38th European Photovoltaic Solar Energy Conference and Exhibition (EU PVSEC 2021)

AA006031

Online
6-10 September 2021

Volume 1 of 3

Editors:

J.M. Almeida Serra
R. Kenny
P. Helm

ISBN: 978-1-7138-5213-1

Printed from e-media with permission by:

Curran Associates, Inc.
57 Morehouse Lane
Red Hook, NY 12571

proceedings.com

Some format issues inherent in the e-media version may also appear in this print version.

Printed with permission by Curran Associates, Inc. (2022)

For permission requests, please contact WIP – Renewable Energies
at the address below.

WIP – Renewable Energies
Sylvensteinstr. 2
81369 Munchen
Germany

Phone: +49 89 72012735
Fax: +49 89 72012791

wip@wip-munich.de

Additional copies of this publication are available from:

Curran Associates, Inc.
57 Morehouse Lane
Red Hook, NY 12571 USA
Phone: 845-758-0400
Fax: 845-758-2633
Email: curran@proceedings.com
Web: www.proceedings.com

38th European Photovoltaic Solar Energy Conference and Exhibition

Proceedings of the International Conference

06 September – 10 September 2021

Edited by:

J.M. ALMEIDA SERRA
University of Lisbon - FCiencias.ID
Lisbon, Portugal

R. KENNY
European Commission Joint Research Centre
Ispra, Italy

P. HELM
WIP Renewable Energies
Munich, Germany

Coordination of the Technical Programme:

European Commission Joint Research Centre
Via E. Fermi 1
21020 Ispra (VA)
Italy

Institutional Support:

European Commission

UNESCO – United Nations Educational, Scientific and Cultural Organization, Natural Sciences Sector

WCRE – World Council for Renewable Energy

Supporting Organisations:

APVI – Australian PV Institute

EREF – European Renewable Energies Federation

ESA – European Space Agency

ESMC – European Solar Manufacturing Council

EUFORES –The European Forum for Renewable Energy Sources

EUREC – The Association of European Renewable Energy Research Centres

ETIP PV – European Technology & Innovation Platform PV

GÜNDER – The Turkish Solar Energy Society

IEA PVPS – International Energy Agency, Photovoltaic Power Systems Programme

REN21 – Renewable Energy Policy Network for the 21st Century

Institutional PV Industry Cooperation:

SolarPower Europe

SOLARUNITED

Supporting Associations:

AIE – European Association of Electrical Contractors

AREA – African Renewable Energy Alliance

CABA – Continental Automated Buildings Association

EU PVSEC realised by:

WIP Renewable Energies
Sylvensteinstr. 2, 81369 München, Germany
Tel: +49 89 720 12 735, Fax: +49 89 720 12 791
Email: pv.conference@wip-munich.de
www.photovoltaic-conference.com
www.wip-munich.de

Proceedings produced and published by:

WIP
Sylvensteinstr. 2, 81369 München, Germany
Tel: +49 89 720 12 735, Fax: +49 89 720 12 791
Email: pv.conference@wip-munich.de
www.photovoltaic-conference.com
www.wip-munich.de

38th EUROPEAN PHOTOVOLTAIC SOLAR ENERGY CONFERENCE AND EXHIBITION
06 SEPTEMBER – 10 SEPTEMBER 2021

EU PVSEC COMMITTEES

INTERNATIONAL SCIENTIFIC ADVISORY COMMITTEE (ISAC)

Chair
P. Szymanski, European Commission Joint Research Centre, Director of Energy, Transport and Climate, Petten, The Netherlands

Committee Members
P. Frankl, Head of the Renewable Energy Division, International Energy Agency, France
M. Getsiou, European Commission, DG RTD, Brussels, Belgium
S.W. Glunz, Head of Division "Solar Cells – Development and Characterization", Fraunhofer ISE, Freiburg, Germany
R. Kenny, European Commission Joint Research Centre, Directorate for Energy and Transport and Climate, Ispra, Italy
P. Malbranche, General Director, CEA INES, France
P. Menna, ret. European Commission, DG ENER, Brussels, Belgium
S. Nowak, Managing Director of NET Nowak Energy & Technology, St. Ursen, Switzerland
W.C. Sinke, Principal Scientist Solar Energy, TNO Energy Transition, The Netherlands
M. Topič, Chairman ETIP PV, Head of Laboratory of Photovoltaics and Optoelectronics of the University of Ljubljana, Slovenia
P. Verlinden, Director at Amrock Pty Ltd, Visiting Professor at Sun Yat-Sen University, Guangzhou, China
A. Grassi, Managing Director, ETA - Renewable Energies, Florence, Italy
P. Helm, Scientific & Managing Director, WIP Renewable Energies, Munich, Germany

CONFERENCE EXECUTIVE COMMITTEE

Conference General Chair
J.M. Almeida Serra, University of Lisbon- FCiencias.ID, Portugal

Technical Programme Chair
R. Kenny, European Commission Joint Research Centre, Directorate for Energy and Transport and Climate, Ispra, Italy

Committee Members
H. Ossenbrink, ret. European Commission Joint Research Centre, Italy
P. Helm, Scientific & Managing Director, WIP Renewable Energies, Munich, Germany
W.C. Sinke, Principal Scientist Solar Energy, TNO Energy Transition, The Netherlands
S. Nowak, Managing Director of NET Nowak Energy & Technology, St. Ursen, Switzerland
A. Grassi, Managing Director, ETA - Renewable Energies, Florence, Italy
J. Bergmiller, Managing Director Events, WIP Renewable Energies, Munich, Germany
J. de Gregorio, Head of Unit, Scientific Services and Cooperation, WIP Renewable Energies, Munich, Germany

2021 SCIENTIFIC COMMITTEE
Topic Organisers and Paper Review Experts

Topic 1 – New Materials and Concepts for Photovoltaic Devices

A. Martí Vega, UPM, Madrid, Spain
P. Pearce, University of Cambridge, United Kingdom
J. Poortmans, imec, Leuven, Belgium
B. Rech, HZB, Berlin, Germany

J.P. Connolly, GeePs, Gif sur Yvette, France
N.J. Ekins-Daukes, UNSW Australia, Sydney, Australia
J. Govaerts, imec, Genk, Belgium
M.A. Hernandez-Fenollosa, UPV, Valencia, Spain
J.C. Hummelen, University of Groningen, The Netherlands
J.-P. Kleider, CNRS/GeePs, Gif sur Yvette, France
D. König, UNSW Australia, Sydney, Australia
I. Konovalov, University of Applied Sciences Jena, Germany
C. Lévy-Clément, CNRS, Thiais, France
Y. Okada, University of Tokyo, Japan
I. Ramiro, NOVA University Lisbon, Caparica, Portugal
F. Roca, ENEA, Portici, Italy
M. Rusu, HZB, Berlin, Germany
R.E.I. Schropp, Driebergen, The Netherlands
T. Tayagaki, AIST, Tsukuba, Japan
W. Warta, Fraunhofer ISE, Freiburg, Germany

Topic 2 – Silicon Materials and Cells

K. Bothe, ISFH, Emmerthal, Germany
D. Bredemeier, ISFH, Emmerthal, Germany
J. Cárabe, CIEMAT, Madrid, Spain
M. Despeisse, CSEM, Neuchâtel, Switzerland
F. Ferrazza, ENI, San Donato Milanese, Italy
S.W. Glunz, Fraunhofer ISE, Freiburg, Germany
G. Hahn, University of Konstanz, Germany
M. Hermle, Fraunhofer ISE, Freiburg, Germany
S. Peters, Hanwha Q CELLS, Bitterfeld-Wolfen, Germany
S. Riepe, Fraunhofer ISE, Freiburg, Germany
P. Roca i Cabarrocas, LPICM-CNRS, Palaiseau, France
M. Topic, University of Ljubljana, Slovenia
A.W. Weeber, TNO Energy Transition, Petten, The Netherlands
P. Wohlfart, Singulus Technologies, Kahl am Main, Germany

O. Anspach, PV Crystalox Solar, Erfurt, Germany
A. Augusto, Arizona State University, Tempe, USA
N. Azkona, UPV/EHU, Bilbao, Spain
D.L. Bätzner, Meyer Burger Research, Hauterive, Switzerland
M.P. Bellmann, SINTEF, Trondheim, Norway
M. Benmansour, CEA, Le Bourget-du-Lac, France
M. Boccard, EPFL, Neuchâtel, Switzerland
R. Brendel, ISFH, Emmerthal, Germany

T.M. Bruton, Progress in Photovoltaics, Woking, United Kingdom
A. Ciesla, UNSW Australia, Sydney, Australia
A. Danel, CEA, Le Bourget-du-Lac, France
W. Dauksher, Arizona State University, Tempe, USA
P. Delli Veneri, ENEA, Portici, Italy
A. Descoeudres, CSEM, Neuchâtel, Switzerland
S. Dubois, CEA, Le Bourget-du-Lac, France
T. Dullweber, ISFH, Emmerthal, Germany
M. Ernst, ANU, Canberra, Australia
P. Fath, RCT-Solutions, Konstanz, Germany
W. Favre, CEA, Grenoble, France
S. Gall, HZB, Berlin, Germany
C. Hagendorf, Fraunhofer CSP, Halle (Saale), Germany
K. Hesse, Wacker Chemie, Burghausen, Germany
J. Horzel, Fraunhofer ISE, Freiburg, Germany
J. John, imec, Leuven, Belgium
A. Meeder, Geo-En Energy Technologies, Berlin, Germany
J. Meier, Meier Technologies, Berlingen, Switzerland
A. Metz, Alzenau, Germany
D. Muñoz, CEA, Le Bourget-du-Lac, France
R.C.G. Naber, Hanwha Q CELLS, Vaassen, The Netherlands
H.-J. Nonnenmacher, Meyer Burger, Hohenstein-Ernstthal, Germany
Y. Ohshita, Toyota Technological Institute, Nagoya, Japan
A. Otaegi, UPV/EHU, Bilbao, Spain
K. Petter, Hanwha Q CELLS, Bitterfeld-Wolfen, Germany
R. Preu, Fraunhofer ISE, Freiburg, Germany
J. Rentsch, Fraunhofer ISE, Freiburg, Germany
J. Schmidt, ISFH, Emmerthal, Germany
M.C. Schubert, Fraunhofer ISE, Freiburg, Germany
L. Serenelli, ENEA, S. Maria di Galeria - Rome, Italy
K. Sporleder, Fraunhofer CSP, Halle, Germany
B. Terheiden, University of Konstanz, Germany
R. Turan, METU, Ankara, Turkey
P.J. Verlinden, AMROCK, McLaren Vale, Australia
M. Vetter, IPHT, Jena, Germany
Q. Wang, Jinko Solar, Haining, China
W. Wu, Tongwei Solar, Chengdu, China
A. Yamada, Tokyo Institute of Technology, Japan
G. Yang, Delft University of Technology, The Netherlands
D.L. Young, NREL, Golden, USA
J. Zhao, CSEM, Neuchâtel, Switzerland

Topic 3 – Perovskites and Other Non-Silicon Materials and Devices, Multijunctions/Tandems

G. Bardizza, European Commission JRC, Ispra, Italy
G. Flamand, imec, Leuven, Belgium
J.C. Goldschmidt, Fraunhofer ISE, Freiburg, Germany
S. Nicolay, CSEM, Neuchâtel, Switzerland
S. Paetel, ZSW, Stuttgart, Germany
R. Po, eni, Novara, Italy
A. Redinger, University of Luxembourg, Luxembourg
G. Siefer, Fraunhofer ISE, Freiburg, Germany
A.N. Tiwari, EMPA, Dübendorf, Switzerland
W. Tress, Zurich University of Applied Sciences, Winterthur, Switzerland
S.C. Veenstra, TNO Energy Transition, Eindhoven, The Netherlands

S. Albrecht, HZB, Berlin, Germany
N. Barreau, University of Nantes, France
V. Bermudez Benito, QEERI, Doha, Qatar
M.J. Brites, LNEG, Lisboa, Portugal
R. Campesato, CESI, Milan, Italy
G. Coletti, TNO Energy Transition, Petten, The Netherlands
S. Cros, CEA, Le Bourget-du-Lac, France
T. Dalibor, Avancis, Munich, Germany
B. Dimmler, NICE Solar Energy, Schwäbisch Hall, Germany
D. Dimova-Malinovska, Bulgarian Academy of Sciences, Sofia, Bulgaria
C.J. Fell, CSIRO, Mayfield West, Australia
I. Gordon, imec, Leuven, Belgium
S. Huang, Macquarie University, Sydney, Australia
S.J.C. Irvine, Swansea University, St. Asaph, United Kingdom
S. Ishizuka, AIST, Tsukuba, Japan
R.R. King, Arizona State University, Tempe, USA
P. Kratzert, SOLARWATT, Dresden, Germany
C.-F. Lin, National Taiwan University, Taipei, Taiwan
D. Lincot, CNRS, Palaiseau, France
M. Lira-Cantú, CIN2, Barcelona, Spain
M.Ch. Lux-Steiner, HZB, Berlin, Germany
M. Meuris, imec, Diepenbeek, Belgium
U.W. Paetzold, Karlsruhe Institute of Technology, Germany
A. Romeo, University of Verona, Italy
V. Sittinger, Fraunhofer IST, Braunschweig, Germany
M. Theelen, TNO/Solliance, Eindhoven, The Netherlands
G. Timò, RSE, Piacenza, Italy
E. Unger, HZB, Berlin, Germany
A. Virtuani, EPFL, Neuchâtel, Switzerland
A. Walter, CSEM, Neuchâtel, Switzerland
S. Yerci, METU, Ankara, Turkey

Topic 4 – Photovoltaic Modules and BoS Components

C. Buerhop-Lutz, HI ERN, Erlangen, Germany
R. Gottschalg, Fraunhofer CSP, Halle (Saale), Germany
W. Herrmann, TÜV Rheinland Energy, Cologne, Germany
N.M. Pearsall, Northumbria University, Newcastle upon Tyne, United Kingdom
A. Wade, Viessmann, Berlin, Germany
K. Wambach, Wambach-Consulting, Petersdorf, Germany

S. Arancón, WIP Renewable Energies, Munich, Germany
P. Bauwens, Ghent University, Belgium
G. Beaucarne, Dow Silicones Belgium, Seneffe, Belgium
N. Bosco, NREL, Golden, USA
C. Camus, LayTec, Berlin, Germany
S. Capaccioli, ETA - Florence Renewable Energies, Italy
R. Einhaus, Apollon Solar, Saint Priest, France
G. Friesen, SUPSI, Canobbio, Switzerland
V. Fthenakis, Columbia University, New York, USA
W. Gambogi, DuPont, Wilmington, USA
G. Graditi, ENEA, Portici, Italy
Y. Hishikawa, AIST, Tsukuba, Japan
U. Jahn, VDE Renewables, Alzenau, Germany
A.R. Lagunas, CENER, Sarriguren-Navarra, Spain
J. Lopez-Garcia, European Commission JRC, Ispra, Italy
M. Mittag, Fraunhofer ISE, Freiburg, Germany
C. Monokroussos, TÜV Rheinland, Shanghai, China
J. Moschner, KU Leuven, Belgium
M.A. Muñoz-García, UPM, Madrid, Spain
H. Nagel, Fraunhofer ISE, Freiburg, Germany
H. Nussbaumer, ZHAW, Winterthur, Switzerland
S. Oktik, Kadir Has University, Istanbul, Turkey
G. Oreski, PCCL, Leoben, Austria
W. Palitzsch, LuxChemtech, Freiberg, Germany
M. Pander, Fraunhofer CSP, Halle (Saale), Germany
J.N. Roy, IIT Kharagpur, India
T. Sample, European Commission JRC, Ispra, Italy
R. Sinton, Sinton Instruments, Boulder, USA

Topic 5 – PV Systems – Modelling, Design, Operation and Performance

J.M. Almeida Serra, University of Lisbon - FCiencias.ID, Portugal
F. Al-Shakarchi, Entech Smart Energies, Quimper, France
I. Antón Hernández, UPM, Madrid, Spain
A.M. Gracia Amillo, European Commission JRC, Ispra, Italy
P. Lechner, ZSW, Stuttgart, Germany
G. Mütter, Enery, Vienna, Austria
H. Ossenbrink, ret. European Commission Joint Research Centre, Italy
S. Taylor, European Space Agency, Noordwijk, The Netherlands
W.G.J.H.M. van Sark, Utrecht University, The Netherlands

C. Alonso-Tristán, UBU, Burgos, Spain
P. Blanc, MINES ParisTech, Paris, France
M. Grottke, WIP Renewable Energies, Munich, Germany
D. Guida, ENEL Green Power, Roma, Italy
P. Hacke, NREL, Golden, USA
V. Khorenko, Azur Space, Heilbronn, Germany
K. Kiefer, Fraunhofer ISE, Freiburg, Germany
K. Lappalainen, Tampere University, Finland
J. Lin, PV Guider, Taipei, Taiwan
A. Louwen, Eurac Research, Bolzano, Italy
O. Mayer, Bayern Innovativ, Nürnberg, Germany
A. Migan-Dubois, GeePs, Gif-sur-Yvette, France
D. Moser, Eurac Research, Bolzano, Italy
K. Nishioka, University of Miyazaki, Japan
C. Nyman, Soleco, Porvoo, Finland
C. Protogeropoulos, EEPS, Athens, Greece
A.H.M.E. Reinders, University of Twente, Enschede, The Netherlands
J. Remund, Meteotest, Bern, Switzerland
M. Richter, 3E, Brussels, Belgium
S. Rodrigues, Forschungszentrum Jülich, Germany
J.A. Ruiz-Arias, University of Malaga, Spain
M. Sengupta, NREL, Golden, USA
J.S. Stein, Sandia National Laboratories, Albuquerque, USA
D. Stellbogen, ZSW, Stuttgart, Germany
J. Tavares Pinho, UFPA, Belém, Brazil
M. Theristis, Sandia National Laboratories, Albuquerque, USA
M. Victoria, Aarhus University, Denmark

Topic 6 – PV Applications, Integration and Storage

G. Adinolfi, ENEA, Portici, Italy
F.P. Baumgartner, Zurich University of Applied Sciences, Winterthur, Switzerland
F. Dolci, European Commission JRC, Petten, The Netherlands
F. Frontini, SUPSI, Canobbio, Switzerland
K.-P. Kairies, ACCURE Battery Intelligence, Aachen, Germany
U. Muntwyler, Bern University of Applied Sciences, Burgdorf, Switzerland
R. Pestana, R&D Nester, Sacavém, Portugal
R. Schlatmann, HZB, Berlin, Germany
A. Scognamiglio, ENEA, Portici, Italy
E. Voroshazi, CEA, Le Bourget-du-Lac, France
I. Weiss, WIP Renewable Energies, Munich, Germany

P.-J. Alet, CSEM, Neuchâtel, Switzerland
P. Alonso, BWP Baumanagement, Munich, Germany
K. Baert, KU Leuven, Heverlee, Belgium
G. Barchi, Eurac Research, Bolzano, Italy
P. Bonomo, SUPSI, Mendrisio, Switzerland
R. Bründlinger, AIT, Vienna, Austria
F. Carigiet, ZHAW, Winterthur, Switzerland
M Centeno Brito, University of Lisbon, Portugal
J. da Costa Fernandes, University of Applied Sciences Offenburg, Germany
K. Ding, Forschungszentrum Jülich, Germany
M. Dörenkämper, TNO Energy Transition, Eindhoven, The Netherlands
G.C. Eder, OFI, Vienna, Austria
V. Efthymiou, University of Cyprus, Nicosia, Cyprus
W. Folkerts, TNO, Eindhoven, The Netherlands
B. Gaiddon, HESPUL, Lyon, France
I.B. Hagemann, Architekturbüro Hagemann, Aachen, Germany
C. Helmke, Clake Energy, Berlin, Germany
B. Herteleer, KU Leuven, Gent, Belgium
B. Kämpfen, kämpfen zinke + partner, Zürich, Switzerland
J.-M. Klein, CEA, Le Bourget-du-Lac, France
P. Lalanne, HydroClapeyron, Dinard, France
B. Lim, ISFH, Emmerthal, Germany
A.C. Neves Foles, University of Évora, Portugal
T. Reijenga, BEAR-iD, Gouda, The Netherlands
E. Román Medina, Tecnalia, Derio, Spain
E. Schüpbach, BFH, Bern, Switzerland
K. Sinapis, Trespa International, Weert, The Netherlands
L.H. Slooff-Hoek, TNO Energy Transition, Petten, Netherlands
L. Tautenhahn, Avancis, Torgau, Germany
R.M.E. Valckenborg, TNO, Eindhoven, The Netherlands

Topic 7 – Finance, Markets and Policies

C. Breyer, Lappeenranta University of Technology, Finland
M. Getsiou, European Commission DG RTD, Brussels, Belgium
P. Malbranche, CEA, Le Bourget-du-Lac, France
G. Masson, Becquerel Institute, Brussels, Belgium
T. Nordmann, TNC Consulting, Feldmeilen, Switzerland
S. Nowak, NET Nowak Energy & Technology, St. Ursen, Switzerland

G. Arrowsmith, EUREC, Brussels, Belgium
S. Caneva, WIP Renewable Energies, Munich, Germany
J. Donoso Alonso, UNEF, Madrid, Spain
I. Kaizuka, RTS Corporation, Chuo-ku, Japan
D. Polverini, European Commission DG GROW, Brussels, Belgium
N. Taylor, European Commission JRC, Ispra, Italy
E. Vartiainen, Fortum Growth, Finland

Awards Coordinators

Student Awards Coordinator
A.H.M. Smets, Delft University of Technology, The Netherlands

Poster Awards Coordinator
J. Cárabe, CIEMAT, Madrid, Spain

SUBJECT INDEX

Topic Code

1 New Materials and Concepts for Photovoltaic Devices
2 Silicon Materials and Cells
3 Perovskites and Other Non-Silicon Materials and Devices, Multijunctions/Tandems
4 Photovoltaic Modules and BoS Components
5 PV Systems – Modelling, Design, Operation and Performance
6 PV Applications, Integration and Storage
7 Finance, Markets and Policies

Session Type

P = Plenary Session
O = Oral Session
V = Visual Session

Day Codes

A = Monday, 06 September 2021
B = Tuesday, 07 September 2021
C = Wednesday, 08 September 2021
D = Thursday, 09 September 2021
E = Friday, 10 September 2021

e.g. 1AO.1 ⇒ 1= New Materials and Concepts for Photovoltaic Devices, A=Monday, O=Oral session, 1=Session 1

TABLE OF CONTENTS
of EU PVSEC 2021 Proceedings Papers

Oral PRESENTATIONS 4BO.5 Interconnects and Soldering

Oral PRESENTATIONS 4CO.1 Polymers and Encapsulation of PV Modules

Oral PRESENTATIONS 4CO.2 PV Module Backsheets

Oral PRESENTATIONS 4CO.3 Performance of BOS Components

Oral PRESENTATIONS 4CO.4 Recycling, Repair and Reuse of PV Modules

Oral PRESENTATIONS 4DO.11 Sustainability Aspects of PV

Visual PRESENTATIONS 5CV.2 Operation, Performance and Maintenance of PV Systems

FOREWORD

The EU PVSEC 2021, as the second online edition of the European PV Solar Energy Conference and Exhibition was a great success and has shown once again that a sustainable energy supply is becoming ever more important and only together, through scientific endeavor and cooperation and political commitments and investments that this goal can be achieved. The Conference provided once again a stage for the dissemination of work that will help ensure ever increasing quantities of PV energy and its integration into the current and future energy network. New discussion sessions focused highly relevant topical questions, provided key lessons for upcoming developments. This year 1,300 participants from 62 countries took part in this most inspiring platform. In the exhibition 16 companies from all parts of the world welcomed online visitors and presented their products and services.

The EU PVSEC 2021 Proceedings give a comprehensive overview of the photovoltaic solar sector, its current status and future prospects in science, research, innovation, development and deployment on almost 3,800 pages. Selection for inclusion in the conference was made by paper review experts and topic organisers (see the listing on pages III-V), to whom we express our sincere gratitude for their comprehensive review work and overall contribution to the success of the conference. In addition to the 364 submitted papers, the proceedings include 103 presentations (slides) shown during the plenary and oral presentations as well as 195 poster files of the visual presentations. In total this amounts to 662 publications.

The 2021 EU PVSEC Proceedings maintain our commitment to providing quick and open access to high quality scientific results. They constitute a powerful tool for targeted and quick information search and retrieval, enabling you to search by keyword, paper title and authors. We are sure that these features will help to simplify the use and exploit the full potential of this extensive source of information.

The Conference Proceedings are published as downloadable DVD file and are also fully accessible online. A DOI code (Digital Object Identifier) has been assigned to each paper. This ensures unequivocal and permanent identification and full citability. The papers can be viewed and downloaded in a full free open access from the EU PVSEC Proceedings website www.eupvsec-proceedings.com.

We are confident that these Proceedings will play an important role in providing a comprehensive overview of the current actors and activities in the global PV sector and that they will disseminate information on the state-of-the-art of technologies and applications. This can generate further research, add momentum to innovation and promote interest in PV worldwide.

We would like to cordially thank all authors and participants of the EU PVSEC 2021 for their contributions and look forward to welcoming you in 2022 at the WCPEC-8, the 8th World Conference on Photovoltaic Energy Conversion in Milan, Italy.

The Editors

LIST OF EXHIBITORS
(in alphabetical order)

Atotech Deutschland GmbH	Germany
Avalon ST Sarl	Switzerland
Bergfeld Lasertech GmbH	Germany
Eaton Electric GmbH	Germany
ECM Greentech	France
exateq GmbH	Germany
h.a.l.m. elektronik GmbH	Germany
HQTS Group Limited	China
John Wiley & Sons Ltd.	United Kingdom
KAUST Solar Center	Saudi Arabia
Nines Photovoltaics	Ireland
pv magazine group GmbH & Co.KG	Germany
SICK AG	Germany
SINGULUS TECHNOLOGIES	Germany
Solaires Enterprises Inc.	Canada
WIP Renewable Energies	Germany

We thank the EU PVSEC 2021 Sponsors

ATOTECH

Avalon ST

BERGFELDLASERTECH
laser for PV and battery cells

EATON
Powering Business Worldwide

exateq
experience aided techniques

halm.

HQTS
Quality Solutions With You Every Step

جامعة الملك عبدالله للعلوم والتقنية
King Abdullah University of Science and Technology
KAUST SOLAR CENTER

Nines Photovoltaics

SINGULUS

Solaires™

ACKNOWLEDGEMENTS

EU PVSEC
European Photovoltaic Solar Energy Conference and Exhibition

The highlights featured in this presentation are largely a compilation of feedback from the Oral and Visual Session Chairs, whom I thank for their invaluable contribution both for the successful running of the sessions as well as their contributions to these highlights.

The first 3 slides contain selected conference statistics, followed by summaries of the four topical panel discussions. Then follow the highlights of selected sessions grouped by Topic.

Thanks are also due to all the members of the Scientific Committee and the Executive Committee for their dedication in putting together the technical programme, as well as of the International Scientific Advisory Committee for its strategic steering.

Thanks to the WIP team for the technical support before and during the conference week.

And last but not least, thanks to all the authors who have produced such excellent results in trying times.

Robert Kenny
EU PVSEC Technical Programme Chair

EU PVSEC
European Photovoltaic Solar Energy Conference and Exhibition
More than 800
live presentations
in Conference Sessions, Parallel Events and the Industry Summit
More than 750
abstracts received
3000+ views of
the Opening during the event
1300+
participants...
...from 62
countries
Scientific Committee of
249 top experts

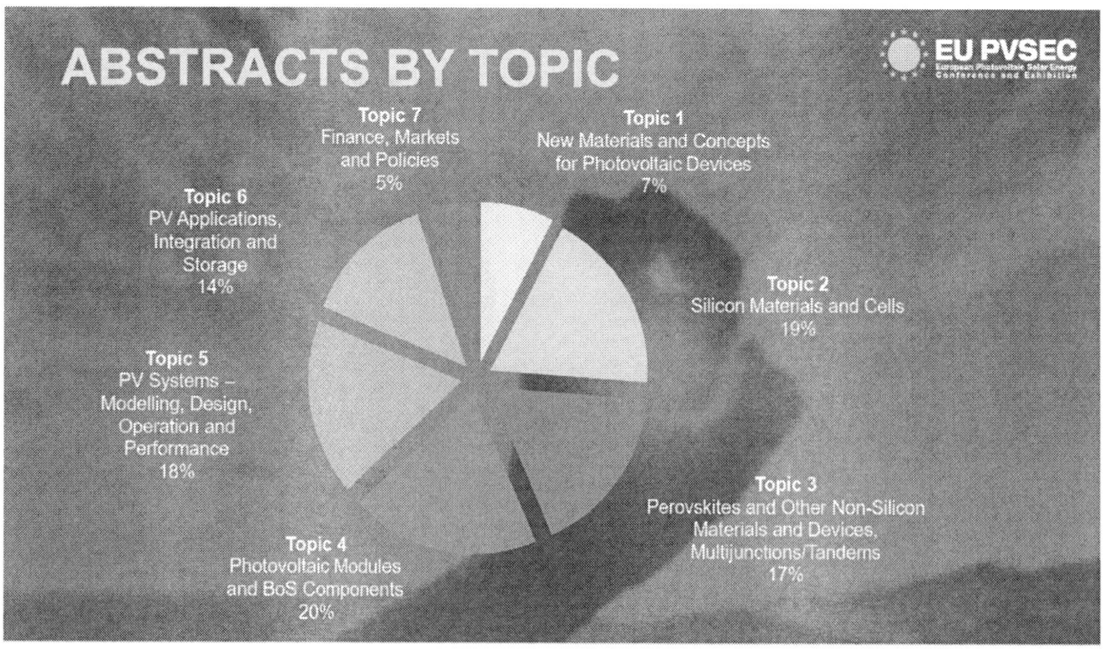
ABSTRACTS BY TOPIC
EU PVSEC
European Photovoltaic Solar Energy Conference and Exhibition
Topic 7
Finance, Markets and Policies
5%
Topic 1
New Materials and Concepts for Photovoltaic Devices
7%
Topic 6
PV Applications, Integration and Storage
14%
Topic 2
Silicon Materials and Cells
19%
Topic 5
PV Systems – Modelling, Design, Operation and Performance
18%
Topic 3
Perovskites and Other Non-Silicon Materials and Devices, Multijunctions/Tandems
17%
Topic 4
Photovoltaic Modules and BoS Components
20%

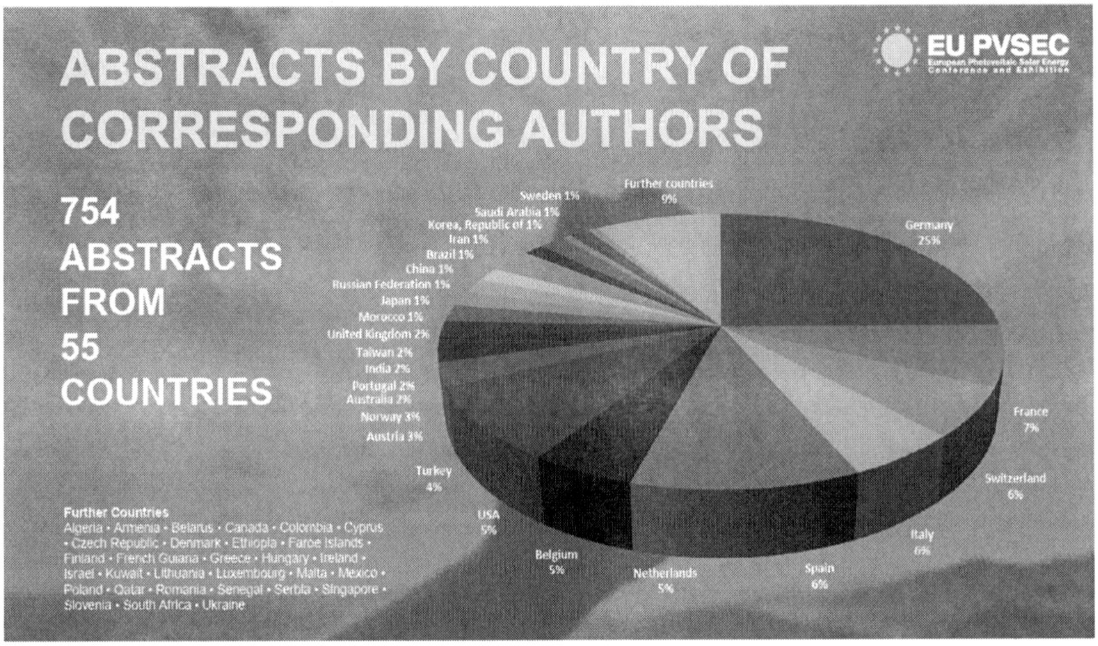
ABSTRACTS BY COUNTRY OF CORRESPONDING AUTHORS
EU PVSEC
European Photovoltaic Solar Energy Conference and Exhibition
754 ABSTRACTS FROM 55 COUNTRIES
Further countries 9%
Germany 25%
France 7%
Switzerland 6%
Italy 6%
Spain 6%
Netherlands 5%
Belgium 5%
USA 5%
Turkey 4%
Austria 3%
Norway 3%
Australia 2%
Portugal 2%
India 2%
Taiwan 2%
United Kingdom 2%
Morocco 1%
Japan 1%
Russian Federation 1%
China 1%
Brazil 1%
Iran 1%
Korea, Republic of 1%
Saudi Arabia 1%
Sweden 1%
Further Countries
Algeria • Armenia • Belarus • Canada • Colombia • Cyprus • Czech Republic • Denmark • Ethiopia • Faroe Islands • Finland • French Guiana • Greece • Hungary • Ireland • Israel • Kuwait • Lithuania • Luxembourg • Malta • Mexico • Poland • Qatar • Romania • Senegal • Serbia • Singapore • Slovenia • South Africa • Ukraine

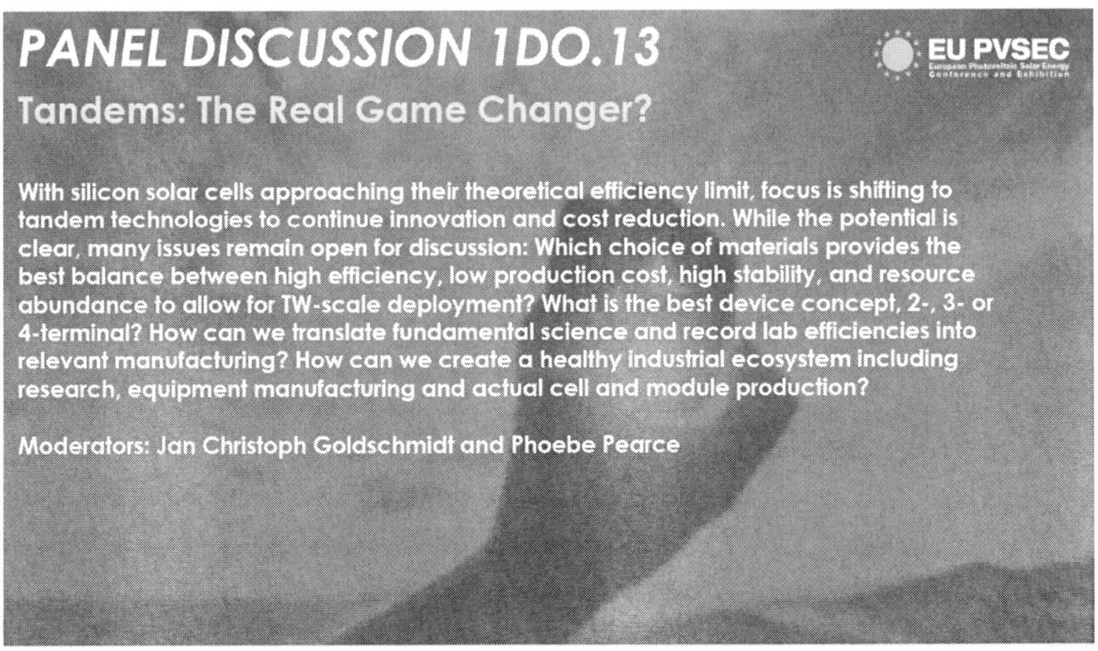
PANEL DISCUSSION 1DO.13
Tandems: The Real Game Changer?
EU PVSEC
European Photovoltaic Solar Energy Conference and Exhibition
With silicon solar cells approaching their theoretical efficiency limit, focus is shifting to tandem technologies to continue innovation and cost reduction. While the potential is clear, many issues remain open for discussion: Which choice of materials provides the best balance between high efficiency, low production cost, high stability, and resource abundance to allow for TW-scale deployment? What is the best device concept, 2-, 3- or 4-terminal? How can we translate fundamental science and record lab efficiencies into relevant manufacturing? How can we create a healthy industrial ecosystem including research, equipment manufacturing and actual cell and module production?
Moderators: Jan Christoph Goldschmidt and Phoebe Pearce

PANEL DISCUSSION 2BO.13

EU PVSEC
European Photovoltaic Solar Energy Conference and Exhibition

Towards Ultimate Single-Junction Silicon Cells – Industry Perspective

Fantastic and quite dynamic panel discussion with five experts from the PV industry and research labs exchanging their vision. The industry is at a historical juncture. Not only we have to continue improving efficiency and reduce LCOE, we cannot develop not-sustainable technologies. The PV community needs to go one step forward in sustainability. Silver, must be rapidly and significantly reduced to less than 5mg/W, before the production level reaches 1TW p.a. (just in a few years). Indium cannot be used. Indium-free TCO's must be developed. And that will be probably more difficult than reducing silver. 25% module efficiency will be the ultimate efficiency for single-junction silicon technology, on G12 (210mm wafers) technology! G12 is already used by several players and certainly imminent for PERC technology. However for Heterojunction, the question is still open though other considerations as wafer thickness can play a major role in LCOE. Which technology would dominate the market in 2030? The question is still open. It would probably be balanced between all the technologies, with demand depending on the location and application."

Moderators: Pierre Verlinden and Delfina Muñoz

PANEL DISCUSSION 4/5AO.7

EU PVSEC
European Photovoltaic Solar Energy Conference and Exhibition

Managing Performance and Sustainability at the TW Scale

Part 1: Computational PV

Almost 10% of presentations at this EUPVSEC deal with advanced computing. Algorithms, big data, machine learning, artificial intelligence, image recognition and satellite data processing appear in the titles of contributions which range from solar cell modeling, manufacturing, preventive maintenance, site selection to forecasting, nowcasting, user behavior and modeling of entire electricity systems.

Part 2: Resource Efficiency of PV in Production, Use and Disposal

The latest IPCC report has been published recently and there is an urgent need for action to mitigate our emissions and save our limited resources. PV systems do contribute significantly to the solution for nature conservation. Modern eco-friendly technologies and long lasting, repairable products are required in combination with sound circular economy approaches to process the huge stock of valuable resources at the end of life.

Moderators: Heinz Ossenbrink and Harry Lehmann

PANEL DISCUSSION 6/7CO.14

EU PVSEC
European Photovoltaic Solar Energy Conference and Exhibition

PV: The Key Element Towards 100% Renewables. How to Make it Happen?

As the cheapest energy source in history, PV is redefining the way we think about energy. Will everything else have to adapt to it? This panel discussion looked into the technical and non-technical changes ahead, including grid integration, markets, policy and finance.

Moderators: Kai-Philipp Kairies and Marion Perrin

PLENARY AP.1

EU PVSEC
European Photovoltaic Solar Energy Conference and Exhibition

Devices in Evolution: Pushing the Efficiency Limits and Broadening the Technology Portfolio

Silicon PV on its way to perfection: Mathieu Boccard of EPFL showed how tackling the biggest remaining loss factor in silicon solar cells (related to the contacts) paves the way to 23-24% full-size modules.

Electrification without limitation: Stephan DeLuca of EMC, USA, argued that very fast roll-to-roll manufacturing of perovskite solar modules can bring up to 20 terawatt-peak installed capacity of PV within reach in time to reach climate targets.

PLENARY CP.1

EU PVSEC European Photovoltaic Solar Energy Conference and Exhibition

Industry and Applications: PV Going Everywhere

Henry Snaith show how high luminescence efficiency in thin perovskite opens way to 34-35% devices. With respect to industrialisation he showed how PK/Si modules can now go through many of the key reliability tests and how Oxford PV targets a market entrance with cells which should reach close to 27%, and with a premium PV module at 23.7%.

Ulrike Jahn, stressed on the importance of combined testing for PV modules, as well as on the emergence of multiple mobile techniques for field testing of PV modules, illustrating this with many data on, performance ratio and failure analyses performed in the field.

Peter Hacke showed the opportunites and challenges for inverters, with an urgent need for international standards in many of the grid supporting functions. Inverters' increasing tasks and complexity, combined with cost pressure, can create risks for reliability and safety/security.

Alexandra Scognamiglio showed some of the prospects for agrivoltaics, where clever light management by various PV configuration can even promote growth of plants. It can also contribute, through an holistic approach, to maintaining the nature of the visual aspect of the landscape, with an analogy to the case of BIPV.

Marko Topic introduced the new strategic research and innovation agenda prepared by ETIP PV, which he is heading. This important document adresses the challenges, from technology, to manufacturing to social aspects which will have to be adressed for solar energy to further develop and contribute massively to climate change mitigation (while creating jobs), in line with the EU CO2 emission reduction goals.

PLENARY EP.1

EU PVSEC European Photovoltaic Solar Energy Conference and Exhibition

Sustainability and Social Acceptance Preparing for the TW Era

Sustainability of PV Manufacturing

A Framework for Implementing Requirements on the Carbon Footprint of Photovoltaic Modules under the Ecodesign Policy

True Cost of Solar Hydrogen

Social Networks and Digital Gamification for Solar Literacy and Photovoltaic Communities - the SOLIS Solar Platform of Lisbon

How Photovoltaics started to Change Architecture over the Last Two Decades

1 New Materials and Concepts

EU PVSEC
European Photovoltaic Solar Energy Conference and Exhibition

1BO.16 Fundamental Studies in the Forefront of PV Many interesting results from novel computational and experimental techniques; a highlight was (1BO.16.2) Juan Camilo Ortiz Lizcano's talk on optic filters, showing an electro-optical model to design and simulate coloured solar cells which showed excellent agreement with experimental measurements.
In (1BO.16.5) an extremely high diode factor in solar cells is discussed from the novel viewpoint of the multiple quasi-Fermi levels.

1AO.3 Innovative Approaches for Module Concepts In (1AO.3.3) High-Resolution Electroluminescence Study of Alternative Patterning P1 Strategies for CIGS Modules

1 New Materials and Concepts

EU PVSEC
European Photovoltaic Solar Energy Conference and Exhibition

1AO.1 Advanced and Novel Approaches for Transparent Layers and Metal Contacting A joint work by researchers at Osazda Energy, University of New Mexico, NREL and Georgia Tech, showed how carbon nanotubes addition enhances silver particle sintering and increases resilience to finger cracking in silicon solar cells by a factor of nearly 7.

1AO.2 Innovative Approaches for Solar Cells and Photovoltaic Materials On the level of materials, the preparation and properties of thin films of a new emerging thermally stable in air inorganic perovskite-type BaZrS3 material for application in PV have been demonstrated. The material consists of environment-friendly and earth-abundant elements. It has a very high absorption coefficient >105 cm-1, thus being suitable for a solar cell absorber with a thickness of few hundreds of nanometers, hence requesting a low material consumption. On the conceptual level, the presentation on HBT-like solar cells was interesting as a demonstration of this new concept by the use of III-V based materials.

2 SILICON

2BO.15 TCO and Metallisation for Silicon Heterojunction Cells Two hot topics towards sustainability of heterojunction solar cells taking into account production constraints: reduction of (or free-) Indium consumption and alternative techniques for silver reduction in metallization (inkjet and parallel dispensing); New understanding of contact issues between metallization and TCO and TCO and amorphous silicon; Edge management of heterojunction solar cells and how to minimize cutting losses using S-ALD to make heterojunction competitive.

2BO.11 Poly-Silicon Passivated Contacts Approaches for local deposition of polySi for IBC cells a) with a mask (Verena Mertens) b) with a laser (Florian Buchholz)

2CO.10 Characterisation of Crystalline Silicon Devices Even in topics that have been worked on for years, there is a continuously growing understanding and new findings. Particularly remarkable is the required measurement accuracy of spectroradiometers to determine the A+ classification of solar simulators. Another highlight of this session was the presentation on the positive effects on Voc, FF and efficiency by switching to high resistivity silicon.

2 SILICON

2A0.8 Manufacturing and Production of Silicon Cells (2A0.8.2) Showcasing of a 1.35 GWp Solar Power Plant presented as a model to reactivate European industry. Machine learning algorithms have the potential to be used in the optimization of the solar cells' industrial production process towards higher efficiencies. In (2A0.8.1) Pietro Altermatt (Trina Solar) presented a very interesting roadmap for Silicon solar cells, from PERC to i-TOPERC, p-TOPCon and finally n-TOPCon using ZnO as a transparent passivating contact. He also presented the new PERC cell record of 23.56% on 210mm wafers and 23.03% (aperture area) efficiency for Vertex modules.
2AO.9 Production Processes of Silicon Solar Cells (2AO.9.5) potential of increased throughput of PERC cells, thus cost reduction; (2AO.9.6) Laser contact enhancement process --> can improve considerably out of class or reject cell, thus improves yield and reduces costs."
2BO.12 Advanced Process Technologies for High-Efficiency Silicon Solar Cells One of the highlights was the detailed study presented (2BO.12.1) on the influence of H on n-type and p-type polySi passivating contacts, where n-type polySi structures seem to suffer significantly more from excess H, causing an increase in recombination current.

2 SILICON

EU PVSEC European Photovoltaic Solar Energy Conference and Exhibition

2BO.14 Wafer-Based Silicon Heterojunction Solar Cells (2BO.14.1) "Temperature-Dependent Performance of Silicon Heterojunction Solar Cells with Molybdenum Oxide as a Hole-Selective Contact"; Detailed study of temperature dependence of transition metal oxide based carrier selective contacts (both MoOx and TiOx for hole resp. electron contact). High efficiencies obtained and the temperature coefficient of cells with TM-oxide based carrier selective contacts is better than of any other silicon cell architecture. Even better than 'conventional' SHJ and TOPCon!

2EO.3 Feedstock and Wafer Processing / Thin-Film and Foil-Based Silicon Cell Three presentations on feedstock and wafer processing and three looking at thin-film and foil-based silicondevices.

2 SILICON

EU PVSEC European Photovoltaic Solar Energy Conference and Exhibition

2CV.1 Characterisation and Manufacturing of Crystalline Silicon Solar Cells Characterization and simulation methodologies of devices and materials for high efficiency solar cells, including optical and in-line approaches. Visual Presentation (2CV.1.4) had a particularly creative short video.

2DV.3 Technologies for High Temperature Passivating Contacts and Homo Junction Silicon Solar Cells / Low Temperature Routes for Silicon Cells Excellent session with several posters addressing fundamental research as well as industrial application of passivating contacts and heterojunctions going towards simplified and more cost-effective solar cell designs.

2CV.1.4 short video

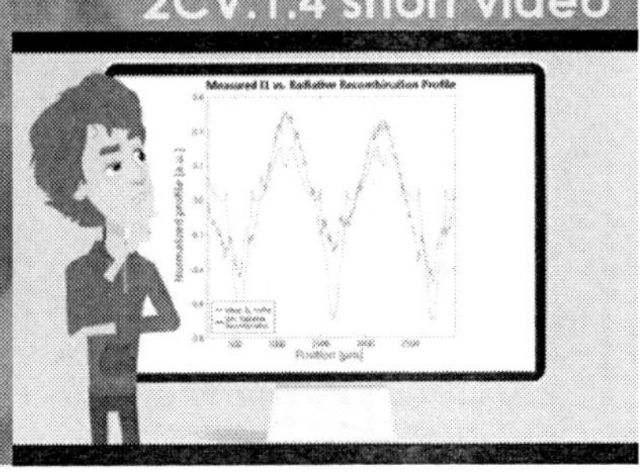

3 PEROVSKITE

EU PVSEC European Photovoltaic Solar Energy Conference and Exhibition

3CO.5 Advances in Perovskite Materials and Process Engineering for High Efficiency Cells (3CO.5.1) Use ethylenediamine to decrease the open-circuit voltage loss and combining FTO-2PACz-MPA, the efficiency achieves over 23%. (3CO.5.4) Use Bromine that penetrates into perovskite bulk to enlarge the bandgap and then the open-circuit voltage. Voc of 1.24 V, corresponding to Eg of 1.6eV, is achieved. The efficiency is above 22%. "• Low stability material can be removed to achieve high efficiency (23%) tin-based perovskite solar cell• Efficient (10%) flexible tin-based perovskite solar cell• Inorganic perovskite-based photodetector by scalable fabrication techniques• Bromide passivation for high voltage perovskite solar cell• Interlayer and additive engineering can be a route to the upscaling of inverted perovskite solar cell.

3CO.7 Large Area Perovskite Solar Cells and Modules Scalability of perovskite solar cells: challenges and techniques to increase the efficiency and performance of the large modules Techniques: blade, vapor deposition, ink Optimization processes in air conditions Future of scalable devices: comparison of nip and pin devices and how that affects the scalability of the cells

3 PEROVSKITE

EU PVSEC European Photovoltaic Solar Energy Conference and Exhibition

3CO.8 Standardisation and Outdoor Performance of Perovskite Solar Cells and Modules
3CO.8.4 Long-Term Outdoor Stability of Perovskite Solar Cells and How It Fits with Indoor Accelerated Aging presented by M. Khenkin, from HZB, Berlin, Germany; First complete results on outdoor performances of perovskite technology over almost 1 year. Very interesting comparison and discrepencies between outdoor and indoor testing with problem of metastability of perovskite technology."

3BV.1 Perovskite Solar Cells and Modules Advanced characterization techniques, originally developed for other PV materials, like PL imaging, EL imaging, DLIT mapping as well as capacity loss mapping, being applied to perovskite cells and mini-modules. Clear examples have been presented where these techniques lead to a better understanding of the investigated perovskite devices (e.g. 3BV.1.27). (3BV.1.26) includes data obtained on CIGS and perovskite absorber layers and aims to extract unifying observations. This may lead to a better understanding of PV materials in general.

3 TANDEMS

EU PVSEC
European Photovoltaic Solar Energy Conference and Exhibition

3A0.4 Tandem Upscaling towards Industrialisation
3AO.4.4 Peter Fiala from EPFL PV Lab showed an efficiency of 28.1% for a textured 2T perovskite-on-silicon tandem solar cell
3AO.4.6 Petra Manshanden from TNO reported an efficiency of 29.2% on 9mm2 for a 4T perovskite-on-silicon tandem solar cell utilizing a BC-SHJ bottom cell
3AO.4.3 Janosch Goldschmidt showed a proof of concept for Cu plating on perovskite solar cells and reported an independently confirmed high current of 19.6 mA/cm2 for a planar 2T perovskite-on-silicon tandem solar cell
3AO.5 Tandem Solar Cells Process and Performance
3AO.5.1 Two-Terminal III-V//Si Triple-Junction Solar Cells with One-Sun Conversion Efficiencies of 35.9 % - impressive efficiency achievement
3BO.10 III-V Solar Cells and Related Compounds
3BO.10.1 Lightweight, flexible, triple junction III-V based space cells, with a power-to-mass ratio of 3 W/g, under AM0 illumination have been demonstrated. BOL efficiency of 30.2% and EOL efficiency of 25.4%, after an electron fluence of 1×1015 cm-2, have been measured.

3 *CIGSe, CdTe and Kesterites/OPV/ III-V and Related Compounds/Tandems*

EU PVSEC
European Photovoltaic Solar Energy Conference and Exhibition

3BO.8.1 Introductory Oral: Improved CIGSSe Absorber Homogeneity with Cd Free Sputtered ZnOS Buffer Layer at World Record Efficiency on 30x30 cm2 Laminated Module
Patrick Eraerds from Avancis reported about the process optimization for the achievement of the CIGSSe module record efficiency of 19.6% on 30x30cm^2
3BV.2 CIGSe, CdTe and Kesterites / OPV / III-V and Related Compounds / Tandems
Progress in tandem solar cells architectures and materials is constant.
Nice contributions on the characterization of tandem solar cells - both the precise calibration at STC as well as the evaluation of the impact of varying spectrum were discussed. Improved methods for accurate tandem solar cells characterization.
3AO.6 Organic and Dye-Sensitised Photovoltaics
Organic solar cells are moving more and more from the lab to the fab. A talk from CSEM in Brazil (3AO.6.1) discussed major challenges towards upscaling of thin-film organic and perovskite solar cells.

4 Modules

EU PVSEC European Photovoltaic Solar Energy Conference and Exhibition

4BO.1 PV Module Design and Materials Highlights were the presentation of a new interconnect scheme using aluminium foil to create shingled modules (4BO.1.6). An innovative investigation of packaging related failures of glass-glass modules demonstrated the dependence of reliability on different encapsulation materials (4BO.1.4).

4BO.5 18:30 Interconnects and Soldering This session presented a wide range of experimental and computational results for characterizing the performance reliability, and ageing of adhesive and solder pastes of different composition.

4CO.2 PV Module Backsheets An innovative measurement technique based on Near-Infrared Spectroscopy was presented and demonstrated, which enables identifying and analyzing backsheets, not only in the lab but importantly also in the field (small, light measurement tool). This is not only useful for the identification of backsheet failures, but will also help for recycling. (4CO.2.2) and (4CO.2.6).

4 Modules

EU PVSEC European Photovoltaic Solar Energy Conference and Exhibition

4BO.3 Induced Degradation in PV Modules "In (4BO.3.1) a comprehensive overview on degradation mechanisms in crystalline silicon solar cells: LID, LeTID and PID; influences of the injection level and the influence of hydrogen on the degradation; measures to avoid degradation for each type of degradation presented. Esther Fokuhl concluded that current test standards for LeTID are a good compromise in between meaningfulness and time consumption for testing. PID under negative bias was shown by Thomas Weber to be a relevant degradation mechanism of current modules.

4BO.4 PV Module Testing and Characterisation
The session highlighted the point that multiple modelling and characterisation methods are needed and are being developed to characterise and understand module degradation. (4B0.4.4) presented the good agreement of calibration laboratories, test laboratories and equipment manufacturers using different approaches for the measurement of high capacitance cells.

4BO.2 PV Module Outdoor Performance and Energy Rating Resilience of PV modules for partial shade was discussed, and advantage of shingled module structure was quantified (4BO.2.6) "Quantitative Evaluation of the Shading Resilience of PV Modules"

4 Modules/Sustainability

EU PVSEC European Photovoltaic Solar Energy Conference and Exhibition

4CO.4 Recycling, Repair and Reuse of PV Modules
there is significant work going on enabling the re-use of Modules and the re-use of module materials including the re-use of former module-polymers in new modules. In order to improve recyclability, solutions on polymer backsheets have been presented pointing out that today it is the most difficult component to recycle. Also, there was a focus on metals demand that will come up in the next few decades (terawatt scale).

4DO.11 Sustainability Aspects of PV
Comprehensive overview on challenges and work in progress towards PV systems that are currently becoming more and more sustainable. - Importance of tools for ecodesign and labelling - Life Cycle Assessment as a generalized method to evaluate environmental sustainability of PV systems- Interesting emerging technologies with good potential for improvement- Need for holistic view, including wide range of issues, both from an environmental and an economic point of view.
Analyses showed that the overall environmental impact of perovskite tandem cells will be small compared to the other material contributions.

4 Modules/Sustainability

EU PVSEC European Photovoltaic Solar Energy Conference and Exhibition

4AV.1 PV Module Design, Components and Ageing There were quite a few contributions on using ECAs in module manufacturing and on shingle interconnection, which is clearly reaching a more mature stage of development. A remarkable milestone towards achieving the full potential of these technologies was presented during the session, namely a full size bifacial module based on shingling heterojunction cells, with a power of 412W (4AV.1.31)

4AV.2 PV Module Characterisation, Testing and Outdoor Performance The session highlights were new module characterization methods. Two examples: Poster 4AV.2.23 showed a new approach for double sided characterization of bifacial modules. And poster 4AV.2.14 showed a test setup for an in situ monitoring of the electrical properties of a PV module under mechanical stress

4DV.2 Components / Sustainability and Recycling of PV Modules A single crystalline Si ingot was manufactured by the use of up to 30% recycled silicon. The bottom part of the ingot had a purity of 99.999998%. From that ingot, 6-inch wafers with 280um thickness were manufactured and the resulting p-type solar cell showed an efficiency of 20.05% (4DV.2.22).

5 SOLAR RESOURCE

EU PVSEC
European Photovoltaic Solar Energy Conference and Exhibition

5BO.7 Forecasting Solar Radiation and PV Power 5BO.7.4: The German Aerospace Center, DLR described a hybrid approach combining all-sky imagers with a persistence method which lead to improved forecast accuracy, for clear sky, overcast, and broken cloud weather. This was developed and validated at the CIEMAT test site in Almeria.

5BV.4 Solar Radiation and Forecasting (5BV.4.1) The International Energy Agency Task 16 recently published the third edition of the ""Best Practices Handbook for the Collection and Use of Solar Resource Data for Solar Energy Applications"", which covers all aspects of solar energy and solar energy applications. Besides the third edition includes significant updates on measurements uncertainties or forecasting, for example. (5BV.4.13) This session also contains a presentation of an open source tool (Photovoltaics Geographic Raster Image Processor) to estimate solar resource and power output considering the effect of surrounding terrain in detail. An example of VIPV simulation is provided. "

5 PV SYSTEMS

EU PVSEC
European Photovoltaic Solar Energy Conference and Exhibition

5BO.6 Solar Radiation Modelling and Instrumentation The PVGIS database with millions of users will be updated to contain data from 2017-2020 with new satellite data. A new radiative transfer model FARMS-DNI was developed by NREL and can predict Direct Normal Irradiance (DNI) significantly more accurately under cloudy conditions than current models from satellite data.Two presentations on uncertainty analysis for photodiode based and reference cell based solar measurements. These are important for improving measurement characterization using cheaper sensors. New method for site adaptation of satellite data using ground measurements.

5DO.2 PV Systems Operation and Field Inspection 1 greateyes presented the development of PL-setup with a crawler moving over the installed PV-modules; 2 Aasreth from Norway presented an interesting step-based algorithm to detect defective bypass diodes in monitoring data; 3 Dow introduced a promising silicon-based material to repair degraded BS.

5EO.2 Concentrators and PV for Space Applications (5EO.2.1) "Suitable Silicon Solar Cell Technologies for Use in Space Application" examined in detail the degradation of a range of Si cell technologies.

5 PV SYSTEMS

EU PVSEC European Photovoltaic Solar Energy Conference and Exhibition

5CO.9 Advanced Monitoring and Fault Detection in PV Systems
In (5CO.9.1) a nice overview of automatic failure detection algorithms applied to real world PV monitoring data from commercially operated PV systems. Atse Louwen (Eurac Research) then presented results from the application of Neural Networks to detect and predict failures. Andreas Livera (University of Cyprus) applied Failure Diagnosis Routines and Trend-based performance Losses Routines to detect PV underperformance due to failures. Gaute Otnes (Institute for Energy Technology) showed how Bayesian Optimization can be used to systematically obtain optimal values for the filter parameters to calculate performance loss rates for PV systems. In the last talk Robert Höller presented the results of a long-term degradation assessment of different Photovoltaic technologies in Austria."

5CO.12 PV Systems: Planning, Plant Optimisation Tools, Advanced Installation Criteria, Construction Issues (5CO12.3) "High Resolution 3D Solar Photovoltaic (PV) Potential Map for TU Delft Campus and Real Estate". This method has a great potential for increasing yield and thus economic performance of PV Systems in the built environment. It also shows how today's computing power, but also sophisticated measurement tools like LIDAR influence the economy of PV."

5 PV SYSTEMS

EU PVSEC European Photovoltaic Solar Energy Conference and Exhibition

5DO.2 PV Systems Operation and Field Inspection A range of innovative imaging techinques were presented directed at O&M improvements

5DO.3 Performance of Field and BIPV Systems (5DO.3.2) PV system owners upload data from their PV systems for a fleet-scale analysis of PV systems in France. Data split into climatic zones to look at degradation for specific areas of the country. The talks (5DO.3.1) and (5DO.3.4) on bifacial PV systems were both enlightening.

5DO.4 The Effect of Soiling and Snow Cover on PV System Performance Soiling sensors can be unreliable and expensive. Progress towards more cost-effective solutions, data availability using PV outdoor test benches were presented. Models can be used to avoid the cost of sensors: however a lot of work needs to be done to solve some issues related to concurring effects. Snow and soil modelling are very challenging as they require high quality input. Nevertheless there are improvements and they can be used in yield assessments.

5 PV SYSTEMS

5CV.2 Operation, Performance and Maintenance of PV Systems
A modelling framework was presented by Austrian researchers (at SAL and AIT), which couples EL imagery with statistical and electrical/physics/chemical modelling. Preliminary results of such deep-learning modelling approach have shown that it is possible to predict degradation/failures propagation and service-lifetime from field-exposed PV modules, under different climatic profiles.
5CV.3 PV Systems: Planning, Plant Optimisation Tools, Advanced Installation Criteria, Construction Issues / Concentrators and PV for Space Applications **PV simulations quality is improving over the years especially for complex system design (bifacial, LSC, etc) including also special attention in increasing computational speed**
5DO.1 Application of Machine Learning & Modelling in PV Systems **The session underpinned the high value of Machine Learning and data based modelling techniques in the operation of PV installations. • The range of applications is broad: from defect classification and fault detection to digital twins, performance modelling and forecasting.• Methods for self-learning and auto-parametrising to minimize the requirement of manual input to the models are particularly promising."**

6 APPLICATIONS, INTEGRATION AND STORAGE

6DO.6 BIPV Implementation: Design, Assessment and Performance **BIPV technology and solutions are no longer the real challenge for the BIPV market, but rather the correct assessment of photovoltaic potential and contribution in urban contexts seems to be very relevant together with the energy simulation. Accurate analysis and simulation models were presented able to consider different colours, technologies, locations and shading.**
6DO.7 BIPV Development and Customisation: Approaches and Experimental Results **The session highlighted that BIPV is a key-strategy for solar exploitation, research sector is very dynamic and many technologies are ready to cover the last km for market introduction to support the energy transition. BIPV topic confirmed as an established research field with a quite mature technological and implementation level in many cases (e.g. coloured and customizable glasses for claddings), as well as with new concepts aimed at satisfying the aesthetic and energy request (e.g. active components with natural materials like stone). Ongoing research on the technologies with a higher TRL are investigating performances, durability and reliability towards commercialization through test campaigns, indoor and outdoor, pilot installations and modeling approaches. Some new technologies (e.g. LSC, perovskite) are under investigation to be used as part of building envelope (especially in windows), so that aspects like aspects like scaling-up, lifetime and competitiveness become crucial for a future outlook.**

EU PVSEC European Photovoltaic Solar Energy Conference and Exhibition

6 APPLICATIONS, INTEGRATION AND STORAGE

6DO.8 Vehicle Integrated PV: Potential Energy Yield Simulations "The level of modelling/simulations has for sure increased in comparison with previous year. Experimental data is now also more available from cars and trucks that are driving around for a full year and collecting irradiance and PV-performance continuously. Many presentations show the enhanced range by making use of VIPV.Difficult to find one specific highlight. Maybe the remark from Bonna Newman that payback time of a PV-system on the top part of a diesel truck is now between 2 and 5 years, which makes it interesting for companies to buy that system; today already.
6DO.9 Vehicle Integrated PV and Floating PV Applications Glass-free solutions for crystalline silicon PV modules for integrated PV applications are being demonstrated and are maturing.
6CV.4 PV on/in Buildings / PV in Infrastructure, on Water and on Vehicles; PV and Agriculture New concepts for building skin systems (prefab, standardized), some mature cases with implemented technologies (shadings, coloured claddings), and different ongoing researches on novel components (half-laminate concept, modules with flexible solar cells, etc.) were presented. BIPV characterization and performance assessments were also presented by reporting results of test campaigns and modelling approaches as well as some economic analysis to support cost-competitiveness. Among the other PV applications, the topics of agrivoltaics, floating and vehicle integrated PV appeared in different contributions. Some new investigation aspects like designing nature inclusive solar parks and the analysis of solar potential for vehicle integrated PV highlighting that, along with BIPV as key-strategy for solar exploitation, also PV integration on infrastructures, urban and landscape areas are emerging sectors that can support the energy transition."

EU PVSEC European Photovoltaic Solar Energy Conference and Exhibition

6 APPLICATIONS, INTEGRATION AND STORAGE

6CO.11 Solar PV in Energy System
In this session, six high quality oral presentations demonstrated that Solar PV can contribute to a large amount to the transformation of the energy system as part of the Energy Transition to achieve the 100% RES target by 2050 in different locations in Europe: First, Solar PV is predictable and new prediction methods improve the quality of short term power data such as machine learning (6CO.11.2 Optimized Machine Learning Method for PV Power Prediction). As such, Solar PV plays a significant role in Energy Transition scenarios at regional level associated with heat pumps, storage and grid interconnections to export and import electricity (6CO.11.3 Energy Model for a Rural Region in Germany - Results and Balancing of Electricity Production and Consumption) and also at European level in association with wind energy, hydro power and biomass (6CO.11.5 Sector Coupled Energy Model for the European Electricity, Heat and Transport Sectors - Methodology). The amount of electricity that can be produced by PV on buildings, land or water can be assessed at different scales with sophisticated algorithms (6CO.11.1 Future PV Supply in the Netherlands: Spatially Resolved Profiles for Building, Land and Water-Bound Installations Considering Three Energy Transition Scenarios). Also, Solar PV can also be a solution for the mobility sector. It can bring benefits when used to power EV charging stations for e-bus for instance (6CO.11.6 Energy Management System for Electric Bus Charging Hub with Local Storage and PV Energy Integration) or for any other types of e-vehicles with also the possibility to produce hydrogen (6CO.11.4 Hybrid PV-Systems for Electrification and Sector Coupling of Road Transport Sector in Norway). To achieve the 100% RES target by 2050 it was demonstrated that in many regions of Europe it also requires sector coupling with the heat and transport sector and often energy storage is necessary."

EU PVSEC
European Photovoltaic Solar Energy Conference and Exhibition

6 APPLICATIONS, INTEGRATION AND STORAGE

6CO.13 Solar Hydrogen There are technologies to expand for the PV sector and enough potential for growth respecting the needs of future generations.• Electrolysers remained a unique technology to eco-friendly produce hydrogen as emission-free fuel. Perovskite and PVT come closer and join to expand the bandwidth. Talking in numbers:Solar to Hydrogen (STH) efficiency of 20 % was demonstrated with a Perovskite / Silicon Tandem solar cell in one device with the electrolyser by the University in Canberra, Australia and followed by a 17 % efficient STH eff using CIGS thin film tandem solar cell systems with integrated electrolyser by the European Joint Project PECSYS presented by the Helmholtz Center Berlin. Efficiency seems to be good enough to go on for devices as PV power is reducing in price and availability for utility grid as well as remote applications.• In 2030 from the demand side the electricity price could fall well below today's 4 reaching 1 Eurocent per kWh leading to storage and Solar Hydrogen solutions driven at daytime with excess of PV electricity. Combined with Desalination the human issue of solving water supply in more and more aridizing zones there are solutions with or without supporting or decoupling storage systems possible. Finally the technical solutions are there, let´s go towards the funds to introduce and establish them to societal dimension.

EU PVSEC
European Photovoltaic Solar Energy Conference and Exhibition

7 APPLICATIONS, INTEGRATION AND STORAGE

7DO.12 PV Business Opportunities Today and Tomorrow (7DO.12.1) PV markets are growing strongly with very soon 200 GW/yr annual installations in the world, maybe in 2027. (DO.12.2) BIPV shows substantial potential, which needs to be activated and is influenced by renovation rates in the building stock. (7DO.12.4) c-Si/perovskite tandem modules with 30% efficiency and competitive module pricing are expected to enter the market by end of this decade7DO.12.6 PV-based hydrogen is increasingly a topic as competitiveness is approaching very fast. Dynamics of all components involved can be handled well. Applications to be expected can be off-grid and on-grid."

7EO.1 Policies, Scenarios and Instruments for Large Scale PV Deployment (7EO.1.2) An early transformation pays off; Building Criteria for the Energy Labelling of Photovoltaic Modules and Residential Systems (7EO.1.1)

7DV.1 Economics, Markets and Education The profitability (and sustainability) of PV systems from small scale integrated to large scale plants is already demonstrated and can be further optimised.

This is a short version of the paper submitted to the journal Progress in Photovoltaics.
It has been selected by the Executive Committee of the EU PVSEC 2021 for submission to Progress in Photovoltaics.

AMORPHOUS SILICON-FREE HETEROJUNCTION CRYSTALLINE SILICON SOLAR CELLS EMPLOYING MoO_X AS HOLE-SELECTIVE AND PASSIVATING CONTACT

Suren Patwardhan[1,2] , Sandeep Maurya[1] , Akash Kumar[1], and Balasubramaniam Kavaipatti[1]
[1]Department of Energy Science and Engineering Indian Institute of Technology Bombay, Mumbai, India
[2]K J Somaiya College of Engineering, Mumbai India

ABSTRACT: Dopant-free heterojunction silicon solar cells are fabricated using sub-stoichiometric molybdenum trioxide (MoO_x) as passivating as well as hole-selective layer for n-type silicon. The average carrier lifetime obtained for 50 nm MoO_x film directly evaporated on both sides of a Si wafer was nearly 100 μs with corresponding implied V_{OC} of 634 mV. In practice, the cells exhibited V_{OC} in the range 590-608 mV and J_{SC} in the range of 30-34 mA/cm^2 without texturization and antireflecting coating. The best cell reached efficiency of 13.6% (14.02% effective efficiency). This is almost 6% absolute rise from our earlier report and close to the best efficiencies reported so far employing direct heterojunction between MoO_x and Si without any passivation layer. However, post-deposition protection of MoO_x layer from ambient conditions holds the key for maintaining high V_{OC} values of these cells as exposure to air showed drastic reduction in their carrier lifetime and accordingly low values of V_{OC}. Although the passivation quality of MoO_x alone does not match to that of intrinsic amorphous silicon, this route significantly reduces the processing steps and avoids use of hazardous gases making the process truly cost-effective.
Keywords: see the list of keywords c-Si, Heterojunction, Lifetime, Passivation

1 INTRODUCTION

In recent years, sub-stoichiometric molybdenum oxide (MoO_x, $2 < x < 3$) has emerged as a prospective dopant free hole-selective layer for Si [1]. Reducing parasitic absorption losses was the primary reason of using MoO_x in place of front boron-doped amorphous Si [a-Si(p^+):H] layer in the typical Heterojunction Intrinsic Thin layer Si solar cell (HIT cell) [2]. In addition to the higher transparency due to its larger band gap (~ 3 eV) compared to a-Si (1.7 eV), use of MoO_x is also advantageous as it can be deposited using simpler and cost-effective deposition techniques. Physical and chemical vapor deposition techniques such as evaporation, sputtering, and solution processing, to name a few have been utilized for this purpose. The typical Carrier-Selective Contact (CSC) solar cell architecture with a-Si(i):H passivation layer on both front and back surfaces of Si has indeed achieved very high efficiencies when MoO_x is used on the front side [3, 4] retaining the n-doped electron selective layer [a-Si(n^+):H] at the rear side. Incorporation of these a-Si layers, even though necessary to obtain highly efficient MoO_x/a-Si(i)/n-Si (uni-functional MoO_x) solar cells, constitutes additional procing steps. Along with the increased fabrication complexity, they are also a harbinger of various detrimental effects such as, degradation of hole selectivity of MoO_x due to H-effusion from a-Si:H layer/s, restriction on annealing temperature, higher risks (and hence maintenance) involved due to use of hazardous gases [5, 6] etc. These issues are liable to hamper expansion of uni-functional MoO_x solar cells to a broader scale. On the other hand, MoO_x itself works as good passivating layer for Si [7, 8]. A large band offset exists between MoO_x and n-type Si due to very high work function (> 5.5 eV) of the former, resulting in barrier for electrons in the conduction band of Si providing a kind of field-effect passivation. When this barrier is carefully engineered, MoO_x can emerge as excellent passivating layer for Si in addition to its use as hole selective-layer [8]. Still, there is a limited literature on using MoO_x for bi-functionality (hole-selective and passivating) and all the attempts made so far have reported efficiencies much lower than the uni-functional MoO_x solar cells. The major focus of this work is to understand the Si-MoO_x interface and identify factors that affect performance. An over-arching goal would be to attain efficiencies comparable to the best reports of uni-functional MoO_x solar cells. We presented a double-sided dopant-free Hetero Junction CSC (HJ-CSC)-Si solar cell having MoO_x and TiO_x films as hole and electron selective contacts respectively reaching 8% efficiency at the 35th EU-PVSEC [9]. The current work is an extension of the same except for the replacement of TiO_x layer by n^+ Si homojunction as back contact in order to focus on optimisation of one interface at a time namely that of Si-MoO_x.

2 EXPERIMENTAL DETAILS

1-3 Ω-cm (1 0 0) oriented 180 μm thick as-cut n-type Fz-Si wafers were used for the experiments. Their average bulk lifetime was about 800 μs as measured using quinhydrone in methanol solution. These 125 mm x 125 mm wafers were pre-conditioned to get n/n^+ high-low homojunction as the back contact and deposited on both sides by 100 nm silicon nitrite (Si_3N_4) to protect the surfaces. The devices are made on small area (about 1.5 cm x 1.5 cm) samples scribed out from these pre-conditioned wafers. Cells were made by evaporating MoO_3 powder source (99.97%) directly on n-side of the Si samples using golve box integrated thermal evaporator (M-BRAUN, MB 20G). After MoO_3 deposition, samples were immediately loaded into sputter chamber for deposition of 80 nm RF sputtered ITO over-layer at room temperature. The Si_3N_4 layer was etched just before MoO_3 evaporation. After ITO deposition, both front and back contacts were made by thermally evaporating silver on full area at the back surface (n^+ Si side) and through a shadow mask at the front surface (ITO coated side). Finger-busbar pattern accounted for about 12% shadow losses. Carrier lifetime measurements were performed using quasi-steady-state photoconductance (QSSPC) based method (Sinton Instruments, WCT-120). A solution of quinhydrone ($C_{12}H_{10}O_4$) dissolved in methanol was prepared for bulk lifetime measurement. Dark and light I-V measurements were carried with class AAA solar simulator (ABET, Sun 3000) using standard AM 1.5G filter and spectrum was calibrated using WPVS reference cell (Fraunhofer). Quantum Efficiency measurements were done using QE tool (Bentham Instruments, PVE 300). Transmission electron microscopy (TEM) was performed in Cross-section TEM mode with EDS line-scan profile for elemental distribution on 300 kV FEI Tecnai G2, F30 instrument having ultimate resolution of 2 Å.

3 RESULTS AND DISCUSSION

In this section, we reslts of experiemtnal work at material level for MoOx thin film parameters ans at device level for MoOx/n-Si diodes/solar cells are presented and discussed

3.1 Passivation study of thin films of MoO_x deposited directly on Si by thermal evaporation

The carrier lifetime data was extracted using photo conductance method using QSS or transient mode wherever applicable. The plot of effective carrier lifetime as a function of minority carrier density for different MoO_x film thicknesses is given in Fig. 1. An improvement in the the effective carrier lifetime (τ) with MoO_x film thickness can be observed from the figure.

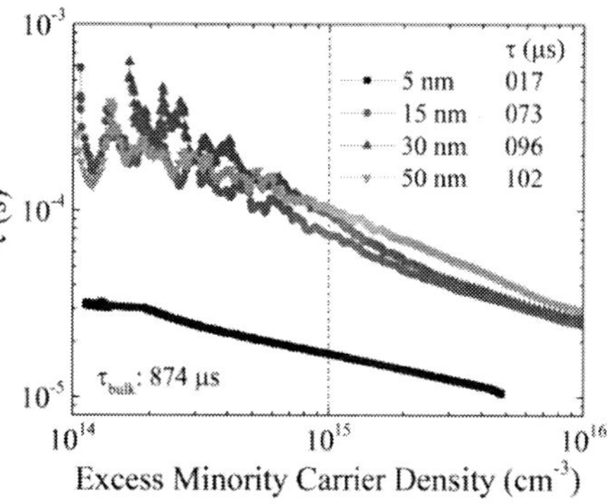

Figure 1: Variation of effective minority carrier lifetime (τ) with excess photo-generated minority carrier density for different MoO_x film thicknesses.

As these samples were prepared by thermally evaporating MoO_x films on Si, for very low thickness (5 nm), the Si surface is not completely covered with MoO_x film, which possibly resulted in poor passivation giving lower carrier lifetime (τ) values. For relatively thicker films, the rise in τ is attributed to complete passivation of dangling bonds at the Si surface. However, the observed variation of carrier lifetime with thickness of MoO_x film is not just a consequence of issues with uniform deposition of evaporated films. The passivation of Si by MoO_x is partially field effect and partially chemical but for thinner films, it is the chemical passivation, which decides the effective surface recombination and hence, τ [10]. However, with increasing thickness, there is a contribution of field effect also as for thicker films, the bulk effect comes into picture. This transition from chemical passivation dominant phase to the combination of chemical and field effect passivation seems to occur for MoO_x thickness of 10-15 nm onward as a significant rise in τ takes place for MoO_x thickness going up from 5 to 15 nm but beyond 15 nm, τ increases only marginally. This is also apparent from the variation in implied open circuit voltage (iV_{OC}) with MoO_x thickness given in Fig. 2. Inset of Fig. 2 gives corresponding variation in minority carrier lifetime measured at 10^{15}/cc excess Minority Carrier Density (MCD). Inset of Fig. 2 also indicates carrier lifetime obtained with 50 nm Si_3N_4 film used for comparison of results. The effectiveness of field effect passivation primarily depends upon work function of MoO_x (higher the work function, better the passivation), which basically corresponds to higher O:Mo ratio in the film [10]. This ratio tends to decrease towards Si surface due to higher affinity of Si towards oxygen and the interface can consist of intermixed Mo-SiO_x states. When the film thickness is 10-15 nm or beyond, the higher O:Mo ratio of the region of MoO_x film that is sufficiently away from the interface assists in setting field effect passivation.

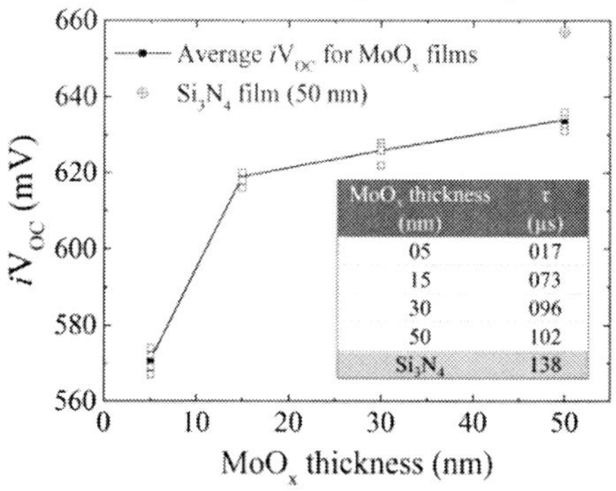

Figure 2: Variation of implied V_{OC} with MoO_x thickness. Also indicated iV_{OC} for 50 nm Si_3N_4 film. Table (inset) gives corresponding carrier lifetime values.

In order to test the performance of MoO_x film to passivate Si surface, we also measured τ with other materials. Variation of τ with MCD for passivation using different materials is given in Fig. 3. We compared the passivation quality of MoO_x films with the passivation using the solution of quinhydrone ($C_{12}H_{10}O_4$)/methanol and SiN (Si_3N_4) films. This data can be used as a reference for bulk lifetime of Si (mainly τ_{SRH}). SiN passivation of Si surface, particularly, n-type Si is also an established surface passivation scheme adopted in Si solar cell fabrication. MoO_x, on the other hand, is still under investigation about its true potential as a passivating layer on its own for Si [11].

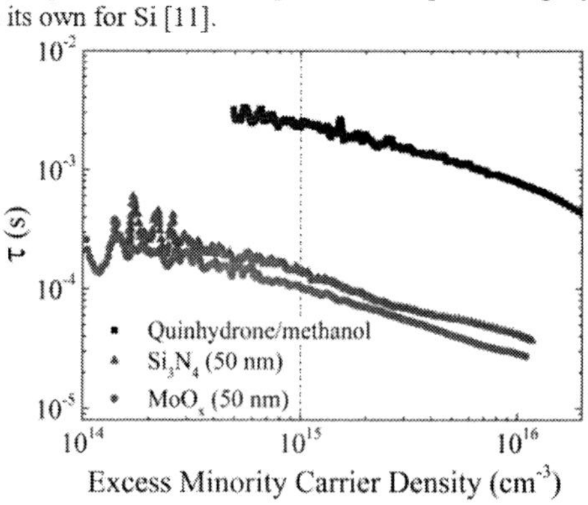

Figure 3: Carrier lifetime as a function of excess minority carrier density for different passivation schemes.

The iV_{OC} obtained from Si wafers immersed in quinhydrone/methanol solution offered the best possible value of 720 mV at 1 Sun (AM 1.5G radiation). This value is a direct reflection of the quality of Si wafer. SiN passivation provided iV_{OC} of 656 mV, which might be limited due to the quality of SiN PECVD deposition. MoO_x films could reach reasonable values of 634 mV for 50 nm film and 623 mV for 30 nm film, which are comparable with the best reported values in the literature involving passivation of Si using only MoO_x [12, 6]. The effective lifetime obtained with MoO_x films is controlled by the surface recombination for high bulk lifetime wafers such as those used in our case. The effective carrier lifetime is given as per the equation [13]:

$$\tau = \frac{1}{\tau_{SRH}^{-1} + \tau_{Radiative}^{-1} + \tau_{Auger}^{-1} + \tau_{Surface}^{-1}}$$

For Si, radiative processes are usually neglected. Further, for low doping densities in Si (< 10^{17}/cc) and for low-level injection, the Auger recombination can also be neglected. This makes $\tau \approx \frac{1}{\tau_{SRH}^{-1}+\tau_{Surface}^{-1}}$. As τ_{SRH} is large for our Si wafers, τ can be expressed as $\tau \approx \frac{1}{\tau_{Surface}^{-1}} = \frac{W}{2S}$ [14]. where, W is wafer thickness and S is the surface recombination velocity. The best value of carrier lifetime measured with MoO_x film (50 nm) was around 102 µs at 10^{15}/cc MCD. The wafer thickness was 150 µm. This gives surface recombination velocity of 79 cm/s, which is a (fairly) low value (< 100 cm/s) considering the condition of Si surface that we were using. Hence, it can be concluded that MoO_x offers a good passivating for as-cut Si wafers (after saw damage removal).

3.2 Interface study of Si-MoO_x junction

Presence of spontaneously formed thin silicon suboxide (SiO_x) layer between MoO_x and Si has been previously reported [15]. As the hole selection from Si by MoO_x has to happen through this inter-layer, a detailed structural analysis of this thin SiO_x inter-layer is required to understand the exact nature of current transport through it. Results of HRTEM imaging and EDS spectra were combined to determine elemental distribution over the inter-layer. This EDS is performed along with HRTEM only as supportive data. The film edges are confirmed by TEM images only as EDX has a spot size of 90 nm and it is not as sensitive enough for alalysing a 20 nm film. For this analysis, n-type (1 0 0) polished Si wafers were used and sequentially deposited with a thicker (20 nm) MoO_x film and about 50 nm ITO. These samples were stored in desiccator for almost a week with the intention that the intermixing layer will reach a stable configuration with its maximum thickness across the Si-MoO_x interface.

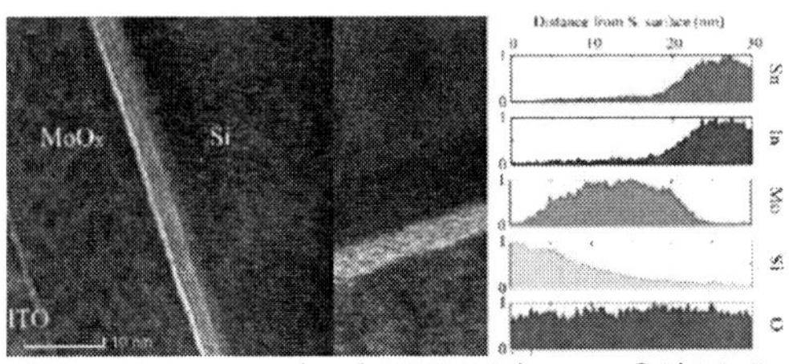

Figure 4: Cross-sectional HRTEM image of Si-MoO_x-ITO interface (leftmost). Figure in middle is the same image magnified. EDS profile is on the rightmost side.

The top ITO layer in a sense protects the Si-MoO_x interface by isolating the whole MoO_x layer from exposure to ambient and other contaminants. Samples for TEM imaging were freshly cut from these main samples. The HRTEM image of the interface along with EDS scan is shown is Fig. 4, which shows distribution of elements in the selected interface area. It also includes parts of Si substrate and ITO layer.

It is noticed that the inter-layer between MoO_x and Si has a spread about 3-4 nm, which is wider than that reported in the literature, which show it around 2 nm when TEM imaging is done either immediately or in-situ [15]. It was mentioned that about a week's time was given before the sample preparation and imaging for TEM. This 3-4 nm grown interlayer thickness can be considered as the stabilized thickness with full intermixing of both layers. A very important observation is about 5 nm of the inter-layer towards Si surface contains not only Si and O signals but significant Mo intensity also. So the inter-layer cannot be SiO_x with only oxygen vacancies, which is usually speculated to support tunnelling through it. Its thickness is too large to allow a large tunnelling through the oxide barrier between Si and MoO_x in the form of hole selection/electron injection. There could be high density of other defects such as Mo interstitial in SiO_x and Si interstitial in MoO_x networks apart from oxygen vacancies, which collectively aid the current transport. This inter-layer can be termed as intermixed a-SiO_x(Mo) instead of simply SiO_x. However, other supporting experiments such as STEM-HAADF analysis and XPS depth profiling are required to make conclusive statements.

3.3 J-V characteristics with variation in MoO_x thickness

Results of illuminated J-V characteristics of diodes fabricated with different thicknesses of MoO_x film are shown in Fig. 5.

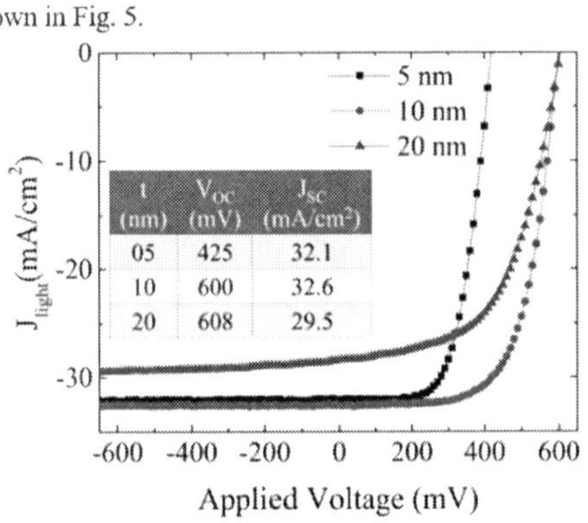

Figure 5: Effect of variation of MoO_x layer thickness on light J-V of MoO_x/n-Si didoes. The values given in the table (inset) are averages.

In the dark (not shown here), typical diode characteristics were obtained for film thickness in the range of 10-15 nm, which showed higher knee voltage exceeding 500 mV and current density around 2 µA/cm² at 1000 mV reverse voltage. These diodes showed higher forward to reverse current ratio as compared to diodes with film thickness of 5 nm. Diodes with 5 nm film thickness suffered from low knee voltage. On the other hand, diodes with film thickness of 30 nm showed heavy series series resistance beyond 500 mV. Accordingly, similar behaviour is reflected in the J-V characteristics under illumination as shown in Fig. 5. Solar cells fabricated with 5 nm MoO_x layer exhibited low open circuit voltages (V_{OC}) ~ 400-425 mV while for thicker MoO_x films (≥ 10 nm), V_{OC} reached or exceeded 600 mV. MoO_x/n-Si didoes with 5 nm MoO_x layer show behaviour similar to a low work function Schottky diode. One possibility of the formation of Schottky diode is lack of uniform or complete coverage of Si surface by MoO_x layer. This would lead to formation of Schottky contact between n-type Si substrate and the metal layer evaporated after MoO_x (we used Ag) at least at some locations over the area of Si wafer selected. However, the Schottky behaviour was exhibited irrespective of the designated area of diode fabricated for J-V measurements (diode area was varied from 2 mm × 2 mm to 10 mm × 10 mm). Moreover, in the corresponding light J-V data of cells, the difference in V_{OC} between different cells fabricated over a total area of 20 mm × 20 mm area Si sample was small (20-30 mV). These two observations suggest some different reason for low V_{OC} at low thickness rather than only the non-uniformity in thickness playing the major role. A possibility of formation of intermixed a-SiO_x(Mo) layer is mentioned in earlier subsecion. Role of this intermixed layer is important in the hole transport. There are various models proposed such as due to oxygen vacancies, deep lying defect states-assisted electron suppression or hole transport via Mo at Si cites in the a-SiO_x layer [16, 17]. However, the effect of this inter-layer will be the same irrespective of thickness of MoO_x layer deposited. From TEM imaging it was revealed that this inter-layer grows up to 4 nm. For thinner MoO_x layers (~ 5 nm or less), almost the entire MoO_x layer would turn into the intermixed state and the hole extraction from Si is then dependent only upon this layer. On the other hand, for thicker MoO_x films, part of the MoO_x layer, which is sufficiently away from away from Si interface (> 5 nm) is only sparsely reduced (x in $MoO_x \approx 2.7$-2.8). This region provides sufficient barrier for blocking electrons as compared to region closer to Si surface. The electron barrier helps in enhancement of hole extraction and increase in V_{OC}. Further, extraction of holes from Si depends upon the presence of sub-band-gap states in MoO_x near the VB of Si. For MoO_x films more than 5 nm thick, density of these states in nearly independent of thickness. Transport of holes from Si to MoO_x occurs via multi-step tunnelling through the thin a-SiO_x(Mo) layer. Once these holes are in MoO_x, the transport takes place via MoO_x sub-band gap defect states, which are almost continuous throughout the MoO_x (Like an acceptor level in doped Si). Therefore, hole extraction remains efficient. However, by increasing thickness, current decreases due to increased lateral resistance offered by MoO_x [18]. Typically, for MoO_x layers beyond 15 nm thickness, the series resistance of MoO_x itself becomes dominating parameter to alter J-V characteristics. Based on these results, we fin that MoO_x thickness in the range of 10-15 nm is ideal for bifunctional MoO_x in MoO_x/n-Si solar cells.

3.4 Analysis of 14% efficient MoO_x/n-Si solar cell with area > 2 cm^2

In this section, we present the partially optimized solar cell of area > 2 cm^2 employing MoO_x as hole selective contact to n-Si without using any interlaying passivation layer. The 5-inch as-cut wafers used in this study are subjected only to indigenously developed chemical etching. We did not texturize these wafers. Despite of this, we could reach nearly 14% efficiency, which is rarely reported for a direct heterojunction between MoO_x and Si having cell area > 2 cm^2. The cell could reach J_{SC} of 33.9 mA/cm^2, V_{OC} of 594 mV with effective efficiency reaching 14.1% using integrated J_{SC} of 35.2 mA/cm^2 as calculated from the EQE curve. The EQE curve along with dark and light J-V curve is shown in Fig. 6. The J_{SC} value of 35 mA/cm^2 obtained from EQE is significant considering the fact that the surface was un-textured. Moreover, we did not perform any rigorous simulation to fix the exact thicknesses of top MoO_x and ITO layers.

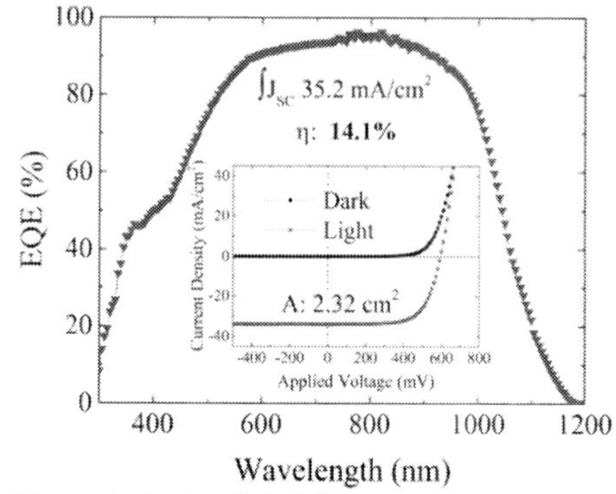

Figure 6: EQE and dark/light J-V characteristics of the best-performing solar cell in this work.

From Fig. 6, it can be noted that EQE stayed close to 90-95% in the wavelength range 600- 900 nm, which typically corresponds to the bulk of Si wafer. This high EQE is an indication of good bulk quality of the Si wafers used, which resulted in maximum photogeneration of carriers. However, EQE shows rapid reduction below 700 nm. The losses are particularly more in the range of 400-550 nm. These losses are critical as the peak of AM1.5G radiation used for determination of efficiency and EQE has a maximum at 550 nm. This leads to higher losses in terms of integrated J_{SC} as the major contribution to J_{SC} is chopped-off due to lower EQE in this region. Using the AM 1.5G standard radiation data, we estimated that a loss of about 3 mA/cm^2 to J_{SC} has occurred in this region. We speculate at least three reasons for this loss. First is due to the poor surface conditioning, second is due to un-passivated back contact (n^- region and metal) and third is the higher reflective losses from the front surface for wavelengths 400-600 nm. Out of these, the major reason for the loss of J_{SC} is the reflection losses at the front surface. Un-textured surface and/or improper ITO+MoO_x thickness seem to be main reasons for this higher reflection at the short wavelengths (< 500 nm). V_{OC} exhibited by this cell stayed at 594.4 mV, lower than the highest V_{OC} obtained from cells within the same batch, which is 608 mV. These differences are small and they mainly arise from variation in surface conditioning during the fabrication steps. The fill factor of the cells could not improve beyond 67%. This is low as compared to similar configuration of direct heterojunction between MoO_x and Si reported in the literature [6]. The ITO layer resistivity in our samples is fair enough with sheet resistance of 57 Ω/square for a 100 nm film.The main reason for low fill factor in our case is mostly due to un-optimized metalization. In order to understand the overall performance of the solar cell, we used two-diode equivalent model of solar cell to extracct diode parameters. These are given in Table I. The recombination factor J_{01} extracted from J-V measurements is close to that extracted from carrier lifetime data. This factor corresponds to recombination of one type of charge carrier (excess minority holes in n-type Si), which depends upon the bulk lifetime of Si wafers used. The second diode with ideality factor n_2 corresponding to recombination factor J_{02} represents recombination due to both types of charge carriers. It can take place in the thin a-SiO_x(Mo) region. Due to oxide traps and various defect levels in this region, once the excess electrons from the CB of Si get into here it would lead to increased J_{02} factor.

Table I: Extracted parameter from dark I-V data of 14% MoO_x/n-Si solar cell.

Parameter	Extracted Value
J_{01}	2.25 x 10^{-12} A/cm^2
J_{02}	1.44 x 10^{-6} A/cm^2
n_1	1.09
n_2	2.32
R_S	650 mΩ-cm^2
R_{SH}	15 kΩ-cm^2

Defect states present at the Si surface due to inefficient passivation or impurities present due to surface contaminants may provide such pinholes for excess (or photo-generated) CB electrons from Si to escape into the a-SiO_x(Mo) region. n_2 exceeds two if there are additional recombination mechanisms apart from recombination in this a-SiO_x(Mo) layer. One cause is the edge recombination. Another is due to high surface roughness. In our case, the latter reasons seems to play a major role. Overall, the cell performance is limited by two main reasons: hight J_{02} factor and lower fill factor.

4 CONCLUSIONS

We demonstrated that MoO_x layers of thickness 10-15 nm are required for its usage as passivating as well as carrier selective layer for Si. The samples showed iV_{OC} in the range 618-623 mV and J_0 factor close to 1 pA/cm^2 for 10 nm thick MoO_x film deposited via thermal evaporation on Si. MoO_x/n-Si solar cells could reach maximum V_{OC} of 608 mV and maximum J_{SC} of 33.94 mA/cm^2. The best-performing cell exhibited an integrated EQE-effective efficiency of 14%. For direct heterojunction between Si and MoO_x, this high V_{OC} and efficiency are rarely reported by using only chemically etched as-cut Si wafers. The efficiency was however, limited by low fill factor due to ineffective contact annealing and high front reflection due to the use of un-textured wafers. The HRTEM images showed thicker (3-4 nm) interlayer at the Si-MoO_x interface, which is composed of sub-stoichiometric SiO_x. Ths interlayer controls the passivation and hole selectivity. Very thin (≤ 5 nm) MoO_x layers predominantly consists of such intermixed state and result in lower V_{OC}. Relatively thicker (10-15 nm) layers possess sub-stoichiometric MoO_x states, which 324 enhance V_{OC} by 175-200 mV. This is attributed to the field effect passivation provided by the sub-stoichiometric MoO_x region in thicker layers.

5 ACKNOWLEDGEMENTS

This work was carried under National Centre for Photovoltaic Research and Education (NCPRE), a project funded by the Ministry of New and Renewable Energy (MNRE), Government of India. We are thankful to the Centre for Excellence in Nanotechnology (CEN), Sophisticated Analytical Instrumentation Facility (SAIF) and Industrial Research and Consultancy Center (IRCC), IIT Bombay for providing various fabrication and characterization facilities.

6 REFERENCES

[1] Martin Bivour, Jan Temmler, Heiko Steinkemper, and Martin Hermle. Molybdenum and tungsten oxide: High work function wide band gap contact materials for hole selective contacts of silicon solar cells. Sol. Energy Mater. Sol. Cells, 142(SI):34-41, 2015.

[2] Corsin Battaglia, Silvia Martin de Nicolas, Stefaan De Wolf, Xingtian Yin, Maxwell Zheng, Christophe Ballif, and Ali Javey. Silicon heterojunction solar cell with passivated hole selective MoO_x contact. Appl. Phys. Lett., 104(11):113902, 2014.

[3] Julie Dreon, Quentin Jeangros, Jean Cattin, Jan Haschke, Luca Antognini, Christophe Ballif, and Mathieu Boccard. 23.5%-efficient silicon heterojunction silicon solar cell using molybdenum oxide as hole-selective contact. Nano Energy, 70, 2020.

[4] Jonas Geissbuehler, Jeremie Werner, Silvia Martin de Nicolas, Loris Barraud, Aicha Hessler-Wyser, Matthieu Despeisse, Sylvain Nicolay, Andrea Tomasi, Bjoern Niesen, Stefaan De Wolf, and Christophe Ballif. 22.5% efficient silicon heterojunction solar cell with molybdenum oxide hole collector. Appl. Phys. Lett., 107(8), 2015.

[5] Stephanie Essig, Julie Dreon, Esteban Rucavado, Mathias Mews, Takashi Koida, Mathieu Boccard, Jeremie Werner, Jonas Geissbuhler, Philipp Loper, Monica Morales-Masis, Lars Korte, Stefaan De Wolf, and Christophe Ballif. Toward annealing-stable molybdenum-oxide-based hole-selective contacts for silicon photovoltaics. Solar RRL, 2(4), 2018

[6] Mrutyunjay Nayak, Krishna Singh, Sapna Mudgal, Sourav Mandal, Sonpal Singh, and Vamsi K. Komarala. Carrier-selective contact based silicon solar cells processed at room temperature using industrially feasible Cz wafers. Phys. Status Solidi A, 216(16), 2019.

[7] Luis G. Gerling, Cristobal Voz, Ramon Alcubilla, and Joaquim Puigdollers. Origin of passivation in hole-selective transition metal oxides for crystalline silicon heterojunction solar cells. J. Mater. Res., 32(2):260-268, 2017..

[8] Zhimin Liang, Mingze Su, Yangyang Zhou, Li Gong, Chuanxi Zhao, Keqiu Chen, Fangyan Xie, Weihong Zhang, Jian Chen, Pengyi Liu, and Weiguang Xie. Interaction at the silicon/transition metal oxide heterojunction interface and its effect on the photovoltaic performance. Phys. Chem. Chem. Phys., 17(41):27409-27413, 2015

[9] Suren Patwardhan, Sandeep Maurya, Akash Kumar, Balasubramaniam Kavaipatti, "Amorphous Silicon-Free Metal-Oxides Based Carrier Selective Contacts to Crystalline Silicon Solar Cells," in 35th EU-PVSEC, 2018, pp. 657–659.

[10]Swasti Bhatia, Irfan M. Khorakiwala, Pradeep R. Nair, and Aldrin Antony. Influence of post deposition fabrication steps and quantitative estimation of band diagram of Si/MoO_x heterojunction for carrier selective solar cells. Sol. Energy, 194:141-147, 2019.

[11]Haider Ali, Celine Maynau, Luc Lajaunie, Geoffrey Gregory, Lijun Wu, John Patrick Looney, Yimei Zhu, Matthew Schneider, Winston V. Schoenfeld, and Kristopher O. Davis. Transmission electron microscopy and electron energy-loss spectroscopy studies of hole-selective molybdenum oxide contacts

in silicon solar cells. ACS Appl. Mater. Interfaces, 11(46):43075-43080, 2019.
[12]G. Gerling Luis, Mahato Somnath, Morales Vilches Anna, Masmitja Gerard, Ortega Pablo, Cristobal Voz, Alcubilla Ramon, and Puigdollers Joaquim. Transition metal oxides as hole selective contacts in silicon heterojunctions solar cells. Sol. Energy Mater. Sol. Cells, 145:109-115, 2016.
[13]D.K. Schroder. Carrier lifetimes in silicon. IEEE Trans. Electron Devices, 44(1):160-170, 1997.
[14]Andres Cuevas and Daniel Macdonald. Measuring and interpreting the lifetime of silicon wafers. Sol. Energy, 76(1):255-262, 2004.
[15]Haider Ali, Geoffrey Gregory, Martin Bivour, Matthew Schneider, Martin Hermle, and Kristopher O. Davis. TEM studies of hole-selective molybdenum oxide contacts in silicon heterojunction solar cells. Microsc. Microanal., 24(S1):1508-1509, 2018.
[16]Ming Gao, Dongyun Chen, Baichao Han, Wenlei Song, Miao Zhou, Xiaomin Song, Fei Xu, Lei Zhao, Yonghua Li, and Zhongquan Ma. Bifunctional hybrid a-SiO_x(Mo) layer for hole-selective and interface passivation of highly efficient MoO_x/a-SiO_x(Mo)/n-Si heterojunction photovoltaic device. ACS Appl. Mater. Interfaces, 10(32):27454-27464, 2018.
[17]Joongoo Kang, Yong-Hoon Kim, Junhyeok Bang, and K. J. Chang. Direct and defect-assisted electron tunneling through ultrathin SiO_2 layers from first principles. Phys. Rev. B, 77:195321, 2008.
[18]Jinyoun Cho, Neerja Nawal, Afshin Hadipour, Maria Recaman Payo, Arvid van der Heide, Hariharsudan Sivaramakrishnan Radhakrishnan, Maarten Debucquoy, Ivan Gordon, Jozef Szlufcik, and Jef Poortmans. Interface analysis and intrinsic thermal stability of MoOx based hole-selective contacts for silicon heterojunction solar cells. Sol. Energy Mater. Sol. Cells, 201, 2019.

Optimization of Carbon-Nanotube-Reinforced Composite Gridlines Towards Commercialization

Andre Chavez[1,2], Brian Rummel[1,2], Nicolas Dowdy[1], and Sang M. Han[1,2]
Nick Bosco[3]
Brian Rounsaville[4] and Ajeet Rohatgi[4]

[1]University of New Mexico, Albuquerque, NM 87131
[2]Osazda Energy, Albuquerque, NM 87102
[3]National Renewable Energy Laboratory, Golden, CO 80401
[4]Georgia Institute of Technology, Atlanta, GA 30332

Abstract — **Solar cells in PV modules crack during field operation, which exposes the modules to environmental stressors, including extreme weather events, such as hailstorms and hurricanes. These cracks gradually lead to power degradation. To directly address the cell-crack-induced degradation, we have reinforced the traditional screen printable silver paste, commonly used throughout the industry for front side contacts, with low-cost multiwalled carbon-nanotubes. We have shown in previous work that these metal matrix composites have little to no effect on the cell's efficiency while enhancing the metallization's fracture toughness and electrical gap-bridging capability. In this work, we focus on optimizing the composite metallization to achieve the best possible performance. We discover that reducing carbon nanotube agglomerations in the paste not only improves the printability for narrow gridlines, but also increases the modulus of toughness of the metallization by over 500%.**

I. INTRODUCTION

The PV industry continues to grow, and U.S. solar installations in the next five years are expected to exceed 107 GWdc [1]. The growth is attributed primarily to large utility-scale solar projects. These projects increasingly demand low-cost, high-performance, high-reliability modules whose useful lifetime exceeds well over 25 years.

The durability of a PV module is measured by their degradation rate, a slow gradual loss of performance over time [2]. This gradual loss is an important factor in the cost model of any PV project. One of the most commonly reported degradation modes in PV, with high severity, is hot spots [3], in which the affected modules show significantly higher power loss. The suspect root cause of hot spots includes solder bond failure and cracked cells [4, 5]. PERC cells are one of the most popular cell architectures used today and can easily crack in a module when subjected to environmental stressors. These cracked cells can limit the current flow through the cell, resulting in hot spots within the module.

The PV field owners and insurers are often responsible for replacing modules that are underperforming due to cell cracks. To directly address this cell-crack-induced module power loss, we have developed a method to incorporate low-cost, multi-walled carbon-nanotubes (MW-CNTs) into commercial, screen-printable silver pastes for terrestrial PV [6, 7]. The CNT-enhanced metal matrix composite (MMC) silver pastes match fineness of grind (FOG) and viscosity of commercial pastes, making them an easily integratable solution within the preestablished manufacturing process flow. The screen printing and firing schedule of the MMC metallization are virtually identical to the commercial paste without compromising cell performance [8].

We have previously shown that our MMC pastes provide increased fracture toughness, electrical gap-bridging, and "self-healing" [9-11] that would enhance module lifetime against cell cracks. In this work, we focus on optimizing the MMC formulation, such that there are no material defects such as CNT agglomerations (see Fig.1). We begin with optimizing the surface functionalization chemistry and mixing techniques to promote CNT adhesion to the surrounding silver matrix while dispersing CNTs well throughout the paste. We then investigate different CNT lengths to see the effects on the overall mechanical strength of the MMC at various concentrations and compare to the conventional metal lines after firing. We discover that uniformly distributing CNTs into commercial silver paste is critically important for consistently obtaining enhanced mechanical properties from MMCs. We use a combination of

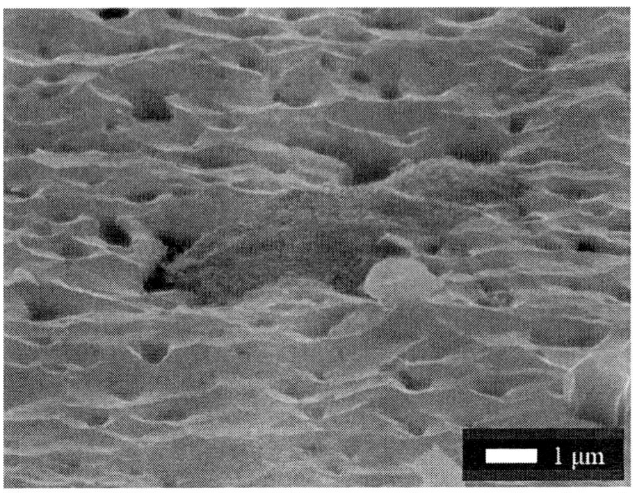

Figure 1: SEM capturing CNT agglomeration protruding through surface of fired silver metallization.

This material is based upon work supported by the U.S. Department of Energy DuraMAT (RGJ-8-82224), DOE SETO (DE-EE0009013), DOE CINT (2019AU0023), and NMSBA.

surface functionalization, a planetary mixer, and a three-roll mill to achieve this uniform mixing that eliminates CNT agglomerates.

II. Experimental Results and Discussion

A. CNT Dispersion Techniques

We have selected a surface functionalization chemistry that would help prevent the CNTs from becoming entagled while promoting adhesion to the silver particles found in commercial silver paste. The specialized CNTs are then suspended in the same solvent used in the commercial product to ensure limited change in the rheological proerties of the paste. The CNT-solvent suspension is then mixed with the commercial paste, using a planetary mixer similar to the popular Thinky mixers used for solder paste. The MMC paste is then processed through a three-roll-mill designed specifically for nano dispersion applications.

Once the MMC paste is processed through the various mixers, fineness of grind (FOG) and viscosity are measured, and we match the commecial paste viscosity after the addition of CNTs to ensure identical printing performance. We analyze the dispersion of CNTs, using an optical microscope and compare the results before and after optimal mixing. As seen in Fig. 2, we can reduce the CNT agglomerates, shown as large black clusters in the image, that would likely cause printability issues in large-volume manufacturing and adversely impact fracture toughness.

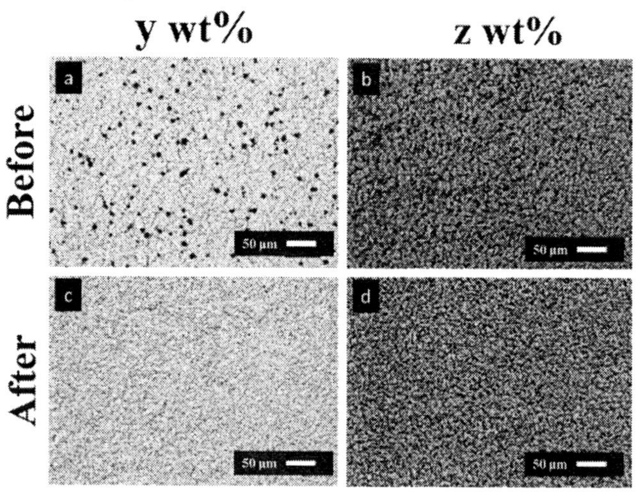

Figure 2: Optical images of MMC before and after optimal dispersion at a low concentration (a and c) and at a higher concertation (b and d).

B. Dynamic Mechanical Analysis

Tensile analysis is performed to extract the elastic modulus as well as the modulus of toughness of various MMC formulations and to compare them to the baseline values from commercial silver paste. The samples are prepared, using a dog-bone stencil, to focus the stress on the center of the sample, followed by the same drying and firing schedule recommended by the commercial paste manufacturer.

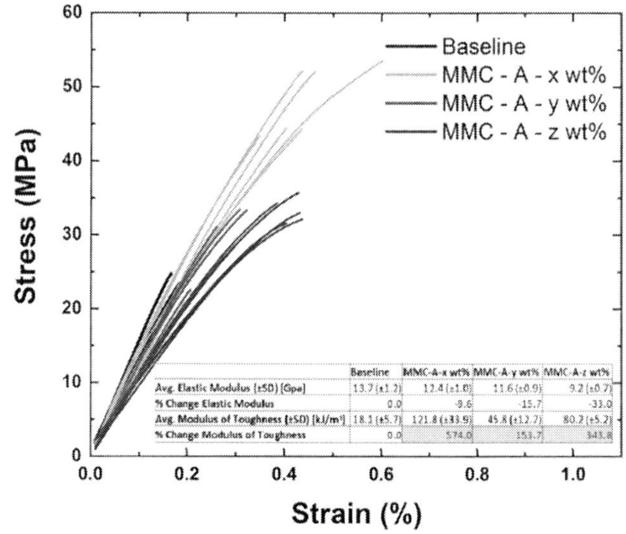

	Baseline	MMC-A-x wt%	MMC-A-y wt%	MMC-A-z wt%
Avg. Elastic Modulus (±SD) [Gpa]	13.7 (±1.2)	12.4 (±1.0)	11.6 (±0.9)	9.2 (±0.7)
% Change Elastic Modulus	0.0	-9.6	-15.7	-33.0
Avg. Modulus of Toughness (±SD) [kJ/m³]	18.1 (±5.7)	121.8 (±33.9)	45.8 (±12.7)	80.2 (±5.2)
% Change Modulus of Toughness	0.0	574.0	153.7	343.8

Figure 3: Stress vs. strain curves of MMC at three different CNT concentrations against the baseline response.

The inclusion of CNTs after optimal dispersion improves the modulus of toughness by up to 574% for the best performing formulation. The data suggest that within the range of CNT concentrations we are working with, any addition of CNTs improves the materials ability to absorb energy before they fracture. The small decrease in elastic modulus suggests that the inclusion of CNTs slightly reduces the stiffness of the material. These results demonstrate that an optimal loading of CNTs produces a stronger material, which can improve the ability to withstand fatigue-like fracture [12].

The increased fracture toughness and reduced stiffness, in comparison to the results from conventional metal gridlines, are expected to enhance the crack-tolerance of MMC gridlines. Our Resistance Across Cleaves and cracKs (RACK) testing [6], where we measure the resistance along each gridline until it fails

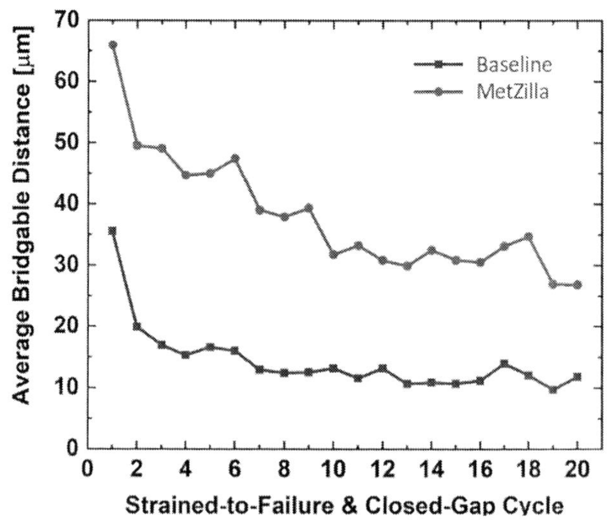

Figure 4: Average trend line of MMC under RACK testing. Each data point represents the average of 8 gridlines being strained to failure.

under strain, reveals that the electrical gap-bridging also improves in its magnitude and statistical repeatability as shown in Fig. 4.

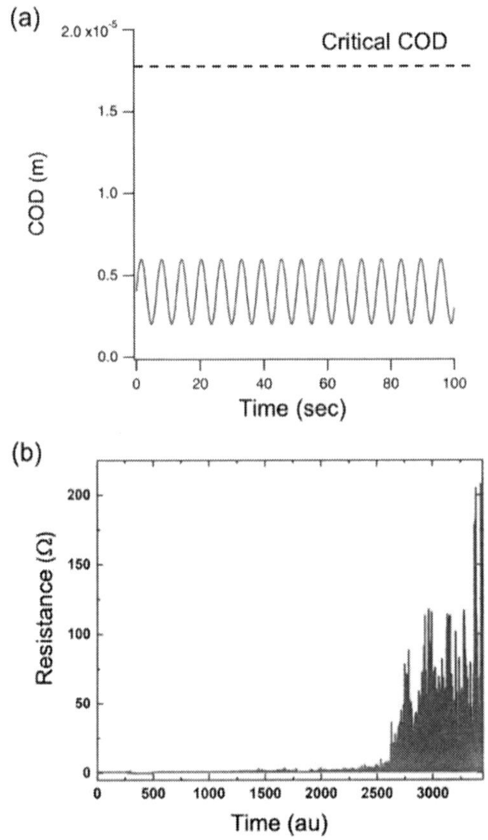

Figure 5: (a) Cyclic strain testing of metal gridlines and (b) corresponding resistance measurements.

We previously reported that metal gridlines undergo fatigue-like failure mode and that the electrical gap bridging behavior decays exponentially [12]. Fig. 5 demonstrates how we characterize the fatigue-like failure mechanism by straining the gridlines in a cyclic fashion below the critical crack opening displacement (COD) and measuring the number of cycles to failure (N_f). N_f is then used to assess the probability of failure for gridlines.[12] In this presentation, we will further discuss the fatigue-like failure of our composite metal gridlines compared to the conventional gridlines.

III. Conclusions

We have optimized the CNT surface functionalization and mixing techniques to eliminate agglomerations of CNTs within silver paste used for front contacts on PERC cells. The DMA strain tests have shown a 5 × improvement in the materials modulus of toughness while becoming slightly less stiff. This improved MMC material will undergo further verification steps, including cyclic strain test, cell performance comparison, mini-module performance comparison, and mini- and full-module-level stress testing. We will include these latter results in the presentation.

IV. Acknowledgment

This material is based upon work supported by the U.S. Department of Energy DuraMAT (RGJ-8-82224), DOE SETO (DE-EE0009013), DOE CINT (2019AU0023), and NMSBA.

References

[1] Wood-Mackenzie and SEIA, "Solar Market Insight Report 2020 Q4," December 2020 Q4 2020.

[2] D. C. Jordan, T. J. Silverman, B. Sekulic, and S. R. Kurtz, "PV degradation curves: non-linearities and failure modes," *Progress in Photovoltaics: Research and Applications,* vol. 25, no. 7, pp. 583-591, 2017, doi: 10.1002/pip.2835.

[3] D. C. Jordan, T. J. Silverman, J. H. Wohlgemuth, S. R. Kurtz, and K. T. VanSant, "Photovoltaic failure and degradation modes," *Prog. Photovolt: Res. Appl.,* vol. 25, no. 4, pp. 318-326, 2017.

[4] E. Bellini. (2020, January 7) UK researchers confirm correlation between micro-cracks and hot spots in polycrystalline cells. *PV Magazine.*

[5] A. G. Tristan Erion-Lorico, Lou Trippel. (2020, January 22) Cracking the case: quantifying the impact of cell cracking. *PV Magazine.*

[6] O. K. Abudayyeh, A. Chavez, J. Chavez, and S. M. Han, "Development of Low-Cost, Crack-Tolerant Metallization Using Screen Printing," *Proc. 46th IEEE PVSC,* pp. 1-5, June 20 2019.

[7] O. K. Abudayyeh, A. Chavez, S. M. Han, B. Rounsaville, V. Upadhyaya, and A. Rohatgi, "Silver-carbon-nanotube composite metallization for increased durability of silicon solar cells against cell cracks," *Sol. Energy Mater. Sol. Cells,* p. under review, 2021.

[8] A. Chavez *et al.*, "Electromechanical Characterization of Crack-Tolerant, Carbon-Nanotube-Reinforced Composite Gridlines Using In Situ Scanning Electron Microscope Strain Test," in *2020 47th IEEE Photovoltaic Specialists Conference (PVSC)*, 15 June-21 Aug. 2020 2020, pp. 1689-1693, doi: 10.1109/PVSC45281.2020.9300651.

[9] O. K. Abudayyeh *et al.*, "Development of Low-Cost, Crack-Tolerant Metallization Using Screen Printing," *Proc. 46th IEEE PVSC,* 2018, doi: 10.1109/PVSC.2018.8547386.

[10] O. K. Abudayyeh, C. Nelson, G. K. Bradshaw, S. Whipple, D. M. Wilt, and S. M. Han, "Crack-Tolerant Metal Composites as Photovoltaic Gridlines," *IEEE Journal of Photovoltaics,* vol. 9, no. 6, pp. 1754-1758, 2019, doi: 10.1109/jphotov.2019.2939096.

[11] O. K. Abudayyeh, N. D. Gapp, C. Nelson, D. M. Wilt, and S. M. Han, "Silver–Carbon-Nanotube Metal Matrix Composites for Metal Contacts on Space Photovoltaic Cells," *IEEE Journal of Photovoltaics,* vol. 6, no. 1, pp. 337-342, 2016, doi: 10.1109/jphotov.2015.2480224.

[12] N. Bosco, A. Chavez, V. Upadhyaya, and S. M. Han, "Fatigue-Like Behavior of Silver Metallization Gridlines and Proposed Damage Mechanics Model," in *Proceedings of the 47th IEEE Photovoltaic Specialist Conference*, Calgary, Canada (Virtual), June 14-19, 2020 2020.

Metal Grid Finger Design Optimization for Cell to Module Ratio Using the Configurable Current Cell Technology

1st Benjamin Damiani and 2nd Abasifreke Ebong

1st Solar Inventions, LLC, Atlanta, GA 30308, USA

2nd University of North Carolina Charlotte, Charlotte, NC 28223, USA

Abstract—**The continued reduction in the amount of silver used to decrease the cost of solar energy is applied to a novel grid finger screen printed metallization design to improve the cell to module ratio for similar efficiency cells with different electrical parameters like open circuit voltage, short circuit current, and fill factor. Just as the current design of busbars has changed to incorporate segmented busbars, floating busbars or no busbars where a multi-wire module is used, the design of grid fingers needs to also evolve to eliminate redundant or unnecessary function within the solar cell device. Using a technology like the Configurable Current Cell shown in this paper will provide additional methods to reduce the amount of any metal needed to optimize the performance of a solar module by coupling the current collected by grid fingers to a particular busbar.**

Keywords— ***silver, grid finger, shading, metallization, cell to module, bus bar***

I. INTRODUCTION (*HEADING 1*)

Crystalline silicon solar cells are rapidly approaching their efficiency limit in high volume production. Each part of the crystalline silicon photovoltaic device is being continually optimized for performance and cost. The International Technology Roadmap for PhotoVoltaics is an annual publication that tracks technology trends for performance and cost [1]. The ITRPV follows trends such as the transition from a "standard" emitter to a "selective" emitter, a full back surface field to a local back surface field, the number of busbars in a module as well as the cost trends for photovoltaic parts and components. This work is focused on exploring a novel application of the cell metallization steps by optimizing the cell design for improved module performance using the Configurable Current Cell, C3, technology. The C3 technology optimizes module performance by modifying the grid finger design to use less silver while taking advantage of the inherent coupling of the grid fingers to a particular busbar in the module for each solar cell. In a study by Green in 2011, the importance of changing the cell design to reduce the amount of silver used per cell was found as a necessary requirement to avoid unnecessary stress on the finite supply of silver [2]. A number of publications in the photovoltaics industry cite the need to reduce the silver consumption per cell, in fact, the International Technology Roadmap for Photovoltaics 2019 [1], reports the silver used per solar cell each year and forecasts half the amount of silver per cell by 2029. Some of the promising technologies such as fine line printing, alternative segmented busbar designs, multi-wire busbar designs have numerous publications that discuss how to optimize the performance and cost [3]. A tremendous amount of research and development has been dedicated to shrinking the printed Ag grid finger width, lowering the height of the printed grid finger/busbar, and optimizing the shading losses from the printed silver pattern while accounting for the electrical losses. Crystalline silicon solar cell devices, including mono facial, bifacial, Metal Wrap Through (MWT), and Interdigitated Back Contact (IBC) solar cells all use metal and silver is most common.

A typical solar module used in 2021 is the mono facial crystalline silicon with the "H-Bar" front silver pattern. Current industrial solar cells may have less than 5% silver shading. State of the art laboratory solar cells do not experience electrical conductivity issues when the metal to silicon interface is less than 2% [Green PERL]. Over the past 5 years there has been a shift in the design of the "H-Bar" pattern busbar to change from a simple solid rectangle that stretches the length of a solar cell to a modified design that balances solder adhesion, edge stress, the amount of silver consumed, lighted IV curve measurement pads, the silver to silicon interface, etc. The design of the busbar is being optimized for each purpose it serves within the cell design. The metal grid finger serves many purposes in the solar cell device. Grid fingers need to reduce contact resistance, transport collected carriers to the busbars, minimize optical shading, minimize electrical shading, and optimize conductivity for generated power transport in both a solar cell and solar module as part of a string.

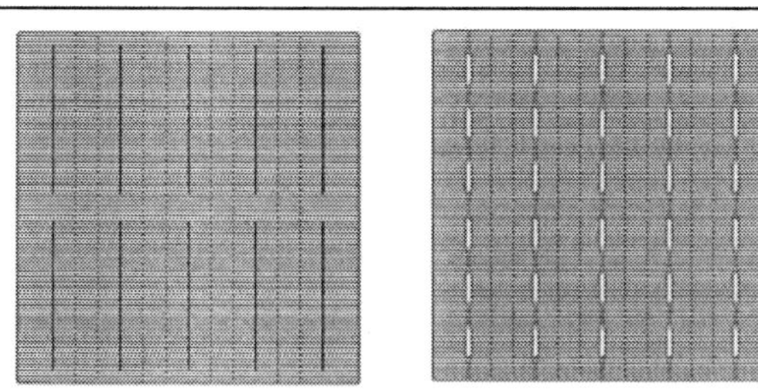

1. Standard "H-bar" front/rear pattern for a bifacial solar cell.

Optimization of the rectangular 2-D design of the busbar into a segmented, or hollow, or stripped design has been shown to improve the Jo-metal resulting in an increase in open circuit voltage while decreasing cost and improving performance while maintaining tabbing/conducting properties. By understanding the functions played by a busbar, the pattern is improved and the solar cell benefits. When considering metal grid fingers, a great deal of development has been made on fine line printing, penetration depth, silver content, deposition layer uniformity, etc. The approach discussed in this work considers the benefits of an alternative metallization that can be used to increase the performance, decrease the amount of silver used, and add new electrical functionality to a solar cell and improves the cell to module ratio of a similar watt standard cell. Using the so called Configurable Current Cell (C3) technology the design of the solar cell is configured to improve the cell to module ratio even compared to a similar efficiency "standard" H-Bar pattern device. This is accomplished by improving the open circuit voltage by 0.5mV - 5mV per cell, by improving the short circuit current proportional to reduced optical shading coupled to the number of busbars, while maintaining the module fill factor (FF).

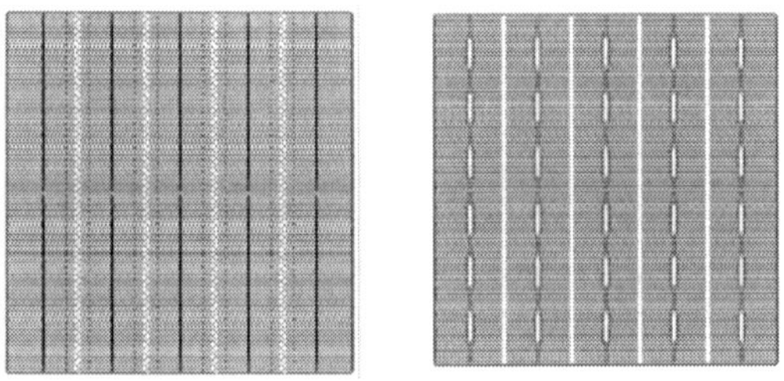

2. Modified "H-bar" zipper front and rear metallization pattern using so called Configurable Current Cell (C3) technology.

A systematic analysis of each process layer involving current state of the art processing will help unlock key solar cell performance enhancements that achieve specific function rather than historical application of solar cell processing. Consider the continuous development of the emitter with a selective emitter or the back surface field with a local back surface field. Metallization of solar cells needs to conduct electricity, fire through a passivation layer, make contact to the silicon surface, minimize shadowing, minimize metal laydown, etc. The goal of this work is to describe a method to further optimize the performance of a solar cell using metallization that is able to extract more power in the module compared to the conventional H-Bar pattern design.

II. DESIGN DISCUSSION

A. Design Optimization for Standard "H-Bar" Pattern

For the purpose of this optimization, a simplified discussion focused on an alternative metal "H-Bar" grid finger design for the front and rear of a bifacial solar cell is investigated, applying the C3 technology. Only metallization alternatives are described. Standard cell processing for texturing, cleaning, diffusion, and passivation are compatible. The role of the traditional "H-Bar" pattern has slightly different responsibilities depending on whether it is used on the emitter or on a surface field layer (such as an aluminum back surface field) where a high-low junction is formed. The metal on an emitter layer is understood to aid in the 2-D lateral conduction of absorbed photons while a surface field layer lowers the resistance of the bulk region and aids a more 3-D movement of collected photons.

The generated current is evenly divided across the total number of busbars. For an "H-Bar" pattern with 5 busbars and a max power point current of 10 amperes, each busbar would carry 2 amperes. Each grid finger provides a lower resistance path for a photon collected in the bulk of the solar cell and typically runs continuously laterally between each busbar, see Figure 1. Creating a region between adjacent busbars where the grid fingers are not continuous would still allow for the 10 amperes of current to be divided equally across the busbars while not significantly impacting the series resistance, $R_s$cell, and while not significantly lowering the fill factor, FF, because the distance traveled by a collected photon does not depend on direction, only the shortest path to a metal grid finger. Figure 2 shows a modified "H-Bar" pattern for the front and rear of common solar cell with a non-continuous grid finger design. The non-continuous grid finger design creates a "gap" in the region located between two busbars where no metal exist. This design increases the amount of light absorbed by decreasing the metal shaded percent of the active area on the solar cell as well as the exposed back surface region capable of absorbing light. The reduction in metal coverage also lowers the Jo-metal providing a potential boost in open circuit voltage. By limiting the metal to silicon interface a number of good things are realized if proper attention is given to maintain the fill factor.

	Control	Δ Voc	ΔIsc	ΔFF	ΔVoc, ΔIsc, ΔFF
Voc	0.680	0.683	0.68	0.68	0.682
Isc	10.300	10.3	10.35	10.3	10.35
FF	81.00%	81.00%	81.00%	80.70%	80.85%
Watts	5.673	5.698	5.701	5.652	5.707
Efficiency	23.22%	23.32%	23.33%	23.13%	23.36%

Table 1. Solar cell parameters used to determine efficiency. A 3mV higher Voc or a 0.05 amperes increases of short circuit current increases the efficiency by 0.1% absolute.

Consider the theoretical data shown in Table 1 for a modeled solar cell that has a gain in short circuit current, or a gain in open circuit voltage, or a loss in fill factor for an M2 wafer size (156.75mm each side). It is straight forward to calculate the efficiency of the solar cells by multiplying the open circuit voltage by short circuit current, by fill factor, and dividing by the surface area. Table 1 shows that by increasing the open circuit voltage by 3 mV or increasing the short circuit current by 0.05 amperes a 0.1% absolute in efficiency is achieve. Similarly, a reduction of fill factor from 81% to 80.7% will reduce the efficiency by 0.1% absolute.

	Cell #1	Cell #2
Voc	0.6720	0.6735
Isc	10.0	10.05
FF	81.00%	80.42%
Watts	5.443	5.443
Efficiency	22.28%	22.28%

Table 2. Similar efficiency solar cells with different electrical parameters for cell #1 and cell #2.

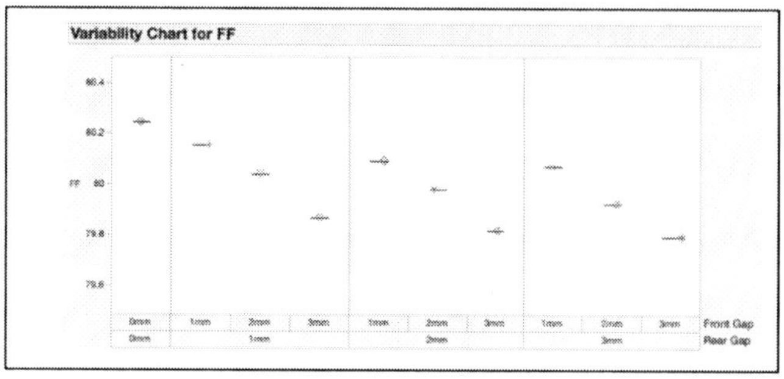

3. Griddler model of bifacial 5 busbar mono-PERC solar cell with a varying grid finger gap on the front and rear between 0mm - 3mm.

Using the Configurable Current Cell design with the properly optimized "gap" sizes for the metal H-Bar patterns on the front and rear of the solar cell will limit or minimize the change in fill factor while gaining extra light absorption from reduced metal shading and increase the open circuit voltage by reducing the Jo-metal component for both the front and rear metallization depositions. Figure 3 shows a 5 busbar PERC solar cell structure modeled in Griddler to determine the impact of the gap size on FF performance, on both the front and rear of the solar cell. Using up to a 3mm rear side gap maintains the cell fill factor within 0.2%. Even a 2mm front gap is shown to maintain a 80.04% FF compared to the 0mm control at 80.24% Thus, by coupling the grid finger design to match to a particular busbar for current transport may result in the increase of 0.15% absolute in efficiency for the solar cell provided the gain in Voc and Isc are achieved. Figure 4 shows production line Voc data of a 5 busbar mono-PERC solar cell pilot run where all samples were processed on the same day through the same equipment where only the metallization screens were changed.

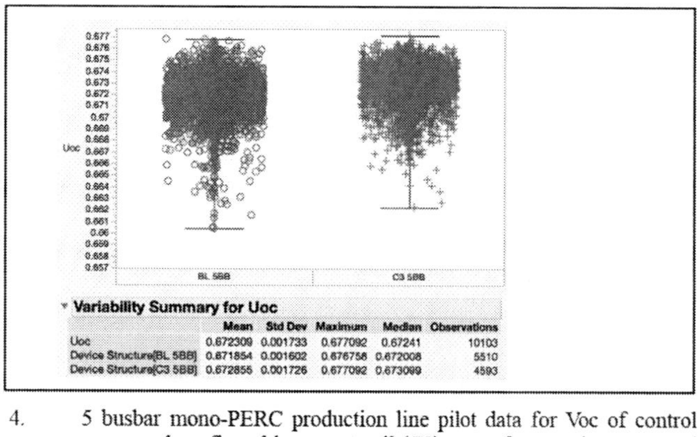

4. 5 busbar mono-PERC production line pilot data for Voc of control group and configurable current cell (C3) group for samples processed exactly the same except for the metallization screens.

III. Application and Results

Now consider two similar efficiency solar cells at 22.3% efficiency where the combination of open circuit voltage, short circuit current and fill factor are different but produce the same efficiency. Table 2 shows two such combinations. Now consider the same two solar cells as a bin of cells assembled into a 60 cell module. The series resistance for a module is described in equation (1) from the literature[5].

$$R_s mod = R_s cell + R_s tab + R_s tabext + R_s cables + R_s bus \quad (1)$$

where $R_s$mod is the module series resistance, $R_s$cell is the cell series resistance, $R_s$tab is the tabbing resistance, $R_s$tabext is the tab extension between two cells resistance, $R_s$cables is the cable resistance, and $R_s$bus is the bus ribbon resistance. The cell series resistance is the only variable that changes between the two modules so the impact on the module FF is suppressed. Experimental data is shown in Figures 5, 6, and 7 where the module FF, Voc, and Power with binned solar cells having electrical parameters similar to Table 2. Figure 5 shows the difference in module fill factor supports little to no difference between the two groups of standard cells and Configurable Current Cells despite the lower cell fill factor. Figure 6 shows the difference in module voltage supports an >1mV per cell advantage for the Configurable Current Cells. Figure 7 shows an improved module power ~ 1 watt between two groups of 50 modules without any optimized binning for the same cell efficiency bin.

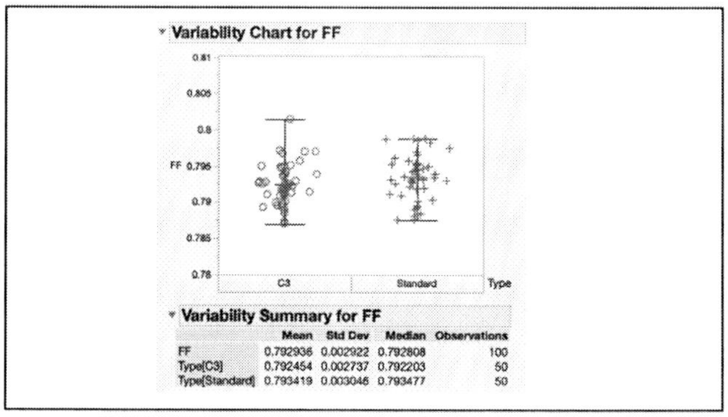

5. Similar module fill factor for a 1mm gap on the front and rear metallization between busbars despite a lower cell fill factor.

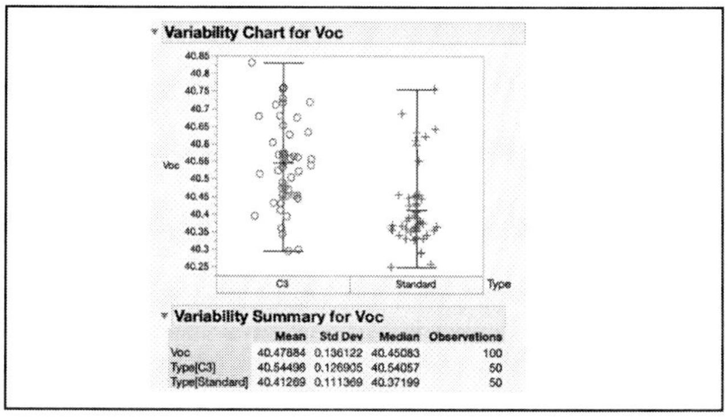

6. Improved module open circuit voltage by 0.14V for a 1mm gap on the front and rear metallization between busbars showing ~2 mV per cell improvement.

Incorporating the modified "H-Bar" pattern on the front and rear of a solar cell results in electrically semi-isolated busbars. During solar cell testing and sorting the electrical parameters are measured effectively in parallel so the behavior is similar to a metallization scheme with continuous metal on the front and rear of a solar cell which enables conventional testing/binning. Additional benefits within a module using alternative electrical architecture will be covered in a future publication for the Configurable Current Cell, C3, technology.

Experimental data using 5,000 standard "H-Bar" solar cells and 5,000 modified "H-Bar" pattern solar cells with 4 gaps, each 1mm, on a 5 busbar design were processed on Cz-Si

Passivated Emitter and Rear Contact (PERC) solar cells. The batch of solar cells were controlled to be processed on the same day, on the same material, by the same equipment with only a change of the metallization screens for the front and rear metal step. No process optimization was considered for the new front and rear metallization steps. Some results for multiple bins from an second module production line are shown in Figure 8 further supporting an improved cell to module ratio for the Configurable Current Cell technology module with similar efficiencies.

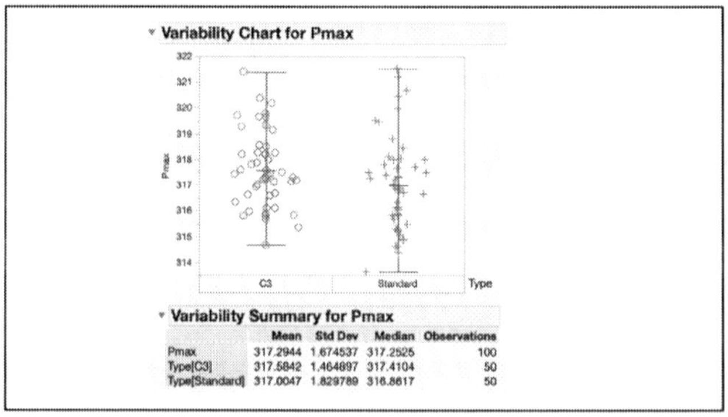

7. Improved module power by ~1 watt for a 1mm gap on the front and rear metallization between busbars even thought the binned cell efficiency was equal.

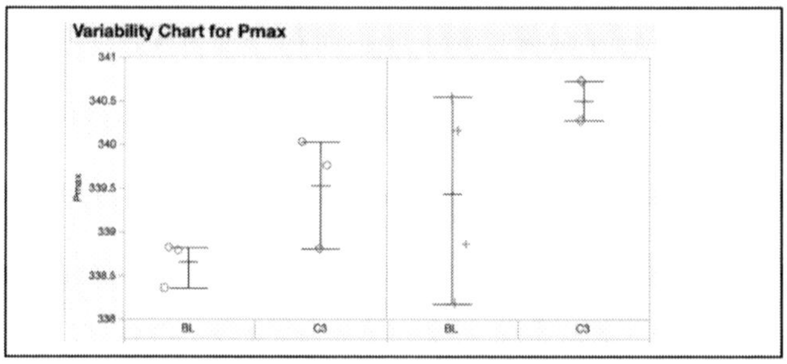

8. Improved solar module power for two separate power bins with a 1 watt power increase for the modified "H-Bar" patterned solar cells.

IV. CONCLUSION

Modification of grid finger design has been shown to simultaneously reduce shading, increase open circuit voltage, increase short circuit current, and increase performance while reducing the amount of metal needed on the front and rear of a solar cell using the Configurable Current Cell technology. Even though the cell fill factor is reduced compared to a standard cell the module fill factor is maintained while keeping the improved open circuit voltage and increased short circuit current resulting in an improved cell to module ratio of ~0.25% for a similar efficiency standard "H-Bar" pattern design. Demonstrating an improved CTM from 97.5% to 97.75%. By optimizing the amount of metal needed on the emitter for each busbar using electrical constraints of the emitter and reducing the amount of metal needed on the rear side of a solar cell by using the constraints pertinent to the solar cell structure, a higher performance, at a lower cost using less material has been demonstrated. Just as the current design of busbars has changed to incorporate segmented busbars, floating busbars or no busbars where a multi-wire module is used, the design of grid fingers needs to also evolve to eliminate redundant or unnecessary function within the solar cell device. Using a technology like the Configurable Current Cell shown in this paper will provide additional methods to reduce the amount of silver or any metal needed to optimize the performance of a solar module by improving the cell to module ratio. In addition, C3 technology provides added flexibility in module circuitry that is discussed in future publications.

ACKNOWLEDGMENT *(Heading 5)*

Thank you to Tomoya Tajima for his hard work and the American Made Solar Prize contest for accelerating this work.

REFERENCES

1. ITRPV, 10th edition, 2019
2. M. A. Green, "Ag requirements for silicon wafer-based solar cells" Progr. Photovoltaics: Res. Appl, vol. 19, no. 8, pp. 911–916, July 2011.
3. C. D. Rodriguez-Gallegos, O. Gandhi, J. M. Yacob Ali, V. Shanmugam, T. Reindl, and S. K. Panda, "On the Grid Metallization Optimization Design for Monofacial and Bifacial Si-Based PV Modules for Real-World Conditions", IEEE Journal of PV, vol. 9, no. 1, pp. 112-117, Jan. 2019
4. Y. S. Khoo, T. Walsh, and A. Aberle, "Optimizing the Front Electrode of Silicon-Wafer-Based Solar Cells and Modules", IEEE Journ. Of Photovoltaics, vol. 3, no. 2, pp.716–722, April 2013.
5. A. Ebong, N. Chen, V. Unsur, A. Chowdhury, and B. Damiani, "Innovative Front Grid Design, Four-Streets and Five-Busbars (4S-5BB), for High Efficiency Industrial Al-BSF Silicon Solar Cell," IEEE EL. Dev. Letters, vol. 37, No. 4, pp. 459–462, April 2016

PROGRESS IN THREE-TERMINAL HETEROJUNCTION BIPOLAR TRANSISTOR SOLAR CELLS

Elisa Antolín [1,*], Marius H. Zehender [1], Simon A. Svatek [1,2], Myles A. Steiner [3], Mario Martínez [1], Iván García [1], Pablo García-Linares [1], Emily L. Warren [3], Adele. C. Tamboli [3] and Antonio Martí [1]

[1] Instituto de Energía Solar, Universidad Politécnica de Madrid, Avda. Complutense 30, 28040 Madrid, Spain
[2] Departamento de Sistemas Informáticos, Universidad Politécnica de Madrid, C/ Alan Turing s/n, 28031Madrid, Spain
[3] National Renewable Energy Laboratory (NREL), 80401 Golden, CO, USA
* Corresponding author: elisa.antolin@upm.es

ABSTRACT: Conventional solar cells, including multijunction solar cells, are based on *pn* junctions as building blocks. In contrast, the three-terminal heterojunction bipolar transistor solar cell (3T–HBTSC) explores the use of a bipolar transistor structure to build a solar cell. The limiting efficiency of this transistor structure equals that of a double-junction solar cell. However, since the 3T–HBTSC does not require tunnel junctions, its minimal structure has only three semiconductor layers, while the minimal structure of a double junction solar cell has six. This work reviews the fundamental theory of this solar cell and the steps carried out towards its practical implementation.

Keywords: see the list of keywords: tandem, high-efficiency, multijunction solar cell, heterojunction, novel concepts

1 INTRODUCTION

The three-terminal heterojunction bipolar transistor solar cell (3T-HBTSC) was proposed as the minimal building block to realize a tandem solar cell [1]. Its fundamental structure consists of three semiconductor layers (*npn* or *pnp*) which are called emitter, base and collector, following the convention for bipolar transistors as represented in Figure 1a. The emitter and base regions are made of wide-bandgap semiconductors and form the *top* junction of the stack. The collector is made of a narrow-bandgap semiconductor. The *bottom* junction is formed by the base and collector regions. Thus, the base region is shared by the *top* and *bottom* subcells. One electric contact is applied to each region, leading to a three-terminal architecture. In contrast, the minimal structure of a conventional two-terminal tandem solar cell based on *pn* junctions, represented in Figure 1b, requires a minimum of six semiconductor layers, two of them dedicated to the implementation of the tunnel junction.

The term "heterojunction" in the name of the device refers to the fact that at least two different semiconductors are needed in the structure (three if emitter and base are not made of the same material). In principle, any combination of semiconductors can be chosen with bandgap energies appropriate to build an efficient double-junction solar cell [2]. Since it is a three-terminal device, the bandgap energy ranges that can produce high efficiencies are significantly wider than in the case of a current-matched device [3]. The concept is especially useful for emerging technologies in which the fabrication of a tunnel junction can be challenging, such as nanowire devices, or when low-cost epitaxial growth techniques are employed. Practical implementations of the 3T-HBTSC have been proposed based on III-V compounds [4], earth-abundant kesterites [5], nanowires [6], [7] and perovskite/silicon [6], [8], [9].

What makes the transistor solar cell concept different from other three-terminal architectures [10]–[12] is the fact that the minority carrier transport through the base is not interrupted, meaning that a majority carrier injected from the emitter into the base (and therefore, becoming a minority carrier in this region) has the potential of reaching the collector, as it is described in the theoretical framework of bipolar transistors [13]. This carrier injection from the emitter and collection by the collector is usually referred to as *transistor effect*. Then, the question arises as whether the transistor effect enhances or worsens the performance of the solar cell with respect to a traditional double-junction device. The answer to this question is given in Section 2 in this paper along with a short review of the theoretical 3T-HBTSC model. The issue of stringing is reviewed in Section 3, and possible base terminal designs in Section 4. In reference [14] the reader can find a longer version of this paper which includes the characterization of a GaInP/GaAs 3T-HBTSC prototype with interdigitated base/collector contacts at the rear of the cell.

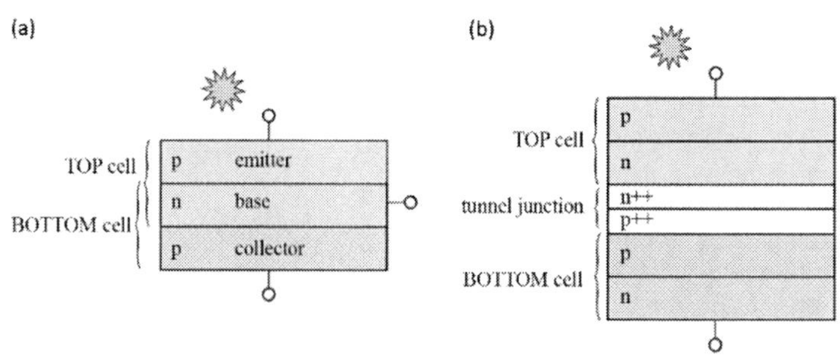

Figure 1. (a) Minimal structure of a 3T-HBTSC. (b) Minimal structure of a current-matched tandem solar cell.

2 IMPACT OF THE TRANSISTOR PARAMETERS ON THE OPERATION OF THE 3T-HBTSC

Figure 2 represents the band diagram of a *pnp* 3T-HBTSC solar that we shall use to illustrate the rationale behind the operation of the 3T-HBTSC. The device is biased at its maximum power point and the dashed lines represent the electron and hole quasi-Fermi (q-F) levels ($E_{F,e}$ and $E_{F,h}$, respectively). The first thing to note in this diagram is that the minority carrier (min-C) q-F level in the base layer, $E_{F,h}$, bends. This is necessary or, otherwise, the output voltage of top and bottom cell (determined by the q-F level split at the base-emitter junction and base-collector junction, respectively) would be the same. In that case, the output voltage of the bottom cell would limit the output voltage of the top cell, and the device would be bound to the efficiency limit of single-gap solar cells.

To enable a bending in the base min-C q-F level, the base min-C mobility must be finite, but this seems to contradict our notion of ideal solar cell. The two main idealizations applied to calculate the efficiency limit of conventional *pn* junction solar cells are the absence of non-radiative recombination and infinite carrier mobility [15]. The second condition rules out the appearance of ohmic losses and implies that there cannot be gradients in the q-F levels. Contrarily, in a solar cell with transistor structure a min-C q-F level gradient is necessary to reach maximum efficiency. Therefore, we need to eliminate the ohmic losses associated with min-C transport across the base with a different condition. Once infinite min-C mobility has been ruled out, the remaining option is to suppress the min-C current density itself across the base.

The min-C current density in the base can be minimized by preventing electron-hole pair generation and recombination in that region, which, in the radiative limit, can be approached by increasing its bandgap energy. When the gap increases, the energy of photons is not enough to excite an electron from the valence band to the conduction band, and carrier generation is inhibited. Furthermore, due to the fundamental Roosbroeck-Shockley relationship [16] that relates band-to-band absorption and radiative recombination in semiconductors, inhibiting band-to-band generation implies that radiative recombination is also suppressed. This argument can also be understood by considering that the min-C current density is proportional to the min-C concentration, which approaches zero as the bandgap energy gap increases. Not exploiting photon absorption in the base has no impact on the ideal device because it can be assumed that the emitter and the collector absorb all photons with energy above their bandgaps.

It is also possible to reduce the min-C concentration in the base region by increasing the dopant concentration as long as high doping effects do not become detrimental to the device performance. A high base doping level can, on the other hand, facilitate the implementation of the metal-semiconductor contact on the base and prevent ohmic losses associated to the lateral extraction of photocurrent in that region [4].

Within the theoretical framework of bipolar transistors, the suppression of the min-C current density results in a transistor with zero emitter and collector carrier injection efficiencies. On the other hand, the increase in the base bandgap leads to a min-C radiative lifetime approaching infinity or, equivalently in the radiative limit, to an infinite min-C diffusion length in the base, that is, a transistor in which the base transport factor approaches unity. In a transistor, the combination of *a low injection efficiency* with a *high transport factor* leads to a *bad* transistor, that is, to a transistor without amplification properties [17]. Therefore, we conclude that the transistor effect, which enables amplification in bipolar transistors, must be avoided in transistor solar cells to maximize efficiency.

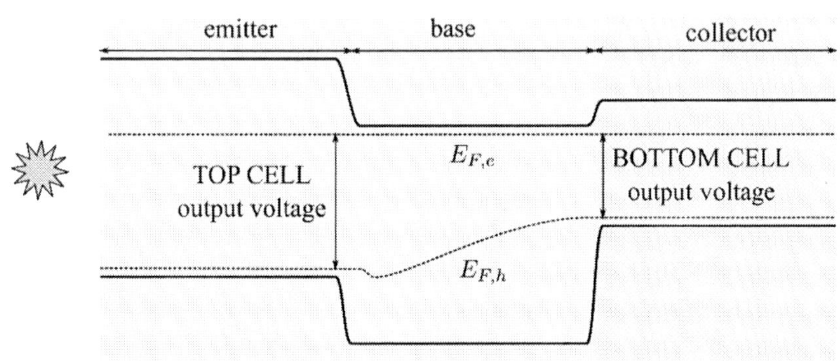

Figure 2. Simplified band diagram of a *pnp* 3T-HBTSC showing qualitatively the shape of the electron ($E_{F,e}$) and hole ($E_{F,h}$) quasi-Fermi levels for the cell biased at maximum-power point. Finite hole mobility is assumed in the base. For simplicity, carrier mobilities in other regions are assumed to approach infinity. Drawing not to scale.

The arguments above are useful to understand the optimum configuration of ideal 3T-HBTSCs, but in practical cells it is necessary to find compromises. For example, pursuing a base with an extremely wide bandgap could require low quality materials or materials that cannot be effectively *n* doped, for a *pnp* 3T-HBTSC, or effectively *p* doped, in the case of an *npn* structure. Also, if the emitter of the cell does not work in the radiative limit, it might be advisable to reduce its thickness and implement a base region with a moderate bandgap, so that the base absorbs a portion of the light providing better collection efficiency than a non-optimal emitter. It has been demonstrated that in realistic 3T-HBTSC designs a base with a moderately high doping and moderately large bandgap energy – suitable to contribute to photon absorption - can have a negligible minority carrier concentration and prevent the appearance of the transistor effect.[18]

3 THE ISSUE OF MODULE STRINGING

The use of a three-terminal solar cell poses the problem of how to interconnect the terminals of several devices to achieve a module output voltage that is sufficient for practical applications [19], [20]. In this respect, Gee [21] provided a partial solution that sacrifices two solar cells in the array and demands the output voltage of the top cell to be an integer multiple of the output voltage of the bottom cell. This approach was experimentally verified using AlGaAs/GaAs 3T-HBTSCs by Zehender *et al.* [22]. The voltage-matching configuration of three-terminal devices has been extensively studied by McMahon *et al.* [23].

An attractive solution to the stringing problem is to stack a 3T-HBTSC on top of a silicon cell [24], as illustrated in Figure 3, interconnecting the two sub-cells by a conductive transparent adhesive such as the one described by Klein *et al.* [25]. Connecting the top contact (T) of the 3T-HBTSC to the rear contact of the Si cell leads to a two-terminal triple-junction solar cell using voltage matching. Remarkably, if the 3T-HBTSC is made of GaInP/GaAs, the resulting triple-junction device has a limiting efficiency of 43% (AM1.5 G spectrum) [24]. Also, the efficiency is tolerant against spectral changes thanks to the combination of a parallel interconnection and a series interconnection (see the circuit diagram on the left side of Figure 3). Using GaInP/GaAs and silicon poses the advantage of combining two mature solar cell technologies.

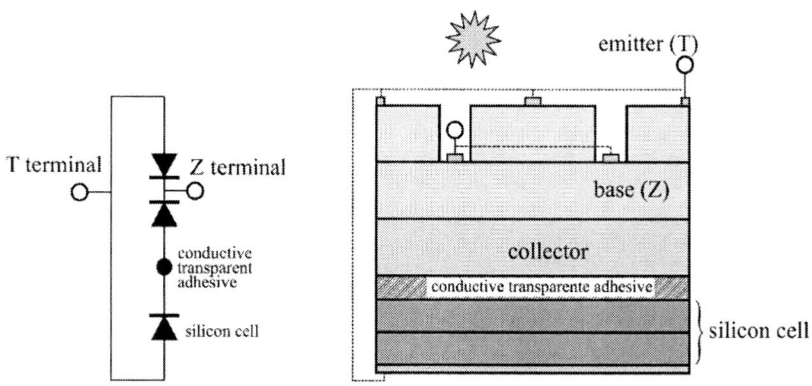

Figure 3. Schematic representation of a GaInP/GaAs 3T-HBTSC on top of a silicon cell to produce a 2-terminal triple-junction solar cell. The equivalent circuit shown at the left illustrates the contacting scheme between the cells.

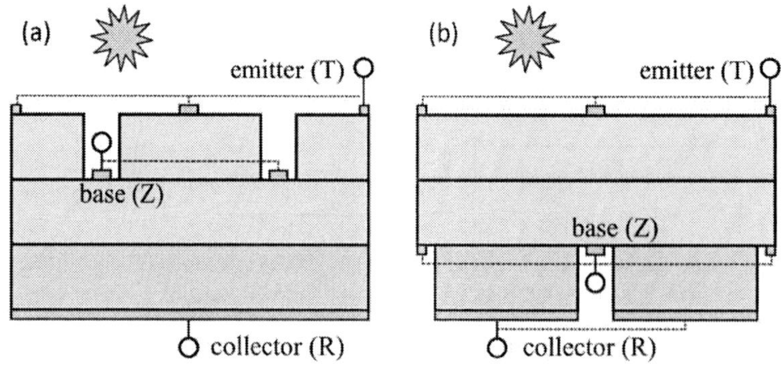

Figure 4. Illustration of interdigitated contact scheme in which the base is contacted (a) from the top (TIC scheme) and (b) from the bottom (BIC scheme).

4 DESIGN OF ELECTRIC CONTACTS

Following the general taxonomy for three-terminal solar cells proposed by Warren *et al.* [26] the three terminals on a 3T-HBTSC are denoted: T (emitter), Z (base), and R (collector). An important practical question in the implementation of this device is the design of the Z terminal. At this respect, two approaches have been considered. In the first one (Figure 4a) the base is accessed from the top with a top interdigitated contact (TIC). In the second approach, the base is accessed from the bottom (Figure 4b) using a bottom interdigitated contact (BIC). In principle, the BIC approach is expected to enable a higher efficiency because of its lower shadowing factor, since in this design the base metal fingers can be aligned underneath the emitter metal fingers. However, the fabrication of a BIC 3T-HBTSC can be difficult in some cases, for instance when the device is intended to be part of a triple-junction stack as the one shown in Figure 3. With either TIC or BIC configuration, having a high doping level in the base – besides eliminating the transistor effect – reduces ohmic losses associated to the lateral flow of carriers in the base and facilitates the implementation of the base terminal. Practical details for the implementation of the base metal contact have been reported by Martínez *et al.* [27].

5 CONCLUSIONS

In this paper the theoretical model of the 3T-HBTSC is reviewed, with emphasis on the importance of preventing the transistor effect to achieve a high efficiency. A possible combination of a 3T-HBTSC with a silicon bottom cell to form a two-terminal triple-junction solar cell has been discussed as a strategy to increase the conversion efficiency and circumvent the stringing problem. Two possible device designs have been proposed for the 3T-HBTSC in which the base contact is either interdigitated with the emitter contact on the top surface or interdigitated with the collector contact on the rear surface.

ACKNOWLEDGEMENTS

This work has been partially supported by the projects MULTIPLIER (Grant N. RTI2018-096937-B-C21, funded by the Spanish Ministry of Science and Innovation) and GRECO (Grant 787289, funded by the European Union's Horizon2020 Research and Innovation Programme). E.A. acknowledges a Ramón y Cajal Fellowship (RYC-2015-18539) and M.M. acknowledges a Formación del Personal Investigador Predoctoral Fellowship (PRE2019-087894), both funded by the Spanish Ministry of Science and Innovation. M.H.Z. is grateful to Universidad Politécnica de Madrid for funding from 'Programa Propio para Ayudas Predoctorales de la UPM'. Partial funding provided by the U.S. Department of Energy, Office of Energy Efficiency and Renewable Energy, Solar Energy Technology Office under contracts DE-00034911 and DE-00034358. The views expressed in the article do not necessarily represent the views of the DOE or the U.S. Government. The U.S. Government retains and the publisher, by accepting the article for publication, acknowledges that the U.S. Government retains a nonexclusive, paid-up, irrevocable, worldwide license to publish or reproduce the published form of this work, or allow others to do so, for U.S. Government purposes.

REFERENCES

[1] A. Martí and A. Luque, "Three-terminal heterojunction bipolar transistor solar cell for high-efficiency photovoltaic conversion," *Nat. Commun.*, vol. 6, no. 1, p. 6902, Apr. 2015, doi: 10.1038/ncomms7902.

[2] A. Martí and G. L. Araújo, "Limiting efficiencies for photovoltaic energy conversion in multigap systems," *Sol. Energy Mater. Sol. Cells*, vol. 43, pp. 203–222, 1996, doi: 10.1016/0927-0248(96)00015-3.

[3] Z. (Jason) Yu, M. Leilaeioun, and Z. Holman, "Selecting tandem partners for silicon solar cells," *Nat. Energy*, vol. 1, p. 16137, Sep. 2016, doi: 10.1038/nenergy.2016.137.

[4] M. H. Zehender *et al.*, "GaInP/GaAs three-terminal heterojunction bipolar transistor solar cell. Under review in Nat. Commun. The preprint can be found in:", doi: 10.21203/rs.3.rs-113488/v1.

[5] U. Saha and Md. K. Alam, "A heterojunction bipolar transistor architecture-based solar cell using CBTSSe/CdS/ACZTSe materials," *Sol. Energy*, vol. 184, pp. 664–671, May 2019, doi: 10.1016/j.solener.2019.04.044.

[6] P. G. Linares, E. Antolín, and A. Martí, "Novel heterojunction bipolar transistor architectures for the practical implementation of high-efficiency three-terminal solar cells," *Sol. Energy Mater. Sol. Cells*, vol. 194, pp. 54–61, Jun. 2019, doi: 10.1016/j.solmat.2019.01.027.

[7] M. H. Zehender, Y. Chen, E. Barrigón, A. Martí, M. T. Borgström, and E. Antolín, "Design study of a nanowire three-terminal heterojunction bipolar transistor solar cell."

[8] G. Giliberti, A. Martí, and F. Cappelluti, "Perovskite-Si solar cell: a three-terminal heterojunction bipolar transistor architecture," in *2020 47th IEEE Photovoltaic Specialists Conference (PVSC)*, Jun. 2020, pp. 2696–2699. doi: 10.1109/PVSC45281.2020.9300949.

[9] S. Wahid and Md. K. Alam, "Efficiency Enhancement of Perovskite Solar Cells Using Heterojunction Bipolar Transistor Configuration," *IEEE Trans. Electron Devices*, vol. 67, no. 2, pp. 552–557, Feb. 2020, doi: 10.1109/TED.2019.2961005.

[10] C. Flores, "A three-terminal double junction GaAs/GaAlAs cascade solar cell," *IEEE Electron Device Lett.*, vol. 4, pp. 96–99, Apr. 1983, doi: 10.1109/EDL.1983.25662.

[11] M. A. Steiner *et al.*, "A monolithic three-terminal GaInAsP/GaInAs tandem solar cell," *Prog. Photovolt. Res. Appl.*, vol. 17, no. 8, pp. 587–593, 2009, doi: 10.1002/pip.913.

[12] M. W. Wanlass, T. J. Coutts, J. S. Ward, K. A. Emery, T. A. Gessert, and C. R. Osterwald, "Advanced high-efficiency concentrator tandem solar cells," in *The Conference Record of the Twenty-Second IEEE Photovoltaic Specialists Conference - 1991*, Oct. 1991, pp. 38–45 vol.1. doi: 10.1109/PVSC.1991.169179.

[13] W. Shockley, "The theory of p-n junctions in semiconductors and p-n junction transistors," *Bell Syst. Tech. J.*, vol. 28, no. 3, pp. 435–489, Jul. 1949, doi: 10.1002/j.1538-7305.1949.tb03645.x.

[14] E. Antolín *et al.*, "Progress in three-terminal heterojunction bipolar transistor solar cells," *Rev. Prog Photovol*, Jul. 2021.

[15] G. L. Araújo and A. Martí, "Absolute limiting efficiencies for photovoltaic energy conversion," *Sol. Energy Mater. Sol. Cells*, vol. 33, no. 2, pp. 213–240, Jun. 1994, doi: 10.1016/0927-0248(94)90209-7.

[16] W. van Roosbroeck and W. Shockley, "Photon-Radiative Recombination of Electrons and Holes in Germanium," *Phys. Rev.*, vol. 94, no. 6, pp. 1558–1560, Jun. 1954, doi: 10.1103/PhysRev.94.1558.

[17] J. J. Ebers and J. L. Moll, "Large-Signal Behavior of Junction Transistors," *Proc. IRE*, vol. 42, no. 12, pp. 1761–1772, Dec. 1954, doi: 10.1109/JRPROC.1954.274797.

[18] E. Antolin, M. H. Zehender, P. Garcia-Linares, S. A. Svatek, and A. Marti, "Considerations for the Design of a Heterojunction Bipolar Transistor Solar Cell," *IEEE J. Photovolt.*, vol. 10, no. 1, pp. 2–7, Oct. 2019, doi: 10.1109/JPHOTOV.2019.2945914.

[19] H. Schulte-Huxel, D. J. Friedman, and A. C. Tamboli, "String-Level Modeling of Two, Three, and Four Terminal Si-Based Tandem Modules," *IEEE J. Photovolt.*, vol. 8, no. 5, pp. 1370–1375, Sep. 2018, doi: 10.1109/JPHOTOV.2018.2855104.

[20] J. F. Geisz, W. E. McMahon, J. Buencuerpo, M. Rienäcker, A. C. Tamboli, and E. L. Warren, "Unique Characterization of Three-Terminal Photovoltaic Devices," *Rev. Joule*, 2021.

[21] J. M. Gee, "A comparison of different module configurations for multi-band-gap solar cells," *Sol. Cells*, vol. 24, no. 1, pp. 147–155, May 1988, doi: 10.1016/0379-6787(88)90044-0.

[22] M. Zehender *et al.*, "Module interconnection for the three-terminal heterojunction bipolar transistor solar cell," *AIP Conf. Proc.*, vol. 2012, no. 1, p. 040013, Sep. 2018, doi: 10.1063/1.5053521.

[23] W. E. McMahon *et al.*, "Homogenous Voltage-Matched Strings Using Three-Terminal Tandem Solar Cells: Fundamentals and End Losses," *IEEE J. Photovolt.*, vol. 11, no. 4, pp. 1078–1086, Jul. 2021, doi: 10.1109/JPHOTOV.2021.3068325.

[24] E. Antolin, M. H. Zehender, S. A. Svatek, P. Garcia-Linares, and A. Marti, "III-V-on-silicon triple-junction based on the heterojunction bipolar transistor solar cell concept," in *47th IEEE Photovoltaics Specialist Conference (PVSC)*, 2020, pp. 1226–1231. doi: 10.1109/pvsc45281.2020.9300681.

[25] T. R. Klein *et al.*, "Transparent Conductive Adhesives for Tandem Solar Cells Using Polymer-Particle Composites," *ACS Appl. Mater. Interfaces*, vol. 10, no. 9, pp. 8086–8091, 2018, doi: 10.1021/acsami.8b00175.

[26] E. L. Warren *et al.*, "A Taxonomy for Three-Terminal Tandem Solar Cells," *ACS Energy Lett.*, vol. 5, no. 4, pp. 1233–1242, Apr. 2020, doi: 10.1021/acsenergylett.0c00068.

[27] M. Martínez *et al.*, "Compensated contacts for three-terminal transistor solar cells," *48th IEEE PVSC*, 2021.

STRUCTURAL AND OPTICAL STUDY OF EUROPIUM DOPED ZnO FILMS GROWN ON DIFFERENT SUBSTRATES

V.F. Gremenok[1,2], E.P. Zaretskaya[1], A.V. Stanchik[1], A.V. Mudryi[1], V.V. Khoroshko[2], A.N. Pyatlitski[3], V.A. Saladukha[3], T.V. Piatlitskaya[3]
[1]State Scientific and Production Association "Scientific-Practical Materials Research Centre of the National Academy of Sciences of Belarus", 220072, Minsk, P. Brovka Str. 19, Republic of Belarus
Phone: + 375 17 240 02 49, Fax: + 375 17 215 15 58, E-mail: gremenok@physics.by
[2]Belarusian State University of Informatics and Radioelecrtonics, 220013, Minsk, P. Brovka str., 6, Republic of Belarus
[3]JSC «INTEGRAL» – «INTEGRAL» Holding Managing Company, 220108, Minsk, Kazintsa I.P. str., 121A, Republic of Belarus, E-mail: office@bms.by

ABSTRACT: It is well known that rare-earth (RE) ions (erbium, terbium, europium, thulium and so on) are a special kind of photoactive centers with narrow emission lines and long emission lifetimes in various semiconductor materials, including ZnO. In the present work, the influence of europium ions (Eu^{3+}) on the physical properties of ZnO films was investigated. Thin films were fabricated on quartz and *p*-Si (100) substrates by magnetron sputtering method at 25 °C. The samples have been annealed at temperature from 400 to 600 °C. X-ray measurements were performed to confirm the crystallinity and phase composition of the films. Morphology and chemical composition of the films were studied using scanning electron microscopy and energy dispersive x-ray spectroscopy, respectively. Particularly, the ZnO:Eu films on silicon substrates exhibit large-scale grain size about of ~ 0.1 - 0.3 µm. The optical characteristics of the samples on glass substrates were obtained from the transmission and reflection spectra. Using photoluminescence (PL) and photoluminescence excitation (PLE) methods are found that ZnO:Eu films exhibit bright red emission which may be useful for the additional generation of nonequilibrium charge carriers in the films. The results presented in this work are important to develop energy efficient solid state lighting devices.
Keywords: Thin Films, ZnO, Doping, Rare Earths, Annealing, Structural Properties, Optical properties

1 INTRODUCTION

The undoped and doped ZnO films are used for various applications, including gas sensors, solar cells, surface acoustic wave devices, photodetectors, light-emitting diodes and UV lasers [1-6]. These applications are connected with the advantageous ZnO properties such as a wide band gap semiconducting oxide, non-toxic, abundant, and inexpensive in comparison to In_2O_3. It crystallizes at relatively low temperatures, easily processed utilizing chemical etching; and having tunable properties by doping to achieve high optical transmittance and low resistance [1].

The trivalent rare earth (RE) elements have been tested as dopants resulting in high ZnO film conductivity and specific lumines cence properties [7]. As a new energy source in the field of lighting, ZnO doped with rare earth ions is utilized for advanced full-color display and white LEDs lighting applications. Among them, the europium (Eu^{3+})-doped ZnO (ZnO:Eu) raised a lot of interest because of the red emission from the Eu^{3+} [8,9]. Pure ZnO generally emits blue–green light, because of its inherent defects. However, due to the considerable difference of structure properties between Eu^{3+} (1.07 Å) and Zn^{2+} (0.74 Å), it is hard to substitute Eu ions for Zn ions in ZnO [8]. In recent years, many researchers indicate that Eu ions can incorporate into the ZnO matrix under suitable conditions because they observe the efficient red emissions [7]. It has been demonstrated that Eu acts as an optically active center if it is surrounded by oxygen forming a pseudo-octahedron structure [9]. This means that Eu replacing Zn in the ZnO matrix forms Eu_2O_3 and does not act as an optically active center. Therefore, an annealing treatment is required to change Eu local structure, forming clusters either in the ZnO matrix or at the grain boundaries.

Thin film growth parameters are largely affected by the deposition techniques and the process parameters. There are various deposition techniques available for preparing ZnO thin film [1,10,11]. One of the methods of ZnO film fabrication is radio frequency (RF) magnetron sputtering. It results in formation of columnar ZnO film with a preferred orientation of c-axes perpendicularly to the substrate surface However, comparison of the results obtained for the films grown on different substrates, as far as we know, sometimes are contradictory. At the same time, the type of substrate can affect the structure as well as Eu incorporation into ZnO grains. Another important factor is the growth and annealing temperature for thin film formation.

In this work, the influence of substrate materials on structural, optical and photoluminescence properties of Eu-doped ZnO thin films after deposition and thermally treated at 400 – 600°C was investigated. The aim of these investigations was to obtain the intense Eu-related red emission of ZnO thin films prepared by magnetron sputtering.

2 EXPERIMENTAL

ZnO films were fabricated on quartz and *p*-Si (100) substrates by reactive magnetron sputtering of Zn target with additives of EuCl in an argon atmosphere with oxygen (20% Ar and 80% O_2) at a pressure of 5×10^{-3} Torr and 25 °C temperature [12]. The deposited ZnO film contained 0.5 at.% - 2.0 at.% of Eu dopant element. Power density applied to the cathode was 2.0 W/cm^2 and the deposition time was 1 h. The samples have been annealed for 30 – 60 min hour in a quartz tubular furnace at temperature from 400 to 600 °C.

Prior to deposition, all substrates were submitted to the cleaning procedure in ultrasonic bath during 5 min to remove mechanical pollution, and then sent to cleaning in propanol for 5 min and drying with nitrogen flow. After cleaning all the substrates were placed in load lock chamber of deposition unit. The thicknesses of the films grown on Si and qartz substrates were in the range 300 - 1100 nm. The difference in the film thickness testifies to different growth rates.

The structural characteristics of films were analysed using Ultima IV X-ray diffractometer (Rigaku) in grazing incidence X-ray diffraction (GIXD) geometry at 1° of incident X-rays with CuKα radiation source scanned in the range of 10 - 80°. The evaluation of the XRD spectra of the films was conducted using JCPDS data cards. Chemical composition and the depth profile of elements in the films were determined by energy dispersive X-ray analysis (EDX) and Auger electron spectroscopy (AES) using a CAMECA SX-100 and Perkin Elmer Physics Electronic 590, respectively. An Ar ion beam with an energy of 5 keV was used for simultaneous sputter etching. Morphology was analyzed by scanning electron microscopy (SEM) using the JEOL 6400. The transmittance (T) and reflectance (R) spectra of the ZnO thin films onto quartz substrates were measured in the spectral range of 200 – 3000 nm using a Carry 500 Scan UV-Vis-NIR (Varian, USA) spectrophotometer. Photoluminescence (PL) and photoluminescence excitation (PLE) measurements were carried out by employing a 1000 W Xe lamp as an excitation source combined with a grating monochromator (600 grooves/mm, focal length ~ 0.3 m).

3 RESULTS AND DISCUSSION

The typical XRD pattern of ZnO and ZnO:Eu thin films is shown in Fig. 1. The scans for the films grown onto different substrates demonstrate two peaks at 2Θ ~ 34.5° and 72.4°, respectivily. These peaks are caused by reflection from (002) and (004) planes of hexagonal phase crystalline ZnO (JCPDS 36–1451 card).

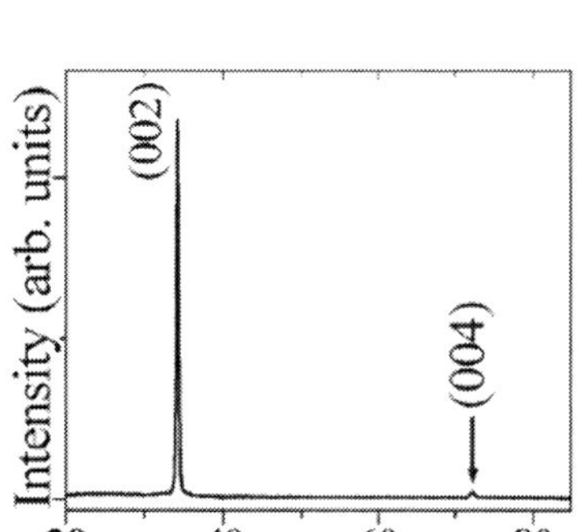

Figure 1: Typical X-ray of powder diffraction pattern of ZnO thin films on quartz substrate

The greater intensities of (002) peaks in respective patterns indicated that crystallites grow preponderantly oriented with *c*-axis normal to the substrate. This preferential growing feature was correlated with the self-ordering effect determined by the film trend to lower its surface energy during growth. As it is shown in [12] the film grains, after their coalescence, grow mainly in the direction normal to the substrate surface. In the case of hexagonal crystalline structure, this direction will be one. Most of the closely packed structures have the lowest free surface energy in the (002) plane and crystallization favorably occurs in this direction.

The observation of this peak testifies to film texture with preferential orientation of *c*-axis perpendicularly to the substrate surface (column-like structure). Furthermore, we can observe the absence of any second crystalline phase, such as free Eu or europium oxide, in particular Eu_2O_3. Experimental data indicate that single phase ZnO:Eu layers without any second crystalline phase, such as free Eu or europium oxide, inparticular Eu_2O_3 may be fabricated under technological conditions. The as-deposited ZnO films did not show any feature related to the Eu_2O_3 phase, suggesting that Eu atoms are either substitutionally replacing Zn in the ZnO lattice or segregated to the non-crystalline region in grain boundaries [13-15]. Films had *c* parameter values slightly higher than that of ZnO powder material ($c = 0.521$ nm), indicating that the unit cells of thin films are elongated along the *c*-axis and the compressive forces were predominant as usual in RF sputtered ZnO thin films. The internal compressive stress in the films is assigned to the bombardment of energetic particles during the deposition and not to the thermal stress originating from the difference between the thermal expansion coefficients of the film and the substrate [1].

On the other hand, when the samples were annealed, the microstructure of the films was not changed by the plausible oxidation of Eu and presented the Eu_2O_3 phase in the XRD patterns. The presence of the Eu_2O_3 phase was not observed in the sample, associated either with the relatively small amount of Eu atoms incorporated into the film (low doping level) or due to the absence of those phases for the employed deposition conditions. After the annealing, the peaks were shifted to higher diffraction angles and the films showed lattice parameters (*a* and *c*) slightly lower than the ideal values for undoped ZnO films. This suggested that the stress was changing from compressive to tensile. It is possible that the annealing temperature produced tensile stress due to the mismatch between thermal energy coefficients when the films cooled down [13]. It was also observed that the annealing of films produces variation in the intensity of the main diffraction peak (002) and showed an increase in its intensity, which unequivocally indicates an enhancement of the film's crystalline arrangement. The average crystallite size in the direction of normal to the reflecting planes was increased after the annealing process.

The composition of the thin films before and after the annealing was estimated from the EDX measurements and appeared to be invariable after the heat treatment. The chemical composition was determined by averaging the concentration values from 10 different points on the surface of the same film. The EDX spectrum of films shows the signals of Zn, O, Eu, Si, C and Ca where the signals of Si, C, Ca are from the substrates. A similar elemental composition (Zn, O, Eu in atomic %) of ZnO thin films has been confirmed by AES method. The spectra were analyzed over a range of kinetic energies from 100 to 1400 eV using the primary electron beam of energy 5 keV. As an example, Fig. 2 shows depth profile of elements for as-grown non-annealed ZnO:Eu films. The homogeneous distribution of all elements Zn, O and Eu along the depth of films is seen from Fig. 2. The similar homogeneous elements distribution is also revealed for all investigated films on different substrates. This fact confirmed the high-quality of deposited films with nearly ideal stoichiometry of Zn and O chemical elements. The measurements show that Eu atomic concentration in ZnO is not exceeded 2.0 at. % at magnetron sputtering.

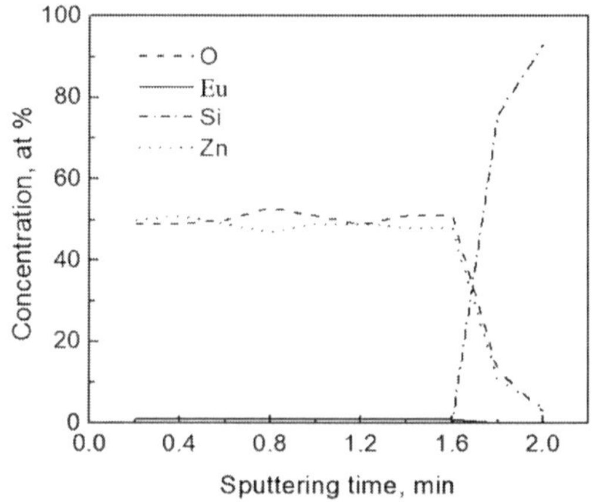

Figure 2: AES depth profile of chemical elements of ZnO:Eu films on silicon substrate

The surface morphology and cross-section images of the thin films were characterized by SEM. According to crystal growth mechanisms the growing faces of crystallites correspond to the crystal shape at equilibrium and are determined by the orientation of the crystal. A growth competition can start among the neighboring crystals according to their orientation. The faster growing crystals will grow over slower growing ones. Once the competition proceeds towards formation of same type of crystal faces, they form the free surface. This competitive growth mode represents an orientation selection resulting in the competitive growth texture [1]. For ZnO, stably preferential orientation is along to *c*-axis. Usually, II–VI binary compound semiconductors crystallize in either cubic zinc blend or hexagonal wurtzite structure where each anion is surrounded by four cations at the corners of a tetrahedron, and vice versa. Thermodynamically stable phase of ZnO is wurtzite symmetry. From the SEM images (Fig. 3), these films exhibit good flatness and the crystal grain size is dependent on the substrate (silicon, quartz) and the deposition conditions. As evident from SEM images the large well-faceted grains without any porosity with dense character of the material may be fabricated on used substrates. Particularly, the ZnO:Eu films on silicon substrates exhibit large-scale grain size about of - 0.1 - 0.3 μm. The difference in the morphology of as-grown and Eu doped ZnO films after the annealing is reflected by the size and shape of grains.

Both films with undoped or doped could present fewer grain boundaries due to the absence of Eu ions at the grain boundaries. The annealing might have helped the grains to grow much bigger, since high-temperature annealing stimulates the migration of grain boundaries and causes the coalescence of more grains. Good crystallinity should help improve the optical and electrical properties of the films.

With the aim of the band gap energy determination of different ZnO:Eu films fabricated on quartz substrates the measurements of absorption and reflectance spectra at room temperature have been performed. As an example Fig. 4 shows transmission spectra of a ZnO:Eu thin films prepared on these substrates (3 ZnO as-grown and 4 ZnO annealed at 400 C).

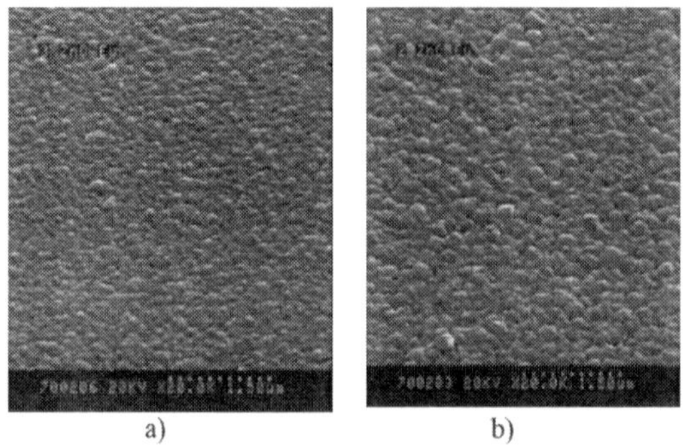

Figure 3: SEM images of ZnO:Eu films on quartz (a) and silicon (b) substrates

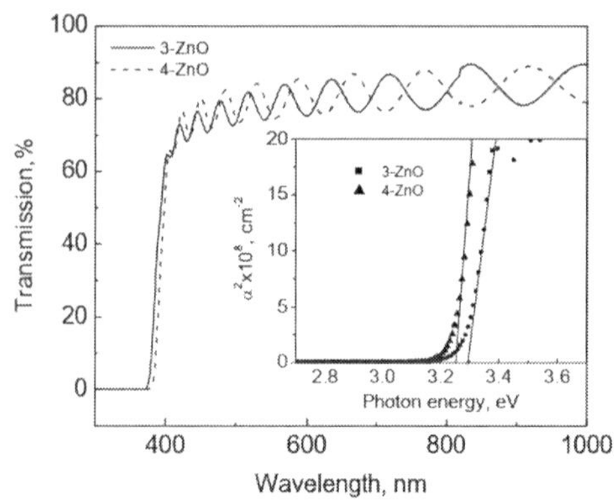

Figure 4: Transmission spectra and the dependence of α^2 vs $h\nu$ for ZnO:Eu films on quartz substrates

As seen all films have high value of transmittance ~ 70 – 80 % in wide spectral range 370-2500 nm, intensive interference fringes, and relatively sharp edge of the intrinsic absorption which starts at less than 390 nm These experimental data confirmed high quality of ZnO:Eu material grown on quartz substrates. Reflectance was measured in the photon energy region 1.0 -5.0 eV and was about 0.17 - 0.20 for both samples. The shifting of intrinsic band edge in ultraviolet spectral region and decreasing transmittance in near infrared region are observed for no annealed films in contrast to i-ZnO films. The absorption in spectral region 1.0 – 2.5 μm and high-energy shift of absorption edge are indicating strong doping effect of ZnO films by Eu atoms. The similar behavior of optical spectra is observed for n-type highly doped semiconductors such as ZnO:Al compounds [1]. This phenomenon may be explained by Burstein-Moss effect. As seen from Fig.4, isochronal annealing at temperatures higher than 400 °C leads to low-energy shift of intrinsic absorption edge and decreasing of infrared absorption.

The absorption coefficient of the films has been calculated using T and R data and following equation:

$$\alpha = -\frac{1}{d}\ln\frac{\sqrt{(1-R)^4+4T^2R^2}-(1-R)^2}{2TR^2}, \quad (1)$$

where α is the absorption coefficient, d is the thickness of the film, T and R are the transmittance and reflectance respectively. For the allowed direct transitions, the absorption coefficient has the following spectral dependence:

$$\alpha = A(h\nu - E_g)^{1/2}, \qquad (2)$$

where A is a constant and E_g is the corresponding semiconductor optical band gap.

The insert in Fig. 4 shows absorption spectra calculated by using relations (1) and (2). Using dependence $\alpha^2 = f(h\nu)$ the band gap energies for ZnO:Eu films E_g~3.28 eV (as-grown) and E_g~3.24 eV (annealing at 400 ºC) by extrapolating of the linear part of the spectrum to the photon energy axis have been determined. This value E_g is in a satisfactory agreement with other optical measurements for ZnO thin films with similar chemical composition.

The main attention was concentrated on the photoluminescence and photoluminescence excitation measurements. To determine the nature of the emission, the experiments with varying excitation and temperature have been performed. Fig. 5 shows the typical PL spectra of ZnO:Eu films on quartz substrates for different annealing temperature.

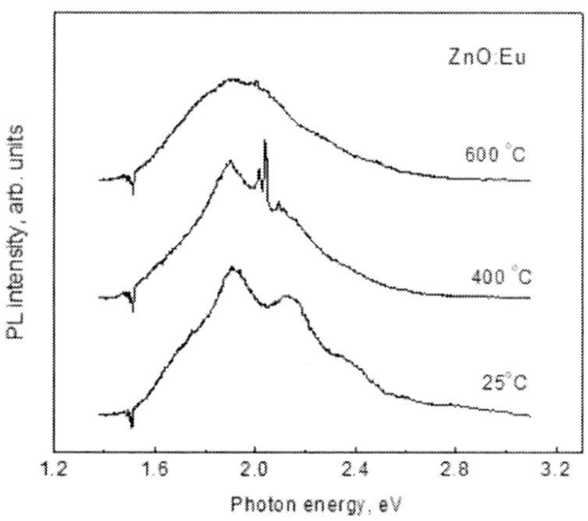

Figure 5: Room temperature photoluminescence spectra of ZnO:Eu films for different annealing temperature

Three broad bands at 2.38, 2.13 and 1.91 eV are presented in photoluminescence spectrum at room temperature for as-grown ZnO:Eu films on quartz substrate. A small shift to higher energy about 5 meV is observed for these PL emission bands with increasing measurement temperature in range of 78 – 300 K. No noticeable spectral shifts for these bands were observed when the excitation power was varied from 1 up to 10 W/cm^2. Therefore, we can conclude that these bands are due to energy band to deep level optical transitions. The experiments show that the relative intensity of these broad bands was depended on the growth process and annealing temperature. In particular, the intensity of green emission band at 2.38 eV decreases with annealing temperature (Fig. 5). The similar behavior of the intensity dependence on annealing temperature has been found for the band at 2.13 eV. Only one broad band at 1.91 eV can be found in PL spectra at annealing temperature higher than 600 ºC. The band at 2.38 eV may be related with single ionized oxygen vacancies (V_O^-). The bands at 2.13 eV and 1.91 eV probably are due to optical transitions on oxygen atoms occupied Zn positons (O_{Zn}) and oxygen interstitial (O_i), respectively. The most interesting experimental result is an appearance sharp emission peaks at about 2.086, 2.031, 2.006 and 1.894 eV correspond to transitions of Eu^{3+} from $^5D_0$ to ^7F_j (j = 1, 2, 3, 4) electronic states for ZnO:Eu thin films annealed at temperatures in range of 400 – 600 ºC. The more intense red emission related with Eu^{3+} optical transitions observed for ZnO:Eu films after annealing at 400 ºC. High-resolution PL spectrum of Er^{3+} related red emission is shown in Fig. 6.

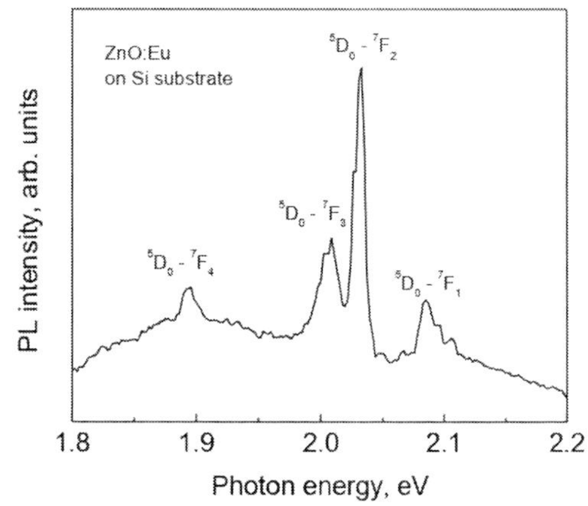

Figure 6: Room temperature high-resolution photoluminescence spectrum of ZnO:Eu thin films on silicon substrate annealed at 400 ºC

Fig. 7 shows the excitation spectra of ZnO:Eu films for monitor emission at about 2.086 eV ($^5D_0 - {}^7F_2$ optical transition).

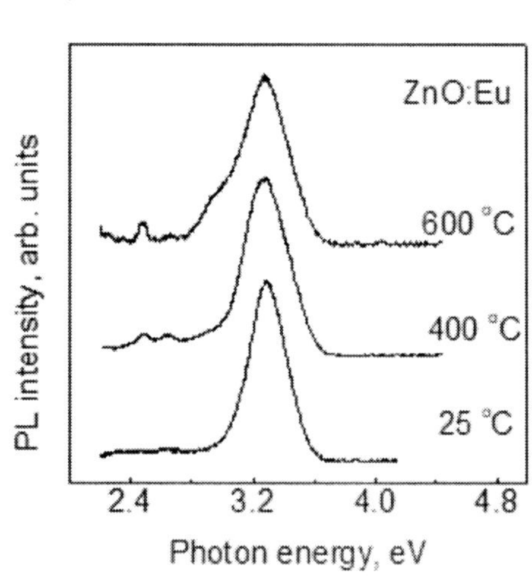

Figure 7: Room temperature photoluminescence excitation spectra of ZnO:Eu films on quartz substrate for different annealing temperature and at E_{det} ~ 2.03 eV (near maximum $^5D_0$ to $^7F_2$ optical transition.

It is clear seen that photoluminescence excitation spectra shows the existence one broad band at 3.28 eV which is respond to the band gap energy of ZnO:Eu material. It is necessary to note that this band gap value is coincide with that found from optical absorption measurements.

4 CONCLUSIONS

The effects of substrate material, quartz and *p*-Si, and of different atomic contents of Eu on structural, optical and photoluminescence properties of ZnO films were studied by different experimental techniques. C-axis oriented ZnO films have been deposited by magnetron sputtering method and have been thermally annealed. All the films showed a (002) preferential orientation with the c-axis perpendicular to the substrate surface. As annealing temperature increased, some changes of the physical properties have been observed. All films thermally treated in temperature range of 400 – 600 ºC show intense red emission related with Er^{3+} intra-shell optical transitions. This result suggests that Eu^{3+} ions are incorporated in ZnO lattice during magnetron sputtering process. Intense red emission of Eu^{3+} dopant in ZnO films are issued by the band-to-band excitation and energy transfer from the host ZnO to europium ions. Thus direct semiconductor material ZnO doped with Eu ions exhibit bright red emission at low temperature annealing. Our results indicate that Eu^{3+} ions are effective co-doping element for ZnO films which can be used for application in lasers, optical amplifiers, thin film solar cells and display technologies. In general, this knowledge will help in the development of devices using semiconductor films as sensitisers to excite RE ions.

5 ACKNOWLEDGMENTS

These investigations was supported by the Belarusian Republican Foundation for Fundamental Research grant № T20UK-022. The part of the work has been finased by the Belarusian State Programme for Research «Physical Material Science, New Materials and Technologies».

6 REFERENCES

[1] V.A. Coleman, A. Victoria, C. Jagadish, Basic Properties and Applications of ZnO, Zinc Oxide Bulk, Thin Films and Nanostructures, Elsevier Science Ltd, Amsterdam (2006) 1.
[2] Y.K. Mishra, G. Modi, V. Cretu, V. Postica, O. Lupan, T. Reimer, L. Kienle, R. Adelung, ACS Appl. Mater. Interfaces. 7 (2015) 14303.
[3] A. Hastir, R.L. Opila, N. Kohli, Z. Onuk, Yuan, K. Bo, Jones, V. Ravi, C. Singh, J. Mater. Sci. 52 (2017) 8502.
[4] C.R. Gorla, N.W. Emanetoglu, S. Liang, W.E. Mayo, Y. Lu, J. Appl. Phys. 85 (1999) 2595.
[5] Y.K. Mishra, G. Modi, V. Cretu, V. Postica, O. Lupan, T. Reimer, I. Paulowicz, V. Hrkac, W. Benecke, L. Kienle, R. Adelung, ACS Appl. Mater. Interfaces 7 (2015) 14303.
[6] D. Punetha, S.K. Pandey, IEEE Sens. J. 19 (2019) 7.
[7] A.N. Ökte, Appl. Catal. A Gen. 475 (2014) 27.
[8] Y. Tan, Z. Fang, W. Chen, P. He, J. Alloys Compd. 509 (2011) 6321.
[9] Vinod Kumar, Vijay Kumar, S. Som, M. M. Duvenhage, O. M. Ntwaeaborwa, H. C. Swart, Appl. Surf. Sci. 308 (2014) 419.
[10] O.M. Ntwaeaborwa, S.J. Mofokeng, V. Kumar, R.E. Kroon, Spectrochim. Acta A. 182 (2017) 42.
[11] R. Kumar, M.R. Sekhar, R. Laha, S.K. Pandey, Applied Physics A. 126 (2020) 859.
[12] V. Malyutina-Bronskaya, A.M. Saad, V. Zalesski, T. Leonova, A. Mudryi, V. Fedorov, Optical Materials 88 (2019) 718.
[13] G. Hodes, Chemical Solution Deposition of Semiconductor Films, Marcel Dekker, Inc, New York, (2005).
[14] E.M Bachari, G. Baud, S. B. Amor, M. Jacquet, Thin Solid Films 348 (1999) 165.
[15] H. Shahroosvand, M. Ghorbaniasl. J. Lumin, 144 (2013) 223.

THE POTENTIAL OF GLASS-FIBRE-REINFORCEMENT: (THERMO-)MECHANICAL TESTING OF LIGHT-WEIGHT PV MODULES

Jonathan Govaerts[1], Kishan Moliya[1], Bin Luo[1,2], Tom Borgers[1], Rik Van Dyck[1,2], Arvid van der Heide[1], Loic Tous[1], Arnaud Morlier[3], Fabiana Lisco[4], Lorenzo Cerasti[5], Marco Galiazzo[5], Jef Poortmans[1,2,6]
[1]imec-Energyville, Thorpark 8320, 3600 Genk, Belgium
[2]Also at KULeuven
[3]ISFH, Am Ohrberg 1, 31860 Emmerthal, Germany
[4]EPFL-PV lab, Rue de la Maladière 71b, 2000 Neuchâtel, Switzerland
[5]Applied Materials Italia, via Postumia Ovest 244, Olmi di San Biagio di Callalta (TV), Italy
[6]Also at UHasselt

ABSTRACT: In this work we report on our approach on integrating c-Si PV into lightweight structures, in particular towards vehicle integration. To this end we want to get rid of the (bulk weight of the) glass but seek a suitable replacement in terms of mechanical stability. First we elaborate on the most basic standards and norms that VIPV products should relate to in terms of (thermo-)mechanical testing.
Then, for the experimental part, 2 concepts are investigated. In a first approach, we reinforced the encapsulant with glass fibre material, while in a second one we applied a dedicated glass-fibre-reinforced sheet as a replacement of the backsheet. In both cases we stay as close as possible to using commercially available material.
Extending our previously reported work, we then elaborate for each approach the testing that has been carried out so far, and what is still ongoing and being planned in the short-term future (before the conference): indicating the promising results so far on thermal cycling and vibration testing, the hail impact testing is a challenge to be tackled still.

Keywords: VIPV, light-weight materials, fibre reinforcement

1 (THERMO-)MECHANICAL TESTING

In the field of VIPV, we identified 4 relevant norms that describe testing related to mechanical and thermomechanical failure modes.

- IEC 61215 for PV modules: ***thermal cycling*** (10.11), (static) mechanical load (10.16), ***hail test*** (10.17)
- IEC TS 62782 for PV modules: Cyclic (dynamic) mechanical load
- ECE R43 for laminated glass: fragmentation, ball impact 227g (A13/5 and A7/4), headform (A13/4, A7/3 and A11/3), resistance to abrasion (A13/6.1, A6/5.1 inner and A9/2 outer), resistance to temperature change (A3/8), flexibility (A3/12)
- ISO 16750-3 for environmental testing of electronics for vehicles: ***vibration***, mechanical shock (door slamming), free fall, surface strength/scratch and abrasion resistance, gravel bombardment

In first instance we focused on a number of them, indicated above in ***bold italic red***, as a first evaluation in the different types of mechanical tests, although we will also consider the other ones carefully in the future.

2 EXPERIMENTAL MODULE FABRICATION

In order to benefit from the continuously advancing PV cell and module technology, we follow as close as possible the developments in this area. Such an approach allows the highest potential of addressing performance and cost requirements in an initial stage. To this end, we consider conventional tabbing and stringing of homojunction cells, and multi-wire interconnection [1,2], as well as shingling of heterojunction cells [3,4,5]. These last concepts of multi-wire and shingling are, apart from providing an increased performance, both also very much interesting from an aesthetics point-of-view.
While for the encapsulation we also aim to stay as close as possible to the existing and proven materials and technologies, the buildup is somewhat adapted to address weight requirements. As the glass front- and/or backside is the main showstopper for these requirements (although very thin glass may still be an option, but at a significant increase in cost), we look into alternatives, based on glass fiber reinforcement, either within the encapsulant (GFRE) or within the cover material (in first instance at the back, GFRB). Figure 1 indicates the weight implications for these different buildup options compared to standard glass-glass (GG) and glass-backsheet (GBS) references, with resp. (2x) 2 mm and (1x) 3.2 mm thick glass.

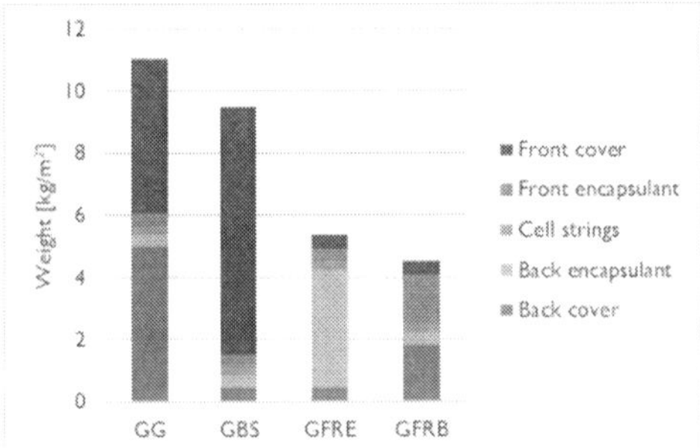

Figure 1: impact of the different options on overall weight of the module; the different buildups for GFRE and GFRB are explained in the following respective sections

For the lamination process, we apply typical lamination profiles with an industrial-type laminator, albeit at slightly elevated temperatures, to reduce the risk of cracking cells during the pressure step.

3 REINFORCED ENCAPSULANT (GFRE)

Here, we replace the front glass with a polymer frontsheet, and provide additional strength by thickening and implementing glass fibres (GF) into the encapsulant. This results in a total thickness slightly below 5 mm.

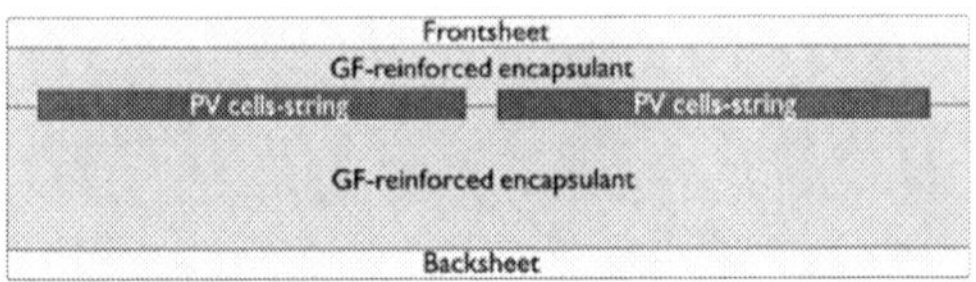

Figure 2: schematic buildup of a lightweight module with reinforced encapsulant

In first instance, we made 2x2-cell laminates based on tabbing and stringing of 3BB cells where each string could be measured separately, and we varied in buildup between a commercial EVA and thermoplastic PO encapsulant, including glass fibre or not and frontsheet or not. If a frontsheet is included, it is applied at both the front and the back. Each sample is IV-measured and EL-inspected throughout the thermal cycling they are subjected to.

The CTE mismatch, in combination with the absence of the front glass, causes significant stresses on the interconnects, as has been reported previously [6,7]. However, implementing chopped glass fibres in the encapsulant clearly ameliorates this impact, as shown in Figure 3.

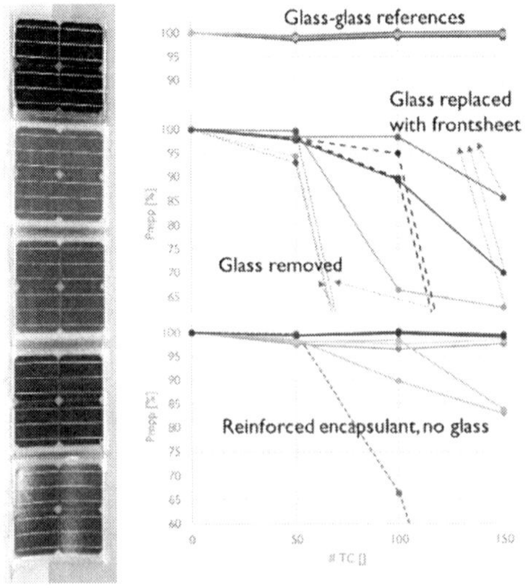

Figure 3: in thermal cycling (TC) of 2x2 laminates, degradation increases when the glass is replaced or removed (center) from a glass-glass buildup (top); however, reinforcing the encapsulant with glass fibre can reduce this degradation (bottom); in all cases the failure mode is interconnect ribbon breakage [6,7]

Getting rid of the ribbons, and implementing shingling as interconnection technology, with 1/6th silicon heterojunction cells [4], also allows to meet thermal cycling demands in the IEC standards and beyond. For this, we prepared laminates with a similar buildup, using shingled strings of 2-cell equivalent length (12 shingles), as well as 1m-long strings. Figure 4 shows that both lengths can sustain >400 cycles with limited degradation. Interesting to note (in the EL image) is that even the significant initial cracks in 3 shingles, in one of the strings, do not result in major P_{mpp} degradation during this cycling.

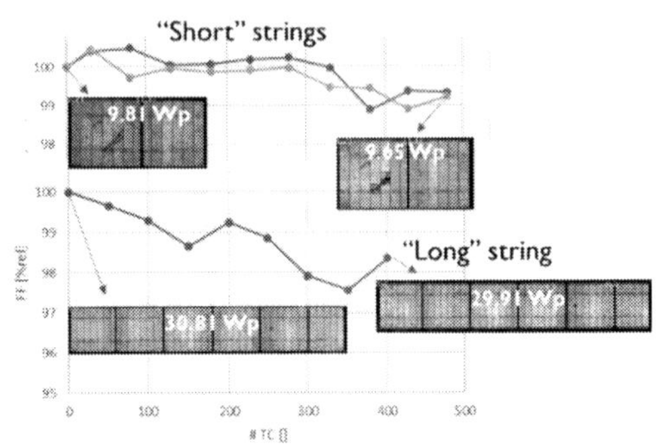

Figure 4: evolution of relative FF throughout thermal cycling for 2 short strings and 1 long string based on shingling

Additionally, we also looked at testing the resilience of such a buildup under dynamical mechanical stress. Sample modules were submitted to random vibrations as experienced by spring mounted vehicle body parts in accordance with the standard for vehicle electronic components ISO FDIS 16750-3 [7]. The vibrations range from 10 Hz to 1000 Hz with an r.m.s. acceleration value of 27,8 m/s² and are applied in a direction perpendicular to the module surface during 8 hours. Although notable deflections of the sample middle point have been observed during the vibration sequences, both the buildup with reinforced PO and the glass-glass references did not show significant damage, as illustrated in Figure 5.

Figure 5: EL images of a sample buildup with reinforced PO and a glass-glass reference before (left) and after (right) the random vibration testing carried out with a shaker (central image)

Despite the above promising results, an important challenge is the mechanical stability at higher temperatures, when the encapsulant becomes too soft and can no longer sufficiently fixate the reinforcement fibres. This deformation is illustrated in the picture in Figure 6, after long-term exposure to damp heat conditions.

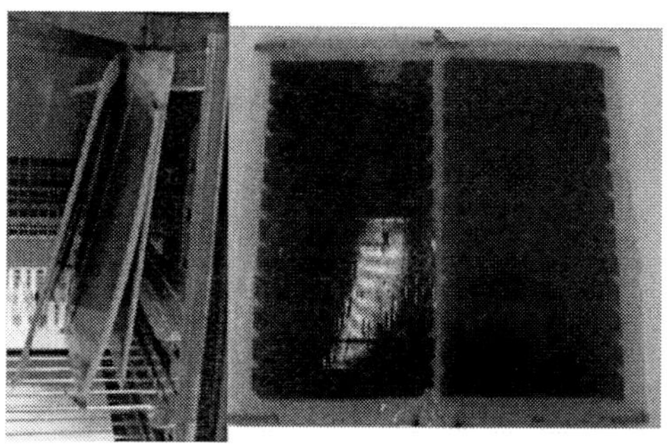

Figure 6: deformation of an "encapsulant-reinforced" laminate, due to vertical loading in damp heat conditions

4 REINFORCED "BACKSHEET" (GFRB)

To ensure stability of the laminate also at higher temperatures, the buildup can be modified. Instead of using the encapsulant as polymer matrix (EVA/PO) for the (chopped) glass fibres, we implement so-called unidirectional (UD) tapes to provide long (unidirectional) glass fibre reinforcement into a polymer matrix with a higher melting temperature at the backside. Of course, to provide stability not only in one direction, tapes are integrated in perpendicular orientations. The resulting total thickness lies around 3 mm.

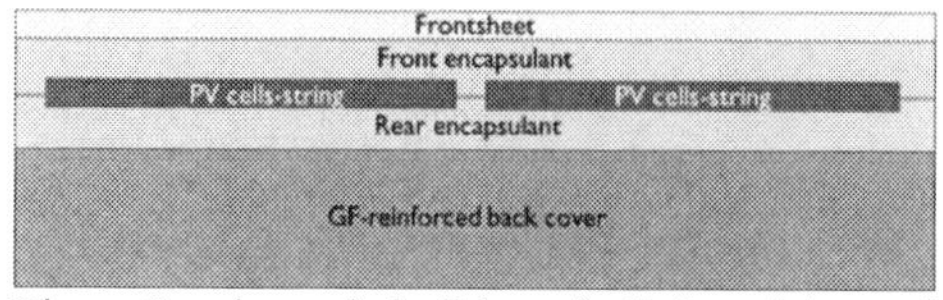

Figure 7: schematic buildup of a lightweight module with reinforced encapsulant

In this way, we can similarly reduce the weight, as indicated in Figure 1.

In first instance, we assembled 1-cell laminates using tabbing and stringing of 3BB cells and subjected them to thermal cycling. The results so far are shown in Figure 8, with negligible losses up until 300 cycles.

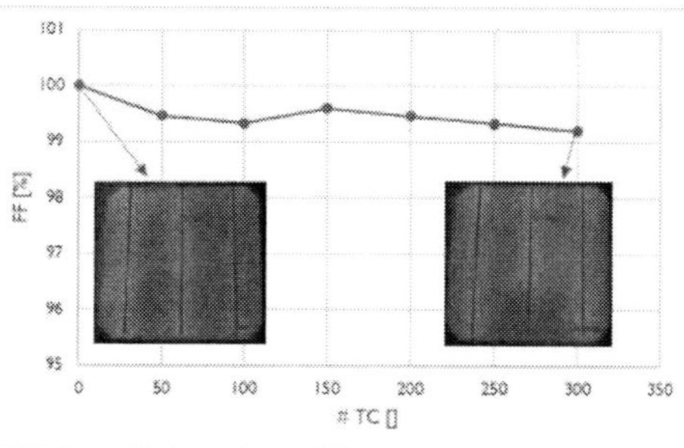

Figure 8: (Absence of) degradation of 1-cell GFRB laminate throughout thermal cycling (300 so far)

Secondly, we prepared also laminates using multi-wire interconnection, as well as shingling (similar to the ones in the previous paragraph). For the test, we also fabricated a reference glass-glass laminate. Thermal cycling is ongoing with initial results given in Figure 9.

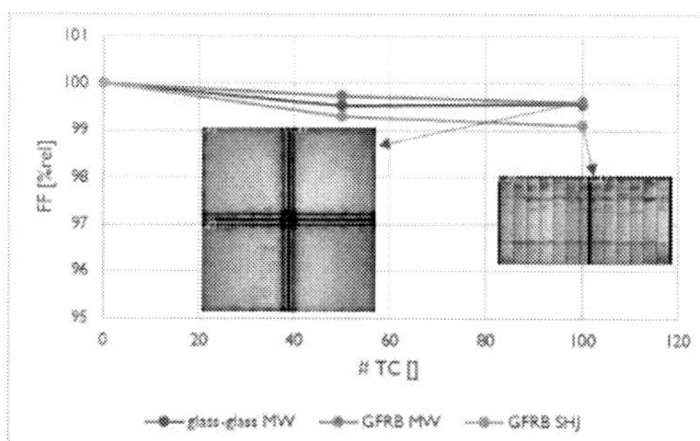

Figure 9: Initial thermal cycling (100 cycles) indicates only limited degradation so far

Thirdly, an initial screening was also conducted to check on the hail impact [9]. For this, again a batch of 2x2-cell laminates was fabricated, based on both multi-wire as well as shingling interconnection, each time including also a reference glass laminate. To check its impact, also a variation of the encapsulant thickness was included. While the glass-glass laminates did not sustain any significant damage that could be observed with EL, all other laminates did not pass the test, with significant losses in power. An impression of the results is shown in Figure 10, seemingly indicating that increasing the front encapsulant thickness somewhat reduces the impact. A similar trend is observed for the multiwire laminates, and in both cases correlates with an increasing power loss for lower thickness.

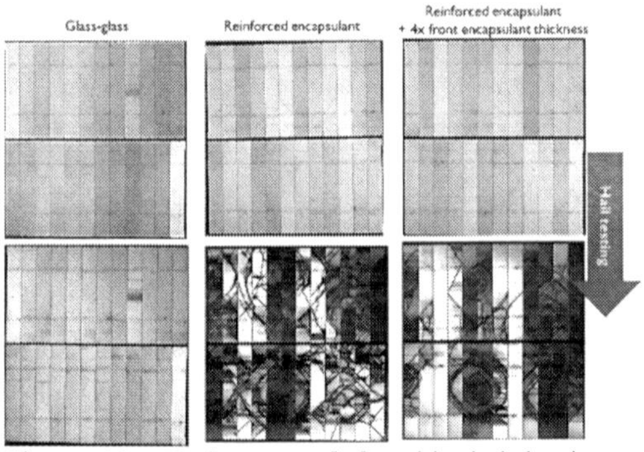

Figure 10: EL images of the shingled laminates before and after hail testing

The hypothesis based on our current understanding is that a stiff front cover combined with a soft encapsulant will help to distribute and also to absorb the mechanical impact stress, while the back cover with the back encapsulant should be stiff enough, and placed close to the cells, to provide the required support during the impact.

At this moment we are preparing additional variations, illustrated in Figure 11, to determine which layers are most critical in this hypothesis. Further progress is reported elsewhere [11].

Figure 11: impression of additional laminate variations in progress for testing the various hypotheses (front cover stiffness, front encapsulant softness, backside stiffness)

6 CONCLUSION AND OUTLOOK

In this work we elaborate on the potential of glass reinforcement for PV modules, replacing the glass to reduce their weight. In 2 encapsulation approaches, either reinforcing the encapsulant or reinforcing the back cover, we perform thermo-mechanical tests to determine challenges and opportunities. For its increasing relevance, we include in these tests, apart from industrial tabbing and stringing, state-of-the-art multi-wire and shingling interconnection.

With promising results so far on thermal cycling and vibration testing, more impact-related testing as hail and shock testing is a challenge to be understood (and hopefully tackled) next. Of course apart from the mechanical impact, also other considerations are important and will be assessed, including e.g. damp heat testing, UV exposure and optical characterization [10].

ACKNOWLEDGMENT

The authors gratefully acknowledge imec's SiPV industrial affiliation programme and its partners, the support of the European Union through the funded H2020 project HighLite under Grant Agreement no. 857793, and the partial funding by the Kuwait Foundation for the advancement of Sciences under project number P115-15EE-01.

REFERENCES

[1] T.Borgers et al., EUPVSEC 2017
[2] T.Söderström et al., EUPVSEC 2013
[3] D.Tonini et al.,Energy Procedia 150, 2018,36-43
[4] S.Harrison et al., EUPVSEC 2020
[5] L.Tous et al., EUPVSEC 2021 (submitted)
[6] J.Govaerts et al., EUPVSEC 2020
[7] C.Kutter et al., EUPVSEC 2020
[8] M.Köntges et al. PiP 2016, DOI: 10.1002/pip.2768
[9] F.Lisco et al., EUPVSEC 2020
[10] S.Wendlandt, EUPVSEC 2021
[11] J.Govaerts et al., EPJ-PV (submitted)

SOLDER PASTE FOR INTERCONNECTING STRUCTURED RIBBONS ON THE BACK SIDE OF THE C-SI CELLS

Narahari S Pujari[1*], Krithika PM[1], Siuli Sarkar[1] and Carl Bilgrien[2]
1: Macdermid Alpha Electronics Solutions Vaishnavi Bhavana 89/1 Industrial Suburb II stage Yeshvanthpur Bangalore-560094 (India)
2: Macdermid Alpha Electronics Solutions, 109 Corporate Blvd South Planfield, New Jersey (USA) 07080
*: Phone:+91-80-4244312, Narahari.Pujari@macdermidalpha.com

ABSTRACT: We report tin-lead (Sn-Pb) and low temperature lead free solder pastes for interconnecting structured ribbons. Lead free option was tin-bismuth (Sn-Bi) based engineered non eutectic alloy HRL1. We found an acceptable wettability and printability of both the solder pastes. These pastes show excellent stability, long stencil life and consistent transfer efficiency (>90 %). Solder joints were strong with peel strength more than 2N/mm and voiding less than 10 %, indicating bonds have better contacts and complete wetting. We found that solder paste based assemblies yield uniform intermetallic compound (IMC) formation and no loss in power output. Both pastes could be fast reflowed and laminated with EVA. The modules assembled using these pastes pass thermal-cycling and damp heat reliability testing according to IEC61215. Our results suggests that solder pastes could be an attractive alternative to expensive electrically conductive adhesives for interconnecting structured ribbons.

Keywords: solder paste, structured ribbon, soldering, tabbing and stringing

1 INTRODUCTION

Structured ribbons help optimize the grid design of the front contacts of a photovoltaic cell. They offer up to 2% power gain by reflecting the incident light that falls on the ribbon [1]. Structured ribbons, also known as light capturing ribbons are copper ribbons patterned with shallow grooves which have sidewalls that are typically 30° from the horizontal. To achieve total internal reflection, copper ribbon is coated with thin silver or solder layer. Typically, the cell interconnection with the ribbon is achieved by electrically conductive adhesives (ECA) or conductive films using production tools [2]. The conductive films are composed of preformed conductive adhesive in foil shape. Although the technology is in the market for some years, long term outdoor performance data proving their stability against environmental influences is not yet available. ECA joints can be easily attacked by moisture [3]. Silver containing adhesives cannot typically be completely cured at low temperature. Accordingly, when a circuit or conductive layer or interconnection is formed on a substrate using such conventional pastes, damage to the substrate or failure in reliability of device may occur.

We present two novel solder pastes as alternative materials for interconnecting structured ribbons. Solder pastes offer many advantages over ECAs such as [4]: 1. Formation of reliable joints 2. Can be dispensed or printed 3. Stable viscosity and higher shelf life compared to ECA 4. Low voiding 5. Better thermal and electrical conductivity compared to ECA, 7. Resistant to moisture and 8. Cheaper than ECA. In this work, we investigated application of tin-lead and low temperature lead free solder pastes in structures of ribbon interconnects on the back side of the c-Si cell. Our aim is to examine solder paste as a joining material and evaluate its bond strength and reliability by assembling minipanels and doing accelerated aging tests. We also studied the role of the solder paste on the electrical and mechanical properties of soldered cells. Our study included evaluation of printing, voiding, microstructure, fast reflowing and snap curing properties of solder pastes. The primary objective of this study was to investigate applicability of solder pastes for interconnecting structured ribbons in an existing automated machines or modified tabbing machines.

2 EXPERIMENTAL

Two types of solder pastes, tin lead and low temperature lead free were used in this study. They were based on eutectic (tin-lead, Sn-Pb) and non-eutectic tin-bismuth pastes, namely, R-276 and OM550® HRL1 from Macdermid Alpha respectively (Table I).

Table I: Solder paste properties

Solder Paste	*Alloy*	*Solder type*	*Liquidus* °C	*Yield Strength* MPa	*Reflow Temp* °C
R276	Sn-Pb (Eutectic)	T4	183	60.1	225
OM550	HRL1 (non eutectic)	T4	138	37.5	175

Fig 1 shows the back side soldering of structured ribbons using solder paste. Typically, paste is printed on the structured ribbon. The printing can be done by automated printer or, dispensing techniques can be used. Based on melting temperature of the alloy, peak temperature for reflow was chosen and cells were exposed for 2-3 sec at peak temperature. A 5 cm x 1 cm cell assembly was prepared for each experiment. An automated tabbing machine from Sunnywell was used prepare the strings. The machine uses IR (infrared) soldering. Other process parameters are given in Table II. After the assembly, bus ribbons were soldered, and lamination was done at 140°C using ethyl vinyl acetate (EVA) as lamination material and polyester white back sheet. A Sinton I-V curve tester was used to evaluate electrical performance of the minimodules. As a part of study, printing evaluation and transfer efficiency, profile optimization, X-ray analysis for voiding and FESEM (field emission scanning electron microscopy) were performed. Panels were also analyzed for contact resistivity and Electroluminescence (EL). Finally, damp heat and thermal cycling (TC) tests were conducted as per IEC61215. These tests were further extended up to 1500 h (damp heat) and 400 cycles (TC) and are continuing. Solder pastes were compared to an ECA which was based on acrylate chemistry. This study explored process optimization for solder paste and demonstrates a cheaper assembly strategy than existing ECA based gluing. One of the solder pastes is low temperature, with reflow around 175°C.

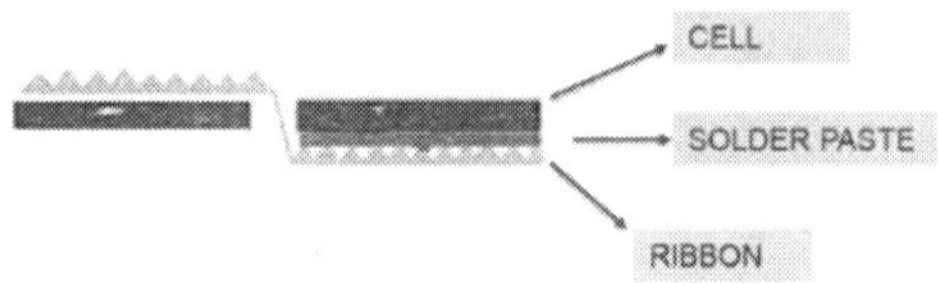

Figure 1: Schematic of back side soldering of structured ribbon using solder paste

Table II: Solder paste assembly process parameters

Attributes	*Value*
Paste	R276 (Sn-Pb) and OM-550 HRL1 (lead free)
Preheating	100-130°C
Soldering time	2-3 sec
Soldering	IR
Solar power	60%
Cell	5BB mono PERC
Soldering Side	Back
Ribbon	Structured Sn-Pb, 1x0.2 mm

3 RESULT AND DISCUSSION

Both, ALPHA® R276 and ALPHA® OM550 HRL1 have a melting point significantly lower than standard SAC305 (tin-silver-copper) alloy. A peak temperature of only 220 and 175°C respectively is also equivalent to most ECA curing temperature ranges. This reduces energy consumption in the PV application process and stress build up on the cells. OM-550 HRL1 is lead free which mean no environment hazard and it can be snap reflowed like ECAs. These two pastes and alloys are specifically designed to improve reliability of the solder joint. In structured ribbon, due to the grooved pattern, soldering is difficult and standard soldering fluxes may not work. To strengthen and maintain the electrical and mechanical properties of the interconnections, including the bulk of the material and the solder-pad interfaces, solder paste is best suited. Low temperature solder joints, specifically with HRL1 alloy, can withstand cyclic contractions and expansions that can deteriorate and weaken the solder joints.

A 3 mil stainless steel stencil was fabricated for the experiments. A DEK03Xi printing machine was used to do automated printing. Just enough paste was printed on the busbar to avoid any spreading. Similarly, pastes could also dispensed on the ribbon just before the assembly. The paste volume was optimized by printing/dispensing paste pads from 300 to 900 microns on the 1 mm wide ribbon. We found the optimized height to be 65 micron which reduces down to around 55 microns after fast reflow. The paste spreads completely to fill the width of the structured ribbon Fig. 2.

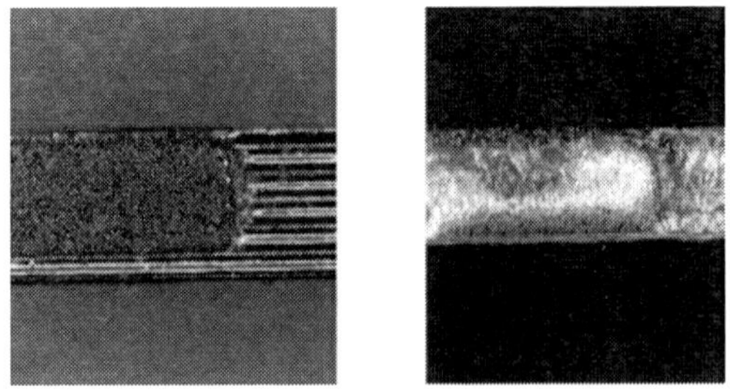

Figure 2: As printed solder pastes and after fast reflow.

Solder paste has excellent stability, and the metal content remained almost unchanged even after 12 h of continuous printing (Fig. 3A). The binder and solvent system are formulated in such a way that viscosity buildup is minimized, and the paste can be printed for hours without cleaning the stencil. In addition, the high print definition and speed works in favor of paste. Even at higher printing speed, the observed transfer efficiency with solder paste was over 90% (Fig. 3B). Generally, ECAs do not have stencil/screen life more than 4 hours. After that, the viscosity of an ECA increases rapidly and often insufficient transfer and screen pore clogging is observed. Material waste is also a problem with ECAs. ECAs additionally need higher pressure to print and through cut stencils do not work with ECAs.

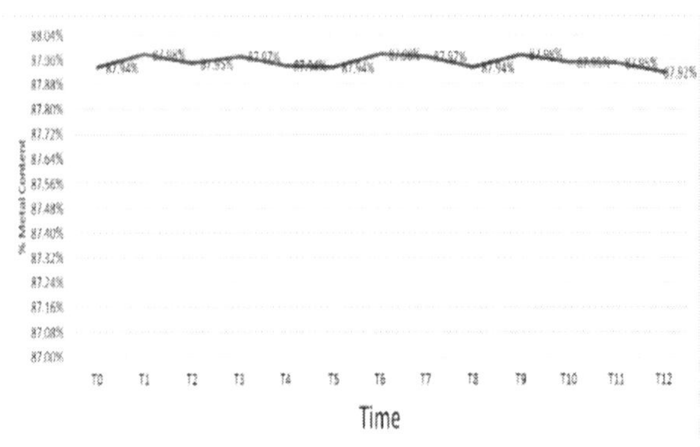

A]

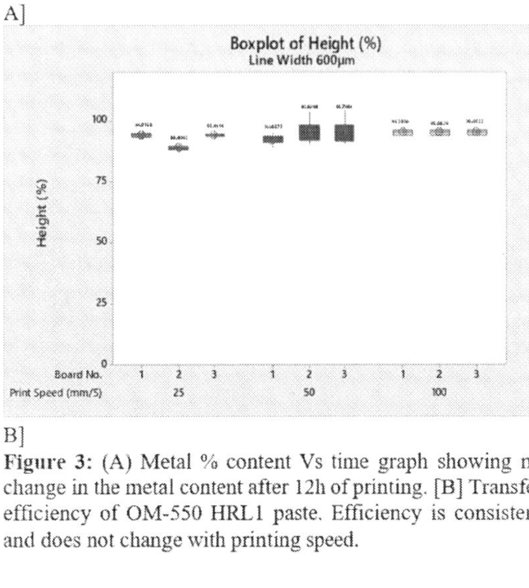

B]

Figure 3: (A) Metal % content Vs time graph showing no change in the metal content after 12h of printing. [B] Transfer efficiency of OM-550 HRL1 paste. Efficiency is consistent and does not change with printing speed.

Tabbing and stringing of the 5 x 1 cells were achieved at rapid rate. Both the pastes could be IR reflowed along with standard fluxing operation on the front side of the cells. The reflow time was around 2-3 sec. Microscopic analysis revealed complete wetting, continuous IMC (intermetallic compounds) formation and no microcracking (Fig. 4A) with solder paste. The wetting angle was found to be 13°. No silver leaching was observed. ECA, on the other hand, does not form any IMC and there is pseudo bonding between silver pad and silver flakes in the ECA, Fig.4B.

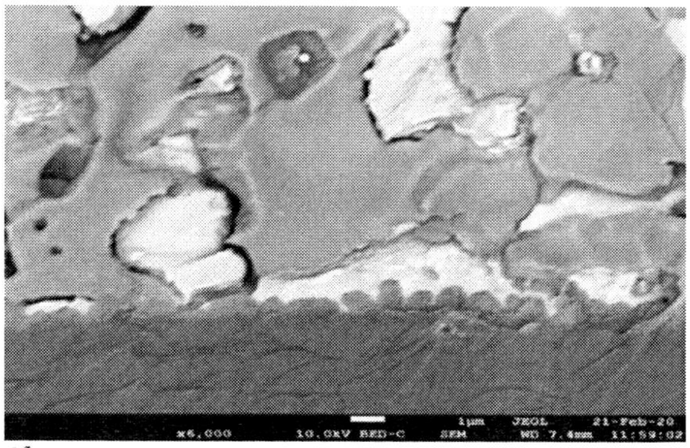

A]

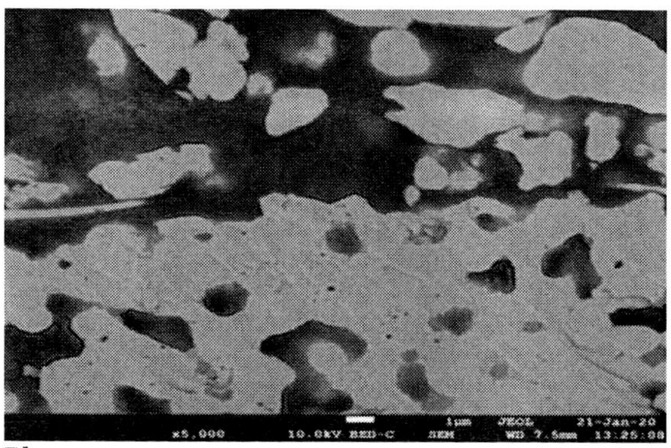

B]

Figure 4: (A)SEM analysis showing IMC formation in the solder paste assembly (B) No IMC and diffusion bonding observed with ECA based assemblies

The quality of the joints was also analyzed by X-ray. This determines voiding in the joints. The solder paste spread well over the bus bar and no solder balls were observed. The voiding was less than 10 % in each assembly which mean there is good thermal contact between the joining surfaces. A large number of voids reduce the solder joint reliability. Voids also reduce the thermal conductivity of the solder joints and can cause solder bridges and solder transfer between neighboring solder joints during the reflow soldering process [5]. In small solder joints, voids can significantly reduce their current carrying capacity. Soldered cells were analyzed for peel strength at 180° using an Imada peel tester. The results are presented in Fig. 5. Both pastes yielded acceptable peel strength, much above the required 1 N/mm as per DIN EN 50461.

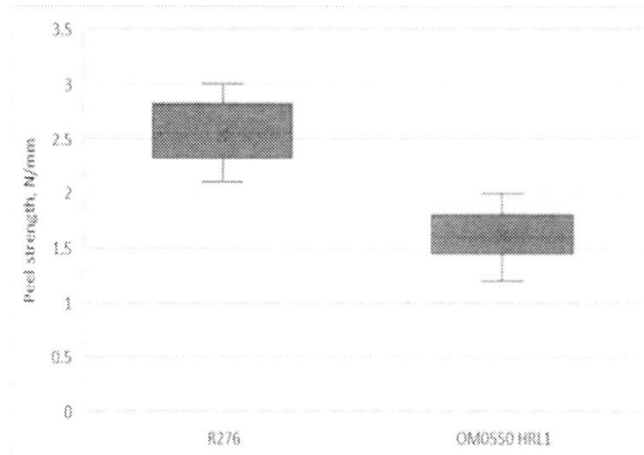

Figure 5: Peel test result of R276 and OM-550 HRL1 assembled soldered cell

Laminated minipanels were exposed to climatic testing, conforming to IEC 61215, at 1000 h of Damp Heat (DH) and 200 temperature cycles (200 TC). Power measurements and electroluminescence imaging (EL) were performed before and after the tests. No degradation in module performance or other functional properties was noted Fig. 5. EL analysis also shows no evidence of cracking or other defects. The maximum power (P_{max}) loss permissible according to IEC 61215 is ±5% after the test. None of the samples reached this value. The maximum change in P_{max} was less than 1 % for both R276 and OM-550 HRL1 assemblies. These results demonstrate that both R276 and OM550 pastes are resistant to oxidation and are resistant to changing weathering conditions. These results further confirm that the standard tin-lead and lead free low temperature alloy-pastes can be used for interconnecting structured ribbons. Solder joints can withstand the expansion and compression forces produced during thermal cycling.

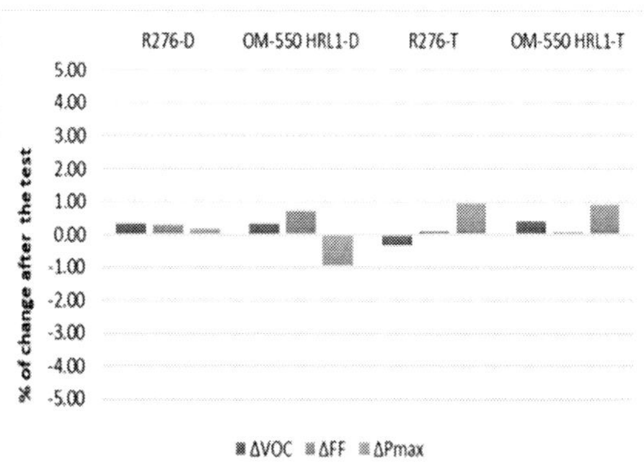

Figure 6: Damp heat and thermal Cycling of R276 and OM550 assembled minipanels. D: 1000 h damp heat test and T: 200 cycles thermal cycling test

3 CONCLUSION

We report tin-lead and low temperature lead free solder pastes that are suitable for interconnecting structured ribbons. We found an acceptable wettability and printability of both solder pastes. These pastes show excellent stability, long stencil life and consistent transfer efficiency (>90 %). Solder joints were strong with peel strength more than 2N/mm on each side and voiding less than 10 %, indicating bonds have better contacts and complete wetting. Both pastes could be fast reflowed and laminated with EVA. The modules assembled using these pastes pass thermal-cycling and damp heat reliability testing according to IEC 61215. Based on these results we conclude that these solder pastes can be used to make solar panel interconnections and other advanced interconnection assemblies, and they offer an attractive alternative to expensive electrically conductive adhesives.

4 REFERENCES

[1] E. M. Sachs, J. Serdy, A. M. Gabor, F. van Mierlo, and T. Booz, "Light capturing interconnect wire for 2% module power gain," in Proc. 24th Eur. Photovolt. Solar Energy Conf., 2009, pp. 3222–3225.

[2] I. Chung, L. Won-jae, E. Cho, and I. Moon, "Light Capturing Film for power gain of silicon PV modules," in 2014 IEEE 40th Photovoltaic Specialist Conference (PVSC), pp. 2689-2692.

[3] M. Estruga, L. Theunissen, A. Ardizzone, B. Willems, and A. Henckens, "Electrical conductive adhesives for cell interconnection in shingled module technology: the impact of material properties on mini-module performance", Proceedings of the 35th European Photovoltaic Solar Energy Conference and Exhibition, Brussels, Belgium, (2018) 67.

[4] N.S. Pujari, PM. Krithika, P. Vishwanath, S. Sarkar, and C. Bilgrien, "Lead free low temperature solder pastes for shingling interconnection," 37th European Photovoltaic Solar Energy Conference and Exhibition., Lisbon, Portugal, 2020, pp. 29-32.

[5] M. Ribas, S. Sarkar, C. Bilgrien, and T. Hunsinger, "Effect of voids on thermomechanical reliability of solder joints" , Proceedings of SMTA International, (2017) 667.

ENHANCING THE PERFORMANCE OF LUMINESCENT SOLAR CONCENTRATOR PHOTOVOLTAIC DEVICES USING MULTIPLE ORGANIC DYES AND BIFACIAL SILICON SOLAR CELLS

N. Desai[1,2], M. Aghaei[1]*, A.H.M.E. Reinders[1,2]
[1]Eindhoven University of Technology, 5612 AE Eindhoven, The Netherlands
[2]University of Twente, 7522 NB Enschede, The Netherlands

* Corresponding author, Email: m.aghaei@tue.nl, Phone: +31402472816

ABSTRACT: This study presents a novel configuration for luminescent solar concentrator photovoltaic (LSC-PV) devices using a mixture of multiple organic luminophores namely, Perylene Green, Lumogen Red 305, and Lumogen Orange 240. This LSC-PV device includes a cubical lightguide with a dimension of 10× 10×10 mm^3 and made of poly (methyl methacrylate) (PMMA), containing a mixture of two or three perylene dyes. The mono-crystalline silicon bifacial solar cells (mono c-Si) are mounted on the edges and bottom sides of this device. This study evaluates the optical efficiency (OE) and power conversion efficiency (PCE) of four different configurations of LSC-PV devices with variations of mixed luminescent dyes in order to obtain the highest performance and optimal structure for the combination of dyes. For this purpose, a Monte Carlo ray tracing simulation is conducted in LightTools software to explore the effect of multiple luminescent dyes on the optical performance of LSC-PV devices and analyse the incident irradiance on the front and rear sides of edges- and bottom-mounted bifacial solar cells. The comparison results show that the highest theoretical optical efficiency is 88.2% under standard test conditions (STC) for an LSC-PV device containing a mixture of Lumogen Orange 240 and Perylene Green dye at 10 ppm. The theoretical power conversion efficiency is then 17.7%.

Keywords: Luminescent Solar Concentrator Photovoltaics (LSC-PV), Organic Luminophores, Ray tracing Simulation, Bifacial Solar cells

1 INTRODUCTION

Luminescent Solar Concentrators Photovoltaic (LSC-PV) as a promising technology aims to reduce the cost of energy generated by photovoltaic (PV) by minimizing the use of silicon materials in PV modules. Moreover, it provides further flexible architectural designs for the PV integration systems into the built environment and transportation [1]. The first generation of LSC has been developed for improving the spectral response of the solar cells in the 1970s [2], [3]. This concept has dominated the challenge of using conventional concentrators that work only with direct radiation namely require sun trackers and probably cause the hot spot on PV cells in long term [4], [5]. LSC-PV includes three main components namely, transparent host material as the lightguide usually made of glass or polymer (e.g. poly(methyl methacrylate) (PMMA) or polycarbonate (PC)) which mixed with luminescent species (i.e. organic dyes [6], [7], quantum dots (QDs) [8], [9] and rare-earth ions [10], [11]) and solar cells which attached to the edge or and back sides of lightguides [12], [13]. LSCs can collect the direct and diffuse irradiance and can be designed in different geometric shapes and colors. The incident photons are entered into the LSC device and absorbed by the luminophores and re-emitted in desire and longer wavelengths. Subsequently, the waveguided photons are trapped in the lightguides and concentrated by total internal reflection (TIR) to the edge-mounted and bottom-mounted PV cells [14], [15]. Conventional silicon PV modules are usually black or dark blue in color and have a rectangular flat shape.
In contrast to this, LSCs offer various beneficial design opportunities such as color, shapeability and transparency leading to the enhancement of the general functionality of PV applications [16], [17].

Optical efficiency (OE) and power conversion efficiency (PCE) of LSC-PV highly depend on the design configuration (i.e. host materials, luminescent species, shape) and also the spectral response of PV cell technology [18].

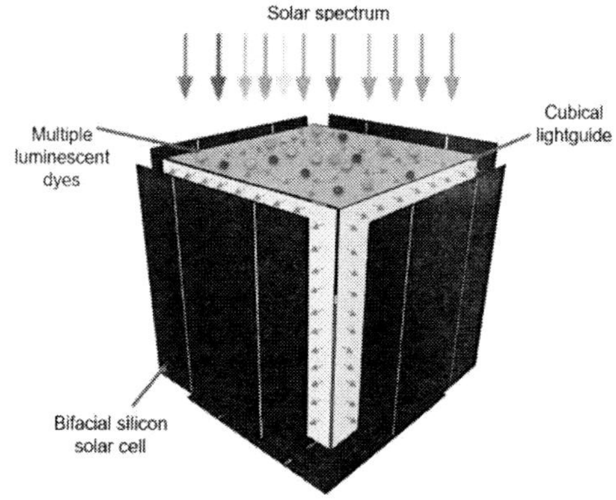

Figure 1: Schematic of a cubical LSC-PV device containing multiple mixed luminophore with edge- and bottom-mounted bifacial solar cells

To date, numerous efforts have been devoted to enhancing the efficiency and lifetime of the LSC-PV devices through investigating luminescent materials with the ideal optical properties namely, high absorption and quantum yield efficiency, large Stokes shift, broad spectral absorption, good solubility in the host matrix materials and also a good matching the emitted photons to the spectral response of the type of solar cells desired to be attached to the LSCs [19], [20].

Organic dyes were investigated as the first luminescent materials for the first generation of LSCs and also currently still used for the fabrication of LSC-PV devices due to their large absorption coefficient, good solubility, and high quantum yield efficiency. Organic dyes as fluorophores for lightguides are categorized into different classes, such as bipyridines, coumarins, dicarbocyanine iodides, dicyano ethylenes, lactones, naphtalimides, oxazines, perylenes and perylenebisimides, erylenebisimidazoles, rhodamines, sulforhodamines, BASF K1 and etc.[20].

To increase the spectral absorption width of organic luminophores, several dyes can be combined whether through a tandem structure of lightguides [21], [22], or a mixture of different organic dyes in a single host matrix [13], [23]. In the latter approach, the dyes absorb the incident photons and transfer to the neighbor dyes with the smallest spectral absorption breadth through radiative or non-radiative transferring namely, Förster resonance energy transfer (FRET) [20]. In this concept, when several dyes are mixed in a lightguide, a cascaded process of absorption and emission occurred may occur in the LSC. Thereby, the incident photons are absorbed and reemitted significantly by the dyes with longer wavelengths [24].

This study presents a novel configuration for LSC PV devices using a mixture of multiple perylene luminophores i.e. Perylene Green, Lumogen Red 305, and Lumogen Orange 240 inside a PMMA cube (10× 10×10 mm^3) and by attaching bifacial silicon solar cells to the edges and rear sides of LSC, see Figure 1 [21]. To evaluate the performance of the proposed LSC-PV devices, ray tracing simulations have been conducted in LightTools software.

2 MATERIALS AND SIMULATION SETUP

In this research, different configurations of LSC-PV devices will be modelled and simulated using the LightTools package which is containing ray tracing simulation software based on Monte Carlo methods [21]–[23] in order to evaluate the effect of cascaded absorption and emission of multiple selected dyes on the optical and electrical performance of this LSC-PV device. The LightTools performs tracing of millions of light rays and analyses the photon-phosphor interaction.

2.1 Design configuration

The design configurations cover four mixtures of two and three organic luminophores namely, Lumogen Red 305 and Lumogen Orange 240 (OR), Perylene Green and Lumogen Red 305 (GR), Lumogen Orange 240 and Perylene Green (OG), Lumogen Red 305 and Lumogen Orange 240 (OR) and Perylene Green (ROG), see table 1.

Table 1: The configurations of the LSC-PV devices with different dye mixtures

Case	Configuration	Number of Organic Dyes	Cubical Dimensions (cm^3)	PV Cell Dimensions (cm^3)
1	Lumogen Red 305/Lumogen Orange 240 (RO)	2	1 × 1 × 1	1 × 1 × 0.02
2	Perylene Green/Lumogen Red 305 (GR)	2	1 × 1 × 1	1 × 1 × 0.02
3	Lumogen Orange 240 and Perylene Green (OG)	2	1 × 1 × 1	1 × 1 × 0.02
4	Lumogen Red 305/ Lumogen Orange 240/ Perylene Green (ROG)	3	1 × 1 × 1	1× 1 × 0.02

2.2 Ray tracing simulation setup

Ray tracing simulations were executed for the LSC PV devices in LightTools software under standard test conditions (STC: 1000 W/m^2, AM 1.5 spectrum, and 25 °C) with 70% direct and 30% diffuse solar irradiance, see Figure 1.

In this study, we have calculated the mean free path (MFP) (as one of the important input parameters in LightTools) for the luminophores according to the Beer–Lambert law and based on the wavelength-dependent optical cross-section as well as molar absorptivity coefficient. The MFP is expressed through Equation (1):

$$l = \frac{1}{n\sigma_{eff}} \qquad (1)$$

where l is the Mean Free Path [mm], n is the density of particles [mm^{-3}], and σ_{eff} is the effective cross-sectional area [m^2]. The optical cross-section is calculated via Equation (2):

$$A = -log_{10}(e^{-\sigma(\lambda) d\, N}) \quad (2)$$

where A is the absorbance, d is the plate thickness [m], N is the dye concentration [moles/m^3] and $\sigma(\lambda)$ is optical cross-section area [m^2]. The Absorbance is calculated through Equation (3):

$$A = -log_{10}T \qquad (3)$$

where T is the transmittance. The Beer-Lambert Law formula is computed by Equation (4):

$$A = \varepsilon c l \qquad (4)$$

where ε is the molar absorption coefficient (M^{-1}cm^{-1}), c is the molar concentration (M), and l is the optical path length (cm).

These simulations aim to evaluate the optical performance as well as power conversion efficiency (PCE) of the LSC-PV devices. The incident photons are traced from entry into the LSC PV until absorbed, re-emitted, and escaped from the devices, or collected by solar cells attached [24].

The optical efficiency (OE) of cubical LSC is computed using Equation (5).

$$\eta_{opt} = \frac{\sum P_{out}}{P_{in}} \qquad (5)$$

where P_{out} is the output power of received irradiance on the surfaces of different sides of the lightguide, and P_{in} is the input power of incident irradiance measured at the top surface of the LSC-PV device. The light rays from the sun source directly incident on LSC PV device's top surface and are considered for measuring the input power. The power conversion efficiency (PCE) of LSC-PV devices is calculated through Equation (6):

$$\eta_{LSC\ PV} = \frac{P_{PV}}{(G \times A)} \qquad (6)$$

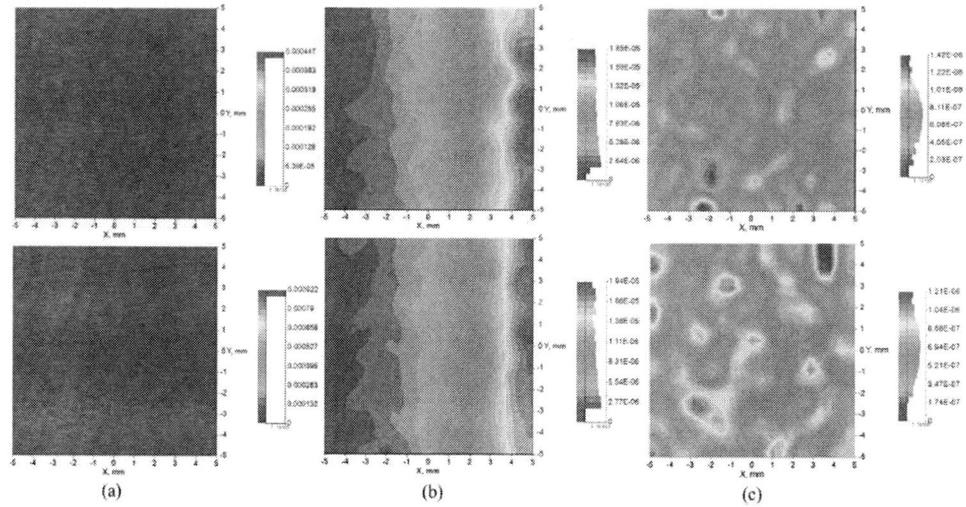

Figure 3: Irradiance distribution plots: (a) on the top surface of the lightguide (Top) and on the bottom-mounted bifacial solar cell (Bottom), (b) on the front, and (c) rear sides of edges-mounted bifacial solar cells.

where P_{PV} (W) is the output power of the solar cells, G is the irradiance (1000 W/m^2) and A is the aperture area (cm^2) of the lightguide.

In this study, the 3D model of the LSC-PV device (see Figure 1) was designed in LightTools and the material properties have been defined in this software for ray tracing simulation. The PMMA host materials with a refractive index of 1.49 at 589 nm were mixed with the combination of two or three perylene dyes. The mono c-Si bifacial solar cells with an STC efficiency of 20% have been modeled as receiver planes with a refraction index of 3.4 at 550 nm. The number of rays varies between 1 million and 5 million. The dye concentration has been increased from 10 ppm to 80 ppm for perylene dyes.

Figure 2 shows the geometry setup of the cubical LSC-PV device with the receiver positions in the model in LightTools software.

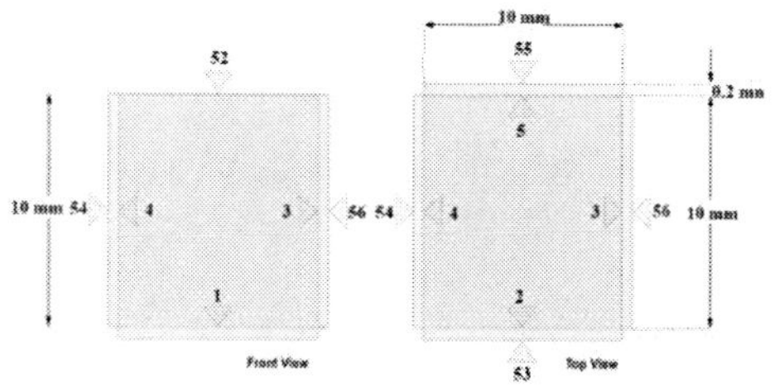

Figure 2: Geometry model of LSC-PV device with front and top view of this device with the receiver positions.

3 RESULTS

Figure 3 depicts the irradiance distribution on the top surface of LSC-PV devices as well as on the front and rear sides of bifacial solar cells.

Figure 3 (a) shows the highest irradiance distribution on the top and bottom surface of LSC-PV device, meaning that a maximum number of photons reaching the top surface and concentrated and waveguided through TIR toward the bottom side where the PV cell-attached and as well as to the edges of lightguide.

Figure 3 (b) illustrates the irradiance distribution on the inner surfaces of cubical LSC and the front side of bifacial solar cells mounted on the edges. As observed in Figure 3 (c), the lowest irradiance falling on the outer surfaces of lightguide. Therefore, the rear side of edges-mounted bifacial solar cells receiving only the diffuse irradiance.

The results of ray tracing simulation have been compared based on optical performance (OE) and power conversion efficiency (PCE) in order to obtain the optimal mixture variation of perylene dyes, see Figure 4 and Table 2.

As shown in Figure 4, the increase in the dye concentration after saturation point of 40ppm, led to a decrease in the incident irradiance for the bifacial PV cells mounted on the edges and bottom sides and consequently, in the efficiency of LSC-PV devices containing the RO, OG and GR dyes. However, the OE and PCE of LSC-PV device with a mixture of Lumogen red and Perylene green exponentially increased after 30ppm due to the scattering of incident irradiance.

Table 2: Modelled OE and PCE for the four differently configurated LSC-PV devices

Case	Optical Efficiency (%)	Power Conversion Efficiency (%)
1	84.9	17
2	67	13.4
3	**88.2**	**17.7**
4	83.7	16.8

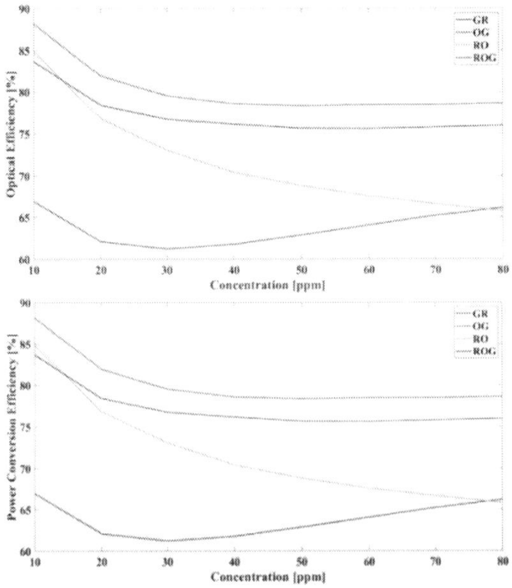

Figure 4: OE and PCE of LSC-PV devices with different combinations of selected organic dyes as a function of the dye concentration.

4 DISCUSSION AND CONCLUSION

In this study, we simulated the optical and electrical performance i.e. OE and PCE, with LightTools software, for four different configurations of LSC-PV devices with different dye mixtures, see Table 2. The highest OE of 88.2% and PCE of 17.7% were obtained, under STC, with 70% direct and 30% diffuse irradiance for the LSC-PV device containing a mixture of Lumogen Orange 240 and Perylene Green dye at 10 ppm. These can be explained by a minimal overlap between the absorption and emission spectrum of mixed luminophores which reduces the reabsorption losses in the device, and consequently results in a higher theoretical PCE of this type of LSC-PV devices.

REFERENCES

[1] F. Meinardi, F. Bruni, and S. Brovelli, "Luminescent solar concentrators for building-integrated photovoltaics," *Nature Reviews Materials*. 2017, doi: 10.1038/natrevmats.2017.72.

[2] W. H. Weber and J. Lambe, "Luminescent greenhouse collector for solar radiation," *Appl. Opt.*, 1976, doi: 10.1364/ao.15.002299.

[3] A. Goetzberger, "Fluorescent solar energy collectors: Operating conditions with diffuse light," *Appl. Phys.*, 1978, doi: 10.1007/BF00885865.

[4] J. Yoon *et al.*, "Flexible concentrator photovoltaics based on microscale silicon solar cells embedded in luminescent waveguides," *Nat. Commun.*, 2011, doi: 10.1038/ncomms1318.

[5] D. D. Nguyen, B. Lehman, and S. Kamarthi, "Performance evaluation of solar photovoltaic arrays including shadow effects using neural network," in *2009 IEEE Energy Conversion Congress and Exposition, ECCE 2009*, 2009, doi: 10.1109/ECCE.2009.5316451.

[6] P. Della Sala *et al.*, "First demonstration of the use of very large Stokes shift cycloparaphenylenes as promising organic luminophores for transparent luminescent solar concentrators," *Chem. Commun.*, 2019, doi: 10.1039/c8cc09859j.

[7] Y. Li, J. Olsen, K. Nunez-Ortega, and W. J. Dong, "A structurally modified perylene dye for efficient luminescent solar concentrators," *Sol. Energy*, 2016, doi: 10.1016/j.solener.2016.07.051.

[8] P. Moraitis, R. E. I. Schropp, and W. G. J. H. M. van Sark, "Nanoparticles for Luminescent Solar Concentrators - A review," *Opt. Mater. (Amst).*, 2018, doi: 10.1016/j.optmat.2018.07.034.

[9] Y. Zhou, H. Zhao, D. Ma, and F. Rosei, "Harnessing the properties of colloidal quantum dots in luminescent solar concentrators," *Chemical Society Reviews*. 2018, doi: 10.1039/c7cs00701a.

[10] V. T. Freitas *et al.*, "Eu3+-based bridged silsesquioxanes for transparent luminescent solar concentrators," *ACS Appl. Mater. Interfaces*, 2015, doi: 10.1021/acsami.5b01281.

[11] A. R. Frias *et al.*, "Transparent luminescent solar concentrators using Ln 3+ -based ionosilicas towards photovoltaic windows," *Energies*, 2019, doi: 10.3390/en12030451.

[12] L. H. Slooff, A. R. Burgers, and M. G. Debije, "Reduction of escape cone losses in luminescent solar concentrators with cholesteric mirrors," in *High and Low Concentration for Solar Electric Applications III*, 2008, doi: 10.1117/12.794522.

[13] L. H. Slooff *et al.*, "A Luminescent Solar Concentrator with 7.1% power conversion efficiency," *Phys. Status Solidi - Rapid Res. Lett.*, vol. 2, no. 6, pp. 257–259, Dec. 2008, doi: 10.1002/pssr.200802186.

[14] A. Reinders, M. G. Debije, and A. Rosemann, "Measured Efficiency of a Luminescent Solar Concentrator PV Module Called Leaf Roof," *IEEE J. Photovoltaics*, 2017, doi: 10.1109/JPHOTOV.2017.2751513.

[15] M. Rafiee, S. Chandra, H. Ahmed, and S. J. McCormack, "An overview of various configurations of Luminescent Solar Concentrators for photovoltaic applications," *Opt. Mater. (Amst).*, vol. 91, no. January, pp. 212–227, 2019, doi: 10.1016/j.optmat.2019.01.007.

[16] A. Reinders, J. C. Diehl, and H. Brezet, *The Power of Design: Product Innovation in Sustainable Energy Technologies*. 2012.

[17] A. Reinders, R. Kishore, L. Slooff, and W. Eggink, "Luminescent solar concentrator photovoltaic designs," *Jpn. J. Appl. Phys.*, vol. 57, no. 8, 2018, doi: 10.7567/JJAP.57.08RD10.

[18] C. Tummeltshammer, A. Taylor, A. J. Kenyon, and I. Papakonstantinou, "Losses in luminescent solar concentrators unveiled," *Sol. Energy Mater. Sol. Cells*, vol. 144, pp. 40–47, 2016, doi: 10.1016/j.solmat.2015.08.008.

[19] W. G. J. H. M. van Sark *et al.*, "Luminescent Solar Concentrators - A review of recent results," *Opt. Express*, 2008, doi: 10.1364/oe.16.021773.

[20] M. G. Debije and P. P. C. Verbunt, "Thirty years of luminescent solar concentrator research: Solar energy for the built environment," *Advanced Energy Materials*, vol. 2, no. 1, pp. 12–35, Jan.

2012, doi: 10.1002/aenm.201100554.
[21] M. Aghaei, M. Nitti, N. J. Ekins-Daukes, and A. H. M. E. Reinders, "Simulation of a novel configuration for luminescent solar concentrator photovoltaic devices using bifacial silicon solar cells," *Appl. Sci.*, 2020, doi: 10.3390/app10030871.
[22] B. Vishwanathan *et al.*, "A comparison of performance of flat and bent photovoltaic luminescent solar concentrators," *Sol. Energy*, 2015, doi: 10.1016/j.solener.2014.12.001.
[23] J. Shu *et al.*, "Monte-Carlo simulations of optical efficiency in luminescent solar concentrators based on all-inorganic perovskite quantum dots," *Phys. B Condens. Matter*, 2018, doi: 10.1016/j.physb.2018.08.021.
[24] J. Shu *et al.*, "Monte-Carlo simulations of optical efficiency in luminescent solar concentrators based on all-inorganic perovskite quantum dots," *Phys. B Condens. Matter*, vol. 548, no. June, pp. 53–57, 2018, doi: 10.1016/j.physb.2018.08.021.

EXPERIMENTAL TEST OF HEAT RECOVERY IN SILICON SOLAR CELLS WITH THERMOELECTRIC MATERIALS

Kenji Kamide[1], Toshimitsu Mochizuki[1], Jun Sakuma[2], Hidefumi Akiyama[2,3], and Hidetaka Takato[1]

[1]National Institute of Advanced Industrial Science and Technology (AIST), Koriyama, Fukushima, 963-0298, Japan,
[2]The University of Tokyo, Kashiwa, Chiba, 277-8581, Japan,
[3]OPERANDO-OIL, Kashiwa, Chiba, 277-8581, Japan

ABSTRACT: Installation environments of solar cells are surrounded by heat, which are also exhausted from the cells themselves and degrade the device performance normally. Here we report an experiment to show that hybrid devices of silicon solar cells and commercial thermoelectric materials, bismuth telluride, can recover heat to exhibit a positive temperature characteristic, i.e., the output power increases with the solar cell temperature. The result implies the hybrid structure can be another type of realization for our concept of "heat recovery solar cell", namely the type II (HERC-II).

Keywords: Solar Cell, Fundamentals, Heat Recovery, Thermoelectric Effect

1 INTRODUCTION

The research level silicon (Si) solar cells have become highly efficient, approaching the theoretical limit in the power conversion efficiency around 29 % [1, 2]. On the other hand, even in such efficient solar cells, large amount, around half of the solar energy is exhausted as heat. Heat exists everywhere around normal installation environments of solar cells e.g. house/car roofs and photovoltaic (PV) power plants, heating up the solar cells and degrading the performance especially in summer. If part of the heat could be recovered as electricity, it will be possible to improve efficient solar cells further. However, heat is nothing but a strong degradation factor to solar cells, mainly due to the negative temperature coefficient of the open-circuit voltage (≡ V_{OC}) [3].

Some concepts for heat recovery in solar cells have been known from the early stage of PV research, e.g. hot carrier solar cells [4] which are not yet realized due to the high technical hurdles. Our recent proposal of "heat recovery (HERC) solar cell [5]" is also along this line of thinking and relaxes the requirement of hot carrier solar cells, ultrafast carrier extraction before thermalization. These solar cells have a common feature that the conversion efficiency increases with the temperature of carriers in the light absorber, i.e., "*positive temperature dependences*," which further attributes to a thermoelectric (TE) or Seebeck effect produced at the energy selective layers, by which hot carriers in the absorber can be extracted to the cold electrodes at high voltage. The energy loss as heat is greatly reduced by the thermoelectric effect in these solar cells.

Following the above, there is another and more direct way for heat recovery in solar cells, that is by using TE materials instead of the energy selective layers. PV-TE hybrid devices have also been studied well [6,7]. A series-connected 2-terminal PV-TE hybrid structure shown in Fig. 1 is a direct analog of HERC solar cells. In this device, heat generated in solar cells is used to gain Seebeck voltage in the TE part (the transmitted subbandgap photons can also be absorbed and used as heat by installing an infrared (IR) absorber). We shall call the device of this structure "heat recovery solar cell type II (HERC-II)."

Despite the economic advantage [7], a small number of papers study the 2-terminal PV-TE hybrids while many papers focus the electrically disconnected 4-terminal PV-TE hybrids [6]. Indeed, a small number of experimental reports are found on the 2-terminal hybrid based on Si solar cells [8]. This may be due to a fact that Si solar cells have large negative temperature coefficients compared to those made of other materials, such as thin-film [9,10], dye-sensitized [11] and perovskite solar cells [12], giving an impression that having the merit of heat recovery with Si solar cells will be difficult [7] when placed at the hot side of the hybrid devices. There is also another concern on the PV-TE devices; the typical temperature difference required for the TE generation to work efficiently is above 100 K, which is beyond the applicable range of PV cells. On the other hand, a simple model simulation shows a potential merit of hybridizing PV and TE, even with presently commercial devices [13].

Here we report an experimental test to find whether or not Si solar cells can be improved by heat recovery with commercial TE materials within an applicable temperature range. We fabricate HERC-II devices based on Si (PERC) solar cells and bismuth telluride (Bi_2Te_3) TE devices, and performed temperature controlled *I-V* measurements within a realistic range of the temperatures.

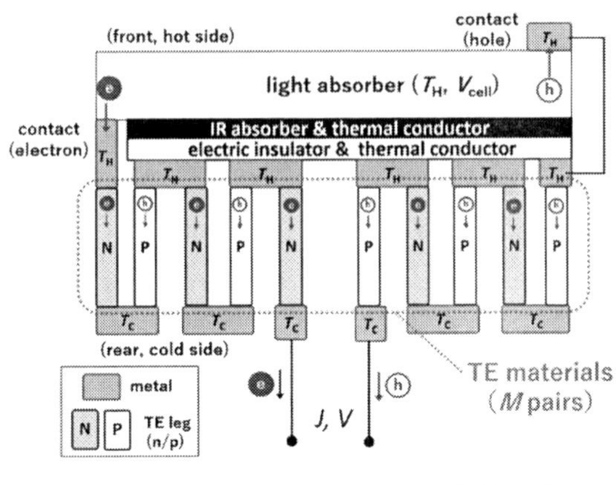

Figure 1: Design of a 2-terminal PV-TE hybrid device (HERC-II), with M pairs of TE legs: T_H = temperature of the PV cell and the hot sides of TE legs, T_C = temperature of the cold sides of TE legs and output electrodes = the room temperature.

2 EXPERIMENTS

We fabricate HERC-II devices with commercial Si

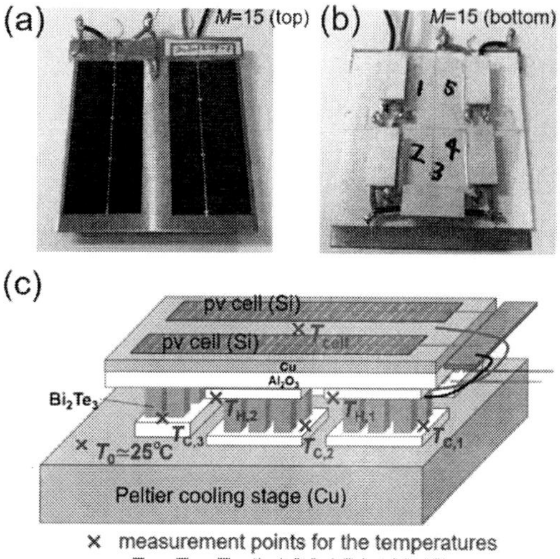

Figure 2: Photographic images of the HERC-II device with $M = 15$, taken from the (a) top and (b) bottom sides. (c) Measurement points for the temperatures in the device (all lead wires are not shown for simplicity).

solar cells (PERC cells) and commercial TE devices of Bi_2Te_3. The TE part consists of plural TE mini modules (Ferrotec Corporation, Product Number: 2003A/003/150A/C01) each of which contains 3 p-n pairs of TE legs with the same dimensions: cross section of 2.5 mm×2.5 mm and 2 mm in height. The devices under test do not contain an IR absorber shown in Fig. 1. In this paper, we present the results for the HERC-II devices with different numbers of TE pairs, $M = 3$ and 15. PERC cells with power conversion efficiencies of 20-21 percent at 25 deg C and total areas of A_{pv}=9 cm^2 and 18 cm^2, are installed on the HERC-II devices with M=3 and 15, respectively. As an example, Figure 2 shows the fabricated device with $M = 15$ viewed from the (a) top and (b) bottom sides.

The I-V measurements (1 sun, AM1.5G) for these devices are performed under temperature control. The hot-side and cold-side temperatures are controlled using a heat gun (with digital temperature control) and a water-cooled Peltier cooling system. The heat gun is used in this experiment, which mimics a situation in real application under the presence of external heat source. The HERC-II devices are placed on a Peltier cooling stage whose temperature is kept at $T_0 = 25$ deg C (=298 K), almost the same as the room temperature, during the whole measurements. This condition is chosen because it is an ideal case realized by passive cooling with a well-designed heatsink.

Figure 2 (c) shows the measurement points of the temperatures to which thermocouples are attached for the device with M=15. As shown in Section 3 (Fig. 4 (b)), thermal gaps open between the solar cells and hot sides of the TE materials and also between the cooling stage and cold sides of the TE materials, due to the imperfect thermal contacts and finite thermal resistances, as being different from Fig. 1. Thus, the temperatures of the PV cells, T_{cell}, and of the hot and cold sides of the TE legs, $T_{H,j}$ and $T_{C,j}$ (j=1,2,3,4,5 for M=15) are measured carefully. We also evaluate the average temperatures, $T_H^{avg.}$ and $T_C^{avg.}$ since they should determine the device characteristics given that all the TE legs have the same material parameters.

3 RESULTS AND ANALYSIS

Figure 3 shows the J-V curves ($J(\equiv I/A_{pv})$ is the current density per acceptance area) of the HERC-II device with $M = 15$ (Fig. 3 (a)) and PV cells mounted on it (Fig. 3 (b)), measured at the PV cell temperature, T_{cell}, of around 25 deg C and 100 deg C. In the HERC-II device, the V_{OC} and the max output power per unit area, P_{max} increases significantly at high temperature, showing a positive temperature dependence. On the other hand, in PV cell only, they decrease significantly at high temperature, showing a negative temperature dependence.

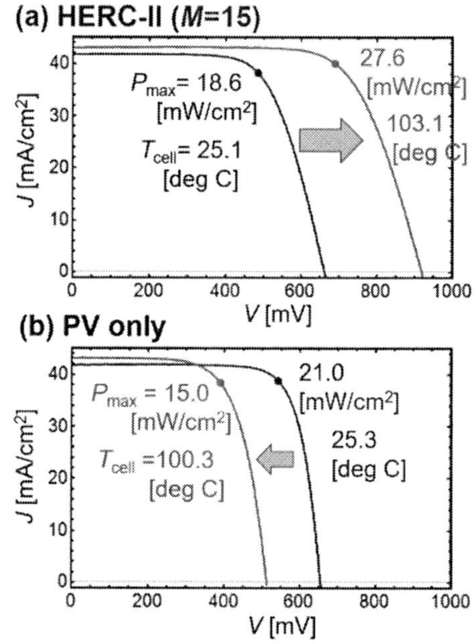

Figure 3: J-V curves of (a) the HERC-II device with M=15 and (b) PV cells mounted on it, measured at the cell temperature of around 25 deg C and 100 deg C.

Figure 4 (a) highlights our main results, where P_{max}, for the HERC-II devices (M=3 and 15) and of the PV cells installed on them are shown as a function of T_{cell}:

- P_{max} of the PV cells, P_{max}^{PV} (dotted lines), decreases with the temperature. The relative temperature coefficient, $\beta_{PV} = 100(\mathrm{d}P_{max}^{PV}/\mathrm{d}T_{cell}/P_{max}^{PV}) \approx -0.4$ %/K, is a typical value for the Si solar cells.
- The temperature coefficient of HERC-II devices with M=15 and 3 (black and red solid) improved significantly compared to those of the PV cells alone. The positive temperature coefficient is found for M=15, while it is negative for M=3.
- At high temperatures, T_{cell}>40 deg C, P_{max} of the HERC-II devices, $P_{max}^{HERC\text{-}II}$, exceeds those of PV cells, P_{max}^{PV}. However, $P_{max}^{HERC\text{-}II}$ is lower than P_{max}^{PV} at low temperature T_{cell}< 30 deg C.
- Experimental data for $P_{max}^{HERC\text{-}II}$ are qualitatively explained by the theory, but quantitatively lower than the theoretical data (by electrical and thermal modeling for

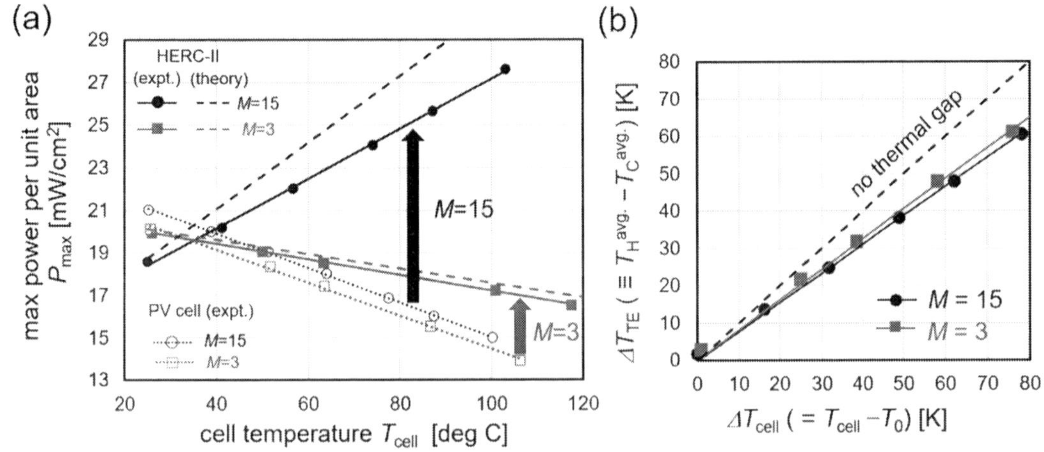

Figure 4: (a) The max output power per unit area, P_{max}, for the HERC-II devices with $M = 3$ and 15 (solid lines) and for the PV cells installed on them (dotted lines), shown as a function of the cell temperature, T_{cell}. The theoretical lines for the HERC-II devices (dashed lines) are obtained by electrical and thermal modeling with no thermal gaps (same as Fig. 1) and the cold-side temperature, $T_C = 25$ deg C, based on 1 diode model fitting to the experimental T_{cell}-dependent *I-V* curves for PV parts and with typical TE parameters of Bi_2Te_3 for the TE parts: Seebeck coefficient, $\alpha = 200$ μV/K, electrical conductivity, $\sigma = 1/\rho = 10^5$ S/m, thermal conductivity $\kappa = 1$ W/mK. (b) Correlation between the average difference between the hot and cold side temperatures of TE legs, ΔT_{TE} ($\equiv T_H^{avg} - T_C^{avg}$), and the temperature rise of the PV cells, ΔT_{cell} ($\equiv T_{cell} - T_0$).

the HERC-II device under the presence of an external heat source with no thermal gap same as Fig. 1, but without an IR absorber).

The second point is the most important findings in this experimental test, which clearly shows that HERC-II devices with proper designs recover heat and exhibits a positive temperature characteristic in the output power under the presence of an external heat source. All the above results can be understood by the following analysis. P_{max} of the HERC-II devices are approximately given by:

$$P_{max}^{HERC\text{-}II}(T_{cell}) \simeq P_{max}^{pv}(T_{cell}) + 2M(\alpha \Delta T_{TE} - R_{TE} I)J, \quad (1)$$

where α and R_{TE} are the Seebeck coefficient and electric resistance of a TE leg, and ΔT_{TE}($\equiv T_H^{avg} - T_C^{avg}$) is the temperature difference inside a TE leg. From Eq. (1), we find the difference between $P_{max}^{HERC\text{-}II}$ at T_{cell} and P_{max}^{pv} at T_0 (=25 deg C) is given by

$$P_{max}^{HERC\text{-}II}(T_{cell}) - P_{max}^{pv}(T_0)$$
$$= (dP_{max}^{pv}/dT_{cell})\Delta T_{cell} + 2M\alpha\Delta T_{TE} J - 2MR_{TE} IJ. \quad (2)$$

The temperature coefficient of HERC-II devices is determined by the first and second terms of the right-hand side of Eq. (2), meaning that the positive temperature dependence is found if the total Seebeck coefficient, $2M\alpha$, exceeds the large negative coefficient of the PV cells i.e. for large M. This explains the second point of the findings. At low temperature, $T_{cell} \simeq T_0$ ($\Delta T_{cell} \simeq \Delta T_{TE} \simeq 0$), we find that the difference $P_{max}^{HERC\text{-}II} - P_{max}^{pv}$ is given by the ohmic loss in the TE legs (the third term of the rhs of Eq. (2)) which increases with M. At the higher temperatures, Seebeck voltage dominates other loss terms for large M. This explains the third point of the findings. Moreover, the effect of thermal gaps is also found in Eq. (2). As was clearly found by the experiments (Fig. 4 (b)), ΔT_{TE} is smaller than ΔT_{cell} by the presence of the gaps. The thermal gap ($=\Delta T_{cell} - \Delta T_{TE}$) reduces Seebeck voltage, and the reduction in the voltage should increase with M. This explains why the experimental $P_{max}^{HERC\text{-}II}$ is smaller than the theoretical $P_{max}^{HERC\text{-}II}$ (without the gaps) and why the degradation found for M=15 due to the thermal gaps is stronger than that for M=3 despite the thermal gaps are similar in the two devices. This tells us that a careful thermal design to eliminate the thermal gaps is necessary, especially for the device with large M.

4 SUMMARY

In this paper, we have shown that well-designed HERC-II devices made of Si solar cells and commercial TE materials, bismuth telluride, can recover heat to exhibit a positive temperature characteristic under the presence of external heat source. Our next target is to observe similar positive temperature characteristics of HERC-II devices by using the heat generated from the solar cells themselves, without relying on the external heat source. Application of HERC-II device to concentrator PV is also an interesting subject while the thermal management especially in the heat extraction part will be a challenging issue.

ACKNOWLEDGEMENTS

This work is partly supported by Japan Society for the Promotion of Science (JSPS KAKENHI Grant Number JP19K04523).

REFERENCES

[1] W. Shockley and H. J. Queisser, "Detailed balance limit of efficiency of p-n junction solar cells," J. Appl. Phys. 32, 510–519 (1961).
[2] K. Yoshikawa, et al. "Silicon heterojunction solar cell with interdigitated back contacts for a photoconversion efficiency over 26%," Nat. Energy, 2, 17032 (2017).

[3] P. Würfel and U. Würfel, "Physics of Solar Cells: From Basic Principles to Advanced Concepts, 3rd ed." (Wiley-VCH, 2016).
[4] R. T. Ross and A. J. Nozik, "Efficiency of hot-carrier solar energy converters," J. Appl. Phys. 53, 3813–3818 (1982).
[5] K. Kamide, T. Mochizuki, H. Akiyama, and H. Takato, "Heat-recovery solar cell," Phys. Rev. Appl. 12, 064001 (2019).
[6] G. Li, S. Shittu, T. M. Diallo, M. Yu, X. Zhao, and J. Ji, "A review of solar photovoltaic-thermoelectric hybrid system for electricity generation," Energy, 158, 41-58 (2018).
[7] D. Narducci and B. Lorenzi, "Challenges and Perspectives in Tandem Thermoelectric–Photovoltaic Solar Energy Conversion," IEEE Transactions on Nanotechnology 15, 348-355 (2016).
[8] K. T. Park, S. M. Shin, A. S. Tazebay, H. D. Um, J. Y. Jung, S. W. Jee, M. W. Oh, S. D. Park, B. Yoo, C. Yu, and J. H. Lee, "Lossless hybridization between photovoltaic and thermoelectric devices," Sci. Rep. 3, 2123 (2013).
[9] A. Virtuani, D. Pavanello, G. Friesen, Overview of temperature coefficients of different thin film photovoltaic materials. In Proceedings of the 25th European Photovoltaic Solar Energy Conference and Exhibition, 4248–4252 (2010).
[10] D. T. Cotfas, P. A. Cotfas, O. M. Machidon, "Study of Temperature Coefficients for Parameters of Photovoltaic Cells", Int. J. Photoenergy, vol. 2018, Article ID 5945602, (2018).
[11] S. R. Raga and F. Fabregat-Santiago, "Temperature effects in dye-sensitized solar cells", Phys. Chem. Chem. Phys. 15, 2328-2336 (2013).
[12] F. Fu, T. Feurer, T. P. Weiss, S. Pisoni, E. Avancini, C. Andres, S. Buecheler, and A. N. Tiwari, "High-efficiency inverted semi-transparent planar perovskite solar cells in substrate configuration," Nat Energy 2, 16190 (2017).
[13] W.G.J.H.M. van Sark, "Feasibility of photovoltaic – Thermoelectric hybrid modules," Applied Energy. 88, 2785-2790 (2011).

ON THE NON-IDEAL DIODE FACTOR IN SOLAR CELLS AND THE EXISTENCE OF MULTIPLE QUASI-FERMI LEVELS

Iñigo Ramiro*, Pablo G. Linares and Antonio Martí
Instituto de Energía Solar
Universidad Politécnica de Madrid, 28040 Madrid, Spain
*i.ramiro@upm.es

ABSTRACT: The diode ideality factor is a key parameter for understanding the recombination mechanisms that dominate in a solar cell. This factor is usually assumed to be in the range 1-2. Values greater than 2 are generally considered anomalous since there is not a clear explanation for them. In this work, we develop a simple *m*-level expansion of the Shockley-Read-Hall model for recombination in a *p-n* junction. Our model can account for ideality factors as high as *m*+1. In the light of this model, we analyze and explain the temperature dependence of the ideality factor of a GaSb/GaAs quantum-dot intermediate band solar cell.

Keywords: recombination, quasi-Fermi level, Shockley-Read-Hall, intermediate band

1 INTRODUCTION

The intermediate band solar cell (IBSC) is a high-efficiency PV-device concept which seeks to exceed the Shockley and Queisser limit for single-gap solar cells by reducing the transmission losses inherent to conventional single-gap semiconductors [1]–[3]. For this, they rely on semiconductor materials that have a narrow third electronic band, called the intermediate band (IB), located in the energy gap between the valence and conduction bands (VB and CB), as shown in Figure 1. The presence of the IB allows for absorption of below-band-gap photons and the generation of extra electron-hole pairs, via electron excitation from the VB to the IB and from the IB to the CB in a two-step process. Consequently, the photocurrent of the solar cell can be increased, thus reducing the transparency or transmission. Materials with an IB can be engineered using different technological approaches (see [2] for a review). So far, the most employed approach is the use of semiconductor quantum dots (QDs) to introduce confined electronic levels in the band gap of a wider-band-gap semiconductor. These confined levels play the role of the IB.

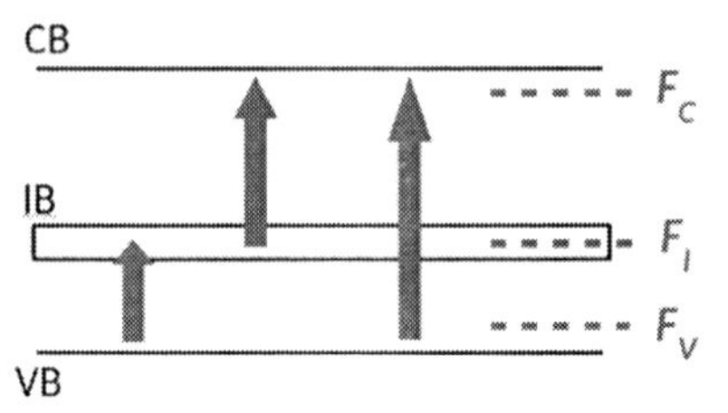

Figure 1: Simplified band diagram of an IB material in an IBSC under operation. The arrows illustrate interband electronic transitions resulting from absorption of photons of different energies.

The IBSC theory framework requires the existence of an electrochemical potential or quasi-Fermi level (QFL) associated to each of the bands, F_C, F_I and F_V, for the CB, IB and VB, respectively (see Figure 1). In the past two decades, different groups have tried to demonstrate the existence of three QFLs in an IBSC [4]–[6] with either negative or inconclusive [7] results. Recently, we have demonstrated the existence of more than 2 QFLs in a QD-based IBSC operating at 9 K [8].

With the aim of studying whether the existence of multiple QFLs has an impact on the recombination properties of our device, we measured its dark *J-V* characteristics at different temperatures (Figure 2). We have fitted the low-current region of the curves to Equation 1:

$$J = J_0\left[\exp\left(V/_{nkT}\right) - 1\right] \tag{1}$$

where J_0 is the reverse saturation current of a diode, V is the applied voltage, n is the diode ideality factor, k is the Boltzmann's constant, and T is temperature. We have fitted the low-current range of the curves to ensure that the devices were operating in the low-injection regime and to avoid series-resistance effects that would mask the actual recombination behavior of the junction. From room temperature (RT) down to 100 K, n increases slowly, going from 1.8 to 2.5. For lower temperatures, n increases more rapidly, reaching 6.2 at 20 K. An additional *J-V* measurement taken at 9 K under illumination (Figure 3) yields $18<n<22$, confirming the rapid increase of n at low temperatures.

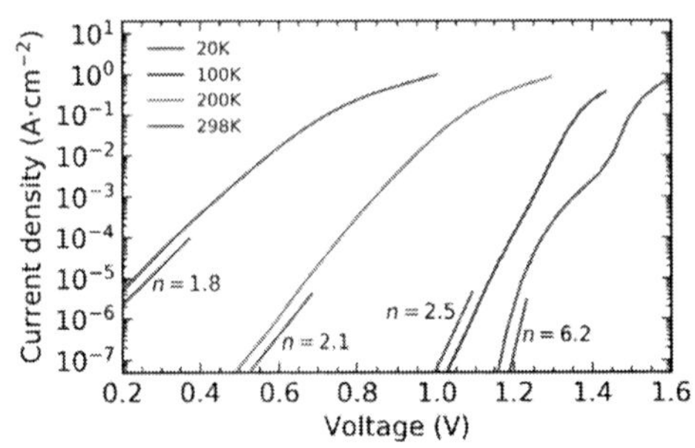

Figure 2: Dark *J-V* characteristic of the studied sample as a function of *T*. The values of *n* obtained by fitting to Equation 1 are indicated.

At low temperatures, the obtained n values are much higher than 2, which is considered the high limit the Shockley-Read-Hall (SRH) recombination [9]. Different explanations for anomalously high values of n can be found in the literature. Some of these are related to the structure of the device. Shah et al. [10] describe how the existence of more than one junction may lead to high n values. Zhu et al. [11] point to the presence of triangular band profiles in devices containing quantum barriers as the cause for high n values. Other explanations add modifications to the conventional SRH model, such as the coexisting contribution of SRH recombination through different recombination centers [12], the contribution of trap-assisted tunneling [13], or the case of a very high density of recombination centers [14]. However, none of these explanations accommodates our case of study, nor do they justify the measured strong temperature dependence of n. In this work, we develop a simple expansion of the SRH model to the case of m recombination levels and show that such a model can account for values of n as high as $m+1$, provided that each recombination center has its own QFL. We use this model to explain qualitatively the temperature dependence of the ideality factor in our QD-based IBSC.

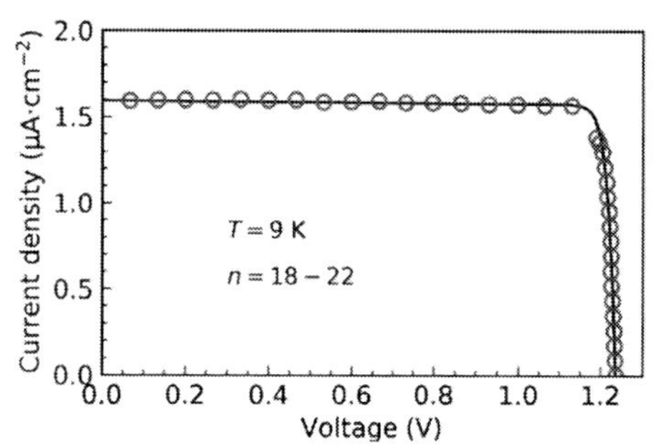

Figure 3: *J-V* characteristic (red circles) of the studied sample under illumination with a 1.93 eV diode laser. The excitation power density was approximately 7.8 μW/cm^2 over a sample area of 6.3 mm^2. The measurement was taken at 9 K. The solid line is a fit to Equation 1 with n=20. Values of n between 18 and 22 resulted in equally good fits.

2 *m*-LEVEL EXPANSION OF THE SRH MODEL

Before expanding the SRH model to the m-level case, it is necessary to briefly review the basics of the conventional SRH theory [9], [15]. It describes the recombination of electrons and holes, in a p-n junction in steady state conditions, via a recombination center that introduces a level inside the band gap. For this, several assumptions are established. The most relevant ones for our work are: (I) The p-n junction is symmetric (in doping) and the capture cross sections of the recombination centers for both carriers (holes and electrons) are the same. (II) There is a single-level recombination center homogeneously distributed in the p-n junction. (III) The level of the recombination centers lies at the energy of the intrinsic Fermi level of the semiconductor. This energy level is called E_t. (IV) The level at Et is associated to its own QFL Ft, which is different from F_C and F_V. Under these assumptions, it is found that the SRH recombination for carriers in the space charge region (SCR) occurs with ideality factor n=2.

Figure 4a illustrates the CB→VB de-excitation of an electron via a SRH process. The conventional equivalent recombination circuit, consisting of one diode with n=2 is shown in Figure 4b. The voltage V applied to the junction, which is equal to the separation between F_C and F_V, drops across the diode. In Figure 4c we present an equivalent two-diode recombination circuit for the SRH model. In this circuit, we make use of the fact the level E_T has its own QFL. Therefore, we can establish three voltage points in the circuit and two diodes connected in series. We highlight that the decomposition of the diode in Figure 4b into the two diodes in Figure 4c emerges naturally from the fact that, in the SRH statistics, the electron and hole capture rates (by the recombination center) are the same [15]. Because of assumptions I and III, the two diodes must be equal, which implies that the voltage drop across each of them is $V/2$. This, in turn, results in n=1.

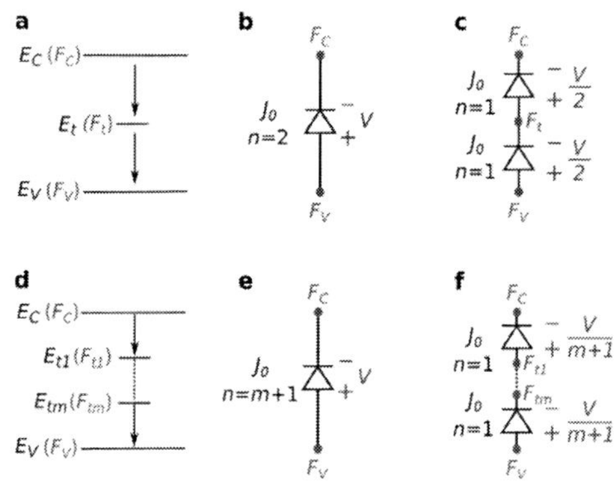

Figure 4: The illustration of the SRH recombination process, the equivalent one-diode recombination circuit, and the equivalent multi-diode recombination circuit, for the conventional 1-level case (a, b and c) and the expanded m-level case (d, e and f).

We can now easily proceed to the m-level generalization of the SRH model (Figure 4d, e and f). We will assume m recombination centers equally distributed in energy, located at energies E_{ti}, i=1, …, m. We will also assume that each level E_{ti} has its own distinct QFL F_{ti}. Analogously to the 1-level case, we can now construct an equivalent recombination circuit containing m+1 diodes, each of them with n=1 and a voltage drop across terminal equals to $V/(m+1)$. It is then straightforward to construct a 1-diode equivalent circuit with $n=m+1$ and a voltage drop of V. This means that the expanded m-level SRH model can account for ideality factors as high as m+1.

3 DISCUSSION

We will now analyze qualitatively the results presented in Section 1 using the model developed in Section 2. First, let us briefly describe the sample under study. The QD-based IBSC consisted of a GaAs p-n junction containing 10 layers of GaSb QDs within the n side of the junction. The whole semiconductor structure was grown by MBE in the University of Michigan (USA). The reader is referred to [16] for a detailed description of the device structure and fabrication. The GaSb/GaAs QDs exhibit a type-II (staggered) band alignment with confinement of holes (Figure 5). For simplicity, only 3 confined levels are illustrated in the figure. However, both theoretical and experimental results indicate that there is a high density of confined hole states (> 100) with an average energy separation of 3 meV [16]. This fact will be key for the following analysis.

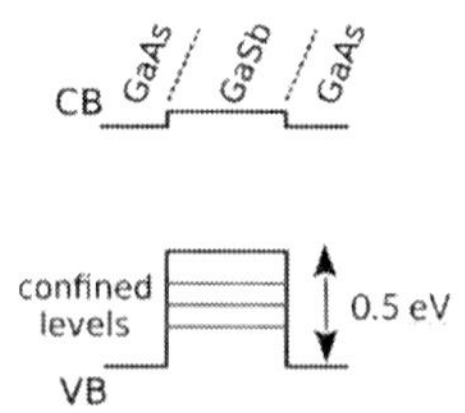

Figure 5: Simplified band diagram of a GaSb/GaAs QDs. The band alignment is type-II with confinement of holes only. The hole potential barrier is approximately 0.5 eV [16].

Table 1 gathers the ideality factors measured in our device at different temperatures. We have included in the table the value of the thermal energy kT for each T. Note that, for the temperature range at which n has values well above 2 (20 K and 9 K), the values of kT are lower than the found average energy separation between confined levels (3 meV). This means that, at these low temperatures, the thermal coupling among the confined levels is weak. In the limit of no thermal coupling, or when the thermal coupling between levels is much lower than the radiative coupling, each level is associated to each own quasi-Fermi level. Hence, it is reasonable to assume that at the lowest temperatures the hole populations of the (many) QD confined states are defined by a large number of QFLs. This is consistent the previous findings [8] mentioned in Section 1.

Table 1: Fitted n values from Figures 2 and 3.

T (K)	kT (meV)	n
298	26	1.8
200	17	2.1
100	9	2.5
20	1.7	6.2
9	0.8	18-22

We are now able to explain the 'anomalously' high ideality factors measured and their temperature dependence (see Figure 6). At 'high' temperatures, the confined levels of the QDs are strongly thermally coupled. Hence, there is essentially one hole population, or one hole gas, described by a single QFL F_I. This is consistent with ideality factors lowers than 2. At 'low' temperatures, the thermal coupling among confined levels is weak and, consequently, multiple QFLs emerge (F_{Ii}, F_{Ij}, F_{Ik}, in the figure). The large number of confined levels introduced by the GaSb QDs and the existence of a large number of QFLs explain the high measured values of n, in the light of the m-level expansion of the SRH model presented in Section 2.

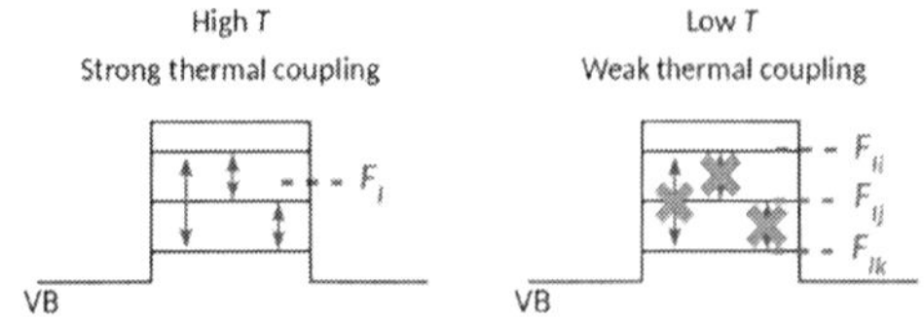

Figure 6: Illustration of the effect of temperature in the interlevel coupling and the existence of multiple QFLs.

4 CONCLUSIONS

We have presented a simple m-level expansion of the SRH recombination model that can account for ideality factors as high as m+1. For this, it is necessary to assume that each of the recombination centers has its own QFL. We have used this model to explain the very high ideality factors measured in an IBSC at low temperatures. We encourage the study of the ideality factor and its dependence with temperature in IBSCs as another tool for analyzing the existence of multiple QFLs.

5 ACKNOWLEDGEMENTS

This work was supported in part by the Project MADRID-PV2-CM (P2018/EMT-4308) funded by the Comunidad de Madrid supported with FEDER funds. I. Ramiro acknowledges funding of the European Commission through a H2020 Marie Skłodowska-Curie Action (GA 891686). The authors acknowledge Andrew Martin, Jinyoung Hwang, Joanna Millunchick, and Jamie Phillips for their work at the University of Michigan providing the solar cell used in this work.

6 REFERENCES

[1] A. Luque and A. Martí, "Increasing the efficiency of ideal solar cells by photon induced transitions at intermediate levels," *Physical Review Letters*, vol. 78, no. 26, pp. 5014–5017, 1997.

[2] I. Ramiro and A. Martí, "Intermediate band solar cells: Present and future," *Prog Photovolt Res Appl*, vol. 29, no. 7, pp. 705–713, 2021.

[3] Y. Okada *et al.*, "Intermediate band solar cells: Recent progress and future directions," *Applied Physics Reviews*, vol. 2, no. 2, p. 21302, 2015.

[4] A. Luque *et al.*, "Experimental analysis of the quasi-Fermi level split in quantum dot intermediate-band solar cells," *Applied Physics Letters*, vol. 87, p. 083505, 2005.

[5] T. Tayagaki, Y. Hoshi, and N. Usami, "Investigation of the open-circuit voltage in solar cells doped with quantum dots," *Scientific reports*, vol. 3, p. 2703, 2013.

[6] T. Aihara, T. Tayagaki, Y. Nagato, Y. Okano, and T. Sugaya, "Investigation of the open-circuit voltage in wide-bandgap InGaP-host InP quantum dot intermediate-band solar cells," *Jpn. J. Appl. Phys.*, vol. 57, no. 4S, p. 04FS04, 2018.

[7] A. A. Abouelsaood, M. Y. Ghannam, and J. Poortmans, "On the reported experimental evidence for the quasi-Fermi level split in quantum-dot intermediate-band solar cells,"

Progress in Photovoltaics: Research and Applications, vol. 21, no. 2, pp. 209–216, 2013.
[8] I. Ramiro *et al.*, "Demonstration of a GaSb/GaAs Quantum Dot Intermediate Band Solar Cell Operating at Maximum Power Point," *Physical Review Letters*, vol. 125, p. 247703, 2020.
[9] C. T. Sah, R. N. Noyce, and W. Shockley, "Carrier generation and recombination in pn junctions and pn junction characteristics," *Proceedings of the IRE*, vol. 45, no. 9, pp. 1228–1243, 1957.
[10] J. M. Shah, Y.-L. Li, Th. Gessmann, and E. F. Schubert, "Experimental analysis and theoretical model for anomalously high ideality factors (n≫2.0) in AlGaN/GaN *p-n* junction diodes," *Journal of Applied Physics*, vol. 94, no. 4, pp. 2627–2630, 2003.
[11] D. Zhu *et al.*, "The origin of the high diode-ideality factors in GaInN/GaN multiple quantum well light-emitting diodes," *Appl. Phys. Lett.*, vol. 94, no. 8, p. 081113, 2009.
[12] J. Beier and B. Voss, "Humps in dark IV-curves-analysis and explanation," in *Photovoltaic Specialists Conference, 1993., Conference Record of the Twenty Third IEEE*, 1993, pp. 321–326.
[13] A. Kaminski, J. J. Marchand, and A. Laugier, "Non ideal dark I—V curves behavior of silicon solar cells," *Solar Energy Materials and Solar Cells*, vol. 51, p. 221—231, 1998.
[14] S. Steingrube, O. Breitenstein, K. Ramspeck, S. Glunz, A. Schenk, and P. P. Altermatt, "Explanation of commonly observed shunt currents in c-Si solar cells by means of recombination statistics beyond the Shockley-Read-Hall approximation," *Journal of Applied Physics*, vol. 110, no. 1, p. 014515, 2011.
[15] W. Shockley and W. T. Read Jr, "Statistics of the recombinations of holes and electrons," *Physical Review*, vol. 87, no. 5, p. 835, 1952.
[16] I. Ramiro *et al.*, "Analysis of the intermediate-band absorption properties of type-II GaSb/GaAs quantum-dot photovoltaics," *Physical Review B*, vol. 96, no. 12–15, p. 125422, 2017.
[17] A. Martí, L. Cuadra, N. López, and A. Luque, "Intermediate band solar cells: Comparison with shockley-read-hall recombination," *Semiconductors*, vol. 38, no. 8, pp. 946–949, 2004.

TANDEM LUMINESCENT SOLAR CONCENTRATORS: OPTIMIZING THE NUMBER OF STACKED PLATES

I.O. Sokolovskyi[1], M.R. Kulish[1], A.V. Sachenko[1], V.P. Kostylyov[1], A.I. Shkrebtii[2]*
[1] V. Lashkaryov Institute of Semiconductor Physics (ISP), NAS of Ukraine,
45 Pr. Nauky, Kyiv 03028, Ukraine
[2] Ontario Tech University, 2000 Simcoe St. N., Oshawa, ON, L1G 0C5, Canada
*Corresponding authors: anatoli.chkrebtii@ontariotechu.ca

ABSTRACT: Using the Monte Carlo formalism, the maximum efficiency of the luminescent photons transport to the plates edges in the tandem luminescent solar concentrator (LSC) is calculated for the stack of plates with different luminophore's parameters. It is shown that increasing the number of plates in the stack decreases thermalization losses and increases reflection losses. The efficiency dependence on the number of plates saturates after 8 plates for the optimal distribution of photon flux between the plates. The thermalization and reflection losses for this case become equal and close to the escape cone losses. On the other hand, for nonoptimal photon flux distribution, the efficiency has a maximum below that one for the optimal flux, and is reached at a smaller number of plates. The higher refractive index of the plates' material, the efficiency maximum is reached for smaller numbers of plates.
Keywords: luminescent concentrator, stack, reflection, thermalization, escape cone

1 INTRODUCTION

The tandem luminescent solar concentrators (LSCs) have recently attracted a lot of interest (see, e.g., [1-3]). To the best of our knowledge, however, there are no investigations on how many LSC plates (also called waveguides or matrices) should be combined in the stack to maximize the tandem photoconversion efficiency. To address this problem theoretically and ensure maximum efficiency of LSC photoconverters as well as acceptable photoconverter cost, we developed a formalism that not only allows to select the optimal number of stacked waveguides, but also allow choosing the quantum dot (QD) types and characteristics of the solar cells (SCs), attached to the waveguides' edges.

Even though the multiple LSC plates in the tandem reduce the thermalization losses, stacking them increases both the reflection and escape cone losses. The formalism developed focuses on the interplay between competing mechanisms: losses of the sunlight photons due to reflection, interaction with the waveguides of the concentrators, and thermalization of high-energy charge carriers, as well as losses in the escape cone. The research presented is aimed to provide the optimal design for this nonconventional photovoltaic device.

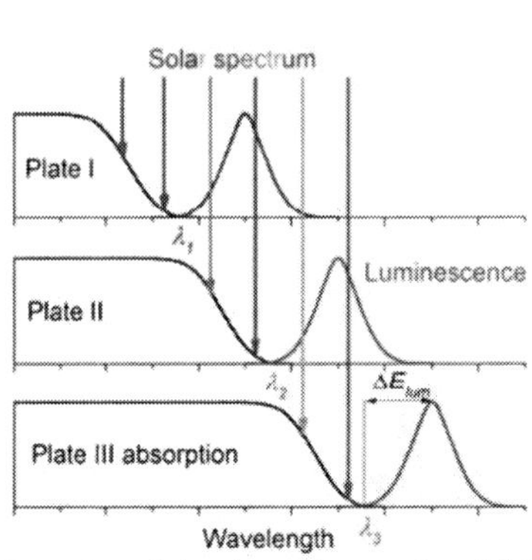

Fig. 1. Schematics of the three-plate tandem LSC. The upper plate fully absorbs higher-energy photons and transmits lower-energy photons to the second and third plates to be absorbed below.

It is predicted theoretically that the LSC efficiency can reach 21.3%, 29.5% and 33.6% for one, two, and three stacked waveguides respectively [4]. Efficiency increase for a few stacked LSC plates in the tandem has been also demonstrated experimentally [5]. However, theoretical or experimental investigations on optimizing the number of solar concentrators plates in the stacked LSCs are still lacking.

With increasing number of the waveguides in the tandem stack, the reflection losses of the incoming sunlight increase while the thermalization losses decrease. Therefore, there is the optimal number of concentrators at which the interplay between thermalization, reflection and other losses maximizes the efficiency. To prove this, using the Monte Carlo method we calculated dependence of the LSC efficiency on the number of plates for several materials with different refractive indexes n of the matrix (n = 1.45 ÷ 1.82). We found that increasing refractive index of the matrix, the optimal number of the concentrators decreases. This is due to increase in the reflection losses in the stack of the incoming sunlight, even though the escape cone losses decreases. If we consider, for instance, the LSC plates, made of fused quartz and doped with quantum dots (QDs), the optimal number of waveguides in the tandem is eight, while exceeding this number decreases the tandem LSC efficiency and increases the device cost.

2 MODEL

Intensity of the incoming solar radiation at the tandem LSC surface, as schematically shown for three-plate LSC in Fig. 1 (see also, [1,3,6]) can be written as:

$$I_0 = \int_{\lambda_0}^{\lambda_{max}} I_{ph}(\lambda)\, d\lambda,$$

where $I_{ph}(\lambda)$ is the intensity spectral density at λ, being the incoming light wavelength, with the shortest wave-length λ_0 and the longest λ_{max} wavelength in the solar photoactive spectral range. Consider that all the incoming sun radiation is absorbed by the luminophores (quantum dots (QDs) are used here) in the plates of the tandem LSC stack. Consider tandem LSC with n plates. Starting from the longest (from the photon photoabsorption edge) wavelength $\lambda_n = \lambda_{max}$ for the n-th cell, while neglecting the absorption at longer than λ_n wavelengths, the solar spectrum can be divided into regions of absorption by the plates, doped with luminophores:

$$I_0 = \int_{\lambda_0}^{\lambda_{max}} I_{ph}(\lambda)\,d\lambda = \int_{\lambda_0}^{\lambda_1} I_{ph}(\lambda)\,d\lambda + \int_{\lambda_1}^{\lambda_2} I_{ph}(\lambda)\,d\lambda + \ldots + \int_{\lambda_{n-1}}^{\lambda_{max}} I_{ph}(\lambda)\,d\lambda = \sum_{l=1}^{n} I_l.$$

Here I_l is the intensity of the light, absorbed by the l-th plate. λ_1, λ_2... are defined by the absorption edge of corresponding QDs, see Fig. 1.

Several mechanisms are responsible for the LSC efficiency η decrease, which can be separated into two main types:

(1) This includes fundamental intrinsic losses, permanent and unavoidable, they cannot be completely eliminated, but only be reduced to their fundamental limits. Such irreversible losses are due to three main basic mechanisms: (i) the reflection from the surfaces of the plates, (ii) the escape cone losses, and (iii) the thermalization losses, all three of them are included in our formalism.

(2) The second type of losses can be mitigated by the LSC design, material choice, etc., and can often be practically eliminated. Several, but not all of them include the light absorption by the matrix [1], reabsorption losses [2], below unity quantum yield of luminescence, and others.

We focus here only on the fundamental (first type, that is irreversible) losses. Two of them, the losses due to reflection and escape cone losses are significantly dependent on the refractive index of the matrix material n_m. If the reflection coefficient (a ratio of the incoming light intensity to that one of the light intensity, reflected from each surface) is $R(\lambda)$, the transmission coefficient $k(\lambda)$ (fraction of the incident energy that is transmitted through the interface), then $k(\lambda) = (1 - R(\lambda))$. In this case, using Fresnel formula, the reflection coefficient $R(\lambda)$ at the normal incidence can be written as:

$$R(\lambda) = \left[\frac{1-n_m(\lambda)}{1+n_m(\lambda)}\right]^2.$$

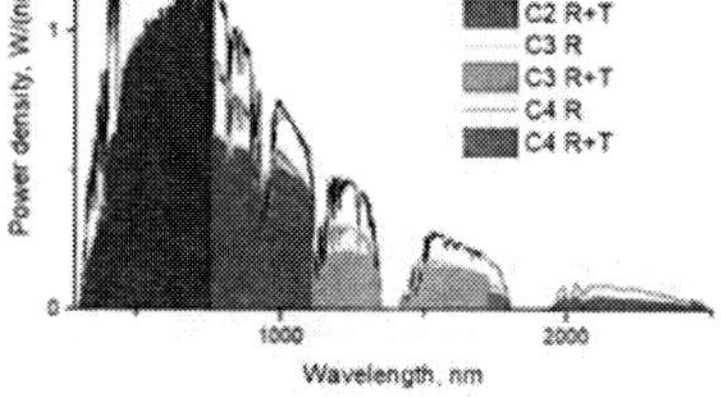

Fig. 2. Power separation for 4-plate (C1-C4) structure. Blackline for total Sun power, color lines for the power on the concentrators after reflection (R) and filled part for the power after reflection and thermal relaxation (R+T).

For the typical matrix materials [9], the reflectivity $R(\lambda)$ for the incoming solar radiation is of order of several percents. For every LSC plate the light undergoes reflection from both top (air-glass) and bottom (glass-air) boundaries, and this is repeated for each subsequent plate. Then the transmission coefficient for the light, which reaches l-th plate k_R^l can be written as:

$$k_R^l = [1 - R(\lambda)]^{1+2(l-1)}.$$

The sunlight intensity, penetrated into the l-th plate and absorbed in it, equals to $I_l k_R^l$, while its part, lost due to the reflection in the l-th part of the spectrum is $I_l(1 - k_R^l)$. It should be noted that for the whole spectrum from λ_0 to λ_l the reflection losses are complete at this point – everything is either reflected or delivered to the plate's edge or radiated into the escape cone.

As we have shown [3], the escape cone losses are determined by the refractive index of the plate at the luminescence wavelength λ_{lum} of the luminophore, which dopes the plate. Thus, the ratio of the luminescent photons that propagate inside the matrix toward the SCs to the total number of photons, emitted by the QDs is:

$$k_{cone} = \sqrt{1 - 1/n_m(\lambda_{lum})^2}\ .$$

Note that the photons, that leak through the escape cone, are also reflected back from the surfaces of the upper boundaries, while the leaked photons, that propagate down, are absorbed by subsequent plates below.

The thermalization energy losses for the photogenerated carriers are determined by the difference between the solar energy, absorbed by the luminophore (QD) inside of each plate and the photon energy that the QD emits. Assuming that all the light, starting from the luminophore wavelength λ_n, is absorbed in the n-th plate, the ratio of this energy conversion can be written as:

$$k_{th}^l = \frac{\left[{hc}/{\lambda_l} - \Delta E_{lum}\right] \int_{\lambda_{l-1}}^{\lambda_l} qN_{ph}(\lambda)d\lambda}{\int_{\lambda_{l-1}}^{\lambda_l} I_{ph}(\lambda)d\lambda}.$$

Here q is the elementary charge, h is the Planck constant, c is the speed of light in vacuum, ΔE_{lum} is the shift between the QD absorption edge and its emission energy, $N_{ph}(\lambda)$ is the density of photons in the solar spectrum.

The power, transmitted to the solar cells by n plates of the tandem LSC can be written as:

$$I_{LSC} = \sum_{l=1}^{n} (I_l k_R^l k_{th}^l + I_{l-1}^{cd}) k_{cone}.$$

Here I_{l-1}^{cd} is the fraction of escape cone losses from the previous plate directed downward. Such additional photon flux is accumulated from plate to plate.

The key issue in the analysis of the efficiency of light conversion by the plates of the luminescent solar concentrator is the choice of absorption energies of the luminophores. Overall efficiency strongly depends on the division of the solar spectrum by the absorption edges of QDs in the plates. Here the Monte Carlo method was applied for this optimization problem and special software has been developed. By randomly selecting the energies of all the luminophores, the software selects the best guess for distribution, which provides the highest efficiency of the tandem LSC under corresponding restrictions. (To highlight a versatility of Monte Carlo formalism for LSCs, we have to note that the same method has been used to approach very different problem in the one-plate LSC [4]).

We show in Fig. 2 how the fractions of spectral density of the solar radiation at AM1.5 (upper black solid line) is adsorbed in each matrix of four-plate tandem LSC. Reflection losses increase with every plate penetrated by the light. The light, absorbed in the upper plates no longer takes part in the losses. On the other hand, relaxation losses are lower for the structures with a higher number of plates.

3 RESULTS

We demonstrate an application of our Monte Carlo formalism to optimize the tandem LSC efficiency η on the number of plates n. We plot the simulated $\eta(n)$ dependence in Fig. 3 for two cases: the red line corresponds to $\eta(n)$ for the unoptimized tandem LSC with equal fraction of the solar intensity, delivered to each plate. The efficiency peaks at 3 to 4 plates due to nonoptimal energy distribution and, therefore, the prevalence of reflection losses. The blue line corresponds to $\eta(n)$ for the optimized tandem LSC (with unequal fractions of the intensity at each plate, defined by Monte Carlo method setting absorption edges of QDs for highest efficiency), for which the efficiency saturates at after n=8. (We have to note that a relatively high theoretical efficiency values calculated (compared to the experimental ones) are because we only treat the three types of inevitable and unavoidable fundamental losses. Adding other losses such a reabsorption, below unity quantum yield of the luminophores, *etc*., will further reduce the efficiency.

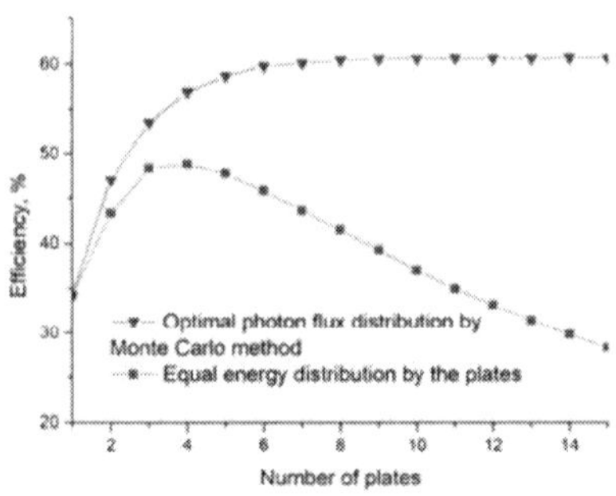

Fig. 3. Transport efficiency for optimized and power-equal distribution of power vs. the number of plates.

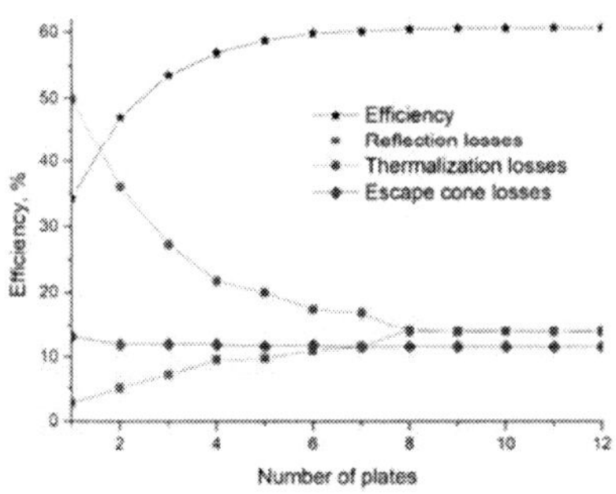

Fig. 4. Transport efficiency (stars) and losses (squares for reflection, circles for thermalization and diamonds for escape cone) vs. the number of plates.

The photoconversion efficiency of solar cells is also set to 100% for the correct estimation of maximal parameters for every plate. It should be noted, that the majority of losses are conducted in the plates of the concentrator and the efficiency of the optimized for this case solar cells itself can be much greater than for the full spectrum and can exceed 50 %.

Figure 4 illustrates the detailed efficiency and losses analysis for the LSC optimized without restrictions for the spectrum partition. Even the endpoint can differ from λ_{max}. One can see that after 8 plates the losses are saturated.

The change in maximal transport efficiency for different plate materials is shown on Fig. 5. With the increase in refractive index escape cone losses decrease, but reflectance losses increase which is critical for large numbers of plates.

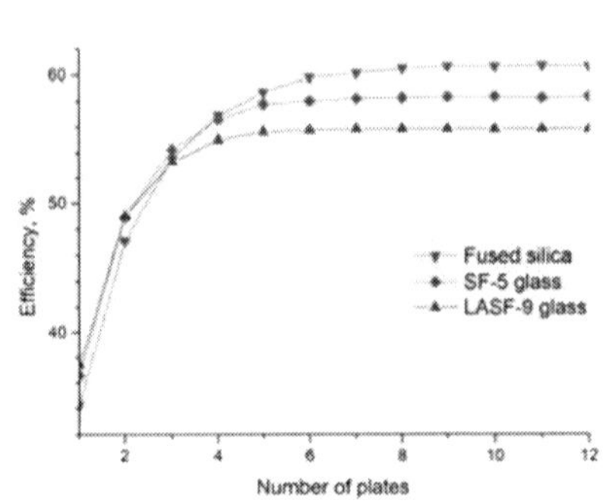

Fig. 5. Transport efficiency for different matrix materials vs. the number of plates.

4 CONCLUSIONS

Monte Carlo analysis of the dominant fundamental losses (which are irreversible) in the tandem luminescent solar concentrators (LSCs) allowed us to determine the optimal number of waveguides, depending on the refractive index of the concentrator matrices and luminophores (quantum dots) absorption/emission properties. In particular, we modelled the interplay between the thermalization losses, reflection from the stacked concentrator plates and escape cone losses to find the optimum number n of the concentrator plates. Depending on the LSC parameters, the number of the plates n, that optimizes the efficiency, ranges from six to eight. Further n increase does not provide performance gain, but makes tandem LSC only more expensive. The agreement of our accurate Monte Carlo results in the tandem LSCs with estimates from a simpler model (which considers equal energy for every plate) has been also demonstrated, as shown in Fig. 3.

Although the quantum dots (QDs) have been considered here as the luminophores, the formalism can be adapted for other luminophore types in LSCs

5 ACKNOWLEDGEMENTS

This work was partially supported by dedicated program of basic research of the National Academy of Sciences of Ukraine "Prospective basic research and innovative development of nanomaterials and nanotechnologies for the needs of industry, health care and agriculture."

6 REFERENCES

[1] K. Wu, H. Li, and V. I. Klimov, Nature Photonics 12 (2018) 105.
[2] D. R. Needell, et al, IEEE Journal of Photovoltaics 8 (2018) 1560.
[3] G. Liu, H. Zhao, F. Diao, Z. Ling, and Y. Wang, J. Mater. Chem. C6 (2018) 10059.
[4] H. Hernandez-Noyola, D. H. Potterveld, et al, Energy Environ. Sci. 5 (2012) 5798.
[5] L. H. Slooff, et al., physica status solidi (RRL) – Rapid Research Letters 2 (2008) 257.
[6] Y. You, X. Tong, W. Wang, J. Sun, P. Yu, H. Ji, et al. Advanced Science 6 (2019) 1801967.
[7] M.R. Kulish, A.I. Shkrebtii, V.P. Kostylyov, I.O. Sokolovskyi, and A.V. Sachenko. *Minimizing Reabsorption in Multilayered Luminescent Solar Concentrators with Quantum Dots*, IEEE Photovoltaic Specialist Conference (PVSC-48) June 20-25, 2021, Miami, Florida, USA, submitted
[8] A. I. Shkrebtii, A. V. Sachenko, I. O. Sokolovskyi, V. P. Kostylyov, M. R. Kulish, D. V. Khomenko, and M. A. Evstigneev, in 7th Forum on New Materials - Part B, Washington DC, USA, June 25-30, 2017 (Trans Tech Publications, 2017), p. 201.
[9] M.R. Kulish, V.P. Kostylyov, A.V. Sachenko, I.O. Sokolovskyi, A.I. Shkrebtii. Semiconductor physics, quantum electronics and optoelectronics, 22 (2019) 80.
[10] M.R. Kulish, A.I. Shkrebtii, V.P. Kostylyov, A.V. Sachenko, and I.O. Sokolovskyi, *Transparent Matrix Materials for Efficient Luminescent Solar Concentrators,* in Proceedings of 37th European Photovoltaic Solar Energy Conference and Exhibition, Sep. 6-11, 2020, p. 68.
[11] S. W. Leow, C. Corrado, M. Osborn, and S. A. Carter, in *High and Low Concentrator Systems for Solar Electric Applications VIII.* Ed. A.P. Plesniak (International Society for Optics and Photonics; SPIE, 2013), p. 24.

DEFECT TRAPPING IN THIN FILMS PROBED BY FREQUENCY DOMAIN PHOTOLUMINESCENCE

Baptiste Bérenguier, Alban Asseko , Jean François Guillemoles

UMR-IPVF 9006, CNRS, Ecole Polytechnique, ENSCP,
IPVF, 18 Boulevard Thomas Gobert, 91120 Palaiseau

Keywords: Characterization, defects, photoluminescence.

Time Resolved Photoluminescence (TRPL) is considered as a standard technique to measure carrier lifetime in thin film semi-conductors. Nevertheless, the excitation of TRPL is sudden, short and therefore the method is focused on fast recombination processes. I case of non mono-exponential decays, the tail of the decays may contain information about shallow traps and slow mechanisms but is quite often noisy and not always considered appropriately. Another approach based on Modulated Photoluminescence, also called Frequency Domain Photoluminescence (FDPL) has been developed previously and consists in the measurement the PL signal under a modulated illumination source. Coupled to mathematical modelling, it allows for extraction of information about the recombination mechanisms including both slows time constants. Here, we present a significant upgrade to the conventional FDPL method by extending the modulation frequency range from 500 Hz to 200 MHz using a Time Correlated Single Photon Counting device, the setup being able to scan both slow and fast mechanisms. We applied FDPL to thin passivated GaAs layers, and observe V-shapes in the phases curves, which are a characteristic of shallow trapping. We also attempt to fit measurement on triple cation perovskites using the two first harmonics of the FDPL signal.

1 INTRODUCTION

Frequency domain methods have been well known for a long time in carrier dynamics probing [1], [2]. Up to know they continue to be developed, mostly for silicon lifetime probing [3], [4] or photo-carrier radiometry [5]. In a recent development [6], the non-linearity of the recombination kinetics was used to perform multi-harmonics fitting allowing to recover the recombination parameters of a silicon wafer. However, in the thin film community, dealing with several order of magnitude shorter lifetimes, Time Resolved Photoluminescence (TRPL) using Time Correlated Single Photon Counting (TCSPC) dominates the contactless optical measurements methods of photogenerated carrier lifetime. The frequency domain methods decay can be explained by several issues. The Frequency Domain Photoluminescence (FDPL) is traditionally performed with lock-in detection and the gain-bandwidth ratio reduce the sensitivity of fast detectors. Photomultiplier tubes and sometimes fast photodiodes can be used successfully [7]–[9] to overcome this difficulty. The main limitation will be the complexity of the observed phase and amplitude response due to unexpected mechanisms. It is often tempting when interpreting TRPL decays to consider only the decay slope of normalized decays: a faster decay is attributed to a stronger recombination rate. On can also fit the decay with several exponential functions and attempt to find a physical meaning to the multiples lifetimes. In a previous article [10], we discovered than the first harmonics phase curves extracted from our high frequency FDPL setup were exhibiting V-shaped patterns instead of classical monotonic arc-tangent shape, which was preventing the direct extraction of a lifetime. Curiously, the first group observing V-shapes [8] concluded to the probable presence of a remnant laser "in phase" signal disturbing the luminescence signal. In our case, the appearance and disappearance of this V-Shapes depending of the material was attributed to the presence of shallow traps in the material. We developed a simulation code to model this behavior.

During the past years we developed a fast modulated setup to perform FDPL. In an opposite way to previous setups, our system uses a TCSPC to record one period of the signal. Initially designed to probe fast dynamics, it was successfully applied to surface recombination characterization of InP wafer [11] using a [40kHz-200MHz] frequency range, and so called High-Frequency MPL. Then, it has been extended to the [500Hz-200MHz] range in order to probe the low frequency shape which appeared in perovskites, CIGS and other thin films. We present some of the obtained results.

2 THEORY

2.1 Simplified theory of the V-shape appearance

As our setup is able to perform both FDPL and TRPL experiments, it was soon pointed out than the V-shape appearing in our FDPL phase curves were correlated with bi-exponential or more complex TRPL decays. During our illumination power dependent study, these behaviors were observed at the lowest fluencies and fluxes. An interesting explanation of bi-exponential decays appearance is given by Maiberg [12] using a dynamic version of the Schockley-Read-Hall recombination equations in the cases of shallow trapping. Figure 1 present the Maiberg's simplified case of a p-type semiconductor with a direct recombination path and a shallow trap communicating only with one band. Furthermore, the illumination is sufficiently low so that the trap filling remains low in comparison to its density.

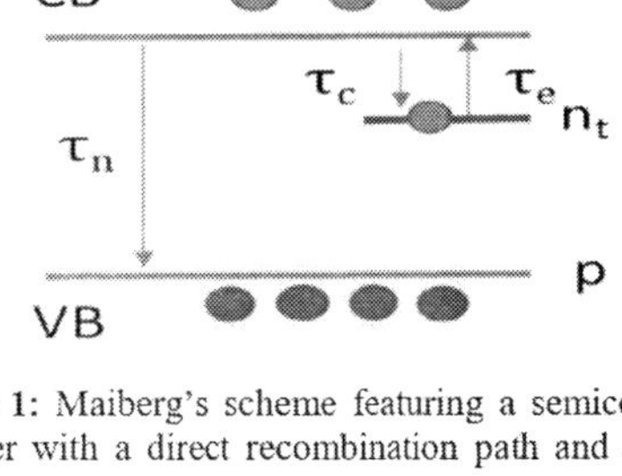

Figure 1: Maiberg's scheme featuring a semiconductor absorber with a direct recombination path and a purely

trapping/detrapping shallow level.
Then three lifetimes govern the carrier density kinetics, the recombination time τ_n, the capture and emission times τ_c and τ_e. We will have:

$$\frac{dn}{dt} = -\frac{n}{\tau_n} - \frac{n}{\tau_c} + \frac{n_t}{\tau_e}$$

$$\frac{dn_t}{dt} = \frac{n}{\tau_c} - \frac{n_t}{\tau_e}$$

In this case the TRPL response become a bi exponential with the apparent lifetimes:

$$\frac{1}{\tau_{short}} = \frac{1}{2}(\frac{1}{\tau_e} + \frac{1}{\tau_c} + \frac{1}{\tau_n} + \Delta)$$

$$\frac{1}{\tau_{long}} = \frac{1}{2}(\frac{1}{\tau_e} + \frac{1}{\tau_c} + \frac{1}{\tau_n} - \Delta)$$

$$\Delta = \sqrt{\left(\frac{1}{\tau_c} + \frac{1}{\tau_c} + \frac{1}{\tau_c}\right)^2 - \frac{4}{\tau_e \tau_n})}$$

According to the same formalism, the small signal modulated electron density would be:

$$\tilde{n} = \frac{\tilde{G}\left(i\omega + \frac{1}{\tau_e}\right)}{\left(i\omega + \frac{1}{\tau_{short}}\right)\left(i\omega + \frac{1}{\tau_{long}}\right)}$$

Where omega is the angular modulation velocity and G the modulated part of the generation rate. Considering the PL signal to be proportional to the minority carrier density, the phase of and the amplitude of the modulated PL would get the typical V-Shape as presented in Figure 2.

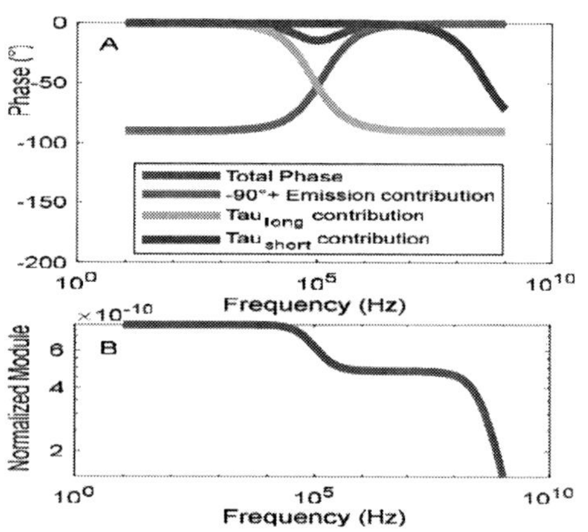

Figure 2: A: Phase of the theoretical modulated PL (blue) with the contribution of Emission, long and short lifetime. B: Corresponding amplitude presenting two kinks.

Of course this is a simplified case. A more complete analytical theory including communication with the valence band will be given soon [13]. Please note that in this system TRPL would lead to the measurement of two lifetimes against three fundamental unknowns when FDPL give three values, allowing the recovery of the initial recombination, emission and capture lifetimes.

3 EXPERIMENTAL

3.1 Setup

The light sources are a 638 nm fast modulated laser for the FDPL and a 532 nm pulsed laser with 80 ps pulse duration for the TRPL. The modulation is based on sinusoidal variations of the light intensity around a given working point. In this study, the ratio between modulated and continuous part of the modulated light is close to 60%. The beam diameter on the sample is 200 µm: it ensures an excitation spot larger than the diffusion length avoiding lateral transport which might influence the carrier dynamics. To detect the PL signal with an optimal sensitivity, we used a single photon avalanche detector. Then, a TCSPC and the in house LabVIEW code allow reconstructing one decay in TRPL mode or one period of the modulated signal. In such a configuration, we obtain a unique experimental setup able to perform FDPL and TRPL at the same point of the samples. Using sixty percent of modulation ratio could create some high order harmonics in the signal. Continuous part, first, second and third harmonics of the modulated signal are recorded which can give a help to identification of carrier dynamics.

3.2 Samples.

The first sample was fabricated at the RCAST laboratory and is constituted by 100 nm of undoped InGaP / 2 µm undoped GaAs /100 nm InGaP / 200nm undoped GaAs / GaAs substrate. During the MPL experiment, we also observed that the PL spectrum was only constituted by the GaAs signature.

The Second sample is a triple cation perovskite with composition $Cs_{0.05}(FA_{0.83}MA_{0.17})_{0.95}Pb(I_{0.83}Br_{0.17})_3$. It was fabricated at the IPVF as presented in [14].

4 RESULTS AND DISCUSSION

4.1 GaAs Layer.

Figure 3 represent the phase of the first harmonics and the ratio between first harmonics amplitude and mean value (continuous part) of the PL at several illumination fluxes starting from $3x10^{16}$ photons/cm^{-2}/s. One can note four V-shapes on the phase at respectively $2x10^4$Hz, $4x10^5$Hz, $1.5x10^7$ and $8x10^7$Hz. Kinks are also present in the modulation ration 1H/Continuous. This is an evidence that several traps are included into this material. The fact that the modulation ratio is always below or above the laser modulation ratio of 0.6 indicates the nonlinearity of the system.

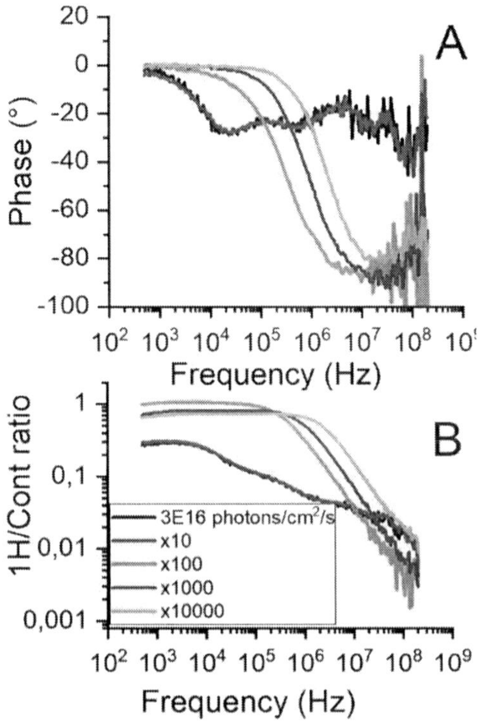

Figure 3: Experimental results on the GaAs sample. A: Phases of the first harmonic at several illumination fluxes. B: ratio between first harmonic amplitude and continuous part of the signal.

4.1 Perovskite absorber.

In this section we perform the estimation of the traps parameters of a perovskite layer via a fitting procedure. The mathematical rate equations are extracted from [15]:

$$\frac{dn}{dt} = -B_{rad}np - R_{pop}(N_t - n_t)n + R_{detrap}n_t$$

$$\frac{dn_t}{dt} = R_{pop}(N_t - n_t)n - R_{detra}\ n_t - R_{depop}n_tp$$

$$\frac{dp}{dt} = -B_{rad}np - R_{depop}n_tp$$

This describe radiative recombination, the electron capture and reemission and hole capture via a trap. The trap is implicitly supposed to be closer from the conduction band so that the hole detrapping is neglected. In opposite to figure 1 the trap is also capturing holes from the valence band.

The preceding equations are solved by a numerical procedure. At each iteration, the TRPL and FDPL at all injection levels are simulated simultaneously with the same parameter set. The fit is thus performed on 9 experimental curves at the same time. This also take into account the repetition rates for both MPL and TRPL as in [10], [16]. The fluencies and fluxes are respectively for TRPL of $5x10^{10}$ photons/pulse/cm2 x1 x13 x233 x1960 and for FDPL of $2.3x10^{16}$ photons/cm2/s x 1 x10 x100 x1000 x10000. Figure 4 present the experimental and numerically calculated curves. All values are normalized by the value of the continuous part et the lowest frequency. One can see that the TRPL shape and the amplitude and phase of first and second harmonics are well simulated. The continuous part has the good shape but is slightly deviating from the fit. The extracted parameters are presented in table I. The detrapping coefficient is low but sufficient to create a small V-Shape at 10^5 Hz at low fluxes.

Table I: Results of the fitting procedure

Brad(cm^3/s)	Nt(cm^{-3})	Rpop(cm^3/s)	Rdepop(cm^3/s)	Rdetrap(1/s)
$6.7x10^{-11}$	$3.3x10^{16}$	$3.2x10^{-10}$	$5x10^{-11}$	$6.4x10^3$

5 CONCLUSION

A novel setup has been built allowing to perform both TRRPL and FDPL experiments with high sensitivity. Using this setup for the characterization of thin films lead to the discovery of V-shapes in the phase shift. It was attributed to the carrier detrapping via shallow traps. Several sample were studied exhibiting multiple V-patterns in the first harmonic phase and kinks in the amplitude curves at low fluxes. Finally as successful fit was performed allowing to recover the trap parameters of a perovskite layer.

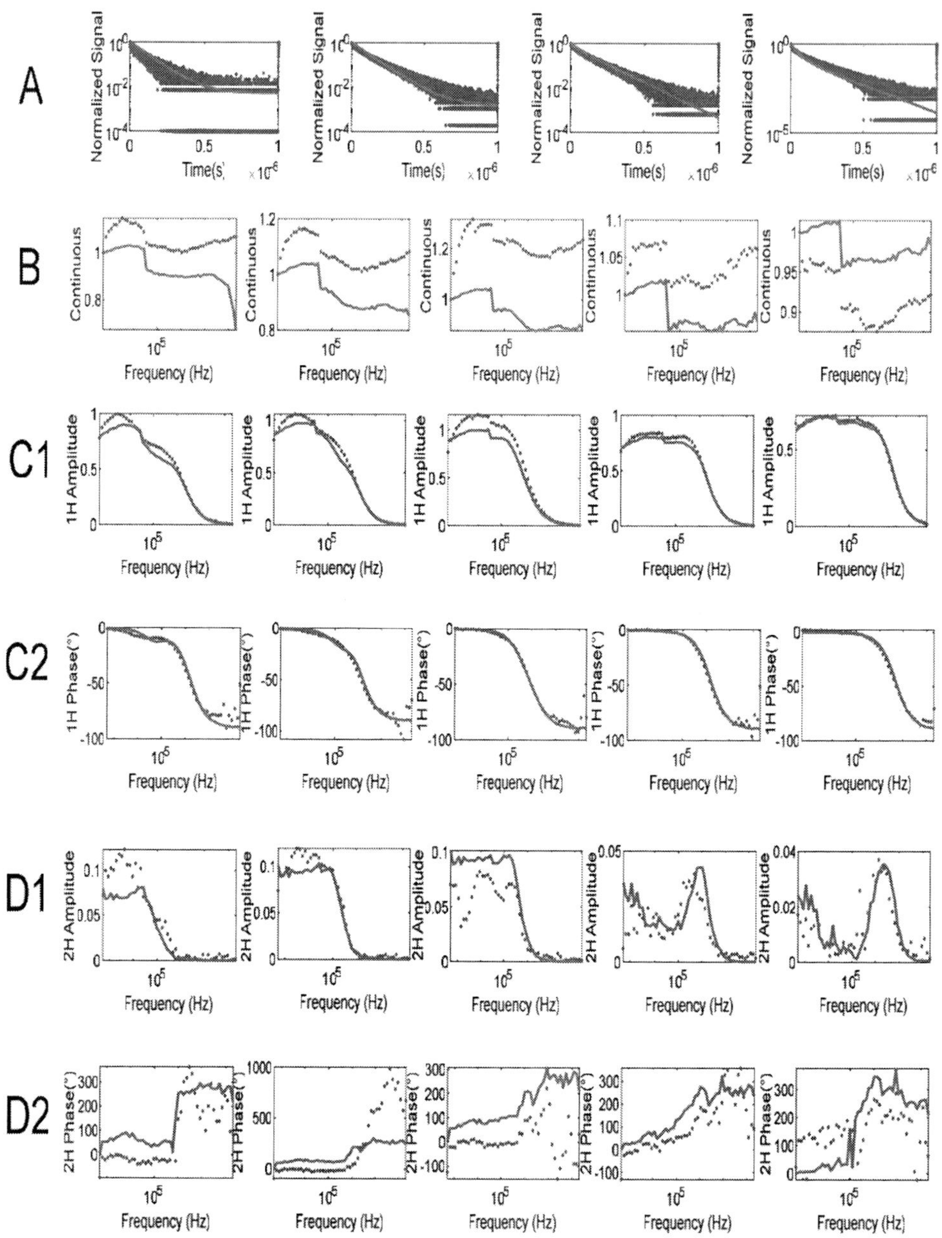

Figure 4: A: experimental (blue) and simulated (red) TRPL decays recorded on the perovskite sample with fluencies from right to left $5x10^{10}$ photons/pulse/cm2 x1 x13 x233 x1960. B: experimental (blue) and simulated (red) continuous part of the recorded FDPL at from right to left $2.3x10^{16}$ photons/cm2/s x 1 x10 x100 x1000 x10000. C1 and C2: corresponding relative amplitude and phase of the first harmonics of the FDPL. D1 et D2: Corresponding amplitude and phase of the second harmonic.

6 AKNOWLEDGMENTS

This work was supported by the French government in the frame or the program of investments for the future (Programme d'Investissement d'Avenir ANR-IEED-002-01).
We would like to thanks for the samples: The RCAST lab Japan, and in at IPVF institute the EdF R&D team (IPVF program II), Armelle Yaiche, Jean Rousset, The ANR InHyMat-PV team: J. Hajhemati, V. Dufoulon, C. Darin Bapaume, P. Schulz, The IPVF program VI, Jackson Lontchi and Daniel Lincot.

7 REFERENCES

[1] M. Orgeret and J. Boucher, "Caractérisation d'un substrat semiconducteur par technique micro-onde et injection photonique," *Revue de Physique Appliquée*, vol. 13, no. 1, pp. 29–37, 1978, doi: 10.1051/rphysap:0197800130102900.

[2] D. Guidotti, J. S. Batchelder, A. Finkel, P. D. Gerber, and J. A. Van Vechten, "Modeling of luminescence phase delay for nondestructive characterization of Si wafers," *Journal of Applied Physics*, vol. 66, no. 6, pp. 2542–2553, Sep. 1989, doi: 10.1063/1.344218.

[3] J. A. Giesecke, M. C. Schubert, F. Schindler, and W. Warta, "Harmonically Modulated Luminescence: Bridging Gaps in Carrier Lifetime Metrology Across the PV Processing Chain," *IEEE Journal of Photovoltaics*, vol. 5, no. 1, pp. 313–319, Jan. 2015, doi: 10.1109/JPHOTOV.2014.2362304.

[4] M. Sreng, "Development of in-situ photoluminescence characterization tools for the study of semiconductors for photovoltaics application," These de doctorat, Institut polytechnique de Paris, 2019. Accessed: Sep. 06, 2021. [Online]. Available: http://www.theses.fr/2019IPPAX003

[5] M. Pawlak and K. Strzałkowski, "Identification of the photoluminescence response in the frequency domain modulated infrared radiometry signal of ZnTe:Cr bulk crystal," *Infrared Physics & Technology*, vol. 78, pp. 190–194, Sep. 2016, doi: 10.1016/j.infrared.2016.08.005.

[6] Q. Sun, A. Melnikov, A. Mandelis, and Y. Song, "Fully nonlinear photocarrier radiometry / modulated photoluminescence dynamics in semiconductors: Theory and applications to quantitative deconvolution of multiplexed photocarrier density wave interference and recombination processes," *Journal of Luminescence*, vol. 236, p. 118075, Aug. 2021, doi: 10.1016/j.jlumin.2021.118075.

[7] I. Reklaitis *et al.*, "Differential carrier lifetime in InGaN-based light-emitting diodes obtained by small-signal frequency-domain measurements," *Journal of Applied Physics*, vol. 121, no. 3, p. 035701, Jan. 2017, doi: 10.1063/1.4973903.

[8] I. Reklaitis *et al.*, "Photoluminescence Decay Dynamics in Blue and Green InGaN LED Structures Revealed by the Frequency-Domain Technique," *Journal of Elec Materi*, vol. 45, no. 7, pp. 3290–3299, Jul. 2016, doi: 10.1007/s11664-016-4557-7.

[9] D. Donetsky *et al.*, "Analysis of Recombination Processes in 0.5-0.6 eV Epitaxial GaInAsSb Lattice-matched to GaSb," p. 10.

[10] B. Bérenguier *et al.*, "Defects characterization in thin films photovoltaics materials by correlated high-frequency modulated and time resolved photoluminescence: An application to Cu(In,Ga)Se2," *Thin Solid Films*, vol. 669, pp. 520–524, Jan. 2019, doi: 10.1016/j.tsf.2018.11.030.

[11] W. Zhao *et al.*, "Coupled time resolved and high frequency modulated photoluminescence probing surface passivation of highly doped n-type InP samples," *Journal of Applied Physics*, vol. 129, no. 21, p. 215305, Jun. 2021, doi: 10.1063/5.0033122.

[12] M. Maiberg, T. Hölscher, S. Zahedi-Azad, and R. Scheer, "Theoretical study of time-resolved luminescence in semiconductors. III. Trap states in the band gap," *Journal of Applied Physics*, vol. 118, no. 10, p. 105701, Sep. 2015, doi: 10.1063/1.4929877.

[13] N. Moron, B. Bérenguier, J. Alvarez, and J.-P. Kleider, "Analytical model of the modulated photoluminescence in semiconductor materials," *to be published*.

[14] S. Cacovich *et al.*, "Light-Induced Passivation in Triple Cation Mixed Halide Perovskites: Interplay between Transport Properties and Surface Chemistry," *ACS Appl. Mater. Interfaces*, vol. 12, no. 31, pp. 34784–34794, Aug. 2020, doi: 10.1021/acsami.0c06844.

[15] M. J. Trimpl *et al.*, "Charge-Carrier Trapping and Radiative Recombination in Metal Halide Perovskite Semiconductors," *Adv. Funct. Mater.*, p. 2004312, Aug. 2020, doi: 10.1002/adfm.202004312.

[16] E. V. Péan, S. Dimitrov, C. S. D. Castro, and M. L. Davies, "Interpreting time-resolved photoluminescence of perovskite materials," *Phys. Chem. Chem. Phys.*, vol. 22, no. 48, pp. 28345–28358, Dec. 2020, doi: 10.1039/D0CP04950F.

OPTIMIZATION OF RANDOM SUBWAVELENGTH STRUCTURES WITH ANTI-REFLECTIVE PROPERTIES ON PHOTOVOLTAIC GLASSES

Cristina L Pinto, Iñaki Cornago, Eugenia Zugasti, Jaione Bengoechea
National Renewable Energy Center, CENER
C/Ciudad de la Innovación 7, 31621 Sarriguren (Navarra) Spain
Telephone: + 34 948 25 28 00
E-mail: cpinto@cener.com

ABSTRACT: Subwavelength structures (SWSs) have the ability to suppress Fresnel reflections, for that reason, random SWSs have begun to be fabricated on glass. However, random structures can hardly be simulated and therefore, they cannot be easily computationally studied. In this work, using a mathematical model to simulate random SWSs and the Effective Medium Theory (EMT), we present a parametric analysis of two different SWSs in order to understand the influence that their main parameters have in the spectral reflectance. Moreover, we analyze the possibility of implanting these structures on photovoltaic glass to reduce its reflectance, and therefore, increase the solar cell efficiency. For this reason, we have made an optimization of these structures maximizing the theoretical photogenerated short-circuit current density that a solar cell would produce laminated with these structured glasses. Finally, we have analyzed the omnidirectional property of these SWSs obtaining promising results.
Keywords: Random subwavelength structures, Photovoltaic glass, Parametric analysis, Anti reflective

1 INTRODUCTION

Glass has been widely used in photovoltaic industry in order to provide mechanical strength and protection against outdoor conditions to solar photovoltaic (PV) modules. However, due to the refractive index mismatch between air and glass, approximately 4% of the incident light is reflected at this first interface. In order to reduce this reflection, anti-reflective coatings based on different technologies are, nowadays, commonly added to the PV glass external surface, increasing sunlight transmission and therefore enhancing efficiency [1].Nevertheless, most of these coatings are only effective for specific wavelengths and at normal incidence, as they use the quarter-wave principle to suppress reflection on a certain wavelength by means of destructive interferences [2].

With the aim of solving this problem, new broadband anti-reflective structures, some of them bioinspired, such as moth-eye or glasswing butterfly configurations, are receiving a lot of attention due to their high broadband anti-reflective and omnidirectional properties [3]–[5]. These structures are characterized by subwavelength sizes and therefore, no light diffraction is produced. However, SWSs are usually be randomly distributed, which hinder its simulation and computational analysis. Even so, the EMT can be applied to analyze these random SWSs, considering them as a layer with a refractive index (n) gradient gradually ranging from n_{air} to $n_{substrate}$ [5] (**Fig. 1**), which noticeable suppresses the Fresnel reflection at this interface.

In this work, in order to understand the spectral effect of the main nanostructures topological parameters, such as average height, randomness and shape, the EMT implemented on GD-Calc [6] computational tool has been used. The influence of these structures on the reflectance of glass-air interface, and consequently, the influence they would have in the photogenerated short circuit current density of a commercial solar cell laminated with that structured glass has been analyzed. These theoretical calculations provide valuable insight into the spectral response of random SWSs and allow to perform an optimization on the main parameter configuration avoiding the need of multiple and costly fabrications.

2 METHODS

The EMT has been used to perform the optical modelling of random SWSs on PV grade soda lime glass (optical data obtained from Vogt et al. [7]). Using the optical model of random nanocylinders with and without tapered pedestals, the glass volume fraction at any elevation has been computed. Secondly, Maxwell-Garnett formula has been applied to obtain the effective refractive index of the composite at any elevation (z), being $z = 0$ nm the glass substrate, $z = 1$ nm located at the SWS base and $z = L$ the top of the highest SWS as is depicted in Fig. 1.

The Maxwell-Garnett formula *(1)* can be written as:

$$n_{eff}^2 = n_{glass}^2 \frac{2fn_{glass}^2 + (3-2f)n_{air}^2}{(3-f)n_{glass}^2 + fn_{air}^2} \qquad (1)$$

where n_{glass} and n_{air} are glass and air refractive index respectively, f is the glass volume fraction, and n_{eff} is the effective refractive index of the each layer [8].

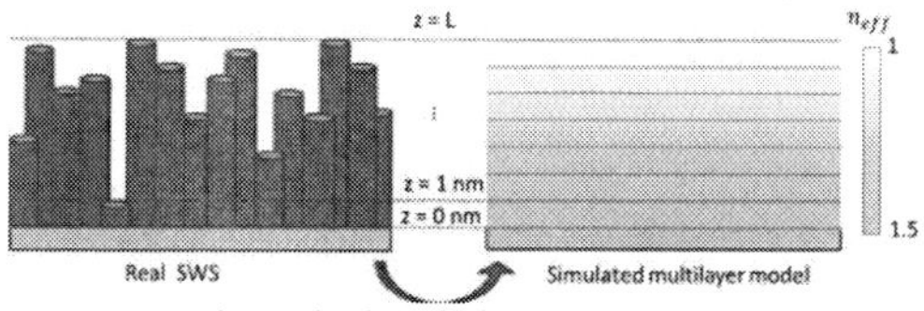

Figure 1: Schematic description of the effective medium theory model.

After, with the calculated effective refractive index, the spectral reflectance and transmittance have been computed using GD-Calc program. Finally, the spectral transmittance, defined as $T = 1 - R$, of the random structures has been weighted by the solar spectral irradiance ($E_0(\lambda)$) as described in standard IEC 60904-3 Ed. 4.0 Feb2019 [9], and with the spectral response ($SR(\lambda)$) of a commercial passivated emitter rear cell (PERC). With these values, the theoretical photogenerated short-circuit current density (J_{sc}) that a solar cell would generate laminated with this structured glass has been calculated for each SWS using the next equation *(2)*:

$$J_{sc} = \sum_{\lambda=300nm}^{\lambda=1186nm} SR(\lambda) \times E_0(\lambda) \times T(\lambda) \qquad (2)$$

The obtained value of the photogenerated short circuit current density has been used to compare between SWSs response and to optimize their main parameters.

2.1 SWSs' volume fraction

First of all, glass volume fractions have been computed using the following formulas developed by Siddique et al [3]. To calculate the glass volume fraction, the structure average height $\bar{h}$ and standard deviation σ_h have to be obtained by means of a Gaussian statistical analysis.

For cylinders, the glass volume fraction is computed according to the next equation:

$$f_c = \begin{cases} 1 & \text{for } z \le 0 \\ \dfrac{\pi r_c^2}{\sqrt{3}d^2}\operatorname{erfc}\left(\dfrac{z-\bar{h}}{\sigma_h\sqrt{2}}\right) & \text{for } z > 0 \end{cases} \qquad (3)$$

and for cylinders with tapered pedestals

$$f_{c-p} = \begin{cases} 1 & \text{for } z \le 0 \\ \dfrac{2\pi}{\sqrt{3}\,d^2}\left[r_p\left(1-\dfrac{z}{h_p}\right)+r_c\right]^2 & \text{for } 0 < z \le h_p \\ \dfrac{\pi r_c^2}{\sqrt{3}d^2}\operatorname{erfc}\left(\dfrac{z-\bar{h}}{\sigma_h\sqrt{2}}\right) & \text{for } z > h_p \end{cases} \qquad (4)$$

Where r_c, d, $\bar{h}$, σ_h and z are the cylinder radius, the distance between cylinders, the average cylinders height, the standard deviation of the cylinder's height distribution and the elevation. For cylinders with pedestals, h_p, r_p are the pedestals height and its radius. For simplicity, the structures have been organized in hexagonal lattice and the fill factor (x) which establish the cylinder radius respect to d has been defined as: $x = \frac{2}{\sqrt{3}d^2}\pi r_c^2$.

These parameters are depicted in Fig. 2.

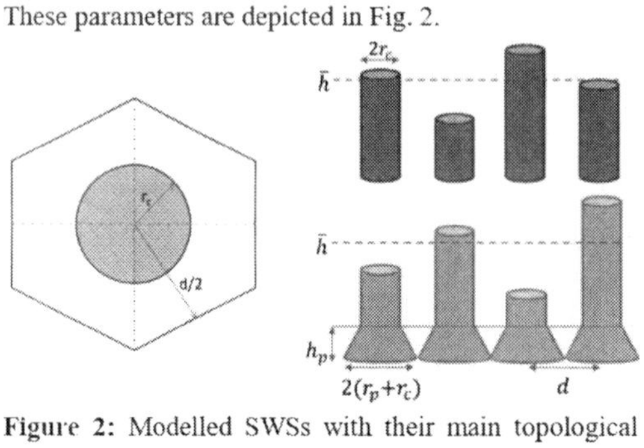

Figure 2: Modelled SWSs with their main topological parameters.

From these equations *(3)(4)* the volume fraction that is used in Maxwell-Garnett formula *(1)* has been obtained. The computed effective refractive indices in function of the relative elevation ($z^* = z/\bar{h}$) are shown in **Fig. 3**. Additionally, a quintic profile, $n_{eff} = 1 + (n_{glass} - 1)\left(10\left(^{z}/_{h_{max}}\right)^3 - 15\left(^{z}/_{h_{max}}\right)^4 + 6\left(^{z}/_{h_{max}}\right)^5\right)$, has been depicted, which is close to the ideal effective refractive index profile to achieve near-zero reflectance [5].

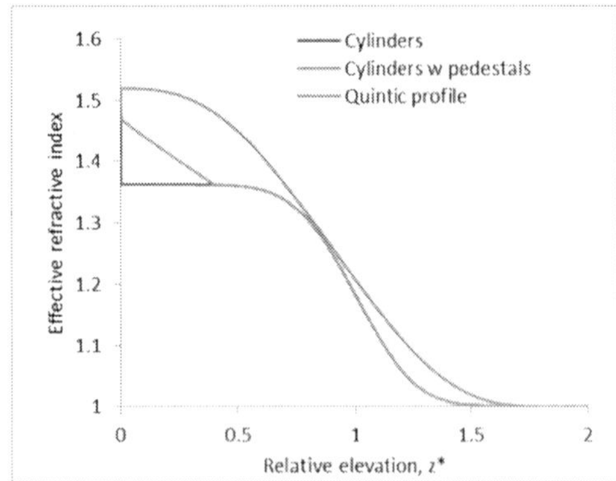

Figure 4: Effective refractive indices of SWSs with exemplary average height ($\bar{h}$) of 250 nm, $\sigma_h = 50$ nm, $d = 100$ nm, $r_c = 44$ nm, $r_p = 6$ nm and $h_p = 125$ nm, together with a quintic profile.

3 RESULTS AND DISCUSSION

First, we have analyzed and discussed how the main significant parameters of these SWSs model influence the effective refractive index. We have performed the analysis by systematically varying the magnitude of different parameters. After, the effect of these changes on the spectral reflectance has been analyzed. Finally, the theoretical J_{sc} values have been computed to find the optimum parameter configuration that maximizes it.

3.1 Parametric analysis

3.1.1. Cylinders

The behavior of the effective refractive index of SWSs when varying several topological parameters such a fill factor, standard deviation and average height is shown in next Fig. 5.

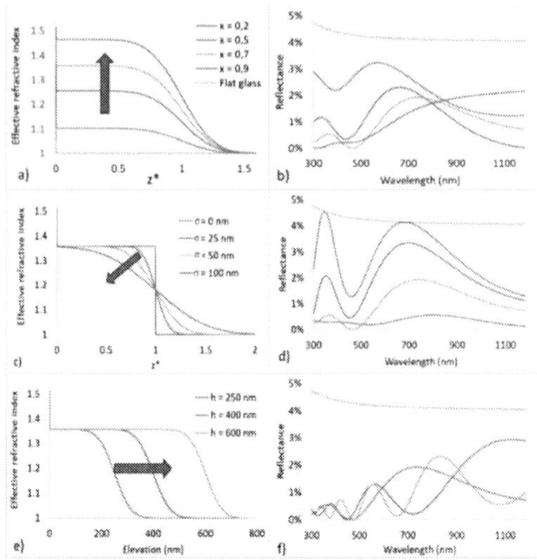

Figure 5: Parameter variation of the random nanostructure model. Subwavelength cylinders' effective refractive index variation in function of the a) fill factor ($\bar{h} = 250$ nm, $\sigma_h = 50$ nm), c) standard deviation ($x = 0.7$, $\bar{h} = 250$ nm), and d) average height ($x = 0.7$, $\sigma_h = 50$ nm). Their corresponding response on spectral reflectance (b, d, f).

As it is depicted in Fig. 4, the effective refractive index that changes like the quintic profile will show the best anti-reflective properties. In Fig. 5a), as the fill factor increases, less air is left between cylinders, and therefore a more gradual variation is achieve in the effective refractive index, minimizing the steep rise at $z^* = 0$. Therefore, the spectral reflectance decreases as the fill factor increases (Fig. 5b).

In Fig. 5c), the randomness of the nanocylinders has been changed via the standard deviation σ_h. The value of $\sigma = 0$ nm corresponds to periodic nanocylinders, all with the same heights and equally separated, and therefore, at the top of the nanostructures ($z = \bar{h}$), a big change in the refractive index takes place. However, as the randomness increases, this gap becomes smoother. This feature improves the anti-reflective property as can be seen in Fig. 5d), where the oscillations amplitude decreases. In consequence, it is favorable to increase the standard deviation in order to obtain better anti-reflective properties.

Fig. 5e) shows the behavior of the effective refractive index when the average height of the nanostructures increases. Unlike the previously analyzed behaviors, increasing the height does not decrease spectral reflectance, but rather shows an impact on its shape. As the nanostructure's average height increases, the number of oscillations that appear on the spectral reflectance increases (Fig. 5f). This is due to the width of the plateau at low z* has increased, creating regions of constant n_{eff} and therefore allowing the appearance of constructive and destructive interferences following the quarter wave principle [10]. However, a certain value of average height is needed to get a low reflectance response.

3.1.2. Cylinders with tapered pedestals

The behavior of the effective refractive index and the spectral reflectance of cylinders with tapered pedestals when the main topological parameters vary is shown in Fig.6. The pedestal heights in function of the cylinders' average height and the pedestal radius have been varied.

As can be appreciated from Fig. 6, the effective refractive index profile of cylinders with pedestals resembles more closely the quintic profile and therefore, it shows better anti-reflective properties. The pedestal makes the constant plateau that appears at low value of z* disappear, making more gradual the change in n_{eff} from air to glass. As the pedestal height increases, the plateau decreases (Fig. 6a). On the other hand, keeping constant the pedestal height and varying the pedestal radius, the gap at low z^* values decreases until it reaches its minimum, that happens when all pedestals bases touch each other (Fig. 6c), however, the impact of these parameters on the spectral reflectance is lower than the other parameters.

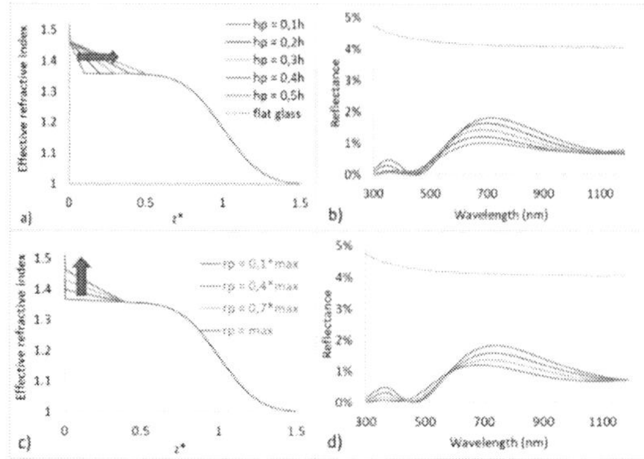

Figure 6: Parameter variation of the cylinders with tapered pedestals ($x = 0.4$, $\bar{h} = 250$ nm and $\sigma_h = 50$ nm). Nanostructure's effective refractive index and spectral reflectance variation with the pedestal height ($r_p = max$) (a, b), and pedestal radius ($h_p = 0.4 * \bar{h}$) (c, d). The maximal radius is stablished by the hexagonal geometry of the lattice.

3.2 SWSs optimization

With respect to the previous results, some parameter configuration creates refraction maximus and minimums, which are more noticeable in cylinder's spectral reflectance. These reflectance maxima can match the maximum absorption of the solar cell (Fig. 7) and therefore decreasing their efficiency. For this reason, the optimization has been done using the theoretical photogenerated short circuit current density that a PERC solar cell would generate laminated with these structured glasses.

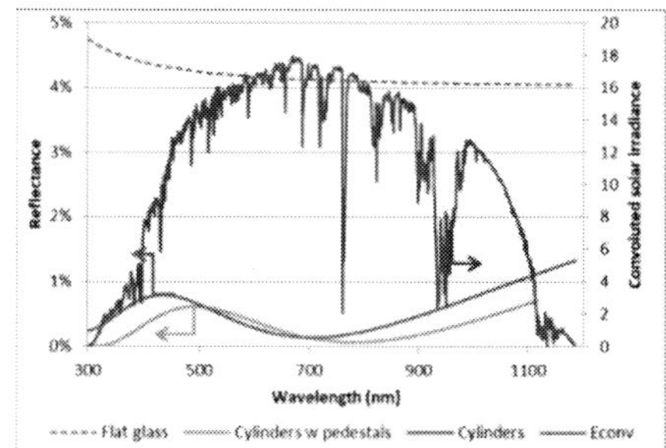

Figure 7: Convoluted solar irradiance defined as $SR(\lambda) \times E_0(\lambda)$ with the spectral reflectance of flat glass, cylinders and cylinders with tapered pedetals

First of all, based on experimental evidences, to be published yet, the standard deviation of different SWSs takes values around 50 nm, and therefore, this value has been kept constant. After, the fill factor has been varied and the best J_{sc} values have been obtained of 0.7 for cylinders and 0.4 for cylinders with pedestals. Next, pedestal height and radius have been analyzed, achieving the best J_{sc} values for the maximal h_p and r_p values ($h_p = 0.5 * \bar{h}$, and $r_p = max$). With all these parameters stablished, the average height has been varied. The results is shown in Fig. 7, where the best average height for cylinders is 140 nm, giving a J_{sc} value of 39.31 mA/cm^2, and 200 nm for cylinders with pedestals, giving a J_{sc} of 39.37 mA/cm^2. With higher $\bar{h}$ values, the J_{sc} decreases due to the different oscillations on the spectral reflectance. The optimal parameters for cylinders and cylinders with pedestals are summarized in Table I.

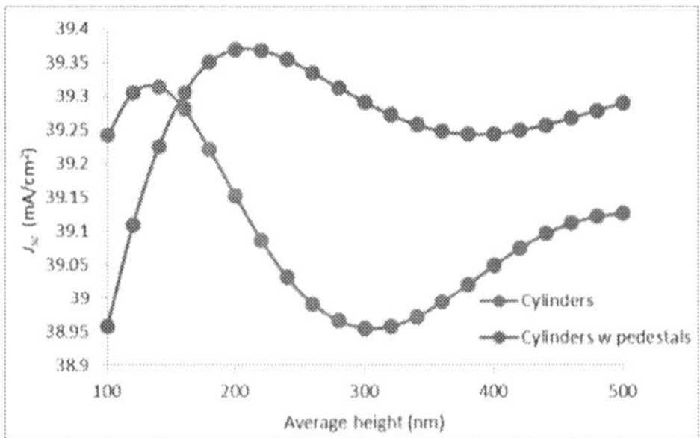

Figure 8: Theoretical photogenerated short circuit current density of SWSs in function of the average height.

Table I: Optimal parameters of nanocylinders and nanocylinders with pedestals.

	Cylinders	Cylinders w pedestals
$\bar{h}$ (nm)	140	200
σ_h (nm)	50	50
h_p (nm)	-	100
r_p (nm)	-	16.8
x	0.7	0.4

3.3 SWSs angle dependence

In this section, the omnidirectional property of the SWSs has been analyzed. The optimum SWSs of each shape has been chosen and computed for every 5 degrees of light incidence angle. The results are depicted in Fig. 8 where all investigated SWSs show an improved behavior in terms of the omnidirectional anti-reflective property with respect to flat glass. Cylinders with pedestal present the best J_{sc} value, producing 39.97 mA/cm^2, 1.52 mA/cm^2 more than flat glass. At normal incidences, both SWSs give almost the same J_{sc} value. However, for angles greater than 70° the difference between them increases, until cylinders with pedestals reach 20.00 mA/cm^2 and cylinders 18.15 mA/cm^2 respectively. These values are summarized on Table II for three different angles.

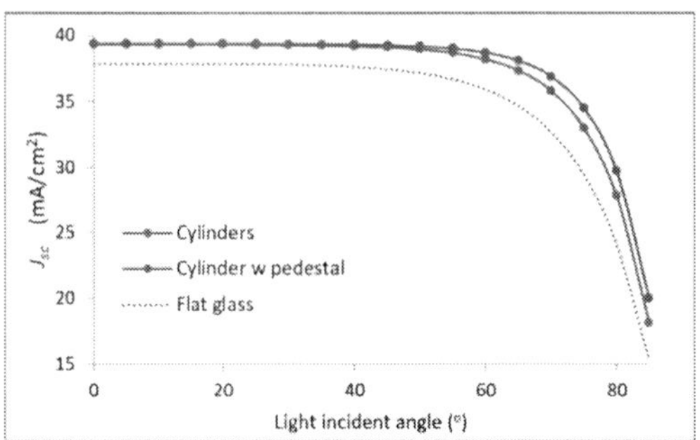

Figure 9: Theoretical J_{sc} in function of the light incident angle for different SWSs and for flat glass.

Table II: Photogenerated short circuit current density at different light incidence angles.

	0°	45°	85°
Flat glass	37.85	37.44	15.24
Cylinders	39.31	39.09	18.15
Cylinders with pedestals	39.37	39.26	20.00

5 CONCLUSION

The broadband omnidirectional anti-reflective property of random sized nanocylinders and nanocylinders with tapered pedestals has been mathematically modelled and parametrically analyzed. A simulation tool, which allows calculating the reflectance of random subwavelength nanostructures has been implemented in order to better understand the effect of the main topological parameters.
According to our simulation results, the height distribution standard deviation has a clear impact on the reflectance reduction, i. e. the more randomly the height is distributed, the lower the reflectance is. This also holds true for the average height to a certain extent, until it reaches a certain value where the reflectance does not decrease further due to the appearance of quarter-wave interferences. The pedestal height and pedestal radius have also been analyzed, concluding that maximizing these parameters leads to an effective refractive index gets closer to the quintic profile, therefore showing a better anti-reflective property. Also, for each SWS, an analysis of the effect of the fill factor has been computed, concluding that a values up to 0.7 present better optical response.
Secondly, the optimum SWS topology has been found whit respect to maximizing the theoretical photogenerated short circuit current density. These J_{sc} values have been used to perform the comparison between different parameter configurations and the best parameters have been obtained.
Moreover, according to our research, the omnidirectionallity is improved with any investigated nanostructure, cylinders with pedestals showing the best J_{sc} result, producing 4.76 mA/cm^2 more than flat glass at 85° of incident light.

6 ACKNOWLEDGEMENTS

Cristina Pinto gratefully acknowledges the Department of University, Innovation and Digital Transformation of the Government of Navarra for the grant for hiring doctoral students by companies, research centres and technological centres: Industrial Doctorates 2020, with file number 0011-1408-2020-000003, received to carry out this work.

7 REFERENCES

[1] VDMA, *International Technology Roadmap for Photovoltaic*, 11th ed. 2020.

[2] N. Shanmugam, R. Pugazhendhi, R. M. Elavarasan, P. Kasiviswanathan, and N. Das, "Anti-reflective coating materials: A holistic review from PV perspective," *Energies*, vol. 13, no. 10, pp. 1–93, 2020, doi: 10.3390/en13102631.

[3] R. H. Siddique, G. Gomard, and H. Hölscher, "The role of random nanostructures for the omnidirectional anti-reflection properties of the glasswing butterfly," *Nat. Commun.*, vol. 6, pp. 1–8, 2015, doi: 10.1038/ncomms7909.

[4] B. M. Phillips and P. Jiang, "Biomimetic Antireflection Surfaces," in *Engineered Biomimicry*, Gainesville: Elsevier Inc., 2013, pp. 305–331.

[5] S. Chattopadhyay, Y. F. Huang, Y. J. Jen, A. Ganguly, K. H. Chen, and L. C. Chen, "Anti-

reflecting and photonic nanostructures," *Mater. Sci. Eng. R Reports*, vol. 69, no. 1–3, pp. 1–35, 2010, doi: 10.1016/j.mser.2010.04.001.

[6] K. C. Johnson, "Grating Diffraction Calculator (GD-Calc)." pp. 1–80, 2008, doi: 10.24433/CO.7479617.v3.

[7] M. R. Vogt *et al.*, "Measurement of the optical constants of soda-lime glasses in dependence of iron content and modeling of iron-related power losses in crystalline si solar cell modules," *IEEE J. Photovoltaics*, vol. 6, no. 1, pp. 111–118, Jan. 2016, doi: 10.1109/JPHOTOV.2015.2498043.

[8] V. A. Markel, "Introduction to the Maxwell Garnett approximation: tutorial," *J. Opt. Soc. Am.*, vol. 33, no. 7, p. 1244, 2016, doi: 10.1364/josaa.33.001244.

[9] "IEC 60904-3:2019 | Photovoltaic devices - Part 3: Measurement principles for terrestrial photovoltaic (PV) solar devices with reference spectral irradiance data," Feb. 2019. https://webstore.iec.ch/publication/61084 (accessed Jun. 18, 2021).

[10] H. A. Macleod, *Thin-Film Optical Filters*, Third Edition. .

ACCURATE DETERMINATION OF CONTACT RESISTIVITY WITH FULLY METALLIZED TEST STRUCTURES

Konstantin Tsoi[1,2,*], Deniz Turkay[3], Selcuk Yerci[1,2,3]
[1]ODTU-GUNAM, Middle East Technical University, Ankara, Turkey
[2]Electrical and Electronics Engineering Department, Middle East Technical University, Ankara, Turkey
[3]Department of Micro and Nanotechnology, Middle East Technical University, Ankara, Turkey
Email: syerci@metu.edu.tr, Tel:+90 312 210 23 44
*knstntn.ts@gmail.com, Tel:+90 531 607 11 76

ABSTRACT: Among the widely used methods such as transfer length or Cox and Strack methods to determine the contact resistivity of various interfaces, contact resistivity can also be extracted using fully metallized test structures. In this structure, both sides of a sample are covered with metal without any patterning, and the resistance is measured between the top and bottom sides. The contact resistivity with such samples is calculated by subtracting the resistance of the bulk, conductive (if present), and metal layers from the total resistance of the structure. However, it is possible that the contribution of contact resistivity to the total resistance may be overestimated due to the metal's finite sheet resistance and partial contacting scheme of the sample. These effects result in current traveling through a limited area inside the sample, increasing the contribution of the contact resistivity. In this work, we use a recently proposed spreading resistance model to accurately determine contact resistivity of the fully metallized test structures. We demonstrate a relative error between the proposed formulation and finite element method simulations below 1.3 %. Moreover, we demonstrate the applicability of the formulation experimentally with Ag/MoO_x/(p)c-Si/MoO_x/Ag structures.
Keywords: contact resistivity, spreading resistance

1 INTRODUCTION

In the pursuit of driving silicon solar cells to higher levels of performance and reduced fabrication costs, heterojunction topologies with passivating carrier-selective layers have demonstrated great potential [1]–[3]. Devices employing transparent conductive oxides with thin-film silicon layers [4], [5], highly doped polysilicon [6]–[8], metal oxides [1], [9]–[11] have been shown to yield efficiencies over 25 % [12]. While there is an abundance of materials to be used as carrier-selective layers, they all require careful investigation of their optoelectronic properties prior to their use in solar cells. One of the crucial properties that should be accurately determined is the contact resistivity (ρ_c, in $\Omega.cm^2$) of the formed interface between the carrier-selective layer and silicon bulk. Thin interlayers can also be inserted ([8], [13], [14]) to further enhance the quality of the interface. In that case, the combined resistive effects of the carrier-selective layers, interlayers, and the silicon substrate are included in ρ_c. Further discussion on this topic can be found in [15]. Some of the widely used methods for the determination of ρ_c are the transfer length method (TLM) [16] and Cox and Strack (CS) method [17]. Alternatively, one can use a fully metallized structure (FMS), shown in Fig. 1a, where both sides of the sample are metallized without patterning of the carrier-selective and metal layers as in the TLM or CS methods. Like in the CS test structures, the voltage is applied between the top and bottom electrodes in FMS. To extract ρ_c using FMS, the resistive contributions of the substrate (typically silicon), metal, and other conductive layers (if present) are subtracted from the total measured resistance, and the remaining resistance is multiplied by the area of the sample. While this approach seems straightforward and simple, the extracted ρ_c is susceptible to overestimation. The reason lies in the experimental procedure of obtaining the current-voltage (I-V) characteristics. It is common to have a measurement setup where the bottom side of the FMS is fully contacted by means of a metallic plate, while the top side is partially contacted with a needle probe with a tip diameter of a few hundred μm. Partially contacting the sample with a small probe from one side may result in current traveling through only a portion of a sample's area (perpendicular to the current direction). If the metal is not thick enough (or, in other words, its sheet resistance is not small), the current will not be able to laterally reach and travel through the regions close to the edges of the sample (as shown in Fig. 1b), resulting in a smaller cross-sectional area through which the carriers pass through the metal-substrate interface (Fig. 1b). Thus, the measured resistance becomes larger than it is with an infinitely conductive metal layer, and the calculation of ρ_c will be faulty.

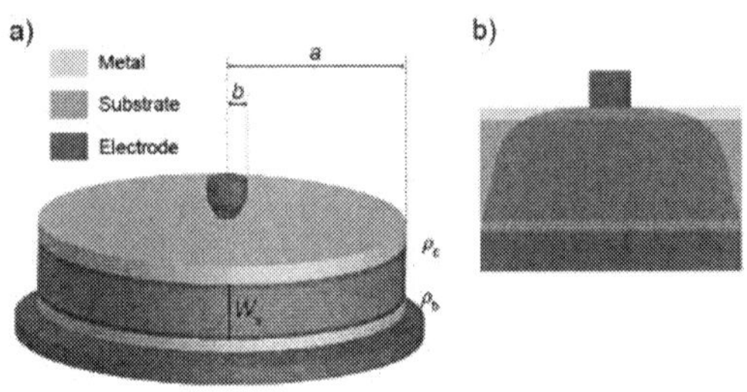

Figure 1: (a) Schematic of the fully metallized structure (FMS) contacted with a small probe from the top and full area electrode from the bottom (blue and purple regions indicate contact layers). (b) Side view of FMS. The light purple region demonstrates the current spreading.

To account for the current spreading in a conductive layer fully covering the substrate, we recently proposed an analytical formulation for the CS test structures [15]. In this work, we adapt this model [15] and use the proposed formulation to calculate the resistance of the FMS's with different metal thicknesses. First, we quantify the accuracy of the analytical formulation by comparing the results with the finite element method simulations. Then, we demonstrate the viability of the formulation by extracting ρ_c using fabricated FMS's.

2 NUMERICAL CALCULATIONS

Following the same procedure as in [15], the distributed resistance network (Fig. 2) can be constructed (in cylindrical coordinates) for structure in Fig. 1a. Here, we define the top metal as the conductive layer and assume that the contact resistivity between the electrode (or probe) is negligible. Solving the set of equations arising from the resistance network in Fig. 2, we arrive at the expression for the measured resistance of the structure,

$$R = \left(\frac{\pi b^2}{\rho_c + 2\pi b \dfrac{\rho_b}{\pi (b+W_s)^2} + \pi R_{spr} b^2} + \frac{2\pi k b}{R_m} Y \right)^{-1}, \quad (1)$$

with

$$R_{spr} = \frac{\rho_s}{2b\pi} \arctan\left(\frac{4}{2b} W_s\right),$$

$$Y = \frac{I_1(ka)K_1(kb) - I_1(kb)K_1(ka)}{I_0(kb)K_1(ka) + K_0(kb)I_1(ka)},$$

$$k = \sqrt{R_m / (\rho_c + \rho_s W_s + \rho_b)},$$

and ρ_b is the contact resistivity between the substrate and the bottom metal, ρ_s (in Ω.cm) is the resistivity of the substrate, R_m is the sheet resistance of the metal, I_0 and K_0 are the zeroth-order modified Bessel equations of the first and second kind, respectively, and I_1 and K_1 are the first-order modified Bessel equations of the first and second-order, respectively.

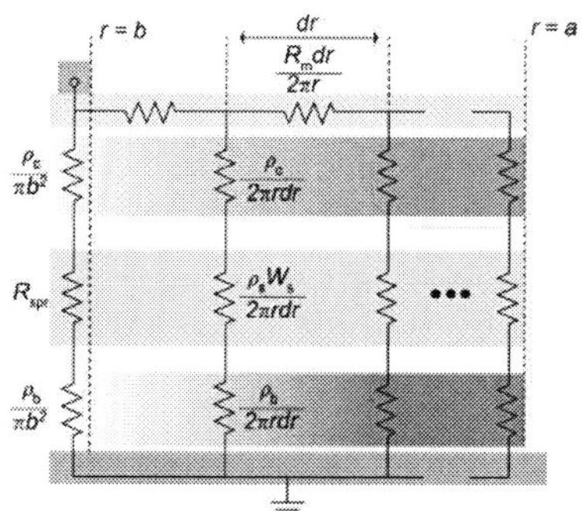

Figure 2: Distributed resistance network representation of the test structure in Fig. 1a.

To investigate the accuracy of Eq. 1, the structure in Fig. 1a was simulated using the 3D finite-element method (FEM), and the results were compared in terms of the relative error (δ_{rel})

$$\delta_{rel} = \frac{R - R_{FEM}}{R_{FEM}},$$

where R is the resistance determined by Eq. 1 and R_{FEM} is the resistance determined by FEM simulations. In simulations, a was 1 cm, ρ_s was as 1 Ω.cm, ρ_c and ρ_b were between 10^0 and 10^3 mΩ.cm^2, respectively, and b was varied between 25, 100, and 200 μm. The effect of the substrate's shape was also investigated by simulating square and circular structures. In the simulations, the edge length of the square substrate ($a_{square} = (\pi)^{1/2} \approx 1.77$ cm) was set to match the area of the circular one.

Fig. 3 shows δ_{rel} for the investigated cases. δ_{rel} is below 1.3 % for the chosen parameter space. The error tends to be larger for the cases with smaller b, thinner metals, and large ρ_c (10^3 mΩ.cm^2), yet the δ_{rel} is still quite small. Moreover, substrate's shape has a negligible effect on the accuracy of Eq. 1 for a = 1 cm (a_{square} = 1.77 cm), demonstrating the formulation's applicability not only for circular substrates (which is assumed in the derivation of Eq. 1) but also for square substrates which are common in experiments involving Si wafers.

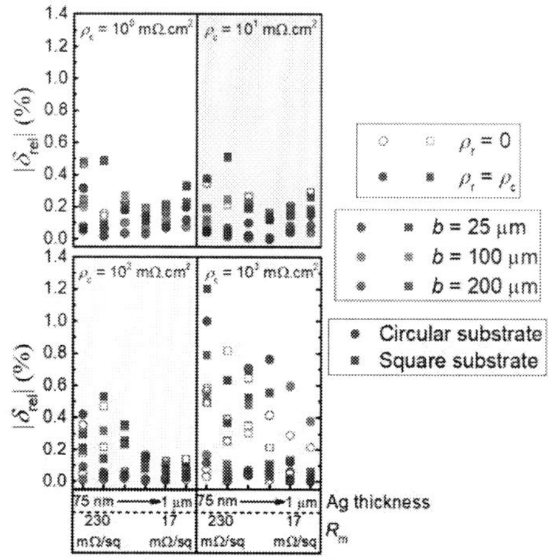

Figure 3: $|\delta_{rel}|$ versus Ag thickness (R_m) with the parameters varied in factorial fashion. b (25, 100, 200 μm), b, ρ_c (10^0, 10^1, 10^2, 10^3 Ω.cm^2), ρ_b (0 or = ρ_c), Ag thickness (75, 150, 250, 500, 750, 1000 nm) and substrate shapes (circular, square). The bulk resistivity of Ag was ~1.7×10^{-6} Ω.cm, and the corresponding R_m were 230, 110, 70, 30, 20, and 17 mΩ/sq, respectively

3 EXPERIMENTS

To investigate the applicability of the spreading resistance model in Eq. 1, we fabricated test structures in Fig. 4. As a substrate, ~200 μm-thick, 1.6×1.6 cm^2 (square-shaped) pyramid textured, p-type monocrystalline silicon [(p)c-Si] wafer was used. Note that the size of the samples used in the experiments is slightly smaller than those analyzed in simulations (by ~ 0.3 cm^2), yet the analytical model is expected to have similar accuracy to the one presented in the previous section. Firstly, the samples were cleaned with RCA-2 and were then dipped into a diluted HF solution (5 vol.%) to remove the chemical oxide. Then, MoO_x/Ag stack was thermally evaporated on both sides of the samples under vacuum (~10^{-6} Torr). All samples received 6 nm MoO_x on both sides. On the bottom side (to be contacted with the full-area chuck), a 200 nm of Ag layer (resistivity ~ 1.7×10^{-6} Ω.cm) was deposited. On the top side (to be contacted with a needle probe), the thickness of the Ag was varied between 100, 150, and 250 nm to investigate the effect of the metal's sheet resistance. Note that all mentioned thicknesses are measured on planar surfaces, and they are expected to be ~1.5 times smaller on the textured surfaces. Due to the modified effective thicknesses of the deposited layers, R_m was calculated by dividing Ag's resistivity by the expected thickness on the textured surface.

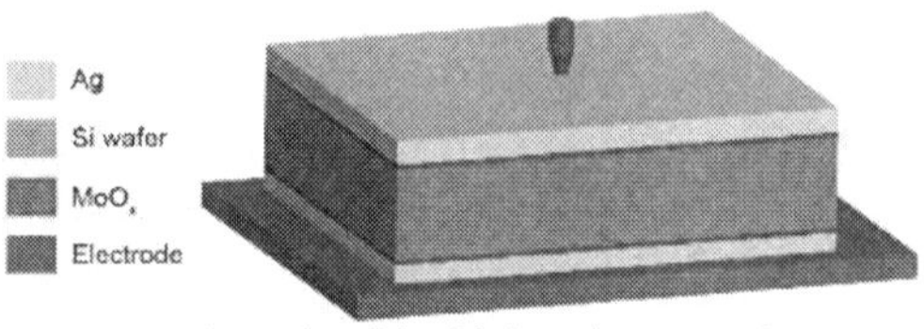

Figure 4: Schematic of the fabricated test sample.

R values were determined by performing a linear fit on the I-V characteristics (voltage was varied between -200 and +200 mV) obtained by Keithley 2400 sourcemeter in a four-wire configuration. Note that while this configuration is necessary to avoid the effect of contact resistivity of the electrode/sample interface experimentally [18], in the simulation domain two-wire configuration is sufficient because the electrode/sample interface is ideal (no contact resistivity). The side with a fixed Ag thickness was contacted via a gold-plated full area chuck (typically used for measuring the performance of the solar cells under illumination), and the other side was contacted with a probe with a tip diameter of ~ 380 µm as measured using a microscope (Fig. 5). Then ρ_c (in the text will be referred to as 'ρ_c with correction') was extracted using MATLAB [19], where the error between experimentally measured *R* (R_{exp}) and *R* calculated by Eq. 1 was minimized by fitting ρ_c. To calculate ρ_c without the correction, the following relation was used

$$\rho_{c\ \text{not corrected}} = \left(R_{exp} - \frac{\rho_s W_s}{a_{square}^2} \right) / 2 , \qquad (2)$$

where a_{square} = 1.6 cm. Note that for all calculations, we assume that $\rho_c = \rho_b$ since both sides of the samples have a 6 nm-thick MoO_x capped with Ag.

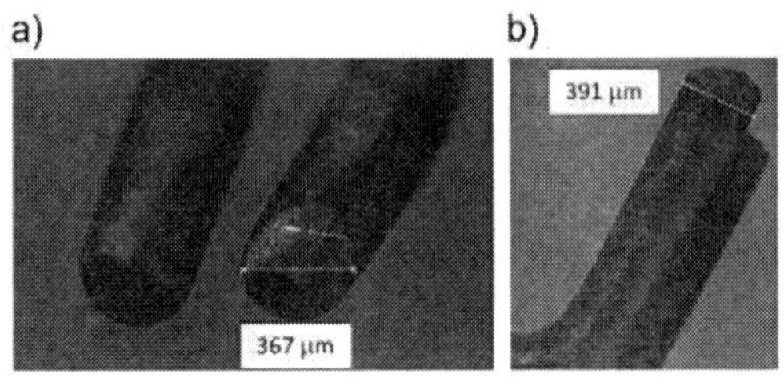

Figure 5: (a) Top and (b) side view of the optical microscope images of the probe used in the experiments.

It is also worth noting that the diameter of the used probe measured with an optical microscope might not correspond to the actual probe contact area on the sample. The contacted area depends on the probe tip shape (e.g. pyramidal, spherical) and the applied pressure. As the applied pressure increases, the contact radius can increase [20]. These effects are extensively investigated in the study of the mechanical properties of the materials via indentation [21], [22]. In our experiments and calculations, we assume that the contacting area is equal to the diameter of the probe measured by a microscope, which can be regarded as the maximum contacting area when the probe touches the surface of the sample.

4 RESULTS AND DISCUSSION

Fig. 6 depicts measured R_{exp} of the fabricated samples versus Ag thickness. R_{exp} decreases for thin Ag sharply as the thickness increases and then tends to saturate at larger thicknesses. This behavior is expected and is related to more pronounced current spreading in the metal as the thickness of the metal increases.

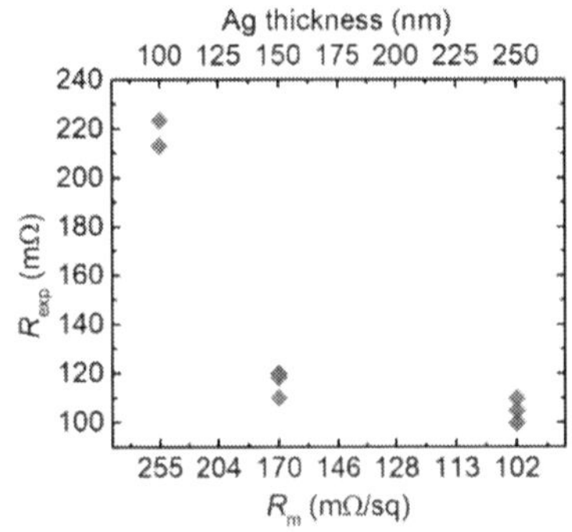

Figure 6: Measured R_{exp} of the fabricated samples versus R_m (bottom axis) and Ag thickness of the side contacted with a probe (top axis).

From the R_{exp} values in Fig. 6, ρ_c was extracted using Eq. 1, assuming that *b* is equal to the maximum contacting radius of 190 µm (Fig. 7). As the thickness of Ag increases (or R_m decreases), ρ_c with the correction remains constant, whereas ρ_c without correction (i.e. using Eq. 2) follows the exponential trend with *R* in Fig. 6. The overestimation in ρ_c can be as large as 100 % when the current spreading in the metal is not considered. As Ag becomes thicker, ρ_c's extracted with and without correction approach one another, indicating that the metal's finite sheet resistance has a lesser effect on the R_{exp} of the samples.

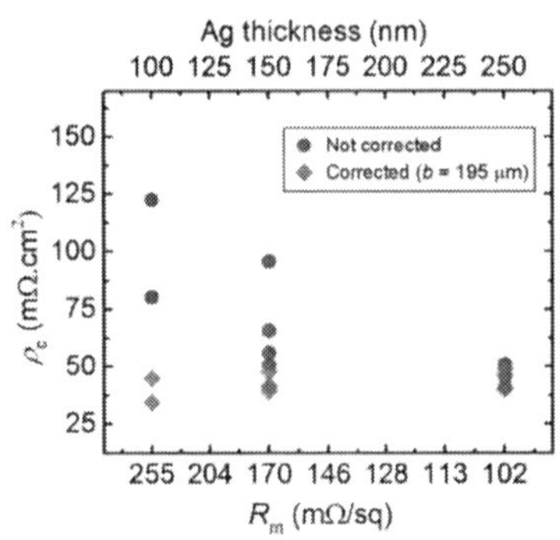

Figure 7: ρ_c extracted using Eq. 2 (black circles) and Eq. 1 with *b* = 195 µm versus R_m (bottom axis) and Ag thickness (top axis).

Finally, we note the importance of knowing the contact area (i.e. contacting radius) between the probe and surface of the sample as it can significantly affect the extracted ρ_c. To see the extent of this effect, we generated *R* values for R_m = 70 and 230 mΩ/sq with *b* = 10 µm and ρ_c = 50 mΩ.cm^2 using Eq. 1. Then, using generated results, we fitted them to find ρ_c while changing the assumed radius ($b_{assumed}$) from 10 (true value) to 10^3 µm. In Fig. 8, it can be seen that for small $b_{assumed}$, the change in the extracted ρ_c is sharp and saturates as $b_{assumed}$ becomes larger. Moreover, the effect is more pronounced for larger R_m (thinner Ag). Although there is a significant dependence of the extracted ρ_c on $b_{assumed}$, assuming that the contact radius is equal to the maximum probe radius (measured by optical or scanning electron microscope) provides an upper bound for the extracted ρ_c, and further analysis can be performed to determine the lower bound.

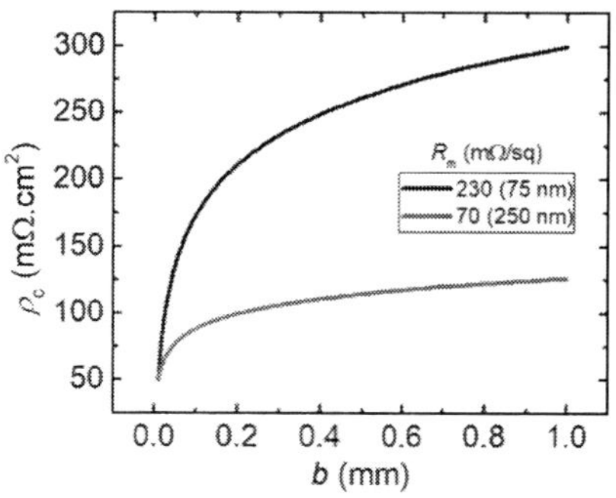

Figure 8: ρ_c extracted Eq. 1 versus $b_{assumed}$ for different R_m's using R generated with b = 10 μm and $\rho_c = \rho_b$ = 50 mΩ.cm^2.

5 CONCLUSION

In this work, we modified a recently proposed analytical formulation [15] to model the spreading resistance in fully metallized test structures. Firstly, we compared the accuracy of the proposed formulation with finite element method simulations, where we obtained a relative error smaller than 1.3 %. Moreover, we have demonstrated the validity of this model experimentally by determining the contact resistivity of (*p*)c-Si/MoO_x/Ag interface with samples with different metal thicknesses.

ACKNOWLEDGEMENT

This work was financially supported by TUBITAK under the grant number: 315M360.

REFERENCES

[1] J. Bullock *et al.*, "Efficient silicon solar cells with dopant-free asymmetric heterocontacts," *Nat. Energy*, vol. 1, no. 3, pp. 1–7, 2016, doi: 10.1038/nenergy.2015.31.

[2] T. G. Allen, J. Bullock, X. Yang, A. Javey, and S. De Wolf, "Passivating contacts for crystalline silicon solar cells," *Nat. Energy*, vol. 4, no. 11, pp. 914–928, 2019.

[3] S. W. Glunz and F. Feldmann, "SiO2 surface passivation layers – a key technology for silicon solar cells," *Sol. Energy Mater. Sol. Cells*, vol. 185, pp. 260–269, Oct. 2018, doi: 10.1016/j.solmat.2018.04.029.

[4] S. DeWolf, A. Descoeudres, Z. C. Holman, and C. Ballif, "High-efficiency silicon heterojunction solar cells: A review," *Green*, vol. 2, no. 1, pp. 7–24, 2012, doi: 10.1515/green-2011-0018.

[5] M. Lozac'h, S. Nunomura, H. Sai, and K. Matsubara, "Passivation property of ultrathin SiOx:H / a-Si:H stack layers for solar cell applications," *Sol. Energy Mater. Sol. Cells*, vol. 185, no. March, pp. 8–15, 2018, doi: 10.1016/j.solmat.2018.05.004.

[6] Y. Xu *et al.*, "Enhanced open-circuit voltage in p-type passivated emitter and rear cell by doped polysilicon layer as passivation contact," *Sol. Energy*, vol. 207, pp. 436–440, Sep. 2020, doi: 10.1016/j.solener.2020.07.002.

[7] D. Yan, A. Cuevas, J. Bullock, Y. Wan, and C. Samundsett, "Phosphorus-diffused polysilicon contacts for solar cells," *Sol. Energy Mater. Sol. Cells*, vol. 142, pp. 75–82, Nov. 2015, doi: 10.1016/j.solmat.2015.06.001.

[8] U. Römer *et al.*, "Recombination behavior and contact resistance of n+ and p+ poly-crystalline Si/monocrystalline Si junctions," *Sol. Energy Mater. Sol. Cells*, vol. 131, pp. 85–91, 2014.

[9] M. Nayak, K. Singh, S. Mudgal, S. Mandal, S. Singh, and V. K. Komarala, "Carrier-selective contact based silicon solar cells processed at toom temperature using industrially feasible Cz wafers," *Phys. Status Solidi Appl. Mater. Sci.*, vol. 216, no. 16, pp. 1–8, 2019, doi: 10.1002/pssa.201900208.

[10] X. Yang, Q. Bi, H. Ali, K. Davis, W. V. Schoenfeld, and K. Weber, "High-performance TiO2-based electron-selective contacts for crystalline silicon solar cells," *Adv. Mater.*, vol. 28, no. 28, pp. 5891–5897, 2016, doi: 10.1002/adma.201600926.

[11] S. Avasthi, W. E. McClain, G. Man, A. Kahn, J. Schwartz, and J. C. Sturm, "Hole-blocking titanium-oxide/silicon heterojunction and its application to photovoltaics," *Appl. Phys. Lett.*, vol. 102, no. 20, 2013, doi: 10.1063/1.4803446.

[12] K. Masuko *et al.*, "Achievement of more than 25% conversion efficiency with crystalline silicon heterojunction solar cell," *IEEE J. Photovoltaics*, vol. 4, no. 6, pp. 1433–1435, Sep. 2014, doi: 10.1109/JPHOTOV.2014.2352151.

[13] K. Tsoi, D. Turkay, M. Ghasemi, E. Dönerçark, and R. Turan, "Heterojunction Silicon Solar Cells Incorporating Unpatterned MoOx and LiF / Al Layers on Wet-Chemically Processed Silicon Surfaces," in *37th European Photovoltaic Solar Energy Conference (EUPVSEC)*, Oct. 2020, vol. 147, no. 03, p. 4, doi: 10.4229/EUPVSEC20202020-2DV.3.51.

[14] M. Taguchi, "Review—Development history of high efficiency silicon heterojunction solar sell: From discovery to practical use," *ECS J. Solid State Sci. Technol.*, vol. 10, no. 2, p. 025002, Feb. 2021, doi: 10.1149/2162-8777/abdfb6.

[15] D. Turkay, K. Tsoi, E. Donercark, T. Rasit, and Y. Selcuk, "Spreading resistance modeling for contact resistivity extraction in multilayer samples with circular electrodes," 2021.

[16] W. Schottky, "Zur halbleitertheorie der sperrschicht-und spitzengleichrichter," *Zeitschrift für Phys.*, vol. 113, no. 5, pp. 367–414, 1939.

[17] R. H. Cox and H. Strack, "Ohmic contacts for GaAs devices," *Solid State Electron.*, vol. 10, no. 12, pp. 1213–1218, 1967.

[18] "Four-terminal sensing - Wikipedia." https://en.wikipedia.org/wiki/Four-terminal_sensing (accessed Sep. 09, 2021).

[19] MATLAB, "R2021a." Natick, Massachusetts, 2021.

[20] L. Charleux, V. Keryvin, M. Nivard, J. P. Guin, J. C. Sanglebœuf, and Y. Yokoyama, "A method

for measuring the contact area in instrumented indentation testing by tip scanning probe microscopy imaging," *Acta Mater.*, vol. 70, pp. 249–258, May 2014, doi: 10.1016/j.actamat.2014.02.036.

[21] A. J. Bushby, "Nano-indentation using spherical indenters," *Nondestruct. Test. Eval.*, vol. 17, no. 4–5, pp. 213–234, Jan. 2001, doi: 10.1080/10589750108953112.

[22] S. Suresh, T. G. Nieh, and B. W. Choi, "Nano-indentation of copper thin films on silicon substrates," *Scr. Mater.*, vol. 41, no. 9, pp. 951–957, Oct. 1999, doi: 10.1016/S1359-6462(99)00245-6.

INCREASE OPTICAL PERFORMANCE OF SILICON BASED HETEROJUNCTION SOLAR CELLS WITH TiO_2 NANOROD STRUCTURES

Betül ŞEKERTEKİN[1], Abdullah YILDIZ[2]
1-KalyonPV Research and Development Center, Kalyon Güneş Teknolojileri Üretim A.Ş.,06909 Ankara, TURKEY
bsekertekin@kalyonpv.com
ORCID: 0000-0002-9594-9635
2-Ankara Yildirim Beyazit University, 06010 Ankara, TURKEY
yildizab@gmail.com,
ORCID: 0000-0002-0250-3731

ABSTRACT: TiO_2 metal oxide layers are used for photovoltaic studies and diode applications because of their excellent properties like chemical resistance, high refractive index, and high dielectric constant. TiO_2 can be operated as an anti-reflecting coating (ARC) and presents an n-type behavior because of the presence of donor-type defects, and titanium interstitials atoms. One-dimensional nanostructures such as nanorods, nanowires, etc. of TiO_2 have been studied due to their ability to increase the charge separation rate and acting as a pathway for charge carrier transport. In this study, the TiO_2 layer as a seed layer was spin-coated on a p-Si wafer and glass. Then, TiO_2 nanostructures were synthesized by hydrothermal method with lower reaction temperature which facilitates a reduction in agglomeration between particles. The morphology of nanostructures as physical shape, length, diameter, and alignment is controlled and changed during hydrothermal synthesis by setting reaction time, temperature, and solution rate. The structural, morphological, and optical properties of the nanostructures were determined utilizing X-Ray Diffraction (XRD), scanning electron microscopy (SEM), and UV-Vis spectrometer measurements. Current-voltage (I-V) characteristics of the p-n heterojunction device based on Ag/ TiO_2 nanostructures/TiO_2 seed layer/p-Si/Ag were investigated under dark and light illumination. The results revealed that the performance parameters of the device were significantly improved with TiO_2 nanostructures.
Keywords: *Hydrothermal method, Heterostructures, TiO_2 Nanostructures*

1 INTRODUCTION

Metal oxides such as ZnO, CeO_2, TiO_2, SnO_2, and Fe_2O_3 are used as a catalyst because of their chemical, electrical and optical properties. Titanium dioxide (TiO_2) has received a considerable attention from scientific researchers because of the semiconductor properties with wide range of applications because of its exceptional chemical stability, abundance, durability, chemical inertness, large band gap, high refractive index and bio-compability. The interest in TiO_2 was mainly because of non-toxicity and good stability in various environments.

Titania (TiO_2) is an important photoactive material with existing and potential uses in a wide variety of environmental and energy applications, such as photocatalytic oxidation of contaminants [1]-[4], photochemical water splitting under ultraviolet (UV) illumination [5]-[6], and dye synthesized photovoltaic cells. [7]-[9]. These applications can be divided in environment and energy categories which depend on properties and modifications of TiO_2 material with another materials. Underground water purification plants based on TiO2 catalyzed photo-degradation reactions are already in work in many countries to decompose environmental pollutants [10]-[14]. TiO_2 an excellent n-type semiconductor that has 3.2 eV band gap. Owing to this property, TiO_2 has a variety application in transparent electrodes, anti-reflecting coatings, photonic crystals, photo-electrochemical solar cells, photocatalysts and sensors [15]-[16].

TiO_2 thin films are prepared using various techniques such as laser deposition, spray pyrolysis, chemical vapor deposition (CVD), physical vapor deposition (PVD), successive ionic layer adsorption and reaction (SILAR) [17], hydrothermal, solvothermal and sol-gel spin coating method.

Wang et al. [18] have reported the optical properties of TiO2 films prepared by sol–gel spin coating technique. The using of sol-gel spin coating and spray pyrolysis techniques to deposit TiO_2 films have several advantages such as low cost, simple apparatus set up and easy controlling the properties of the thin film through the solution composition and reaction conditions.

Generally, TiO_2 exhibits an n-type conductivity behavior and this type conductivity is due to donor type defects, including oxygen vacancies and titanium interstitials atoms [21]. TiO_2 is used in solar cell applications especially heterojunction structures that can be prepared under typical laboratory conditions from a precursor. Specifically, TiO_2 can be employed as an antireflection coating (ARC) in solar cells for silicon surface passivation and as an emitter layer in p-n heterojunction-based devices [22]. Nanostructured TiO_2 thin films have been under intense study for many years Furthermore, one-dimensional (1-D) nanostructures such as nanowires (NWs), nanorods (NRs), nanotubes (NTs) have been demonstrated for potential technological applications due to their high surface to volume ratio and inherent quantum confinement effects [23].

In this study, the n-TiO_2 film was spin-coated with different cycles of coating (x1, x2, x3) on a p-Si wafer and glass. The optical properties of the films were determined utilizing UV-Vis spectrometer.

Current-voltage (I-V) characteristics of the p-n heterojunction device based on Ag/TiO_2/p-Si/Ag were investigated in the dark.

2 EXPERIMENTAL DETAILS

2.1 Preparation TiO_2 thin film

In order to prepare TiO_2 solution, firstly 1.2 ml of titaniumtetraispropoxide [$Ti(OC_3H_7)_4$, Merck]was added in 15 ml of ethanol [C_2H_6O, Merck] and the solution was kept in a magnetic mixture for 1h. Then 5 ml of glacialaceticacide [$C_2H_4O_2$, Merck] and 10 ml of ethanol were added in the solution and after each additive component was added, it was mixed in the magnetic mixture for 1h. As a final step, 1.5 ml trietilamine [$(S_2H_5)_3N$, Merck] was added in the solution and the final solution was subjected to the magnetic mixture for 3h [24].

The substrates (glasses and p-Si wafers (500 um)) were cleaned according to standart cleaning procedures as given below.

Prior to deposition, p-Si wafer was diced to form Si pieces of 1.5cm*1.5cm. Then, the standard Piranha and HF cleaning procedure were applied to eliminate surface contamination. In this work, TiO_2 sol (0.6 M) was prepared by using the precursor of TTIP (Titanium Isopropoxide). TiO_2 metal oxide thin films with three different cycles of coating (x1, x2, x3) were deposited on p-Si (100) substrate via spin coating technique. The spin-coating process was achieved at 2500 rpm, 5.0 acc, and for 30 seconds at room temperature with spin coater. During sol-gel coating process, heat treatment is applied at 300°C for 5 minutes. For the metallization process, Ag paste was applied onto the back and front surface of heterojunction to form metal contacts.

UV-Vis spectrometer is used to determine the optical properties of n-type layers. I-V measurement set up (made up of a Keithley 2400, a computer and a probe station) is used to determine the electrical characteristics of the devices.

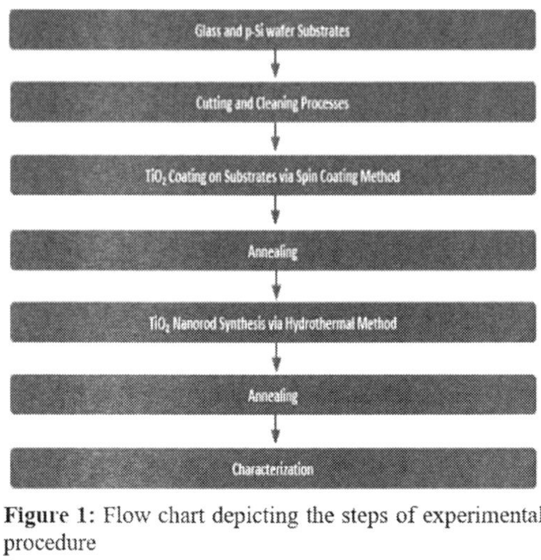

Figure 1: Flow chart depicting the steps of experimental procedure

2.2 Hydrothermal synthesis

For the typical hydrothermal synthesis, 15 mL distilled water was mixed with 15 mL HCl via magnetic stirring in a Teflon-lined stainless-steel autoclave. After stirring for 15 min, 0.5 mL $Ti[O(CH_2)_3CH_3]_4$ was added into the mixture as small droplets.

Then, coated and annealed substrates was placed into the Teflon-liner against the wall with an angle and the coated side was kept face down. The autoclave containing the substrate was sealed and annealed inside an oven at 175°C for 4 h. After process, the autoclave was cooled to room temperature under tap water. The substrate was finally washed with distilled water, followed by air-drying in a nitrogen gas flow.

3 RESULTS AND DISCUSSION

The schematic of n-TiO_2/p-Si heterojunction is shown in *Figure 2*. In the present design, the n-TiO2 layer is depicted as an emitter layer, while p-Si substrate is employed as an absorber layer. In this study, the n-TiO2 is deposited by sol-gel spin coating.

In addition, screen printing method is used to obtain back and front metal contacts.

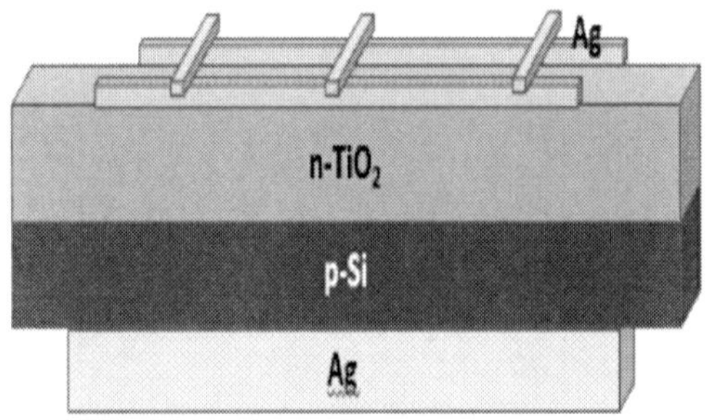

Figure 2: The schematic of n-TiO2/p-Si heterojunction

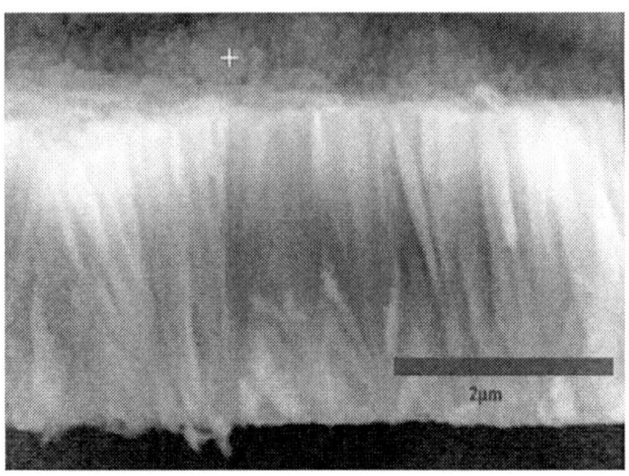

Figure 3: SEM image of TiO_2 nanorods

As seen in Figure 3, the recipe that used for the hydrothermal coating process yielded successful result. TiO_2 nanorods which occurred via hydrothermal method at 175°C for 4 hours in autoclave. Successful formation of approximately 1.86 µm nanorods were observed.

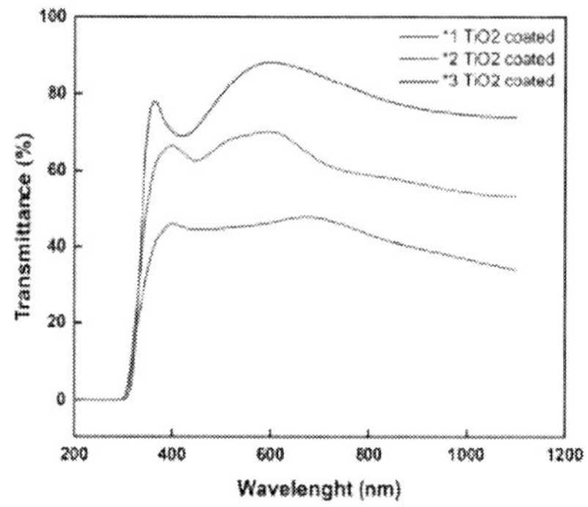

Figure 4: Transmittance spectra of multilayer TiO_2 thin films.

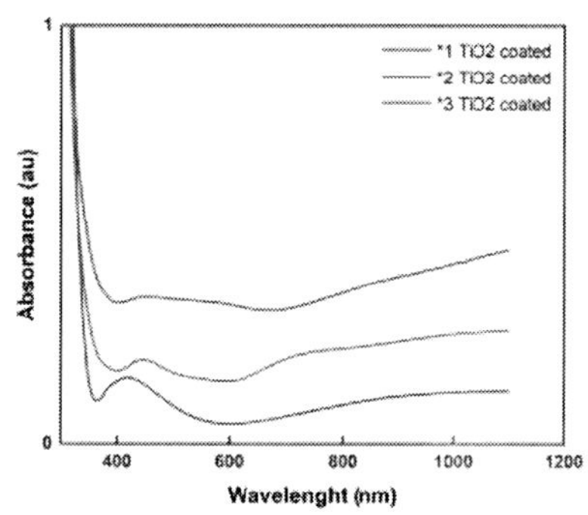

Figure 5: Absorption spectra of multilayer TiO_2 thin films.

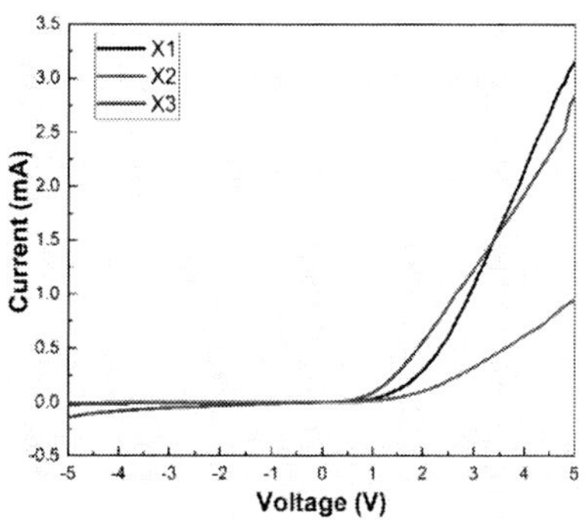

Figure 6: I-V graph of *1-*2-*3 layers of TiO_2 coated p-Si substrates

Figure 6 shows the dark I-V characteristics of the devices. It is expected to obtain diode structure in cell production. For this reason, the diode characteristics of device were measured by making I-V test in the dark condition. According to I-V results, the fabricated devices exhibit good diode behavior.

To analyze the optical properties of TiO_2 structure UV-Vis test is applied and transmittance/absorbance results are given in Figure 4 and Figure 5. According to test results, highest transmittance value observed for thinnest film which has 1-layer TiO_2 thin film. In addition, highest absorbance value observed for thickest film which has 3 layers thin film. High optical and electrical properties make this device highly suitable for electronic and optoelectronic applications.

In conclusion, the TiO_2 thin film layer was successfully coated on the p-Si substrate. It was concluded that the optical properties of the coated thin film were good. The diode characteristics of the devices were determined by I-V measurement. It was found that the best performance was with the device fabricated with 1 cycle of coated TiO_2 film. Optical properties, metallization and coating processes will need to be improved for solar cell fabrication. In future studies, it is aimed to produce solar cell by focusing on these points.

4 REFERENCES

[1] Z. Liu, X. Zhang, S. Nishimoto, T. Murakami, and A. Fujishima: Efficient photocatalytic degradation of gaseous acetaldehyde by highly ordered TiO2 nanotube arrays. Environ. Sci. Technol. 42, 8547 (2008).

[2] A. Fujishima, X. Zhang, and D.A. Tryk: TiO2 photocatalysis and related surface phenomena. Surf. Sci. Rep. 63, 515 (2008).

[3] X. Chen and S.S. Mao: Titanium dioxide nanomaterials: Synthesis, properties, modifications, and applications. Chem. Rev. 107, 2891 (2007).

[4] L. Chen, M.E. Graham, G. Li, and K.A. Gray: Fabricating highly active mixed TiO2 photocatalysts by reactive DC magnetron sputter deposition. Thin Solid Films 515, 1176 (2006).

[5] A. Fujishima and K. Honda: Electrochemical photolysis of water at a semiconductor electrode. Nature 238, 37 (1972).

[6] A. Fujishima and K. Honda: Studies on photosensitive electrode reactions. Bull. Chem. Soc. Jpn. 44, 1148 (1971).

[7] B. O'Regan and M. Gratzel: A low cost, high-efficiency solar-cell based on dye-sensitized colloidal TiO2 films. Nature 353, 737 (1991).

[8] G.K. Mor, K. Shankar, M. Paulose, O.K. Varghese, and C.A. Grimes: Use of highly ordered TiO2 nanotube arrays in dyesensitized solar cells. Nano Lett. 6, 215 (2006).

[9] A. Kongkanand, K. Tvrdy, K. Takechi, M.K. Kuno, and P.V. Kamat: Quantum dot solar cells. Tuning photoresponse through size and shape control of CdSe-TiO2 architecture. J. Am. Chem. Soc. 130, 4007 (2008).

[10] A. Fujisjima, K. Hashaimoto, T. Watanabe, TiO2 Photocatalysis: Fundamentals and Applications, Bkc, Inc., Tokyo, 1999.

[11] M. Murabayshi, K. Itoh, K. Okamura, H. Yamazaki, T. Yoshida, Environmental aspects of electrochemical technology, Electrochemical Society Proceedings, 99–39, The Electrochemical Society, Inc., USA, 2000, pp. 85.

[12] R. Doherty, J. Environ. Forensics 1 (2000) 89.

[13] N. Itoh, S. Kutsuna, T. Ibusuki, Chemosphere 28 (1994) 2029.

[14] J.F. Kenneke, J.L. Ferry, W.H. Glaze, Photocatalytic Purification and Treatment of Water and Air, Elsevier Science Publisher, New York, 1993, 179.

[15] G.K. Mor, M.A. Carvalho, O.K. Varghese, M.V. Pishko, C.A. Grimes, A room tem-perature TiO2nanotube hydrogen sensor able to self-clean photoactively fromenvironmental contamination, J. Mater. Res. 19 (2004) 628–634.
[16] C. Chunlan, H. Chenguo, W. Xue, W. Shuxia, T. Yongshu, Z. Hulin, UV sensorbased on TiO2nanorod arrays on FTO thin film, Sens. Actuators B 156 (2011)114–119.
[17] U.M. Patil, K.V. Gurav, O.-S. Joo, C.D. Lokhande, Synthesis of photosensitivenanograined TiO2thin films by SILAR method, J. Alloys Compd. 478 (2009)711–715.
[18] Z. Wang, U. Helmersson, P.O. Kall, Optical properties of anatase TiO2thin filmsprepared by aqueous sol–gel process at low temperature, Thin Solid Films 405(2002) 50–54.
[19] N. Martin, C. Rousselot, D. Rondot, F. Palmino, R. Mercier, Microstructure mod-ification of amorphous titanium oxide thin films during annealing treatment,Thin Solid Films 300 (1997) 113–121.
[20] K.K. Saini, S.D. Sharma, M. Chanderkant, D. Kar, C.P. Singh, Sharma, Structuraland optical properties of TiO2thin films derived by sol–gel dip coating process,J. Non Crystal. Solids 353 (2007) 2469–2473.
[21] M. K. Nowotny, T. Bak, and J. Nowotny, Electrical Properties and Defect Chemistry of TiO2 Single Crystal. I. Electrical Conductivity, 2006, 110, 16270-16282
[22] Ali, K., Khan, S. A., & Jafri, M. M. Effect of double layer (SiO2/TiO2) anti-reflective coating on silicon solar cells. 2014, 9(12), 7865-7874.
[23] Welte, A., Waldauf, C., Brabec, C., & Wellmann, P. J. Application of optical absorbance for the investigation of electronic and structural properties of sol–gel processed TiO2 films. 2008, 516(20), 7256-7259.
[24] O. Pakma, N.Serin, T.Serin, Ş Altındal. On the energy distribution profile of interface states obtained by taking into account ofseriesresistanceinAl/TiO2/p–Si(MIS)structures.

SURFACE MODIFICATION OF COVER GLASS FOR SOLAR PANELS BY CREATING A NANOSTRUCTURE ON THE SURFACE BY MASKLESS PLASMA TEXTURING WITH FLUOROCARBON

Alena Okhorzina, Norbert Bernhard
Anhalt University of Applied Sciences
Bernburger Str. 55, Köthen, Germany

ABSTRACT: The aim of this work is to increase the optical transmission of the front glass of solar modules. Dry plasma etching is being studied to reduce surface reflection from solar glass, especially borosilicate glass. For solar glass is no optimal coating material to create an anti-reflective coating with a suitable refractive index. The alternative could be a micro- or nano-structuring of the surface by means of plasma etching, whereby a surface layer with the required effective refractive index is generated. Plasma texturing was used as the method for producing this effective layer. Different types of plasmas and the mechanisms of their operation during SiO_2 etching are analyzed. CF_4 and O_2 were chosen as working gases. The etching was performed with capacitive (CCP) and inductive (ICP) coupled plasmas or a combination of both and the effects of their use in the texturing process are described. The changes in the optical properties were determined with a UV / VIS spectrometer.
Keywords: Antireflection Coating (ARC), Plasma Etching, Plasma Texturing, Light Trapping, Solar Glass, Borosilicate Glass.

1 INTRODUCTION

Solar energy is a recognized, used, and promising technology for converting the energy of sunlight into the energy needed by humans. Now it is becoming widespread, not only on the ground in the form of fields [1], but also as part of buildings or floating construction. Until now, a disadvantage of a silicon solar module is its moderate efficiency in comparison to other energy conversion techniques. Several methods can be used to increase efficiency, some important of them are listed here [2, 3]:

- An increase in the transmitted power of solar radiation due to radiation concentrators. But this method leads to an increase in the temperature of the modules, which negatively affects the generation of electricity;
- Solar tracking;
- Use of non-silicon cells. Which often leads to an increase of the production cost of the module (as for GaAs based cells);
- Reduction of reflectivity.

All methods are now used to varying degrees. Let us focus on changing the optical properties to reduce the reflection of the module. This method can be applied to both silicon and cover glass [4, 5, 6].

An anti-reflective or anti-reflective (AR) or anti-reflective coating (ARC) is a type of optical coating that is applied to the surface of lenses and other optical elements to reduce reflection. For the reduction of the glass pane refection in a solar module, also the coating with an anti-reflective layer is beneficial. For optimal results, glass needs a quarter-wave thickness antireflection layer coating in order to reduce reflection losses at the air/glass interface [7]. Here is the problem, that ideally the refractive index of the square-root of the refractive index of glass would be needed, i.e. $\sqrt{1.5} = 1.22$. Unfortunately, there is no material available, which can be easily deposited by the standard techniques of sputtering, evaporation, or CVD for an acceptable price.

Alternatively, the glass surface can be micro- or nano-textured (or made porous) to achieve a thin surface layer with an effective medium refractive index of about 1.22. If the pores or textured structures are smaller than the light wavelength, the light sees only an effective medium with a refractive index, which is a weighted mean of the refractive indices of glass and air/vacuum.

In case of texturing (or varying pore density) this gives a gradually varying refractive index, which can be approximated by a multiple layer sequence with gradually changing refractive indices. [8]. When the surface texture has dimensions comparable or even larger than the light wavelength, the effective medium approximation is no longer valid, but still a reflection reducing effect can be observed and can be calculated numerically by the FDTD (Finite Difference in the Time Domain) method. [9]. Such an effect is also observed when studying the eyes of moths[7, 10]. Physically, it is correctly described as a smooth change in the refractive index or approximated by the theory of an effective mean refractive index.

Dry plasma etching is a well known technique for texturing the Si surfaces for reflectivity reduction in solar cells. This method is sometimes called the black-Si method and is well known from microelectronics form some time [11]. Here this technique is transferred to texture solar glass surfaces.

Plasma etching can be applied to silicon and its compounds used in solar technology silicon nitride, silicon dioxide, and glass.

2 PLASMA ETCHING

A plasma will etch an exposed material whenever the interaction with the surface results in the formation of a stable volatile compound and the removal of small particles. A fluorine plasma easily etches elements in column 14 of the periodic table [12]. The typical materials concerned by fluorine plasma etching are silicon and silicon dioxide. The most widely used, SF_6, CHF_3, CF_4, and C_4F_8, are low toxic or non-toxic gases. The main issue of using these gases is high global warming potential and long atmospheric lifetime. As an example, CHF_3 is a greenhouse gas, about 1.7 times weaker than SF_6 [13]. Therefore using CF_4 or CHF3 is an eco-friendlier approach than using SF_6, assuming that leakage of both gases to the atmosphere during or at the end of the plasma process is the same.

In this study several working gases for etching silicon and silicon dioxide (the main constituent of glass) will be investigated.

SF-plasma:

Silicon (Si) and silicon dioxide (SiO_2) can be plasma-textured by the so-called black-silicon method with SF_6/O_2. The black-silicon method is based on opposing mechanisms, which are only qualitatively understood until now. These mechanisms are (I) an isotropic silicon etching by F radicals, (II) a silicon sidewall passivation by SiOF, which locally reduces the F etching rate, and (III) a local opening of the SiOF layer by accelerated ions from the plasma [4].

CF-plasma:

Due to the presence of CFx radicals that tend to produce fluorocarbon films, fluorocarbon-based plasmas behave very differently from SF_6, for instance, when interacting with silicon and SiO_2 materials. The first feature is (I) the predominance of chemical reactions in reactive plasma etching processes. The second important feature is that both Si and SiO_2 etching in CF-plasma require applying a (II) sufficient bias (typically above 100 V) to be effective. This is especially true for SiO2, which can usually only be etched with a sufficient ion bombardment. Third feature is the significantly larger etching rate of SiO_2 as compared to Si because of the difference in material composition: in the case of Si, the etching proceeds through (III) a thin (few nanometres) fluorocarbon layer that takes part in the etching mechanism [12].

In the case of SiO_2, oxygen atoms released by energetic ion bombardment react with incoming CFx radicals to form stable CO, CO_2, COF_2 particles as volatile etching products and prevent the formation of such a layer. This situation makes it possible to achieve selective etching of SiO_2 with respect to Si [8]. The operator can tune that process by changing the gas mixture: the addition of O_2 will balance the tendency for fluorocarbon deposition, while the addition of H_2 will trap fluorine and promote the formation of fluorocarbon deposits [12].

A schematic representation of some of the processes in a plasma chamber is shown in Fig. 1.

Etching options in SF_6/O_2 [4,5] and CHF_3/H_2 [6] were used earlier, now it is considered to focus on etching with CF_4/O_2.

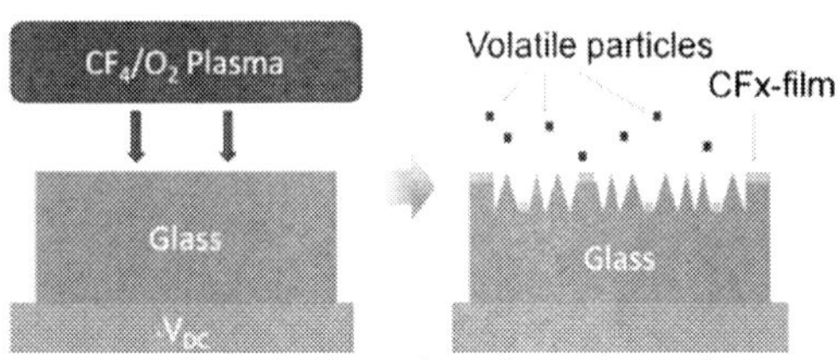

Figure 1: Schematic illustration of texturing surface formation with self-masked dry etching

3 EXPERIMENTAL SETUP

The etching chamber "PlasmaLab System 100" from Oxford Instruments is used to create the textured surfaces. The system consists of a capacitively coupled power generator and an inductively coupled power generator, to create a capacitively coupled plasma (CCP) and an inductively coupled plasma (ICP) or superpose both of them, which allows an independent variation of ion energy (over the self-bias) and plasma density, which is not possible in a pure CCP system. The system is equipped with a load-lock, such that substrates can be loaded without vacuum interruption. Borosilicate glass (81 % SiO_2, 13 % B_2O_3, 4 % Na_2O/K_2O, 2 % Al_2O_3) is used as a substrate, because tests on soda lime glass showed only very small etch rates, probably due to a high concentration of non-volatile alkali flourides in the etch products.

The experimental routine is:

- cleaning the samples surface (IPA, DI);
- weight measurement;
- maskless plasma texturing with CF_4/O_2;
- weight measurement for etch rate calculation;
- cleaning with IPA;
- transmission measurement.

A set of glass etching tests is carried out with various process parameters such as pressure, power and gas content. Test results show a change in light transmission in the visible range.

The process parameters are CF_4/O_2-flow = 50 sccm (ratio 4:1), CCP= 50-150 W, ICP = 100-600 W (the plasma process with CCP+ICP is unstable at values ICP less than 100 W), process pressure = 50-70 μbar.

4 RESULTS FOR GLASS PLASMA TREATMENT

Then, the etching rate (fig 2) will be calculated and its dependence on the process parameters is considered.

Etched samples were separated into three groups according to the parameters to the plasma source used: only CCP (1-3), high ICP-low CCP (4-6), low ICP-high CCP (7-9).

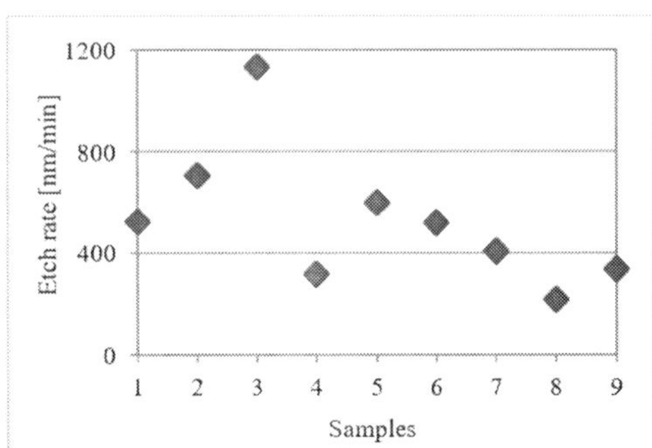

Figure 2: CF_4/O_2 etch rate for different recipes

Figure 3 shows the change in the etching rate when using two types of plasma generators. The greatest increase of etching rate is observed at a power ratio of CCP:ICP=1:2.

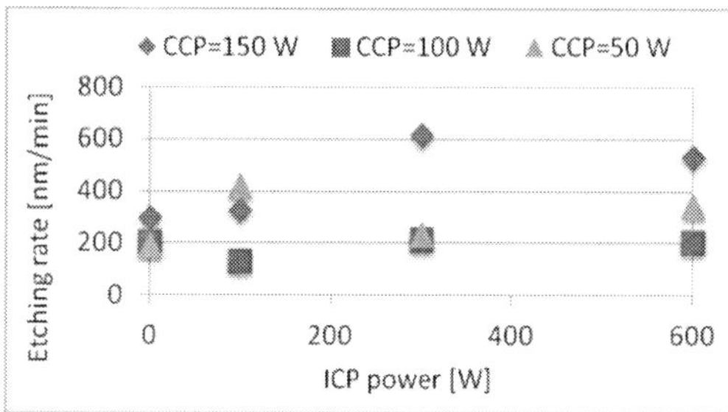

Figure 3: Comparison of glass etching rate at different values of ICP and CCP power. Other process parameters are pressure 65 μbar, etching time 20 minutes, CF_4/O_2-Flow = 50 sccm (ratio 4:1)

Figure 4 shows the variation in self-bias voltage (V_{DC}) with the addition of ICP power.

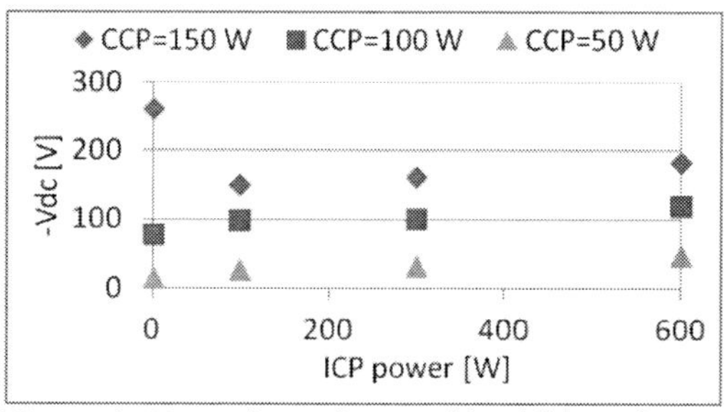

Figure 4: V_{DC} voltage change with addition of ICP power

Probably, the addition of ICP power makes it possible to increase the electron density in the plasma, which leads to an increase in the negative charge accumulated on the sample surface. Accordingly, to compensate for it, a higher bias voltage is required.

Further, the optical properties of the etching samples have been analyzed (fig. 5). The transmission was measured with Boehringer Ingelheim microParts UV/VIS spectrometer in the range between 400 and 850 nm.

Etched samples were separated into three groups according to the parameters to the plasma source used: only CCP (1-3), high ICP-low CCP (4-6), low ICP-high CCP (7-9).

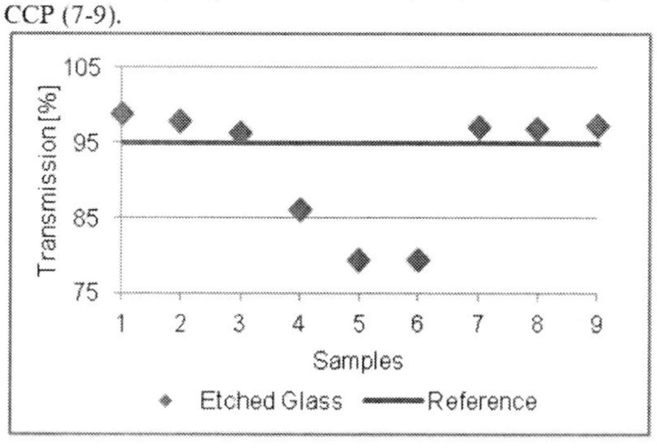

Figure 5: Global transmission of etch glass (400-850 nm)

Figure 6 shows the transmission measurement results for three samples compared to the reference sample.

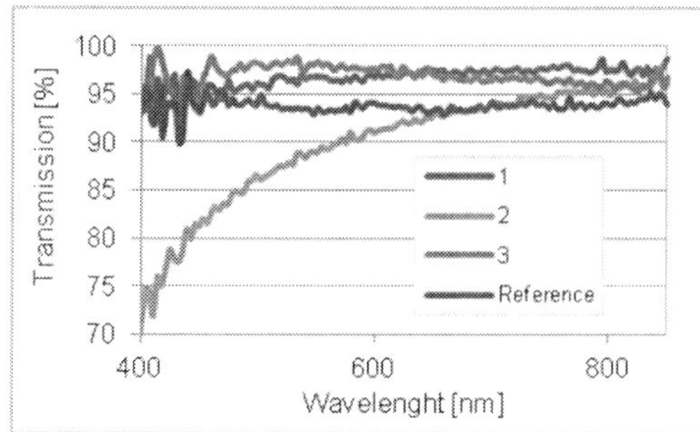

Figure 6: Transmission of etching glass as a function of the wavelength of light (400-850 nm)

The plasma surface treatment can be said to have achieved its goal of increasing transmission. The transmission increase for samples 1 and 3 are 8%, for sample 2 is 3%.

The samples 1 and 3 show approximately the same improvement in light transmission. The sample 2 has a different curve: in the violet-blue region of the spectrum, it shows reflection worse than the reference sample; after the transition to green, it rises, reaching the values of the reflections of other textured samples in the red part of the visible light.

The figure 7 shows how the optical properties of the sample 1 changed qualitatively. The sample is borosilicate glass (10 x 10 cm) with an etched area (5 cm) in the center; the light source is a window on a cloudy day and sunny day. The figure 7 clearly shows how the etched area in the center decreases the reflection relative to the non-etched area.

Figure 7: A qualitative change of the optical properties of the sample 1; the light source is a window on a cloudy day (top) and sunny day (bottom)

5 SUMMARY

It was shown that plasma etching can be used to improve the transmission of borosilicate glass by creating a textured surface on the glass which acts as an effective anti-reflective coating.

The use of a CF_4/O_2 plasma in comparison with CHF_3/H_2 allows achieving a more significant increase in transmission of borosilicate glass. This is because a large amount of H contributes to preventing a film on the surface being etched while O-radicals inhibit the formation of the film, what is agreement with the theoretical knowledge [12, 14].

Such a plasma process could be used for improving the performance of solar modules which use borosilicate glass (e.g. some thin-film modules). For applying the process to standard solar glass (soda lime glass with low Fe content) the problem of low etch rates due to a high concentration of alkali atoms remains to be solved.

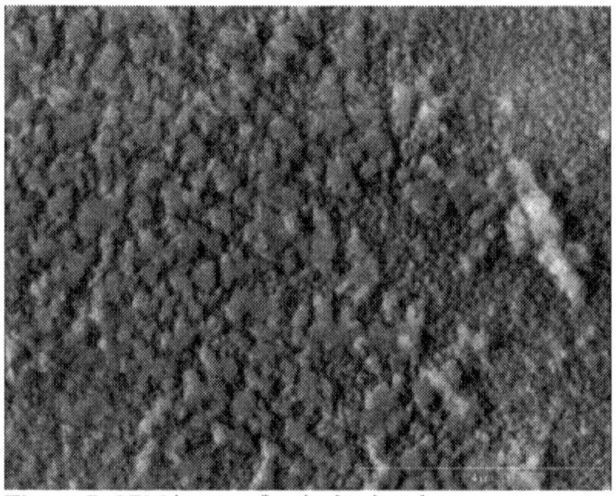

Figure 7: SEM image of etched solar glass

Figure 7 is showing the etched textured surface of glass. This demonstrates that the proposed technique of plasma etching is perspective after some optimization.

REFERENCES:

[1]. Eurostat News Release, 23 January 2020
[2]. D. T. Cotfas and P. A. Cotfas. International Journal of Photoenergy (2014)
[3]. M. Rajvikram, S. Leoponraj. Beni-Suef University Journal of Basic and Applied Sciences 7 (2018) 705–708
[4]. Gaudig M., J. Hirsch J. et al. J. Vac. Sci. Technol. A 33, 05E132 (2015)
[5]. J. Hirsch et al. / Applied Surface Science 374 (2016) 252–256
[6]. Okhorzina A., Bernhard N., Proceedings 37th European Photovoltaic Solar Energy Conference (2020)
[7]. Z. Yu, H. Gao, W. Wu, H. Ge, and S. Y. Chou. J. Vac. Sci.Technol. B 21, 2874–2877 (2003)
[8]. M. Born and E. Wolf, Principles of Optics (Pergamon, 1980)
[9]. M. John, J. Hirsch, N. Bernhard, Proceedings 36th European Photovoltaic Solar Energy Conference (2019)
[10]. P. B. Clapham and M. C. Hutley, Nature 244, 281–282 (1973)
[11]. H. Jansen, M. Boer, R. Legtenberg, M. Elwenspoek, J. Micromech. Microeng., Vol. 5, (1995), 155-120
[12]. Cardinaud C., C. R. Chimie (2018)
[13]. Forster P., Ramaswamy V. et al. Cambridge University Press, Cambridge und New York (2007)
[14]. G.S. Oehrlein, Y. Zhang, D. Vender, M. Haverlag, J. Vac. Sci. Technol.A 12 (1994)

MODIFICATION OF PEDOT:PSS LAYER PROPERTIES BY SWCNT AND Ag NANOPARTICLES

S.V.Mamykin, I.B.Mamontova, T.S.Lunko, O.S.Kondratenko, V.R.Romanyuk
V. Lashkaryov Institute of Semiconductor Physics National Academy of Sciences of Ukraine
41, prospect Nauky, Kyiv 03680, Ukraine, Tel: +38 (044) 5256546; E-mail: mamykin@isp.kiev.ua

ABSTRACT: Conductive and transparent PEDOT:PSS-based composites with carbon nanotubes and silver nanoparticles have been prepared on a flat Si surface using the simplest technologies of layer-by-layer deposition, photochemical growth and spin-coating.

From analysis of optical parameters it was obtained that these composite films in the spectral region of Si photosensitivity reveal high transmittance of light into Si substrate. Together with electrical DC conductivity investigations using a 4-point probe method, unambiguously it was observed that introduction of silver nanoparticles into the composite increases the plasma frequency of free carriers E_p of the film as a whole and film conductivity.

Solar cells have been manufactured on a flat Si surface with these heterostructures. The introduction of metal nanoparticles (Ag) and SWCNT increases the efficiency of photoconversion in PEDOT:PSS/Si heterostructures by reducing the series resistance and increasing the fill factor of solar cells.

Keywords: PEDOT:PSS composite, carbon nanotubes, nanoparticles, photovoltaic heterostructure, silicon

1 INTRODUCTION

The development of modern photovoltaics requires the search and use of new composite organic-inorganic materials with improved electrical (electrical conductivity) and optical (transmittance) characteristics. Poly(3,4-ethylenedioxythiophene) polystyrene sulfonate (PEDOT:PSS) is a leading *p*-type conductive organic polymer mixture widely used to create photosensitive organic/inorganic heterostructures. PEDOT:PSS films is easy to manufacture thanks to low-temperature and vacuum-free technology and promising characteristics. Such films in photovoltaic structures play several roles. These are transparent conductive front electrode, antireflective and passivating coating, charge-selective layer that reflects electrons and extracts holes, thereby significantly reducing recombination [1]. Currently, PEDOT:PSS/Si hybrid silicon-organic solar cells (SC) have been obtained with high solar energy conversion efficiency: 14.8% [2], 15.5% [3], 14.3% [4].

Transparency of PEDOT:PSS thin films is comparable with ITO film transparency, but the conductivity typically is lower. Therefore improving the optical and electrical properties of PEDOT:PSS electrode in hybrid PEDOT:PSS/Si solar cells is an important task for efficiency increasing. In [5], it is shown that the conductivity of PEDOT:PSS thin films can be increased by increasing the conductive PEDOT and decreasing the non-conductive PSS component contents. Performance improvement can be achieved also by creating a composite with adding to the polymer layer (PEDOT:PSS) metal nanoparticles, carbon nanotubes or their mixture [6-10]. In this work, we used the simplest technologies of layer-by-layer deposition of single walled carbon nanotubes (SWCNT), photochemical growth of Ag nanoparticles (NPs), and PEDOT:PSS from appropriate solutions to fabricate transparent conductive composite layer on the *n*-Si surface. Investigation of the optical and electrical properties of such composite films made it possible to create silicon-organic solar cells with good photovoltaic characteristics.

2 EXPERIMENTAL DETAILS

For the deposition of composite PEDOT:PSS films modified with Ag nanoparticles and single walled carbon nanotubes the *n*-Si (100) phosphorus-doped wafers with a specific resistivity of 0.5-2.4 Ohm·cm were used. Before organic polymer layer deposition, native oxide of Si surface was removed in 10% HF solution.

Carbon nanotubes were deposited on substrates by physical adsorption from a colloid solution of single walled carbon nanotubes (SWCNT) in dimethylformamide (DMF) (SWCNT:DMF=1:20 by weight), which was uniformly dispersed in an ultrasonic bath for 13-15 minutes. The deposition time was 1 minute. The thickness of the nanotube layer varied depending on the number of immersions (1, 2 and 4 times) [11]. Photoinduced chemical deposition of silver nanoparticles (Ag NPs) was carried out from an aqueous solution of $AgNO_3$:HF:H_2O (HF:H_2O = 1:4 and an $AgNO_3$ concentration of 1 mM) for 10 s.

To prepare the films, an aqueous PEDOT:PSS (Sigma-Aldrich) emulsion with a concentration of 2 wt% was used. Additionally, 5 wt.% of dimethyl sulfoxide (DMSO) was added to increase conductivity [2]. To ensure the wettability of the hydrophobic surface of Si, the TRITON X100 [12] was added in an amount of 2 wt.% to prevent a decrease in conductivity. The organic conductive layer PEDOT:PSS was deposited on substrates with decorated and undecorated intermediate layer of carbon nanotubes by spin-coating technique at rotation speed ω=3000 RPM, time t=10-30 s [11]. As a result, the composite films with a thickness of 120-170 nm were obtained. These films were annealed at T=130-170°C in air for 5 min. Finally, to complete fabrication of solar cells heterostructures, the Ag semitransparent (27 nm) contact grids have been deposited by thermal evaporation through the masks of corresponding geometry. The general design of the solar cells is shown in Fig. 1.

The thickness of the composite films and their optical properties were obtained from spectroscopic ellipsometric measurements (SE-2000, Semilab) in the spectral range of 0.6-5.0 eV at three angles of light incidence (65°, 70° and 75°). The effective dielectric function of the composite films was calculated for the model of a single-layer effective film on a silicon substrate.

Electrical DC conductivity of the films was determined using a 4-point probe method with CHI-660E

(CH Instruments) workstation.

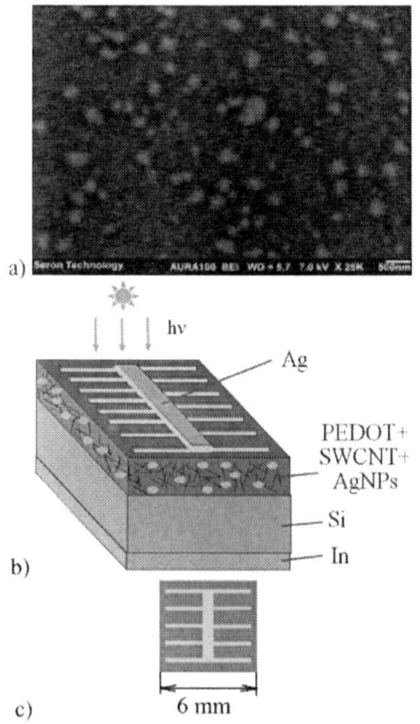

Figure 1: SEM image of the PEDOT:PSS layer with Ag NPs on the flat surface of *n*-type silicon (*a*). Schematic representation of the SCs structure design based on PEDOT:PSS composite on the flat surface of Si (*b*). Top-side image of the structure used in the experiments (*c*)

The dark direct and light current-voltage (I-V) characteristics measured under AM0 simulated illumination with a radiation power of P=136 mW/cm^2.

3 RESULTS AND DISCUSSION

The effective dielectric function of the composite PEDOT:PSS-(AgNPs)-SWCNT films ε(E) in the wide spectral range from 250 to 2000 nm was approximated as a sum of Cauchy and Drude components, $\varepsilon=\varepsilon_C+\varepsilon_D$, and presented in Fig.2.

The classical Drude term describes the optical properties of metals and semiconductors in the long-wavelength free carriers absorption region [13]:

$$\varepsilon_D(E)=\varepsilon_\infty-\frac{E_p^{\ 2}}{E^2-iE\,\Gamma_p}, \quad (1)$$

where E_p and Γ_p are the free carriers plasma oscillation energy and damping parameter, *E* is the energy of incident quantum of light.

Cauchy component, $\varepsilon_C(E) = n^2$ (Eq. 2), is used often to describe the optical properties of dielectrics [13] and makes a major contribution to the refractive index of PEDOT:PSS in transparency region:

$$n(\lambda)=A+B/\lambda^2+C/\lambda^4, \quad (2)$$

where *A*, *B* and *C* are the empirical coefficients, and λ is the wavelength.

From the spectra of obtained refractive index *n*(E) and extinction coefficient *k*(E) (Fig. 2) the free-carrier parameters of that models have been obtained and shown in Table 1.

Taking into account the data that PEDOT: PSS films have a globular structure, as well as the data of [5] that, as a result, the conductivity of PEDOT:PSS films depends on their thickness, it is advisable to analyze the relative change in conductivity for different compositions of composites based on the investigated polymer.

The spectral dependencies of the optical parameters of composite films with embedded nanotubes (Figs. 2 a and 2 b) are almost identical for samples with different content of nanotubes, which indicates a very small volume fraction of nanotubes in the composite. The introduction of silver nanoparticles into the composite, as can be seen from Table 1, increases the plasma frequency of free carriers E_p of the film as a whole. Also Ag NPs in composite increase the light absorption coefficient of the composite mostly in the short-wavelength region of the silicon photosensitivity spectrum.

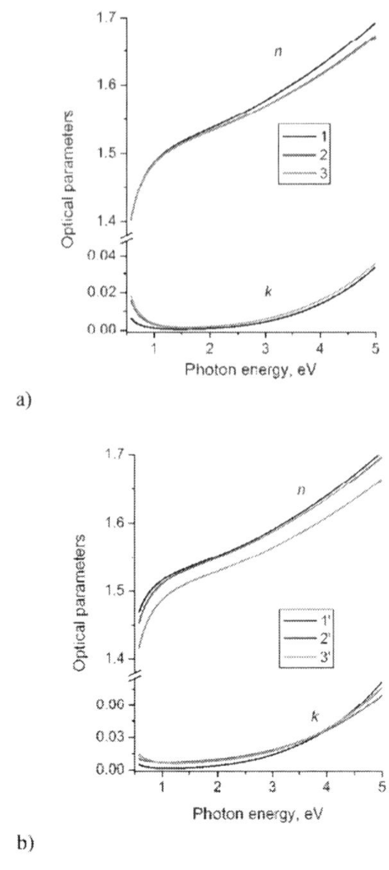

Figure 2: Spectral dependencies of the refractive index and the extinction coefficient of PEDOT:PSS-SWCNT (1-3) – (*a*), and PEDOT:PSS-AgNPs-SWCNT (1′-3′) – (*b*) composite films

Table I: Optical and electrical parameters of PEDOT:PSS composite films modified with SWCNT and Ag NPs

Sample number	Components of the PEDOT:PSS composite film	Total film thickness, d (nm)	E_p (eV)	Γ_P (eV)	Electrical DC conductivity, σ (S/cm)	Short-circuit current, I_{sc} (mA/cm^2)	Open circuit voltage, U_{oc} (mV)
1	1 layer SWCNT	169.8	0.414	0.229	0.6	0.060	450
2	2 layers SWCNT	129.3	0.367	0.135	12.1	0.085	300
3	4 layers SWCNT	128.9	0.359	0.098	2.0	0.050	340
1′	1 layer SWCNT + Ag NPs	141.8	0.282	0.137	2.7	0.070	460
2′	2 layers SWCNT + Ag NPs	123.5	0.291	0.218	11.1	0.155	410
3′	4 layers SWCNT + Ag NPs	138.3	0.374	0.144	31.7	0.185	450

The calculation of the optical transparency of such composite films (Fig. 3), based on the refractive index data obtained from the ellipsometric measurements, showed that, despite some change in the conductivity of the composite film, the total transmittance of light into Si substrate through the composite film in the spectral region of Si photosensitivity changes insignificantly, by 5-10% in average. Also, the film transmission can be tuned by changing the thickness of the film, which will entail a shift in the interference pattern in the spectrum.

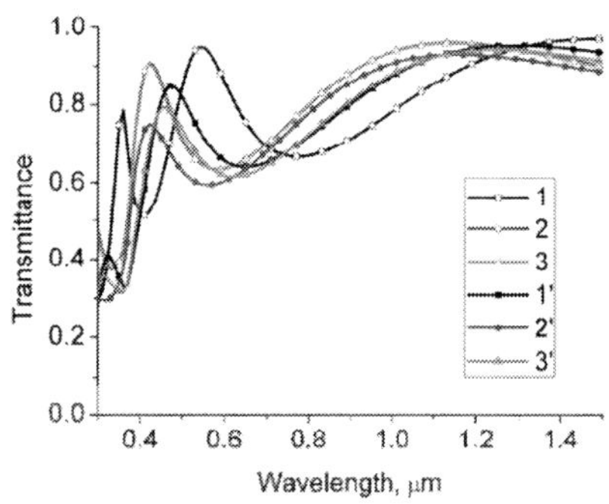

Figure 3: Calculated transmittance of light through the composite film into the semiconductor for PEDOT:PSS-SWCNT/Si (1-3) and PEDOT:PSS-AgNPs-SWCNT/Si (1′-3′) heterostructures. Film thicknesses are represented in Table 1

The dark direct and light current-voltage (I-V) characteristics of the developed SCs are presented in Fig. 4. Improvements in the parameters of I-V characteristics (primarily in the short circuit current I_{sc}) are observed with an introduction of nanotubes and Ag NPs into PEDOT:PSS composite by reducing the series resistance and increasing the fill factor of solar cells.

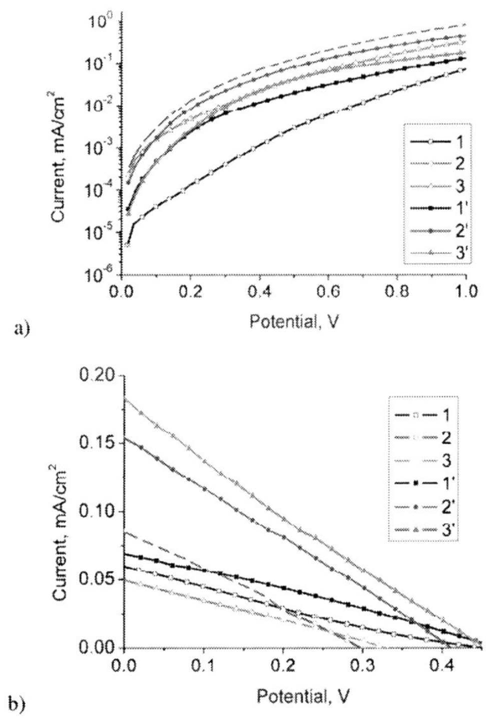

Figure 4: The dark direct (*a*) and light current-voltage (*b*) characteristics of SCs PEDOT:PSS-SWCNT/Si (1-3) and PEDOT:PSS-AgNPs-SWCNT/Si (1′-3′) hetero-structures

4 CONCLUSIONS

Solar cells based on PEDOT:PSS composite on a flat Si surface have been manufactured. Electrical DC conductivity of the films was measured using a 4-point probe method.

From analysis of composite film effective optical parameters it was obtained that introduction of silver nanoparticles into the composite increases relatively the plasma frequency of free carriers E_p of the film as a whole and film conductivity. The total transmittance of light into Si substrate through the composite film with SWCNT and Ag NPs inclusions in the spectral region of Si photosensitivity changes insignificantly, by 5-10% in average.

The introduction of metal nanoparticles (Ag) and SWCNT increases the efficiency of photoconversion in PEDOT:PSS/Si heterostructures by reducing the series resistance and increasing the fill factor of solar cells.

5 References

[1] P. Gao, Z. Yang, J. He, J. Yu, P. Liu, J. Zhu, J. Ye, Advanced Science 5(3) (2017) 1700547.
[2] S. Jäckle, M. Liebhaber, C. Gersmann, M. Mews, K. Jäger, S. Christiansen, K. Lips, Scientific Reports 7 (2017) 2170.
[3] Q. Liu, R. Ishikawa, S. Funada, T. Ohki, K. Ueno, H. Shirai, Advanced Energy Materials 5 (2015) 1500744.
[4] J. He, P. Gao, Z. Ling, L. Ding, Z. Yang, J. Ye, Y. Cui, ACS Nano 10 (2016) 11525.
[5] O.S. Kondratenko, S.V. Mamykin, T.S. Lunko, I.B. Mamontova, V.R. Romanyuk, Molecular Crystals and Liquid Crystals 717(1) (2021) 92.
[6] A.V. Kukhta, A.E. Pochtenny, A.V. Misevich, I.N. Kukhta, E.M. Semenova, S.A. Vorobyova, E. Sarantopoulou, Physics of the Solid State 56(4) (2014) 827.
[7] H. Yun, J. Hun Han, S. Kim, Polymer Journal 44(12) (2012) 1238.
[8] P.C. Ma, B.Z. Tang, J.-K. Kim, Carbon 46(11) (2008) 1497.
[9] K. Choi, C. Yu, PLoS ONE 7(9) (2012) e44977.
[10] N.-R. Shin, S.-H. Choi, J.-Y. Kim, Synthetic Metals 192 (2014) 23.
[11] S.V. Mamykin, I.B. Mamontova, T.S. Lunko, O.S. Kondratenko, V.R. Romanyuk, Semiconductor Physics, Quantum Electronics & Optoelectronics 24(2) (2021) 148.
[12] Y. Jiang, X. Gong, R. Qin, H. Liu, C. Xia, H. Ma, Nanoscale Research Letters 11 (2016) 267.
[13] H. Fujiwara, Spectroscopic ellipsometry: principles and applications, John Wiley & Sons, 2007.

COPPER OXIDES AS BASE MATERIAL FOR SOLAR CELLS

Katarzyna Gawlińska-Nęcek[a], Zbigniew Starowicz[a], Robert P. Socha[b], Mateusz Wlazło[b], Grzegorz Kołodziej[b], Grzegorz Putynkowski[b], Piotr Panek[a]
[a]Institute of Metallurgy and Materials Science PAS, Reymonta 25, 30-059 Krakow, Poland
[b]CBRTP – Research and Development Center of Technology for Industry, Ludwika Waryńskiego 3A, 00-645 Warsaw, Poland

ABSTRACT: In the experiment the thermal copper oxide with sodium doping was elaborated. A simple bar coating technique was used for sodium deposition, while Cu_xO:Na layers were produced in IR belt furnace in only one temperature treatment. The morphology and composition of the manufactured coatings were analyzed. The presence of sodium atoms in the layer was confirmed by the EDS and SIMS techniques. The work function (ϕ) of the layers was calculated from potential difference (CPD) measured by using Kelvin probe. The ϕ values dependence on the process temperature was found. Based on the obtained layers photovoltaic structure was prepared and its dark I-V was measured.
Keywords: Cuprous oxide, cupric oxide, heterojunction solar cell, zinc oxide, sodium dopant

1 INTRODUCTION

Copper in oxidized form occurs most often in the + I or + II oxidation state, forming respectively Cu_2O or CuO. Both copper oxides are p-type semiconductors with direct band gaps which are equal of nearly 2.17 eV for copper (I) oxide and 1.5 eV for copper (II) oxide. The details of the oxidation mechanism of the Cu film with a probable schematic model in terms of thermal diffusion as well as the chemical reactivity are presented in many works [1]. The p-type conductivity of natural copper oxide results from the occurrence of natural defects in the form of copper vacancies (V`Cu) in the lattice [2] which are responsible for the main mechanism of copper oxide conduction. There are two mechanisms for the formation of V`Cu - removing the copper atom from the lattice or split copper vacancy, in which, after the Cu exit, the adjacent copper atom is displaced. The split of the vacancy also favors the generation of a highly localized hole, which may lead to a change in the oxidation state of the shifted Cu atom from Cu^{+} to Cu^{2+} [3]. The negatively charged V`Cu leads to hole occurrence in the valence band.

Cu_2O obtained at temperatures up to 300°C is characterized by a low resistivity of about 100 Ωcm at a charge density of 10^{16} cm^{-3} [4]. As a result of increased temperature from 250 to 350°C, copper I oxide easily transforms into copper II oxide, which is manifested by a color change from orange-red to black. In our previous work we reported that when oxidation temperature increases the crystallinity and stoichiometry of the layer improves. Nevertheless, the CuO share in Cu_2O layer is also increasing what causes layer resistivity growth [5]. To improve the conductivity of metal oxides, additional dopant atoms are introduced into their structure. The most commonly used Cu_2O admixture is sodium which allows to obtain resistivity of about 10^{-2} Ωcm [6]. Other possible dopants which reduce copper oxides resistivity to even 1 Ωcm are nitrogen [7] and strontium [8]. When used Mg admixture the Cu_2O layer resistivity reaches 7 Ωcm due to the increase of the carrier density to $8.1 \times 10^{17} cm^{-3}$ [9]. The presence of Mg also reduces the amount of split vacancies [10].

2 EXPEIMENTAL

2.1 Doped copper oxide preparation

Sodium doped copper oxide was obtained by simple oxidation of the industrially available 200 μm thick ETP R220 copper sheet in IR belt furnace in dry air atmosphere. The furnace heating zone length is 76 cm and it is divided into 3 temperature sectors. In the experiment the same temperature set was applied in all heating zones. Before oxidation copper substrates were cleaned in hot acetone and isopropanol and polished by polishing paste for 5 minutes. Sodium dopant from sodium compound (ComNa) was deposited on a copper sheet by a simple bar coating technique. The exact name of the doping compound cannot be given due to the open patent process. Nevertheless, it should be highlighted that proposed sodium compound was not reported so far. The temperature range from 480 to 760°C and the speed belt 50 cm/min. were applied.

2.2 Photovoltaic structure preparation

The Cu ETP R220 substrate with a thickness of 200 μm has been covered with ComNa dopant precursor by a bar coating technique. Subsequently, the substrate was heated in an IR furnace at 720 °C with a belt speed of 50 cm/min. Secondly, the 200 nm of ZnO was applied on the Cu ETP R220/Cu_xO:Na layer by the atomic layer deposition (ALD) method on a Beneq P400A industrial scale reactor. The Zn precursors was diethyl zinc $Zn(C_2H_5)_2$ (DEZ, CAS 557-20-0, Lanxess, Cologne, Germany). Finally, 400 nm of AZO was deposited in the same process in ALD reactor. Trimethylaluminum $Al(CH_3)_3$ (TMA, CAS 75-24-1, Lanxess, Cologne, Germany) was used as aluminum precursor. The size of the structure was 30 x 30 mm.

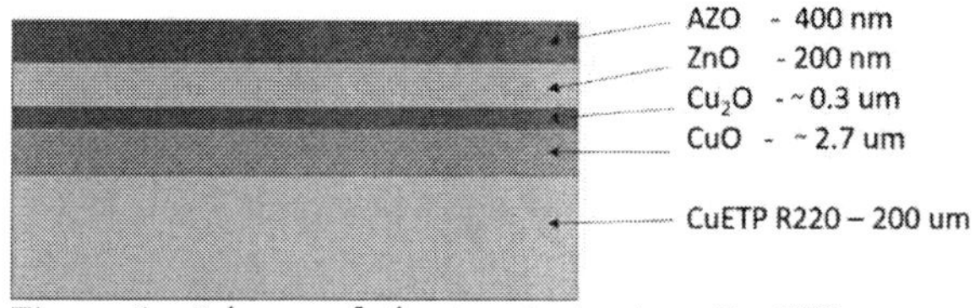

Figure 1: Scheme of the target structure Cu ETP R220/CuxO:Na/ZnO/AZO

3 RESULTS

Fig.2 shows the temperature dependence of the series resistance of undoped Cu_xO/Cu/Cu_xO structures with different sheet thickness (100 and 300 μm). Three temperature ranges can be distinguished in the chart. In the first 300 - 450°C a very slow increase in resistance for both 100 and 300 µm sheets can be observed. In the second range of 500 - 850°C Rs increases linearly, and its value is about two times higher for thinner (100 µm) sheet compared to the 300 µm substrate. The last range 850 - 950°C is resistance stabilization.

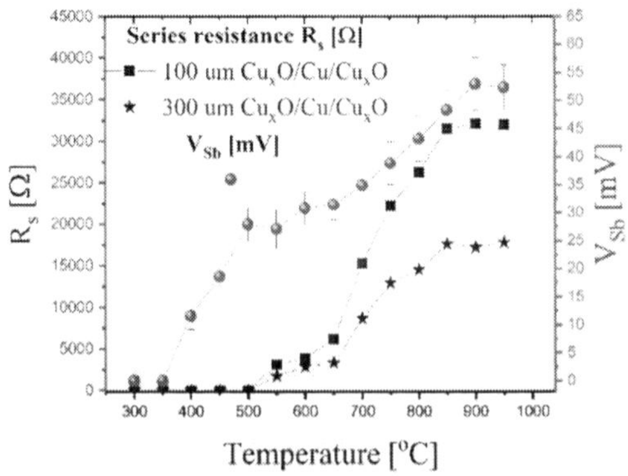

Figure 2: Temperature dependence of series resistance and Seebeck voltage of the $Cu_xO/Cu/Cu_xO$ structure for different copper sheet thicknesses (100 µm and 300 µm) with an area of 6.25 cm^2

The analysis of Seebeck voltage as a function of temperature corresponds to the change of Rs. This observation confirms that with the increase in temperature, the crystallinity and stoichiometry of the layer improved. Layer ideality also improves which can be observed as the growth in thermal voltage.

The obtained Rs values are in good agreement with data presented by Valladares et al. [11] who reported an increase in the oxide resistivity with process temperature.

Based on the above results, the temperature range from 480°C to 760°C was selected for further doping investigation due to the high Rs value of manufactured structure, which favors the observation of the doping effect. Moreover, the structures are also characterized by high Seebeck voltage and good adhesion to the substrate.

The process of manufacturing sodium-doped copper oxide layers by thermal methods on copper sheet was carried out in an IR belt furnace in a natural atmosphere using the elaborated source of sodium atoms admixture (ComNa). For further experiments only 200 µm thick ETP R220 copper sheet was used as the base material. It is an intermediate thickness between the two initially tested 100 µm and 300 µm. Such an approach allow to obtain an increase in resistance of copper oxide layer with the oxidation temperature increasing. On the other hand, it minimize the risk of layer cracking and its detachment what could be observed for too high oxidation speed as for 100 µm Cu sheet. After the oxidation process, homogeneous and adhesive Cu ETP R220/Cu_xO:Na structures were obtained.

Oxide layers manufactured in the process of copper oxidation are continuous and homogeneous at the macroscopic scale (Fig.3). The increase in the annealing temperature is accompanied by a change in the color of the copper surface, reflecting the composition of the newly formed layer. At 480°C the layer is thin and orange which suggests a predominance of Cu_2O. As the temperature increases, the layer darkens due to the increasing amount of low bandgap CuO. Finally, at temperatures above 750°C, the surface layer is completely black.

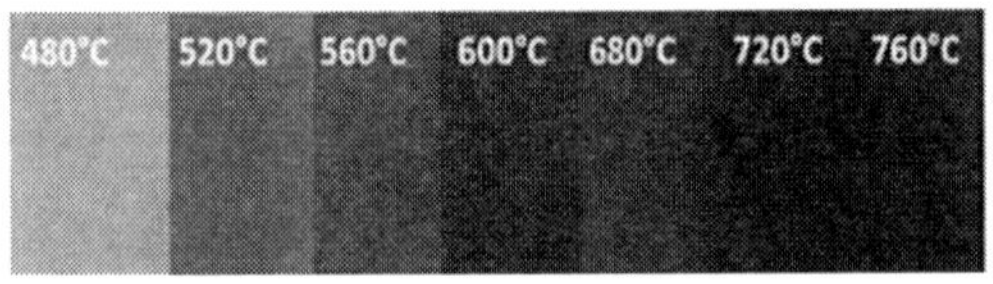

Figure 3: Color scale of the Cu_xO:Na on the 200 µm Cu ETP R220 substrate oxidized in the temperature range 480 - 760°C

For further detailed analysis the 720°C Cu ETP R220/Cu_xO:Na was selected due to the significant Rs reduction, from few kΩ to several Ω after dopant introduction.

Microstructure and chemical composition analysis of Cu ETP R220/Cu_xO:Na manufactured at 720°C was carried out using the FEI E-SEM XL-30 scanning electron microscope, equipped with the EDAX X-ray energy dispersion spectrometer (EDS) (GEMINI 4000) with accelerating voltage 20 kV. The surface morphology analysis confirms that the obtained layer is homogeneous, rough and free from cracks.

Figure 4: The SEM surface microstructure of the Cu ETP R220/Cu_xO:Na

Table I: The content of elements in the Cu_xO:Na layer obtained at 720 °C determined by the EDS method

	Cu	O	Na
[% at.]	67.7	29.7	2.6

The atomic distribution of the elements Cu and O indicates almost stoichiometric character of the Cu_2O with a small amount of CuO on the top of the layer. While the determined value of sodium 2.6%, although near to the detection limit of EDS method, confirms the presence of a dopant in the investigated layer.

To study the sodium dopant profile in the Cu_xO:Na layer (720°C) the SIMS method was applied using Cameca ISM 6f system. The Cs^+ atoms with an energy of 1.5 kV were used as primary ions. The ion current was 110.79 nA while the diameter of the removed surface layer was 150 µm. The dopant distribution determined by the SIMS method is related to the total amount of admixture. This means that it cannot provide precise information about the separation between the substitution and interstitial atoms. It is one of the disadvantages of this method in photovoltaic applications.

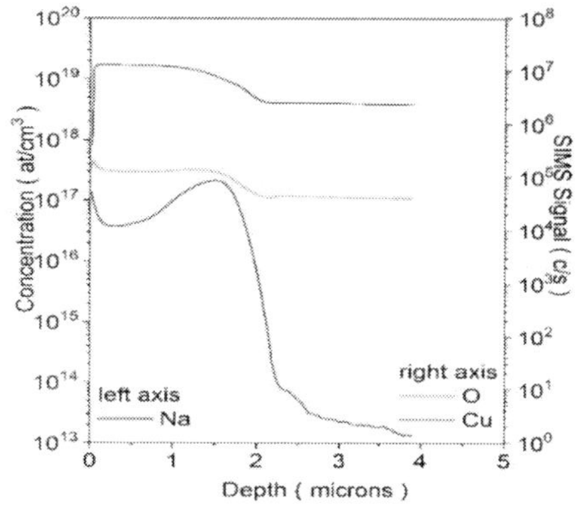

Figure 5: SIMS distribution of the Na admixture in copper oxide Cu ETP R220/Cu_xO:Na

Figure 5 shows the distribution of atoms in the manufactured layer obtained by the secondary ion mass spectrometry (SIMS). Results revealed that the produced layer is 2 μm thick. The concentration of copper atoms (green line) is two orders of magnitude greater than the concentration of oxygen atoms (blue line). The concentration of sodium admixture (red line) is about 10^{17} at/cm^3. Its maximum is located close to the copper substrate, in a place of admixture deposition. In the middle part of the layer number of Na decreases and finally again grows up on the oxide surface. What is more, the results suggest no sodium diffusion into the copper sheet.

Measurements of the contact potential difference (CPD) were made on a Kelvin Probe model 2.0 IF-0008 made by Instytut Fotonowy. The work function (ϕ) of the Cu_xO:Na layers were calculated. The results revealed significant dependence of ϕ from the process temperature for sodium doped copper oxide which increases with temperature (Fig. 6).

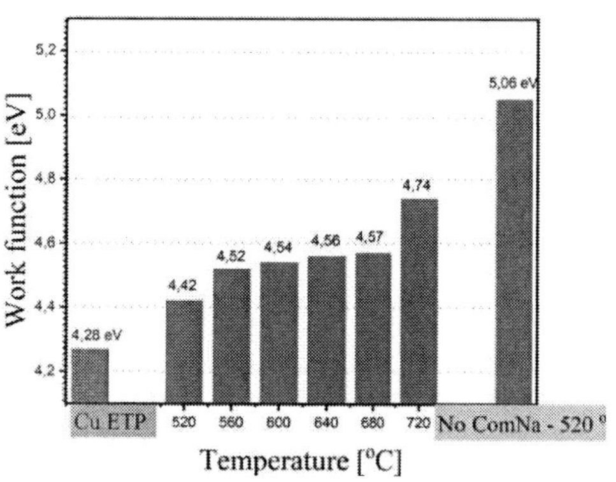

Figure 6: Dependence of the work function value from the oxidation temperature of the Cu ETP R220 sheet with the layer of ComNa precursor

As expected, the presence of sodium admixture in Cu_xO causes a reduction of ϕ which is 5.06 eV for reference copper oxide with no dopant. For comparison, the work function of pristine Cu ETP R220 sheet before oxidation was also calculated and it is 4.28 eV.

The usability of the two different oxide layers as photovoltaic absorbers has been evaluated by APSYS simulation (Fig. 7). Both oxides show good prospects to be used as the absorber in a junction with n-ZnO. At doping concentrations of around $3x10^{15}$ cm^{-3} and higher, CuO shows markedly better performance over Cu_2O. Nevertheless, both stoichiometries improve with increased carrier density.

The solar cell using the Cu_2O absorber shows much higher open circuit voltage than the CuO cell. However, the short circuit current of the Cu_2O cell is generally less than half of CuO. At lower hole concentrations, the Cu_2O absorber appears to suffer significant fill factor losses. The fill factor of the Cu_2O cell matches CuO only at the highest concentrations. However, the current is low across the whole range of carrier concentrations which leads to generally poorer performance of Cu_2O compared to CuO in the Cu/Cu_xO/ZnO/ZnO:Al cell architecture.

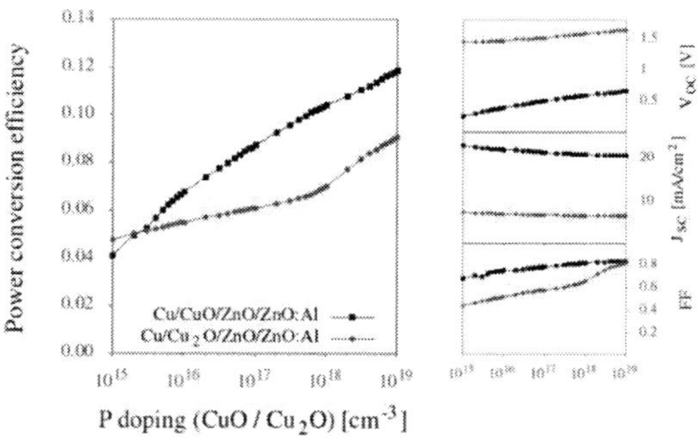

Figure 7: Left: Power conversion efficiency of simulated solar cells with Cu substrate, CuO or Cu_2O as absorbers, ZnO as the n-type partner, and ZnO:Al as the transparent conductor. N-type doping concentration of ZnO was $2.2x10^{19}$ cm^{-3}; Right: Open circuit voltage (V_{OC}), short circuit current (I_{SC}) and fill factor (FF) of the simulated solar cells.

The bright field TEM cross-section analysis (Fig.8) of Cu ETP R220/Cu_xO:Na/ZnO/AZO was made on a microscope manufactured by FEI using a Tecnai F20 200 kV field emission gun (FEG), equipped with an energy dispersive X-ray spectroscopy detector (EDS). Thin foils for TEM investigations were prepared by the focused ion beam (FIB) technique, using a Quanta 200 3D Dual Beam (FIB/SEM) system by FEI equipped with an OmniProbe (Oxford Instruments, High Wycombe, UK) micromanipulator.

Figure 8: The bright field TEM cross-section of R220/CuxO:Na/ZnO/AZO

As seen in Fig. 8, the growth of copper oxide Cu_xO:Na at 720°C occurs in a columnar crystals perpendicular to the CuETP R220 substrate. The diameter of the columns reaches 400 nm and their height is 2 µm. The ZnO layer consists of round shaped crystals and it is 200 nm thick. On top 500 nm of AZO is located.

The dark I-V characteristics of manufactured structure were measured (Fig.9). Analysis of the curve progression indicates that the heterojunction (Cu_xO:Na/ZnO) has been created.

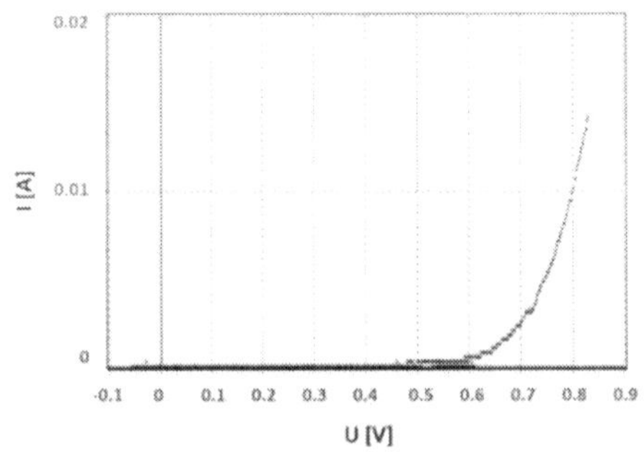

Figure 9: Original printout of the dark I-V characteristic of the Cu ETP R220/Cu_xO:Na/ZnO/AZO photovoltaic structure

4 CONCLUSIONS

The results show that the produced ComNa dopant precursor can be used to obtain sodium-doped thermal copper oxide layers by using simple bar coating method followed by one step thermal treatment. The produced layers are crystalline, columnar in nature and they are characterized by high homogeneity and continuity. Their thickness reaches 2 µm. The presence of sodium atoms in copper oxide layer obtained at 720°C was confirmed using the SIMS method and their concentration is about 10^{17} at/cm^3. What is more, no Na diffusion to Cu substrate was observed. The dependence of the work function of Cu_xO:Na from the process temperature was found. In general, the ϕ value of the doped copper oxide is lower than for the reference sample with no dopant. The theoretical analysis performed by APSYS simulation indicated that both CuO and Cu_2O are suitable as absorbers in the Cu_xO/ZnO p-n junction. On the basis of the developed results, a photovoltaic structure Cu ETP R220/Cu_xO:Na/ZnO/AZO was manufactured and its dark I-V was measured that confirmed the formation of a p-n junction.

5 ACKNOWLEDGEMENTS

The research was supported by the project entitled "Development of technology for manufacturing of functional materials for application in non-silicon photovoltaic cells" co-financed by The National Centre for Research and Development under the contract TECHMATSTRATEG2/409122/3/NCBR.

The authors gratefully thank to DSc Łukasz Major for TEM investigation and PhD Agnieszka Bigos for SEM imaging.

6 References

[1] S. Choudhary, J. V. N. Sarma, S. Pande, S. Ababou-Girard, P. Turban, B. Lepine, S. Gangopadhyay, AIP ADVANCES 8 (2018) 055114.
[2] F. Biccari, C. Malerba, A. Mittiga, Ph.D. thesis. Sapienza – University of Rome 688774 (2009) 1–28.
[3] D. O. Scanlon, A. Walsh, G. W. Watson, Chem. Mater. 21 (19) (2009) 4568–4576.
[4] M. Nolan, S. D. Elliott, Phys. Chem. Chem. Phys. 8 (45) (2006) 5350–5358.
[5] Z. Starowicz, K. Gawlinska-Necek, R.P.Socha, T. Plocinski, J. Zdunek, M. Szczerba, P. Panek, Materials Science in Semiconductor Processing 121 (2021) 105368.
[6] T. Minami, Y. Nishi, T. Miyata, Solar Cells. Appl. Phys. Express 9 (5) (2016) 52301.
[7] C. Malerba, C. L. Azanza Ricardo, M. Dincau, F. Biccari, P. Scardi, A. Mittiga, Sol. Energy Mater. Sol. Cells 105 (2012) 192–195.
[8] L. Bergerot, C. Jiménez, O. Chaix-Pluchery, L. Rapenne, J.-L. Deschanvres, Phys. Status Solidi 212 (8) (2015) 1735–1741
[9] J. Resende, C. Jiménez, N. D. Nguyen, J.-L. Deschanvres, Phys. Status Solidi 7 (2016) 1-7.
[10] K. Kardarian, D. Nunes, P.M. Sberna, A. Ginsburg, D.A. Keller, J.V. Pinto, J. Deuermeier, A.Y. Anderson, A. Zaban, R. Martins, E. Fortunat, Solar Energy Materials & Solar Cells 147 (2016) 27–36.
[11] L. De Los Santos Valladares, D. Hurtado Salinas, A. Bustamante Dominguez, D. Acosta Najarro, S.I. Khondaker , T. Mitrelias, C.H.W. Barnes, J. Albino Aguiar, Y. Majima, Thin Solid Films 520 (2012) 6368-6374.

HYDROGENATION MECHANISMS IN PHOTOVOLTAICS: THE UNCONVENTIONAL IN_2SE_3 NANOMATERIAL AS AN EXAMPLE

R. Minnings, A.I. Shkrebtii*, and G. Perinparajah

Ontario Tech University, Oshawa, ON, L1G 0C5, Canada

Corresponding author: Anatoli.Chkrebtii@uoit.ca

ABSTRACT: Hydrogenation is a versatile tool in photovoltaic technology. To characterize several hydrogenated 2D In_2Se_3 nanomaterials and to reveal important hydrogenation features, we simulated from first principles, the structural, dynamical, electron and optical properties of the hydrogen induced systems by depositing H atoms on a quintuple layer (QL) α-In_2Se_3 film. One-sided hydrogenation (when either the film's top or bottom H is attached), substantially modifies the QL systems initial interatomic separation. Two-sided hydrogenation, by contrast, splits the pristine QL, forming two semiconducting InSeH and $InSe_2H$ nanofilms, both stable at 0K. Both the band structure and optical response of the split hydrogenated films significantly depend on the bonding type of the split film, namely InSeH or $InSe_2H$. From a point view of technology, the two-sided H configuration allows for possible QL etching. The higher temperature MD results indicate a possible instability in the hydrogenated split structures analyzed here, which require further investigation.
Keywords: 2D, Thin films, Hydrogenation, Optical properties, Simulation, Nanoparticles

1 INTRODUCTION

Since the demand for renewable energy is rapidly growing, new materials and technologies are necessary to harvest clean energy from the sun in a more efficient way. Recently, 2D materials have attracted considerable attention as promising systems for the next-generation solar cells [1-3], and hydrogenation is an important component in many photovoltaic technologies [4-6]. For example, the integration of a 2D semiconducting indium selenide α-In_2Se_3 nanofilm, with a hydrogenated Si(111) surface, for solar cell applications, has been demonstrated [7].

A comprehensive understanding of the multiple phases for this exotic In_2Se_3 nanomaterial has been reached [8]. However, contrary to the one-atomic thick flat graphene-like materials, the elemental '2D' α-In_2Se_3 layer forms a quintuple layer (QL) with five alternating Se and In atoms, as shown in Figure 1. Based upon our results on pristine In_2Se_3 nanofilms [9], we theoretically investigated the hydrogenation mechanisms of the α-In_2Se_3 nanolayers.

Due to the asymmetry of QL bonding, namely the Se–In–Se=In=Se, (here In - tetrahedrally coordinated (–) and In- octahedrally coordinated (=) configuration), , it is expected that the interaction with hydrogen will substantially modify not only the elemental α-In_2Se_3 QL but also the weak van der Waals (vdW) interaction between the QLs. It has recently been discovered that various two-dimensional (2D) materials can be derived from traditional III-V, II-VI, and I-VII compound semiconductors [10]. Such materials usually form 2D hexagonal nanofilms that are only a few atoms thick, primarily the single-atom layer honeycomb (SLHC) structure or the double-atom layer honeycomb (DLHC) structure. However, contrary to the one- or two-atomic thick flat graphene-like materials, 2D semiconducting indium selenide α-In_2Se_3 nanofilms, contain five alternating Se and In atoms in the elemental 2D In_2Se_3 layer, referred to as a quintuple layer (QL).

The basic In_2Se_3 layer consists of five alternating Se and In atoms that form the so-called quintuple layer (QL) [9], as shown in Figure 1. Numerous structural variations of the QLs are possible, with the two most energetically stable layered structures belonging to the so-called alpha and beta phases [8]. Both are semiconducting with an energy bandgap in the optical range.

Hydrogen, which can be introduced experimentally into a semiconductor in many ways, such as an atomic H plasma or an H_2 gas, reacts with the target material and its defects in various ways [11]. Hydrogenation is used extensively in the semiconductor industry, for example, to saturate Si dangling bonds [12].

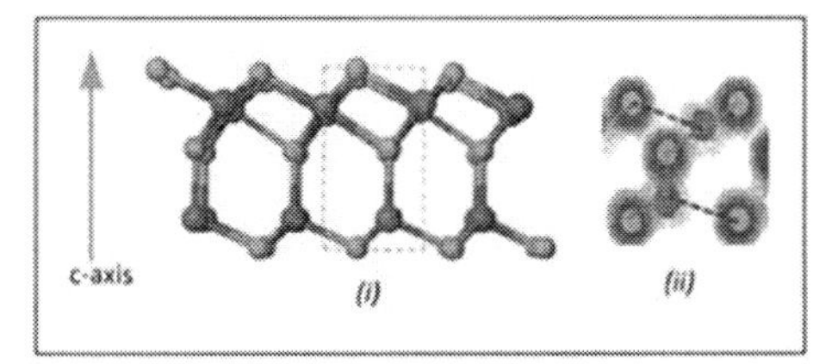

Figure 1: Pristine In_2Se_3 nanofilm. (i) One quintuple layer (QL) of In_2Se_3 quasi two-dimensional film, side view. Yellow spheres correspond to Se atoms, purple spheres to In atoms. The blue dashed line outlines the primitive 2D unit cell that contains one QL. c-axis ('2D' film height) shown, a, b axis (not shown) are in the plane of the film.
(ii) Charge density plot of one QL, [Se–In–Se=In=Se] where (–) indicates a tetrahedral coordination and (=) an octahedral coordination (for details see Figure 3).

The hydrogenation of the alpha (α) phase will be considered here, and further mentions of In_2Se_3 should be assumed to refer to α-In_2Se_3 unless otherwise mentioned. We demonstrate combinations of the possible In_2Se_3 hydrogenation processes by considering the simplest case of one QL of In_2Se_3 and three main hydrogen modified structures, all compared to the pristine single QL case, as illustrated in Table 1, i.e., H at the tetra-bonded side only, H at the octahedral side only, and two-side hydrogenation. The cases of an H atom at either the top or bottom of the QL might be used to interpret the experimentally observed results for the photovoltaic system in [7]. Two-sided hydrogenation (last column, Table 1) allows for a thin film etching possibility, which is important in solar cell technology.

2 TECHNICAL DETAILS

The term hydrogenation used in this paper, describes a theoretical process, whereby H atoms are manually added with a bond length of 1.5 Å to each or either of the Se atoms at the top and bottom of the single QL In_2Se_3, to give the structures depicted in Figure 2, see also Table 1.

We show theoretically that when one QL of In_2Se_3, specifically ZB'- α-In_2Se_3 [8], used for the first principles calculations [9] is hydrogenated at both sides of this QL, and then relaxed, two new 2D films are formed, In-Se-H and In-Se_2-H, where the H atom is bonded with the Se atom in both cases, as in Figure 2.

It is important to note that the single pristine QL structure In_2Se_3 that we hydrogenate, and then relax, has previously been relaxed without hydrogen, next we manually add H atoms (at 1.5Å bond length) to either side of a relaxed In_2Se_3 and then relax this new structure again to observe computationally the impact of the hydrogenation on the In_2Se_3 QL

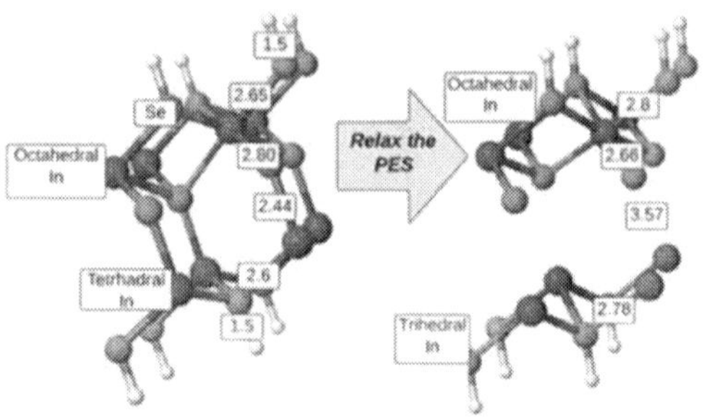

Figure 2: One quintuple layer (QL) of H-In_2Se_3-H before and after relaxation [10]. After relaxation the hydrogenated QL splits into two layers of $InSe_2H$ and InSeH. Bond lengths are shown in Angstrom. The Tetrahedrally and Octahedral coordinated, In atoms are identified in the before relaxation hydrogenated QL. After relaxation, the In in the InSeH layer becomes trihedral coordinated.

After relaxation of the manually created H-In_2Se_3-H structure, Figure 2, it splits spontaneously at 0K into two new layers of $InSe_2H$ and InSeH, separated by approximately 3.5Å. We mostly treat the two new layers, InSeH and $InSe_2H$, separately in our theoretical calculations in this paper for optical response, band gap, and phonon calculations.

However, the two layers, InSeH and $InSe_2H$, after relaxation, appear to form a single film with possible vdW (Van der Waals) forces binding these two new layers together. The energy difference for the relaxed structure with both layers is slightly less (-0.1 Ry) than the sum of the relaxed energy for the separately relaxed InSeH and $InSe_2H$, indicating the formation of weak van der Waals bonding between the two layers, , and discussed later in this paper.

3 COMPUTATIONS AND EXPERIMENT

For a simulation of hydrogenated In_2Se_3 nanolayers we used the DFT-LDA [14] plane-wave pseudo-potential Quantum Espresso package [15] along with the repeated slab geometry with a weak vdW (van der Waals) interaction. Different H bonding configurations with a base one QL α-In_2Se_3 were considered. The initial asymmetry of pristine In_2Se_3 (i.e., [Se–In–Se=In=Se], results in different In–Se bond lengths and of the bonds strength, as well as different ferroelectric dipole orientations [9]. Interaction with H atoms substantially modifies the electron charge density distribution and weakens or strengthens the initial bonds depending on the H adsorption site. For the nanofilm considered (see Table 1) we calculated how the electron band structure, band gap, linear optical response and phonons are modified relative to the non-hydrogenated single pristine QL In_2Se_3.

All calculations for each structural combination used the repeated slab geometry, with the Quantum Espresso [15] software package and the DFT-LDA framework [14], with pseudopotentials plane-waves and a 60 Ry cut-off.

The main focus of this paper was calculations for the structures, shown in column 3 and column 6 of Table 1. However, all structures in Table 1 were first relaxed [14] with a vdW correction. The optical response, band structures and the phonon dispersion along the irreducible Brillouin zone (IBZ) in the plane of the film, were calculated for one QL of In_2Se_3, InSeH, and $InSe_2H$.

	Bond lengths	1. One QL No H	2. QL H bottom	3. QL H on top	4. $InSe_2$-H + InSe-H
	6 Se - H 5 Se - In 4 In - Se 3 Se - In 2 In - Se 1 Se - H	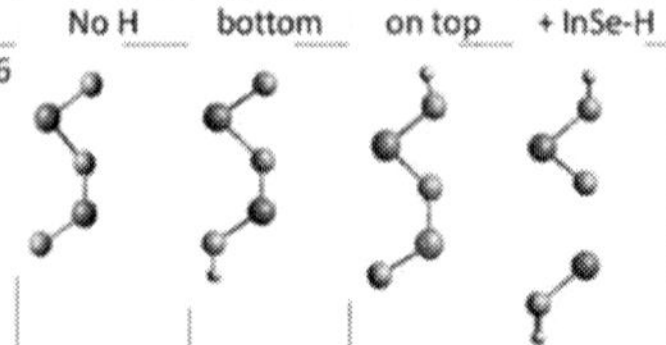			
1	H-Se	N/A	N/A	1.50	1.50
2	Se-In	2.65	2.62	2.75	2.80
3	In-Se	2.80	2.81	2.72	2.66
4	Se-In	2.44	2.51	2.47	Broken
5	In-Se	2.60	2.67	2.57	2.78
6	Se-H	N/A	1.51	N/A	1.51

Table 1. In_2Se_3 and In_2Se_3-H films bond lengths

The relaxed structures and bond lengths used in the computations in this paper. See the *c-axis* orientation in Figure 1 for reference to the terms 'top/bottom' used in the list below.

1. (*Column 3*) One QL of pristine In_2Se_3 with the bonding configuration Se–In–Se=In=Se
2. (*Column 4*) One QL of pristine In2Se3 with an H bond to the bottom Se (i.e., Se bonded to the tetrahedral In)
3. (*Column 5*) One QL of pristine In2Se3 with an H bond to the top Se (i.e., Se bonded to the octahedral In)
4. (*Column 6*) H-In_2Se_3-H after relaxation, demonstrating splitting of the QL into two

Born-Oppenheimer Molecular Dynamics (BOMD) at room temperature was performed on three rectangular supercells, namely In_2Se_3, InSeH and $InSe_2H$, each contains four primitive unit cells. Using the Velocity-Autocorrelation Technique [16] on the BOMD output, the phonon spectra have been determined.

In relation to the bandgap calculations presented here it should be noted that DFT-LDA [14] underestimates the band gap compared to the experimental values. It is important to note that despite the underestimation of the energy gap, DFT-LDA produces the correct electron wave functions for both filled and empty states. Therefore, we show here only DFT-LDA results, while the many-particle effects will be considered in the next phase of the research.

4 RESULTS AND DISCUSSION

We have modelled here from first principles (DFT-LDA [14]) the structural, optical properties, phonon dispersions and the Molecular Dynamics of the hydrogenated 2D In_2Se_3 single QL films.

Table 1 depicts the atomic structures used in the calculations. The column 3 shows the pristine QL In_2Se_3, and column 6 shows the separated hydrogenated InSeH and $InSe_2H$.

Referring to Table 2 and the accompanying figure at the top of this table, the octahedral Indium (purple) can be seen towards the top of this figure and likewise the tetrahedral Indium(purple) towards the bottom of the figure.

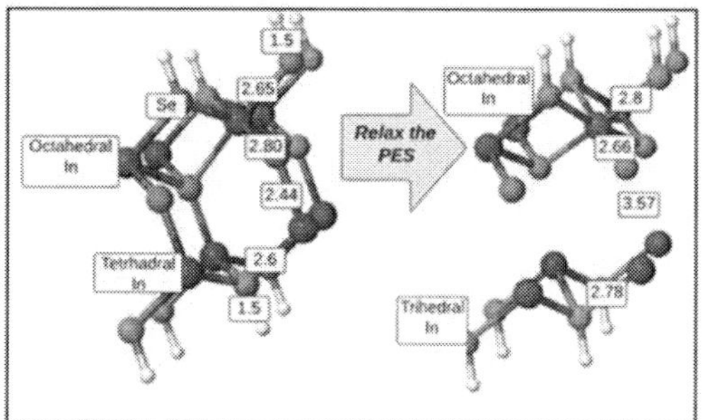

Atom in the single hydrogenated QL $H\text{-}In_2Se_3\text{-}H$	Bond Length (A) to previous atom before relaxation	Bond Length (A) to previous atom after relaxation	In atom coordination before relaxation	In atom coordination after relaxation
H (top)				
Se	*1.5*	*1.5*		
In	*2.65*	*2.8*	Octahedral	Octahedral
Se	*2.8*	*2.66*		
In	*2.44*	*3.57 - broken*	Tetrahedral	Trihedral
Se	*2.6*	*2.78*		
H (bottom)	*1.5*	*1.5*		

Table 2: Bond lengths and In coordination of $H\text{-}In_2Se_3\text{-}H$ before and after relaxation. Note that prior to relaxation of the manually hydrogenated $H\text{-}In_2Se_3\text{-}H$ QL, the pristine In_2Se_3 QL has already been relaxed. The In-Se bond lengths shown for the pre-relaxed hydrogenated structure in this table are therefore for a relaxed In_2Se_3 QL prior to hydrogenation (i.e., these are the bond lengths expected to be seen experimentally for pristine In_2Se_3 QL). In atoms shown as purple, Se atoms shown as yellow in the figure at the top of this table.

It is seen from Table 2 that after relaxation, the $InSe_2H$ structure still has an octahedrally coordinated In, whereas InSeH now has a trihedral In, as the bond that is broken during relaxation is the Se bond to the tetrahedral In, in the initial unrelaxed hydrogenated $H\text{-}In_2Se_3\text{-}H$. The addition of an H atom to the top (and bottom) Se atoms in the one QL In_2Se_3 film, i.e., $H\text{-}In_2Se_3\text{-}H$, modifies the Se-In bond lengths of the resulting two structures, $InSe_2H$ and InSeH, formed after relaxation. The [H-1.5-Se=2.65=In=2.8=Se] top section of the pre-relaxed hydrogenated QL, $H\text{-}In_2Se_3\text{-}H$, (where the numbers indicate bond length in Angstrom, and '=' indicates the bond is to an octahedrally coordinated In), after relaxation, results in the [H-1.5-Se=2.8=In=2.66=Se] structure. The addition of the H to the top Se dangling bond results in the lengthening of the Se-In bond below (down the *c*-axis) and the shortening of the In-Se bond at the bottom of the structure. This effect is also depicted in Table 3. This effect has also been observed in calculations for this paper by taking just the [Se=2.65=In=2.8=Se] structure and alternately adding one H to either Se at either end. The resulting relaxed structure always has the same form, namely [H-1.5-Se=2.8=In=2.66=Se]. A similar effect is seen at the InSeH layer where the addition of the H to the Se dangling bond increases the corresponding Se-In bond length by the same amount as the $InSe_2H$ case.

Atoms	H-Se	Se-In	In-Se
Bond Lengths Pre-Relaxed	*1.5*	*2.65*	*2.8*
Bond Lengths After-Relaxed	*1.5*	*2.8*	*2.66*

Table 3: Bond length modifications for H-Se-In-Se after relaxation. The bond length of the Se-In increases for the Se that has the H attached.

The final relaxed energy of the separated $InSe_2H$ and InSeH, summed together is slightly greater (+0.1Ry) than the final energy of the $H\text{-}In_2Se_3\text{-}H$ relaxed as one structure.. The structural difference that could cause the small energy discepancy between the two sums, is the In-Se layer seperation bewteen the $InSe_2H$ and InSeH films, that results from a relaxation of the full hydrogenated $H\text{-}In_2Se_3\text{-}H$ structure. This layer separation is approximately 3.5A, Table 2. This compares to an approximate 2.9A distance between different layers of multiple QL pristine In_2Se_3 [9] (i.e., un-hydrogenated pristine multiple QL layers of In_2Se_3, the layers between Se atoms in each QL) where these multiple layered pristine QLs of In_2Se_3 have been shown to exist experimentally with vDW forces existing across the 2.9A layer between them [17]. The energy difference is small however, and may be within the error of computation. Additionally, as discussed later, when Molecular Dynamics is performed upon this hydrogenated structure it appears to be unstable and the 3.5A seperation layer collapses. The two sructures where an H atom is added either to the top Se dangling bond or the bottom Se dangling bond in the pristine In_2Se_3 QL (i.e. H- In_2Se_3 or In_2Se_3-H), Table 1 column 4 and 5, appear to exhibit some instabilities (stretching) in the resulting relaxed structure although the bonds are not broken. The tendency of the added H atom to the Se dangling bond to increase the length of the Se-In bond underneath is apprent in both these relaxed structures, but the resulting relaxed unstable (stretched) looking structure needs further examination before further conclusions are drawn.

We calculated the optical response (calculation of the epsilon dielectric function ε_2 [18]) and band structures for the two separated $InSe_2H$, InSeH structures, and the pristine single QL In_2Se_3 for reference, see Figure 3. Here, as previously mentioned, we did not consider calculations with GW self-corrections [19] for the optical response and band structures, due to the computational cost. These corrections will be applied for a later paper. The band structure for the two separated structures, $InSe_2H$, InSeH, and the pristine single QL In_2Se_3 was calculated for high symmetry *k*-points of the IBZ, created from the hexagonal full Brillioun zone in the plane of the '2D' film (*a*, *b*, axis plane [9]), the *k*-points marked along the x-axis of the band structure plots in Figure 3.

The horizontal dashed lines on the band structure plots represent the valence band energy maximum and the conduction band energy minimum respectively. The energy scale (y-axis) is zeroed at the maximum valence band energy for convenience.

Figure 3 also shows the calculated imaginary part of the dielectric function ε_2 for the three structures. The red lines represent the averaged dielectric response for light polarised in the (*a*, *b* axis Figure 1) plane (in the plane of the 2D film), and the blue lines represent the dielectric response for light polarised along the *c*-axis (perpendicular to the plane of the 2D film). The optical bandgap can be seen as the point on the x-axis where the gradient of the dielectric response line starts to increase from zero. It is seen that the two hydrogenated structures are semiconducting, the resulting band gaps (without GW-corrections) are tabulated in Table 4. The band gap figures between the optical response and the band structure calculations cannot be compared directly or with experiment, as the GW correction has not been applied here. However, the trend in the tabulated figures suggests that the two hydrogenated structures $InSe_2H$, InSeH have a higher indirect band gap than the pristine single QL In_2Se_3.

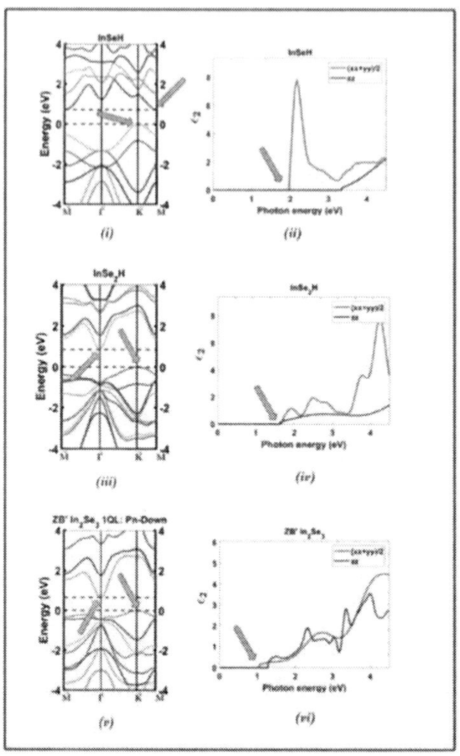

Figure 3: Band structure (i), (iii), (v) and dielectric function plots (ii), (iv), (vi) for InSeH, $InSe_2H$ and In_2Se_3 respectively. The imaginary part of the dielectric function ε_2 is depicted on the y-axis of the optical response plots, for the average of the x and y polarisation (polarisation in the plane of the '2D' film, *color red*) and for the z polarisation (c-axis, perpendicular to the '2D' film, *color blue*). The orange arrows highlight the indirect bandgap in each case, tabulated in Table 4.

Quantum confinement is relevant to this modified bandgap as Quantum confinement plays a role in reducing the bandgap of stacked QL layers relative to the bandgap for a single QL [9]. The motion of electrons in one or multiple QLs is confined to the direction perpendicular to the QL plane, while the carriers can move freely along the films. This confinement of the charge carriers can be qualitatively explained by considering that the In_2Se_3 flakes act as 1D quantum wells [21]. In terms of the electron band structure, if the corresponding 3D layered system is semiconducting with bandgap E_g, the corresponding free-standing or deposited 2D layers (especially on an insulating substrate) should have a larger bandgap due to this confinement effect. The band gap value increases proportional to the inverse of the square of the film thickness [21]. In this case the two films $InSe_2H$, and InSeH created from the pristine single QL In_2Se_3 have a smaller *c*-axis dimension (i.e., smaller height dimension or thickness of the '2D' film) and it would be expected that their bandgap would be larger than the pristine single QL In_2Se_3, if only quantum confinement effects are taken into account. Similarly, the InSeH film would therefore, if quantum confinement effects are taken into account only, be expected to have a larger bandgap than the $InSe_2H$ film. This is the case for the ε_2 derived bandgap in Table 4, but for the band structure derived bandgap in the same table, the InSeH film bandgap is smaller than the $InSe_2H$ bandgap. There are other factors present, namely the hydrogenation, which may impact the bandgap, not investigated here, and the possible instability of these separated films, these factors of which, require further investigation.

In reference to the band structure plots, Figure 3, the indirect bandgap for the pristine single QL In_2Se_3 is between the Γ point and K point for the *k*-point wavevector. This is also the case for the pristine double QL In_2Se_3 [9]. For $InSe_2H$ the indirect bandgap is also between the Γ point and K point, but the InSeH film has the indirect bandgap between the Γ point and M point.

Type	Optical gap from ε_2 (eV)	Bandgap from the band structure (eV)
In_2Se_3	1.1	0.66
InSeH	2.0	0.71
$InSe_2H$	1.8	0.84

Table 4: The DFT-LDA bandgap obtained from the band structure for InSeH, $InSe_2H$ and In_2Se_3 respectively. As noted in the text the DFT-LDA underestimates the band gap compared to the experimental values. However, the trend shown in this table does show that InSeH, $InSe_2H$ have larger band gaps than the pristine In2Se3 single QL.

A phonon dispersion relationship [22] for the pristine single QL, ZB' and WZ' In_2Se_3, calculated for high symmetry *k*-points of the IBZ in the plane of the film (*a*, *b*, axis plane [9]) is plotted in Figure 4 (*iii*) and (*iv*). An independent calculation was done in [8] and the plotted result can be seen in the [8- supplemental paper]. There is good agreement between the calculation for this paper, plotted in Figure 4, and that plotted in [8].

Figure 4 (i) and (ii) also shows the 2nd order phonon dispersion relationship for the two structures $InSe_2H$, InSeH. For InSeH it can be seen that that there is a higher frequency mode, starting at approximately $250cm^{-1}$ at the Γ point, peaking at approximately $350cm^{-1}$ near or at the M point, relative to the pristine single QL's In_2Se_3 (ZB' and WZ'), which have a single highest frequency mode, starting at approximately $250cm^{-1}$ at the **M** point. Similarly, $InSe_2H$ has two higher frequency modes starting above $300cm^{-1}$ at the **M** point. The *H-Se* vibrational mode is not displayed on the Phonon dispersion plots due to scale considerations, but these modes are at $2140cm^{-1}$ and for InSeH and at $2219cm^{-1}$ for $InSe_2H$ respectively.

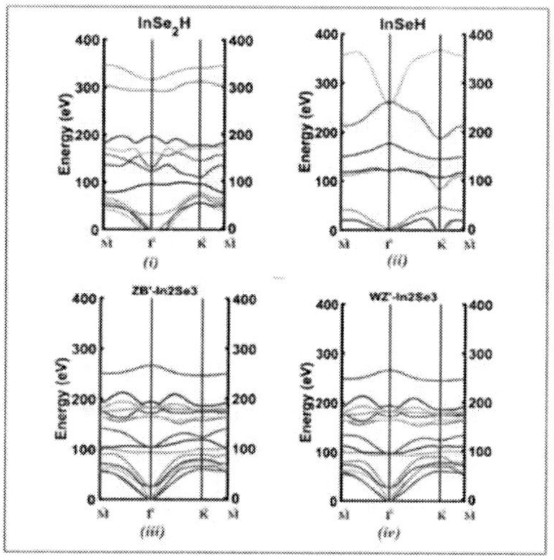

Figure 4: Phonon 2nd order dispersion relation around the 1st IBZ in the x-y plane (i.e., in the plane of the '2D Film), for (i) InSeH, (ii) $InSe_2H$, (iii) ZB'-In_2Se_3 and (iv) WZ'-In_2Se_3 respectively. The ZB' and WZ' In_2Se_3 shown in (iii) and (iv) can be directly compared with the theoretical results in [8-supplement] for the same two structures, (Note: the order of the k-point path traversed in [8] is different to that depicted here). Note: The narrow H bond vibration mode in the $2000cm^{-1}$ range is not shown here in (i), (ii). Thus, theoretically, for InSeH there should be 3×3 = 9 modes in total, 8 modes shown in (ii), and for $InSe_2H$ 3×4 = 12 modes in total, 11 modes shown in (i).

Figure 5 shows the results of analysis of Molecular Dynamics (MD) [23] at approximately room temperature for the pristine single QL In_2Se_3, two hydrogen induced split $InSe_2H$, and InSeH films. Molecular Dynamics was carried out, and the velocity autocorrelation formalism [16] was applied to the atomic velocities at each time step from the MD output, to produce the relative amplitudes and frequencies of the modes of vibration.

For the pristine single QL In_2Se_3, the MD results correlate well to the 2nd order Phonon dispersion relationship shown in Figure 4. (Note: The respective amplitude absolute values between the two techniques cannot be directly compared, as a unit cell is used to calculate the Phonon dispersion relationship, and a supercell is used for the MD calculations.). Experimental Raman scattering results [20], have vibrational peaks at $104cm^{-1}$, $144\ cm^{-1}$, $180cm^{-1}$, $182cm^{-1}$, $187cm^{-1}$, $193cm^{-1}$, $203cm^{-1}$, and $237cm^{-1}$. These experimental peaks apply to specific wavevectors ***k*** and polarization of the wave, whereas the velocity autocorrelation analysis of the MD results presented here apply to all wavevectors and polarizations. Given that, various peaks can be identified from the MD results that correspond to the experimental results in [20]. Namely, and marked up in Table 6, $104cm^{-1}$, and a cluster of peaks in the range $184cm^{-1}$-$200cm^{-1}$.

The MD results for InSeH and $InSe_2H$ shown in Figure 5 (*i*) and (*ii*), show some higher frequency modes in the *500*cm^{-1} region relative to the pristine single QL In_2Se_3. These higher frequency modes do not appear on the equivalent 2nd order Phonon dispersions for InSeH and $InSe_2H$ shown in Figure 4. During the MD runs, the two separated structures InSeH and $InSe_2H$ appear to exhibit instability in that the structure in both cases starts to stretch, and in both cases one bond is broken and does not appear to re-bond. Similarly, MD (*the results not pictured here*) for the relaxed H-In_2Se_3-H structure (i.e., both InSeH and $InSe_2H$ in one supercell separated by 3.5*A*), the structure shown in Figure 2, which has a layer of approximately *3.5***A** between the two InSeH and $InSe_2H$ layers, shows quite rapid instability even at low temperatures, such as 50***K***, with the layer between the InSeH and $InSe_2H$ film disappearing.

Blue background = Experimental	Approximate Energy/Energy range of peaks (cm^{-1})> $100cm^{-1}$							
Raman Scattering (20) α-In_2Se_3	*104 Fig 2(20)*	*144 Fig 3(20)*	*180 Fig 3(20)*	*182 Fig 2(20)*	*187 Fig 3(20)*	*193 Fig 3(20)*	*203 Fig 2(20)*	*237 Fig 3(20)*
Phonon 2nd Order α-In_2Se_3 *Figure 5*	*100-150*	*180-220*						*250-275*
Phonon 2nd Order InSeH *Figure 5*	*80-100*	*120-180*					*200-251*	*250-351*
Phonon 2nd Order $InSe_2H$ *Figure 5*	*100-200*							*300-351*
MD α-In2Se3 *Figure 6*	*90-110*	*180-220*						
MD InSeH *Figure 6*								*240-401*
MD InSe2H *Figure 6*								*450-551*

Table 5: The experimental (blue highlight) results from Raman Scattering [20], first row, compared to the 2nd order Phonon results (Figure 4) and the velocity Auto-correlation of the Molecular Dynamics (MD) results (Figure 5), compared for In_2Se_3, $InSe_2H$, InSeH respectively. The rounded theoretical figures in this table are approximate, taken visually from Figure 4 and Figure 5. Given the possible instabilities of the $InSe_2H$, InSeH structures under MD, the MD result for the pristine In_2Se_3 single QL is of more interest at this time when compared to experiment. (*Note: This table can be zoomed without loss of resolution*)

Further investigation is required into the hydrogenation of pristine single QL α-In_2Se_3, and the MD results of the resulting structures along with calculations on other possible hydrogenated structures.

5 CONCLUSIONS

We present the simulation results on the structure, dynamic and optics of the hydrogenated quasi-2D In2Se3 quintuple layer (QL) nano-material. We show that the hydrogenated In_2Se_3 nanolayers, that is, the In_2Se_3-H systems, offer a variety of structural modifications, which could be attractive for use in photovoltaic technology.

One-side hydrogenation (only at the top or bottom of QL) produces modified QL systems with the initial interatomic separation substantially altered. They demonstrate both semiconducting or metallic behaviour, depending on which side of the QL H atom is located.

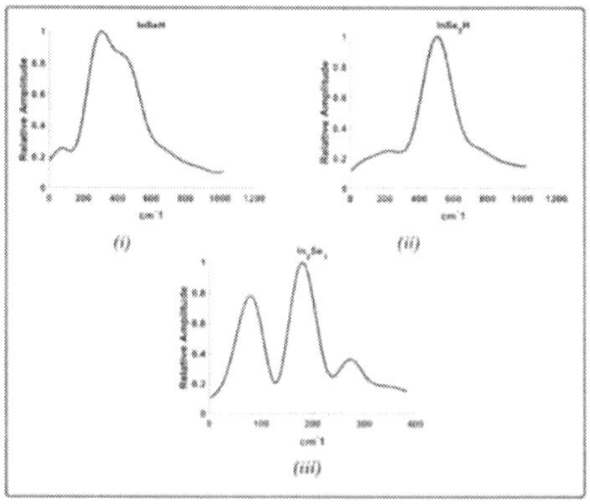

Figure 5: Frequency peaks from the velocities auto-correlation, extracted using the output from Molecular Dynamics at room temperature for InSeH, $InSe_2H$, and pristine In_2Se_3 films respectively. The InSeH, $InSe_2H$ structures investigated appear to exhibit instabilities during the MD runs (broken bonds, stretching of the structure), further investigation is required for these structures. The main vibrational peaks of the pristine In_2Se_3 (iii) agree reasonably well with Raman Scattering experiments [20] (they are tabulated in Table 5).

Two-side hydrogenation, in contrast, splits the pristine QL, forming two semiconducting InSeH and $InSe_2H$ nanofilms, stable at 0K. Both the band structure and optical response of the split hydrogenated films significantly depend on the bonding type, InSeH or $InSe_2H$. From a point view of technology, the two-side H configuration allows for possible QL etching.

The MD results indicate on a possible instability in the hydrogenated split structures analyzed, which requires further investigation. Other possible hydrogenated structures will be studied.

6 REFERENCES

[1] S. Das, D. Pandey, J. Thomas, and T. Roy, Adv Mater 31 (2019) 1802722.
[2] J. Wang, H. Yu, C. Hou, and J. Zhang, ACS Appl. Mater. Interfaces 12 (2020) 26543.
[3] D. Jariwala, A. R. Davoyan, J. Wong, and H. A. Atwater, ACS Photonics 4 (2017) 2962.
[4] F. Xue, J. Zhang, W. Hu, W. Hsu, A. Han, S. Leung, et al., ACS Nano 12 (2018) 4976.
[5] B. J. Hallam, P. G. Hamer, A. M. Ciesla, et al., Prog Photovoltaics Res Appl 28 (2020) 1217.
[6] J. Wang, H. Yu, C. Hou, and J. Zhang, ACS Appl. Mater. Interfaces 12 (2020) 26543.
[7] H. Li, W. Ren, G. Wang, L. Gao, R. Peng, H. Li, et al., J. Phys. D 49 (2016) 145108.
[8] W. Ding, J. Zhu, Z. Wang, Y. Gao, D. Xiao, et al., Nature Communications 8 (2017) 14956.
[9] R. Minnings, A. I. Shkrebtii, G. Perinparajah, N. Arzate, et al., in Proceedings of 37th European Photovoltaic Solar Energy Conference and Exhibition, Sep. 6-11 (2020) 51.
[10] M. C. Payne, M. P. Teter, D. C. Allan, T. A. Arias, and J. D. Joannopoulos. Rev. Mod. Phys. 64 (1992) 1045
[11] R Jones, Philosphical Transactions: Physical Sciences and Engineering, 350 (1995) 1693.
[12] C. Van de Walle, Phys Rev Lett. 60 (1988) 2761.
[13] M. Ishikawa and T. Nakayama, et al., 1997, Japanese Journal of Applied Physics 36 (1997) L1576.
[14] Parr, R. G.; Yang, W. (1989). Density-Functional Theory of Atoms and Molecules. New York: Oxford University Press. ISBN 978-0-19-504279-5.
[15] P. Giannozzi, O. Andreussi, T. Brumme, O. Bunau, M. B. Nardelli, M. Calandra, R. Car, et al, Journal of Physics: Condensed Matter 29 (2017) 465901.
[16] Z. Zhong, et al., Numerical Heat Transfer Fundamentals 46 (2004) 429.
[17] Y. Cho, S. M. Anderson, B. S. Mendoza, S. Okano, R. Carriles, N. Arzate, A. I. Shkrebtii, et al, "Opto-electronic properties of alpha-In2Se3: single-layer to bulk" (https://arxiv.org/abs/1911.01784, 2019).
[18] Z. A. Ibrahim, A. I. Shkrebtii, M. Lee, et al., Phys. Rev. B 77 (2008) 125218.
[19] P. Cudazzo, C. Attaccalite, I. V. Tokatly, and A. Rubio, Phys. Rev. Lett. 104 (2010) 226804.
[20] R. Lewandowskaa, et al., Materials Research Bulletin 36 (2001) 2577
[21] J. Quereda, R. Biele, G. Rubio-Bollinger, N. Agraït, R. D'Agosta, and A. Castellanos-Gomez, Advanced Optical Materials 4 (2016) 1939.
[22] P. Giannozzi, S. de Gironcoli, P. Pavone, and S. Baroni, Phys. Rev. B 43 (1991) 7231.
[23] M. E. Tuckerman, Journal of Physics: Condensed Matter 14 (2002) R1297.

LUMINESCENT SOLAR CONCENTRATOR PHOTOVOLTAICS DEVICES: IMPROVING THE POWER CONVERSION EFFICIENCY BY GEOMETRIC DESIGN MODIFICATIONS

X.Zhu[a*], M. Aghaei[a], and A.H.M.E. Reinders[a,b]
[a] Eindhoven University of Technology, 5612 AE Eindhoven, Netherlands
[b] University of Twente, 7522 NB Enschede, Netherlands
[*] Corresponding author, Email: x.zhu@tue.nl, Phone: +31633678359

ABSTRACT: Luminescent solar concentrator (LSC) devices have been in development for nearly fifty years, usually aiming at enhanced combinations of luminophores and photovoltaic solar cells. In this research, the focus is on the development of new geometric designs of LSC PV devices to improve the performance. The research has been executed by LSC device ray-tracing simulation using software LightTools for LSC PV configurations with a PMMA lightguide doped with Lumogen Red 305 dye. The set of LSC models evaluated have different sizes, different dye concentrations and silicon PV cells on the backside and along all four edges. The simulation results show that with changes in size from 10 to 1,000 mm and dye concentration from 5 to 100 PPM, a 5 mm thick LSC planar light guide can have a total optical efficiency in the range of 77% to 90%. Moreover, a curved LSC lightguide with a dye concentration of 100 PPM also provides a good performance with provided 81% optical efficiency at the backside.

Keywords: Luminescent solar concentrator (LSC), ray-tracing simulation, reabsorption, optical efficiency

1 INTRODUCTOPM

A luminescent solar concentrator photovoltaic (LSC PV) device for which the main application is to realize a transparent photovoltaic system, reducing the cost of the photovoltaic conversion and improving the efficiency of photovoltaics under diffuse light [1], [2]. Although LSC PV technology has a great potential, the power conversion efficiency of LSC PV devices has been limited to a maximum of 8% over the past forty years [3]– [7], and almost all the research has been limited to planar, small-scale LSC PV devices. In order to find methods to improve the efficiency of LSC PV devices and increase the application scenarios of LSC PV devices, it is necessary to conduct research on different geometries of LSC PV devices. The traditional research method of LSC PV device is based on the creating and measuring laboratory-scale prototypes, which has the disadvantages of a long production cycles and high trial and error costs. Therefore, it is more efficient to use ray-tracing simulation as a first stage method for LSC device research.

The simulation of fluorescent particles and their reabsorption properties is crucial for the accuracy of the results. As the first goal, the simulations of different fluorescent dye concentrations were carried out using hypothetical fluorescent particles, and the simulation results were analyzed. The rationality of the re-absorption phenomenon in LightTools, a commercially available ray-tracing simulation software package, is verified, which lays a foundation for the subsequent simulation steps with mixed fluorescent particles or special fluorescent particle LSC devices.

As the second goal, a number of square planar LSC PV configurations based on PMMA 5 mm thick lightguides doped with Lumogen Red 305 dye were simulated (Figure 1), with edge lengths from 10 to 1,000 mm, and dye concentrations from 5 to 100 PPM. Under an AM 1.5G simulated light source of 1000 W/m2, the total optical efficiency namely the number of photons reaching an edge or bottom as a fraction of the number of incidents, of these LSC PV devices is in the range of 77% to 90%. The change of size and dye concentration have separate effects on the optical efficiency.

In the third sub-research, a basic curved 5 mm thick PMMA LSC PV model with the shape of a car engine hood has been simulated (see Figure 2) with mean curvature about 0.0015, a top area of about 1.5 square meters and a dye concentration from 10 to 100 PPM. Under the simulated AM 1.5G solar light source of 1000 W/m2, the backside optical efficiency of this car engine hood is in the range of 81% to 91%. This preliminary work supports the application potential of LSC PV devices as a large-area curved surface.

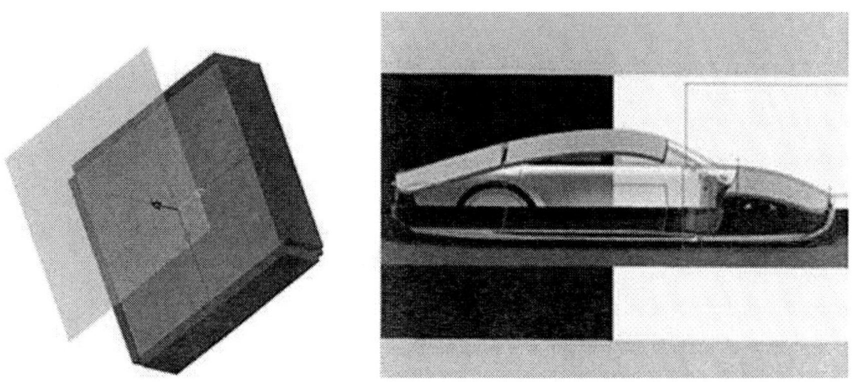

Figure 1: Basic planar LSC PV model in produced in LightTools; (left) model of Curved LSC device application (right).

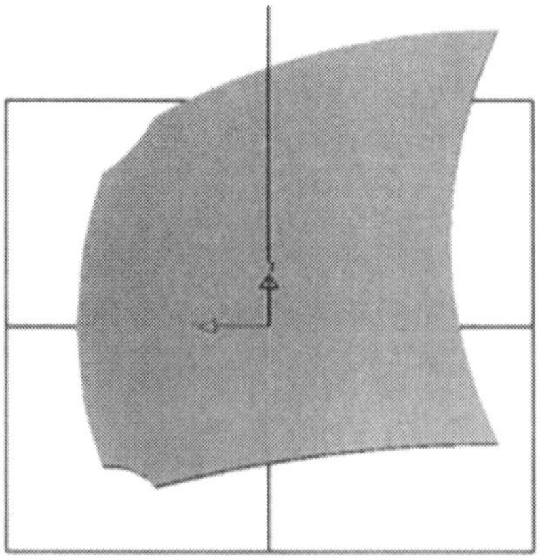

Figure 2: LSC car engine hood model

2 SIMULATION AND RESULT

In this research, we used the ray-tracing software LightTools to simulate the LSC PV devices. All the LSC PV devices used PMMA as the lightguide material and simulated under standard test conditions (STC: 1000 W/m2, AM 1.5 spectrum, and 25 °C). In the first step of the simulation, to verify the re-absorption phenomenon in LightTools (Figure 3a) the hypothetical fluorescent particle was set with two absorption and two emission peaks, the ranges of the second absorption and the first emission regions of the hypothetical fluorescent particles overlap. In the second and third steps, we used Lumogen Red 305 dye (BASF)[8] with seven different sizes from 10 mm to 1,000 mm, and for the car engine hood model we used Solidworks to build a mean curvature about 0.0015 with a top surface area of about 1.5 square meters. In the ray-tracing progress, we traced 1,000 rays per square millimeter for all models. The output data was used to calculate optical performance of each model, and the optical efficiency used to analyze the effects of lightguide size and dye concentration on the LSC-PV devices.

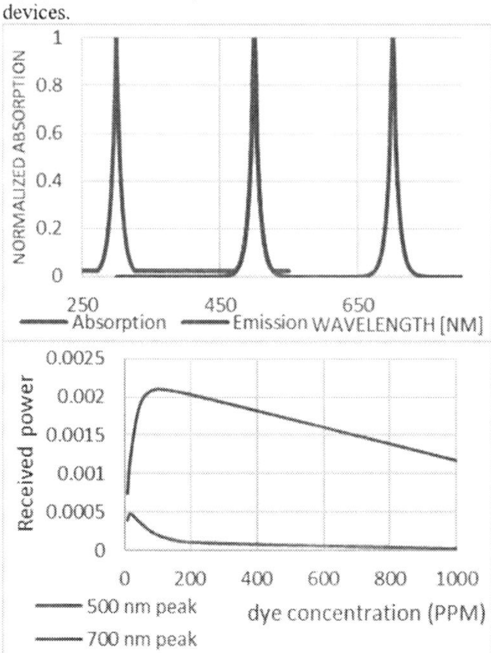

Figure 3: a) Absorption and emission curve of the hypothetical fluorescent particle, b) the emission irradiance comparation between two emission peaks

For the re-absorption verification simulation, we chose to analyze the edge emissions because the edge can only receive irradiance from dye emission. It can be seen from the Figure 3b that the emission irradiance in the first emission peak is less than the second emission peak, and with increasing dye concentration the second irradiance peak decreases. This shows that LightTools ray-tracing simulations can reasonably simulate the phenomenon of light reabsorption by fluorescent particles.

For the square, planar PMMA LSC-PV models, seen in Figure 4, using Lumogen Red 305 dye in size range of 10 to 1,000 mm and dye concentrations in the range of 5 to 100 PPM, the total optical efficiency is in the range of 77% to 90%, both of the increasing of size and dye concentration results in total optical efficiency decreasing. Focusing on the optical efficiency at the edge, the smaller lightguide size with higher dye concentration would have better performance, which also conforms to the results of experiments of some past research [2], [5].

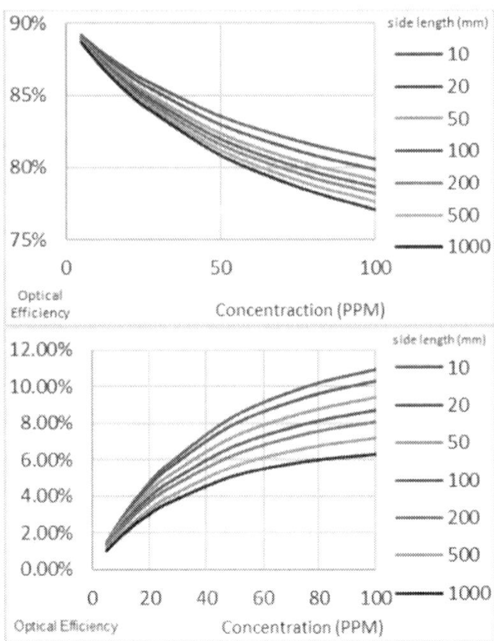

Figure 4: a) the total optical efficiency of planar LSC-PV models, b) the edge optical efficiency of planar LSC-PV models

For the curved LSC car engine hood simulation, the backside optical efficiency of this model is in the range of 80.7 - 90.9%; with increasing dye concentration the edge irradiance would increasingly shift to the red region. As common PV cells always have better Quantum Efficiencies (QE) in the red and near infrared wavelength ranges, using the LSC cover layer for solar cars should be a feasible method to supply more color choices without dramatically hindering performance.

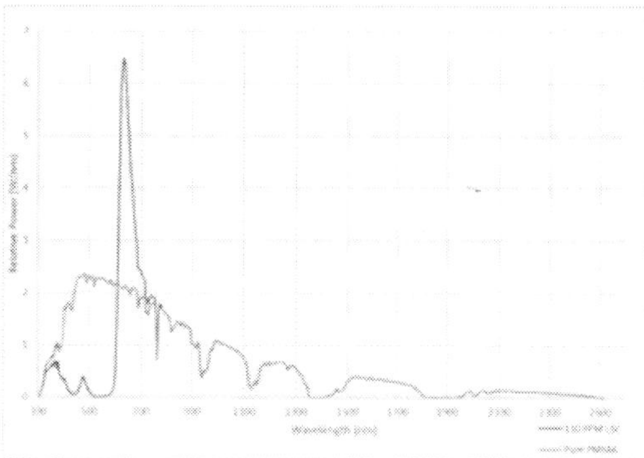

Figure 5: Spectral Distribution received by the LSC (orange line) and PMMA (blue line) car engine hood backside

3 CONCLUSIONS

As the first stage of improving LSC-PV devices, three sub- projects have been executed. Through design simulation, the rationality of the reabsorption phenomenon when LightTools is used in LSC device simulation is verified. The simple simulation of the planar LSC device with different sizes and dye concentrations shows that the total optical efficiency of the planar LSC device can be controlled between 77% and 90% with the dye concentration and size changes. The PV cell attachment area and position would be the method to improve the geometric concentration ratio and power conversion efficiency. For curved LSC applications, a simple car hood model has a backside optical efficiency of more than 80%, and the received light has a significant redshift, which will improve the power conversion efficiency of the attached PV cell.

Reference

[1] A. Goetzberger, "Fluorescent Planar Collector-Concentrators: a Review," Sol. Cells, vol. 4, pp. 3–23, 1980.

[2] V. Wittwer, W. Stahl, and A. Goetzberger, "Fluorescent planar concentrators," Sol. Energy Mater., vol. 11, no. 3, pp. 187–197, 1984.

[3] J. S. Batchelder, A. H. Zewail, and T. Cole, "Luminescent solar concentrators 2: Experimental and theoretical analysis of their possible efficiencies," Appl. Opt., vol. 20, no. 21, p. 3733, 1981, doi: 10.1364/ao.20.003733.

[4] K. Sakuta, S. Sawata, and M. Tanimoto, "Luminescent concentrator module of a practical size," Conf. Rec. IEEE Photovolt. Spec. Conf., vol. 1, pp. 1115–1118, 1994, doi: 10.1109/wcpec.1994.520157.

[5] M. G. Debije and P. P. C. Verbunt, "Thirty years of luminescent solar concentrator research: Solar energy for the built environment," Adv. Energy Mater., vol. 2, no. 1, pp. 12–35, 2012, doi: 10.1002/aenm.201100554.

[6] A. Reinders, M. G. Debije, and A. Rosemann, "Measured Efficiency of a Luminescent Solar Concentrator PV Module Called Leaf Roof," IEEE J. Photovoltaics, vol. 7, no. 6, pp. 1663–1666, 2017, doi: 10.1109/JPHOTOV.2017.2751513.

[7] M. Rafiee, S. Chandra, H. Ahmed, and S. J. McCormack, "An overview of various configurations of Luminescent Solar Concentrators for photovoltaic applications," Opt. Mater. (Amst)., vol. 91, no. March 2018, pp. 212–227, 2019, doi: 10.1016/j.optmat.2019.01.007.

[8] G. Seybold and G. Wagenblast, "New perylene and violanthrone dyestuffs for fluorescent collectors," Dye. Pigment., 1989, doi: 10.1016/0143-7208(89)85048-X.

LOW TEMPERATURE CO-SELENISED ANTIMONY SELENIDE (SB_2SE_3) BASED SOLAR CELLS BY VACUUM EVAPORATION

Vikash Kumar, Elisa Artegiani, Prabeesh Punathil and Alessandro Romeo.
LAPS-Laboratory for Applied Physics, Department of Computer Science, University of Verona, Ca' Vignal 1, Strada Le Grazie 15, 37134 Verona, Italy, tel.:+39-045-8027974, e-mail address: alessandro.romeo@univr.it

ABSTRACT: In this work, we study an alternative path for Sb_2Se_3 preparation by thermal evaporation for low temperature substrate deposition. Thermal evaporation delivered, up to now, lower efficiency devices compared to close space sublimation based solar cells. One of the drawbacks of thermal evaporation for Sb_2Se_3 is the re-evaporation of Se on the substrate, for this reason we present an alternative route to adjust the stoichiometry of selenium deficient Sb_2Se_3 films by adding an excess of selenium (Se) for the deposition of 800 nm thin film by co-evaporation of Se and Sb_2Se_3. Thin films were deposited on Glass/FTO/SnO_2/CdS substrate and their structural and electrical properties have been studied, together with the analysis of the finished devices.

Keywords: Thin Film; Polycrystalline; Heterojunction; Sb_2Se_3; Selenium

1 INTRODUCTION

Antimony Selenide (Sb_2Se_3) is a promising absorber material [1], with elements that are environmentally friendly and abundant in the earth crust. Moreover, the crystals are structured with a 1D crystal system, which limits the charge carriers loss that occurs in polycrystalline absorbers. With a band gap of around 1.2 eV, its theoretical efficiency is 30%, while 9.2% has been reported at laboratory till date [2]. All the top efficiency cells are generally made with close space sublimation as well as vapor transport deposition that allows for a limited re-evaporation of the selenium during growth. However thermal evaporation, with its lower substrate temperature, is interesting for a variety of applications, among them the possibility of deposition on flexible polymers.

Sb_2Se_3 thin film solar cells were made in superstrate configuration by low-temperature fabrication process based on vacuum evaporation (VE). Sb_2Se_3 and Se have been evaporated from special graphite crucibles, and their evaporation rates have been controlled by quartz crystal balance and by monitoring the crucible temperature.

2 EXPERIMENTAL

Sb_2Se_3 based devices were fabricated by thermal evaporation of absorber and buffer layer on commercial fluorine doped SnO_2 (FTO) and SnO_2 (TO) stack (TEC12D). Transparent conductive oxide (TCO) bilayer assures an increased open circuit voltage (Voc) compared to single FTO (TEC10).

Previously, ITO/ZnO bilayer deposited in-house by RF sputtering have been successfully used [3], however for industrial application a commercial TCO is preferable.

The coated glasses are washed with detergent and rinsed in deionized water, successively they are put with a mix of acetone and isopropanol in ultrasonic bath. Subsequently the substrates are put in a chamber with a vacuum of 10^{-4} Pa and annealed at 450 °C for 30 minutes to prepare the layers to the successive depositions. In the same chamber 100 nm thick CdS is deposited by thermal evaporation at a substrate temperature of 150 °C, the rate is controlled by a quartz thickness monitor from Intellemetrics. After a post deposition-annealing the Glass/FTO/TO/CdS stack is transferred to the Sb_2Se_3 evaporation unit.
The Sb_2Se_3 layer is grown under a pressure of $3.6x10^{-4}$ Pa and a substrate temperature of 300 °C. The Sb_2Se_3 graphite crucible is brought to 500-600 °C for an evaporation rate of 0.75 nm/sec, again the thickness of the grown layer is controlled by a quartz balance, typically 800 nm is the optimized thickness for our absorber.

In this work, as previously mentioned, we have added a co-selenisation of Se in order to control the phase of the Sb_2Se_3 compound, the crucible containing Se powder is kept at a temperature in the range of 300-340 °C. The evaporation rate of selenium is regulated between 1.5 ($R_{Se}=2R_{Sb2Se3}$) and 0.75 nm/sec ($R_{Se}=R_{Sb2Se3}$).

The solar cells can be completed with a 30 nm thick Au layer as back contact, deposited by thermal evaporation at $1.1x10^{-3}$ Pa on top of the absorber layer, the described structure of the finished sample is shown in Fig. 1.

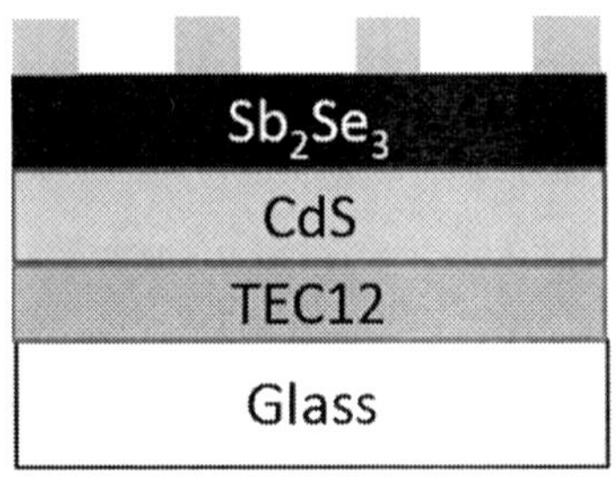

Figure 1. Schematic view of the superstrate configuration.

The morphology of the samples is analysed by atomic force microscopy (AFM) with a NT-MDT Solver Pro at room temperature, in air. The structural properties are analysed by X-ray diffraction (XRD) and by Micro-Raman analysis. XRD is carried out with a Thermo ARL X'TRA powder diffractometer in Bragg-Brentano geometry, equipped with a Cu-anode X-ray source (Kα, λ=1.5418 Å) and a Peltier Si (Li) cooled solid state detector. Micro-Raman spectroscopy is performed in backscattering geometry by using Thermo Scientific DXR2 Raman Microscope, over the Stokes-shifted region between 100 cm^{-1} to 350 cm^{-1} by a diode laser with λ_{exc}=633 nm.

3 RESULT AND DISCUSSION

3.1. Structural Characterization

In Fig.2 the morphologies of Sb_2Se_3 layers are shown and compared: (a) as deposited, (b) co-selenised at (R_{Se}=0.5 R_{Sb2Se3}), (c) co-selenised at (Rse= 1 R_{Sb2Se3}), (d) co-selenised at (R_{Se}= 2 R_{Sb2Se3}). The results show that the absorbers have a very different grain structure: for co-selenised samples at double rate the grains are agglomerated, bigger as well as denser when compared to those co-selenised at 1:1 rate. The films at smaller selenisation rate exhibit a less compact and smaller grain structure.

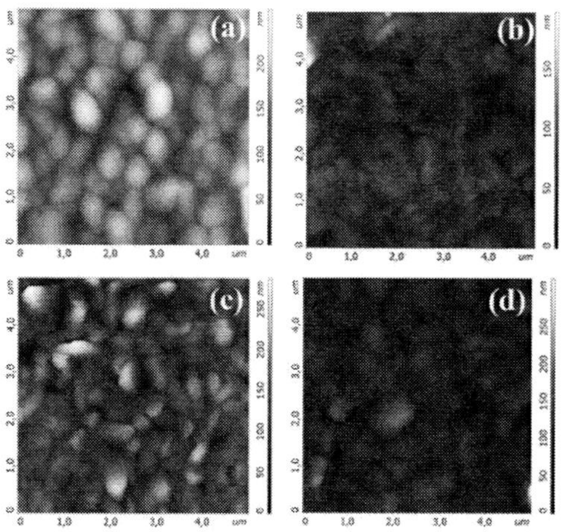

Figure 2. Schematic view of (a) as deposited; (b,c,d) co-selenised at R_{se}= 0.5 R_{Sb2Se3}, R_{se}= R_{Sb2Se3} and R_{se} = 2 R_{sb2se3} respectively.

In Fig. 3, the diffraction patterns show the reflections from the co-selenised samples at different rates. The peaks correspond to the orthorhombic Sb_2Se_3 structure (JCPDS 15-0861) [4][5]. All samples exhibit strong (221), (141), (211), (041), (061) and (321) peaks. The co-selenised samples with lower rate (R_{Se}=R_{Sb2Se3}) show a higher intensity for (021), (322) and (422) when compared to the samples deposited at twice the rate (Rse=2R_{Sb2Se3}). The films exhibit both [hk1] and [hk0] preferred orientation peaks, the (360) peak is very pronounced for the 2 R_{Sb2Se3} and the 0.5 R_{Sb2Se3}, it is known that the [hk1] orientations are crucial for high efficiency.

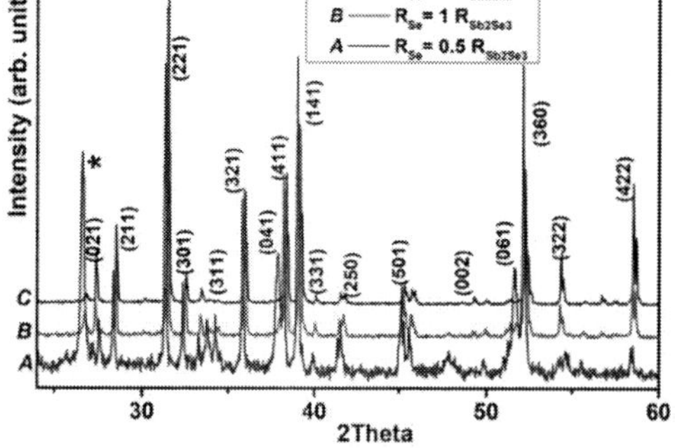

Figure 3. X-ray diffraction patterns of co-selenised Sb_2Se_3 samples.

In figure 4, detailed structural analysis of the co-selenised Sb_2Se_3 thin films, performed at room temperature using Raman spectra with an excitation source of 633 nm, are shown. All samples have Raman bands (centred at 120, 150, 190 and 210 cm^{-1}) associated to Sb_2Se_3, confirming the orthorhombic structure [6]. Films grown at a lower selenisation rate (Rse= 1 Rsb_2se_3) present a better crystalline stoichiometry. The 188 cm^{-1} and 210 cm^{-1} peaks are connected with the Sb-Se vibration mode [7], while the 253 cm^{-1} peak is given by the Se-Se stretching vibration mode [7]. Compared to the as-deposited case, the co-selenised samples show a clear shift of the 189 cm^{-1} peak to the right, demonstrating that the co-selenisation changes the Sb-Se bonds. Paradoxically, the 253 cm^{-1} disappears for the co-selenised samples, so the bonds between Se atoms are removed after selenisation, promoting the 1D-structure made of Sb-Se bonds. This suggests that additional selenium helps to promote the correct stoichiometry without having excess of selenium.

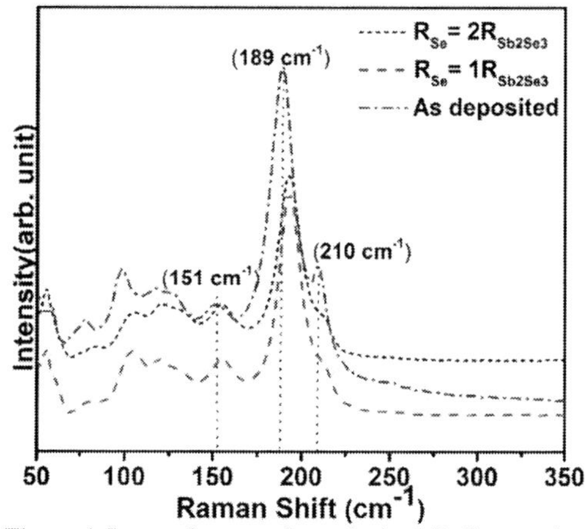

Figure 4. Raman Spectra of co-selenised Sb_2Se_3 samples.

Finally, the finished devices have been completed and their electrical properties studied. The as-deposited sample, without any type of treatment, show quite poor efficiencies. Previously we have demonstrated an efficiency of 3.5%, but in that case the absorbers were annealed at high temperature and with copper insertion in the lattice [3]. In this case we want to analyse the effect of selenium addition by co-evaporation, for this reason we have kept the as-deposited sample as is, without additional improvement treatments.

The co-selenised samples, instead, show a clear improvement of the efficiency, particularly for the open circuit voltage and short circuit current. The fill factor shows a reduction which can be justified by the large increase in the current density, nevertheless for the Sb_2Se_3 fabricated with higher flux of selenium the series resistance is very high, suggesting a reduced overall conductivity of the layer.

The most performing cells result to be the ones with lower Se flux (R_{Se}= 1 R_{Sb2Se3} and R_{Se}= 0.5 R_{Sb2Se3}), they have similar efficiencies but with an improved fill factor and a larger current for the one with R_{Se}=0.5 R_{sb2se3}.

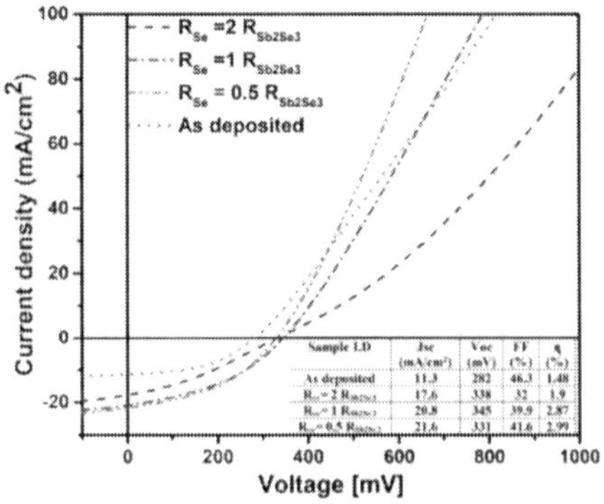

Figure 5. J-V characteristics of the as deposited and selenised samples.

4 CONCLUSIONS

Co-selenisation at varying rates have shown compact films with no or fewer pinholes and cracks, as well as the grains are larger as compared to previously reported as deposited samples, these are one of the major contributing factors in device performance. The X-ray diffraction analysis suggests the formation of crystalline films still with the preferred [hk1] orientations (as in the single evaporation), these planes consist of one –dimensional $(Sb_4Se_6)_n$ ribbon standing on the substrate with a certain angle make it easy for the carriers to be transported when compared to [hk0] planes which are parallel to the substrate [6]. Efficiencies up to 3% have been obtained, with co-selenisation, despite no particular treatment as well as no additional copper was inserted, demonstrating the co-evaporation of Se could be a good path for improving efficiencies in low substrate temperature Sb_2Se_3 made devices.

5 REFERENCES

[1] Y. Zhou, L. Wang, S. Chen, S. Qin, X. Liu, J. Chen, et al., Thin-film Sb2Se3 photovoltaics with oriented one-dimensional ribbons and benign grain boundaries, Nat. Photonics. (2015). doi:10.1038/nphoton.2015.78.

[2] Z. Li, X. Liang, G. Li, H. Liu, H. Zhang, J. Guo, et al., 9.2%-Efficient Core-Shell Structured Antimony Selenide Nanorod Array Solar Cells, Nat. Commun. 10 (2019) 1–9. doi:10.1038/s41467-018-07903-6.

[3] V. Kumar, E. Artegiani, A. Kumar, G. Mariotto, F. Piccinelli, A. Romeo, Effects of post-deposition annealing and copper inclusion in superstrate Sb2Se3 based solar cells by thermal evaporation, Sol. Energy. 193 (2019) 452–457. doi:10.1016/j.solener.2019.09.069.

[4] L. Wang, D.B. Li, K. Li, C. Chen, H.X. Deng, L. Gao, et al., Stable 6%-efficient Sb2Se3 solar cells with a ZnO buffer layer, Nat. Energy. 2 (2017) 1–9. doi:10.1038/nenergy.2017.46.

[5] X. Liu, J. Chen, M. Luo, M. Leng, Z. Xia, Y. Zhou, et al., Thermal evaporation and characterization of Sb2Se3thin film for substrate Sb2Se3/CdS solar cells, ACS Appl. Mater. Interfaces. 6 (2014) 10687–10695. doi:10.1021/am502427s.

[6] A. Shongalova, M.R. Correia, B. Vermang, J.M.V. Cunha, P.M.P. Salomé, P.A. Fernandes, On the identification of Sb2Se3 using Raman scattering, MRS Commun. 8 (2018) 865–870. doi:10.1557/mrc.2018.94.

[7] J. Tao, X. Hu, J. Xue, Y. Wang, G. Weng, S. Chen, et al., Investigation of electronic transport mechanisms in Sb 2 Se 3 thin-film solar cells, Sol. Energy Mater. Sol. Cells. 197 (2019) 1–6. doi:10.1016/j.solmat.2019.04.003.

COPPER DOPED ZNO AS TRANSPARENT METAL OXIDE FOR THIN FILM SOLAR CELL APPLICATION

Mohammed Kamal Hossain*, and Masoud Al-rasheidi
Interdisciplinary Research Center for Renewable Energy and Power Systems (IRC-REPS), Research Institute, King Fahd University of Petroleum & Minerals (KFUPM), Dhahran 31261, Saudi Arabia.
*Correspondence to: kamalhossain@kfupm.edu.sa

ABSTRACT: Zinc oxide (ZnO) is one of the widely studied and promising candidates in multidisciplinary research areas including, but not limited to optoelectronics and advance materials physics. However, the unique characteristics can be further twisted using alien acceptors or donors doping. Here in this work, we have devised a simple and one-step process to fabricate Copper (Cu)-doped ultrathin ZnO film through a co-sputtering technique. The weight percentage of Cu was varied from 50% to 100% to 150% with respect to the reference specimen. A set of topographical, optical and electrical characterizations was carried out to understand the characteristic of Cu-doped ZnO (ZnO:Cu) thin films using various techniques, such as FESEM, UV-absorption and transmission. The same layer was integrated with a standard solar cell and the performances of the overall solar cell were evaluated using a solar simulator. Further to understand the sintering effect, the specimens were annealed at 200, 400 and 600 °C for 2 hours and a detailed comparative study has been carried out. Such systematic and in-depth investigations guide one to understand the potential of modified ZnO on various applications such as thin-film solar cell applications.

Keywords: Sputtering; TCO Transparent Conducting Oxides; Thin Film; ZnO; Doping

1 INTRODUCTION

Transparent conducting oxide (TCO) plays a key role to collect and transport photo-generated excitons available in photovoltaic (PV) solar cells [1-5]. Zinc oxide (ZnO) is well known as the TCO layer of PV solar cells due to its wide band-gap and reasonably high electron mobility in addition to abundance and nontoxicity in nature [6-9]. Metal-doped ZnO has been emerged as an efficient TCO to increase the conductivity at low absorption loss [10-14]. The unique properties of ZnO can be further tuned according to the needs of multidisciplinary applications. However optoelectronic properties of metal-doped ZnO depend on various factors such as deposition techniques, oxygen incorporation during deposition, deposition temperature and post-deposition treatment [15-20].

Here in this work, we have devised a simple and one-step process to fabricate Copper (Cu)-doped ultrathin ZnO film through a co-sputtering technique. The weight percentage of Cu was varied from 50% to 100% to 150% with respect to the reference specimen. A set of topographical, optical and electrical characterizations was carried out to understand the characteristic of Cu-doped ZnO (ZnO:Cu) thin films using various techniques, such as FESEM, UV-absorption and transmission. The same layer was integrated with a standard solar cell and the performances of over solar cell were evaluated using a solar simulator. Further to understand the sintering effect, the specimens were annealed at 200, 400 and 600 °C for 2 hours and a detailed comparative study has been carried out.

2 MATERIALS AND METHODS

A co-sputtering technique was utilized to prepare ZnO:Cu thin film on a glass substrate using a dual-beam sputtering system (model: Nanomaster NSC3000, USA). ZnO and Cu targets with a size of 2" diameter (99.99% purity) purchased from ACI alloy, Inc. was used as received without any further treatment. RF beam was used for ZnO deposition where DC was used simultaneously for Cu deposition. The vacuum in the sputtering chamber (base pressure) and working pressures were about 1.7×10^{-5} torr and 2.8×10^{-3} torr, respectively. The deposition powers were 150 W for RF (ZnO) and 10 W for DC (Cu) Ar flow rate of 30 SCCM respectively. The weight percentage of Cu content within ZnO was controlled by changing the DC (Cu) power. Following the exactly same deposition parameters, 2nd, 3rd and 4th set of specimens were prepared for DC (Cu) power of 15, 20 and 25 W respectively. Afterward, the deposited films were heated in tube furnace (model MTI Corporation OTF-1200X) at 200, 400 and 600 °C for 2 hrs each with a heating rate of 20 °C/min. Systematic and detailed analyses were carried out using various topography, optical and electrical characterization techniques such as FESEM, UV-absorption and transmission. The as-fabricated films and the treated ones were integrated with a standard solar cell with known performances and overall performances were evaluated with a solar cell simulator.

3 RESULTS AND DISCUSSION

Right after the fabrication of specimens, a visual inspection has been carried out for quick screening as shown in Fig. 1. Fig. 1a-d shows CCD images of the specimens with an increasing weight percentage of Cu in ZnO:Cu thin films prepared in Ar. The reference specimen as shown in Fig. 1a represents the Cu-doped ZnO film that was found nearly transparent. For the specimens with increasing Cu content of 50%, 100 and 150% of the reference Cu content were shown in Fig. 1b-d respectively. It was noticed that the films appeared to be darker with increasing content of Cu. Further details in morphology have been carried out by high resolution FESEM and will be demonstrated in the latter part of the texts. To the extent as mentioned in the experimental section, the specimens were annealed at 200, 400 and 600 °C for 2 hrs each. Detailed morphologies of such treated samples were characterized as well. As for the treated ZnO:Cu films, most of the samples were found transparent compared to those fabricated under Ar without annealing. CCD images of such specimens with an increasing weight percentage of Cu in ZnO:Cu thin films prepared in Ar and treated at 600

°C for 2 hrs are shown in Fig. 1e-h.

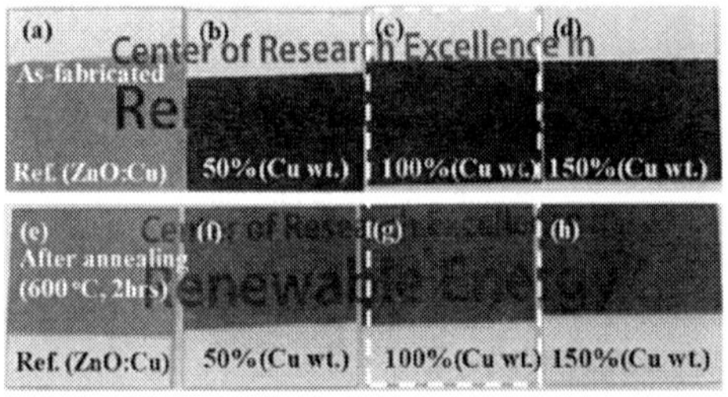

Figure 1: (a)-(d) CCD images of as-fabricated ZnO:Cu samples in Ar with different wt% of Cu as mentioned therein, (e)-(h) the same images of the samples prepared and treated at 600 °C for 2 hrs. White dashed rectangle has been amended to indicate that only these two specimens have been demonstrated in this abstract.

The reference specimen as shown in Fig. 1e represents the Cu-doped ZnO film that was found more transparent with reference to that fabricated under Ar without treatment. Fig. 1f-h represents the specimens fabricated with increasing Cu content of 50%, 100 and 150% of the reference Cu content and treated at 600 °C for 2 hrs respectively. The specimens were found to be a lit bit darker, but more transparent compared to those fabricated under the same condition in Ar without annealing. It is to be noted that the specimen fabricated with 100% increase of Cu content and annealed at 600 °C for 2 hrs will be further analyzed under this work, although detailed topographical, optical and electrical characterizations of all the sets of specimens have been carried out and will be elaborated elsewhere once available.

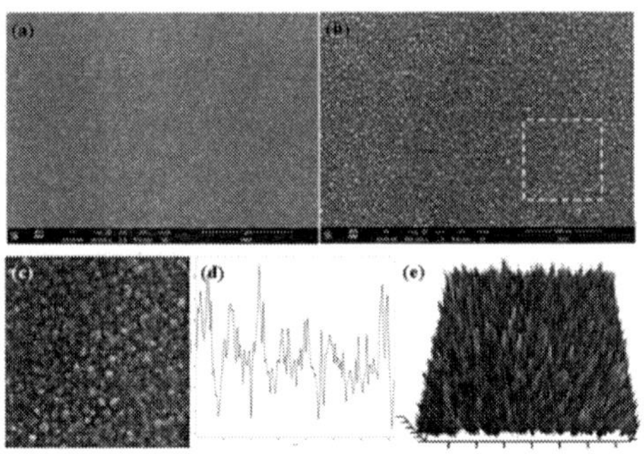

Figure 2: (a) FESEM micrograph of ZnO:Cu film doped with 100% increase in Cu content of the reference sample, (d) high resolution FESEM micrograph of the same, (c) Zoom in view of the selected area (500×500 nm^2) as marked by white dashed square in Fig. 2b, (d) line profile along the white dashed line as shown in Fig. 2c, (e) a hawk-eye view of the same selected area as shown in Fig. 2b.

As mentioned in the experimental section, in-depth and systematic investigations were carried out to understand the detailed morphology, as well as optical, electrical and structural characteristics of such specimens. The specimens, ZnO:Cu film doped with 100% increase of Cu content and fabricated under Ar without and with treatment (Fig. 1c and Fig. 1g respectively) are demonstrated herewith as an example. The same specimen was integrated to standard solar cell and performances were evaluated under the same conditions. Fig. 2a shows FESEM observation of the specimen, ZnO:Cu film doped with 100% increase of Cu content. A high-resolution FESEM shown in Fig. 2b revealed that the surface of the as-fabricated specimen was rough and full of small grains. A selected area (500×500 nm^2) as marked by a white dashed square in Fig. 2b was further zoomed in and shown in Fig. 2c. Small grain was observed whereas the SEM-aided EDS (data not shown here) confirmed the presence

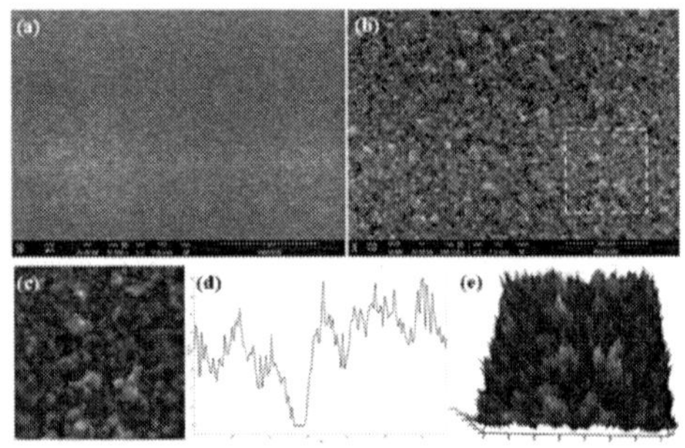

Figure 3: (a) FESEM micrograph of ZnO:Cu film doped with 100% increase in Cu content of the reference sample, (d) high resolution FESEM micrograph of the same, (c) Zoom in view of the selected area (500×500 nm2) as marked by white dashed square in Fig. 3b, (d) line profile along the white dashed line as shown in Fig. 3c, (e) a hawk-eye view of the same selected area as shown in Fig. 3b.

of Zn, Cu and O in EDS peaks (data not shown herewith). It was very hard to identify any Cu grains separately and isolated from those of ZnO. However, a line profile along the line mentioned by a white dashed line in Fig. 2c was shown in Fig. 2d. The dips and hills as observed in Fig. 2d confirmed that the surface was consisted of small grains rather than smooth film. A hawk-eye view of the same selected area further reconfirmed the rough surface of this particular specimen.

A specimen ZnO:Cu film doped with 100% of Cu content and fabricated under Ar followed by annealing at 600 °C for 2 hours has been found more transparent and explained in Fig 3. A FESEM micrograph with low-resolution of the same specimen is shown in Fig 3a. A high-resolution FESEM shown in Fig. 3b revealed that the surface of the as-fabricated specimen was having small pores along with a larger grain size. A selected area (500×500 nm^2) as marked by a white dashed square in Fig. 3b was further zoomed in and shown in Fig. 3c. Few holes and densely populated larger grains were observed clearly whereas the SEM-aided EDS (data not shown here) confirm the presence of Zn, Cu and O in EDS peaks. It was very hard to identify any Cu grains separated from ZnO. However, a line profile along the line mentioned by white dashed line in Fig. 3c was shown in Fig. 3d. The dips and hills as observed in Fig. 3d confirmed the surface was consisted of holes and larger grains. A hawk-eye view of the same selected area further reconfirmed the rough surface of this particular specimen.

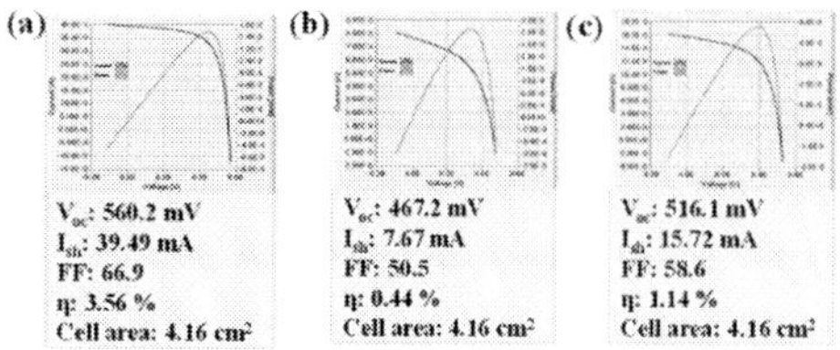

Figure 4: (a)-(c) Performances of typical solar cell integrated with ZnO film only, ZnO:Cu (doped with 100% increase of Cu content) and ZnO:Cu (doped with 100% increase of Cu content followed by annealing at 600 oC for 2 hrs.) respectively. Open circuit voltage, short circuit current, Fill factor and efficiency for each specimen have been mentioned therein.

The specimens, ZnO film as a reference, ZnO:Cu film doped with 100% of Cu content and fabricated under Ar and the same specimen after annealing were integrated with typical Si solar cell and the performances were carried out as shown in Fig. 4. Fig. 4a-c represents performances of typical solar cell integrated with ZnO film only, ZnO:Cu (doped with 100% increase of Cu content) and ZnO:Cu (doped with 100% increase of Cu content followed by annealing at 600 °C for 2 hrs.) respectively. Important characteristics such as open-circuit voltage, short-circuit current, Fill factor and efficiency have been extracted for each scenario. Open circuit voltage was found to increase ~10% whereas open circuit current was increased more than double. As a result the efficiency was observed more than double as well.

4 CONCLUSION

We have reported a simple and one-step process to fabricate Copper (Cu)-doped ZnO (ZnO:Cu) ultrathin layer through co-sputtering technique. The weight percentage of Cu was varied from 50% to 100% to 150% with respect to the reference specimen. Morphological, optical and electrical characteristics of ZnO:Cu thin films with and without annealing were investigated using various techniques, such as FESEM, UV-vis absorption and transmission. The same layers were integrated with a standard solar cell and the performances of over solar cell were evaluated using a solar simulator. The overall performances of the ZnO:Cu thin films were observed to be different with reference to those of pristine ZnO. Open circuit voltage was found to increase ~10% whereas open circuit current was increased more than double. As a result the efficiency was observed more than double as well. The specimens were annealed further at 200, 400 and 600 °C for 2 hours and a detailed comparative study has been carried out.

Acknowledgment: Authors at KFUPM would like to acknowledge the support received under Interdisciplinary Research Center for Renewable Energy and Power Systems (IRC-REPS) Grant INRE2114.

References

1. Afre, Rakesh A., Nallin Sharma, Maheshwar Sharon, and Madhuri Sharon. "Transparent conducting oxide films for various applications: A review." Reviews on advanced materials science 53, no. 1 (2018): 79-89.
2. Park, Hyeongsik, Youngkuk Kim, Donghyun Oh, Duy Phong Pham, Jaechun Song, and Junsin Yi. "Current Status of Low-temperature TCO Electrode for Solar-cell Application: A Short Review." New & Renewable Energy 17, no. 1 (2021): 1-6.
3. Fleischer, K., E. Arca, and I. V. Shvets. "Improving solar cell efficiency with optically optimised TCO layers." Solar energy materials and solar cells 101 (2012): 262-269.
4. Park, Hyeongsik, Youn-Jung Lee, Jinjoo Park, Youngkuk Kim, Junsin Yi, Youngseok Lee, Sangho Kim, Chang-Kyun Park, and Kyung-Jin Lim. "Front and back TCO research review of a-Si/c-Si heterojunction with intrinsic thin layer (HIT) solar cell." Transactions on Electrical and Electronic Materials 19, no. 3 (2018): 165-172.
5. Krc, J., B. Lipovsek, M. Bokalic, A. Campa, T. Oyama, M. Kambe, T. Matsui, H. Sai, M. Kondo, and M. Topic. "Potential of thin-film silicon solar cells by using high haze TCO superstrates." Thin Solid Films 518, no. 11 (2010): 3054-3058.
6. Capper, Peter, Safa Kasap, and Arthur Willoughby. Zinc oxide materials for electronic and optoelectronic device applications. John Wiley & Sons, 2011.
7. Vyas, Sumit. "A short review on properties and applications of ZnO based thin film and devices." Johnson Matthey Technology Review (2020).
8. Yu, Xinge, Tobin J. Marks, and Antonio Facchetti. "Metal oxides for optoelectronic applications." Nature materials 15, no. 4 (2016): 383-396.
9. Djurišić, A. B., Alan Man Ching Ng, and X. Y. Chen. "ZnO nanostructures for optoelectronics: Material properties and device applications." Progress in quantum electronics 34, no. 4 (2010): 191-259.
10. Ismail, Siti Sarah Mohd, and Ainun Rahmahwati Ainuddin. "Effect of annealing temperature on the properties of transition metal doped zinc oxide–A review." In AIP Conference Proceedings, vol. 2068, no. 1, p. 020101. AIP Publishing LLC, 2019.
11. Jeong, S. H., B. N. Park, S-B. Lee, and J-H. Boo. "Metal-doped ZnO thin films: synthesis and characterizations." Surface and Coatings Technology 201, no. 9-11 (2007): 5318-5322.
12. Zhao, J. L., X. W. Sun, H. Ryu, and Y. B. Moon. "Thermally stable transparent conducting and highly infrared reflective Ga-doped ZnO thin films by metal organic chemical vapor deposition." Optical Materials 33, no. 6 (2011): 768-772.
13. Joshi, Kanchan, Mukesh Rawat, Subodh K. Gautam, R. G. Singh, R. C. Ramola, and Fouran Singh. "Band gap widening and narrowing in Cu-doped ZnO thin films." Journal of Alloys and Compounds 680 (2016): 252-258.
14. Shohany, Boshra Ghanbari, and Ali Khorsand Zak. "Doped ZnO nanostructures with selected elements-Structural, morphology and optical properties: A review." Ceramics International 46, no. 5 (2020): 5507-5520.
15. Ueda, Kenji, Hitoshi Tabata, and Tomoji Kawai.

"Magnetic and electric properties of transition-metal-doped ZnO films." Applied Physics Letters 79, no. 7 (2001): 988-990.

16. Patil, Ashokrao B., Kashinath R. Patil, and Satish K. Pardeshi. "Enhancement of oxygen vacancies and solar photocatalytic activity of zinc oxide by incorporation of nonmetal." Journal of Solid State Chemistry 184, no. 12 (2011): 3273-3279.
17. Mosbah, A., and M. S. Aida. "Influence of deposition temperature on structural, optical and electrical properties of sputtered Al doped ZnO thin films." Journal of Alloys and Compounds 515 (2012): 149-153.
18. Cheng, Yung-Chen. "Effects of post-deposition rapid thermal annealing on aluminum-doped ZnO thin films grown by atomic layer deposition." Applied surface science 258, no. 1 (2011): 604-607.
19. Ravichandran, C., G. Srinivasan, Craig Lennon, S. Sivananthan, and J. Kumar. "Influence of post-deposition annealing on the structural, optical and electrical properties of Li and Mg co-doped ZnO thin films deposited by sol–gel technique." Superlattices and Microstructures 49, no. 5 (2011): 527-536.
20. Chen, Xin, Wenjie Guan, Guojia Fang, and X. Z. Zhao. "Influence of substrate temperature and post-treatment on the properties of ZnO: Al thin films prepared by pulsed laser deposition." Applied Surface Science 252, no. 5 (2005): 1561-1567.

EVALUATION OF THE INKJET PRINTABLE DOPANT MATERIAL FOR SOLAR CELL APPLICATIONS

Sergiu Pop, SCP SYS, LLC San Francisco, CA United States
Kent Ridgeway, Desert Silicon, Inc. Tempe, AZ United States
Phil Yin, Desert Silicon, Inc. Tempe, AZ United States

ABSTRACT: In order to deploy silicon PV products at competitive prices one of the most important efforts throughout the industry is to investigate and develop processes and device features aimed to improve the current product performance, reliability, and reduce fabrication costs. In this work we evaluate a novel approach that could alter the traditional fabrication process of passivating poly-Si contacts by employing the inkjet technology of printable dopant (PDI) material to replace the traditional phosphorus ion implant process.
Keywords: Silicon, PV cells, Inkjet printing, SoG material

1 INTRODUCTION

First attempts to use inkjet technology in PV were in 1987 where the aim was the metallization of the solar cells. Since then, the inkjet technology had evolved quite a bit. For example, the number of nozzles per print head and the jetting frequencies increased up to 100 kHz along with many other features which were also improved: variable drop size and/or piezo sensing, all accounting for superior image/shape quality. Also, the recent improvements resulted in print heads which are resistant to a large range of chemicals opening more opportunities for this technology.

Industrial scale inkjet technology is now available from several companies and allows production speeds of 2000 - 4000 wafers/hour, similar to current production standards [1].

In the present collaborative effort, we plan to employ the inkjet technology where the proprietary dopant material developed by Desert Silicon Inc. will be deposited on amorphous silicon, aiming at replacing some of the traditional ion implant processes [2]. The inkjet deposited layers of dopants can be printed on the wafer directly in any shape without the need for masking, followed by an especially designed thermally curing process [3].

If successful inkjet printable dopant (IPD) technology could be implemented as an inline process, opening up further opportunities for yield and process time improvements.

2 RESULTS

2.1 Inkjet printing technology versus traditional screen printing

In this work we are evaluating the performance of a particular dopant ink formulation developed by Desert Silicon Inc.

There are several advantages of inkjet printing over the more common screen printing, especially when it comes to fine features and improvements in mechanical yield due to contactless process. Overall, inkjet process needs digital masks only (no lithography required), which can reduce the costs of the process significantly. Prints with a 5 microns tolerance accuracy can be reached while screen printing can reach only 10 microns tolerance. The features obtained by screen printing at industrial speeds is still beyond the capability of inkjet printing. However, the inkjet process is currently under development at several equipment manufacturers.

2.2 Equipment used and the design of experiments (DoE)

The equipment used throughout this study is a PiXDRO LP50 piezo inkjet printer[1] which has capability of using many heads and an Anton Parr rheometer for evaluation of ink viscosity. All processes and characterizations were performed at Fraunhofer ISE laboratories.

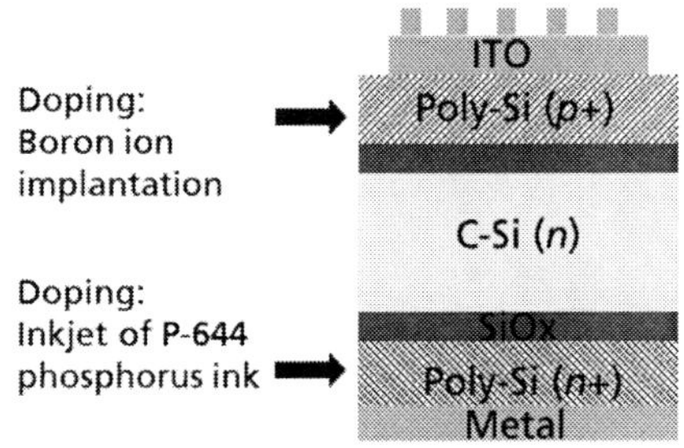

Figure 1: Cross section of cell architecture fabricated for evaluation of poly-Si passivating contact doped using inkjet printing of P-644

The architecture of experimental samples is shown in Figure 1. Float-zone silicon wafers with n-type base doping are used where a layer of silicon oxide (SiOx) with a thickness of approx. ~1.3 nm was grown. It was followed by an LPCVD deposition of amorphous Si with a thickness of about ~ 50 nm. The boron layer was applied on the sunny side of the cell by ion implantation while the phosphorus layer on the rear side was applied by inkjet printing. Upon dopant application, they were cured on a hot plate at a temperature of about 200 °C for about 10 minutes followed by the anneal step in a tube furnace at a temperature between 925 °C and 930 °C for 10 and 30 minutes.

Residual ink was removed in a 20% HF solution for about 5 minutes. Passivation was completed by a reactive plasma hydrogen passivation at a temperature of 400 °C for 30 minutes. The ITO was sputtered followed by photolithography and metallization. QSSPC characterization was carried out at various steps throughout the fabrication process (e.g., upon ink residual removal, RPHP, ITO sputter, etc.) and finalized with I-V sweeps evaluation.

[1] Now SÜSS MicroTec

2.3 Viscosity measurement of phosphorus ink

Prior to inkjet printing, the phosphorus ink was subjected to evaluation using viscosity measurements.
The rheology of the ink was characterized by an Anton Parr rheometer (Figure 2) where the characterization measurement was a shear rate controlled rotational test.

Figure 2: Anton Parr rheometer rotational rheometer at Fraunhofer ISE

The phosphorous ink exhibits a constant viscosity over a wide range of shear rate at 25 °C. This implies that the ink is Newtonian in this range, and hence stable jetting of nozzles can be expected in this range. Samples were measured to check the consistency and the results at low shear rate are plotted in Figure 3.

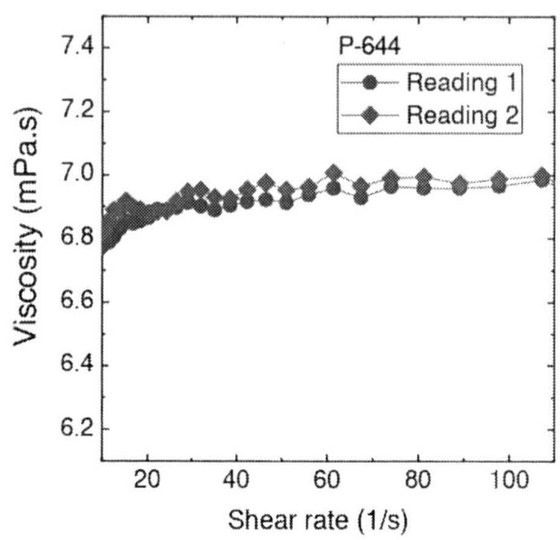

Figure 3: Viscosity of the P-644 phosphorus ink versus shear rate between 20 and 100 1/s measured at 25 °C

2.4 Printing quality

A comprehensive analysis of the droplets and print view was carried out. Droplet analysis for the P-644 ink (Figure 4) illustrates that droplets fly toward the substrate with similar speed (2.85 – 3.10 m/s) and flying angle (represents deviation from the perpendicular direction to the substrate) of -0.56 – 4.48°. Moreover, the volume of all flying droplets is close to the nominal size of the nozzles. These properties of flying droplets ensure a homogenous deposition of the ink over the substrate.

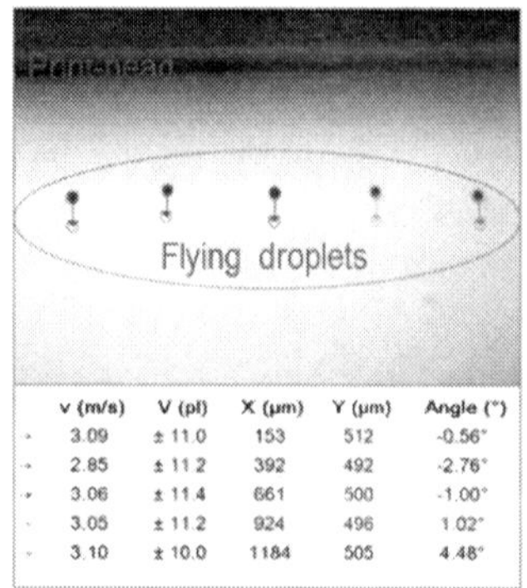

v (m/s)	V (pl)	X (µm)	Y (µm)	Angle (°)
3.09	± 11.0	153	512	-0.56°
2.85	± 11.2	392	492	-2.76°
3.06	± 11.4	661	500	-1.00°
3.05	± 11.2	924	496	1.02°
3.10	± 10.0	1184	505	4.48°

Figure 4: Droplet analysis for P-644

Dots of the ink were inkjet printed on a smooth surface of a shiny etched Si wafer with a printing resolution of 100 dpi in both printing directions using a 10 pl printhead. The dimeter of printed dots was measured to be about ~33 microns (Figure 6). This size is already suitable for application in different solar cell structures, however, it can be reduced by tunning printing parameters or the printhead size (e.g. 1 pl).

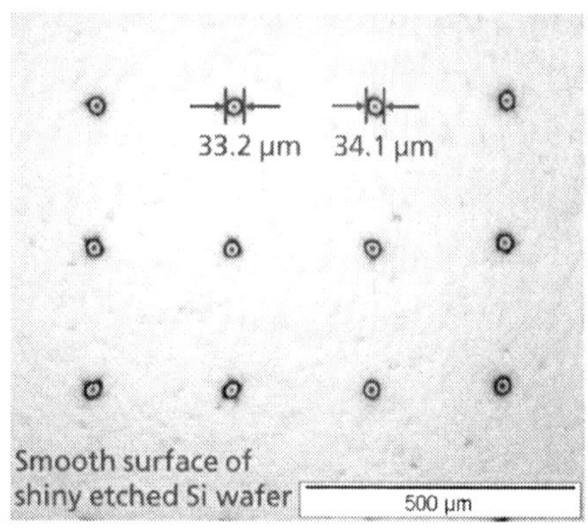

Figure 5: Inkjet printing on textured surface

2.5 Evaluation of electrical properties at device level

The entire spectrum of electrical parameters was investigated. The control sample used has a similar architecture with the experimental samples, the only difference is the phosphorous layer which was applied by ion implantation.
Upon performing I-V sweeps at 1.5 AM spectrum, the main electrical parameters of the experimental samples were extracted and compared against the control.

2.5.1 Photoluminescence Imaging

After the device metallization step, PL images of the control and experimental cells were collected for comparison.

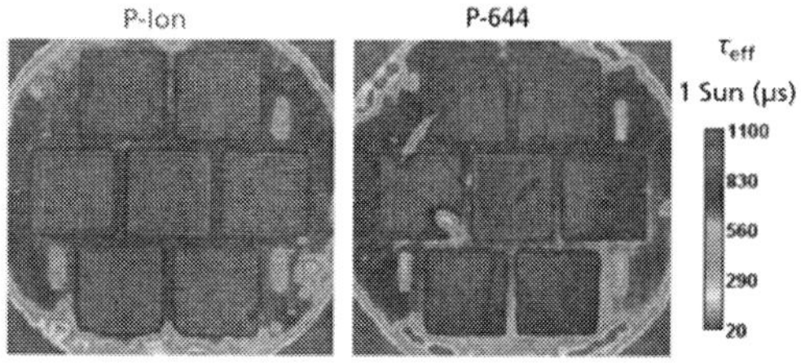

Figure 6: PL images of the control and experimental sample

In Figure 6 the charge carrier lifetime measurements of the control (in the left) and experimental sample (on the right) can be seen. Both samples are very similar in terms of lifetime. The experimental sample saw more manual handling and displays some inhomogeneities. This is likely to be reduced by automated fabrication.

2.5.2 Open Circuit Voltage

The open circuit voltage parameter was extracted and evaluated for both, control and experimental sample and the plotted results can be seen in the Figure 7 below.

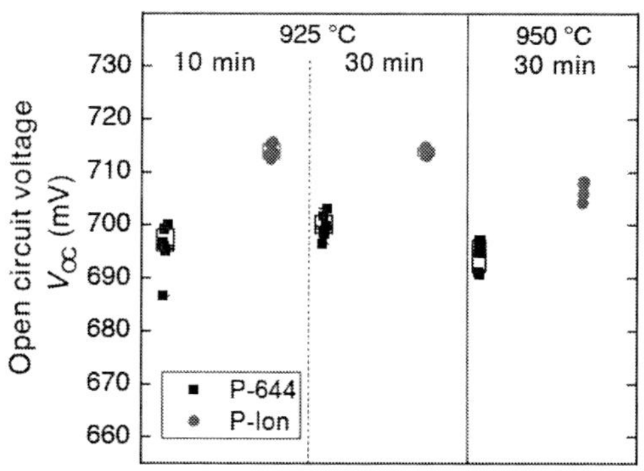

Figure 7: Open circuit voltage for control (red circle) and experimental sample (black square)

The black (square) plot represents the Voc of experimental sample and the maximum recorded value was about 703 mV while the maximum Voc value recorded for the ion implanted sample (control) was about 714 mV. Therefore a high quality of the passivated contact was shown, yet there is still some room for further process optimization.

2.5.3 Fill Factor (FF)

Fill factor was calculated for both samples at various process intervals and the data is plotted in the graph below (Figure 8).

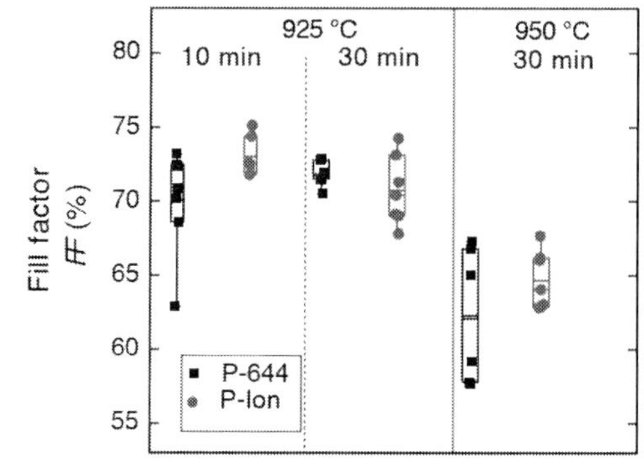

Figure 8: FF results for control (red circle) and experimental sample (black square)

The ion implanted sample (control) exhibits a maximum FF value of about 75 % while the experimental sample show a value of 73.2 % which is slightly lower. However, both boxes overlap largely, which shows a similar level of the FF.

2.5.4 Short Circuit Current

Short circuit current density in shown in Figure 10 below. Please note that all samples did not feature a front side texture, to simplify the experiments. There isn't much difference between the maximum current density measured at ion implanted device 36 mA/cm² compared to 35 mA/cm² for the experimental one.

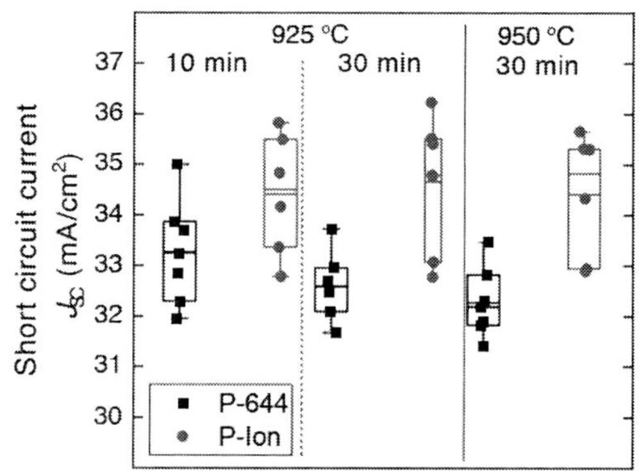

Figure 9: Current density for control (red circle) and experimental sample (black square)

As observed in the plot, the distribution of short circuit current density values is not uniformly distributed, with quite large standard deviation. This is most likely attributed to the non-optimized ITO layer.

2.5.5. Efficiency

All in all, the maximum efficiency obtained at the implanted sample was about 18.2 % while the one measured at the experimental sample was not too far off, at about 17.1 %.

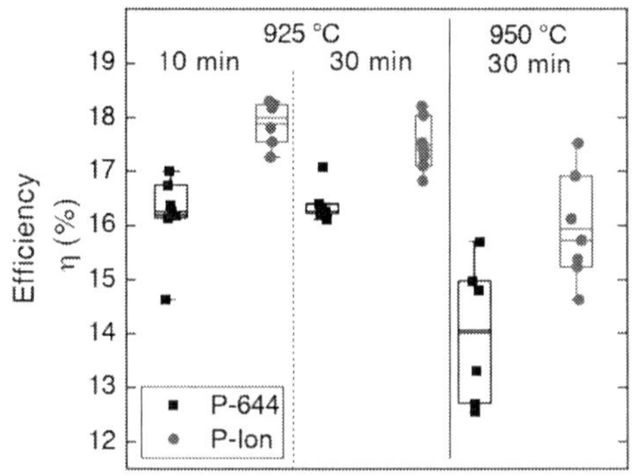

Figure 10: Efficiency

Standard deviation of the efficiency measured for experimental sample shows quite a good standard deviation as well at optimal annealing conditions.

3. CONCLUSIONS.

In this work a novel phosphorous doped ink formulation was evaluated. The viscosity of the ink was found to be 6.6 to 6.8 cP and the obtained minimum feature size was about ~33 microns. Using passivated contacts doped by the ink, a fairly high Voc of about 703 mV was obtained which makes the process quite promising for future solar cell applications. Testing some variations of the annealing process, some possible optimization paths were indicated.

Further work will include the optimization of the printing process such as evaluation of ink-specific correlation between substrate wetting and substrate temperature during printing for example, qualification of annealing profiles by establishing correlations with recombination, resistive losses, etc. A full spectrum of reliability tests will be considered as well where fully functional mini modules will be used to look for potential reliability issues that may arise. The findings resulting from this work will provide important information of how the new materials and processes will perform over time providing feedback for manufacturing process and bill of material optimization.

4. REFERENCES.

[1] David S. *et al.* "Inkjet technology for crystalline silicon photovoltaics", Adv. Mater. 2014.

[2] Jinchao S. *et al.* "Mass production and modeling of high efficiency n-PERT solar cells with ion implanted BSF/selective-BSF", Solar Energy 2017.

[3] https://doi.org/10.1016/j.solmat.2020.110926.

Acknowledgement/Cooperative Effort

Desert Silicon would like to thank and acknowledge the cooperative effort of Dr. Roman Keding and Dr. Jonas Huyeng of Fraunhofer, Germany for data and processing the solar wafers with Desert Silicon InkJet printable dopants.

FROM UPSCALING PERC TO THE NEXT TECHNOLOGY CYCLE: TRANSPARENT PASSIVATING CONTACTS MAY MERGE N- AND P-TYPE CELL TECHNOLOGY

Pietro P. Altermatt, Guanchao Xu, Xueling Zhang, Daming Chen, Yifeng Chen, Zhiqiang Feng

Trinasolar Inc, No 2 Trina Road, Xinbei District, Changzhou, Jiangsu Province, 213031 China

ABSTRACT: Recent progress in the Chinese PV industry will be explained, particularly at Trinasolar. An outlook on cell design leading to the next technology cycle is given. As urgent R&D topics are suggested: how thermal history and injection density affect LeTID degradation of Ga-doped cells; material and processing sequence to produce transparent passivating contacts, both for PERC and TOPCon cells and HJT cells; it is important that the cells and transparent passivating contacts are free of scarce elements such as indium, otherwise the cell type cannot contribute significantly to meeting the Paris Climate Agreement due to planetary boundaries; possibly replacing silver with copper or aluminum is the most urgent task ahead of PV scale-up and much needed CO_2 mitigation.

Keywords: silicon wafer cells, passivating contacts, capex, PERC, TOPCon, HJT

1 AVOIDING DISRUPTION

Figure 1 shows an update of the efficiency of the fastest selling modules (not the highest efficiencies) over the years. It shows that the efficiency jumps during the technology cycles. Such a jump can be expected for the next technology cycle, whether it is to TOPerc (a PERC with a passivating contact on the front), directly to TOPCon, or even by a shift of manufacturing tools to HJT. N the recent past, the technology cycle from standard to PERC was quite disruptive. Not only did it lead to a rapid decline in multi-Si material, but many old production lines were closed, even though standard and PERC processing are compatible. However, the technology cycle also brought new opportunities: many new production lines, better technology, more automation, and last but not least, much improved cleanliness. Predicting a feasible and likely roadmap for the coming technology cycle can align R&D and investment to mitigate the next disruption as much as possible.

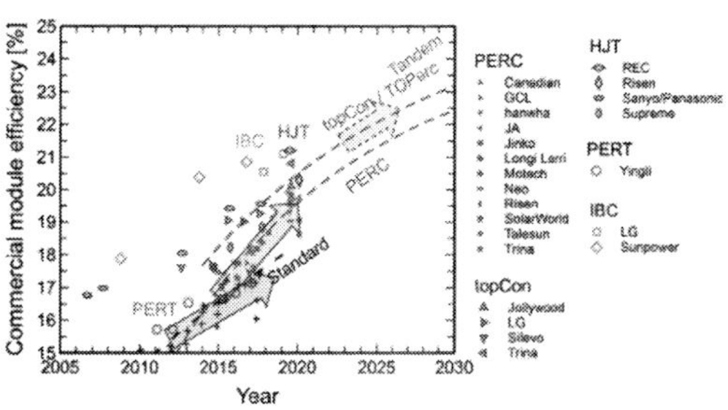

Figure 1: Efficiency of the best-selling modules with c-Si cells produced by the indicated companies (collected by the authors over the years). Dashed lines: least-squares fits of different cell types using the Goetzberger curve.

2 CELL MANUFACTURING

2.1 Ga-doped wafers

Ga-doped wafers took over B-doped ones swiftly. For two main reasons: better lifetime and less LeTID degradation.

Figure 2 shows an update of ingot excess scarier lifetime. Each color represents an ingot at different positions. The Ga-doped ingots (olive dots) match well with the n-type ingots (red dots). This does not mean that the advantage of n-type over p-type material disappears. With optimal cell design, the n-type base region needs less doping than the p-type base region, resulting in higher lifetimes in n-type cells. Closer inspection also shows that for continuous Ga-doped Si Cz pull, the resistance varies only by a factor of 2 along the ingot, which is why many manufacturers order ingots in the range of 0.5 - 1 Ωcm. And observation of the different shades of olive-green markers shows that the 3rd pull is just as good as the 1st pull.

Ga-doped Si does not show light-induced degradation (LID) compared to B-doped Si, because the recombination-enhanced reaction of the Ga_sO_2 complexes from the trapping configuration to a recombination active configuration is suppressed [1], most likely because less strain is induced in the Si lattice (the B atom is smaller than the Ga atom). Light- and elevated-temperature induced degradation (LeTID) occurs, but significantly weaker than in B-doped material [2]. However, it is not well understood yet how LeTID depends on the thermal history and on injection density [3, 4]. There is an urgent need for more research on this topic to give manufacturers strategies to avoid LeTID as much as possible.

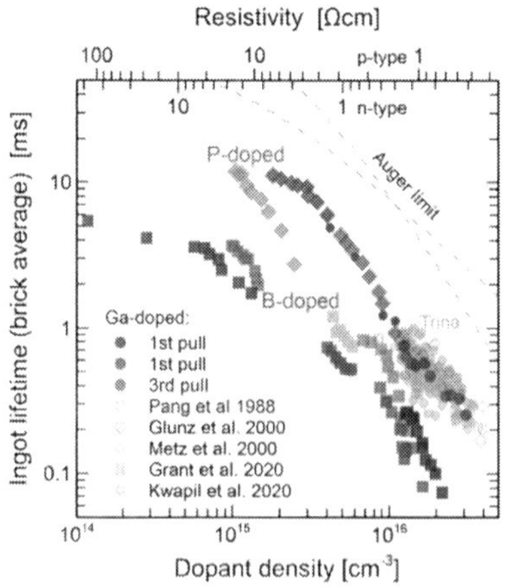

Figure 2: Ingot lifetimes (brick averages) in typical examples of boron- and phosphorus-doped Cz ingots from Longi, China, in 2017 (the boron-doped ingot below a dopant density of 2×10^{15} cm^{-3} was custom-made) and Ga-doped ingots (olive markings) from recent years from various manufacturers.

What are the maximum lifetimes that can be expected from Ga-doped material? Experiments with p-type float-zone material showed an effective lifetime τ_{eff} of about 4 ms at processing temperatures of about 300°C [5]. For comparison, we etched back finished PERC cells from mass production, re-passivated them, and measured the lifetime with a Sinton tester, as shown by crosses in Figure 3. In a Shockley-Read-Hall (SRH) analysis, we used interstitial iron (Fe_i) as the only defect type [6] and a saturation current of both surfaces of 10 fA/cm^2. The effective lifetimes can be reasonably fitted with an iron concentration of about 1.2×10^{10} cm^{-3}. This is reasonable considering that the new factories are very clean, and the ingots are very pure nowadays. This means that, at open-circuit voltage, an almost intrinsic bulk lifetime is achieved (about 3 ms), while at maximum power point it is about 1.2 ms. If more iron contamination is encountered, the lifetime at mpp decreases, lowering V_{mpp} and hence the fill factor FF. With even more iron contamination, V_{oc} also starts to decrease. Therefore, it is very important to keep the new fabs clean.

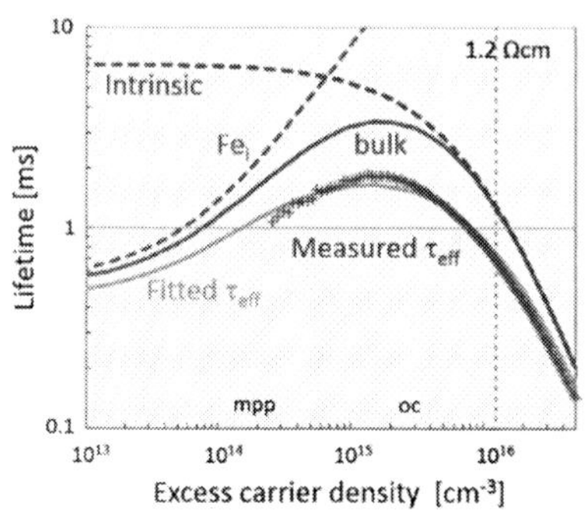

Figure 3: Measured effective lifetime of a mass-produced PERC cell (crosses), after etching the front and rear and re-passivating, and bulk lifetimes from a SRH analysis.

2.2 Mass production of PERC cells

The V_{oc} and efficiency values achieved in typical new factories today are shown in Figure 4. Here, we considered about 31,000 PERC cells produced in a 12-hour shift on a line (in China, work shifts usually go from 8 to 8). Considering the lifetime of the previous figure, the V_{oc} can be significantly improved only by reducing the losses on the front and back sides, not by improving the ingot lifetime, while the efficiency depends on the degree of iron contamination (among other factors such as shading, etc.). We took some of the best 210 mm (M12) sized cells from mass production and did not change anything on them (no triple anti-reflective coating, no annealing treatment or metallization change) and had them accredited by the National Center of Supervision and Inspection on Solar Photovoltaic Product Quality in Wuxi. The IV parameters of the best cell are shown in Table I. Currently, our cell is a world record of 23.56%, but since many manufacturers are close in terms of efficiency, another manufacturer may surpass this record at this conference. The main contributions to the high efficiencies are: (i) less chemical processing and more laser processing instead, as chemicals sometimes led to contamination in the past (high-purity chemicals are expensive to buy); (ii) all handling is done by robots (some cell manufacturers employ about 100 to 150 people per GW); (iii) and cleaner factories, making equipment almost as expensive as production lines.

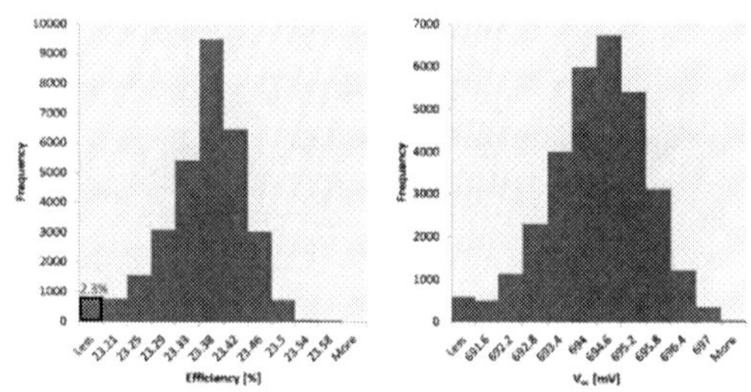

Figure 4: Histogram of efficiency and open-circuit voltage of 31'500 PERC cells, manufactured during a 12-hour work-shift.

Table I: IV and derived parameters of a PERC champion cell from mass production, 210 mm (M12) in size, with no additional features added to it.

J_{sc}	40.65	mA/cm^2
V_{oc}	695.1	mV
FF	83.38	%
Eff	23.56	%
R_s	0.181	Ωcm^2
pFF	84.42	%
pEff	23.85	%

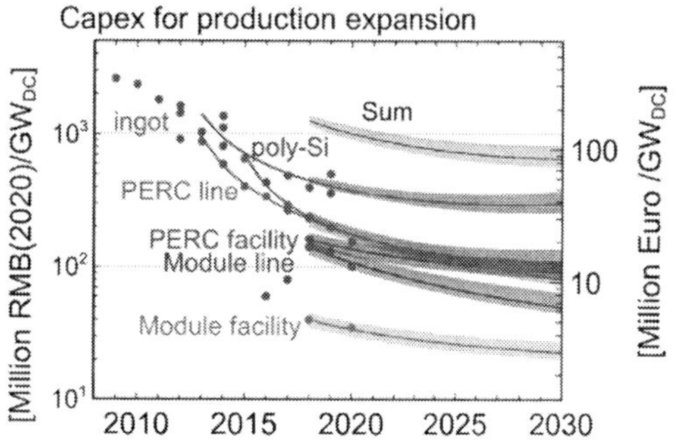

Figure 5: Capital expenditures (capex) for silicon PV production expansion. Historical data (symbols), collected by the authors mainly in China, show the steady and sharp decline fitted with a hyperbolic function to predict future trends (lines) and estimated uncertainties (areas).

Figure 5 shows that capital expenditures for PERC lines have fallen below 15 million € per GW, for module assembly lines below 10 million € per GW, and for the entire value chain from Si purification to finished module, including all equipment, below 100 million € per GW, all considered in China.

The data in Figure 5 has advanced so smoothly that they can be approximated and extrapolated with reasonable confidence. If Si PV fabrication capacities expand by about 20% per year, the required capex will be about 400 billion RMB, totaled from now to 2030. At today's currency exchange rates, this is equivalent to about US$ 55 billion or about 50 billion €. This may seem like an ambitious sum of money. However, a few big investors may secure a large market share in order to get returns once the industry's expansion slows in 10 to 15 years' time. Market share may be more important to investors than fabrication price. We therefore do not expect that upscaling the PV manufacturing capacities in Asia will be the bottleneck in enacting the Paris Climate Agreement.

2.3 PERC design limitations

To analyze the losses, we prefer to plot the recombination current versus bias voltage, as this gives a dynamic picture (see Figure 6). In the example shown here, recombination is active at the cell's edge at mpp, but recovered towards open-circuit condition (red line). This lowered the FF. Only when all recombination losses are lined up, as in Figure 7, is the usual bar graph of Figure 8 used (it is plotted in units of recombination current; if power dissipation is preferred, these currents can be multiplied by V_{mpp}). The bar chart clearly shows that for the usual PERC cells, recombination losses dominate at the front: at the metal contact, at the front surface, and Auger in the emitter dopant profile. The usual simultaneous optimization machinery, shown in Figure 9 (top panel), leaves little room for improvement. It is likely to be a tough job to exceed the average efficiency of 24% without making changes to the manufacturing technology, such as metallization.

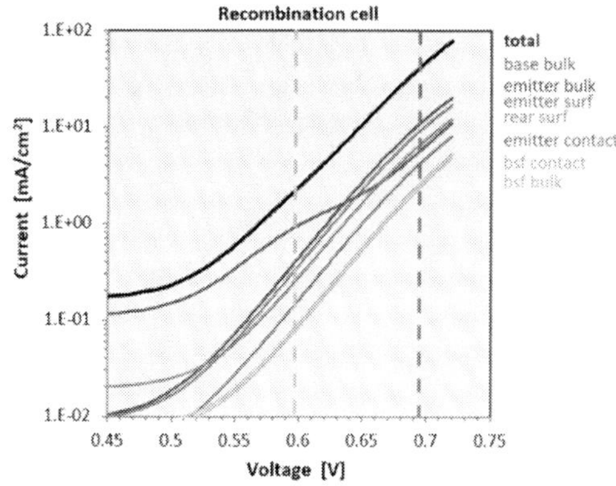

Figure 6: Plotting the amount of recombination in each device region as a function of bias reveals dynamics of perimeter recombination (red line).

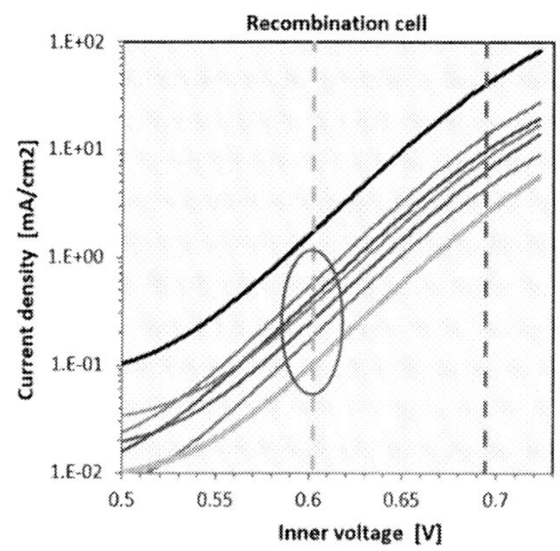

Figure 7: Recombination losses as a function of bias in an optimized PERC cell, with the same colors as in the previous figure.

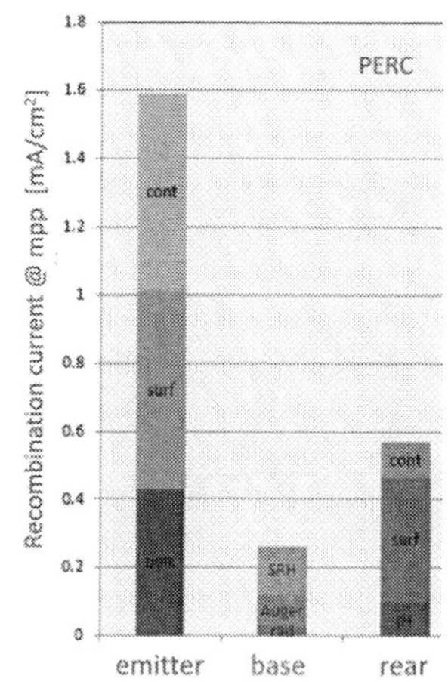

Figure 8: Same as in Figure 7, but depicted as bars at the maximum power point.

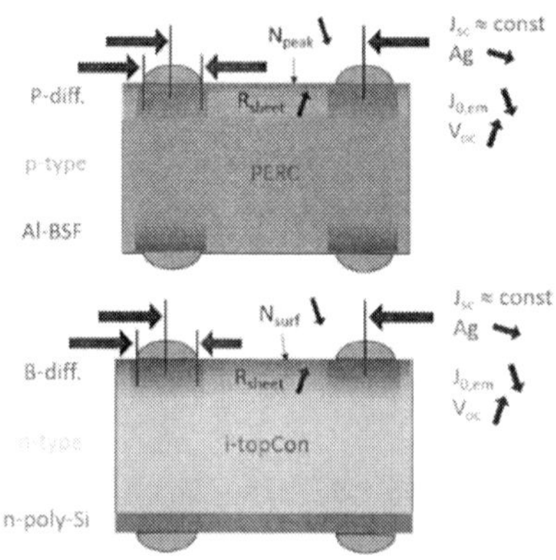

Figure 9: The usual mechanism for simultaneous optimization of the front of PERC and TOPCon cells is basically the same, but with a larger margin for TOPCon than for PERC.

We plot the resistance losses separately from the recombination losses because the resistance losses require different optimization strategies. For space reasons, we do not show them here. However, Figure 10 shows one aspect of resistance losses that manufacturers are very interested in. Does varying the lumped series resistance R_s cause many cells to fall into a lower bin during cell sorting for module assembly? Thanks to the Sinton Cell Flash Tester FCT-450 or FCT-750, we can select [7] a narrow range of measured pseudo efficiencies, pEff, in Figure 10, which means that the recombination losses at mpp are the same in all these selected cells. Plotting these values gives the standard deviation of Eff, which is smaller than a bin width, so that variations in R_s lead to a change in bin assignment in only a few cells.

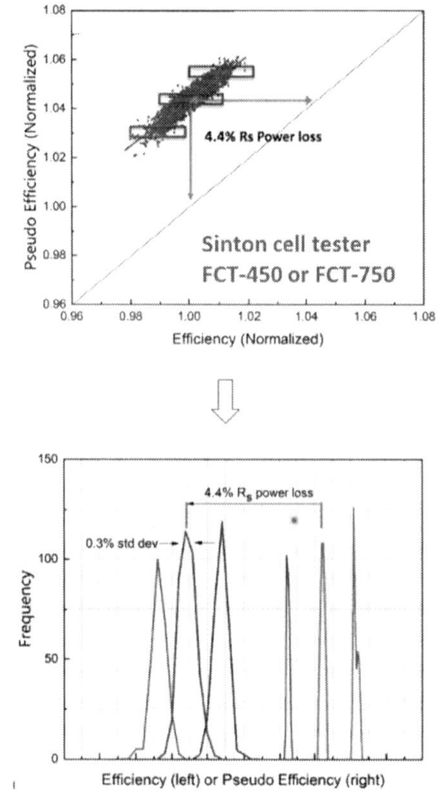

Figure 10: A method from Sinton Consulting to quantify the influence of variations in series resistance on bin placements after IV measurements.

2.4 Module assembly

Trinasolar went directly to 210-mm wafers (M12), the largest size possible with today's Cz pullers. Tool manufacturers adapted very quickly, and fewer problems than expected were encountered in cell and module production. At the end of cell manufacturing, the cells are cut in half and then the IV curves of both halves are measured in parallel. For highly efficient cells, the perimeter losses can be relatively large, so we use non-destructive cutting: A focused laser spot creates a compressive stress, while a de-focused laser spot creates a tensile stress before and after the focused laser spot at a temperature between 200°C and 300°C. This cleaves the cell without melting silicon, avoiding microcracks and producing very smooth cell edges. Nevertheless, FF is reduced, mainly due to inhomogeneities of recombination along the cut surface.

The efficiency of the modules in a designated area (excluding the edge area) reached 23.03% and was certified by TÜV Rheinland and TÜV Nord. This is a world record, but since some manufacturers are very close in terms of efficiency, this record could be surpassed by them at this conference. We routinely use measurements of the designated area to monitor our technical progress (the width of the module edge area is dictated by regulations and is therefore not optimized). As Figure 1 shows, the typical full-area module efficiency in the industry is between 21 and 22%.

2.5 TOPCon cell design

Figure 11 shows that TOPCon cells, similar to PERC cells, are also limited by recombination losses on the front surface. The same simultaneous optimization techniques can be applied as for PERC (see Figure 9), but since front metallization is an issue, there is more room for improvement than for PERC. Nevertheless, this is not a favorable situation. The acronym PERC stands for "passivated emitter *and* rear cell," which means that the main reason for the past technology cycle was that both surfaces were passivated in a similar way. As PERC efficiencies have increased, the front surface has fallen behind and now dominates the overall losses. The passing on of this disadvantage to TOPCon makes it difficult to initiate another technology cycle. Because TOPCon has a passivating contact on only one side, it is sandwiched between cheap PERC and HJT, which has passivating contacts on both sides.

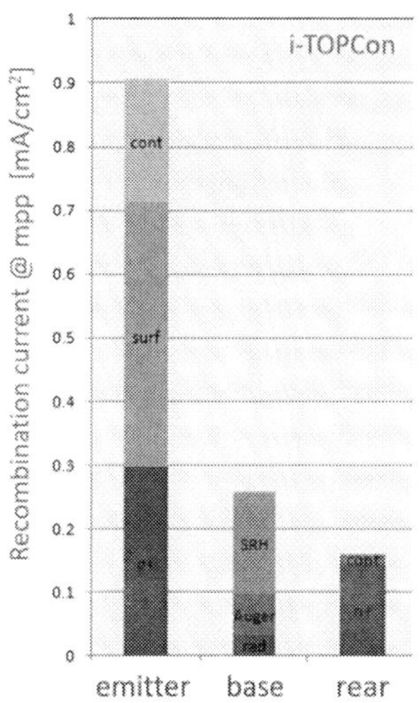

Figure 11: Recombination losses in typical TOPCon cells.

2.6 Benefits from a transparent passivating contact

The development of a transparent passivating contact can provide a significant advantage. Figure 12 shows that the transition to improved cell types would be smoother: Adding a transparent passivating contact to the front of a PERC would turn it into a TOPerc, and replacing the rear Al-BSF with a passivating contact would further increase efficiency. This would then be a p-type TOPCon. Finally, the switch to n-type wafers would be almost trivial, as only the phosphorus diffusion on the front would need to be adapted to its new role as an FSF instead of an emitter.

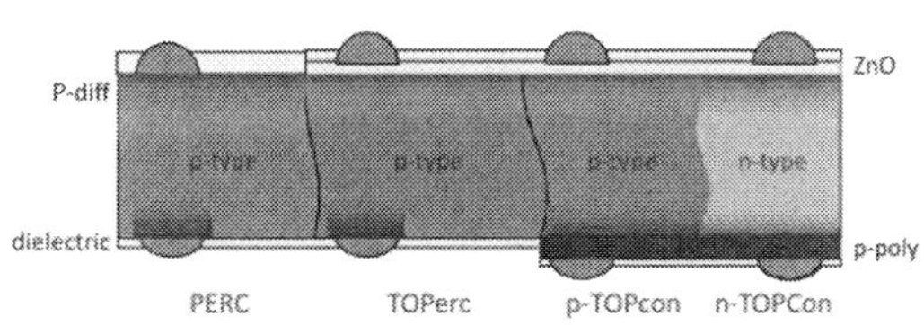

Figure12: Introducing a transparent passivating contact into PERC mass production, and then a backside passivating contact would allow a relatively smooth transition from PERC to TOPCon. However, research and development for such a transparent passivating contact is urgently needed.

Such a transparent layer could be incorporated into mass production via PERC, as it would significantly increase efficiency, as quantified in Figure 13 using a realistic device model for mass production. The introduction of a new process into mass production is usually more expensive at the beginning and may lead to a lower yield for a while, so a significant increase in efficiency is needed.

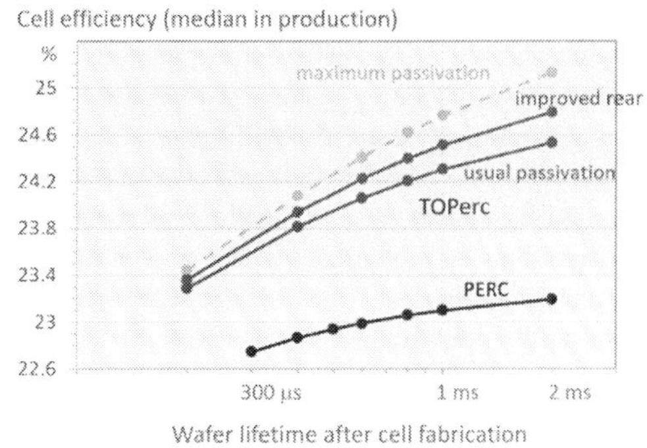

Figure 13: Realistically modeled efficiency as a function of wafer lifetime of PERC and TOPerc cells.

The preferable properties of such a transparent passivating contact are:

- It should be deposited onto the tunnel oxide like an anti-reflective layer.
- It should be low doped and therefore transparent, except around the front metal fingers where doping keeps the contact resistivity low, because
- the main lateral conduction takes place in the Si wafer.
- It should be able to be screen printed, but for plating it can act as a buffer for copper diffusion into the silicon, so no buffer layer needs to be plated.
- It must passivate the underlying tunnel oxide.
- It must not contain scarce elements to gain significant market share in a TW market.

There is a range of materials that can fulfill all these properties. ZnO does not contain a rare element and has been used in photovoltaics for a long time. However, it should be noted that there are two different applications for such transparent conductive coatings (TCO). TOPerc and TOPCon require high transparency but good passivation of the tunnel oxide. In HJT and most thin-film applications, the TCO layer does not need to passivate the interface with the silicon, but it must provide lateral conduction of electrons to the front metal fingers. According to the Kramers-Kronig relations [8] in fundamental optics, the higher the density of free charge carriers, the less transparent a medium is due to the absorption of free charge carriers. This can be better understood quantitatively by combining two equations [9]. First, the equation relating conductivity/resistance to carrier density N and mobility μ,

$$\rho = \frac{1}{\sigma} = \frac{1}{eN\mu}$$

Second, the equation relating the plasma frequency ω to the carrier density:

$$\omega_{ps}^{2} = \frac{e^{2}}{\varepsilon_0 \varepsilon_r m^{*}} N$$

where m* is the effective electron mass. A crucial parameter is the charge carrier mobility: the higher, the more transparent the medium is in the absorption region of silicon. But the higher the free carrier density required for lateral conduction, the lower the mobility due to Coulomb scattering, another fundamental phenomenon. These fundamental properties generally leave little room for optimizing TCOs.

2.7 Relationship between TOPCon and HJT

The two different applications of TCOs mentioned above point to a general difference between the HJT design and the TOPCon design. HJT went to high V_{oc}'s, which is a strong point because there is a lot of room for V_{oc} improvements from a general device physics perspective. HJT did this at the expense of J_{sc}. TOPCon has lower V_{oc} values because it relies on dopant diffusion, but has higher J_{sc} values where there is not much margin. Therefore, TOPCon requires a highly transparent passivating contact.

In Figure 14, these differences are considered from a general point of view: efficiency as a function of saturation current J_0. Efficiency increases exponentially as J_0 decreases, primarily due to increasing V_{oc} (if J_0 were plotted logarithmically, this curve would be a straight line). This is overlooked by some analysis methods that quantify the remaining losses up to the Shockley-Queisser limit. HJT cells have a very low J_0 value, they are in the very steep part of the red line in Figure 14, so any attempt to increase J_{sc} must not be too detrimental to J_0. Thus, it can be said that HJT cells are in a sense fixed to a high J_0 value. In contrast, TOPCon cells are in some sense fixed to a high J_{sc}. Attempts to deposit thin strips of p-poly under the front metal fingers, for example, lower J_{sc} too much and often do not increase V_{oc} enough to compensate, since J_0 is in a flatter part of the red curve in Figure 14.

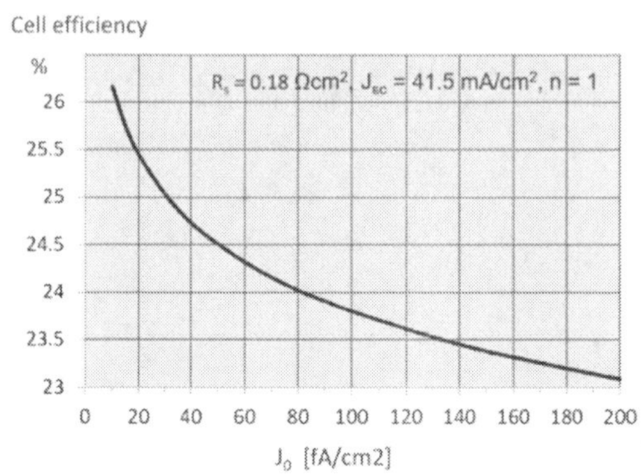

Figure 14: Cell efficiency as a function of saturation current J_0, computed with the one-diode model.

Currently, the Chinese PV tool industry is making great efforts to make HJT manufacturing tools cheaper. Currently, an HJT plant costs about 60 million € per GW (compared to 15 million € for PERC lines). If this capex decreases as steadily and steeply as it did for PERC (see Figure 5), it would take at least 5 to 7 years to approach PERC. Note that the current shortages of silicon and glass affect both PERC and HJT in a similar way. However, the capital cost of HJT could come down very sharply, as it is likely to be over-engineered at present, leaving room for leaner engineering. However, another aspect is that a-Si deposition of HJT cells must be area-based and cannot be boat-based, so the cost reduction due to larger wafers cannot be fully realized, wafer handling is more complex, and throughput may remain lower than PERC and TOPCon. HJT requires diverse gasses, TOPCon diverse chemistry. Another issue is module assembly, where the temperature must remain below 190°C, which may require more silver and/or bismuth. When weighing the advantages and disadvantages of HJT and TOPCon, it is difficult to predict which aspects will have the greatest impact on their market share.

ZnO or another TCO may be introduced into HJT as well to make HJT free of indium. This is necessary to gain significant market share in TW deployment of PV (indium is mainly used by the electronics industry, which can pay much higher prices for it than PV). Additionally, imagine that J_{sc} could be increased, e.g., by very narrow front metal fingers close together, so that lateral conduction of electrons can occur mainly in the silicon wafer. In this way, the advantages of the SHJ concept could be fully exploited.

Considering all these aspects, a transparent passivating contact would indeed support all these routes of cell design.

3 NEED FOR R&D

As outlined above, it is not well understood how LeTID of Ga-doped cells depends on thermal history and injection density. There is an urgent need for further research on this topic to provide manufacturers with strategies to avoid LeTID as much as possible.

There is an urgent need for a material and a process to manufacture transparent passivating contacts, for PERC and TOPCon cells as well as for HJT cells.

It is important that the transparent passivating contact is free of scarce elements such as indium, otherwise the cell type cannot contribute significantly to meeting the Paris climate agreement due to planetary boundaries.

At the conference, audience members of a panel discussion on single-layer cells were asked how much the use of silver will be reduced by 2030 in mainstream technologies. 59% of the participants voted for 50% to 75% and only 14% voted for more than 75%. Considering that PV must be ramped up to a few TW per year by then to meet the Paris Agreement, reducing silver use to half is far too little. Replacing silver with copper and/or aluminum is perhaps the most urgent task ahead of photovoltaics and the urgent need to reduce CO_2.

REFERENCES

[1] J. A. T. de Guzman *et al.*, "Minority carrier traps in Czochralski-grown p-type silicon crystals

doped with B, Al, Ga, or In impurity atoms," in 47th IEEE Photovoltaic Specialists Conference, (2020).

[2] N. E. Grant, J. R. Scowcroft, A. I. Pointon, M. Al-Amin, P. P. Altermatt, and J. D. Murphy, "Lifetime instabilities in gallium doped monocrystalline PERC silicon solar cells," Solar Energy Materials and Solar Cells 206 (2020), https://doi.org/10.1016/j.solmat.2019.110299.

[3] W. Kwapil, J. Dalke, R. Post, and T. Niewelt, "Influence of Dopant Elements on Degradation Phenomena in B- and Ga-Doped Czochralski-Grown Silicon," Solar RRL 2100147 (2021), https://onlinelibrary.wiley.com/doi/10.1002/solr.202100147

[4] N. E. Grant *et al.*, "Gallium-Doped Silicon for High-Efficiency Commercial Passivated Emitter and Rear Solar Cells," Solar RRL 5 (2021) 2000754, https://onlinelibrary.wiley.com/doi/10.1002/solr.202000754

[5] N. E. Grant, F. E. Rougieux, D. MacDonald, J. Bullock, and Y. Wan, "Grown-in defects limiting the bulk lifetime of p-type float-zone silicon wafers," Journal of Applied Physics 117 (2015) 055711, http://aip.scitation.org/doi/10.1063/1.4907804.

[6] F. E. Rougieux, C. Sun, and D. Macdonald, "Determining the charge states and capture mechanisms of defects in silicon through accurate recombination analyses: A review," Solar Energy Materials and Solar Cells 187 (2018) 263, https://doi.org/10.1016/j.solmat.2018.07.029

[7] K. Dapprich, R. Sinton, H. Wilterdink, W. Dobson, C. Sainsbury, and J. Dinger, "Using Doping Variation to Determine J_{0s} at I-V Test," in *Siliconpv2021*, 2021.

[8] P. P. Altermatt, "Optics for PV, 2.6 The Kramers-Kronig relations between n and k", PV Lighthouse,https://www.pvlighthouse.com.au/cms/lectures/altermatt/optics/kramers-kronig

[9] H. C. M. Knoops *et al.*, "Optical modeling of plasma-deposited ZnO films : Electron scattering at different length scales", Journal of Vacuum Science & Technology A 021509 (2015), https://doi.org/10.1116/1.4905086.

This is a short version of the paper submitted to the journal Solar RRL.
It has been selected by the Executive Committee of the EU PVSEC 2021 for submission to Solar RRL.

EXPLAINING THE EFFICIENCIES OF MASS-PRODUCED P-TYPE CZ-SI SOLAR CELLS BY INTERPRETABLE MACHINE LEARNING

S. Wasmer*, K. Hübener and B. Klöter
Hanwha Q CELLS GmbH, Sonnenallee 17-21, 06766 Bitterfeld-Wolfen, Germany
*Phone: +49-3494-6699-51224; e-mail: s.wasmer@q-cells.com

ABSTRACT: We show how at Q CELLS, interpretable machine learning algorithms are used to understand the energy conversion efficiencies of mass-produced *p*-type Cz-Si solar cells. For this end, we use the data of a one week ramp-up of over half a million cells acquired utilizing a single-wafer tracking system. These consist of over 300 single information per cell and feature inline measurements, path- and tool related information, process data as well as the final *I-V* characteristics. In the analysis using state-of-the-art machine learning algorithms, we focus on understanding how the many different input features influence the solar cell efficiency over time by using additive feature impacts. As sources of variation are spread over several features, we pay special attention to correlated features in a hierarchical clustering approach. Finally, after ensuring the model validity at the hands of a hold out test data set (Pearson's coefficient of correlation of 0.84 between true and predicted values), subtle hourly cell efficiency fluctuations in the order of 0.01%$_{abs}$ are explained. We identify features which are relevant for short-term changes in the efficiency and other which influence the efficiency on a general level and do not show temporal changes.
Keywords: PERC, silicon solar cell, manufacturing and processing, machine learning

1 INTRODUCTION

Increasing digitalisation and accompanying data acquisition paves the way into industry for modern machine learning algorithms. Especially the photovoltaics industry is predestined for applying these, as every product is assigned a unique quality related parameter at the end of its production cycle, i.e. for solar cell production its energy conversion efficiency. Hence, the problem definition and model building is rather easy given enough predictive information of each solar cell is acquired during the process chain.

However, a well-fit model alone is not helpful in understanding the cell efficiencies and their time trends, as the learned relations are rather hidden in a black box surrounding the algorithm. This issue will be addressed here in the following by applying additive feature impacts, sample wise assigning each input feature a value that drives the model away from the mean. As we encounter that features and hence their impacts can be correlated, thus distributing source of variation among many inputs, a hierarchical feature agglomeration is proposed, grouping the impacts for a fair comparison.

This work is distinct from the previous works [1-5] as the approach does not need numerical device simulations and fully utilizes the data of our manufacturing execution system. Furthermore, we improve on Reference [6] as we utilize more data (sample- and feature wise) and introduce a mitigation of the issue of highly correlated features by a hierarchical clustering approach. Also, by investigating the time trends of solar cell efficiency and its impacts, we explicitly emphasize the differentiation between features which are relevant for short-term changes in the efficiency and others which influence the efficiency on a general level and do not show temporal changes.

2 APPROACH

2.1 Data acquisition and model

We use the data of over 500.000 cells based on *p*-type Cz-Si wafers of a one week ramp-up acquired utilizing a single-wafer tracking system within one of our production lines. These data consist of inline measurements, path- and tool-related information, process data as well as the final *I-V* characteristics. Since we want to explain the temporal evolution of cell efficiencies, all time related information such as logging times or virtual IDs are removed, resulting in a total of 329 single features per cell.

The model is then trained on 80% of the data using as target variable the standardized (zero mean, unit variance) solar cell conversion efficiency $\eta_{\mathrm{cell,std}}$ and tested on the remaining 20% unseen data.

2.2 Additive feature impacts

Since we do not focus on improving the model accuracy further and further, but rather on understanding what the model has learned, we use additive feature impacts. These explain a model prediction by assigning for each sample and input feature i a distinct value driving the model away from the mean: $y_{\mathrm{pred}} = y_0 + \sum \Delta y_i$, where y_{pred} is the model prediction, y_0 the expected value or the mean of the model and the Δy_i are the impacts depending on the input feature values. They basically average out all other influences and hence provide the impact of a single feature alone, uncovering non-linear relations as well as interactions.

2.3 Hierarchical clustering

We find in our analysis that the aforementioned features and hence their impacts can be highly correlated, possibly distributing sources of variation among many input features, examples of the many different relations can be seen in Figure 1. In order to alleviate this, we propose to group features according to their similarity with respect to their impact (as motivated by Figure 2) using hierarchical clustering [8], iteratively merging the most similar features or clusters, respectively. We then only analyse the impacts of the formed feature clusters, which here contain between one and up to 11 features.

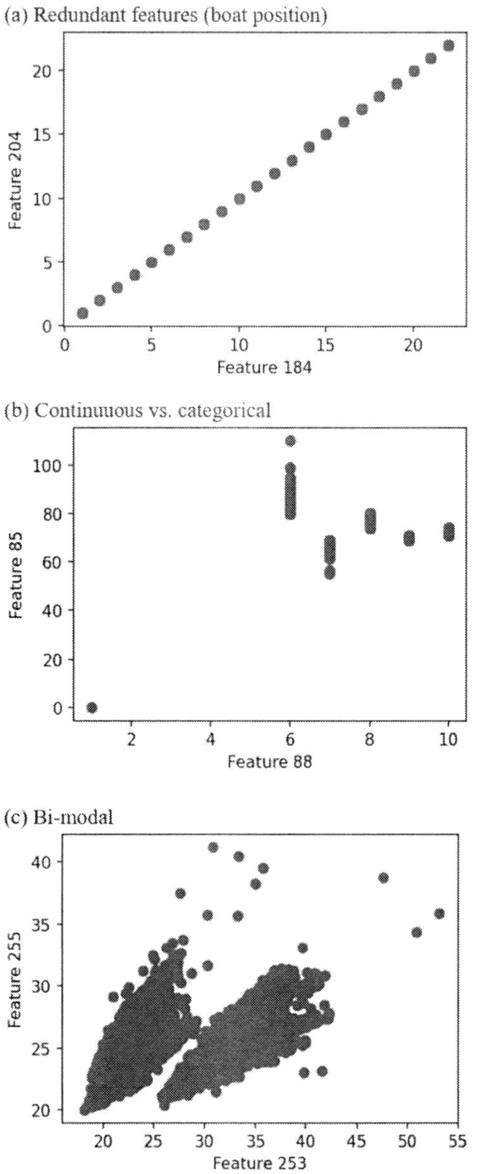

Figure 1: Examples of inter-dependent features, it is hardly possible to find a measure of correlation that can be applied to all relations:
(a) Redundant features (boat position).
(b) Continuuous vs. categorical.
(c) Bi-modal.

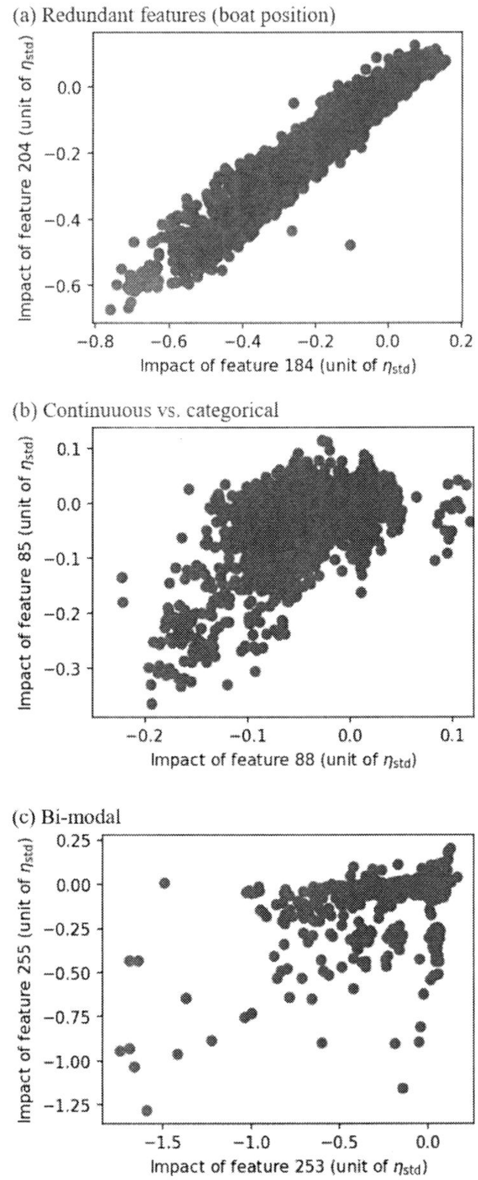

Figure 2: Same scatter plots as in Figure 1, but using the impacts of the features instead of the raw values. For each, a continuous distribution can be seen, and the usual measures of correlation can be applied to define a distance between the features.

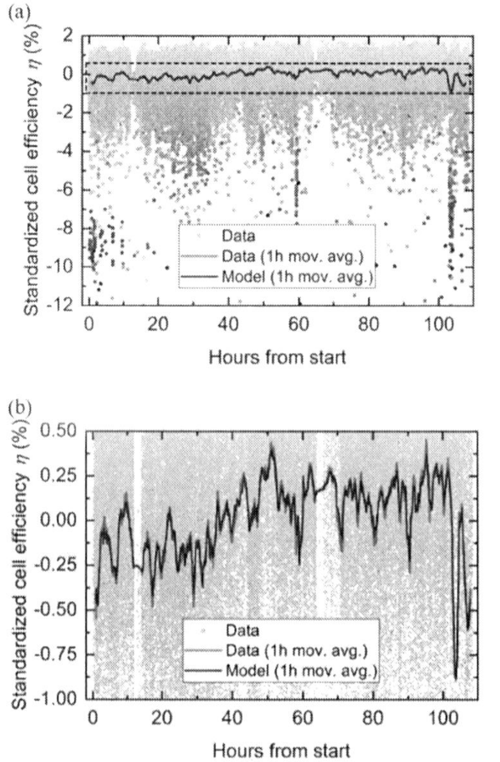

Figure 3: **(a)** Time resolved development of the standardized cell efficiency during the ramp up for the hold out test set (dots) as well as their 1 h moving averages and those of the model predictions (lines). As greyscale color bar serve the model predictions in the same range as the raw data from black to white, showing that even the really poor performing cells are explained by the model.
(b) Zoomed in plot showing the excellent model accuracy on an hourly basis.

3 RESULTS

The temporal evolution of the standardized cell efficiencies can be seen in Figure 3(a) & (b). In summary, the model fares quite well on the unseen data with a Pearson coefficient of correlation of 0.84 and a mean absolute error of 0.33 between true and predicted values. The high prediction accuracy is especially remarkable as all time-related quantities are removed beforehand and as there is only little material-related influence (compare e.g. References [3, 7] for the huge material-related impacts in mc-Si solar cells). As can be seen from the greyscale color bar of the dots showing the model prediction, even the really poor performing cells are explained by the model.

Figure 4(a) shows the Pareto chart of the 20 most important impacts. We highlight the necessity of grouping features for a fair comparison of the impacts, as the chart is dominated by clusters. In Figure 4(b), the temporal evolutions of the most important impacts on $\eta_{\text{cell,std}}$ regarding time trends are shown. It can be clearly seen that the subtle ups and downs of Figure 3(b) can often be assigned to only a few quantities or processes,

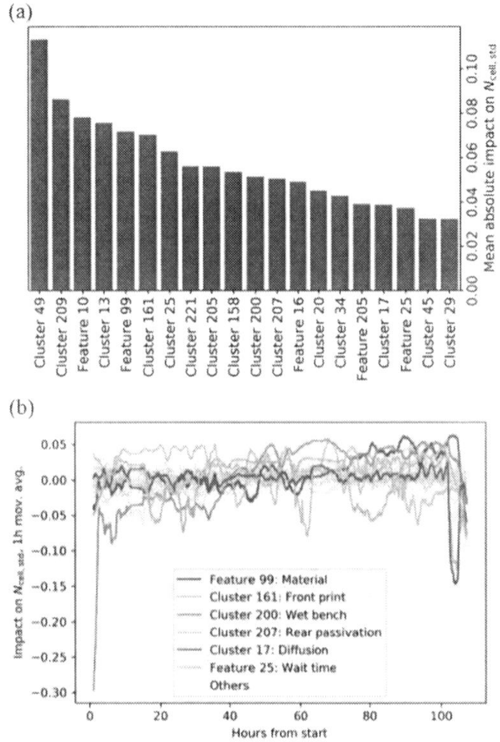

Figure 4: **(a)** Mean absolute impact of the 20 most important impacts.
(b) Time resolved plot of the most important impacts (regarding time trends) on the standardized cell efficiency.

and especially we can hold responsible a wet bench process for the general trend of cell efficiencies.

Interestingly, the top most "Cluster 49" of Figure 4(a) does not show up prominently here, meaning that its high impact is distributed rather homogeneously over time. We also find that a feature relating the wait time between two processes had a negative influence on the cell efficiency during the time period after around 70 hours with a low cycle time of the production line, helping us to determine the critical inter-process wait times.

4 CONCLUSION

Concluding, we showed the valuable benefit of using interpretable machine learning algorithms for explaining the efficiencies of mass-produced *p*-type Cz-Si solar cells including their time trends. As for the model explanation, care has to be taken regarding correlated features, which we addressed with a hierarchical clustering approach. The shown methods help to optimize a running production or to quickly troubleshoot, e.g. during a ramp up of production of new cell technologies such as ones using passivating contacts. They are also applicable to understand off-spec cells featuring for instance hot spots or optical issues.

ACKNOWLEDGEMENT

The authors would like to thank the R&D and production staff at Hanwha Q CELLS for data acquisition.

REFERENCES

[1] R. Evans and M. Boreland, "Multivariate Data Analytics in PV Manufacturing—Four Case Studies Using Manufacturing Datasets," IEEE Journal of Photovoltaics, 8 (1), 38-47 (2018).
[2] P. P. Altermatt, Y. Yang, Y. Sheng, D. Chen, Y. Chen, Z. Feng, and P. J. Verlinden, "A method for optimizing PERC cells in industrial production lines using final IV parameters, statistical procedures and numerical device modeling", AIP Conference Proceedings, 1999 (1), 110001 (2018).
[3] S. Wasmer, J. Greulich, H. Höffler, N. Wöhrle, M. Demant, F. Fertig, and S. Rein, "Impact of material and process variations on the distribution of multicrystalline silicon PERC cell efficiencies", IEEE Journal of Photovoltaics, 7 (1), 118-128 (2017).
[4] H. Wagner-Mohnsen and P. P. Altermatt, "A Combined Numerical Modeling and Machine Learning Approach for Optimization of Mass-Produced Industrial Solar Cells", IEEE Journal of Photovoltaics, 10 (5), 1441-1447 (2020).
[5] Y. Buratti, C. Eijkens, and Z. Hameiri, "Optimization of Solar Cell Production Lines Using Neural Networks and Genetic Algorithms", ACS Applied Energy Materials, 3.11, 10317 10322 (2020).
[6] S. Wasmer and B. Klöter, "Interpretable Machine Learning for Production Optimization", Proc. 36th European Photovoltaic Solar Energy Conference and Exhibition, 272 274 (2019).
[7] M. Demant, S. Rein, J. Haunschild, T. Strauch, H. Höffler, J. Broisch, S. Wasmer, K. Sunder, O. Anspach, and T. Brox , "Inline quality rating of multi-crystalline wafers based on photoluminescence images", Progress in Photovoltaics: Research and Applications, 24 (12), 1533-1546 (2016).
[8] D. Müllner, "Modern hierarchical, agglomerative clustering algorithms", https://arxiv.org/abs/1109.2378v1 (2011).

PRECISE AND LOW-COST I-V CURVE MEASUREMENT OF INDUSTRIAL BUSBAR-LESS SILICON SOLAR CELLS BY USING FLEXIBLE SPRING SUSPENSION (FSS) PROBE BARS

K.Kamatani , H.Kitamura , H.Kojima , Y.Nakamichi , Y.Fujita , K.Shibamoto , S.Kojima
KOPEL (Kyoshin Electric Co ., Ltd.)
18 Shichijo-goshonouchi-nishimachi, Shimogyo-ku, Kyoto 600-8865, Japan

ABSTRACT: A novel contacting system with Flexible Spring Suspension (FSS) has been developed. The system is optimized for the current-voltage (I-V) curve measurement of busbar-less silicon solar cells and is also suitable for other structures such as multi-busbar and shingled solar cells. It is equipped with independent current and voltage contacts on the fingers, in order to achieve ideal 4-wire connection. The contacting structures of the FSS4 consist of multiple small reed-shaped strips, thereby realize more secure and reproducible contact to the fingers with low contact resistance than conventional pin probes. Excellent reproducibility of relative ±0.05% or better in the Isc, Voc, Pmax, and FF has been confirmed for 20 times of dynamic testing of a 6-inch busbar-less cell, which includes mounting/unmounting of the cell to the sample stage and connecting/removing of the FSS4 system to the cell. Undesirable damages to the fingers and cells during the measurements are minimized by the FSS4, because of its innovative structure where each strip has perpendicular contact to the fingers by monolithic plate spring suspentions, and also its line contact to the fingers. It also has an excellent mechanical durability of over 10 million times. In the actual production line conditions, reliability of 4 million times is expected, which enabling low-cost and reproducible production.

Keywords: Contact , Characterization , Solar Cell , Busbar less , Probe bar

1 INTRODUCTION

Recent efforts for high-efficiency and low-cost solar cells have led to various types of electrode structures of silicon solar cells, such as busbar-less, multi-busbar and shingling. Since the busbar-less cells have only thin finger electrodes of several tens of microns in width, their accurate current-voltage (I-V) curve measurements were not possible by using conventional contacting systems with multi pin probes. Efforts have been made to develop suitable contacting methods for the busbar-less cells [1]-[3]. We have recently developed a novel contacting system with Flexible Spring Suspension (FSS) [1], which has enabled very reproducible and stable I-V curve measurements of the busbar-less cells. This study describes the features of the newest and most advanced development of FSS (FSS4) and investigates the precision and accuracy of the I-V curve measurements for mass production and R&D purposes. The excellent durability of the FSS4, which is of crucial importance for the production line, is also discussed [8].

2 FEATURES OF FSS4

The FSS4 is designed to realize precise and durable contacting system for the busbar-less solar cells with an innovative structure, which is based on many years of experience of the cell contacting systems [1] [4]. Its structure is optimized for busbar-less solar cells, and is also suitable for multi-busbar and shingled structures.

As shown in Fig. 1, the FSS4 is thinner (1.2 mm) than conventional pin-probe bars, thereby reducing its shadow on the cell during the measurement. It includes independent current and voltage probes. The current and voltage probes are insulated, and are separated by only 0.2 mm, in order to achieve ideal 4-wire connection to the finger electrodes. Each probe comprises of many reed-shape strips arranged in a straight line, each of which is supported by plate spring suspentions and contacts the finger electrodes. About 200 contacting strips are monolithically integrated in an FSS4 for 6-inch cells, and achieve flexible contact to the finger electrodes. Therefore, good contact is achieved for all types of the finger electrodes, and also for other structures such as multi-busbar and shingled solar cells. The number of the contacting strips is dependent on the size of the cell. A cell settter with twelve FSS4 is shown in Fig. 2.

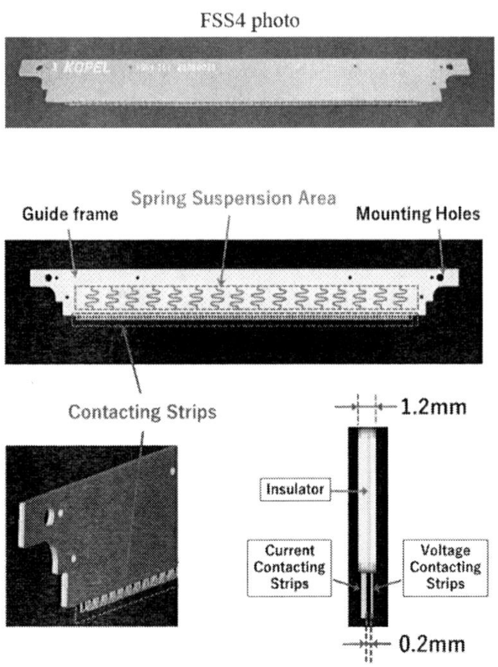

Fig. 1 Schematic structure of FSS4

Fig. 2 A cell setter with twelve FSS4 probe bars. KOPEL KSJ-200SP

The FSS4 has the following advantages over conventional contacting systems.

◆ Lower and more stable contact resistance between the contacting strips and the fingers than conventional pin probes are realized, because of their line contact with the finger and the arrangement in which each finger is contacted by two or more strips.

◆Ideal 4-wire connection is achieved for busbar-less and other kinds of cells. The voltage difference between the current and voltage probes are very small, because the probes are insulated, and make contact on very close positions, i.e., 0.2 mm apart, on the finger electrodes.

◆ Mechanical damages to the fingers and cells during the measurement are much lower than other contacting systems such as those with pin-probes, since each strip makes perpendicular and line contact to the fingers, which is supported by the monolithic plate spring suspention.

◆ The FSS4 has excellent reproducibility and durability in mass production conditions.

◆ The FSS 4 is compatible with conventional probe bars, which enables quick and low-cost replacement to the existing systems.

3 EXPERIMENTAL RESULTS AND DISCUSSION

3.1 Reproducibility and precision of I-V measurements

Examples of the I-V curve measurement results by using the FSS4 is shown in Fig. 4. A 6-inch busbar-less HJT solar cell was repeatedly measured by using twelve FSS4 mounted on KSJ-200SP cell setter (Fig.2) under IV test system KOPEL KSXC-2400RD (Fig.3). The measurements were conducted by using the PDA method [7] under 50 ms pulsed light. Figures 4 (a)-(e) show examples of the reproducibility of Isc, Voc, Pmax, FF and Rs of a 6-inch busbar-less HJT solar cell under repeated measurements by using the FSS4. The I-V curves were dynamically measured, i.e., by mounting and dismounting the cell, and contacting and removing the FSS4 for each I-V curve measurement. Good reproducibility better than ±0.05% for the FF and Pmax, ±0.02% for the Isc and Voc and ±0.012 mΩ for the Rs were obtained. These results confirm the good reproducibility of the contact of FSS4 on busbar-less solar cells. And the results also indicate that the mechanical damages to the finger electrode and cell by the FSS4 are very low even after 20 times repeated dynamic measurements, as well as the basic reproducibility of the FSS4.

Fig.3 IV Test System KOPEL KSXC-2400RD

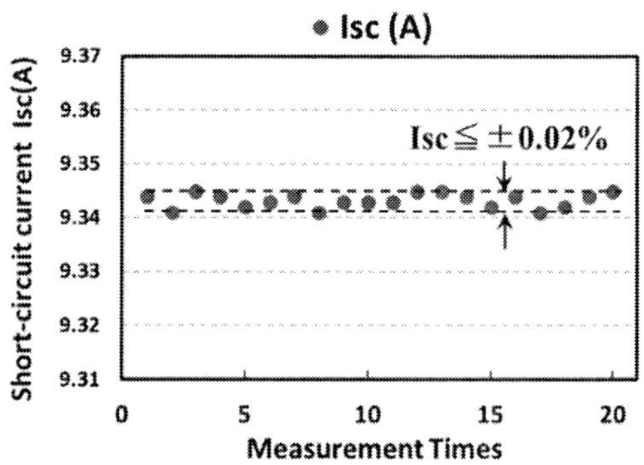

(a)

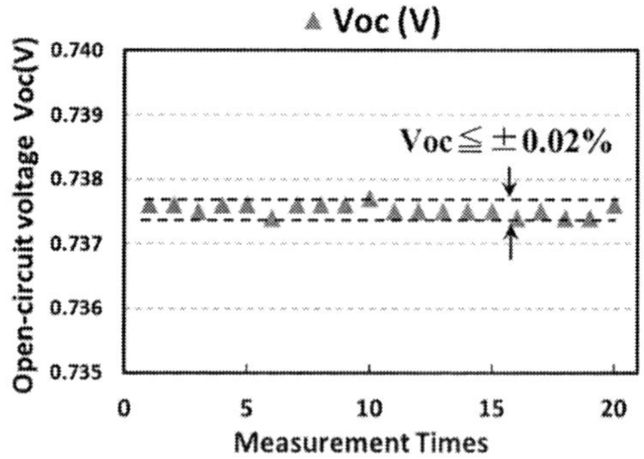

(b)

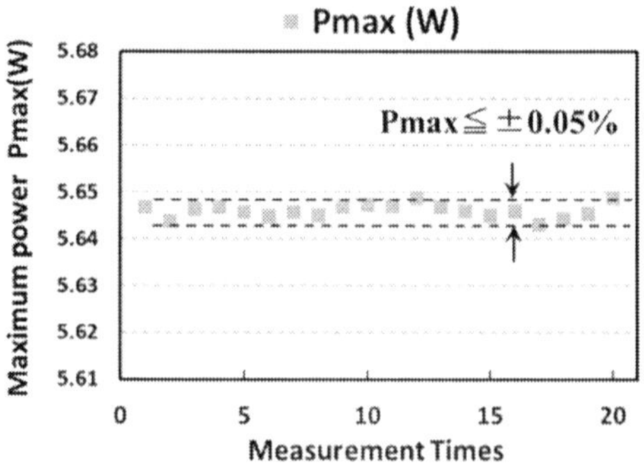

(c)

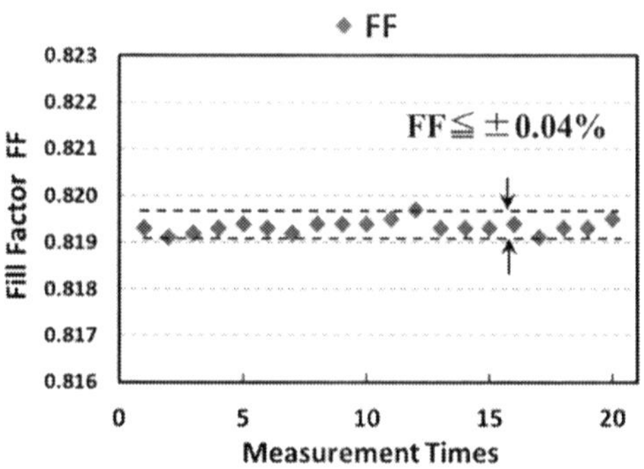

(d)

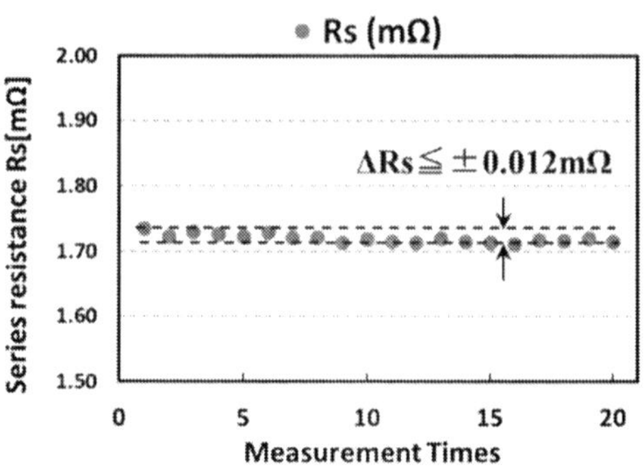

(e)

Fig.4 Examples of the reproducibility of Isc, Voc, Pmax, FF and Rs of a 6 inch busbar-less HJT solar cell under repeated measurements by using the FSS4.

3.2 Comparison with pin-probe

The applicability of FSS4 for conventional cells with busbars was also investigated. The I-V measurements of a cell with five busbars by using the FSS4 and pin probes were compared as shown in Fig. 5. The measurement results of Isc, Voc, Pmax and FF agreed within ±0.05%. This means that the FSS4 is also usable for the I-V measurements of conventional cells with busbars.

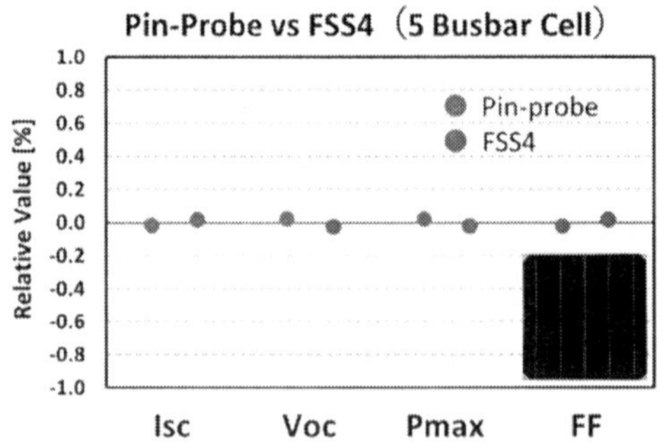

Fig. 5 Comparison of I-V measurement results by using FSS4 and pin-probe for a conventional cell with five busbars.

3.3 Durability in Mass Production Conditions

The FSS4 achieves excellent reproducibility and durability in mass production conditions, because of its unique design in which the contacting strips are supported by monolithically integrated plate springs. They expand and contract within the elasticity limit, and have no sliding parts in the current path. As shown in Fig.6, basic mechanical durability of over 10 million strokes has been confirmed for the previous type of FSS[1] is confirmed by in-house prolonged mechanical tests (1 mm stroke, 1 stroke per second). In actual mass production conditions, the FSS4 is designed to have comparable stability to the previous type of FSS, which has demonstrated stable operation for more than 4 million times. The outstanding durability is one of the major advantages of the FSS4 over conventional contacting systems.

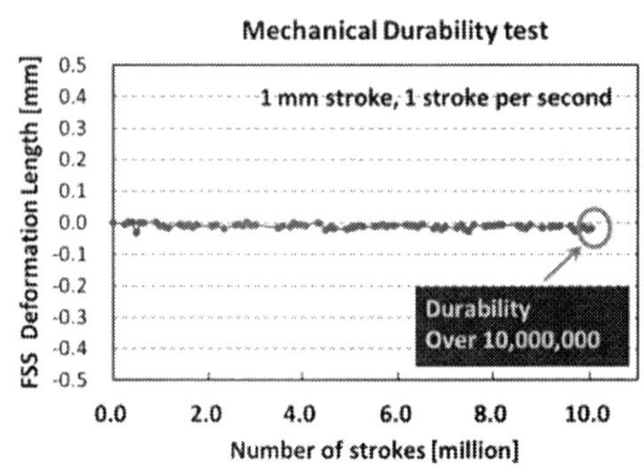

Fig.6 Mechanical durability test result of FSS

4 CONCLUSION

An advanced contacting system with Flexible Spring Suspensions (FSS4) has been developed for the I-V curve measurement of busbar-less, multi-busbar and shingled silicon solar cells. The FSS4 is composed of independent current and voltage probes and has unique features which enable reproducible 4-wire connection to the finger electrodes, while minimizing the damage to the fingers and cells during the operation. Good reproducibility better than ±0.05% for the FF and Pmax and better than ±0.02% for the Isc and Voc was confirmed for a dynamic tests, which include mounting/dismounting of the cell to the sample stage and contacting/removing of the FSS system. Excellent reproducibility of Rs better ±0.012mΩ has been also confirmed. The features are realized by an innovative structure in which multiple small reed-shaped strips contact the finger electrodes. Each strip has perpendicular contact to the fingers by a monolithic plate spring suspension, and also makes line contact to the fingers. The FSS4 is designed for excellent mechanical durability of over 10 million times, and for stable operation for more than 4 million times in actual mass production conditions. The present study focused on the measurement of busbar-less solar cells, for which the FSS4 is optimized. It is also suitable for other structures such as multi-busbar and shingled solar cells. The FSS4 has good compatibility with conventional pin-probe bars, which enables low-cost replacement to the existing measurement systems.

REFERENCES

[1] K. Kamatani et al., Proceedings of the 36th European Photovoltaic Solar Energy Conference and Exhibition, Brussels, 2019, 384-387

[2] V. Fakhfouri et Al., "Measurement accuracy and precision : the Grid TOUCH contacting system™", 2014, www.meyerburger.com

[3] K. Bothe et al. "Contacting of Busbarless Solar Cells for Accurate I-V Measurements", 37th European Photovoltaic Solar Energy Conference and Exhibition, online, 2020, 277-281

[4] patent pending

[5] C. N. Kruse et al. "Impact of contacting geometries when measuring fill factors of solar cell current voltage characteristics", IEEE J. Photovolt. 7-3 (2017) 747-754

[6] IEC60904-1:2020 " Photovoltaic devices - Part 1: Measurement of photovoltaic current-voltage characteristics"

[7] H.Kojima et al., "Accurate and Rapid Measurement of High-Capacitance PV Cells and Modules Using a Single Short Pulse Light", 2014 IEEE 40th Photovoltaic Specialist Conference (PVSC)

[8] https://youtu.be/k54wcsWWFBA

Effects of Plasma Etching on Dopant Compensation Between *p*- and *n*-type *poly*-Si Fingers in Passivated Interdigitated Back Contact Solar Cells

Matthew B. Hartenstein,[a,b] Steve Harvey,[b] William Nemeth,[b] Vincenzo LaSalvia,[b] Matthew Page,[b] David L. Young,[b] Sumit Agarwal*,[a] and Paul Stradins*[b]

***Abstract*— Efficiencies surpassing 26% have been achieved for single-junction Si solar cells based on the interdigitated back contact (IBC) architecture. However, the possibility of lateral shunting between the doped fingers has limited industrial adoption thus far. To avoid this possibility of shunt, complicated patterning techniques of the doped rear fingers have been developed, but spreading can still occur through multiple pathways during cell processing. Patterning can be simplified by using masked plasma-enhanced chemical vapor deposition (PECVD), but spreading during deposition will also lead to contamination of the isolation region between doped fingers. We significantly reduce the effect of this spreading through a short, gentle plasma etching step, which enables a trap-assisted compensation mechanism to take effect more easily than without the plasma etch. These two effects combined allow for the use of a simple patterning technique while still maintaining a highly resistive region between the doped fingers, and will reduce the complexity of IBC cell fabrication overall.**

***Keywords*— Silicon solar cell, passivating contacts, interdigitated back contact, simulations, atom probe**

I. Introduction

Solar cells based on the interdigitated back contact (IBC) solar cell architecture can reach high efficiencies > 26%, holding the current world records for single junction Si solar cells based on Si heterojunction (SHJ) [1] and polysilicon on silicon oxide (*poly*-Si/SiO_x) [2] passivating contacts. However, complicated processing required for rear finger patterning increases cost and reduces industrial feasibility. This patterning needs to be very precise to prevent shunt losses as an insulating region is required between the *p*- and *n*-type fingers (see Fig. 1). Unfortunately, this region can become contaminated during processing [3] due to dopant spreading. We suggest that a simple patterning technique using direct plasma deposition of doped hydrogenated amorphous silicon (*a*-Si:H) through shadow masks can reduce process complexity and thus, cost of IBC manufacturing. We focus specifically on *poly*-Si/SiO_x IBC cells (see Fig. 1) patterned with masked plasma-enhanced chemical vapor deposition (PECVD) of the doped fingers.

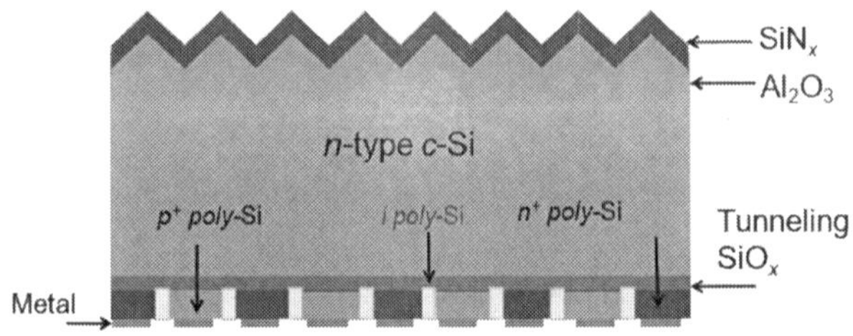

Fig. 1. Schematic of an IBC cell based on *poly*-Si/SiO_x passivating contacts with an insulating *poly*-Si region separating the *p*- and *n*-type fingers.

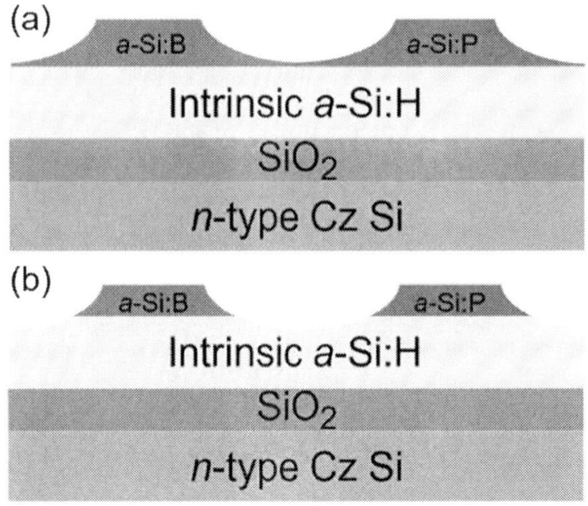

Fig. 2. Cross sectional depiction of adjacent *a*-Si:H fingers in an IBC structure prior to crystallization. (a) The as-deposited fingers have long tails of *doped a*-Si:H extending into the isolation region, resulting in contact of doped regions from opposing fingers. (b) Following plasma etching, the thin tails between the fingers have been removed and an insulating intrinsic region remains.

Though it is simple, this patterning process leads to contamination during the plasma deposition as the B- and P-containing radicals generated in the plasma spread through the thin gap between the shadow mask and the substrate. We show that this contamination can be mitigated through a short, uniform etch in SF_6 plasma following doped *a*-Si:H deposition. Alternatively, it is possible to clear this contamination through subtraction of the entire isolation region through chemical or plasma etching using the metal as a mask after metal deposition, but such processing leads to additional complications such as the requirement to repassivate the isolation region [4]. We measured lateral dopant profiles using secondary ion mass spectrometry (SIMS), predicted the resistance across the isolation region through simulations, and measured local spreading resistance between the interdigitated dopant fingers on samples which received no mitigation strategy. At the conference, we will show how this compares to samples which received the short plasma etch and the complete isolation region subtraction and repassivation.

II. Experiment and Simulation Details

Test structures were formed by first plasma depositing intrinsic *a*-Si:H onto *n*-type Czochralski (Cz) Si wafers insulated from the intrinsic film by a thick SiO_x layer. Shadow masks made by patterning ~200 μm thick *c*-Si wafers were used to pattern doped *n*- or *p*-type fingers deposited by PECVD. Following deposition, some samples underwent dry-plasma etching in an SF_6 plasma for 3-5 s to remove a few nanometers of the deposited *a*-Si:H films and reduce the extent of dopant spreading (see Fig. 2), followed by a crystallization anneal at 850 °C for all samples._The *poly*-Si films were hydrogen passivated with a ~15 nm Al_2O_3 layer grown by atomic layer deposited (ALD) followed by annealing in forming gas at 400 °C. Time-of-flight secondary ion mass spectrometry (TOF-SIMS) was used to obtain lateral dopant concentration profiles extending from doped finger edges. For electrical measurements, samples were metallized with thermally evaporated silver. Scanning spreading resistance microscopy (SSRM) performed to determine local resistivity across the region between doped fingers. Lateral dopant profiles measured with SIMS were used to predict the resistivity in the isolation region between the *p*- and *n*-type doped fingers from net doping concentration and an assumed grain boundary defect density. A more advanced diode model of the system accounting for

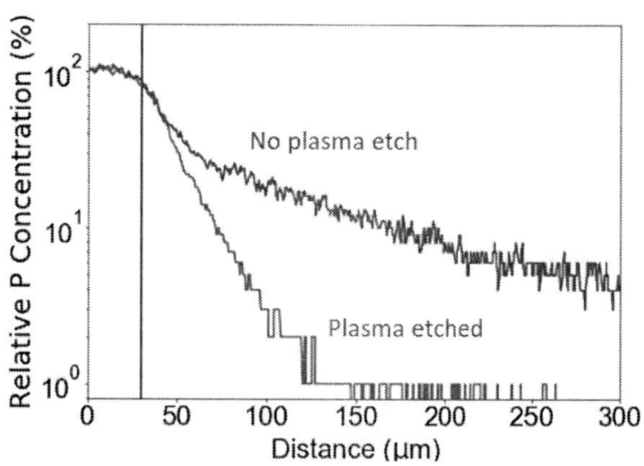

Fig. 3. Measured TOF-SIMS profiles of *n*-type *a*-Si:H tails extending past the PECVD mask edge (black). The blue curve represents the tail extending from an as-deposited doped *a*-Si:H IBC finger, while the red curve corresponds to a doped *a*-Si:H finger following a short SF_6 plasma etch. Profiles are normalized to their initial concentrations for simplicity.

This work was funded US DOE EERE under the PVRD program.

Poisson's equation, drift-diffusion currents of carriers, capture and emission of carriers by defects, and SRH and Auger recombination was used to simulate the current-voltage behavior of model systems [5].

III. Results and Discussion

A. *SIMS measurements confirm plasma etching reduces dopant contamination*

Figure 3 shows SIMS data for two phosphorus-doped tails extending from the edge of the IBC finger. The blue curve represents data from a tail which had received no plasma etching after deposition, while the tail shown by the red curve did receive a short plasma etch following doped *a*-Si:H deposition and quickly drops beneath the measurement threshold of SIMS. Plasma etching drastically reduces the level of contamination caused by dopant radical spreading during the plasma deposition through the shadow masks and

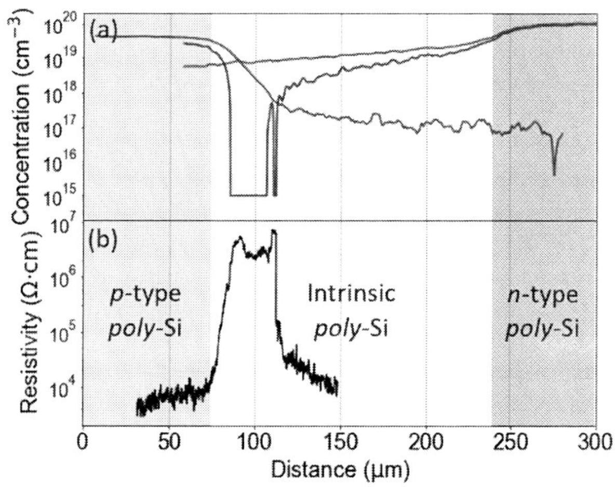

Fig. 4. (a) Overlapping lateral SIMS profiles of boron- (green) and phosphorus-doped (red) tails. The blue curve shows the calculated effective carrier concentration across this region. (b) Local resistivity of the *poly*-Si measured by SSRM. A wide region of high resistivity exists where the overlapping profiles compensate one another.

significantly steepens the rate at which the concentration drops over the distance past the mask edge. This confirms that by only removing a thin layer of a few nanometers of *a*-Si:H following deposition, the contamination can be nearly eliminated.

B. *Simulations predict much wider compensation region following plasma etching*

Figure 4a shows two lateral SIMS profiles of boron (green) and phosphorus (red) overlapping to replicate an IBC structure with a finger separation of ~175 μm. The blue curve shows the effective carrier concentration n_{eff} calculated by applying dopant activation proportions to each dopant concentration profile, determining net doping from compensation, and subtracting an assumed concentration of trap states. The dopant activation proportion accounts for the fact that not all dopant atoms contribute to carrier concentration and that SIMS only counts atomic

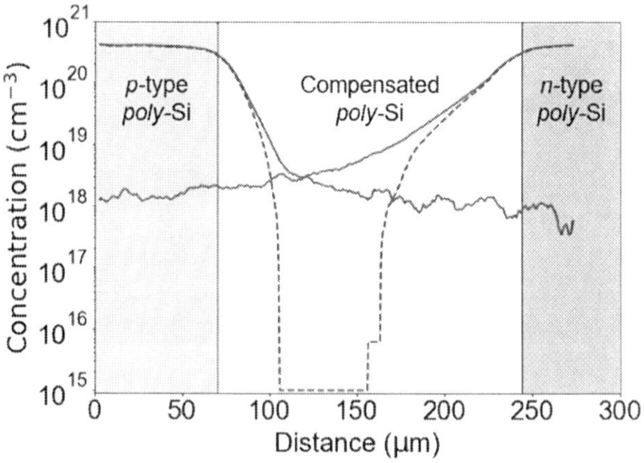

Fig. 5. Overlapping lateral SIMS profiles of boron-doped (green) and plasma-etched phosphorus-doped (red) tails. The dashed blue curve shows the calculated effective carrier concentration across this region.

concentration. Compensation between opposing dopant types results in a decreased net doping density across the overlap regions, but this effect is further enhanced by $\sim 10^{18}$ cm^{-3} defect states within the *poly*-Si, which trap charge carriers and prevent conduction [6]. Effective carrier concentrations $n_{eff} < 10^{15}$ cm^{-3} are shown as 10^{15} cm^{-3} for simplicity. This significantly reduces the conductivity of the *poly*-Si and creates a wide region of high resistivity.

The high resistivity is clearly shown in Fig. 4b, where spreading resistance measurements demonstrate a sharp increase in resistivity at the point when n_{eff} drops sharply. This compensation region of high resistivity is sufficiently wide to prevent shunting across the isolation region between doped fingers. Figure 5 shows a similar analysis as Figure 4a applied to overlapping profiles where the phosphorus-doped finger has been subjected to the SF_6 plasma etching. While the low n_{eff} region for the overlap without plasma etching is ~25 μm, this is much wider for that shown in Fig. 5. When a short plasma etch is applied to the doped *a*-Si:H fingers before crystallization, the width of the isolating compensation region is predicted to be ~60 μm.

Simulations using the diode model described in [5] were performed for pairs of dopant fingers of differing characteristic decay lengths (250 μm, 200 μm, and 167 μm for phosphorus and 143 μm, 125 μm, and 111 μm for boron. Figure 6 shows the simulated current-voltage plots of these systems. For systems with a low degree of spreading (i.e. smaller decay lengths), there is higher shunting current at low voltages, but decreased current at operating voltages of approximately 0.6V, confirming that decreasing the degree of dopant spreading, as the short plasma etch does, will reduce the shunting current between the doped fingers. This effect will occur to an even greater degree when the dopant concentration within the gap is decreased to a level beneath the concentration of trap defects, enabling the trap assisted compensation mechanism to take strong effect. While the isolation region widths studied here are much wider than those typically used in IBC cell fabrication, the concept could easily be applied to much narrower isolation region widths. This would allow for simple plasma deposition of doped fingers to be used as patterning method as opposed to complicated methods such as photolithography and laser processing, with short plasma etching as a quick and easy solution to the spreading that is inherent to masked plasma deposition.

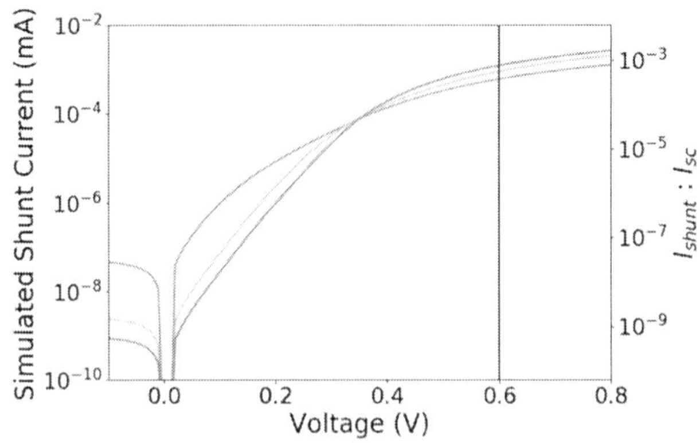

Fig. 6. Simulated current-voltage curves of model *p-i-n* diode systems with differing degrees of boron and phosphorus contamination within the intrinsic isolation region. Degree of dopant spreading increases from the orange curve to the green curve.

IV. CONCLUSIONS AND ACKNOWLEDGMENTS

We have measured dopant profiles spreading from plasma-deposited IBC fingers through SIMS. This isolation region contamination has shown to be significantly reduced by use of a short etch in in SF_6 plasma. Simulations predict the existence of a highly resistive region as a result of dopant compensation enhanced by charge carrier trapping at *poly*-Si defects and these predictions were confirmed through spreading resistance measurements Additionally, we predict that application of the plasma etching step will result in a significantly wider isolation region. We expect that these findings can lead to the application of simple masked plasma deposition to much more precise IBC finger patterning.

This research was supported by US DOE EERE under the Photovoltaic Research and Development (PVRD) program of the Solar Ener-gy Technology Office under Award Number DE-EE0007553 as well as US DOE EERE contract SETP DE-EE00030301 (SuNLaMP)This work was authored in part by the National Renewable Energy Laboratory, operated by Alliance for Sustainable Energy, LLC, for the U.S. Department of Energy (DOE) under Contract No. DE-AC36-08GO28308. Funding provided by the U.S. Department of Energy Of-fice of Energy Efficiency and Renewable Energy Solar Energy Technologies Office. The views expressed in the article do not necessari-ly represent the views of the DOE or the U.S. Government. The U.S. Government retains and the publisher, by accepting the article for publication, acknowledges that the U.S. Government retains a nonexclusive, paid-up, irrevocable, worldwide license to publish or re-produce the published form of this work, or allow others to do so, for U.S. Government purposes.

REFERENCES

[1] K. Yoshikawa *et al.*, "Exceeding conversion efficiency of 26% by heterojunction interdigitated back contact solar cell with thin film Si technology," *Solar Energy Materials and Solar Cells,* vol. 173, pp. 37-42, 2017/12/01/ 2017, doi: https://doi.org/10.1016/j.solmat.2017.06.024.

[2] C. Hollemann, F. Haase, S. Schäfer, J. Krügener, R. Brendel, and R. Peibst, "26.1%-efficient POLO-IBC cells: Quantification of electrical and optical loss mechanisms," *Progress in Photovoltaics: Research and Applications,* 2019, doi: 10.1002/pip.3098.

[3] M. B. Hartenstein *et al.*, "Isolating p- and n- Doped Fingers with Intrinsic poly-Si in Passivated Interdigitated Back Contact Silicon Solar Cells," *IEEE Journal of Photovoltaics,* 2020.

[4] C. Reichel, R. Müller, F. Feldmann, A. Richter, M. Hermle, and S. W. Glunz, "Influence of the transition region between p- and n-type polycrystalline silicon passivating contacts on the performance of interdigitated back contact silicon solar cells," (in English), *Journal of Applied Physics,* vol. 122, no. 18, Nov 14 2017, doi: 10.1063/1.5004331.

[5] M. B. Hartenstein *et al.*, "Trap-Assisted Dopant Compensation Prevents Shunting in Poly-Si Passivating Interdigitated Back Contact Silicon Solar Cells," *ACS Applied Energy Materials,* 2021/09/01 2021, doi: 10.1021/acsaem.1c01775.

[6] B. Lombos, "Deep levels in semiconductors," *Canadian Journal of Chemistry,* vol. 63, pp. 1666-1671, 02/05 2011, doi: 10.1139/v85-279.

EQUIPMENT DEVELOPMENT FOR SIMULTANEOUSLY BIFACIAL PLATING METALLIZATION IN BIFACIAL SOLAR CELLS

Yuan-Chih Chang[1], Sisi Wang[1], Rong Deng[1], Shaoyuan Li[1,2], Jingjia Ji[1]

[1]Shool of Photovoltaic and Renewable Engineering, the University of New South Wales, Sydney, NSW 2052, Australia

[2]Faculty of Metallurgical and Energy Engineering, Kunming University of Science and Technology, Kumming 650093, China

* Corresponding Author: Yuna-Chih Chang (E: y.chang@unsw.edu.au | M: +61 403 692 006)

ABSTRACT: For silicon solar cells, screen-printing has long been favored by the manufacturers due to its simplicity and high throughput. Over decades of technology advancement, screen-printing has almost pushed to its material and equipment limits (e.g. high-aspect ratio of finger, material impacts on cell etc.). However, in recent years, the increasing silver prices, reducing wafer thickness, and emerging new cell designs continue introducing challenges to this mature technology. Those issues have provided incentives to seek metallization alternatives. Among various metallization techniques, Ni/Cu plated contacts could be the most promising candidate due to its scaling up potential. However, there has been a delay of market share increase versus the ITRPV predication. The lack of readily available appropriate equipment and process standard could be the key reason to cause this. Here, we presented the developing UNSW bifacial plating technology. The special designed industrial prototype has greatly improved the plating uniformity and product stability. Moreover, double sides of the cell could be simultaneously plated thereby effectively reducing the processing time. In this paper, the design consideration and engineering solutions were plainly reviewed. The robust cell metallization process has been demonstrated on 6-inch industrial bifacial PERC precursors and primary intergraded with hydrogenation technology successfully. Despite that, this plating prototype could also be applied on other cell structures such as TOPCon, HJT and tandem structure. This innovative technology would remarkably optimize the processing engineering of plating metallization and may promote the utilization of plated contacts in solar cells manufacturing.

Keywords—plated contacts, bifacial plating, plating equipment

1. INTRODUCTION

Screen printing has long been the dominant metallization technology in PV industry due to its simplicity and high throughput. Over the last decade, the continuous refinement has made it become a quite mature technology. For example, the finger aspect ratio has been greatly improved and selective emitters (SE) technology has been well developed to resolve large surface "dead layer" issues. Besides screen printing, the design structure and other section in solar cells manufacturing are also continuously advancing, and some of them may cause processing difficulties when screen printing metallization is used. For example, silicon wafer is consistently becoming lighter and larger. Higher risk of wafer bowling/microcracks due to Ag/Al paste weight are expected. Moreover, cell design structures concepts with passivated contacts using tunnel oxide passivation stacks (i.e. TOPCon, HJT etc.) are getting popular. However, high temperature screen printing firing (~700-800° C) may destroy this ultrathin layer. Those challenges have provided more motivations to seek metallization alternatives[1]–[3].

Besides screen printing, there are several metallization techniques commonly used in PV. Physical Vapor Deposition (PVD) (e.g. sputtering, evaporation etc.) combined with lithography techniques are often applied in high efficiency record cells due to its precise controllability, but the complex processing sequence and high material waste make it less ideal for mass production. In contract, contacts formation via electroplating or/and Light Induced Plating (LIP) would be relatively simple and inexpensive, which may be more suitable[2]. In fact, plated contacts have been around in PV industry since BP solar use it in Buried Contact Solar Cells in 1980s and those cells even showed a better durability than screen printed solar cells after 20 years use[4]. However, the market share of plated contacts is remained low due to appropriate manufacturing equipment has yet been made ready[5]. Currently, large-scale PV plating generally adopts vertical plating method, which immersing whole wafer into the plating bath. Due to non-uniform conductivity, plated contacts have many problems such as peel-off at edges and dielectric layer stacks over etch. Although the adoption of sputtered/evaporated seed layer would help, this would also introduce large amounts of material waste, additional cost, and process complexity. In this paper, we will present a special designed inline plating equipment suitable for PV manufacturing. It not only improves the plating uniformity but also significantly reduces the plating time, which would highly benefit the development of plated contacts use in conventional solar cells.

2. EXPERIMENTAL DESIGN

2.1 Troubleshooting

The design of new plating equipment for PV manufacturing should not only resolve the problems existing in current techniques but also maintain or even enhance the processing fluency and contact performance. This require optimizing quite a few parameters together and then different design components would need to be carefully balanced. Therefore, we started from analyzing the reasons causing undesired plating results, then design new prototype tools based on improving the corresponding parameters first. This is to ensure stable contact properties to be reliably delivered. After that, tool tunning could be more focused on enhancing the contact/cell performance and product reliability.

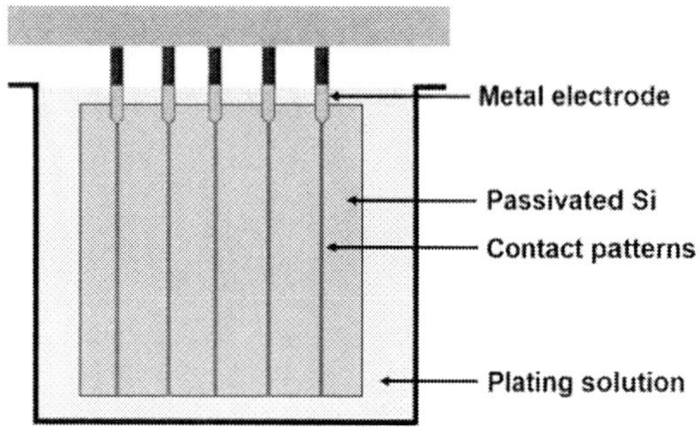

Figure 1: Scheme of typical vertical plating used in PV. There metal electrode only contacts few points of the exposed Si (contact patterns) and the whole wafer is immersed into the plating bath.

The key issues we aimed to resolve from vertical plating is the edge peeling off and potential over-etching of the passivation stack layers. Figure 1 shows the scheme of vertical plating setup, in which we can see that patterns wished to be plated are only points contacting with the electrodes. Since the metal electrodes are much more conductive than Si, the areas close to electrodes having a much faster plating deposition rate thereby resulting the peeling off issues. Moreover, non-uniform deposition rate would also delay the overall required plating time, which means wafers needing to be soaked in plating bath longer thereby increasing the over-etching risks. Therefore, it is required to improve the uniform deposition happening in everywhere needing to be plated on. To achieve this, we firstly need to increase the contact area between the electrode and exposed silicon patterns. However, it would be difficult to fast algin the contact patterns perfectly by adding more metal pins. Moreover, when vertical plating is used, the mechanical stress from both sides of the wafer may enhance the risk of wafer cracking, especially for very thin wafers. Furthermore, it was found that uniform plating is more easily achievable via one of our available plating tools from Aerotech, which adopts horizontal plating and square electrode to contact with the exposed Si. This suggested that increasing the electrode size and using horizontal plating should be the direction to work on. To further allow inline processing possible, line contacts via rollers and horizontal plating were chosen to use in the prototype equipment.

2.2 Test sample structure

Test sample used for plating equipment development in this study is a 6-inch square PREC structure precursor. As illustrated in Figure 2, it has p-type base with front side n-type emitter (120-150ohm/sq), textured front side coated with a ~80nm PECVD SiN_x ARC layer and non-texture rear side coated with a ~10nm PECVD alumina oxide (AlO_x) plus a ~130nm PECVD SiN_x passivation layer stacks. The PREC precursor firstly underwent a passivation layer firing via Camini belt furnace with a set peak temperature at 830°C. After firing, the implied open circuit voltage (iVoc) was within the range of 635±3 mV. Then laser doping was used to open the passivation stack layers and create contact patterns on both sides of the precursor.

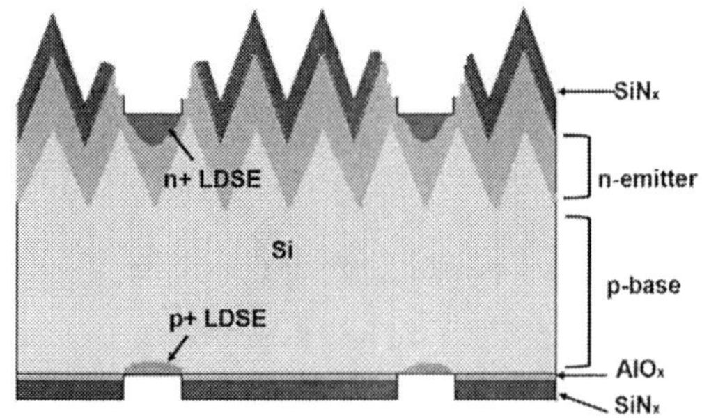

Figure 2: Scheme of the PERC precursors used in this study. The dome-shape colored region at the Si/Ni interface indicates the Laser Doped Selective Emitters (LDSE) (n+ and p+) aiming to be plated on.

2.2 Process flow

As the functions of developing plating equipment is the main study subject, reference samples were firstly tested on Aerotech single side plating equipment to ensure there is no problem associated with other parameters (e.g. laser doping)[6]. By compare the plating results, contact properties, and cell performance fabricated by two plating equipment, the key design components of developing prototype tool would be more clearly to be identified then to be optimized. Figure 3 shows the design scheme of the processing flow, and the processing details are described as below (specified setup parameters in plating sections are data from the reference group). However, only Cu plating trails (Group A) has completed within the current time frame and shown in this paper

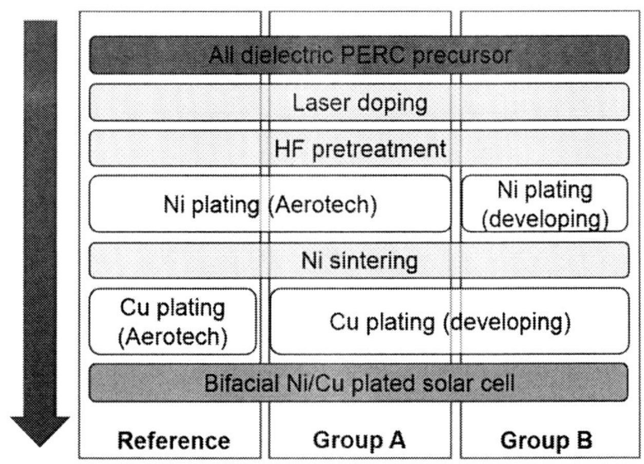

Figure 3: Scheme of the process flow running in this study. There are two groups for optimizing the developing plating equipment.

i. HF pretreatment: A low concentration (~1%v/v, unless specified) HF(aq) dip was applied (60s) right before plating. After dipping, samples were firstly rinsed in DI water and spin-dried for 30s, then immediately transferred to the plating step. HF pretreatment can remove the oxide at the Si exposed area, where is aimed to be plated on, thereby enhancing the Si surface conductivity[7].
ii. Ni plating: electroplating was applied to the rear p-type surface and bias-assisted LIP was applied to the front n-type surface. Constant current (~120mA) was supplied throughout the electroplating and bias-assisted LIP process. A relatively short plating duration (~30s) was used here to avoid thick Ni deposition that could greatly increase the contact resistance[1], [2], [7].
iii. Ni sintering: Ni plated samples were fired via Camini belt furnace with a set peak temperature at 385°C under nitrogen environment.

This is in order to form nickel-silicide as ohmic contact and Cu^{+} diffusion barrier[1], [2].

iv. Cu plating: electroplating was applied to the rear p-type surface and contactless LIP was applied to the front n-type surface. Constant current (~200mA) was supplied throughout the electroplating process.

3. RESULTS AND DISSUSION

3.1. Approach

The designed equipment would combine both electroplating and LIP together, and allow both functions to operate simultaneously. To achieve this, both cases were firstly reviewed separately. Figure 4 and 5 shows the scheme of corresponding electrochemical circuit setup in horizontal plating of electroplating and LIP, respectively. The sun symbol in the top of Figure 4 represents LED light source used in bias-assisted LIP. The light blue background in both figures indicate the area soaking/contacting with the plating solution. For electrode reactions involving charge spices, an electric double layer exists at the surface. Randles equivalent circuit model ($R_{CT,\ cathode}$, $C_{DL,\ cathode}$, $R_{CT,\ anode}$ and $C_{DL,\ anode}$) are used to illustrate the electrical effects of this layer [1], [8]. The ohmic resistance of the solution is $R_{\Omega,\ sol}$ and the rest, including wiring, contacts etc. is R_{Ω}, respectively. Solar cell is presented by using one diode model. Two electrochemical circuits have to be linked together in the designed equipment; however, the plating solutions should be well separated. As cells are transferring horizontally, it is unlikely to have a "floating plating bath" on the top of the wafer without connecting the plating bath underneath the cell at all. Instead, "faucet-like" design was used to constantly dripping certain amount of plating solution. Dripped solution firstly passes through the top anode then reaches to the n-type surface. In addition, the plating bath underneath the cell was separated to allow n-type plating solution to leak from the gap in between. Secondly, to link both electrochemical circuits, conductive and non-conductive rollers were alternatively arranged. The non-conductive rollers were set within the p-type plating bath, only for transportation use. Besides cell transportation, the conductive rollers in between p-type plating bath were also used as cathodes for both sides of the cell. A detail scheme of the designed prototype equipment is shown in Figure 6.

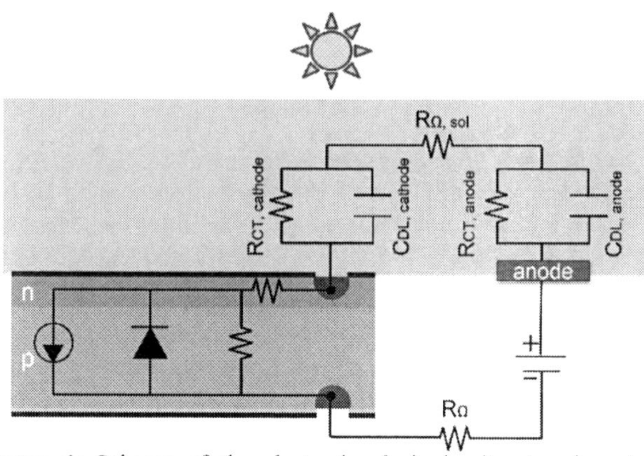

Figure 4: Scheme of the electrochemical circuit setup in n-type plating by using bias-assisted LIP.

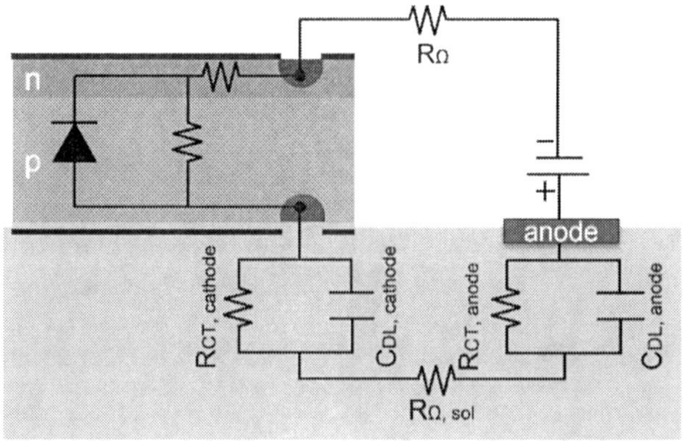

Figure 5: Scheme of the electrochemical circuit setup in n-type plating by using electroplating.

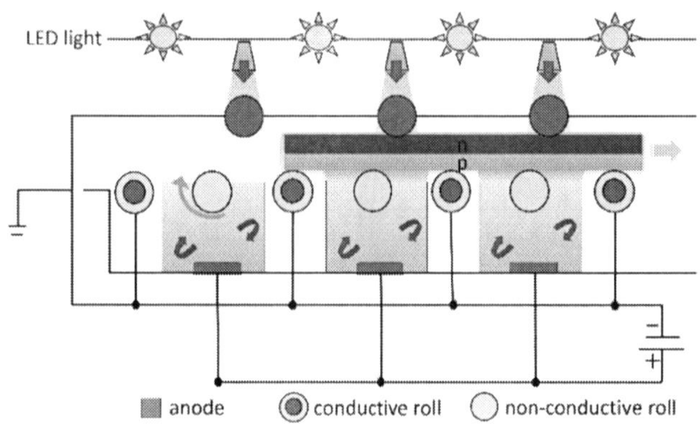

Figure 6: Scheme of developed plating equipment in this study. The conductive roll was used as cathode for both n-type LIP and p-type electroplating.

Figure 7 shows the demonstration images of the 1st gen plating tool for simultaneously bifacial plating. From the top view (Fig. 7b), we can see the design scheme shown in Fig.6 is equally disturbed throughout the cell surface area to ensure best plating uniformity. Although the top plating solution "tube" may partially block the light, this area only takes a relatively low portion. Moreover, as cell is constantly moving on the production line, it was found such temporary light block issues does not reflect on the plating uniformity at all.

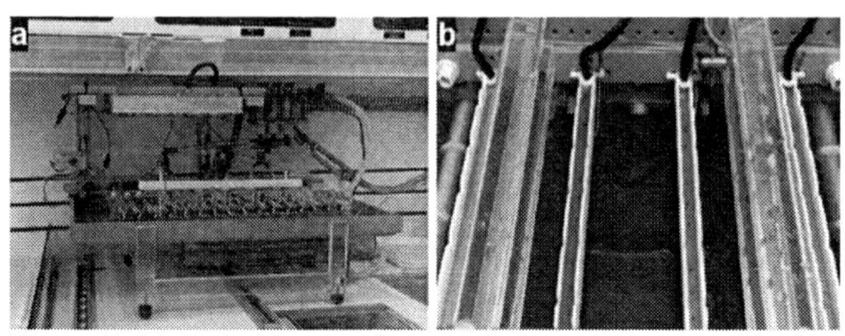

Figure 7: (a) Optical image of 1st gen plating tool developed in this study. (b) top view of 1st gen prototype tool during Cu plating in operation.

3.2. Plated contacts on bifacial PERC solar cells

Figure 8 shows representative top view images of finished cell from both the n-type and p-type side. It is clearly to see there is no ghost plating issues at all throughout the cell surface. Figure 9 shows a close look of both Cu plated contacts under optical microscope. The shinning metallic color is clearly observable, which indicates the well crystallization. However, the inequal intensity also hints there still has rooms for improvement. The average height and width of the contacts are within the range of 20-22μm. Sufficient strong contact adhesion was obtained under stick tape test.

Figure 8: Representative top view optical image of (a) front n-type and (b) rear p-type surface, respectively.

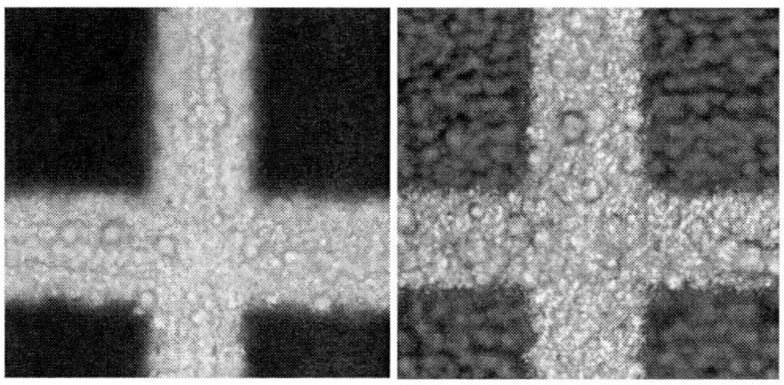

Figure 9: Representative 100x microscope image of (a) front n-type and (b) rear p-type surface, respectively.

Up to ~20% cell efficiencies were measured from as-Cu-plated samples by the developing prototype tool. The series resistance is typically at ~3-4-mΩ and the fill factor is at ~ 0.75. A characterization summary is shown in table I. Furthermore, we trailed integrating UNSW hydrogenation technology into the plated contact cells fabrication process. It was found the cell efficiency could be further enhanced by 0.5-1% and the improvement is stable after >48 hours industry standard light soaking test. Although the achieved cell performance has yet reached the level of current industrial PERC cells with screen printed contacts due to lack of optimization in process not directly related to plating tool development, this result has already shown the viability of simultaneously bifacial plating technology for rapid plated contacts fabrication.

Table I: Characterization summary of finished cells (Group A)

Isc (A)	Voc(mV)	Rs (mΩ)	FF(%)	η(%)
9.52±0.11	647±15	3.39±0.42	75.45±0.03	19.31±0.69

3.3 Further work and potentials

More challenges are expected to encounter when it comes to direct Ni plating as it has stricter requirement in conductivity uniformity. Furthermore, safety, process fluency and quality control are important in terms of up scaling this technology. Therefore, 2nd prototype equipment was built as a light version of industrial equipment as shown in Figure 10. Besides PERC cells, the plating process of other cell structures (e.g. HJT, TOPCon and tandem cells) are also required to be examined to fulfill any possible needs in the future.

Figure 10: Optical image of 2nd gen plating equipment developed in this study, which is a light version of industrial standard equipment.

5. CONCLUSION

In this paper, we demonstrated an innovate plating technology which allows double sides of solar cell to be plated at the same time. By adopting inline horizontal plating approach, the plating uniformity could be greatly improved. This also minimizes the mechanical impacts on the sample thereby effectively reduces microcrack risks. The special arrangement of plating solution supply and electrodes makes simultaneously bifacial plating viable. So far, satisfied Cu plating results on bifacial PERC cells have been achieved by the prototype plating tool. The next step will be to investigate the direct Ni plating on bifacial PERC cells, and then further extend the optimization procedure to other cell structures. This work provides a practical solution to significantly increase the throughput of plated contacts thereby promoting its commercial utilization and further reducing the LCOE of solar energy.

6.ACKNOWLEDGEMENT

This study was supported by the Australian Government through the Australian Renewable Energy Agency (ARENA) through RND2017/004. The Australian Government, through ARENA, is supporting Australian research and development in solar photovoltaics and solar thermal technologies to make solar energy become cost competitive with other energy sources. Responsibility for any views, information, or advice expressed herein is not necessarily accepted by the Australian Government.

7. REFERENCE

[1] A.Lennon, Y.Yao, andS.Wenham, "Evolution of metal plating for silicon solar cell metallisation," *Prog. Photovoltaics Res. Appl.*, vol. 21, no. 7, pp. 1454–1468, Nov.2013.

[2] A.ur Rehman andS. H.Lee, "Review of the potential of the Ni/Cu plating technique for crystalline silicon solar cells," *Materials (Basel).*, vol. 7, no. 2, pp. 1318–1341, 2014.

[3] M.Tao, H.Hamada, T.Druffel, J.-J.Lee, andK.Rajeshwar, "Review—Research Needs for Photovoltaics in the 21st Century," *ECS J. Solid State Sci. Technol.*, vol. 9, no. 12, p. 125010, 2020.

[4] M.Alonso-Abella, F.Chenlo, A.Alonso, andD.Gonzalez, "Toledo PV Plant 1 MWp - 20 Years of Operation," in *29th European Photovoltaic Solar Energy Conference and Exhibition*, 2014, pp. 2728–2733.

[5] ITRPV, "International Technology Roadmap for Photovoltaic," 2019.

[6] Y.-C.Chang, S.Wang, S.Li, R.Deng, J.Ji, andC.-M.Chong, "The Effect of Different Laser Wavelengths on Bificial Plated Laser Doped Selective Emitter Solar Cells," in *37th European Photovoltaic Solar Energy Conference and Exhibition*, 2020, pp. 383–387.

[7] G.Cimiotti, J.Bartsch, A.Kraft, A.Mondon, andM.Glatthaar, "Design Rules for Solar Cells with Plated Metallization," *Energy Procedia*, vol. 67, pp. 84–92, 2015.

[8] J. E. .Randles, "Kinetics of rapid electrode reaction," *Discuss. Faraday Soc.*, vol. 1, pp. 11–19, 1947.

APPLICATION OF ARTIFICIAL INTELLIGENCE TECHNIQUES FOR OPTIMIZATION OF METALLIZATION PROCESS

Enrico Boscolo Marchi, Alessandro Dalla Lana, Stefano Visintin, Marco Galiazzo, Alessandro Voltan*
Applied Materials Italia Srl, Via Postumia Ovest 244, Olmi di San Biagio di Callalta (TV) Italy
*Corresponding author: Alessandro_voltan@amat.com

ABSTRACT: Metallization of PV cells (PERC, HJT, IBC) involves multiple production processes such as screen printing, drying, firing, etc that must be accurately optimized and kept under control in order to guarantee the best IV performances of cells and at the same time the maximum achievable OEE of automation lines.
At Applied Materials Italia, we are increasing the number of sensors and metrology tools that are embedded into metallization lines in order to provide big quantity of real time data to be used to [i] monitor the status of the production tool [ii] predict possible failure and help predictive maintenance [iii] correlate process data with IV performance of solar cells [iv] avoid defects into final product. In this abstract we will focus on application of Artificial Intelligence to Optimization of Fabrication Processes and Reconstruction of Measurement Data reporting 3 examples of recent development:

1) Application of Neural Network to Thermal Processes Optimization
2) DL Approach for Defects detection during Electro-luminescence analysis
3) Data Management for OEE improvement and Preventive Maintenance

1 INTRODUCTION

Metallization of PV cells (PERC, HJT, IBC) involves multiple production processes such as screen printing, drying, firing, etc that must be accurately optimized and kept under control in order to guarantee the best IV performances of cells and at the same time the maximum achievable OEE of automation lines.
Following the recent trends in all the major industrial fields, digitalization is step by step increasing also in the PV field. In particular at Applied Materials Italia, we are increasing the number of sensors and metrology tools that are embedded into metallization lines in order to provide big quantity of real time data to be used to [i] monitor the status of the production tool [ii] predict possible failure and help predictive maintenance [iii] correlate process data with IV performance of solar cells [iv] avoid defects into final product.
In order to process the big quantity of data (*Big Data*), Artificial Intelligence techniques are commonly used for creation of prediction models.
As reported in [1], Machine Learning techniques can be used for

- Prediction of Material Properties
- Optimization of Device Structures
- Optimization of Fabrication Processes
- Reconstruction of Measurement Data

For all these applications, the availability of big data and the dedicated algorithm for data analysis can shorten the development phase avoiding the traditional trial and error approach.
In this abstract we will focus on Optimization of Fabrication Processes and Reconstruction of Measurement Data reporting 3 examples of recent development:

1. Application of Neural Network [2] to Thermal Processes Optimization;
2. Deep Learning approach for Defects detection during Electroluminescence analysis;
3. Data Management for OEE improvement and Preventive Maintenance.

2 THERMAL PROCESSES OPTIMIZATION

The heat treatment on photovoltaic cells represents a fundamental step in order to obtain a product with high quality and efficiency. The thermal process, after the serigraphic printing phase, involves the use of a drying oven which has the main purpose of evaporating the solvents contained in the conductive paste (Ag, Al..).
The definition of a correct drying process passes through the optimization of the control parameters of the oven itself on order to obtain the "optimal" temperature on the cell.
The main oven control parameters are i) Belt speed (m/s), ii) Power lamp/resistance (Watt), iii) Air flow rate (m^3/h). In addition to the parameters listed above, the cell temperature can be influenced by the type of material (mono or multi silicon), wafer size, presence or absence of conductive paste, amount of paste used, type of paste.
It is evident that mapping all possible combinations of parameters, empirically obtaining the temperature of the cell, becomes a very time-consuming activity.
The use of the Deep Learning algorithms and Neural Networks (N.N.), suitably trained with previously collected data, can lead to predicting the temperature curve for a new drying process without having to experimentally measure the temperature (obtained with different process parameters) with thermocouples or IR pyrometers or other empirical techniques (dedicated oven thermal profiling systems).
Fig.1 shows how it is possible to structure a neural network having the main process parameters as inputs and the temperature curve on wafer as output: the numbers of inner layers as well as the number of neurons per layer is a function of the quantity of input data.

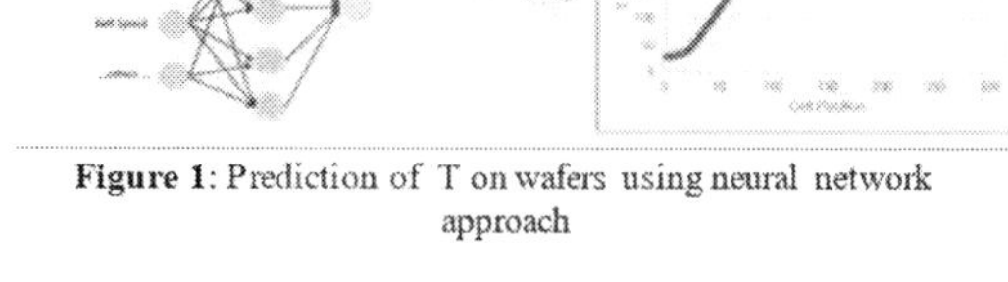

Figure 1: Prediction of T on wafers using neural network approach

Once the network has been trained, it will be possible to make a prediction of the temperature profile, asking the

network to perform the calculation, simply by changing one or more process parameters, without the use of external hardware (dedicated oven thermal profiling systems or Pyrometers) but simply by using the ability to predict of N.N. itself.
The neural network can also be used in other types of oven in photovoltaic fields, such as firing, or as a feedback (closed loop) in order to keep the oven always in the same operating condition.
This approach can bring a significant advantage as it reduces the time in identifying new process parameters to obtain a determined thermal profile on the cell, but it can be used also to identify the control limits for all the input parameters.
In the following, the results of the experimental investigation about the suggested approach are presented.

2.1 Thermal processes optimization: experimental investigation

Data collection represents a fundamental step for the learning/Validation phase of a neural network.
An adequate number of Input and Output (target) data is necessary to ensure that the network is able to learn and, subsequently, properly predict the temperature profiles on the cell.
In the tests carried out, nine input process parameters were identified and used: fig.2 (left) shows the parameters and the sketch on the right highlights the location of the hardware relating to the parameters selected in the oven.

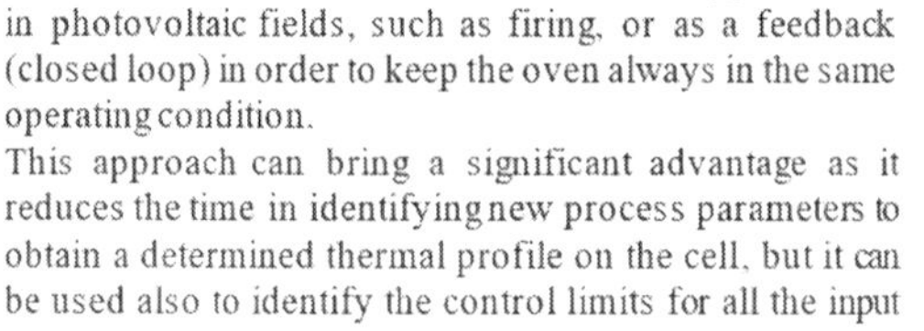

Figure 2: Input parameters

Starting from a nominal set of input parameters a series of 120 curves has been generated changing the input data using DOE matrix within the range reported in Table I, trying to map as many significant configurations as possible even far from the normal conditions of use.

Table I: nominal values and range of variation of input data

Parameter	Unit	Nominal value	Range of variation
R.1	[%]	75	70-90
R.2	[%]	80	70-90
R.3	[%]	75	70-90
R.4	[%]	60	35-60
R.5	[%]	0	0-30
R.D	[°C]	130	130-170
FanIn Speed	[%/rpm]	30	30-40
FanOut Speed	[%/rpm]	90	75-90
Belt Speed	[mm/s]	150	130-230

All the 120 thermal profiles have been collected using thermocouples placed on cells and connected with oven PLC taking approximately 4 working days. Each temperature profile collected is made up of 30 values, this means an acquisition approximately every 10 cm (oven length about 300cm), as a fair compromise between maintaining the information of the trend of the T. curve and the amount of data to be used in the neural network. Curves with a greater number of points could be used but this doesn't lead to any real advantage other than an increase in the computational times of the learning phase in N.N.
Once all the thermal profiles have been collected, a C # routine has been implemented to perform the learning and testing phase of prediction model. We have not used external libraries, therefore all the calculation functions have been developed internally.
The collected Temperature (profiles) data have been divided into three sets:

1) Train Data: set used by N-N. for learning;
2) Validation Data: set of data used to give an estimate of model skill while tuning model's parameters
3) Test Data: used only to assess the N.N. performance after learning

Different learning parameters (activation functions, learning rate, epochs etc..) were tested up to those that guaranteed good convergence in a reasonable time between training and validation curve (Fig. 3) and minimize the cost function (difference between training and validation curves).
The image below (on the left) shows the typical decreasing learning curve (black) and the validation curve (red).
Both curves decrease until about 650000 epochs, after which the validation curve tends to quickly grow (right green square). The learning process was stopped when the validation started to grow to avoid overfitting phenomena.

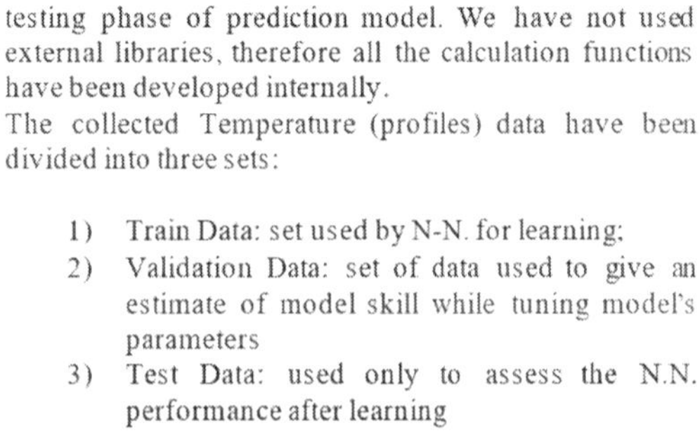

Figure 3: Learning process of Neural Network

2.2 Thermal processes optimization: results

In order to assess the precision of the prediction model, a verification trial has been performed comparing real and predicted thermal profiles using 8 test conditions (reported in table II) chosen randomly and not included in the learning or validation sets.
The goodness of the predicted curves was evaluated simply by taking the maximum absolute and the average deviation between the predicted curve and the real one (Table III). Considering that maximum deviation could be related to a single point, it makes more sense to consider the average deviation estimating a ±3°C as prediction error.
Comparison between predicted and real curves is reported in Fig. 4.
Beside the maximum/average deviation between the predicted and actual curves an important question to answer concerns the analysis of uncertainty and errors related to the described approach.

Table III provides an indication with a good degree of accuracy of the predictive capabilities of the neural network, but, the data shown in the table does not take into account (in any way) the uncertainties on the temperature profiles (used in learning phase) due to multiple factors such, as for example, inexperience of the operator or stability of the oven over time etc.
For the neural network, the learning data are basically error-free data, in fact N.N. doesn't know if the temperature profiles are affected by uncertainties. It is therefore important to analyze all other causes of error that will be added to those deriving from the predictive capacity of the neural network.
To make this simple analysis, we introduce the concept of stability of the temperature curve over time: by stability we mean that, given the same oven input parameters, the Temperature curves obtained at different times have the same characteristics (peak temperature, descent and rise gradients values, etc.).

Table II: input parameters used for validation curves

	1	2	3	4	5	6	7	8
R.1 [%]	85	85	75	70	70	90	90	90
R.2 [%]	85	85	75	70	70	90	90	90
R.3 [%]	85	85	75	70	70	90	90	90
R.4 [%]	45	45	35	30	30	60	60	60
R.5 [%]	15	15	5	0	0	30	30	30
R.D [°C]	150	170	130	150	150	150	130	170
FanIn [%/rpm]	30	30	30	30	30	40	30	30
FanOut [%/rpm]	90	90	90	90	80	90	90	90
Belt [mm/s]	130	230	190	190	230	130	190	230

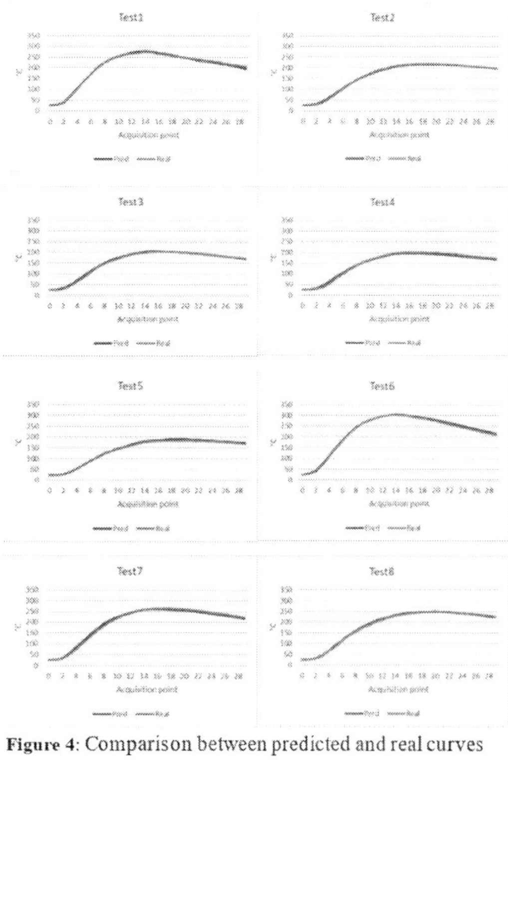

Figure 4: Comparison between predicted and real curves

Table III: deviation between predicted and real curves

	1	2	3	4	5	6	7	8
Average deviation [°C]	0.8	0.9	-1.4	1.9	-2.8	-2.2	0.4	-2.6
Max absolute dev. [°C]	9.7	5.8	6.8	12.1	5.0	7.8	9.5	8.3

To estimate the curve stability and uncertainty of the test, 4 Input Conditions were selected, and, for each condition, 5 temperature profiles were collected. Stability of the curve has been measured to be in a range of about ±5deg which we consider acceptable given the numerous factors involved.
The tests carried out so far demonstrated the goodness of the approach being the average prediction error comparable with curve stability and acquisition uncertainty.

3 DEEP LEARNING APPLIED TO EL IMAGES AND RELATED DATA MINING

As second case study, the application of DL applied to Electro-Luminescence will be described with particular focus to data mining strategies.

Electro-Luminescence (EL) is an analysis technique gaining more and more importance and diffusion in the quality assessment for the solar cell production. It is a fast and sensitive method to highlight the presence of defects and can be applied regardless to the cell's type, size and layout.

Multiple and different defects can be present simultaneously on the same wafer. A few years ago, a Good/Not Good approach was acceptable; at the time being, the quality and process control are becoming more and more demanding and the system is asked to extract a lot of additional information, such as the types of the defects, their number and positions, their size and contrast level and so on.

Deep Learning (DL) has been proved to be a successful approach in detecting both standard (cracks, scratches, finger defects, ...) and non-standard (cloud-foggy, fork-mark, sucker-mark, ...) defects.

The large amount of information outcoming from DL analysis in a production environment must be treated as Big Data, and proper algorithms for Data Mining shall be applied in order to extract information for the process control and its adjustment.

For the purpose of this paper, data have been collected from a site currently running Hetero-Junction (HJT) cells mass production.

3.1 Deep Learning approach

Deep Learning is a subset of Machine Learning algorithms using complex and deep neural networks to analyze an input image.

Four main tasks can be identified in the DL image analysis contest; they're listed below in increasing order of outcoming information (and computational cost):

- Classification
- Classification & Localization
- Object Detection
- Instance Segmentation

Classification considers the image as a whole and can be used in a Good/Not Good approach.

Localization and Object Detection consider each defect as an entity and are applied when limited information are required to be extracted (position and rough size of the defects).

Instance Segmentation is the most complete analysis technique and has been used to process the data reported in this paper. Each pixel of the input image is processed and properly assigned to a defect class or to the background (if it doesn't belong to a defect). The pixels belonging to the same class are clustered into a region considered as a defect entity; each region is then processed to extract the desired information related to the defect (position, size, contrast etc.).

For the reported analysis, 8 defect classes have been defined (big grey areas, black corners, cracks, dark areas, finger defects, fork marks, scratches and sucker marks) and one Convolutional Neural Network (CNN) model has been created for each defect type.

Each CNN model has been trained separately, using a supervised learning approach: training and validation sets have been created from a total of 435 EL images, manually marked and labelled accordingly to the defect classes above.

A loss function has been defined over the training set (about 80% of the collected images) and minimized, epoch by epoch, during the training phase.
The goodness of the model's prediction has been evaluated defining a metric function and evaluating it over the validation set (20% of the remaining images, never seen before from the model itself).
If compared with traditional detection approaches, based mainly on thresholding and contrast gradient analysis, the DL algorithm showed the capability of detecting and classifying complex patterns and not well-defined defects (e.g. the so-called cloud-foggy areas).

Eventually, the overall detection performances of the models have been validated on-site, running the system at production cycle time and volumes (see Table IV).

Two important ratios have been evaluated over 6 days and are reported in the table:

- Overkill ratio: it represents the number of true good cells (from a production batch containing both true good and not good cells) judged as not good by the system
- Underkill ratio: it represents the number of true not good cells (from a batch containing only true not good cells) judged as good by the system.

Underkill and overkill ratio can be balanced adjusting the labelling of the defects for the models' creation and the classification recipe. This operation allows to meet the thresholds set by production requirements.

Table IV: On-site system validation

	Overkill ratio	**Underkill ratio**
Day 1	2.54% (245/9632)	0.56% (5/900)
Day 2	1.24% (262/21142)	0.56% (9/1600)
Day 3	1.58% (312/19809)	0.64% (9/1400)
Day 4	3.80% (403/10619)	0.25% (2/800)
Day 5	4.58% (612/13356)	0.67% (6/900)
Day 6	2.43% (293/12034)	0.56% (5/900)
Total	**2.46% (2127/86592)**	**0.55% (36/6500)**

3.2 Data Mining

Thanks to the large amount of information derived from DL analysis, several considerations can be made about production trends. Starting from this data, the defects' features and their possible causes can be investigated. Given a cell, the main information extracted from the data is the defect type, location, and area. Given a batch of cells, the increment of found defects is considered and the questions, "where" and "when" a defect occurs, can be answered.

Three types of graphs are useful for the analysis: Distribution Charts, Heatmaps and Time-Plots.

Distribution Charts (see Figure 5) give information about the location where defects occur. It can be seen how Sucker Marks are defects localized in restricted regions while Scratches can be frequently found almost everywhere on the cell's surface.

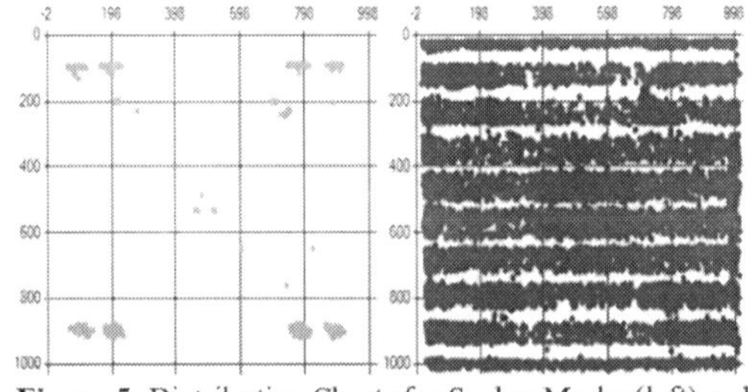

Figure 5: Distribution Charts for Sucker Marks (left) and Scratches (right); units are pixels.

Heatmaps (see Figure 6) highlight the regions having a higher defects' spatial density. In order to take into account, the spatial extension of each defect, it is considered the minimum rectangle containing it. The final plot is obtained superposing all the rectangles representing the defects, coming from all the cells. The overlapping of multiple rectangles generates whiter zones, corresponding to a higher defects' spatial density.

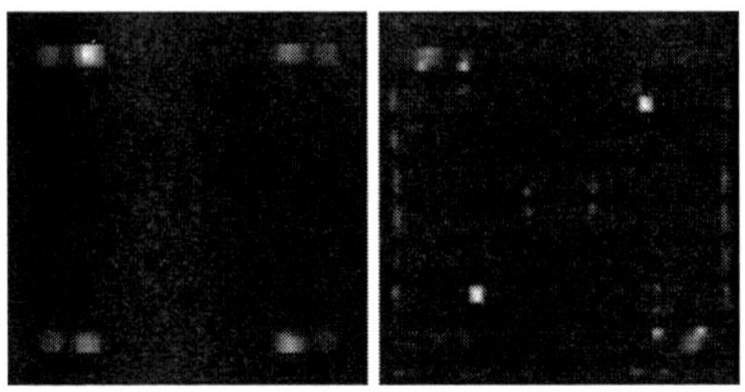

Figure 6: Heatmaps for Sucker Marks (left) and Scratches (right)

Given a batch of cells, the Time-Plots represent the total number of defects identified as a function of the time. A change in the slope of the curve reflects an increase or a decrease in the defect rate.

Heatmaps and Time-plots investigate two different aspects: the heatmaps represent the spatial accumulation of defects while the time-plots represent the temporal accumulation of defects.

A spatial and temporal analysis can be performed using these tools: the goal is to identify clearly and automatically the most critical areas of the cell and to track the changes in the temporal curve's slope.

For the purpose of this article, data from one day's production of Hetero-Junction (HJT) cells were collected, analyzing a total of 35,449 wafers. Some interesting results are reported by analyzing some cell batches during production.

Considering the Sucker Mark defect for a batch of 6,507 cells, a spatial analysis is carried out by dividing the cell into four regions. From the Heatmap and the Time-Plot, it is observed that the defects are mainly located in the Top-Left region of the cell (see Figure 7 and Table V).

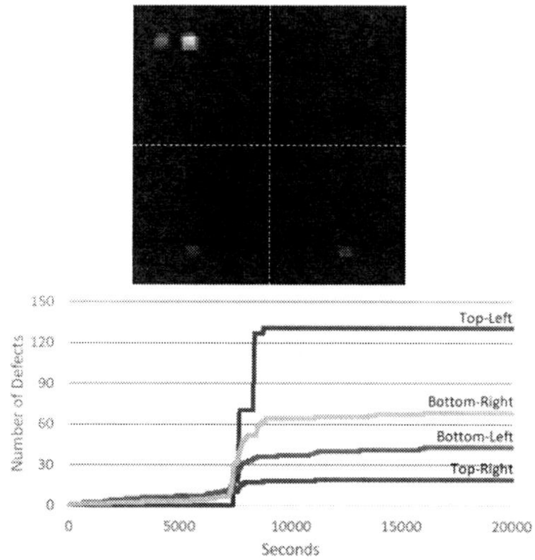

Figure 7: Heatmap (top) and related time-plots (bottom) for Sucker Marks, referred to the four regions

Table V: Sucker Mark spatial analysis

Region	Sucker Marks
Top-Left	131
Bottom-Right	68
Bottom-Left	43
Top-Right	19

Considering the Scratches in a batch of 127 cells, a temporal analysis is performed to track the rate of detected defects (see Figure 8 and Table VI). It can be observed that the slope of the curve increases from a value of 3.7 defects/sec to a value of 21.4 defects/sec. That increase in the temporal defects' rate is confirmed by observing the distribution of defects on the heatmap.

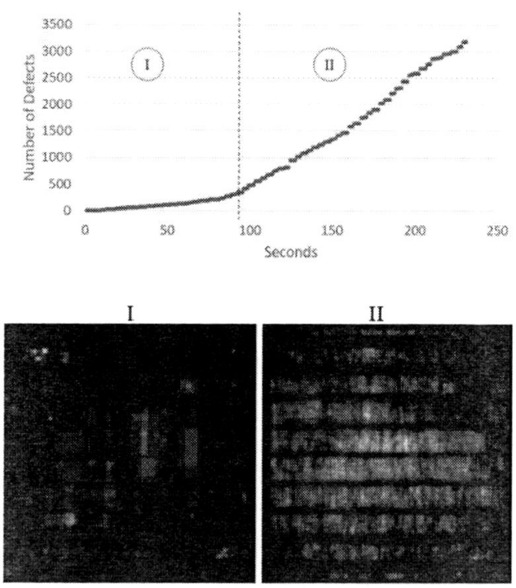

Figure 8: Time-plot (top) and related heatmaps (bottom) for Scratches, referred to the two different time slots (I and II)

Table VI: Scratches temporal analysis

Time Slot	Cells	Defect Rate
I	51	3.7 defects/sec
II	76	21.4 defects/sec

Both heat-maps and time-plots can be properly correlated with process or hardware using proper IoT platform as described in the next paragraph.

4. INDUSTRY 4.0 AND IoT

Today data are pervasive in PV as in any industrial application. Each component of a manufacturing tool, being a PLC, a sensor or a metrology instrument, generates huge amount of information. But raw information is not enough, indeed it can be overwhelming. Value added is coming from proper analysis: useful insight and timely presentation is mandatory.

Traditionally this has been performed through MES systems (Manufacturing Enterprise System): popular solutions for communication protocols are SECS/GEM and OPC/OPC-UA. In both cases an additional level of data integration, analysis and plotting is needed.

We have developed at Applied Materials Italia an embedded software platform which performs the data integration from different sources (PLC, sensors, instrumentations, etc.): more than 200 different variables per printer and per oven can be collected simultaneously from multiple lines using mySQL database and correlated with the identification code of every cell (Fig. 9).

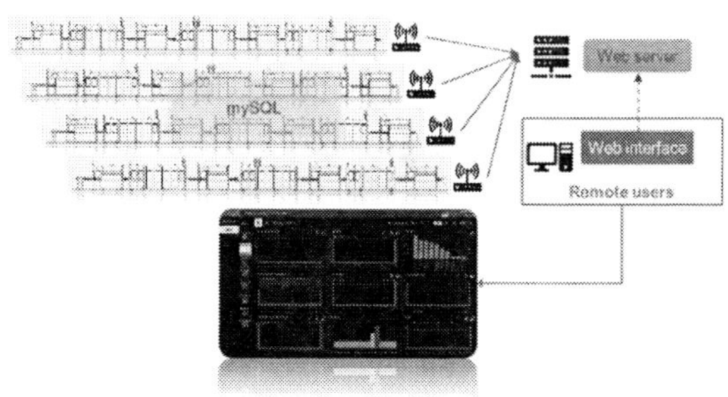

Figure 9: scheme of communication architecture

Example of collected data are process parameters (pressure, speed, snap-off of printers), environmental variables (temperature and humidity), tools parameters (currents absorption) and data coming from metrology systems (AOI, solar simulator, 3D morphology).

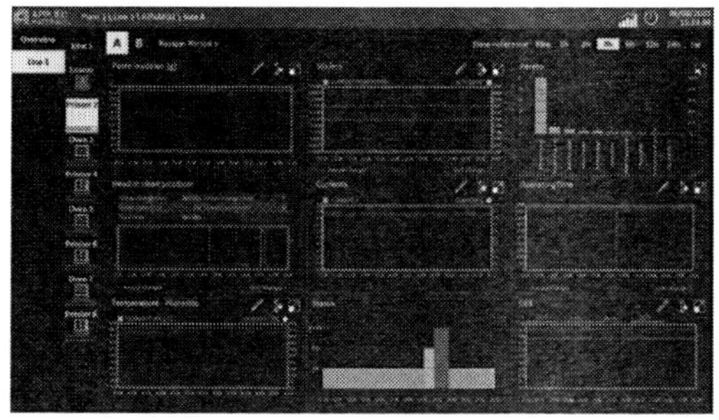

Figure 10: example of dashboard for the printer module

All those variables are processed to automatically calculate critical KPIs (Key Process Indicators):

throughput, process yield, OEE (Overall Equipment Effectiveness), breakage rate, cell efficiency. Furthermore, it can generate charts of trends for printing defects or paste laydown.
KPIs, raw data and trends are organized in dedicated dashboards that can be navigated from fab level (all connected tools) down to each module and side (Fig. 10: example of dashboard).
Final aim of the IoT framework is to build a robust dataset of data including all the critical parameters and measurements needed to train a predictive model that correlates IV results with process inputs. Using this type of approach, is possible to detect critical drifts and react in real time reducing process variability.
On the other hands, real time monitoring of signals coming from electrical and mechanical components is used for predictive maintenance and reduction of production losses.

4.1 Pareto Analysis
Embedded within IoT framework, a special software tool has been developed which automatically performs the Pareto analysis of tool stop reasons and shows dedicated chart (example reported in fig. 11).
In order to optimize metallization line yield and increase uptime, one of the most critical steps is to retrieve objective data identifying the actual causes of production losses. At the moment, this type of data collection is delegated to process engineers with time consuming activities. Level of detail is generally not good enough to understand and classify the root-causes and limited to short time interval.

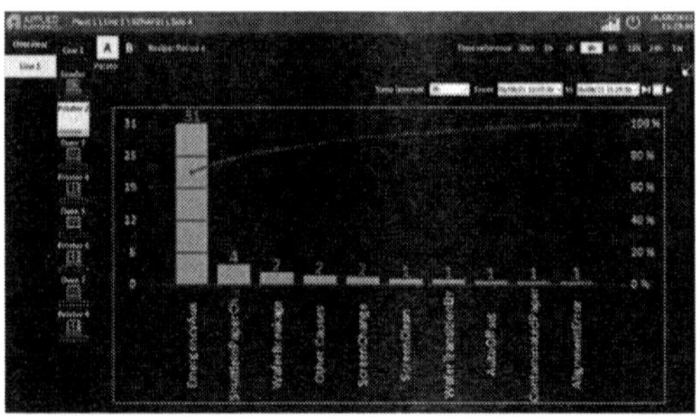

Figure 11: example of Pareto chart for tool stop reasons

The tool uses both hardware/software triggers and machine learning algorithm to properly assign stop labels. The automatic solution has shown to be more reliable than manual mapping and to be an helpful instrument to identify difference of performances between similar lines. Data collection is still ongoing for further improvements.

5 CONCLUSIONS

As reported in this articles, Artificial Intelligence techniques can be widely used for optimization and monitoring of metallization process of PV cells.
As examples, three different types of applications have been reported and described:
1. Application of Neural Network to Thermal Processes Optimization;
2. Deep Learning approach for Defects detection during Electroluminescence analysis;
3. Data Management for OEE improvement and Preventive Maintenance.

As main takeaway, it has been demonstrated that AI techniques can be used both for prediction and modelling of physical phenomena knowing the input data, and for discovering pattern or relationship hidden within big data (example of data mining used with EL).
In any case, the main objective of AI applied to metallization process, will remain the creation of a unique model able to correlate model IV results with all process inputs and machine variables. Using this type of approach, it will be possible to detect critical drifts and react in real time reducing process variability due to human actions.
We are currently proceeding in this direction, firmly convinced that AI will be more and more applied in the next future.

REFERENCES

[1] F. Li et al., "Machine Learning (ML)-Assisted Design and Fabrication for Solar Cells", DOI: 10.1002/eem2.12049, Energy Environmental Materials. 2019, 2, 280–291

[2] https://www.verypossible.com/insights/machine-learning-algorithms-what-is-a-neural-network

HIGH THROUGHPUT SOLAR CELL PROCESSING BY OXIDATION OF WAFER STACKS

M. Meßmer, S. Lohmüller, F. Braun, J. Weber, A. Wolf
Fraunhofer Institute for Solar Energy System (ISE), Heidenhofstr. 2, 79110 Freiburg, Germany
Phone: +49 761 4588-2287; e-mail: marius.messmer@ise.fraunhofer.de

ABSTRACT: In this work, we investigate two approaches for high throughput emitter formation and thermal oxidation for passivated emitter and rear cells (PERC). The low temperature stack oxidation (LoTSOx) approach uses wafer stacks in the typical low temperature thermal oxidation process in the PERC manufacturing sequence. With the LoTSOx approach, we reach similar energy conversion efficiencies of $\eta = 22.2\%$ in comparison to state-of-the-art PERC processing but with 3.1 times higher throughput. The second approach, using a shortened low pressure $POCl_3$ diffusion and a high temperature stack oxidation (HiTSOx approach) with laser-doped selective emitter, yields a peak conversion efficiency of $\eta = 22.0\%$ and low specific contact resistances $\rho_c \approx 0.6$ mΩcm^2. We further investigate different diffusion and oxidation processes for the HiTSOx approach resulting in a wide range of emitter sheet resistances R_{sh}. We reach low emitter dark saturation current densities of $j_{0e} = 12$ fA/cm^2 at $R_{sh} = 389$ Ω/sq while stacking the wafers during the oxidation process.
Keywords: High throughput, wafer stacks, LP-$POCl_3$ diffusion, thermal oxidation, HiTSOx

1 INTRODUCTION

The passivated emitter and rear cell (PERC) [1, 2] is the mainstream solar cell device structure. According to ITRPV (international technology roadmap) expert survey it is expected to remain in this leading market share position for the coming years [3]. In addition, a significant increase in the production tool throughput is predicted [3]. Several process steps in the fabrication of PERC solar cells must be adapted to reach high throughputs. For this, several approaches are under investigation [4]. The PERC sequence usually incorporates a phosphorus oxychloride ($POCl_3$) diffusion after texturing to form the emitter as well as a low temperature oxidation / annealing process between chemical edge isolation and plasma enhanced chemical vapor deposition (PECVD) for passivation [2]. Newest generations of tube furnaces can process up to 1,600 wafers per process tube.

The aim of this work is to strongly increase the production tool throughput of the thermal processes while maintaining a high solar cell performance. We increase the potential throughput by a factor up to 3.1 by implementation of horizontally stacked wafers during the oxidation process. Two approaches are investigated: (i) the low temperature stack oxidation (LoTSOx) which is based on stacking wafers in the low temperature thermal oxidation process typically used in the PERC sequence as surface passivation, and (ii) the high temperature stack oxidation (HiTSOx) [5, 6] for simultaneous emitter drive-in and surface passivation. For the latter, the diffusion process is adapted. Both approaches are implemented in the PERC fabrication sequence.

2 HIGH THROUGHPUT APPROACHES

2.1 Low temperature stack oxidation LoTSOx

The first approach investigated in this paper is the low temperature stack oxidation LoTSOx. This approach follows the typical fabrication sequence for PERC solar cells including thermal oxidation like the baseline process at Fraunhofer ISE [2]. Figure 1 depicts the state-of-the-art diffusion process using phosphorus oxychloride ($POCl_3$) as liquid dopant precursor. The subsequent thermal oxidation process features low temperature as mentioned in Ref. [2, 7] and thus hardly affects the doping profile and grows a thin silicon dioxide SiO_2 as a passivation layer. Both processes usually capture 1,600 wafers (M6 configuration). In the LoTSOx approach, a highly increased throughput of more than 5,000 wafers per batch can be achieved by horizontally stacking of the wafers on top of each other, resulting in a 3.1 times higher throughput. It can be used for solar cells with homogeneous or selective emitter.

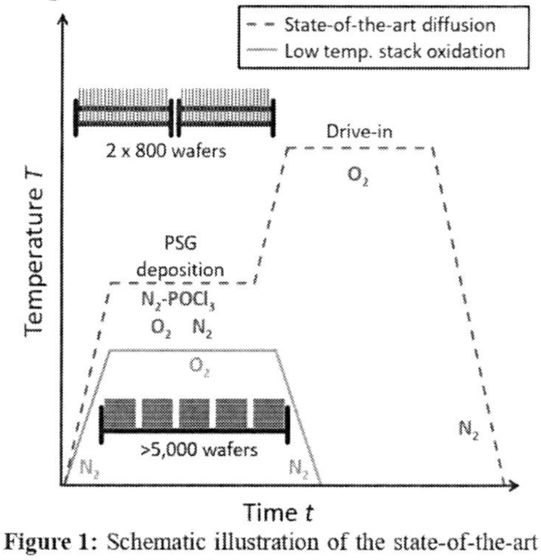

Figure 1: Schematic illustration of the state-of-the-art tube furnace $POCl_3$ diffusion process as a function of the temperature T over time t with a typical process throughput of 1,600 wafers (M6 configuration). Further, the low temperature stack oxidation (LoTSOx) with a throughput of more than 5,000 wafers is illustrated.

2.2 High temperature stack oxidation HiTSOx

The high temperature stack oxidation HiTSOx has already been introduced in Ref. [5, 6]. It uses adapted $POCl_3$ diffusion shortened to only the deposition phase. Thereby, the throughput of the diffusion process is increased by a factor of 1.8. Laser doping from this phosphosilicate glass (PSG) layer [8, 9] shows high suitability [5] for the formation of a laser-doped selective emitter (LDSE) and is used for solar cell processing in this work. Homogeneous emitter formation with the HiTSOx approach is also possible but not investigated here. The drive-in of the phosphorus atoms is shifted to the thermal oxidation process where again horizontally stacking of the wafers increases the throughput to more than 5,000 wafers per batch, increasing the throughput by a factor of 2.4. This stack oxidation process features plateau temperatures in the range of typical drive-in temperatures between $850°C < T_{plateau} < 900°C$ [6]. During this process, the phosphorus incorporated in the silicon is redistributed and diffuses in further. At the same time, a silicon dioxide SiO_2 grows on the wafer surface, which serves as a passivation layer in combination with a hydrogen rich capping layer.

A cost of ownership (COO) calculation for the HiTSOx approach shows a COO reduction of 41% compared to the state-of the art approach [4]. In this calculation, M6 sized wafers and a possible throughput of 1,600 wafers per process tube for state-of-the-art diffusion and thermal oxidation are taken into account.

3 EXPERIMENT

In this investigation, two solar cell batches are manufactured, as depicted in Figure 2. Group A1 and B1 represent the respective reference groups with common PERC fabrication sequence featuring a homogeneous emitter. Group A2 is meant for investigating the low temperature stack oxidation LoTSOx approach as described in section 2.1 and Group B2 the high temperature stack oxidation (HiTSOx) approach as described in section 2.2. For the manufacturing, we use M2 *p*-type boron-doped Cz-Si wafers with a base resistivity of $\rho_b \approx 0.8$ Ωcm. An alkaline texturing process (including saw-damage etching, texturing and surface cleaning) is performed for all groups prior to the diffusion processes. An atmospheric pressure (AP) $POCl_3$ diffusion process forms the homogeneous emitter for the reference groups as well as for the LoTSOx approach in Group A2. For the HiTSOx approach in Group B2, an adapted low pressure (LP) $POCl_3$ diffusion with only the deposition step is performed [6]. Laser diffusion from the phosphosilicate glass (PSG) forms the highly doped region of the selective emitter [8, 9] in Group B2. Then, for all groups, a chemical edge isolation process removes the emitter on the rear side, including the PSG removal on both sides. Subsequently, a wet chemical surface cleaning oxidizes a very shallow part of the emitter surface. The grown oxide is then removed in diluted hydrofluoric acid (HF) prior to the thermal oxidation processes. For the reference groups, a conventional low temperature oxidation process [2, 7] is used (one wafer per slot in vertical orientation), whereas for Group A2, a similar process is used but in this case with horizontally stacked wafers in an experimental setup, see Figure 3. For Group B2, an adapted high temperature stack oxidation is performed. Subsequently, the rear side of all groups is deposited with a passivation stack of $AlO_x/SiO_xN_y/SiN_z$ using plasma enhanced chemical vapor deposition (PECVD). For the front side, silicon nitride (SiN_x) is also deposited by PECVD, serving as anti-reflection coating and surface passivation. For Group A1/2 and B1, the "standard" SiN_x layer is used. For Group B2, a layer adapted in thickness is used to maintain low reflectance. After passivation, some samples (fabricated in a different batch) are fired and measured by quasi-steady-state photoconductance (QSSPC) to determine the implied open-circuit voltage iV_{oc}. For the cells, an infrared laser process locally removes the rear stack layers to form a dot-shaped local contact opening (LCO) for the remaining samples. The metallization is applied using screen-printing. For the rear side a full-area aluminum paste is used. The front side contains 120 fingers without busbars printed in a silver grid. After screen-printing, contact firing is performed in an industrial conveyor belt furnace with variation in the peak temperature. Then, the PERC cells are regenerated to mitigate boron-oxygen (BO) related light-induced degradation [10]. Finally, current-voltage measurements are performed in an industrial inline cell tester. For characterization, 1 cm-wide stripes are cut from selected solar cells to determine the specific contact resistance ρ_c for the front contact according to the transfer length measurement (TLM) [11].

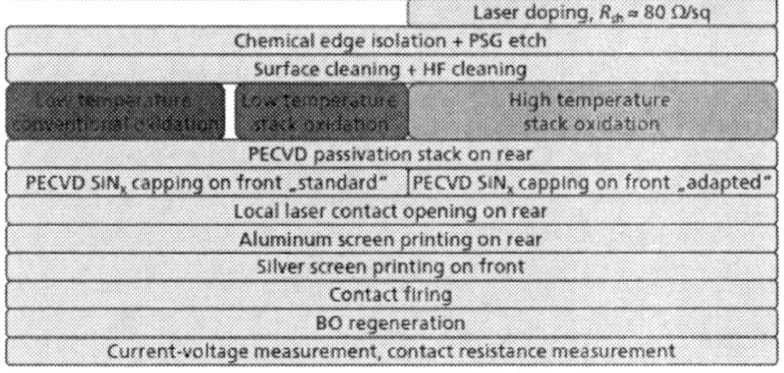

Figure 2: Experiment plan for two solar cell batches A and B, whereby the reference groups A1 and B1 are fabricated similarly. Group A2 represents the low temperature stack oxidation approach (LoTSOx) and Group B2 the high temperature stack oxidation (HiTSOx) approach.

Besides the PERC cell fabrication, symmetric lifetime samples are fabricated to investigate the emitter dark saturation current density j_{0e} of different combinations of diffusion and oxidation processes. Therefore, *n*-type Cz-Si wafers with a base resistivity of $\rho_b \approx 6$ Ωcm are used. After the texturing sequence, a variation in $POCl_3$ diffusion is performed (four different LP $POCl_3$ processes and the reference AP $POCl_3$ process). Then, the PSG is wet chemically etched on both sides. The surfaces are wet chemically cleaned. The thermal oxidation is varied using five different high temperature stack oxidations and the reference low temperature oxidation. The variations in diffusion and oxidation aim to cover a wide range of

Figure 3: Experiment setup for stack oxidation processes. In this quartz boat, wafer groups can be separated by quartz rings.

emitter sheet resistances R_{sh} which is measured by means of inductive coupling [12]. Subsequently, both sides are passivated with either the standard SiN_x for the reference group or the adapted SiN_x for the HiTSOx approach applied by PECVD and the samples are fired. These symmetrical test samples are then characterized using QSSPC measurements to determine the effective lifetime at five spots per sample. Then, j_{0e} is obtained by the procedure described in Ref. [13].

4 RESULTS

4.1 LoTSOx

Table 1 shows the current-voltage characteristics for the LoTSOx approach (group A2 in Figure 2) and the reference process sequence (group A1 in Figure 2) measured in an industrial cell tester after BO regeneration. Comparing the median results of the measurement shows that the LoTSOx approach as well as the reference process sequence result in PERC solar cells with an energy conversion efficiency η of $\eta \approx 22.2\%$. The short-circuit current density j_{sc} and the open-circuit voltage V_{oc} are measured to be $j_{sc} \approx 40.4$ mA/cm^2 and $V_{oc} \approx 673$ mV, respectively, while the fill factor is about $FF \approx 81.7$ %. The best PERC cells in this investigation yield $\eta = 22.2\%$ for both process sequences. In the investigation in Ref. [5], we found that free-standing wafers in a thermal oxidation process show an about 3 nm thicker silicon dioxide in comparison to stacked wafers. Apparently, the different behavior of the silicon dioxide growth has no significant influence on the solar cell performance as the results are independent on the loading configuration that was used for the thermal oxidation process.

To summarize, the LoTSOx approach enables high throughput thermal oxidation processes without energy conversion efficiency losses due to stacking.

Table 1: Current-voltage characteristics of the PERC solar cells fabricated by LoTSOx approach (A2) as well as for the reference group (A1). The median and the best cell for each group are shown. The number of cells per group is shown in brackets. The measurements are performed by an industrial cell tester after boron-oxygen regeneration.

	Process sequence	j_{sc} / (mA/cm^2)	V_{oc} / mV	FF / %	η / %
Median	Ref (6)	40.4 ± 0.1	673 ± 1	81.7 ± 0.1	22.2 ± 0.1
	LoTSOx (4)	40.4 ± 0.1	673 ± 1	81.6 ± 0.1	22.2 ± 0.1
Best	Ref	40.4	673	81.7	22.2
	LoTSOx	40.4	674	81.7	22.2

4.2 HiTSOx

The emitter sheet resistance R_{sh} and the emitter dark saturation current density j_{0e} measured using the symmetric lifetime samples are shown in Figure 4 for the variation of LP $POCl_3$ diffusion processes and high temperature stack oxidation processes in the HiTSOx approach.

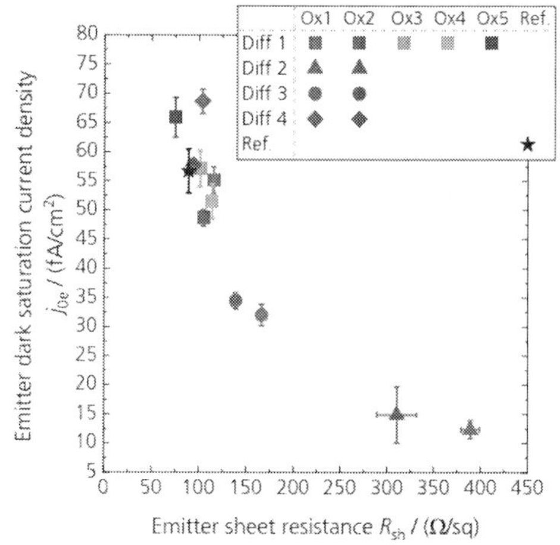

Figure 4: Emitter dark saturation current density j_{0e} on textured surfaces after SiO_2/SiN_x passivation and firing as a function of the emitter sheet resistance R_{sh}.

This investigation shows j_{0e} between 12 fA/cm^2 < j_{0e} < 69 fA/cm^2 within a wide range of emitter sheet resistance of 389 Ω/sq > R_{sh} > 73 Ω/sq. With stacking the wafers during high temperature oxidation, j_{0e} down to j_{0e} = (12 ± 2) fA/cm^2 at R_{sh} = (389 ± 10) Ω/sq is reached. For the combination of "Diff 4" and "Ox 2", R_{sh} and j_{0e} values similar to the ones for the reference manufacturing sequence (j_{0e} = (57 ± 4) fA/cm^2 at R_{sh} = (89 ± 4) Ω/sq) are achieved. Compared to the reference sequence, the combinations in the HiTSOx approach result in up to 15 nm thicker oxide thicknesses [5]. This is the reason why the SiN_x thickness needs to be somewhat thinner to maintain low reflectance at 600 nm wavelength. However, the different ratios between SiO_2 thickness and SiN_x thickness for the reference group and the HiTSOx combination Diff 4 and Ox 2 does not influence the passivation quality.

The combination of Diff 1 and Ox 1 with j_{0e} = (55 ± 2) fA/cm^2 at R_{sh} = (115 ± 1) Ω/sq is chosen for the PERC fabrication in Group B2 (compare Figure 2). By the laser doping process, 150 µm-wide finger structures are formed resulting in a local sheet resistance of $R_{sh} \approx 80$ Ω/sq. The energy conversion efficiency for these PERC cells as well as for the reference group (see Group B1 in Figure 2) are shown in Table 2 (as also reported in Ref. [6]). The cells of the reference process using AP $POCl_3$ diffusion for homogeneous emitter formation yield a peak energy conversion efficiency of $\eta = 22.2\%$, while the cells with the HiTSOx approach and selective emitter peak at $\eta = 22.0\%$. Comparing median results, the reference process reaches $\eta \approx 22.2\%$ with $V_{oc} \approx 675$ mV while the HiTSOx approach results in $\eta \approx 21.9\%$ with the main loss in V_{oc} ($V_{oc} \approx 670$ mV). As the implied open-circuit voltages iV_{oc} for the reference sequence and the HiTSOx approach are almost equal (fabricated in a different batch) with iV_{oc} = 682 mV and iV_{oc} = 681 mV, respectively, the recombination in the contact area with the dark saturation current density in the metallized area and the dark saturation current density in the laser-doped area may be the reason for the V_{oc} loss. This needs to be further investigated.

We also determined the specific contact resistance ρ_c of the front contacts of the cells fabricated with the HiTSOx approach with selective emitter via TLM measurements. Table 3 shows the results for three

Table 2: Current-voltage characteristics of PERC solar cells fabricated by HiTSOx approach as well as for the reference group. The median and the best cell for each group are shown. The number of cells is shown in brackets. The measurements are performed by an industrial cell tester after boron-oxygen regeneration.

	Process sequence	j_{sc} / (mA/cm^2)	V_{oc} / mV	FF / %	η / %
Median	Ref (4)	40.3 ± 0.1	675 ± 1	81.6 ± 0.1	22.2 ± 0.1
	HiTSOx (6)	40.2 ± 0.1	670 ± 1	81.1 ± 0.1	21.9 ± 0.1
Best	Ref	40.3	675	81.7	22.2
	HiTSOx	40.3	672	81.1	22.0

Table 3: Specific contact resistances ρ_c for the HiTSOx approach with selective emitter obtained by transfer length method (TLM) measurements on alkaline textured and phosphorus-doped surfaces after contact firing with the respective set peak firing temperatures $T_{set,peak}$ (SiO_2 and PECVD SiN_x passivation layer).

Set peak firing temperature $T_{set,peak}$ / °C	Specific contact resistance ρ_c / mΩcm^2
820	0.63 ± 0.45
840	0.61 ± 0.51
860	0.57 ± 0.50

different set peak firing temperatures $T_{set,peak}$. We reach low specific contact resistances around $\rho_c \approx 0.6$ mΩcm^2, almost independent from $T_{set,peak}$. The ρ_c in our investigation is as low as for the reference process reported in Ref. [7], which also points out the successful implementation of the new HiTSOx approach.

As higher efficiencies are expected with selective emitter in comparison to homogeneous emitter, more investigations are necessary. Therefore, variations in diffusion, oxidation and laser processing need to be performed on cell level with the HiTSOx approach. However, this first cell experiment shows the promising potential of this high throughput application.

5 SUMMARY AND CONCLUSION

In this work, we fabricated PERC solar cells with two approaches for thermal oxidation enabling processing of more than 5,000 wafers by stacking the wafers which means up to 3.1 times higher throughput. The first one is the low temperature stack oxidation (LoTSOx) approach. Here, the commonly used low temperature thermal oxidation process is performed in a tube furnace while the wafer are horizontally stacked in the quartz boat. The median energy conversion efficiency is not influenced due to stacking of the wafers and reaches $\eta = 22.2\%$ with homogeneous emitter, which is the same as for the reference group with vertically loaded wafers (one wafer per slot).

The second approach uses a shortened $POCl_3$ diffusion with only the deposition step and a subsequent high temperature stack oxidation. The shortened diffusion process enables an additional 1.8 times higher throughput Together with the increased throughput in the oxidation process, it reduces the cost of ownership by 41% compared to the state-of-the-art PERC processing. PERC solar cells with laser-doped selective emitter fabricated with the HiTSOx approach exhibit a peak energy conversion efficiency of $\eta = 22.0\%$ and a median of $\eta = 21.9\%$ which is already high in comparison to the reference sequence with homogeneous emitter and conventional moderate thermal oxidation. Further, the HiTSOx approach yields low specific contact resistances of the front contact around $\rho_c \approx 0.6$ mΩcm^2 independent from the set peak firing temperature.

A variation in the diffusion and stack oxidation processes for the HiTSOx approach allows a wide range of emitter sheet resistances R_{sh} to be covered. Emitter dark saturation current densities as low as $j_{0e} = (12 \pm 2)$ fA/cm^2 are reached at $R_{sh} = (389 \pm 10)$ Ω/sq.

Promising results are reached with our approaches with highly increased throughputs. Further work will investigate the HiTSOx approach with a wide variation in the laser processing as well as homogeneous emitter formation to further improve high throughput solar cell processing.

ACKNOWLEDGEMENTS

The authors would like to thank all colleagues at the Fraunhofer ISE PV-TEC. This work was funded by the German Federal Ministry for Economic Affairs and Energy within the research project "CUT-A PLUS" under contract number 0324282 and "GENESIS" under contract number 0324274C.

REFERENCES

[1] A. W. Blakers, A. Wang, A. M. Milne, J. Zhao, and M. A. Green, "22.8% efficient silicon solar cell," *Appl. Phys. Lett.*, vol. 55, no. 13, pp. 1363–1365, 1989.

[2] R. Preu, E. Lohmüller, S. Lohmüller, P. Saint-Cast, and J. M. Greulich, "Passivated emitter and rear cell—Devices, technology, and modeling," *Applied Physics Reviews*, vol. 7, no. 4, p. 41315, 2020.

[3] ITRPV consortium, "International Technology Roadmap for Photovoltaic (ITRPV): Results 2021," 2021.

[4] M. Meßmer, A. Wolf, M. Zimmer, F. Meyer, G. Hoppe, A. Lorenz, S. Tepner, D. Ourinson, G. Emanuel, S. Nold, B. Goraya, and F. Clement, "Increasing Throughput of Industrial PERC Production," *Photovoltaics International*, 2021, to be published.

[5] M. Meßmer, S. Lohmüller, J. Weber, A. Piechulla, S. Nold, J. Horzel, and A. Wolf, "High Throughput Low Energy Industrial Emitter Diffusion and

Oxidation," in *37th EU PVSEC*, online, 2020, pp. 370–377.
[6] M. Meßmer, A. Wolf, S. Lohmüller, and J. Weber, "Industrial High Throughput Emitter Formation and Thermal Oxidation by the HiTSOx Approach," *Sol. RRL (Solar Rapid Research Letter)*, submitted 2021.
[7] E. Lohmüller, J. M. Greulich, P. Saint-Cast, S. Lohmüller, S. Schmidt, U. Belledin, T. Fellmeth, S. Mack, G. Emanuel, K. Krieg, M. Zimmer, R. Kunert, F. Zobel, M. Linse, J. Horzel, M. Meßmer, A. Wolf, and R. Preu, "Front side optimization on boron- and gallium-doped Cz-Si PERC solar cells exceeding 22% conversion efficiency," in *37th EU PVSEC*, online, 2020.
[8] L. Ventura, A. Slaoui, and J. C. Muller, "Realization of selective emitters by rapid thermal and laser assisted techniques," in *13th EU PVSEC*, Nice, 1995, pp. 1578–1581.
[9] U. Jäger, M. Okanovic, M. Hörteis, A. Grohe, and R. Preu, "Selective emitter by laser doping from phosphosilicate glass," in *24th EU PVSEC*, Hamburg, Germany, 2009, pp. 1740–1743.
[10] A. Herguth and G. Hahn, "Boron-oxygen related defects in cz-silicon solar cells degradation, regeneration and beyond," in *24th EU PVSEC*, Hamburg, Germany, 2009, pp. 974–976.
[11] H. H. Berger, "Contact Resistance and Contact Resistivity," *J. Electrochem. Soc.*, vol. 119, no. 4, p. 507, 1972.
[12] M. Spitz, U. Belledin, and S. Rein, "Fast inductive inline measurement of the emitter sheet resistance in industrial solar cell fabrication," in *22nd EU PVSEC*, Milano, 2007, pp. 47–50.
[13] A. Kimmerle, J. Greulich, and A. Wolf, "Carrier-diffusion corrected J0-analysis of charge carrier lifetime measurements for increased consistency," *Solar Energy Materials and Solar Cells*, vol. 142, pp. 116–122, 2015.

LOCAL PECVD SiO_XN_Y/N-POLY-SI DEPOSITION THROUGH A SHADOW MASK FOR POLO IBC SOLAR CELLS

Verena Mertens*[1], Sören Schäfer[1], Maximilian Stöhr[1], Anja Mercker[1], Adrian Köhler[1], Larissa Mettner[1], Rolf Brendel[1,2], Norbert Ambrosius[3], Thomas Pernau[4], Helge Haverkamp[4], and Thorsten Dullweber[1]

[1]Institute for Solar Energy Research Hamelin (ISFH), Am Ohrberg 1, 31860 Emmerthal, Germany
[2]Institute of Solid-State Physics, Leibniz Universität Hannover, 30176 Hanover, Germany
[3]LPKF Laser & Electronics AG, Osteriede 7, 30827 Garbsen, Germany
[4]centrotherm international AG, Württemberger Str. 31, 89143 Blaubeuren, Germany

*corresponding author: Tel.: +49 (0) 5151 999-315, mertens@isfh.de

ABSTRACT: We present a lean POLO IBC solar cell process flow using *p*-type Cz silicon wafers, where a local SiO_xN_y/*n*-poly-Si emitter is formed by PECVD-deposition of SiO_xN_y/*n*-a-Si through a glass-based shadow mask followed by a subsequent thermal anneal. Light microscope measurements confirm that the PECVD SiO_xN_y/*n*-a-Si pattern deposited on the Cz wafer matches the IBC glass shadow mask layout. A parasitic deposition extending ~60 µm per edge below the mask can occur. However, by an appropriate post-treatment, the parasitic deposition can be reduced to below 5 µm across the whole wafer. On IBC solar cell precursors with two-sided polished surfaces fabricated using the local PECVD-deposition of SiO_xN_y/*n*-a-Si we obtain a median implied open circuit voltage (iV_{oc}) of 709 mV which is comparable to parallel processed samples using a wet-chemical interfacial oxide grown in DI/O_3 plus PECVD n-a-Si with a median iV_{oc} value of 705 mV. After further process optimization including introduction of a post-anneal cleaning step for improved AlO_x/SiN_x passivation the IBC cell precursors with textured front exhibit an excellent median implied V_{oc} of 738 mV. These results show the principal feasibility of the POLO IBC process flow presented.

Keywords: PECVD, shadow mask, IBC solar cell

1 INTRODUCTION

Passivating contacts using polysilicon on oxide (POLO) show excellent electrical surface passivation of silicon enabling highest conversion efficiencies on lab-type IBC solar cells and front and back side-contacted solar cells [1, 2]. When combining an *n*-POLO layer with a local p^+-Al-BSF contact formed by Al finger screen-printing as applied in PERC+ solar cells [3] in an IBC cell [4, 5], these so-called POLO IBC solar cells exhibit a simulated conversion efficiency potential of 25.5% [6] with first experimental devices yielding 23.7% efficiency on 4 cm^2 independently confirmed by ISFH CalTeC [7]. However, for POLO IBC cells the *n*-POLO layer needs to be structured when starting from a full area deposition which so far requires several separate process steps such as masking and etching.

In this paper, we present a lean POLO IBC solar cell process flow using *p*-type Cz silicon wafers, where a local SiO_xN_y/*n*-poly-Si emitter is formed by PECVD-deposition of SiO_xN_y/*n*-a-Si through a glass-based shadow mask followed by a subsequent thermal anneal [8, 9, 10]. The interfacial oxide is grown in a PECVD-N_2O plasma [9]. In accordance to literature we assume the incorporation of some nitrogen into the silicon oxide formed and therefore use the notation SiO_xN_y [8, 9]. The target process flow for the POLO IBC solar cell with shadow mask is shown in Fig. 1 in comparison to an industry-typical PERC+ process flow in its variant for selective emitter and thermal SiO_2 passivation on the front side. Concerning the number of process steps, the "POLO IBC with shadow mask" process flow has the potential of being competitive to the PERC+ variant shown.

In literature, shadow masks were used earlier for local deposition of passivation layers, however, to our knowledge made from other materials [11, 12, 13]. The glass shadow masks used in this paper are fabricated by LPKF AG using their proprietary Laser-Induced Deep Etching glass structuring technology [9] and applied to the POLO IBC solar cell layout designed by ISFH. A Photograph is shown in Figure 2.

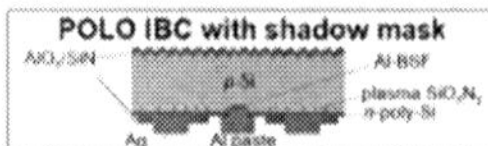

POLO IBC with shadow mask	PERC+ (SE + SiO_x front)
Texture + clean	Texture + clean
	Emitter diffusion ($POCl_3$)
	Laser doping SE
Rear polishing + clean	Rear polishing + clean
local PECVD SiO_xN_y/ *n*-a-Si with shadow mask	
Thermal anneal	Thermal SiO_x growth
AlO_x/SiN_x rear passivation	AlO_x/SiN_x rear passivation
	SiN_x front side
AlO_x/SiN_x front side	
LCO for Base	LCO rear
Screen printing	Screen printing
Firing	Firing

Fig. 1: POLO IBC target process flow applying a local PECVD SiO_xN_y/*n*-a-Si deposition through a glass shadow mask compared to a typical $PERC^+$ process flow with Selective Emitter (SE) and thermal oxide passivation. Differing process steps are written in blue, common process steps in black.

The glass shadow mask enables a confined PECVD deposition and significantly facilitates the local application of poly-Si [9]. The local *n*-a-Si deposition can also be combined with a wet chemical interfacial oxide. Figure 2 shows a photograph of a glass shadow mask featuring a POLO IBC layout consisting of 12 busbars for module interconnection as well as openings for the interdigitated SiO_xN_y/*n*-a-Si fingers. The mask size is 156.75x156.75 mm² pseudo square matching the industrial M2 Cz silicon wafer size.

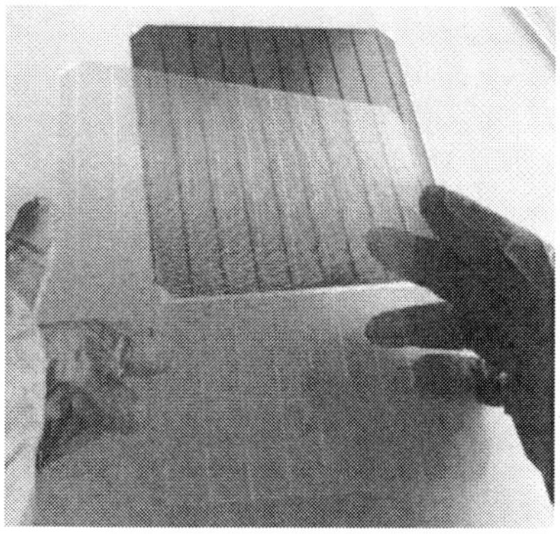

Fig. 2: Photograph of a M2-sized glass shadow mask fabricated by LPKF AG according to the POLO IBC layout designed by ISFH (colors of the image processed for better visualization).

In the following we first give some insight into the local deposition of PECVD *n*-a-Si through the shadow mask covering the image sharpness of the IBC pattern. We also show the current status of our "POLO IBC with shadow mask" process with implied open circuit voltage results on cell precursors without metallisation: comparison of wet chemical interfacial oxide/ *n*-poly-Si stacks and local PECVD-SiO_xN_y/*n*-poly-Si, an optimization example for IBC process flow and iV_{OC} data for optimized "POLO IBC with shadow mask" precursors with textured front and wet chemical interfacial oxide.

2 EXPERIMENTAL

For the investigation of local deposition of *n*-a-Si through the shadow mask and the extend of parasitic deposition we use 1.5 Ωcm boron-doped *p*-type Cz-Si wafers with a thickness of (180±10) µm in M2 format.
The wafers receive a saw damage etch using KOH and a subsequent clean. Then a 100 nm thick thermal silicon dioxide is grown using H_2O vapor as oxidizing agent. Afterwards we deposit 100 nm PECVD-*n*-a-Si through a glass shadow mask made by LPKF AG on one side of the wafers and perform a post-treatment step to reduce the parasitic deposition.
The oxide layer underneath the n-*a*-Si layer enables a good optical contrast and enables visual inspection of parasitic depositions in an optical microscope. We use an automated microscope from MicroVu for visual inspection at 120 points distributed over the wafer.
For all implied V_{oc} and lifetime samples in this paper we use 1.5 Ωcm gallium-doped *p*-type Cz-Si wafers with a thickness of (165±10) µm in M2 format. The lifetime data were measured using a Sinton lifetime tester.
For experiment 1 – the comparison of wet chemical interfacial oxide/ *n*-poly-Si and local PECVD-SiO_xN_y/*n*-poly-Si stacks – the experimental plan and process flow for the POLO IBC cell iV_{oc} precursors are shown in

Fig. 4. Group W (wet-chemical oxide interface) and P (PECVD SiO_xN_y interface) are subdivided into two subgroups 1 and 2 for different PECVD *n*-a-Si layer thicknesses of 140 nm and 210 nm. All other process steps within W and P are identical. The wafers receive an alkaline texture (random pyramids) using KOH and the additive Monotex F from RENA. Afterwards the wafers are etched in an acidic edge isolation process. After porous silicon removal and cleaning for all groups, we do a wet chemical oxidation in ozonated DI water to form a thin interfacial oxide with thickness of ~1.7 nm measured by spectral ellipsometry for the groups W1, W2. Afterwards, the samples are transferred to a PECVD-system using a conductively coupled plasma source. For groups W1 and W2, PECVD *n*-a-Si layers are locally deposited through the glass shadow mask. The Groups P1, P2 (without wet chemical interfacial oxide) receive an HF-Dip and afterwards an in situ plasma oxynitride grown in an N_2O plasma, which is also locally formed through the shadow mask right before the PECVD *n*-a-Si deposition [9, 14, 15]. The PECVD SiO_xN_y has a thickness of ~1.5-2 nm. More process details on the PECVD SiO_xN_y/*n*-a-Si deposition are revealed in [14] and a separate paper at this conference [15]. We deposit two different *n*-a-Si layer thicknesses: ~140 nm for the groups W1, P1 and ~210 nm for the groups W2 and P2. All following process steps are performed identically for all groups: The *n*-a-Si is converted into *n*-poly-Si by a thermal annealing process in a tube furnace in N_2-atmosphere. Afterwards we deposit a stack of 10 nm PECVD-AlO_x and 100 nm SiN_x on the front and on the rear of the wafers. At last the wafers are fired in a fast firing furnace at a peak set temperature of 830°C and a belt speed of 5600 mm/min.
For experiment 2 – process flow optimization example – we prepare lifetime samples representing the p-type base region of the POLO IBC solar cell passivated with an AlO_x/SiN_x stack on both sides. The samples receive an alkaline texture followed by an acidic polishing step including cleaning on front and rear side of the wafers. Group 1/ "without anneal" is passivated with of 10 nm PECVD-AlO_x and 100 nm SiN_x on both sides directly after cleaning, group 2/ "with anneal" receives the inert poly anneal process in N_2 atmosphere at 850°C as in experiment 1 and is afterwards passivated with the AlO_x/SiN_x stack as group 1. The wafers of Group 3/ "with anneal+clean" experience the poly anneal step plus an additional clean after anneal and prior to the deposition of the AlO_x/SiN_x passivation stack, respectively. Finally, all groups are fired in a fast firing furnace at 810°C and a belt speed of 5600 mm/min.
For experiment 3 – iV_{OC} data for optimized "POLO IBC with shadow mask" precursors –the samples receive the process flow shown in Fig. 7: after texture and clean all groups are polished on the rear side, cleaned and receive afterwards a wet chemical interfacial oxide. Then 120 nm *n*-a-Si are deposited on the rear of the samples through a

shadow mask. After an additional clean, group T1 and T2 receive a poly-anneal process 1 with higher start temperature and higher peak temperature, group T3 receives poly-anneal step 2 which has a lower start temperature and lower peak temperature. After anneal group T2 and T3 are treated with the post-anneal clean including an HF-Dip. Group T1 is not cleaned after anneal. Finally, all groups are passivated with 10 nm AlO_x/ 100 nm SiN_x stacks on the front and the rear and afterwards fired in a fast firing furnace at 810°C and a belt speed of 5600 mm/min.

3 RESULTS AND DISCUSSION

The position accuracy of the local *n*-a-Si deposition through a glass shadow mask is shown in Fig. 3. The test wafers without *n*-a-Si deposition have a dark blue/ violet color, so regions with no *n*-a-Si deposition appear dark blue in an optical microscope. After local PECVD deposition plus an appropriate post-treatment, a finger broadening Δw of less than 5 µm per edge is observed which we attribute to a parasitic *n*-a-Si deposition below the shadow mask edges. This post-treatment is also applied to the POLO IBC solar cell precursors discussed below.

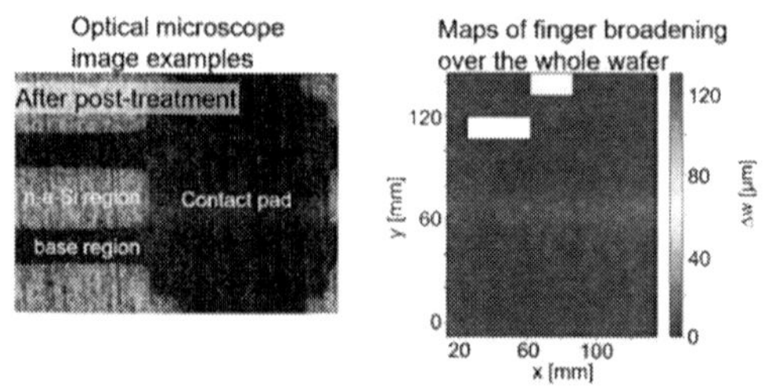

Fig. 3: Exemplary optical microscope images and mappings of n-poly-Si finger broadening measured in an optical microscope after PECVD deposition through the shadow mask plus post-treatment. The images show the IBC finger pattern with poly emitter (grey) and base fingers (blue/violet) and a base contact pad region (blue/violet). The average finger broadening is < 5 µm per edge.

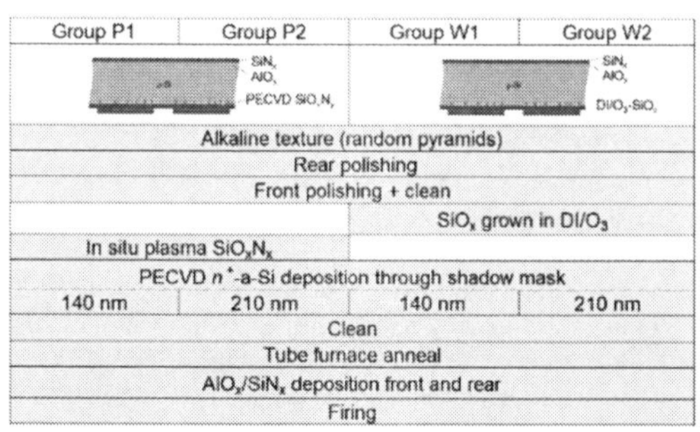

Fig. 4: Experimental plan for experiment 1 – the POLO IBC solar cell iV_{oc} precursors with different interfacial oxides and n-a-Si thicknesses.

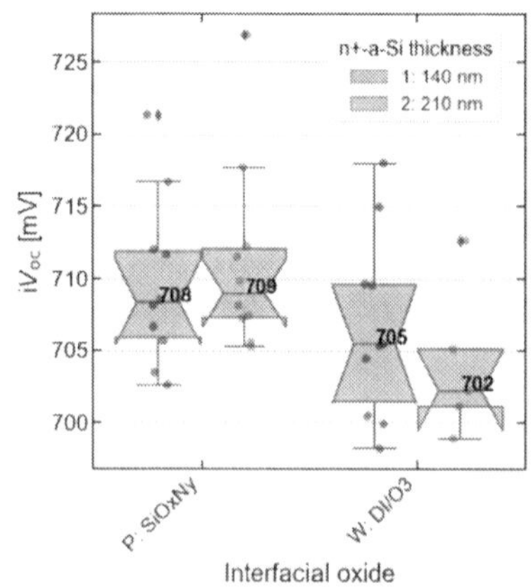

Fig. 5: Implied V_{oc} values of experiment 1 – for IBC solar cell precursors with polished front and rear side and local deposition of the n^+-a-Si. Group W is with wet chemical oxide from DI water/O_3 and *n*-poly Si, P is with PECVD SiO_xN_y, and *n*-poly Si. Subgroup 1 uses a 140 nm n-Poly-Si layer, subgroup 2 uses 200 nm.

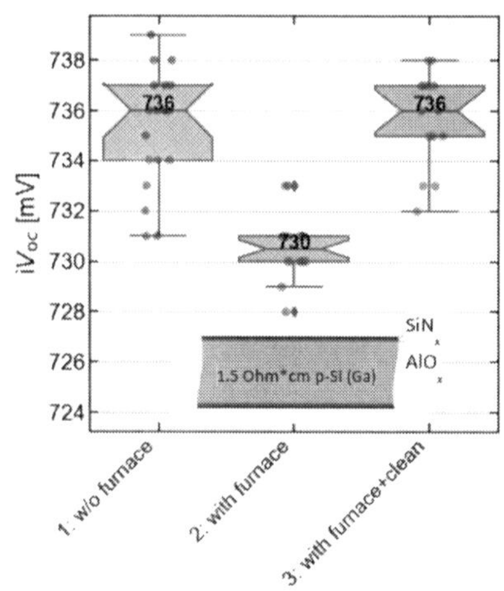

Fig. 6: Implied V_{oc} values for symmetric lifetime samples with polished surfaces and AlO_x/SiN_x passivation on front and rear after firing: the groups 1/ without annealing and 3/ with annealing plus clean show similar values. Group 2 without additional clean after the anneal step is on a lower level presumably due to an interfacial oxide from the anneal process between silicon bulk and AlO_x which is thicker than the usual interfacial SiO_x layer which forms during AlO_x deposition.

Figure Fig. 5 shows the implied open circuit voltages iV_{oc} for the experiment 1 – the comparison of local PECVD-SiO_xN_y/*n*-poly-Si and wet chemical interfacial oxide/ *n*-poly-Si stacks. The corresponding process flow is given in

Fig. 4. POLO IBC cell precursors with polished surfaces from groups P1, P2 and W1, W2. For P1 and P2 with PECVD SiO_xN_y/n-a-Si through a shadow mask, median implied V_{oc} values of 708-709 mV and with a best value of 726 mV are achieved which proofs the feasibility of this local deposition concept concerning its passivation quality. The local n-a-Si deposition combined with a DI water/O_3 interfacial SiO_x gives median implied V_{oc} values of 702 and 705 mV and 717 mV at maximum. The different n-a-Si layer thicknesses of 140 nm and 210 nm show comparable iV_{oc} values within experimental error.

In order to further improve the quality of our IBC cell precursors we perform several optimization steps. One example is given with experiment 2 – where we focus on symmetrical lifetime samples with polished surfaces and AlO_x/SiN_x passivation which represent the base fingers of the IBC solar cell precursors.

Group T1	Group T2	Group T3
Alkaline texture (random pyramids)+clean		
Rear polishing + clean		
SiO_x grown in DI/O_3		
PECVD-n-a-Si deposition through shadow mask (120 nm)		
Clean		
Tube furnace anneal 1		
		Tube furnace anneal 2
Clean		
AlO_x/SiN_x deposition front and rear		
firing step @810°C		

Fig. 7: Experimental plan for the POLO IBC solar cell iV_{oc} precursors with textured front and polished rear: implementation of post-anneal clean and test of two different annealing conditions with different peak temperature and temperature at process start.

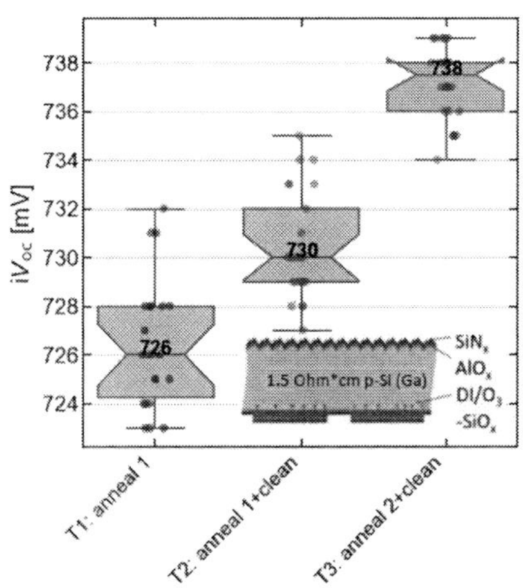

Fig. 8: Experiment 3 – Implied V_{oc} data for IBC cell precursors with textured front. The best median implied V_{oc} is 738 mV for group T3 which is processed with post-anneal clean and lower start temperature and anneal temperature for the poly anneal.

The result of this experiment is shown in

Fig. 6. Group 1 which is passivated directly after texture, polishing and a cleaning step gives a median implied V_{oc} of 736 mV after firing. Group 2 which in addition to group 1 experiences after the clean a poly anneal step in a tube furnace prior to passivation plus firing is with 730 mV on a lower level. However, if an additional cleaning step is introduced after the poly anneal step as for group 3, then the original quality with median iV_{oc} of 736 mV is restored. We assume that during poly anneal, an oxide layer is grown which is thicker than the usual interfacial SiO_x layer which forms during AlO_x deposition. So presumably also an HF-Dip only is sufficient as treatment to restore the original passivation quality. We observed earlier a similar passivation behavior in optimization of our PERC+ process flow comparing the passivation quality of SiO_x/AlO_x/SiN_x with ~6 nm thermal SiO_x intentionally grown and AlO_x/SiN_x stacks plus firing. We attribute this lowered passivation quality of our SiO_x/AlO_x/SiN_x to a reduction or possibly even suppression of negative fixed charge by a SiO_x layer thicker than ~1 nm as investigated in [16] for SiO_x/AlO_x stacks with varying SiO_x thickness.
Finally, we test the post-anneal cleaning step including an HF-Dip to iV_{oc} samples of IBC cell precursors with textured front side.

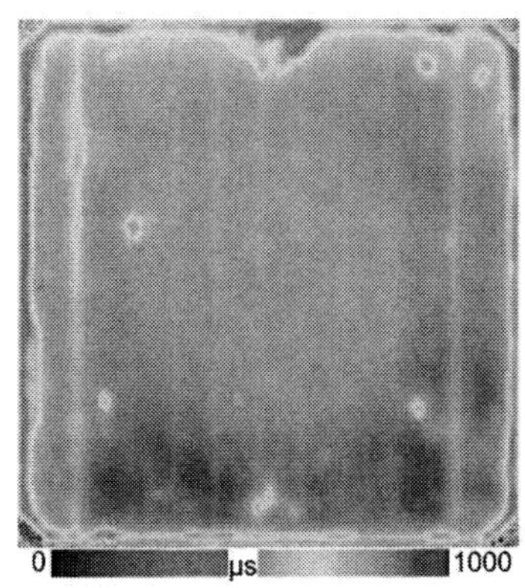

Fig. 9: Infrared lifetime image taken at 1 Sun illumination intensity of one wafer from the best batch T3 of experiment 3 – IBC cell precursors with textured front and improved process flow shown in Fig. 8.

Additionally, we compare two different poly-Si annealing steps. Anneal 1 with higher start temperature and higher peak temperature and anneal 2 with lower start temperature and lower peak temperature. Other optimizations like improved cleaning after texture and polishing process and improved n-a-Si process control not detailed in this paper are also included in the improved process for all batches. The process flow of this experiment is given in Fig. 7. The results of this experiment are shown in Fig. 8. Group T1 without post-anneal clean shows median iV_{oc} of 726 mV. By application of post-anneal cleaning step the iVoc can be improved to 730 mV, so the beneficial effect of this step is also detectable on solar cell precursors. The group T3 with post-anneal step and additionally improved poly anneal 2 with lower start temperature and lower peak temperature gives the best result with median implied V_{oc} of 738 mV. More details on the temperature optimization of n-a-Si deposition and the following poly anneal are given in [14, 15]. An example of the sample homogeneity is given in Fig. **9** where an infrared lifetime mapping at 1 Sun illumination intensity is shown. The regular vertical lines visible are the base busbars of the IBC pattern on the rear of the sample. The image shows no negative impact of the shadow mask on lifetime.

4 SUMMARY

In this paper we present a lean process flow for POLO IBC solar cells using a glass shadow mask for structured deposition of in situ grown PECVD-SiO_xN_y and *n*-a-Si layer. The parasitic deposition of *n*-a-Si is below 5 µm applying an appropriate post-treatment. IBC solar cell precursors with PECVD-SiO_xN_y as interfacial oxide and PECVD *n*-a-Si give similar implied V_{oc} results as PECVD *n*-a-Si combined with a wet chemical interfacial oxide grown in DI water/O_3. Applying further process flow improvements for instance introduction of a post poly-anneal cleaning step which improves the passivation quality of the AlO_x/SiN_x passivated base fingers, results in IBC solar cell precursors with textured front and polished rear side with median iV_{oc} of 738 mV. These data show the principal feasibility of the POLO IBC process flow with shadow mask.

5 ACKNOWLEDGMENTS

This work is funded by the German Federal Ministry for Economic Affairs and Energy under grant FKZ 0324294C (UltraPERC), 0324274B (Genesis), 03EE1102E (Popei) and by the state of Lower Saxony, Germany.

6 REFERENCES

[1] F. Haase et al., *Solar Energy Mat. And Solar Cells*, **186**, 184 (2018).
[2] A. Richter et al., *Solar Energy Mat. And Solar Cells*, **173**, 96 (2017).
[3] T. Dullweber et al., *Prog. Photovolt. Res. Appl.*, **24**, 1487-1498 (2016).
[4] Bende EE., 7th Metallization and Interconnection Workshop, October 23rd and 24th, Konstanz, Germany (2017).
[5] F. Haase et al., Proc. of the 46th IEEE PVSC, Chicago, IL, USA, 2200-2206 (2019). doi: https://doi.org/10.1109/PVSC40753.2019.8980960
[6] C.N. Kruse et al., *Scientific Reports*, **11**, 996 (2021).
[7]. F Haase, private communication.
[8] D. R. Lee et al., *J. Vacuum Science & Technology A*, **13**, 1671-1675 (1995).
[9] M. Stöhr et al., *Proc. 37th EUPVSEC*, 521-524 (2020).
[10] T. Dullweber, SNEC Conf. (2021).
[11] K. Ohdaira et al., Proc. *21st EUPVSEC*, 1716, (2006).
[12] M. Tucci et al., Proc. *IEEE Conference on Optoelectronic and Microelectronic Materials and Devices COMMAD*, 242, (2008).
[13] M. Scherff, *Proc. 26th EUPVSEC*, 2125, (2011).
[14] M. Stöhr et al., *Proc. this conference.*, 2DV.3.2.
[15] M. Stöhr et al., ACS Appl. Energy Mater., 4 (5), 4646–4653 (2021).
[16] G. Dingemans et al., *J. Appl. Phys.* **110**, 093715 (2011).

LOCAL PASSIVATING CONTACTS FROM LASER DOPED P+ POLYSILICON

Florian Buchholz*, Jonathan Linke, Jan Hoss, Haifeng Chu, Valentin Mihailetchi, Aditya Chaudhary, Jayaprasad Arumughan, Jan Lossen, Radovan Kopecek, Eckard Wefringhaus
International Solar Energy Research Center Konstanz e.V.
Rudolf-Diesel-Straße 15, D-78467 Konstanz, Germany
*Corresponding author: +49 7531 36 183 553, florian.buchholz@isc-konstanz.de

ABSTRACT: A simple, mask-less process for the fabrication of localized p+ poly-Si structures is presented based on etch selectivity of highly doped p+ poly-Si. We show the optimization steps that were necessary to obtain a reliable, uniform and reproducible process, namely optimization of the doping source layer, laser parameters, alkaline etching and texturing step. Furthermore, the dopant profiles were tuned by different BBr_3 emitter diffusions in combination with tunnel oxide thickness variation. The suggested process flow allows for simple integration in the our n-type passivated contact cell concept TOUCAN™ as selective poly-Si emitter layers below the metal fingers on the front side.

Keywords: polysilicon, p+, passivating contacts, mask-less structuring, TOPCon

1 INTRODUCTION

Polysilicon (poly-Si) based passivating contacts have gained widespread interest of the scientific community and the industry in the last couple of years. The reason for this is the compatibility with traditional high temperature processing technology and the potential for significantly reduced surface recombination in both metalized and non-metalized areas. So far, mostly n+ poly-Si has found its way into solar cell mass production [1] while p+ poly-Si has been used to far lesser extent. The main reason for this is that for front and rear side contacted n-type TOUCAN™ (tunnel oxide under contacts on n-type wafers) cells, also known as TOPCon, a full area polysilicon layer on the front side would lead to excessive parasitic absorption cancelling out all advantages in surface and contact passivation.

To profit from the excellent TOUCAN™ passivation, one option is to apply the poly-Si on the front side only in the regions below the metal fingers and passivate the regions in between by a classical diffused emitter, as recently demonstrated for n+ poly-Si [2]. Since all common poly-Si deposition methods cover the full wafer area, an additional patterning step is required to allow local removal of the poly-Si between the metal fingers. So far, this is only achieved with substantial effort such as masked ion implantation, photolithography or laser mask ablation [2,3].

For the purpose of simplifying this patterning we have developed a process employing localized laser doping of intrinsic polysilicon from a full area boron silicate glass (BSG). The laser treatment creates an etching resistant p+ poly-Si layer [4] with an etching selectivity with respect to un-doped silicon that is estimated to be in the range of 500-700. Such layers can be used as masking layer against alkaline etching solution attack, e.g. during random pyramid texturing. A resulting local p+ poly-Si finger surrounded by a textured surface is shown in Fig. 1. This fast and simple process can be inserted into the baseline TOUCAN™ process sequence [5,6] without much additional equipment following the saw damage removal[1].

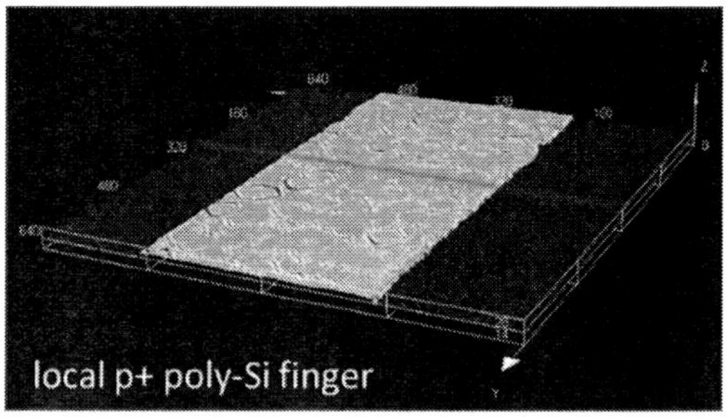

Figure 1: Laser scanning microscope image of local p+ poly-Si finger surrounded by textured silicon surface.

In the following, the optimization procedures for all crucial process steps necessary for the formation of such localized p+ poly-Si contacts are presented. This includes the BSG deposition, laser crystallization and simultaneous doping, etch-back and texturing of untreated poly-Si regions and subsequent emitter diffusion.

2 EXPERIMENTAL

2.1 Sample preparation

The process sequence to obtain localized passivating p+ poly-Si regions as sketched in Fig. 2 was applied to all samples if not stated otherwise. Czochralski grown silicon (Cz) wafer (n-type, 1-7 Ωcm) were saw-damage etched and piranha cleaned before a thin thermal oxide was grown with a thickness of ~2.3 nm. Subsequently, intrinsic a-Si was deposited in a conventional low-pressure chemical vapor deposition (LPCVD) furnace with a nominal thickness of 150 nm, followed by a BBr_3 diffusion for BSG deposition. After a short dip in hydrofluoric acid (HF) for thinning of the BSG, the front side of the wafers was locally laser-treated in order to crystallize and simultaneously dope the a-Si, resulting in a pattern of eight 3.7x3.7 cm^2 squares of etch-resistant p+ poly-Si. The layer stack on the non-laser-treated regions, i.e. the rear side, was removed by etching the remaining BSG in HF and the underlying background doped poly-Si in alkaline solution. The now exposed bare silicon surfaces were textured in alkaline solution and finally passivated by a second BBr_3

[1] The process of mask-less structuring has been applied for patent by the authors.

diffusion for emitter formation in the non-lasered regions. In some cases instead of BBr_3 diffusion, nitrogen anneal was used to study the properties of the layer without the second doing step. Hydrogenation was performed by a silicon nitride (SiN_x) deposition and subsequent firing step.

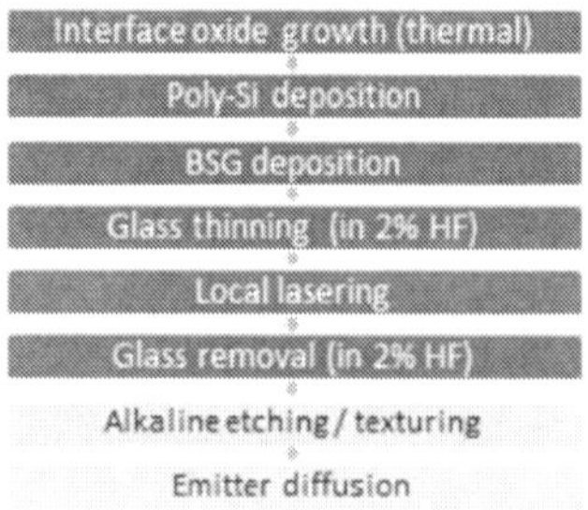

Figure 2: Investigated process sequence for formation of local p+ poly-Si fingers.

All the process equipment used was industrial or industrial-like equipment, such as Centrotherm tube furnaces for interface oxide growth and annealing, for LPCVD deposition of poly-Si, and BBr_3 diffusion furnace for the deposition of BSG as well as emitter diffusion. Hydrofluoric acid etching was performed in 2% HF and alkaline etching in a batch type RENA wetbench. Local laser treatment was performed in an industrially feasible high throughput tool. Where necessary, lifetime samples were produced by depositing SiN_x in a Centrotherm PECVD system and firing in a Centrotherm fast firing furnace.

2.2 Characterization

For each process step optimization suited characterization methods were applied. The passivation quality was determined by photoconductive decay measurements (PCD) in terms of recombination current density. Doping profile and sheet resistance of the p+ poly-Si layers as well as of the diffused emitter were determined by electrochemical capacitance voltage profiling (ECV) and four-point-probe measurements (4PP), respectively. Assessment of the smoothness of the poly-Si surface after laser doping and the texture quality was done by optical microscopy and visual inspection.

3 RESULTS AND DISCUSSION

3.1 BSG deposition optimization

First of all, a uniform BSG layer grown in a BBr_3 diffusion as doping source was required to ensure uniform doping in the poly-Si layer across the whole wafer and thus uniform etching stability. Those regions which show insufficient doping get etched to a larger extend. This may, in an extreme case yield a fully textured surface. In less pronounced cases the sheet resistance increases strongly, indicating a locally thinner layer.

Fig. 3 shows the p+ poly-Si sheet resistance of three reference wafers for several deposition recipes placed in the gas zone (GZ) where the gas is injected, center zone (CZ) and load zone (LZ) of the diffusion boat. After deposition, the wafers were laser treated on one side (full area), followed by etching in 2% HF to yield a target layer thickness of 5-25 nm. Afterwards the samples were subjected to the alkaline etching process and annealing.

The baseline recipe shows the strongest variance, with locally up to 1000 Ω/sq strongly missing the target sheet resistance range between 200 and 300 Ω/sq on 150 nm poly-Si layers.

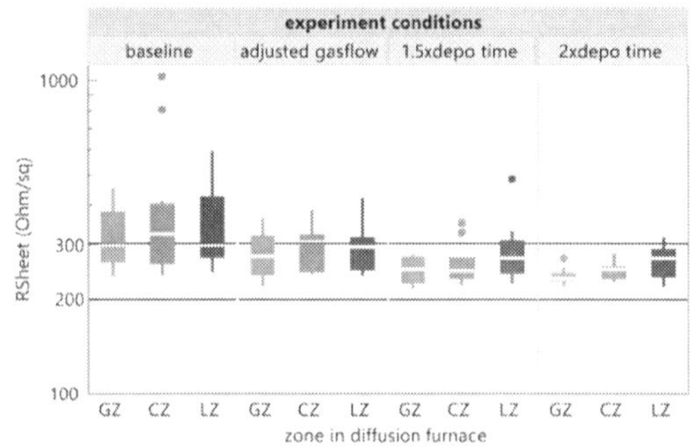

Figure 3: Sheet resistance of p+ poly-Si after laser treatment as function of the BBr_3 diffusion recipe. GZ (gas zone), CZ (center zone) and LZ (load zone) indicate the wafer position in the diffusion boat.

For the second set of data in Fig. 3, the BBr_3 concentration in the gas flow was increased, showing some improvement in lowering the data spread. The third and fourth set of data show the effect of increasing the deposition time by a factor of 1.5 and 2. Doubling the deposition time substantially improves the uniformity and lowers the sheet resistance as desired.

3.2 Laser process optimization

The laser output power P and pulse frequency f were adjusted to yield different pulse energy $E = P/f$. The samples used in this experiments were only treated on one side and textured on the other. Due to the inferior passivation quality on the textured side (SiN_x passivation only) the shown $J_0/2$ values do only partly reflect the good passivation quality of the lasered regions. Since the passivation of the textured side can be assumed to be constant for all samples, optimal laser parameters can be determined anyway.

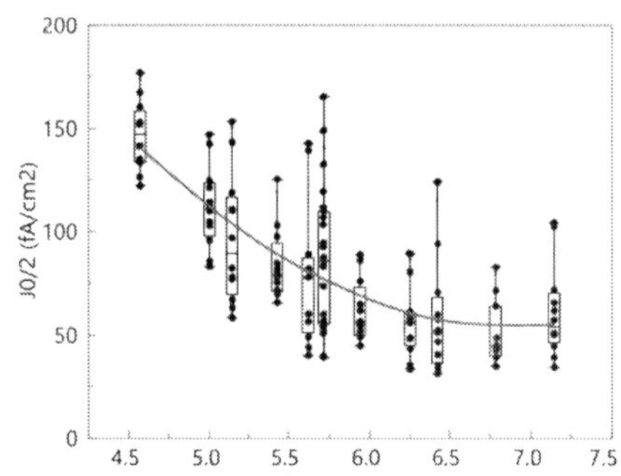

Figure 4: Recombination current of an asymmetrical, single side textured test structure with SiN_x passivated p+ poly-Si on the other side as function of the employed laser pulse energy.

Fig. 4 clearly shows that the recombination is reduced towards higher pulse energies. However, excessive pulse energies lead to severe cracking of the layer as it is shown on the left side of Fig. 5. A compromise between crack avoidance and passivation quality was found for a pulse energy of 5.9 (Fig. 5 right).

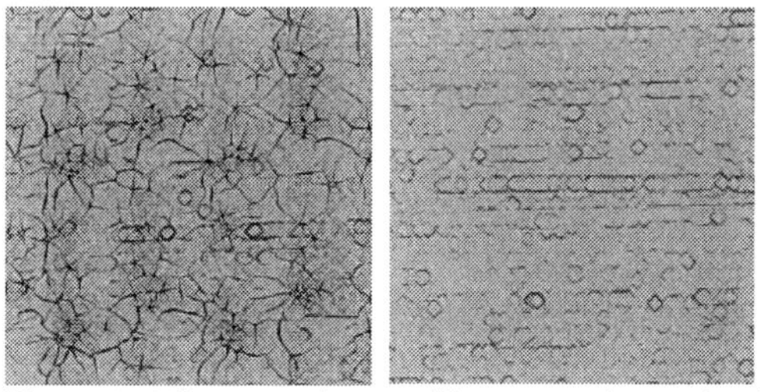

Figure 5: Microscope image of laser treated p+ poly-Si after etching, annealing and SiN_x passivation with excessive pulse energy leading to crack formation (left) and optimized crack-free pulse energy (right).

3.3 Alkaline etching/texturing optimization

The third topic in the optimization is the alkaline etching step to remove the non-lasered poly-Si. In fact, the laser treated area shows enough stability to survive a full texturing process with removal of 4.5 µm of silicon in about 12 min. However, non-lasered regions showed visual non-uniformity after texturing. To overcome this, we performed a dedicated poly-etching step for up to 160 s. Fig. 6 shows the resulting sheet resistances after poly-etch only and poly-etch plus texture of initially 150 nm as well as 200 nm poly-Si thickness and subsequent anneal. The increase of the sheet resistance with poly-etch duration originates from a thickness reduction. For the initially 200 nm poly-Si layer the sheet resistance increase is stronger compared to 150 nm poly-Si and accordingly the layer is less etch resistant.

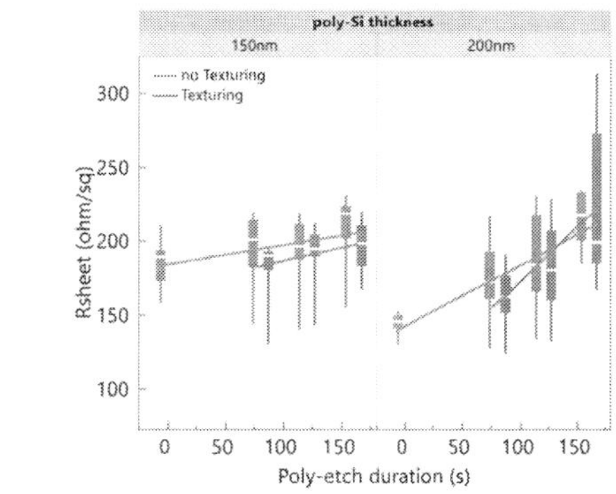

Figure 6: Sheet resistance of laser treated p+ poly-Si regions in relation to initial poly-Si layer thickness and poly-etching duration with and without subsequent texturing.

This can be confirmed by ECV measurements as shown in Fig. 7. The initially 150 nm poly-Si layer is hardly consumed during 160 s poly-etch and subsequent alkaline texturing compared to an un-etched reference. In contrast, the initially 200 nm poly-Si layer clearly shows a reduction of the thickness, thus reduced etch resistance. This observation indicates non-complete saturation of the layer with boron, even though no clear difference is observable in the measured active boron concentration of the ECV profiles. The large spread in sheet resistance observed after etching on the 200 nm indicates that the etching resistance is not uniform across the wafer.

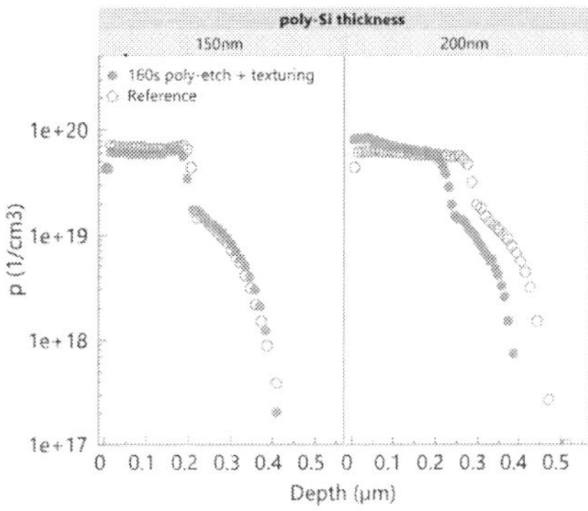

Figure 7: ECV measured boron profiles of laser-treated p+ poly-Si before (Reference) and after poly-etching/ texturing for 150/200 nm poly-Si layer thickness.

3.4 Diffusion profile tuning

In order to fabricate TOUCAN™ solar cells, the textured regions between the localized p+ poly-Si fingers require an emitter diffusion step. Since the open poly-Si fingers already have been doped and an in-diffused tail below the interfacial oxide will develop, the effect of the additional thermal budget and additional dopants needs to be studied. In addition to an emitter diffusion process variation, also the interfacial oxide thickness was varied, as it is well known that it also has a strong effect on the in-diffused tail profile. All four tested diffusions as summarized in Tab. I were applied to samples with initially 200 nm poly-Si after laser doping and alkaline etching/texturing. Three different oxide thicknesses (2.3, 2.5 and 2.7 nm) were used, which were obtained by varying the oxidation temperature.

Table I shows the recombination current densities of the diffusion profiles on textured regions with recipe B14 yielding the best passivation. The corresponding sheet resistances range from 200 – 400 Ω/sq.

Table I: Diffusion recipes relative to the B11 baseline diffusion used for emitter formation in the textured regions and corresponding recombination current densities.

	Depo. temp.	Drive-in temp.	Drive-in time	J_{0e} (text.) in fA/cm²
B11	baseline	baseline	baseline	32.5 ±2.8
B12			-20 min	43.8 ±3.4
B13		-25 K	-15 min	38.9 ±8.4
B14	-30 K	-25 K	-15 min	25.3±5.7

The resulting sheet resistance values of p+ poly-Si regions after laser treatment, poly-etching, texturing, and emitter diffusion are shown in Fig. 8. Decreasing the thermal budget from B11 to B14 increases the sheet resistance of the layers, which may indicate incomplete crystallization and activation of the dopants. An even stronger increase of the sheet resistance can be seen for thicker interface oxide layers.

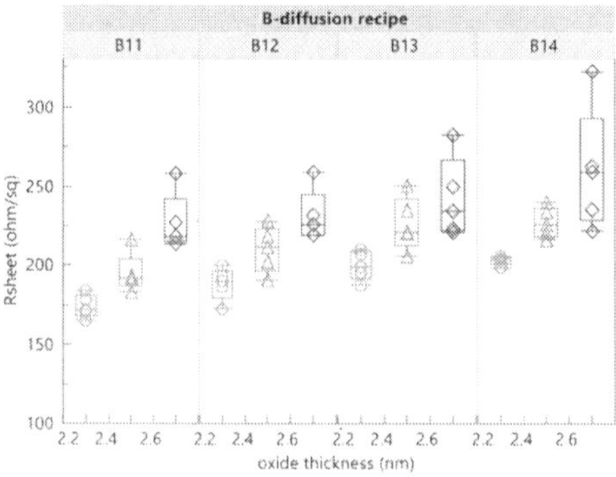

Figure 8: Sheet resistance of laser doped p+ poly-Si after poly-etching and texturing in relation to different subsequent emitter diffusions (B11-B14) and interfacial oxide thicknesses.

Overlaying the ECV measured boron profiles of the p+ poly-Si regions in Fig. 9 reveals that the plateau concentration of the poly-Si layers is hardly affected by the thermal budget or the thermal oxide thickness. In contrast, lower thermal budget and thicker oxides decrease the tail depth, which consistently explains the simultaneous increase in sheet resistance (Fig. 8). Thus, the in-diffused tail depth plays an important role in the overall active carrier concentration and transport properties.

The shown means (diffusion recipe and interface oxide) can now be used to co-optimize passivation and contact properties. It is to be presumed that a certain tail depth will be required for a low poly-Si to c-Si resistance. A low thermal profile diffusion will be favorable for good passivation quality of the textured regions.

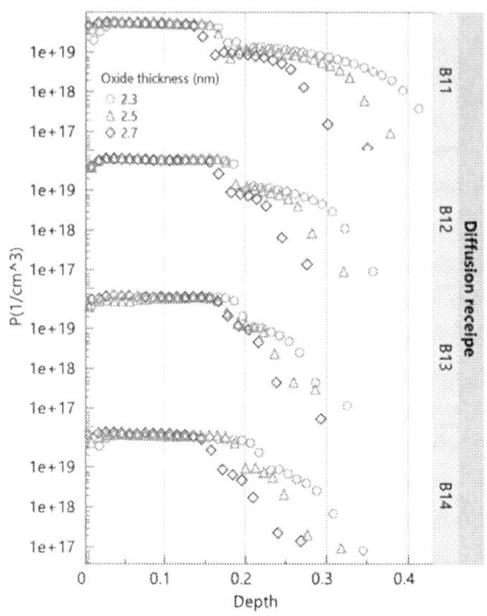

Figure 9: ECV measured boron profiles of laser doped p+ poly-Si after poly-etching and texturing in relation to different subsequent emitter diffusions (B11-B14) and interfacial oxide thicknesses.

4 SUMMARY AND OUTLOOK

We presented the optimization steps that were necessary to develop a process that can be used to produce local p+ poly-Si fingers as selective poly-Si emitter regions below the metal fingers on the front side of TOUCAN™ solar cells. A sufficient amount of boron, uniformly distributed over the wafer surface was accomplished and ensured uniform etching stability after crack-free laser crystallization and doping. Applying a poly-etch step prior to alkaline texturing lead to uniformly distributed pyramids in the non-laser-treated regions, while on the laser-treated p+ poly-Si fingers hardly any etching was observed. A second boron diffusion step for emitter formation in the non-laser-treated regions enabled in combination with an adjustment of the interfacial oxide thickness fine tuning of the final doping tail depth under the p+ poly-Si towards the required needs for reliable contacting. The presented process flow allows for simple integration in the TOUCAN™ process sequence.

5 ACKNOWLEDGEMENTS

This work was supported in part by the project that has received funding from the European Union's Horizon 2020 research and innovation program under grant agreement No 857793.

6 REFERENCES

[1] Chen D, Chen Y, Wang Z, Gong J, Liu C, Zou Y, He Y, Wang Y, Yuan L, Lin W and Xia R. 24.58% total area efficiency of screen-printed, large area industrial silicon solar cells with the tunnel oxide passivated contacts (i-TOPCon) design. Solar Energy Materials and Solar Cells 2020; 206:110258.

[2] Yuan, Shengzhao, et al. "Passivated Emitter and Rear Cell Silicon Solar Cells with a Front Polysilicon Passivating Contacted Selective Emitter." physica status solidi (RRL)–Rapid Research Letters (2021): 2100057.

[3] Yang, G., Guo, P., Procel, P., Limodio, G., Weeber, A., Isabella, O., & Zeman, M. (2018). High-efficiency black IBC c-Si solar cells with poly-Si as carrier-selective passivating contacts. Solar Energy Materials and Solar Cells, 186, 9-13.

[4] Charavel, R., J. Laconte, et al. (2003). "Advantages of p++ polysilicon etch stop layer versus p++ silicon." Proc SPIE, p. 5116.

[5] Kafle, B., Goraya, B. S., Mack, S., Feldmann, F., Nold, S., & Rentsch, J. (2021). TOPCon–Technology options for cost efficient industrial manufacturing. *Solar Energy Materials and Solar Cells*, *227*, 111100.

[6] Chaudhary, A., Hoß, J., Lossen, J., van Swaaij, R., & Zeman, M. (2019, August). Screen printed Ag contacts for n-type polysilicon passivated contacts. AIP Conference Proceedings,Vol. 2147, No. 1, p. 040002.

NOVEL POLY-SI:GA/SIO$_X$ PASSIVATING CONTACTS THROUGH NON-EQUILIBRIUM DOPING

Kejun Chen[1,2], Enrico Napolitani[3], Matteo De Tullio[3], Chun-Sheng Jiang[2], San Theingi[2], William Nemeth[2], Matthew Page[2], Paul Stradins[2], Sumit Agarwal[1,2], and David L. Young[2]
[1]Colorado School of Mines, 1500 Illinois St, Golden, CO, 80401, USA
[2]National Renewable Energy Laboratory, 15013 Denver W Pkwy, Golden, CO, 80401, USA
[3]University of Padova, Padova, PD, 35122, Italy
Email: kejunchen@mines.edu; kchen@nrel.gov

ABSTRACT: *Poly*-Si/SiO$_x$ passivating contacts are one of the key enablers for high-efficiency, low-cost *c*-Si solar cells. In recent years, record devices have reached efficiencies of 26% in the laboratory and 25% in the industry. The current cell structure utilizing a tunneling oxide passivating contact structure with a back phosphorus-doped *poly*-Si/SiO$_x$ passivating contact and a front boron diffused emitter suffers from large emitter recombination. Thus, replacing the front B diffused emitter with a *p*-type passivating contact is a route to overcome this deficiency. This gives rise to front/back *poly*-Si based passivating contacts. To address the low passivation performance of the B-doped *poly*-Si passivating contacts, we replace B with Ga as a novel *p*-type dopant to avoid dopant accumulation in the tunneling oxide, which is known to lead to large degradation loss in passivation quality. Here, we introduce a non-equilibrium method via pulsed laser melting to thermally melt and recrystallize the *poly*-Si and redistribute the dopants, achieving doping concentrations above the solid solubility limit (~4E19 cm^{-3}). We demonstrate a good passivation quality with an iV_{oc} of 721 mV with an active Ga doping concentration in *poly*-Si >10^{20} cm^{-3}. Furthermore, we show a low contact resistivity of 33.2 ± 9.3 mΩ· cm^2 using a diode model calculation. Finally, cross-section scanning spreading resistance microscopy was performed to determine the resistance profile across the non-homogeneously doped *poly*-Si layer.

Keywords: c-Si, polysilicon based passivating contacts, pulsed laser melting, contact resistivity

1 INTRODUCTION

Recently, *poly*-Si based passivating contacts have been attracting large research interests in the monocrystalline Si PV field. The record laboratory cell has reached an efficiency of 26% with a tunnel oxide passivating contact (TOPCon) structure, which features a back *poly*-Si/SiO$_x$ passivating contact with a front boron diffused emitter [1]. The adoption of n^+ passivating contacts is largely owed to the excellent passivation quality of the phosphorus-doped contacts. Boron-doped *poly*-Si contacts, however, have an inferior performance with a lower iV_{oc} and a much higher J_0 (~20 fA/cm^2) compared to the P-doped *poly*-Si contacts (< 5 fA/cm^2) [2]. Although the reason behind this is still under debate, one cause is the accumulation of B dopants at the SiO$_x$/*c*-Si interface, which can create defects within the oxide and degrade passivation [3,4]. In our previous work, we carried out a study exploring Ga as an alternative dopant for *p*-type *poly*-Si/SiO$_x$ to overcome the limitation of the poor passivation quality of B-doped passivating contacts [5]. We show that Ga dopants do not accumulate at the SiO$_x$/*c*-Si interface after high-temperature annealing, due to the high segregation coefficient (k=C_{Si}/C_{SiOx}) and high diffusivity of Ga in the oxide. Despite a good iV_{oc} (> 730 mV) that was demonstrated using the *poly*-Si:Ga/SiO$_x$ contacts, a high metal-to-semiconductor contact resistivity informed us that hyperdoping of the *poly*-Si is needed due to its relatively low solid solubility limit in Si (4×10^{19} cm^{-3}) [6]. Good *poly*-Si/SiO$_x$ passivating contacts generally require *poly*-Si doping >1×10^{20} cm^{-3} to ensure a low resistivity contact with the metal. To achieve a doping concentration above the Ga solid solubility limit in Si, nonequilibrium methods, such as pulsed laser melting, is needed [7]. A ns laser pulse can result in rapid melting and recrystallization of the material, often called liquid phase epitaxial regrowth [8]. During this process, dopant atoms diffuse faster in the melt due to the higher diffusion coefficient in the molten Si and incorporate into the Si lattice upon rapid recrystallization.

Here, we present novel *poly*-Si:Ga/SiO$_x$ contacts on *n*-type polished Cz Si wafers using ion implantation of Ga and pulsed laser melting. Fig. 1 shows the schematic of the *poly*-Si surface after being subjected to an excimer laser. The color indicates the heat distribution from the surface into the bulk *poly*-Si. The molten Si is then rapidly recrystallized after the nanosecond pulse, freezing the Ga dopants within the *poly*-Si lattice. We demonstrate a good passivation quality with an iV_{oc} of 721 mV with an active Ga doping concentration in *poly*-Si >10^{20} cm^{-3}. Furthermore, we demonstrate a low contact resistivity of 33.2 ± 9.3 mΩ· cm^2 using a diode model calculation. To explore its applicability for large-area cells, we applied the laser process on a large area sample (40 mm × 30 mm) and show that a high passivation quality is obtained, similar to the one for the small area sample. Finally, scanning spreading resistance microscopy (SSRM) was performed to analyze the relationship between the lowly doped *poly*-Si region and the obtained low contact resistivity.

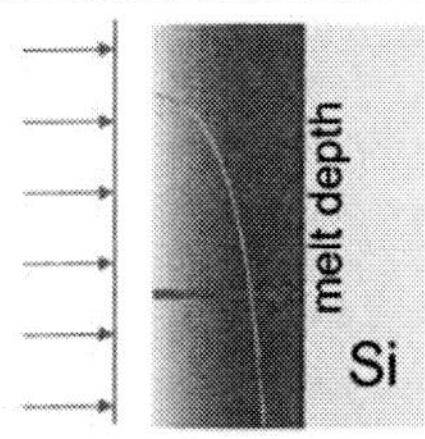

Figure 1: Schematic of excimer laser melting heat distribution profile and the movement of melt front.

2 EXPERIMENTAL DETAILS

Single-side polished *n*-Cz wafers with symmetric tunneling SiO$_x$ (~1.5 nm) and 250 nm thick intrinsic *poly*-Si layers were deposited via low-pressure chemical vapor deposition (LPCVD). All samples were then ion-implanted with 10 keV of Ga-69 at a dose of 6×10^{15} cm^{-2}.

Afterwards, the wafers were laser scribed into 30 × 50 mm test samples and were cleaned with RCA-1 and 2. The samples then received passivation via atomic layer deposition (ALD) of Al_2O_3 and forming gas annealed (FGA) at 400°C for 60 min. Next, all samples were processed with an excimer laser using a KrF laser with a wavelength of 248 nm and a pulse duration of 22 ns. The laser spot had a dimension of 5 × 5 mm^2. Simulations were performed to visualize the maximum melt depth distribution with various laser energy densities (ED) between 300-1100 mJ/cm^2 and different pulses (1-6). A series of different EDs and pulses were tested on a 30 × 50 mm sample (shown in Fig. 2). Following a second passivation using Al_2O_3 and FGA, the samples were characterized for their passivation qualities using photoluminescence imaging (PL) based on a comparison with iV_{oc} values measured on a witness sample using a PCD (WCT-120) lifetime tester. The doping distribution profiles and the electrical characterization of the samples were analyzed using secondary ion mass spectrometry (SIMS) and Van der Pauw Hall measurements, respectively. The contact resistivities were measured and calculated following the expanded Cox and Strack method in Wang *et al* [9]. Scanning spread resistance microscopy (SSRM) was performed to test the local spreading resistance of the sample.

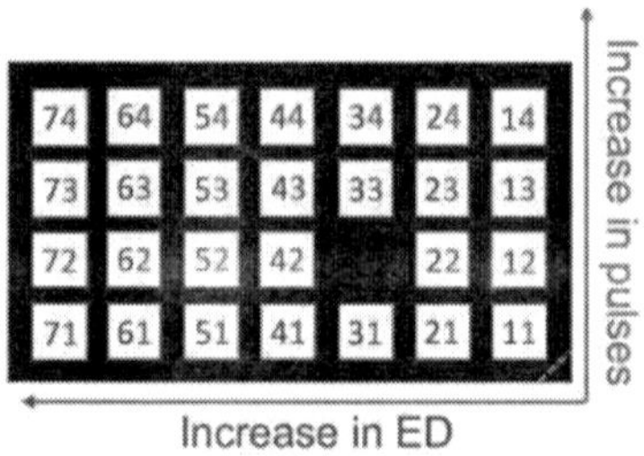

Figure 2: Schematics showing 27 laser processing spots on a 30 mm × 50 mm polished Si wafer. In each laser spot, the numbers in blue scale with the laser energy density (1-7), and the numbers in green indicate the number of the laser pulse (1-4).

3 RESULTS AND DISCUSSION

3.1 Passivation and electrical properties

As described in the experimental section, the samples were subjected to various laser annealing conditions, and Fig. 3 shows that the PL image of one sample after laser processing and the second passivation. The laser spots appear bright in PL and the PL intensity was converted into iV_{oc} values (numbers in Fig. 3, mV units). Based on Fig. 2, we see that the PL brightness increases with increasing laser EDs and pulses. In Fig 3 we see three field regions that did not receive any laser annealing. Generally speaking, the passivation quality of the laser annealed region is better than the nearby field regions. The highest iV_{oc} value measured is 721 mV, with a laser ED of 900 mJ/cm^2 and 4 pulse. Note that the back-side passivation quality of these single-side polished samples was not optimum, which lowered the measured passivation of the laser annealed regions. To explore the ability to process large area samples using the excimer laser, we conducted a test using overlapping laser spots on a 30× 40 mm test sample and show that a similar iV_{oc} value of 721 mV can be achieved with no sign of degradation from the overlapping corners and edges of the laser spots.

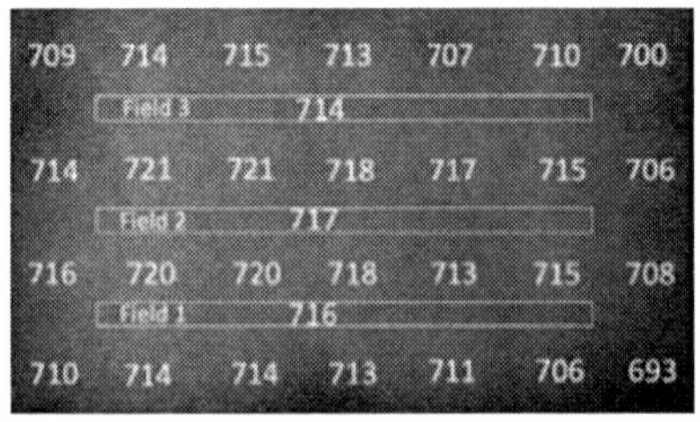

Figure 3: PL image of a polished Ga-doped *poly*-Si/SiO_x test sample after laser annealing and the second passivation step. The numbers indicate the iV_{oc} for each laser spot in mV units. The iV_{oc} values for the three field regions that were unexposed to the laser are also shown.

We also performed SIMS profile for a laser annealed *poly*-Si:Ga/SiO_x sample with varying laser EDs. Ga shows a doping concentration above its solid solubility limit in the first 50 - 200 nm of *poly*-Si. This result shows that the pulsed laser melting is capable of achieving non-equilibrium doping of Ga in *poly*-Si. Moreover, it is clear that the Ga concentration increases with respect to depth for increasing laser EDs. This promising result shows that we can achieve precise doping profiles by controlling the laser conditions. Furthermore, Hall measurements (not shown) indicate that even for the lowest ED, the bulk activated Ga concentration reaches above 10^{20} cm^{-3}, with the highest concentration of 3×10^{20} cm^{-3} realized by the highest ED (1100 mJ/cm^2) and pulse condition (6 pulse).

3.2 Contact resistivity through diode measurement

To evaluate the contact resistivity of the *poly*-Si:Ga/SiO_x contacts on *n*-type wafers, we employed the transfer length method (TLM). The metal-to-*poly*-Si contact resistivity is determined to be 0.9 mΩ·cm^2. However, to accurately extract the contact resistivity contributed from *poly*-Si-to-*c*-Si, we evaporated TiAg contacts onto the samples using a series of variously sized dot patterns through a shadow mask onto each laser spot. The circular electrode diameters were 35 μm, 70 μm, 0.125 mm, 0.5 mm, and 1 mm. Following a reactive ion etch in an SF_6 plasma, mesa structures, isolating the metal contacts, were produced. An ohmic contact was formed on the back of the sample by etching off the back *poly*-Si/SiO_x and contacting directly with a GaIn eutectic alloy. Current-voltage data across the diode were measured by contacting the circular electrode at the front and the GaIn eutectic at the back.

The analysis for extracting contact resistivity follows Equation (1) in Wang *et al* [9].

$$\frac{d(V)}{d(lnI)} = R_T I + \frac{nkT}{q} \quad (1)$$

The total resistance (R_T) can be obtained from the slope of dV/dlnI. The diode resistance (R_{diode}) can be calculated by subtracting the spreading resistance (R_s) and the resistance of the bottom contact (R_0) from R_T. R_s was determined according to Denhoff [10] and R_0 was measured by bypassing the diode region of the sample. For an ED of 900 mJ/cm^2 and 4 pulse (laser spot 53 in Fig. 2), the slope shows that the final calculated diode contact resistivity is 33.2 ± 9.3 mΩ·cm^2. This analysis revealed that despite the lowly doped *poly*-Si region that exists between the heavily doped *poly*-Si and the tunneling oxide (Fig. 4), the contact resistivity remains low. Additionally, we performed cross-sectional SSRM measurements to probe the local spreading resistance as a function of the distance from the top electrode. The results show that there exists a high resistance region with low doping concentration within the *poly*-Si, between the highly doped *poly*-Si and the tunneling oxide, and this region spans about 100 -150 nm. This result matches the SIMS profile showing the lowly doped region near the SiO_x layer. We are still investigating whether the doping concentration needs to be uniformly distributed in *poly*-Si and extend into *c*-Si to obtain a lower contact resistivity.

4 CONCLUSIONS

In this work, we demonstrated the novel *poly*-Si:Ga/SiO_x passivating contacts through non-equilibrium pulsed laser melting and obtained iV_{oc} as high as 721 mV on *n*-Cz Si, with a contact resistivity value of ~33.2 ± 9.3 mΩ·cm². This result addresses our previous work with very high contact resistivity of ~10 Ω· cm². Using the 250 nm of *i-poly*-Si on *n*-Cz samples, we showed that the SiO_x layer was not impacted by the laser heat distribution profile. Additionally, we explored large area laser melting and showed good passivation quality with no damage from the overlapping laser spots. Finally, we show that despite a lowly doped region existing within the *poly*-Si, a low contact resistivity can still be obtained.

5 ACKNOWLEDGEMENTS

This work was authored in part by the National Renewable Energy Laboratory, operated by Alliance for Sustainable Energy, LLC, for the U.S. Department of Energy (DOE) under Contract No. DE-AC36-08GO28308. Funding provided by the U.S. Department of Energy Office of Energy Efficiency and Renewable Energy Solar Energy Technologies Office. The views expressed in the article do not necessarily represent the views of the DOE or the U.S. Government. The U.S. Government retains and the publisher, by accepting the article for publication, acknowledges that the U.S. Government retains a nonexclusive, paid-up, irrevocable, worldwide license to publish or reproduce the published form of this work, or allow others to do so, for U.S. Government purposes. The work at the University Padova was supported by the University of Padova (grant UNIPD-ISR 2017 'SENSITISE').

6 REFERENCES

[1] A. Richter *et al.*, *Nature Energy*, vol. 6, no. 4, pp. 429-438 (2021).
[2] F. Feldmann *et al.*, *Solar Energy Materials and Solar Cells*, vol. 159, pp. 265-271 (2017).
[3] K. Inoue *et al.*, *Applied Physics Letters*, vol. 95, no. 4, p. 043502 (2009).
[4] T. Yamamoto, K. Uwasawa and T. Mogami, IEEE Transactions on Electron Devices **46** (5), 921-926 (1999).
[5] D. L. Young *et al.*, IEEE Journal of Photovoltaics 7 (6), 1640-1645 (2017).
[6] F. A. Trumbore, *The Bell System Technical Journal*, vol. 39, no. 1, pp. 205-233 (1960).
[7] J. M. Poate, *Laser Surface Treatment of Metals*, pp. 597-609 (1986).
[8] P. Baeri, S *et al.*, *Journal of Applied Physics*, vol. 50, no. 2, pp. 788-797 (1979).
[9] W. Wang *et al.*, *IEEE Journal of Photovoltaics*, vol. 9, no. 4, pp. 1113-1120 (2019).
[10] M. W. Denhoff, *Journal of Physics D: Applied Physics*, vol. 39, no. 9, pp. 1761-1765 (2006).

CONTROLLING DOPING DENSITY IN DC-SPUTTERED IN-SITU PHOSPHOROUS-DOPED POLYSILICON LAYERS FOR PASSIVATING CONTACTS

Lasse Nasebandt[1,2], Byungsul Min[1], Simon Hübner[3], Torsten Dippell[3], Peter Wohlfart[3], Robby Peibst[1,4], and Rolf Brendel[1,2]
[1]Institute for Solar Energy Research Hamelin (ISFH), Am Ohrberg 1, 31860 Emmerthal, Germany
[2]Department of Solar Energy, Institute of Solid-State Physics, Leibniz University Hannover, Appelstr. 2, 30167 Hanover, Germany
[3]Singulus Technologies AG, Hanauer Landstraße 103, 63796 Kahl am Main, Germany
[4]Institute of Electronic Materials and Devices, Leibniz University Hannover, Schneiderberg 32, 30167 Hanover, Germany
Corresponding author: phone +49-5151-999-634, fax: +49-5151-999-400, email: nasebandt@isfh.de

ABSTRACT: This article reports on using direct current (DC) sputter deposition of Si for passivating polysilicon on oxide (POLO) contacts. We measure the doping density in DC-sputtered in-situ n^+-type poly-Si layers and study their passivation quality using symmetric lifetime samples. We compare in-situ doped samples with samples that are additionally ex-situ doped by a $POCl_3$ diffusion. The best sputtered and solely in-situ doped sample shows excellent passivation quality with an implied open circuit voltage of 734 mV and a median surface recombination current density of (2.1 ± 0.4) fA/cm² after a hydrogenation step.
Keywords: polycrystalline, sputtering, passivation

1 INTRODUCTION

While passivating poly-Si on oxide contacts (POLO) were mainly reported to use CVD-based methods for the deposition of poly-Si on top of a thin interfacial oxide [1]–[5], other deposition techniques such as physical vapor deposition (PVD) and in particular sputtering is an attractive alternative. In contrast to the CVD-based methods, sputtering lacks the use of hazardous gases such as silane and phosphine and is a strictly single-sided deposition. This spares a subsequent single-sided etching that becomes necessary to remove the deposition in the wrong side (LP-CVD) or for removing a possible wrap around deposition (PE-CVD).

However, achieving high level in-situ P-doping with densities > 5×10^{19} cm^{-3} by sputtering is challenging because that highly phosphorous-doped targets are not readily available. To overcome this challenge an intrinsic sputter target may be used. The dopants are provided by an ex-situ $POCl_3$ diffusion. This allows surface recombination parameters J_0 < 10 fA/cm² [6], [7]. By using a single target without ex-situ doping, a J_0 value of around 35 fA/cm² was reported with lab-scale sputter equipment [8].

However, a passivation quality of J_0 = 5 fA/cm² is typical for CVD processes [9]–[11]. This quality was not yet achieved by sputtering and in-situ doping in an industrial tool.

In this work, we demonstrate the fabrication of passivating contacts via single target DC-sputtering using a phosphorous-doped sputtering target as the only doping source with a median J_0 value of (2.1 ± 0.4) fA/cm². We use an industrial type of sputtering tool.

2 SAMPLE FABRICATION

We fabricate symmetrical lifetime samples with 156.75 mm × 156.75 mm (M2) P-doped Czochralski-grown monocrystalline silicon wafers with a specific resistivity of 4 Ωcm. Figure 1 depicts the process flow and the cross-section of the samples. All samples receive a saw damage etch and cleaning. Subsequently, a wet-chemical oxidation in ozonated de-ionized water (DI/O_3) grows a

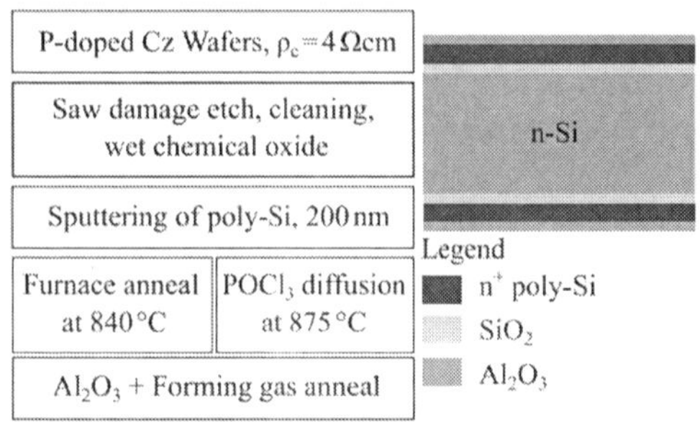

Figure 1 Process flow and cross-section of the symmetrical lifetime samples with chemical interfacial oxide and Al_2O_3 as hydrogenation source

1.5 nm-thin interfacial oxide layer (layer thickness determined via ellipsometry on polished float zone samples). We deposit the silicon layer with a thickness of 200 nm on both sides via DC-sputtering on a SINGULUS sputter coater (Vistaris600) using a silicon target with a purity of 6N and a P-dopant concentration of 4.7×10^{19} cm^{-3}. The deposition pressure during sputtering is 0.75 Pa.

The first group of samples is annealed in a conventional tube furnace at 840 °C for 30 minutes. The second group receives instead of the tube furnace anneal an ex-situ $POCl_3$ diffusion at 875 °C with a $POCl_3$ gas flow of 600 sccm. Secondary-ion-mass spectrometry (SIMS) measures the depth profile of the dopant density and electrochemical capacitance voltage (ECV) profiling measures the electrically active dopants.

An initial QSSPC measurement with a Sinton WCT-120 lifetime tester determines the carrier lifetime. We then coat all samples on both sides with 10 nm Al_2O_3 by plasma assisted atomic layer deposition and annealed in forming gas at 400 °C for 30 minutes to hydrogenate the sample. We then measure the lifetime of the samples again and extract the implied open-circuit voltage iV_{OC} from the irradiance-voltage curve and use the well-known slope method from Kane and Swanson for extracting the prefactor J_0 of the recombination current density [12], [13].

3 RESULTS

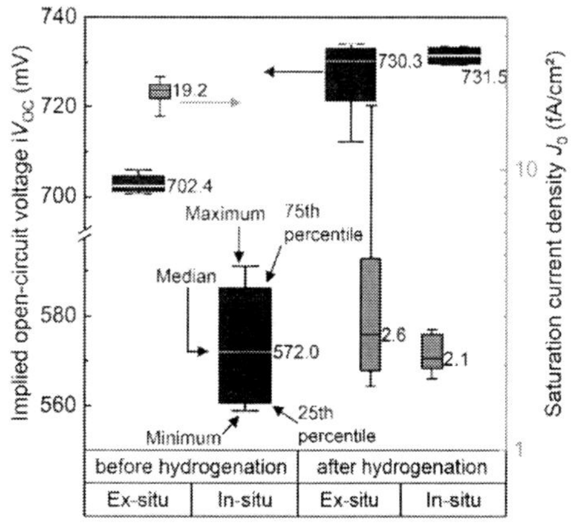

Figure 2 Implied open-circuit voltage and recombination parameter J_0 at the surface before and after hydrogenation with in-situ and ex-situ doped sputtered poly-Si layers.

3.1 Impact of the hydrogenation step on iV_{OC} and J_0

Figure 2 shows the implied open circuit voltage iV_{OC} and the recombination parameter J_0.

Before hydrogenation, the in-situ doped samples show a median iV_{OC} value of 572 mV, indicating poor passivation quality. It was not possible to extract J_0 values because the injection-dependent lifetime curves were nearly flat, hence it was not possible to apply the slope method.

In contrast, the ex-situ $POCl_3$-diffused samples show a good passivation quality already before hydrogenation with a median iV_{OC} value of 702 mV and J_0 of 19 fA/cm².

After hydrogenation, samples from both groups show equally high passivation quality. The improvement in surface passivation caused by the hydrogenation is obviously larger for in-situ doped samples than for ex-situ diffused samples. This increases the median iV_{OC} by

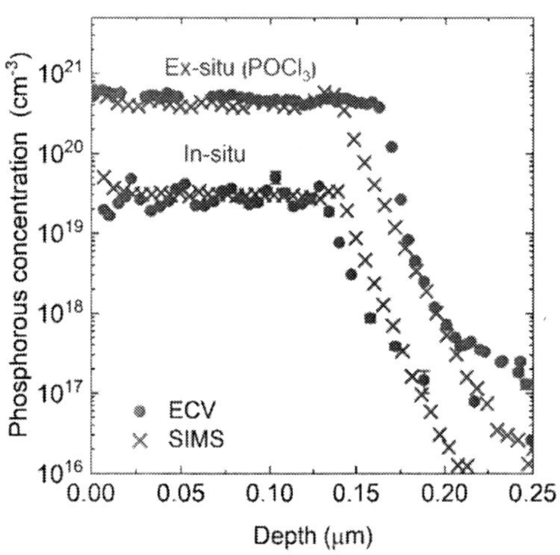

Figure 3 Depth profiles of the P-impurity and active dopant concentration concentrations of ex-situ and in-situ doped samples as measured by ECV (circles) and SIMS (crosses).

160 mV to 732 mV for the in-situ doped samples whereas the ex-situ doped samples only increase by 28 mV. J_0 increases to median values of 2.6 fA/cm² and 2.1 fA/cm² for ex-situ doped and in-situ doped samples, respectively. The best J_0 values for the in-situ doped samples is 1.8 fA/cm².

3.2 Dopant density depth profiles

The total phosphorous concentration in the as-deposited state is 4.2×10^{19} cm^{-3} (not shown here) which agrees well with the specified phosphorous concentration of the silicon target.

Figure 3 shows the SIMS and ECV profiles after anneal and diffusion, respectively. The agreement of the SIMS and ECV measurements indicates a complete activation of dopants within the measurement uncertainty.

The ex-situ doped sample and in-situ doped have an electrically active phosphorous concentration in the poly-Si of 4.9×10^{20} cm^{-3} and 2.8×10^{19} cm^{-3}, respectively. Despite having a doping density more than one magnitude lower than the ex-situ doped samples, the doping density in the in-situ doped samples is apparently high enough for excellent surface passivation quality if a sufficient amount of hydrogen is provided.

4 CONCLUSIONS

In this paper, we achieve a record low J_0 value of 1.8 fA/cm² for passivating contacts with single target, in-situ phosphorus-doped, sputtered poly-Si layers using industrial sputtering tool. The hydrogenation step is more essential for the in-situ doped sample than the ex-situ doped sample due to different dopant density levels in the poly-Si layers as shown in Figure 2.

Our results demonstrate that the dopant density level in commercially available phosphorus-doped sputtering targets is sufficiently high to fabricate excellently passivating contacts by DC-sputtering without an additional doping source such as a $POCl_3$ diffusion.

Our future work will concentrate on screen-printing sputtered poly-Si layers and incorporating these sputtered poly-Si passivating contacts into a solar cell batch using a POLO back junction cell structure [14], [15].

Acknowledgements

The authors would like to thank Bianca Gehring, and Magalie Pollmann (both from ISFH) for their help with sample processing. This work was financially supported by the German Federal Ministry for Economic Affairs and Energy (BMWi) under contact number 03EE1012A (NanoPERC). The content is the responsibility of the authors.

6 REFERENCES

[1] F. Haase, B. Min, C. Hollemann, J. Krügener, R. Brendel, and R. Peibst, "Fully screen-printed silicon solar cells with local Al-p + and n-type POLO interdigitated back contacts with a V_{OC} of 716 mV and an efficiency of 23%," *Prog. Photovolt. Res. Appl.*, vol. 29, no. 5, pp. 516–523, May 2021, doi: 10.1002/pip.3399.

[2] F. Feldmann, M. Simon, M. Bivour, C. Reichel, M. Hermle, and S. W. Glunz, "Carrier-selective

contacts for Si solar cells," *Appl. Phys. Lett.*, vol. 104, no. 18, p. 181105, 2014, doi: 10.1063/1.4875904.

[3] M. K. Stodolny *et al.*, "N-Type polysilicon passivating contact for industrial bifacial n-Type solar cells," *Sol. Energy Mater. Sol. Cells*, vol. 158, pp. 24–28, 2016, doi: 10.1016/j.solmat.2016.06.034.

[4] D. Chen *et al.*, "24.58% total area efficiency of screen-printed, large area industrial silicon solar cells with the tunnel oxide passivated contacts (i-TOPCon) design," *Sol. Energy Mater. Sol. Cells*, vol. 206, p. 110258, 2020, doi: 10.1016/j.solmat.2019.110258.

[5] A. Merkle *et al.*, "Atmospheric Pressure Chemical Vapor Deposition of In-Situ Doped Amorphous Silicon Layers for Passivating Contacts," *35th Eur. Photovolt. Sol. Energy Conf. Exhib. 785-791*, p. 7 pages, 5610 kb, 2018, doi: 10.4229/35THEUPVSEC20182018-2DV.3.49.

[6] J. Steffens, J. Rinder, G. Hahn, and B. Terheiden, "Compensation of the sputter damage during a-Si deposition for poly-Si/SiOx passivating contacts by ex-situ p-Doping," *AIP Conf. Proc.*, vol. 2147, no. 1, p. 040018, 2019, doi: 10.1063/1.5123845.

[7] J. Hoß *et al.*, "Sputtering of silicon thin films for passivated contacts," in *15th International Conference on Concentrator Photovoltaic Systems (CPV-15)*, 2019, p. 040007. doi: 10.1063/1.5123834.

[8] D. Yan, A. Cuevas, S. P. Phang, Y. Wan, W. Yang, and D. Macdonald, "Passivating contacts for silicon solar cells based on physical vapour deposition," presented at the PVSEC-29, Shanxi Guesthouse, Xian, China, 2019.

[9] J. Schmidt, R. Peibst, and R. Brendel, "Surface passivation of crystalline silicon solar cells: Present and future," *Sol. Energy Mater. Sol. Cells*, vol. 187, pp. 39–54, 2018, doi: 10.1016/j.solmat.2018.06.047.

[10] R. Peibst *et al.*, "Implementation of n+ and p+ Poly Junctions on Front and Rear Side of Double-Side Contacted Industrial Silicon Solar Cells," *32nd Eur. Photovolt. Sol. Energy Conf. Exhib. 323-327*, p. 5 pages, 4490 kb, 2016, doi: 10.4229/EUPVSEC20162016-2BO.3.2.

[11] D. Yan, A. Cuevas, J. Bullock, Y. Wan, and C. Samundsett, "Phosphorus-diffused polysilicon contacts for solar cells," *Sol. Energy Mater. Sol. Cells*, vol. 142, pp. 75–82, 2015, doi: 10.1016/j.solmat.2015.06.001.

[12] R. A. Sinton, A. Cuevas, and M. Stuckings, "Quasi-Steady-State Photoconductance, A New Method for Solar Cell Material and Device Characterization," in *Proceedings of the 25th IEEE Photovoltaic Specialists Conference*, [New York, N.Y.] and Piscataway, N.J., 1996, pp. 457–460.

[13] D. E. Kane and R. M. Swanson, "Measurement of the Emitter Saturation Current by a Contacless Photoconductivity Decay Method," in *The Conference Record of the Eighteenth IEEE Photovoltaic Specialists Conference-1985*, 1985, pp. 578–583.

[14] B. Min *et al.*, "716 mV Open-Circuit Voltage with Fully Screen-Printed p -Type Back Junction Solar Cells Featuring an Aluminum Front Grid and a Passivating Polysilicon on Oxide Contact at the Rear Side," *Sol. RRL*, vol. 5, no. 1, p. 2000703, Jan. 2021, doi: 10.1002/solr.202000703.

[15] B. Min *et al.*, "Approaching 23% with p-Type Back Junction Solar Cells Featuring Screen-Printed Al Front Grid and Passivating Rear Contacts," presented at the 38th EU PVSEC 2021, Sep. 07, 2021.

This is a short version of the paper submitted to the journal EPJ Photovoltaics.
It has been selected by the Executive Committee of the EU PVSEC 2021 for submission to EPJ Photovoltaics.

Hydrogenation of Sputtered ZnO:Al Layers for Double Side Poly-Si/SiO$_x$ Integration

Charles SERON[1,3]; Thibaut DESRUES[1]; Frédéric JAY[1]; Adeline LANTERNE[1] ; Gaël BORVON[2] ; Frank TORREGROSA[2] ; Quentin RAFHAY[3]; Anne KAMINSKI[3] & Sébastien DUBOIS[1]

[1] *Univ. Grenoble Alpes, CEA, LITEN, Campus INES, 73375 Le Bourget-du-Lac, France*
[2] *Ion Beam Services (IBS), Rue Gaston Imbert Prolongée, ZI Rousset-Peynier, 13790 Peynier, France*
[3] *Univ. Grenoble Alpes, Univ. Savoie Mont-Blanc, CNRS, Grenoble INP, IMEP-LAHC, 38000 Grenoble, France*

ABSTRACT

This work reports on the development of hydrogenation treatments on indium-free transparent conductive oxide (TCO) layers. The final purpose is to integrate those TCO layers on solar cells featuring double-side thin polysilicon on oxide based passivated contacts. More precisely, ZnO:Al (AZO) layers were developed with in-situ incorporation of hydrogen during their deposition by DC magnetron sputtering. The AZO films feature optical and electrical properties compatible with the elaboration of high efficiency devices. They were integrated in complete solar cells with promising performances, with a maximum conversion efficiency reaching 22.4%. These results show the possibility to move towards indium-free alternatives for contacting poly-Si on oxide based passivated contacts.

I. CONTEXT AND APPROACH

The double side integration of polysilicon on oxide based passivated contacts is a promising way to improve the performances of crystalline silicon (c-Si) solar cells [1]–[3]. With such a double-side passivated contacts configuration, the collecting layers (i.e., poly-Si films) must be as thin as possible to limit parasitic absorption of incident light [4]. Thus, a full surface contact by a transparent conductive oxide layer (TCO) is usually required. If indium-based TCO are known for their excellent electrical and optical properties, issues about indium cost and scarcity are pushing the PV community to investigate new solutions [5], [6]. Compared to standard silicon heterojunction solar cells using hydrogenated amorphous silicon (a-Si:H) layers, the high temperature route with polysilicon on oxide structures presents several assets for using alternative TCO. For instance, poly-Si layers features higher lateral conductivity values (in comparison with a-Si:H), which could reduce the constraints on the TCO properties. It is especially the case in rear emitter configuration, as presented on Figure 1 [7]. However, the formation of poly-Si based structures implies a high temperature (> 800°C) annealing for crystallization and dopants activation. In the case where hydrogen atoms, which can passivate interfaces, crystal defects and impurities [8], were previously incorporated at the surface/bulk during deposition or doping treatments, this annealing step causes the effusion of hydrogen out of the silicon wafer. It is the case in this study where Plasma Immersion Ion Implantation (PIII) with hydrogen-rich PH_3 and B_2H_6 precursors are used for the respective phosphorus and boron doping of polysilicon layers. After dopants activation annealing, this loss of hydrogen affects the effective carrier lifetime due to recombination at unpassivated defects (surface and bulk) [9]. To reach high passivation levels, hydrogen must be incorporated again in the cell structure after this high temperature annealing.

In this work, in-situ addition of hydrogen into aluminum-doped zinc oxide ZnO:Al (AZO) layers is developed as an alternative to indium based TCO and to improve solar cell performances. This in-situ incorporation of hydrogen prevents the need for additional hydrogenation process steps. For TCO, hydrogenation treatments can improve their electrical and optical properties [10]. Moreover, thanks to its high diffusivity, hydrogen atoms initially present within TCO layers could migrate towards the underneath materials and interfaces, especially with thermal treatments, thus lowering the carrier recombination rate. This possibility will be investigated with effective lifetime measurements. Finally, double-side passivated contacts solar cells will be fabricated. This will allow to assess the impact of the TCO layer modifications as well as the TCO hydrogenation, on the performances of complete photovoltaic devices.

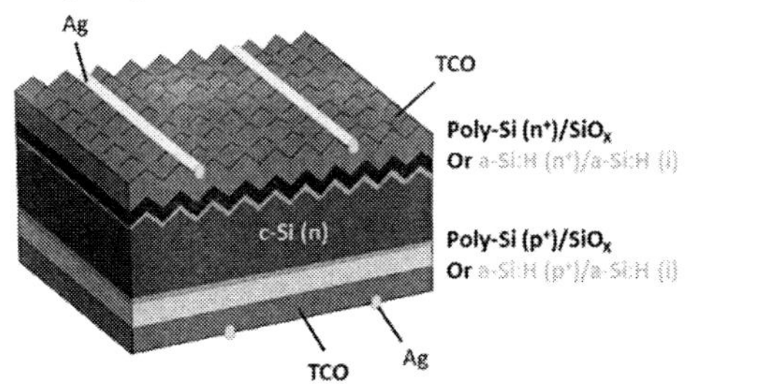

Figure 1: c-Si solar cell structure (rear emitter configuration) with double side passivated contacts elaborated either by low (a-Si:H) or high (poly-Si/SiO$_x$) temperature routes.

III. EXPERIMENTAL RESULTS

A. TCO deposition and hydrogenation process

ZnO:Al layers were deposited by magnetron sputtering with a Jusung equipment. The deposition chamber atmosphere is composed of a mix between pure Ar or Ar:H_2 (98:2%) and O_2 gases. The total gas flowing in the chamber is fixed at 600 sccm and the O_2 flow is also fixed at 1% of the chamber composition. Repartition between Ar and Ar:H_2 converging to the deposition chamber is modified to reach total compositions in hydrogen in the chamber of 0%, 0.5%, 1% and 1.5%. The different AZO:H layers will be distinguished by the H_2 amount used during deposition.

For material comparisons, a reference ITO layer is also deposited by magnetron sputtering. It is worth noting that this TCO is not optimized for contacting poly-Si passivated contacts stacks, and will be used only for material characterizations. For effective lifetime measurements and cell integrations, another ITO layer, hydrogenated and optimized for integration on passivated contacts solar cells is deposited with an in-line Meyer-Burger Physical Vapor Deposition (PVD) tool.

B. Electrical results after AZO:H deposition

Corning glass samples of 2 cm² are used as substrates for electrical and optical characterizations of TCO materials. For these measurements, a TCO layer thickness of 100 nm is targeted. The sheet resistance (R_{sheet}) is then measured by 4-point probes method on a Napson equipment. The charge carriers mobility and concentration values are determined by Hall effect analyses using a Van der Pauw configuration with an Ecopia tool. Finally, the transparency of TCO is assessed with a Perkin-Elmer spectrophotometer on a large range of wavelengths (from 280 nm to 2400 nm). The TCO-covered samples experienced successive low temperature annealing steps from 250°C to 400°C during 5 minutes under N_2 flow in a Carbolite Gero oven. The previously presented characterizations are conducted before and after each annealing.

The values of R_{sheet}, of the charge carrier mobility and concentration for different annealing temperatures and TCO compositions, are presented in Figure 2, Figure 3 and Figure 4, respectively.

As shown by Figure 2, similar R_{sheet} are obtained after deposition for standard AZO and ITO. As deposited, a slight decrease of AZO R_{sheet} layer is measured when hydrogen is added to the TCO layer, for both "0.5% H2" and "1% H2" AZO. Indeed with 1% H_2, post-TCO R_{sheet} is just below 100 Ω/sq, whereas the value is around 120 Ω/sq for a standard AZO deposited without hydrogen flow.

Annealing at 350°C allows to decrease R_{sheet} of standard ITO and of hydrogenated AZO with 0.5% and 1% H_2. As shown by Figure 3 and 4, the R_{sheet} reduction in the case of ITO is mainly due to a strong charge carrier mobility increase, the mobility exceeding 40 cm²/(V.s) after the 350°C annealing. For AZO:H, the main factor of R_{Sheet} reduction is an increase of the charge carrier concentration, approaching 6×10^{20} cm^{-3} ("1% H2", 350°C annealing). 350°C annealing steps on standard AZO and hydrogenated AZO with 1.5% H_2 result in a decrease of carrier mobility causing a slight increase of R_{sheet} values. Notice that this increase in the charge carrier concentration could enhance free carrier absorption (FCA). Therefore, controlling the TCO effective transmission will be particularly important for future device integrations.

The effective transmission (T_{eff}) of different AZO layers are presented on Figure 5. These values are obtained by integrating transmission response as a function of the wavelength in silicon absorption range.

T_{eff} values are higher for AZO layers until 350°C compared to ITO. Moreover, the maximum T_{eff} are obtained after annealing steps at 350°C for hydrogen concentrations of 0.5% and 1%, reaching values higher than 80%. For all the studied AZO layers, T_{eff} values are significantly affected by 400°C annealing steps. On the contrary, for ITO, T_{eff} monotonically increases on the whole range of investigated temperatures.

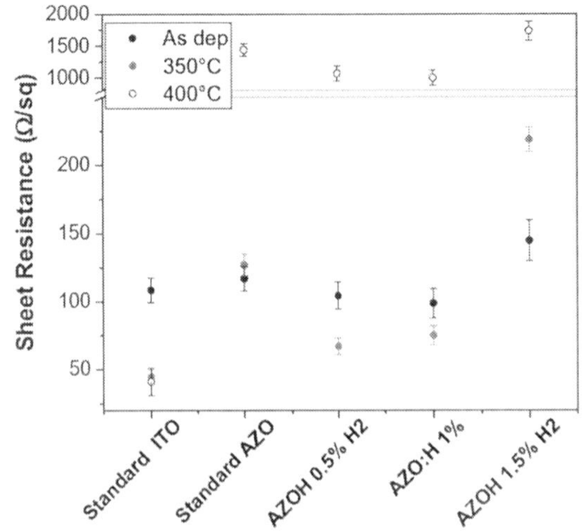

Figure 2: Variation of R_{sheet} with the TCO composition and the annealing temperature.

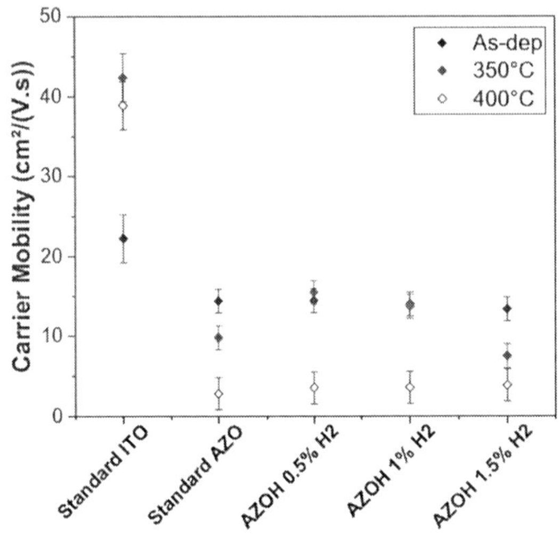

Figure 3: Variation of charge carrier mobility with the TCO composition and the annealing temperature.

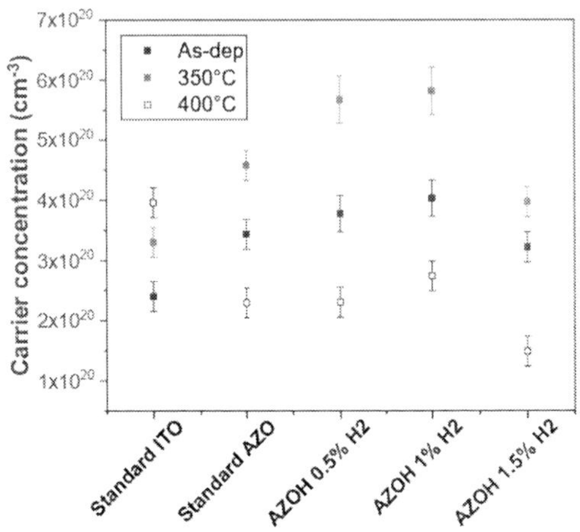

Figure 4: Variation of charge carrier concentration with TCO composition and annealing temperature.

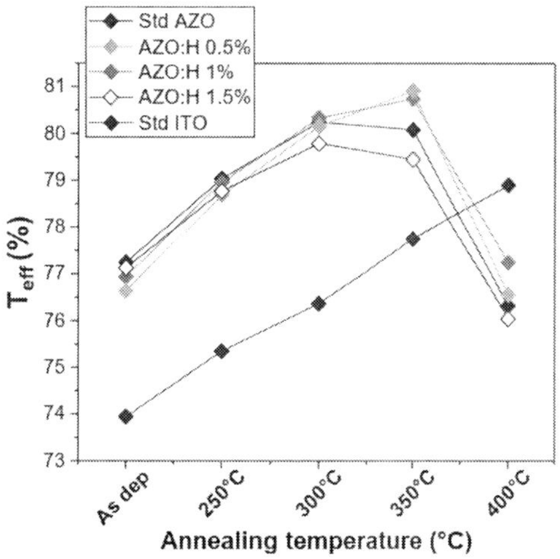

Figure 5: Variation of the effective transmission values of different TCO with the annealing temperature.

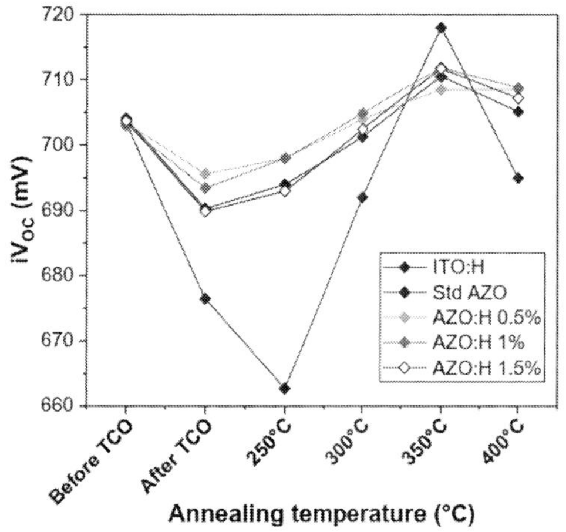

Figure 6: Variation of iV_{OC} of symmetrical samples with TCO layers composition and with the annealing temperature.

C. Passivation behaviour of different TCO layers on n^+ polysilicon/SiO_x stacks

For effective lifetime measurements, n-type (ρ = 2.1 Ω.cm) M2-Cz-silicon wafers, with a double side KOH-texurization, were used. A 1.5 nm thick thermal oxide was grown followed by the LPCVD double-side deposition of a 15 nm-thick intrinsic polysilicon film on a Hortus platform from Semco Smartech company. Symmetrical phosphorus doping is then conducted by Plasma Immersion Ion Implantation (PIII) on a Pulsion Solar® equipment from Ion Beam Services (IBS) company. Acceleration voltage of 4 kV and PH_3 dose of 7.5×10^{15} at/cm² were used for implantation. Then dopant activation and crystallisation are performed by an annealing step at 875°C during 30 minutes under N_2 flow in a conventional Centrotherm tube furnace. Finally, TCO layers are deposited on both surfaces. The surface passivation level is assessed by extracting the implied open circuit voltage (iV_{OC}) from Photo-Conductance Decay measurements with a Sinton equipment in transient mode (the samples featured high carrier lifetime, above 200 µs). Successive TCO annealing steps from 250°C to 400°C under N_2 flow are conducted between each photoconductance measurement.

The evolutions of iV_{OC} in the course of the successive annealing steps for the different samples are presented in Figure 6. For standard AZO, an iV_{OC} decrease of 15 mV after TCO deposition is highlighted. This decrease is likely to be due to sputtering damages during the deposition process [11]. For standard AZO, a complete recovery is obtained with annealing at 350°C. With the low H_2 concentrations (i.e. 0.5% and 1%), a lower drop due to sputtering damage is observed. iV_{OC} values slightly increase for those two concentrations until 350°C. The sample with the 1.5% H_2 concentration during the AZO deposition presents an important sputtering damage. All conditions feature a maximum iV_{OC} around 710 mV, after a 350°C annealing.

D. Cell integration

Cell precursors with double-side polysilicon on oxide passivated contacts were prepared to assess the compatibility of the developed indium free TCO with this structure. Wafers similar to those used for the "passivation study" were prepared, except that the rear side was here KOH-polished. The same LPCVD deposition was conducted, followed by front side phosphorus doping, still using the same conditions, and boron doping was performed at the rear side using an acceleration voltage of 6 kV and a dose of 7×10^{15} at/cm². The dopant-activation annealing step was similar than the treatment used in the previous part. The AZO layers were integrated only on the front side of the cell. Several front TCO were studied: standard AZO, hydrogenated AZO with 1% H_2 and an optimized hydrogenated ITO. The same hydrogenated ITO was used for the rear side contact, for all the samples. For both surfaces a TCO thickness of 80 nm was targeted. After TCO deposition, a busbarless electrode-pattern is screen-printed, using non-firing through pastes. Co-annealing steps of TCO and metal electrode were then conducted at 350°C under N_2 flow during 5 minutes before IV measurements on a Pasan tool.

Average results of the photovoltaic parameters are presented in Table I. For each front TCO, the parameters of the best cell are given in brackets. Similar V_{OC} are obtained with the different TCO, probably due to the recovery of the post-deposition damages when annealing at 350°C. A better J_{SC} is obtained for AZO:H than with AZO. It might partly be due to better effective transmission observed in usable part. Notice that due to the use of hydrogenated ITO on the solar cells, we cannot conclude on J_{sc} difference between AZO and AZO:H with ITO yet. Significant differences are observed for the fill factor values, which are higher (82.1%) for cells with standard AZO, leading to an average conversion efficiency of 22.2%. At the cell level, no benefits are clearly established for hydrogenated AZO compared to standard AZO. Further investigations are on-going to explain this fill factor difference, probably caused by a better contact resistivity of standard AZO on polysilicon, or a more important shunt resistance with AZO:H. Notice that with a front AZO, a champion cell at 22.4% has been obtained.

TABLE I: Cell parameters for devices with various front TCO

Front TCO	V_{OC} (mV)	J_{SC} (mA/cm²)	FF (%)	Efficiency (%)
ITO:H	702.75 *(703.3)*	38.88 *(38.84)*	80.48 *(80.80)*	22.00 *(22.09)*
AZO	702.5 *(703.5)*	38.52 *(38.55)*	82.11 *(82.66)*	22.23 *(22.43)*
AZO:H	702.2 *(701.5)*	38.73 *(38.69)*	81.30 *(82.15)*	22.12 *(22.31)*

CONCLUSION

In this work, the compatibility of indium-free TCO layers with the efficient contacting of polysilicon on oxide structures has been demonstrated. In-situ hydrogenation of ZnO:Al layers allows improvements of the material properties (TCO, interfaces), without additional hydrogenation treatment. It also demonstrates that hydrogen concentration is a key factor to control for improving layer performances. Indeed, low hydrogen concentrations in the deposition chamber, such as 0.5% and 1% led to better material properties of the AZO film, whereas an higher concentration of 1.5% does not provide positive effects. Moreover, annealing of AZO:H at 350°C allows to reach an optimum point for the electrical, optical and passivation properties at the same time. The implementation of these layers (i.e., AZO films with various H contents and reference ITO:H layers) has been conducted on the front side of double side poly-Si on oxide passivated contacts solar cells. The best results were obtained with front AZO layers, with a champion cell reaching 22.4% efficiency. This result demonstrates the possibility to reduce indium consumption for the industrial fabrication of high efficiency double-side passivated contacts silicon solar cells.

ACKNOWLEDGEMENTS

The team acknowledges the French National Research Agency for funding the Oxygen project (ANR-17-CE05-0035), the European Union's Horizon 2020 Programme for supporting the HighLite project (Grant Agreement no. 857793), and the French Environment and Management Agency (ADEME) for co-funding Charles Seron's PhD.

REFERENCES

[1] S. W. Glunz *et al.*, « The Irresistible Charm of a Simple Current Flow Pattern – 25% with a Solar Cell Featuring a Full-Area Back Contact », *31st Eur. Photovolt. Sol. Energy Conf. Exhib. 259-263*, p. 5 pages, 5146 kb, 2015, doi: 10.4229/EUPVSEC20152015-2BP.1.1.

[2] Y. Larionova *et al.*, « Ultra-Thin Poly-Si Layers: Passivation Quality, Utilization of Charge Carriers Generated in the Poly-Si and Application on Screen-Printed Double-Side Contacted Polycrystalline Si on Oxide Cells », *Sol. RRL*, p. 2000177, août 2020, doi: 10.1002/solr.202000177.

[3] R. Peibst *et al.*, « Building Blocks for Industrial, Screen-Printed Double-Side Contacted POLO Cells With Highly Transparent ZnO:Al Layers », *IEEE J. Photovolt.*, p. 1-7, 2018, doi: 10.1109/JPHOTOV.2018.2813427.

[4] S. Reiter *et al.*, « Parasitic Absorption in Polycrystalline Si-layers for Carrier-selective Front Junctions », *Energy Procedia*, vol. 92, p. 199-204, août 2016, doi: 10.1016/j.egypro.2016.07.057.

[5] J.-P. Niemelä *et al.*, « Rear-emitter silicon heterojunction solar cells with atomic layer deposited ZnO:Al serving as an alternative transparent conducting oxide to In2O3:Sn », *Sol. Energy Mater. Sol. Cells*, vol. 200, p. 109953, sept. 2019, doi: 10.1016/j.solmat.2019.109953.

[6] M. Lokanc, R. Eggert, et M. Redlinger, « The Availability of Indium: The Present, Medium Term, and Long Term », *Renew. Energy*, p. 90, 2015.

[7] L.-L. Senaud *et al.*, « Aluminium-Doped Zinc Oxide Rear Reflectors for High-Efficiency Silicon Heterojunction Solar Cells », *IEEE J. Photovolt.*, vol. 9, n° 5, p. 1217-1224, sept. 2019, doi: 10.1109/JPHOTOV.2019.2926860.

[8] B. Hallam *et al.*, « The role of hydrogenation and gettering in enhancing the efficiency of next-generation Si solar cells: An industrial perspective », *Phys. Status Solidi A*, juin 2017, doi: 10.1002/pssa.201700305.

[9] M. Lozac'h, S. Nunomura, H. Umishio, T. Matsui, et K. Matsubara, « Roles of hydrogen atoms in p-type Poly-Si/SiO *x* passivation layer for crystalline silicon solar cell applications », *Jpn. J. Appl. Phys.*, vol. 58, n° 5, p. 050915, mai 2019, doi: 10.7567/1347-4065/ab14fe.

[10] N. Juneja *et al.*, « Effect of hydrogen addition on bulk properties of sputtered indium tin oxide thin films », Fes, Morocco, 2019, p. 040008, doi: 10.1063/1.5123835.

[11] L. Tutsch *et al.*, « Implementing transparent conducting oxides by DC sputtering on ultrathin SiOx / poly-Si passivating contacts », *Sol. Energy Mater. Sol. Cells*, vol. 200, p. 109960, sept. 2019, doi: 10.1016/j.solmat.2019.109960.

Rear side dielectrics on interdigitating p^+-(i)-n^+ back-contact solar cells – hydrogenation vs. charge effects

M. Rienäcker[1], Y. Larionova[1], J. Krügener[2,3], S. Wolter[1], R. Brendel[1,3,4], R. Peibst[1,2]

[1]Institute for Solar Energy Research Hamelin (ISFH), Am Ohrberg 1, 31860 Emmerthal, Germany
[2]Institute of Electronic Materials and Devices, Leibniz Universität Hannover, Schneiderberg 32, 30167, Germany
[3]Laboratory of Nano and Quantum Engineering, Leibniz Universität Hannover, Schneiderberg 39, 30167, Germany
[4]Institute for Solid State Physics, Leibniz Universität Hannover, Appelstraße 2, 30167, Germany

ABSTRACT: Polysilicon-on-oxide (POLO) passivating contacts and interdigitated back-contact (IBC) cell technologies have recently attracted a lot of interest as candidates for the implementation in the next generation of solar cells. An IBC cell with POLO junctions for both polarities – a POLO²-IBC cell – has to electrically isolate the highly defective p^+ and n^+ poly-Si regions on the rear side of the cell to avoid parasitic recombination. Inserting an initially undoped, intrinsic (i) region between the p^+ and n^+ poly-Si regions was demonstrated to successfully prevent the parasitic recombination in the transition region of ISFH's 26.1%-efficient POLO²-IBC cell. In order to further improve the conversion efficiency towards 27%, we apply hydrogen-donating dielectric layer stacks to the p^+-(i)-n^+ POLO interdigitating rear side to enhance the passivation quality of the POLO junctions. We indeed show a significant improvement of POLO junctions on symmetrical full-area homogenously doped reference samples, but when we apply a hydrogen-donating layer stack on the p^+-(i)-n^+ POLO interdigitating rear side, we observe a strong degradation in the performance of the POLO²-IBC cell. We attribute this to the formation of a conductive channel between the p^+ and n^+ poly-Si regions due to the strong negative charge density of the hydrogen-donating layer stack.

Keywords: POLO/IBC/hydrogenation/charge/recombination

1 INTRODUCTION

The photovoltaic (PV) market demands for ever higher PV module efficiencies. Polysilicon-on-oxide (POLO) passivating contacts and interdigitated back-contact (IBC) cell technologies have recently attracted a lot of interest as candidates for the implementation in industrial production in the near future. However, the realization of an IBC cell with POLO junctions for both polarities – a POLO²-IBC cell – requires a separation of the highly defective p^+ and n^+ poly-Si regions on the rear side of the cell to avoid parasitic recombination. Beside the trench separation [1] and local oxidation of the poly-Si [13], inserting an initially undoped, intrinsic (i) region between the p^+ and n^+ poly-Si regions – as reported by several research groups [6, 16, 3, 10] – was demonstrated to successfully prevent the parasitic recombination in the transition region [3, 5]. In 2018, a POLO²-IBC cell with such a p^+-(i)-n^+ POLO interdigitated rear side achieved an efficiency of 26.1% [3], even without an optimized hydrogenation scheme for the POLO junctions. Thus, we recently applied hydrogen-donating layer stacks on symmetric POLO junction samples to demonstrate a significant improvement of the passivation quality compared to that implemented in the 26.1%-efficient cell from 2018 [8]. Moreover, we developed the required laser ablation process for the hydrogen-donating layer stack to be able to create laser contact openings without damaging the POLO junction underneath [12]. In this contribution, we aim to implement the hydrogen-donating layers on the p^+-(i)-n^+ POLO interdigitated rear side to improve our POLO²-IBC cell towards an efficiency of 27%. Since the typical hydrogen-donating layers like Al_2O_3 or SiN_y accommodate a high positive or negative charge, we investigate the influence of such a charge on the POLO²-IBC cell performance. We find that any strong charge density at the p^+-(i)-n^+ POLO interdigitated rear side leads to enhanced non-ideal recombination and diminishes the performance of POLO²-IBC solar cells.

2 EXPERIMENTALS

2.1 POLO²-IBC solar cell precursor with different rear side dielectric layers

Figure 1(a) shows the structure of the precursor stage of our POLO²-IBC solar cell fabricated similar to refs. [3, 5]. One characteristic feature of this cell is the p^+-(i)-n^+ POLO interdigitated rear side, where the p^+-type poly-Si and n^+-type poly-Si regions are separated by an initially intrinsic (i) poly-Si region. The term "initially" refers to the fact that during high-temperature annealing, strong lateral diffusion of dopants in the poly-Si (on the length scale of several micrometres) yields an inter-diffused lateral junction [5].

In brief, the following fabrication process yields the cell precursor in Fig. 1(a): A 2.2 nm-thin interfacial oxide layer is thermally grown onto a 1.3 Ωcm, p-type FZ wafer, which is subsequently capped by a low pressure chemical vapor deposited (LPCVD) intrinsic amorphous Si layer. We perform masked ion implantation of boron and phosphorous into the amorphous Si on the rear side and a blanket phosphorous implantation on the front side. For the former, the PECVD-deposited SiO_x implant barriers are patterned by photolithography, such that a gap of 30 µm i poly-Si remains between the n^+-type and p^+-type poly-Si fingers. Then, the amorphous Si recrystalizes during a pyrogenic oxidation at 900°C to grow a ~200-nm-thick SiO_2 layer and the POLO junction forms in a subsequent higher temperature process at 1035°C. This step also yields a lateral interdiffusion of dopants on the length scale of several micrometers, resulting in an inter-diffused junction within the initially 30µm wide intrinsic poly-Si region [5], which still isolates the charge carrier in the n^+ poly-Si from that in the p^+ poly-Si region. Hartenstein *et al.* have recently confirmed experimentally that the compensation of dopants at the p^+-(i)-n^+ poly-Si junction provides a highly resistive narrow region, which separates n^+ poly-Si and p^+ poly-Si electrically [4].

wafers with full-area doped POLO junction are fabricated and characterized by using the Sinton lifetime tester.

2.2 Advanced hydrogenation of POLO junctions

While the ~200nm-thick thermally grown SiO_2 in Fig. 1(a) was necessary in ref. [3] to enable a damage-free local laser ablation to create contact openings, it hinders the hydrogenation of the POLO junctions from hydrogen-

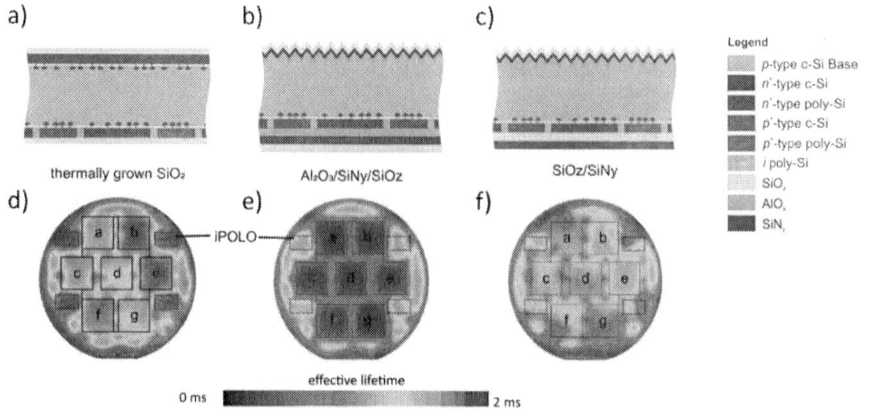

Fig 1: (a-c) Schematic drawing of the structure of a POLO²-IBC solar cell with (a) a thermally grown SiO_2 layer, (b) $Al_2O_3/SiN_y/SiO_z$ or (c) PECVD-SiO_z/SiN_y stack as the rear side dielectric layer. (d-f) Spatially resolved carrier lifetime of the cell structures shown in (a-c) at ~0.25 suns measured by the ILM method. The active cell areas of the seven solar cells are marked with squares a-g.

As in refs. [3, 5], the POLO junctions on the rear side are hydrogenated by depositing a sacrificial hydrogen-containing, silicon-rich SiN_y layer on the thick thermal SiO_2 layer and by subsequently annealing for 30 min at 425°C. Eventually, the SiN_y layer is removed in hot phosphoric acid.

Starting with a cell precursor as in Fig. 1(a), which has resulted in a cell efficiency of 26.1% in ref. [3], an advanced hydrogenation scheme from subsection 2.2 is applied to obtain the cell precursor in Fig. 1(b) with improved POLO junctions on the rear side. For this purpose, we remove the SiO_2 layer from the front side and texturize it. Then, we remove the SiO_2 layer from the rear side and passivated the front side. On the rear side, we deposite a layer stack of a 20-nm-thick Al_2O_3, a 30-nm-thick hydrogen-containing silicon-rich SiN_y and a 200nm-thick PECVD-SiO_z, which is subsequently annealed for 30 min at 425° C.

Alternatively, we remove the layer stack from the rear side of the cell precursor in Fig. 1(b) by a single-sided HF etching and replace it by a 20nm-thick PECVD-SiO_z and a 30-nm-thick SiN_y to obtain the cell precursor in Fig. 1(c). Note that in Fig. 1(c) the 200nm-thick SiO_z is missing for simplicity, but has to be included in the final cell for optical reasons. The minority carrier lifetime of each cell precursor is characterized by spatially resolved infra-red lifetime mapping (ILM) [9] at different illumination intensities. We calculate the illumination-dependent implied open-circuit voltage Suns-iV_{OC} characteristic of the cell precursor and derive an implied pseudo-efficiency.

Furthermore, using the same front-end process of the cell precursors in Fig. 1(a) and (b), symmetrical reference

donating SiN_y layer to some extent. We replace the SiO_2 by a layer stack of Al_2O_3/hydrogen-rich SiN_y/Al_2O_3, which is known to be an efficient hydrogen-donating layer stack [15] to estimate the upper limited for the passivation quality of the POLO junctions used for the 26.1%-efficient POLO²-IBC cell. To study the effect of the improved hydrogenation scheme, symmetric p^+ POLO, n^+ POLO and *i*POLO samples are prepared following the procedure in subsection 0. After POLO junction formation, the thermally grown SiO_2 was replaced by Al_2O_3/hydrogen-rich SiN_y/Al_2O_3 and annealed at temperatures above 425° C.

3 RESULTS

Figure 1(d) shows the lifetime map of the cell wafer from Fig. 1(a), which contains seven cell regions (a-g) and four reference regions for *i*POLO. The lifetime within the cell region b and e reach values of 1.8ms due to the good passivation of the n^+ POLO and p^+ POLO junctions. In this stage, the full-area POLO references reveal a saturation current density of 4 fA/cm² and 5.5 fA/cm² for n^+ POLO and p^+ POLO junctions, respectively. However, as already reported in ref. [5], the intrinsic reference region (*i*POLO) in Fig. 1(d) is almost unpassivated and therefore yields low lifetimes, which is in contrast to properly hydrogenated *i*POLO junctions reported elsewhere [16]. This is a strong evidence that the thick SiO_2 hinders hydrogen from reaching the thin interfacial SiO_z of the *i*POLO junctions of the 26.1%-efficient cell in ref. [3] and that the hydrogenation of n^+ and p^+ POLO junctions is not optimal. Thus, it seems straight forward to apply a better hydrogenation scheme to improve the POLO²-IBC cell efficiency of 26.1% towards 27%.

Indeed, when replacing the thick SiO_2 layer by the Al_2O_3/hydrogen-rich SiN_y/Al_2O_3 layer stack and performing an annealing above 425°C, we were able to improve the saturation current density of our n^+ POLO and p^+ POLO junctions on symmetric lifetime samples down to below 0.5±03 fA/cm² and 3.3±0.7 fA/cm² [8], respectively. For the *i*POLO junction sample, an effective surface recombination velocity S_{eff} of approximately 12 cm/s is achieved after hydrogenation, which is more than an order of magnitude lower compared to the *i*POLO junction of the 26.1%-efficient cell [5]. The improved hydrogenation of the n^+ and p^+ POLO junctions from 4 fA/cm² and 10 fA/cm² to 0.5 fA/cm² and 3.3 fA/cm² was estimated by device simulations to result in an efficiency improvement of ~0.4%$_{abs.}$ [8].

To prove the benefit of an improved hydrogenation scheme on cell level, we deposit a similar hydrogen-donating Al_2O_3/SiN_y/SiO_z layer stack as evaluated above on the cell precursor. This results in the structure shown in Fig. 1(b). Fig. 1(e) depicts the lifetime map of the cell precursor. As expected from the hydrogenation experiments above, the *i*POLO reference region of the wafer improves significantly upon hydrogenation and exhibits a lifetime of 1.5 ms. After hydrogenation, the full-area POLO references of the cell precursor in Fig. 1(b) reveal a saturation current density of 2.5 fA/cm² and 4 fA/cm² for n^+ POLO and p^+ POLO junctions, respectively. Regardless of the improvement of the POLO junctions, the lifetime within cell regions drops from previously 1.5-1.8 ms for the cell precursor as shown in Fig. 1(a) to about 200-300 µs for that in Fig. 1(b). We explain this surprising observation as the result causes by the high density of negative charges of the Al_2O_3 layer in contrast to the almost neutral SiO_2 layer. The charge density in the dielectric layer or at its interfaces to poly-Si seems to establish a conductive channel at the interface of the p^+-(*i*)-n^+ POLO interdigitated rear side to the dielectric layer [2]. The conductive channel connects the electrons and holes within the highly defective poly-Si regions and results in a strong recombination. A similar enhanced recombination was already observed for POLO²-IBC cells without an initially intrinsic poly-Si between p^+ and n^+ poly-Si regions [14, 11, 10].

We confirm our hypothesis by systematically manipulated the charge density on the rear side by depositing a corona charge density on top of the SiO_2 surface of the cell precursor in Fig. 1(a) (for details see ref. [7]).
Given the results in ref. [7], it can be understood why the cell precursor in Fig. 1(b) with Al_2O_3 on the rear side with a typical charge density of $-4e12$ cm^{-2} and the cells in ref. [10] exhibits a worse recombination behavior and a compromised cell performance – even if the POLO junction's passivation quality is improved upon hydrogenation. Moreover, it is expected that a cell with a strongly positively charged SiN_y hydrogen-donating layer would cause the same effect. Therefore, an IBC cell with a p^+-(*i*)-n^+ POLO interdigitated rear side requires an almost uncharged hydrogen-donating dielectric layer on the rear side.
Replacing the strongly charged Al_2O_3 layer by a weakly charged – optimally hydrogen-containing – PECVD/ALD-SiO_z or SiO_zN_y on the rear side of the cell precursor in Fig. 1 (b) poses one possibility for such a dielectric layer stack, which could also be compatible with the laser contact opening process in ref. [12].

To demonstrate that a PECVD-SiO_z meets the requirement of being weakly charged and that it is suitable for the application on the rear side of POLO²-IBC cells, we remove the Al_2O_3/SiN_y/SiO_z dielectric layer stack from a cell precursor as shown in Fig. 1(b) and subsequently deposit a non-optimized PECVD-SiO_z and a SiN_y on the rear side. The resulting cell structure is depicted in Fig. 1(c). The cell precursor with a PECVD-SiO_z /SiN_y layer stack on the rear side shows a strongly improved performance in Fig. 1(f) compared to that with an Al_2O_3/SiN_y/SiO_z layer stack in Fig. 1 (e). The cell precursors in Fig. 1(f) exhibit an implied pseudo-efficiency of up to 26.3%. This is qualitatively comparable with the cell precursor from Fig. 1(d) without corona charging. However, the front side passivation of the cell precursor in Fig. 1(a) deviates from that in Fig. 1(c) and the meaningfulness of a comparison of both cell precursors on an absolute efficiency scale is limited. However, the successful implementation of the PECVD-SiO_z on the rear side of a POLO²-IBC cell precursor represents a good starting point for future optimization towards 27%-efficient POLO²-IBC cells.

5 CONCLUSION

In this contribution, we aim at implementing hydrogen-donating layers on the p^+-(*i*)-n^+ POLO interdigitated rear side to improve our POLO²-IBC cell with an efficiency of 26.1% towards an efficiency of 27%. Since the typical hydrogen-donating layers like Al_2O_3 or SiN_y accommodate a high positive or negative charge, we investigate the influence of such a charge on the POLO²-IBC cell performance. We fabricate POLO²-IBC cell precursor with three different dielectric layers/layer stacks on the rear side: (i). a thermally grown SiO_2 as has been used for the 26.1% cell, (ii) a hydrogen-donating Al_2O_3/SiN_y/SiO_z stack to improve POLO junction passivation quality, (iii) a hydrogen-donating PECVD-SiO_z/SiN_y stack. The implied pseudo-efficiency $i\eta$ of each precursor is deduced from the illumination-dependent implied open-circuit voltage Suns-iV_{OC} of the cell precursor.
While the cell with thermally grown SiO_2 shows the expected $i\eta$ of 26.4%, it is evident from a comparison with the second cell precursor that the thick SiO_2 prevents hydrogen diffusion from a hydrogen-donating layer on top of the SiO_2 to the POLO junction to some extent. The second cell precursor with Al_2O_3/SiN_y/SiO_z stack efficiently hydrogenates the POLO junction, but yields a cell precursor with worse recombination behavior and low performance. We attribute this surprising observation to the presents of a strong negative charge induced by the Al_2O_3 layer.
To confirm this hypothesis, we study the influence of charges at the p^+-(*i*)-n^+ POLO interdigitated rear side on the POLO²-IBC cell performance. We find that strong positive or negative charge densities cause an enhanced non-ideal recombination and degrade the cell performance significantly, which is most likely due to the formation of a conductive path for electrons or holes within the highly defective poly-Si.

We conclude that it is vital to control the charge density of the dielectric layers applied to the p^+-(*i*)-n^+ POLO interdigitated rear side to avoid a strong increase of non-ideal recombination at the p^+-(*i*)-n^+ poly-Si junction. While this effect is highly pronounced for IBC cells with poly-Si, it may also apply – to some extend – to all types of IBC cells with charged dielectric layers on the interdigitated rear side.

Finally, we present a weakly charged hydrogen-donating layer stack comprising a 20-nm-thin and a potentially weakly charged PECVD-SiO_z at the interface with the poly-Si capped by a hydrogen-donating SiN_y layer. This stack may provide an efficient hydrogenation scheme and maintains the isolation between p^+ and n^+ poly-Si regions. A first cell precursor with the weakly charged hydrogen-donating layer stack yields an $i\eta$ of 26.3%.

ACKNOWLEDGEMENTS
The authors thank the state of lower Saxony and the Federal Ministry of Economic Affairs (BMWi) for funding this work, which was performed in the framework of the research project "27Plus6" (FKZ03EE1056A). We are grateful to Hilke Fischer, Sarah Spätlich and Renate Winter (all from ISFH) as well as Raymond Zieseniss and Guido Glowatzki (both from Institute of Electronic Materials and Devices) for sample processing. We thank Felix Haase and Christina Hollemann for fruitful discussions.

REFERENCES

[1] De Ceuster, Denis and Cousins, Peter John and Smith, David D. Trench process and structure for backside contact solar cells with polysilicon doped regions, 2010.

[2] AS Grove and DJ Fitzgerald. Surface effects on pn junctions: Characteristics of surface space-charge regions under non-equilibrium conditions. *Solid-State Electronics*, 9(8):783–806, 1966.

[3] Felix Haase, Christina Hollemann, Sören Schäfer, Agnes Merkle, Michael Rienäcker, Jan Krügener, Rolf Brendel, and Robby Peibst. Laser contact openings for local poly-Si-metal contacts enabling 26.1%-efficient POLO-IBC solar cells. *Solar Energy Materials and Solar Cells*, 186:184–193, nov 2018.

[4] Matthew B Hartenstein, Steve Harvey, Matthew Page, David Young, Paul Stradins, and Sumit Agarwal. Effect of dopant compensation on the conductivity of the intrinsic poly-si isolation region in passivated ibc silicon solar cells. In *2020 47th IEEE Photovoltaic Specialists Conference (PVSC)*, pages 2751–2753. IEEE, 2020.

[5] C. Hollemann, F. Haase, M. Rienäcker, V. Barnscheidt, J. Krügener, N. Folchert, R. Brendel, S. Richter, S. Groß er, E. Sauter, J. Hübner, M. Oestreich, and R. Peibst. Separating the two polarities of the POLO contacts of an 26.1%-efficient IBC solar cell. *Scientific Reports*, 2020.

[6] J. Choi M. Choi, H. Park. Solar cell and method for manufacturing the same, 2013.

[7] J. KrÃ¼gener S. Wolter R. Brendel R. Peibst M. RienÃ¤cker, Y. Larionova. Rear side dielectrics on interdigitating p+-(i)-n+ back-contact solar cells â€“ hydrogenation vs. charge effects. *EPJ Photovoltaics, under review*, 2021.

[8] R. Peibst, M. RienÃ¤cker, Y. Larionova, N. Folchert, F. Haase, C. Hollemann, S. Wolter, J. KrÃ¼gener, P. Bayerl, J. Bayer, M. Dzinnik, R. Haug, and R. Brendel. Towards 28 %-efficient Si single junction solar cells with better passivating POLO junctions. *Sol. Energ. Mat. Sol.*, 2021.

[9] K. Ramspeck, S. Reissenweber, J. Schmidt, K. Bothe, and R. Brendel. Dynamic carrier lifetime imaging of silicon wafers using an infrared-camera-based approach. *Applied Physics Letters*, 93(10):102104, 2008.

[10] Christian Reichel, Ralph MÃ¼ller, Frank Feldmann, Armin Richter, Martin Hermle, and Stefan W. Glunz. Influence of the transition region between p- and n-type polycrystalline silicon passivating contacts on the performance of interdigitated back contact silicon solar cells. *Journal of Applied Physics*, 122(18):184502, 2017.

[11] M Rienäcker, A Merkle, U Römer, H Kohlenberg, J Krügener, R Brendel, and R Peibst. Recombination Behavior of Photolithography-free Back Junction Back Contact Solar Cells with Carrier-selective Polysilicon on Oxide Junctions for Both Polarities. *Energy Procedia*, 92:412–418, 2016.

[12] Michael RienÃ¤cker. *Three-terminal tandem solar cellsenabled by back-contacted bottom cellsfeaturing passivating, carrier-selectivepolysilicon based junctions*. PhD thesis, submitted to Gottfried Wilhelm Leibniz UniversitÃ¤t Hannover, 2021.

[13] Udo RÃ¶mer Robby Peibst. Solar cell and a method for producing a solar cell with oxidised intermediate regions between polysilicon contacts4, 2015.

[14] Udo Römer, Robby Peibst, Tobias Ohrdes, Bianca Lim, Jan Krügener, Tobias Wietler, and Rolf Brendel. Ion implantation for poly-si passivated back-junction back-contacted solar cells. *IEEE Journal of Photovoltaics*, 5(2):507–514, 2015.

[15] M. K. Stodolny, J. Anker, C. J. J. Tool, M. Koppes, A. A. Mewe, P. Manshanden, M. Lenes, and I. G. Romijn. Novel Schemes of p+ poly-Si Hydrogenation Implemented in Industrial 6" Bifacial Front-and-Rear Passivating Contacts Solar Cells. In *35th European Photovoltaic Solar Energy Conference and Exhibition*, pages 414–417, 2018.

[16] D. L. Young, W. Nemeth, V. LaSalvia, R. Reedy, S. Essig, N. Bateman, and P. Stradins. Interdigitated back passivated contact (ibpc) solar cells formed by ion implantation. *IEEE Journal of Photovoltaics*, 6(1):41–47, Jan 2016.

APPROACHING 23% WITH p-TYPE BACK JUNCTION SOLAR CELLS FEATURING SCREEN-PRINTED Al FRONT GRID AND PASSIVATING REAR CONTACTS

B. Min[1], N. Wehmeier[1], H. Schulte-Huxel[1], R. Witteck[1], T. Brendemuehl[1], T. Daschinger[1], F. Haase[1], Y. Larionova[1], L. Nasebandt[1], K. Tsuji[2], M. Dhamrin[2, 3], R. Peibst[1, 4] and R. Brendel[1, 5]
[1]Institute for Solar Energy Research Hamelin (ISFH), Am Ohrberg 1, 31860 Emmerthal, Germany
[2]Toyo Aluminium K.K., 341-14 Higashiyama, Ohtani, Hino-Cho, Gamo-Gun, Shiga 529-1608, Japan
[3]Graduate School of Engineering, Osaka University 1-1 Yamadaoka, Suita 565-0871, Japan
[4]Institute of Electronic Materials and Devices, Leibniz University of Hannover, Schneiderberg 32, 30167 Hannover, Germany
[5]Dep. Solar Energy, Inst. Solid-State Physics, Leibniz University of Hannover, Appelstr. 2, 30167 Hannover, Germany

ABSTRACT: This paper reports the recent progress of p-type back junction (BJ) solar cells featuring passivating polysilicon on oxide (POLO) rear contacts and Al front contacts. The current best cell efficiency on full M2 Cz wafers is 22.9 % as independently confirmed by ISFH CalTeC in Germany. To our knowledge, this is the highest efficiency for a p-type solar cell with screen-printed Al grid on textured front side. We improve the passivation quality beneath rear metal fingers by increasing the SiN_y capping layer thickness. This prevents spiking through the poly-Si layers by screen-printed silver during the firing step and results in a record V_{oc} of 716 mV for a fully screen-printed large-area p-type solar cell with Al-alloyed local contacts. The Al front grid is improved by an upgraded Al paste which enables narrow metal fingers down to 64 µm with higher aspect ratio up to 0.5. We investigate the morphology of the of p^+-type layers beneath the Al fingers by scanning electron microscopy. In addition, the UV resistance of the AlO_x/SiN_y passivation stack on undiffused textured front side is investigated by applying UV transparent encapsulant materials. As a possible upgrade option, we present our first proof-of-concept Ag-free POLO BJ solar cell with an efficiency of 17.9 %.
Keywords: Polycrystalline, Passivating Contacts, Back Junction

1 INTRODUCTION

The PV industry is currently introducing poly-Si based passivating contacts to achieve the cell efficiencies beyond 24 % with crystalline silicon solar cells. For its implementation into large-scale solar cell manufacturing many cell manufactures choose n-type solar cells featuring a boron emitter at the front side and electron-collecting n^+-type passivating contacts at the rear side [1, 2].

In this paper, we present recent results of our alternative POLO back junction (BJ) cell concept [3, 4] for that avoids the boron diffusion. Figure 1 shows the cell structure: The p-type cell features an electron-collecting polysilicon on oxide (POLO) contact on the rear. This cell concept allows higher efficiency than today's PERC cells while utilizing many innovations of these mainstream p-type solar cells. Only minor modification is required to transfer existing PERC production lines to fabricate POLO BJ solar cells because the process flow of POLO BJ cells is quite similar to that of PERC cells.

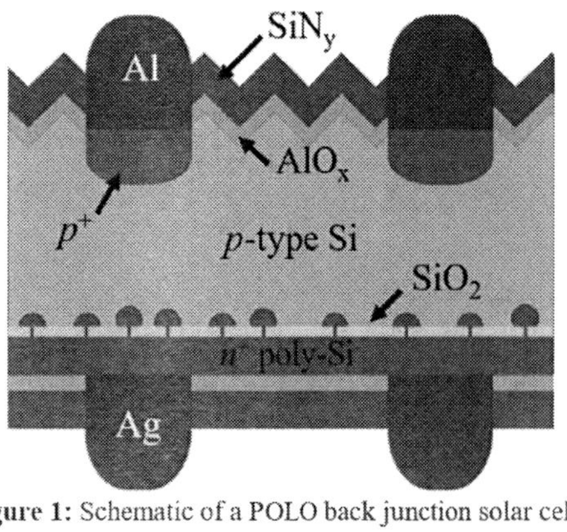

Figure 1: Schematic of a POLO back junction solar cell.

2 EXPERIMENTAL

We start with a boron-doped p-type Cz wafer (M2 size) with a resistivity of 1 Ωcm. An (according to ellipsometry) 1.5 nm-thick interfacial oxide layer grows wet-chemically in de-ionized water with diluted ozone, and is then capped by an in-situ phosphorous-doped poly-Si layer deposited with a Low Pressure Chemical Vapor Deposition (LPCVD) tool. During a high temperature step in a tube furnace, the interfacial oxide layer breaks up and forms the POLO contacts. Simultaneously, 150 nm-thick oxide layers grow on the surface of the poly-Si layers on both sides. The oxide layer on the cell front side is then fully removed. A subsequent KOH-based solution textures the front side. After oxide removal from the rear and cleaning, we deposit AlO_x/SiN_y stacks on the cell front and rear side. We locally open the dielectric layers by a laser on the front side to form contact openings. The contacts are realized by aluminum-paste screen printing at the front side and silver-paste screen printing at the rear side followed by a co-firing process.

3 RESULTS

3.1 Improved passivation quality beneath screen-printed Ag rear fingers

During a rapid firing step in a conveyor belt furnace, the screen-printed silver forms ohmic contact with poly-Si layers. Depending on the penetration depth of the silver crystallites, it is possible that a significant deterioration of the passivation quality occurs [5-7].

In order to reduce the recombination losses beneath the silver fingers, we control the penetration depth of silver by increasing the SiN_y capping layer thickness [8]. Figure 2 shows a 20 mV gain in V_{oc} when increasing the SiN_y thickness from 40 nm to 60 nm. The FF decreases because the contact resistivity between poly-Si and screen-printed silver increases by about one order of magnitude.

However, increasing the peak firing temperature from 800 °C to 810 °C decreases the contact resistivity to below 2 mΩcm² while keeping V_{oc} unchanged. Within this experiment, we achieve 22.6 % a cell efficiency. This best cell from this run has a V_{oc} of 716 mV with a J_{sc} of 39.3 mAcm² and an *FF* of 80.3% as independently confirmed by ISFH CalTeC [8].

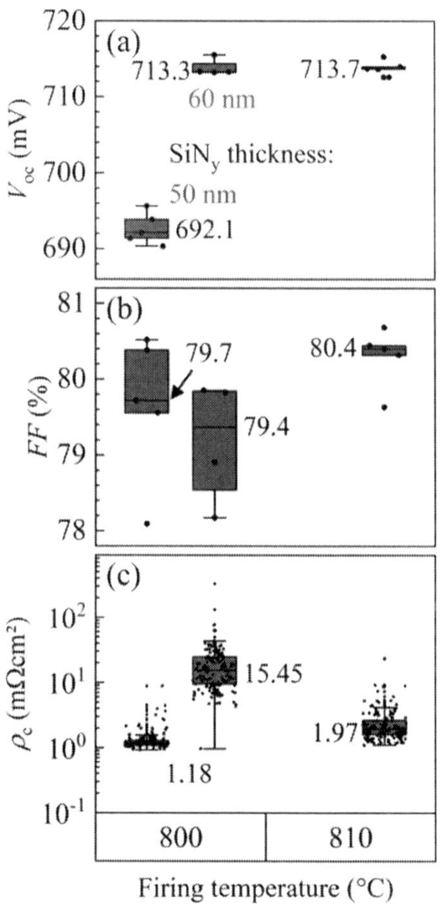

Figure 2 Measured (a) V_{oc} and (b) *FF* of POLO BJ solar cells by different peak firing temperatures, and (c) corresponding contact resistivities between poly-Si and screen-print silver. The box colors indicate different SiN_y layer thicknesses at the cell rear side.

3.2 Improvements in screen-printed Al front grid

The upgraded Al paste leads to narrower fingers with higher aspect ratio compared to the previous Al paste [8, 9]. As shown in Figure 3, the median finger width decreases from 70.1 µm to 63.7 µm while the line resistance is decreases from 4.7 Ω/cm to 2.0 Ω/cm. The median p^+-type layer thickness increases from 1.6 µm to 2.7 µm because more aluminum is provided. The upgraded Al paste seems to avoid regions with very thin p^+-type layers with thickness values below 1 µm. We therefore expect that this upgraded Al paste improves V_{oc} and J_{sc} significantly.

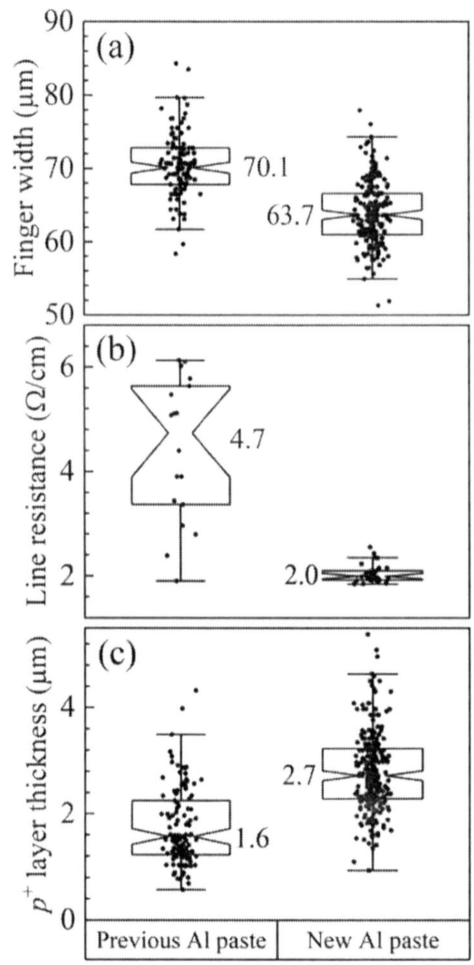

Figure 3 (a) Finger width as measured with optical microscopy, (b) line resistance as measured with the TLM-SCAN tool from pvtool and (c) p^+ layer thickness as measured with scanning electron microscopy. The numbers next to the boxes give the median values.

3.3 22.9 %-efficient POLO BJ solar cell

Table I summarizes the independently confirmed *JV* parameters of our currently best POLO BJ solar cell. The busbarless cell is contacted with 12 contact bars using a grid-resistance neglecting contacting scheme [10]. Compared to the cell results with previous Al paste, the main improvement is the increase in J_{sc} due to narrowed finger width and the increase in *FF* due to an increased p^+ layer thickness. Despite the improved Al front fingers, this best cell has a V_{oc} value of 714 mV that is lower than the best cell from our previous publication [8]. This is because we had to use a different tool for depositing the AlO_x layers in this cell batch.

Table I: *IV* parameters of the best POLO BJ solar cell

	V_{oc} [mV]	J_{sc} [mA/cm²]	*FF* [%]	η [%]
Best cell*	714	39.6	80.9	22.9

*independently confirmed by ISFH CalTeC

3.4 Module integration of POLO BJ solar cells and the first UV degradation tests

Figure 4 shows an electroluminescence (EL) image of a demonstrator module with a POLO BJ half-cell. We apply multi-busbars concept with 9 busbars and use Ag soldering pads for the interconnection wires. Its proof-of-concept is successfully demonstrated as shown with the homogeneously bright EL image in Figure 4.

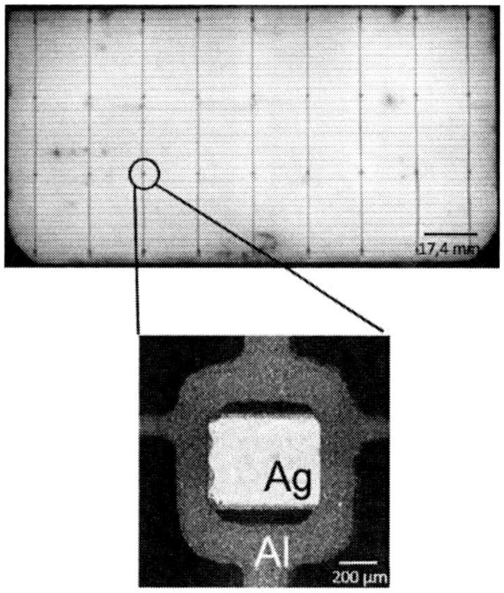

Figure 4 EL image of demonstrator module and optical microscopy image of a soldering pad.

It is known in the literature that the passivation quality of SiN_y on the phosphorus-diffused textured surface degrades significantly during the UV illumination while AlO_x/SiN_y passivation stack on the boron-diffused textured surface shows no degradation [11]. We therefore investigate the UV stability of the AlO_x/SiN_y passivation stack on undiffused textured surface. We first halve a POLO BJ cell which is connected with wires for *IV* measurements. One half cell is encapsulated by ethylene vinyl acetate (EVA) and the other half cell by polyolefin elastomer (POE). Both materials are transparent for UV light. The samples are then laminated between a front side glass and a rear side polymeric backsheet. The samples are then exposed to a UV light, facing the cell front side to the UV lamp. The *IV* measurement is repeated after 24, 48, 112, 208, 325, 512, 942, 1664 hours. In this work, an UV exposure time of 1664 hours equals an UV dose of 285 kWh/m². In order to verify the measurement uncertainty, we fabricate a reference sample with UV-transparent EVA using an additional POLO BJ cell. The reference sample is stored in the dark and not exposed to UV light. Figure 5 shows the measured V_{oc} as a function of UV dose. In both test samples show a decrease of 6 mV in V_{oc} and it stays above 700 mV for a UV dose of 285 kWh/m². This dose corresponds to 8.4 years outdoor exposure in a moderate climate (Potsdam, Germany).

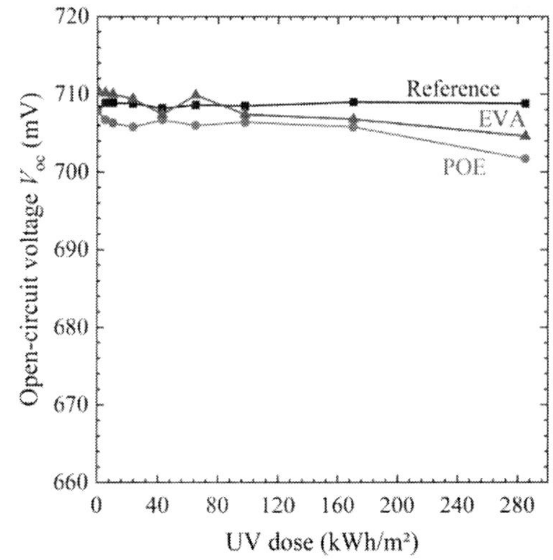

Figure 5 Measured V_{oc} as a function of applied UV dose of different samples encapsulated with ethylene vinyl acetate (EVA) or polyolefin elastomer (POE).

4 SILVER-FREE POLO BJ SOLAR CELLS

Figure 6 shows a POLO BJ cell with the Ag fingers on the rear being replaced by Al fingers.

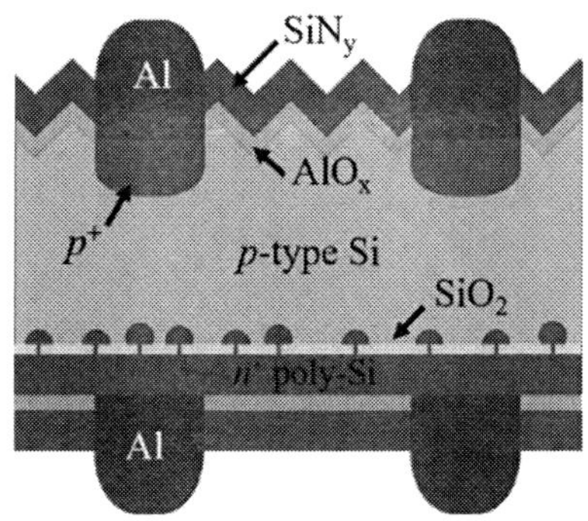

Figure 6 Schematic of a fully Ag-free metallized POLO back junction solar cell.

Figure 7 shows that we achieve median contact resistivities beneath 3.5 mΩcm² with the TTC-series provided by Toyo Aluminium K.K. The best cell efficiency of our first batch with Ag-free POLO BJ solar cells is 17.9 % with a V_{oc} of 675 mV as measured in-house with LOANA tool from pvtools.

5 SUMMARY AND OUTLOOK

In this paper, we reported our recent results on POLO BJ solar cells: (i) We improved the passivation quality beneath the Ag rear fingers and (ii) we improved the shape of the front-side Al fingers to 60 µm and a high aspect ratio. With the first improvement, we achieved 716 mV as the highest V_{oc}. By applying the second improvement, we reached 22.9 % as the best POLO BJ cell efficiency. We are currently running cell batches that utilize both improvements simultaneously to surpass 23 % efficiency.

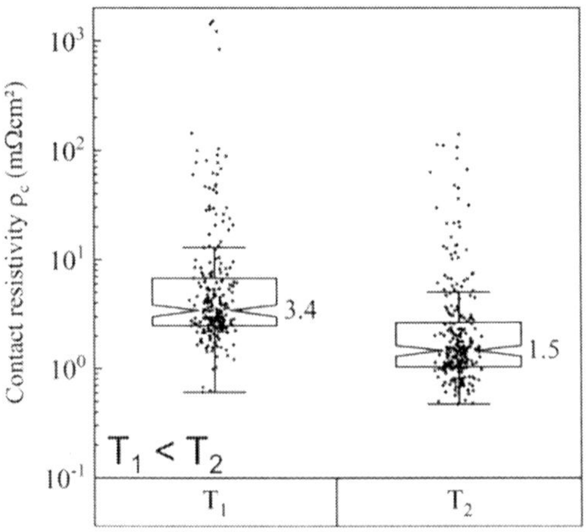

Figure 7 Measured contact resistivities between screen-printed aluminum and phosphorus-doped poly-Si

We gave a proof-of-concept for module integration of POLO BJ solar cells. The V_{oc} of a test module with enhanced UV transparency degrades by less than 6 mV and stays abvoe 700 mV after an UV expose corresponds to 8.4 years outdoor exposure in a moderate climate.

Finally, we presented the first fully Ag-free metallized POLO BJ solar cell by applying screen-printed Al contacts on phosphorus-doped POLO contacts.

6 ACKNOWLEDGEMENTS

The authors thank to M. Pollmann, B. Gehring, R. Winter, S. Spaetlich, T. Neubert and D. Sylla for processing solar cells and K. Bothe, M. Wolf, D. Hinken, V. Mertens, T. Dullweber (all ISFH) for fruitful discussions and continuous support. We also thank S. Huebner, T. Dippell, B. Wattenberg and P. Wohlfart from Singulus Technologies AG for constructive discussions. This work was financially supported by the German Federal Ministry for Economic Affairs and Energy (BMWi) under contact number 03EE1012A (NanoPERC) and a part of this work was supported under contact number 0324171C (Nextstep).

7 REFERENCES

[1] D. Chen, Y. Chen, Z. Wang, J. Gong, C. Liu, Y. Zou, Y. He, Y. Wang, L. Yuan, W. Lin. "24.58% total area efficiency of screen-printed, large area industrial silicon solar cells with the tunnel oxide passivated contacts (i-TOPCon) design", Sol. Energ. Mat. Sol. 206, 110258 (2020).

[2] W. Wu, J. Bao, L. Ma, C. Chen, R. Liu, Z. Qiao, J. Chen, Z. Liu. "Development of Industrial n-Type Bifacial TOPCon Solar Cells and Modules", Proc. of the 37th European Photovoltaic Solar Energy Conference and Exhibition, Marseille, France, 2019, pp. 100-102.

[3] R. Brendel, B. Lim. "Silicon cell concepts - trends in research", Proc. of the PV Module Technology & Applications Forum 2018, TÜV Rheinland, Cologne, Germany, 2018.

[4] R. Brendel, C. Kruse, A. Merkle, H. Schulte-Huxel, F. Haase, R. Peibst. "Screening Carrier Selective Contact Combinations for Novel Crystalline Si Cell Structures", Proc. of the 35th European Photovoltaic Solar Energy Conference and Exhibition, Brussels, Belgium, 2018, pp. 39 - 46.

[5] H.E. Çiftpınar, M.K. Stodolny, Y. Wu, G.J.M. Janssen, J. Löffler, J. Schmitz, M. Lenes, J.-M. Luchies, L.J. Geerligs. "Study of screen printed metallization for polysilicon based passivating contacts", Energy Procedia 124, 851-861 (2017).

[6] A. Chaudhary, J. Hoß, J. Lossen, R.v. Swaaij, M. Zeman. "Screen printed Ag contacts for n-type polysilicon passivated contacts", AIP Conference Proceedings 2147, 040002 (2019).

[7] P. Padhamnath, J. Wong, B. Nagarajan, J.K. Buatis, L.M. Ortega, N. Nandakumar, A. Khanna, V. Shanmugam, S. Duttagupta. "Metal contact recombination in monoPoly™ solar cells with screen-printed & fire-through contacts", Sol. Energ. Mat. Sol. 192, 109-116 (2019).

[8] B. Min, N. Wehmeier, T. Brendemuehl, F. Haase, Y. Larionova, L. Nasebandt, H. Schulte-Huxel, R. Peibst, R. Brendel. "716 mV Open-Circuit Voltage with Fully Screen-Printed p-Type Back Junction Solar Cells Featuring an Aluminum Front Grid and a Passivating Polysilicon on Oxide Contact at the Rear Side", Solar RRL 5, 2000703 (2021).

[9] B. Min, N. Wehmeier, T. Brendemuehl, A. Merkle, F. Haase, Y. Larionova, L. David, H. Schulte-Huxel, R. Peibst, R. Brendel. "A 22.3% Efficient p-Type Back Junction Solar Cell with an Al-Printed Front-Side Grid and a Passivating n+-Type Polysilicon on Oxide Contact at the Rear Side", Solar RRL 2000435 (2020).

[10] K. Bothe, C. Kruse, D. Hinken. "Contacting of busbarless solar cells for accurate I-V measurements", Proc. of the 37th European Photovoltaic Solar Energy Conference and Exhibition, Online event, 2020.

[11] R. Witteck, B. Min, H. Schulte-Huxel, H. Holst, B. Veith-Wolf, F. Kiefer, M.R. Vogt, M. Köntges, R. Peibst, R. Brendel. "UV radiation hardness of photovoltaic modules featuring crystalline Si solar cells with AlOx/p+-type Si and SiNy/n+-type Si interfaces", physica status solidi (RRL) – Rapid Research Letters 11, 1700178 (2017).

EVALUATION OF DIFFERENT APPROACHES FOR HJT CELLS METALLIZATION BASED ON LOW TEMPERATURE AG PASTES

Marco Galiazzo, Nicola Frasson
Applied Materials Italia Srl
Via Postumia Ovest 244, Olmi di San Biagio di Callalta (TV) Italy
*Corresponding author: marco_galiazzo@amat.com

ABSTRACT: One key concern when implementing HJT cells is related to the metallization process, where several limitations are present when using low temperature Ag pastes, curable at 200°C to preserve the a-Si structure. These limitations are reduced printing speed and limited printing ability on fine lines, increased logistics costs, solderability. In this work we investigate two possible alternative routes to the mainstream paste models: the evolutional is based on room temperature paste, not requiring refrigeration and with increased printing performances and the revolutionary, using nano-Ag particles, is surpassing the usual printing speed by three times, with half the deposit if compared to the others.
Keywords: heterojunction, screen-printing, solar pastes, silver paste, nanoparticles, solar cells

I. INTRODUCTION

Heterojunction cells are an industrially proven technology to optimize crystalline Silicon conversion efficiency [1], with notable examples being: the efficiency record of 25.05% absolute set by Maxwell on large wafers [2] and the standing record of 26% for the Interdigitated Heterojunction cells by Kaneka [3].

With the increased R&D and piloting activities involving this technology, both in China and in Europe it is realistic that the learning curve for the processing steps will become steeper and the manufacturing costs will decrease accordingly [4].

One important contributor to the cost structure of the HJT cell is the metallization, and this is linked to the maximum allowed processing temperature. In fact, low temperature Ag based paste systems have 2-3x lower conductivities if compared to fire-through Ag pastes commonly used for mainstream PERC cells manufacturing. This translates into significantly higher pastes deposits per cell, and direct impact on the cost of ownership.

Besides conductivity, several other limitations should be considered for low temperature Ag pastes:
- Reduced printing and flooding speed (2 or 3 times lower than PERC pastes).
- Constraints in terms of shipping and storage temperatures.
- Reduced capability of printing fine lines (<30 µm).
- Limited peel force values for direct soldering of ribbons.
- Longer drying and curing times, overall higher power consumption in the line.

All these elements become even more detrimental for equipment productivity when the development roadmap for HJT is considered, with forecasted adoption of large (up to 210mm in size) and thinner wafers (120 µm and below).

II. AIM AND APPROACH USED

In this work we will compare the reference HJT paste (Paste A) with three alternative pastes:

- Paste B: room temperature storage and marginally increased printing speed
- Paste C: room temperature storage, PERC-like printing speed and reduced curing times
- Paste D: nano-Ag based chemistry suitable for low deposits.

A comprehensive table (Table 1) with specifications for all pastes is reported below:

	Paste A	Paste B	Paste C	Paste D
Storage (° C)	< 10	25	25	25
Ag-content (%)	89-93	90-95	86-94	60-80
Drying/Curing	F/STD	F/STD	1 step, 2.5min	VF/STD
Resistivity (µOhm cm)	5	3.5	4.4	3.2
Printing speed (mm/s)	120	150	350	400

Table 1. STD = standard, 20min @200° C; F = fast, between 30-60s @170° C; VF = very fast, < 30s @170°C. Paste A = reference/commercial paste.

The comparison is based on a head-to head process characterization (Figure 1) including a preliminary phase that includes a test pattern printing and rheological analysis followed by the testing phase that otherwise consists in printing optimization, cell metallization, IV measurements, and finally morphological and electroluminescence (EL) analysis. Additionally, solderability for all pastes using a manual setup and standard Sn-coated Cu ribbons has been tested, showing that no material is suitable for direct soldering. With this respect, possible routes for module processing include: 1) ribbon connection to the cell metallization with Electrically Conductive Adhesive (ECA), 2) implementation of dual printing process, with dedicated paste on busbars optimized for adhesion and alternative interconnection technologies such as ECA-based shingling or low temperature soldering (SWCT). Due to its different chemistry, Paste C has been investigated also in terms of reliability at minimodule level (TC, DH).

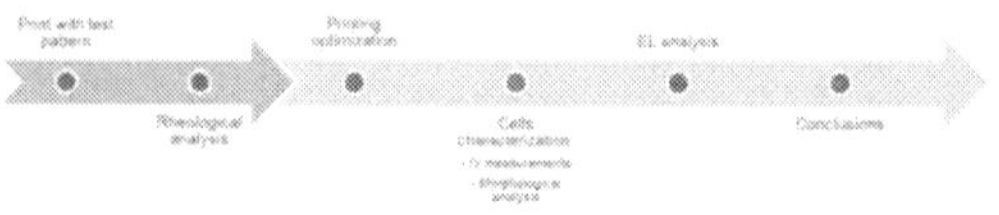

Figure 1. Process characterization workflow applied for the pastes comparison.

III. SCIENTIFIC INNOVATION AND RELEVANCE

Industrial applicability of low temperature Ag pastes has always been limited by multiple constraints, that ultimately lead to overall higher metallization costs. In view of increased adoption of HJT, several paste manufacturers are proposing alternative approaches (systems and chemistries) which will progressively improve process performances

The three alternative pastes proposed in this paper show promising capabilities (see also Table 2) because they can be stored at room temperature, while the reference material is shipped and stored below 10°C. It is also important to note that low temperature Ag pastes are relevant for different PV architectures (OPV, tandem) and will see increased adoption in Printed Electronics (display, medical, packaging).

HJT cell printing process characteristics	
Pros	**Cons**
Lead free pastes	Low conductivity
High conversion efficiencies	Pastes are conserved at < RT
Low temperature	Solderability
Simpler alignment	Printing speed and line width

Table 2. Heterojunction cell printing advantages and disadvantages at the current state-of-art.

IV. RESULTS AND CONCLUSIONS

Since the processing conditions for rheology testing from manufacturers are not always comparable, a direct analysis on our TA Instruments rheometer (plate-plate geometry) has been carried out both for shear-sweep testing and for the oscillatory testing in accordance with [5]. Figure 2 shows how the different pastes change their rheological properties according to the increasing shear rate-stress applied. When low shear rates are considered, the viscosity of the pastes is significantly different as represented by the gap between the nano-based paste and the other three pastes. Although this gap is reduced at higher shear rates, the nano-based ink remains more viscous and this can be also seen during the print: deposits are much lower on both front and back and ad-hoc parameters have been used to fill the gap and obtain an almost comparable process.

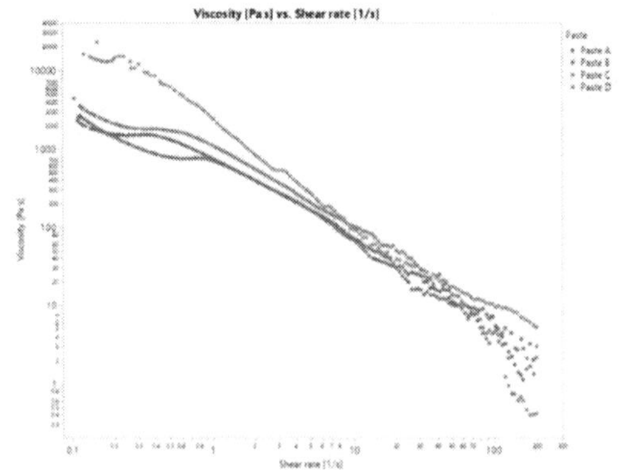

Figure 2. Flow test analysis, viscosity values at increasing shear rates/stress.

Furthermore, this is also confirmed by Figure 3 in which there is a comparison of the different pastes during a print-simulation reproduced on the rheometer (oscillatory test). The nano-based paste reaches delta values around 45deg during the print process that indicate a partial change only in the viscous component while micro-size pastes become less viscous during print (high delta values around 90deg) and then recover the initial state very shortly. This is crucial for the paste transfer trough the screen and limits the deposit as already introduced by the first part of the rheological analysis.

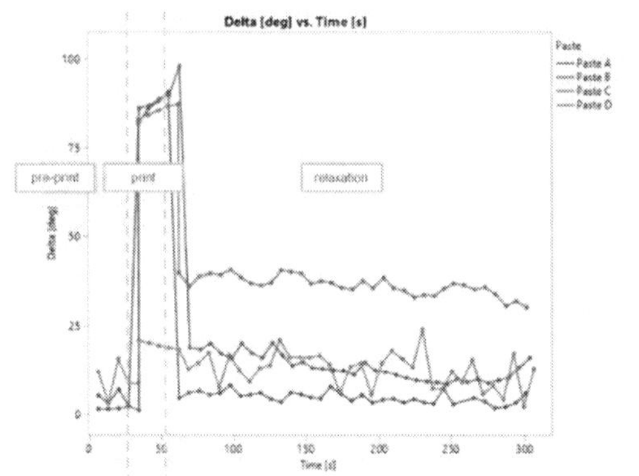

Figure 3. Oscillatory test, simulation of the different phases of a screen-printing process.

The outcome of the rheology analysis is substantially confirmed by the printing trials, where the same screen configuration is adopted for all pastes: stainless steel screen with 380/14 mesh, 15µm EOM, tension 24N/cm on 15" frame size and 32µm opening (squeegee side), single printing (SP) process and 5 busbars layout, 99 fingers on the front side and 199 fingers on the rear side. Industrial precursors used are M2 in size (156.75x156.75 mm), with an initial lifetime range of 1 to 1.5 sec (Sinton Lifetime tester).

Processing results and printing parameters used during the testing phase are reported in the table (Table 3) below:

	Paste A	Paste B	Paste C	Paste D
Finger width (μm)	54.3	53.2	45.2	56.3
Finger height (μm)	17.6	16.6	15.6	5.1
Weight total (mg)	284.1	298.3	241.5	105.7
Print/flood v. (mm/s)	150/200	100/200	300/350	400/1000
Force (N)	90	80	60	100
Snap-off (mm)	1.5	1.7	2.0	1.7

Table 3. Printing parameters and results from the first printing test.

The optimal parameters have been selected according to the rheology analysis and the test pattern printing: in order to evaluate the printability of the different pastes, we use a specific test pattern screen with different openings and geometries (lines, curves, arrays) coupled with a printing parameters Design of Experiment (DoE) to better explore the range of combinations between paste properties and tool setup.

The electrical results (Figure 4), referred to batches of 25 cells, suggest that Paste A and B are substantially comparable in terms of final efficiency, while keeping in mind that Paste B has the great advantage of being stored at room temperature. Paste C cells have instead a slightly lower conversion efficiencies due to substantially reduced fill factor (FF). This limitation on the FF and the consequent gap with respect to the reference is probably related to the lower deposit of paste so an optimization on the paste formulation that allows a higher deposit could be the solution. Finally, Paste D has the lowest efficiency values, and this is mainly related to shading effects of the printed lines on the substrate and the printing characteristics, which strongly limit the deposit. All the IV measurements have been performed on a HALM Solar Simulator and using the same calibration setup for all the measurements.

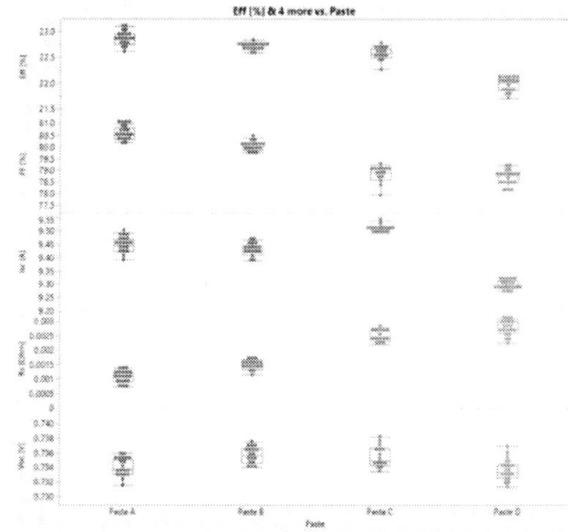

	Eff [%]	FF [%]	Isc [A]	Rs [Ohm]	Voc [V]
Paste	Mean	Mean	Mean	Mean	Mean
Paste A	22.87	80.58	9.45	0.0011	0.734
Paste B	22.72	80.03	9.43	0.0014	0.736
Paste C	22.56	78.81	9.51	0.0025	0.736
Paste D	21.97	78.7	9.3	0.0027	0.734

Figure 4. IV results, electrical performance comparison between the different solar cells. HALM solar simulator for data acquisition.

The different morphological characterizations (Figure 5, Figure 6) have been performed on a Vertex 420 Metrological Microscope for what concerns the line width analysis (15 points for each cell, 5 lines of 3 points equally distributed along the substrate) and on a Cyberscan Vantage 2 metrology system for thickness, roughness and 3D images data acquisition. Data reported are from front sides only and our standard characterization procedure has been applied. Another microscope Olympus BX51 has been also used for real images acquisition (Figure 7).

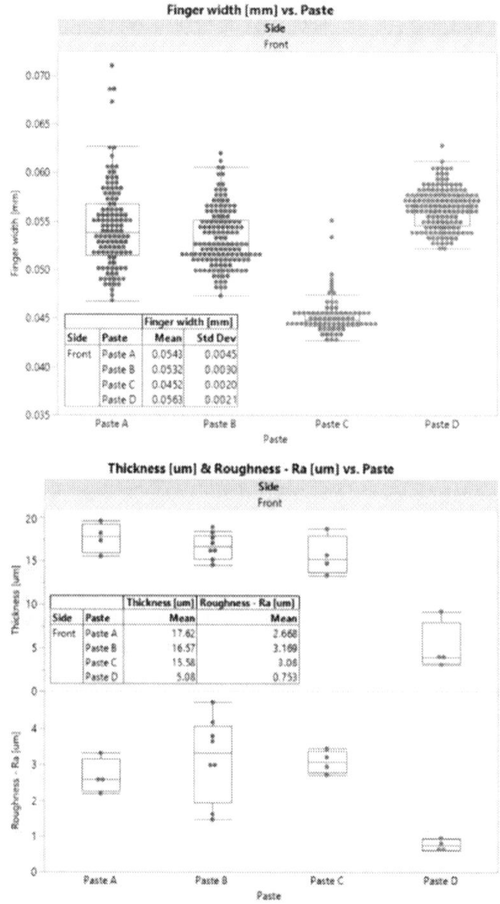

Figure 5. Morphological data and results. Thickness and roughness values from 3 points each cell, front sides only. 100x1000μm of analysis area for finger, 1000x1500μm for busbars.

Paste A and Paste B look very similar in terms of aspect ratio. On the contrary, Paste C allows a much narrower printing because of the limited smearing of the paste (32μm openings, 13μm spreading, total 45μm line width) but the thickness values remain below the reference. Paste D has instead the largest spreading of paste during the print while keeping very low thickness values. However, the paste properties allow much smoother lines as indicated by the roughness measurements and by the images taken at the Olympus microscope (Figure 7).

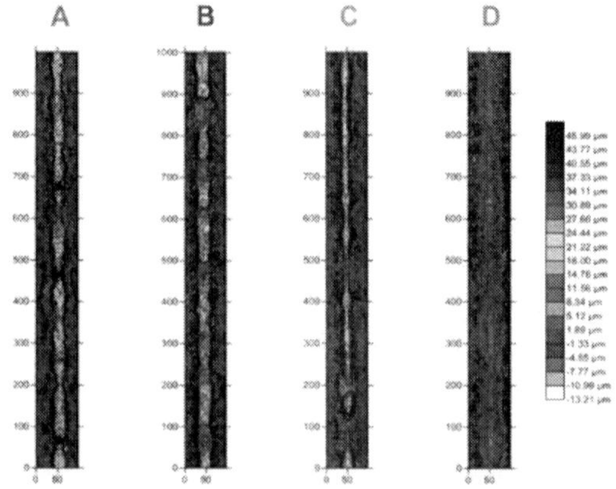

Figure 6. Printed lines and colormap applied to enhance the differences in thickness and roughness.

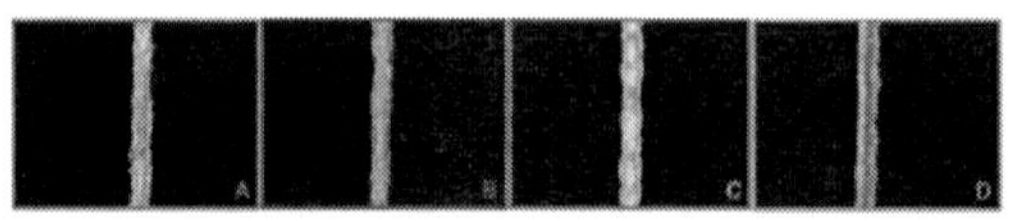

Figure 7. Printed lines captured on the Olympus Microscope.

Specifically for Paste D, 12 minimodules with 2x2 cells have been realized based on a 4 busbars metallization design, POE as encapsulant and using ECA-based interconnection for the busbars in the series, as shown in Figure 8. The modules have been tested on TC200 and DH1000 and the electroluminescence (EL) images have been taken at the end of each step.

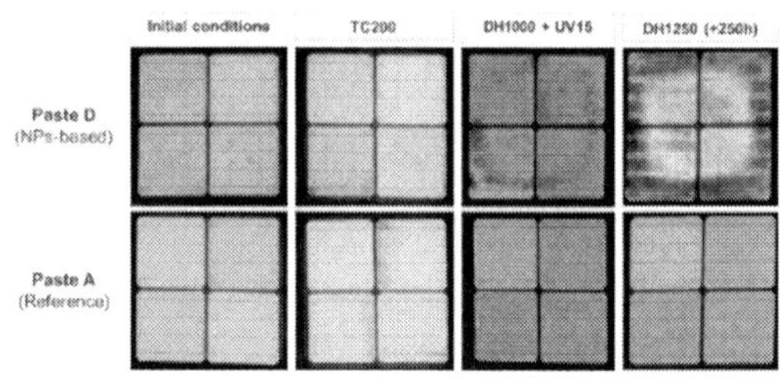

Figure 8. EL images of the minimodules tested during the different phases. Paste A has been used as a reference paste also during the reliability test.

Minimodules made with solar cells printed with Paste A have been also added to the analysis as a reference and assembled under the same conditions and with the same materials. Through the analysis of the EL images it is possible to evaluate the behavior of the modules at different stress conditions and it is clear that for modules made of Paste D cells something went wrong and an halo appears after the DH1000 test: after preliminary investigations on the darker areas a possible interaction between the NPs-based paste and the encapsulant has been considered as the primary cause of the effect. More detailed characterizations are needed for a better understanding and EVA encapsulant will be evaluated. This reliability test will also be extended to the other formulations.

V. CONCLUSIONS

In this work we show the potential that alternative low temperature Ag pastes have in reaching high efficiencies for HJT solar cells. In fact, the results reported above show that the Paste A is confirmed to be our reference paste in terms of overall performance but as already anticipated it suffers of different limitations. Paste B is also confirmed to be one of our best performing pastes with the great advantage of being stored at room temperature while keeping good printing properties. Paste C is even more interesting since allows not only an easier storage but also fast printing speeds and very fast curing processes that are 10-times faster than usual processes. Finally, Paste D is considered an under-development paste and still have different aspects to be implemented but it looks very promising since notable cell efficiencies can be reached even with low deposits (half of the Paste A deposit).

VI. ACKNOWLEDGMENTS

This work has been developed within national Best4U project and has received funding from MIUR within the program Progetti PON ricerca e innovazione 2014-2020.

VII. REFERENCES

[1] G. Condorelli et al., "High Efficiency Hetero-Junction: From Pilot Line To Industrial Production," 2018 IEEE 7th World Conference on Photovoltaic Energy Conversion (WCPEC) (A Joint Conference of 45th IEEE PVSC, 28th PVSEC & 34th EU PVSEC), Waikoloa Village, HI, 2018, pp. 1970-1973, doi:10.1109/PVSC.2018.8548197.

[2] https://www.maxwell-gp.com/en/news_15/421.html, May 2021.

[3] Kenji Yamamoto et al, "High-efficiency heterojunction crystalline Si solar cells" 2018 Jpn. J. Appl. Phys. 57 08RB20.

[4] S. K. Chunduri and M. Schmela, "Heterojunction solar Technology," TaiyangNews, 2020.

[5] Hoornstra, J, Weeber, A W, De Moor, H H.C., and Sinke, W C. The importance of paste rheology in improving fine line, thick film screen printing of front side metallization. Netherlands: N. p., 1997. Web.

LOW TEMPERATURE POST-PROCESS REPASSIVATION FOR HETEROJUNCTION CUT-CELLS

S. Harrison[a], B. Portaluppi[a], P. Bertrand[a], V. Giglia[a], B. Martel[a], A. Sekkat[b], D. Muñoz-Rojas[b]
a Univ Grenoble Alpes, CEA-LITEN, Campus-INES, F-73375 Le Bourget Du Lac, France
b-Univ Grenoble Alpes, CNRS, Grenoble INP, LMGP, F.38000 Grenoble, France

ABSTRACT: Tremendous progresses have been recently made for all cell technologies, and single junction solar cells are currently approaching their theoretical limits. However, to further decrease the Levelized Cost Of Energy, and keep increasing the interest for PV installations, higher module power/efficiencies are still required. Most common approach is thus to work with cut-cells to reduce the resistive losses, but more and more module manufacturers are also considering "zero- gap" or shingling integrations, as we can maximize the active area of silicon within the panels. Such modules with very high power are thus currently gaining market shares, and may soon become the norm for all technologies. However, new challenges also arise, especially with the new cutting-step, which generates additional defectivity and affects quite significantly the cell performances. This is particularly true for Heterojunction (SHJ), because of the very high passivation values needed to maximize its performances. It is thus mandatory to be able to properly characterize the cut- edges generated, and minimize their impact either with adapted cuting-processes or with post-process repassivation. Although several studies already propose different edge repassivation approaches, most of them show only partial performance recovery or unstable final precursor characteristics. Furthermore, contrary to other structure such as PERC or TOPCon, SHJ is manufactured at low-temperature, which brings additional constraints for the optimization of compatible processes. In this paper, we intend to provide insights on cut-edge repassivation capability for SHJ, showing in particular that even in optimized conditions, high passivation is not straightforward to achieve.
Keywords: Heterojunction, Shingle, Low-temperature, AlO_x, Simulation

1 INTRODUCTION

Cut-cells are already becoming the norm in the PV industry. If so far, the power gain observed thanks to the lower current flowing through the cells when interconnected in modules was significant, this may not be still true when high efficiency cells are considered (Heterojunction TOPCon, PERC+...etc), or if small cell dimensions are integrated like for shingle concepts. It seems mandatory to explore thus cut-edge repassivation options. More and more communications arise on this topic [1,2], showing the increased interest of the PV community on this new technological challenge.

This edge passivation is thus especially needed for example for Heterojunction (SHJ) architecture [3], relying on the excellent passivation brought by the use of amorphous layers. These devices are indeed highly sensitive to the defective cut edge(s) generated [4,5], and significant performances losses are observed on the separated devices, especially for the smallest dimensions such as shingle tiles. To fully exploit the high potential of this technology, adapted edge passivation processes may probably be needed, with quite a number of constraints to be taken in account to propose an industry compatible solution:

- Low-temperature deposition/activation
- Sufficiently high passivation values achievable
- Low-cost and high throughput

For the industry compatibility, we thus chose to work first on well-known and already widely used layers in most of the PV production factories: Aluminum Oxide (AlO_x) and hydrogenated amorphous layers (aSi:H). More insights on each layer particularities will be detailed in each dedicated section. To reduce even further the additional costs, it is also clever to combine the additional process steps needed with already existing processes (single final anneal for metal curing, light soaking and passivation layer activation for example). We can also take profit of the deposited passivation layer to improve as well other properties of the device (passivation layer can be used as well as anti-reflective layer and may allow to increase cell current and/or reduce the usage of underneath TCO). These aspects have been all taken into account for the choice of final processes tested.

In the following sections, we will first describe the results obtained with the use of aSi layers, their high passivation obtained at low-temperature allowing us to understand better the different limitations and constraints linked to the cut-edge passivation. Indeed, up to now, only partial repassivation has been demonstrated by the different institutes/companies working on the topic [2,3]. We thus need to understand if this is linked to the difficulty to integrate high quality layers at low-temperature and at the end of the cell integration scheme, or if this is directly linked to the nature of the defects generated during the cut. Then we will focus on AlO_x layers as they seem naturally better suited to large scale industrialization as large size batches can be integrated simultaneously, and it may probably not be needed to localize the deposition only on the cut-edge(s).

Finally, first passivation results generated on SHJ cells will be presented, demonstrating the high potential of the layers developed, with significant performance gain after proper edge passivation processes are applied.

2 CELLS / SAMPLES DETAILS

All cells and samples used in this work were generated on the SHJ production pilot-line located at INES [6]. Standard layers and metallization pastes were used for the different experiments, except for the specific passivation layers tested, and the associated activation anneal step.

Samples used were all mechanically cleaved to avoid excessive defect creation associated to heating (as for laser-based cut for example). Extensive J-V, PL and microscopy analysis were conducted to understand better the inhomogeneity seen on the treated samples. Adapted methodology was applied when needed to properly evaluate the effect of several integration parameters, such as HF-Last, textured or polished final surface morphology, or impact of major cleavage defects revealed by SEM.

Finally, best process developped are tested on 5x5cm² samples generated again by mechanical cleavage out of mother wafers, thanks to the specific use of 45° rotated ingot wafers developped at INES [7], allowing full mechanical cut parallel to the printed metallic busbars used for current collection.

3 HETEROJUNCTION CUT-CELLS: ELECTRICAL OUTPUTS

As previously mentioned, SHJ architecture relies on highly passivating amorphous layers, and such devices have already demonstrated the ability to reach very high efficiencies [8]. This is still true when more agressive configurations such as shingle cells are considered, where the impact of the metal resistances is greated because of the deported busbars. More than 24% efficiency have been obtained on such cells, which process will be used as reference for the different experiments conducted.

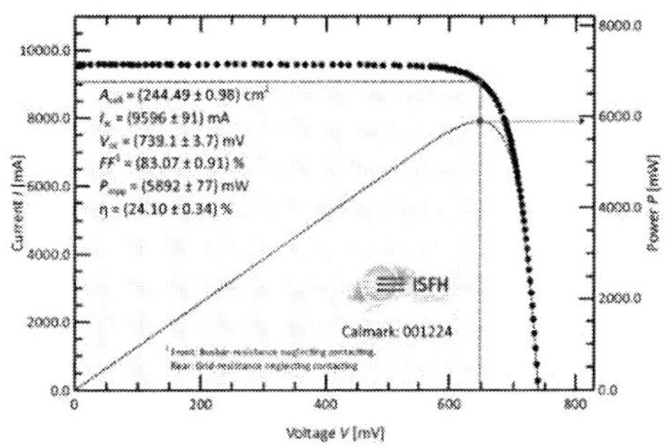

Figure 1: Certified SHJ shingle full-cell (before cut) certified at 24.1%

But SHJ cells are also highly sensitive to the cut-step and the naked defective edge(s) generated. The local high parasitic recombination occurring near the edge can significantly degrade the device performances essentially by modifying around the cut-edge the minority carrier diffusion and collection. This effect can be clearly evidenced by both light IV and dark IV measurements. Main impacted cell parameter is the FF and pFF, directly related to the ability to properly collect the photo-carriers within the cell. By plotting dark J-V curve for different cell geometries (from full cell to shingle), we can also quantify the impact of cut losses on the diffusion diode parameters (J02, n2), and use these values for accurate comparison of different samples considered. These data will also be needed to compute the proper architecture in dedicated simulation tools.

Finally, the different measurements conducted demonstrate the great impact of the final sample size, as edge defectivity impact is greatly increased when the final sample or cell dimension is reduced.

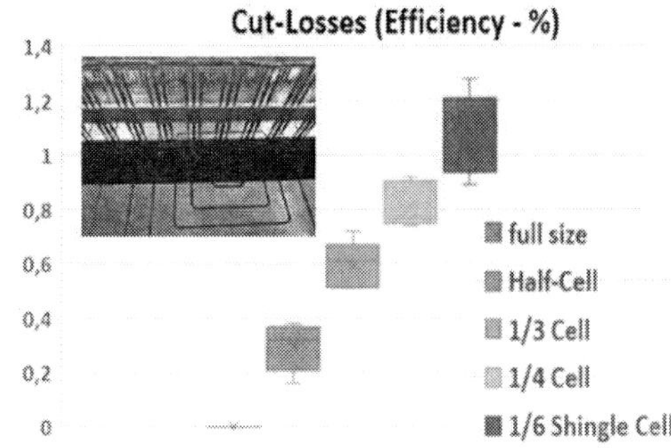

Figure 2: Light IV measurements showing the increased impact of cut-edge defectivity for the smallest devices.

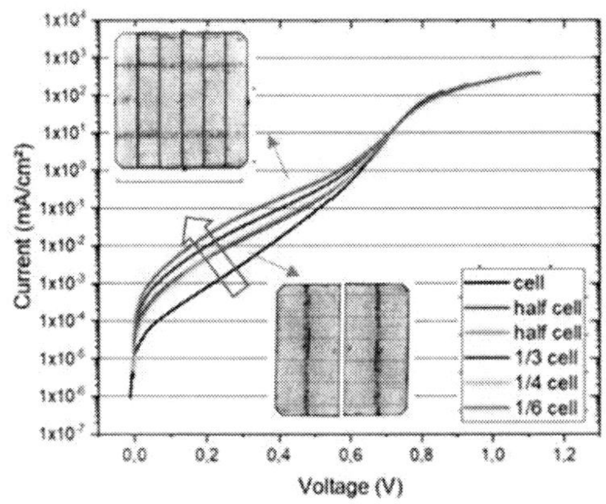

Figure 3: Dark IV measurements showing how the edge defectivity modifies the minority carrier collection within the cell. Again, the visible impact on the diffusion diode is greater for the smallest samples analyzed.

4 EDGE PASSIVATION WITH AMORPHOUS LAYERS

We firstly chose to assess the effect of a-Si:(H) layer passivation on dedicated samples. a-Si:(H) layers cannot be fully considered for post-process repassivation because of the high optical absorption leading to additional integration complexity not compatible with industrial requirements. However, the use of such layers is perfectly adapted to the scope of our study, as very high passivation levels can be achieved at low-temperature. It is thus possible to assess the capability of edge-passivation, without any limitation from the layer itself. Note that only mechanical cleavage is considered in this study. Indeed, as shown in several publications, repassivation of lasered-edges is nearly impossible due to the excessive silicon degradation caused by the local heating occurring during the cut [4]. Several dedicated test samples (N-Type commercial CZ wafers) have finally been used for the study, mixing different process conditions: HF-Last or no HF-Last (cut is then done before and between the clean and the a-Si:(H) deposition) before the a-Si:(H) deposition; textured or as-cut (referred as polished in this paper) final edge morphology. Finally, 20 nm of intrinsic a-Si:(H) is deposited by PECVD on both sides of the wafers and extensive PL characterization is systematically performed.

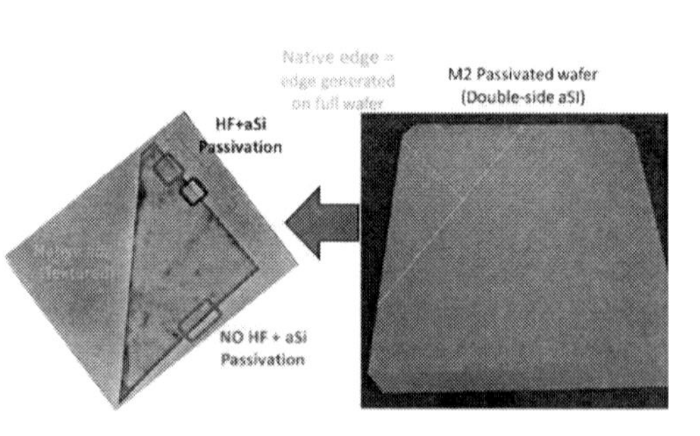

Figure 4: Typical samples generated for this study, allowing proper comparison of different edge configurations (with and without pre-cleaning, textured and polished).

Typical PL signal obtained on most of the samples is presented in figure 2. Despite the near ideal passivation conditions created on several samples (HF-Last + a-Si:(H)), we show that it is still not possible to obtain perfect and uniform passivation all along the wafer edge. We clearly see a random alternation of different areas: highly passivated (black square), partially passivated (blue square) and badly passivated (red square) for a given process. This is clearly visible on the PL scans extracted from the pictures for the three areas considered, with very different PL intensity signals observed close to the cell edge. It is worth to note that in all cases, PL edge defectivity extends up to more than 1 mm in the silicon bulk volume, in-line with different findings previously published [4]. However, it remains difficult to get an objective quantification of the global improvements or impact of the different tested configurations because of the high non-uniformity systematically observed. An in-house dedicated methodology has thus been applied, with in particular the introduction of a specific new metric parameter (labelled Y0 in this paper). For each considered area, an automatic image treatment is performed, with the extraction of the highest PL signal observed for the first pixel located next to the defective edge. Such uniform extraction method allows proper comparison of the different samples, lowering the data dispersion obtained otherwise.

As expected, Y0 results presented in figure 2, show clear differences between the different samples studied. Important differences between well passivated, and unpassivated edges can be clearly illustrated with this new metric, but we can also demonstrate with the reduced data dispersion that slight better passivation level can be reached on polished edge surface when compared to textured edges. We also demonstrate again that despite the ideal conditions created, post- process still shows passivation levels lower than what can be observed on native edge configuration (cell without cut). To understand further this effect, complementary SEM analysis is conducted in the different edge areas already identified (from well passivated to not passivated) and on the best samples only (HF-Last and a-Si:(H) deposition).

As illustrated in Figure 2, in well passivated areas, very neat and smooth surface is generally observed. On the contrary, remaining defects can still be spotted in defective areas despite the precautions taken during the mechanical cell breakage: layer delamination, silicon wrench or tearing are the most common defects seen. However, the presence of such defects is not systematic, and thus cannot fully explain the low-passivation levels measured locally. Results presented in this section thus seem to demonstrate that full passivation of cut-edge remain very challenging, at least if chemical passivation only is considered.

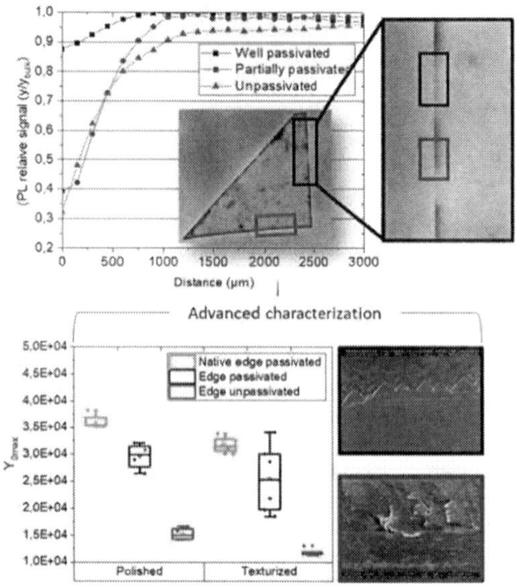

Figure 5: PL analysis of the different passivated edges generated clearly show that only partial and non uniform edge passivation could be achieved despite the high passivation potential of the aSi/H layers used. Lower passivation observed can be partly related to either morphological defects appearing despite proper mechanical cut, or specificities linked to the edge surface (texturation).

Alternative layers, combining both electrical and field effect must thus be considered (such as AlO_x layers discussed in next section), and main results obtained will be presented in next section.

5 DEVELOPMENT OF LOW-TEMPERTATURE AlO_x PASSIVATING LAYERS

Use of AlO_x layers is quite challenging at low-emperature, as thermal anneal over 350°C are generally needed to activate the fixed charges and obtain adequate layer properties. However, use of such layers for our specific repassivation needs seems particularly adapted. These layer are already widely used in the industry, they have high transparency rate which allows to consider full wafer deposition and simpilfied integration, batch deposition and low-cost are also very well adapted to high volume production constraints. But above all that, they combine the very promising impact of both chemical and field effect passivation, which may benefit to the specific cut-edge configuration. Two different deposition techniques are evaluated: thermal tradionnal ALD, and Spatial ALD technique (SALD) [9], allowing relaxed atmosphere and localized layer deposition if required [10].

We firstly tried to evaluate if simple low-temperature thermal anneal (hot plate or oven) could be sufficient to at least partially activate the passivation properties of the AlO_x layers developped. Despite the different deposition conditions tested (temperature, precursors..), no or very poor passivation was observed for all the layers evaluated.

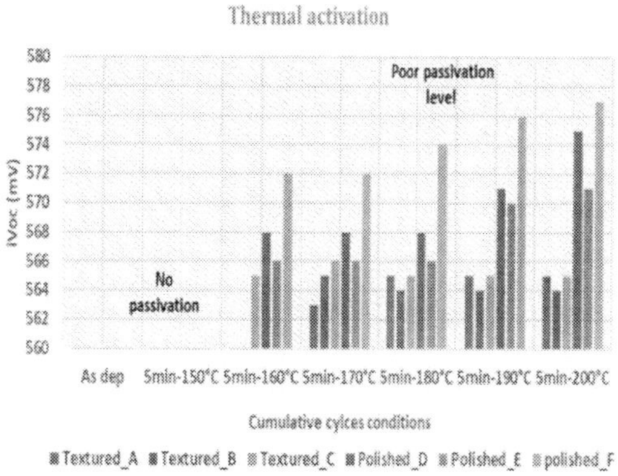

Figure 6: No or poor passivation observed for the AlO_x layers developped if only traditionnal thermal activation is considered.

Taking these first learnings into account, a second step of experiments was conducted, focusing on specific AlO_x layer deposition conditions, and introducing an alternative annealing condition. This new anneal process, still under development at INES is particularly interesting, as it may be used not only for this layer activation, but be considered for other applications such as final metal curing, or light soaking. With this innovative process optimization, very promising passivation results could be obtained for the two ALD and SALD techniques developped, with more than 710mV measured on symetric samples with layers deposited on both sides.

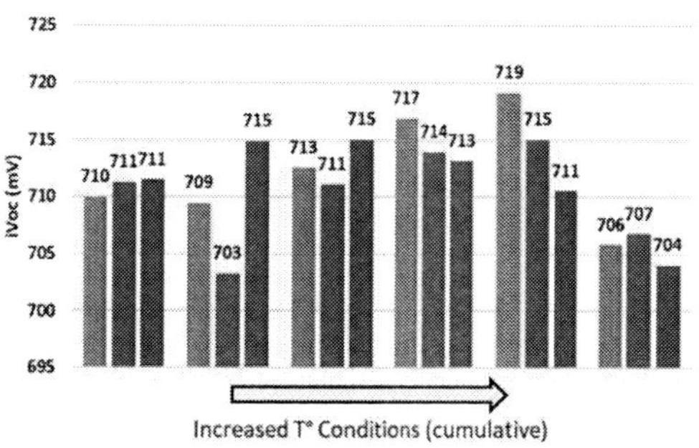

Figure 7: Optimized thermal ALD layer combined with alternative anneal conditions show very promising passivation outputs, with i-Voc close to 720mV for the best layers developped.

We thus successfully demonstrated that high quality low-temperature AlO_x layers could be achievable and thus be compatible with SHJ constraints. However, we still need to work on process stability and repeatability, especially for the SALD deposition technique, to be able to reproduce and stabilize the layers developped.

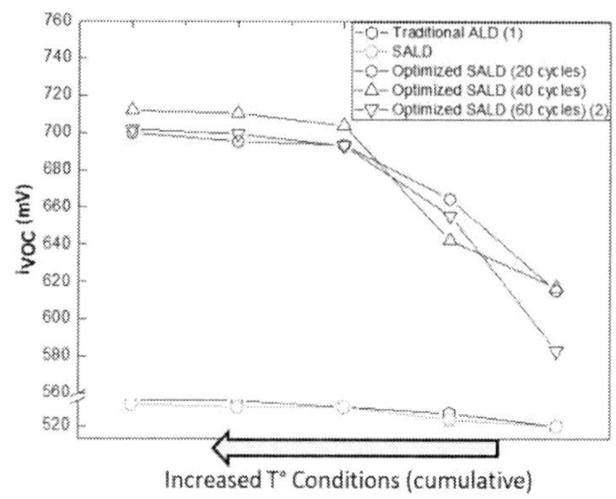

Figure 8: Optimized SALD layer combined with alternative anneal conditions also show very promising passivation outputs, with i-Voc close to 710mV for the best layers developped.

6 SIMULATION: EXPECTED LAYER IMPACT ON DEVICE PERFORMANCES

To evaluate the expected impact of the different layers developped in this work, we ran specific 2D-Silvaco simulations on SHJ Half-Cell and Shingle configurations. Cut-edge defectivity density is linked to local recombination, and finally local i-Voc. Details on simulation optimization can be found in [11]. We thus clearly show the impact of the edge defectivity on cell performances, but also how a proper edge repassivation could improve significantly the device outputs. Interestingly, we show that even moderate passivation levels can lead to great efficiency recovery on final device, and AlO_x layers developped in this work should thus be perfectly adapted to the repassivation scope considered.

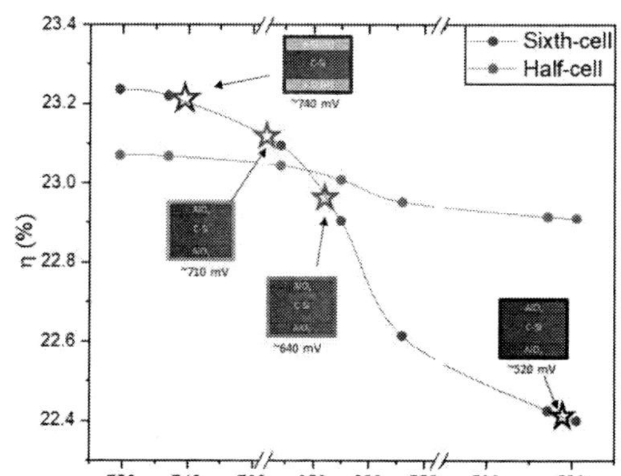

Figure 9: 2D-Simulations are run for Half-Cell and Shingle SHJ devices. Cut-edge defectivity is related to local i-Voc values estimated at the naked edge generated during the cut. Impact of different repassivation layers as developped in this work are included on the graph to estimate the potential recovery that we could expect with the use of such layers.

7 FIRST CELL RESULTS

5x5cm² SHJ cells were prepared to evaluate on real cell configuration the cut-edge repassivation potential of the AlO_x layer developped. Only SALD layer could be tested up to now, as the chamber is fully compatible with the metallized samples we need to use, contrary to the thermal ALD chamber which is dedicated up to now to front-end processes only. On this dedicated run, the SALD reference layers showed a slightly lower passivation level (up to ~670mV) than what was demonstrated previously.
Full wafer deposition was performed in order to cover properly the 4 cut-edges simulteaneously, with an AlO_x layer of approximatively 60nm. The cell pFF is measured for each integration step, as this parameter is the most adapted to edge defectivity follow-up. Indeed, as shown in previous section, FF is the cell paramter the most impacted by the cut edges. Moreover, pFF do not take into account the series resistances, so direct analysis of the extracted values obtained is possible. As seen on the following graph, pFF is degraded after cut, and no recovery is observerved right after the AlO_x deposition. However, when the alternative anneal is applied, we see a clear pFF improvement for the passivated devices, which we attribute to proper edge repassivation. Indeed, reference samples, without the AlO_x layers, going through similar successive annealings show only limited performance recovery, which confirms the successful integration of the passivating layer.

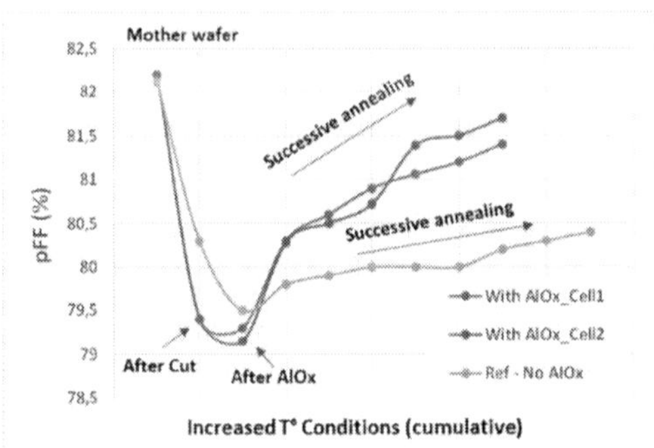

Figure 10: pFF improvement with AlO_x layer and adapted annealing condition shows clear efficiency improvement, demonstrating the effective edge repassivation with the layer developped.

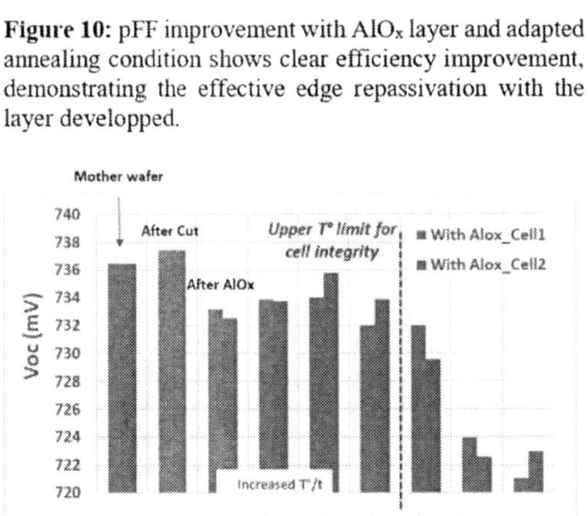

Figure 11: Voc follow-up of the 5x5cm² devices evaluated in figure 10. Globally stable Voc is observed for the first process steps, demonstrating good compatibility with SHJ and defining acceptable process window for the process.

However, as shown on Figure 10, only partial recovery is again observed with the AlO_x layers. This is in-line with the amorphous layer conclusions, and may indicate again that full recovery may not be possible. But we need to take into account the lower passivation layer potential than expected. Moreover, it is more than likely that real passivation value en cell edge may be lower than the values measured on plane surface on the reference samples. These first results are nevertheless very promising and fully validate the approach developped. New integration experiments will soon be conducted on higher volume of samples, as well that on other cell configurations (Shingle, Half-Cell) to fully assess the passivation process considered. Dedicated thermal AlO_x layer will also be soon tested, as a dedicated back-end chamber will be accessible soon for our experiments.

8 CONCLUSION

Cut-edge passivation remains a very tough challenge. Indeed, even with optimal conditions, combining in particular a-Si:(H) layers and appropriate cleaning steps, only partial edge passivation can be achieved. PL associated signal show high level of inhomogeneity all along the considered edge. We thus show that purely chemical passivation approaches may not be sufficient to recover for the initial cut-losses. Alternative layers, where passivation mechanisms rely also on the presence of electrical field might be necessary, and recent developments done on AlO_x low-temperature layers have been initiated in that purpose. Promising passivation values have been recently demonstrated with such layers, and first experiments conducted on SHJ cells confirm the high potential of the developped approach.

REFERENCES:

[1] P.Baliozian et al, "Postmetallization passivated edge technology for separated silicon solar cells", IEEE Journal of Photovoltaics, Vol 10, No 2, March 2020.
[2] D.Tune et al, "Measuring and mitigating edge recombination in modulesemploying cut cells", Proc of the 37th EUPVSEC, 2020.
[3] M. Taguchi et al, "24.7% Record Efficiency HIT Solar Cell on Thin Silicon Wafer", IEEE J. of Photovoltaics, Vol. 4-1 (2014), pp. 96-99.
[4] S.Harrison et al, "400W in Shingle SHJ configuration: promising optimization path for high power modules", proc of the 48th IEEE PVSC, 2021.
[5] F.Gerenton et Al, "Silicon heterojunction and half-cell configuration: optimization path for increased module power", Proc.of the 46th IEEE PVSC, Chicago, 2019.
[6] A.Danel et al, Silicon Heterojunction cells with open circuit voltage over 750mV, 35th EUPVSEC, Belgium (2018).
[7] JF.Lelievre et al, "Alternative CZ ingot squaring and half-cell cutting methodology for low-temperature PV cell and module technologies", proc of the 37th EUPVSEC, 2020.
[8] W.Favre et al, "25% efficienct large area silicon solar cell", proc of the 37th EUPVSEC, 2020.
[9] D. Munoz-Rojas et al, "Spatial atmospheric atomic layer deposition: a new laboratory and industrial tool for low-cost photovoltaics", Mater. Horiz 1(3), 314, 2014.
[10] C.A.Mass de la Huerta, "Gas-Phase 3D Printing of Functional Materials", advanced meterials technologies, volume 5, issue 12, Dec 2020.
[11] V.Giglia et al, "Influence of edge recombinations on the performance of half-, shingled-, and full silicon heterojunction solar cells", Proc of the 37th EUPVSEC

Authors would like to gratefully acknowledge the support of the European Union through the funded H2020 project **HighLite** under GA no. 857793"

ANALYSIS AND CORRECTION OF SYSTEMATIC DEVIATIONS IN MEASUREMENTS OF THE SPECTRAL IRRADIANCE OF SOLAR SIMULATORS

C. Schinke[1,2], D. Hinken[1], M. Wolf[1], K. Bothe[1], I. Kröger[3], M. Bagusch[3], S. Nevas[3] and S. Winter[3]
[1]Institute for Solar Energy Research Hamelin (ISFH), 31860 Emmerthal, Germany
[2]Dep. Solar Energy, Inst. Solid-State Physics, Leibniz University of Hannover (LUH), Appelstr. 2, 30167 Hannover, Germany
[3]Physikalisch-Technische Bundesanstalt (PTB), Bundesallee 100, 38116 Braunschweig

ABSTRACT: Measuring the spectral irradiance of solar simulators is required, among other applications, for rating the simulators with respect to spectral match according to the IEC 60904 standard series. Many laboratories therefore routinely operate spectroradiometers. In this contribution, we discuss the impact of systematic measurement deviations related to the measuring procedure, which have the potential to hamper accurate measurements. We show that deviations of the order of 5% may occur due to distance adjustment and background signal measurement during the calibration of spectroradiometers, which might even sum up in the worst case. Mitigation strategies are discussed and best practices identified: The position of the spectroradiometer's measuring head's optical reference plane should be examined and external stray light should be measured and corrected using a beam block method.
Keywords: Characterisation, Qualification and Testing, Solar Radiation

1 INTRODUCTION

The rating of solar simulators for solar cell characterization and calibration according to the IEC 60904 standard series considers the temporal instability, the spatial nonuniformity and the spectral match with the AM1.5G reference spectrum. For a long time, solar simulators have been rated from A (best) to C in each of these categories. The overall rating corresponds to the lowest of the three individual category ratings. Recently, edition three of the standard was published [1], which introduces a new class A+ for solar simulators with a spectral match better than ± 12.5%. In order to confirm such a rating, the spectral irradiance measurement must provide a significantly higher accuracy, meaning that there are no significant deviations between the measured spectral irradiance values (best estimates) and its actual (true) values. However, various effects can cause systematic measurement deviations of the order of several percent, which might even sum up in the worst case. This may prevent the correct rating of the solar simulator if the deviations remain uncorrected. Systematic deviations in measurements of a solar simulator's spectral irradiance typically result from the following effects:

- Distance adjustment during calibration of the spectroradiometer,
- Background signal measurement during calibration of the spectroradiometer,
  - Detector temperature effects,
  - Detector nonlinearity,
  - Spectral stray light and
  - Spectral bandwidth.

The first two effects refer to the measuring procedure, whereas the other effects are related to the spectroradiometer and thus depend on the specific device. In this contribution, we focus on the first group, which is independent from the instrumentation and thus of general interest. Before discussing the effects and their correction, we give a brief review of the measuring procedure.

2 MEASURING PROCEDURE

Determining the spectral irradiance $E_\lambda^{\mathrm{sim}}(\lambda)$ of a solar simulator as a function of wavelength λ usually requires two measurements. In a first step, the radiometric correction function

$$K(\lambda) = \frac{E^{\mathrm{ref}}(\lambda)}{S^{\mathrm{ref}}(\lambda)} \qquad (1)$$

for the spectroradiometer is determined by measuring the known spectral irradiance $E_\lambda^{\mathrm{ref}}(\lambda)$ of a reference lamp, which leads to a detector signal $S^{\mathrm{ref}}(\lambda)$ (in digital counts per time). This step is often referred to as *reference measurement*. It is also common to denote it as *calibration of the spectroradiometer* (with respect to spectral irradiance), although – strictly speaking – the determination of the radiometric correction function is only the first part of a calibration. The second part is the evaluation of the measurement's uncertainty budget. For many practical applications, however, the determination of the radiometric correction function is sufficient. The evaluation of measurement uncertainty is briefly discussed in section 4. The spectral irradiance $E_\lambda^{\mathrm{ref}}(\lambda)$ is usually given in the calibration certificate of the reference lamp and holds for a specific distance between its reference plane (often defined by a jig) and the measuring head of the spectroradiometer, which must be ensured during the reference measurement. Typical measuring distances are 300 mm when using 250 W halogen lamps and 700 mm when using 1000 W halogen lamps as reference. Having determined the radiometric correction function, the spectral irradiance of the solar simulator is obtained in a second step via

$$E_\lambda^{\mathrm{sim}}(\lambda) = S^{\mathrm{sim}}(\lambda)\, K(\lambda), \qquad (2)$$

where $S^{\mathrm{sim}}(\lambda)$ denotes the spectroradiometer signal (in digital counts per time) when the measuring head is illuminated by the solar simulator (*test measurement*).
Please note that, normally, different integration times are used for the reference and test measurement, since the irradiance levels of reference lamp and solar simulator differ. The detector signals $S^{\mathrm{ref}}(\lambda)$ and $S^{\mathrm{sim}}(\lambda)$ are therefore normalized to the respective integration times.

3 DISCUSSION OF SYSTEMATIC DEVIATIONS

3.1 Distance adjustment during reference measurement

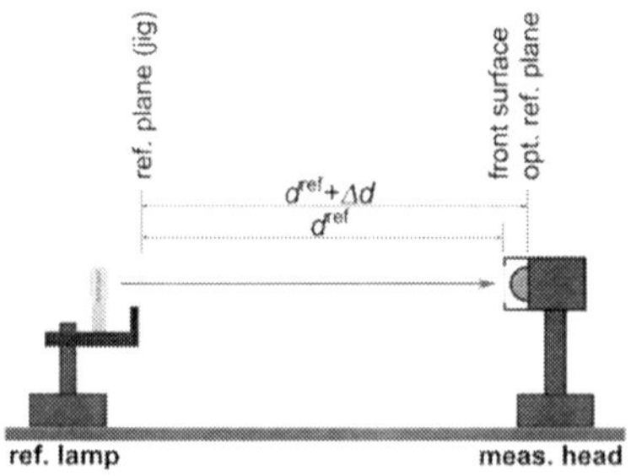

Fig. 1: Position of the optical reference plane of a transmission diffuser measuring head.

For the reference measurement, the distance between the reference lamp and the measuring head of the spectroradiometer must be adjusted to the distance specified in the reference lamp's calibration certificate. This requires a sufficiently accurate measurement of the distance. Inside micrometers or gauge blocks are therefore often used for this purpose. However, the optical reference plane of the measuring head does not necessarily coincide with its front surface, which is used as the touch point for the distance measurement. This holds true especially for the widely used transmission diffuser optics, where the optical reference plane can be located some millimeters behind the front surface, as sketched in Fig. 1. The consequence of the distance offset Δd between front surface and optical reference plane is an underestimation of the detector signal $S^{\mathrm{ref}}(\lambda)$ and consequently an overestimation of $E_{\lambda}^{\mathrm{sim}}(\lambda)$ according to Eqs. (1) and (2). The order of magnitude of this effect can be estimated by considering the inverse square law for the irradiance E of point light sources in a distance d from the light source,

$$E(d) \propto \frac{E_0}{d^2}. \qquad (3)$$

From this relation, the ratio of $E(d^{\mathrm{ref}})$ and $E(d^{\mathrm{ref}} + \Delta d)$ follows as

$$\frac{E(d^{\mathrm{ref}})}{E(d^{\mathrm{ref}}+\Delta d)} = \frac{(d^{\mathrm{ref}}+\Delta d)^2}{(d^{\mathrm{ref}})^2} \approx 1 + 2\frac{\Delta d}{d^{\mathrm{ref}}}, \qquad (4)$$

indicating that $E_{\lambda}^{\mathrm{sim}}(\lambda)$ is overestimated by a factor of roughly $2\frac{\Delta d}{d^{\mathrm{ref}}}$. The approximation in the latter equation is derived by Taylor series expansion around $\Delta d = 0$.

In order to apply a correction, the distance offset Δd must be known. It can be determined by varying the distance d between reference lamp and measuring head and fitting the measured (background and stray light corrected) detector signal $S(d)$ with the expression

$$S(d) = \frac{S_0}{(d+\Delta d)^2}, \qquad (5)$$

where S_0 and Δd are fit parameters [2-4]. Please note that for this measurement, d refers to the distance to the light source (filament of the lamp) and not the jig. Having determined the position of the optical reference plane, there are basically two options for applying a correction:

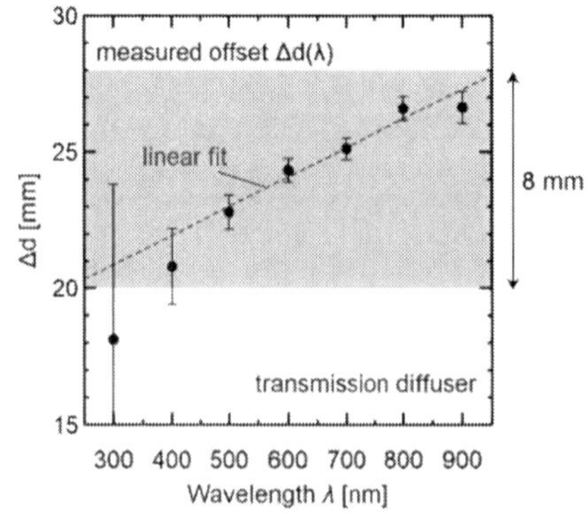

Fig. 2: Wavelength dependence of distance offset Δd observed for a typical transmission diffuser measuring head.

The term $(d^{\mathrm{ref}} + \Delta d)^2/(d^{\mathrm{ref}})^2$ in Eq. (4) corresponds to a correction factor that can be used for a mathematical correction of the effect. The second option is adjusting the distance such that $d^{\mathrm{ref}} + \Delta d$ corresponds to the distance specified in the reference lamp's calibration certificate (i.e., $d = d^{\mathrm{ref}} - \Delta d$).

The situation becomes more complex if the offset Δd turns out to be wavelength-dependent, which is not unusual and results in a spectral tilt of the measured spectral irradiance. Figure 2 exemplarily shows the observed wavelength dependence for a standard transmission diffuser measuring head. As can be seen, the position of the optical reference plane varies by about 8 mm over the wavelength range covered by a typical silicon detector-based spectroradiometer. This variation directly translates into the spectral tilt of the measured spectral irradiance shown in Fig. 3 for a reference measurement using $d^{\mathrm{ref}} = 300$ mm, which results in a maximum deviation of ± 5% compared to the actual values. Note that this deviation solely results from the reference measurement.

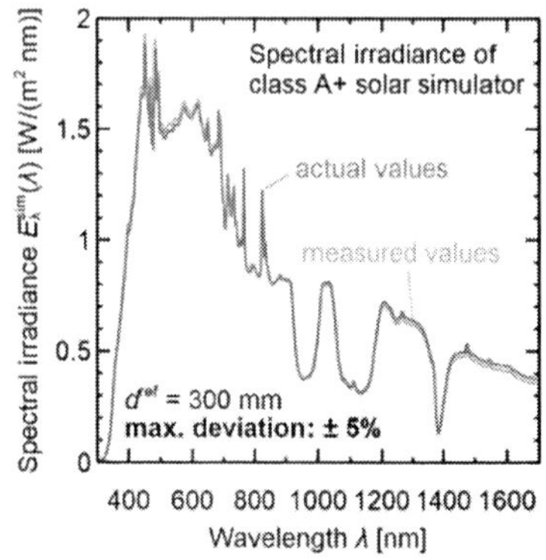

Fig. 3: Spectral tilt of the measured spectral irradiance (yellow) compared to the actual spectral irradiance (grey) following from the wavelength dependence of Δd.

There are essentially two options for circumventing measurement deviations due to the described effect: The first option is using an entrance optics that does not have an offset between its front surface (touch point) and its optical reference plane. This is usually the case for integrating spheres or reflection diffusers; however, using such entrance optics may entail other challenges and they are not always available off-the-shelf, especially for low-cost spectroradiometers. The second option is measuring the offset and applying a mathematical correction. If there is only a slight wavelength dependence, a practical approach might be to correct the distance measurement for a mean offset and to consider the remaining wavelength dependence in the uncertainty budget of the measurement result. If only relative spectral irradiance is of interest (e.g., for spectral mismatch correction), a correction can be neglected as long as there is no significant wavelength dependence of Δd. In any case, an investigation of the optical reference plane's position is advisable to make sure that the measured spectral irradiance is accurate.

Please note that the wavelength dependence of Δd also affects the test measurement. For the test measurement, however, the impact depends on the properties of the specific light field / solar simulator. Thus, the deviations can be increased or partly compensated, depending on the solar simulator under test.

3.2 Background signal measurement during reference measurement

The output signal $S^{\text{total}}(\lambda)$ of spectroradiometers contains a contribution $S^{\text{lamp}}(\lambda)$ due to illumination by the reference lamp and a contribution $S^{\text{dark}}(\lambda)$ from the detector's dark signal, which needs to be corrected. For this purpose, many spectroradiometers have a built-in shutter which allows to measure the dark signal. Additionally, the output signal may contain a contribution $S^{\text{straylight}}(\lambda)$ by stray light from the surroundings, which requires an additional correction. Hence, the output signal can be written as

$$S^{\text{total}}(\lambda) = S^{\text{lamp}}(\lambda) + S^{\text{background}}(\lambda), \quad (6)$$

where the background signal is

$$S^{\text{background}}(\lambda) = S^{\text{dark}}(\lambda) + S^{\text{straylight}}(\lambda). \quad (7)$$

It is common practice to minimise the impact of stray light from the surroundings by using low-reflective (black) materials in the measurement setup, so that $S^{\text{background}}(\lambda) \approx 0$ and $S^{\text{total}}(\lambda) \approx S^{\text{lamp}}(\lambda) + S^{\text{dark}}(\lambda)$. The dark signal is then measured using the built-in shutter of the spectroradiometer and subtracted in order to obtain the output signal due to illumination by the lamp.

Following the procedure described above requires a negligible stray light contribution. However, materials that visually appear black may show an increased reflectivity at invisible wavelengths, which may lead to a non-negligible stray light contribution. This is illustrated in Fig. 4, which shows measured reflectivities of several black materials that are commonly used in optical measurement setups.

A strategy for mitigating stray light effects during calibration of spectroradiometers is depicted in Fig. 5. It is based on two elements, namely: An overall reduction of stray light reaching the measuring head by placing an aperture between lamp and measuring head, and a combined background (dark and stray light) measurement using a removable beam block. Ideally, the beam block should only block the light coming directly from the light source, such that the detector signal only comprises dark signal and stray light signal. The advantage of a combined background signal measurement compared to a separate dark signal and stray light signal measurement is the reduction of measurement noise, since fewer measurements are involved in the calculation of the signal $S^{\text{total}}(\lambda)$. Please note that the stray light contribution discussed here must not be confused with spectral stray light, which refers to light being scattered inside the spectroradiometer and cannot be corrected by the beam block method.

Figure 6 exemplarily examines the impact of different configurations for the background signal measurement when choosing black molton fabric for the housing of our measurement setup, which is a typical material used for this purpose. The black curve represents the reference case for $S^{\text{lamp}}(\lambda) = S^{\text{total}}(\lambda) - S^{\text{background}}(\lambda)$ when using an

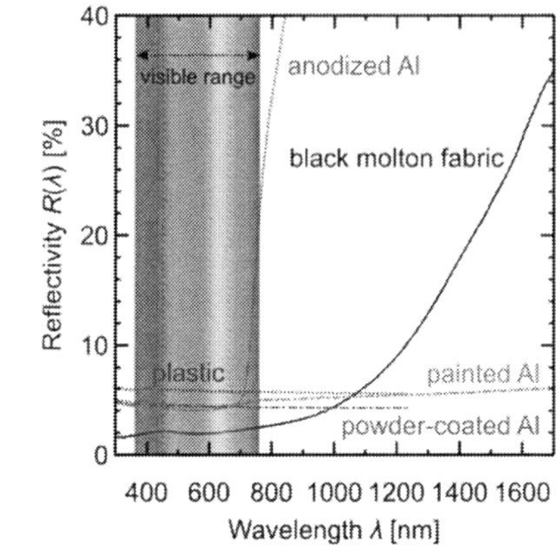

Fig. 4: Measured reflectivities of typical black materials that are used in optical measurement setups.

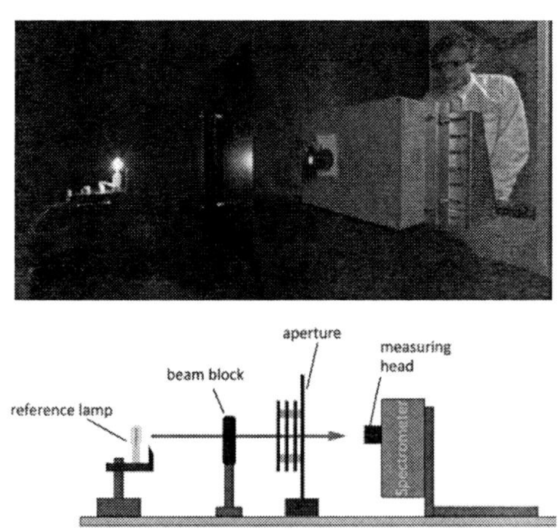

Fig. 5: Photo (top) and schematic drawing (bottom) of measurement setup for a combined background (dark and stray light) signal measurement. The beam block is not shown in the photo, which refers to the illuminated measurement.

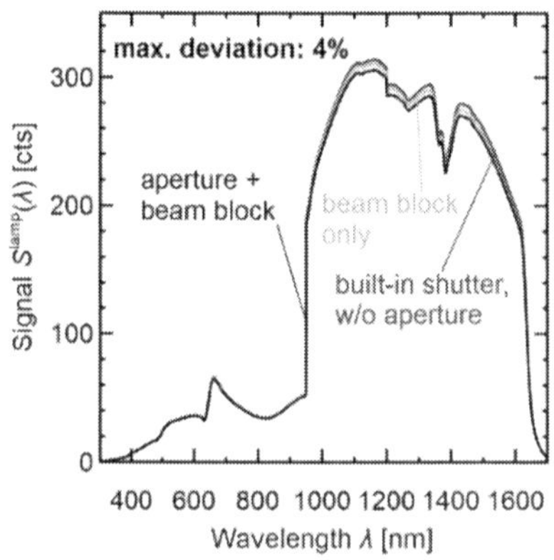

Fig. 6: Impact of different configurations for the background signal measurement (black: reference using aperture and beam block; yellow: beam block only; blue: built-in shutter without aperture).

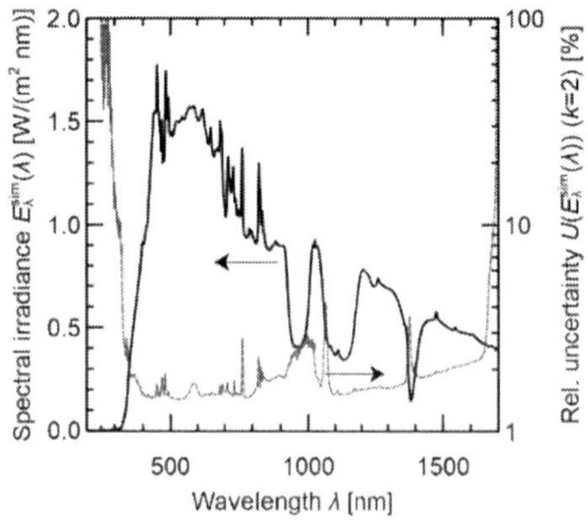

Fig. 7: Measured spectral irradiance (black curve) and corresponding measurement uncertainty (blue curve) for ISFH CalTeC's solar simulator.

aperture and a beam block as described above. The yellow curve represents the measurement results which are obtained when using only a beam block but no aperture for stray light reduction. Finally, the blue curve represents the worst case, where stray light is completely ignored and the dark signal of the detector is measured using the built-in shutter. In this case, deviations of up to 4% are observed in the near-infrared region due to the increased reflectivity of the black molton fabric.

At first glance, it may appear striking that omitting the aperture leads to an increased signal $S^{\text{lamp}}(\lambda)$, although stray light is measured using the beam block (yellow curve). This effect is a consequence of the fact that in reality, not all stray light contributions can be taken into account using the beam block method. In particular, stray light that is reflected in the back of the measurement setup near the optical axis contributes to the detector signal during the illuminated measurement (where the beam block is not present) but is blocked during the background signal measurement. Our simulation studies show that it is mainly this stray light contribution that leads to the increased signal $S^{\text{lamp}}(\lambda)$ compared to the reference case, where its contribution is attenuated by the presence of the aperture.

4 MEASUREMENT UNCERTAINTY ANALYSIS

Uncorrected systematic deviations would result in the measured value not corresponding to the best estimate for the measurand. The identification and correction of systematic measurement deviations is thus a prerequisite for a measurement uncertainty analysis, which quantifies the remaining uncertainty about the exact value of the measurand due to unknown effects. Our procedure for analysing the uncertainty of spectral irradiance measurements has already been published elsewhere [4]. The analysis considers 16 sources of measurement uncertainty and applies a Monte-Carlo calculation for modelling the impact of the uncertainty contributions on the measurement result. Figure 7 exemplarily shows the results of this analysis for ISFH CalTeC' solar simulator, which is a WACOM WXS-156 S-L2 two-lamps system featuring a quadratic light field of 170 x 170 mm². According to IEC 60904-9 Ed. 3 [1], this solar simulator is rated as class A+ in terms of spectral match. The black curve represents the measured spectral irradiance $E_\lambda^{\text{sim}}(\lambda)$ (including corrections for systematic deviations). The blue curve represents the relative uncertainty of the results as obtained from the uncertainty analysis for a coverage factor $k = 2$, which corresponds to a probability of about 95% for the true value of the measurand to be located within the indicated interval.

5 CONCLUSION

Systematic deviations in spectral irradiance measurements for solar simulators originate from effects related to the measuring procedure and effects related to the instrumentation. Whereas effects by the instrumentation can in principle be circumvented by using high-end spectroradiometers, for which the impact of these effects is sufficiently small, effects from the first group may generally affect the results of such measurements. In this work, we discuss the distance adjustment and the background signal measurement during calibration of spectroradiometers. We show that deviations of the order of 5% may occur, which might require correction when attempting to use the measurement results for rating solar simulators according to the IEC 60904 standard series. Moreover, the correction of these deviations is a prerequisite for performing a measurement uncertainty analysis. The following best practices are identified:

First, the position of the optical reference plane of the spectroradiometer's measuring head should be considered for the evaluation of spectral irradiance. This can either be achieved by using a measuring head where the optical reference plane coincides with the front surface for physical reasons (e.g., integrating spheres or reflection diffusers) or by measuring the position of the optical reference plane and applying a correction. Problems may arise for measuring heads that exhibit a significant wavelength dependence of the optical reference plane's position, which leads to a spectral tilt of the measured spectral irradiance.

Second, impact of stray light should be reduced to the extent possible. This can be achieved by using an aperture between light source and measuring head. Remaining stray light should then be taken into account appropriately during the background signal measurement, which can be achieved by using a beam block that only blocks the light coming directly from the light source. Furthermore, light which is directed backwards should be trapped or reflected away from the optical axis because stray light contributions on the optical axis cannot be taken into account by the beam block method.

ACKNOWLEDGEMENTS

The authors would like to thank H. Sträter, S. Riechelmann and P. Sperfeld (PTB) as well as H. Müllejans (JRC) for fruitful discussions about the measurement procedure and the uncertainty analysis.

The projects 16ENG02 (PV-Enerate) and 19ENG01 (Metro-PV) leading to this publication have received funding from the EMPIR programme co-financed by the Participating States and from the European Union's Horizon 2020 research and innovation programme. Parts of this work were funded by the scholarship program of the german national environmental foundation (Deutsche Bundesstiftung Umwelt).

REFERENCES

[1] International Electrotechnical Commission, *International Standard IEC 60904-9*, Ed. 3.0 (2020)

[2] J. Hovilla, M. Mustonen, P. Kärhä and E. Ikonen, *Determination of the diffuser reference plane for accurate illuminance responsivity calibrations*, Appl. Opt. 44, 28 (2005)

[3] K. Gröbner, *Experimental determination of the reference plane of shaped diffusers by solar ultraviolet measurements*, Opt. Lett. 32, 1 (2007)

[4] C. Schinke, H. Pollex, D. Hinken, M. Wolf, K. Bothe, I. Kröger, S. Nevas and S. Winter, *Calibrating spectrometers for measurements of the spectral irradiance caused by solar radiation*, Metrologia 57 (2020)

STABLE REVERSE BIAS OR INTEGRATED BYPASS DIODE IN HIP-MWT+ SOLAR CELLS BASED ON DIFFERENT INDUSTRIAL REAR PASSIVATION

Tadeo Schweigstill[1], Alma Spribille[1], Jonas D. Huyeng[1], Florian Clement[1], Stefan W. Glunz[1,2]
Fraunhofer Institute for Solar Energy Systems ISE
[1] Heidenhofstraße 2, 79110 Freiburg, Germany
Telefon: +49 761 4588-0, Fax: +49 761 4588-9000, E-Mail: info@ise.fraunhofer.de
[2] Department for Sustainable Systems Engineering, University Freiburg
Emmy-Noether-Straße 2, 79110 Freiburg, Germany
Telefon: 0761 / 203 – 96783, E-Mail: info@inatech.uni-freiburg.de

The Metal Wrap Through+ (HIP-MWT+) solar cell is based on the PERC concept but features two additional electrical contacts, namely the Schottky contact between *p*-type Si bulk and Ag *n*-contact and the metal-insulator-semiconductor (MIS) contact on the rear side of the cell below the *n*-contact pads. To prevent hotspots under reverse bias, both contacts shall either restrict current flow or allow a homogenous current flow at low voltage. In this work we present both options. First the stable reverse bias characteristics up to – 15 V with a MIS contact using industrially manufactured SiON passivation and second, an integrated by-pass diode using AlO_X as insulator in the passivation stack allowing current flows at approximately V_{rev} = – 3.5 V depending on the chosen screen-print paste. The examined Schottky contacts break down at around V_{rev} = – 2.5 V. Reverse bias testing of the cells proof a solid performance of the cells under reverse bias and an average conversion efficiency of η = 21.2 % (AlO_X) and η = 20.7 % (SiON), respectively.

Key words: MWT Solar Cell, PERC, Silicon Solar Cell, Passivation, back contact

1 INTRODUCTION

The passivated emitter and rear cell (PERC) [1] technology is now the standard technology in industrial production [2]. An advanced version of the PERC concept is the High-Performance Metal Wrap Through+ (HIP-MWT+) cell where only one additional process step (drilling of vias) is required to create a PERC with *p*- and *n*-contact located on the rear side of the cell [3]. This back-contact configuration results in several intuitive benefits such as a rear side module interconnection and reduced front side shading.

However, wrapping the metal through the Si bulk and placing the external *n*-contact on the rear also introduces new electrical contacts, namely the Schottky contact between *p*-type Si bulk and Ag *n*-contact in the vias and the metal-insulator-semiconductor (MIS) contact on the rear side of the cell below the *n*-contact pads. A schematic cross-section of the cell and the resulting electrical contacts are illustrated in **Figure 1**.

These additional contacts offer the following customization potential. If the contacts are designed for electrical insulation, a cell with a stable reverse bias characteristic is created, comparable to a regular PERC. This means that reverse current flow under partial shading in the module is restrained and thus hotspots are prevented. A standard module integration of such cells should then pass the IEC certification.

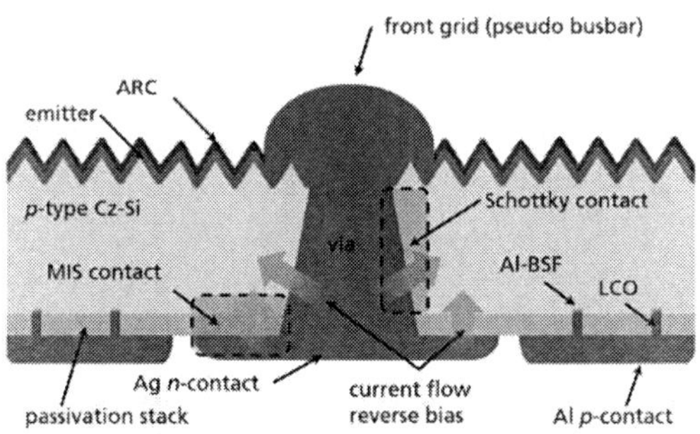

Figure 1: Cross sections of HIP-MWT + cell. Showing location of Schottky and MIS contact and potential current flows under reverse bias (yellow arrows).

On the other hand, one can also exploit the contacts by designing dielectric layers which insulate under forward bias but allow a current flow through the cell under reverse bias across a large area. This way an integrated bypass diode is built into the cell with only little additional production effort. This means that the module assembly is significantly simplified since no parallel bypass diodes are required in the module and a single shaded cell is not leading to an outage of a whole cell-string, which would reduce the power output of a typical module by 33 percent.

To find suitable passivation stacks for the insulating and conducting properties in the MIS contacts, different passivation principles can be compared. First, a chemical passivation using SiO_2 appears suitable to create an isolating passivation stack. Second, a passivation based on the electrical field effect using AlO_X seems appropriate to function as a rectifying contact [4].

2 EXPERIMENTAL

For the experimental realization of different contacts, two solar cell precursor types are used. The precursors are sourced from different industrial partners, after the front and rear surface passivation has been applied. Both precursor types are made from *p*-type Cz-Si and all passivation stacks are deposited using Plasma Enhanced Chemical Vapor Deposition (PECVD). The first stack is relying on the chemical passivation with a total thickness of $d \approx 160$ nm consisting of a hydrogenated siliconoxynitride (SiON) stack ($d \approx 75$ nm) capped with SiN_X ($d \approx 85$ nm) [5]. The second stack, relying on the electrical field effect passivation with a total thickness of $d \approx 150$ nm, is consisting of an AlO_X stack ($d < 20$ nm) and again a SiN_X capping ($d \approx 130$ nm) [6]. The metal side in the Schottky and MIS contact is created by a screen-printing process. While formally referred to as Ag, it is no pure metal due to the additives in the screen-printing paste. As the screen-printing paste contributes to the electric contacts, two screen-printing pastes with slightly different compositions were applied. The following manufacturing of test structures and solar cells was carried out using industrial production equipment located at the PV-TEC laboratory of Fraunhofer ISE [7].

Based on the described precursors, two types of dedicated test structures are manufactured: First a set of test structures to reveal the individual *IV*-characteristics of the MIS and Schottky contact, which show the reverse bias breakdown of the two contacts for different precursors and passivation stacks. Therefore, the different precursor types were metallized via screen printing. While the frontside (no anti-reflection coating) was fully metalized with an Al-paste to create a well-conducting contact, the rear side was metallized with circular Ag-pads with diameters between ⌀ = 0.5 mm and ⌀ = 2.5 mm. To create test samples with MIS as well as Schottky contacts, 50 % of the test samples had the rear side passivation stack removed prior to metallization. After metallization, all samples were then submitted to a contact firing process at T_{peak} = 850 °C. The samples were measured using an industrial 4-Point-Probe *IV*-setup. Two measurement pins were contacted on the front and two on the rear side, thus the *IV*-characteristics of the contacts were determined. The front side contact was proven to be ohmic, hence not influencing the measured rectifying behavior of Schottky and MIS contact. For the measurements, voltage sweeps from – 15 V to + 15 V were conducted with a step size of 10 mV, a delay of 50 ms and a current compliance of 1000 mA. Additionally, ten consecutive voltage sweeps from 10 V to – 10 V were performed (interjacent cooldown phases of 5 s) to show the *IV*-characteristics of the contacts over several reverse bias cycles.

Second, a set of test structures for lock-in thermography imaging (LIT) of the spatial breakdown was processed. These samples ran through the same process steps as the regular HIP-MWT+ cells (see paragraphs below). However, to further investigate the contact characteristics using LIT, the rear side metallization layout and the underlying LCOs were adapted. Instead of the regular rear side *n*-pads, a pad geometry variation was carried out with circular pads reaching from ⌀ = 2 mm to ⌀ = 20 mm each with one via in the center of the pad. After screen printed metallization, the samples were contact fired at T_{peak} = 850 °C and measured in a commercial solar cell analysis system (LOANA, pv-tools [8]). Each sample was measured seven times with set voltages from – 1 V to – 10 V, a LI-frequency of 18.75 Hz and a measurement duration of up to 300 s for low voltages.

Finally, HIP-MWT+ cells with a 6 pseudo-busbar layout in M2 format were manufactured on the industrial precursors with the named industrial process tools at Fraunhofer ISE. The HIP-MWT+ cells were manufactured in five process steps.

In the first process step the precursors were processed in an industrial laser system (ILS500X, Innolas Systems [9]). Here, the HIP-MWT+ cells receive local contact openings (LCOs) (⌀ = 40 μm, pitch of 500 μm) and 48 vias ($⌀_{rear}$ = 148 μm, $⌀_{front}$ = 100 μm) across the whole cell. In the following rear side metallization process, the via-filling is achieved by a controlled suction process of the vacuum chuck [10]. The following screen printing of the Al contact and front side grid were realized in parallel to standard PERC processes using an automated screen printer (XH2, ASYS TECTON GmbH). After contact firing the cell performance was measured in a commercial solar cell analysis system (customized, h.a.l.m. Elektronik GmbH [11]). Additionally, the manufactured cells were used for reverse bias cycling to demonstrate the long-term stability of the cells and the electrical contacts for 20 consecutive cycles at – 15 V (Forward bias at standard testing conditions for 100 ms and reverse bias for 40 ms with a compliance current of I_{rmax} = – 0.5 A).

3 RESULTS AND DISCUSSION

3.1 *IV*-Characteristics of Schottky and MIS Contact

The Schottky contact on the AlO_X based precursor (AlO_X removed) breaks at initial reverse bias at approximately V_{rev} = – 1.5 V to V_{rev} = – 2.5 V depending on the applied screen-printing paste as shown in **Figure 2**. The Schottky contact for the SiON based precursor (SiON removed) breaks at approximately V_{rev} = – 1 V [4]. The different breakdown voltages for the Schottky contacts both consisting of *p*-type Si and Ag (with paste additives) is related to the varying bulk doping of the precursor material. While the SiON based precursor has lower bulk doping of $N \approx 7.3\times10^{15}$ atoms/cm³, the AlO_X based precursor has a significantly higher bulk doping of $N \approx 1.7\times10^{16}$ atoms/cm³, resulting in a reduced width of the Schottky barrier and thus a lower breakdown voltage.

Over several reverse bias cycles the breakdown voltage is decreasing for all measured samples independent from the precursor and utilized screen-printing paste. The effect is the strongest between the first and second bias with the most significant right shift of the *IV*-characteristic. For all following reverse bias cycles the effect is incrementally decreasing, meaning that the contact tends to stabilize over several cycles. The increased contact formation can be explained by the percolation model which has been reported in electrical contacts before [12].

The same effect is observed even more significantly for the MIS contact using the AlO_X passivation as insulator as shown in **Figure 2**. Furthermore, the choice of the screen-print paste seems to have a major impact on the *IV*-characteristics of the MIS and Schottky contact. While the contacts using paste 1 shows a constantly increasing

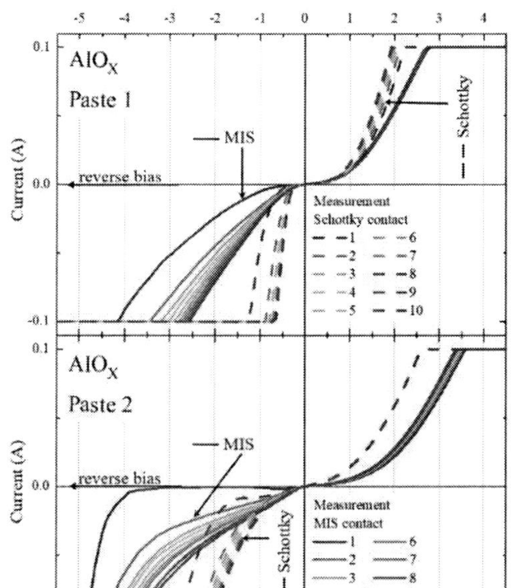

Figure 2: Repeated *IV*-characteristics of quantitative test structures for Schottky and MIS contact of the AlO_X based precursor with two different screen-print pastes.

reverse current towards higher voltages, the *IV*-characteristics using of the contact using paste 2 shows a more distinct knee in the curve during the voltage sweep at negative bias. This means that the reverse current increase is more sudden and thus the breakdown is more abrupt especially for the first bias.

We suggest that the different breakdown behaviors originate from different paste ingredient. The abrupt breakdown of the contacts using paste 2 are closer to the *IV*-characteristic of an ideal diode and suggest that the present contact is more homogenous than contacts using paste 1 with the creeping breakdown. The paste ingredient responsible for the inhomogeneous contact formation is the glass fritt [13] which has a weight content of approximately 5 – 10 $_{wt}$% in paste 1 and approximately 0 – 1 $_{wt}$% in paste 2.

The MIS contact using the SiON passivation is not showing any breakdown in the range from – 15 V to + 15 V independent from the chosen screen-print paste. The described chemical passivation is thus considered stable under reverse bias in a regular solar cell assembly.

3.2 Lock-in Thermography Imaging Results

The Lock-in Thermography (LIT) images of the dedicated test structures confirm the findings of the *IV*-characteristics for the Schottky contact and MIS contact on cell level. As shown in **Figure 3,** the SiON based precursors only allows a small current flow under reverse bias (- 3 V to - 10 V) at the vias where the Schottky contact is located. The MIS contact underneath the *n*-pads is completely restricting the current flow.

The LIT images for the AlO_X based precursor show a different behavior. At a reverse bias voltage of V_{rev} = – 3 V a current flow is essentially observed at the vias as shown in **Figure 3**. At a voltage of V_{rev} = – 5 V accessory currents are flowing from the *n*-pads to the *p*-type Si bulk, meaning that the MIS contact's breakdown voltage is exceeded. At V = – 10 V the currents are spread more homogenously among the whole MIS contact across the whole test wafer.

The LIT image at V = – 10 V is also revealing a higher current flow at the *n*-pad edges (ring structure). The increased current flow at the edges is correlated to an increase pad height. We suggest that the increased paste application at the edges leads to a higher glass frit concentration resulting in an increased contact formation. Therefore, the mass of applied paste during screen-printing seems suitable, among other parameters, to manipulate the breakdown voltage of the MIS contact.

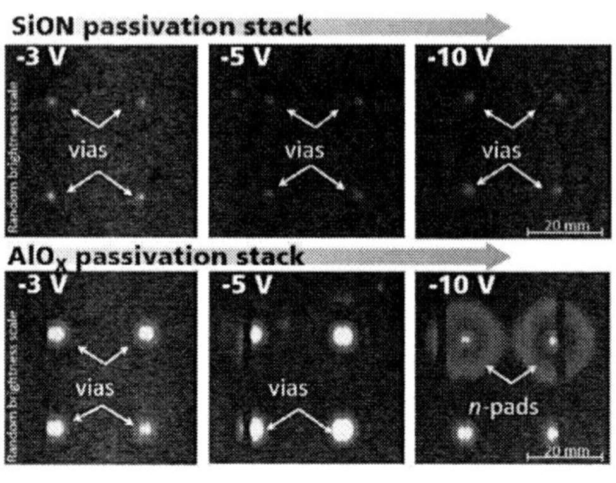

Figure 3: LIT images of test structures at different voltages, showing the spatially resolved reverse current for the SiON and AlO_X based precursor.

3.4 Cell Results and Degradation by Reverse Bias Cycling

The cell results of the manufactured HIP-MWT+ cells suffer from technical issues during the screen-printing process which caused narrowing in the fingers and thus an increased series resistance. However, the cells still showed a solid performance with an average conversion efficiency η = 21.2 % for the the AlO_X based precursor and η = 20.7 % for the the SiON based precursor (20 cells per group). The deviation in the conversion efficiency is related to several precursor parameters such as base resistivity (AlO_X based precursor 0.94 ± 0.03 Ω cm and SiON based precursor 1.97 ± 0.14 Ω cm), and passivation. The open-circuit voltage (V_{oc}) and short-circuit current (j_{sc}) are reflecting the differences in the precursor parameters and can be obtained from **Table I.**

In terms of long-term stability, the HIP-MWT+ cells show a solid performance for the first 20 cycles. While the cells with the SiON passivation stack are not affected by the reverse cycling, the cells using the AlO_X passivation stack show a slight decrease in the conversion efficiency of $\Delta_{\eta abs}$ = 0.2 %, reducing the average conversion efficiency from η_{abs} =21.2 % to η_{abs} =21.0 %.

Table I: Average cell results for HIP-MWT+ cells. Per cell group 20 cells were measured.

Rear passivation group	η (%)	V_{OC} (mV)	j_{sc} (mA/cm^2)	*FF* (%)
AlO_X initial	21.2	671.4	40.2	78.65
AlO_X 20th cycle	21.0	671.9	40.1	77.64
SiON initial	20.7	665.5	39.6	78.71
SiON 20th cycle	20.7	666.2	39.6	78.76

Furthermore, the HIP-MWT+ cells using the AlO_X passivation show acceptable heat dissipation under reverse bias with a power dissipation < 40 W at V_{rev} = – 10 V. This is still significantly below the upper limit of the 80 W limit a module encapsulation is able to cope with without being damaged [14]. This leads to the conclusion that the HIP-MWT+ cells using the SiON passivation stack, as well as the ones using the AlO_X passivation stack are suited for module assembly.

4 CONCLUSION

The *IV*-measurements of the Schottky contacts show a breakdown in the test structures under reverse bias at voltages between V_{rev} = – 1.0 V to V_{rev} = – 2.5 V depending on several factors such as the Si bulk doping (precursor material) and the screen-printing paste. Furthermore, a decrease in the breakdown voltages is observed when several reverse biases are applied. This percolation effect is most significant with the first reverse bias and gradually decreasing with each following applied voltage.

The MIS contact using the AlO_X passivation breaks in a similar manner as the Schottky contacts at slightly higher reverse voltages, while the MIS contact using the SiON passivation shows no breakdown down to V_{rev} = – 15 V.

The LIT images on cell-level confirm the findings from the *IV*-characteristic. The cells using the AlO_X passivation allow a reverse current from via metallization to Si bulk at small voltages (V_{rev} ≈ – 5 V). At higher voltages the current also flows through the MIS contact. The repeated reverse bias testing on cell level leads to marginal cell degradation resulting in a decrease of conversion efficiency of approximately $\Delta_{\eta abs}$ = 0.2 % (from η_{abs} = 21.2 % to η_{abs} = 21.0 %).

Cells using the SION passivation stack only allow small nonhazardous currents from via metallization to Si bulk and no current through the MIS contact even at higher voltages (V_{rev} = – 10 V). The repeated reverse bias testing for these cells is not significantly affecting the cell performance. However, further data is required to determine long-term effects on the MIS and Schottky contacts which must therefore be subject of future research activities.

REFERENCES

[1] A. W. Blakers, A. Wang, A. M. Milne, Appl. Phys. Lett. 55, 13 (1989)
[2] ITRPV, International Technology Roadmap for Photovoltaic (ITRPV) - Results 2018, 2019.
[3] B. Thaidigsmann, Ph.D. thesis, University of Tübingen, 2013
[4] T. Schweigstill, A. Spribille, J. D. Huyeng, F. Clement, S. W. Glunz, Proceedings 38th European Photovoltaic Solar Energy Conference and Exhibition, 2CO.10.2, (2021)
[5] C. Schwab, M. Hofmann, R. Heller, J. Seiffe, J. Rentsch, R. Preu, physica status solidi (a) 210, 11 (2013)
[6] P. Saint-Cast, D. Kania, M. Hofmann, J. Benick, J. Rentsch, R. Preu, Applied Physics Letters, 95, 15 (2009)
[7] Fraunhofer ISE, Neue Technologien und Solarzellen-wirkungsgrade aus dem Fraunhofer ISE PV-TEC press release 26, 2019.
[8] pv-tools GmbH, [Online]. Available: http://www.pv-tools.de/products/loana-system/loana-start.html. [Accessed 17 06 2021].
[9] InnoLas Solutions, [online]. Available: https://www.innolas-solutions.com/laser-machines-products/linexo-linear-table-laser-machine/. [Accessed 17 06 2021].
[10] E. Lohmüller, B. Thaidigsmann, S. Werner, F. Clement, A. Wolf, D. Biro, R. Preu, Proceedings 27th European Photovoltaic Solar Energy Conference and Exhibition, 2AO.2.6, (2012)
[11] h.a.l.m. Elektronik GmbH, [online]. http://www.halm.de/en/products/production/solar-cell.html [Accessed 17 06 2021].
[12] Z. Gingl, C. Pennetta, L. B. Kiss, L. Reggiani, In Semicond. Sci. Technol. 11, 12 (1996)
[13] Hilali, M. M., Sridharan, S., Khadilkar, C., Shaikh, A., Rohatgi, A., & Kim, S. (2006). Effect of glass frit chemistry on the physical and electrical properties of thick-film Ag contacts for silicon solar cells. Journal of electronic materials, 35(11), 2041-2047.
[14] I. Geisenmeyer, F. Fertig, W. Warta, S. Rein, M. C. Schubert, (2012): Proceedings 27th European Photovoltaic Solar Energy Conference. Frankfurt, pp. 870–874.

This is a short version of the paper submitted to the journal EPJ Photovoltaics.
It has been selected by the Executive Committee of the EU PVSEC 2021 for submission to EPJ Photovoltaics.

38th European Photovoltaic Solar Energy Conference and Exhibition

A Round Robin - HighLiting on the Passivating Contact Technology

Tobias Fellmeth[1*], *Frank* Feldmann[1], *Bernd* Steinhauser[1], *Henning* Nagel[1], *Sebastian* Mack[1], *Martin* Hermle[1], *Frank* Torregrosa[2], *Andrea* Ingenito[3], *Franz-Josef* Haug[3], *Audrey* Morisset[3], *Florian* Buchholz[4], *Aditya* Chaudhary[4], *Thibaut* Desrues[5], *Felix* Haase[6], *Byungsul* Min[6], *Robby* Peibst[6], *Loic* Tous[7]

[1]Fraunhofer Institute for Solar Energy Systems (ISE), Heidenhofstr. 2, 79110 Freiburg, Germany
[2]Ion Beam Services (IBS), Rue Gaston Imbert Prolongée, ZI Rousset-Peynier, 13790 Peynier, France
[3]Ecole Polytechnique Fédérale de Lausanne (EPFL), Bätiment 3316 Station 1, 1015 Lausanne, Switzerland,
[4]International Solar Energy Research Center e.V (ISC), Rudolf-Diesel-Str. 15, 78467 Konstanz, Germany
[5]Univ Grenoble Alpes, CEA, LITEN, DTS, INES, 50 avenue du lac Léman, 73375 Le Bourget du Lac, France
[6]Institute for Solar Energy Research Hamelin (ISFH), Am Ohrberg 1, 31860 Emmerthal, Germany,
[7]Interuniversitair Micro-Electronica Centrum vzw (IMEC), Kapeldreef 75, 3001 Leuven, Belgium

Abstract. The aim of this work is to demonstrate the maturity of the TOPCon technology by conducting a round-robin on symmetrically processed lifetime samples in the leading European PV institutes EPFL, ISC, CEA-INES, ISFH, IMEC and Fraunhofer ISE within the H2020 funded project called HighLite. For all layers, dark saturation current-densities ranging between 2-10 fA/cm² can be reported. Simultaneously, no metal induced recombination for the two lower sintering temperatures have been observed pointing towards a true passivated contact. Furthermore, contact resistivities below 10 mΩcm² have been achieved. It seems that the industrial passivating contact matured to a fully passivated and conducting contact enabling full efficiency potential. The fact that this can be realized using either PECVD or LPCVD from various manufacturer is expected to drive costs down and contribute to the increased adoption of the TOPCon technology.

1 Introduction

The poly-Si based surface passivation using a thin tunnel oxide also known as "TOPCon" [1] or "POLO" [2] (for simplicity, in the following only the expression "TOPCon" is used) is subjected to an increasing interest from the industry. In the last years, increasing conversion efficiencies were reported ending so far in record values of 24.6 % reached by Trinasolar [3] and recently of 25.25 % reported by Jinkosolar both in lab scale production.

The aim of this work is to demonstrate the maturity of the TOPCon technology by conducting a round-robin in the leading European PV institutes EPFL, ISC, CEA-INES, ISFH, IMEC and Fraunhofer ISE within the H2020 funded project called HighLite. Deposition tools from various companies using plasma enhanced chemical vapour deposition PECVD and low-pressure chemical vapour deposition LPCVD were used. Properties of the layers and information of the tools are given in Table 1.

2 Approach

N-type Cz-Si wafers were saw damage etched by ISC Konstanz and distributed to the mentioned institutes to use their best-known deposition technology to create n-type TOPCon layers on both sides of the wafers. Despite the differences in the deposition technology, all layers consist of a thin tunnel oxide layer between the c-Si absorber and a phosphorus doped poly-Si layer, which was annealed or $POCl_3$-diffused to transform a- to poly-Si and subsequently capped with a hydrogen-rich SiN_x layer. Then, the samples were sent to Fraunhofer ISE for single side screen-printing of a parallel finger pattern with varying finger pitch consisting of nine individual fields (Fig. 1) [4], contact firing and characterization. The dark saturation current-density for a single TOPCon layer J_{0T} as well as the total dark saturation current-density for each field J_0 both for metallized and unmetallized areas (Fig. 2) were extracted by a combination of the QSSPC [5] J_{0e} analysis from Kimmerle et al. [6] and the modulated photoluminescence Mod-PL [7] method. Finally, transition line method TLM [8] line structures

* Corresponding author: tobias.fellmeth@ise.fhg.de

(equidistant lines) were cut off the wafers to obtain the contact resistivity ρ_c (Fig. 3).

All poly-Si layers investigated here exhibiting a thickness ranging from 150-200 nm being in a typical range for screen-printed contacts with sufficiently low parasitic absorption, if embodied as rear electrode [9].

Table 1. Overview of the round-robin members, their technology and TOPCon layer properties.

Description	ISE	IMEC	ISFH	ISC Konstanz	INES
Technique	PECVD	LPCVD	LPCVD	LPCVD	LPCVD
Deposition tool	Centrotherm c.PLASMA	Tempress SPECTRUM	Centrotherm Europe 2000	Centrotherm	Semco Smartech
Deposition side	single	both	both	both	both
Doping process	in-situ	ex-situ, $POCl_3$ diffusion	in-situ	in-situ	ex-situ (PIII)
Doping tool (only ex-situ)		Tempress SPECTRUM			IBS Pulsion Solar
Interfacial oxide	thermal	thermal	chemical	chemical	thermal
Poly-Si thickness	150 nm	160 nm	200 nm	150 nm	200 nm
Doping (cm^{-3})	$1.7 x 10^{20}$	$2-3 x 10^{20}$	$2 x 10^{20}$	$2 x 10^{20}$	$4-5 x 10^{20}$
Anneal Temperature	900 °C	860°C	860°C	825 °C	975°C
Capping layer	SiN_x PECVD (75 nm)	SiN_x PECVD (85 nm)	AlO_x ALD, SiN_x PECVD (10 nm, 80 nm)	SiN_x PECVD (70nm)	SiN_x PECVD (70 nm)

In Table 1, an overview of the round robin members, their technology and layer properties are given. As can be seen, two CVD technologies from three different companies on five different tools are part of the round-robin.

In Fig. 1, a PL image is shown at the top exhibiting nine individual fields F1-F9 [4]. The fields F1 and F5 serve as reference without metallization, their size being compatible to allow QSSPC measurements. The remaining fields exhibit a parallel allocated finger pattern with nominal finger width of 40 µm and as indicated varying finger pitch in "mm".

The middle image of Fig. 1 shows a selected iV_{oc} calibrated PL image to visualize a challenge the authors observed for some samples during the characterization process. At the bottom, a line scan as indicated in the middle image show an increasing iV_{oc} from field F1 to F3. However, one expects a decreasing iV_{oc} considering the increasing metallized area potentially inducing additional recombination. That seems not to be the case, which points toward a vanishing contact recombination leading to other contrast drivers such as process inhomogeneities.

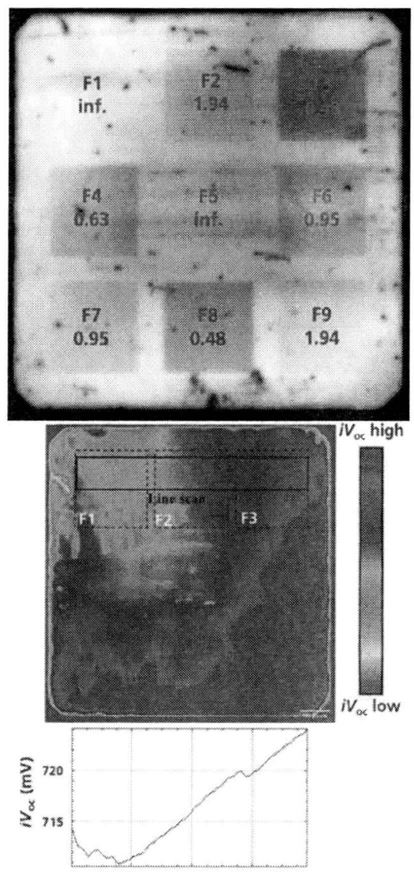

Fig. 1. TOP: A PL image of a over fired wafer, indicating nine individual fields F1-F9 with varying Finger pitches in "mm". On the edges TLM structures are located. Middle: An implied iV_{oc} Mod-PL image of a selected sample showing no field pattern, but other significant inhomogeneities. The iV_{oc} as a result from the indicated line scan area is shown in the bottom graph.

3 Results

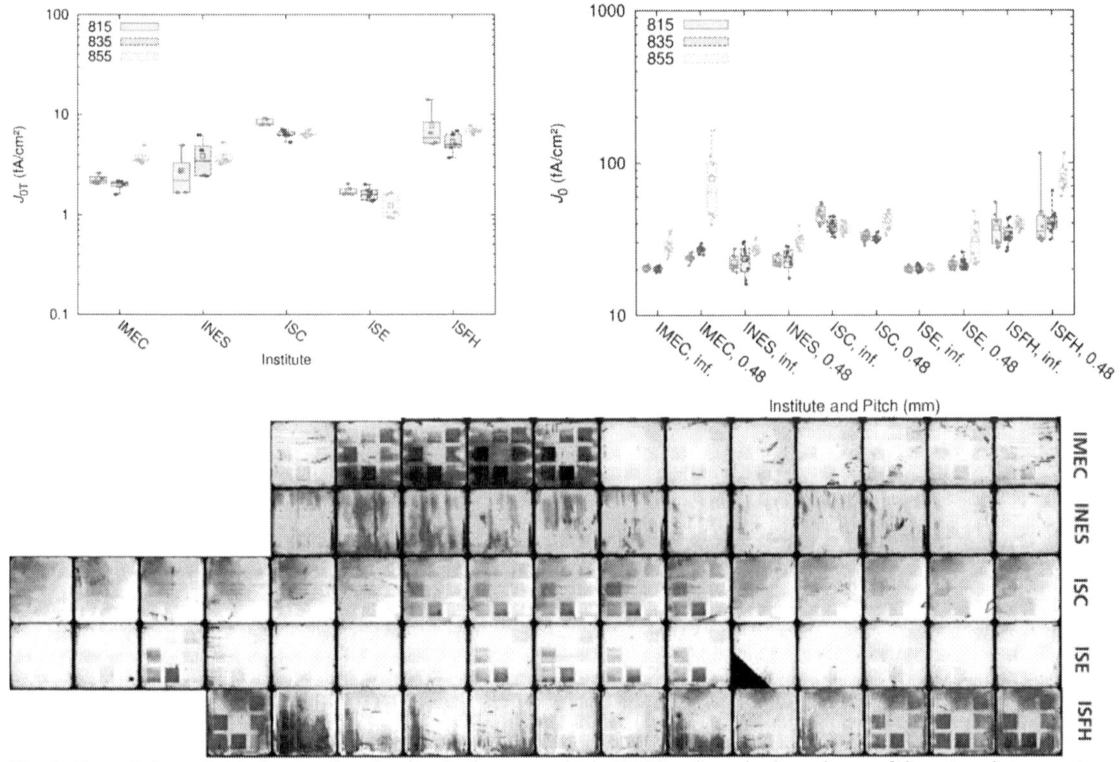

Fig. 2. Upper left graph: Dark saturation density J_{0T} from one single TOPCon layer in dependence of the set peak temperature. Upper right graph: Total dark recombination current-density J_0 either consisting of a bulk and two unmetallized TOPCon layers (infinite pitch) or a bulk and two TOPCon layers one side being metallized (0.48 mm pitch). Lower part: PL images of all specimens from all institutes showing partly no increased recombination or significant inhomogeneities of which both preventing a local $J_{0T,met}$ analysis.

In Fig. 2 upper left, the J_{0T} extracted by QSSPC from field F5 (see Fig. 1) is displayed showing excellent values ranging from 2-10 fA/cm², indicating a wide spectrum of tools and manufacturers being capable of producing high quality layers and therefore excellent surface passivation. It must be noted that such low absolute values as measured by ISE and others are significantly prone to uncertainties or inhomogeneities such as systematic errors from the measurement tool (WCT-120 from Sinton Instruments). Therefore, on the upper right side the corresponding J_0 from the unmetallized reference fields F1 and F5 (infinite pitch) are shown, indicating more realistic values between 20-50 fA/cm². Also, the J_0 from the metallized fields F3 and F8 with a constant pitch of 0.48 mm (smallest pitch) show comparable values to the unmetallized fields, pointing towards a real passivating contact for industrial relevant deposition and metallization technologies.

Eventually, at highest set peak temperature of 855°C, the contact starts to depassivate, due to significant etching by the Ag paste, which is confirmed by SEM analysis. The peak set temperature range was determined with a thermal element in a way that the specimens see the same temperature profile as the final solar cells would experience.

Due to the vanishing contact recombination and frequently occurring inhomogeneities, no meaningful local $J_{0T,met}$ can be determined.

Finally, Fig. 3 completes the picture indicating a conducting contact for most of the samples with specific contact ρ_c below 10 mΩcm². The ISFH samples feature an additional AlO_x layer tremendously increasing ρ_c, if metallized with the ISE paste ("ISE-Ag"). Obviously, that paste is not optimized for an AlO_x layer. Considering this by using a different Ag paste (ISFH-Ag), excellent mean values for the best group of 2 mΩcm² is reached. The ISFH values are obtained on-site from a different experiment on comparable layers [10].

ISC Konstanz completely finalized a part of the round-robin samples internally, and the results are in good agreement to the ones obtained from ISE although different firing conditions are used.

Additionally, the electrochemical voltage (ECV) measured profiles for each TOPCon layer are shown. The profiles were recorded individually by each institute and provided to the main author.

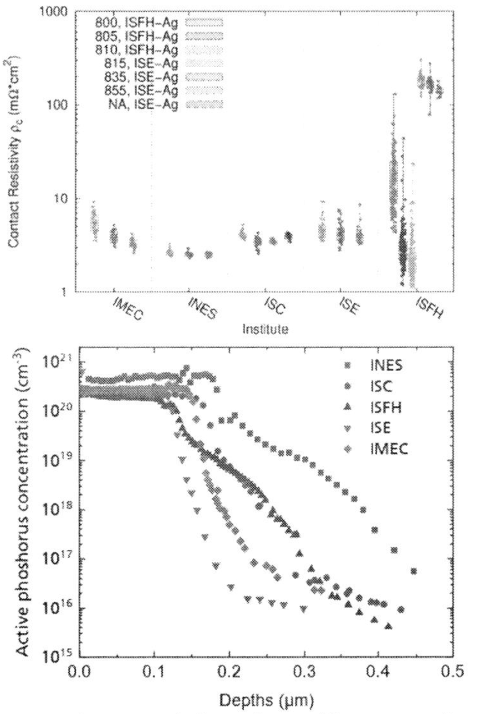

Fig. 3. The top graph shows the specific contact resistance ρ_c measured with a TLM-Scan+ device from PV-Tools for varying peak set temperature in "°C". The bottom graph shows the corresponding ECV profiles for each TOPCon layer including both the phosphorus in the poly-Si (constant part) and the c-Si tail (strongly decreasing part).

4 Conclusions

J_{0T} ranging from 2-10 fA/cm² for a single n-type TOPCon layer and J_0 ranging from 20-50 fA/cm² for the total symmetrical device were reported. Also, for the contacted area similar J_0-values are shown pointing towards a true passivated contact. Simultaneously, low specific contact resistance well below 10 mΩcm² allow a high fill factor potential, if applied as rear side electrode.

All this is realized on a wide technology base using PECVD and LPCVD technology with tools from various leading equipment manufacturers such as Centrotherm, Tempress, and Semco being available in the participating institutes EPFL, CEA-INES, ISC, ISFH, IMEC and ISE. This, together with a general trend towards higher efficient solar cells considering TOPCon, clearly underlines its relevance for the PV community and industry.

It seems that the industrial passivating contact matured to a fully passivating and conducting contact enabling full efficiency potential. The fact that this can be realized using either PECVD or LPCVD from various manufacturer is expected to drive costs down and contribute to the increased adoption of TOPCon technology. This comparable to what happened earlier with the PERC technology with capital expenditures for new production lines coming down by a factor of 5 in less than 10 years due to increased competition between equipment vendors [11].

5 Acknowledgements

The authors would like to thank all people from all institutes not mentioned as authors who have been involved in the process of assembling and characterization, as well as the logistic.

This work was funded by the European Union's Horizon2020 program for research, technological development and demonstration under grant agreement no. 857793.

6 References

1. Feldmann F, Bivour M, Reichel C et al. (2014) Tunnel oxide passivated contacts as an alternative to partial rear contacts. Sol Energy Mater Sol Cells 131:46–50. https://doi.org/10.1016/j.solmat.2014.06.015
2. Hollemann C, Haase F, Schäfer S et al. (2019) 26.1%-efficient POLO-IBC cells: Quantification of electrical and optical loss mechanisms. Prog Photovolt: Res Appl 26:3. https://doi.org/10.1002/pip.3098
3. Chen D, Chen Y, Wang Z et al. (2020) 24.58% total area efficiency of screen-printed, large area industrial silicon solar cells with the tunnel oxide passivated contacts (i-TOPCon) design. Sol Energy Mater Sol Cells 206:110258. https://doi.org/10.1016/j.solmat.2019.110258
4. Chaudhary A, Hoß J, Lossen J et al. (2019) Screen printed Ag contacts for n-type polysilicon passivated contacts. In: 15th International Conference on Concentrator Photovoltaic Systems (CPV-15). AIP Publishing, p 40002
5. Sinton RA, Cuevas A (1996) Contactless determination of current–voltage characteristics and minority-carrier lifetimes in semiconductors from quasi-steady-state photoconductance data. Appl Phys Lett 69:2510–2512. https://doi.org/10.1063/1.117723
6. Kimmerle A, Greulich J, Wolf A (2015) Carrier-diffusion corrected J0-analysis of charge carrier lifetime measurements for increased consistency. Sol Energy Mater Sol Cells 142:116–122. https://doi.org/10.1016/j.solmat.2015.06.043
7. Giesecke JA, Schubert MC, Schindler F et al. (2015) Harmonically modulated luminescence: Bridging gaps in carrier lifetime metrology across the PV processing chain. IEEE J Photovoltaics 5:313–319. https://doi.org/10.1109/JPHOTOV.2014.2362304
8. Meier DL, Schroder DK (1984) Contact resistance: Its measurement and relative importance to power loss in a solar cell. IEEE Trans Electron Devices 31:647–653. https://doi.org/10.1109/t-ed.1984.21584

9. Feldmann F, Nicolai M, Müller R et al. (2017) Optical and electrical characterization of poly-Si/SiO x contacts and their implications on solar cell design. Energy Procedia 124:31–37. https://doi.org/10.1016/j.egypro.2017.09.336
10. Min B, Wehmeier N, Brendemuehl T et al. (2021) 716 mV Open-Circuit Voltage with Fully Screen-Printed p -Type Back Junction Solar Cells Featuring an Aluminum Front Grid and a Passivating Polysilicon on Oxide Contact at the Rear Side. Sol RRL 5:2000703. https://doi.org/10.1002/solr.202000703
11. Chen Y, Altermatt PP, Chen D et al. (2018) From Laboratory to Production: Learning Models of Efficiency and Manufacturing Cost of Industrial Crystalline Silicon and Thin-Film Photovoltaic Technologies. IEEE J Photovoltaics 8:1531–1538. https://doi.org/10.1109/JPHOTOV.2018.2871858

ELECTRONIC PROPERTIES AND STRUCTURE OF BORON-HYDROGEN COMPLEXES IN CRYSTALLINE SILICON

Joyce Ann T. De Guzman[1], Vladimir P. Markevich[1], José Coutinho[2], Nikolay Abrosimov[3], Matthew P. Halsall[1], Anthony R. Peaker[1]
[1] Photon Science Institute and Department of EEE, University of Manchester, Manchester M13 9PL, United Kingdom
[2] Department of Physics and I3N, University of Aveiro, Campus Santiago, 3810-193 Aveiro, Portugal
[3] Leibniz-Institut für Kristallzüchtung (IKZ), Berlin, Germany
Phone: +44 161 306 4746, e-mail: joyceann.deguzman@manchester.ac.uk

ABSTRACT: The significance of hydrogen in the passivation of doping impurities and recombination active defects in silicon is well-established. It has been suggested recently that complexes of hydrogen with boron can be involved in the so-called "light and elevated temperature-induced degradation" (LeTID) in B-doped Si solar cells but the details of the relevant hydrogen-related reactions are not understood. In this work, the subject of hydrogen-boron interactions in crystalline silicon is revisited. We have carried out ab-initio modelling of the structure, binding energy, and electronic properties of complexes incorporating a boron and one or two hydrogen atoms. Further, the electrically active defects have been studied with the use of junction capacitance techniques in n-type Cz-Si samples co-doped with phosphorus and boron and subjected to hydrogenation by different methods. In DLTS spectra of the hydrogenated Si:P+B crystals subjected to reverse bias annealing at 100 °C, an electron emission signal with an activation energy of ~0.175 eV has been detected. The trap is a donor with electronic properties close to those predicted by ab-initio modelling for boron-dihydrogen (BH_2) complex. It is argued that BH_2 can be a very efficient recombination center for minority carriers and may be responsible for LeTID in B-doped p-type Si crystals.

Keywords: defects, silicon, hydrogen, deep-level transient spectroscopy, LeTID

1 INTRODUCTION

Hydrogen passivation of boron acceptors has been known to occur since 1983 [1]. Although there has been much debate about the mechanism of passivation and structure of the resulting complex, it is now generally accepted that hydrogen resides at the bond center between the neighboring silicon and boron atoms. This structure has a strong support from several experimental studies and theoretical calculations [2-8].

The common opinion is that the BH pair does not introduce energy levels into the Si band gap. It has also been suggested that multiple trapping of hydrogen atoms at substitutional boron atoms is possible [7]. Multi-trapping of hydrogen at boron is an old idea which, although postulated previously, has never been demonstrated experimentally [7, 8].

In some recent studies from the literature, it has been suggested that hydrogen is responsible for the so-called light- and elevated temperature-induced degradation (LeTID), which can produce a degradation of efficiency of solar cells between 2.5 and 16% relative and is most significant in multi-crystalline silicon Passivated Emitter and Rear Cells (PERC) [9-12]. It has been speculated that complexes of hydrogen with boron can be involved in the LeTID phenomenon [12, 13].

In view of the above findings, we revisited the subject of hydrogen-boron interactions in crystalline silicon by carrying out ab initio calculations and junction capacitance measurements.

From hybrid density functional calculations, we have determined atomic configurations, electronic properties and binding energies of boron-hydrogen complexes. Using junction capacitance techniques, we have investigated the electrically active defects in n-type Cz-Si samples co-doped with phosphorus and boron during growth and then subjected to two different processes of hydrogenation. The use of Si:P+B crystals has allowed us to monitor the B-H related electron traps in the upper half of the Si band gap and to study their electronic properties and thermal stability. A sequence of boron-hydrogen reactions, which can be related to the LeTID effect in Si:B, is proposed.

2 EXPERIMENTAL DETAILS

Phosphorus and boron co-doped Si wafers used in the present study were cut from an ingot grown at the Leibniz-Institut für Kristallzüchtung (IKZ) in Berlin by the Czochralski technique. The intended phosphorus concentration in the wafers was about 3×10^{16} cm^{-3}, and boron concentration about 2×10^{16} cm^{-3}. Current-voltage and capacitance-voltage (C-V) measurements have been used to assess the quality of the diodes and determine the uncompensated donor concentration $[N_d^+]$ and the width of the probed depletion regions.

Oxygen and carbon concentrations in the wafers have been determined from absorption coefficients, α, at maxima of infrared absorption bands due to vibrations of interstitial O and substitutional C atoms in silicon [14]. The infrared absorption spectra have been recorded at room temperature, and the following values of calibration coefficients for the determination of $[O_i]$ and $[C_s]$ have been used: $[O_i] = 3.14\times10^{17} \times \alpha_{Oi}$ (cm^{-3}), $[C_s] = 0.94\times10^{17} \times \alpha_{Cs}$ (cm^{-3}) [14]. In the Cz-Si:P+B wafers, the oxygen concentration is found to be $(6.5\pm0.5)\times10^{17}$ cm^{-3}, and carbon concentration to be $(4.5\pm1)\times10^{16}$ cm^{-3}.

Prior to hydrogenation processing, the wafers were cleaned using trichloroethylene, acetone, and methanol for five minutes at each step and lightly etched in a solution of hydrofluoric and nitric acids with a 1HF:7HNO_3 ratio. Hydrogen was introduced into the samples either by wet chemical etching at room temperature using the 1HF:7HNO_3 solution or by a treatment in remote hydrogen plasma at room temperature. The parameters of the plasma used were 50 W of RF power with 1-2 mBar chamber pressure and 200-250 cc/min gas flow.

Schottky barrier diodes (SBDs) were fabricated on samples cut from the wafers. The SBDs on the hydrogenated Cz-Si samples were prepared by thermal evaporation of gold through a shadow mask. Subsequently, aluminum was deposited at the back of the sample to serve as an Ohmic contact. The diode area for all the samples is about 0.79 mm^2 with a recorded leakage current of < 0.1 mA at 10 V reverse bias.

To investigate the electrically active defects in hydrogenated Cz-Si:P+B wafers, deep-level transient spectroscopy (DLTS) and Laplace DLTS (L-DLTS) were used to detect and characterize deep level traps [15, 16].

Total energy calculations were carried out from first-principles, within hybrid density functional theory. We used the range-separated functional of Heyd, Scuseria and Ernzerhof (HSE06), as implemented by the Vienna Ab initio Simulation Package (VASP) [17-20]. The projector-augmented wave method was used to treat the core electrons, whereas valence states were expanded in plane-waves with a cut-off energy of 400 eV [21].

Defects were inserted in 216-atom and 512-atom supercells with cubic shape, constructed by replication of 3×3×3 and 4×4×4 conventional cells of silicon. The calculated lattice constant was a = 5.4318 Å. Forces acting on atomic nuclei were obtained within the generalized gradient approximation (GGA) to the exchange correlation potential [22].

Minimum energy structures were found by energy minimization with respect to the atomic coordinates, until the maximum residual force was lower than 0.01 eV/Å. For the calculation of forces and structural relaxation of defects in 216-atom and 512-atom cells, the band structure was sampled over a 2×2×2 **k**-point grid and Γ-point, respectively. For total energy calculations (HSE06-level), the sampling was done at the Γ-point, irrespectively of the cell size. All total energies of defects with non-zero charge include a periodic charge correction, calculated here according to the recipe of Freysoldt, Neugebauer, and Van de Walle [23].

3 EXPERIMENTAL RESULTS AND DISCUSSION

3.1 Ab initio Modelling

We know from early local density functional calculations that atomic hydrogen in B-doped Si effectively roams around four equivalent Si-B bonds and connects directly to the boron atom [5, 6]. These calculations suggested the removal of the boron acceptor level upon formation of a B-H complex.

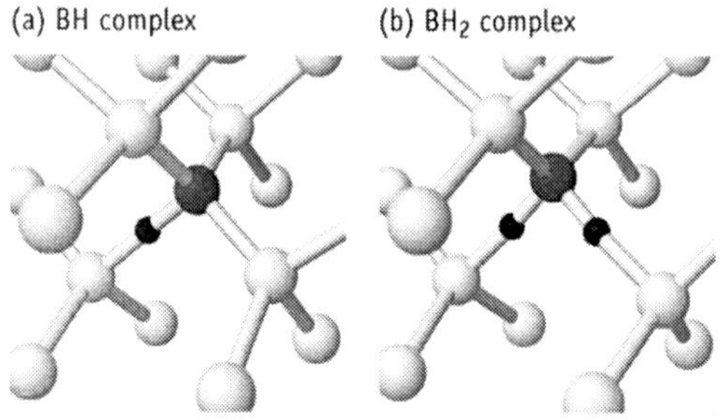

Fig. 1. Atomistic models of (a) BH and (b) BH_2 complexes in silicon. Si, B, and H atoms are shown in white, green and black, respectively.

It has been argued in an ab initio modelling study that hydrogen in the form of BH_2 complexes is more energetically favorable state of hydrogen in B-doped silicon compared to other H-related defects, including H_2 molecules [7]. However, the mechanisms that could explain a sequential or simultaneous reaction of several hydrogen atoms with boron, as well as the electrical activity of the resulting complexes have not been explored.

We inspected several configurations of BH and BH_2 complexes in Si and found that the formation of direct B- H bonds notably lead to the most stable structures, which are shown in Fig. 1 The sequential reaction of two protons with ionized B leads to neutral BH and positively charged BH_2 complexes with respective binding energies E_b = 0.73 eV and 0.4 eV. Molecular hydrogen is also found to effectively bind to B^- (E_b = 0.8 eV). However, in this reaction the donor nature of the resulting BH_2 complex implies the capture of two holes in p-type Si. We confirm that BH is electrically inert. On the other hand, BH_2 has a donor level close to the conduction band edge (E_c– 0.24 eV), reminiscent of the donor state of atomic hydrogen. Being positively charged in p-type material, BH_2 is potentially an attractive center for electrons, and eventually an important recombination center.

3.2 Junction capacitance measurements on hydrogenated Cz-Si:P+B crystals

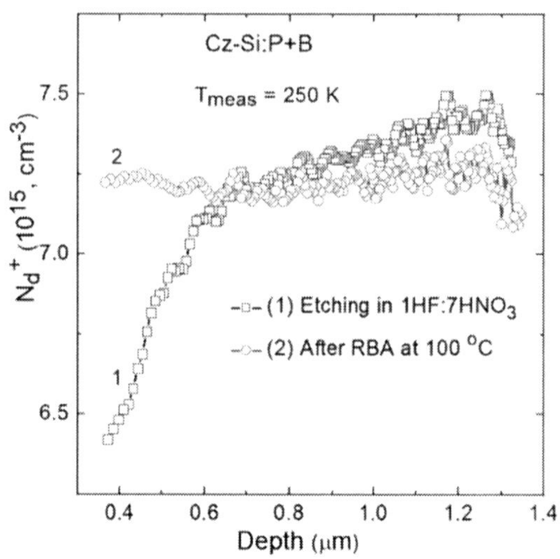

Fig. 2. Spatial profiles of concentration of uncompensated shallow donors in a Cz-Si:P+B sample which was subjected to hydrogenation by etching in 1HF+7HNO_3 solution for 3 min and then to annealing at 100 °C for 20 min with the reverse bias U_b= -9 V applied during annealing.

Fig. 2 shows spatial profiles of uncompensated shallow donors in a diode on a hydrogenated Si:P+B sample in an as-manufactured state and after annealing at 100 °C for 20 min under reverse bias of U_b= -9 V. A decrease in the N_d^+ values in the region from 0.4 to 1.0 μm in the as-manufactured diode is related to passivation of phosphorus donors by negatively charged hydrogen atoms, which were introduced into the bulk regions during wet chemical etching [24, 25]. It is known that the P-H pair is not very stable and dissociates upon annealing at temperatures around 100 °C [25, 26]. Indeed, after annealing of the sample at 100 °C for 20 min a recovery of the N_d^+ values in the subsurface region occurred. We have carried out this annealing with the reverse bias of U_b= -9 V applied to the diode. The reason for the application of the reverse bias was to transform hydrogen atoms from the negatively charged state to the positively charged one in the space charge region of the reverse-biased diodes and initiate interactions of H^+ species with negatively charged boron atoms.

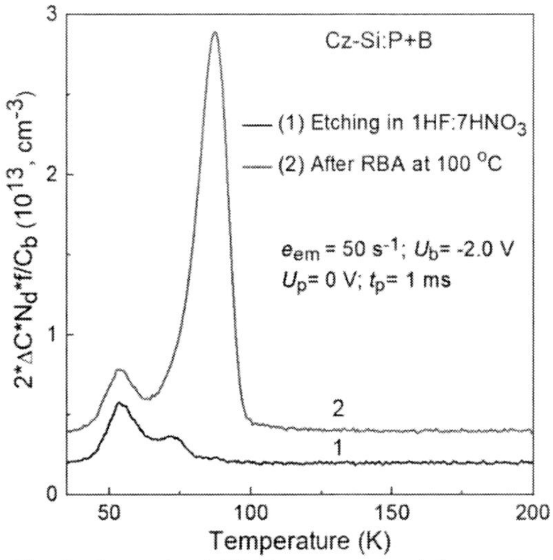

Fig. 3. Conventional DLTS spectra recorded on a Cz-Si:P+B sample which was subjected to (1) hydrogenation by etching in $1HF+7HNO_3$ solution for 3 min and then to (2) annealing at 100 °C for 20 min with the reverse bias U_b= -9 V applied during annealing.

Fig. 3 shows conventional DLTS spectra for a diode, which $N_d^+(W)$ dependencies are shown in Fig. 2. In the hydrogenated Cz-Si:P+B sample, two relatively weak electron emission related peaks with their maxima at about 55 and 73 K are observed in the spectrum recorded just after the hydrogenation. These peaks can be related to oxygen-related thermal double donors, carbon-hydrogen, and carbon-oxygen-hydrogen complexes [24, 26-30].

It is found that annealing under reverse bias has resulted in the appearance in the DLTS spectrum of a strong electron emission signal with its maximum at about 87 K. Concentration values $Y = 2\times(\Delta C/C_b)\times N_d\times f$ are plotted in Fig. 3, where ΔC is the magnitude of a capacitance transient, C_b is the bias capacitance, and f is the correction function, which takes into account depletion widths at the bias and pulse voltages and the so-called "lambda" layer [15]. The Y values at the peak maxima are close to the average trap concentrations in the probed region. As shown in Fig. 3, the RBA-induced trap has an average concentration of 2.5×10^{13} cm^{-3}.

Fig. 4 shows Arrhenius plots of electron emission rates (e_{em}) obtained for the RBA-induced trap at different values of electric field (E) in the space charge region of the diodes. The E-zero-field e_{em} values have been obtained from fitting the experimental $e_{em}(E)$ dependencies according to the Hartke's three-dimensional extension of the standard Poole-Frenkel model [31]. It is clear from a comparison of the Arrhenius plots that the trap has a very strong electric-field dependence of electron emission rates, which is characteristic for a trap with donor properties. The activation energy of electron emission from the corresponding energy level has been determined as ~0.175 eV relative to conduction band edge for E = 0 V/cm. Therefore, the trap is hereafter referred to as $E_{0.175}$.

Based on the results of ab initio calculations, we have tentatively assigned $E_{0.175}$ to a BH_2 complex. It is likely that the trap can be a very efficient recombination center of minority carriers in boron-doped p-type Si crystals.

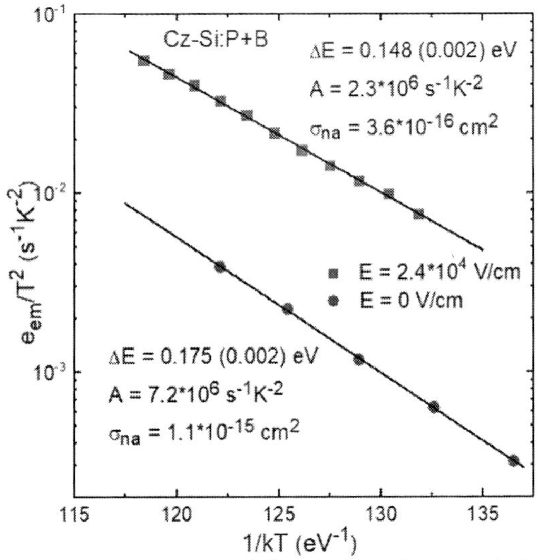

Fig. 4. Arrhenius plots of T^2-normalized electron emission rates recorded in the temperature range 85-100 K at different values of electric field on diodes on Cz-Si:P+B samples, which were subjected to hydrogenation and annealing at 100 °C for 20 min during which a reverse bias of U_b = -9 V was applied. The emission rates at the average electric field of $E \approx 2.4\times10^4$ V/cm have been obtained from measurements with the use of L-DLTS technique with the reverse bias fixed at U_b = -0.5 V and U_p = 0 V. The emission rates at E = 0 V/cm have been obtained from fitting the experimental $e_{em}(E)$ dependencies according to the three-dimensional Hartke model [31]. The $e_{em}(E)$ measurements have been carried using double L-DLTS technique with the reverse bias fixed at U_b = -5 V and changeable values of voltage for both filling pulses with the difference of 0.2 V between them.

Fig. 5 shows spatial profiles of uncompensated shallow donors in a diode on a plasma-hydrogenated Si:P+B sample in as-hydrogenated state and after annealing at 100 °C for 20 min under reverse bias of U_b= - 9 V. The treatment at 100 °C under reverse bias has resulted in a significant increase of concentration of uncompensated shallow donors ($\Delta N_d^+ > 2\times10^{15}$ cm^{-3}) in the sub-surface region of the diode (W< 0.3 μm) with respect to the bulk N_d^+ value.

Fig. 6 shows conventional DLTS spectra for a diode, $N_d^+(W)$ dependencies of which are shown in Fig. 5. The RBA treatment has resulted in the appearance of a very strong electron emission signal, the electronic signatures (emission activation energy, pre-exponential factor, and electric-field enhancement of emission) of which are identical to those for the $E_{0.175}$ donor trap detected in SBDs on samples with hydrogen introduced by wet chemical etching. In the RBA-treated sample the DLTS spectrum represents electron emission from in a region from about 0.2 to about 0.6 μm from the surface. A comparison of magnitudes of the peak with its maximum at 87 K in the DLTS spectra shown in Fig. 3 and Fig. 6 shows that the average concentration of the RBA-induced donor traps in the plasma-hydrogenated samples ($N_{E0.175} \approx 3\times10^{14}$ cm^{-3}) exceed that in the wet chemically etched samples ($N_{E0.175} \approx 2.5\times10^{13}$ cm^{-3}) by more than one order of magnitude. The obtained concentration ratio is consistent with the higher amount of hydrogen introduced into the samples by the H-plasma treatment compared to wet etching.

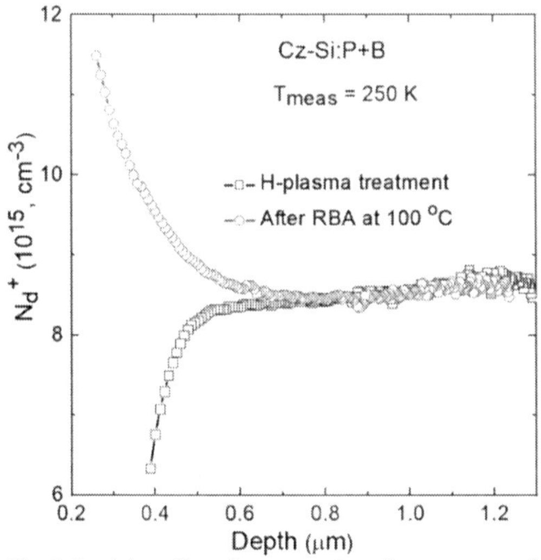

Fig. 5. Spatial profiles of concentration of uncompensated shallow donors measured at 250 K in a diode on Cz-Si:P+B sample, which was subjected to the following subsequent treatments: (i) hydrogenation in H plasma at room temperature for 30 min; (ii) Annealing at 100 °C for 20 min during which a reverse bias of U_b = -9 V was applied.

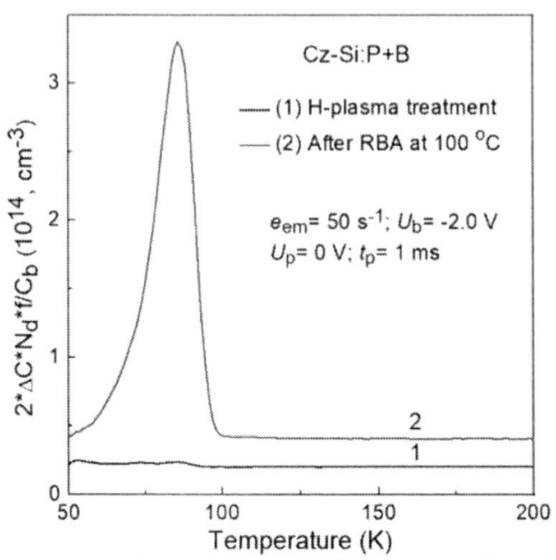

Fig. 6. DLTS spectra recorded for a diode, $N_d^+(W)$ dependencies of which are shown in Fig. 5. The spectra are shifted along the vertical axis for clarity. Measurement parameters are given in the graph.

We have further studied a range of hydrogenated n- type Cz-Si and FZ-Si samples, which were not co-doped with boron, before and after the RBA treatments identical to those used in this work. In some of the samples, the RBA treatments resulted in the appearance of a DLTS signal with its maximum at a temperature close to that for the maximum of the $E_{0.175}$ trap, however, concentrations of the corresponding traps never exceeded 10^{13} cm^{-3}. So, these findings indicate that the $E_{0.175}$ trap and the RBA-induced changes in the $N_d^+(W)$ values in the P+B co-doped samples are related to the presence of boron in these samples and its interactions with hydrogen atoms.

4 CONCLUSIONS

We have investigated the introduction of energy levels in the upper half of the Si band gap due to formation of boron-hydrogen defects. The study was carried out by first-principles calculations and junction capacitance measurements in Schottky diodes on n-type Cz-Si crystals co-doped with phosphorus and boron. Hydrogen was introduced into the Si:P+B crystals either by wet chemical etching or by treatments in remote H-plasma.

We have shown that in the DLTS spectra of hydrogenated Si:P+B crystals, which were subjected to heat-treatments at about 100 °C under reverse bias, a strong electron emission signal with the activation energy of about ~0.175 eV relative to the conduction band edge has been detected. It has been found that the detected electron trap has a very strong dependence of electron emission rate on electric field in the reverse-biased diodes. This indicates a donor character of the trap. We argue that the trap can be identified as a donor energy level of the BH_2 defect.

The RBA treatment at 100 °C results in the appearance of mobile positively charged H atoms in the depletion region of the diodes. The H^+ atoms effectively interact with negatively charged substitutional boron atoms and create the BH defect, which is neutral. In the case of high hydrogen concentration, the BH defect can trap another H atom that results in the formation of the BH_2 defect, which has a donor level at about $E_c - 0.2$ eV according to our ab initio calculations.

The donor character of the BH_2 complex indicates that it can be a very efficient recombination center of minority carriers in boron-doped p-type Si crystals. We suggest that the BH_2 complex could be responsible for the LeTID effect in B-doped p-type Si crystals. It has been shown in a recent study that the kinetics of resistivity changes in hydrogenated B-doped Si samples due to the passivation of boron atoms with hydrogen and changes in lifetime of minority carriers upon dark annealing of the samples at about 160 °C nearly coincide [12]. The resistivity changes in the Si:B+H samples upon initial stages of dark annealing at 160 °C can be related to the interaction of H_2 molecules with the substitutional boron atoms and the appearance of the recombination active BH_2 defects, which may be responsible for the lifetime degradation.

ACKNOWLEDGEMENTS

We would like to thank EPSRC (UK) for funding this work via grant EP/TO25131/1. J.A.T. De Guzman would like to thank the Government of the Philippines through the Department of Science and Technology (DOST) for her PhD funding. JC thanks the support of the i3N projects, Refs. UIDB/50025/2020 and UIDP/50025/2020, financed by the FCT in Portugal.

REFERENCES

[1] J. I. Pankove, et al., Phys. Rev. Lett. 51 (1983) 2224.
[2] J. I. Pankove, P. J. Zanzucchi, C. W. Magee, G. Lucovsky, Appl. Phys. Lett. 46 (1985) 421.
[3] N. M. Johnson, Phys. Rev. B 31 (1985) 5525.
[4] M. Stutzmann, Phys. Rev. B 35 (1987) 5921.
[5] K. J. Chang, D. J. Chadi, Phys. Rev. Lett. 60 (1998) 1422.
[6] P. J. H. Denteneer, C. G. Van de Walle, S. T. Pantelides, Phys. Rev. B 39 (1989) 10809.
[7] L. Korpas, J. W. Corbett, S. K. Estreicher, Phys. Rev. B 46 (1992) 12365.
[8] J. T. Borenstein, J. W. Corbett, S. J. Pearton, J. Appl. Phys. 73 (1993) 2751.
[9] W. Kwapil, J. Schon, T. Niewelt, M. C. Schubert, IEEE J. Photovolt. 10 (2020) 1591.
[10] D. Chen, P. Hamer, M. Kim, C. Chan, A. Ciesla nee Wenham, F. Rougieux, Y. Zhang, M. Abbott, B. Hallam, Sol. Energy Mater. Sol. Cells 207 (2020) 110353.
[11] D. Chen, M. Vaqueiro Contreras, A. Ciesla, P. Hamer, B. Hallam, M. Abbott, C. Chan, Progr. Photovolt. Res. Appl. (2020) DOI: 10.1002/pip.3362.
[12] B. Hammann, L. Rachdi, W. Kwapil, F. Schindler, M. C. Schubert, Phys. Status Solidi RRL 15 (2021) 2000584.
[13] D. C. Walter, D. Bredemeier, R. Falster, V. V. Voronkov, J. Schmidt, in Proc. 37th EU PVSEC, (2020) 140-144.
[14] B. Pajot, B. Clerjaud, Optical Absorption of Impurities and Defects in Semiconducting Crystals: Electronic Absorption of Deep Centers and Vibrational Spectra, New York, Springer-Verlag, 169 (2013).
[15] A. R. Peaker, V. P. Markevich, J. Coutinho, J. Appl. Phys. 123 (2018) 161559.
[16] L. Dobaczewski, A. R. Peaker, K. Bonde Nielsen, J. Appl. Phys. 96 (2004) 4689.
[17] J. Heyd, G. E. Scuseria, M. Ernzerhof, J. Chem. Phys. 118 (2003) 8207.
[18] J. Heyd, G. E. Scuseria, M. Ernzerhof, J. Chem. Phys. 124 (2006) 219906.
[19] G. Kresse, J. Furthmüller, Phys. Rev. B 54 (1996) 11169.
[20] G. Kresse, J. Furthmüller, Comp. Mat. Sci. 6 (1996) 15.
[21] P. E. Blöchl, Phys. Rev. B 50 (1994) 17953.
[22] J. P. Perdew, K. Burke, M. Ernzerhof, Phys. Rev. Lett. 77 (1996) 3865.
[23] C. Freysoldt, J. Neugebauer, C. G. Van de Walle, Phys. Rev. Lett. 102 (2009) 016402.
[24] R. Stübner, Vl. Kolkovsky, J Weber, J. Appl. Phys. 118 (2015) 055704.
[25] A. R. Peaker, V. P. Markevich, L. Dobaczewski, in Defects in Microelectronic Materials and Devices (Eds.: D. M. Fleetwood, S. T. Pantelides, R. D. Schrimpf) CRC Pres, Boca Raton, USA (2009) Chapter 2.
[26] M. Stavola, in Properties of Crystalline Silicon, EMIS Datareviews Series No 20 (ed.: R. Hull) INSPEC, London, United Kingdom (1999) Ch. 9.9.
[27] R. Stübner, Vl. Kolkovsky, J Weber, N. V. Abrosimov, C. M. Stanley, D. J. Backlund, S. K. Estreicher, J. Appl. Phys. 127 (2020) 045701.
[28] L. C. Kimerling, J. L. Benton, Appl. Phys. Lett. 39 (1981) 410.
[29] M. Vaqueiro-Contreras, V. P. Markevich, M. P. Halsall, A. R. Peaker, P. Santos, J. Coutinho, S. Öberg, L. I. Murin, R. Falster, J. Binns, E. V. Monakhov, B. G. Svensson, Phys. Status Solidi RRL 11 (2017) 1700133.
[30] M. Vaqueiro-Contreras, V. P. Markevich, J. Mullins, M. P. Halsall, L. I. Murin, R. Falster, J. Binns, J. Coutinho, A. R. Peaker, J. Appl. Phys. 123 (2018) 161415.
[31] J. L. Hartke, J. Appl. Phys. 39 (1968) 4871.

APPLICATION OF Δn NORMALISED TIME, Γ: LINEAR INJECTION-LEVEL DEPENDENCE ON LETID AND THE RECOVERY IN CRYSTALLINE SILICON

Moonyong Kim[1], Alison Ciesla[1], Chang Sun[2], Daniel Chen[1,3], Malcolm Abbott[1], and Brett Hallam[1]
[1]School of Photovoltaic and Renewable Energy Engineering, University of New South Wales, Sydney, NSW, 2052 Australia
[2]Research School of Electrical, Energy and Materials Engineering, The Australian National University, Canberra, ACT, 2601 Australia
[3]Sundrive Solar, Kirrawee, NSW, 2232 Australia

ABSTRACT: The impact of the injection-level (Δn) on the kinetics of light- and elevated temperature-induced degradation (LeTID) is investigated. Higher illumination intensities (I) are known to increase the LeTID and recovery rates. In this work, the Δn normalisation approach is demonstrated. A similar time scale is observed after normalisation to the Δn at the given I to time. The evidence suggests the recovery rate of LeTID is linearly proportional to Δn and can potentially be used to predict long-term degradation, which can incorporate operating conditions and insolation level, based on an accelerated degradation testing using Δn dependence. The experiment results demonstrated the difference in LeTID and the recovery trends and the overall rates. The variation of Δn during the cycle leads to a slow rate during the degradation since both Δn and the amount of available defect precursors are reduced. However, during the recovery, the Δn is being recovered as the defect states are reduced, which appears to show compressed exponential recovery. The normalisation approach demonstrated here showed very close agreement among the results with different illumination intensities. The implication eliminated both influences due to variation of Δn and the acceleration of the rate due to high Δn or illumination intensity.
Keywords: LID, Silicon Solar Cell, multicrystalline silicon, LeTID, defects, degradation

1 INTRODUCTION

Many studies have been done to identify an approach to reduce the light-induced degradation testing time. Using LeTID as an example, these approaches involve increasing temperature [1] or illumination intensity [2]. A dependence of excess carrier concentration or injection levels (Δn) exists in LeTID [3] and recovery rates [4], where the recovery rate dependence was based on the change with different circuit conditions. For the degradation rate, it is linearly proportional to Δn. For the recovery rate, evidence for this dependence was based on the change in rates under different circuitry operating conditions. The capture cross-section ratio (k) of the defects due to LeTID is typically in the range of 20 and 40 [5,6]. Therefore, increasing the defect concentration during degradation under illumination can lead to more significant changes in lifetime at low Δn or lower illumination intensity. Provided that no other lifetime changes occur during LeTID testing, the three main parameters that can influence the Δn in a given sample throughout the degradation and recovery cycle for LeTID at a constant illumination intensity during testing are: 1) initial effective lifetime of the sample ($\tau_{\mathrm{eff,i}}$); 2) maximum degradation extent; and 3) the intensity of illumination (I). Testing at high I can allow accelerated testing where timescales can be reduced by over three orders of magnitude [2]. Our recent study demonstrated the difference in LeTID under the standard light soaking condition and the accelerated light soaking condition [7]. The evidence alluded to the link between the testing condition and the rates of LeTID and the recovery, which was linked to Δn dependence.

This study investigates the impact of Δn on the behaviour of LeTID and the recovery rates on multi-crystalline silicon (mc-Si) passivated emitter rear cell (PERC) precursor samples. While the industry is transitioning to mono-crystalline silicon, the result still shows significance in scientific findings, given that LeTID can also affect mono-crystalline PERC solar cells. An experimental result is shown to demonstrate the variation of Δn due to increases in recombination rate with the presence of LeTID and under different I. Based on the effective lifetime curves, the corresponding Δn at given I is calculated and used for the normalisation. The normalisation showed a close agreement among the results despite large differences in I. The approach provided new insight into LeTID, allowing us to study LeTID whilst eliminating Δn dependence. The evidence suggests the potential application of Δn normalisation that can help to develop an accelerated testing approach for commercial use.

2 EXPERIMENTAL PLANS

Industrial 6-inch mc-Si with a resistivity of 1.6 Ω·cm ($N_{\mathrm{dop}} \sim 9.1 \times 10^{15}$ /cm^3) PERC precursor samples with phosphorus emitter, silicon nitride (SiN_x:H) front layer and aluminium oxide and SiN_x:H rear passivation were used. Samples then underwent rapid firing at an actual peak temperature of 740 ± 2 °C using a metallisation belt firing furnace. They were then cleaved into 39 mm × 39 mm tokens using a pulsed laser with frequency at 20 kHz, 1064 nm, 15 W. For the light soaking condition, an accelerated testing condition was used. The specification of the light source for the testing was a fibre-coupled diode laser with the wavelength of 938 nm operated in quasi-continuous wave mode with a repetition frequency of 2 kHz and a pulse width of 0.5 ms [2] with different illumination intensities (equivalent to the generation rate, G of 1.2×10^{19} /cm^2/s per sun) at the temperature of ~ 150 °C. The injection-dependent lifetime was measured periodically using the quasi-steady state photoconductance tool (ex-situ PC) from Sinton Instruments (WCT-120TS QSSPC) at 25°C [8] and analysed using the generalised method [9]. The lifetime measurement was also done at the higher temperature, but it showed a variation of less than a factor of 2 at all Δn range (not shown) such that it is safe to assume that the Δn

value measured at 25 °C is comparable and can be used as a reference. In the final paper, both temperature measurements will be compared.

3 EXPERIMENT RESULTS AND DISCUSSION

The LeTID behaviour was different under the different testing conditions. The τ_{eff} at $\Delta n = 0.1 \times N_{dop}$ as a function of accumulated light soaking time, t under different I is shown in Figure 1a. While the initial lifetimes were different among the samples, the minimum τ_{eff} fell to similar values between 15-25 μs. Increasing I from 0.6 suns to 49 suns showed faster rates by close to 3 orders of magnitude. Moreover, samples treated with lower I showed a more stretched exponential-like degradation rate and more compressed exponential-like recovery rate. This observation typically occurs in studies that follow standard testing conditions with illumination intensity around one sun [5,10]. In comparison, higher I showed mono-exponential trends for both degradation and recovery rate.

The injection-level dependent effective minority carrier lifetime (τ_{eff}) measurements of the samples prior to testing and at t_{max} under I of 0.6 suns and 49 suns, as examples, are shown in Figure 1b. The similar trends in lifetime curves after degradation gives an indication that the root cause may be identical. Moreover, in order to estimate the Δn and its variation over time from different I, the curves are converted to Δn as a function of I, as shown in Figure 1c. Note that the measurement was not reached the maximum I of 49 suns. Therefore, Δn at that I was estimated based on the linear extrapolation. The variation of Δn at $t = t_{max}$ comparing to the initial is larger at the lower I (×7 for 0.6 suns and ×1.2 for 49 suns), which hints the non-constant value of Δn under the low intensity.

From each measurement, the Δn for each I as a function of t is shown in Figure 1d. Tests that were operating at higher I tended to have both higher Δn and smaller variation in Δn over the period of time. For $I = 49$ suns, the error bar is based on the extrapolated value and the highest measured value, approximately 30%.

Considering the dependence between Δn and the rate of LeTID and the recovery, we can relate Δn and LeTID and the recovery rates. The normalised of the result with time respect to the corresponding Δn using the results from ($\Gamma = t \times \Delta n$) from Figure 1a and Figure 1c are shown in Figure 2a. While the difference in I vastly varied the rates from Figure 1a, the normalised result showed little or no difference in the result. The close agreement highlights the approach to exclude the Δn dependence in the result. Furthermore, the stretched exponential degradation and the compressed exponential recovery trends with low I were not shown after the normalisation. The t_{max} and Γ_{max} ($t_{max} \times \Delta n$ ($t = t_{max}$)) as a function of I is shown in Figure 2b. The error bar is based on the two intermediate measurements before and after the t_{max}. While t_{max} was reduced by orders of magnitude, the Γ_{max} values were similar, demonstrating the correction from Δn. The matching Γ_{max} value implies that the expected behaviour at different I can be predicted based on Δn. Furthermore, the recovery rate is likely to be linearly proportional to Δn.

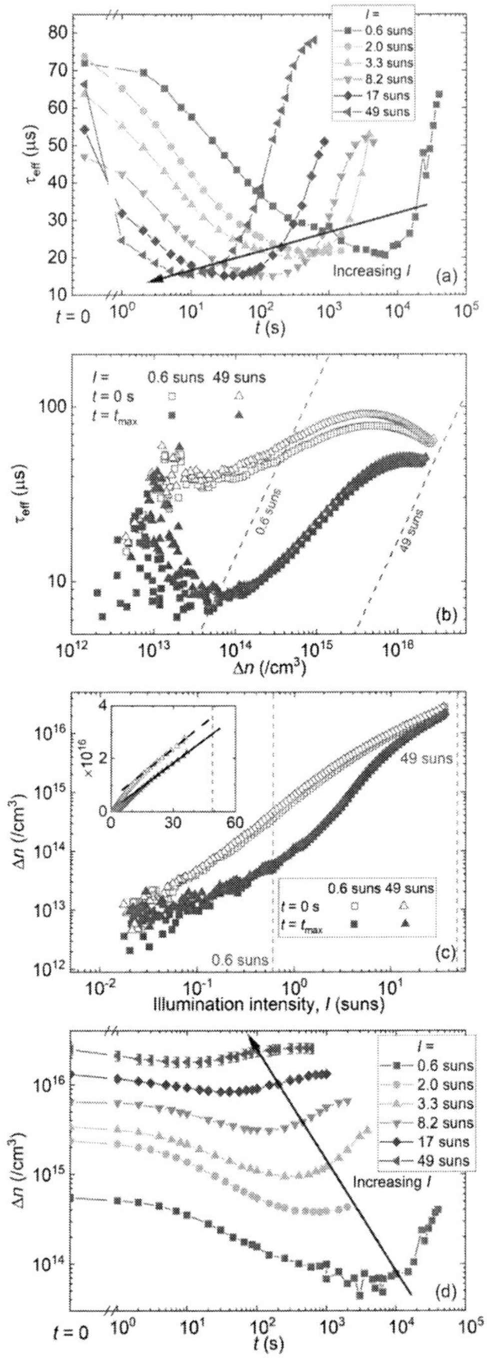

Figure 1: (a) The effective lifetime, τ_{eff} at $\Delta n = 0.1 \times N_{dop}$ with background doping density, $N_{dop} = 9.1 \times 10^{15}$ /cm³ as a function of time, t under different illumination intensity, I. **(b)** effective lifetime (τ_{eff}) as a function of Δn at $t = 0$ and t = maximum degradation extent time, t_{max}. **(c)** Δn as a function of I. The inset shows the results with $I = 49$ suns in a linear scale. The black solid and dashed lines show the linear extrapolation. **(d)** Δn from the lifetime measurement as a function of t. Error bars for $I = 49$ suns is from the measurement at the maximum I and the extrapolated value.

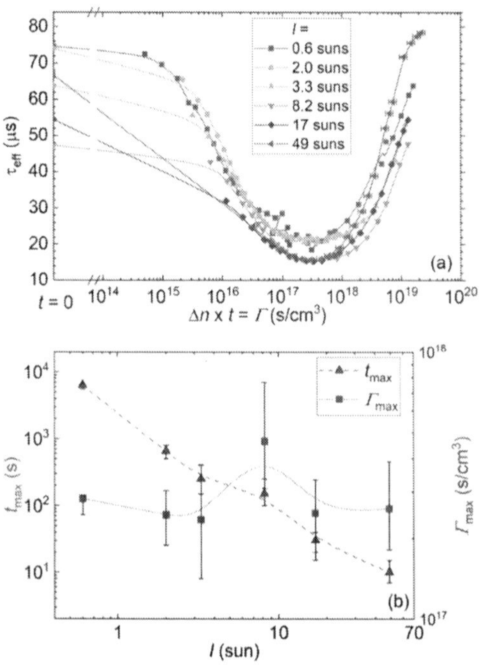

Figure 2: **(a)** Effective lifetime, τ_{eff} as a function of injection level normalised time, Γ. **(b)** t_{max} and Γ_{max} as a function of I. Dashed lines are guided to the eye.

4 CONCLUSIONS

In this study, the Δn dependence of LeTID was investigated. Increasing the illumination intensity to induce LeTID led to a significant increase in the overall rates and more mono-exponential like degradation and recovery rates. Under lower I (0.6 suns), the rates were slower by three orders of magnitude in comparison to the one at the highest I (49 suns). This observation can be explained by relating I and recombination rate to Δn. An increase in Δn with higher I not only reduce the recombination rate due to LeTID but also accelerates the degradation and regeneration rates due to Δn dependence in LeTID. However, after the normalisation of Δn to the light soaking time, t ($\Gamma = \Delta n \times t$), the results under different I showed very similar results. The similarity suggests that the difference is due to Δn dependence and that the recovery rate is also linearly proportional to Δn. This approach has a potential application for rapid light-induced degradation testing.

5 ACKNOWLEDGEMENTS

The support of the Australian Government through the Australian Renewable Energy Agency (ARENA, 2017/RND005, 1-SRI001, 1-060) and the Australian Centre for Advanced Photovoltaics (ACAP) is acknowledged. The views expressed herein are not necessarily the views of the Australian Government, and the Australian Government does not accept responsibility for any information or advice contained herein. The authors acknowledge the Solar Industrial Research Facility (SIRF) for the provision of facilities and equipment used for characterisation. B.H. would like to acknowledge the support of the Australian Research Council (ARC) through a Discovery Early Career Researcher Award (DE170100620) and PhD scholarship for M.K. M.K would like to also acknowledge the support of ACAP for his fellowship.

6 REFERENCES

[1] A. Herguth, C. Derricks, P. Keller, B. Terheiden, Energy Procedia 124 (2017) 740–744.

[2] D.N.R. Payne, C.E. Chan, B.J. Hallam, B. Hoex, M.D. Abbott, S.R. Wenham, D.M. Bagnall, Phys. Status Solidi - Rapid Res. Lett. 10 (2016) 237–241.

[3] W. Kwapil, T. Niewelt, M.C. Schubert, Sol. Energy Mater. Sol. Cells 173 (2017) 80–84.

[4] F. Kersten, P. Engelhart, H.C. Ploigt, A. Stekolnikov, T. Lindner, F. Stenzel, M. Bartzsch, A. Szpeth, K. Petter, J. Heitmann, J.W. Müller, Sol. Energy Mater. Sol. Cells 142 (2015) 83–86.

[5] T. Niewelt, M. Selinger, N.E. Grant, W. Kwapil, J.D. Murphy, M.C. Schubert, J. Appl. Phys. 121 (2017) 185702.

[6] D. Chen, M. Kim, B. V. Stefani, B.J. Hallam, M.D. Abbott, C.E. Chan, R. Chen, D.N.R. Payne, N. Nampalli, A. Ciesla, T.H. Fung, K. Kim, S.R. Wenham, Sol. Energy Mater. Sol. Cells 172 (2017) 293–300.

[7] U. Varshney, M. Kim, M.U. Khan, P. Hamer, C. Chan, M. Abbott, B. Hoex, IEEE J. Photovoltaics 11 (2021) 65–72.

[8] R.A. Sinton, A. Cuevas, M. Stuckings, in: Conf. Rec. Twenty Fifth IEEE Photovolt. Spec. Conf. - 1996, IEEE, 1996, pp. 457–460.

[9] H. Nagel, C. Berge, A.G. Aberle, J. Appl. Phys. 86 (1999) 6218.

[10] D. Bredemeier, D. Walter, S. Herlufsen, J. Schmidt, Energy Procedia 92 (2016) 773–778.

CARRIER LIFETIME MEASUREMENTS OF MONOCRYSTALLINE SILICON INGOTS USING UPGRADED LASER E-PCD TECHNIQUE

D. Krisztián[1,4,*], F. Korsós[1], I. Saegh[1], G. Paráda[1], P. Tüttő[1], X. Dong[2] and H. Deng[3], S. Wang[3], X. Chen[3]
[1]Semilab Co. Ltd. 4/A Prielle K. str., H-1117 Budapest, Hungary
[2]Semilab Trade Shanghai Co., Ltd., Pudong, Shanghai (201210), P.R. China
[3]LONGi Green Energy Technology Co., Ltd., Xi'an, Shaanxi (710100), P.R China
*Address: Prielle Kornélia str. 4/A. H-1117 Budapest, Hungary,
email: david.krisztian@semilab.hu, phone: +36 1 505 4690

ABSTRACT: The photoconductance decay (PCD) method is a standard characterization technique for carrier lifetime characterization of silicon ingots. We analyzed the precision of the eddy current PCD technique using monochromatic light focusing on the impact of the properties of the laser excitation.
The PCD measurement simulations showed that the impact of the surface recombination in the measured lifetime and the injection level is highly influenced by the wavelength and the pulse duration of the laser excitation source. Considering the time evolution of carrier profile as well, an optimal wavelength range between 1050 and 1070 nm has been found. The eddy-current detected PCD (e-PCD) measurements using the original 980 nm and a rather optimal 1064 nm laser was compared experimentally and by simulation in details. In the injection independent lifetime model, the effective lifetime approaches the bulk lifetime asymptotically within a reasonable time, and both laser source provides accurate result as experimentally confirmed. However, if carrier lifetime varies significantly with injection level, a pronounced difference between the asymptotic lifetimes recorded using the different lasers is observed. Lifetime values from the longer wavelength setup approach the saturation value of bulk lifetime even in this case. This is an additional important advantage of using the longer wavelength laser for ingot PCD measurements.
Comparing the laser e-PCD results to flash lamp based PCD systems, a good correlation is found in the long lifetime regime. However, the significant discrepancies found between transient and Quasi-Steady-State (QSS) modes of the flash lamp PCD system indicate that performing the measurement in transient mode for the entire lifetime range is very beneficial, which is realized using the laser based PCD method.

Keywords: Monocrystalline, Silicon, Recombination, Lifetime, Ingot

1 INTRODUCTION

Since the unbroken increase of the monocrystalline silicon's market share is expected in the coming years [1], widely accepted quality inspection methods are very important for the industry. Measurement of the carrier lifetime at ingot level using photoconductance (PCD) based method can qualify the purity of the material and may indicate the presence of harmful defects [2].

The radiofrequency eddy-current sensor and modulated laser excitation based PCD method (laser e-PCD) is suitable for the rapid and contactless measurement of the carrier lifetime on bulk materials [3].

Due very rapid transition of the dopant material from boron to gallium in the dominating p-type wafers for PERC cells [1], both the resistivity and the carrier lifetime varies in a very broad range. The growth in the production of premium quality n-type wafers for high efficiency cell further extends the carrier lifetime range should be properly qualified using the same carrier lifetime measuring system in the industry.

The aim of our laser e-PCD tool development is to realize such a measurement routine at ingot level without any sample preparation. The PCD measurement of thick samples is challenging due to the continuously evolving impact of the surface recombination rate [4]. The dependence of the bulk lifetime τ_b and the carrier mobility on the injection level Δn further makes this task even more complex. From the other side, overcoming these challenges enables to access bulk lifetime data from ingots or thick slabs without any surface preparation.

This requires a sufficiently deep photogeneration and the appropriate detection of charge carriers, which are realized in our commercial product (Semilab WT-1200I/IL) by using an infrared laser and an inductively coupled radio-frequency coil [4] (see Fig. 1).

In contrast to the alternative PCD technique, using flash lamps, the laser source can be switched on and off very rapidly. Thus, recorded PCD curves can be evaluated in purely transient mode over the whole τ_b range of industrially relevant mono-Si samples (2μs – 10ms).

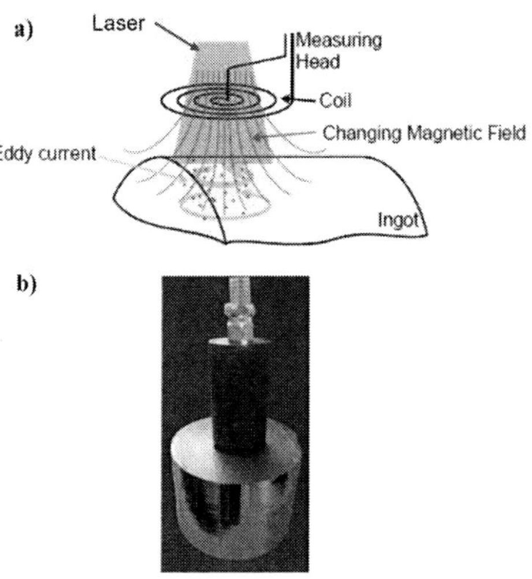

Figure 1: *The schematic illustration of the laser e-PCD measurement technique (a) and the photo of the hand-held WT-1200I/IL system used for the measurements (b).*

The accuracy of the flash-lamp system has been

investigated in several studies for both its transient (for long τ_b) [5] and quasi-steady-state (QSS, for low τ_b) mode [6].

Despite of the thorough optimization, the experimental results indicated up to 70% difference between the transient and QSS modes in their common range.

It clearly implies that a pure transient mode operation of the laser e-PCD system is very beneficial.

The aim of this study is to investigate the dependence of the lifetime on the excitation laser parameters in the entire relevant lifetime range. The complex behavior of surface recombination phenomena and its evolution in time is studied by computer simulation. After the realization of the optimized setup, results are compared to the simulation and the flash lamp based PCD technique.

2 OPTIMIZATION OF LASER EXCITATION

Two major properties have been considered in order to optimize the wavelength of the excitation laser: the steady state injection level Δn_{ss} and the impact of the surface recombination during the transient decay. The goal was to measure the effective lifetime τ_{eff} close to the bulk lifetime τ_b while reaching the standard 1e15 cm^{-3} excess carrier density in a wide bulk lifetime range.

The steady state injection level calculated using the method proposed by Bowden and Sinton [7] in the wavelength range of 900 to 1100 nm is depicted In Fig. 2a.

corresponding to different τ_b levels assuming the same excitation intensity.

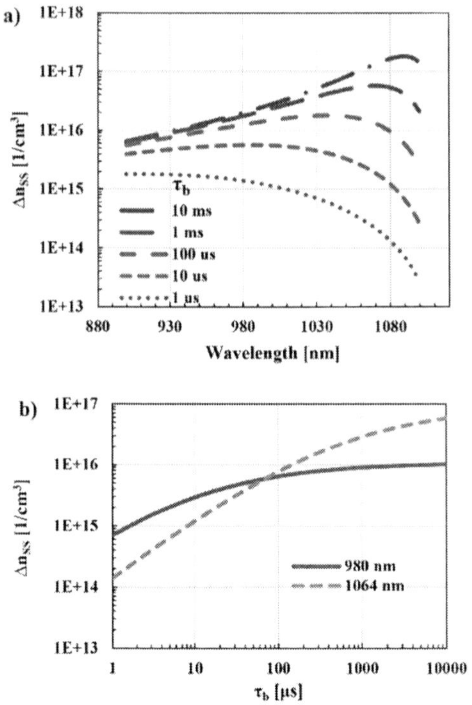

Figure 2: *Steady state injection levels in the versus excitation wavelength for different bulk lifetime samples. $\lambda^{max}_{\Delta n,ss}$ shifts toward larger wavelengths if bulk lifetime increases. Based on the detailed comparison at two commercially available typical wavelengths (b) 1064nm is proposed instead of the earlier standard 980nm [3].*

The maximum injection level position $\lambda^{max}_{\Delta n,ss}$ shifts toward larger wavelength values if τ_b. In the short wavelength range the front surface recombination plays the key role, while for long wavelengths the depth of the excess carrier profile is the limiting factor.

On the other hand, the deeper penetration introduced by longer wavelengths reduce the impact of the surface recombination. The time evolution of excess carrier profile during the decay and the corresponding instantaneous effective lifetime $\tau_{eff}(t)$) - calculated from the derivation of the decay curve related to the areal density of excess carriers - was examined using a 1D computer simulation. After the rapid termination of the laser excitation, in most cases $\tau_{eff} \ll \tau_b$. However, after a certain time it approaches the bulk lifetime as the excess carrier density in the near-surface volume decreased. The dynamics of this convergence is a strong function of τ_b and λ.

Considering these factors as well, the wavelength range between 1050 nm and 1070 nm seems to be the most suitable for measuring silicon ingots. Due to the availability of commercial laser units, 1064 nm was selected to implement.

The Table I. shows the duration in which $\tau_{eff}(t)$ reaches the 80% of the bulk lifetime. We call this quantity as "run-out time" $t_{run-out}$. „SS" means that this condition is satisfied even in steady state, so already in the first moment during the transient decay.

Table I: *After the illumination is terminated the impact of surface recombination decreases clearly demonstrated by $t_{run-out}$. values. The difference in the 10-100 µs range deserves special attention.*

	Time in which τ_{eff} reaches 80% of τ_b ($t_{run-out}$)				
	1 µs	10 µs	100 µs	1 ms	10 ms
980 nm	0.1 µs (10%)	9 µs (90%)	140 µs (140%)	1.5 ms (150%)	15 ms (150%)
1064 nm	SS (0%)	SS (0%)	20 µs (20%)	1.1 ms (110%)	14 ms (140%)

Analyzing the dynamics of the measurement accuracy, the conclusions regarding injection levels (Fig. 2b) must be reconsidered. Although Δn_{ss} is higher in the low lifetime range using the 980 nm laser, close to τ_b lifetimes can be evaluated only from the more decayed parts of the PCD transient, thus at lower injection level. These findings further support the conclusion regarding the optimal wavelength range.

3 EXPERIMENTAL RESULTS

3.1 Comparison of results from the 980 nm and 1064 nm lasers

We investigated the benefits arising from the longer wavelength by performing laser e-PCD measurements at different excitation powers on ingots covering a wide lifetime range. A Semilab WT-1200I system was used for the measurements using the original 980 nm and the optimized 1064 nm laser sources.

Furthermore, the eddy-current sensor was also tuned to expand the ingot resistivity range down to 0.1 Ωcm to follow the industry trends. Therefore, the sensor can report results on larger injection levels as well.

First, we examined a high purity n-type sample featuring negligible injection level dependence of lifetime

(ILD) in Fig. 3a. There is no remarkable difference between the results measured at different laser powers and wavelengths. It confirms, that ILD is indeed not significant. Furthermore, the increasing decay constant of the measured signal $\tau_{eff}(t)$ during the decay is related to the diminishing role of the surface recombination. The limitations of the 1D model are probably responsible for the minor differences.

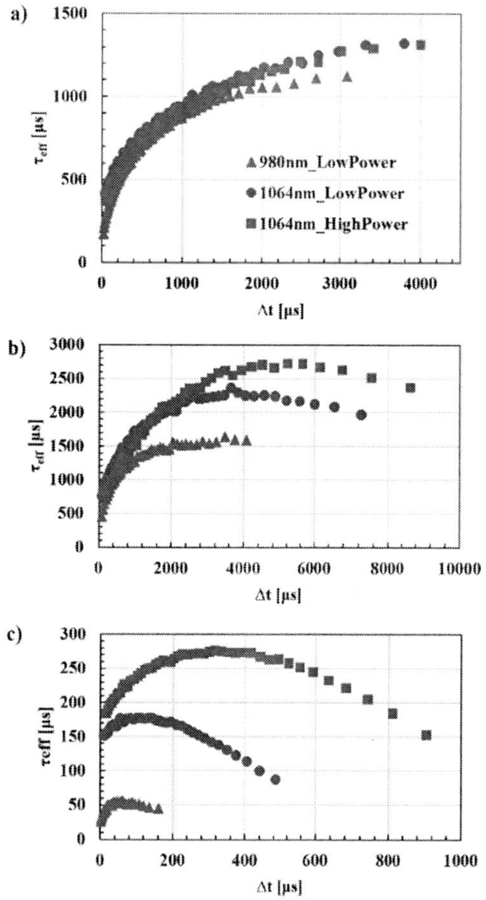

Figure 3: *Measured $\tau_{eff}(t)$ curves during the PCD decay. The n-type ingot featuring minor injection level dependence (ILD) shows a similarly increasing $\tau_{eff}(t)$ regardless of laser wavelength and power (a). In the case of stronger ILD (b and c) longer wavelength and elevated laser power makes possible to determine the carrier lifetime close to the saturation value of τ_b.*

Results from other bulk samples with moderate (Fig. 3b) and strong (Fig. 3c) ILD clearly demonstrate the importance of laser properties regarding measurement quality. Both samples show a pronounced increase in the top lifetime value τ_{eff}^{max} at longer wavelength and higher laser power due to the increased injection level. The larger the ILD, the larger the gap between the results recorded by the 980 nm and 1064 nm lasers, indicating that the saturation value of τ_b can be nicely determined using the proposed optimized laser properties.

3.2 Comparison of results from the optimized laser e-PCD and the flash lamp PCD systems

The comparison of results from the optimized setup of laser e-PCD (1064 nm, elevated power) and lifetime values reported at $1e15\ cm^{-3}$ injection level by a Sinton BCT-400 system using flash lamp is depicted in Fig. 4. For the laser e-PCD measurement τ_{eff}^{max} values are reported.

In most cases, the τ_{eff}^{max} values agree very well to transient mode BCT (>100 µs) results.

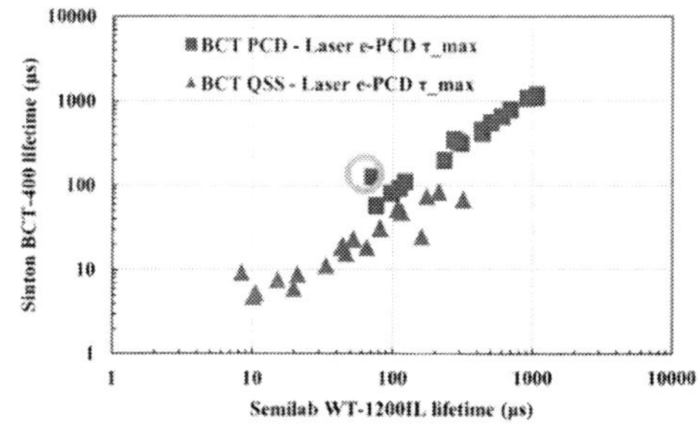

Figure 4: *Correlation to flash-lamp PCD is excellent in its transient mode regardless the evaluation method. The circled out-of-trend value (in τ_{eff}^{max} data) correspond to especially strong ILD. The large discrepancy in BCT results between its QSS and transient mode evaluation indicates the importance of the capability of laser e-PCD providing unified transient mode evaluation in the entire lifetime range.*

However, very large discrepancy is found in the lifetime values between QSS and transient mode of the same flash-lamp PCD system in the overlapping lifetime range (might be over a factor of 3) originated from the very different evaluation method. It indicates the importance of the unique capability of laser e-PCD with unified pure transient mode operation and evaluation in the entire lifetime range naturally avoiding such discrepancy issues.

4 CONCLUSIONS

The flexibility of lasers used in e-PCD measurement opens new paths to boost to accuracy of the lifetime measurement on bulk materials. The combined effect of the injection level dependent bulk lifetime and the continuously changing surface recombination is discussed in the first time considering the parameters of the excitation source as well. Despite this complex behaviour, very clear conclusions are drawn regarding the optimal laser excitation source and the achievable measurement accuracy.

The experimental results show a convincing similarity to the simulated curves and confirms the important benefits if increasing the laser wavelength and intensity. Thus, a commercially available 1064 nm laser unit is proposed to replace the original 980 nm version. The optimized laser setup provides carrier lifetime results much closer to the bulk lifetime, and the accuracy is more immune to issues arising from injection level dependencies. The good correlation with the flash lamp based PCD system (in its transient mode) method confirms the precision of the laser e-PCD method but extending the reliability into the low lifetime ranges as well. This benefit arises from the advantages of using lasers for PCD measurement: rapid switching, variable pulse length and power, simple carrier profile calculation.

ACKNOWLEDGEMENT

We thank Ferenc Simon for useful discussions.

D. Krisztián acknowledge the support of the doctoral student scholarship program of the co-operative doctoral program of the Ministry for Innovation and Technology from the source of the National Research, Development and Innovation Fund.

4.3 References

[1] 12th edition of International Technology Roadmap for Photovoltaic (IRTPV): 2020 results (2021).
[2] F. Korsós, et al., Solar Energy Materials and Solar Cells, Vol. 186: 217–226 (2018).
[3] G. Paráda, et al., "Transient method for lifetime characterization of monocrystalline Si ingots", 29th EU-PVSEC, pp. 1264 - 1266 (2014).
[4] J. S. Swirhun, et al., "Contactless measurement of minority carrier lifetime in silicon ingots and bricks", Prog. Photovolt: Res. Appl. 19:313–319 (2011)
[5] M. Goodarzi, et al., "Improving transient photoconductance lifetime measurements on ingots with deeper photogeneration", AIP Conference Proceedings 1999, 020008 (2018)
[6] M. Goodarzi, et al., "Quasi-steady-state photoconductance bulk lifetime measurements on silicon ingots with deeper photogeneration", AIP Advances 9, 015128 (2019)
[6] S. Bowden and R.A. Sinton, Journal of Applied Physics, Vol. 102., 124501 (2007).

EFFICIENCY POTENTIAL ANALYSIS OF P- AND N-TYPE EPITAXIALLY GROWN SI WAFERS

C. Rittmann, J. Dalke, M. Drießen, C. Weiss, F. Schindler, R. Sorgenfrei, M. C. Schubert, S. Janz
Fraunhofer Institute for Solar Energy Systems ISE
Heidenhofstrasse 2, 79110 Freiburg, Germany

ABSTRACT: In this paper, we investigate the quality of epitaxially grown, 110 µm thick, n- and p-type silicon (Si) layers deposited in a CVD batch reactor of microelectronic standard. Two types of wafers are characterized: 'EpiRef' grown on chemically and mechanically polished reference substrates and 'EpiWafer' grown on substrates with a porous silicon detachment layer. EpiRef wafers exhibit excellent minority carrier lifetimes of 2.5 ms for n-type and 1.3 ms for p-type. EpiWafers show reduced, but still promising lifetimes of 0.5 ms (in local areas up to 1 ms) for n-type and 40 µs for p-type. For EpiWafers, we found that quality limitations are due to stacking faults as well as interstitial iron contamination for p-type. An efficiency limiting bulk recombination analysis (ELBA), allows for an assessment of the corresponding efficiency potential assuming a TOPCon cell model with a cell limit of 25.9 % for n-type and a TOPCoRE cell model with a cell limit of 26.5 % for p-type. In selected 1 cm² areas, potential efficiencies of EpiRef wafers are only -0.4 $\%_{abs}$ below the theoretical cell limit for n-type and -1.3 $\%_{abs}$ for p-type whereas EpiWafers feature losses of -1.2 $\%_{abs}$ for n-type and even -5.4 $\%_{abs}$ for p-type.
Keywords: Epitaxy, Porous Silicon, Photoluminescence Imaging, Efficiency Potential

1 INTRODUCTION

In a silicon photovoltaic module, the CZ wafer contributes to ≈30% of the total costs and to ≈40% of the global warming potential (GWP) [1, 2]. The transfer from crystallized CZ wafers to epitaxially grown Si wafers on reusable substrates with a porous silicon detachment layer (EpiWafers) is therefore a sustainable approach to significantly reduce material and energy consumption and thereby costs and GWP. However, in the EpiWafer community, the origin of quality limitation compared to crystallized CZ Si wafers is still under discussion [3–8]. Hence, we analyze quality limitations of n- and p-type epitaxial layers on chemically and mechanically polished (CMP) substrates as reference 'EpiRef' in comparison to n- and p-type EpiWafers. We specifically highlight differences between EpiRef wafers and EpiWafers as well as differences between n-type and p-type epitaxially grown silicon.

The main parameter to monitor the material quality is the minority carrier lifetime measured by means of lifetime calibrated photoluminescence imaging (PLI) and quasi steady state photoconductance (QSSPC) [9, 10]. Lifetime losses originate potentially from metal contamination which has its origin in the gas phase or in the sample holder during epitaxial growth as well as from thermally induced stress during the process. Furthermore, the growth template for epitaxy plays a major role as it can be particle contaminated or not perfectly flat. Especially the electrochemically etched and reorganized templates for EpiWafers can still show unclosed areas or even stressed areas within the detachment layer of porous silicon leading to crystal defects during growth [11]. In order to investigate all these lifetime limitations for n- and p-type epitaxially grown Si, we use a two-step approach: First, we characterize EpiRef wafers since their limitations can be directly attributed to contaminations of the gas phase or the sample holder, or to thermally induced stress during the process. Second, these EpiRef wafers are compared to EpiWafers since the additionally arising quality limitations can be attributed to properties of the reorganized template.

In addition to the lifetime measurements, efficiency limiting bulk recombination analysis (ELBA) allows to discuss the impact of quality limiting effects on solar cell efficiency. For the spatially resolved efficiency potential analysis based on infection dependent PLI measurements, we assume a TOPcon model for n-type epitaxial layers and a TOPCoRE model for p-type epitaxial layers [12].

2 EXPERIMENTAL PROCEDURE

2.1 Epitaxial growth of silicon

The epitaxial growth of Si is conducted in a chemical vapor deposition (CVD) batch reactor from LPE (PE 2061S), called 'PEpi', by using $SiHCl_3$ and H_2 as precursor gases. For adjustable p- and n-type doping, B_2H_6 and PH_3 gases can be added to the precursors. The gas flows are adjusted to grow Si layers with thicknesses of 130 µm after 140 min of deposition (growth rate ≈ 1 µm/min) at process temperatures between 1090 °C and 1120 °C. The chosen doping concentration of the n-type layers is approximately 3.5×10^{15} cm^{-3} (≈ 1.4 Ωcm) and for p-type layers it is 4.7×10^{15} cm^{-3} (≈ 3.0 Ωcm). Figure 1 shows a schematic of the PEpi pointing out the gas inlets, the induction coil and the silicon carbide coated graphite carrier holding the wafers in a vertical position. The excellent thermal and gas flow stabilities lead to a spatially homogeneous thickness and a low stress level of the epitaxial layer as well as to adjustable and homogeneous doping concentrations over the whole layer depths [13].

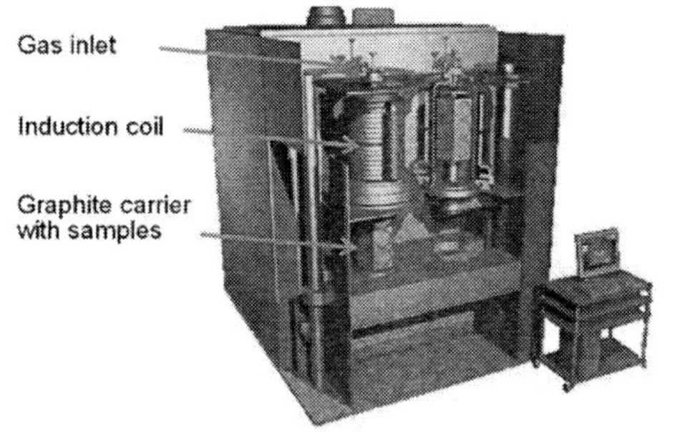

Figure 1: Schematic of the CVD-reactor PE2061 ("PEpi") used for high quality epitaxy.

2.2 Preparation of EpiRef wafers

For obtaining EpiRef wafers, n-type, respectively p-type Si is deposited on 6" CZ Si p-type substrates with chemically mechanically polished (CMP), 'epi-ready' surfaces on microelectonic quality level supplied by Siegert. The substrate's thickness is 675 µm and its resistivity is (1-10) Ωcm. After epitaxial growth, the CZ substrate is locally removed through Taiko-grinding (by the external service provider DISCO). Grinding residues are taken off by KOH etching resulting in an exclusively epitaxial layer with a wafer thickness of 110 µm and a CZ-rim for stability as demonstrated in

Figure 2 For further lifetime measurements, the EpiRef wafers are cleaned by RCA procedure, passivated with Al_2O_3 and forming gas annealed at 425 °C.

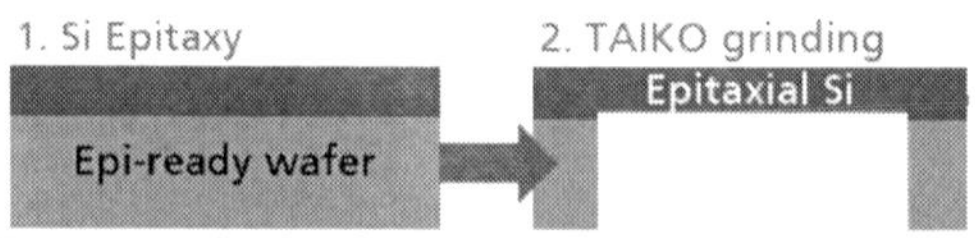

Figure 2: EpiRef wafer scheme: 1. Silicon deposition on an epi-ready wafer in PEpi reactor. 2. Taiko-grinding of the substrate so that only the epitaxial silicon remains in the inner, circular area.

2.3 Preparation of EpiWafers

The EpiWafers are epitaxially grown on Si seed wafers with porous silicon layer stacks supplied by IMS. These seed wafers are 6" CZ Si p^+-type substrates with a resistivity of (10-20) mΩcm and a thickness of 650 µm. Figure 3 shows the whole process starting from the seed wafer with an electrochemically etched sponge-like silicon layer underneath the surface. This layer is reorganized during a 5 min high-temperature step at 1120°C under H_2 atmosphere in the PEpi reactor forming a high porous detachment layer and a closed surface acting as template for epitaxial growth. Then, Si is deposited with the same setting as for the EpiRef samples. Finally, after edge definition with a chip saw, 5x5 cm² pseudo-square EpiWafers can be lifted off the seed wafer by using a vacuum lift-off tool. For the characterization, the EpiWafers are prepared by KOH etching to remove residuals of the porous silicon seed wafer, again leading to 110 µm wafer thickness. Finally, RCA cleaning, passivation with Al_2O_3 and forming gas anneal at 425 °C follows.

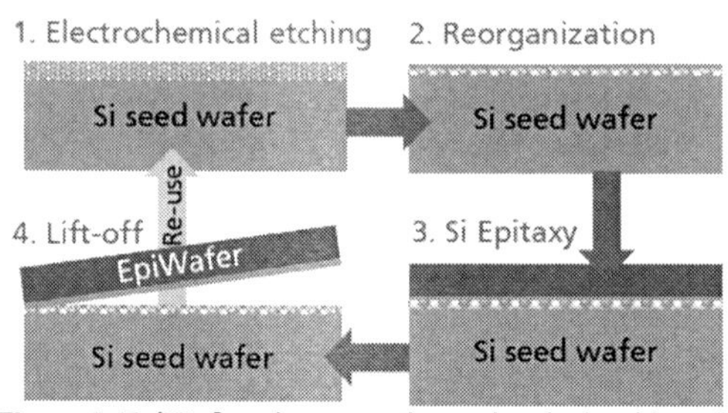

Figure 3: EpiWafer scheme: 1. Electrochemical etching of Si seed wafer. 2. Reorganization process to create porous detachment layer. 3. Silicon epitaxy. 4. Lift-off of EpiWafer from seed wafer.

2.4 Lifetime and efficiency potential characterization

Spatially resolved lifetime images were determined by photoluminescence imaging (PLI) calibrated with modulated photoluminescence at a range of illumination intensities from 0.001 to 2.5 suns [9, 10]. Thus, injection levels of all relevant solar cell operation conditions are covered. At an injection level of 0.05 suns (approximately injection level at maximum power point) the lifetime value τ_{eff} of the whole wafer was obtained by arithmetic averaging. Additionally, injection-dependent, but not spatially resolved, lifetimes were detected by quasi steady state photoconductance (QSSPC).

In a semi-simulative approach the cell efficiency potential was quantified by efficiency limiting bulk recombination analysis (ELBA) based on injection dependent and spatially resolved lifetime data combined with pixelwise numerical cell simulations with Quokka 3 applying a 2D unitcell model [14, 15]. The selected solar cell models are further described in chapter 2.5. Finally, the global solar cell parameters can be calculated by applying the concept of Isenberg *et al.* on the local electrical quantities [16]. The efficiency evaluation was applied to specific regions of interest: an area covering almost the whole wafer and a smaller best performance area with the size of 1 cm².

For further insights into the origin of differences in lifetimes between the wafer types, we performed iron imaging and defect counting considering stacking faults as well as decorated stacking faults with poly-crystalline inclusions leading to a further reduction in lifetime than ordinary stacking faults [17].

2.5 Cell Models for ELBA

For ELBA analysis, we chose two state-of-the-art high efficiency cell models which are depicted in Figure 4 and fully described in detail by Richter *et al.* [12]. In our case, the n-type TOPCon cell can reach an efficiency cell limit of $\eta_{Limit} = 25.9$ % and the p-type TOPCoRE an efficiency cell limit of $\eta_{Limit} = 26.5$ %. These limits assume only intrinsic recombination without any defect related recombination in the base material. Both cell models use similar fabrication technologies with a TOPCon stack at the rear surface and highly doped p^{++}-regions underneath the front metal contact. The main difference is the location of the junction: p^+-doped front junction (FJ) at the n-type Si cell and TOPCon back junction (BJ) without a front surface field (FSF) at the p-type Si cell. Thus, the back junction requires higher material quality (diffusion length) to avoid significant efficiency losses within a p-type TOPCoRE cell.

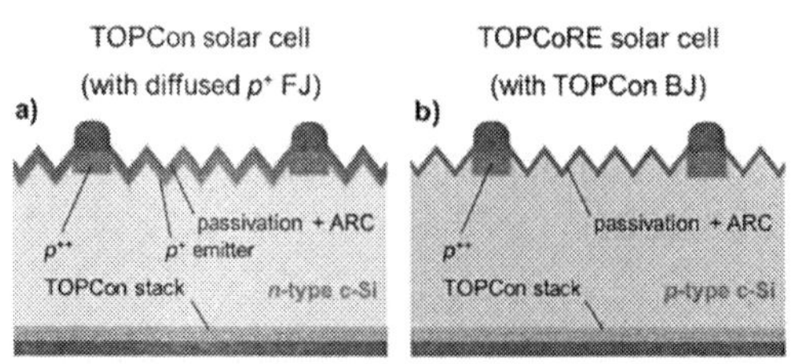

Figure 4: Schematic of cell models used in ELBA: a) TOPCon at the rear surface and n-type Si cell with an Al_2O_3-passivated, boron-doped p^+ FJ, b) TOPCoRE p-type Si cell with an Al_2O_3 passivation of the bare c-Si front surface without a full-area FSF, with highly doped p^{++} regions underneath the front metal contact and a TOPCon BJ [12].

3 LIFETIME RESULTS

3.1 n-type epitaxial layer

In this section, the focus lies on the electrical characterization by means of PLI and QSSPC of n-type EpiRef grown on a substrate with CMP surface and n-type EpiWafer grown on substrate with porous silicon detachment layer.

In Figure 5, lifetime mappings detected by PLI of EpiRef and EpiWafer at an illumination of 0.05 suns are shown. The blue frames in both images mark the areas in which we calculated a lifetime mean value τ_{eff} of (2.5±0.2) ms for the EpiRef wafer and (0.5±0.3) ms for the EpiWafer. The lifetime distribution of the EpiRef wafer is very homogeneous. Only some stacking faults (dark spots) and some slip lines (horizontal and vertical lines) are visible. These minor areas of lower lifetimes originate mainly from gas phase contamination, such as oxygen residuals, and from local thermal inhomogeneities. In contrary, the EpiWafer shows an inhomogeneous spatial lifetime distribution with many spots of locally deviating lifetime manifesting itself also in the large standard deviation of the mean lifetime. The lifetime reduction in comparison to EpiRef is mainly attributed to the increased amount of stacking faults (> 100 cm^{-2} for the EpiWafer vs. < 1 cm^{-2} for the EpiRef). The lower crystal quality can be explained by the epitaxial growth of the EpiWafer on the non-ideally reorganized porous silicon template with surface steps or even holes in the template. However, it is remarkable that the EpiWafer has local areas with very promising lifetime up to 1 ms without any additional gettering process step which shows the potential of EpiWafers.

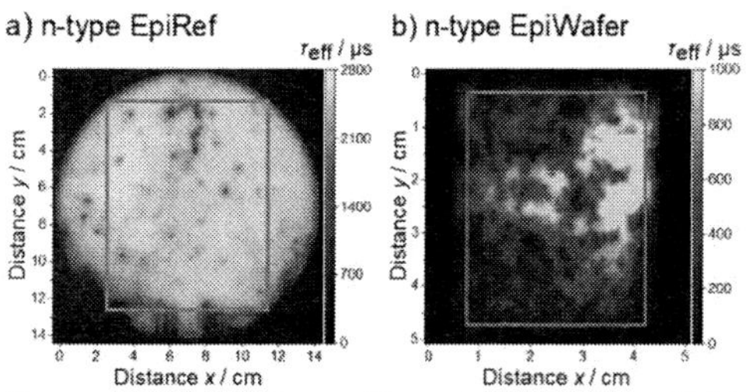

Figure 5: Lifetime detected by PLI at 0.05 suns with blue frames marking the area for the average lifetime τ_{eff} and for efficiency potential analysis. a) n-type reference wafer with 2.5 ms average lifetime. b) n-type EpiWafer with 0.5 ms average lifetime.

QSSPC measurements in Figure 6 confirm that the EpiRef shows higher lifetimes than the EpiWafer although the lifetime for the EpiWafer is overestimated caused by an overweighting of regions with excellent lifetimes on wafers with inhomogeneous lifetimes. Both wafers exhibit relatively stable lifetimes over the hole carrier density range despite of large carrier densities where the lifetime is additionally reduced due to Auger recombination. Further, detected lifetimes are more than ten times below the Auger-limit implying recombination due to contamination in both, EpiRef and EpiWafer as discussed in detail for the p-type wafers in Chapter 3.2.

Reasons for the lifetime reduction of the EpiWafer in comparison to the EpiRef are still discussed and investigated. Besides the non-ideally reorganized surface of the seed wafer, the p^{+}-doping of the seed wafers potentially further reduces the quality of the EpiWafer. We measured a severe reduction of lifetime by about 50% of EpiRef wafers grown on highly doped p^{+}-type epi-ready substrates compared to p-type epi-ready substrates. This reduction is either attributed to an increased metal contamination due to the higher boron content of the substrate, as the boron source is less pure than the silicon source, or attributed to a lattice mismatch between the p^{+}-substrate and the p-type EpiWafer due to different atomic radii of dopants and silicon [18]. Furthermore, we cannot exclude that the EpiWafer is exposed to additional contamination from more handling steps of the seed wafer before epitaxy compared to the template of the EpiRef as every additional handling step poses a source of potential contamination and surface damage.

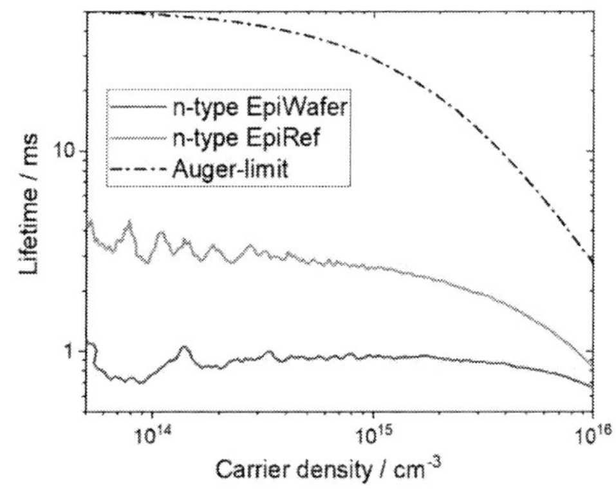

Figure 6: Lifetime curves measured by QSSPC of n-type EpiRef wafer and EpiWafer as well as the Auger-limit for 1.4 Ωcm n-type silicon [19].

3.2 p-type epitaxial layer

In this section, we analyze p-type EpiRef wafer and EpiWafer and point out differences between n- and p-type epitaxial silicon. Similar to the n-type epitaxial layer, there is a significant reduction of lifetimes from p-type EpiRef to EpiWafer. The PLI measurements in Figure 7 exhibit mean lifetime τ_{eff} within the area marked by the blue frame of (1.3±0.2) ms for EpiRef and (0.04±0.01) ms for EpiWafer.

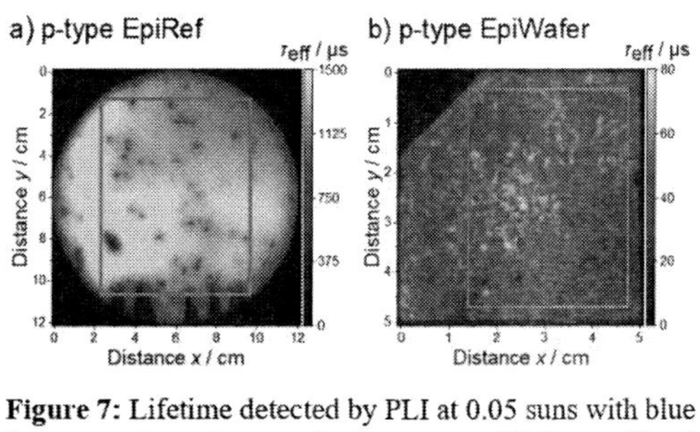

Figure 7: Lifetime detected by PLI at 0.05 suns with blue frames marking the area for the average lifetime τeff and for efficiency potential analysis. a) p-type EpiRef with 1.3 ms average lifetime. b) p-type EpiWafer with 0.04 ms average lifetime.

Again, the EpiRef has a reasonably smooth lifetime distribution over the whole wafer area with little amount of slip lines and stacking faults whereas the EpiWafer appears more spatially inhomogeneous. Same as for n-type wafers, defect counting unveils a large difference in stacking fault density of > 100 cm^{-2} for the EpiWafer and < 1 cm^{-2} for the EpiRef. Hints for decorated stacking faults, which show even higher recombination activity, on EpiWafers were found on microscope images of the layer surface.

Same as for the n-type layers, results of QSSPC measurements on p-type layers in Figure 8 are consistent with PLI. Again, the detected lifetimes depend only little on the carrier density, which means that the PLI results are valid for a wide range of injection levels. For p-type, the lifetimes are even more than 100 times below the Auger-limit, indicating again recombination through contamination. For the p-type EpiWafer, iron imaging confirms an enhanced metal contamination showing an interstitial iron concentration Fe_i in the range of 10^{10} cm^{-3}. In contrast, the interstitial iron concentration of the EpiRef appears to be below the detection limit of 5 x 10^9 cm^{-3}. However, the interstitial iron contamination only makes up for less than 10 % of the total recombination within the p-type EpiWafer. Therefore, other metallic contaminations are assumed to contribute to the recombination, especially for the EpiWafers.

Generally, lifetimes are reduced for p-type wafers compared to n-type wafers. For the EpiRef wafer the lifetime decreases by a factor of two and for the EpiWafer even by a factor of ten as summarized in Table I in chapter 4.1. Several reasons for lower lifetimes in p-type Si epitaxial layers than n-type Si epitaxial layers exist. For example, iron contamination has a larger impact on minority carrier lifetimes in p-type silicon because the capture cross section of interstitial iron is much larger for electrons than for holes [20]. Since the detected interstitial iron content does not fully explain the lifetime reduction, further research on other metallic contamination is ongoing. However, the lower minority carrier lifetimes in p-type silicon does not necessarily have to lead to lower cell efficiencies since the lifetime dependent diffusion length of p-type EpiRef with $D = 2.0$ mm is in the same order as the even slightly smaller diffusion length of n-type EpiRef with $D = 1.7$ mm due to the larger mobility of electrons in p-type silicon [21].

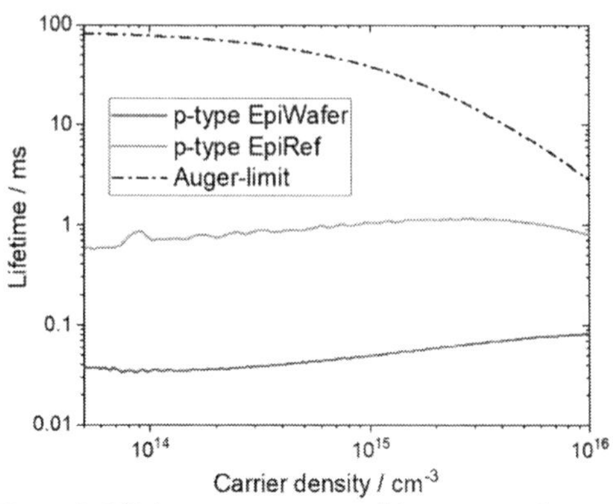

Figure 8: Lifetime curves measured by QSSPC of p-type EpiRef wafer and EpiWafer as well as the Auger-limit for 3 Ωcm p-type silicon [19].

4 EFFICIENCY POTENTIAL ANALYSIS

4.1 ELBA simulation results

In Figure 9, we show the cell efficiency potential of the epitaxially grown layers according to method ELBA described in section 2.4. To compare all wafer types with each other, we define two areas of interest: a large rectangular area (full depicted area) and a second smaller 1 cm² area covering the best minority lifetimes. The n-type EpiRef wafer reaches 25.4 % efficiency on a large scale and 25.5 % in the best spot whereas the n-type EpiWafer reaches 23.8 % efficiency on a large scale and 24.7 % in the best spot. Slightly lower results are obtained for the p-type EpiRef wafer with 25.0 % efficiency at large size and 25.2 % for the small 1 cm² spot. A tremendous reduction in efficiency potential is observed for the p-type EpiWafer with 20.6 % on large area and 21.1 % efficiency at the best location. These results are listed in direct comparison to the lifetime results in Table I.

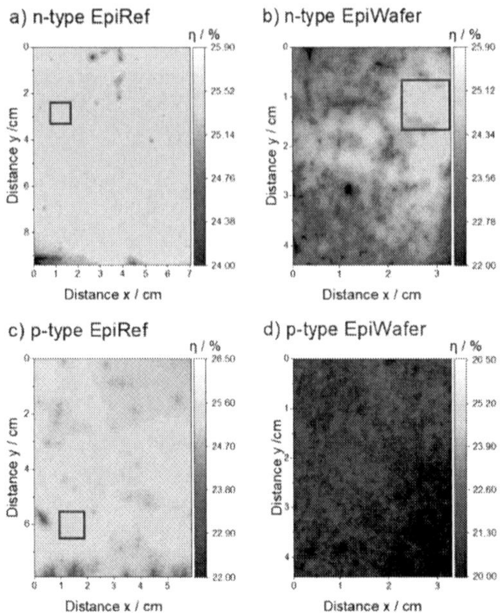

Figure 9: ELBA simulation results predicting the lateral distributed cell efficiency η for different wafers. Further evaluated areas are depicted: η_{Best} best performance area (1 cm²) indicated by the blue box, η_{Wafer} at the full depicted wafer area (indicated by the blue box in Figure 5 and Figure 7).

Table I: Evaluation results of material and cell performance parameters. PLI at 0.05 suns intensity for mean lifetimes τ_{eff} (arithmetic mean) and diffusion lengths D. ELBA simulation results for cell efficiencies η evaluated at a large area and a best 1 cm² area.

	τ_{eff} ms	D mm	η_{Limit} %	η_{Best} %	η_{Wafer} %
n-type EpiRef	2.5	1.7	25.9	25.5	25.4
n-type EpiWafer	0.5	0.8	25.9	24.7	23.8
p-type EpiRef	1.3	2.0	26.5	25.2	25.0
p-type EpiWafer	0.04	0.4	26.5	21.1	20.6

The efficiency loss η_{loss} is given in Figure 10 by comparing the simulated solar cell efficiency η with the specific solar cell limit η_{Limit} = 25.9 % for n-type and η_{Limit} = 26.5 % for p-type material. The inhomogeneity within a wafer is represented in the difference between the bars (blue for the best area vs. gray for the large area). This difference is very prominent for the n-type EpiWafer compared to the n-type EpiRef. Again, a tremendous efficiency loss is visible for the p-type EpiWafer with over -5 $\%_{abs}$. In general, there are small efficiency losses between the large area and the best area in the group of EpiRef wafers corresponding to the reasonably good lifetime homogeneity over the whole EpiRef wafers. Further, the p-type material is on a reduced quality level compared to n-type material for these cell models despite of the longer diffusion length for p-type EpiRef than for n-type EpiRef. However, especially the potential efficiencies for EpiRef wafers indicate, that epitaxially grown n- and p-type silicon layers in the 'PEpi'-reactor are promising candidates for base material for high efficiency cell models. Their calculated efficiencies suggest that high efficiency cells out of our epitaxial layers could exceed efficiencies of industrially produced cells out of crystallized CZ-material [2]. Efficiencies of high-end high efficiency cells out of crystallized FZ-material cannot fully be reached with our epitaxial layers yet [12].

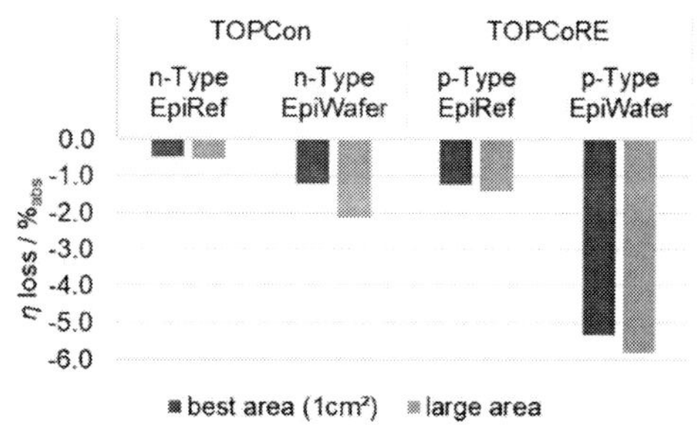

Figure 10: Evaluation summary of efficiency losses based on ELBA simulations (shown in Figure 9). The losses are given in absolute values, referenced to the corresponding cell limits with ηLimit = 25.9 % for n-type and ηLimit = 26.5 % for p-type material.

4.2 Limitations of the conducted efficiency analysis

As the observed material is in an intermediate processing state, there is an uncertainty relative to the final material quality which could be influenced by further treatment steps such as high temperature processes, chemical etching, emitter diffusion, and metallization. These treatment steps could influence the material quality in a positive but also negative way. The selection of an appropriate cell model is another challenge: to fully lift the potential of highly efficient cell models, in specific TOPCoRE, high-quality bulk material is required. Further the material's doping concentration and layer thickness influences the cell limit as well as the potential cell efficiency. In this contribution, both parameters were not optimized for the applied cell model which might offer room for further improvements.

5 CONCLUSION

In this contribution, qualities of n- and p-type EpiRef wafers and EpiWafers were compared. For EpiRef wafers, we achieved remarkable minority carrier lifetimes of 2.5 ms and a predicted cell efficiency potential of 25.4 % for n-type Si (TOPcon model with cell limit of 25.9 %) and 1.3 ms mean lifetime, with 25.0 % efficiency for p-type Si (TOPCoRE model with cell limit of 26.5 %). This confirms the high quality of the epitaxially grown layers for both, n- and p-type Si, and the excellent level of purity in the CVD-reactor 'PEpi'. EpiWafers feature lower minority carrier lifetimes than EpiRef wafers which is attributed to the influence of the highly doped seed wafer with a porous silicon detachment layer used for growing EpiWafers. The lifetime reduction for EpiWafers corresponds to a higher amount of stacking faults and a detectable iron contamination compared to EpiRef wafers grown on epi-ready surfaces. Furthermore, other metallic contaminations are assumed and will be investigated in the future. These results underline the urgent need of seed wafers with low contamination level and a smooth and homogeneous porous silicon detachment layer as well as the importance of further research on the influence of lattice- and doping concentration mismatch between epitaxial layer and seed wafer. In direct comparison between n-type and p-type silicon, n-type EpiWafers and EpiRef wafers exhibit longer lifetimes and higher efficiency potentials than p-type wafers. The explanation for the noteworthy difference is not yet found in detail. Generally, p-type silicon is more prone to certain metallic contaminations than n-type silicon such as interstitial iron. However, lower lifetimes for p-type material can be compensated by longer diffusion lengths due to higher minority carrier mobility and by adjusted cell models. Hence, we achieved encouraging potential cell efficiencies for both, n- and p-type material, for the high efficiency solar cell models TOPCon and TOPCoRE.

6 ACKNOWLEDGMENTS

The authors would like to express their gratitude to Philipp Barth, Andreas Fell, Melanie Frettlöh, Baris Gündogan, Elke Gust, Philipp Häuber, Robin Hintzen, Tom Meyer, Pascal Meßmer, Rainer Neubauer, Armin Richter, Bernd Steinhauser, Michaela Winterhalder, Karin Zimmermann and many more colleagues at Fraunhofer ISE for their support with processing, measurements, and input in many valuable discussions.

This work was funded by the German Federal Ministry of Economic Affairs and Energy (FKZ 03EE1082C and 0324290 B).

7 REFERENCES

[1] S. Riepe, S. Nold *et al.*, Proceedings 37th European Photovoltaic Solar Energy Conference and Exhibition (2020) 121.

[2] ITRPV, *International Technology Roadmap for Photovoltaic (ITRPV): 2020 Results*. 12. Edition, April 2021. Accessed on: Aug. 22 2021.

[3] H. S. Radhakrishnan, R. Martini *et al.*, *IEEE Journal of Photovoltaics* 4 (2014) 70.

[4] C. Gemmel, J. Hensen *et al.*, *Energy Procedia* 92 (2016) 29.

[5] N. Milenkovic, T. Rachow *et al.*, *Energy Procedia* 77 (2015) 613.

[6] D. M. Powell, J. Hofstetter *et al.*, *Applied Physics Letters* 103 (2013) 263902.

[7] M. Drießen, P. Beu *et al.,* Proceedings 35th European Photovoltaic Solar Energy Conference and Exhibition (EU PVSEC) (2018) 350.

[8] C. Gemmel, J. Hensen *et al.*, *AIP Conference Proceedings* 1999 (2018) 130005.

[9] T. Trupke, R. A. Bardos *et al.*, *Applied Physics Letters* 89 (2006) 44107.

[10] J. A. Giesecke, M. C. Schubert *et al.*, *Solar Energy Materials and Solar Cells* 95 (2011) 1011.

[11] M. Drießen, T. Fehrenbach *et al.,* Proceedings 36th European Photovoltaic Solar Energy Conference and Exhibition (2019) 135.

[12] A. Richter, R. Müller *et al.*, *Nature Energy* 6 (2021) 429.

[13] C. Weiss, W. Schreiber *et al.,* Proceedings 37th European Photovoltaic Solar Energy Conference and Exhibition (2020) 444.

[14] A. Fell, J. Schön *et al.*, *Solar Energy Materials and Solar Cells* 173 (2017) 128.

[15] B. Michl, M. Rüdiger *et al.*, *Solar Energy Materials and Solar Cells* 98 (2012) 441.

[16] J. Isenberg, J. Dicker *et al.*, *Journal of Applied Physics* 94 (2003) 4122.

[17] S. Janz, D. Amiri *et al.,* Proceedings 33rd European Photovoltaic Solar Energy Conference and Exhibition (EU PVSEC) (2017) 343.

[18] V. Boureau, J. M. Hartmann *et al.*, *Materials Science in Semiconductor Processing* 87 (2018) 65.

[19] A. Richter, S. W. Glunz *et al.* (en), *Physical Review B: Condensed Matter* 86 (2012).

[20] J. Schmidt, B. Lim *et al.*, *IEEE Journal of Photovoltaics* 3 (2013) 114.

[21] F. Schindler, B. Michl *et al.*, *Energy Procedia* 77 (2015) 633.

LASER CUTTING OF SOLAR CELLS BY USING THE STRESS CUT APPROACH

Stephan Krause[1], Stephan Hensel[1], Manuel Meusel[1], Stefan Eiternick[1], *Marko Turek[1]
[1] Fraunhofer Center for Silicon Photovoltaics, Otto-Eißfeldt-Straße 12, 06120 Halle (Saale), Germany

*Corresponding author: Tel. +49 (0) 345/5589-5121, marko.turek@csp.fraunhofer.de

ABSTRACT: Half-cell modules have entered the mass production market as one route to high-efficiency PV modules with reduced costs. Recent developments such as shingled modules also require reliable solar cell separation processes on a broad variety of solar cell technologies. In this work, we present a new "stress-cut" laser process that leads to optimized laser processing and reduced cell damages. It can be implemented for either ablative or non-ablative processes. Key element of the stress-cut is the laser treatment of a slightly bent solar cell which results in an automatic, intrinsic optimization of the cell separation process. We show, that for both PERC and heterojunction solar cells, well-optimized cell separation is achieved. Furthermore, results are presented showing that our stress-cut approach can be universally applied and transferred to a wide range of laser sources. Going beyond the half-cell layout, it is finally shown that the "stress-cut" process can also be applied to manufacture solar cell stripes for shingled modules.

Keywords: heterojunction solar cells, laser separation, lock-in thermography, efficiency analysis

1 INTRODUCTION

PV-modules with half-cell technologies show a significantly increased cell-to-module power ratio. This can be related to a reduced series resistance together with optical gains [1-4]. However, besides these gains, some additional losses due to recombination within the cell might be caused by the solar cell separation process. At the newly cut surface, the pn-junction crosses a non-passivated surface, which leads to additional dark current losses [1, 5-7].

Current half-cell processes are either based on an ablative, i.e. laser-scribe-cleavage (LSC), approach or on some non-ablative laser processes that induce local mechanical stress due to thermal gradients, e.g. thermal laser separation TLS [8, 9] or laser-direct-cleavage (LDC). In the case of LSC-processing, two separate steps – first laser-scribe, then cleavage – are required. While half cells with low electrical losses can be achieved using LSC, the adjustment of the laser process is rather difficult as an optimum between a sufficiently large scribe depth for a proper cleavage and the required minimal edge damages has to be found. For the TLS or LDC processes, a major challenge lies in achieving a straight-line edge in particular at the beginning and end of the line-cut. This might require special cell-chuck designs or additional cooling. Also, the presence of metallization structures or other cell inhomogeneities can lead to less robust processing results.

In our work, we introduce a newly developed "stress-cut" laser process. This process is a one-step process and can rely on either ablative or non-ablative processes. During the "stress-cut", the solar cells are subject to some mechanical stress, i.e. by using a curved cell chuck. This mechanical stress imposed on the cell already during the laser treatment leads to an automatic termination of the laser impact on the sample. The cell breaks as soon as the laser scribe achieves the minimum depth required for cell separation avoiding any unnecessary further cell damage due to the laser. In the case of a TLS or LSC process, the stress-cut approach can furthermore guide the cell crack into a straight line. Thus, the "stress cut" approach leads to an intrinsic optimization of the actual cell separation. This process has been tested on various cell types and the corresponding electrical losses have been analyzed.

2 EXPERIMENTAL APPROACH

For our experiments, we have designed a variety of cell chuck geometries that lead to a bending of the solar cell while applying the laser. During an ablative laser process, the laser is repeatedly scanning the cell. The material starts to melt and evaporate inducing vertical cracks into the cell. The cell bending leads to a process-specific scribing depth related to a minimal number of required laser scribes initializing the cleavage over the complete cell thickness. This stress-cut laser technique depends on the pre-bending load in combination with the deposited laser intensity. The bending load is related to the geometry of the cell chuck, i.e. its curvature as shown in the working principle of Fig.1. Here, we present optical-microscopic and electrical cell results of different curvature radii (smaller curvature X3 and larger curvature X6) on several solar cell types, e.g. PERC and HJT, as well as special separation layouts and configurations for further applications beyond half-cells.

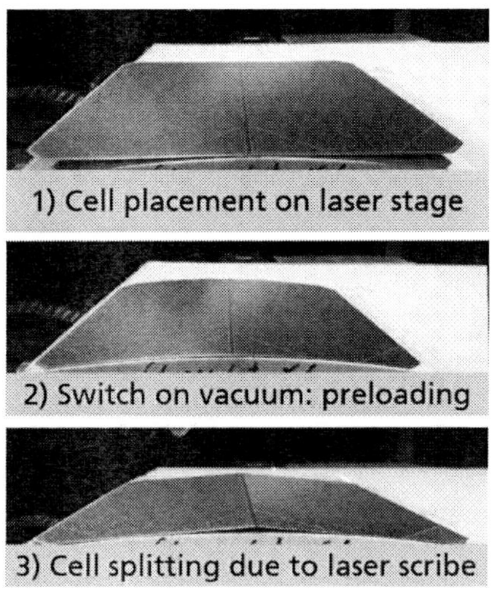

Figure 1. HJT solar cell on a curved base before 1) and after preloading 2) and finally after laser scribing 3). The cell separation seen in 3) has occurred automatically after the required minimal number of laser cycles making the

"stress-cut" approach a one-step process.

For the laser sources, we used short pulses for reduced thermal damage at the scribe region. In our experiment, a ns-laser and ps-laser with 532 nm wavelength at 5 ns and 10 ps, respectively, with a maximum power of 5 W and repetition frequencies up to 100 kHz was used to scribe the cells. The scanning velocities have been modified such that a vertical crack formation occurred after a minimum number of scans. With suitable laser sources and scanner units, process times of less than a second can be achieved. Hence, the self-optimizing stress-cut process is transferable to industrial applications.

In order to assess the different process configurations, we have monitored the number of applied laser scribe iteration, measured the depth of the laser scribe and inspected the quality of the newly-formed cell edge. Furthermore, the electrical losses have been quantified by electrical characterization of the untreated full cells and the laser "stress-cut" half cells. Using the LOANA tool by PV tools GmbH and a SINUS 220 solar simulator by WAVELABS, we have analyzed I-V-parameters and electroluminescence- and lock-in-thermography-images.

3 EXPERIMENTAL RESULTS

The "stress cut" approach leads to an intrinsic optimization as the laser scribing depth influences the electrical losses. To show this effect for our specific laser process with the ns-laser, we implemented laser receipts to processes different scribing depth (as shown in the top of Fig. 2) and measured the electrical behavior of the half cells in comparison to the full cells. The results of the reduction of the fill-factor in dependence on laser-scribe depth are shown in lower part of Fig. 2.

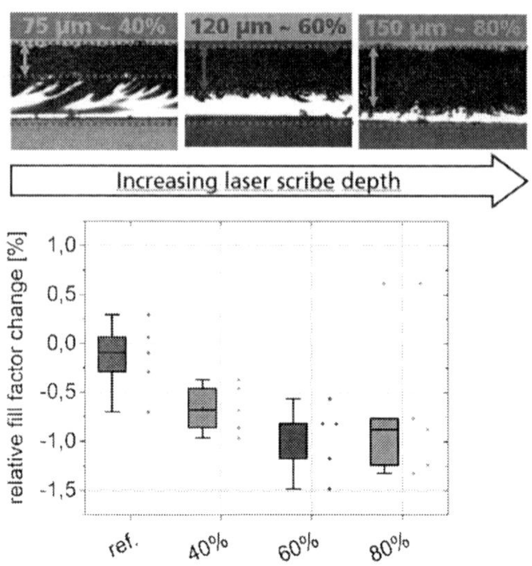

Figure 2: (top) light-microscopy cross-view of the cell edge after laser processing forthree different scribe depth. (bottom) Relative Fill factor in dependence of the laser-scribe depth.

It can be observed that there is a trend of increasing electrical losses with higher scribe depth. These losses can be related to an increase in the J_{02}, due to additional recombination currents due to the laser edge are induced by the laser process [1]. Regarding process stability requirements, we found that a very small scribe depth in the range of 30% or below of the cell thickness results in low production-yield as the cell cleavage cannot be performed reliable in these cases. Thus, laser processing requires an optimal scribe depth for low electrical losses and high production yield.

To establish a self-optimized laser process by the stress-cut approach, we add a mechanical pre-loading of the solar cell before laser treatment. As shown in Fig. 3, we start the stress-cut characterization on hetero-junction solar cells on different pre-loading stages followed by multiple fast laser scribes applied to the curved cells until cell separation takes place. When this cell separation occurs, the cut cell edges slightly open up so that no further laser damage is imposed to the edges, see Fig. 3 (lower left image). The cross-section optical microscopy views on the laser edge after processing show different laser scribe depths caused by the smaller and larger pre-loading.

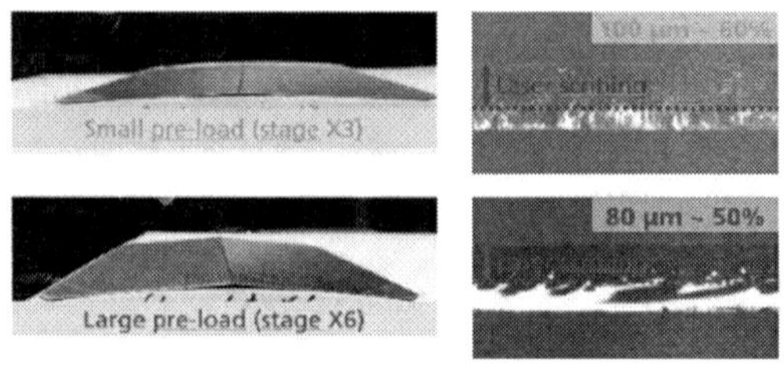

Figure 3: Left: Small (top) and large pre-loading stages (bottom) for different scribe depth during stress-cut process. Right: Microscopic imaging of the laser-cut cell edges showing the achieved laser scribe depth of 60% for small pre-loading and 50% for large pre-loading, respectively.

Moreover, the electrical cell parameters of the stress-cut processed HJT solar cells batch (of about 15 cells) were analyzed, see Fig. 4. The values for the short circuit current and open circuit voltage do not exhibit any significant losses beyond the measurement uncertainty. However, some outliers can be observed which indicate that a further process stabilization might be required. Regarding the fill factor, we observe a larger spread due to the cell separation. Again, this indicates that the process might not yet be completely stable and some further optimization is needed. Also, it is observed that there is some small overall increase in fill-factor. However, this is caused by the specific half-cell contacting scheme which had to be adapted due to the busbar-less cell technology and is thus a measurement artifact. In addition, we investigated the direction of the laser treatment on the cells, indicated in Fig. 4 as "front" (laser hits sunny side of the cell) and "rear" allocation. So far, no significant differences for either process version is observed in the electrical data. Hence, the stress-cut process appears to be robust on the HJT cells for both directions. In general, the electrical characterization does not show any major losses for the HJT sample series.

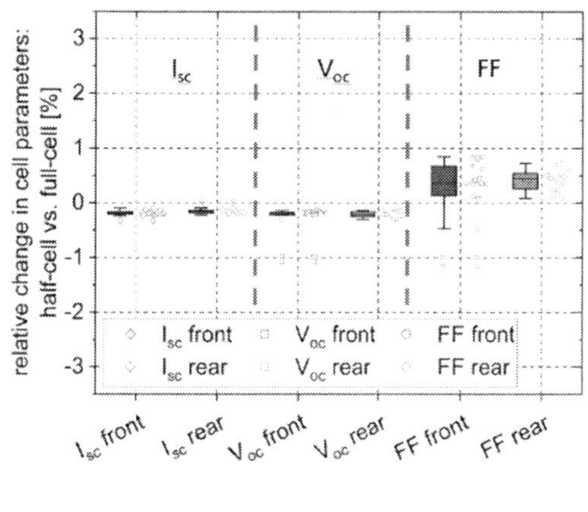

Figure 4: Relative changes of short-circuit currents I_{SC}, open circuit voltages V_{OC} and fill factors FF: comparison between half-cell and full-cell for the stress-cut (stage X6) processed from front and rear side of the HJT cells.

The same electrical characterization procedure was also applied to PERC solar cells for further checking of the process flexibility regarding different cell technologies. Fig. 5 displays the electrical data on the PERC solar cells. The short circuit current and open circuit voltage do not change within measurement errors. They are thus similar to the HJT results. For the fill factor and efficiency, losses of about 0.5%$_{rel}$ - 1 %$_{rel}$ were detected. These losses can be attributed to the recombination behavior at the laser edge and are comparable to previous results for other laser processes [1].

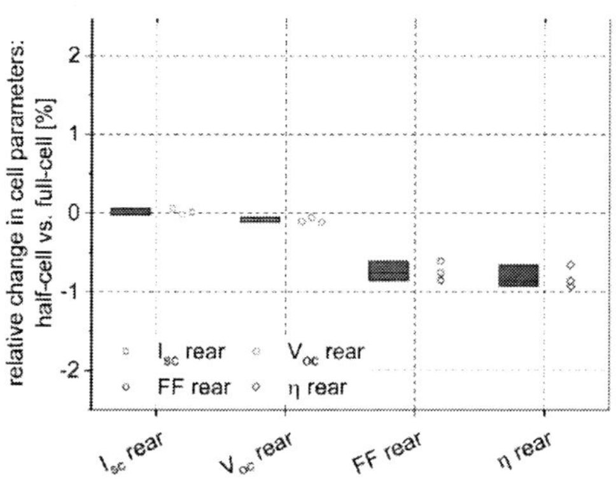

Figure 5: Relative changes of short-circuit currents I_{SC}, open circuit voltages V_{OC}, fill factors FF and efficiency η comparison between half-cell and full-cell for the stress-cut (stage X6) processed PERC cells.

Furthermore, we investigated the stress-cut process regarding its flexibility regarding the laser source. Therefore, a ps-laser source is investigated in addition to the ns-laser source. It has comparable laser parameters to the ns-laser process which was used for the results presented so far. In detail, the scan velocity, mean laser power and wavelength was chosen identically, whereas the repetition frequency was higher by a factor of two due to limits on the setup specification. Nevertheless, the laser impact during both, ns- and ps-laser processing on the sample is quite comparable resulting in a similar laser scribe depth of about 33% of the cell thickness, as shown in Fig. 6. Moreover, Fig. 6 illustrates the electrical parameters of the rear-side stress cut ns- and ps-processed on PERC solar cell samples. The measurements show no relative changes in the parameters when the two different laser sources are compared.

Finally, the stress-cut approach was tested for various layouts and special geometries beyond half-cell cutting. First, a shingle layout for the laser separation to six cell stripes was investigated. A successful run with the X6-stage setup for that multi-cut approach is shown in the upper part of Fig. 7. Here, we found that also a multi-cut process can be established using the stress-cut by means of the integrated laser scanner unit leading to well defined cell stripes.
As shown in the lower part of Fig. 7, another test regarding an even more flexible cell design beyond half-cells was done for IBC cells. Again, the stress-cut could be successfully applied. Additionally, we found that the smaller cell design and metal grid on the rear-side of this design has no significant impact on the pre-loading. Also, the different sizes of the cell stripes did not limit the applicability of the stress-cut approach.

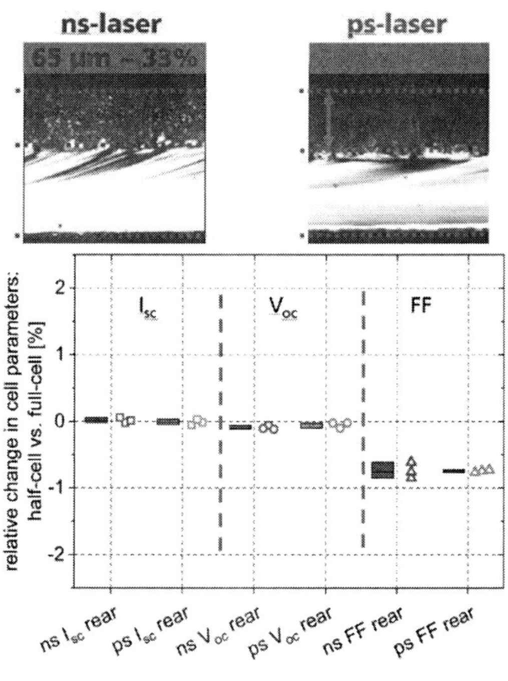

Figure 6: (top) Light-microscopic cross-section view of the ns- and ps-laser edge after stress cut. (bottom) Relative changes of short-circuit currents I_{SC}, open circuit voltages V_{OC} and fill factors FF: comparison between half-cell and full-cell for the stress-cut (stage X6) processed from front and rear side of the HJT cells.

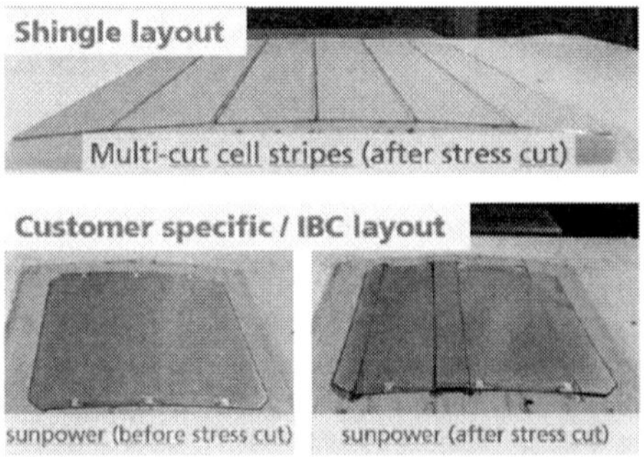

Figure 7. (Top) Stress-cut shingle layout on the pre-loading stage X6 for the laser separation to six cell stripes. (bottom) Flexible configuration of cell stripe separation on IBC cells for customer specific layouts.

4 SUMMARY AND CONCLUSIONS

We presented a new laser process approach for the separation of silicon solar cells. Our stress cut approach relies on a mechanical pre-loading of the sample and a simultaneous laser treatment. The process can be implemented on any laser tool without significant additional upgrade efforts. Due to the pre-loading by the mechanical bending of the cells it can be used as a self-optimized laser process: The laser scribe depth becomes as low as possible but as large as necessary for cell separation. No additional cell treatment, i.e. mechanical cell cleavage after laser scribing, is needed and the stress cut works at one-step process. An optimum curvature level for minimum scribe depth, and thus minimal electrical losses, but high production yield was found for different cell technologies. The process can be applied to a large variety of laser sources, such as ns- or ps-lasers. Furthermore, we provided some examples on the flexibility of the process, by successfully applying it to many variations in geometries, e.g. half-cells and multi-cut shingles, and several different cell technologies, e.g. PERC, HJT and IBC cells.

Acknowledgments

The authors acknowledge funding from the Federal ministry for Economic Affairs and Energy within the HJT-project (FKZ0324172B) and the HoSSa-project (FKZ 03EE1014A).

References

[1] S. Eiternick, K. Kaufmann, J. Schneider, M. Turek. *Loss analysis for laser separated solar cells*. Energy Procedia 55, 326 – 330 (2014).
[2] S. Guo, J. Schneider, F. Lu, H. Hanifi, M. Turek, M. Dyrba, I.M. Peters. *Investigation of the short-circuit current increase for PV modules using halved silicon wafer solar cells*. Solar energy materials and solar cells 133 240 – 247 (2015).
[3] H. Hanifi et al. *Optimized Tab Width in Half-cell Modules*. Energy Procedia, vol. 92, pp. 52–59 (2016).
[4] H. Hanifi, D. Dassler, M. Turek, and J. Schneider. *Evaluation and Comparison of PV Modules With Different Designs of Partial Cells in Desert and Moderate Climates*. IEEE J. Photovoltaics, vol. 8, no. 5, pp. 1266–1273 (2018).
[5] O. Breitenstein. *The Role of Inhomogeneities for Understanding Current–Voltage Characteristics of Solar Cells*. IEEE Journal of Photovoltaics, vol. 8 1429-1435 (2018).
[6] O. Breitenstein. *Nondestructive local analysis of current–voltage characteristics of solar cells by lock-in thermography*. Sol. Energy Mater. Sol. Cells Vol. 95 2933-2936, (2011).
[7] O. Breitenstein. *Local efficiency analysis of solar cells based on lock-in thermography*. Sol. Energy Mater. Sol. Cells 107, 381-389 (2012).
[8] F. Kaule, M. Pander, M. Turek, M. Grimm, E. Hofmueller, S. Schoenfelder. *Mechanical damage of half-cell cutting technologies in solar cells and module laminates*. AIP Conf. Proc. 1999, 020013 (2018).
[9] S. Eiternick, F. Kaule, H.-U. Zühlke, Th. Kießling, M. Grimm, S. Schoenfelder, M. Turek. *High quality half-cell processing using thermal laser separation*. Energy Procedia 77, 340-345 (2015).

14% EFFICIENCY ULTRATHIN SILICON SOLAR CELLS WITH IMPROVED INFRARED LIGHT MANAGEMENT ENABLED BY HOLE-SELECTIVE TRANSITION METAL OXIDE FULL-AREA REAR PASSIVATING CONTACTS

Hisham Nasser[a,*], Mona Zolfaghari Borra[a,b], Emine Hande Çiftpınar[a,b], Basil Eldeeb[a,c], Raşit Turan[a,b,c]
[a]The Center for Solar Energy Research and Applications (ODTÜ-GÜNAM), 06800, Ankara, Turkey
[b]Micro and Nanotechnology Graduate Program of Natural and Applied Sciences, Middle East Technical University (METU), 06800, Ankara, Turkey
[c]Department of Physics, Middle East Technical University (METU), 06800, Ankara, Turkey
*Corresponding author: nasser@metu.edu.tr; hisham.nasser@yahoo.com; Phone: +90-312-210-5095

ABSTRACT: We report efficient 22.5 μm-thick p-type crystalline silicon (c-Si) solar cells with hole-selective transition metal oxide (TMO) layers stacked with silver as full-area rear contacts. The TMO layers, namely molybdenum oxide (MoO_x), vanadium oxide (V_2O_x), and tungsten oxide (WO_x), are prepared by thermal evaporation at room temperature. These films have large work function relative to p-type c-Si creating strong accumulation layer at the interface with high hole concentration which reduces the interfacial recombination current density (J_0) and contact resistivity (ρ_c). Among the TMO p-type Si contacts considered, the MoO_x/p-type Si contact achieves the lowest ρ_c of ~27 mΩ.cm^2 relative to WO_x/p-type Si (~42 mΩ.cm^2) and V_2O_x/p-type Si (~190 mΩ.cm^2). On the other hand, V_2O_x/p-type Si contact yields the lowest J_0 of 65.5 fA/cm^2 compared to MoO_x (79.5 fA/cm^2) and WO_x (264.5 fA/cm^2) based p-type Si contacts. Furthermore, the presence of TMO/Ag contact at the rear resulted in an enhanced external quantum efficiency (EQE) response at long wavelengths compared to reference solar cells with Al at the rear. The champion solar cell with efficiency value reaching 14% was realized from the solar cell with MoO_x/Ag at the rear.
Keywords: Ultrathin c-Si, Hole-selective contacts, Dopant-free, Passivating contacts, Asymmetric heterocontacts

1 INTRODUCTION

Crystalline silicon (c-Si) photovoltaic (PV) dominates more than 90% of the PV market share [1] driven by their stability, good reliability, and high reported efficiencies approaching the Shockley-Queisser limit of 29-32% [2]. On the other hand, c-Si material accounts for 20% of module cost [1]. In this regards, ultrathin c-Si solar cells with thicknesses ≤25 μm featuring low material usage, flexibility, and light weight have received enormous attention in recent years [3–5]. Besides, thin absorber layer has the inherent advantages over thicker counterparts such as enhanced carrier diffusion and collection eliminating the requirement of high Si quality [6]. Thinner wafers have also the potential to attain improved V_{oc} if the wafer is well-passivated [7]. However, the performance of ultrathin c-Si solar cells is limited by: i) weak light absorption of Si; due to indirect band gap nature of c-Si, ii) pronounced parasitic light absorption by the rear metal contact, and iii) carrier recombination losses by heavily diffused and directly metallized Si [8–10]. To overcome these limitations, ultrathin solar cells are usually fabricated with advanced architectures. For instance, to enhance the light absorption, advanced light trapping schemes such as nanowires [11], nanosphere concentrator [12], or combination of nanocones and nanopillars arrays [5] resulted in efficiencies ≤12.2%. Higher efficiencies values were obtained by further incorporating adequate surface passivation. Tan et al., reported an impressive 16.4% efficiency from 20 μm-thick p-type absorber with PERC structure at the rear and small pyramids as front side light management interface [13]. Recently, He et al., have reported 15% from 25 μm-thick n-type c-Si solar cells with periodic nanopyramid arrays where the absorber is passivated by fluorine-doped TiO_x and PEDOT:PSS on the rear and front side, respectively [3]. However, all demonstration of ultrathin c-Si solar cells are based on advanced light trapping schemes with complex fabrication processes, including various types of lithography which are not industry feasible, and high temperature doped regions which imposes undesired fundamental limitations [14].

To eliminate the abovementioned limitations and to reduce the impact of doping-related limitations, dopant-free carrier-selective contacts are ideal candidates for ultrathin c-Si solar cell applications. Several studies have demonstrated competitive efficiency values from dopant-free Si (170-180 μm-thick) heterojunctions featuring hole-selective MoO_x, V_2O_x, or WO_x and/or electron-selective LiF_x, MgF_x, or TiO_x [15–22].

In this work, we prove that the application of hole-selective TMO layers stacked with Ag at the rear of ultrathin (22.5 μm) Cz p-type c-Si solar cells has the potential to passivate the Si surface, enhance the current density response, and improve the overall solar cell performance.

2 EXPERIMENTAL DETAILS

Fig. 1 shows the fabrication flow, of ultrathin p-type c-Si solar cells with hole-selective MoO_x, V_2O_x, or WO_x. As a proof-of-concept, 22.5 μm-thick c-Si slices (2×2 cm^2) were prepared by thinning down 525 μm-thick p-type (100) Si wafers (Cz, 1-3 Ω.cm) through photolithographically defined SiN_x windows in 50 wt% KOH at 85°C. Thin films of metal oxides (~15 nm) are deposited directly at the rear by thermal evaporation at room temperature followed by ~500 nm-thick Ag deposited subsequently in the same chamber without breaking the vacuum [23]. Set of reference solar cells with only aluminum at the rear were also fabricated. Performance of the TMO/ultrathin p-type c-Si solar cells and reference solar cells were evaluated from light *I–V* measurements in a four-wire electrical configuration under standard 1-sun condition (1000 mW/m^2, AM 1.5 spectrum), at an ambient temperature of ~24 °C. The measurement set-up was calibrated using a solar cell with certified PV characteristics. The EQE and total reflection were collected using a photovoltaic characterization system with an integrating sphere (Bentham Instruments, PV300).

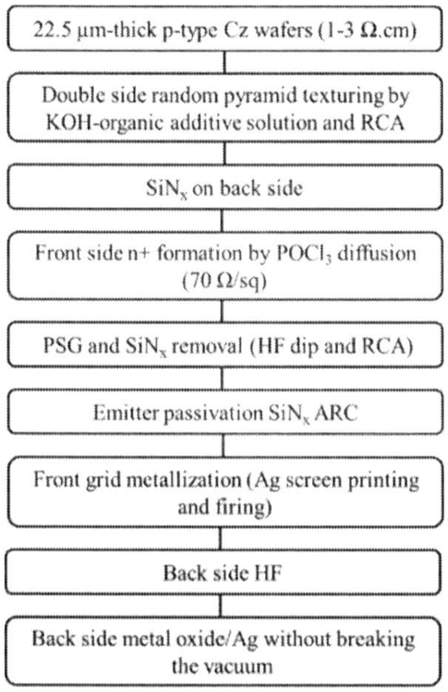

Figure 1: Process flow diagram of 22.5 µm-thick p-type c-Si with full-area hole-selective TMO/Ag rear contact.

3 RESULTS

The measured J_0; from lifetime measurements of symmetrical samples (p-type Cz, DSP, 1-3 Ω.cm) by PCD using Sinton lifetime tester according to Kane and Swanson technique [24], is shown in Fig. 2. The contribution of TMO/p-Si single surface (J_0), displayed in Fig. 2, was calculated by dividing the total measured J_0 of the TMO/p-Si/TMO by two. It can be observed that best passivation quality is achieved from V_2O_x while the highest J_0 is recorded from WO_x. Whilst the passivation quality of these TMO is clearly not as good as the widely used thin films to passivate Si, it shows a certain degree of surface passivation. This passivation is due to the band alignment that the TMO induces at the p-type Si interface which results in formation of an accumulation layer at the interface where electrons are repelled away as well as the formation of a thin SiO_2 (<2 nm) interfacial layer between Si and these TMO during thermal evaporation [25,26].

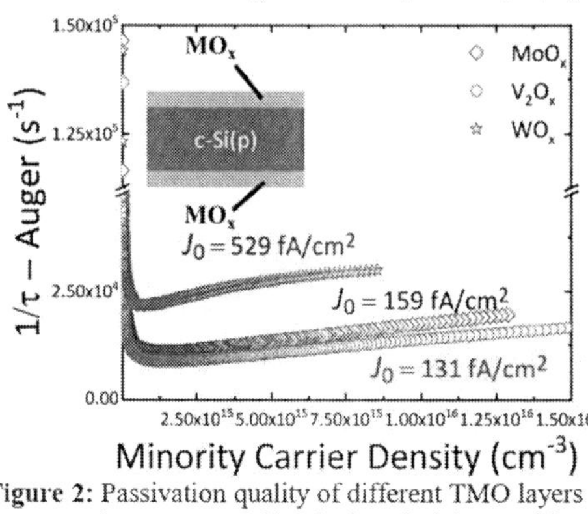

Figure 2: Passivation quality of different TMO layers on p-type wafers represented by the interfacial recombination current density (J_0) extracted at $\Delta n = 2 \times 10^{15}$ cm^{-3}.

Table I shows detailed cell parameters obtained at 1-sun, of the champion ultrathin p-type c-Si solar cells from each TMO. The highest V_{oc} of 566 mV is realized from the cell with V_2O_x (lowest J_0 of 65.5 fA/cm^2). The cell with MoO_x has achieved a reasonable V_{oc} of 555 mV (J_0=79.5 fA/cm^2), while WO_x (J_0=264.5 fA/cm^2) showed 532 mV. The ultrathin cell with directly metallized rear surface (Al-contact) suffers from high recombination rates at the rear resulted in V_{oc} of only 511 mV. FF values show strong correlation with the ρ_c (Ag/TMO/p-type Si); measured on p-type (DSP, 1-3 Ω.cm) Si according to the method devised by Cox and Strack by considering the recently developed approximations [27], as displayed in Fig. 3.

Table I: Photovoltaic parameters of 22.5 µm-thick p-type c-Si solar cells with full-area rear passivating hole-selective TMO/Ag contact.

Cell Rear Structure	V_{oc} (mV)	J_{sc} (mA/cm^2)	FF (%)	η (%)
MoO_x	555	33.2	76.0	14.0
V_2O_x	566	30.7	66.3	11.5
WO_x	532	31.1	73.4	12.1
Al-contact	511	29.7	71.6	10.9

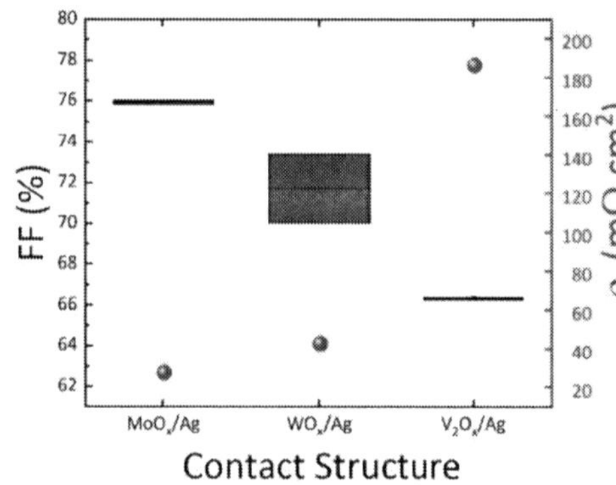

Figure 3: FF relation with ρ_c of TMO/Ag/p-type c-Si.

The cells with TMO/Ag at the rear show noticeable increase in J_{sc} compared to Al-contact counterpart. This is largely due to an improved long-wavelength response (Fig. 4) caused by increased rear side reflectivity (Fig. 5b) into Si combined with low ρ_c and excellent surface passivation. Fig. 4 shows that EQE response increases for wavelength ≥630 nm by inserting TMO/Ag at the rear compared to the cell with only Al. For wavelength <630 nm, all solar cells exhibit identical EQE response. The EQE response at short wavelengths is too weak despite the relatively low front side total reflection (Fig. 5a). It is noteworthy that the total reflection at short wavelengths is directly governed by the front side texturing and SiN_x recipe. Also, the charge carriers generated due to short wavelength-light absorption are sensitive to the emitter quality as well as to the front surface recombination losses. Therefore, the observed poor EQE response at short wavelengths, clearly indicates that the fabricated solar cells suffer from imperfect front side emitter/SiN_x/Ag recipe. On the contrary, the improved EQE response of TMO/Ag based solar cells at longer wavelengths relative to the cell with only Al, further demonstrates the benefits of using TMO/Ag at the rear of ultrathin solar cells.

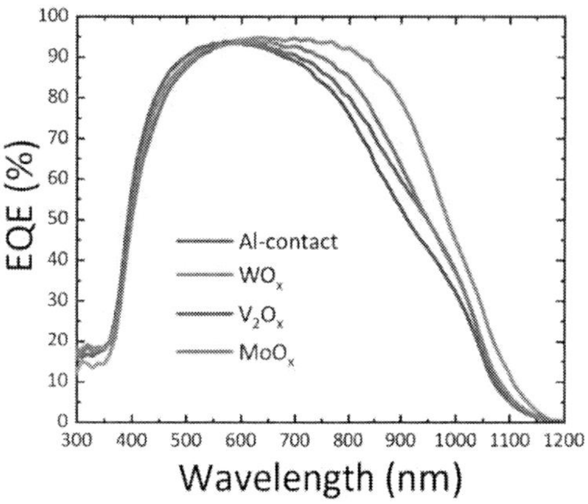

Figure 4: External quantum efficiency (EQE) response of the champion solar cells from each TMO/Ag set.

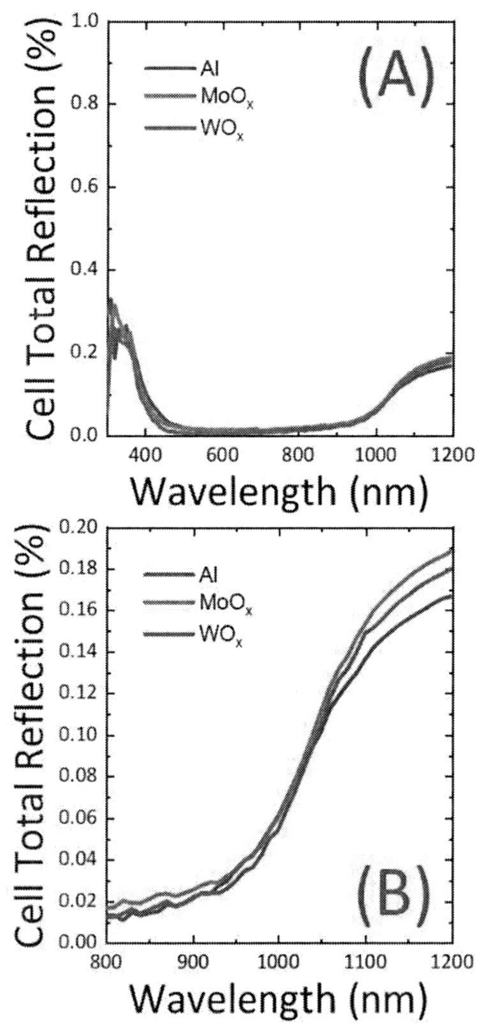

Figure 5: (a)-(b) Measured total reflection of the fabricated solar cells with Al, MoO_x/Ag, and WO_x/Ag rear contacts.

A remarkable high J_{sc} exceeding 33 mA/cm^2 is obtained from MoO_x/Ag without the use of advanced light trapping interface at the front side of the solar cells. The cell with V_2O_x suffers from severe resistive losses (large ρ_c) while the cell with WO_x suffers from low V_{oc} due to large J_0 induced from WO_x/p-Si. Consequently, the J_{sc} and overall cell characteristics obtained from V_2O_x and WO_x cells are lower relative to MoO_x counterpart. Nevertheless, all solar cells with TMO/Ag rear contact exhibit superior infrared EQE response; and thus higher J_{sc}, compared to Al-contact counterpart.

4 CONCLUSION

The efficiency of 22.5 μm-thick solar cells reaches 14% with MoO_x/Ag, which is remarkable for such novel c-Si solar cells concept at these thicknesses. This efficiency is achieved because the rear undoped-Si/metal contact is passivated by the TMO, even without the use of dedicated passivating interlayers. Light absorption in thin Si is greatly enhanced without utilizing advanced light trapping interface at the front thanks to the developed TMO/Ag back reflector which also mitigates parasitic light absorption in the infrared; free-carrier absorption, induced by Al-contact/highly absorbing Al. Further advancement of the obtained results will require the utilization of a proper front side emitter/SiN_x/Ag grid contacts recipe and the use of suitable passivation interlayer between the TMO and Si which will not only improve the V_{oc} but the overall cell performance.

5 REFERENCES

[1] VDMA, International Technology Roadmap for Photovoltaic (ITRPV) 2020 Results, n.d. https://itrpv.vdma.org/en.

[2] W. Shockley, H.J. Queisser, Detailed balance limit of efficiency of p-n junction solar cells, Journal of Applied Physics. 32 (1961) 510–519. https://doi.org/10.1063/1.1736034.

[3] J. He, M.A. Hossain, H. Lin, W. Wang, S.K. Karuturi, B. Hoex, J. Ye, P. Gao, J. Bullock, Y. Wan, 15% Efficiency Ultrathin Silicon Solar Cells with Fluorine-Doped Titanium Oxide and Chemically Tailored Poly(3,4-ethylenedioxythiophene):Poly(styrenesulfonate) as Asymmetric Heterocontact, ACS Nano. 13 (2019) 6356–6362. https://doi.org/10.1021/acsnano.9b01754.

[4] M. Xue, K.N. Nazif, Z. Lyu, J. Jiang, C.Y. Lu, N. Lee, K. Zang, Y. Chen, T. Zheng, T.I. Kamins, M.L. Brongersma, K.C. Saraswat, J.S. Harris, Free-standing 2.7 μm thick ultrathin crystalline silicon solar cell with efficiency above 12.0%, Nano Energy. 70 (2020) 104466. https://doi.org/10.1016/j.nanoen.2020.104466.

[5] J. He, Z. Yang, P. Liu, S. Wu, P. Gao, M. Wang, S. Zhou, X. Li, H. Cao, J. Ye, Enhanced Electro-Optical Properties of Nanocone/Nanopillar Dual-Structured Arrays for Ultrathin Silicon/Organic Hybrid Solar Cell Applications, Advanced Energy Materials. 6 (2016). https://doi.org/10.1002/aenm.201501793.

[6] J. Zhang, T. Song, X. Shen, X. Yu, S.T. Lee, B. Sun, A 12%-efficient upgraded metallurgical grade silicon - Organic heterojunction solar cell achieved by a self-purifying process, ACS Nano. 8 (2014) 11369–11376.

https://doi.org/10.1021/nn504279d.

[7] S.Y. Herasimenka, W.J. Dauksher, S.G. Bowden, >750 mV open circuit voltage measured on 50 μm thick silicon heterojunction solar cell, Applied Physics Letters. 103 (2013). https://doi.org/10.1063/1.4817723.

[8] A. Richter, M. Hermle, S.W. Glunz, Reassessment of the limiting efficiency for crystalline silicon solar cells, IEEE Journal of Photovoltaics. 3 (2013) 1184–1191. https://doi.org/10.1109/JPHOTOV.2013.2270351.

[9] P. Kowalczewski, L.C. Andreani, Towards the efficiency limits of silicon solar cells: How thin is too thin?, Solar Energy Materials and Solar Cells. 143 (2015) 260–268. https://doi.org/10.1016/j.solmat.2015.06.054.

[10] L. Tous, M. Aleman, R. Russell, E. Cornagliotti, P. Choulat, A. Uruena, S. Singh, J. John, F. Duerinckx, J. Poortmans, J. Poortmans, R. Mertens, Evaluation of advanced p-PERL and n-PERT large area silicon solar cells with 20.5% energy conversion efficiencies, Progress in Photovoltaics: Research and Applications. 23 (2015) 660–670. https://doi.org/10.1002/pip.2478.

[11] P. Aurang, R. Turan, H.E. Unalan, Nanowire decorated, ultra-thin, single crystalline silicon for photovoltaic devices, Nanotechnology. 28 (2017) 405205. https://doi.org/10.1088/1361-6528/AA81B9.

[12] B. Wang, T. Gao, P.W. Leu, Broadband light absorption enhancement in ultrathin film crystalline silicon solar cells with high index of refraction nanosphere arrays, Nano Energy. 19 (2016) 471–475. https://doi.org/10.1016/j.nanoen.2015.10.039.

[13] X. Tan, W. Yan, Y. Tu, C. Deng, Small pyramidal textured ultrathin crystalline silicon solar cells with double-layer passivation, Optics Express. 25 (2017) 14725. https://doi.org/10.1364/oe.25.014725.

[14] J. Bullock, M. Hettick, J. Geissbühler, A.J. Ong, T. Allen, C.M. Sutter-Fella, T. Chen, H. Ota, E.W. Schaler, S. De Wolf, C. Ballif, A. Cuevas, A. Javey, Efficient silicon solar cells with dopant-free asymmetric heterocontacts, Nature Energy. 1 (2016). https://doi.org/10.1038/nenergy.2015.31.

[15] H. Nasser, F. Es, M. Zolfaghari Borra, E. Semiz, G. Kökbudak, E. Orhan, R. Turan, On the application of hole-selective MoOx as full-area rear contact for industrial scale p-type c-Si solar cells, Progress in Photovoltaics: Research and Applications. (2020) 1–13. https://doi.org/10.1002/pip.3363.

[16] W. Wang, J. He, D. Yan, C. Samundsett, S.P. Phang, Z. Huang, W. Shen, J. Bullock, Y. Wan, 21.3%-efficient n-type silicon solar cell with a full area rear TiOx/LiF/Al electron-selective contact, Solar Energy Materials and Solar Cells. (2019) 110291. https://doi.org/10.1016/J.SOLMAT.2019.110291.

[17] J. Dréon, Q. Jeangros, J. Cattin, J. Haschke, L. Antognini, C. Ballif, M. Boccard, 23.5%-Efficient Silicon Heterojunction Silicon Solar Cell Using Molybdenum Oxide As Hole-Selective Contact, Nano Energy. 70 (2020). https://doi.org/10.1016/j.nanoen.2020.104495.

[18] X. Yang, K. Weber, Z. Hameiri, S. De Wolf, Industrially feasible, dopant-free, carrier-selective contacts for high-efficiency silicon solar cells, Progress in Photovoltaics: Research and Applications. 25 (2017) 896–904. https://doi.org/10.1002/pip.2901.

[19] X. Yang, H. Xu, W. Liu, Q. Bi, L. Xu, J. Kang, M.N. Hedhili, B. Sun, X. Zhang, S. De Wolf, Atomic Layer Deposition of Vanadium Oxide as Hole-Selective Contact for Crystalline Silicon Solar Cells, Advanced Electronic Materials. 6 (2020) 1–8. https://doi.org/10.1002/aelm.202000467.

[20] H. Mehmood, H. Nasser, T. Tauqeer, R. Turan, Simulation of silicon heterostructure solar cell featuring dopant-free carrier-selective molybdenum oxide and titanium oxide contacts, Renewable Energy. 143 (2019) 359–367. https://doi.org/10.1016/j.renene.2019.05.007.

[21] Y. Wan, C. Samundsett, J. Bullock, T. Allen, M. Hettick, D. Yan, P. Zheng, X. Zhang, J. Cui, J. McKeon, A. Javey, A. Cuevas, Magnesium Fluoride Electron-Selective Contacts for Crystalline Silicon Solar Cells, ACS Applied Materials and Interfaces. 8 (2016) 14671–14677. https://doi.org/10.1021/acsami.6b03599.

[22] J. Bullock, Y. Wan, M. Hettick, X. Zhaoran, S.P. Phang, D. Yan, H. Wang, W. Ji, C. Samundsett, Z. Hameiri, D. Macdonald, A. Cuevas, A. Javey, Dopant-Free Partial Rear Contacts Enabling 23% Silicon Solar Cells, Advanced Energy Materials. 9 (2019) 1–6. https://doi.org/10.1002/aenm.201803367.

[23] O. Akdemir, M. Zolfaghari Borra, H. Nasser, R. Turan, A. Bek, MoOx/Ag/MoOx multilayers as hole transport transparent conductive electrodes for n-type crystalline silicon solar cells, International Journal of Energy Research. (2020) 1–12. https://doi.org/10.1002/er.5145.

[24] D.E. Kane, R.M. Swanson, Measurement of the Emitter Saturation Current by a Contactless Photoconductivity Decay Method, in: Conference Record of the IEEE Photovoltaic Specialists Conference, IEEE, 1985: pp. 578–583.

[25] L.G. Gerling, C. Voz, R. Alcubilla, J. Puigdollers, Origin of passivation in hole-selective transition metal oxides for crystalline silicon heterojunction solar cells, Journal of Materials Research. 32 (2017) 260–268. https://doi.org/10.1557/jmr.2016.453.

[26] H. Nasser, G. Kökbudak, H. Mehmood, R. Turan, Dependence of n-cSi/MoOx Heterojunction Performance on cSi Doping Concentration, Energy Procedia. 124 (2017) 418–424. https://doi.org/10.1016/j.egypro.2017.09.267.

[27] W. Wang, H. Lin, Z. Yang, Z. Wang, J. Wang, L. Zhang, M. Liao, Y. Zeng, P. Gao, B. Yan, J. Ye, An Expanded Cox and Strack Method for Precise Extraction of Specific Contact Resistance of Transition Metal Oxide/n-Silicon Heterojunction, IEEE Journal of Photovoltaics. 9 (2019) 1113–1120. https://doi.org/10.1109/JPHOTOV.2019.2917386.

OPTIMIZATION OF THE CONDUCTIVITY AND CRYSTALLINE FRACTION OF P-TYPE μc-SiO_x:H FILMS FOR SILICON HETEROJUNCTION SOLAR CELLS

Antonio J. Olivares[1], Gurleen Kaur[2], Mateusz Poplawski[1,2], Anatole Desthieux[3,1,2] and Pere Roca i Cabarrocas[1,2]
[1]LPICM, CNRS, Ecole Polytechnique, Institut Polytechnique de Paris, 91128 Palaiseau, France.
[2]Institut Photovoltaïque d'Ile-de-France (IPVF), 18 Bvd Thomas Gobert, 91120 Palaiseau, France.
[3]EDF R&D, 18 Boulevard Thomas Gobert, 91120 Palaiseau, France
antonio-de-jesus.olivares-vargas@polytechnique.edu

ABSTRACT: We report on the development of low-temperature PECVD deposition of p-type hydrogenated microcrystalline silicon oxide films (p-type μc-SiOx:H) as a hole transport layer for heterojunction solar cells. While the production of n-type μc-SiOx:H layers has been widely reported on, p-type ones are far more challenging. The tradeoff between optical properties, passivation quality, and electrical transport requires fine-tuning of the process parameters and post-deposition annealing. We show how the crystalline fraction (Fc) and dark conductivity (σ) depend on RF power, trimethylboron (TMB) gas flow rate, and total pressure during the deposition process at 150 °C, and how these properties evolve with annealing at 250 °C and 300 °C. P-type μc-SiOx:H films were obtained with a conductivity around 10^{-6} S/cm in their as-deposited state and 10^{-5} S/cm after annealing. The layers with the best conductivity on glass substrates are then implemented in solar cells having the structure n-type μcSiOx:H/ a-SiOx:H/ n-type FZ c-Si/ a-SiOx:H/ p-type μcSiOx/p-type μcSi:H where all the layers are deposited by PECVD. These films improved the minority carrier lifetime and field effect, but the solar cell characteristics are still limited by low FF.
Keywords: Microcrystalline silicon oxide thin films; Multicrystalline silicon; p-type; μc-Si:H; Silicon solar cells

1 INTRODUCTION

Microcrystalline silicon oxide (μc-SiOx:H) films are mixed-phase materials consisting of silicon nanocrystals embedded in an amorphous network which itself consists of a mixture of a-SiOx:H and a-Si:H phases (see Fig. 1) They represent an evolution from microcrystalline silicon (μc-Si:H) films, to which the a-SiOx:H phase has been added [1–3]. The addition of CO_2 to the SiH_4-H_2 mixture generally used to grow μc-Si:H is responsible for the incorporation of the a-SiOx:H phase which has a strong impact on the optical properties and thermal stability of the films.

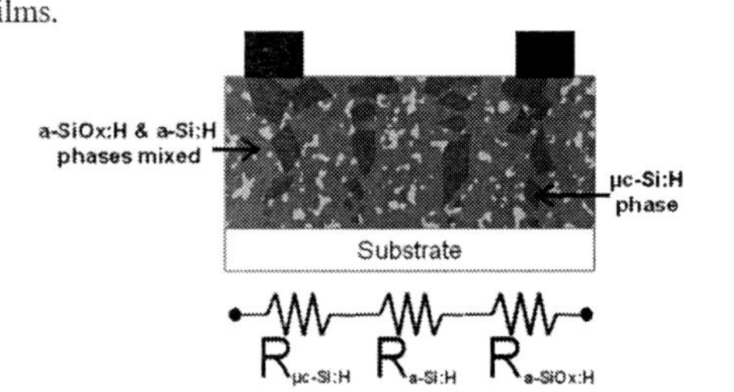

Figure 1: Sketch of μc-SiOx:H films structure and equivalent circuit of the resistances found across the film.

Indeed, μc-SiOx:H films present higher transparency, lower refractive index, and low absorption coefficient compared to a-Si:H and μc-Si:H films, and therefore they reduce the reflectivity and parasitic absorption in c-Si cells [1-2]. Moreover, the higher doping efficiency of the c-Si phase compared to a-Si:H can provide better field-effect passivation than a-Si:H films, thus improving the carrier selectivity in silicon heterojunction (SHJ) solar cells. While n-type μc-SiOx:H films with high conductivity are easy to obtain [4], p-type doping is much more challenging, in particular achieving a crystalline fraction (Fc) higher than 30 %, which can ensure a dark conductivity (σ) of the order of $1x10^{-5}$ S/cm or above [5]. With these objectives in mind, we have studied the effect of various process parameters: RF power, trimethylboron (TMB) gas flow rate, and total deposition pressure, on the optical and electronic properties of μc-SiOx:H films in view of their application as a p-type material and passivated contact in silicon heterojunction (SHJ) solar cells. All the samples have been deposited at a relatively low temperature of 150 °C, which enhances the grown of the c-Si phase but can lead to the dopants atoms (boron atoms) forming B-H complexes [6]. Therefore, post-deposition annealing at 250 and 300 °C for 15 minutes in air each has been applied to determine its impact on the coplanar conductivity of the films and the performance of the SHJ solar cells.

2 EXPERIMENTAL DETAILS

For optoelectrical characterization, 70-nm-thick single layers of p-type microcrystalline silicon oxide were deposited on Corning glass substrates by PECVD with parallel electrodes of 176 cm^2 area at 150°C substrate temperature [7]. Modeling spectroscopic ellipsometry measurements using BEMA approximation allowed us to determine the thickness and crystalline fraction (Fc) of the films [8]. Also, Fc was determined from Raman scattering spectroscopy by calculating the ratio of integrated peaks attributed to crystalline and amorphous regions [3]. The dark conductivity of the films was evaluated using coplanar aluminum electrodes in dark at room temperature.

For SHJ solar cells fabrication, we used double side polished n-type FZ-Si wafers of 260-300 μm thickness with a resistivity of 1-5 Ω cm. The wafers were passivated on both sides with an ultrathin a-SiOx layer (~1 nm). We used on the rear side, an n-type μc-SiOx:H layer (~60 nm) as a passivating contact layer. On the front side in addition to the films studied in this work (p-type μc-SiOx:H (~20 nm)) a p-type μc-Si:H layer (~60 nm) was used to reduce the contact resistance with the front Indium-Tin-Oxide (ITO) electrode. The cells were completed with sputtered ITO on the front and rear sides (using only Al on the rear side resulted in an S-shape in the I-V curve) and Ag contact on the rear side and Ag grid on the top. Minority carrier lifetime and implied V_{oc} (iV_{oc}) were measured on a

Sinton instrument. Finally, the cells were characterized using a solar simulator Oriel Sol 3A under standard test conditions.

3 RESULTS AND DISCUSSION

3.1 Films characterization

The influence of the RF power on the properties of p-type μc-SiOx:H films was investigated in a narrow range from 8 to 12 W while keeping all other process parameters constant (2 Torr, 1.4 sccm SiH_4, 500 sccm H_2, 4 sccm TMB, and 1 sccm CO_2). Fig. 2 shows the Fc and dark conductivity as functions of the RF power.

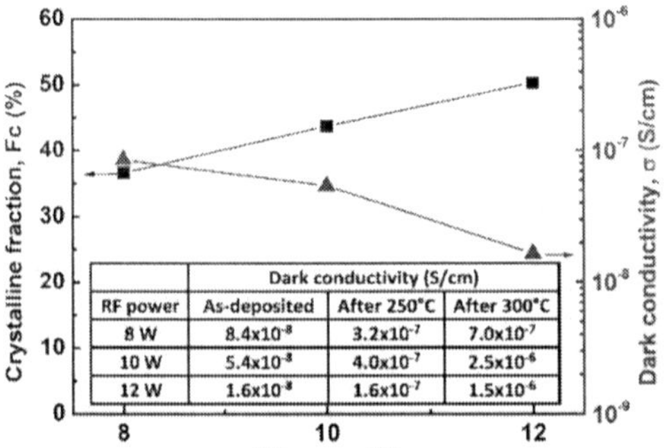

	Dark conductivity (S/cm)		
RF power	As-deposited	After 250°C	After 300°C
8 W	$8.4x10^{-8}$	$3.2x10^{-7}$	$7.0x10^{-7}$
10 W	$5.4x10^{-8}$	$4.0x10^{-7}$	$2.5x10^{-6}$
12 W	$1.6x10^{-8}$	$1.6x10^{-7}$	$1.5x10^{-6}$

Figure 2: Crystalline fraction (Fc) and dark conductivity (σ) of as-deposited p-type μc-SiOx:H films as a function of RF power. The lines are provided as guides to the eye.

Fc of p-type μc-SiOx:H films increased from 36 to 50 % when the RF power was increased from 8 to 12 W as shown in Fig 2. The trend may be linked to the increase of atomic hydrogen flux to the growing surface leading to the formation of the microcrystalline phase [3], and more likely, given our process conditions, to the formation of silicon nanocrystals in the plasma and their incorporation in the film [9]. In contrast with the increase of the Fc, the dark conductivity in the as-deposited state decreased from $8.4x10^{-8}$ to $1.6x10^{-8}$ S/cm when increasing the RF power (see Fig. 2). These results are unexpected since several studies have reported that the conductivity of the μc-Si:H films is controlled by the Fc [10-11]. Our results suggest that the conductivity of the p-μc-SiOx:H films is limited by the a-Si:H and a-SiOx:H phases (see the equivalente circuit in Fig. 1). Previous results from our group [6] suggest that with an increase of the hydrogen content in the plasma, the formation of B-H pairs is enhanced, causing the deactivation of boron and reducing the conductivity of the films. Thus, the decrease of the conductivity of the films with increasing the RF power could be attributed to the increase of hydrogen content in the layers and the deactivation of the boron in the film. The conductivity after both anneals at 250 and 300 °C (shown in the table inserted in Fig. 2) is increased (up to $2.5x10^{-6}$ S/cm for the sample deposited at 10 W). This can be related with the boron thermal activation of both crystalline and amorphous phases.

In order to observe the effect of TMB gas flow rate on the properties of p-type μc-SiOx:H films and increasing the conductivity of the amorphous phases, four samples were deposited under various TMB gas flow rates from 2 to 10 sccm, while keeping all other process parameters constant (2.2 Torr, 10 W, 1.4 sccm SiH_4, 500 sccm H_2 and 1 sccm CO_2). Fc and dark conductivity at different TMB gas flow rates are shown in Fig. 3.

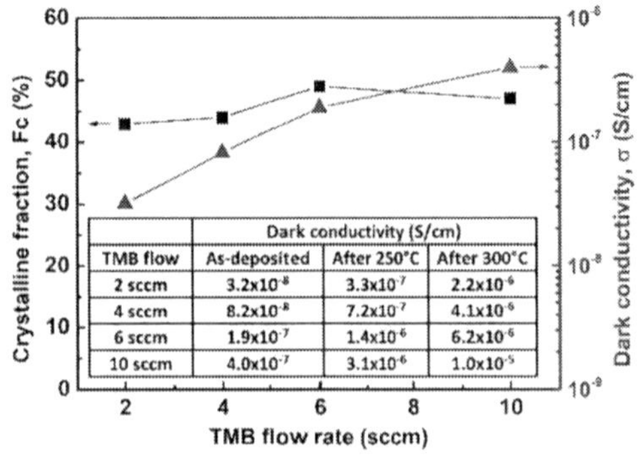

	Dark conductivity (S/cm)		
TMB flow	As-deposited	After 250°C	After 300°C
2 sccm	$3.2x10^{-8}$	$3.3x10^{-7}$	$2.2x10^{-6}$
4 sccm	$8.2x10^{-8}$	$7.2x10^{-7}$	$4.1x10^{-6}$
6 sccm	$1.9x10^{-7}$	$1.4x10^{-6}$	$6.2x10^{-6}$
10 sccm	$4.0x10^{-7}$	$3.1x10^{-6}$	$1.0x10^{-5}$

Figure 3: Crystalline fraction (Fc) and dark conductivity (σ) of as-deposited p-type μc-SiOx:H films as a function of TMB gas flow rate. The lines are provided as guides to the eye.

In the literature [10-11], it has been reported a decrease in the Fc of p-type μc-SiOx:H films deposited at 200 °C when the doping gas flow rate is increased, due to the inhibition of crystalline phase growth by boron incorporation. This is not the case for our films. Indeed, as shown in Fig. 3, Fc shows a small dependence on the TMB gas flow rate, with a minimum of 43 % and a maximum of 49 %. These results suggest that the growth process of the μc-SiOx:H films was not strongly affected by the TMB gas flow rate. On the other hand, the conductivity of p-type μc-SiOx:H films increased from $3.2x10^{-8}$ to $4.0x10^{-7}$ S/cm when increasing the TMB gas flow rate (Fig. 3). The increase in dark conductivity is mainly attributed to the variation of the active doping concentration and due to the incorporation of boron into the film both in the crystalline and amorphous phases. However, the fact that the conductivity increases suggests that the main effect of increasing the TMB gas flow rate is to increase the conductivity of the amorphous silicon and silicon oxide phases. Like in the previous series, the conductivity of the films increased after annealing at 250 and 300 °C (see table inserted in Fig 3). The increase was up to $1.0x10^{-5}$ S/cm for the sample deposited with 10 sccm of TMB gas flow rate. This can be attributed to the enhanced incorporation of boron into the films and the increased conductivity of the amorphous phases in the as-deposited films and, consequently, after anneals, due to the thermal activation of boron.

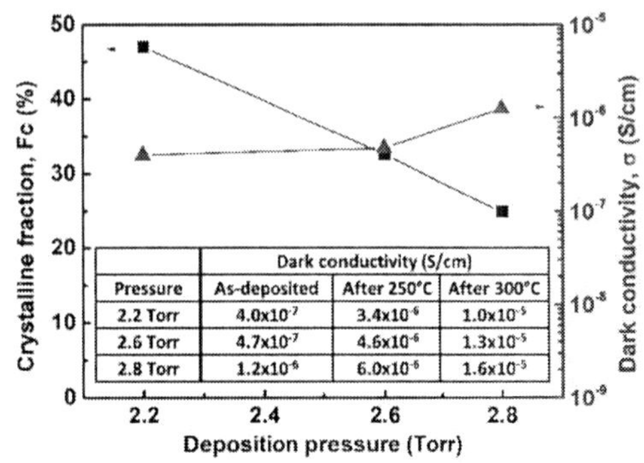

	Dark conductivity (S/cm)		
Pressure	As-deposited	After 250°C	After 300°C
2.2 Torr	$4.0x10^{-7}$	$3.4x10^{-6}$	$1.0x10^{-5}$
2.6 Torr	$4.7x10^{-7}$	$4.6x10^{-6}$	$1.3x10^{-5}$
2.8 Torr	$1.2x10^{-6}$	$6.0x10^{-6}$	$1.6x10^{-5}$

Figure 4: Crystalline fraction (Fc) and dark conductivity of as deposited p-type μc-SiOx:H films as a function of deposition pressure. The lines are provided as guides to the eye.

The dependence of Fc and dark conductivity of p-type μc-SiOx:H films on the deposition pressure was studied from 2.2 to 2.8 Torr while keeping all other process parameters constant (10 W, 1.4 sccm SiH_4, 500 sccm H_2, 10 sccm TMB, and 1 sccm CO_2). Fc and dark conductivity at different deposition pressures are shown in Fig. 4.

Fig. 4 shows that as the deposition pressure increases, the Fc decreases from 47 to 25 %. This trend is similar to that described in the literature by O. Gabriel et al [12], in which it is suggested that in this deposition regime comprising low power and high pressure, the flux of hydrogen towards the surface decreases, in particular H_3^+ ions [13]. Otherwise, the conductivity in the as-deposited state increased from $4.0x10^{-7}$ to $1.2x10^{-6}$ S/cm increasing the total deposition pressure from 2.2 to 2.8 Torr (Fig. 3). These results support our hypothesis of the conductivity by also being controlled by that of the a-SiOx:H and a-Si:H phases. This samples series is no exception and by annealing at 250 and 300 °C, the conductivity is also increased up to $1.6x10^{-5}$ S/cm for the sample deposited at 2.8 Torr owing to the boron activation.

Once p-type μc-SiOx:H films with conductivities around 10^{-5} s/cm are obtained, it is thought to be sufficient for use as a p-type and passivated contact in SHJ solar cells. It is expected to improve the passivation due to the high hydrogen content in the p-type μc-SiOx:H films and to reduce the parasitic resistances due to the conductivity values found.

3.2 Crystalline fraction post-annealing

In the previous section, the improvement of the conductivity of all μc-SiOx:H films after annealing at 250 and 300 °C has been shown and discussed. However, the effect of annealing on the Fc requires a more detailed discussion. It is important to note that the results after the annealing treatment showed the same trend in all the sample series mentioned above, but for simplicity, only the results of the deposition pressure series are presented in this analysis.

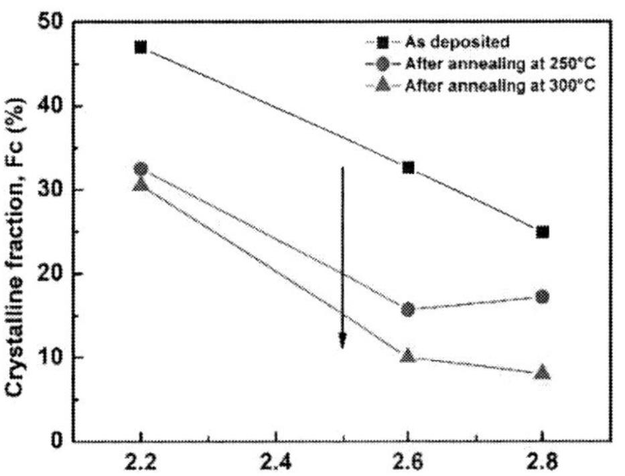

Figure 5: Crystalline fraction (Fc) of p-type μc-SiOx:H films as-deposited (black squares), after annealing at 250°C (green circles) and 300°C (red triangles) for the sample of the deposition pressure series. The lines are provided as guides to the eye.

Fig. 5 shows the Fc obtained by ellipsometry data of p-type μc-SiOx:H films as deposited (black squares), after annealing at 250 °C (green circles) and after annealing at 300 °C (red triangles). Fc decreased from 47 to 32 % after annealing at 250 °C and down to 30 % after annealing at 300 °C for the sample deposited at 2.2 Torr. For the sample deposited at 2.6 Torr, Fc decreased from 32 to 15 % after annealing at 250 °C and down to 10 % after annealing at 300 °C. Finally, for the sample deposited at 2.8 Torr, Fc decreased from 24 to 17 % after annealing at 250 °C and down to 8 % after annealing at 300 °C. The trend of decreasing Fc with respect to deposition pressure previously discussed is maintained. And, in general, after both anneals the Fc is reduced. This is unexpected because we remember that the conductivities are increased after both anneals.

In order to verify the changes in the Fc during annealing, Raman spectra of the best conductivity BC sample (deposited at 2.8 Torr from the deposition pressure series) in the as deposited state and after annealing at 300 °C are shown in Fig 6.

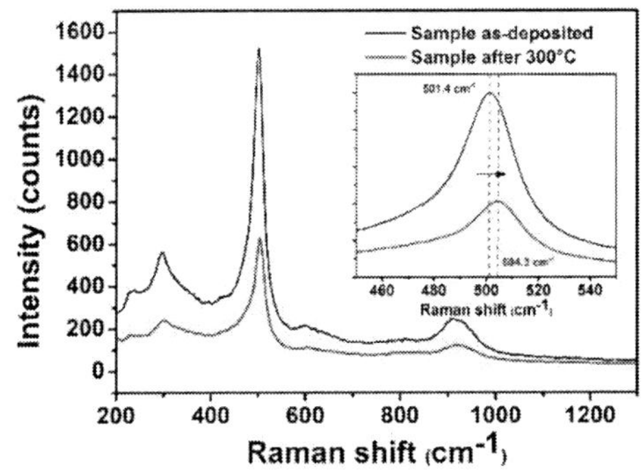

Figure 6: Raman spectroscopy of p-type μc-SiOx:H films as deposited and after annealing at 300°C for the sample with the best conductivity (BC) from the pressure series, measured under the same conditions.

It can be seen in the Raman spectra the main peak at 500-505 cm^{-1}, which is assigned to the presence of silicon nanocrystallites (nc-Si) [14-15]. The reduction of the peak intensity is consistent with the reduction of Fc after annealing. On the other hand, there is a shift of the peak of about 3 cm^{-1} towards higher wavenumbers (see the insert image in Fig. 6). The peak position is related to i) temperature, ii) strain and iii) grain size. In the present case, the shift to the right of the peak can be related to an increase in grain size or to a decrease in tensile stress [16]. However, the effects of deformation cannot be easily differentiated from those of nanocrystallinity in the Raman spectrum, so further studies are required.

3.3 p-type μc-SiOx:H on top of silicon heterojunction (SHJ) solar cells

In the last part p-type μc-SiOx:H materials have been implemented as p-type and passivated contacts of two identical SHJ solar cells (see the structure in Fig 6). In the first structure, a p-type μc-SiOx:H film from the RF power series (referred to as the REF sample) is used. And, in the second structure, the p-type μc-SiOx:H film with the best conductivity from the deposition pressure series (referred to as the BC sample) is used.

Table 1: Minority carrier lifetime at 1.0E+15 Δn and corresponding iVoc.

Sample	Lifetime (μs)	iVoc (V)
REF as deposited	331	0.64
REF after 300°C	431	0.64
BC as deposited	393	0.64
BC after 300°C	455	0.65

Lifetime and implied V_{oc} (iV_{oc}) (shown in table 1) were measured in the as-deposited state. It can see that there is a bit improvement of in the lifetime for the BC sample as deposited and after annealing compared with the REF sample. The higher lifetime for the BC sample could be due to the higher conductivity and field effect passivation provided by the p-μc-SiOx:H. These changes on the lifetime also reflect on the iV_{oc} which is 0.64 V for REF sample and slightly increases to 0.65 V for the BC sample after annealing.

Fig. 6 shows the I-V characteristics measured after deposition electrodes and after annealing at 300°C under AM1.5 illumination. The REF sample exhibits a V_{oc}=0.63 V, J_{sc}=31.2 mA/cm^2, FF=72.7%, and a η=14.3%. In the case of the BC structure, the first thing that should be noted is that the sample was not damaged after annealing at 300°C. Also, there is a gain of current of almost 1 mA/cm^2, but a low FF=63.9% owing to there is an increase in series resistance (Rs) from 3.3 to 4.8 Ω cm^2. Mazzarella et. al. [2] reported that the series resistance losses can be due to i) a very low Fc, ii) a poor contact between a-Si:H/μc-SiOx:H because the incubation layer is too thick and/or, iii) a poor tunnel recombination contact to the ITO.

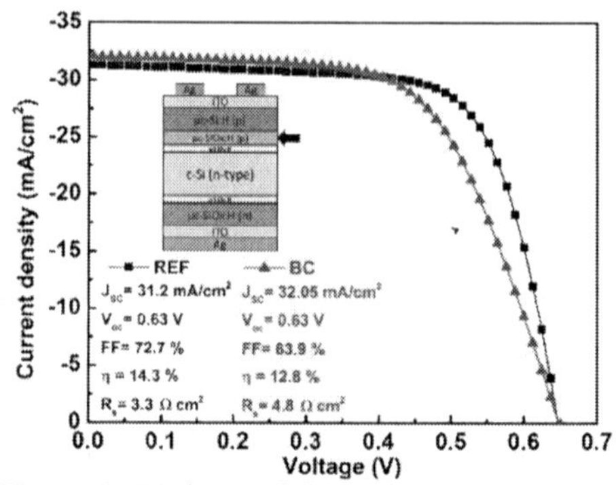

Figure 6: IV-characteristics under AM1.5 illumination and schematic cross-section for SHJ solar cells (after annealing at 300°C) with two different p-type μc-SiOx:H films.

Our structures were fabricated in the same way, except for the p-type μc-SiOx:H film. Therefore, the series resistance losses can be attributed to the increased incubation layer thickness of the p-type μc-SiOx:H film in the BC structure. This increased incubation layer thickness is because the growth rate is higher (1.0 Å/s) than in the REF structure (0.7 Å/s) as deduced from optical modelling of the ellipsometry spectra. After all, the p-type μc-SiOx:H film in the BC structure was deposited at high pressure and high TMB gas flow. Therefore, in our opinion, there must be a tradeoff between the Fc and the conductivity value of the film. Thus, p-type μc-SiOx:H films with similar conductivities but deposited at lower pressure and/or lower gas flow (lower deposition rate) could be needed to reduce the thickness of the incubation layer.

4 CONCLUSIONS

In this work, p-type μc-SiOx:H layers were fabricated with PECVD by varying the RF power, TMB flow rate, and total deposition pressure. This allowed to optimize their crystalline fraction (Fc) in order to obtain a dark conductivity around 10^{-6} S/cm in the as-deposited state and 10^{-5} S/cm after annealing. Surprisingly, Fc was found to decrease after annealing, despite the increase of conductivity. It is concluded that the electrical properties of p-type μc-SiOx:H films are not determined by the crystalline fraction (Fc) but by the conductivity of the amorphous phases (a-SiOx:H and a-Si:H). Furthermore, the films with the best conductivity have been used as p-type material and passivated contact in SHJ solar cells observing that they increase the minority carrier lifetime and due to the conductivity values improve the field effect without degrading the passivation. Although, the solar cell characteristics are limited by poor FF owing to possible low Fc or too thick incubation layer. More interestingly, these structures are improved with annealing up to 300 °C.

ACKNOWLEDGEMENTS

This work was partly funded by the Solar-era.net project "CUSTCO" ANR-19-SOL2-0002-04 of the French Research National Agency (ANR).

REFERENCES

[1] A. Lambertz *et al.*, "Microcrystalline silicon-oxygen alloys for application in silicon solar cells and modules," *Sol. Energy Mater. Sol. Cells*, vol. 119, pp. 134–143, 2013, doi: 10.1016/j.solmat.2013.05.053.

[2] L. Mazzarella, S. Kirner, B. Stannowski, L. Korte, B. Rech, and R. Schlatmann, "P-type microcrystalline silicon oxide emitter for silicon heterojunction solar cells allowing current densities above 40 mA/cm2," *Appl. Phys. Lett.*, vol. 106, no. 2, 2015, doi: 10.1063/1.4905906.

[3] L. Houben, M. Luysberg, P. Hapke, R. Carius, F. Finger, and H. Wagner, "Structural properties of microcrystalline silicon in the transition from highly crystalline to amorphous growth," *Philos. Mag. A Phys. Condens. Matter, Struct. Defects Mech. Prop.*, vol. 77, no. 6, pp. 1447–1460, 1998, doi: 10.1080/01418619808214262.

[4] D. Qiu *et al.*, "Function Analysis of the Phosphine Gas Flow for n-Type Nanocrystalline Silicon Oxide Layer in Silicon Heterojunction Solar Cells," *ACS Appl. Energy Mater.*, 2021, doi: 10.1021/acsaem.1c00654.

[5] S. N. Abolmasov *et al.*, "Substrate and p-layer effects on polymorphous silicon solar cells," *EPJ Photovoltaics*, vol. 5, pp. 0–5, 2014, doi: 10.1051/epjpv/2014007.

[6] M. Chrostowski, J. Alvarez, A. Le Donne, S. Binetti, and P. R. i Cabarrocas, "Annealing of boron-doped hydrogenated crystalline silicon grown at low temperature by PECVD," *Materials (Basel).*, vol. 12, no. 22, pp. 1–8, 2019, doi: 10.3390/ma12223795.

[7] P. Roca i Cabarrocas, J. B. Chévrier, J. Huc, A. Lloret, J. Y. Parey, and J. P. M. Schmitt, "A fully automated hot-wall multiplasma-monochamber reactor for thin film deposition," *J. Vac. Sci. Technol. A Vacuum, Surfaces, Film.*, vol. 9, no. 4, pp. 2331–2341, Jul. 1991, doi: 10.1116/1.577318.

[8] Harland Tompkins Eugene A Irene, *Handbook of Ellipsometry*. William Andrew, 2005.

[9] P. Roca i Cabarrocas, T. Nguyen-Tran, Y. Djeridane, A. Abramov, E. Johnson, and G. Patriarche, "Synthesis of silicon nanocrystals in silane plasmas for nanoelectronics and large area electronic devices," *J. Phys. D. Appl. Phys.*, vol.

40, no. 8, pp. 2258–2266, 2007, doi: 10.1088/0022-3727/40/8/S04.

[10] V. Smirnov, A. Lambertz, S. Tillmanns, and F. Finger, "P-and n-type microcrystalline silicon oxide (μc-SiOx:H) for applications in thin film silicon tandem solar cells1," *Can. J. Phys.*, vol. 92, no. 7–8, pp. 932–935, 2014, doi: 10.1139/cjp-2013-0640.

[11] K. Ding *et al.*, "Impact of doped microcrystalline silicon oxide layers on crystalline silicon surface passivation1," *Can. J. Phys.*, vol. 92, no. 7–8, pp. 758–762, 2014, doi: 10.1139/cjp-2013-0627.

[12] O. Gabriel, S. Kirner, M. Klingsporn, F. Friedrich, B. Stannowski, and R. Schlatmann, "On the plasma chemistry during plasma enhanced chemical vapor deposition of microcrystalline silicon oxides," *Plasma Process. Polym.*, vol. 12, no. 1, pp. 82–91, 2015, doi: 10.1002/ppap.201400114.

[13] and T. N. Tinghui Zhang, Jean-Maxime Orlac'h, Monalisa Ghosh, Vincent Giovangigli, Pere Roca i Cabarrocas, "'Role of H3+ ions in deposition of silicon thin films from SiH4/H2 discharge: modeling and experiments'.," *Plasma Sources Sci. Technol.*, 2021, doi: https://doi.org/10.1088/1361-6595/ac0da2.

[14] J. I. Son, H. J. Nam, and N. H. Cho, "Nanostructural, chemical, and mechanical features of nc-Si:H films prepared by PECVD," *Int. J. Photoenergy*, vol. 2012, pp. 1–9, 2012, doi: 10.1155/2012/643895.

[15] W. Z. GUANG-XU CHENG, HUA XIA, KUN-JI CHEN and and X.-K. ZHANG, "Raman Measurement of the Grain Size for Silicon Crystallites," *Phys. Stat. Sol. A*, vol. 118, pp. K51–K54, 1990.

[16] D. Tuschel, "Stress, Strain, and Raman Spectroscopy," vol. 34, no. September, pp. 1–9, 2019.

This is a short version of the paper submitted to the journal Solar RRL.
It has been selected by the Executive Committee of the EU PVSEC 2021 for submission to Solar RRL.

38th European Photovoltaic Solar Energy Conference and Exhibition

The Empirical Digital Twin: Representation Learning on Solar Cell Images and Efficient Defect Detection with Human-in-the-Loop

Philipp Kunze[1], Stefan Rein[1], Thomas Mueller[1], Marc Hemsendorf[2], Klaus Ramspeck[3], Matthias Demant[1]

[1]Fraunhofer Institute for Solar Energy Systems ISE, Heidenhofstr. 2, 79098 Freiburg, Germany
[2]GP Solar GmbH, Gottlieb-Daimler-Str. 5, 78467 Konstanz, Germany
[3]h.a.l.m. elektronik GmbH, Friesstr. 20, 60388 Frankfurt am Main, Germany

ABSTRACT: Measurement images of solar cells provide information beyond current-voltage characteristics regarding process and material quality in a spatially resolved manner. However, this information is only partially used because algorithms search for human-defined defects and structures. These labels can be inaccurate and incomplete, a relevance in terms of electrical quality is not necessarily given. Thus, we propose a purely data-based approach to derive a comprehensive representation from the measured images that is meaningful in terms of electrical quality and show how it can be used for efficient defect detection. We call this representation the empirical digital twin. For its calculation, we design a convolutional neural network combining multiple measurement images by correlating them with quality variables. The digital twin is an intermediate representation of the network and summarizes quality-related defect signatures that are visible in the images. We show how this representation can be used to derive sorting criteria for quality inspection within an efficient human-in-the-loop approach detecting defects such as finger interruptions, shunts, etc. The human-in-the-loop method not only needs fewer training samples and thus fewer labels but also improved the $F_1 - Score$ detection rate by about 2% on average.

Keywords: Characterization, Deep Learning, Digital Twin, Representation Learning

1 INTRODUCTION

Imaging measurement methods such as electroluminescence (EL), photoluminescence (PL) or infra-red thermography (IR) images are broadly used in production to reveal material and process losses [1–5]. They allow defects such as microcracks, shunts, poor contact formation, etc. to be spatially visualized. Because of this property, they extend the opportunities available through analysis of the current-voltage curve (IV curve) by drawing conclusions about the defect origins and positions.

The interpretation and analysis of the measurement images is challenging even for experts due to overlapping defect structures and high complexity of the data. So far, different approaches have been presented for defect detection:

On the one hand, classical image processing algorithms are developed by experts. These usually combine filtering techniques to extract defect-typical structures and features with machine learning (ML) models such as support vector machines (SVMs) [6] to process those structures and features. For instance, microcracks and finger interruptions in PL images [7,8] or in EL images [9–16] are found by these kinds of techniques.

On the other hand, so-called end-to-end approaches have been pursued in recent years, which are mostly based on convolutional neural networks (CNNs). These are supposed to derive the defects directly from the images. The difference to classical image processing is that here the filters are optimized empirically with a large dataset and not developed by humans. In many works, for example, microcracks, dislocation structures or finger interruptions in EL and IR images are detected using CNNs [17–39].

Both techniques, be it based on image processing or CNNs, have disadvantages in practice limiting their application and making them costly. Especially for CNNs but also for image processing algorithms, annotations of the searched defects are needed before the networks can be trained and evaluated. Firstly, this process of labeling is both time-consuming and costly, as well as error prone. Greulich et al. have shown that even experts cannot reliably find microcracks in EL images and also vary in their detection rate [40]. Secondly, the transferability to other processes is limited, so that re-labeling for CNNs or algorithmic adaptions may be necessary. Here, purely data-based techniques, as for example proposed by Demant et al. and Kovvali et al. regarding as-cut wafer rating, can help overcome these drawbacks [41–43].

Following these ideas, we propose the empirical digital twin, which is learned from measurement images without label overhead, purely data-driven, and can be used for efficient defect detection.

Our contributions are:

1. We present an approach to learn the empirical digital twin describing the quality status of the solar cell by combining the information from multiple measurement images.
2. We utilize the empirical digital twin for efficiently including expert knowledge into defect detection within an iterative "Human-in-the-Loop"-Approach.

This approach, with additional research and explanation, is described in our journal paper [44].

2 APPROACH

We propose a sequence of algorithms for solar cell inspection. First, we derive a meaningful representation of the solar cell from measurement images by semantically compressing measurement images with respect to quality-describing quantities. The representation, which we call empirical digital twin, and its computation is explained in Section 2.1. Following this, we present an efficient way to use the empirical digital twin to incorporate expert knowledge into defect detection. We describe this iterative human-in-the-loop approach in Section 2.2.

2.1 Learning the empirical digital twin

We compute a comprehensive representation of solar cell measurement images by deriving features from them

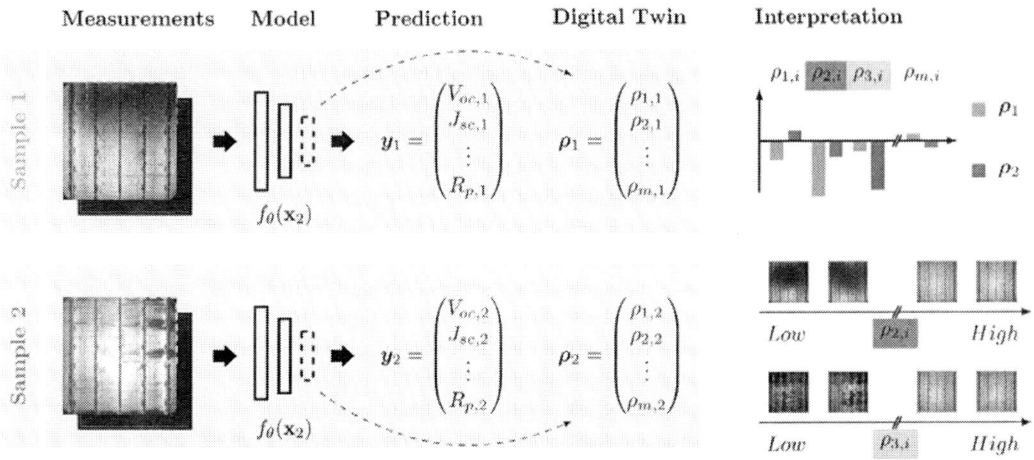

Figure 1: Schematic representation of the calculation of the empirical digital twin for two example cells. The measurements (first column) are passed to a CNN model (second column), which predicts quality variables (third column). The model is used to derive the empirical digital twin (fourth column), whose entries represent quality-describing image features (fifth column).

that are expressive in terms of quality-describing quantities such as IV parameters. The procedure is shown in Figure 1 for two samples (one per row). The measurement images $\boldsymbol{x}_i$ of the i-th sample are given to a CNN model $f_\theta(\boldsymbol{x})$ which predicts the quality parameters in $\boldsymbol{y}_i$, θ being the model's parameters to be optimized.

The CNN consists of a sequence of some convolutional and pooling layers as well as rectified linear unit (ReLU) activation functions. This gradually reduces the spatial resolution of the measurement images while increasing the semantic expressiveness with respect to the target parameters. The last neurons of the network form a vector $\boldsymbol{\rho}_i$ describing the connection between the images and the quality variables. In the following, we call the vector empirical digital twin (EDT).

The EDT offers the possibility to compare measurement images quantitatively. Column 5 in Figure 1 shows the distribution of $\boldsymbol{\rho}_1$ and $\boldsymbol{\rho}_2$ in the form of a bar diagram. Below this, the extreme examples of the dataset studied are shown for features $\rho_{2,i}$ and $\rho_{3,i}$ revealing which image features they represent. Feature 2 (blue) represents poor contact formation during the firing process and feature 3 (yellow) represents finger interruptions. These two process defects occur in the example samples, highlighted by the blue and yellow outlines in the measurement, respectively. Accordingly, these entries in the corresponding EDTs are particularly low as shown in the bar diagram.

In the proposed approach, several different measurement images can be passed to the model and the model can predict various quality parameters. We use EL as well as IR images in conjunction with reflectance values at 390nm and 950nm wavelength, scaled up to match the images' spatial dimensions. In Table 1, columns 1 and 2 show the 18 parameters that the model should predict. These include typical IV parameters such as the open circuit voltage V_{oc}, the short circuit current density J_{sc}, the fill factor FF and the efficiency η, as well as more specific parameters such as the pseudo fill factor pFF, the ideal fill factor FF_0 and their difference $FF_0 - pFF$. We have chosen a large set of parameters so that the EDT can be a comprehensive representation of many properties. However, it is an open question which and how many parameters are best suited for a meaningful EDT.

The EDT vector $\boldsymbol{\rho}$ consists of 720 entries in our implementation. For each of the predicted parameters, 40 entries are provided in $\boldsymbol{\rho}$ within our realization. This adds up to 720 values for 18 parameters. We found this number effective in terms of prediction accuracy, however, it can be optimized in future works.

2.2 Efficient defect detection with Human-in-the-Loop

The digital twin is well suited for quality inspection because it contains the essential measurement image information in a compressed form. Quality-relevant defect structures are already learned indirectly through the regression of the quality parameters. In addition, EDTs have significantly fewer dimensions than their original measurement images and correspondingly less complexity, making it easy to optimize a classifier model based on them.

To use the EDT efficiently and with low label overhead for defect detection, we propose an iterative human-in-the-loop approach, which is schematically outlined in Figure 2. It consists of four steps:

1. The EDTs are calculated for all measurement images of the dataset as described in Section 2.1. From these, a small, representative selection is taken for labeling.
2. This compact dataset is labeled by an expert based on the corresponding measurement images.
3. An ML model based on the labeled compact dataset is trained.
4. The model is applied to the remaining unlabeled dataset and an uncertainty value is calculated for each sample. The most uncertain samples are again given to the expert to be labeled.

Steps 2-4 can be iteratively repeated so that only the most uncertain samples are used for training, since it is assumed that these can give the most information to a classifier model.

The individual steps are described in more detail below. For the representative selection of the of samples for a compact dataset (step 1), we use a k-means [6] with n_{cl} clusters. The n_{cl} samples whose EDTs have the shortest distance to the cluster centroids are selected. These are sorted into defect classes by experts considering the corresponding measurement images (step 2). As a model,

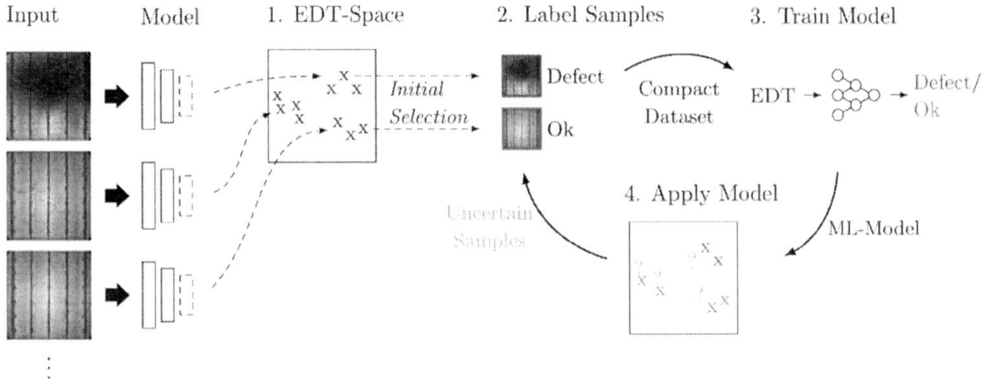

Figure 2: Schematic representation of the human-in-the-loop approach. (1) Measurement images are mapped into the space of empirical digital twins, (2) from which a representative selection based on the measurement images are labeled. (3) Based on the compact dataset, a sorting model is trained. (4) After application to the rest of the dataset, the most uncertain samples are labeled.

we use a small neural classification network (NN) consisting of a few linear layers and ReLU activation functions, which is trained with the n_{cl} EDTs as input and expert labels as targets (step 3).

Regarding the uncertainty, we consider the NN as Bayesian approximation by applying dropout [45] multiple times, as described in [46,47]. The NN is described as $g: \boldsymbol{\rho} \mapsto p_d \in [0,1]$, which maps an EDT $\boldsymbol{\rho}$ to a defect probability p_d. Based on this, we can calculate the entropy H as described in Equation (1):

$$H(\boldsymbol{\rho}) = -p_d \, log_2 \, p_d - (1 - p_d) \, log_2(1 - p_d) \qquad (1)$$

By applying dropout, i.e. with a certain probability some activations of the NN are randomly set to 0 (are dropped), and multiple entering of the same EDT $\boldsymbol{\rho}_i$ to the NN, an entropy distribution results. Equation (2) describes that as uncertainty value U the mean value of this distribution is used:

$$U(\boldsymbol{\rho}_i) = \frac{1}{T}\sum_{t=1}^{T} H(\boldsymbol{\rho}_i) \qquad (2)$$

T describes the number of inputs.

3 EXPERIMENTS

The dataset consists of 1600 Cz-Si industrially processed passivated emitter and rear cells (PERCs) of size $156 \times 156\text{mm}^2$. They were sorted out due to electrical or optical defects. Therefore, they contain various defects such as finger interruptions, microcracks, shunts, edge isolation defects, poor contact formation, etc. Each of the cells was measured by electroluminescence and thermography by a system from h.a.l.m. elektronik GmbH. For the EL measurements, the cells were excited with a current of 20A and the integration time of the Si CCD camera was 50ms. In addition, the reflectance at 390nm and 950nm wavelengths and the parameters from Table 1 were measured.

As CNN we use a variation of the DenseNet [48]. Here, the EL and IR images and the reflectance measurements are first processed in separate CNN branches. Then the results of these are concatenated and processed together. Finally, this path splits to predict the individual quality parameters and compute the EDT. A detailed description of the architecture can be found in [44].

The network is optimized to predict the quality parameters in Table 1, as described in Section 2.1. The input data (EL, IR, and reflectance) are scaled to a size of

Table 1: Predicted parameters, their absolute errors and correlation coefficients with respect to the measured values.

Name	Parameter	Absolute Error	Correlation Coefficient
Open circuit voltage	V_{oc}	0.69 mV	0.93
Short circuit current density	J_{sc}	0.06 mA cm^{-2}	0.62
Fill factor	FF	0.55 %	0.88
Efficiency	η	0.16 %	0.88
Pseudo fill factor	pFF	0.52 %	0.86
Ideal fill factor	FF_0	0.02 %	0.91
Difference: Pseudo & fill factor	$pFF - FF$	0.16 %	0.98
Difference: Ideal & pseudo fill factor	$FF_0 - pFF$	0.53 %	0.86
Suns open circuit voltage	$sunsV_{oc}$	0.65 mV	0.94
Pseudo Efficiency	pη	0.15 %	0.86
Saturation current density D1	J_{01}	0.01 pA	0.96
Saturation current density D2	J_{02}	2.6 nA	0.94
Reverse current density @$-12V$	$J_{rev,1}$	1.05 mA cm^{-2}	0.88
Reverse current density @$-15V$	$J_{rev,2}$	1.15 mA cm^{-2}	0.89
Grid Resistance Front	$R_{grid,fr}$	3.56 Ω m^{-1}	0.93
Grid Resistance Rear	$R_{grid,re}$	1.05 Ω m^{-1}	0.53
Series Resistance	R_s	0.02 Ωcm^2	0.98
Parallel Resistance	R_p	0.29 Ωcm^2	0.83

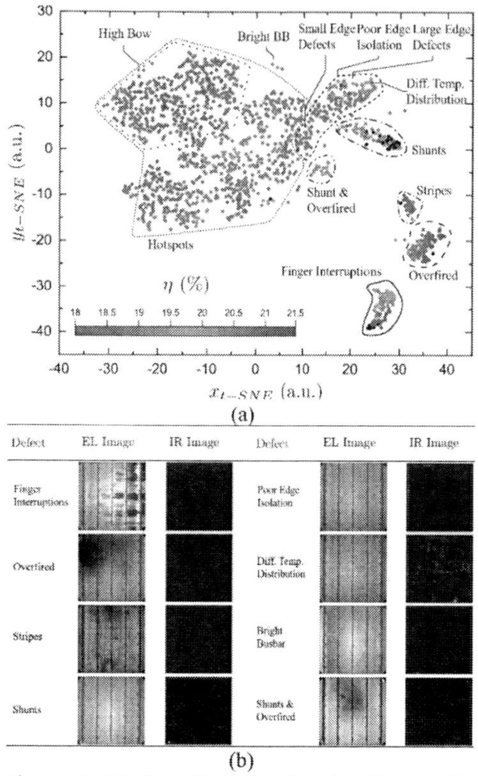

Figure 3: (a) Low-dimensional embedding of the empirical digital twins with cluster annotations and (b) example images for each of the annotated clusters.

$224 \times 224\text{px}^2$, standardized using mean and standard deviation, and randomly rotated and flipped during training. The model is trained on an Nvidia GeForce RTX 2080 Ti GPU for 250 epochs with a batch size of 40 on the training dataset. The adam-optimizer [49] is used with an L1 loss function. The best model is selected based on the validation dataset and tested on the test dataset.

3.1 Analysis of the empirical digital twins

To investigate the meaningfulness of the EDTs, an exploratory analysis of the representation space is performed. For this purpose, all measurements were examined by an expert, so that the main losses of each cell are known. Since the EDTs have 720 dimensions, they cannot be examined directly. Therefore, the t-SNE [50] algorithm is applied for the analysis, which reduces the high-dimensional EDTs to two dimensions. Here, neighboring points in the high-dimensional space are also close to each other in the two-dimensional space. The resulting clusters are compared with the expert labels.

3.2 Defect detection with Human-in-the-Loop

In this experiment, the possibilities of process defect detection and the human-in-the-loop approach are investigated. All cells were sorted according to the defect classes finger interruptions, poor edge isolation, overfired regions and hotspots and shunts by an expert.

Two approaches to defect detection are compared. First, based on the entire labeled dataset, an NN for defect detection is trained with a 5-fold cross-validation. This NN serves as a supervised reference to the human-in-the-loop approach. Here, the same NN is also trained with a 5-fold cross-validation as described in Section 2.2. Starting with $n_{cl} = 100$ samples, in each iteration, the 100 most uncertain samples are added to the training dataset until 1000 samples are reached. Uncertainty is calculated using Equation (2) with $T = 200$. Results are compared using $Precision$, $Recall$ and $F1 - Score$. These quantities are described in the appendix.

4 RESULTS

4.1 Results - Analysis of the empirical digital twins

The quality parameters of the cells can be accurately predicted by the trained CNN. In Table 1, columns three and four show the respective absolute error and correlation coefficient between predicted and measured values. In most cases the correlation coefficient is above 0.8 and sometimes even up to 0.98. Exceptions are the grid resistance on the rear side $R_{grid,re}$ and the short circuit current density J_{sc}, which have lower correlation coefficients of 0.53 and 0.62, respectively.

The EDTs contain information regarding quality-relevant process properties and defects. Figure 3 (a) shows the low-dimensional embedding of the EDTs, the points are colored according to their efficiency. Clusters are formed which contain samples with specific defects. Based on our previous expert labels the groups are bordered and annotated. Typical examples for each cluster are shown in Figure 3 (b).

In Figure 3 (a), four smaller clusters with reduced efficiency are observed on the right side. They contain finger interruptions, overfired regions, and stripe structures and shunts. Example EL and IR images are shown in Figure 3 (b) in the left column. Cells with poor edge isolation and diffuse temperature distribution are positioned in the region above. The example images in Figure 3 (b) in the right column show an increased temperature in the IR image at the edges of the cell, and a larger area of increased temperature, respectively. The rest of the large cluster contains only minor losses, but splits into areas of different bows. In addition, there are two smaller clusters which show a combination of shunts and overfiring, or a radiating region around the busbar contact. Example images of this can be found in Figure 3 (b) in the right column in rows three and four.

Table 2: Detection results of the supervised reference and human-in-the-loop approach with respect to the process defects finger interruptions, poor edge isolation, hotspots & shunts and firing defect.

	Supervised Reference			Human-in-the-Loop		
Process Defect	$F_1 - Score$	$Recall$	$Precision$	$F_1 - Score$	$Recall$	Precison
Finger Interruptions	0.972	0.946	1	0.991	0.982	1
Poor Edge Isolation	0.861	0.812	0.919	0.897	0.874	0.922
Hotspots & Shunts	0.959	0.970	0.949	0.963	0.965	0.961
Firing Defect	0.762	0.643	0.950	0.782	0.689	0.884

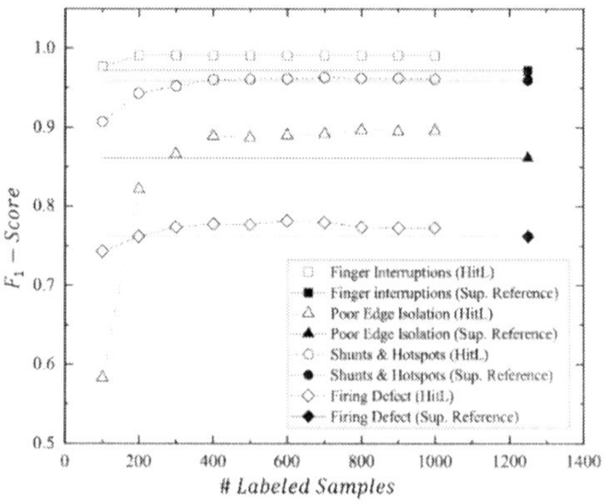

Figure 4: Comparison of $F_1 - Scores$ of the supervised reference (Sup. Reference) and the human-in-the-loop approach with respect to different defect classes.

4.2 Results – Defect detection with Human-in-the-Loop

As a reference to the human-in-the-loop approach, four NNs were trained with respect to the defects finger interruptions, poor edge isolation, hotspots & shunts, and firing defect based on 1250 samples each within a 5-fold cross-validation. They can detect the defects well based on the EDTs. The detection results can be seen in Table 2 in the columns 2-4. In terms of $F_1 - Score$, the best detection results are finger interruptions (0.972), followed by hotspots & shunts (0.959), poor edge isolation (0.861) and firing defect (0.762).

The human-in-the-loop approach can achieve slightly improved results compared to the supervised reference while only needing few labeled samples. In Figure 4, the $F_1 - Score$ is plotted against the number of samples required for the defect classes mentioned. The filled symbols represent the supervised reference (Sup. Reference) and have a horizontal auxiliary line. The empty symbols represent the human-in-the-loop (HitL) approach. Depending on the defect class, this approach achieves the same or slightly better $F_1 - Score$ already with about $200 - 400$ samples.

Figure 4 also shows that the $F_1 - Score$ of the human-in-the-loop approach stabilizes at higher $F_1 - Score$. The corresponding values can be seen in Table 2: Detection results of the supervised reference and human-in-the-loop approach with respect to the process defects finger interruptions, poor edge isolation, hotspots & shunts and firing defect. in columns 5-7. On average, the $F_1 - Score$ can be increased by about 2%.

5 DISCUSSION

It was shown that quality parameters can be derived from EL and IR images as well as reflectance values. The individual influence of the three inputs remains to be investigated. The empirical digital twins of the solar cells, derivable from the CNN, cluster into visible process defects. This indicates that the EDTs are meaningful in terms of the quality parameters and different defects. Thus, they are suitable to compare measurement images quantitatively. It is desirable to investigate the influence of the input data, the predicted quantities, and the size of the EDT in future studies.

The digital twin is suitable for deriving fast and efficient sorting criteria from process defects. The human-in-the-loop method is an effective way to derive classification criteria. With little effort, expert knowledge can be incorporated to derive a classification scheme, enabling user-specific defect detection. Slightly better detection results compared to the reference can be achieved in our experiments. In this context, the label effort was reduced by a factor of 4-6. Due to the low effort, the method promises to detect self-defined defects and process properties. Due to the iterative approach, it can also be trained parallel to the production process.

In addition to the mentioned properties, the digital twin promises high transferability between different cell lines and concepts and could also be suitable for process optimization. Since the measurement images are correlated with measured quality variables, no time-consuming and error-prone human labels are necessary. Therefore, they can be quickly adapted to other cell lines. Furthermore, due to the quantitative representation of the sample with regard to process defects, it is also conceivable to utilize the EDTs for process optimization.

6 CONCLUSION

We introduced the empirical digital twin of the solar cell, which contains quality-describing features with respect to the electrical quality from any number of measured images. For this purpose, a deep neural network was trained to correlate high-dimensional image data with IV parameters, allowing the digital twin to represent features in terms of these quantities. As an example, we combine EL and IR images as well as reflectance measurements to predict in total 18 quality variables. The empirical digital twin is derived from the activations of the network in the form of a vector, which allows a quantitative comparison of one or more measurement images.

The digital twin is suitable for quality inspection. Using a human-in-the-loop approach, we show how expert knowledge can be efficiently incorporated into learning of a classification scheme with 4-6 times less label effort. It was demonstrated that the F1 score of detecting the considered defect types could be increased by about 2% on average. As a result, F1 scores of 0.99 for finger interruptions, 0.96 for hotspots and shunts, 0.90 for edge isolation problems, and 0.78 for inhomogeneous contact formation due to the firing process could be achieved. A detailed and comprehensive investigation of empirical digital twin with additional applications can be found in [44].

7 ACKNOWLEDGEMENTS

This work was funded by the German Federal Ministry for Economic Affairs and Energy within the project "NextTec" (03EE1001A). Philipp Kunze would like to thank the "Studienstiftung des deutschen Volkes" for support within the scope of his scholarship. Furthermore, many thanks to the PV-TEC Team at Fraunhofer ISE for support during the experiments.

References

[1] K. Ramspeck, S. Schenk, D. Duphorn, A. Metz, M. Meixner, In-line thermography for reliable hot spot detection and process control, Energy Procedia 55 (2014) 133–140.

[2] P.K. Kuo, T. Ahmed, H. Jin, R.L. Thomas, Phase-locked image acquisition in thermography, in:

Automated Inspection and High-Speed Vision Architectures II, 1989, pp. 41–47.
[3] T. Fuyuki, H. Kondo, Y. Kaji, T. Yamazaki, Y. Takahashi, Y. Uraoka, One shot mapping of minority carrier diffusion length in polycrystalline silicon solar cells using electroluminescence, in: Conference Record of the Thirty-first IEEE Photovoltaic Specialists Conference, 2005, 2005, pp. 1343–1345.
[4] D. Hinken, C. Schinke, S. Herlufsen, A. Schmidt, K. Bothe, R. Brendel, Experimental setup for camera-based measurements of electrically and optically stimulated luminescence of silicon solar cells and wafers, Review of Scientific Instruments 82 (2011) 33706.
[5] K. Bothe, P. Pohl, J. Schmidt, T. Weber, P. Altermatt, B. Fischer et al., Electroluminescence imaging as an in-line characterisation tool for solar cell production, in: 21st European Photovoltaik Solae Energy Conference, Dresden, 2006, pp. 597–600.
[6] C.M. Bishop, Pattern recognition and machine learning, springer, 2006.
[7] M. Demant, T. Welschehold, M. Oswald, S. Bartsch, T. Brox, S. Schoenfelder et al., Microcracks in silicon wafers I: Inline detection and implications of crack morphology on wafer strength, IEEE Journal of Photovoltaics 6 (2015) 126–135.
[8] I. Zafirovska, M.K. Juhl, J.W. Weber, J. Wong, T. Trupke, Detection of finger interruptions in silicon solar cells using line scan photoluminescence imaging, IEEE Journal of Photovoltaics 7 (2017) 1496–1502.
[9] D. Stromer, A. Vetter, H.C. Oezkan, C. Probst, A. Maier, Enhanced crack segmentation (eCS): a reference algorithm for segmenting cracks in multicrystalline silicon solar cells, IEEE Journal of Photovoltaics 9 (2019) 752–758.
[10] S.A. Anwar, M.Z. Abdullah, Micro-crack detection of multicrystalline solar cells featuring shape analysis and support vector machines, in: 2012 IEEE International Conference on Control System, Computing and Engineering, 2012, pp. 143–148.
[11] D.-M. Tsai, S.-C. Wu, W.-C. Li, Defect detection of solar cells in electroluminescence images using Fourier image reconstruction, Solar Energy Materials and Solar Cells 99 (2012) 250–262.
[12] S. Spataru, P. Hacke, D. Sera, Automatic detection and evaluation of solar cell micro-cracks in electroluminescence images using matched filters, in: 2016 IEEE 43rd Photovoltaic Specialists Conference (PVSC), 2016, pp. 1602–1607.
[13] B. Su, H. Chen, Y. Zhu, W. Liu, K. Liu, Classification of manufacturing defects in multicrystalline solar cells with novel feature descriptor, IEEE Transactions on Instrumentation and Measurement 68 (2019) 4675–4688.
[14] D.-C. Tseng, Y.-S. Liu, C.-M. Chou, Automatic finger interruption detection in electroluminescence images of multicrystalline solar cells, Mathematical Problems in Engineering 2015 (2015).
[15] A.M. Karimi, J.S. Fada, M.A. Hossain, S. Yang, T.J. Peshek, J.L. Braid et al., Automated pipeline for photovoltaic module electroluminescence image processing and degradation feature classification, IEEE Journal of Photovoltaics 9 (2019) 1324–1335.
[16] A.M. Karimi, J.S. Fada, J. Liu, J.L. Braid, M. Koyutürk, R.H. French, Feature extraction, supervised and unsupervised machine learning classification of PV cell electroluminescence images, in: 2018 IEEE 7th World Conference on Photovoltaic Energy Conversion (WCPEC)(A Joint Conference of 45th IEEE PVSC, 28th PVSEC & 34th EU PVSEC), 2018, pp. 418–424.
[17] B. Su, H. Chen, Z. Zhou, BAF-Detector: An Efficient CNN-Based Detector for Photovoltaic Cell Defect Detection, IEEE Transactions on Industrial Electronics (2021).
[18] Y. Jiang, C. Zhao, W. Ding, L. Hong, Q. Shen, Attention M-net for Automatic Pixel-Level Micro-crack Detection of Photovoltaic Module Cells in Electroluminescence Images, in: 2020 IEEE 9th Data Driven Control and Learning Systems Conference (DDCLS), 2020, pp. 1415–1421.
[19] J. Balzategui, L. Eciolaza, N. Arana-Arexolaleiba, J. Altube, J.-P. Aguerre, I. Legarda-Ereño et al., Semi-automatic quality inspection of solar cell based on Convolutional Neural Networks, in: 2019 24th IEEE International Conference on Emerging Technologies and Factory Automation (ETFA), 2019, pp. 529–535.
[20] B. Su, H. yong Chen, P. Chen, G.-B. Bian, W. Liu, others, Deep Learning-based Solar-Cell Manufacturing Defect Detection with Complementary Attention Network, IEEE Transactions on Industrial Informatics (2020).
[21] M. Mayr, M. Hoffmann, A. Maier, V. Christlein, Weakly Supervised Segmentation of Cracks on Solar Cells Using Normalized L p Norm, in: 2019 IEEE International Conference on Image Processing (ICIP), 2019, pp. 1885–1889.
[22] Y. Zhao, K. Zhan, Z. Wang, W. Shen, Deep learning-based automatic detection of multitype defects in photovoltaic modules and application in real production line, Progress in Photovoltaics: Research and Applications 29 (2021) 471–484.
[23] E. Sovetkin, E.J. Achterberg, T. Weber, B.E. Pieters, Encoder-Decoder Semantic Segmentation Models for Electroluminescence Images of Thin-Film Photovoltaic Modules, IEEE Journal of Photovoltaics (2020).
[24] J. Balzategui, L. Eciolaza, N. Arana-Arexolaleiba, Defect detection on Polycrystalline solar cells using Electroluminescence and Fully Convolutional Neural Networks, in: 2020 IEEE/SICE International Symposium on System Integration (SII), 2020, pp. 949–953.
[25] M.R.U. Rahman, H. Chen, Defects inspection in polycrystalline solar cells electroluminescence images using deep learning, IEEE Access 8 (2020) 40547–40558.
[26] P. Kunze, J. Greulich, S. Rein, K. Ramspeck, M. Hemsendorf, A. Vetter et al., Efficient Deployment of Deep Neural Networks for Quality Inspection of Solar Cells Using Smart Labeling, in: Presented at the 37th European PV Solar Energy Conference and Exhibition, 2020 % detection but not with cnn, p. 11.
[27] S. Deitsch, C. Buerhop-Lutz, E. Sovetkin, A. Steland, A. Maier, F. Gallwitz et al., Segmentation of photovoltaic module cells in electroluminescence images, arXiv preprint arXiv:1806.06530 (2018).
[28] M. Alt, S. Fischer, S. Schenk, S. Zimmermann, K. Ramspeck, M. Meixner, Electroluminescence

imaging and automatic cell classification in mass production of silicon solar cells, in: 2018 IEEE 7th World Conference on Photovoltaic Energy Conversion (WCPEC)(A Joint Conference of 45th IEEE PVSC, 28th PVSEC & 34th EU PVSEC), 2018, pp. 3298–3304.
[29] R. Pierdicca, E.S. Malinverni, F. Piccinini, M. Paolanti, A. Felicetti, P. Zingaretti, Deep Convolutional Neural Network for Automatic Detection of Damaged Photovolatic Cells, International Archives of the Photogrammetry, Remote Sensing & Spatial Information Sciences 42 (2018).
[30] A. Bartler, L. Mauch, B. Yang, M. Reuter, L. Stoicescu, Automated detection of solar cell defects with deep learning, in: 2018 26th European signal processing conference (EUSIPCO), 2018, pp. 2035–2039.
[31] Z. Ying, M. Li, W. Tong, C. Haiyong, Automatic Detection of Photovoltaic Module Cells using Multi-Channel Convolutional Neural Network, in: 2018 Chinese Automation Congress (CAC), 2018, pp. 3571–3576.
[32] S. Deitsch, V. Christlein, S. Berger, C. Buerhop-Lutz, A. Maier, F. Gallwitz et al., Automatic classification of defective photovoltaic module cells in electroluminescence images, Solar Energy 185 (2019) 455–468.
[33] H. Chen, S. Wang, J. Xing, Detection of Cracks in Electroluminescence Images by Fusing Deep Learning and Structural Decoupling, in: 2019 Chinese Automation Congress (CAC), 2019, pp. 2565–2569.
[34] C. Dunderdale, W. Brettenny, C. Clohessy, E.E. van Dyk, Photovoltaic defect classification through thermal infrared imaging using a machine learning approach, Progress in Photovoltaics: Research and Applications 28 (2020) 177–188.
[35] A. Ahmad, Y. Jin, C. Zhu, I. Javed, A. Maqsood, M.W. Akram, Photovoltaic cell defect classification using convolutional neural network and support vector machine, IET Renewable Power Generation 14 (2020) 2693–2702.
[36] A. Chindarkkar, S. Priyadarshi, N.S. Shiradkar, A. Kottantharayil, R. Velmurugan, Deep Learning Based Detection of Cracks in Electroluminescence Images of Fielded PV modules, in: 2020 47th IEEE Photovoltaic Specialists Conference (PVSC), 2020, pp. 1612–1616.
[37] J.S. Fada, M.A. Hossain, J.L. Braid, S. Yang, T.J. Peshek, R.H. French, Electroluminescent image processing and cell degradation type classification via computer vision and statistical learning methodologies, in: 2017 IEEE 44th Photovoltaic Specialist Conference (PVSC), 2017, pp. 3456–3461.
[38] M.W. Akram, G. Li, Y. Jin, X. Chen, C. Zhu, X. Zhao et al., CNN based automatic detection of photovoltaic cell defects in electroluminescence images, Energy 189 (2019) 116319.
[39] M.Y. Demirci, N. Beşli, A. Gümüşçü, Defective PV Cell Detection Using Deep Transfer Learning and EL Imaging, Proceedings Book (2019) 311.
[40] J.M. Greulich, M. Demant, P. Kunze, G. Dost, K. Ramspeck, A. Vetter et al., Comparison of Inline Crack Detection Systems for Multicrystalline Silicon Solar Cells, IEEE Journal of Photovoltaics 10 (2020) 1389–1395.
[41] M. Demant, P. Virtue, A. Kovvali, X.Y. Stella, S. Rein, Learning quality rating of as-cut mc-si wafers via convolutional regression networks, IEEE Journal of Photovoltaics 9 (2019) 1064–1072.
[42] M. Demant, P. Virtue, A. Kovvali, X.Y. Stella, S. Rein, Visualizing material quality and similarity of mc-Si wafers learned by convolutional regression networks, IEEE Journal of Photovoltaics 9 (2019) 1073–1080.
[43] A.S. Kovvali, M. Demant, S. Rein, Spatially Resolved Material Quality Prediction Via Constrained Deep Learning, in: 2019 IEEE 46th Photovoltaic Specialists Conference (PVSC), 2019, pp. 3059–3062.
[44] P. Kunze, S. Rein, M. Hemsendorf, K. Ramspeck, M. Demant, Learning an Empirical Digital Twin from Measurement Images for a Comprehensive Quality Inspection of Solar Cells, Solar RRL.
[45] N. Srivastava, G. Hinton, A. Krizhevsky, I. Sutskever, R. Salakhutdinov, Dropout: a simple way to prevent neural networks from overfitting, The journal of machine learning research 15 (2014) 1929–1958.
[46] Y. Gal, Z. Ghahramani, Dropout as a bayesian approximation: Representing model uncertainty in deep learning, in: international conference on machine learning, 2016, pp. 1050–1059.
[47] A. Kendall, Y. Gal, What uncertainties do we need in bayesian deep learning for computer vision?, arXiv preprint arXiv:1703.04977 (2017).
[48] G. Huang, Z. Liu, L. van der Maaten, K.Q. Weinberger, Densely connected convolutional networks, in: Proceedings of the IEEE conference on computer vision and pattern recognition, 2017, pp. 4700–4708.
[49] D.P. Kingma, J. Ba, Adam: A method for stochastic optimization, arXiv preprint arXiv:1412.6980 (2014).
[50] L. van der Maaten, G. Hinton, Visualizing data using t-SNE, Journal of machine learning research 9 (2008).

APPENDIX

If a defect is correctly predicted, it is called true positive (TP), if a defect is incorrectly predicted, it is called false positive (FP). If a non-defect is correctly predicted, it is called true negative (TN), if a non-defect is incorrectly predicted, it is called false negative (FN). Some quantities can be derived from this. The $Precision$ is defined in Equation (3) and indicates how many defect predictions were defects.

$$Precision = \frac{TP}{TP + FP} \quad (3)$$

The $Recall$, defined in Equation (4), is a measure of how many defects were found of all defects.

$$Recall = \frac{TP}{TP + FN} \quad (4)$$

The $F_1 - Score$, defined in Equation (5), is the harmonic mean of Precision and Recall.

$$F_1 - Score = 2\frac{Precision \cdot Recall}{Precision + Recall} \quad (4)$$

3D FINITE ELEMENT MODELLING OF MICRO TRANSFER LENGTH MEASUREMENTS IN CONTACT LAYERS FOR SILICON SOLAR CELLS USING MULTIPHYSICS SIMULATION

Stefan Lange*[1], Mikhail Rumiantcev[1,3], Gao Yiding[1], Gabriel Batista Caldeira[1], Christoph Luderer[2], Alexander Sprafke[3], Volker Naumann[1], Christian Hagendorf[1]
[1]Fraunhofer Center for Silicon Photovoltaics CSP, Otto-Eissfeldt-Str. 12, 06120 Halle (Saale), Germany
[2]Fraunhofer Institute for Solar Energy Systems ISE, Heidenhofstr. 2, 79110 Freiburg, Germany
[3]Martin-Luther-University Halle-Wittenberg, Institute of Physics, Von-Danckelmann-Platz 3, 06120 Halle (Saale), Germany

*corresponding author: phone +49 345 5589-5323 | Stefan.Lange@csp.fraunhofer.de

ABSTRACT: Carrier-selective contacts with low minority carrier recombination and efficient majority carrier transport, i.e. low contact resistivity, play a decisive role in next generation solar cell technologies like silicon heterojunction technology (SHJ), tunnel-oxide passivated contacts (TOPCon), perovskite solar cells or tandem devices thereof. Assessing the contact resistivities between transport layers in thin film systems poses a challenge due to high differences of conductivities of involved layers. The micro transfer length measurement method (μ-TLM) is employed to extract even smallest contact resistivities and also sheet resistances of buried layers in multilayer systems. However, the TLM measurements are not straight forward due to non-homogeneous current densities at the contact pads and within the layers, i.e. current crowding and spreading effects. In this work, we present a finite element model to accurately simulate the current flow in three dimensions. We validate the simulation model by comparison with the analytical model of an ideal single layer system with 1D current flow, where excellent agreement is achieved. Inhomogeneous current injection at the metal pad contacts is studied and deviations of the analytical correction factors accounting for 2D current flow are identified from the simulation, when low contact resistivities on highly conductive substrates are involved and the geometries are not chosen appropriately. Horizontal current spreading due to finite mesa edge insulation is investigated. The simulation model is able to not only qualitatively describe the observed non-linear resistance-distance behavior in SHJ-based μ-TLM samples with insufficient pattern insulation, it can also be used for a simulation-assisted extraction of contact resistivities in an inverse modelling approach even with full 3D current flow. The findings help to understand limitations of TLM and finally allow the evaluation and interpretation of TLM measurements at multilayer stacks with a wide range of geometries and resistivities.

Keywords: Buried Contacts, Heterojunction, Simulation, Laser Processing, Thin Film, TLM, Resistivity Measurement

1 INTRODUCTION

Carrier-selective contacts with low minority carrier recombination (i.e. passivation of surface defect states) and efficient majority carrier transport, i.e. contact resistivity, are necessary to achieve the theoretical efficiency limit of next-generation silicon solar cells or related technologies like perovskite/silicon tandems [1]. Assessing the contact resistivity of involved interfaces is no trivial task and was done in the past by fabrication of discrete test structures for e.g. silicon heterojunction (SHJ) solar cells, where either the same n (p) contact stack is deposited symmetrically on both sides of a n (p) doped silicon wafer or a low-resistive back contact is used, whose contribution is subtracted during data evaluation [2]. This method is easy to perform and the contact resistivity extraction is straight-forward due to the vertical current flow with homogeneous current density distribution perpendicular to the contact areas. The drawback, however, is that this approach is not applicable to fully-processed and finished devices, where resistivities shall be measured independent of the sample backside and the post-processing treatments it received, like plasma exposure or encapsulation. In any case, this method requires a conductive substrate, so it is not applicable to layer stacks with glass substrates often encountered in thin film photovoltaics. Also, it does not enable to measure and observe changes in the one-digit mΩcm² range or even below, since the current transport will be dominated by the c-Si bulk resistivity.

Another method based on transfer length measurements (TLM) at the microscale (μ-TLM) was recently shown by us to provide reliable contact and bulk resistivity values [3]. Thereby, planar contact structures on only one side of the wafer are prepared by means of selective layer removal through ultrashort pulse laser ablation. The so-called current crowding under the metal contacts can be described by the well-known TLM model introduced by Berger and Murrmann/Widmann [4, 5], which basically is a single layer, one-dimensional model. The only allowed in-plane current flow is assumed to be within the conductive bottom layer between the metallic electrodes with all field lines parallel to each other and perpendicular to the metal pads. Vertical current flow is assumed to happen only directly at the interface between metal and conductive layer, represented by the contact resistivity ρ_c.

Shortcomings of this idealized model, when applied to real device structures, were already identified and correction factors based on analytical expressions were proposed, e.g. dealing with finite TLM pattern widths [6] or considering the vertical current flow in the wafer bulk [7]. Apart from that, the microscale dimensions of the structures in combination with non-discrete current paths causes effects such as non-linear resistance-distance relations that do not allow evaluations based on analytical calculations. The 3D model we developed here not only allows a qualitative understanding of the underlying current flow phenomena, it can also be used inversely as electrical parameter (i.e. contact resistivities and sheet resistances) calculation program from experimental data with the real geometries from measurements as input parameters. As such, geometrical influences and correction factors are automatically implemented.

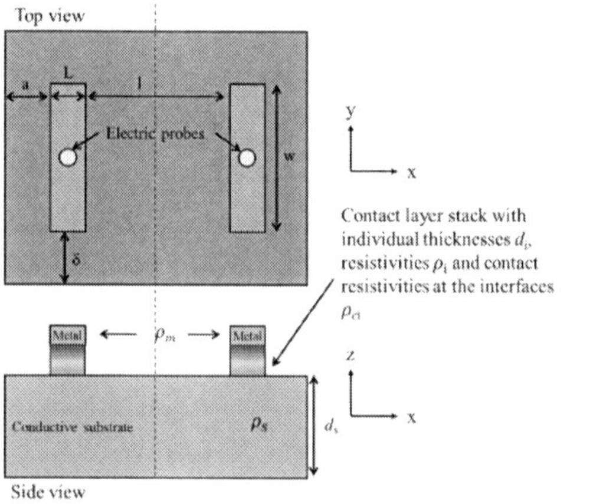

Figure 1: 3D μ-TLM model as implemented in COMSOL. Geometrical and electrical parameters are indicated.

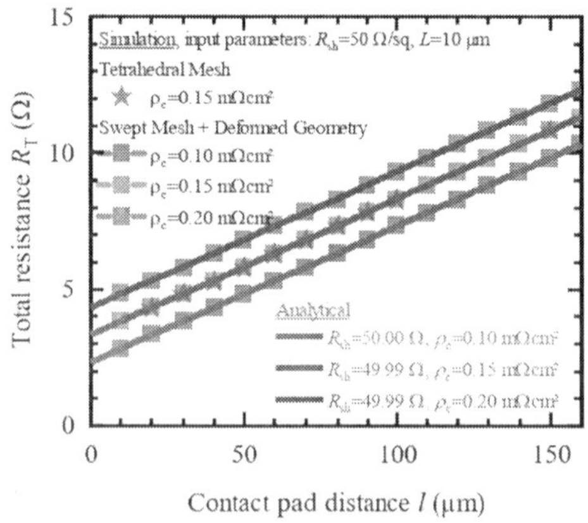

Figure 2: Comparison of analytical and simulation data for an ideal metal/substrate contact.

2 EXPERIMENTAL

2.1 Simulation model

The 3D model based on finite elements was set up in COMSOL Multiphysics, incorporating the AC/DC, the Deformed Geometry and the Optimization modules. The basic model geometry is shown in Figure 1. It consists of two metal contacts (with thickness d_{m} and resistivity ρ_{m}) with dimensions L and w and a separation of l. The current is injected (extracted) in a circular area of 5 μm radius directly in the center of the surfaces of the left (right) contact. The contacts are connected through an arbitrary stack of n layers with a conducting substrate of dimensions $(2\delta + w) \times (2a + 2L + l)$, where a and δ are the mesa insulations in x- and y-direction. Each layer is parameterized by a thickness d_{i} and a resistivity ρ_{i}. The contact resistivities of the resulting interfaces are given by ρ_{ci}, which are manifested by the voltage drop across the (infinitesimally thin) interfaces and modelled by the boundary condition

$$\vec{n} \cdot \vec{J_1} = {}^{1}/_{\rho_c} \cdot (V_1 - V_2)$$
$$\vec{n} \cdot \vec{J_2} = {}^{1}/_{\rho_c} \cdot (V_2 - V_1), \qquad (1)$$

where $\vec{n}$ is the normal vector of the respective interface, $\vec{J}$ is the current density and V the voltage. The subscripts 1 and 2 denote either side of the interface. Voltage probes are placed on both metals, a current of $I = 1$ A was always injected on the left contact, the right contact is grounded. All other boundaries are electrically isolated. The μ-TLM measurement is then simulated by varying the parameter l, which changes the distance between the contacts, and recording the difference of both voltage probes.

2.2 Sample and test structure preparation

300 nm ZnO:Al thin films were magnetron sputtered in pulsed DC mode from ceramic 8-inch targets at room temperature with 300 W power in pure Ar atmosphere on a 10 x 10 cm² glass substrates, followed by 100 nm Al metallization sputtered from an Al target in DC mode with 1000 W power within the same instrument. Furthermore, single-side coated SHJ precursors consisting of metal (400 nm)/ indium tin oxide (ITO) (70 nm)/ n-doped amorphous Si (a-Si) (12 nm)/intrinsic a-Si (6 nm)/n-type crystalline Si (c-Si) (200 μm) were used for experimental verification of our simulation results.

μ-TLM test structures were fabricated by means of ultrashort pulse laser micro machining for selective layer removal. The preparation of the laser trenches was done by ultrashort pulses at a wavelength λ = 532 nm from a picosecond Nd:YAG laser system (neoMOS from neoLASE GmbH), which provides pulses of τ < 10 ps duration at a maximum repetition rate of fmax = 100 MHz and maximum average power of 5 W at this wavelength. A lateral pulse-to-pulse distance of 10 μm was chosen. In addition, a diffractive optical element (DOE) was implemented in the optical setup regarding incoming Gaussian laser beam parameters for a controlled beam shaping to a rectangular intensity profile. For the selective ablation experiments, the ultrashort pulses were focused with a f-theta lens having a focal length of about 255 mm, in connection with a tunable xyz-scanner for positioning and motion control of substrates by dimensions up to 300 x 300 mm².

3 RESULTS

3.1 Optimization of meshing and computation time of ideal system

Finding a good meshing routine and thus reasonable calculation times of μ-TLM structures is challenging due to the high geometrical aspect ratios, $l{:}L \approx 10 \cdots 100$ and especially $l{:}d \approx 10^3 \cdots 10^4$. Therefore, in the beginning, two different routes were tested and evaluated by comparison with the analytical solution of the ideal single layer one-dimensional TLM problem. In the simplest case, the simulated μ-TLM pattern consists of a single conductive layer - in this case a transparent conductive oxide (TCO) - with structured metal pads on top. There is only one interface with the associated contact resistivity ρ_c and the patterns have ideal insulation $a = \delta = 0$. The total resistance R_{T} in dependence of l is given as

$$R_{\mathrm{T}} = \frac{R_{\mathrm{sh}} \cdot l}{w} + \frac{2}{w}\sqrt{R_{\mathrm{sh}} \cdot \rho_c} \cdot \coth\left(L\sqrt{\frac{R_{\mathrm{sh}}}{\rho_c}}\right), \qquad (2)$$

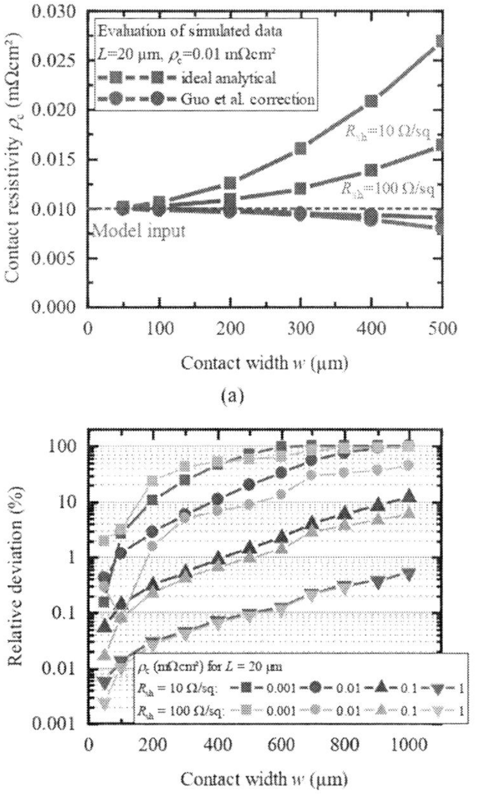

Figure 3: (a) Dependence of contact resistivities on contact width as extracted by the ideal analytical model (squares) and with metal resistivity correction (circles) for a sheet resistance R_{sh} of 10 Ω/sq (black) and 100 Ω/sq (red). (b) Relative deviation between simulated and corrected contact resistivities for different contact resistivities, contact widths and sheet resistances.

where $R_{sh} = {}^{\rho_s}/_{d_s}$ is the sheet resistance of the conductive substrate. Here, the metal resistivity is neglected and there is no voltage drop along the contact pad. The results are shown in Figure 2. Contact resistivities of $\rho_c = 0.10, 0.15, 0.20$ mΩcm² were used as input parameter for the simulation and a sheet resistance $R_{sh} =$ 50 Ω/sq was chosen. The agreement between numerical simulations and analytical calculations are excellent. The simulated data was evaluated with the 1D ideal TLM model by linear regression, which exactly restores the simulation input parameters. The standard meshing technique involves a tetrahedral mesh. The software has to re-mesh the geometry every time l is varied, which is not efficient. Thus, another approach was used for all following investigations where a swept mesh is employed and the increment of l is translated to the time-domain with the Deformed Geometry module. Therefore, re-meshing is not necessary and the calculation is a lot less time-consuming. This method yields the same results as the tetrahedral mesh, see Figure 2.

3.2 Inhomogeneous current injection as a function of contact geometry

In a real scenario, the metal contact pad has a finite

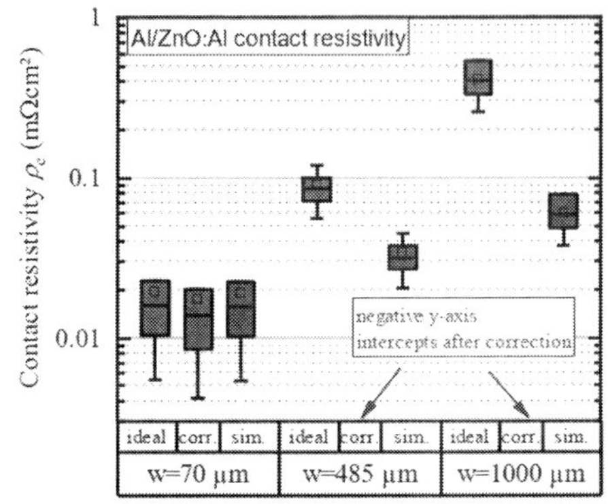

Figure 4: Specific contact resistivities of the Al/ZnO:Al junction evaluated with the ideal analytical expression (black), with the Guo et al. correction (red) and with numerical correction factors (blue) calculated from our COMSOL Multiphysics simulation model.

resistivity $\rho_m > 0$ and typically $a, \delta \neq 0$. It shall be assumed here that the influence of a can be neglected and the current is injected (extracted) in a circular area of 5 µm radius directly in the center of the surfaces of the left (right) contact. Guo et al.[6] proposed a correction factor taking the metal line resistivity into account. However, their model is based on a uniform and parallel current density at any point within the conductive substrate such that basically the whole structure in y-direction is divided into equal slices that start with a different potential on one contact side determined by the voltage drop along the metal. This can clearly be not the case for small R_{sh} and ρ_c, because then there is significant coupling between metal and TCO. In our simulations, for a given metal resistivity $\rho_m = 7 \times 10^{-6}$ Ωcm, metal thickness $d_m =$ 100 nm and contact resistivity $\rho_c = 0.01$ mΩcm², with increasing contact width w this leads to an overestimation of the contact resistivity when the ideal TLM model is used (squares in Figure 3(a)) and to an underestimation if the correction factor is used (circles in Figure 3(a)). The relative deviations of the contact resistivities as obtained by the Guo et al. correction and the simulation input parameters are shown in Figure 3(b). One can see that the deviation is almost negligible for small contact widths $w < 100$ µm or comparably high contact resistivities of $\rho_c > 0.1$ mΩcm². The deviation grows faster for higher contact widths and/or lower contact resistivities. The effect is stronger for lower R_{sh}. The explanation is as follows: if the metal sheet resistance is of the same order of magnitude as the conductive layer sheet resistance and $\sqrt{\frac{\rho_c}{R_{sh}}} \ll L$ and $l < w$, there is good current injection from the metal to the TCO and the current density transfer will be mainly confined to an area with width $< w$. Thus, the uniform current density distribution assumed by Guo et al. is not valid anymore and their model overestimates the metal contribution.

Figure 4 shows experimental results of real Al/ZnO:Al/glass contact resistivities evaluated with different models. As expected, the specific contact resistance calculated using the analytical model (black boxes) increases with increasing contact width. This is not correct because the specific contact resistance should not depend on the geometry of the sample. Applying the Guo

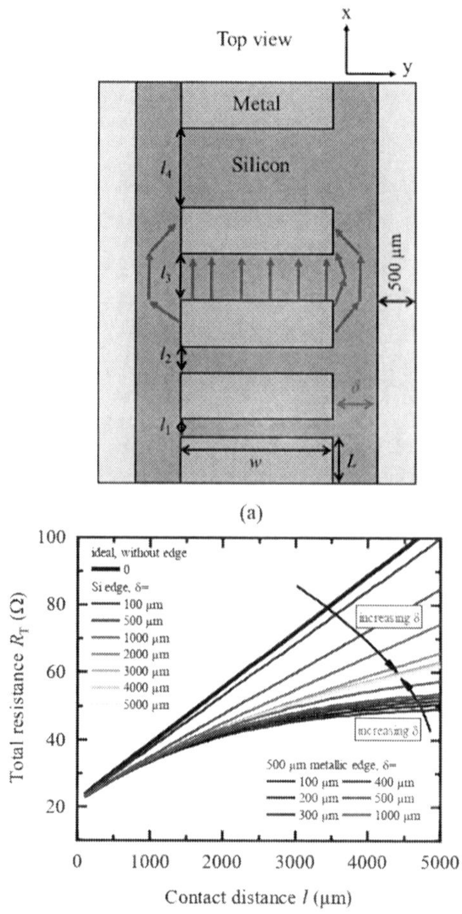

Figure 5: (a) Sketch of the sample model used in the simulation. The non-ideal mesa isolation defined by the parameter δ was varied. (b) Simulated data of μ-TLM measurements in ideal (straight black line) and non-ideal metal/TCO/Si samples. The bluish curves illustrate the case of non-ideal mesa isolation with Si at the pattern edges. The black-to-reddish curves additionally have a 500 μm metal/TCO stripe on top.

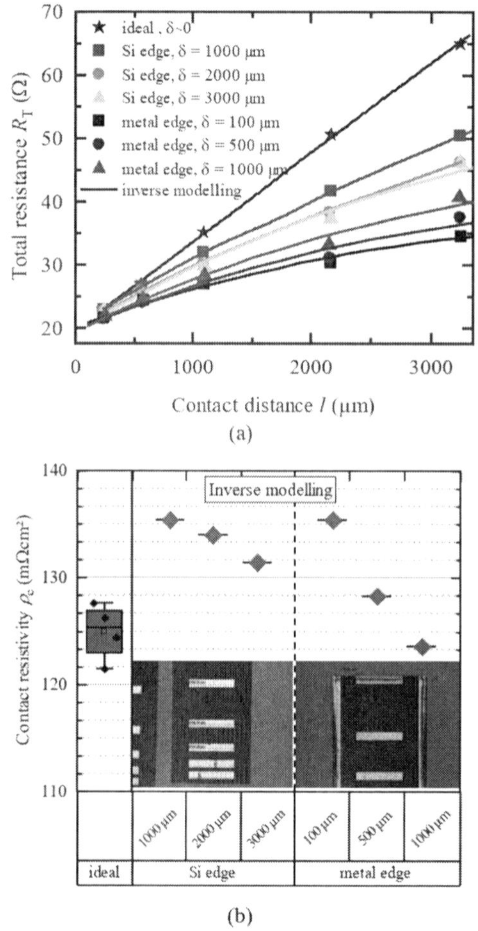

Figure 6: (a) Experimental data points (symbols) of μ-TLM patterns with non-ideal mesa insulation defined by the parameter δ. The data and related geometrical parameters were fed into the FEM simulation for inverse modelling (solid lines). This can be used to estimate the contact resistivity as presented in (b). The insets show pictures of the sample specimen used for the measurements.

et al. correction (red boxes) is only feasible for the w = 70 μm μ-TLM patterns. It yields negative y-axis intercepts for higher contact widths and thus cannot be used to evaluate the contact resistivity. Using numerically calculated correction factors (blue boxes) instead still enables to evaluate the data and to extract reasonable contact resistivities within the measurement uncertainty.

3.3 Horizontal current spreading as a function of mesa conductivity

Consider a SHJ multilayer stack, where the largest contribution to the contact resistivity originates from the TCO/a-Si interface and the intrinsic a-Si layer [8, 9], and only negligible lateral current flow happens in the a-Si layer due to its low conductivity with respect to the TCO or metal. The material system can thus effectively be reduced to consist of metal/TCO/Si layers having metal/TCO and TCO/Si interfacial resistivity contributions. Metal and TCO shall be removed between the contacts, but there remains either (1) a small stripe of Si with width δ at the pattern edges or (2) a 500 μm wide stripe of the full layer stack at the pattern edges, which is separated by the mesa insulation δ from the TLM contacts (see sketch in Figure 5(a)). This situation could happen if ideal wafer breaking of the fabricated μ-TLM structures was not successful. The parameter δ is varied within the same order of magnitude as the contact distances l. Additionally, vertical current spreading in the wafer bulk generally has to be considered, since contact distances are of the same order of magnitude as the wafer thickness. However, this will not be treated in detail here, just note that the impacts of this effect are automatically implemented in the simulation model.

As one can see from the simulated data in Figure 5(b), even in the case of only Si remaining (bluish lines), the edge defined by δ leads to a large deviation compared to the ideal behavior, since the current density field lines are not confined to the μ-TLM pattern anymore. They can spread to the outside and around the contact windows, as indicated by the red arrows in Figure 5(a). This decreases the effective sheet resistance contribution and therefore influences the slope of the μ-TLM graph. The situation gets even worse when one offers a low resistive interface with highly conductive layers at the edges (black to reddish curves). Then there are nonlinearities at even smaller contact distances and the charge carriers tend to stay in the edge regions, eliminating the whole purpose of the TLM methodology.

Samples with non-ideal geometry were prepared to verify the simulation results. The measured μ-TLM data points (symbols in Figure 6(a)) manifest curvatures which agree with the simulated behavior. The experimental data can be used together with the concrete geometrical parameters as input in an inverse modelling approach to obtain material parameters, in particular the contact resistivity. This was done for all sample types and is presented in Figure 6(b). The contact resistivity of the TCO/a-Si interface was determined to be approximately between 120 and 130 mΩcm² from the unperturbed "ideal" μ-TLM patterns. Inverse modelling of the non-ideal structures enables to evaluate the data despite being not linear. The procedure leads to similar values between 124 and 137 mΩcm² and thus proves the validity and applicability of the simulation model.

4 CONCLUSION

A 3D model based on finite elements was set up in COMSOL Multiphysics to simulate μ-TLM measurements. The model was validated with μ-TLM measurements of an ideal single layer system, where an excellent agreement between simulation data and the analytical model was achieved. Two different meshing routines were tested, and a swept mesh has been found to yield superior calculation times.

Inhomogeneous current injection at the metal pad contacts was studied with the simulation model. Deviations of the analytical correction factors accounting for 2D current flow could be identified from our simulation, when low contact resistivities on highly conductive substrates were involved and the geometries were not chosen appropriately.

Additionally, horizontal current spreading due to finite edge insulation has been investigated. Our simulation was able to qualitatively describe the non-linear behavior reported in Lange et al. [3] that one faces, if the TLM pattern insulation is not ideal and charge carriers are allowed to flow around the contact edges. Furthermore, a validation of the simulation results with experimental data was conducted. In conclusion, the model can be used for a simulation-assisted extraction of contact resistivities in an inverse modelling approach even with full 3D current flow in TLM structures of SHJ-type samples, that varies greatly from ideal assumptions.

ACKOWLEDGEMENT

This work was supported by the State of Saxony Anhalt under grant number 1904/00038 ("Neue Halbleiter-Materialien 3") and Fraunhofer lighthouse project "MaNiTU".

REFERENCES

[1] F. Feldmann, M. Simon, M. Bivour, C. Reichel, M. Hermle, and S. W. Glunz, "Carrier-selective contacts for Si solar cells," *Appl. Phys. Lett.*, vol. 104, no. 18, p. 181105, 2014, doi: 10.1063/1.4875904.

[2] C. Messmer, M. Bivour, C. Luderer, L. Tutsch, J. Schon, and M. Hermle, "Influence of Interfacial Oxides at TCO/Doped Si Thin Film Contacts on the Charge Carrier Transport of Passivating Contacts," *IEEE J. Photovoltaics*, vol. 10, no. 2, pp. 343–350, 2020, doi: 10.1109/JPHOTOV.2019.2957672.

[3] S. Lange *et al.*, "Contact and Bulk Resistivity Screening for Advanced Crystalline Silicon Solar Cell Concepts by an Economical and Reliable Transfer Length Measurement Method Based on Laser Micro-Patterning," *Phys. Status Solidi A*, vol. 218, no. 2, p. 2000520, 2021, doi: 10.1002/pssa.202000520.

[4] H. Murrmann and D. Widmann, "Current crowding on metal contacts to planar devices," *IEEE Trans. Electron Devices*, vol. 16, no. 12, pp. 1022–1024, 1969, doi: 10.1109/T-ED.1969.16904.

[5] H. H. Berger, "Contact Resistance and Contact Resistivity," *J. Electrochem. Soc.*, vol. 119, no. 4, p. 507, 1972, doi: 10.1149/1.2404240.

[6] S. Guo, G. Gregory, A. M. Gabor, W. V. Schoenfeld, and K. O. Davis, "Detailed investigation of TLM contact resistance measurements on crystalline silicon solar cells," *Solar Energy*, vol. 151, no. 2, pp. 163–172, 2017, doi: 10.1016/j.solener.2017.05.015.

[7] S. Eidelloth and R. Brendel, "Analytical Theory for Extracting Specific Contact Resistances of Thick Samples From the Transmission Line Method," *IEEE Electron Device Lett.*, vol. 35, no. 1, pp. 9–11, 2014, doi: 10.1109/LED.2013.2290602.

[8] C. Luderer, L. Tutsch, C. Messmer, M. Hermle, and M. Bivour, "Influence of TCO and a-Si:H Doping on SHJ Contact Resistivity," *IEEE J. Photovoltaics*, vol. 11, no. 2, pp. 329–336, 2021, doi: 10.1109/JPHOTOV.2021.3051206.

[9] C. Luderer, C. Messmer, M. Hermle, and M. Bivour, "Transport Losses at the TCO/a-Si: H/c-Si Heterojunction: Influence of Different Layers and Annealing," *IEEE J. Photovoltaics*, vol. 10, no. 4, pp. 952–958, 2020, doi: 10.1109/JPHOTOV.2020.2983989.

"DO NOT BLAME THE BUTTER FOR WHAT THE BREAD DID" OR HOW THE OPTICAL PROPERTIES OF IBC SOLAR CELLS AFFECT THE RESULTS OF SPATIALLY RESOLVED CHARACTERIZATION METHODS

Miha Kikelj[1], Benjamin Lipovšek[1], Matevž Bokalič[1], Florian Buchholz[2], Marko Topič[1]
[1]University of Ljubljana, Faculty of Electrical Engineering
Tržaška cesta 25, SI-1000 Ljubljana, Slovenia
[2]International Solar Energy Research Center Konstanz e.V.
Rudolf-Diesel-Straße 15, D-78467 Konstanz, Germany
E-mail: miha.kikelj@fe.uni-lj.si

ABSTRACT: Interpreting results of spatially resolved characterization methods can prove quite a challenge and it may often happen that a phenomenon takes credit for something it is not responsible for. This is especially important when imaging laterally inhomogeneous devices. We present a derivation, application and validation of a combined opto-electrical model, which sheds light on the roles electrical and optical effects play when it comes to interpretation of EL measurements. We show not only that the device optics and imaging optics cannot be neglected but that they in fact play a major role in shaping the EL measurements.

1 INTRODUCTION

Solar cells are opto-electronic devices with a tightly bound interplay of optical and electrical phenomena. Spatially resolved characterization methods such as electroluminescence (EL) and light beam induced current (LBIC) rely on the assumption that the device under test (DUT) is optically homogeneous in lateral directions. Therefore, when trying to explain measurement results which exhibit spatial inhomogeneities, electrical phenomena usually take the credit. While this may be the case with solar cells with homogeneous back contacts, the situation is vastly different in the case of highly efficient interdigitated back contact solar cells (IBC SCs). Here, the inherent structure with dynamic vertical and particularly lateral variations of materials and textures, can significantly influence the measurement results also from the optical point of view, which could stir up problems and mislead assumptions when it comes to their interpretation.

Close-up EL Image and Profiles of the Zebra IBC

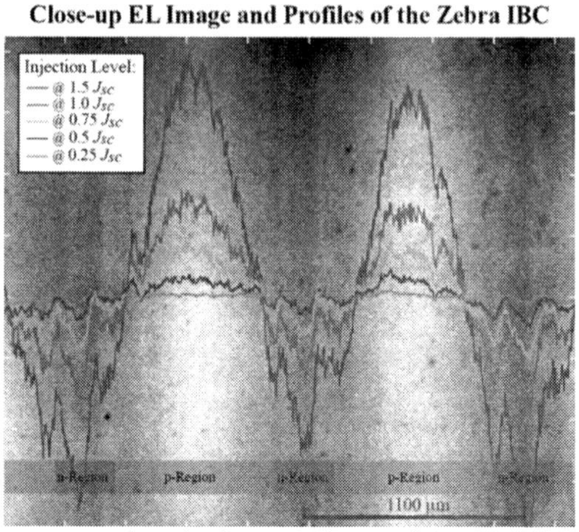

Figure 1: Electroluminescence image of the Zebra IBC cell and the measured EL profiles under different injection levels.

In Fig. 1 one can observe the measured EL profiles of a healthy Zebra IBC [1], [2] solar cell under injection, which exhibit significant lateral fluctuations. In our contribution, we will present a full-3D opto-electrical modelling approach that will shed light on the roles that optical and electrical parameters of the IBC SCs play and how they influence the measurement results obtained by EL imaging. Firstly, we will briefly present the calibration of the electrical model and its use in the combined opto-electrical model. Next, we will describe the derivation and the calibration of the optical model used to simulate the outcoupling of light (in the case of EL) and elaborate on how each distinct geometrical and material feature of the cell influences the outcoupling profiles. As a last step, we will join the two models and evaluate the extent to which each separate part affects the measurement results, namely the EL image of a sample IBC SC under injection. We will demonstrate that neither device optics nor the optical imaging system may be neglected when interpreting the EL measurements of laterally varying structures.

2 MODELLING OF THE IBC DEVICE

2.1 Electrical Modelling

The electrical model of the DUT is based on a drift-diffusion solution of the device's characteristic period in a slightly simplified 2D environment within Sentaurus TCAD [3], where both front and back interfaces are treated as ideally flat, even though in reality they might be textured. The majority of the model's parameters are taken from literature [4]–[7] while dimensions, the doping of bulk, and the doping profiles of the FSF, BSF and the emitter are those of the studied physical Zebra IBC solar cell.

We calibrated the electrical model with the measurements of the solar cell under illumination at standard test conditions (STC), until we achieved a satisfactory match in the key parameter (V_{OC}, I_{SC}, FF, MPP). We then simulated the device at various injection levels, recreating the conditions the EL profiles were measured under, in order to extract the spatial distribution of the radiative recombination rates within the photoactive bulk of the device. These simulated radiative recombination profiles were later used in the combined opto-electrical model to correctly scale the intensities of individual internal light sources of the optical model.

2.2 Optical Modelling

We repurposed our previously developed and calibrated full-3D coupled ray-wave optical model [8] to allow for the simulation of optical outcoupling. In this process, we paid special attention to the interpretation and the effects of the free-carrier absorption [9], [10], realistic reduction of the reflection from the metallic contact fingers [11], and non-negligible light scattering at rear, imperfectly flat interfaces; we found that these three effects have the most influence on the outcoupled optical profile.

To model the light outcoupling profiles which would be easily used in the combined opto-electrical model we established a selection of evenly distributed (spaced 1 μm apart) internal optical sources across the bulk of the device at three major vertical positions: at the back contacts, mid-bulk, and near the top surface. Each optical source was treated as a spherical ray source of approximately 75000 rays with random, but statistically homogeneous ray distribution. Given the number of optical sources this totalled to approximately 42 million starting rays per light profile. The number of rays was chosen as a trade-off between simulation time and noise in the final simulated EL profile.

At this stage we kept the power of each optical source constant and equal to the other ray sources. This allowed us to directly compare the effect of the source position on the outcoupling profile, and secondly it also enabled us to reuse the results in the combined opto-electrical model without the need to recalculate the optical solution each time we updated the electrical model, which in terms changed the profiles of the radiative recombination.

We took note of the most important properties (position, direction, and intensity) of the rays which managed to escape out of the device at the top and bottom interfaces, effectively forming the outcoupled EL profiles of a homogeneous radiative recombination profile. We saved all the aforementioned rays' properties which were later used in post-processing within the combined opto-electrical model.

3 COMBINED OPTO-ELECTRICAL MODEL

In the combined opto-electrical model we seamlessly coupled three aspects which determine the lateral dynamics of the measured EL profile of the IBC cell under injection. Two of them are the aforementioned electrical and optical aspects taken into account in the respective models, while the third is the influence of the optical imaging system.

Because of the rejection of incident light at higher angles [12] due to the camera's limited aperture, only a small amount of incident rays actually reach the sensor. In the case of our imaging setup, all rays, incident at angles above approximately ±2° are rejected. This results in only roughly 1‰ of all simulated rays reaching the sensor and actually contributing to the simulated outcoupling profile. This vastly increases noise of the outcoupled profile but inevitably shapes the profile in a way that cannot be neglected and, therefore, needs to be accounted for also in the scope of device modelling.

To obtain the final results – the simulated EL profile of the IBC solar cell that should accurately recreate the measured one – the three aspects were combined in the following manner: First, the electrical model and the optical model were evaluated separately to obtain (i) the radiative recombination profile and (ii) the optical outcoupling profile of homogeneous radiative recombination, respectively. Because we saved the results of the optical simulations on a source-by-source basis, we were then able to scale each source's contribution according to the actual simulated radiative recombination intensity at its location, thus linking the electrical and the optical models. As the last step, we finally applied the characteristic numerical aperture of the optical imaging system by rejecting rays, incident at angles larger than ±2°.

3.1 Comparison to Measurements of the Zebra IBC SC

We applied the described opto-electrical model to study the electroluminescence profiles of a Zebra IBC cell under injection. We chose an injection of 1.5 J_{SC} as a reference for our simulations because the contrast between light and dark regions of the EL image was the largest in this case, even though the same fluctuations can also be seen at lower injections, albeit they appear less pronounced. Our goal was to systematically investigate how and to what extent each of the aforementioned aspects influences the EL profile of the studied device.

First, in Figs. 2.a and 2.b we show the comparison of only the electrical and only the optical model results to the measurements, respectively. Even though both electrical and optical models' results exhibit fluctuations, which can be easily mapped to the characteristic period of the cell, none of them reliably reproduces or explains the lateral dynamics of the measured EL profile.

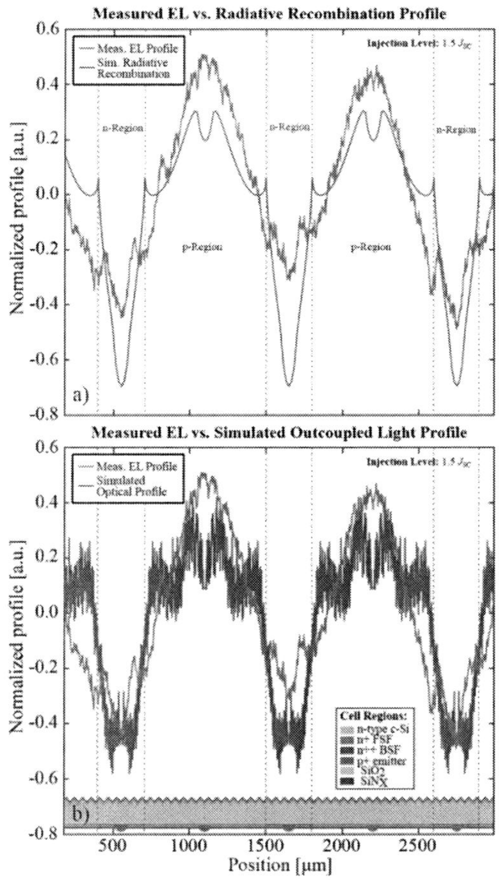

Figure 2: Comparison of the a) Profile of radiative recombination within the bulk, and b) Optical outcoupling profile to the EL measurements at 1.5 J_{SC} injection level. The profiles are mean subtracted and normalized for easier comparison.

We next harnessed the developed opto-electrical model by assuming two ways in which the intensity of the internal sources are determined: first by assuming a homogeneous radiative recombination profile, and second by assuming the actual profile simulated by the optical model. When looking at the results compared in Fig. 3, however, one can observe that the influence of the radiative recombination profile – and thus the electrical phenomena of the studied device – plays only a minor role in contributing to the final EL profile. The only region where it has a significant impact is directly above the n-region contact finger, where there is a noticeable drop in radiative recombination intensity. Apart from that, it is evident that the source of the lateral dynamics of the EL profile is clearly in the domain of the optical effects that determine the outcoupling profile of the generated light.

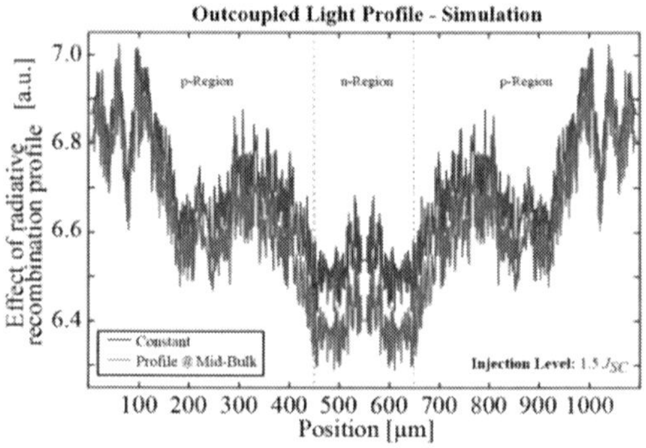

Figure 3: Simulated outcoupling profile – effect of radiative recombination profile.

In the final step, the aspect of the limited aperture of the imaging system was also taken into account, and thus the combined, final EL result determined by joint contributions of all three key players can be seen in Fig. 4. The results accurately recreate the dynamics of the measured EL profile as well all the dips and peaks as seen e.g., around the n-region contact finger and the n-region edges. The noise in the results could be further improved by including more rays in the simulations and increasing the number of simulated optical outcoupling profiles, albeit this was out of scope for the present contribution.

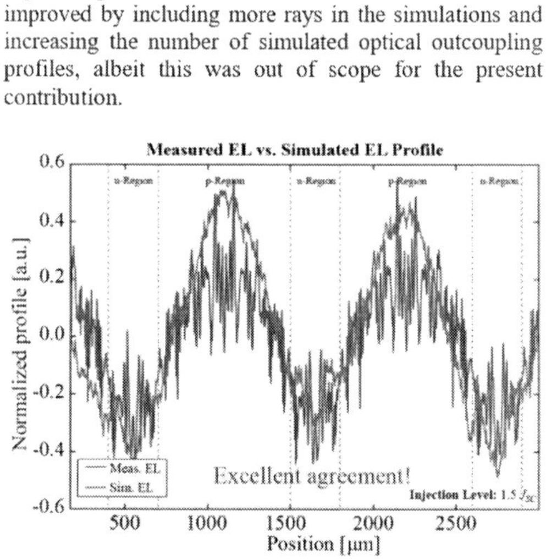

Figure 4: Comparison between the measured EL profile and the EL profile obtained by the presented combined opto-electrical model at 1.5 J_{SC} injection level. The profiles are mean subtracted and normalized for easier comparison.

4 CONCLUSIONS AND OUTLOOK

In this contribution we developed a detailed 3D opto-electrical modelling approach which empowers us with the capability to accurately model complex laterally varying PV structures with unprecedented detail. Using the model, we were able to accurately recreate the experimentally determined EL profile of a state-of-the-art IBC solar cell. We have learned that, counterintuitively, the main aspect, which influences the shape of the EL profile of the DUT are its laterally varying optical properties and *not* the underlying electrical phenomena, which are the usual suspect in interpretation of EL characterization results. This important finding even further emphasises the importance of the presented detailed modelling approach because it showcases the possibility that an electrical phenomenon takes the blame for what is actually of optical origin. With the ability to avoid the pitfall of erroneous result interpretation due to dominant optical phenomena, the model, therefore, proves a valuable tool in further research and optimization of laterally varying highly efficient photovoltaic devices.

5 ACKNOWLEDGEMENTS

The authors thank ISC Konstanz for providing IBC SC samples with a set of simulation parameters The authors acknowledge co-funding from the European Union's Horizon 2020 research and innovation program under grant agreement No 857793 and funding from the Slovenian Research Agency (ARRS) under the research program P2-0197 and project J2-1727. M. Kikelj acknowledges ARRS for his PhD funding.

6 REFERENCES

[1] R. Kopecek *et al.*, 'ZEBRA technology: low cost bifacial IBC solar cells in mass production with efficiency exceeding 23.5%', in *2020 47th IEEE Photovoltaic Specialists Conference (PVSC)*, Jun. 2020, pp. 1008–1012. doi: 10.1109/PVSC45281.2020.9300503.

[2] J. Libal *et al.*, 'The Zebra Cell Concept - Large Area n-Type Interdigitated Back Contact Solar Cells and One-Cell Modules Fabricated Using Standard Industrial Processing Equipment', in *27th European Photovoltaic Solar Energy Conference and Exhibition*, Oct. 2012, pp. 567–570. doi: 10.4229/27thEUPVSEC2012-2AO.2.1.

[3] 'Sentaurus Device'. https://www.synopsys.com/silicon/tcad/device-simulation/sentaurus-device.html (accessed May 14, 2020).

[4] A. Fell *et al.*, 'Input Parameters for the Simulation of Silicon Solar Cells in 2014', *IEEE Journal of Photovoltaics*, vol. 5, no. 4, pp. 1250–1263, Jul. 2015, doi: 10.1109/JPHOTOV.2015.2430016.

[5] P. Procel *et al.*, 'Opto-electrical modelling and optimization study of a novel IBC c-Si solar cell', *Progress in Photovoltaics: Research and Applications*, vol. 25, no. 6, pp. 452–469, 2017, doi: https://doi.org/10.1002/pip.2874.

[6] P. P. Altermatt, F. Geelhaar, T. Trupke, X. Dai, A. Neisser, and E. Daub, 'Injection dependence of spontaneous radiative recombination in c-Si: experiment, theoretical analysis, and simulation', in *NUSOD '05. Proceedings of the 5th International Conference on*

Numerical Simulation of Optoelectronic Devices, 2005., Sep. 2005, pp. 47–48. doi: 10.1109/NUSOD.2005.1518128.

[7] A. Richter, F. Werner, A. Cuevas, J. Schmidt, and S. W. Glunz, 'Improved Parameterization of Auger Recombination in Silicon', *Energy Procedia*, vol. 27, pp. 88–94, Jan. 2012, doi: 10.1016/j.egypro.2012.07.034.

[8] M. Kikelj, B. Lipovšek, M. Bokalič, F. Buchholz, and M. Topič, 'Detailed 3D Optical Modelling of Interdigitated Back Contact Solar Cells', in *2021 IEEE 48th Photovoltaic Specialists Conference (PVSC)*, Jun. 2021, pp. 0997–1000. doi: 10.1109/PVSC43889.2021.9518850.

[9] S. C. Baker-Finch, K. R. McIntosh, D. Yan, K. C. Fong, and T. C. Kho, 'Near-infrared free carrier absorption in heavily doped silicon', *Journal of Applied Physics*, vol. 116, no. 6, p. 063106, Aug. 2014, doi: 10.1063/1.4893176.

[10] M. Nedeljkovic, R. Soref, and G. Z. Mashanovich, 'Free-Carrier Electrorefraction and Electroabsorption Modulation Predictions for Silicon Over the 1–14- um Infrared Wavelength Range', *IEEE Photonics Journal*, vol. 3, no. 6, pp. 1171–1180, Dec. 2011, doi: 10.1109/JPHOT.2011.2171930.

[11] E. Cabrera, S. Olibet, J. Glatz-Reichenbach, R. Kopecek, D. Reinke, and G. Schubert, 'Experimental evidence of direct contact formation for the current transport in silver thick film metallized silicon emitters', *Journal of Applied Physics*, vol. 110, no. 11, p. 114511, Dec. 2011, doi: 10.1063/1.3665718.

[12] M. Bokalic and M. Topic, *Spatially Resolved Characterization in Thin-Film Photovoltaics*. Springer International Publishing, 2015. doi: 10.1007/978-3-319-14651-5.

QUANTITATIVE CONTACTLESS DETERMINATION OF THE SERIES RESISTANCE OF SILICON SOLAR CELLS

H. Höffler[1], W. Wirtz[1], J. M. Greulich[1], S. Rein[1]
[1]Fraunhofer Institute for Solar Energy Systems ISE, Heidenhofstrasse 2, 79110 Freiburg, Germany

ABSTRACT: In this contribution we present a method which *quantitatively* determines the lumped series resistance r_S of silicon solar cells in a *contactless* manner. The method is one essential step on the way to a completely contactless *IV*-measurement of silicon solar cells. The method is based on the evaluation of photoluminescence (PL) signals emitted from the cell. PL signals are recorded in a homogeneously illuminated state and in a partially shaded state. A quantitative value of the lumped series resistance is calculated from the corresponding signal ratio. r_S-values determined via the proposed method are compared to r_S-values determined via a conventional reference method. The comparison is carried out for 20 Cz-PERC solar cells featuring a wide range of r_S-values. A correlation coefficient of 0.87 and a root mean square error of 0.17 Ωcm^2 is obtained. The observed correlation indicates very good agreement.

Keywords: c-Si; Characterization; Photoluminescence; Silicon; Contactless *IV*-Measurement

1 INTRODUCTION

After the production of silicon solar cells in R&D laboratories and production lines, the current voltage curve (*IV*-curve) has to be measured in order to determine the *IV*-parameters efficiency η, short-circuit current j_{sc}, open-circuit voltage V_{oc}, fill factor FF and pseudo fill factor pFF. Conventional *IV*-measurements are usually performed under standard testing conditions defined in the corresponding IEC norm [1]. During conventional *IV*-measurements the solar cells need to be electrically contacted by a suitable contacting unit. Contacting units usually use massive contacting bars with spring-loaded brass pins in case of cells with busbars or brass wires in case of busbarless cells. In any case, the external contacting unit is potentially a source of mechanical damage for the solar cell. Moreover, a certain amount of time is needed to actually bring cell and contacting unit in touch. This amount of time potentially limits the throughput of production lines [2].

In order to circumvent mechanical damage and to increase the throughput of the *IV* tester, a contactless *IV*-measurement would be of great benefit. A contactless measurement of all *IV*-parameters is a highly non-trivial task. Especially the determination of j_{sc} and pseudo fill factor pFF involve certain problems which are not sufficiently solved up to know. Nevertheless, basic ideas are found in literature proposing the contactless determination of V_{oc} [3], pFF [4, 5] and j_{sc} [6]. The determination of these parameters would yield a contactlessly determined pseudo-*IV*-curve. Assuming that the remaining problems can be circumvented, and that a pseudo-*IV*-curve can be measured contactlessly the last missing piece to a completely contactless *IV*-measurement would be the quantitative and contactless determination of the lumped series resistance R_S. This allows to shift the pseudo-*IV*-curve to the actual *IV*-curve, which finally allows the contactless determination of the solar cell's efficiency.

There are different methods which extract the lumped series resistance from *IV*-curves measured in a contacted manner. An overview is given in [7]. The IEC norm proposes to compare the *IV*-curves obtained under dark conditions and at two different irradiations. Another commonly used method which is used as a reference in this work is the comparison of the Suns-V_{oc}-curve and the actual *IV*-curve.

A *qualitative* contactless determination of R_S in a spatially resolved manner has been proposed by Kasemann et *al.* in [8]. The method is based on the idea that partially shaded photoluminescence (PL) images carry information about the local series resistance.

In this contribution, we propose a novel method which determines the lumped series resistance in a *contactless* and *quantitative* way.

2 METHOD

The key idea and the mathematical deviation are laid down in this section.

2.1 Key idea

The key idea of the method is based on the fact that the PL signal of a homogeneously illuminated solar cell ϕ_{hom} and the PL signal in the illuminated part of a partially shaded solar cell ϕ_l are different and that this difference is connected to the cell's series resistance. To visualize that, Figure 1 shows two PL images of the same cell. One has been taken under homogeneous illumination. The other one has been taken at the same illumination intensity, but with the cell being partially shaded during the measurement. f denotes the illuminated area fraction.

It can clearly be seen that the PL signal ϕ_l of the partially shaded image is much lower than the PL signal ϕ_{hom} of the image taken under homogeneous illumination. That arises from the fact that excess charge carriers are flowing from the illuminated part to the shaded part. These excess charge carriers are 'missing' in the illuminated part and do not contribute to radiative recombination. Hence, a lower PL signal compared to the homogeneously illuminated case will be observed. The efficiency with which excess charge carriers flow to the shaded region is determined by the cell's series resistance. We hence expect a quantitative relation between the signal ratio ${}^{\phi_{\text{hom}}}/_{\phi_l}$ and the lumped series resistance r_S.

It should be noted that PL images recorded with a silicon CCD camera and a setup described in [9] are used in this work in order to demonstrate the method in an instructive way. In principle, PL signals can be recorded by a fast IR-sensitive detector (e.g. an InGaAs-diode) which enables integration times of less than 1 ms.

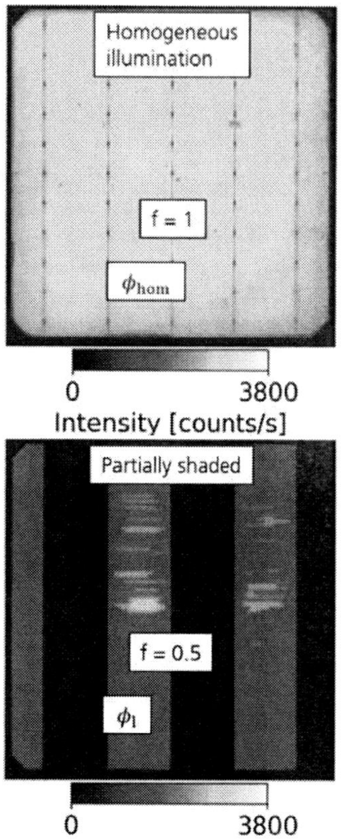

Figure 1: Two PL images of the same PERC cell under the same irradiation. The upper one is taken under homogeneous illumination. The lower one is partially shaded during the measurement.

2.2 Constant-illumination approach

For the quantitative determination of lumped series resistance, the partially shaded scenario is represented by an equivalent circuit according to Figure 2, where the left diode and resistance represent the shaded part and the right resistance, diode and current source represent the illuminated part of the device.

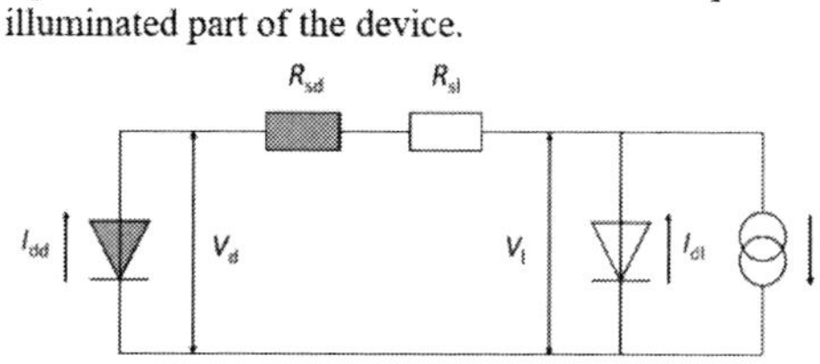

Figure 2: Representation of the scenario of a partially shaded solar cell by an equivalent circuit.

For this equivalent circuit Kirchhoff's voltage law writes as:

$$V_l - I_{dd} \cdot R_{sl} = V_d + I_{dd} \cdot R_{sd} \qquad (1)$$

The illuminated area fraction f is defined as the ration between the illuminated cell area A_l and the total cell area A.

$$f := {}^{A_l}/_{A} \qquad (2)$$

In the next step, the resistances R_{sd} and R_{sl} and the current I_{dd} need to be expressed via the area-related lumped series resistance r_s of the solar cell and the corresponding current density j_{dd}:

$$I_{dd} = j_{dd} \cdot A \cdot (1 - f)$$
$$R_{sl} = \frac{r_s}{A \cdot f} \qquad (3)$$
$$R_{sd} = \frac{r_s}{A \cdot (1 - f)}$$

Note that it might seem counterintuitive that R_{sl} and R_{sd} do *not* add up to the total lumped series resistance R_s if $f \neq 0.5$. Nevertheless, equation (3) describes the correct area-weighting of series resistances. Combining equations (1), (2) and (3) results in an expression for the area related lumped series resistance:

$$r_s = \frac{V_l - V_d}{j_{dd}} \cdot f \qquad (4)$$

Since the voltages V_l and V_d in equation (4) are not accessible via a contactless measurement the general relation between PL signal ϕ and voltage V is needed:

$$\phi = C \cdot e^{V/V_t} \qquad (5)$$

where V_t is the thermal voltage and C is a proportionality constant. The current density j_{dd} is expressed by the voltage V_d using the one-diode equation:

$$j_{dd} = j_0 \cdot e^{V_d/V_t} \qquad (6)$$

The two constants j_0 and C are still not accessible experimentally in a contactless way. In order to eliminate j_0, the homogeneously illuminated case comes into play and we express j_0 via

$$j_o = j_{gen} \cdot \frac{C}{\phi_{hom}} \qquad (7)$$

where j_{gen} is the generation current density during the PL measurement. j_{gen} can be calculated via the impinging photon flux j_γ which is measured by the PL imaging setup and an additional reflectance measurement which is obtained contactlessly. If the irradiation during both PL measurements is equal, we can combine equations (1) - (7) resulting in an expression for the lumped series resistance where all quantities are experimentally accessible in a contactless way:

$$r_s^{con.ill.} = \frac{V_t}{j_{gen}} \frac{1-f}{1 - {}^{\phi_l}/_{\phi_{hom}}} \ln\left(\frac{1-f}{f} \frac{{}^{\phi_l}/_{\phi_{hom}}}{1 - {}^{\phi_l}/_{\phi_{hom}}}\right) \qquad (8)$$

2.3 Constant-signal approach

For the derivation of equation (8), we silently used the assumption that the lateral current flow of excess charge carriers into the shaded part of the cell is the *only* reason for the signal difference between ϕ_{hom} and ϕ_l. Nevertheless, the injection dependence of the effective minority carrier lifetime could also significantly affect the signal difference. In order to circumvent the fact that different injection levels are present during the homogeneously illuminated measurement and the partially shaded measurement, a constant-signal approach is presented. In this approach, the irradiation during the measurement is adopted such that the signals ϕ_{hom} and ϕ_l match. In this case, the difference between homogeneously illuminated and partially shaded case is naturally not found in PL signals but rather in generation current densities j_{gen}^{l} and j_{gen}^{hom} (representing the different irradiation levels). Following an analogous derivation as in section 2.2, we find the lumped series resistance as a function the generation current densities:

$$r_s^{\text{con.sig.}} = \frac{V_t \cdot ln\left(\frac{1-f}{f}\cdot\frac{1}{j_{gen}^{l}/j_{gen}^{hom} - 1}\right)}{j_{gen}^{l} - j_{gen}^{hom}} \cdot (1-f) \qquad (9)$$

3 EXPERIMENTAL RESULTS AND DISCUSSION

We select 20 PERC solar cells fabricated from M2-sized Czochralski grown material. The cells feature a 5-busbar grid layout on the front and full metal coverage on the rear. Every cell is tested in a commercially available *IV*-tester which measures the Suns-V_{oc}-curve and the illuminated *IV*-curve. Every cell is measured in a Modulum PL imaging setup described in [9] which measures the PL signals and the impinging photon flux. PL images and photon fluxes are acquired in a homogeneously illuminated state and in two partially illuminated states using two masks to cover the cell regions between busbar 1 and 2, busbar 3 and 4 and busbar 5 and the right cell edge (see **Figure 1** for visualization) for the first partially illuminated image and cell regions between the left cell edge and busbar 1, busbar 2 and 3 and busbar 4 and 5 (not visualized in **Figure 1**) for the second partially illuminated image. In each state, PL images are acquired at different irradiations. In addition, the injection-dependent minority carrier lifetime of one test sample is measured using the modulated luminescence feature of the Modulum setup.

The reference lumped series resistance (reference r_s) is calculated from the comparison of the Suns-V_{oc}-curve and the illuminated *IV*-curve according to [7]. The data acquired by the PL imaging setup is used to calculate two values for a lumped series resistance determined in a contactless manner (contactless r_s). One is calculated using the constant-illumination approach (equation (8)) and the other using the constant-signal approach (equation (9)). The results for the 20 cells are shown in Figure 3.

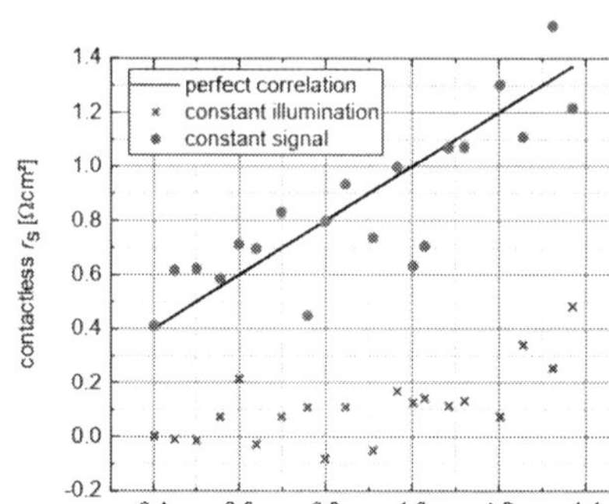

Figure 3: Lumped series resistances of 20 Cz-PERC solar cells. Correlation between reference values and values determined by the two contactless methods proposed in this work.

In the first place the results in Figure 3 show that the constant-illumination method (red crosses in Figure 3) systematically underestimates the lumped series resistances of the cells under test. In some cases, even negative values are obtained, which are obviously artificial. The systematic underestimation can be explained by the injection dependence of effective carrier lifetime. Figure 4 shows one example measurement of an injection-dependent effective carrier lifetime curve of one of the test cells. The green circle indicates the injection level of the partially shaded case. The red cross indicates the injection level of the homogenously illuminated case. A lower minority carrier lifetime is found in the partially shaded case. The luminescence signal ϕ_l in the partially shaded case is hence not only lower than ϕ_{hom} because of the lateral current flow into the shaded region but it is *additionally* lower because of the difference in minority carrier lifetimes at the two states. The ratio ϕ_l/ϕ_{hom} is hence 'smaller than it should be'. According to equation (8) this leads to an underestimation of the value obtained for $r_s^{\text{con.ill.}}$, which is reflected by the data shown in Figure 3.

In contrast to the constant illumination method, values obtained by the constant signal method (green circles in Figure 3) show very good agreement to reference values. The correlation analysis results in a correlation coefficient of $R = 0.84$ and a root mean square error ($RMSE$) of 0.17 Ωcm^2.

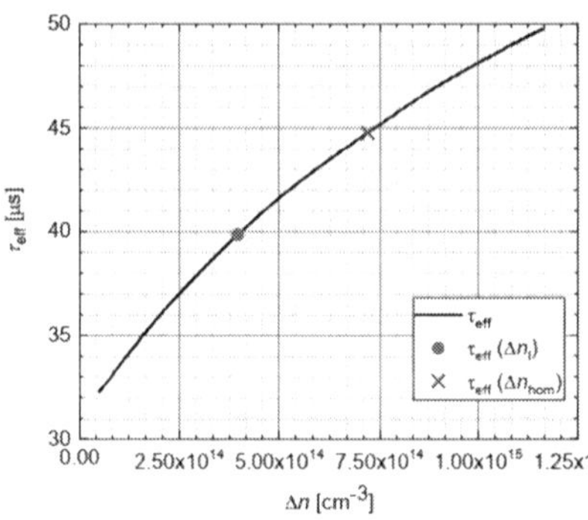

Figure 4: Injection dependent minority carrier lifetime of one sample under test.

4 SUMMARY AND CONCLUSION

A method that quantitatively determines the lumped series resistance of silicon solar cells in a contactless manner has been introduced. Two variations of that method, the constant-illumination and the constant-signal approach, were specified. Both approaches have been applied to 20 test solar cells featuring a wide range of lumped series resistances. The constant-illumination approach led to an underestimation of lumped series resistances. It has been explained and understood that the injection-dependence of minority carrier lifetime is the reason for the underestimation. The constant-signal approach successfully circumvented that problem and led to a very good correlation between reference values and contactlessly determined values of the lumped series resistance. A root mean square error of 0.17 Ωcm^2 was observed when the constant signal approach was applied to the 20 test cells.

5 ACKNOWLEDGMENTS

This work was supported by the German Federal Ministry of Economic Affairs and Energy within the project "NextTec" (03EE1001A).

References

[1] IEC, *Photovoltaic devices - part 3: measurement principles for terrestrial photovoltaic (PV) solar devices with reference spectral irradiance data.*, 2nd ed.: International Electrotechnical Commission, 2008.

[2] K. Ramspeck, S. Schenk, L. Komp, A. Metz, and M. Meixner, "Accurate Efficiency Measurements on Very High Efficiency Silicon Solar Cells Using Pulsed Light Sources," 2014, doi: 10.4229/EUPVSEC20142014-2BV.8.49.

[3] B. Hallam, B. Tjahjono, T. Trupke, and S. Wenham, "Photoluminescence imaging for determining the spatially resolved implied open circuit voltage of silicon solar cells," *J. Appl. Phys.*, vol. 115, no. 4, p. 44901, 2014, doi: 10.1063/1.4862957.

[4] T. Trupke, R. A. Bardos, M. D. Abbott, and J. E. Cotter, "Suns-photoluminescence: contactless determination of current-voltage characteristics of silicon wafers," *Applied Physics Letters*, vol. 87, no. 9, p. 93503, 2005.

[5] B. Michl, D. Impera, M. Bivour, W. Warta, and M. C. Schubert, "Suns-PLI as a powerful tool for spatially resolved fill factor analysis of solar cells," *Prog. Photovolt: Res. Appl.*, vol. 22, no. 5, pp. 581–586, 2014, doi: 10.1002/pip.2293.

[6] A. Paduthol, M. K. Juhl, and T. Trupke, "Addressing limitations of photoluminescence based external quantum efficiency measurements," *J. Appl. Phys.*, vol. 123, no. 2, p. 23105, 2018, doi: 10.1063/1.5004193.

[7] D. Pysch, A. Mette, and S. W. Glunz, "A review and comparison of different methods to determine the series resistance of solar cells," *Solar Energy Materials and Solar Cells*, vol. 91, no. 18, pp. 1698–1706, 2007, doi: 10.1016/j.solmat.2007.05.026.

[8] M. Kasemann, L. M. Reindl, B. Michl, W. Warta, A. Schütt, and J. Carstensen, "Contactless Qualitative Series Resistance Imaging on Solar Cells," *IEEE J. Photovoltaics*, vol. 2, no. 2, pp. 181–183, 2012, doi: 10.1109/JPHOTOV.2012.2184524.

[9] H. Höffler *et al.*, "Review and Recent Development in Combining Photoluminescence- and Electroluminescence-Imaging with Carrier Lifetime Measurements via Modulated Photoluminescence at Variable Temperatures," *37th European Photovoltaic Solar Energy Conference and Exhibition*, 2020, doi: 10.4229/EUPVSEC20202020-2CO.14.2.

SPECTROSCOPIC INLINE CHARACTERISATION OF PARTIALLY PROCESSED EPI WAFERS AFTER POROSIFICATION

S. Al-Hajjawi[1]*, H. Vahlman[1], J. Haunschild[1], S. Kumar[1], H. Schremmer[2], L. Jablonka[3], S. Rein[1]
[1]Fraunhofer Institute for Solar Energy Systems (ISE), Heidenhofstr. 2, 79110 Freiburg, Germany
[2]Meyer Burger (Germany) GmbH, An der Baumschule 6-8, 09337 Hohenstein-Ernstthal, Germany
[3]Nexwafe GmbH, Hans-Bunte-Straße 19, 79108 Freiburg, Germany
Contact email: saed.al-hajjawi@ise.fraunhofer.de

ABSTRACT: With the potential of reduced energy consumption and kerf loss, epitaxial wafer growth is a promising technological innovation compared to the standard ingot growth and wafering techniques. Transferring the technology to mass production requires a fast and robust inline control of the relevant quality parameters in different production stages to ensure high production yield and wafer quality. In this work, we present an inline characterisation method for the quality control after porosification of the substrate wafer. We have integrated an inline spectrophotometer in our automatic wafer inspection system (AWIS) to address this aim. The installed setup measures the reflectance spectra of the wafer surface after electrochemical porosification. Since the following stages of EpiWafer fabrication crucially depend on the quality of the porous layers, monitoring their physical properties is essential and can be achieved by optical modelling of the reflectance spectra. As modelling precision strongly depends on spectrum quality, we made a measurement comparison of the integrated inline spectrophotometer with other inline and offline tools. After appropriate calibration, a qualitative as well as a quantitative correlation with a maximum mean percent error (MPE) of 8% between inline AWIS and offline spectra is found. Over a timespan of almost 11 months, reference measurements with a special porous silicon (PorSi) wafer have been performed. From these results a quality control metric has been defined to ensure automatic reliable results in production.
Keywords: characterization, inline, porosification, EpiWafer.

1 INTRODUCTION

Using the current state-of-the-art wafering and ingot growth techniques, manufacturing the silicon wafer constitutes 60% of the production costs of solar cells [1]. Importantly, 40% of the silicon material is lost as kerf during wafer sawing [2]. As epitaxial wafer (EpiWafer) technology is free from kerf loss, it is superior in terms of both fabrication energy consumption and material efficiency.

Achieving the inherent benefits of the EpiWafer technology requires that these wafers are mass-produced in an inline manufacture environment. Therefore, fast, reliable, and non-destructive inline quality control methods are necessary. One promising approach for this is the inline monitoring of the optical properties of porous silicon (PorSi) layers before epitaxial growth through inline reflectance measurements. Analyzing the optical properties of the PorSi layers is very important because the layer parameters and uniformity are important for subsequent epitaxial layer growth.

Homogeneity, absolute thickness, and porosity of the layers are the key parameters to be controlled by inline characterization. It is noteworthy that ellipsometry has earlier been used to extract such layer parameters [10, 11]. However, ellipsometry is not an inline-compatible method at present. Importantly, the thickness and the porosity of the PorSi layers have earlier been evaluated with laboratory spectrophotometers by analyzing the experimental data based on thin-film optical models, the effective medium theory [12], and Mie scattering [13] [14]. However, inline applicability of this approach is yet to be demonstrated.

In this work, we compare the reflectance spectra of PorSi layers as measured with several different commercially available inline and offline spectrophotometers. Subsequently, we calculate the mean percent error between the different spectra to quantify deviations between the measurement results of the different spectrophotometers. To confirm the stability of the results, the inline spectrophotometer measurements are evaluated for a reference PorSi wafer over a span of 11 months.

Figure 1 shows the manufacturing steps of the EpiWafer. They start with the electrochemical etching process of a highly-doped monocrystalline silicon wafer placed in hydrofluoric acid (HF) electrolyte [3, 4]. The structure of the porous layers depends on the electrical potential, etching time, current density and finally the substrate resistivity [5–8].

In standard fabrication, an etch parameter variation is carried out to yield stacked PorSi layers, where the first one is a high-porosity layer and the second one on top of it is a low-porosity layer. Afterwards, the two porous layers are reorganized during performing a high-temperature annealing process, which causes the pores in the high-porosity layer to further expand and thus form a separation layer. Conversely, the surface of the low-porosity layer reorganizes to a closed surface serving as a template for the epitaxial silicon growth in the subsequent stage.

Finally, the EpiWafer is detached from the silicon substrate. Afterwards, the substrate wafer is reused, maintaining the material cost saving benefit.

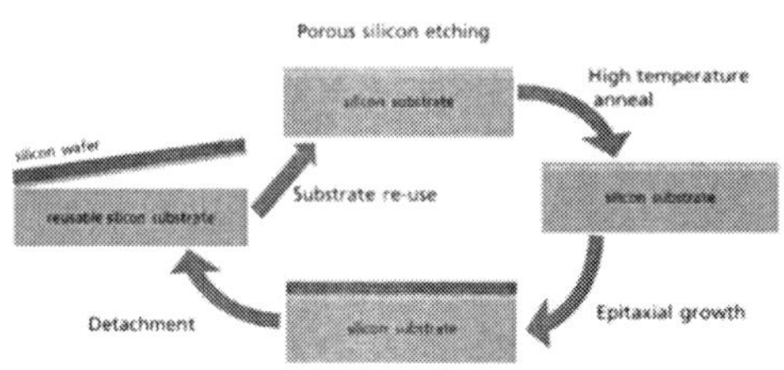

Figure 1: Manufacturing steps of the EpiWafers, extracted from [9].

2 EXPERIMENTAL DETAILS

2.1 Inline measurement setup

A schematic of the inline spectrophotometer setup that is integrated into our automatic wafer inspection system (AWIS) is shown in Figure 2. The spectrophotometer consists of a halogen lamp as a light source, an integrating sphere, and finally two detectors; the first one detects the reflected light from the sample, whereas the second monitors the intensity inside the integrating sphere to eliminate the dependence of the reflectance measurement on the corresponding intensity variations.

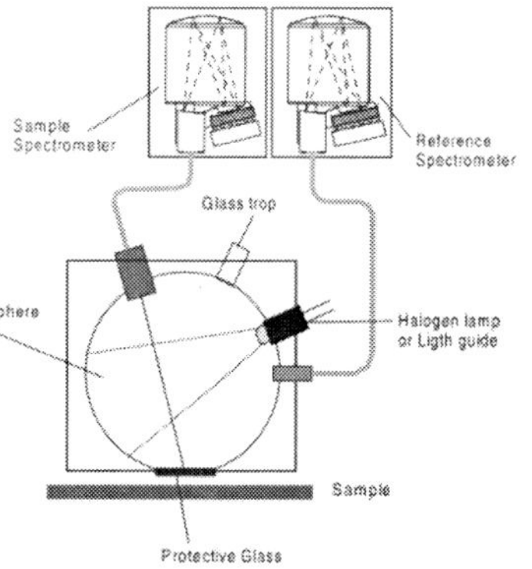

Figure 2. Measurement setup of the inline spectrophotometer [taken from Zeiss product information].

The spectrophotometer setup uses silicon diode array detectors which have 256 pixels and a spectral range of (300-1150) nm. The setup is equipped with a halogen lamp as a light source. It has the advantages that it has a monolithic design in which all components (entrance slit, diode array detector and optical grating) are firmly attached to each other. Furthermore, it provides a high measuring speed, where the wafer is measured on-the-fly and 50 measurement spots on the wafer in transport direction are conducted, i.e., one measurement point every 3 mm for a 156 mm x 156 mm wafer. The number of measurement points is limited by the acquisition time and the belt speed.

Figure 3 shows examples of the spectrophotometer measurements of a Cz-Si as-cut wafer and a PorSi wafer, respectively. The reflectance spectrums at spots 10, 30 and 50 on both wafers are plotted. A strong noise in the UV range of the reflectance spectrum is seen as the intensity of the excitation light decreases below 450 nm. Above 950 nm, the sensitivity of the silicon detector material used decreases sharply, which also results in strong noise in that spectral range in both wafers. In the as-cut wafer (Figure 3(a)), the light is absorbed in the wavelength range of (400-1100) nm. Texturing and antireflection coating will further improve the reflection values. In the PorSi wafer (Figure 3(b)), thin film interference causes the oscillations in the reflection spectra at the three different spots.

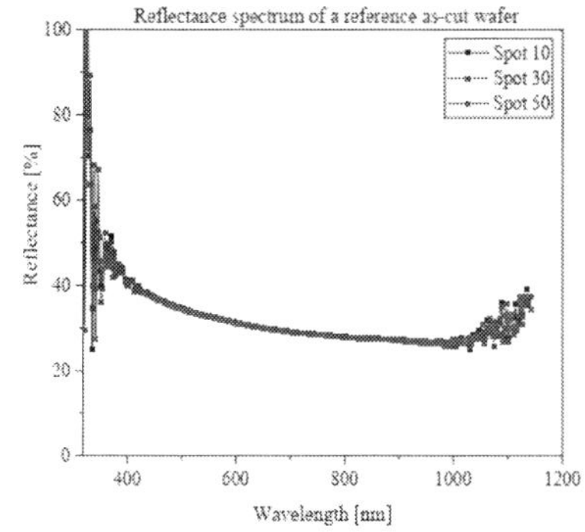

a)

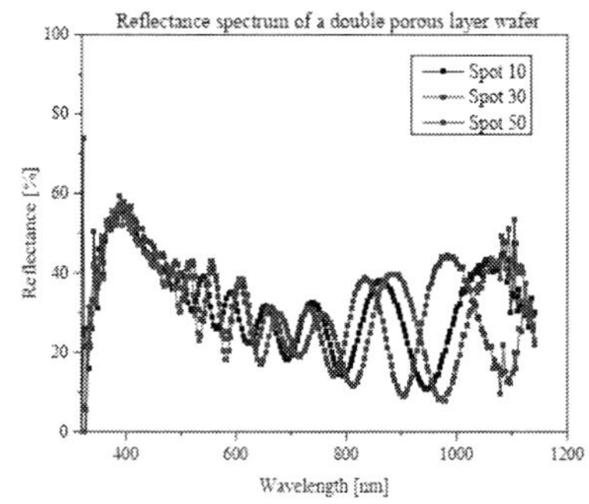

b)

Figure 3 An example reflectance spectrum of: (a) A Cz-Si as-cut wafer. (b) A substrate wafer after porosification. For each wafer, 50 measurement spots are conducted in the spectral range of (300-1150) nm. The as- cut wafer does not show oscillations, unlike the PorSi wafer where such oscillations are caused by optical interference within the porous layer stack.

2.2 Inline and offline Spectrophotometers

To evaluate the quality of the inline spectrophotometric measurements, the reflectance spectra of the inline device are compared to other spectra recorded with three different spectrophotometers: the offline Cary, the offline R-SCAN and finally the inline Zeiss installed in our cell tester.

The offline Cary has the largest spot diameter of 37 mm, where it covers a wide range of wavelengths and has the highest spectral resolution resolution as well as the least noise.

The offline R-SCAN has a small spot diameter of 5 mm and is used for automatic reflection mapping of wafers.

Finally, the inline Zeiss installed in the cell tester has a spot diameter of 30 mm and allows inline reflection measurements of solar cells. Measurements are done at a high resolution of 2 nm in the wavelength range of (360 – 1140) nm. Table 1 shows the specifications of the above-mentioned spectrophotometers.

Table I: Technical data of the offline and inline spectrophotometers used in this investigation.

Name	Spot diameter [mm]	Wavelength Range [nm]	Resolution [nm]
Cary 5000 UV-Vis (Agilent Technologies)	37	250 – 1200	1
R-SCAN (pv-tools GmbH)	5	300 – 1100	5
AWIS inline spectrophotometer (Meyer Burger)	28	300 – 1150	3.3
Cell tester inline spectrophotometer (Zeiss ThinProcess)	30	360 - 1140	2

3 RESULTS

3.1 Comparison of spectrophotometers

Figure 4 shows a comparison between the different tools using a reference Cz-Si as-cut wafer. A clear correlation between AWIS, Cary and R-SCAN is seen in the span of 350-1000 nm. The measurement spot size has a key impact on the comparison as well because averaging over neighboring areas can alter the results if the wafers are inhomogeneous. Conversely, the inline cell tester shows a deviation in the amplitude of reflectance spectrum. The cell tester spectrophotometer is calibrated for coated wafers, which have reflection values less than 10%: This could explain the deviation seen in the inline cell tester compared to other tools.

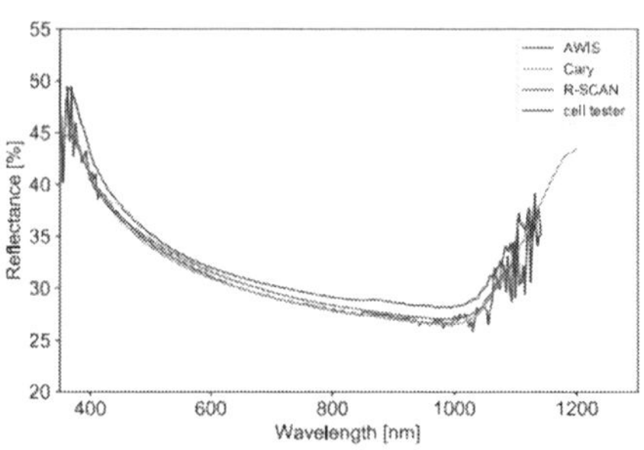

Figure 4. Reflectance spectra measured with the inline spectrophotometer of AWIS, the spectrophotometer of the inline cell tester, the Cary and finally the R-SCAN for a reference as-cut wafer. A good correlation in the wavelength span of 350-1000 nm is seen.

Figure 5(a) shows reflectance spectra of a single-layer PorSi wafer as measured with the different tools used in this work. This wafer has only a single low porosity layer. It is noteworthy that there is a slight difference between the reflectance spectrum of Cary compared to the other spectrometers, which may be explained by a larger measurement spot size and consequent averaging over a larger area in the former case. The difference could also derive from a differing measurement position. Furthermore, Thin film interference has an impact on the reflection spectrum [10]. Figure 5(b) shows the reflectance spectra of a PorSi wafer with two porous layers, and in this case a larger deviation than in Figure 5(b) is observed.

(a)

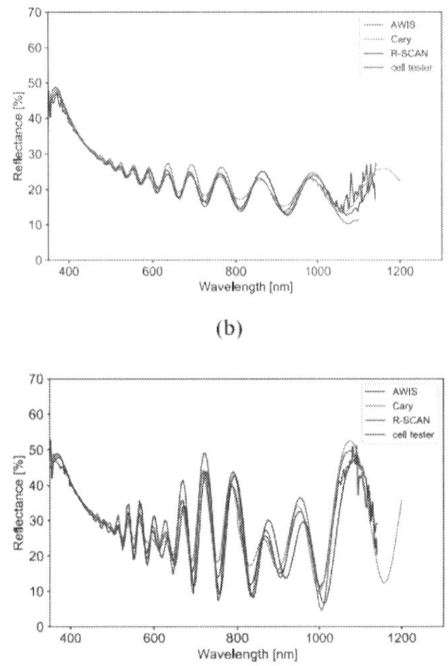

(b)

Figure 5. Reflectance spectra measured with the AWIS inline spectrophotometer, inline spectrophotometer of the cell tester, Cary, and R-SCAN for: (a) A PorSi wafer having only a single low porosity layer. (b) A PorSi wafer that has a double stack of low and high porosity layers. A good correlation is seen, especially between AWIS inline spectrophotometer and the reference laboratory spectrophotometer Cary.

Based on the above comparison between the various tools, the inline spectrophotometer integrated within AWIS shows good agreement with the laboratory reference tool Cary.

To examine the two PorSi wafers used above further, contour plots of the reflectance at a wavelength of 800 nm as measured by the R-SCAN are displayed in Figure 6. Due to the non-uniformity of the reflectance distribution on these PorSi wafers, choosing the same measuring point for comparison between different spectrophotometers becomes more significant.

To quantify the deviation error of the reflectance values between all tools and our reference high-precision laboratory spectrophotometer Cary, we calculate the mean percent error (MPE) between each tool and the offline reference tool Cary. Note that an interpolation is required to enable a quantitative comparison at the same wavelengths for all the tools.

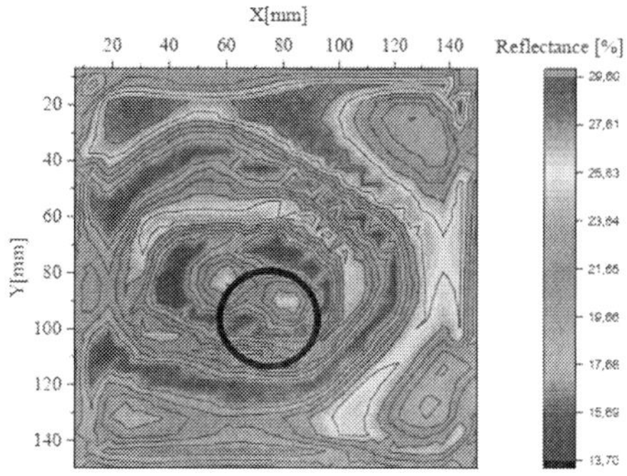

(a)

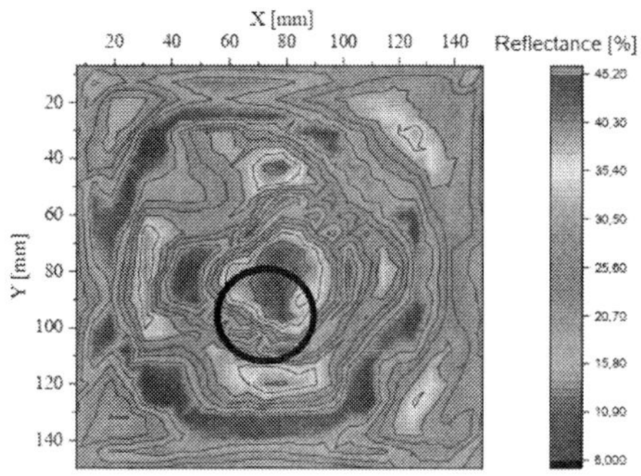

(b)

Figure 6 R-SCAN reflection maps at a wavelength of 800 nm for: (a) The PorSi wafer with one porous layer. (b) The PorSi wafer with two porous layers. The circle shape represents the approximate position of measurements. Due to the non-uniform reflection maps of such wafers, it could be concluded that choosing the exact measurement point is crucial for the comparison of different spectrophotometers, especially with considering the different measurement spot sizes of these tools

Figure 7 shows the MPE for the reference as-cut wafer, the PorSi wafer with one layer and finally the PorSi wafer with two layers. In all wafers, the AWIS inline spectrophotometer shows the best correlation with the Cary (i.e. quantified by the lowest MPE values of 1 % for the reference as-cut wafer). A MPE value of 4% of is calculated for the PorSi wafer with one layer. Finally, a MPE value of 8% is recorded for the PorSi wafer with a stack of two porous layers.

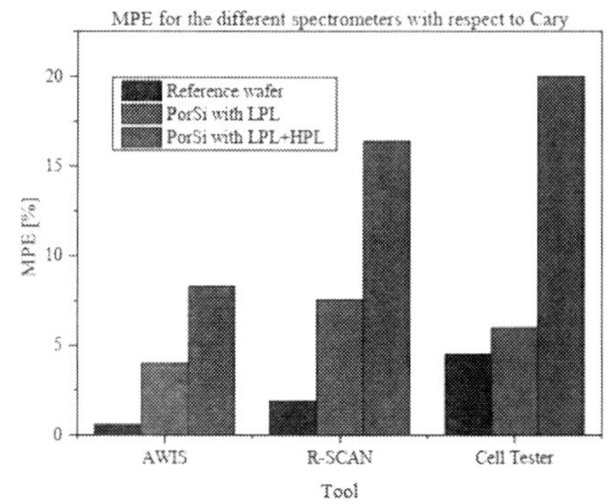

Figure 7. MPE for the different offline and inline spectrophotometers with respect to the reference offline spectrophotometer (Cary) for three different wafers.

3.2 Inline reflectance measurements for PorSi wafers from production

To quantify the stability of the inline spectrophotometer, regular calibration and monitoring is needed. To address this call, calibration of the inline spectrophotometer is done before each PorSi reflection measurement, where a reference PorSi wafer is measured at the middle point at a wavelength of 600 nm after each calibration. This regime was applied repeatedly over the course of almost 11 months with the inline spectrophotometer, resulting in 74 measurements.

Figure 8 presents the spectra at the middle of the reference PorSi wafer at different days. It could be clearly seen that some spectra deviate. Apart from the error in measurement position, such deviations can be caused by forgotten or improper calibration or by changes in the environmental conditions.

a)

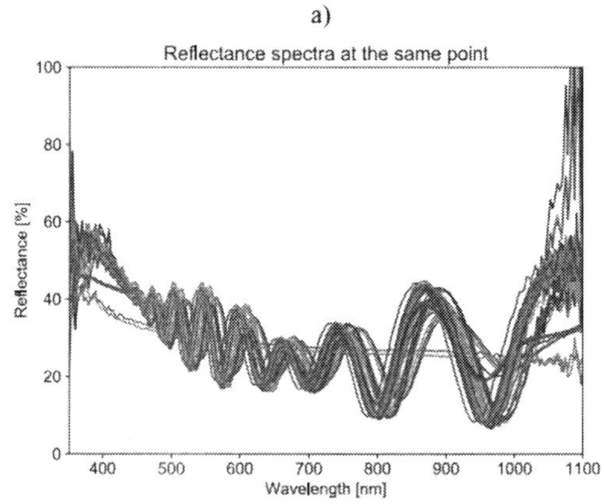

Figure 8. The reflectance spectra at the same point of a reference PorSi wafer at a wavelength of 600 nm each time after calibration during almost 11 months.

To improve the accuracy of measurements and define a proper quality control approach, three regions are defined for accepting reflectance values at 600 nm after each reference measurement, where the accepted values are within the green region. Figure 9 shows the 74 reference measurements after defining the quality control regime. Only 12 measurements are in the red region. A standard deviation of only 0.5% is calculated for measurements within the green region.

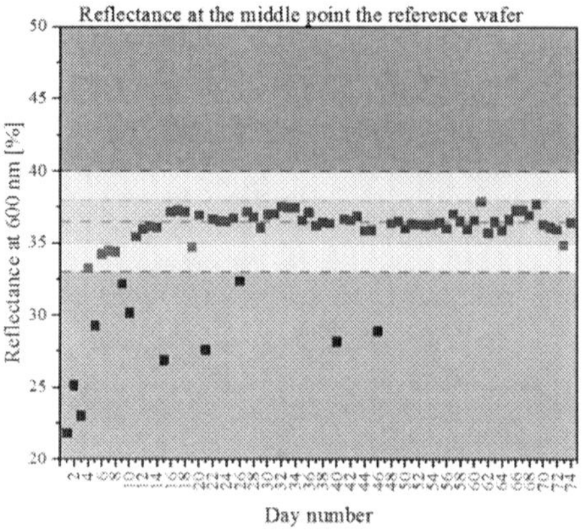

Figure 9. The reflectance values at the middle point of a reference PorSi wafer at 600 nm in the last 11 months after calibration.

Figure 10 represents the reflectance spectra in the defined green region. The aim is to maintain a proper calibration and ensure that the reference wafer reflection values are within the accepted green region.

Accordingly, and for future measurements, a reflection value of the reference wafer at 600 nm should always lie in the green region before conducting PorSi reflection measurements from production, leading to more precise results.

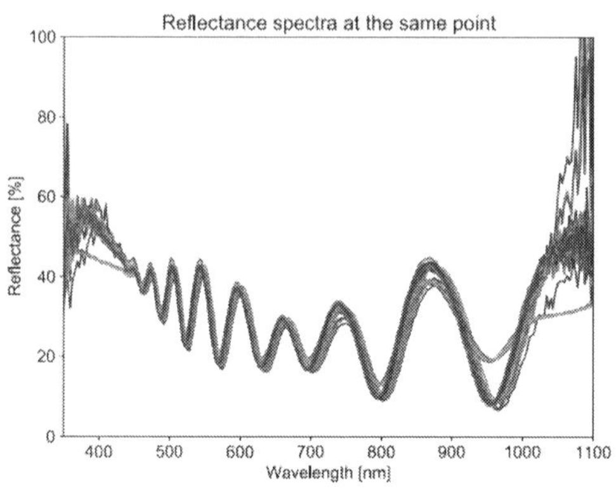

Figure 10. Reflectance values at the middle point of a reference PorSi wafer at 600 nm in the last 11 months after calibration in the defined green region.

6 CONCLUSIONS

Inline reflectance spectroscopy is a promising method for the quality control of porous silicon wafers used for epitaxial Si wafer production. For this, accurate and stable inline measurement of reflectance spectra will be required. To support this goal, this work introduces a fast, accurate and stable device that is usable for industrial application.

Comparing the reflectance spectra of a porous silicon wafer between inline and offline spectrophotometers shows a good quantitative correlation with a maximum percent error (MPE) of 8% between the best inline device and a laboratory spectrophotometer used as a reference device. After monitoring the reflectance values at a wavelength of 600 nm for a reference porous silicon wafer over a timespan of almost 11 months, a quality control approach for evaluating the stability of the inline spectrophotometer is implemented. 76% of the measurements are within the green region, with a standard deviation of only 0.5%.

For future reflection measurements, the reflection value of the reference PorSi wafer at 600 nm should lie in the green region to ensure a precise measurement of the other PorSi wafers from production afterwards.

ACKNOWLEDGEMENTS

This work was supported by the German Federal Ministry for Economic Affairs and Energy BMWi and by the industry partners within the research project Epi-Inspec under contract No. 03EE1043C. The authors would like to acknowledge this support.

REFERENCES

[1] K. Hachichi *et al.*, "Silicon Recovery from Kerf Slurry Waste: a Review of Current Status and Perspective," *Silicon*, vol. 10, no. 4, pp. 1579–1589, 2018, doi: 10.1007/s12633-017-9642-x.

[2] S. A. Sergiienko, B. V. Pogorelov, and V. B. Daniliuk, "Silicon and silicon carbide powders recycling technology from wire-saw cutting waste in slicing process of silicon ingots," *Separation and Purification Technology*, vol. 133, pp. 16–21, 2014, doi: 10.1016/j.seppur.2014.06.036.

[3] A. Uhlir, "Electrolytic Shaping of Germanium and Silicon," *Bell System Technical Journal*, vol. 35, no. 2, pp. 333–347, 1956, doi: 10.1002/j.1538-7305.1956.tb02385.x.

[4] Dennis R. Turner, "Electropolishing Silicon in Hydrofluoric Acid Solutions," *Journal of The Electrochemical Society*, vol. 105, no. 7, p. 402, 1958, doi: 10.1149/1.2428873.

[5] P. Kumar and P. Huber, "Effect of Etching Parameter on Pore Size and Porosity of Electrochemically Formed Nanoporous Silicon," *Journal of Nanomaterials*, vol. 2007, pp. 1–4, 2007, doi: 10.1155/2007/89718.

[6] D. Q. Liu and D. J. Blackwood, "The Role of the Flat-Band Potential in Porous Silicon Formation," *J. Electrochem. Soc.*, vol. 159, no. 12, H909-H911, 2012, doi: 10.1149/2.022212jes.

[7] S. Lust and C. Lévy-Clément, "Chemical Limitations of Macropore Formation on Medium-Doped p-Type Silicon," *J. Electrochem. Soc.*, vol. 149, no. 6, C338, 2002, doi: 10.1149/1.1475688.

[8] V. Lehmann, R. Stengl, and A. Luigart, "On the morphology and the electrochemical formation mechanism of mesoporous silicon," *Materials Science and Engineering: B*, 69-70, pp. 11–22, 2000, doi: 10.1016/S0921-5107(99)00286-X.

[9] Nena Milenkovic, "Kerfless Wafering: Porous Silicon Reorganization and Epitaxial Silicon Growth," PhD Thesis, university of Konstanz, university of Konstanz, konstanz, 2016.

[10] Wikipedia contributors, *Thin-film interference*. [Online]. Available: https://en.wikipedia.org/w/index.php?title=Thin-film_interference&oldid=1010382555 (accessed: Sep. 2 2021).

DETERMINATION OF EFFECT OF LASER CUTTING METHOD ON THE PERFORMANCE OF BIFACIAL PERC CELLS

*Melis Çetmeli[1], Betül Şekertekin[1], Meriç Çalışkan[1], Fırat Es[1]
[1] KalyonPV Research and Development Center, Kalyon Güneş Teknolojileri Üretim A.Ş., 06909 Ankara, Turkey
*Corresponding author: Phone: +905072639703, mail: mcetmeli@kalyonpv.com

ABSTRACT: With the latest developments at photovoltaics such as increasing wafer size and domination of bifacial modules at the market, resistive losses due to increased current became more important. As a result, use of half-cut, triple-cut or shingle technologies are anticipated to be the industry standard in near future where the series resistance losses are reduced by decreasing the generated current and increasing the voltage. However, despite the increase in the module performance, some mechanical damage and electrical losses caused by the cutting process may also occur. Thus, the type of cutting method is important to minimize these drawbacks. In the scope of this study, the effect of laser cutting method on the performance of bifacial PERC cells were investigated.
Keywords: PERC, half-cut, laser, bifacial, thermal laser separation

1 INTRODUCTION

Half-cut cells are preferred in the photovoltaic industry thanks to their several advantages. The market share of half-cut cells is 90% according to the 2021 ITRPV report [1]. In the coming years, it is estimated that the production of both half-cut and triple-cut cells will increase with the spread of large wafer sizes such as M6 and M10 [1]. With half-cut cells, the module power can be increased by 5-8% [2].

First advantage of half-cut cells is that they provide low resistance losses. With the half cut of the solar cells, the current produced from each cell is halved and the resistance losses are reduced thanks to the low current [3]. The other advantage is high shading tolerance over conventional cells. In series-connected full-cell modules, if one cell in a row is shaded and does not produce power, the entire row of cells will stop producing power. Standard modules typically have three separate rows of cells connected to each other; thus, a shadow in one cell of a row will eliminate one-third of the module's power generation [3]. Half-cut cells are also connected in series, but since modules with half-cut cells have twice the number of cells, half-cut modules have twice as many cell rows as modules with full cells. This type of connection causes modules built with half-cut cells to lose less power when a single cell is shaded because a single shaded cell can only remove one-sixth of the total module power output [3]. Another advantage of half-cut technology is high resistance to hot spot damage. When a solar cell in a cell array in a module is shaded, all previous unshaded cells can discharge the energy they have produced into the first shaded cell as heat. Prolonged exposure creates a potentially hot spot that could damage the solar module. In half-cut modules, since the shaded cell has only half area of a full cell, the resultant heat causes less damage. As a result, half-cut modules show an improvement in resistance to hot spot damage [3].

All commercially available silicon solar cells in half-cell size are produced in a two-step manufacturing process. Firstly, standard full-size solar cells are produced. Except for a possible adaptation of the metallization pattern, there is no change in the required process. In a second step, cell cutting takes place. Different laser cutting methods can be used in this step [3]. The first method is "laser scribing and cleavage (LSC)" method. This method is based on a laser ablation process that creates a full-length scratch along the half-cell edge. In some cases, this scratch does not separate the cell completely, but results in a groove with a depth of about half of the cell thickness. Then, mechanical rupture of the cell, guided by the laser, takes place [3]. Another cell cutting method is the "thermomechanical induced cell separation (TMC)" method. This method is based on a non-ablative process in which a small crack is first induced, then propagated by applying a highly localized thermal gradient to the material along the half-cell edge. This thermal gradient causes some local mechanical stress within the cell, causing the crack to propagate [3].

In this paper, in order to investigate the effect of laser cutting method on the performance of bifacial PERC cells, two methods were used to obtain half-cut cells: Laser Scribe and Break method and Thermal Laser Separation (TLS). The latter method offered by 3D Micromac, is a type of TMC method. This TLS method occurs at two steps. In the first step, a break point is created in the wafer to initiate crack formation. Then, separation is carried out by inducing compressive stress by heating and tensile stress by cooling [4]. With this study, these two methods were compared in case of module power, solar cell mechanical strength and cross-section SEM analysis.

2 EXPERIMENTAL

2.1 Production of PERC Cells and Modules

Standard PERC production process was applied to p-type Cz-wafers (Figure 1). G1 size (158.75 mm *158.75 mm), 5 busbar, PERC bifacial solar cells with %22,1 efficiency were selected for the study. The cells were divided into three groups. Group 1 (800 cells) and Group 3 (800 cells) were cut with Autoway SLC-4000 equipment with laser scribing and mechanical breaking technology that is present at Kalyon PV module manufacturing line. Group-2 cells were sent to 3D Micromac and cut by Thermal Laser Separation (TLS) method. Half-cut cells that were processed with in house laser tool at Group-3 was also packed and sent to 3D Micromac and collected back without going through any further process. Group-1 half-cut cells were kept as reference to complete the full set of comparison. Bifacial glass-glass modules were produced at Kalyon PV module manufacturing line by using all groups.

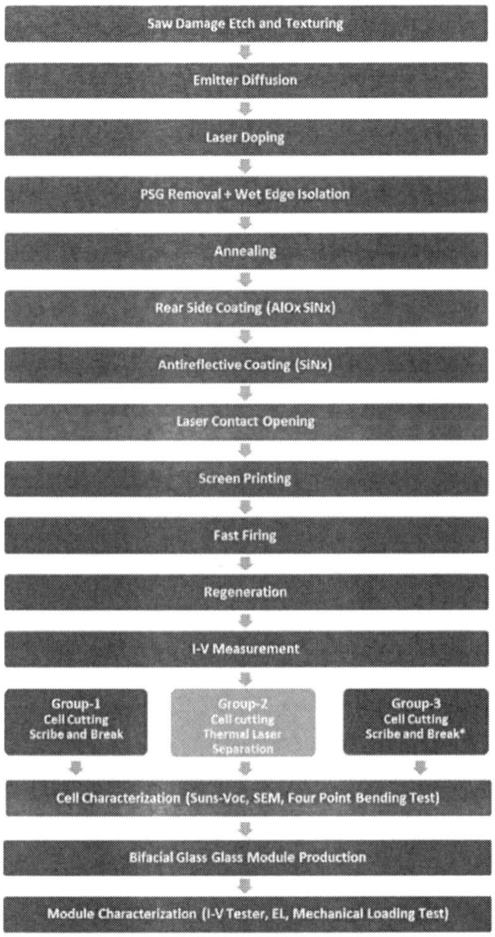

Figure 1: Process flow for cell and module production. *In order to observe the effects of transportation which may cause micro cracks on the cells, Group-3 (1.600 half-cut cells) that was processed with in house laser tool was also packed and sent to 3D Micromac and collected back without going through any further process.

2.2 Cell and Module Characterization

Four point bending test was applied to obtain mechanical strength of half-cut cells. Cross-sections of cells were analyzed by Scanning Electron Microscopy (SEM). Sinton WCT-120TS was used for Suns-Voc analysis of half-cut cells.

Electrical parameters of glass-glass modules were extracted by I-V flash testing. Mechanical loading test (TS EN 61215-2:2017-12) performed to modules. Electroluminescence measurements were also done before and after mechanical loading tests.

3 RESULTS AND DISCUSSION

3.1 I-V Results of Solar Modules

In order to determine the electrical parameters of the produced modules, the modules were put to the test in the I-V tester. Figure 2 illustrates the comparison of average module power, open circuit voltage and short circuit current of the groups. Group 2, which 3D Micromac company use Thermal Laser Separation method to cut the cells, gave the highest average power with 394,84 Watts among three groups. The average power values of Group 1 and Group 3 were determined as 392,55 W and 392,47 W, respectively.

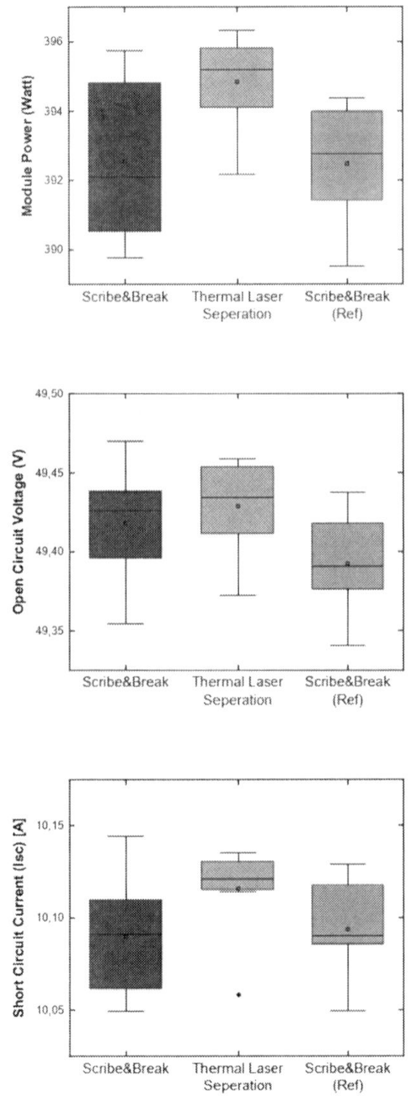

Figure 2: Comparison of module I-V parameters of three groups. Group 1: Half-cut cells were obtained by laser scribe and break technology. Group 2: Thermal Laser separation method is applied to cut the cells in half. Group 3: processed with in house laser tool was also packed and sent to 3D Micromac and collected back without going through any further process.

3.2 Mechanical Loading Test and EL Images

TS EN 61215-2: 2017-12 Mechanical Load Test was performed for the modules. Three modules from each group were randomly chosen and tested.

All modules have passed the test, withstanding mechanical loads of 1.800, 2.100 and 2.400 Pa. When the EL images were examined, no fracture was observed in the cells before and after the test. Figure 3 depicts the EL images before and after the loading test.

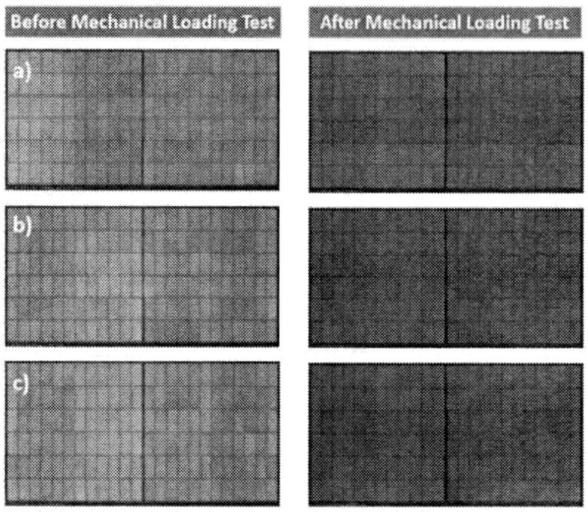

Figure 3: Electroluminescence images of modules before and after mechanical loading test. a) Group-1 b) Group-2 c) Group-3

3.3 Four Point Bending Test

The mechanical strength of solar cells is a significant parameter to provide high returns in the production of solar cells and modules. Every critical mechanical damage during solar cell production affects downstream processes and applications. Also, a decrease in mechanical cell strength causes an increase in cell breakage rates during operation [5].

In studies on half-cut cells in the literature, it has been observed that different cell division processes lead to similar electrical performance parameters, while providing very different mechanical strength properties [5]. A four-point bending test was applied to examine the mechanical strength of half-cut cells in this study.

Four point bending test results are shown in Figure 4. The average mechanical strength of the cells cut by the scribe and break method (Groups 1 and 3) were found to be 113,6 and 126,9 MPa respectively. The average mechanical strength of Group 2 (Thermal Laser Separation) cells was found to be 165,4 MPa. It is seen that Group 2 cells provided higher mechanical strength than those of scribe and break technology.

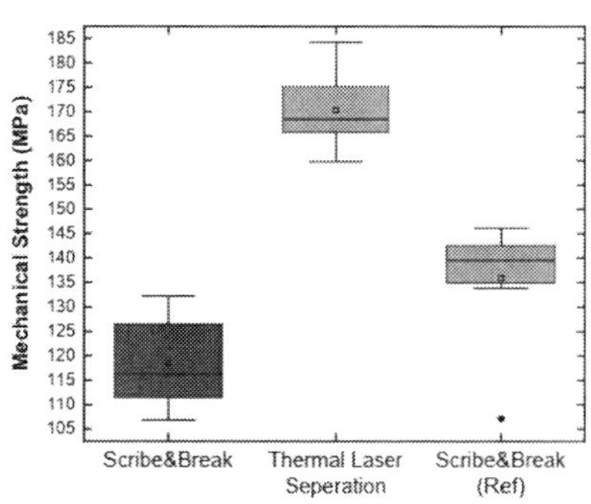

Figure 4: Four point bending test results of three groups. Group 1: Half-cut cells were obtained by laser scribe and break technology. Group 2: Thermal Laser separation method is applied to cut the cells in half. Group 3: processed with in house laser tool was also packed and sent to 3D Micromac and collected back without going through any further process.

3.4 Scanning Electron Microscopy Cross Section Analysis

The cross-sectional areas of half-cut cells were analyzed by SEM analysis to compare the cutting methods. SEM images of the three groups are presented in Figure 5. When the cross-sectional area of the cells cut with the Thermal Laser Separation method is examined, it is seen that it provides a smoother cut compared to the laser scribe and breaking method.

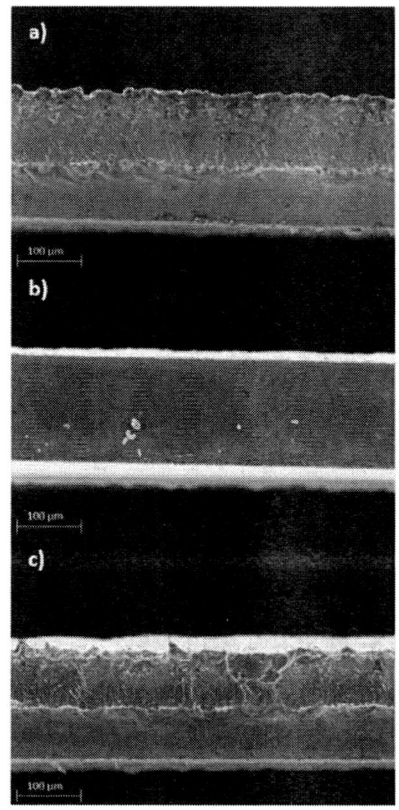

Figure 5: SEM images of cross sections of half cut cells. a) Group 1: Half-cut cells were obtained by laser scribe and break technology. b) Group 2: Thermal Laser separation method is applied to cut the cells in half. c) Group 3: processed with in house laser tool was also packed and sent to 3D Micromac and collected back without going through any further process.

3.5 Suns-Voc Measurement

Randomly selected twenty-five half-cut cells from each group have been analyzed by Sinton WCT-120. Open circuit voltage and pseudo fill factor distribution of the cells measured are given in Figures 6. As can be seen from Figure 6, no significant difference was detected between the open circuit voltage distribution of three groups. Similarly, fill factor values are similar with 83,81% for Group 1, 83,44% for Group 2 and 83,81% for Group 3. Only a minor difference was observed for group 2 cells.

Shunt resistance was also determined from the Suns-Voc measurement. The average shunt resistances of the cells were measured as 150.000, 127.961 and 150.000 ohm.cm^2 for Groups 1, 2 and 3, respectively. It was observed that the mean shunt resistance for group 2 was lower. It was concluded that some damage to the cells may have occurred due to the logistic effect.

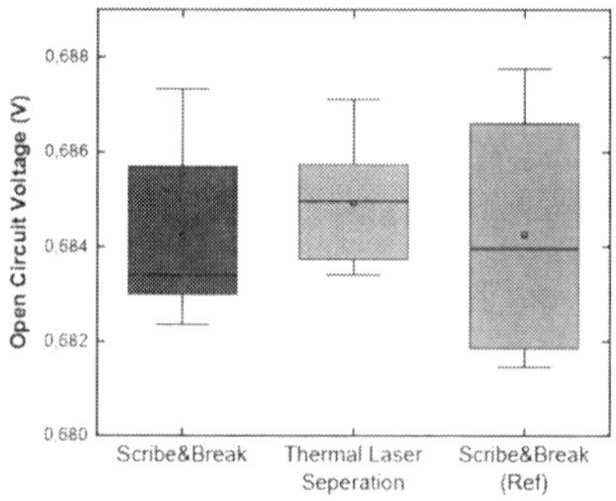

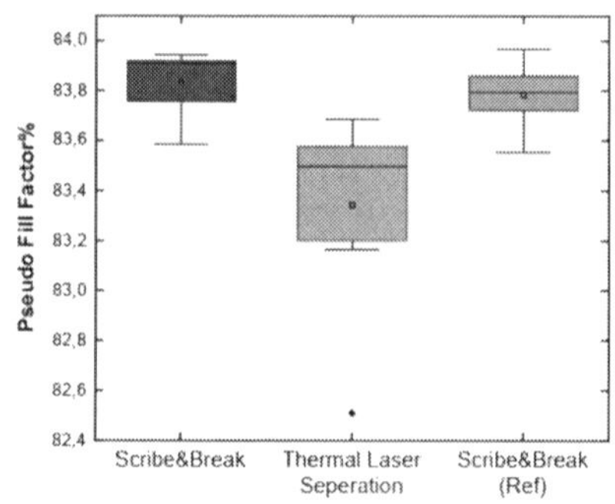

Figure 6: Open circuit voltage and pseudo fill factor parameters of half cut cells obtained by Suns-Voc measruement. Group 1: Half-cut cells were obtained by laser scribe and break technology. Group 2: Thermal Laser separation method is applied to cut the cells in half. Group 3: processed with in house laser tool was also packed and sent to 3D Micromac and collected back without going through any further process.

4 CONCLUSIONS

Table I: Summary of Results

	Group 1	Group 2	Group 3
Number of Modules Produced	8	8	6
Average Module Power (Watt)	392,6	394,8	392,5
Cell Mechanical Strength (MPa)	113.6	165,5	126,9
Module Mechanical Loading Test (1800, 2100 and 2400 Pa)	Pass	Pass	Pass
SEM Cross-section Image	Rough surface	Smooth surface	Rough surface

In this study, three experimental groups were prepared to extract the effects of laser cutting technology and logistics. After the cutting process, bifacial frameless glass-glass modules were produced in the Kalyon PV module production line. All results are summarized in Table I. It was determined that the mechanical strength of the cells cut with the Thermal Laser Separation method was ~40% higher than those of laser scribe and break technology. Moreover, module I-V tester results indicated average power of bifacial frameless glass-glass modules produced with cells cut by Thermal Laser Separation method is 2 W more than those produced according to scribe&break technology. SEM results shown that Thermal Laser Separation method provide a smoother cut than the laser scribe and breaking method. In conclusion, the study revealed that the cutting method has a significant impact on the performance of solar modules.

5 REFERENCES

[1] "International Technology Roadmap for Photovoltaic (ITRPV) 2021." 2021.

[2] T. M. Schneider Jens, Hanifi Hamed, Dassler David, Pander Matthias, Kaule Felix, "Half-cell solar modules: The new standard in PV production?," *Photovoltaics Int.*, vol. 42, 2019.

[3] "How do half-cut solar cells improve panel performance?," *Eco Green Energy - EGE PV News*, 2020.

[4] M. C. Barreto, G. Roeder, M. Steinhoff, M. Schellenberger, and A. Bauer, "Advances in thermal laser separation: Process monitoring in a kerf-free laser-based cutting technology to ensure high yield," *Procedia CIRP*, vol. 74, pp. 645–648, 2018, doi: 10.1016/j.procir.2018.08.056.

[5] F. Kaule, M. Pander, M. Turek, M. Grimm, E. Hofmueller, and S. Schoenfelder, "Mechanical damage of half-cell cutting technologies in solar cells and module laminates," *AIP Conf. Proc.*, vol. 1999, no. August, 2018, doi: 10.1063/1.5049252.

DENOISING OF IMAGE DATA FOR DWS WAFER CHARACTERIZATION USING GENERATIVE ADVERSARIAL NETWORKS

L.Kurumundayil[1], H.Schremmer[2], S.Rein[1], M.Demant[1]
[1]Fraunhofer Institute for Solar Energy Systems, Heidenhofstraße 2, 79110 Freiburg, Germany
[2]Meyer Burger (Germany) GmbH, An der Baumschule 6-8, 09337 Hohenstein-Ernstthal

ABSTRACT: Diamond wire sawing is the wafering technique which recently replaced slurry-cut wafering, due to reduced kerf-loss, lower process expenditures and increased speed. While this technology is successfully used to wafer the ingots, it introduces a strong macroscopic saw damage on the surface of the wafers. These saw marks introduce artifacts in images recorded by multiple characterization methods which lead to poor quality inspection of silicon wafers and solar cells. Instead of a cost intensive update of the implemented measurement systems for the inspection of saw-damaged wafers, we propose a novel denoising technique of the images of diamond-cut wafers using generative adversarial networks.

Keywords: Characterization, Multi-Crystalline, Grain, Experimental Methods, Machine Learning,

1 INTRODUCTION

According to ITRPV 2018 [8], diamond wire sawing (DWS) [1] is a unique approach to wafering technology that has almost replaced slurry-based wafering for both monocrystalline and multi-crystalline silicon wafers. Diamond wire sawing has several advantages over slurry cutting, which include: (i) higher throughput, (ii) lower cost, and (iii) cleaner/environmentally friendly slurry system. However, these advantages occur only if the quality of the cut is as good as or better than slurry-cut [6]. Diamond wire sawn wafers show strong saw damage on their surface because of the wafering process. Differences or problems in the sawing process can cause strong inhomogeneities and variations in reflection on the wafer surface, which disturb the imaging methods and can cause problems in the detection algorithms[3]. In this work, we inspect the grain boundary imaging (GBI) system for mono-cast and multi-crystalline silicon wafers. Due to the inhomogeneities, the grain boundaries are not wholly visible which affects the overall characterization. Figure~1 shows an image example of a diamond wire sawn wafer captured by the grain boundary imaging system. Here, one can see the artifacts caused by the saw marks on the wafer.

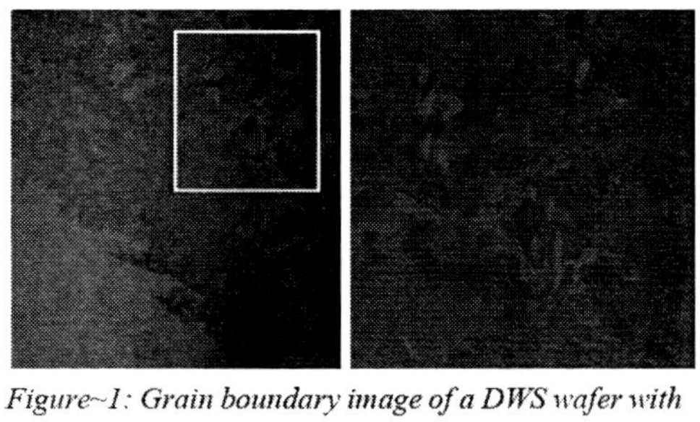

Figure~1: Grain boundary image of a DWS wafer with zoomed in patch (right) shows the artifacts due to sawing damage

Hence our goal is to optimize the grain boundary imaging methods for diamond wire sawn wafers that are implemented for slurry-cut wafers. We approach this problem by digitally removing the saw marks using deep learning technologies. Since the noise patterns are heterogeneously distributed over each wafer, classical image processing techniques are not applicable for denoising the diamond-cut wafer images.

Most machine learning tasks call for supervised learning, i.e., training with paired data which consists of an input and its true target label. In our case, we would need a noisy image with artifacts and a clean image without the artifacts of the same wafer. This is unattainable because a silicon brick can only be cut into wafers by one wafering technology. Hence, we have unpaired data of diamond-wire-sawn and slurry-cut wafer images. An approach using adversarial training with the help of Generative Adversarial Networks (GANs) [2] is used to denoise the diamond wire sawn wafer images. In addition to the unpaired data, we use an experimental strategy to generate paired datasets to aid the adversarial denoising.

2 APPROACH

We use the generative power of deep learning models to denoise images. Therefore, we develop two approaches to overcome the challenges of unpaired data: (a) create pairs of noisy and clean wafer images, (b) implement machine learning models that work for both, paired and unpaired data. For the first approach, two paired datasets are generated, a synthetic dataset by inducing fake saw marks onto slurry-cut wafer images and an experimental dataset by removing the saw damage from the DWS wafers by undergoing an acidic texture process. For the second approach, a network called "ResidualGAN" is implemented which uses adversarial learning to tackle the lack of paired data. The generative power of GANs helps the network to estimate the heterogenous noise patterns and produce noise-free measurements.

2.1 Dataset creation

An informative dataset is crucial to train a machine learning model. We generate three different datasets: one unpaired and two paired, for our denoising task. The experimental flow of obtaining the data is shown in Figure~2. To create the different datasets, we collected 400 diamond-cut wafers and 400 slurry-cut wafers which are of mono-cast [7] and multi-crystalline material. These wafers were sorted out from different points along the brick height to provide a variety in data, e.g., different grain structures along the height of the brick. The data acquisition is explained below based on the complexity, from the easiest to the most difficult achievable data.

Unpaired dataset. To create an unpaired dataset, each DWS wafer and slurry-cut wafer were measured using a grain boundary imaging system. This system records a wafer from eight different angles using an LED illumination and therefore produces eight images of the same wafer. This dataset contains about 2900 images of both diamond-cut and slurry-cut wafers each of 400 wafers. As this dataset is an unpaired dataset, it contains a batch of DWS wafer images as the noisy images batch and a batch of slurry-cut wafer images as the clean images batch as shown in Figure~4 (see Unpaired data).

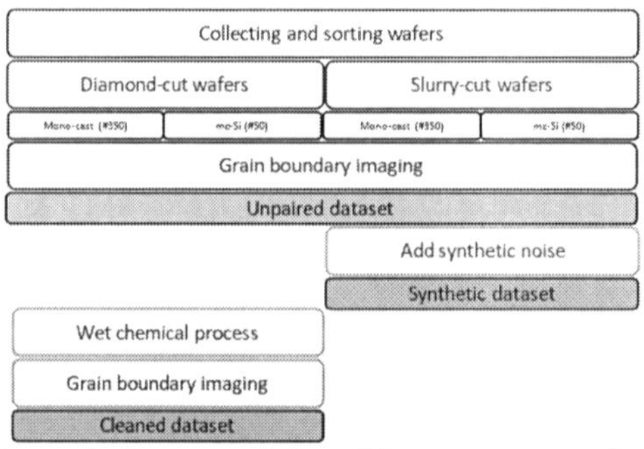

Figure~2: Experimental flow of dataset creation with unpaired (marked in gray), synthetic (marked in orange) and cleaned (marked in blue) data. The processes marked in black are common irrespective of the dataset.

Synthetic dataset. The second dataset is a paired data by adding synthetic noise to the slurry-cut wafer images. It is challenging to synthesize data in such a way that the data correspond to the real data. In our case, it is difficult to reproduce such saw marks due to its unique and very noisy structures across the DWS wafer. To be able to replicate the saw marks accurately, the saw marks on the DWS wafer were analyzed and reproduced on slurry-cut wafer images. Here, we use slurry-cut wafer images because of their artifact-free wafer images. An example pair is shown in Figure~4 (see Synthetic data). Some sample images contain artificial saw mark structures that look more curved than the real saw marks. This allows the model to denoise not only uniform but also uniquely structured saw marks from the DWS wafer images.

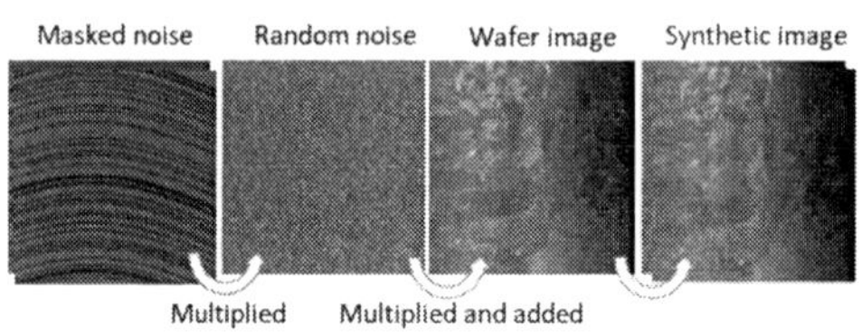

Figure~3: Process flow of noise synthetization on slurry-cut wafer images

Figure~3 shows how the addition of synthetic to the slurry-cut wafer images is performed. First, a mask is created that imitates the line-like structures of the saw marks. To include some variety in the saw marks for better denoising of the noisy wafer images, the lines are slightly curved in randomly selected wafer images. The mask is multiplied with random normal noise. This masked noise is then added and multiplied with the slurry-cut wafer images to create a synthetic DWS wafer image. The original slurry-cut wafer image that is used to create the synthetic DWS wafer image is the corresponding clean label.

Cleaned dataset. An additional paired dataset, called the cleaned dataset, is generated within an experimental set-up. After recording grain boundary images of diamond-cut wafers, the wafers undergo an acidic texture that etches away the saw marks from the wafer surface. Once the saw marks have been removed, the wafers are measured again with the grain boundary imaging system to generate the cleaned diamond-cut wafer images. In Figure~4 (see Cleaned dataset), one can see that the grains on the DWS wafer image look slightly different from the wafer image after surface cleaning. This is due to the change in the optics of the wafer surface related to the texture process.

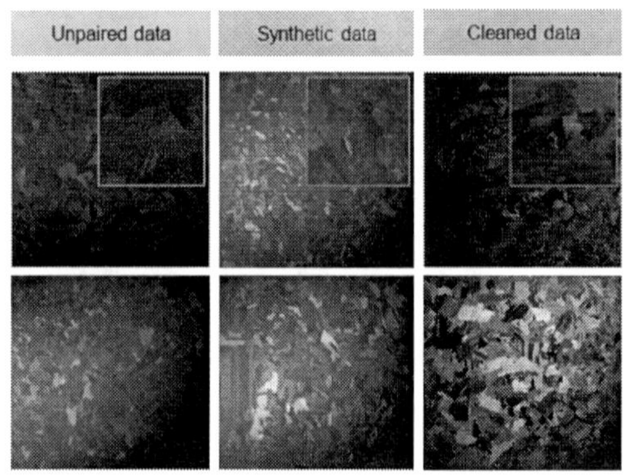

Figure~4: Example pairs of unpaired (left), synthetic (middle) and cleaned (right) datasets

2.2 Machine learning models

The diamond-cut wafers are denoised using an advanced deep learning approach. A generative adversarial network (GAN) is a network used for training with unpaired data by using adversarial training. It consists of a generative and a discriminative network that try to outplay the other during training. The generator tries to fool the discriminator by generating realistic images while the discriminator tries to classify the generated image as real or fake. In our work, we investigate our adversarial model called ResidualGAN shown in Figure~5. The goal of this model is to generate a clean DWS wafer image, which contains the content of the DWS wafer but has a slurry-like surface.

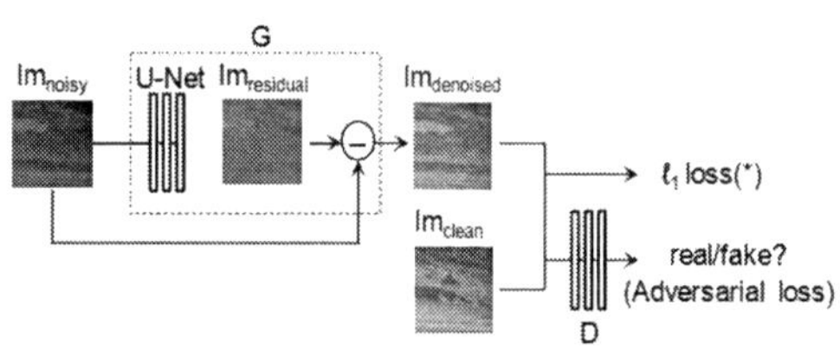

Figure~5: Schematic of our Residual GAN training architecture; (*) ℓ_1 loss is optional: included for paired datasets, excluded for unpaired datasets

ResidualGAN uses a convolutional neural network, namely, a U-Net [3] in the generator network. U-Net was initially developed for image segmentation but has been extended to other tasks. The generator network of the ResidualGAN learns the noise, i.e., the saw marks and subtracts them from the input DWS wafer image to generate the denoised image. To optimize the generator in this model, a discriminator loss is used. The discriminator is trained on both generated "clean" DWS wafer images as well as slurry-cut wafer images. This allows the discriminator to distinguish between a generated and a real image.

Given a generated input from the generator, the discriminator categorizes the image as real or fake. This is fed back to the generator which tries to improve depending on the discriminator's outcome. If the discriminator correctly classifies the generated image as a fake one, the generator will try to create more realistic images than the previous one. On the other hand, if the discriminator classifies the image as real, it implies that the generator is successful at creating slurry-like DWS wafer images. The generator gets better with time at generating slurry-like DWS wafer images. When training this network with paired data, an additional loss function, here l_1 loss, is used along with the discriminator loss to optimize the generator.

3 EXPERIMENTAL

To achieve the goal, different experiments have been conducted. The implemented model is trained with each dataset and tested on real data, i.e., grain boundary images of diamond-wire-sawn wafers. Following the original GAN paper [2], the discriminator D and generator G are updated once in every epoch. To reduce model oscillation [2], we follow Shrivastava et al's strategy [5] and update the discriminator using a history of 50 previously generated images rather than the ones produced by the latest generators [9]. Both the generator and discriminator are optimized by using the Adam optimizer [4] with their learning rates set to 0.0002 for a stable training. In every experiment, the model is trained for a total of 100 epochs. An epoch indicates a round through the entire dataset by the machine learning algorithm.

The results from each experiment are shown in Figure~6 and discussed below. For each test, eight DWS wafer images recorded with the grain boundary imaging system are selected. These images are denoised with each trained model. To understand the results better, the grain boundaries are extracted from the denoised images using a grain boundary detection model that was trained on slurry-cut wafer images. Each column shows three of the eight denoised images and their detected grain boundaries.

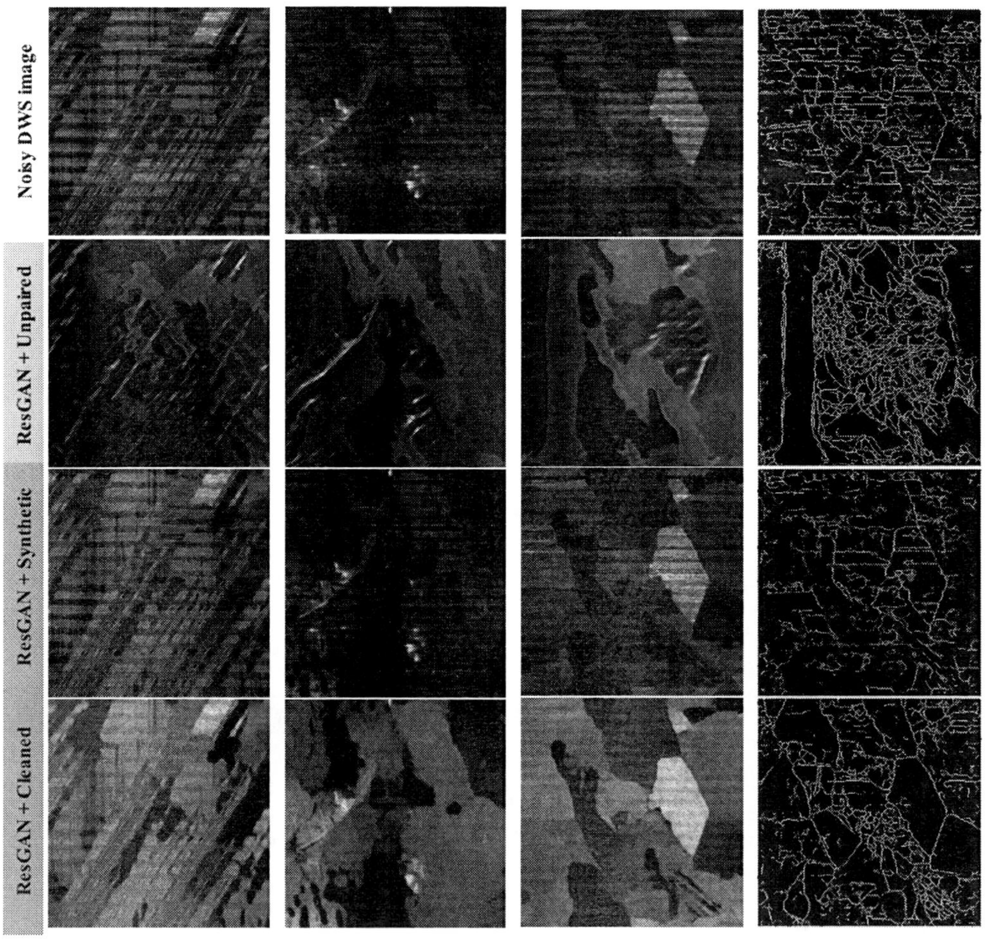

Figure~6: Results of our ResidualGAN trained on unpaired data (second row), synthetic data (third row) and cleaned data (last row) with noisy DWS wafer images (first row); last column shows the detected grain boundaries of the third denoised DWS wafer images

ResidualGAN trained with unpaired data. In the first experiment, the ResidualGAN is trained with unpaired data using an adversarial loss. The network gets a noisy DWS wafer image as the input and is expected to denoise the DWS wafer images. Instead of creating clean DWS wafer images, it generates unrealistic slurry-like images without the grain structures from the DWS wafer. This is due to a phenomenon called mode collapse. Mode collapse occurs when the generator learns how to generate samples from a few modes of the data distribution but misses many other modes, even though samples from the missing modes occur throughout the training data [4]. An example of the generated result is shown in Figure~6 (second row).

ResidualGAN trained with synthetic data. By using an additional direct comparison loss function, the ResidualGAN is trained with the synthetic dataset. The results obtained from this model are better than the ones from adversarial training. The artifacts are less intense and therefore the detection of their grain boundaries partially improves (see third row in Figure~6) when compared to the raw diamond-cut wafer images.

ResidualGAN trained with cleaned data. In the experiment with cleaned data, ResidualGAN did a better job at denoising the wafer images shown in Figure~6 (see ResGAN+Cleaned data). The grain boundaries are sharper than the ones from the previous test. The reason behind it could be that the model was trained on real data instead of synthetic or unpaired data which makes it easier for the model to transfer to real data.

To evaluate the models, a similarity metric called SSIM (Structural Similarity Index Measure) [5] was used. It measures the similarity between the real and generated clean images based on luminance, contrast and structure. An SSIM value of 0 means that the real and generated images do not have any similarity, whereas an SSIM value of 1 represents 100% similarity between the real and generated images. Our model trained with synthetic data achieved an SSIM value of 0.8337 whereas the ResidualGAN trained with cleaned data achieved a value of 0.9779 which indicates that our model performed comparatively better with cleaned data.

4 DISCUSSION

Our model ResidualGAN was trained on three datasets: unpaired, synthetic and cleaned data. In our first experiment, ResidualGAN was trained using unpaired data using an adversarial network. The results obtained from this training scheme turned out to be unrealistic slurry like images, i.e., images with slurry like surfaces but no content from the input image. For the second and third experiments, we used an additional supervised loss function which helped our model to denoise the DWS wafer images. The training with synthetic data aided in a better grain boundary detection than the noisy DWS wafer images. Better results were achieved by training our ResidualGAN on the cleaned data. The results show that most of the saw marks were removed from the wafer images and therefore enables to have a better grain boundary detection with these denoised images. Unlike unpaired data, the training with the synthetic dataset was successful at avoiding mode collapse and generated realistic wafer images. On the other hand, the training with the cleaned data was successful at denoising the DWS wafer images better.

5 CONCLUSION AND OUTLOOK

In this work, we evaluate our model ResidualGAN to denoise the images captured with the grain boundary imaging system DWS wafer of mono-cast and multi-crystalline silicon material. For this approach, we have successfully generated three kinds of datasets. The easiest available dataset is the unpaired data, as no extra effort was required to create this dataset. Additionally, we generated two paired datasets: a synthetic by adding artificial noise to the "clean" slurry-cut wafer images and a cleaned dataset by performing a wet chemical process on the DWS wafers. Our model generates a denoised image of DWS wafers using a combined adversarial and supervised loss mechanism. A purely unsupervised training using adversarial loss led to a failure called mode collapse which led to unrealistic images.

Within the scope of this work, we generated training data for additional imaging systems: photoluminescence and infrared transmission imaging. The denoising task can be extended to these datasets. Additional methods are being investigated to overcome the challenge of mode collapse when training with unpaired data.

6 REFERENCES

[1] A. Bidiville, K. Wasmer, R. Kraft, C. Ballif. Diamond Wire-Sawn Silicon Wafers – From The Lab To The Cell Production.

[2] Ian Goodfellow, Jean Pouget-Abadie, Mehdi Mirza, Bing Xu, David Warde-Farley, Sherjil Ozair, Aaron Courville, And Yoshua Bengio. Generative Adversarial Nets.

[3] J. Haunschild, N. Bergmann, T. Hammer, K. Krieg, T. Kaden, O. Anspach, H. Schremmer, And S. Rein. Inline Characterization Of Diamond Wire Sawn Multicrystalline Silicon Wafers.

[4] Kingma, D. P. And Ba, J. 2014. Adam: A Method For Stochastic Optimization.

[5] Shrivastava, A., Pfister, T., Tuzel, O., Susskind, J., Wang, W., And Webb, R. 2016. Learning From Simulated And Unsupervised Images Through Adversarial Training.

[6] Sopori, B., Devayajanam, S., Shet, S., Guhabiswas, D., Basnyat, P., Moutinho, H., Gedvilas, L., Jones, K., Binns, J., And Appel, J. 2013 - 2013. Characterizing Damage On Si Wafer Surfaces Cut By Slurry And Diamond Wire Sawing. In 2013 Ieee 39th Photovoltaic Specialists Conference (Pvsc). Ieee, 945–950. Doi=10.1109/Pvsc.2013.6744298.

[7] Stoddard, N., Wu, B., Witting, I., Wagener, M. C., Park, Y., Rozgonyi, G. A., And Clark, R. 2007. Casting Single Crystal Silicon: Novel Defect Profiles From Bp Solar's Mono2 Tm Wafers. Ssp 131-133, 1–8.

[8] Vdma. International Technology Roadmap For Photovoltaic: Results 2018 - 10th Edition 2019.

[9] Zhu, J.-Y., Park, T., Isola, P., And Efros, A. A. 2017. Unpaired Image-To-Image Translation Using Cycle-Consistent Adversarial Networks.

STABILITY STUDIES AND CHARACTERISATION OF SILICON SOLAR CELLS VIA IN-SITU TEMPERATURE AND LIGHT DEPENDENT SUNS-VOC MEASUREMENTS

Michelle Vaqueiro-Contreras[1], Arastoo Teymouri[1], Arman Mahboubi Soufiani[1], Catherine Chan[1], Alison Ciesla[1], Harrison Wilterdink[2] and Brett Hallam[1]
[1]Photovoltaic & Renewable Energy Engineering, University of New South Wales, Sydney NSW 2052, Australia
[2]Sinton Consulting, Inc. 4720 Walnut St. Suite 102, Boulder, CO 80301 USA

ABSTRACT: Understanding solar cell performance and stability under field-relevant operation conditions is of paramount importance for long lasting solar energy generation. Diverse forms of degradation can occur during solar cell exposure to high temperatures and illumination. These can originate from bulk-related degradation mechanisms caused by interactions of impurities within the silicon crystal, or from changes at the dielectric or metallic interfaces, resulting in surface-related degradation. In this work, we report on the application of in-situ temperature- and illumination-controlled Suns-V_{OC} measurement analysis for a range of commercial cells of different architectures undergoing boron-oxygen (BO) defect stability testing using a newly developed Suns-V_{OC} *elite* tester capable of extended light-soaking on four industrial size solar cells simultaneously at elevated temperature. It is shown that room temperature current-voltage measurements and those extracted via in-situ Suns-V_{OC} are in excellent agreement with each other giving precise values within ± 1 mV variation. Similarly, the extracted minority carrier density dependent lifetime data using the quasi-steady-state V_{OC} method by the *elite* tool is compared to that of the one measured by the photo-conductance decay method using lifetime structures yielding to effective values of ±5 µs of precision. It is also demonstrated that the application of a BO-related light induced degradation mitigation processes on finished cells as well as unmetallized precursors improves their V_{OC}, avoids degradation, and can be thoroughly characterized using this new batch processing Suns-V_{OC} tool.

Keywords: Solar cells, Suns-V_{OC}, Characterization, Silicon, Recombination

1 INTRODUCTION

The promise of an affordable and environmentally friendly future when it comes to solar energy generation is based on the premise of having long lasting devices delivering the highest possible power. Therefore, a direct dependance on stability of the solar cells under field-operation conditions which include working temperatures as high as 90 °C [1], and the levelized cost of electricity, exists. Mainstream technologies are prone to suffer from various kinds of light induced degradation mechanisms [2, 3], which are still not fully understood, and even high-efficiency n-type device architectures are not necessarily exempt from them [4]. It is therefore desirable for both the research community and industry to have tools available, which are capable of quick but thorough characterization of the cells under the wide temperature and illumination conditions with minimal human-caused error. In this work, a custom designed Suns-V_{OC} *elite* tool by Sinton Instruments is used and the applicability of in-situ, prolonged and continuous Suns-V_{OC} measurements [5] for the characterization of a range of PERC silicon solar cells from different vendors is analyzed. Measurements were performed under different temperature conditions and the results were validated against 1-Sun calibrated current-voltage (I-V) and photoconductance tools. We further highlight the significance of semi-continuous data collection, possible using the in-situ Suns-V_{OC} *elite* tool, as compared to intermittent operator-reliant measurements, particularly important for long-term degradation investigation experiments.

2 EXPERIMENTAL DETAILS

2.1 Materials

For this study we have selected a range of B-doped Czochralski and multi-crystalline silicon solar cells from three different commercial vendors. Details of the sample sets are provided in Table I.

Table I: Solar cell characteristics

	SUPPLIER A	SUPPLIER B	SUPPLIER C
SILICON TYPE	Cz	Cz	Multi
RESISTIVITY	1.3 Ω·cm	1.5 Ω·cm	1.5-6.8 Ω·cm
ARCHITECTURE	PERC	PERC precursor and PERC cell	Bifacial PERC
SIZE	156.75 mm (semi-square)	156.75 mm (semi-square)	158.75 mm (square)

2.2 Characterization and processing tools

We use a custom designed temperature- and illumination-controlled Suns-V_{OC} equipment (Suns-V_{OC} *elite*) by Sinton Instruments. It allows Suns-V_{OC} measurements for up to four cells simultaneously, while providing controlled steady illumination between 0.1- and 1.1-Suns equivalent photon flux and controlled temperature in the range 25 °C (LEDs OFF) to 90 °C. Square wafers of up to 161.7 mm (M4) can be readily mounted for characterization with a single probe for the front contact (Figure 1A). For the stability tests, we also exposed some cells to calibrated 1-Sun equivalent photon flux, home-built illumination set-ups using halogen lamps and LED arrays (Solstice tester). Both set ups had an operation temperature of 75±2 °C even though only the solstice tester is equipped with an active cooling system (Figure 1C). We validate the obtained results with those measured by a calibrated LOANA solar cell analysis system by pv-tools, quasi-steady-state photoconductance (QSS-PC) measurements on a WCT-120 by Sinton Instruments, as well as with photoluminescence data acquired using a BTi R3 luminescence imaging system. An automatic and custom-made DR Laser tool with a 980 nm laser (10±0.2 W power measured at the sample) and a controlled temperature stage at 250 °C was used to process some samples and mitigate the effects of the boron-oxygen light-induced degradation (BO-LID) [6].

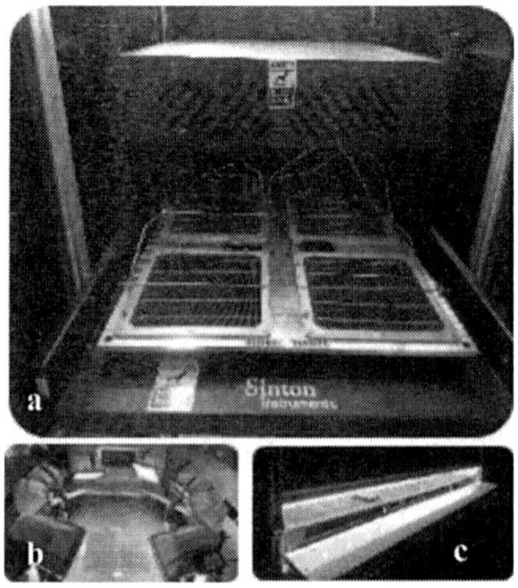

Figure 1: Photo of the (a) Suns-V_{OC} *elite* tool, (b) halogen test station and (c) In-house built Solstice test station.

3 RESULTS

3.1 Current-Voltage validation and BO-LID testing

We have carried out continuous in-situ measurements of a sample cell from supplier A (Sup A) while exposed to 1 sun equivalent illumination at 75 °C for 24 h. To validate the extracted data, intermittent I-V measurements of the same sample were also taken at room temperature on a calibrated I-V tester. Figure 2 shows the degradation of the V_{OC}, pseudo-efficiency and J_{01} during the test. We have found that the error of the V_{OC} measurements taken at 75 °C when compared to the room temperature measurements falls within ±1 mV. However, the calculation of other cell electrical characteristics such as the recombination parameter J_{01} from the fitting of a double-diode equation and the pseudo-efficiency do not lead to accurate values at these high temperatures (see Figures 2b and 2c). On the one hand there is the evident degradation of the short-circuit current density (J_{SC}) that is not accounted for during the in-situ measurements i.e it relies solely on the user-input value. This degradation, although not too relevant for the calculation of parameters such as V_{OC}, V_{mp}, pFF and R_{shunt}, can lead to errors on the calculation of pseudo-efficiency (Figure 2b). On the other hand, the fitted J_{01} value at high temperatures seems to overestimate the room temperature value by a factor of 10 (Figure 2b). An important factor causing this can be the increase in the intrinsic carrier density (ni) at higher temperatures, given that J_{01} is proportional to ni^2. Furthermore, non-radiative recombination coefficient could have increased. Yet, the relative changes of the tracking parameters seem to closely follow those of the I-V measurements at room temperature which is crucial for stability and kinetics studies as shown in Figure 3. This was further validated in another set of samples from Sup-A and Sup-B. Hence, on a second experiment we have exposed three sets of four PERC cells from Sup-A to 1-Sun equivalent illumination at 75°C for approximately 40 hours. Two cells of each of the three sets were subjected to an optimized temperature and illumination process for BO-LID mitigation whereas the other two were tested in their "untreated" state. Four cells were then mounted on each of the three systems shown in Figure 1.

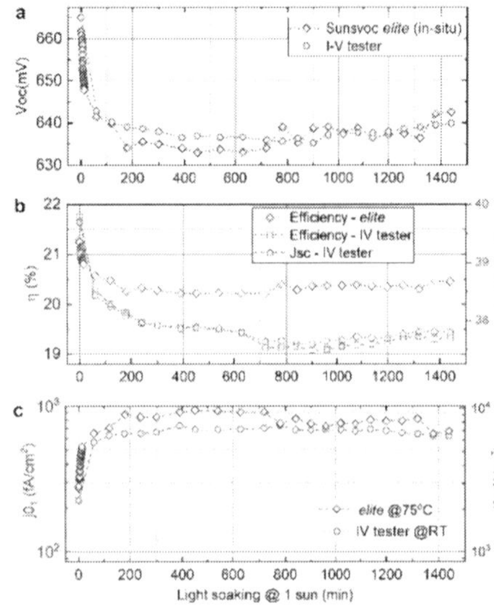

Figure 2: Degradation tracking of Sup-C cells under 1-Sun equivalent at 75 °C. Blue data points were taken by the Suns-V_{OC} *elite* tool at 75 °C and were T-corrected, whereas green data points correspond to room temperature IV measurements.

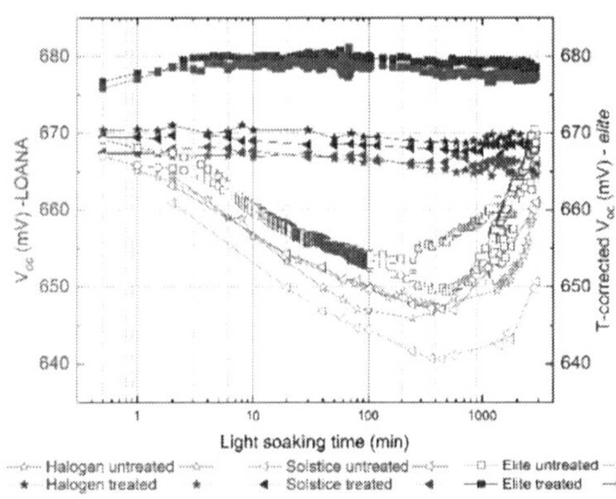

Figure 3: Open-circuit voltage tracking of Sup-C cells under 1 sun eq. at 75 °C. Measurements corresponding to cells tested on the halogen and solstice testers (left axis) were taken at RT with the calibrated solar I-V tester. Measurements on the Suns-V_{OC} *elite* were carried out in-situ at 75 °C and were temperature corrected.

Intermittent measurements were taken at room temperature (RT) with the calibrated solar I-V tester and PL R3 tool from the samples placed on the halogen lamp set up and Solstice tool, whereas continuous in-situ measurements with the Suns-V_{OC} *elite* were carried out. We have then compared the difference in degradation and recovery on the various sample sets.

Figure 3 shows the results of the in-situ Suns-V_{OC} *elite* measurements taken at 75 °C and corrected for temperature using 0.0022 V/°C as the temperature coefficient for the V_{OC}. On the same figure, intermittent measurements taken at RT after illumination on the halogen and Solstice testers are shown. It is observed that the relative degradation of these samples are in very good agreement with each other, irrespective of the system used. A maximum degradation of 20 mV was measured on the cells, which is in line with the commonly observed extent

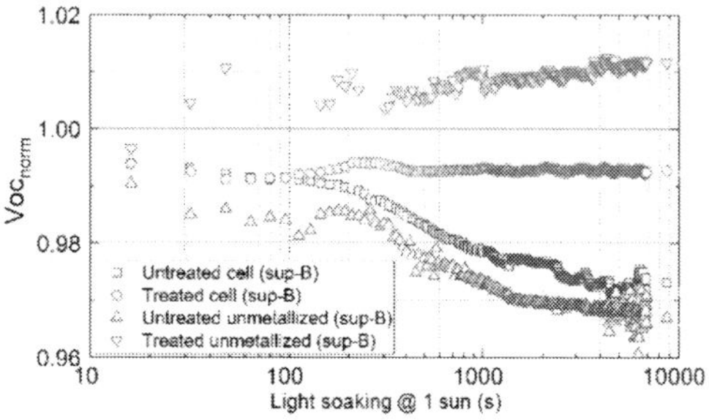

Figure 4: Temperature corrected and normalized V_{OC} measurements of cells from Sup-B taken in-situ at 75 °C.

of degradation produced by the BO defect. Figure 3 also shows the characteristic two-stage degradation (fast and slow component) resulting from BO-LID.

Real-time measurements acquired at *time intervals* as fast as every *one second* have been realized. Time-intervals that are hardly achievable through ex-situ characterization, because besides the challenges resulting from fast handling of hot and thin wafers, there are unavoidable changes in sample brittleness due to the thermal stress as well as unintentional quenching. Depending on the nature of the defects present on the samples, the later may be transforming or altering charge states of metastable defects leading to changes in observed kinetics. Thus, one of the key benefits of this tool to allow for thorough kinetics characterization of relatively fast degradation processes.

In addition to the finished cells data shown in Figure 3, Figure 4 illustrates V_{OC} data collected from PERC precursors (Sup-B), which did not have front metallization. From these measurements we observed again a very similar response from both sets. We have compared the degradation response from these cell precursors (i.e. unmetallized) with a set of finished cells, including the front grid. It was evident from the unmetallized sample which was subjected to the BO-mitigation treatment that the V_{OC} performance surpassed that of the rest of the cells, which could be the result of enhanced defect passivation and/or the non-existent degradation at the front contacts usually observed in cells after the BO-mitigation processing [7].

3.2 Injection dependent effective lifetime of bifacial cells

It has been previously shown that Suns-V_{OC} data can be effectively used to determine the minority carrier lifetime with the application of the so-called QSS-V_{OC} method[8] according to equations:

$$\Delta_n[N_A + \Delta_n] = n_i^2 \exp\left(\frac{V_{oc}}{kT/q}\right) \quad eq.\ (1)$$

$$\tau_{eff} = \frac{n_i^2 \exp\left(\frac{V_{oc}}{kT/q}\right)}{J_{ph}(N_A + \Delta_n)/qW} \quad eq.\ (2)$$

In this work we have used this method on a range of bifacial PERC cells of different base doping concentrations (N_{dop}) and compared the results to those obtained by the QSS-PC measurements from unmetallized wafers of similar N_{dop}. The results are shown in Figure 4. Despite the fact that these wafers differ slightly in N_{dop} and correspond to different wafer growth runs by the supplier, we find very good agreement between the data sets. The combination of these methods therefore can be used in parallel and can provide insights on the material quality and defect analysis.

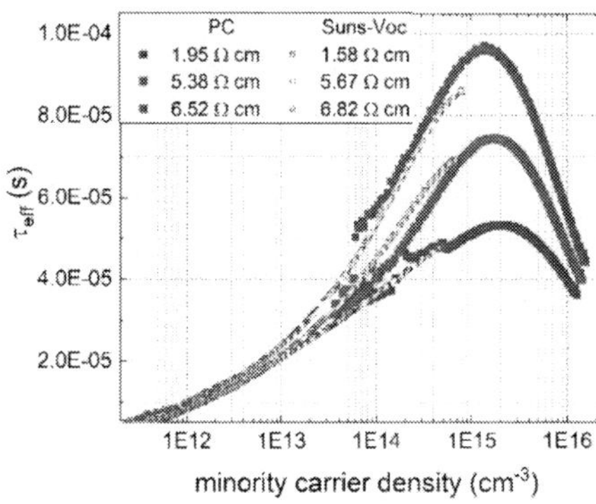

Figure 4: Minority carrier density dependent effective lifetime comparison from QSS-PC and QSS-V_{OC}.

4 CONCLUSIONS

We demonstrate the application of in-situ temperature- and illumination-controlled suns-V_{OC} measurements technique on a variety of finished and unmetallized solar cells. It is shown that this technique allows for an accurate real-time analysis of solar cell performance under heat and illumination which is of the most relevance for photovoltaic applications. Key benefits of the technique, and in particular of this tool, are identified as: reliable parameter in-situ extraction of V_{OC}, V_{mp}, pFF and τ_{eff} at high temperatures, simultaneous characterization of up to 4 cells, increased data-acquisition for more accurate reaction rate studies on finished cells, flexibility of sample sizes and architectures, low injection range lifetime analysis without trapping effects. We confirm the validity of the results extracted by the batch processing tool by comparing them with room temperature measurements of the samples with a calibrated I-V tester and photoconductance decay lifetime measurements of sister wafers.

REFERENCES

1 Kurtz, S., Whitfield, K., TamizhMani, G., *et al.*: 'Evaluation of high-temperature exposure of photovoltaic modules'*Prog. photovoltaics Res. Appl.*, 2011, **19**, (8), pp. 954–965.

2 Niewelt, T., Sch, J., Warta, W., Glunz, S.W., Schubert, M.C.: 'Degradation of Crystalline Silicon Due to Boron – Oxygen Defects'*IEEE J. Photovoltaics*, 2016, **preprint**, (1), pp. 1–16.

3 Chen, D., Vaqueiro Contreras, M., Ciesla, A., *et al.*: 'Progress in the understanding of light- and elevated temperature-induced degradation in silicon solar cells: A review'*Prog. Photovoltaics Res. Appl.*, 2020, (June), pp. 1–22.

4 Wright, B., Madumelu, C., Soeriyadi, A., Wright, M., Hallam, B.: 'Evidence for a Light-Induced Degradation Mechanism at Elevated Temperatures in Commercial N-Type Silicon Heterojunction Solar Cells'*Sol. RRL*, 2020, **2000214**, pp. 1–5.

5 Sinton, R.A., Swanson, R.M.: 'Recombination in Highly Injected Silicon'*IEEE Trans. Electron Devices*, 1987, **34**, (6), pp. 1380–1389.

6 Hallam, B.J., Chan, C.E., Chen, R., *et al.*: 'Rapid mitigation of carrier-induced degradation in commercial silicon solar cells'. *J. Appl. Phys.*, 2017.

7 Chan, C., Hamer, P., Bourret-Sicotte, G., *et al.*: 'Instability of Increased Contact Resistance in Silicon Solar Cells Following Post-Firing Thermal Processes'*Sol. RRL*, 2017.

IMPACT OF CUT EDGE RECOMBINATION IN HIGH EFFICIENCY SOLAR CELLS - MEASUREMENT AND MITIGATION STRATEGIES

Ning Chen*, Daniel Tune, Florian Buchholz, Andreas Halm, Valentin Mihailetchi
International Solar Energy Research Center (ISC) Konstanz, Rudolf-Diesel-Str. 15, D-78467 Konstanz, Germany
*Corresponding author: ning.chen@isc-konstanz.de

ABSTRACT: Many new module designs feature cut cells, especially now that cell and module manufacturers are introducing larger wafer formats (166mm, 182mm, and even 210mm in edge length). However, the cutting procedure introduces extra power losses. This recombination occurs when the cut edge is damaged by a high power laser or exposed to the air without passivation. In this paper, we employ high resolution photoluminescence to investigate the effect of edge recombination on interdigitated back contact solar cells. Edge recombination losses are compared by comparing various photoluminescence profiles obtained from the samples with different structures and cut techniques. First, we compare various laser cutting techniques at various locations. Two questions are addressed: where to cut the cells and how to cut them. Last, we compare the repassivation ability of the samples with different cutting techniques and structures. Repassivation can be used to mitigate edge cut losses with specific cutting techniques.
Keywords: silicon solar cell, edge recombination, photoluminescence, edge repassivation

1 INTRODUCTION

More and more solar modules are using cut cells, because of their lower series resistance loss. Half-cut cell modules, triple-cut cell modules, and even shingle modules are becoming popular in the PV market. Furthermore, with the trend of introducing big wafer formats, it is become more important to cut cells to small pieces to reduce the series resistance losses.

However, the cut process also introduces additional edge recombination losses. The edges after cutting are usually damaged by laser, and unpassivated edges are exposed to the air after cutting. The more pieces the cell cut, the more losses introduced. For high efficiency solar cells, the losses become more critical.

In this paper, we will assess the losses at the cell edges, and we will also evaluate the strategies to reduce them. Our study is focused on Interdigitated Back Contact (IBC) solar cells, and the cell structure is based on the latest ZEBRA cells [1] – IBC technologies developed by ISC Konstanz. High resolution photoluminescence will be used to assess cut edge recombination, and the ability of different samples to repassivate will be investigated to mitigate cut edge losses.

2 METHODS

2.1 PL for edge recombination evaluation

Photoluminescence (PL) can provide both spatial and temporal information about the passivation of wafers, including wafer edges. It is an effective tool for studying edge recombination. PL has been reported to be used for characterization of edge combination losses at ISC [2] and other research institutes [3, 4].

The setup of PL includes,

- Silicon charge-coupled device (CCD) with the cooling system
- Macro lens with a GaAs filter, resolution up to 1024 × 1024 pixels
- Two 30-watt lasers with the wavelength of 808 nm as the optical excitation source

The system is capable of producing high resolution PL images. Using the high resolution PL pictures, we can inspect the PL profile from well passivated areas to the wafer edge. The PL profile curve provides information on the quality of edge passivation/recombination.

2.2 Data analysis

As mentioned, the cross section PL profile from a sample with a cut edge can be obtained using the high resolution PL image. Figure 1 shows one example of converting a PL image to a cross section PL profile. By checking the cross section profile of the image, we get a cross section PL profile. From the PL profile, we can see the passivation quality decreases from the wafer centre to the wafer edge.

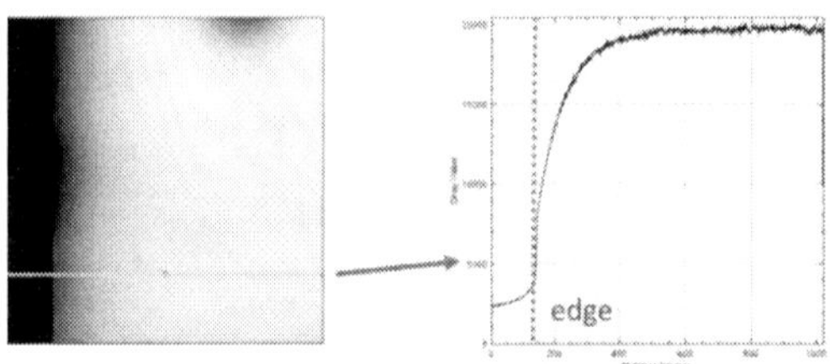

Figure 1: From a PL image (left) to a cross section PL profile (right)

Two methods are used to reduce the measurement error. First, either multi curves or average curves of an area are checked, to eliminate the low lifetime area that may be related to wafer defects or process defects. Second, because lifetime varies from sample to sample, a normalized PL curve is used to eliminate the lifetime influence.

3 EXPERIMENTS

In the experiments, we have two groups of samples.

- Cut samples without any further treatment, the samples include various structures and cut techniques. This group was formed to investigate the effects of cutting techniques and structure on edge recombination.
- Cut samples with additional treatment, to investigate the repassivation to mitigate the edge recombination losses.

3.1 Cut samples without any further treatment

We have prepared samples of various structures to

study the cutting losses. The samples are based on the *Gen2* ZEBRA cells with Front Surface Field (FSF), as shown in Figure 2.

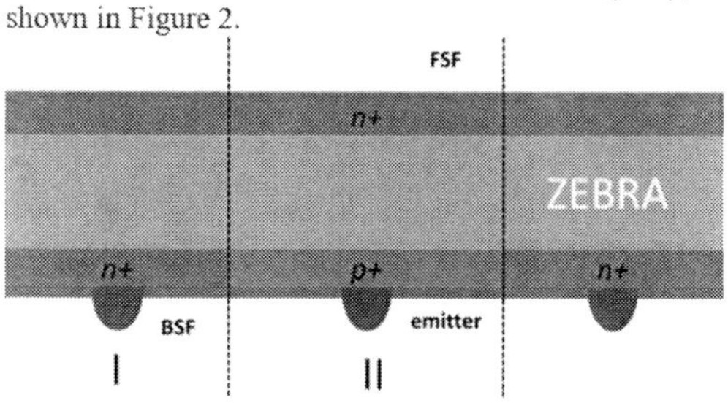

Figure 2: Schematic structure of ZEBRA *Gen2* solar cell

As illustrated in Figure 2, we can cut from several locations on IBC cells.

- On region I, we can cut from the $n+$ area, from either FSF or Back Surface Field (BSF) side
- On Region II, we can cut through $n+$ and $p+$, namely the FSF and emitter sides

We can also select from a variety of cutting techniques. There are three possible cut methods:

- Laser scribing and cleaving (L&C). The laser creat a depth at least of 30% in the depth of a wafer, and followed by mechanical cleaving
- Thermal laser separation (TLS) [5], advanced laser cutting technology with much less laser damage than L&C
- Mechanical cleaving along with the crystal orientation without any laser damage

Based on the two types of structure and different cut techniques, we have prepared two groups of samples, as shown in Figure 3 and Figure 4. There are mainly two types of damages depicted in the figures. The first type is the high-power laser introduces laser damage to the silicon wafers, either on $n+$ or $p+$. The second type is cleaving damage. After cleaving, the silicon edges expose to the air without passivation.

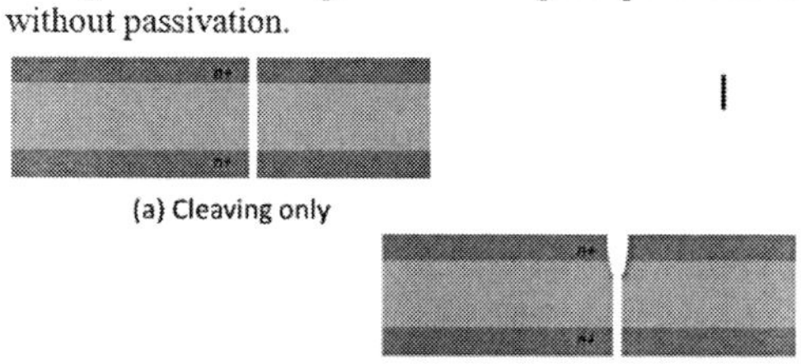

Figure 3: Schematic structure of $n+/n/n+$ samples, (a) cleaving and (b) laser damage on $n+$ side - laser scribing and cleaving or TLS.

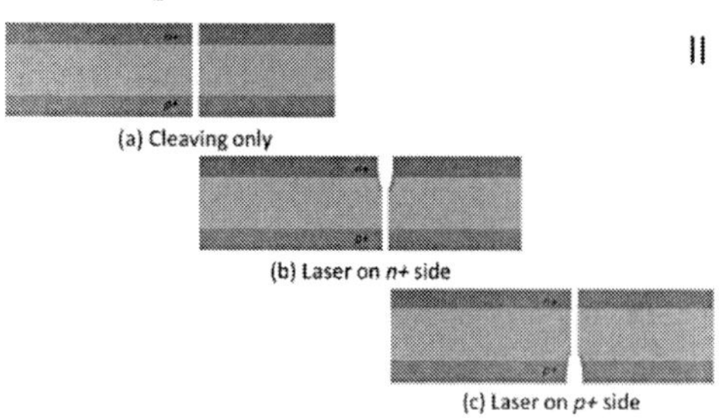

Figure 4: Schematic structure of $n+/n/p+$ samples, (a) cleaving, (b) laser damage on $n+$ side, and (c) laser damage on $p+$ side.

Detailed samples list as shown in Table I.

- Samples from G1 to G6 were cut from n+ with different techniques
- G1, G3, and G5 are samples with the structure of region I; G2, G4, and G6 are samples with region II structure
- L&C 5 passes introduce laser damage ~30% of wafer thickness, while L&C 15 passes introduce laser damage ~60% of wafer thickness
- G7 and G8 are samples that were mechanical cleaved
- G17 is the reference group, and the edge was naturally passivated during the cell process
- G18 and G19 are type II samples cut from the emitter by L&C

Table I: Samples list for PL evaluation.

Group no.	*Cut region*	*Sample str. (f | b | r)*	*Cut type*	*Cut side*
1	I	$n+ \| n \| n+$	TLS	$n+$
2	II	$n+ \| n \| p+$	TLS	$n+$
3	I	$n+ \| n \| n+$	L&C (5 passes)	$n+$
4	II	$n+ \| n \| p+$	L&C (5 passes)	$n+$
5	I	$n+ \| n \| n+$	L&C (15 passes)	$n+$
6	II	$n+ \| n \| p+$	L&C (15 passes)	$n+$
7	I	$n+ \| n \| n+$	Cleave	n.a.
8	II	$n+ \| n \| p+$	Cleave	n.a.
17	reference	$n+ \| n \| p+$	No cut, passivated edge	n.a.
18	II	$n+ \| n \| p+$	L&C (15 passes)	$p+$
19	II	$n+ \| n \| p+$	L&C (5 passes)	$p+$

3.2 Repassivation samples

After laser cutting, repassivation can be utilized to reduce edge recombination losses. The Atomic layer deposition (ALD) of aluminium oxide (AlO_x) was reported [6, 7] for repassivation. This research focuses on a polymer-based passivation approach that is simple to implement in our lab. In our test, we employed Nafion®, for passivation. Researchers have employed Nafion to passivate wafers and achieved good passivation [8, 9].

Our optimized repassivation process flow is as follows. First, samples are cleaned with 10 wt.% Hydrofluoric acid (HF) for 10 seconds. Then the samples are washed with deionized water for 10 seconds and followed by paper drying. After drying, the samples are repassivated by dipping into a 5 wt.% Nafion® solution. Finally, samples are dried with nitrogen.

Four samples with different structures and cutting techniques were prepared, to test their repassivation ability. Sample details as below.

- $n+/n/p+$ sample (region II), TLS cut from $n+$, with minimum laser damage
- $n+/n/p+$ sample (region II), L&C cut from $n+$, with laser damage of 60% of wafer thickness
- $n+/n/p+$ sample (region II), cleaving, with no laser damage
- $n+/n/n+$ sample (region I), TLS cut

The stability of repassivation was also examined. Continuous PL pictures were taken over time. The repassivation stability can be assessed by comparing the PL curves taken at different times.

4 RESULTS

4.1 Sample without further treatment

Normalized PL profiles of different samples are shown in Figure 5. The PL curves in the figure were measured at various light intensities. Figure (a) was taken under 1-sun, and Figure (b) was measured under 0.3-sun. There are two main takeaways from the figures. To begin, the PL curves can be classified into two splits. Higher PL curves are found in the split of the blue-like colours (groups G1, G3, G5, and G7). The other split in different colours has more edge recombination than the first split, and it can be further subdivided into three sub-groups. This observation will be discussed in greater detail in the following paragraphs. Second, we can see a clear difference between the 1-sun and 0.3-sun PL curves. The edge recombination effects are further from the edge under low light intensity, and the difference between groups is becoming larger. The conclusion is consistent with the previous report [10]. At low illumination, the PL curves near the edges become more sensitive to recombination.

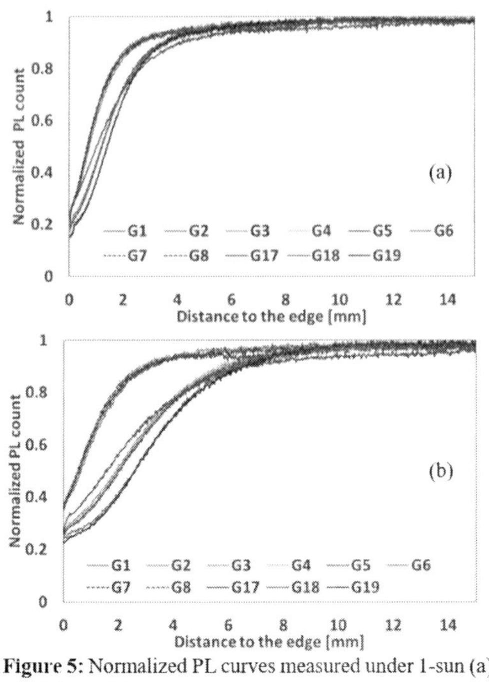

Figure 5: Normalized PL curves measured under 1-sun (a) and 0.3-sun (b).

Let's take a closer look at the different groups. Figure 6 depicts the outcomes of G1, G3, G5, and G7. The four curves are all from the region I and have an *n+/n/n+* structure. G1 is the TLS cut sample, G3 is the L&C 5 passes cut sample, G5 is the L&C 15 passes cut sample, and G7 is the cleaved sample. The *n+/n/n+* structure has less edge recombination losses than the *n+/n/p+* structure. There is no discernible difference between TLS cuts, L&C 5 passes, L&C 15 passes, and cleaving. Under low light, the results are consistent.

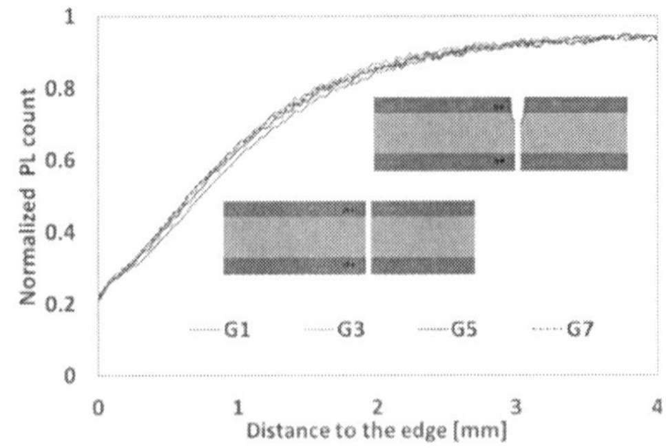

Figure 6: Normalized PL curves measured on *n+/n/n+* samples

The normalized curves of region II with *n+/n/p+* structure are shown in Figure 7. The curves can be further subdivided into three groups. The first sub-group with only G17. G17 with the passivated edge shows the best edge passivation. The second sub-group, which includes Groups G2, G4, G6, and G8, has medium losses. G2 received TLS, G4 received L&C 5 passes, G6 received L&C 15 passes, and G8 was cleaved only. In this sub-group, the laser damage is on the *n+* side or there is no laser damage and only cleaving damage (G8). Laser damage to the *n+* side results in recombination losses similar to cleaving damage. The last sub-group includes G18 and G19. Both groups have laser damage on the *p+* side. Despite the laser damage depths in the two groups differ, the PL profiles appear to be similar.

In summary, on the *n+/n/p+*, samples with FSF cut and the sample was cleaved only showed similar edge recombination, and both had less edge losses than samples cutting from the emitter.

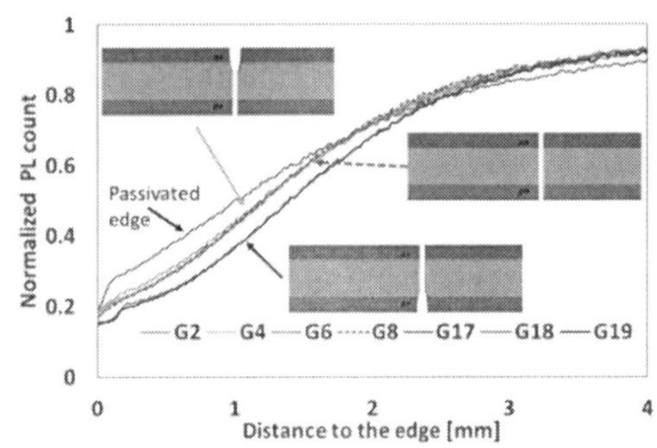

Figure 7: Normalized PL curves measured on *n+/n/p+* samples

Based on the comparison, we can conclude that the cut regions are significant for edge losses. The best way to minimize cut losses for IBC solar cells is to cut at *n+/n/n+* regions. If you need to cut from the *n+/n/p+* region, cut from the *n+* side rather than the *p+* side. Furthermore, no difference in cut techniques was observed in the experiments.

4.2 Repassivation

The repassivation results for the four different samples are shown in Figure 8.

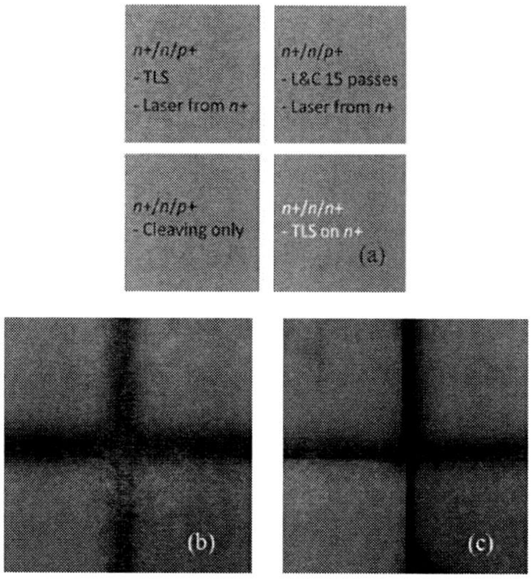

Figure 8: Repassivation samples locations (a), before repassivation (b) and after repassivation (c)

For *n+/n/p+* samples, both TLS cut and cleaved samples were repassivated – the edges after repassivation (shown in figure (c)) are brighter in colour than the edges before repassivation. The cleaved edges exhibit consistent repassivation for both edges. The TLS cut sample exhibits one edge that is better repassivated than the other, and it may be related to the process that needs further investigation. No repassivation for the samples with 60% laser damage.

For the *n+/n/n+* sample with TLS cut, the edges were also repassivated.

The repassivation stability of the cleaved sample shows in Figure 9. Good repassivation from Nafion at the beginning. Then, without encapsulation, the repassivation quality begins to deteriorate over time; after 100 minutes, the repassivation remains 50% of the initial repassivation value. After 16 hours, the repassivation remain ~25% of the initial value.

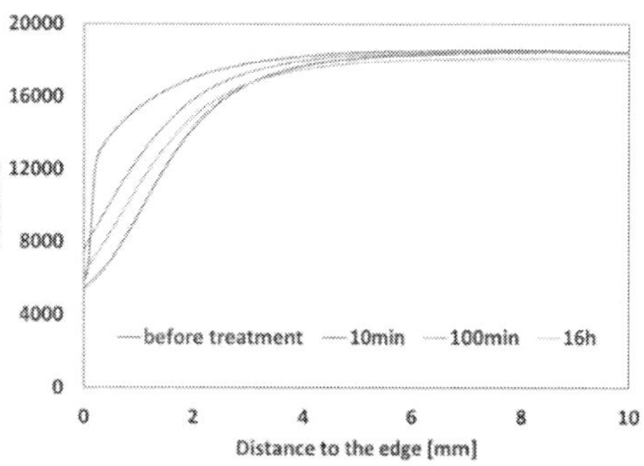

Figure 9: Results of repassivation on the cleaved edge. PL curves were measured before repassivation, and after repassivation at various times.

5 SUMMARY AND OUTLOOK

High resolution PL was used to investigate edge recombination in IBC solar cells. The results for the most recent ZEBRA cell structure show that the losses are primarily determined by the cutting regions, with no influence from cut techniques. It is preferable to cut IBC cells from *n+/n/n+* regions to reduce edge cut losses. As a next step, we'll confirm the findings at the solar cell level.

The repassivation experiment revealed differences in the cut techniques. While the TLS cut and cleaved edge can be repassivated, the sample with 60% laser damage cannot. If repassivation is performed later, the cut techniques should be carefully chosen. Finally, the repassivation in our experiment is not stable, and notable degradation is observed over time. We will continue to improve the stability in the future.

ACKNOWLEDGEMENTS

This work was supported by the German Federal Ministry for Economic Affairs and Energy (BMWi) as part of the HoSSa project with funding reference number 03EE1014D, and has also received funding from the European Union's Horizon2020 Programme for research, technological development and demonstration under Grant Agreement no. 857793.

REFERENCES

[1] R. Kopecek et al., "ZEBRA technology: Low cost bifacial IBC solar cells in mass production with efficiency exceeding 23.5%," Conf. Rec. IEEE Photovolt. Spec. Conf., vol. 2020-June, pp. 1008–1012, 2020

[2] D. Tune, F. Buchholz, I. Ullman, and A. Halm, Measuring and Mitigating Edge Recombination in Modules Employing Cut Cells. 37th Eur. Photovolt. Sol. Energy Conf., no. October, pp. 322–328, 2020

[3] H. Stolzenburg et al., "Edge recombination analysis of silicon solar cells using photoluminescence measurements," AIP Conf. Proc., vol. 2147, no. 1, p. 20017, Aug. 2019

[4] K. C. Fong et al., "Perimeter recombination characterization by luminescence imaging," IEEE J. Photovoltaics, vol. 6, no. 1, pp. 244–251, 2016

[5] F. Kaule, M. Pander, M. Turek, M. Grimm, E. Hofmueller, and S. Schoenfelder, "Mechanical damage of half-cell cutting technologies in solar cells and module laminates," AIP Conf. Proc., vol. 1999, no. August, 2018

[6] P. Baliozian et al., "Postmetallization 'Passivated Edge Technology' for Separated Silicon Solar Cells," IEEE J. Photovoltaics, vol. 10, no. 2, pp. 390–397, 2020

[7] B. Portaluppi, S. Harrison, V. Giglia, A. Sekkat, and D. M. Corresponding, "Insights on cell edge defects impact and post-process repassivation for heterojunction," 37th Eur. Photovolt. Sol. Energy Conf., no. October, pp. 504–507, 2020

[8] D. D. Tune, N. Mallik, H. Fornasier, and B. S. Flavel, "Breakthrough Carbon Nanotube–Silicon Heterojunction Solar Cells," Adv. Energy Mater., vol. 10, no. 1, p. 1903261, Jan. 2020

[9] D. Biro and W. Warta, "Low temperature passivation of silicon surfaces by polymer films," Sol. Energy Mater. Sol. Cells, vol. 71, no. 3, pp. 369–374, 2002

[10] M. Hermle, J. Dicker, W. Warta, S. W. Glunz, and G. Willeke, "Analysis of edge recombination for high-efficiency solar cells at low illumination densities," Proc. 3rd World Conf. Photovolt. Energy Convers., vol. B, pp. 1009–1012, 2003

PERFORMANCE ENHANCEMENT OF PERC SOLAR CELL WITH SIO_XN_Y BACK SURFACE PASSIVATION BY LOW TEMPERATURE ANNEALING PROCESS

Ahmet Emin Keçeci[1,2*], Gence Bektaş[1,2], Emine Hande Çiftpınar[1,2], Sümeyye Koçak Bütüner[1], Hasan Asav[1,2], Gamze Kökbudak[2], Hasan Hüseyin Canar[1,3], Bülent Arıkan[1], Raşit Turan[1,3]
[1]ODTÜ-GÜNAM, Çankaya, 06800 Ankara, Turkey
[2]Micro and Nanotechnology Graduate Program, Middle East Technical University, Çankaya, 06800 Ankara, Turkey
[3]Department of Physics, Middle East Technical University, Çankaya, 06800 Ankara, Turkey

ABSTRACT: Phosphorus doped emitter surfaces have been widely passivated by PECVD deposited silicon nitride (SiN_x) layers which also act as anti-reflection coating (ARC). However, the following high temperature processes applied in the regular p-based cell fabrication sequences might result in significant degradation in open circuit voltage (V_{oc}) values. In this study, an additional heat treatment called as low temperature annealing (LTA) process after SiN_x deposition is applied to improve implied open circuit voltage (iV_{oc}) values of symmetrical p-type samples with symmetrically diffused emitter. Around 15 mV increase in iV_{oc} is obtained with the optimized LTA process conditions for symmetrical samples. We also demonstrate the effect of LTA process on solar cell performance by approximately 10 mV gain in open circuit voltage and approximately 0.5%abs gain in cell efficiency.

1 INTRODUCTION

Silicon (Si) solar cell technology is the dominant photovoltaic technology due to low cost, superior stability and conversion efficiencies over 20% in mass production[1]. P-type Si wafers have a higher market share than n-type wafers in the photovoltaic industry due to ease in manufacturing of uniformly doped Si ingot and phosphorus diffusion to form emitter on p-type Si wafer. However, the passivation of p-type wafers can be challenging due to low tolerance to metal impurities (Fe, Ni, Co, Al) and defects such as B-O complexes [2], [3]. Amongst the existing cell structures which are interdigitated back contact (IBC) [4], heterojunction technology (HJT) [5], tunnel oxide passivated contact (TOPCON) [6], passivated emitter rear contact (PERC) [7], aluminum back-surface field (Al-BSF),. PERC technology is the most dominant one due to the simplicity of the process flow, low cost, high production rate and an average conversion efficiency of 22% [8]. With the help of the rear side passivation and local contacts, recombination in the rear side of the cell is decreased dramatically compared to the Al-BSF solar cells [9]. Emerging metallization technologies provide better metal-semiconductor contacts and low recombination under metal contacts. Considering all these aspects, the main loss in a PERC solar cell comes from the highly doped emitter region. To achieve a low recombination rate in that area, emitter passivation quality is an important factor as much as emitter itself. It has been shown that LTA after $POCl_3$ deposition can be used for redistribution of inactive phosphorus atoms in the emitter region [10]–[12] or increasing gettering effect of interstitial iron (Fe_i) [12], [13]. This paper shows the optimization of LTA process after SiN_x deposition to improve emitter passivation.

2 EXPERIMENTAL DETAILS

In this study, p-type, textured, industrial size Cz-Si wafers with a resistivity of 1-3 Ω.cm and thickness of 180 µm were used as substrate. For the optimization of the LTA process, symmetrical samples were fabricated by depositing a SiN_x layer on both sides of the wafers after the phosphorus diffusion with a resulting 90 Ω/□ emitter sheet resistance and 0.3 µm junction depth by using $POCl_3$ as the dopant source. Then, five different temperatures and two different durations were applied for LTA processes in a tube furnace under N_2 flow. Following the LTA, a fast-firing step was applied using an industrial conveyor belt furnace with a peak set temperature of 910°C. Quasi-steady-state photo conductance (QSSPC) method [14] was used to characterize the symmetrical samples. After optimizing the LTA condition, PERC cells are fabricated. For the cell fabrication, textured wafers underwent through a phosphorus diffusion followed by the single side etching process for rear side emitter removal. While silicon oxynitride (SiO_xN_y) layer capped with SiN_x was used for passivation of the rear surface, SiN_x layer was used for the front surface. To see the effects of the optimized LTA process on cell performance, wafers were divided into two groups. One group of wafers was exposed to LTA while the other group was used as the control group. The rear side of the samples was ablated by a picosecond pulsed laser prior to metallization. Finally, metallization and firing processes were performed for contact formation. For the cell characterization, current-voltage (I-V) measurements were conducted with a class AAA solar simulator.

3 RESULTS AND DISCUSSION

3.1 Symmetrical samples

To reach the optimum condition for LTA, symmetrically phosphorus diffused and SiN_x passivated samples were processed in two different batches. For the first batch, as illustrated in Figure 1, two different annealing temperatures (600°C and 700°C) were applied for two different time periods (17 and 60 minutes). By comparing the iV_{oc} values "After diffusion" (black boxes) and "After SiN_x deposition" (blue boxes), it was observed that the samples were almost identical as expected. In the control group, samples were directly fast fired after SiN_x deposition (no LTA). The results show that if LTA process is conducted at 600°C, no significant change is observed in the after fast firing. On the other hand, LTA process at 700°C significantly improves iV_{oc} values of the fired samples (named "After Firing") by approximately 10 mV regardless of duration. Hydrogen might be playing a role for the improvement of iV_{oc} since diffusion of H from SiN_x to Si depends on the temperature [15]. It is also deduced from the improvement of iV_{oc} values from "After SiN_x" to "After LTA" (green boxes) results. Moreover, the behavior between "After LTA" and "After Fast Firing" (red boxes) results should be taken into consideration. For the firing stability of the symmetrical samples, it is necessary to apply LTA process at a higher temperature and for a longer time. In the case of LTA at 600°C, iV_{oc} values decrease with a process time of 60 minutes. However, for 700°C and 60 minutes duration, iV_{oc} values or the passivation quality improves.

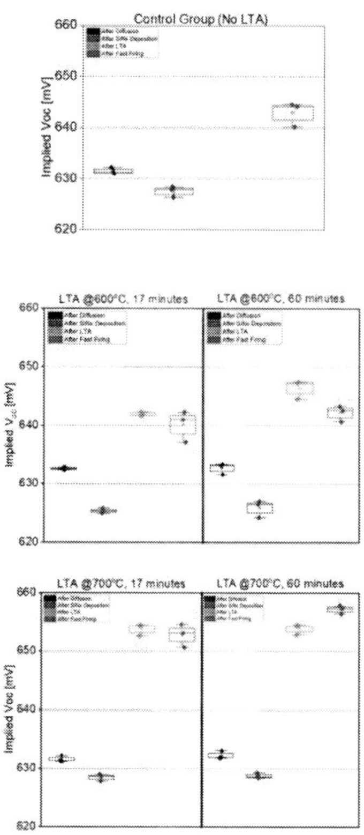

(a)

(b)

(c)

Figure 1: iV_{oc} results of the first batch of symmetrical emitter samples (a) samples with no LTA treatment (control group) (b) samples with LTA at 600°C for 17 and 60 minutes (c) samples with LTA at 700°C for 17 and 60 minutes

Comparing with the control group, LTA condition at 700°C for 60 shows more than 10 mV improvement for the finished samples. According to the results of the first batch, increasing LTA temperature enhances the iV_{oc} of the samples. Therefore, to investigate the higher temperatures above 700°C, we fabricated a second batch of symmetrical samples. In the second batch, three different LTA temperatures are applied to the group of samples as 700°C, which was the best LTA condition in the first batch and used as a reference, 725°C and 750°C for 17 minutes and 60 minutes. We also included reference samples without LTA.

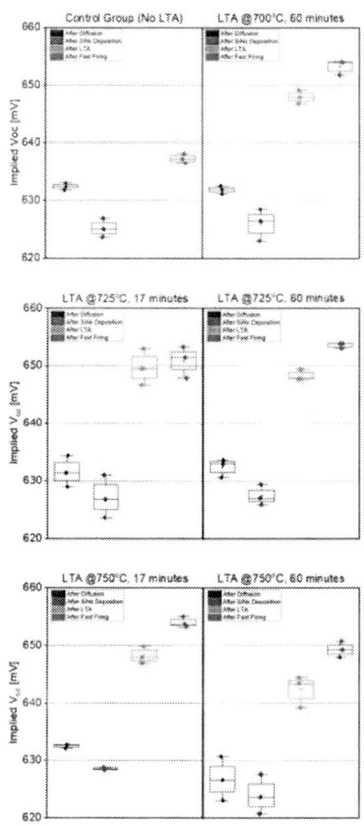

(a)

(b)

(c)

Figure 2: iV_{oc} results of the second batch of symmetrical emitter samples (a) samples with no LTA treatment (control group) and LTA at 700°C for 60 minutes as a reference (b) samples with LTA at 725°C for 17 and 60 minutes (c) samples with LTA at 750°C for 17 and 60 minutes

Figure 2 demonstrates the results of the second batch of samples. Regardless of the process time and temperature, more than 10 mV of iV_{oc} gain is observed and the same passivation quality is obtained for all LTA conditions. At 750°C, for 60 minutes of process time, the results are lower than the other LTA conditions. This may be due to lower quality emitter, which can be deduced from the "After Diffusion" measurements results. Also, 17 minutes of process time for LTA at higher temperatures (725°C and 750°C) can be enough to reach almost the maximum iV_{oc} values.

3.2 PERC cells

Figure 3: Schematics of the PERC solar cell

Process condition of LTA was chosen based on the results of symmetrical samples. Since 700°C for 60 minutes results in maximum iV_{oc} values in both batches, this condition is applied for the fabrication of PERC solar cells. The study with the symmetrical emitter samples demonstrated that LTA has a major impact on the passivation of the emitter. For the next step, a new set of samples were fabricated to make PERC cells, the structure can be seen in Figure 3. Samples are divided into two groups to observe the effect of LTA. The first group of samples were fabricated without LTA process as a reference set while the second group of samples was exposed the LTA at 700°C for 60 minutes. For the rear surface passivation, silicon oxynitride (SiO_xN_y) was grown on the surface by using plasma enhanced chemical vapor deposition (PECVD) due to its thermal stability so that no degradation will occur during long LTA processes [16]. Figure 4 demonstrates the results of the solar cells in terms of V_{oc}, J_{sc}, FF and efficiency. V_{oc} of the PERC cells with LTA treatment shows an approximately 10 mV increase, which is consistent with iVoc results of symmetrically passivated emitter samples. Along with an increase in V_{oc}, we obtained 0.25 mA improvement in average J_{sc} thanks to LTA process. Moreover, LTA does not affect the fill factor values of the device.

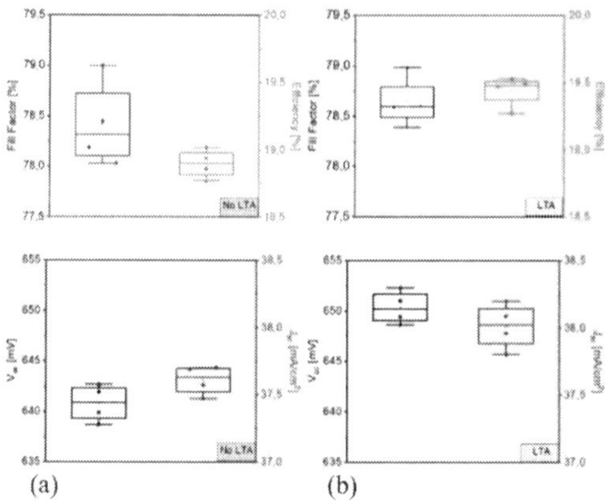

Figure 4: PERC cell results from I-V curve showing the effect of LTA on V_{oc}, J_{sc}, FF and efficiency on (a): PERC cells fabricated without LTA and (b): PERC cells fabricated with LTA (at 700°C for 60 minutes)

4 CONCLUSIONS

In this study, we showed an optimized LTA condition to increase device performance on PERC solar cells with SiO_xN_y passivation layer at the rear side. LTA process after PECVD:SiN_x resulted in 10 mV gain in V_{oc} along with 0.25 mA increase in J_{sc}. The result of the symmetrical samples implies that this increase in the cell performance is related to the enhanced passivation of the emitter.

5 REFERENCES

[1] S. Ma, H. B. Tang, Z. P. Li, X. Y. Kong, and W. Z. Shen, "Application of SiOxNy films in industrial bifacial PERC solar cells," *Solar Energy Materials and Solar Cells*, vol. 230, Sep. 2021, doi: 10.1016/j.solmat.2021.111199.

[2] A. Herguth, G. Schubert, M. Kaes, and G. Hahn, "A new approach to prevent the negative impact of the metastable defect in boron doped CZ silicon solar cells," in *Conference Record of the 2006 IEEE 4th World Conference on Photovoltaic Energy Conversion, WCPEC-4*, 2006, vol. 1, pp. 940–943. doi: 10.1109/WCPEC.2006.279611.

[3] N. Nampalli, B. J. Hallam, C. E. Chan, M. D. Abbott, and S. R. Wenham, "Influence of Hydrogen on the Mechanism of Permanent Passivation of Boron-Oxygen Defects in p-Type Czochralski Silicon," *IEEE Journal of Photovoltaics*, vol. 5, no. 6, pp. 1580–1585, Aug. 2015, doi: 10.1109/JPHOTOV.2015.2466457.

[4] M. D. Lammert and R. J. Schwartz, "The Interdigitated Back Contact Solar Cell: A Silicon Solar Cell for Use in Concentrated Sunlight," *IEEE Transactions on Electron Devices*, vol. ED-24, no. 4, pp. 337–342, 1977, doi: 10.1109/T-ED.1977.18738.

[5] M. Taguchii\ *et al.*, "Special Issue HIT TM Cells*High! Ef_ciency Crystalline Si Cells with Novel Structure."

[6] F. Feldmann, M. C. Bivour Reichel, M. Hermle, and S. W. Glunz, "A PASSIVATED REAR CONTACT FOR HIGH-EFFICIENCY n-TYPE SILICON SOLAR CELLS ENABLING HIGH V OC S AND FF>82%," 2013.

[7] A. Wang, J. Zhao, and M. A. Green, "22.8 % efficient siHcon solar cell," 1363.

[8] B. Min *et al.*, "A Roadmap Toward 24% Efficient PERC Solar Cells in Industrial Mass Production," *IEEE Journal of Photovoltaics*, vol. 7, no. 6, pp. 1541–1550, Nov. 2017, doi: 10.1109/JPHOTOV.2017.2749007.

[9] S. Gatz, J. Müller, T. Dullweber, and R. Brendel, "Analysis and optimization of the bulk and rear recombination of screen-printed PERC solar cells," in *Energy Procedia*, 2012, vol. 27, pp. 95–102. doi: 10.1016/j.egypro.2012.07.035.

[10] A. Peral *et al.*, "Electrically-inactive phosphorus re-distribution during low temperature annealing," *Journal of Applied Physics*, vol. 123, no. 16, Apr. 2018, doi: 10.1063/1.5002627.

[11] A. Dastgheib-Shirazi, A. Peral, M. Steyer, J. Rinder, H. Wagner, and G. Hahn, "Dissolution of Electrically Inactive Phosphorus by Low Temperature Annealing," in *Energy Procedia*, 2015, vol. 77, pp. 286–290. doi: 10.1016/j.egypro.2015.07.040.

[12] H. Li, M. Abbott, B. Hallam, and S. Wenham, "Low Temperature Thermal Treatments for n-type Emitters on Si Solar Cells," in *Energy Procedia*, Aug. 2016, vol. 92, pp. 466–473. doi: 10.1016/j.egypro.2016.07.128.

[13] A. Jiménez, C. del Cañizo, C. Cid, and A. Peral, "Phosphorus Diffusion Gettering Efficacy in Upgraded Metallurgical-Grade Solar Silicon."

[14] R. A. Sinton and A. Cuevas, "Contactless determination of current-voltage characteristics and minority-carrier lifetimes in semiconductors from quasi-steady-state photoconductance data," *Applied Physics Letters*, vol. 69, no. 17, pp. 2510–2512, Oct. 1996, doi: 10.1063/1.117723.

[15] S. Wilking, S. Ebert, A. Herguth, and G. Hahn, "Influence of hydrogen effusion from hydrogenated silicon nitride layers on the regeneration of boron-oxygen related defects in crystalline silicon," *Journal of Applied Physics*, vol. 114, no. 19, Nov. 2013, doi: 10.1063/1.4833243.

[16] F. H. P M Habraken *et al.*, "Properties of PECVD SiO Ny Films as Selective Diffusion Barrier," 1986.

DEVELOPMENT AND IMPLEMENTATION OF A REFINED MODEL FOR COMPREHENSIVE CHARACTERIZATION AND OPTIMIZATION OF HIGHLY EFFICIENT SILICON SOLAR CELLS

A.V. Sachenko[1], V.P. Kostylyov[1], V.M. Vlasiuk[1], I.O. Sokolovskyi[1],
G. Perinparajah[2] and A.I. Shkrebtii[2*]
[1] V. Lashkaryov Institute of Semiconductor Physics, NAS of Ukraine, 03028 Kyiv, Ukraine
[2] Ontario Tech University, Oshawa, ON, L1G 0C5, Canada

*Corresponding author: anatoli.chkrebtii@ontariotechu.ca

ABSTRACT: Crystalline Silicon (c-Si) remains a dominant photovoltaics material in solar cell industry. Currently, scientific and technological advances make possible of producing the c-Si solar cells (SCs) efficiency close to the fundamental limit. Therefore, combining the experimental results with modeling becomes crucial to further improve the efficiency and reducing the photovoltaics systems cost. We carried out the experimental characterization of the highly-efficient textured c-Si SCs and compared the results with the analytical modeling formalism [1] and a *preliminary* modeling, using the commercial solar cell simulation software [2]. The analytical model accounts for all recombination mechanisms, including nonradiative exciton recombination and recombination in the space-charge region (SCR). To compare the theoretical results with the experiment, we proposed empirical formula for the external quantum efficiency (EQE), which describes its experimental spectral dependence near the absorption edge. The approach used allows a comparative modeling of the short-circuit current and photoconversion efficiency in the textured crystalline silicon solar cells. The theoretical results, both from analytical formalism and computational tools, compared to the experimental measurements, allowed: (i) Validating the formalism developed, which was then used to optimize the key parameters of the SCs, such as the base thickness, doping level and others. (ii) We demonstrated several advantages of the application of our analytical formalism, compared to the computational software tools.
Keywords: silicon, solar cell, texture, efficiency, recombinations, optimization

1 INTRODUCTION

To reduce the cost of solar cells (SCs) while maximizing their photoconversion efficiency η, surface texturing is used to reduce the optical reflection losses and maximize the number of photoactive photons absorbed. Therefore, surface texturing in photovoltaics systems is currently drawing much attention [3-5]. Recent record-breaking efficiency of 26.6% for Si solar cells have been achieved on such textured systems [6-8], while the theoretical limit of the efficiency for Si SCs is about 29%. The important component of the photon management is the texturing, an artificial roughening of the surface, which reduces its reflection by increasing the chances for reflected light bouncing back onto the semiconductor surface, rather than out of it to the surrounding and increasing the photon propagation path inside the semiconductor. This is important for making the solar cells thinner and even flexible [5], [9]. The commonly accepted model for the textured surface from the reflection reduction point of view is a completely randomized Lambert surface [10], although different textured patterns are experimentally used [5]. This complicates the theoretical, both analytical and computational, descriptions of the textured SCs.

To advance even more in improvement of the SCs efficiency the available optimisation approaches should be further developed and applied. Currently, further increase of the silicon SC efficiency becomes more and more difficult, and requires better consideration of the physical processes in the SCs. In particular, to further increase the SCs efficiency it is necessary to take into account all contributing recombination mechanisms in silicon. This has to include a non-radiative exciton recombination via deep impurity centers [11] and recombination in the space-charge region (SCR) [12] . Still, the above two mechanisms are not included in the existing SC optimization formalisms and software [2]. However, as we previously demonstrated [13], these recombination processes are detrimental in the high-efficiency silicon-based SCs. They may significantly influence the SCs efficiency as compared to the mechanisms that are already considered, such as, in particular, the radiative and band-to-band Auger recombinations. In order to model and optimize textured high-efficiency silicon SCs we first measured and analyzed experimentally the external quantum efficiency (EQE). For the analytical formalism the main input parameter required is the experimentally measured short-circuit current density J_{SC}. Next, the short-circuit current and the photoconversion efficiency under AM1.5 conditions, both dependent on the base thickness d and its doping level are derived, and used to optimize the characteristics of the textured silicon SCs. Important is that the theoretical approach, presented here, is applicable to any type of textured SCs surface: The experimentally measured short-circuit current density J_{SC} is a required parameter in the analytical formalism [1]. Apart from being dependent on the standard solar cell parameters such as recombination, doping level, etc., J_{SC} also includes implicitly influence of the surface texture type, antireflective coating (ARC), and angular distribution of the photoactive photon flux in the semiconductor. The last three factors cannot be very accurately accounted for in the solar cell simulations software.

As an example, to demonstrate a considerable variation of the SC reflectivity, we used the online simulator Opal 2 [14] to calculate the change in the Si solar cell reflectivity when the texture was created using regular pyramids, random pyramids and spherical caps, as well as the planar surface. The results are shown in Fig. 1. For this example, we also used a single layer TiO_2 ARC with thickness of 78 nm (the technical details of the reflectivity calculations are discussed in the next sections)

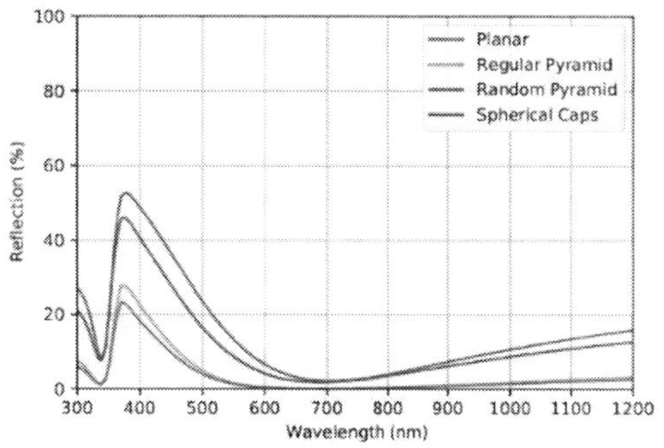

Figure 1: Calculated silicon solar cell surface reflectivity under AM1.5 illumination for different texture types *vs.* flat (planar) surface: regular pyramids (orange line), random pyramids (green) and spherical caps (red) *vs.* planar (blue)). The single 78 nm thick layer TiO_2 ARC has been used. Texturing combined with ARC substantially lower the reflectivity losses in SCs.

Using the solar cell simulation Quokka software [2] we also optimized the solar cells considered, the preliminary results of the simulation are presented and compared to the experimental results as well as the analytical formalism outcome.

To develop the analytical formalism, it is established that the experimental spectral dependencies of the external quantum efficiency $EQE(\lambda)$ in textured silicon SCs near the absorption edge can be described by an empirical formula $EQE(\lambda)=[1+b/(4n_r(\lambda)^2\cdot\alpha(\lambda)d)]^{-1}$, where $\alpha(\lambda)$ is the light absorption coefficient, d is the base thickness, n_r is the refractive index, and b is a non-dimensional coefficient, which characterizes the texturing profile impact. It is usually greater than 1, however, when b=1 the empirical formula transforms into the well-known Yablonovitch formula [15]. Physical meaning of the parameter b is the ratio of the photon path length in the case of a perfectly randomized surface, considered in [10], to the photon path length in a particular sample with nonrandomized flat surface. This formula allows to analytically model and optimize the photo-conversion efficiency of the solar cell in terms of base thickness d.

For the textured silicon SCs considered, the experimental $EQE(\lambda)$ dependence was compared to both empirical and simulated dependencies. It is shown that close to the absorption edge $EQE(\lambda)$ behavior can be accurately described by the empirical formula with the parameter b, ranging from 1.6 to 4. Using the experimental values $EQE(\lambda)$ and the proposed formula, we calculated the dependence of the short-circuit current density J_{SC} on the base thickness d for the textured SCs.

2 SUMMARY OF THE ANALYTICAL FORMALISM

We model the silicon p^+- n - n^+ type SCs, for details of the analytical formalism [1]. When the bulk minority-carriers diffusion length $L_d = \sqrt{D_p\tau_{eff}}$ is much bigger than $d/4$ (d is the base width), and combined surface and SCR recombination velocity $S_{SC} + S_{S0} << \frac{2D_p}{d}$, the excess carriers' concentration is practically uniform in the base region. (Here D_p is the carriers' diffusion coefficient and τ_{eff} is their effective recombination time). In this case within the narrow-base approximation the I-V characteristic under the illumination can be written as:

$$I(V) = I_L - I_r(V) + \frac{V+IR_s}{R_{SH}}, \quad (1)$$

where the first term is the photocurrent, the second term is the recombination current, A_{SC} is the SC area, and R_S and R_{SH} are its series and shunt resistances. The excess carrier concentration depends on the applied voltage V by a modified law of mass action as:

$$(n_0 + \Delta n)(p_0 + \Delta n) = n_{i0}^2\, exp\left(\frac{\Delta E_g + q(V-IR_S)}{kT}\right), \quad (2)$$

where n_{i0} is the intrinsic concentration at low injection level [7], and $\Delta E_g(n_0, \Delta n)$ is the bandgap narrowing in Si [8]. Eq. (2) can be used to write the excess concentration as:

$$\Delta n = -\frac{n_0}{2} + \sqrt{\frac{n_0^2}{4} + n_{i0}^2 e^{\Delta E_g/kT}\left(e^{q(V-IR_s)/kT} - 1\right)}. \quad (3)$$

Equations (1) – (3) have to be solved numerically. The open-circuit voltage and the short-circuit current are obtained by setting in (1) I and V to zero, respectively. The expression for $V_{OC}(V)$, obtained from (3) has the following form:

$$V_{OC} = \frac{kT}{q} ln\left(\frac{\left(\Delta n_{OC}+\frac{n_0}{2}\right)^2 - \frac{n_0^2}{4}}{n_{i0}(T)^2 e^{\Delta E_g/kT}} + 1\right). \quad (4)$$

The excess electron density Δn_{OC} at open-circuit conditions when $L_d >> d$, can be found from the balance equation in the form:

$$I_L/q = A_{SC}\left[\frac{d}{\tau_{eff}(\Delta n_{OC})}\right]\Delta n_{OC}, \quad (5)$$

where the effective recombination time $\tau_{eff}(\Delta n_{OC})$ is determined by expression (6). The total effective recombination time in the SC depends on the intrinsic mechanisms (that cannot be decreases below the well-defined fundamental limits) and extrinsic recombination mechanisms (that can be modified, *e.g.*, in the process of the sample preparation, *etc.*)

$$\tau_{eff}^{-1} = \tau_{intr}^{-1} + \tau_{extr}^{-1}, \quad (6)$$

with the lifetime for intrinsic recombination, which includes radiative and interband Auger recombinations:

$$\tau_{intr} = \left(\tau_{rad}{}^{-1} + \tau_{Auger}{}^{-1}\right)^{-1}. \quad (7)$$

The extrinsic recombination lifetime (which can be reduced by sample bulk preparation, surface passivation, etc.) depends on Shockley-Reed-Hall lifetime, non-radiative exciton recombination lifetime due to Auger mechanism (recombination via deep level [5]), lifetime due to surface recombination and lifetime for recombination in the SCR [6]. In general

$$\tau_{extr} = \left(\tau_{SRH}^{-1} + (\tau_{exc}^{n})^{-1} + \left(\tau_{eff}^{s}\right)^{-1} + (\tau_{scr})^{-1}\right)^{-1}. \quad (8)$$

The photoconversion efficiency η, photovoltage V_m,

photo-current I_m, and output power P_m in the maximum-power output regime are found by setting the derivative of $P = IV$ with respect to the voltage to zero.

This self-consistent approach for textured Si solar cells efficiency η optimization has been applied and validated experimentally. Dark and light *I–V* characteristics as well as light intensity dependence of the open-circuit voltage were measured for a number of highly efficient silicon *p-n* junction SCs with the photoconversion efficiency $\eta \geq 20\%$. The spectral dependencies of $EQE(\lambda)$ under AM1.5 conditions were also investigated experimentally.

3 NUMERICAL SIMULATIONS

For the numerical simulations we considered c-Si solar cells with both the planar and top surfaces The top surface reflection was calculated using the SunSolve Software provided by PV Lighthouse [14], as was shown in Fig. 1. In this simulation, a 78 nm thick TiO_2 ARC film was deposited on top of a 500 μm silicon substrate [16]. The first simulation uses a film with a planar texture and the minimum reflection is 2.2% at 705 nm. The second simulation uses a film with a regular upright pyramid texture with a characteristic angle of 54.74° and has a minimum reflection of 0.2% at 690 nm. The third simulation uses a film with a random upright pyramid texture with a characteristic angle of 54.74° and has a minimum reflection of 0.2% at 685 nm. The last simulation uses a film with a spherical cap texture with a characteristic angle of 54.74° and has a minimum reflection of 1.8% at 690 nm.

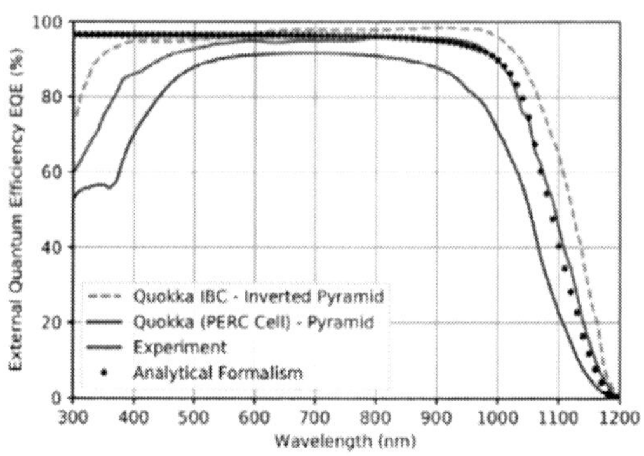

Figure 2: Experimental $EQE(\lambda)$ for commercial 21.3% efficient SunPower SC (green line). Black circles are for *EQE* using empirical formula [1]. Cyan-dashed line is the calculated $EQE(\lambda)$ dependence for the 25.6% efficient IBC SC [17], and, for comparison, the blue line is for the PERC solar cell.

Preliminary simulations of *EQE*, *I–V* characteristics, and the efficiency were carried out using the Quokka 2 software with the *standard* parameter set from [17], including the bulk recombination parameters, for Auger bulk recombination, the radiative recombination, etc. The External Quantum Efficiency was calculated using a thickness of 165 μm and a doping level of $9{\cdot}10^{14}$ cm^{-3}. The IBC external quantum efficiency was obtained from [4]. For figure 3, we calculate the *I–V* current density as a function of voltage using a thickness of 165 μm and a doping level of $9{\cdot}10^{14}$ cm^{-3}.

4 COMPARISON: THEORY *VS.* EXPERIMENT

In Fig. 2 we present the photon wave length dependent external quantum efficiency $EQE(\lambda)$. This compares the experimentally measured and theoretical results, calculated using the empirical formula [1], as well as simulated using Quokka software. The EQE for 25.6% efficient IBC SC is from [17], and that one from the PERC solar cell, for comparison.

Next Fig. 3 compare the measured and calculated light *I–V* characteristics. Circles are the experimental values for the commercial 21.3% efficient SunPower SC. The solid lines are from the theory: the red line is from the analytical formalism [1]; green line is from Quokka simulations for the IBC SCs.

As can be seen from Figs. 2 and 3, the analytical formalism [1] provides very good agreement with the experimentally measured results. The agreement between the preliminary Quokka 2 simulations with the experiment, however, is not as good for the analytical formalism

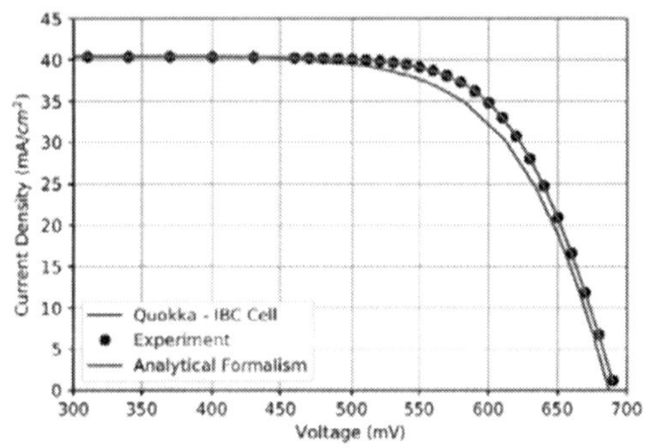

Figure 3. Measured (circles) and calculated (line) light *I–V* characteristics. Circles are the experimental values for the commercial 21.3% efficient SunPower SC. The solid lines are from the theory: the red line is from the analytical formalism; green line is from Quokka simulations for the IBC SC.

As it is clear from Figure 3, the open circuit voltage V_{OC}, calculated using both theoretical approaches are in a perfect agreement with the experimental results, while below V_{OC} the simulated (using Quokka) current deviates from the experimental values. We attribute, however, the disagreement with the experiment mostly due to the use of the standard [17] parameter set in the Quokka input, especially series and shunt resistance. In the future simulations the focus will be on the optimization of series and shunt resistances as well as other parameters.

5 CONCLUSIONS

We compared the results of the comprehensive analytical formalism [1] that models and optimizes the high-efficiency textured silicon-based solar cells with the experimental characterization and the *preliminary* results from the numerical simulation software Quokka 2 [2]. It is demonstrated in [1] that the contribution of nonradiative exciton recombination, as a rule, exceeds the radiative recombination contribution and substantially affects the photo-conversion efficiency, stronger than the gap narrowing effect. Such important mechanism, however, cannot be readily modelled within the numerical simulation software.

The calculated SCs parameters such as the short-circuit current density J_{SC}, open-circuit voltage V_{OC} and photo-conversion efficiency η have been determined for several SCs types. Following [1], a very good agreements of the calculated characteristics with experimental ones has been achieved, better than using Quokka 2 [2] simulation. We believe that using the accurate and specific, to the SC considered, parameter sets will allow better agreement of the simulation and the experiment. Using the analytical formalism [1] allows to optimize the SCs base thickness and the base doping level. Further Quokka 2 modeling is in progress.

6 ACKNOWLEDGEMENTS

This work was partially supported (V. Kostylyov and V. Vlasiuk) by National Research Foundation of Ukraine by the state budget finance (project 2020.02/0036 "Development of physical base of both acoustically controlled modification and machine learning–oriented characterization for silicon solar cells") and (G. Perinparajah) by Ontario Centres of Excellence (OCE) and IntriEnergy Canada Labs Inc.

7 REFERENCES

[1] A.V. Sachenko, V.P. Kostylyov, V.M. Vlasiuk, I.O. Sokolovskyi, A.I. Shkrebtii, et al, Proceedings 2021 IEEE 48th Photovoltaic Specialists Conference (PVSC), p. 0544 – 0550.
[2] Quokka 2 simulation software https://www.pvlighthouse.com.au/cms/simulation-programs/quokka2.
[3] J. Liu, Y. Yao, S. Xiao, and X. Gu, J. Phys. D 51 (2018) 123001.
[4] A. Augusto, J. Karas, P. Balaji, S. G. Bowden, and R. R. King, J. Mater. Chem. A 8 (2020) 16599.
[5] H. Sai, T. Oku, Y. Sato, M. Tanabe, T. Matsui, et al., Prog Photovoltaics Res Appl 27 (2019) 1061.
[6] K. Yamamoto, K. Yoshikawa, H. Uzu, et al., Japanese Journal of Applied Physics 57 (2018) 08RB20.
[7] K. Yoshikawa, W. Yoshida, T. Irie, H. Kawasaki, et al., Solar Energy Mater. Solar Cells 173 (2017) 37.
[8] M. Green, E. Dunlop, J. Hohl-Ebinger, M. Yoshita, et al., Prog Photovoltaics Res Appl 29 (2021) 3.
[9] R. Saive, Prog Photovoltaics Res Appl (2021) 1.
[10] M. A. Green, Prog. Photovolt: Res. Appl. 10 (2002) 235.
[11] A. V. Sachenko, Diamond and Related Materials 141 (2016) 53.
[12] A. V. Sachenko, V. P. Kostylyov, V. M. Vlasiuk, R. M. Korkishko, I. O. Sokolovs'kyi, and V. V. Chernenko, Ukr. J. Phys. 61 (2016) 917.
[13] A. Sachenko, V. Kostylyov, I. Sokolovskyi, B. Arzhang, and M. Evstigneev, 47th IEEE Photovoltaic Specialists Conference (PVSC) 0715 (2020) 0715.
[14] ARC simulator, OPAL 2 calculator OPAL 2 (pvlighthouse.com.au), https://www2.pvlighthouse.com.au/calculators/OPAL%202/OPAL%202.aspx, 2020).
[15] T. Tiedje, E. Yablonovitch, G. D. Cody, and B. G. Brooks, IEEE Trans. Electron Devices 31 (1984) 711.
[16] M. A. Green, Solar Energy Mater. Solar Cells 92 (2008) 1305.
[17] A. Fell, K. R. McIntosh, P. P. Altermatt, G. J. M. Janssen, R. Stangl, A. Ho-Baillie, et al, IEEE Journal of Photovoltaics 5 (2015) 1250.

LIGHT TRAPPING IN SILICON NANOWIRE: CORRELATED ABSORPTION DEPTH PROFILE, EM-FIELD DISTRIBUTION AND EXCITON GENERATION RATE DISTRIBUTION

Mohammed Kamal Hossain[1,*] , and Ayman Wajeeh Mukhaimer [2]
[1]Interdisciplinary Research Center for Renewable Energy and Power Systems (IRC-REPS), Research Institute, King Fahd University of Petroleum & Minerals (KFUPM), Dhahran 31261, Saudi Arabia.
[2]School of Engineering, RMIT University, GPO Box 2476, Melbourne VIC 3001, Australia.
*Correspondence to: kamalhossain@kfupm.edu.sa

ABSTRACT: Silicon nanowires (Si-NWs) have been able to draw huge attention in solar cell research particularly for their inherent and unique characteristics of light confinement. Here, we have reported correlated absorption depth profile, electromagnetic field distribution and exciton generation rate distributions for different diameters of the vertically aligned Si-NWs (coverage, ca. $6.5\times10^8/cm^2$ with diameter 50-200 nm) at different spectral wavelengths. Although Si-NWs were observed to be different in size and shape in experiment investigations, for simplicity, a single Si-NW of cylindrical shape was modeled and optimized to correlate the aforementioned characteristics. Light absorption and energy flow distribution confirmed that Si-NW facilitates to confine photon absorption of several orders of enhancement whereas the energy flow is also distributed along the wire itself. On the other hand, electric field distributions and excitons generation distributions highlighted the higher probability of strongest sites compared to those of flat silicon wafer.
Keywords: c-Si; Silicon Solar Cell; Simulation; Spectral Response; Thin Film.

1 INTRODUCTION

Silicon (Si) is a well-studied and unprecedented material in the microelectronics era including solar energy harvesting using photovoltaic (PV) technologies [1-5]. It has many superior advantages such as stability, non-toxicity, low cost, abundance, etc. Conventional Si-based PV solar cells are not that efficient because of relatively low light absorption [6-10]. Nanometric features add a unique flavor and therefore Si nanowires (Si-NWs) have been emerged as an extremely attractive candidate [11-13]. Si-NWs show strong optical absorption compared to that of conventional planer PV solar cell. Apart from absorption, it has been reported that radial p–n junctions Si-NW solar cell provides short collection length for charge carriers, thus allowing the use of lower-quality Si materials and a way to reduce production cost substantially [14-15].

It has been reported earlier that vertically aligned Si-NWs can be fabricated using gold-catalyzed chemical vapor deposition (CVD) technique. High resolution scanning electron microscope (SEM) revealed the diameter of the as-grown Si-NWs to be in the range of 50 to 200 nm [16]. Here in this work, based on the experimental observations, a detailed simulation has been carried out to understand the mechanism how Si-NWs interact with solar spectral wavelengths and contribute to absorption depth profile, EM-field distribution and exciton generation rated distribution. Cross-sectional analyses as well as aforementioned distributions for different sizes of Si-NWs have been investigated. Such insights are not only indispensable to understand the mechanism behind but also open up new possibilities in various applications.

2 METHODOLOGYS

Considering two specific objectives in mind (i.e. cost reduction in processes involved and nanostructure features for light trapping), a simple and hand-on strategy had been devised to grow Si-based vertically aligned nanowires. Silicon wafer of resistivity <1Ωcm was well-cleansed following standard procedure (i.e. 3:1 concentrated H_2SO_4/30% H_2O_2 for 15 min at 80 °C) and was transferred to the sputtering chamber for ultrathin metallic layer deposition followed by subsequent sintering to achieve metal nanoparticles. Then the specimen was transferred to CVD reactor to grow Si-NWs. Morphology of as-fabricated Si-NWs was confirmed by SEM and high-resolution SEM investigations [16]. A typical model, Si-NW on c-Si wafer, was designed in finite deference time domain (FDTD) analysis as shown in Fig. 1. Excitation generation rate distributions were extracted for different sizes of the Si-NWs at different spectral wavelengths. Additional features such as absorption profile, energy transfer, and electromagnetic field distribution were carried out in the same simulation.

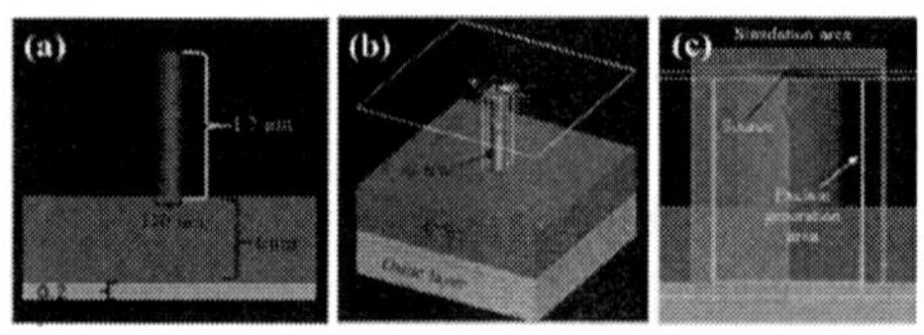

Figure 1: (a)-(b) Modeling dimension, hawk-eye view of the same and monitors for extracting optically simulated characteristics.

3 RESULTS AND DISCUSSION

Optical characteristics are interrelated to each other and therefore visualization of absorption depth profile, EM near-field distribution and corresponding exciton generation rate distribution can be extracted for a particular geometry. Figure 2 shows the absorption depth profile of vertically designed c-Si NWs modeled above the c-Si slab. The dimensions are mentioned in the experimental section. It is to be noted that the photons are mostly confined within the Si-NW at 700 nm of solar spectral wavelength whereas at lower and higher wavelengths such absorptions are mostly found confined

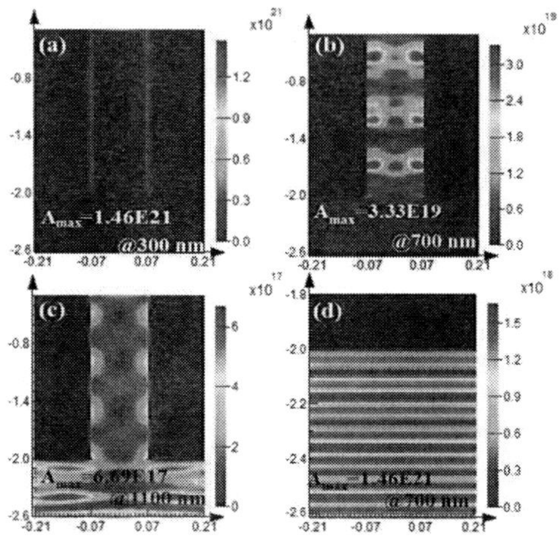

Figure 2: (a)-(c) Absorption depth profile distributions of vertically aligned Si-NW of 150 nm diameter at 300, 700 and 1100 nm solar spectral wavelengths respectively, (d) Absorption depth profile distribution of c-Si slab only at 700 nm solar spectral wavelength as reference.

along the wall as shown in Fig. 2a-c. The absorption depth profile of a c-Si slab without Si-NW was shown in Fig. 2d as reference.

In photovoltaic, the processes happen simultaneously; 1. photon absorption within in the absorbing layer, 2. exciton generation by photoactive materials and 3. exciton collection through the TCO. Therefore, efficient absorption of photons is the initial step to have an efficient solar cell. In brief, photons absorbed from solar radiation excite free electron-hole pairs or excitons. Therefore, electric field distribution defines how effectively excitons will be generated within the absorbing layer. Figure 3 displays EM near-field distribution of the same model as shown in Fig.1. The dimensions are mentioned in the experimental

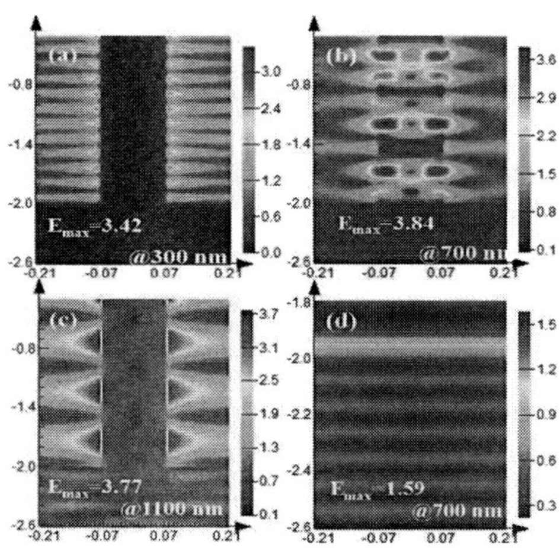

Figure 3: (a)-(c) EM near-field distributions of vertically aligned Si-NW of 150 nm diameter at 300, 700 and 1100 nm solar spectral wavelengths respectively, (d) EM near-field distribution of c-Si slab only at 700 nm solar spectral wavelength as reference.

section. It is noteworthy to mention that the EM near-field confinement effectively occurs within the Si-NW at 700 nm of solar spectral wavelength as shown in Fig. 3b. However, at lower and higher wavelengths such distributions are mostly found confined along the wall as shown in Fig. 3a and Fig.3c. EM near-field distribution of a c-Si slab without Si-NW was shown in Fig. 2d as reference.

Figure 4 shows exciton generation rate distributions of vertically aligned Si-NWs of 50 and 200 nm diameters at 700 and 1100 nm solar spectral wavelengths. Fig. 1a-b shows Exciton generation rate distributions of a vertically aligned Si-NW of 50 nm diameters at 700 and 1100 nm solar spectral wavelengths respectively. Excitation rate distribution was found to be localized near the edges in both scenarios. At 700 nm wavelength confined spots were observed in repeated fashion along with lower intensities, whereas at 1100 nm wavelength such spots increased in intensity and lower in repetition. The same distributions of a vertically aligned Si-NW of 200 nm diameter at 700 and 1100 nm solar spectral wavelengths were depicted in Fig. 1c-d respectively. Unlike those observed for Si-NW of 50 nm diameter, excitation rate distribution was found to be localized near the center with higher intensity at 700 nm solar wavelength, whereas at 1100 nm wavelength such distribution disappeared.

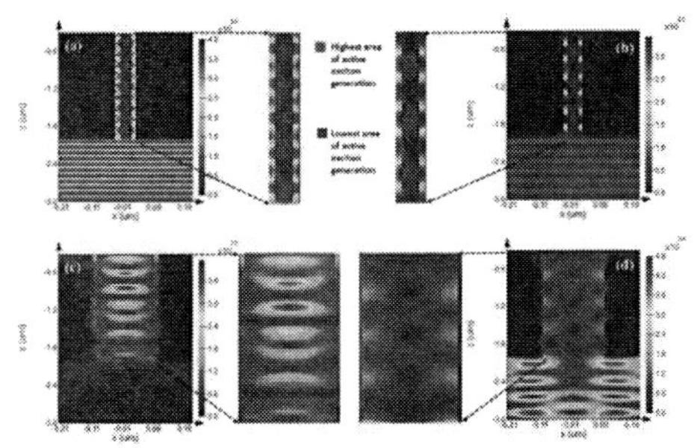

Figure 4: (a)-(b) Exciton generation rate distributions of vertically aligned Si-NW of 50 nm diameters at 700 and 1100 nm solar spectral wavelengths respectively, insets: corresponding magnified area confirming areas confirming active portion of exciton generation and (c)-(d) that of vertically aligned Si-NW of 200 nm diameters at 700 and 1100 nm solar spectral wavelengths respectively, insets: corresponding magnified areas confirming active portion of exciton generation.

4 CONCLUSION

Based on the experimental observations; a detailed simulation has been carried out to understand the mechanism of how Si-NWs interact with solar spectral wavelengths and contribute to absorption depth profile, EM-field distribution and exciton generation rated distribution. It was found that the solar spectrum of 700 nm wavelength performs the best for Si-NW based model system. Detailed cross-sectional analyses as well as aforementioned distributions for different size of the Si-NW have been investigated. Such insights are essential to facilitate high efficient nanostructures-based solar cell applications as well as pave away new and emerging routes to reduce process costs and possibilities in various applications.

Acknowledgment: The leading author at KFUPM would like to acknowledge the support received under Interdisciplinary Research Center for Renewable Energy and Power Systems (IRC-REPS) Grant INRE2114.

References

1. De Wolf, Stefaan, Antoine Descoeudres, Zachary C. Holman, and Christophe Ballif. "High-efficiency silicon heterojunction solar cells: A review." green 2, no. 1 (2012): 7-24.
2. Green, Martin A. "Thin-film solar cells: review of materials, technologies and commercial status." Journal of Materials Science: Materials in Electronics 18, no. 1 (2007): 15-19.
3. Yamaguchi, Masafumi, Kan-Hua Lee, Kenji Araki, and Nobuaki Kojima. "A review of recent progress in heterogeneous silicon tandem solar cells." Journal of Physics D: Applied Physics 51, no. 13 (2018): 133002.
4. Bhopal, Muhammad Fahad, Doo Won Lee, Atteq ur Rehman, and Soo Hong Lee. "Past and future of graphene/silicon heterojunction solar cells: a review." Journal of Materials Chemistry C 5, no. 41 (2017): 10701-10714.
5. Menna, P., G. Di Francia, and V. La Ferrara. "Porous silicon in solar cells: a review and a description of its application as an AR coating." Solar Energy Materials and Solar Cells 37, no. 1 (1995): 13-24.
6. Sharma, Shruti, Kamlesh Kumar Jain, and Ashutosh Sharma. "Solar cells: in research and applications—a review." Materials Sciences and Applications 6, no. 12 (2015): 1145.
7. Neuhaus, Dirk-Holger, and Adolf Münzer. "Industrial silicon wafer solar cells." Advances in optoelectronics 2007 (2007).
8. Imamzai, Mohammadnoor, Mohammadreza Aghaei, Y. Hanum Md Thayoob, and Mohammadreza Forouzanfar. "A review on comparison between traditional silicon solar cells and thin-film CdTe solar cells." In Proceedings of National Graduate Conference (Nat-Grad), pp. 1-5. 2012.
9. Lee, Youngseok, Cheolmin Park, Nagarajan Balaji, Youn-Jung Lee, and Vinh Ai Dao. "High-efficiency silicon solar cells: A review." Israel Journal of Chemistry 55, no. 10 (2015): 1050-1063.
10. De Wolf, Stefaan, Antoine Descoeudres, Zachary C. Holman, and Christophe Ballif. "High-efficiency silicon heterojunction solar cells: A review." green 2, no. 1 (2012): 7-24.
11. G. Otnes and M. T. Borgström. "Towards high efficiency nanowire solar cells." Nano Today 12, 31-45 (2017).
12. L. Yu, S. Misra, J. Wang, S. Qian, M. Foldyna, J. Xu, Y. Shi, E. Johnson, and P. R. Cabarrocas. "Understanding light harvesting in radial junction amorphous silicon thin film solar cells." Scientific reports 4, 4357-4364 (2014).
13. P. Krogstrup, H. I. Jørgensen, M. Heiss, O. Demichel, J. V. Holm, M. Aagesen, J. Nygard, and A. F. Morral. "Single-nanowire solar cells beyond the Shockley–Queisser limit." Nature Photonics 7(4), 306-310 (2013).
14. L. Tsakalakos, J. Balch, J. Fronheiser, B. A. Korevaar, O. Sulima, and J. Rand. "Silicon nanowire solar cells." Applied physics letters 91(23), 233117 (2007).
15. G. Jia, M. Steglich, I. Sill, and F. Falk. "Core–shell heterojunction solar cells on silicon nanowire arrays." Solar Energy Materials and Solar Cells 96, 226-230 (2012).
16. Hossain, Mohammad Kamal, Billel Salhi, and Ayman Mukhaimar. "Fabrication and Spectral Characteristics of Silicon Nanowires for Efficient Solar Energy Harvesting." Plasmonics 16, no. 1 (2021): 1-8.

THE EFFECT OF DOPING PROFILE ON CONTACT RESISTIVTY BETWEEN SCREEN PRINTED AG/AL AND IMPLANTED BORON EMITTER

Ege ÖZMEN [1,2], Selin SEYREK [1,2], Gence BEKTAŞ [1,2*], Hasan Hüseyin CANAR [1,3], Sümeyye KOÇAK BÜTÜNER [1], Hasan ASAV [1,2], Ahmet Emin KEÇECİ [1,2], Raşit TURAN [1,2,3]

[1]ODTÜ-GÜNAM, Çankaya, 06800 Ankara, Turkey
[2]Micro And Nanotechnology Department, Middle East Technical University, Çankaya, 06800 Ankara, Turkey
[3]Department of Physics, Middle East Technical University, Çankaya, 06800 Ankara, Turkey

ABSTRACT: In this work, the contact resistivity between fire through metal contacts and boron implanted n-type textured silicon wafer with various doping profiles is studied. The metal contacts are screen printed on boron emitter passivated with Al_2O_3/SiN_x stack and subsequently underwent a fast-firing process by a conveyor belt furnace. Electrochemical capacitance-voltage (ECV) measurements show that increasing the activation temperature and time results in a decrease in dopant peak concentration and a deeper junction. The contact resistivity has generally shown a decreasing trend in TLM results as the activation temperature and duration increase due to the increasing junction depth. The sample activated at 1050°C for 60 minutes shows a slight increase in contact resistivity which may be caused by decreased surface doping concentration. The lowest contact resistivity is measured as 1.4 $m\Omega.cm^2$ for the emitter activated at 1050°C for 30 minutes.
Keywords: boron, ion implantation, contact resistivity, screen printing

1 INTRODUCTION

Even though p-type solar cells dominate the solar cell market today, n-type solar cells attract attention due to their superior material properties such as the absence of boron-oxygen related defects, and greater tolerance to metal impurities like Fe, Al, Co and Ni [1]. Ion implantation is an alternative method for doping which is favored because of reducing the fabrication steps thanks to single side doping and creating a controllable profile [2]. Contact resistivity between metal contacts and the emitter underneath them crucially affects solar cell efficiency [3]. The contact resistivity, defined as the contact resistance per unit area, is determined by various factors such as firing temperature, electrical properties of interacting materials which are Ag/Al and Si in our case. The doping profile is known to have an impact on the emitter quality; however, its effect on the contact resistivity has also been reported in the literature.

In this work, we doped n-type silicon wafers by the ion implanter with boron species and activated dopants at different temperatures for varying durations. The doping profiles of the wafers were measured by electrochemical capacitance-voltage (ECV) tool and the sheet resistances of the samples were measured with four-point probe (4PP). Later, we screen printed Ag/Al metal contacts and fired them. Then, we measured their contact resistivity with transmission line measurement (TLM) to observe the dependence of contact resistivity on doping profile.

2 EXPERIMENTAL

Emitter was formed by implanting boron on textured n-type Si wafer which was performed at 7.5 keV. After that the samples went through the annealing process at different temperatures of 950 °C, 1000 °C and 1050 °C for dopant activation under N_2 flow. After the annealing process, the samples were HF dipped to remove unintentionally grown oxide. Then, the samples were divided into two groups. One group was reserved for sheet resistance and doping profile measurements by 4PP and ECV methods, respectively. The surfaces of the emitters in the second group were passivated with Al_2O_3 by atomic layer deposition (ALD) system and subsequently capped with SiN_x deposited by plasma enhanced chemical vapor deposition (PECVD) tool. Following the dielectric deposition processes, Ag/Al paste was screen printed on boron implanted emitter and a fast-firing process was performed. The contact resistivity between fire through metals and implanted boron emitters was measured with TLM setup.

The process flow and the schematic of the samples fabricated for TLM is shown in Figure 1.

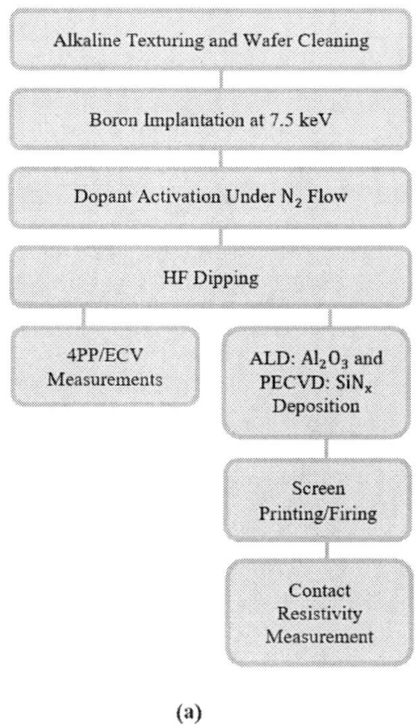

(a)

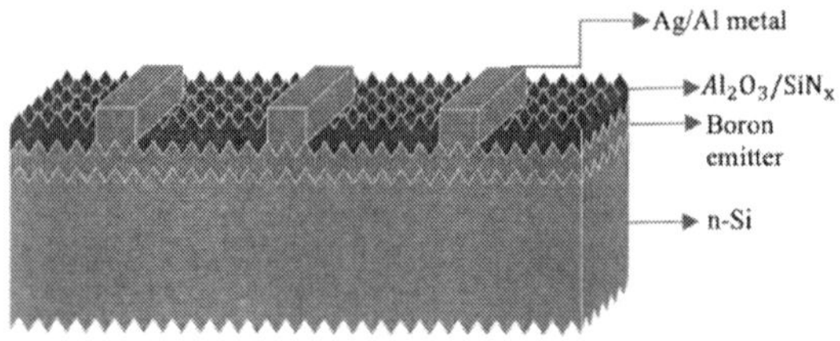

(b)

Figure 1: (**a**) Process flow (**b**) Schematic of the samples prepared for TLM measurements

3 RESULTS AND DISCUSSION

3.1 ECV Results

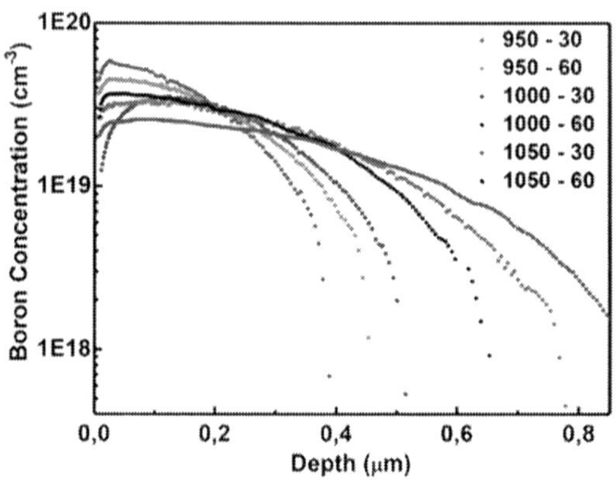

Figure 2: ECV graphs of the samples activated at 950 °C, 1000 °C, 1050 °C for 30 min and 60 min

ECV results show that an increase in the activation temperature and duration results in a decrease in peak dopant concentration while the junction depth increases. It should be noted that the result for the sample which was activated at 1000 °C for 30 minutes should be hold separated because N_2 supply was interrupted during activation process.

3.2 TLM Results

Sample Name	Activation Temperature(°C)	Activation Time (min)	R_{sheet} (Ω/sq)
950-30	950	30	93.39
1000-30	1000		90.91
1050-30	1050		75.83
950-60	950	60	85.75
1000-60	1000		75.77
1050-60	1050		71.21

Table I: R_{sheet} values of samples measured by 4PP

It is observed that the R_{sheet} values decrease with increasing activation temperature and time since junction gets deeper.

As it can be seen in Figure 3, the average contact resistivity between screen printed Ag/Al and implanted emitter decreases as the doping profile gets deeper, except for the samples 1000-30, which is the problematic sample, and 1050-60. A similar trend for the contact resistivity of diffused boron emitter is reported in the literature [4].

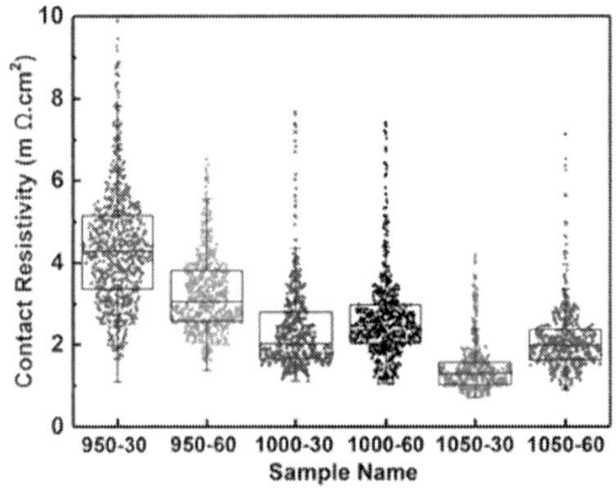

Figure 3: Contact resistivity between screen printed Ag/Al and implanted boron emitters

For shallower emitters, a higher contact resistivity results from the local spikes of fire-through metal paste in the emitter [5,6].

The contact resistivity increases as the surface doping concentration decreases, which may be the reason for the slight increase in contact resistivity of sample 1050-60 [7], since increasing doping level means reducing Si resistivity [8].

4 CONCLUSION

The variant activation temperature and duration was shown to change the junction depth of the implanted boron emitter and corresponding sheet resistance values. The R_{sheet} values decreased as the activation temperature and time increased in consequence of creating a deeper junction. Generally, the average contact resistivity decreased as the doping profile got deeper. The sample 1000-30 was annealed under ambient air by accident and the sample 1050-60 got higher contact resistivity results which may be caused by the relatively lowest peak concentration. The samples named 1000-30, which was annealed under ambient air by accident, and 1050-60, which has the lowest peak concentration, have relatively higher contact resistivity values than the expected trend.

5 REFERENCES

[1] B, A., MM, K., MI, H., OE, D., AA, A., F, A., & N, T. (2015). Emerging frontiers of N-Type silicon material for photovoltaic applications: The impurity-defect interactions. *Frontiers in Nanoscience and Nanotechnology*, *1*(1), 2–12

[2] Bektaş, G., Kökbudak, G., Keçeci, A. E., Bütüner, S. K., Asav, H., Çiftpınar, E. H., Canar, H. H., Arıkan, B., & Turan, R. (2021). Simplified process flow for the fabrication of PERC solar cells with

ion implanted emitter. *Renewable Energy*, *179*, 2295–2299

[3] Gabor, A. M., Gregory, G., Payne, A. M., Janoch, R., Anselmo, A., Yelundur, V., & Davis, K. O. (2017). Dependence of solar cell contact resistivity measurements on sample preparation methods. *2017 IEEE 44th Photovoltaic Specialist Conference, PVSC 2017*, 777–780

[4] Çiftpinar, H. E., Stodolny, M. K., Wu, Y., Janssen, G. J. M., Löffler, J., Schmitz, J., Lenes, M., Luchies, J. M., & Geerligs, L. J. (2017). Study of screen printed metallization for polysilicon based passivating contacts. *Energy Procedia*, *124*, 851–861

[5] Bao, J. (2020). Towards 24% Efficiency for Industrial N-Type Bifacial Passivating-Contact Solar Cells With Homogeneous Emitter. *37th European Photovoltaic Solar Energy Conference and Exhibition*, *December*, 3–6

[6] Padhamnath, P., Wong, J., Nagarajan, B., Buatis, J. K., Ortega, L. M., Nandakumar, N., Khanna, A., Shanmugam, V., & Duttagupta, S. (2019). Metal contact recombination in monoPoly^{TM} solar cells with screen-printed & fire-through contacts. *Solar Energy Materials and Solar Cells*, *192*(December 2018), 109–116

[7] Moki, A., Shenoy, P., Alok, D., Baliga, B. J., Wongchotigul, K., & Spencer, M. G. (1995). Low resistivity as-deposited ohmic contacts to 3C-SiC. *Journal of Electronic Materials*, *24*(4), 315–318

[8] Lei, S. (2008). Contact Resistance Study on Polycrystalline Silicon Thin-Film Solar Cells on Glass by Lei Shi. *Master's Thesis*, *June*.

SELECTIVE EMITTER FORMATION VIA LASER DOPING WITH PICOSECOND PULSED LASER FOR HIGH-EFFICIENCY PERC SOLAR CELLS

Hasan Asav[1,2], Ahmet Emin Keçeci[1,2], Gence Bektaş[1,2], Sümeyye Koçak Bütüner[1], Gamze Kökbudak[1,2], Hasan Hüseyin Canar[1,3], Bülent Arıkan[1], Raşit Turan[1,2,3]

[1] ODTÜ-GÜNAM, Çankaya, 06800 Ankara, Turkey
[2] Micro And Nanotechnology Department, Middle East Technical University, Çankaya, 06800 Ankara, Turkey
[3] Department of Physics, Middle East Technical University, Dumlupinar Blvd. 1, 06800 Ankara –Turkey

ABSTRACT: In this study, we develop a $POCl_3$ diffusion recipe so that we obtain a thicker phosphosilicate glass layer for laser doping. The formation of a laser doped selective emitter is achieved by utilizing an industrial-type picosecond pulsed laser operating at 532 nm (green) wavelength (λ) and pulse repetition rate (f_{rep}) of 1 MHz. To demonstrate the selective emitter concept, sheet resistance reduction of the laser processed region after tube furnace diffusion is provided. In addition, produced samples are metalized and the difference of contact resistivities of both homogenous emitter and laser doped emitter is shown. By applying a moderate amount of heat accumulation on the surface, specific contact resistivity of the laser doped region is reduced to around 3 mΩ.cm^2. We report process details and results in this paper.
Keywords: PERC, selective emitter, laser doping, picosecond pulsed laser

1 INTRODUCTION

Selective emitter concepts have gained great interest since energy conversion efficiency can be increased by more than 0.3%$_{abs}$ in the passivated emitter and rear solar cell (PERC) technology compared to homogenous emitter [1-2]. The gain in the selective emitter is mainly due to the reduced emitter dark saturation current density of passivated emitter regions $j_{0e,pass}$, which is possible with lower surface doping concentration [3-4]. On the other hand, the emitter regions under the metal contacts should be heavily doped for a lower saturation current density of the metalized region, $j_{0e,met}$ [2-4-5]. Also, heavily doped areas yield minimal specific contact resistivity (ρ_c), which is desirable for low series resistance in a solar cell. In the current standard process, the emitter is formed by the diffusion of phosphorus atoms at high temperatures on p-type Si wafers. By adding phosphorous at much higher concentrations than the background boron doping level, the surface region is inverted from p into n-type silicon and a p-n junction is formed [6]. On the other hand, selective emitters are made of a lowly doped emitter and local heavily doped regions under metal contacts, which can be efficiently done by laser scanning [5-7-8]. The nanosecond laser is commonly used for local heating of PSG layer for heavy diffusion of P [9]; however, there is hardly enough study conducted for picosecond lasers.
In this study, we investigated the contact resistivity between screen-printed Ag metal and heavily doped emitter which was formed by applying a picosecond pulsed laser on the PSG layer. For an efficient heavy diffusion of P by the picosecond pulsed laser, a thicker PSG was formed by an additional $POCl_3$ deposition step.

2 METHODOLOGY

In this study, we used alkaline textured and cleaned p-type Cz-Si wafers as a substrate. A recipe was developed for tube furnace diffusion with phosphorus oxychloride ($POCl_3$) as a liquid dopant precursor. The diffusion recipe contains a second $POCl_3$ deposition after drive-in. This additional deposition step provides a thicker PSG layer on top of the emitter compared to the standard tube furnace diffusion recipe that contains only one deposition step. Since tube furnace diffusion creates a stacked layer that is composed of PSG and SiO_2, analysis of both layers is a preliminary step of achieving laser doping. Following the diffusion process on p-type Cz-Si polished wafer, PSG and SiO_2 stack layer thickness was measured by spectroscopic ellipsometry as approximately 29 nm. The wafer was cut into small pieces and each piece was dipped into the diluted HF solution for different time intervals. Since the PSG etching rate is higher than the etching rate of SiO_2, the reduction in individual thicknesses was different. As the SiO_2 layer acts as a barrier during laser doping, our major aim is to minimize the SiO_2 thickness while keeping the PSG thickness maximum. After the analysis, we employed an industrial-type picosecond pulsed laser system to obtain laser-doped selective emitters. Half of the alkaline textured and cleaned p-type Cz-Si wafer surface was completely scribed by laser and the other half of the surface was left intact. Finally, laser-processed samples were dipped into the HF solution to remove the residual PSG and SiO_2 layers. By using the four-point probe (4PP) tool, sheet resistance of both surfaces was measured. Then, an anti-reflection coating (ARC) was deposited by using a plasma-enhanced chemical vapor deposition (PECVD) system. The front surface of the samples was then metalized with Ag paste and the fast-firing process was applied by a conveyor belt furnace.

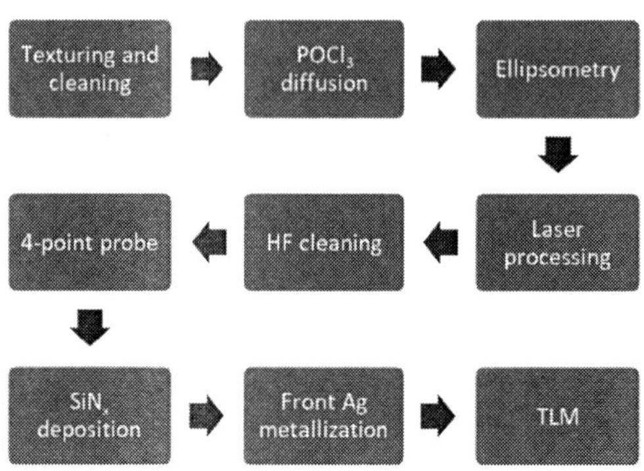

Figure 1: Process flow of the produced samples

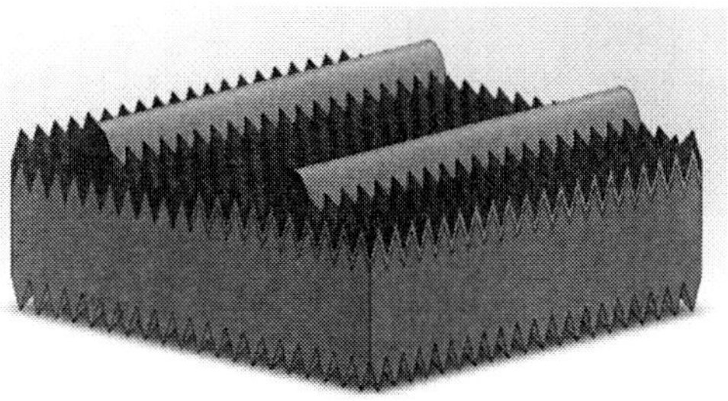

Figure 2: 3D visualization of the produced samples

Schematic representation of the produced samples can be seen in Figure 2. The red region denotes the homogenous emitter and the yellow region denotes the laser doped emitter. Specific contact resistivity of both emitters was measured with the transmission line measurement (TLM) tool.

3 RESULTS

Figure 3 shows the results of the stepwise etching experiment with diluted HF. It is seen that there are two different etch rates corresponding to PSG and SiO_2 layers. From the slopes of the two regions in Figure 3, it is depicted that the PSG layer is around 25 nm and the SiO_2 layer is around 4 nm. These results indicate that our diffusion recipe is convenient for subsequent laser doping processes [10].

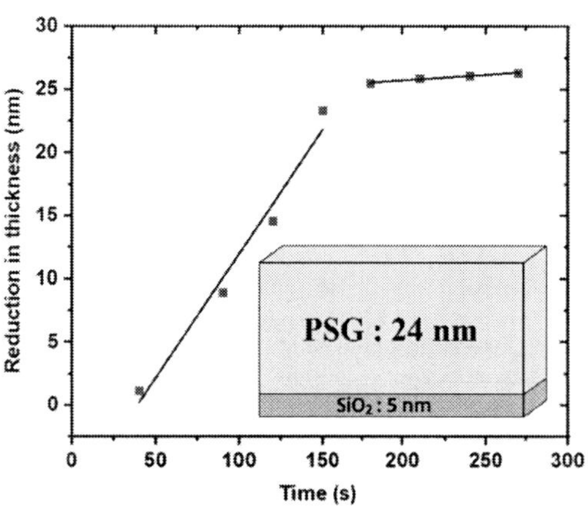

Figure 3: Reduction in thickness of PSG and SiO_2 layers with time

As a next step, the laser power of the system was optimized for the formation of laser doped regions.
Table 1 displays the sheet resistance values measured in the homogenous emitter and laser doped emitter regions of the wafer. We see that laser doping provides a significant reduction in the sheet resistance (R_{sheet}) of the homogenous emitter.

Table I: Sheet resistances of homogenous and laser doped emitters

Emitter type	*Sheet resistance (Ω/sq.)*
Homogenous emitter	135±5.00
Laser doped emitter	85±5.00

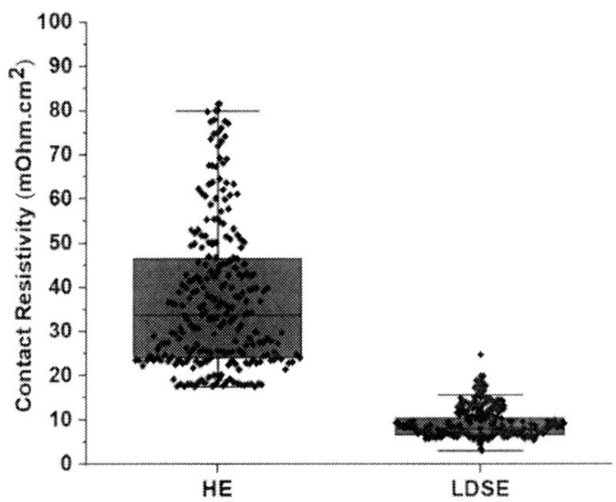

Figure 4: Contact resistivity of the homogenous emitter (HE) and laser doped selective emitter (LDSE)

Finally, metalized samples underwent the TLM measurements for the determination of contact resistivity of both homogenous emitter and laser doped emitter. Eight stripes were obtained from one wafer and each stripe contains a homogenous emitter and laser doped emitter. TLM results are shown in Figure 4. A contact resistivity value as low as $\rho_c = 3.19$ mΩ.cm^2 has been obtained after the laser doping, which is sufficiently low for good electrical contact to the emitter region.

4 CONCLUSIONS

This study determined the thicknesses of PSG and SiO_2 by taking advantage of two different etch rates of the layers as 24 nm and 5 nm, respectively. Picosecond pulsed laser doping provides a significant reduction in the sheet resistance of homogenous emitter from 135 Ω/□ to 85 Ω/□ on average. Also, the specific contact resistivity of the metal junctions with the underlying laser doped substrate is reduced from 35 mΩ.cm^2 to 7 mΩ.cm^2 on average via laser doping. Picosecond pulsed laser, which is commonly used for contact openings at the rear side of the PERC cells in production lines, can be utilized for laser doping for high-efficiency PERC solar cells with the selective emitter.

5 REFERENCES

[1] M. Müller. G. Fischer, B. Bitnar, et al., "22.61% efficient fully screen printed PERC solar cell," Energy Procedia, vol. 124, pp. 131-137, 2017.

[2] J. Weber, S. Lohmüller, E. Lohmüller et al., "Simulations on laser-phosphorous-doped selective emitters," in 7 th World Conference on Photovoltaic Energy Conversion, to be published, 2018.

[3] A. Wolf, A. Kimmerle, S. Werner, et al., "Status and perspective of emitter formation by POCl3- diffusion," in 31st European Photovoltaic Solar Energy Conference and Exhibition, 2015, pp. 414- 419.

[4] S. Werner, E. Lohmüller, S. Maier, et al., "Challenges for lowly-doped phosphorus emitters in silicon solar cells with screen-printed silver contacts," Energy Procedia, vol. 124, pp. 936-946, 2017.

[5] ITRPV, "International technology roadmap for photovoltaic (ITRPV) - 2017 Results - Ninth edition, March 2018," available online: http://www.itrpv.net/Reports/Downloads/, 2018.

[6] Bultman J, Cesar I, Geerligs B, Komatsu Y, Sinke W, Method of Emitter formation for crystalline silicon solar cells, The Journal of Photovoltaic International;2010; 8:69-81

[7] U. Jäger, "Selective emitter by laser doping from phosphosilicate glass" 24th European Photovoltaic Solar Energy Conference, 21-25 September 2009, Hamburg, Germany, pp. 1740-1743.

[8] L. Ventura, A. Slaoui, and J. C. Muller, "Realization of selective emitters by rapid thermal and laser assisted techniques," in 13th European Photovoltaic Solar Energy Conference and Exhibition, 1995, pp. 1578-1581.

[9] S. Werner, E. Lohmüller, P. Saint-Cast, J. M. Greulich, J. Weber, S. Schmidt, A. Moldovan, A. A. Brand, et al., "Key aspects for fabrication of p-type Cz-Si PERC solar cells exceeding 22% conversion efficiency," in 33rd European Photovoltaic Solar Energy Conference and Exhibition, 2017, pp. 406- 412.

[10] Werner, S.; Mourad, S.; Hasan, W.; Wolf, A. Structure and composition of phosphosilicate glass systems formed by POCl3 diffusion. Energy Procedia 2017, 124, 455–463.

INVESTIGATION OF THE POTENTIAL INDUCED DEGRADATION FOR PERC SOLAR CELLS

Min-An Tsai[1], Yean-San Long[1], Teng-Chun Wu[1]
[1]Center for Measurement Standards, Industrial Technology Research Institute, Taiwan

ABSTRACT: In the past few years, potential induced degradation (PID) has received considerable attention due to its catastrophic impact on the reliability of PV modules. The transportation of sodium ions triggered by a system voltage across PV modules was suspect to be the root causes and was further supported by findings of sodium ions on PID-affected regions in silicon solar cells. Here we use consistent lower volume resistivity packaging materials to test high-efficiency cells fabricated by different manufacturers. Under 192 hour test, Some samples have slightly obvious defects.

Keywords: PID, LID, Accelerated Degradation Test

1 Introduction

Until now, the Potential induced degradation (PID) is still a great challenge for the solar industry, due to it could lead to dramatic failure. Potential-induced degradation (PID) of PV modules is an important degradation mechanism caused by a high system voltage that drives ion drift towards the solar cell. These ions affect the proper functioning of the silicon solar cells and have been shown to induce rapid and significant performance losses of up to 50% at module level within one year in the field. The main external factors affected the PID phenomenon of the module are the environment temperature, the environment humidity and the voltage potential. The internal factors are the cells, encapsulation film, and glass and so on.

A number of researchers have analyzed the effect of temperature and humidity, and the results indicate that temperature increases from 25°C to 85°C, leading to increased degradation, while an increase in humidity from 20% to 85% will cause severe potential induced degradation problems that range from 5% to as much as 95%. The effect of the voltage stress on the solar module affects the PID phenomenon in terms of a system factor which is determined by comparing the generated module voltage to the ground level stress. System grounding techniques, as well as the module's position within the photovoltaic array, are major determinants of the system-level factor. Other factors affecting PID are the type of materials used within the solar modules, in the front sheet; the encapsulating material, in the back sheet; and the overall module design, which are also called as module-level factors. Regarding cell-level factors, the antireflective coating, emitter depth, and type of base doping are three major parameters that also affect PID.

In this work use consistent lower volume resistivity packaging materials to test high-efficiency cells fabricated by different manufacturers. Under 192 hour test, to observe the behavior of anti-PID by different fabricated processes.

2 Experiments and details

Each manufactories provided 3 pieces high efficiency P-type solar cell. Then the mini-module were packaged by using consistent lower volume resistivity packaging materials. Before the PID test, all sample were test by light soaking 60 kWh/m^2. PID test was performed in the environment chamber. the relative humidity set 85 %, temperature set 85 , the stress voltage set +/- 1000 V. The test flow chart for the solar cell and the electric connection for the PID test is shown as figure 1. Using the EL image to observe the defect before and after PID test.

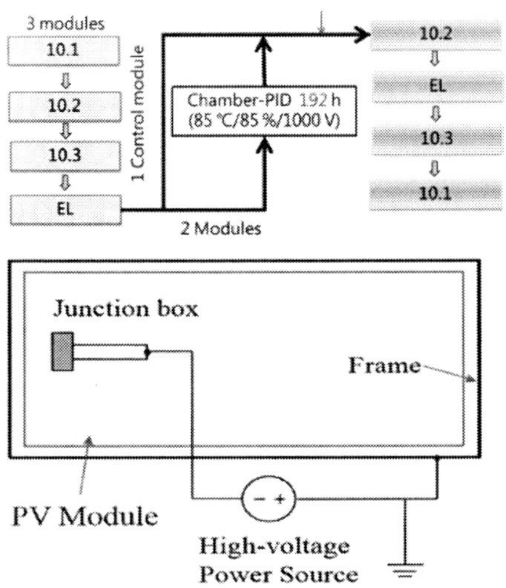

Figure 1: The test flow chart for the solar cell and the electric connection for the PID test.

3 Results

As the figure 2, the sample A reveals a about 2 % power degration, other sample (BCD) is less than 1 % degradation after PID(+1000 V) test. The EL image for the PID(+1000 V) as shown in figure 3, after the PID 192 h it can obviously see the defect in the edge of the cell (Sample A). And for the sample C, it show a slight defect in the top of the cell.

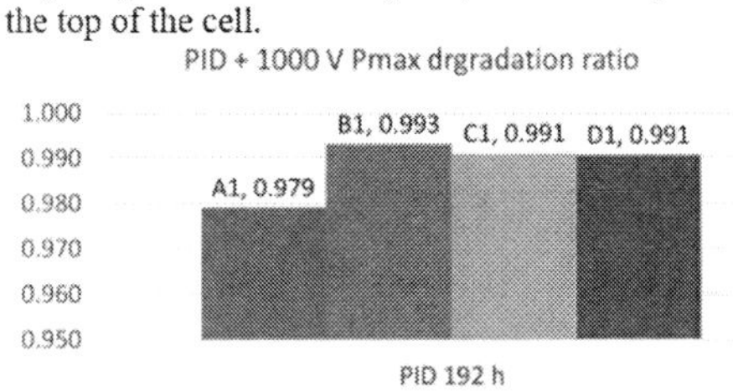

Figure 3: The normalized maximum power for the PID +1000 V.

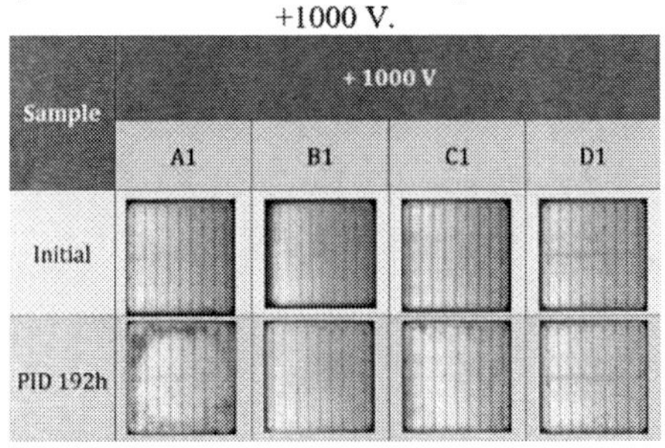

Figure 3: The EL image for the PID +1000 V.

As the figure 4, the sample A and B reveals a more than 3 % power degration, other sample (CD) is less than 2 % degradation after PID(-1000 V) test. The EL image for the PID(-1000 V) as shown in figure 5, after the PID 192 hr it can obviously see the defect in the edge of the cell (Sample A). For the sample B, it shows an obvious point defect in the center of the cell. For the sample D, it shows a circle defect in the center of the cell.

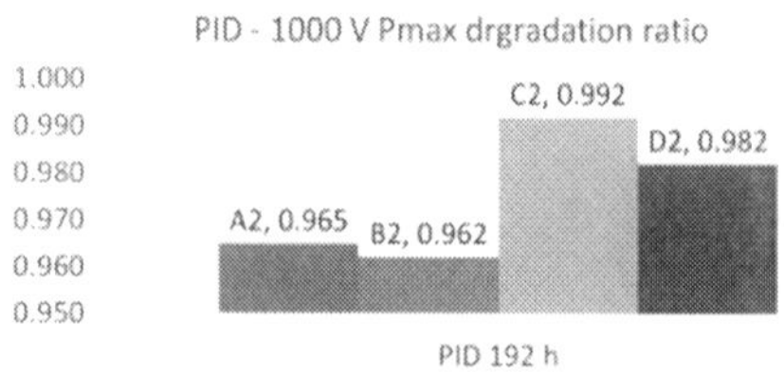

Figure 4: The normalized maximum power for the PID -1000 V.

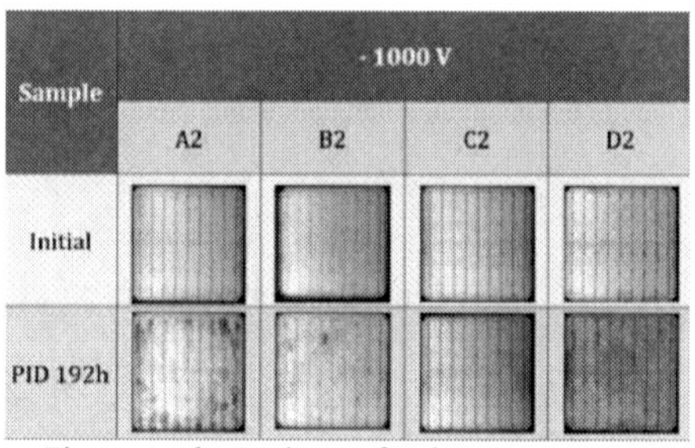

Figure 5: The EL image for the PID -1000 V.

4 Conclusions

In this work, it demonstrated that the different high efficiency PERC solar cell by different companies show the different resistance behavior of PID. For consistent lower volume resistivity packaging materials, the sample A show a little weak compare to others for PID +1000V condition. And the sample C show a best resistance of PID for PID -1000 V condition. the sample C show a best anti-PID effect for +/- 1000 V stress voltage.

5 References

[1] S. C. Akcaoğlu,1 G. Martinopoulos ,1 and C. Zafer, " Experimental Analysis of the Potential Induced Degradation Effect on Organic Solar Cells", International Journal of Photoenergy 2017; Volume 2017 |Article ID 2101932.

[2] J. A. Del Cueto and T. J. McMahon, "Analysis of leakage currents in photovoltaic modules under high-voltage bias in the field," Progress in Photovoltaics: Research and Applications, vol. 10, no. 1, pp. 15–28, 2002.

[3] P. Hacke, K. Terwilliger, S. Glick et al., "Interlaboratory study to determine repeatability of the damp-heat test method for potential-induced degradation and polarization in crystalline silicon photovoltaic modules," IEEE Journal of Photovoltaics, vol. 5, no. 1, pp. 94–101, 2015.

[4] J. Berghold, O. Frank, and H. HoehneS. Pingel, B. Richardson, and M. Winkler, "Potential induced degradation of solar cells and panels," in Proceeding 25th EUPVSEC, pp. 3753–3759, Valencia, Spain, 2010.

[5] P. Hacke, K. Terwilliger, R. Smith et al., "System voltage potential-induced degradation mechanisms in PV modules and methods for test," in 2011 37th IEEE Photovoltaic Specialists Conference, pp. 000814–000820, Seattle, WA, 2011.

[6] Oh, J.; Bowden, S.; TamizhMani, G. Potential-induced degradation (PID): Incomplete recovery of shunt resistance and quantum efficiency losses. IEEE J. Photovolt. 2015, 5, 1540–1548.

[7] Islam, M.A.; Hasanuzzaman, M.; Abd Rahim, N. A comparative investigation on in-situ and laboratory standard test of the potential induced degradation of crystalline silicon photovoltaic modules. Renew. Energy 2018, 127, 102–113

INFLUENCE OF DIAMOND WIRE SAWING PROCESS VARIATION ON COMMERCIAL MONO PERC SOLAR CELL PRODUCTION PARAMETERS

M. Raval[1,3*], B. Gumus[2,3], E. Toker[2,3], M. Ender[3], E. Muti[3], W. Jooss[1], F. Es[2], P. Fath[1]
1-RCT Solutions GmbH, Line - Eid - Str. 1, D-78467 Konstanz, Germany.
2- KalyonPV Research and Development Center, Kalyon Günes Teknolojileri Üretim A.,S., 06909 Ankara, Turkey.
3 - KalyonPV Solar Factory, Kalyon Günes Teknolojileri Üretim A.,S., 06909 Ankara, Turkey.
*-Corresponding Author, E-mail: mehul.raval@rct-solutions.com.

ABSTRACT: Diamond wire sawing method has become a standard for sawing mono c-Si bricks. There are established criteria based on parameters like total thickness variation and saw mark depth for acceptance of the sawn wafers. Based on such criteria as specified in the wafer datasheets, the wafers are classified as A grade, B grade and unqualified. However, variations in the sawing process that can still possibly classify the wafers as good wafers could adversely influence the cell production parameters. In this work, we report the observations from an integrated factory relating the quality of manufactured wafers on PERC solar cell production parameters as compared to the standard wafer quality parameters. It is observed that wafers with non-uniform saw damage can lead to decrease in the A+/A grade cells by up to 6% relative due to increase in the B grade and unqualified cells. In addition, the breakage rate was also observed to be more than 500% higher compared to the standard production. The observations point to the need to adjust the wafer inspection system criteria to minimize such wafers being classified as A grade and optimization of the sawing process to reduce the non-uniform sawing.

Keywords: c-Si, PERC, sawing, saw damage.

1 INTRODUCTION

Diamond Wire Sawing (DWS) was introduced few years ago as an alternative technology to Multi-Wire Slurry Sawing (MWSS) process and is being used for sawing mono c-Si and multi c-Si wafers since 2018 [1]. The DWS method has reduced kerf loss, uses water-based coolant (against PEG), has higher productivity, reduced cost of waste treatment and cheaper recyclability of silicon from discharge compared to the MWSS method [2]. The wet chemistry processes for saw damage removal and texturing of mono c-Si wafers have also been quickly adapted for the DWS process [1]. Studies have indicated that parameters like wire feed speed and wire cutting speed for the DWS process have an influence on microcrack formation and the surface roughness of the sawn wafers [3]. Hence qualification and quantification of the saw damage is important for determining the wafer quality [4].

There are established criteria for classification of DWS wafers based on parameters like wafer thickness, total thickness variation, resistivity, saw mark depth, micro cracks and chipping. Based on the industrially accepted criteria, the wafers are classified as A grade, B grade and unqualified. However, there are some variations in the sawing process that can possibly classify the wafers as good wafers, but their properties can adversely influence the cell production parameters.

2 OBJECTIVE OF WORK

The aim of this study is to understand the influence of the DWS process variation on commercial mono PERC solar cell production parameters like efficiency distribution and yield. Given that maintaining consistently high production yield for a PERC factory is critical, understanding the influence of variations in the incoming wafer surface properties becomes vital, especially in an integrated PV factory (Ingot to module production).

3 EXPERIMENTAL DETAILS

The study is carried out in the cell section of Kalyon PV, which is an integrated 1 GW bifacial p-type mono PERC PV factory. P-type (Boron doped) mono c-Si ingots are manufactured from poly-Si, which are then cropped, squared and chamfered into bricks. The bricks are then sliced into wafers using 55 μm diamond wire into wafers of 180 +20/-10 μm thickness. The wafers are fed to the cell section to make bifacial p-type mono PERC cells, which are in turn used to make glass-glass bifacial modules with half-cut cells. The cell section is equipped with characterization equipment like QSSPC lifetime tester, photoluminescence (PL) measurement tool, 3D microscope, quantum efficiency (QE) system, in line and offline electroluminescence (EL) tester for detailed characterization of the produced solar cells.

Trial group of non-uniformly sawn wafers (as shown in figure 1) were processed through the complete production flow and compared with the baseline wafers. The I-V parameters of the trial groups were compared with the baseline production. In addition, quality control (QC) aspects like A+/A grade, B grade, scrap rate, rework rate, unqualified rate were compared to give insights on the influence of wafer quality on such parameters.

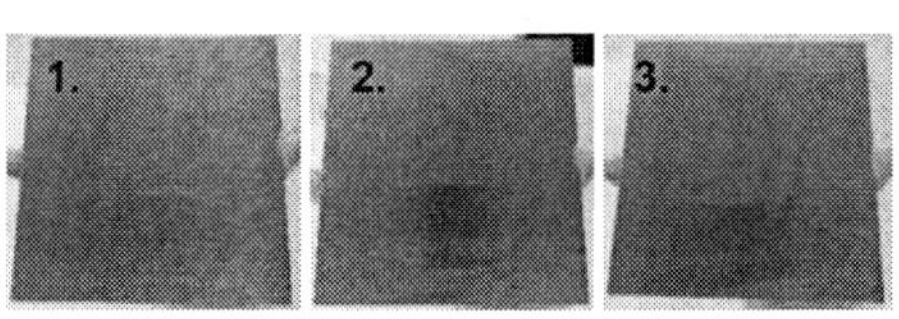

Figure 1: Representative images of incoming p-type wafers with non-uniform saw damage.

4 RESULTS AND DISCUSSIONS

4.1 Wafer surface morphology studies

Figure 2 shows the surface profile for the left-side regions of wafer 1 (in figure 1) in which a distinct wire saw mark (Width = 50 μm) with corresponding depth of 2.2 μm is evident. The smoother profile in the right-side region of wafer 1 is evident in figure 3.

Figure 2: Microscopic image and corresponding surface profile for left-side region of type 1 wafer in figure 1.

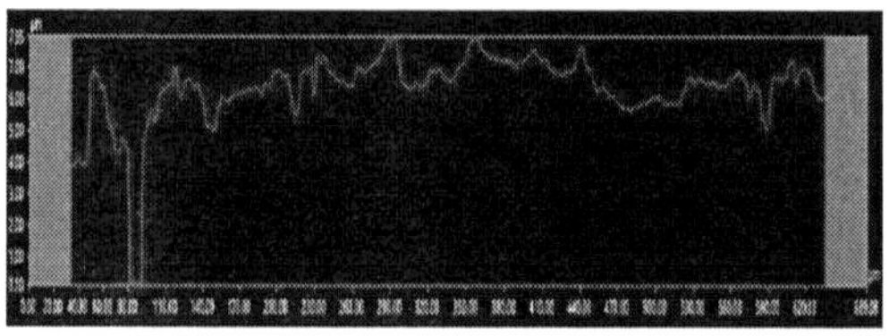

Figure 3: Surface profile for right-side region of type 1 wafer (in figure 1).

A microscopic view of the interface region in wafer 2 (in figure 4) indicates that the alternations between smoother and saw-damage region of 3.6 μm. The left-side region of wafer 2 is smoother compared to the interface region as can be observed in figure 5, while the rougher surface on the right-side is evident in figure 6.

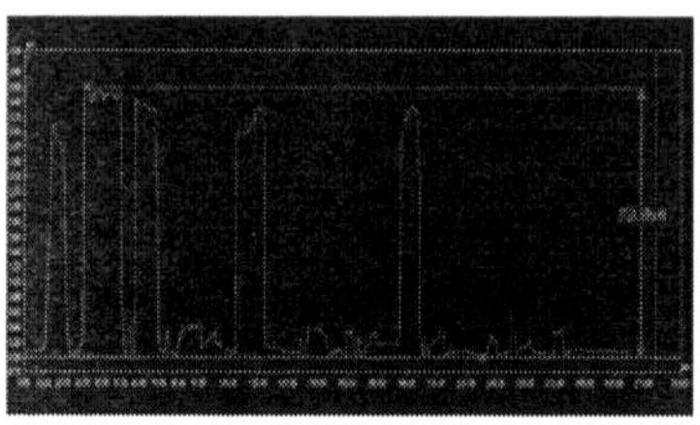

Figure 4: Surface profile of type 2 wafer (in figure 1) at interface region.

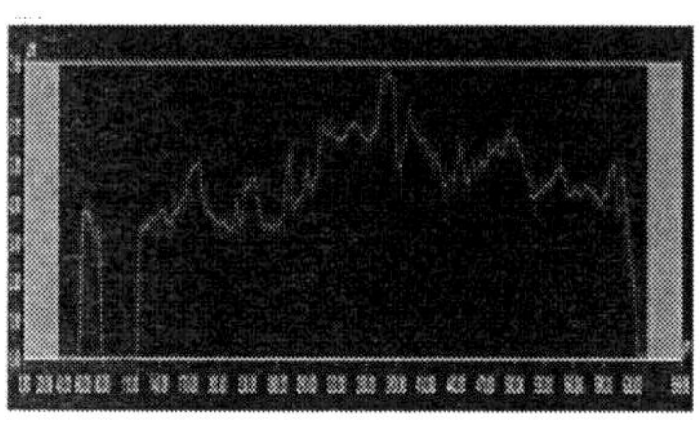

Figure 5: Surface profile of type 2 wafer (in figure 1) on the left-side region.

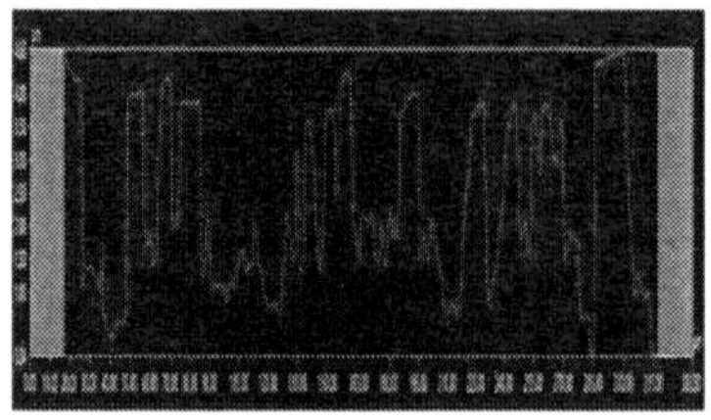

Figure 6: Surface profile of type 2 wafer (in figure 1) on the right-side region.

The left-side and right-side regions of wafer 3 exhibits a pulse-like profile as can be observed in figure 7 and 8, respectively. Figure 9 shows the profile at the interface region which clearly shows height variation in the wafer. The data from the wafer inspection system indicated that such wafers have average saw damage of 17.0 μm compared to standard production wafers which have saw damage of 4.5 μm. It can be inferred that the surface morphology of such wafers is not uniform due to DWS process variations.

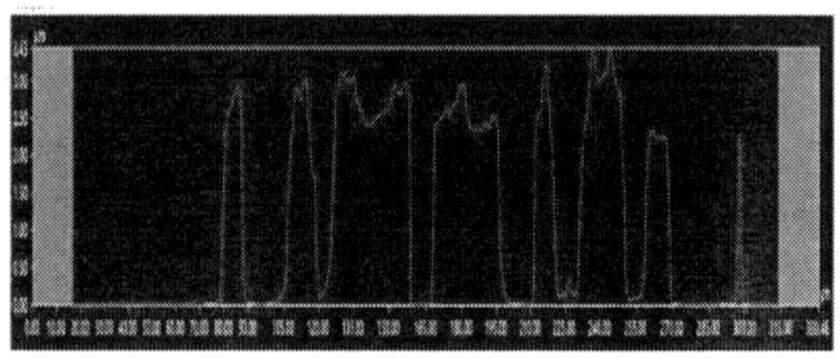

Figure 7: Surface profile of type 3 wafer (in figure 1) on the left-side region.

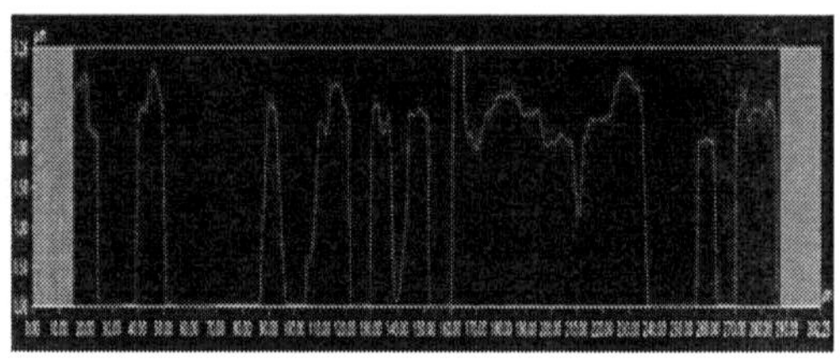

Figure 8: Surface profile of type 3 wafer (in figure 1) on the right-side region.

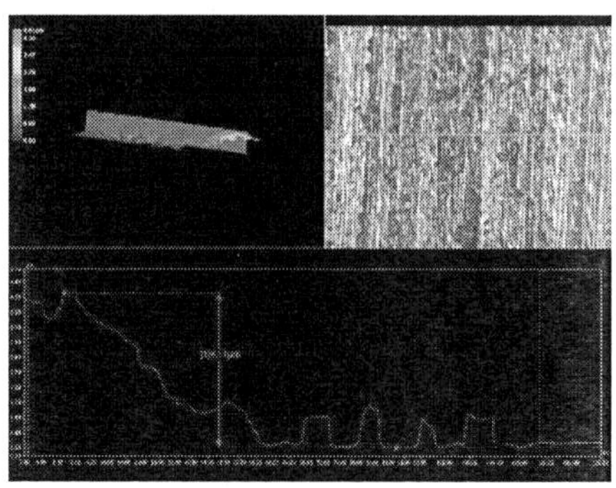

Figure 9: Surface profile of type 3 wafer (in figure 1) at the interface region.

4.2 PL & EL studies

The PL signal images of standard incoming wafers after texturing and passivation (AlO_x + SiN_x) are shown in figure 10. It can be observed that the high intensity region in center is circular with lower intensity region on the outer region. The PL signal images of the trial wafers after texturing and passivation are shown in figure 11. It can be observed that the high intensity region in center is elliptical with lower intensity in the outer region. The signature is different as compared to the standard production wafers indicating a possible correlation of the cell performance and EL image of the fabricated cells.

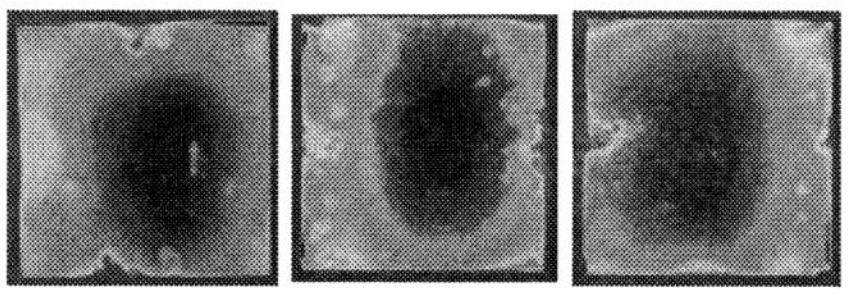

Figure 10: PL images of regular production wafers after texturing, RS PECVD deposition on both sides and fast firing (in that order). Blue region corresponds to higher PL signal strength, while green, yellow and red correspond to lower PL strength in that order.

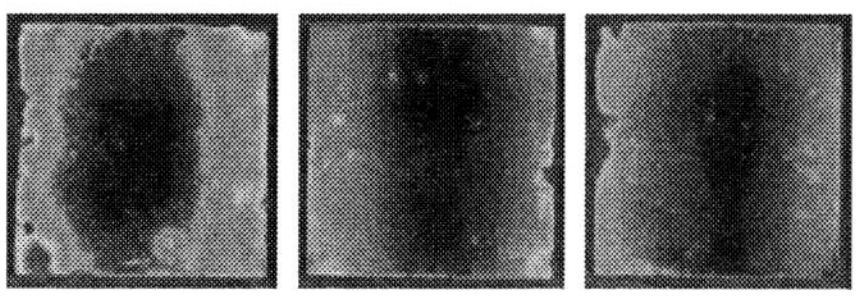

Figure 11: PL images of non-uniformly sawn wafers after texturing, RS PECVD deposition on both sides and fast firing (in that order). Blue region corresponds to higher PL signal strength, while green, yellow and red correspond to lower PL strength in that order.

Representative PL images of standard production wafers processed till front-side (FS) PECVD with uniform intensity are shown in figure 12. The PL images for representative trial wafers are shown in figure 13. It can be observed that the left-side wafer has lower PL intensity region where the saw mark lines are more clearly visible (which are possibly not completely removed during the texturing process). For the PL image on the right-side such saw marks seem to have influenced a significant area of the wafer.

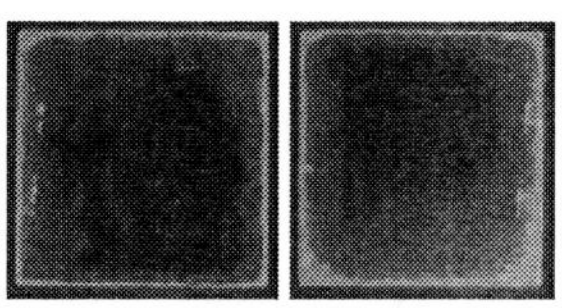

Figure 12: PL images of baseline production wafers after texturing, diffusion, Laser SE, edge isolation, RS PECVD, FS PECVD deposition and fast firing (in that order). Blue region corresponds to higher PL signal strength, while green, yellow and red correspond to lower PL strength in that order.

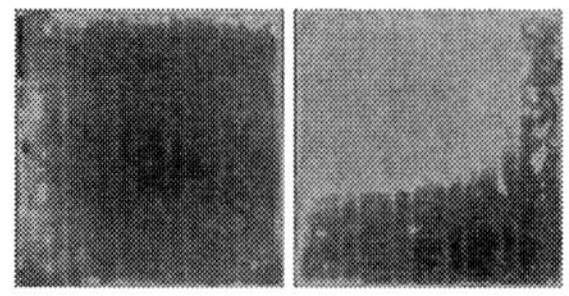

Figure 13: PL images of non-uniformly sawn wafers after texturing, diffusion, Laser SE, edge isolation, RS PECVD, FS PECVD deposition and fast firing (in that order). Blue region corresponds to higher PL signal strength, while green, yellow and red correspond to lower PL strength in that order.

4.3 I-V data comparison

The I-V data comparison for the baseline production and trial group cells are shown in table 1. The average efficiency difference between the two groups is 0.04%-0.08% absolute. The difference is mainly coming from FF decrease of ≈ 0.1% absolute for both groups. The series resistance and shunt resistance values were comparable for both groups.

Table I: Comparison of the I-V data for the baseline production with the trial groups.

Test	Group	Nos	V_{oc} (mV)	I_{sc} (A)	FF (%)	η (%)	I_{rev12} (A)
1	Baseline	21,115	683.4	10.01	81.15	22.04	0.053
	Trial	1,166	682.1	10.01	81.02	21.96	0.056
2	Baseline	23,641	682.4	10.07	81.25	22.18	0.047
	Trial	10,800	682.2	10.07	81.16	22.14	0.044

From the cell efficiency distribution shown in figure 14 & 15 for the first test, it can be observed that there is a wider efficiency distribution for the trial group compared to a tighter distribution of the regular production. The percentage of solar cells with efficiency between 22.0-22.5% for the baseline production and trial group are 67.2% and 53.3%, respectively.

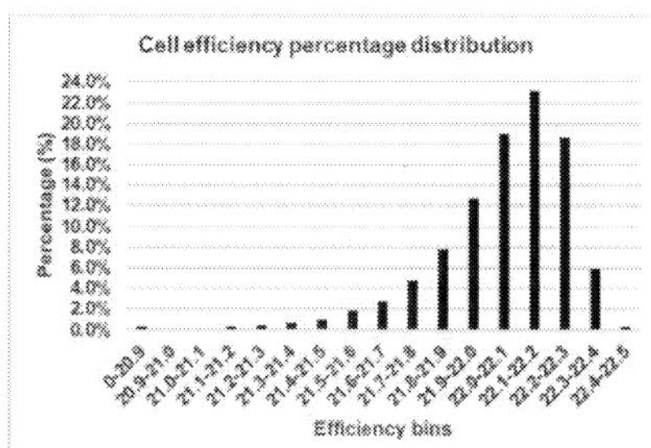

Figure 14: Efficiency distribution for the baseline production in the 1st test.

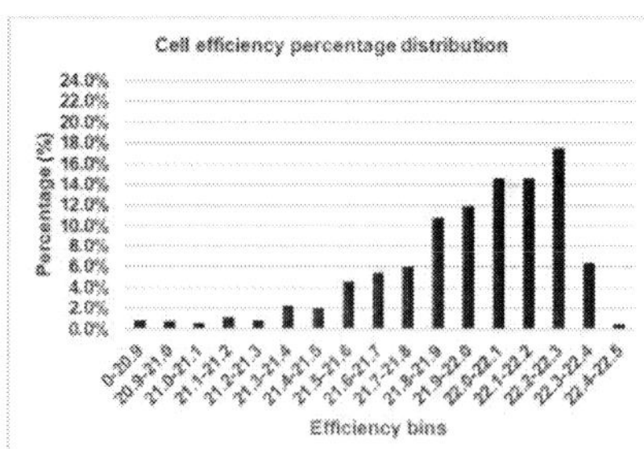

Figure 15: Efficiency distribution for the trial group in the 1st test.

The efficiency distribution for the second test (with a larger trial batch size) are shown in figure 16 and 17. The percentage of solar cells with efficiency between 22.0-22.5% for the baseline production and trial group are 90.1% and 73.1%, respectively. This supports the decrease in the average efficiency for the two groups with non-uniform saw damage. It can be inferred that wafers with such non-uniform saw damage also contribute to a wider spread of cell efficiency during production.

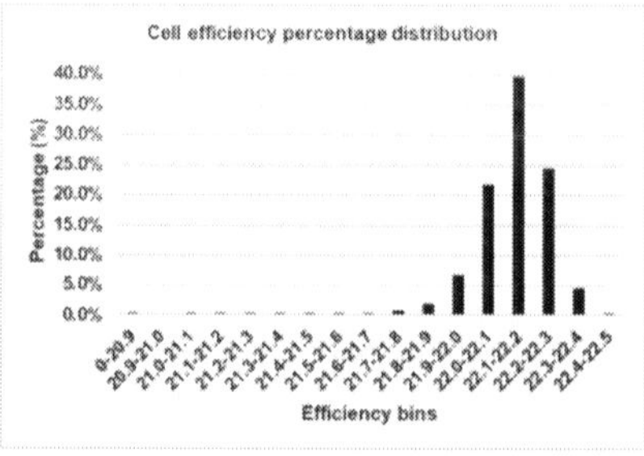

Figure 16: Efficiency distribution for the baseline production in the 2nd test.

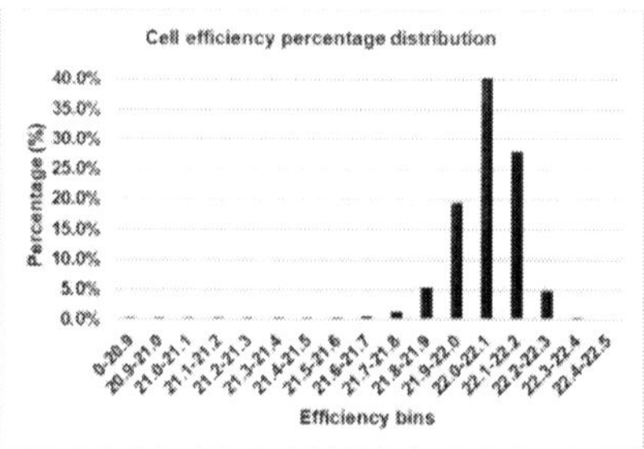

Figure 17: Efficiency distribution for the trial group in the 2nd test.

4.4 Comparison of cell classification

The comparison of classification of the baseline and trial groups are shown in table 2 and table 3 for the two test batches. It can be observed that the percentage of B grade and unqualified cells is higher for the trial groups compared to the standard production. This leads to a decrease in the over-all yield of A & A+ grade cells by up to 6.0% relative compared to the regular production. It is also important to observe that the scrap rate and rework wafers are also significantly higher in the trial group of the second test. In addition to decreasing the A+/A grade cells, the breakage rate and rework rate is also adversely affected by wafers with non-uniform saw damage.

Table II: Comparison of classification for the 1st run of the baseline production and the trial group with batch sizes of 21,115 and 1,166, respectively.

Parameter	Baseline production	Trial group relative difference (%)
A+ & A	Reference	-2.2
B grade	Reference	+32.8
0-20.9%	Reference	+235.0
Low R_{sh}	Reference	-22.8
High I_{rev12}	Reference	0.0
Unqualified	Reference	+155.6

Table III: Comparison of classification for the 2nd run of the baseline production and the trial group with batch sizes of 23,641 and 10,800, respectively.

Parameter	Baseline production	Trial group relative difference (%)	Observations
A+ & A	Reference	-6.0	Lower compared to baseline for both runs
B grade	Reference	+41.0	Higher for both runs
0-20.9%	Reference	-9.0	Lower for larger batch size run.
Low R_{sh}	Reference	+64.0%	Higher for larger size batch run.
High I_{rev12}	Reference	0.0%	Not critical for both runs.
Unqualified	Reference	+237.0%	Higher for both runs.
Scrap	Reference	+553.0%	Significantly higher for larger batch size run.
Rework	Reference	+77.0%	Higher for larger batch size run.

5 SUMMARY

In our study it was observed that non-uniformly sawn wafers can reduce the efficiency of produced mono PERC solar cells by ≈ 0.1% absolute and more significantly, lead to yield loss of up to 6.0% relative compared to the regular production. The reduction in average efficiency is reflected in wider efficiency spread to lower values for the trial groups. The percentage of B grade and unqualified cells for the trial groups are higher by 30% and 150%, respectively relative compared to the regular production wafers. In addition, the breakage rate and rework rate were also observed to be > 500% and > 70%, respectively higher compared to the baseline production. The study points to the detrimental influence of non-uniformly sawn wafers which can be classified as A grade wafers on the solar cell production parameters. It is recommended to finetune the criteria of the wafer inspection system to minimize the classification of such wafers into A grade and also optimize the DWS process to minimize such non-uniformly sawn wafers.

ACKNOWLEDGEMENTS

The authors would like to thank the colleagues Mert Kahraman and Metehan Gulaydın for support in performing the trials in production and related analysis.

REFERENCES

[1] International Technology Roadmap for Photovoltaic (2020) 11th Edition. October, 2020.
[2] Arkadeep Kumar and Shreyes N. Melkote. "Diamond Wire Sawing of Solar Silicon Wafers: A Sustainable

Manufacturing Alternative to Loose Abrasive Slurry Sawing", Procedia Manufacturing 21 (2018) 549–566.

[3] Erick Cardoso Costa, Caroline Piesanti dos Santos, Fabio Antonio Xavier, Walter Lindolfo Weingaertner "Influence of diamond wire sawing parameters on subsurface microcracks formation in monocrystalline silicon wafer", Proceedings of 25th ABCM International Congress of Mechanical Engineering (2019).

[4] Srinivasamurthy Devayajanam "Experimental and theoretical evaluation of in-depth damage distribution in sawn silicon wafers", Doctor of Philosophy, Materials Science and Engineering, New Jersey Institute of Technology, Newark, NJ (2016).

COMPARISON OF ELECTRICAL PERFORMANCES OF FULLY ION IMPLANTED N-PERT SOLAR CELLS MADE OF DIFFERENT N-TYPE WAFER FEEDSTOCKS

Gence Bektaş[1,2*], Hasan Asav[1,2], Hasan Hüseyin Canar[1,3], Ozan Aydın[4], Ahmet Emin Keçeci[1,2], Samet Ozdemir[1,], Sümeyye Koçak Bütüner[1,2], Selin Seyrek[1,2], Fırat Es[4], Raşit Turan[1,2,3]

[1]ODTÜ-GÜNAM, Çankaya, 06800, Ankara, Turkey
[2]Micro And Nanotechnology Department, Middle East Technical University, Çankaya, 06800, Ankara, Turkey
[3]Department of Physics, Middle East Technical University, Çankaya, 06800, Ankara, Turkey
[4]KalyonPV Research and Development Center, Kalyon Güneş Üretim Teknolojileri A.Ş, 06909, Ankara, Turkey

ABSTRACT: This work compares the electrical performances of fully ion-implanted bifacial PERT solar cells made of n-type wafers from Kalyon PV and Longi Solar. The cell performances are presented with comparative emitter and BSF characterization results. The highest front photo-conversion efficiencies are measured as 19.3% and 19.2% with bifaciality factors (η_{rear}/η_{front}) of 86.8% and 85.1% for the solar cells made of Longi and Kalyon wafers, respectively. The V_{oc} values of the solar cells and iV_{oc} values of the symmetrical emitter and BSF samples fabricated on n-Si wafers from the two companies are almost equal, implying similar wafer qualities from the device point of view.
Keywords: n-PERT, wafer feedstock, boron, phosphorus, ion-implantation

1 INTRODUCTION

N-type wafers take advantage of a higher minority carrier lifetime than p-type wafers and lack of light-induced degradation due to the absence of boron-oxygen (B-O) complexes and their resistance to metal impurities. [1]. Although the price of n-type wafers has been estimated 8% more expensive than p-type due to the difference in brick cost [2], it is projected that the market share of n-type solar cells will increase in the solar energy market [3]. This might require manufacturers to improve their n-type wafer production capacity in the near future. Besides, the bifacial solar cells with a so-called "H-pattern" grid at the rear surface can provide an additional power generation by utilizing the albedo of the surroundings [4].

Here, we fabricated n-type bifacial PERT solar cells in ODTÜ-GÜNAM to compare different n-type wafer feedstocks. The ion implantation method is used for the emitter and back surface field (BSF) formation since it simplifies the process steps by enabling single side doping [5]. In this context, we also prepare symmetrical boron implanted emitter and phosphorus implanted back surface field (BSF) samples passivated with Al_2O_3/SiN_x stack layer to measure their implied open circuit voltage (iV_{oc}) values.

2 EXPERIMENTAL

The industrial n-type Cz-Si solar wafers with base resistivity of 1-3 Ω.cm from Longi Solar and Kalyon PV were alkaline textured and ozone cleaned. Then, the samples were divided into two groups; the first one is for the symmetrical emitter and back surface field (BSF) sample fabrication and the latter one is for the solar cell fabrication. The first group of wafers was double-side implanted with either boron at 7.5 keV with a dose of D1 or phosphorus at 10 keV with a dose of D2. Boron (B) and phosphorus (P) implanted wafers were activated at 1050 °C and 875 °C, respectively. Afterward, the samples were dipped into the HF solution to remove unintentionally grown oxide during the activation process. The double sides of the symmetrical emitter and BSF samples were passivated with 2 nm and 4 nm ALD: Al_2O_3, respectively. Finally, all the samples were symmetrically capped with a 75 nm PECVD: SiN_x layer and subsequently fired by a conveyor belt furnace to mimic the cell process.

The second batch of wafers was used for bifacial n-PERT solar cell fabrication, as the process flow is shown in Figure 1.

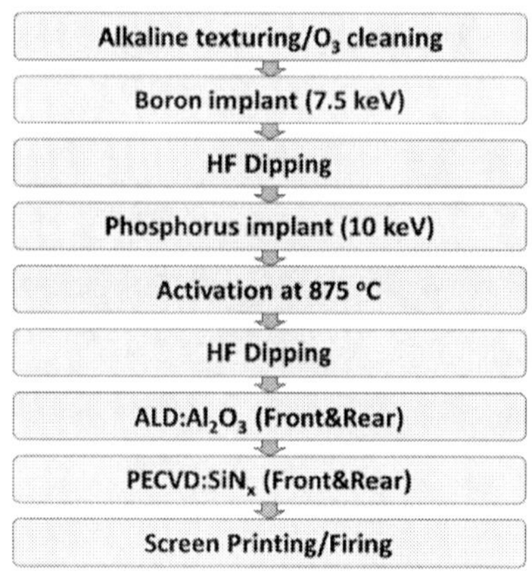

Figure 1: Process flow for the cell fabrication.

The schematic representation of the resulting bifacial n-PERT solar cells in this study is given in Figure 2.

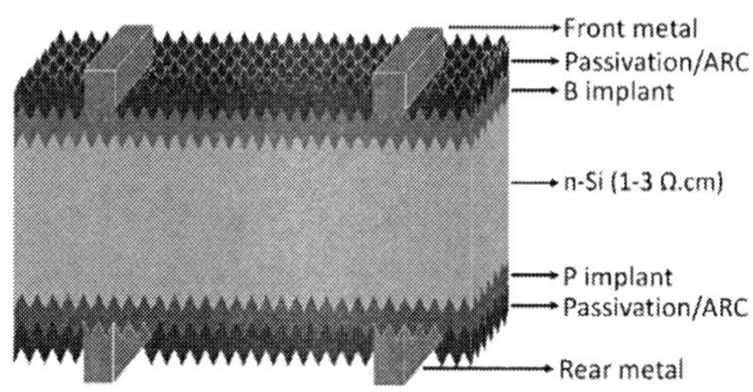

Figure 2: The schematic representation of n-PERT solar cells fabricated in this study.

The doping profiles and the sheet resistances (R_{sheet}) for the emitter and BSF were determined on the wafers with the structure of $p^{++}/n/n^{++}$ by the electrochemical capacitance-voltage (ECV) and the four-point probe (4PP) methods, respectively. iV_{oc} values on the symmetrical samples were measured with the photoconductance decay (PCD) method by a Sinton lifetime tester. The current-voltage (I-V) characteristics of the fabricated solar cells were measured by a class AAA solar simulator under standard test conditions.

3 RESULTS AND DISCUSSION

The averages of measured sheet resistances of the emitter and BSF on n-Si wafers were measured as 100.6 Ω/sq and 33.4 Ω/sq, respectively. The corresponding ECV graphs are given in Figure 3. Active dopant doses were calculated as 1.16E15 cm^{-2} for B and 1.76E15 cm^{-2} for P.

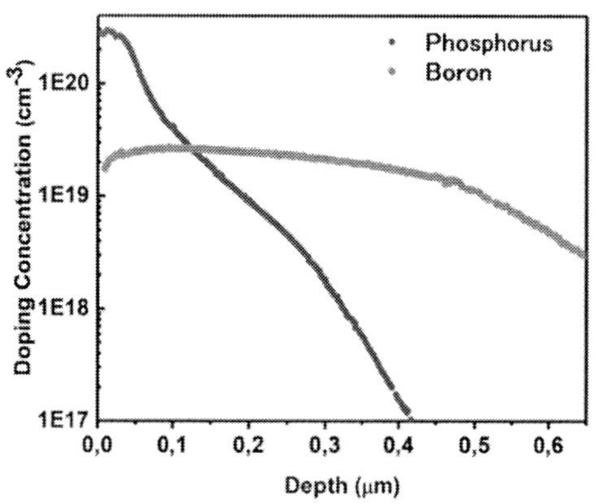

Figure 3: ECV graphs of implanted emitter (green) and BSF (red)

The iV_{oc} values measured on the symmetrically implanted and passivated n-type wafers from the two companies are shown in Figure 4. The average iV_{oc} for the emitter samples is around 624 mV and 623 mV for Kalyon (named K-B) and Longi (named L-B) wafers, respectively. Additionally, the average iV_{oc} values of the BSF samples are measured as 641 mV for Kalyon and 642 mV for Longi wafers.

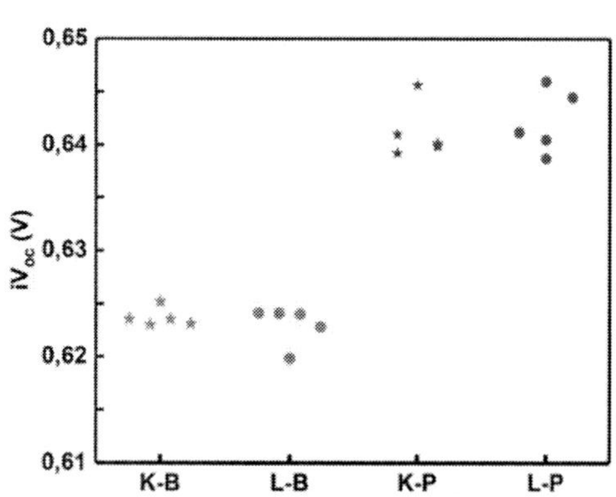

Figure 4: iV_{oc} of symmetrically implanted and passivated emitter (green) and BSF (red) formed on Kalyon (star) and Longi (square) wafers.

The solar cell parameters obtained by I-V measurement are given in Table 1.

Table I: The solar cell parameters measured by the solar simulator on fabricated bifacial n-PERT cells from Longi (named L-1 and L-2) and Kalyon (named K-1 and K-2) wafers

Sample Name	Measured Side	η (%)	J_{sc} (mA/cm^{-2})	V_{oc} (mV)	FF (%)
L-1	Front	19.3	38.1	629.5	80.4
	Rear	16.8	33.5	625.4	80.0
L-2	Front	**19.3**	38.2	628.7	80.5
	Rear	16.8	33.5	625.0	80.1
K-1	Front	**19.2**	38.1	629.5	80.1
	Rear	16.3	32.8	625.7	79.8
K-2	Front	19.1	37.9	630.4	79.8
	Rear	16.4	33.0	626.0	79.2

V_{oc} values are measured as around 630.0 mV at the front side of the solar cells made of Kalyon and Longi wafers. This implies a similar wafer quality for the two brands. The front efficiency of the champion cells is around 19.2% and 19.3% for the fabricated cells from Kalyon and Longi wafers, respectively. The 0.1% difference in the efficiency of the solar cells from the two wafer brands is due to a difference in fill factor (FF). This can be due to unalike metallization quality for the cells.

4 CONCLUSIONS

Herein, we investigated the performances of the fully ion-implanted bifacial n-PERT solar cells made of wafer feedstocks of Kalyon PV and Longi Solar. The measured V_{oc} values of the cells and iV_{oc} values of the symmetrical emitter and BSF samples were almost equal for the wafers from the suppliers, which implies similar wafer qualities from the device perspective. Efficiencies of the fabricated solar cells slightly differ due to the variance in FF, which may be due to the difference in resistive losses of the cells. The highest efficiencies obtained at the front sides of the cells from Kalyon and Longi wafers were 19.20% and 19.30%, respectively.

5 REFERENCES

[1] A. B *et al.*, "Emerging frontiers of N-Type silicon material for photovoltaic applications: The impurity-defect interactions," *Front. Nanosci. Nanotechnol.*, vol. 1, no. 1, pp. 2–12, 2015.

[2] N. L. Chang, M. Wright, R. Egan, and B. Hallam, "The Technical and Economic Viability of Replacing n-type with p-type Wafers for Silicon Heterojunction Solar Cells," *Cell Reports Phys. Sci.*, vol. 1, no. 6, p. 100069, 2020.

[3] VDMA, "Update ITRPV 2020 Results 2019 including maturity report 2020." 2020.

[4] A. Lanterne, S. Gall, Y. Veschetti, R. Cabal, F. Milési, and A. Tauzin, "High efficiency fully implanted and co-annealed bifacial n-type solar cells," *Energy Procedia*, vol. 38, pp. 283–288, 2013.

[5] G. Bektaş *et al.*, "Simplified process flow for the fabrication of PERC solar cells with ion implanted emitter," *Renew. Energy*, vol. 179, pp. 2295–2299, 2021.

TEXTURIZATION OF MONOCRYSTALLINE SILICON BY METAL-ASSISTED CHEMICAL ETCHING: ANALYSIS OF REACTION DYNAMICS

David Pera, Ivo Costa, Filipe Serra, Afonso Guerra, Killian Lobato, João M. Serra, José A. Silva
Instituto Dom Luiz - Faculdade de Ciências Universidade de Lisboa
Edifício C1, Campo Grande, 1749-016 Lisboa, Portugal

ABSTRACT: We present here recent results of an optimized metal-assisted chemical etching (MACE) based process. The process consists of a simple maskless etching based on the use of a solution of hydrogen peroxide (H_2O_2), hydrofluoric acid (HF), and silver nitrate ($AgNO_3$). H_2O_2 and HF work as etching agents, while $AgNO_3$ donates silver ions that locally catalyze the etching reaction, creating steep and anisotropic structures on the surface and sharply reducing its reflectivity. In this study Czochralski monocrystalline silicon wafers both p-type and n-type were used as substrates and the etching time was swept between 30 s and 270 s. It was observed that, although for optimal conditions p-type and n-type wafers achieved similar reflectance values after etching, the reaction's dynamic showed to be dependent on the doping type and concentration level (resistivity). The lowest effective reflectivity (R_{eff}) reached during this study was 3.0%, obtained for a p-type wafer.
Keywords: Crystalline, silicon, texturization, light-trapping

1 INTRODUCTION

Improving the efficiency of photovoltaic systems is a goal increasingly pursued by the photovoltaic industry. One of the most effective ways to improve the efficiency of photovoltaic cells and systems is by increasing the number of photons that reach and are absorbed in the solar cell active area, thus boosting the solar cell photocurrent. Also, by enhancing the solar cells' light-capture efficiency, thinner and lower quality substrates can be used, allowing significant reductions in the device's production costs [1].

The standard method to reduce front surface reflectivity in crystalline silicon solar cells is to perform a chemical etching with an alkaline or acidic bath, followed by the deposition of a hydrogenated silicon nitride film by PECVD [2]. In the last years, several innovative strategies to improve light capture in solar cells have been proposed, such as the use of plasmonic structures [3]. However even if the introduction of plasmonic structures can produce a significant increase absorption, for certain regions of the electromagnetic spectrum, the overall light absorption gain, averaged in the AM1.5 spectrum, is not weighty. The use of reactive ion etching (RIE) to produce nanostructures that enhance solar cells light-capture is seen as very promising approach [4]. But until know a cost-efficient RIE based process, to be used in industrial solar cell production has not been demonstrated. The interest in the use of metal-assisted chemical etching (MACE) [5] based methods to reduce the reflectivity of the solar cells' front surface has increased significantly in the last years. Due to its simplicity, scalability, and potential cost-effectiveness, MACE is one of the most promising light-trapping techniques to be applied in the industrial production of high-efficiency crystalline silicon solar cells. In fact, the capability of this technique to increase the solar cell efficiency has already been demonstrated [6].

In our study we use a simple maskless MACE process based on the use of a solution of hydrogen peroxide (H_2O_2), hydrofluoric acid (HF), and silver nitrate ($AgNO_3$). The process has been characterized and optimized in previous studies [7,8]. Here we analyze the reaction dynamics and compare results for p-type and n-type silicon.

2 METHODS

2.1 Etching process

The mechanics of the MACE texturization process is well known [7], the substrates are dipped in the etching solution. H_2O_2 acts as oxidation agent while HF removes the silicon oxide formed, according to the equation:

$$Si+H_2O_2 + 6HF \rightarrow 2H_2O + H_2SiF_6 + H_2 \qquad (1)$$

$AgNO_3$ is dissociated in the solution, donating holes to the reaction thus catalyzing it. Consequently, the presence of silver originates faster etching reactions and deeper texturizations.

During this study three groups of Czochralski monocrystalline silicon wafers were used: (i) 1-3 Ω.cm p-type (boron doped); (ii) 10-20 Ω.cm p-type (boron doped); (iii) n-type (phosphorous doped) 3-7 Ω.cm. For all these wafers the surface orientation was (100).

During this study the concentrations of the three reactants were fixed (Table I) and the etching time was varied between 30 and 270 seconds.

Table I: Reactants' concentrations used

	Concentration (mol/l)
HF	12
H_2O_2	1.1
$AgNO_3$	6×10^{-3}

Since the etching reaction is highly exothermic and the reaction's kinetics is strongly dependent on the temperature. To adequately characterize the reaction's dynamics, controlling the reaction temperature is mandatory. For this purpose, the reaction beaker was placed in a water buffer bath kept at 30°C and the evolution of the reactants' temperature was carefully monitored. After removal from the etching bath, the wafers were rinsed in deionized water. To remove the silver deposit, the wafers were then dipped in a nitric acid solution, and again rinsed in deionized water.

2.2 Characterization

To analyze the mass variation during the MACE process, the wafers' mass was measured before etching and after silver removal. The spectral reflectance of the etched surfaces was measured in the range 350-1100 nm with an UV-vis Spectrophotometer Shimadzu UV – 2600 PC. The effective reflectivity R_{eff} was calculated by weighting the reflectance values with the reference AM1.5 spectrum [9] according to equation (2), where R_λ and n_λ are respectively the reflectance measured and the number of photons in the AM1.5 spectrum for the wavelength λ.

$$R_{\text{eff}} = \frac{\sum_\lambda R_\lambda n_\lambda}{\sum_\lambda n_\lambda} \qquad (2)$$

Moreover, the structural characterization of the textured surfaces was carried out through scan electron microscope (SEM) observations.

3 RESULTS

3.1 Structural characterization

In Figure 1 a SEM crosscut image of a MACE etched p-type silicon wafer is presented. The disordered nature of the nanostructures created is noticeable, also the typical height of the nanostructures can be estimated to be around 2 nm. The observations made on n-type silicon wafers etched in analogous conditions, showed similar surface morphology.

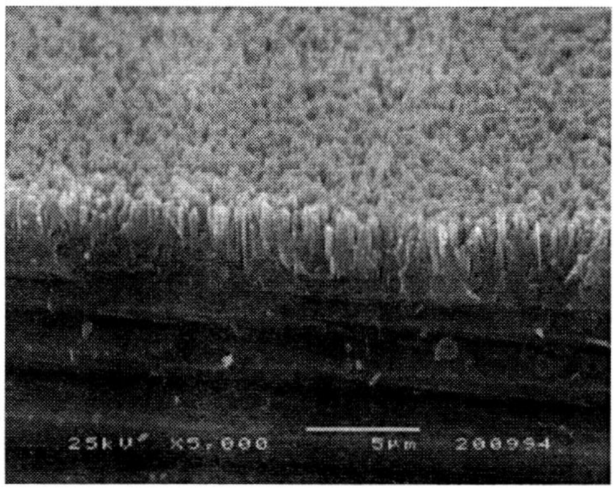

Figure 1: SEM image of p-type silicon wafer, 1-3 Ω.cm etched during 120 s.

3.2 Mass variation

The variation of the wafers' mass during etching is presented in Figure 2. It can be observed that for the three groups of wafers texturized, the mass is linearly reduced with the etching time.

We can also observe that, for the etching times tested, the reduction of mass over time is very similar for the samples type n (3-7 Ω.cm), and type p II (10-20 Ω.cm).

When comparing the mass reduction for wafers p-type I (1-3 Ω.cm) with the remaining groups we can conclude that the mass reduction rate is slightly lower. In fact, for p-type I wafers the Δm/m (t) line slope is $m=1.6\times10^{-3}$, compared to 1.8×10^{-3} for the other two series of wafers. The results suggest that for n- and p-type wafers with comparable resistivities, the etch rate is approximately 10% higher for n-type ones. It must be mentioned that such difference in the reaction dynamics, for the two types of silicon, is lower than the values previously reported in the literature for MACE based processes [10].

On the other hand, the results obtained for p-type wafers with different levels of resistivity, suggest etching rate increases with the doping level, agreeing with previous results [10].

Finally, it is clear that for the p-type I wafers' group, the mass reduction as function of time is significantly delayed nearly 30 s when for the two-remaining wafers' groups (i.e., the series is shifted to the left). Such result might be attributed to a different initial surface condition for the wafers comprised in this group.

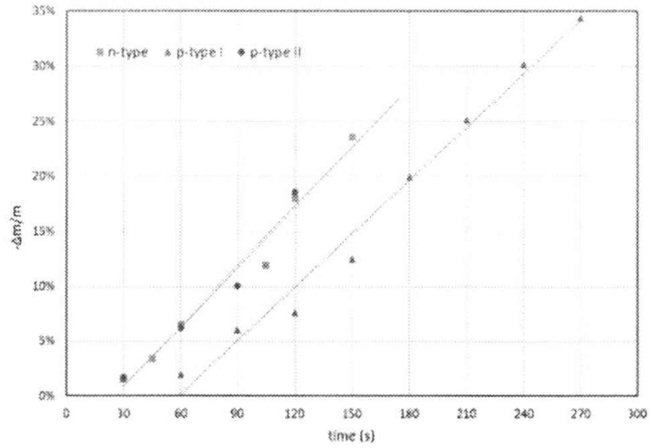

Figure 2: Mass variation as function of etching time for silicon wafers n-type (3-7 Ω.cm), p-type I (ρ ~ 1-3 Ω.cm) and p-type II (ρ ~ 10-20 Ω.cm).

3.2 Reflectance

In Figure 3 the reflectance curves after the etching process, for a n-type and a p-type wafer are compared. The reference spectrum AM1.5 is also presented (secondary axis). It can be observed that for the most spectral range the reflectance is very similar for both wafers. The only noticeable differences are observed in the ultraviolet and near-infrared regions, where the n-type wafer slightly higher reflectance values. It must be mentioned that such differences are inside the typical variability of the reflectance measurement process and were not considered significant.

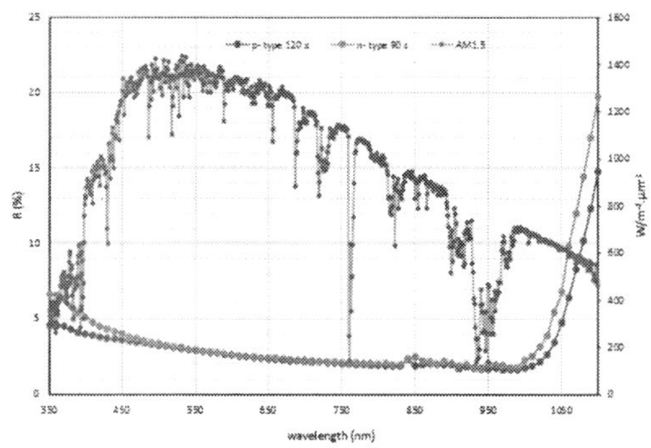

Figure 3: Reflectance curves for a p-type wafer (ρ ~ 1-3 Ω.cm) and n-type wafer (3-7 Ω.cm), in the range 350-1100 nm.

Based on the reflectance curves obtained and using equation (2) the R_{eff} values were obtained. The lowest R_{eff} obtained for n-type wafers was R_{eff}=3.0%, while for n-type the minimum was R_{eff}=3.3%.

4 CONCLUSIONS

The aim of the study here presented was to analyze sensibility of the reaction dynamics of a MACE based process, on doping type and concentration, by performing etchings of n- and p-type wafers with several etching times.

It was observed that the mass removal rates were similar for the three groups of samples, showing a linear behavior of the MACE process. For the same etching conditions and similar resistivities mass removal in n-type samples showed to be slightly faster than for p-type wafers. When comparing p-type samples with resistivity values with one order of magnitude difference, results showed that the mass removal increases for higher resistivity.

The reflectance measurements showed no significant difference between the three sets of samples. The lowest R_{eff} obtained for p-type was 3.0% while for n-type was 3.3%.

5 REFERENCES

[1] Y. Liu, W. Zi, S. Liu, and B. Yan, 'Effective light trapping by hybrid nanostructure for crystalline silicon solar cells', Solar Energy Materials and Solar Cells 140 (2015) 180-186.

[2] S. W. Glunz, R. Preu, and D. Biro. 'Crystalline silicon solar cells: state-of-the-art and future developments', Comprehensive Renewable Energy 1 (2012) 353-387.

[3] F. Enrichi, A. Quandt and G. C. Righini, 'Plasmonic enhanced solar cells: Summary of possible strategies and recent results', Renewable and Sustainable Energy Reviews 82 (2018) 2433-2439.

[4] H. S. Lee, J. Suk, H. Kim et al., 'Enhanced efficiency of crystalline Si solar cells based on kerfless-thin wafers with nanohole arrays', Science Reports 8 (2018) 3504.

[5] K. Tsujino, M. Matsumura and Y. Nishimoto, 'Texturization of multicrystalline silicon wafers for solar cells by chemical treatment using metallic catalyst', Solar Energy Materials and Solar Cells 90 (2006) 100-110.

[6] J. Oh, H.C. Yuan, H.M. Branz, 'An 18.2%-efficient black-silicon solar cell achieved through control of carrier recombination in nanostructures', Nature Nanotechnology 7 (2012) 743-748.

[7] I. Costa, D. Pera, J. A. Silva, 'Improving light capture on crystalline silicon wafers', Materials Letters 272 (2020) 127825.

[8] D. Pera, et al. 'Advanced light-trapping structures for back-contact solar cells produced by metal-assisted chemical etching.' Proceedings 37th European Photovoltaic Solar Energy Conference (2020) 354 - 357.

[9] ASTM, 'Reference Solar Spectral Irradiance: Air Mass 1.5', American Society for Testing and Materials (ASTM) Terrestrial Reference Spectra for Photovoltaic Performance Evaluation, 2003. [Online]. Available: http://rredc.nrel.gov/solar/spectra/am1.5/.

[10] Z. Huang, N. Geyer, P. Werner, J. De Boor, and U. Gösele, 'Metal-assisted chemical etching of silicon: A review', Advanced Materials. 23.2 (2011) 285-308.

6 AKNOWLEDGEMENTS

The authors acknowledge the financial support of the Portuguese Fundação para a Ciência e Tecnologia (FCT) through the projects TaCIt / PTDC/NAN-OPT / 28837 / 2017 and UIDB / 50019 / 2020 – IDL.
We would also like to thank Dr. Olinda Monteiro for the use of the UV-Vis Spectrophotometer and Telmo Nunes for performing the scan electron microscope observations.

OPTIMIZATION OF PHOSPHORUS IMPLANT DOSE AND ACTIVATION TEMPERATURE FOR EMITTER FORMATION OF P-TYPE SILICON SOLAR CELLS

Gence Bektaş[1,2*], Ahmet Emin Keçeci[1,2], Sümeyye Koçak Bütüner[1], Raşit Turan[1,2,3]

[1]ODTÜ-GÜNAM, Çankaya, 06800, Ankara, Turkey
[2]Micro And Nanotechnology Department, Middle East Technical University, Çankaya, 06800, Ankara, Turkey
[3]Department of Physics, Middle East Technical University, Çankaya, 06800, Ankara, Turkey

ABSTRACT: Today, p-type solar cells such as monofacial and bifacial Passivated Emitter and Rear Cell (PERC) with phosphorus emitter dominate the photovoltaic industry. For emitter formation on p-type Si, ion implantation is an attractive doping method because it allows a simplified solar cell process flow. Here, we investigate the effect of phosphorus implant dose and activation temperature on the iV_{oc} of the emitter of p-type textured Si solar wafers, which are passivated with 5 nm of thermal SiO_2 capped with PECVD:SiN_x. Our results show that the implanted emitter with a relatively lower dose is severely affected by activation temperature. We demonstrate iV_{oc} of 648 mV for implanted emitter with optimum implant dose and activation temperature.
Keywords: phosphorus, ion implantation, emitter, implied open circuit voltage

1 INTRODUCTION

Although the share of n-type solar cells is predicted to increase, p-type solar cells still dominate the photovoltaic industry (PV) industry [1]. In particular, monofacial and bifacial PERC solar cells with diffused emitter are mainstreams in the solar markets. On the other hand, the ion implantation method for emitter formation has been applied by several groups in recent years [2] since it can provide process simplicity and has precise control of the doping profile [3]. Moreover, the ion implantation method can be more advantageous for bifacial p-Si PERC cells. In the case of a diffused emitter, the rear surface of the wafer should be exposed to a single side etching process to remove n^{++}; however, it is not necessary in the case of ion implantation since it allows singe side doping. This will bring about flexibility to optimize the surface morphology in terms of optical and electrical performance. It has been shown that the roughness of the rear surface of bifacial p-Si PERC cells affects the efficiency depending on the amount of albedo [4].

Optimizing phosphorus implantation dose and subsequent activation processes is crucial for a high quality emitter formation, thus a solar cell efficiency. Here, we investigate the impact of doping profile, which was varied by ion implantation dose and activation temperature, on implied open-circuit voltage (iV_{oc}) of the symmetrical samples.

2 EXPERIMENTAL

The industrial p-type Cz-Si solar wafers with bulk resistivity of 1-3 ohm.cm and thickness of 180±20 µm underwent potassium hydroxide (KOH) texturing to form random pyramids and ozone cleaning process to remove organic and metallic contamination. Then, the wafers were double side phosphorous (P) implanted with two different doses (D1<D2) by the so-called Bosphorus implanter system manufactured by Dong Guan Plasma Ltd. The implanted samples were HF dipped and activated at various annealing temperatures of 850 °C, 875 °C, and 900 °C under nitrogen flow. In order to ensure the same surface passivation quality, the samples were dipped into HF to remove unintentionally grown thermal oxide during activation, and approximately 5 nm of thermal oxide layers were grown by dry oxidation in an annealing furnace. After that, all the samples were coated with PECVD: SiNx layers and fired by a conveyor belt furnace to mimic solar cell fabrication steps. Finally, the samples were annealed under forming gas for further improvement of iVoc of the samples. The sheet resistances (R_{sheet}) and active dopant profiles of the samples were measured by four-point probe (4PP) and electrochemical capacitance-voltage (ECV) methods, respectively. For the determination of the implied open circuit voltage (iVoc) values, the symmetrical emitter samples were measured by the quasi-steady state photoconductance (QSSPC) method using a Sinton WCT-120 instrument. The schematic of our process flow and characterization steps are given in Figure 1.

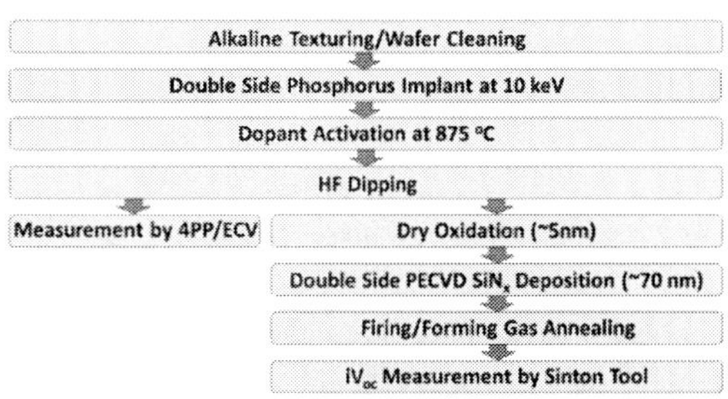

Figure 1: Process flow for the sample preparation

The schematic of our symmetrical samples prepared for iV_{oc} measurement is shown in Figure 1.

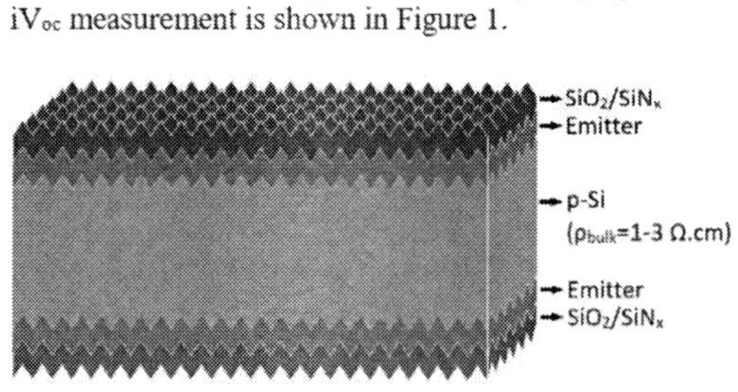

Figure 2: The schematic of the symmetrical emitter samples.

3 RESULTS AND DISCUSSION

R_{sheet} values of the samples with various implantation doses and activation temperatures are depicted in Table 1. R_{sheet} values decrease as the implantation dose and activation temperature increase.

Table I: R_{sheet} values of the samples measured by 4PP

Sample Name	*Dose*	*Activation Temperature (°C)*	*R_{sheet} (Ω/sq.)*
D1-850		850	120.4±9.5
D1-875	D1	875	105.6±6.3
D1-875		900	93.6±3.6
D2-850		850	90.7±3.9
D2-875	D2	875	77.4±3.0
D2-900		900	73.7±4.6

Figure 3 illustrates the active phosphorus doping profiles of the samples measured by the ECV technique. A higher activation temperature results in a decrease in peak concentration and an increase in junction depth. The trend is similar for both of the emitters but the peak concentrations are higher for larger dose named, D2.

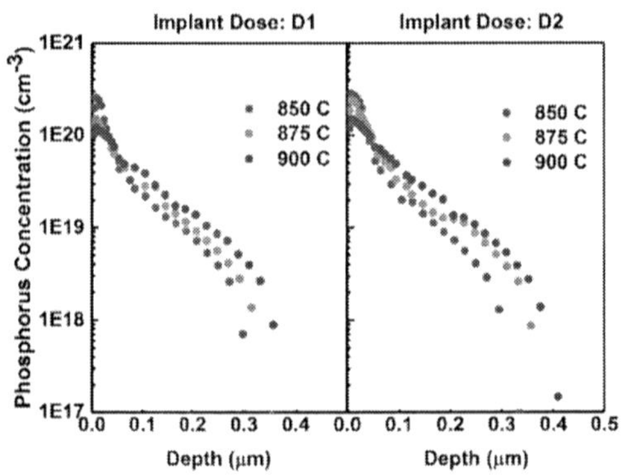

Figure 3: Phosphorus doping profile of the emitters with doses of D1 (left) and D2 (right), and activation temperature of 850 °C (red), 875 °C (green), and 900 °C (blue) measured by ECV.

Figure 4 shows the iV_{oc} variation of the symmetrical samples with activation temperature. iV_{oc} of the emitters with a relatively lower dose named D1 significantly decreases with increasing activation temperature. On the other hand, it is slightly dependent on activation temperature for a higher dose, D2. The optimum value was obtained for the emitter with a dose of D2 and activation temperature of 875 °C.

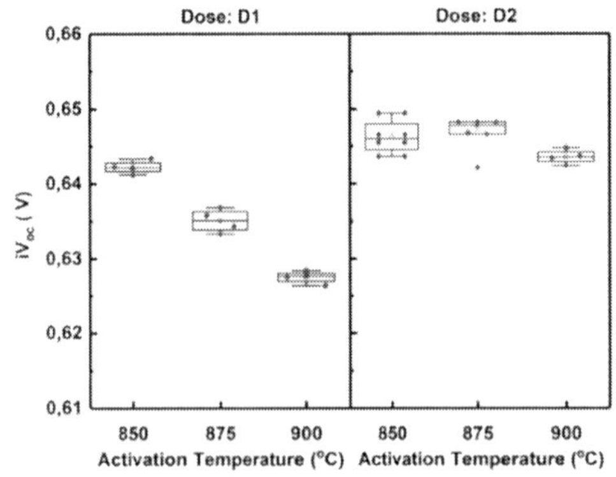

Figure 4: iV_{oc} values for the symmetrical emitter samples with implant dose of D1 (blue) and D2 (red) activated at 850 °C, 875 °C, and 900 °C.

4 CONCLUSIONS

In conclusion, we achieved to control the doping profile of the implanted emitter by changing the implant dose and subsequent activation temperature. Our results indicated that the iV_{oc} of the emitter implanted with a relatively lower dose, D1, is very sensitive to the activation temperature. The highest average iV_{oc} was 648 mV for the emitter implanted with dose D2 and activation temperature of 875 °C.

5 REFERENCES

[1] "International Technology Roadmap for Photovoltaic (ITRPV), 11th edition, April 2020."

[2] C. S. L. Mandrell, H. Huang, Jb. Wang, Jl. Wang, Z. Dong, J. Lv, R. Zhu, J. Sullivan, I. Latchford, B. Adibi, "Extending Solar Cell Efficiency Beyond 20% Using Energi Ion Implantation," in *29th European Photovoltaic Solar Energy Conference and Exhibition*, 2019, vol. 1, pp. 11–28.

[3] G. Bektaş *et al.*, "Simplified process flow for the fabrication of PERC solar cells with ion implanted emitter," *Renew. Energy*, vol. 179, pp. 2295–2299, 2021.

[4] H. B. Tang, S. Ma, Y. Lv, Z. P. Li, and W. Z. Shen, "Optimization of rear surface roughness and metal grid design in industrial bifacial PERC solar cells," *Sol. Energy Mater. Sol. Cells*, vol. 216, no. July, 2020.

Investigation of the Rear Side Passivation layer on Bifacial PERC Solar Cells

Cheng-Wen Kuo, Ta-Ming Kuan, Wei-Lo Chueh, Li-Guo Wu, Shih-Chieh Lin and Cheng-Yeh Yu
TSEC Corporation, No.85, Gaungfu N. Rd., Hsin-Chu 30351, Taiwan.

ABSTRACT: In this work, we demonstrated that the p-type bifacial PERC solar cells can also be improved front side performance based on rear side SiO_xN_y passivation structure. In the rear side passivation experiment, It was used the previous best conditions as the basic conditio
ns to change the thickness. The original thickness was the reference, and the thickness was increased by 1.2 times and 1.4 times. From the experimental results, a significant average efficiency gain for front side efficency with 1.4 time thickness passivation layer was achieved. Finally, by optimizing the rear structure, the bifacial PERC solar cells achieved the best performance with average front side efficiency of 22.57% and the bifaciality of bifacial PERC solar cells was about 50%.
Keywords: PERC, bifacial, SiO_xN_y, aluminum finger width

1 INTRODUCTION

In recent years, the concept of bifacial solar cells has received more attention. This is because in photovoltaic power plants, the use of bifacial solar modules instead of monofacial solar modules can increase power generation. However, industrial bifacial solar cell is mainly applying n-type passivated emitter and rear totally diffused (PERT) solar cells. The challenge of these bifacial PERT cells is that they must require single sided boron and phosphorus doping which requires additional or other process steps compared to p-type PERC cells [1-3]. It means that the manufacturing cost will increase.

Previous literature [2] also reported a bifacial PERC solar cell called PERC+. The PERC+ could be fabricated using monofacial PERC process by changing full area printing to local area printing on the rear surface. However, most of the rear side passivation layer for p-type PERC+ used the Al_2O_3/SiN, and there is very little discussion on SiON/SiN passivation layer for p-type PERC+.

In order to achieve the best bifaciality for PERC+ solar cell, the front side efficiency will be slightly degraded. The current sales method is to calculate the front side power for the module industry. It means that the rear side power generation for consumers is without payment. Therefore, we want to develop a PERC+ solar cell that has the best fornt side efficiency.

In this work, we demonstrated that the p-type bifacial PERC solar cells can also be improved front side performance based on rear side SiOxNy passivation structure. The bifacial PERC solar cells from optimizing the rear structure achieved the best performance with front side average efficiency of 22.57%.

2 EXPERIMENTAL

Figure 1 showed the PERC and PERC+ structure in this work. The 6-inch (158.75 mm x 158.75 mm) monofacial and bifacial PERC solar cells were fabricated on 175 μm thick p-type boron doped Cz-silicon wafers with a resistivity of 0.4~1.1 Ω-cm.

The process sequence in this work was shown in Figure 2. Chemical wet etching methods were used for saw damage removal and surface texturing, which include both acidic etching for multi-crystalline silicon wafers and alkaline etching for mono-crystalline silicon wafers. The solution composition of acidic etching was nitric acid (HNO3) : hydrofluoric acid (HF) : deionized water (DI) = 1 : 2 : 1.5, and the solution composition of alkaline etching was potassium hydroxide (KOH) : deionized water (DI) = 1 : 50. Conventional phosphorous diffusion was employed by furnace to form the n+ emitters with a sheet resistance of 80 - 100 ohm/sq. The diffusion process also improved to uniformity for this work.

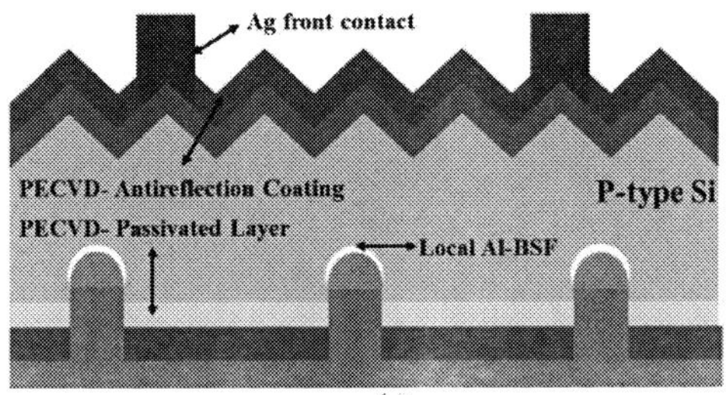

(a)

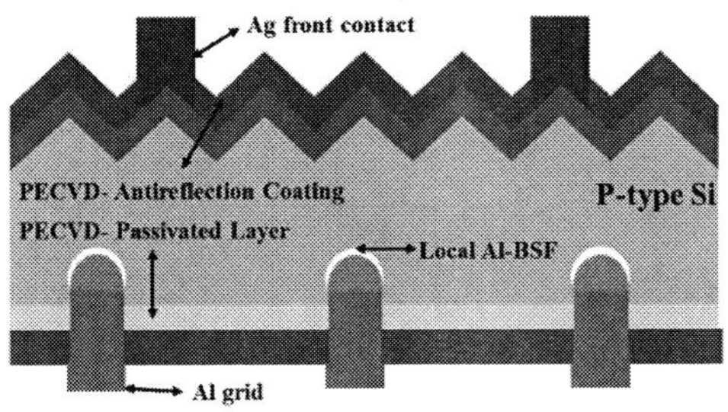

(b)

Figure 1: (a)PERC and (b)PERC+ structure as investigated in this work.

The phosphosilicate glass (PSG) was then removed by HF-contained solutions, formed edge isolation in the meantime. Then, the rear surface is polished which removes 5 μm and 10 μm of silicon in this work, in order to reduce the roughness and its impact on the effective charge carrier lifetime [4]. In order to decrease PID effect and increase efficiency, we optimized the CVD process parameter to improve. An antireflection coating (ARC) layer of silicon nitride (SiN_x: H) and/or SiN_x/SiO_x dual ARC was deposited by plasma enhanced chemical vapor deposition (PECVD) with thickness of 90 nm. The different rear side passivation was carried out on rear side also using a PECVD. The thickness of rear side passivation thickness changed from 280 to 200 nm. Afterward, the rear side laser opening ratio is also optimized [5]. All PERC solar cells have the laser-ablated opening widths in the rear-side passivation stacks which were about 45 μm, as shown in Fig 3. The rear Al finger width is 150 μm in this work.The detail parameters were listed in Table 1. The front side silver and rear side aluminum metallization was formed by screen printing technique and followed by co-firing process, which resulted in formation of front silver grid contact and local back-surface-field. Finally, the current-voltage measurements of solar cells under AM 1.5 illumination were performed.

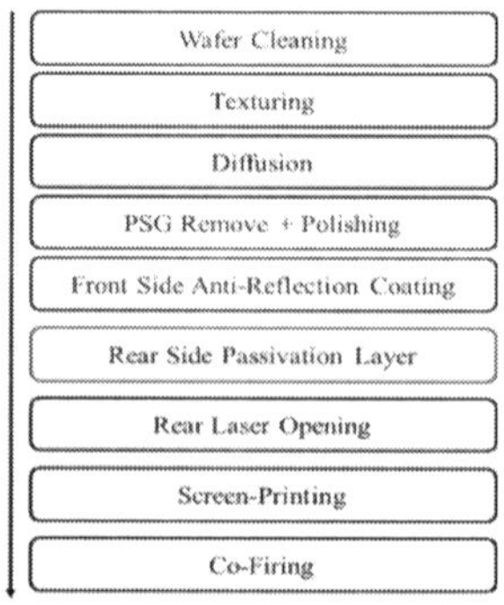

Figure 2: Process flow for the fabrication of PERC solar cell as developed in this work

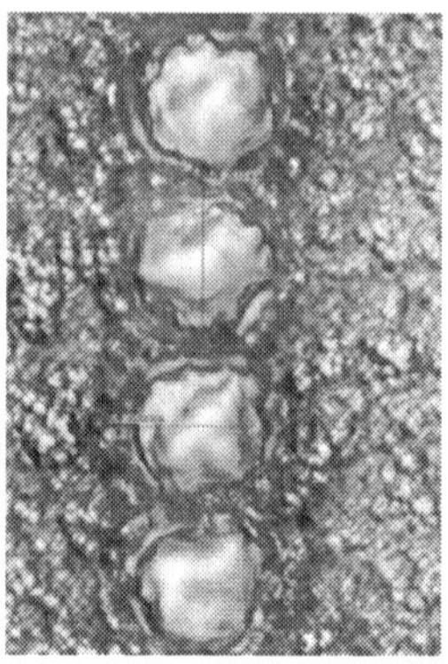

Figure 3 : OM image shown the laser ablated opening width in this work.

3 EXPERIMENTAL RESULTS AND DISCUSSION

Figure 4 shows scanning electron microscope (SEM) images of the different rear side passivation thickness for PERC and PERC+ solar cells. It could be found that the rear side passivation thickness is similar to the content of table I.

Table 1: The rear structure parameters of the PERC and PERC+, include the rear side passivation thickness and rear Al finger width

	Rear side layer thickness (nm)	Rear Al Finger Width (μm)
Mono PERC	180	-
Bifacial PERC REF (1)	200	150
Bifacial PERC (1.2)	240	150
Bifacial PERC (1.4)	280	150

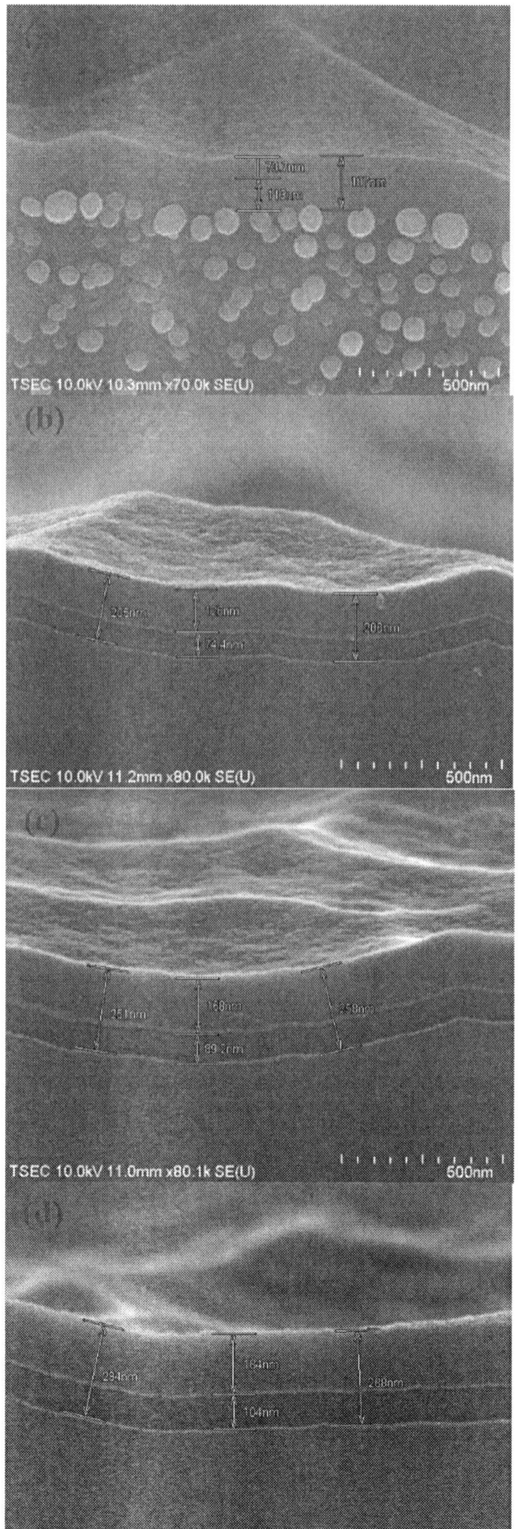

Figure 4: Scanning electron microscope images of rear side passivation thickness for the (a) PERC, (b) bifacial PERC REF (1), (c) bifacial PERC (1.2), and (d) bifacial PERC (1.4)

Table 2: The average electrical results of mono PERC and bifacial PERC cells.

	EFF (%)	FF (%)	VOC (V)	ISC (A)
Mono PERC	22.43	81.02	0.6813	10.242
Bifacial PERC REF (1)	22.40	80.25	0.6837	10.288
Bifacial PERC (1.2)	22.51	80.41	0.6850	10.297
Bifacial PERC (1.4)	22.57	80.41	0.6851	10.322

Table 2 summarized the average electrical results of PERC and bifacial PERC solar cells. It could be observed that the front side efficiency of Mono PERC and Bifacial PERC REF (1) are similar. However, the conversion efficiency on the rear side of the bifacial solar cell is about 70% of that the front side. The photo current of Bifacial PERC (1.4) solar cell is higher than Mono PERC and other bifacial PERC solar cells. This may be caused by the thickness of the rear passivation layer being too thick. However, it also caused the bifaciality to be only 50% for Bifacial PERC (1.4) solar cell. On the other hand, it could be found that the open circuit voltage (V_{OC}) of bifacial PERC solar cells is higher than Mono PERC solar cells. Figure 5 shows the SEM images of the local aluminum contacts for the PERC solar cell as well as for the PERC+ solar cell. The depth of the local Al-BSF of the PERC solar cell is 3.3±1.0 μm as average value of four measured local Al contacts (Figure 5a), the bifacial PERC solar cell exhibits the local Al-BSF depths of 4.5±1.2 μm as shown in Figure 5b. The increased local Al-BSF depth of the bifacial PERC solar cell is due to the Al finger layout. During furnace firing, silicon from the wafer diffuses through the laser contact opening into the Al paste layer. During cool down, the liquid silicon in the aluminum layer epitaxially re-grows at the silicon wafer surface incorporating Al up to the solid solubility forming the local Al-BSF[1]. The deeper local Al-BSF lowers the contact recombination and hence potentially enables higher VOC of the bifacial PERC solar cells [2], this is consistent with our experimental results. In addition, It can be found that the VOC of Bifacial PERC (1.2) and Bifacial PERC (1.4) is almost the same, and this is due to the SiON passivation effect has reached its limits in this experiment.

4 CONCLUSIONS

In this work, we have demonstrated the p-type bifacial PERC solar cells could be completed using rear side SiO_xN_y. In the rear side passivation experiment, it could be observed that the front side efficiency of bifacial cells which thicken the rear passivation thickness has improved, but the rear side efficiency has decreased significantly. Finally, the bifacial PERC solar cells achieved the best performance with front side average efficiency of 22.57% and the bifaciality of bifacial PERC solar cells was about 50%.

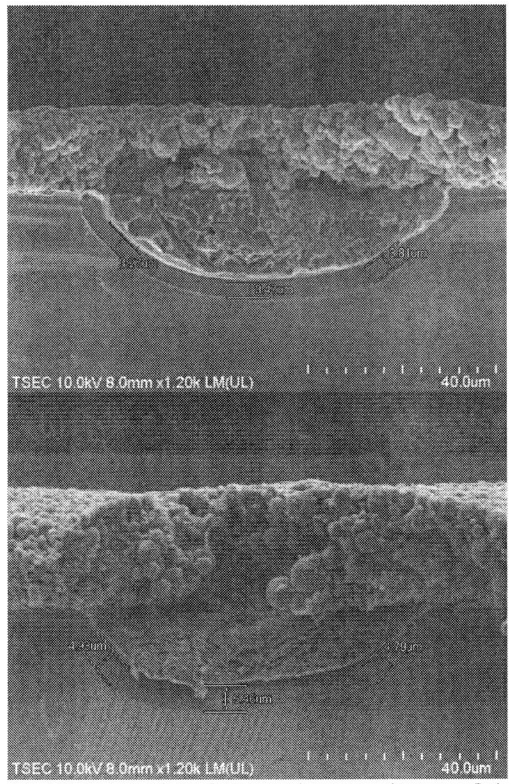

Figure 4: Scanning electron microscope images of the local aluminum contacts (a) for the PERC and (b) for the PERC.

5 REFERENCES

[1] T. Dullweber, C. Kranz, R. Peibst, U. Baumann, H. Hannebauer, A. Fülle, S. Steckemetz, T. Weber2, M. Kutzer, M. Müller, G. Fischer, P. Palinginis, D.H. Neuhaus, "The PERC+ Cell: A 21%-Efficient Industrial Bifacial PERC Solar Cell," 31st EUPVSEC, pp. 341, 2015.

[2] T. Dullweber, C. Kranz, R. Peibst, U. Baumann, H. Hannebauer, A. Fülle, S. Steckemetz, T. Weber, M. Kutzer, M. Müller, G. Fischer, P. Palinginis, and H. Neuhaus, "PERC+: Industrial PERC solar cells with rear Al grid enabling bifaciality and reduced Al paste consumption", *Prog. Photovolt: Res. Appl.*, vol. 24, pp. 1487, 2016.

[3] J. Dong, X. Chen, Y. Zhang, J. Lv, Z. Shen, J. Li, Q. Ye, W. Wei, W. Wang, J. Sheng, C. Zhang, "82% Bifaciality Industrial p-Type mc-Si Bifacial PERC Solar Cell," 35th EUPVSEC, pp. 778, 2018

[4] E. Urrejola, R.Petres, J. Glatz-Reichenbach, K. Peter, E. Wefringhaus, H. Plagwitz, G. Schubert., "High Efficiency Industrial PERC Solar Cells with all PECVD-Based Rear Surface Passivation" 26th EUPVSEC, pp. 2233, 2011..

[5] C.-W. Kuo, T.-M. Kuan, C.-C. Liu, C.-C. Huang, L.-G. Wu, C.-Y. Yu, "Impact of Laser Opening Ratio for Mass Production High Efficiency Mono-Crystalline Silicon Solar Cells" 29th EUPVSEC, pp. 1392, 2014

EFFECT OF SURFACE MORPHOLOGY ON PASSIVATION QUALITY OF AL_2O_3/SiN_x STACK LAYER FOR PERC SOLAR CELL

Sümeyye Koçak Bütüner[1*], Gence Bektaş[1,2], Gamze Kökbudak[2,] Mona Zolfaghari Borra[1,2], Hasan Asav[1,2],
Ahmet Emin Keçeci[1,2], Bülent Arıkan[1], Raşit Turan[1,2,3]
[1]ODTÜ-GÜNAM, Çankaya, 06800 Ankara, Turkey
[2]Micro And Nanotechnology Department, Middle East Technical University, Çankaya, 06800 Ankara, Turkey
[3]Department of Physics, Middle East Technical University, Çankaya, 06800 Ankara Turkey

ABSTRACT: In this study, we analyze the passivation quality of the Al_2O_3/SiN_x stack layer on various surface morphologies for the rear side of PERC solar cells. For this purpose, four types of samples that have different surface morphologies were produced symmetrically. Surface characterization of the samples is conducted by using SEM and AFM. The impact of surface morphology on passivation quality is evaluated by implied open-circuit voltage (iV_{oc}) and minority effective carrier lifetime (τ_{Eff}) values measured by Sinton WCT – 120 device. Our analysis reveals that the smoother surface improves passivation or (that passivation is enhanced when surface roughness decreases) unless the surface is polished. The highest iV_{oc} and τ_{Eff} values are obtained for the surface with root mean square roughness (Rq) value of 280 nm as 700.6 mV and 156.47 µs, respectively.
Keywords: PERC, chemical etching, surface morphology, rear side passivation

1 INTRODUCTION

Passivated emitter and rear cell (PERC) dominates today's PV industry[1]. Rear emitter removal and subsequent passivation processes are two essential and correlated steps in the PERC cell production lines [2]. The surface obtained after removal of the rear emitter affects the quality of passivation. In general, random pyramids which form during texturing process cause recombination problems compared to the smoother surfaces [3]. This phenomenon is because the amount of dangling bonds for surfaces [111] is higher than for surfaces [100]. [4]. Accordingly, smoothening of the pyramids before the passivation process is necessary to improve the final cell performance. This study aims to reveal the effect of surface morphology on the Al_2O_3/SiN_x stack passivation layer.

2 EXPERIMENTAL

2.1 Preparation of the samples

Symmetrical lifetime samples with various surface morphologies are prepared as in Fig.1.

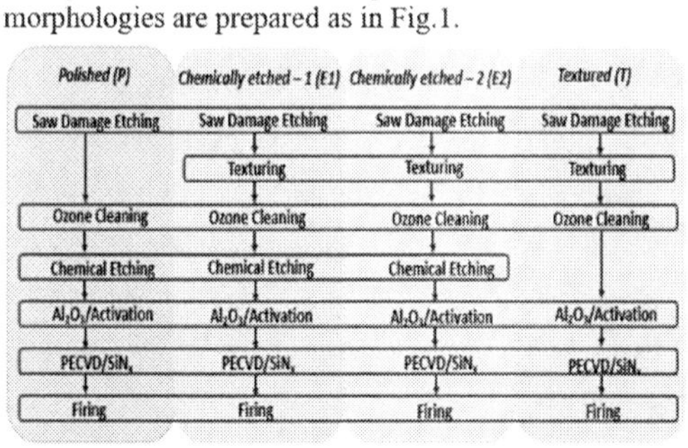

Fig.1. Process flow for the preparation of symmetrical lifetime samples

Monocrystalline p-type silicon wafers were exposed to saw damage etching solution and directly sent to the cleaning process to prepare polished samples (P). Etching of the samples was done by an industrial scale RENA Inpilot chemical etching tool. Then, for the surface passivation, aluminum oxide (Al_2O_3) by an atomic layer deposition (ALD) system and SiN_x layer by a plasma-enhanced chemical vapor deposition (PECVD) system were symmetrically deposited with thicknesses of 5 nm and 120 nm, respectively. Finally, the samples were subjected to a fast-firing process by a conveyor belt furnace to mimic the cell fabrication. Chemically etched samples, denoted as E1 and E2, underwent the same procedure as polished samples but were exposed to an additional texturing step following the saw damage removal.

It should be noted that_the etching time for E2 was three times longer than E1. For the fabrication of textured samples, the wafers were subjected to all process steps except for chemical etching. At the end of these processes, four sample types with various surface morphologies were produced.

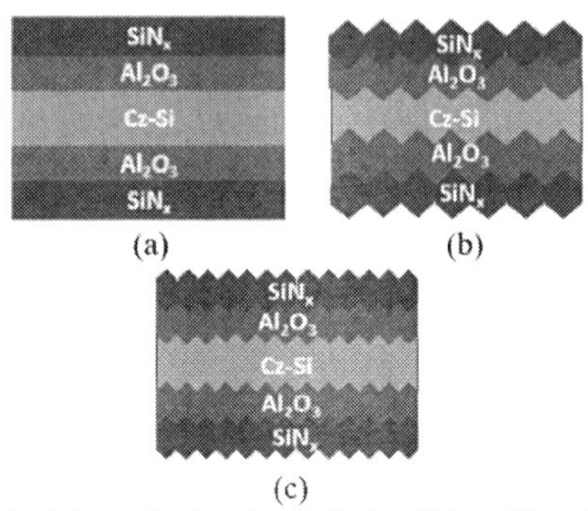

Fig.2. Schematic drawings of (a) polished (b) etched and (c) textured samples

2.2 Characterization of the samples

Surface morphologies of the samples were characterized by scanning electron microscopy (SEM) and atomic force microscopy (AFM).

Passivation quality was evaluated by implied open-circuit voltage (iV_{oc}) and effective minority carrier lifetime (τ_{Eff}) values measured by Sinton WCT – 120 tool.

3 RESULTS

3.1 Surface Morphology

Fig. 3 demonstrates SEM and AFM images of the various surface morphologies produced in our study.

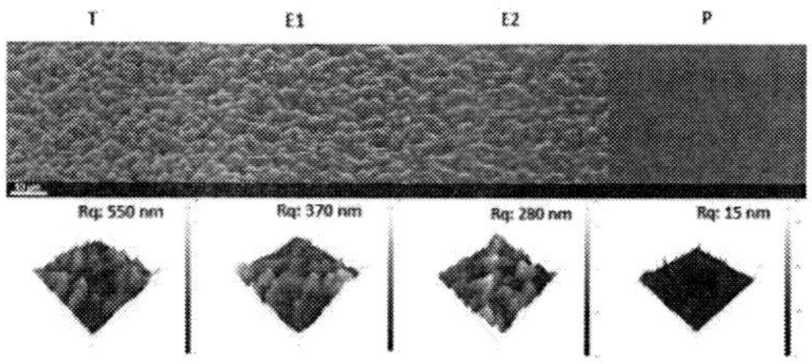

Fig.3. SEM and AFM images of the samples with different surface roughnesses

Rq values with 10μm x 10μm scan scale images were measured for the T, E1, E2, and P as 550, 370, 280, and 15 nm, respectively. As we go from sample T to sample P, surface morphologies become smoother. Moreover, the etching time of E2 is longer than that of E1. As a result, samples specified as E2 have a smoother surface among chemically etched samples. As the etching time increases, the surface morphology becomes flatter.

3.2 Passivation Quality

According to Fig. 4, the highest iV_{oc} and τ_{Eff} were achieved for sample E2 with 700.6 mV and 156.47 μs, respectively. Textured samples gave the lowest values, and polished samples showed poor performance than the chemically etched samples E1 and E2.

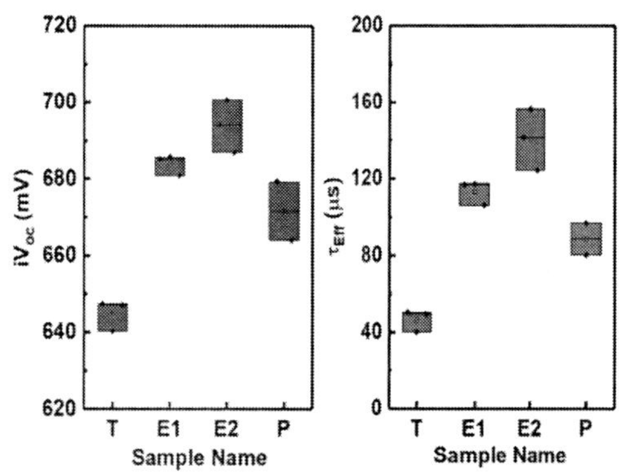

Fig.4. iVoc(left) and τeff (right) of symmetrical samples

4 CONCLUSIONS

In this paper, obtained results indicate that passivation quality improves as the surface becomes smoother unless the surface is fully polished. This outcome is also consistent with the literature [5]. Besides, the chemical etching time influences the surface morphology. The more exposure to the etching chemicals causes the more smooth surfaces. The highest iV_{oc} of 700.6 mV and τ_{Eff} of 156.47 μs was obtained with sample E2 with the surface roughness of 280nm.

5 References

[1] M. Fischer, "International Technology Roadmap for Photovoltaic (ITRPV) - Results 2018," *Itrpv*, no. 10th Edition, pp. 1–38, 2019, [Online]. Available: http://www.itrpv.net/.

[2] T. Dullweber and J. Schmidt, "Industrial Silicon Solar Cells Applying the Passivated Emitter and Rear Cell (PERC) Concept-A Review," *IEEE J. Photovoltaics*, vol. 6, no. 5, pp. 1366–1381, 2016, doi: 10.1109/JPHOTOV.2016.2571627.

[3] K. R. McIntosh and L. P. Johnson, "Recombination at textured silicon surfaces passivated with silicon dioxide," *J. Appl. Phys.*, vol. 105, no. 12, 2009, doi: 10.1063/1.3153979.

[4] C. Gong, E. Simoen, N. E. Posthuma, E. Van Kerschaver, J. Poortmans, and R. Mertens, "Study of silicon-silicon nitride interface properties on planar (1 0 0), planar (1 1 1) and textured surfaces using deep-level transient spectroscopy," *J. Phys. D. Appl. Phys.*, vol. 43, no. 48, pp. 1–6, Dec. 2010, doi: 10.1088/0022-3727/43/48/485301.

[5] J. Horzel *et al.*, HOW MUCH REAR SIDE POLISHING IS REQUIRED? A STUDY ON THE IMPACT OF REAR SIDE POLISHING IN PERC SOLAR CELLS," 2012, doi: 10.4229/27thEUPVSEC2012-2AO.1.6.

PROCESS FABRICATION OF COST-EFFECTIVE IBC SOLAR CELLS USING APCVD GLASS LAYERS

V. Kuruganti[a*], R. Kopecek[a], S. Seren[b], V. D. Mihailetchi[a]
[a]ISC Konstanz, Rudolf-Diesel-Straβe 15, 78467 Konstanz, Germany
[b]SCHMID, Robert-Bosch str. 32-36, 72250 Freudenstadt, Germany
Telephone: +49 7531 36183 305, e-mail: vaibhav.kuruganti@isc-konstanz.de

ABSTRACT: With PERC technology approaching its efficiency limits, more companies and research institutes are interested in developing industrially feasible TOPCon and IBC cell architectures. Though the highest efficiency solar cells are realized on IBC cell structure, making them industrially viable is challenging. In this work, the process sequence for obtaining cost-effective IBC solar cells was studied. We investigate boron-doped emitter and phosphorus-doped FSF formation APCVD BSG and PSG layers, respectively. Lifetime samples with different dopant concentrations were exposed to different high-temperature co-annealing recipes. Also, the co-annealing was optimized to perform dual tasks: diffusion of dopants of both polarities and, at the same time, growing an in-situ SiO_2 at the glass/Si interface for passivation purposes. To further optimize the APCVD IBC process sequence, a single wet-bench process after co-annealing was used to remove the dopant glasses using the in-situ grown SiO_2 as the etch barrier. The passivation quality of lifetime test samples was lower than anticipated, and a systematic study pointed out that the source of low passivation quality is metal contamination.

1 INTRODUCTION

The photovoltaic domain has been one of the most promising and rapid growing industries in the energy sector, with 0.5-0.6%$_{abs}$ improvement in cell efficiency every year since the beginning of the 20th century. However, due to its structural constraints and unable to meet the growing efficiency demands, cell technologies like Al-BSF have perished by early 2015 [1]. Now PERC technology is approaching its efficiency limits at 23%; hence more companies and research institutes are interested in high-efficiency n-type cell technologies like TOPCon and IBC. A few advantages of n-type substrate cell concepts include the lack of light-induced degradation (LID), higher tolerance to metal impurities & favoring bifacial cell designs [2].

In IBC solar cells, the trade-off between electrical and optical losses on the front side of the solar cell is eliminated as all the contacts are moved to the rear non-illuminated side. Since all the contacts are moved to the rear side, losses from the optical shading by metal finger and bus bars is eliminated, thereby improving the J_{sc} by typically 5-8 %. Some other advantages of IBC architecture include easy module fabrication and lower series resistance due to higher metal fraction on the rear side [3]. Kaneka Corporation has reported a conversion efficiency of 26.7% [4] for its heterojunction IBC, ISFH has reported a conversation efficiency of 26.1% on its POLO solar cells [5], SunPower has reported an efficiency of 25% [6] on the SunPower X-Series technology, and SPIC has reported efficiency in excess of 23.5% on its low-cost bifacial IBC ZEBRA technology [7].

Although IBC cell technology yields very high conversion efficiencies, the process complexity and production costs make them not so industrially viable. Firstly, making an IBC cell requires three different doped areas: a front surface field (FSF), a back surface field (BSF), & an emitter layer, which would require several high-temperature steps, increasing the thermal budget of process fabrication. Secondly, since both polarities are located on the rear side of the IBC structure, it would mean several masking and wet bench processing to prevent cross-contamination. Finally, the passivation of the IBC cell structure is sophisticated as both p+, and n+ regions are present on the rear side. It is well established in the literature that SiN_x (Al_2O_3) passivates well in n+ [8] (p+ [9]) regions, which would again mean multiple deposition and masking steps.

An extensive study has been carried out to investigate various viable technologies for doping semiconductors, such as ion implantation [10] [11], PECVD [12], APCVD [13] etc. APCVD is a viable alternative to conventional gas tube diffusion to make cost-effective IBC solar cells. It is a single-sided process with a very minimal wrap-around, eliminating the need for additional process steps like edge isolation. APCVD depositions take place at atmospheric pressure and hence would require lower equipment and operational costs. In addition, it yields higher throughput as no homogenous gas flow between the wafers is not needed [14]. Another advantage of CVD layers over gaseous diffusion sources is the separation between the deposition and diffusion step, which gives the flexibility to perform laser processing before the high-temperature step [15]. One of the approaches to reducing the process complexity and production cost of IBC solar cells is using co-annealing, which means forming the FSF and the emitter simultaneously using a single high-temperature process [16].

In this work, the process sequence of a cost-effective industrially viable solar cell is done using APCVD glass as a solid dopant source, co-annealing to drive-in dopants of both polarities using a single high-temperature step. The co-annealing was optimized such that, apart from dopant diffusion in bulk, an in-situ SiO_2 was grown at the glass and silicon interface for passivation purposes. Due to the difference in etch selectivity of SiO_2 w.r.t BSG and PSG in HF, it is used as an etch barrier. Finally, the SiO_2/SiN_x layer stacks are used for the passivation of p+ and n+ regions.

2 EXPERIMENT

Symmetric lifetime samples were processed using industrial-type production equipment from ISC Konstanz and SCHMID. We use the M2 (156.75×156.75 mm) wafer format, pseudo-square n-type wafers with a base resistivity of 4 Ωcm. The samples were sawed damage etched (SDE, up to 8 μm on each side) and then cleaned in piranha solution to remove all the organic and inorganic impurities. All the wet-bench processes were done using the RENA inline tool.

Later, the samples were sent to SCHMID for deposition of APCVD dopant glass. The APCVD equipment has multiple deposition chambers, making it capable of depositing a single doped layer or a stack of different layers. PSG layer deposition is carried out by injecting silane (SiH_4), nitrogen diluted phosphine (PH_3) and oxygen O_2 through a chemical vapour deposition (CVD) injector. In the case of BSG, diborane (B_2H_6) is used in place of phosphine. Different boron and phosphorus concentrations in PSG/BSG can be controlled by regulating the dopant gas flow. This work used three different doping concentrations of both BSG and PSG glass (5, 6, and 7 wt. %). Due to the hygroscopic nature of PSG, an additional undoped silicate glass (USG) was deposited as a capping layer.

Following the dopant glass deposition at SCHMID, the samples were shipped back to ISC Konstanz for co-annealed. Two different co-annealing recipes were used, hereafter AN1 & AN2. Both recipes have similar properties except for the oxidation time. For both the recipes, the minimum and maximum temperatures are 810 °C and 995 °C, respectively. Both recipes have a temperature ramp-up and ramp-down of 12 °C/min and four °C/min, respectively. The oxidation time in AN1 and AN2 was 8 and 22 minutes, respectively.

A hydrofluoric acid-based (HF) solution is used for etching the doped glasses. A systematic study was done to determine the HF etching time such that about 15-20nm of the in-situ grown oxide remains on the Si surface. Consequently, the samples have undergone two PECVD steps for the SiN_x deposition on both sides to obtain lifetime samples. Both the oxidation and PECVD deposition were performed on c.PLASMA by Centrotherm AG. The firing of the samples was done according to the best-known method (BKM) in our lab.

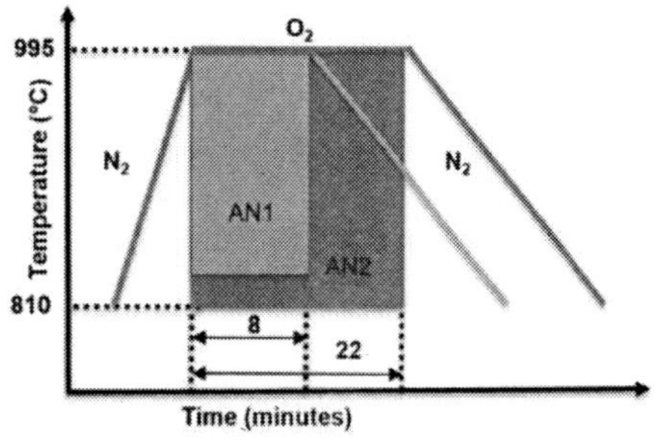

Figure 1: Schematic representation of co-annealing recipe, namely AN1 and AN2.

3 RESULTS

3.1 Sheet resistance and ECV profiles

The sheet resistance of the samples was measured using the well-established four-point probe (FPP) contacting technique, and the active dopant carrier concentration profiles after annealing were measured using an electrochemical capacitance-voltage (ECV) technique.

The difference between the two co-annealing recipes used in this study is only in the oxidation time. The Deal-Grove model clearly explains the relationship between oxidation time (t) and oxide thickness (X) for a given temperature. The film thickness X is proportional $\sqrt{t}$ because it is related to the depth of the oxide diffusion in silicon [17]. Hence, in the case of AN1 and AN2, which have an oxidation time of 8 and 22 minutes, we have in-situ SiO_2 of 16nm and 32nm, respectively.

In Fig. 2, as the diffusion coefficient of phosphorous is higher than boron at 995 °C, for the same dopant concentration, we observe lower R_{SH} in the case of phosphorous than in the case of boron [18]. As anticipated, the increase in dopant concentration led to a decrease in the R_{SH} for both types of dopants. The average R_{SH} of AN2 co-annealing is always higher by around 50 Ω/□ compared to AN1. This phenomenon is attributed to the difference in the in-situ grown SiO_2 thickness. During co-annealing, the in-situ grown SiO_2 acts as a diffusion barrier and reduces the surface concentration of dopant in the c-Si [19]. Having a thicker oxide makes it harder for the dopant to reach the silicon interface.

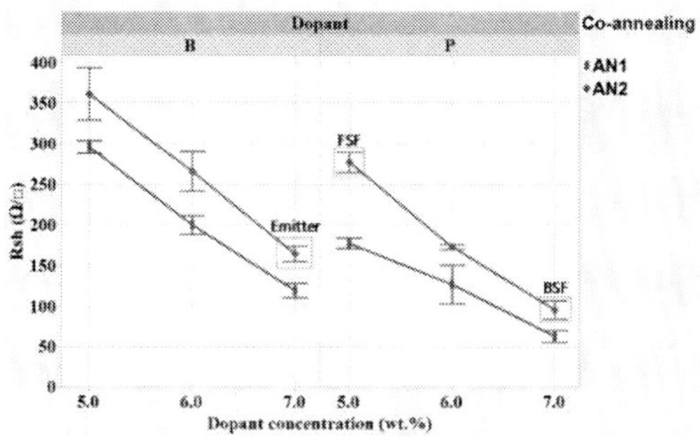

Figure 2: Sheet resistance (Ω/□) vs doping concentration (cm^{-3}) as a function of the co-annealing recipes and dopant type.

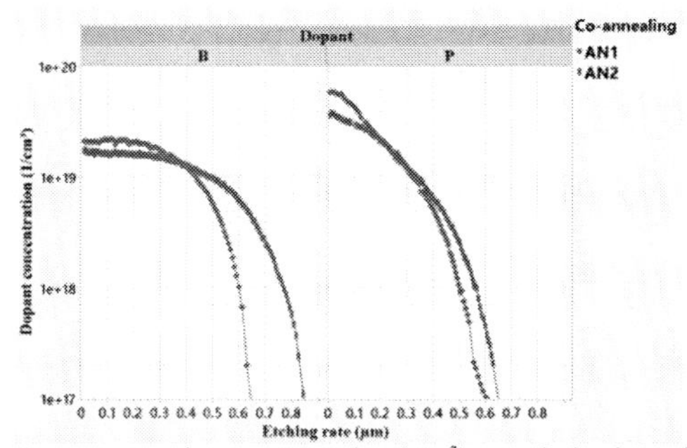

Figure 3: Doping concentration (cm^{-3}) vs etch depth (µm) as a function of the co-annealing recipes and dopant type.

In Fig. 3, the ECV profile of 7 wt. % BSG and 7 wt. % PSG is depicted for different co-annealing recipes. As mentioned above, the higher the diffusion coefficient of phosphorous, the higher the dopant surface concentration than boron [18]. The surface dopant concentration for BSG was found to be 2×10^{19} cm^{-3} for AN1, whereas, for AN2, it was found to be 1.5×10^{19} cm^{-3}. Similarly, in the PSG layer, AN1 had a peak surface concentration of 5×10^{19} cm^{-3}, and AN2 had 5×10^{19} cm^{-3}. The lower surface concentration and deeper junction for AN2 in comparison to AN1 are due to the longer oxidation time [20].

Based on ECV profile & sheet resistance values, we concluded that the AN2 co-annealing recipe was more suitable for our process. Also, having a thicker in-situ oxide provides a higher process window for the following wet bench processing. Finally, BSG 7 wt. % and PSG 5 wt. % were shortlisted to function as emitter and FSF of the APCVD-based ZEBRA cell process.

3.2 Etching rate analysis

Instead of using different passivation schemes for the p+ and n+ regions of the IBC solar cell, we have used a single SiO_2/SiN_x stack layer. However, optimizing the SiO_2/SiN_x stack layer properties is imperative to achieve effective passivation of p+ regions. Therefore, Barbora et al. [21] carried out a systematic study to find the right thickness of SiO_2 and SiN_x needed to provide good V_{oc} and J_{sc}. The results showed that optimum passivation is achieved by a stack of a thermal SiO_2, with a thickness of at least 10 nm, and a SiN_x layer with a low refractive index.

In this work, the etch selectivity of SiO_2 w.r.t BSG & PSG is used to have a controlled single glass removal step post-co-annealing such that a SiO_2 layer of about 10-25 nm remains on the Si surface, which will be eventually capped with a suitable thickness and low refractive index SiN_x using PECVD to obtain the passivation for doped regions on the front and rear side.

The thickness measurement of different glass layers was carried out on the SE800-PV ellipsometer. A simple recipe was used to measure the thickness of BSG, PSG and SiO_2 assuming the refractive index of n = 1.46.

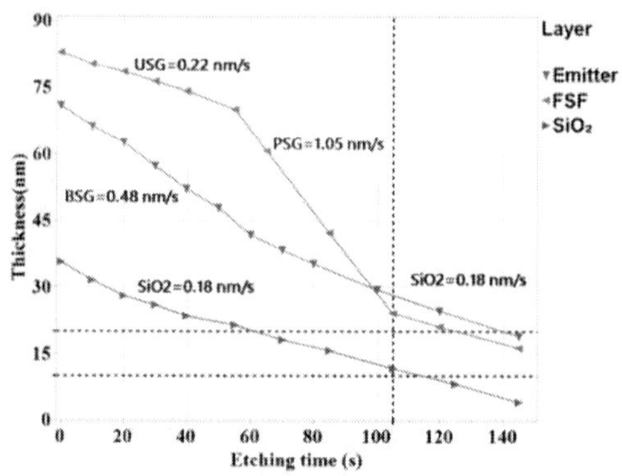

Figure 4: Etch rate analysis of emitter, FSF and thermal grown SiO_2 layers

Etching analysis was carried out in a 2% HF solution currently used in the PV industry for glass removal. Obtained etching rates were compared to literature values and were observed to be in good agreement. In Fig. 4, the etching analysis was carried out for three layers after co-annealing: (i) control sample with thermal grown SiO_2, (ii) 7 wt. % BSG layer, and (iii) 5 wt. % PSG layer. As mentioned in section 2, we have an additional USG on top of the PSG to protect it from reacting with the environment. The etching rate of BSG and PSG was much higher than SiO_2 because the breakage rate of B-O and P-O bonds is much higher than Si-O bonds [19]. The difference in the etch selectivity of SiO_2 w.r.t BSG and PSG makes it a viable option to act as an etch barrier [22]. The etching rate of 0.18 nm/s (green square in Fig. 4) was measured for the thermally grown in-situ oxide. For the seven wt. % BSG layer (blue triangle in Fig. 4), the etching rate of 0.48 nm/s was observed until 100 sec, and after that, the etching rate dropped to 0.18 nm/s, which is a clear indication of the presence of in-situ SiO_2. In the case of the PSG layer (red rhombus in Fig. 4), we observe an etching rate of 0.22 nm/s for the first 50 seconds, which corresponds to the USG layer etching, and later the etching rate increases drastically to 1.05 nm/s until 85 seconds, corresponding to 7 wt. % PSG, and finally, the etching rate drops to 0.18 nm/s due to the in-situ SiO_2. The difference in the etch rate of USG, which is deposited using APCVD, and thermal grown SiO_2 is due to the difference in layer composition and porosity. One hundred fifty seconds of HF etching would result in a thin thermally-grown SiO_2 (10-25 nm) remaining on the surface, which a SiN_x could cap to form the passivation layer.

3.3 Passivation quality study

The passivation stack of SiO_2/SiN_x was fired according to our lab's best-known method (BKM). The passivation quality of the lifetime samples was measured using the commercially available tool WCT-120 from Sinton Instruments. The measured iV_{oc} of the emitter layer was found to be 630 mV, and that of FSF was found to be around 645 mV which was much lower than the anticipated values.

A systematic study was carried out to find the source of the passivation loss. All the samples from groups A, B and C had similar process fabrication until co-annealing as mentioned in the experiment section, and group D consisted of a gas-tube diffused layer (Fig5). Different passivation schemes were realized on the APCVD layers and were compared to the gas tube diffused layers with similar electrical properties. The four different groups studied are as follows: (A) - previously used SiO_2/SiN_x passivation stack, (B) - complete etching of in-situ SiO_2 and then thermally regrown a new SiO_2/SiN_x passivation stack, (C) - complete etching of in-situ SiO_2 and then use AlO_x/SiN_x passivation stack and (D) - SiO_2/SiN_x passivation stack for gas tube diffused layer.

According to the literature, AlO_x passivation is much better than SiO_2/SiN_x stack due to fixed charge passivation for p+ regions [9]. From Fig. 5, it can be seen that APCVD layers (A, B & C group) have iV_{oc} around 645-650 mV irrespective of the passivation scheme, whereas the tube diffused layer has shown around 675 mV. Also, in PSG glass for APCVD layers (A & B group), the iV_{oc} of 650 mV is observed, whereas the tube diffused layer has shown about 695 mV. Hence, irrespective of the passivation scheme, the passivation quality of APCVD LT samples are much lower than tube diffused LT samples with similar electrical properties. Alongside n-type samples, p-type samples were also processed into the four different groups as mentioned above. Using WCT-120 from Sinton Instruments, we can determine the interstitial iron concentration [24] on the p-type samples using the approach developed by Zoth and Bergholz [23].

Table I: Interstitial iron (Fe_i) cm^{-3} concentration of different passivation scheme groups

Group/ Fe_i (cm^{-3})	BSG	PSG
Group A	$6.0*10^{11}$	$1.2*10^{11}$
Group B	$1.0*10^{12}$	$2.3*10^{11}$
Group C	$8.0*10^{11}$	
Group D	$5.5*10^{10}$	$4.0*10^{10}$

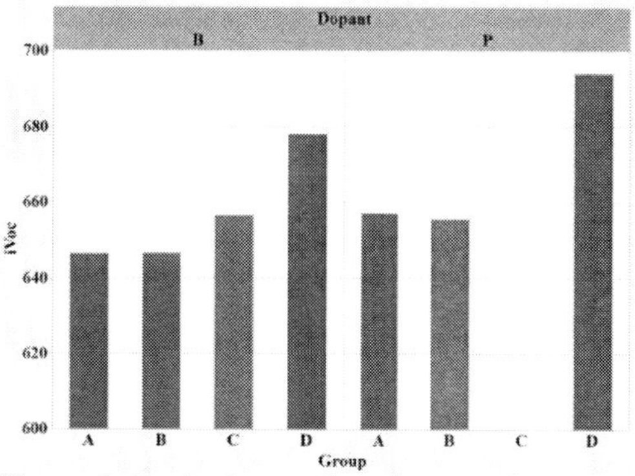

Figure 5: Passivation quality of group A) APCVD SiO_2/SiN_x passivation stack, B) Thermal grown SiO_2/SiN_x passivation stack, c) ALD $AlO_x/SiNx$ passivation stack and D) tube diffused SiO_2/SiN_x stack of passivation for both types of dopants.

From Table I, it can be inferred that all the APCVD groups (group A, B&C) have an order of magnitude higher Fe_i compared to tube diffused layers (group D) for both BSG and PSG groups. During the annealing process, the impurities are driven into the bulk and hence irrespective of the final passivation; all the APCVD groups show low iV_{oc}. Process optimization is in progress to obtain a contaminant-free APVCD layer. Once the desired passivation quality is attained, the process fabrication mentioned in this paper would be used to make the APCVD IBC solar cells.

4 CONCLUSIONS

The use of APCVD in tandem with co-annealing has been shown to be a more straightforward and cost-effective way to fabricate IBC solar cells. After investigating a wide range of dopant concentrations and annealing recipes, Co-annealing recipe AN2 on BSG 7wt. %, PSG 7wt. % and PSG 5wt. % were best suited to function as emitter, BSF, and FSF, respectively, in the APCVD IBC solar cell. Co-annealing recipe AN2 has also been proven to be better suited for the subsequent wet-bench processes.

The etch selectively of SiO_2 w.r.t BSG and PSG in 2% HF was used to perform a controlled etching of co-annealed dopant glass. For the shortlisted process parameters, a systematic study has shown that 150 seconds of etching in 2% HF is best suited to remove both the dopant glasses and leave behind a thin SiO_2 of 10-20nm, which could be capped with SiN_x for passivation purposes.

In the passivation study it was observed that passivation schemes involving APCVD layers had shown at least an order of magnitude higher interstitial iron (cm^{-3}) concentration w.r.t tube diffused layers pointing out the source for the low passivation quality. Therefore, work is in progress to deposit contaminant-free APCVD layers. Once that has been achieved, the APCVD IBC solar cells would be processed using the process sequence mentioned in the paper.

5 ACKNOWLEDMENT

This work was funded by the German Federal Ministry for Economic Affairs and Energy within the research project "RALPH" (contract no. 03EE1018C).

6 REFERENCES

[1] M. Hermle, F. Feldmann, M. Bivour, J. C. Goldschmidt, and S. W. Glunz, "Passivating contacts and tandem concepts: Approaches for the highest silicon-based solar cell efficiencies," Appl. Phys. Rev., vol. 7, no. 2, 2020, doi: 10.1063/1.5139202.

[2] A. Ur Rehman and S. H. Lee, "Advancements in n-type base crystalline silicon solar cells and their emergence in the photovoltaic industry," Sci. World J., vol. 2013, 2013, doi: 10.1155/2013/470347.

[3] P. Verlinden, "Interdigitated Back Contact Solar Cells1.1 A Different Concept for High-Efficiency Silicon Solar Cells," 2017.

[4] K. Yoshikawa et al., "Exceeding conversion efficiency of 26% by heterojunction interdigitated back contact solar cell with thin film Si technology," Sol. Energy Mater. Sol. Cells, vol. 173, no. June, pp. 37–42, 2017, doi: 10.1016/j.solmat.2017.06.024.

[5] C. Hollemann et al., "Separating the two polarities of the POLO contacts of an 26.1%-efficient IBC solar cell," Sci. Rep., vol. 10, no. 1, pp. 1–15, 2020, doi: 10.1038/s41598-019-57310-0.

[6] D. D. Smith, G. Reich, M. Baldrias, M. Reich, N. Boitnott, and G. Bunea, "Silicon solar cells with total area efficiency above 25 %," Conf. Rec. IEEE Photovolt. Spec. Conf., vol. 2016-Novem, pp. 3351–3355, 2016, doi: 10.1109/PVSC.2016.7750287.

[7] R. Kopecek et al., "ZEBRA technology: low cost bifacial IBC solar cells in mass production with efficiency exceeding 23.5%," Jan. 2021, pp. 1008–1012, doi: 10.1109/pvsc45281.2020.9300503.

[8] J. F. Lelièvre et al., "Efficient silicon nitride SiNx:H antireflective and passivation layers deposited by atmospheric pressure PECVD for silicon solar cells," Prog. Photovoltaics Res. Appl., vol. 27, no. 11, pp. 1007–1019, 2019, doi: 10.1002/pip.3141.

[9] J. S. D. Zielke, J.H. Petermann, F. Werner, B. Veith, R. Brendel, "21.7 % EFFICIENT PERC SOLAR CELLS WITH ALOx TINNELING LAYER," 26th Eur. Photovolt. Sol. Energy Conf. Exhib., no. August 2020, pp. 1115–1119, 2011, [Online]. Available: http://www.eupvsec-proceedings.com/proceedings?paper=10815.

[10] G. Yang, A. Ingenito, O. Isabella, and M. Zeman, "IBC c-Si solar cells based on ion-implanted poly-silicon passivating contacts," Sol. Energy Mater. Sol. Cells, vol. 158, pp. 84–90, Dec. 2016, doi: 10.1016/j.solmat.2016.05.041.

[11] R. Müller, C. Reichel, J. Benick, and M. Hermle, "Ion implantation for all-alumina IBC solar cells with floating emitter," Energy Procedia, vol. 55, pp. 265–271, 2014, doi: 10.1016/j.egypro.2014.08.078.

[12] A. Fallisch et al., "Analysis of phosphorus-doped silicon oxide layers deposited by means of PECVD as a dopant source in diffusion processes," IEEE J. Photovoltaics, vol. 2, no. 4, pp. 450–456, 2012, doi:

10.1109/JPHOTOV.2012.2200455.
[13] Y. Schiele, F. Book, C. Demberger, K. Jiang, and G. Hahn, "Co-Diffused APCVD Boron Rear Emitter with Selectively Etched-Back FSF for Industrial n-Type Si Solar Cells," Proc. 29th EU PVSEC, no. September, pp. 821–824, 2014, doi: 10.4229/29thEUPVSEC2014-2AV.2.4.
[14] P. Rothhardt, S. Meier, C. Demberger, A. Wolf, and D. Biro, "Codiffused bifacial n-type solar cells (CoBiN)," Energy Procedia, vol. 55, pp. 287–294, 2014, doi: 10.1016/j.egypro.2014.08.084.
[15] G. H. Felix Book, Holger Knauss, Carsten Demberger, Florian Mutter, "Phosphorous doping from APCVD deposited PSG," in Proceeding of the 34th European Photovoltaic Solar Energy Conference and Exhibition, Hamburg, Germany, 2016, pp. 2–5.
[16] F. C. J. D. Huyeng, A. Spribille, S. Nold, R. Efinger, R. Keding, O. Doll, "A cost driven reserach strategy towards industrially feasible high-efficiency back-contact back-junction silicon solar cells," Proceeding 33rd Eur. Photovolt. Sol. Energy Conf. Exhib. Hamburg, Ger., vol. 6, no. September, pp. 5–9, 2017.
[17] B. E. Deal and A. S. Grove, "General relationship for the thermal oxidation of silicon," J. Appl. Phys., vol. 36, no. 12, pp. 3770–3778, 1965, doi: 10.1063/1.1713945.
[18] J. S. Christenserr, H. H. Radamson, A. Y. Kuznetsov, and B. G. Svensson, "Phosphorus and boron diffusion in silicon under equilibrium conditions," Appl. Phys. Lett., vol. 82, no. 14, pp. 2254–2256, 2003, doi: 10.1063/1.1566464.
[19] V. D. Mihailetchi, H. Chu, J. Lossen, and R. Kopecek, "Surface Passivation of Boron-Diffused Junctions by a Borosilicate Glass and in Situ Grown Silicon Dioxide Interface Layer," IEEE J. Photovoltaics, vol. 8, no. 2, pp. 435–440, 2018, doi: 10.1109/JPHOTOV.2018.2792422.
[20] P. Rothhardt, C. Demberger, A. Wolf, and D. Biro, "Co-diffusion from APCVD BSG and POCl3 for industrial n-type solar cells," Energy Procedia, vol. 38, pp. 305–311, 2013, doi: 10.1016/j.egypro.2013.07.282.
[21] B. Mojrová et al., "A comparison study of boron emitter passivation by silicon oxide and a PECVD silicon nitride stack," Energy Procedia, vol. 124, pp. 288–294, 2017, doi: 10.1016/j.egypro.2017.09.301.
[22] G. A. C. M. Spierings, "Wet chemical etching of silicate glasses in hydrofluoric acid based solutions," J. Mater. Sci., vol. 28, no. 23, pp. 6261–6273, 1993, doi: 10.1007/BF01352182.
[23] G. Zoth and W. Bergholz, "A fast, preparation-free method to detect iron in silicon," J. Appl. Phys., vol. 67, no. 11, pp. 6764–6771, 1990, doi: 10.1063/1.345063.
[24] D. Macdonald, T. Roth, P. N. K. Deenapanray, K. Bothe, P. Pohl, and J. Schmidt, "Formation rates of iron-acceptor pairs in crystalline silicon," J. Appl. Phys., vol. 98, no. 8, 2005, doi: 10.1063/1.2102071.

INVESTIGATION OF EFFECT OF POLY-SI IMPURITIES ON CZ-INGOT TO SOLAR MODULE PRODUCTION

Fatma Çambay Kuban[1*], Emre Uçar[1], Nurhayat Yıldırım[1], Fatma Sena Yıldırım[1], Fırat Es[1]
[1] KalyonPV Research and Development Center
Kalyon Güneş Teknolojileri Üretim A.Ş., 06909 Ankara, Turkey

ABSTRACT: Achieving precise purity control by analysis methods is a crucial step in the development of semiconductor devices because impurities greatly change the basic properties (mechanical, optical, electrical, thermal) of semiconductors. This paper presents and analyses the photovoltaic conversion performance of cell and modules from ingots grown by Czochralski (Cz) method using two poly-Si suppliers (Supplier1 and Supplier2) with different purity levels. Effects of raw material quality, ingot and wafer lifetime, cell efficiency, module power and light induced degradation (LID) mechanism are investigated. It has been observed that Supplier2 poly-Si has lower Cr impurity than for Supplier1 and also does not contain Ni impurities. Although it was expected that the impurities originating from Ni and Cr should decrease the yield, trial production showed that the effect of metallic impurities on productivity at these levels is insignificant.
Keywords: Impurities, Silicon Solar Cell, Czochralski, Electrical Properties

1 INTRODUCTION

In the photovoltaic sector, silicon-based solar cells contribute more than 90% of the total PV module supply. Therefore, the technology that will most likely dominate the PV supply in the coming years is expected to be based on crystalline silicon [1], [2] .

Photovoltaic modules, which produce electricity from solar energy, which is one of the leading renewable energy sources, are expected to have an efficient operating life of at least 25 years [3]. One of the Light Induced Degradation (LID) mechanisms, the B-O complex, has a great influence on module lifetimes [4], [5]. If boron-oxygen are present in the growing ingot; the bulk lifetime and cell efficiency are decreased due to the activation of the boron-oxygen complex. This phenomenon is observed in silicon ingots grown by the boron doped Czochralski method [6].

To reduce Si-PV production costs, Cz-Si ingot producers use scrap under the quartz crucible and silicon used in remelting as low-grade feedstock. The purity of the generated ingot is likely to be compromised by such low-cost ingot production technologies. According to the segregation coefficients into the melt, impurities in the raw material are transported to the crystal during ingot growth [7]. The segregation coefficient describes the relationship between the concentration of impurity atoms in the developing crystal and the concentration of melt atoms in thermodynamic equilibrium. Usually, this coefficient is much lower than 1 because impurity atoms tend to remain in the melt. The solubility of impurity atoms in the melt is greater than in the solid. Thus, slow cooling process while growing ingot can provide a "getter effect" [7], [8]. The positive side of this situation is that the crystal will be cleaner than liquid. In other words, crystal growth can also be considered as a purification method provided that the last part of the crystal in which all impurities are concentrated is discarded. For all these reasons, the determination and control of impurities are critical in the manufacturing of poly-Si ingots.

Ingot growth process parameters are crucial step to decreasing the lifetime loss depends on the boron-oxygen concentration in the ingot. Factors such as rotational speed, crystal crucible diameter, volume of silicon melt, process time, process temperature, crucible lift rate, oxygen content of the inert gas all influence the mono-crystalline silicon ingots produced [9]. All these parameters and production conditions affect the oxygen-carbon content, resistance and minority charge carrier life of the ingot and determine the efficiency and quality of the modules to be produced precisely at these stages.

The effects of impurities in poly-Si on the solar cell base and junction site must be considered since the defects can spread to different parts of the cell. Since the recombination properties of metals in silicon are different, among the metallic impurities; Figure 1 shows that elements Zr, W, Ti and V have the highest effect on solar cell efficiency, while Cr, Mn, Fe and Co have a moderate effect, Al and Ni have less effect, and element Cu has an even lower effect [1], [8], [10]–[14]. However, these studies mostly remained at laboratory scale, therefore industrial scale studies will contribute to the literature. It is necessary to carry out an industrial level study that will cover all stages of production from raw material to final product.

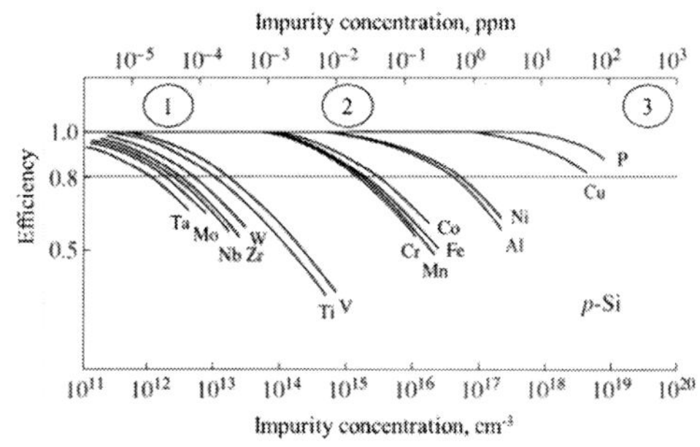

Figure 1: Limits on impurity concentrations in p-type silicon for impurities determining the degradation threshold of solar cells: 1.semiconductor grade, 2.solar grade, 3.metallurgical grade silicon [15].

In this study, commercial Cz-ingots were grown using poly-Si (different impurity levels) from two different suppliers. These ingots were used to fabricate wafers, cells, and modules in an integrated factory environment where all steps of production from poly-Si to module are carried out in the same factory. The effect of impurity level of raw material on solar cell and module performance has been investigated.

2 METHODS

2.1 NAA and VPD-ICPMS Method

The two most common analytical methods used for impurity control in semiconductors are neutron activation analysis (NAA) and vapor phase decomposition inductively coupled plasma mass spectrometry (VPD-ICPMS). The measurement of gamma rays emitted from a specimen exposed by NAA. The basic principle of this method is based on the measurement of the heating emitted as a result of excitation of a stable isotope by exposing it with neutrons or charged particles. The fractionation half-life of the radioactive nucleus generated as a result of the bombardment is used to calculate the amount of impurities [16], [17]. In the VPD-ICPMS analysis, the oxide layer on the surface as well as other impurities are etched with the acid vapor formed on the wafer surface. After etching process, acid vapor condensate is collected and analyzed for trace metals using ICP-MS [18], [19].

2.1 Device Fabrication

For ingot fabrication, approximately 220 kg of poly-Si raw material from each supplier were melted at 1420 °C inside quartz crucibles of two different Cz-growers. A rotating seed crystal with the desired crystallographic orientation was dipped into the solution and was pulled up with rotation. The ingot was allowed to grow to the desired diameter by adjusting the drawing rate and melting temperature. 180-micron thick, 158.75 mm x 158.75 mm sized full square Cz-wafers were produced with resistivity of 1,5 Ω.cm. Bulk resistivity values were measured with Semilab Resistivity Tester RT-1000, the minority carrier lifetime of each ingot was measured with the Semilab WT-1200iand VPD-ICPMS analysis was performed to determine the surface impurities in poly-Si.

The wafers were used to fabricate over 10.000 industrial p-type bifacial PERC solar cells in the cell fabrication line. Cells went through anti-LID treatment with electrical injection. Performance of the cells was determined with the flash I-V tester. LID measurements were conducted via respective test tool. Finally, glass/glass 144 half-cut cell modules were produced in the module line. Module performances were determined with I-V tester and LID tests were also performed on module level.

3 RESULTS and DISCUSSION

3.1 NAA and VPD-ICPMS Measurements

The impurity level is a major factor for solar cell yield optimization because metallic impurities are common contaminants in PV Si and can greatly impact the performance of the final devices by generating metal silicides and/or functioning as recombination sites. It is clearly seen from Table I the impurity values are lower than the literature [20], [21].

Table I. NAA Results of Supplier1 and Supplier2

	Bulk Impurities (ppt) – NAA Results						
	Cr	As	Ni	Sb	W	Re	Au
Supp.1	9,41	<DL	53,30	3,71	3.64	0,93	0,04
Supp.2	3,89	1,99	<DL	0,85	3,65	<DL	<DL

DL: Detection Limit

The ICP-MS analysis results of the wafers produced with Supplier1 and Supplier2 poly-Si are given in Table II. When the elements except Ni element are examined, it is seen that the impurities are higher in Supplier1 wafer than in Supplier2 wafer. Considering the wafer production process (ingot squaring, brick cutting, grinding) from raw material; it is expected that metal impurities in the wafer increase during the processes.

Table II. VPD-ICPMS Results of Supplier1 and Supplier2

	Element Name	Supplier1	Supplier2
Surface Metal Impurities (ppb) - VPD-ICPMS	Na (23)	8,05	3,22
	Mg (24)	8,11	3,57
	Al (27)	14,99	10,23
	Ti (47)	1,00	0,11
	Cr (52)	0,59	0,46
	Fe (56)	13,76	4,67
	Ni (60)	7,15	34,09
	Cu (63)	0,47	<LOD
	Zn (66)	1,93	<LOD
	Pb (208)	<LOD	<LOD

DL: Detection Limit

3.2 Impact on the Solar-Ingot Lifetime

In order to see the effect of impurities on the ingot lifetime, the ingot was produced, and the resistivity values of the ingots were within the desired resistivity range (0.5 – 1.5 Ω.cm) as seen in Table III. Top and tail minority lifetime values of ingots produced with different poly-Si suppliers with similar impurity levels were found to be like each other.

Table III. Resistivity and minority carrier lifetime measurement results of Supplier1 and Supplier2

	Top minority carrier lifetime (μs)	Tail minority carrier lifetime (μs)	Top resistivity (Ω.cm)	Tail resistivity (Ω.cm)
Supplier1	115.5	257.7	1.21	0.65
Supplier2	125.4	300.4	1.17	0.81

3.3 Impact on the Solar-Cell Parameters

In order to investigate the effect of impurities on efficiency, V_{OC}, I_{SC} and fill factor (FF) values in solar cells over 10,000 cells produced by Supplier1 and Supplier2 were examined, and these results are given in Figure 2. It's clear that there is no significant difference between suppliers even though the used raw material has different purity levels.

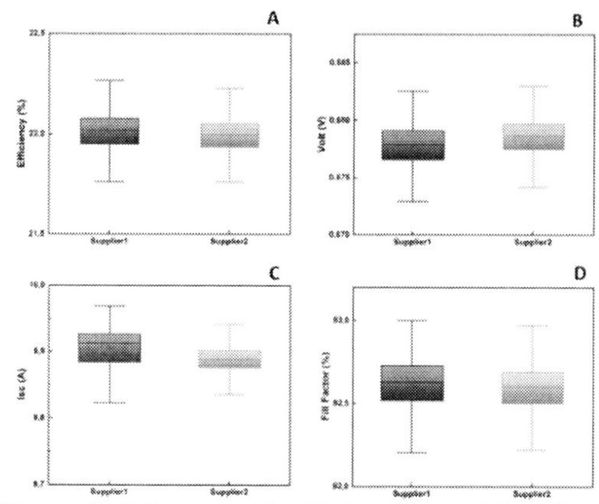

Figure 2: Cells produced with Supplier1 and Supplier2 poly-Si, A. efficiency graph, B. open circuit voltage graph, C. short circuit current graph, D. fill factor graph.

3.4 Impact on the Solar-Module Parameters

To observe the effect of impurities on solar modules, glass/glass 144 half-cut 62 bifacial Supplier1 modules and 81 bifacial Supplier2 modules were produced in the module line. Module performances were measured with an I-V tester. Comparative module results are given in Figure 3. Similarly, the difference in performance is again not very significant.

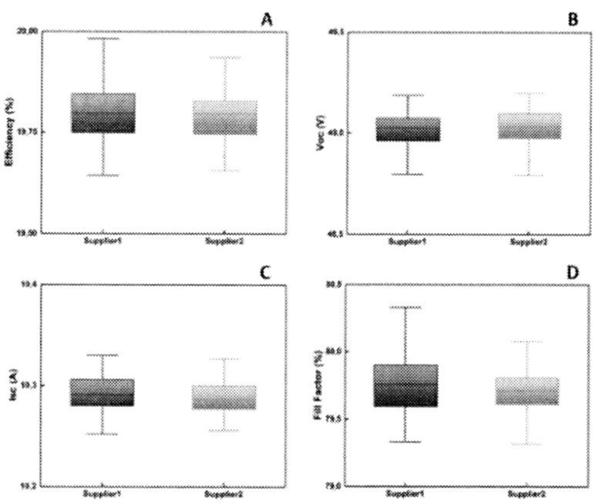

Figure 3: Modules produced with Supplier1 and Supplier2 poly-Si, A. efficiency graph, B. open circuit voltage graph, C. short circuit current graph, D. fill factor graph.

3.5 Impact on the LID Mechanism

The LID test of the cells was carried out under 1000 W/m^2 irradiation for 24 hours at 35°C to investigate the effects of impurities on the LID mechanism. In-house quality standard LID acceptance value is "<3%" and when the Figure 4 LID results are examined, the efficiency loss in Supplier2 cells is 1.1%, while the efficiency loss in Supplier1 cells is 2.0%. Cells produced from both suppliers were found to meet the acceptance criterion value.

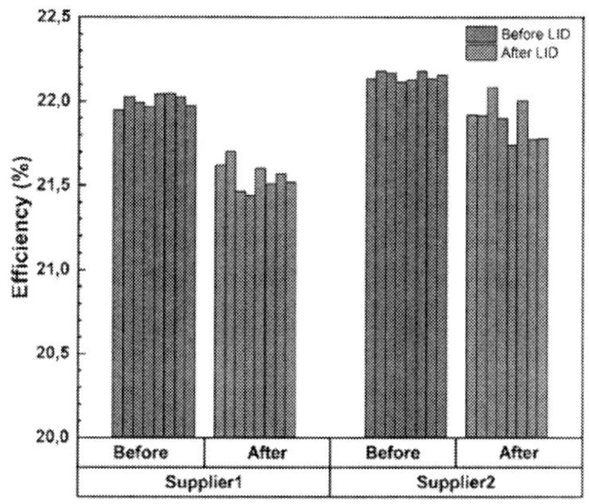

Figure 4: Efficiency graph of Supplier1 and Supplier2 cells before/after LID

4 CONCLUSIONS

The efficiency differences in solar cell and module produced with two polysilicon raw materials with different metal impurities were investigated. According to NAA result, Cr, Ni, Sb and Re impurities for Supplier2 are lower than Supplier1. Considering the literature, it is seen that impurities originating from Ti and Fe elements shall cause more efficiency decrease than impurities originating from Ni [11], [12]. For this reason, the efficiency of cells and modules produced with Supplier2 poly-Si is expected to be higher than those produced with Supplier1 poly-Si. However, there is no significant difference in efficiency between the cells and modules produced with both suppliers. This study shows that the effect of metallic impurities at these levels on the efficiency is insignificant. Similarly, LID mechanism at this impurity level is not crucial.

4.1 References

[1] G. Coletti, "Sensitivity of state-of-the-art and high efficiency crystalline silicon solar cells to metal impurities," *Prog. Photovoltaics Res. Appl.*, vol. 21, no. 5, pp. 1163–1170, 2013.

[2] A. F. B. Braga, S. P. Moreira, P. R. Zampieri, J. M. G. Bacchin, and P. R. Mei, "New processes for the production of solar-grade polycrystalline silicon: A review," *Sol. energy Mater. Sol. cells*, vol. 92, no. 4, pp. 418–424, 2008.

[3] E. D. Dunlop, "Lifetime performance of crystalline silicon PV modules," in *3rd World Conference onPhotovoltaic Energy Conversion, 2003. Proceedings of*, 2003, vol. 3, pp. 2927–2930.

[4] T. Luka, C. Hagendorf, and M. Turek, "Multicrystalline PERC solar cells: Is light-induced degradation challenging the efficiency gain of rear passivation?," *Photovoltaics Int.*, vol. 32, pp. 37–44, 2016.

[5] B. Damiani, A. Ristow, A. Ebong, and A. Rohatgi, "Design optimization for higher stabilized efficiency and reduced light-induced degradation in boron-doped Czochralski silicon solar cells," *Prog. Photovoltaics Res. Appl.*, vol. 10, no. 3, pp. 185–193, 2002.

[6] R. C. Newman, "Defects in silicon," *Reports Prog.*

Phys., vol. 45, no. 10, p. 1163, 1982.
[7] B. Sopori, "Silicon solar-cell processing for minimizing the influence of impurities and defects," *J. Electron. Mater.*, vol. 31, no. 10, pp. 972–980, 2002.
[8] I. Périchaud, "Gettering of impurities in solar silicon," *Sol. Energy Mater. Sol. Cells*, vol. 72, no. 1–4, pp. 315–326, 2002.
[9] B. Gao, S. Nakano, and K. Kakimoto, "Reduction of oxygen impurity in multicrystalline silicon production," *Int. J. Photoenergy*, vol. 2013, 2013.
[10] T. Buonassisi *et al.*, "Engineering metal-impurity nanodefects for low-cost solar cells," *Nat. Mater.*, vol. 4, no. 9, pp. 676–679, 2005.
[11] B. S. Xakalashe and M. Tangstad, "Silicon processing: from quartz to crystalline silicon solar cells," *Chem Technol*, no. March, pp. 6–9, 2012.
[12] J. Schmidt *et al.*, "Impurity-related limitations of next-generation industrial silicon solar cells," in *2012 IEEE 38th Photovoltaic Specialists Conference (PVSC) PART 2*, 2012, pp. 1–5.
[13] G. Coletti *et al.*, "Impact of metal contamination in silicon solar cells," *Adv. Funct. Mater.*, vol. 21, no. 5, pp. 879–890, 2011.
[14] J. Safarian, G. Tranell, and M. Tangstad, "Processes for upgrading metallurgical grade silicon to solar grade silicon," *Energy Procedia*, vol. 20, pp. 88–97, 2012.
[15] B. R. Bathey and M. C. Cretella, "Solar-grade silicon," *J. Mater. Sci.*, vol. 17, no. 11, pp. 3077–3096, 1982.
[16] G. D'Agostino, L. Bergamaschi, L. Giordani, G. Mana, and M. Oddone, "Instrumental neutron activation analysis of an enriched 28 Si single-crystal," *J. Radioanal. Nucl. Chem.*, vol. 299, no. 1, pp. 277–282, 2014.
[17] G. Revel, N. Deschamps, C. Dardenne, J. Pastol, B. Hania, and H. Nguyen Dinh, "Characterization of solar grade silicon by neutron activation analysis," *J. Radioanal. Nucl. Chem.*, vol. 85, no. 3, pp. 137–150, 1984.
[18] A. Krushevska, S. Tan, M. Passer, and X. R. Liu, "Advances in trace element analysis of silicon wafer surfaces by vapor phase decomposition (VPD) and inductively coupled plasma mass spectrometry (ICP-MS) Presented at the 2000 Winter Conference on Plasma Spectrochemistry, Fort Lauderdale, FL, USA, January 10–15, 2000.," *J. Anal. At. Spectrom.*, vol. 15, no. 9, pp. 1211–1216, 2000.
[19] J. TAKAHASH, "Determination of metal impurities present in solar-grade bulk silicon by ORS-ICP-MS," *Am. Lab.*, vol. 41, no. 7, pp. 28–30, 2009.
[20] B. G. Gribov and K. V Zinov'Ev, "Preparation of high-purity silicon for solar cells," *Inorg. Mater.*, vol. 39, no. 7, pp. 653–662, 2003.
[21] L. P. Hunt, "Compositional analysis of silicon for solar cells," *J. Electrochem. Soc.*, vol. 131, no. 8, p. 1891, 1984.

INVESTIGATION OF THE CURRENT INDUCED DEGRADATION FOR GALLIUM DOPED SILICON SOLAR CELLS

Min-An Tsai[1], Yean-San Long[1], Cheng-Wen Kuo[2], Ta-Ming Kuan[2], Cheng-Yeh Yu[2], Teng-Chun Wu[1]
[1]Center for Measurement Standards, Industrial Technology Research Institute, Taiwan
[2]TSEC Corporation, No.85, Gaungfu N. Rd., Hsin-Chu 30351, Taiwan.

ABSTRACT: For the high efficiency solar cell technology, the light-induced-degradation (LID) and the light-and-elevated temperature-induced-degradation (LETID) are still an important issue need to be solved by the PV industries. In this work, We design the different solar cell to see the LETID phenomenon, one is using with/without Gallium doped solar cell, another is using 4 different fabricated process by different companies. Using the Gallium doped solar cell reveals an antiresistance for LETID, but form another result show that it still influenced by fabricated process.

Keywords: LeTID, CIT, Accelerated Degradation Test

1 Introduction

LeTID is a form of solar cell degradation seen in the field and is accelerated by high irradiance at higher temperatures after hundreds of hours of light exposure. Indeed, a drop in efficiency of up to 10% can be seen if steps are not taken to adequately prevent it. As the name suggests, LeTID is exacerbated by higher operating temperatures and higher light intensity. Unlike LID, LeTID can occur over a much longer period of time, showing that the development rate of LeTID in the cell is slower than any boron/oxygen combination process

Previous work, we treated the cells under 1 SUN at 130°C and obtained that the required time for recovery shrank down to 4hrs compared with 6hrs for 80°C. Moreover, the enhanced illumination of 2 SUN accelerated the LID effect much faster within 60mins, indicating the light intensity would be a dominant factor for the acceleration. Yet, we employed an LED light source of 396nm and 969nm and obtained that both maximum degradations of P_{max} were quite different. The consequence implied that the behavior of LID might be attributed to the particular position (penetration depth) where carriers located. In this work, we design the experiment of the current injected test (CIT) In this work, We design the different solar cell to see the LETID phenomenon, one is using with/without Gallium doped solar cell, another is using 4 different fabricated process by different companies.

2 Experiments and details

In this work, the design test method is reference with the test protocol which was added in 2019 to a separate Technical Specification (TS) for LETID by the TC82 committee as a module quality LETID detection test. In the test process, a current I_{SC}-I_{MPP} is injected in the dark at 75 °C module temperature. The operation mode at this current value is referred to as Maximum Power Point mode, because the excess charge carrier density is similar to MPP at irradiance with 1 sun. The injection level is therefore in a typical range for field conditions. However, this approach requires timescales of several weeks to reach maximum degradation, which makes it time- and cost-intensive. Increasing the temperature and the injection level could accelerate the test significantly.

The test condition for temperature was 75 °C. The injection current two time of I_{sc}-I_{mpp}. The detail test flow is shown as figure 1. The experiment I is using the solar cell with/without Gallium doping. And anther is using the solar cell fabricated by four different companies.

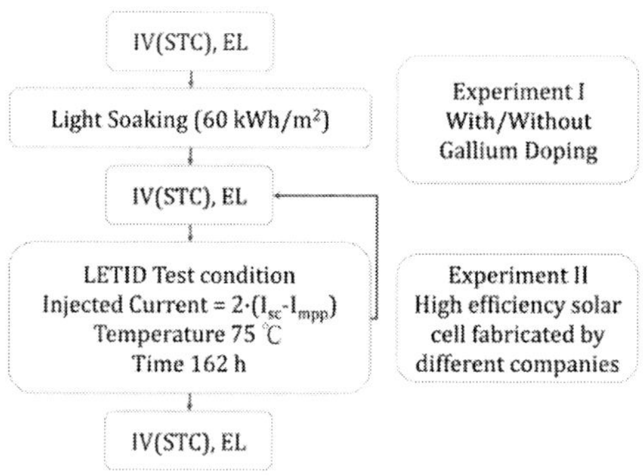

Figure 1: the test flow chats for the solar cell.

3 Results

The result of the experiment I shows in figure 2 and figure 3. Each bar graph represents a sample, and the purple line represents the average of the three samples. It shows that both types solar cell with/without Gallium doping has the first degradation by after the light soaking process (60 kWh). After CIT 486h the without doped Ga solar cell reveals a continued degradation, but the other one recovers to the initial value.

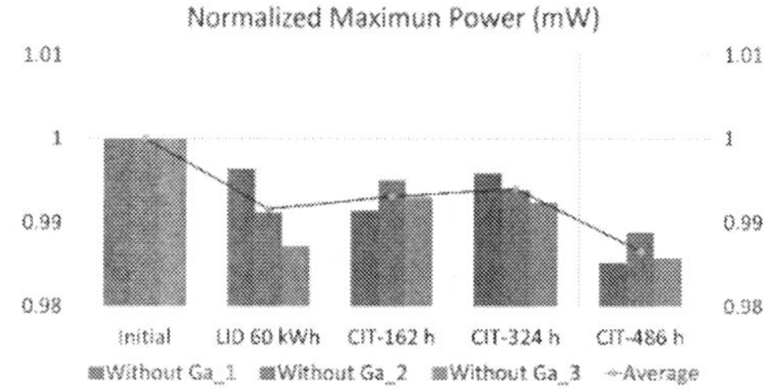

Figure 2: the normalized maximum power for without Gallium doping.

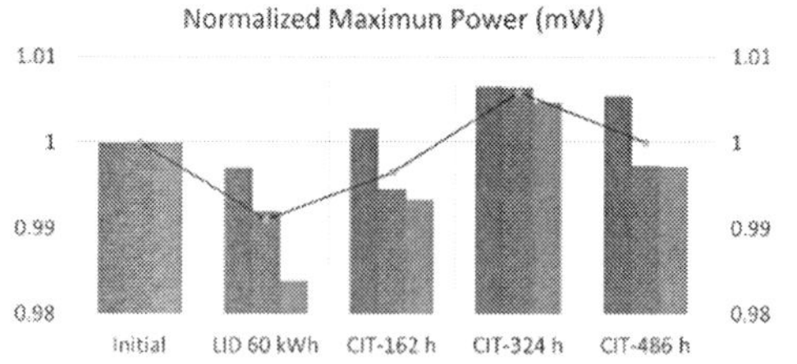

Figure 3: the relative power degradation ratio for different injection current.

The result of the experiment II shows in figure 4. This test using different fabrication solar cells (high efficiency P-type PERC solar cells) from different companies, . Here using the injection current (two time of Isc-Impp) is want to see the normal degradation effect for the different process solar cells. The result reveals the different fabricated process have different degradation behavior. All sample (ABCD) reveal an antiresistance for LETID after 300 hour CIT. But when increase the CIT time to 462h, Sample D reveals a dramatic power degradation of about 4.5 %, and Sample ABC show slight power degradation. Figure 5 is the EL image of Sample D. it show some obvious defects in the edge of the cell and the around of the busbar.

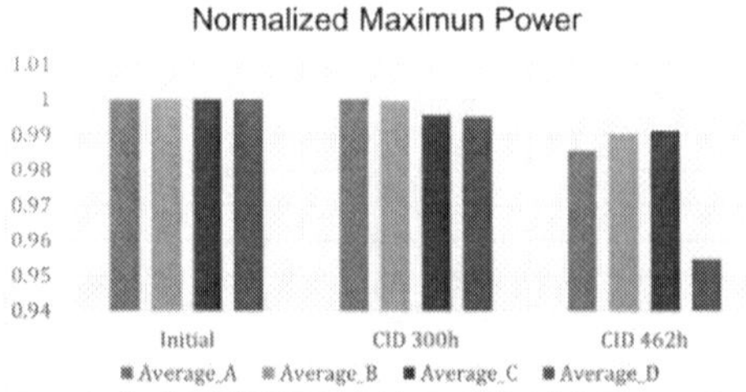

Figure 4: the relative power degradation ratio for different fabricated process solar cell.

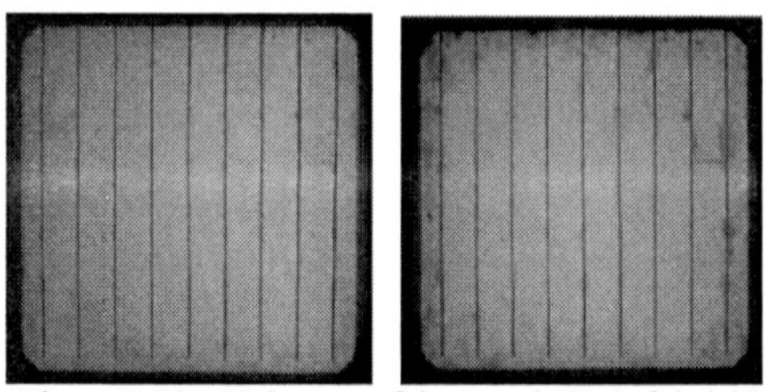

Figure 5: The EL images of the sample D, Left is initial condition, Right is after CID 462 hour.

4 Conclusions

So far, the LeTID for a solar cell or the module still needs to deeply discuss. In this work, experiment I demonstrates by using the Gallium doped solar cell reveal a good resistance for LETID test. Then the experiment II demonstrates the same high efficiency solar cell by different companies/fabricated processes reveal different degrees of degradation when CID test time lager than 462 hour.

5 References

[1] F. Kersten, P. Engelhart, H-C. Ploigt, A. Stekolnikov, T. Lindner, F. Stenzel, M. Bartzsch, A. Szpeth et al. Degradation of multicrystalline silicon solar cells and modules after illumination at elevated temperature. Solar Energy Materials and Solar Cells 2015;142:83–86.

[2] Esther Fokuhl, Tayyab Naeem, Alexandra Schmid, Paul Gebhardt, Torsten Geipel, Daniel Philipp, "Letid A Comparison Of Test Methods On Module Level" EU PVSEC 2019.

[3] Ching-Wen Wu ; Yen-Chun Lee ; Li-Chieh Yu ; Min-An Tsai ; Hung-Sen Wu ; C.W. Kuo ; T.M. Kuan ; C.Y. Yu, Peichen Yu, "Influence on light-induced degradation in Cz-Si PERC cells under light soaking of variant wavelength and intensity", 2018 IEEE 7th World Conference on Photovoltaic Energy Conversion (WCPEC).

[4]F. Fertig*), F. Kersten, K. Petter, M. Bartzsch, F. Stenzel, A. Mette, B. Klöter, J. W. Müller," Light and Elevated Temperature Induced Degradation of Multicrystalline Silicon Solar Cells and Modules", 26th NREL Workshop 2016.

[5]F. Fertig*), F. Kersten, K. Petter, M. Bartzsch, F. Stenzel, A. Mette, B. Klöter, J. W. Müller," System performance loss due to LeTID ", Energy Procedia 124 (2017) 540–546.

[6] Daniel Chena,, Phillip G. Hamerb, Moonyong Kima, Tsun H. Funga, Gabrielle Bourret-Sicotteb, Shaoyang Liua, Catherine E. Chana, Alison Cieslaa, Ran Chena, Malcolm D. Abbotta, Brett J. Hallama, Stuart R. Wenhama," Hydrogen induced degradation: A possible mechanism for light- and elevated temperature- induced degradation in n-type silicon ", Solar Energy Materials and Solar Cells 185 (2018) 174–182.

[7] C.-W. Kuo, T.-M. Kuan, Y.-C. Lee, M.-A. Tsai, H.-H. Hsieh, W.-L. Chueh, Y.-H. Chao, L.-G. Wu, C.-Y. Yu, " Investigation of the Accelerated Light Soaking Testing for p-Type PERC Cell with and without Laser LIR Technology", " EU PVSEC 2019.

SOL-GEL METHOD FOR DOUBLE LAYER COATED COLORED SILICON SOLAR CELLS

M. Rudzikas[1,4], A. Šetkus[1], N. Curmei[2], D. Serban[2], J. Donėlienė[3], J. Ulbikas[4], A.G. Ulyashin[5]

[1] Center for Physical Sciences and Technology, Sauletekio av. 3, LT-10257 Vilnius, Lithuania;
[2] The Institute of Applied Physics, 5 Academiei str. Chisinau, MD-2028, Moldova;
[3] The Applied Research Institute for Prospective Technologies.
[4] UAB "Modern E-Technologies", Vismaliukų str. 34, Vilnius LT-10243, Lithuania;
[5] SINTEF Industry Forskingsveien 1, 0314 Oslo, Norway;

ABSTRACT: In this work, few cost effective solutions are proposed for improving solar cell aesthetic features in a more efficient way than it is usually done in industry. We demonstrated a method of color adding to the cell by deposition of ITO or $TIO_2/SiO_2/TiO_2$ films on commercial monocrystalline Si cell and glass cover. Simple and cheap sol-gel technology combined with spray pyrolysis or dipping were used for the metal oxide films. Colors are intentionally tuned by the thicknesses of the layers to fit the design concept, e.g. for building integrated photovoltaics (BIPV).
Keywords: sol-gel, BIPV, thin films, colored PV, Bragg lattice, DLARC.

1 INTRODUCTION

BIPV sector is expanding due to the strong request for solutions contributing to the energy efficient buildings with autonomous power generating systems tailored to architectural needs of particular façade. However the PV industry is yet not ready to meet the requests of architects in the terms of the cost and the aesthetics value as well. Therefore the applied BIPV growth is still relatively slow.

Most of the Si-cell market is based on blue or black colored modules and colored solutions are rare and expensive. These colors can be modified by changing the SiN_x:H layer thickness. However, this typical approach is not very convenient because the modifications of the ARC interfere with other technological processes in the solar cell manufacturing line reducing line overall throughput and cell efficiency [1].

More efficient approach can be realized by additional metal oxide coatings on the top of a commercial solar cell that actually results in a double layer anti-reflective coating (DLARC) [1–3].

2 METHODOLOGY

Commercial monocrystalline silicon solar cells with efficiency of 20% and size 15.675x15.675 cm^2 were cut into pieces of 3x3 cm^2 by nanosecond laser system.

The sol-gel solutions for spraying were prepared by mixing $InCl_3$ and $SnCl_4$ in proportion of 9:1 and diluted in ethanol. Then the solution was deposited by vapor phase pyrolysis using compressed oxygen as the carrier gas. Dedicated spray system was developed for better dispersion and purification from large size droplets using glass tube with a labyrinth of specific obstacles. Substrate temperature was set to 450 °C. The duration of deposition was varied from 1 to 5 minutes for each sample.

TiO_2 and SiO_2 sol-gel solutions for preparing Bragg lattice structures by dipping technique on soda lime glass plates were made from: titanium butoxide, HCl, H_2O, ethanol and tetraethyl orthosilicate, HCl, H_2O, ethanol (respectively).

Bentham PVE 300 spectrometer was used for recording reflection spectra. Wavelength was varied from 300 to 1200nm and beam was directed between fingers, to avoid shading.

OPAL 2 was used for the numerical simulation, which is made by Australian company - "PV Lighthouse".

3 RESULTS: SPRAY PYROLYSIS

A set of the produced double layered ARC solar cells is visualized in Figure 1. The blue cell in the 1st row is the reference cell and the next 8 cells had individual thickness of the ITO layers on the top of SiN_x:H ARC. Vivid colors from blue to purple can be clearly seen. Some variations in the color were noticeable due to the technology based non uniformity in the ITO.

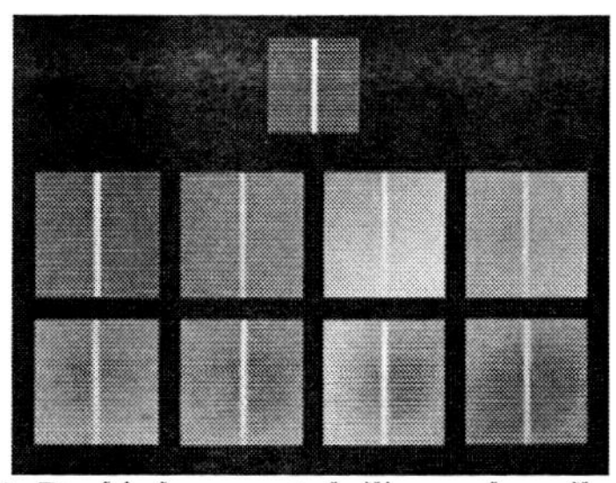

Figure 1: Double layer coated silicon solar cells with ITO sprayed on the top. Upper blue cell is a reference cell and further cells are with increasing spraying duration from 1 to 5 minutes.

The reflectance spectra were measured for the produced cells and are plotted in Figure 2 (a-d). Theoretical curves were also shown (using OPAL2). In figure (a) the experimental data graphically show the shift of the reflection maxima with increasing spray duration (the arrow direction). The peak position and the width of maximum were directly related to the color that is perceived as a specific color by human eye. In Figs (b-d) the simulations were compared with the experiments. The experimental and numerical results coincide well. Some discrepancies can be explained by the technology imperfections in the development stage and by the discrepancies between the reference and the real refractive indexes of the ITO layer. Likely, the reflection in red and IR region can be caused by the reflections from backside of cell that was not evaluated during the simulation. There simulated ITO layer thickness was 26nm and 103nm for 1min and 5min spray durations, respectively.

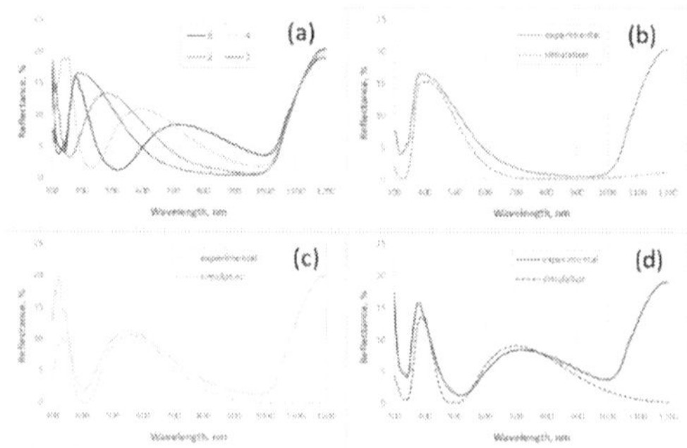

Figure 2: Reflectance spectra of double layer coated solar cells, simulation curves drawn besides experimental data (b-d).

The reflectance spectra were used in the CIE diagrams for qualitative representation and comparison of colors (Figure 3 (a)). The black rectangle is the point with 0nm ITO thickness whereas the subsequent points visualize the color of the thicker samples increasingly till 5 min deposition duration (approx. 103nm) sample is reached (see arrow). For the comparison, in figure (b), the results from our previous work [3] are shown for the magnetron coated cells. It can be stated that significant differences between obtained diagrams does not exist. That indicates suitability to use spraying technique for the ITO layer formation.

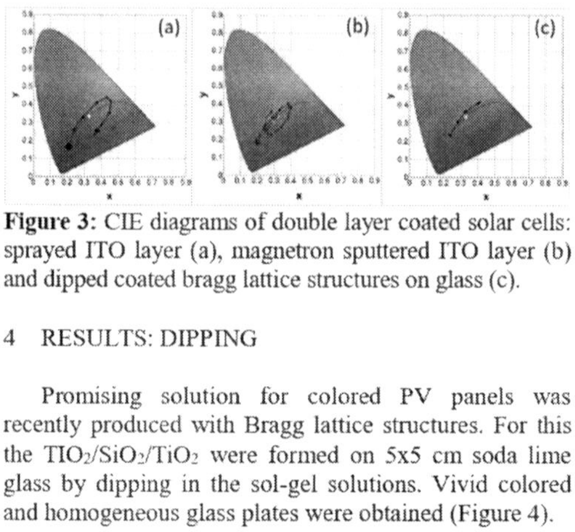

Figure 3: CIE diagrams of double layer coated solar cells: sprayed ITO layer (a), magnetron sputtered ITO layer (b) and dipped coated bragg lattice structures on glass (c).

4 RESULTS: DIPPING

Promising solution for colored PV panels was recently produced with Bragg lattice structures. For this the $TiO_2/SiO_2/TiO_2$ were formed on 5x5 cm soda lime glass by dipping in the sol-gel solutions. Vivid colored and homogeneous glass plates were obtained (Figure 4).

Figure 4: Bragg lattice structures formed on soda lime glass plates with dipping technique with different thicknesses. 3 Layers structures were formed: $TiO_2/SiO_2/TiO_2$.

CIE diagrams for these Bragg lattice colored glass plates are drawn in Figure 3 (c). Color purity and hues are similar compared to the both double layer ARC coated structures. It must be noted that, in figure (b), the colors of the encapsulated cells were also shown for 2 samples. The arrow indicated the change of CIE parameters after the encapsulation. Points were shifted towards D65 illuminant, meaning that color purity slightly decreased. Therefore, it can be stated, that colors obtained with Bragg lattice structure are more pure, than with encapsulated solar cells.

5 CONCLUSION

It was proved that the color of the commercial monocrystalline solar cells can be effectively changed by easy upscalable spray pyrolysis technique. Vivid colors were produced with intentional change of the ITO thickness of ITO/SiN_x:H DLARC. This approach can be used for any type of cells. Large area technology for 15.675x15.675 cm and bigger cells requires strict process control for higher ITO film homogeneity.

The experiments and simulations proved that the ITO thicknesses must be from 26 to 103nm that can be produced by the spray duration from 1 to 5min.

The Bragg lattice coated glass plates are vivid colored and more homogeneous compared to the DARC coated solar cells.

6 ACKNOWLEDGEMENTS

Part of research was supported by the H2020 Programme under the project INFINITE-CELL (H2020-MSCA-RISE-2017-777968).

Part of research was supported by the programmes "J05-LVPA-K Intelektas" and "LVPA-K Eksperimentas" under the projects "CORAL CELLS" (Nr. J05-LVPA-K-04-0030) and "BUILDinCOLOR" (Nr. 01.2.1-LVPA-K-856-01-0029) respectively.

7 REFERENCES:

[1] L. Zeng, M. Li, Y. Chen, H. Shen, A simplified method to modulate colors on industrial multicrystalline silicon solar cells with reduced current losses, Sol. Energy. 103 (2014) 343–349. https://doi.org/10.1016/j.solener.2014.02.012.

[2] P.P. Altermatt, Y. Chen, Y. Yang, Z. Feng, H. Shen, Color modulation of c-Si solar cells without significant current-loss by means of a double-layer anti-reflective coating, in: Eur. Photovolt. Sol. Energy Conf., 2012. https://doi.org/10.4229/27thEUPVSEC2012-2CV.7.12.

[3] M. Rudzikas, A. Šetkus, M. Stange, J. Ulbikas, A. Ulyashin, Simple interference based colorization of Si based solar cells and panels with ITO/SiNx:H double layer antireflective coatings, Sol. Energy. 207 (2020) 218–227. https://doi.org/10.1016/j.solener.2020.06.091.

[4] K.R. McIntosh, S.C. Baker-Finch, OPAL 2: Rapid optical simulation of silicon solar cells, in: Conf. Rec. IEEE Photovolt. Spec. Conf., 2012: pp. 265–271. https://doi.org/10.1109/PVSC.2012.6317616.

FIRING-STABLE PECVD SiO_xN_y/N-POLY-SI PASSIVATING CONTACTS FOR HIGH-EFFICIENCY SILICON SOLAR CELLS

Maximilian Stöhr[1], Johannes Aprojanz[1], Rolf Brendel[1,2], Thorsten Dullweber[1]
[1]Institute for Solar Energy Research Hamelin (ISFH), Am Ohrberg 1, 31860 Emmerthal, Germany
[2]Department of Solar Energy, Institute of Solid-State Physics, Leibniz Universität Hannover, Appelstraße 2, 30167 Hannover, Germany

ABSTRACT: Passivating contacts based on SiO_x/poly-Si enable very high solar cell conversion efficiencies. In this paper, we investigate and optimize the plasma-enhanced chemical vapor deposition (PECVD) of SiO_xN_y/n-a-Si stacks, their subsequent annealing to SiO_xN_y/n-poly-Si stacks followed by PECVD SiN_x deposition and firing. Whereas the J_0 of thermal SiO_x/n-poly-Si stacks degrade after firing, PECVD SiO_xN_y/n-poly-Si stacks exhibit an excellent firing stability with J_0 values down to 1.3 fA/cm² after firing which corresponds to an outstanding implied V_{OC} of 744 mV. The application of different hydrogenation processes to the thermal SiO_x/n-poly-Si and PECVD SiO_xN_y/n-poly-Si layers reveals, that both stacks achieve excellent passivation properties with J_0 = 1.5 fA/cm² after maximum hydrogenation. However, only the PECVD SiO_xN_y/n-poly-Si stack maintains this excellent surface passivation after firing possibly due to a superior capability to retain the hydrogen at the c-Si/SiO_xN_y interface during firing, and thus demonstrates the potential as a future manufacturing process sequence.
Keywords: Silicon, Polycrystalline, Passivation

1 INTRODUCTION

Passivating contacts based on SiO_x/poly silicon (SiO_x/poly-Si) layer stacks are a promising technology for next-generation high-efficiency industrial silicon solar cells [1-3]. An outstanding passivation quality in combination with a good contact resistivity are the key advantages of SiO_x/poly-Si contacts which significantly reduce the recombination losses at the metal contacts and hence, exhibit a higher efficiency-potential compared to industrial PERC and PERC+ solar cells [4]. The technology is commonly denoted either as "poly-Si on oxide" (POLO) or "tunneling oxide passivated contact" (TOPCon). In terms of process technology, intrinsic or doped amorphous Si (a-Si) layers are deposited by low-pressure chemical vapor deposition (LPCVD) [5] or plasma-enhanced chemical vapor deposition (PECVD) [6] on top of thermally, wet-chemically or PECVD-grown SiO_x [7-9]. The conversion from amorphous to poly-crystalline silicon (poly-Si) is done during a high-temperature annealing process.

An industrial application of SiO_x/poly-Si passivating contacts in silicon solar cells requires the development of lean and hence, cost-effective manufacturing process sequences. PECVD offers single-sided deposition and the growth of a complete layer stack in a single process tool [8]. Furthermore, shadow masks can be applied for the deposition of local SiO_x/a-Si fingers [10] thereby enabling very lean manufacturing process sequences for local passivating contacts as e.g. in development for the POLO IBC solar cell [11]. In addition, PECVD is a proven low-cost and high-throughput mass production process for the deposition of AlO_x and SiN_x layers of industrial PERC and PERC+ solar cells.

In literature, the combination of PECVD oxide by plasma-assisted N_2O oxidation and n-type poly-Si (n-poly-Si) resulted in excellent implied V_{OC} (iV_{OC}) values up to 747 mV and corresponding J_0 values down to 2.0 fA/cm² [9]. An additional hydrogenation by Al_2O_3/SiN_x capping and subsequent low-temperature annealing was applied in order to improve the surface passivation properties. However, industrial solar cells like TOPCon, POLO-BJ and POLO-IBC apply a short high-temperature firing process for alloying the metal contacts. Hence, it is mandatory that the passivation quality of the PECVD SiO_x/n-poly-Si stack is maintained after the PECVD SiN_x deposition and subsequent firing.

In this paper, we investigate in detail how to obtain an excellent passivation quality of fired PECVD SiO_x/n-poly-Si stacks capped with SiN_x. We apply optimized PECVD of n-type a-Si (n-a-Si) and subsequent annealing recipes to minimize blistering by enabling hydrogen effusion through a cavity-rich n-a-Si morphology as published previously in Ref. 12. After additional PECVD SiN_x capping and firing, we obtain J_0 values down to 1.3 fA/cm² corresponding to an iV_{OC} of 744 mV for symmetrically passivated Cz silicon test wafers. We apply the same PECVD n-a-Si, annealing, SiN_x, and firing process sequence to a thermally-grown SiO_x interface, which however, degrades after firing to J_0 values up to 6.8 fA/cm².

2 EXPERIMENTAL

Figure 1 shows the processing sequences of four split groups T1, T2, P1 and P2 all resulting in symmetrically processed silicon test wafers that we used to determine the passivation quality of the SiO_x/poly-Si layer by carrier lifetime measurements. Groups featuring thermal SiO_x or PECVD SiO_xN_y are labeled with "T" or "P", respectively. The integer "1" is for groups with hydrogenation by PECVD SiN_y-deposition plus forming gas anneal (FGA) whereas the split groups T2 and P2 were processed without FGA relying on hydrogenation by SiN_x and firing, only. All experiments are conducted on planar phosphorous-doped Czochralski (Cz) silicon wafer (156.75 x 156.75 mm²) with (100) orientation, 3 Ωcm base resistivity and 170 µm thickness. After an initial saw damage etch (SDE) in KOH solution, a RCA clean and a phosphorous diffusion is applied to improve the charge carrier bulk lifetime by gettering. A dip in a solution of 1% hydrogen fluoride (HF) diluted in de-ionized water removes the phosphorus-silicate glass followed by a KOH etch to remove the phosphorous doping and a RCA clean. Afterwards, group T1 and T2 (c.f. Figure 1) receive an approx. 1.7 nm thin SiO_x thermally grown in a tube

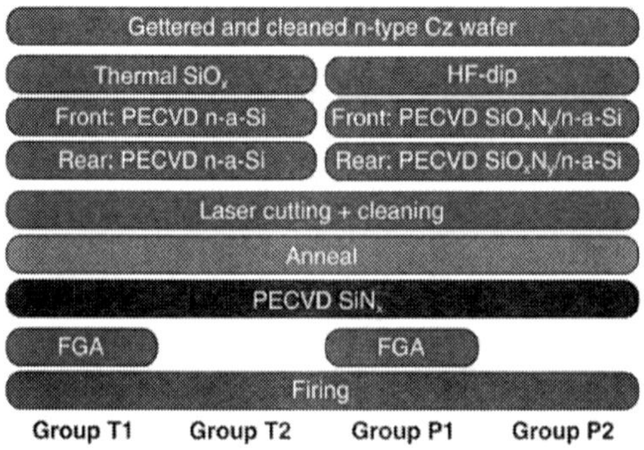

Figure 1. Processing sequence of Cz wafers with symmetrical SiO_x/n-poly-Si passivation. Split groups are labeled with "T" for thermal SiO_x and "P" for PECVD SiO_xN_y, while "1" indicates hydrogenation including FGA and "2" without FGA. Split group P2 is the shortest and cheapest process sequence and hence targeted for industrial application.

furnace. The topping n-a-Si layers are deposited by PECVD in a von Ardenne CS400P clustertool using a capacitively coupled plasma (CCP) source. 100 nm-thick in-situ P-doped a-Si is deposited, applying 20 sccm SiH_4, 50 sccm H_2 and 50 sccm PH_3 (3% in H_2) to a high frequency generator operating at 13.56 MHz, 20 W plasma power and a substrate temperature of 220 °C. A wafer flip is performed outside the vacuum chamber after the front and before the rear side deposition.

In contrast to T1 and T2, the sample groups P1 and P2 in Figure 1 receive an additional HF-dip (0.25 % in deionized water) to remove any natively grown SiO_x before PECVD deposition. Subsequently, the samples are rinsed with deionized water, dried with a nitrogen gun and immediately loaded into the clustertool for PECVD deposition of both the SiO_xN_y (100 sccm N_2O, 400 s) layer and n-a-Si layer. The latter was grown with the same deposition parameters as in T1 & T2.

After another wet-chemical clean, the crystallization of the n-a-Si to n-poly-Si is done in an inert high-temperature anneal using a tube furnace with nitrogen atmosphere. The temperature profile consists of an initial 10 minute anneal at 400 °C followed by a ramp-up to the plateau temperature $T_{plateau}$ of 800, 850 or 900 °C with a heating rate of 10 °C/min. The crystallization temperature remains for 30 minutes and is then subsequently reduced to 600 °C with a cool down rate of 10 °C/min. Afterwards, both wafer sides are capped with SiN_x (thickness d=100 nm, refractive index n=2.05) with a PECVD tool. Groups T1 and P1 receive further hydrogenation by a 30 min FGA at 450 °C utilizing a dedicated tube furnace. The process is concluded for all samples with firing in a belt-furnace. We use a firing profile typically used in PERC+ solar cell production and similarly applied to POLO and TOPCon solar cells for contact formation with a set peak temperature of 830 °C and a belt speed of 5.6 m/min.

All samples are applied to photo-conductance lifetime measurements (WCT-120, Sinton Instruments). Effective charge carrier lifetime values τ_{eff} are extracted at an excess carrier density of Δn = $1*10^{15}$ cm^{-3} and the saturation current density J_0 attributed to a single wafer side is extracted at $\Delta n = 5*10^{15}$ cm^{-3} using the Kane and Swanson method [13]. The implied open circuit voltage iV_{OC} is determined at 1 sun illumination. For all split groups, we perform the QSSPC measurement after several consecutive process steps in order to monitor the impact of the process step on the passivation quality, namely after anneal, PECVD SiN_x, FGA if applied, and firing.

3 RESULTS AND DISCUSSION

3.1 PECVD SiO_xN_y/n-poly-Si

We utilize a N_2O plasma oxidation to form SiO_xN_y on the surface of a Cz silicon wafer in a self-limiting process [14], which enables precise control over the SiO_xN_y thickness on the nanometer scale by tailoring the PECVD process parameters. According to Ref. 14, N_2O plasma oxidation results in the incorporation of nitrogen in the vicinity of the interface and therefore we name the oxide SiO_xN_y. We grow SiO_xN_y on float zone Si wafer with 10 to 30 W plasma power and measure the oxide thickness with spectral ellipsometry achieving values in the range of 1.5 to 2.2 nm. Figure 2 (a) displays the τ_{eff} values of SiO_xN_y/n-

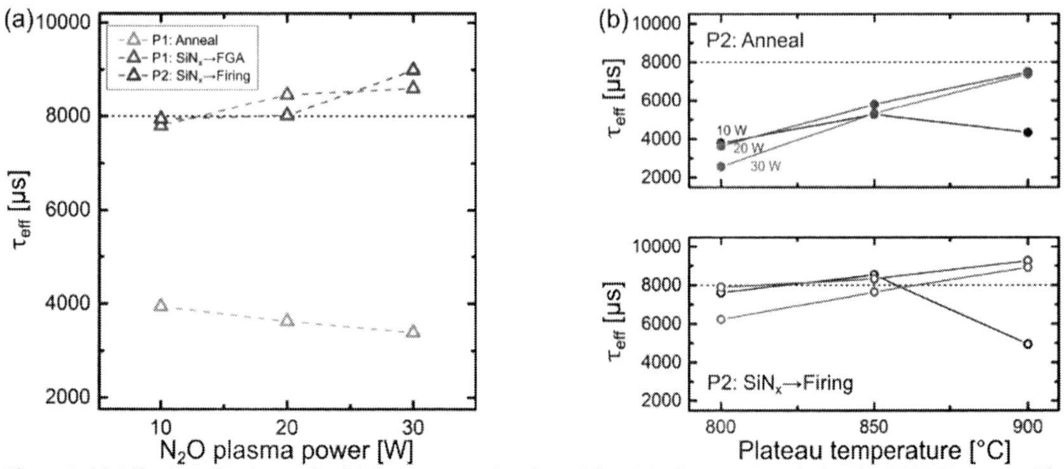

Figure 2. (a) Effective minority carrier lifetimes τ_{eff} as a function of the N_2O plasma power during the PECVD process. The samples were subjected to a two-stage anneal with an initial temperature of 400 °C followed by a ramp-up to 800 °C. The color code highlights different stages of hydrogenation. (b) Dependence of τ_{eff} on the furnace crystallization temperature and the N_2O plasma power. The top panel depicts results acquired right after annealing, while the samples analyzed in the bottom panel were subjected to a subsequent hydrogenation by SiN_x deposition and fast firing at 830 °C.

poly-Si stacks as a function of the N_2O plasma power after different process steps according to the process sequence P1 and P2 in Figure 1. We obtain high τ_{eff} values between 8-9 ms after both, hydrogenation by SiN_x and FGA (P1 without firing) as well as SiN_x with subsequent firing (P2). Furthermore, there is a slight increase of τ_{eff} with increasing N_2O plasma power reaching a highest τ_{eff} of 9.0 ms after firing for SiO_xN_y grown at 30 W indicating that the final passivation quality slightly increases with the SiO_xN_y thickness. Further optimization can be achieved by the adjustment of the thermal budget of the annealing process. Figure 2 (b) depicts the dependence of τ_{eff} after annealing on $T_{plateau}$. For all N_2O plasma powers, the lowest anneal with $T_{plateau}$ of 800 °C results in the lowest effective lifetimes ranging between 2.5-3.8 ms. For SiO_xN_y with t > 1.5 nm (20, 30 W), τ_{eff} approximately doubles by increasing the peak temperature to 900 °C, which translates to an J_0 of 3 fA/cm² even without further hydrogenation of the interface. The combination of 10 W plasma power and 900 °C shows lower τ_{eff} after anneal, most likely due to parasitic diffusion of phosphorous through the ultra-thin oxynitride, which translates into reduced τ_{eff} due to enhanced Auger-recombination. Hydrogenation with PECVD SiN_x and firing but without FGA, see economically preferred process sequence P2 in Figure 1, enhances the passivation quality of all stacks (cf. lower panel of Figure 2 (b)) resulting in a highest lifetime of 9.3 ms corresponding to an excellent iV_{OC} = 740 mV and J_0 = 3.8 fA/cm².

3.2 Comparison of thermal SiO_x and PECVD SiO_xN_y

In order to compare the passivation quality of thermal SiO_x with PECVD SiO_xN_y, we prepare samples following the process flows T1, P1, T2 and P2 in Figure 1. Figure 3 (a) shows the τ_{eff} (top panel) and J_0 (bottom panel) values for both interface oxides along the process sequences of groups T1 and P1, which include a FGA anneal after the deposition of SiN_x before firing (cf. Figure 1). As can be seen, the passivation quality of the thermal SiO_x/n-poly-Si/SiN_x stack (group T1) significantly improves after FGA with a mean τ_{eff} = 10.8 ms and J_0 = 1.5 fA/cm², which corresponds to a reduction by 7.5 fA/cm² as compared to the mean J_0 value after anneal. The PECVD SiO_xN_y/n-poly/SiN_x stack (group P1) shows a similar improvement after anneal to τ_{eff} = 9.7 ms and J_0 =1.5 fA/cm² suggesting that the hydrogenation by SiN_x and FGA supplies sufficient hydrogen to achieve excellent surface passivation irrespective of the interface oxide.

However, remarkable differences are evident after firing. In the case of thermal SiO_x, τ_{eff} decreases to 4.3 ms and the J_0 increases to 6.8 fA/cm² reversing part of the gain in passivation quality provided by the previous hydrogenation with SiN_x and FGA. Contrary to that, the PECVD SiO_xN_y/n-poly Si/SiN_x stacks show negligible impact of firing with average J_0 values of 1.6 fA/cm² post firing. Finally, we compare τ_{eff} as well as J_0 obtained with the shorter process sequences T2 and the economically most attractive sequence P2 as shown in Figure 3 (b). Interestingly, τ_{eff} increases rather steadily for both interface oxides with further process steps. The mean τ_{eff} of the PECVD SiO_xN_y/n-poly Si/SiN_x stacks rises up to 9.5 ms after firing, while the thermal SiO_x/n-poly Si/SiN_x only reaches a mean value of 4.6 ms after firing similar to the one obtained for group T1 after FGA and firing shown in Figure 3 (a). The bottom panel of Figure 3 (b) depicts the corresponding J_0 values. The mean J_0 of the thermal SiO_x/n-poly Si stack reaches its minimum after SiN_x deposition at 4.9 fA/cm² and subsequently increases to 6.3 fA/cm² after firing. In contrast, the mean J_0 of PECVD SiO_xN_y/n-poly Si/SiN_x stacks decreases steadily with further process steps down to 1.9 fA/cm² in average and with a best value of 1.3 fA/cm² after firing, corresponding to an outstanding iV_{OC} = 744 mV.

Three main observations can be derived from Figure 3, which clearly demonstrate that the passivation quality is crucially influenced and determined by the nature of the interfacial oxide:

1. The mean τ_{eff} of PECVD SiO_xN_y/n-poly-Si stacks after annealing is about 2 ms higher and the mean J_0 about 4.5 fA/cm² lower as compared to thermal SiO_x/n-poly-Si.
2. Maximum hydrogenation by SiN_x and FGA without firing results in excellent passivation properties for both interface oxides.
3. Only the PECVD SiO_xN_y/n-poly-Si/SiN_x stacks retain the passivation quality induced by FGA after firing and also enable excellent τ_{eff} and J_0 values after firing without the use of FGA.

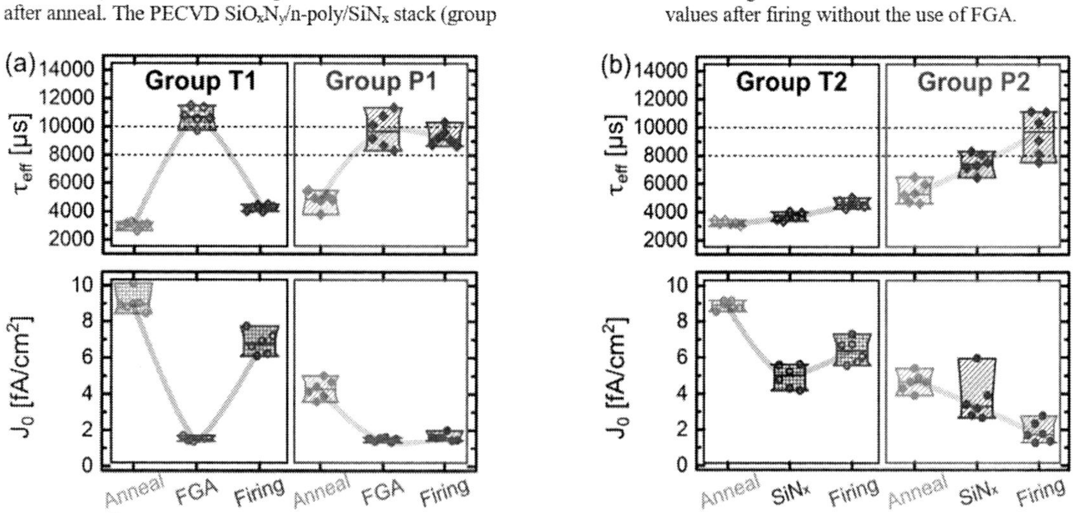

Figure 3. Effective minority carrier lifetime τ_{eff} (top panel) measured for SiO_x/n-poly-Si (black frame) and PECVD SiO_xN_y/n-poly-Si (blue frame) after different process steps including FGA (group T1 and P1) in (a) and without FGA in (b). The corresponding J_0 values are shown on the bottom panels. See Figure 1 for further information about the processing sequences of the groups T1, T2, P1, and P2.

We suggest the following aspects to be considered as part of a future c-Si/$SiO_x(N_y)$ interface hydrogenation model. The properties of ultra-thin dielectrics are highly determined by the chemical configuration in close vicinity to the substrate interface. It is known that thermally grown SiO_x feature a mixed composition of suboxides at the SiO_x/c-Si interface, which comes along with interface defects acting as potential recombination centers [15-17]. For both oxides, the saturation of trap states can be achieved by providing a sufficient amount of hydrogen with SiN_x deposition and a subsequent FGA. Moreover, the robustness of SiO_xN_y/n-poly Si/SiN_x stacks against firing suggest the existence of a mechanism, which retains a sufficient amount of hydrogen at the interface thereby maintaining the excellent surface passivation. In literature, the oxidation of Si surfaces by N_2O plasma is reported to incorporate nitrogen at the Si/SiO_xN_y interface [14, 18]. In addition, FTIR and hydrogen effusion measurements showed that an increasing concentration of N in PECVD SiN:H layers shifts the Si-H bonds towards higher wavenumbers and therefore, increases the Si-H bond energy and reduces the hydrogen effusion during firing [19]. Hence, we suggest that the incorporation of nitrogen in the SiO_xN_y/n-poly-Si stacks increases the Si-H bond energy thereby reducing hydrogen out diffusion and maintaining the passivation quality during firing. However, it should be noted that firing-stable passivation with $J_0 \leq 3$ fA/cm² has been demonstrated for a wet-chemical interface oxide and AlO_x/SiN_x capping [20]. Thus, the presence of nitrogen at the interface might promote thermal resilience to firing but may not be required. In future experiments, we aim on demonstrating low resistivity metal contacts while maintaining the excellent passivation of the PECVD SiO_xN_y/n-poly-Si stacks. From a technological point of view, the presented results pave the way for a cost-effective integration of PECVD SiO_xN_y/n-poly-Si passivating contacts in solar cell manufacturing.

4 CONCLUSION

We obtain blister-free PECVD poly-Si layers after annealing with mean $\tau_{eff} > 9$ ms for both interface oxides after sufficient hydrogenation. The J_0 of thermal SiO_x/poly-Si stacks decreases down to 1.5 fA/cm² after maximum hydrogenation with SiN_x deposition and FGA suggesting a sufficient saturation of dangling bonds at the c-Si/SiO_x interface. However, after firing the J_0 of thermal SiO_x/poly-Si stacks degrades to around 6.8 fA/cm², which indicates a change of the c-Si/SiO_x interface microstructure. In contrast, PECVD SiO_xN_y/poly-Si stacks exhibit an excellent firing stability enabling excellent J_0 values down to 1.3 fA/cm² after firing which are similar to the values obtained with maximum hydrogenation and which correspond to an outstanding iV_{OC} = 744 mV. This finding suggests a superior capability to retain the hydrogen saturation of dangling bonds at the c-Si/SiO_xN_y interface. As possible root cause, we propose that when using N_2O as oxidizing gas nitrogen is incorporated at the c-Si/SiO_xN_y interface thereby increasing the Si-H bond energy. The in-situ PECVD SiO_xN_y/poly-Si stack developed in this work in combination with annealing, PECVD SiN_x deposition and firing represents an industrially attractive process technology for implementing poly-Si based passivating contacts into mass production of future high-efficiency silicon solar cells.

5 REFERENCES

[1] Brendel, R.; Dullweber, T.; Gogolin, R.; Hannebauer, H.; Harder, N.-P.; Hensen, J.; Kajari-Schröder, S.; Peibst, R.; Petermann, J. H.; Römer, U.; Schmidt, J.; Schulte-Huxel, H.; Steckenreiter, V. Recent progress and options for future crystalline silicon solar cells. in: Proc. 28th Eur. Photovolt. Sol. Energy Conf., 2013, 676–690.

[2] Peibst, R.; Römer, U.; Larionova, Y.; Rienäcker, M.; Merkle, A.; Folchert, N.; Reiter, S.; Turcu, M.; Min, B.; Krügener, J.; Tetzlaff, D.; Bugiel, E.; Wietler, T.; Brendel, R. Working principle of carrier selective poly-Si/c-Si junctions: Is tunnelling the whole story?. Sol. Energy Mat. Sol. Cells, 2016, 158, 60 – 67.

[3] Feldmann, F.; Bivour, M.; Reichel, C.; Hermle, M.; Glunz, S. W. A passivated rear contact for high-efficiency n-type Si solar cells enabling high Voc's and FF > 82%. In Proc. 28th Eur. Photovolt. Sol. Energy Conf. 2013, 988-992.

[4] Dullweber, T.; Stöhr, M.; Kruse, C.; Haase, F.; Rudolph, M.; Beier, B.; Jäger, P.; Mertens, V.; Peibst, R.; Brendel, R. Evolutionary PERC+ solar cell efficiency projection towards 24% evaluating shadow-mask-deposited poly-Si fingers below the Ag front contact as next improvement step. Sol. Energy Mat. Sol. Cells, 2020, 212, 110586.

[5] Stodolny, M.; Lenes, M.; Wu, Y.; Janssen, G.; Romijn, I.; Luchies, J.; Geerligs, L. n-Type polysilicon passivating contact for industrial bifacial n-type solar cells. Sol. Energy Mat. and Sol. Cells, 2016, 158, 24 – 28.

[6] Temmler, J.; Polzin, J.-I.; Feldmann, F.; Kraus, L.; Kafle, B.; Mack, S.; Moldovan, A.; Hermle, M.; Rentsch, J. Inline PECVD Deposition of Poly-Si-Based Tunnel Oxide Passivating Contacts. physica status solidi (a), 2018, 215, 1800449.

[7] van der Vossen, R.; Feldmann, F.; Moldovan, A.; Hermle, M. Comparative study of differently grown tunnel oxides for p-type passivating contacts. Energy Procedia, 2017, 124, 448 – 454.

[8] Gao, T.; Yang, Q.; Guo, X.; Huang, Y.; Zhang, Z.; Wang, Z.; Liao, M.; Shou, C.; Zeng, Y.; Yan, B.; Hou, G.; Zhang, X.; Zhao, Y.; Ye, J. An industrially viable TOPCon structure with both ultra-thin SiO_x and n+-poly-Si processed by PECVD for p-type c-Si solar cells. Sol. Energy Mat. and Sol. Cells, 2019, 200, 109926.

[9] Liao, M.; Huang, Y.; Wang, Z.; Guo, X.; Jiang, C.; Yang, Q.; Yuan, Z.; Huang, D.; Yang, J.; Zhang, X.; Wang, Q.; Jin, H.; Al-Jassim, M.; Shou, C.; Zeng, Y.; Yan, B.; Ye, J. Ultrathin silicon oxide prepared by in-line plasma-assisted N_2O oxidation (PANO) and the application for n-type polysilicon passivated contact. Sol. Energy Mat. and Sol. Cells, 2020, 208, 110389.

[10] Stöhr, M.; Beier, B.; Aprojanz, J.; Vogt, A.; Ambrosius, N.; Brendel, R.; Dullweber, T. PECVD Shadow Mask Deposition of Amorphous Silicon – a Shortcut to Local Passivating Contacts. Proc. 37th Eur. Photovolt. Sol. Energy Conf., 2020, 521–524.

[11] Mertens, V.; Schäfer, S.; Stöhr, M.; Mercker, M.; Köhler, A.; Brendel, R.; Ambrosius, N.; Pernau, T.; Haverkamp, H.; Dullweber, T. Local PECVD SiO_xN_y/n-Poly-Si deposition through a shadow mask for POLO IBC solar cells. Proc. 38th Eur. Photovolt. Sol. Energy Conf., 2021

[12] Stöhr, M.; Aprojanz, J.; Brendel, R.; Dullweber, T. Firing-Stable PECVD SiO_xN_y/n-Poly-Si Surface Passivation for Silicon Solar Cells. ACS Appl. EnergyMaterials 2021 4 (5), 4646-4653

[13] Kane, D.; Swanson, R. Measurement of the emitter saturation current by a contactless photoconductivity decay method. IEEE Photovoltaic Specialists Conference, 1985, 18, 578-583.
[14] Masuda, A.; Morimoto, A.; Kumeda, M.; Shimizu, T.; Yonezawa, Y.; Minamikawa, T. Novel oxidation process of hydrogenated amorphous silicon utilizing nitrous oxide plasma. Appl. Physics Letters, 1992, 61, 816-818.
[15] Grunthaner, F. J.; Lewis, B. F.; Zamini, N.; Maserjian, J.; Madhukar, A. XPS Studies of Structure-Induced Radiation Effects at the Si/SiO_2 Interface. IEEE Transactions on Nuclear Science, 1980, 27, 1640-1646.
[16] Fukuda, H.; Yasuda, M.; Iwabuchi, T.; Kaneko, S.; Ueno, T.; Ohdomari, I. Process dependence of the SiO_2/Si(100) interface trap density of ultrathin SiO_2 films. Journal of Appl. Physics, 1992, 72, 1906-1911.
[17] Green, M.; Gusev, E.; Degraeve, R.; Garfunkel, E. Ultrathin (<4 nm) SiO_2 and Si-O-N gate dielectric layers for silicon microelectronics: Understanding the processing, structure, and physical and electrical limits. Journal of Appl. Physics, 2001, 90, 2057-2121,
[18] Lee, D. R.; Lucovsky, G.; Denker, M. S.; Magee, C. Nitrogen-atom incorporation at Si-SiO_2 interfaces by a low-temperature (300°C), pre-deposition, remote-plasma oxidation using N_2O. Journal of Vacuum Science & Technology A, 1995, 13, 1671-1675.
[19] Gatz, S.; Dullweber, T.; Mertens, V.; Einsele, F.; Brendel, R. Firing stability of SiN_y/SiN_x stacks for the surface passivation of crystalline silicon solar cells. Sol. Energy Mat. and Sol. Cells, 2012, 96, 180-185.
[20] Hollemann, C.; Rienäcker, M.; Soeriyadi, A.; Madumelu, C.; Haase, F.; Krügener, J.; Hallam, B.; Brendel, R.; Peibst, R. Firing stability of tube furnace-annealed n-type poly-Si on oxide junctions. Progress in photovoltaics, 2021, DOI: 10.1002/pip.3459.

Interplay of IBC cell's front surface doping, passivation quality, and stability under ultraviolet light exposure

Haifeng Chu, Vaibhav Kuruganti, Razvan Roescu, Christoph Peter, Valentin D. Mihailetchi

International Solar Energy Research Center Konstanz e. V., Rudolf-Diesel Str.15, D-78467 Konstanz, Germany
haifeng.chu@isc-konstanz.de

ABSTRACT: We study the UV stability of lightly doped *n*- type Si surfaces with a sheet resistance of 154 Ohm/sq, 935 Ohm/sq and 1480 Ohm/sq. These surfaces are passivated by a layer stack consisting of thermal SiO_2 and PECVD SiN_x and the resulting J_0 is in the range of 10 to 40 fA/cm^2. We applied these lightly diffused surfaces for the FSFs of IBC cells, and for symmetric $n^+/n/n^+$ lifetime samples. We found out that although the 6-inch IBC cells with lighter doped FSFs lead to better V_{oc}, they showed more degradation after the UV exposure test (dose of 45 kWh/m^2, equivalent to three times of IEC test). The degradation is likely caused by the increase of front surface recombination. Lastly, we summarize several UV degradation test results for both *p*- and *n*-type surfaces, based on which it seems that it is advisable to check the UV stability for not only the IBC cells but also other cell concepts which have a lightly doped or even undoped front surface.

Keywords: Si solar cells, UV degradation, surface passivation, IBC

1 INTRODUCTION

To further increase the efficiency of industrial wafer-based crystalline Si solar cells to > 25%, it is important to minimize the recombination current prefactor J_0 [1] for the front surface, e.g. <15 fA/cm^2 (J_0 is also called as "saturation current density" in many papers). Decrease of front surface J_0 leads to better spectral response in ultraviolet range (280 nm to 400 nm), and further gain in module power can be obtained by utilizing UV-transmitting encapsulants. However, a potential problem associated with such a well-passivated front surface is its susceptibility to ultraviolet exposure. Previous studies show that many popular high efficiency solar cell structures such as PERC, PERT, HJT, and IBC show different extent of ultraviolet-induced degradation (UVID), depending on the surface doping concentration, the material, and process that are used for the passivating dielectric [2–4].

The reduced front surface passivation quality was reported to be responsible for the UVID. For instance, when using SiO_2 as the passivating layer, the UV light may excite the electrons from the conduction band of the Si to SiO_2 and create recombination centers at the Si-SiO_2 interface [5]. But the exact physical mechanism, especially under non-concentrated sunlight, is still under debate [6, 7] and the UV stability problem becomes more complicated as the intensity of the UV light may also influence the degradation rate [3].

In this study we investigate UVID for IBC cells. The front surface of an IBC silicon solar cell is free of metal contacts, and therefore the phosphorus doped front-surface-field (for *n*-type bulk) can be solely designed to obtain minimum surface recombination without having significant loss due to high metal-Si contact resistance when using a lightly doped front surface. Numerous studies have shown that the recombination current prefactor J_0 of a phosphorus doped surface decreases with decreasing doping concentration [8]. It has also been shown that by applying a lighter doped front-surface-field (FSF), the cell's V_{oc} and efficiency increase [9]. Unfortunately, the UV stability appears to show an opposite trend. The IBC cell without a doped front surface degrades significantly after being exposed to ultraviolet (UV) light [5, 9, 10], whereas having a doped front surface (e.g. FSF) improves the UV stability [5, 9] but significant degradation is still present [3].

We fabricated IBC cells with three different doping profiles for the front surface, and using the same diffusion recipes we fabricated symmetric $n^+/n/n^+$ lifetime samples as well. The resulting J_0 is in the range of 10 to 40 fA/cm^2. The bare cells and lifetime samples (not covered with any encapsulant or glass) were placed in a UV chamber to study their UV stability. Before and after UV exposure, *IV*, QSSPC, spectral response, and PL measurements were performed. Lastly we discuss the implication of this work for rear-junction solar cells.

2 EXPERIMENT

The IBC cells and J_0 samples were fabricated from 175 μm thick 6-inch *n*-type Cz wafers with a base resistivity of 3.6 Ω·cm. We followed our baseline process [11] except for a modification of the FSF region. BBr_3 and $POCl_3$ diffusions were used to form all the diffused regions, whereas the front surface is passivated by a SiO_2/SiN_x layer stack. The SiO_2 layer is 15 nm thick and it was grown *in situ* during the diffusion process [12], while the PECVD SiN_x layer (40kHz, direct plasma) is 54 nm thick. The rear alternating p^+ and n^+ doping lines were realized by laser ablation. Screen-printed and firing-through Ag paste was utilized to contact the rear p^+ and n^+ doped surfaces. All processes can be considered to be industrially feasible.

Furthermore, we have produced lifetime samples by using *n*-type Cz wafers with a base resistivity of 3.0 Ω·cm; these samples do not go through the phosphorus diffusion. They are passivated with a 14 nm thick wet SiO_2 and 54 nm thick PECVD SiN_x. These undiffused samples are designed to stand for the case of IBC cells without FSF. The front surfaces of our IBC cells and the two types of lifetime samples are pyramids textured as depicted in Fig. 1, whereas Fig. 2 shows ECV measurements of the three FSF doping profiles used in this study.

After the initial characterization such as *IV* and QSSPC, we randomly selected in total of 23 samples to start the UV exposure test, and the front sides of them were not cover with any encapsulants or glass. The UV light sources are low pressure mercury vapor fluorescent lamps with a major peak at approximately 350 nm. This kind of lamp is also used in the solarium. The test samples received a total UV irradiance of 253 W/m^2, where 96.1 % of the total irradiance belongs to UVA (320 nm to 400 nm) and the rest belongs to UVB (280 nm to 320 nm). Temperature in the UV test chamber was set at 60 °C. Our test condition almost meets the IEC 61215-2 (2016) standard, except that the UV irradiance is slightly higher than the maximum allowed irradiance of 250 W/m^2 as defined in the standard. The total UV dose in this test is 45 kWh/m^2, which is triple of that is required in the standard. This UV dose corresponds to about 970 hours of UV exposure from the AM1.5 Global spectrum. As a comparison, the annual UV dose for a moderate climate (Potsdam, Germany) reaches 34 kWh/m^2 [13], while for a desert climate (Atacama, Chile), it reaches to 160 kWh/m^2 [14]. Therefore, we consider our test sufficient to find out early failure but too short to assess long-term reliability.

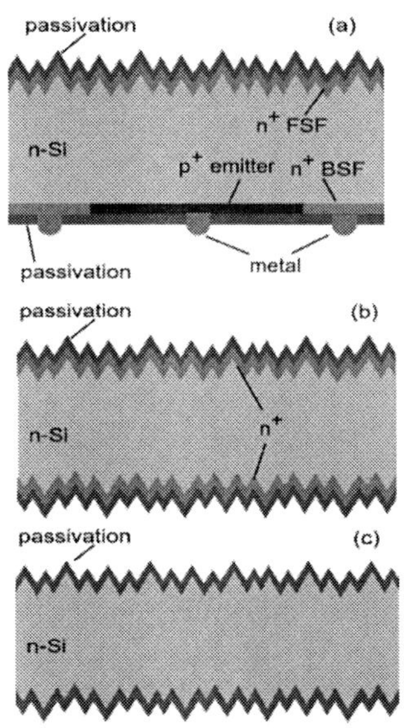

Figure 1: Simplified sample structure of the IBC cell (a), the lifetime sample to monitor J_0 of FSF (b), and the intrinsic *n*-type lifetime sample (c)

3 SIMULATION OF FSF SHEET RESISTANCE

First, we performed simulation to study the change in cell efficiency along with the change in sheet resistance of FSF by using the simulation tool Quokka 3 [15]. In the simulation we fix all the cell parameters except for the sheet resistance of the front FSF and the bulk resistivity. The cell pitch is set as 1.1 mm; the sheet resistance of the rear emitter and BSF is set as 150 Ohm/sq and 90 Ohm/sq, respectively. The Shockley-Read-Hall bulk lifetime of 2 ms is set for all bulk materials, whereas the J_0 for the FSF is fixed at 30 fA/cm^2. The results as presented in Figure 3 thus show the pure influence that is caused by the reduced lateral conductivity at the front surface when increasing the FSF sheet resistance. For the substrate with a bulk resistivity of 16 Ω·cm, by varying the FSF sheet resistance from 50 Ohm/sq to intrinsic, FF and Eta drop 0.16% and 0.06% absolute, respectively. However, in reality by reducing the doping for the front surface, the J_0 of FSF probably reduce (not the case as simulated in Fig. 3) and thus it could compensate the slight FF drop and improve the cell efficiency. Our simulation shows that the decreased front lateral conductivity when using high ohmic FSF will have little negative impact on the cell performance.

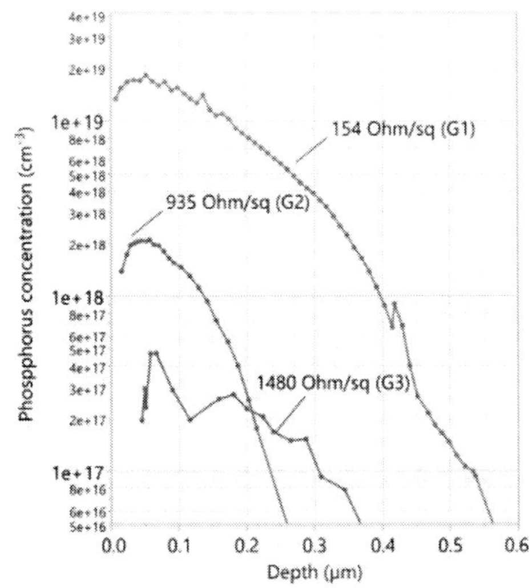

Figure 2: ECV measurement of three FSF doping profiles used for the IBC cells and J_0 samples

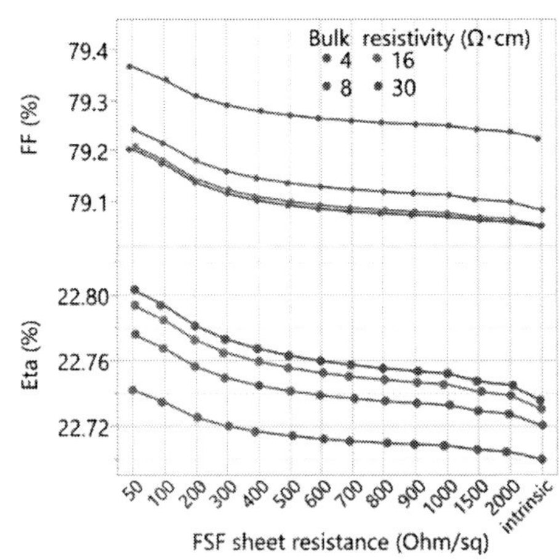

Figure 3: Simulated performance of IBC cells with respect to various FSF doping. A fixed J_0 for the front surface was used in the simulation

4 RESULTS OF UV EXPOSURE TEST

The recombination current prefactor J_0 of the lifetime samples as shown in Fig. 1 (b) and (c) was extracted from the quasi-steady-state photoconductance (QSSPC) measurements [16] at an injection level of 5×10^{15} cm^{-3}. The only exception is the undiffused group, where the injection level after the UV exposure test was set at 3×10^{15} cm^{-3}, and this was done because the fitting at 5×10^{15} cm^{-3} is not good anymore. Note that before the QSSPC measurement, fast firing with a set peak temperature at 820 °C has been performed for all lifetime samples to activate the hydrogen passivation of the SiO_2/ SiN_x layer stack. Furthermore, the J_0 of the front surface after UV exposure test is calculated by using the following formula (same as in Reference [17]):

$$J_0\text{ (front surface)} = J_0\text{ (both surfaces after test)} - 1/2 \times J_0\text{ (both surface before test)}$$

Fig. 4 shows that the intrinsic *n*-type surface has the lowest initial J_0, whereas the second lowest J_0 is obtained from G2 (935 Ohm/sq), while G1 (154 Ohm/sq) has the highest initial J_0. However, after UV exposure G1 (154 Ohm/sq) has the least increase in J_0; the median J_0 from G1 after the UV exposure test increases slightly from 33 fA/cm^2 to 38 fA/cm^2. By contrast, G2 (935 Ohm/sq) shows a moderate increase from 19 fA/cm^2 to 37/fA/cm^2. By contrast, J_0 of the sample with an undiffused surface increases significantly from 12 fA/cm^2 to 287 fA/cm^2. Using the online tool EDNA2 [18] we have extracted the surface recombination velocity (SRV) by fitting J_0 of the front surface, and the obtained initial SRV are 2210 cm/s, 285 cm/s, and 169 cm/s for G1 to G3, respectively.

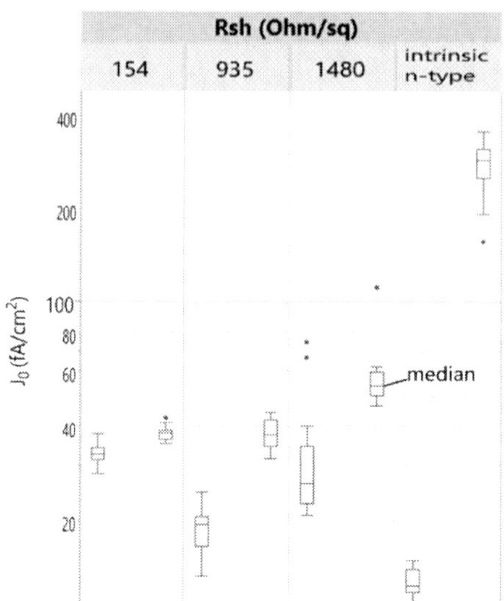

Figure 4: J_0 measurement of the lifetime sample's front surface before and after the UV exposure test. QSSPC measurement was made on nine spots of each sample. Group size: three samples each group, except for G3 (1480 Ohm/sq) where only two samples were used.

IV data measured under standard test condition are shown in Fig. 5, from which one can observe that although G2 (935 Ohm/sq) and G3 (1480 Ohm/sq) have higher initial V_{oc} than G1 (154 Ohm/sq), all *IV* parameters including FF, V_{oc}, J_{sc} and Eta degrade faster than G1. The trend of the *IV* data is in line with the change of J_0 (see Fig. 4). It is plausible that the efficiency of an IBC cell with a lower front surface J_0 or SRV is more sensitive to the increase of surface recombination [19].

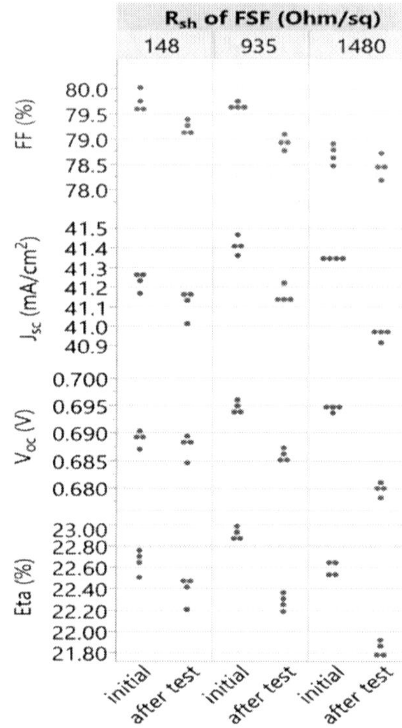

Figure 5: *IV* data before and after the UV exposure test. IBC cells with three different FSF doping are compared. Group size: four samples each group.

Fig. 6 presents the PL images taken before and after the UV exposure test for one solar cell with an initial efficiency of 22.9% from G2 (935 Ohm/sq). One can see that the degradation takes place over the entire wafer, instead of only in several local spots. The same observation is made for the samples from other groups. The IQE (internal quantum efficiency) measurement of the same cell as shown in Fig. 6 is presented in Fig. 7. Decrease of IQE can be found from 280 nm to 1000 nm, which is likely caused by the increase front J_0. The change of reflection is found to be negligible before and after UV exposure (not shown) and therefore the EQE follows the trend of IQE (not shown).

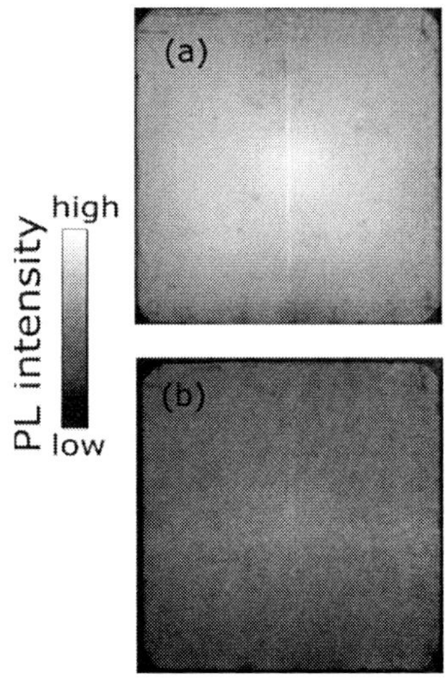

Figure 6: PL images of a 22.9% cell from G2 taken before (a) and after (b) the UV exposure test. Degradation takes place over the entire cell

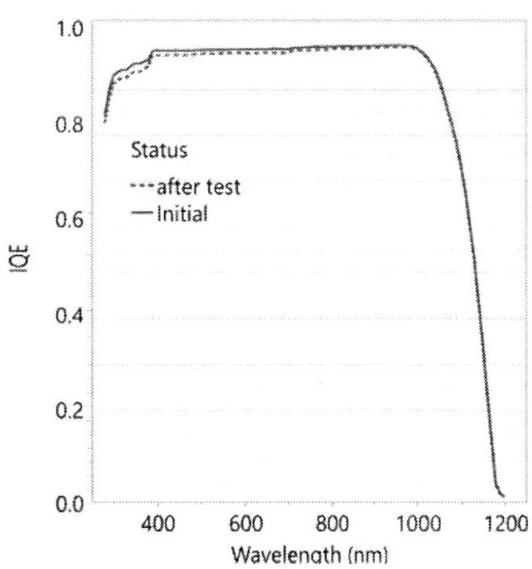

Figure 7: Comparison of IQE measurements before and after the UV exposure test.

5 REVIEW OF LITERATURE

We summarize the results from several papers about UV degradation in Table I (*n*-type surface) and Table II (*p*-type surface). In these two tables, the results of UV exposure test in each of the cited papers are categorized into three groups: "A" stands for zero degradation or even gain; "B" stands for < 100% (relative) increase of J_0, or < 7.5 % (relative) decrease of cell efficiency; "C" stands for cases which are worse than class "B".

It is challenging to compare results from different studies due to the difference in cell process, passivation, lamp spectrum, and test duration. Nevertheless based on Table I, it seems that for *n*-type surfaces, SiO_2 or SiO_2/ SiN_x stack are more stable under UV exposure as compared with SiNx. If the SiO_2/ SiN_x stack is used to passivate the front surface of an IBC cell, significant UVID (class "C") may start to take place for an FSF with a J_0 of about 25 fA/cm^2, or when the surface concentration is in the range of 10^{18} cm^{-3} to 10^{19} cm^{-3}. Cells with lighter front side doping tend to be more susceptible to UVID.

Fewer studies exist for the *p*-type surface. However, it seems that Al_2O_3/ SiN_x provides very stable passivation for the intrinsic *p*-type surface as well as the p^+ emitter. The increase of negative charges in the Al_2O_3 layer after UV exposure is assumed to be the cause [20]. It was also found out that proper thermal treatment after the formation of Al_2O_3 (e.g., firing) is important to have good UV stability [20].

6 SUMMARY AND OUTLOOK

Reducing the phosphorus concentration for the front surface of an IBC cell helps to obtain a lower J_0 and higher V_{oc}, but the cell with a lightly doped FSF (initial J_0 in the range of 10 to 30 fA/cm^2) degrades more as compared with the sample with a heavier doped FSF. Therefore when optimizing the front side doping of an IBC cell, it is worth to check not only the *IV* data but also the UV stability, especially when J_0 of the front surface is below 30 fA/cm^2 or surface concentration is below 1×10^{19} cm^{-3}.

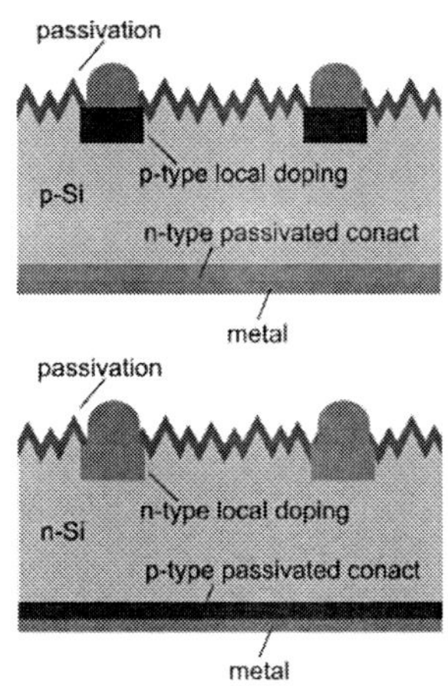

Figure 8: Novel rear-junction cell concepts which have undoped regions between the front fingers.

Our study may also be relevant to other new rear-junction cell concepts as depicted in Fig. 8. Efficiency as high as 26% has been demonstrated for these cell concepts [21], but the non-metallized part of the front side of these cell concepts is intrinsic, and therefore it might be prone to UVID, just like the IBC cell without an FSF.

The UV stability of high efficiency solar cells should therefore be considered when choosing the module encapsulant. Nowadays both UV transparent and UV opaque encapsulants are available [7, 22]. Higher J_{sc} for the module can be obtained by using the UV transparent encapsulant, but it has a transmittance of ~80% at 350 nm, which may cause significant degradation for some solar cell concepts. Therefore, the ultimate solution is to develop a passivation dielectric that provides good passivation as well as UV stability.

Table I: Summary of UVID studies for *n*-type doped surfaces (e.g. phosphorus doped)

surface properties				results of various surface passivation		
Rsh (Ohm/sq)	surface concentration (cm^{-3})	structure & surface morphology	initial J_0 (fA/cm^2)	SiO_2	SiN_x	SiO_2/SiN_x
90	N. A.	lifetime sample, planar	70		C [20]	
125	2.0×10^{20}	p-PERC, pyramids textured	67		C [17]	
150	2.0×10^{19}		36			B [17]
160	N. A.	n-IBC (concentrated sunlight), planar	N. A.	B [23]		
intrinsic *n*-type Si	N. A.		N. A.	C [5]		
150	3.0×10^{18}	n-IBC & lifetime sample, pyramids textured	34			B [19]
350	4.0×10^{19}		29			B [19]
intrinsic *n*-type Si	1 Ω·cm		48			C [19]
440	4.5×10^{17}	n-IBC, pyramids textured	10			C [3]
154	1.3×10^{19}	n-IBC & lifetime sample, pyramids textured	33			B [this work]
935	1.4×10^{18}		19			B [this work]
1480	1.9×10^{17}		26			C [this work]
intrinsic *n*-type Si	3 Ω·cm	lifetime sample, pyramids textured	12			C [this work]

Table II: Summary of UVID studies for *p*-type doped surfaces (e.g. boron doped)

surface properties					results of various surface passivation	
Rsh (Ohm/sq)	surface concentration (cm^{-3})	structure & surface morphology	initial J_0 (fA/cm^2)	initial SRV (cm/s)	SiN_x	Al_2O_3/ SiN_x
90	N. A.	lifetime sample, planar	16			A [20, 24]
intrinsic *p*-type Si, Gallium doped, 0.8 Ω·cm				30 (SiN_x) / 39 (Al_2O_3)	C [20, 24]	A [20, 24]
intrinsic *p*-type Si		p-PERT, pyramids textured	N.A.			B [25]
650	2.0×10^{18}	n-IBC, pyramids textured	7			B [3]

ACKNOWLEDGEMENTS
This work was supported by the German Federal Ministry for Economic Affairs and Energy in party under Contract 03EE1005A (Zquadrat), and in part under Contract 03EE1018C (Ralph).

REFERENCES
[1] A. Cuevas and Di Yan, "Misconceptions and Misnomers in Solar Cells," *IEEE J. Photovoltaics*, vol. 3, no. 2, pp. 916–923, 2013.
[2] D. C. Jordan *et al.*, "Silicon Heterojunction System Field Performance," *IEEE J. Photovoltaics*, vol. 8, no. 1, pp. 177–182, 2018.

[3] R. Müller *et al.*, "Ion Implantation for All-alumina IBC Solar Cells with Floating Emitter," *Energy Procedia*, vol. 55, pp. 265–271, 2014.

[4] A. Sinha *et al.*, "UV-Induced Degradation of High-Efficiency Solar Cells with Different Architectures," in *2020 47th IEEE Photovoltaic Specialists Conference (PVSC)*: IEEE, 6142020, pp. 1990–1991.

[5] P. E. Gruenbaum *et al.*, "Photoinjected hot - electron damage in silicon point - contact solar cells," *Journal of Applied Physics*, vol. 66, no. 12, pp. 6110–6114, 1989.

[6] L. E. Black and K. R. McIntosh, "Defect Generation at Charge-Passivated Si-SiO2 Interfaces by Ultraviolet Light," *IEEE Trans. Electron Devices*, vol. 57, no. 8, pp. 1996–2004, 2010.

[7] R. Witteck *et al.*, "UV-induced degradation of PERC solar modules with UV-transparent encapsulation materials," *Prog. Photovolt: Res. Appl.*, vol. 25, no. 6, pp. 409–416, 2017.

[8] A. Kimmerle *et al.*, "Precise parameterization of the recombination velocity at passivated phosphorus doped surfaces," *J. Appl. Phys.*, vol. 119, no. 2, p. 25706, 2016.

[9] F. Granek, "High-efficiency back-contact back-junction silicon solar cells," Ph.D. dissertation, University of Freiburg, Freiburg, 2009.

[10] P. E. Gruenbaum *et al.*, "Light - induced degradation at the silicon/silicon dioxide interface, " *Appl. Phys. Lett.*, vol. 52, no. 17, pp. 1407–1409, 1988.

[11] G. Galbiati *et al.*, "Latest results in screen-printed IBC-ZEBRA solar cells," in *Proc. 7th World Conference on Photovoltaic Energy Conversion (WCPEC-7)*, 2018, pp. 1540–1543.

[12] V. D. Mihailetchi *et al.*, "Surface passivation of boron-diffused junctions by a borosilicate glass and in situ grown silicon dioxide interface layer," *IEEE J. Photovoltaics*, vol. 8, no. 2, 435--440, 2018.

[13] U. Feister and K.-H. Grasnick, " Solar UV radiation measurements at potsdam (52° 22′ N, 13° 5′ E)," *Solar Energy*, vol. 49, no. 6, pp. 541–548, 1992.

[14] A. Marzo *et al.*, "Comparison of Atacama Desert Solar Spectrum vs. ASTM G173-03 Reference Spectra for Solar Energy Applications," in *Proceedings of EuroSun2016*, 2016, pp. 1–13.

[15] A. Fell *et al.*, "The concept of skins for silicon solar cell modeling," *Solar Energy Materials and Solar Cells*, vol. 173, 128--133, 2017.

[16] R. A. Sinton *et al.*, "Quasi-steady-state photoconductance, a new method for solar cell material and device characterization," in *Proc. 25th IEEE Photovoltaic Spec. Conf.*, 1996, 457--460.

[17] R. Witteck *et al.*, "UV-stable surface passivation for crystalline silicon cells in solar modules with UV light transmitting encapsulation materials," in *2019 IEEE 46th Photovoltaic Specialists Conference (PVSC)*, 2019, pp. 2238–2242.

[18] K. R. McIntosh *et al.*, "An examination of three common assumptions used to simulate recombination in heavily doped silicon," in *Proc. 28th Eur. Photovoltaic Sol. Energy Conf.*, 2013, 1672--1679.

[19] F. Granek and C. Reichel, "Back-contact back-junction silicon solar cells under UV illumination," *Solar Energy Materials and Solar Cells*, vol. 94, no. 10, pp. 1734–1740, 2010.

[20] B. Veith-Wolf *et al.*, "Spectra-Dependent Stability of the Passivation Quality of Al2O3/c-Si Interfaces," *IEEE J. Photovoltaics*, vol. 8, no. 1, pp. 96–102, 2018.

[21] A. Richter *et al.*, "Design rules for high-efficiency both-sides-contacted silicon solar cells with balanced charge carrier transport and recombination losses," *Nat Energy*, vol. 6, no. 4, pp. 429–438, 2021.

[22] P. Ferrada *et al.*, "Potential for photogenerated current for silicon based photovoltaic modules in the Atacama Desert," *Solar Energy*, vol. 144, 580--593, 2017.

[23] P. E. Gruenbaum *et al.*, "Stable passivations for high-efficiency silicon solar cells," in *IEEE Conference on Photovoltaic Specialists*, 1990, 317-322 vol.1.

[24] B. Veith-Wolf *et al.*, "Effect of UV illumination on the passivation quality of AlOx/c-Si interfaces," in *2016 IEEE 43rd Photovoltaic Specialists Conference (PVSC)*, 2016, pp. 1173–1178.

[25] B. Min *et al.*, "Approaching 23% with p-Type Back Junction Solar Cells Featuring Screen-Printed Al Front Grid and Passivating Rear Contacts," in *EUPVSEC 2021*.

NOVEL AG-PASTE FOR SIMULTANEOUS CONTACTING n^+ and p^+ EMITTERS THROUGH CONTACT-SUPPORTIVE WELL-PASSIVATING DOPED APCVD LAYERS FOR PERT AND IBC SOLAR CELLS

Fabian Geml[1*], Benjamin Gapp[1], Melanie Mehler[1], Christian Ebert[2], Jonathan Booth[3], Patsy Sutton[3], Simon Johnson[3], Beatriz Cela[4], Heiko Plagwitz[1] and Giso Hahn[1]
[1]University of Konstanz, Department of Physics, 78457 Konstanz, Germany
[2]SCHMID Group, Gebr. SCHMID GmbH, 72250 Freudenstadt, Germany
[3]Johnson Matthey Technology Centre, Sonning Common, Reading, RG4 9NH, UK
[4]Jonson Matthey AGT, Fregatweg 38, 6222 NZ Maastricht, Netherlands
*Tel: +49 7531 88 4995, Fax: +49 7531 88 3895, fabian.geml@uni-konstanz.de

ABSTRACT: The European SolarEraNet project HEAVENLY has the goal to jointly develop simultaneously contacting novel silver pastes and chemical vapor deposited (CVD) layers which are used for both, surface passivation and formation of emitter and back surface field regions. Within the project, pastes have been produced that contact n^+ and p^+ doped regions simultaneously over a wide firing-temperature range, achieving contact resistivities below 0.01 mΩcm^2 on a 100 Ω/sq n^+ emitter. The doped regions are formed by using the high-throughput, cost efficient atmospheric pressure chemical vapor deposition (APCVD) technique which enables the single-sided deposition of doped glass layers, to allow for an efficient co-diffusion step for the formation of both, emitter and back surface field regions. The glass layers remain on the wafer surface after the diffusion process, providing very good surface passivation ($S_{eff} \leq 2.5$ cms^{-1}) and also supporting the contact formation during the subsequent firing process of the Ag pastes. As pastes and passivation layers are developed together, the production of high-efficiency low-cost PERT (Passivated Emitter and Rear Totally diffused) and IBC (interdigitated back contact) solar cells can benefit in three ways from the project's results: by low contact resistivities, high quality surface passivation, and an efficient processing sequence.
Keywords: APCVD, PERT, IBC, Ag

1 INTRODUCTION

Doped Si-based glasses have a variety of applications in photovoltaics, such as dopant sources and/or passivation layers. Nowadays, many solar cell concepts (e.g., passivated emitter, rear totally diffused (PERT) cells) rely on chemical vapor deposited (CVD) layers due to the advantage of the high parameter variability during deposition [1].

Another big advantage of the CVD doping source compared to conventional gas tube diffusion (e.g., $POCl_3$ or BBr_3 [2]) is the possibility of single-side drive-in of dopants into the substrate's surface. Especially for cell architectures with a pn-junction on one side and a high-low junction on the other side, this enables process simplification and more cost effective co-diffusion in which B and P diffuse from different sources simultaneously using single-side deposited doped glasses [3].

Atmospheric pressure CVD (APCVD) tools provide a feasible process with high throughput, low operational costs as well as high uptime and provide the required high layer quality for high-efficiency solar cell production. Based on diborane/phosphine, mono-silane and oxygen, boron silicate glasses (BSG) and phosphorus silicate glasses (PSG) are deposited. These silicate glass layers have shown to provide excellent surface passivation [4].

Until today boron doped emitters are often contacted with Al-containing Ag pastes, which creates the risk of spike formation through the emitter that is detrimental to the fill factor and thus efficiency of a solar cell [5]. However, p^+ regions are also reported to be contacted with pure Ag pastes [6]. A simultaneously n^+ and p^+ contacting Ag paste allows cost-efficient single-print of IBC solar cells or process simplification for bifacial PERT solar cells. Preliminary studies and the literature indicate that doped glass layers support contact formation, resulting in lower contact resistivity and, thus, enabling higher cell efficiencies [7].

This work concentrates on the one hand on the investigation of the newly developed paste and its firing temperature range. On the other hand, the impact of doped CVD layers with various phosphorus content on the contact formation behavior is analyzed in a wide range of firing peak temperatures. For these temperatures, the surface passivation quality of the doped CVD layers is determined.

2 EXPERIMENTAL

During sample preparation, the monocrystalline n-/p-type Czochralski (Cz) Si substrate (5 Ωcm) is etched and cleaned. The samples are either textured using a KOH-based alkaline solution with alcohol-based additives or chemically polished.

BSG and PSG are deposited by an APCVD roller system. Based on these glass layers, the emitters and back/front surface fields are formed within a co-diffusion step for one group. For investigations of the emitter profile and emitter sheet resistance, the samples are etched in diluted HF and measured using a four-point probe setup and the electro-chemical capacitance voltage technique. One group of samples is coated again with APCVD PSG to investigate the impact of the P concentration in the glass layer on the contact formation on the same emitter. For these samples and a second sample group (lifetime samples without a deep highly doped region) a high-temperature annealing step at 850°C for a few minutes is performed to re-form the glass. Thus, the glass layers behave more like as-diffused glass but without an (additional) emitter formation.

The samples are then coated with plasma-enhanced CVD (PECVD) SiN_x:H, after which the contact samples are screen-printed with Ag paste. All samples are (co-) fired in a belt furnace at various peak temperatures. The contact samples are analyzed by applying TLM measurements and the lifetime samples are measured by photoconductance decay (PCD). The general process flow is schematically shown in Fig. 1.

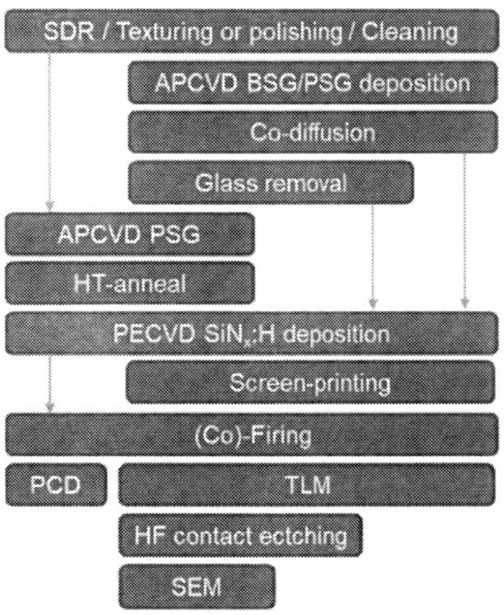

Figure 1: Schematic process flow of samples.

3 RESULTS AND DISCUSSION

3.1 Paste characteristics and contact behavior

Johnson Matthey has developed a novel metallization solution in which the conductive Ag paste does not contain glass particles. This patented technology consists of a Pb-free mixture of crystalline metal compounds, including tellurium oxide, milled and used as substitute flux for the typically used glass powders.

The use of a simple mixture of finely milled crystalline metal compounds, rather than a glass frit, has several advantages including:

- No glass frit forming step for paste manufacture so lower energy demand, more environmentally friendly, lower production costs/complexity
- Potential for more flexibility in component mixtures, improving performance and speed/flexibility for developing new recipes tailored for new solar cell designs

The contact formation mechanism of a paste containing crystalline metal oxide mixture is different to that observed using paste containing a glass made with the same composition of the crystalline metal oxides mixture.

In the systems containing the crystalline mixture it is observed that the reactions between TeO_2 and SiN_y:H starts at lower temperatures than in systems using glasses with similar composition. At higher firing temperatures conductive pastes containing glasses seem to tend to form larger Te-rich particles and additional crystalline phases in comparison to when using the crystalline system.

The pastes used in this study were substantially glass-free and comprised 1-3wt% of a crystalline metal oxides mixture; 88-90wt% of one or more metallic Ag powders and 9wt% of an organic vehicle.

3.2 Contact resistivity and firing range

The contact resistivities of this screen-printed paste on n^+ and p^+ emitters diffused from PSG/BSG is shown in Fig. 2 on 5 Ωcm Cz-Si. The contacted emitters have sheet resistances of $R_{sheet} \approx 85$ Ω/sq (p^+) and $R_{sheet} \approx 105$ Ω/sq (n^+) and are either passivated by single SiN_x:H or by SiN_x:H stacked on the as-diffused BSG/PSG. The SiN_x:H thickness is chosen to fulfill the anti-reflection condition.

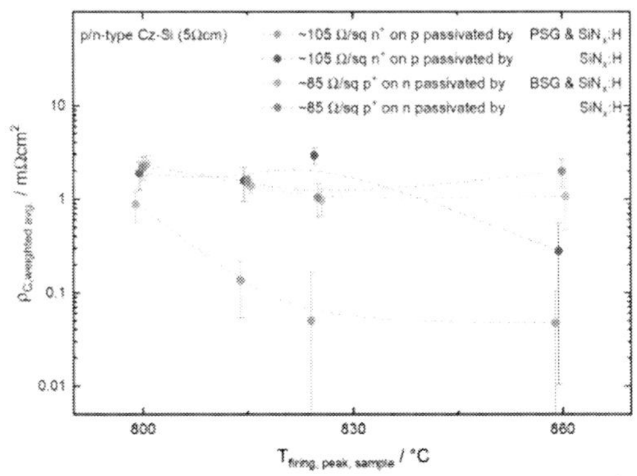

Figure 2: Contact resistivities of different passivated n^+ and p^+ emitters contacted by the novel developed Ag paste depending on the sample peak firing temperature.

For both emitter types and passivation layers, respectively, contact resistivity values below 2 mΩcm^2 can be reached. For the contacts supported by the PSG layer, even values below 0.1 mWcm^2 are achieved. For P containing glass, this proves that the layer supports the contact formation. The applied paste shows these very low contact resistivities in a wider firing range, measured up to 860°C sample peak firing temperature, than for other industrially used silver pastes.

3.3 Impact of glass dopant on contact formation

In order to investigate the influence of the CVD layers (especially their phosphorus content) on contact formation, PSG layers with different amounts of phosphine in the total gas flux (4.3%, 7.2%, 10.1%) are deposited on the already diffused emitter. The glass is re-formed in a high-temperature annealing step and coated by SiN_x:H to form an anti-reflection layer stack. Additionally, a reference group is produced by depositing a single SiN_x:H layer onto the bare emitter surface, without an intermediate PSG film. The samples are screen-printed with an industrially used Ag paste.

Fig. 3 shows the specific contact resistance of the contacts formed by these layers on textured 5 Ωcm Cz-Si as a function of the sample peak firing temperature.

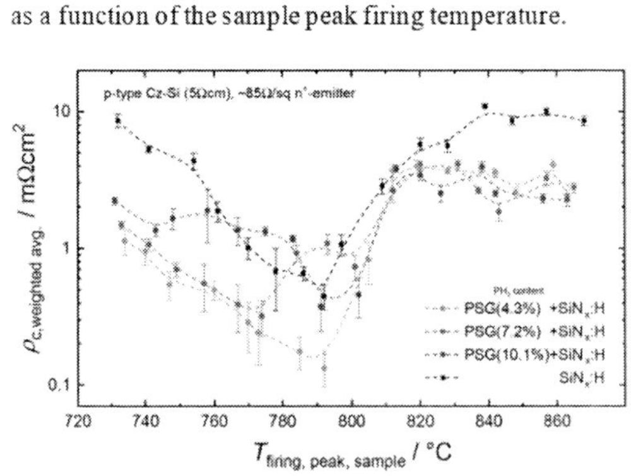

Figure 3: Contact resistivities of differently passivated sample groups depending on the sample peak firing temperatures.

Overall, samples with PSG/SiN_x:H stacks show lower contact resistivities. Even though the minima of the groups with the medium and highly doped PSG are close to the reference group, especially the lowly doped PSG enables a significant improvement of the specific contact resistance.

Except for the group with the medium doped PSG, which appears to form a kind of double minimum, all groups including the reference group show a minimum at the same sample peak firing temperature, which can be attributed to the paste. SEM images show a significantly higher number of Ag crystallites on the pyramid flanks for the contacts that had to etch through the doped glass layers during their formation. An example of this is shown in Fig. 4, where the HF-etched contacts, i.e. only the crystallites, can be seen. The images of the contacts at the minimum of the reference group (792°C, (top)) and of the PSG deposited at 10.1% phosphine content (794°C, (B)) were selected because the contact resistivity values are almost the same.

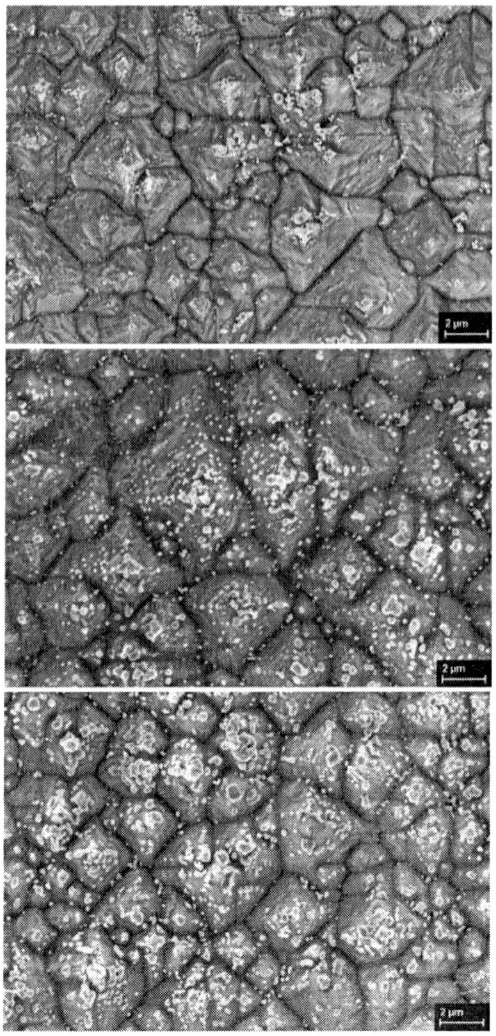

Figure 4: SEM pictures of HF-etched contacts of the samples passivated with SiN_x:H (top), a PSG(10.1%)/SiN_x:H stack (middle) and a PSG(4.3%)/SiN_x:H stack (bottom).

In Fig. 4 (top), it can be observed that the crystallites form preferentially on the top whereas in Fig. 4 (middle), with almost the same contact resistivity, many small crystallites grow on the flanks of the pyramids. This behavior could also be observed for B doped glass layers in [7]. Further SEM images (e.g. in Fig. 4 (bottom)) show that for the samples with lower doped PSG the amount of crystallites remains nearly the same at the same peak firing temperature, but their size increases. The latter may lead to the lower contact resistivity observed in Fig. 3.

3.4 Temperature dependency of passivation

Chemically polished 5 Ωcm n-type Cz-Si is coated by different PSG layers with varying phosphine content. With a subsequent high-temperature annealing step, the glass is re-formed and subsequently covered by SiN_x:H to investigate the (nearly) undiffused samples regarding their passivation quality as a function of the peak firing temperature.

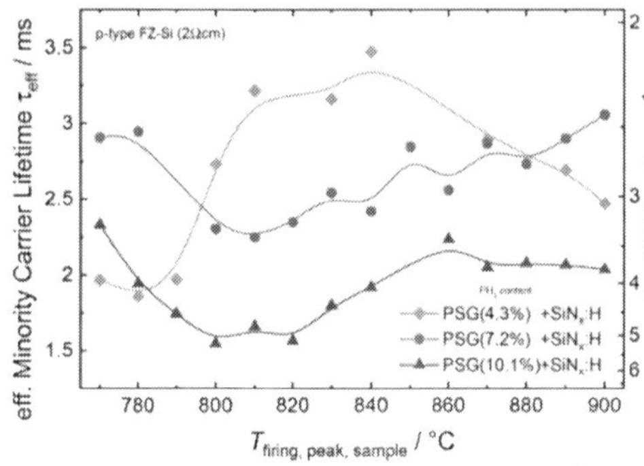

Figure 5: Passivation quality expressed by the effective minority carrier lifetime evaluated at an injection density of 10^{15} cm^{-3} as well as the effective surface recombination velocity in dependence of sample peak firing temperature.

Fig. 5 shows the passivation quality, i.e. the effective minority carrier lifetime τ_{eff} (evaluated at an injection of 10^{15} cm^{-3}) as well as therefrom calculated effective surface recombination velocity S_{eff}, depending on the sample peak firing temperature. The curves indicate a drop in passivation quality for all sample groups with increasing temperature until 800-820°C, and a subsequent increase of passivation quality until 840-860°C. At higher temperatures, samples with highly doped PSG show a saturation, while samples with 7.2% even show a further increase of τ_{eff}. The lowly doped PSG leads to a distinctive optimum between 810°C and 840°C peak firing temperature, achieving surface recombination velocities of $S_{eff} \leq 2.5$ cms^{-1}.

The observed increase is within the firing range of an industrially used silver paste. Together with the wide firing range of the newly developed Ag paste (see Fig. 2), cell efficiency can be optimized by adjusting the peak firing temperature for the highest passivation quality and lowest specific contact resistance.

As a next step, the above findings will be implemented into an industrial solar cell process.

4 CONCLUSIONS

In conclusion, the use of doped glass layers can be beneficial for metal contact formation. It could be shown that an as-diffused PSG layer can reduce the contact resistivity by supporting the Ag crystallite formation on the pyramid flanks. Furthermore, the PSG films enable very low surface recombination velocities below $S_{eff} \leq 2.5\ cms^{-1}$. The firing temperature shows an impact on this passivation quality, especially in the typical industrially applied firing range. With a novel paste established in the HEAVENLY project, the firing temperature range is widened enabling the optimization for such CVD layers. Additionally, this paste contacts both, n^+ and p^+, regions which makes it a cost-efficient alternative for industrial usage in PERT and IBC solar cell development.

5 REFERENCES

[1] D. Chen, Y.Chen, Z. Wang, J. Gong, C. Liu, Y. Zou, Y. He, Y. Wang, L. Yuan, W. Lin, R. Xia, X. Zhang, G. Xu, Y. Yang, H. Shen, Z. Feng, P. P. Altermatt, P. J. Verlinden, Sol. Energy Mater. Sol. Cells 206 (2020) 110258.

[2] A. Wang, J. Zhao, M. A. Green, Appl. Phys. Lett. 57 (1990) 602-604.

[3] H. P. Yin, K. Tang, J. B. Zhang, W. Shan, X. M. Huang, X. D.Shen, Sol. Energy Mater. Sol. Cells 208 (2020) 110345.

[4] N. Wehmeier, B. Lim, A. Nowack, J. Schmidt, T. Dullweber, R. Brendel, Phys. Status Solidi RRL 10(2) (2015) 148.

[5] A. Frey, S. Fritz, J. Engelhardt, G. Hahn, B. Terheiden, Energy Procedia 92 (2016) 919.

[6] S. Fritz, J. Engelhardt, S. Ebert, G. Hahn, Phys. Status Solidi RRL 10(4) (2016) 305.

[7] J. Engelhardt, A. Frey, S. Gloger, G. Hahn, B. Terheiden, Appl. Phys. Lett. 107 (2015) 042102.

NOVEL APPROACH FOR SELF-ALIGNED LOCAL POLYSILICON LAYER FORMATION

Yalun Cai, Ran Chen, Brett J. Hallam and Fiacre. E. Rougieux

School of Photovoltaic and Renewable Energy Engineering, University of New South Wales, Sydney, NSW 2052, AUSTRALIA

Correspondence: Yalun Cai yalun.cai@unsw.edu.au Ph: +61 414 034 720

ABSTRACT: In this work, we investigate the use of localised laser recrystallisation on full area amorphous silicon layers to enable an approach for local polysilicon contact formation. We show that laser recrystallisation paired with selective etching of amorphous silicon enables the formation of localised polysilicon layers with high aspect ratios. We demonstrate that a 32 wt% potassium hydroxide solution is suitable to achieve selective etching on hydrogenated amorphous silicon with a comparatively higher etch rate of about 100 nm per minute, while avoiding significant etching of the polysilicon regions, with a vertical etch rate of up to 50 nm per minute on a <100> c-Si surface. We confirm that a 2 nm interfacial oxide can survive laser processing via the existence of its characteristics FTIR peak at 1079 cm^{-1}. We propose a cell process using this method to fabricate selective tunnel-oxide polysilicon contacts.

Keywords: a-Si/μ-Si, Laser Processing, Etching

1 INTRODUCTION

In 2020, monocrystalline silicon (mono-Si) wafers accounted for than 75% of the total solar cell market share , surpassing that of multicrystalline silicon (mc-Si) [1]. Passivated emitter and rear cell (PERC) structure enabled the growing dominance of mono-Si [2,3]. The PERC structure is replacing the conventional full-area aluminium back surface field (Al-BSF) with a localized Aluminium-alloyed back surface field by laser ablation and, as a result, PERC structures have achieved efficiency of 22% [4], 22.8% [2] and 23.83% [5]. Laser doped selective emitter (LDSE) was one important step to unlock the efficiency potential of PERC in an industry setting, by forming highly doped emitters locally, directly underneath the metal and reducing resistive loss within the front emitters of a solar cell [5-6]. Both PERC and LDSE technologies function as a solid step towards the efficiency limit of 29.4% for a single-junction silicon solar cell [8,9].

To further improve the efficiency of a single-junction silicon solar cell, passivating contacts are developed on both front and rear sides of a single-junction solar cell. Polysilicon on oxide contact is a successful passivating contacts that has been reported to lower the power loss from recombination [10] and maintaining the same low-level resistive loss. For example, a conversion efficiency of 25.7% is achieved with a full-area polysilicon contact on the rear side and LDSE on the front side of the sample cell in [11]. However, it would be unwise to replace the front LDSE with a full-area polysilicon contact because the resulting parasitic absorption on the front side would be problematic. To reach the efficiency goal of 27%, predicted achievable for a front-and-rear-contact solar cell [12], a localised polysilicon contact is needed as an upgrade for front LDSE.

Polysilicon contact has two components: (1) a highly doped polysilicon layer that is in direct contact with the metal and front-surface passivation and (2) an ultra-thin oxide between the polysilicon layer and the bulk of a silicon solar cell. Together, they provide carrier tunnelling and passivation. By replacing LDSE with polysilicon contact directly underneath the metal at the front of a silicon solar cell, the loss from recombination can be significantly reduced while a low contact resistivity can be maintained. There have been several reported attempts to achieve polysilicon contacts at the front of a silicon solar cell. The first method involves a shadow mask [13] during deposition of Potential-enhanced Chemical Vapor Deposition (PECVD) amorphous silicon (a-Si). The method can lead to localised a-Si fingers but whether the usage of a shadow mask would be successful in industrial settings is a uncertainty. The second method uses metal tabs as protection islands for selected regions in the underlying full-area polysilicon contact during Reactive-ion Etching (RIE) [14]. The second method can yield self-aligned local polysilicon contact under metal, but the etching duration and processing cost would be two concerns for industry. The third method employs a UV laser to locally oxidize both the front and rear surfaces of a pre-fabricated full-area Tunnel oxide passivating contact (TOPCon) sample to form a protective oxide layer against KOH etching; the region without UV laser exposure is etched off and the remaining region becomes a local polysilicon contact [15].

Given the importance of local polysilicon contact formation as a potential passivated contact structure for the front surface of the next-generation of PERC solar cells, this study aims at the fabrication of self-aligned polysilicon layer are ready for $POCl_3$ diffusion.

There are two key steps in the proposed proof-of-concept method to achieve local polysilicon contact formation. The first step is the laser recrystallization of a-Si:H in local regions; the second step is to selectively etch off a-Si:H with respect to laser processed intrinsic-polysilicon region using KOH. It has been reported that the etch rates of c-Si, and SiO_2 in KOH are three and one thousand higher than that of a-Si:H [16,17], respectively. We use Raman spectroscopy to confirm local recrystallization of the a-Si:H layer, FTIR to measure the integrity of the interfacial oxide after laser processing and Dektak profiling to measure the aspect ratio and selectivity of the etched a-Si:H.

2 EXPERIMENTAL METHODS

The samples were double-side-polished phosphorus-doped n-type Czochralski (Cz) c-Si wafers with a thickness of 275 μm. The samples were RCA cleaned with and HF dip followed by the growth of a 2 nm oxide via dry thermal oxidation and the deposition of 300 nm a-Si:H on both sides using and Oxford PECVD. Recrystallisation was then achieved via a Newport continuous wave (CW) laser with a wavelength of 532 nm, a power of 2W, a scan speed of 1 m·s and a gaussian beam width of about 25 μm at full-width half maximum (FWHM). The pitch of laser lines was set to 5 μm when processing large areas.

Recrystallization of a-Si:H by laser was characterized by Raman spectroscopy. The power and exposure time of Raman laser was kept low to minimize the recrystallization of a-Si:H by Raman laser. The InVia2 Raman tool has a 532 nm laser, implying a measurement depth of around 300 nm (95% of the laser is absorbed in a 300 nm layer of a-Si:H).

Selective etching of a-Si:H against poly-Si by KOH solution was confirmed by Dektak profiler. After laser processing, the sample was immersed in a solution of 32 wt% KOH and 13 wt% isopropanol (IPA) to selectively etch off the a-Si:H without removing too much of the laser-recrystalized poly-Si. The etch lasted around 90 seconds for an a-Si:H layer of ≈ 300 nm. We then used Dektak profiling to quantify the height differences between regions with and without laser processing and etching.

The integrity of the interfacial oxide pre- and post-laser processing was investigated via FTIR. The FTIR signal of the interfacial SiO_x was obtained by subtracting the absorbance spectrums of the oxide amorphous deposited sample (a-Si/SiO_x/c-Si/SiO_x/a-Si) with the amorphous deposited sample (a-Si/c-Si/a-Si). The FTIR signal of the a-Si:H layer was obtained by subtracting the absorbance spectrums of amorphous deposited sample (a-Si/c-Si/a-Si) sample by bulk silicon sample (c-Si) sample. The native oxide of all samples was removed by HF dip immediately before FTIR measurements. The stretching mode of Si-O-Si bond in the bulk of the sample results in a peak at around 1079 cm^{-1} [18–20], which is also the indicator of interfacial oxide existence.

The steps of the experiment in this study are summarized in Fig. 1. The characterization step comprises of Raman spectroscopy, Dektak profiling and FTIR spectroscopy.

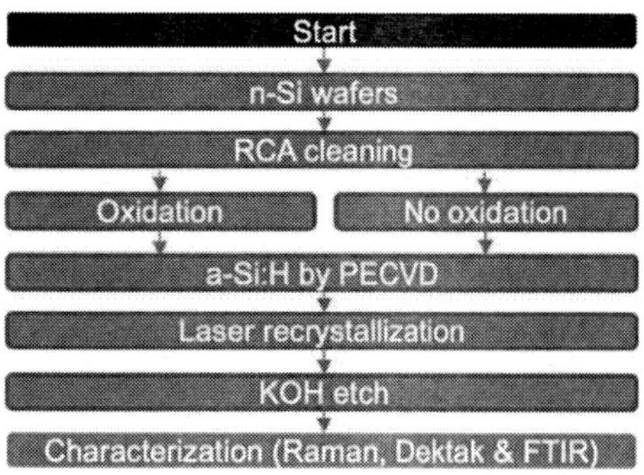

Figure 1: Process flow of experimental steps in this study. The a-Si:H deposition, laser recrystallization and the KOH etch were all symmetrically took place on both sides of the samples.

3 RESULT AND DISCUSSION

3.1 Targeted sample structure and Quokka3 simulations

The characterization step confirmed the sample structure in Fig. 2. The laser-processed intrinsic poly-Si fingers are on top of the ultra-thin interfacial oxide by etching. The samples are then ready for diffusion that turns the poly-Si (i) to either n-type or p-type highly doped poly-Si contact.

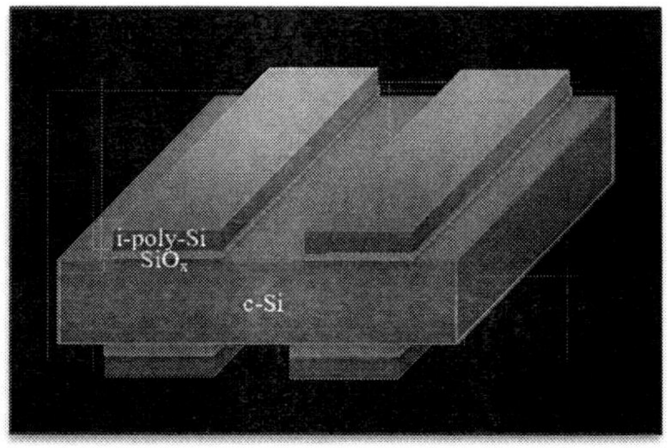

Figure 2: Targeted sample structure on a c-Si wafer featuring local polysilicon contact.

The efficiency of solar cells with front local polysilicon front contacts is simulated in Quokka3. We vary the recombination current density (J_0) to obtain the efficiency, as demonstrated in Fig. 3. By reducing the J_0 of the front contacts from 500 to 35 fA·cm^{-2} and keeping the contact resistivity (ρ_c) unchanged the overall cell efficiency can increase by roughly 0.15%. The results were obtained from with p-type substrate, p-PERC on the rear and n-LDSE on the front.

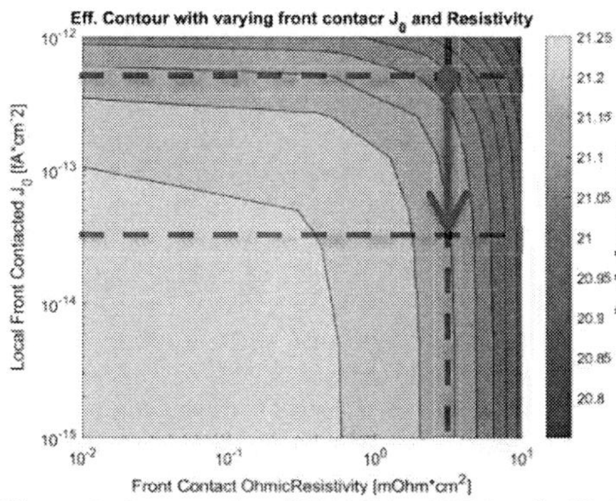

Figure 3: Contour plot of 2018 Trina PERC cell efficiency over a various combinations of recombination current density (J_0) and contact resistivity (ρ_c).

3.2 Laser recrystallization of a-Si:H and characterization by Raman

In this section, we demonstrate a laser process that leads to recrystallization of the amorphous silicon. Raman spectroscopy has been widely used to differentiate crystalline silicon from amorphous silicon. The spectrum of laser recrystalized silicon and the corresponding fitting is presented in Fig. 4. The peak positions for a-Si, nc-Si, and c-Si are around 475 cm^{-1}, 510 cm^{-1}, and 520 cm^{-1}, respectively. The fraction of a-Si, nc-Si and c-Si were calculated by calculating the integrated peak intensities of the three characteristic Raman peaks of a-Si, nc-Si, and c-Si, using Eq. 1.

$$X_a = \frac{\gamma * A_a}{\gamma * A_a + A_{nc} + A_c} \qquad \text{Equation 1,}$$

where γ is a pre-factor whose value was chosen to be 0.8, from a commonly-accepted range of 0.8 to 0.9 [21]. X_a is the fraction of amorphous silicon, and A_i (i=a-Si, nc-Si, or c-Si) are the respective integrated peak intensities.

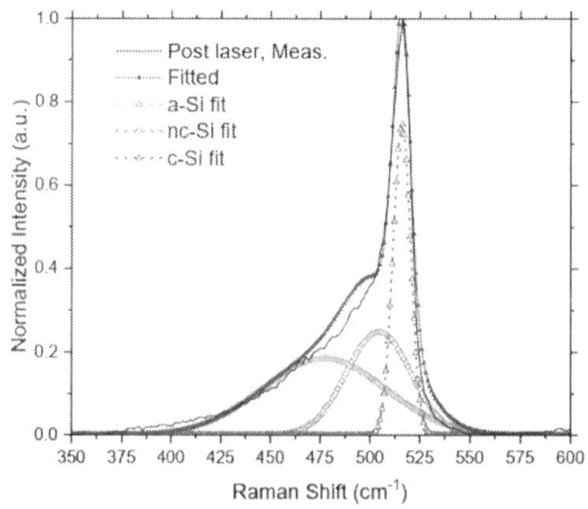

Figure 4. The experimental spectrum of a sample after laser recrystallization. The spectrum was deconvoluted into three characteristic gaussian peaks. The a-Si:H (i) layer was turned into a layer that contains a-Si (pink dashed line), nanocrystalline silicon (green dashed line), and c-Si (purple dashed line).

An optical image of the laser scan pattern is shown in Fig. 5 (a). The fit shown in Fig. 4 is performed at every position. This allows to obtain the profiles of a-Si:H crystallised fraction via Raman mapping as shown in Fig. 5 (b)-(d). The maximum recrystallization on a 300 nm a-Si:H on c-Si sample is about 60%, at the centre of the laser line. Full recrystallization is not needed, as the purpose of the laser recrystallization step is to make designated regions on the sample to become more resistant to the subsequent etch by KOH. After recrystallisation, the a-Si:H inside the selected regions becomes laser-processed poly-Si. Fig. 1 (b)-(d) shows the profile of the fraction of crystalline, nanocrystalline and amorphous silicon along the laser scan line, measured via Raman spectroscopy (before and after laser process). The reduction in the ratio of a-Si:H and the increase in nc-Si and c-Si can be observed between the two profiles in Fig. 5 (b) and (c). Therefore, successful partial recrystallisation via laser is demonstrated. Furthermore, Fig. 1 (d) shows that a-Si:H in laser-free regions disappears completely after etching. This validates the success of the selective etching process.

(a)

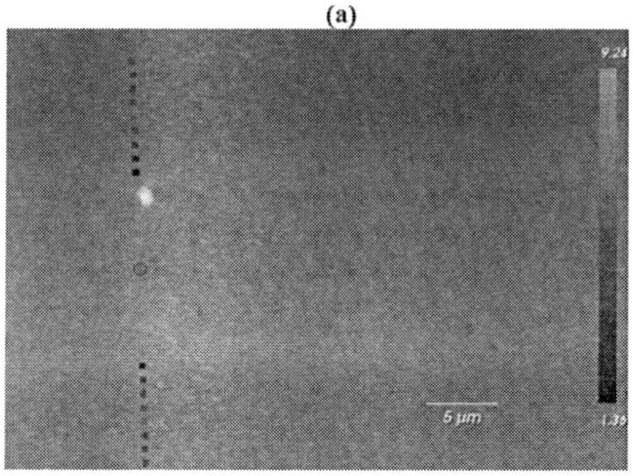

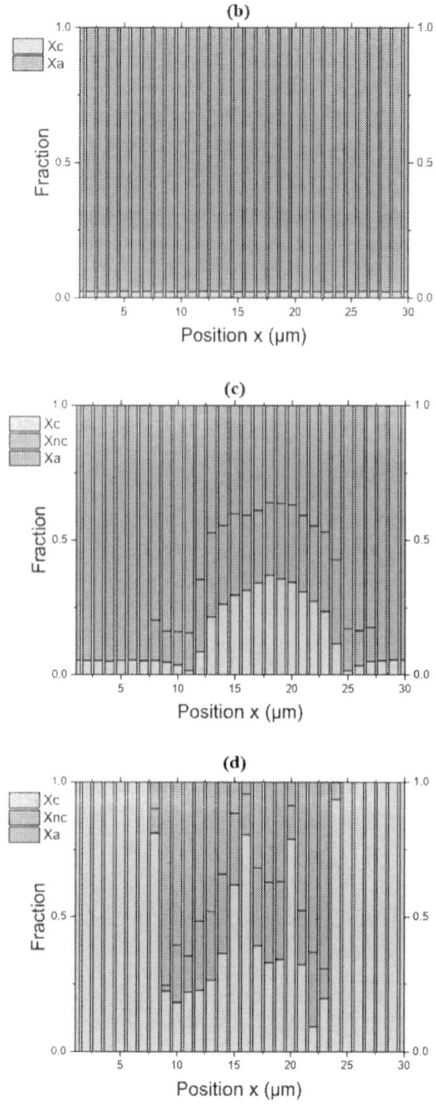

Figure 5: (a) Optical-microscopic view of the recrystallization of a-Si:H deposited on a sample with a single-line laser scan. The square-shaped dots were the line of mapping with a colour bar showing the c-Si peak area, indicating the ratio of c-Si and large-grain poly-Si combined, at each spot of measurement. The green dot was the spot of Raman laser on focus. (b) is the Raman profile of a-Si:H layer before laser exposure showing no sign of laser recrystallization. (c) is the profile of laser recrystallization across the single laser line, showing different levels of partial recrystallization. (d) is the profile across the laser line after receiving 90 seconds of KOH etch, showing existence of a-Si:H in the laser line.

3.3 Selective Etching by KOH and characterization by Dektak

The next step is to immerse the sample in the KOH solution to remove the a-Si:H in regions that did not receive laser recrystallization process, while the mixture of a-Si:H, nc-Si and c-Si are retained in the laser processed region. As reported, the KOH solution has a preference to etch a-Si, especially a-Si:H much faster than c-Si and SiO_2.

the result of the Dektak profiling session on the KOH etched sample is shown in Fig. 6 (a) and (b) and Table 1. The selectivity of KOH solution is confirmed by the result of the Dektak profiling session, shown in Fig. 6 (b).

(a)

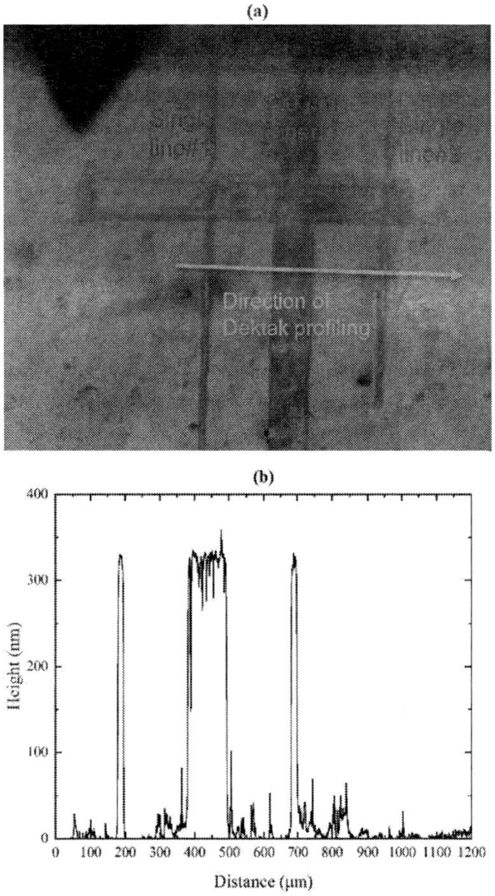

Figure 6: (a) Optical image of Dektak profiling and local features on the sample, including the direction of profiling; and (b) the resulting height profile from the Dektak profiling session, clearly showing the height of single line#1, #2 and the finger.

As shown in Table 1, the two single lines and the 100-μm-wide finger that were the result of recrystallization induced by the doping laser all had a height of 300 nm above surroundings. The selective etching of a-Si by KOH solution can be confirmed by the height difference between the local features and their surroundings. Considering the change in thickness of a-Si:H during the KOH etching process, the etch rate was estimated to be 100 nm·min^{-1}, at least twice the vertical etch rate that KOH would have on c-Si surfaces with the similar etching conditions [22].

Table I: Heights and widths of the two single-line laser fingers and the 100-μm finger after laser recrystallization and KOH etching.

	Height	Approx. Width
Single line #1	360 nm	25 μm
100 um wide finger	329 nm	100 μm
Single line #2	339 nm	25 μm

3.4 Oxide existence during laser recrystallization by FTIR

The existence of the interfacial oxide is essential for the successful fabrication of the desired sample structure in Fig. 1. We use FTIR spectroscopy to investigate whether the oxide remains post-laser recrystallisation process. All samples received a HF dip to remove the surface oxide immediately before FTIR measurements. The FTIR signal of the thermally grown interfacial silicon oxide after laser processing is shown in Fig. 7, by subtracting the absorbance signal of samples with and without dry thermal oxidation. The absorbance spectrums were converted from transmission spectrum. The peak around 1079 cm^{-1} suggests that the interfacial oxide remains after laser recrystallization. The microstructure of the interfacial oxide requires further investigation. The purpose of performing FTIR spectroscopy has been achieved—to confirm that the interfacial oxide exists after laser processing.

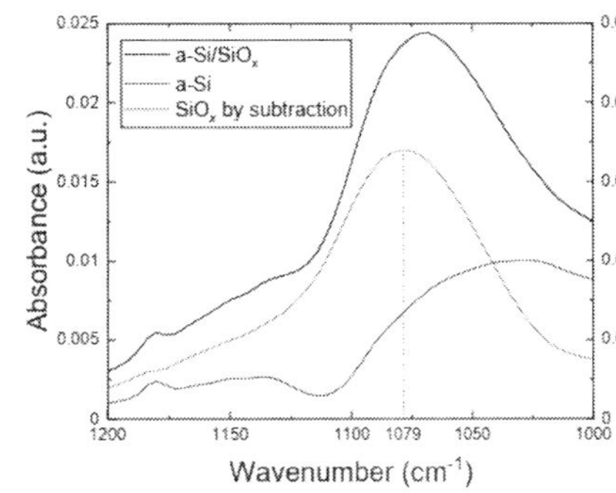

Figure 7: The FTIR spectrum of a-Si/SiO$_x$ was obtained by subtraction of c-Si sample from a laser-processed a-Si/SiO$_x$/c-Si/SiO$_x$/a-Si sample. In a similar fashion, the FTIR spectrum of a-Si was obtained by subtraction of the same c-Si sample from a laser-processed a-Si/c-Si/a-Si sample. By subtracting the spectrum of a-Si/SiO$_x$ with a-Si spectrum, the spectrum of interfacial SiO$_x$ was obtained. The characteristic peak from the stretching vibrational mode of Si-O-Si at 1079 cm^{-1} was marked by a dashed line.

4 CONCLUSIONS

We have demonstrated partial recrystallization of a-Si:H by laser, through Raman spectroscopy. The FTIR measurements suggest that the interfacial oxide still exist after laser processing. Via Dektak profiling, we show successful selective etching of a-Si:H against laser-recrystalized poly-Si. Together our new process and results could pave the way for the advent of industrial, local poly-Si contacts. Future work will explore the doping of such structures.

5 REFERENCE

[1] VDMA, Update ITRPV 2020 Results 2019 including maturity report 2020, (2020).

[2] A.W. Blakers, A. Wang, A.M. Milne, J. Zhao, M.A. Green, 22.8% Efficient Silicon Solar Cell, Appl. Phys. Lett. 55 (1989) 1363–1365.

[3] M.A. Green, The Passivated Emitter and Rear Cell (PERC): From conception to mass production,

Sol. Energy Mater. Sol. Cells. 143 (2015) 190–197.
[4] F. Fertig, R. Lantzsch, A. Mohr, M. Schaper, M. Bartzsch, D. Wissen, F. Kersten, A. Mette, S. Peters, A. Eidner, J. Cieslak, K. Duncker, M. Junghänel, E. Jarzembowski, M. Kauert, B. Faulwetter-Quandt, D. Meißner, B. Reiche, S. Geißler, S. Hörnlein, C. Klenke, L. Niebergall, A. Schönmann, A. Weihrauch, F. Stenzel, A. Hofmann, T. Rudolph, A. Schwabedissen, M. Gundermann, M. Fischer, J.W. Müller, D.J.W. Jeong, Mass production of p-type Cz silicon solar cells approaching average stable conversion efficiencies of 22 %, Energy Procedia. 124 (2017) 338–345.
[5] R. Chen, H. Tong, H. Zhu, C. Ding, H. Li, D. Chen, B. Hallam, C.M. Chong, S. Wenham, A. Ciesla, 23.83% efficient mono-PERC incorporating advanced hydrogenation, Prog. Photovoltaics Res. Appl. 28 (2020) 1239–1247.
[6] B. Hallam, S. Wenham, A. Sugianto, L. Mai, C. Chong, M. Edwards, D. Jordan, P. Fath, Record large-area p-type CZ production cell efficiency of 19.3% based on LDSE technology, IEEE J. Photovoltaics. 1 (2011) 43–48.
[7] B. Hallam, A. Urueña, R. Russell, M. Aleman, M. Abbott, C. Dang, S. Wenham, L. Tous, J. Poortmans, Efficiency enhancement of i-PERC solar cells by implementation of a laser doped selective emitter, Sol. Energy Mater. Sol. Cells. 134 (2015) 89–98.
[8] A. Richter, J. Benick, F. Feldmann, A. Fell, B. Steinhauser, J.-I. Polzin, N. Tucher, J.N. Murthy, M. Hermle, S.W. Glunz, Both Sides Contacted Silicon Solar Cells: Options for Approaching 26% Efficiency, 36th Eur. Photovolt. Sol. Energy Conf. Exhib. (2019) 9–13.
[9] W. Shockley, H.J. Queisser, Detailed balance limit of efficiency of p-n junction solar cells, J. Appl. Phys. 32 (1961) 510–519.
[10] F. Feldmann, J. Schön, J. Niess, W. Lerch, M. Hermle, Studying dopant diffusion from Poly-Si passivating contacts, Sol. Energy Mater. Sol. Cells. 200 (2019) 109978.
[11] A. Richter, J. Benick, R. Müller, F. Feldmann, C. Reichel, M. Hermle, S.W. Glunz, Tunnel oxide passivating electron contacts as full-area rear emitter of high-efficiency p-type silicon solar cells, Prog. Photovoltaics Res. Appl. 26 (2018) 579–586.
[12] A. Richter, R. Müller, J. Benick, F. Feldmann, B. Steinhauser, C. Reichel, A. Fell, M. Bivour, M. Hermle, S.W. Glunz, Design rules for high-efficiency both-sides-contacted silicon solar cells with balanced charge carrier transport and recombination losses, Nat. Energy. 6 (2021) 429–438.
[13] T. Dullweber, M. Stöhr, C. Kruse, F. Haase, M. Rudolph, B. Beier, P. Jäger, V. Mertens, R. Peibst, R. Brendel, Evolutionary PERC+ solar cell efficiency projection towards 24% evaluating shadow-mask-deposited poly-Si fingers below the Ag front contact as next improvement step, Sol. Energy Mater. Sol. Cells. 212 (2020) 1–8.
[14] D.L. Young, K. Chen, S. Theingi, V. LaSalvia, D. Diercks, H. Guthrey, W. Nemeth, M. Page, P. Stradins, Reactive ion etched, self-aligned, selective area poly-Si/SiO2 passivated contacts, Sol. Energy Mater. Sol. Cells. 217 (2020) 110621.
[15] S. Sören, M. Anja, M. Verena, N. Tobias, K. Adrian, M. Larysa, B. Rolf, P. Robby, Laser-induced modification of doped poly-Si surface for Si solar cells with structured passivated contacts, EU PVSEC Proc. (2020).
[16] I. Haller, Y.H. Lee, J.J. Nocera, M.A. Jaso, Selective Wet and Dry Etching of Hydrogenated Amorphous Silicon and Related Materials, J. Electrochem. Soc. 135 (1988) 2042–2045.
[17] ETCHING PROPERTIES OF HYDROGENATED AMORPHOUS SILICON Y.S. Tsuo, Y. Xu, D.W. Baker, and S.K Deb Solar Energy Research Institute, Golden, Colorado 80401-3393, 219 (1991) 805–810.
[18] D.B. Mawhinney, J.A. Glass, J.T. Yates, FTIR study of the oxidation of porous silicon, J. Phys. Chem. B. 101 (1997) 1202–1206.
[19] I. Prasad, A.N. Chandorkar, Spectroscopy of silicon dioxide films grown under negative corona stress, J. Appl. Phys. 94 (2003) 2308–2310.
[20] B. Shokri, M.A. Firouzjah, S.I. Hosseini, FTIR analysis of silicon dioxide thin film deposited by metal organic-based PECVD, Proc. 19th Int. Plasma Chem. Soc. (2009) 1–4.
[21] R. Tsu, J. Gonzalez-Hernandez, S.S. Chao, S.C. Lee, K. Tanaka, Critical volume fraction of crystallinity for conductivity percolation in phosphorus-doped Si:F:H alloys, Appl. Phys. Lett. 40 (1982) 534–535.
[22] H. Seidel, L. Csepregi, A. Heuberger, H. Baumgärtel, Anisotropic Etching of Crystalline Silicon in Alkaline Solutions: II . Influence of Dopants, J. Electrochem. Soc. 137 (1990) 3626–3632.

STABLE COPPER PLATED METALLIZATION ON SHJ SOLAR CELLS & INVESTIGATION OF SELECTIVE Al/AlO$_x$ LASER PATTERNING

Thibaud Hatt*, Jonas Bartsch, Stefan Schellinger, Jale Schneider, Andreas A. Brand, Sven Kluska and Markus Glatthaar
Fraunhofer Institute for Solar Energy Systems ISE, Heidenhofstr. 2, 79110 Freiburg, Germany

ABSTRACT: Copper plating metallization is growing in importance to replace silver and to enable growth of photovoltaic to terawatt-scale. Besides better performance of the plated Cu contacts on solar cells, the processing needs to be less complex and more cost effective. The „NOBLE“ metallization responds to cost savings for bifacial silicon heterojunction solar cells. This study shows that the processing with inkjet-printing for patterning of the top Al masking layer is yet mature enough for first tests on industrial pilot. Efficiencies comparable to the Ag screen-printed reference on the same precursors grade are reached. To even improve the patterning throughput a selective laser ablation patterning of the Al/AlO$_x$ layer is further investigated. There, laser-induced damaging is avoided by choosing adapted laser settings and first large area SHJ solar cells are manufactured with this method yielding an encouraging 21.4% efficiency.

Keywords: Silicon Heterojunction Solar Cell, Metallization, Copper Electroplating, Laser Processing, Inkjet-printing, Cost reduction

1 INTRODUCTION

Next generation high efficiency silicon solar cells rely on passivating contacts such as the silicon heterojunction (SHJ) technology [1]. Improvement of the technology enabled to reach record efficiencies above 25% on large area cells in the past years [2–5]. To maintain the exponential growth of photovoltaic (PV) towards terawatt (TW) scale, material supply is an important concern [6–8]. Especially silver supply will soon become a bottleneck and is already causing 10% of the costs for a typical solar module. The metal contacts on bifacial SHJ solar cells are manufactured on both-sides by screen-printing low-temperature silver paste. This causes even higher silver consumption and costs for this technology. Unfortunately, the silver saving smart wire technology (SWCT) suitable for interconnection and module integration of SHJ solar cells will be no longer available to all manufacturers [9].

Thus, intrinsic low-temperature copper plating metallization featuring high performance is growing in importance as alternative for SHJ solar cells [10]. In different parts of the globe an industrially viable electroplating route using a sacrificial organic resist-mask (on a PVD metal-seed deposited on the transparent conductive oxide (TCO)) is under development and outstanding efficiencies around 25% were reported [2, 11].

Our resist-free native oxide barrier layer for selective electroplating („NOBLE“) metallization aims at reducing the plating process complexity and processing cost [12], making this route attractive for industrial implementation. Instead of using an organic mask for electroplating, the „NOBLE“ metallization implements a conductive self-passivated Al layer as detailed below. The first proof-of-concept was achieved on lab-scale SHJ solar cells [13] before being transferred on large area recently yielding up to 22.7% similar to the screen-printed reference on the same SHJ precursors [14].

This study shows that the „NOBLE“ metallization is already running stabile on large area medium grade SHJ precursors when using the optimized NaOH$_{aq}$ inkjet-printing patterning of the Al layer [15]. In addition, a selective laser ablation patterning route of the Al/AlO$_x$ layer is investigated to facilitate the processing sequence, improve the patterning resolution and increase the throughput even more.

2 APPROACH

Our „NOBLE“ metallization consists in using a conductive self-passivated Al layer as plating mask while providing homogeneous current distribution on the wafer surface during simultaneous bifacial electroplating. The complete process is depicted in **Figure 1**.

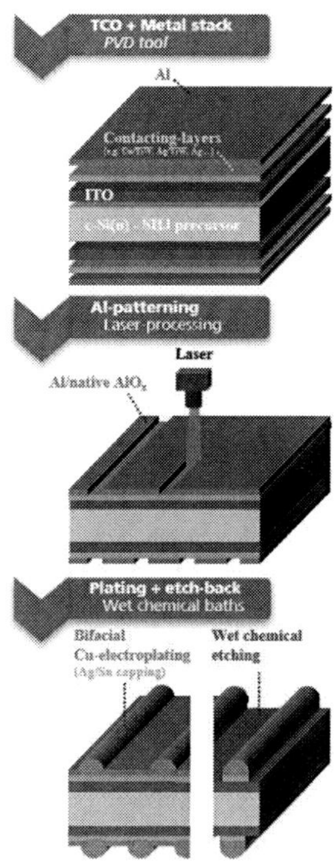

Figure 1: Schematic sequence of the „NOBLE“ metallization with laser patterning

Following the physically vapor deposition (PVD) of the TCO layer, an additional thin PVD metal stack (with Al on top) is deposited on full area. At the grid position,

Al is locally removed from the underlying PVD-Cu or -Ag layer. This patterning is was until now performed by $NaOH_{aq}$ inkjet-printing but a new method using selective laser ablation of the Al/AlO_x layer is being developed as shown in the schematic. The exposed PVD-Cu or -Ag layer act as seed-layer for selective Cu electrodeposition. Finally, the PVD metal stack is etched back in non-grid positions.

The „NOBLE" plated Cu contacts on SHJ solar cells achieved superior electrical and mechanical performance compared to typical Ag screen-printed ones as shown in **Table** 1. It is noteworthy that by reducing the finger width, contact resistivities ρ_c below 1 $m\Omega \cdot cm^2$ are required to avoid impacting the series resistance of the solar cell. This requirement is well met with the „NOBLE" metallization.

First part of this work consists in a demonstration of a first pilot run with a higher amount of SHJ solar cells with the optimized $NaOH_{aq}$ inkjet-printing patterning of the Al. Therefore, industrial SHJ precursor (M2) with ITO were covered on both-sides by different PVD metal stacks (e.g. TiW/Ag, Ag…) with a 50 nm thin Al layer on top. After complete „NOBLE" processing, the I-V curves were measured in a sun-simulator.

Table 1: Comparison of the metal finger characteristics on SHJ solar cells with ITO based on [12, 16]

Process	Width [µm]	ρ_L [$\mu\Omega \cdot cm$]	ρ_c [$m\Omega \cdot cm^2$]	Peel-force [N/mm]
Plated Cu (NOBLE)	25-30	≈ 2	0.1 - 1	≥ 2
Screen-printed Ag	30-40	≥ 5	≥ 2	≤ 0.5

Secondly, a new patterning method is investigated using a femtosecond UV laser (at 343 nm) for selective ablation of the top Al layer on the SHJ precursors. The laser ablation is optimized for the respective layer stack and characterized by photoluminescence imaging (PL) at 1-sun. The thickness of the PVD metal stack (Cu_{500nm}/Al_{100nm} and Ag_{100nm}/Al_{50nm}) is varied to evaluate the laser-induced damage of the passivation on such SHJ solar cells (same batch as above). The complete „NOBLE" process is finally performed to produce the first large area SHJ solar cells using the selective laser patterning of Al.

3 RESULTS

3.1 Stabilized sequence with inkjet-printing patterning

The large area SHJ solar cells were metallized with the „NOBLE" approach using different PVD metal seed-layers below the Al (e.g. TiW/Ag, Ag…). The I-V performances presented in **Figure 2** highlight the reproducibility of the cell efficiency around 22.6%, which is very close to the Ag screen-printed reference. The inkjet-printing enables to keep the passivation quality of the SHJ solar cells at the highest level above 735 mV. Last improvements of the inkjet-printing process are still necessary to benefit fully from the narrow plated contacts and to reach higher fill-factor FF and current density j_{sc} above 38.5 mA/cm² on this precursors type (no additional ARC, just ITO).

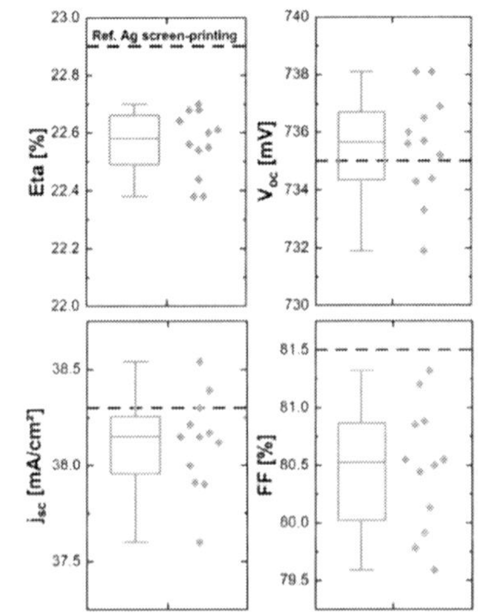

Figure 2: Reproducible I-V characteristics reached on 12 large area SHJ solar cells metallized with the „NOBLE" approach (inkjet patterning) compared to the best screen-printed reference (black dashed line) on the same industrial SHJ precursor material

These promising I-V results confirm that the „NOBLE" metallization with inkjet-printing patterning yields stable results which is required to soon highlight the full potential of plating on high quality materials.

3.2 Patterning by selective laser ablation of Al/AlO_x

First developments of the new selective laser ablation of the thin Al layer show two possible ways to avoid damaging the SHJ cell passivation. One is using a thicker underlying PVD metal layer (contacting layer to plate) as buffer during lasering with high laser pulse energy to locally remove the Al layer with a single laser pulse. The impact of pulse energy on the damage is shown by photoluminescence imaging in **Figure** 3 **(Left)**. Except for the highest pulse energies (4.3 and 4.7 µJ), this buffer layer enables to prevent laser-induced passivation damaging. Furthermore, the Al layer opening seems to be compatible with selective copper electroplating.

The second way is to use low laser pulse energy but multiple laser pulses on the same spot to remove the Al layer bit by bit. As illustrated in **Figure 3 (Right)**, the process window that allows avoiding laser-induced damaging of the passivation is smaller. The Al opening seems less efficient but selective copper plating also worked with this approach for defined parameters and an encouraging contact resistivity ρ_c down to 1.6 ± 0.3 $m\Omega \cdot cm^2$ was measured.

Further investigations will refine the damage free process window and explain the mechanismus of degradation resulting from ablation.

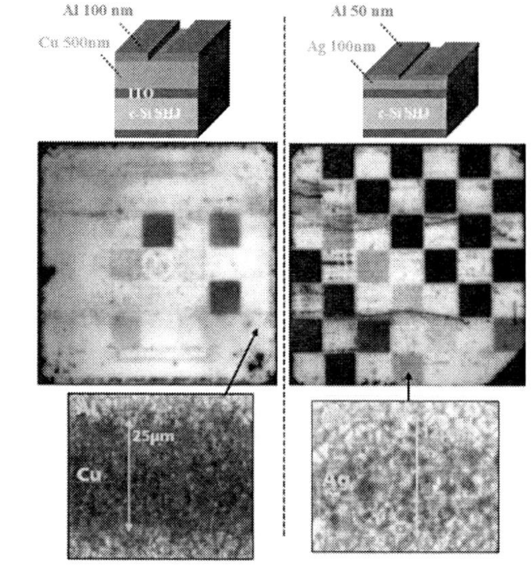

Figure 3: Photo-luminescence (PL) imaging at 1-sun and respective microscopy pictures after femtosecond pulse laser ablation (at 343 nm) of the Al layer depending on the PVD metal stack thickness. (Left) High pulse energies between 2.2 and 4.7 µJ. (Right) Low pulse energies between 0.2 and 0.7 µJ with pulse repetitions per laser dot from 1 to 35 times

3.3 SHJ solar cells manufactured by laser patterning

This early development of the selective laser ablation patterning of the Al/AlO_x masking layer enabled to manufacture first large area SHJ solar cells with the modified „NOBLE“ route. The high laser pulse energy with the adequate thicker "buffer layer" was prioritized. The best laser settings were used to metallize the front-side grid in the Al layer. After full processing of the front-side, the rear-side of the SHJ solar cell was metallized by Ag-paste screen-printing. The I-V characteristics reported in **Table** 2 are quite promising as proof-of-concept and not too far from the reference with inkjet patterning. The limited efficiency is explained by a loss of about 10 mV in V_{oc} probably due to passivation damages with such an early stage laser process. The FF is not as high as expected due to finger interruptions resulting from inhomogeneities in the PVD Al layer thickness and/or the focusing of the laser beam. It can be noticed that the pFF of the SHJ cell also seems to be affected. However, a gain in j_{sc} is achieved thanks to the narrower contacts ≤ 25 µm enabled by the laser patterning.

Table 2: I-V characteristics of the best large area SHJ solar cells after „NOBLE“ metallization with laser ablation or $NaOH_{aq}$ inkjet-printing patterning of the Al top layer

Patterning process	Eta [%]	V_{oc} [mV]	j_{sc} [mA·cm^{-2}]	FF [%]	pFF [%]
Laser	21.4	726.8	38.6	76.3	83.4
Inkjet-p.	22.7	735.7	38.3	80.5	84.4

The low pulse energy and pulse overlaps laser patterning approach also enabled to manufacture first large area SHJ solar cells with the most promising laser settings. Despite of the non-optimized opening of the Al layer, selective plating was successful and narrow Cu contacts ≤ 25 µm wide were produced after complete processing as shown in **Figure** 4. The PL characterization of a quarter of the solar cell after complete sequence revealed an encouragingly low damage of the passivation as observed in **Figure 4**.

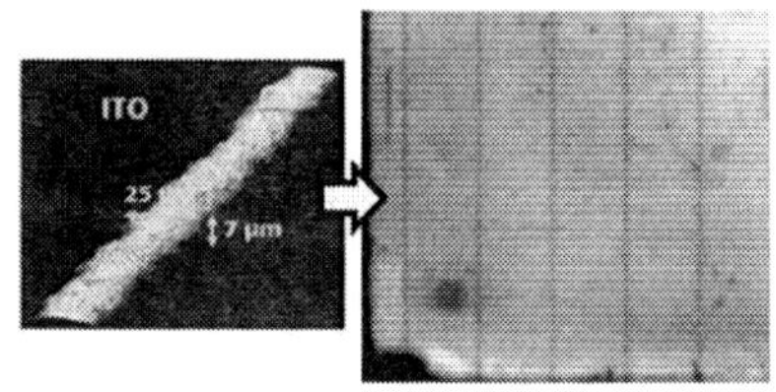

Figure 4: SHJ solar cell after metallization with laser patterning (low pulse energy and pulse repetitions per laser dot). (Left) Microscopic picture of a narrow plated Cu finger. (Right) Enlarged PL picture at 1-sun from the grid-side

4 CONCLUSION

This work shows SHJ solar cells fabricated with an optimized plating process sequence resulting in stabilized I-V performances of 22.6% efficiency comparable to the screen-printed reference. The so-called „NOBLE“ approach with inkjet patterning is considered to be mature enough to be tested on industrial pilot line.

Furthermore, a new selective laser ablation process for the top Al layer is demonstrated on textured SHJ solar cells. An encouraging efficiency of 21.4% in this first experiment confirm the potential of the fast contactless patterning method. Further work should therefore enable a fast improvement of this laser-based method for industrial implementation of our resist-free copper metallization.

ACKNOWLEDGEMENTS

This work was supported by the Fraunhofer society within the research project „LEO“ (contract n°840190) and the German Federal Ministry for Economic Affairs and Energy under the project „PaSoDoble“ (contract n°03EE1031A). The authors acknowledge the colleagues at the Fraunhofer ISE for fruitful discussion and technical assistance. The authors especially thank T. Wenzel for Ag screen-printing, J. Löffler for solar cell *I–V* measurements and I. E. Elias Ilosvay from Fraunhofer EMFT for metal sputtering.

REFERENCES

[1] G. M. Wilson, M. Al-Jassim, W. K. Metzger, S. W. Glunz, P. Verlinden, G. Xiong, L. M. Mansfield, B. J. Stanbery, K. Zhu, Y. Yan, J. J. Berry, A. J. Ptak, F. Dimroth, B. M. Kayes, A. C. Tamboli, R. Peibst, K. Catchpole, M. O. Reese, C. S. Klinga, P. Denholm, M. Morjaria, M. G. Deceglie, J. M. Freeman, M. A. Mikofski, D. C. Jordan, G. TamizhMani, and D. B. Sulas-Kern, "The 2020 photovoltaic technologies roadmap," (en), *J. Phys. D: Appl. Phys.*, vol. 53, no. 49, p. 493001, 2020.

[2] D. Adachi, J. L. Hernandez, and K. Yamamoto, "Impact of carrier recombination on fill factor for large area heterojunction crystalline silicon solar cell with 25.1% efficiency," *Appl. Phys. Lett.*, vol. 107, no. 23, p. 233506, 2015.

[3] W. Favre, M. Sciuto, A. Danel, R. Varache, A. Ragonesi, A. Di Matteo, D. Nicotra, F. Rametta, D.

Iuvara, M. Foti, C. Coletti, M. Vandenbossche, A.-S. Ozanne, F. Jay, F. Pernoud, A. Bettinelli, D. Muñoz, C. Roux, A. Jouini, and C. Gerardi, "25% efficient large area silicon solar cell: paving the way for premium PV manufacturing in Europe," 37th EUPVSEC, Virtual, 2020.

[4] X. Ru, M. Qu, J. Wang, T. Ruan, M. Yang, F. Peng, W. Long, K. Zheng, H. Yan, and X. Xu, "25.11% efficiency silicon heterojunction solar cell with low deposition rate intrinsic amorphous silicon buffer layers," *Solar Energy Materials and Solar Cells*, vol. 215, p. 110643, 2020.

[5] K. Yoshikawa, H. Kawasaki, W. Yoshida, T. Irie, K. Konishi, K. Nakano, T. Uto, D. Adachi, M. Kanematsu, H. Uzu, and K. Yamamoto, "Silicon heterojunction solar cell with interdigitated back contacts for a photoconversion efficiency over 26%," *Nat. Energy*, vol. 2, no. 5, p. 17032, 2017.

[6] N. M. Haegel, H. Atwater, T. Barnes, C. Breyer, A. Burrell, Y.-M. Chiang, S. de Wolf, B. Dimmler, D. Feldman, S. Glunz, J. C. Goldschmidt, D. Hochschild, R. Inzunza, I. Kaizuka, B. Kroposki, S. Kurtz, S. Leu, R. Margolis, K. Matsubara, A. Metz, W. K. Metzger, M. Morjaria, S. Niki, S. Nowak, I. M. Peters, S. Philipps, T. Reindl, A. Richter, D. Rose, K. Sakurai, R. Schlatmann, M. Shikano, W. Sinke, R. Sinton, B. J. Stanbery, M. Topic, W. Tumas, Y. Ueda, J. van de Lagemaat, P. Verlinden, M. Vetter, E. Warren, M. Werner, M. Yamaguchi, and A. W. Bett, "Terawatt-scale photovoltaics: Transform global energy," (eng), *Science (New York, N.Y.)*, vol. 364, no. 6443, pp. 836–838, 2019.

[7] P. J. Verlinden, "Future challenges for photovoltaic manufacturing at the terawatt level," (en), *Journal of Renewable and Sustainable Energy*, vol. 12, no. 5, p. 53505, 2020.

[8] E. Gervais, S. Shammugam, L. Friedrich, and T. Schlegl, "Raw material needs for the large-scale deployment of photovoltaics – Effects of innovation-driven roadmaps on material constraints until 2050," *Renewable and Sustainable Energy Reviews*, vol. 137, p. 110589, 2021.

[9] Int. Wirtschaftsforum Regenerative Energien (IWR®), *Meyer Burger prüft Bau einer Solarfabrik in Deutschland*, 2020.

[10] J. Yu, J. Li, Y. Zhao, A. Lambertz, T. Chen, W. Duan, W. Liu, X. Yang, Y. Huang, and K. Ding, "Copper metallization of electrodes for silicon heterojunction solar cells: Process, reliability and challenges," *Solar Energy Materials and Solar Cells*, vol. 224, p. 110993, 2021.

[11] A. Lachowicz, A. Descoeudres, J. Champliaud, A. Faes, J. Geissbühler, M. Despeisse, S. Nicolay, and C. Ballif, "Copper Plating Processes for Silicon Heterojunction Solar Cells: an Overview," *Proceedings of the 36th EUPVSEC Conference, Marseille, France*, pp. 568–571, 2019.

[12] T. Hatt, J. Bartsch, S. Kluska, S. Nold, S. W. Glunz, and M. Glatthaar, "Low-cost Cu-plated metallization on TCOs for SHJ Solar Cells - Optimization of PVD Contacting-layer," *Proceedings of the IEEE 47th Photovoltaic Specialists Conference (PVSC)*, pp. 397–400, 2020.

[13] T. Hatt, S. Kluska, M. Yamin, J. Bartsch, and M. Glatthaar, "Native Oxide Barrier Layer for Selective Electroplated Metallization of Silicon Heterojunction Solar Cells," *Sol. RRL*, vol. 3, no. 6, p. 1900006, 2019.

[14] T. Hatt, J. Bartsch, S. Kluska, and M. Glatthaar, "Resist-free and Busbar Compatible Plating Route on TCOs Developed on SHJ Solar Cells," 9th Metallization and Interconnection Workshop, 2020.

[15] T. Hatt, J. Bartsch, V. Davis, A. Richter, S. Kluska, S. W. Glunz, M. Glatthaar, and A. Fischer, "Hydrophobic AlOx surfaces by adsorption of a SAM on large areas for application in solar cell metallization patterning," (eng.), *ACS applied materials & interfaces*, vol. 13, no. 4, pp. 5803-5813, 2021.

[16] D. Erath, S. Pingel, R. Khotimah, A. de Rose, D. Eberlein, T. Wenzel, S. Roder, A. Lorenz, and F. Clement, "Fast screen printing and curing process for silicon heterojunction solar cells," *AIP Conference Proceedings*, vol. 2367, no. 1, p. 20006, 2021.

EDGE PASSIVATION OF HETEROJUNCTION SOLAR CELLS FOR RESEARCH PURPOSES

Virginia Boldrini[1], Marica Canino[1], Rita Rizzoli[1], Emanuele Centurioni[1], Filippo Bonafè[1],
Salvatore Lombardo[2], Alessandro Di Mauro[3], Marcello Sciuto[3], Claudio Colletti[3], Caterina Summonte[1]
[1]Institute for Microelectronics and Microsystems (CNR-IMM Bologna), Via Gobetti 101, 40129 Bologna, Italy
[2]Institute for Microelectronics and Microsystems (CNR-IMM Catania), Strada VIII, 5, 95121 Catania, Italy
[3]Enel Green Power, Catania, Italy

Corresponding author: Phone: +39 051-6399127, E-mail: canino@bo.imm.cnr.it

ABSTRACT: Silicon heterojunction (HJ) solar cell is one of the leading technologies for single junction photovoltaic devices and it could be successfully integrated into a silicon based multijunction solar device, in conjunction with a large band gap cell. To be used as a bottom cell, the HJ device needs to adapt to the current technology for top cells, which are typically far less mature and characterized by smaller areas. However, if the HJ device is cut from a finished larger cell, the original edge passivation gets lost, causing the introduction of surface recombination paths and an overall decrease of the device performance. In this context, the development of a proper edge passivation procedure to be applied to high quality HJ solar cells becomes of major importance. We tried two different strategies of edge passivation based either on wet processes, such as the mesa etching of the device, or plasma treatment of the vertical edges. The latter consisting in the deposition of 1-10 nm of hydrogenated amorphous Si on the edges. In order to characterize the recombination losses and quantify the effect of different passivations, we did current-voltage measurements in light and dark conditions and quasi-steady-state photoconductivity measurements.
Keywords: Edge passivation, Heterojunction, dark JV, lifetime, mesa structure.

1 INTRODUCTION

Silicon heterojunction (HJ) solar cell is one of the leading technologies for single junction photovoltaic devices. Moreover, given a set of peculiarities such as the relatively simple fabrication process, the low thermal budget required, and the maturity of the technology at the industrial level, where large areas devices are demonstrated [1], it also represents the best option for the bottom junction in silicon based multijunction devices, in conjunction with a large band gap device, for which several approaches are presently being proposed.

However, to be used as the bottom cell in a multijunction device, the HJ device needs to adapt to the current technology for the top cell device, which is typically far less mature, and characterized by smaller areas. For this reason, smaller areas HJ bottom cells must be used. If the device is cut from a finished larger cell, edge passivation of the original device gets lost [2], which causes the introduction of new recombination paths, and an overall decrease of the device performance [3-6]. The impact of edge recombination consequent to a cut is also relevant in the case of solar modules that englobe shingled solar cells, where each single cell is halved into two and the two halves are connected in series [7]. This confirms that the development of a proper passivation procedure to be applied to finished, high quality HJ solar cells is of importance. The issue is in fact addressed by several laboratories [3-5, 7, 8].

In this contribution, we started from standard industrial HJs and performed edge passivation on small cleaved areas (about 1 cm^2), because this is the case in which the recombination losses across the unpassivated edges significantly affect the performance of the device. Two different procedures were employed, based either on plasma treatment of the base edges, or on mesa etching applied to the transparent conducting oxide (TCO) of the device. For the sake of compatibility with HJ technology, the maximum processing temperature was limited below 200°C. For this reason, approaches such as thermal oxide passivation could not be considered.

In order to make the result of the experiment detectable, all major carrier recombination paths need to be eliminated. In this contribution, the experiment could be done thanks to the extremely high quality of the starting material (more details in Experimental section), which was supplied by Enel Green Power (Catania, Italy) [9].

The obtained cells were characterized by current-voltage (JV) measurements carried out both in dark and under AM1.5G. The quality of the passivation on the borders was evaluated by quasi-steady-state photoconductivity (QSSPC) [10] measurements.

2 EXPERIMENTAL

The starting material consisted of extremely high quality bifacial, p+/n rear junction HJ solar cells, which were supplied by Enel Green Power (Catania, Italy). These cells, based on M2 size wafers (>244 cm^2) and terminated with screen-printed metallization, are characterized by 22.5% average efficiency and a V_{oc} as high as 739 mV [9]. We cleaved them in samples of small areas (about 1 cm^2). Several processes were tested to suppress the leakage in V_{oc} and efficiency that arises from this procedure: a mesa etching, which could make more resistive, long and disadvantageous the carrier path across lateral surfaces, and different recipes of a-Si:H plasma deposition, to passivate the just exposed (111) surface edge.

2.1 Mesa etching

As a first strategy to suppress edge recombination, we considered the formation of a double-sided partial mesa structure, whose layout is outlined in Figure 1 with active region of 0.84 cm^2 area. The idea behind this structure was that even if the HJ is left untouched, the lateral surface leakage is reduced if a certain area of the transparent conductive oxide (TCO) and metal fingers all around the borders both on the front and back side are removed. , The cleaved sample was cleaned with hot acetone and 2-propanol. A photoresist mask (PR), consisting in a double layer of AZ®10XT for a total thickness of about 20 μm, has been spinned on both sun-facing and rear side to protect the device region during wet etching. The baking stages were optimized for a double side processing. A good PR definition could be obtained by exposing the PR only on one side, while leaving the other side entirely protected during etching. The TCO etching was performed in a wet solution of $HCl:HNO_3:H_2O$ (4:1:5) for 40 min, while the metal fingers were removed with the action of ultrasonic bath for 5 more minutes. At this point, the PR mask was stripped off and the mesa structure is formed on one face. The same procedure was repeated on the opposite face of the sample, in order to create the mesa structure also on the other side.

Figure 1: Sketch of the double-sided partial mesa structure, obtained by removing TCO and metal fingers near the edges.

2.2 a-Si:H plasma deposition

The second strategy adopted to suppress edge recombination on small cleaved cells was the deposition of an a-Si:H passivating layer directly on the lateral surfaces. For this experiment we started from a 10×10 cm^2 area of the starting HJ cell, which was cleaned and covered with two layers of AZ® 10XT photoresist, both on front and back surfaces in order to protect them during the following a-Si:H depositions. Then, the sample was cleaved into four cells, of area about 1 cm^2 (see measured areas in Table I). In this way the cell surface was protected on both sides while the edges were uncovered. The samples were then inserted into the PECVD reactor chamber (details in Fig. 2) Among all possible deposition conditions, four different PECVD processes were tested: one was done at very high frequency (VHF), 100 MHz and 22.5 W power, while the others were done at 13.56 MHz and 4 W power (the details of all processes are reported in Table I). Moreover, 5×5 cm^2 textured c-Si substrates coming from the same batch as the ones used to fabricate the solar cells used in this work and intended for QSSPC characterization were inserted in the PECVD together with the devices, just after the removal of the native oxide with a dip in HF 0.5% for 5 min. Identical PECVD process was then repeated on the other side of the textured substrates to have symmetrical a-Si:H layers on both faces. QSSPC measurements on such wide-area substrates are dominated by recombination at the front and rear surfaces, and are indicative of the passivation properties of the deposited a-Si:H material on the (111) faces of the pyramids on both sides.
The thickness deposited on the vertical edges has not been evaluated, however, it is estimated to be of the order of 1-10 nm. All the other experimental details are accurately depicted and described in Figure 2.

Table I: Details of PECVD recipes for a-Si:H deposition. The deposition temperature is indicated in the process name. The electrode area and distances are 144 cm^2 and 20 mm respectively.

Process name	Time [s]	Press. [mbar]	SiH_4 [sccm]	H_2 [sccm]	Area [cm^2]
VHF170°C-dil	30	0.40	20	20	1.00
RF200°C	90	0.60	20	-	1.36
RF200°C-dil	90	0.75	20	20	1.25
RF170°C	90	0.60	20	-	1.14

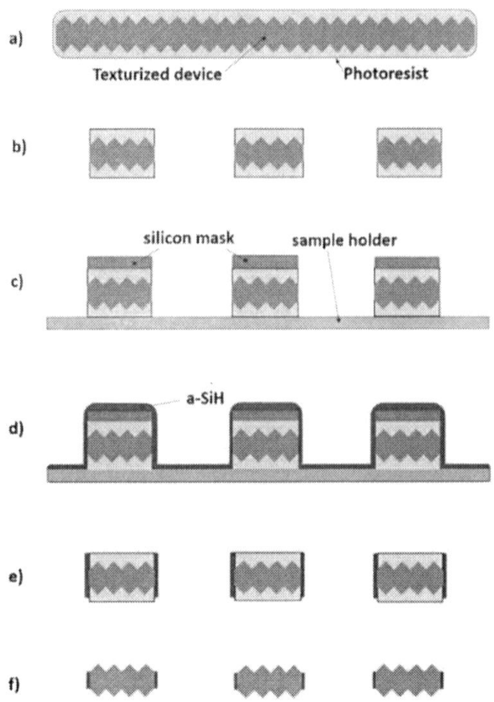

Figure 2: Simplified sketch describing the process used to deposit a-Si:H on the edges of the devices. a): Deposition of 20 μm photoresist on both sides. The thickness is dictated by the need of covering metal fingers, omitted in the figure. b): Cleavage of the circa 1×1 cm^2 devices. This step leaves exposed the non passivated vertical edges. c): Stick the devices on the sample holder, and cover the devices with 1x1 cm^2 silicon masks. The masks are needed to avoid a-Si:H deposition on the photoresist, which would hinder its removal. Sticking of the devices on the sample holder, and of the masks on the devices, is required in our PECVD system given its sample face-down configuration, omitted in the figure. The sticking tape is also omitted in the figure. d): Deposition of a-Si:H, which then covers also the vertical edges of the devices. The thickness of the vertical deposition has not been measured. e): Mechanical removal of the silicon masks from the top of the devices, and the devices themselves from the sample holder, assisted by the immersion in a wet solution (microstrip or acetone). f): Removal of photoresist. The final samples have a-Si:H passivated vertical edges.

2.3 Characterization techniques

The effect of the two passivation processes that were applied, mesa formation and a-Si:H deposition, was characterized by the use of different techniques: QSSPC and the measurement of JV characteristics under dark and light conditions. QSSPC was performed on textured c-Si substrates of the same batch as the ones used for cell fabrication, with symmetrical a-Si:H deposition on both sides, and is intended to determine the passivation properties of the deposited a-Si:H material.

3 RESULTS AND DISCUSSION

Figures 3 and 4 show the minority carrier lifetime measured at 3×10^{15} cm^{-3} carrier concentration, and the implied V_{oc} at 1 sun (iV_{oc}), obtained by QSSPC.

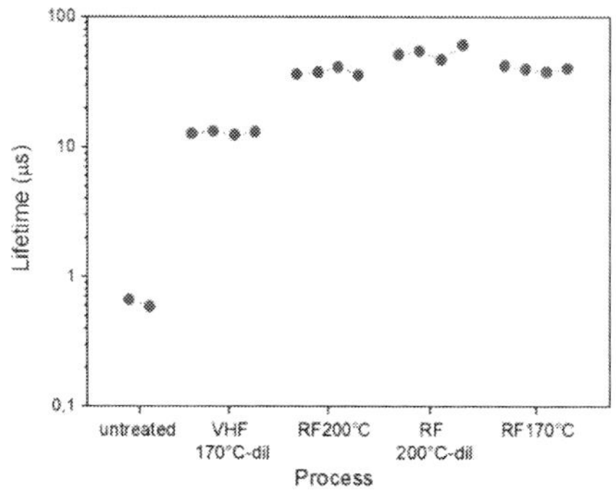

Figure 3: Minority-carrier lifetime for the untreated substrate and for different surface passivation treatments.

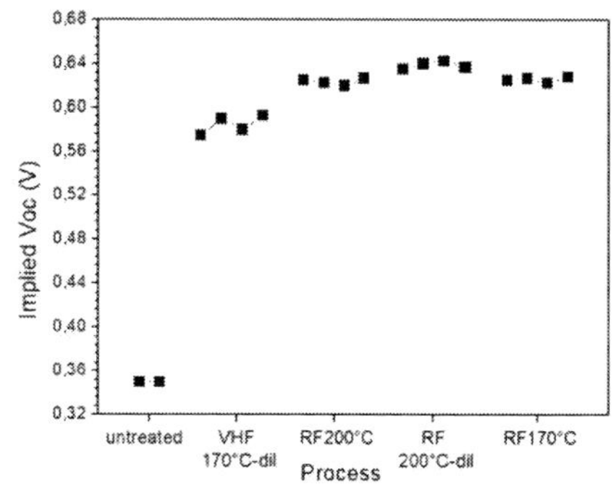

Figure 4: The implied V_{oc} at 1 sun, obtained from QSSPC measurements, is reported for the untreated substrate and for different passivation treatments.

The untreated (as-cleaved) sample exhibited very high recombination and was practically unmeasurable. Among the remaining processes, the best result (highest value for both quantities) is obtained by using the 'RF200°C-dil' process.

The effect of the mesa etching process is shown in Figure 5, which reports the V_{oc} measured on the as-cleaved and processed devices, of area 1.06 and 0.84 cm^2 respectively. It is seen that the mesa etching process is effective in improving the V_{oc}, indicating a reduction in the recombination path after the treatment. The figure also shows that the process on the n-side (front surface) is already effective, in spite of the rear-junction character of the devices.

Figure 6 shows dark JV characteristics for all the samples. From a qualitative observation of the graph, different considerations can be done. Reverse current densities, indicative of the different amount of leakage, are different from sample to sample. All treated cells have lower J_{REV} than the as-cleaved one. Specifically, the 'RF200°C-dil' cell and the mesa one are those with the lowest J_{REV}.

In Figure 7 the local ideality factor as function of voltage is plotted for all measured cells. In three curves, those relating to the 'as-cleaved', 'VHF170°C-dil' and 'RF200°C' cells, a high hump at about 0.45 V is visible. This feature is ascribed to a deviation from the equivalent 2-diode circuit model caused by resistance-limited enhanced recombination [11]. Such recombination is plausible to occur at the edges, associated to unsaturated bonds that are present after cleavage.

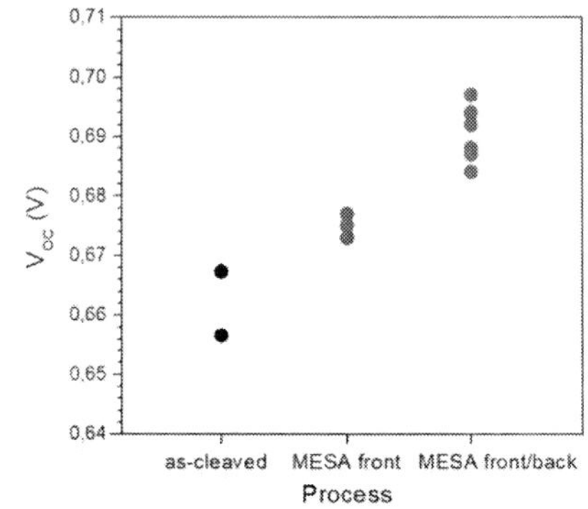

Figure 5: Open circuit voltage measured on as-cleaved and on mesa-etched 1x1 cm^2 devices.

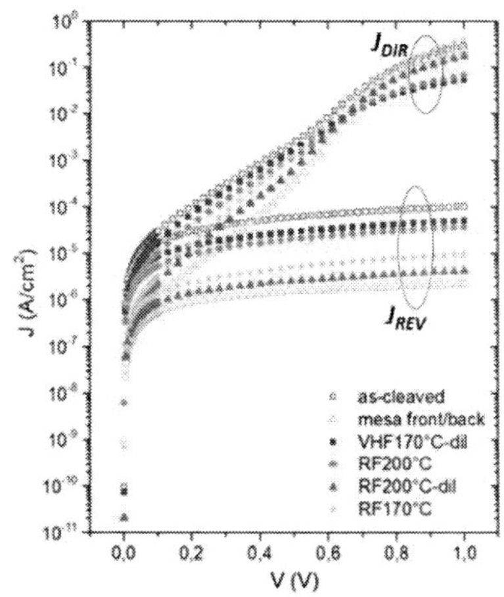

Figure 6: Comparison between the dark JV characteristics of all the cells under study.

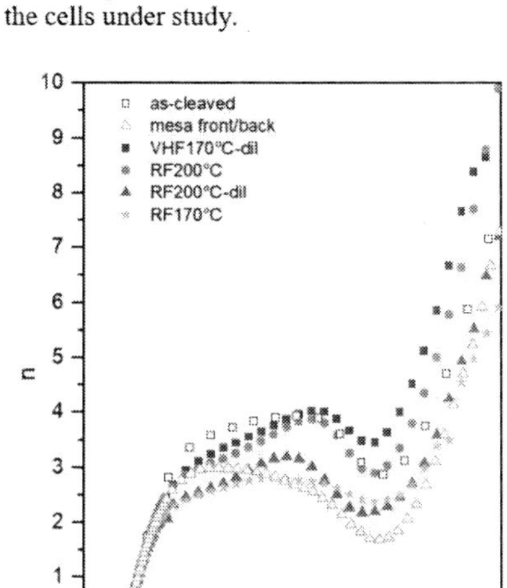

Figure 7: Experimental local ideality factors.

The reduction of the hump in the other curves ('RF200°C-dil', 'RF170°C' and 'mesa fr/back') is indicative of better edge passivation. The results on plasma-passivated samples correlate with QSSPC measurements. In fact, lower J_{REV} and hump signal are associated to higher lifetime and iV_{oc} data measured on textured substrates. This is indicative of the validity of the applied technique, also considering the well known potentiality of a-Si:H passivation properties under proper fabrication conditions [12 - 15].

As a summary of this section, we have shown that the examined methods are effective in reducing the cleavage-induced recombination in the 1×1 cm^2 devices. In fact, the passivation properties of a-Si:H are well known, and, for research purposes, TCO mesas are already used to isolate HJ devices fabricated on the same substrate.

However, both methods show some drawbacks. Mesa etching implies a reduction of the useful aperture of the final device, whereas plasma passivation implies the application of thermal treatments to an already optimized device. Both methods require a certain amount of manipulation, here included the use of photolithography, which limits the application to laboratory purposes. In spite of this, no fundamental limitations have been observed, and both methods can be further improved and adapted to specific needs.

4 SUMMARY AND CONCLUSIONS

We propose two different methods to circumvent the problem of edge-induced reduced efficiency in small (1×1 cm^2) solar cells cut from a high quality large device. In one case, the current path along the edge is made highly resistive by means of a mesa-etching applied to front and rear TCO. In the second method, a passivating a-Si:H layer is deposited on the edges of the devices. Using the TCO mesa etching results in an improvement of V_{oc} by about 4 mV with respect to the as-cleaved device and in the elimination of the edge-related hump visible in the local ideality factor. Concerning a-Si:H depositions, RF 170°C conditions result in the elimination of the edge-related hump in the curve n(V) as well. Although both methods are open to refinements, the results can be applied to the fabrication of small area, non-monolithic 4 terminal tandem devices based on high quality HJ.

ACKNOWLEDGMENTS
This work has been partially supported by the PON Ricerca e Innovazione 2014-2020 project BEST4U, funded under Decreto Direttoriale di concessione dell'agevolazione del 21-05-2019 prot. n. 991, contract code ARS01_00519.

REFERENCES

[1] W. Favre et al., "25% efficient large area silicon solar cell: paving the way for premium PV manufacturing in Europe", EUPVSEC 2020.

[2] A. Danel, "Recent Progress On The CEA-INES Heterojunction Solar Cell Pilot Line", EUPVSEC 2016, pp. 279–283.

[3] S. Eiternick, "Loss analysis for laser separated solar cells", Energy Procedia 55 (2014) 326–330.

[4] D. Bertrand et al. "Modeling of Edge Losses in Al-BSF Silicon Solar Cells", IEEE Journal of Photovoltaics 7 no. 1 (2017) 78–84.

[5] K. Rühle, M. K. Juhl, M. D. Abbott, L. M. Reindl, and M. Kasemann, "Impact of edge recombination in small-area solar cells with emitter windows," IEEE Journal of Photovoltaics, vol. 5, no. 4, (2015) 1067–1073.

[6] A. Fell et al., "Modeling edge recombination in silicon solar cells", IEEE Journal of Photovoltaics, vol. 8, no. 2 (2018) 428–434.

[7] F. Gérenton et al., "Analysis of edge losses on silicon heterojunction half solar cells", Sol. En. Mater. Sol. Cells 204 (2020) 110213.

[8] J. Choi et al., "Schottky-type edge passivation of silicon solar cells", Sol. En. Mater. Sol. Cells 159 (2017) 20–25.

[9] C. Colletti et al., "AMPERE: The European PV Manufacturing Ready To Compete In The Premium High Efficiency Market", EUPVSEC 2020, pp.256–260.

[10] R. A. Sinton and A. Cuevas, "Contactless determination of current–voltage characteristics and minority-carrier lifetimes in semiconductors from quasi-steady-state photoconductance data", Appl. Phys. Lett. 69, (1996) 2510.

[11] K.R. McIntosh, "Lumps, Humps, and Bumps: Three Detrimental Effects in the Current-Voltage Curve of Silicon Solar Cells", PhD thesis, University of New South Wales (2001).

[12] S. De Wolf, C. Ballif, and M. Kondo, "Kinetics of a-Si:H bulk defect and a-Si:H/c-Si interface-state reduction", Phys. Rev. B 95 (2012) 113302.

[13] L. Zhao, W. Zhang, J. Chen, H. Diao, Q. Wang, W. Wang, "Plasma enhanced chemical vapor deposition of excellent a-Si:H passivation layers for a-Si:H/c-Si heterojunction solar cells at high pressure and high power", Front. Ener. 11 (2017) 85–91.

[14] Xiaoning Ru, Minghao Qu, Jianqiang Wang, Tianyu Ruan, Miao Yang, Fuguo Peng, Wei Long, Kun Zheng, Hui Yan, Xixiang Xu,"25.11% efficiency silicon heterojunction solar cell with low deposition rate intrinsic amorphous silicon buffer layers", Sol. En. Mater. Sol. Cells 215 (2020) 110643.

[15] Hitoshi Sai, Hung-Jung Hsu, Po-Wei Chen, Pei-Ling Chen, and Takuya Matsui, "Intrinsic Amorphous Silicon Bilayers for Effective Surface Passivation in Silicon Heterojunction Solar Cells: A Comparative Study of Interfacial Layers", Phys. Status Solidi A 218 (2021) 2000743.

POURING THE REMAINING MELT AS A METHOD TO REDUCE THE RED-ZONE IN THE TOP REGION OF MC-SILICON INGOTS

T. Bähr*, M. Ghosh*, M. Hamacher*, C. Kranert+, C. Reimann+

* ACCESS e.V., Intzestraße 5, D-52072 Aachen, Germany
+ Fraunhofer IISB, Schottkystr. 10, D-91058 Erlangen, Germany

ABSTRACT: Completing our investigations on different melt removal techniques presented at previous PVSEC we installed an appliance in our SCU-G1 furnace for pouring the impurity-enriched silicon melt at the top of the ingot at the end of crystallization. With this method any amount of silicon melt between 0 and 25% (compared to the original mass of the G1 ingot) can be removed completely within a few seconds. Thus, the top surface of the growing crystal can be freed from the impurity enriched melt immediately and completely before the end of the regular crystallization process. This can be detected by optical investigation of the ingot's top surface. However, in all of our experiments a top red-zone of at least 1.7 cm thickness in the mc-Si ingot remained. Compared to reference ingots, where no melt was removed, no significant thickness reduction was achieved. A deeper analysis of the minority charge carrier lifetime by a diffusion model [9] revealed that, typically, the Fe_i content in the ingot's top region was reduced by a factor of about 3 compared to reference ingots. Yet, this seems to be not enough to affect the thickness of the top red zone. We suppose additional contamination of the ingot via the gas phase to be a reason for this behavior.
Keywords: multicrystalline silicon, crystallization, minority carrier lifetime, red-zone, diffusion

1 INTRODUCTION

Multicrystalline silicon ingots produced by directional solidification exhibit a reduced minority charge carrier lifetime at their top, bottom and edge regions, known as red-zone [1, 2]. Specifically, the use of silicon from the top and bottom red-zone in solar cells would result in a strongly reduced conversion efficiency. The reduced minority carrier lifetime at the ingot's top is assumed to be caused by impurities. Their distribution at the top is determined by segregation during and the in- and back-diffusion after the crystallization process.

Several concepts to remove the impurity-enriched melt from the top of the growing mc-Si crystal have been reported and claimed in patents [4, 5], but no reports of testing or realization of these concepts are publicly available. In previous PVSEC contributions, we presented experimental test results for the techniques of sponge absorption and pipe suction, applied in our G1 sized lab furnace [6, 7].

In this paper we report on

- the realization of a pouring device for the use on G1-sized mc-Si ingots,
- the efficiency of this method compared to the other melt removal techniques,
- specific results of reducing the red zone and the impurity level at the ingots top region.

2 TECHNICAL IMPLEMENTATION OF MELT POURING

2.1 Requirements

In order to reduce the ingot's top red-zone contaminated with segregated impurities, the following three requirements have to be satisfied by any melt removal technique:

Fast:
The duration of the melt removal should be quick enough such that the concurring crystal growth during this time can be neglected. Therefore, the targeted removal process time was to be no longer than a few seconds.

Complete:
If some melt were to remain after the removal procedure, the impurities present therein would (after solidification) still be able to eventually extend the red-zone by back-diffusion at high temperatures during the cool-down of the ingot.

Sufficient Amount:
Regarding the goal to increase the yield of PV-useable ingot material, the maximum meaningful amount of melt to be removed is given by the silicon amount forming the typical top red-zone, which is less than 15% of our G1-ingots (16.5 kg). The minimum amount of melt to be removed can be estimated using the Scheil equation [3] describing the impurity height distribution caused by segregation. Assuming an iron concentration of $5 \cdot 10^{15}$ atoms per cm^3 in the initial melt, which is a reasonable estimate for the ingots in question, and setting a red-zone carrier lifetime threshold limit of 10 µs, at least 5% of the melt should be removed.

2.2 Rating of tested melt removal methods

After a considerable amount of test experiments with the three proposed techniques we can classify them according to the requirements as follows:

The sponge absorption technique can take up a sufficient amount of silicon under favorable circumstances, but the removal duration was far too long to suppress additional crystal growth during the absorption process. Additionally, the absorption process is not easy to control, and there is a risk of crystal contamination through substances from the sponge material.

The pipe suction technique is fast enough to end the crystal growth process immediately after suction, but its efficiency to set free the whole crystal top surface strongly depends on the position of the suction pipes and is affected by the flatness of the crystal surface. So, typically a part of the surface remains covered by melt.

Only the pouring technique fulfills all the stated requirements, as we will show in this paper. Table I gives a summary of these results.

Table I: Properties of tested melt removal techniques

	Absorption by Sponge	Pipe Suction	Pouring
Number of experiments	15	12	12
Duration	few min	few sec	few sec
Silicon removed	≤ 12%	≤ 15%	≤ 15%
Phase boundary set free	Only locally	Typ. 25 – 60%	80-100%

2.3 Realization of the pouring technique in our G1-sized furnace

Preliminary pouring experiments have already been reported in a previous PVSEC contribution [7]. In that experiment a G0 sized crucible placed inside an empty G1 crucible was tilted at the end of the crystallization process, the G1 crucible easily being able to hold the poured melt. To realize the G1 pouring technique the question of where to put the poured melt was to be answered. The solution chosen was to tilt the crucible and allow the melt to drain into a collecting container attached to the appropriate side of the crucible. Fig.1 shows the principle, figs. 2 and 3 the realization.

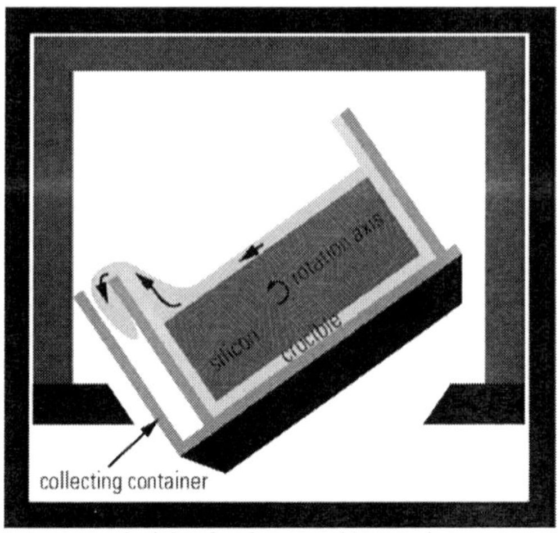

Figure 1: Principle of melt removal by pouring.

Figure 2: Filled Crucible with collecting container in the G1 furnace at process start

Figure 3: Collecting container filled with solidified silicon after pouring process

The typical process when applying this technique runs as follows:

- Filling the crucible with prime poly Si
- Mounting the collecting container at one side of the crucible
- Melting process under standard conditions
- Crystallization process while monitoring the height of the growing mc-Si crystal
- Pouring the remaining melt by tilting when the crystal has reached a predefined height
- Tilting back the crystal to the initial position after complete removal of the melt
- Standard cool down process

3 CHARACTERISATION METHODS

To characterize our experiments, we used the same techniques as reported in previous contributions [6, 7]. The evaluation of the ingot's top surface was made with regard to the following differences:

- Ingot surface permanently freed from melt by tilting.
- Ingot surface freed by tilting, but covered again with melt after tilting back.
- Surface crystallization already completed before tilting

So we can estimate the effectiveness of the melt removal process.

The other characterization quantity is the minority charge carrier lifetime, measured on a center vertical slab of the mc crystal. The samples were measured in the as-cut state using an MDP map tool from Freiberg Instruments, which is based on the principle of microwave-detected photocurrent (MDP). Because the regions of high minority carrier lifetime were not of interest for our investigations, we used a high-injection mode with a 4.8 W laser focused to a spot size of 1 mm. In this text all minority carrier lifetime mappings are scaled according to figure 4.

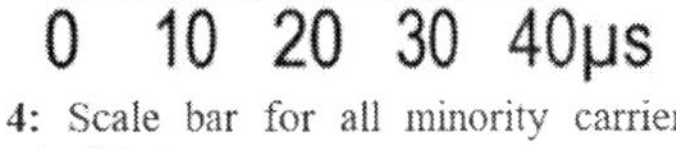

Figure 4: Scale bar for all minority carrier lifetime mappings in this text.

4 PROCESS PARAMETERS AND REFERENCE INGOTS

The typical dimensions of the G1 ingots are 22.5 cm in both horizontal directions and 14.4 cm in height for a mass of 16.5 kg.

For feedstock, we used Wacker prime poly-silicon in the form of nuggets (size 3 and 4) and granules (size 1). Boron dopant was added in the form of highly doped electronic grade silicon to reach a final resistance of ~2 Ωcm. During the standard process, the furnace atmosphere was Argon (purity 5.0) at a pressure of 600 mbar and a flow rate of 10 NL/min.

All ingots were grown under nearly the same process conditions. This comprises a crystallization time of 16-17 h for the ingot without melt removal, an average propagation speed for the solid/liquid interface of about 0.9 cm/h, and a cool-down time of 10 h. The solid/liquid interface was slightly convex, especially in the upper region of the ingot.

Under these process conditions, the typical MDP values of the minority carrier lifetime are as depicted in Figure 6 with a red-zone thickness at the top of slightly less than 2 cm.

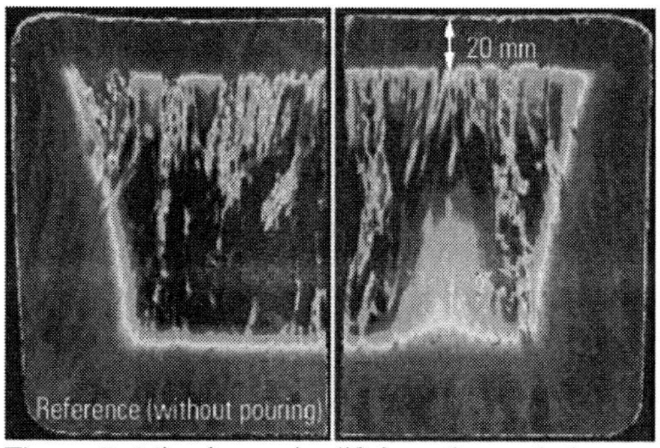

Figure 5: Minority carrier lifetime of a vertical cross-section of the reference G1-ingot.

5 RESULTS

5.1 Melt Removing Efficiency

In previous contributions to PVSEC Conferences, we reported about the efficiency of melt removal by sponge absorption [6] and pipe suction [7]. While sponge absorption is a relatively slow process, pipe suction can set free the ingot's surface within a few seconds. Yet, both method suffer in the reliability of removing the remaining melt from the complete ingot surface. In most of our experiments, after the process we identified areas on the ingot surface where the melt was not removed, but covered some "valleys" on the surface. Additionally, no real process control could be performed in order to know how far the crystal already was solidified. In a few preliminary pouring experiments on G0 size, we were able to ensure that the ingot surface is completely melt-free. For that reason we realized the pouring device for G1 crucibles. In all pouring experiments that we carried out, a clear and complete melt removal was achieved.

So, from this point of view, the technique of melt removal from the top surface of the growing ingot has been realized successfully and the requirements given in chapter 2.1 are fulfilled. Figure 6 gives an overview of all results of melt removal experiments concerning the amount of removed melt and the removal duration. Clearly, the pouring technique fulfills all the requirements given in chapter 2.1.

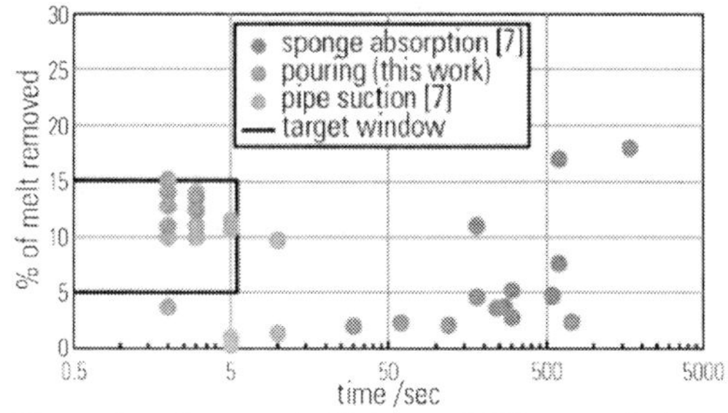

Figure 6: melt removal efficiency of different techniques

5.2 Characterization of Ingot Surfaces

To give an impression of the melt removal efficiency, two examples of ingot surfaces are discussed in this section. As the surface of the crystal growing within the silicon melt is not completely flat but somewhat wavy, there are always zones where residual melt can accumulate and solidify. This is the case in the following example of pipe suction. As the pipes are "concentrated" at certain positions which are typically not at the lowest points of the grown crystal, the melt suction ends when the melt level is falling under this level. This is the case in fig. 7, where a significant amount of the surface is covered with remaining melt which crystallized after the suction process.

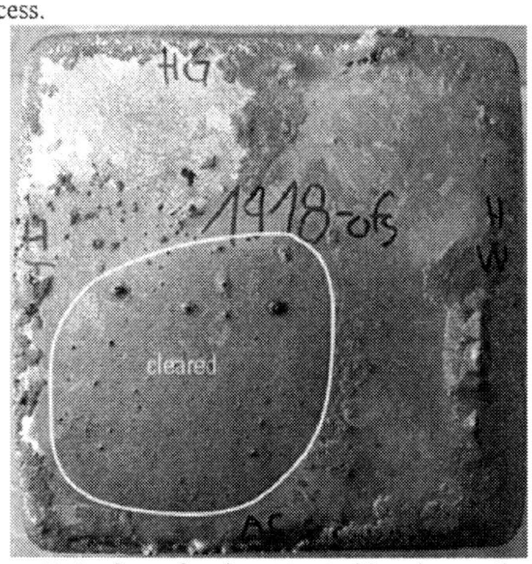

Figure 7: Surface of an ingot treated by pipe suction [7]. Due to a non-optimal position of the sucking pipes only one third of the surface was cleared from melt.

On the other hand, the pouring process does not distinguish between "hill and valley" as long as the tilt angle allows the melt to flow over the crucible edge. So, in fig.8, despite the ingot surface not being flat but having a hill at one point in the middle, the melt has been removed almost completely by the tilting procedure. Only a remnant stayed in a narrow section on the left side and solidified there.

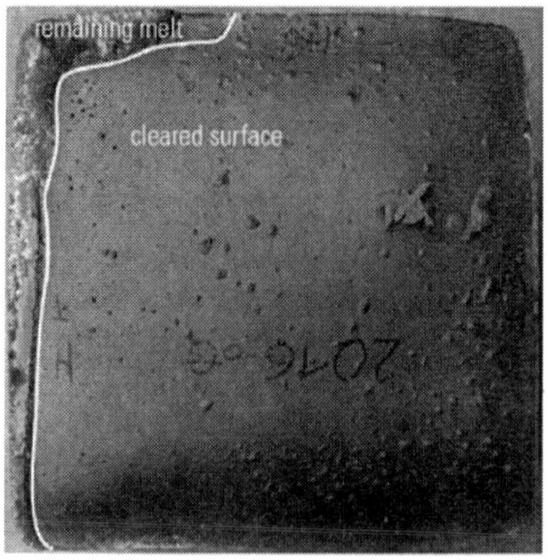

Figure 8: Surface of an ingot treated by pouring. The surface was cleared from melt almost completely. Only on the container side a few cm² are covered due to an incomplete tilting by less than 90°.

5.3 Minority Carrier Lifetime – Top Red Zone Reduction

The original purpose of our project was to eliminate (or at least strongly reduce) the top red zone in mc-Si ingots. In order to test this, we characterized the charge carrier lifetime by the method described in chapter 4. Contrary to the hopes associated to the original patents, the top red zone never vanished in any experiment. Sometimes its height was reduced by up to 10% compared to reference ingots grown under similar conditions. Figs. 8 and 9 show typical results. We also must state that an increase of the volume fraction with higher charge carrier lifetime was never achieved.

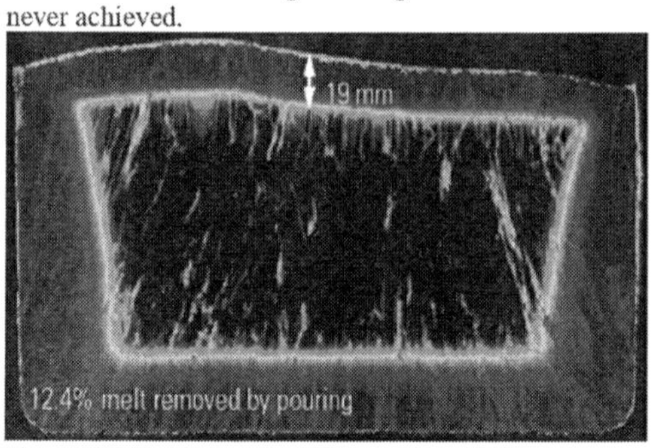

Figure 8: Minority carrier lifetime map of an ingot, where the top melt mas removed by tilting

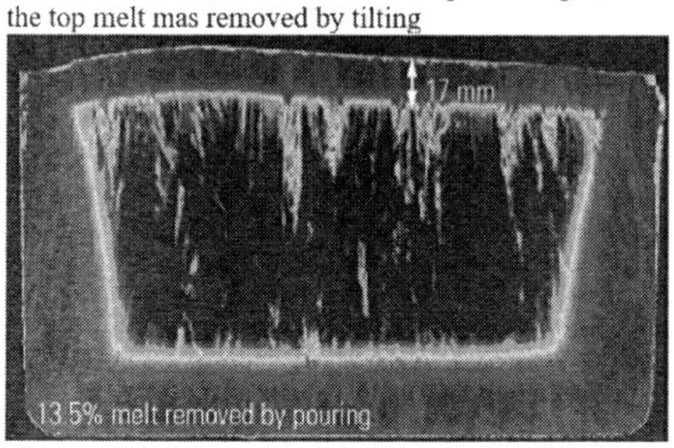

Figure 9: Minority carrier lifetime map of another ingot, where the top melt mas removed by tilting

6 DISCUSSION

A priori it is not clear if the slightly reduced red-zone extension is indeed related to the removal of impurities by the pouring process or if the melt removal had an effect on the cool-down behavior, which may also affect the extension of the top red-zone. In order to distinguish between these two effects, we introduce a simplified diffusion model to which the measured charge carrier lifetime profile along the crystallization axis can be fitted.

6.1 Diffusion Model for Fe_i in Top Zone of Ingot

We assume Fe_i to be the electrically dominant species in the top region. Our model [8] includes the back diffusion according to Fick's law from a limited source concentrated at an infinitively thin layer at the top surface of the ingot. If we assume the back-diffusion to be mostly dependent on the temperature history of the cooldown process, we can describe the height(x)-dependent concentration by

$$c(x) = \frac{c_0}{\sqrt{\pi} L_{\text{eff}}} \exp(-\frac{x^2}{L_{\text{eff}}^2})$$

with

$$L_{\text{eff}} = 2\sqrt{\int D(T(t))\,\mathrm{d}t}$$

and T(t) being the temperature history of the ingot during cool down. The height dependent concentration c(x) can be converted to a charge carrier lifetime by the SRH recombination of FeB pairs to

$$\tau = 10\mu s\ c_{crit}/c$$

with $c_{crit} = 3 * 10^{12}$ cm^{-3} [9]. The measured height-dependent charge carrier lifetime can be fitted to the free variables c_0 and L_{eff}. The following figure 11 (left) gives an example for such a fit. Fig.11 (right) shows the correlation between the fitted L_{eff} and the "expected" diffusion length integrated from the temperature history L_{int}. The fit and correlation are in good agreement.

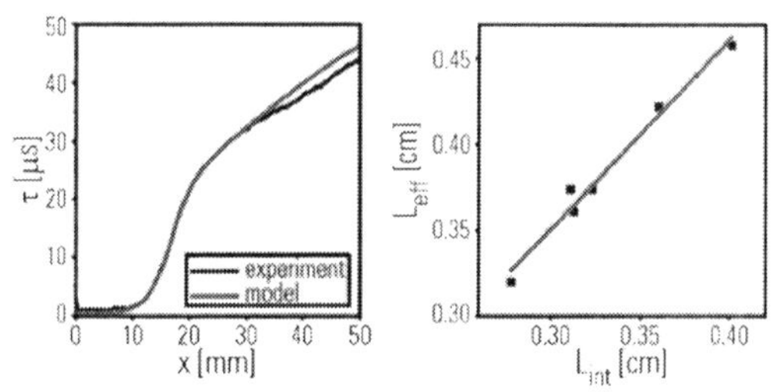

Figure 11: Diffusion model for Fe_i, example of the resulting charge carrier lifetime τ fitted/measured (left). Correlation of fitted L_{eff} to L integrated from the temperature history (right).

6.2 Application of the model

We fitted c_0 and L_{eff} for 2 reference ingots and various experiments with melt removal. The result is shown in fig.12. Melt-removed and regular ingots clearly differ in the derived c_0, while their diffusion length is similar. Ingots where the remaining melt was removed by pouring have a Fe_i concentration at their top of a factor of 2-3 lower than reference ingots. Thus, the slight reduction of the top red-zone extension is indeed a consequence of the reduction of the metal contamination by the melt removal and is not related to a variation of the cool-down conditions. However, the contamination clearly did not drop to zero as would be expected from the fast and complete melt removal. Thus, the remaining contamination is expected to originate from a different source, namely from in-diffusion from the furnace atmosphere, a phenomenon which we investigated in another project under similar conditions [10].

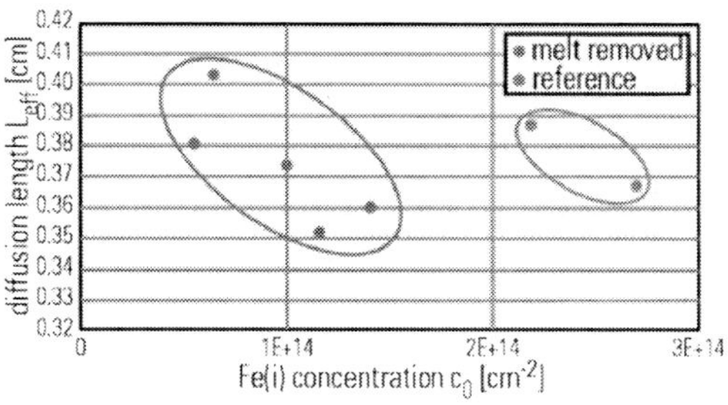

Figure 12: Fitted c_0 and L_{eff} for different Si ingots. When remaining melt was removed, the calculated Fe_i concentration is significantly lower.

7 CONCLUSIONS

An appliance for melt removal by pouring was realized in our SCU-G1 crystallization furnace and used for the crystallization of several mc-Si crystals. A stable process could be established. The use of tilting to pour out the contaminated Si resulted in complete melt removal. Compared to other melt removal techniques the pouring turned out to be the most effective, the ingot surface could be freed from remaining melt in all cases. Nevertheless, the top red zone was nearly unaffected by melt removal. A deeper look into the height distribution of the minority charge carrier lifetime revealed that the Fe_i concentration was reduced by a factor of 2-3, which is not enough to suppress the top red zone completely.

REFERENCES

[1] F. Ferrazza: Large size multicrystalline silicon ingots. Solar Energy Materials and Solar Cells. Volume 72, Issues 1–4, p 77-81 (2002)

[2] Y. Boulfrad, G. Stokkan, M. M'hamdi, E. J. Øvrelid, L. Arnberg: Modeling of lifetime distribution in a multicrystalline silicon ingot. Solid State Phenomena, Volume 178-179. p. 507-512 (2011)

[3] E. Scheil: Bemerkungen zur Schichtkristallbildung. Zeitschrift für Metallkunde 34, 70-72 (1942)

[4] Solar World Innovations GmbH, Patent Specification DE 10 2009 039070 B4

[5] Solar World America Inc., Patent Specification US 00000 893 6652 B2

[6] T. Bähr et al., 2AV.1.11, proc. of 35. PVSEC Bruxelles (2018)

[7] T. Bähr et al., 2AV.2.16, proc. of 36. PVSEC Marseille (2019)

[8] C. Kranert et al.: Assessment of residual melt removal as approach to reduce the top redzone of cast silicon ingots. J. Cryst. Growth 571 (2021), 126261

[9] T. Isobe, H. Nakashima, K. Hashimoto: Diffusion coefficient of interstitial iron in silicon, Jap. J. Appl. Phys. 28 (part 1, vol.7) (1989), 1282-1283

[10] C. Kranert et al.: Metal contamination of silicon from the furnace atmosphere after crystallization. J. Cryst. Growth Volume 559 (2021), 126026

ACKNOWLEDGEMENTS

This project was funded by the German Ministry of Economics and Energy under contract 0324089 A and 0324089 B, which is gratefully acknowledged.

Supported by:

Federal Ministry
for Economic Affairs
and Energy

on the basis of a decision
by the German Bundestag

INFLUENCE OF AN ACTIVE CRYSTAL COOLING DEVICE ON THE SHAPE OF THE PHASE BOUNDARY IN MONO INGOTS GROWN BY THE CZOCHRALSKI TECHNIQUE

F. Mosel[1], A.V. Denisov[1], K. Hess[1], B. Klipp[1], N. Sennova[1], C. Kranert[2], M. Trempa[2], C. Reimann[2], J. Friedrich[2]
[1] PVA Crystal Growing Sytems GmbH, Im Westpark 10-12, 35435 Wettenberg, Germany
[2]Fraunhofer-Institut für Integrierte Systeme IISB, Schottkystraße 10, 91058 Erlangen, Germany
e-mail: frank.mosel@pvatepla.com
tel: +49 64168690-125, fax: +49 64168690-822

ABSTRACT: In Cz-Si the shape of the crystallization front, which is essentially controlled by the interaction of the various convection phenomena, determines the radial distribution of the crystal properties and is the subject of numerous publications. In our former investigations [1] on the limits of the pull speed in the Czochralski process, we were able to determine a clear influence of an active crystal cooling on the extent of the deflection of the phase boundary in the growing crystal. In order to investigate the influence of the crystal cooling element on the shape of the interface in more detail, two different series of crystal growth experiments were carried out and interpreted with 2D numerical simulations. For this purpose the shapes of the phase boundaries of the grown crystals were made visible by means of LPS measurements and compared with the calculated interface shapes. In one of the two series, the influence of a certain crystal cooling element in connection with the growth configuration was investigated. In the other series, the influence of different crystal cooling elements on the phase boundary was examined. Based on the heat flux balance at the crystallization front, the essential parameters for stable crystal growth conditions were evaluated. For all experiments performed, the cooling rate of the growing crystal in its growth environment were calculated from the point of view of the distribution of intrinsic defects.
Keywords: crystal cooling, pull speed, melt/crystal interface shape, Czochralski process

1 INTRODUCTION

The motivation for the development of an active crystal cooling was the increase of the average pull speed for a higher productivity of the Czochralski process. The first experiments with a water-cooled coil as an active cooling element were reported by Samuel N. Rea in 1977 in a research report on the development of "low-cost solar arrays" [2]. In the meantime, the Czochralski technology has undergone a rapid development for both, the production of crystals for photovoltaics and for microelectronics. At the same time, the use of active crystal cooling has also become established, albeit with different objectives. In this paper, the effects of active crystal cooling are investigated from the point of view of an increase in pull speed and from the point of view of the influence on the shape of the solidification front and on the distribution and concentration of the intrinsic defects in the crystal. The interface shape has an influence on the thermo-elastic stresses at the solidification front. Strongly W-shaped phase boundaries are considered to be one of the root causes of dislocation formation [3]. For the studies in this report, 8-inch crystals were grown in different growth configurations with and without active crystal cooling. The growth conditions have been simulated with the commercially available 2D software code CGSIM [4] and compared with experimental results.

2 EXPERIMENTAL

2.1 Crystal growth

The crystal growth experiments presented in this paper were performed in different growth configurations in a SC22 and a SC24/26 Czochralski-puller from PVA Crystal Growing Systems GmbH. The SC24/26 puller can be equipped optionally with a 24-inch or a 26-inch hotzone. The different crystal growth configurations are sketched in Fig.1 and summarized in Tab.1.

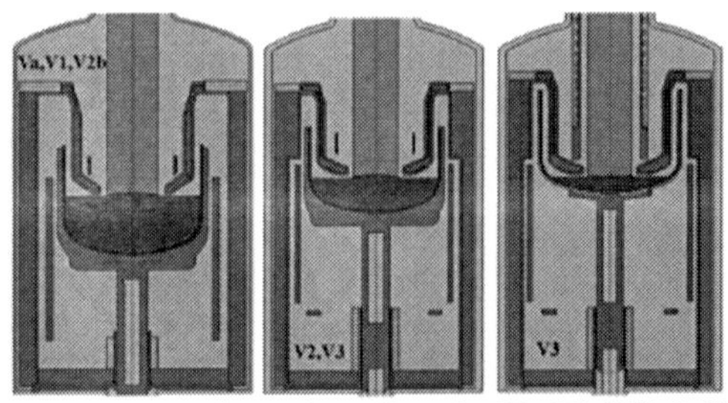

Figure 1: Crystal growth configuration Va, V1, V2b (left), V2, V3 (middle), V3 (right).

growth configuration	Cz-puller type	crucible dimension [inch]	bottom heater	cooling element
Va	SC24	24	no	-
V1	SC22	22	no	A
V2	SC24/26	24, 26	yes	A
V2b	SC24	24	no	-
V3	SC26	26	yes	A, B, C, D, E

Table I: Main parameter of the different growth configurations.

For our investigations, we used five different cooling devices which differed in their geometry and position within the hotzone. The water-containing elements were made of heat-resistant stainless steel embedded in different radiation absorbers.

Two series of crystal growth experiments were realized. The first one (Series A) was performed from the standpoint of maximizing the average pull speed (Tab.II).

A second series of crystal growth experiments (Series B) was carried out (Tab.III) to investigate the influence of the different cooling elements (geometry and position) on the shape of the crystallization front. These growth experiments were all executed in the same crystal growth configuration with the almost identical growth parameters. Vertically cut slices were prepared from all crystals, and the solid liquid interface shapes were made visible by means of LPS measurements [5]. To ensure the comparability between the growth experiments, the LPS measurements as well as the numerical simulations of the crystal growth conditions were all done at the growth stage of the body of 500 mm.

2.2 Numerical simulation

All crystal growth runs were simulated using the commercially available CGSIM software [4]. The results presented here are derived from 2D simulations which depict the axially symmetric formation of the calculated interface shape. The situation of the heat fluxes at the solidification front (Eq. 1) is given by the heat flow balance:

$$v_g = \frac{\lambda_s G_s - \lambda_l G_l}{\rho_s L} \quad (1)$$

in which λ is the thermal conductivity, G is the axial temperature gradient, v_g is the constant growth velocity, ρ_s is the crystal density, L is the latent heat of fusion and s,l are the subscripts for the solid and liquid phase.

In the melt, heat transport by conduction and convection is simulated, taking buoyancy, Marangoni and forced convection into account. The crystal is assumed to be opaque, and heat transport from the interface in the crystal takes place through conduction. Heat exchange with the environment is driven by radiation and convection, and the gas atmosphere is assumed to be transparent for radiation under the growth conditions. The main results of the numerical simulations are the deflection and the shape of the interface, the heat flows and the axial temperature gradients at the crystallization front as well as the distribution of the isotherms and the von Mises-stresses in the growing crystal.

Based on the comparison of the calculated and experimentally determined interface shapes, the boundary conditions for the simulation calculations were adjusted in such way that experiment and simulation showed satisfactory agreement.

3 RESULTS AND DISCUSSION

3.1 Characterization of the crystals

The shape of the phase boundary can be described by two key figures, one by the deflection in the crystal center H (Fig. 2) and the other by the W-shape factor ϕ (Eq. 2), according to Noghabi et al. [6].

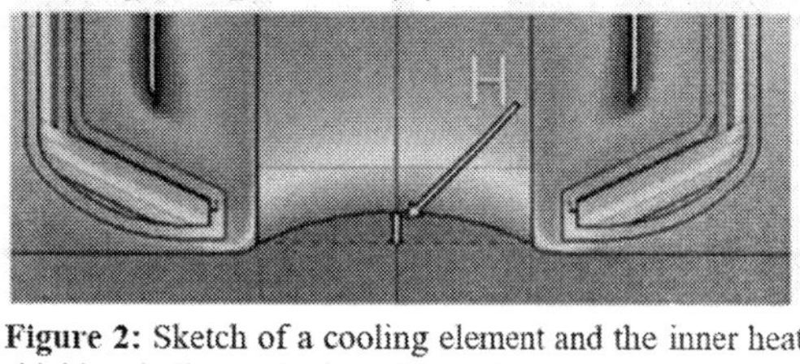

Figure 2: Sketch of a cooling element and the inner heat shield, H indicates the interface deflection.

$$\Phi[\%] = \frac{(\delta_{max} - \delta_{center})}{\delta_{max}} \quad (2)$$

δ_{max} and δ_{center} describe the distance between the two extreme values of the deflection and the value of the deflection in the crystal center. A W-factor of 0 means a concave, convex or planar interface. A positive sign of H denotes a concave deflection (Fig.2) and a negative sign a convex deflection of the interface. The results from the crystal growth experiments are listed in Tab.II and Tab.III. Examples of the LPS-measurements are shown for the growth condition CC227V2 in Fig.3 and CC262V3 in Fig.4.

Ingot/growth configuration	active cooling element	average pull speed [mm/min]	deflection H [mm] measured	deflection H [mm] calculated	W-shape ϕ [%] measured	W-shape ϕ [%] calculated
CC212 Va	-	0.9	12	12	4	0
CC213 V2b	-	1.25	16	20	0	0
CC227 V2	A	1.8	20	22	0	0
CC246 V3	A	1.6	13	13	0	0
CC249 V1	A	1.2	1 to 2	1	35	68

Table II: Interface parameters of Series A.

Ingot/growth configuration	active cooling element	average pull speed [mm/min]	deflection H [mm] measured	deflection H [mm] calculated	W-shape ϕ [%] measured	W-shape ϕ [%] calculated
CC262 V3	E	0.9	-2	-3	55	33
CC263 V3	B	0.9	-5 to -7	-5	0	15
CC265 V3	B	0.9	-6 to -8	-5	0	18
CC266 V3	D	0.9	-0.9 to -1.8	-4	60	25
CC267 V3	C	0.9	-3.1 to -4.1	-3	38	21

Table III: Interface parameters of Series B.

Figs. 6-10 show the phase boundary shapes (thin lines) derived from the LPS-measurements together with the phase boundary calculated for a body length of 500 mm (blue line).

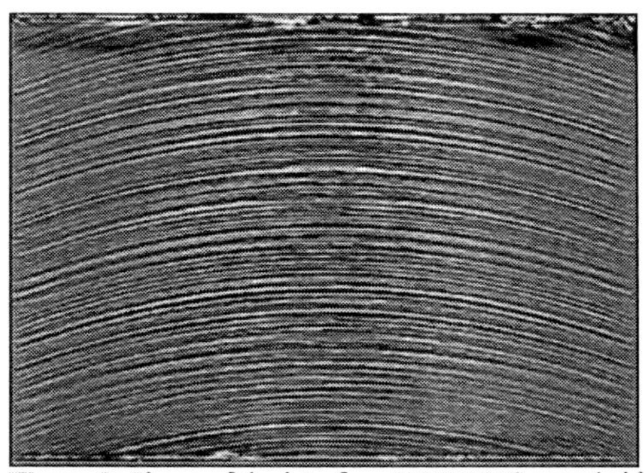

Figure 3: Shape of the interface (CC227V2) revealed by LPS-measurement, image width equals 8 inches.

Figure 4: Shape of the interface (CC262V3) revealed by LPS-measurement, image width equals 8 inches.

3.1.1 Growth experiments of Series A

The main growth parameters of Series A are summarized in Tab. IV.

ingot/growth configuration	average pull speed [mm/min]	crystal rotation [rpm]	crucible rotation [rpm]	gap [mm]	bottom heater	cooling element
CC212 Va	0.9	10	10	25	no	-
CC213 V2b	1.25	10	10	17	no	-
CC227 V2	1.8	10	6	17	yes	A
CC246 V3	1.6	10	6	17	yes	A
CC249 V1	1.2	10	10	25	no	A

Table IV: Main growth parameters of Series A, gap denotes the distance between the melt surface and the lower edge of the inner heat shield.

The calculated interface deflection H versus the average pull speed for the different growth configurations at a body length of 480 mm - 520 mm of the growing ingot is shown in Fig.5. The symbols in this figure represent individual results of the simulations. A linear relationship between the average pull speed and the deflection of the phase boundary is evident and indicated by the inserted lines. The figure represents a stability diagram in which three areas are marked. In the stable growth region (green) the system is insensitive to changes in the average pull speed. This region is robust and suitable for industrial production. In the metastable growth region (yellow) all growth parameters have to be well tuned. Small changes can lead to unstable growth with loss of shape, i.e. spiral growth. In the unstable growth region (red) no regular crystal growth is possible. The theoretical results were confirmed by LPS-measurements and are indicated in the graph by the open red circles. In the growth configuration V2 an average pull speed of 1.8 mm/min was maintained over the entire body length (CC227V2).

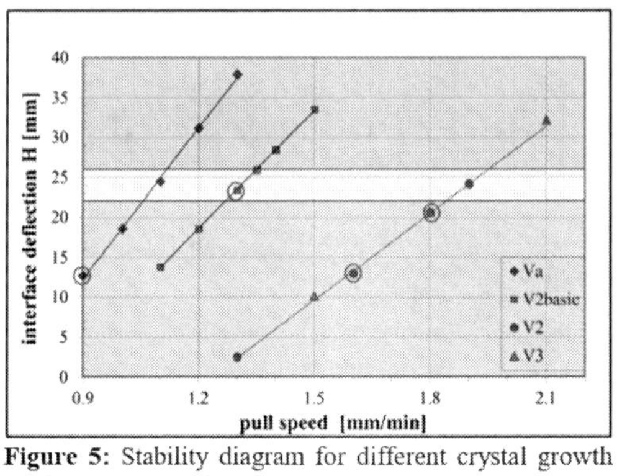

Figure 5: Stability diagram for different crystal growth configurations (Series A).

Fig.5 is an empirically determined diagram and does not represent an analytical model. Fig.6 and fig.7 show the concave phase boundaries of the two experiments CC212Va and CC227V2 as extreme examples. The calculated shapes agree well with the measured ones as indicated by the symbols in fig.5. For the shapes of the phase boundary of series B the simple linear relationship from Fig.5 holds no longer. The growth configuration of CC249V1 (Fig.8) shows the transition from a purely concave to a convex shaped interface with a wavelike formation.

The coloured thin lines in Figs. 6-10 represent different LPS- measurements on a sample. These lines are shown as an example in Fig.4.

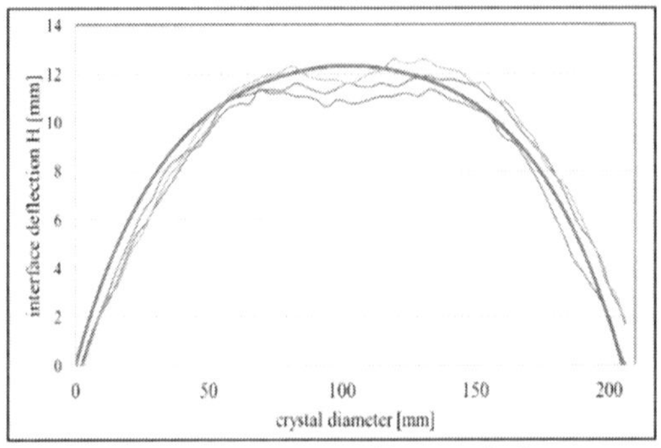

Figure 6: Calculated (blue line) and from LPS-measurement deduced interface shape (coloured thin lines) of CC212Va.

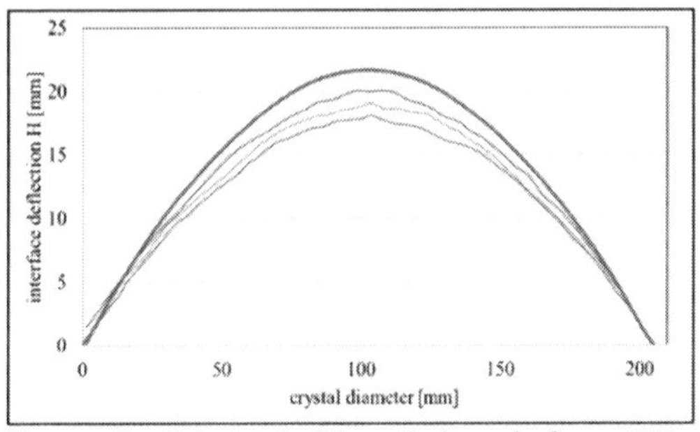

Figure 7: Calculated (blue line) and from LPS-measurement deduced interface shape (coloured thin lines) of CC227V2 (see also Fig. 3).

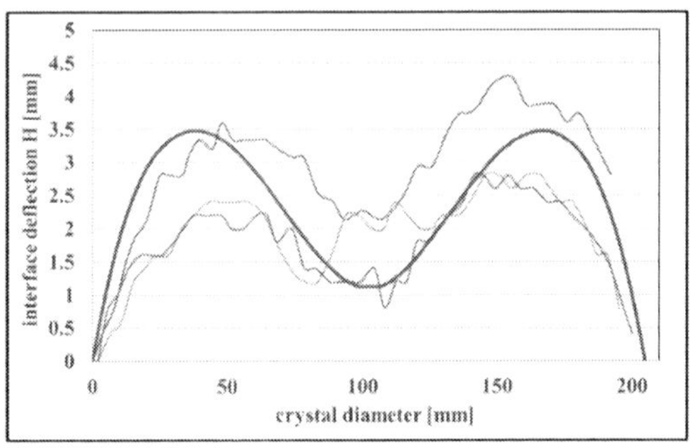

Figure 8: Calculated (blue line) and from LPS-measurement deduced interface shape (coloured thin lines) of CC249V1.

3.1.2 Growth experiments of Series B

The main growth parameters of Series B are summarized in Tab.V.

ingot/growth configuration	average pull speed [mm/min]	crystal rotation [rpm]	crucible rotation [rpm]	gap [mm]	bottom heater	cooling device
CC262 V3	0.9	10	6	25	yes	E
CC263 V3	0.9	10	6	25	yes	B
CC265 V3	0.9	10	6	25	yes	B
CC266 V3	0.9	10	6	25	yes	D
CC267 V3	0.9	10	6	25	yes	C

Table V: Main growth parameters of Series B, gap denotes the distance between the melt surface and the lower edge of the inner heat shield.

All interface shapes of series B show a more or less similar wave-shaped formation of the phase boundary (W-shape). Two examples of the series are given in Fig.9 and Fig.10.

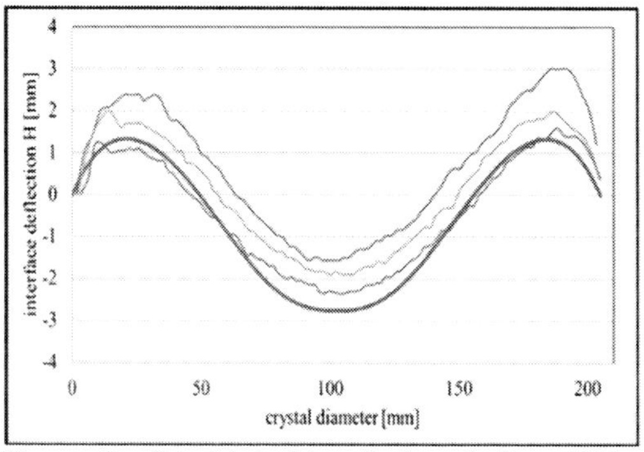

Figure 9: Calculated (blue line) and from LPS-measurement deduced interface shape (coloured thin lines) of CC262V3.

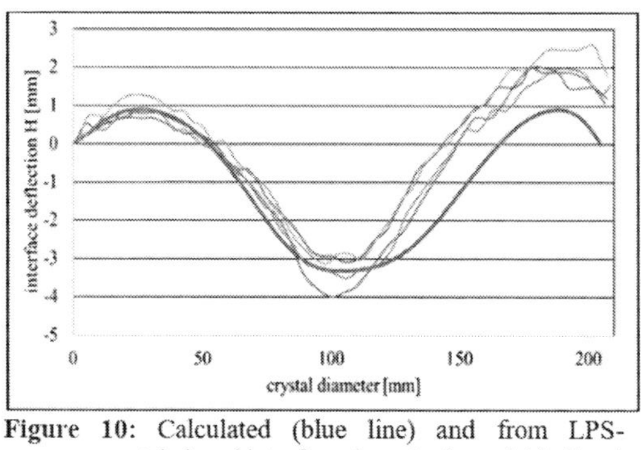

Figure 10: Calculated (blue line) and from LPS-measurement deduced interface shape (coloured thin lines) of CC267V3.

3.2 Influence of active crystal cooling on the heat fluxes and temperature gradients at the interface of the growing crystal

The temperature gradients at the phase boundary derived from the calculated heat flows are given in Tab.VI and Tab.VII.

Ingot/growth configuration	latent heat of fusion [W]	heat flow crystal [W]	heat flow melt [W]	G_l~Q/(Aλ) [K/cm]	G_s~Q/(Aλ) [K/cm]	G_{cr} [K/cm]
CC212 Va	2033	2990	960	4.8	45.3	3.7
CC213 V2b	2847	3470	625	3.1	52.6	1.3
CC249 V1	2727	4510	1780	8.9	68.3	7.9
CC246 V3	3639	4860	1220	6.1	73.6	5.6
CC227 V2	4102	4780	680	3.4	72.4	0.8

Table VI: Thermal boundary conditions at the crystallization front of Series A.

Ingot/growth configuration	latent heat of fusion [W]	heat flow crystal [W]	heat flow melt [W]	G_l~Q/(Aλ) [K/cm]	G_s~Q/(Aλ) [K/cm]	G_{cr} [K/cm]
CC262 V3	2053	3528	1475	7.3	53.5	7.8
CC263 V3	2046	3894	1848	9.2	59.0	9.5
CC265 V3	2047	3794	1746	8.7	57.5	8.4
CC266 V3	2043	3641	1604	8.0	55.2	7.5
CC267 V3	2057	3450	1400	7.0	52.3	6.4

Table VII: Thermal boundary conditions at the crystallization front of Series B.

At this point, it should be pointed out that, when estimating the melt-side temperature gradient by means of the thermal conductivity λ_l in the liquid phase, the convective contribution to the heat transport at the phase boundary is neglected. However, this fact seems to be justified within certain limits from the point of view of impulse momentum boundary layer, which is in the order of the meniscus height.

The radial temperature distribution on the melt meniscus at the crystal edge is an important criterion for the stability of the growth process, especially with regard to the limitations of the growth rate [5]. Therefore, the radial temperature gradient G_{cr} at the crystal edge was determined from the simulation calculations as shown in Figure 11.

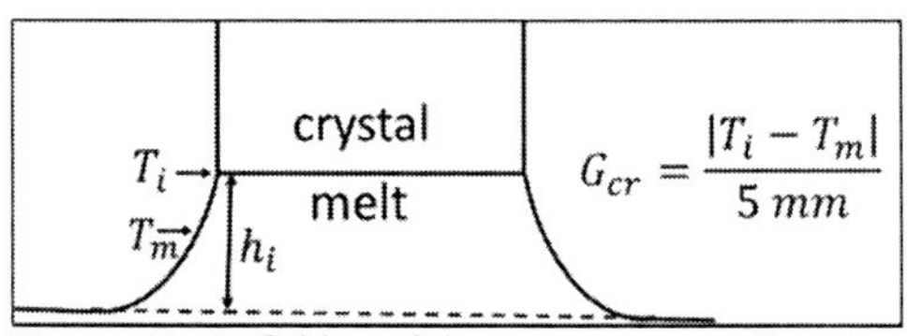

Figure 11: Definition of the temperature gradient G_{cr} at the crystal rim.

Here, T_i is the solidification temperature and T_m is the temperature on the melt meniscus at a distance of 5 mm from the crystal edge. T_m is therefore still within the thermal boundary layer which corresponds to the meniscus height h_i (~ 6.9 mm). This key figure among others is used to judge the process stability.

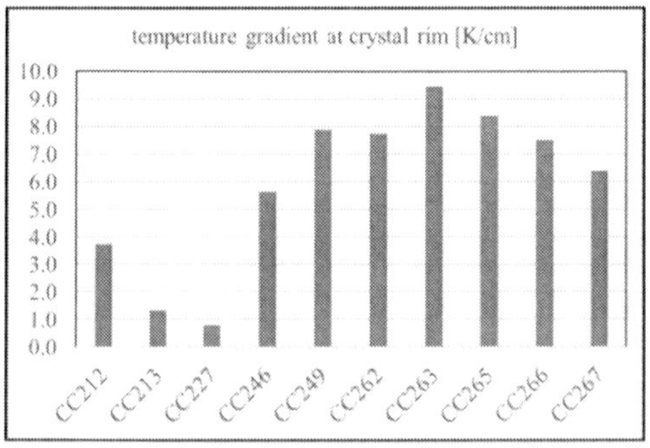

Figure 12: Calculated temperature gradient G_{cr} at the crystal rim.

From the experimental results of CC213V2b and CC227V2, which represent the stability limit of the average pull speed in the body, it can be concluded that a minimum temperature gradient in the melt G_l and at the crystal edge G_{cr} must be ensured for stable growth conditions.

3.3 Influence of active crystal cooling on the distribution of isotherms and von-Mises stresses in the growing crystal

Due to the presence of non-linear temperature gradients during the cooling of the growing ingot, thermal stresses are induced inside the crystal, which play an essential role as a source for defect formation. The thermal stress state can be represented by a single value, the von-Mises stress.

As expected, the calculated von-Mises stresses show a clear dependence on the growth rate (Fig.13). For crystals with a concave interface shape of Series A, the maximum of the von Mises stresses is located at the crystallization front in the crystal center. The exception in series A is the crystal growth configuration CC249V1. In this experiment no pronounced maximum appears. The stress distribution is rather uniform and decreases towards the crystal edge.

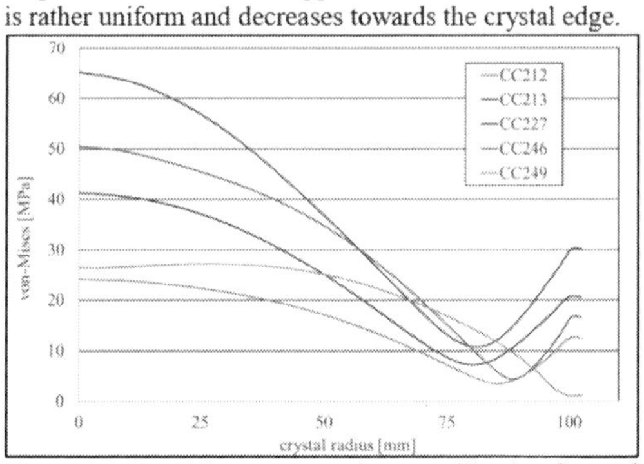

Figure 13: Von-Mises stress at the crystallization front of Series A.

In the growth configurations of series B the maxima of the stresses on the crystallization front are approximately at the turning point (half radius) of the interface (Fig.14). The maximum stress levels in the crystals (Fig.15) are located at the crystal edge at some distance from the interface (Figs.16-19).

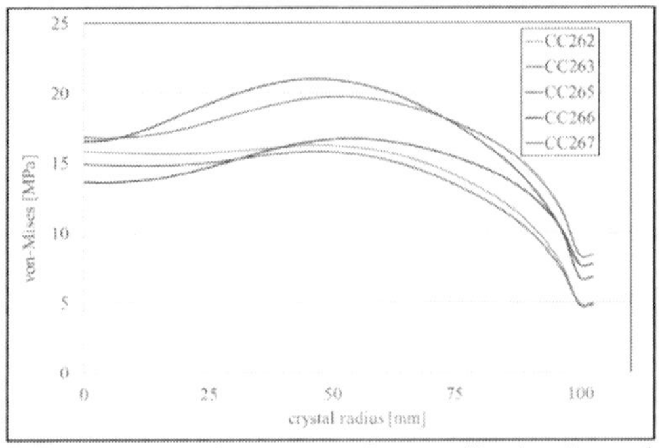

Figure 14: Von-Mises stress at the crystallization front of Series B.

The crystals CC212 and CC262 as well as the crystals CC213 and CC249 were produced under comparable conditions regarding the growth rate. CC212 und CC213 were grown without crystal cooling in contrast to CC249 and CC262 which were grown with crystal cooling. The comparison of the growth experiments show that the maxima of the stresses are comparable (dashed lines in fig.15). The main difference is the local distribution of the stresses. In the two crystals grown with crystal cooling the maxima of the von-Mises stresses is moved towards the crystal edge with a certain distance to the interface whereas the stress maxima of the two crystals grown without crystal cooling are situated at the crystal center on the crystallization front (Fig.13).

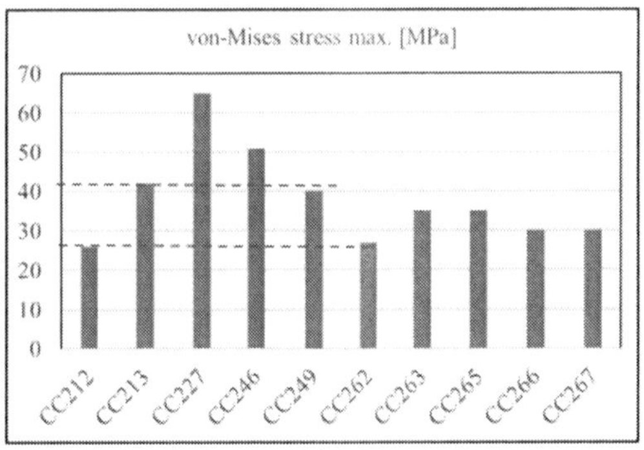

Figure 15: Calculated maximum von-Mises stress at a body length of 500 mm of the growing ingot.

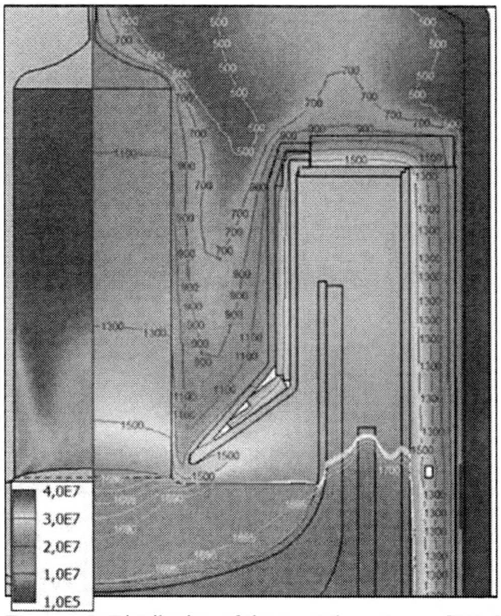

Figure 16: Distribution of the von-Mises stresses [MPa] (left) and isotherms (right) of CC212Va.

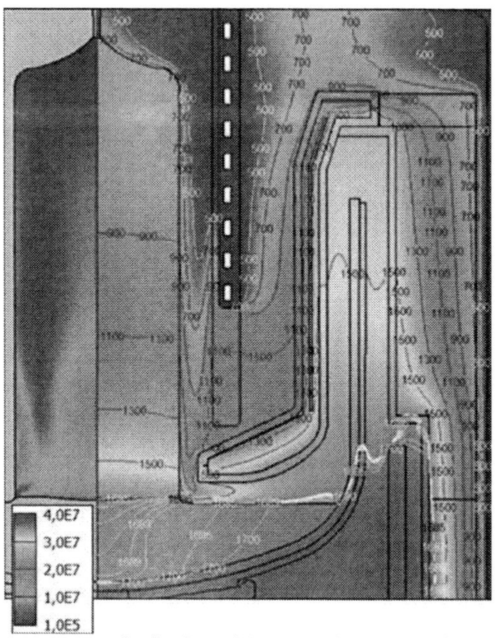

Figure 17: Distribution of the von-Mises stresses [MPa] (left) and isotherms (right) of CC262V3 (cooling element E).

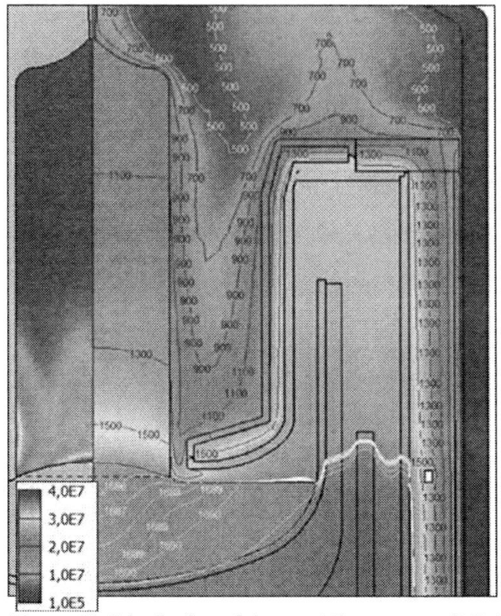

Figure 18: Distribution of the von-Mises stresses [MPa] (left) and isotherms (right) of CC213V2b.

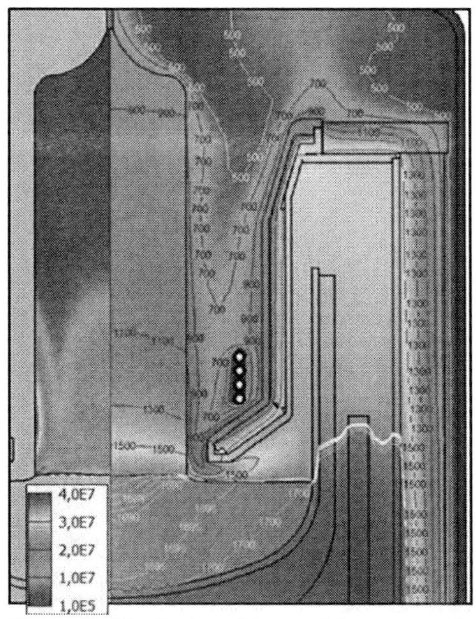

Figure 19: Distribution of the von-Mises stresses [MPa] (left) and isotherms (right) of CC249V1 (cooling element A).

3.4 Influence of active crystal cooling on the axial temperature profile in the growing crystal

While the temperature gradient on the solid side of the interface G_s in connection with the growth rate v_g determines the concentration and radial distribution of the intrinsic defects in the crystal, another important factor for the controlled incorporation of the intrinsic defects is the cooling rate of the growing crystal. The dwell time of the growing crystal in the temperature range between 900 °C and 1050 °C affects the density and size of the defects. Thus, the formation of vacancy clusters (voids) is strongest in this temperature range. The average size of the voids increases with the dwell time in this temperature range at the expense of their density. A simple empirical relation can be given for the density of the void clusters N_{Void} as a function of time with A as a constant [7].

$$N_{void} = Ae^{-0.011min^{-1}t} \quad (3)$$

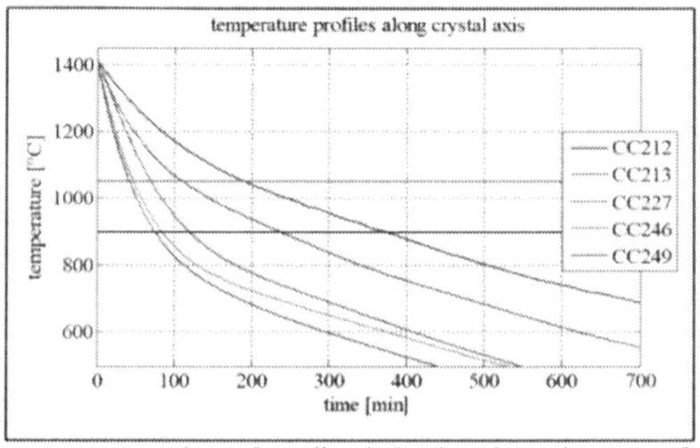

Figure 20: Estimated cooling behavior along the axes of the ingots of Series A.

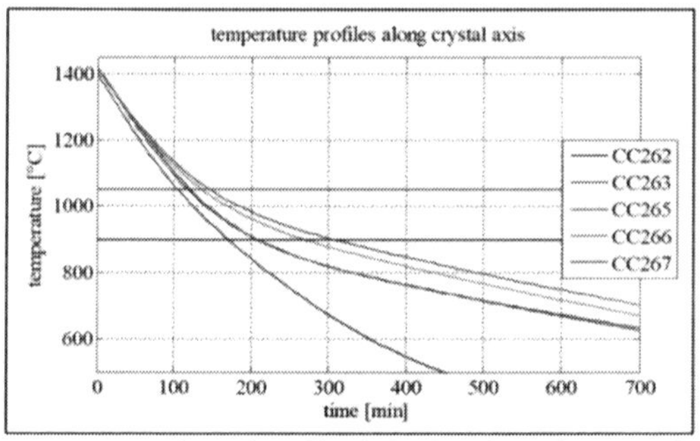

Figure 21: Estimated cooling behavior along the axes of the ingots of Series B.

The numerical simulation was also used to estimate the temporal temperature profile in the growing crystals. The cooling trends were obtained from the calculated temperature profiles along the central axis of the growing crystal versus body length.
The results for series A and B are plotted separately in Fig.20 and Fig.21.
The diagram in Fig.20 clearly shows the influence of the average pulling speed on the dwell time of the growing crystal in the critical temperature range between 900 °C and 1050 °C. The graph in Figure 21 shows the influence of the different cooling elements on the cooling behavior of the growing crystal. The time span of the growing crystal in the critical temperature range is by far the shortest when cooling element E is used.

3.5 Influence of active crystal cooling on the Voronkov parameter v/G

The radial distribution of the heat flux density from the melt into the crystal depends not only on the heat transport by convection and conduction in the melt, but also on the transport in the crystal and the heat release from the crystal to its surroundings. The interface shape is determined by this interplay, which includes the heat of crystallization released by the imposed growth rate. The temperature gradients resulting from the heat flow balance determine the intrinsic defect distribution in connection with the growth rate. The Voronkov parameter v_g/G_s takes into account the convective (v_g) and diffusive (G_s) transport processes in the crystal. For the critical value of the Voronkov parameter, which is given with different values in the literature, the solid side of the crystallization front is considered.
Another point of view results also from the one-dimensional heat balance (Eq. 1) at the solidification front:

$$\frac{G_l}{G_s} = \frac{\lambda_s}{\lambda_l} - \Gamma_{crit}\frac{\rho_s L}{\lambda_l} \quad mit \quad \Gamma = \frac{v_g}{G_s} \quad (4)$$

For a literature value of the critical Voronkov parameter $\Gamma_{crit}=1.34*10^{-3}cm^2/K\,min$ [12], a certain ratio of the temperature gradients at the phase boundary is obtained, in the example given here of Gs/Gl=5.3.
The gradient ratios of the studied crystal growth configurations are listed in Tab.VIII and Tab.IX. A ratio of the temperature gradients Gs/Gl > 5.3 is obtained in all listed growth configurations. Thus, under the assumptions made, all crystals were grown under vacancy rich conditions.

Ingot/growth configuration	CC212 Va	CC213 V2b	CC249 V1	CC246 V3	CC227 V2
G_s/G_l	9.49	16.91	7.72	12.13	21.41

Table VIII: Ratio of the temperature gradients G_s/G_l at the crystallization front averaged over the interface of Series A.

Ingot/growth configuration	CC262 V3	CC263 V3	CC265 V3	CC266 V3	CC267 V3
G_s/G_l	7.29	6.42	6.61	6.91	7.5

Table IX: Ratio of the temperature gradients G_s/G_l at the crystallization front averaged over the interface of Series B.

4 CONCLUSIONS

The essential prerequisite for increasing the pull speed is to increase the temperature gradient G_s at the crystallization front in the crystal. With the application of an active cooling element during crystal growth, the effective radiation exchange with the crystal surface and thus the heat flux in the crystal can be increased. If the pulling speed exceeds a critical value, cylindrical growth is no longer possible. So-called spiral growth occurs. The limit of the pulling speed correlates in certain limits with the deflection of the phase boundary (Fig.5). Other growth parameters that are critical for high growth rates are the temperature gradient at the liquid side of the interface and the temperature gradient in the meniscus region of the crystal edge. Numerical simulations can obtain these mentioned parameters. From the calculated heat fluxes at the phase boundary a temperature gradient in the melt of 3 K/cm and a temperature gradient at the crystal rim G_{cr} of 1 K/cm can be estimated, which of course depends on its definition.
Another key feature of the active crystal cooling is the ability to influence the shape of the interface at moderate growth rates. Crystal cooling causes an increased heat exchange of the crystal with its environment, resulting in an increased heat conduction in the crystal center. As a consequence, the phase boundary undergoes a convex deflection resulting in a wave-shaped phase boundary. By additionally adjusting the rotation rates of the crystal and crucible, it should be possible to establish an approximately flat phase boundary. Fig.22 shows a theoretical example obtained by numerical simulation.

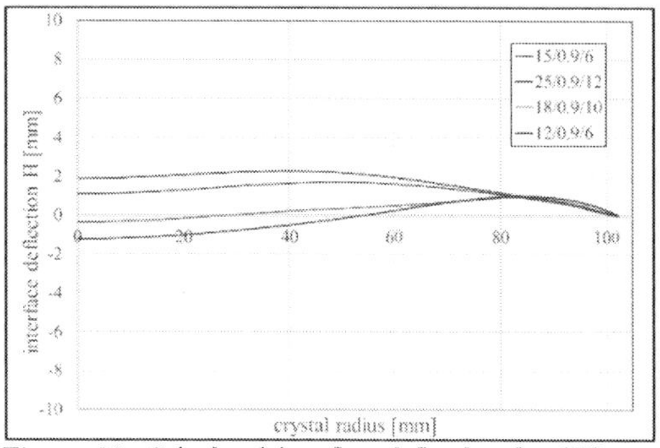

Figure 22: Calculated interface deflection for different rotation conditions at an average pull speed of 0.9 mm/min and active crystal cooling. The legend gives crystal rotation/pull speed/crucible rotation, opposed.

The type of the cooling element does not seem to have any noticeable influence on the shape of the phase boundary but on the cooling behavior of the growing crystal. Thus, cooling can also be used for shorter cooling times of the crystal, which in turn leads to an increase in productivity (Fig.23).

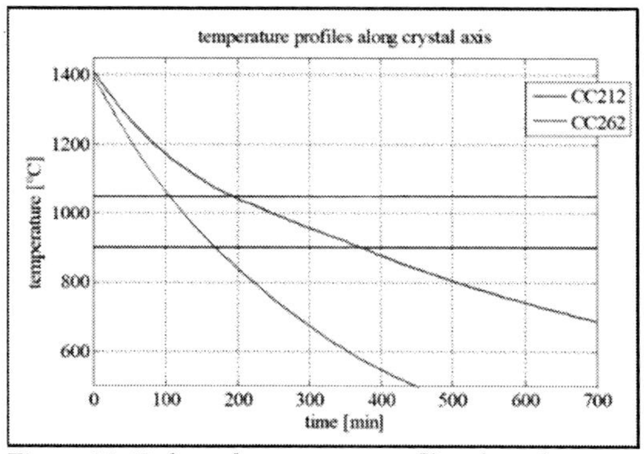

Figure 23: Estimated temperature profiles along the axes of the ingots CC212 grown without crystal cooling and CC262 grown with cooling element E. Both crystals were grown with an average pull speed of 0.9 mm/min.

By tailoring the temperature-time profile in the growing crystal, it should be also possible to influence the development of the defect structure in the growing crystal.

5 ACKNOWLEDGEMENTS

This work was supported by the German Federal Ministry for Economy Affairs and Energy under contract number 0324281B and 0324357B

6 REFERENCES

[1] J. Friedrich, T. Jung, M. Trempa, C. Reimann, A.V. Denisov, A. Mühe, Journal of Crystal Growth **524** (2019) 125168

[2] S.N. Rea, Final Report, ERDA/JPL 954475 (April 1977)

[3] O.V. Smirnova, N.V. Durnev, K.E. Shandrakova, E.L. Mizitov, V.D. Soklakov, Journal of Crystal Growth **310** (2008) 2185-2191

[4] CGSim package, STR Group,Ltd.

[5] H.J. Schulze, A. Lüdge, H. Riemann, Electrochem. Soc. **143**, 4105 (1998)

[6] Omidreza Asadi Noghabi, Moez Jomaa, Mohammed M'hamdi, Journal of Crystal Growth **362** (2013) 77-82

[7] W. von Ammon, E. Dornberger, P.O. Hansson, Journal of Crystal Growth **198-199** (1999) 390-398

COMPARISON OF THE OXYGEN CONCENTRATION IN CZOCHRALSKI SILICON CRYSTAL OBTAINED BY A SIMPLE LUMPED-PARAMETER MODEL AND SOPHISTICATED 2D-3D SIMULATIONS

J. Friedrich[1], M. Trempa[1], H. Koch[1], F. Mosel[2], A. Mühe[2]
[1]Fraunhofer IISB, Schottkystr. 10, 91058 Erlangen, Germany
[2]PVA Crystal Growing Sytems GmbH, Im Westpark 10-12, 35435 Wettenberg, Germany
Email: jochen.friedrich@iisb.fraunhofer.de

ABSTRACT: In the present work a simple boundary layer model for the oxygen transport during Cz silicon crystal growth is revisited and extended by introducing classical scaling laws for the thicknesses of the different boundary layers at the crucible wall, at the free melt surface and at the solid-liquid interface. This lumped model is used to calculate the influence of crystal and crucible rotation on the oxygen transport during Cz growth of 100kg heavy silicon crystals with 210mm diameter in a 24" hot zone geometry. The results of this reduced order model are compared to results from sophisticated, coupled 2D-3D, global numerical simulations of the whole Cz puller. It is found that at a first glance the accuracy of the boundary layer model can be sufficient for certain coarse case studies. However, the boundary layer model hits quickly the wall, when a precise analysis of the oxygen transport is required. The root cause is believed to be the complex flow structure, which results from the combination of crystal and crucible rotation in the vicinity of the solid-liquid interface. For an accurate analysis of oxygen transport sophisticated simulations are indispensable.
Keywords: Convection; Computer simulation; Czochralski; Single Crystal Growth; Silicon

1 INTRODUCTION

Numerical modeling has become an essential tool for optimization and further development of the Cz process of Si mono-crystals for photovoltaic applications with respect to low oxygen concentrations. This is particularly true for large diameter growth processes as respective growth experiments are very expensive. This precise modeling approach is typically achieved by a combination of a global, quasi-stationary, axisymmetric model of the whole Cz puller with a local three-dimensional, time-dependent model of the melt-crystal-crucible region. Thereby, the usage of Large Eddy Simulation (LES) turbulence model is well established [1-9].

However, such numerical calculations, giving precisely high-fidelity spatiotemporal profiles of important process parameters such as oxygen distribution, temperature distribution, and flow profiles, take typically several weeks on a high performance cluster to complete. To get a quick estimation on the influence of process parameters on the crystal properties with reasonable accuracy, a reduced-order model would be desired where results can be obtained more or less within real time. Such lumped parameter model would also allow understanding easier the phenomena e.g. occurring in the melt, if the main physical mechanisms are considered.

Such reduced order models were typically developed in the 1970ies and 1980ies when numerical simulation of the Cz process was in an infancy stage. One example is the boundary layer model for the oxygen transport published by Carlberg et al. [10] and Hoshikawa et al. [11]. With increasing accuracy of numerical simulations these reduced-order models have slided into obscurity. Therefore, it is unclear whether lumped-parameter models are able to predict the influence of different process parameter on the oxygen transport for large diameter Cz silicon crystal growth with almost the same accuracy than highly sophisticated, time consuming 2D-3D simulations.

Therefore, in this work we used a further developed version of the boundary layer model of [10, 11] to calculate the influence of crystal and crucible rotation on the oxygen transport. The results of this reduced-order model will be compared to results obtained by coupled 2D-3D simulations.

2 COUPLED 2D-3D NUMERICAL SIMULATIONS

We performed coupled 2D-3D simulations of the heat and oxygen transport during growth of Si crystals with 210mm diameter and a weight of 190kg in a 24" (diameter of the silica crucible) hot zone. In the 2D-3D model the convective heat and oxygen transfer are computed by a time-dependent 3D-LES model [5, 6]. Temperature, melt velocities and oxygen concentration are averaged azimuthally and in time, and a turbulent heat flux is calculated form the observed fluctuations of temperature and velocity. These average values plus the turbulent heat fluxes are then imported into a Reynolds averaged form of the temperature equation inside the 2D-model, in which the temperature field as well as the SiO evaporated from the melt surface in the whole Cz puller geometry is calculated. The details of the 2D-3D model can be found in [5, 6, 12]. The 2D-3D model was already validated by comparison to data from Direct Numerical Simulation as well as by comparison to experimental data with respect to the shape of the solid-liquid interface and oxygen distribution in large diameter Si Cz crystals [5, 6].

3 THE REDUCED-ORDER MODEL

Our approach in developing a reduced-order model begins with the assumption of a well-mixed melt region, thus eliminating spatial variation of dissolved oxygen from the crucible. Due to strong flows from the combination of Rayleigh-Bernard convection with azimuthal crystal and/or crucible rotation, adequate mixing from turbulence is assumed. Fluxes of oxygen in and out are estimated as simple linear Fickian diffusion across boundary layers of defined width as originally proposed by [10, 11] (see Figure 1).

The basic variables of interest are defined in Table 1. In general, subscripts of c, s, m, and x will refer to crucible, free melt surface, melt volume, and crystal, respectively. The material properties of silicon used in the present work can be found in [5].

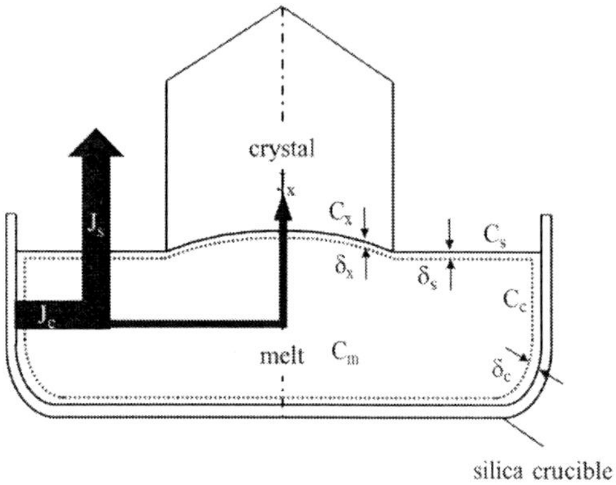

Figure 1: Cartoon of basic mass flows.

Vari-able	Description
C_c	Equilibrium concentration at crucible wall
C_m	Solute concentration in the melt volume
C_s	Solute concentration at the free melt surface
C_x	Solute concentration at the s/l interface
A_c	Contact area between melt and crucible wall
A_s	Area of free melt surface
A_x	Area of s/l interface
D	Diffusion coefficient
r_c	Crucible radius
r_x	Crystal radius
δ_i	Solute boundary layer thickness
$\dot{n}_i$	Molar flow rate
J_i	Molar flux
ω_i	Angular velocity

Table 1: Basic variables of the boundary layer model. In general, subscripts of c, s, m, and x will refer to crucible, free melt surface, melt volume, and crystal, respectively.

3.1 Mass fluxes J_i at crucible wall, free surface and solid/liquid interface

We define the molar oxygen fluxes from the crucible into the melt, out of the free melt surface, and into the crystal as $\dot{n}_c$, $\dot{n}_s$, and $\dot{n}_x$, respectively. Thus, the flux of oxygen into the melt from the crucible wall is given by

$$\dot{n}_c = J_c A_c \qquad \text{eq. 1}$$
$$J_c = \frac{D}{\delta_c}(C_c - C_m) \qquad \text{eq. 2}$$

At the melt surface the flux of oxygen from the melt into the gas ambient is defined as

$$\dot{n}_s = J_s A_s \qquad \text{eq. 3}$$
$$J_s = \frac{D}{\delta_s}(C_m - C_s) \qquad \text{eq. 4}$$

We assume that we have an ideal transport of SiO from the melt surface into the ambient gas. Therefore, all SiO formed at the surface is immediately removed and the oxygen concentration at the melt surface C_s approaches zero. This assumption is also used in our 2D-3D model, because it has been proven that the interaction of the gas flow with the melt flow can be neglected for the given puller geometry and process parameters.

At the solid-liquid interface the flux of oxygen out of the melt is given by

$$\dot{n}_x = J_x A_x \qquad \text{eq. 5}$$
$$J_x = k_{eff} V C_m \qquad \text{eq. 6}$$

Here we use the Burton-Prim-Schlichter (BPS) model to derive the effective segregation coefficient k_{eff}.

$$k_{eff} = \frac{k}{k+(1-k)\exp(-\frac{V\delta_x}{D})} \qquad \text{eq. 7}$$

δ_x is the solute boundary layer at the solid-liquid interface, V the pull rate. δ_x can be approximated by the model for a rotating disk, which gives [13]:

$$\delta_x = 1.6 D^{1/3} \upsilon^{1/6} \omega_x^{-1/2} \qquad \text{eq. 8}$$

3.2 Mass balance

Because of slow growth rates, we assume a pseudo-steady concentration inside the melt volume. The mass balance then becomes:

$$\frac{dn_m}{dt} = 0 = J_c A_c - J_s A_s - J_x A_x \qquad \text{eq. 9}$$

And solving for C_m gives

$$C_m = \frac{1}{1+\frac{A_s\delta_c}{A_c\delta_s}+\frac{A_x\delta_c}{A_c D}V k_{eff}} C_c \qquad \text{eq. 10}$$

If we neglect the bending of the solid-liquid interface, then the areas A_i are all known from the crystal and crucible geometry and the initial melt mass. Also the equilibrium oxygen concentration at the crucible wall C_c is known. Figure 2 shows the temperature dependence of the oxygen equilibrium concentration published from several authors [14-17]. Most papers within the last ten years are using the curve from Togawa et al. [14] as boundary condition in their numerical simulations. We apply also this curve in our coupled 2D-3D approach as well as in our lumped model. However, in the lumped model we neglect completely the temperature dependence and assume a constant value of C_c=2x10^{18}cm^{-3}.

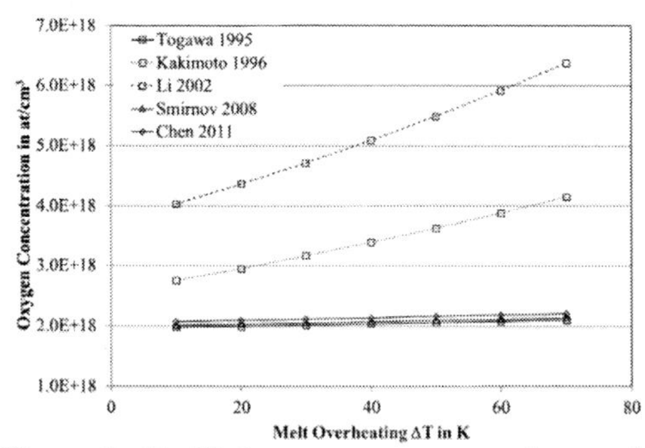

Figure 2: Equilibrium oxygen concentration at the crucible wall C_c versus melt overheating ΔT published by several authors [14-17].

A priori, the solute boundary layer thickness at the crucible, at the melt surface and at the solid-liquid interface are unknown, but can be estimated from transport scaling relationships and process parameters for the momentum boundary layers. Thereby, the Schmid number Sc = (ν/D) is used to approximate the thickness of the solutal boundary layer δ_i from the momentum boundary layer δ_i' [18] which is marked by a prime further on.

$$\delta_i = 2.4 * Sc^{-1/3}\delta_i' \qquad \text{eq. 11}$$

To estimate $\delta c'$, we turn to an analysis of King et al. [19]. They investigated the different flow regimes in rotating Ralyeigh-Benard configurations in dependence on the Ekman number Ek and Rayleigh number Ra which are defined for the present work as follows:

$$Ek = \frac{\nu}{2\omega_c r_c^2} \qquad \text{eq. 12}$$

$$Ra = \frac{\beta g \Delta T r_c^3 (\frac{h}{r_c})}{\kappa\nu} = Gr * Pr \qquad \text{eq. 13}$$

with volumetric expansion coefficient β, melt height h, crucible radius r_c, crucible rotation ω_c, melt overheating ΔT, kinematic viscosity ν, thermal diffusivity κ, Grashof number Gr, and Prandtl number Pr= κ/ ν.

King et al. [19] found that for geostrophic conditions ($Ra \cdot Ek^{3/2} < 10$) the momentum boundary layer thickness follows a scaling relationship with the Ekman number Ek:

$$\delta_c' = 3Ek^{1/2}r_c \qquad \text{eq. 14}$$

This criterion for geostrophic flow is fulfilled for the present Cz set-up ($0.01 < Ra \cdot Ek^{3/2} < 0.1$), therefore eq. 14 should be valid. However, to check this assumption we used also the alternate estimation for non-rotating Rayleigh-Bernard convection [19], which is possible by the Grashof number, Gr.

$$\delta_c' = Gr^{-1/4}r_c \qquad \text{eq. 15}$$

For the free surface, we utilize the analysis of [20]. The thickness of momentum boundary layer associated with Marangoni flow over the free surface is found to follow the following relationship with the Marangoni number Ma. The relevant length scale here is the distance between the crucible wall and the crystal, or $r_c - r_x$. The Marangoni number is defined as

$$Ma = \frac{-(\frac{\partial\sigma}{\delta T})\Delta T(r_c - r_x)}{\rho\nu^2} \qquad \text{eq. 16}$$

with density ρ, and temperature dependence of surface tension $\frac{\partial\sigma}{\delta T}$. Accordingly, the momentum boundary layer thickness at the free melt surface becomes:

$$\delta_s' = Ma^{-1/3}(r_c - r_x) \qquad \text{eq. 17}$$

Finally, we get the following expression for the oxygen concentration by using the momentum instead of the solute boundary layer thicknesses.

$$C_m = \frac{1}{1 + \frac{A_s\delta_c'}{A_c\delta_s'} + \frac{A_x}{A_c(D/V)}\frac{\delta_c'}{} \frac{2.4 k_{eff}}{Sc^{1/3}}} C_c \qquad \text{eq. 18}$$

4 RESULTS AND DISCUSSION

4.1 Influence of melt overheating

In eq. 18 all parameters are known except the melt overheating ΔT, which is needed to calculate the Grashof- and Marangoni-number. Figure 3 and Figure 4 show the dependence of the overheating ΔT on the crucible rotation ω_c and crystal rotation ω_x for different body lengths (resp. solidified fractions g), obtained from our 2D-3D simulations. As already known from literature [21] the overheating increases with increasing crucible rotation, because the convective heat transport in the lateral direction is suppressed due to the higher Coriolis force for higher crucible rotation. The influence of crystal rotation is minor.

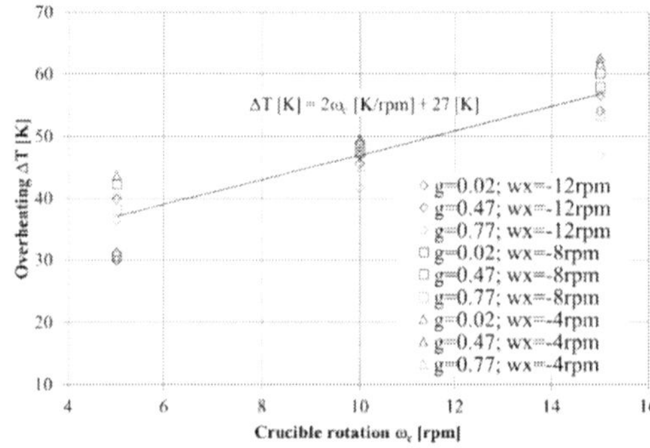

Figure 1: Melt overheating ΔT versus crucible rotation ω_c for different solidified fractions g and crystal rotations ω_x obtained from 2D-3D simulations.

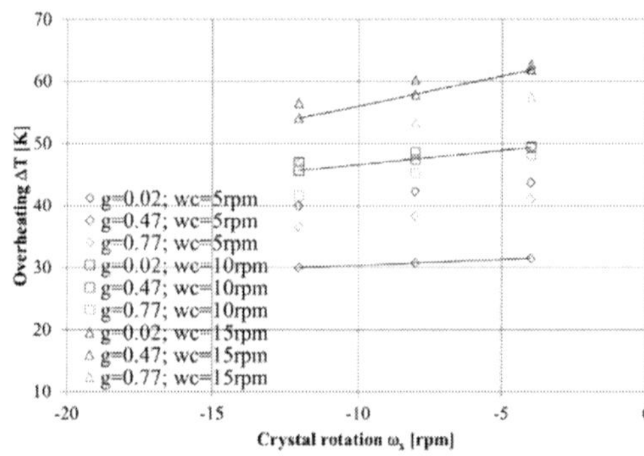

Figure 4: Melt overheating ΔT versus crystal rotation ω_x for different solidified fractions g and crucible rotations ω_c obtained from 2D-3D simulations.

In a first step, let us neglect J_x and consider J_c and J_s only. For a given crucible rotation the overheating decreases by less than 10K with increasing the crystal rotation, which means that the momentum boundary layer $\delta s'$ increases, the term $A_s\,\delta c' / A_c\,\delta s'$ decreases and therefore, the oxygen concentration increases by less than 3% according to the lumped model. Therefore, we consider only the dependence of the overheating ΔT on the crucible rotation, which can be expressed as follows when we apply a linear trend line to the data in Figure 3:

$$\Delta T \approx 27K + 2\omega_c[\frac{K}{rpm}] \qquad \text{eq. 19}$$

4.2 Importance of mass flux J_x

In a second step, we analyze the importance of the mass flux J_x. A priori, it is expected that most of the oxygen entering the melt over the crucible wall will evaporate as SiO from the melt surface [22], that means that the oxygen concentration depends strongly on the contact area A_c between melt and crucible and that J_x is small compared to J_s. The latter is equivalent that the third term in the denominator of eq. 18 is negligible. Figure 5 shows indeed that the oxygen concentration decreases with increasing body length, because the contact area A_c decreases and thus, the denominator of eq. 18 increases. It is also obvious that the oxygen concentration will increase by less than 5% only when the mass flux J_x is not considered.

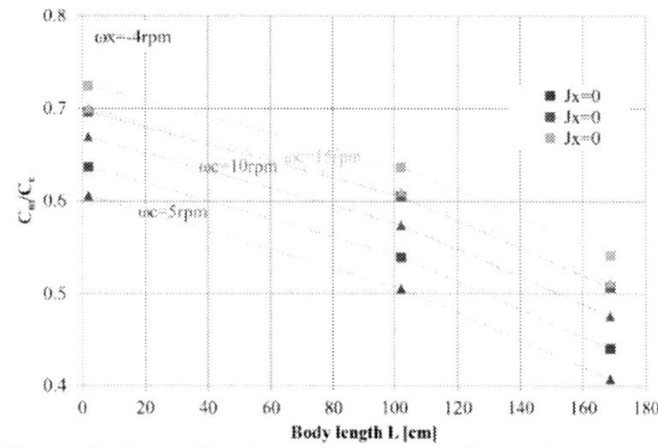

Figure 5: Normalized oxygen concentration C_m/C_c versus body length of the crystal for different crucible rotations ω_c at crystal rotation ω_x=-4rpm calculated by eq. 18 with eqs. 14, 17 and 19 w/o consideration of the mass flux J_x.

4.3 Influence of crucible rotation

In a further step, we compare the results of the boundary layer model with coupled 2D-3D simulations for varying crucible rotation and constant crystal rotation. In the boundary layer model, eq. 19 was used to calculate the melt overheating ΔT. Figure 6 shows the oxygen concentration in the crystal calculated by the boundary layer model and obtained from the 2D-3D simulations. The results of the boundary layer model are in qualitative agreement to the 2D-3D simulations when the Ekman number is used to determine the boundary layer thickness at the crucible wall (eq. 14). In this case, the well-known increase of oxygen concentration with increasing crucible rotation [22] is well represented. The root cause is the decreasing boundary layer thickness δ_c at the crucible wall with increasing crucible rotation, which results in an increase of the oxygen flux into the melt.

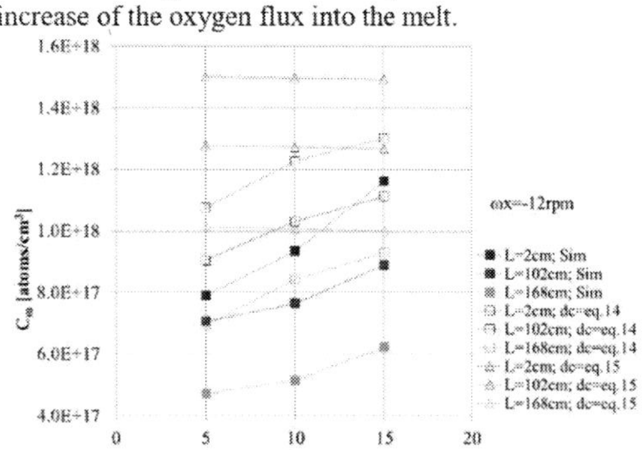

Figure 6: Oxygen concentration versus crucible rotation at constant crystal rotation (-12rpm) for different body lengths. Full squares represent the results of the simulations, open squares the results of the boundary layer model using eq. 14, open triangles the results of the boundary layer model using eq. 15.

On the other hand, when the boundary layer at the crucible wall is calculated by using the Grashof number (eq.15), no dependence of the oxygen concentration on the crucible rotation is found. This proofs that convection in the silicon melt represents a geostrophic flow under these conditions.

4.4 Influence of crystal rotation

In a last step, we compare the results of the boundary layer model with coupled 2D-3D simulations for constant crucible rotation and varying crystal rotation.

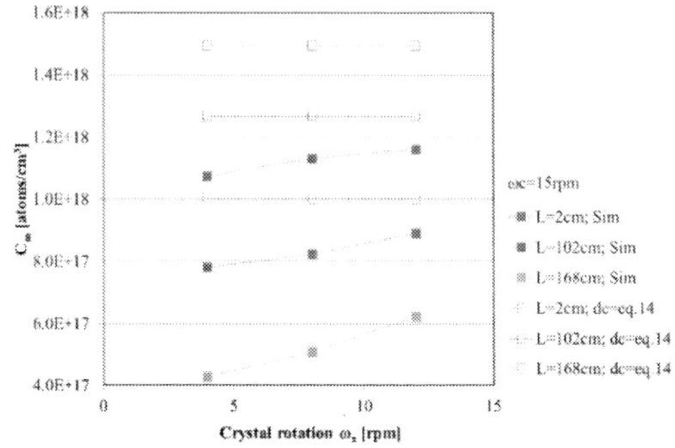

Figure 7: Oxygen concentration versus crystal rotation at constant crucible rotation (15rpm) for different body lengths. Full squares represent the results of the simulations, open squares the results of the boundary layer model using eq. 14.

From Figure 7 is obvious that the slight increase of the oxygen concentration with increasing crystal rotation in the 2D-3D simulation is not reproduced by the boundary layer model. Obviously, the influence of the crystal rotation is underestimated in the boundary layer model. In order to check whether this could be caused by the assumption of a flat interface in the boundary layer model we replaced the ideal area of the flat interface by the real area of the curved interface A_x which we extracted from the 2D-3D simulations. In addition, we replaced the approximated melt overheating (eq. 19) by the values obtained from the simulation. However, there is almost no difference between these three cases. Thus, the complex flow structure itself is contributing much to the real oxygen transport, which is not represented in this very simple lumped parameter model.

5 CONCLUSIONS

The boundary layer model allows the calculation of the oxygen concentration at the touch of a button on a simple PC, whereas the coupled 2D-3D simulations take several weeks on a PC cluster. The comparison to the 2D-3D simulations shows that at a first glance the accuracy of the boundary layer model is sufficient for certain coarse case studies. Therefore, it can contribute to a quick and improved understanding about the oxygen transport during silicon Czochralski crystal growth, especially for beginners in crystal growth. Extensions of the model to take into account magnetic field effects or the influence of the interaction of melt and gas flow can be carried out easily as already shown in literature.

However, the boundary layer model hits quickly the wall, when a precise analysis of the oxygen transport is required. The root cause is believed the complex flow structure, which results from the combination of crystal and crucible rotation in the vicinity of the solid-liquid interface. For a better accuracy, it might be useful to consider several melt regions beneath the crystal in the model. However, it is assumed that too much data have to be extracted from the 2D-3D simulations which serve as input parameter for such improved boundary layer model.

This contradicts the charm of a lumped model.

Therefore, only first rough estimations can be made by using the boundary model. For more accurate treatments coupled 2D-3D simulations of the oxygen transport are indispensable.

6 ACKNOWLEDGEMENT

Our former colleague Dr. Thomas Jung is highly acknowledged for carrying out the 2D-3D simulations. We thank also Prof. Jeff J. Derby and Kerry Wang from the University of Minnesota for the fruitful discussion about the applicability of the lumped model. Research and development published in this work were part of the *CZ3003* project, which is funded partly by the German Federal Ministry for Economic Affairs and Energy (BMWI) under contract number 0324357A.

7 REFERENCES

[1] I. Yu. Evstratov, V. V. Kalaev, A. I. Zhmakin, Yu. N. Makarov, A. G. Abramov, N. G. Ivanov, E. M. Smirnov, E. Dornberger, J. Virbulis, E. Tomzig, W. von Ammon, Journal of Crystal Growth 230 (2001) 22-29

[2] D. P. Lukanin, V. V. Kalaev, Yu. N. Makarov, T. Wetzel, J. Virbulis, W. von Ammon, Journal of Crystal Growth 266 (2004) 20-27

[3] D. Smirnov, V.V. Kalaev, Journal of Crystal Growth 310(12) (2008) 2970-2976

[4] A. Krauze, N. Jekabsons, A. Muiznieks, A. Sabanskis, U. Lacis, Journal of Crystal Growth 312 (2010) 3225-3234

[5] T. Jung, J. Seebeck, J. Friedrich, Journal of Crystal Growth 368 (2013) 72–80

[6] J. Friedrich, T. Jung, M. Trempa, C. Reimann, A. Denisov, A. Muehe, Journal of Crystal Growth 524 (2019) 125-168

[7] J. Zhang, J.-C. Ren, D. Liu, Results in Physics 13 (2019) 102127

[8] R. Yokoyama, T. Nakamura, W. Sugimura, T. Ono, T. Fujiwara, K. Kakimoto, Journal of Crystal Growth 519 (2019) 77–83

[9] V. Kalaev, Journal of Crystal Growth 532 (2020) 125413

[10] T. Carlberg, T. B. King, A. F. Witt, J. Electrochem. Soc. 129/1 (1982) 189–193

[11] K. Hoshikawa, X. Huang, Materials Science and Engineering B 72(2–3) (2000) 73-79

[12] J. Fainberg, D. Vizman, J. Friedrich, G. Mueller, Journal of Crystal Growth 303(1) (2007) 124-134

[13] A. G. Ostrogorsky, G. Müller, Journal of Crystal Growth 121(4) (1992) 587-598

[14] S. Togawa, X. Huang, K. Izunome, K. Terashima, S. Kimura, Journal of Crystal Growth 148 (1–2) (1995) 70-78

[15] K. Kakimoto, K. Yi, M. Eguchi, Journal of Crystal Growth 163 (3) (1996) 238-242

[16] M. Li, Y. Li, N. Imaishi, T. Tsukada, Journal of Crystal Growth 234 (1) (2002) 32-46

[17] J. C. Chen, Y. Y. Teng, W. T. Wun, C. W. Lu, H. I. Chen, C. Y. Chen, C. Y., W. C. Lan, Journal of Crystal Growth 318(1) (2011) 318-323

[18] B. Vartak, A. Yeckel, J. J. Derby, Journal of Crystal Growth 283 (2005) 479–489

[19] E. M. King, S. Stellmach, and B. Buffett, Journal of Fluid Mechanics 717 (2013) 449-471

[20] Y. Okano, A. Hatano, A. Hirata, J. Electrochem. Eng. Jpn.22(4) (1989) 385-388

[21] D. Vizman, O. Gräbner, G. Müller, Journal of Crystal Growth 233 (2001) 687–698

[22] J. Friedrich, W. von Ammon, G. Müller, Czochralski Growth of Silicon Crystals, in: Handbook of Crystal Growth Vol. II. (Eds. T. Nishinaga, P. Rudolph) Elsevier (2015) 45–104

PROCESSING OF A HIGH-QUALITY SINGLE CRYSTAL SILICON: OPTIMIZED PARAMETERS FOR THE CZOCHRALSKI METHOD

Emre Uçar[1]*, Nurhayat Yıldırım[1] , Fatma Sena Yıldırım[1], Neval Dilan Yıldırım[1], Melis Çetmeli[1], Mehmet Konyar[1], Fırat Es[1]
Kalyon PV Research and Development Center,
[1] KalyonPV Research and Development Center, Kalyon Güneş Teknolojileri Üretim A.Ş.,06909 Ankara, TURKEY
* Corresponding Author: TEL: +90 5319210038, E-mail: eucar@kalyonpv.com

ABSTRACT: Optimization of hot zone parts and increasing resistance is the key to increase minority carrier lifetime, thereby, efficiency. In this study, Boron-doped p-type monocrystalline silicon ingots were grown by the Czochralski (CZ) method simply by employing necessary hot zone parts and 28" inch crucibles to obtain higher minority carrier lifetime values. Defects, minority carrier lifetime and resistivity were evaluated for the optimized growing processes. Results revealed that utilization of larger parts for the CZ pullers yielded higher values for the top part of the ingot.

Keywords: Czochralski, Dislocation, Impurities, Lifetime, Silicon

1. INTRODUCTION

Czochralski, float zone or directional solidification methods can be utilized for production of c-Si ingots. Czochralski method is commonly preferred to produce single crystal silicon ingots due to its advantages. The float-zone technique is also advantageous due to its lower impurities (e.g. oxygen and carbon) and higher purity levels, but this is an additional step which requires an almost produced ingot. Therefore, many photovoltaic applications do not require this additional step because of higher costs [1, 2].

Ingots grown by the CZ-technique are susceptible to oxygen and carbon contamination which are incorporated by the polysilicon, crucible and hot zone parts. Relevant parts (i.e. coating layer of the crucible) of the CZ pullers are also the source of the contamination which decrease the values of important parameters (resistivity, lifetime values etc.) to obtain an efficient solar cell [3, 4].

In this study, larger hot zone parts and 28'' crucibles were employed to produce Boron-doped p-type monocrystalline silicon ingots. Recipes of ingots with changing resistivities (0.2 Ωcm, 1.5 Ωcm, 143 Ωcm) were processed to discuss the effect of employing larger parts on minority carrier lifetime, resistivity and impurities.

2. EXPERIMENTAL

Boron doped p-type monocrystalline silicon ingots at 228 mm diameter were produced by the Czochralski (CZ) method at (100) orientation. A CZ puller (Jingsheng TDR-115P) equipped with a 28'' crucible and necessary hot zone components (Meijing Material Co. Ltd.) adaptable to this size was employed to melt the polysilicon (9N grade, Wacker Chemie AG) at 1450°C. Boron doping was applied to obtain p-type structure at the level of 10^{16} at/cm^3. A p-type seed with (100) orientation was dipped to the homogeneous melt and solidification was applied at a pulling speed of 1.6 mm/min. Solidification was progressed under a counterclockwise rotation of 10 rpm where crucible was rotated vice versa at 8 rpm to reduce defects and thermal gradience. Argon gas was purged at 75 min/L to prevent excess oxygen incorporation. 4 consecutive runs were conducted by semi-continuous feeding.

All ingots were characterized by their crystallinity by IR-Imaging technique (Semilab PLB-55R). Then, each was subjected to cropping before they are characterized by their resistivity, minority carrier lifetime and impurity content. Top and tail parts are removed, and all ingots are cut to sections to form bricks before obtaining wafers. Disks with 2 mm thickness were cut between each section to analyze oxygen and carbon content by FT-IR Spectroscopy (Thermo Fisher Scientific Nicolet iS20). Minority Carrier Lifetime and resistivity measurements were conducted from top and tail parts of sections by Ingot testing (Semilab WT 1200I) and a mobile resistivity measurement device (Semilab RT – 1000), respectively.

3. RESULTS AND DISCUSSION

3.1 Effect of larger crucible and enhanced hot zone parts

Ingots were produced successfully by employing larger hot zone parts and 28'' crucible. Processing of at least 500 ingots were evaluated for this study to compare both designs. Minority carrier lifetime value results of the ingots grown by employing smaller and larger furnace parts were given in Figure 1a. Higher lifetime values were obtained for the ingots produced by the utilization of larger parts. However, similar resistivity values almost duplicating each other was obtained for both ingot types (Fig. 1b). Oxygen concentration was less than expected for ingots produced by larger parts (Fig. 2a) and carbon concentration remained almost at same level (Fig. 2b). These results could be attributed to the thermal gradient which could be less by employing larger parts. These parts should yield less impurity content to increase lifetime values. As Colletti et. al. mentioned, these parts are closely correlated with determining solar cell efficiency [3].

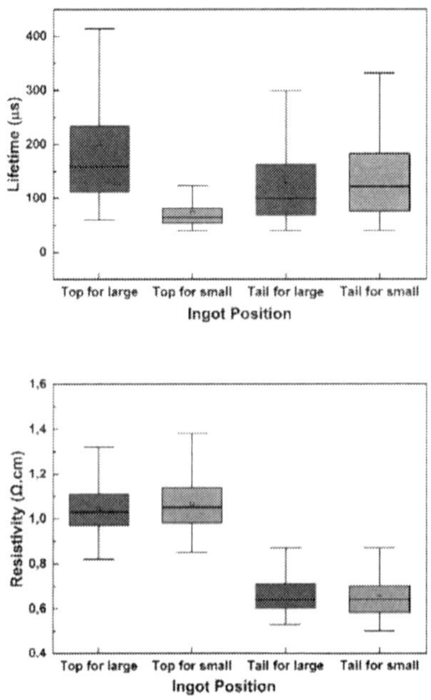

Figure 1. The mean minority carrier lifetime (a) and resistivity (b) values for ingots employing small and large parts.

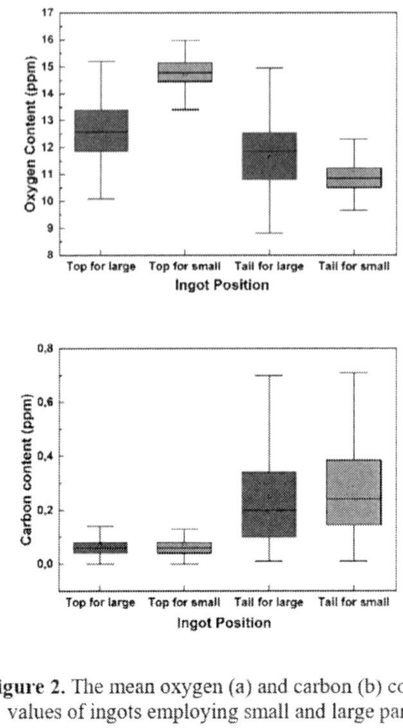

Figure 2. The mean oxygen (a) and carbon (b) content values of ingots employing small and large parts.

3.2 Effect of enhanced resistivity

Ingots with varying resistivities (0.2 Ωcm, 1.5 Ωcm, 143 Ωcm) were produced by changing the dopant concentration of the original recipe. Ingots with a resistivity value of 143 Ωcm were obtained by using without any dopant. Minority carrier lifetime values were increased tremendously as dopant concentration was decreased (Fig. 3). This could be attributed to the impurity level of the structure; however, this is not yet successful to produce an efficient solar cell as a p-n junction is required [5]. Impurity level of the ingots remained almost at the same level for all types of resistivities except for 0.2 Ωcm (Fig. 4). An oxygen incorporation was analyzed for the top part which could be eliminated because of a processing mistake.

Table 1. Minority carrier lifetime values of ingots by their resistivities.

Resistivity (Ωcm)	M. C. Lifetime Top (µs)	M. C. Lifetime Tail (µs)
0.2	6,8	4,8
1.5	288,3	91,3
143	9656	9842,1

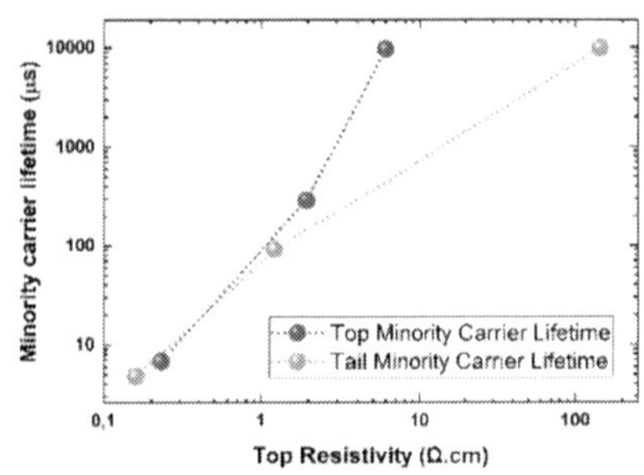

Figure 3. Lifetime variation amongst different resistivity of ingots.

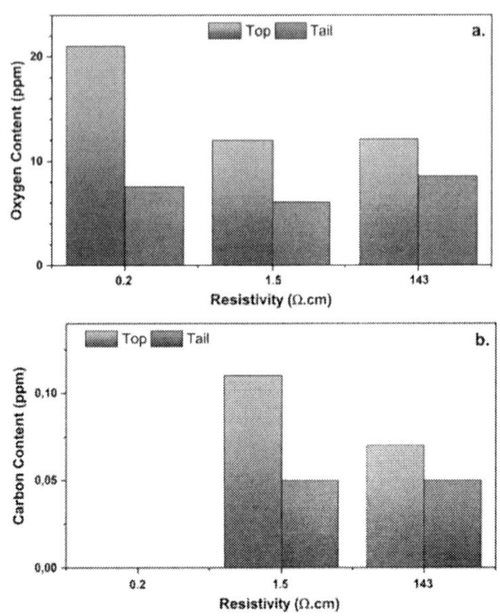

Figure 4. Oxygen (a) and carbon content of the ingots by their resistivities.

4. CONCLUSIONS

Ingots with varying resistivity values were produced successfully. Reducing the amount of dopant concentration yielded higher minority lifetime values. However, as building a p-n junction enhances efficiency of a solar cell and maintains a prolonged carrier lifetime, it is important to adjust dopant concentration of ingots before they are produced to wafers. Utilization of larger parts for furnaces also improved minority lifetime values for the same recipes. This also maintained an easier way of mass production of photovoltaic ingots.

5. REFERENCES

[1]Silicon Ingot Production: Czochralski- and Float - Zone Technique, https://www.microchemicals.com/products/wafers/silicon_ingot_production.html

[2] Crystal Growth Technique, http://www.jiwaji.edu/pdf/ecourse/physics/M.Sc.%20Physics%20on%20Crystal%20Growth%20Technique.pdf

[3] G. Coletti, Impurities in Silicon and Their Impact On Solar Cell Performance; Utrecht University: Utrecht, The Netherlands, (2011).

[4] Mohammad ZiaurRahman, Modeling Minority Carrier Recombination Lifetime in Silicon Solar Cell", International Journal of Renewable Energy Research, Vol.2 (1), pp. 449- 454, (2012)

[5] C-Y. Wu, J-F. Chen, Doping and Temperature Dependences of Minortiy-Carrier Diffusion Length and Lifetime Deduced From the Spectral Response Measurements of p-N Junction Solar Cells, Solid-State Electronics Vol. 25, No. 7 pp. 679-682, (1982)

TOWARDS ULTRA-THIN SILICON SOLAR CELLS FOR HIGH SPECIFIC POWER APPLICATIONS

Yuchi Lan[1, 2], Maryknol Delos Santos[2], Donny Lai[2], Xia Yan[2], Aaron Danner[1, 2]
[1]Department of Electrical and Computer Engineering, National University of Singapore, Singapore.
[2]Solar Energy Research Institute of Singapore, National University of Singapore, Singapore.
Contact: yuchi.lan@u.nus.edu, adanner@nus.edu.sg

ABSTRACT: This paper introduces a detailed study between carrier lifetime and wafer thickness in Czochralski-grown (Cz) crystalline silicon (c-Si) wafers which will later be processed into solar cells. Wafers are thinned to different thicknesses with a minimum of 54.8 μm via wet chemistry etching as carrier lifetime drops from around 3000 μs to lower than 500 μs. A negative correlation appears between the carrier lifetime and wafer thickness.
Keywords: c-Si, etching, lifetime, thin film

1 INTRODUCTION

Ultra-thin silicon solar cells have recently attracted increased attention due to their properties of being thin, light, and bendable thus suitable to be used in applications where weight is a significant factor like aerospace and wearable electronics. However, as silicon wafers become thinner and thinner, they usually have a decrease in overall quality (for example, in carrier lifetime and optical absorption) which results in a trade-off in solar cell performance. This has multiple causes, including increased effects of surface recombination on overall lifetime. Here in this paper, we introduce a detailed study between carrier lifetime and wafer thickness in Cz c-Si wafers which will later be processed into solar cells. Understanding physical mechanisms of decreased efficiency in thin silicon solar cells will be increasingly important in applications requiring high specific power and may later motivate the need for improved surface passivation, or improved light trapping for instance.

2 EXPERIMENTAL DETAILS

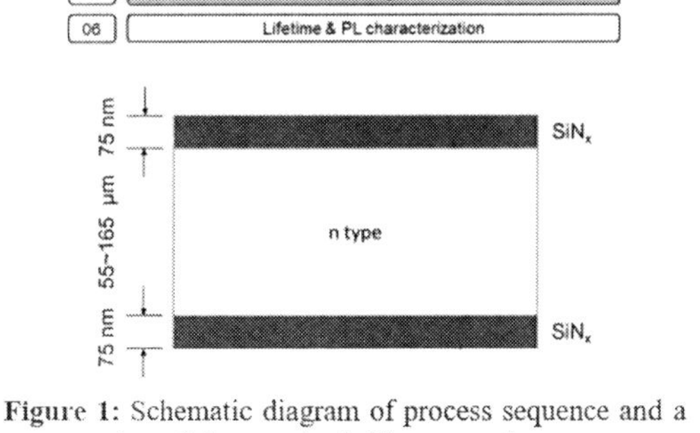

Figure 1: Schematic diagram of process sequence and a cross section of characterized silicon samples.

Industrial n-type Cz-Si wafers of as-cut thickness 190 μm are first thinned to various thicknesses prior to characterization. To avoid having the lifetime being influenced by scratches during processing, wafers are divided into two sets—one for thickness measurement and one for lifetime characterization. The method of etching here is saw damage etching (SDE), which uses a chemical bath of 20% KOH and 80% deionized (DI) water. Wafers in each set are separated into nine groups of 10-15 wafers, and each group receives an SDE time between 5 and 90 min. The nominal bath temperature is 78 °C and it is worth noting that the SDE process is exothermic and during the process the actual bath temperature rises up to 83 °C. After SDE, standard RCA cleaning processes are applied to remove organic and metallic contaminations, followed by a 1% HF dip to remove native oxide. Wafers are dried in a surface tension gradient (STG) dryer. Afterwards, a layer of 75 nm SiN_x is deposited on both sides of each wafer via plasma-enhanced chemical vapor deposition (PECVD), which provides the surface passivation. Then wafers are transferred to a fast-firing furnace with a stepwise temperature profile up to 725 °C.

For the thickness measurement wafer set, thickness is measured before and after SDE with an MX6012 eddy current sensor from E+H. The lifetime characterization for the other set is carried out by the quasi-steady-state photoconductance (QSSPC) method via WCT-120 from Sinton Instruments. Photoluminescence (PL) images of wafers with surface passivation are characterized via LIS-R2 from BT Imaging. Wafers without surface passivation are characterized for surface roughness via an AXIO Imager A1m microscope from Zeiss.

3 RESULTS AND DISCUSSION

As wafers become thinner and thinner, it is observed that they initially become more bendable, but below 70 μm the wafers appear to be brittle and break easily during processing. This phenomenon is particularly evident after the etching time reaches 70 mins, where the wafer thickness is around 55 μm. In the 70-min group, all wafers in both sets survived the SDE but during the drying process only 2 wafers in the thickness measurement set survived. In the 80-min group, all 10 wafers broke in the chemical bath, so this represents our experimental boundary without process alterations. Our thinnest intact wafer has a 54.8-μm thickness.

3.1 Characterization of thickness and lifetime

As shown in Figure 2, the wafer thickness decreases from the original thickness of 190 μm almost linearly as

SDE time increases. However, as longer SDE is carried out, the lifetime increases first with a peak value around 3000 μs and then drops. That is due to the diamond-wire induced saw-damage on the surface of as-cut wafers and a 5-min SDE is not enough to remove it, so the first group has a lower lifetime. Otherwise, the resulting wafer thickness and lifetime both have a negative correlation with SDE time. The lowest lifetime in this experiment is around 500 μs, which is achieved in 70-min group.

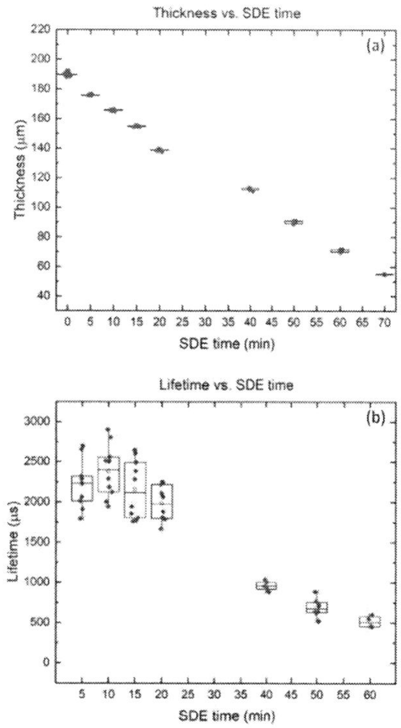

Figure 2: Relationship between wafer (a) thickness, (b) lifetime and SDE time.

Moreover, the standard deviation of carrier lifetime in shorter SDE groups is larger than that in longer SDE groups. This phenomenon is probably caused by the difference in surface morphologies from wafer to wafer due to saw-damage, which will be discussed more in section 3.3. When longer SDE is performed to wafers, this surface damage is more thoroughly removed thus the surface passivation via PECVD can be better carried out. As a consequence, the values of carrier lifetime of wafers in the same long SDE group are more concentrated.

3.2 Photoluminescence

Photoluminescence (PL) images of wafers with surface passivation are characterized to observe their surface uniformity. Figure 3 shows PL images of wafers etched for 10/20/40/60 mins, respectively. The color of the center and the corners all have similiar lightness in PL images, which indicates that the chosen wafers are fairly uniform. However, more scratches and water marks appear on the surface of wafers etched for 40 and 60 mins due to the challenges in dealing with ultra-thin silicon wafers during the wet processes. As mentioned before, ultra-thin wafers are more bendable and thus tend to stick to each other during wet processes. Hence, double spacing between neighboring wafers in the cassette is necessary and effective to handle this problem.

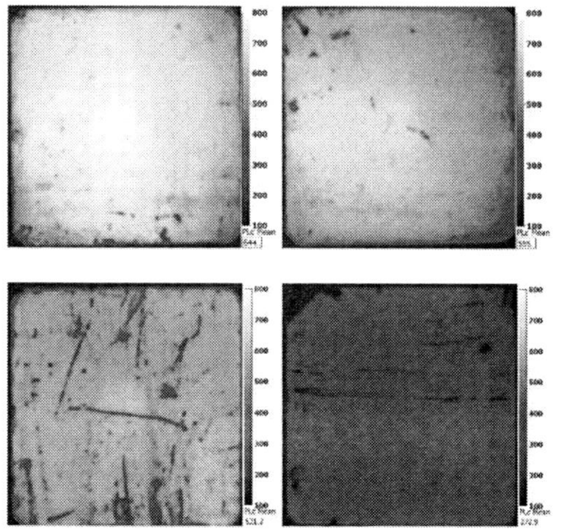

Figure 3: PL images of wafers etched for 10/20/40/60 mins, respectively.

From PL images, it can be clearly seen that as the SDE time increases the images turn darker (i.e., a lower PL intensity), which represents a lower lifetime. In this case, it is a negative correlation between lifetime and SDE time. In other words, when a wafer is thinner, its lifetime is lower.

3.3 Surface roughness

Wafers after SDE and without surface passivation are characterized for surface roughness. Figure 4 shows microscopic images of wafer surface in the 10/20/40/60-min groups. Under the microscope, a difference of surface roughness is obvious, and this will further influence the lifetime.

Figure 4: Microcopy images of wafers etched for 10/20/40/60 mins, respectively.

Horizontally arranged square patterns are observed, which are parallel to the saw-damage on as-cut wafers. Alkaline solutions like KOH and NaOH etch <111> planes of Si wafers with a much slower etching rate compared to other planes, especially <100> planes. The ratio of etching rate of alkaline solutions on these two planes can reach 10 to 35. That explains why each square pattern is located around 45° to the diamond-saw lines and the longer the SDE, the bigger the patterns. Meanwhile, the depth of these etched patterns are more uniform when a longer SDE is applied, which causes the difference in the standard deviation of lifetime discussed in section 3.1.

3.4 Lifetime v.s. thickness

The effective lifetime of Si is related to its bulk lifetime and surface lifetime.

$$\frac{1}{\tau_{eff}} = \frac{1}{\tau_b} + \frac{1}{\tau_s}$$

where τ_{eff} is effective lifetime, τ_b is bulk lifetime, and τ_s is surface lifetime. Bulk lifetime is unrelated to thickness and comprises radiative recombination, Auger recombination, and Shockley-Read-Hall (SRH) recombination, with respective lifetimes τ_{rad}, τ_A, τ_{SRH}.

$$\frac{1}{\tau_b} = \frac{1}{\tau_{rad}} + \frac{1}{\tau_A} + \frac{1}{\tau_{SRH}}$$

Surface lifetime is a function of surface recombination velocity S (in our case, both wafer sides have identical velocities), wafer thickness d, and minority carrier diffusivity D. The solution for τ_s is a transcendental equation.

$$\sqrt{\frac{1}{D\tau_s}} \cdot \tan\left(\frac{d\sqrt{\frac{1}{D\tau_s}}}{2}\right) = \frac{S}{D}$$

In two limiting cases, $S \ll D$ and $S \gg D$, it can be simplified to $\tau_s \approx \frac{d}{2S}$ and $\tau_s \approx \frac{d^2}{\pi^2 D}$, respectively. The solution can thus be written as,

$$\tau_s \approx \frac{d}{2S} + \frac{d^2}{\pi^2 D}.$$

According to the equation, surface lifetime decreases as the wafer gets thinner. Therefore, again a negative correlation appears between the effective lifetime of silicon wafers and SDE time, which corresponds well to the experimental data.

4 CONCLUSION & PERSPECTIVE

In summary, we have undertaken a detailed study of the relationship between silicon carrier lifetime and wafer thickness. These results shall be correlated with solar cell efficiency and other parameters and will ultimately prove useful in the design of solar cells for high specific power applications.

Since wafers thinner than 50 µm cannot survive the SDE, the etching of ultra-thin wafers becomes challenging. Some reports exist where silicon was first deposited on another substrate before plasma etching. One of the challenges of this method is the difficulty to find a suitable substrate material. Meanwhile, the substrate should be easy to peel from a silicon wafer without breakage.

Here, we propose a window framing method – depositing a material inert to KOH such as SiN_x on the periphery of a silicon wafer. Such a patterned SiN_x deposition could be achieved by using a hard mask during PECVD or by depositing a full-area layer of SiN_x and then patterning it by inkjet masking and chemical etching. Both approaches are under investigation. Using such a SiN_x mask, only the center of each wafer is etched. The window frame will then survive the SDE and provide structural support for wafers against lateral force caused by the chemical solution. Solar cells constructed with wafers of various thicknesses will be compared in detail and the results presented at the next stage.

5 ACKNOWLEDGEMENT

Solar Energy Research Institute of Singapore (SERIS) is a research institute at the National University of Singapore (NUS). SERIS is supported by NUS, the National Research Foundation Singapore (NRF), the Energy Market Authority of Singapore (EMA) and the Singapore Economic Development Board (EDB).

6 REFERENCES

[1] E. Abdur-Rahman, I. Alghoraibi, and H. Alkurdi, "Effect of isopropyl alcohol concentration and etching time on wet chemical anisotropic etching of low-resistivity crystalline silicon wafer," *International journal of analytical chemistry*, vol. 2017, pp. 1-9, 2017.

[2] W. Shockley, "The theory of p-n junctions in semiconductors and p-n junction transistors," *Bell System Technical Journal*, vol. 28, no. 3, pp. 435-489, 1949.

[3] P. Auger, "Sur les rayons β secondaires produits dans un gaz par des rayons X," *Comptes rendus de l'Académie des Sciences*, vol. 177, pp. 169–171, 1923.

[4] W. Shockley and W. Read Jr, "Statistics of the recombinations of holes and electrons," *Physical Review*, vol. 87, no. 5, pp. 835-842, 1952.

[5] R. N. Hall, "Electron-hole recombination in germanium" *Physical Review*, vol. 87, no. 2, p. 387, 1952.

[6] K. L. Luke and L. J. Cheng, "Analysis of the interaction of a laser pulse with a silicon wafer: Determination of bulk lifetime and surface recombination velocity," *Journal of Applied Physics*, vol. 61, no. 6, pp. 2282-2293, 1987.

[7] T. Horanyi, T. Pavelka, and P. Tüttő, "In situ bulk lifetime measurement on silicon with a chemically passivated surface," *Applied Surface Science*, vol. 63, no. 1-4, pp. 306-311, 1993.

[8] A. Lochtefeld et al., "15%, 20 Micron thin, silicon solar cells on steel," in *2013 IEEE 39th Photovoltaic Specialists Conference (PVSC)*, 2013: IEEE, pp. 1364-1365.

[9] T.-Y. Wang, C.-H. Chen, J.-C. Shiao, S.-Y. Chen, and C.-H. Du, "Fabricating 40 µm-thin silicon solar cells with different orientations by using SLiM-cut method," *Journal of Micromechanics and Microengineering*, vol. 27, no. 10, p. 105017, 2017.

DYNAMIC HW-CVD PROCESS DEVELOPMENT FOR VERY HIGH-RATE THIN-FILM SILICON DEPOSITION

S. Leszczynski*[1], B. Leszczynska[2], C. Strobel[1], M. Albert[1], F. Stahr[3], J. W. Bartha[1]
1 Technische Universität Dresden, Institut für Halbleiter- und Mikrosystemtechnik, 01062 Dresden, Germany, *phone: +49 351 463 32448, e-mail: sebastian_sylwester.leszczynski@tu-dresden.de
2 Leibniz Institute for Solid State and Materials Research Dresden
3 Forschungs- und Applikationslabor Plasmatechnik GmbH, Gostritzer Str. 67b, 01217 Dresden, Germany

ABSTRACT: The catalytic chemical vapor deposition technology (CAT-CVD), also called Hot-Wire CVD (HW-CVD), is a commonly used technique for the production of various thin-film materials in many branches of the thin-film industry. The focus of this work has been placed on the process development of the hydrogenated amorphous silicon thin-film (a-Si:H) deposition used for fabrication of highly efficient heterojunction solar cells with intrinsic thin layer (HIT). The advantages of the HW-CVD technique are high deposition rates and a lack of plasma damage which is a significant feature in the production of the thin-film layers. The main subject of this work is the development of a high-rate dynamic deposition technique, which enables the realization of a process on large area substrates and in roll-to-roll systems.
Keywords: HW-CVD, a-Si:H, Manufacturing and Processing, High Deposition Rate, Thin Film

1 INTRODUCTION

In our previous work, an integration of plasma enhanced chemical vapor deposition (PECVD) and Hot-Wire CVD (HW-CVD) in a single in-line deposition system was presented [1].

In this work, a study of the HW-CVD process development for the fabrication of hydrogenated amorphous silicon thin-films (a-Si:H) is presented. Especially the high-rate dynamic deposition of very thin a-Si:H layers was analyzed. Moreover, a suitable solution for the extension of the filament lifetime has been applied. Additionally, our standard gas feed system has been improved to increase the process efficiency and consequently the static and dynamic deposition rates.

1.1 In-line combined deposition system

In this work, the process development on an in-line deposition system containing PECVD and HW-CVD chambers is presented (Figure 1). The PECVD method used here is based on the linear plasma source concept developed by Dresden University of Technology and FAP GmbH Dresden [2]. Additionally, the deposition system has been extended by an HW-CVD module. This unique deposition equipment enables a direct comparison of static and dynamic processes for both techniques. A high-rate deposition, combined with a carrier movement (speed up to 0.5 mm/min), is essential for the development of a cost competitive silicon thin film production.

During the dynamic deposition the substrates are placed on the horizontally moved carrier. Using this technique, the growth of wide thickness range (form 3-5 nm, through tens of nm, up to hundreds of nm, or even micrometers) can be achieved at very high deposition rates.

Design of the equipment enables deposition of different combinations of i-, p- and n- layers using PECVD and HW-CVD during a single in-line dynamic process, without vacuum interruption [1]. This work is focused on development of intrinsic layers. Nevertheless, the described procedure can be extended to p- and n- doped layers.

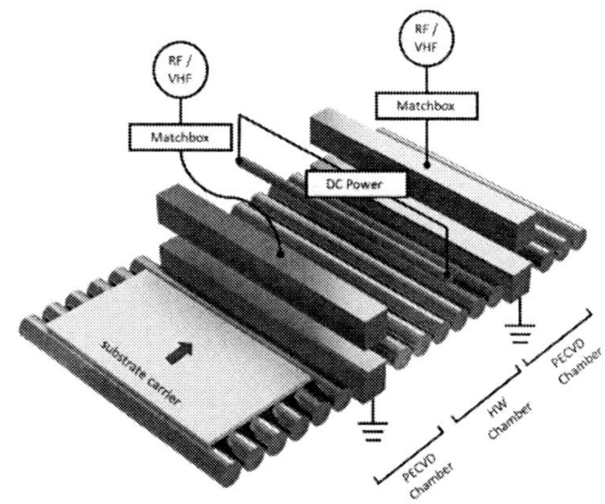

Figure1: In-line deposition system with integrated HW-CVD module [2].

2 EXPERIMENTAL

First, to characterize the deposition process, the influence of the process parameters such as the filament temperature, distance from the substrate, gas flow, gas pressure, deposition time and the substrate velocity, on the deposition rate has been analyzed. Furthermore, the static and dynamic deposition were compared, whereby the relationship between them was established. Additionally, the deposition homogeneity in a single filament setup using static and dynamic technique was analyzed. In the next step, the problem of the filament lifetime was considered, and a suitable solution was developed.

To investigate the material properties, the intrinsic and doped thin silicon films were fabricated on 1.1 mm thick Float glass substrates with the surface area of 2,5x2,5 mm^2. Additionally, double sided polished c-Si wafers were used to obtain the passivation properties of intrinsic materials.

To investigate the deposition rates the thickness of the layers was measured by a surface profilometer from Veeco.

The passivation properties of thin silicon films deposited by HW-CVD on 300 μm thick n-type double sided polished float zone silicon wafers were investigated on the structure shown in Figure 2. The passivation quality was obtained by the effective minority carrier lifetime measurements using microwave detected photoconductivity (MDP) [3].

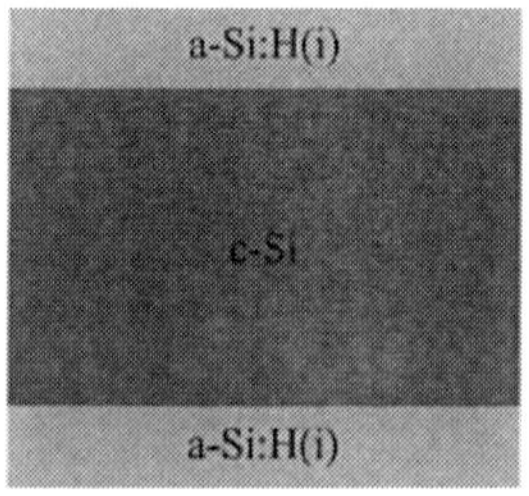

Figure2: Passivation scheme for silicon intrinsic thin films deposited using HW- CVD technology.

To investigate the growth homogeneity on a larger area (15 x 15 cm^2), the static deposition on aluminum foil was performed. To achieve more precise measurement of the layer thickness distribution, the ellipsometry on the 6-inch wafers was applied using step by step measuring in one dimensional raster.

The hot filament was driven using a constant direct current power source. The temperature of the filament was measured using a pyrometer applied on the center of the wire.

During this study a 0,3 mm thick tungsten, as well as 0,3 mm thick tantalum wire was used.

3 RESULTS AND DISCUSSION

3.1 Temperature of the hot filament

In the first step in the process development of HW-CVD the influence of the driving power on the filament temperature was established (Figure 3). The degradation of the wire during the deposition changes the driving characteristics over time. Therefore, the wire temperature has to be used as a process parameter, and has to be kept constant by monitoring the pyrometer data and accordingly adjusting the DC-power source.

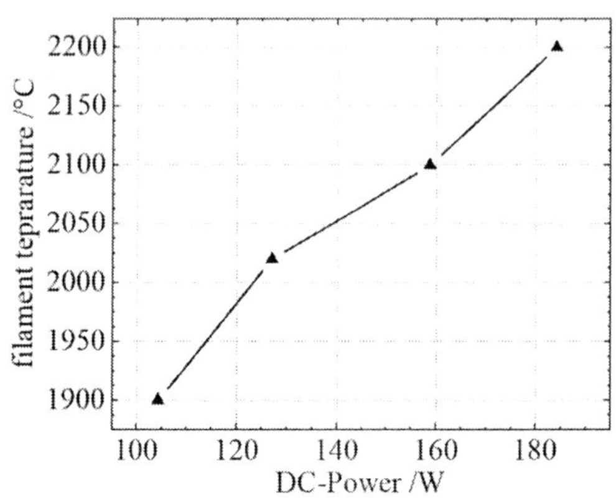

Figure 3: Influence of the DC-power on the filament temperature

In the next step the influence of the wire temperature on the dynamic deposition rate was established. (Figure 4).

The deposition rate can be driven by the filament temperature, nevertheless there are some significant limitations, that make this approach less practical.

The minimum filament temperature is limited to 1500 °C by the formation of the Si on the surface of the filament since it is lower than the Si melting point [4]. At the same time, the maximum wire temperature is limited by the filament lifetime.

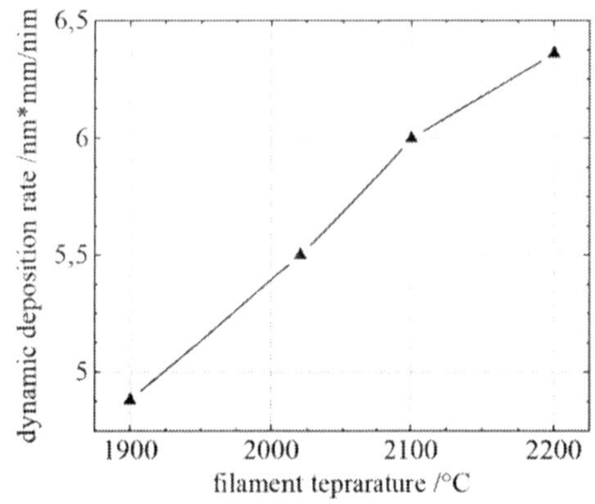

Figure 4: Influence of the filament temperature on the deposition rate.

3.2 Silane flow and gas pressure

Depending on the demands considering the properties of the deposited layers, a different process working point has to be chosen. As an example, in our previous work [1], for the fabrication of highly-effective heterojunction solar cells with intrinsic thin layer (HIT) the carrier lifetime of the intrinsic a-Si:H layers deposited on silicon wafers was the priority.

First, the variation of the silane flow rate was investigated to achieve the maximum deposition rate (Figure 5). The optimal value of 80 sccm was established, and used in further process development.

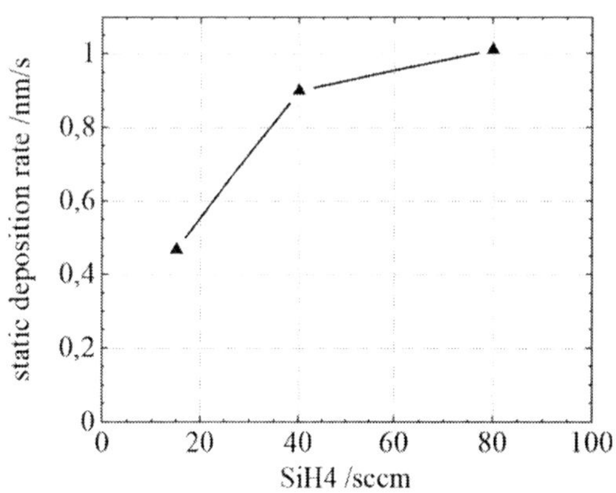

Figure 5: Static deposition rate as the function of silane gas flow

To optimize the passivation properties of a-Si:H layers, the variation of the gas pressure was investigated (Figure 6). The value of 3 Pa was established for the maximum of the carrier life time, maintaining very high deposition rate. The life time value of around 3 ms was achieved during the initial process development. In our recent work [1] further optimizations were performed, including fabrication of various layer stacks, where the value of 14 ms was achieved.

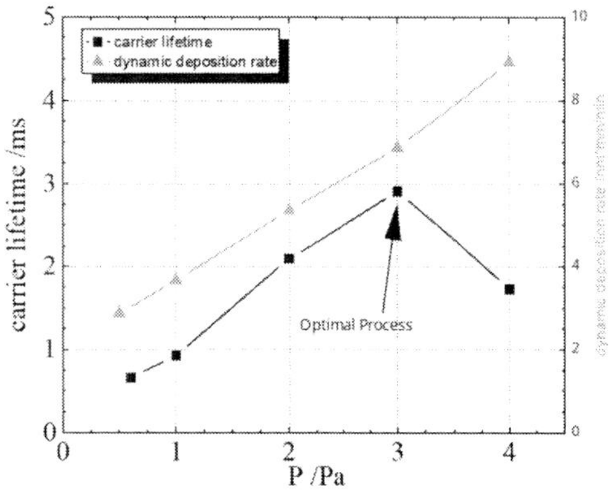

Figure 6: Dynamic deposition rate and passivation properties of a-Si:H layers in the function of silane gas pressure

3.3 Extension of the filament lifetime

The filament life time is a crucial property for long-time process flows, as the replacement of the damaged wire is time consuming, and can require recalibration of the deposition process. There are already solutions for extending the lifetime of the filaments [5,6], but these solutions require specially designed holder system.

The degradation of the filaments is caused by the silicide formation on the surface. In a conventional HW system, it occurs at the colder ends of the catalyzer during film deposition, in the region where the wires are electrically contacted. The thermal-mechanical load leads to failure of the wire after several hours of operation.

In this work a cost-effective solution with a simple holder was applied. Using a specially designed wiring and suspension of the filament (Figure 7), mechanical stress at the suspension was reduced. Moreover, the resistivity (Figure 8) and consequently the heat distribution was flattened, extending the lifetime to several tenth of hours. The resistance of the active part was higher than that of the twisted part. These values are not reproducible, since the electrical contact between twisted wires occurs only at the crossings (Figure 9). Nevertheless, the variations of the exchanged filaments were acceptable, and can be easily neglected by adjusting the DC-power according to the temperature measurements on pyrometer.

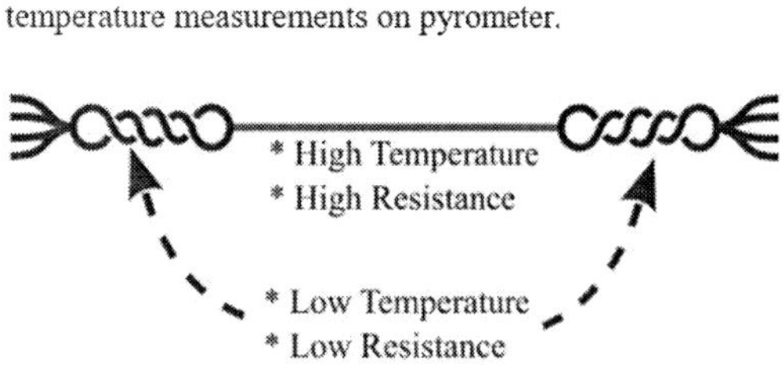

Figure 7: Schematic of the filament wiring system

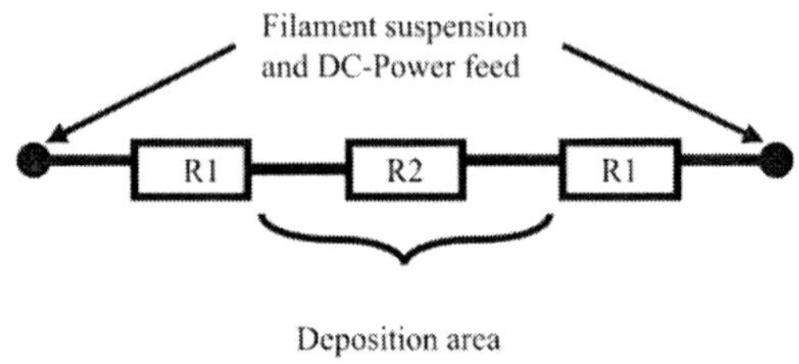

Figure 8: Electrical principle of the filament wiring system used in the HW-CVD chamber, where R1 < R2.

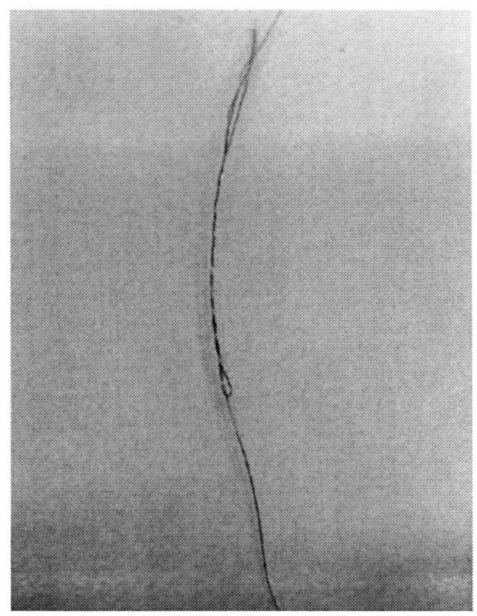

Figure 9: Filament wiring system with twisted wires.

To even further increase the wire lifetime, especially for non-continuous process, where the filament is being cooled down between, and heated up before new runs a pre- and post- treatment scheme was developed, where the tempering cycles in vacuum and hydrogen atmosphere were performed. Before starting the deposition process, the wire is heated at the operational temperature in high vacuum for several minutes. In the next step, the silane gas flow is introduced, and the process begins. After the deposition, the silane is evacuated and the hydrogen flow is introduced. After several minutes the DC-power is disconnected, and the wire cools down in the presence of the hydrogen.

3.4 Gas feed system

To increase the process efficiency and consequently the static and dynamic deposition rates, a gas feed system has been modified. The silane mixture is introduced into the deposition chamber through two parallel arranged pipes with periodically arranged nozzles placed along the and above the wire (Figure 10a.). Due to the angular orientation of the nozzles, the gas outflow is aimed directly at the wire (Figure 10b).

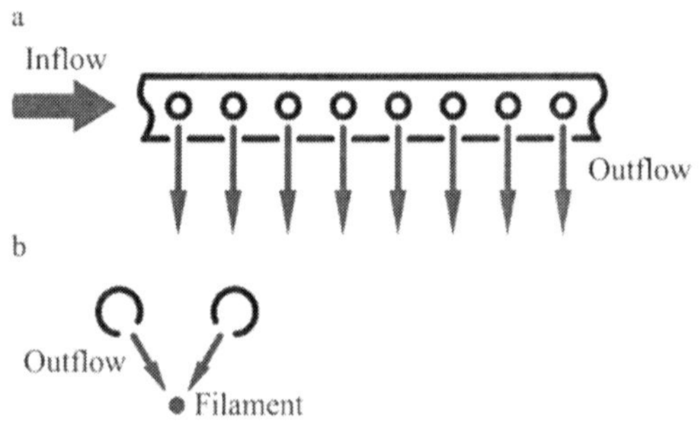

Figure 10: Gas feed pipes with periodically arranged nozzles. Nozzles aimed directly at the wire (angular orientation of the nozzles)

3.5 Dynamic deposition

A very significant property of each fabrication process is the homogeneity of the deposited layers, especially during the static deposition in the single-wire setup. To ensure the homogenous preparation of the thin-films different solutions, for example multi-wire arrangement [6, 7], can be used.

In this work the dynamic deposition technique is proposed. In order to keep the optimal layer properties, the process parameters have to be maintained at optimized levels. In order to achieve desired film thickness the velocity of the substrate carrier has to be obtained. Since this dependency is not linear, a mathematical model was established.

After the static deposition, the deposition rate distribution profile $R(x)$ along the axis x, perpendicular to the filament is significantly inhomogeneous (Figure 11).

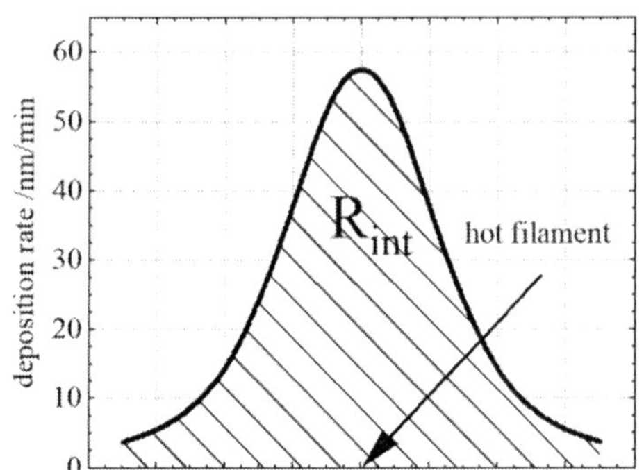

Figure 11: Deposition profile and integrated deposition rate.

During the dynamic deposition, the substrate moves across this deposition profile with a constant velocity V. At every single small distance Δx covered during a very small time Δt, the layer thickness increases of a small amount Δd. Integrating the partial growth ΔR over the whole distance x during the whole deposition time t gives the final layer thickness d after a single run (Eq. 1).

$$d = R_{int} \cdot V^{-1} \qquad \text{Equation 1}$$

The integrated deposition rate R_{int} can be calculated from the deposition profile, or directly from the eq. 1 after a single test deposition at freely chosen velocity.

Then, the estimated curve is used to predict carrier velocities for desired sheet thicknesses (Figure 12).

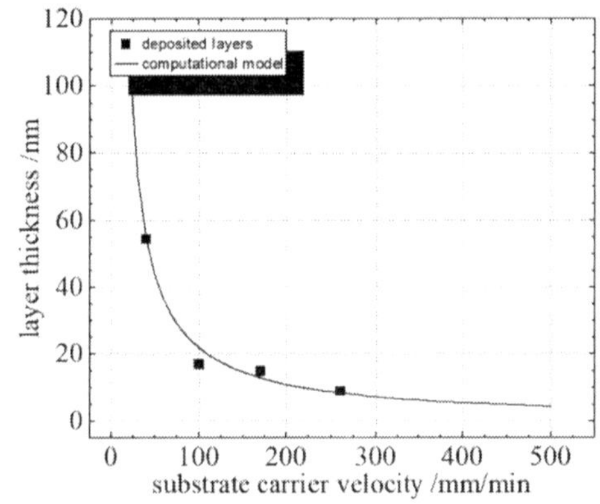

Figure 12: Dependence of layer thickness as a function of substrate carrier velocity calculated on the basis of equation 1.

4 SUMMARIES

In this work, a study of the HW-CVD process development for the fabrication of hydrogenated amorphous silicon thin-films was presented. The main process features of the HW-CVD such as the controlling and maximizing the deposition rate, extending the lifetime of the filaments and homogenization of the layers were investigated, and suitable solutions were presented. To improve the control of the dynamic growth, a suitable mathematical toll was proposed.

5 REFERENCES

[1] S. Leszczynski, B. Leszczynska, C. Strobel, M. Albert, F. Stahr, J. Kuske, J.W. Bartha, Plasma Enhanced Chemical Vapor Deposition (PECVD) and Catalytic CVD (Hot-Wire CVD) for High-Rate Fabrication of Thin-Film Silicon Layers, in: 37th Eur. Photovolt. Sol. Energy Conf. Exhib., 2020: pp. 290–294. https://doi.org/10.4229/EUPVSEC2020-2DO.10.4.

[2] J. Rüdiger, H. Brechtel, A. Kottwitz, J. Kuske, and U. Stephan, "VHF plasma processing for in-line deposition systems," Thin Solid Films, vol. 427, no. 1–2, pp. 16–20, Mar. 2003.

[3] F. Benner, P. M. Jordan, M. Knaut, I. Dirnstorfer, J. W. Bartha, and T. Mikolajick, "Investigation of the c-Si/Al2O3 Interface for Silicon Surface Passivation," in 27th European Photovoltaic Solar Energy Conference and Exhibition, 2012, pp. 1793–1796.

[4] Matsumura, H., 1998. Formation of Silicon-Based Thin Films Prepared by Catalytic Chemical Vapor Deposition (Cat-CVD) Method. Jpn. J. Appl. Phys. 37, 3175–3187. https://doi.org/10.1143/JJAP.37.3175

[5] Laukart, A., Harig, T., Höfer, M., Schäfer, L., 2015. Strategy for silicon based hot-wire chemical vapor deposition without wire silicide formation. Thin Solid Films 575, 38–41. https://doi.org/10.1016/j.tsf.2014.10.026

[6] Ishibashi, K., 2001. Development of the Cat-CVD

apparatus and its feasibility for mass production. Thin Solid Films 395, 55–60. https://doi.org/10.1016/S0040-6090(01)01207-X
[7] Ishibashi, K., Karasawa, M., Xu, G., Yokokawa, N., Ikemoto, M., Masuda, A., Matsumura, H., 2003. Development of Cat-CVD apparatus for 1-m-size large-area deposition. Thin Solid Films 430, 58–62. https://doi.org/10.1016/S0040-6090(03)00071-3

QUANTIFYING THE PERFORMANCE GAIN OF 6-INCH BIFACIAL 4T PK-SI TANDEM MODULES

P. Manshanden[1*], G. Coletti[1], V. Rosca[1], M.J. Jansen[1], K. de Groot[1], G.J. de Graaff[1], M. Creatore[2], K.Datta[2], R. Janssen[2], L. Simurka[3], D. Zhang[3], M. Najafi[3], V. Zardetto[3], I. Dogan[3], H. Fledderus[3], S.C. Veenstra[3],
[1] TNO Energy Transition, Westerduinweg 3, 1755 LE Petten, the Netherlands
[2] Department of Applied Physics, Eindhoven University of Technology, P.O. Box 513, 5600 MB Eindhoven, The Netherlands
[3] TNO Energy Transition, High Tech Campus 21, 5656 AE Eindhoven, the Netherlands

ABSTRACT: This short paper aims to show the application of four terminal perovskite-silicon tandem modules in indoor and outdoor applications with a six inch area. We demonstrate a 24.5 mW/cm² bifacial tandem module which shows a 3 mW/cm² gain with respect to monofacial use of the same tandem. In addition we show that bifacial tandems outperform identical monofacial tandems by over 25% in a 10% albedo environment. Lastly, we show our way forward with a new record tandem of 29.2%.
Keywords: bifacial, 4 terminal tandem, perovskite

1 INTRODUCTION

In order to improve the performance of photovoltaic panels above the practical limit of crystalline silicon, tandem module technology offers an attractive prospect.[1] In addition, the application of bifaciality in solar panels offers an effective path to increase performance and decreases sensitivity to irradiation direction.[2] Combining tandem module technology with bifacial technology is expected to give the highest yield by combining optimal spectral utilization with high module power.

Our choice is to combine 4 terminal (4T) tandem modules with bifacial application.

2 SAMPLE DESCRIPTION

Perovskite modules were created on six inch glass substrates using scalable deposition methods such as sputter coating, slot die coating and ALD. A laser scribed monolithic interconnection process was used to create the 31 cells for a total effective area of 92.5 cm².

For the bottom devices commercial bifacial PERC+ silicon solar cells were chosen with size matched to the perovskite modules. Top and bottom devices were encapsulated using a commercial TPO. The edges were sealed using a commercial edge sealant to prevent moisture ingress. One of the samples was made monofacial by application of a black rear foil, as shown in figure 1.

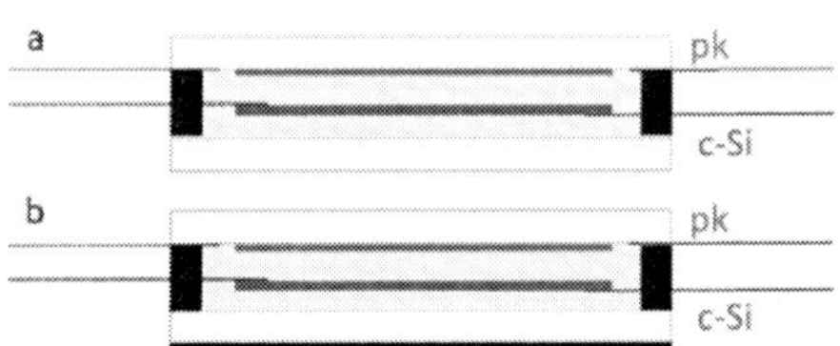

Figure 1: Sample schematic for bifacial (a) and monofacial (b) samples.

3 INDOOR RESULTS

3.1 Full size tandem modules

The indoor results are measured with a class AAA solar simulator using a spectrally matched reference and an adapted IEC procedure as explained in [3].

The PERC+ solar cell had a bifaciality of 70%. The efficiency of the tandem is listed in table I.

Table I: Indoor power conversion efficiency

	Power density (mW/cm²)
Perovskite	14.6
PERC+ monofacial	20.6
Monofacial tandem	21.4
Bifacial tandem Bifi100	23.2
Bifacial tandem Bifi200	24.5

3.2 Mini tandem modules

Higher efficiencies for tandem devices can be reached by improving the transparency and efficiency of the perovskite devices, as well as the efficiency of silicon devices. We attempted this on 9 mm² samples using similar materials as for the large devices, which yielded the results depicted in table II.

Table II: Record power conversion efficiency

	Conversion efficiency
Perovskite	17.8 %
BC-HIT (filtered)	11.4 %
Tandem (monofacial)	29.2 %

4 OUTDOOR RESULTS

4.1 Outdoor setup

Bifacial and monofacial tandem devices were placed outdoors at 52.5 degrees northern latitude between 18 December 2020 and 18 August 2021. The angle of the modules was fixed at 30 degrees. Ambient and module temperature was monitored. The incident light at the front and rear of the samples was monitored. The albedo of the location was approximately 10% (bitumen). Every 10 minutes an IV sweep of 10s was performed. Between the IV sweeps, the samples were kept at the last maximum power point.

4.2 Stability of the samples

In order to draw conclusions on the eight month outdoor stability of the samples, first we must establish that the data quality is sufficient for analysis.

Silicon solar cells are well established to be extremely stable [4]. However, the perovskites constitute a newer technology which still has to prove it's long term stability. Within the context of tandems, perovskite transparency is one of the factors which needs to be established. This can be monitored over the entire measurement period by comparing the short circuit current at 500 W/m^2 over the monitoring time. This irradiance level was chosen as it is present over the entire monitoring period.

In figure 2, the short circuit current of monofacial and bifacial bottom devices of tandems is shown together with the module temperature. The module temperature is of the monofacial device; however the bifacial device remains within 1 °C of the monofacial device at all times. The bifacial device shows more variance because the rear illumination has a variance which does not at all times depend on the front illumination. The decrease of current over the year can be ascribed to the increase in temperature of the devices.

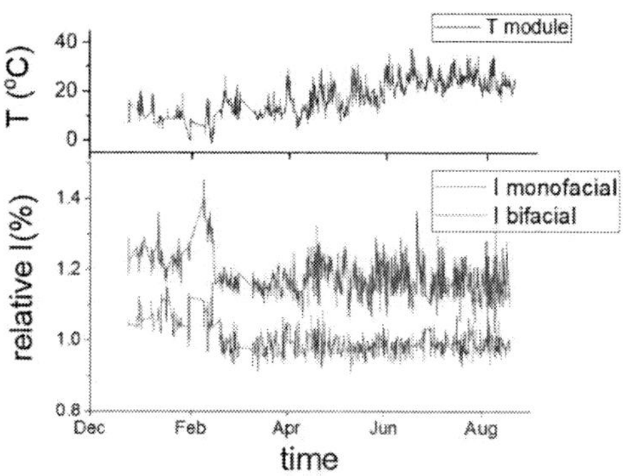

Figure 2: Short circiut current of bottom cell of bifacial and monofacial tandems

The performance of the top devices was not entirely stable. For the purposes of this analysis, the relative performance outdoors is more important than the absolute performances. Therefore the relative performance is shown in figure 3.

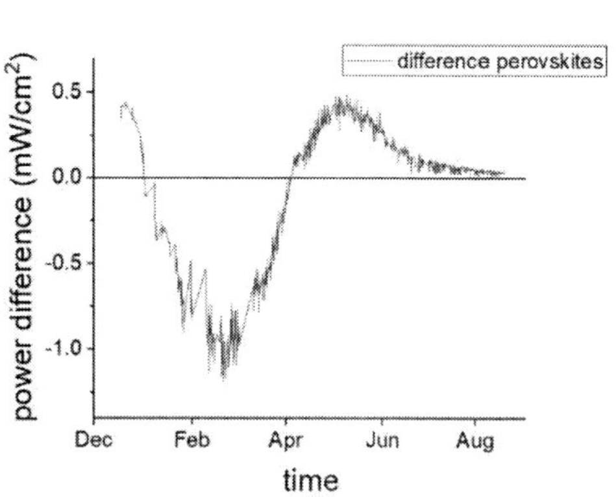

Figure 3: difference in performance of the perovskites over the year.

The differences in the power of the perovskite top devices are not insurmountably large, and variable over the monitoring time. Therefore, we can compare the performance of the monofacial and bifacial tandems outdoors with reasonable accuracy.

4.3 Outdoor performance

The difference in outdoor performance of bifacial and monofacial modules depends on the absolute irradiance level and the proportion of irradiance between the front and rear of the modules. This difference is shown in figure 4.

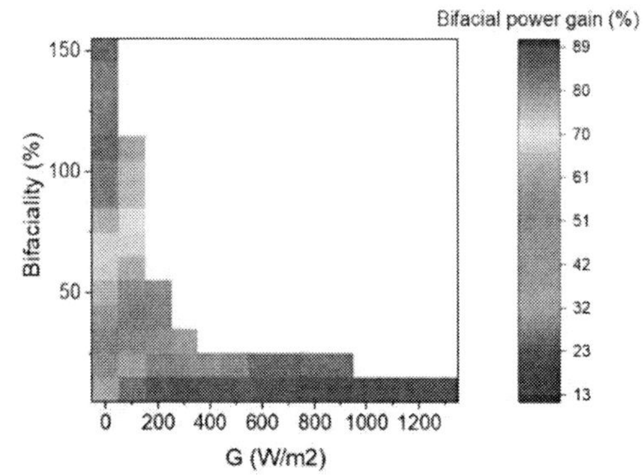

Figure 4: bifacial power gain as a function of front irradiance and bifacial irradiance.

Bifacial proportion exceeds 100% in summer during sunrise and sunset at 52.5 degrees latitude. In addition, certain weather conditions like snow can change the bifaciality of the irradiance as well. Similarly, front injection can exceed 1000 W/m^2 because of effects of surroundings and latitude.

Average gain for bifaciality is 28% for an average bifacial irradiance proportion of 17%. On average this constitutes a power gain of 60 mW over the entire monitoring period.

5 DISCUSSION AND CONCLUSIONS

Based on the indoor performance, a gain in power from monofacial to bifacial performance of 8 to 14 % is expected. This is less than the observed outdoor performance gain over the course of eight months. This discrepancy can be explained by the fact that the indoor observation is at one specific irradiance point. Indeed, when limited to the range of 1000 W/m^2, the prediction nearly exactly coincides with the outdoor measurements.

The second observation which can be made from the outdoor data is that the bifacial irradiance is not a constant at any given location. Rather, it depends on angle of the direct light as well as the proportion between specular and diffuse incident light, caused by time of day, year and weather conditions. This implies that optimization for a specific bifacial irradiance proportion will yield suboptimal results at best, which has implications for the applicability of 2 terminal bifacial tandems.

Improving power conversion efficiency of our devices can be achieved by simultaneously improving the efficiency of the top and bottom devices, as well as improving the transparency of the top device. We show this can be achieved, resulting in a tandem efficiency of 29.2%.

6 REFERENCES

[1] R.M. Swanson. Approaching the 29% limit efficiency of silicon solar cells. Conference Record of the Thirty-first IEEE Photovoltaic Specialists Conference. 2005.
[2] G. Coletti et al. Bifacial Four-Terminal Perovskite/Silicon Tandem Solar Cells and Modules ACS Energy Lett. 2020, 5, 5, 1676-1680.
[3] P.Manshanden *et al.*, Quantifying the performance gain of 6 inch bifacial 4T pk-Si tandem modules, EDJPV 2021.
[4] VDMA, 2020. 2020 International Technology Roadmap for PV.

TWO-TERMINAL III-V//SI TRIPLE-JUNCTION SOLAR CELL WITH POWER CONVERSION EFFICIENCY OF 35.9 % AT AM1.5G

Patrick Schygulla, Ralph Müller, Oliver Höhn, Hubert Hauser, Benedikt Bläsi, Felix Predan, Jan Benick, Martin Hermle, Frank Dimroth, Stefan W. Glunz, and David Lackner
Fraunhofer Institute for Solar Energy Systems ISE
Heidenhofstr. 2, 79110 Freiburg, Germany

ABSTRACT: III-V//Si multi-junction solar cells offer a pathway to increase the power conversion efficiency beyond the fundamental Auger limit of silicon single-junctions. In this work we demonstrate how the efficiency of a two-terminal wafer-bonded III-V//Si triple-junction solar cell is increased from 34.1 % to 35.9 % under an AM1.5g spectrum, the highest reported efficiency for silicon based multi-junction solar cell technologies. This improvement was accomplished by two main factors. First, the integration of a GaInAsP absorber in the middle cell increased the open-circuit voltage by 51 mV. Second, a better current matching of all subcells enhanced the short-circuit current by 0.7 mA/cm².
Keywords: MOVPE, III-V/Si, silicon based tandem solar cells, multijunction solar cells, GaInAsP, wafer bond

1 INTRODUCTION

Single-junction crystalline silicon solar cells have a theoretical efficiency limit between 29.4 % and 29.5 % [1, 2]. To make better use of limited areas and to continue reducing levelized costs of electricity in terrestrial applications, it is vital to explore new solar cell concepts that can exceed this limit. In multi-junction solar cells thermalisation and non-absorption losses are reduced and thus the efficiency limit is enhanced. The current record efficiency of all solar cell concepts under one-sun AM1.5g illumination was demonstrated for an inverted metamorphic sextuple-junction solar cell and amounted to 39.2 % [3].

Such multi-junction solar cells are based on III-V compound semiconductors. The need of rare elements for wafers and precursors, the expensive facilities that are required for epitaxial growth and cell processing, and the low numbers of produced units result in large cell costs of around 60 $/W_{DC}$ [4] which is more than two orders of magnitude larger compared to single-junction silicon (Si) cells. Strong efforts are being undertaken to reduce the cost of III-V solar cell fabrication. One important breakthrough towards this goal are demonstrated growth rates above 120 µm/h [5, 6]. The substrate costs amount to around a third of the total costs for III-V multi-junction solar cells. Substrate removal and reuse approaches can thus also help to reduce these costs [7]. Here, the expensive substrate is used for epitaxial growth and then the solar cell is transferred to a carrier wafer consisting of a cheaper material.

Another possibility is to use a less expensive substrate material, for instance silicon, as the bottom cell of a multi-junction architecture. The III-V cell layer stack can be grown directly on the Si cell using a metamorphic buffer structure [8, 9] to overcome the mismatch in lattice constants between Si and the III-V cells. The highest power conversion efficiency was reported for a triple-junction GaInP/GaAs/Si solar cell with a value of 25.9 % [10]. This cell suffers from a voltage loss due to non-radiative recombination at the threading dislocation defects introduced by the metamorphic buffer growth. This loss mechanism can be avoided if the III-V top structure is grown lattice matched on GaAs and then either mechanically stacked on the Si cell (four-terminal configuration) [11] or monolithically connected to the Si cell (two-terminal configuration). The latter can be achieved by using either a direct wafer bond [12] or a transparent conductive adhesive [13].

With a four-terminal design the maximum conversion efficiency demonstrated amounted to 35.9 % for a triple-junction GaInP/GaAs-Si solar cell [11]. This has so far been the highest conversion efficiency of any silicon based multi-junction solar cell [14]. The two-terminal configuration allows for a direct integration into modules and the exploitation of existing technology for Si solar panels. The previous wafer bonded triple-junction GaInP/AlGaAs//Si champion device, however, achieved a lower maximum conversion efficiency of 34.1 % [15]. Other promising developments for cost reduction in multi-junction solar cells include the combination of a perovskite/Si dual-junction. This technology has recently obtained a record conversion efficiency of 29.2 % [16].

Hence, combining the high-efficiency multi-junction concept with the mature and cost-effective silicon technology is a promising route for redefining the practical solar cell efficiency limit. In this work, we explore the experimental efficiency potential of monolithic wafer-bonded III-V//Si triple-junction solar cells in a two-terminal configuration.

2 EXPERIMENTAL METHODS

2.1 Solar Cell Design

The ideal band gap combinations of the upper two junctions were calculated to be 2.01 eV and 1.50 eV [17]. For the top junction only AlGaInP or AlGaAs can be grown at the required band gap lattice matched to GaAs. Both exhibit severe disadvantages, though. $Al_{0.55}Ga_{0.45}As$ is an indirect semiconductor and hence unsuitable for maximising the voltage. $Al_{0.11}Ga_{0.40}In_{0.49}P$ is known to have a rather low carrier mobility and an increased incorporation probability for parasitic oxygen resulting in reduced diffusion lengths, reduced open-circuit voltage, and increased sheet resistance, which is detrimental for the uppermost junction [18]. Choosing GaInP with a direct band gap of 1.90 eV our group can at present reach higher absolute voltages in practice. Because of a higher minority carrier lifetime and reduced recombination in the depletion region resulting in higher voltages and a higher filling factor a rear-heterojunction cell architecture with a thin AlGaInP base was implemented [19]. Choosing a lower band gap top junction absorber does not alter the ideal middle cell band gap, though, which remains at 1.50 eV. This band gap can be realised either by $Al_{0.06}Ga_{0.94}As$ [20] or $Ga_{0.93}In_{0.07}As_{0.87}P_{0.13}$ [21] as shown in Figure 1.

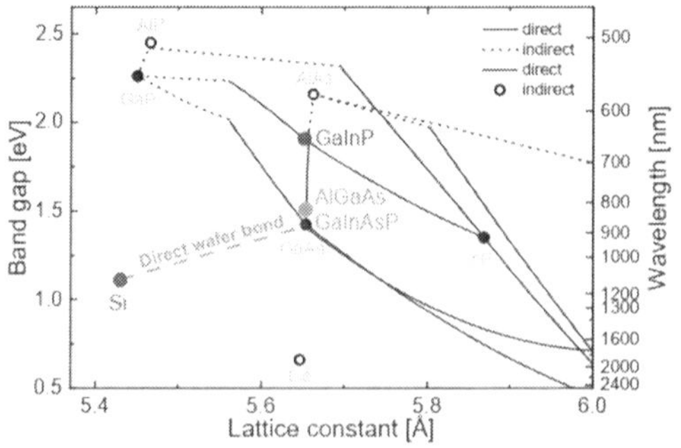

Figure 1. The target band gaps for the top III-V-dual-junction at 1.90 and 1.50 eV can be realised by GaInP and GaInAsP or AlGaAs. To overcome the large lattice mismatch between the III-V semiconductors and silicon a direct wafer bond is used.

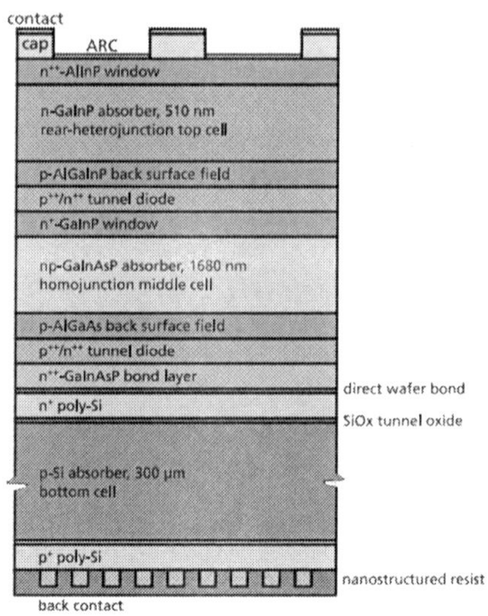

Figure 2. Schematic layer stack of the III-V//Si triple-junction solar cell design including a double-layer antireflection coating (ARC), a highly doped n-GaAs cap layer below the contacts, a GaInP-rear-heterojunction top cell, a GaInAsP homojunction middle cell, a silicon bottom cell with tunnel oxide passivating contacts (TOPCon), and a nanostructured diffractive rear-side grating for light path enhancement.

2.2 Epitaxial Growth of the III-V Top Structure

The III-V top junction solar cells were grown using metalorganic vapor phase epitaxy (MOVPE) in a commercial AIXTRON AIX2800G4-TM reactor. GaAs wafers of 4" size with a 6° miscut of the (001) surface towards the (111)-B direction were employed as substrates. The precursors for the group-III elements were trimethylgallium (TMGa), trimethylindium (TMIn), and trimethylaluminium (TMAl). For the group-V elements phosphine (PH_3) and arsine (AsH_3) were used. The doping agent precursor for n-type absorber layers was silane (SiH_4). For p-type material dimethylzinc (DMZn) was used. Epitaxial growth was performed at standard growth temperatures ranging from 550 – 680 °C with a preference for lower temperatures where possible to reduce the thermal load and interdiffusion effects. The V/III ratios were set between 20 and 40 for arsenide-based layers and between 60 and 140 for phosphide-based layers. For the top and middle junction absorber layers the V/III-ratio has been optimised in terms of the open-circuit voltage of single-junction component cells [22]. Hydrogen was used as the carrier gas for the precursors.

2.3 Silicon Bottom Cell

Float-zone grown p-type silicon wafers with a bulk resistivity of 4 Ω cm and a thickness of 280 µm, that had been polished on both sides, were used for the fabrication of the silicon bottom cells. Tunnel-oxide passivating contacts (TOPCon) [23] were formed on the n-type front and p-type back side. These layers passivate the surface very well and thus allow for a high voltage. A thin oxide was grown in HNO_3. A 100 nm thick intrinsic amorphous silicon was deposited by liquid phase chemical vapor deposition (LPCVD) on both sides. This layer was then doped by ion implantation of phosphorus at the front and of boron monofluoride at the back side. The amorphous layer was eventually annealed to polysilicon at 850 °C. Last, the samples were exposed to a remote hydrogen plasma at 425 °C. The front side was polished by chemical mechanical treatment (CMP) to remove particles and thin the poly-Si layer to around half its thickness.

2.4 Solar Cell Processing

The samples were first temporarily bonded to a sapphire carrier wafer. Then, the GaAs substrate was etched away, and the uncovered bond surface was smoothened by chemical mechanical polishing (CMP). Next, a direct wafer bond between the III-V layer stack and the silicon bottom cell was performed in an Ayumi SAB100 high vacuum direct wafer bonder. The native oxides of the surfaces were first removed by argon ion sputtering at an energy of 0.3-0.4 keV. Then, the wafers were pressed together with a force of 2.4 – 10 kN for 5 min. Further details on the wafer bonding process can be found elsewhere [12, 24]. The sapphire carrier wafer was removed by thermal slide at 190 °C.

The front contacts were deposited with the help of photolithography and metal evaporation. The cell area of 4.028 cm^2 was defined by wet chemical mesa etching. A Ta_2O_5/MgF_2 antireflection coating (ARC) was deposited on the front side by evaporation. On the rear side an optical grating for light path enhancement in the silicon cell was implemented by nanoimprint lithography of a SU-8 photo resist [25]. Finally, a 1 µm thick silver contact layer was deposited on the nanostructured grating.

2.5 Solar Cell Characterisation

The triple-junction solar cells were characterised in the Fraunhofer ISE CalLab under calibrated conditions. The external quantum efficiency (*EQE*) was measured with a grating monochromator and adjustable bias illumination in lock-in mode. Current-voltage (*IV*) characteristics were acquired for one-sun conditions using the AM1.5g spectrum (IEC90604-3, ed. 2 with 1000 W/m^2). A spectrally adjustable sun simulator with one xenon and two halogen lamp fields at a cell temperature of 25 °C was used to set the correct illumination conditions [26]. For the analysis of the current-voltage data the two-diode model including resistances was employed [27].

Electroluminescence (EL) spectra were acquired under different injection conditions. Applying the reciprocity relations [28], the EQE and EL could be combined to calculate the contribution of each junction to the overall open-circuit voltage as a function of the short-circuit current density [29]. The band gaps of the III-V semiconductors were determined from the intersection of two exponential fits to the EQE [30].

3 RESULTS AND DISCUSSION

3.1 Performance under One-Sun Illumination

The external quantum efficiencies (*EQE*) of the three subcells of a triple-junction solar cell with upright grown top cell structure and their sum are shown in Figure 3. Integrating the AM1.5g spectrum under consideration of the quantum efficiency of each junction, one obtains the photocurrent beyond which the respective junction becomes limiting. The GaInP top junction is slightly too thick for a perfect current match and the Si bottom cell limits the total current density that is produced by the triple-junction. An additional solar resource of 0.67 mA/cm² or 0.22 mA/cm² per subcell could still be distributed by adjusting the absorber layer thicknesses. Note that the sum of the *EQE*s is constant at a high level ranging between 95 % and 98 % in the wavelength interval from 470 nm to 910 nm.

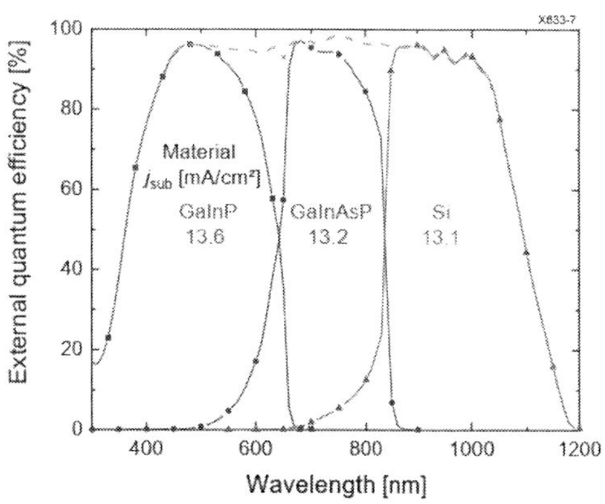

Figure 3. External quantum efficiencies of the top (blue, squares), middle (green, circles), bottom (red, triangles) subcells, and the overall sum (grey, dashed) in the two-terminal record III-V//Si solar cell X633-7. The current density that each subcell could contribute to the photocurrent is given below the absorber material for each junction. The external quantum efficiencies were calibrated using the short-circuit current density of the triple-junction solar cell.

The current-voltage characteristics of the champion cell is presented in Figure 4. Compared to the previous III-V//Si record device [15] with a conversion efficiency of 34.1 % the performance could be improved by 5.3 $\%_{rel}$ or 1.8 $\%_{abs}$. The two main reasons for this improvement were an increase in the voltage and in the current density. The open-circuit voltage (V_{oc}) was improved by 2.2 % or 71 mV. The origin of this voltage gain will be discussed in more detail in the following section. The current density could be enhanced by 5 % or 0.7 mA/cm². The previous cell was strongly current limited by the second junction with an AlGaAs absorber layer with a maximum difference in subcell current of 1.2 mA/cm² [15]. This imbalance could be reduced to 0.6 mA/cm² in the current champion device thanks to an increased quantum efficiency and thus current generation in the GaInAsP based middle junction.

From fitting the *IV*-curve with a two-diode model the series resistance was found to be around 1 Ω cm². Since the solar cell is designed to work under one-sun condition, this results in a rather low voltage and filling factor loss. The low series resistance is a confirmation of the electrical functionality of the wafer bond and of a sufficiently high conductivity in the n-GaInP top layer. No impact of a finite parallel resistance could be observed in the modelling.

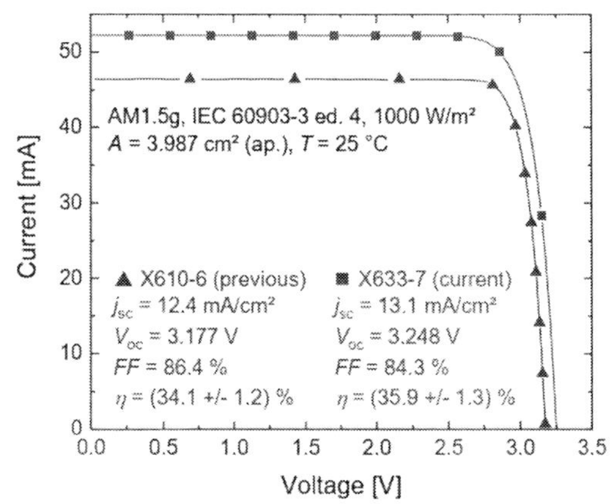

Figure 4. Current-voltage characteristics of the current two-terminal record III-V//Si solar cell X633-7 (red, squares) compared to the previous champion device X610-6 (blue, triangles) under the AM1.5g spectrum.

3.2 Voltage Gain by Improved Subcell Absorber Material

The improvement in open-circuit voltage compared to the previous champion device [15] can be understood from the subcell voltages shown in Figure 5. The voltage increases logarithmically with increasing injection current density because of the higher resulting charge carrier density. At injection conditions that correspond to AM1.5g illumination, 13.1 mA/cm² in this case, the open-circuit voltage can be broken down into a contribution of the three subcells. Comparing these subcell voltages to the ones of the previous champion device X610-6 the improvements can mostly be attributed to one development step. The biggest voltage gain resulted from replacing the AlGaAs absorber by a GaInAsP absorber in the middle junction as shown in Figure 5. As both materials had a similar band gap to within 10 meV, the gain is due to an increased minority carrier lifetime resulting in a lower loss to the radiative limit on substrate of 55 mV, or 125 mV for a cell with perfect rear-side reflector. This corresponds to an external radiative efficiency on substrate of 11.1 %. Given that this is a value of a junction inside a multilayer structure this compares well to the highest performing GaAs single-junction solar cell which has a rear-side mirror and exhibits an external radiative efficiency on mirror of 35.7 % [31]. A second smaller voltage gain originates in the GaInP top cell. Possible reasons for this increase could be fluctuations in the solar cell processing which damage the top cell most, e.g. from the ARC deposition or cap etching inhomogeneities that generate local shunts.

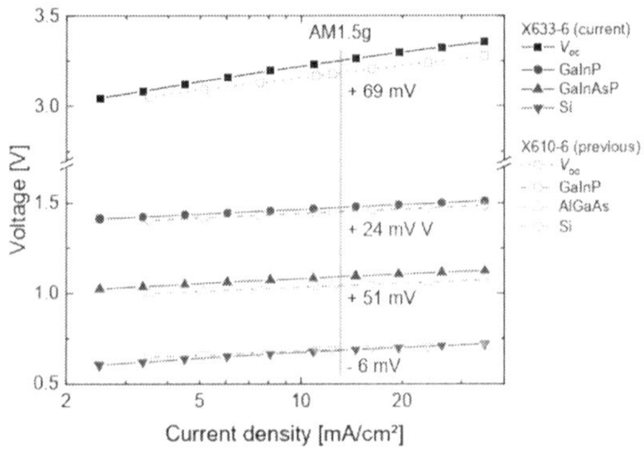

Figure 5. Open-circuit voltages (black, squares) and subcell voltages (coloured, squares, triangles) of the cell X633-6 from the current record III-V//Si wafer (solid, filled symbols) and of the previous champion device X610-6 (dotted, open symbols) at different injection current densities.

4 CONCLUSIONS

We have shown how the efficiency potential of silicon based multi-junction solar cells can be further exploited. By choosing III-V semiconductor materials, that can be deposited with a high crystal quality at the right band gap, and improving the MOVPE growth conditions for these materials we were able to present a two-terminal wafer-bonded triple-junction solar cell which sets a new efficiency record with 35.9 % under an AM1.5g spectrum. In particular, the employment of a GaInAsP middle junction resulted in a significant voltage gain.

It was demonstrated that a monolithic, two-terminal silicon based multi-junction can achieve at least the same efficiency as the currently best performing four-terminal device. No separate interconnection of the mechanically stacked subcells is required for the two-terminal configuration as it is the case for the four-terminal configuration. Hence, installation costs of modules made from such cells are lower. If the costs can be reduced in the future, the presented results may therefore increase the market entrance chances for III-V/Si solar cells.

ACKNOWLEDGEMENTS

The authors would like to thank S. Stättner and S. Meier for help with epitaxial growth, F. Schätzle, A. Leimenstoll, C. Reichel, R. Freitas, R. Koch, P. Barth, N. Brändlin, and R. Neubauer for sample processing, M. Bauer from Freiburg University for a-Si deposition, J. Krügener from Hannover University for ion implantation, E. Paz Alpuche and T. Kroyer for technical support, F. Martin, E. Schäffer, M. Schachtner, D. Chojniak, G. Siefer, E. Fehrenbach, and A. Wekkeli for solar cell measurements, and H. Helmers and J. Ohlmann for helpful discussions. This project received funding from the German Federal Ministry for Economic Affairs and Energy (FKz. 0324247- PoTaSi). P. Schygulla acknowledges his PhD scholarship from the Heinrich-Böll Foundation.

REFERENCES

[1] A. Richter, M. Hermle, and S. W. Glunz, "Reassessment of the Limiting Efficiency for Crystalline Silicon Solar Cells," *IEEE J. Photovolt.*, vol. 3, no. 4, pp. 1184–1191, 2013.

[2] B. A. Veith-Wolf, S. Schäfer, R. Brendel, and J. Schmidt, "Reassessment of intrinsic lifetime limit in n-type crystalline silicon and implication on maximum solar cell efficiency," *Solar Energy Materials and Solar Cells*, vol. 186, pp. 194–199, 2018.

[3] J. F. Geisz, R. M. France, K. L. Schulte, M. A. Steiner, A. G. Norman, H. L. Guthrey, M. R. Young, T. Song, and T. Moriarty, "Six-junction III–V solar cells with 47.1% conversion efficiency under 143 Suns concentration," *Nat. Energy.*, vol. 5, no. 4, pp. 326–335, 2020.

[4] Kelsey A. W. Horowitz, Timothy Remo, Brittany Smith, Aaron Ptak, "A Techno-Economic Analysis and Cost Reduction Roadmap for III-V Solar Cells," National Renewable Energy Laboratory (NREL), Golden, Colorado NREL/TP-6A20-72103, 2018.

[5] R. Lang, F. Habib, M. Dauelsberg, F. Dimroth, and D. Lackner, "MOVPE growth of GaAs with growth rates up to 280 µm/h," *J. Cryst. Growth*, vol. 537, p. 125601, 2020.

[6] W. Metaferia, K. L. Schulte, J. Simon, S. Johnston, and A. J. Ptak, "Gallium arsenide solar cells grown at rates exceeding 300 µm h-1 by hydride vapor phase epitaxy," (eng), *Nat. Commun.*, vol. 10, no. 1, p. 3361, 2019.

[7] N. Jain, D. Crouse, J. Simon, S. Johnston, S. Siol, K. L. Schulte, C. E. Packard, D. L. Young, and A. J. Ptak, "III–V Solar Cells Grown on Unpolished and Reusable Spalled Ge Substrates," *IEEE J. Photovoltaics*, vol. 8, no. 5, pp. 1384–1389, 2018.

[8] J. T. Boyer, A. N. Blumer, Z. H. Blumer, D. L. Lepkowski, and T. J. Grassman, "Reduced Dislocation Introduction in III–V/Si Heterostructures with Glide-Enhancing Compressively Strained Superlattices," *Crystal Growth & Design*, vol. 20, no. 10, pp. 6939–6946, 2020.

[9] M. Feifel, D. Lackner, J. Ohlmann, J. Benick, M. Hermle, and F. Dimroth, "Direct Growth of a GaInP/GaAs/Si Triple-Junction Solar Cell with 22.3% AM1.5g Efficiency," *Sol. RRL*, vol. 3, no. 12, p. 1900313, 2019.

[10] M. Feifel, D. Lackner, J. Schön, J. Ohlmann, J. Benick, G. Siefer, F. Predan, M. Hermle, and F. Dimroth, "Epitaxial GaInP/GaAs/Si Triple-Junction Solar Cell with 25.9% AM1.5g Efficiency Enabled by Transparent Metamorphic Al x Ga 1− x As y P 1− y Step-Graded Buffer Structures," *Sol. RRL*, p. 2000763, 2021.

[11] S. Essig, C. Allebé, T. Remo, J. F. Geisz, M. A. Steiner, K. Horowitz, L. Barraud, J. S. Ward, M. Schnabel, A. Descoeudres, D. L. Young, M. Woodhouse, M. Despeisse, C. Ballif, and A. Tamboli, "Raising the one-sun conversion efficiency of III–V/Si solar cells to 32.8% for two junctions and 35.9% for three junctions," *Nat. Energy.*, vol. 2, 17144, 2017.

[12] R. Cariou, J. Benick, F. Feldmann, O. Höhn, H. Hauser, P. Beutel, N. Razek, M. Wimplinger, B. Bläsi, D. Lackner, M. Hermle, G. Siefer, S. W. Glunz, A. W. Bett, and F. Dimroth, "III-V-on-silicon solar cells reaching 33% photoconversion efficiency in two-terminal configuration," *Nat. Energy.*, vol. 3, no. 4, pp. 326–333,

2018.

[13] U. Heitmann, J. Bartsch, S. Kluska, H. Hauser, O. Hohn, R. Hermann, D. Lackner, S. Janz, and S. W. Glunz, "Pathways and Potentials for III–V on Si Tandem Solar Cells Realized Using a ZnO-Based Transparent Conductive Adhesive," *IEEE J. Photovoltaics*, vol. 11, no. 1, pp. 85–92, 2021.

[14] M. Green, E. Dunlop, J. Hohl-Ebinger, M. Yoshita, N. Kopidakis, and X. Hao, "Solar cell efficiency tables (version 57)," *Prog. Photovolt: Res. Appl.*, vol. 29, no. 1, pp. 3–15, 2021.

[15] D. Lackner, O. Höhn, R. Müller, P. Beutel, P. Schygulla, H. Hauser, F. Predan, G. Siefer, M. Schachtner, J. Schön, J. Benick, M. Hermle, and F. Dimroth, "Two-Terminal Direct Wafer-Bonded GaInP/AlGaAs//Si Triple-Junction Solar Cell with AM1.5g Efficiency of 34.1%," *Sol. RRL*, p. 2000210, 2020.

[16] A. Al-Ashouri, E. Köhnen, B. Li, A. Magomedov, H. Hempel, P. Caprioglio, J. A. Márquez, A. B. Morales Vilches, E. Kasparavicius, J. A. Smith, N. Phung, D. Menzel, M. Grischek, L. Kegelmann, D. Skroblin, C. Gollwitzer, T. Malinauskas, M. Jošt, G. Matič, B. Rech, R. Schlatmann, M. Topič, L. Korte, A. Abate, B. Stannowski, D. Neher, M. Stolterfoht, T. Unold, V. Getautis, and S. Albrecht, "Monolithic perovskite/silicon tandem solar cell with 29% efficiency by enhanced hole extraction," (eng), *Science (New York, N.Y.)*, vol. 370, no. 6522, pp. 1300–1309, 2020.

[17] G. Létay and A. W. Bett, "EtaOpt - a program for calculating limiting efficiency and optimum bandgap structure for multi-bandgap solar cells and TPV cells," in *Proceedings of the 17th European Photovoltaic Solar Energy Conference and Exhibition*, Munich, Germany, 2001, pp. 178–181.

[18] E. E. Perl, J. Simon, J. F. Geisz, W. Olavarria, M. Young, A. Duda, D. J. Friedman, and M. A. Steiner, "Development of High-Bandgap AlGaInP Solar Cells Grown by Organometallic Vapor-Phase Epitaxy," *IEEE J. Photovolt.*, vol. 6, no. 3, pp. 770–776, 2016.

[19] J. F. Geisz, M. Steiner, I. Garcia, S. R. Kurtz, and D. J. Friedman, "Enhanced external radiative efficiency for 20.8% efficient single-junction GaInP solar cells," (English), *Appl. Phys. Lett.*, vol. 103, no. 4, 041118 (5 pp.), 2013.

[20] S. Adachi, *GaAs and Related Materials: Bulk Semiconducting and Superlattice Properties*: World Scientific, 1994.

[21] R. L. Moon, G. A. Antypas, and L. W. James, "Bandgap and lattice constant of GaInAsP as a function of alloy composition," *JEM*, vol. 3, no. 3, pp. 635–644, 1974.

[22] P. Schygulla, F. D. Heinz, F. Dimroth, and D. Lackner, "Middle Cell Development for Wafer-Bonded III-V//Si Tandem Solar Cells," *IEEE J. Photovoltaics*, vol. 11, pp. 1264–1270, 2021.

[23] F. Feldmann, M. Bivour, C. Reichel, M. Hermle, and S. W. Glunz, "Passivated rear contacts for high-efficiency n-type Si solar cells providing high interface passivation quality and excellent transport characteristics," *Sol. Energy Mater. Sol. Cells*, vol. 120, pp. 270–274, 2014.

[24] S. Essig and F. Dimroth, "Fast Atom Beam Activated Wafer Bonds between n-Si and n-GaAs with Low Resistance," *ECS Journal of Solid State Science and Technology*, vol. 2, no. 9, Q178-Q181, 2013.

[25] B. Bläsi, O. Höhn, H. Hauser, R. Cariou, J. Benick, F. Feldmann, P. Beutel, D. Lackner, M. Hermle, G. Siefer, S. W. Glunz, A. W. Bett, F. Dimroth, and N. Tucher, "Photonic structures for III-V//Si multijunction solar cells with efficiency >33%," in *Photonics for Solar Energy Systems VII*, Strasbourg, France, 2018, p. 2.

[26] M. Meusel, R. Adelhelm, F. Dimroth, A. W. Bett, and W. Warta, "Spectral Mismatch Correction and Spectrometric Characterization of Monolithic III-V Multi-junction Solar Cells," *Prog. Photovolt: Res. Appl.*, vol. 10, no. 4, pp. 243–255, 2002.

[27] S. Suckow, T. M. Pletzer, and H. Kurz, "Fast and reliable calculation of the two-diode model without simplifications," *Prog Photovolt Res Appl*, vol. 22, no. 4, pp. 494–501, 2014.

[28] U. Rau, "Reciprocity relation between photovoltaic quantum efficiency and electroluminescent emission of solar cells," *Phys. Rev. B, Condens. Matter*, vol. 76, 085303-1 - 8, 2007.

[29] T. Kirchartz, U. Rau, M. Hermle, A. W. Bett, A. Helbig, and J. H. Werner, "Internal voltages in GaInP/GaInAs/Ge multijunction solar cells determined by electroluminescence measurements," *Appl. Phys. Lett.*, vol. 92, no. 12, p. 123502, 2008.

[30] H. Helmers, C. Karcher, and A. W. Bett, "Bandgap determination based on electrical quantum efficiency," *Appl. Phys. Lett.*, vol. 103, no. 3, 032108 (3pp.), 2013.

[31] M. A. Green and A. W. Y. Ho-Baillie, "Pushing to the Limit: Radiative Efficiencies of Recent Mainstream and Emerging Solar Cells," *ACS Energy Lett.*, pp. 1639–1644, 2019.

CALIBRATION OF A DYE-SENSITIZED PHOTOVOLTAIC LARGE AREA MODULE

G. Bardizza*, D. Pavanello, H. Müllejans, E.D. Dunlop
European Commission, Joint Research Centre, Ispra (VA) - Italy
* E-mail: giorgio.bardizza@ec.europa.eu - Tel: +39 0332785529

ABSTRACT: Emerging photovoltaic (PV) technologies like organic (OPV) and dye-sensitized (DSC) devices are rapidly evolving and considerable research effort has addressed various approaches to increase their efficiency, reaching in the last years promising values. Examples of large area DSC modules for outdoor installation have been showcased by different manufacturers and proposed for multiple applications.
Here we present an indoor calibration of a large area DSC PV device, at standard test conditions (STC), through I-V measurements and the determination of the spectral responsivity following the procedures described in IEC 60891 and 60904 standards. Initial measurements have focused on the analysis of the module time response to light stimulus and on the necessary pre-conditioning step to ensure a reliable calibration at steady state conditions. Then the electrical performance was measured at STC, at irradiances from 100 W/m^2 to 1000 W/m^2 (at 25 °C) and for temperatures from 10 °C to 50 °C (at 1000 W/m^2).

1 INTRODUCTION

The rapid evolution of PV technologies calls for innovative approaches to increase the devices' efficiency and lifetime [1]. For a reliable assessment of the technology and a correct evaluation of the potential of new devices, the scientific community must agree on a method in order to make accurate performance measurements at standard test conditions (STC) and under real operating conditions, in compliance with standards IEC 60904 [2], IEC 60891 [3] and IEC 61853-1 [4].

There is a clear need of guidelines and protocols on how to perform a reliable calibration of emerging PV devices as these new technologies are rapidly evolving. In this work we present a practical example of a calibration of a large area DSC module performed under a steady-state solar simulator. The procedure applied followed as much as possible the method described in the international standards, but critical issues specific to this emerging technology (like the relatively slow response time) are considered and discussed.

2 DEVICE UNDER TEST

The large area DSC module involved in this study was provided by the company H.Glass under a non-disclosure agreement (ESTI CODE TV801). The module has an active area of 390 x 935 mm^2 and it is formed by 50 cells serially connected in W configuration. Alternating narrow (12 mm wide) and large (16 mm wide) cells are connected along the module. All of them have a length of 310 mm. For the calculations reported in this study we considered an average cell width of 14 mm resulting in a cell area of 4340 mm^2.

Standard MC4 type connectors are available for electrical connections. A picture of TV801 is shown in Figure 1.

3 CHARACTERIZATION AT STANDARD TEST CONDITIONS

In order to characterize the module at standard test conditions (STC) as defined in IEC TS 61836 [5] we followed the procedure implemented previously in our laboratory [6] [7].

3.1 Time response analysis

In the setup of the time response analysis [6], the electrical signals generated by the device under test TV801 (DUT) and by the calibrated c-Si reference cell PX302C (REF) illuminated by the chopped white light from the Xenon lamp were simultaneously acquired while the chopping frequency was scanned from 0.2 Hz to 80 Hz. The irradiance of the incident white light measured with REF on the test plane was 60 W/m^2.

Figure 1 Picture of the device under test TV801.

The DUT module shows a slow response compared to the c-Si REF cell. Different bias lights have been used: halogen lamp, green LEDs, white LEDs. In all the conditions, the DUT signal showed distortion when the chopping frequency was higher than 1.4 Hz and consequently the peak-to-peak DUT signal amplitude started to decrease. This is visible in Figure 2 where the normalized DUT/REF amplitude variation with chopping frequency is shown. Hence we selected 1 Hz as chopping frequency for the SR measurement.

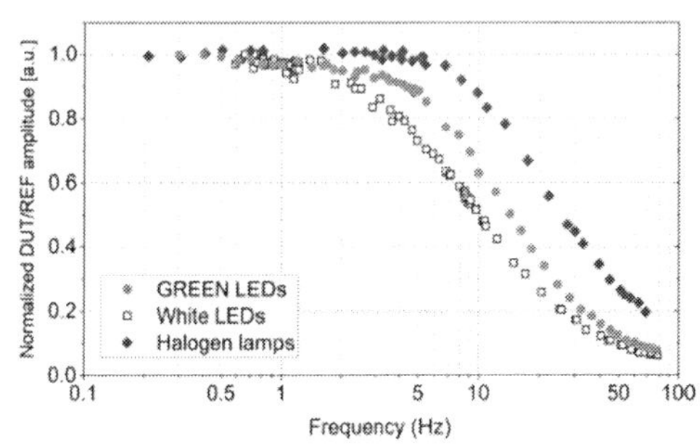

Figure 2 Time response analysis data: normalized DUT/REF amplitude variation with chopping frequency is shown for different bias lights: green LEDs, white

LEDs and halogen lamps. Distortion of the DUT signal is visible above 1.4 Hz, hence a chopping frequency smaller needs to be chosen to ensure measurements at steady state conditions.

3.2 Spectral responsivity

The TV801 spectral responsivity (SR) was measured with the Oriel system. The Oriel setup for SR measurements basically consists of a continuous illumination from a Xenon lamp, band-pass filters covering the interested wavelength range, a mechanical chopper and lock-in amplifiers. An exhaustive description of the setup is available in [8].

The device was mounted horizontally and different bias lights (LEDs and halogen lamps) were adopted for the SR measurements (Figure 3).

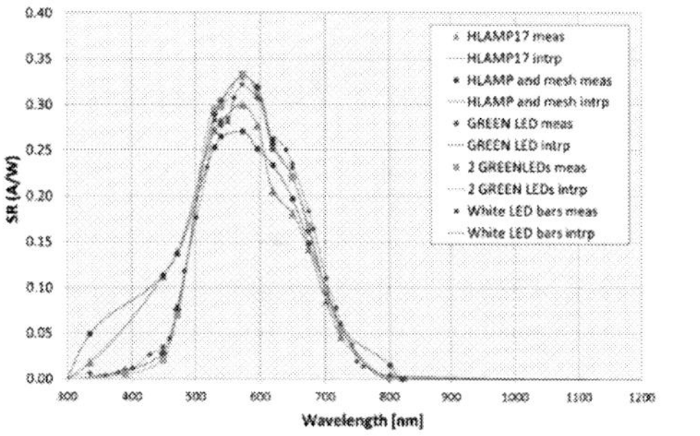

Figure 3 SR data of TV801 measured with Oriel setup using different bias lights: halogen lamp (HLAMP), halogen lamp filtered with meshes (HLAMP and mesh), green LEDs (GREEN LED), white LED bars (White LED bars).

The corresponding SR curve measured with white LED bias light after optimization of the parameter is reported in Figure 4.

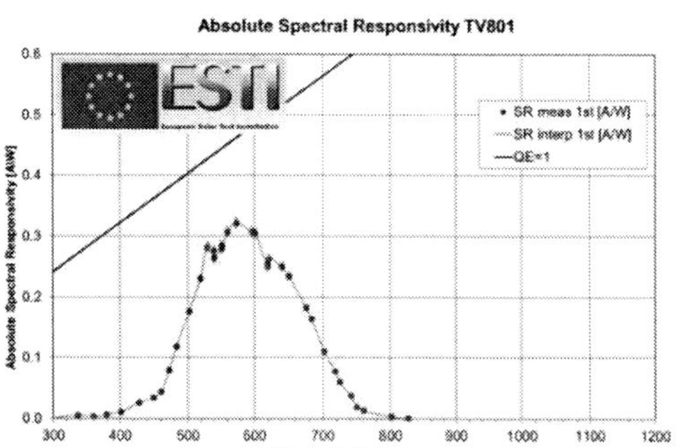

Figure 4 Final SR data of TV801 measured with Oriel setup

3.3 I-V measurements

The I-V characterization of TV801 was performed with the Apollo solar simulator (large area steady-state solar simulator) [9]. The device was mounted vertically in the test plane. As the DUT (time) response during the I-V sweeps was unknown, I-V curves for different (total) sweep times ranging from 0.1 s to 8 s were measured first. The extracted electrical parameters and their associated uncertainties are reported in Figure 5.

A device pre-conditioning was performed keeping the module under light at about P_{max} before I-V measurements. The procedure described in IEC 61215 was adopted to select the correct pre-conditioning time. Following MQT 19 protocol of IEC 61215 the DUT underwent a light soaking (LS) with a constant illumination of about 760 W/m^2, while controlling the DUT temperature to about 40 °C. The LS lasted 64 hours corresponding to an irradiation of about 49 kWh. The extracted relevant parameters are averaged and reported in Figure 6 and Table 1. The superimposed error bars are fully dominated by the uncertainties of the system, associated to each set of measurements.

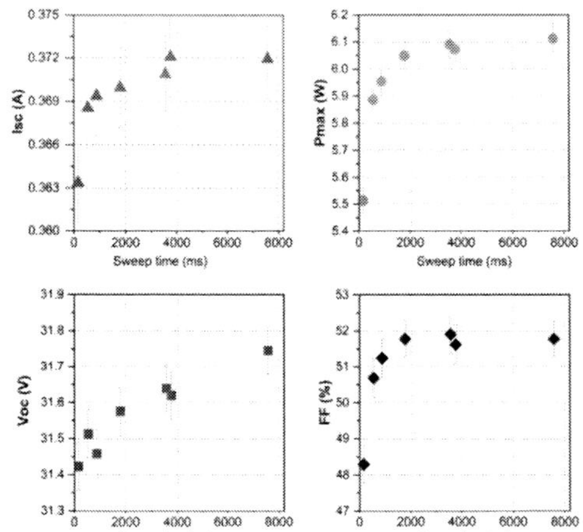

Figure 5 Absolute values of the relevant electrical parameters extracted from I-V curves measured at different sweep times in the range 0.11 s – 8 s. Error bars are the calculated uncertainties. We concluded that sweep times larger than 2 seconds are appropriate for this device. Finally 5 s sweep time was selected for the I-V characterization of TV801.

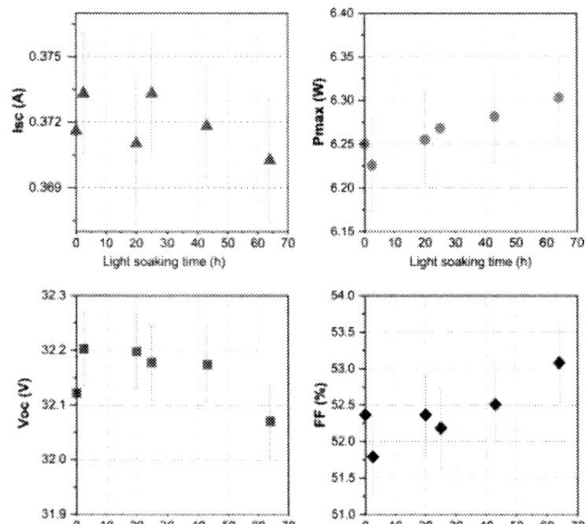

Figure 6 Light soaking (LS) experiment: the electrical parameters are extracted from at least four I-V curves performed periodically during LS. The error bars are dominated by the associated uncertainties.

Table 1 Stability test following MQT 19 protocol of IEC 61215.

LS	LS hours	Dose (kWh)	P_{max} (W)	$(P_{max}-P_{min})/P_{avg}$
0	0	0	6.250	
1	2.5	2	6.226	0.385%
2	20	15	6.255	0.465%
3	25	19	6.268	0.674%
4	43	33	6.282	0.423%
5	64	49	6.303	0.557%

Four I-V curves were measured at STC after the pre-conditioning (Figure 7) and the corresponding electrical parameters extracted (Table 2).

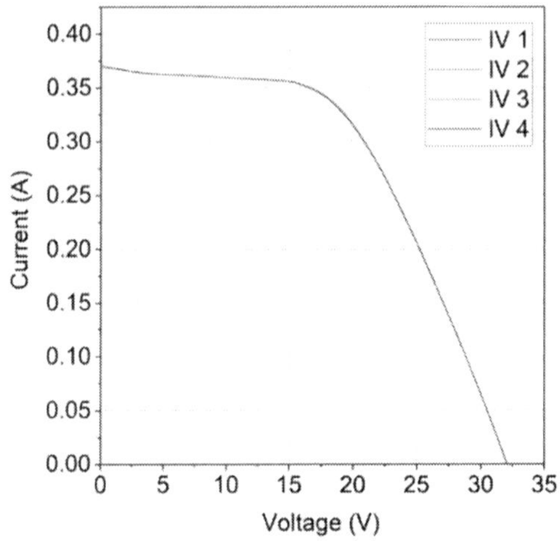

Figure 7 Four I-V curves measured at STC after pre-conditioning (5 s sweep time). The four curves overlay almost perfectly and are indistinguishable in this plot.

Table 2 Electrical parameters extracted from I-V curves of **Figure** 7.

Measurement	Temperature (°C)	Isc (A)	Voc (V)	Pmax (W)	FF (%)	Imp (A)	Vmp (V)
1	24.91	0.3705	32.09	6.313	53.11	0.3170	19.92
2	24.90	0.3703	32.08	6.303	53.05	0.3165	19.91
3	24.93	0.3702	32.04	6.296	53.07	0.3162	19.91
4	24.97	0.3700	32.07	6.301	53.10	0.3164	19.92

4. MEASUREMENTS vs IRRADIANCES

In order to investigate the behaviour of TV801 at different irradiances, measurements have been performed with the Apollo solar simulator varying the incident irradiance by gradually opening the adequate number of lamps [9]. The power of each lamp has been kept fixed at maximum level. The irradiance from each lamp in the test plane was about 100 W/m². The results of the experiments are shown in Figure 8. The evolution of the electrical parameters extracted from the measured I-V curves at the different irradiances are reported in the plots.

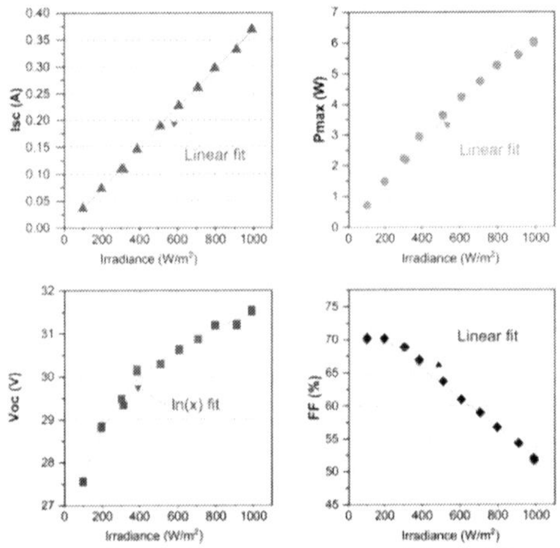

Figure 8 Electrical parameters extracted from I-V curves measured at different irradiances.

5. MEASUREMENTS vs TEMPERATURES

In order to investigate the behaviour of TV801 at different temperatures, two temperature coefficient measurements (TCO) were performed at 1000 W/m² with the Apollo solar simulator whose lamps were used both to illuminate during I-V characterization and to heat up the device in between the measurements [9]. A climatic chamber set to 5 °C has been used to cool down the device prior I-V characterization.

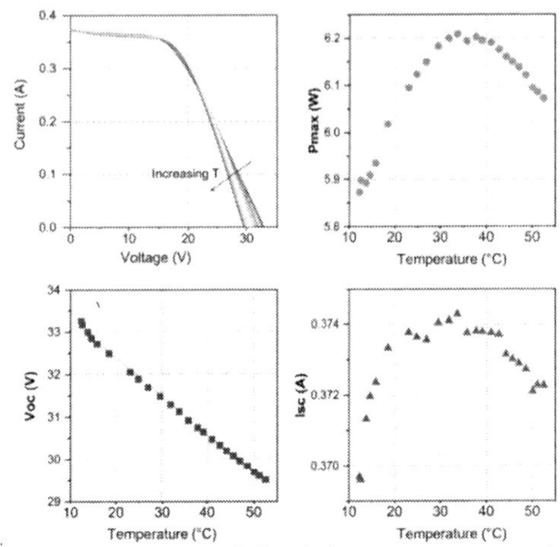

Figure 9 I-V curves and electrical parameters extracted from I-V curves measured at different temperatures during the warm up of the TV801 module.

The two experiments performed showed comparable results. The results of one experiment are shown as example in Figure 9. While V_{oc} decreases linearly with increasing temperature, the other parameters clearly do not vary linearly with temperature. Therefore, the linear fit usually made to extract temperature coefficients is not an appropriate approximation and cannot be used. A more complex analysis will be required, which will be presented elsewhere.

6 CONCLUSIONS

We have presented the electrical characterization of a large area DSC module. After determining the appropriate measurement parameters for SR and IV through time response analysis and after preconditioning, the performance at STC and a range of irradiances and temperatures was measured. The dependence on irradiance was found to be similar to those known from traditional c-Si PV modules, whereas the variation of the electrical parameters with temperature was found to be non-linear (except for open circuit voltage). Further analysis of this behaviour and energy rating studies will be published elsewhere.

REFERENCES

[1] M. A. Green, E. D. Dunlop, J. Hohl-Ebinger, M. Yoshita, N. Kopidakis e X. Hao, «Solar cell efficiency tables (Version 58),» *Prog Photovolt Res Appl*, pp. 1-11, 2021.

[2] IEC 60904 Series - Photovoltaic devices - ALL PARTS, International Electrotechnical Commission, 2020.

[3] IEC 60891 (Ed.2) Photovoltaic devices - Procedures for temperature and irradiance corrections to measured I-V characteristics, 2009.

[4] IEC 61853 (Ed. 1) Photovoltaic (PV) module performance testing and energy rating - Part 1: Irradiance and temperature performance measurements and power rating, 2011.

[5] IEC TS 61836 Solar photovoltaic energy systems - Terms, definitions and symbols, 2016.

[6] G. Bardizza, D. Pavanello, H. Müllejans e T. Sample, «Spectral responsivity measurements of DSSC devices at low chopping frequency (1Hz),» *Progress in Photovoltaics: Research and Applications*, n. 24, pp. 428-435, 2016.

[7] G. Bardizza, D. Pavanello, R. Galleano, T. Sample e H. Müllejans, «Calibration procedure for solar cells exhibiting slow response and application to a dye-sensitized photovoltaic device,» *Sol. Energy Mater. Sol. Cells*, vol. 160, pp. 418-424, 2017.

[8] M. Pravettoni, A. Komlan, R. Galleano, H. Müllejans e E. and Dunlop, «An alternative method for spectral response measurements of large-area thin-film photovoltaic modules,» *Progress in Photovoltaics: Research and Applications*, vol. 20, pp. 416-422, 2012.

[9] H. Müllejans, E. Salis, D. Pavanello, G. Bardizza, J. Lopez-Garcia, W. Zaaiman, D. Shaw, T. Sample e E. D. Dunlop, «Apollo large-area steady-state solar simulator as a versatile tool for PV device performance measurements,» in *Proc. 37th Eur. PVSEC, online*, 2020.

Study of $SnO_2/CdSe_xTe_{1-x}/CdTe$ solar cells fabricated by selenium treatment of the absorber layer

Elisa Artegiani, Vikash Kumar, Prabeesh Punathil, Solidea Zanetti, Matteo Bertoncello*, Matteo Meneghini* and Alessandro Romeo

LAPS—Laboratory for Photovoltaics and Solid State Physics,
Department of Computer Science, University of Verona, Ca' Vignal 1, Strada Le Grazie 15, 37134 Verona, Italy
e-mail address: alessandro.romeo@univr.it

* Department of Information Engineering, University of Padova, via Gradenigo 6/B, 35131, Padova, 35131, Italy

ABSTRACT: The reboot after the global Covid-19 crisis will go through the investment of large economic resources around the world. To take advantage of the opportunity to improve the world situation, this money should be invested extensively for the technological and green energy revolution. Photovoltaics is nowadays a mature technology, capable of providing green energy at competitive prices compared to fossil fuels. Regarding manufacturing costs, CdTe thin film devices have a great advantage over traditional crystalline silicon cells, since they can be fabricated in a single production line. Furthermore, CdTe panels nowadays exceed an efficiency of 19 %, close to that of silicon technology. The efficiency improvement of the last few years is mainly due to the introduction of a $CdSe_xTe_{1-x}$ layer in the absorber. This reduces the band gap increasing the absorption in the long wavelength region. Moreover, it allows a better band alignment with wider band gap buffers, enabling the removal of the CdS window layer. In this work we study a $SnO_2/CdSe_xTe_{1-x}/CdTe$ solar cell where the band gap grading of the absorber has been achieved through an innovative method: the selenization of the CdTe layer.

Keywords: CdTe, $CdSe_xTe_{1-x}$, band gap grading, selenization

1 INTRODUCTION

CdTe commercial panels have reached a conversion efficiency of 19 %, very close to mono-crystalline silicon wafer based technology [1]. Moreover, CdTe thin film devices are fabricated in a single production line with high throughput and simple fabrication process. For these reasons, CdTe solar cells are increasing their share in the PV market.

Laboratory scale CdTe cells have reached efficiencies up to 22.1 % [2], with a large increase in the last few years. The efficiency improvement has been mainly achieved by increasing the short circuit current of the devices. This has been obtained primarily by introducing a $CdSe_xTe_{1-x}$ layer in the absorber, narrowing its band gap and thus increasing the absorption at long wavelengths. Secondly, by replacing the CdS window layer with a wider band gap buffer such as Mg_xZnO_{1-x} (MZO) or SnO_2, thus avoiding the parasitic absorption of CdS in the short wavelength region.

The grading of the band gap has been achieved by depositing CdSe [3–5] or directly $CdSe_xTe_{1-x}$ [6] compound at the junction and then promoting the inter-diffusion by performing the activation treatment.

In our laboratories, the $CdSe_xTe_{1-x}$ compound is formed through an innovative method: the selenization of CdTe layer [7].

In this work we study $SnO_2/CdSe_xTe_{1-x}/CdTe$ solar cells and compare them with standard CdS/CdTe devices.

2 EXPERIMENTAL

2.1 Preparation of the devices

In our laboratories, CdTe cells are deposited in superstrate configuration by low-temperature fabrication process. The front contact of the standard CdS/CdTe cells is an ITO/ZnO bilayer deposited on soda lime glass by sputtering [8]. This is subsequently annealed at 450 °C in vacuum. Then a 150 nm thick CdS window layer is deposited by evaporation and annealed at 450 °C in vacuum, followed by the deposition of 4 μm thick CdTe layer.

Instead, the front contact of the $CdSe_xTe_{1-x}/CdTe$ devices consists of a fluorine doped tin oxide (FTO) layer followed by a high resistivity transparent (HRT) un-doped tin oxide (SnO_2) layer deposited on soda lime glass (NSG TEC 12D). The application of SnO_2 instead of ZnO is due to the better suitability of this buffer layer with the $CdSe_xTe_{1-x}$ compound. On this bilayer a thin (from 150 up to 300 nm) CdTe layer is deposited by vacuum evaporation (VE). The stack is then put in a tubular oven in argon atmosphere and annealed at 400 °C in a Se+Ar atmosphere. Then a 4 μm thick CdTe layer is deposited by VE [7].

In both configurations, after deposition of the thick CdTe layer a typical CdTe activation treatment is applied. This consists of a wet deposition of a $CdCl_2$ solution and subsequent annealing in air at 380 °C.

The sample surface is cleaned from the $CdCl_2$ residues generating a Te-rich region to improve the performance and stability of the back contact, by a bromine/methanol etching [9].

In this case, the back contact is formed by a 2 nm thick copper layer followed by a 30 nm thick gold layer deposited by VE, followed by an annealing at 200 °C in air to form Cu_xTe_{1-x} compounds in the contact [9]. However we have recently presented a different insertion of copper by depositing a $CuCl_2$-methanol solution on the surface, to better control the Cu content and homogeneity [10].

2.2 Characterization

Current density-voltage (JV) characteristics are measured with a Keithley Source Meter 2420, under an AM 1.5 spectrum at 100 mW/cm^2, using a LOT Quantum Design Europe solar simulator LS0306.

The external quantum efficiency (EQE) is obtained using a commercial LOANA solar cell analysis system, calibrated with a silicon reference sample with known EQE using an incident spotlight of 1 mm × 2 mm area.

3 DEVICES CHARACTERIZATION

3.1 External Quantum Efficiency

Fig. 1 displays the EQE of $SnO_2/CdSe_xTe_{1-x}/CdTe$ and CdS/CdTe cells. The right shift at long wavelengths of the red curve compared to the black one indicates that in these devices the absorber has a narrower band gap. This demonstrates that the selenization of the first thin CdTe layer, subsequently mixed with a thicker one by the activation treatment, is successful in the formation of a suitable $CdSe_xTe_{1-x}$ compound in the absorber. Moreover, from the EQE it is possible to calculate the bang gap of the absorber, which corresponds to the maximum negative derivate of the EQE curve at long wavelengths [11]. This, in case of band gap grading in the absorber, corresponds to the value at the bottom of the bowing. For $SnO_2/CdSe_xTe_{1-x}/CdTe$ samples the calculated band gap is 1.43 eV compared to the 1.49 eV of standard CdS/CdTe cells. Confirming the formation of a suitable $CdSe_xTe_{1-x}$ compound in the absorber.

The presence of a $CdSe_xTe_{1-x}$ compound at the junction favours a better band alignment with high band gap buffers such as SnO_2 or MZO, compared to pure CdTe. This allows to remove the CdS window layer from the structure, avoiding its parasitic absorption at short wavelengths. In fig.1 the EQE gain is evident in the short wavelengths' region (red curve compared to black one) due to the absence of the CdS layer in $SnO_2/CdSe_xTe_{1-x}/CdTe$ samples.

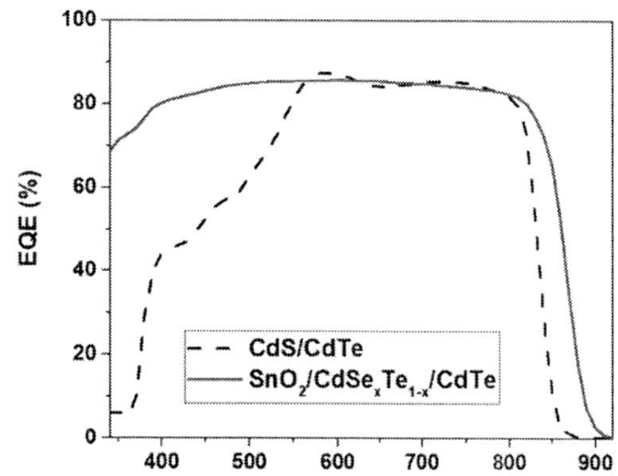

Figure 1: External Quantum Efficiency of CdS/CdTe (black line) and $SnO_2/CdSe_xTe_{1-x}/CdTe$ (red line) solar cells.

To quantitatively assess the current gain due to the introduction of the $CdSe_xTe_{1-x}$ compound and the removal of the window layer, the Jsc of the solar cells was extracted from the EQE data. The standard CdS/CdTe cell, made with the same process parameters used for the new configuration, has a calculated Jsc of 21.8 mA/cm^2, while the $SnO_2/CdSe_xTe_{1-x}/CdTe$ configuration leads to 26 mA/cm^2, a current gain of more than 4 mA/cm^2.

3.2 Current density - voltage characteristic

To evaluate and compare the overall performance of the cells due to the different configurations, in Figure 2 we present the current-voltage characteristic of CdS/CdTe and $SnO_2/CdSe_xTe_{1-x}/CdTe$ devices. Their efficiency parameters are reported in Table I.

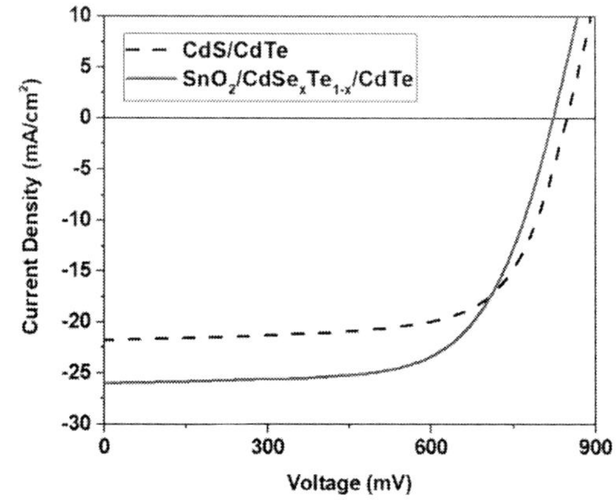

Figure 2: Current density - voltage characteristic of CdS/CdTe (black line) and $SnO_2/CdSe_xTe_{1-x}/CdTe$ (red line) solar cells.

Table I: Efficiency parameters of the CdS/CdTe and $SnO_2/CdSe_xTe_{1-x}/CdTe$ solar cells, whose current - voltage characteristics are showed in Figure 2.

	FF [%]	Voc [mV]	Jsc [mA/cm^2]	Eff [%]
CdS/CdTe	68	852	21.8	12.6
$SnO_2/CdSe_xTe_{1-x}/CdTe$	66	824	26.0	14.1

Fig. 2 clearly emphasizes the current gain calculated from the EQE in Fig. 1. The introduction of the $CdSe_xTe_{1-x}$ compound causes a slight decrease of the open circuit voltage, due to the lower band gap energy compared to CdTe. On the other hand, this is more than offset by the substantial current gain. In fact, overall, the $SnO_2/CdSe_xTe_{1-x}/CdTe$ structure leads to an efficiency gain of 1.5 percentage points compared to standard CdS/CdTe made with the same fabrication process (higher efficiencies have been obtained but with a different configuration in the past [12]).

4 CONCLUSIONS

An innovative method to form a $CdSe_xTe_{1-x}$ compound in order to produce a band gap grading of the absorber in CdTe solar cells has been developed with the application of selenization of a thin CdTe layer.

The external quantum efficiency and the current-voltage characteristics show that the selenization process is suitable for the formation of a $CdSe_xTe_{1-x}$ compound. This also allows the removal of the CdS window layer from the structure, promoting the current gain both in the short and long wavelength regions. Overall, the new configuration leads to a current gain of more than 4 mA/cm^2, and a consequent conversion efficiency increase of 1.5 percentage points.

5 REFERENCES

[1] I. Fraunhofer Institute for Solar Energy Systems, with support of P.P. GmbH, Photovoltaics Report, (2020). www.ise.fraunhofer.de.

[2] S. Krum, S. Haymore, First Solar Achieves Yet Another Cell Conversion Efficiency World Record, First Sol. (2016). http://investor.firstsolar.com/releasedetail.cfm?ReleaseID=956479.

[3] J.D. Poplawsky, W. Guo, N. Paudel, A. Ng, K. More, D. Leonard, Y. Yan, Structural and compositional dependence of the CdTe x Se 1-x alloy layer photoactivity in CdTe-based solar cells, Nat. Commun. 7 (2016) 1–9. https://doi.org/10.1038/ncomms12537.

[4] M. Lingg, A. Spescha, S.G. Haass, R. Carron, S. Buecheler, A.N. Tiwari, Structural and electronic properties of CdTe1-xSex films and their application in solar cells, Sci. Technol. Adv. Mater. (2018). https://doi.org/10.1080/14686996.2018.1497403 .

[5] T. Baines, G. Zoppi, L. Bowen, T.P. Shalvey, S. Mariotti, K. Durose, J.D. Major, Incorporation of CdSe layers into CdTe thin film solar cells, Sol. Energy Mater. Sol. Cells. 180 (2018) 196–204. https://doi.org/10.1016/j.solmat.2018.03.010.

[6] A.H. Munshi, J. Kephart, A. Abbas, J. Raguse, J.N. Beaudry, K. Barth, J. Sites, J. Walls, W. Sampath, Polycrystalline CdSeTe/CdTe absorber cells with 28 mA/cm2 short-circuit current, IEEE J. Photovoltaics. (2018). https://doi.org/10.1109/JPHOTOV.2017.2775139.

[7] E. Artegiani, A. Gasparotto, P. Punathil, V. Kumar, M. Barbato, M. Meneghini, G. Meneghesso, F. Piccinelli, A. Romeo, A new method for CdSexTe1-x band grading for high efficiency thin-absorber CdTe solar cells, Sol. Energy Mater. Sol. Cells. 226 (2021) 111081. https://doi.org/10.1016/j.solmat.2021.111081.

[8] E. Artegiani, D. Menossi, A. Salavei, S. Di Mare, A. Romeo, Analysis of the influence on the performance degradation of CdTe solar cells by the front contact, (2016). https://doi.org/10.1016/j.tsf.2016.09.052.

[9] I. Rimmaudo, A. Salavei, E. Artegiani, D. Menossi, M. Giarola, G. Mariotto, A. Gasparotto, A. Romeo, Improved stability of CdTe solar cells by absorber surface etching, Sol. Energy Mater. Sol. Cells. (2017). https://doi.org/10.1016/j.solmat.2016.12.044.

[10] E. Artegiani, D. Menossi, H. Shiel, V. Dhanak, J.D. Major, A. Gasparotto, K. Sun, A. Romeo, Analysis of a novel $CuCl_2$ back contact process for improved stability in CdTe solar cells, Prog. Photovoltaics Res. Appl. (2019) pip.3148. https://doi.org/10.1002/pip.3148.

[11] J.M. Raguse, Electroluminescence of Thin-Film CdTe Solar Cells and Modules (PhD), Color. State Univ. (2015).

[12] E. Artegiani, J.D. Major, H. Shiel, V. Dhanak, C. Ferrari, A. Romeo, How the amount of copper influences the formation and stability of defects in CdTe solar cells, Sol. Energy Mater. Sol. Cells. 204 (2020) 110228. https://doi.org/10.1016/j.solmat.2019.110228.

III-V MULTIJUNCTION SOLAR CELLS ON ULTRATHIN GE|SI VIRTUAL SUBSTRATES GROWN AT LOW TEMPERATURE BY RF-PECVD

Iván García[1*], Monalisa Ghosh[2], Víctor Orejuela[1],
Pere Roca i Cabarrocas[2], Ignacio Rey-Stolle[1]
[1]Instituto de Energía Solar, Universidad Politécnica de Madrid, 28040 Madrid, Spain
[2]LPICM-CNRS, Ecole Polytechnique, 91128, Palaiseau, France

ABSTRACT: The application of Ge|Si virtual substrates with an ultrathin Ge layer to the development of high efficiency III-V solar cells is analyzed. The potential to achieve high efficiencies in GaInP/GaInAs/Si triple-junction cells, which is affected by the optical absorption in the Ge layer, is theoretically demonstrated under realistic scenarios of Ge|Si substrate characteristics and III-V cell configurations. The first dual-junction GaInAs/Ge|Si III-V cells fabricated on these substrates demonstrate the concept with functional devices and show the route to high efficiency implementations by using appropriate Si substrates and optimized Ge deposition.

Keywords: germanium virtual substrate, low-temperature RF-PECVD, III-V multijunction solar cell

1 INTRODUCTION

Virtual substrates fabricated by direct deposition of Ge on Si have gained attention over the last years due to the high quality obtained, i.e., low threading dislocation densities (TDD) and surface roughness, offering a high potential as low-cost templates for the implementation of all kinds of GaAs-based electronic devices, including high efficiency multijunction solar cells. We have shown GaInP/GaInAs/Ge 3-junction cells (3JSC) on active (thick Ge) Ge|Si virtual substrates in previous conferences [1], [2]. The formation of cracks due to the mismatch in thermal expansion coefficients between the Si substrate and the rest of Ge and III-V layers poses an important challenge to this approach. Thinning the III-V structure has been demonstrated to prevent cracking at the expense of some efficiency loss [3].

In this work we are presenting a parallel approach consisting on using Ge|Si virtual substrates with an ultrathin Ge layer (20-140 nm) fabricated by low-temperature radio-frequency plasma-enhanced chemical vapor deposition (RF-PECVD), with a reported TDD below $5 \cdot 10^6 cm^{-2}$ [4]. This thin Ge layer bridges the lattice constant difference between Si and GaAs and thus would enable the deposition of high efficiency III-V solar cell structures on top without surpassing the critical thickness for crack formation. In this case, the Ge layer is too thin to function as a subcell, but the Si substrate can perform the role of bottom cell. This method is simpler and potentially lower cost than designs using GaP nucleation and GaAsP-based buffer layers [5], or wafer-bonding [6]. However, the thin Ge layer introduces a parasitic optical absorption that affects the efficiency potential. Therefore, the potential to obtain high efficiency triple-junction solar cells (3JSC) with a Si bottom junction is first analyzed theoretically, considering the effect of the parasitic absorption in the Ge layer, the design of the upper junctions, etc, showing that efficiencies approaching 34% can be obtained. Then, we show our first experimental results on GaInAs solar cells grown on ultrathin Ge|Si virtual substrates with a Ge layer as thin as 20 nm, aiming to test the stability of this thin Ge layer and its suitability as epitaxy template, resulting in a GaInAs/Si dual-junction solar cell, which shows no cracking and offers promise to obtain high efficiencies in next implementations using higher quality Ge|Si templates.

2 METHODS

The theoretical calculation of the external quantum efficiency, photocurrent and dark currents is performed using 1D models based on the Hovel approach and the generalized matrix method. The Ge|Si virtual substrates are fabricated by RF-PECVD using p-type Si substrates with a resistivity of 1-10 ohm·cm. The results presented in this abstract are obtained for non-epiready Si substrates and with a miscut of 5° towards <110>. Epiready substrates with a more appropriate miscut of 6° towards <111> are being used for the next developments. 20 nm Ge films are deposited in a standard 13.56 MHz capacitively coupled RF-PECVD, using the process parameters explained elsewhere [4]. The virtual substrates used are not optimized in terms of Ge layer quality, and they show a high TDD. III-V structures were grown by MOVPE (details can be found in [3]) and solar cell devices were fabricated using standard photolithography and wet etching techniques, with appropriate metal contacts for GaAs (front) and Si (back). No passivation of the Si back surface was applied. External quantum efficiency (EQE) and I-V curves were taken using conventional methods.

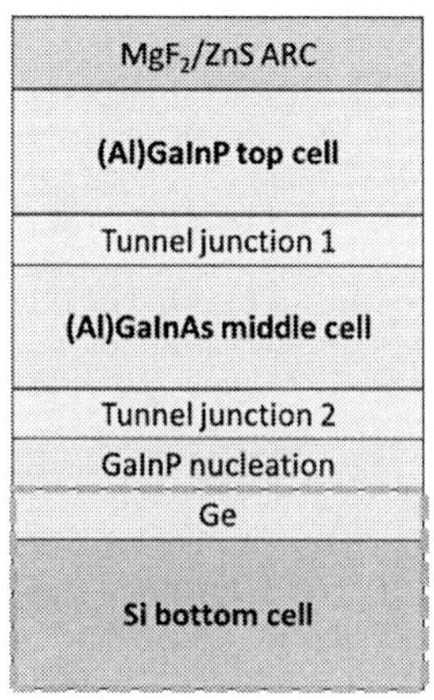

Figure 1: Baseline structure of the 3JSC on Ge|Si virtual substrate studied

3 ANALYSIS OF PERFORMANCE POTENTIAL

The suitable Ge layer thickness to achieve the highest quality template by low temperature RF-PECVD in terms of TDD and surface roughness has been found to lie between 20 and 140 nm. Therefore, the parasitic light absorption in this layer can be significant. Thus, the potential of these virtual substrates was analyzed towards the achievement of high efficiency two-terminal, monolithic (Al)GaInP/(Al)GaInAs/Ge|Si 3JSC. The baseline structure is shown in Fig. 1 (further details can be found elsewhere [7]).

Table I: Summary of the minority carrier parameters used in the analysis of potential.

Layer		τ (ns) TDD: $1{\cdot}10^4$ (cm^{-2})	τ (ns) TDD: $5{\cdot}10^6$ (cm^{-2})	μ (cm^2/ V·s)	SRV (cm·s)
Top cell	Emitter	1	0.1	40	1000
	Base	10	1	1100	100
Middle cell	Emitter	2	0.2	100	1000
	Base	20	0.5	3100	10
Bottom cell	Emitter	$1{\cdot}10^5$		500	1000
	Base	$1{\cdot}10^6$		1000	2

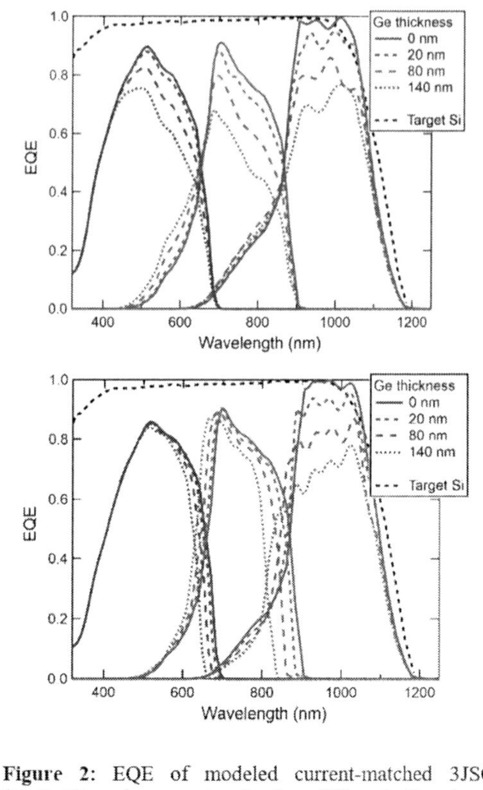

Figure 2: EQE of modeled current-matched 3JSC (AM1.5G solar spectrum) for different Ge layer thicknesses in the virtual substrate. Current matching for cells will Ge layer is achieved by: top: thinning the upper junctions; bottom: increasing the bandgap of the upper junctions using AlGaInP and AlGaAs. The dashed black line corresponds to the target EQE of the Si subcell.

Concerning the Si bottom cell, we assume the typical light trapping factor and back surface passivation achieved in III-V on Si designs [5]. We use typical experimental minority carrier parameters, as detailed in Table 1. We consider two scenarios: negligible TDD and the best value obtained in previous Ge|Si virtual substrates fabricated by RF-PECVD [4], which has a significant effect on the minority carrier properties [8]. Current developments on such virtual substrates are expected to improve the TDD, so with this analysis we are exploring the upper and lower performance limits that can be projected. We assume the same minority carrier parameters in the Al-containing compounds of the high bandgap design analyzed, since the Al compositions needed are below 10%.

Figure 2 shows the modeled EQE of current matched 3JSC for different thicknesses of the Ge layer in the virtual substrate. The graph on the top corresponds to a design (low-bandgap design) that achieves current matching as the Ge layer is made thicker by thinning the upper junctions. In the graph on the bottom (high-bandgap design) current matching is achieved by increasing the bandgap in these junctions, by using Al-containing compounds. It can be clearly observed how the EQE of the Si bottom junction is affected by the Ge layer, which forces the upper junctions to be thinned and lowering their EQE too. In the case of the high bandgap upper junctions, current is also lost as the Ge layer is made thicker, but the increased bandgap increases the junctions' voltage, assuming that their minority carrier parameters is preserved. It is important to point out that the total thickness of the III-V structure ranges from 1.5 to 2.2 μm, for the span of Ge layer thicknesses explored. These thicknesses are well below the critical thickness for crack propagation, a clear advantage of this 3JSC design to achieve crack-free III-V solar cells on Si substrates. Also, the EQE of the Si subcell is assumed to match the state-of-the-art, which implies using an efficient back passivation, light trapping and prevention of impurity diffusion during deposition of the III-V stack.

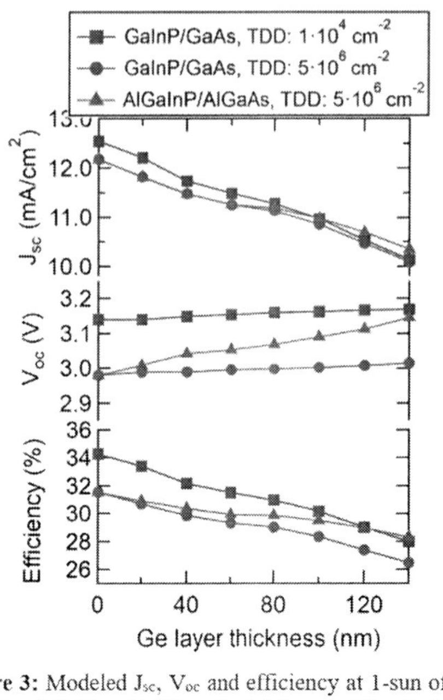

Figure 3: Modeled J_{sc}, V_{oc} and efficiency at 1-sun of the optimized 3JSC on Ge|Si vs the thickness of the Ge layer, for three design scenarios

Figure 3 shows a summary of the modeled electrical performance for the low and high bandgap cases with a TDD of $5\cdot10^6 cm^{-2}$. The case of negligible TDD for the low bandgap design is also included. The J_{sc}, V_{oc} and efficiency of the cells at 1-sun is plotted versus the Ge layer thickness for the 3 cases described above. The effect of the TD in the performance is clearly observed. As the Ge layer thickness is increased and the upper junctions thinned, the minority carrier diffusion length becomes more resilient to the TD and so the J_{sc} of the high and low TDD cases converge. The thinning of the upper junction has also an effect on the V_{oc}, which slightly, but noticeably, increases as the Ge layer thickness is increased. The expected significant increase in the V_{oc} for the high bandgap case is also observed. Finally, the efficiency drop with Ge layer thickness is less steep for the high bandgap design, thanks to the effect of the V_{oc} increase. This efficiency chart shows that values between 30-34% can be achieved for Ge layer thicknesses below 60 nm. This must be considered when optimizing the Ge/Si templates. Besides, the significant effect of the TDD, despite the thin III-V junctions used, also encourages the pursue of lower TDD virtual substrates.

4 EXPERIMENTAL GAAS/GE|SI SOLAR CELLS

The preliminary experimental work presented here aimed to test the stability of the ultrathin Ge layer and its suitability as template for epitaxy. For this, we used virtual substrates with Ge layer thicknesses in the lower range, of just 20 nm. The test structure is depicted in Fig. 4. For this, we grew GaInAs cells using virtually the same process as in the case of standard Ge substrates. A GaInP nucleation layer was grown first, followed by a GaInAs overbuffer and a tunnel junction before the GaAs cell structure. In the fabrication of the devices, we did not attempt to modify the Si rear side to achieve any improved passivation or light trapping. Besides, we expect that some degradation of the Si substrate occurs during the MOVPE process, due to in-diffusion of impurities from the graphite susceptor, as described before [9]. All these issues will be addressed to achieve a proper EQE in the Si subcell in the future. Moreover, as commented in the Methods section, the Si substrates used were not fully epi-ready and they have not the optimum surface miscut. The reflectance of these virtual substrates was found to be virtually identical to standard Ge, but macroscopic defects developed during the III-V deposition affected the performance of the cells. The EQE and J-V curves obtained are shown in Fig. 5. Firstly, these measurements demonstrate functional devices with a reasonable performance for a first attempt using ultrathin Ge|Si virtual substrates. A well-defined response of the GaInAs and Si subcells is observed. An unusually high EQE for devices without ARC was measured, due to the antireflecting effect of the surface roughness. By measurement of the reflectance and comparison with that of similar devices on Ge substrates we could find a correction factor that was applied to the EQE curves. The modeling of the EQE data indicates a Si bottom cell with a back surface recombination around $1\cdot10^4$ cm/s and base lifetime over 1 µs, as expected for the fabrication conditions used. The GaInAs subcell shows an EQE limited by the diffusion length in the base, below 600 nm, as evidenced also by the virtually identical curves for 2500 nm and 4000 nm thickness in the base layer of the devices (comparison not shown in Figure 5). The J-V curves show a fine shape and low variability among the set of devices fabricated (J_{sc} varies due to different shadowing factor in the devices). The ideality factor of 3 in the 2JSC indicates that the GaInAs top cell is operating at n=2 regime, with a high J_{02} saturation current. The origin of the low minority diffusion length and low V_{oc} observed is mainly ascribed to the high TDD of the test virtual substrates used. Current experiments on new Ge|Si substrates show no macroscopic defects and a surface roughness comparable to standard Ge substrates that, together with optimized Ge layer deposition and thickness, are expected to enable a significant boost in the material quality and solar cell performance in the short term.

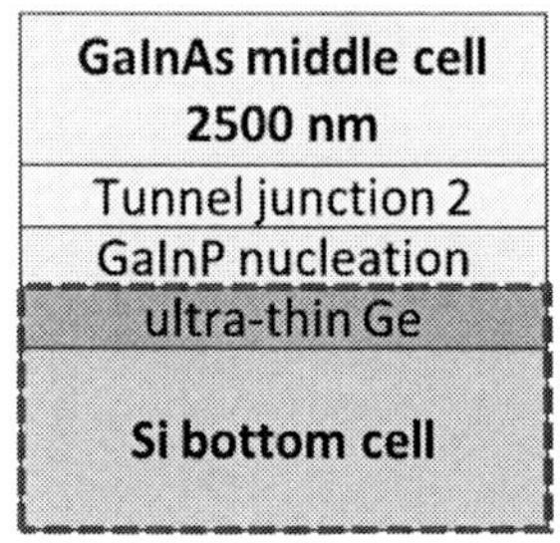

Figure 4: GaInAs/Si tandem solar cell structure used for the experimental test of the virtual substrates.

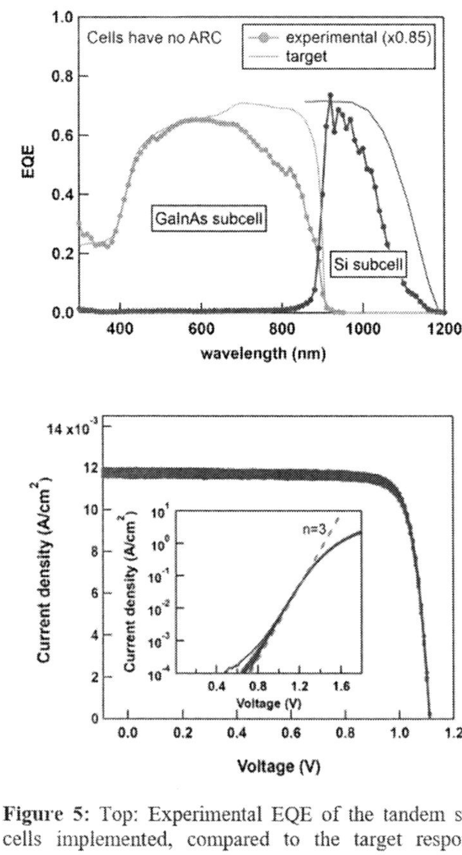

Figure 5: Top: Experimental EQE of the tandem solar cells implemented, compared to the target response. Bottom: uncalibrated light I-V curve of a set of devices. The inset shows the dark I-V curves.

5 SUMMARY AND CONCLUSIONS

Efficiencies approaching 34% are attainable with 3JSC on ultrathin Ge|Si substrates with an active Si bottom junction. For this, the Ge layer used must be kept below ~60 nm, and it is essential that a state-of-the art Si solar cell performance is achieved in the bottom cell. The first III-V cells on ultrathin Ge|Si have been implemented and show functional devices with EQE and J-V results demonstrating the potential of these virtual substrates. An improved performance is expected as soon as the appropriate Si substrates are used to fabricate the Ge|Si templates with optimized Ge layers.

6 ACKNOWLEDGEMENT

We acknowledge technical support from J. Bautista and L. Cifuentes for device processing. Funding from AEI (project RTI2018-094291-B-I00) is gratefully acknowledged. Parts of this work were done in the framework of the TOTAL-LPICM joint research team. V. Orejuela is funded by the Spanish Ministerio de Ciencia e Innovación (MICINN) through a FPI grant (PRE2019-088437).

REFERENCES

[1] I. García, M. Hinojosa, I. Lombardero, L. Cifuentes, I. Rey-Stolle, C. Algora, H. Nguyen, Edwards, S, Morgan, A, and Johnson, A, "Ge virtual substrates for high efficiency III-V solar cells: applications, potential and challenges," in *46th IEEE PVSC*, Chicago, USA, 2019, pp. 1444–1451, doi: 10.1109/PVSC40753.2019.8980914.

[2] I. García, I. Rey-Stolle, M. Hinojosa, I. Lombardero, L. Cifuentes, C. Algora, H. Nguyen, A. Morgan, and A. Johnson, "Space III-V multijunction solar cells on Ge/Si virtual substrates," in *2019 European Space Power Conference (ESPC)*, Juan-Les-Pins, France, Oct. 2019, pp. 1–6, doi: 10.1109/ESPC.2019.8932026.

[3] I. García, L. Barrutia, S. Dadgostar, M. Hinojosa, A. Johnson, and I. Rey-Stolle, "Thinned GaInP/GaInAs/Ge solar cells grown with reduced cracking on Ge|Si virtual substrates," *Sol. Energy Mater. Sol. Cells*, vol. 225, p. 111034, Jun. 2021, doi: 10.1016/j.solmat.2021.111034.

[4] R. Cariou, R. Ruggeri, X. Tan, G. Mannino, J. Nassar, and P. Roca i Cabarrocas, "Structural properties of relaxed thin film germanium layers grown by low temperature RF-PECVD epitaxy on Si and Ge (100) substrates," *AIP Adv.*, vol. 4, no. 7, p. 077103, Jul. 2014, doi: 10.1063/1.4886774.

[5] M. Feifel, D. Lackner, J. Ohlmann, K. Volz, T. Hannappel, J. Benick, M. Hermle, and F. Dimroth, "Advances in Epitaxial GaInP/GaAs/Si Triple Junction Solar Cells," in *2020 47th IEEE Photovoltaic Specialists Conference (PVSC)*, Jun. 2020, pp. 0194–0196, doi: 10.1109/PVSC45281.2020.9300594.

[6] R. Cariou, J. Benick, F. Feldmann, O. Höhn, H. Hauser, P. Beutel, N. Razek, M. Wimplinger, B. Bläsi, D. Lackner, M. Hermle, G. Siefer, S. W. Glunz, A. W. Bett, and F. Dimroth, "III–V-on-silicon solar cells reaching 33% photoconversion efficiency in two-terminal configuration," *Nat. Energy*, vol. 3, no. 4, pp. 326–333, Apr. 2018, doi: 10.1038/s41560-018-0125-0.

[7] L. Barrutia, I. Garcia, E. Barrigón, E. Barrigón, M. Ochoa, I. Lombardero, M. Hinojosa, P. Cano, J. Bautista, L. Cifuentes, I. Rey-Stolle, and C. Algora, "Development of the Lattice Matched GaInP/GaInAs/Ge Triple Junction Solar Cell with an Efficiency Over 40%," in *2018 Spanish Conference on Electron Devices (CDE)*, Nov. 2018, pp. 1–4, doi: 10.1109/CDE.2018.8596996.

[8] N. Jain and M. K. Hudait, "Impact of Threading Dislocations on the Design of GaAs and InGaP/GaAs Solar Cells on Si Using Finite Element Analysis," *IEEE J. Photovolt.*, vol. 3, no. 1, pp. 528–534, Jan. 2013, doi: 10.1109/JPHOTOV.2012.2213073.

[9] E. García-Tabarés, J. A. Carlin, T. J. Grassman, D. Martín, I. Rey-Stolle, and S. A. Ringel, "Evolution of silicon bulk lifetime during III–V-on-Si multijunction solar cell epitaxial growth," *Prog. Photovolt. Res. Appl.*, vol. 24, no. 5, pp. 634–644, 2016, doi: https://doi.org/10.1002/pip.2703.

200 MM GERMANIUM WAFERS FOR EPITAXIAL GROWTH OF III/V SPACE GRADE SOLAR CELLS

Tim Kubera[1], Jan Strate[1], Kristof Möller[1], Victor Khorenko[1], Steffen Sommer[1], Chérubin Noumissing Sao[1], Johannes Vanpaemel[2], Kristof Dessein[2]

[1)] AZUR SPACE Solar Power GmbH, Theresienstraße 2, 74072 Heilbronn, Germany,
[2)] UMICORE Electro-Optic Materials, Watertorenstraat 33, 2250 Olen, Belgium

ABSTRACT: In the framework of the H2020 project 'RadHard' we demonstrate for the first time high efficiency triple-junction space solar cells on 200 mm Ge wafers with excellent efficiencies exceeding 30% (AM0, 25°C). Crystal growth of dislocation free germanium ingots with sufficient diameters is achieved and a pilot-line for wafer manufacturing is implemented successfully at UMICORE. The surface quality obtained on the 200 mm wafers is close to the grade of commercially available 150 mm substrates. A comprehensive growth study on different III/V materials comparing MOVPE growth on 150 and 200 mm substrates shows good to excellent workability of the 200 mm substrates at AZUR SPACE. The epitaxy process of the AZUR standard 3G30 triple-junction structure has been performed in an AIXTRON planetary reactor using an industrially feasible MOVPE recipe, whereas due to the wafer format >11% wafer area is treated in a single epitaxial run by using 200 mm instead of 150 mm wafers. For solar cell manufacturing the standard 30.18 cm² cell design is applied, using a cell arrangement that allows the assessment of the epitaxy homogeneity across the 200 mm wafer surface by means of cell performance. Engineering tests are performed on a test coupons built up with these cells verify the device quality.
Finally, the growth of solar cell structures on two separate 200 mm Ge substrates and their subsequent merging by the direct semiconductor bonding technology (SBT) is demonstrated, which is known to enable ultra-high efficient tandem solar cells but is cost-intensive when using smaller wafers.

Keywords: 200 mm germanium wafers; Ge substrates, MOVPE; high-efficiency space solar cells, direct semiconductor bonding technology (SBT)

1 INTRODUCTION

Space solar cell development is traditionally driven by demands for high conversion efficiency and high radiation tolerance for the harsh in orbit conditions. The recent rise in various space based applications – in particular constellation programs – is leading to the need of an increased cost efficiency for space crafts, which is broken down to the single component level. For some opto-electronic components, cost reduction is achievable by substitution with similar components produced in e.g. for automotive or industrial applications in higher numbers (so-called components-off-the-shelf; 'COTS'). However, this approach is not applicable to tandem solar cells, as to date there is no sufficiently large terrestrial PV market for such cells (this may change, when HCPV using III/V cells is going to be implemented for large-scale power plants).

Another way to improve cost effectiveness is to use processes and equipment widely implemented in other semiconductor markets for mass production. Therefore, we consider the upscaling of the Ge wafer format to 200 mm as a major step towards compatibility with low-cost integrated circuit (IC) processes mainly using silicon wafers. The today's standard format for epitaxy of III/V multi-junction cells is changing from wafers with 100 mm to 150 mm diameter, whereas both formats are currently utilized in roughly equal share. The limited volume of the actual space market and high germanium raw material costs restrain a further upscaling of the wafer size and therefore a less dynamic trend is observed compared to the silicon IC industry several years ago. Nonetheless, the next step towards 200 mm Ge substrates becomes increasingly attractive as this opens new processing opportunities. E.g., the direct semiconductor bonding technology (SBT) is a well-developed process and used in industrial scale for other products, but optimized for 200 mm or even larger substrates.

In this contribution, we present the results achieved in the project 'RadHard' [1] on the developments on 200 mm germanium wafers as well as epitaxial growth of III/V semiconductor devices on these. For the first time we demonstrate state-of-the-art multi-junction 3G30 space solar cells [2] realized on this wafer format.

Moreover, we demonstrate a first direct semiconductor bonded solar cell structure conducted on 200 mm wafer format. Such SBT tandem solar cells are known to enable highest conversion efficiencies of >35% at AM0 [3, 4] but the high surface preparation efforts when using smaller wafers inhibit a corresponding product development.

2 EXPERIMENTAL

2.1 Germanium ingot and wafer manufacturing

The first step in making a 200 mm germanium wafer is pulling a sufficiently large dislocation-free germanium crystal. In order to achieve this, the crystal growth recipe was optimized to make sure the internal stresses in the large diameter crystal were minimized during the pulling process. As a result, a full 160 kg, 200 mm dislocation-free germanium crystal was produced.

The crystal was then sawn into individual wafers using a wire-saw and afterwards, these wafers followed a similar flow as standard 150 mm Ge wafers comprising grinding, etching and polishing steps with some hardware and software modifications to the tools. The biggest challenge in developing the 200 mm germanium wafers however was to ensure that these wafers were "epi-ready". This feature is typically accomplished during the final clean and packaging processes. Hence, a semi-automatic procedure to clean and dry the germanium wafers was implemented resulting in a germanium surface that is on par with the quality of our smaller standard wafers. Figure 1 shows a Surfscan image of a 200 mm germanium wafer. The LPD (light point defect) counts are around 500 or 1.53 cm^{-2} and matches that of standard 150 mm wafers. In addition, the area counts that are mostly linked to the presence of crystal-originated defects are similar with respect to standard products.

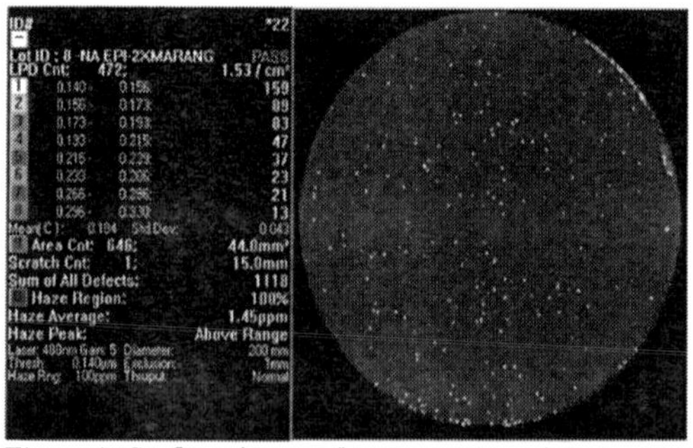

Figure 1: Surfscan image of a 200 mm Ge substrate.

2.2 Epitaxy on 200 mm germanium wafers

Due to the adapted tooling and wafer arrangement in the reactor, as a first step we performed a comprehensive series of experiments to determine the most important epitaxial growth parameters such as temperature, gas flows, material composition and doping homogeneities and compared these to the 150 mm standard process. Figure 2 shows the used epitaxy reactor from AIXTRON loaded with five 200 mm Ge substrates.

Figure 2: MOVPE reactor loaded with 200 mm wafers.

As baseline for this study AZUR's standard 3G30 structures & material were selected. In figure 3, the relative growth rates of $In_{0.50}Ga_{0.50}P$ (top cell material) and $In_{0.01}Ga_{0.99}As$ (middle cell material) are illustrated. The growth rates are measured optically in-situ (solid line) and are crosschecked after growth by thickness measurements using a profilometer at different positions across the wafer radius (each mark represents a thickness measurement result transformed into the growth rate). Below 35 mm radius, the in-situ measurements of the 200 mm epitaxy runs are at a constant level down to the center (0 mm) as the position of the measurement tool does not yet allow measurements in this area for the 200 mm setup.

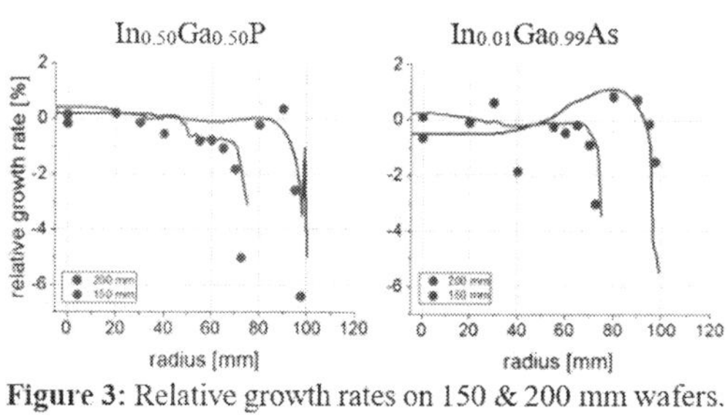

Figure 3: Relative growth rates on 150 & 200 mm wafers.

To achieve the right band-gap energies and high-quality crystal growth, the material compositions have be controlled precisely. Figure 4 shows the relative indium-content as an indicator to assess the material composition of both sub-cell materials. Again, a number of measurement positions along the wafer radius is evaluated and a comparison of 150 and 200 mm wafers is made. The measurement method used to determine the Indium content is XRD (X-ray diffraction).

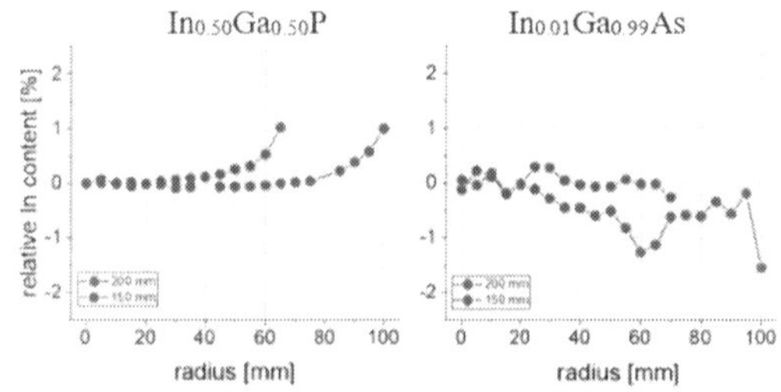

Figure 4: Relative In-content on 150 & 200 mm wafers.

Except for a slightly lower Indium content towards the outer diameter for the InGaAs material on 200 mm wafers, all relevant parameters are on a high level. Therefore, in the next steps we performed growth of full triple-junction space solar cell structures. In terms of defect density and general surface quality, none of the post-epitaxy inspection tests showed irregularities for all four runs conducted within the test campaign.

2.3 Solar cell manufacturing

The ideal evaluation of an epitaxial 3J cell structure quality is by means of the electrical performance of fully processed solar cells. In order to be as close as possible to a standard process flow as a reference we re-sized the 200 mm wafers after epitaxy into two 100 mm standard wafers according to figure 5 using a laser process.

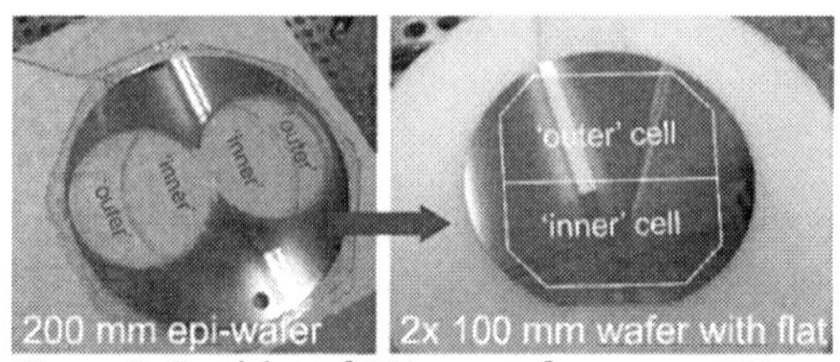

Figure 5: Re-sizing of 200 mm wafers to two 100 mm standard wafer format and used solar cell arrangement.

These resulting substrates with 100 mm wafer format were processed using standard processes in the AZUR cell manufacturing line. On each of these substrates, we arranged two standard solar cells (30.18 cm^2) in a manner that one cell is located at the flat and the other one on the opposite side. Therefore, a comparison between 'inner' and 'outer' areas in terms of epitaxy homogeneity and quality on the 200 mm is possible. As the initial thickness of the 200 mm substrates is significantly higher than for the smaller wafers, additionally a thinning process was applied to achieve the standard cell thicknesses for this cell type (150μm). In total, we fabricated 60 solar cells from 19 epitaxy wafers in 200 mm format.

2.4 Solar cell assembly (SCA) & coupon manufacturing

Finally, these bare cells were further processed to SCA's and coupons to allow first engineering test on higher integration levels as well. Figure 6 shows a space coupon that was built up at AIRBUS Defense & Space comprising 10 of the described cells / SCA's and subsequently was subjected to thermal cycling testing.

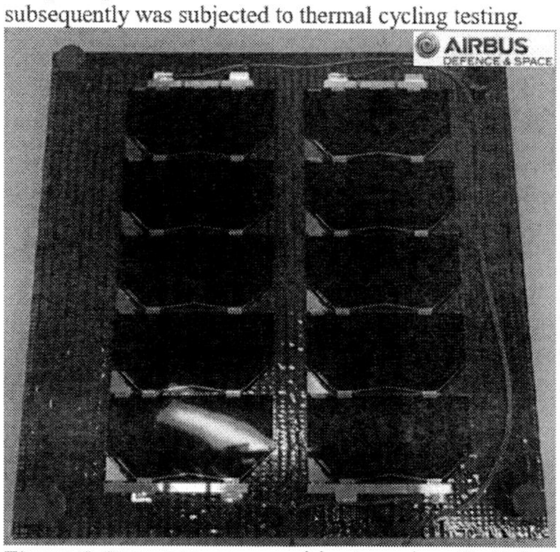

Figure 6: Space coupon comprising ten solar cells / SCAs originating from 200 mm Ge wafers (before thermal cycle testing).

In parallel to the 3G30 epitaxy, we additionally performed a small number of particular epitaxy runs to demonstrate feasibility of a SBT tandem solar cell structure on 200 mm wafer format. To our knowledge, this is the first time a SBT tandem cell structure is performed using 200 mm wafers. Figure 7 illustrates the epitaxial structure and targeted cell built-up including the location of the SBT interface. After epitaxy of the two different stacks (top & bot), both 200 mm wafers were polished (CMP) at III/V-Reclaim as preparation for the direct semiconductor bond. The bonding process was performed at EVG. To enable this on top of the InGaP and InGaAs surface we integrated an additional thin and highly doped dedicated bonding layer. These after defined polishing form the needed transparent and conductive bond interface.

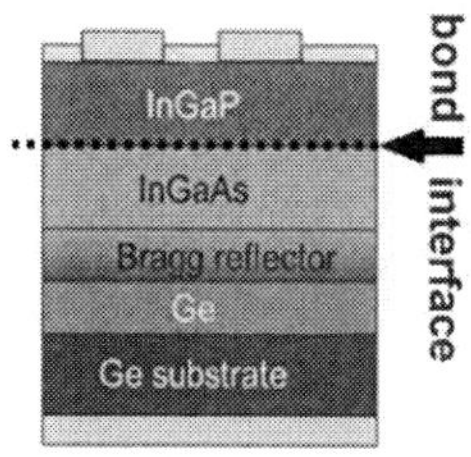

Figure 7: Target SBT cell structure for evaluation of the first 200 mm SBT cell demonstrators.

Figure 8 shows the scanning acoustic microscope (SAM) image of the 200 mm bond after post-bond annealing. In the following we again laser-separated two smaller 100 mm wafers out of this 200 mm sample to perform the further cell processing (100 mm wafer arrangement is indicated in blue in fig. 8). An exemplary 100 mm wafer sample dissected from the 200 mm bond sample after removal of the top Ge wafer material prepared for solar cell manufacturing is shown in figure 9.

As the cell processing sequence was not finalized at the time this paper was created the performance data cannot be presented here. We will finalize processing later in 2021 and report on the results separately.

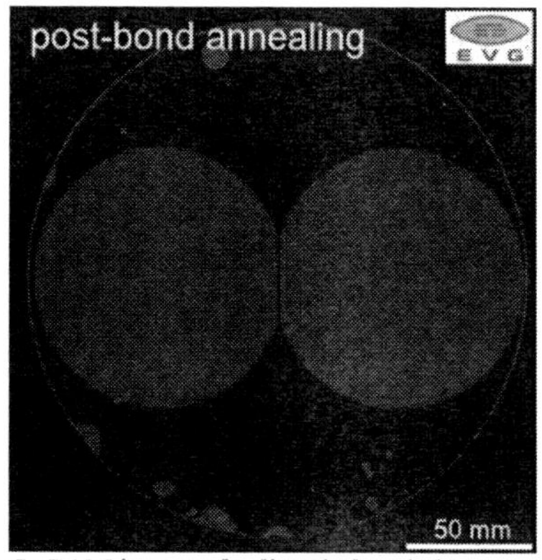

Figure 8: SAM image of a directly bonded 200 mm wafer pair composing a triple-junction high efficiency cell structure; the two 100 mm wafers laser-separated for cell manufacturing afterwards are marked in blue.

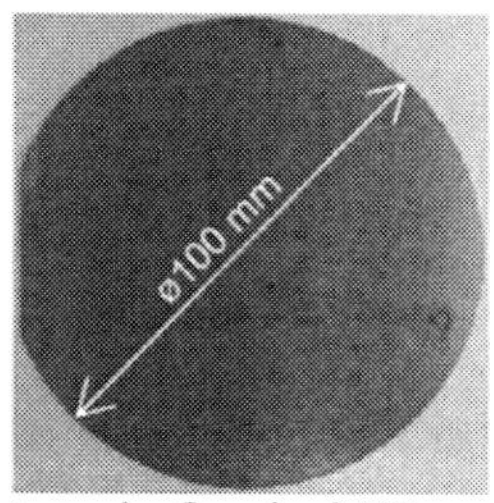

Figure 9: Photograph of re-sized 100 mm wafer sample comprising 3J SBT solar cell epitaxy stack after top wafer removal and ready for cell processing.

3 RESULTS AND DISCUSSION

3.1 Solar cell results

In total, we deposited the 3J space cell structure on 19 wafers (200 mm) within four epitaxy runs. 30 wafers were fully processed resulting in 60 solar cells with an active area of 30.18 cm² (std. 4x8 design). In table I the LIV results of these cells are summarized whereas the two lowest performing cells were excluded for this evaluation.

Table I: Light-IV performance data of 58 cells manufactured from 200 mm wafers (AM0; 1367 W/m²).

	I_{SC}[A]	V_{OC}[V]	FF[%]	eta[%]
Mean	533	2.716	85.6	30.0
Median	533	2.716	85.8	30.1
Min.	530	2.687	84.3	29.5
Max.	535	2.728	86.1	30.3

In Figure 10 the electrical performance values for I_{SC}, V_{OC}, FF and efficiency of all 60 cells are presented in standard box plots, split into 'inner' and 'outer' cells to assess the epitaxy homogeneity in terms of performance also. Quantum efficiency measurements indicate a slightly too high top-to-mid cell current ratio for all cells, which is not optimal for EOL performance but resulting in a higher BOL efficiency (approx. +8 mA at BOL). Apart from that, all parameter deviations are in the usual range for this product type. However, the voltage values (V_{OC}) indicate an actual difference of approx. +15 mV for the 'inner' cells. This, we assume is related to a doping level gradient along the radius and is currently under further investigation.

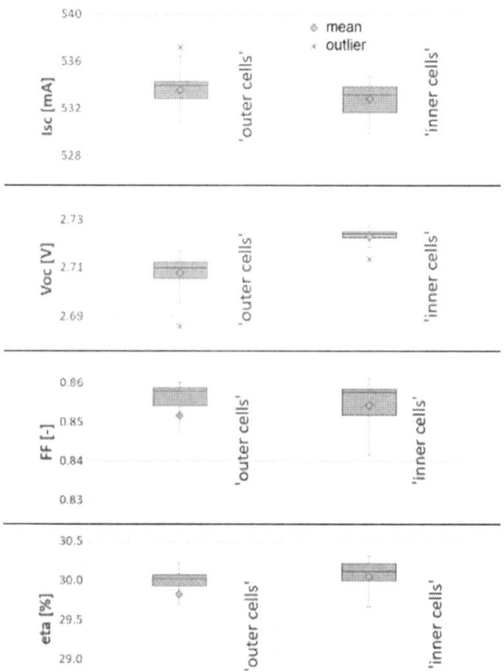

Figure 10: AM0 cell performance parameters for 'inner' and 'outer' cells originating from the 200 mm wafers (the outlier cells are not shown in all diagrams, as their values are partly outside the display range)

3.2 Results on coupon level

The electrical performance of the coupon was tested before and after thermal vacuum (5 cycles) and thermal cycling testing (+95 cycles) within a temperature range from -175°C to +130°C. Table II summarizes the relevant current and voltage values and in figure 6 the pre- and post-test IV characteristics of string 1 is shown. No relevant degradation is detected.

Thales Alenia Space built up a second similar coupon also comprising ten solar cells. This coupon was put to a thermal cycling test as well and again no notable degradation was observed.

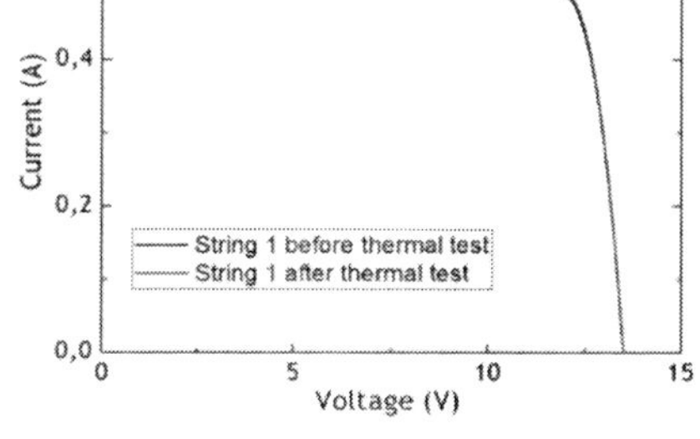

Figure 6: IV-curve of string 1 before and after testing.

Table II: Current and V_{OC} values of both test coupon strings with 5 cells each before and after thermal testing.

	String	I_{OP}[A]	I_{SC}[A]	V_{OC}[V]	$\|\Delta I_{OP}\|$
Before test	1	0.513	0.517	13.50	-
After TV/TC	1	0.513	0.518	13.50	<1%
Before test	2	0.513	0.519	13.47	-
After TV/TC	2	0.514	0.517	13.47	<1%

4 CONCLUSIONS

We demonstrated manufacturing of 200 mm Ge wafers achieving high surface quality and epi-readiness for industrial MOVPE growth on these. We succeeded in growing a 160 kg, 200 mm in diameter, dislocation-free germanium crystal. Furthermore, we executed an epitaxy homogeneity study on relevant III/V materials for tandem space solar cells on the 200 mm wafers. Therefore, we evaluated the growth homogeneity based on multiple measurements across the wafer radii and in addition, by solar cell performance values comparing cells from the 'inner' and 'outer' areas of the 200 mm substrates.

Prototype cells in standard 4x8 format (30.18 cm²) made of the 200 mm wafers show no irregularities compared to cells manufactured on smaller wafer format and exceed 30% efficiency (AM0, BOL). Further processing and thermal tests up to coupon level confirm the feasibility of using 200 mm substrates for growth of high efficiency space solar cells.

In addition, we present first 200 mm epitaxial grown solar cell structures merged by the direct semiconductor bonding technology (SBT). To our knowledge, this is the first time a SBT solar cell structure is fully applied on 200 mm, which will enhance cost effectiveness for this technology approach due to the larger wafer format.

5 ACKNOWLEDGEMENTS

The authors would like to thank the team at III/V-Reclaim for polishing the wafers and EVG for enabling and realizing the direct semiconductor bonding process on these novel samples. Furthermore, we thank AIRBUS Defense & Space and Thales Alenia Space for building up the space PV coupons and performing the thermal tests. Moreover, we acknowledge the contributions of all colleagues, and especially the MOVPE team, T. Lehmann, M. Shirnow and T. Bergunde.

This work was supported by the European Commission in the framework of the Horizon 2020 Project 'RadHard' (Grant Agreement #776362).

6 REFERENCES

[1] H2020 Project 'RadHard'; URL1: https://radhard.org/; URL2: https://cordis.europa.eu/project/id/776362/de
[2] W. Guter et al., "Space Solar cells – 3G30 and next generation radiation hard products"; European Space Power Conference – ESPC 2016
[3] G. Siefer et al., "Four-Junction Wafer Bonded Solar Cells for Space Applications"; European Space Power Conference – ESPC 2019
[4] P. Chiu et al., „35.8% space and 38.8% terrestrial 5J direct bonded cells"; Photovoltaic Specialist Conference – PVSC 2014

ELECTROLUMINESCENCE INTENSITY STABILIZATION IN PEROVSKITE SOLAR CELLS

Matevž Bokalič, Marko Jošt, Kristijan Brecl, Marko Topič
University of Ljubljana, Faculty of Electrical Engineering
Tržaška 25, SI-1000 Ljubljana, Slovenia
E-mail: Matevz.Bokalic@fe.uni-lj.si

ABSTRACT:
We have used electroluminescence to evaluate short-term stabilization process of perovskite solar cells. The cells were kept in the dark between periodic characterization sessions. In a characterization session, a cell was forward biased with a quarter of short circuit current density, while voltage was continuously logged and electroluminescence (EL) images were acquired. When a constant current forward bias was applied, the voltage on all cells immediately jumped to a certain level and then it gradually stabilized to a voltage around 1 V over the course of one to several minutes. Similar stabilization was observed for intensity of EL images. Based on the observed behaviour we differentiated four different stabilization types: positive exponential decay, overshoot, exponential rise continued with linear rise, and deficit. During the first three, the temporal EL emission can be predicted from applied voltage using reciprocity relation. However, during the deficit stabilization, the EL intensity, calculated from applied voltage, initially misses the measured EL intensity forming a kind of EL deficit. The observed stabilization processes were both, positive and negative. Spatial information from EL images reveals that when there is no EL deficit, the images mostly just change in intensity, but when EL deficit is present, the (in)homogeneity of the images also changes.
Keywords: Electroluminescence (EL), Stabilization, Perovskite, Solar Cell

1 INTRODUCTION

Characterization of perovskite solar cells is of key importance for understanding operation principles and evaluating the performance throughout their lifetime. The characterization techniques applied have, however, often proved challenging due to varying response of the perovskite material and cells. Some of the most known effects are their hysteresis and light-soaking [1] changes in I-V measurements, which vary temporally with repeating measurements. Thus, it is not trivial to select the correct measurement.

A recent report has dealt with reliable extraction of performance parameters [2], while another study has been published on long-term stability [3]. However, an in-depth study on short-term stabilization dynamics is missing. Electro- and photoluminescence were introduced in perovskite solar cell characterization [4], followed by several studies on spatial inhomogeneity [5,6], and on the dynamics of luminescence emission from very short time scales [7], to long term changes [5,8], however none of them reports on variability of responses from cell to cell.

With this paper we report on the electroluminescence stabilization of perovskite solar cells as observed throughout the experiment.

2 EXPERIMENTAL

2.1 Cells

The cells used in the experiment were fabricated at Hemholtz Zentrum Berlin during a research stay of M. Jošt. Each cell has an area of 1 cm^2. The cells are made on the TCO (120 nm ITO) glass. A 1 nm MeO-2PACz and 550 nm Pervoskite layers were spin coated in nitrogen atmosphere and annealed at 100°C for 10 and 40 minutes, respectively. Further structure consists of 23 nm layer of thermally evaporated C_{60}, 20 nm layer of SnO_2 deposited by thermal ALD, and a 100 nm layer of evaporated copper. The cell is encapsulated with another sheet of glass.

2.2 The experiment

Characterization sessions were performed periodically during the experiment. At first the period was 3 days and was gradually elongated to 14 days towards the end of the experiment. Between the characterization sessions, cells were kept in the dark and some were connected to high potential. A characterization session consisted of electroluminescence and IV characterization.

Electroluminescence inspection consist of two steps. First, during a stabilization phase, a forward bias injection of a quarter of short circuit current was applied to the cell, while the voltage was monitored and EL images were acquired periodically. Stabilization phase in general lasted until the voltage and mean EL intensity stabilized. Second, during a regular EL phase, a current injection was increased to one half and one short circuit current, while the voltage was still monitored and EL images were still acquired. It must be noted that during the EL inspection (i.e. stabilization and regular EL) the current was constantly applied and was not interrupted between the images nor between the current increase.

After acquisition, the EL images were processed by background subtraction, automatic region of interest (ROI) detection, rotation, cropping and finally time normalization to provide comparable results between characterization sessions and between cells.

The results are presented in **Figure 1**. The figure shows temporal dependence of the voltage (blue line, left y axis) and the mean EL intensity of the ROI in the images (horizontal red lines, right y axis). The mean EL intensity is fitted with the reciprocity relation [9]:

$$EL(V) = C_{EL} \cdot e^{\frac{V}{n \cdot V_{th}}} \tag{1}$$

and is plotted with a pink colour.

3 RESULTS AND DISCUSSION

3.1 Stabilization types

The most common stabilization type is **positive**

exponential decay presented in **Figure 1**, a. In this stabilization type, both voltage and electroluminescence follow the positive exponential decay function (similar to charging a capacitor over a resistor to a certain voltage):

$$V(t) = V_{stab} \cdot \left(1 - e^{-\frac{t-t_{off}}{\tau}}\right) + V_{off} \qquad (2)$$

Temporal dynamics (initial delay and time constant) and voltage levels (stabilization voltage and offset voltage) vary between cells, and between characterization sessions. In case of larger initial delay and smooth initial voltage rise, the first part can also be fitted with a logistic function:

$$V(t) = V_{amp} \cdot \frac{1}{\left(1+e^{-stp\cdot(t-t_{mid})}\right)} + V_{off} \qquad (3)$$

In this stabilization type, EL obeys the reciprocity relation and the EL images do not change spatially (only the intensity changes).

We called another common stabilization type an **overshoot**. It is presented in **Figure 1**, b. Both, voltage and electroluminescence reach a peak shortly after the start of stabilization process, and then gradually fall off towards the stabilized value. Initial rise obeys either positive exponential decay, or logistic function, while the fall off can be described by a negative exponential decay. The voltage response may resemble the response of the second order system, but the attempt to fit it with the second order system equation failed. EL obeys the reciprocity relation and EL images do not change spatially.

Not so common stabilization type is an **exponential rise continued with linear rise**. In this stabilization type, EL intensity starts to rise exponentially and then transforms to a linear rise, as shown in **Figure 1**, c. The main property of this type is that it does not stabilize within a "reasonable" time, e.g. 5 minutes in this particular case. The EL intensity should eventually stabilize in order to stay within the physical limits. EL still obeys the reciprocity relation and EL images do not change spatially.

We have noticed another rare stabilization type and named it **deficit**. It is presented in **Figure 1**, d, and as the name implies, the characteristics of this type is that the EL intensity does not completely obey the reciprocity relation. Based on the reciprocity relation and measured voltage, there should be a spike in EL signal in the beginning of the stabilization, but the EL signal is missing at that time, hence the name deficit. Characteristic to this stabilization type are also spatial changes in the corresponding EL images (note the central part of the device becoming slightly brighter in normalized EL images at the bottom of **Figure 1**, d).

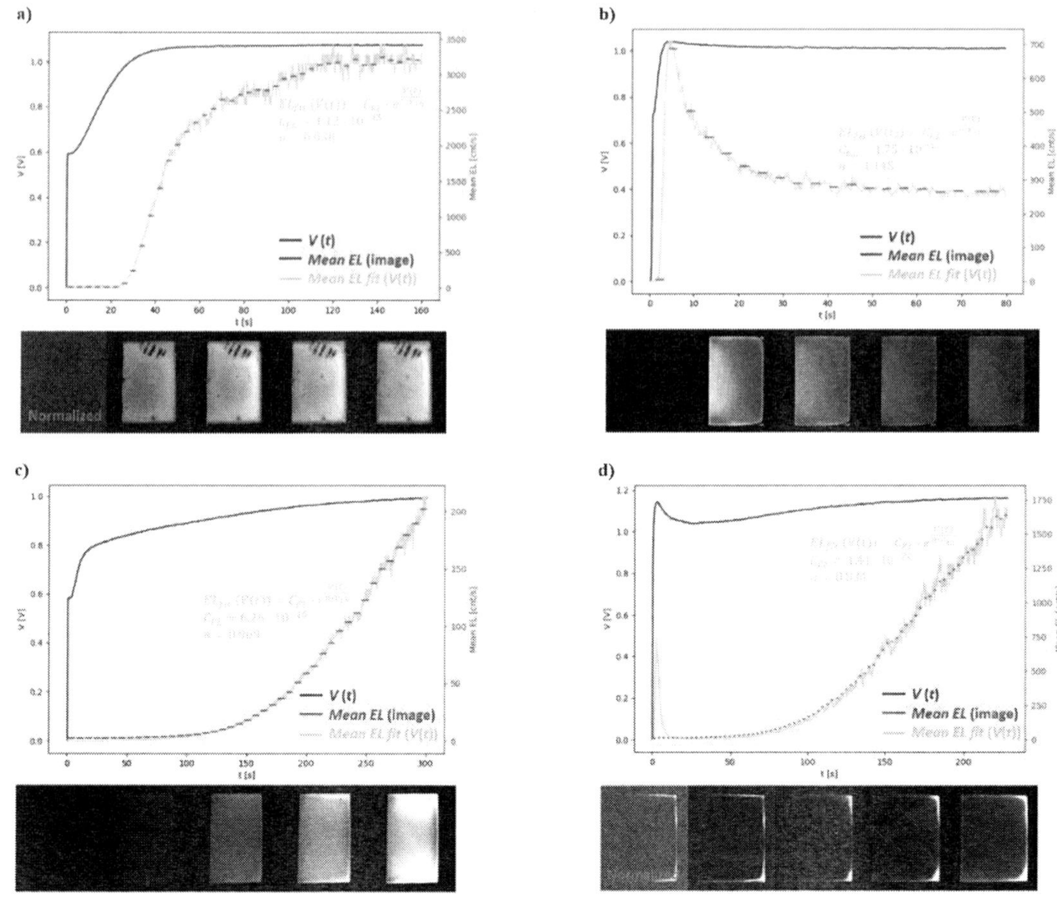

Figure 1: Different stabilization types observed throughout the experiment. Pink plot presents the EL fit based on the reciprocity relation and measured voltage. a) positive exponential decay, b) overshoot, c) exponential rise continued with linear rise, and d) deficit. Below each chart there are several EL images, which are scaled absolutely for subfigures b) and c), and individually normalized for subfigures a) and d).

4 SUMMARY AND CONCLUSION

We presented how perovskite solar cells made in the same batch behave differently during the initial stabilization at constant forward current bias. We show four different trends of initial stabilization: positive exponential decay, overshoot, exponential rise continued with linear rise, and deficit. Electroluminescence stabilization is most often a consequence of temporal voltage stabilization. The stabilization process can take from 10 seconds up to 10 minutes and possibly even longer.

Awareness of temporal luminescence dynamics and of variability of dynamics between samples from the same batch is crucial to perform correct stabilization steps before acquiring representative spatial luminescence image and even more important when extracting quantitative parameters. While it is probably too early to set standardized procedures for luminescence characterization, we want to emphasise the importance of stabilization. Furthermore, the variability in stabilization is most probably caused by mechanisms in the perovskite solar cells, such as ion migration, charge trapping, etc. Thus, by confirming a link between the stabilization mode and its key root mechanism, EL stabilization may provide a simple method to identify mechanisms in perovskite solar cells.

5 ACKNOWLEDGMENTS

The authors acknowledge the Young Investigator Group of Prof. Steve Albrecht at HZB for cell fabrication and research stay of Marko Jošt. Authors gratefully acknowledge the Slovenian Research Agency for funding the research programme P2-0197 and project J2-1727, and Hemholtz association for funding the project TAPAS within the EU partnering program.

6 REFERENCES

[1] H. Tsai, R. Asadpour, J.-C. Blancon, C. C. Stoumpos, O. Durand, J. W. Strzalka, B. Chen, R. Verduzco, P. M. Ajayan, S. Tretiak, J. Even, M. A. Alam, M. G. Kanatzidis, W. Nie, A. D. Mohite, *Science* **2018**, *360*, 67.

[2] G. Bardizza, H. Müllejans, D. Pavanello, E. D. Dunlop, in *37th European Photovoltaic Solar Energy Conference and Exhibition*, WIP, **2020**, pp. 1–13.

[3] M. V. Khenkin, E. A. Katz, A. Abate, G. Bardizza, J. J. Berry, C. Brabec, F. Brunetti, V. Bulović, Q. Burlingame, A. Di Carlo, R. Cheacharoen, Y.-B. Cheng, A. Colsmann, S. Cros, K. Domanski, M. Dusza, C. J. Fell, S. R. Forrest, Y. Galagan, D. Di Girolamo, M. Grätzel, A. Hagfeldt, E. von Hauff, H. Hoppe, J. Kettle, H. Köbler, M. S. Leite, S. (Frank) Liu, Y.-L. Loo, J. M. Luther, C.-Q. Ma, M. Madsen, M. Manceau, M. Matheron, M. McGehee, R. Meitzner, M. K. Nazeeruddin, A. F. Nogueira, Ç. Odabaşı, A. Osherov, N.-G. Park, M. O. Reese, F. De Rossi, M. Saliba, U. S. Schubert, H. J. Snaith, S. D. Stranks, W. Tress, P. A. Troshin, V. Turkovic, S. Veenstra, I. Visoly-Fisher, A. Walsh, T. Watson, H. Xie, R. Yıldırım, S. M. Zakeeruddin, K. Zhu, M. Lira-Cantu, *Nature Energy* **2020**, *5*, 35.

[4] Z. Hameiri, A. Mahboubi Soufiani, M. K. Juhl, L. Jiang, F. Huang, Y.-B. Cheng, H. Kampwerth, J. W. Weber, M. A. Green, T. Trupke, *Progress in Photovoltaics: Research and Applications* **2015**, *23*, 1697.

[5] A. M. Soufiani, J. Kim, A. Ho-Baillie, M. Green, Z. Hameiri, *Advanced Energy Materials* **2018**, DOI 10.1002/aenm.201702256.

[6] M. C. Schubert, L. E. Mundt, D. Walter, A. Fell, S. W. Glunz, *Advanced Energy Materials* **2020**, *10*, 1904001.

[7] R. Gegevičius, M. Franckevičius, J. Chmeliov, W. Tress, V. Gulbinas, *J. Phys. Chem. Lett.* **2019**, *10*, 1779.

[8] O. A. Jaramillo-Quintero, R. S. Sanchez, M. Rincon, I. Mora-Sero, *J. Phys. Chem. Lett.* **2015**, *6*, 1883.

[9] U. Rau, *Phys Rev B* **2007**, *76*, 85303.

ON THE EQUILIBRIUM ELECTROSTATIC POTENTIAL AND LIGHT-INDUCED CHARGE REDISTRIBUTION IN HALIDE PEROVSKITE STRUCTURES

Davide Regaldo[a,b,c]*, Aleksandra Bojar[a,b,c], Sean P. Dunfield[d,e], Pilar Lopez-Varo[a], Mathieu Frégnaux[a,f], Vincent Dufoulon[a,g], Shan-Ting Zhang[a], José Alvarez[a,b,c], Joseph J. Berry[d], Jean-Baptiste Puel[a,h], Philip Schulz[a,g], Jean-Paul Kleider[a,b,c].

[a] IPVF, Institut Photovoltaïque d'Ile-de-France, 18, Boulevard Thomas Gobert, 91120 Palaiseau France; [b] Université Paris-Saclay, CentraleSupélec, CNRS, Laboratoire de Génie Electrique et Electronique de Paris, 91192, Gif-sur-Yvette, France; [c] Sorbonne Université, CNRS, Laboratoire de Génie Electrique et Electronique de Paris, 75252, Paris, France, [d] National Renewable Energy Laboratory, 15013 Denver West Pkwy, Golden, CO 804011;[e] Materials Science & Engineering Program, 027 UCB, University of Colorado Boulder, Boulder, CO 803032;[f] Institut Lavoisier de Versailles, Université de Versailles Saint-Quentin-en-Yvelines, Université Paris-Saclay, CNRS, UMR 8180, 45 avenue des Etats-Unis, 78035 Versailles Cedex, France7; [g] CNRS, École Polytechnique, IPVF, UMR 9006, 18, Boulevard Thomas Gobert, 91120 Palaiseau, France; [h] EDF R&D, 91120 Palaiseau, France;

Address: * davide.regaldo@ipvf.fr

ABSTRACT: Lead halide perovskites are a class of semiconductor materials which are employed as non-intentionally doped absorbers inserted between two selective carrier transport layers (SCTL), realizing a p-i-n or n-i-p heterojunction. In our study, we have developed and investigated a lateral device, based on methylammonium lead iodide ($MAPbI_3$) in which the p-i-n heterojunction develops in the horizontal direction. Our research suggests that the effective doping level in the $MAPbI_3$ film should be very low, below 10^{12} cm^{-3}. Along the vertical direction, this doping level is not enough to screen the electric field of the buried heterojunction with the SCTL. The perovskite work function is therefore affected by the work function of the SCTL underneath. From drift-diffusion simulations we show that intrinsic perovskite-SCTL structures develop mV range surface photovoltages (SPVs) under continuous illumination. However, perovskite-SCTL structures can develop SPVs of hundreds of mV, as confirmed by our measurements. We therefore analyzed the compatibility between low doping and low defect densities in the perovskite layer and such high SPV values using numerical modelling. It is shown that these high SPV values could originate from electronic processes due to large band offsets in the buried perovskite-SCTL heterojunctions, or at the SCTL - transparent conductive oxide (TCO) buried heterojunction. However, such electronic processes can hardly explain the long SPV persistence after switching off the illumination, which could be a sign that light-induced physicochemical processes are at work.

Keywords: Perovskite solar cell, surface photovoltage, drift-diffusion modelling, photoemission spectroscopy, defects.

1 INTRODUCTION

Metal halide perovskite (MHP) solar cells (PSC) have experienced a rapid increase in efficiency in the last decade. The quality of the perovskite absorber (e.g. high absorption coefficient, low recombination and high diffusion length in the domain of polycrystalline materials) and of the heterointerfaces (selectivity by carrier-blocking band offsets, low interface recombination, transparency of the selective carrier transport layers (SCTLs)) are considered as key factors for the high power-conversion efficiency (PCE) achieved [1][2].

However, despite the record improvements in PCE of PSCs, the community still faces an ongoing debate about the nature and concentration of defects in the absorber layer. Several studies report the presence of native defects in MHPs, which are dependent on the MHP deposition process [3]. Moreover, upon external stimuli, such as illumination, an oxidizing or reducing environment, or a bias voltage, defects could be created or destroyed [4][5]. Polycrystalline MHP thin films show defects both in bulk and on the grain boundaries [6]. Related to the crystal lattice properties, the perovskite's shallow defect levels have very low formation energies and could induce a nonzero doping level in the material [6][7].

Direct measurements of the doping level in perovskites, like capacitance-voltage profiling are still a major challenge. Perovskites are generally chemically reactive when brought in contact with a wide variety of metals [8]. In addition, capacitance techniques rely on the presence of a space charge region in the semiconductor, the extension of which should be modulated by the application of a voltage, a condition which is difficult to achieve for lowly doped and thin (sub-micron) semiconducting layers [9].

X-ray and ultraviolet photoemission spectroscopy (XPS, UPS) have been employed to characterize electronic and chemical properties of the perovskite material [10]. Furthermore, transient surface photovoltage (SPV) measurements in fixed capacitor configuration, Kelvin probe and Kelvin probe force microscopy (KPFM) have been employed to measure the charge extraction capabilities of perovskite - SCTL structures [11][12].

In several past studies, a combination of these techniques revealed that the Fermi level position in the bandgap of the perovskite can be changed by interfacing the MHP layer with different substrates [13][14][15][16]. However, to date the exact origin of this effect remains unresolved and could be rooted in subtle chemical interactions between the perovskite film and substrate. This supposed variability, together with the effect of different deposition techniques, produces a very complex system, loosely described by the term "p-i-n heterojunction".

In conventional, vertically stacked n-i-p or p-i-n solar

cells, the absorber is sandwiched between the SCTLs; therefore, only indirect measurements can be employed to study the impact of the SCTL on the electronic properties of the junction. In this work, we used a lateral solar cell layout (see Figure 1a), similar to the design of interdigitated back contact solar cells (IBC). The two SCTLs were deposited prior to the perovskite layer, thus forming two buried heterojunctions. In this way, the absorber properties can be measured with surface sensitive techniques and the contribution of each buried heterojunction can be separated by making the channel between the two SCTLs large enough to contain micrometric space charge regions (SCRs).

This structure was previously characterized using different techniques [17]. We focus here on the XPS, UPS, KPFM and SPV techniques and on full 3D finite-element modelling of experimental data to derive conclusions regarding the doping and defects in the perovskite material, as well as the impact of the SCTLs on the energy profiles and carrier distributions in the dark and under illumination.

2 RESULTS AND DISCUSSION

2.1 XPS surface potential analysis

We consider the perovskite lateral heterojunction device (LHJ) described in Figure 1a. It is composed of an FTO substrate on which two SCTLs, namely TiO_x and NiO_x, which act as electron transport layer (ETL) and hole transport layer (HTL), respectively, were deposited and laser scribed in order to create a channel between them (l_{ch} > 50 μm). Then, a $MAPbI_3$ perovskite layer was deposited on top, leaving the surface available for characterization and illumination.

We employed XPS both in scanning and imaging mode. The first method allowed us to acquire the signal rapidly, minimizing the possible radiation damage during measurement, with a total measurement time of 30 minutes. In this acquisition mode, we swept the probed area across the channel (in the x direction, as illustrated in Figure 1a) and performed local XPS measurements. We measured the variation across the channel of the $I_{3d_{5/2}}$ core level peak energy, BE_I, which is representative of the electrostatic potential variation, as shown in Figure 2. XPS scanning data show that a perceived gradual potential drop is present across the channel [17] (Figure 1b black dashed line).

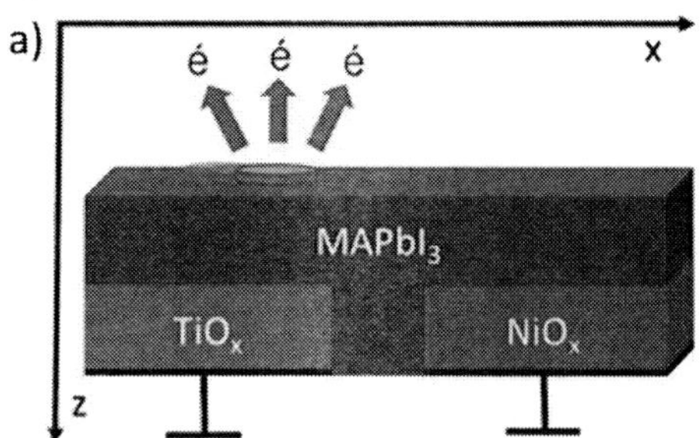

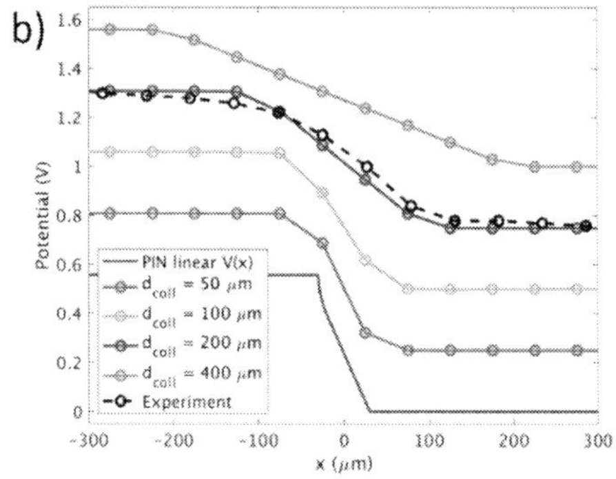

Figure 1: Sketch of the LHJ device with the representation of the XPS electron collection area in the scanning mode (a). In (b) the apparent surface electrostatic potential profile across the channel obtained from XPS experimental data (black dashed line) is compared to the one calculated from modelling using an intrinsic perovskite, under various electron collection diameters (d_{coll}). The blue solid line shows the calculated surface potential profile without convolution effect with the probe area (also corresponding to $d_{coll} = 0$). The reference for the electrostatic potential (0 V) is taken on top of the NiO_x side and the origin of the x axis is taken in the middle of the channel. Curves have been shifted with respect to each other for clarity.

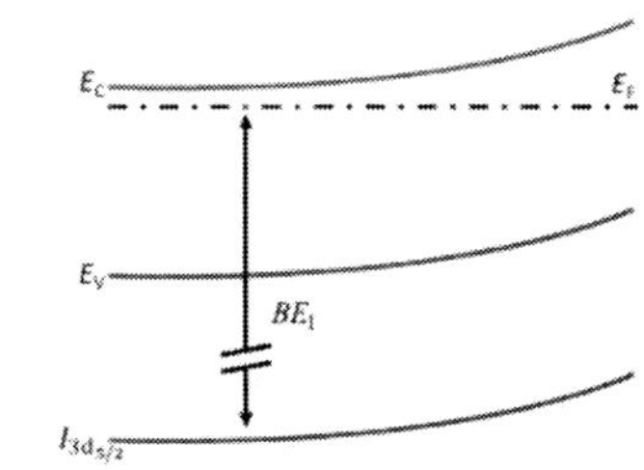

Figure 2: Sketch of the $I_{3d_{5/2}}$ core level, top of the valence band, E_V, and bottom of the conduction band, E_C, in presence of a space-dependent electrostatic potential, $q\varphi(x) = BE_I(x) = E_F - I_{3d_{\frac{5}{2}}}(x)$, q being the unit charge and E_F the Fermi level.

In this scanning XPS mode, each measured point is the average value of the collected electrons arriving from a given probe area. Therefore, to have a proper comparison with the experiments, the surface electrostatic potential values calculated from modeling have been averaged over a simulated collection distance d_{coll}, that we varied between 50 μm and 400 μm. In Figure 2b we compare the corresponding apparent surface potential profiles to the XPS experimental data (dark dashed line) and to the true calculated surface potential profile (blue curve without symbols). On the regions above the HTL and the ETL, far from the channel location, the potential is constant. The linear voltage drop in the channel is due to the perovskite being considered undoped and defect-free.

Convoluting the surface potential with the collection area of the XPS setup shows that, when the area increases, the slope of the surface potential decreases, while the perceived channel length increases. We here emphasize that a good agreement is found between the experimental data and the apparent surface potential profile calculated with a collection diameter of 200 µm, which corresponds to the value in our experimental XPS scan mode. This good agreement was obtained assuming the perovskite to be undoped. Now, we need to study how doping and defects affect the surface potential variation and the corresponding apparent variation obtained from the XPS scan with the 200 µm collection diameter. We thus added a trap level in the bandgap close to the valence band edge. It was considered as an acceptor (E_{tA} - E_V=0.02 eV) in order to induce p-type doping [7].

We simulated the surface potential across the channel for various acceptor trap densities in the bulk perovskite and convoluted the data with the experimental electron collection area (d_{coll} = 200 µm), as shown in Figure 3. As the trap density increases, the effective p-type doping in the perovskite increases, and the potential drifts away from the linear trend, towards a parabolic-like shape. This is due to the presence of a fixed electronic charge inside the traps, which modifies the surface potential curvature. For trap density values above 10^{12} cm^{-3}, two separate SCRs appear close to the buried heterojunctions with the SCTLs along the x-direction, and a constant potential region is formed in the middle of the channel. However, this quasi-neutral region in the middle of the channel is invisible to the measurement probe due to the large electron collection distance (Figure 3).

Above 10^{15} cm^{-3} acceptor trap density in the perovskite, the electrostatic potential at the two extremities (i.e. on top of the SCTLs) starts to decouple from the respective SCTL WF, as SCRs start to form along the z-direction.

For trap density values above 10^{17} cm^{-3}, the parabolic-like shape disappears and the surface potential becomes constant along the x-direction: it does not depend anymore on the WF of the SCTL underneath.

Therefore, we conclude that acceptor trap densities higher than 10^{12} cm^{-3} are not compatible with the experimental data, as they would modify the gradual potential drop as shown in Figure 3.

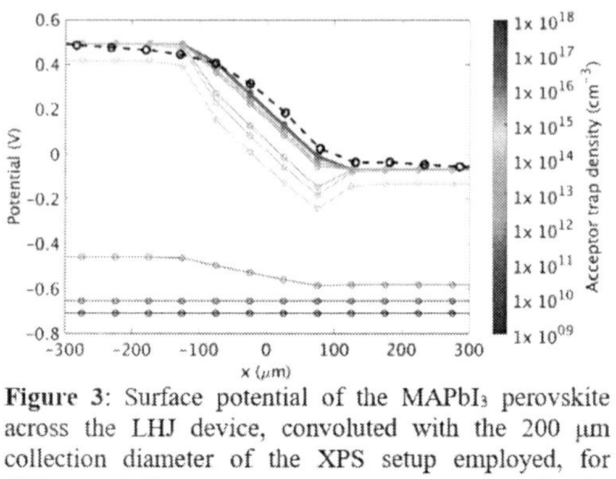

Figure 3: Surface potential of the $MAPbI_3$ perovskite across the LHJ device, convoluted with the 200 µm collection diameter of the XPS setup employed, for different shallow acceptor trap concentrations in the perovskite. The dark dashed line represents the experimental data. The surface potential $V(x)$ is referenced here to the intrinsic Fermi level $E_i(x)$: $V(x) = (E_F - E_i(x))/q$.

In Ref [19] we also considered additional high-resolution imaging mode XPS data (not shown here), that allowed us to rule out n-type doping of the perovskite with donor trap concentrations larger than 10^{12} cm^{-3}.

2.2 KPFM surface photovoltage analysis

KPFM yields the contact potential difference (CPD) between the tip of the AFM and the probed surface. Applying a DC bias between the tip and the sample or submitting the sample to illumination leads a change of CPD that depends on the charge redistribution across the layer stack. This CPD variation is therefore influenced by the presence of surface or bulk defects, or by electric fields present in surface or buried junctions and by band alignments at heterojunctions [11]. Specifically, by subtracting the CPD measured in the dark to that measured under illumination, one obtains the local surface photovoltage (SPV). Preliminary measurements on the $MAPbI_3$ perovskite LHJ device of Figure 1a were reported elsewhere [17]. We observed a large positive SPV (≈ 500 mV) on top of TiO_x, and a smaller but negative SPV on top of NiO_x (≈ -320 mV). Here we want to extend the measurements and analysis to triple cation mixed halide perovskite layers over different transport layers. We thus fabricated perovskite-SCTL structures employing a triple cation mixed halide perovskite and two different SCTLs: TiO_x and NiO_x. We characterized the bilayers with our KPFM setup, in which we illuminated the samples with a 532 nm laser centered on the KPFM tip.

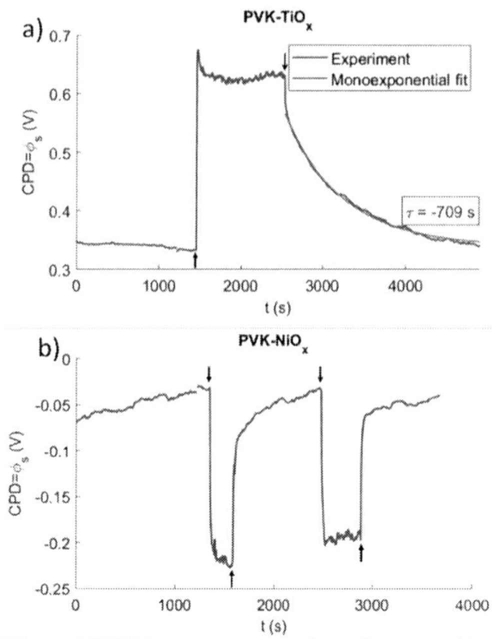

Figure 4: KPFM measurements performed on perovskite-SCTL structures, under 532 nm laser illumination. CPD measurement of (a) perovskite-TiO_x and (b) perovskite-NiO_x. Arrows indicate switch on and subsequent switch off times of the 532 nm laser illumination.

The measured sign of SPV is consistent with the nature of the SCTL underneath the perovskite, i.e. negative for the perovskite-HTL and positive for the perovskite-ETL structure (Figure 4). This is comparable to our previous measurements on $MAPbI_3$ [17]. The sign of the respective SPVs can be attributed to charge separation at the perovskite-SCTL interface. Due to the accumulation of blocked carriers inside the perovskite, a net charge builds up and the surface potential shifts under illumination [18]. The amplitude of the SPV is on the order of 300 mV for the perovskite on TiO_x, and 170 mV for the perovskite on NiO_x. Regarding the perovskite-TiO_x structure, SPVs of

150-200 mV have already been observed, while perovskite-HTL structures showed 2 orders of magnitude lower SPVs. This difference has been ascribed to a higher efficiency of charge extraction at the perovskite-ETL interface, with respect to the perovskite-HTL one [11].

The timescales of the SPV buildup after switch-on and decay after switch-off are signatures of the processes involved in the SPV formation. In our case, the rise of the SPV upon switching on the illumination is at least three orders of magnitude faster than the decay of SPV after turning off the illumination, taking into account the low temporal resolution of the instrument (~4 s).

We also observe that, for the perovskite-TiO_x sample, the photovoltage decay follows a monoexponential trend with a characteristic time constant of about 700 s. (Figure 4a).

We have employed drift-diffusion modelling to unveil the root causes of the SPV and study its dynamics under continuous monochromatic illumination in our perovskite - SCTL bilayers. For comparison with experimental data, we have chosen the same light intensity employed in the KPFM measurements (~8 W/cm^2).

We start with the ideal case, in which the perovskite is intrinsic and without defect states in the bandgap, as suggested by the electrostatic potential measurements in Sec. 2.1, and the SCTL is also without defect states in the bandgap but doped ($N_{\rm dop} = 10^{17}{\rm cm}^{-3}$). In addition, we consider zero band offset between the perovskite and the SCTL for the type of carriers that should be extracted at this interface. At equilibrium, the perovskite being undoped and defect-free, the position of the Fermi level in the perovskite is imposed by the doped SCTL underneath, and only a low built-in voltage of about 100 mV forms across the heterojunction.

Under illumination, electrons and holes are separated at the heterointerface: holes accumulate in the perovskite, while electrons accumulate inside the ETL. The sign of the calculated SPV is therefore positive. The charge redistribution is opposite in the perovskite - HTL structure, where the sign is negative. Noticeably, the amplitude of the SPV in both types of structures is lower than 1 mV (Figure 5).

In conclusion, this simple modelling of the perovskite-ETL and perovskite-HTL structures cannot explain the large amplitudes that are experimentally observed. In addition, the calculated transients are much faster, with stabilization times lower than 1 µs. We therefore studied which other factors could affect rise, decay time and magnitude of the SPV.

a)
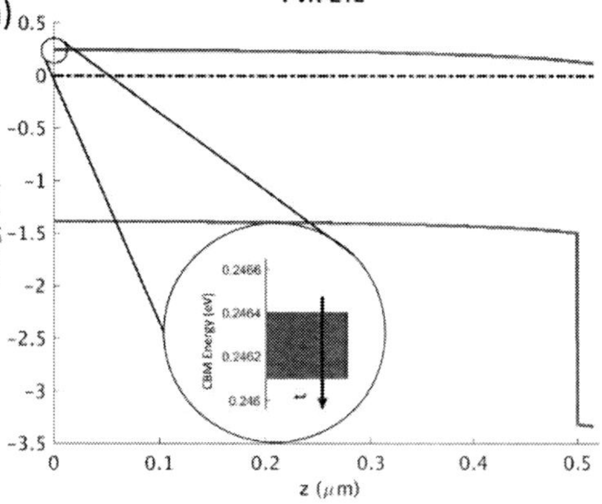

b)
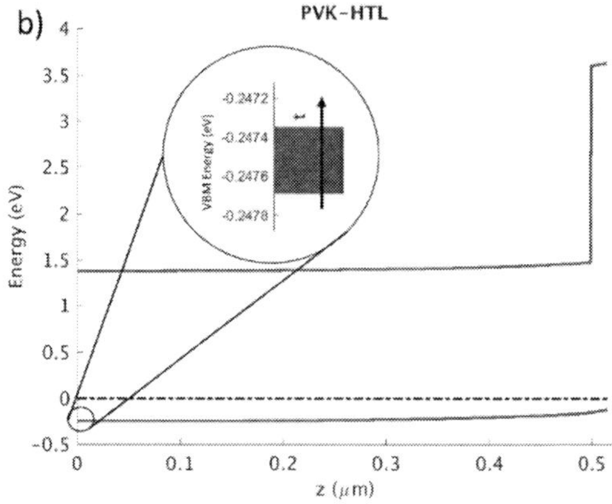

Figure 5: perovskite (PVK)-SCTL band diagrams under illumination; (a) PVK-ETL, and (b) PVK-HTL; the small time-dependent shift of the surface potential after switch-on is magnified in the insets (positive for PVK-ETL, negative for PVK-HTL). The black arrows indicate the direction of the shift of the surface potential as a function of time.

A built-in voltage across a buried (hetero)junction is usually considered as a primary source of surface photovoltage formation. Regarding the perovskite - TiO_x structure, the presence of effective p-type doping in the perovskite would determine a built-in voltage across the heterojunction, since TiO_x has been considered as n-type (with a donor concentration $N_{\rm d} = 10^{17}{\rm cm}^{-3}$). Thus, although the XPS mapping results presented in Sec. 2.1. suggest that high doping of the perovskite is unlikely, we calculated the SPV response under experimental illumination conditions for different doping density values of the perovskite. Results of the transient SPV curves for both switch-on and switch-off are displayed in Figure 6, showing that the magnitude of the SPV and corresponding decay times do indeed increase with the perovskite acceptor doping concentration.

a)
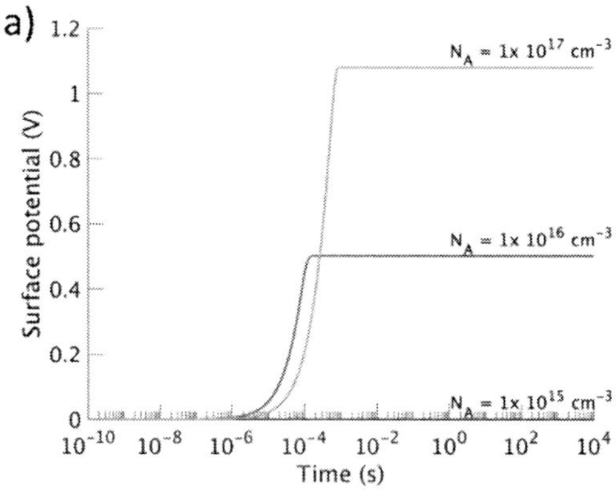

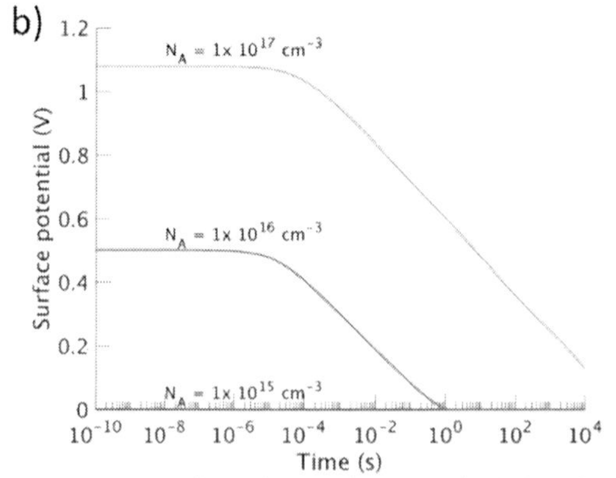

Figure 6: Perovskite - TiO_x transient SPV simulations for different values of acceptor dopant concentration in the perovskite; (a) build-up of SPV after illumination is switched on, and (b) SPV decay after illumination is switched off.

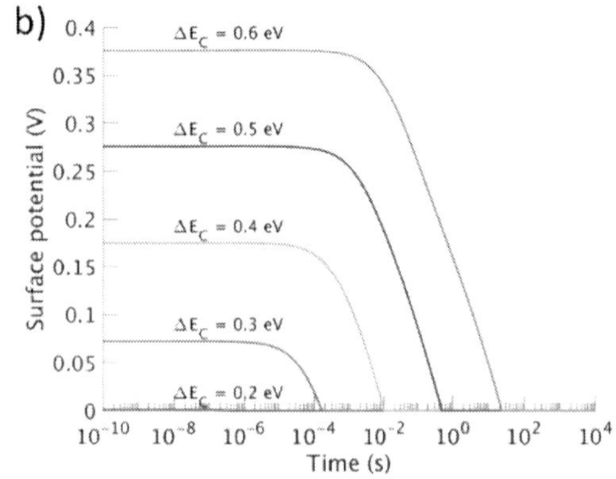

Figure 7: Perovskite - TiO_x transient SPV simulations for different values of conduction band offset, $\Delta E_C = E_{C,\mathrm{pvk}} - E_{C,\mathrm{TiO_x}}$; (a) build-up of SPV after illumination is switched on, and (b) SPV decay after illumination is switched off.

However, the characteristic decay time of the simulated SPV is considerably lower than the 700 s observed experimentally. We also note that the simulated transient after switch-off exhibits a non-exponential decay, in contrast with the exponential behavior observed experimentally for long times (Figure 4a). Therefore, p-type doping of the perovskite would be unable to explain the experimental SPV results, which further confirms the results from XPS data analysis in Sec. 2.1.

For selectivity purposes, large valence band offset at the ETL interface is preferable in order to prevent holes, generated in the perovskite, from diffusing into the ETL. At the same time the conduction band offset should be small in order to facilitate electron collection. Thus, the conduction band offset defined as the difference at the interface between the bottom of the conduction band in the perovskite, $E_{C,\mathrm{pvk}}$, and that in the TiO_x, $E_{C,\mathrm{TiO_x}}$, $\Delta E_C = E_{C,\mathrm{pvk}} - E_{C,\mathrm{TiO_x}}$, should ideally be close to zero. However, we found that a significantly positive value could be a source of large surface photovoltage, as depicted in Figure 7.

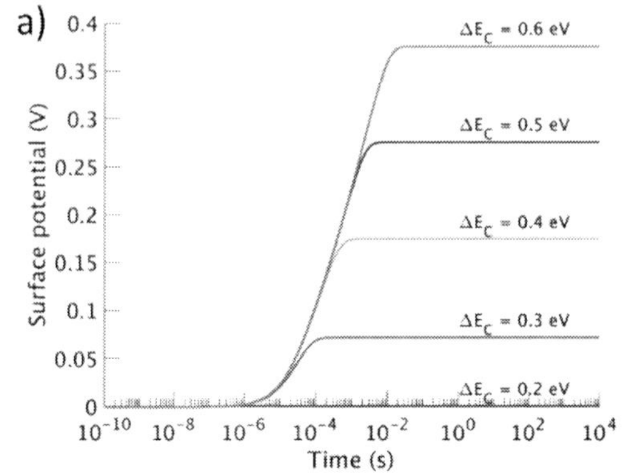

Under illumination, electron-hole pairs are continuously generated inside the perovskite layer. In steady state conditions, however, the net flux of electrons across the perovskite - ETL heterojunction should be zero (open circuit conditions). For positive values of the conduction band offset, the electrons reinjected towards the perovskite from the TiO_x have to overcome a potential barrier in a thermionic emission process. If the barrier height increases, the amount of electrons accumulated in the TiO_x layer has to increase, in order to keep a net zero flux of electrons in steady state, thus increasing the negative charge in TiO_x. An opposite positive charge of holes forms in the perovskite, which shifts the surface potential towards more positive values. This shift is represented in Figure 8a, where the transition from blue to red curves shows the accumulation of holes in the perovskite as a function of time, after illumination is switched on. The carrier separation is also highlighted in Figure 8b: increasing the conduction band offset value leads to larger accumulation of holes in the perovskite, and of electrons in the TiO_x layer.

The same reasoning would be valid for a perovskite - HTL structure with opposite values of the valence band offset with respect to the ones shown in Figure 7.

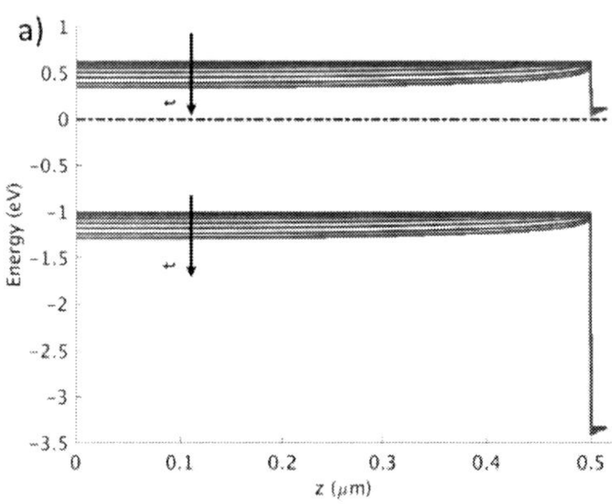

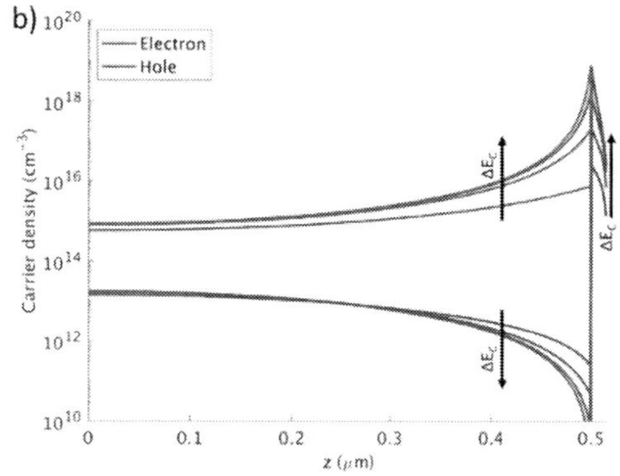

Figure 8: Band diagram evolution after illumination is switched on for perovskite - TiO_x (a), and (b) photogenerated carrier distribution across the same heterostructure under illumination in steady state, for different values of conduction band offset. The interface between perovskite and TiO_x is located at $z = 0.5$ μm .

Conduction band offsets of 0.5 and 0.6 eV provide SPVs of 0.28 and 0.38 V, respectively, which are comparable to the experimental values. However, the characteristic decay time of the SPV related to this process is again considerably lower than experimentally observed, and the simulated transient at switch-off is non-exponential. In addition, we expect that such large values of conduction band offset would limit the efficiency of solar cells. In conclusion, the experimental SPV amplitude and transients cannot be explained solely by the conduction band offset at the perovskite-SCTL heterojunction.

In addition to doping and band offset considerations, we studied how the SPV is affected by the presence of defects in the SCTL and how it could depend on the WF of the contact electrode. Indeed, defects could trap photogenerated carriers selectively injected in the SCTL, generating an SCTL-dependent SPV, while changing the WF of the contact electrode can introduce a Schottky contact at the back of TiO_x, which also affects the SPV.

The full analysis, described in [19], shows that a conduction band offset of 500 meV may explain the amplitude of the SPV but is unable to describe the transient behaviour, while the decay time after switch-off is more compatible with the presence of a modest ($\simeq$ 400 meV) difference between the back contact and the TiOx work functions, which however cannot reproduce the measured SPV amplitude.

3 DISCUSSION AND CONCLUSIONS

In this study, we combined experiments and modelling on specific perovskite-SCTL structures to discuss the equilibrium electrostatic potential distribution and the redistribution of charges upon illumination. On the one hand, our analysis conducted in equilibrium conditions supports the hypothesis of a low effective doping density, smaller than 10^{12} cm^{-3}, inside the perovskite. On the other hand, from drift-diffusion simulations for the bilayers under study, we showed that nearly intrinsic perovskite - moderately doped SCTL structures provide surface photovoltage values lower than 1 mV, while experimental values exceed hundreds of mV under the same continuous illumination conditions. Regarding the perovskite - TiO_x structure, we also excluded the effect of band offsets at the buried heterojunction due to decay time after switching off the illumination. Considering the presence of a contact electrode at the bottom of the structure, we studied the influence of its work function, which affects the SPV. Also in this case, however, varying uniquely this parameter does not allow to fit both the SPV magnitude and decay times.

We can suggest that the large timescale of the photovoltage decays of the perovskite-ETL samples could indicate the occurrence of reversible chemical reactions at the perovskite/SCTL interface, which were not considered in drift-diffusion simulations. Those reactions could be driven by the presence of a net charge inside the perovskite under illumination, due to the presence of the SCTLs and the TCO.

4 REFERENCES

[1] N. G. Park, Mater. Today 18 (2015) 65.
[2] M. I. H. Ansari, A. Qurashi, M. K. Nazeeruddin, J. Photochem. Photobiol. C Photochem. Rev. 35 (2018) 1.
[3] W. J. Yin, T. Shi, Y. Yan, Appl. Phys. Lett. 104 (2014).
[4] D. Meggiolaro *et al.*, Energy Environ. Sci. 11 (2018) 702.
[5] S. G. Motti *et al.*, Adv. Mater. 31 (2019) 1.
[6] H. Jin *et al.*, Mater. Horizons 7 (2020) 397.
[7] J. Euvrard, Y. Yan, D. B. Mitzi, Nat. Rev. Mater. 6 (2021) 531.
[8] L. Zhao *et al.*, ACS Energy Lett. 1 (2016) 595.
[9] O. Almora, C. Aranda, E. Mas-Marzá, G. Garcia-Belmonte, Appl. Phys. Lett. 109 (2016).
[10] S. Béchu, M. Ralaiarisoa, A. Etcheberry, P. Schulz, Adv. Energy Mater. 10 (2020) 1.
[11] B. N. Lee, S. Kirmayer, E. Edri, G. Hodes, D. Cahen, J. Phys. Chem. Lett. 5 (2014) 2408.
[12] J. R. Harwell *et al.*, Phys. Chem. Chem. Phys. 18 (2016) 19738.
[13] T. Gallet, D. Grabowski, T. Kirchartz, A. Redinger, Nanoscale 11 (2019) 16828.
[14] P. Schulz, L. L. Whittaker-Brooks, B. A. Macleod, D. C. Olson, Y. L. Loo, A. Kahn, Adv. Mater. Interfaces 2 (2015).
[15] E. M. Miller *et al.*, Phys. Chem. Chem. Phys. 16 (2014) 22122.
[16] A. Zohar, M. Kulbak, I. Levine, G. Hodes, A. Kahn, D. Cahen, ACS Energy Lett. 4 (2019) 1.
[17] S. P. Dunfield *et al.*, (2021).
[18] L. Qiu, S. He, L. K. Ono, Y. Qi, Adv. Energy Mater. 10 (2020) 1.
[19] D. Regaldo *et al.*, this conference, submitted to Progress in Photovoltaics.

SCALABLE MANUFACTURING OF MULTIPLE CATION PEROVSKITE SOLAR MODULES

Luigi Vesce[1], Maurizio Stefanelli[1], Luigi Angelo Castriotta[1], Jan Philipp Herterich[2], Markus Kohlstädt[2], Uli Würfel[2], Aldo Di Carlo[1, 3]
[1]CHOSE – Centre for Hybrid and Organic Solar Energy, Department of Electronic Engineering, University of Rome "Tor Vergata"
Via del Politecnico 1, 00133 Rome, Italy, vesce@ing.uniroma2.it, aldo.dicarlo@uniroma2.it
[2]Fraunhofer Institute for Solar Energy Systems ISE
Heidenhofstraße 2, 79110 Freiburg, Germany
[3]ISM-CNR, Istituto di Struttura della Materia, Consiglio Nazionale delle Ricerche
Via del Fosso del Cavaliere 100, 00133 Rome, Italy

ABSTRACT: In the last decade, perovskite solar cell (PSC) technology showed an efficiency improvement from 9.7% to 25.2% approaching the Si record of more than 26%. The low-cost perspective of PSCs is achievable only if scalable processes in manufacturing conditions such as pilot line or plant factory are designed and optimized for the full device stack. Here, we demonstrate a semi-automatic and scalable solvent quenching/doublestep method, based on the blade-coating technique, to deposit the triple cation perovskite in ambient air on module size. We assess the homogeneity of the perovskite layer and of the full stack by scanning electron microscopy images and profilometry measurements. Moreover, we compared the efficiency of modules fabricated by different deposition procedures: blade-coating in ambient air, spin-coating in nitrogen air and ambient air. In the first case the efficiency was about 16%, while the spin-coating technique exhibited efficiencies about 15% and 13% in nitrogen and ambient air, respectively.
Keywords: perovskite, module manufacturing, PV module, blade coating

1 INTRODUCTION

The comparable fabrication procedures with organic electronics, dye-sensitized and copper indium gallium selenide solar cells pushed the performance of perovskite (PVSK) solar cell technology till 25.2% [1–5]. The perovskite for photovoltaic technology is formed by an organic/inorganic cation as methylammonium CH3NH3+, formamidinium CH(NH2)2+, Cs or a mixture of them, a divalent IV-A group metal (usually lead) and an inorganic halide anion (I-, Br- or Cl-). The triple cation perovskite (CsMAFA) mixture pushes the device efficiency thanks to the MAFA compound and suppresses the yellow phases impurities thanks to Cs [6]. The maximum efficiency in literature on small area cell is 23.5% by adding a passivation interlayer between the perovskite and the HTM (hole transporting material) to reduce interfacial defects [7]. Generally, the CsMAFA perovskite is deposited in nitrogen glovebox environment at room temperature by spin coating technique with a solvent (chlorobenzene) quenching process [8]. The described process shows some issues for the perovskite technology upscaling route: i) the spin coating technique cannot be considered a scalable deposition method if the aperture area of the device increases [9]; ii) the process is strongly dependent to the user and needs intense manual practice [8]. Moreover, the reference process reported in literature is realized in a nitrogen-filled glove box that is not replicable in a real plant [10]. Based on our knowledge, very few papers reported triple cation small area cells in ambient air by ink-jet printing, blade or meniscus coating with efficiency of 20% [10–13], whilst the fabrication of modules based on stable triple cation perovskite by using scalable coating technique in ambient air without any dipping step is lacking. In this work we show an efficient CsMAFA perovskite module (11 cm^2 aperture area) fabricated in ambient air by adopting a semi-automatic air-assisted blade coating machine (Fig. 1) in a double step quenching method.

Figure 1: Blade coater assembled by Cicci Research in collaboration with the Centre for Hybrid and Organic Solar Energy (Tor Vergata University of Rome).

We demonstrate the deposition homogeneity by scanning electron microscopy images and profilometry measurements. Finally, we compare three deposition procedures, blade-coating in ambient air and spin-coating both in ambient and nitrogen air, in terms of module efficiency. We have not the chance to test the nitrogen air/blade condition, because the machine we used has a big size (more than 1 m^3) since it is designed for large scale production (Fig. 1).

2 MATERIALS AND METHODS

We adopted the mesoscopic n-i-p structure (Fig. 2a).

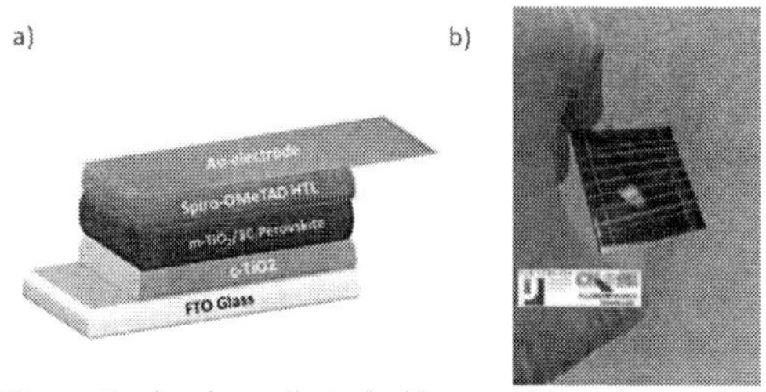

Figure 2: a) n-i-p cell stack. b) perovskite solar module fabricated.

The process to realize the blade-coated module on a 5.6x5.6 cm^2 substrate is as follow [14]. Fluorine-doped tin oxide (FTO)-glass substrates are patterned (P1) by a nanosecond UV raster scanning laser (λ=355 nm, Nd:YVO_4 pulsed at 80 kHz) [15]. We deposit the c-TiO_2 by spray-pyrolysis technique and the diluted mp-TiO_2 ink by blade-coating or spin-coating technique and then sintered at 480 °C for 30 min. The CsMAFA 3C perovskite ($(Cs_{0.05}(MA_{0.17}FA_{0.83})_{0.95}Pb(I_{0.83}Br_{0.17})_3$ in DMF/DMSO) is deposited by blade coating (ambient air, 30% RH) and spin coating (glove box and ambient air environment) techniques. The blade coating deposition process is assisted by a controlled and heated airflow to pre-dry and smooths the perovskite surface. The solvent quenching step is with 2-propanol by blade-coating technique. Blading and drying systems are placed at a precise distance from each other and can be controlled via software by the user. The blade speed for the two steps (10 and 20 mm/s) and the perovskite solution concentration (1.2 M) play a crucial role in the optimization process. The spin-coating in nitrogen air parameters are 1000 rpm, 5 s ramp up, 10 s then 5000 rpm, 2s ramp up, 30 s. Then, we add chlorobenzene 7 s before ending. In both cases, blade and spin, the modules are annealed at 100 °C for 40 min. The ambient air/spin module is fabricated by following the protocol from Singh et al. [12]. The HTL Spiro-OMeTAD solution (Borum 60 mM in chlorobenzene; the molar ratio between the dopants and the Spiro-OMeTAD was 0.5, 3.3, and 0.03 for lithium bis(trifluoromethanesulfonyl)imide, 4-tert-butylpyridine, and cobalt additive, respectively) is coated by blade-coating and spin-coating (nitrogen air) methods. The next step is the removal (laser P2) of the full stack (ETL/mp-TiO2/PVSK/HTL) deposited on FTO from the vertical connection areas to series connect two adjacent cells with the subsequent electrode deposition. Then, 90 nm thick Au counter-electrode is thermally evaporated in a high vacuum chamber. Finally, the counter electrode P3 laser ablation permits the electrical insulation between adjacent cells. The patterned modules show 5 series connected cells with a total active area of 10 cm^2 (Fig. 2b).

The morphological analysis is realized with a field emission scanning electron microscope (Auriga 60 workstation, Zeiss) equipped with a patented Gemini electron gun. Acceleration voltage was set to 5 kV, the working distance was approx. 5 mm. An in-lens detector was used for image generation. The photovoltaic characteristics are measured with a class A sun simulator (Sun 2000, Abet) at AM 1.5 1000W/m^2 calibrated with an SKS 1110 sensor (Skye Instruments Ltd., Llandrindod Wells, UK); the system is equipped with a 2612 source meter (Keithley Instruments Inc., Cleveland, OH, USA) and a LabVIEW interface.

3 RESULTS AND DISCUSSIONS

We realized top view SEM images of the bladed samples (Figs. 3a, b) at the beginning and at the end of the substrates to testify the homogeneity of the perovskite deposition by blade-coating technique.

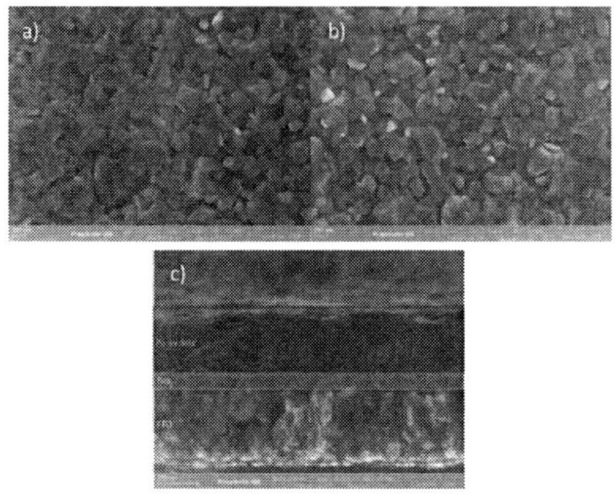

Figure 3: Perovskite layer a) at the beginning and b) at the end of the coating process. c) SEM cross section.

The images reveal the same morphology and grain size distribution at beginning and end of the coating area, indicating a homogeneous coating of the whole module substrate. The grain size was found to be about 200 nm and the perovskite thickness in different spots of the deposition was about 490 nm, being in accordance with the CsMAFA perovskite in literature [6,8]. The SEM cross section (Fig. 3c) shows the perovskite grains avoid stacking on top of each other resulting in high charge transport and FF of the devices.
We compared the reported fabrication procedure in ambient air with the procedure based on the spin-coating technique both in nitrogen and in ambient air (Fig. 4).

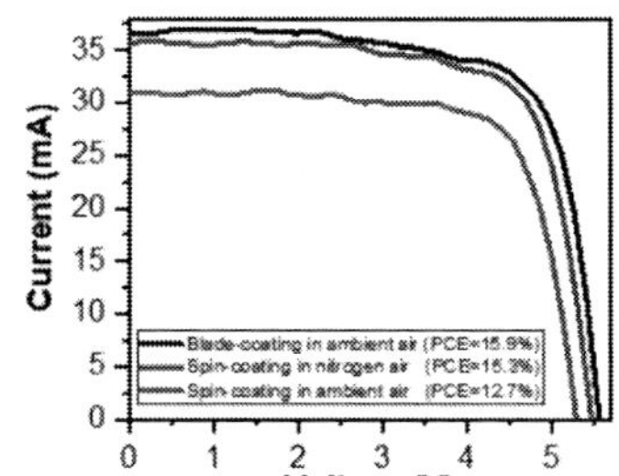

Figure 4: IV curves of modules fabricated by different fabrication procedures: blade-coating in ambient air, spin coating in nitrogen and ambient air.

The scalable technique reached an efficiency of 15.9%, about 5% more than the spin-coated module in nitrogen air and 20% the spin-coated module in ambient air. The reproducibility of the proposed method is about 2% respect to 10% of the spin coating method. The compared methods have different performance because the bladed samples reveal smaller defects in respect to the spinned samples. The spin-coated device in ambient air has mainly a lower density current respect to the reference glove box condition. We believe that further investigations about spin coating conditions out of glove box and solvent optimization are required, but it goes beyond the scope of the present work that focuses on the identification of a scalable coating technique in ambient air. Moreover, the optimized blade-coating procedure has about 70% less material consumption respect to the spin-coating technique.

4 CONCLUSIONS

In few years, perovskite solar cell got efficiency closed to the silicon technology. The scaling up from small area cells to modules is the development step for any Photovoltaic technology [16]. The large area modules manufacturing process should be low cost, reliable and high throughput, with good performance and stability of the devices [17]. One of the advantages of the perovskite technology is the possibility to use the well-known and cheap printing/coating techniques [3]. The CsMAFA perovskite entered in between the different perovskite compositions as a possible way to have efficient and stable devices [6]. In literature many researchers reported several methods to upscale the technology [18], but no one demonstrated the feasibility of efficient triple cation perovskite module in ambient air by using scalable technique for both perovskite and antisolvent. In this work we showed a process based on the air-assisted blade-coating technique in a double step quenching method. The module efficiency at 1SUN illumination is about 16%, 15.3% and 12.7% for blade-coated in ambient air, spin-coated in nitrogen air and spin-coated in ambient air devices, respectively. Further characterizations about failure analysis and stability are in progress.

ACKNOWLEDGEMENTS

Authors were supported by the European Union's Horizon 2020 Framework Program for funding Research and Innovation under grant agreement no. 764047 (ESPResSo). The authors acknowledge the project UNIQUE, supported under the umbrella of SOLAR-ERA.NET_cofund by ANR, PtJ, MUR (GA 775970), MINECOAEI, SWEA, within the European Union Framework Programme for Research and Innovation Horizon 2020 (Cofund ERANET Action, No. 691664). Dr. L. Vesce would like to thank Dr. Ilaria Farina (IFO-IRCSS, Rome, Italy) for interesting discussions about the literature databases. The authors would like to thank Jutta Zielonka (Fraunhofer ISE, Freiburg, Germany) for performing electron microscopy.

REFERENCES

[1] H. J. Snaith, *Nat. Mater.* **2018**, *17*, 372.
[2] P. K. Nayak, S. Mahesh, H. J. Snaith, D. Cahen, *Nat. Rev. Mater.* **2019**, *4*, 269.
[3] P. Mariani, L. Vesce, A. Di Carlo, *Semicond. Sci. Technol.* **2015**, *30*, DOI 10.1088/0268-1242/30/10/104003.
[4] NREL, "No Title," can be found under https://www.nrel.gov/pv/cell-efficiency.html, **2020**.
[5] M. A. Green, E. D. Dunlop, J. Hohl-Ebinger, M. Yoshita, N. Kopidakis, A. W. Y. Ho-Baillie, *Prog. Photovoltaics Res. Appl.* **2020**, *28*, 3.
[6] M. Saliba, T. Matsui, J. Y. Seo, K. Domanski, J. P. Correa-Baena, M. K. Nazeeruddin, S. M. Zakeeruddin, W. Tress, A. Abate, A. Hagfeldt, M. Grätzel, *Energy Environ. Sci.* **2016**, *9*, 1989.
[7] H. Zhu, Y. Liu, F. T. Eickemeyer, L. Pan, D. Ren, M. A. Ruiz-Preciado, B. Carlsen, B. Yang, X. Dong, Z. Wang, H. Liu, S. Wang, S. M. Zakeeruddin, A. Hagfeldt, M. I. Dar, X. Li, M. Grätzel, *Adv. Mater.* **2020**, *32*, 1.
[8] M. Saliba, J. P. Correa-Baena, C. M. Wolff, M. Stolterfoht, N. Phung, S. Albrecht, D. Neher, A. Abate, *Chem. Mater.* **2018**, *30*, 4193.
[9] F. Matteocci, L. Vesce, F. U. Kosasih, L. A. Castriotta, S. Cacovich, A. L. Palma, G. Divitini, C. Ducati, A. Di Carlo, *ACS Appl. Mater. Interfaces* **2019**, *11*, 25195.
[10] H. Eggers, F. Schackmar, T. Abzieher, Q. Sun, U. Lemmer, Y. Vaynzof, B. S. Richards, G. Hernandez-Sosa, U. W. Paetzold, *Adv. Energy Mater.* **2020**, *10*, DOI 10.1002/aenm.201903184.
[11] F. Mathies, H. Eggers, B. S. Richards, G. Hernandez-Sosa, U. Lemmer, U. W. Paetzold, *ACS Appl. Energy Mater.* **2018**, *1*, 1834.
[12] T. Singh, T. Miyasaka, *Adv. Energy Mater.* **2018**, *8*, 1.
[13] M. C. Tang, Y. Fan, D. Barrit, X. Chang, H. X. Dang, R. Li, K. Wang, D. M. Smilgies, S. F. Liu, S. De Wolf, T. D. Anthopoulos, K. Zhao, A. Amassian, *J. Mater. Chem. A* **2020**, *8*, 1095.
[14] L. Vesce, M. Stefanelli, J. P. Herterich, L. A. Castriotta, M. Kohlstädt, U. Würfel, A. Di Carlo, *Sol. RRL* **2021**, *5*, 1.
[15] G. Mincuzzi, L. Vesce, M. Liberatore, A. Reale, A. Di Carlo, T. M. Brown, *IEEE Trans. Electron Devices* **2011**, *58*, 3179.
[16] L. Vesce, A. Guidobaldi, P. Mariani, A. D. I. Carlo, M. L. Parisi, S. Maranghi, R. Basosi, in *World Sci. Ref. Hybrid Mater.*, **2018**.
[17] L. Vesce, M. Stefanelli, A. Di Carlo, *2019 Int. Symp. Adv. Electr. Commun. Technol.* **2020**, *764047*, 1.
[18] L. A. Castriotta, F. Matteocci, L. Vesce, L. Cinà, A. Agresti, S. Pescetelli, A. Ronconi, M. Löffler, M. M. Stylianakis, F. Di Giacomo, P. Mariani, M. Stefanelli, E. M. Speller, A. Alfano, B. Paci, A. Generosi, F. Di Fonzo, A. Petrozza, B. Rellinghaus, E. Kymakis, A. Di Carlo, *ACS Appl. Mater. Interfaces* **2021**, *13*, 11741.

EXPLORING THE USE OF METHYLAMMONIUM IODIDE TO IMPROVE THE UNIFORMITY OF THE MAPBI3 LAYER IN HTM-FREE PEROVSKITE SOLAR CELLS EQUIPPED WITH BACK CONTACTS BASED ON PASTES OF GRAPHITE AND BLACK CARBON IN A POLYVINYLIDENE FLUORIDE IN TOLUENE SOLUTION

C. Montes[1,3], L. Ocaña[1,3], S. González-Pérez[2], B. González-Díaz[3], E. Llarena[1].
[1]Instituto Tecnológico y de Energías Renovables, S. A. (ITER)
Pol. Industrial de Granadilla, s/n, E 38600 Granadilla de Abona, Spain. Ph. +34 922 747 700. Fax +34 922 747 701. E-mail cmontes@iter.es
[2]Departamento de Didácticas Específicas.
C/ Pedro Zerolo, s/n
Edificio Central Planta 2
Apartado 456. Código postal 38200
San Cristóbal de La Laguna. S/C de Tenerife. Spain.
[3]Departamento de Ingeniería Industrial.
Universidad de La Laguna.
Camino San Francisco de Paula, s/n
38206 S/C de Tenerife, Spain

ABSTRACT: A new iteration of a perovskite solar cell has been developed that evolves from architectures previously studied by our group, this time exploring the introduction of a graphite and black carbon with methylammonium iodide (MAI) in polyvinylidene fluoride layer, which is placed over the methylammonium lead triiodide ($MAPbI_3$) one, in order to reduce the heat-induced perovskite decomposition that results from the thermal processes in ambient conditions that are required to deposit the back electrode, which also acts in the cell as hole transport material and encapsulate. A batch of devices has been produced and characterized, obtaining power conversion efficiencies (PCE) up to 1.32%.
Keywords: Perovskites, Contact, Metallization, Graphite

1 INTRODUCTION

The Solar Cell Laboratory from the Instituto Tecnológico y de Energías Renovables (ITER), with the support of the La Laguna University (ULL), has been working for several years on the development of screen printing inks in order to provide back contacts for perovskite solar cells as a low-cost alternative to thermal evaporation [1-5]. Also, recent efforts have been devoted in order to find solutions for encapsulating methylammonium lead triiodide ($MAPbI_3$) solar cells, produced in ambient conditions, and study their subsequent degradation due to exposure. In these trials, the chosen encapsulating material was ethylene-vinyl acetate (EVA), in the form of commercial sheets, which were applied by techniques similar to the ones commonly used for laminating PV modules [6,7]. By using the knowledge acquired with these two initiatives and inspired by the work of Guanglan Liao et Al [8] and of Xingzhong Zhao et Al [9], who developed carbon-based back electrode for Hole Transport Material Free (HTM-free) planar heterojunction perovskite solar cells, an agglomerate of graphite and black carbon in an EVA in toluene solution was developed, in order to be applied as back contact electrode by means of an innovative cell architecture, so as to produce HTM-free perovskite solar cells [10].

The IV curves performed on these cells suggested the existence of uniformity issues in the $MAPbI_3$ layers, in the form of disconnected grains, which allowed unwanted shunting paths that heavily drained the light-generated current.

These non-uniformities could be consequence of the thermal stresses inflicted on the perovskite layers during the electrode deposition stages, due to the formation of PbI_2 on the $MAPbI_3$ grain boundaries, caused by the loss of methylammonium iodide (MAI) when subjected to temperatures above 100 ºC in humid conditions [12, 13].

Thus, further inspired by yet another work of Xingzhong Zhao et Al [14], a modification on the back electrode deposition technique has been developed, so as to introduce MAI in the carbon sublayer that is placed above the $MAPbI_3$, in order to reduce heat-induced perovskite decomposition and improve its uniformity.

2 EXPERIMENTAL

2.1 Obtaining the carbon and MAI with PVDF in Toluene paste.

In order to develop a carbon based electrode, a procedure was developed, which consisted of the following stages:

On the first stage, polyvinylidene fluoride (or PVDF) in N-Methyl-2-pyrrolidone (or NMP) solutions were prepared with a concentration of 10:90% in weight (wt%), by adding 0.2056 g of PVDF [15] and 2 ml of NMP [16] to 7ml capacity lid vials. Initially the mixtures were stirred manually, using a fine spatula, and then they were homogenized using a heated magnetic stirrer for 30 minutes at 75 ° C and 450 rpm. The result of the mixing was approximately 2.2 grams of a solution per vial that have a relatively thick appearance and a yellowish to brownish color.

Then, on a second stage, a mixture of PVDF in NMP with graphite and black carbon was prepared, at a PVDF:NMP:Graphite:BlackCarbon 2:58:30:11% weight percentage (wt%), by adding to a beaker 3,375 g of the solution of PVDF in NMP, as described in the previous paragraph, 6.328 g of graphite powders [17], 2.443g of black carbon [18] and, finally, adding also 9 ml of extra NMP. The mixture was first stirred manually using a spatula for 5 minutes, then it was passed through a three roll mill 5 times. The visual appearance of the result of the mixing process is that of an intensely dark and thick paste.

Next, a quantity of the so produced carbon paste, was spread on a clean glass and then, in order to evaporate all the NMP, it was placed in a stove furnace [19] at 150°C for one hour. Once the glass had cooled, the layer deposited on it was extracted using a glass scraper, and the detached materials were collected in an agate mortar, where they were grinded for several minutes, pouring the contents into a sample cup with lid for storage. The weight percentage (wt%) of the resulting powder was 96:4% (Conductive powders:Binder).

Next, 2.10 g of carbon and PVDF of the powder so obtained, was placed in a beaker, to which 5 ml of Toluene [20] and 0,7327g of Methylammonium iodide (CH_3NH_3I or MAI) [21] were added. These three components were mixed by using a spatula during 5 minutes, resulting on a rather dense paste, with a weight percentage (wt%) of 28:1:10:60% in their main components: Conductive Powders, binder, MAI and Solvent.

2.2 Preparation of the HTM-free Perovskite Solar Cells (PSCs)

The fabrication process of the device was carried out inside an ISO 7 cleanroom in ambient conditions, at relative humidity (RH) ranging from 50 up to 60%, and at a temperature set to 23°C.

First, the Fluorine Tin Oxide (FTO) covered glass substrates (with sizes 2.5 x 2.5 mm^2) were cleaned sequentially by ultrasonication in a 2% soap solution, deionized water, acetone and finally in 2-propanol. Then, the substrates were dried with nitrogen gas flow [22]. Coming up next, an UV Ozone Cleaner was used to remove the contaminants from the surface of the samples [23].

Second, a compact layer of Titanium dioxide (TiO_2) was deposited by spin coating over the FTO layer, at a speed of 2000 rpm for 60 seconds. The titanium oxide precursor solution was prepared by dropwise adding a solution of 175 µl of Titanium isopropoxide ($Ti\{OCH(CH_3)_2\}_4$) in 1.25 ml of Ethanol (EtOH) into a solution of 17.5 µl of 2 M hydrochloric acid (HCl) in 1.25 ml of EtOH [24]. This layer was cured in a furnace at 500 ° C for 60 min. Prior to their deposition, the produced solutions were passed through a 0.22 µm Polytetrafluoroethylene (PTFE) filter.

Third, thin layers of perovskite, from the methylammonium lead iodide kind ($CH_3NH_3PbI_3$ or $MAPbI_3$), were fabricated in ambient conditions. The perovskite precursor solution was produced by mixing Lead iodide (PbI_2) and Methylammonium iodide (CH_3NH_3I) in an equimolar ratio of 1:1, with a solvent mixture of Dimethylformamide (DMF) and Dimethyl Sulfoxide (DMSO), prepared with a 9:1 volume ratio. The solution was mixed in a magnetic stirrer with heating capabilities by set it at 300 revolutions per minute (rpm) and 75 °C during 50 minutes. Then the heating was interrupted, leaving the solution to get to room temperature, while keeping the stirring at 1050 rpm overnight. Prior to its deposition, the produced solution was also passed through a 0.22 µm PTFE filter. The deposition proper was made by spin-coating 100 µl of the perovskite precursor solution at 5000 rpm for 30 seconds so that, 6 seconds after the start of the spinning, 300 µl of a non-polar solvent, in this case chlorobenzene, was dropped over the substrates. Throughout the spin-coating process the RH inside the chamber was maintained at a range of 6-8 %, depending on the existing ambient humidity, by using a constant flow of clean dry air [25]. Finally, after the deposition process took place, the samples were cured at 100 °C for 10 minutes inside a stove furnace and stored in a vacuum desiccator, so as to avoid their degradation [26,27].

In order to incorporate the carbon based layer as the back contact for the HTM-free perovskite solar cell, a new cell architecture was developed (see Figure 3), in which polyimide (kapton) tape was used as a tool to screen and electrically isolate parts of the device, while copper adhesive tape was applied for interfacing with both, front and back electrodes, as a way to transport the charges generated by the device when illuminated.

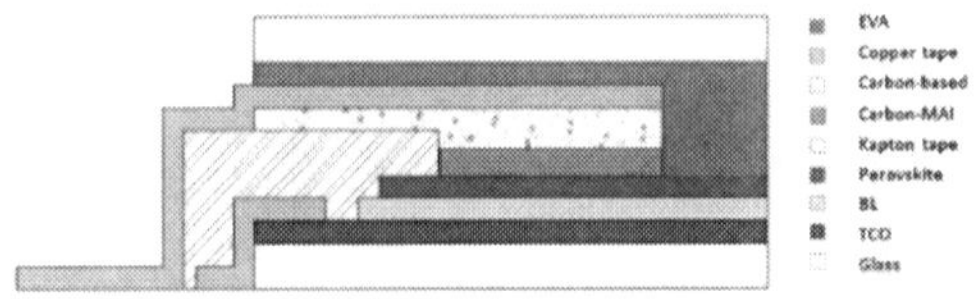

Figure 1: Schematic view of the developed HTM-free PSC architecture, in which the type and arrangement of the different layers that correspond to it can be appreciated.

The back electrode was placed by following a four step procedure: In the first one, a small amount of the carbon, PVDF and MAI in Toluene paste, obtained as described in section 2.2, was applied over the perovskite layer by following a technique similar to cement rendering, in order to obtain a coating as thin and regular as possible. The second step consisted of subjecting the device to a thermal process in which it was placed in a stove furnace during 30 minutes at 100 °C. With this step the electrode was cured, eliminating the excess of Toluene. The third step is where the kapton tape, the copper tape, the EVA sheet and a second glass were applied, each one fulfilling their respective tasks and following the architecture shown at Figure 3. A small amount of carbon in PVDF powder, from the one prepared in section 2.2, was dusted over the carbon-MAI in PVDF layer, in order to facilitate the electrode's bonding.

The fourth and final step was where the device encapsulation took place, in which a thick steel plate was placed over the device, in order to exert uniform pressure during yet a second thermal process, also performed in a stove furnace set at 100 °C, but this time lasting 165 minutes, in order to take into account the thermal inertia of the steel plate. Finally, the device was allowed to cool down for another 125 minutes, before removing the steel plate.

2.3 HTM-free PSCs characterization

First, a visual inspection of the produced samples was carried out by using a digital microscope [28], in order to evaluate their morphology. Special attention was paid to detect encapsulation failures and to look for signs that suggested that damage had been caused to any of the many elements that comprised the devices.

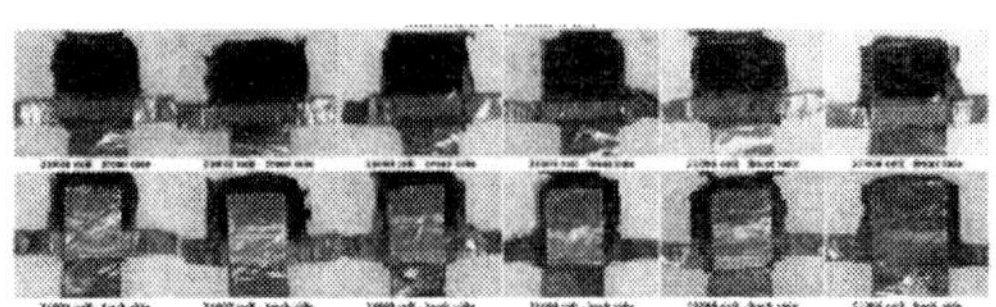

Figure 2: Batch of HTM-free PSCs produced at a clean room at 23°C and relative humidity (RH%) in between 45 and 50%.

Then, the devices were characterized by using a

semiconductor characterization system [29] in combination with a solar simulator [30], illuminating through a mask with an open surface of 0.38 cm^2 (see Figure 5), for determining their electrical properties. Four measurements were made on each device, two fast measurements (forward and reverse) and two stability measurements (forward and reverse). So far, these measurements have been repeated three times: The first one was carried out three days after the devices were manufactured (on the 07/30/2021), the second about one month later (on the 08/31/2021) and the third on the following day (the 09/01/2021), after having subjected the devices to another thermal process, this time of 100°C for 30 minutes in a stove furnace. It is important to indicate that the cooling process was carried out with the devices weighted down by a thick steel plate, in order to avoid any possible delamination.

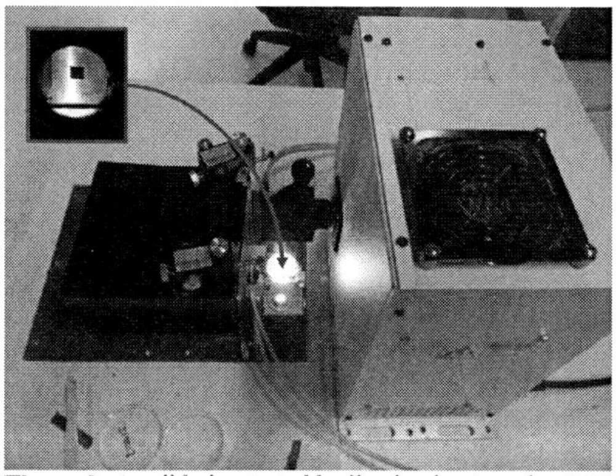

Figure 3: A cell being tested by iluminating trough a 0.38 cm^2 mask by using a solar simulator set at STC conditions.

3 RESULTS AND DISCUSSION

Table I summarizes the most relevant data measured at the champion cell produced in the manufactured batch.

Table I: Summary with the values of short circuit current (Isc), open circuit voltage (Voc), fill factor (FF) and power conversion efficiency (PCE) obtained from the measurements made at the sample 21003.

Measurement	Isc (mA)	Voc (V)	FF (%)	PCE (%)
		30/07/2021		
Cell stability forward	-0.90	0.72	43.65	0.69
Fast forward	-0.37	0.51	26.99	0.12
Fast reverse	-0.45	0.60	31.00	0.22
		08/31/2021		
Cell stability forward	-0.63	0.62	34.78	0.36
Cell stability reverse	-0.47	0.62	24.09	0.19
		09/01/2021		
Cell stability forward	-1.65	0.82	37.07	**1.32**
Cell stability reverse	-0.99	0.62	26.43	0.43

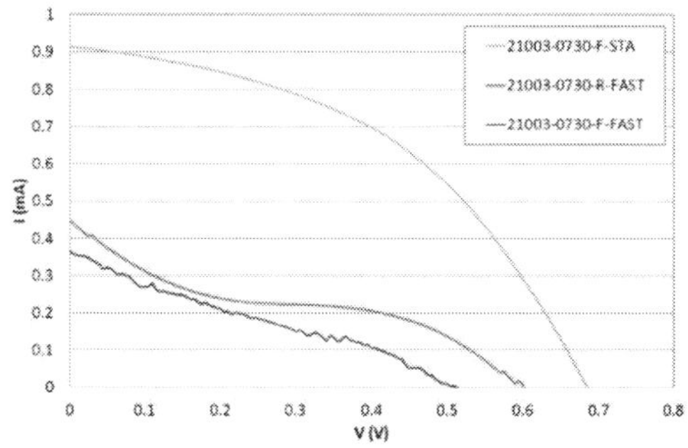

Figure 4: IV curve measured in the forward direction at stability conditions (orange), together with fast IV curves, measured both forward (blue) and reverse (green). All measurements were obtained from cell 21003 on the 30th of July.

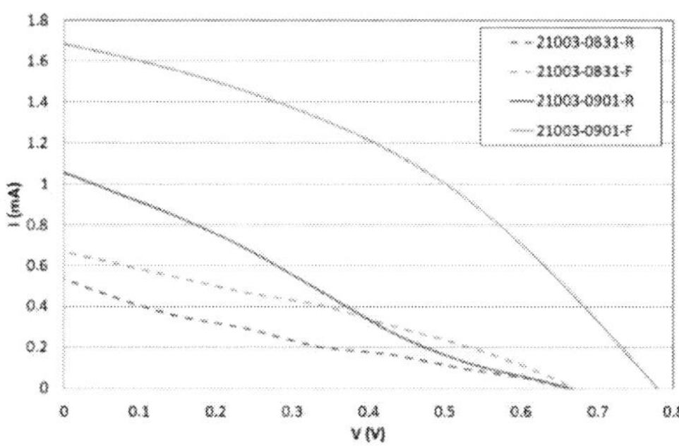

Figure 5: IV curves performed in the forward (orange) and reverse (blue) directions from cell 20030 at stability conditions on the 31st of August (dashed lines) and on the 1st of September (solid lines).

The gathered data indicates that, despite the efforts made to reduce the thickness of the electrode, it is still way too thick for the devices to perform efficiently, providing high series resistance to them.

We hypothesize that the MAI dispersed in the first layer of the carbon electrode, although it does counter the loss of such material in the perovskite layer (which in turn results on an increase in the devices' shunt resistance) during the thermal processes required for the electrode deposition, this effect nevertheless quickly fades when the cells reach room temperature.

This explains why the first measurements, which were performed three days after their fabrication, were already so poor, while the ones performed one month later (on the 31st of August) indicated a degradation process taking place at a slower pace and yet, the measurements taken on the cells right after they were subjected to a new thermal process, provided the best power conversion efficiencies registered for all the cells in the produced batch so far (with the champion cell, the 21003, reaching 1.32%), suggesting that the degradation process had been considerably reversed.

The hysteretic behavior observed in these curves

suggests the existence of a great deal of ionic migration in our cells, which may also cause degradation on the MAPbI3 layer [32-34]. This suggests that incorporate multication mixtures, such as formamidium [FA; $HC(NH_2)_2+$], cesium (Cs) or rubidium (Rb) to act as passivation agents [35-37] could be necessary.

Finally, although the results presented in this paper indicate a slight improvement on this line of research, it still is necessary to perform more test trials with the devised fabrication procedures, without further modifications, in order to properly evaluate them, in terms of quality and consistency. That is, to carry out laboratory work which, in these times of pandemic and crisis, has become more difficult if possible.

4 ACKNOWLEDGEMENTS

This work has been developed within the MACLAB-PV project framework, which has been co-financed by the INTERREG Madeira-Azores-Canarias Territorial Cooperation Programme (MAC) 2014-2020. 2nd Call. Axis 1 – Enhancing research, technological development and innovation.

The authors would like to thank Susana Gómez, Product Manager at Internatio-Müller Chemical Distribution (IMCD) - España Especialidades Químicas, S.A., for providing the black carbon used throughout our experiments.

5 REFERENCES

[1] C. Quinto, A. Linares, E. Llarena, C. Montes, O. González, D. Molina, A. Pío, L. Ocaña, M. Friend and M. Cendagorta, "Screen printing for perovskite solar cells metallization", 31st European Photovoltaic Solar Energy Conference and Exhibition, pp. 1144-1148, 14-18 September 2015, Hamburg, Germany.

[2] C. Montes, A. Linares, E. Llarena, O. González, D. Molina, A. Pío, L. Ocaña, C. Quinto, M. Friend and M. Cendagorta, "Conductive inks with epoxy resin based vehicles for perovskite screen printing metallization", 32nd European Photovoltaic Solar Energy Conference and Exhibition, pp. 1303-1309, 21-24 June 2016, Munich, Germany.

[3] C. Montes, L. Ocaña, C. Quinto, S. González-Pérez, B. González-Díaz, M. Friend and M. Cendagorta, "Conductive inks with epoxy resin based vehicles for perovskite screen printing metallization as a viable and low –cost alternative to thermal evaporation" 33rd European Photovoltaic Solar Energy Conference and Exhibition, pp. 1158-1162, 25-28 September 2017, Amsterdam, The Netherlands.

[4] C. Montes, L. Ocaña, L. De Sousa-Vieira, J.S. Moreno-Ramírez, S. González-Pérez, B. González-Díaz, M. Fiend and M. Cendagorta, "Exploring the use of collodion as a binder in screen printing vehicles for perovskite solar cells metallization" 34th European Photovoltaic Solar Energy Conference and Exhibition, pp. 928-932, 24-28 September 2018, Brussels, Belgium.

[5] C. Montes, L. Ocaña, L. De Sousa-Vieira, S. González-Pérez, B. González-Díaz, M. Friend & M. Cendagorta, 2019, "Studying the use of mixed binders made with epoxy resin and collodion for producing conductive inks for the metallization of perovskite solar cells via screen printing techniques", 36th European Photovoltaic Solar Energy Conference and Exhibition, 742 - 745. DOI: 10.4229/EUPVSEC20192019-3BV.2.17

[6] L. Ocaña, C. Montes, L. De Sousa-Vieira, J.S. Moreno-Ramírez, S. González-Pérez, B. González-Díaz, C. Hernández-Rodríguez, M. Friend & M. Cendagorta, 2019, "On the stability of planar CH3NH3PBI3 perovskite solar cells produced in ambient conditions by using polymer encapsulates", 36th European Photovoltaic Solar Energy Conference and Exhibition, 721 - 727. DOI: 10.4229/EUPVSEC20192019-3BV.2.10

[7] L. Ocaña, C. Montes, S. González-Pérez, B. González-Diaz, M. Friend, M. Cendagorta. "Improvements on the Stability of MAPbI3 Thin Films Against UV Radiation by a Low Temperature Encapsulation Method that Uses Commercial Ethylene-Vinyl Acetate Sheets", 37th European Photovoltaic Solar Energy Conference and Exhibition, 659-662. DOI DOI: 10.4229/EUPVSEC20202020-3BV.1.22

[8] Z.Liu, T. Shi, Z. Tang, G. Liao, (2017) "A large-area hole-conductor-free perovskite solar cell based on a low-temperature carbon counter electrode". Materials Research Bulletin, Volume 96, Part 3, 196-200. DOI: 10.1016/j.materresbull.2017.03.069

[9] Fangjie Li Changlei Wang Pei Liu Yuqing Xiao Lihua Bai Fei Qi Xiaoyi Hou Huijie Zhang Yuan Wang Shaofu Wang Xingzhong Zhao , "Fully Air-Processed Carbon-Based Efficient Hole Conductor Free Planar Heterojunction Perovskite Solar Cells With High Repoducibility and Stability", Solar RRL, 3, December 2018.

[10] C. Montes, L. Ocaña, S. González-Pérez, B. González-Diaz, M. Friend, M. Cendagorta, "Developing an Agglomerate of Graphite and Black Carbon in an Ethylene-Vinyl Acetate in Toluene Solution for Producing Electrodes for HTM-Free Perovskite Solar Cells", 37th European Photovoltaic Solar Energy Conference and Exhibition, 667-672. DOI: 10.4229/EUPVSEC20202020-3BV.1.27

[11] Yu, Z., Chen, B., Liu, P., Wang, C., Bu, C., Cheng, N., Bai, S., Yan, Y. and Zhao, X. (2016), Stable Organic–Inorganic Perovskite Solar Cells without Hole-Conductor Layer Achieved via Cell Structure Design and Contact Engineering. Adv. Funct. Mater., 26: 4866-4873. DOI: 10.1002/adfm.201504564

[12] A. Dualeh , P. Gao , S. Il Seok , M. K. Nazeeruddin , M. Grätzel , "Thermal Behavior of Methylammonium Lead-Trihalide Perovskite Photovoltaic Light Harvesters". Chem. Mater. 2014, 26 , 6160. DOI: 10.1021/cm502468k

[13] Y. Juang, E.J. Juarez-Perez, Q. Ge, S. Wang, M.R. Leyden, L. K. Ono, S. R. Raga, J. Hu, Y Qi, "Post-annealing of MAPbI3 perovskite films with methylamine for efficient perovskite solar cells". Mater. Horiz. 2016. 548-555. DOI: 10.1039/C6MH00160B

[14] Yu, Z., Chen, B., Liu, P., Wang, C., Bu, C., Cheng, N., Bai, S., Yan, Y. and Zhao, X. (2016), Stable Organic–Inorganic Perovskite Solar Cells without Hole-Conductor Layer Achieved via Cell Structure Design and Contact Engineering. Adv. Funct. Mater., 26: 4866-4873. DOI: 10.1002/adfm.201504564

[15] KF Polymer W#1700 by KUREHA

[16] Reference 328634, purchased from Sigma Aldrich.

[17] Graphite powder SC 20 OS from Graphite Kropfmühl GmbH Langheinrichstraße 1, 94051 Hauzenberg Germany.

[18] Black carbon powder. Conductex SC Ultra by

Birla Carbon.
[19] Stove Furnace. CONTERM 19L, J.P.Selecta.
[20] Toluene anhydrous, 99.8% (244511). Sigma aldrich.
[21] Methylammonium iodide, 99%, anhydrous (901434). Sigma Aldrich.
[22] C. Montes, L. Ocaña, L. De Sousa-Vieira, J. Moreno-Ramírez, S. González-Pérez, B. González-Díaz B, M. Friend and M. Cendagorta, "Producing uniform and smooth thin layers of perovskite under ambient conditions by adjusting to the existing levels of moisture". 35th European Photovoltaic Solar Energy Conference and Exhibition. pp. 933–7. 24-28 September 2018. Brussels, Belgium.
[23] UV Ozone Cleaner, model L2002A2-UK by Ossila.
[24] Sandeep Pathak, Alessandro Sepe, Aditya Sadhanala, Felix Deschler, Amir Haghighirad, Nobuya Sakai, Karl C. Goedel, Samuel D. Stranks, Nakita Noel, Michael Price, Sven Hüttner, Nicholas A. Hawkins§, Richard H. Friend, Ullrich Steiner, and Henry J. Snaith. Atmospheric Influence upon Crystallization and Electronic Disorder and Its Impact on the Photophysical Properties of Organic–Inorganic Perovskite Solar Cells. ACS Nano 2015, 9, 3, 2311–2320
[25] C. Montes, L. Ocaña, L. De Sousa-Vieira, S. González-Pérez, B. González-Díaz, J. Moreno-Ramírez, M. Friend and M. Cendagorta, "Fabrication of smooth, mirror-like and PbI2-free thin film perovskite layers in ambient conditions", 36th European Photovoltaic Solar Energy Conference and Exhibition, pp. 717–20. 9-13 September 2019. Marseilles, France.
[26] Xiao, M., Huang, F., Huang, W., Dkhissi, Y., Zhu, Y., Etheridge, J., Gray-Weale, A., Bach, U., Cheng, Y.-B. and Spiccia, L. (2014), "A Fast Deposition-Crystallization Procedure for Highly Efficient Lead Iodide Perovskite Thin-Film Solar
[27] Pablo P. Boix, Kazuteru Nonomura, Nripan Mathews, Subodh G. Mhaisalkar, "Current progress and future perspectives for organic/inorganic perovskite solar cells", Materials Today, Volume 17, Issue 1, January–February 2014, Pages 16-23, ISSN 1369-7021
[28] Digital Microscope U500X, focus range 15mm - 40mm, power range 50X-500X by Unimake.
[29] Semiconductor Characterization System model SCS-4200 by Keithley.
[30] Solar Simulator, model ABET-10500 by ABET Technologies
[31] Choi, J. J., Yang, X., Norman, Z. M., Billinge, S. J. L., and Owen, J. S., "Structure of methylammonium lead iodide within mesoporous titanium dioxide: Active material in high-performance perovskite solar cells," Nano Lett. 14, 127–133 (2014). https://doi.org/10.1021/nl403514x.
[32] Snaith, H. J., Abate, A., Ball, J. M., Eperon, G. E., Leijtens, T., Noel, N. K., Stranks, S. D., Wang, J. T., Wojciechowski, K., and Zhang, W., "Anomalous hysteresis in perovskite solar cells," J. Phys. Chem. Lett. 5, 1511–1515 (2014). https://doi.org/10.1021/jz500113x.
[33] Unger, E. L., Hoke, E. T., Bailie, C. D., Nguyen, W. H., Bowring, A. R., Heumüller, T., Christoforo, M. G., and McGehee, M. D., "Hysteresis and transient behavior in current–voltage measurements of hybrid-perovskite absorber solar cells," Energy Environ. Sci. 7, 3690–3698 (2014). https://doi.org/10.1039/C4EE02465F.
[34] Yuan, Y. and Huang, J., "Ion migration in organometal trihalide perovskite and its impact on photovoltaic efficiency and stability," Acc. Chem. Res. 49, 286–293 (2016). https://doi.org/10.1021/acs.accounts.5b00420.
[35] Yang, Zhibin & Chueh, Chu-Chen & Po Wei, Liang & Crump, Michael & Lin, Francis & Zhu, Zonglong & Jen, Alex. (2016). Effects of Formamidinium and Bromide Ion Substitution in Methylammonium Lead Triiodide toward High-Performance Perovskite Solar Cells. Nano Energy. 22. 10.1016/j.nanoen.2016.02.033.
[36] M. Saliba, T. Matsui, K. Domanski, J.-Y. Seo, A. Ummadisingu, S. M. Zakeeruddin, J.-P. Correa-Baena, W. R. Tress, A. Abate, A. Hagfeldt, M. Grätzel, "Incorporation of rubidium cations into perovskite solar cells improves photovoltaic performance". Science, 206–209 (2016).
[37] M. Saliba, T. Matsui, J.-Y. Seo, K. Domanski, J.-P. Correa-Baena, M. K. Nazeeruddin, S. M. Zakeeruddin, W. Tress, A. Abate, A. Hagfeldt, M. Grätzel, "Cesium-containing triple cation perovskite solar cells: Improved stability, reproducibility and high efficiency". Energ. Environ. Sci. 9, 1989–1997 (2016).

EVAPORATION PRESSURE CONTROL ON HIGH CRYSTALLIZATION PEROVSKITE LAYER VIA SANDWICH EVAPORATION TECHNIQUE

Hui-Hung Shen[1], Chung-Han Chang [1], Wei-Cheng Lo[1], Ching-Fuh Lin[1, 2, 3]*
1. Graduate Institute of Photonics and Optoelectronics, National Taiwan University,
No. 1, Sec. 4, Roosevelt Road, Taipei 106319, Taiwan
2. Graduate Institute of Electronics Engineering, National Taiwan University,
No. 1, Sec. 4, Roosevelt Road, Taipei 106319, Taiwan
3. Department of Electrical Engineering, National Taiwan University,
No. 1, Sec. 4, Roosevelt Road, Taipei 106319, Taiwan
**lincf@ntu.edu.tw*

ABSTRACT: Sandwich evaporation technique (SET) process is used to make the perovskite layer. First of all, CH_3NH_3I (MAI) is spin-coated. Next, PbI_2 is thermally evaporated. Then, MAI is evaporated at low vacuum by the SET process. In the process of the double-interdiffusion of MAI and PbI_2, we adjust the reaction conditions by different pressures in the SET chamber, which can form better morphology of perovskite layer.
Keywords: perovskite, sandwich evaporation technique, double-interdiffusion, pressure

1 INTRODUCTION

Perovskite, showing great potential in the photovoltaic field, can be used to make high conversion efficient solar cells with the advantages of low cost on forming perovskite crystalline thin films and fast fabrication process. In this study, we fabricate perovskite crystals via sandwich evaporation technique (SET) by hand-made chamber. This process, with the high elasticity on adjusting evaporation pressure by our home-made SET chamber, enables to improve the crystallization of perovskite layers.

Among various methods for the preparation of the perovskite layers, SET process is relatively easy and gives good perovskite crystals. So far, thermal evaporation technique and solution method are the most common methods of fabricating perovskite thin films. The solution method can be divided into two kinds of processes: one-step and two-step deposition. One-step deposition, first, is to dissolve raw materials into solvents, and further, spin-coating perovskite precursor on the substrate. Taking organic metal halide perovskite $MAPbI_3$ for instance [1], dimethylformamide(DMF) is commonly used as a solvent. , an anti-solvent often applied to the perovskite thin film. In order to form uniform crystals, chlorobenzene(CB) is chosen to be the anti-solvent for its property of low solubility, with the participation of anti-solvent, perovskite can be uniformly crystallized on the thin film. Hence, one-step deposition has the advantages of low cost, low fabrication time and simple process, but poor coverage. In comparison, in two-step deposition, perovskite is synthesized after reaction with better film coverage. Take $MAPbI_3$ as an illustration, the reactants MAI and PbI_2 are respectively dissolved in the solvent, and then spin-coated on the substrate. Co-evaporation is another way in which MAI and PbI_2 are evaporated simultaneously in a thermal evaporator under high vacuum. Although it has better film coverage and uniformity, time-consuming and high vacuum equipment cost make it difficult to achieve mass production.

Unlike the current perovskite film forming methods, we fabricate perovskite via SET process with our own designed chamber. It costs less time to achieve operating pressure. Besides, it can fabricate large perovskite grain size, compared to that in traditional solution methods. Through the SET process, double interdiffusion enables crystallization to be more complete, which improves the photovoltaic performance of perovskite solar cells. This concept has been proposed in our laboratory and achieves excellent performance. The crystalline film of $MAPbI_3$ can be produced within 30 min, combined with evaporation pressure control, which can further increase the efficiency of perovskite solar cells.

2 EXPERIMENTS

The ITO glass substrates were cleaned by sequential ultrasonic treatment in acetone, isopropanol and methanol for 15 min. Then, the pre-clean ITO glass was spin-coated with PEDOT:PSS. The PEDOT:PSS-coated substrates were annealed at 120 °C for 10 min. A solution of CH_3NH_3I (MAI) and dimethylformamide (DMF) is deposited on the hole transport layer (HTL) by a spin coating process. After annealing, PbI_2 was evaporated on the MAI layer by thermal evaporation. After SET process, IPA was used to clean the remaining MAI to get a smooth surface and dark brown perovskite layer. The $PC_{60}BM$ (Phenyl-C_{60}-methyl butyrate) as the electron transport layer (ETL) was spin-coated on the perovskite layer. Finally, bathocuproine (BCP) and Ag electrodes were deposited by thermal evaporation.

3 RESULTS AND DISCUSSION

Fig. 1. is the appearance of perovskite films under different evaporation pressure and time. In Fig. 1a. PbI_2 is coated on the MAI layer by thermal evaporation with film color yellow. And then, in the SET chamber, MAI is evaporated on the substrates with PbI_2 at 130 °C to form perovskite thin film. In Fig. 1b. Under 0.08 torr, with the increasing evaporation time, the color of film becomes deeper, which represents the reaction between MAI and PbI_2 is more complete. However, compared to Fig. 1c. under 2 torr, 30 min, the color of perovskite thin film is much deeper than that of any conditions under 0.08 torr. Hence, the proper evaporation pressure plays the important role in the formation of perovskite crystals.

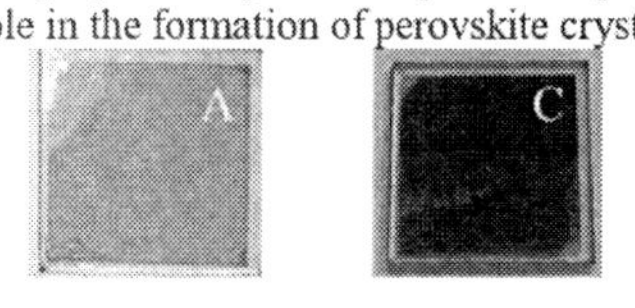

Fig. 1. Appearance of (a)PbI_2 film and perovskite thin films at 130 °C under (b)0.08 torr, 35,45,55,65 min (c)2 torr, 30 min

In Fig. 2. perovskite crystalizes on the surface of PbI_2 film, and the crystals become more obvious when the evaporation time increases. Fig. 2a. is the surface scanning electron microscope (SEM) images of perovskite thin films, for 65 min, the perovskite layer has clearer and more stereoscopic crystals than for 35 min.

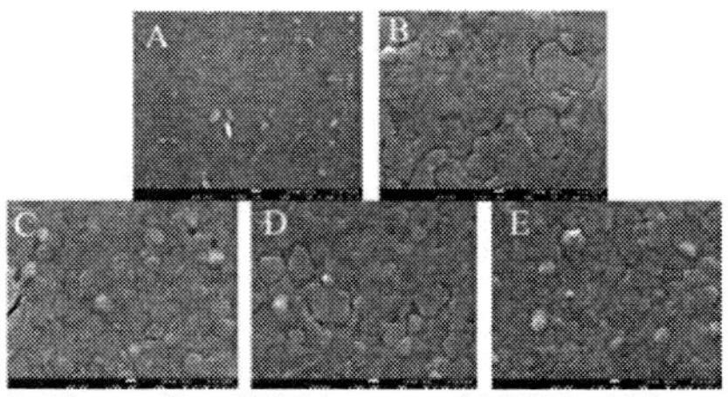

Fig. 2. The surface SEM images of (a)PbI_2, 185nm, after SET process, perovskite thin film is fabricated under 0.08 torr for (b)35min(c)45min(d)55min(e)65min

Fig. 3a. shows the cross-section SEM images of PbI_2 after thermal evaporation, and then MAI is evaporated on PbI_2 surface to form perovskite film via SET process under 0.08 torr. From the Fig. 3c, the thickness of perovskite film achieves 431 nm for the evaporation time 45 min. However, the crystal of perovskite becomes a nonclassical club structure when the evaporation time is more than 55 min as in Fig. 3d, which is very different from Fig. 4, under 2 torr for 30 min. With standard grain structure, the club structure is caused by the excessive MAI. Consequently, we find that the process of the double-interdiffusion via SET process under 0.08 torr is not proper for perovskite crystallization. Besides, perovskite crystal becomes a club structure but not the classical grain structure.

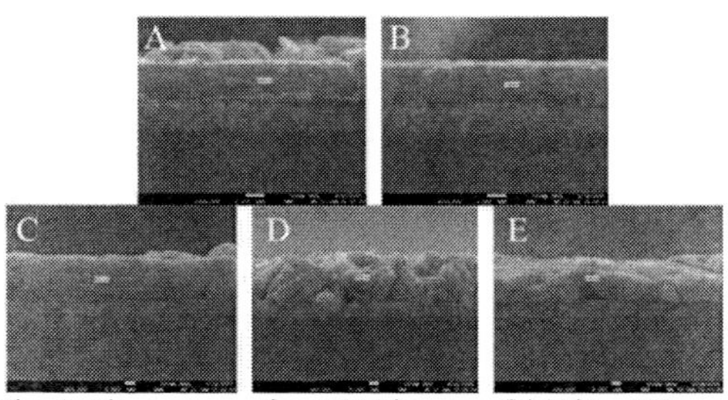

Fig. 3. The cross-section SEM images of (a)PbI_2, 185nm, after SET process, perovskite thin film is fabricated under 0.08 torr for (b)35min(c)45min(d)55min(e)65min

Fig. 4. shows the SEM images of perovskite thin films via SET process under 3 torr for 30 min. In Fig. 4a, the surface SEM images, obvious grains are formed on the perovskite film under 3 torr, and it takes less evaporation time for 30 min, whereas the formation of perovskite crystal is much better than under 0.08 torr. From the Fig. 4b, cross-section SEM images, we can find that the thickness of film is 323 nm under 3 torr for 30 min, thicker and denser, compared to the perovskite film under 0.08 torr for 35 min, showing that the perovskite crystallization is much better under 3 torr than 0.08 torr.

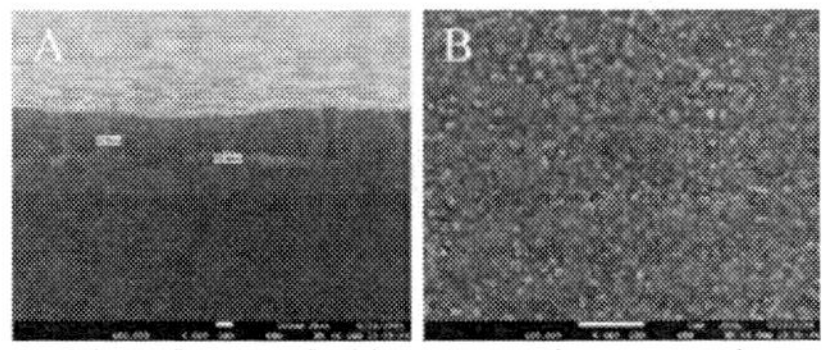

Fig. 4. (a)Surface and (b)cross-section SEM images of perovskite thin films under 3 torr for 30 min

Fig. 5. shows the XRD patterns of perovskite thin films at 130 °C for 35 min under different evaporation pressure. When the SET evaporation pressure is up to 2 torr from 0.2 torr, PbI_2 diffraction peak intensity dramatically decreases. Nevertheless, $MAPbI_3$ diffraction peak intensity changes slightly, which means that the change in pressure helps MAI react with PbI_2, crystal lattice of PbI_2 is broken and combined with MAI to form $MAPbI_3$ crystal. Besides, the evaporation time is proper for 35 min to avoid destroying $MAPbI_3$ crystals by excessive MAI.

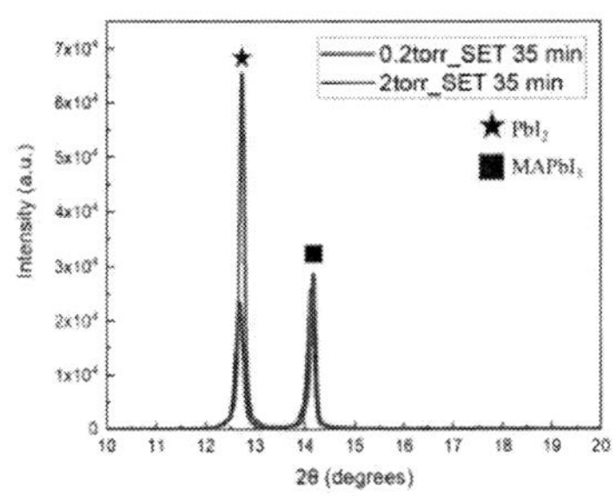

Fig. 5. XRD patterns of perovskite thin films at 130 °C 35 min under different evaporation pressure.

Fig. 6. shows the absorption spectrum of perovskite thin films at 130 °C for 35 min respectively under 0.2 torr and 2 torr. When the SET evaporation pressure is 2 torr, compared to 0.2 torr, the absorption spectrum extremely rises at the range of 500 nm to 700 nm. At 625 nm, the absorption rate rises from the lowest 45% to 60%, which shows the change in evaporation pressure could greatly enhance the absorption rate at visible spectrum.

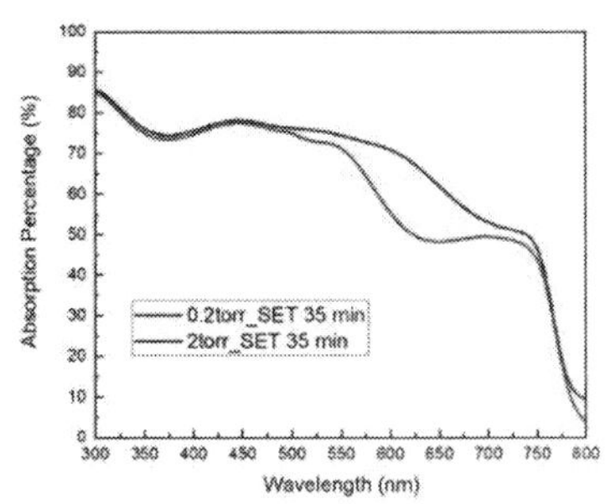

Fig. 6. Absorption spectrum of perovskite thin films at 130 °C for different SET evaporation pressure.

4 CONCLUSION

The SET process under 0.08 torr is not proper for PbI_2 to react with MAI, the formation of perovskite crystals is still dissatisfactory even as the evaporation time increases. However, evaporation pressure is raised properly to 2 torr and could help break PbI_2 crystals to react with MAI to form $MAPbI_3$, which accelerates the formation of $MAPbI_3$ crystals. Moreover, raising evaporation pressure not only fabricates denser perovskite film, but also enhances the absorption rate at visible spectrum. Perovskite solar cells with higher photovoltaic conversion efficiency are expected.

5 ACKNOWLEDGMENT

The authors gratefully acknowledge Ministry of Science and Technology in Taiwan for supporting the research with the contract numbers: MOST 108-3116-F-002-009-CC2, MOST 108-3116-F-002-010-CC2, MOST 107-2221-E-002-172-MY3. Besides, we would like to thank Ms. Yuan-Tzu Lee of the Instrumentation Center at National Taiwan University for their assistance with the SEM operations.

6 REFERENCES

[1] Kojima, A., Teshima, K., Shirai, Y., & Miyasaka, T. , "Organometal halide perovskites as visible-light sensitizers for photovoltaic cells, " Journal of the American Chemical Society, 131(17), 6050–6051 (2009).

[2] Yang, D., Yang, R., Priya, S., & Liu, S. F. , "Recent Advances in Flexible Perovskite Solar Cells: Fabrication and Applications, " Angewandte Chemie (International ed. in English), 58(14), 4466–4483 (2019).

[3] Guo, X., McCleese, C., Kolodziej, C., Samia, A. C., Zhao, Y., & Burda, C. , "Identification and characterization of the intermediate phase in hybrid organic-inorganic $MAPbI_3$ perovskite, " Dalton transactions (Cambridge, England : 2003), 45(9), 3806–3813 (2016).

TESTING ENCAPSULATED PEROVSKITE SOLAR CELLS IN A CLIMATIC CHAMBER BY FOLLOWING THE IEC 61215 AND IEC 61646 STANDARDS FOR THE THERMAL CYCLING TEST

L. Ocaña[1,3], C. Montes[1,3], S. González-Pérez[2], B. González-Díaz[3], E. Llarena[1].
[1]Instituto Tecnológico y de Energías Renovables, S. A. (ITER)
Pol. Industrial de Granadilla, s/n.
E 38600 Granadilla de Abona, Spain.
Ph. +34 922 747 700.
Fax +34 922 747 701.
E-mail lmgonzalez@iter.es
[2]Departamento de Didácticas Específicas.
C/ Pedro Zerolo, s/n
Edificio Central Planta 2
Apartado 456. Código postal 38200
San Cristóbal de La Laguna. S/C de Tenerife. Spain.
[3]Departamento de Ingeniería Industrial.
Universidad de La Laguna.
Camino San Francisco de Paula, s/n
38206 S/C de Tenerife, Spain

ABSTRACT: This document summarizes the work carried out to evaluate the performance of perovskite solar cells, based on $MAPbI_3$ absorbers, with ethyl vinyl acetate (EVA) encapsulations at low temperature, submitted to thermal cycling test established in the standards IEC-61215 for crystalline silicon PV modules and IEC-61646 for thin film PV modules. Three types of samples were studied, those composed of MAPbI3, those composed of MAPbI3 deposited on a compact layer of TTIP, and those composed of MAPbI3, but deposited so that the layer dons not reach the edge of the substrate (aka with setback). The degradation of the samples were characterized by imaging microscope, optical ellipsometry and the electrical performance has been simulated by the SCAPS-1D software. The results indicate the EVA encapsulated samples remained stable after the 200 thermal cycles, keeping the setback samples a 14.84% PCE compared to the 15.48% PCE of the pristine sample.
Keywords: Perovskite, degradation, absorption, encapsulated, Ethylene-vinyl acetate.

1 INTRODUCTION

Nowadays perovskite solar cells (PSCs) have achieved high efficiency values, obtaining 25.5% for perovskite cells [1], 29.5% for perovskite / Si tandem [1] and 17.9% for perovskite modules [2]. However, the stability under atmospheric agents needs to be improved in order to compete with other solar cells technologies in the sector, such as Silicon, Amorphous silicon, GaAs III-V or Chalcogenide. The best values reported concerning the perovskite stability to date are above 1000 hours [3]–[5], highlighting the results of Jang Y. et al, where the encapsulated device retained 94% of its initial efficiency after 1,056 h under the damp heat test (85 °C/85% relative humidity) and 98% after 1,620 h under full-sun illumination [6]. These stability values are intrinsically related with the composition of the device, the encapsulation method and the activating agent of the degradation process [7].

This work summarizes the actual work to enlarge the stability of perovskite solar cells, based on CH3NH3PbI3 absorbers, with ethyl vinyl acetate (EVA) encapsulation at low temperature [8], submitted to thermal cycling test established in the standards IEC-61215 for crystalline silicon PV modules and IEC-61646 for thin film PV modules. The performance of the fabricated perovskites solar cells was analyzed before and after being treated in a climatic chamber, in order to verify that these devices pass the standard conditions. The optical and chemical properties are being characterized. Also, their electrical parameters such as efficiency (η), filling factor (FF), short-circuit intensity (Isc), open-circuit voltage (Voc) and maximum power (Pmax), are studied in order to evaluate the quality of the encapsulation method, by the SCAPS-1D simulation software, developed by the Department of Electronics and Information Systems (ELIS) of the University of Ghent [9]–[15].

2 EXPERIMENTAL

In order to carry out the test to analyze the stability of the perovskite cells submitted to atmospheric agents, three batches of perovskite solar cells were fabricated by spin coating techniques inside an ISO 7 cleanroom in ambient conditions [16]. In the first batch (named as 1-Pk, 2-Pk and 2-Pk), the samples were composed of a layer of perovskite, in order to evaluate the behavior of the perovskite layer against the test. In the second batch (named as 1-TTIP-Pk, 2-TTIP-Pk and 2-TTIP-Pk), the samples were composed of a perovskite thin film deposited over a TiO_2 compact layer formed from Titanium(IV) isopropoxide (TTIP), in order to evaluate the interaction of the layers during the test. In the third and last batch (named as 1-Pk-R, 2-Pk-R and 2-Pk-R), the samples were composed of a layer of perovskite but, this time, deposited so that the layer do not reach the edge of the substrate, but was setback about two millimeters. This was made in order to evaluate if, by doing so, there was any improvement in the performance of the encapsulation.

Subsequently, the solar cells were encapsulated with EVA at a low temperatures [8]. Then, the encapsulated cells were taken to a climatic chamber [17]. Inside the chamber, the effect of atmospheric agents can be recreated by means of automatic temperature control and recirculation of air, with cycling temperatures ranging between -40ºC and 85ºC, while minimizing condensation. Therefore, this kind of equipment guarantee test compliance with the standards established in the IEC-61215 and IEC-61646.

The purpose of the thermal cycling test is to determine the ability of the device to resist thermal imbalances, fatigue or other stresses caused by repeated changes in temperature. The testing process consists of installing the samples inside the chamber at room temperature. Subsequently, the samples are exposed to 200 thermal cycles, where the temperature of the sample is varied from -40ºC±2ºC to +85ºC±2ºC . The temperature rate between the low and high extremes must not exceed 100ºC/h and the temperature of the samples must remain stable for a time of at least 10 min and the duration of the cycle will never exceed 6 hours (see Figure 1).

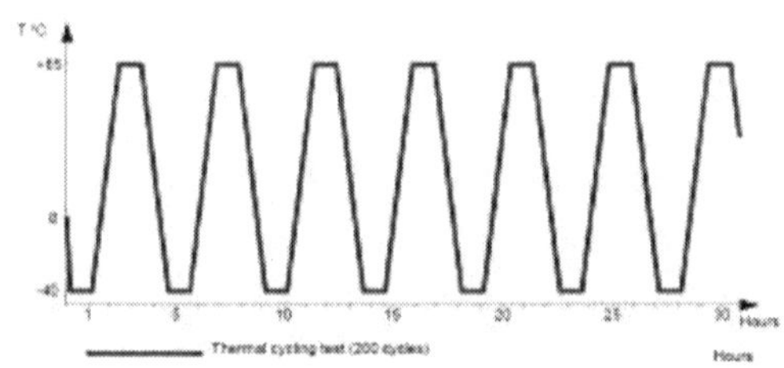

Figure 1. - Thermal cycling test following the standard IEC-61215 for crystalline silicon PV modules and IEC-61646 for thin film PV modules, inside the climatic chamber.

3 CHARACTERIZATION

The characterization process was divided in two phases: On the first one, a set of measurements were performed on the samples before going the climatic chamber, before beginning with the tests. On the second phase, the same set of measurements were performed on the samples at the end of each trial. This approach was aimed to not only assess whether the samples meet the aforementioned standards, but also in order to try to elucidate what caused the degradations that may have happened (if any).

The referred set of measures consists on obtaining macroscopic images by digital microscope [18] in order to observe changes in the appearance of the studied samples, determining the absorbance, degradation percentage and the band gap in the samples by using a spectroscope ellipsometer [19] and obtaining their electrical parameters (η, FF, J_{sc}, V_{oc} and Pmax) by using the SCAPS-1D simulation software [20].

4 RESULTS AND DISCUSSION

4.1 Digital microscope.

Figure 2 depicts images of the samples after the thermal cycling test. The first row corresponds to the samples composed of an encapsulated perovskite layer without setback, the second row shows the samples composed of an encapsulated perovskite thin film deposited on a compact layer formed by TTIP without setback and the third row composed of an encapsulated perovskite layer with setback. In all cases, after 200 thermal cycles, the samples maintained their mirrored brown color and did not show macroscopic defects. Which seems to indicate that the samples remained stable after the test.

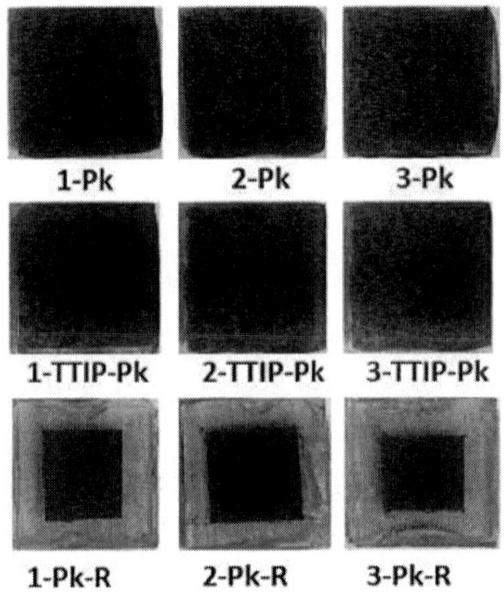

Figure 2. - Images of the samples after de thermal cycling test inside the climatic chamber.

4.2 Spectroscope ellipsometer.

In order to evaluate the effect that the EVA encapsulation had on the perovskite layers during the thermal cycling test, the absorbance curves of the samples were measured before and after their encapsulation.

As it can be seen in Figure 3, the overall shape obtained for both curves was rather similar so, in order to identify where they diverged the most, a percentage reduction of the absorbance was calculated. Thus, the degradation of the samples after the thermal cycling test can be divided into three ranges of the spectrum. At the first range, between 200 nm and 400 nm, the samples remain stable, the reduction in absorbance being very small. The second range, between 400 nm and 600 nm, the reduction in absorbance was more pronounced, varying between 10% and 30%, which some authors associate with the formation of multiiodide plumbate species such as $PbI_5S_3^-$, PbI_6^{4-} and $(Pb_2I_4)_n$ [21]. The third range, from the bandgap value (1.6 eV) on, the samples increase their degradation, generating a lot of noise in the results. This deviation above 750 nm can be addressed as merely scattering effects [21], [22].

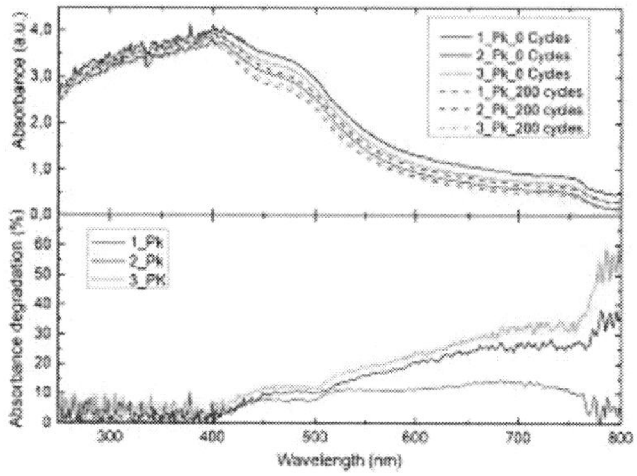

Figure 3. - Above: Absorbance curves of an encapsulated pristine MAPbI3 layer without setback, before thermal cycling test (continuous line) and absorbance curves of an encapsulated pristine $MAPbI_3$ layer, without setback, after thermal cycling test (dash line). Below: Percentage reduction of the absorbance.

Figure 4 depicts the absorbance curves of an encapsulated pristine $MAPbI_3$ layer deposited on a compact layer formed by TTIP, without setback, before and after thermal cycling test. As it can be seen in Figure 4, the behavior of the curves was very similar to that of the samples composed only of perovskite without setback, except between 400 nm and 500 nm, where for two of them, there was an increase in the reduction of absorbance. This seems to indicate that formation of multiiodide plumbate species such as PbI_5S_3-, $PbI6_4$- and (Pb_2I_4)n [21] was higher for these cases.

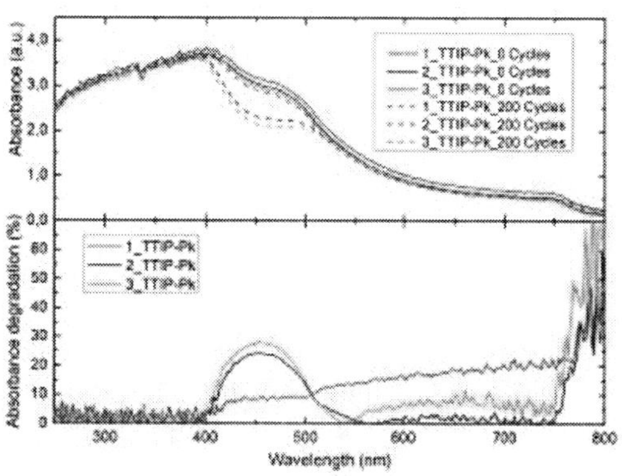

Figure 4. - Above: Absorbance curves of an encapsulated $MAPbI_3$ layer deposited on a compact layer formed by TTIP, without setback, before thermal cycling test (continuous line) and absorbance curves of an encapsulated $MAPbI_3$ layer deposited on a compact layer formed by TTIP, without setback, after thermal cycling test (dash line). Below: Percentage reduction of the absorbance.

Figure 5 depicts the absorbance curves of an encapsulated pristine $MAPbI_3$ layer, with setback, before and after thermal cycling test. As it can be seen in Figure 5, at the first range, the reduction in absorbance was negligible. At the second range, and as in the previous cases, there was an absorbance decrease, but in this time it was lower, around 10%-20% and more uniform. At the third range, as in the rest of the cases, from 760 nm the reduction in absorbance has a notable increase.

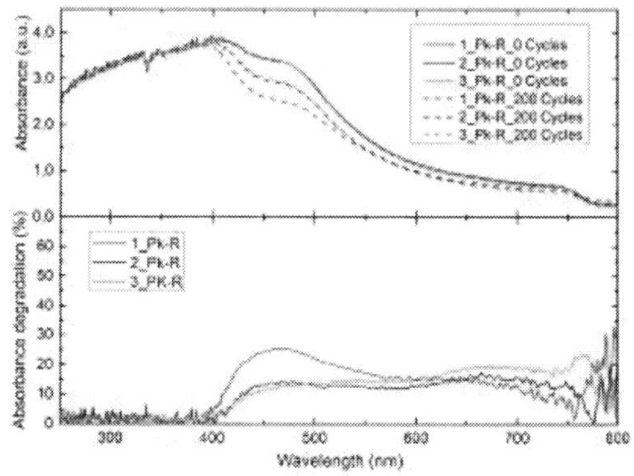

Figure 5. - Above: Absorbance curves of an encapsulated pristine $MAPbI_3$ layer, with setback, before thermal cycling test (continuous line) and absorbance curves of an encapsulated pristine MAPbI3 layer, with setback, after thermal cycling test (dash line). Below: Percentage reduction of the absorbance.

4.2 SCAPS-1D simulation software

In order to evaluate the effect of the degradation of the perovskite layer on the device, the SCAPS-1D simulation software, developed by the Department of Electronics and Information Systems (ELIS) of the University of Ghent, Belgium, has been used (because of being widely reported as a tool for the study of PSCs [9]–[15]). The thickness, the band gap and the curve of the absorbance of the device layers allowed us to extrapolate the degradation of the samples after the thermal cycling test radiation to the SCAPS 1-D software, determining their influence on the short-circuit current (J_{sc}), the open circuit voltage (V_{oc}), the fill factor (FF) and the energy efficiency (PCE).

Figure 6 compares the curve IV of the pristine sample with the best samples composed of a layer of perovskite without setback, a layer of perovskite deposited on TTIP without setback and a layer of perovskite with setback after undergoing the thermal cycling test. As it can be seen both in figure 6 and in table 1, the sample with set-back presents better electrical parameters compared to those of the samples without set-back. Maintaining the power conversion efficiency after the test of 16.24% compared to 16.90% of the original sample

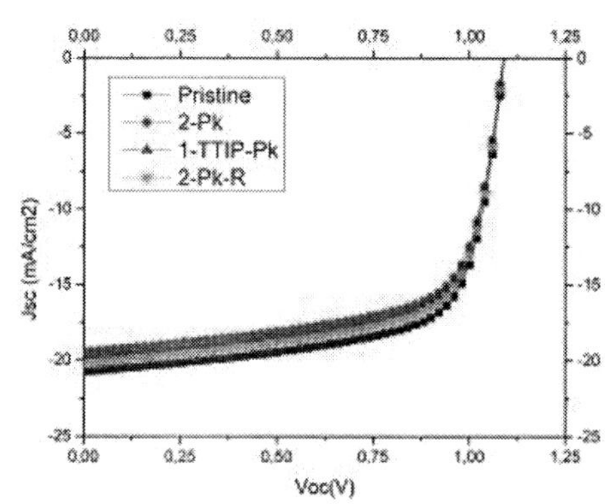

Figure 6. - Comparative of the Curve IV between pristine device and the devices composed by $MAPbI_3$ layer without setback, $MAPbI_3$ layer deposited on a compact layer formed by TTIP, without setback and $MAPbI_3$ layer with setback.

Table I: Electrical parameters

Parameters	Pristine	2-Pk	1-TTIP-Pk	2-Pk-R
V_{oc} (V)	1,09	1,08	1,08	1,08
J_{sc} (mA/cm^2)	20,75	19,45	19,72	20,20
FF (%)	68,33	67,43	67,56	67,39
PCE (%)	15,482	14,27	14,51	14,84
V_{mp} (V)	0,916	0,91	0,91	0,91
J_{mp} (mA/cm2)	16,90	15,63	15,88	16,24

4 CONCLUSIONS

In communications presented by our laboratory on this line of work prior to the present Congress, the chosen encapsulation method allowed perovskite layers to remain stable after 25 thermal cycles, between 15ºC and 85ºC [16], 6 thermal cycles between 23ºC and 150ºC [16] and 350 hours under full solar radiation [8]. We are now in the process to assess what effects a comprehensive set of tests concerning variations in temperature, humidity-freezing and damp heat have on EVA encapsulated perovskite solar cells.

The results of this work reveal that the developed encapsulation method, which was performed at low temperatures using commercial EVA, has passed the thermal cycling test, showing a similar behavior for the three kinds of samples. It has been observed that the absorbance reduction was minimal for the spectrum range between 200 nm and 400 nm, which indicates that the formation of $PbI_3S_3^-$ does not occur [21]. At the range from 400 nm to 600 nm, an absorbance decrease has been observed, being more noticeable for non-setback samples, reaching 30% absorbance decrease, while for the setback samples was between 10% and 20% and more uniform. This degradation may be due to the formation of multiiodide plumbate species such as PbI_5S_3-, $PbI6_4$- and $(Pb_2I_4)n$ [21], indicating that it was produced by the effect of temperature. Regarding the last range of the spectrum, from 760 nm, passed the band gap, the degradation registered was much higher.

The data obtained by the spectroscope ellipsometer (absorbance, band gap and thickness) has been used to simulated the device with the SCAPS-1D software, verifying that the electrical parameters, such as efficiency, fill factor, short circuit intensity and open circuit voltage, remain fairly stable after the thermal cycling test, keeping the setback samples a 14.84% PCE compared to the 15.48% PCE of the pristine sample.

Finally, in the next few months this encapsulation method is expected to be submitted to humidity-freezing test and damp heat test, established in the standards IEC-61215 for crystalline silicon PV modules and IEC-61646 for thin film PV modules.

5 ACKNOWLEDGEMENTS

This work was developed within the MACLAB-PV project framework, which has been co-financed by the INTERREG Madeira-Azores-Canarias Territorial Cooperation Programme (MAC) 2014-2020. 2nd Call. Axis 1. Enhancing research, technological development and innovation.

6 REFERENCES

[1] (NREL) Laboratory National Renewable Energy, "Best Research-Cell Efficiency Chart," *National Renewable Energy Laboratory*, 2020. [Online]. Available: https://www.nrel.gov/pv/cell-efficiency.html.

[2] (NREL) National Renewable Energy Laboratory, "Champion Photovoltaic Module Efficiency Chart," *Natl. Renew. Energy Lab.*, 2020.

[3] Y. S. Kwon, J. Lim, H. J. Yun, Y. H. Kim, and T. Park, "A diketopyrrolopyrrole-containing hole transporting conjugated polymer for use in efficient stable organic-inorganic hybrid solar cells based on a perovskite," in *Energy and Environmental Science*, 2014, vol. 7, no. 4, pp. 1454–1460, doi: 10.1039/c3ee44174a.

[4] T. J. Wilderspin, F. De Rossi, and T. M. Watson, "A simple method to evaluate the effectiveness of encapsulation materials for perovskite solar cells," *Sol. Energy*, vol. 139, pp. 426–432, 2016, doi: 10.1016/j.solener.2016.09.038.

[5] R. Cheacharoen *et al.*, "Encapsulating perovskite solar cells to withstand damp heat and thermal cycling," *Sustain. Energy Fuels*, vol. 2, no. 11, pp. 2398–2406, 2018, doi: 10.1039/c8se00250a.

[6] Y.-W. Jang *et al.*, "Intact 2D/3D halide junction perovskite solar cells via solid-phase in-plane growth," *Nat. Energy*, vol. 6, no. 1, Jan. 2021, doi: 10.1038/s41560-020-00749-7.

[7] C. C. Boyd, R. Cheacharoen, T. Leijtens, and M. D. McGehee, "Understanding Degradation Mechanisms and Improving Stability of Perovskite Photovoltaics," *Chemical Reviews*, vol. 119, no. 5. American Chemical Society, pp. 3418–3451, 13-Mar-2019, doi: 10.1021/acs.chemrev.8b00336.

[8] L. Ocaña, C. Montes, S. González-Pérez, B. González-Díaz, M. Friend, and & M. Cendagorta, "Improvements on the stability of MAPbI3 thin films against UV radiation by a low temperature encapsulation method that uses commercial ethylene-vinyl acetate sheets," in *37th European Photovoltaic Solar Energy Conference and Exhibition*, 2020, doi: 10.4229/EUPVSEC20202020-3BV.1.22.

[9] M. S. Chowdhury *et al.*, "Effect of deep-level defect density of the absorber layer and n/i interface in perovskite solar cells by SCAPS-1D," *Results in Physics*, vol. 16. Elsevier B.V., 01-Mar-2020, doi: 10.1016/j.rinp.2019.102839.

[10] A. Slami and A. B. Belkaid, "Numerical Study of Based Perovskite Solar Cells by SCAPS-1D."

[11] U. Mandadapu, S. V. Vedanayakam, and K. Thyagarajan, "Simulation and Analysis of Lead based Perovskite Solar Cell using SCAPS-1D," *Indian J. Sci. Technol.*, vol. 10, no. 11, pp. 1–8, Mar. 2017, doi: 10.17485/ijst/2017/v10i11/110721.

[12] Institute of Electrical and Electronics Engineers, *2018 International Conference on Advancement in Electrical and Electronic Engineering (ICAEEE)*. .

[13] L. Lin, L. Jiang, P. Li, B. Fan, Y. Qiu, and F. Yan, "Simulation of optimum band structure of HTM-free perovskite solar cells based on ZnO electron transporting layer," *Mater. Sci. Semicond. Process.*, vol. 90, pp. 1–6, Feb. 2019, doi: 10.1016/j.mssp.2018.10.003.

[14] S. Bansal and P. Aryal, "Evaluation of new materials for electron and hole transport layers in perovskite-based solar cells through SCAPS-1D simulations," in *Conference Record of the IEEE Photovoltaic Specialists Conference*, 2016, vol. 2016-November, pp. 747–750, doi: 10.1109/PVSC.2016.7749702.
[15] K. Chakraborty, M. G. Choudhury, and S. Paul, "Numerical study of Cs2TiX6 (X = Br−, I−, F− and Cl−) based perovskite solar cell using SCAPS-1D device simulation," *Sol. Energy*, vol. 194, pp. 886–892, Dec. 2019, doi: 10.1016/j.solener.2019.11.005.
[16] L. Ocaña *et al.*, "On the stability of planar CH3NH3PbI3 perovskite solar cells produced in ambient conditions by using polymer encapsulates," in *36th European Photovoltaic Solar Energy Conference and Exhibition*, 2019, pp. 721–727, doi: 10.4229/EUPVSEC20192019-3BV.2.10.
[17] "CTC 256 MEMMERT, Climatic Chamber, DD BIOLAB, S.L., Manufactured by MEMMERT GmbH + Co. KG, Schwabach, Germany." .
[18] "Digital Microscope U500X, focus range 15mm - 40mm, range 50X-500X by Unimake." .
[19] "J. A. Woollam Co., Inc. M-2000 spectroscopic ellipsometer." .
[20] "SCAPS-1D simulation software, developed by the Department of Electronics and Information Systems (ELIS) of the University of Ghent has been used, Belgium."
[21] C. Aranda, C. Cristobal, L. Shooshtari, C. Li, S. Huettner, and A. Guerrero, "Formation criteria of high efficiency perovskite solar cells under ambient conditions," *Sustain. Energy Fuels*, vol. 1, no. 3, pp. 540–547, 2017, doi: 10.1039/c6se00077k.
[22] Y. Tian and I. G. Scheblykin, "Artifacts in absorption measurements of organometal halide perovskite materials: What are the real spectra?," *Journal of Physical Chemistry Letters*, vol. 6, no. 13. American Chemical Society, pp. 3466–3470, 01-Aug-2015, doi: 10.1021/acs.jpclett.5b01406.

STUDY OF ALD-GROWN SnO_2 AS AN ELECTRON SELECTIVE LAYER FOR NIP PEROVSKITE-BASED SOLAR CELLS

Félix Gayot[1], Elise Bruhat[1], Matthieu Manceau[1], Eric De Vito[2], Stéphane Cros[1]
[1]Univ. Grenoble Alpes, INES, F-73375 Le Bourget du Lac, France - CEA Liten, Department of Solar Technologies, F-73375 Le Bourget du Lac
[2]Univ. Grenoble Alpes, CEA Liten, Departement of Nanomaterial technologies, 38054 Grenoble Cedex, France
felix.gayot@cea.fr

ABSTRACT: Numerous challenges remain unresolved to achieve photovoltaic conversion efficiency above 30% for mainstream application by utilization of PK/Si tandem solar cells. For a 2-Terminal (2T) tandem configuration, it is necessary to adapt the perovskite (PK) top-cell to a crystalline silicon (c-Si) bottom-cell, in terms of size and fabrication processes. One adjustment concerns the ESL deposition. Solvent-based processes mainly employed in single-junction PK cells are not appropriate for the top-cell integration in the tandem case, where the c-Si bottom-cell is a large and textured substrate. Among other possibilities, Atomic Layer Deposition (ALD) seems more suitable for such ESL deposition. To this day, few studies report the use of ALD-grown tin oxide (SnO_2) as an ESL in NIP PK top-cell for 2T tandems. Aiming at a better understanding on how ESL-PK interaction relates to device performances, the present work investigates SnO_2 thin films, then PK films properties and, finally, analyses integration of ALD-grown SnO_2 as ESL in single-junction PK cells.

Keywords: Perovskite, Tandem, ALD, Electron Transport Layer

1 INTRODUCTION

The reduction of thermalization losses enabled by stacking a wide bandgap PK top-cell (~1.65 eV) on a Si bottom-cell (1.12eV) allows PK/Si tandem solar cells to show outstanding theoretical efficiency limits above 35% (*1, 2*), thereby overcoming the standard Si c-Si solar cells efficiency limit of 30% (*3*). Today, PK/Si tandem cells have already reached 29.5% of power conversion efficiency at lab scale (*4*). If c-Si solar cells are the industrial standard that dominates the global market (*5*), PK-based PV on the other hand is less mature, despite an exceptionally quick increase in efficiencies up to 25% at the lab scale over the last decade (*6*). Several technical issues originate from this maturity level difference, as for instance a lack of industrial technique to deposit PK top-cell on large areas c-Si bottom cells.

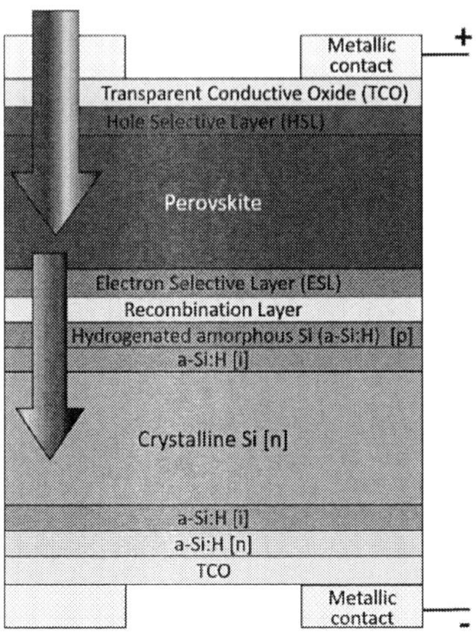

Figure 1 *- Schematic of a 2-T PK/Si tandem solar cell with a PK top-cell in NIP configuration on top of a SHJ bottom-cell.*

This study is part of the development of a monolithic 2T tandem cell, where the PK top-cell is integrated in NIP configuration on top of a c-Si heterojunction (SHJ) bottom-cell (Fig. 1). According to this architecture, the ESL is the first layer of the top-cell being deposited on the recombination layer, which makes it crucial for sub-cells interaction. First, it must allow electrons transfer from the PK layer to the recombination layer while blocking the photogenerated holes. Besides, the ESL transparency over the 750-1100 nm range is essential, as it corresponds to the fraction of light that needs to be converted in the c-Si wafer. Furthermore, the PK must be able to nucleate and grow as a homogeneous film on top of it.

SnO_2 is one of the various inorganic materials matching these requirements and is now commonly used as an ESL in PK devices (*7*). SnO_2 layers are commonly deposited by spin-coating (*8, 9*), a solution process which gives poorly conformal and homogeneous thin films on top of a large and textured substrate such as a SHJ bottom-cell (*10*). Various vacuum-based deposition techniques can be considered, provided they are non-aggressive with respect to the SHJ bottom-cell (*11*). Among them, Atomic Layer Deposition (ALD) seems particularly attractive for ESL deposition (*12*). Thanks to its cyclic nature and the segregation of precursors injections, it allows ultra-conformal, dense and non- aggressive thin film deposition over large areas with fine control over thicknesses (*13*). ALD specific properties have therefore already given rise to effective PK single junction or PK/Si tandem fabrication (*14–17*).

Yet, these studies are few, especially for tandems where SnO_2 ESL are mostly used in PIN configuration. Hence, the direct interaction between ALD-grown SnO_2 (SnO_2^{ALD}) and the overlying PK and how it is related to device performances remains relatively unexplored. Besides, ESL made of SnO_2^{ALD} seem to suffer from

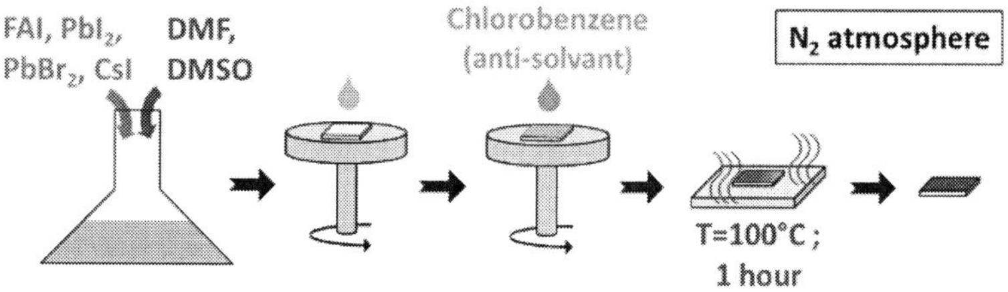

Figure 2 - PK film deposition process

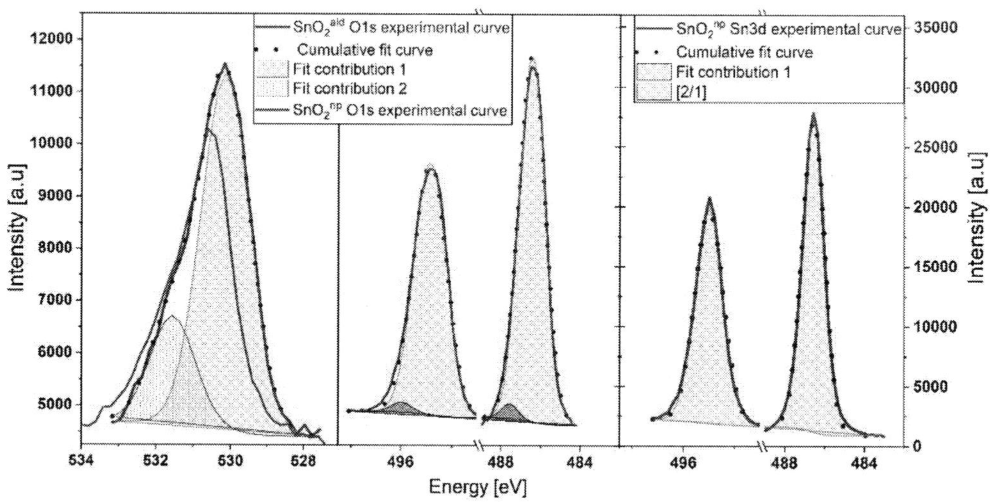

Figure 3 - XPS spectrum of SnO_2^{ALD} and SnO_2^{np} in O1s region (525-535eV) (a.) and XPS spectrum for Sn3d doublet region (480-505eV) of SnO_2^{ALD} (b) and of SnO_2^{np} (c). Fitting curves are shown in grey.

interfacial defects with the PK and would need the adjunction of another material layer to passivate such defects (*14*, *18*). Still, reported performances remain lower than for cells using solution-processed SnO_2 (*19*, *20*). Aiming at a better understanding about SnO_2^{ALD} limitations when employed as single-layer ESL, this work presents a systematic comparison between state-of-the-art SnO_2 nanoparticles thin films (SnO_2^{np}) and SnO_2^{ALD}. Chemical, electrical, optical and topographical properties of each type of film were investigated as well as their integration in PK single junction solar cells.

2 EXPERIMENTAL DETAILS

SnO_2^{ALD} layers are grown Glass ITO substrates at 150°C from Tetrakisdimethylamino-tin(IV) (TDMASn) and hydrogen peroxide (H_2O_2) as tin precursor and oxidizing reactant, respectively. Reference SnO_2 ESL (SnO_2^{np}) were prepared by spin-coating of a SnO_2 water-based nanoparticles dispersion on UV-O_3 treated substrates. Spin-coated samples undergo annealing at 80°C for 1 minute.

PK films are solution-processed under N_2 atmosphere (Fig. 2) on top of SnO_2. In the case of SnO_2^{ALD}, substrates undergo a UV-O_3 treatment during 30 minutes before PK coating to improve surface wettability. 150µL of chlorobenzene is dropped 5s before the spinning of the PK precursors solution ends. Afterwards, PK Films are annealed at 100°C for one hour. The Hole Selective Layer (HSL) is a spin-coated organic thin film made of doped Poly(triaryl amine) (PTAA).

SnO_2 surface chemical environment was examined through X-ray Photoelectron Spectroscopy (XPS) with a standard Al-Kα X-ray source. Binding energies are calibrated according to the contamination carbon C1s peak at 285.0eV and the spectral resolution is of 0.1 eV. Resistivity was recorded via 4-point probe measurements on SnO_2 layers deposited on glass substrates. Transmission and reflection measurements recorded with a UV-vis-IR spectrophotometer enabled to extract PK absorption properties according to the Tauc method (*21*). PK films morphology was observed thanks to Scanning Electron Microscopy (SEM), and PK films X-ray diffraction (XRD) was performed with a Co-Kα X-ray source. The presented diffraction patterns were re-expressed as a function of equivalent 2θ angle for a standard Cu-Kα X-ray source.

All characterization methods were performed on at least three samples.

3 RESULTS AND DISCUSSION

3.1 Checking SnO_2 stoichiometry

Charge selectivity is strongly dependent on film stoichiometry (*22*), thus XPS measurements were performed on 30nm-thick SnO_2^{np} and SnO_2^{ALD} thin films deposited on ITO. The O1s related peak observed in Fig. 3, left panel, is made of two contributions for both SnO_2^{ALD}

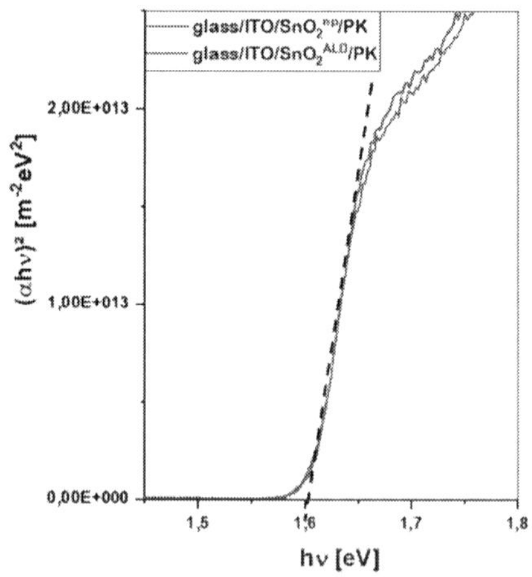

Figure 4 - *Tauc-plots centred on the absorption edge of PK films deposited either on glass/ITO/SnO_2^{np}, either on glass/ITO/SnO_2^{ALD}.*

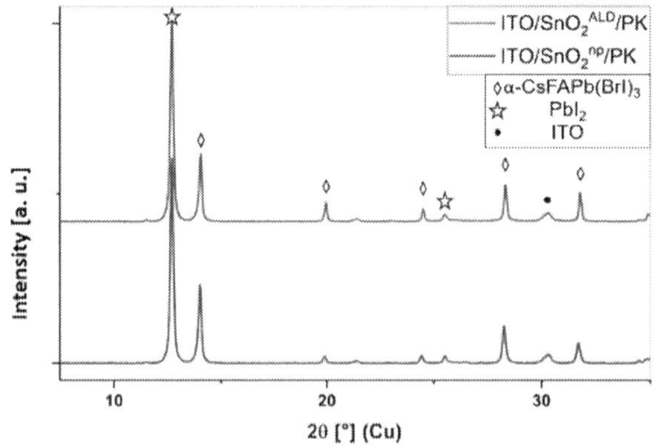

Figure 5 - *XRD patterns of PK films deposited on glass/ITO/ SnO_2^{np} or on glass/ITO/SnO_2^{ALD}.*

and SnO_2^{np}. For the sake of clarity, only the SnO_2^{ALD} fitting curves are shown. Deconvoluted peaks rise at 530.1 eV and 531.7 eV for SnO_2^{ALD} and at 530.4 eV and 531.7 eV for SnO_2^{np}. The low-energy contribution corresponds to O^{2-} for SnO_2 while the high-energy contribution may be caused by O-H groups due to water chemisorption on samples surface (*23*). The shift of about 0.3 eV observed for the main O 1s contribution peak between each SnO_2 type could suggest a more O rich environment in the SnO_2^{np} films. On the opposite, Sn 3d doublet of both studied SnO_2 presents a single main contribution that rises at 486.5±0.1 eV for both SnO_2^{ALD} and SnO_2^{np}. One can however note a small additional contribution at higher energies in the case of SnO_2^{ALD}, but it is difficult to know if it is a fitting articfact or if it comes from few Sn atoms bound to contaminants (ALD precursor residuals for instance). In any case, it appears at too high energy to be related to a small amount of SnO within the film. Indeed, Sn^{4+} related $Sn3d_{5/2}$ peak is reported in the literature in the 486.3-487.5eV range while the Sn^{2+} related $Sn3d_{5/2}$ peak is always reported at about 0.6-1.0eV lower energies (*23–25*). Hence, the presented spectra confirm the SnO_2 stoichiometry of our ALD-grown thin film and that they have a similar chemical environment with SnO_2^{np}.

3.2 Optical and electrical properties of SnO_2^{ALD}

It is essential to check that SnO_2^{ALD} layers have a relatively low resistivity and that they do not prevent transmission of the light that has not been photoabsorbed within the PK. The comparison of electrical resistivity and effective optical transmission mean values of SnO_2^{ALD} and SnO_2^{np} thin films in Table 1 shows that for a very close effective optical transmission to the reference, the SnO_2^{ALD} electrical resistivity is about 3.5×10^4 smaller than the one SnO_2^{np}, for which charge conduction is limited at particles boundaries. This electrical resistivity discrepancy between SnO_2^{ALD} and reference films for a similar optical transmission gives widely extended flexibility on the SnO_2^{ALD} thickness, which may lead to optical gain in devices.

3.3 Perovskite films quality on SnO_2^{ALD}

PK crystallisation depends strongly on the substrate, and PK films could show significant differences whether they grow on SnO_2^{ALD} or SnO_2^{np}. In addition to SnO_2^{ALD} intrinsic properties, another part of this work focuses on analysing the quality of PK films grown onto SnO_2^{ALD} layers.

3.3.a Optical absorption properties

Tauc-plots extracted from UV-vis-IR transmission and reflectivity measurements and presented in Fig. 4 allows one to note that PK films optical bandgap is the same (1.6 eV) when deposited on SnO_2^{ALD} as when deposited on reference SnO_2^{np}. Furthermore, absorption edges seems to be as sharp for both PK films. These observations confirm that PK films grown on SnO_2^{ALD} show an equivalent photoabsorbing behaviour as films grown on SnO_2^{np}.

3.3.b Similar XRD patterns

The evaluation of PK films crystal phases by XRD (Fig. 5) shows diffraction peaks corresponding to pure perovskite as well as peaks corresponding to crystalline PbI_2. For both phases, peaks rise at the same angles for each type of films, and with similar intensity. Therefore, PK films crystallize similarly whether they are deposited on SnO_2^{np} or SnO_2^{ALD}. One can note nevertheless the particularly intense <001> PbI_2 peaks, which supposes a rather large amount of crystalline PbI_2 within the films. It is difficult to conclude on its origin, since PbI_2 is a precursor as well as a degradation product

Table 1 - *Electrical resistivity of 30nm-thick SnO_2^{np} and SnO_2^{ALD} layers deposited on glass and effective optical absorption of 30nm-thick SnO_2^{np} and SnO_2^{ALD} layers deposited on glass/ITO substrates.*

	Electrical resistivity [Ω.m]	**Effective optical absorption [%] (300-1200nm)**
SnO_2^{np}	2.26	10
SnO_2^{ALD}	5.9×10^{-5}	11

of PK. However, in agreement with other work (*4*), it is not believed to have an impact on device performances as we will see later on.

3.3.c SEM morphology

No significant difference as well can be noted between the SEM pictures presented in Fig. 6. Both PK films deposited on SnO_2^{ALD} and SnO_2^{np} show large grains (~100-500nm) as well as pinholes of approximately the same size. Several brighter particles laying in between PK grains are also homogeneously present in both films. These grains might actually correspond to a PbI_2 segregated phase (*26*). Their presence widespread all over the PK films seems consistent with XRD results.

3.4 SnO_2^{ALD} in single-junction PK cells

PK single junction solar cells that integrate 30nm-thick layers of SnO_2^{ALD} as ESL were fabricated. The current-voltage characteristics under illumination shown in Fig. 7 outline how replacing the ESL deposition method can affect the device performance: SnO_2^{ALD}-based cells perform less (efficiencies below 10%) than SnO_2^{np} -based reference cells (averaged efficiency of 16%), this latter result confirming that PbI_2 crystalline phase is not significantly detrimental.

For all SnO_2^{ALD}-based cells, current densities delivered at short-circuit ranged from 18 to 19 mA.cm^{-2}, which is slightly lower than reference values (20.5 to 21 mA.cm^{-2}) but could simply arise from the lower optical transmission of a 30nm-thick layer of SnO_2^{ALD} in the

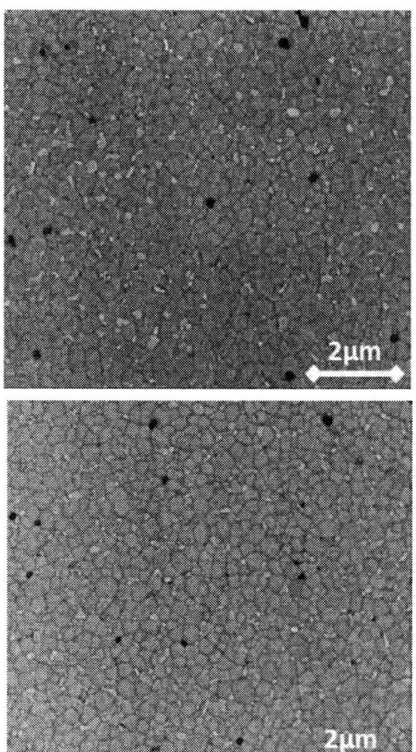

Figure 6 *- SEM pictures of the surfaces of a PK film grown on SnO_2^{np} (a.) and on SnO_2^{ALD} (b.), themselves deposited on glass/ITO substrates.*

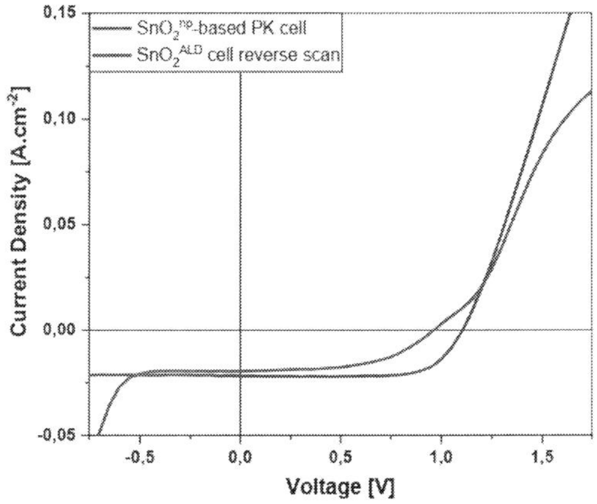

Figure 7 *- Current-voltage characteristics under 1 sun illumination of a reference cell that uses SnO_2^{np} and of a cell using SnO_2^{ALD} as ESL*

visible range. Open-circuit voltage (V_{oc}) however is more deteriorated with a median value of 930mV, which is about 200mV lower than reference values (1050 to 1200 mV). Furthermore, all measured Fill Factors (FF) ranged from 40% to 55%, which is substantially below the reference ones (>70%). The curves show a strong s-shape as well as high series resistance that drastically degrade V_{oc} and FF in SnO_2^{ALD}-based cells. Yet, we have seen that SnO_2^{ALD} electrical conductivity is particularly high with respect to SnO_2^{np}. In addition, transmission line model (TLM) measurements showed us that SnO_2^{ALD} efficiently contacts ITO. Therefore, such large series resistance is more likely to arise from a strongly limited electron extraction at the SnO_2/PK interface or a lack of selectivity of SnO_2^{ALD}. It may be explained by an energetic misalignment of the conduction band (CB) edges between the PK and the SnO_2^{ALD}. Yet, literature often highlights the well-matched CB edge alignment between SnO_2 and PK (*7, 27*). Ultra-violet electron spectroscopy and Kelvin force microscopy could give new insight on this issue. Furthermore, coupling SnO_2^{ALD} ESL with another layer, like C_{60} for instance, might also lead to better electron extraction performances, which would point out a selectivity problem at the SnO_2^{ALD}/PK interface. A better electron extraction for the SnO_2^{np} could also be assumed arguing infiltration of the PK within the nanoparticle-based layer. Regarding this, AFM but also SnO_2 porosity measurements could eventually give more understanding on the effective contact area between SnO_2 and PK for electron extraction.

4 CONCLUSION

The intrinsic properties of SnO_2^{ALD} thin films in regards of their application in PK/Si tandem cells have been examined. PK films grown on SnO_2^{ALD} exhibit similar properties as reference films. Finally, integration of SnO_2^{ALD} ESL in PK cells revealed a strong electron extraction inhibition and/or poor electron selectivity. In this regard, further investigation regarding the SnO_2/PK interface shall lead to more specific insights. Particularly, energetic alignment between PK and SnO_2^{ALD} shall be analysed as well as the effective contact area available for electron extraction at perovskite interface. Besides, SnO_2^{ALD} annealing at several temperatures, for different times and under various atmosphere is likely to have non-negligible impact on electron extraction in devices.

REFERENCES

1. P. Löper *et al.*, *IEEE Journal of Photovoltaics*. **4**, 1545–1551 (2014).
2. Q. Wali, N. K. Elumalai, Y. Iqbal, A. Uddin, R. Jose, *Renewable and Sustainable Energy Reviews*. **84**, 89–110 (2018).
3. W. Shockley, H. J. Queisser, 11.
4. A. Al-Ashouri *et al.*, *Science*. **370**, 1300–1309 (2020).
5. D. D. Smith *et al.*, *IEEE Journal of Photovoltaics*. **4**, 1465–1469 (2014).
6. M. A. Green *et al.*, *Progress in Photovoltaics: Research and Applications*. **28**, 3–15 (2020).
7. Q. Jiang, X. Zhang, J. You, *Small*. **14**, 1801154 (2018).
8. W. Ke *et al.*, *J. Am. Chem. Soc.* **137**, 6730–6733 (2015).
9. K.-H. Jung, J.-Y. Seo, S. Lee, H. Shin, N.-G. Park, *J. Mater. Chem. A*. **5**, 24790–24803 (2017).
10. J. Werner *et al.*, *ACS Energy Lett.* **1**, 474–480 (2016).
11. S. De Wolf, A. Descoeudres, Z. C. Holman, C. Ballif, *green*. **2** (2012), doi:10.1515/green-2011-0018.
12. V. Zardetto *et al.*, *Sustainable Energy Fuels*. **1**, 30–55 (2017).
13. S. M. George, *Chem. Rev.* **110**, 111–131 (2010).
14. Y. Lee *et al.*, *Adv. Sci.* **5**, 1800130 (2018).
15. Y. Kuang *et al.*, *ACS Appl. Mater. Interfaces*. **10**, 30367–30378 (2018).
16. S. Albrecht *et al.*, *Energy Environ. Sci.* **9**, 81–88 (2016).
17. F. Hou *et al.*, *ACS Appl. Energy Mater.* **2**, 243–249 (2019).
18. A. F. Palmstrom *et al.*, *Advanced Energy Materials*. **8**, 1800591 (2018).
19. Q. Jiang *et al.*, *Nature Photonics*. **13**, 460–466 (2019).
20. J. J. Yoo *et al.*, *Nature*. **590**, 587–593 (2021).
21. B. D. Viezbicke, S. Patel, B. E. Davis, D. P. Birnie, *physica status solidi (b)*. **252**, 1700–1710 (2015).
22. M. Batzill, U. Diebold, *Progress in Surface Science*. **79**, 47–154 (2005).
23. Q.-H. Wu *et al.*, *Materials Letters*. **61**, 3679–3684 (2007).
24. L. Y. Liang, Z. M. Liu, H. T. Cao, X. Q. Pan, *ACS Appl. Mater. Interfaces*. **2**, 1060–1065 (2010).
25. B.-E. Park *et al.*, *Applied Surface Science*. **480**, 472–477 (2019).
26. L. Lin *et al.*, *J. Mater. Chem. A*. **7**, 25511–25520 (2019).
27. L. Xiong *et al.*, *Adv. Funct. Mater.* **28**, 1802757 (2018).

CHARACTERIZATION OF CESIUM LEAD BROMIDE IODIDE MIXED PEROVSKITE ($CsPbBr_{3-x}I_x$) PREPARED BY SEQUENTIAL EVAPORATION

G. Gordillo[1], J.C. Peña[1], O. G. Torres[1], and M. C. Abella[1]
[1] Departamento de Física, Universidad Nacional, Bogotá, Colombia

ABSTRACT: Thin films of cesium lead bromide iodide mixed perovskite ($CsPbBr_{3-x}I_x$) with x varying at different molar ratios were synthesized by sequential evaporation of precursors (CsBr, $PbBr_2$, PbI_2), in order to get light absorbers with tunable band gap. Special emphasis was put on studying the effect that the amount of bromine that is substituted by iodine, has on the optical, structural and morphological properties. In particular, the effect caused by the substitution of Br by I in the $CsPbBr_3$ lattice on the optical gap, morphology and the formation of structural defects was studied, by means of transmittance, photoluminescence, scanning electron microscopy (SEM) and energy measurements of Urbach.

1. INTRODUCTION

Solar cells based on hybrid organic-inorganic methylammonium/Formamidinium lead halide $(MA/FA)PbX_3$, where X is a halide Cl, Br, I, or mixture of halides, have accomplished conversion efficiencies over 25% [1,2]. However these types of compound easily degrade to its component salts by humidity [3,4] and also by a combination of UV radiation and oxygen [5–7]; it is also thermally unstable close to its working temperature.

To overcome this instability, several inorganic $CsPbX_3$ (X: I, Br, or mixed halides) perovskites have been developed, which are also tolerant to oxygen exposure and ultraviolet irradiation [8–10]. Among all the inorganic perovskites, the mixed halide $CsPbBr_{3-x}I_x$ has attracted the attention due to its good thermal/light stability and tunability of the band gap through substitution of halide ions yielding optimum performance of solar cells [11]. It has been reported that inorganic $CsPbBr_3$ perovskite have higher thermal stability than hybrid perovskite $MAPbBr_3$ [12] and also that exhibit good electrical transport properties with an electron mobility around 1000 cm^2/Vs [13,14]. On the other hand, solar cells based on $CsPbI_3$ with an efficiency of 19.03% [15] were fabricated, which are characterized by having improved optical and electrical properties [16]. However the $CsPbI_3$ perovskite present structural changes at room temperature. Among the $CsPbX_3$ family, $CsPbI_2Br$ has been also highlighted because has shown to be thermally stable [17,18] and additionally because efficiencies of 16.79%, have been achieved with $CsPbI_2Br$ based solar cells [19].

This work presents a detailed description of the synthesis of thin films of $CsPbBr_{3-x}I_x$ following a route based on the sequential evaporation of the precursors PbI_2, CsBr and $PbBr_2$ and additionally reports results of a study on the influence of the x molar ratio on its optical, structural and morphological properties carried out by transmittance, photoluminescence, Scanning Electron Microscopy (SEM) and Urbach Energy measurements.

2. EXPERIMENTAL DETAILS

The samples were prepared by sequential evaporation of the precursors ($PbI_2/PbBr_2$/CsBr), where the chemical composition of the resulting compound is adjusted by varying the thickness ratio of precursors, which is determined with the help of a thickness monitor. Sequential evaporation was performed following a routine that includes initial evaporation of PbI_2 followed by evaporation of $PbBr_2$ and CsBr, keeping their respective deposition rates constant (around 3 Å/s). The precursors are evaporated at room temperature from Knudsen cell-type evaporation sources and after the deposition, these are annealed in vacuum for 20 minutes at a temperatures of 280°C. Good reproducibility of both the composition and properties of the $CsPbBr_{3-x}I_x$ films was achieved using an equipment with facilities to control electronically with accuracy the deposition rate of precursors, by means of advanced algorithms. Details of equipment used to prepare the $CsPbBr_{3-x}I_x$ films is described in a previously published paper (see Ref. [20]), where a similar setup was used to deposit thin films of $MAPbI_3$ by sequential evaporation of precursors.

The samples prepared were characterized by means of transmittance and reflectance measurements performed using a Varian–Cary 5000 spectrophotometer, as well as by XRD measurements performed with a Philips X'Pert Pro PANalytical diffractometer, using the radiation Cu-Kα (1.540598 Å), an acceleration voltage of 40 KV and a current 40 mA. The SEM images were obtained with a TESCAN scanning electron microscope, model Vega 3, and the film thickness was determined using a Veeco Dektak 150 surface profiler.

3. RESULTS

3.1 Optical characterization

The influence of the molar ratio x on the optical properties of $CsPbBr_{3-x}I_x$ films was studied through spectral transmittance and reflectance measurements. Figure 1 displays typical transmittance and reflectance spectra of $CsPbBr_{3-x}I_x$ films prepared varying x between 0 and 2. The curves of $(\alpha h\nu)^2$ vs $h\nu$ depicted in the insets were built from absorption coefficient (α) values calculated using the relation: $\alpha = -(1/d)\ln\frac{T(\lambda)}{1-R(\lambda)}$ [21]. The Eg value indicated in each figure was obtained from the intercept with the axis $h\nu$ of the curve of $(\alpha h\nu)^2$ vs hν. It is observed that the energy gap Eg of the $CsPbBr_{3-x}I_x$ films decreases from 2.35 eV to 1.91 eV, when the iodine concentration increases from x=0 to x=0.2 in molar ratio.

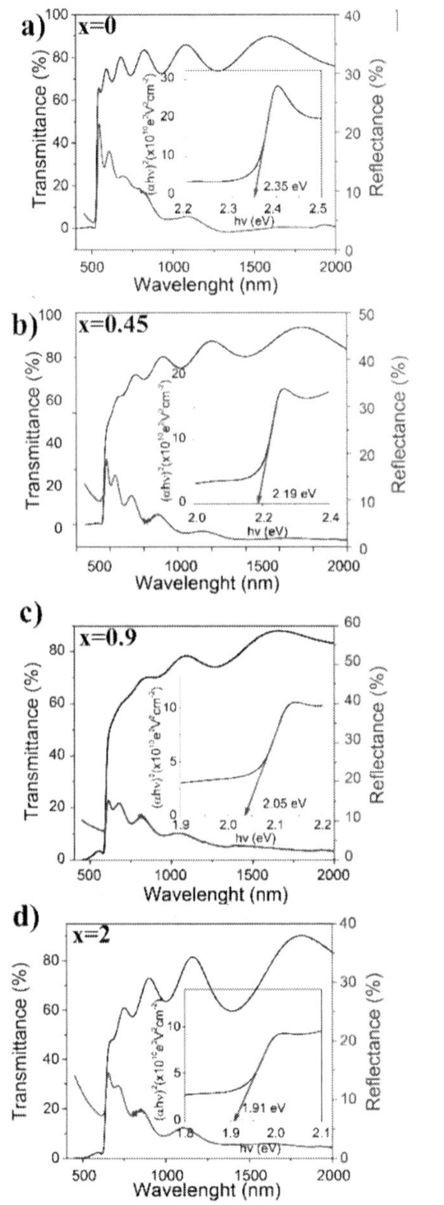

Fig.1: Transmittance and reflectance spectra of $CsPbBr_{3-x}I_x$ films, prepared with different x molar ratios. a) x=0, b) X=0.46, c) X=0.9 and d) X=2. In inset are shown the corresponding Tauc plots.

It is also observed in Fig.1 that the slope of the transmittance curves are significantly affected by the concentration of iodide in the $CsPbBr_{3-x}I_x$ films. This behavior seems to be caused by structural defects which give rise to distortion of the bands, generating tails of states that extend the bands within the gap, causing a decrease of the slope of the transmittance curves. Absorption via transitions between states of band's tails was evaluated through the calculation of the absorption coefficient near the band edge (α_U) that has an exponential dependence on photon energy and Urbach energy [22] ($\alpha_U = \alpha_0 exp\left[\frac{h\nu - E_i}{Eu}\right]$). Where E_U is the Urbach Energy, E_I and α_0 are constant. Thus, a plot of ln(α) vs. hν should be linear and Urbach Energy can be obtained from the slope.

In Fig.2 are depicted curves of α vs $h\nu$ (near the band edge) of samples prepared with different x molar ratio as well as curves of ln(α) vs hν showing the values of E_U calculated from their respective slopes. Considering that Urbach energy is related to density of localized states in the band gap induced by structural disorder (stress and dislocation) [22], the results of Fig. 2 reveal that the increase in the iodine concentration gives rise to an increase in the Urbach energy indicating that an increase in the density of structural defects is induced by the incorporation of iodine in the perovskite structure, apparently due to the fact that the ionic radius of iodine (0.220 nm) is greater than that of bromine (0.196 nm) [23].

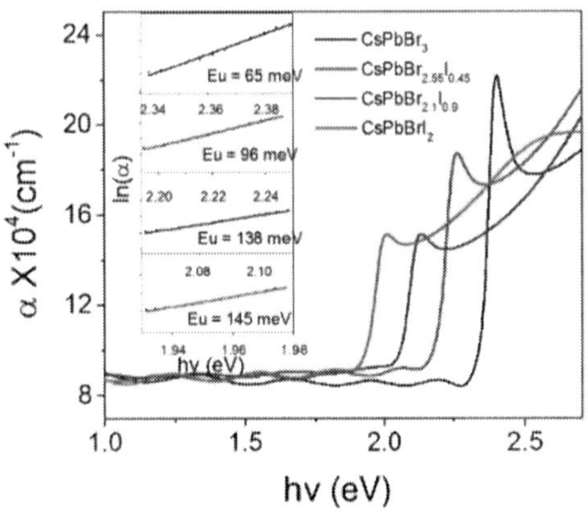

Fig. 2: Curves of α vs hν of $CsPbBr_{3-x}I_x$ films prepared with different x molar ratios. In inset are displayed curves of ln(α) vs hν and values of Eu calculated from their respective slopes

The effect of the x molar ratio on the photoluminescence spectrum emitted by the $CsPbBr_{3-x}I_x$ films, excited with a 400 nm laser, was also evaluated.

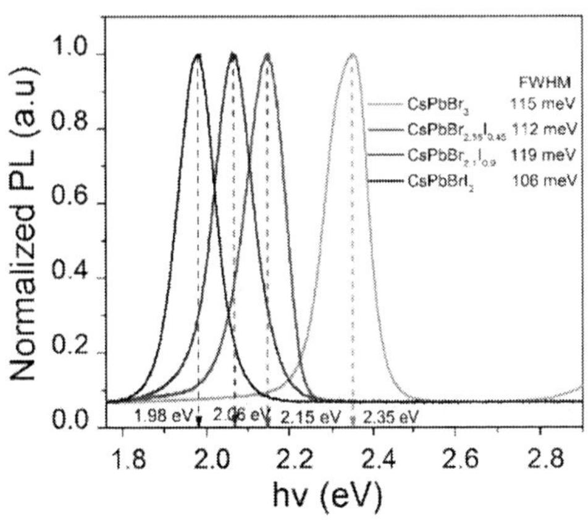

Fig. 3: Curves of normalized photoluminescence spectra of $CsPbBr_{3-x}I_x$ films, where x is 0, 0.46, 0.9, 2 respectively.

Fig. 3 shows typical PL emission spectra of $CsPbBr_{3-x}I_x$ films prepared with molar ratios varying between x=0 and x=2. From the results shown in Fig. 3, the following facts can be highlighted:

When the concentration of iodine in $CsPbBr_{3-x}I_x$ film increases, a reduction in Eg is observed from 2.35 ($CsPbBr_3$), to 1.98 eV ($CsPbBrI_2$), results that are consistent with the reported by other Authors [24,25] and with those obtained from the Tauc curves.

The width at half maximum (FWHM) of PL spectra decreases when the concentration of iodide in $CsPbBr_{3-x}I_x$ film increases. The broadening of the PL emission peaks has been attributed to an increase of the density of excitons thermally ionized [26].

3.2 Structural properties

In Fig. 4 are compared XRD spectra of $CsPbBr_{3-x}I_x$ films prepared varying the molar ratio between x=0 and x=2. These results reveals that the $CsPbBr_3$ films (x=0) exhibit only reflections corresponding to the cubic-$CsPbBr_3$ phase (ICSD 29073) and when the content of iodine in the $CsPbBr_{3-x}I_x$ sample increases, the crystalline structure does not change. However the reflections observed in the XRD pattern shift towards lower 2θ values, when the concentration of iodine increases. This is an expected behavior taken into account that iodine has an ionic radius greater than that of bromine, which leads to an expansion of the $CsPbBr_{3-x}I_x$ crystal lattice when the Br^{-1} anion is replaced by I^-. This expansion of the lattice leads to an increase in the lattice constant, which is manifested in a shift of the diffraction peaks.

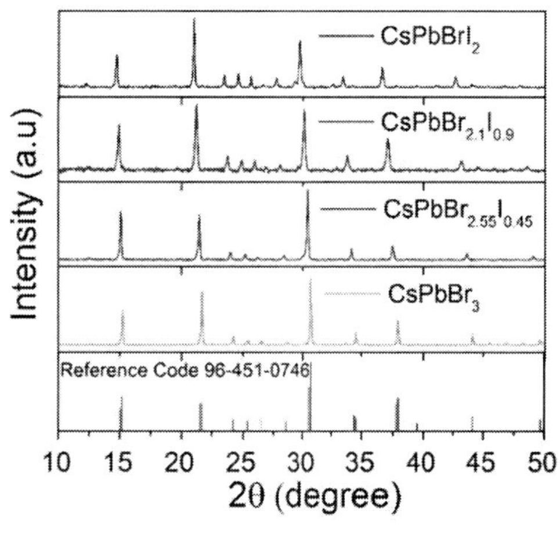

Fig. 4 : XRD spectra of $CsPbBr_{3-x}I_x$ films, with molar ratios corresponding to: x=0, x=0.46,x= 0.9, x=2.

3.3 Morphological characterization

The influence of the iodide content in $CsPbBr_{3-x}I_x$ films on their morphology was evaluated through scanning electron microscopy (SEM) measurements. Fig.5 shows the SEM micrograph (top view) of $CsPbBr_{3-x}I_x$ samples prepared with molar ratios varying between x=0 and x=2. It is observed that all the studied samples exhibits a morphology consisting of compact grains with sub-micron spherical shaped size. On the other hand, when the concentration of iodine in the $CsPbBr_{3-x}I_x$ films increases from x=0 to x=2 in molar ratio, a reduction in the average grain size is observed, from 0.82μm for $CsPbBr_3$ films, to 0.5 μm for $CsPbBrI_2$ films.

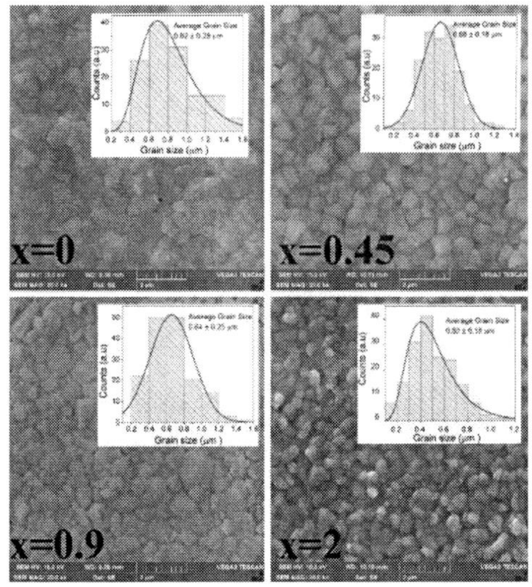

Fig. 5: SEM images of $CsPbBr_{3-x}I_x$ films with molar ratios corresponding to: x=0, x=0.46,x= 0.9, x=2.

4 CONCLUSIONS

The influence of the iodide concentration in mixed perovskite $CsPbBr_{3-x}I_x$ films prepared by sequential evaporation on their optical, structural and morphological properties was evaluated. It was found that regardless of iodine concentration the $CsPbBr_{3-x}I_x$ films grow with cubic perovskite type structure. The study also revealed that the incorporation of Iodine in the $CsPbBr_3$ matrix generates both an expansion of the crystal lattice and an increase in the density of structural defects. On the other hand the energy gap of $CsPbBr_{3-x}I_x$ films decreases from 2.35 eV, to 1.91 eV when the concentration of iodide increase from x=0 ($CsPbBr_3$) to x=2 ($CsPbBrI_2$), indicating that thin films of cesium lead bromide iodide mixed perovskites could be potentially used as active layer for both single junction solar cells, and perovskite-silicon hybrid tandem devices.

From SEM measurements it was found out that in general the $CsPbBr_3$-xI_x films deposited by sequential evaporation grow with a morphology characterized by compact grains with sub-microm size and that the grains size decreases by increasing the content of iodide in the $CsPbBr_3$ matrix. When the concentration of iodine in the $CsPbBr_{3-x}I_x$ films increases from x=0 to x=2 in molar ratio, a reduction in the average grain size is observed, from 0.82μm for $CsPbBr_3$ films, to 0.5 μm for $CsPbBrI_2$ films.

ACKNOWLEDGEMENTS

This work was supported by DIEB-Universidad Nacional de Colombia, Sede Bogotá, Facultad de

Ciencias, Departamento de Física, Grupo GMS&ES, K30 #45-03, Bogotá DC, Colombia.

REFERENCES

[1] National Center for Photovoltaics at the National NREL, Research cell efficiency records, available in https://www.nrel.gov/pv/assets/pdfs/best-research-cell-efficiencies.20200203.pdf

[2] M. Jeong, I.W. Choi, E.M. Go, Stable perovskite solar cells with efficiency exceeding 24.8% and 0.3-V voltage loss. Science. 2020;369 (6511):1615-1620.

[3] J.A. Christians, P. A. Miranda, P.V. Kamat, Transformation of the Excited State and Photovoltaic Efficiency of CH3NH3PbI3 Perovskite upon Controlled Exposure to Humidified Air. J. Am. Chem. Soc. 2015, 137 (4), 1530−1538.

[4] E. Mosconi, J. M. Azpiroz, F. De Angelis, Ab Initio Molecular Dynamics Simulations of Methylammonium Lead Iodide Perovskite Degradation by Water. Chem. Mater. 2015, 27 (13), 4885−4892.

[5] N. Aristidou, I. Sanchez-Molina, T. Chotchuangchutchaval, M. Brown, L. Martinez, T. Rath, S. A Haque, The Role of Oxygen in the Degradation of Methylammonium Lead Trihalide Perovskite Photoactive Layers. Angew. Chem., Int. Ed. 2015, 54 (28), 8208−8212.

[6] N. Aristidou, C. Eames, I. Sanchez-Molina, X. Bu, J. Kosco, M. S. Islam, S. A. Haque, Fast Oxygen Diffusion and Iodide Defects Mediate Oxygen-Induced Degradation of Perovskite Solar Cells. Nat. Commun. 2017, 8, 15218−15227.

[7] D. Bryant, N. Aristidou, S. Pont, I. Sanchez-Molina, T. Chotchunangatchaval, S. Wheeler, J. R. Durrant, S. A. Haque, Light and Oxygen Induced Degradation Limits the Operational Stability of Methylammonium Lead Triiodide Perovskite Solar Cells. Energy Environ. Sci. 2016, 9 (5), 1655−1660.

[8] A. Ho-Baillie, M. Zhang, C. F. J. Lau, F.J. Ma, S. Huang, *Joule* **2019**, *3*, 938.

[9] J. K. Nam, D. H. Chun, R. J. K. Rhee, J. H. Lee, J. H. Park, *Adv. Sci.* **2018**, *5*, 1800509.

[10] Q. D. Tai, K. C. Tang, F. Yan, *Energy Environ. Sci.* **2019**, *12*, 2375.

[11] G. E. Eperon, S. D. Stranks, C. Menelaou, M. B. Johnston, L. M. Herz, H. J. Snaith, Formamidinium Lead Trihalide: A Broadly Tunable Perovskite for Efficient Planar Heterojunction Solar Cells. Energy Environ. Sci. 2014, 7 (3), 982−988.

[12] M. Kulbak, S. Gupta, N. Kedem, I. Levine, T. Bendikov, G. Hodes, D. Cahen, Cesium Enhances Long-Term Stability of Lead Bromide Perovskite-Based Solar Cells. *J. Phys. Chem. Lett.*,167–172 (2016).

[13] C. C. Stoumpos, *et al.* Crystal Growth of the Perovskite Semiconductor CsPbBr3: A New Material for High-Energy Radiation Detection. *Cryst. Growth Des.* **13**, 2722–2727 (2013).

[14] M. Kulbak, D. Cahen and G. Hodes, How Important Is the Organic Part of Lead Halide Perovskite Photovoltaic Cells? Efficient CsPbBr3 Cells. *J. Phys. Chem. Lett.* **6**, 2452–2456 (2015).

[15] Y. Wang, X. Liu, T. Zhang, X. Wang, M. Kan, J. Shi, The Role of Dimethylammonium Iodide in $CsPbI_3$ Perovskite Fabrication: Additive or Dopant?, Angew Chem (Int Ed Engl), 2019 58(46):16691-16696.

[16] T. S. Ripolles, K. Nishinaka, Y. Ogomi., Y. Miyata, and S. Hayase, Efficiency enhancement by changing perovskite crystal phase and adding a charge extraction interlayer in organic amine free-perovskite solar cells based on cesium. *Sol. Energy Mater. Sol. Cells* **144**, 532–536 (2016).

[17] R. J. Sutton, G. E. Eperon, L. Miranda, E. S. Parrott, B.A. Kamino, J. B. Patel, M. T. Hörantner, M. B. Johnston, A. A. Haghighirad, D. T. Moore, H. J. Snaith, Bandgap-Tunable Cesium Lead Halide Perovskites with High Thermal Stability for Efficient Solar Cells. Adv. Energy Mater. 2016, 6, 1502458.

[18] R. E. Beal, D. J. Slotcavage, T. Leijtens, A. R. Bowring, R. A. Belisle, W. H. Nguyen, G. Burkhard, E. T. Hoke, M. D. McGehee, Cesium Lead Halide Perovskites with Improved Stability for Tandem Solar Cells. J. Phys. Chem. Lett. 2016, 7, 746−75

[19] Y. Han, H. Zhao, Ch. Duan, Sh. Yang, Zh. Yang, Zh. Liu, and Sh. Liu, Controlled n-Doping in Air-Stable CsPbI2Br Perovskite, Solar Cells with a Record Efficiency of 16.79%, Adv. Funct. Mater. 2020, 1909972

[20] M. A . Reinoso, C. A. Otálora and G. Gordillo, Improvement Properties of Hybrid Halide Perovskite Thin Films Prepared by Sequential Evaporation for Planar Solar Cells, *Materials* 2019, *12*, 1394

[21] F. Ruske , M. Wimmer, G. Köppel, A. Pflug, B. Rech, Optical characterization of high mobility polycrystalline ZnO:Al films, Proc. SPIE. 2012; 8263; p. 826303.

[22] M.V. Kurik, Review of Urbach's tail. Phys. *Status Solidi A*, vol. 8, No. 9, 1971.

[23] R.D. Shannon, "Revised Effective Ionic Radii and Systematic Studies of Interatomic Distances in Halides and Chalcogenides", *Acta Cryst.* **A32** 751-767 (1976).

[24] Ch. Huey, T. S. Ripolles , K. Hamada, S. H. Teo, H. N. Lim, J. Bisquert and Sh. Hayase, Tunable Open Circuit Voltage by Engineering Inorganic Cesium Lead Bromide/Iodide Perovskite Solar Cells, SCIENTIFIC REPORTS, (2018) 8:2482

[25] Y. Han, H. Zhao, Ch. Duan, Sh. Yang, Z. Yang, Z. L. Shengzhong, F. Liu, Controlled n-Doping in Air-Stable CsPbI2Br Perovskite Solar Cells with a Record Efficiency of 16.79%, Adv. Funct. Mater. 2020, 1909972

[26] J. Li, X. Yuan, P. Jing, J. Li, M. Wei, J. Hua, J. Zhao and L. Tian, Temperature-dependent photoluminescence of inorganic perovskite nanocrystal films, RSC Adv., 2016, 6, 78311

THE OPPORTUNITY OF CADMIUM STANNATE AS TRANSPARENT CONDUCTING OXIDE FOR PEROVSKITE-BASED CONCENTRATED PHOTOVOLTAIC SYSTEM

Maria Khalid,[1] Anurag Roy,[1] Senthilarasu Sundaram,[1] Tapas K. Mallick[1]
[1]Environment and Sustainability Institute, University of Exeter, Penryn Campus, Cornwall TR10 9FE, U.K.

ABSTRACT: The relative scarcity and high cost associated with indium or fluorine, fabrication approaches surface roughness, and low adhesion to polymeric materials are significant drawbacks for indium tin oxide (ITO) and fluorine tin oxide (FTO)-based transparent conductive oxide (TCO) glasses. TCOs are materials that are conductive and, at the same time, exhibit a transparent characteristic that allows light to pass through the material itself. In this regard, as a potential, low-cost semiconducting transparent oxide $CdSnO_3$ (CTO) having excellent electrical properties and stable structure with excellent electrical properties and sturdy structure is a suitable alternative to ITO and FTO. However, the cost of the FTO and fabrication approaches surface roughness and low adhesion to polymeric materials make it less viable to use as TCO. In this study, CTO has been successfully synthesized by a cost-effective solution process and further endeavoured as a TCO candidate for perovskite solar cell application. Interestingly, the CTO-based device shows 5.42% power conversion efficiency while the FTO-based device shows 6.02%. This work outlines a future direction to use low-cost CTO based TCO as an alternate to FTO. In addition to that, this also signifies a great opportunity towards its implementation of perovskite solar cells. Additionally, coupling a concentrating device to boost the device's performance and reduce the production cost appears as another scope of research interest for this type of device.

Keywords: Transparent conducting oxides, Cadmium Stannate ($CdSnO_3$), Perovskite solar cell, CPV

1 INTRODUCTION

Transparent conducting oxides (TCO's) as transparent electrodes plays a significant role in solar cells technology to allow incident light to get to the photoactive layer. TCO's are electrically conductive with low light absorption and are primarily prepared using thin-film technologies in potential optoelectrical applications such as low emissivity windows, thin-film transistors, and semiconductor lasers, flat panel displays and LED etc. TCO's have a wide bandgap of ~3 eV with higher conductivity of ~10^6 s. The high conductivity of TCO's is due to doping of oxygen vacancies or extrinsic dopants. TCO's conductivity can be tuned via semiconducting from insulating, and their transparency can also be adjusted. They have a wide range of scope for power-saving circuitries and other optoelectronic technological applications due to both n-type and P-type conductors. TCO materials are to be used in various applications spanning from photovoltaics to lighting, smart windows, or gas sensors [1][2][3].
However, the ample combination of good conductivity and transparency is difficult to maintain in intrinsic stoichiometric oxides. It is possible to achieve with non-stoichiometric composition and doping appropriate material. In this direction, CdO, ZnO, SnO_2, In_2O_3 and SnO_2 were introduced to process such characteristics without degrading optical transmission [4][5]. The material selection needs careful decision and consideration of essential parameters such as optical and electrical properties and maintaining abrasion resistance, electron work function, environmental stability, a combination of the TCO with other deposited layers, especially in perovskite solar cells application. Moreover, high mobility is another essential aspect to consider, which can be possible by wide lowest energy band in the conduction band of the TCO's. The ITO/ATO/TiO_2 triple-layered TCO was thermally stable and showed a photovoltaic property to comparable FTO because the antimony-doped tin oxide (ATO) layer on ITO functioned as a heat-resistant layer [6].
In the case of binary TCO, the symmetric anion coordination environment is helpful to meet the condition. The second considerable condition for compelling TCO is maintaining conduction or valance by doping suitable material that does not degrade the optical transmission or carrier mobility. TCO's play an integral role in the structure of perovskite solar cells (PSC's), which is an emerging field due to its flexibility in several ways; lightweight, ease of fabrication, inexpensive production and low temperature processability. Moreover, the thin film structure, unique properties and facile manufacturing techniques promise an astonishing forthcoming solar technology field.
The most common TCO is indium tin oxide (ITO) and fullerene doped transparent oxide (FTO) for PSC application due to their transparent and conductive nature. However, ITO is preferred over FTO for a bottom electrode in PSC configuration. Although the cost of indium is high, studies revealed that ITO electrode costing is higher among other deposited layers such as HTL, ETL, photoactive layer, etc. Moreover, stress-strain constraints also lead ITO as TCO to look to other alternates.
Besides the cost, stability is another concern in ITO. ITO was found to diffuse into other top deposited layers even in conventional architecture or generating instability of the device [7][8]. A few studies are presented to use FTO as the bottom electrode as an alternate of ITO. FTO has better stability and high tolerance to physical abrasion as compared to ITO. The more significant transmission percentage at visible wavelength and low resistance rate, chemical inertness, and low cost of FTO are better options than ITO; however, the surface roughness leading to current leakage is another primary concern of FTO [9].
Both the above mentioned TCO's have drawbacks of brittleness and low adhesion to polymeric materials and complications of fabricating thin film to meet the conditions of the transparent electrode. In addition, the expensive fabrication method such as the vacuum sputtering process is commonly used for ITO/FTO layers deposition. Hence, it is imperative to find new feasible material to substitute ITO/FTO as TCO.

Here, we introduce the advantage of using the oxides within PSC fabrication: their simple design, low cost, low weight, and high sensitivity. $CdSnO_3$ has been reported for gas sensing applications using easy manufacturing methods such as coprecipitation [10]. Besides gas sensing applications, $CdSnO_3$ (CTO) has some other potential applications in nanostructure form. Ternary or complex quaternary TCO's can also have electronic configuration $(n-1)d^{10}ns^0$ like binary n-type TCO. CTO attained two stable crystal forms like distorted perovskite crystal form and rhombohedral structure.

In 1977, Shannon et al. surmised that edge shearing of materials such as In^{3+}, Cd^{2+} and Sn^{4+} could form transparent conductors. In their work, they studied the complex structure of $CdSnO_3$, Cd_2SnO_4, $CdIn_2O_4$ [11]. Nano- $CdSnO_3$) As reversible capacity with the stability of ~475 mAhg-[1] for about 40 cycles between 5 mV to 1.0 V at 0.13 C current rate. Additionally, low hysteresis, convenient charge and discharge potential and stable capacity make this compound speedier than other TCO. The combined oxide of $CdSnO_3$ with a bandgap of 0.694 eV has excellent potential to combined electrical, chemical, optical and physical properties more quickly than the alternation of chemical composition. Extensive research has been executed on CTO since Nozik et al. indicated the compound's potential as TCO [12].

The purpose of the paper is to unveil the potential of $CdSnO_3$ as a promising TCO to fabricate PSC using low-cost in a facile, easy colloidal synthesis way. $CdSnO_3$ has shown magnificent prospective in other technological applications, but TCO, no practical approach has been reported yet. CTO is a good choice as an alternative to FTO, compare photovoltaic (PV) performance utilizing as a coating on the glass. The inverted architecture of the device is another novelty of the work. Furthermore, a unique idea of implementation of concentrated light is proposed to enhance the performance of the inverted perovskite based on $CdSnO_3$ based TCO.

2 Materials and Method

2.1 Materials

Cadmium acetate [$Cd(OOCCH_3)_2.2H_2O$, Merck], tin (IV) chloride pentahydrate ($SnCl_4.5H_2O$, Fisher Chemicals), triton X (Sigma-Aldrich) and sodium hydroxide beads (NaOH, Merck) were used to synthesized CTO without any further purification.

Besides, poly(3,4-ethylene dioxythiophene) polystyrene sulfonate or PEDOT: PSS (Ossila UK), PbI_2 (Sigma-Aldrich, >99.99 %), CH_3NH_3I (Sigma-Aldrich, ≥99.5 %), [6,6]-Phenyl-C_{61}-butyric acid methyl ester or $PC_{61}BM$ (Ossila ≥99 %), dimethylformamide or DMF (Sigma-Aldrich, 99.8 %), dimethyl sulfoxide or DMSO (Sigma-Aldrich, 99.8 %), toluene (Fisher Chemical) and isopropanol (Sigma Aldrich, 99.5 %) were purchased from commercial suppliers and used without further purification.

2.2 Synthesis of CTO nanoparticles

CTO sample was synthesized using the precipitating method at a pH ≥ 10 followed by calcination at 800°C to attain the pure phase CTO. The overall synthesis was adapted from the literature by Roy et al. (2019) [13].

2.3 Device fabrication and structure

The CTO films were prepared by the doctor blade method from the CTO powder dispersions with Triton X and isopropyl alcohol on a cleaned glass, followed by heat treatment at 250°C under vacuum. Before using the glass, they were cleaned in an ultrasonic bath in detergent, deionized water, acetone, isopropanol sequentially and treated with UV-ozone treatment for 10 mins.

In inverted PSC structure, conductive polymer PEDOT: PSS layer is commonly used as hole transport layer (HTL) due to its easy processability, ample mechanical flexibility and excellent optical transparent characteristics and in the visible range of light spectrum and appropriate energy band levels. Fullerene derivatives [6], [6]-phenyl C61 butyric acid methyl ester (PCBM) is often used as electron-transport layers (ETL) in PSC devices, mostly in inverted architecture.

The device fabrication process was adopted from the literature [14]. Instead of ITO, synthesized CTO coated glass and commercial FTO glasses (8Ω/cm²) were used for this case. A low-temperature carbon layer has been used as a back electrode, deposited as a screen-printing layer on the top of the $PC_{61}BM$ layer [15].

The device was fabricated following inverted architecture. For device 1, the position of different components is as follows: FTO coated glass was utilised for TCO layer, followed by PEDOT: PSS as HTL, MAPbI3 as photo active layer, PCBM as well ETL and carbon as a back electrode. Whereas device 2 also follows the same pattern, expect CTO rather than FTO as TCO, shown in fig. 1. Therefore, this work's motivation is to scrutinise the photovoltaic performance of CTO-based inverted PSC as a replacement of FTO-based glasses.

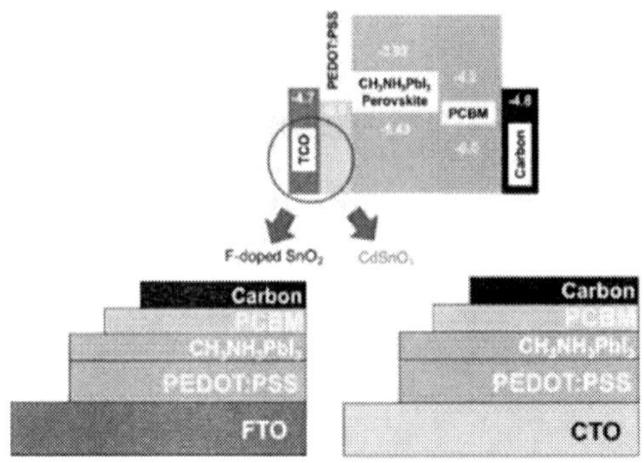

Figure 1: Schematic illustration of energy level alignment of TCO/PEDOT: PSS/perovskite/PCBM/carbon was as a TCO layer FTO and CTO have been used, respectively.

3 CHARACTERIZATION TECHNIQUES

The confirmation of the phase formation was assessed by Bruker D8 Advance X-ray diffractometer on a X'pert pro MPD XRD of PANalytical with Cu K_α radiation (λ = 1.5406 Å) (Billerica, MA, USA). Optical properties (transmittance spectra) of the bare glass and CTO coated glass were recorded using a PerkinElmer Lambda 1050+ UV/Vis/NIR spectrophotometer. The Elemental analysis of CTO film was performed using a scanning electron microscope TESCAN VEGA3 SEM coupled with energy-dispersive spectroscopy (EDS; Oxford instrument elemental analysis). The conductivity and resistivity measurements were carried out using the Ossila (UK)

Four-Point Probe Instrument. The devices' current density–voltage (J-V) measurement was recorded via a AAA+WACOM solar simulator solar having sun condition of AM 1.2G, 1000 mW/cm^2 with attached EKO MP-160i I-V Tracer. The light intensity was calibrated using a reference crystalline silicon cell, providing the light intensity of 1000 mW/cm^2. The device's active area was 0.16 cm^2, and six different cells were measured to understand the best performance of the devices.

4 RESULTS AND DISCUSSION

XRD results of the CTO film indicate the crystalline nature and phase purity of $CdSnO_3$. All the diffraction peaks of the sample indexed to match with the ilmenite crystal structure of $CdSnO_3$. The calculated lattice parameters of the sample also well matched with the standard data JCPDS (Joint Committee on Powder Diffraction Standards, card 34-0758) corresponding to α-$CdSnO_3$ phase, confirming the successful synthesis of well organised CTO film as shown in fig. 2.

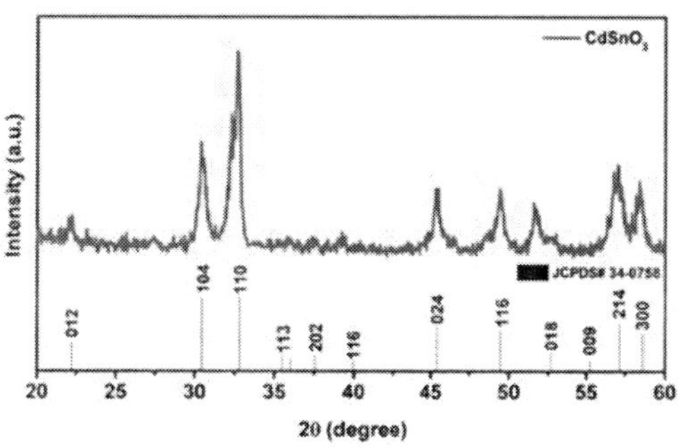

Figure 2: X-ray diffraction patterns of prepared precipitate of $CdSnO_3$ well matched with the JCPDS data (card no. 34-0758) of $CdSnO_3$ phase.

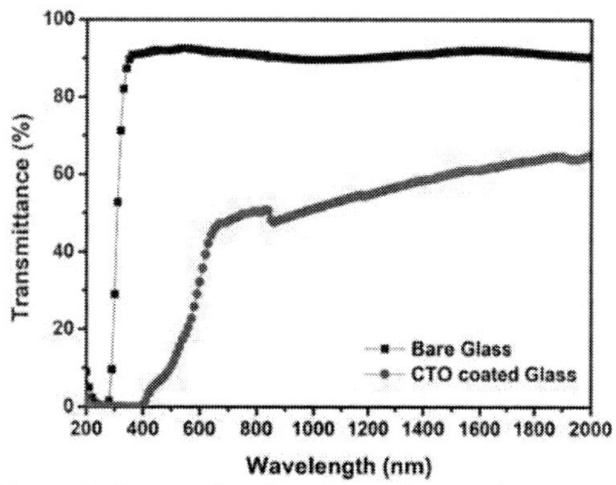

Figure 3: Comparative plot of transmittance of Bare glass and CTO coated glass.

The bare glass and CTO coated glass's transmittance spectral range is between 200 nm to 2000 nm (Fig. 3). The spectral indicates that the glass's transmittance is ~90 %, while a reduction of ~30% for CTO coated glass can be seen. A maximum of 62% of transmittance is shown for CTO coated glass. The decline can be attributed to the coating material.

Figure 4: SEM microstructural image of CTO coated glass coupled with EDS analysis to strengthen the formation of the $CdSnO_3$ nanoparticles.

Morphological studies of the CTO coated glass was done using electron microscopy and the incorporation of EDS to analyse the synthesized $CdSnO_3$ film uniformity and size shown in fig 4. The results reveal the appropriate orientation of the prepared sample in cuboidal shape with closely held formation and length of 200 nm. The cuboidal shape appeared to form in a bundles pattern, and each bundle is closely regulated with some visible voids, which can decrease the device's output. The successful formation of Cd, Sn with an excess of O can also be seen in the figure. The presence of Cu is due to the probe of EDS not being present in the sample.

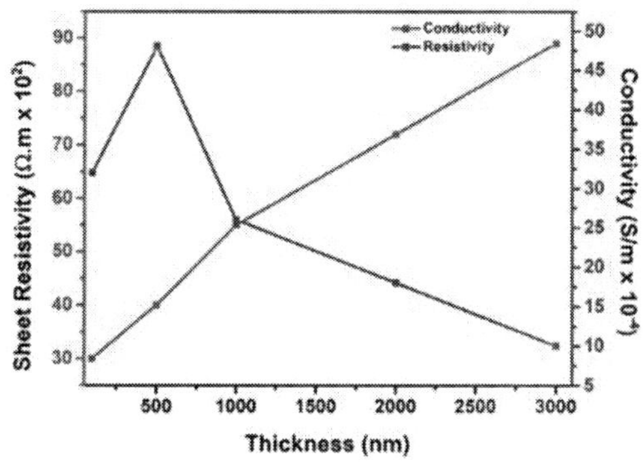

Figure 5: Plot of conductivity and resistivity to optimize the thickness of type sample.

The CTO coating thickness on glass has been varied to optimized its maximum resistivity and conductivity parameters. The number of layer deposition controls the thickness of the coating during their doctor blade method. The conductivity and resistivity study of the device was carried out using four probes, as shown in fig. 5. It can be seen that resistivity increases with the increment of thickness, and conductivity also varies. At 500 nm, thickness is optimised, and we get maximum conductivity with minimum resistance, and the gap is highest among others.

After CTO glass coating, the cross-sectional SEM images were analysed again, as shown in fig. 6. It can be seen that the CTO optimised thickness of the layer is ~550 nm, while for bare glass, it's 90°. For the second sample, the thickness of CTO glass is ~1.82 µm and 90° for bare glass as the first comparison.

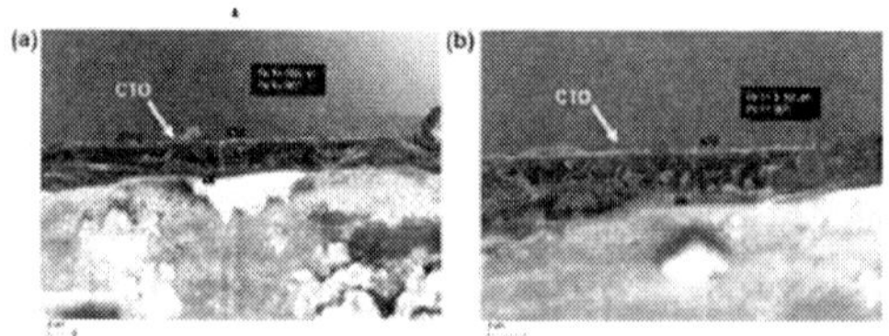

Figure 6: Cross-sectional SEM micrographs of CTO-coated glass with a maximum thickness of (a) 550 nm and (b) 1.82 μm, respectively.

5 PHOTOVOLTIAC PERFORMACE

The excited electrons are diffused through the ETL layer and are transferred to the TCO layer. The conduction band edge (CBE) of the CTO is positioned below the lowest unoccupied molecular orbital (LUMO) of the PEDOT: PSS. Through the external circuit, the photogenerated electrons reach the counter electrode.

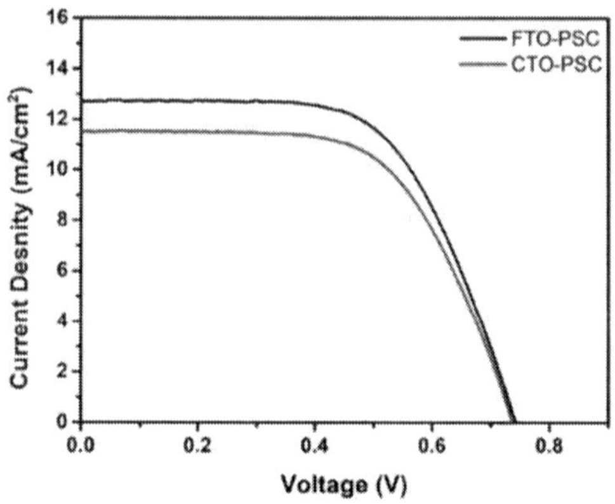

Figure 7: Comparison J–V characteristic plot of CTO and FTO based PSC devices with an active area of 0.16 cm^2 under 1 sun (100 mW/cm^2) light illumination.

Table 1:Summary of the photovoltaic parameters of the devices based on CTO and FTO as TCO-coated glass under the illumination of 100 mW/cm^2.

Sample	J_{SC} (mA.cm^{-2})	V_{OC} (V)	FF	Efficiency (%)
CTO	11.53	0.74	0.61	5.42
FTO	12.72	0.74	0.62	6.02

To evaluate the PV parameters performance of both devices with an active area of 0.16 cm^2, such as short circuit current density (Jsc), open-circuit voltage (Voc), fill factor (FF). Power conversion efficiency (PCE) was also calculated to determine the quality of the devices in ambient conditions (fig. 7). The current density vs voltage (J–V) characteristic was recorded under constant solar irradiation with simulated AM 1.5 (1000 mW/cm^2). The PSC device's maximum photoconversion efficiency (PCE) with CTO as conducting layer was found 5.42% with 11.53 mA/cm^2 of Jsc,0.74V of Voc, and FF of 0.61. In comparison, the FTO based PSC device showed PCE of 6.02%, Jsc of 12.72 mA/cm^2, Voc of 0.74V and FF of 0.62, respectively.

Table 1 summarised the PV parameters of the PSC devices. Interestingly, The results show no significant output difference between CTO and FTO the-based on PSC devices. However, in the fabrication of the devices, CTO-based device is low cost and facile to cope with the big-budget need and excellent replacement to FTO glass. The low efficiency can be attributed to the layers deposited, such as HTL consist of PEDOT: PSS has low conductivity of (<1 Scm^{-1}), which is not enough for efficient hole transportation. Moreover, its hygroscopic nature and incapability to block electrons backflow from the photoactive layer can lead to degradation and consequently restrict the output of the inverted PSC devices. Moreover, inadequate coverage of the transparent conductive layer on the glass substrate surface would also result in current leakage and ultimately poor device performance. The inhomogeneous film formation of the ETL of PCBM can also originate thin region between the photoactive layer and ETL, resulting in direct contact between the layers [16]. Moreover, colossal energy level differences at the photoactive/PCBM layer and low electron mobility can directly restrict the inverted PSC performance. However, the better performance of the device was attributed to the reduction of interfacial defects and improved series and shunt resistance of the PSC. No change in the V_{OC} was consistent, indicating that the thin PEDOT: PSS layer did not affect the charge recombination rate.

However likely, the chemistry of the CTO/PEDOT: PSS interface is also an essential factor. Rapid hole transfer to the polymer where adsorbed water is oxidized decreases the CTO photo corrosion. PEDOT: PSS that acts as a coat of CTO layer plays an essential role in protecting the stability of the CTO-coating [17]. It is well known that high-temperature sintering techniques measure the electronic conductivity and Seebeck Coefficient of semiconducting materials. These procedures may lead to unreliable results such as changing transport properties from p- to n-type and thus affect their performance [18]. However, the charge transport properties are related to the transport resistance in impedance measurement. The higher series resistance (R_S) of the device can also decrease the short-circuit current. The R_S in DSSC appears from the contact resistance, the TCO's resistance, semiconductor film, electrolyte, connected wires in the cell, etc. Evaluation of CTO as a suitable replacement of FTO requires more attention for future study.

6 FUTURE RECOMMENDATION

In PV technology, photovoltaic concentrator systems are made up of optical devices to increase the solar radiation intensity on the PV cell, leading to the reduction of the cost of the system by decreasing the PV material area. A concentrator PV system can employ one or more optics dependent on the type of the system. Some CPV systems are either reflective or refractive or both and may involve different optical stages such as refraction directly within the lens, TIRs within the lens and reflection out of that lens. The output of the CPV system can be increased by concentrating more light on the substrate surface using different optical components.

The most studied and known ideal concentrator, which collects all the incoming rays and acceptance angle into the geometry, is a compound parabolic concentrator (CPC). The design of the 3D CPC is also very efficient. In our recent work, we utilised the idea to use PSC with optical integration using CPC and achieved excellent results [19].

Another work was presented in 2018 by Snaith et al. by integrating high solar concentration to exceed Shockley-Queisser limit and to achieve remarkable voltage close to 1.4 V. They assess solar PV parameters based on theoretical predictions with the trend of light concentration levels. They determined results based on extraction rates and charge carrier recombination assumptions of the device. The obtained results indicate the scope of PSC under solar concentration to exhibit a substantial increment of the output of the PV systems [20]. Hasan et al. extended the field by offering an appealing alternative to achieve high PCE of PSC under concentrated solar light. A PSC device with a 0.81 mm^2 size was used with low concentrated PV, and 23.1% efficiency was achieved to open a new area of research in PV technology [21].

Some other advantages of CPV systems are appreciable such as the existence of CPV systems, which could be related to the tremendous slow fall in the price of the PV systems. The recent innovations regarding solar cells and the immense potential of these systems offer to harness the available solar energy. These concentrator systems are easy to manufacture compared to complicated procedures adopted for manufacturing solar cells, shifting the burden of cost from the expensive solar cells towards the optical concentrator systems. This implies that the more extensive system demands can be meet with CPV instead of PV systems, making CPV technology more critical. The technology is now quite mature. Using the latest innovations in solar cell technology (multijunction cells) can generate more electricity and have a potential market. By motivated all these aspects, we recommend this present PSC device to integrate with CPC to get proficient output at low cost due to the cost of efficient cost compelling TCO to provide a good foundation that can stimulate further research on enhancement of PSC devices. A schematic illustration of the proposed idea is explained in Fig.8.

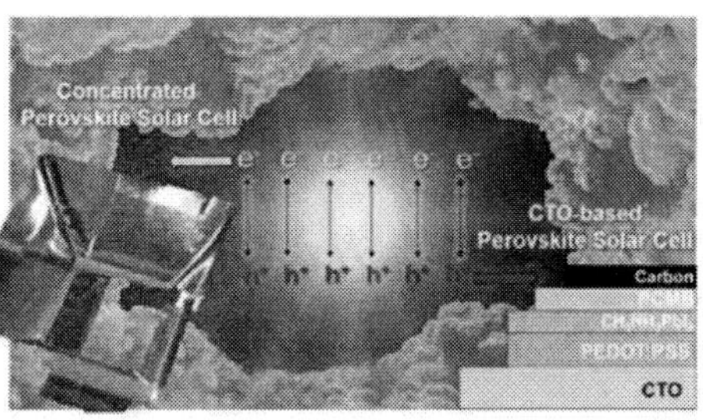

Figure 8: Schematic representation of CTO- based PSC device integrated with a concentrator (CPC) possibly enhances the device's output.

7 CONCLUSIONS

In summary, we have synthesized $CdSnO_3$ film to utilise as TCO in PSC devices to replace other expensive TCO such as FTO or ITO. The sample was successfully prepared using a facile, easy, and fast synthesis spin coating approach. The XRD pattern of the coated CTO glass reveals the successful formation of the rhombohedral ilmenite $CdSnO_3$ well-matched with standard data. SEM images also confirm the synthesised film's cuboidal shape with an average diameter of 200 nm. In contrast, EDS further strengthen the formation of the pure $CdSnO_3$ sample. Transmittance spectra of the sample showed 90% T for bare glass, while a decrement of ~30 % is observed for the $CdSnO_3$ coated sample. Four probe characterizations suggest 500 nm thickness of the film is optimised to get maximum conductivity and minimum resistance. The cross-sectional SEM images were analysed after CTO coating again. The optimised thickness of the CTO film layer is observed to be ~550 nm, whereas, for the second sample, the thickness of CTO glass is ~1.82 µm. The PV parameters of both the devices showed a slight difference between them. CTO coated device obtained % while FTO based device showed % of PCE. However, interestingly FTO based device showed a little higher Isc than CTO based device.

8 REFRENCES

[1] "The Future of TCO Materials: Stakes and Challenges Marie-Isabelle Baraton University of Limoges & CNRS, Faculty of Sciences, SPCTS, 87060 Limoges, FRANCE," *Mater. Res.*, vol. 1209, pp. 1209-P03-06, 2010.

[2] H. Hosono, "Recent progress in transparent oxide semiconductors: Materials and device application," *Thin Solid Films*, vol. 515, no. 15 SPEC. ISS., pp. 6000–6014, 2007, doi: 10.1016/j.tsf.2006.12.125.

[3] H. Hosono, H. Ohta, M. Orita, K. Ueda, and M. Hirano, "Frontier of transparent conductive oxide thin films," *Vacuum*, vol. 66, no. 3–4, pp. 419–425, 2002, doi: 10.1016/S0042-207X(02)00165-3.

[4] K. Bädeker, "Über eine eigentümliche Form elektrischen Leitvermögens bei festen Körpern," *Ann. Phys.*, vol. 334, no. 8, pp. 566–584, 1909, doi: 10.1002/andp.19093340807.

[5] C. Coatings, "TRANSPARENT," pp. 73–94, 1977.

[6] B. Yoo *et al.*, "Enhanced charge collection efficiency by thin-TiO2-film deposition on FTO-coated ITO conductive oxide in dye-sensitized solar cells," *J. Mater. Chem.*, vol. 20, no. 21, pp. 4392–4398, 2010, doi: 10.1039/b926145a.

[7] P. Cheng and X. Zhan, "Stability of organic solar cells: Challenges and strategies," *Chem. Soc. Rev.*, vol. 45, no. 9, pp. 2544–2582, 2016, doi: 10.1039/c5cs00593k.

[8] M. Jørgensen, K. Norrman, and F. C. Krebs, "Stability/degradation of polymer solar cells," *Sol. Energy Mater. Sol. Cells*, vol. 92, no. 7, pp. 686–714, 2008, doi: 10.1016/j.solmat.2008.01.005.

[9] Z. Bashir and S. Rastogi, "The explanation of the increase in slope at the Tg in the plot of d-spacing versus temperature in Polyacrylonitrile," *J. Macromol. Sci. - Phys.*, vol. 44 B, no. 1, pp. 55–78, 2005, doi: 10.1081/MB-200044588.

[10] W. Xing-Hui, W. Yu-De, L. Yan-Feng, and Z. Zhen-Lai, "Electrical and gas-sensing properties of perovskite-type CdSnO3 semiconductor material," *Mater. Chem. Phys.*, vol. 77, no. 2, pp. 588–593, 2003, doi: 10.1016/S0254-0584(02)00116-5.

[11] R. D. Shannon, J. L. Gillson, and R. J. Bouchard, "Single crystal synthesis and electrical properties of CdSnO3, Cd2SnO4, In2TeO6 and Cdln2O4," *J. Phys. Chem. Solids*, vol. 38, no. 8, pp. 877–881, 1977, doi: 10.1016/0022-3697(77)90126-3.

[12] A. J. Nozik, "Optical and electrical properties of Cd2 SnO4: A defect semiconductor," *Phys. Rev. B*, vol. 6, no. 2, pp. 453–459, 1972, doi: 10.1103/PhysRevB.6.453.

[13] A. Roy, P. P. Das, P. Selvaraj, P. S. Devi, and S. Sundaram, "Template free synthesis of CdSnO3 micro-cuboids for dye sensitized solar cells," *J. Photochem. Photobiol. A Chem.*, vol. 380, no. January, p. 111824, 2019, doi: 10.1016/j.jphotochem.2019.04.035.

[14] K. Jiang *et al.*, "Inverted planar perovskite solar cells based on CsI-doped PEDOT:PSS with efficiency beyond 20% and small energy loss," *J. Mater. Chem. A*, vol. 7, no. 38, pp. 21662–21667, 2019, doi: 10.1039/c9ta08995k.

[15] S. Bhandari, A. Roy, A. Ghosh, T. K. Mallick, and S. Sundaram, "Perceiving the temperature coefficients of carbon-based perovskite solar cells," *Sustain. Energy Fuels*, vol. 4, no. 12, pp. 6283–6298, 2020, doi: 10.1039/d0se00782j.

[16] J. Meiss, M. K. Riede, and K. Leo, "Towards efficient tin-doped indium oxide (ITO)-free inverted organic solar cells using metal cathodes," *Appl. Phys. Lett.*, vol. 94, no. 1, pp. 92–95, 2009, doi: 10.1063/1.3059552.

[17] P. T. S. de la Cruz, K. Irikura, A. Lachgar, J. C. Cardoso, H. A. Cavero, and M. V. B. Zanoni, "Preparation of FTO/CU2O Electrode Protected by PEDOT:PSS and Its Better Performance in the Photoelectrocatalytic Reduction of CO2 to Methanol," *Electrocatalysis*, vol. 11, no. 5, pp. 546–554, 2021, doi: 10.1007/s12678-020-00612-z.

[18] S. Singh, P. Singh, O. Parkash, and D. Kumar, "Synthesis, microstructure and electrical properties of Ti doped SrSnO3," *Adv. Appl. Ceram.*, vol. 106, no. 5, pp. 231–234, 2007, doi: 10.1179/174367607X202573.

[19] M. Khalid, A. Roy, S. Bhandari, S. Sundaram, and T. K. Mallick, "Integrating concentrated optics for ambient perovskite solar cells," *Energies*, vol. 14, no. 9, 2021, doi: 10.3390/en14092714.

[20] Z. Wang *et al.*, "High irradiance performance of metal halide perovskites for concentrator photovoltaics," *Nat. Energy*, vol. 3, no. 10, pp. 855–861, 2018, doi: 10.1038/s41560-018-0220-2.

[21] H. Baig, H. Kanda, A. M. Asiri, M. K. Nazeeruddin, and T. Mallick, "Increasing efficiency of perovskite solar cells using low concentrating photovoltaic systems," *Sustain. Energy Fuels*, vol. 4, no. 2, pp. 528–537, 2020, doi: 10.1039/c9se00550a.

ASSESSING THE USE OF POLYVINYLIDENE FLUORIDE AS A BINDER MATERIAL FOR PRODUCING CARBON BASED INKS SUITABLE FOR HTM-FREE PEROVSKITE SOLAR CELLS

C. Montes[1,3], L. Ocaña[1,3], S. González-Pérez[2], B. González-Díaz[3], E. Llarena[1].
[1]Instituto Tecnológico y de Energías Renovables, S. A. (ITER)
Pol. Industrial de Granadilla, s/n, E 38600 Granadilla de Abona, Spain. Ph. +34 922 747 700. Fax +34 922 747 701. E-mail cmontes@iter.es
[2]Departamento de Didácticas Específicas. Universidad de La Laguna.
C/ Pedro Zerolo, s/n, Edificio Central Planta 2, Apartado 456. Código postal 38200
San Cristóbal de La Laguna. S/C de Tenerife. Spain.
[3]Departamento de Ingeniería Industrial. Escuela Superior de Ingeniería y Tecnología, Universidad de La Laguna (ULL).
Avenida Astrofísico Francisco Sánchez, S/N. 38206 La Laguna (Tenerife), Spain.

ABSTRACT: A process has been devised to develop pastes based on the use of polyvinylidene fluoride (PVDF), as a binder material, together with graphite and black carbon, as conductive materials to produce back contacts to HTM-free perovskite solar cells. Although the consistency of the pastes produced was not compatible with their spreading using Dr. Blade techniques, contacts were deposited by an alternative method, similar to cement rendering, as it was done in previous attempts, when ethylene-vinyl acetate (EVA) was being evaluated as binder material, resulting in a more than three-fold reduction in the measured contact resistance.
Keywords: Perovskites, Contact, Metallization, Graphite.

1 INTRODUCTION

The Solar Cell Laboratory from the Instituto Tecnológico y de Energías Renovables (ITER), with the support of the La Laguna University (ULL), has been working on developing HTM-free perovskite solar cells, by means of an innovative cell architecture that incorporates a carbon back electrode based on the use of an agglomerate produced by mixing graphite and black carbon with ethylene-vinyl acetate (EVA) in toluene solutions [1], a work that was inspired by similar efforts carried out by Guanglan Liao et Al [2] and by Xingzhong Zhao et Al [3].

The first step in our procedure for depositing the back electrode over the perovskite substrate begins by spreading a small amount of the carbon agglomerate which, for that purpose, had been further diluted by using toluene. This substance was found not to be cohesive enough for using Dr. Blade techniques in order to be spread, so we had to use an alternative method, similar to cement rendering. It should be noted that this new spreading method produces thicker and uneven layers and, thus, contributes to an increase in the series resistance of the device.

In the work made by Xingzhong Zhao et Al, already mentioned above, a carbon-based electrode was developed by using a commercial carbon slurry, which was modified in order to be deposited over perovskite layers by Dr. Blade techniques. Carbon slurries, which are commonly used for producing electrodes of graphite and graphite with black carbon, are made by mixing these materials with polymers such as polyethylene terephthalate (PET) [4], polyvinylidene fluoride (PVDF) [5], combinations of PVDF and poly(methyl methacrylate) (PMMA), in solvents such as N-Methyl-2-pyrrolidone (NMP) [6], dimethylformamide (DMF), dimethylacetamide (DMA), dimethyl sulfoxide (DMSO) [7] or even water [8].

Thus, based on the studies carried out by A. Bottino et Al [9], in which the solubility parameters of PVDF in 50 different solvents are analyzed, and in the work of Mengyuan Li et Al [10], in which PVDF is dissolved in DMF at temperature, we have developed a procedure to obtain a carbon-based slurry, by mixing graphite and black carbon in a solution of PVDF in NMP so that, by following a technique similar to the one used by Xingzhong Zhao et Al, it can be further modified in order to achieve a material compatible with perovskite thin films that can be spread to produce thin and homogeneous contacts.

2 EXPERIMENTAL

All the experiments carried out in this work have been developed in an ISO 7 clean room environment.

2.1 PVDF in NMP solutions

The PVDF in NMP solutions were prepared with a concentration of 10:90% in weight (wt%), by adding 0.2056 g of PVDF [11] and 2 ml of NMP [12] to 7ml capacity lid vials. Initially the mixtures were stirred manually, using a fine spatula, and then they were homogenized using a heated magnetic stirrer for 30 minutes at 75 ° C and 450 rpm. The result of the mixing was approximately 2.2 grams of a solution per vial that have a relatively thick appearance and a yellowish to brownish color.

Figure 1: Vial with a solution of PVDF in NMP.

2.2 Preparation of carbon pastes based on graphite and black carbon dispersed in PVDF in NMP solutions

A mixture of PVDF in NMP with graphite and black carbon was prepared, at a PVDF:NMP:Graphite:BlackCarbon 2:58:30:11% weight percentage (wt%), by adding to a beaker 3,375 g of the solution of PVDF in NMP, as described in section 2.1, 6.328 g of graphite powders [13], 2.443g of black carbon [14] and, finally, adding also 9 ml of extra NMP. The mixture was first stirred manually using a spatula for 5 minutes, then it was passed through a three roll mill 5 times. The visual appearance of the result of the mixing process is that of an intensely dark and thick paste.

Figure 2: Image of the paste being homogenized in a three roll mill.

2.3 Preparation of the carbon and PVDF powder

A quantity of the carbon paste, obtained as described in section 2.2, was spread on a clean glass and then, in order to evaporate all the NMP, it was placed in a stove furnace [15] at 150°C for one hour. Once the glass had cooled, the layer deposited on it was extracted, using a glass scraper, and the detached materials were collected in an agate mortar, where they were grinded for several minutes, pouring the contents into a sample cup with lid for storage.

The weight percentage (wt%) of the resulting powder was 96:4% (Conductive powders:Binder).

Figure 3: Image of carbon in PVDF powder preparation.

2.4 Preparation of carbon and PVDF in Toluene pastes

Two pastes were prepared as follows: For the first one, 4.14 g of carbon and PVDF powder, obtained as described in section 2.3, was placed in a beaker, to which 6 ml of Toluene [16] was added. These two components were mixed by using a spatula during 5 minutes, resulting on a rather dense paste.

For the second paste, 8.28 g of the first one was spared in yet another beaker, to which further 4 ml of Toluene was added. Then, as in the first case, a 5 minute mixing took place by using a spatula. The resulting product, although more liquid than the first one, could not yet be considered as an ink.

No further attempt was made to dilute the carbon and PVDF powder as the second paste was already showing signs of starting to disperse, denoting binder failure.

Table I: Studied paste recipes, expressed in weight percentage (wt%).

Paste #	Wt %		
	Conductive Powders	Binder	Solvent
1	43	2	56
2	30	1	69

2.5 Contact deposition and curing.

Unfortunately, the pastes produced were not liquid enough to be spread by using Dr. Blade methods. However, in order to deposit contacts whose electric properties could be measured, a technique was devised, in which small quantities of paste were thinly distributed over glass substrates of 175mm x 175mm x 2mm, along narrow straight lines, at which ends pieces of conductive and adhesive copper tape [17] had been previously glued. Then strips of adhesive Kapton tape were placed over the contacts thus produced, in order to prevent them from disaggregating during their characterization. Finally, the samples were cured in a stove furnace at 100°C during 30 minutes.

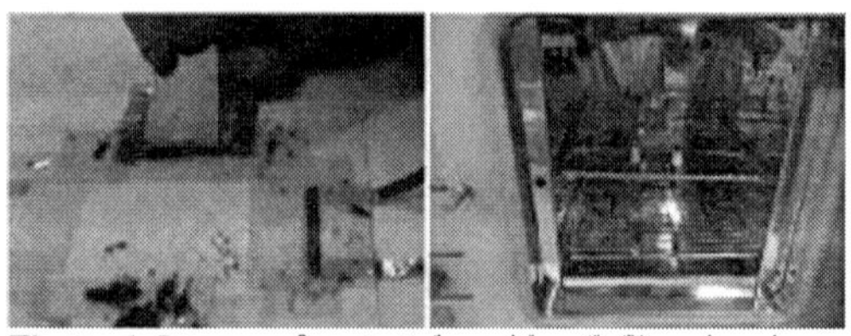

Figure 4: Images of contact deposition (left) and curing (right).

2.6 Characterization

Initially, the purpose was to deposit contacts that could be characterized according to the Standard F 1896 Test Method for Determining the Electrical Resistivity of a Printed Conductive Material [18]. However, the produced contacts did not meet the required 1:60 width to length ratio so, although it was possible to make very precise measurements of the contacts' Resistance (Ω) by using a semiconductor characterization system [19], configured to operate as an ohmmeter, other parameters such as Sheet Resistance (Ω/sq) or Volume Resistivity (Ω·cm) could not be inferred.

3 RESULTS AND DISCUSSION

Although the shape and thickness of the deposited contacts varied considerably, their overall aspect was homogeneous, particularly for the ones deposited with the second paste. However, their adherence to the substrates was not very strong and the contact itself tended to disaggregate. This suggest that the developed binder was successful at keeping the ink together, even after applying relatively high amounts of toluene as thinner material, but not enough to make it solidly stick to the glass, as it was intended.

Table II shows the values of the contacts' Resistance measured. Although these results indicate that both pastes produced similar contacts, it is possible to infer that the second paste resulted on more homogeneous contacts, despite of having higher amounts of Toluene.

Table II: Resistances measured of deposited contacts

Paste #	Electrode (Ω)				Avd. (Ω)	σ (Ω)
	A	B	C	D		
1	687.64	113.94	156.79	87.2	261.47	285.54
2	432.05	104.88	186.41	135.22	214.64	148.79

In order to assess whether these results represent an improvement over previous studies, we have compared them with the ones presented by our group during the 37th European Photovoltaic Solar Energy Conference [1], in which contacts with similar characteristics, in terms of size and thickness, were deposited by using an analogous technique, from an agglomerate of graphite and black carbon in an EVA in Toluene solution. As can be seen in Table III, despite having almost the same composition, a reduction of more than three times in contact resistance has been achieved.

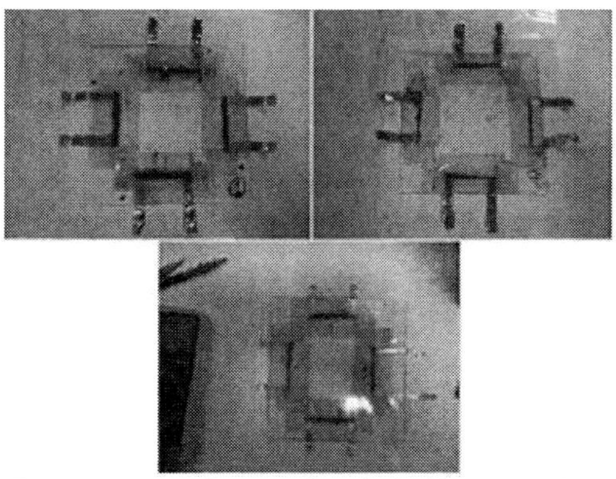

Figure 5 Images of contact deposited with paste #1 (top left), paste #2 (top right) and with the EVA-based one (below).

Table III: Resistances measured in the deposited contacts together with their corresponding composition in weight percentage (wt%) for the cured material.

ID	R (Ω)	Wt%	
		Conductor	Binder
EVA based agglomerate	759.57	95	5
PVDF based paste #1	261.47	96	4
PVDF based paste #2	214.64		

Finally, although it is clear that further research is required in order to achieve an ink-like material that can be spread over perovskite thin layers by using Dr. Blade techniques, the results obtained so far suggest that the developed PVDF and carbon based pastes, are capable of providing back contacts to HTM-free perovskite solar cells, with a reduction in their series resistance.

4 ACKNOWLEDGEMENTS

This work has been developed within the MACLAB-PV project framework, which has been co-financed by the INTERREG Madeira-Azores-Canarias Territorial Cooperation Programme (MAC) 2014-2020. 2nd Call. Axis 1 – Enhancing research, technological development and innovation.

Also, the authors would like to thank Marcos Barraca, Sales Manager at Plastic Agents, S.L. and Susana Gómez, Product Manager at Internatio-Müller Chemical Distribution (IMCD) - España Especialidades Químicas, S.A., for kindly providing the PVDF (Mr. Barraca) and the black carbon (Ms. Gómez) that are in use throughout our experiments.

5 REFERENCES

[1] C. Montes, L. Ocaña, S. González-Pérez, B. González-Diaz, M. Friend, M. Cendagorta, "Developing an Agglomerate of Graphite and Black Carbon in an Ethylene-Vinyl Acetate in Toluene Solution for Producing Electrodes for HTM-Free Perovskite Solar Cells", 37th European Photovoltaic Solar Energy Conference and Exhibition, 667-672. DOI: 10.4229/EUPVSEC20202020-3BV.1.27

[2] Z.Liu, T. Shi, Z. Tang, G. Liao, (2017) "A large-area hole-conductor-free perovskite solar cell based on a low-temperature carbon counter electrode". Materials Research Bulletin, Volume 96, Part 3, 196-200. DOI: 10.1016/j.materresbull.2017.03.069

[3] Fangjie Li Changlei Wang Pei Liu Yuqing Xiao Lihua Bai Fei Qi Xiaoyi Hou Huijie Zhang Yuan Wang Shaofu Wang Xingzhong Zhao , "Fully Air-Processed Carbon-Based Efficient Hole Conductor Free Planar Heterojunction Perovskite Solar Cells With High Repoducibility and Stability", Solar RRL, 3, December 2018.

[4] JUJO Techincal Information (2010). 9000 Series Pet Ink. www.jujo-chemical.co.jp/e/tec_info/yozai/9000_PET.html

[5] KF Polymer - Poly(vinylidene fluoride) (PVDF) leaflet by KUREHA

[6] Yan Wang, Huiyuan Zheng, Qunting Qu, Li Zhang, Vincent S. Battaglia, Honghe Zheng, "Enhancing electrochemical properties of graphite anode by using poly(methylmethacrylate) – poly(vinylidene fluoride) composite binder", Carbon, Volume 92, October 2015, Pages 318-326.

[7] Surface Chemistry and Morphology of Binders in Graphite Anodes of Lithium Ion Batteries

[8] Haoxiang Zhong, Aiqin He, Jidian Lu, Minghao Sun, Jiarong He, Lingzhi Zhang, "Carboxymethyl chitosan/conducting polymer as water-soluble composite binder for LiFePO4 cathode in lithium ion batteries" Journal of Power Sources, Volume 336, 30 December 2016, Pages 107-114

[9] A. Bottino, G. Capannelli, S. Munari, and A. Turturro, "Solubility parameters of poly(vinylidene fluoride)," J. Polym. Sci. Part B Polym. Phys., vol. 26, no. 4, Apr. 1988, doi: 10.1002/polb.1988.090260405.

[10]M. Li et al., "Controlling the microstructure of poly(vinylidene-fluoride) (PVDF) thin films for microelectronics," J. Mater. Chem. C, vol. 1, no. 46, pp. 7695–7702, Dec. 2013, doi: 10.1039/c3tc31774a.

[11]KF Polymer W#1700 by KUREHA

[12]Reference 328634, purchased from Sigma Aldrich.
[13]Graphite powder SC 20 OS from Graphite Kropfmühl GmbH Langheinrichstraße 1, 94051 Hauzenberg Germany.
[14]Black carbon powder. Conductex SC Ultra by Birla Carbon.
[15]Stove Furnace. CONTERM 19L, J.P.Selecta.
[16]Toluene anhydrous, 99.8% (244511). Sigma aldrich.
[17]Copper tape 3M 1181, Antistatic, w. 9.5mm, l. 16m.
[18]F 1896 – 98 (Preaproved 2004). Test Method for Determining the Electrical Resistivity of a Printed Conductive Material. Test method © by ASTM International, 100 Barr Harbor Drive, PO Box C700, West Conshohocken, PA 19428-2959, United States.
[19]Semiconductor Characterization System model SCS-4200 by Keithley.

SYNTHESIS AND DEFECT CHARACTERIZATION OF 2D HYBRID-PEROVSKITES

Gerd Fischer [a,*], Katharina Klose [b], Tomas Kleint [a], Hartmut Stöcker [c],
Jan Beyer [b], Johannes Heitmann [b], Matthias Müller [b]
[a] Hochschule Zittau/Görlitz, Theodor-Körner-Allee 16, 02763 Zittau, Germany;
[b] TU Bergakademie Freiberg, Institute of Applied Physics, Leipziger Str. 23, 09599 Freiberg, Germany
[c] TU Bergakademie Freiberg, Institute of Experimental Physics, Leipziger Str. 23, 09599 Freiberg, Germany;
* corresponding author: gerd.fischer@hszg.de

ABSTRACT: In addition to the now established hybrid-perovskites, materials with a 2D-perovskite structure are promising candidates for the further improvement on the way to industrial multi-junction solar cells. In this work, two synthesis routes for the preparation of 2D hybrid-perovskites have been evaluated. The 2D-perovskite [benzodiimidazolium]PbI_4 was successfully synthesized via the solution method. For other perovskites, we found that the precipitation of PbI_2 is kinetically favored using this method. Instead, the direct synthesis method has been successfully applied for the 2D-perovskite [benzimidazolium]$_2PbI_4$. XRD analysis confirmed the 2-dimensional perovskite structure. The photoluminescence and absorption spectra for both perovskites show a double peak and a shoulder structure, respectively. We attribute this either to an exciton with a binding energy in the range of 100 meV or a defect band which lies at this energy below the band gap. The bandgap energy of 2.1eV was determined by UV-VIS absorption and photoluminescence measurements with an adjusted Tauc model.
Keywords: Perovskite, Defects, Photoluminescence, Band gap

1 INTRODUCTION

For some years now, materials with a perovskite structure have been seen as the most promising options for new types of solar cells [1]. The achieved efficiencies are impressive. We believe the greatest potential for an actual industrial application of these materials is as a top cell on a silicon solar cell. This would enable the path from multi-junction solar cells to industrial production, as predicted by the International Technology Roadmap for Photovoltaics (ITRPV 11th edition) [2]. With such an industrial application, the need for detailed defect characterization will increase enormously.

In addition to the established hybrid-perovskites, materials with a 2D-perovskite structure are promising candidates for the further improvement. Particularly a higher robustness against hydrolysis (moisture) and oxidation (air) can be expected [3].

The intention of this work is to analyze intrinsic and induced defects in 2-dimensional hybrid-perovskites suitable for photovoltaic applications. For this purpose, we develop simple and cheap syntheses for 2D hybrid-perovskites and characterize the resulting material by optical and electron microscopy, UV-VIS spectroscopy, power-dependent and transient photoluminescence (PL) and XRD.

2 SYNTHESIS PROCEDURES

2.1 Solution Method

In the so-called solution method [4] hybrid perovskites are synthesized from a mixture of hydroiodic acid and phosphinic acid. This method can result in very pure and defect-free crystalline material. We were able to synthesize [benzodiimidazolium]PbI_4 according to [3] with the solution method (see Fig. 1). This has resulted in a crystalline material as reported earlier [5]. No PbI_2 could be detected in the XRD measurements of [benzodiimidazolium]PbI_4. For other hybrid-perovskites, we found that the precipitation of PbI_2 is kinetically favored using the solution method [6]. Nevertheless, we continue to pursue promising approaches for an applicable solution method by varying the polarity of the solvent.

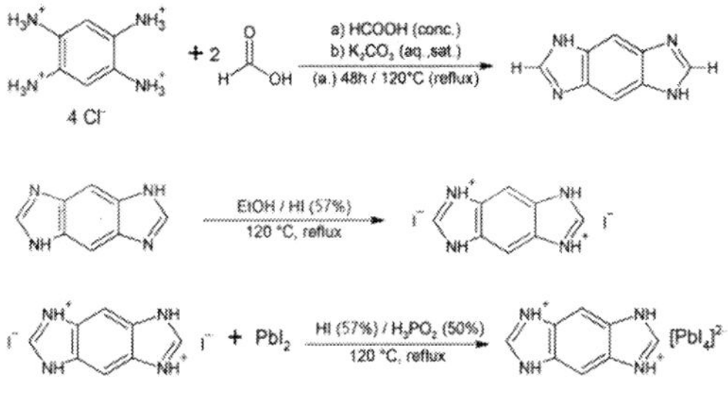

Figure 1: Synthesis route to [benzodiimidazolium]PbI_4

2.2 Direct Synthesis

Since we found the solution method not yet practicable for other 2D hybrid-perovskites due to the precipitation of PbI_2, we evaluated alternative synthesis procedures. Another route for the synthesis of perovskites is the procedure in which the hydroiodide and PbI_2 are mixed in solvents like dimethylformamide (DMF) [4]. This has been successfully applied for the 2D perovskite [benzimidazolium]$_2PbI_4$. The corresponding synthesis route is shown in Fig. 2.

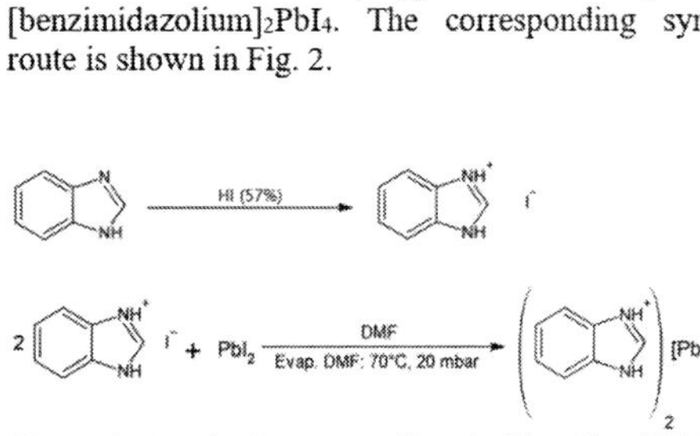

Figure 2: Synthesis route to [benzimidazolium]$_2PbI_4$

The product could be obtained in good crystallinity (see Fig. 3).

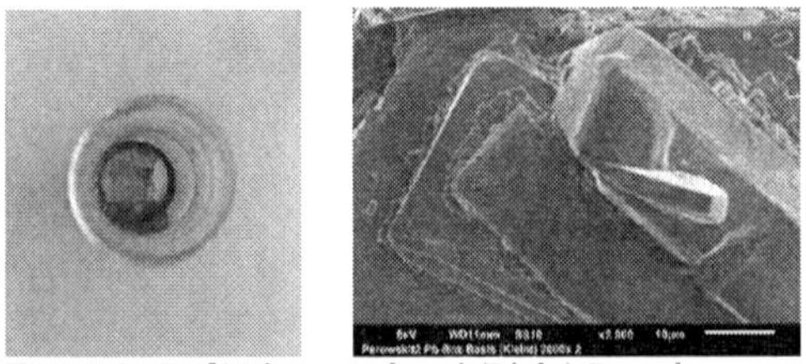

Figure 3: (Left) Photograph and (Right) Scanning Electron Microscopy image of the crystals of [benzimidazolium]$_2$PbI$_4$.

3 CHARACTERIZATION RESULTS AND DISCUSSION

3.1 Powder X-Ray Diffractometry (XRD)

The XRD analysis confirms the reported 2-dimensional perovskite structure for [benzodiimidazolium]PbI$_4$ (see Fig. 4) [3]. A crystallite size of approx. 125 nm is obtained from the refinement. Neither PbI$_2$ nor PbO can account for the additionally observed reflections (marked by * in Fig. 4). Their origin remains to be resolved.

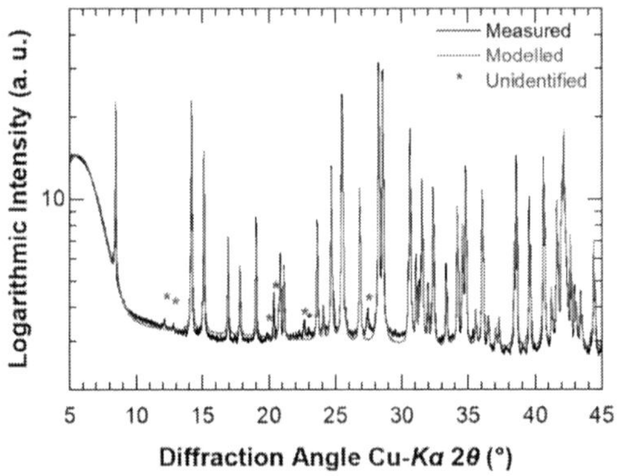

Figure 4: XRD measurement [benzodiimidazolium]PbI$_4$.

The structure of the [benzimidazolium]$_2$PbI$_4$ [7] obtained via direct synthesis in DMF can be identified accordingly (see Fig. 5). Again, the origin of the unidentified reflections (marked by * in Fig. 5) still has to be resolved.

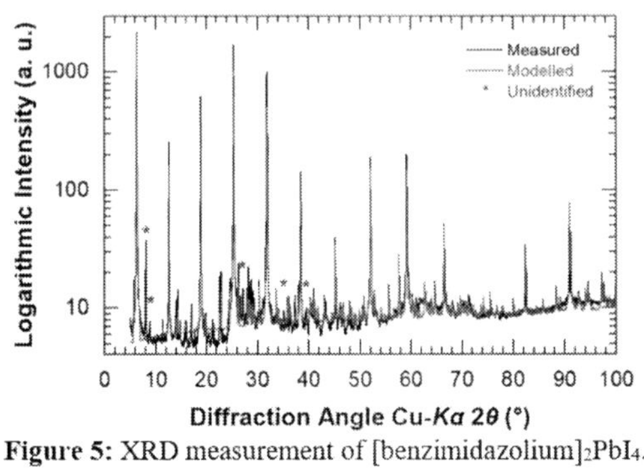

Figure 5: XRD measurement of [benzimidazolium]$_2$PbI$_4$.

3.2 PL and UV-VIS absorption spectra characteristics

Both 2D-perovskites have been analyzed by PL and UV-VIS absorption spectroscopy.

The PL was excited by a solid-state laser emitting at 442 nm (HeCd 442). Typically, employed excitation powers were in the range of 0.5 μW to 30 mW, focused to a diameter of ~150 μm. The spectra were detected by a 0.5 m grating spectrograph (Acton SpectraPro 2560i) equipped with a liquid nitrogen cooled CCD camera (Princeton Instruments SPEC-10:100BR_eXcelon). All spectra were corrected for the spectral response of the detection system. Power-dependent PL spectra have been taken using the same detection system with a 337 nm pulsed nitrogen laser (3 ns pulse width) and a variable attenuator as the PL excitation source.

The UV-VIS spectra were measured in a 150 mm InGaAs integrating sphere from Perkin Elmer.

The PL and absorption spectra show a double peak and a shoulder structure at 2.05 eV and 2.15 eV, respectively (see Fig. 6 and 7). We attribute this to an exciton with a binding energy in the range of 100 meV in 2D-perovskites [8] or a defect band which lies at this energy below the band gap. It is noteworthy that this double-peak structure appears very similar in both perovskites. In our opinion, this indicates that the semiconductor properties are predominantly determined by the 2-dimensional layer structure formed from the [PbI$_6$] octahedrons.

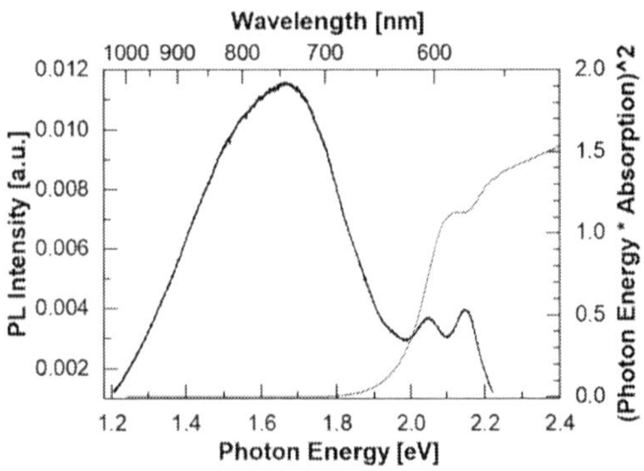

Figure 6: PL (black line) and absorption spectrum (red line) of [benzodiimidazolium]PbI$_4$

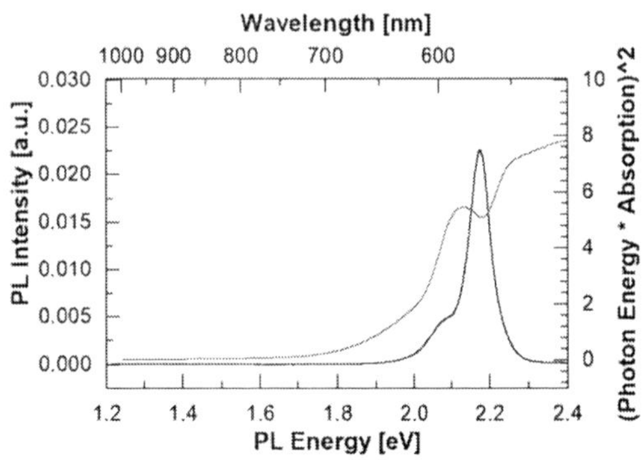

Figure 7: PL (black line) and absorption spectrum (red line) of [benzimidazolium]$_2$PbI$_4$.

Additionally, the photoluminescence spectrum of [benzodiimidazolium]PbI$_4$ shows a broad peak at 1.6 eV. The nature of this peak is attributed to defect states being identified by a typical sublinear behavior in the power-dependent PL (see Fig. 8).

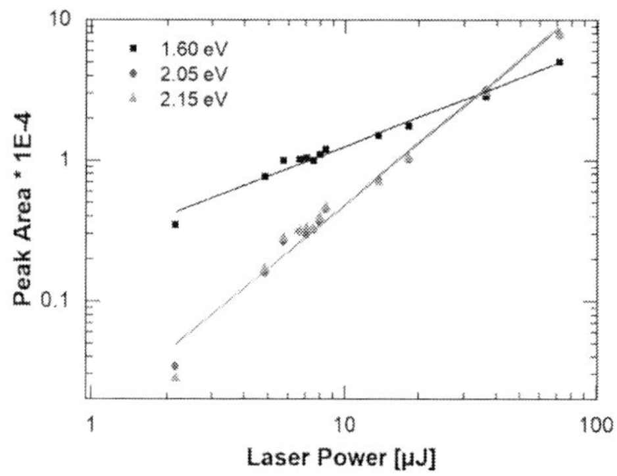

Figure 8: Power- dependent PL of [benzodiimidazolium]PbI_4 at the three PL peaks.

3.3 Determination of the band gap energy

It is assumed that the double peak structure in the absorption spectra (compare Fig. 6 and 7) is caused by an overlap of either an excitonic or a defect band (resulting in the low-energy shoulder) and the direct band-to-band absorption (at higher energies). In order to fit the band gap correctly, the Tauc model must be corrected accordingly. We used a Gaussian function approximation to separate shoulder and direct band gap influence in the spectra (see Fig. 9 and 10, light blue curve).

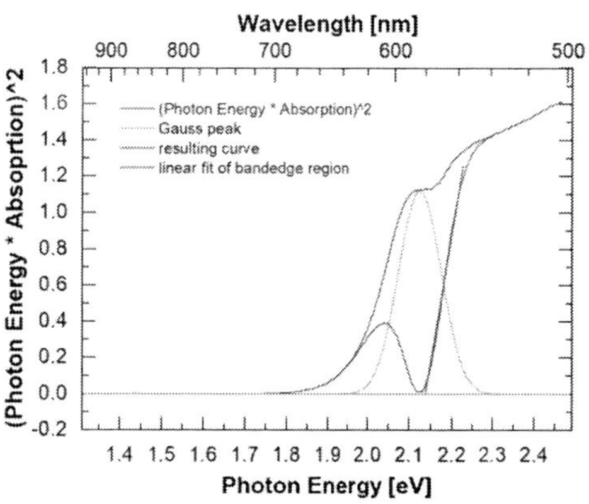

Figure 9: Band gap fitting from the absorption measurement for [benzodiimidazolium]PbI_4

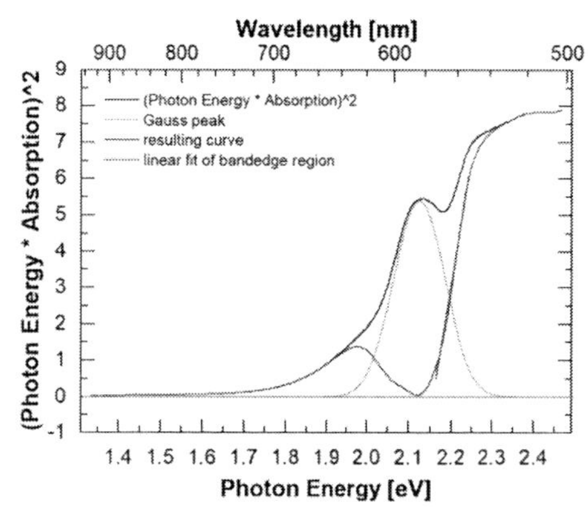

Figure 10: Band gap fitting from the absorption measurement for [benzimidazolium]$_2$$PbI_4$.

Even if the resulting difference curve (dark blue line) still has a low energy tail, which is not reflected by the Gaussian excitonic peak, the band edge can be fitted sufficiently well giving the energetic position of the band gap. The linear fit for a direct band gap (see Fig. 9 and 10, red curve) results in a band gap energy of 2.14 eV and 2.16 eV for [benzodiimidazolium]PbI_4 and [benzimidazolium]$_2$ PbI_4, respectively. Thus, we suggest, that in order to characterize the absorption edge itself, the formed exciton or defect band at about 100 meV lower energy has to be taken into consideration.

4 CONCLUSIONS

We have evaluated two synthesis routes for the preparation of 2D hybrid-perovskites. This allows us an in-depth analysis of these promising materials and a further development of our characterization capabilities for hybrid-perovskite based photovoltaic components.

5 ACKNOWLEDGMENTS

This work was supported by the SMWK (Project DeHyPe). This measure is co-financed with tax funds based on the budget decided by the members of the Saxon State Parliament.

6 REFERENCES

[1] Kazim, S., Nazeeruddin, M. K., Grätzel, M., & Ahmad, S. (2014). Perovskite as light harvester: a game changer in photovoltaics. Angewandte Chemie International Edition, 53(11), 2812-2824.

[2] International Technology Roadmap for Photovoltaic (ITRPV) 11th Edition. (2020). https://itrpv.vdma.org/

[3] Zimmermann, I., Aghazada, S., & Nazeeruddin, M. K. (2019). Lead and HTM Free Stable Two-Dimensional Tin Perovskites with Suitable Band Gap for Solar Cell Applications. Angewandte Chemie, 131(4), 1084-1088.

[4] Stoumpos, C. C., Malliakas, C. D., & Kanatzidis, M. G. (2013). Semiconducting tin and lead iodide perovskites with organic cations: phase transitions, high mobilities, and near-infrared photoluminescent properties. Inorganic chemistry, 52(15), 9019-9038.

[5] Fischer, G., Beyer, J., Müller, L., Jarosz, A., Heitmann, J, & Müller, M. (2020). Synthesis of Mixed-Halide Hybrid Perovskites for Defect Characterization. Presented at the 37th European Photovoltaic Solar Energy Conference and Exhibition (Vol. 7, p. 678-681).

[6] Müller, L., Kleint, T.; assignment work, Hochschule Zittau/Görlitz (2020).

[7] Lermer, C., Harm, S. P., Birkhold, S. T., Jaser, J. A., Kutz, C. M., Mayer, P., Schmidt-Mende, L., & Lotsch, B. V. (2016). Benzimidazolium lead halide perovskites: effects of anion substitution and dimensionality on the bandgap. Zeitschrift für anorganische und allgemeine Chemie, 642(23), 1369-1376.

[8] Han, X., Zheng, Y., Chai, S., Chen, S., & Xu, J. (2020). 2D organic-inorganic hybrid perovskite materials for nonlinear optics. Nanophotonics, 9(7), 1787-1810.

A STUDY OF QUENCHING APPROACHES TO OPTIMIZE ULTRASONIC SPRAY COATED PEROVSKITE LAYERS SCALABLE FOR PV

Joao Silvano[*†§], Jacopo Sala[*†§], Tamara Mercks[*§], Yinghuan Kuang[*§], Tom Aernouts[*§], Bart Vermang[*†§], Wim Deferme[*†]
[*] Imec – Partner in Solliance, Kapeldreef 75, 3000 Leuven, Belgium
[†] Hasselt University, Institute for Materials Research, Wetenschapspark 1, 3590 Diepenbeek, Belgium
[§] EnergyVille, Thor Park 8320, 3600 Genk, Belgium

Email: joao.silvano@uhasselt.be; Tel: +32 (0)11 26 88 74

APPLICATION AREA: 3.1 Perovskites.

JUSTIFICATION: While the main focus of the PhD project is the integration with CIS bottom cells in 2T tandem devices, the work reported here describes the upscaling of perovskite film deposition by ultrasonic spray coating, with a focus on the use of different quenching techniques.

1 INTRODUCTION

With the fast paced and incremental improvements in efficiency and stability, perovskite solar cells have been consolidated as a critical player in the future of photovoltaic energy generation. Aside for its high efficiency, the band gap tunability granted by the diversity of compositions allows the material to be used in alignment with other technologies, such as silicon [1], CIGS [2], and even in devices with two different types of perovskite [3]. This factor gets more relevant as tandem devices are close to reaching the 30% efficiency margin [4].

In the case of dual junction tandems, two cells can be produced independently and mechanically stacked - which is the case of four terminal (4T) devices - or deposited layer by layer monolithically, requiring only two terminals (2T), or electrodes [5]. While 4T fabrication is less challenging, it also limits the performance due to parasitic absorption and requires twice as much external electronics, as well as transparent electrodes [6]. 2T-monolithic devices have the set-back of processing the layers of the second cell directly on top of the first, a surface that can be rough, chemical- and thermo-sensitive. To overcome these challenges it is crucial to develop compatible methods, with material deposition by replicable techniques that do not rely on high temperatures.

The fabrication of a perovskite thin film is usually a two steps process: deposition of a precursor solution, where the components are present as ions in stoichiometric proportions, followed by crystallization, when the solvent evaporates and the octahedral structure is obtained. Post processing, such as annealing, can be used to improve the characteristics of the film by creating larger grains. However, the crystallization step does not necessarily require high temperature, owing to the low crystallization activation energy of the material, typically between 50 and 100 kJ mol^{-1} [7]. In order to produce a high-quality film, compact and without pin-holes, the crystallization step must be well controlled. Initially, this was done in laboratories by solvent quenching, or the anti-solvent technique, which involves the dripping of large amounts of solvents into the precursor solution during spin coating. Problems with replicability, scalability and material waste lead to the development of alternative methods, such as vacuum quenching and later, the more industrially viable, gas quenching [8].

In the race to commercialization, novel processes, *e.g.* plasma quenching [9], are associated with high throughput deposition techniques, for instance ultrasonic spray coating and blade coating, to allow a fully roll-to-roll fabrication of perovskite films. As those processes approach the high efficiencies obtained in laboratory, perovskite photovoltaics stop being viewed as a promise for the distant future, and start gathering the interest of PV companies worldwide.

2 AIM & APPROACH

In this work we aim to optimize the fabrication of perovskite layers via quenching-assisted ultrasonic spray coating to be ultimately combined in monolithic tandem devices with CIS cells. The total efficiency of the device must be compatible with the state-of-the-art (>25% PCE [10]), with a single junction efficiency of at least 15% for the ultrasonically spray coated perovskite. Furthermore, the deposition and quenching processes of the perovskite layer must include scalable processes, focusing on replicability, low cost and low material waste. The device cannot present high risk of environmental hazard and toxicity, and must have high operational stability.

In order to accomplish these goals, the quenching-assisted ultrasonic spray coating was optimized for the fabrication of an homogeneous perovskite thin film (600 nm). The processing was then integrated in the fabrication of single junction perovskite cells. Electronic performance and morphological analysis are used for the iterative improvement of the film deposition and quality.

The selection of perovskite composition was done based on simulations for optimal performance of monolithic 2T devices with CIS (1.0 eV) as bottom cell; as such the perovskite film must have a bandgap between 1.60 and 1.65 eV [11]. Moreover, a composition of 2 cations, with a low content of cesium (Cs) added to formamidinium (FA), is preferred over the presence of methylammonium (MA) due to heightened stability.

Out of the three most frequently studied bottom cell materials, CIGS cells are thin film devices with bandgap tunability - in contrast to silicon - while being a proven technology with substantial market share, a position still to be reached by low bandgap perovskites. In addition, CIS cells show high efficiency with a bandgap of 1.0 eV [10]. As previously mentioned, the 2T configuration is chosen in order to avoid the efficiency-limiting boundaries of 4T tandems, and the monolithic deposition challenges will be tackled by studying the interlayer materials and optimizing the top cell fabrication with CIS-compatible techniques. Ultrasonic spray coating is chosen as it allows the deposition on rough surfaces, while being a scalable process with low material waste.

The preliminary results reported in this paper refer to the first step of this work: the ultrasonic spray coating of perovskite solar cells. First, the technique was studied by depositing perovskite inks of different concentrations onto plain glass. Once a high-quality layer was achieved, single junction solar cells were fabricated and evaluated in regards of their performance. One important step in this process is the study of perovskite crystallization with different quenching techniques. Replicability and scalability are also key factors in this step, so the choice of techniques leads away from batch processes, such as anti-solvent and vacuum quenching, and in favor of large area, roll-to-roll compatible ones. Nitrogen gas quenching is an intensively reported method which matches these criteria, while plasma treatment and near-infrared sintering could provide even higher throughput and cost efficiency. With that in mind, this work will be focused on this category of quenching techniques.

3 SCIENTIFIC INNOVATION AND RELEVANCE

The work presented here contributes to the lab-to-industry transition of perovskite tandem photovoltaic devices. Replacing batch techniques (*e.g.* spin coating, anti-solvent quenching) with high throughput, scalable fabrication, whilst maintaining top performance and stability, is fundamental to the commercialization of novel materials. In addition, the process of perovskite deposition on rough surfaces via ultrasonic spray coating enables tandem applications, not only with CI(G)S solar cells, but also textured silicon and other challenging surfaces.

This rises in importance as perovskite research is on the rise, on the verge of reaching 30% PCE for tandem devices, and at the same time tackling the intrinsic toxicity problems and developing device stability.

4 PRELIMINARY RESULTS & CONCLUSIONS

In the course of this work, the deposition of perovskite by ultrasonic spray coating was optimized, resulting in high quality homogeneous films with low roughness. Single junction cells were fabricated with the same techniques as spin-coated perovskite cells. While the performance is not yet on the same level as lab scale, spray coating shows the potential to approach the high efficiencies of batch, spin coated cells. For that, the search for optimal quenching techniques and controlled conditions will be continued.

Ultrasonic spray coating of perovskite

The initial experiments with the ultrasonic spray coater focused on obtaining an homogeneous perovskite layer, covering the entire surface of the sample (1 inch^2 glass), and comparing it to spin coated layers. Since the scalable technique has significantly less material waste, the precursor concentration could be reduced from 1.33 M (used in spin coating) to 0.25 M. The molecular composition was maintained in order to guarantee the desired band gap of 1.6 eV. To study the effects of evaporation time, two solvents were employed - N,N-Dimethylformamide (DMF) as high volatile, and N-Methyl-2-Pyrrolidone (NMP) as low volatile. Crystallization occurs after the ink reaches the saturation point, so having the wet film long enough to allow the introduction of a quenching technique is desired. In experiments without the presence of NMP, premature crystallization was detected. The long evaporation resulted in a non-uniform film; with crystal agglomerations and fiber-like structures.

Higher temperatures were attempted up to the system limit (100 °C), however this alone was not successful in obtaining an homogenous perovskite film. Even higher temperatures could present a risk of damaging the bottom cell in the case of monolithic tandem fabrication, so alternative methods were employed. With the mixture of NMP and DMF and low temperature (30-40°C), a wide range of spray coating parameters is able to achieve a fully covering wet layer. For the subsequent experiments, the following parameters were repeated: flow rate = 1 mL/min; hot plate temperature = 40 °C; power = 3 W; N2 pressure = 3 psi, nozzle-to-substrate distance = 10 cm; pass speed = 30 mm/sec (single pass).

Vacuum Quenching

As the first attempted quenching technique, vacuum quenching was responsible for a significant improvement in film quality, with significant roughness reduction easily observed by the layer reflectiveness and transparency. Using laser and optical microscopy it is possible to see a fully covered sample with no large agglomeration of material or fiber-like structures, although still with some impurities [Figure 2]. Furthermore, the technique provides an estimated average thickness of 609 nm - similar to the desired thickness of spin coated samples - and roughness lower than 35 nm.

With the perovskite deposition on glass reaching the desired morphology, the next step was fabricating single junction cells with the spray coated layer. The structure of the cell was in p-i-n design, with thermal evaporated copper (Cu) and indium tin oxide (ITO) electrodes, sputtered nickel oxide (NiOx) as hole transport layer (HTL), and the triple layer of thermal evaporated lithium fluoride (LiF), fullerene (C60), and bathocuproine (BCP) as electron transport layer. Each sample (3x3 cm^2 glass) produces 12 0.125 cm^2 cells. After perovskite deposition, the samples were always kept in a controlled N2 atmosphere, where their performance was measured with a solar simulator under standard test conditions (1000 W/m2 of AM 1.5G and a cell temperature of 25°C). The JV curves allow for the determination of the short-circuit current (Jsc), open circuit voltage (Voc), fill factor (FF) and power conversion efficiency (PCE).

Table I: Vacuum quenched perovskite solar cells average performance per sample (standard deviation).

Sample	Voc (mV)	Jsc (mA/cm2)	FF (%)	η (%)
Vac A	675 (62)	18.3 (0.4)	48.5 (5.2)	6.0 (1.1)
Vac B	651 (48)	18.4 (0.5)	46.6 (3.3)	5.6 (0.9)
Vac C	628 (35)	18.1 (0.4)	45.9 (3.8)	5.3 (0.8)
Vac D	605 (34)	17.8 (0.4)	49.2 (3.9)	5.3 (0.6)
Vac E	558 (20)	18.1 (0.3)	47.5 (2.7)	4.8 (0.4)
Vac F	584 (25)	18.2 (0.4)	44.3 (2.8)	4.7 (0.5)

By varying the exposure time to vacuum, a minimum of 2 minutes was determined as necessary for complete quenching of the sample after ultrasonic spray coating. Additional time under vacuum didn't significantly alter the performance of the cells. In this manner 6 samples were tested, for a total of 72 cells. The average PCE was 5.30 % ($\sigma = 0.87$ %), with top cell performance being: PCE = 8.3 %, Voc = 790 mV, Jsc = 18.2 mA/cm², FF = 58.7 % (sample Vac A, cell 6). A summary of the sample results is given in table I. These results are far from the typical values for spin coated perovskites, possibly due to the long time required in the vacuum system, which translates into slow crystallization and generates porous films.

Gas Quenching

Aiming for a more compact layer, as well as better process scalability, attention was shifted to gas quenching. Following the same spray coating procedures, the perovskite precursors are deposited as a wet film in low temperature and immediately exposed to quenching by a flow of gas originating from a N2 gun. In this process the formation of crystals on the sample are easily verified by the color change, from the transparent yellow film to a dark brown reflective thin layer. In spin coated samples, the change starts a few seconds after exposure to the gas flow, and after around 10 seconds no further change is observed. Due to lower nitrogen pressure, the ultrasonically spray coated samples exhibited longer crystallization, taking from 20 to 30 seconds before the entire surface reached a steady state, with the homogeneous brown color. Nevertheless, the crystallization was faster than with vacuum quenching, which led to far superior cell performance.

Table II: Vacuum (control sample) and gas quenched perovskite solar cells average performance per sample (standard deviation).

Sample	Voc (mV)	Jsc (mA/cm2)	FF (%)	η (%)
N2 A	888 (25)	17.2 (0.3)	71.1 (2.6)	10.8 (0.7)
N2 B	868 (6)	15.9 (0.6)	75.7 (1.7)	10.5 (0.4)
N2 C	863 (8)	16.4 (0.4)	74.9 (2.5)	10.6 (0.3)
N2 D	845 (13)	16.9 (0.3)	70.2 (1.7)	10.0 (0.4)
N2 E	868 (21)	17.6 (0.7)	71.6 (1.7)	10.9 (0.6)
N2 F	854 (25)	18.0 (1.0)	68.8 (2.5)	10.6 (1.2)
N2 G	852 (17)	17.6 (0.7)	71.9 (4.2)	10.8 (0.4)
Vac ctrl	658 (44)	17.1 (0.3)	56.4 (4.3)	6.4 (0.9)

The cells were fabricated with the same processes previously described on the vacuum quenching section, with the only difference being the quenching step for the perovskite layer. In the first experiment, 8 samples were produced, for a total of 96 cells. One of the samples was fabricated with vacuum quenching for control over different batches, and had results compatible to the previously seen range. The average PCE for the N2 quenched cells was 10.61 % (σ = 0.66 %). The top performing cell had a PCE = 13.26%, Voc = 900 mV, Jsc = 19.7 mA/cm², FF = 74.74% (sample N2 F, cell 12). The results for each gas quenched sample, as well as for the vacuum quenched control sample, are on table II. After 20 minutes of illumination, the top performing cell showed an increase in PCE and Voc, to 14.30% and 1.0 V, respectively, after which point the values stabilized. Additionally, external quantum efficiency (EQE) measurements verified the bandgap of the cell to be the desired value of 1.6 eV, ideal for tandem applications with CIS bottom cells.

In order to test the limits of the current low pressure nitrogen line, a second experiment used more focused gas flow to optimize individual cells instead of quenching the entire sample simultaneously. With this, the color change happened slightly faster in the focused region of the sample, which leads to faster and more compact crystallization. By using this method, the total of 54 studied cells exhibited improved performance overall, with an average of 12.71 % PCE ($\sigma = 0.67$ %), with the top performing cell having PCE = 13.93%, Voc = 990 mV, Jsc = 17.5 mA/cm² and FF = 80.38%. Moreover, 92% of the cells had PCE higher than 12%, which makes it likely that with further optimization of the N2 pressure would result in even higher efficiencies.

Preliminary conclusions and future plans

With the results obtained in this work so far, it is clear that the ultrasonic spray coating is a viable substitute for the batch process of spin coating perovskite layers. Similar characteristics were archived in regards to complete coverage and thickness, with 600 nm thick perovskite layers being successfully integrated in complete stacks of single junction solar cells. The large-scale technique also allows less material usage, with the more-than-80% lower ink concentration as compared to spin coating, owing to the fact that all the ink being applied to the substrate, while in the traditional technique a large part of it is wasted during the spread by spinning.

The quality and opto-electrical properties of the final perovskite film is associated mostly with the crystallization step, which occurs after the deposition of the precursor ink. By comparing different quenching techniques it was possible to go from non-homogeneous films to above-10% efficient solar cells. The comparison

between vacuum and gas quenched perovskite solar cells fabricated in this study permit the visualization of the effects of evaporation time on final photovoltaic performance, with the faster quenching method resulting in more than twice as efficient devices. While it is possible to improve both methods, vacuum quenching lacks the scalability of gas quenching. Future experiments will focus on the optimization of the N2 flow and the investigation of other industrially viable routes, especially plasma treatment and NIR sintering. These processes represent significant improvements in cost efficiency and consume less raw material, if compared to the anti-solvent technique, still used in laboratory scale. Once the processing of single junction perovskite cells is optimized, the quenching-assisted ultrasonic spray coating will be used to deposit the perovskite layer on rough surfaces, as in monolithic tandem devices with CIS bottom cells.
Alternatively, perovskite-on-kesterite tandems will be developed in parallel. CZTSSe kesterites have achieved > 12% PCE [12], with the advantage of using earth abundant, non-toxic and cheaper materials than CIGS technology. Kesterites can also be solution processed, and are suited for tandem applications with perovskite solar cells [13].

ACKNOWLEDGEMENTS

This study was supported by the Special Research Fund (BOF) of Hasselt University, BOF number: BOF19OWB17. This project has also received funding from the European Union's Horizon 2020 research and innovation programme under grant agreement No 850937.

REFERENCES

[1] A. Al-Ashouri, E. Köhnen and S. Albrecht, "Monolithic perovskite/silicon tandem solar cell with >29% efficiency by enhanced hole extraction," *Science,* vol. 370, no. 6522, pp. 13000-1309, 2020.

[2] A. Al-Ashouri, A. Magomedov and S. Albrecht, "Conformal monolayer contacts with lossless interfaces for perovskite single junction and monolithic tandem solar cells," *Energy Environ. Sci.,* vol. 12, pp. 3356-3369, 2019.

[3] G. E. Eperon, T. Leijtens, M. D. McGehee and H. J. Snaith, "Perovskite-perovskite tandem photovoltaics with optimized band gaps," *Science,* vol. 354, no. 6314, pp. 861-865, 2016.

[4] M. Hutchins, "Oxford PV retakes tandem cell efficiency record," pv magazine, 21 December 2020. [Online]. Available: https://www.pv-magazine.com/2020/12/21/oxford-pv-retakes-tandem-cell-efficiency-record/. [Accessed 10 January 2021].

[5] J. Werner, B. Niesen and C. Ballif, "Efficient Near-Infrared-Transparent Perovskite Solar Cells Enabling Direct Comparison of 4-Terminal and Monolithic Perovskite/Silicon Tandem Cells," *ACS Energy Lett.,* vol. 1, no. 2, pp. 474-480, 2016.

[6] M. T. Hörantner and H. J. Snaith, "Predicting and optimising the energy yield of perovskite-on-silicon tandem solar cells under real world conditions," *Energy Environ. Sci.,* vol. 10, pp. 1983-1993, 2017.

[7] D. T. Moore, U. Wiesner and L. A. Estroff, "Crystallization Kinetics of Organic–Inorganic Trihalide Perovskites and the Role of the Lead Anion in Crystal Growth," *J. Am. Chem. Soc.,* vol. 137, no. 6, pp. 2350-2358, 2015.

[8] A. Babaygit, J. D'Haen, H.-G. Boyen and B. Conings, "Gas Quenching for Perovskite Thin Film Deposition," *Joule,* vol. 2, no. 7, pp. 1205-1209, 2018.

[9] N. Rolston, W. J. Scheideler and R. H. Dauskardt, "Rapid Open-Air Fabrication of Perovskite Solar Modules," *Joule,* vol. 4, pp. 1-18, 2020.

[10] T. Feurer and A. N. Tiwari, "Efficiency Improvement of Near-Stoichiometric CuInSe2 Solar Cells for Application in Tandem Devices," *Adv. Energy Mater.,* vol. 9, p. 1901428, 2019.

[11] T. Feurer, "Narrow band gap Cu(In,Ga)Se2 for tandem solar cell application," ETH Zurich, 2019, p. 15.

[12] W. Wang and D. B. Mitzi, "Device Characteristics of CZTSSe Thin-Film Solar Cells with 12.6% Efficiency," *Adv. Energy Mater.,* vol. 4, pp. 1-5, 2014.

[13] Y. Li, N. Yuan and J. Dind, "Solution-processed perovskite-kesterite reflective tandem solar cells," *Solar Energy,* vol. 155, pp. 35-38, 2017.

MINIMIZING THE INTERCONNECTION WIDTH OF LASER PATTERNED PEROVSKITE SOLAR CELLS

Christof Schultz[a], Markus Fenske[a], Janardan Dagar[b], Andreas Bartelt[a], Rutger Schlatmann[a,c], Eva Unger[b], Bert Stegemann[a]

[a] University of Applied Sciences – HTW Berlin, Wilhelminenhofstr. 75a, D-12459 Berlin, Germany
[b] Helmholtz-Zentrum Berlin für Materialien und Energie, Young Investigator Group Hybrid Materials Formation and Upscaling, Kekuléstraße 5, 12489 Berlin, Germany
[c] PVcomB / Helmholtz-Zentrum Berlin für Materialien und Energie GmbH, Schwarzschildstr. 3, D-12489 Berlin, Germany

ABSTRACT: A key to the production of large-format perovskite solar modules is the monolithic series interconnection of neighboring cells with minimal power losses, which requires the optimization of the cell width and the minimization of the interconnection width. The optimal width of the cell strips was calculated to be 6 mm based on loss analysis, which was also confirmed experimentally by identifying the highest power conversion efficiencies (PCE) of laser-patterned mini-modules. By varying the scribe line spacings (P1/P2 and P2/P3), an optimal interconnection width (yielding the highest PCE) of about 230 μm was determined, corresponding to directly adjacent scribe lines and a geometric fill factor (GFF) of 94.6 %. This is an improvement compared to results achieved so far.
Keywords: Perovskite, Module manufacturing, Laser processing, Dead area

1 INTRODUCTION

Within the last years, lead halide perovskite solar cells (PSC) have shown an impressive evolution resulting in high power conversion efficiencies (PCE) exceeding 23% [1]. The high PCE of these absorber materials are assigned to the outstanding material properties, such as high optical absorption and tuneability of the band gap [2-4]. Beside the improvement of the material properties, several groups started patterning PSCs a few years ago with the goal of establishing monolithic series interconnection, which is common in other thin film solar cells and required for module fabrication.

The monolithic interconnection scheme comprises three laser-scribed lines (P1-P3, see Fig. 1) which enable series interconnection, with the P2 step opening the perovskite absorber layer and the P1 and P3 scribes separating the front and back electrode layer, respectively. The total interconnection width of these three scribe lines is referred as dead area, since it is electrically inactive and does not contribute to charge carrier generation. In order to use as large an area of the solar module as possible, the dead area width (DA) should be as small as possible while still providing full electrical functionality.

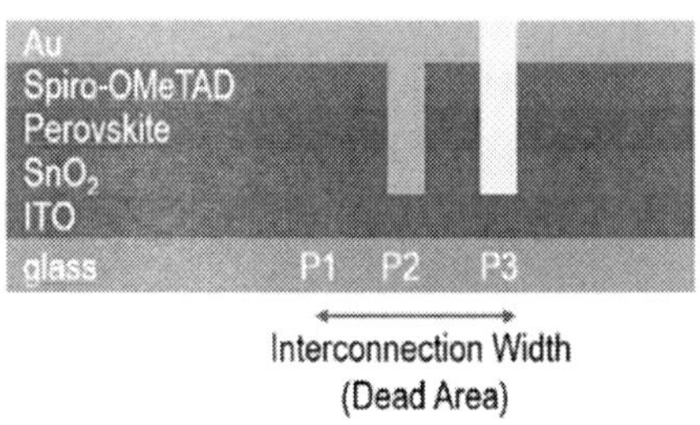

Figure 1: Schematic overview of the perovskite solar cell layer stack and the scribing lines P1-P3 for the monolithic series connection.

In previous studies the laser material interaction behavior was systematically studied [5-7] and parameters for patterning, using ps and ns laser pulses, were developed, that enable low-ohmic series interconnections as well as high isolation resistances between neighboring cells. Interconnection losses can be further reduced by minimizing the DA, which comprises the interconnection width between the respective outer edges of the P1 and P3 lines

However, current approaches are based on P2 and P3 patterning processes where the scribe lines are multiple times patterned and slightly shifted against each other to match the required electrical functionality. The achieved interconnection widths are in the range of 300 μm to 400 μm [8, 9] and, thus, rather large. However, another advantageous property of perovskite helps to reduce the DA: It was shown, that neither significant morphological changes nor increased electronic defect densities occur at the scribe line edges [10]. This seems surprising and is different from the laser patterning of CIGSe solar cells [11], which might be due to the specifics of heat propagation, the passivation of electronic defects at voids, grain boundaries and defects by laser-induced PbI_2 formation [12], as well as self-healing effects in perovskite [13]. Therefore, from the material properties point of view, potentially small DAs are achievable, when both P1 and P3 can be placed very close to the P2.

However, this also requires the consideration of so-called safety areas, which are typically inserted between the scribe lines P1 and P2 as well as between the scribe lines P2 and P3 to prevent line crossings [9]. However, currently high-resolution CCD cameras are available that allow the detection of the sample position with a very high accuracy of one micron. Thus, safety areas are no longer absolutely necessary.

Despite possible reduced interconnection widths, the electrical functionality must be ensured and viable laser processes must be explored in order to take the next step towards industrial production of efficient perovskite solar modules (PSMs). This is the subject of the present investigation.

In our approach, we first determined the optimum cell width, which influences the geometric fill factor (GFF), which is defined as the ratio of the active area (AA) in the module to the aperture ratio (i.e., sum of AA and DA). By calculations based on an equivalent circuit diagram described in detail elsewhere [14], the total losses which are the sum of the electrical losses and the area losses, were determined for different cell stripe widths (at a fixed DA of 500 μm). The results were then verified with PSCs prepared with the corresponding properties.

In the next step, perovskite mini-modules with the optimized cell stripe width were fabricated by fully laser-based series interconnection of three cells with dead area widths gradually reduced from over 500 μm to nearly 100 μm. The DA reduction is achieved by gradually reducing the distances between the P1 / P2 lines and P2 / P3 lines.

The morphology of the scribe lines was characterized by laser scanning microscopy. By analyzing the resulting j-V characteristics of the interconnected mini-modules, the influence of the dead area width on the solar cell parameters was determined and the optimum of efficiency and GFF was derived.

2 EXPERIMENTAL

2.1 Sample preparation

The layer stack of the solar cell samples is schematically shown in Fig. 1 and was prepared following the preparation procedure for perovskite baseline samples described in detail elsewhere [6, 15]. In brief, a standard glass substrate coated with a ITO front contact layer (140 nm) was used for sample preparation. Then, the ITO layer was patterned by laser (ps, 1064 nm) and a SnO_2 electron transport layer (23 nm) was deposited by spin-coating. Curing was carried out under ambient atmosphere. Subsequently, the perovskite ($Cs_{0.05}(FA_{0.83}MA_{0.17})_{0.95}Pb(I_{0.83}Br_{0.17})$) absorber layer was deposited by spin-coating, its resulting thickness was about 650 nm after annealing. Spin-coating was also used to deposit the spiro-OMeTAD hole transport layer (180 nm). Subsequently, P2 patterning was performed by laser (ps, 532 nm). A thin Au film (100 nm) was evaporated as back contact. Finally, in order to electrically separate neighboring cells, a laser (ps, 532 nm) was applied for P3 patterning. The resulting perovskite mini-module (PSM) consists of 3 series interconnected cells (3 × 0.733 cm²) with a total active area of 2.2 cm².

2.2 Laser patterning

The samples utilized in this work were patterned by a customized, commercially available patterning tool (Rofin Baasel Lasertech). This tool is equipped with high-precision motor stages for x-y sample translation with velocities of up to 1.2 m/s and a positioning accuracy below 5 μm. A ps laser source (Time Bandwidth, Duetto) emitting laser pulses at wavelengths of 1064 nm and 532 nm. Maximum pulse energy of about 114 μJ is supplied at 1064 nm; at 532 nm the maximum pulse energy is about 65 μJ. The pulse duration is specified with 10 ps for both wavelengths; the intensity distribution is Gaussian. The optimal patterning parameters have been determined in previous studies and can be found elsewhere [5, 6]. Here, in this work the optimal parameters for sample patterning were applied. Precise positioning of the laser spot on the sample is ensured by camera-based fiducial recognition. As fiducials two crosses (size: 2 mm) located at opposite corners of the sample and patterned in parallel with the P1 step were utilized.

2.3 Characterization techniques

The electrical functionality of the PSM was verified by current density-voltage (j-V) measurements using an AAA-class solar simulator (Wavelabs Sinus 70), operated at 1 sun equivalent (AM 1.5). In addition, the morphology of the scribed lines was examined by laser scanning microscopy (LSM, Keyence VK-X260).

3 RESULTS AND DISCUSSION

Before the minimization of the dead area can be carried out, the optimal width of the solar cell within the mini-module must be determined. Therefore, a common equivalent electric circuit diagram was utilized [14], to calculate the total power losses as a function of the cell width (range: 1 to 10 mm) for a fixed DA of 500 μm. A detailed explanation and an overview of the input parameters can be found elsewhere [14].

In Fig. 2a, the calculated total losses resulting from the electrical losses and the area losses are plotted against the cell stripe width. The minimum of the losses is at a cell stripe width of 6 mm. In Fig. 2b the experimental results of the electrically measured mini-modules, with the corresponding cell stripe widths, are displayed. The diagram shows the data points for the top efficiency and with the mean value of about 50 samples each. Here, a good agreement of the minimum total loss and the highest achieved efficiency of the mini modules can be seen. A cell width of 6 mm was therefore used for the subsequent investigations to minimize the dead area width.

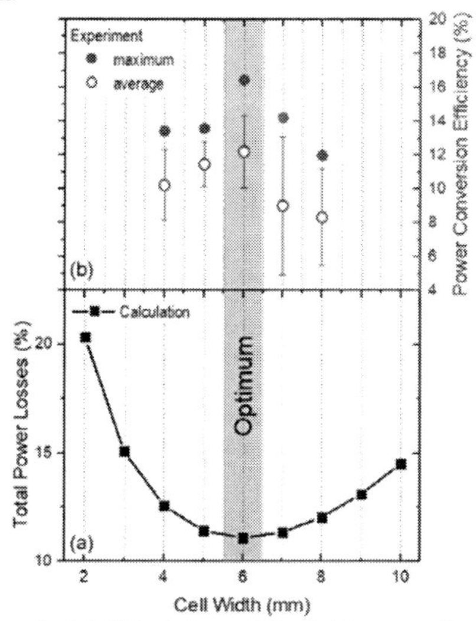

Figure 2: (a) Calculation of the total power losses as function of the cell width. (b) Statistical analysis of the power conversion efficiencies of the prepared mini-modules (average and maximum values of about 50 samples each).

Figs. 3a-d show laser scanning microscopy images of the interconnection regions. There, the total interconnection widths were gradually reduced from 430 μm down to 130 μm, by reducing the interline P1/P2 and P2/P3 spacings. The line widths and the varying interline spacings are marked in the corresponding images of Fig. 3. Since the patterning was carried out with the same laser parameters, the individual widths of the scribe lines P1, P2, and P3 are constant in each case.

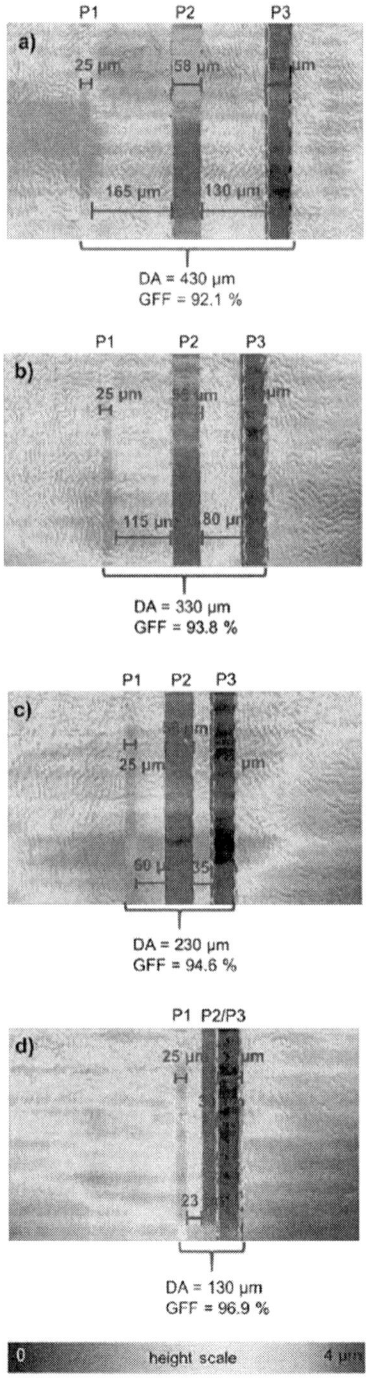

Figure 3: Height images of the interconnection regions, acquired by laser scanning microscopy for total interconnection widths of (a) 430 µm, (b) 330 µm, (c) 230 µm, and (d) 130 µm. The heights correspond to the given scale bar. The scribe line spacings P1/P2 and P2/P3 as well as the P1, P2, P3 line widths are indicated in each image.

The images in Fig. 3 demonstrate that for total interconnection widths of 430 µm, 330 µm, and 230 µm (see Figs. 3a-c) the scribe lines P1, P2, and P3 are clearly separated from each other, while at narrower interconnection widths (see Fig. 3d) the P2 and P3 scribe line start to partially overlap. Furthermore, the results prove that by using a camera based fiducial detection system, to detect the sample position and aligning the patterning layout, unintentional crossing of the adjacent parallel lines can be successfully prevented.

Comparing the different interconnecting regions with the corresponding series resistances and the resulting solar cell efficiencies, a clear dependence can be observed. In Fig. 4, the measured series resistances, approximated at the open-circuit voltage from the j-V characteristics of the laser-patterned perovskite mini-modules (i.e., R_{oc}), are plotted as a function of the total interconnection width.

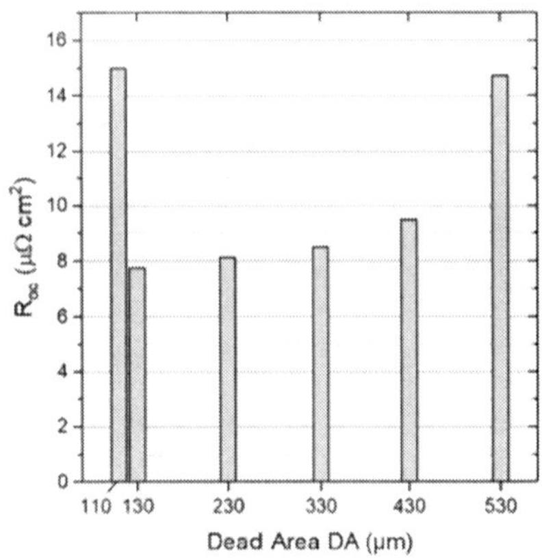

Figure 4: Measured series resistances as function of the dead area width.

It can be seen that with decreasing dead area the series resistance decreases and reaches its minimum at about 130 µm, while it increases again at narrower interconnections width, i.e., at the onset of overlap of the P2 and P3 lines.

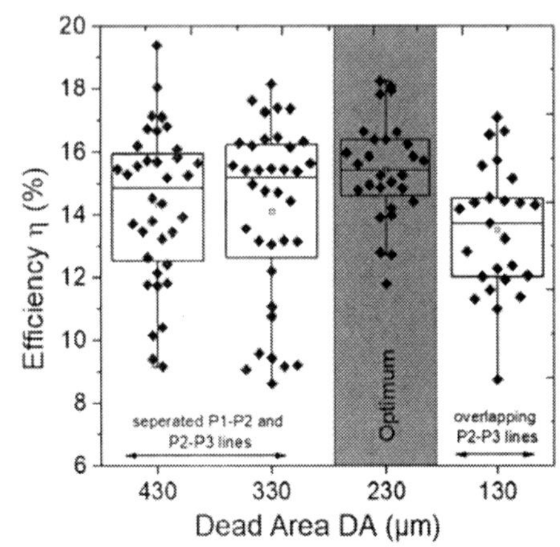

Figure 5: Solar cell efficiencies as a function of the dead area width.

In Fig. 5 the corresponding mini-module efficiencies, are shown. Here, a similar trend can be observed, that with decreasing dead area the mini-module efficiencies slowly increase. But unlike the series resistance, here the maximum is not found at DA = 130 µm but at DA = 230 µm. This discrepancy can be attributed to increased losses in the ITO layer in case of large interconnection widths, while increased contact resistances occur for the overlapping scribe lines due to a reduced surface contact.

Accordingly, the optimal interconnection width is achieved when the P2 and P3 scribe lines are directly adjacent without overlap (i.e. at DA= 230 µm), which is remarkable lower compared to the values reported previously in literature. The corresponding geometric fill factor (GFF) is 94.6%. However, functional mini-modules were also manufactured with interconnection widths of 110 µm and 130 µm (FF: 96.8% and 97.3%, respectively), albeit with the drawback of higher electrical losses.

4 SUMMARY

To minimize the interconnection losses of perovskite solar cells interconnected in series to form a module, the optimum cell stripe width and the optimum of interconnection width and power conversion efficiency were determined. The optimal width of the cell strips was calculated to be 6 mm based on loss analysis, which was also confirmed experimentally by determining the highest PCE of laser-patterned mini-modules. By varying the scribe line spacing (P1/P2 and P2/P3), an optimum interconnection width (yielding the highest PCE) of about 230 µm was determined, corresponding to a geometric fill factor of 94.6%, which is an improvement over the previously obtained results [8, 9].

These results also show that the optimum interconnection width is achieved when the P2 and P3 scribe lines are directly adjacent and do not overlap. However, functional mini-modules have also been produced with smaller interconnection widths of 130 and 110 µm, albeit with the disadvantage of higher electrical losses. But here, too, there is still room for improvement, because a further reduction in the P1-P2 spacing allows interconnection widths below 200 µm to be achieved without scribe line overlap and thus higher efficiencies.

5 REFERENCES

[1] M.A. Green, E.D. Dunlop, J. Hohl-Ebinger, M. Yoshita, N. Kopidakis, X. Hao, Progress in Photovoltaics: Research and Applications 29 (2021) 657.

[2] J.S. Manser, J.A. Christians, P.V. Kamat, Chemical Reviews 116 (2016) 12956.

[3] E.L. Unger, Current Opinion in Green and Sustainable Chemistry 4 (2017) 72.

[4] E.L. Unger, L. Kegelmann, K. Suchan, D. Sörell, L. Korte, S. Albrecht, Journal of Materials Chemistry A 5 (2017) 11401.

[5] M. Fenske, C. Schultz, J. Dagar, F.U. Kosasih, A. Zeiser, C. Junghans, A. Bartelt, C. Ducati, R. Schlatmann, E. Unger, B. Stegemann, Energy Technology (2021) 2000969.

[6] C. Schultz, M. Fenske, J. Dagar, A. Zeiser, A. Bartelt, R. Schlatmann, E. Unger, B. Stegemann, Solar Energy 198 (2020) 410.

[7] C. Schultz, F. Schneider, A. Neubauer, A. Bartelt, M. Jošt, B. Rech, R. Schlatmann, S. Albrecht, B. Stegemann, IEEE Journal of Photovoltaics 8 (2018) 1244.

[8] F. Di Giacomo, L.A. Castriotta, F.U. Kosasih, D. Di Girolamo, C. Ducati, A. Di Carlo, Micromachines 11 (2020) 1127.

[9] A.L. Palma, F. Matteocci, A. Agresti, S. Pescetelli, E. Calabrò, L. Vesce, S. Christiansen, M. Schmidt, A. Di Carlo, IEEE Journal of Photovoltaics PP (2017) 1.

[10] C. Schultz, M. Fenske, J. Dagar, G.A.F. Basulto, A. Zeiser, A. Bartelt, C. Junghans, R. Schlatmann, E. Unger, B. Stegemann, Materials Today: Proceedings (2021.

[11] C. Schultz, A. Bartelt, C. Junghans, W. Becker, R. Schlatmann, B. Stegemann, Applied Surface Science 536 (2021) 147721.

[12] B. Roose, K. Dey, Y.-H. Chiang, R.H. Friend, S.D. Stranks, The Journal of Physical Chemistry Letters 11 (2020) 6505.

[13] Y. Rakita, I. Lubomirsky, D. Cahen, Materials Horizons 6 (2019) 1297.

[14] M.A. Green, Solar cells: operating principles, technology, and system applications (Prentice-Hall, Inc., Englewood Cliffs, NJ, 1992).

[15] J. Dagar, G. Paramasivam, C. Klimm, M. Fenske, C. Schultz, R. Schlatmann, B. Stegemann, E. Unger, Micromachines 12 (2021) 423.

OUTDOOR MONITORING AND ASSESSMENT OF PEROVSKITE MINI MODULES

Vasiliki Paraskeva[1](vparas01@ucy.ac.cy, tel: +35722894321), Matthew Norton[1](matthewn@ucy.ac.cy), Maria Hadjipanayi[1](hadjipanayi.maria@ucy.ac.cy), Afshin Hadipour[2](Afshin.Hadipour@imec.be), Aranzazu Aguirre[2](Aranzazu.Aguirre@imec.be), Rita Ebner[3](Rita.Ebner@ait.ac.at), George E. Georghiou[1](geg@ucy.ac.cy)
[1] FOSS Research Centre for Sustainable Energy, Department of Electrical and Computer Engineering, University of Cyprus, 75 Kallipoleos St., Nicosia, 1678, Cyprus
[2]imec, Kapeldreef 75, 3001 Leuven, Belgium,
[3]AIT Austrian Institute of Technology, Center for Energy, Giefinggasse 2, 1210 Vienna,

ABSTRACT: The long-term performance stability of perovskite-based photovoltaic (PV) devices under real operating conditions remains in question since these devices are sensitive to moisture, illumination, temperature etc. Two groups of perovskite mini modules with different perovskite active layers in each group have been fabricated and exposed outdoors for the purpose of studying their outdoor performance stability and of identifying differences in their long-term performance that can be attributed to their composition. Preliminary results indicated differences in the long-term stability of the two groups of modules, one group exhibiting less degradation than the other. Furthermore, significant current losses have been detected in all modules under test whilst voltage and Fill Factor losses were found to be lower This assessment also studies long-term trends in defect evolution by comparing spatially resolved electroluminescence (EL) images taken indoors before outdoor exposure and after outdoor testing. Significant changes in EL emission have been detected in the modules before and after outdoor testing due to the increase of degradation pathways in the devices.
Keywords: perovskite, characterization, degradation, electroluminescence, PV module

1 INTRODUCTION

Over the last decade, perovskite-based solar cells have shown impressive power conversion efficiencies (PCEs) over 25.2% for single-junction cells [1]. However, the performance stability of those devices remains in question since perovskite-based devices are sensitive to moisture, illumination, temperature etc. [2]. Several indoor testing techniques have been applied to study performance degradation so far [3] while few papers have presented outdoor field-testing results [4]. With the improvement in materials and solar cell fabrication, some perovskite devices are now passing standardized test protocols such as IEC 61215 involving indoor tests in which modules are heated up to 85°C for 1000 hours at 85% relative humidity and undergo temperature cycling from -40°C to 90°C. The IEC tests for perovskite solar cells are summarized in [5]. These measurements and aging procedures are done to qualify to a minimum standard of initial performance, and they do not resemble the outdoor operational conditions where temperature, humidity and irradiance are continuously changing. Only limited experience of long-term outdoor testing of perovskite-based PV devices is available in the literature due to the relative novelty of this technology [6], [7]. Furthermore, the long-term outdoor study of different structure perovskite modules by using a well-defined measurement procedure with numerous samples is essential for the establishment of reliable long-term stability and performance from perovskite samples.

In this context, this paper aims to characterize the output of several perovskite mini modules of different structure under real operating conditions. Alongside these devices, measurements of irradiance in the plane of array were taken in combination with other environmental factors. Such measurements help to investigate degradation processes and their correlation with environmental parameters over the same period. This assessment also studies long-term trends in defect evolution by comparing spatially-resolved electroluminescence images taken indoors before outdoor exposure and at regular intervals after field exposure.

2 EXPERIMENTAL APPROACH

Two different types of perovskite modules (three of type A and four of type B) have been mounted outdoors in a fixed plane (see Figure 1) and current-voltage (I-V) measurements have been collected at regular intervals. The active layer of the type A module is a two-cation perovskite ($Cs_{0.18}Fa_{0.82}PbI_{2.82}Br_{0.18}$) with no additives while type B modules have the same perovskite active layer with formamidinium chloride (FACl) additives inside. The purpose of studying several modules of the same structure is to have better statistical data from samples of the same type. Open-circuit (V_{oc}) loading was applied between the IV scans. A dummy module was placed alongside the 7 modules under study for the collection of temperature at the backside of the modules. Forward and reverse voltage sweeps were applied to the module during each IV measurement. The outdoor testing of the samples started on the 12th of January 2021 and lasted for 17 weeks. Alongside the IV traces from the devices, environmental sensors have been used to collect solar irradiance in the plane of array, ambient and device temperature, wind velocity and humidity/precipitation levels. The electrical measurements have been acquired by a single current-voltage source-meter multiplexed to take sequential measurements from the devices under test. A LabVIEW program was created to record the IV-traces every 5 minutes at high Global Normal Irradiance (GNI) conditions (GNI > 400 W/m^2). Both forward (< 0 V to > V_{oc}) and reverse ((> V_{oc} to < 0 V) voltage sweeps have been applied to the devices. Forward sweeps were performed first on all modules under test. The voltage sweep rate was chosen to be 1V/sec.

For type A perovskite modules studied outdoors, a control module of identical structure was kept indoors under dark, controlled conditions to help distinguish the degradation that occurs outdoors from the aging degradation processes occurring inherently within the device. The current-voltage characteristic of the control modules was measured at AIT with the OAI's TRI-SOL Solar Simulator, Class AAA, Xe Arc Lamp with 100 mW/cm^2 power output. The current at the contacts of the cell was collected by using a Keithley 2651A source meter and by applying the ivRider software. The temperature of the samples was kept at 25°C.

Figure 1: Perovskite mini-modules mounted outdoors on a tracker at the University of Cyprus site.

Alongside the collection of IV traces, spatially-resolved EL measurements were taken from the modules before outdoor exposure to look for changes occurring in EL images after outdoor exposure thus enabling the study of defect evolution at the different perovskite modules. A high-performance CCD matrix camera at -5°C was used during the EL studies. Dedicated software was utilized to collect the EL images at the same testing parameters for comparable results.

3 RESULTS & DISCUSSION

During the whole period of outdoor exposure, no visible degradation due to humidity ingress inside the encapsulant material was observed in the modules. Over the same period significant rainfall was present at the test location but this seems not to have caused any distortion of the encapsulant material indicating the good quality of the encapsulation of these modules.

Figure 2 depicts the reduction in the normalized power conversion efficiency (PCE), Fill Factor (FF), open-circuit voltage (V_{oc}) and short-circuit current (I_{sc}) observed in the first ten weeks of outdoor exposure from all 7 modules under test. These measurements were taken at reverse sweeps.

In Figure 2 modules labelled A, C, and I are of type A while modules O, S, P and Q are of type B. All parameters have been normalized to their initial values obtained outdoors.

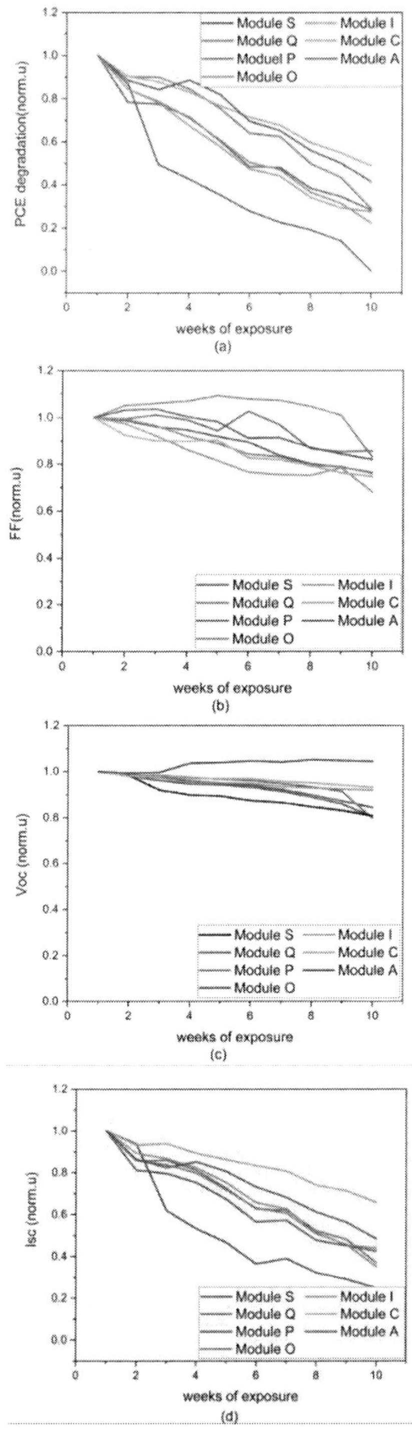

Figure 2: a) Efficiency (b) Fill Factor (c) Open-circuit voltage and (d) Short-circuit current degradation of perovskites mini-modules of type A and B the first 10 weeks of outdoor exposure. All parameters have been normalized to their initial values.

The majority of type A modules present similar degradation rates over the first 10 weeks of outdoor exposure (module A, module C). However, in type B modules more deviations exist in efficiency degradation. Only two out of four type B modules present similar efficiency degradation levels (module O, module P) while the rest of the type B modules present higher (module S) or lower (module Q) efficiency degradation. Having a closer look on the efficiency behavior of the modules the testing period can be classified into two periods: the initial burn-in period which lasts up to 3 weeks after the modules' installation and the period after the 3rd week of operation. During the initial burn-in performance period type B modules degrade more than type A modules while after the 3rd week of operation the degradation rate is the same for the two types of modules under test.

To identify the root cause of efficiency degradation of the modules, all the other electrical parameters (Fill Factor, open-circuit voltage, short-circuit current) have been plotted over the testing period (see Figure 2). Figure 2(c) demonstrates that voltage losses are not significant in all the modules under test. The maximum open-circuit voltage reduction obtained in the devices was 20%, with type A modules presenting slightly lower open-circuit voltage losses than type B modules. The mean voltage losses in type B modules was calculated to be 19.5% while voltage losses in type A modules are 5.2%. It is worth noting that one module of type A (module A) presents a small open-circuit voltage enhancement over the testing period. This result agrees with the indoor data under the solar simulator at standard testing conditions. The origin of the open-circuit enhancement in module A is unclear and more investigation is underway. An investigation of Fill Factor changes over the testing period has been carried out and plotted in Figure 2(b). The maximum Fill Factor reduction of 30% was present. The majority of type A modules present higher Fill Factor losses, but this is not a clear trend since large fluctuations in FF are present in all modules under test. Short-circuit current losses were detected in all modules under study indicating that the major cause of the efficiency degradation is the current (Figure 2d). Current losses up to 75% were detected during outdoor exposure. The mean short-circuit current loss in type B modules was 60.5% while for type A modules it was 47.6%. The fastest short-circuit current degradation rate was obtained in the first two weeks of outdoor exposure in most modules. After that time the degradation rate for current decreases. These results indicate that a 'faster' degradation mechanism occurs in devices just after the exposure to light and temperature levels outdoors and this is replaced by a 'slower' degradation mechanism after some time.

By studying the efficiency behavior of the modules over a day it can be observed that their performance presents changes over a day depending on whether the data was collected in the morning or afternoon. The modules in most cases perform better in the morning hours than in the afternoon on the days under examination. The efficiency drop during the day is mainly attributed to current reduction. In some cases, Fill Factor losses are detected over the day but this was obtained only in some modules. The current drop is most likely related to irradiance since the first hours after following exposure there is a rapid current degradation. The decrease of current over the day could be also attributed to temperature rise of the module, however, after data analysis we have found that the impact of temperature on current reduction is a minor effect. The improved performance in the morning hours suggests that the devices present some form of recovery overnight. This result was obtained in the outdoor assessment of perovskite modules recently [7].

To compare the degradation of the modules tested outdoors with the concurrent degradation of the control modules kept indoors, the control modules were tested regularly under the solar simulator over the timespan of the outdoor studies. Figure 3 illustrates the changes in the current-voltage characteristics of a control module kept indoors and a module located outdoors. The changes in all electrical parameters of the device are more severe in the modules kept outdoors. The efficiency degradation of the module exposed outdoors reaches 58.5% while at approximately the same time interval the control module degradation was 15.7%. It is assumed that this degradation originates from an innate aging process occurring within the devices under the environmental conditions present during storage (humidity, temperature etc.).

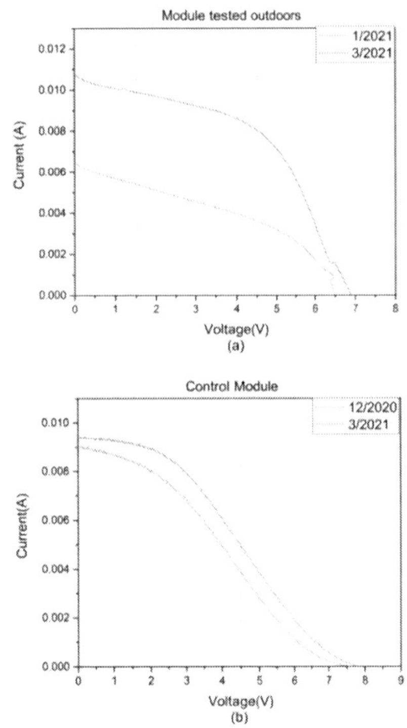

Figure 3: Current-voltage characteristics of a (a) control module and a (b) module exposed outdoors at different times during outdoor campaign.

For the investigation of long-term trends in radiative emission and defect evolution, a spatially-resolved EL imaging technique has been used prior to and after field testing. EL images have been collected from all modules under test before and after outdoor exposure. It is worth noting that all the modules have been left outdoors for in total 17 weeks and the EL images were collected before their installation outdoors and after the 17th week of operation. The EL images before and after outdoor testing from two modules of type A and two modules of type B can be found in Figure 4 and Figure 5 respectively. The same voltage applied in all devices before outdoor testing for comparison purposes.

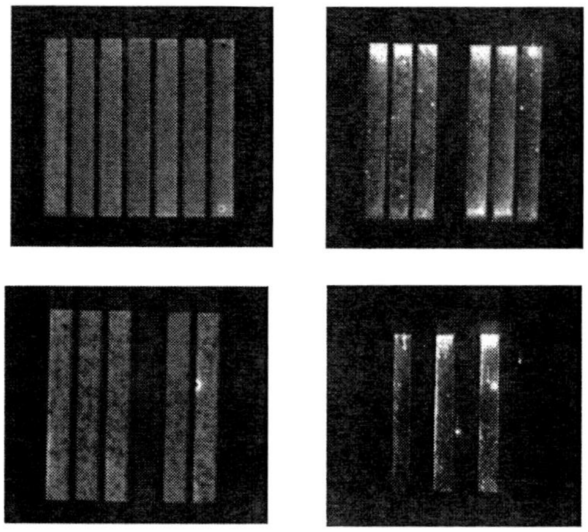

Figure 4: Spatially-resolved EL images before (left) and after 17 weeks of outdoor testing (right) for two modules of type A.

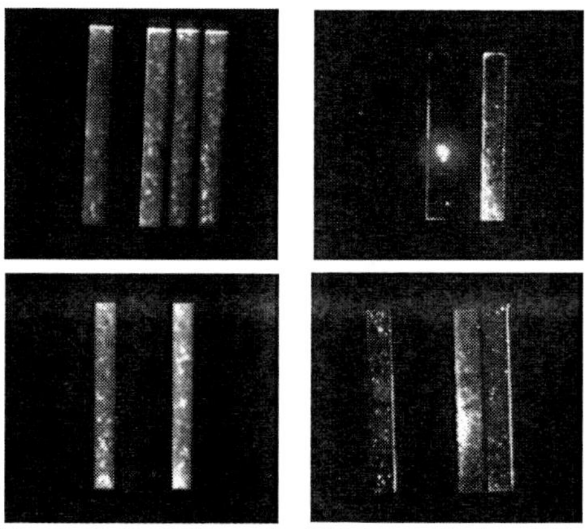

Figure 5: Spatially-resolved EL images before (left) and after 17 weeks of outdoor testing (right) for two modules of type B.

Similar EL image patterns were obtained over the modules of the same type. Some cells within the modules did not luminesce prior to the outdoor testing, indicating a degree of degradation of the modules before their exposure outdoors. The majority of type A modules presents better EL emission compared to emission from type B modules in studies prior to field testing giving an indication of the presence of more non-radiative paths and thus lower material quality in type B modules. EL images after outdoor testing demonstrate lower emission and more inhomogeneities within the modules.

4 CONCLUSIONS

The long-term outdoor performance of different perovskite mini-modules installed side-by-side was studied. Preliminary results indicated differences in the long-term stability of the two groups of modules with less efficiency degradation obtained in modules without additives in the active perovskite layer. The type B modules were shown to degrade over the course of 10 weeks to around 20% of their initial value while type A modules degrade to around 39% of their initial value. The statistical differences in the efficiency degradation of two different types of perovskite cell structure under continuous on-sun exposure suggests that the test methodology used will be successful in revealing which cell type is better suited to long term outdoor use. Significant current losses were obtained in all modules under study while voltage losses were insignificant. Results have indicated that the encapsulation of these devices successfully protects against degradation mechanisms associated with atmospheric exposure. Control modules were tested indoors in parallel with the outdoor campaign and indicated an efficiency degradation of 15% which arise from innate aging processes within the devices. Finally, spatially-resolved EL measurements of the modules were undertaken before and after field testing which demonstrated significant changes in the EL emission of the test devices.

6 AKNOWLEDGEMENTS

This work was co-financed by the European Regional Development Fund and the Republic of Cyprus through the Research & Innovation Foundation (Grant No.: INFRASTRUCTURES/1216/0043).

7 REFERENCES

[1] NREL, "Best efficiency chart Photovoltaic Research," *https://www.nrel.gov/pv/cell-efficiency.html*, 2019.

[2] K. Domanski, E. A. Alharbi, and A. Hagfeldt, "Systematic investigation of the impact of operation conditions on the degradation behaviour of perovskite solar cells.," *Nat. Energy*, vol. 3, pp. 61–67, 2018.

[3] S. Kundu and T. L. Kelly, "In situ studies of the degradation mechanisms of perovskite solar cells," *EcoMat*, vol. 2, no. 2, pp. 1–22, 2020, doi: 10.1002/eom2.12025.

[4] M. Jošt *et al.*, "Perovskite Solar Cells go Outdoors: Field Testing and Temperature Effects on Energy Yield," *Adv. Energy Mater.*, vol. 10, no. 25, 2020, doi: 10.1002/aenm.202000454.

[5] P. Holzhey and M. Saliba, "A full overview of international standards assessing the long-term stability of perovskite solar cells," *J. Mater. Chem. A*, vol. 6, no. 44, pp. 21794–21808, 2018, doi: 10.1039/C8TA06950F.

[6] V. Stoichkov, N. Bristow, J. Troughton, F. De Rossi, T. M. Watson, and J. Kettle, "Outdoor performance monitoring of perovskite solar cell mini-modules: Diurnal performance, observance of reversible degradation and variation with climatic performance," *Sol. Energy*, vol. 170, no. June, pp. 549–556, 2018, doi: 10.1016/j.solener.2018.05.086.

[7] M. Norton, M. Kohlstädt, M. Hadjipanayi, V. Paraskeva, U. Würfel, and G. E. Georghiou, "Experiences of continuous on-sun performance measurements of perovskite mini-modules," 2020.

OPTICAL AND ELECTRICAL CHARACTERIZATION OF PEROVSKITE MINI-MODULES

R. Ebner[1], G. Ujvari[1], A. Mittal[1], M. Hadjipanayi[2], V. Paraskeva[2], M. Hadjikypris[2], G. E. Georghiou[2], A. Hadipour[3], A. Aguirre[3]
[1]AIT Austrian Institute of Technology, Center for Energy, Giefinggasse 2, 1210 Vienna, Austria, T +43 50550-6628, F +43 50550-6390, rita.ebner@ait.ac.at, www.ait.ac.at
[2]University of Cyprus, 1 Panepistimiou Avenue, 2109 Aglantzia, Nicosia, Cyprus, www.foss.ucy.ac.cy
[3]imec, Kapeldreef 75, 3001 Leuven, Belgium, www.imec.be

ABSTRACT: Commercially available solar modules undergo a series of characterization procedures that analyze their properties and ensure quality. However, these procedures cannot simply be applied to perovskite solar cells. Perovskites are inorganic-organic materials for a new generation of solar cells that have only been studied for about eleven years [1]. Perovskite solar cells achieve very high efficiencies (world record for perovskite-silicon tandem cell: 29.52% [2]) and can be applied as extremely thin films at extremely low cost. The main challenge for a successful commercialization of perovskite solar cells is to achieve long lifetimes with good stability at module level [1] [3]. More advanced characterization methods are needed to understand the degradation mechanisms that occur. In this context, optical and electrical measurement methods are effectively used in quality control and development support and are important characterization tools in industry and research.

KEYWORDS: Perovskite, Optical and electrical characterization

1 INTRODUCTION

In the proposed work, the degradation of perovskite solar cells and mini modules was investigated by using optical and electrical characterization methods.
Electroluminescence (EL), photoluminescence (PL) and dark lock-in thermography (DLIT) are non-destructive measurement techniques. These optical measurements provide high-resolution images that show a two-dimensional distribution of the characteristic features of PV cells and allow the investigation of cracks, defects, shunts, and stacking faults in the cells [4]. Measurements of power output (IV) and external quantum efficiency (EQE) measurements help to provide a complete picture of solar cell or (mini) module power conversion efficiency. IV measurements indicate the current generated by the cell under simulated solar irradiance conditions and EQE measurements provide quantification of the current generated at each wavelength wherein the ratio of extracted free charge carriers to incident photons was determined.

2 EXPERIMENTS

A 2-cation perovskite active layer with the composition $\mathbf{Cs_{0.18}FA_{0.82}PbI_{2.82}Br_{0.18}}$ was used. In order to make large area devices, so called mini-modules were produced by laser scribing, to generate 7 sub-cells connected in series. In order to prevent penetration of metallic particles of the top electrode into the soft perovskite layer, ITO was used. ITO was also selected as a top electrode to obtain semi-transparent modules. The module stack was as follows:
glass/ITO/PTAA/Al2O3/550nm 2C Perovskite with bandgap of 1.6 eV/LiF/C60/BCP/ITO/glass. Fig.1 shows the structure of the perovskite sub-cell and Fig. 2 shows the cross-section of the mini-module. Fig.3 shows two images of the front and back sides of one perovskite mini-module (size: 3cm x 3cm). 4 perovskite mini-modules ("B", "D", "E" and "F") were produced and characterized by applying DLIT-, EL- and PL methods alongside IV and EQE measurements. The results are presented in point 4.

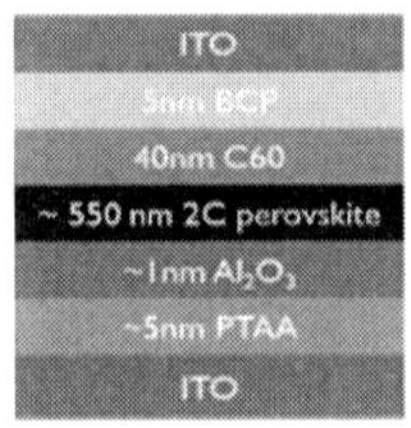

FIG.1 PIN-structure of each sub-cell.

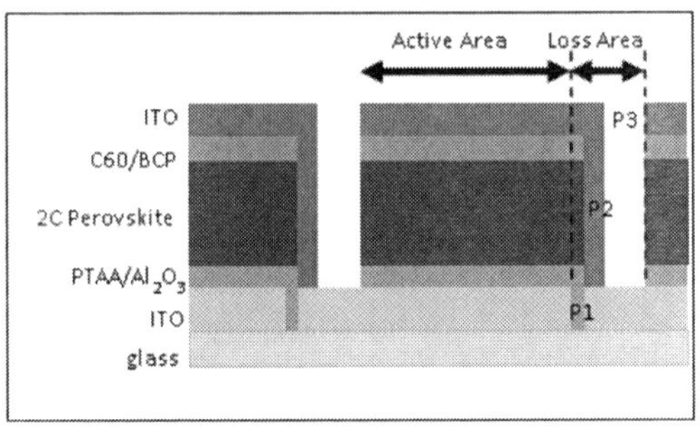

FIG.2 Cross-section of the mini-module.

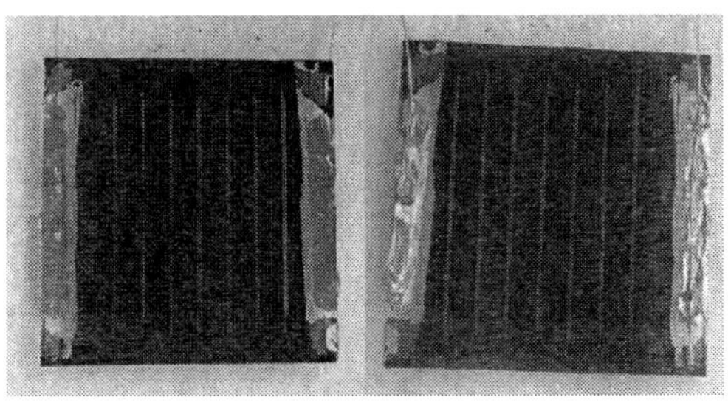

FIG.3 Perovskite mini-module: Front side (left), Back side (right).

3 RESULTS

3.1 Mini-module "B"

Fig.4 shows the EL images (front side and back side) of mini-module "B" with an inactive cell. The IV measurement results of mini-module "B" are shown in Table 1. The EQE measurement results are shown in Fig.5.

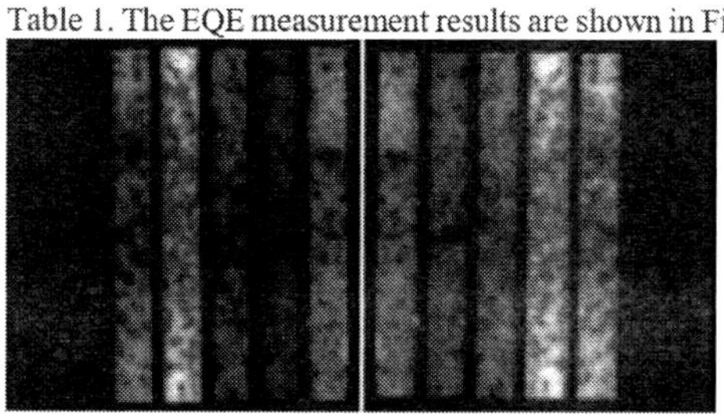

FIG.4 Perovskite mini-module "B": Front side (left) and back side (right).

Table 1:
IV-measurement results, Perovskite mini-module "B"

I_{SC} [A]	V_{OC} [V]	FF [%]	P_{MPP} [W]	Jsc [mA/cm²]
0.0094	6.1879	39.92	0.0233	12.049

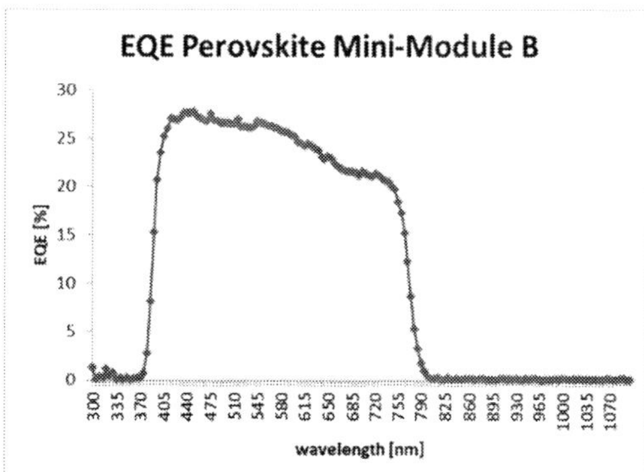

FIG.5 EQE Perovskite mini-module "B"

3.2 Mini-module "D"

Fig.6 shows the EL images of minimodule "D" with one defective and one inactive cell. The IV measurement results of minimodule "D" are shown in Table 2. The EQE measurement results are shown in Fig.7.

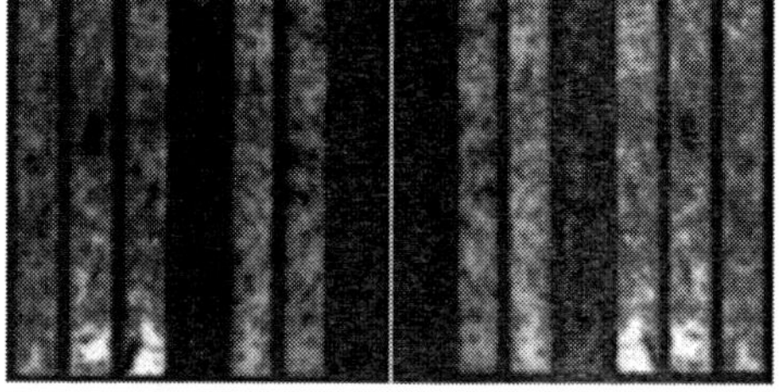

FIG.6 Perovskite mini-module "D": Front side (left) and back side (right).

Table 2:
IV-measurement results, Perovskite mini-module "D"

I_{SC} [A]	V_{OC} [V]	FF [%]	P_{MPP} [W]	Jsc [mA/cm²]
0.0093	7.5501	35.35	0.0247	11.831

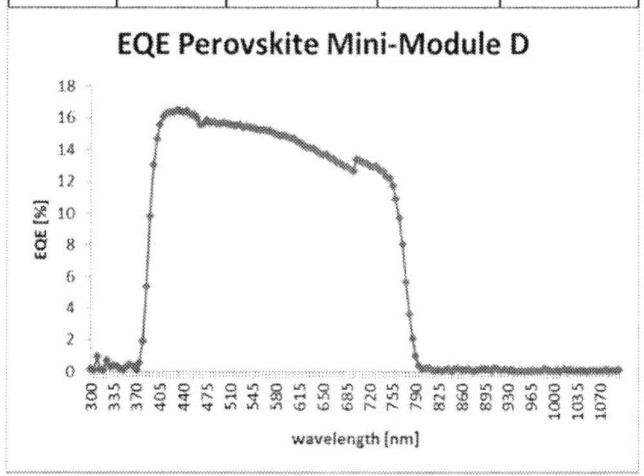

FIG.7 EQE Perovskite mini-module "D"

3.3 Mini-module "E"

Fig.8 shows the EL recordings of the minimodule "E" with a defective cell. The IV measurement results of the minimodule "E" are shown in Table 3. The EQE measurement results are shown in Fig.9.

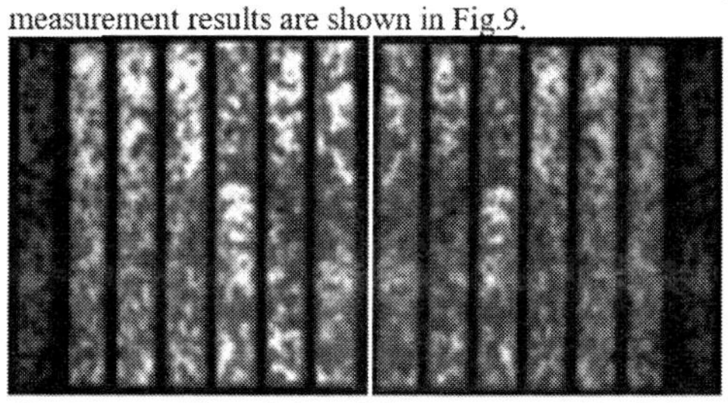

FIG.8 Perovskite mini-module "E": Front side (left) and back side (right).

Table 3:
IV-measurement results, Perovskite mini-module "B"

I_{SC} [A]	V_{OC} [V]	FF [%]	P_{MPP} [W]	Jsc [mA/cm²]
0.0097	7.8345	36.70	0.0280	12.419

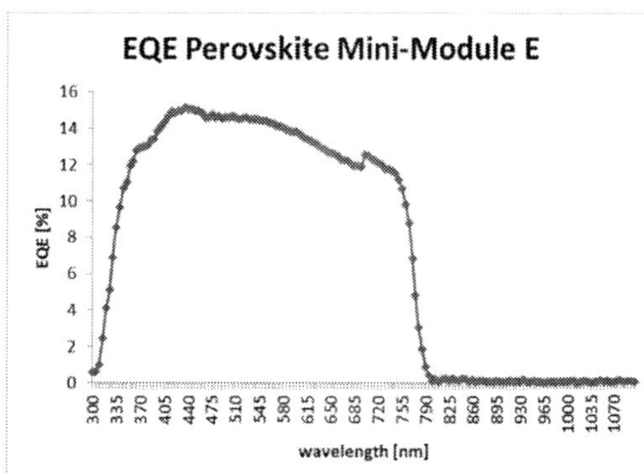

FIG.9 EQE Perovskite mini-module "E"

3.4 Mini-module "F"

Fig.10 shows the EL images of the minimodule "F" with an inactive cell. The IV measurement results of the minimodule "F" are shown in Table 4. The EQE measurement results are shown in Fig.11.

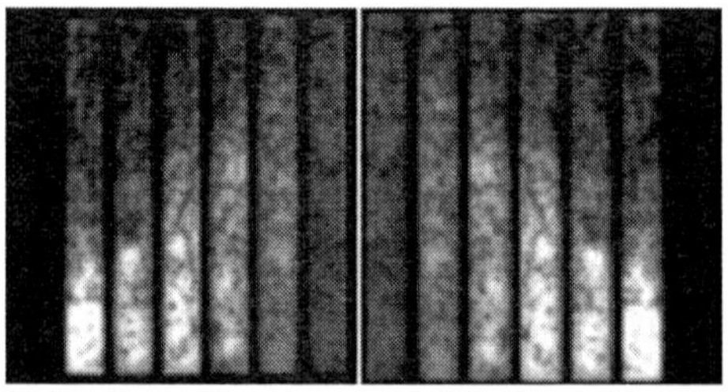

FIG.10 Perovskite mini-module "F": Front side (left) and back side (right).

Table 4:
IV-measurement results, Perovskite mini-module "F"

I_{SC} [A]	V_{OC} [V]	FF [%]	P_{MPP} [W]	Jsc [mA/cm^2]
0.0094	7.7664	34.13	0.0249	12.006

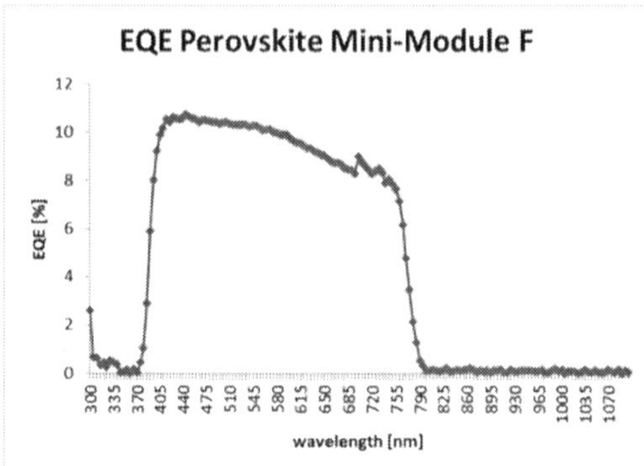

FIG.11 EQE Perovskite mini-module "F"

4 CONCLUSION

First defects and shunts were identified by EL, PL and DLIT measurements. The EQE measurements gave very high values for perovskite cells (e.g. about 90%, cell size:0.13cm^2 see Fig.12), but still too low values for perovskite mini modules (see Fig.5).

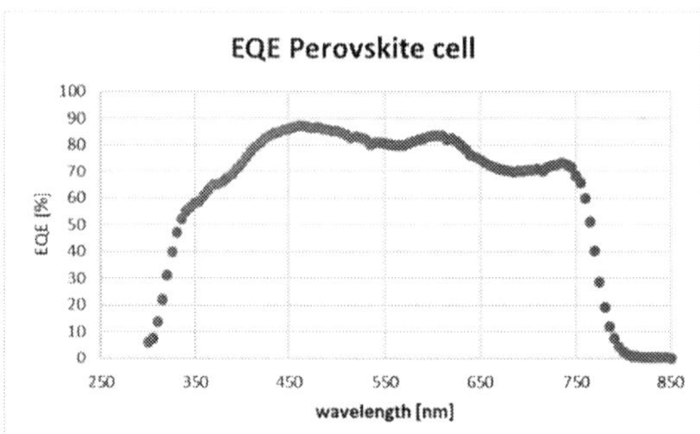

FIG.12 EQE Perovskite cell

Further indoor tests and outdoor tests of differently structured perovskite samples and also perovskite tandem cells have to be performed in order to identify more defects and thus be able to improve the perovskite cell and module structure. In addition, measurement protocols for indoor and outdoor tests of perovskites will be established.

5 REFERENCES

[1] Antonio Urbina 2020 J. Phys. Energy 2 022001
[2] "Oxford PV retakes tandem cell efficiency record",https://www.pv-magazine.com/2020/12/21/oxford-pv-retakes-tandem-cell-efficiency-record/
[3] Enzheng Shi et al., "Two-dimensional halide perovskite lateral epitaxial heterostructures", *Nature*, 2020; 580 (7805): 614
[4] R. Ebner et al., "Non-destructive techniques for quality control of PV modules", 39th Annual

Acknowledgement:
This work was funded through the European Regional Development Fund and the Republic of Cyprus in the framework of the project "DegradationLab" with grant number INFRASTRUCTURES/1216/0043.
http://www.foss.ucy.ac.cy/degradationlab/

A UNIFIED CAPACITANCE LOSS MAPPING FOR SOLAR CELLS DEFECTS

Jacopo Sala*†§, Thierry Kohl*†§, Guy Brammertz*†§, Mehrad Ahmadpour~, Morten Madsen~, Tom Aernouts†§, Michaël Daenen*†§ Bart Vermang*†§,
*IMO-IMOMEC, Hasselt University, Wetenschapspark 1, 3590 Diepenbeek, Belgium
†imec, Kapeldreef 75, 3001 Heverlee, Belgium
§EnergyVille, Thor Park 8310, 3600 Genk, Belgium
~SDU NanoSYD, Mads Clausen Institute, University of Southern Denmark, Alsion 2, 6400 Sønderborg, Denmark

ABSTRACT: With the rise of perovskite-based solar devices, Solar PV has not only reinforced its central position in the renewable energy transition for electricity generation, but it has also reinvigorated the race for untapped integrated power generation. To succeed in producing cheaper and more affordable electricity, capacitive techniques are essential to understand the transitional properties of solar devices and their impact on the short- and long-term performance. In this publication, we present a study on combined admittance spectroscopy measurements at different biases that are commonly used to characterize metal–oxide–semiconductors and pn-junction devices with the same approach being already tested and validated on CIGS devices[1]. This study generalises the use of this technique to other PV technologies such as perovskites and organic solar cells and identifies correlations among the different structures and material properties
Keywords: Capacitance, Multijunction Solar Cell, Perovskite, Cu(InGa)Se2, Organic Solar Cell

1 INTRODUCTION

In recent years, solar photovoltaics has established itself as one of the cheapest forms of electrical energy and it is becoming the new king of electricity[2]. With the rise of thin-film technologies, solar cells are finding new applications starting from Building Integrated PhotoVoltaics (BIPV) to Vehicle Integrated PhotoVoltaics (VIPV), from flexible devices to customisable substrate shapes. As hybrid perovskites have recently demonstrated in their rapid rise from 3.8% to 25.5% efficiency in just a decade[3], there is more than an opportunity for researchers to learn from other technologies. Similarly, admittance spectroscopy has been utilized to investigate the ionic and defective behaviour of perovskite solar cells. So far, the perovskite community has focused on identifying and evaluating the impact of vacancies and ionic diffusion in relation to the hysteretic behaviour of I-V characteristics. Even more interest was raised about the presence of negative capacities that have recently started to find an explanation in the relaxation of the accumulated charges at the interfaces[4–7]. Furthermore, it was possible to relate the detrimental effects of ionic diffusion in the transport layers defining the importance of understanding and controlling this effect.

The most common approaches to these measurements connect devices to a DC voltage bias and to an AC voltage source with a modifiable frequency. The measurement is usually performed at a specific DC bias voltage (Cf – Admittance) or at a specific frequency (CV). The former allows measuring the capacitance as the depletion layer is collapsed, providing insights on the built-in potential, the depletion layer, and the doping density. The latter is more commonly used to identify thermally activated processes or losses, and defect states kinetics[8].

In some cases, the temperature is an additional parameter that is changed to evaluate the activation energy of the presented phenomenon. However, as previously presented on CIGS by Brammertz et al.[1], it is possible to see from a combined variation DC bias voltage and frequency that some defects may be present at a different frequency or DC bias voltage for the same solar cell stack. Therefore, in this study, the same technique is presented and generalised for different PV technologies such as perovskites, silicon, and organic solar cells..

2 AIM AND APPROACH USED

The main objective of this research is to extend this comprehensive capacitance-based technique and data visualization to all the photovoltaic technologies. To do so, the admittance data is acquired with an LCR Meter and measure the values of an equivalent circuit consisting of a capacitance and a conductance in parallel. The presented results were also focused on a frequency range between 1 kHz and 1 MHz and a variable bias voltage from -0.5 V to a few hundreds of millivolts higher than the most common Voc for the analysed technology as measured at 1 sun illumination. The output data is collected in the form of a matrix of values based on the used frequency and DC bias voltage. The results are then elaborated and represented in different contour plots to show the dielectric losses via the relation -fdC/df in the frequency and voltage 2D domain as well as the Mott-Schottky relation and the Nyquist plot. The analysis of these separated graphs can be used to identifies electronic/ionic defects and relate them to different sources of inefficiencies.

Moreover, 2D data visualization is a key element to identify clearly and easily the most common or relevant issues. So far most of the capacitance measurements have focused on fixed frequencies or bias voltages, as described by the red lines in Figure 1.

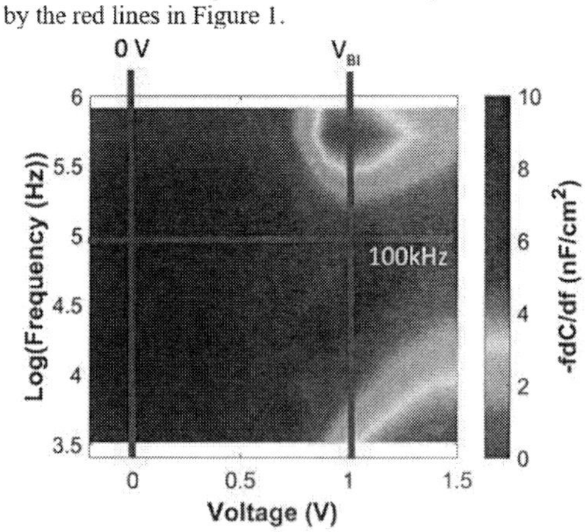

Figure 1: A CVf map of a perovskite solar cell. The red lines show the data usually considered for this kind of measurement

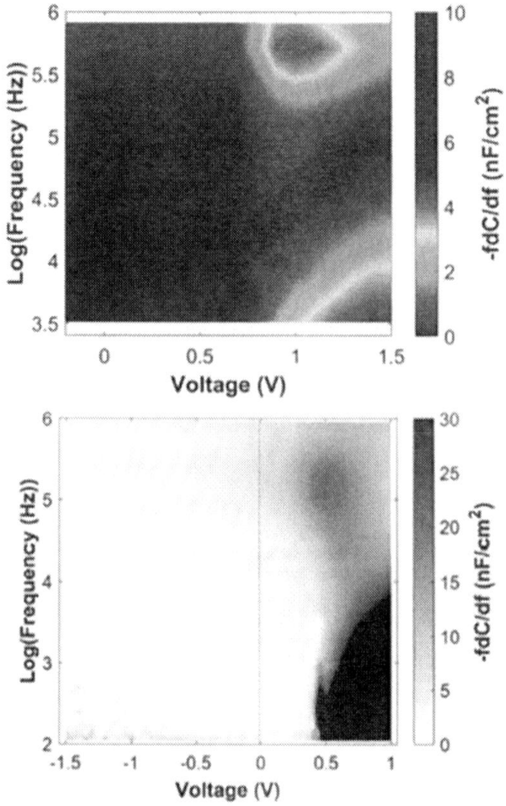

Figure 2: A perovskite (Top) and a CIGS (Bottom) solar cell show similar responses. CIGS map reproduced from [1] with author permission.

However, this can mislead the interpretation of the results as for the defect region that appears in the top-right side of the graph and is both frequency and voltage-dependent.

To validate the relevance of this approach, CIGS, perovskite, and organic solar cells are initially investigated. In a second moment, the insights are extended to other technologies.

3 RESULTS AND DISCUSSION

Building on the previous work of Brammertz et al.[1], the numerical model obtained via SCAPS provides a good 1D drift-diffusion model of a CIGS solar cell and insights on how defect translate in CVf maps. Starting from an ideal device, it was possible to identify the characteristic defects maps associated with interface or bulk defects, resistances, energy barriers or more generic dissipation losses. Similarly, perovskite solar cells demonstrate comparable defect plots.

In this case, both images show the presence of similar nonidealities. The signals present at low frequencies are due to dissipations linked to DC currents passing through the device. On the other hand, the high-frequency signal at around the built-in voltage can be explained as an acceptor-like bulk defect as suggested by SCAPS simulations.

With a different perovskite structure, it was also possible to see that some other similarities are present for other technologies than for perovskites and CIGS. In fact, shows how hybrid perovskites can produce similar responses to organic solar cells if the data is properly related to the device's built-in voltage.

4 SUMMARY AND OUTLOOK

We presented a methodology to analyse solar cell defects and to further extend the shareable knowledge among different technologies. It was possible to show how different materials and structures are affected by similar responses behaviours and how understanding the irregularities of a technology can lead to insights for different devices. This work also targets silicon-based solar cells and will provide further insights into possible sources of losses and their specific traces in the 2D CVf map.

This work is supported by the European Union's Horizon 2020 projects Uniting PV (Grant agreement ID: 715027) and PERCISTAND (Grant agreement ID: 850937). M.M. And M.A. acknowledges Danmarks Frie Forskningsfond, DFF FTP for funding of the project React-PV, grant number 8022-00389B.

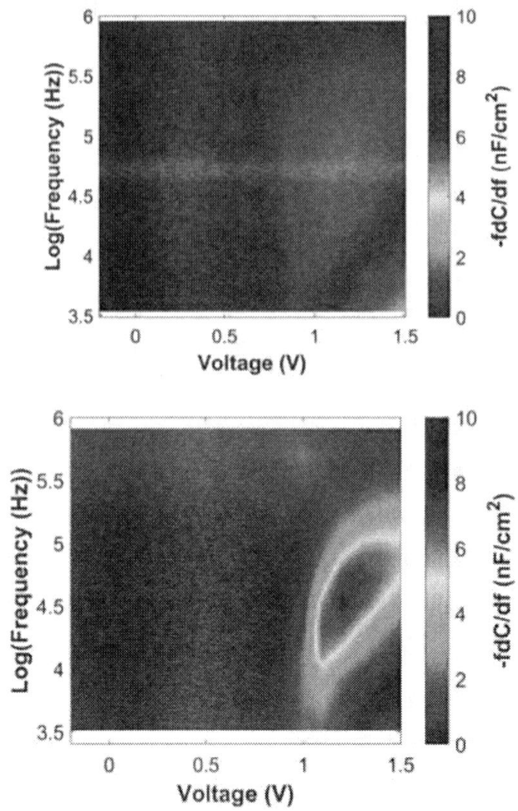

Figure 3: Organic (top) vs Perovskite (bottom) solar cells showing a similar response around their built-in potentials

5 REFERENCES

[1] Brammertz, G.; Kohl, T.; De Wild, J.; Buldu, D. G.; Birant, G.; Meuris, M.; Poortmans, J.; Vermang, B.

Bias-Dependent Admittance Spectroscopy of Thin-Film Solar Cells: Experiment and Simulation. IEEE J. Photovoltaics 2020, 10 (4), 1102–1111.
[2] IEA. World Energy Outlook 2020; Paris.
[3] National Renewable Energy Laboratory. NREL Best Research-Cell Efficiencies Chart. 2021.
[4] Srivastava, P.; Kumar, R.; Bag, M. Discerning the Role of an A-Site Cation and X-Site Anion for Ion Conductivity Tuning in Hybrid Perovskites by Photoelectrochemical Impedance Spectroscopy. J. Phys. Chem. C 2021.
[5] Ebadi, F.; Taghavinia, N.; Mohammadpour, R.; Hagfeldt, A.; Tress, W. Origin of Apparent Light-Enhanced and Negative Capacitance in Perovskite Solar Cells. Nat. Commun. 2019, 10 (1), 1574.
[6] Alvarez, A. O.; Arcas, R.; Aranda, C. A.; Bethencourt, L.; Mas-Marzá, E.; Saliba, M.; Fabregat-Santiago, F. Negative Capacitance and Inverted Hysteresis: Matching Features in Perovskite Solar Cells. J. Phys. Chem. Lett. 2020, 11 (19), 8417–8423.
[7] Jacobs, D. A.; Shen, H.; Pfeffer, F.; Peng, J.; White, T. P.; Beck, F. J.; Catchpole, K. R. The Two Faces of Capacitance: New Interpretations for Electrical Impedance Measurements of Perovskite Solar Cells and Their Relation to Hysteresis. J. Appl. Phys. 2018, 124 (22).
[8] Futscher, M. H.; Gangishetty, M. K.; Congreve, D. N.; Ehrler, B. Quantifying Mobile Ions in Perovskite-Based Devices with Temperature-Dependent Capacitance Measurements: Frequency versus Time Domain. arXiv 2019, 1–7.

TWO LOCAL BUILT-IN POTENTIALS OF CZTSE GE BI-LAYERS DEVICES BY MODULUS SPECTROSCOPY

Sanghyun Lee[a], Kent J. Price[b], Edgardo Saucedo[c]
[a]Department of Electronics and Computer Engineering Technology, Indiana State University, Terre Haute, IN 47809, USA
[b]Department of Physics, Morehead State University, Morehead, KY 40351, USA
[c]Catalonia Institue for Energy Research, Sant Adria de Besos, Barcelona 08930, Spain

ABSTRACT: We fabricated and analyzed CZTSe: Ge bi-layers solar cells to accurately evaluate the electrical structure at hetero and back contact junctions by modulus spectroscopy. Two separate built-in potentials of hetero and back contact junctions are extensively investigated to describe qualitative each junction model of CZTSe: Ge bilayers devices. Fabricated devices are analyzed with an equivalent circuit model, and LEVM/LEMW software resolved two different parallel circuit branch components. Modeling with LEVM/LEMW software accurately fits our model to modulus and impedance spectra in the given testing frequency domain. Using TCAD numerical simulations, we examine a variety of device models with a set of built-in potentials for appropriate parallel circuits with two capacitance and two resistance (R^2=0.99). Our study shows that heterojunction built-in potential is 801 meV, which is ~10 % smaller than a theoretical value, 894 meV. Conversely, the built-in potential of the back contact junction is 428 meV. Total built-in potential is 373 meV

1 INTRODUCTION

Kesterite or CZTSSe ($Cu_2ZnSnS_xSe_{4-x}$) solar cells has been steadily emerging third-generation thin-film solar cells as cost-effective CZTSSe solar cells have shown <12.6 % efficiency [1-3]. In particular, we reported that CZTSe Ge bi-layers devices have shown greatly improved performance with the enhanced grain and grain boundary properties [2-3]. CZTSe: Ge bi-layers devices consist of several different functional parts such as carrier generation and carrier separation, associated with electric field in the depletion region. The current-voltage (I-V) measurements can provide the device performance of photovoltaic devices (DC characteristics), but the electrical property associated with each interface junction of CZTSe: Ge bi-layers devices can not be readily accessed with DC characteristics. Conversely, AC characteristics (modulus and impedance spectroscopy) can provide each interface junction's properties in the frequency domain as a certain range of frequency separately reveals them. [4-6].

Modulus spectroscopy is a powerful technique that can decouple the individual junction's electrical property by setting up equivalent circuit models when combined with impedance spectroscopy methods. Crystalline Si solar cells [7] have been extensively studied by impedance spectroscopy. Regarding thin-film solar cells, however, there are a few reports about CdTe [4], CIGS [6], CZTSSe devices with impedance spectroscopy [8,9]. In contrast, there is no report about Ge alloyed CZTSe nano-layer devices with modulus spectroscopy combined with impedance spectroscopy [10,11].

In this work, we fabricated and analyzed CZTSe:Ge bi-layers devices to accurately evaluate the electrical structure at hetero and back contact junctions by modulus characteristics. Moreover, we studied the dependence of light and voltage biases with respect to impedance and modulus behaviors of hetero and back contact junctions. We used an equivalent circuit model for CZTSe:Ge bi-layers devices that describes the junction modulus and impedance characteristics. This model is utilized to extract the most appropriate device parameters under various voltage bias conditions, which agrees with the numerical simulation result.

2 EXPERIMENTAL METNODS

The multi-layer structure of CZTSe:Ge bi-layers devices consist of soda-lime glass, Molybdenum (Mo) metal, $MoSe_2$ interlayer (100 nm), pre-absorber Ge film (7.5 nm), p-CZTSe absorber (<2 um), post-absorber Ge film (2.5 nm), n-CdS buffer(50 nm), intrinsic ZnO (50 nm), In_2O_3-SnO_2 (ITO, 350 nm) as a transparent conductive window layer, and metal contact (Ni/Al). On the soda-lime glass, Mo metal and $MoSe_2$ were deposited by DC magnetron sputtering. On $MoSe_2$ film, a pre-absorber layer of thin 2.5 nm Ge film is deposited. We used direct current magnetron sputtering to deposit metal precursors (Cu/Sn/Cu/Zn) for the CZTSe absorber formation. For the completion of Ge bi-layers stack deposition, an additional 7.5 nm post Ge film is deposited, followed by subsequent Se and Sn ambient reactive annealing. The composition of thin-film is Cu-poor and Zn-rich CZTSe by maintaining Cu/(Zn+Sn) = 0.78 and Zn/Sn = 1.20. The total absorber film thickness is 1.8 – 2.0 um. To minimize the secondary phases, several chemical etchings, ($KMnO_4$+H_2SO_4, $(NH_4)_2S$, KCN) were done. After 50 nm CdS chemical bath deposition, intrinsic-ZnO (50 nm) and In_2O_3-SnO_2 (ITO, 350 nm) were deposited as a transparent conductive window layer. Devices were laser-scribed and completed with front metal contact (Ni/Al).

The current-voltage of devices is measured with Sun 3000 solar simulator from ABET technologies and Keithley 2400 Sourcemeter. For the impedance spectroscopy characterization, Agilent 4284A precision

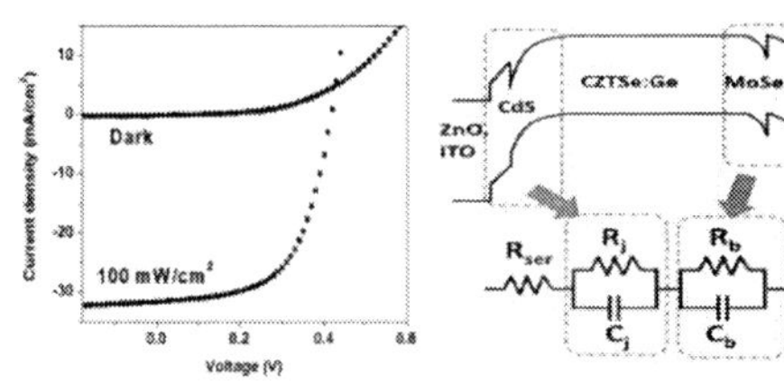

Fig. 1 (a) current-voltage characteristics, (b) band diagram for an equivalent circuit model

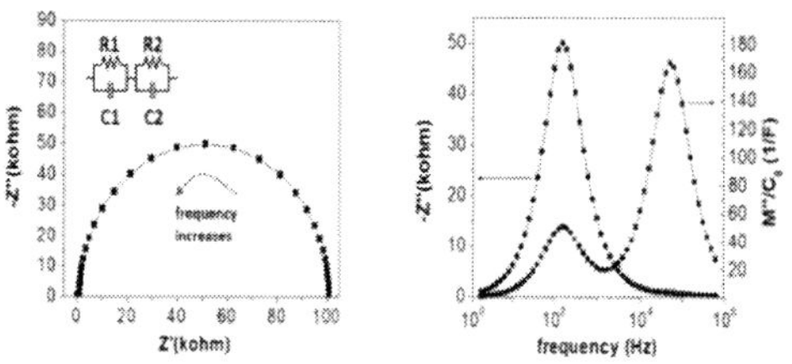

Fig. 2 (a) Simulated impedance plots, (b) Transformed into Modulus format (M"/Co) with two distinctive peaks.

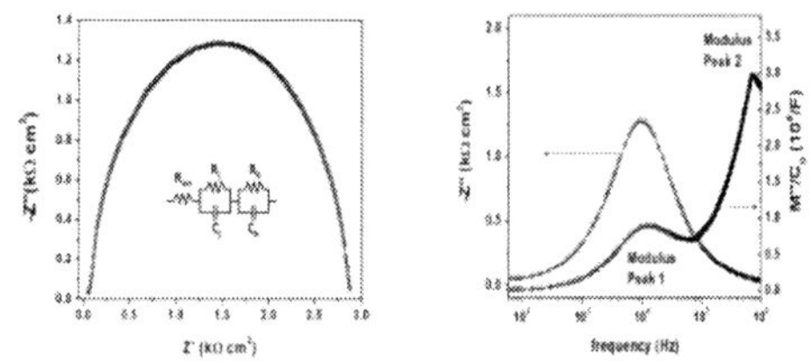

Fig. 4 (a)Measured impedance spectra, (b) Modulus spectroscopy, compared to impedance spectra.

LCR meter was used with frequencies ranging from 20 Hz to 1 MHz. For all measurements, a 10 mV AC signal was applied to the test devices. Measurements were obtained under dark conditions at room temperature. The fitting of the acquired data was conducted with LEVM/LEVMW.

3 RESULT AND DISCUSSION

The overall device performance was characterized under dark and one sunlight conditions by conducting I-V measurements. The short circuit current (Jsc) is 33.1 mA/cm^2, open-circuit voltage (Voc) is 418 mV, fill factor (FF) is 58.8 %, and efficiency (8.1 %) in Fig. 1. For the modulus and impedance analysis, Fig 1 (b) shows an equivalent circuit model for hetero and back contact junctions. For practical impedance spectroscopy with thin-film photovoltaic devices, one of junction impedance spectroscopy response (CdS/CZTSe:Ge heterojunction) tends to dominate the impedance spectroscopy response, and back contact junction is buried in the spectral response, which make it indistinguishable for two different junctions (Fig. 2(a)). However, modulus (M) spectroscopy can be converted from impedance (Z) spectroscopy by the relation, $M(f) = j\omega C_0 Z(f)$, where $j^2=-1$ and ω is angular frequency. Modulus spectroscopy weights to capacitance, but impedance spectroscopy weights to resistance. Hence, impedance plots highlight the most resistive elements in the sample, while modulus plots pick out the smallest capacitive element. In Fig. 2(b), the simulated result clearly demonstrates two different peaks in the frequency domain of an imaginary part of modulus, M", as opposed to that of impedance, Z". Therefore, a combined approach of modulus and impedance spectroscopy tends to provide better results for certain combinations of resistance and capacitance in the testing frequency domain.

Fig. 3(a) demonstrates Nyquist plots under dark by applying a set of DC voltage biases to the devices. These plots are represented on the complex plane with the real part of the impedance, Z' (resistance, x-axis), and the imaginary part of the impedance, Z" (reactance, y-axis). To identify the contribution of each interface to impedance spectra, we define an equivalent circuit model as in Fig. 3(b) inset. By increasing voltage from 0 V to 0.35 V,

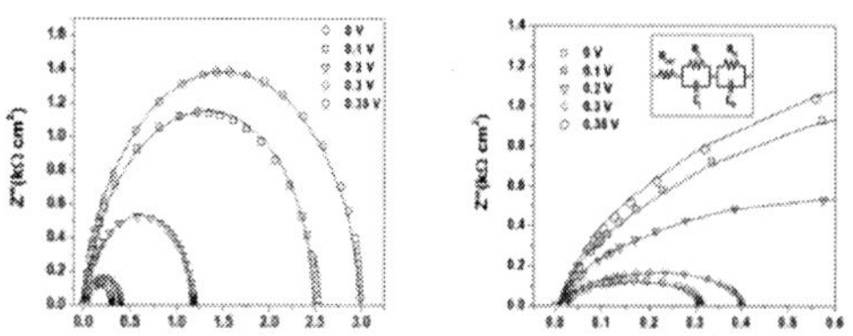

Fig. 3 (a) Impedance spectra, (b) Zoom-ups of (a) and an equivalent circuit model (inset) to which the line is fitted after modeling.

CZTSe:Ge bi-layers devices demonstrate a single semicircle at each voltage on the complex plane. The diameter of a semicircle depends on bias conditions, and all the semicircles intersect the Z' axis near 18 Ω, at high frequencies (lower side Z'-axis intercept).

To decouple two circuit components of hetero and back contact junctions, we measured weak light-biased modulus spectroscopy, combined with impedance spectroscopy. Firstly, we applied -1 V during measurements to minimize the diffusion capacitance, considering an equivalent circuit model. To inflate each parallel circuit branch's response, we apply weak light biases of 0.1 Sun (1 mA/cm^2 UV at 400 nm) and 0.2 Sun (2 mA/cm^2 IR at 1150 nm), which could reduce the resistance of each parallel circuit branch. In Fig. 4(a), there is one typical semicircle for impedance spectra. Fig. 4(b) demonstrates a single peak of impedance spectroscopy in the frequency domain, whereas two modulus peaks are present in modulus spectra, which is consistent with the theoretical study in Fig. 2.

Based on modulus and impedance spectra, each resistance and capacitance at voltage biases are resolved in Fig. 5. Heterojunction resistance (R_j) is almost an order of magnitude larger than back contact junction resistance (R_b) at 0V, whereas series resistance (R_{ser}) is smallest at all voltage biases. As voltage increases, all the resistance and the difference between them decrease. Interestingly, heterojunction capacitance (C_j) increases in Fig. 5(b), whereas back contact capacitance (C_b) decreases and cross over near 0.28 V due to the so-called a back-to-back diode structure as in the band diagram in Fig. 1(b). Since the depletion width of the back contact junction is much smaller than that of the heterojunction, capacitance of the back contact junction is larger at 0 V. As voltage increases, the heterojunction is forward-biased while the back contact junction is reverse-biased, making a crossover due to the increase of diffusion capacitance.

Due to the opposite polarity of the back-to-back diode structure, the total built-in potential, reduced $V_{b,tot}$ is determined by $V_{bi,j}$ - $V_{bi,b}$, where $V_{bi,j}$ and $V_{bi,b}$ are the

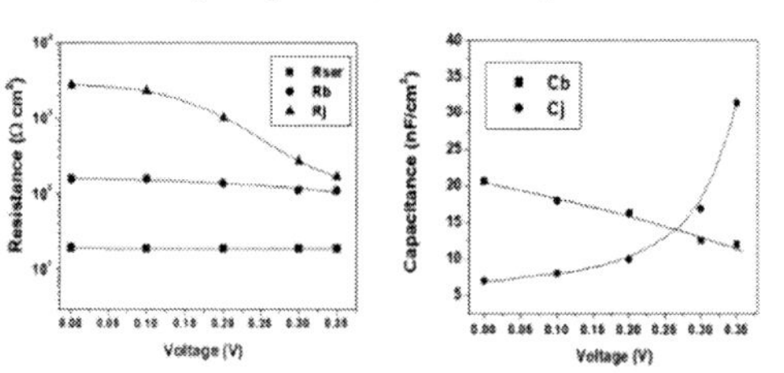

Fig. 5 Resolved (a) Resistance, (b) Capacitance. R_{ser}: series resistance, R_b and C_b: back contact resistance and capacitance, R_j and C_j: heterojunction resistance and capacitance

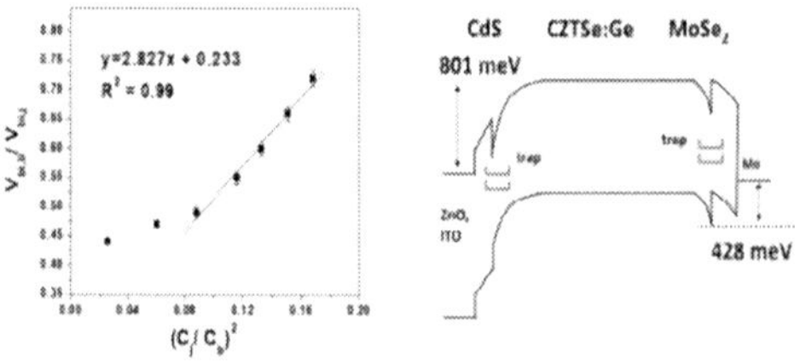

Fig. 6 (a) Linearity between Capacitance squared vs built-in potential ratio, (b) Updated band diagram with adjusted $V_{bi,j}$ and $V_{bi,b}$

built-in potential of hetero and back contact junctions [12]. To understand the impact of the built-in potential on capacitance, we simulated the ratio of each junction capacitance (C_j/C_b) at 0 V as the Schottky barrier height increases based upon our device models. Simultaneously, the ratio of the built-in potential of each junction is recorded. With the assumption of one-sided heterojunction, the ratio of the built-in potential of the device is derived from Mott-Schottky equations without applied biases below.

$$\left(\frac{C_j}{C_b}\right)^2 \propto k \cdot \left(\frac{V_{bi,b}}{V_{bi,j}}\right)$$

, where k the constant, showing the linearity.

From the simulation and equation above, Fig. 6(a) shows the linear correlations between the ratio of each junction's capacitance squared and that of each junction's built-in potential. Based on the results, the built-in potential for heterojunction is estimated 801 meV and the back contact junction is 428 meV in Fig. 6(b). For the comparison, TCAD simulation of an ideal case indicates the ideal built-in potential for the heterojunction is 894 meV, which is 93 meV larger than the estimated built-in potential.

5 CONCLUSION

The built-in potential of hetero and back contact junction of CZTSe: Ge bi-layers devices have been investigated by impedance spectroscopy combined with modulus spectroscopy. Each semicircle is resolved to explain each hetero and back contact junctions with appropriate equivalent circuit models, which tend to respond to different interface junctions at different frequency ranges. The total built-in potential of CZTSe: Ge bi-layers devices is 373 meV from capacitance characterization, which is reduced due to the opposite polarity of the back contact Schottky barrier. Consequently, the built-in potential of the heterojunction is derived as 801 meV, which is ~10 % smaller than the ideal built-in potential of the heterojunction from modeling and numerical simulation. All in all, modulus data combined with impedance spectra agree well with modeling and numerical simulation results.

ACKNOWLEDGEMENT

Authors acknowledges the generous support by the Duke Energy Foundation and CSRC at Indiana State University.

REFERENCES

[1] S. Hood, A. Walsh, C. Persson, K. Iordanidou, D. Huang, M. Kumar, Z. Jehl, M. Courel, J. Lauwaert, S. Lee, "Status of materials and device modelling for kesterite solar cells", *Journal of Physics Energy*, vol. 1, pp. 042004, 2019.

[2] S. Lee, Kent J. Price, Edgardo Saucedo, Sergio Giraldo, "Improved Device Models of CZTSe: nanolayer Ge solar cells with Quantum Efficiency", *Proceedings of IEEE 7th World Conf. on Photo. Ener. Conv.*, vol. 1, pp. 1-3 , 2018.

[3] S. Giraldo, M. Neuschitzer, T. Thersleff, S. Lopez, Y. Sanchez, H. Xie, M. Colina, M. Placidi, P. Pistor, V. Roca, K. Leifer, A. Perez, E. Saucedo, "Large efficiency improvement in $Cu_2ZnSnSe_4$ solar cells by introducing a superficial Ge nano-layer", *Adv. Energy Mater.*, vol. 5, pp. 1501070, 2015.

[4] Y. Proskuryakov, K. Durose, M. Turkestani, I. Mora, G Belmonte, F. Satiago, J. Bisquert, V. Barrioz, D. Lamb, S. Irvine, E. Jones, "Impedance spectroscopy of thin-film CdTe/CdS solar cells under varied illumination", *J. Appl. Phys.*, vol. 106, pp. 044507, 2009

[5] S. Lee, K. Price, E. Saucedo, S. Giraldo, "Improved Quantum Efficiency Models of CZTSe: nanolayer Ge solar cells with linear electric field", *Nanoscale*, 10, 2990 - 2997 (2018)

[6] M. Sugiyama, M. Hayashi, C. Yamazaki, N. Hamidon, Y. Hirose, M. Itagaki, "Application of impedance spectroscopy to investigate the electrical properties around the pn interface of $Cu(In,Ga)Se_2$ solar cells", *Thin Solid Films*, vol. 535 pp. 287-290, 2013

[7] I. Mora-Sero, G. Garcia-Belmonte, P. Boix, M. Vazquez, J. Bisquert, "Impedance spectroscopy characterisation of highly efficient silicon solar cells under different light illumination intensities", *Energy Environ. Sci.*, 2, 678-686, 2009

[8] S. Lee, K. Price, E. Saucedo, "Interface models of CZTSe: Ge bilayers solar cells with Impedance Spectroscopy", *Proc. of IEEE Photovoltaic Spec. Conf., 47th*, 1, 1-3, 2020

[9] S. Lee, K. Price, D-H Kim, "Two local built-in potentials of H_2S processed CZTSSe by complex impedance spectroscopy", *Solar Energy*, 225, 11-18, 2021

[10] F. S. Chien, A. Herawati, M. He, C. Huang, C. Chen, Optical Responses of the Heterojunctions in $Cu_2ZnSn(S,Se)_4$ Solar Cells Studied by Electric Modulus Spectroscopy, *ACS Appl. Electron. Mater.*, 2, 3, 796-801, 2020

[11] S. Lee, K. Price, E. Saucedo, "Estimation of hetero- and back contact junctions of CZTSe:Ge bi-layers solar cells by impedance and modulus spectroscopy", *J. Phys. D: Appl. Phys.*, 54, 335501 (2021)

[12] S. Moore, C. Hages, N. Carter, R. Agrawal, M. Lindstro, "The Physics of V_{bi}-Related IV Crossover in Thin Film Solar Cells: Application to Ink Deposited CZTSSe", *Proc. of IEEE Photovoltaic Spec. Conf., 39th, 1, 3255, 2013*

ANALYSIS OF ENVIRONMENTALLY FRIENDLY AND LOW COST NON VACUUM PROCESS FOR $Cu_2ZnSn(S,Se)_4$ SOLAR CELLS.

Prabeesh Punathil, Solidea Zanetti[1], Elisa Artegiani, Vikash Kumar and Alessandro Romeo
LAPS-Laboratory for Applied Physics, Department of Computer Science, University of Verona, Ca' Vignal 1, Strada Le Grazie 15, 37134 Verona, Italy, tel.:+39-045-8027974, e-mail address: alessandro.romeo@univr.it
[1]also at Isopan S.p.A., via Giona 5, Trevenzuolo, Italy

ABSTRACT: In this work CZTSSe absorber films were prepared by spin coating technique followed by selenization at optimized suitable condition. The structural properties and phase purity of the films were analyzed by XRD and Raman spectroscopy, surface morphology of the films was analyzed by SEM and AFM. Elemental composition was analyzed by energy dispersive x ray spectroscopy. Phase pure CZTSSe films with the required photovoltaic properties was obtained by selenizing the precursor films at 450 °C for 30 min with the optimal precursor drying condition. Prototype CZTSSe solar cell was fabricated and its performance were studied systematically. The highest device efficiency of 4.9% with V_{oc}=321 mV, J_{sc}=30 mA/cm^2, and FF=51% was obtained for spin coated CZTSSe device. It was observed that precursor drying condition significantly influence the performance of devices.

Keywords: kesterite; CZTS; stability; drying temperature; light soaking.

1 INTRODUCTION

Kesterite-based $Cu_2ZnSn(S,Se)_4$ (CZTSSe) solar cell devices has attracted wide attention for its merits of high abundance of elemental constituents and low environmental hazardous [1,2]. To date, highest certified device efficiency of 12.6% was obtained by hydrazine based pure solution-based approach [3]. Which indicates that, controlled solution-processing techniques even overmatches the traditional expensive vacuum based techniques. To obtain CZTSSe solar cell with high photovoltaic performances, researchers are focusing on synthesizing well crystallized phase pure off-stoichiometric CZTSSe films through annealing optimization [4,5]. Most of the solution based routes use a two-step approach using toxic H_2S gas for sulphurization [6]. In this work, prepared CZTS thin-film by spin coating technique from a stable precursor solution that contain sulphur source and selenization was carried out using Se pellets with roughing vacuum.

The present study emphasizes the importance of precursor drying condition on the device performance and propose a solution-based technique for developing cost-effective and device quality kesterite films. The successful development of absorbers having optimal durability and good optoelectronic properties through a cheap fabrication process would certainly lead to cheaper solar cells. In this work phase pure Cu_2ZnSnS_4 thin films were prepared by a low-cost solution-based technique using non-toxic solvents. Influence of drying condition on the device performance were systematically studied with the help of XRD, Raman, AFM and JV characteristics.

2 EXPERIMENTAL METHODS

2.1 Device fabrication

The precursor solution was prepared by dissolving copper (II) acetate monohydrate, zinc (II) acetate dihydrate, tin (II) chloride, and thiourea into 2-methoxy ethanol and diethanolamine. All chemicals were used without further purification. 2-methoxy ethanol was used as a solvent and diethanolamine as a stabilizer. Spin coating was carried out at 2400 rpm for 15 s and dried at 300 °C with three different drying times, namely 3, 5 and 10 minutes. This process is repeated for 5 times to increase the film thickness. The CZTS precursor films were annealed at 450°C in selenium atmosphere to complete the phase formation and to incorporate Se into the CZTS lattice. Finally, the solar cell is fabricated with the following structure:
SLG/Mo/CZTSSe/CdS/ZnO/ITO.

CdS is deposited at 60 °C by chemical bath deposition, in particular 15 ml of Cd $(CH_3COO)_2$ $2H_2O$ (0.025 M) solution, 10 ml of thiourea (0.422 M) with 25 ml of NH_4OH (20%) solution are mixed into a glass with 200 ml of distilled water. ITO and ZnO layers are deposited in the same RF-sputtering machine, with a substrate temperature of 150 °C and with an oxygen flux of 0.5 sccm, respectively with a power of 160 W and 60 W.

2.2 Characterization

The structural properties of the films were analyzed by X-ray diffraction (XRD) (Thermo ARL X'TRA powder diffractometer in Bragg-Brentano geometry, equip- ped with a Cu-anode X-ray source (Kα, λ = 1.5418 Å) and Peltier Si (Li) cooled solid-state detector) and Raman (Horiba Jobin-Yvon LabRam HR800 microprobe setup (in backscattering geometry), He-Ne laser at 632.8 nm) spectroscopies. The surface morphology and composition of the films were analyzed by scanning electron microscopy (SEM) and energy dispersive X-ray spectroscopy (EDXS) (Philips FEI XL30 Scanning Electron Microscope). Atomic force microscopy (AFM) analysis have been made by NT-MDT Solver Pro.
Finally, the J-V characteristics of the solar cells with an active area of 0.16 cm^2 were measured under AM1.5 solar irradiation with 1000 W/m^2 light intensity with a Keithley Source Meter 2420, using a halogen lamp calibrated with a silicon solar cell under an irradiation of 100 mW/cm^2 (AM 1.5).

3 RESULT AND DISCUSSION

3.1 Structural analysis

Structural properties and phase purity of the CZTS thin films were systematically analyzed by XRD and Raman spectroscopy. Very little changes were registered among the three differently dried samples (3, 5 and 10 minutes drying time for the spin coated precursor-as described above). From the XRD pattern, all the films show peaks corresponds to the CZTSSe phase and Mo phase (from the substrate), which were indexed from the ICDD files PDF No. 066-0163 (CZTSe), PDF No 026-0575 (CZTS) and PDF No 42-1120 (Mo). In fig.1 (a) the pattern of the sample dried for 5 minutes is shown. There is very slight trace of possible secondary phases, however it is difficult to confirm the exact phase formation of CZTS just with this because of the similar crystal structure possessed by its secondary and/or ternary phases. Also the Raman spectra of the differently dried samples are very similar.

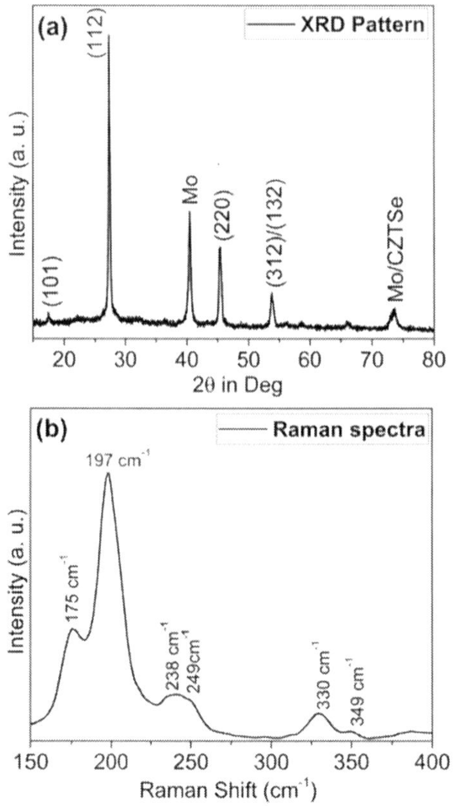

Figure 1: (a) XRD pattern and (b) Raman spectra of CZTS thin films selenized at 450 °C, with precursors dried for 5 minutes.

In Fig. 1(b) only the spectrum of the sample dried for 5 minutes is presented. The Raman spectra of CZTSSe samples showed an intense characteristic peak at 197 cm^{-1} which corresponds to A1 symmetry modes. In addition to this, 6 other CZTSSe peaks were also observed at 175 cm^{-1}, 238 cm^{-1}, 330 cm^{-1}, 349 cm^{-1}, which corresponds to the B, B(LO), A, E(TO) vibrational modes of CZTSSe phase [7–10]. A minor peak at 249 cm^{-1} was observed in the Raman spectra, which corresponds to the $MoSe_2$ phase [11]. The peak corresponds to the $MoSe_2$ phase resulting from the back contact absorber interface. XRD and Raman confirm the phase pure CZTSSe formation at the specified drying condition. A deeper analysis can be found in [12].

Differently the AFM pictures show a very different morphology, according to the different drying times (see Fig. 2). If for 3 and 5 minutes drying time the shape and size of the polycrystals is similar, for 10 minutes drying time the morphology and roughness is completely different. Moreover, the 5 minutes drying time sample results in a slightly larger grain size and improved compactness. Films shows dense morphology with average grain size around ~1 μm. Further surface morphology and elemental composition was studied by SEM and EDXS analysis, here we present the results for the 5 minutes dried sample (see Fig. 3), which corresponds to the optimized absorber as we will see further on.

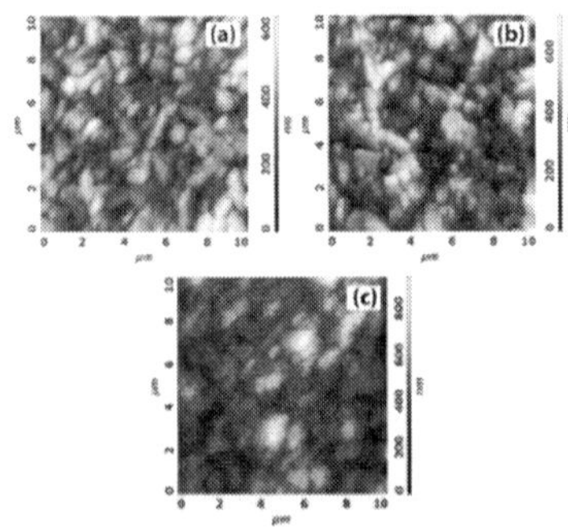

Figure 2: AFM images of the CZTSSe film: a) precursor dried for 3 minutes, b) precursor dried for 5 minutes, c) precursor dried for 10 minutes.

SEM shows a typical CZTS structure, with polycristals of about one micrometer in size, the grains are compact with limited grain boundaries. The EDXS analysis (for the 5 minutes drying time absorber) shows a Sn-rich composition with a Zn/Sn ratio of 0.56. The measurements also confirm a large incorporation of Se in the crystal lattice by selenium treatment (S/Se=0.14).

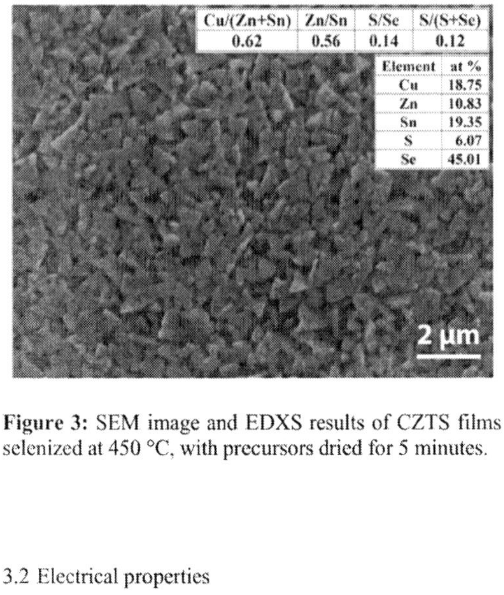

Figure 3: SEM image and EDXS results of CZTS films selenized at 450 °C, with precursors dried for 5 minutes.

3.2 Electrical properties

CZTSSe solar cells were fabricated with the above mentioned standard device structure (SLG/Mo/CZTSSe/CdS/ZnO/ITO), with active area of the cell is 0.16 cm^2, and they have been measured under AM 1.5 (see Fig. 4). As shown in Fig.4 the best performance is given with the 5 minutes drying time, with the following parameters: V_{oc}=321 mV, J_{sc}=30 mA/cm^2, FF=51 % and an efficiency of 4.9 %.

This clearly shows that the optimization of CZTS process can come through process steps that might seem irrelevant such as the drying time of the precursor, more details on this can be found in [12]. This detailed study show also the improvement carried out from the previous work presented in 2020 [13].

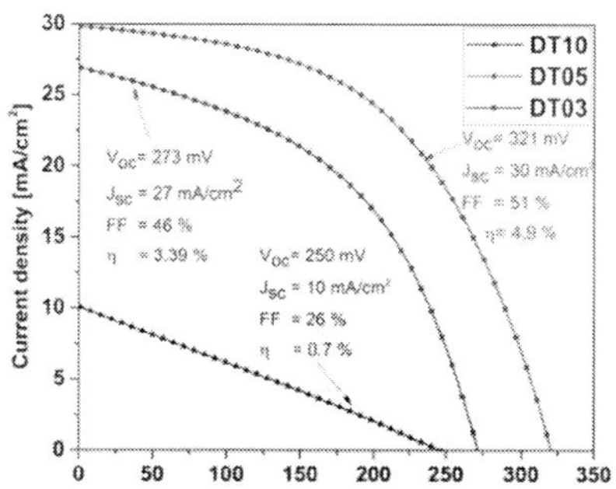

Figure. 4: JV characteristics of CZTS thin films dried at 3 min (DT03), 5 min (DT05) and 10 min (DT10).

Cells have also been tested at 80 °C and one sun in a special box and left there for several hundreds of hours (see fig. 5). This has brought to the demonstration of a light soaking effect which enhances the efficiency. In particular the fill factor strongly enhances, suggesting that many possible carrier traps are getting passivated and, at the same time, this shows the high stability of the devices.

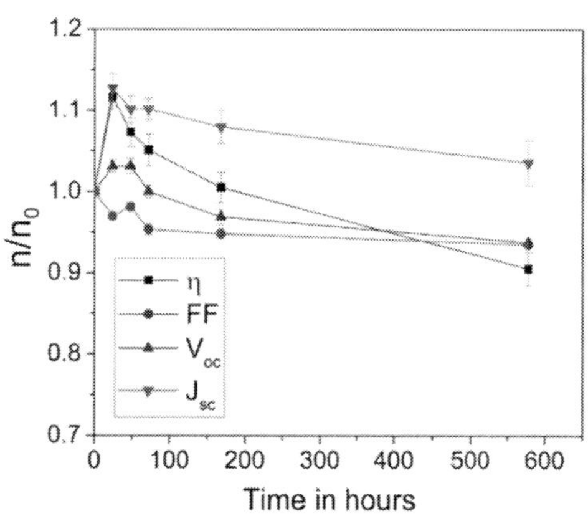

Figure 5: Accelerated lifetime stability tests on the CZTSSe solar cell with 5 min drying duration, the values are normalized to the initial ones.

4 CONCLUSIONS

CZTS thin films were prepared by spin-coating technique with optimal drying condition followed selenization. XRD and Raman confirm the phase purity of the CZTSSe films. In addition to the CZTSSe phase, $MoSe_2$ phase was also observed in the Raman spectra. The presence of $MoSe_2$ interfacial layer gives better ohmic contact between back contact and absorber layer. AFM image of the CZTSSe shows dense morphology with average grain size around ~1 µm. Finally, CZTSSe solar cell device was fabricated with standard device structure and observed an efficiency of 4.9% with V_{oc}=321 mV, J_{sc}=30 mA/cm^2, and FF=51%.

5 REFERENCES

1. Xue, Y., Yu, B., Li, W., Feng, S., Wang, Y., Huang, S., Zhang, C., and Qiao, Z. (2017) Effect of annealing atmosphere on properties of Cu2ZnSn(S, Se)4 thin films *. *Superlattices Microstruct.*, **112**, 311–7.
2. Park, J.Y., Chalapathy, R.B. V, Lokhande, A.C., Hong, C.W., and Kim, J.H. (2016) Fabrication of earth abundant Cu2ZnSnSSe4 (CZTSSe) thin film solar cells with cadmium free zinc sulfide (ZnS) buffer layers. *J. Alloys Compd.*, **4**.
3. Wang W, Winkler MT, Gunawan O, Gokmen T, Todorov TK, Zhu Y, M.D. (2014) Device characteristics of CZTSSe thin-film solar cells with 12.6% efficiency. *Adv. Energy Mater.*, **4** (7), 1301465.
4. Prabeesh, P., Sajeesh, V.G., Packia Selvam, I., Divya Bharati, M.S., Mohan Rao, G., and Potty, S.N. (2020) CZTS solar cell with non-toxic buffer layer: A study on the sulphurization temperature and absorber layer thickness. *Sol. Energy*, **207**, 419–427.
5. Schorr, S., Gurieva, G., Guc, M., Dimitrievska, M., Pérez-Rodríguez, A., Izquierdo-Roca, V., Schnohr, C.S., Kim, J., Jo, W., and Merino, J.M. (2019) Point defects, compositional fluctuations, and secondary phases in non-stoichiometric kesterites. *J. Phys. Energy*, **2** (1), 012002.
6. Prabeesh, P., Saritha, P., Selvam, I.P., and Potty, S.N. (2017) Fabrication of CZTS thin films by dip coating technique for solar cell applications. *Mater. Res. Bull.*, **86**, 295–301.
7. Li, C., Yao, B., Li, Y., Ding, Z., and Zhao, H. (2016) Impact of sequential annealing step on the performance of Cu2ZnSn(S, Se)4 thin film solar cells. *Superlattices Microstruct. J.*, **95**, 149–158.
8. Yang, K., Sim, J., Son, D., Kim, Y., and Kim, D. (2017) Precursor designs for Cu2ZnSn(S, Se)4 thin-film solar cells. *Nano Energy*, **35**, 52–61.
9. Liu, F., Zeng, F., Song, N., Jiang, L., and Han, Z. (2015) Kesterite Cu2ZnSn(S, Se)4 Solar Cells with beyond 8 % Efficiency by a Sol-gel and Selenization Process. *ACS Appl. Mater. Interfaces*.
10. Dimitrievska, M., Gurieva, G., Xie, H., Carrete, A., Cabot, A., Saucedo, E., Pérez-rodríguez, A., Schorr, S., and Izquierdo-roca, V. (2015) Raman scattering quantitative analysis of the anion chemical composition in kesterite Cu2ZnSn(SxSe1-x)4 solid solutions. *J. Alloys Compd.*, **628**, 464–470.
11. Khare, A., Himmetoglu, B., Johnson, M., Norris, D.J., Cococcioni, M., and Aydil, E.S. (2012) Calculation of the lattice dynamics and Raman spectra of copper zinc tin chalcogenides and comparison to experiments. *J. Appl. Phys.*, **111** (8), 2000–2002.
12. Punathil, P., Zanetti, S., Artegiani, E., Kumar, V., and Romeo, A. (2021) Analysis of the drying process

for precursors of Cu2ZnSn(S, Se)4 layers by low cost non vacuum fabrication technique. *Sol. Energy*, **224** (July), 992–999.

13. Punathil, P., Zanetti, S., Artegiani, E., Kumar, V., and Romeo, A. (2020) ANNEALING TEMPERATURE AND POST SULPHURIZATON / SELENIATION EFFECTS ON SOLUTION-BASED CZTS DEVICES. *37th Eur. Photovolt. Sol. Energy Conf. Exhib. ANNEALING*, **37**, 691–694.

ALTERATION OF PV CELL PARAMETERS VIA N-DOPED GRAPHENE QUANTUM DOTS INCORPORATION ON SOLUTION-PROCESSED CIGS THIN FILM-BASED PHOTOVOLTAIC CELLS

Firoz Khan[1], Jae Hyun Kim[2]

[1]Interdisciplinary Research Center for Renewable Energy and Power Systems (IRC-REPS), King Fahd University of Petroleum & Minerals (KFUPM), Dhahran-31261, Saudi Arabia
[2]Division of Energy Technology, Daegu Gyeongbuk Institute of Science & Technology, 333, Techno Jungang-Daero, Hyeonpung-Myeon, Dalseong-Gun, Daegu 42988, South Korea

*Corresponding author: Tel: +966 13 8607019; Fax: +966 13 8607312,
E-mail: firoz.khan@kfupm.edu.sa (Firoz Khan)

ABSTRACT: Copper indium gallium selenide (CIGS)-based photovoltaic (PV) cells have been considered a renowned candidate for thin-film-based PV technology because CIGS materials offer a high absorption coefficient and tunable bandgap. Only a few microns CIGS absorbing layer is sufficient to absorb more than 95% of the suitable range of solar radiations. Nevertheless, due to the parasitic absorption of the short-wavelength (λ) in the top layers, its high efficiency is severely hampered. N-doped graphene quantum dots (NGQDs) are employed as luminescent photon downshifting (LDS) material. Moreover, poly (methyl methacrylate) (PMMA) is used as a polymer matrix to enhance the chemical and mechanical stability of NGQDs. However, the polymer matrix creates photoluminescence (PL) quenching. Here, the NGQDs/PMMA LDS layer is employed on to CIGS solar cell. The short circuit current density is enhanced by 3.40% relative to the control cell (34.44 mA/cm^2). However, a slight improvement (0.68%) in the open-circuit voltage is observed. The overall enhancement in the power conversion efficiency is ~4.50%.
Furthermore, the PV cell parameters are analytically predicted. The photogenerated current density, shunt resistance, series resistance, diode ideality factor, and reverse saturation current density are changed by 3.43%, 18.34%, 22.01%, -5.47%, and -51.92%, respectively.
Keywords: CIGS solar cells, N-doped graphene quantum dots, Photon management, Photon downshifting, Downconversion

1 INTRODUCTION

Due to the excellent optoelectrical properties of copper indium gallium selenide (CIGS), it has gained more attention toward thin film-based photovoltaics (PVs) [1]. It has a tunable bandgap and a wide absorption range. However, there is a large gap between theoretical and experimental performance [2]. The parasitic absorption in the top layers (ZnO and CdS) is the key factor for the decline of the short-wavelength (λ) response, which results in the lessening of the performance of the CIGS solar cells [3]. The short-λ can be improved via photon management.

Numerous luminescent photon downshifting (LDS) materials i.e., organic-inorganic QDs [4-6], dyes [7, 8], GQDs [9-11], and rare-earth ions/complexes [12, 13] are investigated. They can be employed on the top of the solar cell to retrieve the spectral loss. However, most of them are not suitable for the CIGS cell due to one or more obstacles, such that low photoluminescent (PL) quantum yield (QY), insufficient absorption in the short-λ range, low Stokes-shift, chemical permanence, toxicity, complicated synthesis process, etc. Thus, N-doped GQDs are used previously employed as LDS in CIGS solar cells [14]. However, to provide mechanical and chemical stability, a polymer matrix is required. The insertion of the polymer matrix causes PL quenching, lowers the transmittance, and absorbs the photons of higher-λ.

In this work, the NGQDs are employed as LDS material. Moreover, poly(methyl methacrylate) (PMMA) is used as a polymer matrix to enhance the chemical and mechanical stability of NGQDs. To evaluate the efficacy, the NGQDs/PMMA layer is employed on the CIGS solar cell. The performance parameters (short circuit current density (J_{sc}), open-circuit voltage (V_{oc}), fill factor (FF), and efficiency (η) of the cell are experimentally determined. Furthermore, the PV cell parameters (photogenerated current density (J_{ph}), shunt resistance (R_{sh}), series resistance (R_s), diode ideality factor (n), and reverse saturation current density (J_0)) are analytically predicted. The changes in the performance and PV cell parameters were determined to explore the loss reduction via incorporation of NGQDs/PMMA on to CIGS PV cell.

2 EXPERIMENTAL

2.1 Synthesis of NGQDs

First, graphene oxide (GO) powder was synthesized as reported elsewhere [15, 16]. An aqueous solution of GO (2 mg/ml) was made in deionized (DI) water followed by 5 h sonication. Polyethyleneimine is added to GO solution as an N-dopant source followed by stirring at 90 °C for 4 h to get a paste. The paste was transferred to a horizontal furnace and annealed at 300 °C for 1 h. The annealed sample was dispersed in chlorobenzene. The NGQDs were extracted using a syringe filter.

2.2 Synthesis and characterization of NGQD/PMMA

The PMMA (average M_w ~120,000) was dispersed in the NGQDs solution to get a concentrated solution of NGQD/PMMA. The NGQD/PMMA solution was spin-coated on the quartz slide to get a thin layer, which was used as LDS. To study the atomic structure of the NGQD/PMMA, high-resolution transmission electron microscopy (HRTEM: HF-3300/NB5000/S-4800, Hitachi) was used. The other characteristics of the NGQDs are reported elsewhere [14].

2.2 Solar cell characterization

The synthesized NGQDs/PMMA composite was coated onto CIGS solar cells (active area of ~0.43 cm^2). The J-V characteristics of the control (without NGQD/PMMA) and the cell with NGQD/PMMA layer were measured under 1-sun AM1.5G illumination using a source meter (Keithley 2400). The illumination intensity was calibrated using a Si reference solar cell (PV Measurements Incorporated, USA). All measurements were performed at 25 °C.

3 RESULTS AND DISCUSSION

3.1 Morphological and elemental study

The TEM image of the NGQD/PMMA is shown in Figure 1. It can be seen that the NGQDs are embedded in the PMMA matrix. Also, NGQDs are uniformly distributed in the matrix.

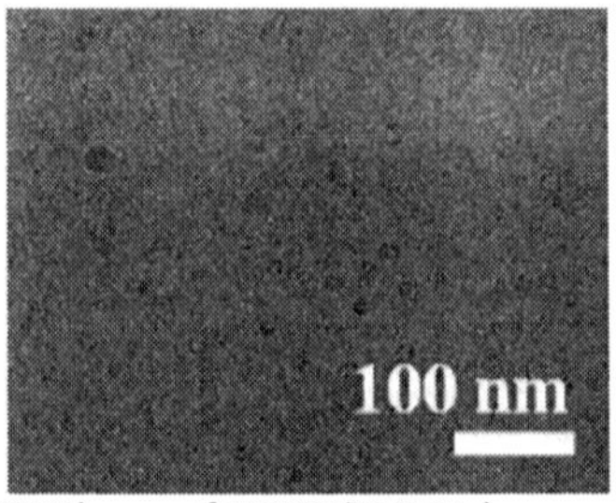

Figure 1: TEM image of NGQD/PMMA layer.

The NGQD/PMMA acts as an LDS layer. The optical images of the bare and NGQD/PMMA coated quartz glass are displayed (under illumination with UV-light of wavelength 365 nm) in Figure 2. The color of the NGQD/PMMA layer indicates that the wavelength of the emitted light is ~490 nm. The color of the NGQD/PMMA coated quartz class can be compared with bare glass. This result confirms that the Stokes-shift of the NGQD/PMMA LDS layer is ~125 nm. The Stokes shift can vary with the excitation wavelength.

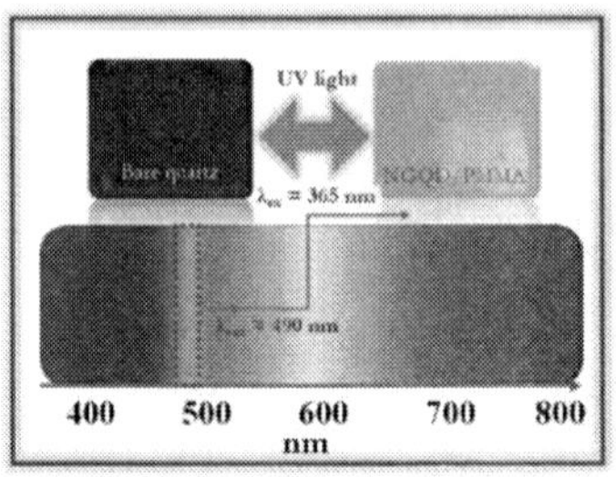

Figure 2: Optical images of (left) bare and (right) NGQD/PMMA deposited quartz glass.

The NGQD/PMMA composite layer is applied on the top of the CIGS solar cell. The cross-sectional image of the CIGS solar cell is displayed in Figure 3, which revealed the structure of the CIGS solar cell. The thickness of the CIGS absorbing layer is ~ 1 μm. The NGQD/PMMA layer lies on the top of the CIGS solar cell, which acts as an LDS layer.

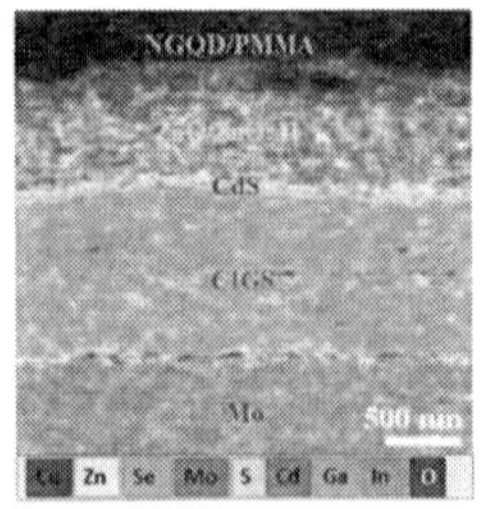

Figure 3: Cross-sectional TEM-EDS mapping the CIGS solar cell.

3.2 Performance analysis

Moreover, the schematic of the photon downshifting is illustrated in Figure 4. The photons of λ > 500 nm absorb directly into CIGS absorbing layer. However, the LDS layer absorbs the photons of short-λ (< 500 nm) and these photons are re-emitted with higher-λ. Thus, the number of photon generation can be increased, which results in enhancing the photogenerated current density.

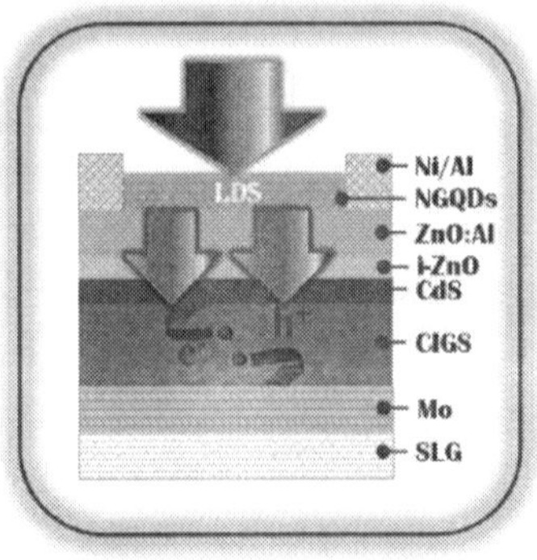

Figure 4: Schematic for the photon downshifting thru NGQD/PMMA LDS layer.

The illuminated J-V curve is shown in Figure 5. The performance parameters are extracted from the illumination *J-V* curves. The performance parameters of the control cell and CIGS solar cell with the LDS layer are listed in Table I.

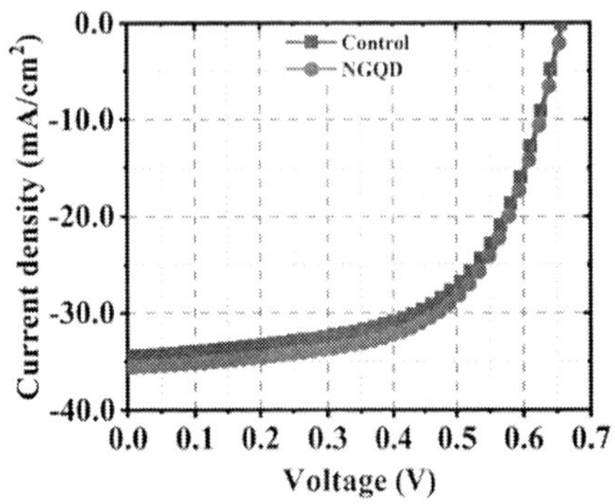

Figure 5: Illuminated J-V characteristics of the control and CIGS with LDS layer.

Table I: Performance parameters of the control and NGQD/PMMA CIGS solar cells.

Cell	J_{sc} (mA/cm^2)	V_{oc} (mV)	FF (%)	η (%)
Control	34.44	655.9	60.16	13.59
NGQD/PMMA	35.61	660.4	60.38	14.20

The changes in performance parameters are illustrated in Figure 6. The changes in the J_{sc}, V_{oc}, and FF are 3.40%, 0.68%, and 0.36%, respectively. It revealed that the overall performance is enhanced by ~4.5% relative to the control cell.

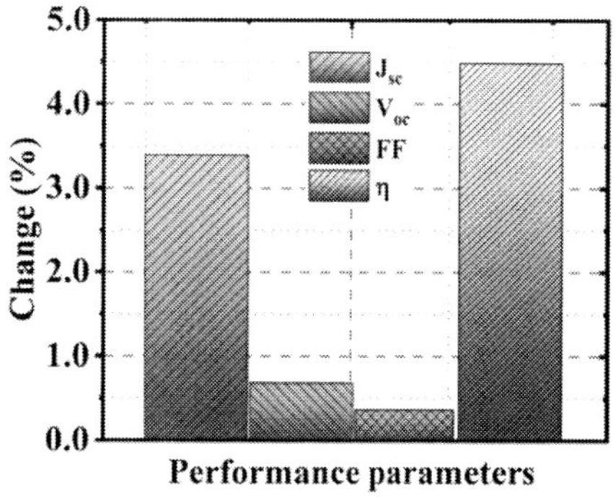

Figure 6: Change in performance parameters via insertion of NGQD/PMMA.

3.3 PV cell parameters analysis

The PV cell parameters were analytically predicted using an analytical method [17, 18]. The obtained values of PV cell parameters are listed in Table II.

Table II: PV cell parameters of the control and NGQD/PMMA CIGS solar cells.

Cell	J_{ph} (A/cm^2)	R_{sh} (Ωcm^2)	R_s (Ωcm^2)	n	J_0 (A/cm^2) x10^{-7}
Control	0.03475	182.21	1.64	2.2964	4.6026
NGQD/ PMMA	0.03594	215.63	2.01	2.1708	2.3562

The changes in the PV cell parameters are illustrated in Figure 7. The obtained changes in J_{ph}, R_{sh}, R_s, n, and J_0 are 3.43%, 18.34%, 22.01%, -5.47%, and -51.92%, respectively. By application of the LDS layer, the number of photons is increased, which results in an improvement in the J_{ph}. Thus, J_{sc} is also increased, which also fills the traps [19]. Moreover, the values of R_{sh}, n, and J_0 are reduced. However, the value of R_s is significantly increased. With further optimization, the performance of the CIGS solar cell can be improved. The results confirm that via employing the LDS, the charge carrier generation is improved, which results in a lessening in the charge recombination.

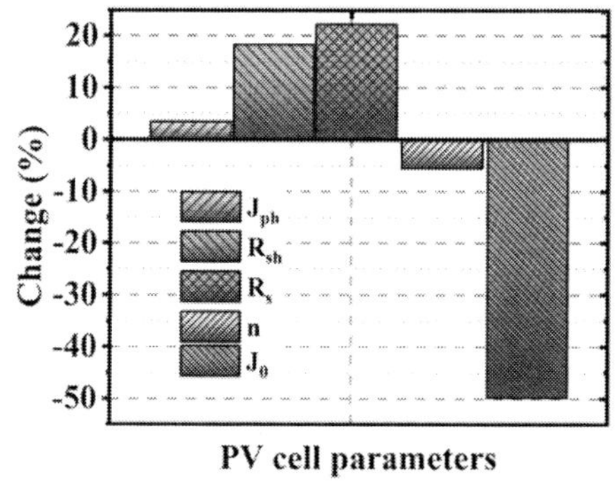

Figure 7: Change in performance parameters via insertion of NGQD/PMMA.

4 CONCLUSIONS

Via incorporation of NGQD/PMMA layer the enhancements in the J_{sc}, V_{oc}, and FF, were 3.40%, 0.68%, and 0.36%, respectively. Thus, overall performance was enhanced by ~4.5% relative to the control cell. However, the observed changes in J_{ph}, R_{sh}, R_s, n, and J_0 are 3.43%, 18.34%, 22.01%, -5.47%, and -51.92%, respectively. The results confirm that via employing the NGQD/PMMA as an LDS layer, the charge carrier generation is improved, which results in a lessening in the charge recombination. However, the leakage current is greatly reduced.

ACKNOWLEDGMENTS

The author gratefully acknowledges funding (Project No. INRE2109) from the Dean Ship Research (DSR), King Fahd University of Petroleum & Minerals (KFUPM), Saudi Arabia.

REFERENCES

[1] F. Khan, H.-J. Lee, M. Oh, J.H. Kim, Solar Energy, 108 (2014) 189-198.
[2] S. Siebentritt, Sol. Energy Mater. Sol. Cells 95 (2011) 1471-1476.
[3] F. Khan, J.H. Kim, Scientific Reports 9 (2019) 1-9.
[4] V. Švrček, A. Slaoui, J.-C. Muller, Thin Solid Films 451 (2004) 384-388.
[5] H.-J. Muffler, M. Bär, I. Lauermann, K. Rahne, M. Schröder, M. Lux-Steiner, C.-H. Fischer, T. Niesen, F. Karg, Sol. Energy Mater. Sol. Cells 90 (2006) 3143-3150.
[6] W. Van Sark, Appl. Phys. Lett. 87 (2005) 151117.
[7] B. Richards, K. McIntosh, Progress in Photovoltaics: Research and Applications 15 (2007) 27-34.
[8] T. Maruyama, A. Enomoto, K. Shirasawa, Sol. Energy Mater. Sol. Cells 64 (2000) 269-278.
[9] D. Qu, M. Zheng, J. Li, Z. Xie, Z. Sun, Light: Science & Applications 4 (2015) e364-e364.
[10] K.D. Lee, M.J. Park, D.-Y. Kim, S.M. Kim, B. Kang, S. Kim, H. Kim, H.-S. Lee, Y. Kang, S.S. Yoon, ACS applied materials & interfaces 7 (2015) 19043-19049.
[11] M.-L. Tsai, W.-C. Tu, L. Tang, T.-C. Wei, W.-R. Wei, S.P. Lau, L.-J. Chen, J.-H. He, Nano Lett. 16 (2016) 309-313.
[12] K. Kawano, N. Hashimoto, R. Nakata, Effects on solar cell efficiency of fluorescence of rare-earth ions, in: Mater. Sci. Forum, 1997.
[13] K. Kawano, K. Arai, H. Yamada, N. Hashimoto, R. Nakata, Sol. Energy Mater. Sol. Cells 48 (1997) 35-41.

[14] F. Khan, J.H. Kim, ACS Photonics, 5 (2018) 4637-4643.
[15] F. Khan, J.H. Kim, Journal of Industrial and Engineering Chemistry 68 (2018) 129-139.
[16] F. Khan, S.-H. Baek, J.H. Kim, Carbon 100 (2016) 608-616.
[17] J. Phang, D. Chan, J. Phillips, Electron. Lett, 20 (1984) 406-408.
[18] F. Khan, S.-H. Baek, Y. Park, J.H. Kim, Energy Convers. Manage., 76 (2013) 421-429.
[19] F. Khan, S. Singh, M. Husain, Sol. Energy Mater. Sol. Cells, 94 (2010) 1473-1476.

CADMIUM SULFIDE THICKNESS REDUCTION IN CIGS SOLAR CELLS BY APPLICATION OF ALD-ZINC MAGNESIUM OXIDE

Dominik Bagrowski, Stefanie Spiering, Thomas Schnabel, and Jan-Philipp Becker
Zentrum für Sonnenenergie- und Wasserstoff-Forschung Baden-Württemberg (ZSW), Meitnerstraße 1, 70563 Stuttgart, Germany
dominik.bagrowski@zsw-bw.de

ABSTRACT: We investigated the potential of reducing the thickness of solution grown CdS buffer in $Cu(In,Ga)Se_2$–based solar cells by substitution of the adjacent radio frequency sputtered i-ZnO high resistive layer with an atomic layer deposited $Zn_{1-x}Mg_xO$ layer. Typically, the standard CdS thickness is about 50 nm combined with a 50-80 nm sputtered i-ZnO layer. Former investigations showed a significant decrease of the device performance by reducing the CdS thickness below 30 nm. We deposited the $Zn_{1-x}Mg_xO$ layer in a standard thermal as well as in a plasma-assisted mode. We compared both processes by varying the Zn/Mg-pulse ratios and the layer thicknesses. We realized best results with a reduced CdS thickness down to 5 nm when combined with 10-30 nm $Zn_{1-x}Mg_xO$ with $0.15 \leq x \leq 0.3$ without efficiency loss compared to reference cell with standard CdS thickness. The cells showed no or very low cold light soak (CLS) effect. A maximum cell efficiency of 18.5 % (V_{OC} = 727 mV, FF = 78.0 %, J_{SC} = 32.7 mA/cm², 10 min CLS) could be achieved comparable to reference cells of 18.6 % efficiency (V_{OC} = 741 mV, FF = 79.9 %, J_{SC} = 31.4 mA/cm²).

Keywords: CIGS, thin-film solar cell, buffer, ALD, ZnMgO

1 INTRODUCTION

For decades, $Cu(In,Ga)Se_2$ (CIGS) based solar devices with a chemical bath deposited (CBD) CdS buffer layer show highest and reliable efficiencies [1,2]. A wide range of alternative buffer materials, mainly Zn- or In-based ternary compounds, have been studied in parallel to enhance the device performance by reducing the parasitic absorption in the short wavelength region as well as to avoid the cadmium compound for environmental reasons. Reviews of these efforts are presented in [3,4]. In particular, atomic layer deposition (ALD) is one of the most interesting methods for the deposition of alternative buffer materials due to the possibility of precisely control of the material composition in particular for ternary compounds, as well as for the deposition of homogeneous ultrathin layers with an excellent step coverage [5].

$Zn_{1-x}Mg_xO$ is one of the promising candidates explored as buffer layer in CIGS devices in combination with other materials including CdS [6], Zn(O,S) [7,8], i-ZnO [9], or also as stand-alone buffer layer [10,11]. In 2019, a new world-record cell with 23.35 % efficiency was presented with a buffer layer of CBD-Zn(O,S) and a magnetron-sputtered $Zn_{1-x}Mg_xO$ on a $Cu(In,Ga)(S,Se)_2$ absorber [6,10]. For $Zn_{1-x}Mg_xO$ several deposition methods were applied, predominantly based on rf-sputtering or ALD. The band alignment in the CIGS solar cell can be tuned by changing the Mg content [12,13], and thereby adapted to the needs of the particular CIGS and buffer system used [13]. One challenge of these solar devices are metastable effects [11]. In some cases, devices need a specific light-soak-treatment to achieve their best performance [14,15].

The thickness of CdS buffer layers is usually 50-60 nm [16]. The reduction of the CdS thickness in CIGS solar cells typically leads to a decrease in the device performance. Contreras et al. also described this effect in [17]. In recent years, a post-deposition alkali metal treatment (PDT) following the CIGS deposition process has become common [18]. Chirila et al. showed a possible reduction of CdS thickness for PDT treated absorbers, but still 30 nm of CdS were used [19].

Here we present the potential for high efficiency CIGS devices with a very thin buffer system of a total thickness of 15 nm consisting of 5 nm CdS and 10 nm $Zn_{1-x}Mg_xO$ showing a negligible light-soak-effect.

2 EXPERIMENTAL

2.1 Cell preparation and structure

The complete CIGS solar cells investigated in this study consist of a soda-lime glass substrate, a sputtered Mo back contact, the CIGS absorber, a double buffer layer of either CBD-CdS and rf-sputtered i-ZnO or CBD-CdS and ALD-$Zn_{1-x}Mg_xO$, a dc-sputtered ZnO:Al front contact, and Ni-Al grid fingers on top. Per sample, 10 cells with a total cell area of 0.5 cm² each were processed. No anti-reflective coating was applied. A schematic overview of the cell structures investigated in this work is shown in Figure 1.

The CIGS absorbers were deposited by means of in-line multi-stage co-evaporation process with an alkali-based post-deposition treatment (PDT) [1].
The thickness of the CdS layers in combination with the ALD buffer was varied between 0-30 nm. The deposition of $Zn_{1-x}Mg_xO$ was performed in a FlexAL ALD system from Oxford Instruments GmbH. The system allows depositions in two different process modes, a thermal (th) and a plasma-enhanced (pl) mode, which were both compared in this study. The ALD cycles were as follows:

th: DEZ/Ar/H_2O/Ar/Mg$(CpEt)_2$/Ar/H_2O/Ar
pl: DEZ/Ar/O_2-Plasma/Ar/Mg$(CpEt)_2$/Ar/O_2-Plasma/Ar

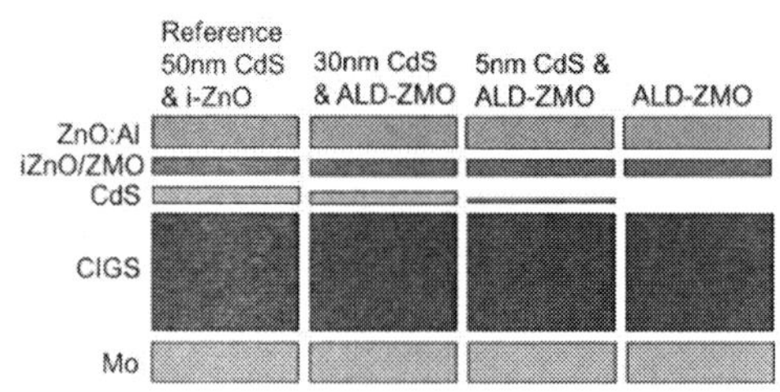

Figure 1: schematic layer structure of the investigated CIGS devices with CdS/sputter-i-ZnO and CdS/ALD-$Zn_{1-x}Mg_xO$.

The substrate temperature was kept constant at 150°C. The thickness of the $Zn_{1-x}Mg_xO$ layer was varied between 5-50 nm. The ratio of Mg/Mg+Zn-pulses (M/MZ-pulse) was varied between 11-25 %. For comparison, standard reference cells with 50 nm CdS and 80 nm sputtered i-ZnO were prepared.

2.2 Analytics

The I-V performance of the CIGS solar cells were analyzed by a WACOM AM1.5G solar simulator with four point geometry at standard testing conditions as-grown (ag) and after cold-light-soaking (CLS) for 30 minutes under AM1.5G conditions. No post-annealing step was applied. Additionally, the external quantum efficiency (EQE) was measured for single devices using a Bentham PVE300 setup.
The composition of ALD-$Zn_{1-x}Mg_xO$ with varying M/MZ-pulse ratio was determined by energy dispersive x-ray spectroscopy (EDX) mapping on silicon wafers with a FEI Tecnai Osiris transmission electron microscope. The thickness was determined using a Woollam M2000X ellipsometer.

3 RESULTS

In the following sections, we present results of completed cell devices processed with several variations of the buffer process conditions with a focus on the influence of the M/MZ-pulse ratio compared for both thermal and plasma-enhanced ALD. Based on former works at ZSW with sputtered zinc magnesium oxide [13] and other groups like [12,13], we focused on M/MZ-pulse ratios ≤ 25 %.

In previous investigations, CIGS standard cells with different CdS thicknesses between 30–120 nm were processed at ZSW. Figure 2 shows the efficiency distribution of the completed cells. Best results were obtained for 60 nm CdS. For the 30 nm CdS buffered cells the open circuit voltage V_{OC} is significantly lower, whereas for the 120 nm CdS buffered devices the short circuit current J_{SC} is reduced due to high absorption. In the following, we explored the best CdS thickness in combination with ALD-$Zn_{1-x}Mg_xO$.

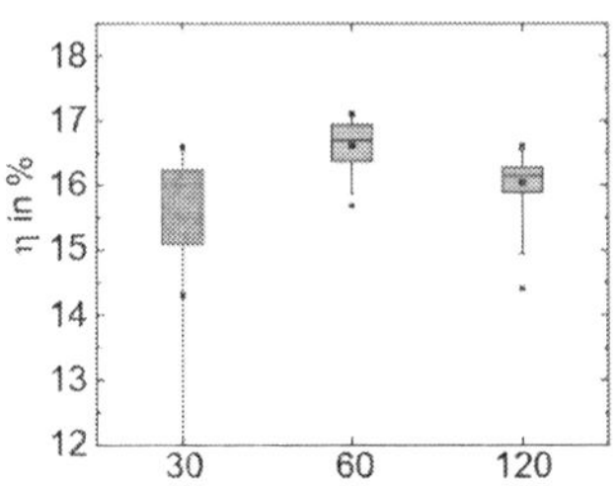

Figure 2: Efficiency distribution for CIGS/CdS/i-ZnO/ZnO:Al cells with various CdS thicknesses.

3.1 CdS thickness reduction

To evaluate the best CdS thickness in combination with ALD-$Zn_{1-x}Mg_xO$, a cell series with either 0 nm, 5 nm or 30 nm CdS layer was processed. The ALD layer was deposited in thermal mode at two different M/MZ-pulse ratios of 14 and 20 %, respectively. Figure 3 shows the efficiency distribution of the completed devices following a cold-light-soak treatment for 30 minutes. The devices without CBD-CdS-layer were all shunted. The best results were obtained for the devices with a 5 nm CdS layer. Thus, in the following experiments 5 nm CdS was used as standard thickness for the underlying buffer.

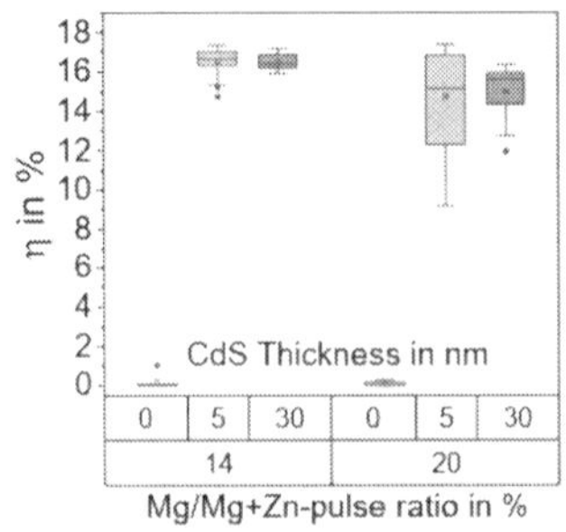

Figure 3: Efficiency distribution for CIGS cells (after CLS) with a combined buffer system of various CdS thicknesses and 30 nm $Zn_{1-x}Mg_xO$ (M/MZ-pulse ratio: 14 and 20 %).

3.2 Cold-Light-Soak (CLS) effects

The I-V performance of the cell devices was measured as grown (ag) and after a cold-light-soak (CLS) treatment for 30 minutes to explore metastable behavior. The results indicated a strong influence of the M/MZ-pulse ratio during the ALD process on the light-soak behaviour. Figure 4 shows the I-V curves of two example cells (ALD layer deposited in plasma-mode) with a M/MZ-pulse ratio of 14 % and 20 %, respectively, before and after light-soak. For the lower M/MZ-pulse ratio of 14 %, we observed a negligible performance change, whereas the cell with 20 % M/MZ-pulse ratio showed a strong efficiency improvement after CLS.

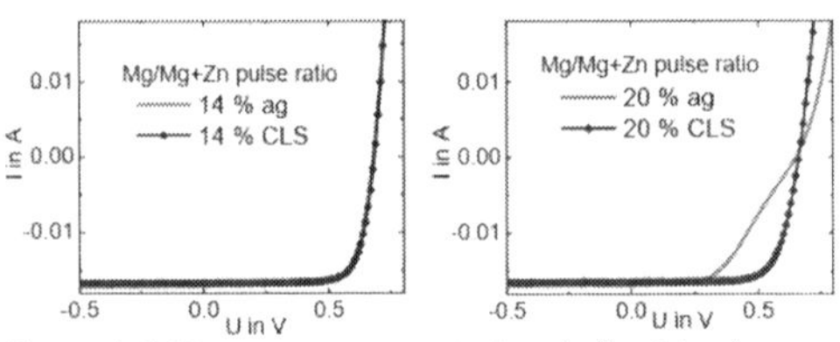

Figure 4: I-V curves as grown (ag) and after 30 minutes cold light-soak (CLS) of CIGS cells with a 5 nm CdS/30 nm $Zn_{1-x}Mg_xO$ (plasma-ALD) buffer system for a M/MZ-pulse ratio of 14 % (left) and 20 % (right).

Overall, all cells with initially low performance showed a significant light-soak-effect. However, their efficiency did not improve by the light-soak treatment towards the same level of cells that initially showed high performance. A CLS of almost 3 hours was necessary to reach a stable performance for the example cell in Figure 5.
Figure 6 presents an overview of the efficiency distribution of different cell series processed with $Zn_{1-x}Mg_xO$ in ALD plasma mode as well as in thermal mode with 14 % and 20 % M/MZ-pulse ratios before and after light-soaking. The efficiencies clearly decreased with increasing M/MZ-pulse ratio. When comparing both ALD processes, cells with plasma-$Zn_{1-x}Mg_xO$ showed a significantly better performance and quite lower light-soak effect compared to cells with thermal-$Zn_{1-x}Mg_xO$.
For a M/MZ-pulse ratio of 14 %, no or only a minor light-soak effect was observed for devices processed in plasma-mode. Upon light-soak, the devices processed in thermal mode with 14 % M/MZ-pulse ratio reached the same efficiency level as in plasma mode.

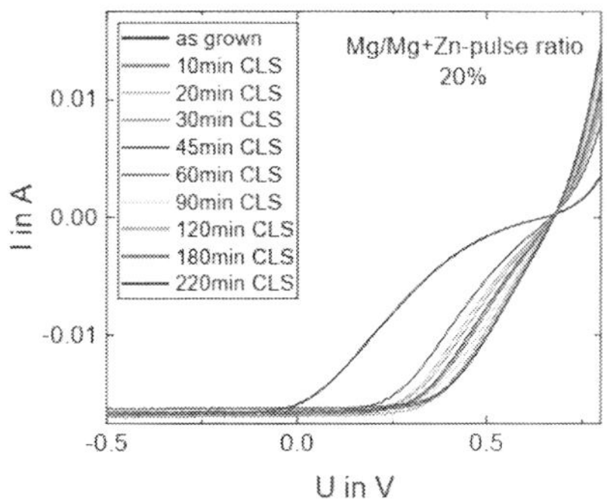

Figure 5: I-V curves as grown (ag) and after 10-220 minutes light-soak (CLS) of CIGS cells with a 5 nm CdS/30 nm $Zn_{1-x}Mg_xO$ (thermal-ALD) buffer system for a M/MZ-pulse ratio of 20 %.

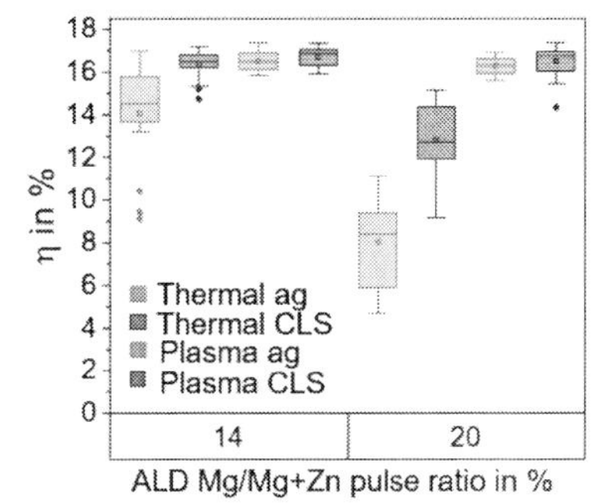

Figure 6: Average efficiency of CIGS cells with a CdS/ ALD-$Zn_{1-x}Mg_xO$ buffer layer (thermal and plasma assisted) before (ag) and after 30 min cold-light-soak (CLS) for 14 % and 20 % M/MZ-pulse ratio.

3.3 Variation of ALD-M/MZ-pulse ratio

Based on the effects shown in section 3.2, another cell series was processed using an extended ALD-M/MZ-pulse ratio variation between 11-25 %. Again, both thermal and plasma-enhanced ALD processes were compared.

Figure 7 shows the I-V data of the completed cell devices after light-soaking. For the short-circuit current density J_{SC} a slight decrease with increasing M/MZ-pulse ratio was observed for cells processed with thermal ALD mode, whereas no effect was observed for the plasma-enhanced devices. Transmission measurements on $Zn_{1-x}Mg_xO$-layers showed no significant dependence on the M/MZ-ratio. The open circuit voltage V_{OC} showed a clear decrease with increasing M/MZ-pulse ratios. The same was observed for the FF, which is more prominent for thermal-ALD samples. Overall, for both ALD processes efficiencies of up to 18 % were achieved for M/MZ-pulse ratios $\leq$ 17 %.

The different behaviour of samples processed with the two different ALD processes may be explained by different Mg contents in the resulting layers. We deposited ALD-$Zn_{1-x}Mg_xO$ layers of various M/MZ-pulse ratios in parallel on silicon wafers and subsequently analysed them using an EDX system.

The EDX measurements revealed a significant higher Mg/Mg+Zn ratio for the thermally deposited layers (see Figure 8). For the plasma-enhanced ALD layers the Mg content was slightly higher than the original pulse ratio, yet it was nearly doubled in the thermal ALD layers.

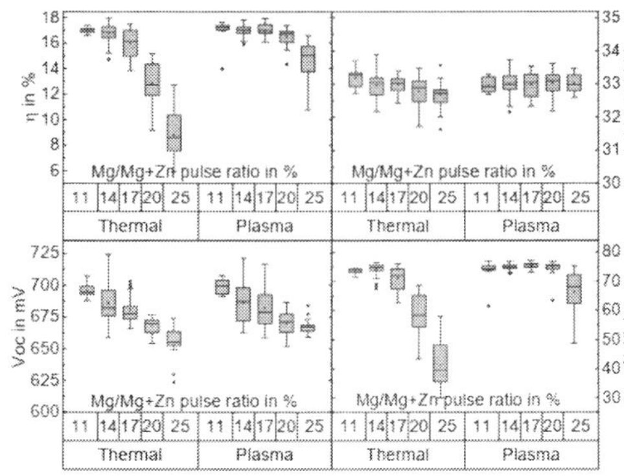

Figure 7: I-V data of CIGS cells with a CdS/$Zn_{1-x}Mg_xO$ buffer layer after cold-light-soak, comparing thermal- and plasma-ALD mode with various M/MZ-pulse ratios.

A Mg/Mg+Zn ratio of ~20 % was verified for the 11 %-thermal process and the 14 %-plasma process. Comparing the I-V results of the corresponding cell devices, a similar performance was obtained.

For future experiments, the ALD process conditions should be adjusted to the targeted Mg content. From the EDX-measurements, it can be assumed, that the best Mg content in the $Zn_{1-x}Mg_xO$ layers is $0.15 \leq x \leq 0.25$. This result is consistent with the work of other groups [14,20,21].

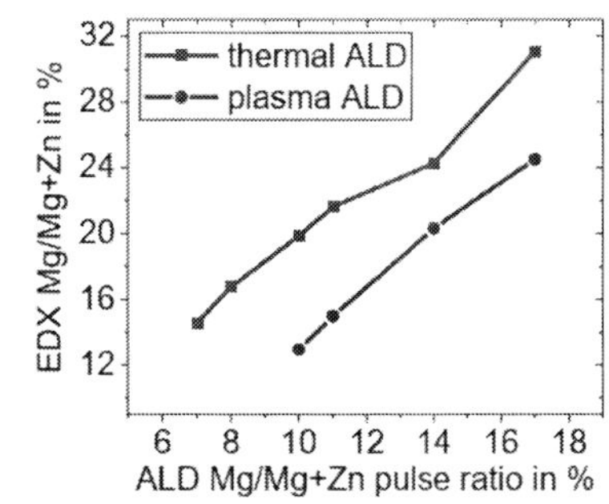

Figure 8: Measured Mg content in $Zn_{1-x}Mg_xO$ layers processed either with ALD thermal or plasma mode from EDX compared to the original ALD-M/MZ-pulse ratio.

3.4 ALD-$Zn_{1-x}Mg_xO$ thickness variation

In section 3.1. CdS thickness for the lower buffer was varied at constant $Zn_{1-x}Mg_xO$ thickness of 30 nm for the upper buffer part. In this section, the CdS thickness was kept constant at 5 nm, whereas the $Zn_{1-x}Mg_xO$ thickness was modified between 5 and 50 nm on substrates of a different CIGS batch. For this deposition series, the M/MZ-pulse ratios for the thermal and plasma ALD processes were adapted according to prior EDX measurements. A Mg/Mg+Zn-ratio of ~15 % was achieved by applying a thermal process with 7 % M/MZ-pulse ratio and a plasma process with 11 % M/MZ-pulse ratio.

The devices processed with thermal ALD mode showed better performance compared to the plasma-processed samples (see Figure 9). On average, best results were achieved using a thickness of 10 nm $Zn_{1-x}Mg_xO$.

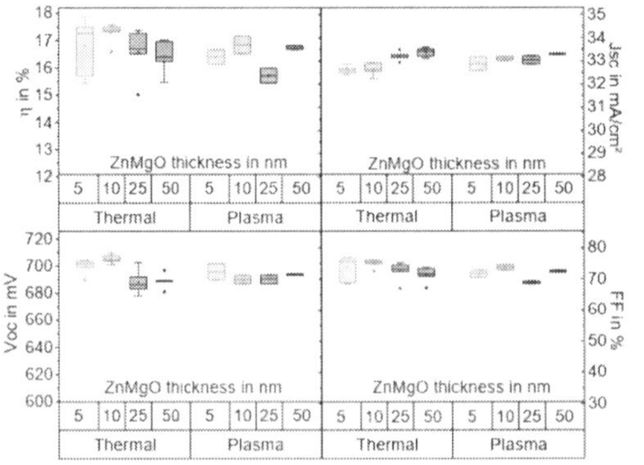

Figure 9: Distribution of average I-V data (multiple samples with 10 CIGS cells each, after CLS) with a 5 nm CdS/ALD-$Zn_{1-x}Mg_xO$ buffer layer (thermal and plasma) for various $Zn_{1-x}Mg_xO$ ($x = 0.15$) thicknesses.

The difference in the current density is due to the fact that the layer thickness was not optimized with regard to interference. The slightly V_{oc} decrease with increasing layer thickness may occur because of the influence of temperature due to longer processes for thicker layers.

3.5 Best devices

The performance of cell devices based on a CBD-CdS/ALD-$Zn_{1-x}Mg_xO$ buffer system with two different thicknesses of the ALD layer (10 nm + 30 nm) was compared to CIGS solar reference cells (see Table I).

Table I: I-V data of best cell devices with thermal and plasma-enhanced ALD-$Zn_{1-x}Mg_xO$ buffer combined with 5 nm CdS in comparison with standard reference cell with 50 nm CdS/80 nm i-ZnO.

Best cell devices		CdS thickness	Eta [%]	V_{OC} [mV]	FF [%]	J_{SC} [mA/cm²]
ZMO Thermal	10 nm	5 nm	18.5	727	78.0	32.7
ZMO Plasma	10 nm	5 nm	18.4	702	78.1	33.6
ZMO Thermal	30 nm	5 nm	17.9	694	76.7	33.7
ZMO Plasma	30 nm	5 nm	18.1	722	76.9	32.7
iZnO-Ref 1	80 nm	50 nm	18.5	747	79.0	31.0
iZnO-Ref 2	80 nm	50 nm	18.6	741	79.9	31.4

Highest efficiencies > 18 % were achieved for cells with a combination of 5 nm CdS + 10 nm $Zn_{1-x}Mg_xO$, similar to reference cells with 50 nm CdS + 80 nm i-ZnO. The I-V curves are presented in Figure 10.

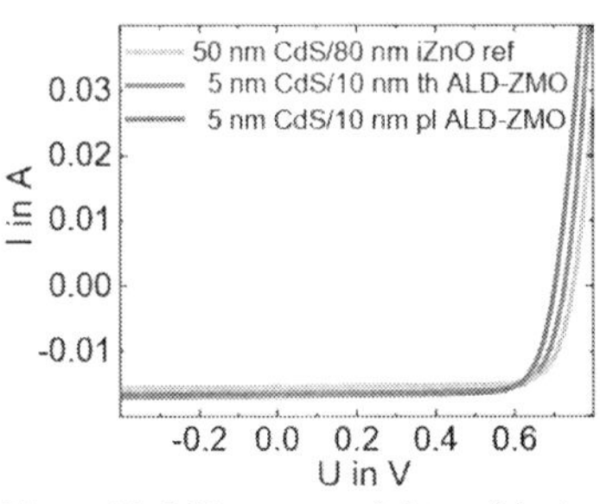

Figure 10: I-V curves and data of the best cell devices (A_{ap} = 0.5 cm²) with a 5 nm CdS/10 nm ALD-$Zn_{1-x}Mg_xO$ (thermal and plasma mode) buffer layer system compared to a reference device with CdS/i-ZnO.

Nevertheless, the V_{OC} and the FF were slightly lower compared to standard cells, which is compensated by a gain in J_{SC} (see also the EQE measurement in Figure 11).

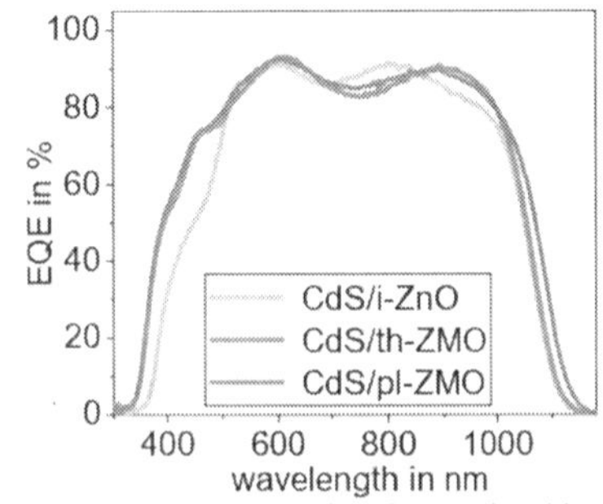

Figure 11: EQE curves of the best cells with CdS/ALD-$Zn_{1-x}Mg_xO$ layer (thermal + plasma mode) compared to a reference device with CdS/i-ZnO.

4 DISCUSSION AND SUMMARY

CIGS solar devices with a combined buffer system of CBD-CdS and ALD-$Zn_{1-x}Mg_xO$ were processed. Two different ALD processes were applied, thermal and plasma enhanced. The influence of the buffer thickness and M/MZ-pulse ratio was explored. The I-V parameter of the completed cells before and after light-soaking were compared.

A direct deposition of the $Zn_{1-x}Mg_xO$ on the CIGS absorber led to shunted cell devices. A very thin CdS layer of 5 nm was sufficient to achieve high performance cells.

With respect to thickness variation of the ALD-$Zn_{1-x}Mg_xO$ layer, best results were obtained with 10 nm thickness. Thus, the total buffer thickness for the best devices was only 15 nm. Consequently, minimizing the parasitic absorption losses, commonly observed for CdS, leads to an increasing current density.

In this study, devices processed with plasma-ALD showed on average better results than the thermally processed devices, which could be related to a significantly higher Mg content in the layers compared to the original Mg/Zn+Mg-pulse ratio during the thermal process. After adjusting the Mg/Mg+Zn-pulse ratios for both the thermal and the plasma-enhanced processes the device performances were comparable. EDX measurements revealed an optimum Mg content range in $Zn_{1-x}Mg_xO$ of $0.15 \leq x \leq 0.25$.

A cold-light-soak of 30 minutes was applied on the completed cells. For cells with initially high performance, there was no or only minor light-soak effects.

In comparison to reference devices with 50 nm CBD-CdS and 80 nm i-ZnO and maximum efficiencies up to 18.6 %, our best device with 5 nm CBD-CdS and 10 nm $Zn_{1-x}Mg_xO$ achieved a similar level with a maximum efficiency of 18.5 %. Although the V_{OC} and the FF were slightly lower compared to the reference, the J_{SC} was higher. In order to further enhance the performance of CdS-reduced and possibly CdS-free CIGS solar cells, future approaches should include a detailed characterization and adjustment of the band alignment to avoid any losses in V_{OC}. In addition to the fine-tuning of the buffer composition, also surface treatments of the CIGS layer might be crucial here.

5 ACKNOWLEDGEMENTS

The authors like to thank the CIGS team for their support in cell preparation, as well as the analytic team at ZSW for EDX and I−V measurements. We thank Dr. Wolfram Witte for providing the results of the CdS thickness variation.
This work was supported by the Bundesministerium für Wirtschaft und Energie within the „ALDELA“ project FKZ 0324352 and the „CISHiTec“ project FKZ 0324179.

6 REFERENCES

[1] P. Jackson, R. Wuerz, D. Hariskos, E. Lotter, W. Witte, M. Powalla, Phys. Status Solidi RRL 10 (2016) 583.
[2] T. Kato, J.-L. Wu, Y. Hirai, H. Sugimoto, V. Bermudez, IEEE J. Photovoltaics 9 (2019) 325.
[3] W. Witte, S. Spiering, D. Hariskos, Vakuum in Forschung und Praxis 26 (2014) 23.
[4] N. Naghavi, D. Abou-Ras, N. Allsop, N. Barreau, S. Bücheler, A. Ennaoui, C.-H. Fischer, C. Guillen, D. Hariskos, J. Herrero, R. Klenk, K. Kushiya, D. Lincot, R. Menner, T. Nakada, C. Platzer-Björkman, S. Spiering, A.N. Tiwari, T. Törndahl, Prog. Photovolt: Res. Appl. 18 (2010) 411.
[5] S. Sinha, D.K. Nandi, P.S. Pawar, S.-H. Kim, J. Heo, Sol. Energy 209 (2020) 515.
[6] M. Nakamura, K. Yamaguchi, Y. Kimoto, Y. Yasaki, T. Kato, H. Sugimoto, IEEE J. Photovoltaics 9 (2019) 1863.
[7] D. Hariskos, P. Jackson, W. Hempel, S. Paetel, S. Spiering, R. Menner, W. Wischmann, M. Powalla, IEEE J. Photovoltaics 6 (2016) 1321.
[8] D. Hariskos, R. Menner, P. Jackson, S. Paetel, W. Witte, W. Wischmann, M. Powalla, L. Bürkert, T. Kolb, M. Oertel, B. Dimmler, B. Fuchs, Prog. Photovolt: Res. Appl. 20 (2012) 534.
[9] C.-S. Lee, Y.M. Shin, B.T. Ahn, ECS Trans. 41 (2011) 213.
[10] J. Chantana, Y. Kawano, T. Nishimura, Y. Kimoto, T. Kato, H. Sugimoto, T. Minemoto, Prog Photovolt Res Appl 28 (2020) 79.
[11] R. Hertwig, S. Nishiwaki, M. Ochoa, S.-C. Yang, T. Feurer, E. Gilshtein, A.N. Tiwari, R. Carron, EPJ Photovolt. 11 (2020) 12.
[12] T. Minemoto, Y. Hashimoto, T. Satoh, T. Negami, H. Takakura, Y. Hamakawa, Journal of Applied Physics 89 (2001) 8327.
[13] I. Gharibshahian, S. Sharbati, A.A. Orouji, Thin Solid Films 655 (2018) 95.
[14] J. Pettersson, C. Platzer-Björkman, M. Edoff, Prog Photovolt Res Appl 17 (2009) 460.
[15] S. Kim, C.-S. Lee, S. Kim, R.B.V. Chalapathy, E.A. Al-Ammar, B.T. Ahn, Phys. Chem. Chem. Phys. 17 (2015) 19222.
[16] D. Abou-Ras, G. Kostorz, A. Romeo, D. Rudmann, A.N. Tiwari, Thin Solid Films 480-481 (2005) 118.
[17] M.A. Contreras, M.J. Romero, B. To, F. Hasoon, R. Noufi, S. Ward, K. Ramanathan, Thin Solid Films 403-404 (2002) 204.
[18] T.M. Friedlmeier, P. Jackson, A. Bauer, D. Hariskos, O. Kiowski, R. Menner, R. Wuerz, M. Powalla, Thin Solid Films 633 (2017) 13.
[19] A. Chirilă, P. Reinhard, F. Pianezzi, P. Bloesch, A.R. Uhl, C. Fella, L. Kranz, D. Keller, C. Gretener, H. Hagendorfer, D. Jaeger, R. Erni, S. Nishiwaki, S. Buecheler, A.N. Tiwari, Nature materials 12 (2013) 1107.
[20] C.-S. Lee, S. Kim, Y.-M. Shin, B.G. Park, B.T. Ahn, H. Kwon, RSC Adv. 4 (2014) 36784.
[21] M. Powalla, G. Voorwinden, D. Hariskos, P. Jackson, R. Kniese, Thin Solid Films 517 (2009) 2111.

A study of Ag paste contacts on various TCO layers for $Cu(In,Ga)Se_2$ thin film modules

Busra Sesli[1,2,3], Jorne Carolus[1,2,3], Jan D`Haen[1,2], Dieter Reenaers[1,2,3], Veronique Gevaerts[4], Sharona Sente[5]
Marc Meuris[1,2,3], Michael Daenen[1,2,3], Bart Vermang[1,2,3]

1 Institute for Material Research (IMO), Hasselt University (partner in Solliance), Wetenschapspark 1, 3590 Diepenbeek, Belgium
2 Imec division IMOMEC (partner in Solliance), Wetenschapspark 1, 3590 Diepenbeek, Belgium
3 EnergyVille, Thorpark, Poort Genk 8310 & 8320, 3600, Belgium
4 TNO (partner in Solliance) , High Tech Campus 21, Eindhoven, 5656 AE, The Netherlands
5 Henkel, Nijverheidsstraat 7, 2260 Westerlo, Belgium

Keywords: Screen printing, Contact resistance, TFPV, Ag paste, TCO, CIGS

1 INTRODUCTION

The power conversion efficiencies (PCE) of Cu(In,Ga)Se2 (CIGS) cells on lab-scale are approaching c-Si cells. The module efficiencies, on the other hand, are considerably lower.[1], [2] Large-area modules show reduced fill factors and power conversion efficiencies. This so-called cell to module efficiency gap exists for all PV technologies. However, the difference in efficiency for CIGS cells modules is considerably higher than for the more mature Si-technology. This efficiency gap is mainly ascribed to the laterally spread electric defects due to non-uniformities in the materials' properties and the longer current paths over the larger area.[3]
The most used modularization technique for CIGS cells is monolithic interconnection (MLI). Typical of this modularization technique is the deposition of the PV stack on a full module scale, instead of making individual cells and interconnecting them to make modules. The individual cells are created by scribing simultaneously during the deposition process. The top electrode for the CIGS modules is a transparent conductive oxide (TCO). Thicker TCO leads to lower resistive losses while decreasing the transparency. This represents a trade-off when optimizing the top electrode. One of the approaches to optimize CIGS cells is to deposit metal grids on top of TCO. Different studies have shown through simulation and experimentally that this can help to reduce the resistive losses while relaxing the TCO requirements.[4]–[8] Another drawback of the monolithic interconnection technique is the scribed area or so-called dead-zone, that does not contribute to the generation of current. By implementing grids on TCO, longer cells are allowed through a decrease in resistive losses, resulting in a relative reduction in the module's dead zone area. However, one of the challenges regarding this interconnection approach is to obtain low-ohmic contacts by contacting metal onto the transparent conductive oxide. When metals are in contact with oxides, electrical and chemical interactions can affect the properties of the contacts [9]. Losses in the contact regions significantly impact the PCE since they contribute to the cell's series losses and reduce the fill factor. Furthermore, the resistivity of the gridlines must be low enough to compensate for the additional shadow losses. For photovoltaic applications, screenprinting is one of the most applied techniques for c-Si because of its relatively low cost and high throughput.[10] CIGS cells are susceptible to high temperatures, therefore, low-temperature cure pastes are investigated to deposit grids on the TCO layer for CIGS modules. This work aims to compare contacting low-temperature silver paste on two different TCO types: Al-doped zinc oxide (AZO) and Indium tin oxide (ITO). ITO is the TCO of preference for most industries because of its high conductivity and optical transparency for visible light. Another often used and the preferred TCO for the CIGS industry is AZO. It is an interesting alternative due to its low cost, non-toxicity, and excellent chemical stability, but it is less conductive than ITO.[11] The screen print process on top of ITO is typical for a-Si: H/c-Si heterojunction (SHJ) solar cell applications. Since the TCO layer for Si-based cells is mostly ITO, different studies can be found on forming a good electrical contact between silver paste and ITO. [10], [12], [13]Although AZO is not the TCO for preference regarding SHJ applications, Khanne et al.[14] characterized contacting of a low-temperature paste (Dupont) to AZO for SHJ application. They measured a contact resistance in the range of 3-5 $mOhm.cm^2$.
This study investigates the contact resistance of a commercially available low cure temperature silver paste on the Al-doped ZnO and ITO layer. Resulting electrical performance and adhesive properties will be shown and discussed.

2 MATERIALS AND METHODS

The contact resistance measurements are carried out by using the principle of the transmission-line method (TLM). TLM structures consist of pads that are apart from each other with increasing distance between the successive pads. By plotting the total resistance measured between two consecutive pads as a function of the distance between these pads, contacts resistance (R_C) and sheet resistance (R_{sh}) of the TCO layer can be extracted, as shown in fig. 1.[15] For ohmic contacts, the I-V measurements between two contact pads will show a linear relationship with the slope representing the total resistance.

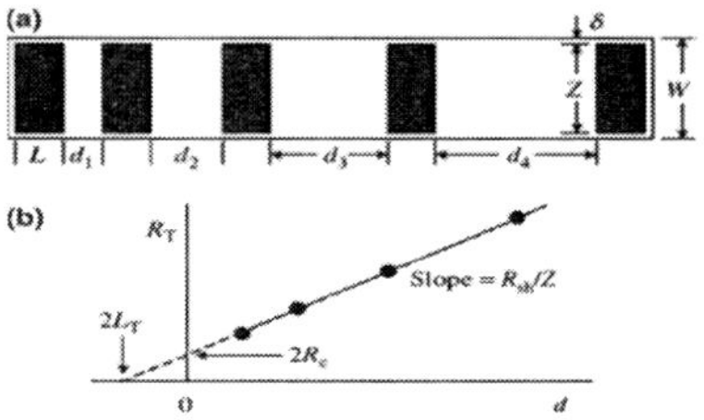

$$R_T = \frac{R_{sh} * d}{Z} + 2\,R_c$$

Figure 1: *Illustrative of the TLM structure for determining the contact resistance.* [13]

The samples are fabricated by screenprinting the TLM pattern on glass samples with TCO on top. First, the glass substrates are cleaned and prepared to deposit the TCO layers on top by RF sputtering. The ITO is deposited with a thickness of 150 nm, while the thickness of AZO was 280 nm. The sheet resistances of the ITO samples and AZO samples were both about 35 as deposited, and measured by the Van Der Pauw method on selected samples before deposition. Next, The the glass samples with TCO are cleaned with isopropanol (IPA) before the printing process. Afterward, the TLM pattern is screenprinted on the glass substrates with TCO. These TLM structures consist of pads sequentially 1, 2, 4, and 8 mm apart, with each an area of 265x3000 µm2. The silver paste that is used for this experiment is a commercially available paste of Henkel: LOCTITE ECI 1011 E&C. After printing, the samples are annealed in ambient air at 150 °C for 10 min, according to specifications. Subsequently, the two sets of TLM structures are isolated by laser cutting through the TCO to avoid edge effects influencing the measurements, and the area is defined. Then, I-V measurements are carried out on the TLM pads under identical conditions for all samples. The cross-sectional microstructures of the grids are examined by field emission gun - scanning electron microscopy (FEG-SEM) after the contact resistance measurements. Lastly, the adhesion of the Ag paste on the ITO and AZO layers are compared.

3 RESULTS

Fig 2. summarizes the results obtained from the TLM measurements for the different sets of samples. The lowest contact resistance is achieved on ITO. We can conclude that the ITO layers' thicknesses do not significantly affect the contact resistance. The contact resistance on AZO had the highest value, and the voltage was drifting during the measurements. Thus the contact resistance measured by the TLM method is only estimated.

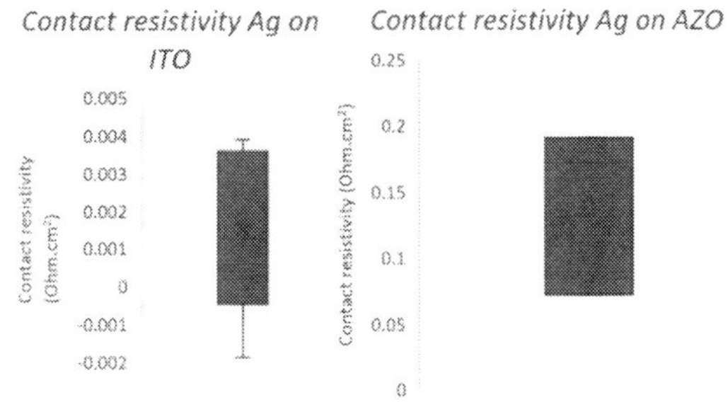

Figure 2: Summary of the TLM measurements

The drift in voltage is shown in fig. 3(b) and (c), together with the measurements on ITO (a) for comparison. The measurements are taken by imposing a constant current of 10 μA and measuring the voltage across two pads, in function of time.

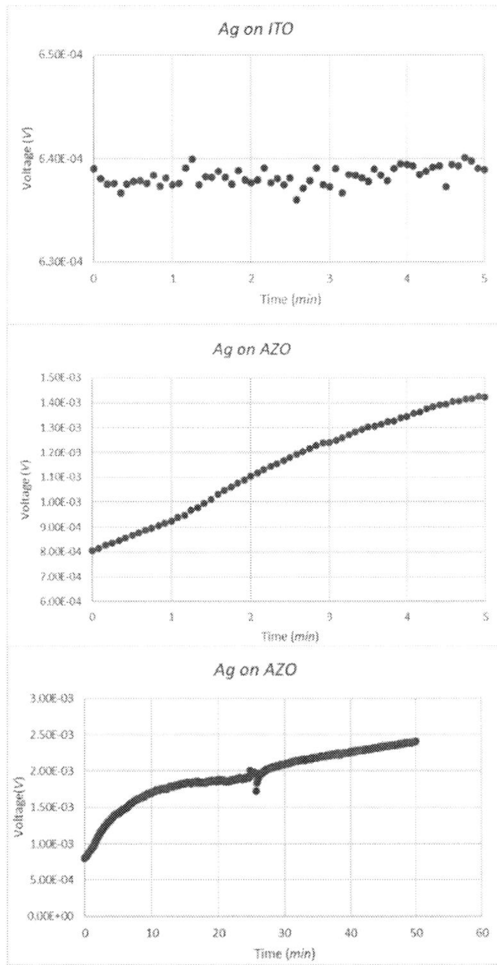

Figure 3: *Measured voltage between two pads as in function of time. The current source is set to 10 µA current and the measurement points are taken every 5 seconds. (a) Ag paste screenprinted contacts on ITO layer, (b) Ag paste screenprinted contacts on AZO layer, (c) same as (b) taken for a longer period (600 measurement points, which is about 50 min).*

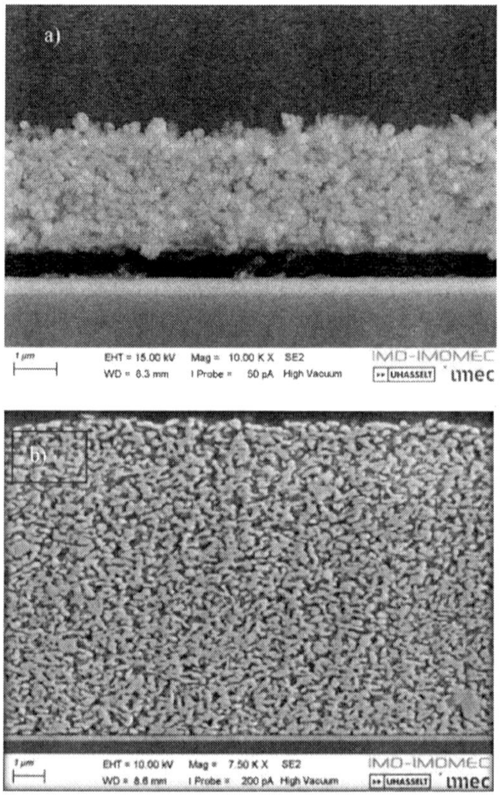

Figure 4: *SEM images of the cross-sections, (a) screenprinted Ag metallization on AZO (b) screenprinted Ag metallization on ITO.*

Fig 4. compares the cross-sections imaged by SEM. The samples' cross-sections with Ag on AZO (fig. 4(a)) exhibit a void at the interface. It can be seen that detachment of the paste from the substrate may have happened during sample preparation. This already indicates a poor adhesion of the paste on the AZO layer. On the contrary, the Ag paste on the ITO layer (fig. 4(b)) showes good contact at the interface. Furthermore, a widely used 'scotch tape' test is carried out to compare the adhesion of the Ag paste on ITO and AZO. Fig.5 shows the result on a sample with AZO layer and ITO layer. The grids could be peeled off from the AZO, while the grids stayed intact on ITO samples.

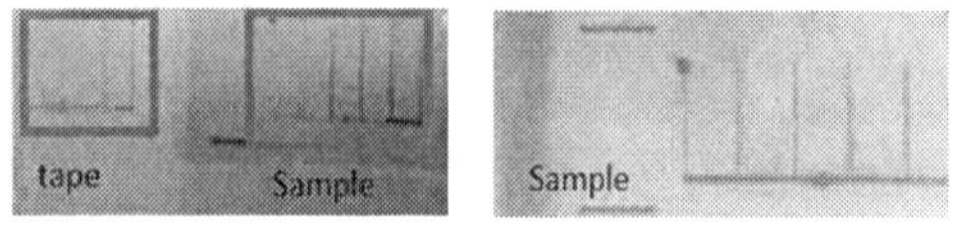

Figure 5: *Scotch tape test on a sample with screenprinted Ag grids on AZO (left) on ITO(right).*

Next contact angle measurements are carried out to determine the wettability of the silver past on the ITO and AZO layer. The results are shown in Fig.6. Lower contact angle en thus better wettability is observed on the ITO layer compared to the AZO layer, before and after curing.

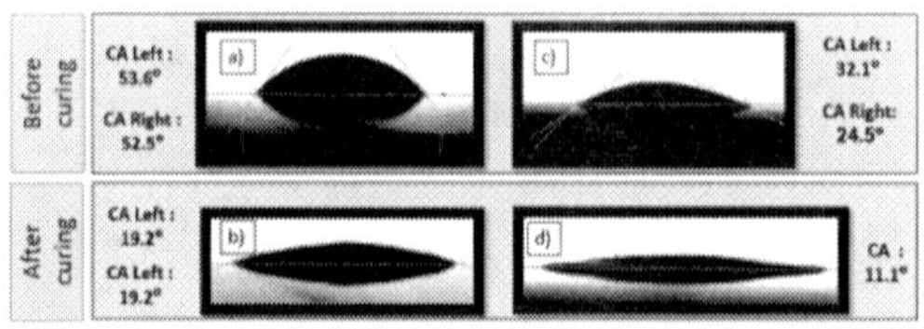

Figure 6: *The contact angle measurements on AZO before curing (a) and after curing (b) on ITO before curing (c) and after curing (d).*

4 DISCUSSION

One reason for the increase in current shown in fig.3 (b and c) might be the poor adhesion between the silver paste and Al: ZnO layer. The lower effective contact area between the paste and the AZO would result in hot spots due to locally heated regions. The results of the `Scotch tape test` confirm that the adhesion on the substrates with ITO was substantially better. This could explain the voltage drift during the TLM measurements, and the higher contact resistance measured.

According to previous studies, the effective contact area has a significant influence on contact resistance. As an example, Yuanming Li et al.[10] found a method to further improve the contact resistance on ITO by increasing the effective contact area between the Ag particles and the ITO. They added a thin Ag contact layer by inkjet printing between the screenprinted Ag gridline and the ITO layer to lower the contact resistance from 6.23 to 0.55 mOhm.cm^2. In a compartive study of Cu-alloy paste for heterojunction solar cells with Ag paste[16], the authors found that binder polymer in the Ag paste caused a thin insulating layer at the interface that worked as tunneling barriers. From the literature, it can be concluded that the Ag particles' direct contact with the TCO is essential to achieve low ohmic contacts.

5 CONCLUSION

The lowest ohmic contact resistance is achieved on ITO. For the AZO layer, besides the higher contact resistances, a voltage drift is observed. The higher contact resistance of the paste on the AZO layer is attributed to poor adhesion.

6 ACKNOWLEDGEMENTS

This work received funding from the European Union's H2020 research and innovation program under grant agreement No. 715027.

7 REFERENCES

[1] NREL, "Best Research-Cell Efficiency Chart.pdf," *Nrel*, p. 1, 2019, [Online]. Available: https://www.nrel.gov/pv/assets/pdfs/best-research-cell-efficiencies.20190802.pdf.

[2] National Renewable Energy Laboratory (NREL), "Champion Photovoltaic Module Efficiency Chart," *Nrel*, p. 1, 2020, [Online]. Available: https://www.nrel.gov/pv/module-efficiency.html.

[3] V. Bermudez and A. Perez-Rodriguez, "Understanding the cell-to-module efficiency gap in Cu(In,Ga)(S,Se)2 photovoltaics scale-up," *Nat. Energy*, vol. 3, no. 6, pp. 466–475, 2018, doi: 10.1038/s41560-018-0177-1.

[4] R. Wächter, G. Kaune, T. Repmann, and K. Orgassa, "NICE SOLAR ENERGY Loss analysis and efficiency potentials for CIGS PV modules without and with metal grid : Experimental results analyzed by simulation," 2020, no. September.

[5] J. van Deelen and C. Frijters, "CIGS cells with metallized front contact: Longer cells and higher efficiency," *Sol. Energy*, vol. 143, pp. 93–99, 2017, doi: 10.1016/j.solener.2016.12.050.

[6] M. GREEN *et al.*, "Solar cell efficiency tables (version 40)," *Ieee Trans Fuzzy Syst*, vol. 20, no. 6, pp. 1114–1129, 2012, doi: 10.1002/pip.

[7] T. R. & K. O. T. Freund, N. Zancan, G. Kaune, W. Bromenne, R. Wächter, "NICE SOLAR ENERGY Large Area Screen-Printed Front Contact Metallization for Thin-Film Solar," in *EU PVSEC 2020*, 2020, no. September, pp. 1–27.

[8] J. Carolus, "PROPOSING AN ELECTRO-THERMAL SPICE MODEL TO INVESTIGATE THE EFFECT OF PARTIAL SHADING ON CIGS PV MODULES," 2020.

[9] M. T. Greiner, L. Chai, M. G. Helander, W. M. Tang, and Z. H. Lu, "Metal/metal-oxide interfaces: How metal contacts affect the work function and band structure of MoO3," *Adv. Funct. Mater.*, vol. 23, no. 2, pp. 215–226, 2013, doi: 10.1002/adfm.201200993.

[10] Y. Li, H. S. Kim, J. Yi, D. Kim, and J. Y. Huh, "Improved electrical performance of low-temperature-cured silver electrode for silicon heterojunction solar cells," *IEEE J. Photovoltaics*, vol. 8, no. 4, pp. 969–975, 2018, doi: 10.1109/JPHOTOV.2018.2834955.

[11] E. Fortunato, D. Ginley, H. Hosono, and D. C. Paine, "Transparent conducting oxides for photovoltaics," *MRS Bull.*, vol. 32, no. 3, pp. 242–247, 2007, doi: 10.1557/mrs2007.29.

[12] Y. Lee *et al.*, "Performance of hetero junction with intrinsic thin-layer solar cell depending upon contact resistivity of front electrode," *J. Photonics Energy*, vol. 4, no. 1, p. 043094, 2014, doi: 10.1117/1.jpe.4.043094.

[13] D. Chen, L. Zhao, H. Diao, W. Zhang, G. Wang, and W. Wang, "Choice of the low-temperature sintering Ag paste for a-Si:H/c-Si heterojunction solar cell based on characterizing the electrical performance," *J. Alloys Compd.*, vol. 618, pp. 357–365, 2015, doi: 10.1016/j.jallcom.2014.08.175.

[14] A. Khanna, "screen printed metallisation for silicon heterjunction solar cells," 2013, vol. 4, pp. 25–28.

[15] P. N. Vinod, "Specific contact resistance measurements of the screen-printed Ag thick film contacts in the silicon solar cells by three-point probe methodology and TLM method," *J. Mater. Sci. Mater. Electron.*, vol. 22, no. 9, pp. 1248–1257, 2011, doi: 10.1007/s10854-011-0295-z.

[16] M. Yoshida, H. Tokuhisa, U. Itoh, T. Kamata, I. Sumita, and S. Sekine, "Novel low-temperature-sintering type Cu-alloy pastes for silicon solar cells," *Energy Procedia*, vol. 21, no. October 2011, pp. 66–74, 2012, doi: 10.1016/j.egypro.2012.05.009.

PV CELL FOR VARIED ANGLE PERFORMANCE UNDER INDOOR LIGHTING SIMULATOR

Yean-San Long[1*], Min-An Tsai[1], Teng-Chun Wu[1]
[1]Center for Measurement Standards, Industrial Technology Research Institute, Taiwan
*Corresponding author (mickeylong88@itri.org.tw, Phone: +(886) 3 5911026, Fax: +(886) 3 5725540)

ABSTRACT: In our study, we used RTOS method for I-V then the results showed better accuracy by eliminating in real time the acceptance effect. We also used this method to compare PV cells with the latter cells exhibited promising power conversion efficiency and no hysteresis behavior for varied angle under indoor lighting simulator. In this study, we measured a varied angle of the PV cell performance will be show different performance trend in indoor lighting application, which depends on device. That's will be shown PV cell have the energy harvesting for varied angle in indoor lighting.

Keywords: Organic Photovoltaic, Real-Time Removing Capacitance Effect, One-Sweep Method.

1 INTRODUCTION

Energy harvesting for IoT (Internet of things) application presents a straightforward solution for easily powering those remote devices using clean energy. The power-generating element must be selected after considering the type of energy to be collected from the surrounding environment, whether vibration, light, or heat. The most common elements used are solar, piezoelectric, or thermoelectric. Therefore, the standard of the emerging PV applied in IoT follow

(1) For Outdoor environment, we meet IEC 60904 series, SEMI PV57, SEMI PV69, SEMI PV76 and IEC 61215.

(2) For Indoor environment, we meet SEMI PV57, SEMI PV76 and SEMI PV80.

In general indoor lighting, a lot of different light sources exist. The International Commission on Illumination (CIE) is responsible for publishing binding standards for different types of light sources, called illuminants. A standard illuminant represents a mathematical table of relative energy versus wavelength, used for colorimetric calculations. It is a (theoretical) source of visible light with a set spectrum, determined by convention and therefore provides a worldwide basis for comparing images under different lighting. Which illuminants are useful standards for indoor illumination ? We consider some typical, widely used, light sources for indoor environments (residential and commercial) and relate them to an appropriate illuminant (standard). First, we will propose a standard for the common incandescent light bulb. An incandescent lamp consists of a balloon of (usually) glass in which a filament is placed that is heated to high temperatures by the Joule's effect. To a first approximation the filament can be considered as a black body radiator. In incandescent lamps, the filament usually consists of tungsten heated to a temperature of 2000 to 3000 K. The CIE has agreed upon a standard for the incandescent light bulb, called "illuminant A". This illuminant is intended to represent typical, domestic, tungsten-filament lighting. Its relative spectral power distribution is that of a Planck radiator at a temperature of approximately 2856 K. In other words, a typical light bulb can be characterized by a black body at temperature T = 2856 K. Technically, illuminant A is only defined over the spectral region from 300 nm to 830 nm. However, solar cells often absorb longer wavelengths. Therefore, we will not use illuminant A itself to represent the spectrum of the incandescent light bulb, but we will use Planck's law for a black body at a temperature of 2856 K. In the 300–830 nm range, the difference between Planck's law at this temperature and illuminant A never exceeds 1%. However, In indoor lighting, the International Commission on Illumination is an organization devoted to international cooperation and exchange of information between its member countries on all matters relating to the science and art of lighting. In accordance with an agreement between ISO and CIE, standards are published as double logo standards by ISO. Standards produced by the CIE are a concise documentation of data defining aspects of light and lighting, for which international harmony requires a unique definition. Therefore, the standard follow that the CIE has defined some lighting spectrums for different indoor colorful applications and situations. However, the requirements for color testing do not apply to related indoor energy harvesting under different lighting uniformity and stability, for performance measurement of emerging PV and device. Furthermore, countless types of indoor lighting patterns are required to do performance characteristics measurement for different application contexts. The most common performance indicator is the PV efficiency under standard test conditions (STC). It is the maximum electrical power divided by the total irradiance. For indoor lighting illuminance values are measured used a lux meter or spectradiometer.

(a) Lux is a typical unit for measuring indoor lighting.

(b) For outdoors, irradiance is measured in W/m^2 and there is an approximate conversion of 0.0079 W/m^2/ lx for the sun.

For indoor lighting, fundamental difference between illuminance and irradiance is the weighting of the spectral response. Irradiance includes the power from all wavelengths weighted equally, whereas illuminance weights the power from each wavelength in proportion to the sensitivity of the human (Fig. 1). Although these conditions are seldom achieved in practice (except in the lab), this characterization provides a reasonable guideline for comparing different solar cell types under outdoor conditions. However, the STC are not relevant for indoor applications. Accurate characterization of emerging PV requires level lighting consideration on each very slow temporal response in the I-V curves of the emerging PV are clearly dependent on the voltage sweep direction, even when the sweep time is the order of seconds [1].

Furthermore, the temporal response is dependent on different level lighting consideration.

In addition, we developed the Indoor lighting simulator, meet SEMI PV80, for energy harvesting under Indoor environment for IoT applications. In our study, RTOS (real-time one-sweep) method can real-time remove the capacity effect under solar simulator for I-V successfully and proposes the forward/backward schematic reaction mechanism for DSC [2]. In this study, RTOS method [3, 4] practices for measuring perovskite solar cell performance showed having better accuracy results by eliminating acceptance effect instantly. Additionally, RTOS method will be useful to measuring cell performance more accurately and rapidly when evaluating emerging PV performance in indoor lighting condition. Therefore, we also defined a performance ratio to compare between indoor lighting and solar simulator.

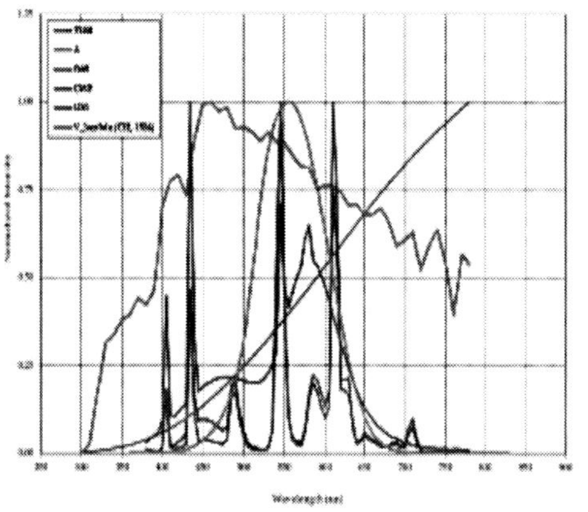

Figure 1: Illumination spectra of CIE Standard Lighting [11, 12]

2 EXPERIMENTS

The samples Sample size ranges from 0.1 cm^2 to 1 cm × 5 cm. These I-V curves were done on a Keithley Source Meter (Model 2460), AM1.5G solar simulator, and Indoor Multi-Light Source System (see Fig. 4). We developed the Indoor lighting simulator, meet SEMI PV80 and follow as:

1) Spectral match, available methods are the use of spectroradiometer. Standard lighting spectrum is defined by the deviation from CIE[12]. For three wavelength intervals of interest, the percentage of total irradiance is specified in Table I. Calculate the spectral match for each wavelength interval, which is the ratio of calculated percentage for the CIE spectrum and the lighting simulator spectrum (Table 1). The data comparison with the solar spectrum shall indicate the spectral match classification as per the following as Table 2.

2) Non-uniformity of irradiance in the test plane, the lux meter is recommended to be used as uniformity detector for determining the non-uniformity of irradiance in the test area of the lighting simulator. The uniformity detector shall have a spectral response appropriate for the simulator. The linearity and time response of the uniformity detector shall conform to the characteristics of the simulator being measured. And the designated test area divided by 25. The area covered by the detector measurements should be 100% of the designated test area. The measurement positions should be distributed uniformly over the designated test area. Therefore, the maximum and minimum irradiance are those measured with the detector(s) over the designated test area, likely in Fig. 2. The calculation of non-uniformity is defined as Eq. (1).

3) Temporal instability of irradiance, the lux meter is recommended to be used as temporal instability detector for determining the temporal instability of irradiance in the test area of the lighting simulator. For lighting simulator the temporal instability is related to the irradiance change of measured data sets during the time of data acquisition (Fig. 3 and Table 3).

Table 1: The ratio of calculated percentage for the CIE spectrum

Wavelength	Bandwidth	Percentage of total irradiance (300 nm – 900 nm)				
Band		A	D65	U30	CWF	TL84
1	300 - 450	3.78	20.31	11.40	17.99	15.18
2	450 - 650	40.15	45.35	81.72	76.73	79.13
3	650 - 900	56.06	34.34	6.88	5.29	5.69

Table 2: Spectral definition of lighting simulator classifications

Classification	Percentage
A	1 %
B	3 %
C	5 %

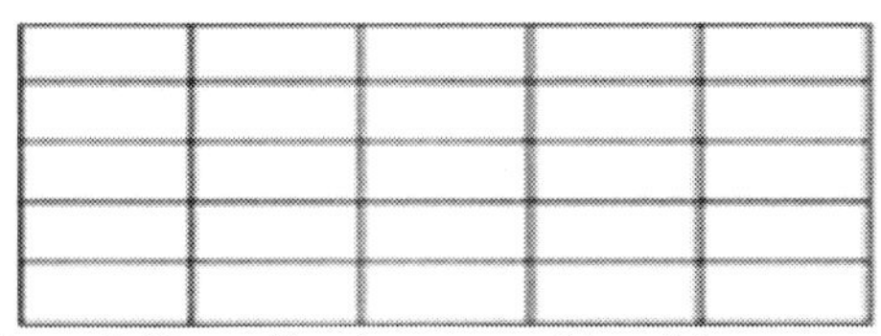

Figure 2: Non-uniformity put a position of detector.

$$Non-uniformity(\%) \equiv \frac{(max-min)}{(max+min)} \times 100\% \quad (1)$$

The Indoor Multi-Light Source included D65 (6500 K, Average North USA sky daylight) / TL84 (4100 K, European shop fluorescent) / CWF (4150 K, cool white fluorescent, shop lighting) / U30 (3000 K, shop lighting) / A (2856 K, typical home lighting), meeting CIE Standard illuminants. The adjustable light intensity (ranging from 0 to 2500 lx) is supplied with an ultraviolet light source for aging testing, with non-uniformity of less than 2% (at 20 cm × 20 cm) and temporal instability of less than 2%. During the I-V measurement in the dark and under Indoor Multi-Light Source System, we followed the testing flowchart in Fig. 5; the scan direction was forward and backward, the sample temperature should be stable at 25 °C with a fluctuation of less than 1 °C, and the irradiance and luminance were determined using a NIST-traceable spectroradiometer (StellarNet) and an NML (National

Measurement Laboratory, Taiwan)-traceable lux meter (HIOKI, model FT3424).

Table 3: Temporal instability of indoor lighting simulator.

Item	Average (lx)	Short-term (lx)	Long-term (%)
U30	2709.33	6.35	0.23
	1492.13	1.64	0.11
TL84	2772.08	10.43	0.38
	1507.82	5.61	0.37
D65	1480.16	5.75	0.39
	663.42	2.35	0.35
CWF	2536.80	0.00	0.00
	1199.10	1.91	0.16
A	1902.85	9.01	0.47
	282.92	3.19	1.13

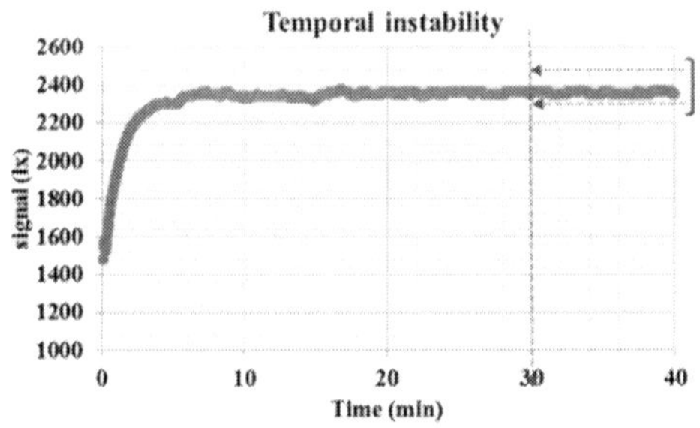

Figure 3: Evaluation of temporal instability.

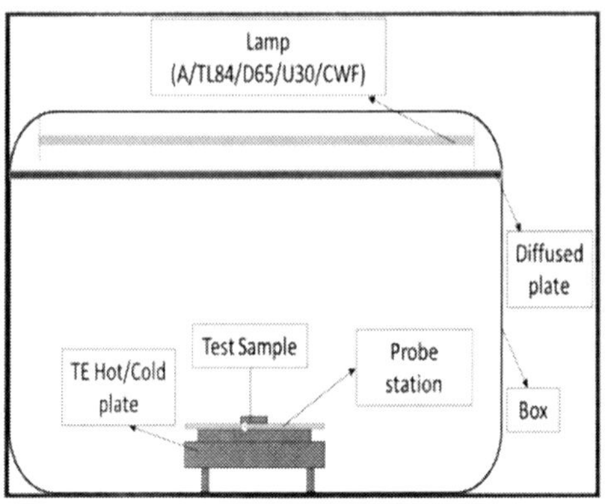

Figure 4: Indoor multi-light source system custom-made by CMS/ITRI for solar cell measurement.

3 ANALYSIS AND RESULTS

In this study, we follow the testing flow [3,4] in the I-V measurement. We used real-time removal of capacity effect; the RTOS method shows the best repeatability (see Fig. 5). A proposed list of steps [2,5] for the measurement of a sample is given in Fig. 8. For most traditional semiconductor PV devices, the current will respond to small changes in bias voltage in less than a millisecond. Sweeping the voltage from open-circuit voltage (Voc) to short-circuit current (Isc) in a few seconds at a constant illumination for such a device will yield accurate currents. Many emerging PV cells respond far more slowly, and consequently, the sweep time must be increased. Fig. 9 shows the I-V for a slowly responding emerging PV cell with different sweep rates.

Therefore, we used the RTOS Method [2] following the testing flowchart (see Fig. 8) in the I-V measurement. applied bias voltage is changed from Isc to Voc (forward scan) or in the reverse direction (backward scan, Voc to Isc). When a capacitance effect occurs, this measurement process produces the phenomenon. In the forward scan, an overshot current appears immediately after an abrupt increase of applied voltage and then the current gradually decreases to an equilibrium state. Sweeping to determine maximum power (Pmax) may not be valid if the device is metastable because of the long times that it sits near Isc or Voc before getting to Pmax shown in Fig. 7(b) to confirm the "near-true" Pmax of emerging PV, especially for samples that depend strongly on sweep direction or rate. This also shows that the asymptotic I-V method may be the only reliable way to obtain efficiency.

For indoor lighting measurement system light source evaluation is performed according to SEMI PV80-0218, the check cycle is one year, and the light source evaluation results are recorded to confirm that the uncertainty of the system light source conditions falls within the scope of this document. Sources of uncertainty for maximum power measurements of the emerging PV include type A uncertainty assessed by statistical methods and type B uncertainty assessed by non-statistical methods:

Table 4: Uncertainty source for indoor lighting simulator.

Type A		Type B	
u_{A1}	Repeatability	u_{B1}	Effective illumination
u_{A2}	Stability	u_{B2}	Temperature effect
		u_{B3}	Accuracy for electrical measuremet equipment
		u_{B4}	Measurement condition correction

In the light intensity under artificial lighting conditions found in offices and factories is less than 5 W/m2 as compared to 100–1000 W/m2 under outdoor conditions, depending on the type of light source and its distance. Moreover, the spectrum can be totally different from the outdoor solar reference spectrum (e.g. AM1.5G). The spectrum depends not only on the type of light source, but also on the presence of reflected and diffused light. Therefore, we shown a definition of performance (η) between indoor lighting and solar simulator below as

$$\eta \equiv (P_{max}/Area)/P_{in}. \qquad (2)$$

Where (a) P_{max} means maximum power of device, (b) Area means area of device, (c) P_{in} means Illuminant power of lighting source. Therefore, we also defined a normalized power density (performance ratio, η_n) to compare between indoor lighting and solar simulator below as

$$\eta_n \equiv \eta_{Indoor} / \eta_{AM1.5G} \qquad (3)$$

Where (a) η_{Indoor} means performance of device at each indoor lighting (A/TL84/D65/U30/CWF), (b) $\eta_{AM1.5G}$ means performance of device at solar simulator.

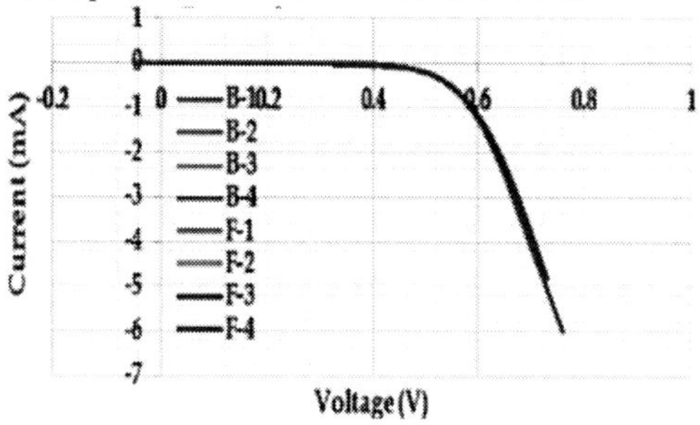

Figure 5: I-V curves measured by RTOS method at dark condition.

4 CONCLUSIONS

The characterization after testing showed that one must measure performance under all typical artificial light sources. Therefore, we propose that each solar cell should be measured varied indoor and outdoor lighting condition for its characterization by RTOS method. In Fig. 8a, we will be compare the Pmax of the different solar cells in light levels in indoor environments to show linearity under near range lighting level, which depends on device. In Fig. 8b shown nearly a constant from 400 lx to 1000 lx, it will be get an average of normalized power density and follow the defined of equation (3). Therefore, the results will be show the different solar cells in light levels in indoor environments to show linearity under near range lighting level, which depends on device. From Fig. 10b, it is also shown the different solar cells in indoor lighting and solar environments to show different performance, ex. DSSC have more 0.26 times for TL84 than D65. We are also known STC under outdoor application, therefore, we also define a indoor lighting-STC for indoor application, that is " A/ 25 °C/ 1000 lx ".

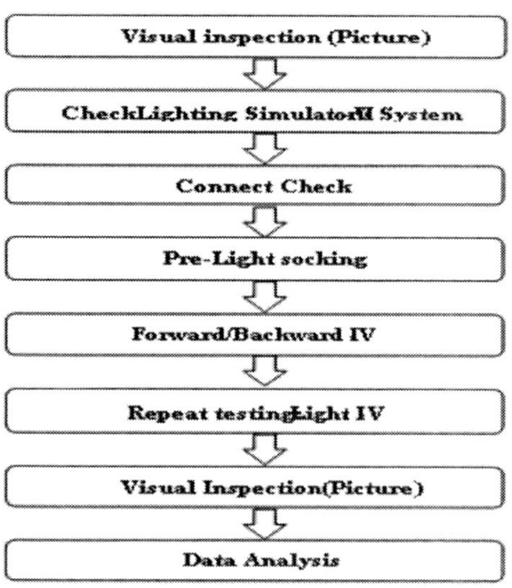

Figure 6: Testing flowchart [2,5].

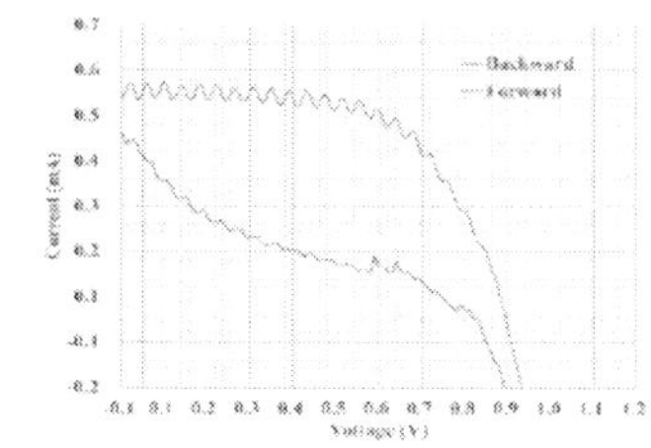

(a) sweeping rate method

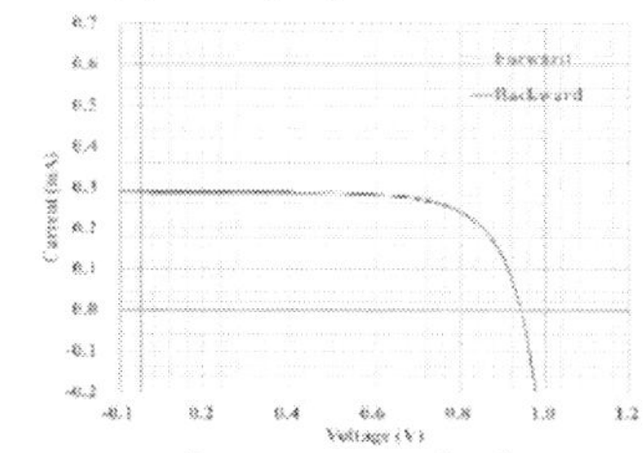

(b) RTOS method

Figure 7: I-V measurement of PSC (at STC) using the (a) sweeping rate method and (b) RTOS method.

In this work, we defined performance and normalized power density, is called performance ratio, shown a comparison for different emerging PV in indoor lighting and solar environments (see Fig. 10). Therefore, the results also find DSSC have more 0.26 times for TL84 than D65, but OPV/a-Si have less 0.5 times for TL84 than D65. That's shown DSSC have more indoor application than others in this case study.

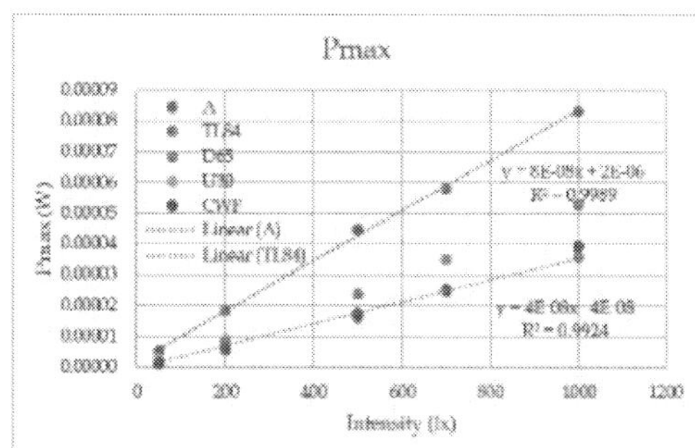

(a) P_{max}

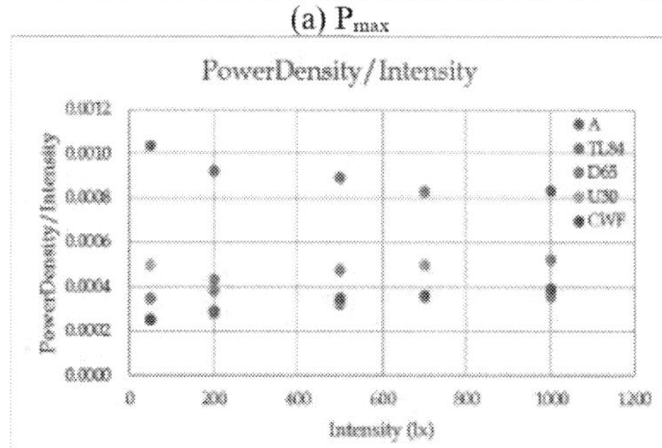

(b) Normalized Power density

Figure 8: Light linearity of DSSC measured by indoor multi-light source system under varied lighting condition (I-V using RTOS method).

The 1000 lx of the system uncertainty is shown as

Table 5: Uncertainty budget for 1000 lx for indoor lighting simulator.

	Uncertainty Sources	I_{sc}	I_{mp}	V_{oc}	V_{mp}	P_{max}	FF
u_{B1}	Effective Irradiance	1.54	1.54	0.0921	0.0921	1.54	0.185
u_{B2}	Temperature	0.0268	0.0268	0.131	0.131	0.134	0.190
u_{B3}	Electrical Accuracy	0.00405	0.00405	0.00247	0.00247	0.00474	0.00670
u_{B4}	Correction to 1000 lx/25 °C	0.0572	0.0572	0.140	0.140	0.152	0.214
u_{A1}	Repeatability	0.0937	0.215	0.0293	0.236	0.0698	0.164
u_{A2}	Stability	0.235	0.653	0.143	0.438	0.829	0.532
Effective degree of freedom		8982	213	52	9	101	11
Confidence Level		95 %					
Coverage factor k		1.96	1.96	2.01	2.26	1.96	2.2
Relative Combination Standard Uncertainty (%)		1.56	1.69	0.258	0.551	1.76	0.653
Expanded Uncertainty (%)		**3.1**	**3.4**	**0.52**	**1.3**	**3.5**	**1.5**

When we will be follow through CIE and SEMI PV80 of the defined some lighting spectrums for different indoor colorful applications and situations. However, the requirements for color testing do not apply to related PV indoor performance measurement under different lighting uniformity and stability, for performance measurement of emerging PV and device, and Its characteristics will be having a classification. Furthermore, countless types of indoor lighting patterns are required to do performance characteristics measurement for different application contexts. We should be define a different STC than STC in AM1.5G, which we will be define a STC in indoor lighting as below as "I-V measurement for indoor lighting-STC condition: 1000 lx/25 °C/Illuminant A". Furthermore, our goal is to determine which light source is relevant for a new international standard test method. Upon the indoor characterization of solar cells, it is the spectrum range of the light source that decides the lighting performance, not the light intensity, which is present and affects the cell equally through different light sources.

REFERENCES

[1] T. C. Wu, S. T. Hsu, and Y. S. Long, "New Set-up Procedures and Integrated Measurement System for Organic Photovoltaic (OPV) Module," *International Photovoltaic Science and Engineering Conference (PVSEC23)*, 2013.

[2] Y. S. Long, S. T. Hsu, and T. C. Wu, "Induction of Internal Capacitance Effect in Performance Measurement of OPV (Organic Photovoltaic) Device by RTOSM (Real-Time One-Sweep Method)," *Journal of Energy and Power Engineering*, Vol 8, pp1059-1066, 2014.

[3] Y. S. Long, S. T. Hsu, and T. C. Wu, "Capacitance Effect from High to Low Level Lighting by Real-Time One-Sweep Method for OPV Device," *IEEE PVSC42*, 2015.

[4] Y. S. Long, S. T. Hsu, and T. C. Wu, "A Study of Capacitance Effect in DSC from High to Low Level Lighting by Real-Time One-Sweep Method," *OPTIC2014*.

[5] SEMI PV57-1214, "Test Method for Current-Voltage (I-V) Performance Measurement of Organic Photovoltaic (OPV) and Dye-Sensitized Solar Cell (DSSC)."

[6] US PTO. 8224598, Method for Forming Optimal Characteristic Curves of Solar Cell and System Thereof.

[7] ASTM E490, "Standard Solar Constant and Zero Air Mass Solar Spectral Irradiance Tables."

[8] Handbook of Photovoltaic Science and Engineering, Chapter 16

[9] IEC 60904-1, "Photovoltaic Devices – Part 1: Measurement of Photovoltaic Current-Voltage Characteristics," 2006.

[10] IEC 60904-3, "Photovoltaic Devices – Part 3: Measurement Principles for Terrestrial Photovoltaic (PV) Solar Devices with Reference Spectral Irradiance Data," 2008.

[11] ISO 10526, "CIE Standard Illuminants for Colorimetry," 2007.

[12] CIE S 014-2, "CIE Standard Illuminants for Colorimetry," 2006.

[13] K. Emery, S. Winter, S. Pinegar, and D. Nalley, "Linearity Testing of Photovoltaic Cells," *4th World PV Conf.*, 2006.

[14] CIE S 014-2, "CIE Colorimetry - Part 2: Standard Illuminants for Colorimetry "

[15] ISO 11664-2:2007, "Colorimetry -- Part 2: CIE standard illuminants"

[16] ASTM D 1729, "Standard Practice for Visual Appraisal of Colors and Color Differences of Diffusely-Illuminated Opaque Materials";

[17] SEMI Doc. 5979, "Specification of indoor lighting simulator requirements for emerging PV "

[18] Handbook of Photovoltaic Science and Engineering, Chapter 16

GROWTH AND STRUCTURAL CHARACTERIZATION OF GASB/GAAS QUANTUM DOTS: PROSPECTIVE APPLICATIONS IN PHOTOVOLTAIC CELLS

C.C. Ahia[1,*], E.L. Meyer[1], N. Tile[2]
[1]University of Fort Hare, Institute of Technology, Private Bag X1314, Alice 5700, South Africa
[2]Department of Physics, Nelson Mandela University, PO Box 77000, Port Elizabeth, South Africa
*Phone: +27(0)833447761; email: achinedu@ufh.ac.za

ABSTRACT: In an attempt to develop high quality GaSb quantum dots (QDs) on GaAs substrate for prospective application in optoelectronic device based on p-n junction, the results of cross-sectional scanning transmission electron microscopy (STEM) measurements, electron energy loss spectroscopy (EELS) elemental mapping analysis and optical transitions of metalorganic vapor phase epitaxy (MOVPE) grown GaSb QDs are presented in this study. Cold capping and growth interruption was found to be favorable for the epitaxial growth of densely populated GaSb QDs while simultaneously preserving the shape of the QDs after capping. The thermal activation (Ea) energy deduced for competing non-radiative processes in the capped samples using applicable deconvolution technique and least-squares fits is = 320.4 ± 1.2 meV. Results from the photoluminescence (PL) spectra shows a quick reduction in intensity of the higher energy wetting layer (WL) peaks at high temperatures compared with the peaks corresponding to QDs at lower energy which can be attributed to improved migration of holes from the WL to the QD at elevated temperature. The growth conditions adopted in this study are promising for the deposition of high quality stack of GaSb QDs, an essential characteristics needed for the development of high efficiency intermediate band solar cells (IBSCs).
Keywords: GaSb, Quantum dots, Substrates, III-V semiconductors, high-efficiency

1 INTRODUCTION

Evolving procedures for photovoltaic technology applications has its primary focus on achieving high efficiency solar cell production at relatively affordable cost. Some of these emerging techniques which are promising for delivering high power conversion efficiency include multi-exciton generation (MEG) and IBSCs. III-V semiconductor nanostructures have attracted significant research interest for their prospective applications in IBSCs and have been extensively investigated as potential candidates for the generation of solar cells with highest efficiencies which exceeds the Shockley-Queisser limit.

Various advanced techniques for the development of III-V photovoltaic technology are widely explored as a result of the intensive demand for an alternative energy sources other than conventional fossil fuels. One of the unique III-V semiconductor material system of interest which is promising for a variety of opto-electronic device applications such as solar cells [1] and photodetector [2] is GaSb/GaAs QDs. They are known to have a type-II, staggered band alignment, whereby the holes are confined and electrons deterred from the area surrounding the QD, giving rise to a weak recombination rate. In order to fabricate a solar cell with high power conversion efficiency, a lot of factors should be taken into consideration. Apart from the choice of well suited semiconductor compounds, the growth conditions and the purity of the precursors can affect the crystal quality and the overall device performance.

A number of deposition methods have been utilized over the years to develop GaSb/GaAs QDs. Some of these techniques are the molecular beam epitaxy (MBE), metalorganic vapour phase epitaxy (MOVPE), and liquid-phase epitaxy (LPE). Nevertheless, the low surface mobility and non-volatility of elemental antimony has made the development of high quality Sb-based compounds challenging regardless of epitaxial deposition method used.

With the aim of investigating suitable growth conditions for the deposition of cap/spacer layers of QDs without significantly altering their morphology after capping, this paper gives a detailed description of the composition, structural and optical properties of GaSb/GaAs QDs which was fabricated at atmospheric pressure (AP) using MOVPE system. Additionally, fundamental approaches for developing high density GaSb QDs which are embedded in GaAs matrix for prospective application in high efficiency IBSCs are presented.

2 THEORETICAL FRAMEWORK

The band alignment of semiconductor material system is one of the essential properties which determines its specific photonic device application. GaSb/GaAs QDs has a type-II (staggered) misalignment of the conduction and valence band which implies that electrons are loosely bound in the host matrix through Coulombic attraction, with strong confinement of hole within the structure. Figure 1 shows the simplified schematic band alignment for type-II structures.

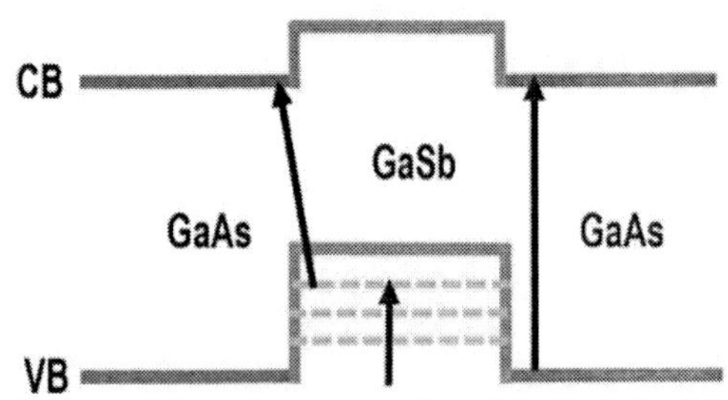

Figure 1: Schematic representation of GaSb/GaAs QDs band setup (type-II) showing confined hole states acting as the intermediate band

Three confined states and possible transitions are specified in the band alignment diagram displayed in Figure 1. The spatial distance between the holes and electrons in structures with type-II band alignment can give rise to extended radiative lifetimes [3].

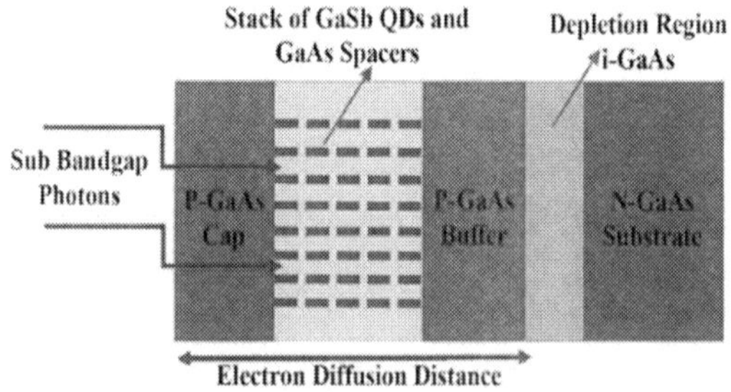

Figure 2: Simplified schematic of proposed GaSb/GaAs QD IBSC

Additionally, it is expected that the process of Auger recombination can be controlled in QD systems with type II configuration, hence counteracting challenges associated with common semiconductor lasers [4]. The extensive radiative lifetimes and the suppression of Auger recombination mechanism in materials with type-II band alignment makes them potential candidates for intermediate band solar cells and alternatives for single gap solar cells. Figure 2 is a schematic diagram of the proposed IBSC with the GaSb QDs implanted between the GaAs buffer and cap regions which are situated outside the GaAs p-n junction depletion region. The arrangement of the IBSC shown in Figure 2 is expected to be promising in proffering high efficiency while preserving simple device structural complexity.

3 EXPERIMENT

3.1 Growth procedure

A Thomas Swan MOVPE reactor, with a horizontal quartz tube operated at atmospheric pressure was used for the deposition of the sample. The capped sample comprise of a ~ 6.5 nm thick GaSb QD layer embedded between a 50 nm and 200 nm thick GaAs cap and buffer layers respectively, while the uncapped sample only contains the unmasked GaSb QDs on GaAs buffer layer without the GaAs cap layer, hence enabling the uncapped sample to be preferable specifically for atomic force microscopy (AFM) studies. In all cases, a (100) semi-insulating GaAs substrate with a 2° misorientation towards (111)B was used. Before the growth procedure, the reactor was purged with H_2 and the molybdenum susceptor was properly cleaned, while and the substrate was annealed at 600 °C for 300 s. These procedures are essential in order to get rid of impurities such as lingering oxides and water vapor from the reactor.

With palladium diffused H_2 carrier, the molar flow rates of the source precursors and interfacial growth interruptions were controlled using calibrated mass flow controllers in order to regulate the chemical activities on the substrate likewise obtaining strained structures and a consistent dot distribution. The precursors which are mixed in the manifold are channeled into the reactor at elevated temperatures, hence enabling the molecules to pyrolyze, therefore depositing adatoms of interest on the substrate. Using a proportional-integral derivative (PID) temperature controller, the temperature of the substrate during deposition is controlled as a function of the melting point and cracking temperatures of the precursors respectively. This was achieved by the use of a type-K thermocouple which is positioned in the molybdenum susceptor.

During deposition, different concepts which are conventionally known as the cold capping and growth interruption techniques were adopted in order to preserve the shape of the QDs after capping.

4 RESULTS AND DISCUSSION

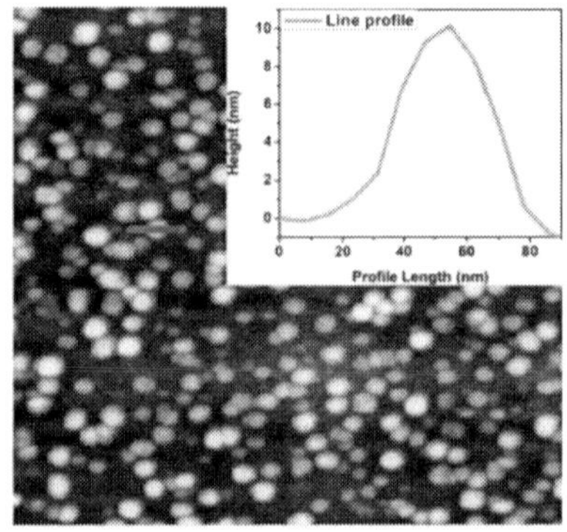

Figure 3: AFM micrograph of GaSb QDs before capping with inset displaying height profile of the QDs

The size distribution and lateral dimensions of the uncapped GaSb QDs are shown in the 1 μm × 1 μm AFM image in Figure 3. The maximum height and lateral dimensions of the QDs are approximately 9.5 nm and 40 nm respectively as seen in the inset in Figure 3. Similarly, a high dot density which is measured to be in the $10^{10}cm^{-2}$ range was obtained from the sample over a scan size of 1 μm^2.

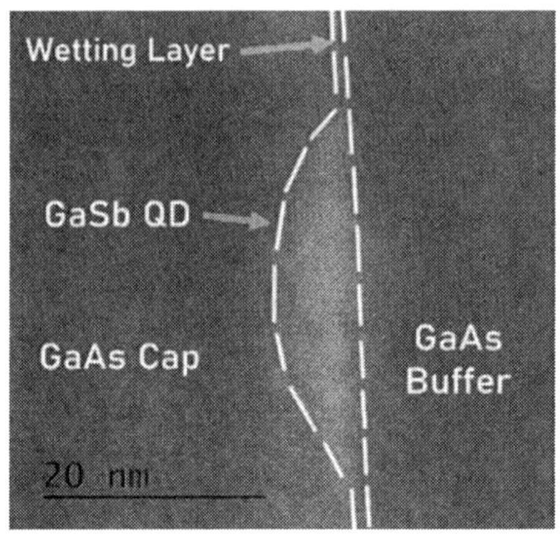

Figure 4: Cross-sectional scanning transmission electron microscopy (STEM) micrograph of capped GaSb/GaAs QDs.

The region where the GaSb/GaAs QD is sandwiched is displayed in Figure 4 in the form of a High Angle Annular Dark Field (HAADF) STEM micrograph. Analysis of the micrograph in Figure 4 shows a close correlation in the value of approximate height and diameter of the dots compared to measurements obtained from AFM for the uncapped GaSb/GaAs QD. The absence of threading dislocations and growth defects from the micrograph is an indication of the quality of the epitaxial layers deposited and the growth conditions.

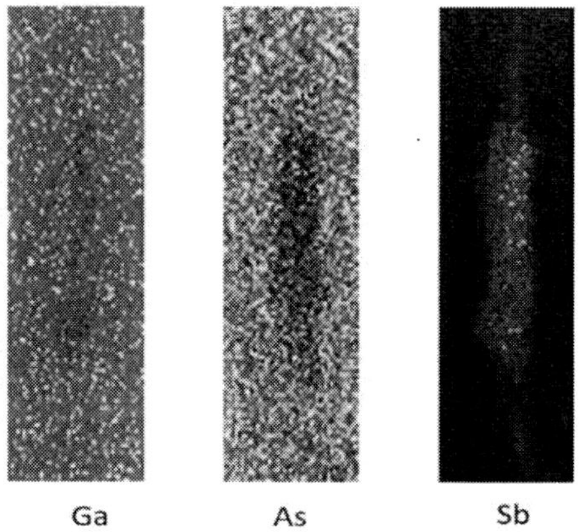

Figure 5: Electron energy loss spectroscopy (EELS) imaging of area around the capped GaSb/GaAs QDs.

EELS elemental mapping analysis were carried out on the capped sample in order to quantify and determine the percentage composition and distribution of elements in the QD. The EELS spectrum image obtained from the capped sample is displayed as a 50 nm × 20 nm micrograph in Figure 5. The micrograph reveals the individual constituents in the dot and the neighboring region. Analysis of the constituents which corresponds to Ga, As and Sb indicates the dominance of Ga which was found to be consistently spread throughout the area surrounding the dot compared to proportion of As and Sb which were found to be inconsistent. However, a slight reduction in percentage composition of Ga was detected at the dot center, while As is more prevalent at the outer region of the dot. The EELS micrograph in Figure 5 suggests a significant proportion of intermixing between Sb/As within the area encompassing the dot.

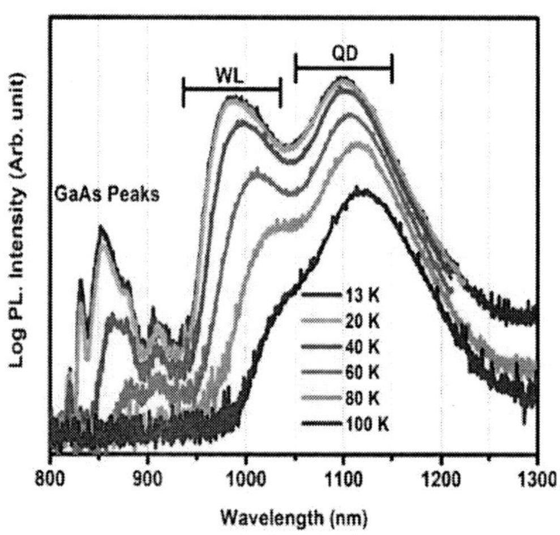

Figure 6: Temperature dependent photoluminescence (PL) spectra of capped GaSb/GaAs QDs

In order to probe the luminescence emission patterns and the discrete energy levels in the QDs, PL spectroscopy investigations at variable temperatures were carried out on the capped sample. Different transitions were observed from the PL spectra at different temperatures as seen in Figure 6. The PL spectrum collected at low temperature is characterized by four different PL lines. The lines at 831 nm and 855 nm are emissions associated with GaAs. The other two lines at higher wavelength (990 and 1100 nm) are ascribed to transitions from the wetting layer (WL) and the QDs respectively. The abrupt reduction in intensity of the GaAs related peaks at elevated temperature range is suggested to be as a result of non-radiative recombination [5] and the increased trapping localization of carriers within the GaSb layer. Similarly, the PL spectra in Figure 6 shows a quick reduction in intensity of the higher energy (WL) peaks at high temperatures compared with the peaks corresponding to QDs at lower energy. This phenomenon can be attributed to improved migration of holes from the WL to the QD at elevated temperature [3], while the decrease in PL intensity of the QDs at higher temperature is due to thermal escape of carriers [6]. At 100 K, the transition line associated with QD shifts to 1135 nm which allows ~ 420 meV hole localization energy, with a GaAs energy gap of 1.515 eV. This result is closely related to the values determined from deep level transient spectroscopy and PL measurements for MOVPE [7] and MBE [8] grown GaSb/GaAs QDs. Further details on the behaviour of the capped sample at different laser excitation power has been reported elsewhere [9].

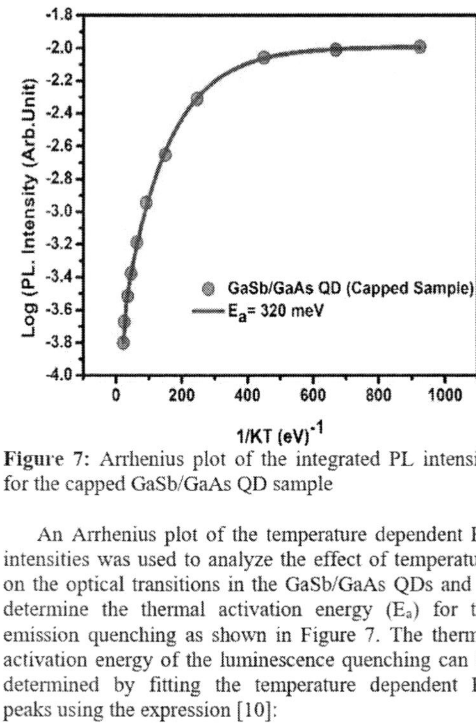

Figure 7: Arrhenius plot of the integrated PL intensity for the capped GaSb/GaAs QD sample

An Arrhenius plot of the temperature dependent PL intensities was used to analyze the effect of temperature on the optical transitions in the GaSb/GaAs QDs and to determine the thermal activation energy (E_a) for the emission quenching as shown in Figure 7. The thermal activation energy of the luminescence quenching can be determined by fitting the temperature dependent PL peaks using the expression [10]:

$$I = \frac{I_0}{\left[1 + C\exp\left({E_a}/{kT}\right)\right]} \tag{1}$$

where I_0 is the initial intensity at 10 K. C is a constant which is often presumed to be temperature independent and derived from the temperature where the emission begins to quench (i.e. referred to as the "knee" in the Arrhenius plot [11]). The thermal activation energy deduced for competing non-radiative processes in the capped samples using applicable deconvolution technique and least-squares fits is E_a =320.4 ± 1.2 meV. The thermal activation energy obtained in this study for the capped sample is higher than the values of 39 meV [12] ,

140 meV [13] and 130 meV [14] which was reported for GaSb QDs grown using AP-MOVPE, low pressure MOVPE and MBE respectively. The 320 meV value of thermal activation energy measured in present work for GaSb/GaAs QDs suggests considerable excitonic localization in the QD even though it is still lower than the estimated values of 450 meV [7] and 500 meV [8] reported by other authors as GaSb/GaAs QD hole localization energy/confinement potential.

5 SUMMARY

From the study conducted, the uncapped sample, the QDs were densely populated with fairly uniform size distribution. Cross-sectional STEM measurements of the capped dots reveals a close correlation in the value of approximate height and diameter of the dots compared to measurements obtained from AFM for the uncapped GaSb/GaAs QD, a phenomenon which indicates good capping conditions used during epitaxial growth. EELS elemental mapping analysis on the capped sample indicates the dominance of Ga which was found to be consistently spread throughout the area surrounding the dot compared to proportion of As and Sb which were found to be inconsistent. Different optical transitions were observed from the PL spectra at different temperatures with PL lines at 831 nm and 855 nm ascribed to be emissions from GaAs, while the lines at higher wavelength (990 and 1100 nm) are transitions from the WL and the QDs respectively. Results emanating from present work presents prospective foundation for optimizing the growth conditions of GaSb/GaAs QDs in order to achieve high quality nanostructures which ultimately enhances the efficiency of solar cells produced using these materials.

6 ACKNOWLEDGMENT

The authors wish to acknowledge the financial support of the National Research Foundation (GUN 93215) and Department of Science and Innovation (DST/CON 0170/2019). The authors also acknowledge the support of Nelson Mandela University and the assistance of Centre for High Resolution Transmission Electron Microscopy at Nelson Mandela University.

7 REFERENCES

[1] Y. Shoji, R. Tamaki, and Y. Okada, "Multi-stacked GaSb/GaAs type-II quantum nanostructures for application to intermediate band solar cells," AIP Adv., vol. 7, no. 65305, pp. 1-7, 2017.

[2] W. H. Lin, C. C. Tseng, K. P. Chao, S. C. Mai, S. Y. Kung, S. Y. Wu, S. Y. Lin, and M. C. Wu, "High-Temperature Operation GaSb/GaAs Quantum-Dot Infrared Photodetectors," IEEE Photonics Technol. Lett., vol. 23, no. 2, pp. 106-108, 2011.

[3] F. Hatami, N.N. Ledentsov, M. Grundmann, J. Böhrer, F. Heinrichsdorff, M. Beer, D. Bimberg, S.S. Ruvimov, P. Werner, U. Gösele, J. Heydenreich, U. Richter, S.V. Ivanov, B. Ya. Meltser, P. S. Kop'ev, andZh. I. Alferov, Appl. Phys. Lett. 67, 656 (1995)., "Radiative recombination in type-II GaSb/GaAs quantum dots," Appl. Phys. Lett., vol. 67, p. 656, 1995.

[4] G.G. Zegrya and A.D. Andreev, "Mechanism of suppression of Auger recombination processes in type II heterostructures," Appl. Phys. Lett, vol. 67, p. 2681, 1995.

[5] Y. Varshni, "Temperature dependence of the energy gap in semiconductors," Physica, vol. 34, no. 1, pp. 149-154, 1967.

[6] G. Bacher, C. Hartmann, H. Schweizer, T. Held, G. Mahler, H. Nickel, "Exciton dynamics in InxGa1-xAs/GaAs quantum well heterostructures: Competition between capture and thermal emission," Phys. Rev. B, vol. 47, p. 9545, 1993.

[7] M. Geller, C. Kapteyn, L. Müller-Kirsch, R. Heitz, D. Bimberg, "450 meV hole localization in GaSb/GaAs quantum dots," Appl. Phys. Lett., vol. 82, p. 2706, 2003.

[8] D. Alonso-Álvarez, Benito Alén, Jorge M. García, José M. Ripalda, "Optical investigation of type II GaSb/GaAs self-assembled quantum dots," Appl. Phys. Lett., vol. 91, p. 263103, 2007.

[9] N. Tile, C.C. Ahia, J. Olivier and J.R. Botha, "Atmospheric pressure-MOVPE growth of GaSb/GaAs quantum dots," Physica B, vol. 535, pp. 20-23, 2018.

[10] D. S. Jiang, H. Jung, and K. Ploog, "Temperature dependence of photoluminescence from GaAs single and multiple quantum-well heterostructures grown by molecular-beam epitaxy," J. Appl. Phys., vol. 64, no. 3, pp. 1371-1377, 1988.

[11] C. C. Ahia, N. Tile, Z. N. Urgessa, J. R. Botha, J. H. Neethling, "An investigation of near-infrared photoluminescence from AP-MOVPE grown InSb/GaSb quantum dot structures," J. Cryst. Growth, vol. 458, pp. 53-59, 2017.

[12]Motlan, E.M. Goldys, L.V. Dao, "Photoluminescence of GaSb self-assembled quantum dot layers grown by metalorganic chemical vapour deposition," J. Vac. Sci. Technol., B, vol. 20, pp. 291-294, 2002.

[13] L. Müller-Kirsch, R. Heitz, A. Schliwa, O. Stier, D. Bimberg, H. Kirmse, W. Neumann, "Many-particle effects in type II quantum dots," Appl. Phys. Lett. , vol. 78, p. 1418, 2001.

[14] K. Suzuki, R.A. Hogg, Y. Arakawa, "Structural and optical properties of type II GaSb/GaAs self-assembled quantum dots grown by molecular beam epitaxy," J. Appl. Phys. , vol. 85, p. 8349, 1999.

MECHANICALLY-STACKED AND ELECTRICALLY-CONNECTED TWO-TERMINAL TANDEM MODULE

Kyotaro Nakamura and Yoshio Ohshita
Toyota Technological Institute
2-12-1 Hisakata, Tempaku-ku, Nagoya 468-8511, Japan
phone: +81-52-809-1830, fax: +81-52-809-1879, e-mail: kyotaro@toyota-ti.ac.jp

ABSTRACT: We would like to propose a new configuration of tandem solar modules. It is a Mechanically-Stacked and Electrically-Connected two-terminal tandem module, that is to say, MSEC tandem module. The basic idea is very simple, just to connect the output terminals of the top and bottom modules of a mechanically-stacked four-terminal tandem module. Just by doing so, you can make high efficiency two-terminal tandem module easily and inexpensively. To validate this concept, a specific MSEC tandem module was designed and its electrical characteristics were calculated by SPICE simulation. The top module fabricated with 71 mm square III-V cells and the bottom module fabricated with 79 mm square Si cells are connected in series to form a MSEC 2-terminal tandem module, which allows the top and bottom currents to be matched. As a result, we were able to minimize the losses associated with the 2-terminal configuration and confirmed that the output power is almost the same as that of a 4-terminal tandem module.
Keywords: Design, Module, Simulation, Tandem

1 INTRODUCTION

As the market for photovoltaic (PV) power generation expands, the demand for higher efficiency and higher power output of solar cells and PV modules continues to increase. At the same time, however, the maximum efficiency of Si solar cells, which are currently the mainstream, has reached 26.7%, which is very close to the expected practical efficiency limit of 27% [1]. This fact strongly suggests that there is little room for efficiency improvement in single junction devices. In recent years, therefore, researchers and developers of solar cells and modules have increasingly turned their attention to multi-junction devices, which consist of two or more stacked sub-cells that each absorb different parts of the solar spectrum. Among the multi-junction devices, Si-based tandem solar cells and modules using Si solar cells as the bottom cell have been actively researched and developed, and the Perovskite/Si tandem cell is expected to be commercialized as soon as possible because it is considered to have a good balance between cost and efficiency, and the highest efficiency of 29.15% (two-terminal) has been reported [1]. According to ITRPV 12th edition, Si-based tandem cells with efficiency over 28% will start mass production after 2023, and their module efficiency will reach 22.5% around 2023 and 26% in 2031 [2].

On the other hand, another promising candidate for Si-based tandem cell, the III-V/Si tandem cell using high-efficiency compound solar cells as the top cell, has already achieved a very high cell efficiency of 35.9% (four-terminal) [3]. This efficiency exceeds both the theoretical limit of conventional Si technology (29.4%) and the efficiency of record III-V dual-junction devices (32.8%), demonstrating the high potential of this type of Si-based tandem solar cell. The key issues for mass production of these III-V/Si tandem solar cells will be to achieve low cost and high productivity.

Usually, tandem solar cells are either monolithic two-terminal tandems, in which the top and bottom cells are physically and electrically connected directly, or mechanical stack four-terminal tandems, in which the cells are simply physically stacked on top of each other but not electrically connected, and the output is extracted from both the top and bottom cells. Fig. 1 shows the schematic views of both types of tandem solar cells and modules. Each tandem cell and module has its own advantages and disadvantages, which are inextricably linked to each other. Monolithic two-terminal tandem has the following disadvantages; (1) It is difficult to manufacture because the top and bottom must be physically stacked and electrically connected with low resistance. This should increase cell cost. (2) It is difficult to obtain high efficiency because the current matching of the top and bottom is required. On the other hand, the mechanical stack four-terminal tandem does not have such disadvantages, so it is easy to manufacture with high efficiency. However, since the output comes from both the top and bottom, the maximum power point (MPP) tracking is required for each output, which increases the burden on the system side. This disadvantage will increase the system cost.

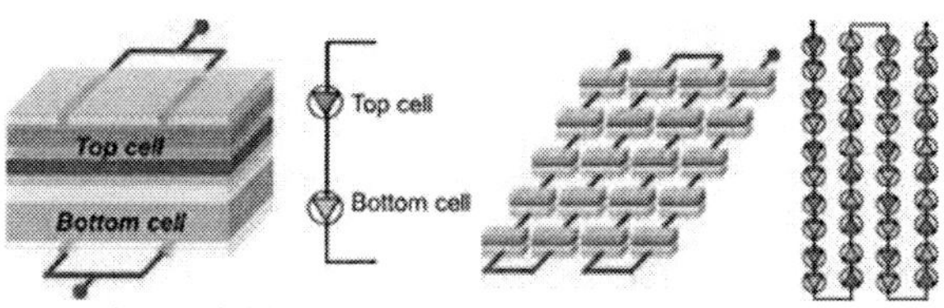

(a) Monolithic two-terminal tandem cell and module

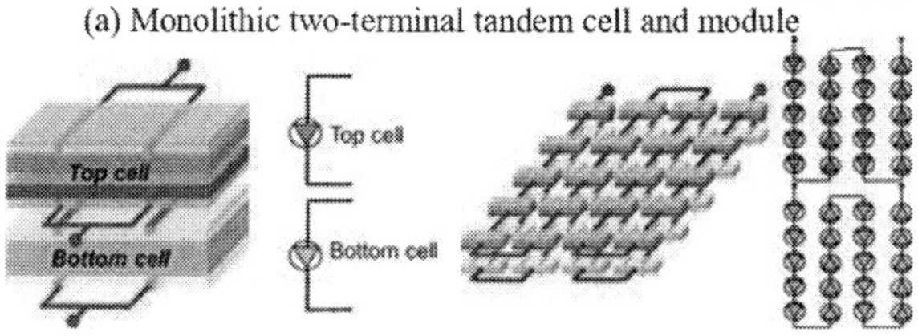

(b) Mechanical stack four-terminal tandem cell and module

Figure 1: Schematic views of tandem cells and modules

The one of the simplest approaches to solve these problems is to connect the output terminals of the top and bottom modules of a mechanically-stacked four-terminal tandem module. However, it is not enough to simply connect them. That would not match the electrical characteristics of the top and bottom, and the output would be significantly reduced. To avoid this problem, a little ingenuity is required.

In this paper, we would like to propose a new configuration of tandem solar modules. It is a Mechanically-Stacked and Electrically-Connected two-terminal tandem module, that is to say, MSEC tandem module. We will also explain the "little ingenuity" to avoid the problem of reduced power output due to the electrical mismatch between top and bottom, using a specific example.

2 MSEC 2-TERMINAL TANDEM MODULE

Figure 2 shows the examples of MSEC two-terminal tandem modules with (a) the top and bottom are connected in series, and (b) the top and bottom are connected in parallel. In this paper, we call (a) MSEC-serial and (b) MSEC-parallel, respectively. In addition, there are previous studies on MSEC-parallel by other researchers [4]. M. Stocks et al. have already investigated this type of module configuration and demonstrated its advantages.

In each configuration, the following considerations should be made when designing the module. In the case of MSEC-serial, it is necessary to match the currents of the top and bottom modules, but this can be done by changing the size of each cell. On the other hand, in the case of MSEC-parallel, it is necessary to match the voltages of the top and bottom modules, but this can be done by changing the number of series of each cell.

The advantages of this configuration are as follows; (1) It is easy to fabricate because it does not require direct deposition or wafer bonding of top and bottom cells as in the case of monolithic tandem. In other words, it can be made at low cost. (2) As a module, it has two terminals, which reduces the burden on the system side compared to a four-terminal module. In other words, it is easier to use and decrease system cost. (3) the electrical matching between top and bottom cell can be done at the module level, which makes matching easier than with monolithic tandem. In other words, the module efficiency can be increased compared to the monolithic tandem.

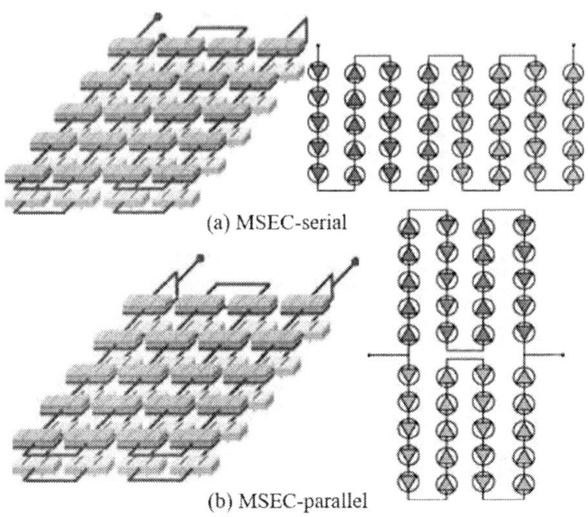

Figure 2: Schematic views of MSEC two-terminal tandem modules

3 PROOF OF CONCEPT

To validate this concept, a specific MSEC tandem module was designed and its electrical characteristics were calculated by SPICE simulation. The following calculations are based on the cell parameters (shown in Table I) of the GaInP/GaAs//Si 4-terminal tandem solar cell described in the paper by S. Essig et al [3].

Table I: Cell parameters assumed in the simulation [3]

	Jsc [mA/cm²]	Voc [V]	FF [%]	η [%]
III-V 2J top cell	13.61	2.520	87.5	30.01
Si bottom cell	11.03	0.6811	78.5	5.90

3.1 Module design

At first, we designed a MSEC-serial tandem module based on the following steps (See Fig. 3);

(1) Cut out a square of the III-V 2J top cell fabricated on 4" wafer ⇒ Top cell size = 71 x 71 mm² ⇒ Top cell area = 50.4 cm² ⇒ Isc_{top} = 0.686 A

(2) Arrange the top cells in 10 rows x 10 columns with 1 mm spacing ⇒ Top module size = 719 x 719 mm² ⇒ Top module area = 0.517 m²

(3) Divide a G1 size Si cell into four ⇒ Bottom cell size = 79 x 79 mm² ⇒ Bottom cell area = 62.4 cm² ⇒ Isc_{bottom} = 0.688 A

(4) Arrange the bottom cells in 9 rows x 9 columns with 1 mm spacing ⇒ Bottom module size = 719 x 719 mm² Bottom module area = 0.517 m²

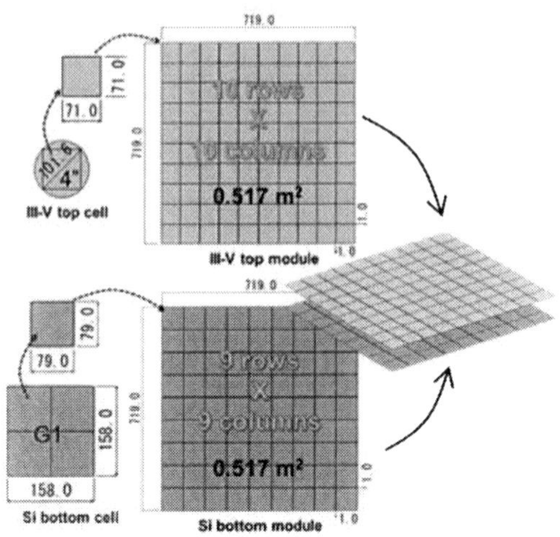

Figure 3: Schematic views of MSEC two-terminal tandem module design

3.2 SPICE Simulation

As mentioned in the previous section, we were able to design top and bottom modules with matching external shapes and almost equal current value, using a typical 4-inch size III-V solar cell and a typical G1 size Si solar cell as starting points. Next, the electrical behavior of these modules when connected in series was investigated using LTspice, a simulation software from Analog Devices. Fig. 4 schematically shows the simulation model of MSEC two-terminal tandem module. Assuming an appropriate connection resistance, a circuit model of a module with 100 III-V top cells and 81 Si bottom cells connected in series was constructed, and its electrical characteristics under the standard testing condition were calculated. Table II shows the simulation result. If the top and bottom modules are operated individually, the output power of the top and bottom modules will be 151.3W and 29.8W, respectively. On the other hand, the output of the MSEC module with the top and bottom modules connected in series is estimated to be 180.3W. This value is only 0.4% smaller than 181.1 W, which is the sum of the outputs of the top and bottom modules.

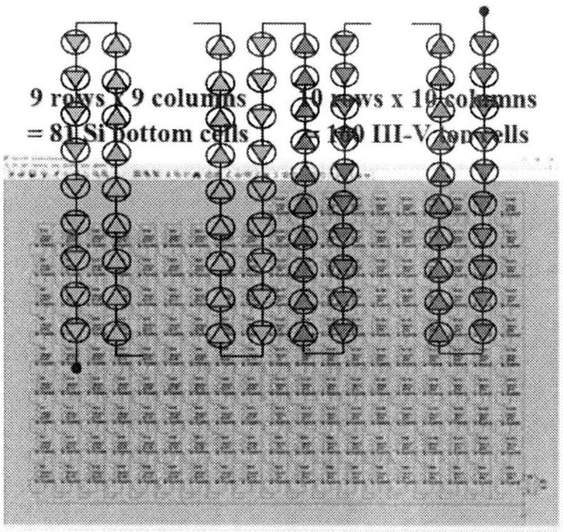

Figure 4: Schematic views of SPICE simulation of MSEC two-terminal tandem module

Table II: Simulated MSEC module parameters

	Isc [A]	Voc [V]	FF [%]	Wp [W]	η [%]
III-V top module	0.686	252.2	87.5	151.3	29.3
Si bottom module	0.688	55.2	78.5	29.8	5.76
MSEC module	0.686	307.4	85.5	180.3	34.9

As described above, the top module fabricated with 71 mm square III-V cells and the bottom module fabricated with 79 mm square Si cells are connected in series to form a MSEC 2-terminal tandem module, which allows the top and bottom currents to be almost matched. As a result, we were able to minimize the losses associated with the 2-terminal configuration and confirmed by simulation that the output power was almost the same as that of a 4-terminal tandem module.

4 CONCLUSIONS

We proposed a new configuration of tandem solar module, i.e., "Mechanically-Stacked and Electrically-Connected two-terminal tandem module". We further validated this concept by designing a specific module and estimating its electrical characteristics using SPICE simulation. The MSEC tandem module is a very simple method of connecting the output terminals of the top and bottom modules of a mechanically stacked 4-terminal tandem module. However, this method has several advantages, as mentioned before, and it is an easy and inexpensive way to make highly efficient 2-terminal tandem modules.

5 ACKNOWLEDGEMENTS

This work was supported by the New Energy and Industrial Technology Development Organization (NEDO) under the Ministry of Economy, Trade and Industry (METI) in Japan.

6 References

[1] Green, M. A. *et. al.* Solar cell efficiency tables (version 57). *Prog. Photovolt. Res. Appl.* **29**, 3–15 (2021)
[2] International Technology Roadmap for Photovoltaic, Edition 2021, retrieved July, 2, 2021. From https://itrpv.vdma.org/
[3] Essig S, Allebé C, Remo T, *et al., "Raising the one-sun conversion efficiency of III–V/Si solar cells to 32.8% for two junctions and 35.9% for three junctions" Nat Energy*. 2017;2:17144. https://doi.org/10.1038/nenergy.2017.144
[4] Matthew Stocks, *et. al.*, "Free the Bandgap! Series-Paralles Connection of Tandem Cells" *Proc. 32nd European Photovoltaic Solar Energy Conference and Exhibition*, Munich, Germany, 2016. pp. 223-227.

BICHROMATIC LIGHT SOURCE FOR SUBCELL PERFORMANCE ANALYSIS IN PEROVSKITE/SILICON TANDEM SOLAR CELLS

Marko Jošt[1], Gašper Matič[1], Eike Köhnen[2], Bor Li[2], Boštjan Glažar[1], Marko Jankovec[1], Steve Albrecht[2], Marko Topič[1]
University of Ljubljana, Faculty of Electrical Engineering, Tržaška cesta 25, 1000 Ljubljana, Slovenia
Helmholtz-Zentrum Berlin, Kekulestrasse5, 12489 Berlin, Germany

ABSTRACT: We present a tailored bichromatic LED array setup for in-depth analysis of tandem solar cells. Two spectrally independent LED arrays are used to selectively bias the two subcells of monolithic perovskite/silicon tandem devices, which facilitates intensity calibration and improves stability of the light source. The developed bichromatic LED setup (BCLED) offers a reliable and spectrally controlled several 1000-hours long maximum power point stability test of tandem solar cells. Additionally, selective biasing of subcells enables investigation of individual subcell performance. We showcase this by successfully disentangling the tandem *J-V* curve into subcell *J-V* curves. A detailed procedure how to use BCLED to obtain parameters for one diode-model of each subell is described, with the parameters later used to reconstruct tandem and both subcell *J-V* curves using simulations. The procedure is validated by simulations and then employed on a high-efficiency perovskite/silicon tandem device, showing great matching with the measured *J-V* curve. Therefore, the BCLED setup can be used for quick subcell quality control and performance analysis as well as for long-term stability testing in maximum-power-point tracking mode with periodic I-V scans.

Keywords: tandem, perovskite, stability

1 INTRODUCTION

Monolithic perovskite/silicon tandem solar promise to enter the photovoltaic market in the near future. Their power conversion efficiency (PCE) has reached almost 30% [1]–[3], with several groups revealing decent stability in the range of a few hundred hours [3]–[6]. One of the reason that there are no longer MPP tracking results are the spectral requirements of the light source. Due to the monolithic connection of the two subcells and their specific absorption regions, long-lived, typically used white LEDs are not suitable for any tandem solar cell measurements.

Additionally, in field operation the devices have to produce high power for a couple of decades at the conditions much more demanding than those in the laboratory, such as temperature changes, different spectral mismatches and irradiances. Due to perovskite ionic structure and stronger sensitivity to environmental factors compared to silicon, such conditions have to be thoroughly analyzed, requiring specialized equipment and analytical techniques. An additional challenge presents the monolithic connection of the two subcells, aggravating the subcell analysis [3], [7], [8], since some of the most common characterization techniques are not capable of distinguish between the two subcells.

To overcome these challenges, we develop a bichromatic light source that serves as a dedicated sun-simulator for tandem solar cells. The tool is primarily designed for long-term MPP tracking measurements [3], [9], but also capable of subcell-dependent analysis. In this contribution, we briefly describe the tool and its applications.

2 BICHROMATIC LED SOURCE

The schematics and a photograph of the BCLED setup are presented in Figure 1. Two arrays of LEDs, blue with a $\lambda = 470$ nm and infrared with a $\lambda = 940$ nm are used to selectively bias top and bottom subcell in monolithic tandem. The uniformly distributed light of BCLED over 7 x 7 cm^2 area has several advantages over classical solar simulators: i) longevity of the LEDs compared to halogen and xenon lamps, especially needed for long-term MPP tracking, ii) easy control of the light intensity and facilitated calibration of the light and iii) subcell dependent analysis. In combination with an in-house developed MPP tracker [10], [11], the setup enables long-term measurements in the order of several hundreds or even thousands of up to 4 tandem devices at the same time. While primary designed for perovskite/silicon tandem solar cell, it can be used for any tandem technology with bandgaps corresponding to the wavelengths of the LEDs used.

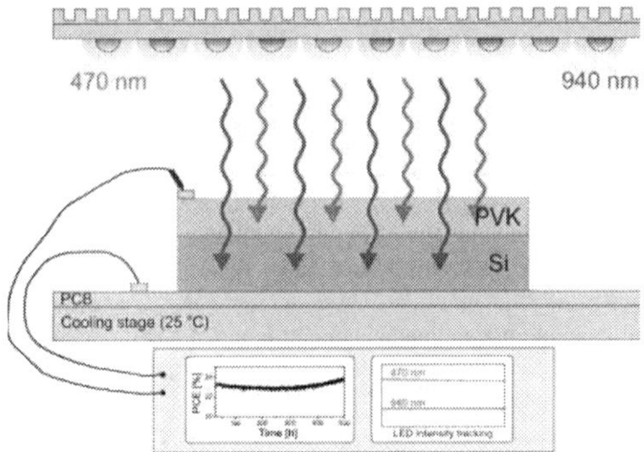

Figure 1: Schematics and a photograph of a bichromatic LED setup. Blue LEDs are used to selectively bias the top cell, while IR LEDs are used to bias the bottom cell. The cooling stage ensures constant temperature during the testing, while MPP tracker is used for long-term stability testing.

3 RESULTS

First, BCLED is ideal for long-term stability testing under various conditions. In Figure 2, a 600 hour long MPP track of a perovskite/silicon tandem solar cell is shown. This result was obtained under equivalent AM1.5 conditions. However, importantly conditions such as bottom subcell limiting or top subcell limiting, can controllably and reliably be tested. This is a regular occurrence in diurnal outdoor operation due to changes in spectrum throughout the day. In the PV community, the focus has mostly been towards current matched tandem solar cells. However, constant temperature and spectral changes, transition from bottom to top cell limitation and vice versa change the current limitation in the device and might degrade perovskite faster compared to silicon [12], [13]. Thus, an initial mismatch might be beneficial for long-term energy yield. In-depth analysis of tandems under spectrally different conditions, for which the bichromatic light source is designed, is therefore necessary for designing a decade-efficient and surviving perovskite-based tandem solar cell.

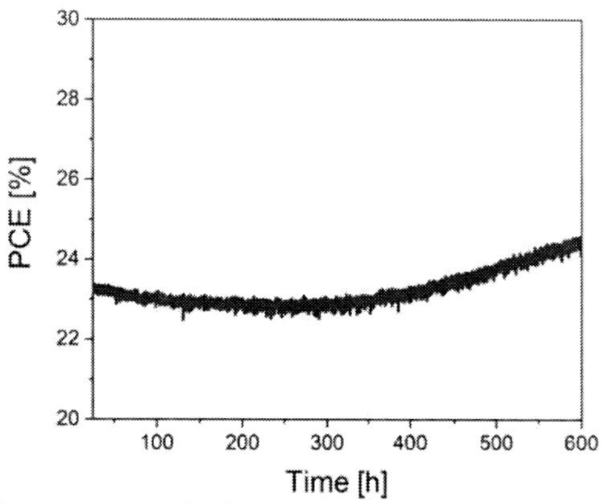

Figure 2: MPP track of perovskite/silicon tandem solar cell under BCLED.

Secondly, with the selective biasing of subcells, BCLED enables subcell-dependent analysis, such as fill factor dependence on current mismatch or as a biasing source for external quantum efficiency and photoluminescence measurements. Significantly, at EU-PVSEC we will present how to disentangle the tandem *J-V* curve into individual *J-V* curves of subcells using BCLED. With this procedure, we can reliably extract subcell PV parameters, such open-circuit voltage (V_{OC}) and fill factor (FF), with especially the latter being otherwise challenging to obtain due to monolithic connection of the subcells. Both V_{OC} and FF possess critical information on the quality of the subcell, successfulness of tandem integration and possible damage in the fabrication. While for the perovskite/silicon it is most likely that the perovskite top cell quality is lower, for other technologies, such as perovskite/CIGS and perovskite/perovskite also the bottom cell could be the reason for weaker performance. There both subcells are prone to poor performance due to shunting. To disentangle this, tools such as BCLED are necessary, since we can easily point to the under-performing subcell and its origin for under-performance.

The subcell *J-V* extraction is based on a one-diode model for each subcell. Thus, following parameters have to be known: series resistance R_S, shunt resistance R_{sh}, saturation current density J_0 and ideality factor n. We obtain those parameters by performing a series of *I-V* measurements under different light intensity for top and bottom subcell. For example, to obtain the data that are relevant for the perovskite subcell, we overbias the silicon cell with IR light and then perform tandem *I-V* measurements at different blue LED instensities (Figure 3a). The R_S and R_{sh} can be obtained from the curve's slope around open-circuit and at short-circuit conditions, respectively. J_0 and n can be extracted plotting the *J-V* data to satisfy the following equation:

$$\ln(J_L) = \frac{1}{n}\frac{q}{kT}V_{OC} + ln(J_0) \quad (1)$$

The procedure is first validated using SPICE simulations. A series of *J-V* curves corresponding to perovskite/silicon tandem solar cells were simulated and used as a base for parameter extraction and *J-V* reconstruction. An excellent match is obtained between originally simulated and reconstructed *J-V* curves, if the Rsh of perovskite was larger than 500 Ω cm². The same approach was then applied to a fabricated monolithic perovskite/silicon tandem solar cell. Again, a successful reconstruction was achieved, with a small discrepancy around MPP was observed (Figure 3b). However, we have revealed high dependence of the ideality factor on V_{OC} (light intensity) in our tandem solar cells. Additional aspects with detailed results are published in [14].

The BCLED setup combined with our reconstruction procedure has therefore proved as a powerful tool for subcell operation analysis, enabling to extract important parameters on the subcell quality.

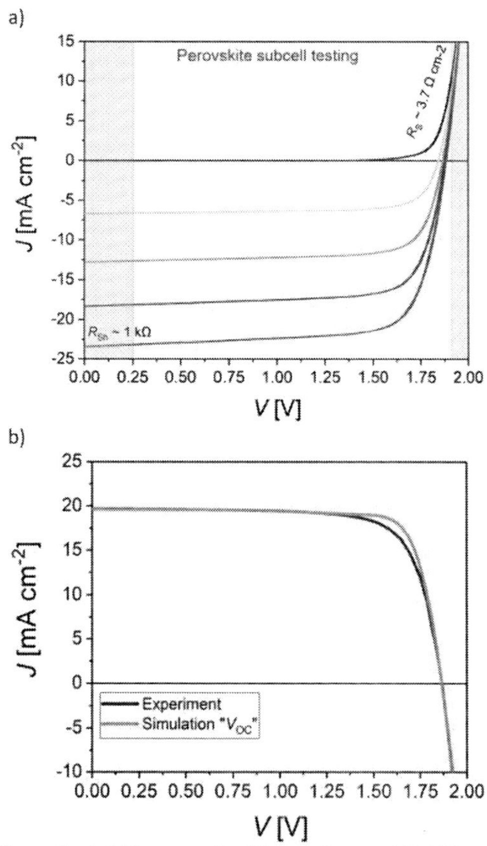

Figure 3: a) *J-V* curves of a fabricated perovskite/silicon tandem solar cell with overbiased silicon bottom cell measured at different bias (blue LED intensity) for the perovskite top cell. b) A comparison between the experimentally measured *J-V* (black) curve of a perovskite/silicon tandem solar cell and the reconstructed (blue) one, using the procedure described in the main text.

4 CONCLUSIONS

We have presented a bichromatic LED light source for advanced characterization of tandem solar cells. Two sets of LEDs (blue and infra-red) enable individual and selective biasing of both subcells, thus facilitating calibration of the light source, allowing testing of different biasing conditions and also subcell dependent analysis. An example of the latter is shown here, where we demonstrate a procedure to extract the *J-V* curves of both subcells.

We get good matching between the experimentally measured and the reconstructed *J-V* of the highly efficient perovskite/silicon tandem solar cells. The slight difference is attributed to the voltage-dependent ideality factor, which shows the current limitation of the devices. Such applications of the bichromatic LED source, together with the possibility of long-term MPP tracking, make it a powerful tool for photovoltaic characterization and analysis.

5 ACKNOWLEDGEMENT

The authors acknowledge the funding by the Slovene Research Agency (ARRS) for research programs P2-0197 and J2-1727, German Federal Ministry of Education and Research (BMBF, grant no. 03SF0540) and the Helmholtz association for funding the project TAPAS (TAndem Perovskite And Silicon solar cells - Advanced optoelectrical characterization, modelling and stability) within the EU partnering program.

6 REFERENCES

[1] M. Jošt, L. Kegelmann, L. Korte, and S. Albrecht, 'Monolithic Perovskite Tandem Solar Cells: A Review of the Present Status and Advanced Characterization Methods Toward 30% Efficiency', *Adv. Energy Mater.*, vol. 10, no. 26, p. 1904102, 2020, doi: 10.1002/aenm.201904102.

[2] 'Best Research-Cell Efficiency Chart'. https://www.nrel.gov/pv/cell-efficiency.html (accessed Feb. 03, 2021).

[3] A. Al-Ashouri *et al.*, 'Monolithic perovskite/silicon tandem solar cell with >29% efficiency by enhanced hole extraction', *Science*, vol. 370, no. 6522, pp. 1300–1309, Dec. 2020, doi: 10.1126/science.abd4016.

[4] Y. Hou *et al.*, 'Efficient tandem solar cells with solution-processed perovskite on textured crystalline silicon', *Science*, vol. 367, no. 6482, pp. 1135–1140, Mar. 2020, doi: 10.1126/science.aaz3691.

[5] F. Sahli *et al.*, 'Fully textured monolithic perovskite/silicon tandem solar cells with 25.2% power conversion efficiency', *Nat. Mater.*, vol. 17, no. 9, Art. no. 9, Sep. 2018, doi: 10.1038/s41563-018-0115-4.

[6] J. Xu *et al.*, 'Triple-halide wide–band gap perovskites with suppressed phase segregation for efficient tandems', *Science*, vol. 367, no. 6482, pp. 1097–1104, Mar. 2020, doi: 10.1126/science.aaz5074.

[7] J. Gilot, M. M. Wienk, and R. A. J. Janssen, 'Measuring the External Quantum Efficiency of Two-Terminal Polymer Tandem Solar Cells', *Adv. Funct. Mater.*, vol. 20, no. 22, pp. 3904–3911, 2010, doi: 10.1002/adfm.201001167.

[8] S. Roensch, R. Hoheisel, F. Dimroth, and A. W. Bett, 'Subcell I-V characteristic analysis of GaInP/GaInAs/Ge solar cells using electroluminescence measurements', *Appl. Phys. Lett.*, vol. 98, no. 25, p. 251113, Jun. 2011, doi: 10.1063/1.3601472.

[9] E. Köhnen *et al.*, '27.9% Efficient Monolithic Perovskite/Silicon Tandem Solar Cells on Industry Compatible Bottom Cells', *Sol. RRL*, vol. n/a, no. n/a, p. 2100244, doi: https://doi.org/10.1002/solr.202100244.

[10] 'LPVO: Cell monitoring'. http://lpvo.fe.uni-lj.si/en/services/pv-monitoring/cell-monitoring/ (accessed Oct. 12, 2020).

[11] M. Jošt *et al.*, 'Perovskite Solar Cells go Outdoors: Field Testing and Temperature Effects on Energy Yield', *Adv. Energy Mater.*, vol. 10, no. 25, p. 2000454, 2020, doi: https://doi.org/10.1002/aenm.202000454.

[12] E. Aydin *et al.*, 'Interplay between temperature and bandgap energies on the outdoor performance of perovskite/silicon tandem solar cells', *Nat. Energy*, pp. 1–9, Sep. 2020, doi: 10.1038/s41560-020-00687-4.

[13] M. De Bastiani *et al.*, 'Toward Stable Monolithic Perovskite/Silicon Tandem Photovoltaics: A Six-Month Outdoor Performance Study in a Hot and Humid Climate', *ACS Energy Lett.*, pp. 2944–2951, Jul. 2021, doi: 10.1021/acsenergylett.1c01018.

[14] M. Jošt *et al.*, 'Subcell Operation and Long-Term Stability Analysis of Perovskite-Based Tandem Solar Cells Using a Bichromatic Light Emitting Diode Light Source', *Sol. RRL*, vol. n/a, no. n/a, p. 2100311, doi: 10.1002/solr.202100311.

THE PERFORMANCE OF FOUR-TERMINAL PEROVSKITE-SILICON TANDEM SOLAR CELLS UNDER DIFFERENT IRRADIANCE LEVELS

A.B. Nikolskaia, M.F. Vildanova, S.S. Kozlov, O.V. Alexeeva, O.K. Karyagina, O.I. Shevaleevskiy
Solar Photovoltaic Laboratory, Institute of Biochemical Physics, Russian Academy of Sciences
Kosygin St., 4, 119334, Moscow, Russia

ABSTRACT: Four-terminal (4-terminal) tandem perovskite-silicon solar cells (PSC/c-Si) were manufactured and studied. Semitransparent PSCs with the device architecture of glass/FTO/c-TiO_2/MO_2/$CH_3NH_3PbI_3$/PTAA/PEDOT:PSS/FTO/glass (M = Zr, Hf) were prepared and used as top solar cells. To fabricate the PSCs nanostructured metal oxides with very large bandgap ($E_g > 5$ eV), such as ZrO_2 and HfO_2, were used as photoelectrodes. A comparative analysis of photovoltaic parameters for the top PSCs, the bottom crystalline silicon (c-Si) solar cells and PSC/c-Si tandem systems was performed. We have demonstrated that the application of very large-bandgap materials allows to improve the performance of the PSCs and to increase the efficiency of PSC/c-Si tandem solar cell under different illumination conditions. The performance of 4-terminal PSC/c-Si tandem solar cell at low irradiation level of 15 W/m^2 has shown a 4-times increase in comparison with a standalone c-Si solar cell.
Keywords: solar cell, perovskite, c-Si, tandem, solar radiation, photoelectrical properties

1 INTRODUCTION

To date, crystalline silicon (c-Si) solar cell performance under standard illumination conditions AM1.5G (1000 W/m^2) has nearly reached its theoretical maximum [1]. Besides the power conversion efficiency (PCE) of c-Si solar cells is drastically reduced with the decrease of light intensity thus limiting their possibilities in real weather conditions [2, 3]. The increase of PCE can be achieved by integrating c-Si solar cells into the tandem system [4]. Partially transparent solar cells, through which a certain part of the solar spectrum passes with minimal losses, should be used as top devices. Low cost and easily manufactured perovskite solar cells (PSCs) look to be the best candidates for this purpose [5, 6]. In contrast to c-Si solar cells the performance of the PSCs is nearly independent on the light intensity drop. More of it the PSCs demonstrate high performance even under diffuse light conditions [3, 6].

Using different types of electron transport layers in PSCs makes it possible to adjust the photoelectrical and optical characteristics of the top solar cell and to optimize the performance of perovskite-silicon (PSC/c-Si) tandem system under varying illumination conditions [7, 8]. In previous works [9, 10] we demonstrated that very wide-bandgap nanostructured materials, like ZrO_2 and HfO_2 with E_g exceeding 5 eV could be successfully used as photoelectrodes in PSCs. The incorporation of mesoscopic ZrO_2 (or HfO_2) layer into the PSCs with the common architecture of glass/FTO/c-TiO_2/ZrO_2(or HfO_2)/$CH_3NH_3PbI_3$/Spiro-MeO-TAD/Au led to the improvement of the photovoltaic (PV) performance [9]. However, these PV devices cannot be used as top elements in tandem systems because of low optical transparency of Spiro-MeO-TAD layer and Au contacts in the visible range of solar spectrum. Alternatively PTAA or P3HT as a hole transporting material and PEDOT:PSS-based counter electrode can be used for fabrication semi-transparent PSCs [11, 12].

In this study high efficiency semi-transparent PSCs based on nanostructured ZrO_2 (or HfO_2) photoelectrodes were fabricated under ambient conditions and used as top elements in four-terminal (4-terminal) PSC/c-Si solar cells. The PV performance of PSCs, standalone c-Si and PSC/Si tandem solar cells was studied under varying illumination conditions. The results obtained demonstrate a new approach for developing high efficiency PV devices for real weather conditions.

2 EXPERIMENTAL

2.1 Materials and samples preparation

ZrO_2 and HfO_2 nanoparticles were obtained by hydrothermal treatment of co-precipitated zirconium and hafnium hydroxides from solutions of the corresponding metal salts [13]. TiO_2 nanoparticles (P25 Aeroxide Degussa) were purchased from Sigma-Aldrich (USA) and were used for fabrication of state-of-the-art TiO_2 photoelectrode [14]. Thick pastes from MO_2 (where M – Ti, Zr or Hf) nanopowders were prepared in organic solvent following the method described in [15]. The pastes were dissolved in ethanol in mass ratio 1:5 and spin-coated onto FTO conductive glass (Solaronix, 2×2 cm) covered by compact layer (c-TiO_2) with subsequent annealing at 500 °C for 30 min. Mesoscopic MO_2 photoelectrodes with the thickness of 180-220 nm were obtained.

Perovskite ($CH_3NH_3PbI_3$) layer was formed on MO_2 using a conventional one-step deposition method [16, 17]. PTAA with Li-TFSI and 4-tert-Butylpyridine additives in toluene was used as a hole transporting material. It was drop-casted onto the perovskite surface and then covered by counter electrode of PEDOT:PSS/FTO according to the procedure described elsewhere [11]. PTAA, PEDOT:PSS, Li-TFSI and 4-tert-Butylpyridine were purchased from Sigma Aldrich. All the steps of PSC fabrication process were provided under ambient conditions with a relative humidity about 40%.

The c-Si solar cells were kindly provided by the research group of G. Untila from Skobeltsyn Institute of Nuclear Physics, Moscow State University. All types of the fabricated PSCs based on mesoscopic MO_2 photoelectrodes were used as top elements in PSC/c-Si tandem solar cells and were mechanically stacked with bottom c-Si solar cells in 4-terminal configuration.

2.2 Characterization Studies

The measurements of PV parameters for PSCs fabricated, for c-Si solar cells and PSC/c-Si tandem solar cells were provided under standard illumination conditions of 1000 W/m^2 (AM1.5G) using Solar Simulator Abet 10500 (Abet Technologies, USA). Neutral optical filters

(Marumi, Japan) providing a uniform decrease of the light flux in the spectral range of 300-1100 nm were used to reduce gradually the illumination intensity from 1000 W/m^2 to 10 W/m^2. PSCs and c-Si solar cells were covered by masks limiting active area to 1 × 1 cm.

The current–voltage (*J–V*) characteristics were measured by Semiconductor Characterization System 4200-SCS (Keithley, USA). Monitoring of solar cell performance under outdoor conditions was carried out at the rooftop of IBCP RAS (Moscow, Russia). During the monitoring process, solar cells were operated at the maximum power point using STEVAL-ISV006V2 MPPT controller (STMicroelectronics), and the current and voltage values generated by the solar cells were recorded. The intensity of solar radiation was measured using the Kimo SL200 solarimeter.

3 RESULTS AND DISCUSSION

3.1 Performance under standard illumination (1000 W/m^2)

Semi-transparent PSCs based on mesoscopic MO_2 photoelectrodes were fabricated under ambient conditions. We prepared solar cell architecture of glass/FTO/c-TiO_2/MO_2/$CH_3NH_3PbI_3$/PTAA/PEDOT:PSS/FTO/glass. Comparative *J–V* curves for all samples recorded under standard illumination (1000 W/m^2, AM1.5G) are shown in Fig. 1a. The PV parameters of PSCs including short circuit current density (J_{SC}), open circuit voltage (V_{OC}), fill factor (*FF*) and power conversion efficiency (PCE) are listed in Table 1.

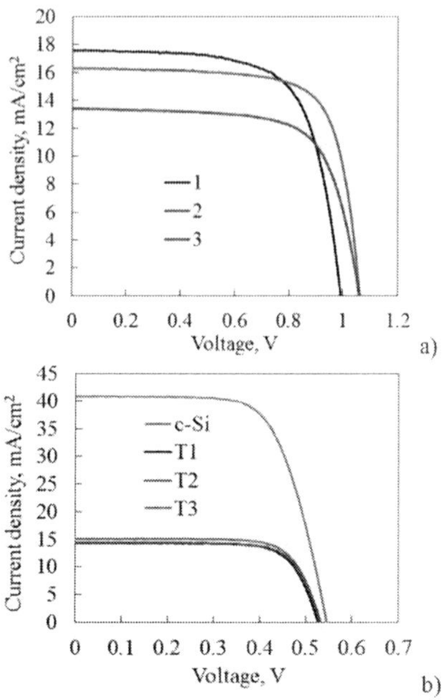

Figure 1: *J-V* curves for (a) PSC devices with mesoscopic MO_2 photoelectrodes (1 – TiO_2, 2 – ZrO_2, 3 – HfO_2) and for (b) bottom c-Si solar cell in tandem systems (T1 – TiO_2-based PSC/c-Si, T2 – ZrO_2-based PSC/c-Si, T3 – HfO_2-based PSC/c-Si)

PSC devices with state-of-the-art TiO_2 photoelectrodes demonstrate the average PCE of 11.94%. Substitution the TiO_2 layer with a very wide-bandgap material leads to the improvement of the V_{OC} value, which is attributed to the difference in the conduction band positions of TiO_2 and ZrO_2 (HfO_2) [9, 10]. However, a drop in J_{SC} values is found with incorporation of ZrO_2 and HfO_2 materials, that can be explained by their insulating nature [10].

Table I: The PV characteristics for PSC devices with mesoscopic MO_2 photoelectrodes (1 – TiO_2, 2 – ZrO_2, 3 – HfO_2), for c-Si solar cell and for bottom c-Si solar cells in tandem systems (T1 – TiO_2-based PSC/c-Si, T2 – ZrO_2-based PSC/c-Si, T3 – HfO_2-based PSC/c-Si)

	J_{SC}, mA/cm^2	V_{OC}, V	*FF*, a.u.	PCE, %
1	17.58±0.17	0.99±0.01	0.69±0.01	11.94±0.15
2	16.30±0.24	1.06±0.02	0.74±0.02	12.72±0.29
3	13.40±0.32	1.06±0.01	0.70±0.02	9.96±0.36
c-Si	40.79±0.23	0.55±0.02	0.68±0.02	15.07±0.11
c-Si in T1	14.32±0.14	0.53±0.01	0.74±0.01	5.61±0.15
c-Si in T2	15.10±0.18	0.54±0.02	0.74±0.01	5.97±0.22
c-Si in T3	14.36±0.19	0.54±0.01	0.74±0.02	5.71±0.22

The highest average performance of 12.72% is demonstrated for PSC samples with ZrO_2 photoelectrode which is 6.5% higher than for TiO_2-based devices (Fig. 1a and Table 1). The observed result can be attributed to the various charge transfer mechanisms realized in different photoelectodes. The charge transfer in TiO_2 layer occurs through the conduction band whereas in ZrO_2 photoelectrode the hopping conduction mechanism is dominated via localized states within band gap [10]. As a result, the recombination losses at the ZrO_2/perovskite interface are lower than for TiO_2/perovskite. The PCE of HfO_2-based PSCs is lower than for others due to lower J_{SC} values. This is more likely attributed to the insufficient layer morphology and less favorable positions of the electronic states in HfO_2 band gap [10].

The fabricated PSC devices based on mesoscopic MO_2 photoelectrodes were used as top elements in 4-terminal PSC/c-Si tandem solar cells. The *J–V* curves measured under standard illumination (1000 W/m^2, AM1.5G) and the PV parameters for standalone c-Si sample and for bottom c-Si elements are shown on Fig. 1b and in Table 1. The bottom c-Si cell performance is found to be nearly independent on the type of photoelectrode in the top PSC that is due to the same thickness of MO_2 photoelectrodes [18].

Comparative charts, illustrating the relative contributions of both top and bottom sub-cells into the resulting PCE of tandem PSC/c-Si solar cells are presented in Fig. 2. The overall conversion efficiency for the tandem systems under investigation is higher than that for standalone c-Si solar cell and strongly depends on the top cell PCE. The best performance under standard illumination conditions (1000 W/m^2, AM1.5G) was achieved for PSC based on mesoscopic ZrO_2 photoelecrode in combination with c-Si. It was found to be 18.69%, which exceeded the PCE of a standalone c-Si solar cell (15.07%) by ~ 24%. The PCE of TiO_2-based PSC/c-Si and HfO_2-based PSC/c-Si was 17.51% and 15.71%, respectively.

The obtained results reveal that PSCs based on very wide-bandgap materials can be successfully used as top elements in combination with the bottom c-Si cells and its allows to increase significantly the overall conversion efficiency of 4-terminal PSC/c-Si tandem systems. In addition, ZrO_2 can serve as an alternative material to the conventional TiO_2 in-photoelectrodes for PSCs.

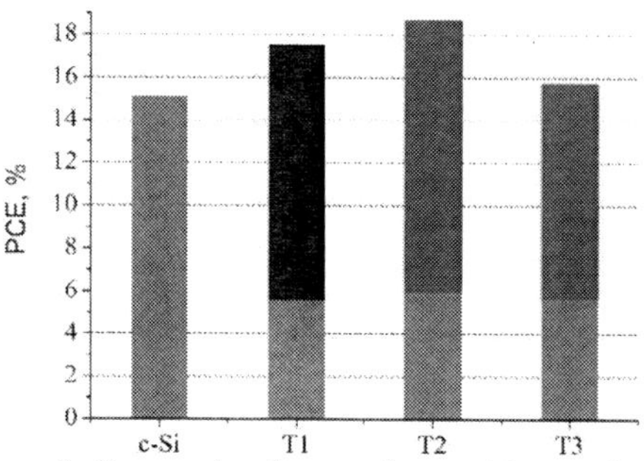

Figure 2: Comparative diagrams for standalone c-Si solar cell and for 4-terminal tandem systems (T1 – TiO_2-based PSC/c-Si, T2 – ZrO_2-based PSC/c-Si, T3 – HfO_2-based PSC/c-Si)

3.2 Performance under different irradiance levels (10-1000 W/m^2)

PV characteristics for PSCs based on mesoscopic MO_2 photoelectrodes were investigated under varying illumination conditions ranging from 10 to 1000 W/m^2. The PCE versus the light intensity plots are presented in Fig. 3a. The data obtained demonstrate the outstanding low light performance for PSCs fabricated.

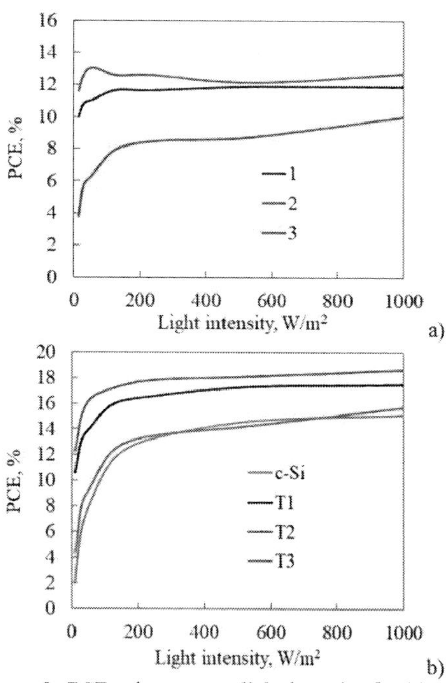

Figure 3: PCE values versus light intensity for (a) PSC devices with mesoscopic MO_2 photoelectrodes (1 – TiO_2, 2 – ZrO_2, 3 – HfO_2) and for (b) standalone c-Si solar cell and tandem systems (T1 – TiO_2-based PSC/c-Si, T2 – ZrO_2-based PSC/c-Si, T3 – HfO_2-based PSC/c-Si)

The efficiency of PV device with mesoscopic TiO_2 photoelectrode was found to slightly decrease (no more than ~ 20%) with the reduction of light intensity that correlated with literature values [6, 19]. At the same time, the PCE for PSC based on mesoscopic ZrO_2 layer declined by only 10% with light intensity drop from 1000 to 10 W/m^2. Moreover small increase (~ 3%) was observed in the range 10-100 W/m^2 with a maximum at 60 W/m^2 (PCE = 13.06%). In contrast, PCE of HfO_2-based PSC has been reduced by more than two times.

The obtained data confirm that due to the very wide ZrO_2 bandgap and the hopping mechanism of electron transfer over localized states within this zone the recombination losses at the ZrO_2/perovskite interface were minimised. In addition, the light scattering effect within the photoelectrode, the most important factor for low light performance, is more effective in mesoscopic ZrO_2 layer. It can be caused by more dense and uniform structure of ZrO_2 photoelectrode than for TiO_2 (or HfO_2) material.

Investigations of PV performance under varying illumination conditions (10-1000 W/m^2) revealed that the efficiency of 4-terminal PSC/c-Si tandem solar cells was significantly higher than for c-Si cell under low light conditions (Fig. 3b). PCE for PSC/c-Si tandem solar cell where PSC was based on mesoscopic ZrO_2 layer at 15 W/m^2 illumination was found to be 12.88% that was ~ 4 times higher than that for a standalone c-Si solar cell (3.17%).

3.3 Outdoor performance

PV performance of ZrO_2-based PSC, standalone and bottom c-Si solar cells, 4-terminal ZrO_2-based PSC/c-Si tandem solar cell was examined in outdoor conditions during sunny and cloudy days. The monitoring results obtained at the rooftop of IBCP are demonstrated in Fig. 4.

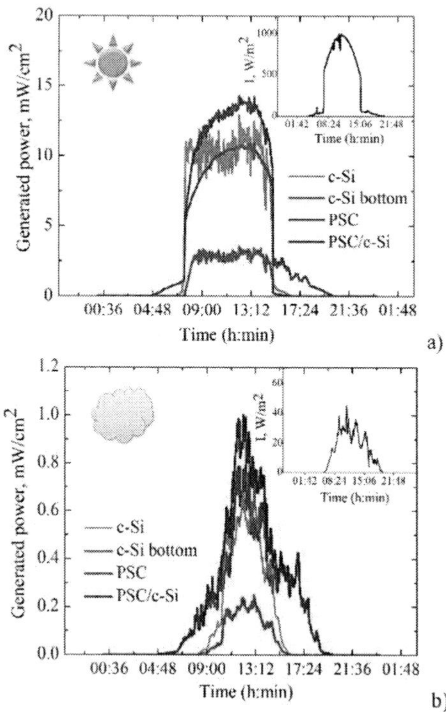

Figure 4: Outdoor performance monitoring during sunny (a) and cloudy (b) days for ZrO_2-based PSC, standalone and bottom c-Si solar cells, 4-terminal ZrO_2-based PSC/c-Si tandem solar cell (built-in graphs in the upper right corner show the daylight intensity (I) falling onto PV devices)

Fig. 4 shows that the generated power for ZrO_2-based PSC/c-Si tandem solar cell is higher than that for standalone c-Si solar cell regardless of weather conditions.

These results demonstate that PSCs with the developed photoelectrodes based on very wide-bandgap ZrO_2 material allow to increase significantly the overall conversion efficiency of tandem solar cells not only under standard illumination conditions but also at low light intensities. In addition, a new approach for developing high efficiency PV devices for real weather conditions is presented.

4 CONCLUSIONS

New high efficiency semi-transparent PSCs based on very-wide nanostructured materials were fabricated and successfully used as top elements in 4-terminal PSC/c-Si tandem solar cells. The best efficiency of 18.69% was obtained for ZrO_2-based PSC/c-Si tandem system under standard illumination (1000 W/m^2, AM1.5G) This result exceeds PCE for state-of-the-art TiO_2-based PSC/c-Si tandem solar cells by ~ 6.5% and the PCE of a standalone c-Si solar cell by ~ 24%. PCE of ZrO_2-based PSC/c-Si tandem solar cells under 15 W/m^2 illumination was found to be 12.88% that is ~ 4 times higher than that for a standalone c-Si solar cell. Outdoor measurements provided have demonstrated that that the generated power from ZrO_2-based PSC/c-Si tandem solar cell was higher than that obtained from a standalone c-Si solar cell regardless of weather conditions. The obtained results demonstrate that the ZrO_2-based photoelectrodes are promising for the manufacturing of high efficiency PSCs as top elements and for high performance PSC/c-Si tandem systems operating under different irradiance levels.

5 ACKNOWLEDGEMENTS

This work was supported by the Russian Foundation for Basic Research (grant no. 19-08-01042).

6 REFERENCES

[1] N.G. Park, Advanced Energy Materials 10 (2020) 1903106.
[2] L. Cojocaru, S. Uchida, K. Tamaki, P.V.V. Jayaweera, S. Kaneko, J. Nakazaki, T. Kubo, H. Segawa, Scientific Reports 7 (2017) 11790.
[3] A.B. Nikolskaia, S.S. Kozlov, M.F. Vildanova, O.I. Shevaleevskiy, Semiconductors 53 (2019) 550.
[4] T. Leijtens, K.A. Bush, R. Prasanna, M.D. McGehee, Nature Energy 3 (2018) 828.
[5] M.I.H. Ansari, A. Qurashi, M.K. Nazeeruddin, Journal of Photochemistry and Photobiology C: Photochemistry Reviews 35 (2018) 1.
[6] A.B. Nikolskaia, M.F. Vildanova, S.S. Kozlov, O.I. Shevaleevskiy, Russian Chemical Bulletin 69(2020) 1245.
[7] M.F.M. Noh, C.H. Teh, R. Daik, E.L. Lim, C.C. Yap, M.A. Ibrahim, N.A. Ludin, A.R. Yusoff, J. Jange, M.A.M. Teridi, Journal of Materails Chemistry 6 (2018) 682.
[8] K. Mahmood, S. Sarwar, M.T. Mehran, RSC Advances 7 (2017) 17044.
[9] L.L. Larina, O.V. Alexeeva, O.V. Almjasheva, V.V. Gusarov, S.S. Kozlov, A.B. Nikolskaia, M.F. Vildanova, O.I. Shevaleevskiy, Nanosystems: Physics, Chemistry, Mathematics 10 (2019) 70.
[10] M.F. Vildanova, A.B. Nikolskaia, S.S. Kozlov, O.I. Shevaleevskiy, Journal of Physics: Conference Series 1697 (2020) 012187.
[11] H. Heo, H.J. Han, M. Lee, M. Song, D.H. Kim, S.H. Im, Energy & Environment Science 8(2015) 2922.
[12] B. Chen, Y. Bai, Z. Yu, T. Li, X. Zheng, Q. Dong, L. Shen, M. Boccard, A. Gruverman, Z. Holman, J. Huang, Advanced Energy Materials 6 (2016) 1601128.
[13] M.F. Vildanova, A.B. Nikolskaia, S.S. Kozlov, O.K. Karyagina, L.L. Larina, O.I. Shevaleevskiy, O.V. Almjasheva, V.V. Gusarov, Doklady Physical Chemistry 484 (2019) 36.
[14] A. Nikolskaia, M. Vildanova, S. Kozlov, N. Tsvetkov, L. Larina, O. Shevaleevskiy, Sustainability 12 (2020) 788.
[15] S. Ito, P. Chen, P. Comte, M.K. Nazeeruddin, P. Liska, P. Péchy, M. Grätzel, Progress in Photovoltaics: Research and Applications 15 (2007) 603.
[16] A.B. Nikolskaia, S.S. Kozlov, M.F. Vildanova, O.I. Shevaleevskiy, Semiconductors 53 (2019) 540.
[17] N. Ahn, D.Y. Son, I.H. Jang, S.M. Kang, M. Choi, N.G. Park, Journal of the American Chemical Society 137 (2015) 8696.
[18] A.B. Nikolskaia, M.F. Vildanova, S.S. Kozlov, O.V. Almjasheva, V.V. Gusarov, O.I. Shevaleevskiy, Nanosystems: Physics, Chemistry, Mathematics 12 (2021) 246.
[19] I. Raifuku, Y. Ishikawa, S. Ito, Y. Uraoka, Journal of Physical Chemistry C 120 (2016) 18986.

ADVANCED LED SOLAR SIMULATOR: FLEXIBLE AND FAST CHARACTERISATION TOOL FOR RESEARCH AND INDUSTRIALISATION OF PEROVSKITE/SILICON TANDEM SOLAR CELLS

B. Mette[1], B. Mitchell[1], M.Scherff[1], P. Fuss-Kailuweit[1], L. Korte[2], E. Köhnen[2], S. Mariotti[2], S. Esefelder[1],T. Brammer[1]
[1]WAVELABS Solar Metrology Systems GmbH, Spinnereistr. 7, 04179 Leipzig, Germany
[2]Helmholtz-Zentrum Berlin für Materialien und Energie GmbH, Young Investigator Group Perovskite Tandem Solar Cells, Kekuléstr. 5, D-12489 Berlin, Germany

ABSTRACT: Advanced LED solar simulators feature flexible and enduring light sources for various applications. The need for current matching of the series connected perovskite and silicon cells presents optimization challenges and tradeoffs for maximum power output. A flexible sun simulator combined with suitable equipment to measure cell characteristics is necessary in order to allow cell-preconditioning (light soaking) and continuous tracking of the maximum power point (MPP) during the characterization process. In this paper, we will demonstrate a routine to presoak and track the tandem cell current as well as a fast method to detect the current matching based on a spectrally sensitive characterization method. Besides the individual setting of single LED channels, the creation of nearly any user definded spectra for e.g. different solar altitudes or any spectral albedo is shown and hence offers the research of multiple junction cells - especially perovskite technology - an uncomplicated and fast characterization tool. A 1cm² bipolar perovskite/silicon tandem solar cell fabricated at Helmholtz-Zentrum Berlin (HZB) was used for detailed investigations. A 4-step measurement cycle was recorded, starting with 15 min of light soaking, followed by an MPP tracking procedure to bring the cell to a stable working state for subsequent measurements and a 1-min forward and finally a 1-min reverse I-V curve. Hysteresis effects as well as time-dependent changes in the electrical behavior (degradation) of the cell were observed. Finally, spectrometric characterization was performed by varying certain wavelength ranges in the UV/VIS and NIR regions to determine the current matching point. Either due to the changing electrical properties of the cells over time or due to the layer structure not being spectrally matched to AM1.5g, no current matching point could be determined.

Keywords: LED solar simulator, tandem solar cells, characterization, light soaking, MPP tracking, spectrometric characterization

1 INTRODUCTION

1.1 PEROVSKITE/SILICON TANDEM SOLAR CELLS

Since their discovery in 2009, the quickly growing know-how on perovskite/silicon tandem solar (PST) cells enabled an unprecedented increase in power conversion efficiencies up top the current world record of 29.5% [1]. According to the European White Paper of 2019 [2] an efficiency potential for PST cells of 35% and for PST modules up to 30% (Perosvkite/c-Si) is estimated. But there are major challenges in measurement of PST cells. Light soaking - time dependent for the individual PST cell technology -, Pmpp tracking and especially the irradiated spectrum play a crucial role in order to determine the cells efficiency. [3]. This is to optimize material combinations and device designs for the purpose of maximizing e.g. the efficiency at standard test conditions (STC), or energy yield for specific sites and their specific environmental conditions. The SINUS sun simulator series by WAVELABS is designed to both preirradiate and characterize the extensive zoo of possible perovskite compositions and its properties when put into both single and especially dual junction devices. Unlike conventional Xenon based solar simulators, the SINUS-300 allows to set exposure times of any length, ranging from a few milliseconds to several hours.

Monolithic two terminal (2T) perovskite/silicon tandem solar cells (PST cells) undergo a steep improvement of their power conversion efficiency and durability. Due to the more complex device structure [4] compared to single junction devices, further detailed research and industrialization requires specialized metrology tools to investigate and control their intricacies accurately and conveniently. Current matching at short circuit (I_{sc}) and maximum power point (I_{mpp}) conditions can be determined by applying various irradiation spectra to keep one sub cell in saturation (light bias), while the other sub cell limits the current. Such analysis requires subsequent stepwise modification of the light spectrum to specifically manipulate current matching [3][4], where conventional methods are time-consuming and laborious.

In tandem structures, it is necessary to study not just the overall I-V curve, but to monitor the individual sub-cells and their influence on the overall cell characteristics. An overarching issue of 2T tandem cells is the need to limit losses by achieving current matching of series connected solar cells for the desired operation conditions. Despite of the series connection, the external quantum efficiency EQE (Figure 2) of the top perovskite and bottom silicon cell must be measured or at least known, to account for the spectra to irradiate with in the subsequent I-V measurement.

Figure 1: Measuremet setup of the 1cm² investigated Perovskite Tandem cell

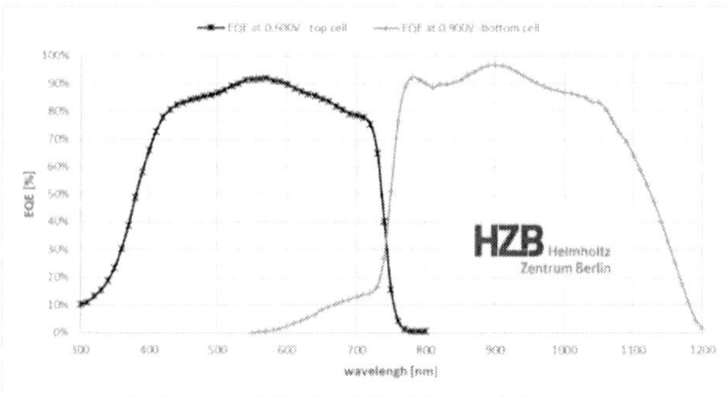

Figure 2: EQE of the top (black) and bottom PST cell (orange)

APPROACH

1.2 CHALLENGES

Even though a 2T tandem solar cell might be current matched at STC, there are many effects which will influence the I_{sc}- or I_{mpp}-matching during operation, such as

1. Shunt resistances: Different values in each sub cell will lead to a mismatch in the individual I_{mpp}, leading to a power loss during operation. Köhnen et al. [4] reported a method for determining the shunt resistance of both sub cells using a SINUS-70 Flasher from WAVELABS and generating several AM1.5g spectra with varying intensities of UV light.
2. Temperature coefficients: Aydin et al. [5] reported different temperature coefficients for both sub cells, which can lead to a current mismatch of about 5 % at field operating conditions, even if the tandem cell shows perfect current matching at STC conditions.
3. Cell degradation: The individual sub cells can degrade differently, which in turn affects current mismatch and cell performance. Therefore, it is important to monitor not only the standard I-V parameters, but also the current mismatch during degradation tests.

1.3 REQUIREMENTS

To understand the above-mentioned cell properties and to optimize the parameters in the manufacturing process, precise and reproducible measurement sequences are required. Those procedures should be flexibly adaptable to the respective measurement task. A sun simulator that allows controlled spectral variations and long irradiation times will enable a methodical search for optimal material combinations and device designs.

Selecting suitable spectral irradiation allows to maximize either STC efficiency or to optimize energy yield for specific sites and their specific environmental conditions. The SINUS simulator series by WAVELABS provides widely adjustable spectra ranging from STC to desert or many other local or regional spectra (Figure 3). This allows to characterize the full zoo of possible perovskite compositions and its properties when configured as both single and, in particular, dual junction devices.

Current matching behaviour at short circuit current (I_{sc}) and at MPP (I_{mpp}) conditions can be determined by measuring the cell while applying a light spectrum which is chosen to keep one sub cell in saturation, while the other sub cell limits the current. This can be achieved by increasing the spectral irradiation in a wavelength range which is only absorbed by the subcell supposed to be saturated.

State-of-the-art analysis such as spectrometric charaterization requires solving complex linear equation systems [6] or the step by step variation of intensity of selected wavelength ranges to manipulate the current matching as described in [4]. The principal approach described in [6] is used for more than 20 years to analyze multi-junction space- and concentrator solar cells or to compare 2 terminal and 4 terminal tandem cells. Advanced LED solar simulators, which feature flexible and enduring light sources, allow for preconditioning, tracking of the maximum power point and accurate I-V measurements of perovskite/silicon tandem solar cells. In the following chapters, we demonstrate a reproducible method for light soaking, MPP tracking and spectrometric charaterization based on a spectral adjustment procedure of the irradiation spectrum.

2 EXPERIMENTAL SETUP

The used LED-based solar simulator SINUS-300 by WAVELABS is a 21-color LED light source emitting across the UV/VIS/NIR part of the solar spectrum to reproduce the desired solar spectrum with great accuracy. Apart from replicating STC spectra, it is possible to reproduce nearly any spectrum within the supported wavelength range (Figure 3). This allows to simulate different terrestrial irradiation conditions like with NIR rich or NIR poor spectrum. In particular it is possible to the investigate tandem solar cell architectures with respect to current matching under different irradiation conditions using the spectrometric characterization method [6].

A lens system distributes the light from the narrow bandwidths LED channels homogeneously over the illuminated area of 210mm x 210mm. By means of an integrated spectrometer and multiple photodiodes, the emitted light is permanently recorded in millisecond intervals to ensure continuous self-correction and light stabilization during irradiationn. This guarantees a stable intensity and a deviation between target spectrum and simulated spectrum below 5% during the full irradiation period.

Unlike conventional Xenon solar simulators, the SINUS-300 allows to set exposure times of any length, from a few milliseconds to continuous illumination. Even at lower light intensities, the spectrum remains stable. This allows to record precise and highly reproducible I-V curves. All settings can be adjusted as needed for the experiment and the results are saved as simple ASCII file format (*.txt,*.csv etc).

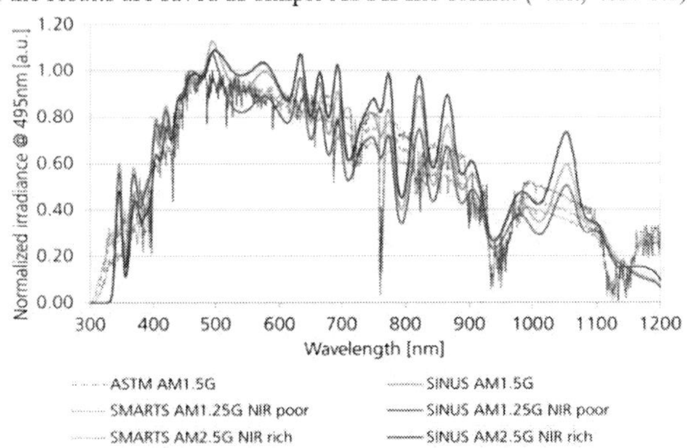

Figure 3: AM1.5g, NIR poor AM1.25g spectrum and a NIR rich AM2.5g spectra (dashed/dotted lines) and the corresponding SINUS-300 LED sun simulator spectra. Reference spectra taken from SMARTS [7].

3 MEASUREMENT AND RESULTS

Our at HZB fabricated test cells were placed on a temperature controlled vacuum chuck to ensure a reliable back side contacting and stable cell temperature. For the cells top-contact manipulator probes were used which were lowered carefully onto the contact area. A 3-point measurement configuration (separate source and sense contacts on the front side and one common contact for the rear side) was used, as is shown in Figure 1.

To achieve sufficient reproducibility, we did the following characterisation steps in the same chronological order, to be described in further detail:

1. Pre light soaking
2. MPP tracking (including initial full IV-sweep)
3. Slow I-V-Sweep
4. Spectrometric characterization

3.1 MPP TRACKING

As shown in [8], light soaking is required to achieve stable working conditions of the PST cell under irradiation. The time which is necessary for an optimal soaking depends on the individual cell and can vary from some milliseconds to several minutes. This necessitates the supply of a stable continuous reference light.

The SINUS-300 cell tester provides two tracking modes to monitor the maximum power point of the solar cell, "Mini-sweep" (a) and "P&O (Perturb and Observe)" (b). Both modes offer a comprehensive selection of configuration parameters for an optimally adapted measurement procedure. For this work, we have selected the mini-sweep method for MPP tracking.

a) Mini-sweep MPP tracking

Selecting the mini-sweep tracking method, a small voltage sweep is performed within each tracking step and the voltage at maximum power within this sweep is determined. The center of the mini-sweep for the next step is set to the point where the maximum power of the current step has been found.

b) Perturb and Observe (P&O) MPP tracking

Choosing the P&O method, the cell under test is first operated at a start voltage setpoint and the power is measured. After a settling time, another setpoint is applied (perturbation) and the power is measured (observation). The change to the next target setpoint depends on whether the power measured in the perturbation step is larger (keep going into the same direction) or smaller than in the previous step (change of direction for the new setpoint).

The MPP tracking is the most adequate technique to simulate the field operation of a cell or module.

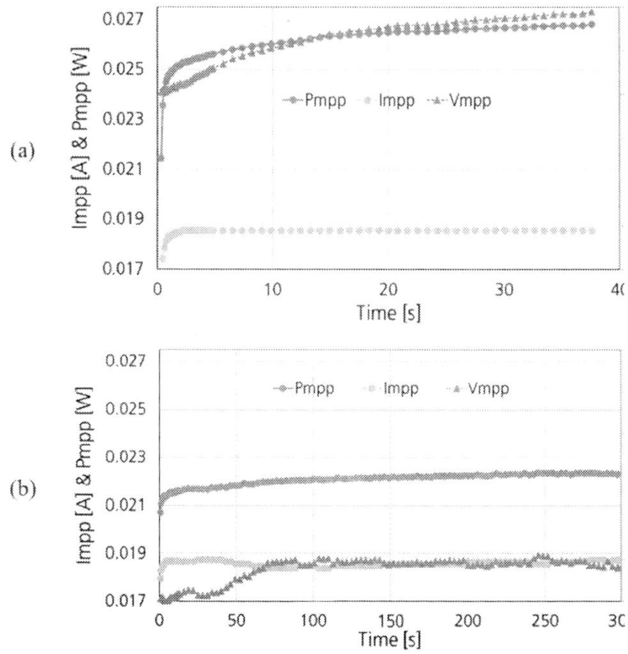

Figure 4: MPP tracing during a 40 second (a) and a 300 second (b) measurement. Current (I_{mpp}), voltage (V_{mpp}) and power (P_{mpp}) in the MPP were tracked using the minisweep method

Figure 4a) & b) show the MPP tracking measurements versus time using the mini-sweep-method. To learn the cells behaviour and demonstrate the solar cell dynamics, we did not perform previous light soaking in this case before this measurement. The MPP was tracked for 40 seconds (a) and 300 seconds (b). One can see a strong continuous power increase in this measurement phase during approximately the first measurement minute, which is mostly correlated with the increase of the MPP voltage (V_{mpp}) that is observed. Afterwards, the voltage stabilizes and only a slight power increase is observed. The latter correlates with the increase of the MPP current (I_{mpp}). The cause for this change of cell parameters is the dark current stabilization, which happens during the first minutes of the measurement and is the reason why we have chosen to perform the measurement cycles in the order shown above. After doing the light soaking, we have used the MPP tracking to verify, that our solar cell under test has stabilized sufficiently to perform the next characterization steps.

3.2 LIGHT SOAKING AND I-V-SWEEPING

After performing the light soaking and a 40s MPP-tracking, we have characterized the solar cells by doing current voltage sweeps on a temperature controlled measurement chuck. First, we have done one minute sweeps under STC in forward and reverse direction (Figure 4). Then we have repeated the same measurement sequence multiple times, always following the same cycle of 15 minutes light soaking, 40 seconds MPP tracking and finally forward and reverse sweep. During the whole measurement, the SINUS-300 tracked the irradiation conditions and ensured a constant spectrum and intensity. Figure 5 shows the characteristics of our cell at the end of a full measurement cycle. We observe a small hysteresis between forward and reverse sweep, which is not unexpected for PST solar cells.

The result of our repeatability measurements is shown in Figure 6. A total of four full measurement cycles of the same solar cell is shown and reveals a moderate variation of solar cell parameters. In particular, we observe a decrease of the cell's short circuit current I_{sc} with each subsequent measurement. In contrast, the open circuit voltage V_{oc} and the voltage at the maximum power point V_{MPP}, increase with each measurement. Except for the second measurement cycle, the Vmpp measurd is probably a measurement artefact. The other cell parameters have a maximum at the second measurement and decrease again afterwards. This is a hint to significant degradation of the top cell while the measurements were repeated. Despite of the cell's electrical parameters change over time, the hysteresis effect resulting out of the difference of forward in reverse I-V-curve is present for all four measurement cycles.

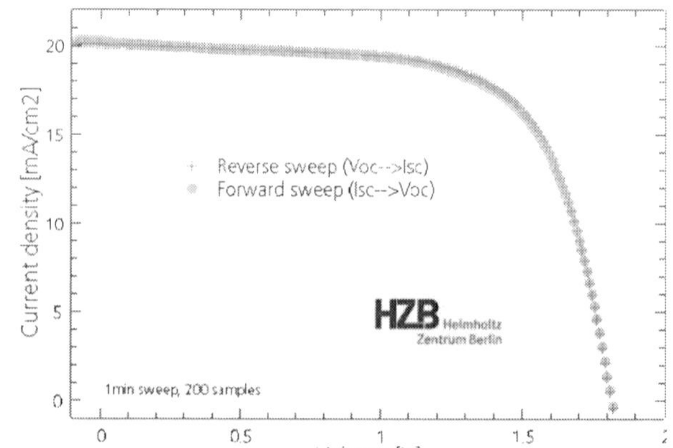

Figure 5: Current-voltage-sweep in forward and reverse direction after 15min soaking at STC with a total sweep time of 1min each

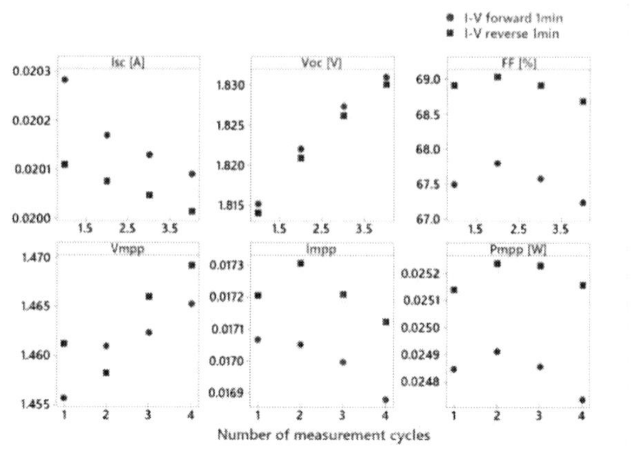

Figure 6: Repeatability of I-V-parameters with each cycle containing 15min soaking at STC, 1min forward, 1min reverse sweep. A total of 4 cycles is shown revealing some moderate variations in cell parameter

3.3 SPECTRAL CHARACTERIZATION

As a last step in our characterization cycles, we have performed a spectrometric characterization to determine the current mismatch. This method is well established for III/V multijunction solar cells [6]. A spectrometric characterization consists of a subsequent measurement of the solar cell characteristics for a series of spectral irradiation. The simulator spectrum is varied between a" blue-rich" spectrum - shifting the bottom solar cell into current limit - to a "red rich" spectrum - shifting the top cell to limit the overall current. Figure 7 illustrates the external quantum efficiency and the simulator spectra we have applied to determine the current match behaviour. The simulator spectra were derived from a standard AM1.5g spectrum. To generate the needed "red"- and "blue rich" spectra, we have changed the set power of individual LED channels located in the blue wavelength range (between 450nm and 530nm) – declared in Figure 7 as range "I" - and the red wavelength range (between 830nm and 915nm) –declared in Figure 7 as range "III". The spectra were adjusted to all provide the same overall irradiance, i.e. when increasing power of the "blue" LED channels, we decreased the power of the "red" LED channels by the same amount in order to keep the integrated total irradiance [W/m²] constant and vice versa. The modified LED channels were selected carefully to be located in wavelength ranges (range "I" & range "III" as shown in Figure 7), where the subcell not meant to be measured has a zero external quantum efficiency. Therefore the range "II" in Figure 7 is kept unchanged during the whole measurement series to avoid parasitic measurement signals originating from the current limiting solar cell.

$$\text{UV/VIS(I) x\%} \uparrow + \text{NIR(III) x\%} \downarrow = \text{Total irradiance [W/m}^2\text{]}$$

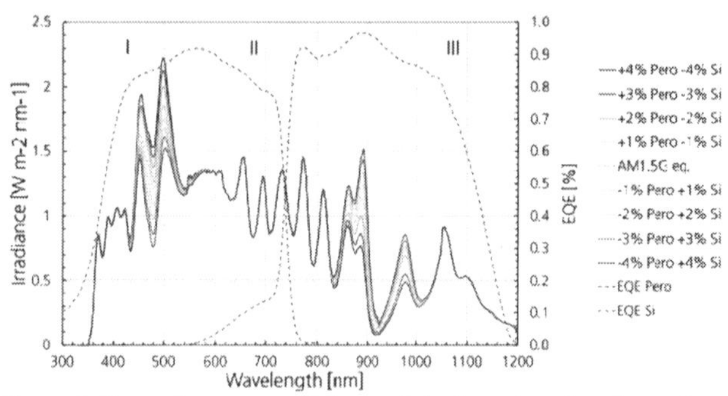

Figure 7: External quantum efficiency of the investigated tandem solar cell (dashed lines) and the different simulator spectra which have been applied to vary the spectral metric. All spectra are designed to have the same total irradiation power. Spectral variations were restricted to regions I and III, where only one of the two subcells has a non zero quantum efficiency.

A spectrometric characterization of our solar cell under test is shown in Figure 8. Short circuit current (I_{sc}), open circuit voltage (V_{oc}), fill factor (FF) and maximum power point (P_{mpp}) are plotted versus spectral deviation of the simulator spectra versus the AM1.5g spectrum. This deviation is derived from the sum of the deviations of the spectral irradiance in the wavelength regions I and III (Figure 7) from the AM1.5g spectrum. Smaller values indicate a blue-rich spectrum (which should bring the bottom cell into current limitation) while larger values indicate a red-rich spectrum (which should make the top cell limit the current). We observe, that the I_{sc} decreases continuously from "blue-rich" to" red-rich" spectra, while the V_{oc} and the maximum power have a maximum close to the original AM1.5g spectrum ~ 0.98). The fill factor decreases continuously from step 1 to step 5 and increases slowly from step 6 to step 9.

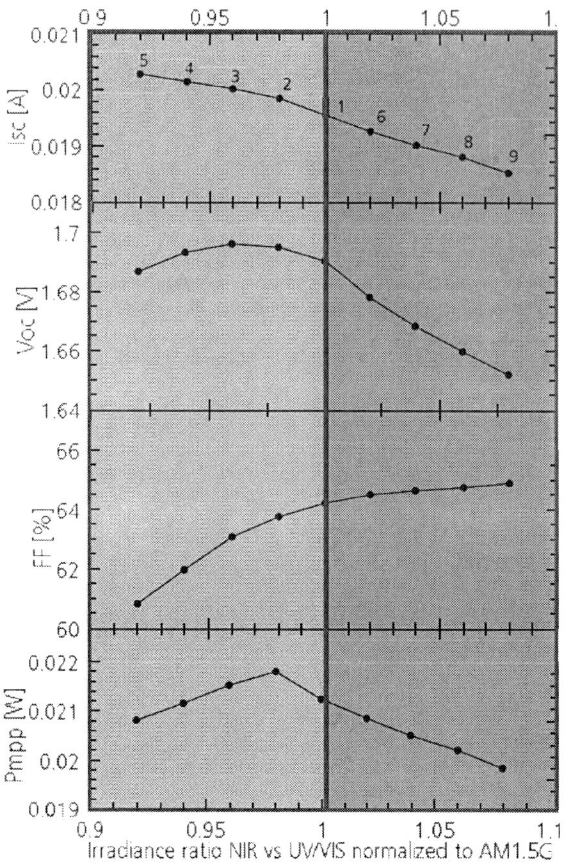

Figure 8: Results of the spectrometric characterization of one measurement cycle. $I_{sc}V_{oc}$, FF and Pmpp are plotted versus the ratio of used irradiance spectrum and the AM1.5g spectrum. Small spectra quotients indicate a "blue-rich" spectrum, while large values indicate a "red-rich "spectrum.

4 DISCUSSION

4.1 SLOW I-V SWEEP:

P_{mpp} tracking combined with a light soak step found to be a suitable method in order to find out the cells stable operating point. These features offer an optimal opportunity for perovskite researches to measure the I-V at the exact right time without complicated separate procedures including the risk of cell degradation between presoak and measurement.

4.2 PRESOAK & PMPP TRACKING:

Figure 6 summarizes the results for alternating sweeps in forward and reverse direction. A small hysteresis was observed which is well known for some type of PST solar cells. Depending on the I-V scan direction (U_{oc}→I_{sc} or I_{sc}→U_{oc}), different I-V characteristics are obtained during the sweep and correspondingly different MPPs are detected. For this type of PST cell, we learned afterwards that the P&O method described in 3.1.b) would have been a better choice in order to avoid hysteresis effects. In the course of further measurements, the cell also showed strong degradation over time. Due to its strong hysteresis, strongly varying cell parameters are measured despite of very slow I-V sweeps of one minute measurement duration. Apart from the hysteresis, degradation already started after the second measurement cycle of 15min presoak and 1min measurement (forward & reverse). The same trend of decreasing FF and J_{SC} was also observed on perovskite single junction cells at in-house measurements.

4.3 SPECTROMETRIC CHARACTERIZATION:

In the spectrometric characterization shown in Figure 8, the irradiance of the spectral ranges is varied such that the"red-rich" NIR range is increased in the same amount [W/m^2] as the irradiation of the "blue-rich" UV/VIS range decreases. For this measurement sequence no maximum in Isc is observed, but a continuous increase which points toward a current limitation of the tandem cell by the perovskite layer thoughout the whole measurement. Two effects could affect the results here:

1) either the investigated cell structure was not matched to the AM1.5g spectrum yet or
2) not anymore, e.g., due to degradation of the Perovskite layer

The measurements were carried out in the step order shown in Figure 8. The open circuit voltage V_{oc} on stable tandem structures is usually observed to be constant throughout the whole spectral variations [6]. Showing small changes from step 1 to step 5 of about 5mV in V_{oc}, a constant voltage drop of ≈ 40mV is meausured from step 6 to step 9, so the spectral variation is probably superposed by the degradation of the top layer. This influences as well the asymmetric characteristics of the fill factor FF not exhibiting any minimum over the varied spectrum as usually observerd on spectrally matched and temporarily stable tandem structures [6].

5 CONCLUSIONS

Using SINUS 300 for characterizations of 2-terminal perowskite/silicon tandem solar cells, we have demonstrated that (i) presoak, (ii) hysteresis measurement parameters and (iii) spectrometric characterisation must be very well adjusted to the specifics of the cell under test. If the presoak time is not long enough, the cell is not activated and if it is too long, the danger of degradation occurs. With SINUS simulator series by WAVELABS necessary measurement sequences can be carried out fast and automated through the SINUS GUI. The user is able to investigate the cells behaviour relevantly, dependably and with ease. We demonstrated modern LED solar simulators are convenient and powerful tools to obtain output characteristics of perovskite silicon tandem solar cells. The ability to perform light soaking, maximum power point tracking and subsequent spectral characterization in few steps helps to systematically investigate effects such as cell degradation and current matching conditions.

ACKNOWLEDGEMENTS

The authors acknowledge R. Holle for technical support and V. Gutewort for software upgrade as well as the financial support by the Federal Ministry for Economic Affairs and Energy of Germany in the project P3T (FKZ 03EE1017E (WAVELABS) & FKZ 03EE1017A (HZB)).

REFERENCES

[1] M. A. Green, E. D. Dunlop, J. Hohl-Ebinger, M. Yoshita, N. Kopidakis, and X. Hao, "Solar cell efficiency tables (Version 58)," *Prog. Photovoltaics Res. Appl.*, vol. 29, no. 7, pp. 657–667, 2021.

[2] EPKI, "Perovskite-based photovoltaics : A unique chance for European PV-industry," *White Pap. from EPKI*, no. September, 2019.

[3] R. Adelhelm and K. Bücher, "Performance and parameter analysis of tandem solar cells using measurements at multiple spectral conditions," *Sol. Energy Mater. Sol. Cells*, vol. 50, no. 1–4, pp. 185–195, 1998.

[4] E. Köhnen *et al.*, "Highly efficient monolithic perovskite silicon tandem solar cells: Analyzing the influence of current mismatch on device performance," *Sustain. Energy Fuels*, vol. 3, no. 8, pp. 1995–2005, 2019.

[5] E. Aydin *et al.*, "Interplay between temperature and bandgap energies on the outdoor performance of perovskite/silicon tandem solar cells," *Nat. Energy*, vol. 5, no. 11, pp. 851–859, 2020.

[6] M. Meusel, R. Adelhelm, F. Dimroth, A. W. Bett, and W. Warta, "Spectral mismatch correction and spectrometric characterization of monolithic III-V multi-junction solar cells," *Prog. Photovoltaics Res. Appl.*, vol. 10, no. 4, pp. 243–255, 2002.

[7] Christian Gueymard, "SMARTS version 2.9.5: Simple Model of the Atmospheric Radiative Transfer of Sunshine." NREL, 2005.

[8] A. Al-Ashouri *et al.*, "Monolithic perovskite/silicon tandem solar cell with >29% efficiency by enhanced hole extraction," *Science (80-.).*, vol. 370, no. 6522, pp. 1300–1309, Dec. 2020.

THE IMPACT OF HIGH SPECTRAL MATCH: USING AN LED SOLAR SIMULATOR FOR TANDEMS

Blagovest Mihaylov, Benjamin C. Duck, Christopher J. Fell, Timothy W. Jones, Kenrick F. Anderson and Gregory J. Wilson
Commonwealth Scientific and Industrial Research Organisation (CSIRO) Energy
10 Murray Dwyer Circuit, Mayfield West, NSW, 2304, Australia

ABSTRACT: Spectral responsivity (SR) measurements on stable multi-junction (tandem) solar cells are never trivial, however the procedure is well-established and methods exist to correct for measurement artefacts. Perovskite/Silicon (PSK/Si) devices present a challenge however, since they typically exhibit metastability and degradation, even during the measurement. This complicates SR measurements, leading to uncertainty in the setup of the solar simulator for the primary current-voltage (I–V) measurement. We demonstrate a way to optimize the spectrum of a LED-based solar simulator so that both the spectral mismatch (SMM) and its sensitivity to the spectrum are small. The method leverages the inherent flexibility in the multichannel light source and combines it with known spectral responsivity data from a range of similar devices. The impact is that tandem solar cells can be measured accurately even with no prior information about their spectral responsivity. We estimate for a PSK/Si device the additional uncertainty in measured efficiency can be as low as 0.32%. This significantly reduces the exposure of the devices to light and voltage bias, thus minimizing metastability and degradation prior to the most important part of the measurement.
Keywords: Solar Cell Efficiencies, Spectral Response, LED, Perovskite, Tandem

1 INTRODUCTION

The race to develop market-ready perovskite (PSK) solar cells, including perovskite-silicon tandem (PSK/Si) cells, is impacted by the difficulty in securing accurate, reproducible and representative measurements of the cell efficiency [1,2]. Despite these difficulties, progress in the field has been rapid, with the latest International Technology Roadmap for Photovoltaic estimating that mass production of PSK/Si tandems will commence as early as 2023 and may grow to ~5% of market share by 2031 [3]. Formal measurements of photovoltaic efficiency have always required determination of the device spectral responsivity (SR), in order to account for spectral mismatch between the laboratory light source and the standard reference spectrum. For perovskite-based photovoltaics however, the prolonged exposure to light and voltage during SR measurements leads to higher uncertainties in measured efficiencies [4]. This is a highly undesirable situation in a field where incremental improvements to world-record efficiencies are given great significance.

In tandem devices, the measurement problem is magnified since SR measurements must be done prior to the efficiency measurements in order to set up the solar simulator correctly [5]. Measurement standards require that the spectral mismatch for each junction in the tandem cell falls below a predefined and reportable threshold before the measurement can be considered valid. It is important to realize that simply achieving this level does not constitute an ideal measurement. Any residual spectral mismatch increases the measurement uncertainty and therefore reduces the confidence in the reported value. In the case of a PSK junction, consideration should be given to the metastable junction response to applied light and voltage bias during the SR measurement (Fig. 1). Differences between the measured SR data and the device response under operating conditions will further reduce the efficiency measurement accuracy.

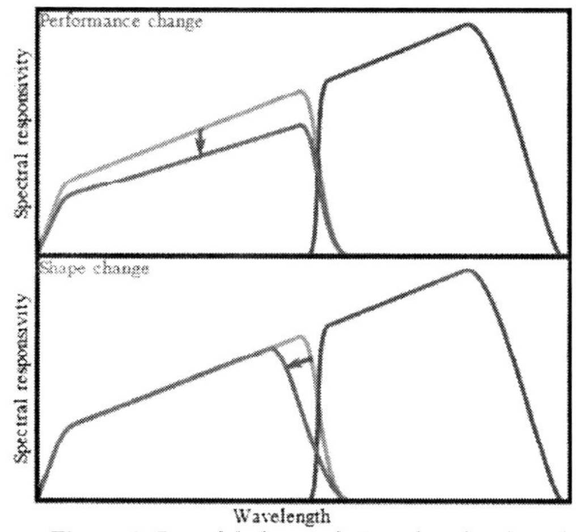

Figure 1: Potential changes in PSK junction (green) spectral responsivity curves relative to Si junction (blue) that may be observed with light and voltage bias exposure.

Tandem SR evaluation is also much more complex and time-consuming, which exacerbates the likelihood of detrimental SR deviations. Particular challenges for SR in PSK/Si tandem cells include: (1) multiple measurements are often required to establish the linearity of SR with bias light, (2) additional testing may be required to ensure measurement artefacts such as luminescence coupling and shunt effects are minimized and corrected for, (3) SR measurements may still be relative, but all junctions must scale equally to establish the correct current ratio between junctions, and (4) the SR is possibly a function of voltage bias [6]. The additional exposure time spent under non-operating bias conditions will drive the device further towards a different operating state.

These issues considerably increase the complicated impact spectral mismatch between the light source and the standard reference spectrum has on efforts to accuratly determine PV cell efficiency for metastable perovskite-based devices, particularly multi-junction configurations [4]. The problems can be avoided however if a laboratory light source could be developed with an ultra-high match to the reference spectrum. With this in mind, CSIRO has developed a prototype LED-based solar simulator using 40 different LED colors, in an attempt to achieve the closest match to the AM1.5G reference spectrum possible with current technology. In this work, we demonstrate how close the prototype is to an ideal reproduction of AM1.5G, and discuss the implications for accurate measurement of tandem solar cells, including perovskites.

2 EXPERIMENTAL METHOD

2.1 Novel LED solar simulator

We have developed a prototype LED-based solar simulator (LEDSS) using 40 different colors, in an attempt to achieve the closest match to the AM1.5G reference spectrum possible with current technology. Each channel is independently controlled and has been characterised at the operating plane using a high-end spectroradiometer set (Instrument Systems CAS 140D and 140CT). These measurements are performed under a full operation equivalent thermal load for the system and at a range of operating powers for the LED channel under test. These measurements form the base inputs for the optimization model discussed later.

The 40 LED types chosen consist of commercially available nominal peak wavelengths spanning the range from 340 nm to 1300 nm. The operation of each has been confirmed to meet the stability requirements of IEC 60904-9 ED3 [7]. Each channel is individually addressable, and its output can be controlled across the entire operational range of the LED type. This allows the flexibility to change the spectrum to attempt matching any arbitrary desired spectral shape.

2.2 Solar simulator output optimization

The IEC 60904-9 ED3 [7] standard describes the classification of solar simulators. It describes the methodology for measuring spectral irradiance, calculating the classification based on matching total irradiance and maximum deviation in a set of wavelength bands each approximately corresponding to 16.6 % of total irradiance. Since these can be met with six discrete LED types, the standard introduces additional metrics for *spectral coverage* (SPC) and *spectral deviation* (SPD) . SPC highlights gaps in the spectrum; SPD is the sum of the absolute differences from 300 to 1200 nm. Only the band classification is used in the measurement and reporting on PV efficiency, however both SPC and SPD are informative metrics for the quality of the light source used. In particular, since the light source we have developed uses 40 different LED types achieving a classification of A+ is trivial, and only the SPC and SPD metrics have any true comparative value.

The flexibility in the light source output due to the large number of LED types introduces an optimization problem for how best to configure the output of each channel. We have approached this problem using a number of different criteria as follows.

The simplest approach to directly optimize the solar simulator spectrum is to match the AM1.5G in the wavelength range of interest, i.e. from 300 to 1200 nm (*LSQ*). For n LEDs:

$$\sum_{i=1}^{n} E_{Li}(\lambda_{300})x_i = E_{Ref}(\lambda_{300})$$

$$\vdots$$

$$\sum_{i=1}^{n} E_{Li}(\lambda_{1200})x_i = E_{Ref}(\lambda_{1200}) \quad \ldots(1)$$

where $E_{Li}(\lambda)$ is the spectral irradiance of each LED and $E_{Ref}(\lambda)$ is the spectral irradiance of the reference spectrum. x_n is the contribution of each LED channel. If we write it in a matrix form $Ax=b$ and regularize the data to integer wavelengths, A is a 900 by n matrix with the spectral irradiance of each LED type, x is the solution vector with the contribution of each LED type and b is the AM1.5G spectrum. The overdetermined linear system of equations in (1) cannot be solved exactly, but can be approximated with ordinary least squares, i.e. we minimize the sum of squares of Ax-b. This is a stronger metric than the SPD since larger deviations between the simulator and reference spectrum are given greater weight, i.e. sum of squares vs absolute difference.

It is important to note that this solution would not necessarily integrate to the same total irradiance as the standard spectrum. As such we can add an extra row to A (multiplied by a weighting factor) and an element to b that corresponds to the sum of all elements in each column, effectively ensuring eq. (2) below holds (*TotalE*):

$$\sum_{i=1}^{n} x_i \int E_{Li}(\lambda)d\lambda = \int E_{Ref}(\lambda)d\lambda \quad \ldots(2)$$

Similarly, we can add rows that ensure the irradiance integrates to the same number for any wavelength range by summing the appropriate rows in A and b. We do this for the wavelength ranges defined in IEC 60904-9 ED3 (*6rangeE*).

Finally, we consider the effective junction current produced by potential candidates for the device under test (DUT) top and bottom junction (*SRs*). These are calculated from a pool of 51 SR curves obtained from digitizing curves found in the literature (Fig. 2). The current each produces under the solar simulator spectrum is matched to the value produced under the AM1.5G reference as shown in eq (3). Each SR dataset will add an additional row to A and corresponding element to b, i.e. an additional 51 rows in this instance.

$$\sum_{i=1}^{n} x_i \int E_{Li}(\lambda)\cdot SR_{DUT}(\lambda)d\lambda = \int E_{Ref}(\lambda)\cdot SR_{DUT}(\lambda)d\lambda \quad \ldots(3)$$

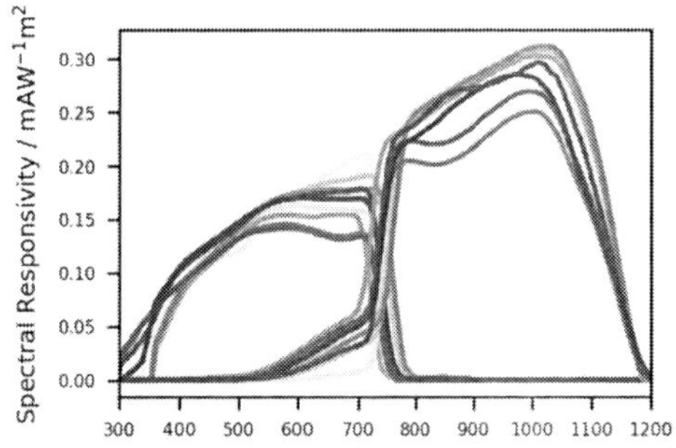

Figure 2: Pool of 51 spectral responsivity curves from Perovskite on Silicon tandem devices.

2.2 Assessment of spectral mismatch uncertainty

The maximum junction mismatch requirements imposed by PV measurement standards (e.g. [6]) imply that a tunable solar simulator will be required, however care needs to be taken when assessing the impact of the light source on the measured result. For example, with a double junction device, a dual light source is sufficient, and there is a unique solution that ensures the current ratio is correct between the two junctions as described in [8]. The concern of ensuring the reference cell (RC) used generates its calibrated value can be addressed by adding another source. Therefore, with three correctly chosen LEDs one can have negligible spectral mismatch (SMM) and ensure junction current balances are maintained under the test illumination. However, the resulting measurement uncertainty would be large and the light source output would be unlikely to meet the required classification standard.

The fact that a small SMM does not mean small overall uncertainty (u(SMM)) was highlighted in [9]. In particular, there will be a large sensitivity to any errors in either the SR measurements or the spectral irradiance of each LED. As the number of LED channels is increased and the light source spectrum approaches a perfect match to the AM1.5G solar spectrum, the SMM would be 1 and its uncertainty should be negligible. Calculating the exact value of u(SMM) is complex, requiring wavelength-dependent uncertainties alongside their correlations for both SR and spectral irradiance measurements [10]. In this work, we assess the sensitivity of the SMM to the LEDSS spectrum using a statistical approach as described in Annex A of IEC 60904-9 ED3. We calculate a large set of SMM values for each of the optimized spectra above using SR data from one pool of 10 RC (Fig. 3) together with the pool of 51 DUTs shown earlier. We have not limited the calculations to matched devices; this is to highlight worse case scenarios and achieve a conservative estimate of u(SMM). Due to the large number of SMMs calculated the u(SMM) estimate approaches the standard deviation of the population as expected.

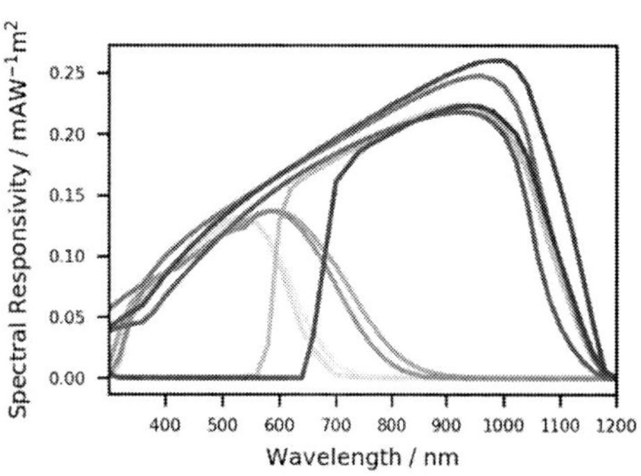

Figure 3: Pool of 10 spectral responsivity curves of available reference cells.

3 RESULTS AND DISCUSSION

The flexibility afforded with the LEDSS allows us to achieve an excellent match between solar simulator output and the AM1.5G reference spectrum. In Fig. 4, we compare the output of the LEDSS against examples of a traditional Xe lamp and an LED solar simulator that uses only 15 color channels. In all cases, the simulators are spectral Class A as defined by IEC 60904-9. Table I shows key comparison metrics for the three sources as described above.

The SPD metric demonstrates the improvement achieved by the LEDSS with a value of more than half of either of the other two sources. While both the 15-channel and LEDSS (40-channel) have excellent average mismatch values, only the LEDSS can achieve a low mismatch uncertainty value.

Table I: Comparison metrics for solar simulator types

Type	SPD	SMM mean	u(SMM)
Xe Lamp	26.39	0.9905	0.86%
15 Channel	19.55	0.9987	1.08%
LEDSS	9.75	1.0036	0.36%

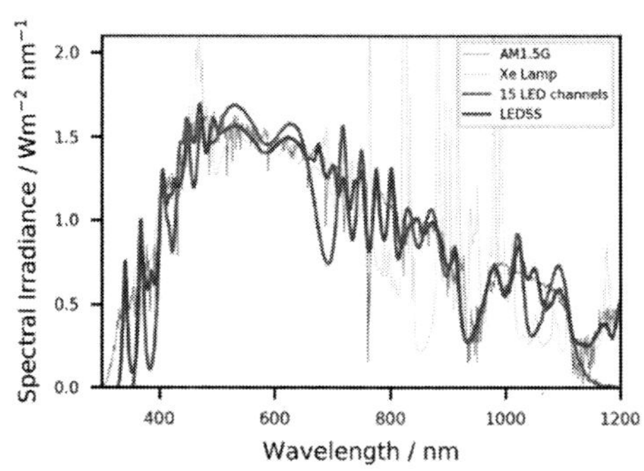

Figure 4: Spectral output of solar simulators. The LEDSS (blue) provides an excellent match to the AM1.5G, particularly compared to traditional Xe lamp output (orange). A simpler 15-channel system would be able to achieve Class A status, but with an inferior spectral match

While the value of u(SMM) for the LEDSS shown in Table I is already reasonable, there exists a potential to lower it further by leveraging the flexibility of the 40 independent channels in the LEDSS. We have explored this space using combinations of the different optimization criteria described in 2.2, above. Fig 5 shows the resulting variation in spectrum when these optimization targets are used.

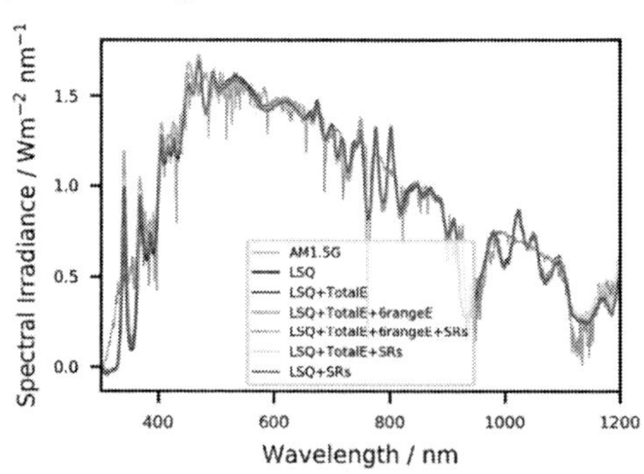

Figure 5: LEDSS output with different combinations of optimization targets.

In all cases, the different optimizations cause shifts to occur to compensate for the gaps in the spectrum. This effect is most clearly observed at shorter wavelengths due to gaps in the LEDSS spectra below 330 nm and at 350 nm. While high power LEDs at 320 nm exist, these would have required that all optical components are made of fused silica. This would have been prohibitively expensive for a negligible contribution to the total irradiance. In contrast, we could not currently identify any high-power LEDs at 350 nm on the market. Manufacturers have indicated that these wavelengths are possible, but currently there is insufficient demand for them. Use of either total irradiance (*TotalE*) or the banded irradiance (*6rangeE*) optimization targets induces a relatively large impact, particularly at these lower wavelengths. The use of the spectral responsivity derived target (*SRs*) serves to dampen any changes induced by the irradiance targets and to place additional focus of the spectral shifts on the region near the Perovskite absorption edge. In all cases, the shifts are subtle; however, the impact on measurement uncertainty can be profound, as shown below.

In Fig. 6 we display the mismatch values calculated by combining the DUT and RC SR pools with each of the optimized spectra. This data is also summarized in Table II. We calculate the additional uncertainty impact of directly using each of these simulator outputs on a new device without knowledge of its exact spectral responsivity curve (u(nSR)). This is achieved by considering the value of u(SMM) as well as the deviation of the calculated mean SMM value from the ideal value of 1. Fig. 7 shows a histogram of the spread of calculated mismatch values for the LSQ+SRs optimization case.

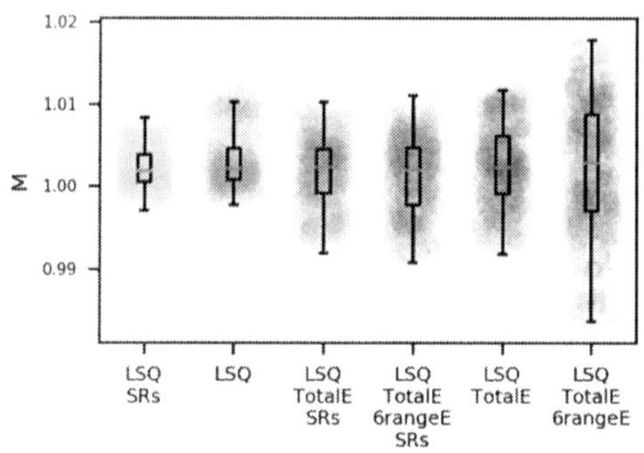

Figure 6: Mismatch values calculated using the pool of DUT and RC SR data for different LEDSS optimization approaches.

Table II: Comparison metrics for optimization approaches. The additional uncertainty from measuring without SR correction is evaluated (u(nSR)).

Optimization	SMM mean	u(SMM)	u(nSR)
LSQ+SRs	1.0022	0.23%	0.32%
LSQ	1.0032	0.33%	0.46%
LSQ+TotalE+SRs	1.0018	0.39%	0.42%
LSQ+TotalE+6rangeE+ SRs	1.0013	0.44%	0.46%
LSQ+TotalE	1.0025	0.48%	0.54%
LSQ+TotalE+6rangeE	1.0025	0.76%	0.80%

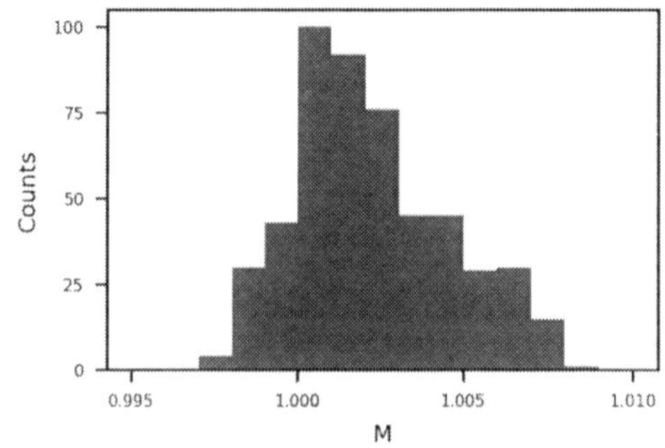

Figure 7: Histogram of mismatch values for the LSQ+SRs optimization case.

Inclusion of the spectral responsivity data into the optimization shows a significant benefit with a reduction of ~30% in additional uncertainty compared to the base case. We estimate that the additional uncertainty in the measured performance of a PSK/Si tandem when no additional SR correction is performed would be 0.32%. Note that the uncertainty in the relative spectral irradiance measurments are included separately in the overall measurement uncertainty. For many laboratories, this value will be dwarfed by other considerations in their uncertainty budget. Importantly it should be recognized that this value is also much lower than the deviation that some authors have observed in SR measurements on metastable devices [4].

4 CONCLUSIONS

We have demonstrated the potential for highly configurable LED-based solar simulators to perform accurate solar PV efficiency measurements in the absence of device-specific spectral responsivity measurement data. We demonstrate the excellent spectral match achieved with our newly prototyped 40-channel LED solar simulator and have shown that its flexibility is key to lowering measurement uncertainties. Direct efficiency measurements of PSK/Si tandem devices without prior spectral responsivity measurement is estimated to lead to only an additional 0.32% uncertainty. Performing measurements in this way avoids any issues around metastability or degradation during the spectral responsivity characterization step that may lead to a worse overall result and greater measurement uncertainty.

5 ACKNOWLEDGEMENTS

This Project was conducted in partnership with Lastek Pty Ltd and received funding from ARENA as part of ARENA's Advancing Renewables Program.

6 REFERENCES

[1] T. Song, D. J. Friedman, and N. Kopidakis, "Comprehensive Performance Calibration Guidance for Perovskites and Other Emerging Solar Cells," Adv. Energy Mater., vol. 11, no. 23, pp. 1–16, 2021.
[2] H. J. Snaith et al., "Anomalous Hysteresis in

Perovskite Solar Cells," no. iii, 2014
[3] VDMA, "International Technology Roadmap for Photovoltaic (ITRPV)", 12 Edition, 2021
[4] M. Bliss, A. Smith et al, "Spectral Response Measurements of Perovskite Solar Cells", IEEE J. Photovoltaics, vol. 9, pp. 220–226, 2019.
[5] IEC 60904-1-1:ED1, "Photovoltaic devices – Part 1-1: Measurement of current-voltage characteristics of multi-junction photovoltaic (PV) devices." 2017
[6] IEC 60904-8-1:ED1, "Photovoltaic devices – Part 8-1: Measurement of spectral responsivity of multi-junction photovoltaic (PV) devices." 2017
[7] IEC 60904-9:ED3, "Photovoltaic devices – Part 9: Classification of solar simulator characteristics." 2020.
[8] M. Meusel, R. Adelhelm, F. Dimroth, A. W. Bett, and W. Warta, "Spectral mismatch correction and spectrometric characterization of monolithic III-V multi-junction solar cells," PIP: Res. Appl., vol. 10, no. 4, pp. 243–255, 2002.
[9] J. Hohl-Ebinger and W. Warta, "Uncertainty of the spectral mismatch correction factor in STC measurements on photovoltaic devices," PIP: Res. Appl., vol. 19, pp. 573–579, 2011
[10] S. K. Reichmuth, G. Siefer, M. Schachtner, M. Muhleis, J. Hohl-Ebinger, and S. W. Glunz, "Measurement Uncertainties in I-V Calibration of Multi-junction Solar Cells for Different Solar Simulators and Reference Devices," IEEE J. Photovoltaics, vol. 10, no. 4, pp. 1076–1083, 2020.

TCO OPTIMIZATION OF C-SI HETEROJUNCTION SOLAR CELLS FOR TANDEM ARCHITECTURE BY OPTICAL SIMULATION

Marica Canino[1], Virginia Boldrini[1], Rita Rizzoli[1], Emanuele Centurioni[1], Alberto Maurizi[1], Salvatore Lombardo[2], Antonio Terrasi[3,4], Caterina Summonte[1]
[1]Institute for Microelectronics and Microsystems (CNR-IMM Bologna), Via Gobetti 101, 40129 Bologna, Italy
[2]Institute for Microelectronics and Microsystems (CNR-IMM Catania), Strada VIII, 5, 95121 Catania, Italy
[3]University of Catania, Dipartimento di Fisica e Astronomia "Ettore Majorana", Via S. Sofia, 64, 95123 Catania, Italy
[4]Institute for Microelectronics and Microsystems (CNR-IMM Catania Università), Via S. Sofia, 64, 95123, Catania, Italy

Corresponding author: Phone: +39 051-6399127, E-mail: canino@bo.imm.cnr.it

ABSTRACT: Silicon heterojunctions (Si HJs) are regarded as a promising structure to develop high efficiency tandem solar cells. However, the Si HJ structure has been optimized for use as single junction device and variations might be made in order to adapt it to tandem applications. In this work we analyse by optical simulations the improvements that could be achieved in the short circuit current of Si HJs used as bottom cell in tandem devices, if the front TCO layer is made thicker. This solution allows to adapt the absorption in the Si HJ to the incident light spectrum, that is rich of high wavelength photons if a wide band gap cell is put on top of the Si HJ. The investigation has been carried out for two top cell technologies: a high efficiency InGaP solar cell and a semitransparent perovskite solar cell. The improvement in J_{SC} is of the order of 1 mA in both cases, and is more pronounced when the edge of transparency of the top cell occurs at higher wavelength.
Keywords: Tandem, Heterojunction, Optical losses, Transparent Conductive Oxide.

1 INTRODUCTION

Tandem solar cells are the solution to obtain higher efficiency with respect to single junction solar cells. Since industrial c-Si solar cells have already exceeded 26% in the heterojunction (HJ) architecture, most of the effort is directed towards the optimization of the top cell [1] and of the interconnection layer [2], given a certain top cell material. The optimization of the bottom cell mainly concerns the alignment of the metal grid to limit shadowing [3] and the texturing of the silicon surface prior to coating with the perovskite layer [4]. However, in the wide landscape of top cell materials, a survey on the impact of each material on the requirements for the bottom cell is still lacking. In this work we perform optical simulations [5] to optimize the anti-reflective coating of the bottom cell and interconnection layer in a 4-terminal configuration with different top cell materials: III-V [3] and perovskite [6]. In an optimized 4-terminal solar cell, the bottom cell needs to be designed in order to optimize the absorption not to the solar spectrum, but to the residual spectrum transmitted by the top cell. In the case of HJ solar cells, the thickness of the transparent conductive oxide (TCO), typically indium tin oxide (ITO), needs to be increased, to lead to a minimum of reflectance shifted to the red. Moreover, if the tandem cell is a 4-terminal, the relaxation of the requirement of current equalization has the consequence that such an optimization does not necessarily coincide with the 2-terminal case. Specific optimization is then needed, given the performance of the top cell. The guiding parameter is the maximization of the current generated in the bottom cell, J_{SC}, by varying the interlayer and TCO composition and thickness, as well as electronic properties of the material, and in particular the electrical conductivity, through the variation of free carrier concentration and absorption, which in turn has an impact on carrier generation in the c-Si absorbing layer. The upper limit for J_{SC} is determined by the overlap integral of the AM1.5G solar spectrum with the absorptance of the c-Si wafer, while the light absorbed in the top cell, in the interlayer, in the TCO and in the selective contacts does not contribute to J_{SC}.

2 METHOD

The procedure employs the software *Optical*, a simulation tool that computes the optical absorption in multilayer structures by the generalized transfer matrix method [5]. Among others, the following quantities can be computed by *Optical*:

- the light absorption in each single layer A_{Layer} from the energy flux across the specific layer [5]; the calculation is rigorously performed and is only valid in the flat case;
- The J_{sc} of a solar cell, which is calculated through the integration over the solar spectrum of the product

$$A_{Layer}(\lambda) \times IL(\lambda) \qquad (1)$$

where the relevant layer is the absorber material (silicon in our case) and $IL(\lambda)$ is the solar spectrum AM1.5G. The calculation is rigorously performed and is valid in the flat case;
- The front reflectivity R_{Flat} of the entire layer stack;
- The front reflectivity R_{Text} of textured c-Si, which is approximated by the product of the reflectivity at the first and second reflection angles:

$$R_{Text} = R(\delta) \times R(180°-3\delta) \qquad (2)$$

where δ is the angle at the base of the pyramid.

From these data, the J_{sc} of a c-Si-based textured solar cell can be computed by integrating over the solar spectrum the following quantity:

$$A_{Layer}(\lambda) \times I(\lambda) \times r(\lambda,t) \qquad (3)$$

where the factor $r(\lambda,t)$ is given by

$$r(\lambda,t) = \frac{1-R_{Text}(\lambda,t)}{1-R_{Flat}(\lambda,t)} \qquad (4)$$

and has been inserted to account for the decreased reflectance in the textured case, and consequent increased absorption in the relevant layer. Equation (4) represents an approximation, as it should contain the absorption instead, which, in the textured case, is not available, and is only valid if the sample is thick enough that transmittance is negligible. The approximation leads to overestimation of J_{sc} with respect to the case of practical wafer thicknesses. The impact of this approximation will be evaluated in section 3.

In this work, in order to simulate the effect of the top cell, the incident light spectrum (*IL*) is calculated by filtering the solar spectrum by the Transmittance (T_{Top}) of the top cell, extracted from literature data:

$$IL(\lambda) = AM1.5G(\lambda) \times T_{Top}(\lambda) \qquad (5)$$

The HJ solar cell structure is constituted by a stack of ITO/a-Si:H/c-Si/a-Si:H/ITO. The thickness of the a-Si:H layers is set to 5 and 10 nm while the ITO is about 65 nm both on the front and on the back side in order to minimize the reflectance at 500 nm. The thickness of the silicon wafer was set to 5 cm for the reason described above. In fact, with reference to fig. 1, we observe that for this thickness and larger, J_{sc} does no longer depend on thickness, indicating complete absorption in the silicon wafer.

The method used for this calculation is described in Figure 1, that shows the absorption spectra in the c-Si layer for different wafer thicknesses (a) and the calculated J_{SC} as a function of the wafer thickness (b).

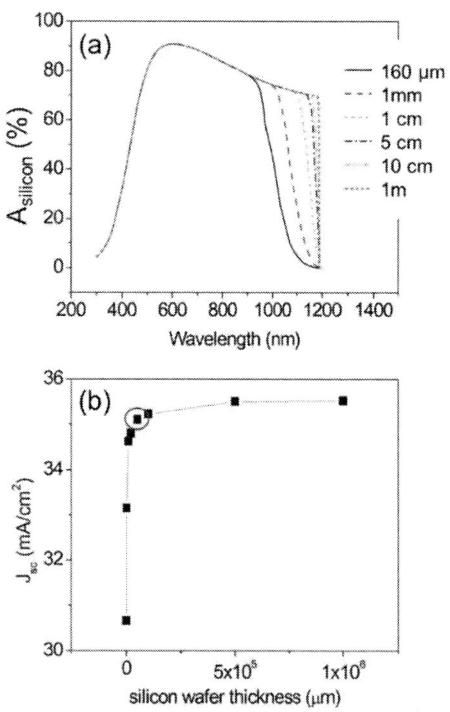

Figure 1: Light absorption spectra in the silicon wafer calculated for different values of wafer thickness (a). J_{SC} extracted from the solar cell as a function of the wafer thickness (b). The structure is set as flat and the incident light is the AM1.5G spectrum.

By using this very simple method, the calculated J_{sc} approaches the value which is reached for perfect light trapping at the back side, similar to what attained by rear side texturing, which is not explicitly calculated by the software. With this approximation, the fraction of light that reaches the layers located at the rear side of the silicon wafer is almost zero. As a consequence, the value of J_{SC} does not depend on the thickness and composition of the rear side of the cell.

We investigated two top cell technologies: a high efficiency III-V solar cell with a GaInP absorber, having a band gap equal to 1.8 eV and 18.1% conversion efficiency (PCE) [3] and a semitransparent perovskite solar cell with 19.8% PCE [6]. Both cells were used as top cells in tandem structures with silicon HJ bottom cells. The PCE values reported for the tandem structures are 29.8% and 28.3%, respectively. The J_{SC} measured in the silicon HJ bottom cells is 22.7 mA/cm² in conjunction with the GaInP top cell [3] and 14.5 mA/cm² in conjunction with the perovskite solar cell [6].

3 RESULTS AND DISCUSSION

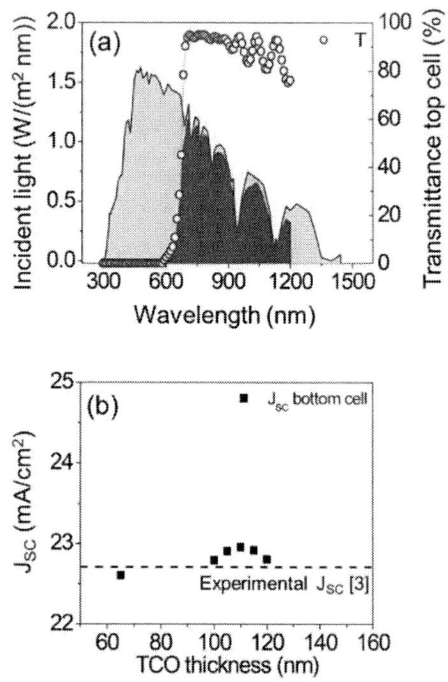

Figure 2: Spectrum of the incident light in the bottom cell (magenta area) calculated by filtering the AM1.5G spectrum (yellow area) by the transmittance of the GaInP top cell [3] (blue circles) (a). Computed J_{SC} in the bottom cell as a function of the thickness of the front ITO (b). The J_{SC} measured in the bottom cell of the structure fabricated in [3] is reported for comparison (black dashed line).

Figure 2a shows the AM1.5G spectrum (yellow area), the light transmitted from the top cell T_{Top} extracted from [3], and portion of the AM1.5G spectrum that is transmitted through the GaInP top cell, *IL* (magenta area) computed according to Eq. (5). The calculation of the J_{SC}, i.e. the maximum current density that can be extracted from the bottom cell, is made by integrating the quantity reported in eq. 3, where $I(\lambda)$ is replaced by $IL(\lambda)$. Figure 2b shows the computed values of J_{SC} as a function of front ITO thickness.

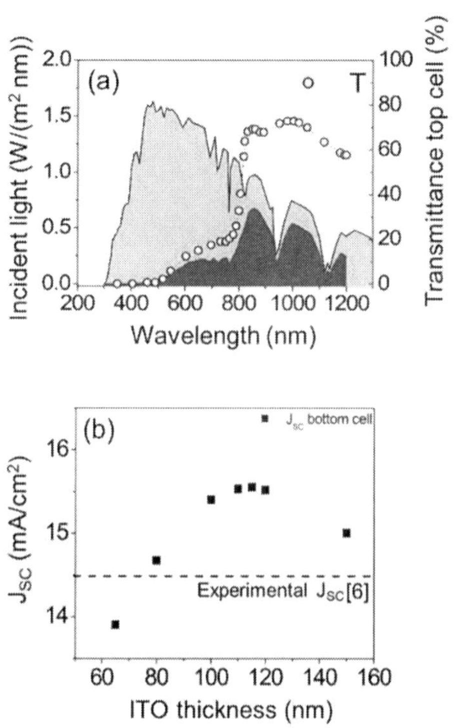

Figure 3: Spectrum of the incident light in the bottom cell (magenta area) calculated by filtering the AM1.5G spectrum (yellow area) by the transmittance of the perovskite top cell [6] (blue circles) (a). Computed J_{SC} in the bottom cell as a function of the thickness of the front ITO (b). The J_{SC} measured in the bottom cell of the structure fabricated in [6] is reported for comparison (black dashed line).

The GaInP top cell is highly transparent with a sharp rise in T around 650 nm. (Fig. 2a). The resulting J_{SC} in the bottom cell (Fig. 2b) is about 22.6 mA/cm^2 if the TCO is as thick as 65 nm (i.e. the typical TCO thickness in texturized HJs). The TCO optimization results in an improvement in J_{SC} around 3.8% of the non-optimized value for 110 nm TCO.

Figure 3a shows the spectra of AM1.5G and *IL* in the case of the perovskite top cell, whose T_{Top} is also plotted. Figure 3b shows the computed values of J_{SC} for different thicknesses of front ITO. The T spectrum of the perovskite solar cell [6] is featured by a two-step rise and its saturated value of about 70% is attained at 830 nm. The J_{SC} of the bottom cell, (Fig. 3b) is impacted by the low T of the top cell. The value of J_{SC} is 13.9 mA/cm^2 with the standard TCO thickness of 65 nm. However, with a thicker TCO this value can improve by 11.2% of its original value, attaining 15.5 mA/cm^2 for 115 nm front ITO.

The improvement of J_{SC} obtained for both configurations by increasing the thickness of the front ITO is due to the higher fraction of high wavelength light in the incident spectrum with respect to the AM1.5G spectrum. In order to increase the absorption in this range, the minimum of the bottom cell reflectance, which depends on the front ITO thickness, must be red-shifted. The enhancement is more pronounced if the region of best transparency of the top cell occurs at longer wavelengths.

It can be seen from Figures 2b and 3b that the computed J_{SC} for the non-optimzed structure is remarkably close to the values of 22.7 mA/cm^2 and 14.5 mA/cm^2 measured in [3] and [6] respectively. This is not surprising in the case of [3], because the bottom cells used in [3] are HJs with a very similar structure as the one used in this calculation. Conversely, the authors of [6] used HJs with a different anti-reflective coating. Nevertheless, the computed and experimental values are remarkably close also in the case of perovskite top cell. These similarities confirm the accuracy of the employed procedure.

These results are independent on the thickness of the rear TCO because the semi-infinite thickness of the wafer eliminates the losses from the rear side of the silicon wafer. In order to understand the effect of this approximation, we compared the J_{SC} resulting from the increase in the rear TCO thickness in a thin wafer (160 μm, even thinner than average Si wafers used in PV technologies) and flat rear surface. This is the worst case which could occur. The J_{SC} undergoes a decrease of 0.2 mA/cm2 (i.e. approximately 2% of its value) for rear TCO increasing from 40 nm to 110 nm. This result excludes any dramatic impact of the rear TCO thickness on the optical losses, and implies that, in monofacial solar cells, the rear TCO could be made as thick as the front TCO in order to prevent mechanical stability problems which could arise in asymmetric structures. The calculation made for a flat structure can differ from a non-optimal texturing by the beneficial effect of reflections from the back side, which are not expected in textured structures. However, these reflections introduce a systematic error which might have a major effect on the absolute value of the calculated J_{SC}, not on the percentage reported here.

4 CONCLUSIONS

We propose an approach to enhance the optical absorption of the bottom cell in a tandem structure by optimizing the thickness of the front TCO of the bottom cell itself. The results are analyzed in terms the maximum output current in the bottom cell, under the hypothesis that all electron-hole pairs are collected. It is shown that a thicker ITO can improve the J_{sc} by about 1mA/cm^2. The extent of the improvement is higher when the edge of the top cell transmittance occurs at higher wavelength. For transparent contacts, this corresponds to lower band gap top cells. The larger ITO thickness relaxes the requirement for the conductivity of the TCO layer, thus allowing for the reduction of free carrier absorption, further improving carrier generation at long wavelengths, and opening the way to TCO materials other than ITO in high efficiency HJ solar cells used as bottom cells in tandem structures.

ACKNOWLEDGEMENTS

This work has been partially supported by the PON Ricerca e Innovazione 2014-2020 project BEST4U, funded under Decreto Direttoriale di concessione dell'agevolazione del 21-05-2019 prot. n. 991, contract code ARS01_00519.

REFERENCES

[1] F. E. Subhan, Aimal Daud Khan, Adnan Daud Khan, N. Ullah, M. Imran and M. Noman, RSC Adv. 10 (2020) 26631.

[2] K. Bittkau, T. Kirchartz, and U. Rau, Opt. Express 26 (2018) A750-A760.
[3] S. Essig, M.A. Steiner, C. Allebé, J.F. Geisz, B. Paviet-Salomon, S. Ward, A. Descoeudres, V. LaSalvia, L. Barraud, N. Badel, A. Faes, J. Levrat, M. Despeisse, C. Ballif, P. Stradins, D. L. Young IEEE Journal of Photovoltaics, vol. 6, no. 4 (2016) 1012.
[4] K. Jager, J. Sutter, M. Hammerschmidt, P. I. Schneider, C. Becker, Nanophotonics 10 (2021) 1991.
[5] E.Centurioni, Applied Optics 44 (2005) 7532.
[6] Dong Yang, Xiaorong Zhang, Yuchen Hou, Kai Wang, Tao Ye, Jungjin Yoon, Congcong Wu, Mohan Sanghadasa, Shengzhong (Frank) Liu, Shashank Priya, Nano Energy 84 (2021) 105934.

WIDE BANDGAP PURE SULFIDE CIGS LAYERS FOR SI/CIGS TANDEM CELLS FROM METAL COEVAPORATION ENGINEERING AND SULFUR ANNEALING

Alexandre Crossay[a], Hugo Gloaguen[b], Davide Cammilleri[c], Jackson Lontchi[c], Amelle Rebai[a,c], Nicolas Barreau[d], Daniel Lincot[a,c].

a CNRS UMR 9006 IPVF, 18 Boulevard Thomas Gobert, 91120 Palaiseau, France; b Université de Nantes, F-44000 Nantes, France; c Institut Photovoltaïque d'Ile de France (IPVF), 18 Boulevard Thomas Gobert, 91120 Palaiseau, France; d Université de Nantes, CNRS, Institut des Matériaux Jean Rouxel, IMN, 2 Chemin de la Houssinière, 44300 Nantes, France.

ABSTRACT: Recent improvements in wide bandgap $Cu(In,Ga)S_2$ (CIGSu) solar cells efficiencies [1, 2] confirm the potential of tandem silicon/CIGSu solar cells. In this work, the deposition of pure sulfide wide gap CIGSu on silicon via a two-step method is investigated. Cu, In and Ga are deposited by the highly versatile coevaporation process, which enables to engineer the properties of the metallic films prior to sulfurization. Power conversion efficiencies of 7.5% with a CdS buffer layer and 6.5% with a Zn(O,S) buffer layer were achieved with this method on standard glass/Mo substrates. On silicon substrates, the effects of (co)deposition sequence of the metals were investigated, and show a strong influence on the morphology of the layers after sulfurization. This study validates the relevance of using this coevaporation-based two-step deposition method for the investigation of CIGSu deposition on silicon for tandem solar cells.
Keywords: CIGS ; $Cu(In,Ga)S_2$; tandem solar cell ; silicon.

1 INTRODUCTION

Recent developments in the field of wide gap CIGS solar cells led to the achievement of several record pure sulfide devices with values of 16.9% at 1.55 eV by a two-step sequential process [1], 15.2% at 1.6 eV by co-evaporation [2] and a value of 14.2% at 1.65 eV by co-evaporation [3]. A 12.2% CIGSu was also recently reported, prepared by evaporation of Cu-In-Ga and sulfurization [4]. Results from [1], [4] show that sequential processes have a strong potential for obtaining high efficiencies with pure sulfide CIGS.

In this context, we report on the investigation of a two-step deposition of pure sulfide CIGS, where precursor films are first deposited by co-evaporation, followed by a reactive sulfur annealing. Co-evaporation is seen as a very versatile technique for designing the metallic precursor layer (sequential, mixed...). Experiments were performed i) on classical glass/Mo substrates for process validation, and ii) on p+ silicon substrates, as a first step towards Silicon/wide-gap CIGS tandem configuration. A further advantage of using silicon substrates is to promote epitaxial growth of wide band gap CIGS, which would be also beneficial for improvement of efficiencies thanks to the reduction of grain boundaries.

Deposition of CIGSu on Si by coevaporation of metal precursors and sulfurization was recently reported by our group [5]. Analyses of morphology, crystalline structure and composition confirmed the formation of a dense CIGSu film on silicon, with an in-depth gallium gradient, compatible with high efficiency solar cells. Following this promising result, the study presented in the current paper aims now at understanding the effects of tuning the precursor metallic layer on the final CIGSu absorbers.

2 EXPERIMENTAL

2.1 CIGSu on glass/Mo

The glass/Mo substrates were standard molybdenum-coated 3mm thick soda lime glass substrates prepared by the company Nexcis. 600 nm thickness Mo layers were deposited by sputtering.

Cu, In and Ga layers were sequentially deposited by evaporation on glass/Mo at 140°C substrate temperature. The metallic precursors were then annealed at 600°C for 10 minutes in presence of sulfur powder in a tubular furnace.

Two solar cells are presented in this paper, with two different architectures of buffer/window layers: CdS/i-ZnO/ZnO:Al and Zn(O,S)/ZnMgO/ZnO:Al. CdS and Zn(O,S) were deposited by chemical bath deposition (CBD). i-ZnO, ZnMgO and ZnO :Al were deposited by sputtering. Current-Voltage (J-V) measurements were carried out on 0.5 cm^2 (cells with CdS) and 0.1 cm^2 (cells with Zn(O,S)) solar cells under A.M. 1.5G illumination using an Oriel Newport class AAA solar simulator. The External Quantum Efficiency (EQE) was measured using a IQE200 Newport instrument. The Zn(O,S)-based solar cell was subjected to a light soaking treatment of 2 hours at 1 sun, prior to J-V and EQE measurements.

2.2 CIGSu on silicon

The Silicon substrates used for this study were (100) p-type doped Si wafers, etched for 30 s in a 5% HF solution. Additional etching of Si substrates was performed by evaporation of NaF prior to the deposition. Cu, In and Ga were deposited on silicon substrates by coevaporation, at rates of 1.5 nm/min for NaF, 6 nm/min for Cu, 5.5 nm/min for In and 8 nm/min for Ga unless specified otherwise, with the following sequences:

- sample A2 : i) NaF 15 minutes, ii) In 120 min, iii) Cu 45 min, iv) Ga 30 min.
- sample B2 : i) NaF 15 min, ii) In 75 min, iii) In and Cu 15 min, iv) In, Cu and Ga 30 min.
- sample C2 : i/ NaF 15 min, ii) Cu 45 min, iii) In 90 min, iv) In and Ga 30 min.

Si/Cu-In-Ga precursors were annealed in a tubular furnace at 600 °C during 10 min in a graphite box with sulfur powder. Microcrystalline structure of the layers was analyzed by X-ray diffraction (XRD) on a PANalytical - DY711 system with Cu Kα radiation.

3 RESULTS AND DISCUSSION

3.1 CIGSu on glass/Mo

Stacked Cu/In/Ga precursors were evaporated on glass/Mo and sulfurized. A typical SEM cross section of these layers is shown on figure 1.a). A dense structure is observed, with a strong gallium gradient detected by GD-OES (figure 1.b). Solar cells were prepared with CIGSu layers originating from two different metallic samples: A1 made from a precursor stack 224 nm Cu / 397 nm In / 128 nm Ga, and B1 made from a stack 224 nm Cu / 490 nm In / 158 nm Ga. Best efficiencies for these two samples were of 7,4% for A1 with a CdS buffer layer, and 6,5% for B1 with Zn(O,S) (figures 1.c). EQE analysis of these two best solar cells is shown on figure 1.d). Despite the EQE of B1 being lower overall, a slight improvement of EQE in the lower wavelength region is detected on sample B1, which can be related to the difference in buffer layer, because of the higher absorption of CdS compared to Zn(O,S) in this energy range. This tends to confirm the benefits of using Zn(O,S) as an alternative of CdS for wide gap CIGSu.

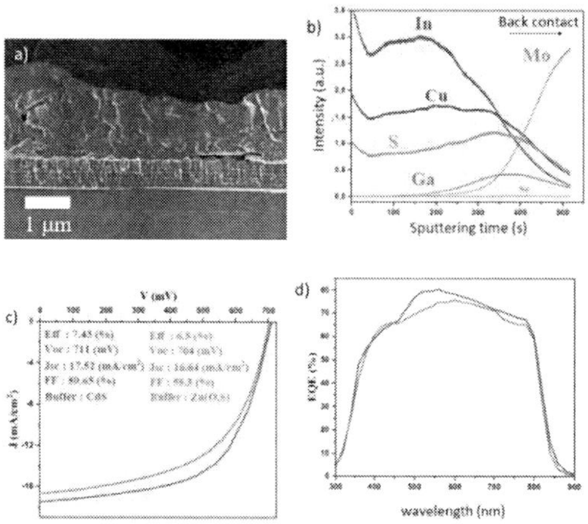

Figure 1: a) typical SEM cross section image, and b) typical GD-OES in-depth analysis of CIGSu on glass/Mo; c) IV and d) EQE measurements of best CIGSu solar cells on glass/Mo. A1 in blue and B1 in green.

These cells results obtained with wide-gap CIGSu on glass/Mo were used as a first validation of the conditions used for our two-step process of metals evaporation and sulfurization.

3.2 CIGSu on Si

One of the main advantages of using an evaporation process for the deposition of metallic precursors on silicon is the versatility of the method, which allows the exploration of various ordering of metallic stacks and/or of co-deposition sequences. It is expected that interdiffusion of metals and formation of alloys can lead to segregation of phases, which will ultimately influence the formation of CIGSu during the sulfurization step.

A first sample, referred to as sample A2 in the following, was deposited by sequential evaporation of In, Cu and Ga. An SEM image of this metallic precursor sample is shown on figure 1. A dense and smooth film is observed, with segregation of 5-10 µm diameter hillocks of indium. Another approach was then investigated, where indium is gradually incorporated in the film, with two different deposition sequences. Sample B2 was deposited by i) indium evaporation, ii) coevaporation of indium and copper and iii) coevaporation of indium, copper and gallium. Sample C2 was deposited by i) copper evaporation, ii) indium evaporation and iii) coevaporation of indium and gallium. Segregation of indium is also detected in samples B2 and C2, but with a much lower surface density of indium hillocks. Very different morphologies are observed by SEM for samples B2 and C2 (figure 2), with the presence of micro-cavities on the surface of sample B2 and the segregation of small crystals on the surface of sample C2.

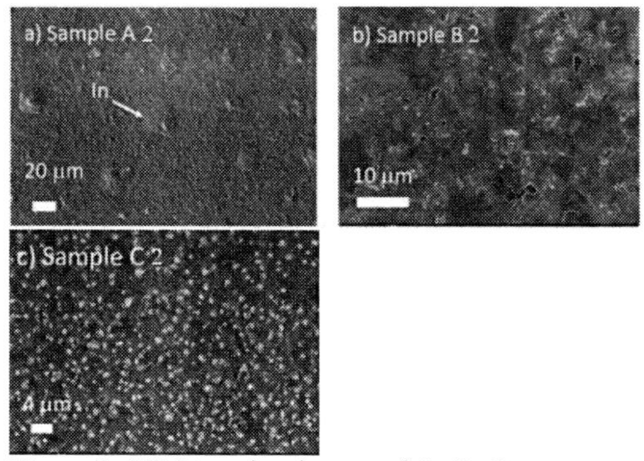

Figure 2: SEM top view images of Cu-In-Ga precursors on silicon. a) sample A2, b) sample B2, c) sample C2.

After sulfurization, strong differences are again observed between the three samples, as can be seen on figure 3. SEM cross section images of the samples show a dense structure for sample B2, and the presence of cavities in sample C2. Both cross section images present very strong in-depth inhomogeneities, where indium segregates at the front on the film and gallium at the back, with the formation of two distinct layers. This phenomenon has already been reported, and is not specific to two-step processes, as a spontaneous gradient of In and Ga has already been reported during the coevaporation of pure sulfide CIGSu which gave a 14% efficient solar cell [3]. XRD analyses in figure 3.f) show the presence of an indium rich and a gallium-rich chalcopyrite CIGS phase, together with a cubic thiospinel $CuIn_5S_8$ compound (XRD references from [6]).

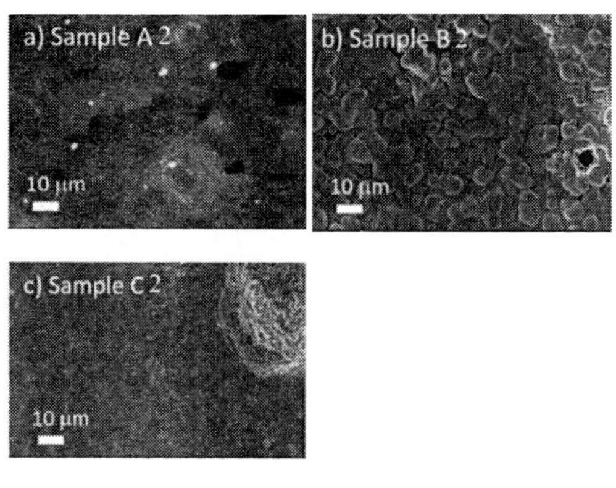

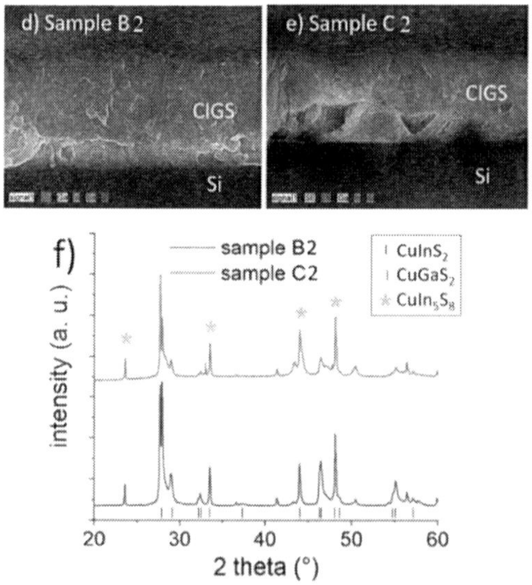

Figure 3 : a), b) and c) SEM top view images of sulfurized Cu-In-Ga films on silicon, d) and e) SEM cross section images colorized with EDX intensities of In in blue and Ga in yellow of sulfurized Si/Cu-In-Ga layers, and f) XRD pattern of samples B and C after sulfurization.

4 CONCLUSION

A two-step deposition method of CIGSu by evaporation of metals and sulfurization was investigated. Solar cells of 7.4% and 6.5% efficiencies were achieved on glass/Mo substrates, with respectively CdS and Zn(O,S) buffer layers. The investigation of this deposition method on silicon substrates shows the importance of having the ability to tune the deposition sequence of the metallic precursor sample, in order to optimize the formation of the CIGSu film in a two-step process. The versatility of metals deposition via a coevaporation process was here validated. Owing to the superior photovoltaic performances of two step processes, using coevaporation may provide a route for further improvement. Furthermore, beneficial epitaxial growth may also be achieved by these two step processes thanks to unique recrystallization paths, which is an ultimate challenge for this study.

5 ACKNOWLEDGEMENTS

This project has been supported by the French Government in the frame of the program of investment for the future (Programme d'Investissement d'Avenir), by the ANR project EPCIS (ANR-20-CE05-0038) and by the ANR project PROOF (ANR-IEED-002-01-PROOF).Dr Frédérique Donsanti is acknowledged for coevaporation support and Sophie Gaillard for silicon etching.

6 REFERENCES

[1] H. Sugimoto, H. Hiroi, Y. Iwata, and A. Yamada, 27th International Photovoltaic Science and Engineering Conference, Japan 2017.

[2] S. Shukla *et al.*, Joule 5, 1–16, 2021.

[3] N. Barreau and A. Thomere, 47th IEEE Photovoltaic Specialists Conference, 2020.

[4] I. Kwon *et al.*, Current Applied Physics, vol. 22, p. 71-76, 2021.

[5] A. Crossay *et al.*, 48th IEEE Photovoltaic Specialists Conference (PVSC), 2021.

[6] A. Thomere *et al.*, Thin Solid Films, vol. 665, 46-50, 2018.

INTERFACE ENGINEERING OF SILICON/PEROVSKITE TWO-TERMINAL TANDEM SOLAR CELLS

Afshin Hadipour[1], Jacopo Sala[*,1,3], Yinghuan Kuang[1], Hariharsudan Sivaramakrishnan Radhakrishnan[1], Yaser Abdulraheem[4], Tom Aernouts[1], and Jef Poortmans[1,2,3]

[1] Imec, Thin-Film Photovoltaics - partner in Solliance and EnergyVille, Thor Park 8320, 3600 Genk, Belgium
[2] ESAT, KU Leuven, Heverlee, 3001, Belgium
[3] University of Hasselt, Diepenbeek, 3590, Belgium
[4] Department of Electrical Engineering, Kuwait University, P.O. Box 5969, Safat 13060, Kuwait

ABSTRACT: To overcome the efficiency limitation of single-junction Si solar cells, tandem structures can be made with a top cell based on perovskite materials, benefitting from their low processing cost, versatility in bandgap tuning and high performance. In 2-terminal (2T) silicon/perovskite (Si/PK) tandem devices controlling the recombination between the two sub-cells is critical to the overall performance of the tandem. Here, we present improved open circuit voltage for a PTAA based recombination junction via double-layered hole transporting stack. Tandem devices based on a dual-cation (2C) perovskite $Cs_y(MA_xFA_{1-x})_{1-y}Pb(I_{1-x}Br_x)_3$ with 1.6 eV bandgap in the top cell and silicon heterojunction bottom cell with open-circuit voltages above 1.8 V and 24.6 % power conversion efficiency with 1 cm^2 monolithic active area have been obtained with reproducible results.
Keywords: perovskite, tandem, interface engineering

1 INTRODUCTION

Multijunction solar cells have demonstrated to be an effective way to overcome the efficiency limitation of single-junction solar cells. Moreover, a perovskite based multijunction device benefits from their low processing costs, versatility in bandgap tuning and high performance to maintain costs competitive. In a silicon/perovskite (Si/PK) 2-terminal (2T) tandem device the control over the recombination junction between the two sub-cells is critical to the overall performance of the tandem. In fact, to achieve high performance in monolithic tandem devices, one of the most important aspects is the recombination layer between two sub-cells allowing charges with opposite polarity to recombine efficiently. Studies in literature have already successfully reported the potential of an ITO/PTAA based hole transport layer[1], [2], even though PTAA is known to hinder perovskite solution deposition due to low wettability. Here, we present the introduction of an inorganic intermediate layer on top of PTAA[3] that leads to improved perovskite formation and results in a more optimal summation of open-circuit voltages of the two sub-cells. Moreover, tandem devices based on a dual-cation (2C) perovskite CsyFA1-yPb(I1-xBrx)3 with 1.6 eV bandgap in the top cell and silicon heterojunction bottom cell have been obtained reproducibly with open-circuit voltages above 1.8 V and 24.6% power conversion efficiency with 1 cm2 monolithic active area. Moreover, the layer thickness of the top cell and front electrode have been fine-tuned to achieve good current matching between the sub-cells. Additionally, the perovskite layer was processed by using gas quenching, which is considered a method also suitable for upscaled, wafer-size processing. Tuning the bandgap of the perovskite layer and the transparency of the front TCO electrode will yield even higher performance for these Si/PK tandem solar cells with such double-layered interface at recombination centre. Therefore, in this work, we implement two main elements, (a): A very thin insulating interlayer between the hole transporting polymer PTAA (5 nm) and the perovskite active layer is realized to enhance the wettability of PTAA. (b): The use of a cost-effective, scalable deposition process for the perovskite active layer, the gas quenching (GQ) method.

2 RESULTS AND DISCUSSION

To achieve high performance in monolithic tandem devices based on silicon and perovskite sub-cells, many different issues must be addressed in the same device. As previously mentioned, one of the most important aspects is the recombination layer between two sub-cells allowing charges with opposite polarity to recombine. Only via efficient charge recombination at the intermediate layer a 2T tandem device can produce high open-circuit voltage equal to the sum of the open circuit voltages of the sub-cells. In addition, for optimum performance and long-term stability, the perovskite layer must be pinhole-free which means there is a need for very good wetting of the perovskite solution on the underlying surface. Figure 1 shows the tandem structure used in these experiments, while reports a 2C perovskite processed onto 3×3 cm2 silicon substrates with PTAA only and with an additional interlayer at the PTAA/Pk interface, with the former producing a defective film compared to the latter. Due to poor wetting of the perovskite solutions on PTAA in general, pinholes similar to the one shown in Figure 2(b) generate potential shunt paths that hinder the tandem performance. These conditions are only made worse on larger area devices. Therefore, it is not possible to use PTAA as a single layer for larger area, wafer-size applications, especially when using other coating methods than spin coating, such as blade and/or slot-die coaters. On the other hand, the introduction of a multi-layered structure with a good energy alignment and passivation properties, enables optimum recombination. In addition, the intermediate layer can have an additional passivation effect that reduces the number of traps at the interface between the PTAA and the perovskite layer. In our experiments, an increase of the open-circuit voltage of the top sub-cell and consequently of the tandem device resulted from this strategy and lead to an open-circuit voltage above 1.8V for the tandem device. Moreover, the results obtained from two runs with a 3 months' time span is shown in **Figure 1**, which confirms the reproducibility of the developed approach and shows to be promising for enhanced performances. Additionally, the use of a gas quenching deposition method can also be scaled on large area coating tools such as blade and slot-die coaters using N_2-knife and is compatible with leading production of full wafer tandem devices. This approach is also suitable for a single junction solar device where Glass/ITO/PTAA is used. The introduction of a double HTL structure led to a champion device with a Voc of 1.11 V, with a Jsc of 22.00 mA/cm2 and a fill factor of 74.32% and a power conversion efficiency of 18.15%.

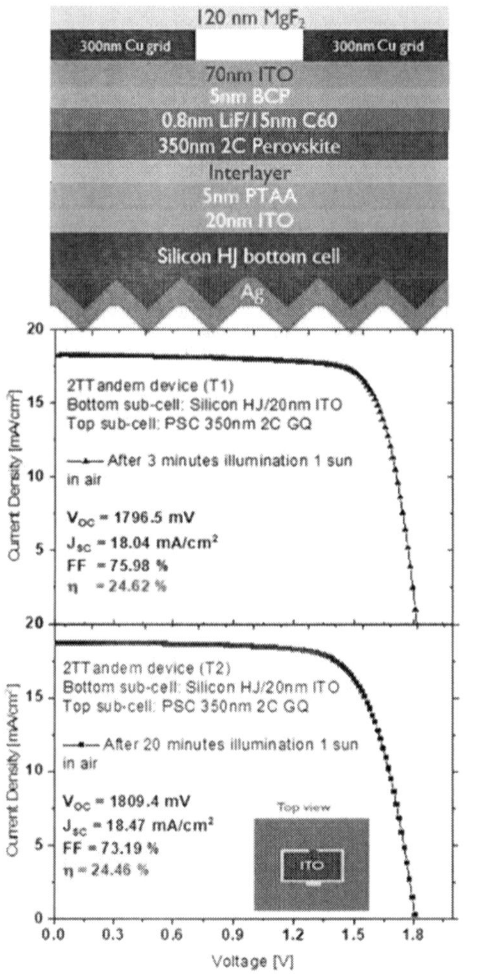

Figure 1: The 2T tandem device structure and JV-measurements of two different runs.

3 NEXT STEP

In order to increase the performance of the tandem device to higher values, wider bandgap perovskites[4], [5] must be used to realize higher V_{OC}. To do so, a triple-cation (3C) perovskite can be used as a platform in which the amount of Cesium (Cs) and Bromine (Br) can be tuned in such a way that different wide bandgaps (here up to 1.71 eV) can be realized. Here, we are demonstrating initial results of bandgap tuning by perovskite solution composition engineering and their properties based on single junction devices all done by gas quenching method as mentioned before.

The corresponding bandgaps of 3C perovskite with different amounts of Bromide (Br) and Cesium (Cs) are mentioned in Table I & II (based on photoluminescence peaks).

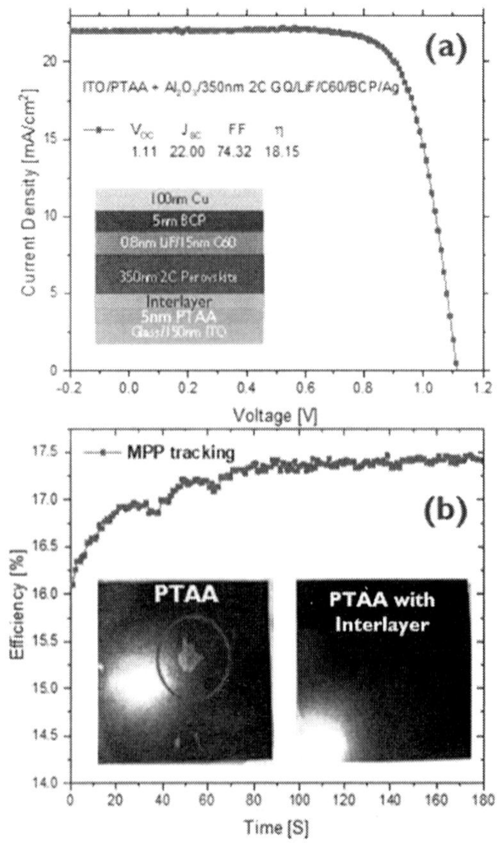

Figure 2: (a) JV measurement of single junction 2C perovskite solar cell and the opaque device structure used. and (b) the MPP tracking of the same device. Inset of (b) demonstrates a big opening of coated perovskite layer onto PTAA only, while using a double layer shows perfectly closed film.

4 CONCLUSION AND SUMMARY

We mention here the comparison of 2C and 3C perovskites engineered for tandem applications with bandgaps of 1.6, 1.65 and 1.67 eV. The EQE and JV-measurement results of opaque single junction devices are shown in Figure 3. The blue shift of the absorption edge of the 3C perovskite solar cells would allow more light to pass through to the bottom cell when used in a Pk/Si tandem cell. These newly developed 3C perovskite will be tested in tandem devices and presented in future work. In conclusion, this work has explored the impact of an insulating thin inorganic interlayer on the performance and reproducibility of a 2-Terminal silicon/perovskite tandem solar cells. It has been found that the introduction of an interlayer between PTAA and the perovskite layer is essential to obtain high efficiency tandem structures. Good tandem efficiency (24.6%), optimum open-circuit voltage (Voc = 1.8 Volt) and reproducibility were realized. For future work, we have developed new 3C based perovskites with different (wider) bandgaps. Gas quenching method for perovskite processing was used in all cases which is compatible with commercialization progress to the development of highly efficient silicon/perovskite tandem devices.

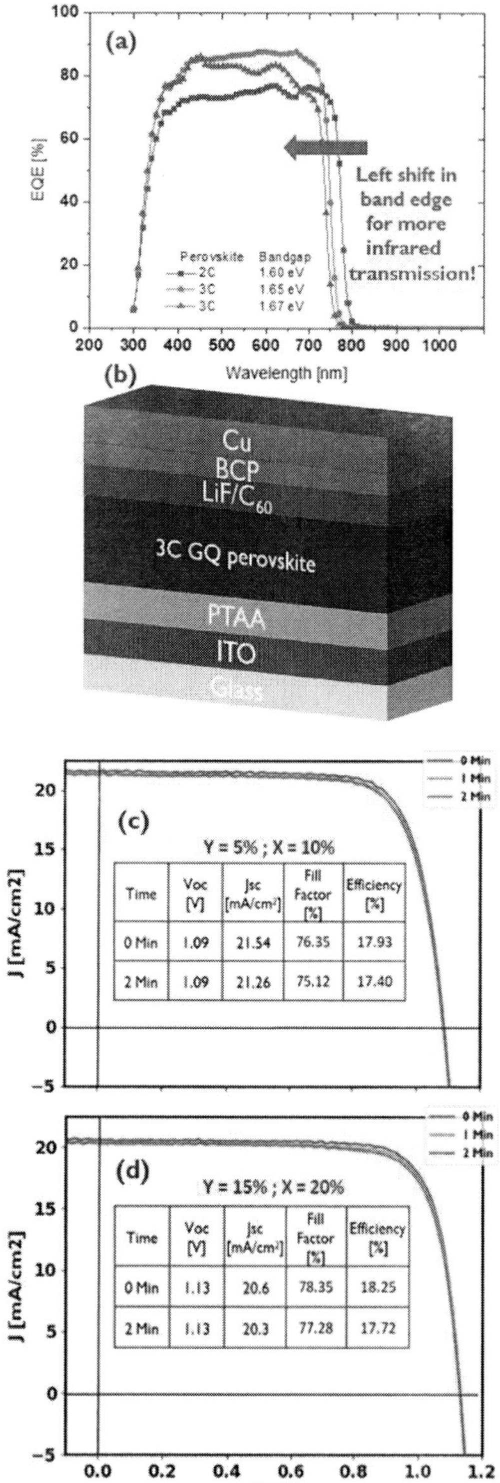

Figure 3: (a) Left shift in EQE spectra for 3C perovskite with wider bandgaps of 1.65 and 1.67 eV are compared with 2C perovskite with bandgap of 1.6 eV, (b) opaque single junction device structure and JV measuremnts under 1 sun illuniation (c) of perovskite with bandgap 1.65 eV (5 mol% Cs + 20 mol% Br), (d) of perovskite with bandgap 1.67 eV (15 mol% Cs + 20 mol% Br).

Table I: Different bandgaps of 3C perovskite $Cs_y(MA_xFA_{1-x})_{1-y}Pb(I_{1-x}Br_x)_3$ by variation of Br contents.

Br Content %	10%	17%	20%	24%	30%
Bandgap (eV) (Cs = 5%)	1.60	1.64	1.65	1.67	1.71

Table II: Different bandgaps of 3C perovskite $Cs_y(MA_xFA_{1-x})_{1-y}Pb(I_{1-x}Br_x)_3$ by variation of Cs contents.

Cs Content %	5%	10%	15%	20%	25%
Bandgap (eV) (Br = 20%)	1.64	1.66	1.67	1.67	1.69

6 REFERENCES

[1] M. De Bastiani, A. S. Subbiah, E. Aydin, F. H. Isikgor, T. G. Allen, and S. De Wolf, "Materials Horizons Recombination junctions for efficient monolithic perovskite-based tandem solar cells : physical principles , properties , processing and prospects," 2020.

[2] E. Köhnen *et al.*, "Highly efficient monolithic perovskite silicon tandem solar cells: Analyzing the influence of current mismatch on device performance," *Sustain. Energy Fuels*, vol. 3, no. 8, pp. 1995–2005, 2019.

[3] V. Zardetto *et al.*, "Atomic layer deposition for perovskite solar cells: research status, opportunities and challenges," *Sustain. Energy Fuels*, vol. 1, no. 1, pp. 30–55, Mar. 2017.

[4] T. C. J. Yang, P. Fiala, Q. Jeangros, and C. Ballif, "High-Bandgap Perovskite Materials for Multijunction Solar Cells," *Joule*, vol. 2, no. 8, pp. 1421–1436, 2018.

[5] J. DM *et al.*, "Reversible Halide Exchange Reaction of Organometal Trihalide Perovskite Colloidal Nanocrystals for Full-Range Band Gap Tuning," *Nano Lett.*, vol. 15, no. 8, pp. 5191–5199, Aug. 2015.

Optimizing Top-TCO for Perovskite-Silicon Tandem Solar Cells

Hunter King[1], Volker Sittinger[1], Tino Harig[1], Daniel Stoll[1]

[1]Fraunhofer Institute for Surface Engineering and Thin Films, Bienroder Weg 54E, 38108 Braunschweig, Germany

Corresponding Author: Hunter King, Hunter.King@ist.fraunhofer.de, +49 (0) 531 2155 641

Abstract: Using reactive sputtering processes, indium zinc oxide (IZO) and hydrogen-doped indium oxide (IO:H) films were developed for use as the top-TCO contact for perovskite-silicon tandem solar cells. Perovskite materials have shown significant promise as an upcoming absorber material for use in tandem solar cell configurations. Both the IZO and IO:H films were deposited at temperatures below 100 °C in order to prevent any possible degradation of the perovskite absorber material. The IO:H films required a post-deposition thermal annealing process of 220 °C for 30 minutes in air, after which they showed comparable conductivities and improved absorptance when compared with the IZO films. Additionally, the IO:H films showed over double (112 vs < 50) the carrier mobilities of the IZO films, showing potential as a future top-TCO candidate.

1. INTRODUCTION

Perovskite-silicon tandem solar cells have shown promise as a new solar technology to provide higher efficiencies and lower cost of energy than either silicon or perovskite single junction cells alone. However, further improvements to the efficiency and stability of these cells is required in order to provide an industry-viable lifetime and to compete with the current highest efficiency cells on the market. While improvements in absorber materials are a central focus, a significant amount of work needs to be done on the other intermediate and contact layers in the cell. While these layers have been optimized for traditional cells, there is still much work to be done on the optimization of these films for perovskite-silicon tandem solar cell designs. IZO is known to enable world record efficiencies in perovskite-silicon tandem solar cells [1]. In this work the aim is to provide a detailed insight into the characteristics of the top-TCO which causes overall cell efficiency improvements.

2. EXPERIMENTAL

The experiments performed in this work were carried out on two separate systems. The deposition of IZO films were made using a reactive serial co-sputtering process in a Systech ZVL 1000 vertical inline coater with a specialized co-sputter environment (as outlined in Figure 1). For these processes the zinc concentration was varied by varying the corresponding target power in the planar cathode (500 x 88 mm^2) towards the indium rotatable target (d = 133 mm, l = 400 mm). The zinc target is protected from oxidation by means of a gas separation wall (also shown in Figure 1).

The process of serial co-sputtering involves the use of a traditional cylindrical rotating primary target with additional secondary target(s). The secondary target(s) sputter onto the surface of the rotatable primary target, while remaining separated from any potential poisoning from reactive gasses.

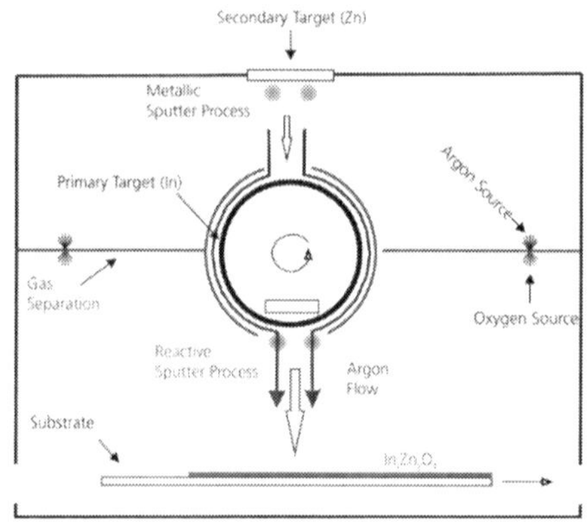

Figure 1: Schematic process of the Systec ZVL1000 serial co-sputter tool

During this process the primary target is implanted with additional element(s) from the secondary target(s) which will then be re-sputtered, along with the primary material, either by means of a metallic or reactive process onto a substrate, leading to the production of a mixed-composition film. This technology is also known as "Megatron™", which involves the separation between the primary and auxiliary chamber volumes during the sputtering process. This system, utilizing a primary indium rotatable target, and a secondary zinc planar target material was used for the preparation of the indium zinc oxide films investigated in this work.

The hydrogen doped indium oxide (IO:H) films were produced using a vertical in-line coater Leybold A700V at Fraunhofer IST (shown in Figure 2). This system utilizes PK750 cathodes with a size of 750 x 88 mm^2. This allows for the investigation into industrially scalable PVD processes. Using this system, IO:H films were produced by a DC reactive sputtering process utilizing a metallic indium target, oxygen as the reactive gas, argon as the carrier gas, and both molecular hydrogen and water vapor as hydrogen sources for doping.

Figure 2: The large-scale A700V inline coater for the IO:H processes at Fraunhofer IST

The specifics of both deposition parameters are outlined in Table I.

Table I: Deposition conditions for IO:H and IZO films produced by magnetron sputtering onto corning glass

A700V with rotatable cathode, Plasma excitation with AE Pinnacle DC (metallic target) and ZVL 1000 serial co sputtering configuration with two metallic targets, a planar PK 500 cathode and a rotatable target powered by AE Pinnacle DC supplies.

		IO:H	IZO
Target substrate distance	d_{ST}	90 mm	
Target material		IO:H 3N	In 3N / Zn 3N
Deposition pressure	p_{tot}	480 mPa	350 mPa
Argon flow	q(Ar)	~100 sccm	100 sccm
Oxygen flow	$q(O_2)$	50 sccm	34 sccm
Substrate temperature	T_s	RT	50-80 °C
Excitation power	P	0.5 – 1.3 kW	0.5 – 1.2 kW
Oxygen partial pressure	$p(O_2)$	40 – 43 mPa	48 - 54 mPa

The system parameters used in this work corresponds to sputtering in the transition regime in order to optimize deposition rate, optical, and electrical properties of the resulting films. The properties of these films were characterized by means of optical measurements (photometry and Ellipsometric measurements) and electrical measurements (4-point resistivity and Hall effect measurements) in order to quantify the optical absorptance of the films in addition to the carrier concentration and mobility. For the hydrogen doped indium oxide films these measurements were performed both before and after a tempering process.

All films on both systems were deposited onto Corning® Eagle XG® glass substrates.

3. RESULTS

Indium-zinc-oxide (IZO) and hydrogenated indium-oxide (IO:H) films were produced using a reactive co-sputtering, and reactive sputtering process, respectively. All films were optimized on Eagle XG glass substrates.

3.1 IZO Film Properties

Figure 3 shows the transmittance of the IZO films (d_{film} ~ 80nm) as a function of oxygen partial pressure and Zn target power. As can be seen, the transmittance of the film can be increased in the low-wavelength regime (l < ~ 1300 nm) by reducing the oxygen partial pressure, however increasing the oxygen partial pressure leads to a significant improvement of the higher wavelength optical properties (green arrows highlight the transmittance trend with increasing oxygen partial pressure).

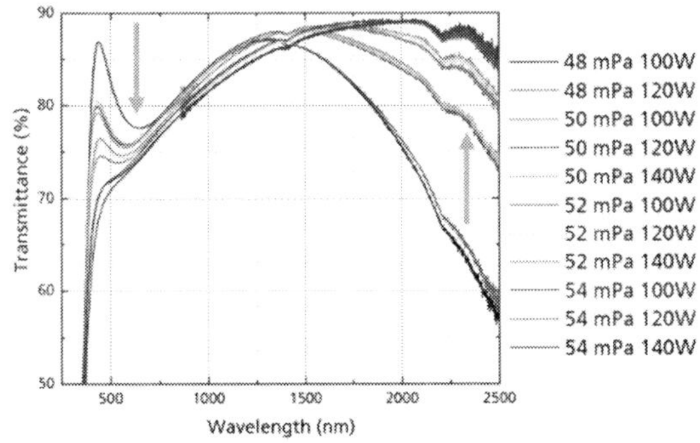

Figure 3: Transmittance of the IZO films as a function of O_2 partial pressure and Zn target power

This highlights the potential for overall cell performance improvements by means of absorber-specific TCO properties.

The developed IZO films showed comparable resistivities to indium tin oxide (ITO) reference films, but significantly improved absorptance values (reduction of more than 1.5 abs % at comparable specific resistivity, see Figure 4).

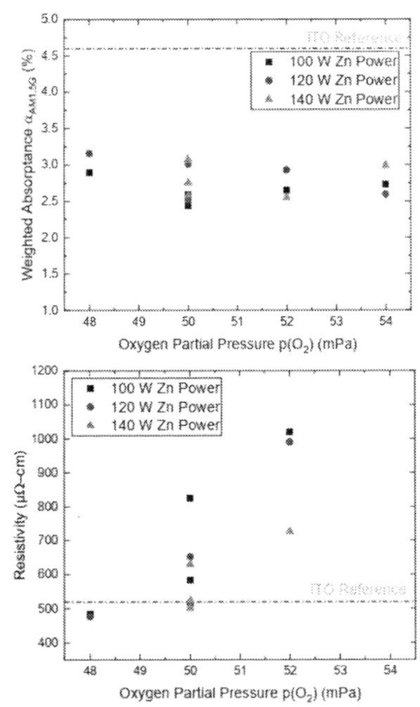

Figure 4: Weighted absorptance (top) and resistivity (bottom) of IZO films as function of Zn target power and oxygen partial pressure. Also shown is the absorptance and resistivity of the ITO reference film (in dotted-blue).

The absorptance values were measured as an absorptance value weighted by AM1.5G spectral data in order to more accurately model the expected influence of absorption improvements on overall cell efficiencies. The films were further quantified by means of Electron Probe Micro

Analyzer (EPMA) to determine film stoichiometries (see Figure 5).

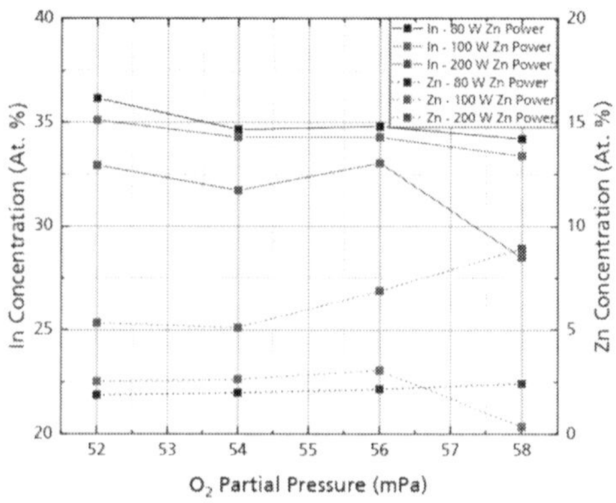

Figure 5: Indium and zinc concentrations of the IZO films (At. %) measured by EPMA as a function of Zn target power and oxygen partial pressure.

As can be seen in figure 5, the zinc concentration of the film is mostly dependant on the Zn target power, and does not appear to have a strong correlation with the oxygen partial pressure. The measured indium concentration is shown to slightly reduce as the oxygen partial pressure in the system is increased.

In addition to EPMA, Hall effect measurements were made (Figure 6). Hole mobilities of up to 50 cm^2/Vs were achieved with the best performing cells.

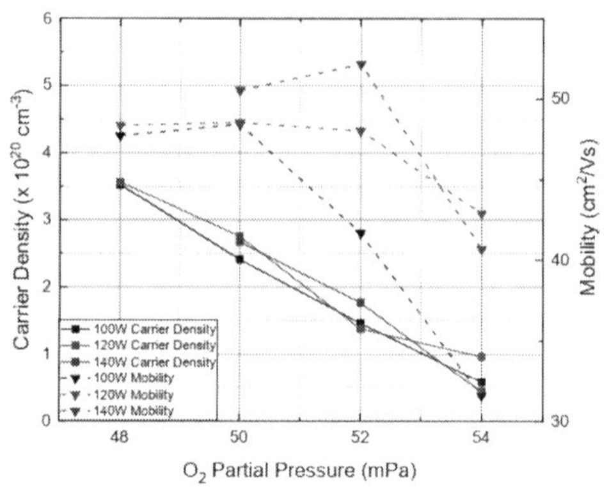

Figure 6: Carrier density (hole concentration) and mobility of the IZO films as a function of O_2 partial pressure and Zn target power

Initial IO:H depositions involved variations in both the oxygen partial pressure in addition to the flow rate of H_2O. Figure 7 and 8 show the resistivity and weighted absorptance of the as-deposited and post-annealing IO:H samples as a function of H_2O flow and O_2 partial pressure, respectively.

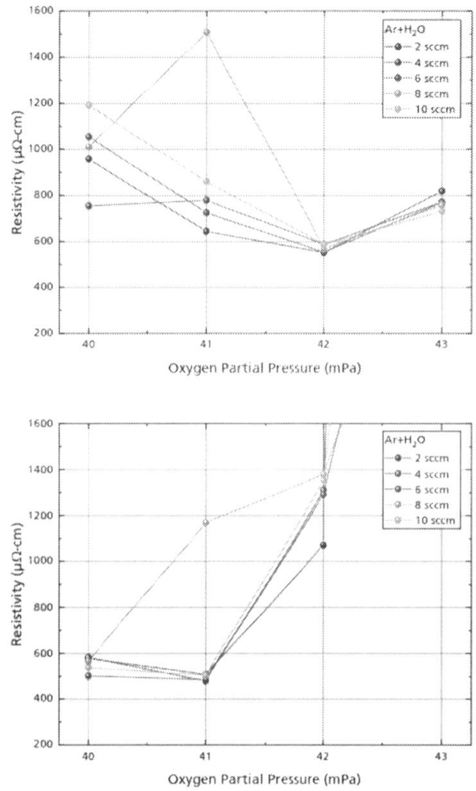

Figure 7: Top: as-deposited, and bottom: post-annealing resistivities of the IO:H samples as a function of H_2O flow and O_2 partial pressure.

The IO:H films showed acceptable resistivity values around 600 μΩ-cm before annealing at 220 °C for 30 minutes. After annealing, the samples showed significant improvement in the more metallic samples (reaching < 530 μΩ-cm), but significant degradation of the more oxidized samples. Significant improvement in the weighted absorptance was also seen in all but the most metallic samples (from > 5.0 abs % before annealing to < 1.5 abs % post-annealing process at 42 – 43 mPa). This significant improvement is likely due to an expecting crystallization of the as-deposited amorphous film [3]. The best post-annealing films were then reproduced and further characterized by means of Hall effect measurement, however due to drift from indium target oxidation [4] the optimum which was previously near 42 mPa was 40.5 mPa, which was then used for the subsequent series (see figure 9).

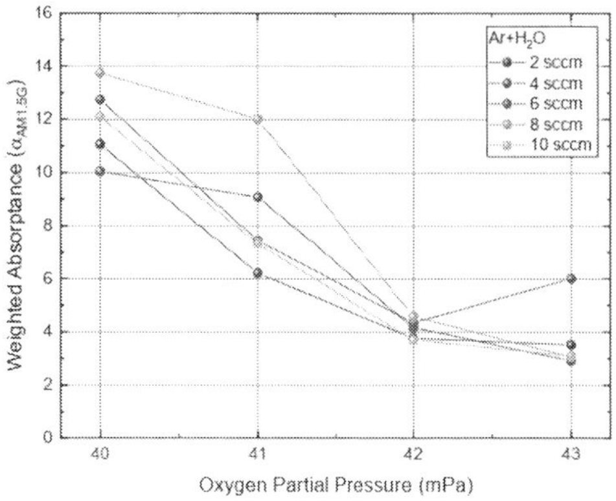

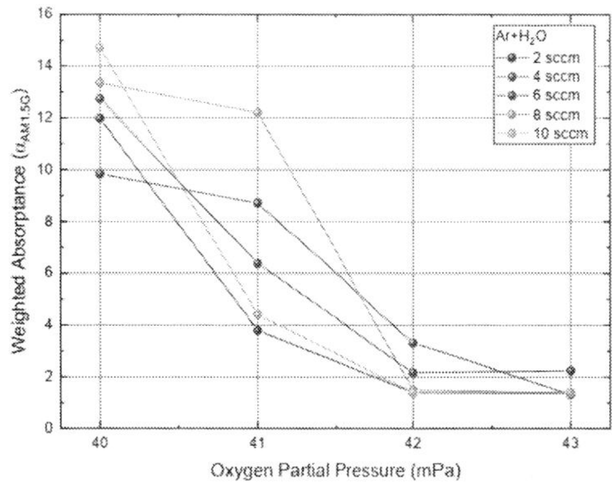

Figure 8: Top: as-deposited, and bottom: post-annealing weighted absorptance of the IO:H samples as a function of H_2O flow and O_2 partial pressure.

Figure 9 shows that while the IZO and the post-annealing IO:H films have comparable conductivities, the IO:H films have over double the carrier mobility of their IZO counterparts, with a value of 112 cm^2/Vs achieved at an Ar/H_2O mix of 4 sccm and an oxygen partial pressure of 40.5 mPa while maintaining a weighted absorption of 2.41 %. This increase of mobility, when implemented into the perovskite-silicon tandem cell should reduce free-carrier absorption, while maintaining comparable conductivities (with both IZO and ITO) [5]. This should help to further improve the overall power conversion efficiency of the cell.

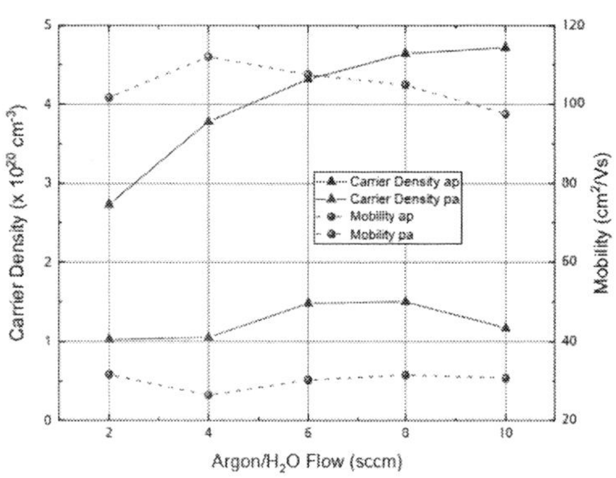

Figure 9: Carrier density and mobility of IO:H films as a function of Argon/H_2O mixed gas flow for 40.5 mPa O_2 partial pressure. Shown in blue: as-prepared (ap), and in red: post-annealing (pa).

4. DISCUSSION / CONCLUSIONS:

While the IO:H films developed in this work show promise as a candidate for top TCO material in perovskite-silicon tandem cells, they are still limited by their required post-deposition annealing treatment. The annealing process used in this work (as originally outlined by Erfurt et al. [3]) is not suitable for IO:H films deposited onto perovskite-silicon cells due to degradation of the perovskite absorber material above 100 °C. For this reason, in order for this material to be a suitable replacement, an alternative process must be developed (e.g. a flash or laser-based annealing process) which can maintain a lower substrate temperature, while still achieving a crystallization of the a-IO:H film.

Acknowledgements

This work was supported by the Fraunhofer Lighthouse project (MaNiTU).

REFERENCES

1. A. Al-Ashouri, E. Köhnen, B. Li, A. Magomedov, H. Hempel, P. Caprioglio, J. A. Marquez (2020) "Monolithic perovskite/silicon tandem solar cell with >29% efficiency by enhanced hole extraction," *Science*, Vol. 370, Issue 6522, pp. 1300-1309,

2. D. Erfurt, M. D. Heinemann, S. Körner, B. Szyszka, R. Klenk, R. Schlatmann, "Improved electrical properties of pulsed DC magnetron sputtered hydrogen doped indium oxide after annealing in air," *Materials Science in Semiconductor Processing*, 89, 170–175, 2019, doi: 10.1016/j.mssp.2018.09.012

3. D. Erfurt, "New TCO for Use as Transparent Front Contact in Chalcopyrite Thin Film Solar Cells," *Doctoral Thesis*, 2019, doi: 10.14279/depositonce-8251

4. V. Sittinger, F. Ruske, A. Pflug, W. Dewald, B. Szyszka, G. Dittmar, "Optical on-line monitoring for the long-term stabilization of a reactive mid-frequency sputtering process of Al-doped zinc oxide films", *Thin Solid Films*, 518 (2010) 3115–3118

5. L. Barraud, Z. C. Holman, N. Badel, N. Reiss, A. Descoeudres, C. Battaglia, S. De Wolf, C. Ballif, "Hydrogen-doped indium oxide/indium tin oxide bilayers for high-efficiency silicon heterojunction solar cells," *Solar Energy Materials and Solar Cells*, 115, 151-156, 2014, doi: doi.org/10.1016/j.solmat.2013.03.024

ENERGY YIELD AND PERFORMANCE RATIO OF III-V ON SILICON DUAL JUNCTION SOLAR CELLS IN DIFFERENT CLIMATE ZONES

Oliver Höhn, Mario Hanser, Marc Steiner, Elke Lorenz, Benedikt Bläsi, Stefan W. Glunz, Frank Dimroth
Fraunhofer Institute for Solar Energy Systems ISE
Heidenhofstr. 2, 79110 Freiburg, Germany

ABSTRACT: The importance of predicting outdoor performance of tandem solar devices is growing, as silicon-based tandem solar cells approach industrialization. A 2-terminal configuration has the advantage of lower parasitic absorption and less electrical wiring at module level as compared to a 4-terminal device. However, a 2-terminal tandem device is sensitive to current-mismatch losses. As a result, there is a strong interest to determine the outdoor energy yield in addition to the STC efficiency in order to evaluate the real impact of current-mismatch losses on the energy yield. We compare the energy yield, harvesting efficiency and performance ratio of a silicon-based dual junction solar cell in 2- and 4-terminal configuration to a silicon single junction device for different locations in the world.
We find that changes due to current mismatch do not lead to a relevant decrease in harvesting efficiency in 2-terminal tandem solar cells and thus are no reason to principally prefer 4-terminal tandem solar cells. The assumption that current mismatch might be a severe issue in 2-terminal tandems does not hold – they show similar performance as silicon single-junction devices. Note that the performance ratio of 4-terminal tandems shows a lower dependence on location than 2-terminal tandems as well as single junction solar cells, and thus for certain locations show an exceptionally high performance ratio.

1 INTRODUCTION

With silicon-based tandem solar cells approaching industrialization, the significance of predicting outdoor performance of tandem solar devices is growing. Compared to a 4-terminal tandem solar cell, a 2-terminal configuration has the advantage of lower parasitic absorption and less electrical wiring at module level. On the other hand, a 2-terminal tandem device design can suffer from current-mismatch losses. As a result, there is a strong interest to determine the outdoor energy yield instead of the standard testing condition (STC) efficiency to evaluate the real impact of current-mismatch losses. Such an analysis of the energy yield is crucial when assessing the potential and the electricity costs of new and emerging PV technologies in comparison to existing technologies.

We compare the energy yield, harvesting efficiency and performance ratio of an $Al_{0.21}Ga_{0.79}As$-on-silicon dual junction solar cell in 2- and 4-terminal configuration to a silicon single-junction device for different locations in the world.

Various models exist to describe energy yield modelling of silicon-based tandem solar cells [1–14]. Those are often limited to one location, idealized solar cells are assumed, temperature effects are neglected, or purely modeled spectra (e.g. by SMARTS2 [15]) are used. Apart from this, most of these investigations model perovskite silicon tandem solar cells.

We use standardized satellite-based data for different climate zones as representatives for different conditions all over the world. With these, the energy yield of 2- and 4-terminal AlGaAs-on-silicon dual junction devices is determined and compared to single-junction devices accounting for temperature effects. Based on this, we investigate if spectral variations lead to significant losses especially in 2-terminal tandem solar cells at different locations.

2 MODELING METHOD

2.1 Spectral Data

We use spectral data and ambient conditions such as temperature from the IEC international standard "Photovoltaic (PV) module performance testing and energy rating – Part 4: Standard reference climatic profiles" [16] that are based on satellite data and represent irradiation in all relevant climate zones (Table 1). We assume fix tilted modules oriented towards the equator (south or north respectively) and a tilt angle that corresponds to the latitude angle. This data is used for performance testing of silicon single-junction solar cells. In [17] the energy yields of AlGaAs-on-Silicon devices between ground-measured and satellite-based data are compared for Golden, Co, USA. It was shown that the difference in yield is below 1%. That study is restricted to just one location and therefore the overall accuracy of energy yield calculations based on satellite data cannot be estimated exactly. However, for technology comparison (e.g. single-junction vs dual-junction) a difference between satellite-based data and ground measured data does not have to be a problem. We do compare energy yields for different technologies for different locations, i.e different spectral irradiance, but the perfectly accurate absolute values are not the focus. Thus, satellite-based data are suitable for this purpose.

Table 1 For the 6 investigated climate zones the total annual irradiance, the average photon energy APE (307-1200nm (1.80 for AM1.5g)) and the mean temperature are summarized together with the latitude at which the spectral data was recorded.

	Irradiance [kWh/m²a]	Latitude	APE [eV]	Mean T [K]
tropical humid	1759	1°S	1.87	298.6
temperate continental	1307	33°N	1.81	281.2
subtropical coastal	1475	33°N	1.84	291.2
subtropical arid	2345	56°N	1.82	295.5
High elevation	2202	34°N	1.8	271.1
temperate coastal	956	57°N	1.82	283.8

2.2 Modeled Device

As modelled device we assume a module stack consisting of a PERC-like silicon bottom solar cell including rear random pyramids with a rear reflector made from a stack of silicon oxide and silver. The silicon solar cell front side is well passivated and planar. On top a highly performing AlGaAs solar cell is assumed to be monolithically integrated (e.g. by direct growth or bonding). On top of the solar cell a single layer tantalum oxide ARC minimizes reflection between the top cell and the front encapsulation (EVA/glass stack). The front glass has an ARC with a refractive index of 1.3 (See **Figure 1**). We call this type of device, where contacts, cell distances, partial shading, cell interconnection and other module-related effects are neglected, a module stack. In other words, we account for the module glass, as this has a relevant optical influence when modelling yield, but for the purpose of technology comparison, we neglect other effects that are not directly linked to the technology.

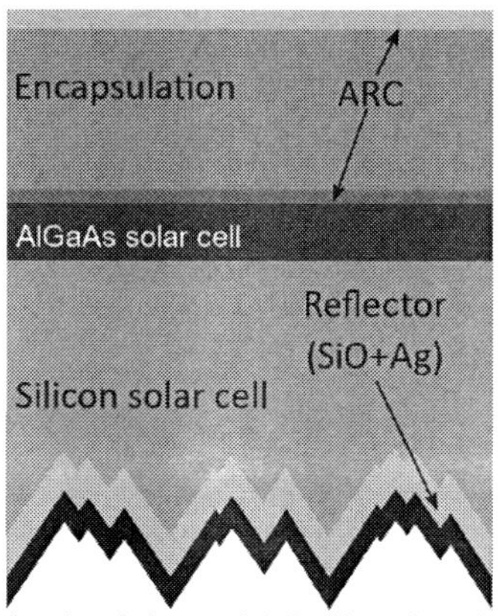

Figure 1 Sketch of the modeled silicon based tandem device.

In the case of the 2-terminal device, the cells are assumed to be interconnected in series. For the 4-terminal result the same sub cell EQEs are assumed. The IVs are modeled separately, and the efficiencies are added up. This results in an overestimation of the 4-terminal efficiency. Usually, the required additional interface will create additional optical losses. The additional contacts in-between will add resistive losses, additional shading and parasitic absorption. As these losses depend severely on the exact design, we decided to neglect them in the quantitative analysis, and account for them qualitatively in the conclusions.

The electrical parameters of the relevant device with $Al_{0.225}Ga_{0.775}As$ are shown in Table 2 and the resulting cell performance at standard test conditions in Table 3

Table 2. Cell parameters of the $Al_{0.225}Ga_{0.775}As$ top junction and the silicon bottom junction used in the temperature dependent two diode model. The values are given for a temperature of $T = 298$ K.

	$Al_{0.225}Ga_{0.775}As$	Si
J_{01} [mA/cm²]	6.75e-21	1.4e-10
J_{02} [mA/cm²]	0	2e-6
R_s [Ω cm²]	0.103	0.05
R_p [Ω cm²]	10 000	30 000
E_g [eV]	1.73	1.12

Table 3 Open circuit voltage V_{oc}, short circuit current J_{sc} as well as fill factor FF and efficiency η_{stc} at standard test conditions are summarized. For the sake of comparison, the silicon cell is also shown as a single junction device assuming front side pyramids. In case of the tandem device all results are additionally shown for the top and bottom junction separately. The STC efficiency of the 4-terminal tandem is given by the sum of the top and bottom cell efficiency.

	Single Junction	Top	Bot	2-terminal-Tandem
V_{oc} [V]	0.68	1.29	0.66	1.95
J_{sc} [mA/cm²]	41.9	19.4	21.7	19.4
FF [%]	84	90	83	90
η_{stc} [%]	23.8	22.5	11.9	34.0

Note that the shown 2-terminal tandem device is not current matched at STC. On average, the spectra of the reference regions are slightly blue shifted, compared to AM1.5g and the operating temperature is significantly higher than 298 K. The investigated device is optimized for these conditions.

The band gap of the optimal 2- and 4-terminal tandem solar cell in terms of yield are very close to each other, which allows a comparison of 2- and 4-terminal configurations with the same top cell. For detailed explanation why this device is optimal see section 3.1.

2.3 Two diode modeling

The electrical model is based on a temperature dependent 2-diode model for each subcell including series and parallel resistance R_s and R_p. The 2-diode equation is an implicit equation for the current density of a solar cell

$$J = J_{ph} - J_{01}\left\{\exp\left(\frac{q(V-JR_s)}{k_B T}\right) - 1\right\} - J_{02}\left\{\exp\left(\frac{q(V-JR_s)}{2k_B T}\right) - 1\right\} - \frac{V-JR_s}{R_p}, \quad (1)$$

depending on the voltage V, the incoming photo current density J_{ph}, the dark saturation currents J_{01} and J_{02} and the Temperature T, where q is the elementary charge and k_B the Boltzmann constant. The saturation currents temperature dependence is described by the Wanlass model [18]:

$$J_{01} = \beta_{01} T^3 \exp\left(-\frac{E_g(T)}{k_B T}\right), \quad (2)$$

$$J_{02} = A_{01} \exp\left(B_{01} E_g(T)\right) T^{5/2} \exp\left(-\frac{E_g(T)}{k_B T}\right), \quad (3)$$

with the Wanlass parameters β_{01} and A_{01} and the band gap $Eg(T)$. The Varshni relation [19]

$$E_g(T) = E_0 - \frac{\alpha T^2}{T+\beta} \quad (4)$$

describes the temperature dependence of the band gap with the material constants $\alpha = 4.73e\text{-}4$ and $\beta = 636$ as well as the band gap at absolute zero $E_0 = 1.166$eV for silicon taken from [20]. In case of the AlGaAs top junction the temperature dependence of the band gap is calculated with binary and tertiary material parameters reported by Vurgaftmann [21]. For both solar cells the Wanlass parameters β_{01} and A_{01} and J_{02} parameter A_{01} and B_{01} are extracted at an initial temperature $T = 298$ K, with the corresponding band gap $E_g(T)$ from the J_{01} and J_{02} values shown in Table 2.

Within this study we assume the following empirical

relation between the ambient temperature T_{amb} and the module temperature T as given by [22]

$$T = I \cdot 0.025 \frac{m^2 K}{W} + T_{amb}, \qquad (5)$$

where I is the global irradiance power on the module plane. All input parameters for the $Al_{0.225}Ga_{0.775}As$ composition as well as the silicon junction are listed in Table 2 for a temperature of $T = 298$ K. The value 0.025m²K/W is an empirical value that represents a ground-mounted PV power plant. We use the same values for both the single junction and the tandem device. In the end the higher efficiency of the tandem device will lead to a lower temperature increase. However, this is a minor effect, which is neglected in this study and will further boost both the 4-terminal and 2-terminal tandem performance ratio.

The temperature coefficient dV_{oc}/dT is calculated as the slope of the V_{oc} curves in Figure 2. For the silicon solar cell the values -2.1 mV/K as part of the tandem device with J_{sc} = 21.7 mA/cm², and -2 mV/K for the single junction under AM1.5g with J_{sc} = 42.1 mA/cm² are in good agreement with earlier publications [23], which reported a temperature coefficient of -2.1 mV/K for the silicon device. Thus, we can assume that the Wanlass model and the J_{02} model that was originally developed for III-V-solar cells reasonably reproduces the temperature dependence of the silicon bottom solar cell.

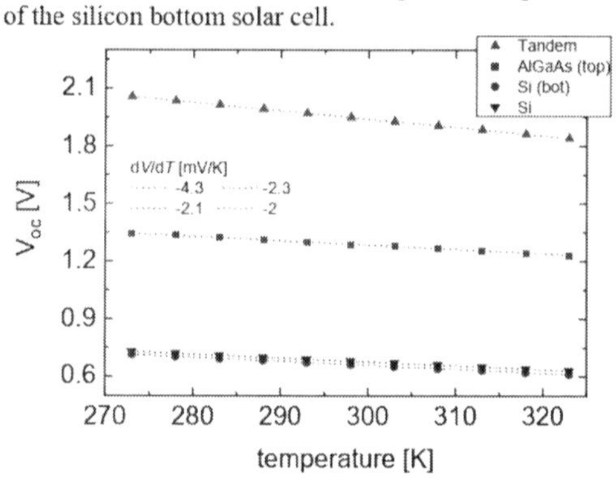

Figure 2 The open-circuit voltage V_{oc} is plotted over the temperature T for the silicon single junction device at a short circuit current of J_{sc} = 41.9 mA/cm², as well as for the top and bottom junction of a tandem device consisting of $Al_{0.225}Ga_{0.775}As$ and silicon. The short-circuit currents in this case are J_{sc} = 19.4 mA/cm² and J_{sc} = 21.7 mA/cm² for the top and bottom junction, respectively. The slope of these curves correlates with the temperature coefficients supporting the chosen model for the temperature depended band gaps.

The temperature and angle of incidence-dependent EQE is modelled using the OPTOS formalism [8, 24, 25]. In this manner, the impact of cell temperature, as well as the incident angle of the solar radiation on the solar cell power output is considered. Effects of the temperature and the incidence angle on the EQE are depicted in **Figure 3** and **Figure 4**, respectively.

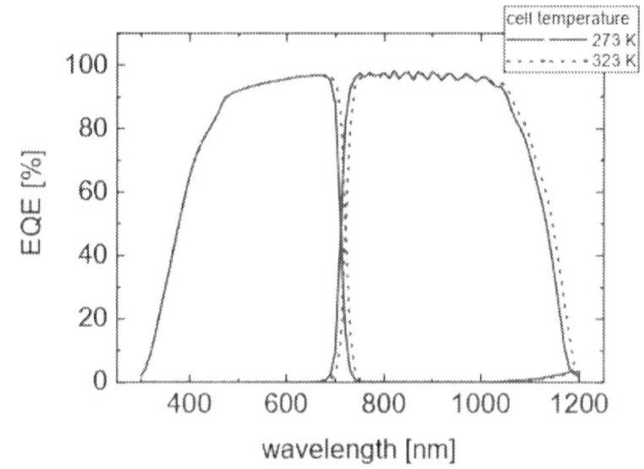

Figure 3 The EQE of the investigated tandem device with a $Al_{0.225}Ga_{0.775}As$ composition as top cell and a silicon bottom cell is plotted for two different temperatures. The top cell EQE's are depicted in blue and the bottom cell EQE's in red. A shift of the total EQE is observed towards higher wavelength for increasing temperature.

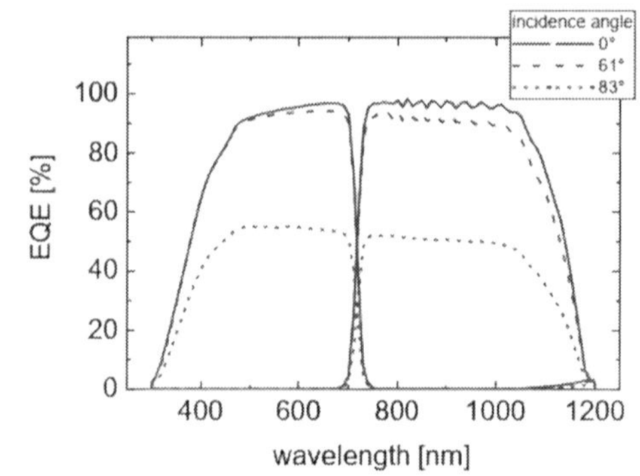

Figure 4 Investigated is the EQE's angle dependence of a tandem device with a $Al_{0.225}Ga_{0.775}$ As composition as top cell and a silicon bottom cell. It was found, that the EQE change with incoming angle is very small for low incoming angles. For higher angles, starting at approximately *60°*, the EQE begins to drop drastically.

2.4 Energy yield modeling

The used model for the determination of the energy yield (YieldOpt) is described in more detail in [5, 8, 17]. It combines the above mentioned spectral and ambient data from the specific locations, the 2-diode-models for the cells and the temperature and angle dependent EQE into one single modelling tool.
The energy yield is calculated in 6 different climate zones. 2- and 4-terminal dual junction solar cells are compared with a silicon solar cell in terms of harvesting efficiency (Annual Energy Output divided by Annual Incident Energy) and performance ratio (harvesting efficiency divided by STC efficiency).

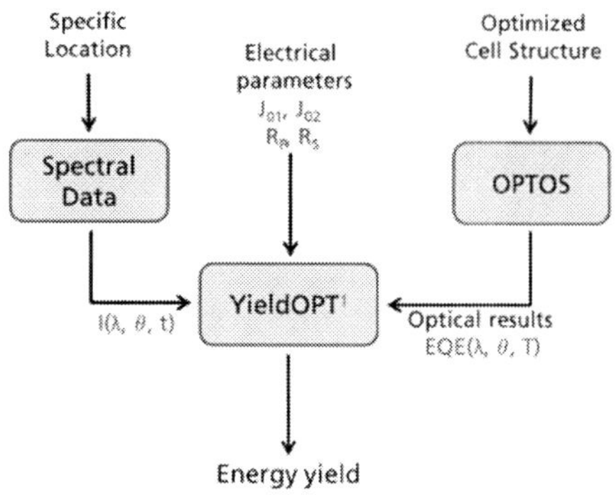

Figure 5 Sketch of the simulation formalism.

3 RESULTS AND DISCUSSION

3.1 Variation of top cell band gap

As a first step the harvesting efficiency was modeled for the locations from the IEC international standard [16] for a 2- and 4-terminal $Al_xGa_{1-x}As$-silicon-tandem solar cell with different Al-content X, to vary the top cell band gap (Figure 6). The Wanlass parameter extracted from the saturation current J_{01} in Table 2 were used for all top cells to simulate a constant cell quality.

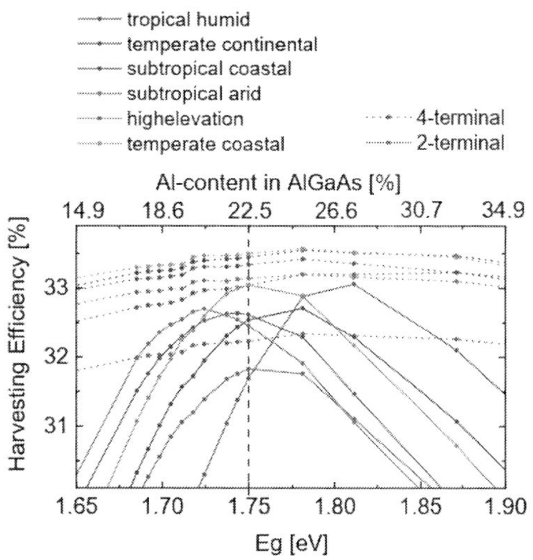

Figure 6 Harvesting efficiency for an $Al_xGa_{1-x}As$-on-silicon solar cell for different Al-content x and different locations.

The harvesting efficiency of the 4-terminal device shows only a very weak dependence on the band gap. Thus, a device with one single Al-content will be perfectly suitable for every location. The 2-terminal tandem shows a dependence of harvesting efficiency on band gap and especially the optimal band gap depends on the location. However, as for an industrialization the aim is, to produce only one single device, we decided to compare 2- and 4-terminal yield for one device with an Al-content of x=22.5%. This composition leads to good harvesting efficiencies considering the 2-terminal performance in all climate zones and is at the same time very close to the optimal working point of the 4-terminal solar cell. Thus, a fair comparison between the 2 device-configurations is possible with the selected top cell.

3.2 Results of the optimal device

In the following we have a more detailed look at the optimal device, build of an $Al_{0.225}Ga_{0.775}As$ top and a silicon bottom solar cell.

The harvesting efficiency (Figure 7) of both tandem devices is significantly higher than the harvesting efficiency of the single junction solar cell. The 4-terminal device shows the highest harvesting efficiency, with the median 0.5% absolute above the 2-terminal cell median value. Note that additional losses in the 4-terminal device as mentioned in section 2.2 are neglected, meaning that this gain will most likely vanish or even be overcompensated depending on the exact design of the 4-terminal interconnection.

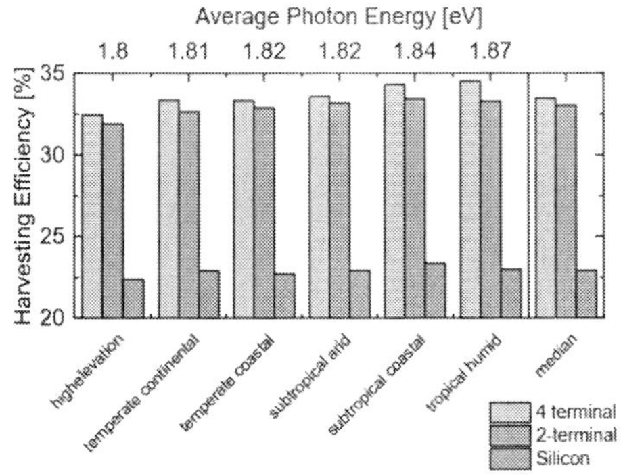

Figure 7 Harvesting efficiency (Annual Energy Output divided by Annual Incident Energy) of the 3 investigated devices for different climate zones.

While the performance ratio PR (PR is given as the harvesting efficiency divided by the STC efficiency) of the 2-terminal and 4-terminal device show similar performance for the locations with APE 1.8-1.82eV, the PR of the 4-terminal device for high APE regions (subtropical costal and tropical humid) is higher (see Figure 8). Thus, the median of the PR of the 4-terminal tandem is higher by 1.0%. This explains the above mentioned slightly higher harvesting efficiency of the 4-terminal device as compared to the 2-terminal configuration. Note that the STC efficiencies of the 4- and 2-terminal device in this optimized system are very similar (2-terminal: 34%, 4-terminal 34.4%).

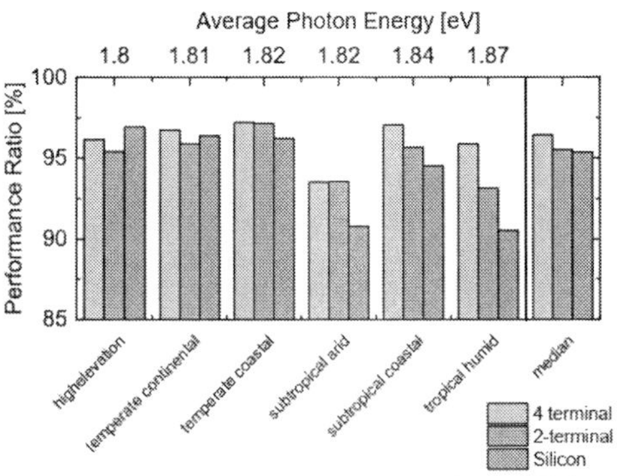

Figure 8 Performance ratio (Harvesting Efficiency divided by STC Efficiency) of the three investigated devices for different climate zones.

One can see that the PR depends on the average photon energy APE (APE describes the overall power in the

spectrum divided by overall number of photons per second) for the 2-terminal tandem and the silicon single junction in almost the same way (see Figure 9). The PR of the 4-terminal tandem device shows a lower dependence on the average photon energy. The PR of a 2-terminal tandem could be further improved by adjusting the top cell band gap to the APE. However, this would significantly increase the effort in production and might not be feasible. Note that the median performance of the 2-terminal tandem device is as good as for the pure silicon cell. Only the 4-terminal tandem performs particularly well in certain locations.
The lower points at APE=1.82 eV belong to the "subtropical arid" climate zone, where very high temperatures occur. As the silicon single-junction solar cell has a larger relative voltage temperature coefficient, it loses more in efficiency than the 2- and 4-terminal tandem devices. Consequently, the PR of the single junction is decreased stronger due to the high temperatures. When comparing the two locations with an APE of 1.82eV, it is observed that the temperature can have a severe effect to the performance ratio. This occurs for 2- and 4-terminal tandem cells in a very similar way.

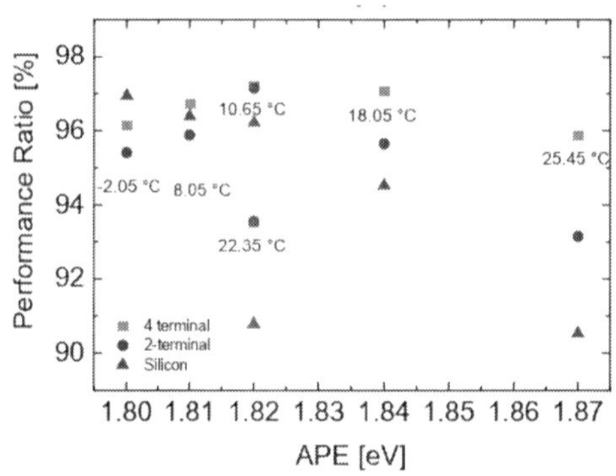

Figure 9 Performance ratio of the three investigated devices depending on the average photon energy The points are labeled with the average ambient temperature derived from the satellite data (only during daytime, as the night temperature has no influence on the device performance).

The calculation was repeated for a constant module temperature of T=298K, to prove that the difference in PR at an APE of 1.82 eV is simply caused by temperature effects and not just by different spectral influences of spectra with the same mean APE. One can see that the devices at the two locations with an APE of 1.82 eV behave very similar, when neglecting temperature effects. This proofs that the difference in the performance ratio of these two locations is mainly caused by temperature effects. It also hints towards APE in combination with ambient temperature being a good indicator for the performance ratio of this type of dual-junction solar cell.

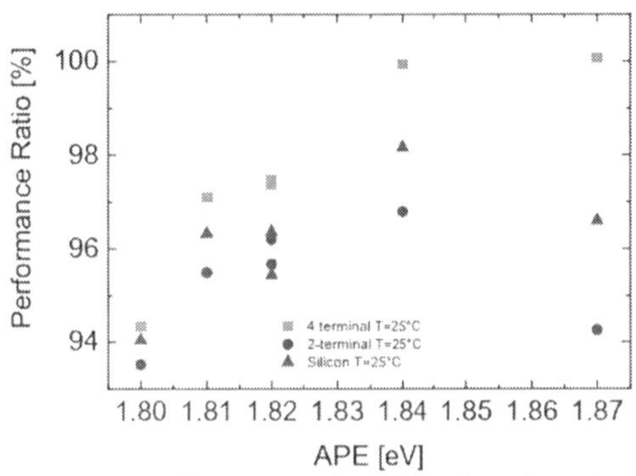

Figure 10 Performance ratio of the three investigated devices depending on the average photon energy for an assumed module temperature of *298* K.

Summarized, we find that the PR of well-designed 2-terminal III-V-on-silicon dual junction solar cells behave like the PR of silicon single junction solar cells. There are no relevant changes in the PR due to spectral variations that only occur in the 2-terminal dual junction but not in the single junction device. We show that this holds for all investigated climate zones. Thus, the STC efficiency gain of well-designed 2-terminal III-V-on-silicon dual junction solar cells will translate to nearly the same gain in energy harvesting efficiency.

It was also found that the PR of 4-terminal tandem solar cells (assuming the same EQE and IVs and no additional losses) is principally higher than the 2-terminal device PR. This is especially true for locations with high APE. However, this gain (~1% in median) will be mostly compensated by the neglected additional losses e.g. due to reflection at additional interfaces and losses due to more complex electrical wiring. Thus, well designed 2-terminal and 4-terminal devices will show similar harvesting efficiencies under outdoor conditions. In such a device the PR of a 2-terminal tandem is well comparable to the PR of a silicon single junction solar cell.

When comparing these results to temperature-independent calculations, i.e. temperature is set to *298* K, one observes that the module temperature can have a relevant effect on the performance ratios. The single-junction solar cell has a larger temperature coefficient than the tandem device. Thus, the performance ratio of the single junction is boosted at very cold locations, while the PR of the tandems behaves better at rather warm locations. At locations with very high APE (>1.84eV), that are typically warm, the slight impairment from current mismatch in 2-terminal tandems is compensated by the better temperature coefficient as compared to single junction solar cells.

4 CONCLUSION

We showed that spectral changes due to current mismatch do not lead to a relevant decrease in harvesting efficiency in 2-terminal III-V-on-silicon dual-junction solar cells. The assumption that current mismatch might be a severe issue in 2-terminal tandems does not hold for the investigated devices – they show similar performance as silicon single-junction devices. Note that the PR of 4-terminal tandems shows a lower dependence on location than 2-terminal tandems as well as single junction solar cells, and thus benefits at certain locations

5 ACKNOWLEDGEMENTS
This work was supported in part by the European Union's Horizon 2020 Research and Innovation Program within the Project SiTaSol under Grant 727497

6 REFERENCES

[1] I. M. Peters, H. Liu, T. Reindl, and T. Buonassisi, "Global Prediction of Photovoltaic Field Performance Differences Using Open-Source Satellite Data," *Joule*, vol. 2, no. 2, pp. 307–322, 2018.
[2] J. Lehr, M. Langenhorst, R. Schmager, S. Kirner, U. Lemmer, B. S. Richards, C. Case, and U. W. Paetzold, "Energy yield modelling of perovskite/silicon two-terminal tandem PV modules with flat and textured interfaces," *Sustainable Energy Fuels*, vol. 8, p. 506, 2018.
[3] M. T. Hörantner and H. J. Snaith, "Predicting and optimising the energy yield of perovskite-on-silicon tandem solar cells under real world conditions," *Energy & Environmental Science*, vol. 10, no. 9, pp. 1983–1993, 2017.
[4] M. Jošt, E. Köhnen, A. Morales Vilches, B. Lipovšek, K. Jäger, B. Macco, A. Al-Ashouri, J. Krc, L. Korte, B. Rech, R. Schlatmann, M. Topic, B. Stannowski, and S. Albrecht, "Textured interfaces in monolithic perovskite/silicon tandem solar cells: Advanced light management for improved efficiency and energy yield," *Energy & Environmental Science*, vol. 11, no. 12, pp. 3511–3523, 2019.
[5] M. Steiner, G. Siefer, T. Hornung, G. Peharz, and A. W. Bett, "YieldOpt, a model to predict the power output and energy yield for concentrating photovoltaic modules," *Prog. Photovolt: Res. Appl.*, vol. 23, no. 3, pp. 385–397, 2015.
[6] H. Schulte-Huxel, T. J. Silverman, M. G. Deceglie, D. J. Friedman, and A. C. Tamboli, "Energy Yield Analysis of Multiterminal Si-Based Tandem Solar Cells," *IEEE J. Photovoltaics*, pp. 1–8, 2018.
[7] R. Schmager, M. Langenhorst, J. Lehr, U. Lemmer, B. S. Richards, and U. W. Paetzold, "Methodology of energy yield modelling of perovskite-based multi-junction photovoltaics," *Opt. Express*, vol. 27, no. 8, A507, 2019.
[8] N. Tucher, O. Höhn, J. N. Murthy, Martinez, J. C. Steiner, M., A. Armbruster, E. Lorenz, B. Bläsi, J. C. Goldschmidt, N. Tucher, O. Höhn, J. C. Martinez, M. Steiner, B. Bläsi, and J. C. Goldschmidt, "Energy yield analysis of textured perovskite silicon tandem solar cells and modules," *Optics Express*, vol. 27, no. 20, A1419, 2019.
[9] E. F. Fernández, D. L. Talavera, F. M. Almonacid, and G. P. Smestad, "Investigating the impact of weather variables on the energy yield and cost of energy of grid-connected solar concentrator systems," *Energy*, vol. 106, pp. 790–801, 2016.
[10] D. Dirnberger, G. Blackburn, B. Müller, and C. Reise, "On the impact of solar spectral irradiance on the yield of different PV technologies," *Solar Energy Materials and Solar Cells*, vol. 132, pp. 431–442, 2015.
[11] Y. Boose, C. Heller, O. Stern, and M. Lynass, "The Effect of Spectral Mismatch on the Annual Energy Output of CdTe-, CIGS- and Si-Based Photovoltaic Modules," (eng), 2012.
[12] M. Alonso-Abella, F. Chenlo, G. Nofuentes, and M. Torres-Ramírez, "Analysis of spectral effects on the energy yield of different PV (photovoltaic) technologies: The case of four specific sites," *Energy*, vol. 67, pp. 435–443, 2014.
[13] H. Liu, Z. Ren, Z. Liu, A. G. Aberle, T. Buonassisi, and I. M. Peters, "The realistic energy yield potential of GaAs-on-Si tandem solar cells: a theoretical case study," *Opt. Express*, vol. 23, no. 7, pp. A382, 2015.
[14] O. Dupré, B. Niesen, S. de Wolf, and C. Ballif, "Field Performance versus Standard Test Condition Efficiency of Tandem Solar Cells and the Singular Case of Perovskites/Silicon Devices," (eng), *The Journal of Physical Chemistry Letters*, pp. 446–458, http://pubs.acs.org/doi/pdf/10.1021/acs.jpclett.7b02277, 2018.
[15] C. Gueymard, "SMARTS2, A Simple Model of the Atmospheric Radiative Transfer of Sunshine," Florida Solar Energy Center. University of Central Florida, Cocoa, FL, USA, 1995.
[16] *Photovoltaic (PV) module performance testing and energy rating – Part 4: Standard reference climatic profiles*, 61853-4, 2018.
[17] O. Höhn, J. N. Murthy, M. Steiner, N. Tucher, E. Lorenz, J. C. Goldschmidt, F. Dimroth, and B. Blasi, "Impact of Irradiance Data on the Energy Yield Modeling of Dual-Junction Solar Module Stacks for One-Sun Applications," *Photovoltaics, IEEE Journal of*, vol. 11, no. 3, pp. 692–698, 2021.
[18] M. W. Wanlass, K. A. Emery, T. A. Gessert, G. S. Horner, C. R. Osterwald, and T. J. Coutts, "Practical considerations in tandem cell modeling," *Solar Cells*, vol. 27, no. 1-4, pp. 191–204, 1989.
[19] Y. P. Varshni, "Temperature dependence of the energy gap in semiconductors," *Physica*, vol. 34, no. 1, pp. 149–154, 1967.
[20] P. Singh, S. Singh, M. Lal, and M. Husain, "Temperature dependence of I–V characteristics and performance parameters of silicon solar cell," *EuroSun2004*, vol. 92, no. 12, pp. 1611–1616, 2008.
[21] I. Vurgaftman, J. R. Meyer, and L. R. Ram-Mohan, "Band parameters for III–V compound semiconductors and their alloys," *J. Appl. Phys.*, vol. 89, no. 11, p. 5815, 2001.
[22] E. Barykina and A. Hammer, "Modeling of photovoltaic module temperature using Faiman model: Sensitivity analysis for different climates," *Sol Energy*, vol. 146, pp. 401–416, 2017.
[23] S. Bensalem, M. Chegaar, and M. Aillerie, "Solar Cells Electrical Behavior under Thermal Gradient," *Energy Procedia*, vol. 36, pp. 1249–1254, 2013.
[24] N. Tucher, J. Eisenlohr, P. Kiefel, O. Höhn, H. Hauser, I. M. Peters, C. Müller, J. C. Goldschmidt, and B. Bläsi, "3D optical simulation formalism OPTOS for textured silicon solar cells," *Opt. Express*, vol. 23, no. 24, pp. A1720, 2015.
[25] N. Tucher, J. Eisenlohr, H. Gebrewold, P. Kiefel, O. Höhn, H. Hauser, J. C. Goldschmidt, and B. Bläsi, "Optical simulation of photovoltaic modules with multiple textured interfaces using the matrix-based formalism OPTOS," *Optics Express*, vol. 24, no. 14, pp. A1083, 2016.

TUNNEL JUNCTION FORMATION ON SILICON P++ EMITTERS BY GAS IMMERSION LASER DOPING

G. Gaspar[1*], A. Guerra[1], F. C. Serra[1], A. S. Viana[2], J. Arumughan[3], I. Costa[1], D. M Pêra[1], J. A. Silva[1], L. Vines[4], J. M. Serra[1], K. Lobato[1]
[1]Instituto Dom Luiz (IDL), Faculdade de Ciências, Universidade de Lisboa, 1749-016, Lisboa, Portugal
[2]Centro de Química Estrutural, Faculdade de Ciências (CQE), Universidade de Lisboa, 1749-016, Lisboa, Portugal
[3]International Solar Energy Research Center (ISC), D-78467 Konstanz, Germany
[4]Department of Physics, Center for Materials Science and Nanotechnology, University of Oslo, N-0316 Oslo, Norway
*e-mail: gmgaspar@fc.ul.pt

ABSTRACT: Monolithically integrated 2-terminal Si/perovskite tandem solar cells have the potential to overcome the efficiency of Si single junction devices, whilst being cost-effective. The formation of the layers interconnecting the two sub-cells is a critical step, with several possible pathways for monolithic integration, such as amorphous Si, TCOs and crystalline silicon tunnel junctions. The latter one has the advantage of transparency, whilst also being more stable under high temperature steps required to deposit the perovskite sub-cell over the silicon sub-cell.
Here we report on the status of our approach at forming tunnel junctions directly on a silicon sub-cell using Gas Immersion Laser Doping (GILD). We believe that this approach is scalable, cost-effective, and simple to integrate into manufacturing lines, unlike tunnel junction formation by eg, ion implantation. Our GILD system has already been used to successfully form phosphorus doped thin (100 to 400 nm) layers on p-type Cz-Si wafers.
We now present the results of laser processing dependence according to wafer type (p^{++} emitter on p^{+} Cz-Si wafer vs p-type Cz-Si wafer without an emitter) and surrounding atmosphere. These influence the resultant surface morphology and the boron/oxygen depth profiles.
Keywords: Tandem Solar Cells; c-Si Tunnel Junction; Laser Doping

1 INTRODUCTION

On monolithically integrated 2-terminal Si/perovskite tandem solar cells, interconnecting the two sub-cells is a critical step that will impact the quality of the final device, since this layer must: (i) efficiently transport each sub-cell carrier types; (ii) have a high vertical conductivity, but low lateral conductivity; and (iii) be optically transparent (low absorption and reflectivity). Several approaches to produce the interconnecting layer have been developed such as amorphous Si (a-Si), transparent conducting oxides (TCOs) and crystalline silicon tunnel junctions [1,2].

An a-Si layer on top of the Si sub-cell restricts the fabrication of the top-cell layers to low temperature processing techniques (below 200 °C) since degradation of a-Si will occur above this temperature [3].

Common TCOs like indium zinc oxide (IZO) and indium tin oxide (ITO), have successfully been used as a recombination layer between the two sub-cells. Lab-scale solar cells using indium zinc oxide (IZO) and indium tin oxide (ITO), achieved power conversion efficiencies of 19.2% and 25.1%, respectively [4, 5]. Despite having good electro-optical properties, these oxides have indium, a rare earth metal, which might impact on future widespread TW PV deployment.

Crystalline silicon tunnel junctions are formed by joining two highly doped layers n^{++}/p^{++}, as the abrupt shift in the band diagram (valence and conduction bands aligned) allows for the electron tunnelling effect between these two layers, without energy loss.

Plasma-enhanced chemical vapor deposition (PECVD) has been used to create nanocrystalline silicon tunnel junctions on top of the silicon heterojunction sub-cell, followed by the formation of the perovskite top-cell, yielding a 25.2% power conversion efficiency device [6]. Silicon tunnel junction have also been produced by ion implantation and crystalline silicon sub-cell devices [7].

We believe that Gas Immersion Laser Doping (GILD) [8] is also a viable option to create a silicon tunnel junction, and unlike the two previously mentioned techniques, it is scalable, cost-effective, and easier to integrate into manufacturing lines. The GILD process consists in the main following steps: (i) sample is immersed in atmosphere containing dopant source; (ii) shallow surface melting occurs by laser pulsing; (iii) surface adsorbed dopant is dissolved into the melt and rapidly diffuses within melt region; (iv) melt recrystallizes; (v) dopant surface re-adsorption; and (vi) melting of adjacent region by another laser pulse.

Previously, we successfully shallow doped p-type Cz-Si wafers with phosphorus at atmospheric pressure, using the GILD technique. Morphological analysis of the processed Si revealed shallow melting of the surface with the formation of pyramidal-like structures. Secondary ion mass spectrometry (SIMS) depth profiling showed phosphorus incorporation into Si, with doping depth and concentration profile depending on laser processing parameters [9].

In this paper, we present the results delivered by our GILD setup for a different silicon wafer, a p^{+} Cz-Si wafer with a highly doped p^{++} emitter on top. Field-emission scanning electron microscopy (FE-SEM) and SIMS show that both atmosphere chemical composition and wafer type have an impact on the laser processed sample properties.

2 EXPERIMENTAL

The experimental setup comprises two main systems, the reaction chamber, and the laser apparatus. It is described in detail elsewhere [9].

A cylindrical reaction chamber (see Fig. 1), made in aluminum with an IR-transparent window, can hold up to 10 x 10 cm^2 silicon wafers. The chamber is connected to a gas feed system, capable of delivering both argon and $POCl_3$ through a bubbler containing the liquid phosphorus precursor [10,11].

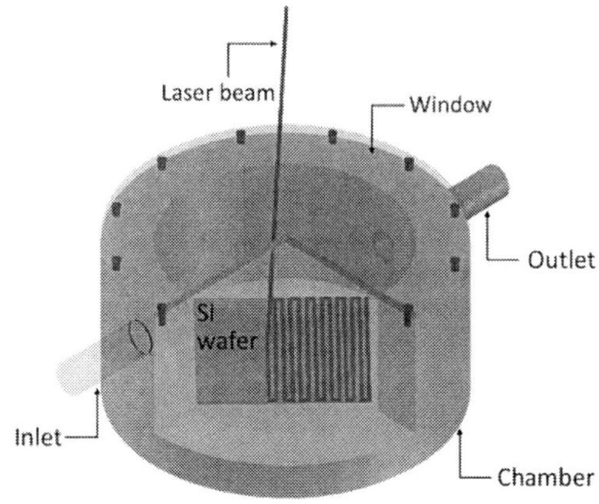

Figure 1: Reaction chamber schematic. A Si wafer placed in a $POCl_3$-rich atmosphere is laser processed.

After placing the silicon wafer and closing the chamber with the window, a series of vacuum and argon filling cycles were performed. Then, with the chamber's inner pressure set, an argon flow at 1.05 bar, passing through the $POCl_3$ bubbling system was maintained until the reaction chamber pressure equals, saturating its atmosphere with $POCl_3$ (vapor partial pressure of 0.037 bar at 293 K [12,13]).

Surface melting of the Si wafer was accomplished by laser rastering sequences with a 1064 nm solid-state pumped laser. Laser properties, such as pulse energy (0.04 mJ), pulse rate (500 kHz), beam diameter (20 μm) and distance between scan lines (10 μm), were kept constant. Laser rastering parameters, such as distance between consecutive laser spots (d = 1.0, 1.5, 2.0, 2.5 and 3.0 μm) and number of rastering sequences (n = 1, 3, 6, 9 times) were varied and combined to obtain a wide range of 10 × 5 mm^2 laser processed areas with different surface morphologies and doping depth profiles. Fig. 2 summarizes the process flow of the laser doping setup.

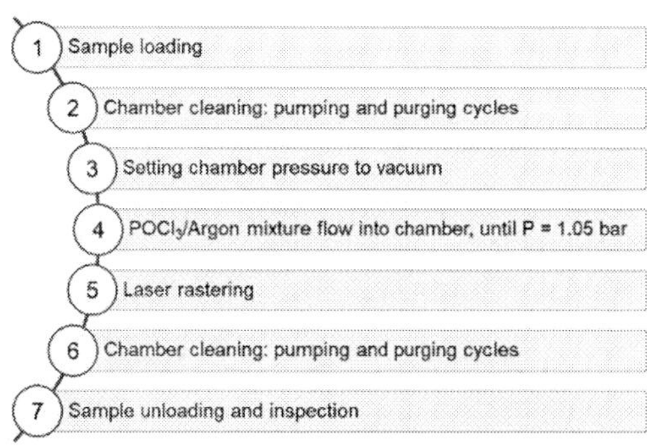

Figure 2: Summarized process flow diagram of the laser doping setup.

p^+ (100-oriented) Cz-Si wafers with a p^{++} emitter were used. The wafers are 100 mm in diameter, 525 μm thick, the base dopant concentration is ~ 10^{16} - 10^{17} cm^{-3} and the emitter side is mirror-like mechanically polished. Emitter peak surface dopant concentration and emitter depth are ~ 10^{20} cm^{-3} and ~ 0.4 μm, respectively, and were produced through a standard boron diffusion in a tube furnace apparatus.

3 RESULTS AND DISCUSSION

3.1 Surface morphology

Wafer type influences the surface morphology of the laser processed samples, as seen in FE-SEM micrographs of Fig. 3. Both p-type Cz-Si wafers and p^{++} emitter on p^+ Cz-Si wafers were laser processed.

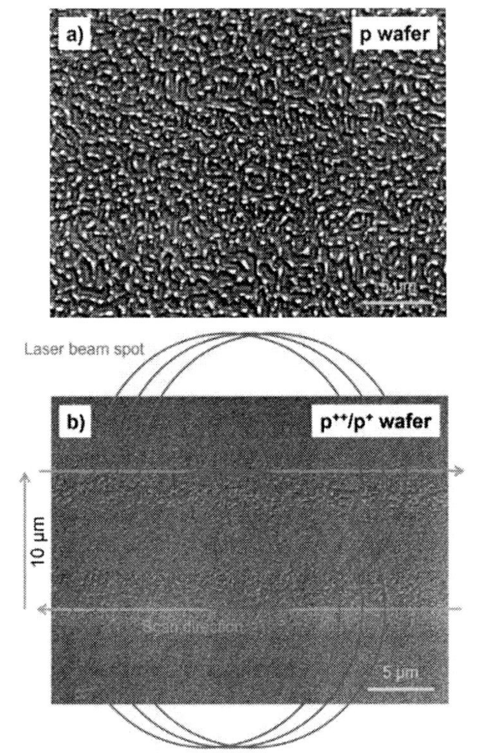

Figure 3: High-resolution FE-SEM micrographs (4000× magnification) of laser processed samples under argon atmosphere, using d = 2 μm, n = 1×: a) p-type Cz-Si wafer; b) p^{++} emitter on p^+ Cz-Si wafer.

The p-type Cz-Si wafer has a surface with a high density of pyramidal-like shape structures with sharp edges and sub-μm size bases, an indication of facet growth mechanism during melt recrystallization. It should be pointed out that while the structure base edges are in the sub-μm size range, their heights are in the 10-100 nm size range [9].

The p^{++} emitter on p^+ Cz-Si wafer has a smoother surface morphology, with both lower density and sized structures. This allows us to visualize the interaction of the successive laser pulses on the sample's surface. As the laser pulse moves in the direction marked by the superimposed green arrows, the rear part of the laser pulse creates consecutive circular recrystallized crusts separated by ca 2 μm (the distance between consecutive laser spots, marked as orange circles). The distance between scan lines is 10 μm, thus the next scan line above the previous one, remelts half the recrystallized surface below the center of the current laser pulse, leaving behind the same consecutive circular pattern, but in the opposite direction. It is also clearly noticeable the formation of structures at the edge of each scanned line distancing apart ca 10 μm. It is possible that these structures are formed as molten silicon is released from the laser spot due to Marangoni forces present in the molten areas [14].

The difference in surface morphology between the p-type Cz-Si wafer and p^{++} emitter on p^{+} Cz-Si wafer is clear evidence of higher laser interaction on the p-type Cz-Si wafer. This may be attributed to the following factors: (i) lower thickness (275 μm), of the p-type Cz-Si wafer, thus heat loss rate by conductivity is lower; (ii) lower boron doping level of the p-type Cz-Si wafer, 10^{15} cm^{-3}; and (iii) different surface reflectivity.

Fig. 4 presents the FE-SEM micrographs of the p^{++} emitter on p^{+} Cz-Si wafer. Samples were laser processed by varying the number of rastering sequences, with both sets processed under argon or $POCl_3$ atmospheres respectively.

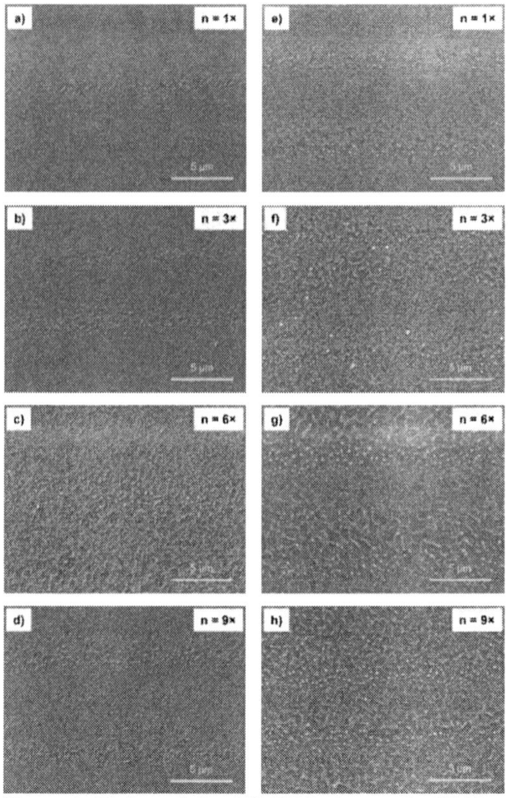

Figure 4: High-resolution FE-SEM micrographs (4000× magnification) of laser processed samples using a fixed laser spot spacing of d = 2 μm and varying number of rastering sequences (n = 1×, 3×, 6×, 9×). Comparison between argon, a) to d), and $POCl_3$, e) to h), atmospheres.

Under argon atmosphere there is a noticeable increase in structure size and density on the sample's surface as the number of rastering sequences increases from 1 to 9. The same behavior is also present on the samples processed under $POCl_3$ atmosphere, with the increase of laser interaction resulting in a rougher surface morphology. For the same number of rastering sequences, samples processed under $POCl_3$ atmosphere always showed higher surface roughness than the equivalent argon processed sample. Adsorbed $POCl_3$ plays a key role in increasing the laser interaction.

3.2 Boron and oxygen depth profiles

Boron and oxygen depth profile concentrations of p^{++} emitter on p^{+} Cz-Si wafer laser processed samples under argon atmosphere were measured with SIMS using a Cs primary ion beam at 15 keV to sputter craters of 100 × 100 μm^2 and obtaining data from a physical aperture of 8 μm diameter. Boron depth profile concentrations are shown in Fig. 5.

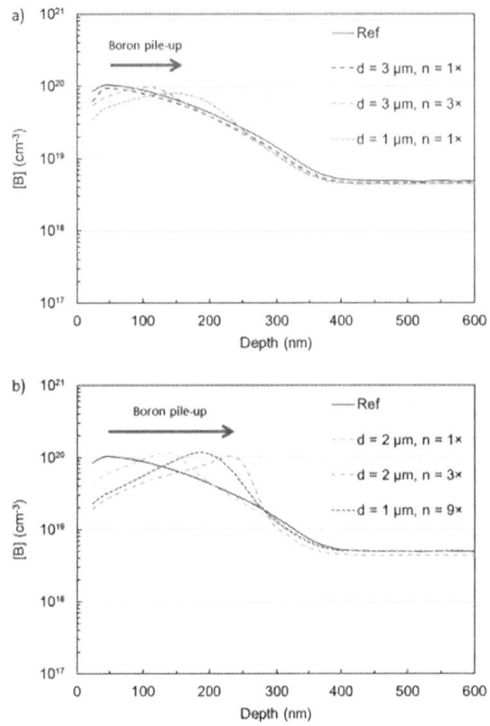

Figure 5: SIMS boron depth profiles for unprocessed reference sample (black solid line), and processed ones under argon atmosphere using: a) lower laser interaction conditions (d = 3 μm, n = 1×; d = 3 μm, n = 3× and d = 1 μm, n = 1×); b) higher laser interaction conditions (d = 2 μm, n = 1×; d = 2 μm, n = 3× and d = 1 μm, n = 9×).

Samples processed with lower laser interaction conditions show a small yet noticeable increase of boron concentration towards the maximum melt depth (taken as the point where the concentration profile matches the Ref), whilst there is also a decrease in concentration closer to the sample surface. This effect is more pronounced in samples processed with higher laser interaction conditions with a distinct "shoulder" becoming apparent.

This phenomenon, known as boron-pile up at maximum melt depth, is due to the segregation coefficient for boron being greater than 1 when high recrystallization velocities are present, as is the case here. As opposed to steady state melt conditions, where the boron segregation coefficient is 0.8 causing segregation of boron into the melt, a higher than 1 value causes the boron to pile-up at greater depths where the silicon begins to recrystallize [15].

Oxygen depth profile concentrations are shown in Fig. 6.

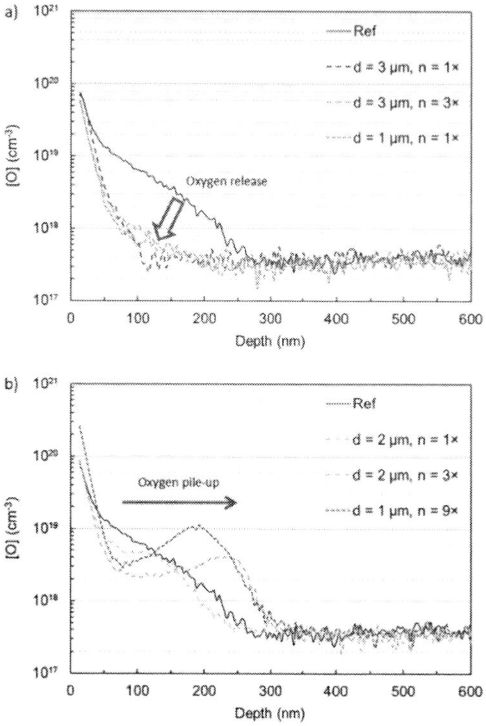

Figure 6: SIMS oxygen depth profiles for unprocessed reference sample (black solid line), and processed ones under argon atmosphere, using: a) lower laser interaction conditions ($d = 3$ µm, $n = 1\times$; $d = 3$ µm, $n = 3\times$ and $d = 1$ µm, $n = 1\times$); b) higher laser interaction conditions ($d = 2$ µm, $n = 1\times$; $d = 2$ µm, $n = 3\times$ and $d = 1$ µm, $n = 9\times$). Nb: the apparent oxygen concentration of ca. 4×10^{17}cm^{-3} is due to the limits of the SIMS system (eg, O_2 gas entering the vacuum chamber) and do not represent the true bulk oxygen content.

Samples processed with lower laser interaction conditions present a decrease in oxygen concentration. The oxygen, due to its high vapor pressure, evaporates from the molten silicon as SiO gas before the rapid recrystallization [16].

Oxygen pile-up occurs when the sample is processed with higher laser interaction conditions, as also observed for boron. This is in-line with the oxygen segregation coefficient being greater than 1 under these rapid recrystallization conditions [17]. However, oxygen concentration remains high at the surface. The maximum melt depths (taken as where the concentration profile returns to the Ref baseline), are in line with those inferred from the boron concentration profiles.

4 CONCLUSIONS

The experimental setup has proven effective in surface shallow melting of p^{++} emitter on p^{+} Cz-Si wafers, in a $POCl_3$ environment at atmospheric pressure.

Wafer surface doping level influences laser interaction, as laser interaction is weaker with higher doping level.

For p^{++} emitter on p^{+} Cz-Si wafers we observe: (i) increasing the number of rastering sequences results in greater laser interaction, as SEM micrographs show rougher surface morphology; and (ii) addition of $POCl_3$ doping atmosphere results in higher surface roughness.

SIMS depth profiles of p^{++} emitter on p^{+} Cz-Si wafer show: (i) for boron, a distinct "shoulder" appears as the laser interaction is more intense (boron pile-up towards the maximum melt depth [5]); and (ii) oxygen is released for low laser interaction, whilst pile-up occurs for higher laser interaction, representing segregation into the bulk.

REFERENCES

[1] M. Jošt *et al.*, Advanced Energy Materials 10 (2020) 1904102.
[2] Y. Ko *et al.*, Advanced Materials 32 (2020) 2002196.
[3] S. Albrecht *et al.*, Energy & Environmental Science 9 (2016) 81.
[4] J. Werner *et al.*, The Journal of Physical Chemistry Letters 7 (2016), 161.
[5] P. Schulze *et al.*, Solar RRL 4 (2020) 2000152.
[6] F. Sahli *et al.*, Nature Materials 17 (2018) 820.
[7] P. Bellanger *et al.*, IEEE Journal of Photovoltaics 8 (2018) 1436.
[8] G. B. Turner *et al.*, Applied Physics Letters 39 (1981) 967.
[9] G. Gaspar *et al.*, Proceedings 37th European Photovoltaic Solar Energy Conference (2020) 765.
[10] H. Li *et al.*, Frontiers in Energy 11 (2017) 42.
[11] H. Ghembaza *et al.*, Silicon 10 (2018) 377.
[12] D. R. Stull, Industrial & Engineering Chemistry 39 (1947) 518.
[13] National Institute of Standards and Technology, NIST Chemistry WebBook, SRD 69 (2021, September 8).
[14] E. Haro-Poniatowski *et al.*, Journal of Applied Physics 115 (2014) 224309
[15] P. C. Lill *et al.*, Materials 10 (2017) 189.
[16] G. Gaspar *et al.*, Journal of Crystal Growth 418 (2015) 176.
[17] M. Ametowobla, Characterization of a laser doping process for crystalline silicon solar cells, PhD Thesis, University of Stuttgart (2010).

ACKNOWLEDGEMENTS

This work was supported by the S-LoTTuSS project through the grant agreement PTDC/CTM-CTM/28962/2017, financed by national funds FCT/MCTES (PIDDAC), and by FCT – Project UIDB/50019/2020 – IDL – Instituto Dom Luiz.
The authors acknowledge the support from The Research Council of Norway (RCN) through Project No. 239895/F20.

DUAL QUANTUM TUNNELING IN A MONOLITHIC N-I-P PEROVSKITE/C-SI TANDEM DEVICE: BOTTOM CELL WITH MODIFIED SQIS STRUCTURE

Z.Q.Ma*, K.J.Wu, Z.X.Lan, Y.L.Wang, F.Xu and L.Zhao
SolarE PV joint laboratory, Department of Physics, Shanghai University, Shanghai 200444, P.R.China
*Corresponding author Dr. Z.Q.Ma, e-mail:zqma@shu.edu.cn

ABSTRACT: An efficient photovoltaic framework has been constructed with metal organic-inorganic halide perovskite upper cell and asymmetric heterojunction silicon-based dual carrier's tunneling as bottom cell, in which a natural recombination layer of about 80 nm ITO film is deposited by r.f magnetron sputtering of hole selected ultra-thin passivating contact a-SiO_x(In) layer (1.3 – 1.8 nm) on silicon substrate coupled with ITO capping layer. At rear side of the Si wafer a wet chemical oxidized SiO_x (<1.6 nm) and subsequently d.c magnetron sputtered TiN_x coupling layer of SiO_x/TiN_x acts as electron selective contact. Thus, the quantum tunneling from both hole and electron in the front or back of silicon wafer has been figured out on both sides, respectively. Besides the quantum tunneling and the pinholes within the silicon oxides playing a crucial role, the delocalized states and negative charged indium (In) are localized between ITO film and Si substrate are significant for hole diffusion and field passivation effects in the device. Finally, the Si-based bottom cell is covered by a structure of merit n-i-p perovskite cell as a monolithic tandem PV system. The photoelectric conversion efficiency of the tandem device has been numerically evaluated to be more than 30 %.
Keywords: Heterojunction, tandem, high-efficiency, n-Si, perovskite

1 INTRODUCTION

In the current and future, a direct way for improving the power conversion efficiency (PCE) of a photovoltaic (PV) device is to assemble multi-junction tandem solar cells, basing on the energy bands matching and the state-of-the art process of materials. Hence, the exploration and innovation-driven significance of science and technology has been far greater than the co-dependence of levelized cost of electricity (LCOE) on manufacturing cost and efficiency of the PV devices and modules [1]. So that the evaluation of the value of efficiency (VOE) together with ideal scenarios project boosts of about 4% in absolute efficiency for perovskite on silicon and 6% for GaAs on silicon solar cells has been reported [1, 2]. But we believe that the balance between efficiency and cost will be the ultimate goal of the photovoltaic community. In this sense, it is necessary to construct new photovoltaic devices to improve efficiency as much as possible. In this way, we try to propose a novel perovskite / c-Si monolithic tandem device that is possible to achieve the above purpose.

For single-junction solar cells, it is well known that the ultimate PCE is governed by the Shockley–Queisser theoretical limit (29.43%) [3]. The primary energy loss of a single-junction solar cell includes the unabsorbed long-wavelength photons and thermalization of high-energy carriers [4]. However, the appropriate overlying design may planarity utilize solar energy, i.e., high-energy photons are absorbed by the upper wide-bandgap subcell, and lower energy photons are absorbed by the bottom narrow-bandgap subcell [5], such as a tandem cell on the monolithic wafer.

In accordance with the PV enhancement principles mentioned above, a realistic option is proposed for the tandem scheme. As we know, high photoelectric conversion efficiency crystalline silicon PV device, so far, besides SHJ (heterojunction) and TOPCon's (homojunction), respectively, other types have been developed with improved technologies in time line. Recently, we cultivate a simplified method to fabricate a dual tunneling Si-based PV device (named as SQIS) with ITO/a-SiOx(In) and SiOx/TiNx passivation contact layers for either hole and electron selectivity, respectively [6-8]. On the other hand, because the overlayer of ITO film, its optical and electronic structures in the front surface of the device are suitable for the connection to a wide-band gap device as an upper subcell.

Therefore, an architecture of metal organic-inorganic halide perovskite solar cell (PSCs) as upper-cell with our silicon based bottom-cell (SQIS) have become the most promising new-generation tandem PV device. To date, halide perovskites also represent the only polycrystalline thin-film absorber technology that has enabled >25 % efficiency for wide-bandgap solar cells, making wide-bandgap PSCs uniquely positioned to enable high efficiency and low-cost tandem solar cell technologies by coupling wide bandgap perovskites with low-bandgap absorbers, such as crystalline silicon, CIGS or itself with lower band gaps [9, 10]. Thus, in this report, an effective tandem scheme is present for the n-i-p perovskite / c-Si solar cell and its efficiency has been evaluated through numerical calculation.

2 MONOLITHIC TANDEM PHOTOVOLTAIC SCHEME

2.1 Preparation of SQIS device

A simplified SQIS (semiconductor-quasi-insulator-semiconductor) device was fabricated by r.f magnetron sputtering of ITO film on n-Si. The detail experimental methods and sample processing can be inferred to our previous report [11]. However, in the present investigation, the sputtering conditions have been changed to forward power of 120 W, plasma pressure of 0.5 Pa, the thickness of ITO films in the range of 80-100 nm, and the ITO ceramic target, being composted of 90 wt. % In_2O_3 and 10 wt. % SnO_2, from BJYAMT (Beijing Jinyuan Advanced Materials Technology Co).

For this PV device, the delocalized states in ultra-thin a-SiO_x(In) layer may result from the In-O-Si bonding like molecule after an efficient doping of indium into a-SiO_x(In) medium during the energetic collision and diffusion of both In and O immerged on the top layers of n-Si substrate. The delocalized states for our photovoltaic device might play an important role for the transport of photo-generated carriers (electron and hole). In the previous investigations [11,12], we suggested a mechanism of oxygen vacancy assisted-tunneling of holes through the ultra-thin a-SiO_x(In) layer, complementarily. Now we propose a view of point that the delocalized states in the a-SiO_x(In) layer also bring into play quantum tunneling by those factors, such as the existence of stable amorphous phase in the ultra-thin a-SiO_x(In) layer; the high probability of forming In-O-Si bonding within the a-SiO_x(In) materials and leading to LUOM and HOMO orbitals; the very low series resistance and higher filled factor (FF>0.73) present for the ITO-SiO_x(In)/n-Si PV device. Therefore, the multi-functions such as good passivation, built-in field, inversion layer and carriers tunneling has been integrated into the a-SiO_x(In) materials, which may be a good candidate for the front hole selective contact of silicon-based high efficient heterojunction solar cells.

Moreover, the co-present of O and In together with Si within the ultra-thin intermediate region between ITO and n-Si are detected by X-ray photoelectron spectroscopy (XPS) measurement. The ternary compound In-Si-O_x is predicted by first-principles calculation together with density functional theory (DFT). It was found that there were two gap states in the a-SiO_x(In) layer induced by two indium atoms and were located at E_v+4.60 eV and E_v+3.85 eV, respectively, acting as "empty levels" for photon-generated hole carriers' tunnel transport. The strong inversion layer in the surface of n-Si substrate is associated with In-Si-O_x buffer layer in which a direct tunneling path of hole is produced. Additionally, a level localized at Ev + 0.30 eV, as the acceptor state, associated with negative indium ion in a-SiO_x(In) layer is found by DFT, which also provides a field passivation effect [12].

For the test device having more than 30% reflectivity in a non-textured planar surface, the relationship of current-voltage with or without SiO_x/TiN_x passivation contact layers is shown in Fig.1. It can be found that a merit PV feature has been illustrated, while the PCE is relative lower owing to the high surface reflection and the non-optimal implied potential (V_i) from the work function difference between ITO film and n-Si wafer for this heterojunction device. However, the ideal SQIS device with back SiO_x/TiN_x passivation contact has an improved conversion efficiency more than 23.0 %, when taking an optimized optical and electric structures through AFORS-HET simulation program [13], and the resulted PV parameters are listed in table I.

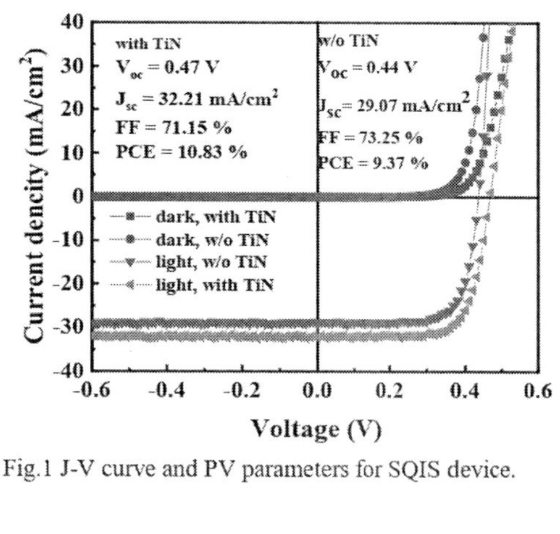

Fig.1 J-V curve and PV parameters for SQIS device.

2.2 Preparation of n-i-p perovskite device PSCs

The state of the art processing for n-i-p metal halide perovskite solar cell (PSCs) is a technological imitation of report [14], in which the composition of $Cs_{0.05}MA_{0.15}FA_{0.8}Pb(I_{0.75}Br_{0.25})$ and the band gap of 1.65 eV for the organic-inorganic halide perovskite material were adopted, as a top metal halogen perovskite cell in the Si-based tandem device. The perovskite device uses TiO_2 film as the electron selective layer, and its conduction band bottom E_c is about 4.26 eV. Organic material Spiro-Me OTAD is as hole selection layer of metal halide perovskite cell with a E_c of 2.05 eV. As the front subcell, the wide-bandgap PSCs simultaneously exhibited a high Voc (low V_{oc} deficit, E_g–V_{oc}: 410-490 V), high fill factor (FF), and sufficient photocurrent density to match the bottom junction. The most potential for this optimal PSCs device was that the incorporation of MA did not necessarily reduce the formation of defects, but the incorporation of the dipolar MA cation in mixed cation–halide wide-bandgap perovskites cured deep trap defects, resulting in a more defect-tolerant material. Thus, the configuration of hybrid materials system is the best choice for our purpose.

2.3 Assembling of SQIS and PSCs devices

The shcematic of monolithic tandem PV device with n-i-p perovskite and n-Si absorbers is shown in Fig. 2. The connected recombination layer of the both cells has been deployed by TiO_2/ITO for electron and hole transport and meeting at the middle intermediate region, in order to make a current matching.

3 PREDICTION OF CONVERSION EFFICIENCY OF THE TANDEM SYSTEM

The prediction of the PCE of the planar n-i-p perovskite/SQIS with stacked passivating contact and transport layers mainly obeys the following three principles:

1) for the entire stack device, silicon-based short circuit current density loses nearly half in that the EQE response by two current matching cells;

2) due to the decrease of short circuit current density and the recombination of interconnection layer, the open-circuit voltage has a little lost (deficit);

3) fill factor uses the comparison of the simulation results and the empirical formula.

3.1 Short-circuit current density Jsc, the relationship between short-circuit current density and external quantum efficiency can be expressed as formula

$$J_{sc} = q\int F(\lambda)IQE(\lambda)[1 - R(\lambda)]d\lambda = q\int F(\lambda)EQE(\lambda)d\lambda \quad \text{(eq. 1)}$$

Where IQE (λ)[1-R (λ)] = EQE (λ), R (λ) is reflectivity, EQE (λ) and IQE (λ) are external quantum efficiency and internal quantum efficiency, respectively, F (λ) is the density of incident photon flow, q is the elemental charge. Metal halogen perovskite and silicon-based subcells should have good current matching by adjusting the thickness of the recombination layer and the band gap, so that the J_{sc} of tandem cells is considered to be half the J_{sc} of silicon-based cell.

So that a value of about 20.0 mA/cm² has been taken as the short-circuit current density of the monolithic two-terminal tandem PV device.

3.2 Open-circuit voltage V_{oc}, the open-circuit voltage of the tandem cell is conceptually the sum of the two subcell's V_{oc}. However, due to the loss of short-circuit current density due to the shelter from top perovskite cell, the open-circuit voltage of silicon-based bottom cell has a certain loss, thus, the total open-circuit voltage is as follows:

$$V_{oc,tandem} = V_{oc,top} + V_{oc,bottom} + \Delta V_{oc} \qquad \text{(eq. 2)}$$

Where $V_{oc,top}$ is open-circuit voltage of perovskite top cell, $V_{oc,bottom}$ is that of bottom cell, while ΔV_{oc} is a loss arisen from the decrease of short-circuit current density and a recombination cause of electron and hole in the interconnection layers of the both cells. The reasons for ΔV_{oc} mainly come from two aspects:

(i) Deficiency from short-circuit current density: The V_{oc} of the subcells in the tandem is estimated by the logarithmic dependence of V_{oc} on the photocurrent density [15]. The deficit of open-circuit voltage (ΔV_{oc}) on the SQIS device in this part is:

$$\Delta V_{oc} = \frac{k_B T}{e}\left[ln\left(\frac{J_{sc,before} + J_0}{J_{sc,after} + J_0}\right)\right] \qquad \text{(eq. 3)}$$

Where k_B is Boltzmann constant, T is absolute temperature, J_o is a recombination current density. Since the order of magnitude of J_o in both cells is a few tens of fA/cm^2 at most, and the order of magnitude of J_{sc} in both cells is in the level of mA/cm^2, so J_o can be ignored. Thus, if $J_{sc,before} = 40.05\ mA/cm^2$, or $J_{sc,after} = 20.00\ mA/cm^2$, it can be derived that the ΔV_{oc} is about 18.05 mV. Nevertheless, for upper perovskite cell, the short-circuit current density decreases slightly, hence the deficit of its open-circuit voltage is negligible.

(ii) Deficit from interconnection layer: The intermediate layer for connecting two subcells inevitably produces a lower carrier recombination rate, resulting in the accumulation of electrons from perovskite cell and holes from silicon-based cell in the intermediate layer [16]. It has been estimated for a tandem system that about 20 mV of ΔV_{OC} is lost mainly due to the electrical loss at the interconnection layer [17]. Thus, the total loss of open-circuit voltage ΔV_{OC} is about 38.05 mV, combined with the loss of open-circuit voltage caused by halving the short-circuit current density mentioned above. It is derived that the open-circuit voltage of planar n-i-p perovskite / SQIS tandem cells with SiO_x-TiN rear contact can reach about 1.892 V.

3.3 Filled factor pFF, ideally, pFF is just a function of open circuit voltage. Owing to the fill factor of any solar cell is dependent on the ratio of carrier's transport velocity

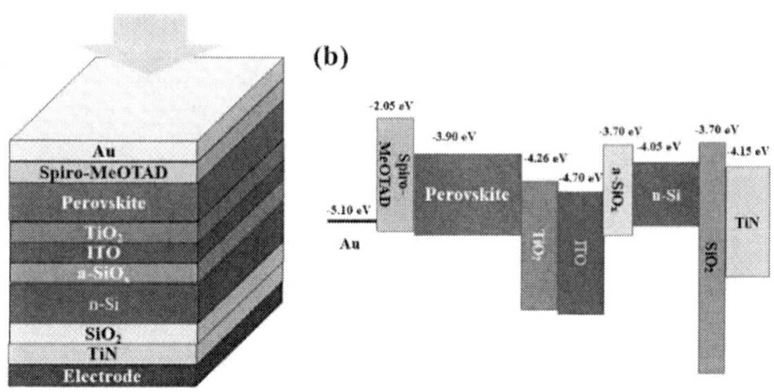

Fig.2 (a) Schematic diagram of n-i-p perovskite / n-Si stacking; (b) Roughly energy band profiles of these photovoltaic materials.

Table I: The photoelectric parameters acquired by experimental and numerical calculation

Individual and tandem cells	V_{oc} (V)	J_{sc} (mA/cm^2)	pFF (%)	η (%)
SQIS with SiO_x/TiN_x	0.710	40.05	81.19	23.08
Planar n-i-p perovskite [14]	1.220	21.20	80.50	20.70
Planar n-i-p perovskite / SQIS with SiO_x/TiN_x	1.892	20.00	81.19	30.72

to recombination velocity, so do the tandem cell. Therefore, the fill factor of the tandem cell should be dominantly determined by the modified SQIS cell, additionally, the following approximate empirical formula 4 can be deployed to estimate the pseudo-fill factor (pFF), because it is only a function of open circuit voltage (V_{oc}) under ideal conditions and without the consideration of parasitic shunt resistance [16, 18].

$$pFF = [\upsilon - \ln(\upsilon + 0.72)]/(\upsilon + 1) \qquad \text{(eq. 4)}$$

Where υ is the normalized open-circuit voltage given by $\upsilon = \frac{q V_{oc}}{n K_B T}$, where q is electron charge, while n is the ideal factor of a diode and in the range of 1-2. Generally, the pFF value is in the range of 75.6 % - 85.1 % for a tandem cell, dependent on the quality of the cells, but in our case, the optimized value of pFF is estimated to be about 81.19 % for the modified SQIS device.

3.4 Photoelectric conversion efficiency η, the conversion efficiency of the tandem cell can also be calculated by the formula as

$$\eta = (J_{sc}V_{oc}\, pFF/P_{in}) \times 100\% \qquad \text{(eq. 5)}$$

Thus, the results of J_{sc}, V_{oc}, pFF and conversion efficiency η are summarized in table I for the individual and tandem cells.

4 CONCLUSION

In this research, a higher prospective conversion efficiency of 30.72 % for the planar n-i-p perovskite/SQIS with SiO_x-TiN tandem cells has been predicted through an available estimation on the maximum values of J_{sc}, V_{oc} and pFF, mentioned in section 3.4 and Table I, where the photovoltaic parameters of both perovskite and n-Si based devices are taken from reference [13,14] and optimized SQIS by means of the state of the art fabrication and numerical simulation, respectively. The near-infrared response and enhancement of EQE after SiO_x-TiN back contact to Si-based SQIS device is crucial for improving the performance of modified SQIS and tandem cells as well.

Moreover, there are three key points to successfully realize in the monolithic tandem photovoltaic device: first, the rational design and preparation of silicon-based SQIS device; second, the band gap of halogen perovskite is more than 1.65 eV; third, the intermediate electron-hole interconnection layer and other auxiliary thin film materials are preferred. Thus, the requirements of the photovoltaic device principle and the goal of low cost and high efficiency can be achieved.

5 ACKNOWLEDGEMENT

This work is mainly supported by the National Natural Science Foundation of China (No. 61874070, 61674099 and 61274067) and the R&D Foundation of the SHU-SOENs PV Joint Laboratory (No. SSE0700601).

6 REFERENCES:

[1] I.M. Peters, C.D.R. Gallegos, S.E. Sofia and T. Buonassisi, The value of efficiency in photovoltaics, Joule 3 (2019) 2732.
[2] Z. J. Yu, J.V. Carpenter and Z.C. Holman, Techno-economic viability of silicon-based tandem photovoltaic modules in the United States. Nat. Energy 3 (2018)747.
[3] W. Shockley and H. J. Queisser, Detailed balance limit of efficiency of p–n junction solar cells. J. Appl. Phys. 32 (1961) 510.
[4] C. A. Nelson, N. R. Monahan and X.-Y. Zhu, Exceeding the Shockley-Queisser limit in solar energy conversion. Energy Environ. Sci. 6 (2013) 3508.
[5] G. E. Eperon, M. T. Hörantner and H. J. Snaith, Metal halide perovskite tandem and multiple-junction photovoltaics. Nat. Rev. Chem. 1 (2017) 0095.
[6] X. M. Song, M. Gao, Z. G. Huang, B. C. Han, Y. Z. Wan, Q. Y. Lei, and Z. Q. Ma, Interface Properties of ITO/n-Si Heterojunction Solar Cell: Quantum Tunneling, Passivation and Hole-Selective Contacts, Solar Energy 173 (2018) 456.
[7] M. Gao, Y.Z. Wan, Y. Li, B.C. Han, W.L. Song, F. Xu, L. Zhao and Z.Q. Ma, An effective passivation and tunneling hybrid a-SiO_x(In) layer in ITO/n-Si heterojunction photovoltaic device, ACS Appl. Mat. Interface 9 (2017) 17565.
[8] K.J. Wu, M. Gao, H.W. Du, D.Y. Chen, L. Zhao and Z.Q. Ma, Improvement of the performance of ITO/a-SiOx/n-Si device by controllable sputtering power and reducible interface states, Materials Science in Semiconductor Processing 105 (2020) 104702.
[9] N. J. Jeon, J. H. Noh, W. S. Yang, Y. C. Kim, S. Ryu, J. Seo and S. I. Seok, Compositional engineering of perovskite materials for high performance solar cells. Nature 517 (2015) 476.
[10] Q. Jiang, Z. Ni, G. Xu, Y. Lin, P. N. Rudd, R. Xue, Y. Li, Y. Gao, J. and Huang, Interfacial Molecular Doping of Metal Halide Perovskites for Highly Efficient Solar Cells. Adv. Mater. 32 (2020) 2001581.
[11] H. W. Du, J. Yang, Y. H. Li, F. Xu, J. Xu, and Z. Q. Ma, Preparation of ITO/SiOx/n-Si solar cells with non-decline potential field and hole tunneling by magnetron sputtering, Appl. Phys. Lett. 106 (2015) 093508.
[12] Y. Z. Wan, M. Gao, Y. Li, H. W. Du, Y. H. Li, H. B. Guo, F. Xu, Z. Q. Ma, Potentiality of delocalized states in indium-involved amorphous silicon oxide, Appl. Phys. Lett 110 (2017) 213902
[13] K.J. Wu, Performance Optimization and Interface Electronic Structure of Silicon-based Indium-Tin Oxide Heterojunction PV Devices, a dissertation for the degree of master in science, Chapter 4, Shanghai University of China, April 2020.
[14] H. Tan, F.L. Che, M.Y. Wei, Y.C. Zhao, M. I. Saidaminov, P. Todorović, D. Broberg, G. Walters, F.R. Tan, T.T. Zhuang, B. Sun, Z.Q. Liang, H.F. Yuan, E. Fron, J. Kim, Z.Y. Yang, O. Voznyy, M. Asta and Edward H. Sargent, Dipolar cations confer defect tolerance in wide-bandgap metal halide perovskites, Nature Communications 9 (2018) 3100.
[15] U. Wurfel, D. Neher, A. Spies and S. Albrecht, Impact of charge transport on current-voltage characteristics and power-conversion efficiency of organic solar cells, Nat Commun. 6 (2015) 6951.
[16] P. Campbell and M.A. Green, The limiting efficiency of silicon solar cells under concentrated sunlight, IEEE Transactions on Electron Devices, 33 (1986) 234.
[17] B.F. Liu, L.S. Bai, Z. Chen, X.D. Zhang, D.K. Zhang, J. Ni, Q. Huang, C.C. Wei, J. Sun, X.L. Chen, H.Z. Ren, G.F. Hou, G.C. Wang and Y. Zhao, High efficiency triple junction thin film silicon solar cells with optimized electrical structure, Progress in Photovoltaics, 23 (2015) 1313.
[18] L. Meng, Y. Zhang, X. Wan, C. Li, X. Zhang, Y. Wang, X. Ke, Z. Xiao, L. Ding, R. Xia, H.L. Yip, Y. Cao and Y. Chen, Organic and solution-processed tandem solar cells with 17.3% efficiency, Science, 361 (2018) 1094.

A FRAMEWORK FOR IMPLEMENTING REQUIREMENTS ON THE CARBON FOOTPRINT OF PHOTOVOLTAIC MODULES UNDER THE ECODESIGN POLICY

Polverini[1], D., Espinosa[2], N., Klos[1], C., Arcipowska[2], A.

[1]European Commission, DG Internal Market, Industry, Entrepreneurship and SMEs, Brussels, Belgium
[2]EuropeanCommission, Joint ResearchCentre, Spain

Emails: davide.polverini@ec.europa.eu; nieves.espinosa@ec.europa.eu; caspar.klos@ec.europa.eu; aleksandra.arcipowska@ec.europa.eu

ABSTRACT: In December 2020, the European Council [1] endorsed a new binding European Union (EU) target for a net domestic reduction in greenhouse gas emissions of at least 55% by 2030, compared to the 1990 level. Renewable energies will certainly play an important role in policies in the short and medium term; to date, the 2018 Renewable Energy Directive [2] establishes a binding renewable energy target for the EU for 2030 of at least 32%. Within this framework, photovoltaics (PV) is expected to be a major contributor, therefore it is crucial to ensure that newly installed PV products in the EU are environmentally-friendly and do not create new future burdens on the environment. The aim of the paper is to assess the feasibility of requirements on the carbon footprint of PV modules under the already existing legal framework of the Ecodesign Directive.

Keywords: photovoltaics, emission reduction, carbon footprint

1 AIM AND APPROACH USED

In December 2020, the European Council [1] endorsed a new binding European Union (EU) target for a net domestic reduction in greenhouse gas emissions of at least 55% by 2030, compared to the 1990 level. Renewable energies will certainly play an important role in policies in the short and medium term; to date, the 2018 Renewable Energy Directive [2] establishes a binding renewable energy target for the EU for 2030 of at least 32%. Within this framework, photovoltaics (PV) is expected to be a major contributor, therefore it is crucial to ensure that newly installed PV products in the EU are environmentally-friendly and do not create new future burdens on the environment.

A preparatory study on the environmental impacts – and related policy approaches - of PV products [3] was finalised by the Joint Research Centre of the European Commission at the end 2019. In the course of the preparatory study, a streamlined life cycle assessment for PV products was performed using the standard methodology for the Ecodesign of Energy Related Products (MEErP) [5]. The assessment of the resulting environmental hotspots was complemented by a literature screening of the life cycle assessments published for this family of products. The findings were subsequently translated into proposals for policy requirements [6].

The study concluded that improvements in energy yield, long-term performance and circularity of PV modules, inverters and systems could be ensured by (mandatory) legal instruments, in particular by implementing measures in the framework of the Ecodesign Directive [4]. The Ecodesign Directive requires manufacturers placing products on the EU market to mandatory minimum energy efficiency requirements and other environmental requirements such as water consumption or material efficiency aspects. Furthermore and in parallel to the implementation of the Ecodesign Directive, at European Commission level there are two ongoing initiatives[1] aimed at exploring the feasibility of requirements concerning the environmental footprint of products; implementing measures to this regard could be expected in the medium term.

The global warming potential (GWP) category was identified as one of the primary parameters that can be optimised. For the PV modules it is within the manufacturing phase where the 'carbon'[2] emissions have shown a significant contribution - depending on the considered end of life (EoL) scenario. These findings are also in line with the kind of requirements that can be established in the framework of Ecodesign policy, i.e. requirements that can be verified on the products when they enter the EU market. Hence, a declaration of the carbon footprint of the photovoltaic modules required by the Ecodesign regulation should target at least this stage and at least carbon emissions. In the future a material passport may be a vehicle for the information on the environmental impacts.

The aim of the paper is to assess the feasibility of requirements on the carbon footprint of PV modules under the already existing legal framework of the Ecodesign Directive, i.e. a policy approach for finalisation in the short term, by means of the research approach devised in the next sections.

2 SCIENTIFIC RELEVANCE

Even though the manufacturing of PV modules is an energy intensive and comprised of high throughput processes, e.g. silicon (Si) ingot manufacturing for Si-based technologies, there is a significant improvement potential that depends on the PV cell technology. The variance of the carbon dioxide equivalent emissions for the different module design options of the preparatory study spans from 28% to 96% compared to a base case (multicrystalline Si back surface field, 14.7% PCE). The lowest CO2e emissions are corresponding to thin-film PV modules. If optimisations are realised at the level of the

[1] See 'Environmental performance of products & businesses – substantiating claims' at https://ec.europa.eu/info/law/better-regulation/have-your-say/initiatives/12511-Environmental-claims-based-on-environmental-footprint-methods and 'Sustainable Products Initiative' at https://ec.europa.eu/info/law/better-regulation/have-your-say/initiatives/12567-Sustainable-Products-Initiative

[2] Carbon' is a shorthand for the carbon dioxide equivalent of all emissions of greenhouse gases, quantified as 'kilograms of carbon dioxide equivalents per kWh' (kg CO2e/kWh).

cell on Si-based technologies, the emissions savings can be of 28% compared to the base case.

In spite of the usefulness of having a carbon footprint declaration for products, so far such a declaration is not legally binding for any product entering the EU markets. In addition, the comprehensive calculations are uneasy to perform and require quality data input.

There are several methodologies, guidance, standards and norms available to perform a greenhouse gas (GHG) accounting – also termed carbon footprint - on PV modules. The general frameworks that can serve as a reference include: the ISO 14067:2018 standard (specifically addressing carbon footprint of products) [7], the GHG protocol product standard [8] and the standard EN 15804 [9] framing environmental product declaration (EPD) requirements for construction products. In Norway, for example the latter standard has been used to develop the product category rules for photovoltaics. Particular environmental category rules have been established by the EU Commission in the PEFCR for photovoltaic products [10].

Table I: Non exhaustive list of standards, documents and guidelines that can be used as reference for the carbon footprint declaration as a potential Ecodesign mandatory requirement for PV products.

Standards and other guidelines/ reference documents	Description	Method
ISO 14067: 2018 [7]	Greenhouse gases — Carbon footprint of products — Requirements and guidelines for quantification	ISO 14067
GHG Protocol Product Standard [8]	Product Life Cycle Accounting and Reporting Standard to evaluate the full life cycle GHG emissions of a product	GHG Protocol Product Standard
EU Member states PCR (Italy, France, Norway, Finland, Netherlands) [9]	Databases and Product Category Rules for construction products/services where PV modules and inverters are part of new and renovated buildings	EN 15804
European PEFCR Guide for PV modules [10]	Guidance for calculating and reporting products' life cycle environmental impacts	PEF method
Italy's LCA legislation Promotion of the Green Economy [11]	Legislation fully based on the Environmental Footprint methods. Voluntary "Made Green in Italy" label	PEF method
NSF/ANSI 457 Sustainability Leadership Standard for PV Modules and PV Inverters [12]	Standard to establish product sustainability performance criteria and corporate performance metrics exemplifying sustainability leadership in the market. Basis of conformity assessment, such as third-party certification.	NSF ANSI 457
France's public tenders for utility scale PV plants [13]	Public tenders include carbon footprint requirements to prioritize projects with low-carbon manufacturing processes[3].	ADEME guidelines

Table 1 summarises existing standards and reference documents that could be potentially used for the carbon footprint Ecodesign requirement. All listed documents ideally recommend adopting a life cycle perspective when addressing PV modules, i.e. from the necessary raw material extraction, through their manufacturing and operations, up to their recycling and disposal. Nevertheless, for the proposed Ecodesign requirement to be verifiable, and since a significant part of the climate change impacts can be attributed to the raw materials and manufacturing stages, the minimum scope of the carbon footprint calculation here should comprise from the resource extraction to the factory gate; the so called cradle to gate assessment.

The feasibility of the following approaches will be therefore explored, in order to establish a methodology to calculate the carbon footprint that could be used in the potential Ecodesign and Energy Label regulations:.

1. Perform a carbon footprint accounting using as a reference methodology the following guidelines:
    a. The PEFCR for photovoltaics
    b. GHG Protocol standard
    c. The ISO 14067:2018
    d. Product Category rules, EN 15804
1. This calculation could be or not third party reviewed.
2. Set a provision of tabulated carbon emission data for different module components, i.e. as specified in French Tender requirements.
3. Use of a streamlined life cycle assessment as the 'Ecoreport tool' that follows the application of the MEErP.

For the necessary calculation of the gross energy requirements (GER) there are two possibilities related to the electricity mix considered in the production of a cell, wafer, ingot block, solar-grade silicon, solar substrates, solar superstrates or other solar grade semiconductor materials. The first one could be based on using electricity grid mix of the relevant production location. A second option, in the case that the local energy mix would be composed of a higher rate of renewable energy, could allow for the use of European Energy Certification or similar schemes. For example, the Norwegian Product Category Rules state that electricity shall be the physical national grid mix and electricity based on guarantees of origin shall not be used to model electricity for the PV value chain, but LCIA results using these may be reported as additional environmental information.

It is foreseen that the carbon footprint reporting will be accompanied by the documentation relative to the general information and scope, inter alia the functional

[3] Cut-off criteria for November 2020 were set on 1150 kgCO2eq/kWc

unit, the type of inventory, cradle-to-grave or cradle-to-gate and its date/version. The quality of the inventory data for conducting the assessment is something dealt in the PEFCR and in the PCR (EN 15804). In some of the listed frameworks, a minimum number of processes have to be reported as well, being specified and differentiated by PV modules technologies.

3 BUILDING ECODESIGN REQUIREMENTS ON THE CARBON FOOTPRINT OF PHOTOVOLTAIC PRODUCTS

In legal terms, the Ecodesign Directive could provide a clear legal basis for the possible introduction of requirements related to the carbon footprint. The methods for setting new Ecodesign requirements are laid out in Annex I and II of the Directive. Annex I, firstly, structures the Commission's assessment of possible requirements. It provides a list of life-cycle stages, environmental aspects and related parameters to be picked as appropriate to the relevant product. Greenhouse gas emissions and consumption of energy are specifically identified as relevant parameters, as well as the manufacturing phase as a separate life-cycle stage. In addition, Annex I Part 2 lists as a possible requirement the provision of information 'on the significant environmental characteristic and performance of a product [...] to allow consumers to compare these aspects [...].' This could provide a basis for ensuring the communication of the outputs calculated and described by the manufacturer in the relevant carbon footprint to the relevant actors, and to allow them to compare PV modules on the basis of this metric. However, to date there are no practical applications of these principles, meaning that not a single Ecodesign Regulation foresees product-specific requirements on the carbon footprint, for a number of reasons already analysed in literature [14].

A regulatory proposal on the carbon footprint of PV modules could be composed of the two following parts:

- An 'holistic' information requirement to report on the (overall) greenhouse gas emissions of the production phase, and
- Targeted information requirement(s) on the most relevant aspect/parameters, inter alia the energy mix of the production phase.

The regulatory proposal should also clearly indicate the assessment method for the calculation of the carbon footprint, identifying the environmental impact categories to declare (here: emissions to air or climate change). Limiting the calculation of the carbon footprint to the production phase seems the most straightforward choice, for two reasons. In first place, the relevance of this phase in environmental impact terms clearly emerges from the analysis of the impacts. In second place, it is noteworthy to highlight that Ecodesign requirements affect manufacturers at the moment of the placing on the market of their products; therefore, the reporting obligation should ideally target phases/processes under their control.

In line with the Ecodesign Directive, and in order to ensure maximum comparability between results, the implementing measure should lay down harmonized rules for assessing the identified life-cycle stages and parameters. This then provides an occasion to address the conceptual challenges identified in the previous section, for example by setting clear rules for data quality or the use of guarantees of origin. The harmonized standard or transitional method chosen for the implementation of the footprint requirement should then follow these rules.

4 CONCLUSIONS AND FUTURE WORK

Following the conclusions of the preparatory study on the environmental impact of PV products carried out by the Joint Research Centre, the paper will elaborate on the technical and policy aspects related to Ecodesign requirements on the carbon footprint of PV modules, in particular by means of compulsory reporting. Potential formulations for these requirements are devised, together with an analysis on verifiability aspects.

The reporting obligations could act as an incentive for manufacturers to improve the environmental performance of their production phase, thus contributing to the 'ecodesign' of their products.

The proposals presented in this paper obviously represent a first attempt of regulating the carbon footprint of products within the Ecodesign Directive. Further analysis would be needed to shed light on the following aspects:

- which kind of verification procedure should be used to assess the compliance of the products with requirements on the carbon footprint? Certainly not the 'traditional' ones implemented in existing Ecodesign Regulations, as these procedures foresee the testing of specific (physical) features of the product at the moment of its placing on the market
- (related to the previous one) how to ensure level playing field between product manufactured within the EU (where market surveillance activities typically take place) and outside the EU?
- the variability (and related verifiability) of the energy mix over time. How to report it?
- how to trace environmental impacts of components/subassemblies throughout the value chain?

From the above, it is clear that refined formulations for requirements on the carbon footprint are needed, to overcome these issues and make the requirements themselves robust enough in view of their potential application.

References:

[1] Council of the European Union. Proposal for a Regulation of the European Parliament and of the Council establishing the framework for achieving climate neutrality and amending Regulation (EU) 2018/1999 (European Climate Law). Brussels, 17 December 2020. Available at: https://data.consilium.europa.eu/doc/document/ST-14171-2020-INIT/en/pdf (accessed on 01.02.2021)

[2] Directive (EU) 2018/2001 of the European Parliament and of the Council of 11 December 2018 on the promotion of the use of energy from renewable sources. OJ L 328, 21.12.2018, p. 82–209

[3] Dodd N, Espinosa N et al. Preparatory study for solar photovoltaic modules, inverters and systems.

https://publications.jrc.ec.europa.eu/repository/handle/JRC122431

[4] Directive 2009/125/EC of the European Parliament and of the Council of 21 October 2009 establishing a framework for the setting of ecodesign requirements for energy-related products, OJ L285, 31.10.2009, p. 10-35

[5] Kemna R. Methodology for ecodesign of energy-related products (MEErP 2011). Publications Office of the European Union. 2011

[6] Polverini D, Dodd N, Espinosa N. Potential regulatory approaches on the environmental impacts of photovoltaics: Expected improvements and impacts on technological innovation. Prog Photovolt Res Appl. 2021; 29: 83– 97. https://doi.org/10.1002/pip.3344

[7] ISO Standard 14067: 2018 Greenhouse gases — Carbon footprint of products — Requirements and guidelines for quantification

[8] GHG Protocol Product Standard - Product Life Cycle Accounting and Reporting Standard. https://ghgprotocol.org/product-standard

[9] EN 15804:2012+A1:2013 Sustainability of construction works - Environmental product declarations - Core rules for the product category of construction products

[10] Product Environmental Footprint Category Rules (PEFCR)—photovoltaic modules used in photovoltaic power systems for electricity generation, 2019. https://ec.europa.eu/environment/eussd/smgp/pdf/PEFCR_PV_electricity_v1.1.pdf

[11] Disposizioni in materia ambientale per promuovere misure di green economy e per il contenimento dell'uso eccessivo di risorse naturali. https://www.normattiva.it/uri-res/N2Ls?urn:nir:stato:legge:2015-12-28;221

[12] NSF/ANSI 457 - 2019 Sustainability Leadership Standard for Photovoltaic Modules and Photovoltaic Inverters. https://greenelectronicscouncil.org/wp-content/uploads/2019/12/NSF-457-2019.pdf

[13] Délibération de la CRE du 3 septembre 2020 portant décision relative à l'instruction des dossiers de candidature à la huitième période de l'appel d'offres sur la réalisation et l'exploitation d'installations de production d'électricité à partir de l'énergie solaire «Centrales au sol» https://www.cre.fr/content/download/22950/289288.

[14] van Rossem, C., Dalhammar, C. and Toulouse, F. (2009). Designing Greener Electronic Products: Building Synergies between EU Product Policy Instruments or Simply Passing the Buck? Report: European Environmental Bureau (EEB), Brussels.

OVERVIEW OF THE LATEST RESULTS ACHIEVED IN THE H2020 FUNDED PROJECT HIGHLITE AIMING FOR HIGH-PERFORMANCE, LOW-COST AND SUSTAINABLE C-SI PV MODULES TAILORED FOR DIFFERENT APPLICATIONS

Loic Tous[1], Jonathan Govaerts[1], Samuel Harrison[2], Carolyn Carriere[2], Florian Buchholz[3], Andreas Halm[3], Antonin Faes[4], Gizem Nogay[4], Andrea Ingenito[5], Franz-Josef Hang[5], Frank Feldmann[6], Daria Raine[6], Tobias Fellmeth[6], Martin Heinrich[6], Max Mittag[6], Dirk Reinwand[6], Felix Haase[7], Arnaud Morlier[7], Matevz Bokalič[8], Kristijan Brecl[8], Marko Topic[8], Josco Kester[9], Stefan Wendlant[10], Marco Galiazzo[11], Alessandro Voltan[11], Giuseppe Galbiati[12], Liesbeth Theunissen[12], Frank Torregrossa[13], Michael Grimm[14], Julius Denafas[15], Tadas Radavicius[15], Povilas Lukinskas[16], Jarno Kaakkunen[17], Tuukka Savisalo[17], Thomas Regrettier[18].

[1] imec-Energyville, Thorpark 8320, 3600 Genk, Belgium
[2] CEA-INES, 50 Avenue du Lac Léman, 73370 Le Bourget-du-Lac, France
[3] ISC Konstanz, Rudolf-Diesel-Straße 15, 78467 Konstanz, Germany
[4] CSEM, Rue Jaquet-Droz 1, 2002 Neuchâtel, Switzerland
[5] EPFL, Route Cantonale, 1015 Lausanne, Switzerland
[6] Fraunhofer ISE, Heidenhofstraße 2, 79110 Freiburg im Breisgau, Germany
[7] ISFH, Am Ohrberg 1, 31860 Emmerthal, Germany
[8] UL FE,Tržaška cesta 25, 1000 Ljubljana, Slovenia
[9] TNO Energy Transition – Solar Energy, Westerduinweg 3, 1755 LE Petten, the Netherlands
[10] PI-Berlin, Wrangelstraße 100, 10997 Berlin, Germany
[11] AMAT Italia SRL, Via Postumia Ovest, 244, 31048 Olmi TV, Italy
[12] HENKEL Belgium NV, Nijverheidsstraat 7, 2260 Westerlo, Belgium
[13] IBS, Rue Gaston IMBERT prolongée, ZI Peynier Rousset, 13790 Peynier, France,
[14] 3D-Micromac AG, Technologie-Campus 8, 09126 Chemnitz, Germany,
[15] Solitek, Mokslininkų g. 6A, Vilnius 08412, Lithuania,
[16] Valoe Cells, Mokslininkų g. 6A, Vilnius 08412, Lithuania,
[17] Valoe Oyj, Insinöörinkatu 5, FI-50150 Mikkeli, Finland,
[18] Voltec Solar, 1 Rue des Prés, 67190 Dinsheim-sur-Bruche, France

ABSTRACT: The EU crystalline silicon (c-Si) PV manufacturing industry has faced strong foreign competition in the last decade. To strive in this competitive environment, the EU c-Si PV manufacturing industry needs to focus on highly performing c-Si PV technologies and on reducing environmental impact (lower CO_2 footprint, improved sustainability, recyclability, etc.). In this work, we give an overview of the latest results achieved within the H2020 funded project called HighLite which addresses the work programme LC-SC3-RES-15-2019: Increase the competitiveness of the EU PV manufacturing industry. We report the progress made when implementing novel layers and processes in existing pilot-lines producing high-efficiency n-type silicon heterojunction (SHJ) cells with a shingle layout and interdigitated back-contact (IBC) cells with passivated contacts. In particular, we present the work done together with several industrial partners towards innovative production equipment for assembly of cut-cells into PV modules. Finally, we explain key learnings made so far when tailoring c-Si PV modules for various applications including building-applied PV (BAPV), building-integrated PV (BIPV) and vehicle-integrated PV (VIPV).

Keywords: HighLite, H2020, industry, SHJ, IBC, BAPV, BIPV, VIPV

1 INTRODUCTION

The EU crystalline silicon (c-Si) PV manufacturing industry has faced strong foreign competition in the last decade. This was further intensified by the EU commission's decision to end the Minimum Import Price (MIP) measures in 2018 and with competitors in Asia very quickly ramping-up multi-GWp production of low-cost PERC cells and modules in the last years. This initially led to a drastic reduction of the EU PV manufacturing capacity in the 2010-2019 period with mainly "small" PV module producers (<500 MW_p/year manufacturing capacity) surviving by importing low-cost PERC cells with a relatively high CO_2 footprint from China. On the other hand, this has spurred demand significantly with the IEA now forecasting annual PV capacity additions in the EU to exceed 25 GWp per year in the period between 2023 and 2025 [1].

Looking at the EU c-Si manufacturing landscape today (see Fig. 1), one can see that the industry is enjoying strong momentum with multiple companies adding new production capacity. This is driven by factors such as the rapid transition to larger format wafers/cells (M6/M10/G12) requiring significant upgrades (or adding new lines altogether), the stronger demand for made-in-EU products, and the facts that imported PV modules currently suffer from very high shipping costs (mainly due to the prolonged impact of the Covid19 pandemic).

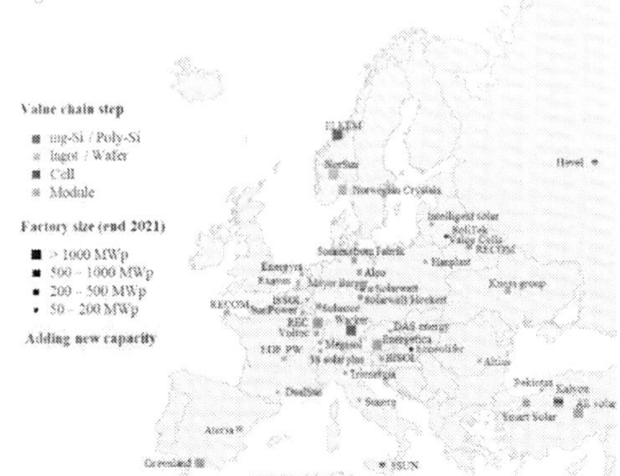

Fig. 1: EU c-Si PV manufacturing landscape in late 2021 with companies adding new capacity marked in red.

To strive in this competitive environment, the EU PV manufacturing industry needs to focus on highly performing PV technologies and products tailored for different applications. Distributed applications such as building-applied PV (BAPV), building-integrated PV (BIPV), and vehicle-integrated PV (VIPV) are particularly fast-growing markets due to the transition towards self-consumption and electro-mobility. The EU PV manufacturing industry also needs to focus on reducing environmental impact (lower CO_2 footprint, improved sustainability, recyclability, etc.) to differentiate itself from the competition and benefit from specific incentives that are already in place (e.g. CRE tenders in France) or being envisaged on a wider scale by the EU commission [2].

In this context, leading EU PV institutes/universities have joined forces with major material, equipment, and module manufacturers under the H2020 funded project called HighLite (GA no. 857793) which started in October 2019 for a duration of 36 months.

2 MAIN OBJECTIVES IN H2020 HIGHLITE

The HighLite project focuses on bringing two n-type c-Si technologies to high technology readiness levels (TRL 6-7) taking profit of the unique strength and expertise of the project partners. The first one is based on shingle assembly of n-type heterojunction (SHJ) cells. The second one is based on the assembly of industrial n-type interdigitated back-contact (IBC) cells cut in half or smaller. Shingle SHJ assembly enables the manufacturing of high-performance lead-free modules (improving sustainability) while modules with IBC cut-cells are interesting for applications demanding the best aesthetics and power density (such as VIPV). Both SHJ and IBC cells can be produced from thinner wafers (100-160 µm vs 140-200 µm for PERC) without compromising cell efficiencies and yield as shown by the project partners in previously funded projects [3-5]. This allows to produce SHJ and IBC modules with much lower CO_2 footprints particularly if one selects poly-Si/wafer suppliers from the low-carbon footprint EU supply chain (ELKEM, NorSun, ...). Thus, the ambition of the HighLite project is to develop PV modules with high-performance, low-cost, and excellent environmental profiles that are tailored for various applications (BAPV, BIPV, VIPV).

As shown in Fig. 2, the main objectives of the HighLite project partners are to: (1) develop innovative manufacturing solutions and processes, (2) develop innovative PV equipment and materials, and (3) implement novel layers and processes in existing pilot-production lines (cells and modules).

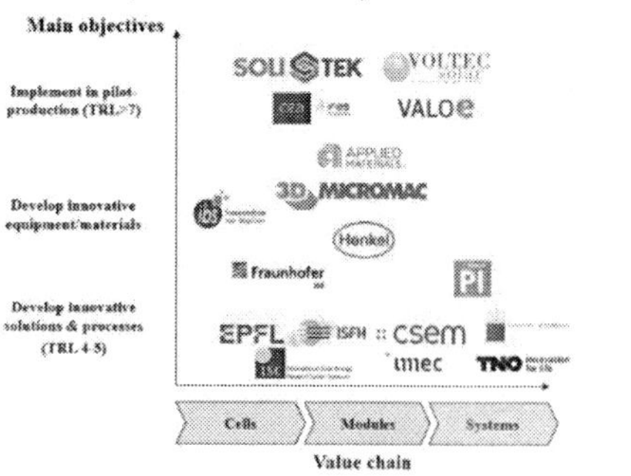

Fig. 2: Main objectives of HighLite project partners along the value chain.

In the following sections we discuss the main results achieved so far in the project.

3 HIGHLITE PROJECT RESULTS

3.1 Advanced processing of SHJ and IBC cells

Shingle assembly requires cells with a dedicated metal design with typically 5-6 deported busbars (BB) because the cells are interconnected near the cut-edges using electrically conductive adhesives (ECA). Performance/reliability/cost optimization of the front side (FS) and back side (BS) metal designs is even more crucial in SHJ cells due to the comparatively higher cost and lower conductivity of low-temperature Ag pastes vs Ag pastes used in PERC cells. This optimization is led by CEA-INES who operates a SHJ pilot-line using M2-size wafers (area= 244.3 cm^2).

Key results in the first 24 months of the project include: the development of a patented busbar design to improve reliability, the implementation of double printing to reduce resistive losses, and significant work on reducing defectivity caused by certain automation steps to produce high-efficiency SHJ cells in shingle configuration [6-7]. CEA-INES has been able to produce over 6000 cells that were distributed to the partners and to gradually increase the average efficiency (η) of small batches from ~22% at the beginning of the project to over 23.5% today (see Fig. 3). The best cell efficiency (full area measurement including busbars shading) from one of those record batches was measured at 24.1% (externally confirmed by ISFH CalTec).

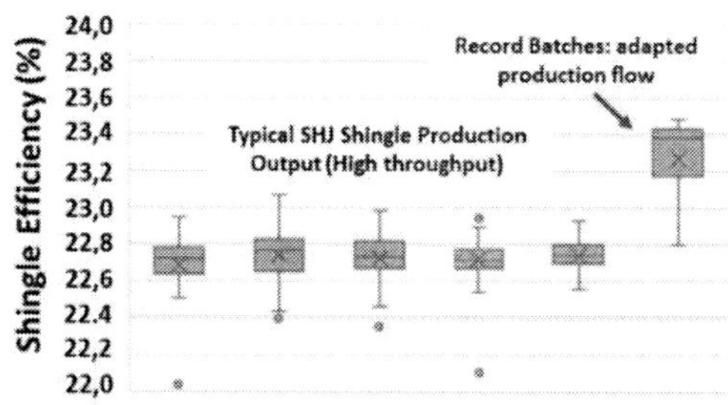

Fig. 3: Typical SHJ shingle production output at CEA-INES pilot-line with record batches producing M2-size cells (area = 244.3 cm^2) with efficiencies well over 23%. Further details are reported at this conference in Ref. [7].

As for IBC cells, one of the key objectives of the project is to reach $\eta \geq 24.5\%$ by the end of 2022. Valoe is currently ramping-up production of IBC cells based on the low-cost and industrially proven ZEBRA technology licensed from ISC Konstanz [5]. In parallel, the HighLite partners have been working on various approaches based on polysilicon (poly-Si) passivating contacts to significantly reduce recombination losses. Key results achieved so far include: (i) the fact that the various poly-Si approaches developed by the partners all offer excellent performance with industrial screen-printed Ag contacts [8] giving companies like Valoe more freedom in terms of upgrade choices and (ii) the development of novel manufacturing methods and equipment to produce low-cost IBC cells with poly-Si based passivating contacts [9-11]. So far, the best result obtained is η=23.7% (4 cm^2 designated area on M2 wafer) using the POLO-IBC concept from ISFH with room for improvement as the efficiency potential of this concept is over 25.5% [9].

3.2 Advanced cell cutting and re-passivation of cut edges

Cutting cells in 2 ("half-cut"), 3 ("tri-cut") or even more segments (typically 5 or 6 for shingling technology) is attractive to improve packing density and reduce interconnection losses at module level but presents multiple challenges. First, it causes significant efficiency losses at cell level [12-13]. Second, it can lead to cell breakage during cutting and subsequent module processing (interconnection, lamination, etc.) causing significant yield losses. Finally, it can also cause long-term reliability issues in the field (hot spots, propagation of micro-cracks, snail trails, etc.).

In HighLite, three different cell cutting approaches are being optimized by the project partners through experiments as well as advanced characterization and modeling [12-17]. Overall, 3D-Micromac's proprietary Thermal Laser Separation (TLS) appears to be most promising as it is industrially proven (with >10 GW of systems running in production) and offers smooth cut edges which are key for effective and stable edge re-passivation [16-17]. Thanks to the progress made within HighLite, 3D-Micromac has been developing a new version of its tool called microCELL MCS that is based on an innovative circulating chuck concept (see Fig. 4). This tool is available as standalone (with load/unload systems), or it can be integrated into an inline system. It is capable of cutting cells (up to M12 format) into 1/6 segments while reaching an impressive gross throughput > 6000 wph [18].

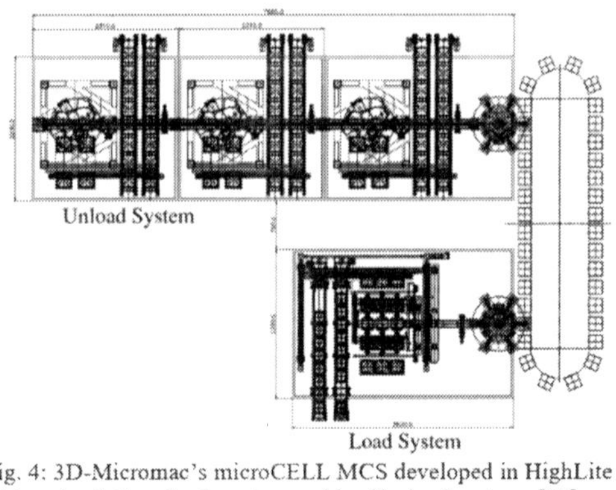

Fig. 4: 3D-Micromac's microCELL MCS developed in HighLite with a circulating chuck system and load/unload systems [18].

As for advanced re-passivation of cut-edges, various techniques are being optimized by the HighLite project partners including thermal atomic layer deposition (ALD) of aluminum oxide (Al_2O_3), hydrogen termination, polymer passivation, or plasma enhanced chemical vapor deposition (PECVD) of silicon nitride (SiN_x) layers. So far, the most promising results have been obtained with ALD of Al_2O_3 [16-17] and with PECVD SiN_x [15].

3.3 Automated assembly of cut-cells

AMAT has been developing a new version of its Sonetto platform that is capable of cutting and assembling full cells (up to M6 format) into shingled strings at a gross throughput of 4000 wph. Key lessons learned together with the project partners include: (i) the reduction of laser cutting losses, (ii) the optimization of ECA formulation and printing for cost, performance, and reliability, (iii) the feasibility of reducing cell overlap down to 0.5 mm (versus the 0.8-2.0 mm typically used in the industry) to improve performance and reduce costs, (iv) the feasibility of handling 120 µm thick cells with high yield and (v) the integration of an automated bussing unit which are all key differentiators on the market. By the end of the project, AMAT plans to integrate 3D-Micromac's microCELL MCS tool into a new version of its Sonetto platform that is compatible with wafers up to M12 formats and a gross throughput of 6000 wph [18].

Various solutions are being explored by the project partners for the assembly of IBC cut-cells. Valoe has developed a fully automated production line for the assembly of IBC modules using conductive backsheet (CBS) technology that can be licensed/sold to third parties [19]. The line features integrated ECA printing and is compatible with wafers sizes up to G1 format (full-size, half-cut or quarter-cut). Valoe is also assembling a more flexible line with integrated cell cutting/cleaving and ECA printing off-line that will be used to produce lightweight modules (<1 kg/m^2) for space applications but also odd-form modules for various application such as VIPV. In parallel, several project partners have been exploring possibilities to interconnect busbarless IBC cells (thereby enabling large materials savings at cell level) using various multi-wire technologies with promising results achieved so far [20-21].

3.4 Tailored PV module designs

Another key objective in HighLite is to develop various PV module designs that are specifically tailored for BAPV, BIPV, or VIPV applications.

For BAPV, the objective by the end of 2022 is to demonstrate PV modules with excellent aesthetics, high efficiency ($\eta \geq 22\%$) and very low carbon footprint ($\leq$250 kg-eq.CO_2/kWp). For comparison, standard PERC modules imported from China typically have efficiencies in the range of 19-21% [22] and carbon footprints typically in the range of 500-1000 kg-eq.CO_2/kWp depending on the calculation method [23]. So far, CEA-INES has been able to demonstrate full-size SHJ shingle modules with excellent all-black aesthetics (see Fig. 5) and with promising results in terms of efficiency and reliability [24]. Of note, shingle strings with 120 µm thick cells (enabling a much lower carbon footprint than with standard 160 µm thick cells) were shown to pass 600 thermal cycles without issues. In parallel, ISC Konstanz has been optimizing the bill-of-materials of BAPV modules with IBC cut-cells towards 35 years lifetime and produced all-black glass-glass IBC modules with P_{max} up to 345 W_p (60 G1-cell equivalent). Finally, Fraunhofer ISE and Voltec Solar have been working on reducing the carbon footprint of BAPV modules using novel wooden or Al frame designs. Based on detailed cell-to-module (CTM) analysis and life cycle analysis (LCA) performed by the project partners, we remain on track to reach the objectives set in HighLite for BAPV modules by the end of 2022.

Fig. 5: 396 Wp (left) and 387 Wp (right) shingled SHJ modules (72 M2-cell equivalent). For further details see Ref. [6].

For BIPV, the objective within HighLite is to realize modules with $\eta \geq 21\%$ and improved shading tolerance. Among the key learnings made so far, we can mention the fact that short shingle strings are preferred for mechanical robustness while block layouts (with series-parallel connections) are preferred for improved shadow tolerance [25]. With respect to the various BIPV applications, Valoe is mainly focusing on developing modular modules with IBC cut cells tailored for balcony applications (see Fig. 6) while Voltec solar has been focusing on shingle SHJ modules for rooftile applications with the support of several project partners. Both BIPV module designs have successfully passed critical reliability tests such as DH1000, TC200, UV (65 kWh/m^2), mechanical load (ML), etc. In the coming months, we plan to install and monitor multiple BIPV prototypes for facade and rooftop applications at various partner locations. Their outdoor performance against commercial products will be monitored regularly with monthly "scorecards" results released publicly on the project website by the end 2022.

Fig. 6: Front side (left) and rear side (right) of BIPV facade module.

Finally, the objective in HighLite for VIPV applications is to demonstrate 3D-curved vehicle-integrated PV modules with $\eta \geq 20\%$ and a weight ≤ 5 kg/m^2. In this context, IMEC has been developing 3D-curved lightweight modules using shingled SHJ strings, glass-fiber reinforced encapsulants, and a composite backsheet with promising results in terms of performance/reliability (TC, DH, UV, vibration, etc.) [26]. Other promising results include 3D lamination of IBC-SWCT strings on various substrates (steel, carbon fire composites, etc.) to form self-bearing modules by ISFH and the development of lightweight IBC modules using IBC cut-cells, conductive backsheet (CBS) technology, and ultra-thin glass (for improved impact and scratch resistance) by Valoe. In the next months, we plan to produce full-size car bonnets that will be used for demonstration and testing purposes.

4 CONCLUSIONS

In this work we gave an overview of the main results achieved within HighLite since the project start in October 2019. We reported about the progress made with advanced processing of high-efficiency SHJ and IBC cut-cells which are key to produce premium c-Si modules with high efficiency, great aesthetics, and low environmental impact. We also reported the excellent progress made by the industrial partners towards novel automated cutting and cell-assembly equipment that offer key benefits in terms of throughput, performance, and flexible manufacturing. Finally, we explained the key results results/learnings made so far when developing modules for various applications such as BAPV, BIPV, and VIPV. Overall, the results obtained within HighLite can already help to increase the competitiveness of the EU PV manufacturing industry across the value chain. In the coming months, we plan to continue our developments and make plans for innovative and potentially large-scale PV manufacturing plants in the EU.

5 AKNOWLEGMENTS

This project has received funding from the European Union's Horizon2020 Programme for research, technological development and demonstration under Grant Agreement no. 857793.

7 REFERENCES

[1] IEA, Renewables 2020: Analysis and forecast to 2025.
[2] D. Polverini, et al, doi: 10.1002/pip.3344
[3] G. Galbiati, et al, Large Area IBC ZEBRA Solar Cells in Pilot Production: the Results of FP7 HERCULES Project, in Proceedings of the 32nd EUPVSEC, pp. 980-982, 2016.
[4] A. Danel et al. Silicon Heterojunction Solar Cells with Open-Circuit-Voltage above 750mV, EUPVSEC, pp. 444-447, 2018.
[5] R. Kopecek, et al, doi: 10.1109/PVSC45281.2020.9300503
[6] S. Harrison, et al, doi: 10.1109/PVSC43889.2021.9518854
[7] S. Harrison, et al, 2BO.15.3 Low Temperature Post-Process Repassivation for Heterojunction Cut-Cells, this conference
[8] T. Fellmeth, et al, 2CO.10.4 A Round Robin - HighLiting on the Passivating Contact Technology, this conference.
[9] R. Peibst, et al, 2AP.1.2 Beyond Single Junction Efficiencies, this conference
[10] F. Buchholz, et al, 2BO.11.3 Local Passivating Contacts from Laser Doped P+ Polysilicon, this conference
[11] F-J. Haug, et al, DOI: 10.1002/pssr.202100272
[12] B.Portaluppi, et al., doi: 10.4229/EUPVSEC20202020-2DV.3.16
[13] V. Giglia et al, doi: 10.4229/EUPVSEC20202020-2CO.15.4.
[14] M. Bokalič, et al, doi: 10.4229/EUPVSEC20202020-2CV.1.9
[15] D.D. Tune et al, doi: 10.4229/EUPVSEC20202020-2CV.1.13
[16] T. Fellmeth, Minimizing Cut Edge Losses in Advanced Silicon Solar Cells, TaiyangNews webinar, 8th

July 2021, https://youtu.be/ZuOi7gcleY0
[17] S. Harrison, et al, 2BO.15.3 Low Temperature Post-Process Repassivation for Heterojunction Cut-Cells, this conference
[18] M. Grimm and L. Cerasti, High throughput shingling equipment for various applications, TaiyangNews webinar, 8th July 2021, https://youtu.be/Evv9EK73NKA
[19] T. Savisalo, Flexible Manufacturing of Integrated PV Products with IBC Technology, TaiyangNews webinar, 8th July 2021, https://youtu.be/jS8jjSW-DoI
[20] JR. Van Dyck, et al, 4AV.1.12 Three-Dimensional Multi-Ribbon Back-Contact Interconnection: Latest Results on Reliability Testing, this conference
[21] A. Morlier, et al, 6DO.7.2 Photovoltaic Modules with Natural Materials for a Seamless Building Integration, this conference
[22] TaiyangNews Report on Very High-Power Solar Modules 2021, June 2021
[23] L. Stoker, Chinese manufacturers slam South Korean solar module carbon emissions results, 22 September 2020, pv-tech.org
[24] C. Carrière, et al, 4AV.1.15 Shingle Interconnection on HJT Solar Cells: Reliability Study and Upscaling for High Power PV Modules, this conference
[25] S. Wendlandt, et al, Indoor Characterization & Reliability Investigations on Shingled c-Si PV Module, TaiyangNews, 8th July 2021, https://youtu.be/r_vDljfhebk
[26] J. Govaerts, et al, AO.3.1 The Potential of Glass-Fibre-Reinforcement: (Thermo-)Mechanical Testing of Light-Weight PV Modules, this conference

NEEDS, CHALLENGES AND APPROACHES FOR NEW SERVICE LIFE ESTIMATION MODELS FOR PV MODULES- RESULTS FROM IEA-PVPS-TASK 13 SUBTASK 1.4

Karl-Anders Weiß[1]*, Ismail Kaaya[1], Gernot Oreski[2], Laura S. Bruckman[3], Roger H. French[3], Tadanori Tanahashi[4]
[1] Fraunhofer ISE, Heidenhofstr. 2, 79110 Freiburg, Germany
Corresponding author: karl-anders.weiss@ise.fraunhofer.de, T: +49 761 4588-5474
[2] PCCL Polymer Competence Center Leoben, Roseggerstraße 12, 8700 Leoben, Austria
[3] Case Western Reserve University, Cleveland, USA
[4] National Institute of Advanced Industrial Science and Technology (AIST), Koriyama, Japan

ABSTRACT: The economic success of photovoltaic (PV) power plants depends crucially on the lifetime energy yield. The degradation effect and the total lifetime directly influence the produced electricity and therefore the cash flow. In most cases, lifetimes and degradation rates, which are used for the estimation of plant behaviour, are nowadays still not system specific, but based instead on average values of evaluations of older systems or data sheets. So, these values unfortunately have no direct correlation with the specific components of the respective PV system. Also, the mathematical models used for calculated power output typically expect linear degradation rates which are not found in real degradation processes found in the field.
Activity 1.4 of the IEA PVPS Task 13 developed an overview on service life prediction and degradation modelling of PV modules since they are the major component of PV systems causing degradation effects and for other components no comparable scientific data is available.

Keywords: PV Module, Reliability, Degradation, Modelling

1 INTRODUCTION

The economic success and environmental impact of photovoltaic (PV) power plants depends crucially on the degradation and service life of the PV modules and other components of the PV power plants. The behaviour of PV modules is especially relevant since they typically show gradual degradation effects over time. The useful service lifetime and degradation of PV modules directly influences the lifetime yield of electricity, and therefore, the levelized cost of the electricity (LCOE) produced [1]. Degradation and service life are influenced not only by the materials used and the quality of module manufacturing, but also by local environmental effects that dictate the exposure conditions of the PV modules. Therefore, the lifetime and degradation of PV modules cannot be determined easily and are not valid for all locations and applications. Since both are dependent on local and operational conditions, the prediction of service lifetime from PV module degradation rates must be taken into consideration all of these factors and incorporate them into mathematical models.

2 PREPARING THE MANUSCRIPT

This report introduces the influencing factors for service life and degradation of PV modules and components as well as the modelling of degradation effects and service life prediction. It describes relevant stresses and load effects in section 3 and different modelling approaches, as well as models which have shown to fulfill the requirements of PV stakeholders in section 4. The descriptions are written in a way to address the needs of readers from all stakeholder groups, so on one hand people with no background in mathematical modelling who are interested in the influencing factors, potential of service life prediction, and interpretation of given data, as well as experts in reliability and degradation modelling. The chapters therefore briefly describe the approaches and background as well as examples and list relevant literature for further reading.

Since the content of the report would exceed the possibilities of such a conference paper, only the executive summary is presented in the following. The readers are encouraged to download the full report which is available on the IEA PVPS Task 13 website.

3 EXECUTIVE SUMMARY

The economic success of photovoltaic (PV) power plants depends crucially on their lifetime energy yield. Degradation effects and the total lifetime directly influence the produced electricity and therefore the cash flow, which also impacts the levelized costs of energy (LCOE) and therefore the profitability of the power plant. In most cases, the lifetimes and degradation rates that are used to estimate the system performance are not system-specific, but are based on average values from the evaluations of older systems or data sheets. So, these values unfortunately have no direct correlation with the specific components of the specific PV system, nor the operational and climatic conditions at the specific location. Also, the mathematical models used for calculating the expected power output typically expect linear degradation rates which are not in line with real degradation processes found in the field, which are typically non-linear.
This report gives an overview on empirical degradation modelling and service life prediction of PV modules since they are the major components of PV systems that are subject to the effects of degradation. For other components, no comparable scientific data is available.
The structure of the document addresses different stakeholders with different backgrounds. Chapter 1 begins with a short introduction including a condensed overview of the state of the art.
Chapter 2 follows with the definition of relevant terms and definitions. Since especially in discussions on lifetime and degradation different terms are not used

coherently in industry or science, the authors try to improve the situation with this dedicated glossary. In addition, the extremely relevant term "end-of-life" is discussed with different definitions, depending on the point of view and perspective of the user and the typical factors impacting the PV module or PV system. For this "end-of-life" term, no definition which is generally applicable in all situations can be given. Since the definition is crucial for the calculated service life, yield, and all related parameters, through to LCOE it is important to be aware of this when evaluating power plants and PV investments.

Climatic factors play a major role in degradation and are by nature location specific. It is precondition for the creation of meaningful service life prediction or degradation data to know about the relevant (climatic) stressors. Therefore Chapter 3 introduces the different relevant climatic stressors as well as classification schemes and methodologies to handle and analyse them. The chapter also describes differences and relations of the so-called macroclimatic stressors, describing the climatic conditions in the ambience of the modules, and the situation at material level, the so called micro-climatic stressors. The latter describes the relevant parameters for degradation processes and so also the mathematical models addressing module degradation and service life prediction. The ambient macro-climatic conditions at specific locations can be estimated using data for the climatic regions or adapted climatic maps and so be classified using climatic classification schemes which exist also specified for PV purpose like the Köppen-Geiger PV scheme. For the determination of microclimatic loads - which are typically input parameters for degradation models, further calculations are necessary. The report presents possible ways to determine the necessary data for the most important micro-climatic parameters which are temperature and humidity. This data is also very important for the definition of accelerated tests, which can deliver module specific parameters for the service life and degradation prediction. Chapter 3 also describes basic accelerated ageing tests, as described in the respective IEC standards, and how they can support degradation and service life prediction and modelling as well as their limitations.

Chapter 4 addresses general degradation and service-life modelling approaches including related issues. It starts in section 4.1 with general issues of empirical modelling one has to be aware of when working on mathematical modelling solutions for service life and degradation prediction and interpreting results. There are very different approaches for empirical modelling of the lifetime performance prediction and service life of products such as PV modules empirical statistical modelling, and empirical physical modelling. Physical empirical models are those that utilize analytic or numerical forms to represent the fundamental physics and chemistry of the phenomena. Statistical models, often referred to as data driven models, use mathematical forms which are able to fit the (measured) data without direct relation to physical or chemical processes. Both approaches use empirical (measured) data to determinate parameters which can be used for predicting future behaviour.

Section 4.2 introduces on one hand models for specific degradation modes or phenomena of modules (e.g., backsheet or cell cracking or electrochemical corrosion). On the other hand, modelling approaches for degradation effects of components and materials are presented. A special focus is here on degradation of polymeric materials since these materials are known to be sensitive to degradation effects causes by typical climatic stressors like high temperature, humidity and UV radiation. The modelling approaches using predictive models and inferential mechanistic models are presented using polyethylene terephtalate (PET) degradation as catchable example. It is shown that different modelling approaches are necessary to describe all degradation effects. Weak points of modules can be identified, and focussed optimization of products can be supported.

Performance degradation models are addressed in Section 4.3 which are the core models for the prediction of degradation of modules over time for specific types and locations. Combined with defined end-of-life conditions, these models can be used for service life prediction. Different approaches which have been specifically developed for PV modules are presented. Starting with an approach focusing on physical and chemical processes and the specific application. An approach to develop performance loss rate (PLR) models following the statistical methodology is presented as well including the processes to determine the relevant parameters from field data.

The modelling approaches are presented including the methodological approach to the problem the used input data, and parameters related to specific module types or local climatic conditions, down to calculations of degradation rates over time or remaining useful lifetime (RUL) or total expected lifetime.

The latest scientific work shows that service lifetime and degradation models for PV modules are of specific use if they combine different modelling approaches and include know-how and modelling parameters of the most relevant degradation effects. Such models can differentiate between the behaviour of different module types and to include the situation at different service locations. For some modules, it is also necessary to use multi-step modelling approaches to enable meaningful results.

Advanced approaches of data analysis and modelling also enable the determination of degradation signatures which can be related to specific degradation effects. This approach is expected to be very helpful in future work to identify failures based on operational data.

Since uncertainties of input parameters can have significant impact on the results but are often not totally avoidable, these topics are addressed in Chapter 4.3.

4 ACKNOWLEDGEMENT

This paper is supported by the German Federal Ministry for Economic Affairs and Energy (BMWi) under contract no. 0324304A and 0324304B.

HOLISTIC DESIGN OPTIMIZATION OF THE PV MODULE FRAME: CTM, FEM, COO AND LCA ANALYSIS

Ammar Tummalieh[1,2], Andreas J. Beinert[1], Christian Reichel[1], Max Mittag[1], Holger Neuhaus[1]
[1] Fraunhofer Institute for Solar Energy Systems ISE, Heidenhofstrasse 2, 79110 Freiburg, Germany
[2] University of Freiburg, IMTEK, Georges-Koehler-Allee 103, 79110 Freiburg, Germany

ABSTRACT: We present a holistic approach for the photovoltaic (PV) module frame optimization that considers technical as well as economic and ecological aspects for different frame designs. This provides insights into a method to reduce frame costs and carbon footprint without compromising mechanical stability as well as module power. Finite element method (FEM) simulations of module and frame are used to assess mechanical stability, cell-to-module (CTM) analysis is used to evaluate power losses affected by frame overlap, a bottom-up cost model is used for the economic analysis of material and process cost shares of frame manufacturing, and a life cycle assessment (LCA) analysis is applied to evaluate the ecological footprint ($CO_{2\text{-eq}}$/kW$_P$). Compared to a reference frame, the exemplary optimized frame design shows 2.6% less deflection, which corresponds to around 0.7 mm. CTM results show that a bigger frame width lightly decreases the cover coupling power gain. By increasing the front frame width from 16 mm to 20 mm the module power is reduced by 0.12 W$_P$. Cost analysis suggests that the optimized frame can save around 30 g aluminum which reduces the module cost by 0.1%. LCA results are directly correlated to the material mass of the corresponding design. Results show that using the optimized frame can save 0.8 kg $CO_{2\text{-eq}}$/kW$_P$ due to the saving in aluminum compared to the reference frame.

Keywords: CTM, FEM, LCA, COO, Frame, PV module, Optimization, Simulation

1. Introduction

As PV technologies evolve rapidly, the PV market expands and becomes more complex with all components of the module being permanently optimized [1, 2]. One of these components is the aluminum frame, which on the one hand is an important structural component of the module and provides the required mechanical stability, and on the other hand affects the performance and overall cost of the PV module.

An analysis of more than 250 technical datasheets of PV modules from 1995 to 2020 shows the historical evolution of the module frame dimensions, where for example the frame height h_{frame} was reduced from 50 mm to 35 mm in the last 30 years.

The frame dimensions play a crucial role in a holistic optimization as frame height and width which are closely related to the necessary aluminum amount, which is related to costs, environmental impact, as well as indirectly to the mechanical stability and electrical performance.

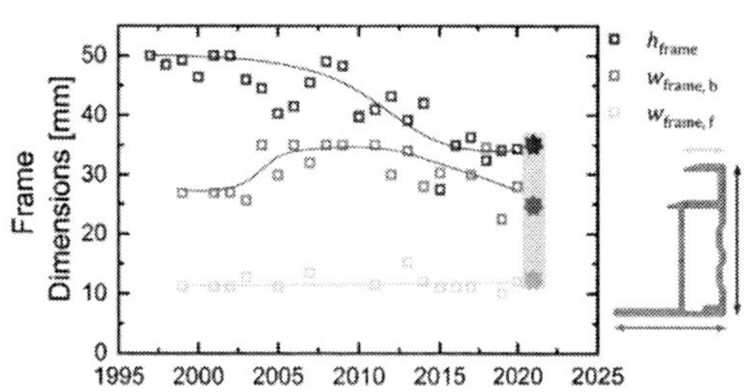

Figure 1: Historic trend of the frame dimensions, including frame height h_{frame} (dark purple), front frame width $w_{\text{frame,f}}$ (purple) and frame base width $w_{\text{frame,b}}$ (light purple). The corresponding lines are guide-to-the-eyes. Dimensions of the reference frame (stars) are marked in purple.

Here, we represent an approach for a holistic PV module frame optimization. Within this study we investigate and optimize exemplary PV module frame design based on mechanical, electrical, economic, and environmental analysis. Concerning the mechanical analysis, previous studies [3–6] investigated the module deformation and stress distribution under mechanical pressure loads. As related previous work studies the mechanical stress of solar cells [7] we propose an approach to optimize the module frame design and apply FEM simulations on the PV module including the aluminum frame with various designs to define the module maximum deflection under 2400 Pa push load according to the IEC 61215 standard. Changing the frame design also affects the geometry of the module and its light-exposed internal areas which have impact on the loss and gain channels in the PV module. For this purpose, we complement the models in a previous related work [8], where loss and gain mechanisms in the PV module are investigated using CTM analysis. Additionally, the module frame forms about 9-12% of the whole module cost [9, 10], which highlights the importance of the design optimization. Therefore, we study the impact of the frame design on the cost of ownership (COO). Lastly, most frames are made of aluminum, which is an energy-intensive material and may significantly contribute to a higher carbon footprint [11–14]. This aspect is investigated by applying detailed LCA analysis. All previous aspects enlarge the importance of optimizing the frame design and finding a balance point between its mechanical, electrical, economic, and ecologic impact. The Approach and results presented in this paper are based on detailed work in [15].

2. Method

We parameterize the frame design and define variables that affect the PV module mechanically and electrically and influence the amount of aluminum used. Relevant parameters that affect the different aspects considered in this study are illustrated in Figure 2. Like common PV module designs, we assume that the rear side frame width $w_{\text{frame,r}}$ is equal or bigger than the front frame width $w_{\text{frame,f}}$ with a fixed frame thickness t_{frame} of 1.8 mm and rubber seal thickness t_{rubber} of 2 mm. Furthermore, we ensure that the distance through insulation d_{ti} between the string connector and the aluminum frame conform to the standard IEC 61730 which is approximately 21 mm for class II PV modules. We set different sets of front and rear frame widths shown in Table 1 to specify a wide realistic and reasonable range of the different frame parameters including standard and most common frame parameters. The cross-section of studied frame designs is shown in Figure 3 and the overlap is calculated using Equation 1:

$$w_{overlap} = w_{frame} - t_{frame} - t_{rubber} \quad \text{(Eq. I)}$$

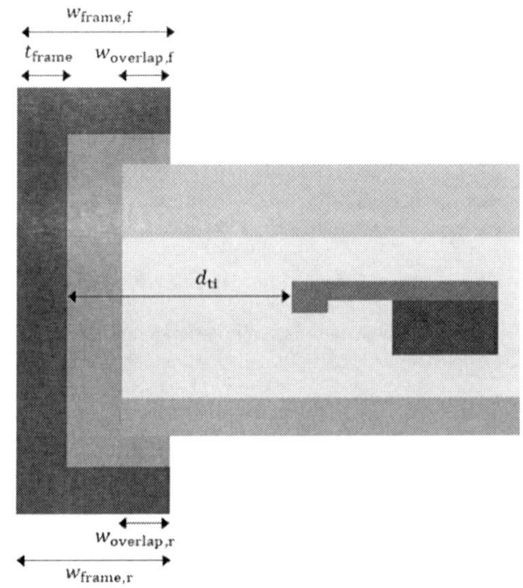

Figure 2: Module layer structure and parameters for frame description.

Table 1: Frame parameters used in the simulations. Reference frame values are sprinted in italic purple.

Front frame width $w_{frame,f}$ [mm]	Rear frame width $w_{frame,r}$ [mm]
10	12
12	14
14	16
16	18
18	*20*

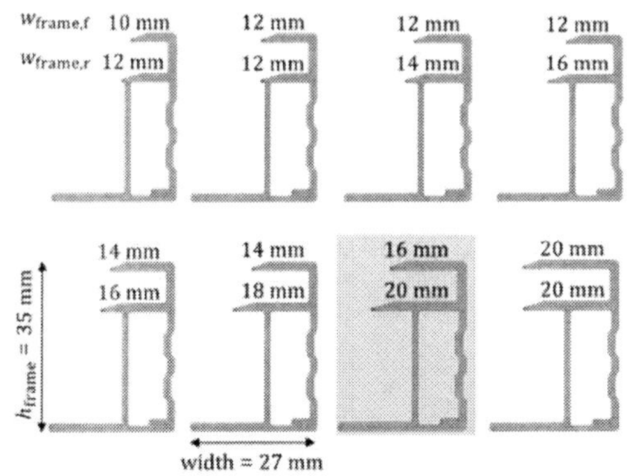

Figure 3: Cross-section of exemplary frame designs with the mesh used in the FEM simulations. Reference frame is marked in purple.

As reference module for the comparison in our study, we choose standard frame parameters of 16 mm front width and 20 mm rear width. After calculating the relevant frame parameters, we use them to define different sets of combinations to represent different frame designs variants for the FEM analysis. In a first step, we simulate the deflection at 2400 Pa push load for the different frame geometries with the parameters in Table 1. Afterwards, CTM, COO and LCA analyses are carried out to simulate the impact on the electrical output of the module, cost, and global warming potential. Table 2 and Table 3 show the solar cell parameters and the used module components in the simulation. The simulated PV module is a 120 M6 half-cells glass-backsheet module with laminate dimensions of 1.76×1.05 m².

Table 2: Solar cell parameters used in the simulations.

Solar Cell	
Format	5 busbar M6 half cell (83×166 mm)
Power	3.10 W

Table 3: Module parameters used in the simulations.

Solar Module	
Laminate Length	1.67 m
Laminate Width	1.05 m
Glass Thickness	3.2 mm AR-coated glass
Encapsulation	2×0.45 mm EVA
Backsheet	0.218 mm white backsheet

2.1. Mechanical Analysis

For the mechanical analysis, we adapt FEM models from previous studies [3, 16] to the investigated PV module topology. The fully parameterized frame is considered in full detail, as depicted in Figure 4. The attachment to the PV laminate is realized by silicone adhesive. For a more detailed description of the FEM model as well as the material properties, we refer to [3]. We simulate the lamination process prior to the mechanical push load according to the IEC 61215 standard. As an improvement to the previously published FEM models, the geometric non-linearity of the deflection is considered in the static simulation as well as the contact to the mounting rail. The module clamp is simulated by a weak constraint, which suppresses the displacement in z-direction on the outer edge of the frame, as depicted in Figure 4 by the purple line. For each frame design using the parameters in Table 1 we evaluate the maximum deflection d_{max} at 2400 Pa push load.

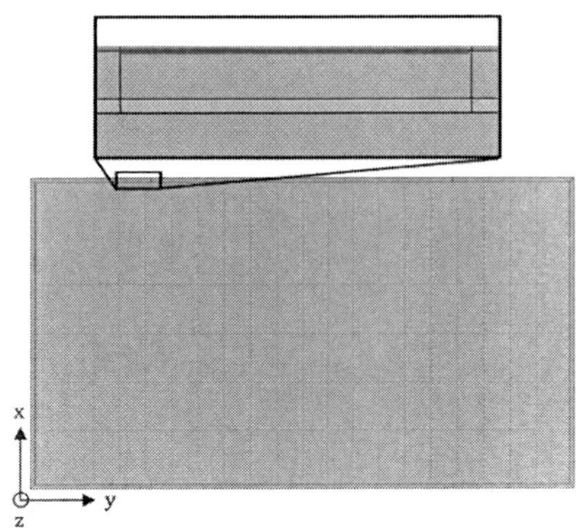

Figure 4: Module topology as modelled in the FEM analysis with the weak constraint on the frame edge (purple line) simulating the frame clamp.

2.2. Cell-To-Module (CTM) Analysis

The CTM simulation is done at standard test conditions (STC) on a 120 M6 monofacial half-cells glass-backsheet PV module with different frame designs as presented above. The models are based on a bottom-up multi-physics approach and are integrated into a consistent framework that allows the direct comparison of different module designs [17–19]. Using the CTM analysis we evaluate the impact of frame design on the loss and gain mechanisms in the PV module. We adapt the CTM models in [8, 20] to incorporate effects of the frame and overlap widths. Changing the parameters of the frame can affect the module margin loss factor k_1 in case of a changed module area as described in Figure 5.

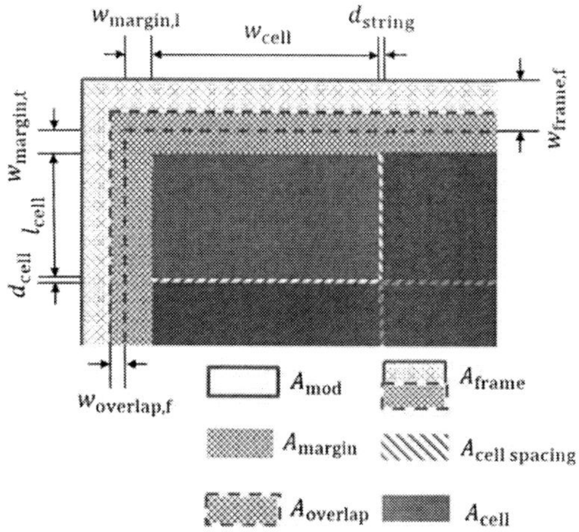

Figure 5: Outer and inner geometry and areas of PV module based on [21].

Furthermore, the frame design has impact on the cover coupling gain factor k_{11} due to changed module internal geometry sown in Figure 5 [20] and therefore optical coupling by light recycling, which is described in Figure 6.

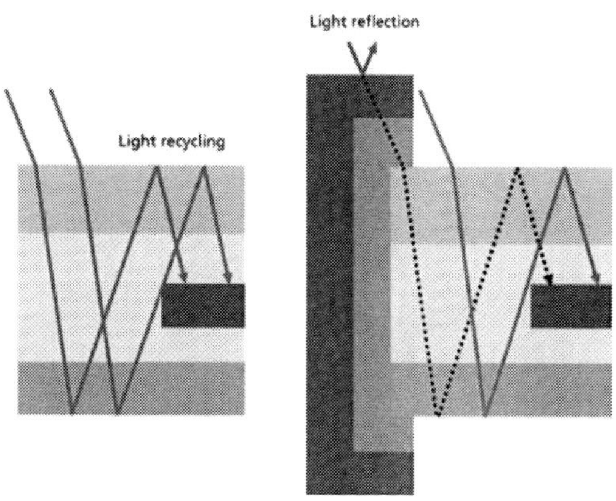

Figure 6: Schematic of the frame effect on the light recycling.

By calculating the additional module area due to the frame, we adjust the loss factor k_1. By defining the frame/glass overlap width, we can calculate the shaded internal area. Shading reduces the short circuit current change ΔI_{SC} used to calculate the gain factor k_{11}. The geometric frame overlap is used to correct the models of [11, 21].

2.3. Cost Analysis

Frames are typically manufactured by extrusion molding of aluminum ingots. Those ingots include the raw material price and the costs of manufacturing the billet used for further extrusion. Depending on the price for raw aluminum (06/2021 ~2 €/kg) [21], the ingots currently account for approximately 70% of the total frame costs. This stresses the importance of reducing frame weight without compromising mechanical stability. We calculate the cost share of the module frame by considering the material price for aluminum and a production share for the variations. The weight of the aluminum frame (kg/m) and the module circumference are taken from the dimensions used in mechanical and CTM analysis. We calculate the costs of the frame and the power loss due to frame shading and combine both to calculate the specific module costs C_{module} (€/Wp). A previously published study for module cost calculation is applied [10].

We calculate the costs of different frame designs based on individual aluminum consumption, which is directly related to the frame parameters. The manufacturing of the reference frame is calculated to account 2.34 €/piece and is assumed constant for all designs. The manufacturing costs of the frame, the framing process costs in module manufacturing (0.30 €) and the costs of the frameless module (62.00 €) are used with the frame material costs and the module power to calculate the module specific cost C_{module}.

2.4. Life Cycle Assessment (LCA)

The life cycle assessment is of great importance to evaluate the potential environmental impact caused by the module and especially the frame design. The carbon footprint or global warming potential GWP of the module, including the studied frames designs, is determined by using SimaPro Analyst v9.0 [22]. The PV foreground processes are based on previous work [12, 23], while background processes are based on Ecoinvent v3.7 database [24]. For the LCA, Germany is assumed as manufacturing location.

3. Results

The results of the different analyses for the various frame designs created from the parameters in Table 1 are presented separately within this section.

3.1. Mechanical analysis

Figure 7 exemplarily shows the deflection of the reference configuration at 2400 Pa as simulated by FEM. The maximum deflection d_{max}, occurring in the PV modules center, is 25.9 mm.

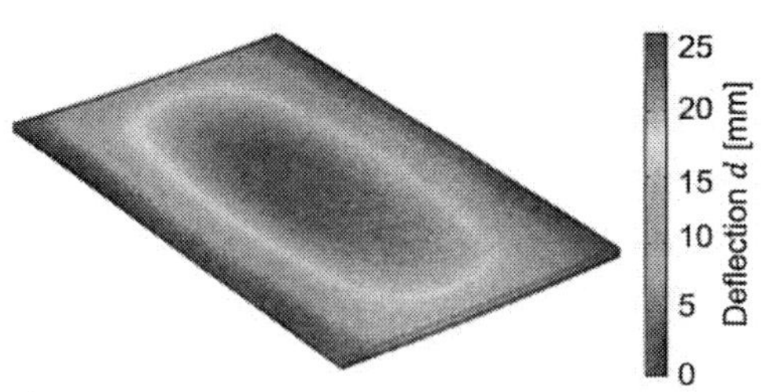

Figure 7: Deflection of the PV module of the reference frame at 2400 Pa push load simulated by FEM.

The maximum deflection d_{max} of all investigated frame geometries is depicted in Figure 8. The variations show a deflection between 25.1 mm and 28 mm by changing the front and rear frame width from 10 mm and 12 mm to 20 mm and 20 mm, respectively. Compared to the reference frame, this is a decrease of 3% for the strongest frame design and an increase of 8.2% deflection for the weakest frame design. Comparing the influence of the front and rear width shows a stronger impact of the front width on the mechanical stability.

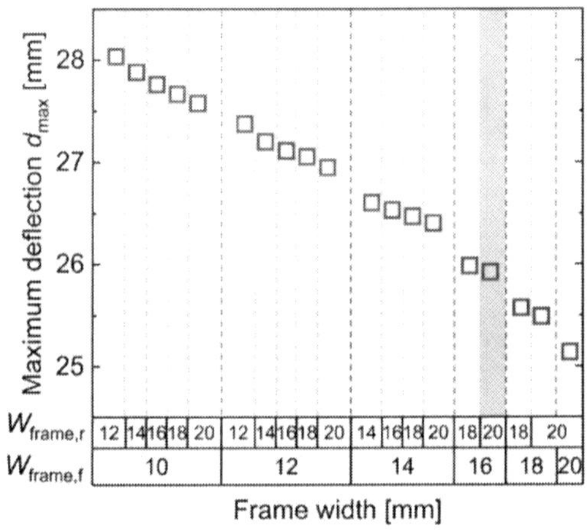

Figure 8: Maximum deflection d_{max} of different frame designs at 2400 Pa push load. The reference module is marked in purple.

3.2. Cell-To-Module (CTM) Analysis

Since no light is photoconverted from the backside (glass-backsheet monofacial PV module), the rear side parameters of the frame do not affect module power or efficiency. Therefore, we only consider changing the front parameters of the frame for this part of the analysis. As the change in the overlap width in our approach equals the change in the frame width, the overall PV module area is constant. This leads to a constant module margin efficiency loss of -1.84% for all frame designs. In general, a smaller front frame overlap leads to a higher cover coupling gain factor k_{11} by allowing for more internal reflection on the white backsheet as described in Figure 6. In Figure 9, the PV module power and efficiency using different frame designs are shown.

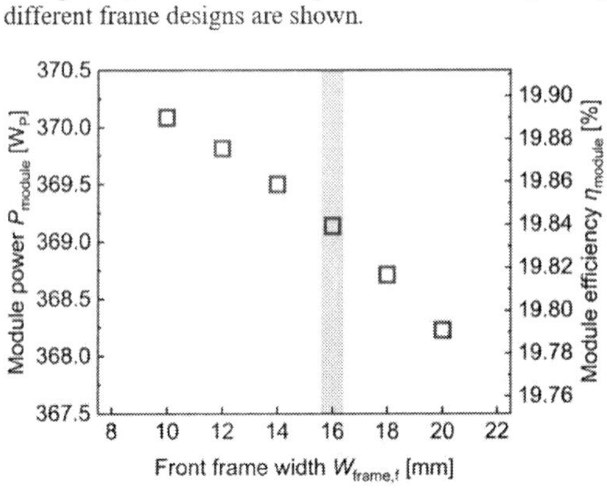

Figure 9: Power P_{module} and efficiency η_{module} of a monofacial 120-halfcell glass-backsheet PV module with different frame designs. The reference module is marked in purple.

It is recognizable that the module power increases due to higher internal reflection when decreasing the front frame width and thereby the overlap width. The module with the lowest front frame width and thereby the lowest overlap has the highest module power of 370.1 Wp and proportionately a module efficiency of 19.89%. This refers to the highest power gain due to cover coupling k_{11} of 7.85 W. Contrariwise, simulating the PV module with a 20 mm front frame width results in 1.86 Wp and 0.1% less module power and efficiency, respectively, due to lower cover coupling gain of 5.96 W. Compared to the PV module with the reference frame design, simulating the module with 10 mm front frame width results in a relative power and efficiency increase of about 0.26%.

3.3. Cost Analysis

We calculate the specific module cost C_{module} based on the different costs defined in the method part, the frame weight, and the PV module output power. Figure 10 shows an expected increasing trend of the module specific cost by increasing the frame width. This refers to the larger amount of aluminum used and to the reduced module power due to smaller coupling gains.

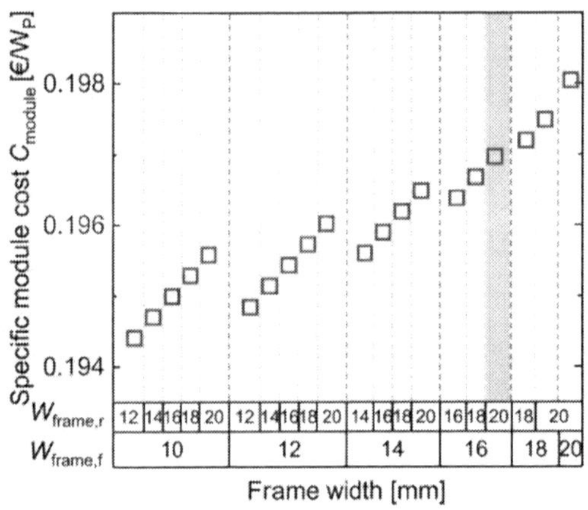

Figure 10: Specific module cost C_{module} using different frame designs. The reference frame design is marked in purple.

We find the module frame to have a share of approximately 10% of total module costs as shown in Figure 11.

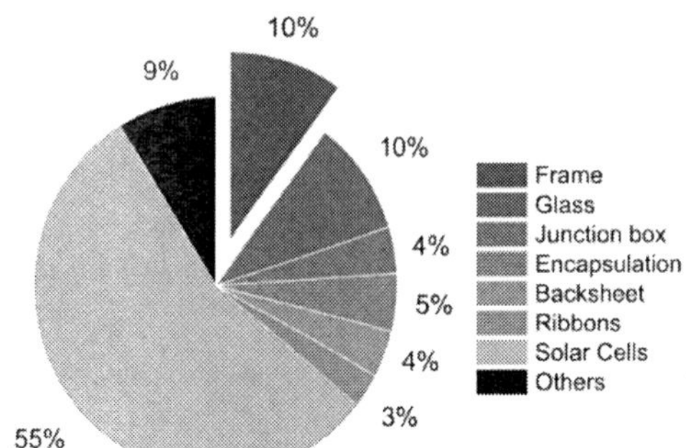

Figure 11: Cost of ownership share of a PV module with reference frame design.

The impact of margin shading due to frame overlap on specific module cost is small due to the small changes in module power. On the other hand, the frame mass has a larger impact on the specific module cost. Due to the cost shares in module and frame manufacturing we find that the module with the lowest specific cost has a 11.9% lower frame mass but leads to lower module costs of only 1.33%, which corresponds to around 1 €. Considering all designs, we find that a reduction of frame weight by about 9% reduces total module costs by 1%.

3.4. Life Cycle Assessment (LCA)

The results of the global warming potential *GWP* per kW_P nominal power of the reference module produced in Germany is shown in Figure 12.

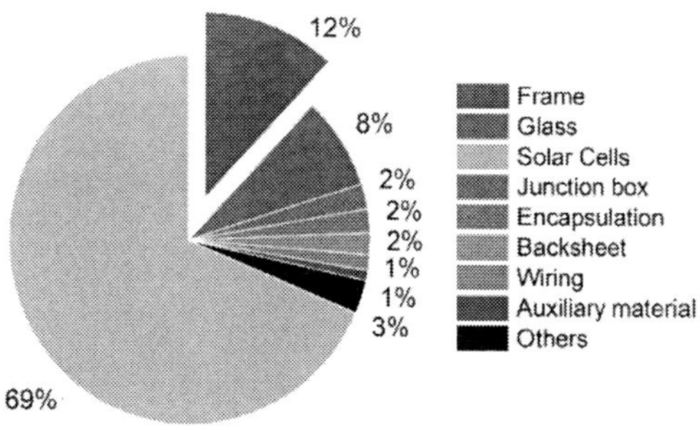

Figure 12: Global warming potential *GWP* share of PV module with the reference frame design.

Based on the results in Figure 12, the module frame has a significant impact on the global warming potential of the PV module. 31% of the total carbon footprint is related to the module materials, whereas the frame with 12% and the glass with 8% have the highest share of the emissions.

Hence, optimizing the frame design of PV modules offers a great potential to reduce the carbon footprint on the module as depicted in more detail in Figure 13.

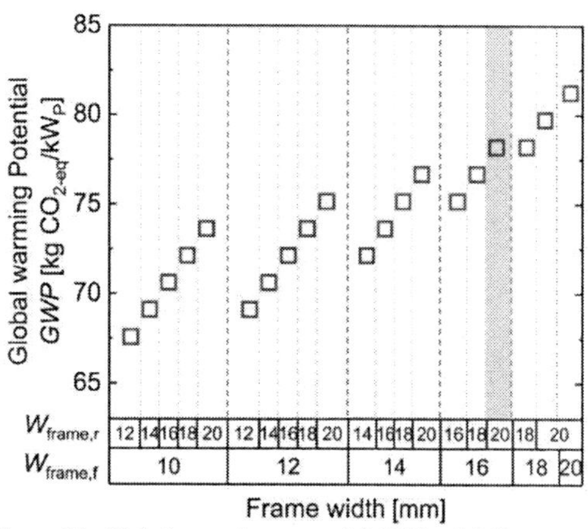

Figure 13: Global warming potential *GWP* of different frame designs. Reference frame is marked in purple.

It is obvious that decreasing the front and rear frame width (and thus the frame mass) has a significant influence on the CO_2 emissions. In comparison with the reference module, a CO_2 amount of up to 12%, corresponding to around 9 kg $CO_{2\text{-eq}}/kW_P$ can be avoided when reducing front and rear frame widths to 10 mm and 12 mm respectively due to the savings in aluminum. Furthermore, it can be observed that a reduction in the rear frame width is almost equally beneficial as minimizing the front frame width.

4. Discussion

The analyses show a potential to optimize the PV module frame. The results show that the mechanical stability is directly related to the frame height and weight which are relevant for costs and global warming potential. Therefore, the frame design optimization must be done with a focus on this aspect.

The results stress the necessity to reduce frame weight without compromising mechanical stability to reduce the module costs. An increase in overlap to achieve that is possible due to the comparatively small impact on the CTM power change. CTM analysis shows that the frame design has a small impact on the power of the PV module. Increasing the front frame width to 20 mm results in a decrement of $0.05\%_{abs}$ in efficiency compared to the reference frame design. LCA analysis shows that an aluminum frame is large contributor to the module carbon footprint and that a reduction in frame weight is crucial for optimization.

Having a closer look to the results and combining them reveal that the frame design with 18 mm front and rear width has a lower cost as well as deflection and the same CO_2 footprint compared to the reference frame. Therefore, in a second step this design is further investigated in terms of weight reduction potential. We simulate ten different variations of this frame by FEM to identify the influence of different parameters. The variations do affect all aspects except the CTM analysis since the frame front and rear widths are constant. Figure 14 shows the change in maximum deflection and frame weight by applying the variations.

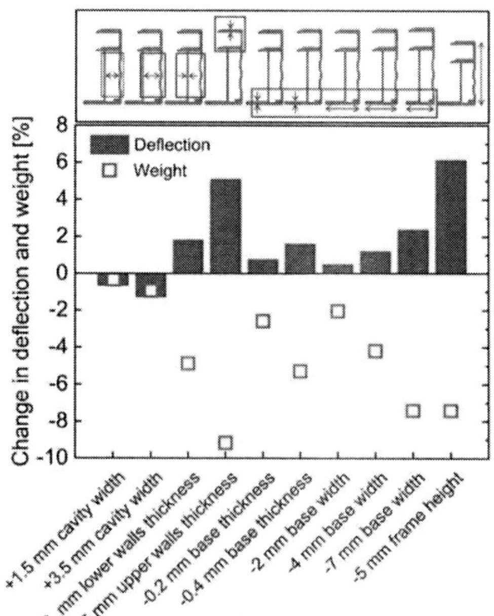

Figure 14: Change of maximum deflection and weight relative to the frame design with 18 mm front and rear width.

Enlarging the cavity decreases the deflection up to 1.3% and simultaneously the weight by 0.9%. All other changes lead to an increase of the deflection. The highest impact is due to decreasing the frame height (+6.1%) followed by decreasing the wall thickness of the upper frame part (+5.1%). Both also have a large impact on the weight (-7.4% and -9.2%). Decreasing the base thickness and length show a smaller impact on the deflection (between 0.8% and 2.5%) while having the same weight reduction potential as the frame height (2-7.4%). Therefore, we identify the cavity enlargement as well as a base thickness reduction as further optimization potential. However, choosing the frame design relies on manufacturer priorities regarding optimization.

5. Summary and Conclusion

By combining different simulation methods and analyses, we defined an approach for a holistic digital prototyping and optimization of the PV module frame. The simulation methods consist of mechanical FEM, CTM, COO and LCA analysis. With these methods, frame designs are investigated, and optimization potential is identified.

The highest optimization potential is identified for the frame front width followed. Figure 15 shows the reference design along with the variations that lead to a balanced optimization regarding mechanical stability, module power, costs, and carbon footprint.

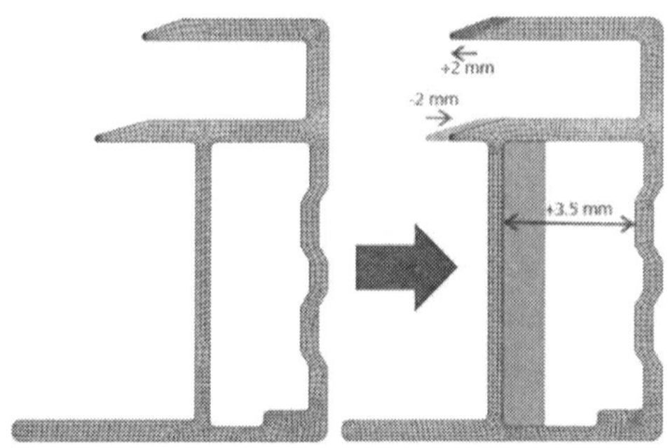

Figure 15: Reference frame design (left) and the variation with the highest optimization potential (right).

Compared to the PV module with the reference frame, the optimized design has a 2.6% lower deflection at 2400 Pa push load while saving 0.9% weight. The design reduces the PV module cost by 0.1% but also the PV module power by about 0.4 Wp. Around 1% of CO_2 emissions can be saved, which corresponds to 0.8 kg $CO_{2\text{-}eq}$/kW$_P$ due to around 30 g savings in aluminum. Other frame designs revealed an even higher reduction in cost and CO_2 footprint, with slight loss in mechanical stability. Therefore, the benefit of the presented holistic approach is that each aspect can be weighted individually to find a design optimized to the individual demand.

6. Acknowledgements

The authors acknowledge the funding of the work by the German Ministry of Economic Affairs and Energy under Grant No. 03EE1028A "CTS1000plus".

7. References

[1] M. A. Green, "Silicon photovoltaic modules: a brief history of the first 50 years," *Progress in Photovoltaics*, vol. 13, no. 5, pp. 447–455, 2005, doi: 10.1002/pip.612.

[2] A. Tummalieh, A. Pfreundt, and M. Mittag, "Trend Tracking of Efficiency and CTM Ratio of PV Modules: 37th European Specialist Conference and Exhibition," 2020.

[3] A. J. Beinert, P. Romer, M. Mittag, and J. Aktaa, "The Effect of Cell and Module Dimensions on thermomechanical Stress in PV Modules," *IEEE J. Photovoltaics*, vol. 10, 2020.

[4] A. J. Beinert, M. Ebert, U. Eitner, and J. Aktaa, "Influence of Photovoltaic Module Mounting Systems on the Thermo-Mechanical Stresses in Solar Cells by FEM Modelling," 2016.

[5] Y. Lee and A. A. Tay, "Stress Analysis of Silicon Wafer-Based Photovoltaic Modules Under IEC 61215 Mechanical Load Test," *Energy Procedia*, vol. 33, pp. 265–271, 2013, doi: 10.1016/j.egypro.2013.05.067.

[6] J. Y. Hartley *et al.*, "Effects of Photovoltaic Module Materials and Design on Module Deformation Under Load," *IEEE J. Photovoltaics*, vol. 10, no. 3, pp. 838–843, 2020, doi: 10.1109/JPHOTOV.2020.2971139.

[7] J.Schicker, Ch.Hirschl, and R.Leidl, "Effect of PV Module Frame Boundaries on Stresses in Solar Cells," *Journal of Energy Challenges and Mechanics*, 2014, doi: 10.13140/2.1.3374.1441.

[8] I. Haedrich, U. Eitner, M. Wiese, and H. Wirth, "Unified methodology for determining CTM ratios: Systematic prediction of module power," *Solar Energy Materials and Solar Cells*, vol. 131, pp. 14–23, 2014, doi: 10.1016/j.solmat.2014.06.025.

[9] Jibran Shahid, Max Mittag, and Martin Heinrich, "A Miltidimentional Optimization Approach to Improve Module Efficiency, Power and Costs," 2018.

[10] Fabian Fertig, Sebastian Nold, Nico Wöhrle, Johannes Greulich, Ingrid Hädrich, Karin Krauß, Max Mittag, Daniel Biro, Stefan Rein, Ralf Preu, "Economic feasibility of bifacial silicon solar cells," *Progress in Photovoltaics*, 2016, doi: 10.1002/pip.2730.

[11] V. Fthenakis and E. Alsema, "Photovoltaics energy payback times, greenhouse gas emissions and external costs: 2004–early 2005 status," *Prog. Photovolt: Res. Appl.*, vol. 14, no. 3, pp. 275–280, 2006, doi: 10.1002/pip.706.

[12] A. Müller, L. Friedrich, C. Reichel, S. Herceg, M. Mittag, D. H. Neuhaus, "A Comparative Life Cylce Assemssment of Silicon PV Modules: Impact of Module Design, Manufacturing Location and Inventory," *Solar Energy Materials and Solar Cells*, vol. 230, 2021.

[13] A. Müller, K. Wambach, and E. Alsema, "Life Cycle Analysis of Solar Module Recycling Process," *MRS Proc.*, vol. 895, 2005, doi: 10.1557/PROC-0895-G03-07.

[14] A. Stoppato, "Life cycle assessment of photovoltaic electricity generation," *Energy*, vol. 33, no. 2, pp. 224–232, 2008, doi: 10.1016/j.energy.2007.11.012.

[15] A. Tummalieh, A. J. Beinert, C. Reichel, M. Mittag, and D. H. Neuhaus, "Holistic Design Optimization of the PV Module Frame: FEM, CTM, COO and LCA Analysis," *Progress in Photovoltaics and Applications*, 2021.

[16] A. J. Beinert and A. Masolin, "Enhancing PV Module Thermomechanical Performance and Reliability by an Innovative Mounting Solution: 37th European Photovoltaic Solar Energy Conference and Exhibition," 2020.
[17] M. Mittag, T. Zech, M. Wiese, D. Bläsi, M. Ebert, and H. Wirth, "Cell-to-Module (CTM) Analysis for Photovoltaic Modules with Shingled Solar Cells: 44th IEEE Photovoltaic Specialists Conference," 2017.
[18] M. Mittag, C. Reise, N. Wöhrle, R. Eberle, M. Schubert, and M. Heinrich, "Approach for a Holistic Optimization from Wafer to PV System: IEEE 7th World Conference on Photovoltaic Energy Conversion," 2018.
[19] M. Mittag, A. Pfreundt, J. Shahid, N. Wöhrle, and D. H. Neuhaus, "Techno-Economic Analysis of Half Cell Modules - the Impact of Half Cells on Module Power and Costs," 2019.
[20] I. Haedrich, D. C. Jordan, and M. Ernst, "Methodology to predict annual yield losses and gains caused by solar module design and materials under field exposure," *Solar Energy Materials and Solar Cells*, vol. 202, p. 110069, 2019, doi: 10.1016/j.solmat.2019.110069.
[21] London Metal Exchange, *LME Aluminium*. [Online]. Available: www.lme.com (accessed: Jun. 13 2021).
[22] *SimaPro Analyst v9.0*: PRé Sustainability. Accessed: Jun. 9 2021. [Online]. Available: https://simapro.com/
[23] L. Friedrich, S. Nold, A. Müller, J. Rentsch, R. Preu, "Global Warming Potential and Energy Payback Time Analysis of Photovoltaic Electricity by Passivated Emitter and Rear Cell (PERC) Solar Modules," *IEEE Journal of Photovoltaics*, 2021.
[24] *Ecoinvent database v.3.7.* [Online]. Available: https://www.ecoinvent.org/

FOILMET®-INTERCONNECT SHINGLING (FIS): ALUMINUM FOIL BASED INTERCONNECTION FOR MODULES WITH SHINGLED SOLAR CELLS

Jan Paschen, Oliver John, Torsten Rößler, Puzant Baliozian, Jan Nekarda
Fraunhofer Institute for Solar Energy Systems ISE
Heidenhofstr. 2, 79110 Freiburg, Germany

ABSTRACT: The investigation of novel cell-to-cell interconnection methods has gained importance with the increase of wafer sizes. Shingling (i.e. overlapping) of solar cells is not only a solution for the interconnection of smaller solar cells but also a chance to increase the output power density by (i) increasing the active cell area within the module, (ii) decreasing shading losses, and (iii) reducing electrical interconnection losses. Replacing the commonly used expensive materials such as the silver in the cell's electrodes and the electrically conductive adhesive increases the attractiveness of a shingled interconnection.

This article introduces the FoilMet®-Interconnect, an approach using laser-welded aluminum foil, for shingling. With the production of aluminum foil-based shingle strings and modules, saving 36.7% of silver, a proof of concept was successful. In subsequent measurements an efficiency of $\eta_{FoilMet}$ = (21.0± 0.2)% was achieved with foil-based strings of three shingled cell.

1 INTRODUCTION

The fabrication of silicon wafers and cells trends towards larger formats that reach to around 210 mm edge length [1]. Large cells, however, result in higher currents, which increase the effect of the series resistance contribution of the modules metallization. One way to reduce the cell interconnection losses is the reduction of string currents by interconnecting separated, i.e. smaller, solar cells such as half cells [2–5] and shingle cells [6–9]. Since increased recombination occurs at the currently non-passivated cell edge, the cell pieces cannot be chosen arbitrarily small and a compromise between the reduction of series resistance and edge recombination must be found. Nevertheless, the market share for modules of separated silicon solar cells has gained relevance [1].

The typical approach to interconnect shingle cells is based on using electrically conductive adhesives (ECA) interconnecting the rear side busbar of one cell to the front side busbar of an adjacent cell [10, 11].

As the commercialization of ECA-based shingle modules just took place in the 2010s (e.g. Sunpower and Solaria), the experience with reliability of this technology is still limited compared to conventional interconnection approaches. A focus of recent research is the understanding of thermo-mechanical stress in the ECA joint. The thermo-mechanical stress is caused by temperature changes and different thermal expansion coefficients of the module materials that mainly lead to shearing and bending at the shingle joint [12, 13]. The right choice of ECA and processing conditions considering aspects such as adhesion strength, mechanical flexibility of the material, conductivity and bond line geometry is believed to result in good thermomechanical long-term stability.

Because of the low price of Al compared to copper or Ag and because laser welding of Al with IR lasers is well understood [14], the usage of Al-foils in solar manufacturing is of high interest [15].

2 MATERIAL AND METHODS

2.1 FoilMet®-Interconnect Shingling (FIS)

FIS, addresses the above problems by means of laser-welded Al-foil. First it presents a cost-effective and flexible solution for the interconnection of separated cells. Second it holds a high potential of conserving resources and third, it has a potentially increased string efficiency due to lowered series resistance in the Al-foil joint.

In our novel concept, we omit the ECA and the screen-printed busbar on both sides and replace it with an 8 µm thick Al-foil (EN AW-8079). The Al-Foil is directly joined to the silicon nitride (SiNX) passivation using the so-called Laser Metal Bond (LMB)-process [16]. The Ag-fingers and the Al-fingers are micro welded to the foil using 300-700 µm long line-shaped joints performed with a single 1 kW infrared continuous-wave laser leaving no necessity for busbar-like structures. The simplicity of FoilMet, which connects the fingers of adjacent cells without more intermediate layers than a single Al-foil, is of great advantage compared to an ECA interconnection, with joints from the Al-finger to the Ag-busbar, to the ECA, to the Ag-busbar, and to the Ag-finger, and therefore consisting of a multitude of layers.

2.2 Implementation for shingling via Al-foil

One way to produce shingle strings with the connections described is to place two cells on top of each other with their front sides facing each other (Figure 1A). Al-foil is placed on both cells with the upper cell being positioned 2 mm off-center and brought into direct contact with the substrate using vacuum. After all laser steps have been performed, the upper cell is folded over and remains next to the lower cell with an overlap of 0.7 to 2.5 mm (Figure 1B). The size of the overlap can be adjusted due to the flexibility of the foil forming a U-shape between the upper and the lower cell.

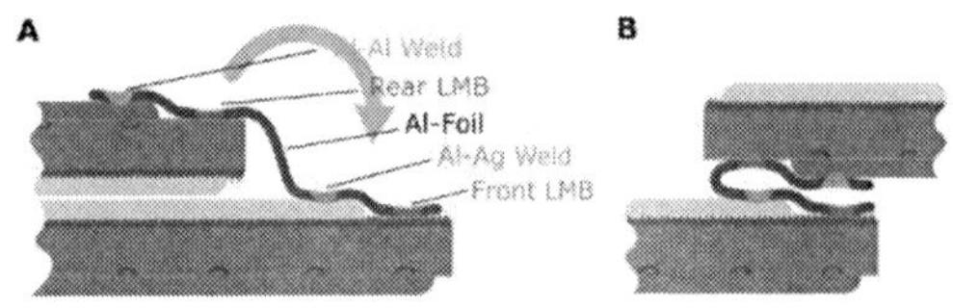

Figure 1: Sketch of different configuration creating a foil shingle string. (A) In the first configuration, the shingles are placed with their front side facing each other with a slight offset. (B) After laser processing, the top cell can be folded over. (not to scale).

Due to the configuration, the Al-foil forms a kink between the two cells which allows the interconnection to be extremely flexible. Depending on the dimensioning of the foil and the configuration, great mobility in almost all directions is possible without exerting forces on the LMB or welds. These flexible connections can absorb the large forces generated during deformation at different temperatures.

3 EXPERIMENTAL DESIGN

3.1 Rating of the Performance of FIS in Comparison to ECA

In this experiment, ten shingle strings were produced (Figure 2), each consisting of three shingle cells. Pre-characterized and within the groups homogeneously distributed bifacial p-type Cz-Si PERC host cells with an edge length of 156.75 mm and 210 mm diameter were used. Each host cell features five rectangular shingle solar cells with a size of 22 mm x 156.75 mm (the two wafer parts with the pseudo-square edges are not used). The two groups were screen-printed in the same run with identical front Ag finger and identical rear Al finger metallization layouts. The FoilMet® group bypassed the screen-printing steps for the front and rear Ag busbars. Six strings were adhered in a conventional way using ECA and standard end connectors, and four were joined using FIS ending with an open lace of Al-foil. The FoilMet® strings have on average (0.89 ± 0.46) mm larger overlap per interconnection than their ECA reference due to manual handling and therefore a smaller active area. The active area consists of all visible cell area in the module including Ag-fingers but excluding areas covered by Al-foil as in the end-connections or stretched interconnections. I-V-measurements were performed for all strings by contacting the end connectors respectively Al-foil laces with kelvin clamps.

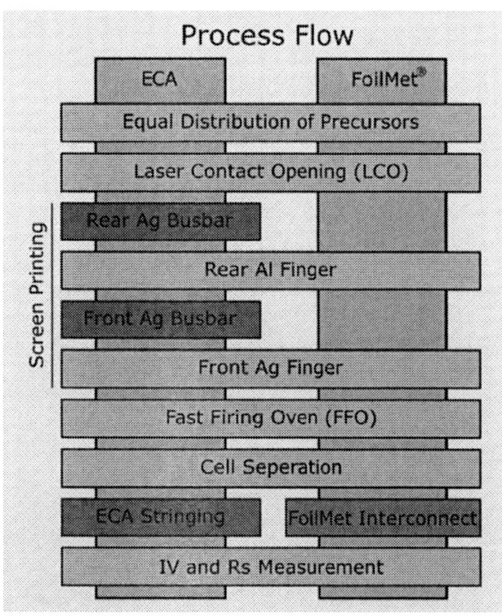

Figure 2: Process flow of the experiments in this work. Two groups of shingle strings were produced within the first experiment of this work. The conventionally processed strings with ECA and our novel approach with Al-foil, FIS. Process steps that differ between the two groups are highlighted. The FoilMet® group does not require a screen-printing step for rear and front Ag-busbar.

4 RESULTS

In the following, we compare our novel FIS to the conventional interconnection using ECAs. In this experiment, the performance of the foil interconnection was benchmarked against a conventional interconnection using ECA. The ECA strings achieved an average efficiency of η_{ECA} = (20.3 ± 0.2)%. The FoilMet® group on the other hand achieved with $\eta_{FoilMet}$ = (21.0± 0.2)% an outstanding efficiency, 0.7%$_{abs}$ higher than the ECA group (Figure 3).

The data for short circuit current (Figure 3B), fill factor (Figure 3C), and open circuit voltage (Figure 3D), imply the efficiency advantage is mainly due to the increased fill factor (+2.5%$_{abs}$). It indicates that the welding of the aluminum foil with the Ag-/Al-finger has excellent electrical properties.

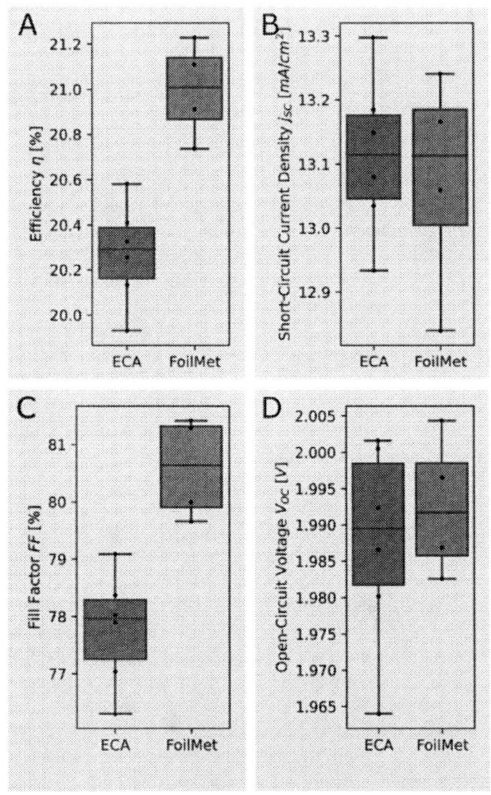

Figure 3: Comparison of strings adhered in the conventional way using ECA (ECA) and strings with the novel FIS (FoilMet). I-V-measurements of shingle strings consisting of three cells each itemized to (A) energy conversion efficiency, (B) fill factor, (C) short-circuit current density, and (D) open-circuit voltage.

The following reasons for the disparity of the fill factor are under discussion. (i) An increased resistance in the interface of the ECA and the screen-printed Ag-busbars would explain the effect, reflected in the measurement. (ii) Also, the interface between the Al-finger and the Ag-busbar could have led to increased resistance contributions. (iii) Due to the manual placement of the cells within the strings, the size of their active area does vary and could lead to a mismatch of the cells within the string. This would result in the fill factor being overestimated and a reduction in string current. Regardless of the cause leading to the disparity in the filling factor, in general good interconnections using ECA can be achieved. FIS, however, demonstrates the ability to reliably create interconnections with low series resistance and strings with high efficiencies. The simplicity of FoilMet, which connects the fingers of adjacent cells without more intermediate layers than a single Al-foil, is of great advantage and might be a way to improved reliability of interconnections.

During the screen-printing process, the wafers were weighed before and after the individual screen-printing steps. Therefore, the exact Ag-consumption of the Ag-paste with an Ag-content of 91% could be determined. Since the FoilMet® cells did not require any screen-printing steps for busbars, 27.5% less Ag-paste was required for their production. A further saving of silver results from the omission of the ECA with an Ag-content of 50% and an average usage of 4 mg per interconnection resulting in a total conservation of 36.7% Ag.

5 SUMMARY

In this article, we introduce FoilMet®-Interconnect Shingling (FIS), a flexible and potentially low-cost shingle interconnection without the need for screen-printed Ag-busbars nor ECA. We use Al-foil to replace the interconnection that is normally established with ECA. Three joining processes are introduced: the Al-foil is welded directly (i) to the fingers of the rear Al-electrode, (ii) to the fingers of the front Ag-electrode, and (iii) to the front and rear passivation. All three joining processes are evaluated in terms of adhesion and conductivity and are examined based on polished cross-sections.

In the first experiment, our novel shingle strings were produced and benchmarked to conventional strings using ECA. The FoilMet® strings achieved an efficiency of $\eta_{FoilMet} = (21.0 \pm 0.2)\%$. This is $0.7\%_{abs}$ higher than its ECA reference, most likely as a result of a low series resistance and leads to an improvement in the fill factor by $2.5\%_{abs}$. During the production in this experiment, 36.7% Ag was conserved due to the unnecessity of busbars and ECA within FIS. In the second experiment, we produced modules with the novel interconnection concept. These modules were subjected to a TC50 test to obtain first insights of the longterm stability of the Al-interconnection. With one exception, all shingle strings withstood the stresses of encapsulation and a TC50 test. In the experiment, it has been shown that FIS might be a way to low resistant and high-efficiency interconnection. Since the Al-foil is bonded to the cell and welded to its electrodes, FIS is potentially applicable to nearly all metallization and cell concepts.

5 ACKNOWLEDGEMENT

The German Federal Ministry for Economic Affairs and Energy is gratefully acknowledged for financially supporting this work within the project "FolMet" (contract number 0325574). The authors also would like to thank all colleagues at Fraunhofer ISE who have contributed to this work.

6 REFERENCES

[1] ITRPV, International Technology Roadmap for Photovoltaic (ITRPV) - Results 2016, http://www.itrpv.net/Reports/Downloads/2019/.

[2] F. Gérenton, *et al.*, "Analysis of edge losses on silicon heterojunction half solar cells," *Solar Energy Materials and Solar Cells*, vol. 204, p. 110213, 2020, doi: 10.1016/j.solmat.2019.110213.

[3] A. Münzer, *et al.*, "Post -Separation-Processing for Industrial High Efficiency Silicon Heterojunction Solar Half Cells with Passivated Edges," *IEEE Journal of Photovoltaics (submitted)*, 2020.

[4] J. Muller *et al.*, "Resistive Power Loss Analysis of PV Modules Made From Halved 15.6 × 15.6 cm 2 Silicon PERC Solar Cells With Efficiencies up to 20.0%," *IEEE J. Photovoltaics*, vol. 5, no. 1, 2015, doi:10.1109/JPHOTOV.2014.2367868.

[5] J. Schneider, *et al.*, "Solar Module with Half Size Solar Cells," 2014, doi: 10.4229/EUPVSEC20142014-1BV.6.48.

[6] P. Baliozian, *et al.*, "PERC-Based Shingle Solar Cells and Modules at Fraunhofer ISE," *Photovoltaics International*, vol. 43, pp. 63–79, 2019.

[7] D. Tonini, *et al.*, "Shingling Technology For Cell Interconnection: Technological Aspects And Process Integration," *Energy Procedia*, pp. 36–43, 2018, doi: 10.1016/j.egypro.2018.09.010.

[8] N. Klasen, A. Mondon, A. Kraft, and U. Eitner, "Shingled Cell Interconnection: A New Generation of Bifacial PV-Modules," in *7th Workshop on Metallization & Interconnection for Crystalline Silicon Solar Cells*, Konstanz, Germany, 2017.

[9] D. Rudolph, *et al.*, "Cell Design Optimization for Shingled Modules," *33rd European Photovoltaic Solar Energy Conference and Exhibition, Amsterdam, The Netherlands*, 2017.

[10] G. Beaucarne, "Materials Challenge for Shingled Cells Interconnection," *Energy Procedia*, vol. 98, pp. 115–124, 2016, doi: 10.1016/j.egypro.2016.10.087.

[11] R. Morad *et al.*, "METHOD FOR MANUFATURING SOLAR CELL MODULES," EP3489848 (A1), EP EP20180215615 20150526, May 29, 2019.

[12] M. Springer and N. Bosco, "Linear viscoelastic characterization of electrically conductive adhesives used as interconnect in photovoltaic modules," *Prog Photovolt Res Appl*, vol. 28, no. 7, pp. 659–681, 2020, doi: 10.1002/pip.3257.

[13] N. Klasen, P. Romer, A. Beinert, and A. Kraft, "FEM simulation of deformations in strings of shingled solar cells subjected to mechanical reliability testing," *AIP Conference Proceedings*, vol. 2156, p. 20016, 2019, doi: 10.1063/1.5125881.

[14] X. Cao, W. Wallace, C. Poon, and J.-P. Immarigeon, "Research and Progress in Laser Welding of Wrought Aluminum Alloys. I. Laser Welding Processes," *Materials and Manufacturing Processes*, vol. 18, no. 1, pp. 1–22, 2003, doi: 10.1081/AMP-120017586.

[15] J. Paschen *et al.*, "FoilMet® - Connect: A new rear metallization upgrade for PERC and other cell concepts," *AIP Conference Proceedings*, vol. 2156, p. 20004, 2019, doi: 10.1063/1.5125869.

[16] O. John *et al.*, "Laser Metal Bonding (LMB) - low impact joining of thin aluminum foil to silicon and silicon nitride surfaces," *Procedia CIRP*, vol. 94, pp. 863–868, 2020, doi: 10.1016/j.procir.2020.09.109.

AUTHOR INDEX

WIP – Renewable Energies
Sylvensteinstr. 2
81369 Munchen
Germany

ISBN 978-1-7138-5213-1

CURRAN ASSOCIATES INC.
proceedings.com

38th European Photovoltaic Solar Energy Conference and Exhibition (EU PVSEC 2021)

Online
6-10 September 2021

Volume 2 of 3

WIP RENEWABLE ENERGIES

38th European Photovoltaic Solar Energy Conference and Exhibition (EU PVSEC 2021)

Online
6-10 September 2021

Volume 2 of 3

Editors:

J.M. Almeida Serra
R. Kenny
P. Helm

ISBN: 978-1-7138-5213-1

Printed from e-media with permission by:

Curran Associates, Inc.
57 Morehouse Lane
Red Hook, NY 12571

Some format issues inherent in the e-media version may also appear in this print version.

Printed with permission by Curran Associates, Inc. (2022)

For permission requests, please contact WIP – Renewable Energies
at the address below.

WIP – Renewable Energies
Sylvensteinstr. 2
81369 Munchen
Germany

Phone: +49 89 72012735
Fax: +49 89 72012791

wip@wip-munich.de

Additional copies of this publication are available from:

Curran Associates, Inc.
57 Morehouse Lane
Red Hook, NY 12571 USA
Phone: 845-758-0400
Fax: 845-758-2633
Email: curran@proceedings.com
Web: www.proceedings.com

38th European Photovoltaic Solar Energy Conference and Exhibition

Proceedings of the International Conference

06 September – 10 September 2021

Edited by:

J.M. ALMEIDA SERRA
University of Lisbon - FCiencias.ID
Lisbon, Portugal

R. KENNY
European Commission Joint Research Centre
Ispra, Italy

P. HELM
WIP Renewable Energies
Munich, Germany

Coordination of the Technical Programme:
European Commission Joint Research Centre
Via E. Fermi 1
21020 Ispra (VA)
Italy

Institutional Support:
European Commission
UNESCO – United Nations Educational, Scientific and Cultural Organization, Natural Sciences Sector
WCRE – World Council for Renewable Energy

Supporting Organisations:
APVI – Australian PV Institute
EREF – European Renewable Energies Federation
ESA – European Space Agency
ESMC – European Solar Manufacturing Council
EUFORES –The European Forum for Renewable Energy Sources
EUREC – The Association of European Renewable Energy Research Centres
ETIP PV – European Technology & Innovation Platform PV
GÜNDER – The Turkish Solar Energy Society
IEA PVPS – International Energy Agency, Photovoltaic Power Systems Programme
REN21 – Renewable Energy Policy Network for the 21st Century

Institutional PV Industry Cooperation:
SolarPower Europe
SOLARUNITED

Supporting Associations:
AIE – European Association of Electrical Contractors
AREA – African Renewable Energy Alliance
CABA – Continental Automated Buildings Association

EU PVSEC realised by:
WIP Renewable Energies
Sylvensteinstr. 2, 81369 München, Germany
Tel: +49 89 720 12 735, Fax: +49 89 720 12 791
Email: pv.conference@wip-munich.de
www.photovoltaic-conference.com
www.wip-munich.de

Proceedings produced and published by:
WIP
Sylvensteinstr. 2, 81369 München, Germany
Tel: +49 89 720 12 735, Fax: +49 89 720 12 791
Email: pv.conference@wip-munich.de
www.photovoltaic-conference.com
www.wip-munich.de

38th EUROPEAN PHOTOVOLTAIC SOLAR ENERGY CONFERENCE AND EXHIBITION
06 SEPTEMBER – 10 SEPTEMBER 2021

EU PVSEC COMMITTEES

INTERNATIONAL SCIENTIFIC ADVISORY COMMITTEE (ISAC)

Chair
P. Szymanski, European Commission Joint Research Centre, Director of Energy, Transport and Climate, Petten, The Netherlands

Committee Members
P. Frankl, Head of the Renewable Energy Division, International Energy Agency, France
M. Getsiou, European Commission, DG RTD, Brussels, Belgium
S.W. Glunz, Head of Division "Solar Cells – Development and Characterization", Fraunhofer ISE, Freiburg, Germany
R. Kenny, European Commission Joint Research Centre, Directorate for Energy and Transport and Climate, Ispra, Italy
P. Malbranche, General Director, CEA INES, France
P. Menna, ret. European Commission, DG ENER, Brussels, Belgium
S. Nowak, Managing Director of NET Nowak Energy & Technology, St. Ursen, Switzerland
W.C. Sinke, Principal Scientist Solar Energy, TNO Energy Transition, The Netherlands
M. Topič, Chairman ETIP PV, Head of Laboratory of Photovoltaics and Optoelectronics of the University of Ljubljana, Slovenia
P. Verlinden, Director at Amrock Pty Ltd, Visiting Professor at Sun Yat-Sen University, Guangzhou, China
A. Grassi, Managing Director, ETA - Renewable Energies, Florence, Italy
P. Helm, Scientific & Managing Director, WIP Renewable Energies, Munich, Germany

CONFERENCE EXECUTIVE COMMITTEE

Conference General Chair
J.M. Almeida Serra, University of Lisbon- FCiencias.ID, Portugal

Technical Programme Chair
R. Kenny, European Commission Joint Research Centre, Directorate for Energy and Transport and Climate, Ispra, Italy

Committee Members
H. Ossenbrink, ret. European Commission Joint Research Centre, Italy
P. Helm, Scientific & Managing Director, WIP Renewable Energies, Munich, Germany
W.C. Sinke, Principal Scientist Solar Energy, TNO Energy Transition, The Netherlands
S. Nowak, Managing Director of NET Nowak Energy & Technology, St. Ursen, Switzerland
A. Grassi, Managing Director, ETA - Renewable Energies, Florence, Italy
J. Bergmiller, Managing Director Events, WIP Renewable Energies, Munich, Germany
J. de Gregorio, Head of Unit, Scientific Services and Cooperation, WIP Renewable Energies, Munich, Germany

2021 SCIENTIFIC COMMITTEE
Topic Organisers and Paper Review Experts

Topic 1 – New Materials and Concepts for Photovoltaic Devices

A. Martí Vega, UPM, Madrid, Spain
P. Pearce, University of Cambridge, United Kingdom
J. Poortmans, imec, Leuven, Belgium
B. Rech, HZB, Berlin, Germany

J.P. Connolly, GeePs, Gif sur Yvette, France
N.J. Ekins-Daukes, UNSW Australia, Sydney, Australia
J. Govaerts, imec, Genk, Belgium
M.A. Hernandez-Fenollosa, UPV, Valencia, Spain
J.C. Hummelen, University of Groningen, The Netherlands
J.-P. Kleider, CNRS/GeePs, Gif sur Yvette, France
D. König, UNSW Australia, Sydney, Australia
I. Konovalov, University of Applied Sciences Jena, Germany
C. Lévy-Clément, CNRS, Thiais, France
Y. Okada, University of Tokyo, Japan
I. Ramiro, NOVA University Lisbon, Caparica, Portugal
F. Roca, ENEA, Portici, Italy
M. Rusu, HZB, Berlin, Germany
R.E.I. Schropp, Driebergen, The Netherlands
T. Tayagaki, AIST, Tsukuba, Japan
W. Warta, Fraunhofer ISE, Freiburg, Germany

Topic 2 – Silicon Materials and Cells

K. Bothe, ISFH, Emmerthal, Germany
D. Bredemeier, ISFH, Emmerthal, Germany
J. Cárabe, CIEMAT, Madrid, Spain
M. Despeisse, CSEM, Neuchâtel, Switzerland
F. Ferrazza, ENI, San Donato Milanese, Italy
S.W. Glunz, Fraunhofer ISE, Freiburg, Germany
G. Hahn, University of Konstanz, Germany
M. Hermle, Fraunhofer ISE, Freiburg, Germany
S. Peters, Hanwha Q CELLS, Bitterfeld-Wolfen, Germany
S. Riepe, Fraunhofer ISE, Freiburg, Germany
P. Roca i Cabarrocas, LPICM-CNRS, Palaiseau, France
M. Topic, University of Ljubljana, Slovenia
A.W. Weeber, TNO Energy Transition, Petten, The Netherlands
P. Wohlfart, Singulus Technologies, Kahl am Main, Germany

O. Anspach, PV Crystalox Solar, Erfurt, Germany
A. Augusto, Arizona State University, Tempe, USA
N. Azkona, UPV/EHU, Bilbao, Spain
D.L. Bätzner, Meyer Burger Research, Hauterive, Switzerland
M.P. Bellmann, SINTEF, Trondheim, Norway
M. Benmansour, CEA, Le Bourget-du-Lac, France
M. Boccard, EPFL, Neuchâtel, Switzerland
R. Brendel, ISFH, Emmerthal, Germany

T.M. Bruton, Progress in Photovoltaics, Woking, United Kingdom
A. Ciesla, UNSW Australia, Sydney, Australia
A. Danel, CEA, Le Bourget-du-Lac, France
W. Dauksher, Arizona State University, Tempe, USA
P. Delli Veneri, ENEA, Portici, Italy
A. Descoeudres, CSEM, Neuchâtel, Switzerland
S. Dubois, CEA, Le Bourget-du-Lac, France
T. Dullweber, ISFH, Emmerthal, Germany
M. Ernst, ANU, Canberra, Australia
P. Fath, RCT-Solutions, Konstanz, Germany
W. Favre, CEA, Grenoble, France
S. Gall, HZB, Berlin, Germany
C. Hagendorf, Fraunhofer CSP, Halle (Saale), Germany
K. Hesse, Wacker Chemie, Burghausen, Germany
J. Horzel, Fraunhofer ISE, Freiburg, Germany
J. John, imec, Leuven, Belgium
A. Meeder, Geo-En Energy Technologies, Berlin, Germany
J. Meier, Meier Technologies, Berlingen, Switzerland
A. Metz, Alzenau, Germany
D. Muñoz, CEA, Le Bourget-du-Lac, France
R.C.G. Naber, Hanwha Q CELLS, Vaassen, The Netherlands
H.-J. Nonnenmacher, Meyer Burger, Hohenstein-Ernstthal, Germany
Y. Ohshita, Toyota Technological Institute, Nagoya, Japan
A. Otaegi, UPV/EHU, Bilbao, Spain
K. Petter, Hanwha Q CELLS, Bitterfeld-Wolfen, Germany
R. Preu, Fraunhofer ISE, Freiburg, Germany
J. Rentsch, Fraunhofer ISE, Freiburg, Germany
J. Schmidt, ISFH, Emmerthal, Germany
M.C. Schubert, Fraunhofer ISE, Freiburg, Germany
L. Serenelli, ENEA, S. Maria di Galeria - Rome, Italy
K. Sporleder, Fraunhofer CSP, Halle, Germany
B. Terheiden, University of Konstanz, Germany
R. Turan, METU, Ankara, Turkey
P.J. Verlinden, AMROCK, McLaren Vale, Australia
M. Vetter, IPHT, Jena, Germany
Q. Wang, Jinko Solar, Haining, China
W. Wu, Tongwei Solar, Chengdu, China
A. Yamada, Tokyo Institute of Technology, Japan
G. Yang, Delft University of Technology, The Netherlands
D.L. Young, NREL, Golden, USA
J. Zhao, CSEM, Neuchâtel, Switzerland

Topic 3 – Perovskites and Other Non-Silicon Materials and Devices, Multijunctions/Tandems

G. Bardizza, European Commission JRC, Ispra, Italy
G. Flamand, imec, Leuven, Belgium
J.C. Goldschmidt, Fraunhofer ISE, Freiburg, Germany
S. Nicolay, CSEM, Neuchâtel, Switzerland
S. Paetel, ZSW, Stuttgart, Germany
R. Po, eni, Novara, Italy
A. Redinger, University of Luxembourg, Luxembourg
G. Siefer, Fraunhofer ISE, Freiburg, Germany
A.N. Tiwari, EMPA, Dübendorf, Switzerland
W. Tress, Zurich University of Applied Sciences, Winterthur, Switzerland
S.C. Veenstra, TNO Energy Transition, Eindhoven, The Netherlands

S. Albrecht, HZB, Berlin, Germany
N. Barreau, University of Nantes, France
V. Bermudez Benito, QEERI, Doha, Qatar
M.J. Brites, LNEG, Lisboa, Portugal
R. Campesato, CESI, Milan, Italy
G. Coletti, TNO Energy Transition, Petten, The Netherlands
S. Cros, CEA, Le Bourget-du-Lac, France
T. Dalibor, Avancis, Munich, Germany
B. Dimmler, NICE Solar Energy, Schwäbisch Hall, Germany
D. Dimova-Malinovska, Bulgarian Academy of Sciences, Sofia, Bulgaria
C.J. Fell, CSIRO, Mayfield West, Australia
I. Gordon, imec, Leuven, Belgium
S. Huang, Macquarie University, Sydney, Australia
S.J.C. Irvine, Swansea University, St. Asaph, United Kingdom
S. Ishizuka, AIST, Tsukuba, Japan
R.R. King, Arizona State University, Tempe, USA
P. Kratzert, SOLARWATT, Dresden, Germany
C.-F. Lin, National Taiwan University, Taipei, Taiwan
D. Lincot, CNRS, Palaiseau, France
M. Lira-Cantú, CIN2, Barcelona, Spain
M.Ch. Lux-Steiner, HZB, Berlin, Germany
M. Meuris, imec, Diepenbeek, Belgium
U.W. Paetzold, Karlsruhe Institute of Technology, Germany
A. Romeo, University of Verona, Italy
V. Sittinger, Fraunhofer IST, Braunschweig, Germany
M. Theelen, TNO/Solliance, Eindhoven, The Netherlands
G. Timò, RSE, Piacenza, Italy
E. Unger, HZB, Berlin, Germany
A. Virtuani, EPFL, Neuchâtel, Switzerland
A. Walter, CSEM, Neuchâtel, Switzerland
S. Yerci, METU, Ankara, Turkey

Topic 4 – Photovoltaic Modules and BoS Components

C. Buerhop-Lutz, HI ERN, Erlangen, Germany
R. Gottschalg, Fraunhofer CSP, Halle (Saale), Germany
W. Herrmann, TÜV Rheinland Energy, Cologne, Germany
N.M. Pearsall, Northumbria University, Newcastle upon Tyne, United Kingdom
A. Wade, Viessmann, Berlin, Germany
K. Wambach, Wambach-Consulting, Petersdorf, Germany

S. Arancón, WIP Renewable Energies, Munich, Germany
P. Bauwens, Ghent University, Belgium
G. Beaucarne, Dow Silicones Belgium, Seneffe, Belgium
N. Bosco, NREL, Golden, USA
C. Camus, LayTec, Berlin, Germany
S. Capaccioli, ETA - Florence Renewable Energies, Italy
R. Einhaus, Apollon Solar, Saint Priest, France
G. Friesen, SUPSI, Canobbio, Switzerland
V. Fthenakis, Columbia University, New York, USA
W. Gambogi, DuPont, Wilmington, USA
G. Graditi, ENEA, Portici, Italy
Y. Hishikawa, AIST, Tsukuba, Japan
U. Jahn, VDE Renewables, Alzenau, Germany
A.R. Lagunas, CENER, Sarriguren-Navarra, Spain
J. Lopez-Garcia, European Commission JRC, Ispra, Italy
M. Mittag, Fraunhofer ISE, Freiburg, Germany
C. Monokroussos, TÜV Rheinland, Shanghai, China
J. Moschner, KU Leuven, Belgium
M.A. Muñoz-García, UPM, Madrid, Spain
H. Nagel, Fraunhofer ISE, Freiburg, Germany
H. Nussbaumer, ZHAW, Winterthur, Switzerland
S. Oktik, Kadir Has University, Istanbul, Turkey
G. Oreski, PCCL, Leoben, Austria
W. Palitzsch, LuxChemtech, Freiberg, Germany
M. Pander, Fraunhofer CSP, Halle (Saale), Germany
J.N. Roy, IIT Kharagpur, India
T. Sample, European Commission JRC, Ispra, Italy
R. Sinton, Sinton Instruments, Boulder, USA

Topic 5 – PV Systems – Modelling, Design, Operation and Performance

J.M. Almeida Serra, University of Lisbon - FCiencias.ID, Portugal
F. Al-Shakarchi, Entech Smart Energies, Quimper, France
I. Antón Hernández, UPM, Madrid, Spain
A.M. Gracia Amillo, European Commission JRC, Ispra, Italy
P. Lechner, ZSW, Stuttgart, Germany
G. Mütter, Enery, Vienna, Austria
H. Ossenbrink, ret. European Commission Joint Research Centre, Italy
S. Taylor, European Space Agency, Noordwijk, The Netherlands
W.G.J.H.M. van Sark, Utrecht University, The Netherlands

C. Alonso-Tristán, UBU, Burgos, Spain
P. Blanc, MINES ParisTech, Paris, France
M. Grottke, WIP Renewable Energies, Munich, Germany
D. Guida, ENEL Green Power, Roma, Italy
P. Hacke, NREL, Golden, USA
V. Khorenko, Azur Space, Heilbronn, Germany
K. Kiefer, Fraunhofer ISE, Freiburg, Germany
K. Lappalainen, Tampere University, Finland
J. Lin, PV Guider, Taipei, Taiwan
A. Louwen, Eurac Research, Bolzano, Italy
O. Mayer, Bayern Innovativ, Nürnberg, Germany
A. Migan-Dubois, GeePs, Gif-sur-Yvette, France
D. Moser, Eurac Research, Bolzano, Italy
K. Nishioka, University of Miyazaki, Japan
C. Nyman, Soleco, Porvoo, Finland
C. Protogeropoulos, EEPS, Athens, Greece
A.H.M.E. Reinders, University of Twente, Enschede, The Netherlands
J. Remund, Meteotest, Bern, Switzerland
M. Richter, 3E, Brussels, Belgium
S. Rodrigues, Forschungszentrum Jülich, Germany
J.A. Ruiz-Arias, University of Malaga, Spain
M. Sengupta, NREL, Golden, USA
J.S. Stein, Sandia National Laboratories, Albuquerque, USA
D. Stellbogen, ZSW, Stuttgart, Germany
J. Tavares Pinho, UFPA, Belém, Brazil
M. Theristis, Sandia National Laboratories, Albuquerque, USA
M. Victoria, Aarhus University, Denmark

Topic 6 – PV Applications, Integration and Storage

G. Adinolfi, ENEA, Portici, Italy
F.P. Baumgartner, Zurich University of Applied Sciences, Winterthur, Switzerland
F. Dolci, European Commission JRC, Petten, The Netherlands
F. Frontini, SUPSI, Canobbio, Switzerland
K.-P. Kairies, ACCURE Battery Intelligence, Aachen, Germany
U. Muntwyler, Bern University of Applied Sciences, Burgdorf, Switzerland
R. Pestana, R&D Nester, Sacavém, Portugal
R. Schlatmann, HZB, Berlin, Germany
A. Scognamiglio, ENEA, Portici, Italy
E. Voroshazi, CEA, Le Bourget-du-Lac, France
I. Weiss, WIP Renewable Energies, Munich, Germany

P.-J. Alet, CSEM, Neuchâtel, Switzerland
P. Alonso, BWP Baumanagement, Munich, Germany
K. Baert, KU Leuven, Heverlee, Belgium
G. Barchi, Eurac Research, Bolzano, Italy
P. Bonomo, SUPSI, Mendrisio, Switzerland
R. Bründlinger, AIT, Vienna, Austria
F. Carigiet, ZHAW, Winterthur, Switzerland
M Centeno Brito, University of Lisbon, Portugal
J. da Costa Fernandes, University of Applied Sciences Offenburg, Germany
K. Ding, Forschungszentrum Jülich, Germany
M. Dörenkämper, TNO Energy Transition, Eindhoven, The Netherlands
G.C. Eder, OFI, Vienna, Austria
V. Efthymiou, University of Cyprus, Nicosia, Cyprus
W. Folkerts, TNO, Eindhoven, The Netherlands
B. Gaiddon, HESPUL, Lyon, France
I.B. Hagemann, Architekturbüro Hagemann, Aachen, Germany
C. Helmke, Clake Energy, Berlin, Germany
B. Herteleer, KU Leuven, Gent, Belgium
B. Kämpfen, kämpfen zinke + partner, Zürich, Switzerland
J.-M. Klein, CEA, Le Bourget-du-Lac, France
P. Lalanne, HydroClapeyron, Dinard, France
B. Lim, ISFH, Emmerthal, Germany
A.C. Neves Foles, University of Évora, Portugal
T. Reijenga, BEAR-iD, Gouda, The Netherlands
E. Román Medina, Tecnalia, Derio, Spain
E. Schüpbach, BFH, Bern, Switzerland
K. Sinapis, Trespa International, Weert, The Netherlands
L.H. Slooff-Hoek, TNO Energy Transition, Petten, Netherlands
L. Tautenhahn, Avancis, Torgau, Germany
R.M.E. Valckenborg, TNO, Eindhoven, The Netherlands

Topic 7 – Finance, Markets and Policies

C. Breyer, Lappeenranta University of Technology, Finland
M. Getsiou, European Commission DG RTD, Brussels, Belgium
P. Malbranche, CEA, Le Bourget-du-Lac, France
G. Masson, Becquerel Institute, Brussels, Belgium
T. Nordmann, TNC Consulting, Feldmeilen, Switzerland
S. Nowak, NET Nowak Energy & Technology, St. Ursen, Switzerland

G. Arrowsmith, EUREC, Brussels, Belgium
S. Caneva, WIP Renewable Energies, Munich, Germany
J. Donoso Alonso, UNEF, Madrid, Spain
I. Kaizuka, RTS Corporation, Chuo-ku, Japan
D. Polverini, European Commission DG GROW, Brussels, Belgium
N. Taylor, European Commission JRC, Ispra, Italy
E. Vartiainen, Fortum Growth, Finland

Awards Coordinators

Student Awards Coordinator
A.H.M. Smets, Delft University of Technology, The Netherlands

Poster Awards Coordinator
J. Cárabe, CIEMAT, Madrid, Spain

SUBJECT INDEX

Topic Code

1 New Materials and Concepts for Photovoltaic Devices
2 Silicon Materials and Cells
3 Perovskites and Other Non-Silicon Materials and Devices, Multijunctions/Tandems
4 Photovoltaic Modules and BoS Components
5 PV Systems – Modelling, Design, Operation and Performance
6 PV Applications, Integration and Storage
7 Finance, Markets and Policies

Session Type

P = Plenary Session
O = Oral Session
V = Visual Session

Day Codes

A = Monday, 06 September 2021
B = Tuesday, 07 September 2021
C = Wednesday, 08 September 2021
D = Thursday, 09 September 2021
E = Friday, 10 September 2021

e.g. 1AO.1 ⇒ 1= New Materials and Concepts for Photovoltaic Devices, A=Monday, O=Oral session, 1=Session 1

TABLE OF CONTENTS
of EU PVSEC 2021 Proceedings Papers

Visual PRESENTATIONS 5CV.3 PV Systems: Planning, Plant Optimisation Tools, Advanced Installation Criteria, Construction Issues / Concentrators and PV for S

FOREWORD

The EU PVSEC 2021, as the second online edition of the European PV Solar Energy Conference and Exhibition was a great success and has shown once again that a sustainable energy supply is becoming ever more important and only together, through scientific endeavor and cooperation and political commitments and investments that this goal can be achieved. The Conference provided once again a stage for the dissemination of work that will help ensure ever increasing quantities of PV energy and its integration into the current and future energy network. New discussion sessions focused highly relevant topical questions, provided key lessons for upcoming developments. This year 1,300 participants from 62 countries took part in this most inspiring platform. In the exhibition 16 companies from all parts of the world welcomed online visitors and presented their products and services.

The EU PVSEC 2021 Proceedings give a comprehensive overview of the photovoltaic solar sector, its current status and future prospects in science, research, innovation, development and deployment on almost 3,800 pages. Selection for inclusion in the conference was made by paper review experts and topic organisers (see the listing on pages III-V), to whom we express our sincere gratitude for their comprehensive review work and overall contribution to the success of the conference. In addition to the 364 submitted papers, the proceedings include 103 presentations (slides) shown during the plenary and oral presentations as well as 195 poster files of the visual presentations. In total this amounts to 662 publications.

The 2021 EU PVSEC Proceedings maintain our commitment to providing quick and open access to high quality scientific results. They constitute a powerful tool for targeted and quick information search and retrieval, enabling you to search by keyword, paper title and authors. We are sure that these features will help to simplify the use and exploit the full potential of this extensive source of information.

The Conference Proceedings are published as downloadable DVD file and are also fully accessible online. A DOI code (Digital Object Identifier) has been assigned to each paper. This ensures unequivocal and permanent identification and full citability. The papers can be viewed and downloaded in a full free open access from the EU PVSEC Proceedings website www.eupvsec-proceedings.com.

We are confident that these Proceedings will play an important role in providing a comprehensive overview of the current actors and activities in the global PV sector and that they will disseminate information on the state-of-the-art of technologies and applications. This can generate further research, add momentum to innovation and promote interest in PV worldwide.

We would like to cordially thank all authors and participants of the EU PVSEC 2021 for their contributions and look forward to welcoming you in 2022 at the WCPEC-8, the 8th World Conference on Photovoltaic Energy Conversion in Milan, Italy.

The Editors

COMPARISON OF DIFFERENT APPROACHES TO DETERMINE THE NOMINAL PV MODULE OPERATING TEMPERATURE (NMOT)

Werner Herrmann[1]*, Christos Monokroussos[2], Kyumin Lee[3]
[1]TÜV Rheinland Group, Cologne, Germany
[2]TÜV Rheinland Group, Shanghai, China
[3]Array Technologies Inc., Albuquerque, U.S.A.
*Corresponding author: werner.herrmann@de.tuv.com, +49.221.806 2272

ABSTRACT: The aim of this work was to re-evaluate the Climate Specific Energy Rating (CSER) methodology with regard to modelling the temperature of PV modules as defined in the IEC 61853 series of standards "Photovoltaic (PV) module performance testing and energy rating". The temperature behaviour of PV modules is described by a simple model in which the effects of light absorption and wind are considered by two modelling parameters u_0 and u_1. These are derived from a series of measurements under natural sunlight, from which only data points that meet certain stability criteria for solar irradiance and wind speed are used for linear regression (Method A). Our work addresses the shortcomings of the IEC data filter reported by test laboratories in the past: a) small number of filtered data points, b) filtered data points are not well distributed in the wind speed range. Both can lead to considerable uncertainty in the modelling parameters. We investigated an alternative data filter that provides a higher number of usable data points and improves the robustness of the regression by wind speed binning (Method 2). Furthermore, we introduce two indicators for the quality of the regression: a) the mean bias error with respect to the differences between measured and modelled temperatures and b) the irradiance-weighted module temperatures. Based on these indicators, a third method for determining u_0 and u_1 was investigated that does not require data filtering (Method 3) and focuses only on the best fit of modelled and measured module temperatures. For a measurement series from New Mexico with mostly clear skies, the application of the three methods resulted in a large scatter of the u_0 and u_1 parameters, but comparable NMOT values. Furthermore, the evaluation of a measurement series from Germany showed a scattering of the modelling parameters due to the measurement season mainly for Methods A and B. Our results confirm clear advantages for Method C. It is easy to apply, the data collection period can be kept short and the parameters u_0 and u_1 are less sensitive to seasonal variations.

Keywords: Energy Rating, PV Module, Energy Performance.

1 INTRODUCTION

The electrical performance of PV modules and thus their energy production depends on the operating temperature, which is determined by the balance between the heat generated in the PV module and the heat loss to the environment. Heat is primarily generated by incomplete light absorption in the solar cells. Several mechanisms are relevant for heat loss to the environment: a) conductive heat loss due to thermal gradients, b) convective heat loss due to wind, c) radiation exchange between the module and the sky or ground. These loss mechanisms depend on the thermal resistances of the module materials, the emission characteristics of the surfaces, and the environmental conditions such as wind conditions and sky temperature.

This makes the modelling of the PV module temperature very complex, as it depends on many variables (**Figure 1**). In particular, fluctuations of solar irradiance will cause dynamic processes with successive heating and cooling phases, that have different time constants (**Figure 2**). Therefore, temperature modelling approaches in the literature commonly refer to constant ambient conditions [2], [3], [4], [5], [6], [7], [8].

The Climatic Energy Rating of PV modules (CSER) according to IEC 61853 [1] uses the Faiman model [4] to estimate the operating temperature of PV modules from incident solar irradiance (G_{POA}), wind speed (WS) and ambient temperature (T_{AMB}). It is calculated according to the following formula

$$T_{MOD} = T_{AMB} + \frac{G_{POA}}{u_0 + u_1 \cdot WS} \qquad (1)$$

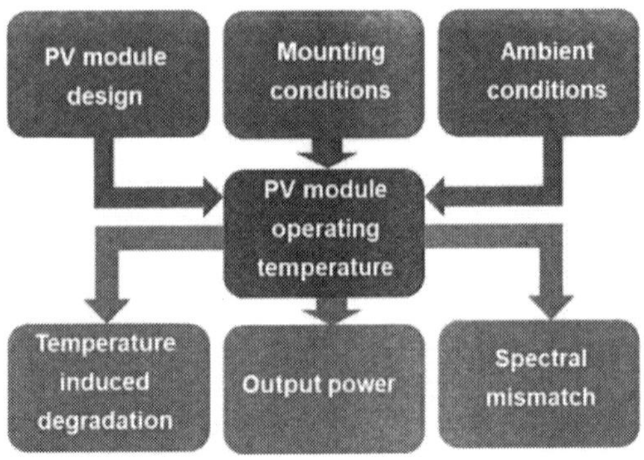

Figure 1: Importance of PV module operating temperature for PV energy yield

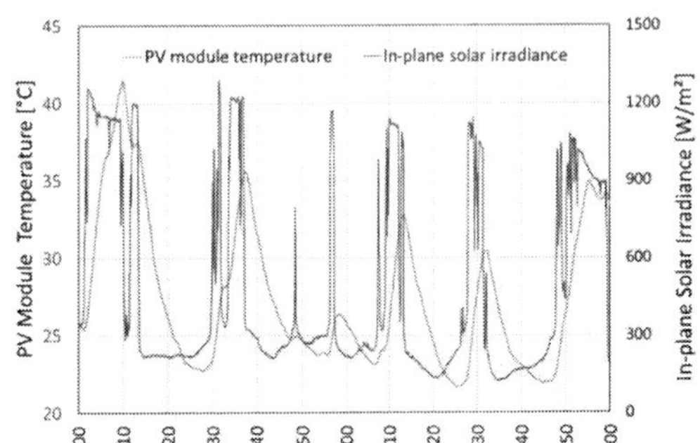

Figure 2: PV module temperature as a function of fluctuations of solar irradiance measured at Braunschweig, Germany.

The coefficients u_0 and u_1 are specific to the module type and mounting conditions and describe the effects of solar irradiance (u_0) and wind speed (u_1). They depend on the PV module design and the properties of the materials used. In order to compare the temperature behaviour of different PV modules, the Nominal PV Module Operating Temperature (NMOT) is specified in the data sheets of manufacturers. This variable is calculated from equation (1) and refers to G_{POA} = 800 W/m², T_{AMB} = 20°C and WS = 1 m/s. Therefore, the determination of the coefficients u_0 and u_1 is of particular importance since it affects the accuracy of NMOT. The NMOT of a PV module is an indicator for temperature induced energy losses.

The coefficients u_0 and u_1 are determined from a series of measurements under natural sunlight. The corresponding test conditions are specified in the standard IEC 61853-2:

- PV module inclination angle = of 37.5° ±2.5°;
- Test sample surrounded by PV modules of the same design;
- Test sample operated at MPP;
- Device temperature (T_{MOD}) measured at 4 positions at the module rear;
- G_{POA}, T_{MOD}, T_{AMB} and WS recorded in intervals 5 s.

The standard also defines a procedure for determining u_0 and u_1 parameters. First, a data filter is applied (see section 3.1), that delivers data points that meet the specific stability criteria. Because transforming equation (1) yields

$$\frac{G_{POA}}{T_{MOD} - T_{AMB}} = u_0 + u_1 \cdot WS \qquad (2)$$

the parameters u_0 and u_1 are then obtained from a linear regression with wind speed corresponding on the x-axis and $G_{POA}/(T_{MOD} - T_{AMB})$ on the y-axis.

Our work addresses the shortcomings of the IEC approach that have been reported by test laboratories in the past: a) the data filter results in a small number of usable data points, b) the filtered data points are not well distributed over the wind speed range. Both of these can lead to a significant uncertainty in the regression and thus in the modelling parameters and NMOT.

The objective of this work was to analyse the practicality of the IEC 61853-2 filter methodology and compare results with alternative approaches for determining the modelling parameters u_0 and u_1. This comparative study was conducted with measurement data recorded at Albuquerque, New Mexico, and Braunschweig, Germany.

2 DATA BASIS USED FOT THE STUDY

Figures 3 and 4 show the 5 s instantaneous values of in-plane solar irradiance G_{POA} and module temperature T_{MOD} plotted against time of the day during the data collection period of one summer month in Albuquerque. The high point density in a narrow band indicates mostly clear skies during the measurement period. The red lines represent the average daily curves. These were used to calculate the average daily insolation and the average module temperature. It should be noted that more than 270.000 data points were collected in the time window between 5.30 am and 7.30 pm.

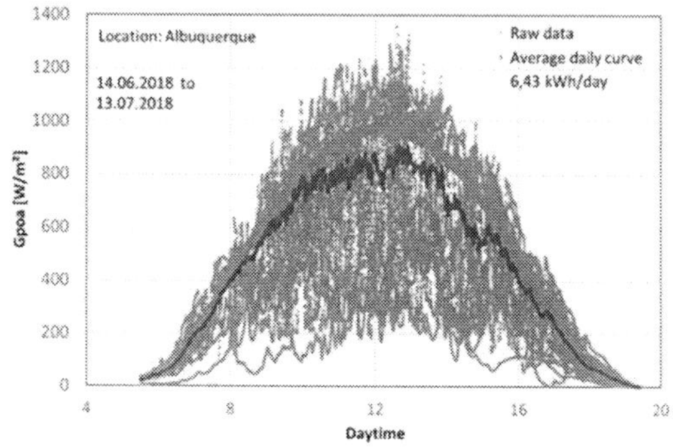

Figure 3: Incident solar irradiance data recorded in time intervals of 5 seconds.

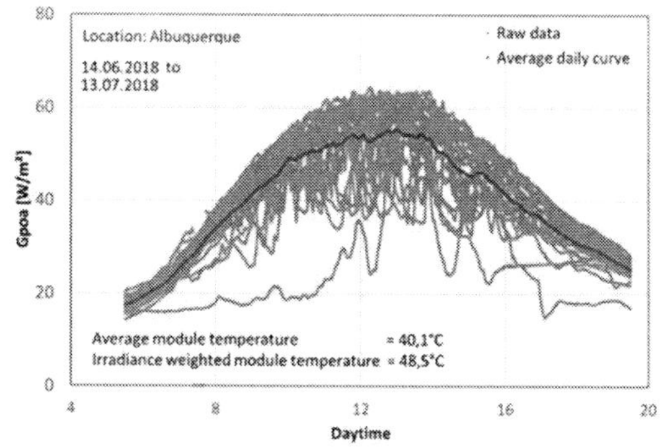

Figure 4: PV module temperature data recorded in time intervals of 5 seconds.

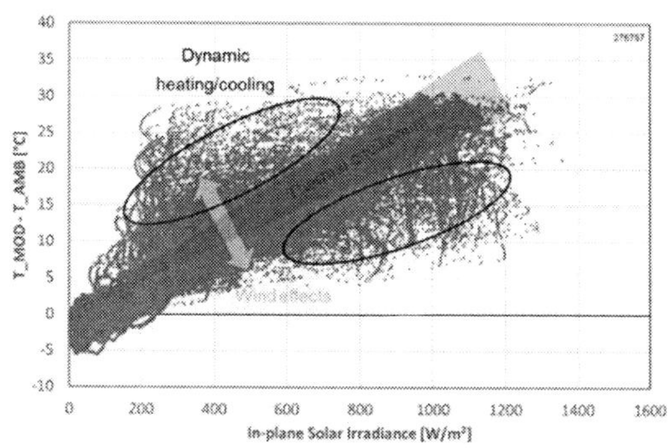

Figure 5: Module surplus temperature above ambient. Temperature differences below 0 K indicate a radiation loss due to sky temperature and errors due to measurement uncertainty.

Figure 5 shows the combination of the two parameters: The surplus module temperature (T_{MOD} – T_{AMB}) is plotted against G_{POA}. This diagram best illustrates the temperature behaviour of a PV module. The cloud of data points can be subdivided into different areas:

a) The triangular region in the centre with a high density of points represents conditions of thermal equilibrium. The upper leg of the triangle represents conditions with WS = 0 m/s, and the opening angle of the triangle is correlated with the effects of wind speed.

b) The areas on either side of the centre with a lower point density. These represent the dynamic operating conditions with heating and cooling phases due to variations in irradiance and wind conditions.

3 QUALITY INDICATORS

A major shortcoming of the current IEC 61853-2 procedure for determining u_0 and u_1 parameters is that no control measure is defined to monitor how well modelled temperatures match the measured module temperatures. To this end, we introduce two quality indicators. These will be used to evaluate the results of the different methods for determining u_0 and u_1 parameters.

3.1 Mean Bias Error

For each data point (G_{POA}, WS) the corresponding module temperature can be calculated according equation (1) and is called modelled temperature. The mean bias error is the average value of temperature differences

$$MBE = \frac{1}{N}\sum_{N}\left(T_{MOD,modelled} - T_{MOD,measured}\right) \quad (3)$$

where N is the total number of raw data points.

3.2 Irradiance Weighted Module Temperature

A suitable indicator for evaluating temperature-related energy yield losses is the irradiance-weighted PV module temperature [9], which is calculated according to equation (4). Assuming a P_{MAX} temperature coefficient of -0.5%/K, the temperature-related PV energy yield losses are -0.5% ·($T_{MOD,Gw}$ – 25°C).

$$T_{MOD,Gw} = \sum(T_{MOD} \cdot G_{POA})/\sum G_{POA} \quad (4)$$

This parameter is used as a second quality indicator for the accuracy of temperature modelling by calculating it for a) the measured module temperatures and b) the modelled module temperatures according to equation (1). The accuracy of modelling is better the smaller the temperature difference $\Delta T_{MOD,Gw}$ between both.

4 METHODOLOGIES FOR DETERMINING THE TEMPERATURE MODELLING PARAMETERS

We studied three methods for determination of u0 and u1 parameters:

Method A: Data filter and regression procedure are applied as defined in IEC 61853-2.

Method B: Data filter is based on fluctuations of GPOA, TMOD and WS in preceding 1-minute intervals combined with a wind speed binning of resulting data points.

Method C: No data filter is applied, but a variation of u_0 and u_1 parameters is conducted until the best fit between measured and modelled module temperatures is achieved.

4.1 IEC 61853-2 approach (Method A)

Method A follows the approach of IEC 61853-2 Ed. 2. The details of the data filtering to extract points for which thermal equilibrium can be assumed are given in **Table 1. Figure 6** shows the results. The blue data points refer to the raw data and the yellow data points are those that meet the stability criteria and are used for linear regression to determine u_0 and u_1 parameters.

The diagram reveals the weaknesses of Method A, even for a site with mostly clear skies. The data filter is too stringent and results in only 0.3% usable data points. In addition, the data points are not well distributed in the wind speed range, which means that ranges with a higher point density have a higher weight in the regression. Both observations contribute to a low accuracy for u_0 and u_1 parameters. Higher regression accuracy can be only achieved by increasing the number of usable data points and a longer data recording interval. However, this is not practical for sites in temperate climates, where only the period April to September is suitable for the measurements anyway.

The summary of Method A results:

- u_0 =29.27 W/(m² °C)
- u_1=5.60 Ws/(m³ °C)
- NMOT=42.9°C
- MBE=0.7 K
- $\Delta T_{MOD,Gw}$=0.3 K

Table 1: Description of the IEC 61853-2 data filter

Minimum G_{POA}: 400 W/m²
Reject data in a following 10 min interval if the G_{POA} variation in the preceding 10 minute period related to instantaneous G_{POA} >10 %
Reject data if the 5-min WS running average is less than 1 m/s and larger than 8 m/s
Reject data in 10 min interval for wind gusts: Instantaneous WS is larger than 200% related to the 5 min wind speed running average
Use the 5 minute wind speed averages as the basis for regression

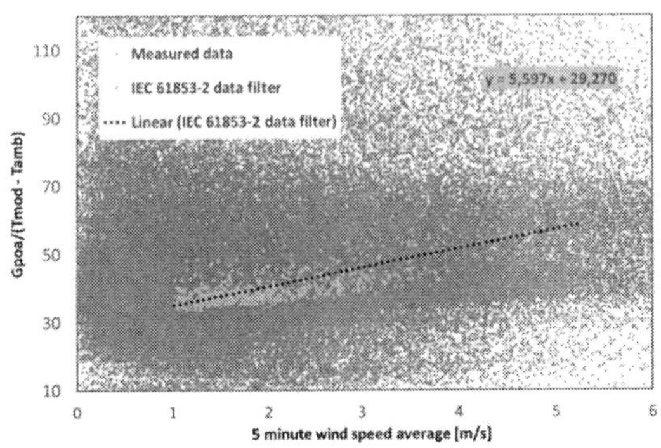

Figure 6: Result for Method A data filter: u_0 =29.27 W/(m² °C), u_1=5.60 Ws/(m³ °C s).

4.2 Alternative Data Filter (Method B)

The approach for data filtering is based on the fluctuations of G_{POA}, T_{MOD} and WS at 1-minute intervals. The details are shown in **Table 2**. The data filter results in a significantly higher number of useful data points that meet the stability criteria (about 17%). However, these are still unevenly distributed over the wind speed range (see **Table 3**). This makes it necessary to introduce another data processing step, namely the binning of wind speed in intervals of 0.5 m/s. For each wind speed interval, the medians of useful 1-minute averages are used for linear regression, so that each wind speed interval is represented by one data point. Another limitation is that only wind speed bins with more than 3% data are accepted.

Figure 7 illustrates this procedure. The yellow data points are the useful data points from the data filter and the red points result from the wind speed binning. It can be seen that the wind speed binning results in a more robust linear regression compared to Method A. However, Method B leads to large deviations in the u_0 and u_1 values, while NMOT deviates only slightly. Both MBE and $\Delta T_{MOD,Gw}$ are 0.5 K higher, indicating poorer agreement between modelled and measured temperatures.

Table 2: Description of the alternative data filter

Minimum G_{POA}: 400 W/m²
Reject data if the G_{POA} fluctuation in the preceding 1-min interval is larger than 10 W/m²
Reject data if the T_{MOD} fluctuation in the preceding 1-min interval is larger than 1 °C
Reject data if the WS fluctuation in the preceding 1-min interval is larger than 2 m/s

Table 3: Distribution of data points that fulfil the stability criteria in wind speed bins

WS bin	Data points		WS median X	$G_{POA}/(T_{MOD}-T_{AMB})$ Y
0 – 0.5	3439	7,1%	0.40	35.18
0.5 – 1	15426	31.8%	0.77	35.44
1 – 1.5	14242	29.3%	1.22	35.81
1.5 – 2	8534	17.6%	1.72	36.36
2 – 2.5	4117	8.5%	2.20	37.15
2.5 – 3	1716	3.5%	2.68	38.79
3 – 3.5	641	1.3%		
3.5 – 4	291	0.6%		

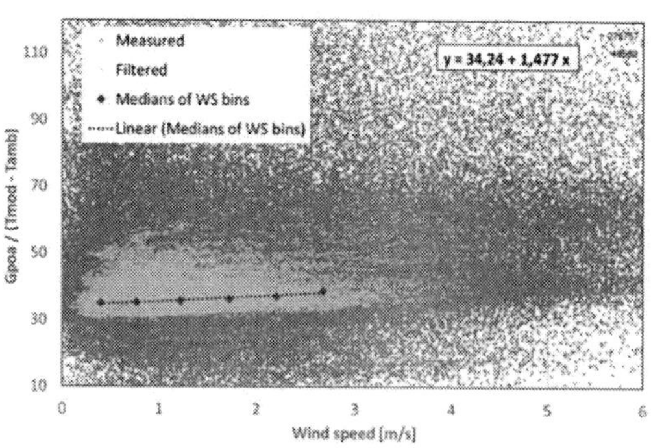

Figure 7: Result for Method B data filter: u_0 =29.27 W/(m² °C), u_1=5.60 Ws/(m³ °C).

The summary of Method B results:

- u_0 =34.24 W/(m² °C)
- u_1=1.48 Ws/(m³ °C)
- NMOT=42.4°C
- MBE=1.3 K
- $\Delta T_{MOD,Gw}$=0.8 K

4.3 Best Fit of Measured and Modelled PV Module Temperatures (Method C)

The previous sections have shown that the use of different data filters may not lead to consistent results for parameters u_0 and u_1. To ensure robust linear regression and high accuracy of u_0 and u_1 parameters, a long monitoring period may be required to obtain a sufficiently large number of data points representing thermal equilibrium of the PV module.

With this in mind, we investigated a third approach to derive the parameters u_0 and u_1. It does not require a data filter, but only a cleanup with implausible data. The method is applied on the entire raw data. The method uses a variation of u_0 and u_1 parameters until the best match of measured and modelled module temperatures is achieved. For this purpose, the quality indicators MBE and $\Delta T_{MOD,Gw}$ are used. The procedure is as follows:

a) Initial settings

The initial parameter settings u_0=35 and u_1=0 are applied and surplus temperatures ($T_{MOD,modelled} - T_{AMB}$) and ($T_{MOD,measured} - T_{AMB}$) are plotted in a diagram against G_{POA}. Since the effects of wind are not considered in the modelling, Eq. 1 transforms into a linear equation with slope $1/u_0$.

b) Optimization of u_0 value

The parameter u_0 is changed in steps of 0.5 until the displayed line coincides with the upper boundary line of measured data points, for which thermal equilibrium and WS=0 m/s exists. This condition can be easily seen by the transition from a high to a low density of data points, provided that there are days with clear sky conditions and high irradiance levels during the measurement period. **Figure 8** shows the resulting diagram for the Albuquerque data set.

c) Optimization of u1 parameter

The parameter u_1 is changed in steps of 0.5 until the absolute values of MBE and $\Delta T_{MOD,Gw}$ are minimal. **Figure 9** shows that the variation of parameter u_1 leads to a scattering of the calculated temperature differences below the line u_1=0. The higher u_1 is chosen, the larger is the scatter.

The distinctiveness of Method C is that it uses specific operating conditions of the PV module as fixing point: Thermal equilibrium for high irradiance levels and WS=0 m/s. The overall results for Method C are shown below. The modeling parameters lie between Methods A and B, but NMOT is lower.

- u_0 =32 W/(m² °C)
- u_1=4.5 Ws/(m³ °C)
- NMOT=41.9°C
- MBE=0.3 K
- $\Delta T_{MOD,Gw}$=0.2 K

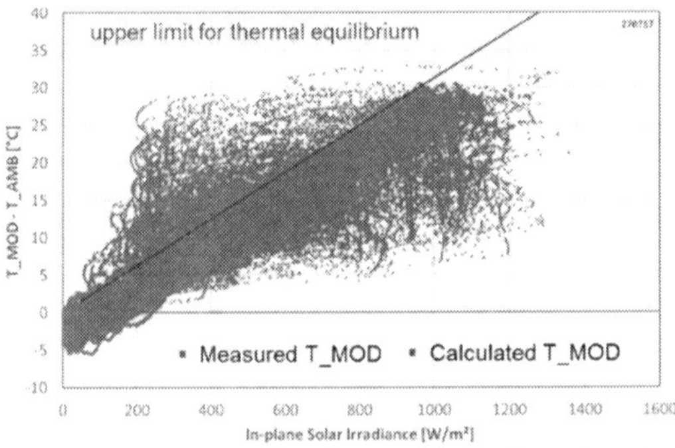

Figure 8: Method C plot for u_1 = 0 Ws/(m³°C) and optimized u_0=32 W/(m² °C).

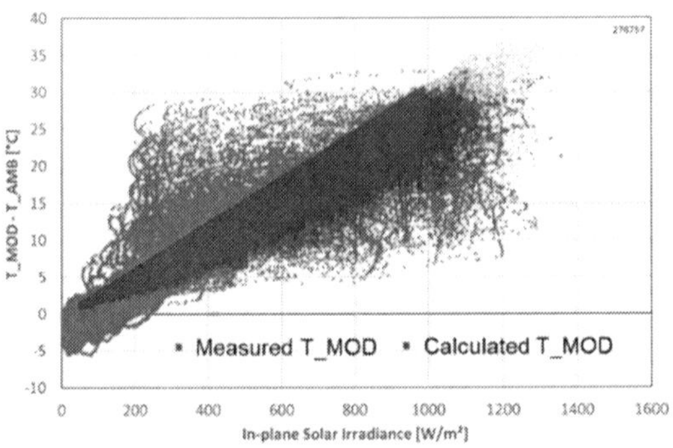

Figure 9: Method C plot for optimized u_0=32 W/(m² °C) and optimized u_1 = 4.5 Ws/(m³°C).

5 DISPERSION OF MODELLING PARAMETERS DUE TO SEASONAL VARIATION

The variation of parameters u_0 and u_1 was studied for a measurement series from Braunschweig, Germany, representing a location with a temperate climate. The data recording period covered 5 months May 2020 to September 2020, and the measured data from each months was separately subjected to Methods A, B and C, respectively. **Tables 4, 5 and 6** summarize the results for each method. They also include information on the dispersion and observed averages for the modelling parameters and NMOT. The main results of this study are:

- Methods A and B show comparable scatter for modelling parameters and comparable NMOT;
- Method B shows worst results for quality indicators;
- Method C gives the best results in terms of sensitivity to seasonal effects. It leads to lower NMOT values compared to Methods A and B, but also the best fits in terms of MBE and $\Delta T_{MOD,Gw}$.

Table 4: Variation of results for Method A

	useable data points	u_0	u_1	NMOT °C	MBE	$\Delta T_{MOD,Gw}$ °C
May 20	220	36,1	4,6	39,6	0,3	0,8
Jun 20	429	31,9	5,8	41,2	0,7	1,3
Jul 20	79	31,5	4,8	42,0	-0,6	-0,6
Aug 20	429	31,9	5,8	41,2	0,6	0,7
Sep 20	556	30,6	7,2	41,2	0,7	0,3
Spread		5,5	2,5	2,4		
Average		32,4	5,6	41,1		

Table 5: Variation of results for Method B

	useable data points	u_0	u_1	NMOT °C	MBE	$\Delta T_{MOD,Gw}$ °C
May 20	14046	35,9	4,3	39,9	0,5	1,1
Jun 20	16024	37,0	2,8	40,1	0,8	1,6
Jul 20	4707	35,1	2,1	41,5	0,9	2,2
Aug 20	18827	32,7	4,7	41,4	0,9	1,1
Sep 20	19301	33,2	4,8	41,0	0,7	0,3
Spread		4,3	2,8	1,6		
Average		34,8	3,8	40,8		

Table 6: Variation of results for Method C

	useable data points	u_0	u_1	NMOT °C	MBE	$\Delta T_{MOD,Gw}$ °C
May 20		37,0	5,0	39,0	0,0	0,5
Jun 20		35,5	5,5	39,5	-0,1	0,3
Jul 20		35,5	4,5	40,0	-0,2	0,2
Aug 20		34,0	5,5	40,3	-0,1	-0,2
Sep 20		37,0	5,0	39,0	0,1	-0,6
Spread		3,0	1,0	1,2		
Average		35,8	5,1	39,6		

6 IMPACT OF FINDINGS ON CSER

At this point, the question arises how much the differences in u_0 and u_1 parameters affect the CSER values calculated for the six climate data sets defined in IEC 61853-4. To answer this question, we calculated the CSER by changing only the u_0 and u_1 parameters as determined by Methods A, B and C for Albuquerque site.

The results are shown in **Table 7**. The absolute effect on CSER is less than 1%. With the exception of tropical climate, where the three CSER values show the least deviation, Method B results in the smallest CSER. For all climates, the results of Method A and Method C well agree and differ by maximum 0.3%.

This study has shown that the CSER is not very sensitive to variations of the modelling parameters u_0 and u_1. Moreover, the CSER variation due to seasonal effects is less than 0.3% when the results of Method C and Braunschweig data are used.

Table 7: CSER calculation for different climates and results for location Albuquerque

	Method A	Method B	Method C	Spread
u0	29,3	34,2	32	4,9
u1	5,6	1,5	4,5	4,1
CSER:				
Temperate coastal	0,979	0,972	0,978	0,7%
Temperate continental	0,971	0,967	0,972	0,5%
Subtropical arid	0,956	0,945	0,955	1,0%
Subtropical coastal	0,907	0,902	0,909	0,6%
Tropical humid	0,915	0,917	0,918	0,3%
High elevation	0,981	0,971	0,981	1,0%

7 CONCLUSIONS

We studied three methodologies to determine the u_0 and u_1 temperature modelling parameters for PV modules, which are required as input variables for the Climate Specific Energy Rating (CSER) of the IEC 61853 series of standards.

The main sources of uncertainty in determination the modelling parameters in the IEC 61853-2 procedure are the small number of useful data points that are unevenly distributed over the wind speed range. These drawbacks were eliminated with an alternative data filter and additional wind speed binning. However, no improvement was achieved with respect to the seasonal variability of the modelling parameters.

To address these shortcomings, a new approach to determining the modelling parameters was investigated that does not require a data filter for selecting data points for which thermal equilibrium can be assumed. This method focuses on the best match of modelled (calculated with u_0 and u_1 parameters) and measured module temperatures considered for the entire series of measurement. For this purpose, it has been proven useful to use specific quality indicators: a) Mean bias error and b) difference of irradiance-weighted module temperatures.

Finally, we answered the question about the influence of u_0 and u_1 uncertainties on the CSER value. Our analysis showed that observed variations of modeling parameters can cause a maximum of 1% CSER uncertainty.

ACKNOWLEDGEMENTS

This work was funded by the European Metrology Programme for Innovation and Research (EMPIR) 19ENG01 "Metro-PV", co-financed by the participating states and from the European Union's Horizon 2020 research and innovation programme. Special thanks to PTB, Braunschweig (Germany), and CFV test laboratory, Albuquerque (New Mexico), for the provision measurement data.

REFERENCES

[1] IEC 61853: Photovoltaic (PV) module performance testing and energy rating, Parts 1 - 4

[2] Evans, D.L.: Simplified method for predicting photovoltaic array output. Sol. Energy 1981, 27, 555–560.

[3] Jakhrani, A.Q.: Comparison of Solar Photovoltaic Module Temperature Models, World Appl. Sci. J. Spec. Issue Food Environ. 2011, 14, 1–8.

[4] Faiman, D.: Assessing the outdoor operating temperature of photovoltaic modules, Progress in Photovoltaics: Research and Applications, vol. 16, no. 4, pp. 307–315, 2008.

[5] Kurtz S., Whitfield K., Miller D., Joyce J., Wohlgemuth J., Kempe M., Dhere N., Bosco N., Zgonena T.: Evaluation of high-temperature exposure of rack mounted photovoltaic modules, 34th IEEE Photovoltaic Specialists Conference (PVSC), Philadelphia, PA, USA, 7–12 June 2009; pp. 2399–2404.

[6] Ross R.G.: Interface design considerations for terrestrial solar cells modules, 12th IEEE Photovoltaic Specialist's Conference, Baton Rouge, LA, USA, 7–10 January 1976; pp. 801–806.

[7] Rauschenbach, H.S.: Solar Cell Array Design Handbook; Van Nosstrand Reinhold: New York, NY, USA, 1980; pp. 390–391.

[8] 16. Risser V., Fuentes M.K.: Linear regression analysis of flat-plate photovoltaic system performance data, 5th Photovoltaic Solar Energy Conference, Athens, Greece, 17–21 October 1983.

[9] Schweiger M., Herrmann W., Gerber A., Rau U.: Understanding the energy yield of photovoltaic modules in different climates by linear performance loss analysis of the module performance ratio, IET Jornals (2016), doi: 10.1049/iet-rpg.2016.0682

ENERGY CONTRIBUTION OF REAR-SIDE IRRADIANCE FOR BIFACIAL PHOTOVOLTAIC MODULES

Giuliano L. Martins[1,2], Sebastian Dittmann[1,3], Carlos D. Rodríguez-Gallegos[4], Hyeonwook Park[5], Soo-Young Oh[5], Woo Kyoung Kim[5], Laurie Burnham[6], Ben Figgis[7], Ricardo Rüther[2], Ralph Gottschalg[1,3]

[1]Anhalt University of Applied Sciences, Bernburger Str. 55, 06366 Köthen, Germany
E-mail: giuliano.luchettamartins@student.hs-anhalt.de, Tel.: +49 3496 67 2354
[2]Universidade Federal de Santa Catarina, Florianópolis, Brasil
[3]Fraunhofer-Center for Silicon Photovoltaics CSP, Halle, Germany;
[4]Solar Energy Research Institute of Singapore (SERIS), National University of Singapore (NUS), Singapore;
[5]Yeungnam University, Gyeongsan, Republic of Korea;
[6]Sandia National Laboratories, Albuquerque, USA;
[7]Qatar Environment and Energy Research Institute BKU, Doha, Qatar;

ABSTRACT: Bifacial Photovoltaic (PV) modules are up taking an important market share of PV industry in the next few years. Rear surface irradiance modeling imposes significant challenges, as it lacks in reliable and standardized datasets. Besides, irradiance contribution and distribution depend highly on location and mounting conditions. A standardized methodology was developed and applied to characterize rear irradiance behavior using on-site measured data. Data from seven test sites within the PV CAMPER collaborative were evaluated including different albedo scenarios. It was found that surplus energy can vary between 6% - 60%, depending on the ground substrate and site location. As expected, latitude has great influence on horizontal overall (front + rear) irradiance distribution, while rear-side Plane-of-Array (POA) tends to concentrate energy contribution in the higher front-side irradiance ranges. In addition, it was observed that idiosyncratic climate conditions, such as desert environment and snow season, have a strong effect in rear irradiance distribution, thus a very relevant aspect of Bifacial PV prospections.

Keywords: PV Modules, Bifacial, Albedo

1 INTRODUCTION

Bifacial Photovoltaic (PV) modules are one of the promising module technologies that will enhance PV energy generation worldwide, and reduce the levelized cost of energy. Some manufacturers pledge up to a 30% energy generation increase. However, there is still a significant lack of information regarding the operating conditions; not only related to on-site assessment, the rear-side irradiance potential, and its seasonal variation, but as well as regarding a proper indoor characterization, which reflects real operation conditions, such as an energy rating for bifacial modules. In 2019, the IEC60904-1 [1] has been updated with a bifacial module measurement procedure (IEC TS 60904-1-2), which describes procedures for the measurement of the current-voltage (I-V) characteristics of bifacial photovoltaic devices in natural or simulated sunlight. With simulated sunlight, the standard allows two types of measurements; (1) measurements by means of a single light source (Ge method) and (2) by means of a simulator with two different light sources to illuminate front and rear simultaneously. However, the existing standard considered only two rear-side irradiance levels, 100 W/m², and 200 W/m². This might not represent the rear-side irradiance in different climate zones [2] or rather the real operation conditions of a bifacial PV module.

Therefore, *Photovoltaic Collaborative to Advance Multi-climate Performance and Energy Research* (PV CAMPER) [3] has created a common technical approach and set of best practices to ensure their global research platform maintains high standards for data quality and availability, enabling data comparisons across multiple sites. Within the PV CAMPER collaboration, a unique measurement setup for albedo and rear-side irradiance has been developed and implanted at seven test sites worldwide. Based on this data, a methodology to characterize the energy contribution of rear-side irradiance was developed. Thus, properly characterizing sites worldwide allows for a better overview of how bifacial PV modules will operate, assisting the development of realistic guidelines for e.g. indoor measurements and energy yield prediction models.

2 MATERIAL & METHOD

This section will discuss the available datasets and the methods proposed to analyse them.

2.1 DATASET

Data was gathered from seven PV CAMPER test sides and quality checks were applied. Since shading can affect measurements as shown in past works [2]. Thus, guidelines for measurement setup were proposed within PV CAMPER. The main aspect of the measurement setup was the positioning of pyranometers at a 1.5m height, without any nearby shading. Since not all partners have full data sets of rear-side Plane-of-Array (POA) irradiance measurements in this study the analysis will be performed using Horizontal irradiance only, apart from Anhalt/GER for which rear-side POA measurements will be analysed.

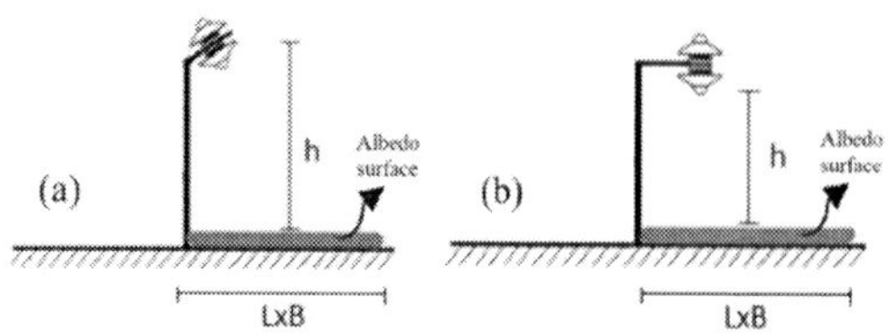

Figure 1: Representative albedometers, with upward and downward-facing pyranometers, one set at a tilt angle representative of plane-of-array irradiance (a); the other one horizontal to the ground (b) (h=1.5m).

Data from partners which were able to comply with guidelines were updated, otherwise, they are tagged with a star (*) in the following table. Information about Sandia and QEERI data can be found in previous work [2].

Two additional data sets from SERIS have been analyzed with a slightly different installation method: rear-side irradiance measurements in a floating PV (FPV) and a rooftop (ROOF) installation with a mounting height of the sensor at 2.5m and 0.7m, respectively. At YU two different substrates are uses, concrete with a light gray color and concrete with a white color.

Table 1 - **Site specifications, relating to Köppen-Geiger** [4] **Climate classification and dataset period.**

Site	*Country*	*Lat./ Long.*	*Clim. Zone*	*Surface*	*Meas. Period*
Anhalt	GER	51.77 N/ 11.76 E	Cfb	Grass/ Straw	Dec/20 - Aug/21
CSP	GER	51.49 N/ 11.93 E	Cfb	Green roof	Mar/21 - Aug/21
QEERI *	Qatar	25.32 N/ 51.43 E	Bwh	White/ Yellow gravel	Jan/18 - Oct/19
SERIS*	Singapore	1.38 N/ 103.67 E	Af	Water & Concrete	Jan/19 - Jan/20
Sandia NM*	New Mexico USA	35.03 N/ 106.54 W	Bsk	Grey gravel	Mar/17 - Feb/18
Sandia VM*	Vermont USA	44.47 N/ 73.10 W	Dfa	Grass/ Snow	Jan/18 - Dec/18
YU	South Korea	35.82 N/ 128.77 E	Cwa	Concrete & White floor	Mar/21 - Jul/21

2.2 METHOD

In this paper, irradiance is referred as "G". Energy is referred as the sum or integral of the irradiances. Overall irradiance will be considered the sum of front (G_{Front}) and rear surface irradiance (G_{Rear}).

$$G_{Overall} = G_{Front} + G_{Rear} \qquad \text{(Eq. 1)}$$

The analysis considered only the energy contribution from Albedo in a horizontal installation or α in a title installation, which is the ratio between G_{Rear} and G_{Front} as described in Eq.2.

$$\alpha = {G_{Rear}}/{G_{Front}} \qquad \text{(Eq. 2)}$$

To observe the variation within energy contribution in relation to front irradiance, the dataset is divided into steps and bins as shown in the following table.

Table 2 - **Bin and step information for rear-side Energy contribution.**

Variable	*Step*	*Bin*
Front Irradiance	100 W/m²	±50 W/m²
α	5%	±2.5 %

Furthermore, surplus energy contribution from rear-side irradiance will be assessed, in relation to the available front-side energy.

Besides, considering Table 2 guidelines for front irradiance, the variation in contribution within every front irradiance step will be calculated. Thus, the percentual increase of energy, due to rear-side for each G_{Front} step.

$$\Delta G = \left[{(\sum G_{Overall} - \sum G_{Front})}/{\sum G_{Front}} \right] * 100 \qquad \text{(Eq. 3)}$$

Each site, fornt and rear, will be treated independently, and the analysis will consider any idiosyncratic conditions inherent to the location.

3 RESULTS & DISSCUSION

This section will discuss the results obtained throughout the available dataset, applying the methods previously discussed.

3.1 Geographical influence of rear-side irradiance

The stratified energy contribution in relation to α is displayed in Figure 2.

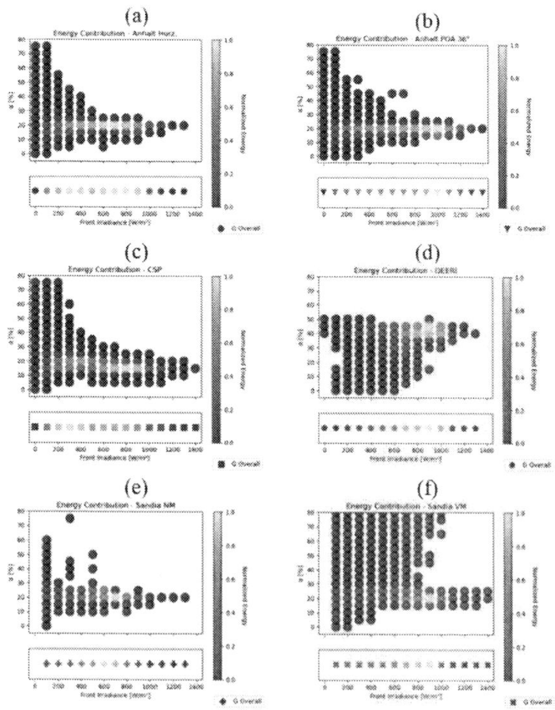

Figure 2: Energy contribution relation between α [%] and Front Irradiance [W/m²] steps. Energy normalized and displayed in colorbar for (a) Anhalt horizontal, (b) Anhalt rear POA 36°, (c) CSP, (d) QEERI, (e) Sandia New Mexico, (f) Sandia Vermont.

The Figure 2 illustrates interesting aspects regarding latitude influence. Even though, it is noticeable that surface has the highest impact on α, latitude was observed to impact its distribution. Higher latitude sites, such as Anhalt and CSP, have shown broader range of α distribution at low magnitude front-side irradiances.

The other sites, in general, presented rectangular distribution shapes. This situation can be explained due to the frequency of higher solar zenith angles throughout the year, in the case of higher latitudes.

QEERI, in Figure 2 (d) , located in a desert climate, has an albedometer at two meters height. Its height can be affecting energy contribution. On the other hand, albedo and irradiance in desert climates are consistently higher, therefore it is reasonable that this specific site would achieve its most energetic point at higher α values, as observed.

Figure 2 (b) (Anhalt, POA) illustrates a different behavior due to the snow in the winter, which values go up to 100% of α. As shown in previous work [1], the snow has a strong effect on albedo distribution; besides, we can observe that it shall favor bifacial PV modules generation, especially at the winter season.

3.2 Surface Impact on rear-side energy contribution

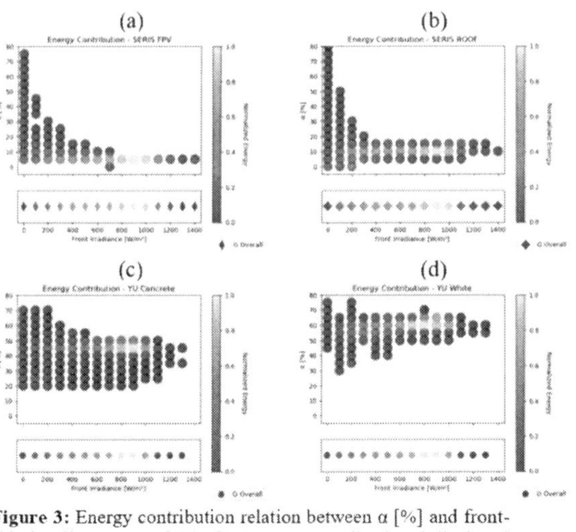

Figure 3: Energy contribution relation between α [%] and front-side irradiance steps [W/m²]. Energy normalized and displayed in color-bars. For (a) SERIS Floating PV, (b) SERIS Rooftop, (c) YU gray concrete, and (d) YU white painted concrete.

The type of surface or substrate has a strong impact in the mean of α [2]. However, Figure 3 (c) & (d) (YU site) demonstrate there is an effect on its distribution, as well. Concrete surface has a broader distribution of α, varying between 20% and 70%; on the other hand, YU white has most energetic points within 45% to 65% of α. Such differences could be relevant when predicting energy for bifacial PV modules.

Floating PV (FPV) is a new trend in PV applications [5]. Due to albedo properties of water, rear-side irradiance is expected to be minimal. From Figure 3 (a) it is clear that rear-side energy contribution is minor, thus bifacial PV modules cost-benefit implementations shall be carefully assessed.

3.3 Energy Distribution

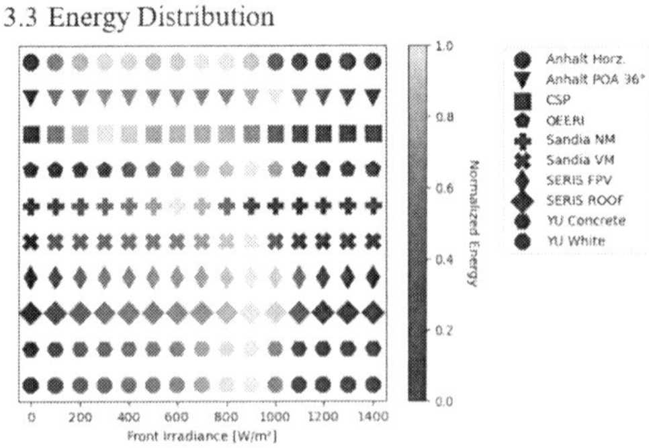

Figure 4: Normalized Overall Energy contribution for each test site in relation to front-side irradiance.

Figure 4 demonstrate that independently of rear-side irradiance, the highest contribution generally stays in the G_{Front} range of 800 W/m² to 900 W/m². On the other hand, lower G_{Front} is vastly populated by different albedo ranges; therefore, generating more impact in bifacial PV generation. Besides, Anhalt's rear POA illustrates that optimized angles (here 36°) are likely to change energy contribution to higher magnitudes and to a more concentrated pattern.

Besides, higher latitudes are shown to have a more evenly distribution of energy. A larger frequency of high solar zenith angles during the year is the main cause for it. Thus, favoring the use of Bifacial PV modules in these regions.

3.3 Energy Surplus due to rear-side irradiance

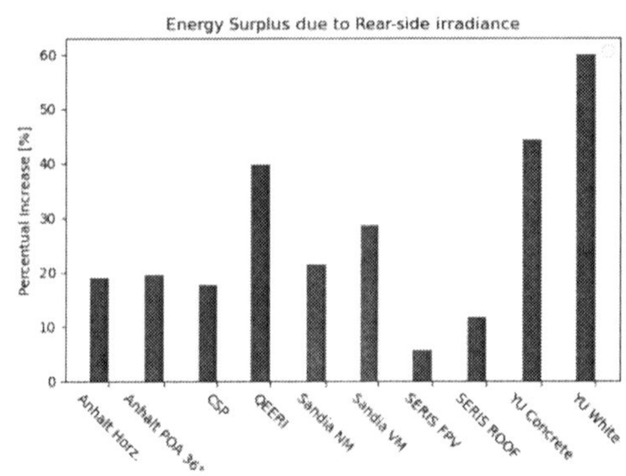

Figure 5: Graph showing surplus energy by site, due to rear-side irradiance contribution.

Figure 5Figure 5 demonstrates relative and not absolute values. The highest energy contribution from α is from YU and QEERI. Thus, showing that idiosyncratic climate conditions, such as desert, favor rear-side energy contribution.

In addition, the artificial implementation of the new surface can increase energy contribution, as is seen in the results of YU. As a matter of fact, an economical evaluation is needed to a proper conclusion of viability in large-scale PV power plants.

Besides, these results demonstrate that some manufacturer's claims of up to 30% more energy generation are not reasonable, for most cases. Since this analysis does not consider losses, bifaciality factor of bifacial PV modules or shading within a PV array; even though rear POA would increase rear-side energy surplus, it would not be enormous, as shows Anhalt's difference of less than 1%.

Another interesting aspect is that in the opposite of energy distribution, the surplus due to rear-side energy is not strongly affected by latitude. Therefore, is related mainly to the α.

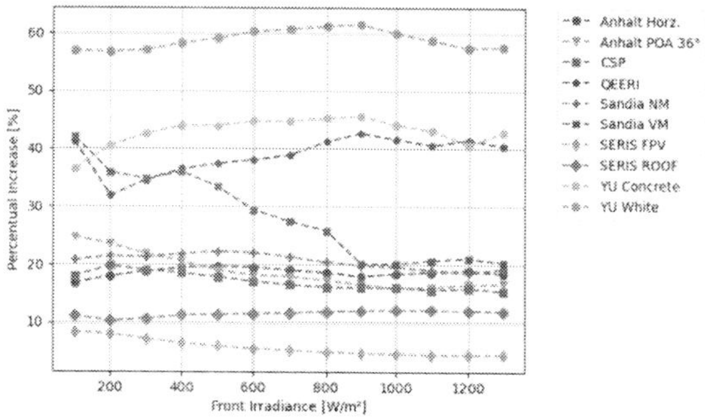

Figure 6: Energy percentual increase per front-side irradiance steps. Relating overall and front-side irradiance.

Figure 6 demonstrates the relative increase per G_{Front} bin.

QEERI and Sandia VM have strong effects, as expected. High insolation is propelled by high α values, in the desert; and rear-side irradiance in lower G_{Front} magnitudes, significantly increases contribution, because of the snow in the winter season at Sandi VM.

In addition, an important remark can be inferred from figure 6 that irradiance percentual increase is higher between 600 W/m² and 900 W/m², representing a higher contribution in these ranges.

Rear POA (Anhalt) demonstrates a high contribution in lower irradiances. However, it can be seen as an effect of a bigger field of view of the pyranometer, thus more effective in lower irradiance due to nearby reflections.

Furthermore, the difference between the German sites can be explained since CSP does not have a full year of measurements, winter months are missing. Therefore, lower front-side irradiances with minor α contributions are mostly not considered, this creates a deviation between both test sites.

4 CONCLUSIONS

This paper proposed and applied a methodology to characterize the rear energy contribution, independently of the test site. Seven different locations were analyzed, mainly using horizontal sensors, besides of one rear POA evaluation.

It is clear that special climate conditions have a strong effect on irradiance distribution. The desert regions show an increase for higher G_{Front} (>800 W/m²). On the other hand, snow events propelled rear-side energy contribution in the lower magnitudes. Both representing an important advantage of bifacial PV modules.

Surplus energy due to different surface/substrate depends on site location; however, mainly on the surface conditions. The cost-benefit of changing surface demands intense research and it is left as future work.

In addition, water in FPV systems has small rear-side surface contributions, thus using bifacial PV modules with higher module cost would probably be unreasonable, thus future work is needed to evaluate this matter.

Besides, energy surplus can range from 6% up to 60%. Thus, bifacial PV modules pledges of higher energy yield must be carefully considered, as it is strongly site and surface dependent.

Finally, rear-side energy contribution is shown in all ranges of G_{Front}. As a matter of fact, climate conditions are illustrated in this work as the main influence in the rear-side irradiance distribution.

Finally, the work showed that assuming constant α for energy prediction is not a reasonable approach. The method developed provides relevant inputs. Therefore, site characterization shall be considered, not only in the discussions, but in future work e.g. extending the current energy rating standards for bifacial PV modules or consider seasonal and geographical effects on α, the rear-front-side ratio.

5 REFERENCES

[1] International Electrotechnical Commission, "IEC 60904-1:2020," 2020.

[2] Dittmann, S.; Sánchez, H.; Burnham, L.; Gottschalg, R.; Oh, S.-Y.; Benlarabi, A.; Figgis, B.; Abdallah, A.; Rodríguez-Gallegos, C. D.; Ruther, R.; Fell, "Comparative Analysis of Albedo Measurements (Plane-of-Array, Horizontal) at Multiple Sites Worldwide", 2019, doi: 10.4229/EUPVSEC20192019-5DO.1.4.

[3] PV CAMPER Annual Report, December 2020, DOI: 10.13140/RG.2.2.15902.46405

[4] H. E. Beck, N. E. Zimmermann, T. R. McVicar, N. Vergopolan, A. Berg, and E. F. Wood, "Present and future Köppen-Geiger climate classification maps at 1-km resolution," *Sci. Data*, vol. 5, no. 1, p. 180214, Dec. 2018, doi: 10.1038/sdata.2018.214.

[5] R. Cazzaniga and M. Rosa-Clot, "The booming of floating PV," *Sol. Energy*, vol. 219, no. June 2020, pp. 3–10, 2021, doi: 10.1016/j.solener.2020.09.057.

6 ACKNOWLEDGEMENTS

SERIS is a research institute at the National University of Singapore (NUS). SERIS is supported by NUS, the National Research Foundation Singapore (NRF), the Energy Market Authority of Singapore (EMA) and the Singapore Economic Development Board (EDB).

For Sandia the work was funded by the U.S. Department of Energy Solar Energy Technologies Office, under Award Number 34362.

QUANTITATIVE EVALUATION OF THE SHADING RESILIENCE OF PV MODULES

[1,2,*]Nils Klasen, [1]Florian Lux, [1]Julian Weber, [1]Daniel Weißer, [1]Torsten Roessler, [1]Achim Kraft, [1]Dirk Holger Neuhaus

[1] Fraunhofer Institute for Solar Energy Systems ISE
Heidenhofstraße 2, 79110 Freiburg i.Br., Germany

[2] Karlsruhe Institute of Technology KIT, Institute for Applied Materials – Material and Biomechanics (IAM-WBM)
Hermann-von-Helmholtz-Platz 1, 76344 Eggenstein-Leopoldshafen, Germany

[*]corresponding author, e-mail to: nils.klasen@ise.fraunhofer.de

ABSTRACT: We present simulation results on the partial shading behavior of four PV module layout containing three different solar cell sizes. Two types of shingle interconnection are compared to the widely used "butterfly" layout for half-cut solar cells and the conventional solar cell interconnection of 60 full size solar cells. For the LT spice simulations, we measured the *I-V* characteristic of PERC shingle solar cells and included the reverse breakdown region into an extended two-diode model. Two basic shading cases, rectangular and random shading, are investigated. Therefore, we created sets of > 1000 scenarios per case by Latin Hypercube Sampling (LHS). The scenarios are transferable for all solar cell sizes and the number of full solar cell equivalents identical in all topologies. We apply a new criterion [1] for evaluation of the shading resilience SR from 0 to 1 to compare and rank the different cell topologies regarding their performance under partial shading. The shading resilience for rectangular shadings of the shingle matrix layout ($SR = 0.692$) excels the one of the shingle string layout ($SR = 0.602, 87\ \%$ of the matrix layout), the butterfly layout ($SR = 0.461, 67\ \%$) and the conventional full cell interconnection ($SR = 0.213, 31\ \%$) scheme. We find minor differences in the absolute values for random shading. However, the ranking of the solar modules stays identical. Finally, we conclude that besides higher power densities and a valuable aesthetical appearance, shingle solar modules feature an additional very important but so far little noticed and discussed property: Overall higher power outputs under partial shading. This makes them especially suited for vehicle and building integrated applications.

Keywords: Partial Shading, Shingle Solar Cell Interconnection, LTspice

1 INTRODUCTION

Shading of photovoltaic modules can cause severe losses in the output power of PV generators. In integrated applications such as buildings and cars this becomes especially relevant. U. Jahn and W. Nasse showed in early studies of roof top installations in Germany that 41 % of the installed systems were affected by shading and shading-related energy yield losses of up to 20 % [2]. C. Schuss *et al.* investigated the shading of car roofs due to vegetation during operation [3, 4]. Poles and antennas cause shading as well as shown in [5, 6]. From these two cases we derive two cases to be considered in this study: Random and rectangular shading of solar modules.

So far bypass-diodes are the established measure to reduce shading losses in PV solar modules. However, in recent years a variety of new interconnection technologies and solar cell geometries allow for another option of shading loss reduction: Adapted module layouts. When shading occurs, the current generation reduction in shaded solar cells and the resulting current mismatch with the unshaded cells is responsible for the loss in power. Parallel interconnection is the solution to avoid these losses since currents add up and current mismatch has no negative effect. However, there are limits to that due to ohmic loss reduction and the need of reasonable voltage levels. Adapted module layouts allow the combination of serial and parallel interconnections of the incorporated solar cells.

Half-cut and shingle solar cells allow the implementation of mixed interconnections while at the same time maintaining similar current-voltage levels compared to conventional full-cell topologies. We investigated four topologies based on full-sized, half-cut and shingle solar cells with respect to their shading resilience under random and rectangular shading. All four topologies are highly relevant for the application in commercial module products.

2 METHODOLOGY

2.1 *I-V* Characteristics

In this study, we use *I-V* characteristics obtained in previous work [7]. An extended two-diode model is used to describe the entire current-voltage curve including the reverse breakdown region. The exponential current increase during reverse breakdown is added to the known two-diode equivalent circuit by another Schottky diode term as proposed by H.S. Rauschenbach [8]. This Schottky diode term requires three additional fit parameters: The reverse breakdown saturation current density J_{Br}, an avalanche factor to be multiplied with the diode ideality factor n to form n_{Br} and a reverse breakdown voltage V_{Br}. Equation (1) describes the entire solar cell characteristic. For more detailed information on the measurement procedure and data obtained for lab-scale and commercial PERC shingle solar cells, we refer to previous work [1]. *I-V* data are obtained for PERC shingle solar cells and scaled to half-cut and full-sized solar cells for better comparability.

$$\begin{aligned} J(V) = J_{\mathrm{ph}} - J_{01}\left(\exp\left\{\frac{e(V+JR_{\mathrm{s}})}{k_{\mathrm{B}}Tn_{01}}\right\}-1\right) \\ - J_{02}\left(\exp\left\{\frac{e(V+JR_{\mathrm{s}})}{k_{\mathrm{B}}Tn_{02}}\right\}-1\right) \\ + J_{\mathrm{Br}}\exp\left\{-\frac{e(V+JR_{\mathrm{s}}-V_{\mathrm{Br}})}{k_{\mathrm{B}}Tn_{\mathrm{Br}}}\right\} - \frac{V+JR_{\mathrm{s}}}{R_{\mathrm{p}}} \end{aligned} \quad (1)$$

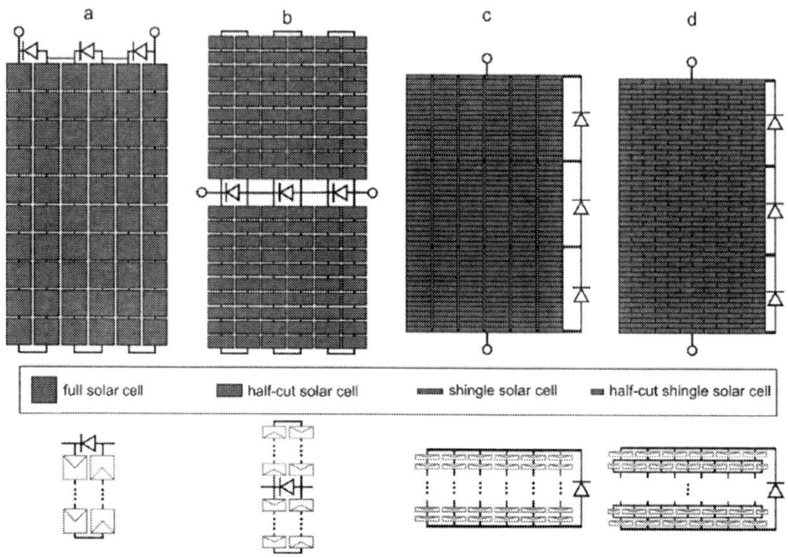

Figure 1: Module layouts investigated in LTspice shading simulations taken from [1]: a) conventional topology full-sized solar cell interconnection b) "butterfly" topology for half-cut solar cell c) shingle string and d) shingle matrix topology for shingle solar cells. For the latter a half-cut shingle solar cell is added in every second row. The electrical interconnection schemes of the different topologies within one string or block, secured by a bypass diode, are given in the bottom.

2.2 Module Layouts

Figure 1 shows a schematic drawing of the four investigated module layouts.

a) The conventional layout for full-size solar cells containing 60 solar cells, divided into strings of 10 solar cells. Two strings are interconnected in series and are bypassed by one diode each. 20 solar cells are in serial interconnection as suggested by the interconnection scheme given in the bottom part of **Figure 1**.
b) Half-cut solar cells are assembled in the so called "butterfly" layout. The upper and the lower block of solar cells are interconnected in parallel. Each block contains again three strings with a bypass-diode each. Note that each bypass-diode is in parallel to a string of the upper and the lower block. When in its conductive state at around -0.4 V both strings are bypassed.
c) The shingle string interconnection contains 6 parallel interconnected strings of 50 solar cells each. Three bypass-diodes split the strings into pieces of 16, 17 and 17 solar cells. Note, that in a real PV solar module, there would be three times 17 thus 51 solar cells, since overlapping of the solar cells allows for additional solar cells on the same area. However, for better comparability, all simulations contain the identical number of full-cell equivalents.
d) The shingle matrix interconnection features an additional lateral overlap of the solar cells. From these results an intrinsic serial and parallel interconnection of each solar cells to its direct neighbours. Note that this requires half-cut shingle solar cells in each second row to form a rectangular active area. Similar to the string interconnection three bypass diodes split the module containing 50 solar cells in the direction of serial interconnection. In an earlier publication we showed that this scheme allows lateral current transport along the joints formed by busbars and electrically conductive adhesives (ECA). These electrode bridges allow the currents to flow past shaded areas and thus reduce the losses due to shading significantly [7].

2.3 Shading Scenarios

We consider two basic cases: rectangular and random shading. Both are shown in **Figure 2** along with exemplary irradiation maps for each case on the right side. Although shading is never absolute and *e.g.* in case of a leafy tree shaded areas are still illuminated with ~200 W/m^2 [9], for better comprehensibility we assume 0 W/m^2 on shaded areas. In case of rectangular shading, the shaded areas are computed as precisely as the machine error. In case of random shading, we use a pixel-based approach to create random shading patterns. Both will be explained in more detail.

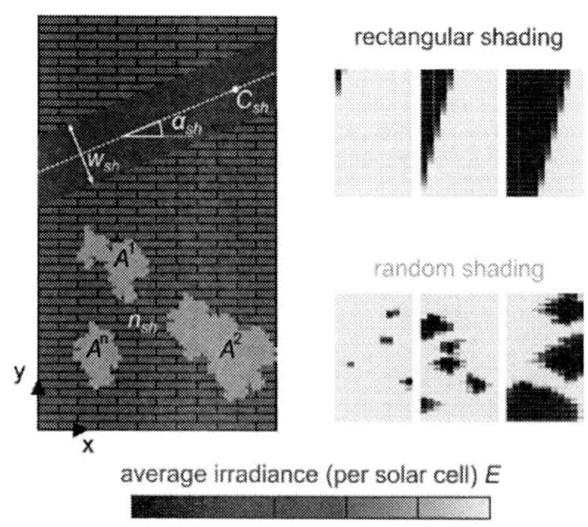

Figure 2: Left: Shading cases: rectangular and random shading. Rectangular shading is defined by three parameters width w_{sh}, angle towards the x-axis α_{sh} and the position of the rotation center C_{sh}: $(x|y)$. Random shading is defined by the shaded area fraction $A_{\mathrm{sh}} = \sum A^n$. Right: Exemplary irradiation maps for both cases based on which the photocurrent of each solar cell is computed and passed to the LTspice circuit simulation. Irradiation levels refer to the average irradiation on one solar cell.

Rectangular shading is defined by three respectively four parameters: The shading width w_{sh}, the angle towards the x-axis α_{sh} and the rotation center C_{sh} given by its x and y coordinate. Perpendicular to the shading width the shadow expands infinitely. We compute the exact shaded area of each solar cell. According to findings of V. Quaschning [10] there is only a small error of ~2% on the MPP of a solar cell, when neglecting the spatial distribution of the irradiation and taking into account only the average irradiation on the solar cell surface. We use this finding to set the photocurrent density of a shaded solar cell $J_{ph,s}$ according to its shaded area fraction A_{sh} and fractional irradiation $\hat{\imath}$ by Eq. (2).

$$J_{ph,s} = J_{ph,0}A_{sh}(1 - \hat{\imath}) \quad (2)$$

$J_{ph,0}$ is the photocurrent density measured under Standard Test Conditions (STC) (25 °C, $AM1.5$, $1000\ \mathrm{W/m^2}$). A_{sh} is given by the ratio of shaded area A to entire area A_0, thus $A_{sh} = A/A_0$. $\hat{\imath}$ is defined similarly as the irradiation on shaded areas E divided by the irradiation on illuminated areas I_0, thus $\hat{\imath} = E/E_0$.

Random shading is defined only by A_{sh} which is distributed on 1 to n_{sh} individual and randomly shaped areas. n_{sh} thereby defines the maximum number of shaded areas and thus is somewhat in between a parameter and a random quantity. When creating the shading scenarios, the number of shading patches is randomly chosen between $1 \dots n_{sh}$ according to the boundary value n_{sh}. A_{sh} is then distributed randomly on the shading patches. Starting from a shadow seed pixel randomly chosen on the module area, each shaded area A^i is built pixel by pixel. We use $25x25$ pixels per solar cell. Exemplary irradiation maps for both base cases are given in **Figure 2**. The color code refers to the average irradiation on the individual solar cells. Note that in **Figure 2** the individual solar cells, not the pixels per solar cell are displayed.

2.4 Shading Resilience

In prior work we defined a factor called the Shading Resilience SR quantifying the ability of a PV module to minimize the partial shading losses. For a more detailed explanation to the derivation of SR we refer to our publication [1].

In **Figure 3** we show the graphical definition of SR. The PV module power is given dependent from A_{sh}. The dotted lines define the minimum possible power output thus for $A_{sh} = 1$. Naturally this value depends on the irradiation of the shaded areas and therefore $P_{min} = \hat{\imath}P_0$ with the initial unshaded power output P_0. The core aspect is the definition of the upper physical limit for the power output under partial shading. The limit functions are given by solid lines. They represent a hypothetical parallel interconnection of all solar cells in the PV solar module. Since in parallel interconnections there is no current mismatch. Assuming ideal interconnection, the solar cells operate almost independently from each other and only the power of the shaded solar cells is lost. This defines an upper physical limit for the power output. Any possible maximum power point for any shading scenario is found between these two limiting functions. A PV module obviously incorporates a higher SR the closer it approaches P_{lim}.

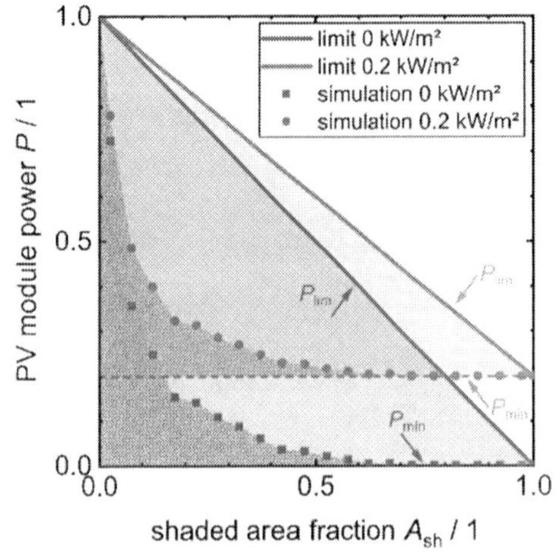

Figure 3: Graphical definition of the shading resilience according to [1]. SR is defined as the ratio of the spotted to the filled areas for two cases of $\hat{\imath} = [0; 0.2]$. The data points are a result of the simulations from the power function $P(A_{sh})$.

To create common and comparable ground, we define SR as the ratio of two areas: first, the area enclosed by a power function $P(A_{sh})$ and P_{min}. Second, the area enclosed by P_{lim} and P_{min}. $P(A_{sh})$ is obtained in circuit simulations based on LTspice with an adequate number of data points from $A_{sh} = 0$ to 1. The area below $P(A_{sh})$ is obtained by integration with the trapezoidal rule. This graphical definition of SR is also given by Eq. (3). Again for more details we refer to our previous work [1].

$$SR = \frac{2}{(1-\hat{\imath})P_0}\int_0^1 P\, dA_{sh} - \frac{2\hat{\imath}}{1-\hat{\imath}} \quad (3)$$

2.5 LTspice Simulations

As discussed in the previous section, we used LTspice to obtain $P(A_{sh})$. Therefore, the equivalent circuit model of the (extended) two-diode model was implemented in a python code creating the netlist for LTspice. Each solar cell is represented by two virtual sub-cells. This is necessary to adequately represent the shingle matrix interconnection where the solar cells overlap laterally on half of their length. In previous work, we discussed this approach and the non-ideal current transport properties of this lateral interconnection [7]. Lateral resistors with absolute values R_{lat} interconnect the solar cells in the shingle matrix layout. For comparability, we transfer the virtual sub-cell concept on all topologies. Values of R_{lat} measured in [7] are also used in this study. We conducted validation experiments and found experimental and computed values to deviate approximately by 2 % over various shading scenarios [7].

3 SIMULATION STUDY

In the last section we introduced the components needed to obtain the shading resilience of varying module topologies. The last missing component is the power function $P(A_{sh})$ for both basic shading cases. The entire process flow from the generation of input to postprocessing and obtaining SR is summarized in **Figure 4**.

We performed parameter studies based on Latin

Hypercube Sampling (LHS) [11], a method to randomly but uniformly cover a multidimensional parameter space.

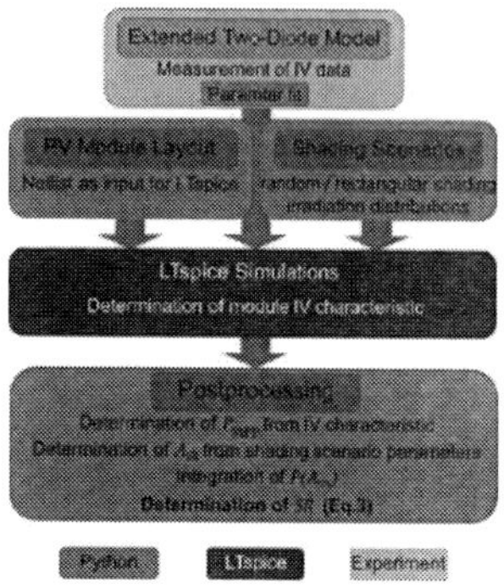

Figure 4: Flowchart of the process performed to obtain the shading resilience *SR* for different module layouts and basic shading cases (rectangular and random). The color code represents the software or experiment used for each step.

Therefore, each parameter is split in equidistant steps between given boundaries where the step size is defined by the chosen number of scenarios. LHS then generates random combinations of all parameters so that each step occurs exactly once.

For rectangular shadings there are four input parameters: w_{sh}, α_{sh} x and y. Their boundary values are given in **Table 1.** A_{sh} is then computed for each individual parameter set. We chose 2000 scenarios, from which 777 result in a shaded area fraction of 1. From the remaining scenarios computed in LTspice the power function $P(A_{sh})$ is obtained.

Table 1: Boundaries of the parameters defining the rectangular shading scenarios

Parameter	Lower limit	Upper limit
x / mm	0	$l_M = 1567.5$
y / mm	0	$w_M = 940.5$
α_{sh} / °	0	90
w_{sh} / mm	0	$2\sqrt{l_m^2 + w_M^2} = 3656.0$

In case of random shading there is only one parameter: A_{sh}. We adjusted the number of computations to rectangular shading. Thus 1250 random scenarios form $P(A_{sh})$ in equidistant steps from $A_{sh} = 0$ to 1 for random shading.

4 RESULTS AND DISCUSSION

Table 2 summarizes the obtained values for the shading resilience of all module layouts and both basic shading cases. We emphasize, that due to the large number of computed scenarios and the capability of LHS to represent multidimensional parameter spaces the *SR* values feature a very high significance. As mentioned before, *SR* can be interpreted as the share of the ideal shading resilient module layout: A parallel interconnection of all solar cells. Thus, the results show that *e.g.* a shingle matrix module under any possible rectangular shading yields close to 70 % of the maximum possible power, while *e.g.* the conventional layout can only generate ~20 %. Hence, in average there is a factor of ~3.5 in the power output under rectangular shading.

Table 2: Shading resilience computed for all module topologies and both basic shading cases.

Module Layout	*SR* rectangular	*SR* random
Shingle Matrix	0.692	0.545
Shingle String	0.602	0.446
Butterfly	0.461	0.319
Conventional	0.213	0.207

We find similar results for random shading. However, the absolute values of SR are reduced by $\Delta SR = 15\ \%_{abs}$ compared to rectangular shading for all module layouts. This is explained by the fact that rectangular shading covers distinct parts of the solar module, which in consequence can be bypassed. Random shading on the other hand is very likely to distribute on the entire surface. Thus, all bypassed substrings are likely to be affected equally. Since besides the conventional layout all others incorporate parallel interconnected solar cells, they are affected more by this resulting in the above-mentioned differences. The conventional layout however has very few cases in which only one or two strings are affected by shading after all.

To further investigate reasons for the differences in *SR* we show *I-V* data of all topologies in **Figure 5**. Each plot shows the currents at the MPP on the left y-axis (colored data) and the voltage at the MPP on the right y-axis (grey data). On the x-axis the shaded area fraction is shown. In the left column the results for rectangular, in the right column the results for random shading are shown. Please note, that there are slight differences in the I_{sc} and V_{oc} between the shingle and the other two topologies which is due to the fact, that we used 1/5th shingle solar cells. This leads to slightly higher currents at lower voltages in the shingle modules. First and most prominent difference is the fact, that both shingle module topologies produce currents over the whole range of A_{sh} from 0 to 1. The half-cell butterfly and the full cell conventional layout on the other hand show a distinct threshold at $A_{sh} \cong 0.9$ and $A_{sh} \cong 0.8$, respectively, where currents drop to 0. For rectangular shading this threshold is located at $A_{sh} \cong 0.7$ and $A_{sh} \cong 0.3$, respectively. This again is caused by the fact, that for a concentrated, rectangular shading it is more likely to affect all strings, while it takes a larger area of randomly shaped, distributed shading to achieve the same effect.

All graphs show distinct horizontal lines in the I_{MPP} data. They correspond to states with active bypass diodes or in case of the butterfly layout at ~4.5 A to a fully illuminated module half. They illustrate how bypass diodes even for large shaded area fractions enable the extraction of the unshaded I_{MPP} current from the solar modules. This becomes especially important, when many modules are interconnected in power plant applications and the individual module MPP becomes less important than the global module string MPP. Again, both shingle topologies show significant more data points at the unshaded I_{MPP} compared to the butterfly and conventional layout. These data points are found even for values of $A_{sh} > 0.5$ where both butterfly and conventional layout in most of the cases are at zero current already.

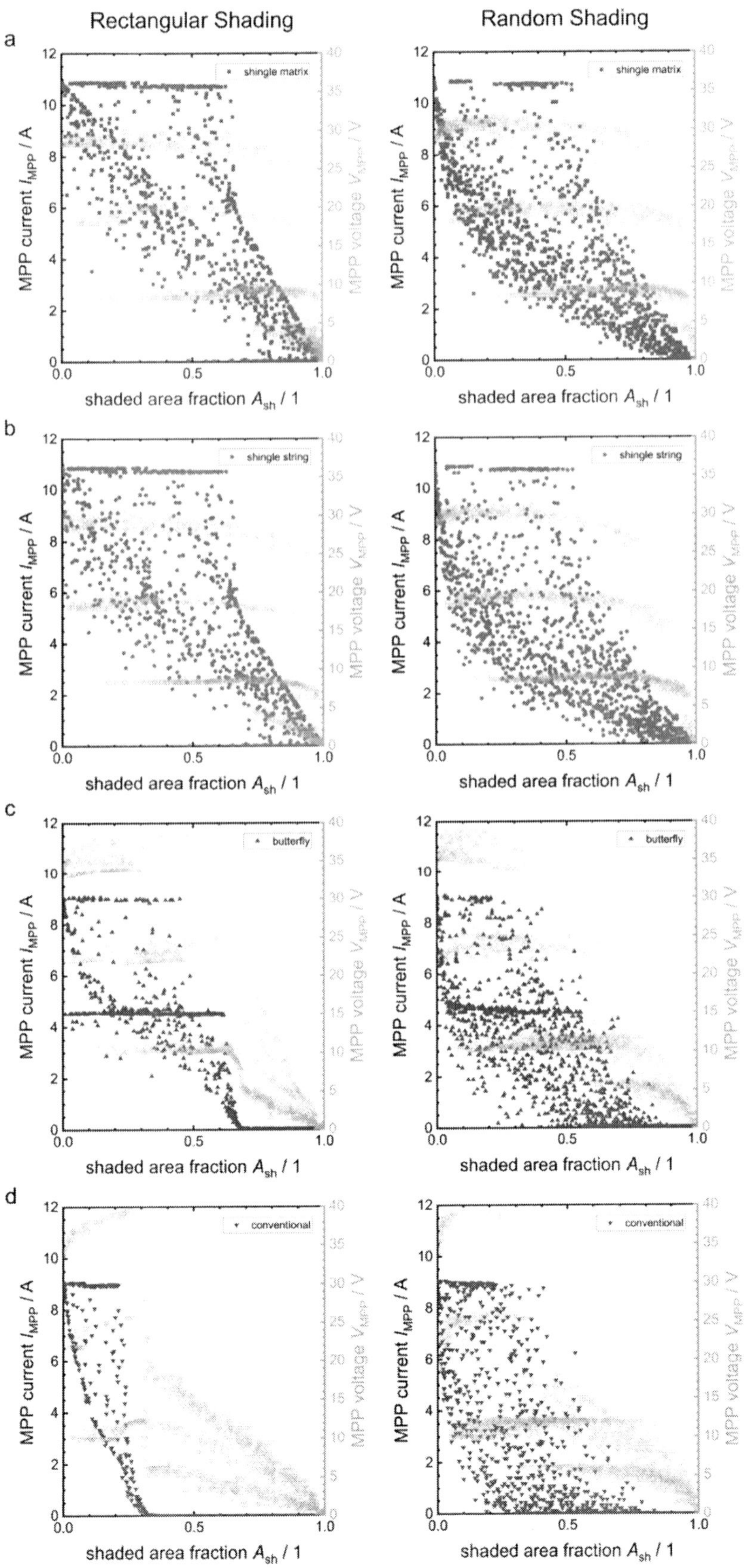

Figure 5: Diagrams of the I_{MPP} and V_{MPP} dependent from the shaded area fraction A_{sh}. The left y-axis shows $I_{MPP}(colored\ data)$, the right y-axis shows the V_{MPP} values (grey data). **a** to **d** correspond to the different module layouts: a) shingle matrix, b) the shingle string, c) the half-cell butterfly and d) the conventional full cell topology. While currents distribute over the entire range from 0 to the unshaded I_{MPP} the V_{MPP} data points form distinct clusters corresponding to the number of conductive bypass diodes. Horizontal lines in the current data also indicate states of conductive bypass diodes or in case of the butterfly module at ~4.5 A one fully operational half of the module.

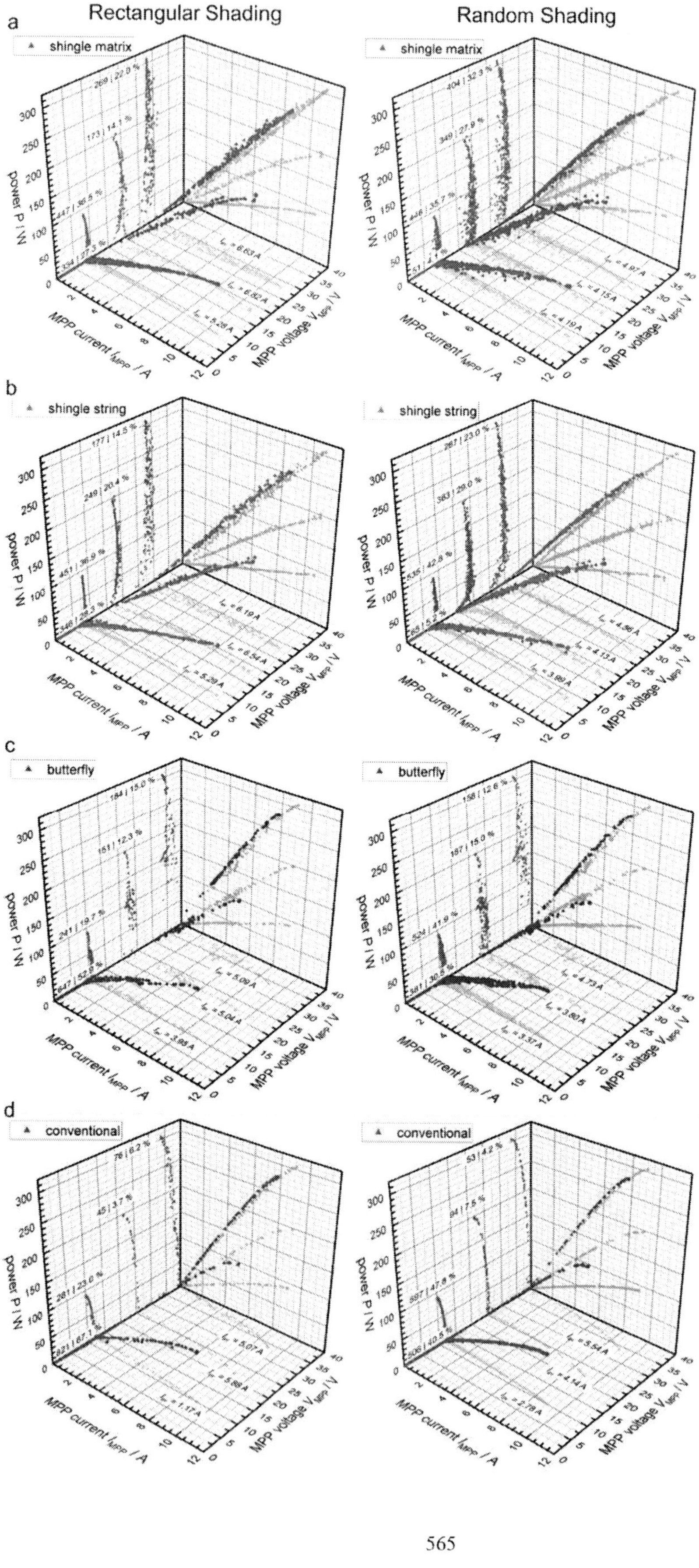

Figure 6: 3D plot of the I_{MPP} and V_{MPP} data shown in **Figure 5**. Additionally, the module power is given on the z-axis. It becomes visible that the data points form distinct groups again correspond to the different states of conductive bypass diodes. For each group the absolute number of data points and percentage of all scenarios are shown on the yz-projection plane. On the xy-projection plane the average current of each group is added.

Figure 6 shows three-dimensional plots of the I_{MPP} and V_{MPP} data from **Figure 5**, adding the module power on the z-axis. In all cases the data show three respectively four distinct groups. The first three represent the different configurations of conductive bypass diodes. The fourth group corresponds to the data points found at zero currents. For all groups absolute numbers of data points and percentages from the entire set of scenarios are shown on the yz-projection plane. In the same way, the xy-projection plane incorporates the average MPP currents I_{av} for all groups. We use the V_{MPP} values to group the data points. For the shingle modules we use 5 V to 15 V, 15 V to 25 V and 25 V to 35 V. For butterfly and conventional layout, we use 10 V to 20 V, 20 V to 30 V and 30 V to 40 V.

We find the before mentioned differences confirmed, especially with respect to current extraction and number of states without conductive bypass diodes. As an example, the groups with no conductive bypass diodes are examined: for the shingle matrix it contains 22.0 % of the data with an average MPP current of 6.63 A. The shingle string contains 14.5 % of the data at $I_{av} = 6.19$ A. The butterfly slightly exceeds the shingle string with 15.0 % of the data. However, its I_{av} is at 5.09 A. The conventional layout only in 6.2 % of the cases is without conductive bypass diodes and has $I_{av} = 5.07$ A.

Random scenarios in general lead to more cases without conductive bypass diodes. Again, this is caused by the distribution of shading on the entire surface leading to less mismatch between bypassed regions.

5 CONCLUSIONS

In this work we present results of partial shading simulations based on an extended two diode model with reverse bias characteristics. We compared four different module layouts: Two modules with shingle solar cells, a half-cut solar cell module with butterfly layout and a conventional solar module with full-sized cells. Rectangular and random shading have been examined to cover the shading by pylons or chimneys and the shading by leaves or bird droppings. We calculated > 1000 scenarios each. From the obtained data, we evaluated the shading resilience SR of the module as defined in [1]. For rectangular and random shading, the shingle modules exceed the half-cell butterfly and the full cell modules significantly. We find the highest values for SR under rectangular shading with $SR = 0.692$ for the shingle matrix, $SR = 0.602$ for the shingle string, $SR = 0.461$ for the butterfly and $SR = 0.213$ for the conventional layout. This means that *e.g.* in any rectangular shading scenarios the shingle matrix module has in average a factor 3.5 $(0.692/0.213)$ higher power output compared to the conventional layout. The large parameter space in combination with Latin Hypercube Sampling ensures a high significance of the obtained SR values.

In random shading scenarios, SR is reduced for all module topologies. This is explained by the distribution of shading on the entire solar module surface, lowering the overall module current which can be seen by comparing the average currents, *e.g.* for the shingle matrix module 6.63 A for rectangular vs. 4.97 A for random shading. Similar differences are found for all other topologies.

In conclusion, shingle solar modules offer a generally increased resilience to partial shading. In both rectangular and random shading cases they are closer to the ideal shading resilient solar module layoutlayout - a parallel interconnection of all solar cells - than the half-cell or full cell topologies. This makes modules with shingle solar cells especially suited for integrated applications such as VIPV and BIPV. Future work aims to not only show the advantage in power output, but also energy yield by outdoor yield experiments under varying partial shading situations.

6 ACKNOWLEDGEMENTS

The authors thank all colleagues from Fraunhofer ISE who contributed to this work and namely Heinz Riesch-Oppermann from KITs IAM-WBM for fruitful discussions on sampling and statistics. This work was supported by the German Federal Ministry for Economic Affairs and Energy (BMWi) under the contract number 03EE1026A, acronym Shirkan. We also thank the German Federal Environmental Foundation (DBU) for supporting our research within their graduate scholarship program.

6 REFERENCES

[1] N. Klasen, F. Lux, J. Weber, T. Roessler, and A. Kraft, “"Shading Resilience" as a General Measure of Shading Tolerability in PV Modules,” *IEEE Journal of Photovoltaics*, tbp.

[2] U. Jahn and W. Nasse, “Operational performance of grid-connected PV systems on buildings in Germany,” *Prog Photovoltaics*, vol. 12, no. 6, pp. 441–448, 2004, doi: 10.1002/pip.550.

[3] C. Schuss, B. Eichberger, and T. Rahkonen, “Impact of solar radiation on the output power of moving photovoltaic (PV) installations,” in *2018 IEEE International Instrumentation and Measurement Technology Conference (I2MTC)*, Houston, TX, May. 2018 - May. 2018, pp. 1–6.

[4] C. Schuss, T. Kotikumpu, B. Eichberger, and T. Rahkonen, “Impact of dynamic environmental conditions on the output behaviour of photovoltaics,” in pp. 993–998.

[5] Deutsche Gesellschaft für Sonnenenergie, *Planning and installing photovoltaic systems: A guide for installers, architects and engineers*, 3rd ed. London: Routledge Taylor & Francis Group, 2013. [Online]. Available: http://site.ebrary.com/lib/hawhamburg/docDetail.action?docID=10737858

[6] D. Götz, D. Hahn, R. Gottschalg, D. Dassler, S. Schindler, and H. Hanifi, “Evaluation of Shading Tolerance of PV Modules with different Module Designs for mobile Applications by Simulation, indoor and outdoor Measurements,” in *Proceedings of the 36th European Photovoltaic Solar Energy Conference and Exhibition*, Marseille, France, 2019.

[7] N. Klasen, D. Weißer, T. Geipel, D. H. Neuhaus, and A. Kraft, “Performance of Shingled Solar Modules Under Partial Shading,” *Progress in Photovoltaics: Research and Applications*, tbp.

[8] H. S. Rauschenbach, *Solar cell array design handbook: The principles and technology of*

photovoltaic energy conversion. New York: Van Nostrand Reinhold Co, 1980.
[9] M. A. Sattler and S. Sharpies, "FIELD MEASUREMENTS OF THE TRANSMISSION OF SOLAR RADIATION THROUGH TREES," in *Advances In Solar Energy Technology*: Elsevier, 1988, pp. 3846–3850.
[10] V. Quaschning, "Simulation der Abschattungsverluste bei solarelektrischen Systemen," Dissertation, Technische Universität Berlin, Berlin, 1996.
[11] M. D. Mckay, R. J. Beckman, and W. J. Conover, "A Comparison of Three Methods for Selecting Values of Input Variables in the Analysis of Output from a Computer Code," *Technometrics*, vol. 21, no. 2, p. 239, 1979, doi: 10.2307/1268522.

REVIEW OF INDUCED DEGRADATION PHENOMENA AFFECTING PV MODULES

Giso Hahn
University of Konstanz, Department of Physics
78457 Konstanz, Germany, Tel. +49 7531 88 3644, Fax. +49 7531 88 3895, giso.hahn@uni-konstanz.de

ABSTRACT: Degradation phenomena of solar modules occurring under operating conditions in the field may cause severe power losses for a PV system. To avoid or at least minimize the negative effects correlated with these phenomena, a good understanding of the nature and origin of the degradation is crucial. In this contribution the state of knowledge of four different important degradation phenomena is reviewed, and mitigation strategies are discussed.
The underlying defects of these induced degradation phenomena form at operating conditions of the solar cell under injection, (elevated) temperature and realistic ambient conditions. Under these conditions, the quasi-Fermi level in p-type based Si wafers changes under injection and may cause defects to change their charge state. This has implications for some of the degradation phenomena, since charge states of species determine the underlying reaction rates and degradation/regeneration kinetics in general.
BO-related light induced degradation seems to be well understood and mitigation strategies are available, although the exact microscopic picture of the defect itself is still unclear. Potential induced degradation of the shunt type is only occurring under specific polarity and high voltage drop in the string of interconnected modules. The formation mechanism seems to be clear and mitigation strategies exist as well. In contrary, for light and elevated temperature induced degradation the underlying defect is still not known, but a lot of empirical findings are available, pointing strongly towards the involvement of H being present in the crystalline Si bulk. Mitigation strategies exist, but seem to cause extra steps/costs in solar cell production. Degradation of surface passivation quality was recently observed for dielectric layer systems being used in new cell architectures allowing high efficiencies. Again, the exact nature of the underlying defect(s) is still unclear, but H may play a role in here, too.
Keywords: c-Si, Degradation, LID, Module

1 INTRODUCTION

Power degradation phenomena in solar cells/modules have been reported for a very long time. They can be separated into phenomena having their origin in the solar cell or in the solar module. As many of them occur only after completion of the cell/module under working conditions of the solar module under illumination and operating temperature over time, they can strongly affect power output and therefore cause warranty issues, as usually which a certain power is guaranteed over 25 or even 30 years of operation.

On cell level, classical examples for cell-based degradation are the Staebler-Wronski degradation in amorphous silicon [1] or light-induced degradation (LID) caused by BO-related defects in crystalline silicon (c-Si) [2] which have been first reported already in the 1970s. Another well-known degradation phenomenon is degradation via FeB-pairing or splitting, respectively [3]. More recently, also Cu was found to form a defect under working conditions of the solar cell in the Si bulk [4]. In 2012, a "new" defect was discovered, which was mainly observed under illumination and elevated temperature >50°C [5]. It was therefore named LeTID (light and elevated temperature induced degradation) [6]. Recently, it was reported that the surface passivation of the solar cell might be prone to degradation phenomena, too (e.g., [7]).

On module level, there exists a multitude of degradation phenomena, too. Examples apart from obvious malfunctions like cracks are delamination issues, hot spots, so-called snail trails and PID (potential induced degradation) [8]. See also [9] for a good overview on PID.

One of the reasons why some of the cell-based degradation phenomena have not been discovered earlier is based on the current shift of solar cell design from a full area Al BSF (back surface field) architecture towards the PERC (passivated emitter and rear cell) solar cell. While in the Al BSF cell the effective charge carrier lifetime τ_{eff} is mainly determined by the limited surface passivation quality of the Al BSF, the PERC design allows to make use of the high quality of the Si bulk by providing a better surface passivation. This enables higher efficiencies on the one hand, but makes the cell more prone to degradation effects on the other hand, as a degradation effect reducing either bulk lifetime or surface passivation quality has a higher impact on power output and can now be detected more easily on cell level.

In this presentation, a review of some of the mentioned degradation phenomena and possible mitigation strategies will be given. The focus of the review is on c-Si technology and covers BO-LID, PID (shunt-type), LeTID and degradation of surface passivation.

2 EFFECT OF INJECTION ON CHARGE STATE OF DEFECTS

To better understand defect reaction kinetics of some of the discussed degradation phenomena, it is helpful to discuss the consequences of injection conditions on the charge state of defects being present in the bulk of the Si wafer.

Under illumination, the generation of excess charge carriers (electrons and holes) causes a splitting of the Fermi level E_F as defined in thermal equilibrium into two quasi-Fermi levels (one for electrons E_{Fn} and one for holes E_{Fp}). In p-type c-Si with E_F close to the valence band edge E_V, this causes E_{Fn} to move upwards within the band gap towards the conduction band edge E_C. Depending on the position of a defect level in the c-Si band gap, this might affect the charge state of the defect.

One prominent example is the charge state of isolated H in c-Si. H in c-Si can act as a donor or acceptor, showing two defect levels in the band gap [10]. While the formation energy favors the positive charge state H^+ for the Fermi level close to E_V and around mid-gap, the negative charge state H^- is favored for E_F lying close to E_C [11]. Interestingly, the neutral charge state H_0 is never the favored one, but its occupation at room temperature shows a relative maximum for E_F positioned slightly above mid-gap [12].

Assuming a positively charged defect, it becomes clear that in thermal equilibrium H in p-type Si might never have the chance to bind to the defect due to Coulomb repulsion. On the other hand, injection of charge carriers can change the charge state of H^+ and therefore allow for a binding of H to the positively charged defect. But the situation might be more complicated as the defect itself might also change its charge state under injection conditions.

3 DEGRADATION PHENOMENA

3.1 BO-LID

BO-LID in c-Si is mainly observed in B-doped Czochralski(Cz)-grown wafers due to their high O concentrations originating from the quartz crucible. The degradation can be described with a so-called annealed state A (not recombination active) and a degraded state B (recombination active) which can be reached by injection at room temperature. An anneal at higher temperature around 200°C in the dark allows for the reverse reaction and state A is populated again. In 2006 it was discovered that the defect can be changed under injection at higher temperatures into a third state C, which shows no recombination activity and is stable under standard degradation conditions (injection at moderate temperature) [13]. This B→C reaction (also termed regeneration) seems to be only possible with the presence of H in c-Si [14,15].

As BO-LID has a severe impact on the efficiency potential after installation of c-Si modules made of B-doped wafers, a lot of research effort went into strategies to avoid or minimize the negative effects of BO-LID. One mitigation strategy is to treat the fully processed solar cell in an extra step after the co-firing to bring ideally all of the BO-related defects into state C. Doing this in an industrial production environment in a cost-effective way means to find solutions that can be applied fast with reliable results. Therefore, it is desirable to be able to manipulate the degradation and regeneration rate to speed up the population of state C.

It could be shown that degradation can be sped up not only by increasing temperature, but also by strong injection conditions [16]. For manipulating the regeneration rate, a multitude of findings are reported in literature. While increase of injection and temperature also speed up regeneration rate [13,17], it can also be manipulated by changes in the dielectric surface passivation layer system and the firing conditions. The introduction of an AlO_x layer between c-Si and SiN_x:H layer slows down the regeneration rate. The regeneration rate is getting slower with thicker AlO_x interlayer, pointing towards AlO_x being a barrier of H diffusing from the SiN_x:H into the c-Si bulk during the firing step [18]. A high peak firing temperature causes a faster regeneration rate and a high belt speed during firing also helps to speed up regeneration [19]. All these results can be explained by the regeneration rate being influenced by the amount of H present in the c-Si bulk during the regeneration process, with high concentration of H being beneficial for a fast regeneration process. As a side note, it has to be mentioned that the H has to be present in a suitable form, as low temperature annealing steps after the firing step can affect the regeneration rate, too [20]. This could be explained by the low temperature anneal changing the form H is present/bound in c-Si [21].

The most obvious technical solution to apply the regeneration procedure is to illuminate the cell after firing at elevated temperature. Depending on the technical solution typical temperatures are around 250°C at 10-20 suns illumination, with treatment times <1 min (e.g. [22]).

Another method is to apply a bias to the solar cell at elevated temperature. This can be elegantly realized by stacking the solar cells after firing and applying a bias to the stack in an environment providing the envisaged temperature for the stack [23]. As the cells are stacked, treatment time can be increased without negatively influencing throughput, and treatment temperatures are in the range of 180°C [24]. This seems to be currently the most cost-effective method to regenerate B-doped solar cells and is widely used in industry.

A third mitigation strategy is to avoid B as a dopant completely by using Ga. For a long time, this has been ruled out as the segregation coefficient of Ga in Si is much smaller than for B, causing an inhomogeneous distribution within the grown ingot. But continuous feeding during Cz crystal growth allowed for the use of Ga and industry has started shifting from B- to Ga-doped material much faster than anticipated in the ITRPV roadmap of 2020 [25,26].

In summary, the problem of BO-LID seems to be solved/solvable on an industrial scale, although the microscopic picture of the defect is not yet fully clear. See also [27] for the current understanding of influencing factors of BO-LID regeneration.

3.2 PID (Shunt Type)

Another frequently observed type of degradation affecting only modules, but not isolated solar cells, is potential induced degradation of the shunt type. Here a reduction in parallel resistance R_P of the module under working conditions (illumination/injection, elevated temperature, realistic ambient conditions) is observed in the field, reducing power output over time. This degradation phenomenon resulted in severe losses of power output of many PV systems worldwide and was therefore heavily researched. It could be shown that this degradation is triggered by a high potential difference between the glass and the embedded cells, with humidity and temperature playing important roles, too.

A closer look revealed that degradation strength depends strongly on polarity of the potential difference, with PID becoming stronger when the cell is on high negative potential compared to the grounded frame/glass [8]. Degradation strength also depends on the position of the cell in the string or module, with cells closer to the grounded frame having a higher probability to be affected by PID [28]. Humidity, soling and temperature are other parameters that have been found to trigger the strength of PID.

An important finding was the detection of Na decorating stacking faults penetrating the space charge region at locations showing a locally reduced R_P in a solar cell affected by PID [29]. With this finding, a model explaining the mechanism behind PID of the shunt type could be proposed. It is based on drift of Na ions from the glass through the encapsulant and the dielectric layers on top of the cell into the emitter. Here the Na decorated extended defects, reducing R_P and causing a local shunt.

The drift is triggered by the high potential difference between the frame (grounded) and the cell lying on high negative potential. This also explains why the opposite polarity (cell on a positive potential compared to the grounded frame) does not provoke PID in standard monofacial cells, due to the positively charged Na^+ ions.

Within this model, the conductivity of the surface of the glass plays an important role, as it determines how far the high potential difference between frame and cell reaches out into the center region of the module. The conductivity can be strongly influenced by humidity, soiling and temperature, and modules in hotter and more humid climate zones show indeed a much higher probability of being affected by PID indeed due to the higher average conductivity of the surface of the glass [28].

Mitigation strategies can be separated into the different regions where action has to be taken. On cell level one possibility is to avoid the high voltage drop over the dielectric layer (SiN_x:H-based anti-reflective coating (ARC) on the front). An increase in conductivity of the layer can, e.g., be reached by using a more Si-rich SiN_x:H layer, showing a higher refractive index. As the ARC/optical conditions might be negatively influenced by that change, frequently multi-layer dielectric layer stacks are in use nowadays. Another measure is the use of a thin SiO_x layer on top of the emitter, although the exact mechanism of how this reduces PID seems to be not clear yet [8].

Mitigation strategies on module level follow the idea to avoid the drift of Na^+ towards the cell. Therefore, an adaption of the encapsulation material showing a higher electrical resistivity or the avoidance of Na in the glass show the wanted effect, although the use of Na-free or Na-lean glass might not be the most economical solution.

Measures can also be taken on system level. The most obvious strategy would be to completely avoid a negative bias between grounded frame and cell. Depending on the inverter used, this might be an option or not. Module level inverters and/or power optimizers might be an option, too. If no other measures to avoid PID from developing can be taken, the last option is to use an opposite biasing of the string to drive the Na out of the cell (e.g. at night). It could be shown that up to a certain level of decoration of the extended defects this is working.

In summary, PID of the shunt type is a quite well understood phenomenon, although some open questions remain. But mitigation strategies exist to avoid its negative impact for future installation of PV systems. See [8] for a very good overview.

3.3 LeTID

In 2012 a degradation mechanism was first described in multicrystalline (mc) p-type material that causes a reduction of bulk lifetime in c-Si under illumination at elevated temperature [5]. As it was not scaling with the B and O concentration and only detectable at elevated temperatures, its origin is different from BO-LID and it was named LeTID [6]. As for BO-LID, also for LeTID the prerequisite is not the presence of photons, but the presence of excess charge carriers, therefore both abbreviations are misleading as the degradation can also be triggered by biasing the solar cell. Similar to BO-LID, a degradation and regeneration phase can be observed, but degradation on realistic timescales is only visible at temperatures $\geq$50°C. LeTID kinetics can be increased by higher treatment temperature and higher carrier concentration [6], indicating that both reaction rates show a strong injection dependency. LeTID can also be observed under realistic outdoor conditions on module level [30], with the extent and kinetics strongly dependent on the geographic location, mainly due to varying solar cell temperature.

Many studies have been carried out to find the root cause of LeTID, as the phenomenon can strongly affect module power output, especially for the PERC design [5]. It could be shown that the degradation affects all areas of the solar cell, independently on the initial material quality, provided that firing temperature is high enough [31]. Higher firing temperatures lead to more severe LeTID, while samples fired at temperatures $\leq$650°C do not show LeTID [32]. Regeneration sets in earlier in areas of initially high τ_{eff}, indicating that regeneration kinetics depend on injection conditions [33]. Thinner samples show less degradation, pointing towards a mobile species involved in the LeTID phenomenon, interpreted in [34] as a possible involvement of a mobile species diffusing towards the surface in the course of the degradation/regeneration cycle. Firing with H-lean dielectric layers or the use of interlayers between H-rich SiN_x:H and c-Si leads to less LeTID defects [35]. Thicker AlO_x interlayers cause less LeTID most probably as they act as a barrier layer for H diffusing from the SiN_x:H through the interlayer into the c-Si bulk [36]. Slow cool-down ramps after firing are also beneficial for a less LeTID [37].

The findings listed above all point towards H being involved in the LeTID phenomenon. But contrary to BO-LID, where H in high concentrations (in a suitable form) is desired for fast regeneration kinetics and therefore cost-effective treatment of BO-LID when going through the degradation/regeneration process within short treatment times, H seems to be a prerequisite for LeTID to occur. The microscopic picture of the defect is still unclear, but some models have been proposed. One of it is the "H in buckets" model which is based on the BO-LID 3-state model, but contains an additional reservoir state [38]. It is based on H diffusing in c-Si being released from dopant-H pairs under LeTID conditions and being recaptured in more stable bonds during the regeneration phase [38]. Most models assume that an additional species is involved, but its nature is not clear yet.

LeTID was also observed in p-type Cz [39,40] and also p-type Floatzone material [41], but with different kinetics than in mc-Si using similar illumination and temperature conditions. As it is partly triggered by the same conditions necessary for BO-LID (carrier injection) and the kinetics are similar in monocrystalline material, usually an overlay of BO-LID and LeTID is observed in B-doped Cz material. One possibility to differentiate between both phenomena are the different ratios of capture cross sections for electron and holes for both phenomena. Using this, it could be shown that at high firing temperature LeTID is the dominating phenomenon, while BO-LID dominates for lower firing temperature [42].

Mitigation strategies are based on controlling the H content in the c-Si bulk, e.g. via adapting firing temperature and ramps or the composition of the dielectric surface passivation layer stack. Going towards thinner wafers is currently not an option for industry due to yield issues. Annealing steps can influence the LeTID kinetics [43] and apparently can influence the defect precursors and/or the reservoir and might be interesting to speed up the LeTID degradation/regeneration cycle.

In summary, LeTID is a phenomenon that remains a little bit mysterious, as not much is known yet about the composition of the defect. But it is known empirically how the kinetics can be influenced by external parameters (temperature, injection, processing steps). LeTID is also occurring in Ga-doped material, and with the change towards Ga doping it remains to be seen how LeTID can be treated in the future.

3.4 Degradation of Surface Passivation

Decrease in surface passivation quality over time is affecting solar cell power output as it reduces the effective lifetime of charge carriers. Various degradation phenomena have been studied, some of them could be tracked down to ultraviolet (UV) radiation cracking Si-H bonds.

Recently, it could be shown that PERC solar cells with SiO_x/SiN_x:H surface passivation stack suffer from loss of surface passivation quality after a treatment at 1 sun and 150°C [7]. A similar loss in surface passivation quality could also be seen in lifetime samples with different passivation stacks and different doping (p- and n-type) [44]. Here the degradation occurred already at 1 sun and 80°C, therefore this effect could really harm solar cell performance in realistic outdoor conditions. It could be tracked down to a loss in chemical surface passivation quality.

The root cause for this type of degradation is not yet clear, but it was speculated that H diffusing towards the c-Si surface might be responsible for the observed degradation. For longer treatment times and/or higher treatment temperatures, again a regeneration phase can be observed after the degradation of surface passivation, indicating that the loss in surface passivation is only temporarily [44,45].

Most probably, loss in surface passivation quality under outdoor performance conditions (illumination, elevated temperature) will have an effect on a variety of solar cell architectures. But not too much published results can be found in that direction of research, although many new cell concepts are currently in development. Nevertheless, intense long-term stability testing under injection and elevated temperature should be done for all of these new concepts to clarify the underlying mechanisms and to avoid surprises.

4 SUMMARY

In this contribution, the state of knowledge for some prominent examples of degradation phenomena has been presented. The shown examples all have in common that they are triggered by typical operation conditions of solar modules (illumination, elevated temperature, realistic ambient conditions).

To understand some of the observed findings, the understanding of the underlying defect reactions is very helpful. The charge state of the species interacting during the degradation/regeneration reactions may change during injection, as it may depend on the position of E_{Fn} in the band gap. As H seems to be involved in some of the described phenomena, especially its change of charge state under operation conditions of the solar cell is key for understanding the underlying mechanisms.

BO-LID seems to be understood up to the degree that satisfying mitigation strategies exist (regeneration procedure on fired solar cells or use of Ga doping).

For PID of the shunt type an accepted model exists, based on drift of Na ions from the glass into extended defects in the space charge region. Here also mitigation strategies are available (on cell, module, and system level), although some of them might increase fabrication costs significantly.

The microscopic picture of the defect causing LeTID is unknown yet. It is only known that H triggers its occurrence. Therefore, mitigation strategies mainly deal with the H management and/or thermal treatments to go through the regeneration process. Most strategies cause extra costs. Therefore, a more detailed understanding would be very helpful to try to better control its occurrence.

Degradation of surface passivation due to injection at elevated temperature was observed during LeTID tests. Not much is known yet about its origin, but H seems to play an important role here, too. More research effort is needed to clarify its possible impact, especially for new cell architectures.

With solar cell efficiencies increasing due to improved c-Si wafer quality and new solar cell architectures with less losses at surfaces, contacts and highly doped regions, degradation effects gain more and more importance. Or in short: with higher efficiencies, there is more to lose.

5 ACKNOWLEDGEMENTS

Part of this work was funded by the German BMWi (0324204B, 03EE1051C, 03EE1052A). The content is the responsibility of the author.

6 REFERENCES

[1] D.L. Staebler, C.R. Wronski, Appl. Phys. Lett. 31(4) (1977) 292.

[2] H. Fischer, W. Pschunder, Proc. 10th IEEE PVSC (1973) 404.

[3] G. Zoth, W. Bergholz, Appl. Phys. 67 (1990) 6764.

[4] J. Lindroos, H. Savin, Sol. Energy Mater. Sol. Cells 147 (2016) 115.

[5] K. Ramspeck, S. Zimmermann, H. Nagel, A. Metz, Y. Gassenbauer, B. Birkmann, A. Seidl, Proc. 27th EUPVSEC (2012) 861.

[6] F. Kersten, P. Engelhart, H.C. Ploigt, A. Stekolnikov, T. Lindner, F. Stenzel, M. Bartzsch, A. Szpeth, K. Petter, J. Heitmann, J.W. Müller, Sol. Energy Mater. Sol. Cells. 142 (2015) 83.

[7] A. Herguth, C. Derricks, D. Sperber, IEEE J. Photovolt. 8 (2018) 1190.

[8] W Luo, Y.S. Khoo, P. Hacke V. Naumann, D. Lausch, S.P. Harvey, J.P. Singh, J. Chai Y. Wang, A.G. Aberle, S. Ramakrishna, Energy Environ. Sci. 10 (2017) 43.

[9] https://iea-pvps.org/wp-content/uploads/2020/01/IEA-PVPS_T13-01_2014_Review_of_Failures_of_Photovoltaic_Modules_Final.pdf

[10] N.M. Johnson, C.G. Van de Walle, Semiconductors and Semimetals 61 (1999) 13.

[11] C.G. Van de Walle, P.J.H. Denteneer, Y. Bar-Yam, S.T. Pantelides, Phys. Rev. B 39 (1989) 10791.

[12] C. Herring, N.M. Johnson, C.G. Van de Walle, Phys. Rev. B 64 (2001) 125209.

[13] A. Herguth, G. Schubert, M. Kaes, G. Hahn, Proc. 4th WCPEC (2006) 940.
[14] K.A. Münzer, Proc. 24th EUPVSEC (2009), 1558.
[15] G. Krugel, W. Wolke, J. Geilker, S. Rein, R. Preu, Energy Procedia 8 (2011) 47.
[16] P. Hamer, B. Hallam, M. Abbott, S. Wenham, Phys. Status Solidi RRL 9(5) (2015) 297.
[17] A. Herguth, G. Schubert, M. Kaes, G. Hahn, Prog. Photovolt. Res. Appl. 16 (2008) 135.
[18] S. Wilking, A. Herguth, G. Hahn, J. Appl. Phys. 113 (2013) 194503.
[19] S. Wilking, S. Ebert, A. Herguth, G. Hahn, J. Appl. Phys. 114 (2013) 194512.
[20] S. Wilking, C. Beckh, S. Ebert, A. Herguth, G. Hahn, Sol. Energy Mater. Sol. Cells 131 (2014) 2.
[21] V.V. Voronkov, R. Falster, Phys. Status Solidi B 254 (2017) 1600779.
[22] C. Derricks, O. Romer, T. Pernau, G. Hahn, A. Herguth,:Sol. Energy Mater. Sol. Cells 195 (2019) 358.
[23] A. Herguth, G. Hahn, Proc. 28th EUPVSEC (2013) 1507.
[24] J. Ye , B. Ai ,J. Jin, D. Qiu, R. Liang, H. Shen, Int. J. Photoenergy (2019) 5357370.
[25] ITRPV 11 (4/2020)
[26] ITRPV 12 (4/2021) https://itrpv.vdma.org/viewer/-/v2article/render/71528684
[27] A. Herguth, B. Hallam, AIP Conf. Proc. 1999 (2018) 130006.
[28] S. Koch, Dissertation, University of Halle-Wittenberg, 2019
[29] V. Naumann, D. Lausch, A. Hähnel, J. Bauer, O. Breitenstein, A. Graff, M. Werner, S. Swatek, S. Großer, J. Bagdahn, C. Hagendorf, Sol. Energy Mater. Sol. Cells 120 (2014) 383.
[30] F. Kersten, F. Fertig, K. Petter, B. Klöter, E. Herzog, M.B. Strobel, J. Heitmann, J.W. Müller, Energy Procedia 124 (2017) 540.
[31] D. Skorka, A. Zuschlag, G. Hahn, AIP Conf. Proc. 1999 (2018) 130015.
[32] D. Bredemeier, D. Walter, S. Herlufsen, J. Schmidt, AIP Advances 6 (2016) 035119.
[33] A. Zuschlag, D. Skorka, G. Hahn, Prog. Photovolt. Res. Appl. 25 (2017) 545.
[34] D. Bredemeier, D.C. Walter, J. Schmidt, Sol. RRL 2 (2018) 1700159.
[35] F. Kersten, P. Engelhart, H.C. Ploigt, F. Stenzel, K. Petter, T. Lindner, A. Szpeth, M. Bartzsch, A. Stekolnikov, M. Scherff, J. Heitmann, J.W. Müller, Proc. 31st EUPVSEC (2015) 1822.
[36] A. Schmid, C. Fischer, D. Skorka, A. Herguth, C. Winter, A. Zuschlag, G. Hahn, IEEE J. Photovolt. 11 (2021) 967.
[37] R. Eberle, W. Kwapil, F. Schindler, M.C. Schubert, S.W. Glunz, Phys. Status Solidi RRL 10 (2016) 861.
[38] T.H. Fung, M. Kim, D. Chen, C.E. Chan, B.J. Hallam, R. Chen, D.N.R. Payne, A. Ciesla, S.R. Wenham, M.D. Abbott, Sol. Energy Mater. Sol. Cells 184 (2018) 48.
[39] F. Fertig, R. Lantzsch, A. Mohr, M. Schaper, M. Bartzsch, D. Wissen, F. Kersten, A. Mette, S. Peters, A. Eidner, J. Cieslak, K. Duncker, M. Junghänel, E. Jarzembowski, M. Kauert, B. Faulwetter-Quandt, D. Meißner, B. Reiche, S. Geißler, S. Hörnlein, C. Klenke, L. Niebergall, A. Schönmann, A. Weihrauch, F. Stenzel, A. Hofmann, T. Rudolph, A. Schwabedissen, M. Gundermann, M. Fischer, J.W. Müller, D.J.W. Jeong, Energy Procedia (2017) 338.
[40] D. Chen, M. Kim, B.V. Stefani, B.J. Hallam, M.D. Abbott, C.E. Chan, R. Chen, D.N.R. Payne, N. Nampalli, A. Ciesla, T.H. Fung, K. Kim, S.R. Wenham, Sol. Energy Mater. Sol. Cells 172 (2017) 293.
[41] D. Sperber, A. Schwarz, A. Herguth, G. Hahn, Phys. Status Solidi A 215(24) (2018) 1800741.
[42] A. Graf, A. Herguth, G. Hahn, AIP Conf. Proc. 2147 (2019) 140003.
[43] C. Chan, T.H. Fung, M. Abbott, D. Payne, A. Wenham, B. Hallam, R. Chen, S. Wenham, Sol. RRL 1 (2017) 1600028.
[44] D. Sperber, A. Graf, D. Skorka, A. Herguth, G. Hahn, IEEE J. Photovolt. 7 (2017) 1627.
[45] D. Sperber, A. Heilemann, A. Herguth, G. Hahn, IEEE J. Photovolt. 7(2) (2017) 463.

THE FRONT AND REAR SIDE CONTRIBUTIONS TO THE POTENTIAL INDUCED DEGRADATION OF BIFACIAL SILICON HETEROJUNCTION SOLAR MODULES

Olatz Arriaga Arruti[1], Luca Gnocchi[1], Fabiana Lisco[1], Alessandro Virtuani[1], Christophe Ballif[1,2]
1 – École Polytechnique Fédérale de Lausanne (EPFL), Institute of Electrical and Microengineering (IEM), Photovoltaics and Thin Film Electronics Laboratory (PV-Lab), 2000 Neuchâtel, Switzerland
2 – CSEM, PV-center, 2000 Neuchâtel, Switzerland
Email: olatz.arriagaarruti@epfl.ch

ABSTRACT: Recent studies showed that silicon heterojunction (SHJ) solar cells may be prone to potential induced degradation (PID), when encapsulated with a low volume resistivity ethylene vinyl acetate (EVA). This, however, can be prevented when using high-volume resistivity encapsulants (e.g. ionomer or polyolefin elastomers (POE)), or an edge sealant in combination with EVA. Here, we perform PID tests in a climatic chamber (85°C/85% RH) on glass/glass one-cell mini-modules, by using different combinations of encapsulants. We combine EVA and POE, interchanging their position between the front- and rear-sides (i.e. POE/EVA and EVA/POE), with the aim of studying the PID mechanism and analyzing the contribution from each side. The SHJ solar cells employed are bifacial rear-emitter cells. We demonstrate that for all combinations, the addition of a POE layer contributes to considerably slowing the degradation kinetics. Although not completely prevented, after 450 hours of PID exposure (corresponding to nearly 5 times the duration of the IEC pre-standard), the deceleration of PID is mostly successful when the POE layer is placed at the rear-side of the module (i.e. EVA/POE encapsulation scheme). After 450 hours of PID test, the POE/POE and EVA/EVA samples degrade by 0.3% and 31%, respectively. If a POE is placed only on the front- or rear-side, the corresponding degradation is, respectively, of 21% and 8%.
Keywords: Potential Induced Degradation (PID), Silicon heterojunction (SHJ), POE, EVA, recombination.

1 INTRODUCTION

The long-term performance of photovoltaic (PV) modules is of utmost importance for investors and to further reduce the cost of solar electricity. Potential induced degradation (PID) has been a hot topic in recent years, due to its potential large impact on PV modules' performance. It is a tricky phenomenon to identify and understand due to its lack of any visual defects and dependency on many factors, such as cell technology [1].

The underlying PID mechanism has been extensively researched in conventional crystalline silicon (c-Si) modules [1]. Developing a method to prevent PID is, however, necessary to ensure their application in the market. A well known method for preventing PID in conventional modules is the use of high-volume resistivity encapsulants [2], [3]. These avoid the diffusion of sodium (Na) ions that cause shunts in conventional c-Si solar cells. The market share of PV technologies has been changing in recent years though, and high efficiency solar cell technologies such as silicon heterojunction (SHJ) are expected to increase their market volume in the coming years [4]. Consequently, preventing and understanding PID in this solar cell technology is of high relevance. In previous works, presented last year, we demonstrated that we can prevent PID in SHJ solar cells applying two different methods.

1. Using high-volume resistivity encapsulants, e.g. ionomer, polyolefin elastomer (POE) and thermoplastic polyolefin (TPO) [5].
2. Applying an edge sealant (ES) at the edges of glass/glass (G/G) mini-modules, making them impermeable to water ingress, while using a low-volume resistivity encapsulant, i.e. ethylene vinyl acetate (EVA) [6].

In this work, we aim to understand the PID mechanism in bifacial SHJ solar cells, studying the contribution from both the front- and rear-sides. We do this by manufacturing bifacial SHJ one-cell mini-modules in a glass/glass scheme. We compare the contribution of each of the sides by combining two different polymers, EVA along with a high resistivity encapsulant (POE) (see Table I). We aim to isolate the contribution of PID to one side of the module at a time, by using one layer of POE on one side, which has shown to be PID-resistant in SHJ solar cells [5].

2 EXPERIMENTAL WORK

We investigate the impact of moisture by performing PID accelerated tests in humid conditions (85°C/85% RH) and employing a glass/glass (G/G) encapsulation design. An aluminum tape was placed at the four edges of the mini-module to simulate the presence of a metallic frame. A voltage bias of -1000 V was applied to the short-circuited module leads with respect to the grounded frame to analyze the effect of voltage polarization. At each testing condition, a couple of samples were kept unbiased as reference.

In previous works we observed that SHJ cells encapsulated with EVA in a G/G module structure can suffer from PID. This however, can be completely prevented by using a high-volume resistivity encapsulant, such as POE. Consequently, we consider that the POE acts as a protective layer against degradation. Here, we study the contribution of each of the sides of the bifacial SHJ cells by protecting one side with a POE, whilst leaving the other one "unprotected", by employing a layer of EVA (see Table I). Furthermore, in order to make a more exhaustive study, these modules were compared at all times to the EVA/EVA and POE/POE ones.

Table I. Encapsulant combinations on the glass/glass mini-modules using SHJ solar cells.

Name	Front encapsulant	Rear encapsulant
EVA/EVA	EVA	EVA
POE/POE	POE	POE
POE/EVA	POE	EVA
EVA/POE	EVA	POE

The mini-modules were characterized during the PID testing by one-sun and dark current-voltage (IV) curves, electroluminescence (EL) and external quantum efficiency (EQE) measurements.

3 RESULTS

In this contribution, we present results that correspond to the PID tests performed on modules encapsulated with a combination of EVA and POE. The comparison in degradation among the mini-modules described in Table I is summarized in Figure 1. These modules were subjected to PID under a bias of -1000V for 450 h. Figure 1 a) summarizes the evolution of the maximum power (P_{max}) throughout the duration of the extended PID test for the front-side of all the module structures mentioned. The IEC TS 62804-1 [7] states the standard test duration for PID tests is of 96 h. However, we extend the tests for longer times in order to accelerate the degradation. Therefore, it is to be considered that the results showed here are for 450 hours of test, almost 5 times the duration of the standard. All module structures degrade less than 5% after 96 hours of test, thus "passing" the IEC TS. In fact, the kinetics for the different structures change after 192 hours (2 times the IEC TS), with very different evolutions.

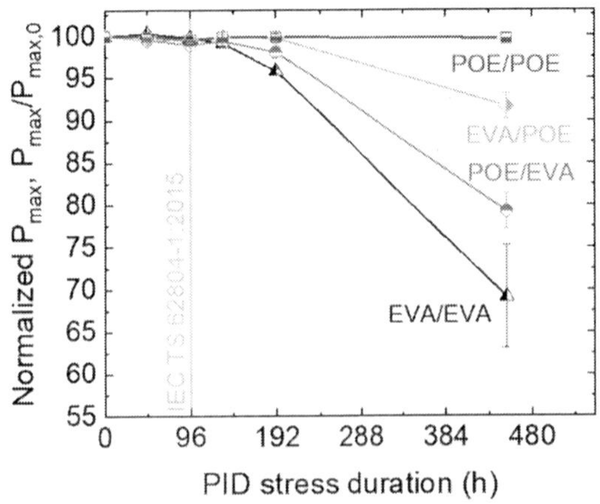

Figure 1. Normalized power of the **front-side** of SHJ 1-cell G/G mini-modules EVA/EVA, POE/POE, POE/EVA and EVA/POE structures, after an extended 450 h of test at 85°C/85% RH and -1000 V conditions. The test conditions foreseen in IEC TS 62804-1 (i.e. 60°C/85% RH for 96h), have increased in terms of relative humidity (85% RH) and test duration (almost 5 times the standard). The values plotted in the chart are the averages of each type of sample.

As we demonstrated in previous works, PID is prevented in SHJ mini-modules when using POE as the only encapsulant (POE/POE), whereas the modules in an EVA/EVA structure degrade around 31% after 450 hours of test **[5]**, **[6]**. Conversely, we observe that having one POE layer in one of the sides of the cell already has an impact on the degradation of the mini-module. We observe that the POE/EVA modules present a power loss of 21% after the PID test, while the EVA/POE modules, with the POE layer at the rear-side, show a power loss of 8%.

The relative variation of electrical parameters for the POE/EVA and EVA/POE module structures is summarized in Figure 2. The degradation in the former is dominated by a considerable loss in the fill factor (FF). Since the bifacial SHJ cell is rear-emitter, this could be caused by defects at the rear-side of the cell, where the p-n junction is located. Conversely, the degradation of the cells with an EVA/POE structure is less pronounced, and driven equally by losses in the short-circuit current density (J_{SC}) and FF.

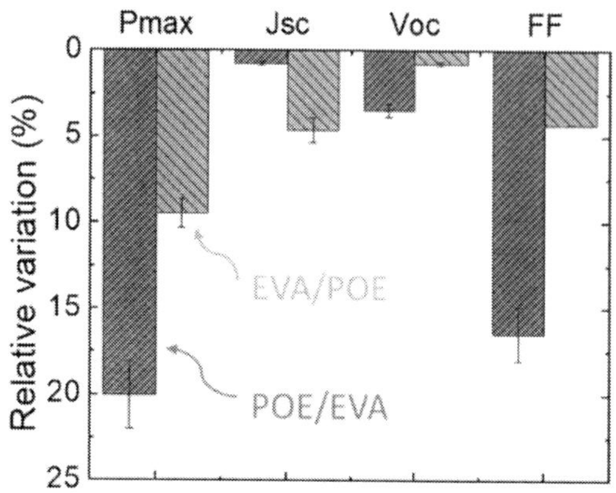

Figure 2. Relative variation of electrical parameters of the front-side for mini-modules encapsulated in the POE/EVA and EVA/POE schemes.

As observed in the relative variation of the modules with the EVA/EVA scheme, the loss of P_{max} at the front-side of these bifacial SHJ solar cells is 10% larger than at the rear-side (see Figure 3). Furthermore, the mechanism itself is different: the front-side is characterized by a loss in Jsc, derived from an increased front-surface recombination [6], whereas the degradation at the rear-side is driven by FF losses, potentially by an increased recombination at the p-n junction of the cell.

These findings lead to believe that protecting the front-side would be more efficacious when reducing the effect of PID in these cells. Nonetheless, we observe the opposite effect. Indeed, protecting the rear-side, where the p-n junction is located (i.e. EVA/POE scheme), improves the performance of the module by 10%, compared to the POE/EVA configuration. Therefore, using POE as the encapsulant to protect the rear-side of a bifacial SHJ mini-module can have a very positive impact in decelerating degradation, albeit it does not completely prevent it.

Figure 4 compares the relative variation of electrical parameters between the front- and rear-sides of the bifacial SHJ mini-modules for the POE/EVA and EVA/POE module structures. We see that in both configurations, the loss of P_{max} at the front-side of the module is considerably minimized with respect to the EVA/EVA structure (as already observed in Figure 1). Conversely, the behavior of the rear-side of the modules depends highly on the encapsulant combination. At the rear-side of the cell, the P_{max} loss in the EVA/POE structure is of 11%, compared to the 21% of the POE/EVA configuration, same as the EVA/EVA scheme.

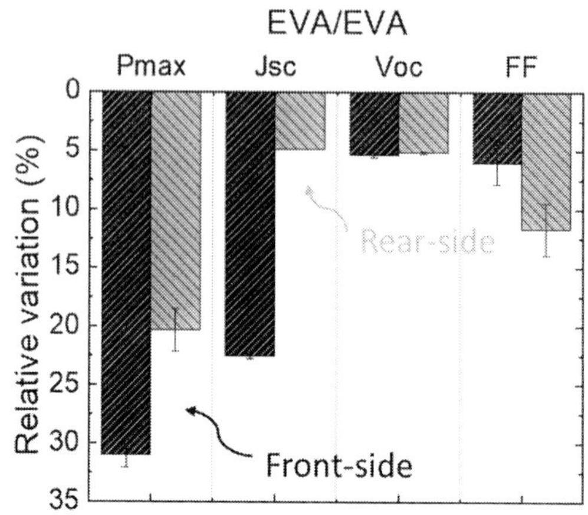

Figure 3. Relative variation of electrical parameters at the front- and rear-sides of mini-modules encapsulated in the EVA/EVA scheme. These are not the standard test conditions (60°C/85% RH for 96h), but extended both in humidity (with a higher relative humidity of 85%) and in test duration (almost 5 times the standard).

The POE layer at the front-side in the POE/EVA scheme does not have an effect at the rear-side of the SHJ cell. Although, as expected, it does at the front-side. Conversely, the module benefits in its entirety when this POE layer is placed at the rear-side, that is, in the EVA/POE module structure. All electrical parameters display a lower degradation in this module scheme. In fact, the degradation of the V_{OC} in the SHJ cell is minimal, showing a stable passivation throughout the extended PID test. Consequently, we can say that the front-surface recombination observed in EVA/EVA module schemes is caused by the contribution of issues generated at both the front- and rear-sides of the cell [6]. Thus, any defects in the p-n junction caused by an "unprotected" rear-side in rear-emitter SHJ solar cells have also a direct effect on the front-surface recombination we have previously observed.

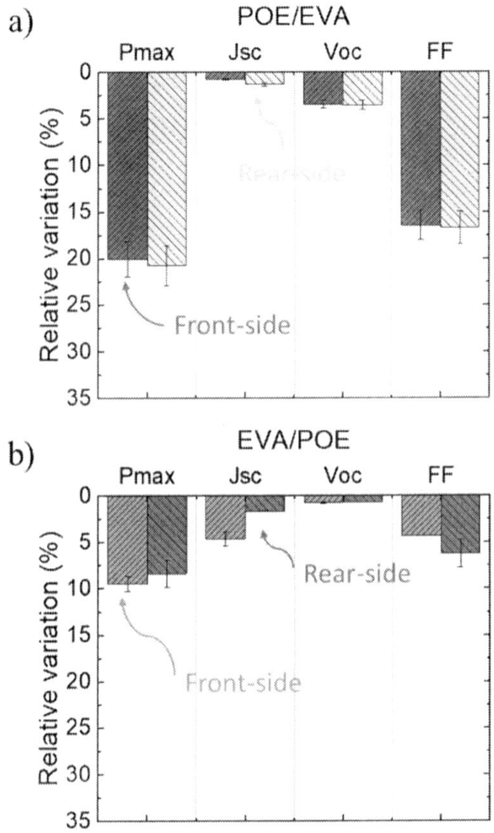

Figure 4. Relative variation of electrical parameters for modules in a) POE/EVA and b) EVA/POE encapsulant structures, comparing the degradation between the front- and rear-sides of the bifacial SHJ solar cells. The results show the maximum power loss after an extended 450 h of test at 85°C/85% RH and -1000 V conditions. These are not the standard test conditions (60°C/85% RH for 96h), but extended both in humidity (with a higher relative humidity of 85%) and in test duration (almost 5 times the standard).

4 CONCLUSIONS

In previous works, we demonstrated that PID can be prevented in bifacial SHJ mini-modules when using solely a high-volume resistivity encapsulant (i.e. ionomer, POE and TPO) [5] or an edge sealant in combination with EVA to prevent moisture ingress [6]. In this work, we aim to understand the contribution of the rear- and front-sides to the PID mechanism. We performed PID on one-cell SHJ mini-modules encapsulated with one layer of POE and another layer of EVA, interchanging their positions at the front- and rear-sides, having POE/EVA and EVA/POE (front/rear encapsulants) structures. Besides showing that only one layer of POE can have a beneficial impact in reducing PID in SHJ cells, we have demonstrated that having this layer of POE at the rear-side decelerates the degradation kinetics more than placing it at the front. Remarkably, despite the hard testing conditions, modules with the EVA/POE encapsulant scheme degrade 10% (at the front-side) after 450 hours of test. We demonstrate that any defects created at the p-n junction on the cell have direct impact at the front, exacerbating the front-surface recombination observed previously [6]. Therefore, protecting just the rear-side has a beneficial impact on the SHJ solar cell as a whole.

In a recently presented work [7], we have demonstrated that one of the origins causing degradation is the diffusion of Na ions through the front transparent conductive oxide (TCO) layer and into the amorphous Si/TCO interface. Therefore, if we prevent this Na diffusion with the use of POE at the rear-side of the cell, we avoid the potential defects where the p-n junction is, thus minimizing the recombination current causing shunts, and allowing a good carrier collection at both sides of the cell.

5 ACKNOWLEDGEMENTS

This work has been funded by the European Commission as part of the Horizon 2020-GOPV Project (grant agreement # 792059). We kindly thank CEA-INES (Chambery, France), in particular Romain Soulas and Stéphane Guillerez for providing testing material (SHJ solar cells). We gratefully acknowledge support from all PV-Lab staff members, in particular from Xavier Niquille.

6 REFERENCES

[1] W. Luo, Y. S. Khoo, P. Hacke, V. Naumann, D. Lausch, S. P. Harvey, J. P. Singh, J. Chai, Y. Wang, A. G. Aberle and S. Ramakrishna, "Potential-induced degradation in photovoltaic modules: a critical review," *Energy & Environmental Science,* vol. 10, no. 1, pp. 43-68, 2017.

[2] S. Yamaguchi, C. Yamamoto, K. Ohdaira and A. Masuda, "Comprehensive study of potential-induced degradation in silicon heterojunction photovoltaic cell modules," *Progress in Photovoltaics: Research and Applications,* vol. 26, no. 9, pp. 697--708, 2018.

[3] A. Virtuani, E. Annigoni and C. Ballif, "One-type-fits-all-systems: Strategies for preventing potential-induced degradation in crystalline silicon solar photovoltaic modules," *Progress in Photovoltaics: Research and Applications,* pp. 1-9, 2018.

[4] E. IEC, "IEC TS 62804-1:2015 Photovoltaic (PV) modules - Test methods for the detection of potential-induced degradation - Part 1: Crystalline silicon," IEC, 2015.

[5] O. Arriaga Arruti, L. Gnocchi, A. Virtuani and C. Ballif, "Encapsulant Selection for PID Resistant Modules made with Silicon Heterojunction Solar Cells," in *EU PVSEC Proceedings*, 2020.

[6] O. Arriaga Arruti, L. Gnocchi, F. Lisco, A. Virtuani and C. Ballif, "Optimized Module Packaging for Silicon Heterojunction Solar Cells and Increased PID Resistance," in *EU PVSEC Proceedings*, 2020.

[7] O. Arriaga Arruti, L. Gnocchi, Q. Jeangros, A. Virtuani and C. Ballif, "Potential Induced Degradation mechanism in rear-emitter bifacial silicon heterojunction solar cells encapsulated in different module structures," in *IEEE PVSC Proceedings*, 2021.

A grey box model for shunting-type potential induced degradation in silicon photovoltaic cells under environmental stress

Arnaud Schils[1], Robbe Breugelmans[2,3], Jorne Carolus[2,3], Julián Ascencio-Vásquez[4], Andreas Wabbes[5], Emmanuelle Bertrand[5], Bader Aldalali[6], Michael Daenen[2,3], Eszter Voroshazi[1], Stijn Scheerlinck[5]
[1] imec (partner inEnergyVille), Thorpark 8320, B-3600 Genk, Belgium
[2] Institute for Materials Research (IMO), Hasselt University (partner in EnergyVille), 3590 Diepenbeek, Belgium
[3] IMOMEC, imec (partner in EnergyVille), 3590 Diepenbeek, Belgium
[4] 3E sa, Quai à la chaux 6, B-1000 Brussels, Belgium
[5] Engie Laborelec, Rodestraat 125, B-1630 Linkebeek, Belgium
[6] Kuwait University, College of Engineering and Petroleum, Khaldiya, Kuwait

Topic 4.1 PV Module Design, Manufacture, Performance and Reliability
PV module design, module manufacturing processes. Type approval testing, degradation, ageing and lifetime; new and improved measurement and characterisation methods, correlation between laboratory testing and field performance, energy yield, energy rating.

ABSTRACT: This paper proposes a grey box model for cell-level simulation of potential induced degradation of the shunting type (PID-s) by extending a physics-based energy yield simulation framework. The leakage current is first computed using a physics-based approach based on an equivalent frame-to-cell electric circuit composed of temperature and humidity dependent resistances. Then, the increase of PID-s degradation due to leakage current is retrieved from measurements of PID-s evolution performed on poly-cSi cells. Finally, IV-curves of cells affected by PID-s are approximated by modifying the IV-curve of a PID-free cell using measurements of short circuit current and open-circuit voltage decrease with PID-s. A first comparison of simulated versus measured final PID-s levels is reported for a PV string after nine years of operation. The advantage of our grey box model over an empirical model is the possibility to investigate changes in material properties without the need for new measurements and to provide quantitative energy yield loss estimates depending on climate and system design.

Keywords: potential induced degradation (PID), reliability, high voltage stress (HVS)

1 INTRODUCTION

Estimating accurately the power produced by photovoltaic (PV) systems is critical for many applications such as financial assessment of PV projects and design of energy management systems [1]. During the past decade, this need motivated the development of new physics-based models for PV systems energy yield simulation [2]. Despite their accuracy over the first year of PV system operation, results over lifetime are often too optimistic: a recent solar risk assessment report states that predicted P90 production is reported to occur more than 1-in-3 years instead of 1-in-10 [3]. One of the causes is the absence of accurate models for degradation mechanisms.

Among possible degradation types, potential induced degradation (PID), in particular of the shunting type (PID-s), is one of the most common and severe reliability issue of crystalline silicon PV cells in the field [4]. PID-shunt occurs due to migration of sodium cations (Na^{+}) from the module tempered glass ($Na_2O.3SiO_2$) into small defects of the cell silicon lattice, creating stacking faults.

When many PV modules are connected in series, a high potential difference builds up between cells and frames of PV modules, which are grounded for safety reasons. Sodium cations drift from the front glass to the silicon nitrate (SiN) layer under the effect of the electric field. If the potential drop across the SiN layer of the cell is higher than 10V, sodium cations goes through the SiN potential barrier and then diffuse into small defects of the cell silicon lattice.

The first and main impact of this degradation on cell electric properties is the decrease of the shunt resistance R_{SH} [4]. At a more severe degradation stage, open circuit voltage (V_{OC}), short-circuit current (I_{SC}), operating point voltage (V_{MPP}), current (I_{MPP}) and power (P_{MPP}) are all drastically reduced.

The current state of the art for PID simulation relies on empirical models [5,6] which have their advantages but also come with the following shortcomings. Firstly, the model requires experimental input of any new PV module to properly fit the parameters of the models. Therefore, these models do not allow to perform "what if" simulations. Secondly, to ease the test matrix the model assumes a quadratic and linear evolution of PID with time and voltage, respectively, while measurements report logistic dependencies [7]. Finally, these models do not consider the cell-level differences in PID evolution due to different frame to cell distances.

This work is a first step to extend the imec's PV energy yield coupled optical, thermal and electrical simulation framework introduced in [2,8-10] with a cell-level model to predict the evolution of PID-s and its impact on cell current-voltage (IV) curve, power and temperature. The model takes as input climatic data (up to 1 min temporal resolution), thermoelectrical and optical module properties from datasheets and geometrical parameters.

The PID-s model is composed of three steps. Firstly, the computation of the Na^+ leakage current (I_{leak}) depending on environmental conditions. For now, only I_{leak} flowing through the front glass surface and then through the cell thickness, reported to be the most detrimental in outdoor conditions [11], is considered. Secondly, the computation of the PID-s level (power loss at STC) increase due to degradation by I_{leak}. Finally, the IV curve of the solar cell depending on its PID-s level is obtained.

2 LEAKAGE CURRENT

The current I_{leak} is computed by solving an equivalent electrical circuit modelling the path of Na^+ cations [12] (see Figure 1). The effect of environmental stress factors on PID-s evolution is modelled by varying, at each timestep and for each cell, the electrical resistances.

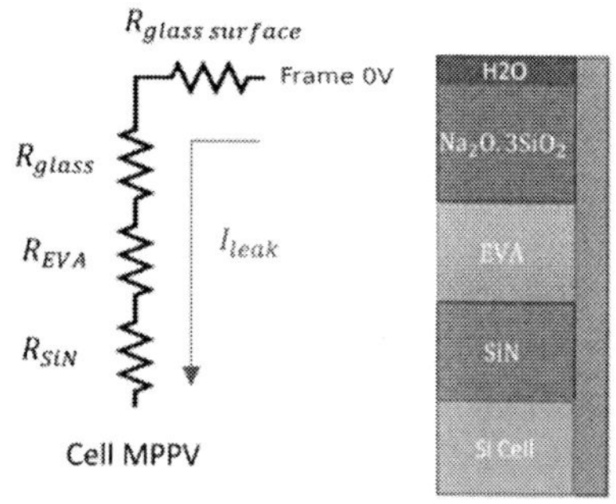

Figure 1: Simplified equivalent electric circuit for the Na^+ path from the grounded frame to a solar cell operating at V_{MPP} at the front side of a PV module [12]. From top to bottom, the four electrical resistances model the glass surface, the tempered glass, the encapsulant (EVA) and the silicon nitrate.

2.1 Frame to cell voltage drop

The frame being grounded, the frame to cell voltage drop can be inferred knowing operating voltages of all cells in the string electric circuit and the position of the zero-voltage point. The imec's PV energy yield framework computes the IV curves and V_{MPP} of the string and each solar cell using cell-level single diode model (see [2,8-10] for details). Module bypass diodes, installed in parallel of substrings, are also modelled by their IV curve in the string electric circuit (see Figure 2). Bypass diodes prevent decrease of the string I_{SC} when only a part of string is affected by PID.

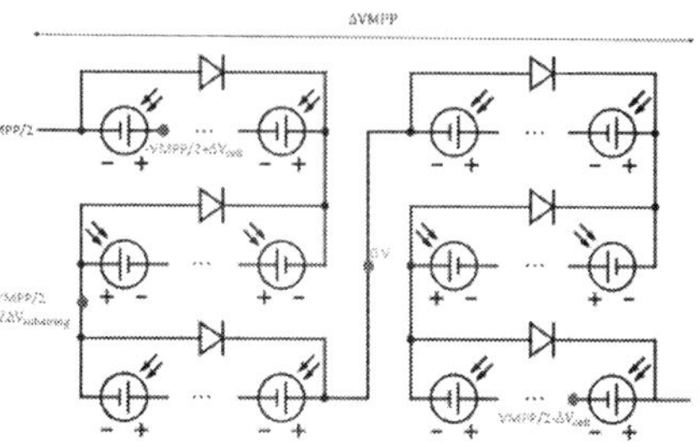

Figure 2: Electric circuit modelling a two modules string. Voltage differences with the frame are indicated in red.

The frame to cell voltage drops are retrieved going through the electric circuit from the negative towards the positive plug and incrementally adding up the voltage drops across each component of the circuit (see Figure 2).

2.2 Glass, EVA and SiN electrical resistances

The ionic electrical resistances of tempered glass and EVA as a function of layer thickness, temperature, water content and humidity are implemented using measurements reported by Takata et al. [13] and Mon et al. [14]. Accurate glass and EVA temperatures are computed by the imec's PV energy yield framework thermal model. The SiN layer being very thin, its ionic electrical resistance can be safely neglected [12]. The consideration of the impact of the SiN electronic conductivity on cell susceptibility to PID, reported by Naumann et al. [15], is a future work.

2.3 Glass surface electrical resistance

The glass surface electrical resistance models the propagation of Na^+ ions through a thin layer of humid air at top of the module front glass. The electrical resistivity of this layer is provided as a function of glass surface humidity by Naumann et al. [12]. The glass temperature T_{glass} is retrieved from the energy yield framework and is used to compute the glass surface humidity from relative ambient humidity [16, 17].

A different glass surface resistance is obtained for each cell by modelling the frame to cell glass surface as a conductor connecting the frame rectangular equipotential ($0V$) to a second rectangular equipotential (V_{stack}) inner to both the cell glass surface and the frame rectangles. This process is performed many times for each cell through a finite difference method.

This approach, even if being an extension of the model introduced by Naumann et al. [12], nevertheless remains an approximation whose accuracy on I_{leak} needs to be assessed [18]. An example of obtained I_{leak} spatial variation is presented in Figure 3. More formal modelling is possible using 3D FEM, at the cost of slower computation.

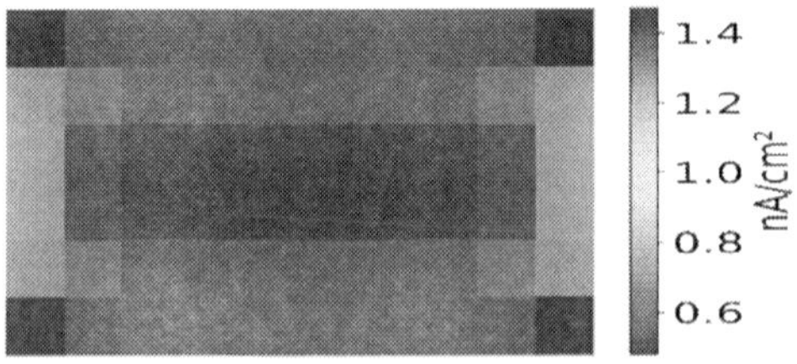

Figure 3: I_{leak} in a 60 cells module. Voltage drop between frame and cells set to 1000 V, temperatures of all layers set to 60°C, glass surface and EVA RH set to 60% and water content in tempered glass set to 0%.

3 LEAKAGE CURRENT IMPACT ON PID-SHUNT DEGRADATION

The PID-s level evolution with time for three different constant voltage V_{stack} (200 V, 600 V and 1000 V) applied with the foil method on multi-crystalline one-cell laminates has been measured by J. Carolus [19-21] at 60°C and 60% ambient temperature and relative humidity (RH), respectively. Since the foil method is used, the voltage V_{stack} is applied around the bulk glass, EVA and SiN resistances in Figure 1 (i.e. no glass surface resistance). Evolution of PID-s with I_{leak} is obtained by dividing the stack voltage by the stack resistance at the lab temperature and RH,

$$R_{stack} = R_{glass} + R_{EVA} + R_{SiN}, \quad (1)$$

$$I_{leak} = \frac{V_{stack}}{R_{stack}(60°C,\ 60\%\,RH)} \quad (2)$$

The measurements are well fitted by a two parameters logistic function [19],

$$PID_s(t_{lab}) = \frac{100\%}{1 + \exp\left(-k * (t_{lab} - t_0)\right)} \quad (3)$$

The parameters $k(I_{leak})$ and $t_0(I_{leak})$ are obtained by linear interpolation over the three I_{leak} points allowing to evaluate Equation (3) for any I_{leak} (see Figure 4). The derivative of Equation (3) w.r.t. t_{lab} multiplied by the timestep of the simulation Δt yields PID-s level increase during Δt due to any constant I_{leak} applied since a time t_{lab} on an initially PID-s free cell.

An approach like the one proposed by Annigoni et al. [6] is used to extend the model to time-varying I_{leak} in outdoor conditions: the effective time t_{lab} is obtained by computing the inverse of Equation (3), i.e. $t_{lab}(PID_s)$, with PID_s being the PID-s level of the cell at the end of the previous timestep.

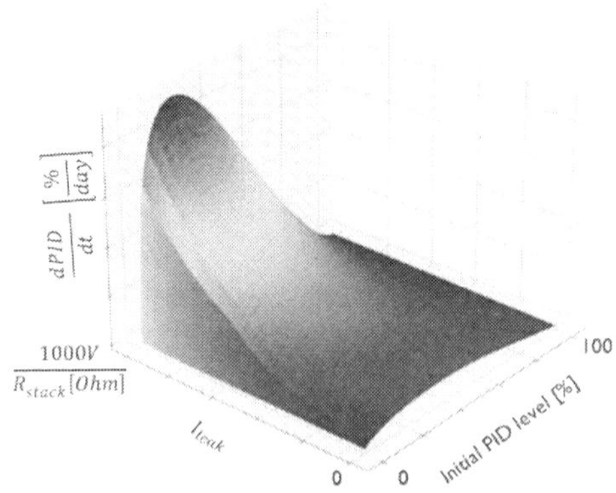

Figure 4: PID-s derivative at any leakage current I_{leak} and for any initial cell PID-s level.

4 IV-CURVES OF PID-SHUNTED CELLS

The estimation of PID-s level is the basis to reconstruct the full IV curve of a PID-s affected cell at specific irradiance and temperature. Because no analytical expression is available for the IV-curve of a PID-s affected cell, this is approximated applying a geometrical algorithm on the IV-curve of a PID-free cell. Operating powers of the full string and cells are then computed using these IV curves.

First, the I_{sc} and V_{oc} are scaled using evolution of these quantities with PID-s level measured by J. Carolus [19-21]. A geometrical algorithm then decreases the distance between each (V,I) point and the straight line passing by I_{sc} and V_{oc} by PID-s level % (see Figure 5): the higher the PID-s level, the closer to a straight line the IV curve will be.

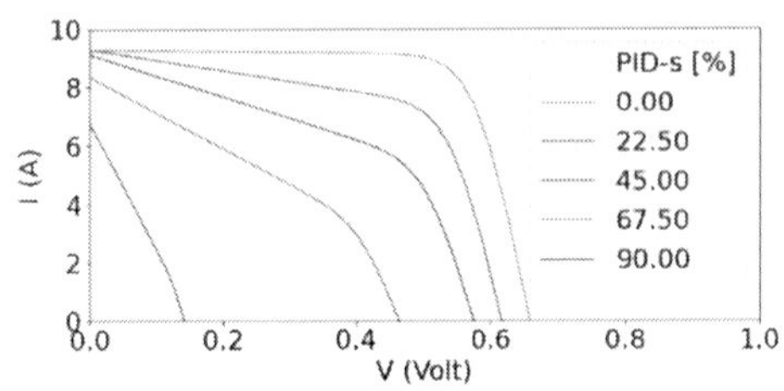

Figure 5: Solar cell IV-curves for different PID-s level.

Accuracy of this approach has been assessed comparing obtained P_{MPP} with expected P_{MPP} in view of the PID-s level (see Figure 6).

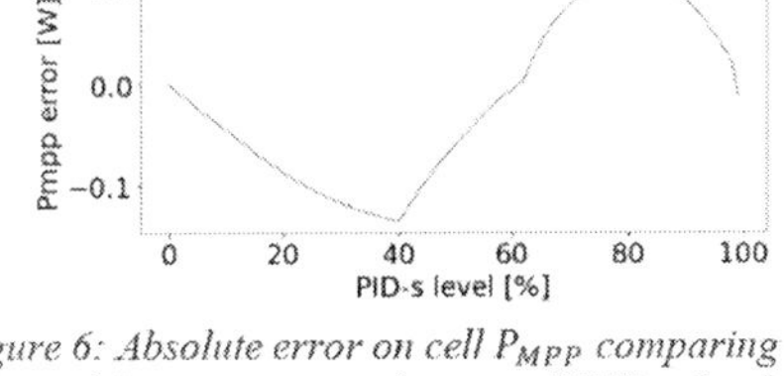

Figure 6: Absolute error on cell P_{MPP} comparing with expected P_{MPP} corresponding to cell PID-s level. The P_{MPP} at STC and 0% PID-s is 4.5W.

5 RESULTS AND VALIDATION

First validation is conducted on a 22 modules string located south of France, oriented south and tilted by 25°. The PV polycrystaline 60-cells modules in the system are rated for 235 Wp and have three bypass diodes distributed among three substrings. The front glass and EVA encapsulant thicknesses are considered to be 3.2 mm and 0.3 mm, respectively.

The time period between the 25/03/2011, i.e. the first day of operation of the string, and the 06/10/2019 was simulated using a 1h time resolution. Input in plane irradiance and ambient temperature were measured onsite, at the exception of 157 days of missing measured data filled with the ERA5 climatic dataset [22]. Wind speed, wind direction and ambient RH are retrieved from the same ERA5 dataset for the nine years time period.

The simulated maximum DC power among one year drops quickly from 5 to 3 kW during the first two years and then keeps slowly decreasing to reach ~1.5 kW at the end of the nine years (see Figure 7).

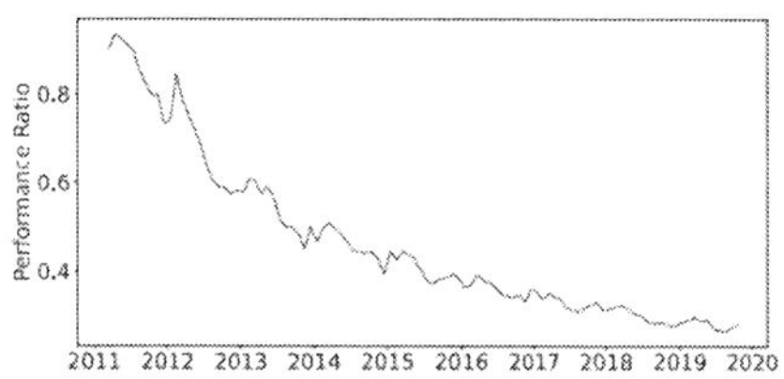

Figure 7: Simulated monthly PR for string DC power decreases from 0.935 to 0.284 due to PID-s.

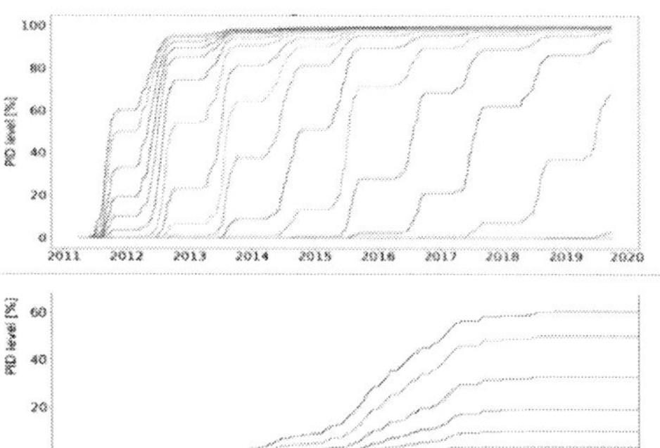

Figure 8: Evolution of per module average of the cells PID-s levels in order of distance from negative polarity of the inverter.

As expected, modules undergoing higher frame to cell voltage drops suffer from faster PID-s evolution (see Figure 8). Modules PID-s levels increase sharply during summer but remain constant during winter. This is explained by the high ambient RH (on average 69% RH during daylight over the nine years) in the region, which makes the glass surface resistance negligible compared to the EVA resistance. Therefore, PID-s evolution is mainly driven by the EVA temperature which increases highly during summer (see Figure 9). Due to higher temperatures, the EVA electrical resistance is up to 10 times lower in summer explaining the fast increase of PID-s during this period.

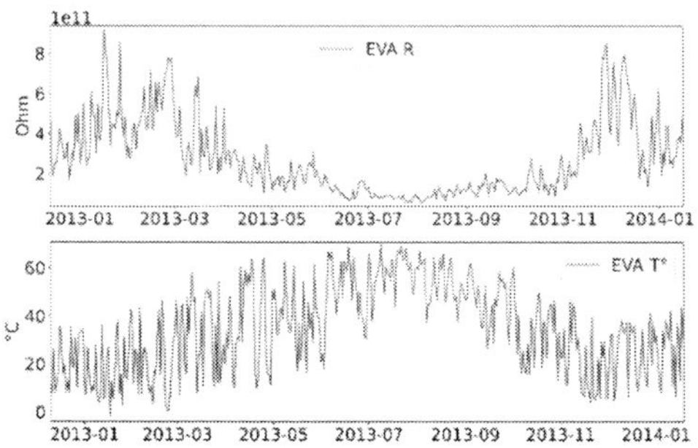

Figure 9: daily maximum of the EVA electrical resistance and temperature in year 2013.

At the end of the nine years simulation, five modules end up not being affected by PID-s, four have a PID-s level between 1% and 99% and 13 reached a 100% PID-s level. A comparison between measured and simulated final modules PID-s levels for six modules is presented in Table 1. From this comparison it appears that simulation overestimates the impact of PID-s degradation. A possible explanation is the lack of model for PID regeneration.

Position	Final PID-s levels	
	Measured	Simulated
Most +	0%	0%
2nd most +	2%	0%
4th most -	53%	100%
3rd most -	78%	100%
2nd most -	100%	100%
Most -	100%	100%

Table 1: Measured versus simulated final PID-s levels for the two and four modules closest to the positive and negative voltage inverter plugs, respectively.

6 CONCLUSION AND FUTURE WORK

An advantage of the presented grey-box approach over fully empirical models is the possibility to investigate the effect of changes in the module layers properties without experimental data to calibrate. First results of obtained PID-s levels for the studied string show interesting correlations between PID-s degradation and environmental stress factors. Future work will be to further validate the model, to implement a water ingress model, spontaneous PID-s recovery and assess the generality of the empirical part of the model for different Si cells (PID-s evolution with I_{leak}, V_{OC} and I_{SC} evolution with PID-s level).

7 ACKNOWLEDGEMENTS

This work was executed partially within the imec.icon project ANALYST PV, partially within the DAPPER projet and received funding from the European Union's Horizon 2020 research and innovation program under grant agreement N952957. ANALYST PV is a research project bringing together academic researchers (imec-PV and imec-IPI-Ghent) and industry partners (3E, Sitemark, AllThingsTalk and Engie Laborelec). The ANALYST PV project was co-financed by imec and received project support from Flanders Innovation & Entrepreneurship (project nr. HBC.2019.0050) and Innoviris. The DAPPER project is financed by Flux50 and Flanders Innovation & Entrepreneurship (project nr. HBC.20202144). The work in this paper was also partially funded by the Kuwait Foundation for the Advancement of Sciences under project number CN18-15EE-01.The authors would like to thank ECMWF for providing the ERA5 dataset generated using Copernicus Climate Change Service Information 2020.

REFERENCES

[1] Viktor Rudolf, Konstantinos D. Papastergiou, Financial analysis of utility scale photovoltaic plants with battery energy storage, Energy Policy, Volume 63, 2013, Pages 139-146, ISSN 0301-4215.

[2] Horvath Imre, Goverde Hans, Manganiello, Patrizio, Schils Arnaud, van der Heide Arvid, Govaerts Jonathan, Voroshazi Eszter, Yordanov Georgi, Moschner Jens et al., Next Generation Tools for Accurate Energy Yield Estimation of Bifacial PV Systems – Best Practices, Improvements and Challenges, 2019.

[3] kWh analytics, Solar Risk Assessment: 2020, Quantitative Insights from the Industry Experts, [https://www.kwhanalytics.com/blog-archive/solar-risk-assessment-2020-quantitative-insights-from-the-industry-experts].

[4] Jorne Carolus, John A. Tsanakas, Arvid van der Heide, Eszter Voroshazi, Ward De Ceuninck, Michaël Daenen, Physics of potential-induced degradation in bifacial p-PERC solar cells, Solar Energy Materials and Solar Cells, Volume 200, 2019, 109950, ISSN 0927-0248.

[5] P. Hacke et al., Accelerated Testing and Modeling of Potential-Induced Degradation as a Function of Temperature and Relative Humidity, in IEEE Journal of Photovoltaics, vol. 5, no. 6, pp. 1549-1553, Nov. 2015, doi: 10.1109/JPHOTOV.2015.2466463.

[6] Annigoni Eleonora, Jankovec Marko, Galliano Federico, Li Heng-Yu, Perret-Aebi Laure-Emmanuelle, Topic Marko, Sculati-Meillaud Fanny, Virtuani Alessandro, Ballif Christophe, Modelling potential-induced degradation (PID) in crystalline silicon solar cells: from accelerated-aging laboratory testing to outdoor prediction, 2016, 10.4229/EUPVSEC20162016-5BO.11.2.

[7] Hattendorf J., Löw R., Gnehr W.-M, Wulff L., Koekten M.C., Koshnicharov D., Blauaermel A., Esquivel J.A., Potential Induced Degradation in Mono-Crystalline Silicon Based Modules: An Acceleration Model, 2012, 10.4229/27thEUPVSEC2012-4BV.2.51.

[8] Imre T. Horváth, Hans Goverde, Patrizio Manganiello, Jonathan Govaerts, Loic Tous, Bader Aldalali, Eszter Vörösházi, Jozef Szlufcik, Francky Catthoor, Jef Poortmans, Photovoltaic energy yield modelling under desert and moderate climates: What-if exploration of different cell technologies, Solar Energy, Volume 173, 2018, Pages 728-739, ISSN 0038-092X.

[9] Goverde H., Herteleer B., Anagnostos D., Köse G., Goossens D., Aldalali B., Govaerts J., Baert K., Catthoor F., Driesen J., Poortmans J., 2014, Energy Yield Prediction Model for PV Modules Including Spatial and Temporal Effects. PVSEC, EU.

[10] Anagnostos, D., Goverde, H., Catthoor, F., Soudris, D., 2014, Presentation of a VerilogAMS model for detailed transient electro-thermal simulations of PV modules and systems. Presented at the 29th EU PVSEC.

[11] Luo, Wei and Khoo, Yong Sheng and Hacke, Peter and Naumann, Volker and Lausch, Dominik and Harvey, Steven P. and Singh, Jai Prakash and Chai, Jing and Wang, Yan and Aberle, Armin G. and Ramakrishna, Seeram, Potential-induced degradation in photovoltaic modules: a critical review, Volume 10 of Energy Environ. Sci., The Royal Society of Chemistry, 2017

[12] Volker Naumann, Otwin Breitenstein, Klemens Ilse, Matthias Pander, Kai Sporleder and Christian Hagendorf, Increase of PID susceptibility of PV modules under enhanced environmental stress, Volume 22 of PV Tech Power, 2020.

[13] Takata M., Tomozawa M., Watson E.B., 1980, Electrical Conductivity of $Na_2O\ 3SiO_2$ Glasses with High Water Content. Journal of the American Ceramic Society, 63: 710-712.

[14] Mon, G. R. and Ross, R. G., Jr. and Whitla, G. and Orehotsky, J., Predicting electrochemical breakdown in terrestrial photovoltaic modules, 17th Photovoltaic Specialists Conference, 1984, pages 682-692.

[15] Naumann, Volker & Ilse, Klemens & Hagendorf, Christian. (2013). On the Discrepancy between Leakage Currents and Potential-Induced Degradation of Crystalline Silicon Modules. 10.4229/28thEUPVSEC2013-4DO.3.2.

[16] Koehl, M., Heck, M. and Wiesmeier, S. (2018), Categorization of weathering stresses for photovoltaic modules. Energy Sci Eng, 6: 93-111. [https://doi.org/10.1002/ese3.189]

[17] David H. Otth and Ronald G. Ross Jr., Assessing photovoltaic module degradation and lifetime from long term environmental tests, 1983, Proceedings of the 1983 Institute of Environmental Sciences 29th Annual Meeting Los Angeles, pages 121-126.

[18] Utah University, Basic Electricity and Magnetism 3910: Current Flow in Ohmic Resistors, [http://www.physics.utah.edu/~rprice/PMT/ohmic.pdf]

[19] Jorne Carolus, A study on potential-induced degradation: from conventional to emerging photovoltaic technologies, Faculty of Engineering Technology, University of Hasselt, Belgium, 2019.

[20] J. Carolus, W. De Ceuninck and M. Daenen, "Irreversible damage at high levels of potential-induced degradation on photovoltaic modules: A test campaign," *2017 IEEE International Reliability Physics Symposium (IRPS)*, Monterey, CA, 2017, pp. 2F-5.1-2F-5.6, doi: 10.1109/IRPS.2017.7936275.

[21] Carolus, Jorne & Govaerts, Jonathan & Voroshazi, E. & De Ceuninck, Ward & Daenen, Michael. (2017). Voltage dependence of potential-induced degradation and recovery on photovoltaic one-cell laminates.

[22] Hersbach, H., Bell, B., Berrisford, P., Biavati, G., Horányi, A., Muñoz Sabater, J., Nicolas, J., Peubey, C., Radu, R., Rozum, I., Schepers, D., Simmons, A., Soci, C., Dee, D., Thépaut, J-N. (2018): ERA5 hourly data on single levels from 1979 to present. Copernicus Climate Change Service (C3S) Climate Data Store (CDS). (Accessed on < 23-11-2020 >), 10.24381/cds.adbb2d47

DEPENDENCY OF IAM LOSSES IN COLORED BIPV PRODUCTS ON THE REFRACTIVE INDEX OF COLORANTS

Markus Babin*, Sune Thorsteinsson, Adrian A. Santamaria Lancia, Peter B. Poulsen, Anders Thorseth, Carsten Dam-Hansen, Michael L. Jakobsen
Department of Photonics Engineering, Technical University of Denmark
Frederiksborgvej 399, 4000 Roskilde, Denmark
*Corresponding Author: marbab@fotonik.dtu.dk

ABSTRACT: Measurements of the incidence angle modifier (IAM) of colored PV modules based on inkjet-printed glass have shown differences between colors, which are uncorrelated with the total transmittance and reflectance of the colored glass. In this paper, the complex refractive index of ceramic inks is estimated based on measurements of reflectance and transmittance spectra. These estimates are used in ray-tracing simulations to determine the IAM of coupon-sized PV modules with printed ceramic inks. Comparisons of the simulation results to measured IAMs show a high correlation between the refractive index of colorants and the IAM.
Furthermore, estimations of the impact of IAM on performance based on TMY data at different locations are performed. The results indicate that changes in IAM due to sample coloration can lead to annual irradiance losses of up to 5% compared to uncolored glass. These losses are caused by additional reflections at the glass-ink interface according to the Fresnel equations, dependent on the refractive index of the ceramic inks.
Keywords: Building Integrated PV (BIPV), Optical Losses, Performance, Ray Tracing, Color

1 INTRODUCTION

Building-integrated photovoltaics (BIPV) present a huge potential market for PV deployment and its costs are already approaching competitiveness with conventional building materials. [1] Furthermore, BIPV eliminates the need for increased land usage and, due to its distributed nature, reduces the need for grid expansion by achieving physical proximity between generators and loads. [2]

One major barrier limiting the deployment of BIPV are aesthetic requirements, e.g. through building codes in city centers or cultural heritage sites. This issue is tackled by a variety of coloration technologies attempting to tailor BIPV appearance to the architectural requirements.

Coloration is usually achieved through spectrally selective absorption and reflection in a material between the glass cover and the PV cell, or through thin-film coatings leading to constructive or destructive interference of reflected light. Absorptive materials, either printed on the glass cover or introduced within the laminate have shown to achieve better color saturation and color constancy independent of incidence or view angle, but a significantly higher transmission loss compared to thin film coatings. [3] Nevertheless, most colored BIPV solutions on the market currently rely on this technology.

In a previous comparison of differently colored, inkjet-printed glasses, a significant difference in angular dependent transmittance – described by the incidence angle modifier (IAM) – was observed. This phenomenon is however correlated with neither the total transmittance at normal incidence, nor the primary wavelength of reflected or transmitted light. [3] As a possible alternative factor, this paper therefore investigates the correlation between the spectrally varying refractive index of the ceramic ink colorant and the measured IAM.

2 METHODOLOGY

For this paper, reflectance and transmittance spectra are measured on printed glass samples and used to estimate the wavelength dependent complex refractive index of the colorants. This data is then used as input in raytracing simulations used to predict the IAM of PV laminates with said colored glass. The simulation results are subsequently compared to measurements of the IAM of laminated samples. Finally, calculations based on TMY data are carried out to estimate the impact of the measured IAM on annual PV yields.

2.1 Sample preparation

In total, seven coupon-sized samples are investigated. The sample cover consists of a 200x300x3.5 mm low-iron float glass with ceramic ink printed on the inner side. The colored layer has a thickness of approximately 20 μm and is printed using a DipTech inkjet printer and thermally fused with the glass. The layer is printed in five primary colors (white, blue, green, red, yellow) and one compound color (terracotta).

The printed glasses (printed side facing the laminate) are used as superstrate for single-cell modules featuring mono-Si cells laminated on a black polymeric backsheet with EVA as encapsulant.

2.2 Measurements of the IAM

The main illuminating light source for all measurement setups is an Energetiq EQ-99X laser-driven light source (LDLS). This light source emits broadband light with high temporal stability. The light is collimated using three off-axis parabolic mirrors and an aperture to filter for divergent light. Furthermore, the setup includes a long-pass UV filter with a cut-off wavelength at 320 nm.

The samples themselves are mounted in a rotary stage with stepper motor, with a rotational axis aligned with the center of the light beam, allowing accurate variations of the incidence angle. I_{sc} of samples is determined through measurements of the voltage across a 0.5 Ω shunt resistor, with measurements averaged over 20 individual acquisitions to reduce noise. Simultaneously the temperature is monitored through a J-type thermocouple affixed to the center of the module backside. A schematic of the measurement setup is shown in Figure 1.

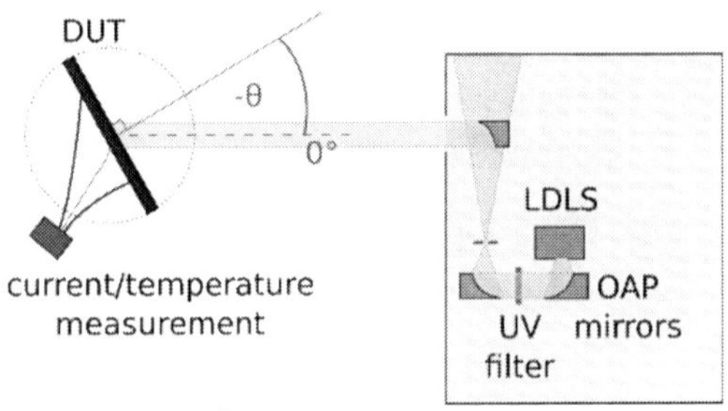

Figure 1: Setup for IAM measurements

For measurements of the IAM, an aperture of 10 mm diameter is placed within the collimated beam, so that the spot of incident light can be contained within the PV cell at all measured angles. This alleviates the need for cosine corrections, as the total incident flux on the sample is kept constant for all angles. Even though the cell is not fully illuminated as prescribed in the IEC 61852-2 standard [4], the measurement method shows high accuracy, as has been validated through inter-laboratory comparison. [5]

Four measurements in 5° steps are carried out per sample and averaged, including averaging over positive and negative incidence angles. In addition to the absolute IAM, a relative IAM is determined for colored samples as a relative measure compared to the unprinted reference:

$$rel.\ IAM_{sample} = \frac{IAM_{sample}}{IAM_{reference}}$$

2.3 Measurements of reflectance

For reflectance measurements, a 2" integrating sphere is used to capture both specular and diffuse reflections from the sample. For this, the collimated light from the LDLS is reflected towards the input port located at 4° compared to the surface normal. The reflected light is collected in the sphere and transferred to an OceanOptics QE65000 spectrometer using a 600 µm optical fiber, as can be seen in Figure 2. This spectrometer allows measurements of the spectral flux between approximately 200 and 980 nm with varying integration times.

The incident light is characterized using a LabSphere Spectralon reflectance standard with known spectral reflectivity. Subsequently, reflectance is measured on samples with the inkjet printed side facing towards the integrating sphere. A black backsheet is placed several mm below the glass surface to avoid additional reflections from transmitted light.

For each measurement, five spectra are acquired with five seconds integration time each and averaged, to reduce measurement noise. Additionally, for each measurement, dark signal spectra are acquired and subtracted.

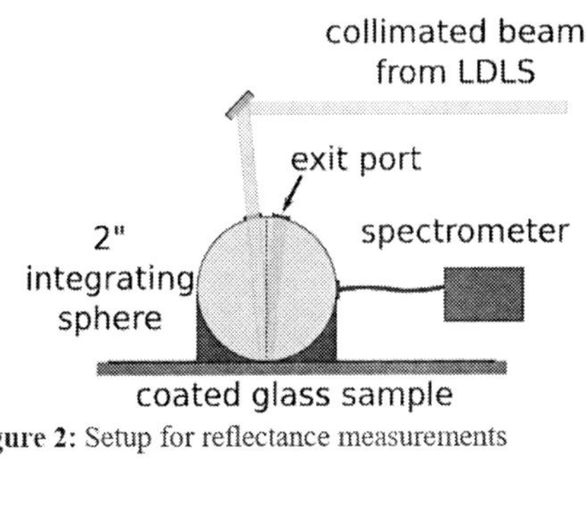

Figure 2: Setup for reflectance measurements

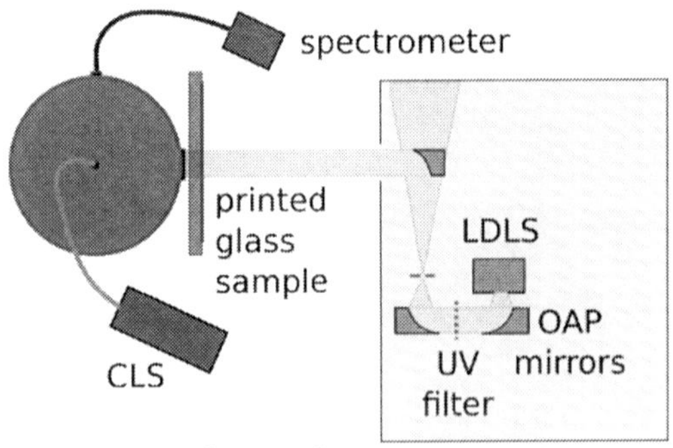

Figure 3: Setup for transmittance measurements

2.4. Measurements of transmittance

Transmittance measurements again use the same light source described in Section 2.2, adding a 6" integrating sphere directly into the collimated light path. Here, the full diameter of collimated light, 1.5", is used. As shown in Figure 3, a correction light source (CLS), an OceanOptics DH2000 halogen-deuterium lamp, is connected using a 400-µm optical fiber to the secondary port of the integrating sphere to allow compensation for back-reflections from the sample. Spectra of the transmitted light are acquired using the aforementioned OceanOptics QE65000 spectrometer, connected via a 600-µm optical fiber to the baffled port of the integrating sphere.

Measurements are carried out for all samples with an integration time of 2.5 seconds and averaging over 5 measurements. In addition to measurements with the LDLS and dark background acquisitions, measurements with the CLS are carried out. Additionally, spectra without any samples mounted are acquired for reference purposes.

The transmittance through measured samples is calculated according to the following equation to compensate for back-reflections from the glass surface to the inside of the integrating sphere [6]. Here ϕ_x refers to measurements with the coated sample in place, while ϕ_0 refers to reference measurements without any samples:

$$T_x = \frac{\phi_{LDLS,x}}{\phi_{CLS,x}} \frac{\phi_{CLS,0}}{\phi_{LDLS,0}}$$

2.5 Refractive index and extinction coefficient estimation

Using the spectral reflectance, the refractive index can be estimated according to the following Fresnel equation for normal incidence, giving the following relationship for a single interface between medium 0 and medium x:

$$R_{0\to x} = \left|\frac{n_{\lambda,0} - n_{\lambda,x}}{n_{\lambda,0} + n_{\lambda,x}}\right|^2$$

While the inkjet-printed glass samples feature several interfaces, for an initial estimate of the refractive index of the ceramic ink, only the initial air-ink interface is considered.

In addition to the interface itself, absorption in the ink layer has to be considered. This absorption can be described by the extinction coefficient k derived from the Beer-Lambert Law. It can be estimated based on transmittance measurements and knowledge of the layer thickness:

$$T_\lambda = e^{-k_\lambda \frac{4\pi d}{\lambda}}$$

In this case, absorption in the glass substrate is neglected and the extinction coefficient is calculated based on the full transmittance of the printed samples.

2.6 Ray-tracing simulations and processing

The optical stack in the colored samples is simulated in OpticStudio by Zemax with the following simplifications: All surfaces are considered to be flat and all objects to have a constant thickness. Additionally, reflections on the detector surface, representing the PV cell or backsheet, are neglected. The model is outfitted with detectors for transmitted light (located within the encapsulant) and reflected light.

Ray-tracing simulations are carried out for incidence angles from 0° to 80° in 1° steps and for wavelengths ranging from 320 to 1100 nm in 10 nm steps. For each configuration, 1000 rays of randomized polarization are generated and traced through the optical stack, splitting rays into reflected and transmitted components at each interface.

The flux incident on each of the detectors is obtained through analysis of the exported ray paths in MATLAB, averaged over wavelengths and compared for different angles, to give a total, angular dependent transmittance analogous to the IAM.

2.7 IAM impact simulations

In order to estimate the impact of IAM differences on the annual yield of BIPV installations, a simulation tool based on weather data is developed in MATLAB.

At a chosen location, the angle of incidence for direct and ground reflected light are calculated, as well as an effective incidence angle for diffuse light based on the installation tilt. [7] Direct and diffuse irradiance data for a typical meteorological year (TMY), obtained from PVGIS (ec.europa.eu/jrc/en/pvgis), are then multiplied by the IAM for each timestamp using a 5th-order polynomial fit to the calculated incidence angle.

The recomposed irradiance is then compared for modules with and without ceramic ink printing on the inside of the glass, resulting in a measure for the additional, color dependent annual reflection losses.

For exemplary purposes, calculations are carried out for Risø, Denmark (55.7°N, 12.1°E) and Syracuse, Italy (37.1°N, 15.2°E) and five different configurations with installation tilt ranging from horizontal to vertical.

3 RESULTS

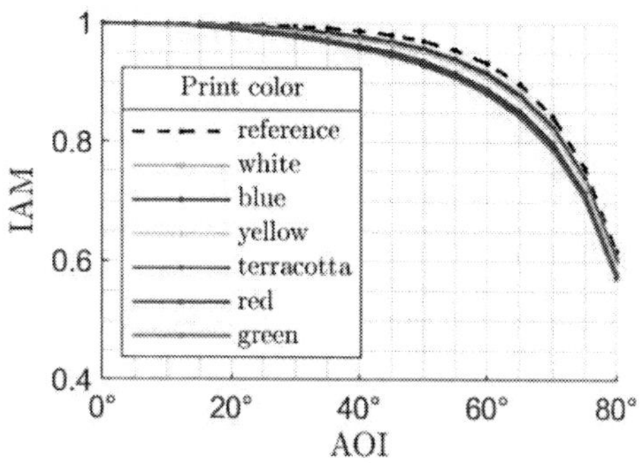

Figure 4: Measured absolute IAM

3.1 IAM Measurements

The results of the IAM measurements can be seen in Figure 4, showing the absolute IAM for incidence angles of 0° to 80°. Most noticeably, all colors show a significantly reduced IAM compared to the unprinted reference already at intermediary incidence angles, with highest losses for the red, green and terracotta colored samples. In contrast, the white, blue and yellow samples show a significantly lower reduction. This difference in colors becomes even more apparent in Figure 5, showing the relative IAM compared to the unprinted reference.

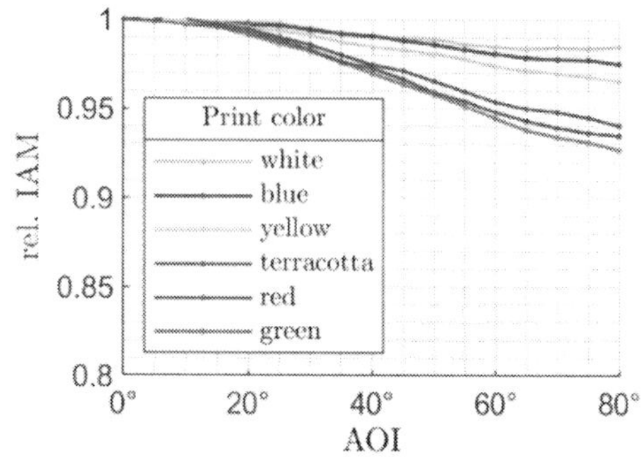

Figure 5: Measured relative IAM

Compared with measurements of the short-circuit current and/or maximum power of the PV cells as published previously [3], no direct correlation between relative IAM and total solar-weighted transmittance losses can be observed, as shown in Figure 6.

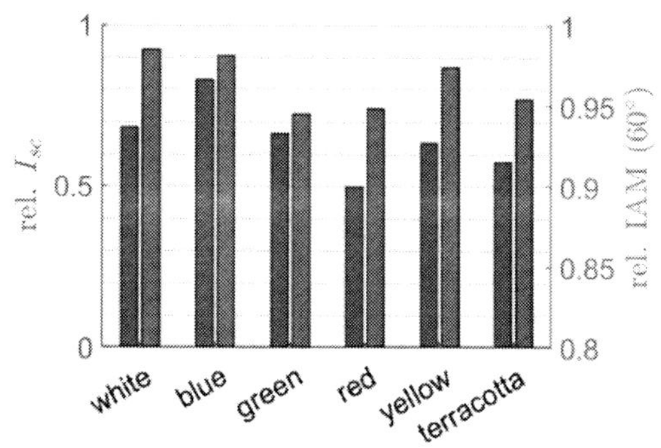

Figure 6: Comparison of relative Isc and relative IAM

3.2 Reflectance and transmittance

Both reflectance and transmittance are measured according to the descriptions in section 2.3 and 2.4. The resulting spectra are shown in Figures 7 and 8 respectively, showing the expected spectral variations for different colors.

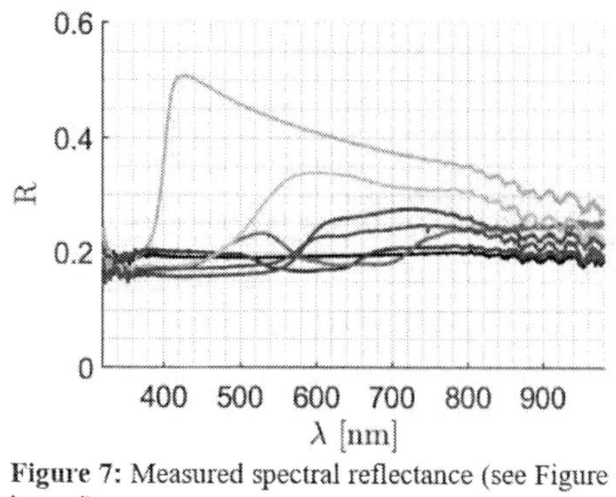

Figure 7: Measured spectral reflectance (see Figure 4 for legend)

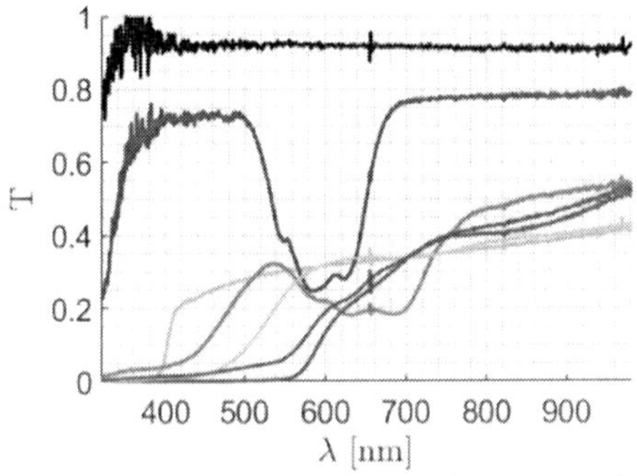

Figure 8: Measured spectral transmittance (see Figure 4 for legend)

Of special note is the blue colored sample, showing vastly increased transmittance compared to all other samples and low reflections over the entire spectrum. This can most likely be attributed to the specific pigment used in the ceramic ink, having different properties from those of the remaining colors.

3.3 Estimated refractive index

Figure 9 and 10 show the refractive index and extinction coefficient estimated from the reflectance and transmittance spectra. Beyond 980 nm, both n and k are extrapolated as constant values.

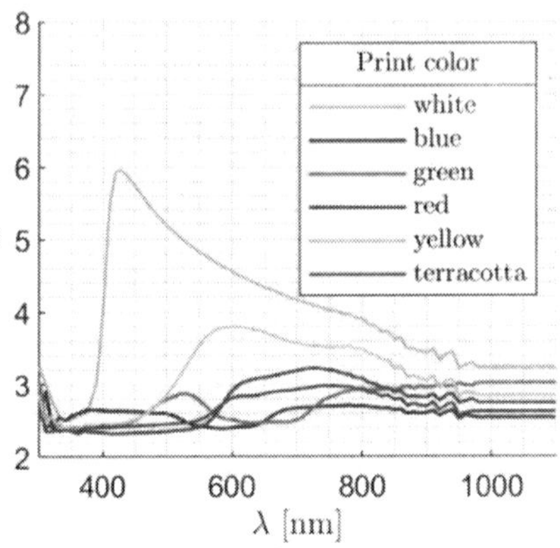

Figure 9: Estimated refractive index

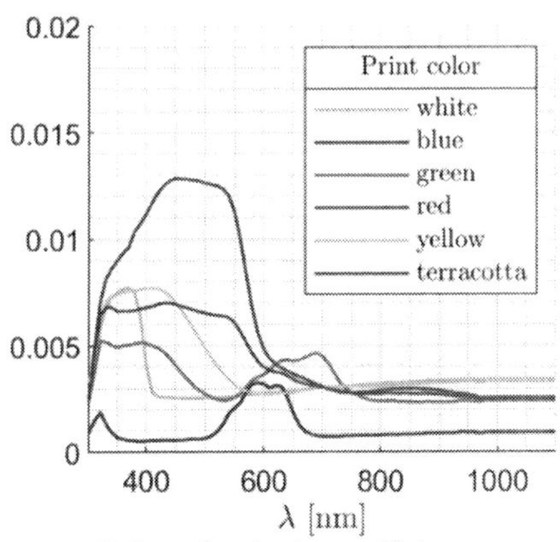

Figure 10: Estimated extinction coefficient

3.4 Raytracing results

The simulated absolute and relative IAM are shown in Figures 11 and 12 respectively. Immediately apparent are the oscillations in IAM over incidence angles, which can be attributed to variations in reflectance and transmittance at interfaces for differently polarized light as well as to interference effects between the wavelength of light and the thickness of the ink layer, differing only by approximately one magnitude. Accordingly, these effects are only visible in the raytracing simulations since no light scattering is considered, as diffuse light normally compensates for any spectral oscillations

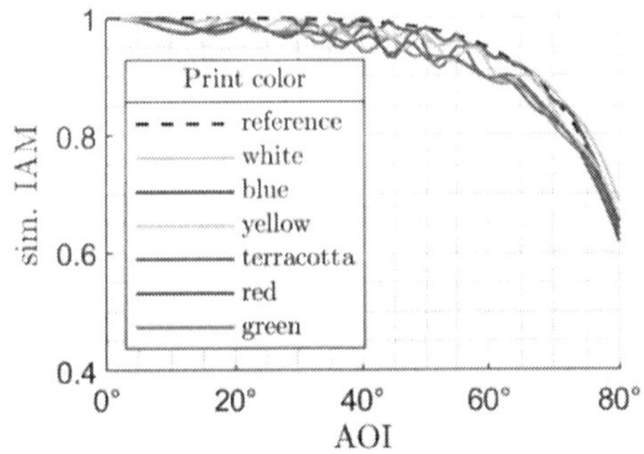

Figure 11: Simulated total IAM

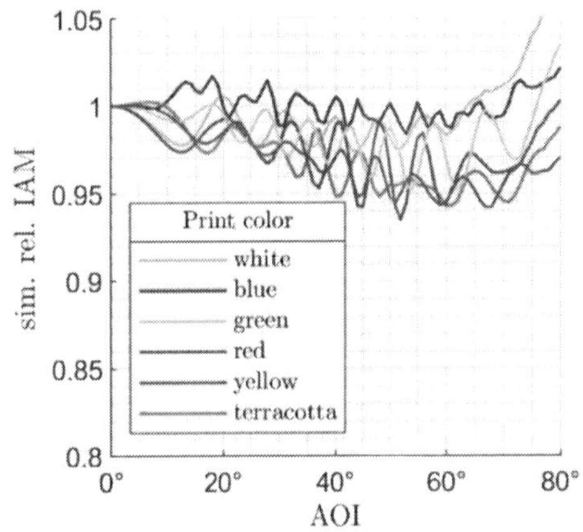

Figure 12: Simulated relative IAM

Apart from the oscillations, the simulated IAM curves generally agree well with the measurements, regarding both magnitude and trends for different colors. Only at high incidence angles does the simulation overestimate the IAM. This could possibly be attributed to changes in the reflectance and/or scattering behavior at high incidence angles, which are not considered in the simulation.

3.5 Yield simulations

For both locations, depending on color and installation tilt, the additional annual IAM losses range between approximately 1-5%, as can be seen in Figures 13 and 14.

For vertical installation, e.g. in façades, the worst performing colors show losses of more than 4%, and for the terracotta color, commonly used to mimic clay roof tiles, losses of up to 3% are evident for common roof tilts. Therefore, coloration dependent IAM losses can significantly affect BIPV system performance.

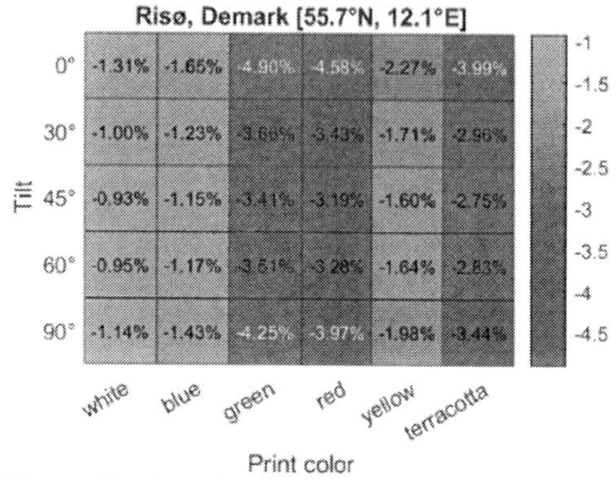

Figure 13: Annual additional reflection losses due to coloration in Risø, Denmark

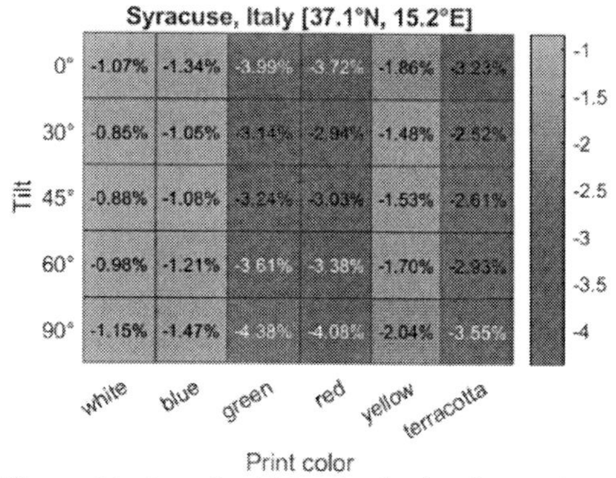

Figure 14: Annual additional reflection losses due to coloration in Syracuse, Italy

4 DISCUSSION

4.1 Other factors influencing the optical transmission

Changes in the IAM due to color could also be attributed to differences in light scattering, both for transmitted and reflected light. The diffusing properties of the ceramic inks may very well differ between colors, leading to different amounts of diffuse and specular components. Measurements of the bi-directional reflectance distribution function (BRDF) indicate a possible correlation between reflectance or transmittance specularity with the IAM. [3]

In addition, fluorescence effects may explain the very high reflectance of the white ink in Figure 6. Since a broadband light source is used for all spectral measurements, fluorescence cannot be identified from the measurements and may lead to errors in the estimated refractive index.

Finally, interactions between the pigments in the ceramic ink and the glass surface may affect the accuracy of results. During the thermal fusing of glass and ink, pigments may diffuse into the glass substrate, creating a refractive index gradient throughout the interface. Furthermore, pigment particle sizes may have an influence through interactions with light, leading to increased absorption or scattering.

4.2 Power yield estimations and PV module testing

As shown in section 3.5, the choice of color can have a significant impact on the annual yield of BIPV installations due to the differences in IAM. This has implications for both estimations and simulations of the power yield as well as for manufacturing and testing of BIPV products:

First of all, the choice of coloration has to be taken into account when estimating BIPV system performance. While the impact of coloration materials on the overall transmittance is well established, these observations indicate a requirement of new IAM measurements for every new color selection.

Secondly, these IAM measurements have to be made accessible for yield simulations, as otherwise BIPV system performance may be significantly overestimated. Furthermore, software used for BIPV yield prediction must accommodate more accurate modelling of the IAM to allow for accurate results.

5 CONCLUSIONS

In this work, the refractive index of ceramic inks used in colored BIPV modules was estimated and used in raytracing simulations to model the IAM. The simulations showed a high correlation between refractive index and IAM, indicating that the refractive index is one of the major driving force for these differences.

As alternative factor leading to different IAM, the diffusing properties of the ceramic inks, leading to light scattering, may be considered. As a next step, light scattering will be implemented in the raytracing model based on BRDF measurements to include these properties as well.

Finally, annual yield simulations have been shown that indicate additional reflection losses in colored BIPV modules of up to 5%, depending on color choice, location and orientation, compared to uncolored modules.

6 ACKNOWLEDGEMENTS

The authors would like to acknowledge the support from EUDP as a funding agency for project "64019-0029 PV-ROCK-ROOF". This work was made possible through the Zemax smart research program with the use of OpticStudio under the license number L112498.

7 REFERENCES

[1] BIPVboost, BIPV Solutions in Europe: Competitiveness Status & Roadmap towards 2030 – White Paper, Report, Becquerel Institute (2021)

[2] M.J. Guillén et al., Solar Skins: An opportunity for greener cities, Report, SolarPower Europe and ETIP PV (2019)

[3] M. Babin, A.A. Santamaria Lancia, S. Thorsteinsson, A. Thorseth, Characterisation of Angular Dependent Optical Properties of Different Coloring Technologies Employed in BIPV Products, 37th EUPVSEC (2020) pp. 1136-1142

[4] Photovoltaic (PV) module performance testing and energy rating – Part 2: Spectral responsivity, incidence angle and module operating temperature measurements, Standard, IEC 61853-2:2016 (2016)

[5] N. Riedel Lyngskær et al., Interlaboratory Comparison of Methodologies for Measuring the Angle of Incidence Dependence of Solar Cells, 35th EUPVSEC (2018) pp. 1034-1039

[6] A. Tirpak, R. Young, Accurate transmission measurements of translucent materials, Report, Optronic Laboratories (2008)
[7] M.J. Brandemuehl , W.A. Beckman, Transmission of diffuse radiation through CPC and flat plate collector glazings, Solar Energy vol. 24, no. 5 (1980) pp. 511-513

DARK IV-CURVES AS A METHOD FOR IN SITU MODULE CHARACTERISATION

Jens Froebel*, Bengt Jaeckel, Matthias Pander, Ulli Zeller
Fraunhofer Center for Silicon Photovoltaics, Otto-Eissfeldt-Str. 12, 06120 Halle (Saale), Germany
phone: +49-345-5589-5211, fax: +49-345-5589-5999, jens.froebel@csp.fraunhofer.de

ABSTRACT: For PV module characterization and intermediate assessment during a long-term stress test, the test has typically to be interrupted and the modules are brought to standard test conditions, which brings a number of disadvantages. The interruption of the test inevitably involves temperature changes and relaxation time. It is not possible to measure several test specimens at the same time, and the measurement is performed without the presence of the load. Therefore, methods are being considered to obtain information in-situ about the condition / health of the sample during the stress test.

Measuring the dark IV curve of PV modules is a method to obtain information about the current state of the module even during an ongoing stress test. Depending on test conditions and setup dark IV curves can be measured easily. However, care must be taken not to influence sample temperature or to induce other degradation due to external applied voltages. Dark IV in-situ measurements generally allow detection of parameter changes (current I and voltage V) and thus in-situ analysis of tests with changing stresses, such as thermal cycling, humidity freeze, damp heat, PID degradation/recovery, LeTID or cyclic mechanical loading tests with a high number of cycles can be performed.

The information gain depends on type of defect and not all defects can be detected as for example cell cracking and current mismatching of strings. Nevertheless, various types of defects found in the light IV curves of a module can be detected by dark IV curve. The change in shunt resistance during PID testing or the changes of the module during LeTID testing can be well tracked utilizing dark IV in-situ measurements. Defective bypass can be analyzed during the stress test while checking reverse voltage dependence of dark IV curve. This procedure is also easily applicable outdoor to assess larger quantities of modules regarding defective bypass diodes. Some defects that are easily visible in light IV curves and Electroluminescence (EL) are not easily detectable by the dark IV curves. Therefor conversion methods for temperature and irradiance should be used with care but can add information overall.

Generally, dark IV curves offer the possibility to obtain data with higher temporal resolution, independent of time-consuming intermediate measurements and allow therefor to monitor stress tests in more detail and possibly detect defects and degradation earlier. From this, improved stress models can be developed, e.g., to more accurately answer questions about the number of maximum stress cycles, i.e., when solder joints break, or whether degradation converges, i.e. during PID- recovery or extended LeTID tests.

Keywords: Dark IV, Defects, PV Module

1 INTRODUCTION

For the assessment of condition of a PV module, the power measurement with a solar simulator is essential. The requirements for the measurements are contained in IEC 60904 [1] and are challenging depending on the desired accuracy and reproducibility. Furthermore, there are time consuming interruptions and workloads for conditioning and intermediate measurements of modules when these are characterized in the course of long-term tests. The interruption of the tests also inevitably adds temperature change und relaxation time. A number of test specimens cannot be measured simultaneously, and the measurement is made without the presence of the stress.

The measurement of PV module IV characteristics in the dark is therefore an interesting option to determine the electrical condition of modules even during a running test [2]. The use of dark IV (DIV) curves in solar cell analysis relies on the principle of superposition. That is, in the absence of resistive effects, that the light IV curve is the dark IV curve shifted by the light generated current [3]. This is a good approximation for most cases, except e.g. for very high series resistance or when carrier lifetime is a function of injection level (e.g. for defective multi-crystalline materials). After removing internal resistances, it is possible to complement and match light and dark IV curves [4]. Several defects that can be found in the light characteristic of a module can also be detected in the dark IV characteristic. A great example is the change in shunt resistance as a reliable indicator for potential induced degradation (PID) [5]-[8]. In the course of PID stress test, observations about the course of degradation can be made already during the application and without having to switch off the test conditions [9]. Since it is already obligatory to perform a current or continuity test during the several IEC tests such as temperature change (TCT) and humidity-freeze (HF), supplementary dark measurements during the test procedure are also feasible here [10]. DIV has also been used as an analytical method for diagnosing low or inconsistent device efficiency of CIGS [11].

In this work, the possibilities of characterizing and monitoring photovoltaic modules by measuring DIV-curves are investigated. Dark IV-curve measurement is used to extract in-situ information about the current status of the module. Methods are investigated to determine module parameters such as series resistance and parallel resistance. Additionally, results are presented to determine the temperature coefficient for open-circuit voltage using DIV-curves, however, the results are non-conclusive.

2 METHODOLOGY

2.1 Experimental Setup

A Keithley Sourcemeter 2651A was used for detailed measurements of the dark IV characteristics. A disadvantage of this device was the limitation of maximum voltage of 40 V (close to V_{OC} of 60 cell equivalent PV Modules) and maximum power of 200 W, which translates to a maximum current of 5 A for most modules. This is unfavorable well below short circuit current of current generation solar cells (approx. 9 - 10 A I_{SC}). For monitoring during climate chamber experiments the internal monitoring of commercially available TDK Lambda Genesys 150-10 power supplies was used. The correctness of the generated values was verified with a calibrated voltmeter.

To compare DIV results all samples were initially, intermediated and final characterized at STC and 200W/m² using a AAA class solar simulator from Berger Lichttechnik followed by EL imaging.

2.2 Simulation and general behavior of parameters R_S, R_{sh} and temperature

The starting point for the theoretical description of a solar cell is the two-diode model [12].

$$I = I_{ph} - I_{o1}\left(e^{\frac{q(V+IR_S)}{n_1 k_B T}} - 1\right) - I_{o2}\left(e^{\frac{q(V+IR_S)}{n_2 k_B T}} - 1\right) - \frac{V + IR_S}{R_{sh}} \quad (1)$$

V – voltage
I_{ph} – photo current
I_{o1}, I_{o2} – saturation currents
n_1, n_2 – ideality factors
k_B – Boltzmann's constant
R_S – series resistance
R_{sh} – parallel resistance

For the behavior in the dark, the photocurrent is zero. The model was simulated in LTSpice. Here the parameters can be adjusted and the effects on characteristic curves can be calculated. We vary the parameters parallel resistance R_{sh}, series resistance R_S, temperature and short circuit current I_{SC}.

However, the simulation also has limitations, such as the missing temperature coefficient at the resistors. In addition, the behavior of the diodes in reverse direction is not modeled.

In the simulation, we can see the effects on the dark characteristic curve of a cell when changing R_S (Figure 1) and R_{sh} (Figure 2) in a linear (top) and semi-logarithmic (bottom) plot. The simulations nicely show where to expect the largest changes in the IV curve by change of R_S (high currents) or R_{sh} (very small currents). Additionally, the region of interest determines the requirements for the measurement resolution.

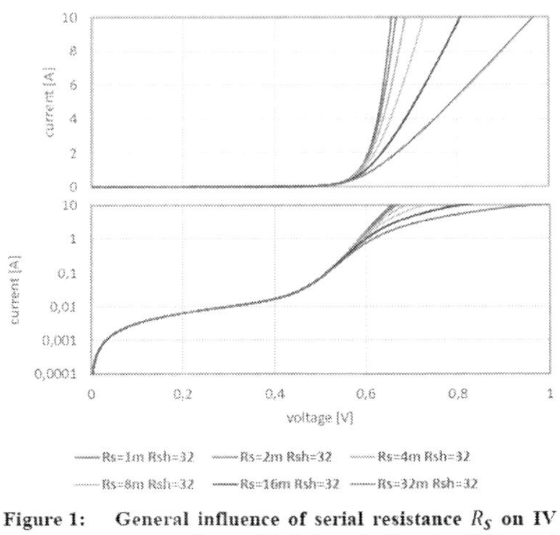

Figure 1: **General influence of serial resistance R_S on IV curves. R_S modified from 1 Ohm to 32 Ohm at constant R_{sh}**

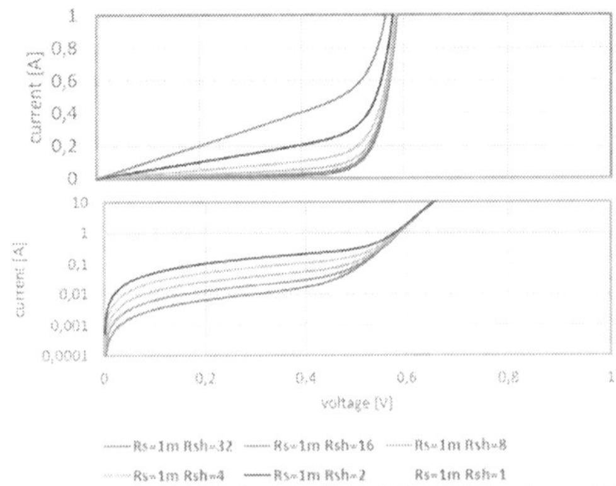

Figure 2: **General influence of parallel resistance R_{sh} on IV curves. R_{sh} modified from 1 Ohm to 32 Ohm at constant R_S**

A change of the parallel resistance can be measured significantly below the open-circuit voltage, for example at 1/3 V_{OC}. The effects of the change at the series resistance can be observed mainly above V_{OC}.

Note: The simulations work generally well for individual solar cells. As typical modules consist of 60 and more cells the measured IV curves are the result of all changes in all cells, both in R_S and R_{sh}. Small changes of an individual cell can't be expected to be detectable.

As PV modules are typically not operated at a constant temperature and also are tested at varying temperatures it is interesting to look for the temperature dependence of IV curves. Limitations exist within LTSpice like implementation of temperature coefficients for resistors. Still the general behavior can nicely be displayed (Figure 3).

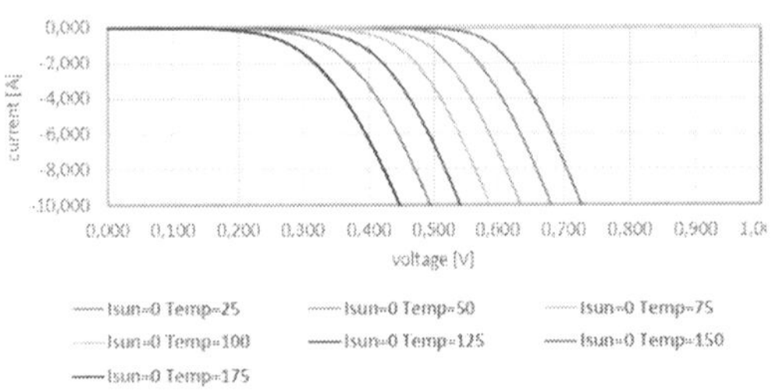

Figure 3: **Simulated dark IV-curves with variation on temperature**

3 RESULTS

3.1 Temperature dependent DIV

Based on the general temperature behavior of IV curves (Figure 3) an experiment was performed to utilize this idea to determine a temperature coefficient of a set PV mini module (see Figure 4-top). The Samples were heated to several temperatures and prior the DIV a wait time was implemented to assure that no significant temperature gradient exist between thermal couple (attached to backsheet) and solar cell. The advantage of this method is that multiple PV modules can be measured simultaneously and, more important, the temperature of the sample can be better determined as the inside and outside of the module are at thermal steady conditions. Standard temperature coefficient measurements done in a Flashtester typically have a temperature offset between solar cell and backsheet surface of a few degree, where the difference get smaller when temperature get close to ambient.

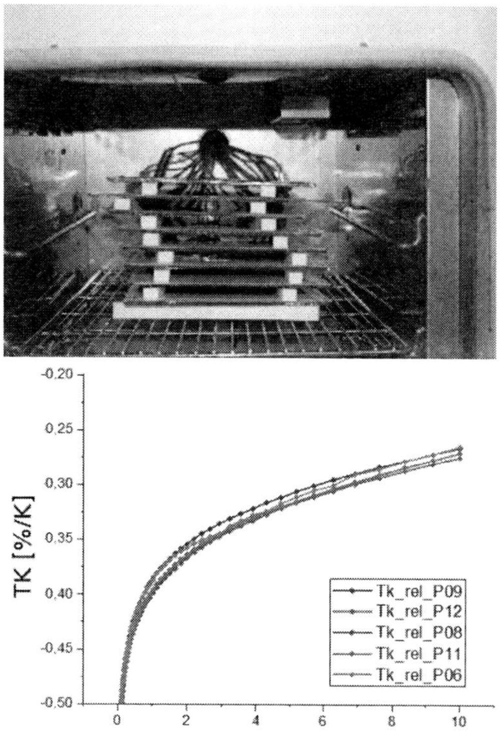

Figure 4: Top: Setup in an oven to modify temperature; bottom: resulting temperature coefficients determined from DIV curves

Form the taken DIV curves voltage readings were taken at varying currents (Figure 4-bottom). Obviously there seems to be a strong current dependent of the determine temperature coefficient. This is not in agreement with results from Light IV curves. Therefor the method seems not to work properly or at least needs further work.

3.2 Cracked cells

During Thermal cycling test (TCT) or mechanical load tests cells break and parts can get disconnected. A monitoring current is applied to check for continuity of the electric circuitry. To better understand cell breakage and modifications during such testing DIV are easy to implement. To validate the idea a severely damaged module was used as reference. Flash tests, DIV and EL image were taken (Figure 5 and Figure 6). The DIV (black) and the shifted DIV curve (pseudo Light IV – dotted black) shows a smooth curve. Both light IV curves show several knees (marked with ①② for 1000 W/m^2 and 200 W/m^2, respectively) that are a result from cell cracking and current mismatch compensated by the bypass diodes. None of these features is visible in the DIV. Based on that result DIV seem not be reasonable to check for cell cracking.

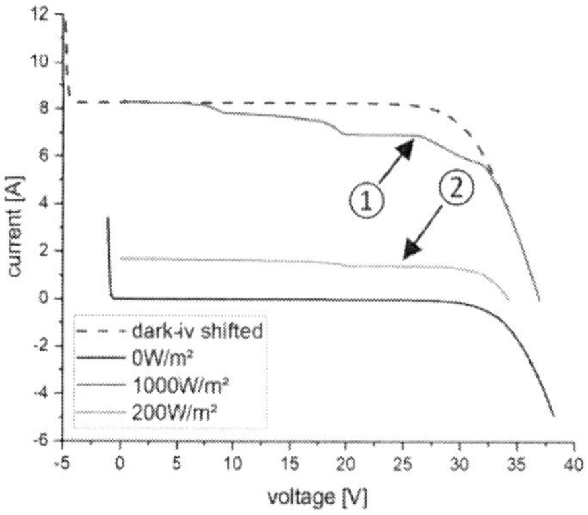

Figure 5: Illuminated (200W/m^2 and 1000W/m^2) and dark IV-curves on PV module with severe cell cracks

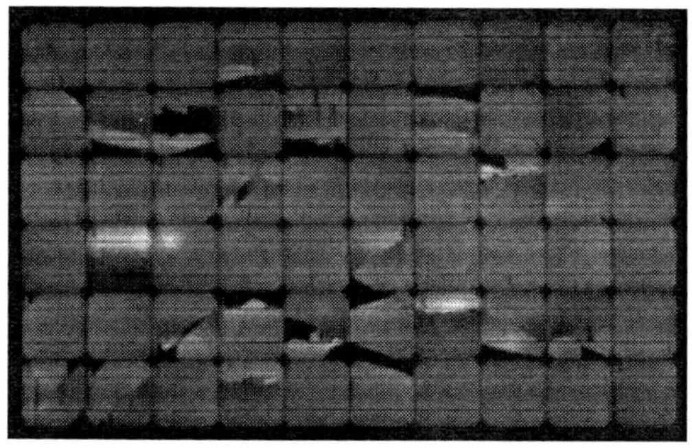

Figure 6: Electroluminescence image of a severely damaged PV-module with a lot of cell cracks including dark areas and ribbon breakage. The module showed significant power loss.

3.3 Track changes during extended stress testing

In most long-term stress tests no in-situ detailed performance monitoring is done. Intermediate sample characterization involves wait time to reach ambient conditions, sample handling and time for characterization. During that time samples can change as for example start to show recovery phenomena well known from PID testing. Therefor it is aimed to have as few test interruptions as possible. However, a method is needed to track changes. DIV are a suitable method to check for several defects.

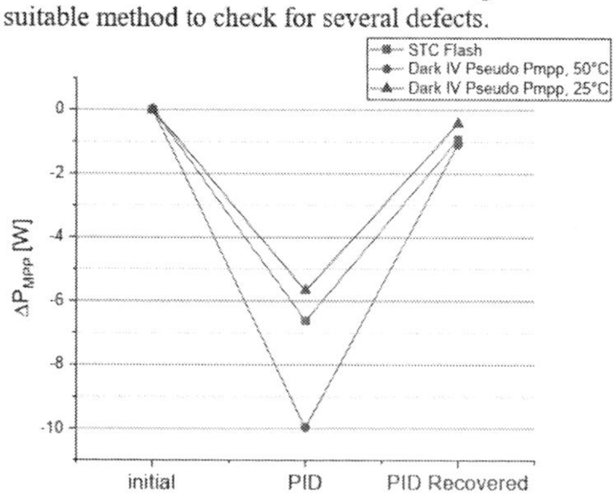

Figure 7: Change of STC power and Pseudo P_{MPP} for PID test

To compare DIV with Flash test results the superposition principle is applied where photocurrent (I_{SC}) is simply added to the DIV curve and a pseudo $P_{MPP,pseudo}$ can be determined. This works quite well for all case were a degradation is not affecting I_{SC}. An example taken from PID testing is shown in Figure 7 where change of P_{MPP} from DIV and STC-Flash are compared. The order of magnitude is well comparable. It can also be assumed from the 50°C data that some recovery may have already taken place during cooling. This method is suitable for tests with a termination condition of 80 % residual power especially during PID [5].

The direct transfer via adding I_{SC} current works only to a limited extend for degradations like LID or LETID were there is beside a loss in V_{OC} also a significant loss of I_{SC} during the degradation ([14],[15]). The total power loss or gain for a regeneration experiment is therefore underestimated. The change in measured DIV during a LETID regeneration experiment is shown in Figure 8. The increase in voltage during the experiment is well observed at constant climate chamber temperature. As stated before the improvement in Pseudo P_{MPP} is underestimated compared to the light STC flash (Figure 9).

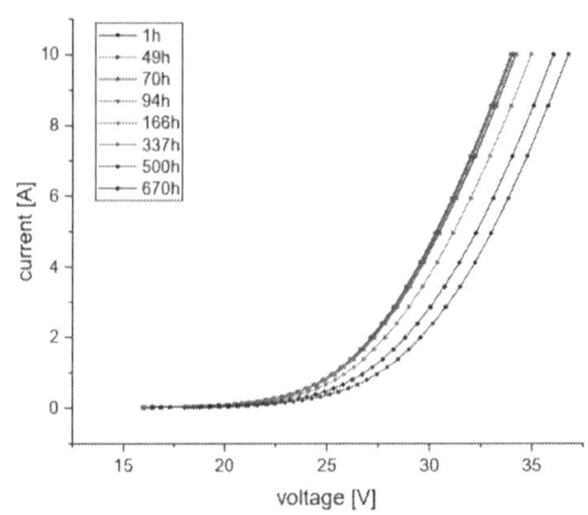

Figure 8: Dark-IV curves during LETID Regeneration experiment at 85°C, 9.5 A current injection

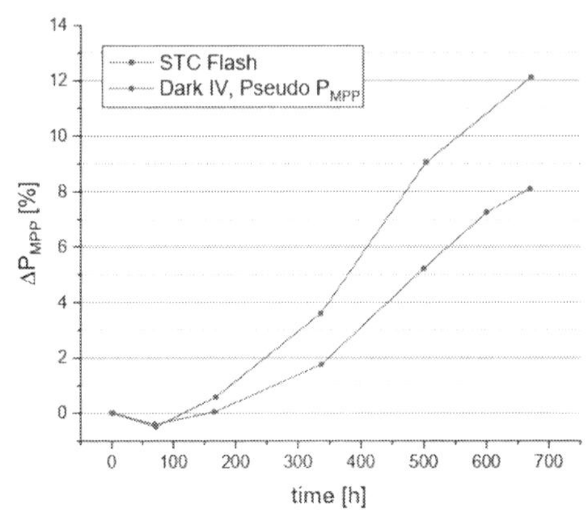

Figure 9: Comparison of Pseudo P_{MPP} from Dark-IV curves in Figure 8 and STC Flash for LETID Regeneration experiment at 85°C, 9.5 A current injection

Tracing changes of PV modules, at least for c-Si modules, via DIV tends to work quite well during such stress tests.

3.4 Monitoring the applied voltage

Taking full DIV curves might lead to heating effects in the sample if IV curve is taken too often, or sweeps take too long. To monitor changes a constant current can be applied to the sample and the required voltage can be measured similar to a continuity check setup.

That way the degradation during LeTID testing can be monitored easily in-situ. The applied voltage to the modules under test can be monitored and can be called Current and elevated Temperature Induced Degradation (CeTiD) test. This way it is possible to follow decreasing voltage as indicator for degradation during the test (Figure 10). The changes in voltage follow nicely the changes of output power from the modules during Flash testing.

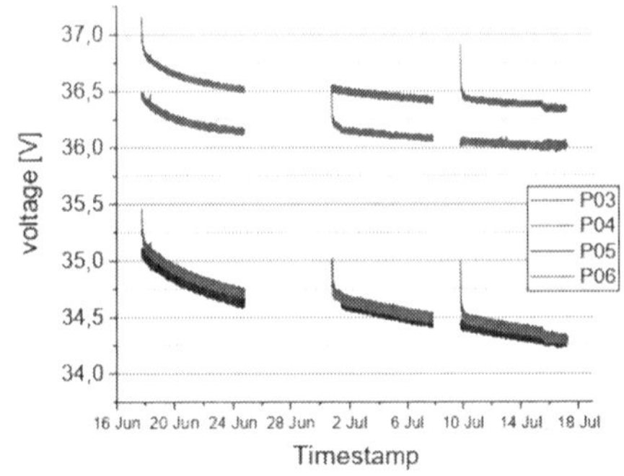

Figure 10: Voltage monitoring during a long-term CeTiD with intermediate interruption for Flash testing.

Monitoring and evaluating voltage dependencies during e.g. CeTID testing is an easy way to follow degradation and recovery (not shown as voltage change, but as DIV trend in Figure 8).

3.5 Checking the bypass diodes

Bypass diode failures are a quite common observation, especially in field return modules of a solar system, but can also occur during indoor testing. The following procedure works for both and is a non-intrusive method to check diodes in PV module junction boxes without the need to have direct access to the diode.

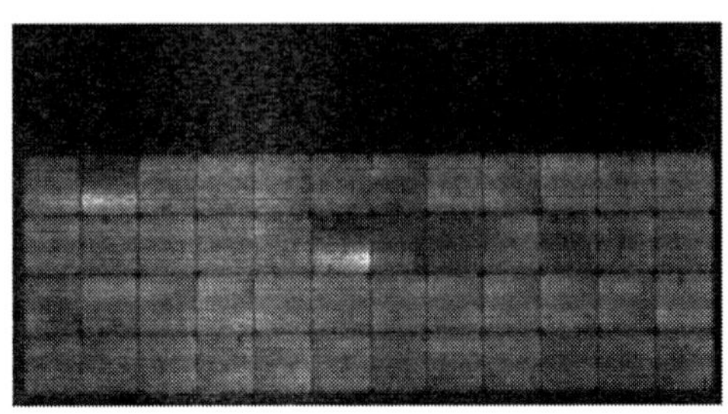

Figure 11: EL image of PV-module with a shorted bypass diode

Figure 11 shows an EL image of a module with a short-circuited bypass diode. Usually Schottky diodes are used, which have a forward voltage of about 0.3 V. At the same time, the reverse breakdown voltage of the bypass diode must be higher than the voltage in the substring, so that a dark characteristic can be measured. Example DIV measurements for modules with defective bypass diodes are shown in Figure 12. Defective bypass diodes can either be short-circuited, in which case the affected substring also fails in the solar simulator ① and the total diode voltage is reduced by the proportion of the defective bypass diode ②. Or there is an open circuit ③, then the protection function is no longer given in case of partial shading or mismatch.

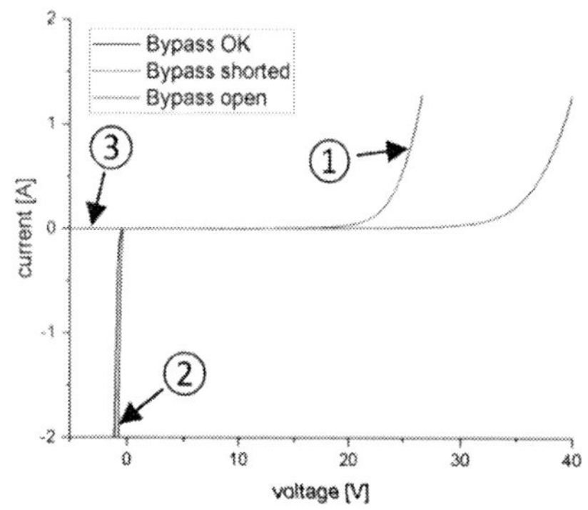

Figure 12: Dark IV-curves with negative voltage showing defective bypass diodes

The measurement of the DIV curve of the PV module in reverse direction shows the sum of the diode voltage when all diodes are functional. Solar simulators with an option for a 4-quadrant measurement can perform this measurement.

4 SUMMARY

In this contribution advantages and difficulties of dark IV-curve measurement are discussed. Several defects that can be found in the light characteristic of a module can also be detected and monitored in-situ with the DIV curves. The influence of series R_S and parallel resistance R_{sh} can be utilized to track certain changes of the PV module.

The monitoring of parallel resistance R_{sh} of the module is for example very valuable for detecting PID and trace changes during degradation and recovery experiments. The calculation of pseudo-module power is most reliable at invariable short-circuit current I_{SC}. It can be used for quantifying PID, and with limitations LID and LETID. Dark IV data therefore offer the possibility of in-situ monitoring and information with higher temporal resolution can be obtained independently of time-consuming intermediate measurements. From these data improved stress models can be developed for instance to more precisely answer questions about the degradation or recovery rate or when solder joints break. Additionally, the functionality of the bypass diodes in the module can be tested by extending the current in the opposite direction.
However, there are also defects that cannot be detected by DIV. These are, among others:

- cell breakage, cell cracks
- power mismatch
- yellowing, discoloration and reduction of light transmission and reflectance
- glass defects, glass breakage
- contamination, shading
- bifacial properties
- defects in laminate or back sheet
- power reduction with decreasing I_{SC}
- temperature coefficient of the I_{SC}

Further investigations are ongoing to work on more details what can be derived in-situ and ex-situ from DIV.

5 REFERENCES

[1] IEC 60904-1 Photovoltaic devices - Part 1: Measurement of photovoltaic current-voltage characteristics
[2] Report IEA-PVPS T13-24:2021 chapter 5 Dark i-v-measurements of pv strings and pv modules
[3] D.L. King et al., Dark current-voltage measurements on photovoltaic modules as a diagnostic or manufacturing tool, 26th PVSC, 1997
[4] H. Kojima, "Accurate and Rapid Measurement of High-capacitance PV Cells and Modules Using Dark Current Characteristics", 29th European Photovoltaic Solar Energy Conference - Amsterdam, 2014
[5] Florides, M. et al.: Early Potential Induced Degradation (PID) Detection in the Field: Solar Cell Shunt Resistance Measurement at different Bias Conditions. In. 35th European Solar Energy Conference and Exhibition, 2018
[6] D. Lausch et al. "Sodium out diffusion from stacking faults as root cause for the recovery process of potential-induced degradation (PID)", SiliconPV, 2014
[7] V. Naumann et al. "Explanation of potential-induced degradation of the shunting type by Na decoration of stacking faults in Si solar cells", Solar Energy Materials & Solar Cells, 2013
[8] Ch. Taubnitz et al.: Potential-Induced Degradation: An Improved Unterstanding of Mechanisms and Influence Factors. In: 32nd European Photovoltaic Solar Energy Conference and Exhibition, 2016
[9] W. Luo et al., "In-Situ Characterization of Potential-Induced Degradation in Crystalline Silicon Photovoltaic Modules Through Dark I–V Measurements", 45th IEEE Photovoltaic Specialists Conference, 2017
[10] IEC 61215-2 Terrestrial photovoltaic (PV) modules - Design qualification and type approval - Part 2: Test procedures, 2016
[11] B. L. Williams et al.: Identifying parasitic current pathways in CIGS solar cells by modelling dark J-V response. Progress in Photovoltaics, Research and Publications, Vol. 23 (2015), pp. 1516-1525
[12] T. Pletzer, "Influence of cracks on the local current - voltage parameters of silicon solar cells", Progress in Photovoltaics: Research and Applications, 2015
[13] IEC 60891 Photovoltaic devices - Procedures for temperature and irradiance corrections to measured I-V characteristics, 2021
[14] F. Kersten et al., Degradation of multicrystalline silicon solar cells and modules after illumination at elevated temperature, Sol. Mat. Ener. Sol. Cells 142 (2015) 83-86
[15] M. Pander et al., Benchmarking Light and elevated Temperature Induced Degradation (LeTID), 36th EUPVSEC, 2018

INTERLABORATORY COMPARISON OF VOLTAGE SWEEP METHODS USED FOR THE ELECTRICAL CHARACTERIZATION OF ENCAPSULATED HIGH-EFFICIENCY C-SI SOLAR CELLS

Christos Monokroussos[1]*, Masahiro Yoshita[2], Kengo Yamagoe[2], Harald Müllejans[3], Diego Pavanello[3], Klaus Ramspeck[4], David Hinken[5], Karsten Bothe[5], Yuji Fujita[6], Gilles Arnoux[7], Filipe Pinto[7], Rajesh Ambigapathy[7], Qiang Shi[8], Harrison Wilterdink[9], YiFeng Chen[10], Yan Ping[10], Qi Gao[1]

[1]TÜV Rheinland Group, [2]AIST, [3] European Commission JRC, [4] h.a.l.m. elektronik GmbH, [5] ISFH, [6] Kyoshin Electric Co., Ltd., [7] Pasan SA (Meyerburger Group), [8] SIMIT CAS, [9] Sinton Instruments, [10] Trina Solar

* Email: Christos.Monokroussos@tuv.com

ABSTRACT: This work presents the comparison of measurement results for four types of encapsulated high efficiency (HE) c-Si solar cells measured by ten laboratories based in Asia, Europe and North America utilizing a wide range of voltage sweeping methods, which include well-established procedures that represent good industry practice, as well as recently introduced ones that have not been verified yet. A proficiency test was employed to examine the consistency of results and their corresponding uncertainties. The results of all participant laboratories generally remained well within [−1; 1], thus indicating consistency between the measured values and the reference values within stated measurement uncertainties. A preliminary analysis revealed that differences remained within ±1.15% in P_{MAX} and within ±0.35% in V_{OC} for all participants and methods applied. Essentially this work forms the basis to validate all applied methods and their stated measurement uncertainties.

Keywords: round-robin, interlaboratory comparison, high-efficiency c-Si photovoltaic module, electrical performance, characterization

1 INTRODUCTION

The power measurements of high efficiency (HE) c-Si technologies have been a challenge for both production lines and calibration laboratories for over two decades [1-3]. HE c-Si devices exhibit measurement deviations due to their slow response causing hysteresis in IV measurements in particular in pulsed solar simulators. Such deviations are attributed to the diffusion capacitance of c-Si devices, which derives from the redistribution of charge carriers in the depletion region under forward bias. HE c-Si have longer diffusion lengths for minority carriers, and hence would also have a higher inert diffusion capacitance than conventional solar cells giving rise to measurement artefacts. Such deviations have been previously reported for both solar simulator and outdoor measurements. [1-6].

Recent developments in HE c-Si technologies (such as *p*-type PERC, *n*-type PERT, HJT and IBC) have popularized these devices in the market and made the accuracy of their calibration increasingly important. In parallel advances in measurement systems have opened new possibilities for characterization [7-11] compared to proven standard procedures involving long-pulse solar simulators, the multi-flash method and steady-state conditions such as those provided by natural sunlight.

The aim of this work is to examine the comparability of measurements of different laboratories utilizing different procedures for the measurement of HE solar cells with particular focus on voltage sweeping methods. The sweeping methods utilized by the participants of the round-robin are: 1. linear sweep rate, 2. non-linear sweep rate, 3. hysteresis measurement and correction, 4. voltage-stepping and correction and 5. voltage-irradiance modulation and correction. Furthermore, the work aims to compare different methods and laboratory setups used between ISO/IEC 17025 accredited calibration and testing as well as industrial laboratories that have processes and procedures in place in accordance with IEC 60904-1. In this context a HE encapsulated solar cell round-robin was initiated. A preliminary analysis is presented in the following paragraphs.

2 METHODOLOGY

2.1 Participant laboratories, measurement uncertainty and setups

A round-robin was realized between ten laboratories based in Asia, Europe and North America. The participants were five ISO/IEC 17025 accredited laboratories (three for calibration and two for testing) and five industrial laboratories. Namely the participant laboratories were: AIST (Japan), Berger Lichttechnik (Germany), JRC ESTI (Italy), h.a.l.m. Electronic (Germany), ISFH (Germany), Kyoshin Electric Co., Ltd. (Japan), Pasan, Meyer Burger (Switzerland), SIMIT, CAS (China), Sinton Instruments (USA), TÜV Rheinland (China). The participant laboratories employed one or more of the above-mentioned sweeping methods to electrically characterise the devices under test (DUTs). Table I lists information related to the accreditation, but also the method and the sweeping procedure that each of the participant laboratories employed.

Measurement uncertainty in HE PV devices are driven by device hysteresis and reference irradiance. As both effects can impact the I_{SC}, P_{MAX} and V_{OC} of the DUTs, it is desirable to design an experiment which can isolate the influence of reference irradiance calibration, so measurement deviations can be directly associated with device hysteresis and consequently the voltage sweeping strategy employed. In order to do so, predetermined I_{SC} values were assigned to each of the DUTs by the initiator and communicated to all participants in advance. The randomly assigned values were within ±3% of the measured I_{SC} value at STC. All laboratories were then asked to perform IV characterization of the DUTs so that the I_{SC} matches the predefined value while controlling device temperature at (25±1) °C. All participants employed four-terminal connections, while the contact points at voltage and current terminals were clearly marked, so one may also assume that deviations due to contacting played a secondary role.

Table I: List of participant laboratories, methods and IV sweeping procedures employed in the round-robin.

Lab. Participants	Accreditation	Illumination (Min. Pulse Length)	IV Sweeping Method
AIST (Japan)	17025 for calibration	Long-pulse solar simulator (>100ms)	Linear sweep
Berger Lichttechnik (Germany)	None	Pulse solar simulator (>10ms)	Linear step-like
CAS, SIMIT (China)	17025 for testing	Long-pulse solar simulator (>100ms)	Linear step-like
JRC ESTI (Italy)	17025 for calibration	Steady-state solar simulator (>1s)	Linear sweep
h.a.l.m. Electronic (Germany)	None	Pulse solar simulator (>110ms)	Advanced Hysteresis [11]: 1. 2×16ms (hysteresis) 2. 2×24ms (hysteresis) 3. 2×50ms (hysteresis) 4. 2×110ms (hyst./multiflash) 5. 2×2500ms (hyst./multiflash)
ISFH (Germany)	17025 for calibration	Steady-state solar simulator (>10s)	Linear sweep
Kyoshin Electric Co., Ltd. (Japan)	None	Pulse solar simulator (>20ms)	1. Multiflash 2. Photo and dark analysis (hysteresis) [10] 3. Fast and versatile (non-linear sweep)
Meyer Burger, Pasan (Switzerland)	None	Pulse solar simulator (>10ms)	1. *Dragonback®* (non-linear voltage stepping) [9] 2. Smart Sweep (non-linear sweep)
Sinton Instruments (USA)	None	Pulse Solar simulator (multiple flashes of >1ms)	Voltage-irradiance modulation [7]
TÜV Rheinland (China)	17025 for testing	1. Pulse solar simulator A (>10ms) 2. Pulse solar simulator B (>80ms)	Dynamic IV (non-linear voltage stepping) [8]

In this context a comparison of P_{MAX} and V_{OC} (as I_{SC} was almost identical for all participants) could reveal deviations arising from device hysteresis, due to the different voltage sweeping strategies employed without influence from irradiance traceability and spectral mismatch. This strategy effectively allows to isolate and quantify measurement artefacts due to device hysteresis. The measurement uncertainty contributions (UCs) for the electrical parameters at predefined I_{SC} values are reported in Table II.

Table II: Laboratory measurement uncertainty contributions (UCs) of electrical parameters at predefined I_{SC} values (k=2). For laboratories, which practiced measurements with multiple methods, the name of the method is indicated next to the laboratory name.

	P_{MAX}	V_{OC}
AIST	0.30%	0.30%
Berger Lichttechnik	0.70%	0.50%
JRC ESTI	0.75%	0.20%
h.a.l.m., 2x16ms	0.60%	0.40%
h.a.l.m., 2x24ms	0.60%	0.40%
h.a.l.m., 2x50ms	0.60%	0.40%
h.a.l.m., Multiflash, 2x110ms	0.60%	0.40%
h.a.l.m., Multiflash, 2x2500ms	0.70%	0.50%
ISFH	0.70%	0.35%
Kyoshin Electric, FAV*	0.59%	0.30%
Kyoshin Electric, Multiflash	0.52%	0.30%
Kyoshin Electric, PDA**	0.55%	0.30%
Pasan, *Dragonback®*	0.70%	0.50%
Pasan, Smart Sweep	1.50%	0.50%
Sinton Instruments	1.20%	0.70%
SIMIT, CAS	1.20%	0.70%
TÜV Rheinland	1.11%	0.40%

* FAV: Fast, Accurate, and Versatile method
** PDA: Photo and Dark Analysis method

2.2 Devices under test (DUTs)

The DUTs were four types of encapsulated HE c-Si solar cells (two types of *p*-type PERC and two types of *n*-type HJT) with initial aperture solar cell efficiencies varying between 20.5% and 23.1%. In total 11 single-cell encapsulated devices were used in the round-robin, seven of which were measured by all participants, while four were kept by the initiator in a controlled environment of temperature, humidity and light and were used as control cells (one per type). More details about the test specimens are given in Table III. Prior to any measurements all samples were subject to (15-25) kWh/m^2 light soaking in accordance with the stabilization procedures listed in IEC 61215-2. Therefore, all devices were sufficiently stabilized prior to the round robin.

2.3 Interlaboratory comparison Approach and Project Timeflow

The interlaboratory comparison took place from June 2018 to July 2020. The round-robin approach was employed here, i.e. the test samples were measured in turn by each participant and shipped to the subsequent laboratory. The agreed test protocol required electrical characterization, so the I_{SC} of the DUT matches its predetermined value at 25°C. Optionally EL measurements were also required for all samples, as long as the laboratory was qualified to do so. A timeflow of the round-robin is shown in Table IV. The identity of participants is not disclosed to ensure the confidentiality of results. Participant laboratories are rather identified with capital letters A-J assigned by order of testing. Due to shipment issues the initiator laboratory (A) repeated its final measurements before laboratory (J) concluded the round-robin.

Table III: List of the encapsulated cells used in the round-robin together with their constructional characteristics.

Sample No.	Use	Cell Technology	Initial Aperture Solar Cell Efficiency*	Soldering / Dimensions
HJT-1 #1	Round-Robin			
HJT-1 #2	Round-Robin	*n*-type HJT	~21.6%	double soldered per +/- terminal; 20 x 20 cm
HJT-1 #3	Control			
HJT-2 #1	Round-Robin			
HJT-2 #2	Round-Robin	*n*-type HJT	~23.1%	double soldered per +/- terminal; 38 x 38 cm
HJT-2 #3	Control			
PERC-1 #1	Round-Robin	*p*-type PERC	~21.6%	double soldered per +/- terminal; 30 x 30 cm
PERC-1 #2	Control			single soldered per +/- terminal; 30 x 30 cm
PERC-2 #1	Round-Robin			double soldered per +/- terminal; 30 x 30 cm
PERC-2 #2	Round-Robin	*p*-type PERC	~20.5%	double soldered per +/- terminal; 30 x 30 cm
PERC-2 #3	Control			single soldered per +/- terminal; 30 x 30 cm

(*) Initial aperture solar cell efficiency is calculated based on the initial nominal power after stabilization assuming the total projected area of the solar cell under test (which were at all cases ~15.6 x 15.6 cm), but excluding its surrounding area. It is given here indicatively of the total conversion solar cell efficiency. It is noted that the actual solar cell total conversion efficiency would be lower. As the surrounding area far exceeded the actual cell dimensions, there is a non-negligible part of light, which is being redirected towards the cell from its surroundings due to multiple internal reflections inside the glass, which rises the aperture solar cell efficiency.

Table IV: Time flow of measurement in the round robin between participant laboratories (A-J) highlighted with different colour code; different measurements sets performed by the initiator laboratory are marked with a number. Inter-laboratory shipment time is indicated with arrows and highlighted in grey.

Year	2018							2019					
Month	6	7	8	9	10	11	12	1	2	3	4	5	6
Lab.	A1			→	B	→ C	→	D	→ E		→ F	→	G

Year	2019					2020							
Month	7	8	9	10	11	12	1	2	3	4	5	6	7
Lab.	→	H		→		I	→		A2		→		J

2.4 Methods for statistical analysis

The consistency of the results was first verified by employing the E_n number proficiency test according to ISO/IEC 17043 against ISO/IEC 17025 calibration laboratories (i.e.. AIST, ESTI and ISFH). Consensus values were then assigned based on the weighted (by measurement uncertainty) average of all values of accredited calibration and qualifying testing laboratories. This approach has been previously used successfully in other interlaboratory comparisons [12, 13] and represents the state of the art, as it provides a higher weight to measurement values, which measured with a lower uncertainty.

The disadvantage of such an approach is that it depends on each laboratory's stated uncertainties, but also on the metastability of the DUTs. A reliable uncertainty estimate requires firstly traceability (i.e. an unbroken chain of prior measurements to international standards), secondly a rigorous analysis of all possible contributions in the measurement itself and thirdly a stable device which would allow the direct interlaboratory comparison of measurements practiced by different laboratories. Furthermore, some devices exhibited metastabilities, which limited the usefulness of the interlaboratory comparison. When a DUT showed metastability, the consensus values were assigned directly by accounting the weighted average of all testing and calibration laboratories omitting the E_n number proficiency test, as the test does not provide useful information for the laboratory method and procedure. These cases will be discussed in more detail in section 3.1.

2.4.1 Electrical performance analysis

A validity and consistency of all results was firstly tested in order to identify and, if necessary, eliminate outliers. For this purpose the values submitted by all participants were compared with a reference value, which was chosen as the weighted (by inverse square measurement uncertainty) average of all participant calibration laboratories according to ISO/IEC 17025. A proficiency test was performed employing E_n numbers in accordance with ISO/IEC 17043 [14], as:

$$E_{n_i} = \frac{x_i - \bar{x}}{\sqrt{{U_{95_i}}^2 + {\bar{U}_{95}}^2}} \quad (1)$$

E_{n_i} is the E_n number of the ith participant laboratory, x_i is the measurement value corresponding to a measurement uncertainty U_{95_i} of the ith participant laboratory. $\bar{x}$ is the reference value with a measurement uncertainty $\bar{U}_{95}$. An agreement between results within stated UCs is obtained when the E_n number is between -1 and 1; the results do not agree with the reference, and therefore the UCs are to be revised (or some cause for an erroneous measurement has to be identified), in case the E_n number is above 1.

The weighted mean for each electrical parameter (either V_{OC}, or P_{MAX}) is used here as its reference value. In addition, corrections were applied to consider multiple results submitted by a single laboratory. In particular, laboratory A measured twice all the modules. Therefore, a formula accounting for both reliability weights (i.e. measurement uncertainty) and frequency weights (i.e. number of independent measurements submitted per laboratory) was employed. Specifically, the weighted mean, $\bar{x}^*$, of measurements was derived as follows [15, 16]:

$$\bar{x}^* = \frac{\sum_{i=1}^{n} w_i x_i}{\sum_{i=1}^{n} w_i} \quad (2)$$

with weights w_i defined as:

$$w_i = \frac{1}{N_{lab} \cdot {U_{95_i}}^2} \quad (3)$$

and x_i is a measurement with expanded measurement uncertainty (k=2), U_{95} [17]. In eq. (2) N_{lab} is the number of independent measurement sets submitted by each laboratory. The notation n represents the total number of independent measurements submitted by all participants in the round-robin. The derived values are considered as the best estimate of the physical true value of the electrical parameters.

The uncertainty of the weighted mean can then be expressed as [18]:

$$\bar{U}_{95} = \frac{1}{\sqrt{\sum_{i=1}^{m} \frac{1}{{U_{95_i}}^2}}} \quad (4)$$

The notation m represents the number of participant testing or calibration laboratories, i.e. five in our case.

Inter-laboratory deviations were then expressed in terms of their deviation from the weighted mean as defined above. This is a more conservative approach as compared to the maximum absolute laboratory differences, because it accounts for the dispersion of measurements, and therefore can provide a more comprehensive estimate of laboratory agreement. The weighted deviation was used rather than the standard deviation, which gives statistical weight to differences measured with lower uncertainty. As the sample size was relative small (<30), the unbiased estimator correction was used for the population variance [19]:

$$\bar{s}_w = \sqrt{\frac{\sum_{i=1}^{n} w_i (x_i - \bar{x}^*)}{\sum_{i=1}^{n} w_i - \frac{\sum_{i=1}^{n} {w_i}^2}{\sum_{i=1}^{n} w_i}}} \quad (5)$$

3 PRELIMINARY RESULTS

A preliminary analysis of the results is discussed in the following paragraphs in terms of the metastability of the reference samples and interlaboratory deviations for the electrical performance parameters. A more comprehensive analysis will be presented elsewhere.

3.1 Drift or metastability of test samples

As device metastability or drift would directly affect the outcome of the round-robin, it is important to minimize its impact. When this is not possible it is equally important to be able to quantify any performance changes, to isolate artefacts deriving from device hysteresis from metastability or drift in general. In the following paragraphs, the metastable behaviour or drift is discussed.

DUT HJT-2 #2 sustained damage in its electrical termination while in laboratory I. The damage was significant, as an electrical terminal was completely cut-off (Figure 1), which changed the measured P_{MAX} on average by approximately 2.0 % (see Figure 2). For this reason, the results of DUT HJT-2 #2 will not be analysed further in the following paragraphs.

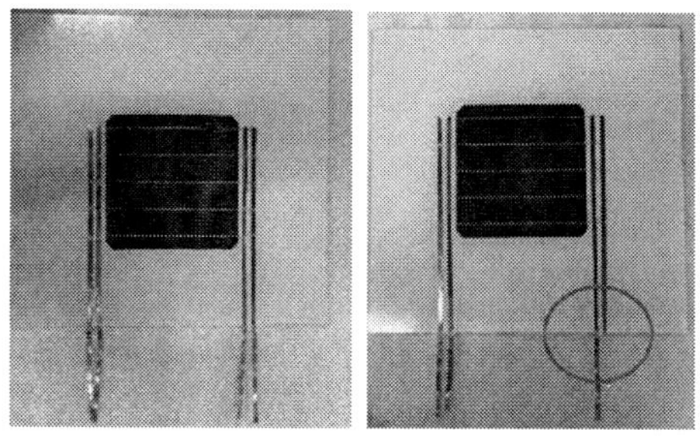

Figure 1: Sample HJT-2 #1 (left) and HJT-2 #2 (right) carrying double soldered electrical terminations. The sample HJT-2 experienced damage in its exterior terminations during the round-robin.

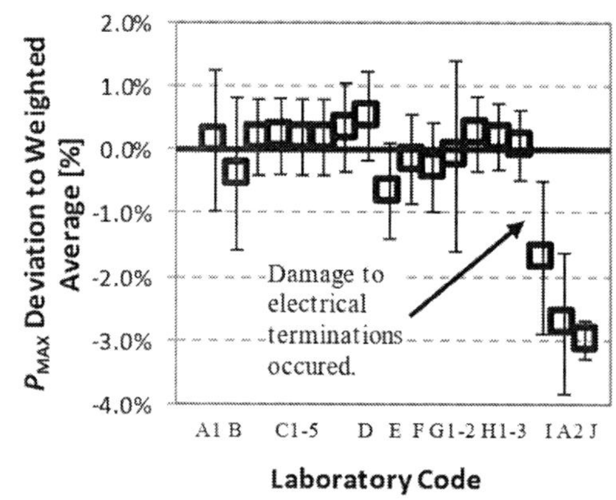

Figure 2: Maximum power deviation for DUT HJT-2 #2 as measured blindly by each laboratory at predefined I_{SC} condition at 25°C in respect to the weighted average of all measurements before any damage occurred. The measurements of the last three laboratories was not accounted in the weighted average calculation. X-axis denotes the test laboratory code in chronological order of testing. The estimated values of uncertainty (k=2) are shown by error bars for each laboratory. The figure indicates the drift shown after an electrical termination was accidentally cut-off.

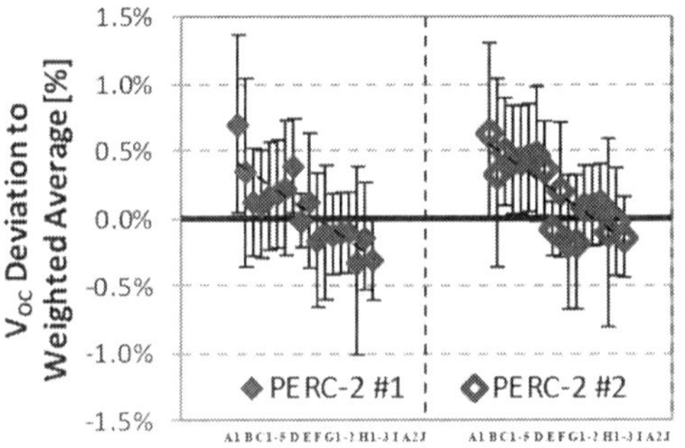

Figure 3: Open-circuit voltage deviation for samples PERC-2 #1 and #2 as measured blindly by each laboratory at predefined I_{SC} condition at 25°C in respect to the weighted average of all measurements. X-axis denotes the test laboratory code in chronological order of testing. The estimated values of uncertainty (k=2) are shown by error bars for each laboratory. The figure shows that samples PERC-2 #1 and #2 exhibited a linear metastable behaviour during the course of round-robin. Dashed lines have been added as a guideline to show the linear regression of open-circuit voltages measured over time.

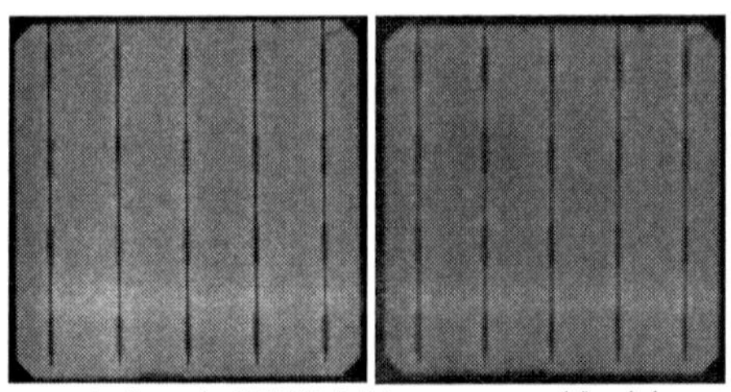

Figure 4: Sample PERC-2 #1 as measured by laboratory B on the 25th October 2018 and laboratory A on 26th April 2020. A minor darkening of the image was observed, which indicates a decrease of minority carrier lifetime.

Samples PERC-2 #1 and #2 also exhibited moderate systematic metastable behaviour, which became notable in P_{MAX} and V_{OC} (as shown in Figure 3). Specifically, for device PERC-2 #1 P_{MAX} degraded approximately 1.10% and V_{OC} approximately 0.84%, while device PERC-2 #2 degraded approximately 0.75% in P_{MAX} and 0.67% in V_{OC}. The estimates provided were based on the differences seen between initial and final measurements carried out by the initiator laboratory at STC (not predefined I_{SC}). The effect was also noticeable when comparing the EL images of this cell taken at the beginning and the end of the round-robin (as shown in Figure 4), which generally indicates an increase of series resistance or a decrease in the minority carrier lifetime. The increase of series resistance was ruled out, because a reduction in V_{OC} was also observed, but also because a comparison of EL images of the cell at higher and lower injection still showed a darkening of these cells. Therefore, it is assumed that minority lifetime changes influenced the metastability of cells PERC-2 #1 and #2. Interestingly, the control samples (which were stored in a dark environment) have not shown similar behaviour, pointing out that exposure to light may have had an impact in the metastability of the samples. Indeed the control sample PERC-2 #3 was remeasured higher at STC by the initiator laboratory and its values changed by +0.23% in I_{SC} and +0.33% in P_{MAX}; suggesting either a minor increase in performance, or an influence from laboratory reproducibility. As the effects seen were rather moderate (approx. within 1%) it was decided to include the solar cells in the round-robin examination, but exclude them from E_n proficiency test. Generally, the results of PERC-2 should be read with a caution.

3.2 Proficiency Test - E_n Numbers

The results of the proficiency verification are expressed with the E_n numbers for P_{MAX} and V_{OC} in tables V and VI respectively. The results presented here correspond to all samples whose behaviour did not show systematic metastabilities, or drift over the course of the round-robin. A single outlier was observed in the E_n numbers among all participants for P_{MAX}. Specifically, the measurement of laboratory E for device PERC-1 #1 is associated with an E_n number of 1.30. No outlier was observed for V_{OC}. The results of all participant laboratories generally remained well within [−1; 1], thus indicating consistency between the measured values and the reference within the declared UCs for all measurement procedures and laboratories.

Table V: E_n numbers for P_{MAX} for the participant laboratories at predefined I_{SC} condition at 25°C.

Laboratory	HJT-1 #1	HJT-1 #2	HJT-2 #1	PERC-1 #1
A1	0.53	0.22	0.59	0.28
B	0.08	0.09	-0.18	0.32
C1	0.29	0.17	0.43	0.54
C2	0.26	0.12	0.41	0.47
C3	0.25	0.09	0.37	0.45
C4	0.21	0.07	0.35	0.49
C5	0.33	0.13	0.39	0.46
D	0.83	0.94	0.84	0.73
E	-0.58	-0.53	-0.77	-1.30
F	0.39	0.38	-0.34	0.99
G1	0.32	0.02	-0.19	0.06
G2	0.09	0.24	-0.18	0.51
H1	0.59	0.97	0.67	0.64
H2	0.39	0.64	0.40	0.54
H3	0.18	0.51	0.42	0.54
I	0.60	0.48	0.99	0.14
A2	0.02	-0.12	-0.04	0.05
J	-0.10	-0.15	-0.04	0.16

3.3 Electrical Performance Parameters Comparison

Interlaboratory deviations in the electrical performance parameters P_{MAX} and V_{OC} are shown in Figure 5 and Figure 6 respectively.

Since I_{SC} values were predefined, the weighted deviation for I_{SC} stayed below 0.06% (k=2) for all modules, which shows that all calibrations have been carried out correctly for all modules. The worst case agreement in I_{SC} has been -0.30%, which was measured by laboratory D for DUT PERC#2-1. This would introduce an error contribution of the same size to P_{MAX}.

The calculated uncertainty of the consensus value of P_{MAX} varied between ±0.63% and ±0.80% (k=2) depending on the DUT. The weighted deviations of DUTs varied between 0.79% and 1.12% (k=2) in P_{MAX}. These

Table VI: E_n numbers for V_{OC} for the participant laboratories at predefined I_{SC} condition at 25°C.

Laboratory	HJT-1 #1	HJT-1 #2	HJT-2 #1	PERC-1 #1
A1	0.35	0.33	0.49	0.56
B	0.13	0.05	0.10	0.33
C1	0.34	0.04	0.55	0.71
C2	0.34	0.08	0.55	0.50
C3	0.40	0.14	0.58	0.36
C4	0.40	0.11	0.55	0.64
C5	0.38	0.25	0.55	0.49
D	0.62	0.74	0.71	0.94
E	-0.46	-0.47	-0.63	-0.53
F	-0.11	-0.01	-0.05	0.46
G1	-0.21	-0.43	-0.65	-0.25
G2	0.21	0.15	-0.31	0.02
H1	0.43	0.66	0.66	0.47
H2	0.41	0.44	0.33	0.47
H3	0.35	0.59	0.62	0.54
I	0.15	0.01	0.20	0.14
A2	0.48	0.38	0.54	0.50
J	0.26	0.18	0.45	0.11

values indicate good agreement between laboratories. Worst case agreement in P_{MAX} was seen in the first measurement of laboratory A for DUT PERC#2-1, where the laboratory deviation to the consensus value was higher by 1.21%. The difference is largely attributed to the metastable behaviour of this device, which drifted in P_{MAX} over time (see also Figure 5). Another noticeable difference is the value measured by laboratory I for sample HJT-2 #1, which deviated by 1.15% in P_{MAX} to the consensus value. Lastly, laboratory E measured sample PERC-1 #1 by -1.04% lower than the consensus value of P_{MAX}. The last two deviations are most likely single events, which are difficult to explain, as the deviations in the associated values of V_{OC} were less than 0.2% of the consensus values. Lastly, sweep time and temperature control are critical parameters as they can affect P_{MAX}, but the errors reported in this work do not show any systematic under- or overestimation of P_{MAX}, which point towards the directions that the employed methods and procedures are valid.

Laboratory weighted deviations in V_{OC} remain within 0.35% to 0.56% (k=2) for all samples, while the uncertainty of the consensus values of V_{OC} stayed between ±0.29% and ±0.48% (k=2). Generally, these values are encouraging and indicate a high level of agreement observed amongst the participant laboratories. Worst case agreement in V_{OC} from the consensus values were the values recorded in the first measurements of laboratory A for devices PERC-2 #1 and #2, which deviate by 0.70% and 0.66% respectively. These errors are consistent with the P_{MAX} measurement of the laboratory, which was also overestimated by 1.21% and 0.86% for devices PERC-2 #1 and #2 respectively. As previously discussed, the PERC-2 devices exhibited a linear degradation in P_{MAX} and V_{OC} during the round-robin. The reported values agree well on the reported linear trend of degradation for devices PERC-2 #1 and #2, which degraded by 1.10% and 0.84% approximately in V_{OC} during the round-robin (see also Figure 4). It is also noted that laboratory D systematically overestimated V_{OC} for all DUTs by 0.21% - 0.37% higher to the consensus values, but the reported values were within the uncertainties stated. Since no significant systematic overestimation or underestimation was recorded, it is assumed that all employed methods and procedures are valid.

Generally, in this round-robin interlaboratory comparison no significant deviations for any of the four devices, which were considered stable (HJT-1 #1, HJT 1 #2, HJT-2 #1 and PERC-1 #1), were observed. In total 144 measurement results were reported, and only one outlier has been reported. The latter implies that the stated uncertainties are more conservative, as statistically one would expect at least five outliers for every 100 measurements practiced at a 95% level of confidence. Specifically, the results obtained utilizing the following methods: linear, non-linear sweep rate, hysteresis measurement and correction, voltage-stepping and correction and voltage-irradiance modulation and correction, are comparable within ±1.00% (k=2) for P_{MAX} and within ±0.40% (k=2) for V_{OC} with the results of established methods such as long-pulse, and multiflash. Therefore, this investigation leads to the conclusion that hysteresis artefacts are countered sufficiently well by all practiced methods and procedures.

It is noted that a round-robin has been conducted in the past, which also examined the magnitude of capacitive artefacts on full-sized PV modules [13], the reported level of agreement has been ±0.90% (k=2) for P_{MAX} and ±0.33% (k=2) for V_{OC}, which are quantitatively similar with the values reported in this round-robin.

4 CONCLUSIONS

A round-robin interlaboratory comparison was realized between ten laboratories based in Asia, Europe and North America. The participants were five ISO/IEC 17025 accredited laboratories (three for calibration and two for testing) and five industrial laboratories. The scope of this work was to examine the comparability of measurements made by different laboratories utilizing different procedures on the measurement of HE solar cells with particular focus on voltage sweeping methods. The sweeping methods examined in this work are: 1. linear sweep rate, 2. non-linear sweep rate, 3. hysteresis measurement and correction, 4. voltage-stepping and correction and 5. voltage-irradiance modulation and correction. Furthermore, the work aimed to compare different methods and laboratory setups used between ISO/IEC 17025 accredited calibration and testing as well as industrial laboratories that have processes and procedures in place in accordance with IEC 60904-1.

A proficiency test was employed to examine the consistency of results and their corresponding uncertainties, but also isolate possible outliers. For all stable devices the results of all participant laboratories generally remained well within [−1; 1], thus indicating consistency between the measured values and the reference within the declared UCs for all measurement procedures and laboratories. A preliminary analysis revealed that the maximum measured deviations varied for all stable samples within ±0.67% to ±1.15% for P_{MAX} and within ±0.21% to ±0.35%, for V_{OC}. The weighted deviations per sample type ranged within ±0.79% to ±1.12% (k=2) for P_{MAX} and ±0.35 to ±0.56% (k=2) for V_{OC}, which can be seen as a conservative estimation of

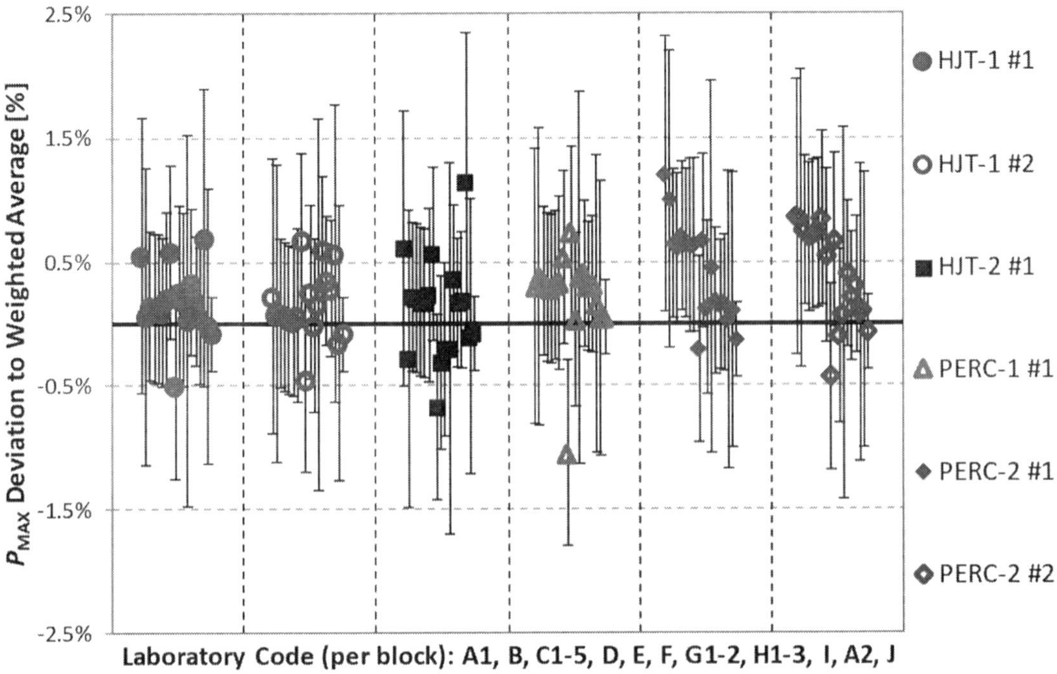

Figure 5: Maximum power point deviations to the weighted mean for all modules as measured blindly by each laboratory at predefined I_{SC} condition at 25°C. The estimated values of measurement uncertainty (k=2) are shown as error bars for each laboratory.

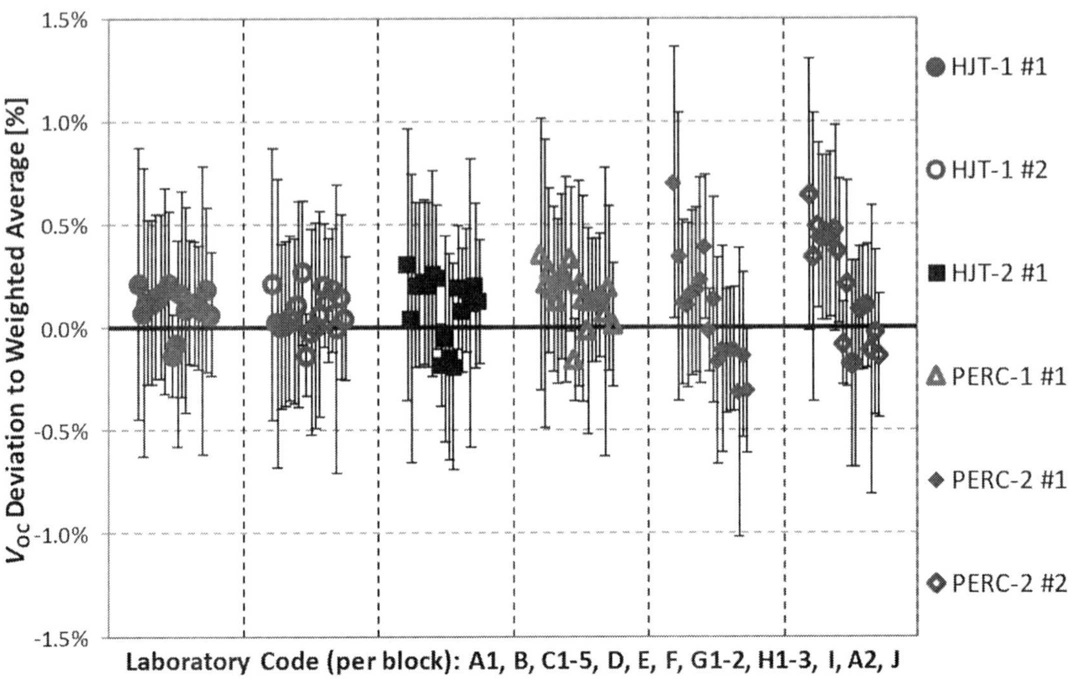

Figure 6: Open-circuit voltage deviations to the weighted mean as measured blindly by each laboratory at predefined I_{SC} condition at 25°C. The representations used are the same as in Figure 5.

interlaboratory agreement of this round-robin. These numbers are encouraging and indicate good agreement for all participants.

Essentially this work validated all applied methods and their stated measurement uncertainties. A more comprehensive analysis of the results will be presented elsewhere.

ACKNOWLEDGEMENTS

The authors would like to acknowledge the contribution of CAS SIMIT, Jinko Solar, LONGi Solar and Meyer Burger Group for providing the DUTs, which were used in the round-robin interlaboratory comparison.

REFERENCES

[1] H. A. Ossenbrink, W. Zaaiman . J. Bishop, "Do multi-flash solar simulators measure the wrong fill factor?" (1993); pp. 1194-1196.
[2] G. Friesen, H.A. Ossenbrink, "Capacitance effects in high-efficiency cells", Solar Energy Materials & Solar Cells, Vol. 48, Issues 1–4, (1997); pp. 77–8
[3] J. Metzdorf, A. Meier, S. Winter, T. Wittchen, "Analysis and correction of errors in current-voltage characteristics of solar cells due to transient measurements", 12th EUPVSEC (1994); pp. 496-499
[4] Y. Hishikawa, "Precise Performance Measurement of High-Efficiency Crystalline Silicon Solar Cells", 4th IEEE World Conference on Photovoltaic Energy Conversion (2006); pp. 1279-1282
[5] C. Monokroussos, R. Gottschalg, A. N. Tiwari, G. Friesen, D. Chianese, S. Mau, "The Effects of Solar Cell Capacitance on Calibration Accuracy When using a Flash Simulator", 4th IEEE World Conference on Photovoltaic Energy Conversion (2006); Vol. 2 (2006); pp. 2231 – 2234
[6] A. Virtuani, H. Müllejans, F. Ponti, E. Dunlop, "Comparisonof indoor and outdoor performance measurements of recent commercially available solar modules", 23rd EU PVSEC (2008); pp. 2713-2718
[7] R. A., Sinton, "A solution to the problem of accurate flash-testing of high-efficiency modules", Proceedings of the 21st European Photovoltaic Solar Energy Conference, Dresden, Germany (2006)
[8] C. Monokroussos, D. Etienne, K. Morita, C. Dreier, U. Therhaag, W. Herrmann, "Accurate power measurements of high capacitance PV modules with short pulse simulators in a single pulse", 27th EUPVSEC (2012)
[9] A. Virtuani, G. Rigamonti, G. Friesen, D. Chianese, P. Beljean, "Fast and accurate methods for the performance testing of highly-efficient c-Si photovoltaic modules using a 10 ms single-pulse solar simulator and customized voltage profiles", J. Meas. Sci. Technol., 23 (2012); doi:10.1088/0957-0233/23/11/115604
[10] H. Kojima, K. Iwamoto, A. Shimono, J. Abe and Y. Hishikawa, "Accurate and Rapid Measurement of High-Capacitance PV Cells and Modules Using a Single Short Pulse Light", IEEE PVSC (2014)
[11] K. Ramspeck, S. Schenk, L. Komp, A. Metz, M. Meixner, "Accurate efficiency measurements on very high efficiency silicon solar cells using pulsed light sources", Proceedings of the 29th EUPVSEC (2014)
[12] D. Dirnberger, U. Kräling, H. Müllejans, E. Salis, K. Emery, Y. Hishikawa, K. Kiefer, "Progress in PV module calibration – results of a world-wide intercomparison between four reference laboratories", Measurement Science and Technology 25 (2014); 105005 (17pp)
[13] C. Monokroussos, E. Salis, D. Etienne, X.Y. Zhang, S. Dittmann, G. Friesen, K. Morita, J. Stang, T. Herbrecht, V. Fakhfouri, N. Rebeaud, D. Pavanello, H. Müllejans, "Electrical characterization intercomparison of high-efficiency c-Si modules within Asian and European laboratories", Prog. Photovolt. Res. Appl., Vol. 27, Iss. 7 (2019); pp. 603-622 (https://doi.org/10.1002/pip.3134)
[14] ISO/IEC 17043:2010, "Conformity assessment - general requirements for proficiency testing"
[15] M.G. Cox, "The evaluation of key comparison data", Metrologia 39 589 (2002)
[16] M.G. Cox, C. Eiø, G. Mana and F. Pennecchi, "The generalized weighted mean of correlated quantities Metrologia", 43 S268 (2006)
[17] ISO/IEC Guide 98–3:2008 (JCGM/WG1/100). Uncertainty of measurement - Part 3: Guide to the expression of uncertainty in measurement (GUM:1995)
[18] H. Müllejans, W. Zaaiman, E. D. Dunlop, "Reduction of uncertainties for photovoltaic reference cells", Metrologia 52 (2015); pp. 646–653, doi:10.1088/0026-1394/52/5/646)
[19] C.F. Dietrich, "Uncertainty, calibration and probability", Ed. 2 (1991); pp. 37–42

EFFECTS OF INHOMOGENEOUS SNOW LOAD ON THE MECHANICS OF A PV MODULE

Pascal Romer[1], Andreas J. Beinert[1]
[1]Fraunhofer Institute for Solar Energy Systems ISE, Heidenhofstraße 2, 79110 Freiburg, Germany.
Corresponding author: Pascal Romer | Phone: +49 (0)761 4588 5044 | E-mail: Pascal.Romer@ise.fraunhofer.de

ABSTRACT: The phenomenon that snow sliding down a tilted PV-module can cause an inhomogeneous load is well known. However, just in 2020 a standard for this load scenario was established with the IEC 62938. This paper compares the difference between a homogeneous and such an inhomogeneous load distribution. Additionally, the influence of different temperatures on this new standard is analyzed. Compared to a homogeneous load distribution the first principal stress is lower for an inhomogeneous distribution. Moreover, it is shown that temperature has a large influence on the stress during mechanical loading. For TPO there is a specific temperature at which both the first principal stress and the probability of cell fracture reach a maximal value.

Keywords: Finite element modelling, FEM simulations, inhomogeneous snow load, photovoltaic module, stress, thermomechanics, mechanical load, virtual prototyping, coefficient of thermal expansion, material characterization.

1 INTRODUCTION

According to the IEC 61215 [1] a homogeneous mechanical load of at least 2400 Pa at 25 °C is one criteria to pass the certification. However, in the field load is often distributed inhomogeneously over the module and occurs mainly at temperatures different from 25 °C. Therefore the IEC 62938 [2] proposes a non-uniform snow load test for photovoltaic modules. A lot of papers investigate a homogeneous mechanical load according to IEC 61215 [3–10]. A few others deal with an non-uniform load due to wind [11, 12] and snow [12], but to the present knowledge none of them investigate the load according to IEC 62938. Hence, this work analyzes the effect of such a non-uniform snow load on the mechanics of a photovoltaic module for TPO (thermoplastic polyolefin) as the encapsulant. Furthermore some experimental works [13, 14] already investigated the influence of the temperature on the homogeneous mechanical load. Therefor this influence is analyzed for the inhomogeneous load using FEM simulations (finite element method).

2 METHOD

2.1 Storage modulus measurement

To analyze the effect of the encapsulants thermal properties, the temperature dependent storage modulus of TPO is measured with a DMA Eplexor from Netzsch Gabo. Due to the melting point of TPO, the storage modulus is only measured from -60 °C up to 80 °C and then extrapolated with a constant value for the FEM simulations. The samples are measured with an oscillation frequency of 1 Hz and a temperature ramp of 2 K/min.

2.2 CTE measurement

The coefficient of thermal expansion (CTE) is measured using a DMA242 C from Netzsch. To ensure that only the sample's thermal expansion is analyzed, a reference measurement using copper, with a known CTE, is carried out. Based on this measurement, thermal expansion of the DMA is calculated and can later be subtracted from that of the material measurement. Three measurements are performed, and a linear function is fitted to evaluate the CTE.

2.3 FEM simulation

The used FEM model is based on previous studies with neglected metallization and ribbons [4, 5]. Unlike to the referenced FEM models, due to the inhomogeneous load, the PV module exhibits only one symmetry plane. Therefore, half of a 120 half-cell PV module is simulated. In a first step, the lamination process is simulated by using a single temperature drop from 150 °C down to 25 °C. The residual stress from the lamination is used as an initial stress state for the later simulation of the mechanical load. The mechanical load simulation considers an aluminum frame and is performed at -40 °C, -20 °C, 0 °C as well as at 25 °C. Both, the homogeneous as well as the inhomogeneous load, are simulated with an equivalent load of 2400 Pa and 5400 Pa, respectively. Due to the slanted mounting of most PV modules, the snow load is not evenly distributed over the entire module but tends to slide downwards to the edge of the module. To reproduce this behavior the IEC 62938 works with a surface load S_A covering the lower 2/3 of the module combined with a linear load S_E as depicted in Figure 1.

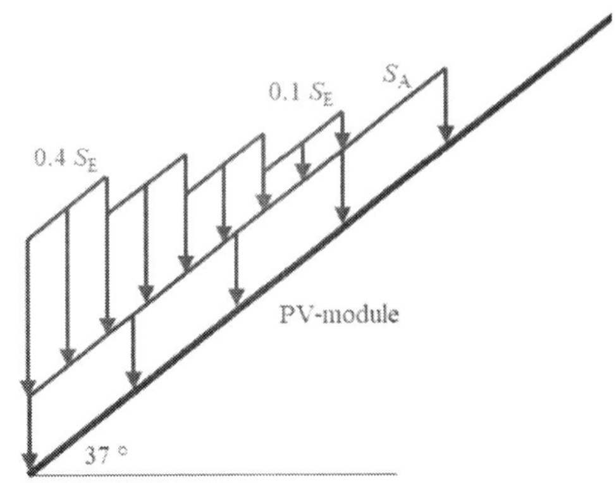

Figure 1: Inhomogeneous load distribution according to IEC 62938 on a PV module, consisting of an uniform surface load (S_A) and a linear load (S_E).

For comparable results between the homogeneous and the inhomogeneous simulations the total load mass must be the same for both cases. To guarantee this for the inhomogeneous simulation, the same load mass, used in the homogeneous case, is distributed unevenly over the pv module according to IEC 62938. The load profiles for the homogeneous as well as the inhomogeneous case that are implemented in the FEM simulation are shown in Figure 2. The material parameters used in the simulation are shown in Table 1.

Table 1: Specifications and material properties of the PV module. *: provided by manufacturer, †: measured.

Layer	Material	Dimension	Density [g/cm³]	Young's modulus [GPa]	Poisson's ratio [-]	CTE [$10^{-6}K^{-1}$]
Front glass	soda-lime glass	3.2 mm	2.5*	73*	0.24*	9.2*
Encapsulant	TPO	460 µm	0.96 [17]	Table 2†	0.22 [17]	293†
Solar cell	Cz-silicon	166 × 83 × 0.180 mm³	2.329 [17]	Elasticity matrix [17]		T-dep. [15, 16]
Backsheet	TPT	218 µm	2.52 [17]	3.5 [17]	0.29 [17]	50.4 [17]
Frame	aluminum		2.7 [18]	70 [18]	0.33 [18]	23 [18]
Frame-inlay	rubber		0.067*	0.00043*	0.49*	769*

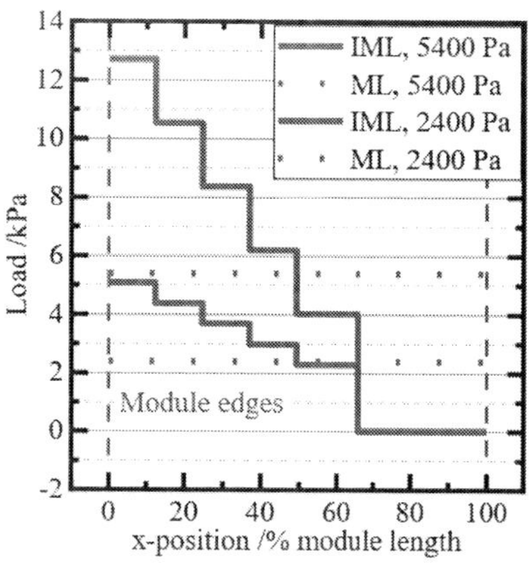

Figure 2: Profiles for the homogeneous as well as the inhomogeneous load distribution, implemented in the FEM simulation.

Silicon is a brittle material, which means that solar cells tend to break under tensile stress. This analysis focuses therefore on the first principal stress σ_I (interpreted as tensile stress) during mechanical load. Furthermore, the probability of cell fracture P_{fail} is calculated using the Weibull distribution [19] considering size effects [20]:

$$P_{fail} = 1 - \exp\left(-\sum_i - A_{eff,i}\left(\frac{\sigma_{I,max}}{\sigma_{0,i}}\right)^{m_i}\right) \text{ with } \quad A_{eff,i} = \int \left(\frac{\sigma_{I,i}(x,y)}{\sigma_{I,max}}\right)^{m_i} dA_i \qquad (1)$$

with the effective area A_{eff}, the maximum first principal stress $\sigma_{I,max}$, the Weibull scale factor σ_0 and the Weibull modulus m. Both parameters σ_0 and m are taken from [21]. The effective area A_{eff} can be interpreted as an equivalent area where the maximum first principal stress $\sigma_{I,max}$ occurs. The first principal stress $\sigma_{I,i}(x,y)$ at the position (x, y) is integrated over the surface $A_i(x,y)$. It should be mentioned that the probability of cell fracture is the probability that at least one crack is present in at least one solar cell of the entire module and does not correlate with the expected loss of power.

3 RESULTS

3.1 Storage modulus measurements

The temperature dependent storage modulus is depicted in Figure 3 and can be found in Table 2. TPO shows a glass transition below 0 °C where the storage modulus increases from about 33 MPa at 0 °C to 217 MPa at -40 °C.

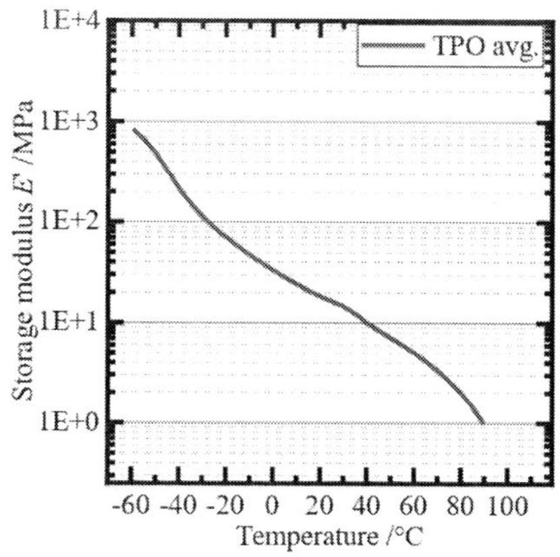

Figure 3: Measured storage modulus for the three different encapsulants EVA, TPO, POE over temperature.

3.2 CTE measurements

For the CTE measurement at least three samples are measured, and a linear model is fitted to the data, where the slope is taken as the CTE. Exemplary one sample with the fitted data is shown in Figure 4. The average CTE for TPO is found to be $293 \cdot 10^{-6} K^{-1}$.

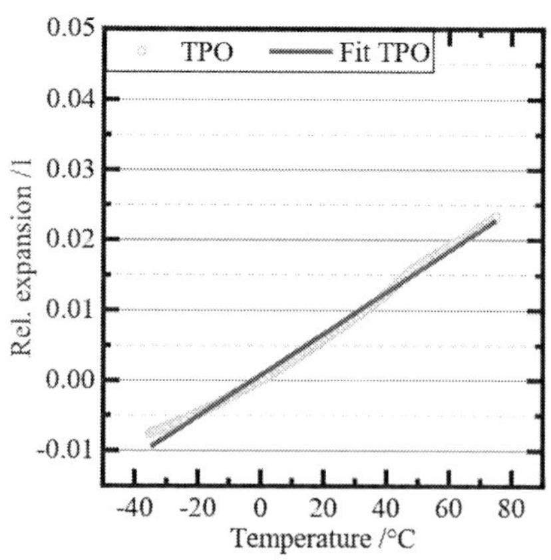

Figure 4: Measured relative expansion for one sample (symbols, light colors) and linear fits to the data (line, dark colors), from which the CTE is taken.

Table 2: Measured storage modulus for the three encapsulants EVA, TPO and POE for temperatures between -55 °C and 90 °C.

Temp / °C	-55	-50	-45	-40	-35	-30	-25	-20	-15	-10	-5	0	5	10	15
E'_{TPO} / MPa	669	479	322	217	153	113	88.0	70.5	57.2	47.2	39.6	33.4	28.4	24.2	20.9
Temp / °C	20	25	30	35	40	45	50	55	60	65	70	75	80	85	90
E'_{TPO} / MPa	18.3	16.4	14.6	12.3	9.98	8.41	7.15	6.06	5.03	4.13	3.34	2.62	2.01	1.48	0.98

3.2 Lamination

The minimum third principal stress $\sigma_{III,min}$ (interpreted as compressive stress) after lamination with TPO as the encapsulant is found to be -93 MPa.

3.3 Mechanical load

In this section the results of the mechanical load simulations will be discussed. First, the homogeneous load test according to IEC 61215 will be compared to the inhomogeneous one out of the IEC 62938, distributed along the module's short side as well as the long side. Second, the temperature dependency of the inhomogeneous load on the resulting first principal stress will be discussed.

Figure 5 exemplary shows the maximal first principal stress as well as the probability of cell fracture for TPO at an equivalent load of 2400 Pa. In all cases the homogeneous and the inhomogeneous ones, the maximal first principal stress is well below the critical value of 177 MPa [21], and accordingly the probability of cell fracture is negligible. Consequently, the following analysis will focus only on the results for an equivalent load of 5400 Pa.

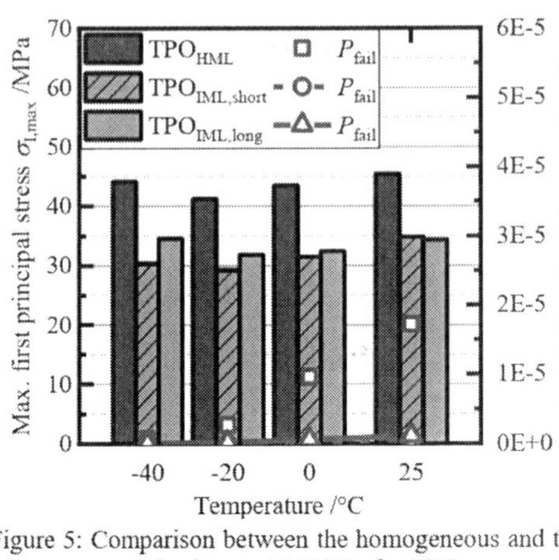

Figure 5: Comparison between the homogeneous and the inhomogeneous load case at 2400 Pa for TPO.

3.3.1 Comparison homogeneous vs. inhomogeneous load

Figure 7 depicts the overall stress distribution at an equivalent load of 5400 Pa and a temperature of 25 °C for a load distributed along the short side (middle) as well as along the module's long side (rigth). Compared to the homogeneous load, the stress distribution of the inhomogeneous load shifts to the loaded PV module side and has a lower overall magnitude on the first principal stress. In Figure 6 the maximal first principal stress as well as the probability of cell fracture is shown for all three load cases. Due to the lower stress during the inhomogeneous loading the cells are less prone to cell fracture.

Even though the homogeneous loading leads to a higher probability of cell fracture, this loading case is well investigated in previous research, therefore the following analysis will focus on the inhomogeneous loading, especially when distributed along the module's long side.

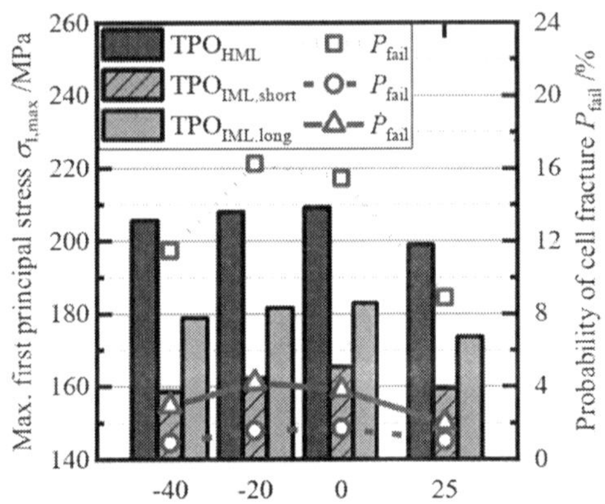

Figure 6: Comparison between the homogeneous and the inhomogeneous load case at 5400 Pa for TPO as a representative for the three encapsulants.

3.3.3 Influence of temperature

In Figure 8 the maximum first principal stress as well as the probability of cell fracture is depicted. Both values depend strongly on the temperature. A maximum in the probability of cell fracture can be seen at -20 °C. To explain this, three different effects must be considered. First, the encapsulant serves as a stress buffer for the solar cells. The colder and therefore the harder the encapsulant the worse the ability to buffer stresses, see Figure 3. Second, the harder the encapsulant the stiffer the whole PV module. Hence, colder temperatures lead to a decrease of the PV module deflection during mechanical loading, shown in Figure 9. Finally, going to colder temperatures, the amplitude of the compressive stress increases due to thermal compression. This can be seen in in the third principal stress in Figure 10, which shows the highest amplitude of the third principal stress in the cells at different temperatures from a FEM simulation without mechanical loading.

The first effect causes an increase of the first principal stress. Opposed to that, the latter two effects lead to a decrease of the first principal stress when going to colder temperatures. Above the glass transition at roughly -20 °C the first effect dominates and below it the latter two become more dominant.

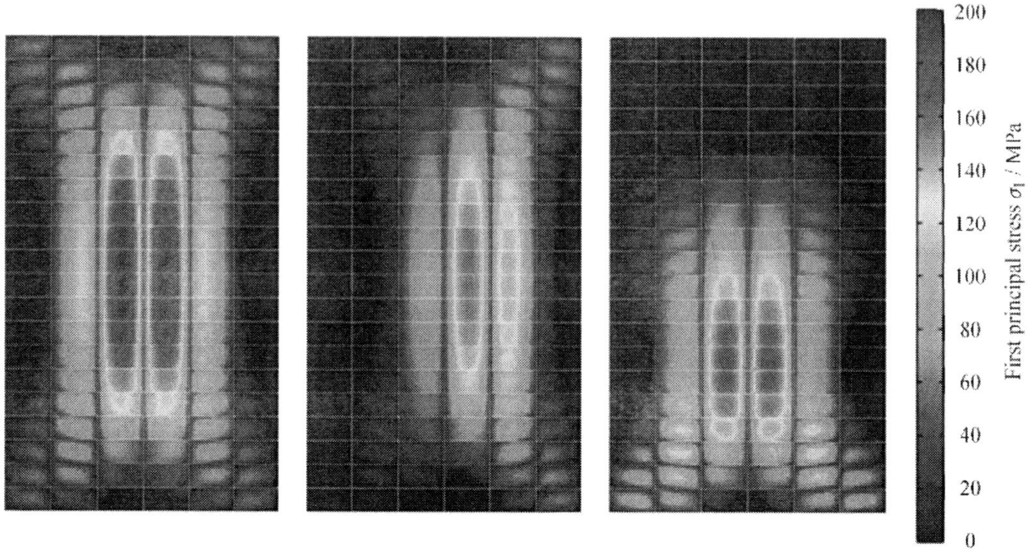

Figure 7: Distribution of the first principal stress σ_I for at an equivalent load of 5400 Pa and 25 °C for a homogeneous load (left) compared to an inhomogeneous loads distributed along the short side of the module (middle) as well as along the module's long side (right).

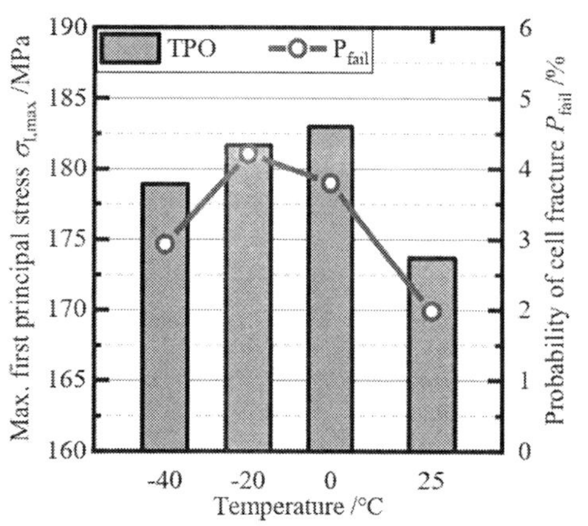

Figure 8: Temperature dependent first principal stress as well as probability of cell fracture for the inhomogeneous distribution with an equivalent load of 5400 Pa

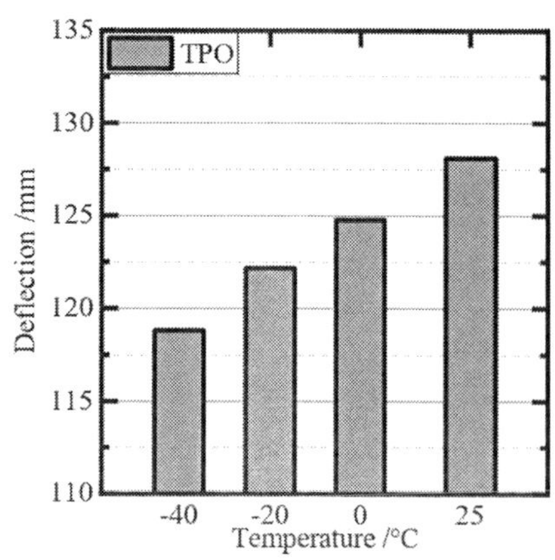

Figure 9: Maximal deflection at an equivalent load of 5400 Pa at different temperatures.

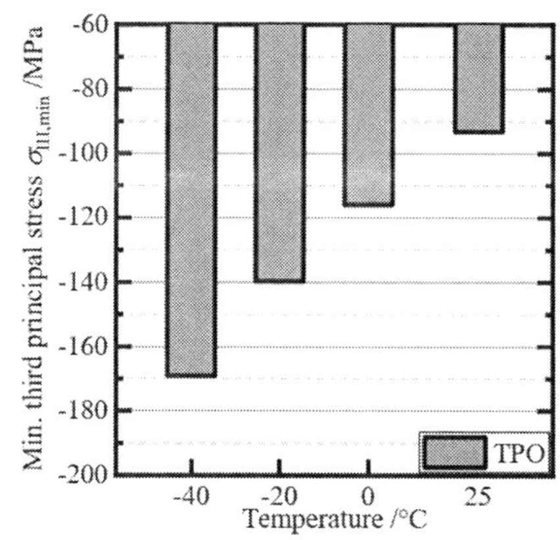

Figure 10: Minimal third principal stress without mechanical loading at different temperatures.

4 CONCLUSIONS

Applying an inhomogeneous load distribution on photovoltaic modules changes the stress distribution in the solar cells. The maximal stress shifts to the lower module edge with higher applied load. Looking at the inhomogeneous load distribution, the maximum first principal stress is significantly lower. Consequently, also the cells are less prone to cell fracture during inhomogeneous loading.

Inhomogeneous loading appears in the field usually not at room temperature, as specified in the IEC 62938, but most often at temperature below 25 °C. Therefore, the influence of the temperature is analyzed. It is shown that three effects determine the PV module's behavior. First, the encapsulant can buffer less stress at colder temperatures. Second, the overall PV module's stiffness increases caused by the increase of the encapsulants storage modulus. This leads to lower deflections at the same load at colder temperatures. Third, the third principal stress, caused by thermal compression increases significantly when going to colder temperatures. The first effect is more dominant at higher temperatures and opposes the latter two effects, that are more dominant at lower temperatures. Therefore, a peak at a specific temperature where the probability of cell fracture is the highest can be seen. Concluding, the inhomogeneous loading leads to a significant change in the solar cell's stress distribution, with a lower overall first principal stress as well as a lower probability of cell fracture. Here it is to mention that also the PV module's frame is loaded differently what may cause other failure pictures and is part of current and future work. Even though the cells are less prone to failure the inhomogeneous load test is a valuable addition to the homogeneous load described in the IEC 61215.

4 REFERENCES

[1] *Terrestrial photovoltaic (PV) modules – Design qualification and type approval – Part 1: Test requirements*, IEC 61215-1:2016, International Electrotechnical Commission (IEC), Geneva, Switzerland, 2016.

[2] *Photovoltaic (PV) modules - Non-uniform snow load testing*, IEC 62938:2020, International Electrotechnical Commission (IEC), May. 2020.

[3] S. Dietrich, "Numerische Untersuchungen zur mechanischen Zuverlässigkeit verkapselter Siliziumsolarzellen," Dissertation, Fakultät für Maschinenbau, Otto-von-Guericke-Universität Magdeburg, Halle (Saale), 2014.

[4] A. J. Beinert, P. Romer, M. Heinrich, M. Mittag, J. Aktaa, and H. Neuhaus, "The Effect of Cell and Module Dimensions on Thermomechanical Stress in PV Modules," *IEEE J. Photovoltaics*, vol. 10, no. 1, pp. 70–77, 2020, doi: 10.1109/JPHOTOV.2019.2949875.

[5] P. Romer, G. Oreski, A. J. Beinert, D. H. Neuhaus, and M. Mittag, "More Realistic Consideration of Backsheets Coefficient of Thermal Expansion on Thermomechanics of PV Modules," in *Proceedings of the 37th European Photovoltaic Solar Energy Conference and Exhibition (EU PVSEC)*, online, 2020, pp. 772–776.

[6] A. J. Beinert, P. Romer, M. Heinrich, M. Mittag, J. Aktaa, and H. Neuhaus, "Thermomechanical evaluation of new PV module designs by FEM simulations," in *Proceedings of the 36th European Photovoltaic Solar Energy Conference and Exhibition*, Marseille, France, 2019, pp. 783–788.

[7] A. J. Beinert, M. Ebert, U. Eitner, and J. Aktaa, "Influence of photovoltaic module mounting systems on the thermo-mechanical stresses in solar cells by FEM modelling," in *Proceedings of the 32nd European Photovoltaic Solar Energy Conference and Exhibition*, Munich, Germany, 2016, pp. 1833–1836.

[8] S. Dietrich, M. Sander, M. Pander, and M. Ebert, "Interdependency of mechanical failure rate of encapsulated solar cells and module design parameters," in *Proceedings of the SPIE Solar Energy + Technology*, San Diego, California, USA, 2012, 84720P-1-84720P-9.

[9] J. Y. Hartley *et al.*, "Effects of Photovoltaic Module Materials and Design on Module Deformation Under Load," *IEEE Journal of Photovoltaics*, vol. 10, no. 3, pp. 838–843, 2020, doi: 10.1109/JPHOTOV.2020.2971139.

[10] M. Aßmus, S. Bergmann, K. Naumenko, and H. Altenbach, "Mechanical behaviour of photovoltaic composite structures: A parameter study on the influence of geometric dimensions and material properties under static loading," *Composites Communications*, vol. 5, pp. 23–26, 2017, doi: 10.1016/j.coco.2017.06.003.

[11] S. Dietrich, U. Zeller, M. Pander, and M. Ebert, "Evaluation of non-uniform mechanical loads on solar modules," in.

[12] M. Aßmus, S. Bergmann, J. Eisenträger, K. Naumenko, and H. Altenbach, "Consideration of Non-uniform and Non-orthogonal Mechanical Loads for Structural Analysis of Photovoltaic Composite Structures," in *Advanced Structured Materials*, vol. 46, *Mechanics for materials and technologies*, H. Altenbach, R. V. Goldstein, and E. Murashkin, Eds., Basel, Switzerland: Springer International Publishing, 2017, pp. 73–122.

[13] A. M. Gabor, R. Janoch, A. Anselmo, J. L. Lincoln, H. Seigneur, and C. Honeker, "Mechanical load testing of solar panels — Beyond certification testing," in *2016 IEEE 43rd Photovoltaic Specialists Conference (PVSC)*, Portland, OR, USA, Jun. 2016 - Jun. 2016, pp. 3574–3579.

[14] G. Muelhoefer, H. Berg, C. Ferrara, W. Grzesik, and D. Philipp, "Influence of mechanical load at low temperatures on cell defects and power degradation in full-scale PV modules," in *Proceedings of the 28th European Photovoltaic Solar Energy Conference and Exhibition*, Paris, France, 2013, pp. 2968–2971.

[15] R. B. Roberts, "Thermal expansion reference data: silicon 300-850 K," *Journal of Physics D: Applied Physics*, vol. 14, no. 10, L163-163, 1981.

[16] R. B. Roberts, "Thermal expansion reference data: silicon 80-280K," *Journal of Physics D: Applied Physics*, vol. 15, no. 9, L119-120, 1982.

[17] U. Eitner, S. Kajari-Schroeder, M. Koentges, and H. Altenbach, "Thermal stress and strain of solar cells in photovoltaic modules," in *Advanced Structured Materials*, vol. 15, *Shell-like Structures: Non-classical Theories and Applications*, H. Altenbach and V. A. Eremeyev, Eds., Berlin/Heidelberg: Springer, 2011.

[18] W. M. Haynes, Ed., *CRC handbook of chemistry and physics*: CRC Press, 2014.

[19] W. Weibull, *A Statistical Theory of the Strength of Materials*. Stockholm: Generalstabens Litografiska Anstalts Förlag, 1939.

[20] D. Munz and T. Fett, *Ceramics: Mechanical properties, failure behaviour, materials selection*, 1st ed. Berlin | Heidelberg: Springer, 2001.

[21] F. Kaule, M. Pander, M. Turek, M. Grimm, E. Hofmueller, and S. Schoenfelder, "Mechanical damage of half-cell cutting technologies in solar cells and module laminates," in *SiliconPV 2018, The 8th International Conference on Crystalline Silicon Photovoltaics: Conference date, 19-21 March 2018 : location, Lausanne, Switzerland*, 2018, p. 20013.

OPTICAL STRAIN AND TEMPERATURE SENSING WITHIN PHOTOVOLTAIC LAMINATES

Nivelle Philippe[1,2,3], Maes :Lowie[1], Poortmans Jef[1,2,3,4], Daenen Michaël[1,3]
[1] University of Hasselt, Martelarenlaan 42, 3500 Hasselt, Belgium
[2]Imec, Kapeldreef 75, 3001 Leuven, Belgium
[3]Energyville, Thor park 8320, 3600 Genk, Belgium
[4]KULeuven, Oude markt 13, 3000 Leuven, Belgium

Summary
Quantifying the lifetime of photovoltaic modules is getting increasingly important with the aim to drive further technological developments, to reduce financial risks and to accelerate the deployment for new photovoltaic applications. Many lifetime limiting factors can be directly or indirect related to temperature and strain acting on the laminate constituents. In this work a novel optical sensing solution is proposed based on Fibre Bragg Grating sensors which enables a direct validation of the thermo-mechanical strain within a PV laminate. A combined sensor package enables a direct monitoring or validation of the thermo-mechanical stress of a PV cell during production, accelerated testing and field testing. Initial testing shows a high potential due to the non-invasive nature, high sensitivity, scalability and optical transparency.

Keywords: thermo-mechanical, Fibre Bragg Grating, in-situ, photovoltaic, modules,

1 Introduction

The lifetime and reliability of Photovoltaic (PV) modules or systems should always be warranted. However, as new module technologies or materials are employed or new applications such as vehicle and building integration are explored, establishing a representative lifetime can become challenging. There are various factors which influence the reliability of PV modules ranging from production-induced stress such as firing or annealing of the metal contacts on the cell, soldering of the cells into strings, lamination of the cells into a module, up to stress induced in the field due to module interactions with the environment (temperature fluctuations, irradiance fluctuations, hail, wind and snow deformations). The various degradation mechanisms, failure modes and their respective impact are generally investigated through a combination of field testing, accelerated aging tests, controlled lab experimentation and simulation. Each variation of testing utilizes specific characterization tools to quantify results.

Field testing largely relies on intermittent characterization using IR imaging, EL-imaging combined with I/V-tracing or power monitoring and visual inspections. As many degradation modes are linked to temperature (module or cell temperatures) thermocouples and PT100's are often used by either laminating them into or attaching them onto the PV module. Although integrating these sensors can be quite invasive on the laminate, continuous thermal monitoring of PV modules is essential as validation or input for simulations (refs).

Standardized stress testing tests such as thermal cycling, according to IEC 61215 and or non-standard combined accelerated stress testing provides a basis for controlled degradation testing [1], [2]. In most cases, similar characterization techniques, as employed during field testing, are used for accelerated stress testing. However, the shorter test-time, controlled conditions and accessibility do allow a more systematic approach to study the performance of various laminate materials and or cell technologies.

During mechanical testing of modules non-contact displacement sensing or 3D scanning [3]–[5] can be used to capture the displacement field of the module. Numerical or analytical simulations can sequentially translate the measured displacements to internal stress and strain levels of the individual constituents. However as argued by , E.H Amalu *et. al.* [6] the accuracy of this approach relies heavily on proper material representation and assumptions used within the models. Advanced characterization methods such as synchrotron X-ray microdiffraction [7], [8] and confocal Raman spectroscopy [9], [10] do allow to directly quantify internal stress and strain levels for small samples. Similarly in-cell sensors have been proposed as an additional tool to directly measure internal cell strain levels and absolute temperatures [11]. In-laminate sensing as the aforementioned technique poses great potential to investigate internal stress and strain values during various loading conditions.

In order to allow for quantization of stressors like strain and temperature on larger scales, physics based simulations are often employed. Simulations pose great potential due to their ability to capture multiple failure mechanisms [12]–[15]. As simulations require assumptions ranging from material models to the abstraction method used, a validation is required. Performing proper validation often proves challenging and in many cases as indicated before rely on indirect measurements.

This work proposes an alternative novel approach to in-laminate sensing which aims to directly quantify thermo-mechanical stress and temperatures during field testing, accelerated stress testing as well as for other lab experimentation. Thereby providing also a valuable tool for the validation of various simulations. This is achieved without the need for adapting any of the laminate constituents and with minimal impact on both the optical and thermo-mechanical behaviour of the module. The proposed sensing solution is optical in nature thereby removing the impact of electromagnetic interference and enabling high accuracy field measurements during operation. Customized optical fibre's containing Fibre Bragg Gratings (FBG) that are integrated into the PV laminate to provide the in-situ thermal and mechanical strain as well the absolute temperature.

2 CONVERTING LIGHT TO AN ABSOLUTE CELL TEMPERATURE

Optical fibres have been employed in telecommunication and sensing applications for many years [16]. Although various types of optical sensors exist, FBG-based sensors have been adopted by many industries and its properties captured our interest given their spatial accuracy, optical multiplexing capabilities and commercial availability.

2.1 Optical principles

An FBG is created by creating a localized periodic change in the refractive index within the core of an optical fibre. This pattern of alternating refractive indexes referred to as a grating will cause a specific wavelength, named the Bragg wavelength, to be reflected back through the optical fibre as indicated in Figure 1. The transmitted spectrum shows an equal drop in power at the Bragg wavelength allowing sensing both in transmission or reflection. Any strain (thermal or mechanical) exerted to the active part of the fibre will cause a shift in this Bragg wavelength. Its sensitivity for thermal and mechanical strain is mainly determined by the integration method used. The wavelengths used are similar to those used in common telecommunication with typical wavelengths ranging between 1520-1580 nm. These specific wavelengths minimize absorption and scattering effects in single mode fibre thereby maximizing attenuation and thus the noise-to-signal ratio

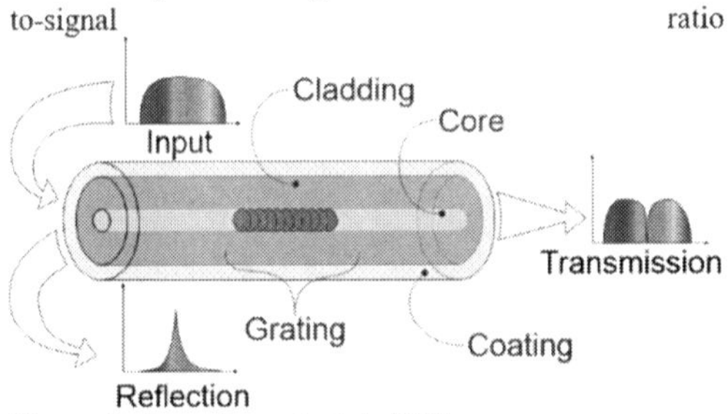

Figure 1: Operating principle FBG sensor

2.2 Temperature sensing packaging

An FBG is by definition a strain sensor. However, by decoupling the fibre from any mechanical influences only the thermal strain is detected and can be converted to absolute temperatures by characterizing the thermal expansion and thermo-optic properties of the fibre.

Decoupling of the fibre from mechanical strain is achieved by placing the fibre in a capillary tube and fixing it only from one side. This leaves one end of the fibre to expand freely within the capillary. Inside the capillary also a lubricating, high density, transparent liquid sits between the fibre and inner walls of the capillary. As indicated in Figure 2 the ends of the capillary are sealed off using a UV curable adhesive. When exiting the laminate, the thin fibre is protected by a sleeve to prevent breakage during manipulation.

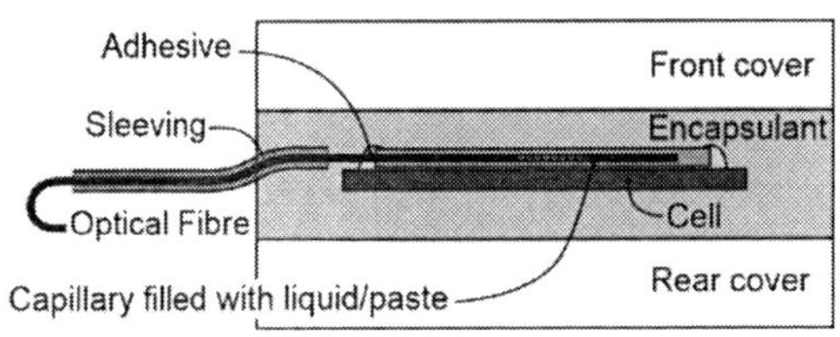

Figure 2: Temperature sensor integration

Typical dimensions used for the quartz capillary is an outer diameter between 300-400 μm and an inner diameter between 200-300 μm. As an encapsulant typically has a thickness of 400 μm or above, the capillary will be fully embedded in the capsulation layer. However, the overall bulk of the packaging can be further reduced to about 100-150 μm without any signifcant changes in processing or approach.

As most materials used in the packaging are optically transparant, there was no measurable impact on the EQE when considering a 5mm square area with the sensor packaging positioned in the middle.

Figure 3: Temperature sensor integration on to the cell

2.3 Accuracy and sensitivity

In order to demonstrate the accuracy of the sensor a single cell PV laminate was prepared using the structure as referred in Figure 2. The sample was exposed to a cyclic temperature profile ranging from -40 °C till 80°C with a steady plateau of at least 30 min at every temperature.

Next the optical parameters were determined based on the stable temperature plateaus in order to translate the measured wavelengths into absolute temperatures. Comparing the temperatures provided by the climate chamber to the ones provided by our sensor a repeatability of ±0.3 °C was achieved as shown in Figure 4. The limiting factor during calibration was found to be the climate chamber accuracy, thus proving that even better thermal accuracy could be achieved if needed.

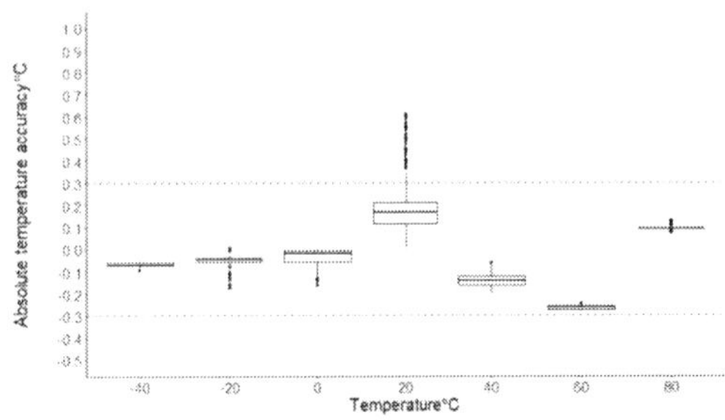

Figure 4: Absolute temperature accuracy as determined by calibration

The temperature sensitivity of the packaging was found to about 9.8pm/°C which provides a relative temperature accuracy of about 0.05°C using an FBG-scan-904 interrogator.

3 THERMOMECHANICAL SENSING

The thermal packaging can be further expanded by including an additional strain fibre next to the temperature sensor as indicated in Figure 5. This allows to decouple mechanical effects from thermal effect which is useful for the validation of various types of simulations. Additional adhesive is applied where the strain sensor is bonded to the cell. The thermo-mechanical sensing package further enhances multiplexing capabilities, as it can be looped throughout the module without the need for additional fibres. With our current setup up to 64 sensing points can be integrated into a single fibre which provides the potential to measure the stress in every cell for an 60 cell module using a single fibre.

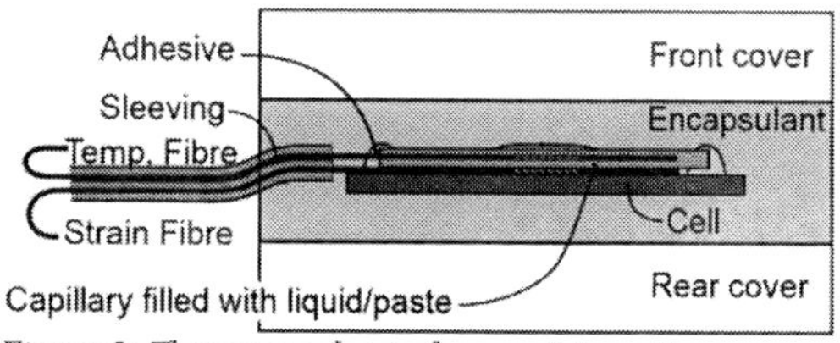

Figure 5: Thermo-mechanical sensor integration

5 APPLICATIONS

Various PV applications and tests can benefit from this technology. However as the sensing solution is already present during production of the laminate, an obvious application is apply it for process optimization.
In Figure 6, a lamination process is shown for a glass-glass laminate using a polyolefin encapsulant. A lamination temperature of 165°C was set and also reached.

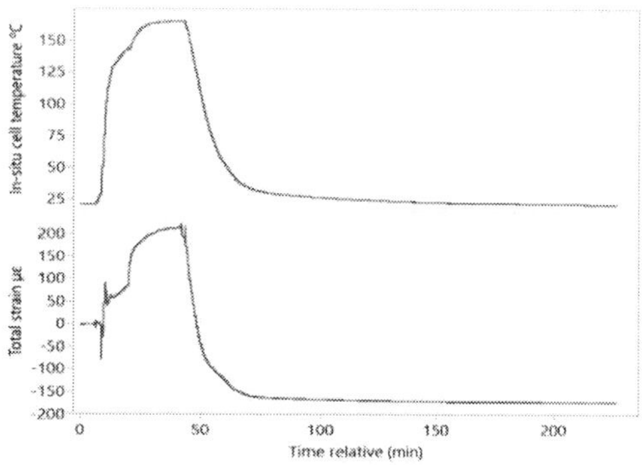

Figure 6 In-situ temperature and strain development during lamination

The strain response reveals important steps in the lamination process, such as the placing of the laminate in the laminator (first peak), starting of the lamination process (second peak), start of the pressing stage (sudden increase in temperature) and end of the process (sudden drop in strain). Due to the thermal mass of the sample, the temperature does not decrease immediately. The sample was naturally cooled down to room temperature. Failed lamination recipes also revealed the ability to detect cell cracks during the lamination process. These were also observed as discrete steps in the lamination strain after the process initiated. For the considered encapsulant a final compressive strain of about 170με was measured.

6 CONCLUSION and OUTLOOK

An novel in-situ optical thermo-mechanical sensing solution is proposed with a demonstrated accuracy of ±0.3 °C. The optical transparency of the solution was found to have no significant impact on the optical performance of the module. This show great potential for its use in field applications. The thermo-mechanical packaging was successfully employed to detect various effects occurring during lamination. The technology and packaging is fully scale-able from lab-scale up to field tests. Future work will focus on applying this technology to demonstrators of emerging applications such as infrastructure-integrated PV and building-integrated PV. Furthermore, various failure modes will be investigated using the technology.

6 ACKNOWLEDGEMENTS

This study has received funding from the projects SUNOVATE, ROLLING SOLAR and DAPPER, financed by the cross border collaboration program Interreg Flanders-Netherlands and EMR with financial support of the European Funds for Regional Development. This work has also been supported by Flanders Innovation & Entrepreneurship and Flux50 under project DAPPER, HBC.2020.2144

[1] "IEC 61215, Crystalline silicon terrestrial photovoltaic (PV) modules: Design qualification and type approval."

[2] M. Owen-Bellini *et al.*, "Advancing reliability assessments of photovoltaic modules and materials using combined-accelerated stress testing," *Prog. Photovoltaics Res. Appl.*, vol. 29, no. 1, pp. 64–82, 2021, doi: 10.1002/pip.3342.

[3] W. H. Coulter, R. C. Czyzewicz, W. E. Farneth, A. Prince, and R. G. Rajendran, "Using finite element methods and 3D image correlation to model solar cell bowing," *Conf. Rec. IEEE Photovolt. Spec. Conf.*, no. 2, pp. 000262–000267, 2009, doi: 10.1109/PVSC.2009.5411682.

[4] P. Yoon, T. Baek, H. Chung, H. Song, and S. Shin, "Numerical simulation of bowing phenomenon in ultra-thin crystalline silicon solar cells," *Sol. Energy*, vol. 105, pp. 705–714, 2014, doi: 10.1016/j.solener.2014.04.027.

[5] T. Geipel, L. C. Rendler, M. Stompe, U. Eitner, and L. Rissing, "Reduction of Thermomechanical Stress Using Electrically Conductive Adhesives," *Energy Procedia*, vol. 77, pp. 346–355, 2015, doi: 10.1016/j.egypro.2015.07.049.

[6] E. H. Amalu, D. J. Hughes, F. Nabhani, and J. Winter, "Thermo-mechanical deformation degradation of crystalline silicon photovoltaic (c-Si PV) module in operation," *Eng. Fail. Anal.*, vol. 84, no. October 2017, pp. 229–246, 2018, doi: 10.1016/j.engfailanal.2017.11.009.

[7] S. K. Tippabhotla *et al.*, "Thermomechanical residual stress evaluation in multi-crystalline silicon solar cells of photovoltaic modules with different encapsulation polymers using synchrotron X-ray microdiffraction," *Sol. Energy Mater. Sol. Cells*, vol. 193, no. January, pp. 387–402, 2019, doi: 10.1016/j.solmat.2019.01.016.

[8] S. K. Tippabhotla, W. J. R. Song, A. A. O. Tay, and A. S. Budiman, "Effect of encapsulants on the thermomechanical residual stress in the back-contact silicon solar cells of photovoltaic modules – A constrained local curvature model," *Sol. Energy*, vol. 182, no. October 2018, pp. 134–147, 2019, doi: 10.1016/j.solener.2019.02.028.

[9] A. J. Beinert, P. Romer, A. Büchler, V. Haueisen, J. Aktaa, and U. Eitner, "Thermomechanical stress analysis of PV module production processes by Raman spectroscopy and FEM simulation," *Energy Procedia*, vol. 124, pp. 464–469, 2017, doi: 10.1016/j.egypro.2017.09.282.

[10] A. J. Beinert *et al.*, "Enabling the measurement of thermomechanical stress in solar cells and PV modules by confocal micro-Raman spectroscopy," *Sol. Energy*

Mater. Sol. Cells, vol. 193, no. January, pp. 351–360, 2019, doi: 10.1016/j.solmat.2019.01.028.

[11] A. J. Beinert *et al.*, "Silicon solar cell–integrated stress and temperature sensors for photovoltaic modules," *Prog. Photovoltaics Res. Appl.*, vol. 28, no. 7, pp. 717–724, 2020, doi: 10.1002/pip.3263.

[12] M. Paggi, M. Corrado, and M. A. Rodriguez, "A multi-physics and multi-scale numerical approach to microcracking and power-loss in photovoltaic modules," *Compos. Struct.*, vol. 95, pp. 630–638, 2013, doi: 10.1016/j.compstruct.2012.08.014.

[13] M. Pander, S. Dietrich, S. H. Schulze, U. Eitner, and M. Ebert, "Thermo-mechanical assessment of solar cell displacement with respect to the viscoelastic behaviour of the encapsulant," in *EuroSimE*, 2011, pp. 5–10, doi: 10.1109/ESIME.2011.5765831.

[14] N. Bosco, T. J. Silverman, and S. Kurtz, "The Influence of PV Module Materials and Design on Solder Joint Thermal Fatigue Durability," *IEEE J. Photovoltaics*, vol. 6, no. 6, pp. 1407–1412, 2016, doi: 10.1109/JPHOTOV.2016.2598255.

[15] P. Lenarda and M. Paggi, "A geometrical multi-scale numerical method for coupled hygro-thermo-mechanical problems in photovoltaic laminates," *Comput. Mech.*, vol. 57, no. 6, pp. 947–963, 2016, doi: 10.1007/s00466-016-1271-5.

[16] G. Allwood, G. Wild, and S. Hinckley, "Fiber bragg grating sensors for mainstream industrial processes," *Electron.*, vol. 6, no. 4, pp. 1–19, 2017, doi: 10.3390/electronics6040092.

RELIABILITY ASSESSMENT FOR INDUSTRIAL SOLDERED AND GLUED BJ-BC SOLAR CELL INTERCONNECTIONS

Authors: Tudor Timofte[1], Stephan Großer[2], Robert Wade[3], Thomas Fischer[3], Dominik Rudolph[1], Johannes Linster[1], Maria Ignacia Devoto[1], Daniel Tune[1], Ruediger Farneda[1], Frank Köhler[3], Andreas Halm[1]
[1] ISC Konstanz, Rudolf-Diesel-Str. 15, 78467 Konstanz, Germany
[2] Fraunhofer-Center für Silizium-Photovoltaik CSP, Otto-Eißfeldt-Str. 12, 06120 Halle (Saale), Germany
[3] teamtechnik Maschinen und Anlagen GmbH, Planck Str. 40, 71691 Freiberg, Germany

ABSTRACT: The metallization scheme of the back-junction back-contact "ZEBRA" cell, which is a configuration based on several metallization and insulation paste layouts, having all busbars located at rear side [1], require only process adaptation at already existing automated mass production interconnection equipment, as well as comprehensive qualification. By means of that, also based on previous research [2], two interconnection technologies were considered for investigation: soldering and gluing with electrically conductive adhesive for PV (ECA), whereby one is epoxy- and the other one acrylic-based, both with silver particle fillers. For these contacting media available industrial equipment was used, a teamtechnik TT2100 stringer for soldering process, with a new interdigitated back contact cell (IBC) process and a teamtechnik TT1600ECA stringer for gluing process, also with process adaptation for IBC cell type. Both these new processes implied trials and several characterization techniques for identification of capability and for optimization. For testing long-term reliability of the joints, a setup for investigation of interconnection susceptibility to ageing factors (UV, temperature cycling and damp heat) was defined for unprotected samples (strings) and for samples protected by encapsulation (laminated). A suitable testing tool for our approach was the peel force measurement, which is a direct measure for the quality of the joint. Main target was to identify the optimal characterization and features specific for an industrial ready process for the IBC "ZEBRA" cells.

Keywords: back contact, solar cell interconnection, qualification and testing

1 INTRODUCTION

While back-junction back-contact solar cell technology, named "ZEBRA" is since 2019 at mass production implementation level [3], further trials are necessary for new processes regarding intensive qualification and understanding of material and process interactions, mainly at cell interconnection. In a previous study [2], it was not possible through these two interconnection technologies: soldering or gluing with conductive adhesive, to reduce the full cell bowing to an acceptable processing level in range of 2 mm (best bowing results for 156.75 x 156.75mm IBC cell were in range of 7 mm). In agreement with market trends [4] and with confirmed consistent IBC cell warpage reduction after interconnection of IBC half-cells (in 2.0 mm range); we used only this format for the tests of this study. For creating the industrial process environment, two standard available stringers (teamtechnik TT2100 and teamtechnik TT1600ECA) were involved, adapting the front-back contacting process to IBC "ZEBRA" cell interconnection pattern. For soldering $Sn_{60}Pb_{40}$ coated PV ribbon was used, whereas for gluing Ag coated PV ribbon was chosen. Two established and market available types of conductive adhesives were selected: an epoxy- and an acrylic-based ECA, both with Ag particle fillers. Automated processes are the highlight of this study, since these depict with correct setup the reproducible industrial environment. By means of that, three directions were projected for this investigation:

1. Initial process characterization
2. Characterization of the created interconnection
3. Investigations on the reliability of the interconnection

For identification of an industry ready adequate interconnection process, this study contains the first step of comprehensive process qualification. To achieve that, cross section, peel force and interconnection fracture nature, differential scanning calorimetry [5], sensibility and reliability of interconnection at small cell batch level, with 20 - 30 half-cells per testing group were investigated, for three soldering process parameter variations and two ECAs with same interconnection stringer parameter setup.

2 EXPERIMENTAL

For our tests, mono crystalline IBC ZEBRA cells with M2 wafer format (156.75 x 156.75 mm), as well as G1 wafer format (158.75 x 158.75 mm) and specific rear side metallization with four busbars per polarity, each with a width of approx. 1.5mm, were used. The metallization of busbars consist of low temperature silver paste, similar to the silver paste used for heterojunction cell type metallization, as uninterrupted pattern for each busbar for half-cells. These cells were produced partly at ISC Konstanz e.V., partly at an industrial production plant. This cell batch variation was considered in our study (no testing groups with cells mixed from two batches). Furthermore, to verify the implication of busbar paste on the interconnection process, for a soldering experiment two cell batches with two different busbar metallization pastes were used. All cells for the experiment were segmented into halves at ISC Konstanz with an LSC (laser scribing and mechanical cleavage) process. For the interconnection of cells to strings, PV ribbons were used, with $Sn_{60}Pb_{40}$ alloy coated copper core for the soldering ribbon and Ag_{100} coated copper core for the ribbons used for ECA interconnection. For soldering process a low residue, halogen free and non-rosin, common industrial flux was used. The typical sample type for soldering or gluing process implied a two half-cell string with 2 mm cell-cell distance, which was used for further testing and characterization, mentioning that at soldering process also a testing group with 4 half-cells and 2 mm cell-cell distance was produced.

Soldering samples in this study were produced exclusively automatically at a teamtechnik TT2100 single-track stringer with a new modified process for IBC ZEBRA cell. The PV ribbon tracks mounted and used at stringer have the rated copper core geometry of 1.00 x 0.20 mm, $Sn_{60}Pb_{40}$ coating thickness in range 16-20 µm and offset yield strength ($Rp_{0.2}$) 68 N/mm^2. Fluxing of cell busbars took place automatically via nozzles. The ribbons were racked by 1.28 % for soldering process, cut at set length prior automatic positioning on cells and mechanical fixation by special down holders during conveyor transportation. Only the modified parameters (Table I) will be referred to in the next pages. The non-referred parameters from the stringer recipe are within the range of manufacturer recommendations and were not target for adjustment in this study.

Table I: Soldering process parameter variation

Soldering Test Recipe	1	2	3
Dwell Time per heating zone (ms)	1680	2200	2200
Setup IR Pre-Soldering (°C)	190	190	180
Setup IR Soldering (°C)	220	220	210
Fluxing Speed (mm/s)	500	500	400

A schematic diagram of the soldering process is represented in the figure 1 below.

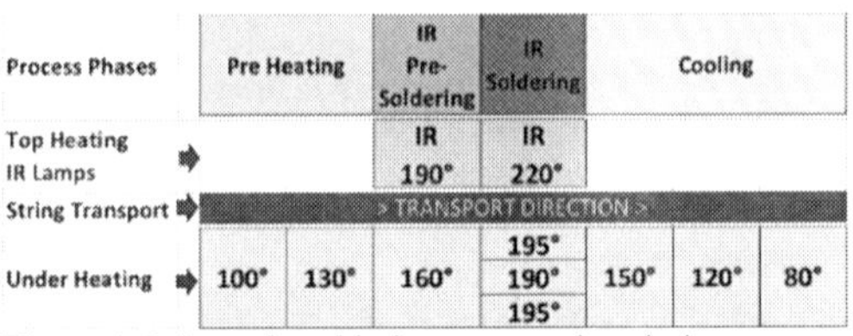

Figure 1. Schematic soldering process description

For half-cells, their dwell time on each of the process heating plates is double as presented in Table I, since cycle time is cell length dependent.

Three half-cell strings (with two half-cells each) per testing group were prepared for peel tests (whereas all available ribbons for testing were peeled / eight per string). An additional testing group with two half-cell strings, with 4 half-cells each, was prepared for testing the dependency of peel force on the position of cell in string and transport direction, whereby at all sample strings only the ribbons from the first and last half-cells were peeled.

With soldering recipe 2 were produced at a later point half-cell strings (PV ribbon rated technical data: 1.20 x 0.24 mm copper core ; 16-20 µm $Sn_{60}Pb_{40}$ coating; $Rp_{0.2}$ 64 N/mm^2) for comparison purposes with glued test groups, since another cell batch was used. Variations of fluxing speed compared to original recipe 2 setup were necessary (new setup: 350 mm/s), since the flux quantity had to be adjusted for adequate processing, meaning no flux excess, however sufficient to wet the entire busbar surface.

Cells interconnected by glued ribbons were produced mostly with an automatic process at a dedicated teamtechnik TT1600ECA single-track stringer. This machine is fitted with an automated screen-printing unit, which was adapted for IBC ZEBRA cell layout. For qualification of manual processing intervention, a semiautomatic mode was tested, where all the processing steps took place automatically, except application of ribbons (stretched prior use by 0.3 % and cut at desired length) on cell's busbars by manual positioning, without contamination of ECA or ribbon surface. The ribbon for automatic process was silver-coated LCR (light capturing ribbon) ribbons, racked by 0.65 % for ECA process with a rated copper core geometry: 1.20 x 0.22 mm and $Rp_{0.2}$ in range 90-100 N/mm^2. For semiautomatic process Ag_{100} coated ribbons were used with rated core dimension 1.50 x 0.20 mm and $Rp_{0.2}$ 77 N/mm^2 . The necessary energy for curing of glue at interconnection is delivered through the heating plates on transport path, which contains one additional heating plate compared to soldering stringer (to increase available curing time) and special thermal insulating lids over heated path for better process heat homogeneity. At figure 2 below is presented schematically the gluing process.

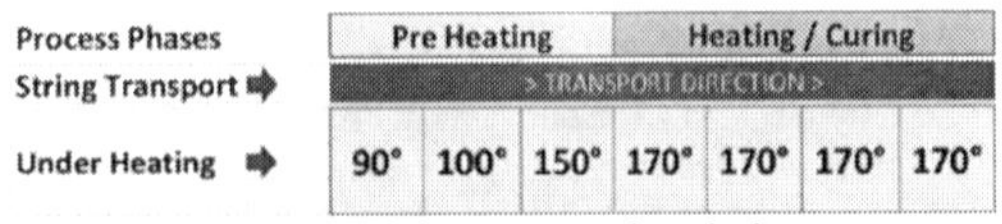

Figure 2. Schematic gluing process description

At the gluing process, two different, however market relevant ECAs were used: one as single-component delivered ECA with acrylic-based polymer matrix, and Ag particle conductive filler, named subsequently ECA1, and one as two-component delivered ECA with epoxy-based polymer matrix, and Ag particle conductive filler, named hereafter ECA2 (the ECA2 was mixed shortly before usage according to producer guidelines). For both ECAs was used the setup presented in figure 2, with a process step time of 2.5 s, meaning, that on each heating zone a half-cell spends 5 s. Furthermore, both ECAs were printed with same screen print parameters: printing speed = 400 mm/s; snap off = 2.0 mm and squeegee pressure = 3 bar.

For this process setup slightly different screens (due to optimization of printing position on busbar and trial for ECA quantity reduction per cell) were used. For ECA1 a screen print of type: polymer mesh 40-70x45° / FL-190 20-32µm EOM was implemented, and for ECA2 a screen print of type: high strength stainless steel mesh 250-36x45° / FL-190 SD+20µm EOM was used, which cause certain limitation for the comparison between the two ECAs. With the above-described process approximately 31-33 mg ECA was printed on a half-cell for ECA1 and approximately 26 – 28 mg for ECA2 (15.6 wt.% less at ECA2 compared to ECA1 process). Both amounts were determined per ECA type by weighing three consecutively printed cells at an analytical balance. For ECA process, two additional features are important to consider, the geometry of the ECA printed pattern (figure 3 for ECA2)

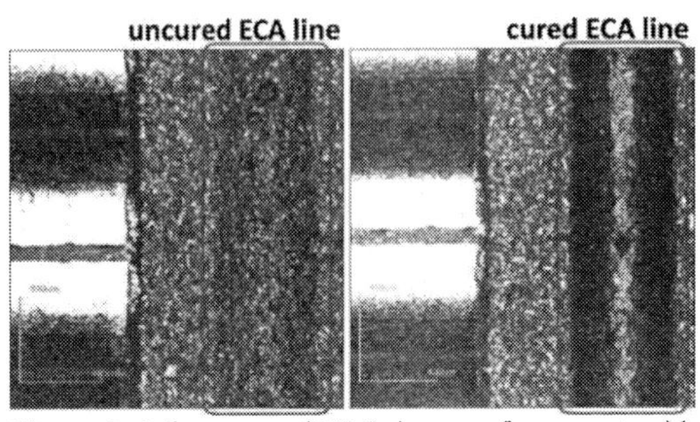

Figure 3. Microscopy (LSM) image of segment with printed ECA2 pattern (left image: uncured ECA pattern / right image: cured ECA pattern)

and the crosslinking or curing of the ECA, since it is essential for reliability purposes to verify if within a process run the ECA is fully cured. For this aspect, the versatile differential scanning calorimetry was deployed [5]. To measure the degree of ECA curing, following approach was used: at first step a fresh uncured ECA sample was measured at DSC (Netzsch 214 Polyma) under nitrogen protective gas stream (protective N_2 flow rate: 40 ml/min; purge N_2 flow rate: 60 ml/min). A defined setup regarding sample weight (10 mg range), heating and cooling ramps (10K/min) and heating up to max. 200°C was used. The exothermal curing peak of each ECA, which corresponds to 100% curing (two further heatings within same measurement confirm full curing), was recorded. In the second step, a sample of same ECA (with weight in same range as for first DSC in order to avoid weight related DSC peak shifting) was introduced in a DSC crucible and sent on a solar half-cell used as carrier through the ECA stringer process. The crucible with cured ECA was measured at DSC: no peak at first heating of this DSC measurement means that ECA is cured completely (or eventual residual curing lies under detection limit of DSC unit) at stringer process; any residual peak which occurs within the range of the peak from the uncured sample means most probably partial curing. Below are presented the combined DSC diagrams of uncured and at process cured ECA samples for both ECAs, in figure 4 for ECA1 and in figure 5 for ECA2.

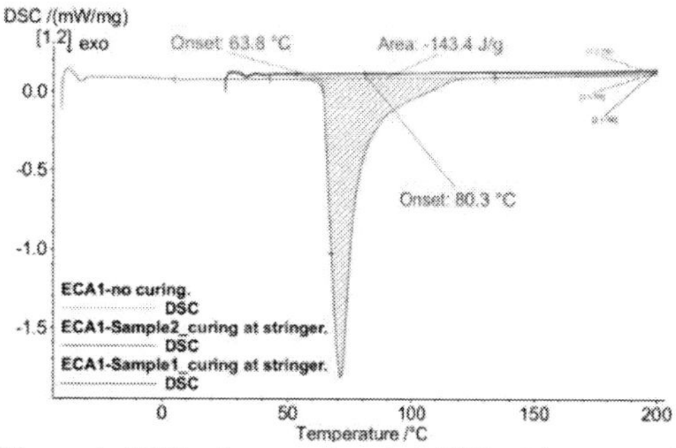

Figure 4. DSC of uncured and at ECA stringer cured samples for ECA1 (all 3 lines represent first heating patterns)

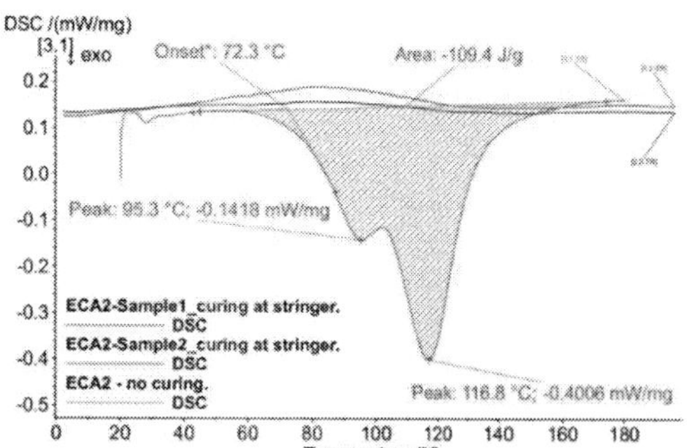

Figure 5. DSC of uncured and at ECA stringer cured samples for ECA2 (all 3 lines represent first heating patterns)

It can be concluded after both inspections that shape and aspect ratio of the printed ECA lines is constant through the entire length of sample with an average line width by 450-470 µm for ECA1 and 480 - 500 µm for ECA2. DSC measurement confirms broad curing (> 99%) for both ECAs with two remarks: for ECA1 a weak DSC peak was detected at one cured sample (marked in fig. 4 with an onset of red line at 80.3°C), which could be correlated to residual curing, corresponding to approx. 0.26 % residually cured ECA1 (ratios of cured /uncured sample peaks). At DSC diagrams of ECA2 samples, additional phenomena can be observed, which is described by a broad, low intensity peak in temperature range 20°C – 130°C. According to literature [6] and to position, shape and behaviour (peak size reduction by drying the second sample for approximately 30 minutes under vacuum: blue line at fig. 5), this peak is assigned to trapped moisture.

Since first process inspection for ECAs reveal good results regarding dispensing pattern and confirmed full curing of ECAs, qualification procedure continues, using the mentioned ECA process setup, with two directions:

1. Initial characterization, which refers to peel force of the soldered or glued interconnection, regarding different variable dependencies and its relation to the associated cross section of samples and
2. proof of interconnection susceptibility towards environmental factors (UV-, temperature cycling and damp heat tests with testing profile according to IEC 61215-2 norm), as well as interconnection reliability study at mini-laminate level by means of accelerated climate chamber tests (temperature cycling and damp heat).

With samples produced at both soldering and ECA processes (schematic representation of sample in the figure 6 below /red arrow show transport direction, interrupted red lines mark the string separation regions) tests were made. Interconnection cross section investigations, as well as comprehensive peel and environmental, respectively accelerated weathering testing were conducted to identify features, dependencies and sensitivities of both processes and involved materials.

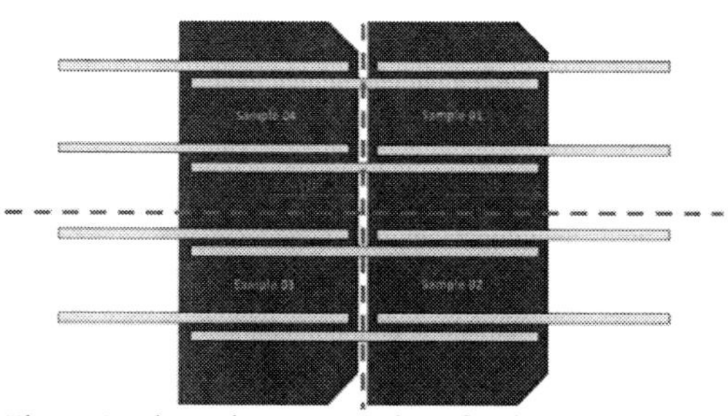

Figure 6. Schematic representation of main sample type of this study: two half IBC cell strings interconnected by means of soldering or ECA through PV ribbons

Cross section analysis was made using contrast mode microscopy (dark field / bright field) at metallographically polished samples from different positions of a two half-cell string. For soldering process only a control, measurement cross section was considered.

For the peel force test quarter cell segments were used. This sample shape was determined by the sample string type (half-cell string) and the fixation possibility of samples at the peel force-testing unit (ZwickRoell BT1-FR0.5TN.D14 testing machine) with special fixture, consisting of string holder and peel clamp for ribbons. This sample type was used for initial characterization of peel force, as well as for investigation of interconnection susceptibility towards environmental factors like UV, temperature cycling and damp heat.

3 RESULTS

The findings of this study will be sorted in three categories for better structure and understanding:

1. Initial peel force analysis for soldering and gluing processes,
2. Susceptibility of stand-alone (no encapsulation or other protection) soldered of glued interconnection toward environmental factors: UV, temperature cycling and damp heat and last,
3. Reliability study of interconnection at laminate level by means of temperature cycling and damp heat test.

Parallel to peel and reliability tests for the ECA interconnections comprehensive cross section analysis was made, considering two aspects: to investigate via microscopy the interconnection interfaces and to determine its geometry (width and thickness). The observed voids at interconnection occurred during sample production at both ECAs, however for the identified joint cracks it is not clear if those were present before or after cross section sample preparation.

For acrylic ECA1 following results were obtained: minor ribbon-busbar offset for automatic and semiautomatic processes, incomplete interconnection through entire available busbar-ribbon common width (due to narrow printed ECA pattern); widening the printed line can be considered as an option to increase adhesion. At different positions, voids and cracks could be noticed in the ECA or at the interface between ECA and ribbon, respectively busbar. Figure 7 reveals several exemplary pictures at cross section of ECA1 with acceptable interconnection (images a, b – at b image only at edge a fine narrow fissure marked with red arrow) and rough interconnection issues, like voids, incomplete interconnection (c) or cracks (d).

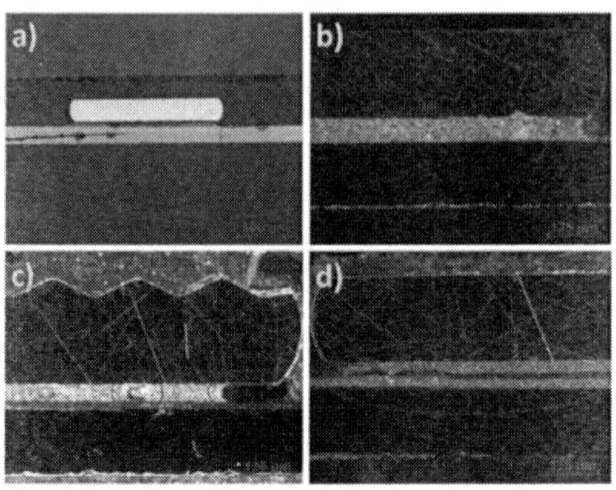

Figure 7. Cross section images of ECA1 interconnection with selected good and bad sections (a, b – acceptable sectors / c, d –sectors with defects)

Epoxy ECA2 reveals a similar picture as ECA1 (acceptable interconnections sectors in pictures a and b from fig.8/ at picture b a small crack is marked by red arrow), with the mention that at ECA2 the voids (pictures c, d at fig. 8) are larger and more common than at samples with ECA1. Since ECA2 is a two component ECA, which was mixed on site before screen-printing and deliberately without degassing prior use, to evaluate the impact of this sample preparation for this specific ECA formulation, this is most probably the cause for the observed increased incidence of holes at the interconnection.

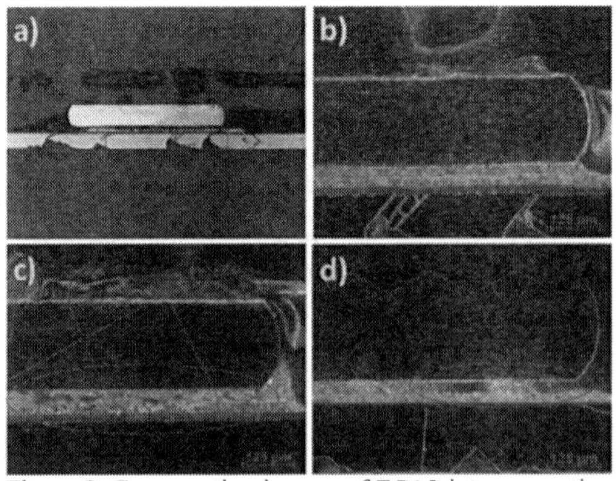

Figure 8. Cross section images of ECA2 interconnection with selected good and bad sections (a, b – acceptable sectors / c, d – sectors with defects)

Thickness and width at cross section were analysed at different positions of the two half-cell string, as marked in the figure 9 below. Sample at the string extremities (pos.1,2,4,5,6 and 8) have approximately 10 mm clearance to wafer edge; otherwise pos. 3 and 7 are at the centre of each half-cell.

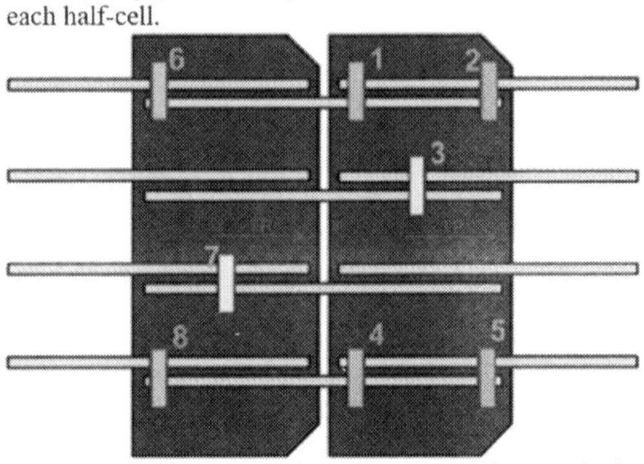

Figure 9. Marking of positions for cross sections analysis

The distributions of thicknesses (fig.10) and of width (fig.11) for ECA2 sample were compared through string surface. For the correlation of measured position, the colour code of sampling in fig.9 must be considered.

Figure 10. Distribution of thickness at ECA2 string sample for automatic and semiautomatic process

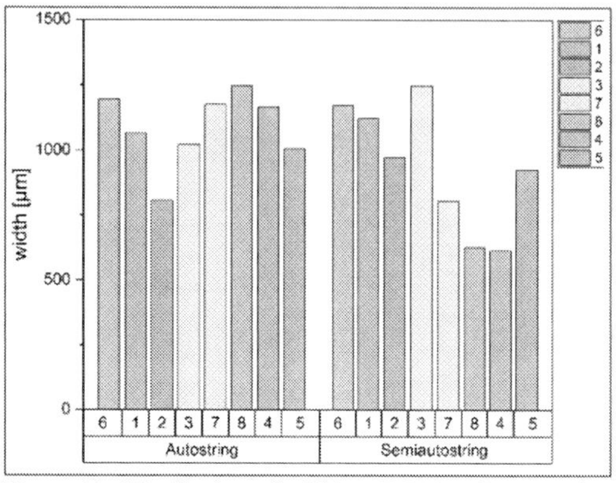

Figure 11. Distribution of width at ECA2 string samples for automatic and semiautomatic process

A clear correlation between thickness, respectively width fluctuations and position at string for both process types could not be identified so far, however the observed variations correlate most probably with the position of down holder blades on ribbons, this effect being under investigation.

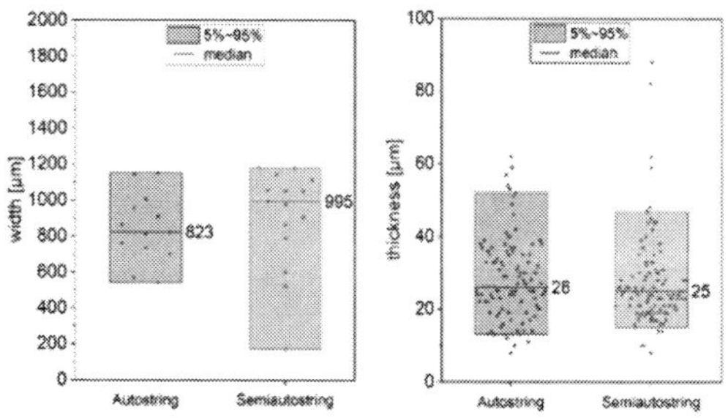

Figure 12. Distribution of width and thickness for ECA1 for automatic and semiautomatic process with median values displayed at boxplots

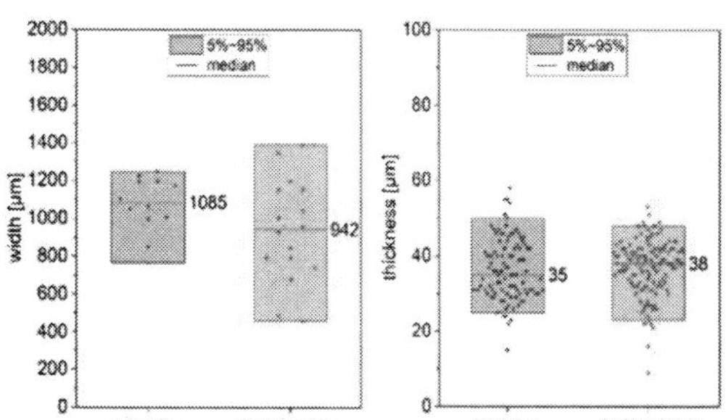

Figure 13. Distribution of width and thickness for ECA2 for automatic and semiautomatic process with median values inserted at boxplots

At one cross section of soldered interconnection with ribbon's rated copper core geometry: 1.20 x 0.24 mm, an effective interconnection width of 1175 µm and thickness of 23 µm were measured (fig. 14). These values will be considered as guideline for the soldered string samples at following peel tests analysis, since this effective interconnection width is almost as wide as the copper core width of ribbon. Furthermore, this complies with observed peel image of soldered samples (broad usage of available width, which is the width of ribbon, since the widest used ribbon is as wide as the busbar width of the cells).

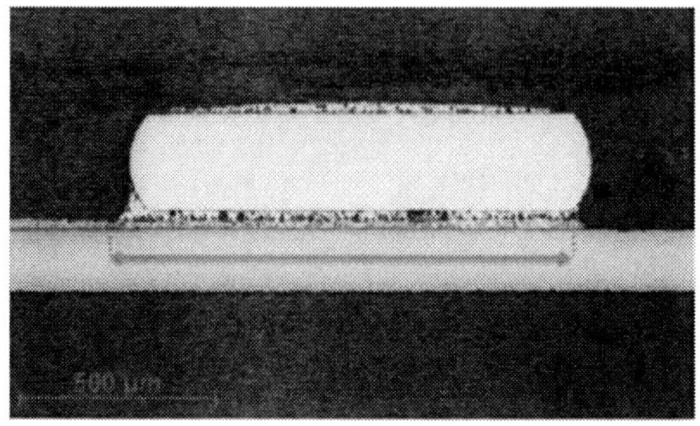

Figure 14. Cross section of a soldering interconnection sample used in this study, revealing interconnection through almost entire ribbon width

3.1 Initial peel force analysis for soldered and glued interconnections

For each peel diagram, the first 10 mm and last 15-20 mm were deliberately removed to increase reliability of analysed data, since initial and final segments of peel diagrams contain measurement artefacts.

For the three testing groups mentioned in experimental section, peel test samples were measured at a ZwickRoell force tester. For a reliable comparison of data sets, only the peel data of the second cell in string for each of all three groups was considered for the results discussed below.

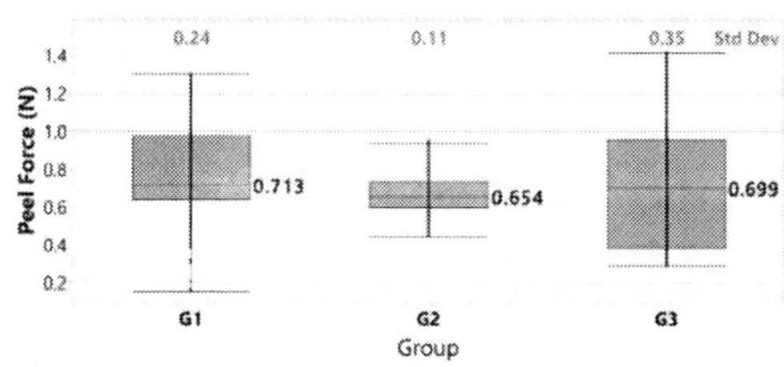

Figure 15. Variation of peel force by the variation of soldering process for three tested groups (median values displayed at boxplots)

Figure 15 reveals a wide spread at test groups G1 and G3, which is an indicator for unstable process. Group G2 has an intermediate median peel force value; however, it has the narrowest distribution from all three tested groups (indicating most stable process). It seems a longer dwell time for the ZEBRA cells on heating zones has a positive effect (this being the only difference between G2 and G1). For further soldering tests only the setup from G2 will be considered (with slight variations at fluxing speed for proper fluxing result: full fluxed busbar area). Having the soldering setup defined (G2), process variations could be evaluated for dependencies.

Neither for soldering process, nor for gluing process could be identified a clear dependency of peel force by busbar polarity, therefore the peel force is in same range for positive and negative busbar polarities, without clear differentiation.

Also between cells with the same material setup (metallization pastes) and processing, but different batches, could not be found a clear differentiation at peel force.

However, it could be identified a correlation of peel force by busbar paste. In figure 16 the clear difference at peel force between two groups with different busbar pastes (different paste producers and different drying processes) is visible. Also at this test, only peel data from second cell in string was analysed for each of both groups.

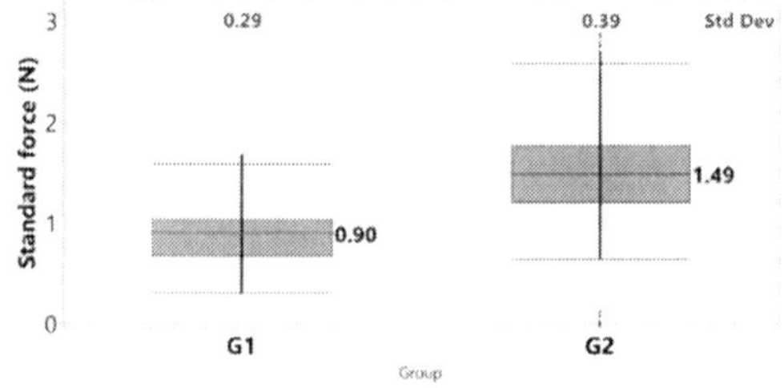

Figure 16. Differentiation of peel force depending on the busbar metallization paste

Group 2 has approximately 65% higher peel force than Group 1. It was noticed that a poor solder alloy wetting (high contact angle of molten solder alloy on busbar, evident in the formation of islands outside of the leading edge of the molten alloy) might be the cause for lower adhesion. None of several tested soldering fluxes could improve clearly the contact between busbar and soldering alloy. Either physical (surface roughness) or chemical features of busbar, or a combination of both lead to the observed differentiation.

At stringing process was identified a peel force gradient depending on the position of cell in string. This is assumed to occur due to the start-stop nature of the R&D trials, which influence the homogeneous temperature distribution through the entire string. We investigated the effect up to the fourth cell in string (fig.17), with cell 1 being the first cell in string and cell 4 cell being the fourth cell in string in transport direction (only peel tests at first, second and fourth cell were measured).

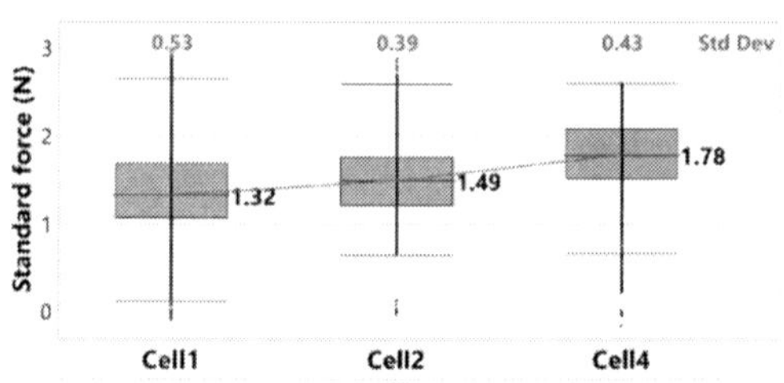

Figure 17. Differentiation of peel force (median value) in string by the position of cell in transport direction

For better visualization, and relevance, only the group 2 (G2) results are presented in figure 17, however this trend was confirmed by test data from group 1 as well.

No clear differentiation or trend could be identified at peel force between the two outer peeled ribbons on each cell and the two inner peeled ribbons, which imply a homogeneous process perpendicular to transport direction.

Regarding glued interconnections, minor differences could be noticed at peel force between both ECAs, which are considered related to voids and poor surface contacting, as well as related to the different printed ECA amount between the two groups (delta of 15.6 wt.%).

For a reliable qualification, 24 ribbons were peeled at ECA1 group, 34 ribbons peeled at ECA2 group and 28 ribbons were peeled at the soldering group.

At figure 18 can be seen the distribution of peel force for three investigated interconnections: ECA1 (acrylic), ECA2 (epoxy) and soldering ($Sn_{60}Pb_{40}$). It is interesting to notice at this graph not only the medians of each interconnection, but also the maximum values for each interconnection medium, as a proof that consistent optimization potential exists (maxima for ECA1 at 1.77 N and for ECA2 at 2.06 N).

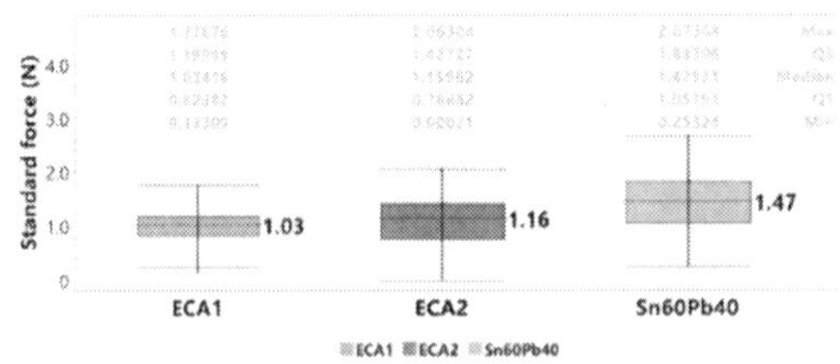

Figure 18. Peel force for the three interconnection media: ECA1, ECA2 and soldering

It was assumed, since no infrared lamp heating is involved at ECA process, that no dependency between peel force and cell position in string exists, which was confirmed at ECA1 batch (fig.19).

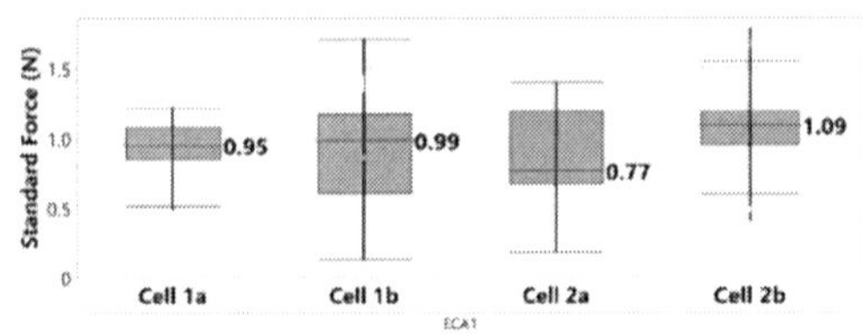

Figure 19. Peel force for ECA2 with differentiation at the cell position in string

Compiling all the acquired data regarding process dependencies and features, a robust comparison method is possible between both ECA interconnections and the soldering process. The basis of the comparison is the peel force data as standard force in N, which is divided for each test group by the corresponding cross section effective interconnection width (overview of results in Table III). For this comparison only peel data from one cell batch and one process type were used.

Table III: Comparison of peel force between interconnections at full automatic processing

Testing Group	Soldering	ECA1	ECA2
Process	Automatic	Automatic	Automatic
Test Volume (Strings)	28	24	34
Data Points	18417	15364	11438
Peel Force (N) (Median)	1.47	0.98	1.18
Eff. Width (mm) (Median)	1.17*	0.82	1.08
Peel Force (N/mm)	1.25	1.20	1.09

*one cross section measurement as guideline

Peel force is by 4 % higher at soldering group compared to ECA1 group, whereas peel force of ECA2 group is by 12.8 % lower than for the same soldering group. Most probably, the observed excessive voids at interconnection of ECA2 influence the peel force result (ECA1 interconnection contains also voids, but clearly less and of smaller dimension); an additional factor for lower adhesion could be the lower ECA printed amount at ECA2 group. Furthermore, a peel force difference could be also determined by ECA material bulk properties, on one side through the cohesive force within bulk ECA, and on the other side, the ability of ECA to wet and build bonds with certain strength with the busbars, respectively the ribbon surfaces.

Below are presented the typical peeled interfaces at cell's busbar for the three tested interconnections: at soldered sample (fig.20 a) the fracture is mostly adhesive, between busbar paste and isolation paste, while in range of finger contacting, the fracture occurs in busbar paste; for ECA1 (fig.20 b) and ECA2 (fig.20 c) fracture of interconnection is cohesive.

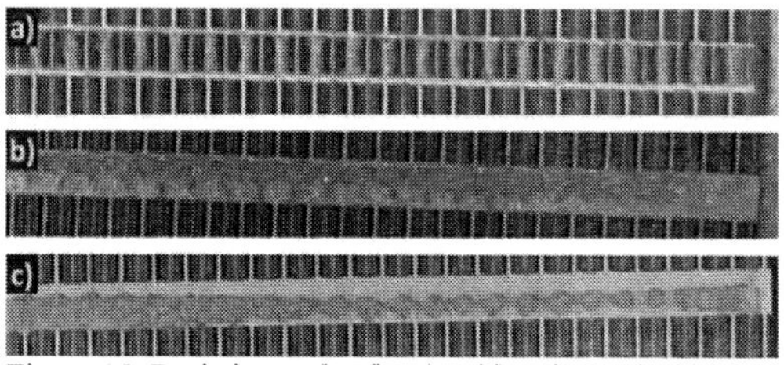

Figure 20. Peeled samples for a) soldered sample, b) ECA1 sample, c) ECA2 sample

3.2 Susceptibility study of stand-alone interconnection to environmental ageing factors

To investigate the susceptibility of the interconnection as stand-alone feature of the string to ageing factors, such as UV, temperature cycling and damp heat, half-cell strings were used. For each string with a certain process variation, the ribbons from one-quarter string were peeled as reference before ageing, the remaining three quarters were used for peel tests at three different ageing intervals. For ageing testing, the samples were exposed to specific ageing factors without any protection or isolation from testing environment.

Tests with fluorescent UV lamps up to 15 kWh UV at 60°C, as well as temperature cycling in range -40° to +85°C (TCT) for up to 55 cycles, reveal no changes to initial for all three interconnection media (ECA1, ECA2 and soldering). At damp heat test (85°C / 85% rel. humidity) ECA1 has a clear drop at peel force compared to initial, up to 71% (fig. 21).

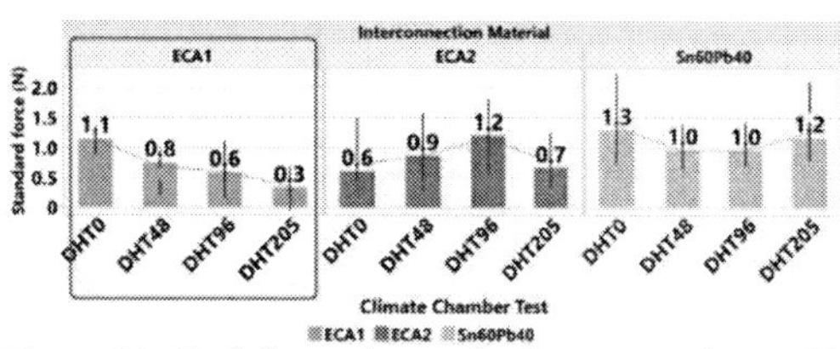

Figure 21. Peel force change for interconnections with ECA1, ECA2, and soldering up to DHT205

The peeled interface is very similar to initial peeled image at busbar level and is an indicator of possible impairment of polymer matrix of ECA1 up to DHT205, since peel rupture seems for initial and after test to be mostly cohesive.

3.3 Investigation of interconnection reliability at laminate level by accelerated ageing tests (temperature cycling and damp heat)

For the investigation of different interconnections close to real operation conditions, sample strings with ECA1, ECA2 and soldering interconnections were embedded in mini-modules, in a glass-EVA-transparent backsheet configuration. For production of all samples, same solar, low iron, float glass (size: 3.2 x 200 x 200 mm), same encapsulation: ethylene vinyl acetate (one EVA layer on each side of cell matrix/ thickness: 450 µm), same backsheet (transparent, multilayer PET backsheet, fluorine free / total thickness: 260 µm), and same bussing (5.0 x 0.3 mm $Sn_{60}Pb_{40}$) were used. Also same lamination process was used for all mini-modules. For each test group 3 mini-modules were made.

As an initial step, cell to modules (CTMs) ratios were determined. SmartCalc v1.2.2 software from Fraunhofer ISE was used to simulated power losses of mini-modules as a function of ribbon cross section, in order to be able to compare mini-modules with different ribbons.

In figure 22 can be seen, that ECA2 has the highest CTM drop values from all tested interconnections, up to 8.58% compared to solder and ECA1 groups.

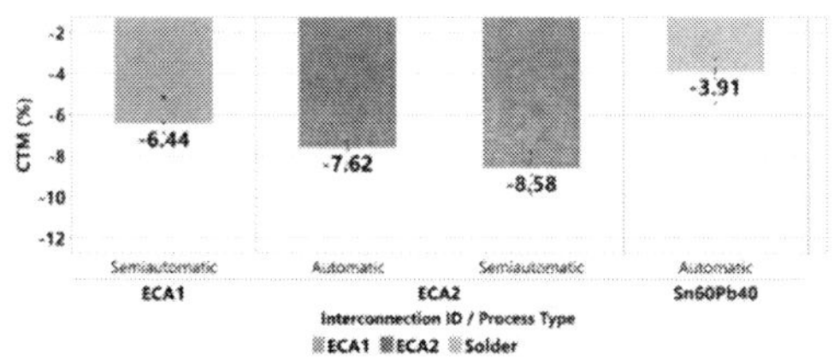

Figure 22. CTM analysis for the mini-modules produced with different interconnections

First hypothesis would be to consider the lower ECA amount per cell for the worse performance of ECA2. However if this should not be the case, result leads to the conclusion that either the conductivity in Z direction of ECA2 is lower than that of ECA1, or that contact resistance of ECA2 is higher than that of ECA1. Furthermore, it is assumed that the observed excessive voids impact also the CTM by affecting the interfaces as well as the bulk of ECA2 interconnection.

All mini-module samples designated for accelerated ageing, despite differences in CTM, were visually inspected, the electrically values proved and classified as acceptable for further testing (free of inactive cells, shunts, defects, which could impair the test purpose).

The mini-modules were measured with an AAA h.a.l.m. GmbH flasher using 4 point contacting at STC (flasher irradiance 1000 W/m², 25°C ±2°C) before and after accelerated ageing test sequences. For the examination of interconnection condition after TCT or DHT it was decided to consider the fill factor as a direct indicator of damaged interconnection or interconnection interface.

At figure 23 can be seen, that mini-modules with ECA1 have a fill factor drop of 2.5 % relative to initial after 400 temperature cycles; however the mini-modules with soldered ribbons, even after 400 cycles reveal a relative fill factor drop of only 0.2 %.

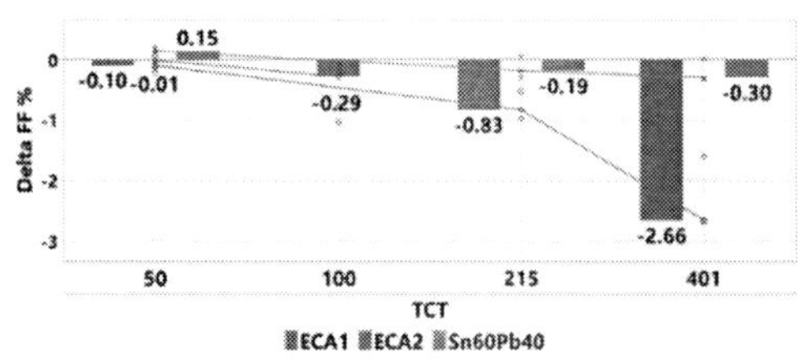

Figure 23. Relative change of fill factor at interconnections test groups during temperature cycling test (TCT)

Contrary to expectations, ECA1 performs very well at damp heat test (fig.24), with a stable relative increase of 0.9 % to initial even after 2000 h damp heat. It is still under investigation what causes this increase, which indicates an improvement of contacting. This effect correlates the opposite trend for peel force at the damp heat test on strings samples.

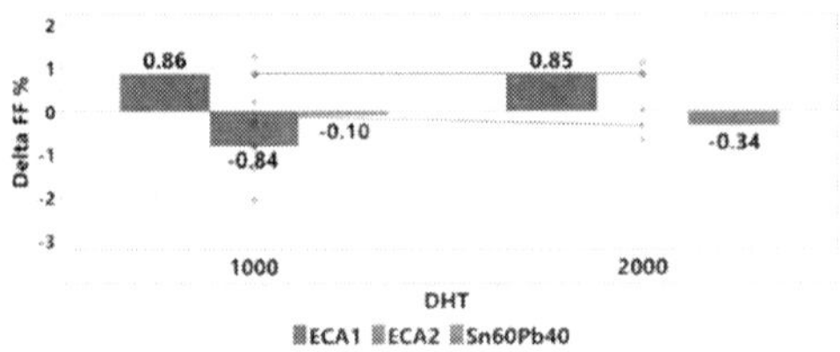

Figure 24. Relative change of fill factor at interconnections during damp heat test (DHT)

For ECA2 no clear trend or dramatic change up to 100 temperature cycles could be detected. At damp heat test however, the relative fill factor change to initial has a different trend compared to ECA1, meaning a worsening of fill factor by almost 1% to initial. A comparison with the results from peel tests after ageing test allow to conclude that the worsening of electrical contacting at ECA2 samples after DHT does not correlate with the peel force, which remains stable at this sample up to the 205h damp heat.

4 SUMMARY

Through this study of interconnection qualification a robust sequence of tests and data analysis for characterization of solar cell joints was developed, based on two market available (epoxy and acrylic based) ECAs and a widely used soldering alloy ($Sn_{60}Pb_{40}$). From testing extent so far, by inspecting peel force, susceptibility to environmental ageing factors and performance at module level, it can be clearly concluded that soldering ensure the best interconnection. However, long-term tests (test to fail) are not concluded and other test aspects like mechanical load tests were not covered by this work. On the other hand, the tested ECAs could be processed at industrial level for ZEBRA cell at acceptable level, considering that no comprehensive optimization was made for the implemented R&D processes. The ECAs had acceptable adhesion values and no signs of mechanical instability at strings before lamination could be noticed. ECA1 manage to pass double the TCT cycles and DHT duration required by the IEC 61215 norm; ECA2 pass 100 TCT cycles and 1000h DHT. Considering the lower adhesion and performance drop after DHT1000 for the group with ECA2, following improvement aspects could be identified for the processing of ECAs:

1. It is a challenge considering cost factors, however optimized quantity of ECA and proper assembly process should ensure a stable reliable and sufficiently wide interconnection between ribbon and cell busbar; low ECA amounts could affect adhesion and module performance;
2. Preparation and processing of ECAs is known to be accompanied by voids. However, the high incidence of voids at the interconnection of ECA2 group is a clear contributing factor to reduced adhesion, as well as impairment of long-term reliability. For reproducible and reliable cell interconnections with adhesives, degassing techniques at preparation of adhesive, as well as at interconnection process, should be deployed;
3. Considering the ECA bulk material, no matter which polymer base or adhesive type was used, could be noticed that cross section samples contain fissures in the interconnection; it is recommended to invest efforts to tune the polymer matrix toward increased ductility.

As a conclusion of this work can be mentioned that none of the tested interconnection materials disqualify, proving all (both ECAs and the soldering) up to the reached testing level, suitable for industrial interconnection of IBC ZEBRA cell. Further testing is necessary to understand and solve certain lower performance at peel tests or impact on reliability.

5 ACKNOWLEDGEMENTS

The authors like to thank German Federal Ministry of Economic Affairs and Energy in the frame of Zquadrat Project (contract nr.: 03EE1005) for support, as well as to TeamTechnik, Fraunhofer CSP and all partners and colleagues, which contributed, for great cooperation.

6 REFERENCES

[1] A.Halm, et.all – doi: 10.4229/27thEUPVSEC2012-2AO.2.1
[2] T.Timofte, et.all – doi: 10.4229/EUPVSEC20202020-4CO.1.4
[3] R. Kopecek, et.all – doi: 10.1109/PVSC45281.2020.9300503
[4] P.Baliozian, et.all – doi:10.4229/EUPVSEC20202020-2CV.1.59
[5] G.W.H. Hoehne, et.all - doi: 0.1007/978-3-662-06710-9
[6] R. Artiga, et.all – doi: 10.1016/j.tca.2004.11.016

CORROSION BEHAVIOR OF THE METALLIZATION INCLUDING TERNARY GLASS FRITS IN THE HIGH-TEMPERATURE AND HIGH-HUMIDITY TEST

Taeko SEMBA and Atsushi MASUDA
Graduate School of Science and Technology, Niigata University, 8050 Ikarashi 2-no-cho,
Nishi-ku, Niigata City, Niigata 950-2181, Japan
Phone: +81-25-262-7856
E-mail: taesemba@eng.niigata-u.ac.jp

ABSTRACT: Ethylene-vinyl acetate copolymer generates acetic acid by hydrolysis, and this acetic acid corrodes the metallization in the PV modules. Metallization corrosion behaves differently depending on the composition of the glass frits contained in the metallization paste. In this study, we investigated the metallization corrosion using ternary composition of $PbO-SiO_2-TeO_2$ system glass frits in a high-temperature and high-humidity test. The electrical characteristics of the module with the metallization dropped rapidly over 2500 h of the test, and the EL dark areas spread from the cell edge and from near the busbars. Metallization corrosion also showed different behaviors at the cell edge and near the busbars. The glass layer corrosion at the interface of Si wafer was observed at the cell edge, and the gap formation between the bulk Ag and Si wafer was found at the vicinity of the busbars. These corrosions are due to the respective degradation of the binary glass frits of $PbO-SiO_2$ system and TeO_2-PbO system.

Keywords: Acceleration test, acetic acid, Ag paste, damp heat, degradation mechanism

1 INTRODUCTION

The photovoltaic module is exposed to various stresses that affect the power generation capacity during its outdoor installation period. Accelerated tests are conducted to confirm the durability to corrosion and degradation of components used in the module. One of the causes of failure of crystalline silicon solar cell modules in a high-temperature and high-humidity test is the loss of electrical contact between the emitters of the cells and the finger metallization. Because acetic acid is generated from the ethylene-vinyl acetate (EVA) copolymer used as an encapsulant and corrodes the metallization [1]-[3].

The composition of the glass frits in the metallization paste has been suggested to be involved in the metallization corrosion in high-temperature and high-humidity tests [4], [5]. The appearance of the electroluminescence (EL) dark area from the cell edge with the finger is due to $PbO-SiO_2$ system glass frits in the metallization. On the other hands, the EL dark area due to TeO_2-PbO systema glass frits in the metallization is thought to spread from the vicinity of the busbars.

In this study, a one-cell module was fabricated using a cell with the metallization paste containing $PbO-SiO_2-TeO_2$ system glass frits. The metallization corrosion caused by glass frits compositions was investigated using this ternary composition glass frits. High-temperature and high-humidity test was conducted under the condition of 95°C/95% relative humidity (RH) and changes in the electrical characteristics and the EL images of the module were confirmed after the test. The cross section of fingers was observed after the test using field-emission scanning electron microscopy (FE-SEM).

2 EXPERIMENT

2.1 Test sample

The 1-cell modules was fabricated using p-type Al-back-surface field (BSF) solar cells. A 6 inch p-type monocrystalline Si solar cell was used. The surface of the cell was textured, and an emitter layer was formed by diffusing phosphorus. A silicon nitride layer was deposited as an antireflection film on the surface of the Si cell by plasma-enhanced chemical vapor deposition. The front-side Ag, back-side Ag, and Al paste were screen printed and co-fired in a high-speed firing industrial furnace. $PbO-SiO_2-TeO_2$ system glass frits was used as the front-side metallization paste.

A solder ribbon was used for tab attachment in the fired cell. The test PV modules were fabricated using tempered cover glass on the front-side, an EVA encapsulant, a solar cell, and a backsheet on the back-side of the module.

2.2. High-temperature and high-humidity test

A high-temperature and high-humidity test was conducted at 95 °C and 95% RH to determine the electrical characteristics and analyze the EL of the module. The initial electrical characteristics of the module was 9.21 A for the short-circuit current (I_{sc}), 0.624 V for the open-circuit voltage (V_{oc}), 73.8% for the fill factor (FF), and 4.24 W for the maximum power (P_{max}).

3 RESULTS

3.1 High-temperature and high-humidity test

Table I shows the rate of change in electrical characteristics of the module. The FF of the module with the metallization containing $PbO-SiO_2-TeO_2$ system glass frits dropped rapidly over 2500 h of the test.

Table I: The rate of change in the electrical characteristics of the test module.

	Rate of change			
Time (h)	I_{sc}	V_{oc}	FF	P_{max}
0	0.00	0.00	0.00	0.00
992	-0.85%	-0.02%	0.01%	-0.87%
1342	-0.28%	0.21%	0.11%	0.04%
2002	-0.40%	0.29%	-0.56%	-0.68%
2398	-0.90%	0.14%	-5.42%	-6.14%
2710	-9.79%	-0.43%	-26.86%	-34.31%

Figure 1 shows the EL images of the module before and after the test. The EL dark areas spread from the vicinity of the busbars, and it is also occurring around the cell edge.

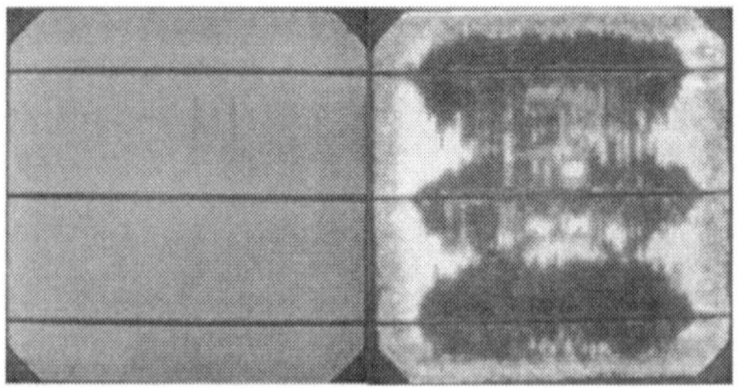

Figure 1: EL images of the module before (left) and after test for 2710 h (right).

3.2 SEM observation

Figure 2 shows the SEM image of the metallization in the fingers located in the EL dark area at the cell edge. Corroded glass layer was seen along the interface between bulk Ag and glass on the Si wafer. Figure 3 shows the glass by the energy dispersive X-ray spectroscopy (EDX) both with and without corrosion. In particular, as shown in Figure 3, the amount on Pb detected was different between the points of dark tone with corrosion and bright tone without corrosion.

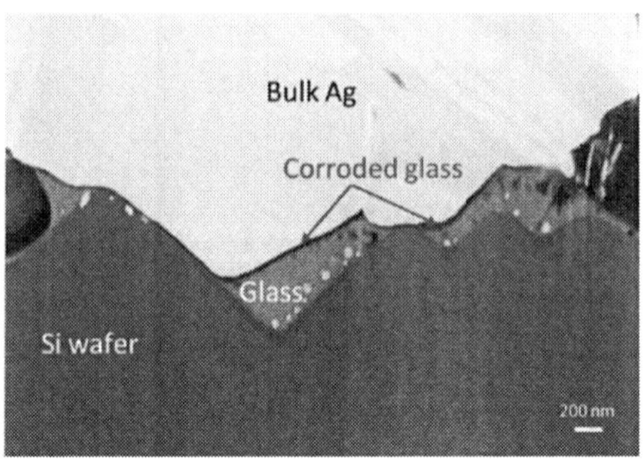

Figure 2: SEM image of the metallization at the cell edge.

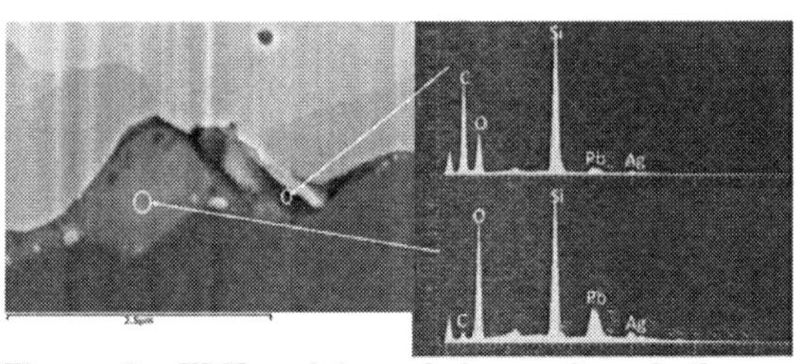

Figure 3: EDX point analysis results of the metallization of the glass layer at the interface of Si-wafer. Upper side graph shows dark tone glass point and lower side graph shows bright tone glass point.

Figure 4 shows the SEM image of the metallization in the fingers located in the EL dark area at the vicinity of the busbars. The gap was formed at the interface between bulk Ag and the Si wafer in the fingers near the busbars.

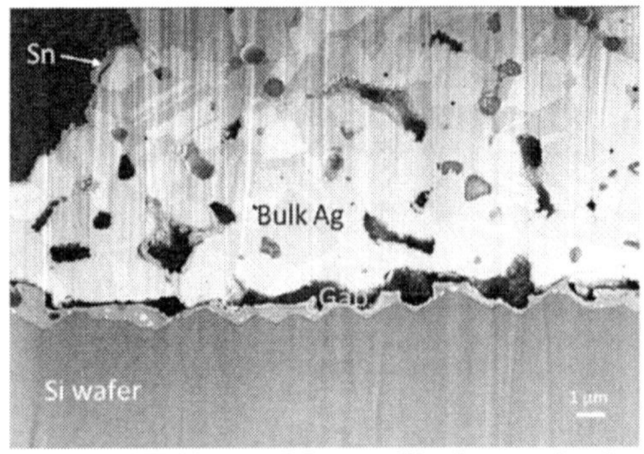

Figure 4: SEM image of the metallization near busbars.

Figure 5 shows the EDX results of the interface between the bulk Ag of the metallization and EVA encapsulant. Sn precipitation was detected on the surface of the bulk Ag at the vicinity of the busbars.

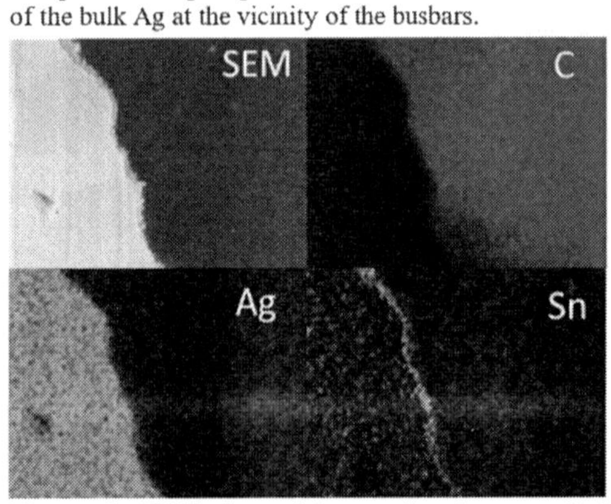

Figure 5: EDX elemental mapping images of the metallization at the interface of EVA encapsulant.

4 DISCUSSION

The EL dark areas of the metallization containing PbO-SiO_2-TeO_2 glass frits showed two modes; one from the cell edge and the other from near the busbars. The EL dark area appearing from the cell edge is found in older types of solar cells, many of which contain PbO-SiO_2 system glass frits in the metallization paste. Corrosion of PbO-SiO_2 system glass by acetic acid causes Pb^{2+} to leach from the glass layer, forming a corroded glass layer on the surface of glass layer [6], [7]. The corroded glass layer is observed to generate at the interface with bulk Ag in the metallization of solar cells [8], [9]. The corroded glass layer at the cell edge shown in Figure 2 is also formed by this mechanism during the high-temperature and high-humidity test.

The EL dark area spreading from the vicinity of the busbars were observed for the metallization containing TeO_2-PbO system glass frits. A gap was formed at the interface between the bulk Ag and the glass layer at the finger in the EL dark area. For the metallization using TeO_2-PbO system glass frits, a discontinuous Te alloy layer was observed at the interface between the bulk Ag and glass layers of the metallization after firing. Te alloy layer is corroded by acetic acid, resulting in a gap [10]. The metallization with reduced adhesion strength due to the dissolution of the Te alloy is considered to delaminate due to residual stress on the solar cell.

Sn precipitation on the metallization surface was only observed in some of the EL dark areas around the bus bars. Sn diffuses into EVA as Sn^{2+} due to corrosion of the interconnector ribbon. However, the EL dark area from the vicinity of the busbars was not observed and no deposition of Sn has been observed in the metallization using PbO-SiO_2 system glass frits. It is suggested that the presence of TeO_2 in the glass frits is related to the Sn precipitation around the busbars. It was suggested that the composition of the glass frits has a significant effect on the corrosion of the metallization.

5 SUMMARY

The corrosion of the metallization containing PbO-SiO_2-TeO_2 system ternary glass frits was investigated. The electrical characteristics of the module were greatly degraded by the high-temperature and high-humidity test for more than 2500 h. The EL dark areas of the module and the corrosion of the metallization showed both the characteristics of metallization containing PbO-SiO_2 system glass frits and TeO_2-PbO system glass frits.

Acknowledgement
This work was supported by the New Energy and Industrial Technology Development Organization, Japan.

References

[1] C. Peike et al., Sol. Energy Mater. Sol. Cells **116**, 49 (2013).
[2] A. Kraft et al., IEEE J. Photovolt. **5**, 736 (2015).
[3] T. Tanahashi et al., Proc. 43rd IEEE Photovoltaic Specialists Conf., 2016, p. 1075.
[4] T. J. Trout et al., Proc. 44th IEEE Photovoltaic Specialists Conf., 2017, p. 2312.
[5] T. Semba et al., Proc. 46th IEEE Photovoltaic Specialists Conf., 2019, p. 1525.
[6] A. Paul, "Chemistry of Glasses", Chap. 6, Chemical durability of glass, Chapman and Hall, London, New York, 2nd ed., 1990.
[7] L. L. Hench and D. E. Clark, J. Non-Cryst. Solids **28**, 83 (1978).
[8] T. Semba, Jpn. J. Appl. Phys. **59**, 054001 (2020).
[9] T. Semba et al., Proc. 7th World Conf. Photovoltaic Energy Conversion, 2018, p. 1333.
[10] T. Semba, Jpn. J. Appl. Phys. **59**, 114001 (2020).

THERMOMECHANICAL FATIGUE OF SOLDER JOINT AND INTERCONNECT RIBBON: IMPACT OF LOW LAMINATION TEMPERATURE

Dag Lindholm[1], Heng Yu Li[2], Gaute Otnes[1], Gianluca Cattaneo[2], Sean Erik Foss[1] and Hallvard Fjær[1]
[1]Institute for Energy Technology, 2007 Kjeller, Norway
[2]Centre Suisse d'Electronique et Microtechnique, Neuchâtel, Switzerland

Thermomechanical loads constitute an important stressor affecting the durability of photovoltaic (PV) modules. Because materials with very different coefficient of thermal expansion are stacked together at a pre-set pressure and temperature in the module lamination process, typically in the range of 140 to 160 °C, stresses are induced in the module assembly after cooldown to room temperature. The purpose of our study was to understand the potential of a lower lamination temperature to reduce thermomechanical fatigue of PV modules, including effects both on the solder bond and the interconnect ribbons. This was done for two different encapsulants, EVA and POE. For this purpose, a three-dimensional FEM model was applied, using a temperature profile that accounted both for lamination and accelerated thermal cycling according to the IEC 61215 standard. With respect to thermomechanical fatigue, results show that a reduced lamination temperature lowers the subsequent thermomechanical stresses experienced both in the interconnect ribbons and the solder joint. Because degradation by means of thermomechanical fatigue of the cell interconnects is typically found among the main degradation modes in photovoltaic modules, our study concludes that a reduced lamination temperature has potential to extend the module lifetime in the field.
Keywords: Modelling, Degradation, Fatigue

1 INTRODUCTION

Degradation of cell interconnects is typically found among the main degradation modes in photovoltaic (PV) modules [1], potentially leading to significant power losses, hence revenue loss, and/or safety concerns [2]. In the lamination process of PV modules, material layers with very different coefficient of thermal expansion (CTE) are stacked together at a pre-set pressure and lamination temperature, which for the latter is typically in the range 140 to 160 °C [3]. Subsequent cooling to room temperature induces stresses in the module assembly, which has potential to influence the module lifetime in the field.

Today EVA is the dominant material for manufacturing of encapsulant films in PV modules. Next-generation encapsulation films based on polyethylene, polyolefin elastomer (POE) or thermoplastic polyolefin (TPO), are expected to increase from about 8 % in 2019 to 30 % within 2030 [4]. Compared to EVA, TPO and POE have the advantage that no acetic acid is produced with aging, which can cause corrosion of the metallization [3].

CSEM in Switzerland has developed a low-temperature processable (LTP) encapsulant based on a POE which can be laminated at temperatures ranging from 85 to 170 °C. This LTP encapsulant foil has been produced successfully with a width of 1.2 m, as shown in Fig. 1. This encapsulant has been shown to meet all basic performance requirements, including but not limited to glass adhesion, damp heat reliability (on mini-module level), and UV stability. Results of these tests will be reported separately.

The purpose of our study was to compare two encapsulants, EVA and POE, and the potential a low lamination temperature has to reduce thermomechanical fatigue of PV modules, including effects both on the solder bond and the interconnect ribbons. For this purpose, a three-dimensional (3D) FEM model was applied, using a temperature profile that accounted both for lamination and accelerated thermal cycling according to the IEC 61215 standard [6].

Figure 1: The LTP encapsulant (cross-linking POE) developed at CSEM.

2 THE COMPUTATIONAL DOMAIN

A 3D computational domain that included glass (rear and front), wafer, encapsulant, rear contact, busbar, solder and ribbon was applied in the study. Because the size of a typical module is large compared to its thickness, and the fact that it contain several thin layers of different materials, it is a well-known challenge to make a mesh that provides sufficient accuracy at acceptable computational time. The strategy chosen in this study was to use a coupled global (coarse mesh) and local model (dense mesh) that represents a quarter of a solar panel, as displayed in Fig. 2. This model, including the boundary conditions, was identical to the one applied by Lindholm et al. [7]. A cross-section of the local model that identifies both the materials and the individual thicknesses of the various material layers is shown in Fig.

3. The gap between adjacent wafers was 2.0 mm in the module lay-up before lamination. Other details are given in the cited reference.

3 MATERIAL PROPERTIES

The material behavior of typical encapsulants is complex with significant softening at high temperatures in the field, implying that the mechanical impact of the glass on the interconnect cells decreases with an increase in temperature. While the melting temperature for typical TPO is 109.4 °C, it is lower for POE and EVA, respectively 81.2 °C and 65.9 °C [5]. Nivelle et al. [13] points out that several authors apply a visco-elastic material model for the encapsulant, accounting for time- and temperatur-dependency. In our study, we have used a simpler approach by applying a thermo-elastic model and temperature-dependent values for Young's modulus [7]. Values provided by Beinert et al. [8], see Fig. 4, were used since these cover the complete temperature window that was considered in our study, -40 °C to 160 °C.

Except for the solder and the copper ribbon, all materials stacked together in the assembly was treated as thermo-elastic materials in the model. The accumulated plastic work in the solder, which was used as metric for fatigue damage, was modelled using Anand's visco-plastic creep model. For copper, the bilinear elastic-plastic model applied by Hasan et al. [9] was adopted. Applied values for the Young's modulus (E) and the CTE are displayed in Table II.

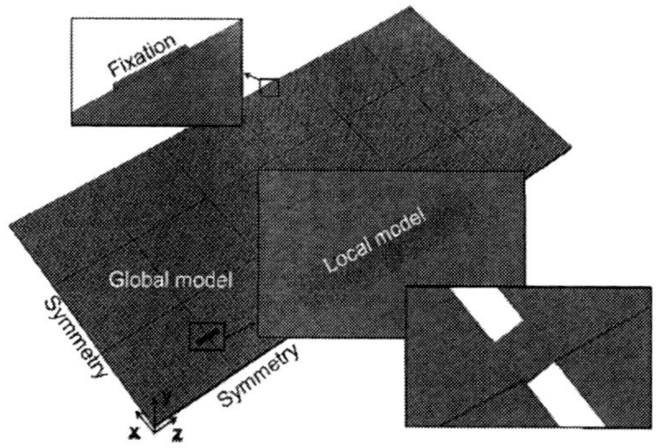

Figure 2: The computational domain consisting of a coupled global and local model.

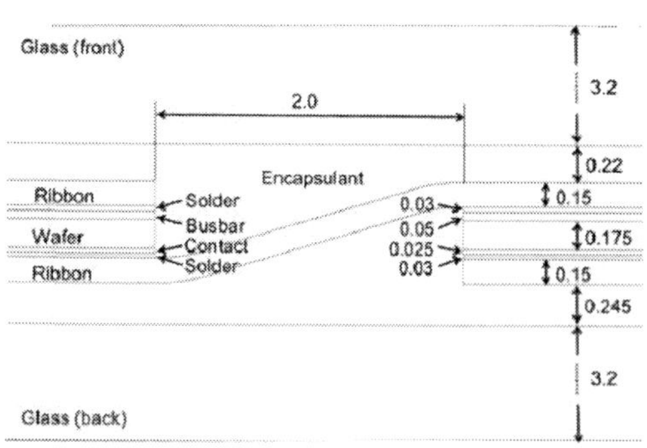

Figure 3: Cross-section in the local model that displays the type of materials and layer thicknesses [7].

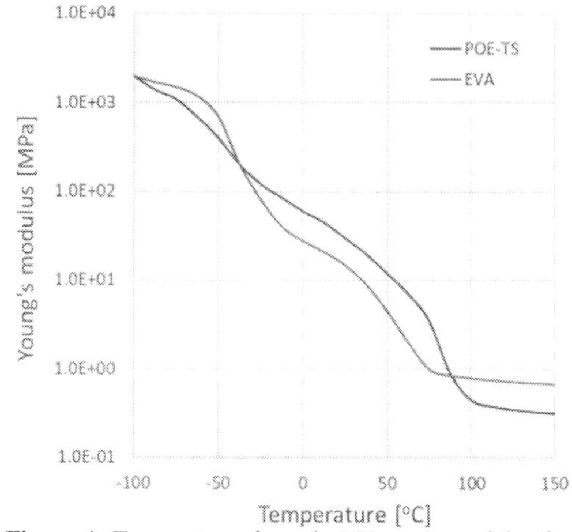

Figure 4: Temperature dependent Young's modulus for POE and EVA. Values obtained by Beinert et al. [8].

Table II: Mechanical material properties.

Layer	E [MPa]	CTE [1/K]	Ref.
Front cover (glass)	73 100	$9 \cdot 10^{-6}$	[10]
Back cover (glass)	73 100	$9 \cdot 10^{-6}$	[10]
Encapsulant (POE)	Fig. 6 (f(T))	$270 \cdot 10^{-6}$	[8]
Encapsulant (EVA)	Fig. 6 (f(T))	$270 \cdot 10^{-6}$	[8]
Contact (Aluminium)	70 000	$23.1 \cdot 10^{-6}$	[11]
Busbar (Silver)	83 000	$18.9 \cdot 10^{-6}$	[11]
Solder (60Sn40Pb)	75 842 - 152T	$24.5 \cdot 10^{-6}$	[11]
Ribbon (Copper)	f(T)	$17 \cdot 10^{-6}$	[9]/[10]
Wafer (Silicon)	170 000	$2.6 \cdot 10^{-6}$	[10]

4 TEMPERATURE CURVES

4.1 Cooling from lamination

With respect to thermomechanical fatigue of the cell interconnects, different lamination temperatures were considered. For POE and EVA, lamination temperatures of 160 °C, 120 °C and 90 °C, and 160 °C and 120 °C were used as lamination temperatures, respectively. While 160 °C represents a typical lamination temperature [3], the lowest value applied for POE is close to the melting temperature, 81.2 °C [5]. The used cooling profiles after lamination are shown in Fig. 5. All three curves in Fig. 5 have adopted the same temperature-dependent cooling rate as the cooling profile of Li et al. [12], which starts at a temperature of 140 °C.

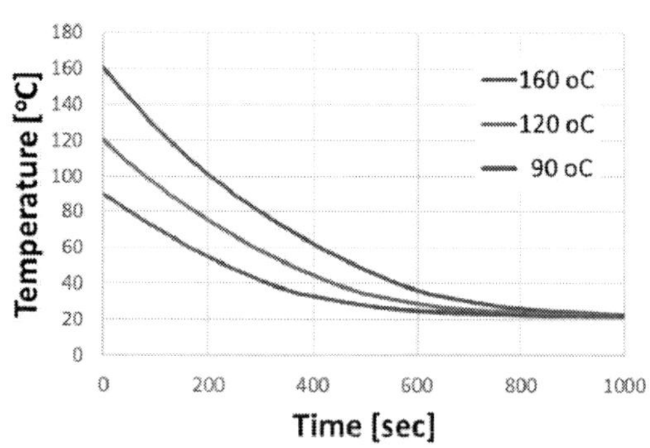

Figure 5: Cooling profiles after lamination.

4.2 Accelerated thermal cycling

After initial computation of the cooling from the lamination temperature down to room temperature, thermomechanical fatigue of the cell interconnects by ATC, according to specifications in the IEC 61215 [6], was studied in the model. The extreme temperatures in the thermal cycles were -40 °C and 85 °C. Ramping and holding times are shown in Fig. 6.

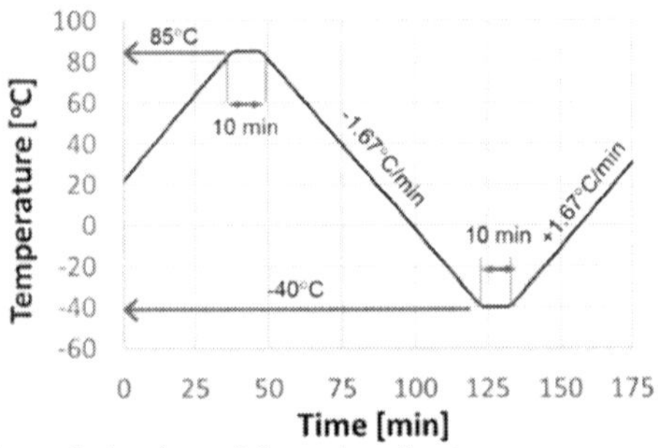

Figure 6: Accelerated thermal cycling temperature profile used in the simulations, compliant with IEC 61215.

5 RESULTS

5.1 Thermomechanical fatigue of interconnect ribbon

In-plane strains of the interconnect ribbons, which were derived from gap size variations between adjacent wafers when the module was exposed to ATC, were used to assess the influence of the lamination temperature on thermomechanical stresses regarding fatigue damage. The in-plane displacements were computed as the difference between the cell gap at the lamination temperature (u_0) and the corresponding gap at the exposed temperature (u), see Fig. 7, i.e.

$$\Delta u = (u - u_0) \qquad (1)$$

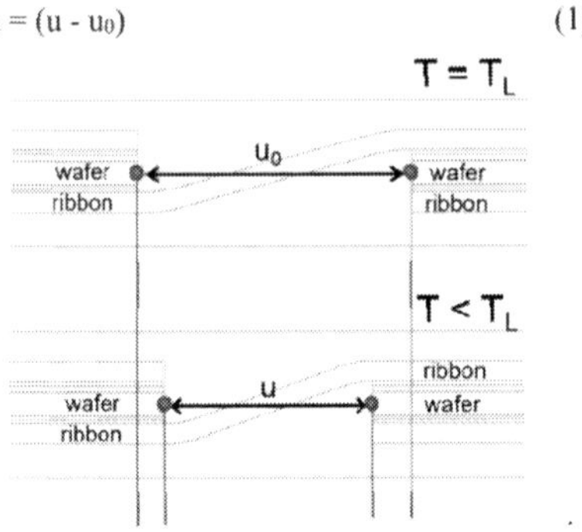

Figure 7: Upper part: In-plane gap size at the lamination temperature (T_L). Lower part: Illustrating a corresponding smaller gap size at an exposed temperature lower than T_L.

At the lamination temperature, the module assembly is assumed to be in a stress-free condition. Because of CTE-mismatch between the various materials, stresses emerge during cooling and are locked in the module laminate at room temperature. In-plane strains were derived from the in-plane displacements as

$$\varepsilon = 100\ \% \cdot \Delta u/u_0 \qquad (2)$$

The in-plane strain amplitude of the repeated thermal cycles was then calculated as

$$\Delta\varepsilon = \varepsilon_{max} - \varepsilon_{min} \qquad (3)$$

Computed in-plane displacements and strains are displayed in Fig. 8. Blue curves represent the module laminate based on POE as encapsulant, red curves are for EVA. While the vertical axis to the left in the figure shows both in-plane displacements and corresponding strains, the axis to the right displays the temperature. At room temperature, but also in the thermal cycles, all five cases exhibit a cell gap that is always smaller than at T_L, hence the ribbon is always in compression. The computations confirm that a lower lamination temperature reduces the in-plane displacements/strains in the cell gap. This is as expected, because in this case there is a smaller temperature difference between the lamination temperature (the stress-free state) and the operating temperature (the stressed state).

For the highest lamination temperature considered in the study, 160 °C, POE turned out better than EVA both with respect to the maximum in-plane displacement (-30.3 µm for POE vs. -36.8 µm for EVA) and the in-plane displacement amplitude in the thermal cycles (13.2 µm vs. 20.3 µm). A lower lamination temperature of 120 °C proved to have significant impact on the in-plane displacement amplitude, which was reduced to 1.6 µm (POE) and 7.3 µm (EVA).

The shape of the curves in Fig. 8 depends strongly on the lamination temperature. This has been explained by the significant softening of the encapsulant with increasing temperature, which reduces the impact of the glass expansion/contraction on the materials inside the encapsulant [7]. For the lowest lamination temperature that was considered in the study (90 °C), the shape of the curve representing POE has an opposite shape to the one based on a lamination temperature of 160 °C. The lowest lamination temperature is close to the largest temperature in the thermal cycle (80 °C), which is close to the stress-free condition of the module laminate with zero in-plane displacement in the wafer gap. For the highest lamination temperature, the cell gap for both POE sand EVA continues to contract with increasing temperature until the temperature exceeds a certain level. Above this level, the contraction ceases, and the cell gap expands towards a condition with zero displacement, i.e. the original gap size at the lamination temperature. In Fig. 8, this temperature level has been exceeded by EVA, but not by POE.

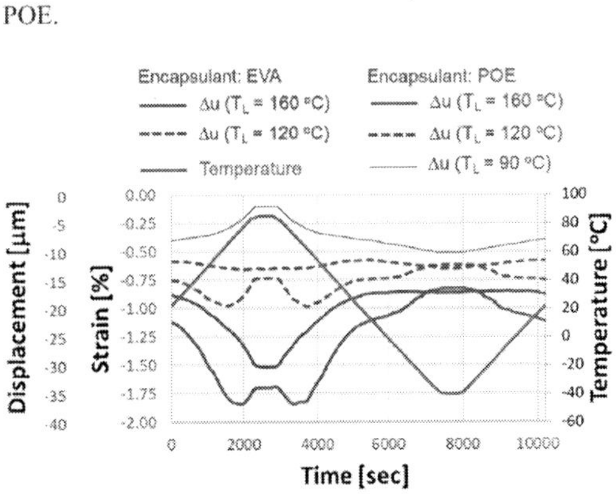

Figure 8: Computed in-plane displacements, in-plane strains and temperatures during a temperature cycle.

Key values for the in-plane strains and the strain amplitudes are summarized in Table III. In the literature, fatigue failure of metals, in our case the interconnect Cu ribbons, is expressed by the number of load cycles until failure (N_f) which is linked to the strain amplitude. Encouraged by Basquin's law of fatigue, Bosco et al. [10] proposed a power law equation to determine high-cycle fatigue failure of interconnect ribbons. This equation reads

$$\Delta\varepsilon = A \cdot N_f^{\,b} \qquad (4)$$

where A and b are fitting constants. Eq. (4) disregards the strain level. As seen, this formula predicts that the number of thermal load cycles until fatigue failure decreases with increased strain amplitude. Eq. (4) was not used to quantify the relative lifetime in the thermal cycles, but rather as a guide to rank the different cases considered our study. Based on the mechanical material properties applied for the encapsulants, the results confirm our hypothesis that a lamination temperature of 120 °C leads to more robust interconnect ribbons with respect to fatigue failure, hence potential extended lifetime of the modules, than a temperature of 160 °C. This conclusion applies to both POE and EVA as encapsulant. For POE, a further reduction of the lamination temperature to 90 °C further lowers the in-plane strain at room temperature, but at the same time increases the strain amplitude compared to lamination temperature of 120 ºC. However, the curve in Fig. 8 that represents POE and a lamination temperature of 120 °C is composed of two sub-cycles which makes a direct comparison to the case with the lowest lamination temperature more complicated. This also apply to the curve based on EVA and a lamination temperature of 120 °C. Nevertheless, the results show that a reduced lamination temperature from the typical lamination temperatures of 140-160 ºC is beneficial to improve the robustness of the interconnect ribbons with respect to subsequent thermomechanical fatigue, as it reduces the contractions of the gap between adjacent cells when the module assembly is exposed to thermal loads.

Table III: Type of encapsulant, lamination temperature, in-plane strain at room temperatur and the in-plane strain amplitude.

Encapsulant (T_L)	ε_0 [%]	$\Delta\varepsilon$ [%]
POE (160 °C)	-0.87	0.67
POE (120 °C)	-0.58	0.08
POE (90 °C)	-0.40	0.41
EVA (160 °C)	-1.12	1.01
EVA (120 °C)	-0.75	0.37

5.2 Solder fatigue damage

The solder layer was assumed to be continuous along the busbars in the model. As a metric for solder fatigue damage, an average plastic work was computed for shorter regions adjacent to the cell gap where the plastic work proved to be largest [7]. A stripped computational domain that only displays the interconnect ribbon and four 1 mm long solder regions facing the cell gap, defined as S1-S4, is shown in Fig 9. The computations give the largest solder fatigue damage in S1, which is attached to the bottom interconnect ribbon.

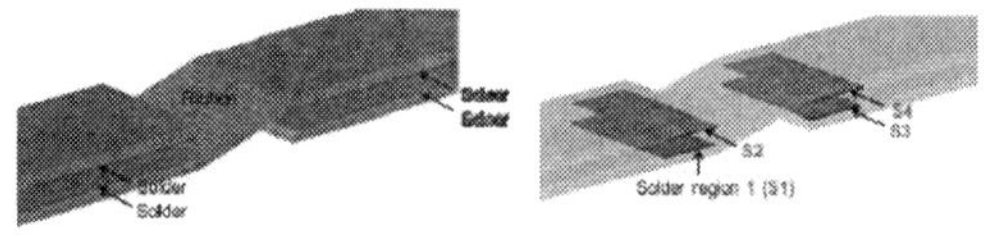

Figure 9: Focused solder regions [7].

The curves in Fig. 10 display the accumulated plastic work in solder region S1, normalized with respect to a value of 27.8 kJ/m^3. While the red curves represent the module based on EVA as encapsulant, blue curves show POE. For the two lamination temperatures of 160 °C and 120 °C, the accumulated plastic work in the solder tip was computed to be about the same for POE and EVA, with a similar development through the thermal cycle. With the lamination temperature reduced from 160 °C to 120 °C, the accumulated plastic work in solder tip S1 was reduced by 20 % for both POE and EVA. For POE a lamination temperature of 90 °C the reduction was 33 % compared to the initial value. These results demonstrate that there is a significant potential to reduce solder fatigue damage by laminating the module at a low lamination temperature.

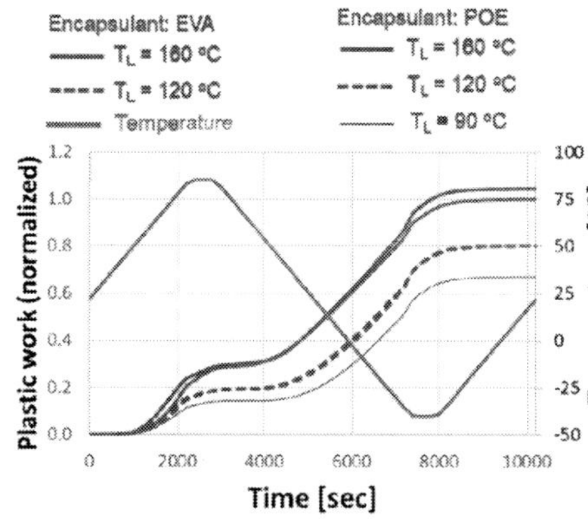

Figure 10: Accumulated plastic work in solder region S1. POE and EVA as encapsulants. Three lamination temperatures, 160 °C and 120 °C and 90 °C (POE only).

6 OUTLOOK

By the time this work was prepared for the EUPVSEC 2021 conference, low temperature data (Young's modulus) for CSEM's LTP encapsulant going down to -40 °C was not available to the analysis. We aim to follow up on the benefits of reduced lamination temperature by also considering material properties for the LTP encapsulant specifically. There are also other factors that influence on the robustness of the cell interconnects with respect to fatigue failure. A review of FEM analyses particularly adressing stresses and strains in module laminates, also the influence of lamination, has been published by Nivelle et al. [13]. Experimental validation for the modelling results could be pursued. Similar simulation could also be performed on other module designs and other interconnection technologies.

7 CONCLUSIONS

Interconnect failure is typically found among the main degradation modes in photovoltaic modules, potentially leading to significant power losses, hence revenue loss, and/or safety concerns. Based on a coupled local-global 3D FEM model, this study aimed to examine the potential of lower lamination temperatures to reduce thermomechanical fatigue of PV module interconnects, including effects both on the solder bond and the interconnect ribbons.

Regarding fatigue damage, the computational results show that low lamination temperature is beneficial to lower thermomechanical stresses in the interconnect ribbons, as it leads to reduce the contractions of the gap between adjacent cells when the module assembly is exposed to thermal loads. Similar for solder fatigue damage: lower lamination temperature decreases the accumulated plastic work in the solder joint. However, our analysis does only consider a single module configuration. To expand the basis for the conclusion, future work should include analyses of other configurations as well.

ACKNOWLEDGEMENT

The authors acknowledge the valuable contribution of all BE-SMART project partners. This project has received funding from the European Union's Horizon 2020 research and innovation programme under grant agreement 818009. The information reflects only the author's view and the Commission is not responsible for any use that may be made of the information it contains.

REFERENCES

[1] M. Köntges et al., "Review of Failures of Photovoltaic Modules," IEA Technical report, 2014.
[2] D.C. Jordan, T.J. Silverman, J.H. Wohlgemuth, S.R. Kurtz, and K.T. VanSant, "Photovoltaic Failure and Degradation Modes," Progress in Photovoltaics: Research and Applications, Volume 25 (4), pp. 318–326, 2017.
[3] "Designing New Materials for Photovoltaics: Opportunities for Lowering Cost and Increasing Performance through Advanced Material Innovations," ISBN 978-3-907281-02-4, Report IEA-PVPS T13-13:2021, April 2021.
[4] "International Technology Roadmap for Photovoltaic (ITRPV)," 11th edition, April 2020
[5] G. Oreski et al., "Properties and degradation behaviour of polyolefin encapsulants for photovoltaic modules", Prog Photovolt Res Appl 28(12), pp. 1062-7995, 2020.
[6] IEC 61215:2016. Crystalline silicon terrestrial photovoltaic (PV) modules – Design qualification and type approval, 2005.
[7] D. Lindholm, G. Otnes, H. Fjær, G. Cattaneo, H.Y. Li, S.E. Foss, "Thermomechanical Fatigue of Solder Joint and Ribbon Interconnect: A Comparison Between Glass-Glass and Glass-Foil Modules," 37th European Photovoltaic Solar Energy Conference (2020).
[8] A.J. Beinert, R. Leidl, P. Sommeling, U. Eitner and J. Aktaa, "FEM-Based Development of Novel Back-Contact PV Modules with Ultra-Thin Solar Cells," 33rd European Photovoltaic Solar Energy Conference (2017).
[9] O. Hasan, A.F.M. Arif, M.U. Siddiqui, "Finite Element Modeling and Analysis of Photovoltaic Modules," ASME International Mechanical Engineering Congress and Exposition (2012).
[10] N. Bosco, T.J. Silverman and S. Kurtz, "Climate Specific Thermomechanical Fatigue of Flat Plate Photovoltaic Module Solder Joints, "Microelectronics Reliability, Volume 62, pp. 124-129, 2016.
[11] E.H. Amalu, D.J. Hughes, F. Nabhani and J. Winter, "Thermo-Mechanical Deformation Degradation of Crystalline Silicon Photovoltaic (c-Si PV) Module in Operation", Engineering Failure Analysis, Volume 84, pp. 229-246, 2018.
[12] H.Y. Li, Y. Luo, C. Ballif and L.E Perret-Aebi, "Effect of Cooling Press on the Optical Transmission Through Photovoltaic Encapsulants," Polymer-Plastics Technology and Engineering, Volume 54, pp. 416-424, 2015.
[13] P. Nivelle, J.A. Tsanakas, J. Poortmans and M. Daenen, "Stress and Strain within Photovoltaic Modules using the Finite Element Method: A Critical Review," Renewable and Sustainable Energy Reviews, Volume 145, 2021, https://doi.org/10.1016/j.rser.2021.111022.

IMPROVED MEASUREMENTS OF THE CONTACT RESISTIVITY OF ECA-BASED JOINTS

María Ignacia Devoto*, Tudor Timofte, Andreas Halm, Daniel Tune
International Solar Energy Research Center (ISC) Konstanz
Rudolf-Diesel-Str. 15, D-78467 Konstanz, Germany
Corresponding author*: +49 (0) 7531 36183341; ignacia.devoto@isc-konstanz.de

ABSTRACT: Electrically conductive adhesives (ECAs) can replace soldering for the interconnection of cells within photovoltaic modules such as those based on IBC-based technology or the shingling interconnection approach. They can also be used for applications in which soldering is no longer appropriate or lacks long-term reliability, such as temperature-sensitive SHJ-based technology or vehicle-integrated photovoltaics. The aim of this work is to show further improvements in the development of a method to characterize the contact resistivity of joints based on these adhesives. The developed method, which is based in the Transmission Line Method, is shortly explained while the treatment of the raw data and the obtained results are clarified in detail. Samples were manufactured using nine different commercially available adhesives that contained different polymer matrices (acryl and epoxy) and conductive filler (silver and copper with silver coating). It is concluded that the introduction of the adhesive significantly increases the complexity of the system under study and increases the number of calculations required to remove non-important resistive components so that the contact resistivity of the joint can be extrapolated. It is also shown that although measurements per sample are statistically significant, a larger number of probes should be manufactured so as to extrapolate a non-misleading value. This means that many measurements per sample with several samples per ECA composition are required.

Keywords: see the list of keywords

1 INTRODUCTION

Rather than the traditional soldering approach, several commercially available alternative module concepts use electrically conductive adhesives (ECAs) as a means of cell interconnection. Typically, concepts such as shingling, conductive backsheets, and heterojunction cells use adhesive because soldering is not suitable for them. The shingling approach requires direct joints between busbars while conductive adhesives need to perform the interconnection process during lamination. Heterojunction cells are sensitive to high temperatures and their passivation can be damaged by soldering. ECAs possess several advantages over soldering; they are lead-free, mechanically flexible, and can better tolerate thermal cycling and vibrating environments. In addition, they can be precisely dispensed through screen/stencil printing or pressure-time/jet dispensing. Among the main disadvantages of ECAs compared to soldering are the higher per-module cost of the material and the initial investment for new tools/machinery.

For ECAs to gain wider acceptance in photovoltaic (PV) manufacturing they need to be further studied. Physical, electrical, and mechanical characterization is necessary to compare different adhesive compositions, test electrical and mechanical performance of ECA-based joints, and to have a means of quantifying the impact of accelerated aging tests at adhesive and joint level. To quantify the electrical performance of conductive joints, the contact resistivity is a useful figure of merit.

Geipel et al. used the Stacked Greek Cross (or Cross Bridge) structure with a four-point probe measurement to extract the contact resistivity, concluding that due to current crowding, contact resistivity is overestimated.[1] The transmission line method (TLM) has proved to be a useful method for electrical characterization of ECA-based joints and similar applications. [2]-[5] Nevertheless, TLM has been revised several times in the last decades and several improvements have been made.[6]-[11] However, although its use is widespread in the microelectronics community and it is also used in the PV field, its usage is neither simple nor straightforward.

In this work, previous manual studies are expanded by performing the measurements in an automated way. Only two contacting events are required to obtain a very large number of measurements from each sample, thereby minimizing any measurement differences due to contact damage. Many measurements are carried out after each contacting event and these are used to reduce the standard deviation of the outcomes, increasing their statistical significance. The sample design is chosen according to what previous work has shown is best. A revised methodology is also used to treat the raw data.

2 SAMPLE DESIGN AND ADHESIVES

2.1 Sample design

In our previous work, several designs were tested in order to find the best sample design, in which the contact behavior is found to be ohmic. The sample design used in this work is shown in Figure 1.

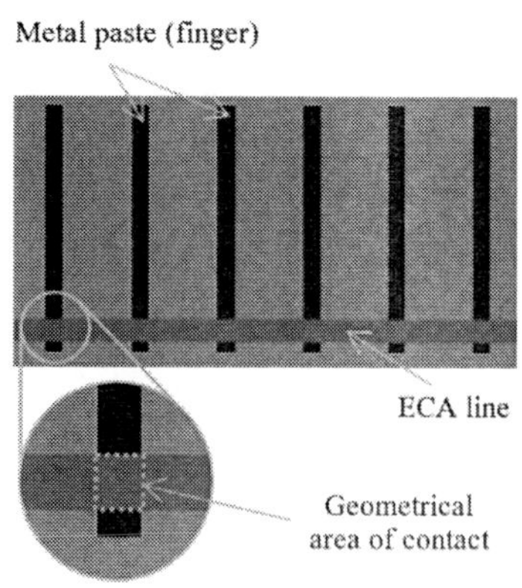

Figure 1. Sample design. The contact under study is highlighted with a green dotted line.

To study different contacting area sizes, six different samples were designed with different finger widths (0.2, 0.4, 0.8, 1.0, 1.2, and 1.5 mm).

2.2 ECA composition under study

Nine commercially available ECAs were tested. Two different polymeric materials were chosen: acryl and epoxy, and two different metals were chosen as fillers: pure silver and copper core with silver coating. A summary of the tested ECAs is shown in Table I.

Table I. ECA composition of adhesives under test.

Name	Polymer material	Filler material	Best performance application
ECA 1	Acryl	Ag	Silicon-to-Silicon
ECA 2	Acryl	Ag	Silicon-to-Silicon
ECA 3	Acryl	Ag	Silicon-to-Silicon Silicon-to-(Sn)Ribbon
ECA 4	Epoxy	Ag	Silicon-to-Ribbon
ECA 5	Epoxy	Ag	Silicon-to-Ribbon
ECA 6	Acryl	Cu(Ag)	Silicon-to-(Sn/SnPb)Ribbon
ECA 7	Epoxy	Cu(Ag)	Silicon-to-Silicon Silicon-to-(Ag/Ni)Ribbon
ECA 8	Epoxy	Cu(Ag)	Conductive backsheet
ECA 9	Epoxy	Cu(Ag)	Conductive backsheet

3 METHODOLOGY

3.1 Equipment

A source–measure unit (SMU, Keithley 2602A) was used to inject current between two pins (high and low source) and to sense the voltage between two pins (high and low sense). A custom contacting fixture with 37 pins was designed to simultaneously contact 2 or 4 fingers for each measurement. To assign a role to each pin, a multiplexer was used between the Keithley 2602A unit and the fixture.

3.2 Types of measurements

Four different types of measurements were performed as illustrated in Figure 2:

- Two pins on whole finger: R@2PWF
- Two pins on half finger: R@2PHF
- Four pins on whole finger: R@4PWF
- Four pins on half finger: R@4PHF

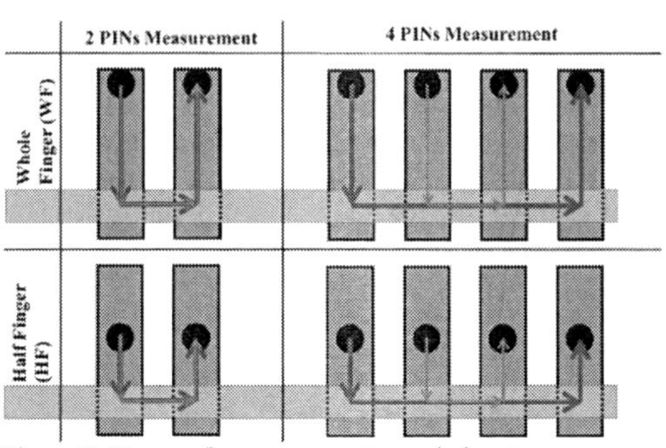

Figure 2. Types of measurements carried out.

Considering the sample circuits shown Figure 2, the following equations are expected to be true:

$$R@2PWF = 2R_c' + 2R_{WF} + 2R_c + R_b \quad (1)$$

$$R@2PHF = 2R_c' + 2R_{HF} + 2R_c + R_b \quad (2)$$

$$R@4PWF = R_b \quad (3)$$

$$R@4PHF = R_b \quad (4)$$

where R_c' is the contact resistance between the pin and the finger (in Ω); R_{WF} and R_{HF} correspond to the resistance measured when contacting on the whole and half finger, respectively (in Ω); R_c is the resistance of the contact (in Ω); and R_b is the bulk resistance of the ECA line (in Ω).

Equations 1 to 4 assume the following:

- The resistance of the pins is negligible.
- For R@4PWF and R@4PHF, the voltage drop along the sensed fingers is negligible (this was tested with Keithley 2602A and is in accordance with manufacture references).

To extrapolate the contact resistivity for a particular sample using TLM, only $2R_c + R_b$ should be plotted. To remove the R_{WF} and R_{HF} components, another TLM is performed using R@2PWF and R@2PHF, as shown in Figure 3. The distance on the x-axis corresponds to twice the distance between the pin positions for whole and half finger measurements. The extrapolated y-axis intercept of this first plot corresponds to the hypothetical case of finger length = 0 and thus yields $2R_c' + 2R_c + R_b$.

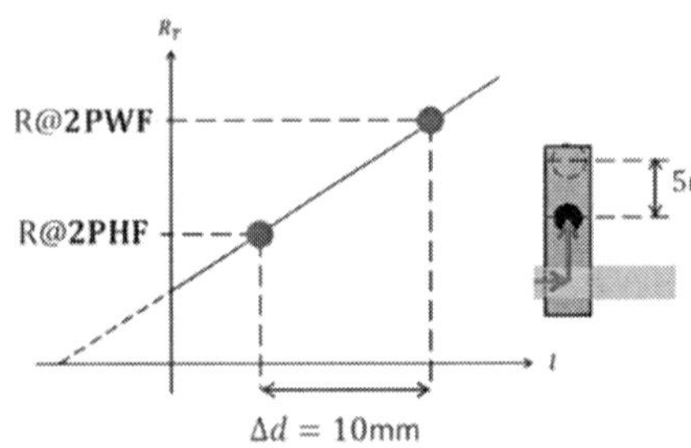

Figure 3. TLM to remove finger resistance component.

To remove the R_c' term, an additional sample was used to perform another TLM along a longer line of the same metallization paste as used in the fingers and with the same , width as the fingers of the sample. As shown in Figure 4, for each sample several measurements were taken at a distance of 4–20 mm. The minimum distance can be measured 10 times (e.g. between fingers 1 and 2 in both directions, 2 and 3 in both directions, etc.) while the maximum distance can only be measured twice (between fingers 1 and 6 in both directions). For each distance the median is calculated.

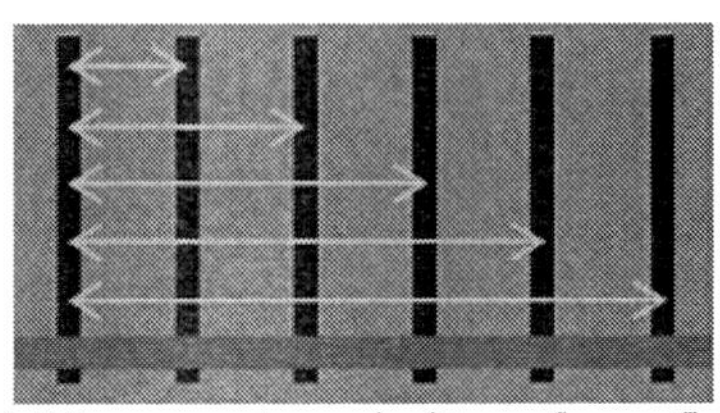

Figure 4. Measurements required to perform a final TLM and extrapolate R_c and R_b.

The distance used to plot R_{xmm} (the resistance between pins with a separation of x mm) corresponds to the distance between the inner edges of the fingers (d_{xmm}). This distance depends on the width of the finger W_F in mm and decreases when the finger is wider. That is, $d_{xmm} = x - W_F$, where x can be from 4 mm up to 20 mm. The resulting TLM plot is shown in Figure 5.

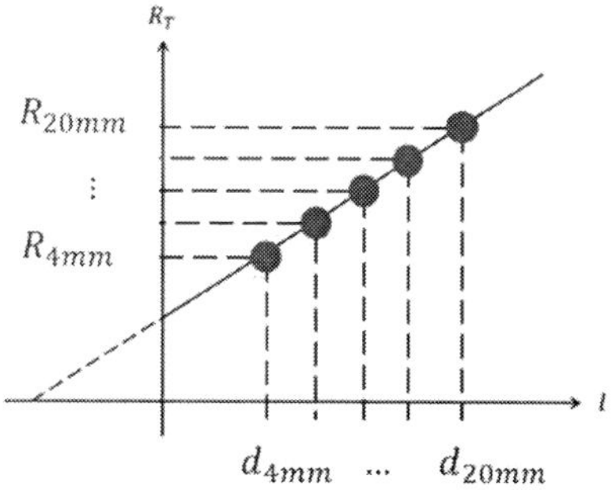

Figure 5. TLM plot using the measurements shown in **Error! Reference source not found.** to extrapolate R_c and R_b.

3.3 Validation of measurements

Because the method is under development, resistance values are not known in advance and literature values can vary significantly for different sample geometries. For this reason, the SMU was configured to sense using auto ranging. To know that a sensed voltage value is valid, the range in which the SMU had performed the measurement is registered. If the range used was equal to the voltage upper limit, the value was excluded. If a whole (half) finger measurement was excluded, then its respective half (whole) finger measurement was also excluded.

Before performing any calculations, the following conditions were required to be true:

- R@2PWF > R@2PHF (condition 1)
- R@4PWF = R@4PHF (condition 2)

Condition 1 was used exactly as stated here. However, condition 2 was relaxed and $\pm 5\,\%$ deviation was accepted.

4 RESULTS

4.1 ECA line resistance versus effective and geometrical contact area

As shown in the methodology, the ECA line resistance (in Ω mm^{-1}) can be calculated from the slope of the TLM plot $(R_{L[TLM]})$ as shown in Figure 5 or directly calculated using either R@4PWF or R@4PHF or the median of both $(R_{L[4PXF]})$. The ratio of these two values is referred to as the ECA line resistance ratio (α). Figure 6 shows the variation of α with the percentage of the geometrical contact area used (β), which depends on the transfer length. The two ratios are calculated as follow:

$$\alpha = \frac{R_{L[TLM]}}{R_{L[4PXF]}} \cdot 100\ [\%]$$

and,

$$\beta = \frac{\text{Effective area}}{\text{Geometrical area}} = \frac{\text{Transfer length}}{\text{Finger width}} \cdot 100\ [\%]$$

As shown in Figure 6, when the contact is overflowed with a high current density ($\beta > 100\,\%$), a condition commonly known as current crowding, TLM tends to underestimate the bulk resistance of the adhesive ($\alpha < 100\,\%$). Hence, the line resistance of the ECA is also underestimated. On the other hand, when the current density is small and it occupies a small fraction of the geometrical area ($\beta < 100\,\%$), TLM overestimates the bulk resistance of the ECA ($\alpha > 100\,\%$). However, when the current occupies somewhat more than the whole geometric contact area ($\beta \sim 100\,\%$, red oval), TLM tends to extrapolate the bulk resistance as expected ($\alpha \sim 100\,\%$). Hence, the extrapolated line resistance of the ECA via TLM under these conditions is comparable to the values measured using four pins.

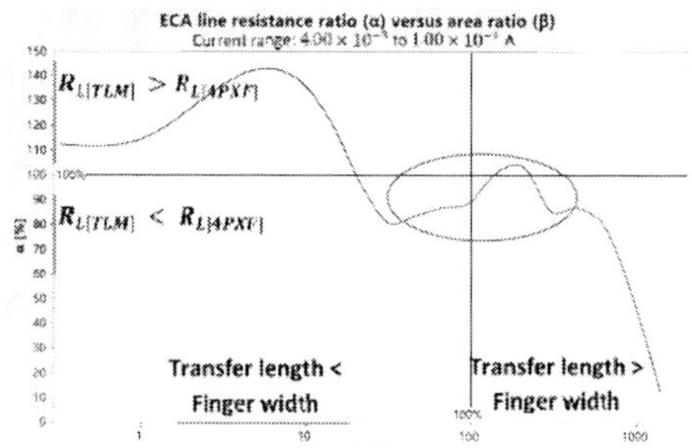

Figure 6. ECA line resistance ratio (α) depending on the percentage of the used contact area (β). Data is shown for a current range of 4.00×10^{-3} to 1.00×10^{-1} A and a TLM squared Pearson correlation coefficient larger than 0.9 ($R^2 > 0.9$).

It is then concluded that in order to calculate an accurate ECA line resistance, the injected current should be selected so that the current flows through the whole area, avoiding current crowding due to currents that are too large as well as underestimation due to current injections that are too small.

4.2 Effective contact resistivity versus current injection

The dependence of the effective contact resistivity, calculated with TLM, on the current injection is shown in Figure 7. So that the area of the contact does not influence the analysis, the curves are separated by their area. The contact resistivity tends to stabilize at the upper current injection levels between 4.00×10^{-3} A and 1.00×10^{-1} A, independent of the area of the contact. For the cases in which the current injection levels are small (see red ovals for ECA 3, 4, and 6 in Figure 7), TLM tends to underestimate the contact resistivity.

4.3 Contact resistivity of all tested ECA

Table II shows the effective contact resistivity and its statistical values. The median and the median absolute deviation are in mΩ-cm^2. All values within the table are in the current range of 4.00×10^{-3} A to 1.00×10^{-1} A and $R^2 > 0.9$. However, if results are filtered to only consider contact resistivities calculated when the usage of the geometrical contact area is 100 % ($\alpha \sim 100\,\%$), then there are only a few ECAs with valid results. For this reason,

Table II shows values that are calculated for any contact area usage. That means, that some contact resistivities are calculated on current crowding conditions and other are not.

As it can be seen in the table, the median deviation is still in the same order of magnitude when compared to the median, and we believe that this is related to sample manufacture and also to the number of measured samples.

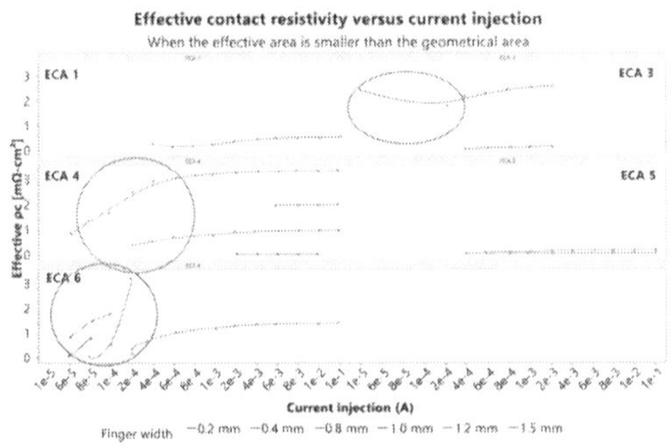

Figure 7. Dependence of the effective contact resistivity on the current injection. Data is shown for effective contact areas that are smaller than the geometrical contact areas and a TLM squared Pearson correlation coefficient larger than 0.9 ($R^2 > 0.9$).

Table II. Statistics (median and median absolute deviation) of effective contact resistivity (in mΩ-cm²) for all ECAs under test.

ECA	N	Median	Median Absolute Deviation
ECA 1	10	2.9587	2.4512
ECA 2	14	7.9956	1.6845
ECA 3	15	5.7412	1.3821
ECA 4	18	1.4631	1.0091
ECA 5	20	0.4439	0.3483
ECA 6	25	3.2895	1.7915
ECA 7	15	2.9285	0.4581
ECA 8	20	10.7865	3.6106
ECA 9	25	6.0917	3.4497

5 CONCLUSIONS

As also shown previously, TLM is a method that can be used to extrapolate the contact resistivity of ECA-based joints. However, the complex sample geometry and microscale variations introduce new levels of complexity into the system, and this added complexity requires more steps to properly treat the raw data to obtain reliable and consistent values with statistical confidence.

In this work, the contact resistivity of several ECA-based joints was extrapolated via TLM. The samples were manufactured by pressure-time dispensing and the measurements were carried out automatically with the use of specially designed samples and measurement equipment.

Many different measurements were performed for each sample, but it remains necessary to increase the number of samples under study. By obtaining very large data sets it is clear that although there are certain combinations of geometry, current injection, and measurement methodology that can produce apparently sensible results, such conditions do not reliably produce sensible results for all samples and there remains significant deviation in the contact resistivity calculated when different contact areas are used.

Now that a measure system has been developed that is fast and automated, future work will focus on further improving and expanding sample manufacture to obtain greater confidence in the statistical significance of calculated contact resistivities independent of sample geometry or current injection.

6 ACKNOWLEDGEMENTS

This work is supported by the German Federal Ministry for Economic Affairs and Energy (BMWi) as part of the HoSSa and Zquadrat projects with funding reference numbers 03EE1014D and 03EE1005A.

7 REFERENCES

[1] T. Geipel, M. Meinert, A. Kraft, and U. Eitner, "Optimization of Electrically Conductive Adhesive Bonds in Photovoltaic Modules," *IEEE J. Photovoltaics*, vol. 8, no. 4, pp. 1074–1081, Sep. 2018.

[2] M. I. Devoto, T. Timofte, A. Halm, and D. Tune, "Measuring the contact resistivity of ECA-based joints," in *37th European Photovoltaic Solar Energy Conference and Exhibition*, 2020, pp. 1001–1008.

[3] M. I. Devoto, T. Timofte, A. Halm, and D. Tune, "Contact resistivity of ECA bonded joints," *AIP Conf. Proc.*, vol. 2367, 2021.

[4] G. Beaucarne *et al.*, "Study of compatibility of silicone-based electrically conductive adhesives and conductive backsheets for MWT modules," *Energy Procedia*, vol. 55, pp. 444–450, 2014.

[5] M. Estruga, L. Theunissen, A. Ardizzone, B. Willems, and A. Henckens, "Electrically Conductive Adhesives for Cell Interconnection in Shingled Module Technology: The Impact of Material Properties on Mini-Module Performance," in *35th European Photovoltaic Solar Energy Conference and Exhibition*, pp. 67–70.

[6] W. Schockley, A. Goetzberger, R. Scarlett, R. Gereth, V. Williams, and N. Zetterquist, "Research and investigation of inverse epitaxial UHF power transistors," 1964.

[7] H. Murrmann and D. Widmann, "Current Crowding on Metal Contacts to Planar Devices," *IEEE Trans. Electron Devices*, vol. 16, no. 12, pp. 1022–1024, 1969.

[8] H. H. Berger, "Contact resistance and contact resistivity," *J. Electrochem. Soc.*, vol. 119, no. 4, 1972.

[9] H. H. Berger, "Models for contacts to planar devices," *Solid State Electron.*, vol. 15, no. 2, pp. 145–158, 1972.

[10] G. K. Reeves and H. B. Harrison, "Obtaining the specific contact resistance from transmission line model measurements," *IEEE Electron Device Lett.*, vol. 3, no. No. 5, pp. 111–113, 1982.

[11] Y. Pan, "Versatile Circular Test Structure for Ohmic Contact Characterisation," no. March, 2015.

Degradation Evaluation of Thermoplastic Polyolefin Encapsulant used in c-Si PV Modules and Laminates

Baloji Adothu[1*], Francis Reny Costa[2], and Sudhanshu Mallick[1]

[1]Metallurgical Engineering and Materials Science, and National Centre for Photovoltaic Research and Education (NCPRE), Indian Institute of Technology Bombay, Mumbai, Maharashtra, India-400076
[2]Borealis Polyolefine GmbH, St.-Peterstraße 25, A-4021, Linz, Austria

ABSTRACT: Reliability and degradation product evaluation is the most important study to predict the long-term performance of new materials used in photovoltaic (PV) module encapsulation. Thermoplastic polyolefin (TPO) is a newly developed non-crosslinking encapsulant. Its reliability and degradation products have been evaluated and compared with the most commonly used ethylene vinyl acetate (EVA) copolymer under UV and damp heat aging conditions. EVA glass-to-glass laminate shows strong fluorescence at the center, whereas TPO did not show any fluorescence during UV-365 aging. It has been found that the fluorescence brim pattern dominated in between the edge and center of solar cell for EVA, whereas TPO dominated only at the edges. This is due to the EVA mini PV laminate has four times higher oxygen diffusion length and coefficient than the TPO PV laminates during the Xenon aging. UV dose of 367.2 and 162 kWh/m^2 is sufficient to detect the early-stage UV-induced discoloration degradation under the hot-dry climatic aging conditions. Electroluminescence images of TPO shows insignificant changes, whereas EVA shows strong metallization corrosion in 4000 h damp-heat tested PV modules. Due to this, overall 5% and & 7% power degradation has been observed for TPO and EVA, respectively. A factor of nine times higher acid index difference and two times higher hydroxyl index have been identified with infrared spectroscopy for EVA as compared to the TPO. A higher yellowness index has been observed for EVA during all three aging conditions. This work will help the PV industry and encapsulant manufacturers to understand the TPO degradation behavior in hot-dry and hot-humid situations.
Keywords: Encapsulant, Ethylene-vinyl acetate, PV modules, thermoplastic polyolefin, Fluorescence.

1 INTRODUCTION

Solar modules technology is fast-growing technology in renewable energy industries. PV modules are composed of many composite layers such as solar glass, solar cells, and polymer materials. An encapsulant layer is a crucial component, and it should provide structural, electrical isolation, and adhesion support to the PV module components [1–3]. EVA is the most commonly used crosslinked encapsulant in the PV industry. EVA has significantly developed in the last few years but still has lots of problems like discoloration, delamination, acetic acid generation, and moisture absorption [2–6]. EVA performance strongly depends upon the lamination time, temperature, and gel content [7–9]. Thermoplastic polyolefin (TPO) encapsulant material is a newly developed non-crosslinking material for PV module lamination in the PV industry to replace the crosslinking encapsulant [10]. Due to the absence of curing reaction, the lamination time is shorter for TPO than the crosslinked encapsulant [11].

The UV pre-conditioning and damp heat tests are standard recommended (IEC 61215) qualifying tests for new modules and encapsulant material manufacturers. An encapsulant discoloration in the PV modules mainly results from the UV radiations. This discoloration results in reducing optical transmittance and short circuit current in the PV modules [12]. The PV module has shown delamination and corrosion of metalization parts (metal grid lines and busbar strings) [13]. UV radiation also degrades the PV modules' adhesion properties between backsheet and solar cells [14]. An encapsulant material degrades very fast under hot-dry and hot-humid conditions with a combination of UV radiation [15–17]. Studying degradation behavior under the hot-dry and hot-humid (damp heat) conditions is very important for a new encapsulant material. High humidity and temperatures accelerate the degradation of encapsulant material, including the corrosion of metals, delamination, and discoloration [18].

Very limited studies are available for TPO encapsulant degradation comparison with the most commonly used EVA. B. Ottersbock et al. stated that EVA had shown a slightly higher discoloration and oxidation index than the TPO during the 2000 h of Xenon UV and damp heat aging. UV aging has not shown significant changes in the morphology of TPO and EVA films than the damp heat aging [5]. López-Escalante et al.stated that no significant degradation was observed for the TPO mini-module compared with EVA during the UV aging (carried out according to the IEC61215 normative)[19]. A strong fluorescence and a factor of nine times slower yellowness were observed for TPO as compared to the EVA after the 50 days of UV-365 aging [20]. C. Barretta et al. explained that the transmittance of EVA decreased during the DH (up to 3300 h) and UV aging (340 nm fluorescent lamps irradiated up to 200 kWh/m^{-2}) is due to the UV absorber loss (between 250 and 350 nm). A slight decrease of transmittance of TPO below 390 nm is due to the formation of chromophores (results in material discoloration) during the UV aging 85 kW h m^{-2} [21].

G.Oreski et al. revealed that the TPO films' shear modulus and shear viscosity properties are significantly higher than the EVA films. These are required beneficial properties for PV module encapsulation. He also explained that EVA shows strong corrosion effects on PV module (single-cell) metalization parts (silver grides and ribbons) due to the strong generation of acetic acid after 3000 h of damp heat aging [22]. G.Oreski et al. stated that the strongest yellowing had been detected for both TPO encapsulant-based PV modules (3 x 2 cell module) prepared with co-extruded PP and PPF backsheets compared to the EVA during the 3000 h of damp heat test [23]. Glass-to-glass EVA laminate showed a significant development of discoloration visualization from 2000 h to 4000 h compared to TPO (3000 to 4000 h). More than two and six times higher carbonyl and hydroxyl functional groups have been found for EVA than the TPO glass-to-glass laminates during the damp heat aging [24].

There are several long-term studies available for the discoloration evaluation of encapsulant materials. The early fluorophore-induced discoloration or early detection of fluorescence patterns for TPO and EVAs are very important than the long-term UV aging studies. The fluorescence patterns are different for the different encapsulant materials [25]. The fluorescence patterns and early-stage identification have been studied only for EVA mini PV laminates [25]. There are missing details of TPO fluorescence patterns and their early-stage identification. Asper our literature survey also observed that the damp-heat test is conducted only on TPO mini-modules and on mini glass-to-glass laminates compared with EVA. In our study, a damp heat test has been conducted up to 4000 h on standard industrial 72 cell TPO PV modules and compared its performance with EVA modules. Degradation products such as corrosive acid and hydroxyl groups have been evaluated for both the encapsulants in both UV and damp heat aging.

This study aims to evaluate the high UV transparent TPO encapsulant degradation behavior under UV and damp heat aging conditions. In this study, the first novelty is to investigate the early-stage identification of UV-induced degradation and fluorescence patterns under the UV-365 and Xenon accelerated hot & dry aging conditions. The second novelty is to examine the extended damp-heat degradation up to 4000 h on standard industrial PV modules. In both UV and damp heat aging studies, degradation products of TPO encapsulant have been evaluated and compared with high UV transparent EVA. This work will help the PV industry and encapsulant manufacturers understand the TPO and EVA degradation behavior in hot-dry and hot-humid conditions.

2 EXPERIMENTAL METHODS

2.1 Materials for PV laminates and modules

The TPO encapsulant (approximate composition ≤25% acrylate, ≤ 5% silane, Quentys BPO-8828F, 0.54 mm thickness) has been supplied by Borealis Polyolefine GmbH. Commercially available fast cure type EVA (approximately 70% ethylene and 30% VA content, 0.45 mm thickness) has been used in this study. Standard 72 cell solar glass (Gujarat Borosil Ltd., ARC coated tempered solar glass, 3.2 mm thickness, 1.6 m^2), mono-crystalline Si solar cells (200 mm thickness, 20% efficiency) a polymer back sheet (Coveme, dyMate KL 50/250, polyvinyl fluoride (PVF or Tedlar)/polyethylene terephthalate (PET)/Primer (i.e., Tedlar/PET/Primer)) has been used. ARC coated annealed solar glass (Gujarat Borosil Ltd., 3.2 mm thickness) and quartz glass (QG) (20 mm × 20 mm, 2 mm thickness) has been used for mini PV and glass-to-glass laminates preparation.

2.2 PV laminates and modules preparation

A glass/two-layer of encapsulant/glass (QG glass-to-glass) structure has been used to visualize fluorescence patterns and changes in transmittance during UV aging. Solar glass /encapsulant/solar cell/encapsulant/backsheet structure has been used for PV module fabrication. In EVA-TT PV laminate, UV transparent EVA has been used as a front-side and rear-side encapsulant. In TPO-TT PV laminate, UV transparent TPO has been used as a front and rear encapsulant, whereas in TPO-TW PV laminate, UV transparent TPO has been used as a front and white TPO a rear encapsulant. EVA-TT, TPO-TT, and TPO-TW laminates are based on single cut cell PV laminates (6.5 cm^2) used to evaluate the TPO and EVA fluorescence pattern and early detection under UV aging. Standard 72 cut cell (130 x 150 mm) modules have been prepared for assessing the transparent TPO and EVA encapsulant corrosion and degradation products under damp heat conditions.

Standard PV modules, glass-to-glass, and mini PV laminates have been prepared at 150 °C for both encapsulants. The evacuation time is set to 282 and 182 seconds, and pressure holding time is set to 482 and 382 seconds for EVA and TPO encapsulants, respectively, during the lamination process in an automatic industrial laminator (Boost Solar, BSL22360AC-II & III laminator). The pressure around 95 and 100 kPa has been maintained during the pressure holding process. Laminates are placed between Teflon sheets during the lamination process and removed after cooling down to room temperature.

2.3 Accelerated aging tests

In order to evaluate the fluorescence pattern and its effect on transmittance, glass-to-glass TPO and EVA laminates have been exposed under the UV-365 aging conditions. UV-365 accelerated aging chamber is based on a 365 nm LED light source (make: Thorlabs, model: SOLIS-365C, with an optical output of 3 Watts). The aging has been conducted with a maximum light intensity of ca. 950 W/m^2 at 90 ° C, and 10-13% relative humidity. Xenon light source will take months to a year to observe the reliability results, whereas UV-365 aging test will take four to five weeks. UV dose of 367.2 kWh/m^2 is sufficient to detect the fluorescence patterns with the UV-365 aging test at the given conditions. Due to the sample size and space limitation, mini PV modules have not been evaluated with UV-365 aging chamber. To assess the fluorescence pattern in mini PV laminates, Xenon aging chamber has been used.

To bring near about 90 °C, all three EVA-TT, TPO-TT, and TPO-TW mini PV laminates have been insulated during the Xenon aging study. A T-type thermocouple has been attached to the backsheet side and center of the PV laminate to check the working temperatures for all three modules. All three laminates have been placed on a 12 cm insulation (glass wool) box and packed with aluminum tape. These packed laminates have been placed inside the Xenon accelerated aging test chamber. The Xenon chamber (make: Q-Lab, model: Q-sun Xe-3) has been operated at a black panel temperature of 60 °C, a chamber air temperature of 40 °C, 20% of chamber relative humidity and a total UV intensity of 75 W/m^2 (from 280 nm to 390 nm). After 162 kWh/m^2 UV dose, a clear difference between the fluorescence patterns has been observed through the fluorescence imaging and samples removed for further investigation of diffusion properties and degradation products of both the encapsulants.

The hot-humid climatic induced degradation of TPO encapsulant-based standard PV modules has been evaluated with EVA modules under damp heat conditions. The test has been conducted at 85% relative humidity & 85 °C for 1000 h as per the IEC-61215 recommendation and later extended to 4000 h. Damp heat test has been conducted through an environmental chamber (Make: ESPEC, Model: EWSH282-5CW).

2.4 Characterization

Early-stage identification of UV-induced degradation and fluorescence patterns of TPO and EVA glass-to-glass and mini PV laminate have been characterized with UV fluorescence imaging and inverted grayscale imaging. The oxygen diffusion length and diffusion coefficients have been measured on inverted grayscale images. For UV fluorescence imaging, UV-365 light has been used as an excitation light source. UV fluorescence images have been captured using a digital camera (Make: Nicon corporation, model: D-3200) under UV light illumination. The changes in optical transmittance of glass-to-glass and UV absorber loss have been measured with UV/Vis spectrophotometer (make: Jasco, model: V-650 Series). The transmittance spectra have been measured from the spectral range of 200 nm to 900 nm, at 1 nm step-size and 0.1 s integrating time. Yellowness index has been recorded with a portable colorimeter (make: Testronix, model: TP110). The electrical current-voltage (I-V) parameters have been measured with Spire Solar Simulator (SPI-SUN 5600 SLP BLUE, Class AAA simulator) at standard AM 1.5G conditions. Electroluminescence (EL) image has been captured with Module/Solar String Electroluminescence (EL) tester (Model: SWT-REL 16M, Make: RENUSCO Image Technology Co. Ltd.) has been used for visual image observation of PV modules.

The degradation products and functional groups of TPO and EVA encapsulant films have been recorded with Fourier transform infrared (FTIR) spectroscopy in an attenuated total reflection (ATR) mode (BRUKER ALPHA ECO-ATR). The FTIR spectrum has been measured with 32 scans, 4 cm^{-1} optical resolution from 600 cm^{-1} to 4000 cm^{-1} spectral range. Acid index and hydroxyl index have been measured after the baseline subtraction and normalization. Acid index and hydroxyl index values have been obtained by integrating the area under the acid and hydroxyl peaks (1730 and 3450 cm^{-1}), and dividing with methylene peak (724 cm^{-1}).

3 RESULTS AND DISCUSSIONS

3.1 UV induced degradation and its early detection

UV-induced fluorescence and its early detection have been studied through the UV-365 and Xenon accelerated aging tests. UV-induced fluorescence is related to the degraded chemical products of encapsulant material. The fluorescence can be easily captured under the illumination of UV light [25,26]. Fig. 1 (a) depicts the fluorescence images of TPO and EVA glass-to-glass laminates captured before and after the UV-365 accelerated aging test. EVA encapsulant has shown strong fluorescence at the laminate center, whereas TPO did not show any fluorescence. This indicates that EVA goes faster discoloration than TPO during the UV-365 accelerated aging test. UV dose of 367.2 kWh/m^2 is sufficient to detect the early discoloration of EVA during the UV-365 accelerated aging test. Due to the oxygen bleaching at the edge, the photobleached region dominated at the edge, and discoloration dominated at the center of EVA laminate. Due to the discoloration, a significant change in transmittance spectra of EVA laminate has been observed in the UV region from 220 nm to 380 nm compared to the TPO shown in Fig. 1 (b). Peaks located at 264 nm and 309 nm belong to the UV absorber i.e., identified as "Cyasorb 531". The changes in transmittance spectra at 264 nm and 309 nm are due to the UV absorber loss, and the change at 398/400 nm is due to the discoloration in EVA[6,27].

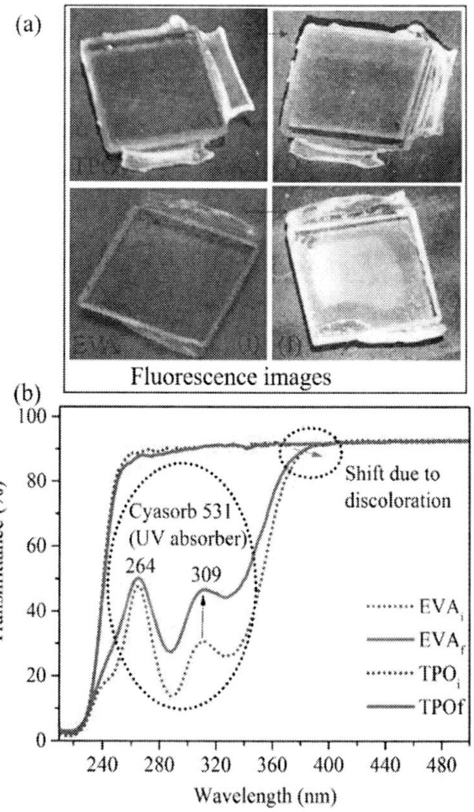

Figure 1: (a) Fluorescence images, and (b) transmittance spectra of TPO and EVA glass-to-glass laminates. Note: (i) and (f) indicates the initial and final laminates under the UV-365 accelerated aging test.

The transmittance measured in the UV region (at 264 nm and 309 nm) and at 400 nm is shown in Table 1. The overall transmittance of EVA has been increased to 8% in the UV region (220 nm to 380 nm). This is due to the UV absorber loss resulting in an increase of transmission in the UV region (around 8%, 385 to 420 nm) and a degrease of transmittance (around 2%) near at 390/400nm. There has no such UV absorber involved in TPO, so insignificant changes have been observed in transmittance spectra. Due to the UV absorber loss-induced fluorescence, the yellowness has been increased. More than three-factor higher increment in yellowness index has been observed for EVA laminate as compared to the TPO. The yellowness index values are shown in Table 2.

Table 1. Optical properties

Sample details	Transmittance (%)			
	264 nm	309 nm	200-380 nm	400 nm
EVA$_i$	46	30	42	93

TPO_i	---	---	81	92
EVA_f	50	46	50	91
TPO_f	---	---	81	92

Note: i and f indicates the initial and final values under the UV-365 accelerated aging test

Table 2. Yellowness index

Sample details	change in YI	Change in YI (%)
EVA_i	9.32	0
TPO_i	9.81	0
EVA_f	10.12	8.58
TPO_f	10.07	2.65

Fig. 2 (a) depicts the fluorescence images of TPO and EVA PV module laminates captured before and after the Xenon accelerated aging test. EVA-TT PV module laminate is based on UV transparent EVA encapsulant (front and rear). The final EVA-TT laminate shows the fluorescence brim square pattern between the photobleached (edge) and the center region of the solar cell. This pattern is almost similar to the recently published article[25]. TPO-TT PV module laminate is based on UV transparent TPO encapsulant (front and rear). In contrast, TPO-TW PV module laminate is based on UV transparent TPO as a front and white TPO as a rear encapsulant. The final TPO-TT laminate shows the fluorescence brim square pattern on the edges of solar cells, whereas the final TPO-TW laminate shows the fluorescence brim square pattern near the edges (inner photobleached region) of solar cells. UV dose of 162 kWh/m^2 is sufficient to find the early-stage fluorescence pattern in PV laminates during the Xenon accelerated aging test. The fluorescence pattern initiated in between the photobleached edge and center, on the edges, and near the edges of solar cell is due to the difference in oxygen diffusion length or diffusion coefficient of laminates.

Fig. 2 (b) shows the inverted greyscale images of the fluorescence image of Fig. 2 (a). Inverted greyscale images show a clear difference between the photobleached (edge white region) and fluorescence square brims (thin black ring) region on the solar cell of EVA and TPO PV laminates. The photobleaching effect on the solar cell edge is due to a rich oxygen environment [1]. A square photobleached/discoloration pattern appears on the solar cell when oxygen diffuses equally from all the sides (assume without cracks/defects). If photobleaching/discoloration is distributed as a square brim, the oxygen diffusion length and oxygen diffusion area (ODA) can be obtained by using ImageJ software analysis. By using the oxygen diffusion length, the oxygen diffusion coefficient (D) of three laminates has been calculated using a simple diffusion model (1) [25].

$$L = \sqrt{Dt} \qquad (1)$$

Where L is the oxygen diffusion length over the 90 days of Xenon accelerating time (t). The calculated oxygen diffusion length (ODL) and oxygen diffusion coefficient (D) values are summarized in Table 3. It has been found that EVA-TT PV laminate has four times and one order of magnitude higher oxygen diffusion coefficient than the TPO-TW and TPO-TT PV laminates, respectively. The diffusion length of EVA-TT has a factor of more than two and four times higher than the TPO-TW and TPO-TT PV laminates, respectively. Due to the higher oxygen diffusion length/coefficient, EVA laminate has shown a higher photobleached area. TPO-TT laminate also shown delamination outside the solar cell edges (dark black regions in the invested grayscale image or milky white regions in fluorescence image). An average of ten YI has been recorded on both the EVA and TPO min PV modules. The average YI, change in YI, increment in YI has been shown in Table 4. About one factor higher increment in YI has been observed for EVA-TT PV laminate. This may be due to the higher fluorescence/discolored area of EVA as compared to the TPO PV laminates.

(a)
(b)
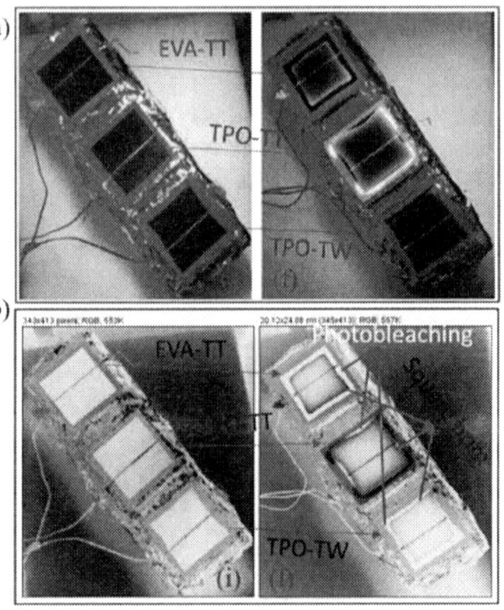

Figure 2: (a) Fluorescence images (b) inverted grayscale images of TPO and EVA PV laminates. (i) and (f) indicates the initial and final images of PV laminates.

Table 3. Oxygen diffusion parameters.

Details	ODL (cm)	D (cm^2/s)
$EVA\text{-}TT_f$	0.63	5.1×10^{-8}
$TPO\text{-}TT_f$	0.13	2.1×10^{-9}
$TPO\text{-}TW_f$	0.32	1.3×10^{-8}

Note: f indicates the final laminate.

Table 4. Yellowness index values.

Sample details	Initial YI	Final YI	Change in YI	Increment in YI (%)
EVA-TT	3.83	5.69	1.86	48%
TPO-TT	2.99	4.15	1.16	38%
TPO-TW	3.12	4.02	0.89	28%

3.2 Damp heat induced degradation

The damp heat-tested TPO and EVA encapsulant-based PV module EL images have been shown in Fig. 3 (a). Before (i: indicates 0 h) and after (f: 4000 h) of the damp heat test, TPO PV modules don't show a significant change, whereas the EVA module shows a significant change in EL image. The interconnect and metallization have been corroded all over solar cells of

EVA-based PV module. This is due to the free-acetic acid and other acid group's formation during the damp heat aging.

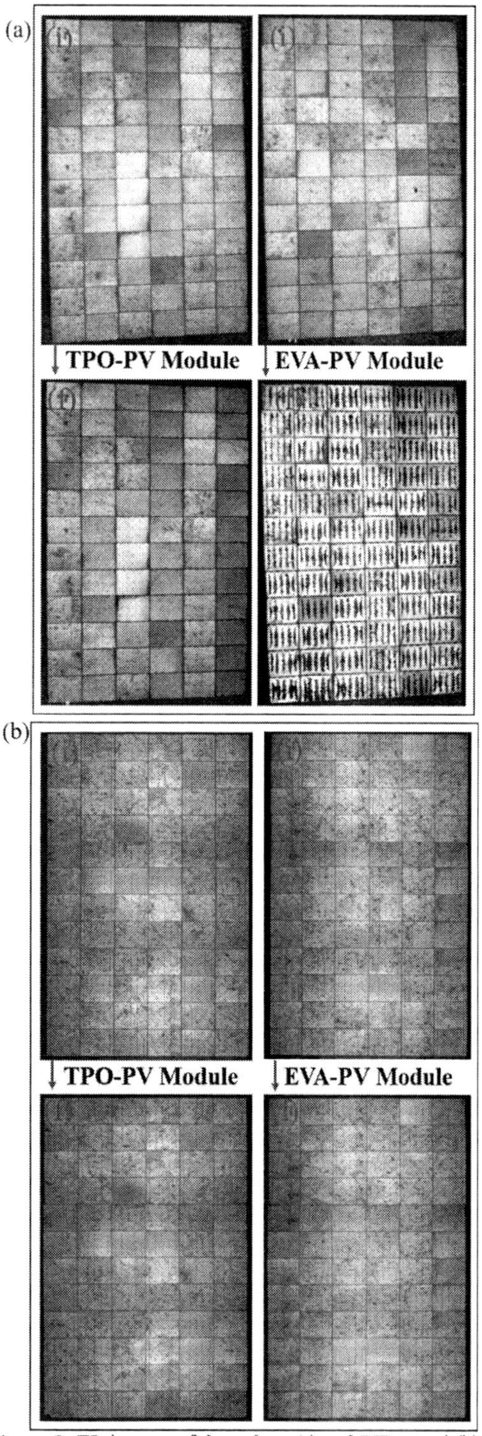

Figure 3: EL images of damp heat (a) and PID tested (b) TPO and EVA PV modules. (i) and (f) indicates the initial and final EL images.

EVA has a higher water vapor transition rate and higher water absorption than the TPO [19]. Thi may leads metalization corrosion in EVA. In EVA, vinyl acetate is weakly bonded to the polymer main chain [1][24]. This leads to an easy liberation of free-acetic acid and causes metallization corrosion during the damp heat aging test. There are no such metallization corrosion signs have been observed in TPO. This may be due to no such a vinyl acetate unit or direct free-acetic acid/other acid-forming groups involved in TPO. PID test also conducted at damp heat condition on TPO and EVA encapsulant-based PV modules shown in Fig. 3 (b). When a PV module is affected by PID, the solar cells appear dark. The dark solar cell indicates the current leakage cell during the PID test. Before (i: indicates 0 h) and after (f: indicates 384 h) the PID test at damp heat condition, TPO and EVA PV modules doesn't show any significant changes in EL images. So only admp heat tested modulus upto 4000 h has been used for further analysis.

Fig. 4 (a) depicts the power degradation of TPO and EVA modules. The EVA module showed 2.7 % power loss until the 3000 h and sharply dropped to 7% from 3000 to 4000 h of damp heat test.

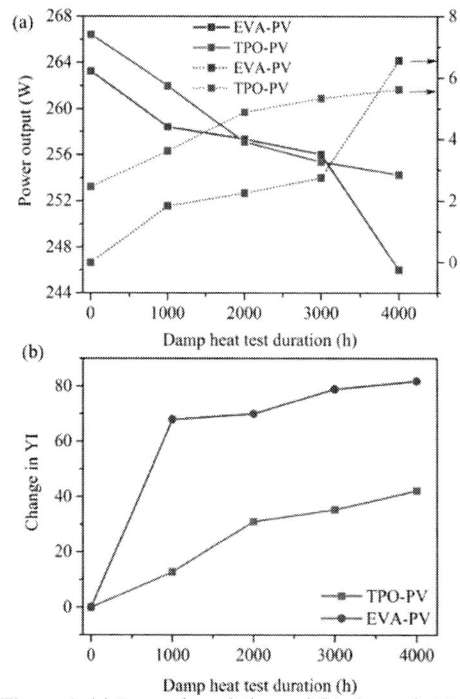

Figure 4: (a) Power degradation and (b) change in YI of damp heat tested TPO and EVA PV modules.

The overall 7% and 5% power degradation were observed for EVA and TPO PV modules, respectively. Higher power degradation in EVA than TPO is due to metallization corrosion. TPO and EVA modules tested more than four times of the IEC 61215 recommended qualification damp heat test. PV module encapsulant is considered as damp heat and PID resilient when its power loss is less than 5%. TPO and EVA modules are four and three times more damp heat resilient than the IEC 61215 recommendations, and both modules are four times more PID resilient than the IEC 62804 recommendations. The electrical I-V parameters of damp heat and PID tested TPO and EVA PV modules are summarized in Table 5 and Table 6. TPO and EVA modules are shown only

2.5% power loss during the PID test. Due to the insignificant changes in EL images and power output, further analysis has not been done on PID-tested TPO and EVA modules.

YI has been measured after every 1000 h of damp heat test. YI values have been measured at seven points on TPO and EVA PV modules, and their average values are summarized in Table 7. Fig. 4 (b) depicts the change in YI of TPO and EVA modules as a function of damp heat aging time. Both the TPO and EVA PV modules showed a significant increment in a change in YI. The rate of change in YI is more than two times higher for EVA than the TPO. This is well-matched with the recently reported values of the rate of change in YI of TPO and EVA glass-to-glass laminates (damp heat tested up to 4000 h)[24]. A higher increment in a change in YI of EVA is due to the higher formation of free-acetic acid or other acid functional groups than the TPO.

Table 5. Electrical I-V parameters of EVA and TPO PV modules during the damp heat test.

Details	I_{sc} [A]	I_{mp} [A]	V_{oc} [V]	V_{mp} [V]	P_{mp} [W]
EVA-0	7.43	7.01	46.12	37.55	263.24
EVA-1000	7.30	6.89	46.06	37.48	258.42
EVA-2000	7.34	6.99	46.13	36.80	257.33
EVA-3000	7.23	6.84	46.34	37.44	256.02
EVA-4000	7.37	6.70	46.54	36.71	245.99
TPO-0	7.49	7.06	46.29	37.75	266.41
TPO-1000	7.43	7.07	45.99	37.05	261.97
TPO-2000	7.28	6.87	46.36	37.45	257.08
TPO-3000	7.23	6.83	45.99	37.38	255.36
TPO-4000	7.30	6.87	45.79	37.01	254.25

Table 6. Electrical I-V parameters of EVA and TPO PV modules during the PID test.

PID Test duration (hr)	**TPO module**			
	V_{oc} (V)	I_{sc} (A)	P_{mp} (W)	P_{mp} loss (%)
0	46.41	9.1	322.01	0.00
192	46.53	9.06	316.45	-1.73
288	46.71	9	314.17	-2.43
384	46.8	9.04	313.8	-2.55
PID Test duration (hr)	EVA module			
	V_{oc} (V)	Isc (A)	P_{mp} (W)	P_{mp} loss (%)
0	46.41	9.12	322.05	0.00
192	46.97	8.89	317.14	-1.52
288	46.67	9.05	316.1	-1.85
384	46.89	9.03	314.63	-2.30

Table 7. Yellowness index values.

DH test Duration	**TPO**	**EVA**
	YI	YI
0	-61.05	-91.3
1000	-48.34	-23.43
2000	-30.07	-21.43
3000	-25.75	-12.41
4000	-18.89	-9.52

3.2 Damp heat induced degradation

UV and damp heat aged TPO and EVA encapsulant degradation products and functional groups have been identified with ATR-FTIR spectroscopy. Fig. 5 depicts the FTIR spectra of EVA and TPO films. The absorption peaks at 2924, 2852, 1638, 1540, 1462, and 724 cm^{-1} belong to the methylene functional group of the ethylene unit in EVA and TPO. The absorption bands at 1730, 1238, 1158, and 1020 cm^{-1} belong to vinyl acetate (VA) in EVA copolymer [28]. While absorption peaks at 1732, 1240, and 1020 cm^{-1} bands belong to acrylates in TPO [22][24]. The ether stretching band at 1158 cm^{-1} and the ester stretching band at 1238 cm^{-1} are the significant differences between the TPO and EVA films. The ether is a non-polar, and ester is a polar functional group in TPO and EVA, respectively. Other more common functional groups have been summarized in Table 8.

Fig. 5 (a) depicts the FTIR spectra of UV-365 aged EVA and TPO films recorded at the edges of glass-to-glass laminate (Fig. 1 (a)). Fig. 5 (b) depicts the FTIR spectra of Xenon aged films recorded at the fluorescence/discoloration brim region of mini PV laminates (Fig. 2 (a)). For Fig. 5 (b), the aged and unaged films have been removed from the EVA and TPO mini PV laminates. For Fig. 5 (c), the damp heat aged and unaged films have been removed from the standard 72 cut cell PV modules.

A significant change has been observed in FTIR spectra of UV-365, Xenon, damp heat aged EVA, and TPO films. Acid (carbonyl, C=O) and hydroxyl (OH) functional groups show significant changes before and after aging. The acid index and hydroxyl index and their difference between the initial and the final have been calculated and summarized in Table 9. Negative and positive difference in index values indicates the loss and formation of acid and hydroxyl functional groups of encapsulants during the aging study. Acid and hydroxyl groups are formed significantly in EVA films during the UV-365 and Xenon aging study. In contrast, TPO shows insignificant changes, a slight loss of hydroxyl groups (negative value) during the Xenon aging.

A severe loss of acid groups and a strong formation of hydroxyl groups have been observed during the damp heat aging. A factor of nine times higher acid index difference has been obtained for EVA. Due to this acid group generation in EVA, all the metallization parts have been strongly corroded in the EVA PV module shown in Fig. 3 a (f) compared to the TPO (Fig. 3 a (i)). A factor of two times higher hydroxyl index difference has been obtained for EVA. This indicates the more amount of moisture absorption in EVA as compared to the TPO. This is well agreed with the recent article[24]. This is due to the more polar function groups content in EVA as compared the TPO. In EVA, VA is a weaker unit and easily splits/generates free-acetic acid through deacetylation[6]. The generation of free-acetic acid and ester groups causes more discoloration (yellowness index) and metal corrosion in EVA modules. In TPO, ester (C-O) or ether (C-O-C), carbonyl acid (C=O) groups of acrylate unit are strongly bonded (C-C bond) with polymer main[24]. So, the generation of free-acetic acid and other functional groups is not easy in TPO compared to the EVA.

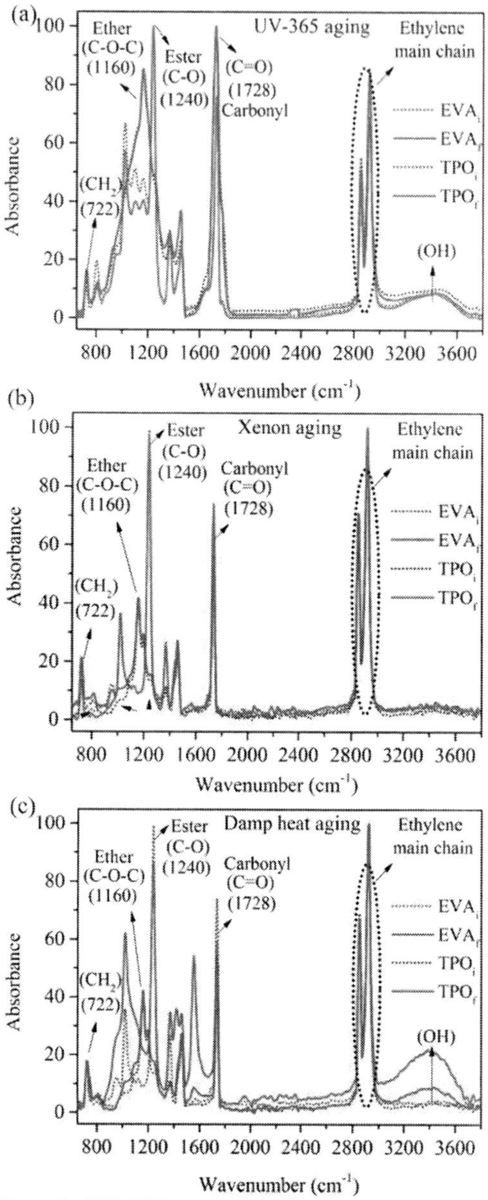

Figure 5: ATR-FTIR normalized absorbance spectra of UV-365 (a), Xenon (b), and damp heat (c) aged TPO and EVA encapsulants.

Table 8. FTIR functional groups.

Wavenumber (cm^{-1})	Functional groups	Band characteristics
3450	Hydroxyl	OH stretching
2924	Methylene	Asymmetric stretching of CH2
2852	Methylene	Symmetric stretching of CH2
1730	Acid carbonyl	Stretching of C=O
1638	Polyenes	Vibration of C=C
1540	Polyenes	Vibration of C=C
1462	Hydrocarbon	-CH2 bending
1372	Hydrocarbon	-CH2 bending
1238	Ester	Stretching of C-O
1158	Ethers	Stretching of COC
1020	Vinyl	Stretching of C-O
724	Methylene	Stretching of CH_2

Table 9. Change in acid index and hydroxyl index during the UV aging

Details	**UV-365 aging**			
	Acid index (AI)		Hydroxyl index (HI)	
	EVA	TPO	EVA	TPO
Initial	14.91	23.94	10.05	2.89
Final	16.47	24.61	11.09	3.94
Difference	1.56	0.67	1.04	1.05
Details	Xenon aging			
	AI		HI	
	EVA	TPO	EVA	TPO
Initial	6.69	3.87	0.15	1.38
Final	5.88	3.66	0.31	1.24
Difference	1.19	-0.21	0.16	-0.14
Details	Damp heat aging			
	AI		HI	
	EVA	TPO	EVA	TPO
Initial	5.76	3.91	0.83	1.07
Final	2.79	3.58	9.54	5.56
Difference	-2.97	-0.33	8.71	4.49

4 CONCLUSION

UV-365, Xenon, and damp heat accelerated aging tests have been successfully conducted to evaluate the TPO encapsulant degradation and compared with the commercially available standard EVA encapsulants. UV dose of 367.2 kWh/m^2 is sufficient to detect the early UV-induced degradation in glass-to-glass laminates under the UV-365 accelerated aging test conditions. EVA glass-to-glass laminate shows strong fluorescence at the center, whereas no fluorescence has been observed for TPO. A significant change in UV region of transmittance spectra has been observed for EVA due to its UV absorber loss during the UV-365 aging study. Due to the UV absorber loss, EVA shows a higher yellowness index as compared to the TPO. UV dose of 162 kWh/m^2 is sufficient to detect the Xenon UV-induced degradation of EVA and TPO mini PV laminates. The different fluorescence patterns have been found in EVA and TPO mini PV laminates during the Xenon aging test. The final EVA-TT shows the fluorescence brim square pattern in between the photobleached and center region, whereas TPO-TT laminates show the fluorescence brim square pattern on edges of the solar cell due to difference in oxygen diffusion length and diffusion coefficients. The higher fluorescence, oxygen diffusion length, and diffusion coefficients have been found for EVA as compared to the TPO mini PV laminates during the Xenon aging test. Damp heat test has been conducted up to 4000 h. TPO PV modules don't show significant change, whereas the EVA module showed significant differences in EL images during the damp heat test. The metallization, busbar, and other metal strings have been corroded all over the solar cells of the EVA-based PV module. Due to the corrosion, EVA module power has been dropped suddenly from 2.7 to 7% power loss. TPO module shows less than 5% power degradation and less change in yellowness index up to the 4000h of damp heat test.

The aged TPO and EVA encapsulant films degradation products and functional groups have been identified with ATR-FTIR spectroscopy. The ether and ester functional groups are the major difference between the TPO and EVA films, respectively. The formation of acid and hydroxyl groups is significant in EVA compared to the TPO films during the UV-365 and Xenon aging study. A severe loss of acid groups and a strong formation of hydroxyl groups have been observed for EVA during the damp heat aging. All the results indicate that TPO shows better performance in hot-dry and hot-humid climatic conations.

ACKNOWLEDGEMENT

The authors would like to acknowledge Borealis Polyolefine GmbH, Austria, and Waaree Energies Ltd. Surat, India, for providing material and DH test support during this study. We are thankful to the SAIF at IIT Bombay for the characterization facilities. This work was supported by the NCPRE funded by the Ministry of New and Renewable Energy of the Government of India through Project No. 31/09/2015-16/PVSE-R&D dated 15th June 2016. The authors confirm that there is no conflict of interest in this study.

REFERENCES

[1] A.W. Czanderna, F.J. Pern, Encapsulation of PV modules using ethylene vinyl acetate copolymer as a pottant: A critical review, Sol. Energy Mater. Sol. Cells. 43 (1996) 101–181. doi:10.1016/0927-0248(95)00150-6.

[2] A. Omazic, G. Oreski, M. Halwachs, G.C. Eder, C. Hirschl, L. Neumaier, G. Pinter, M. Erceg, Relation between degradation of polymeric components in crystalline silicon PV module and climatic conditions: A literature review, Sol. Energy Mater. Sol. Cells. 192 (2019) 123–133. doi:10.1016/j.solmat.2018.12.027.

[3] A. Ndiaye, A. Charki, A. Kobi, C.M.F. Kébé, P.A. Ndiaye, V. Sambou, Degradations of silicon photovoltaic modules: A literature review, Sol. Energy. (2013). doi:10.1016/j.solener.2013.07.005.

[4] F.J. Pern, Luminescence and absorption characterization of ethylene-vinyl acetate encapsulant for PV modules before and after weathering degradation, Polym. Degrad. Stab. 41 (1993) 125–139. doi:10.1016/0141-3910(93)90035-H.

[5] G. Oreski, G. Pinter, B. Ottersb, Comparison of different microclimate effects on the aging behavior of encapsulation materials used in photovoltaic modules, 138 (2017) 182–191. doi:10.1016/j.polymdegradstab.2017.03.010.

[6] N. S. Allen, M. Edge, M. Rodriguez, C. M. Liauw, E. Fontan, Aspects of the thermal oxidation, yellowing and stabilisation of ethylene vinyl acetate copolymer, Polym. Degrad. Stab. 71 (2000) 1–14. doi:10.1016/S0141-3910(00)00111-7.

[7] G.C.E. Ch. Hirschl, L. Neumaier, S. Puchberger, W. Mühleisen, G. Oreski, M.K. R. Frank, M. Tranitz, M. Schoppa, M. Wendt, N. Bogdanski, A. Plösch, Determination of the degree of ethylene vinyl acetate crosslinking via Soxhlet extraction: Gold standard or pitfall?, Sol. Energy Mater. Sol. Cells. 143 (2015) 494–502. doi:10.1016/j.solmat.2015.07.043.

[8] W. Stark, M. Jaunich, Investigation of Ethylene/Vinyl Acetate Copolymer (EVA) by thermal analysis DSC and DMA, Polym. Test. 30 (2011) 236–242. doi:https://doi.org/10.1016/j.polymertesting.2010.12.003.

[9] B. Adothu, A.K. Singh, S. Kumar, S. Zele, S. Mallick, Effect of curing temperature on properties of Ethylene vinyl-acetate used for crytalline silicon solar module encapsulation, in: 36th Eur. Photovolt. Sol. Energy Conf. Exhib., 2019: pp. 1204–1207.

[10] B. Adothu, P. Bhatt, S. Zele, J. Oderkerk, F.R. Costa, S. Mallick, Investigation of newly developed thermoplastic polyolefin encapsulant principle properties for the c-Si PV module application, Mater. Chem. Phys. 243 (2020) 122660. doi:10.1016/J.MATCHEMPHYS.2020.122660.

[11] B. Adothu, P. Bhatt, P. Kartikay, S. Zele, F.R. Costa, J. Oderkerk, S. Mallick, Determination of Crystallinity and Thermal Stability of Newly Developed Thermoplastic Polyolefin Encapsulant for the c-Si PV Module Application, in: Conf. Rec. IEEE Photovolt. Spec. Conf., Institute of Electrical and Electronics Engineers Inc., 2019: pp. 487–490. doi:10.1109/PVSC40753.2019.8980526.

[12] D.C. Miller, D.M. Burns, S. Fowler, L. Peter, C. Christian, N.H. Phillips, K.P. Scott, A.F. Zielnik, M. Köhl, M.D. Kempe, Degradation in photovoltaic encapsulant transmittance : Results of the first PVQAT TG5 artificial weathering study, Prog Photovolt Res Appl. (2019) 391–409. doi:10.1002/pip.3103.

[13] J. Tracy, N. Bosco, R. Dauskardt, Encapsulant Adhesion to Surface Metallization on Photovoltaic Cells, IEEE J. Photovoltaics. 7 (2017) 1635–1639. doi:10.1109/JPHOTOV.2017.2746572.

[14] J. Tracy, N. Bosco, F. Novoa, R. Dauskardt, Encapsulation and backsheet adhesion metrology for photovoltaic modules, Prog. Photovoltaics Res. Appl. 25 (2017) 87–96. doi:10.1002/pip.2817.

[15] F. Pern, Ethylene-vinyl acetate (EVA) encapsulants for photovoltaic modules: Degradation and discoloration mechanisms and formulation modifications for improved photostability, Die Angew. Makromol. Chemie. 252 (1997) 195–216. doi:10.1002/apmc.1997.052520114.

[16] S. Chattopadhyay, R. Dubey, V. Kuthanazhi, J.J. John, C.S. Solanki, A. Kottantharayil, B.M. Arora, K.L. Narasimhan, V. Kuber, J. Vasi, A. Kumar, O.S. Sastry, Visual degradation in field-

aged crystalline silicon PV modules in India and correlation with electrical degradation, IEEE J. Photovoltaics. 4 (2014) 1470–1476. doi:10.1109/JPHOTOV.2014.2356717.
[17] R. Dubey, S. Chattopadhyay, V. Kuthanazhi, A. Kottantharayil, C. Singh Solanki, B.M. Arora, K.L. Narasimhan, J. Vasi, B. Bora, Y.K. Singh, O.S. Sastry, Comprehensive study of performance degradation of field-mounted photovoltaic modules in India, Energy Sci. Eng. 5 (2017). doi:10.1002/ese3.150.
[18] C. Peike, S. Hoffmann, P. Hülsmann, B. Thaidigsmann, K.A. Weiß, M. Koehl, P. Bentz, Origin of damp-heat induced cell degradation, Sol. Energy Mater. Sol. Cells. 116 (2013). doi:10.1016/j.solmat.2013.03.022.
[19] M.C. López-Escalante, L.J. Caballero, F. Martín, M. Gabás, A. Cuevas, J.R. Ramos-Barrado, Polyolefin as PID-resistant encapsulant material in PV modules, Sol. Energy Mater. Sol. Cells. 144 (2016). doi:10.1016/j.solmat.2015.10.009.
[20] B. Adothu, P. Bhatt, S. Chattopadhyay, S. Zele, J. Oderkerk, H.P. Sagar, F.R. Costa, S. Mallick, Newly developed thermoplastic polyolefin encapsulant–A potential candidate for crystalline silicon photovoltaic modules encapsulation, Sol. Energy. 194 (2019) 581–588. doi:10.1016/J.SOLENER.2019.11.018.
[21] C. Barretta, G. Oreski, S. Feldbacher, K. Resch-Fauster, R. Pantani, Comparison of degradation behavior of newly developed encapsulation materials for photovoltaic applications under different artificial ageing tests, Polymers (Basel). 13 (2021). doi:10.3390/polym13020271.
[22] G. Oreski, A. Omazic, G.C. Eder, Y. Voronko, L. Neumaier, W. Mühleisen, C. Hirschl, G. Ujvari, R. Ebner, M. Edler, Properties and degradation behaviour of polyolefin encapsulants for photovoltaic modules, Prog. Photovoltaics Res. Appl. 28 (2020). doi:10.1002/pip.3323.
[23] G. Oreski, G.C. Eder, Y. Voronko, A. Omazic, L. Neumaier, W. Mühleisen, G. Ujvari, R. Ebner, M. Edler, Performance of PV modules using co-extruded backsheets based on polypropylene, Sol. Energy Mater. Sol. Cells. 223 (2021) 110976. doi:https://doi.org/10.1016/j.solmat.2021.110976.
[24] B. Adothu, F.R. Costa, S. Mallick, Damp heat resilient thermoplastic polyolefin encapsulant for photovoltaic module encapsulation, Sol. Energy Mater. Sol. Cells. 224 (2021) 111024. doi:10.1016/j.solmat.2021.111024.
[25] B. Adothu, S. Chattopadhyay, P. Bhatt, P. Hui, F.R. Costa, S. Mallick, Early-stage identification of encapsulants photobleaching and discoloration in crystalline silicon photovoltaic module laminates, Prog. Photovoltaics Res. Appl. (2020). doi:10.1002/pip.3269.
[26] G.C. Eder, Y. Voronko, C. Hirschl, R. Ebner, G. Újvári, W. Mühleisen, Non-destructive failure detection and visualization of artificially and naturally aged PV modules, Energies. 11 (2018) 1–14. doi:10.3390/en11051053.
[27] A. Jentsch, K.J. Eichhorn, B. Voit, Influence of typical stabilizers on the aging behavior of EVA foils for photovoltaic applications during artificial UV-weathering, Polym. Test. 44 (2015) 242–247. doi:10.1016/j.polymertesting.2015.03.022.
[28] R.J. Koopmans, R. van der Linden, E.F. Vansant, Quantitative Determination-of the Vinylacetate Content in Ethylene Vinyl Acetate Copolymer- A Critical Review, Polym. Eng. Sci. 22 (1982) 878–882.

This is a short version of the paper submitted to the journal Progress in Photovoltaics.
It has been selected by the Executive Committee of the EU PVSEC 2021 for submission to Progress in Photovoltaics.

TRANSPARENT BACKSHEETS FOR BIFACIAL PHOTOVOLTAIC MODULES

Xiaohong Gu, Soshana Smith, LaKesha Perry, Stephanie Watson, Stephanie L. Moffitt, Shih-Jia Shen, Stefan Mitterhofer, Li-Piin Sung, Po-Chang Pan and Deborah Jacobs
Engineering Laboratory, National Institute of Standards and Technology, Gaithersburg, MD 20899, USA

ABSTRACT: Bifacial PV modules are promising to achieve a lower levelized cost of electricity (LCOE) compared to monofacial modules because light absorption occurs from both sides of the devices. Compared to the glass/glass bifacial modules, glass/backsheet modules show many advantages including lighter weight and high corrosion resistance. However, research on the weatherability and long-term reliability of transparent backsheet materials and bifacial modules under their service environments is lacking. In this study, accelerated weather testing using the NIST SPHERE *(Simulated Photodegradation via High Energy Radiant Exposure)* was conducted to investigate the durability of three fluoropolymer-based transparent backsheets and their laminated coupon counterparts. The transparent backsheets were exposed at UV/75 °C/50 % relative humidity (RH), while the coupons were exposed at UV/65 °C/50 % RH, both subjected to UV irradiance of approximately 140 W/m^2 for up to 2000 h. Optical, chemical properties and mechanical properties of specimens were characterized. Results indicate that different transparent backsheets perform differently after aging, and potential cracking could occur once films become brittle. The cracking of polyethylene terephthalate (PET) core layer was observed, as well as the cracking of the ethylene vinyl-acetate (EVA) inner layer for a fluoroethylene vinyl ether (FEVE)/ PET/ EVA transparent backsheet. The other two backsheets, polyvinyl fluoride (PVF)/PET/FEVE and polyvinylidene fluoride (PVDF)/PET/FEVE, showed no obvious signs of cracking up to 2000 h of UV exposure. This study suggests that UV stability of PET core layer is critical to the application of transparent backsheets for bifacial PV modules.
Keywords: Transparent backsheets; bifacial modules; NIST SPHERE; accelerated laboratory testing; chemical and optical properties; yellow index

1 INTRODUCTION

Bifacial modules are anticipated to surpass monofacial modules to achieve a lower levelized cost of electricity (LCOE) because of the capability of absorbing light and generating electrical power from both front and rear sides [1]. The additional power yield of bifacial modules can be nearly 30 % by optimization of various configurations [2, 3], such as installation conditions (e.g., height, angle, and direction), ground albedo, tracking, and the bifaciality of solar cells.

Traditional bifacial modules consisting of a double glass structure are resistant to moisture, fire, scratch and sand abrasion. However, these modules can be heavy, resulting in high installation and transportation cost at a large scale [4]. In addition to lightweight, glass/transparent backsheet modules have the advantage of higher light transmittance, lower susceptibility for hail damage, smaller heat capacity, higher stain-resistance, corrosion-resistance to saline alkali and a high throughput manufacturing process [5, 6]. Despite these advantages, there are currently limited studies addressing their reliability and long-term performance. In a survey of mechanical durability of PV backsheets, Kempe et al. [7] found all seven of the transparent backsheets cracked after the standardized IEC 62788-7-2 test followed by a mandrel bend test. Failure of backsheets can lead to concerns of safety and loss of power performance.

In this study, we aim to provide more insight into understanding the weatherability of transparent backsheets for their application in bifacial modules. Material characterization and durability tests of three fluoropolymer-based transparent backsheets films were performed. Corresponding tests for their laminated coupon counterparts as well as glass/glass coupons were also conducted. These multi-layered backsheets consisted of a fluoropolymer based outer layer, a PET core layer and an outer layer consisting of either a fluoropolymer coating or ethylene-vinyl acetate (EVA) polymer layer. Accelerated laboratory weathering was carried out on the NIST SPHERE (*Simulated Photodegradation via High Energy Radiant Exposure*), an integrating sphere-based device with a uniformly high energy light source. The exposure conditions for the materials were UV/75 °C/50 % RH and UV/65 °C/50 % RH for films and coupons, respectively. Optical changes as a function of exposure were characterized by UV-Visible spectrophotometry (UV-Vis). Chemical degradation of the exposed air side and the opposite cell side of the backsheets were examined with the attenuated total reflectance Fourier transform infrared (ATR-FTIR) spectroscopy. Surface and bulk mechanical properties were characterized by nanoindentation and tensile tests, respectively. The crack formation and location identification were studied by laser scanning confocal microscopy.

2 EXPERIEMNTAL*

2.1 Materials

Three types of transparent backsheets (CB1, CB2, CB3) were obtained from industry suppliers. The materials and structures of these backsheets are presented in Table 1.

TABLE 1. Materials and structures of transparent backsheets, glass/transparent backsheet coupons and glass/glass coupons.

Specimens	Materials and Structures	Remarks
CB1	Proprietarily modified PVF/PET/FEVE$_1$	PVF: polyvinyl fluoride PVDF: polyvinylidene fluoride FEVE: fluoroethylene vinyl ether coating POE: polyolefin elastomer (TF4/TF8) EVA: ethylene vinyl acetate (F406S/F806) PET: polyethylene terephthalate
CB2	Proprietarily modified PVDF/PET/FEVE$_2$	
CB3	FEVE$_3$/PET/EVA	
G/CB1	Silica/POE/CB1	
G/CB2	Silica/POE/CB2	
G/CB3	Silica/POE/CB3	
G/POE/G	Silica/POE/Silica	
G/EVA/G	Silica/EVA/Silica	

Silica/encapsulant/transparent backsheet (G/B) coupons were prepared based on the same three backsheets using a 3 mm thick polished fused silica wafer with a transmittance of around 95 % in the UV and visible ranges. The top and bottom encapsulants used for lamination were a UV-transparent polyolefin elastomer (POE, TF4) and a UV-cut off POE (TF8), respectively. Silica/POE/Silica (G/POE/G) coupons were laminated with two fused silica wafers and the same POE combination as G/B coupons. For comparison, Silica/EVA/Silica (G/EVA/G) coupons were prepared using a transparent-EVA (F406S) and a UV-cut-off EVA (F806) for the top and bottom encapsulants. All encapsulants were products of Hangzhou First Applied Material Co.

2.2 Accelerated Laboratory

Laboratory accelerated weathering was performed on the NIST SPHERE at 75 °C/50 % RH and 65 °C/50 % RH for transparent films and coupons, respectively, under approximately 140 W/m^2 of UV irradiance (295 nm to 400 nm). Two pieces of 60 mm x 80 mm free-standing films of each backsheet were mounted in the sample holder with the air side of each backsheet facing to the UV light. Silica/backsheet coupons (G/B) of 75 mm x 100 mm were exposed with the backsheets air side facing to the light. Samples were periodically (approximately biweekly) removed from the chamber for optical, chemical and mechanical characterizations. For G/B coupons, both air side exposure and glass side exposure were performed.

2.3 Characterization

2.3.1 UV-Vis and ATR-FTIR

Optical and chemical degradation of the transparent backsheets were measured as a function of exposure time. UV-Vis spectra of aged and unaged transparent backsheet films, silica/transparent coupons or silica/silica coupons were obtained by a Perkin-Elmer Lambda 900 spectrophotometer equipped with an integrating sphere in the range of 190 nm to 820 nm with a resolution of 1 nm. The yellow index (YI) was calculated from the acquired UV-Vis spectra based on ASTM E313. Chemical degradation was characterized by ATR-FTIR using a Nicolet iS50 FTIR spectrometer (Thermo Scientific). All FTIR spectra were the average of 128 scans recorded at a resolution of 4 cm^{-1} using dry air as a purge gas. A diamond crystal and 45° incident angle were used for all ATR-FTIR measurements. All measurements were done in at least three different locations of the sample. All acquired spectra were baseline corrected and normalized in order to compare peak intensity more accurately over time.

2.3.2 Laser Scanning Confocal Microscope (LSCM)

A Zeiss model LSM800 LSCM (Carl Zeiss Microscopy, LLC) was utilized to image the cross-section and surface of free-standing transparent backsheet films. Images were captured in reflection mode with a laser wavelength of 405 nm using objectives of 5x (Z-step of 3 μm), 20x (Z-step of 0.5 μm) and 50x (Z-step of 0.3 μm). Typical LSCM images presented in this paper are 2D projected images, which is a stack of Z-scan images projected into X-Y plane. Without specification, each frame consisted of 1024 × 1024 pixels. The scan sizes are 2.56 mm x 2.56 mm for 5x, 639 μm x 639 μm for 20x, and 256 μm x 256 μm for 50x.

2.3.3 Raman Microscopy

The Raman spectra were collected on a Senterra II Confocal Raman Microscope (Bruker) with excitation wavelength of 785 nm, 25 mW laser power, and 4 cm^{-1} spectral resolution in the range of 50 cm^{-1} to 3640 cm^{-1}. A dispersive mapping method was used to characterize the cross-section of three transparent multilayer backsheets. All backsheet layers were measured using the 50x objective with spot size 2 μm to ensure accurate representation of the material, particularly for thin layers. A 10x objective was used to capture a broader view of the transition from layer to layer along the cross-section. All spectra were baseline corrected using the OPUS software and normalized using a peak appropriate for each material.

2.3.4 Small Angle Neutron Scattering (SANS)

SANS was performed on the NG-7 30m SANS instrument at the NIST Center for Neutron Research. Free-standing films of three transparent backsheets were measured in transmission geometry at a sample-to-detector distance of 4 meters with a combination of various wavelengths. The scattered intensity was corrected for background and parasitic scattering.

2.3.5 Tensile Test

Backsheet films before and after aging were cut to 70 mm x 5 mm along the traverse direction (TD). The tests were performed at a speed of 50 mm/min and a gauge length of 35 mm on a tensile test device (MTEST Quattro, ADMET). A total of 5 samples were used for each test series.

2.3.6 Nanoindentation

Surface mechanical properties, such as elastic modulus and hardness were measured using a nanoindenter (G200, KLA Corporation) with a 1 μm radius 60° diamond cone indenter in a continuous stiffness mode (CSM) [8]. The CSM oscillation frequency was 45 Hz with 2 nm amplitude. Nanoindentation was conducted at a fixed strain rate of 0.05 s^{-1} and indented to a depth of 2 μm. Reported moduli of the samples are the averaged values obtained between the depths of 500 nm to 1 μm for a minimum of 15 indents.

3 RESULTS AND DISCUSSION

3.1 Backsheet Multilayer Structures and Microstructures

Cross-sectional Raman microscopy was used to identify the various layers of the backsheets. Optical images from 10x objective on the Raman microscope showing the outer, inner and core layers of the clear backsheets are presented in Figure 1. The outer layers of three backsheets consist of different fluoropolymers, namely PVF, PVDF and FEVE for CB1, CB2 and CB3, respectively. The fluoropolymer layer is followed by an adhesive and a PET core layer. The PET core layer makes up the majority of the CB1 and CB2 backsheets (≈ 260 μm) with a subsequent thin FEVE coating. CB3 has a different structure consisting of a thinner PET layer (130 μm) and an EVA inner layer.

SANS patterns of unaged free-standing films of three transparent backsheets are shown in Figure 2. Overall isotropic structural patterns are observed for CB1 and CB3, though CB3 scatters weakly. A clear anisotropy in the long-period spacing is seen along machine direction (MD) for CB2 (modified PVDF/PET/FEVE). Such a scattering pattern in polymeric materials typically corresponds to the ordering of crystalline domains. Thus, SANS results suggest that the PVDF-based clear backsheet contains ordered crystalline structures. Though the bulk mechanical property of a complete backsheet is commonly dominated by the high-strength PET core layer, this anisotropic structure may lead to an anisotropic mechanical property for the PVDF-based outer layer.

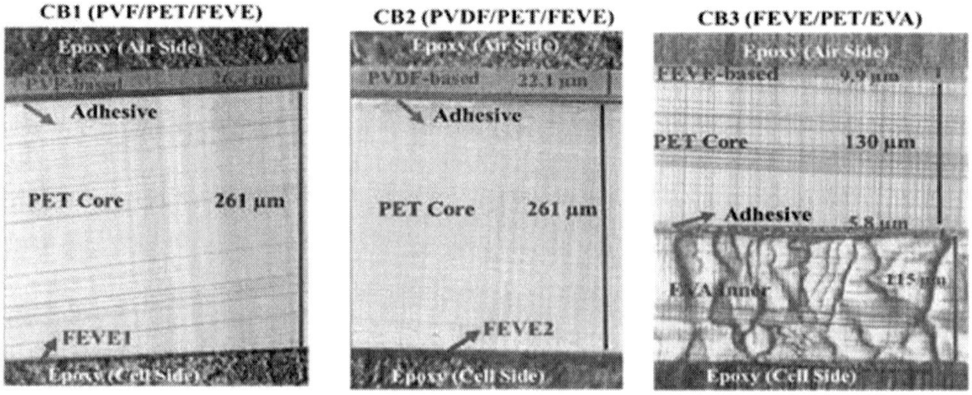

Figure 1: Optical images of cross-sections of three transparent backsheets by the Raman microscope, showing the materials and structures of each transparent backsheet. (CB1: PVF/PET/FEVE; CB2: PVDF/PET/FEVE; CB3: FEVE/PET/EVA)

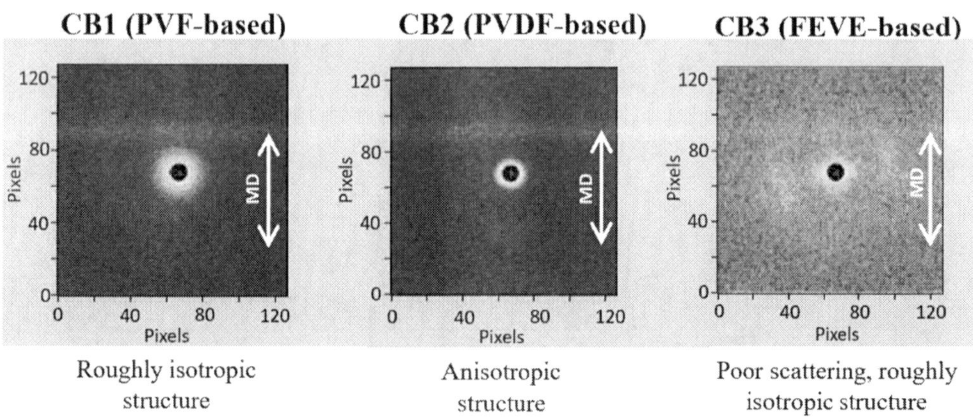

Figure 2. Raw SANS patterns of unaged free-standing films of three transparent backsheets (CB1-Left, CB2-Middle, CB3-Right). MD is the machine direction.

3.2 Optical and Chemical Changes of Transparent Backsheets

Figure 3 shows the UV-vis transmittance spectra of three transparent backsheets before and after SPHERE exposure with accumulated UV doses (295 nm to 400 nm) approximately 600 MJ/m^2 and 1000 MJ/m^2. Given a 12 % albedo for reflected light on the backside of a PV module, these doses are equivalent to about 15 y and 25 y, respectively, of field exposure in Arizona. With increasing UV dose, an increase in yellowing and a shift of UV cut-off wavelength to a lower wavelength are seen. Compared to those of coupon samples (G/G or G/B), increases in yellow index of free-standing films (CB1, CB2, CB3) are more obvious.

However, only CB3 films and their corresponding G/B coupons show significant changes among the three backsheets, seen in Figure 3. The yellow index of CB3 film and G/CB3 coupons increases much faster than those of CB1 and CB2. The increase in the transmittance near 350 nm might be associated with the loss of UV absorbing species in CB3. The drop in the transmittance in low UV range for CB1 could be related to the change in fluorescence after UV exposure. Compared to G/POE/G coupons, G/EVA/G coupons show faster changes in yellowing.

The optical changes (Figure 3) of three transparent backsheets are consistent with the chemical changes shown in Figure 4. Decreases in intensities at 2936 cm^{-1} and 1463 cm^{-1} (C-H stretching and bending, respectively) indicate that chain scission and mass loss have occurred on the air side (FEVE) of CB3 after exposure, and the broadening of peaks around 1690 cm^{-1} (C=O stretching) suggests the formation of photo-oxidative degradation species. Visible chemical changes are also observed on the PVF-based outer layer of CB1 and the PVDF-based outer layer of CB2, mainly due to the degradation of the acrylate polymers mixed in those fluoropolymers.

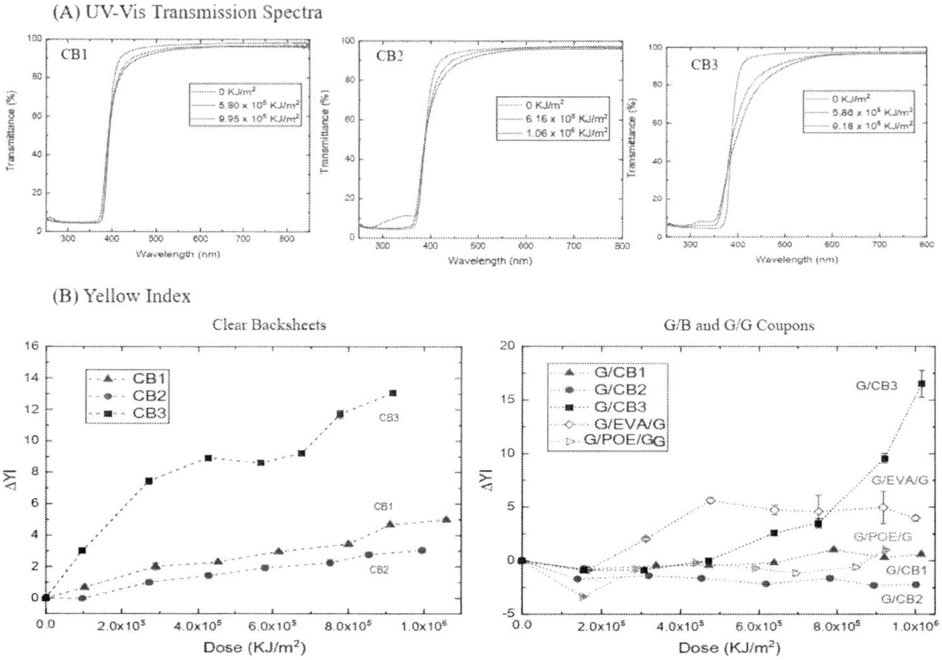

Figure 3. (A) UV-Vis transmission spectra of three transparent backsheets and their G/B coupons (CB1-Left, CB2-Middle, CB3-Right) before and after UV exposure under 75 °C/50 % RH (films) and 65 °C/50 % RH (coupons). (B) Yellow Index of three transparent backsheets and their G/B coupons and G/G coupons (G/EVA/G and G/POE/G). Error bars represent one standard deviation of uncertainty.

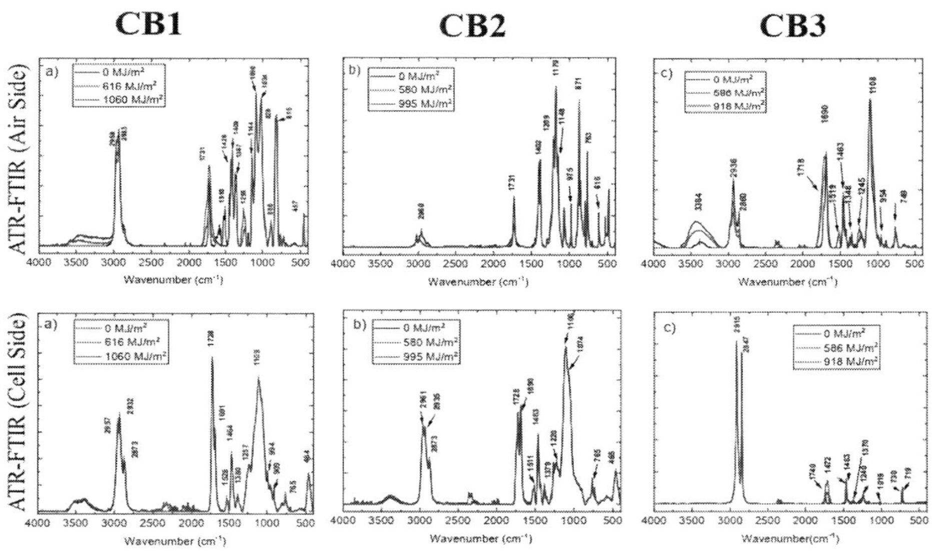

Figure 4. ATR-FTIR spectra of air side (upper row) and cell side (lower row) of three transparent backsheets (a-CB1, b-CB2, c-CB3) before (black) and after UV exposure (red-600 MJ/m^2, blue-1000 MJ/m^2) at UV/75 °C/50 % RH.

3.3 Surface and Bulk Mechanical Changes of Transparent Backsheets

The bulk mechanical properties of three clear backsheets before and after UV exposure were measured by tensile tests, and their results can be seen in Figure 5. CB1 and CB2 show an increase in the yield stress but little changes in the Young's modulus after UV dose up to ≈1000 MJ/m^2, indicating there might be signs of post crystallization during exposure. In contrast, CB3 shows a 50 % increase in the Young's modulus with half the UV dose (568 MJ/m^2), suggesting the CB3 was severely degraded and became stiffer and brittle with UV exposure. The CB3 backsheet exposed with UV dose at 918 MJ/m^2 became too brittle to cut and to measure. Tensile strength-at-break results show even more obvious differences between the backsheet films. Though CB1 and CB2 display similar behavior in terms of Young's modulus after exposure, the decreases in their tensile strengths are different, i.e., 32 % for CB1 while 17% for CB2. However, CB3 has approximately 67 % loss in the tensile strength, showing the most signs of degradation followed by CB1 then CB2. The elongation-at-break data also supports this finding. Noted that an elongation-at-break for CB3 is less than 2 % at 568 MJ/m², breaking as soon as strain was applied to the sample. This almost instantaneous break further indicates that the CB3 backsheet has become very brittle due to the post-crystallization and chain scission after UV exposure.

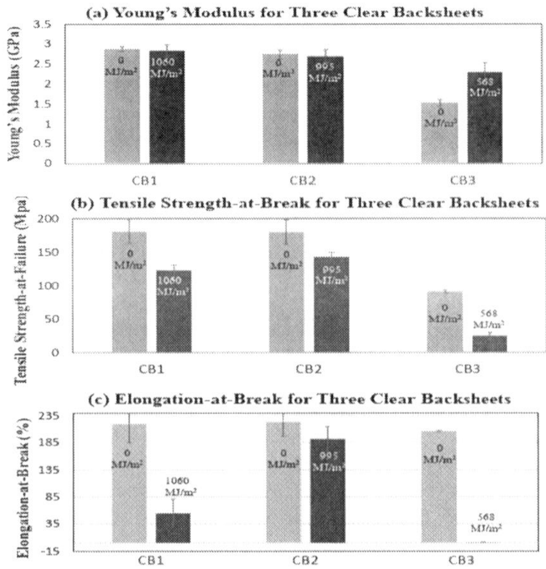

Figure 5. (a) Young's modulus, (b) Tensile strength-at-break, and (c) Elongation-at-break of the unaged and aged backsheets after approximately 1000 MJ/m² UV dose for CB1 and CB2, and approximately 600 MJ/m² UV dose for CB3 at UV/75 °C/50 % RH. Error bars represent one standard deviation of uncertainty.

Compared to bulk mechanical properties, surface mechanical properties are more sensitive to UV-exposed outer layers. Table 2 shows surface modulus and hardness of the air side of backsheets before and after exposure as characterized by nanoindentation. The surface moduli for three unaged clear backsheets are similar. After exposure, modulus and hardness values increase, indicating the exposed fluoropolymer outer layers for three clear backsheets became more brittle. UV degradation within the layer is the most likely reason for this brittleness.

Table 2. Surface modulus (E) and hardness (H) of unexposed and exposed backsheets with UV dose ≈1000 MJ/m². Uncertainties represent the standard deviation of 20 measurements.

		E (GPa)	H (GPa)
CB1	unexposed	2.6±0.1	0.25±0.02
	exposed	3.3±0.1	0.28±0.01
CB2	unexposed	2.4±0.0	0.19±0.01
	exposed	3.3±0.1	0.26±0.01
CB3	unexposed	2.9±0.04	0.22±0.01
	exposed	3.8±0.03	0.30±0.004

3.4 Cracks Formation and Characterization

Cross sectional LCSM images of CB3 after approximately 568 MJ/m² of exposure show the propensity of CB3 to cracks (Figure 6). Figure 6a shows a crack along the length of the PET core layer while Figure 6b shows a perpendicular crack that extends through whole layers of EVA inner, adhesive and PET core, reaching the FEVE outer layer. Delamination was also observed in the adhesive layer between EVA and PET, as well as the interface between FEVE and PET layers. The observation of cracks in the PET layer further supports the notion of the brittleness of the core layer as a result of UV degradation and post-crystallization. Note that caution is needed when preparing these samples because the preparation of cross-sectional samples could induce cracking as a result of cutting.

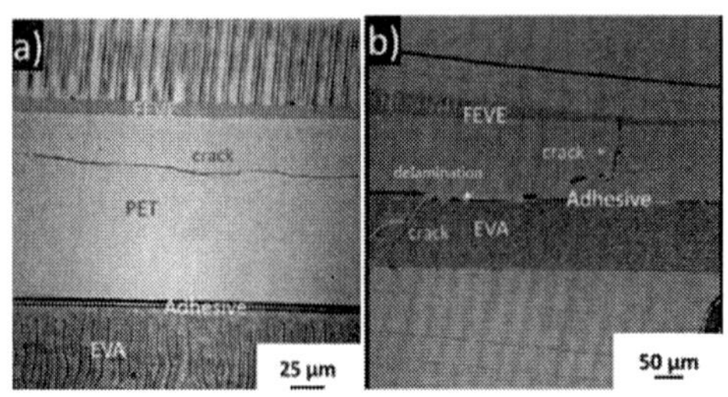

Figure 6. Cross-sectional LSCM images of PET core layer cracking in the aged CB3 after 1223 h (568 MJ/m² UV dose) of exposure under UV/75 °C/50 % RH. (a) The image shows a crack propagating along the PET layer of CB3, and (b) A perpendicular crack was observed across the layers of EVA inner, adhesive and PET core, reaching the FEVE outer layer. Delamination also appeared in the adhesive layer between EVA and PET as well as the interface between FEVE and PET layers.

4 SUMMARY

Durability of three fluoropolymer-based transparent backsheets, along with coupons composed of glass/encapsulant/transparent backsheet and glass/encapsulant/glass were investigated under UV/75 °C/50 % RH and UV/65 °C/50 % RH up to 2000 h for films and coupons, respectively. Results show degradation is most likely to take place in the PET core layer due to UV degradation and post-crystallization of the polymer. Among three types of clear backsheets, CB3 (FEVE/PET/EVA) exhibited the most chemical, optical and mechanical property changes after only 1200 hours (≈ 600 MJ/m^2) of exposure. LCSM images showed some cracks appeared in the PET core layer. The other two backsheets, CB1 (PVF-based) and CB2 (PVDF-based) showed no signs of cracking up to 2000 hours of exposure (≈ 1000 MJ/m^2), despite an increase in yellowing and decreases in mechanical properties were seen. This study indicates the importance of an appropriately UV-stabilized PET core layer for the long-term performance of transparent backsheets. This is an ongoing study. The exposure from the glass side of the G/CB coupons is being conducted and the results will be compared to the data shown in this paper. Furthermore, effects of the light source and the sample temperature on degradation of transparent backsheets will also be studied.

*DISCLAIMER:

Certain commercial products or equipment are described in this paper to specify adequately the experimental procedure. In no case does such identification imply recommendation or endorsement by the National Institute of Standards and Technology, nor does it imply that it is necessarily the best available for the purpose.

5 ACKNOWLEGEMENT

We greatly acknowledge the support from a NIST/Industry consortium on Characterization and Modeling of Polymers for Photovoltaic Systems. We also acknowledge Drs. Paul Butler, Yimin Mao and Yun Liu from NIST Center for Neutron Research for their help in the SANS experiment and useful discussions.

6 REFERENCES

[1] A. Cuevas, A. Luque, J. Eguren, and J. Del Alamo, "50 Percent more output power from an albedo-collecting flat panel using bifacial solar cells," Solar Energy, 29 (1982) 419.
[2] X. Sun, M. R. Khan, C. Deline, and M. A. Alam, "Optimization and performance of bifacial solar modules: A global Perspective," Appl. Energy, 212 (2018) 1601.
[3] G. W. Wilson, et al., "The 2020 photovoltaic technologies roadmap," J. Phys. D: Appl. Phys, 53 (2020) 493001.
[4] T. C. Felder, K. R. Choudhury, L. Garreau-Iles, S. MacMaster, et al., "Analysis of Glass-Glass Modules." In *New Concepts in Solar and Thermal Radiation Conversion and Reliability*; International Society for Optics and Photonics, 10759 (2018) 107590E. https://doi.org/10.1117/12.2321637.
[5] W. Gambogi, M. Demko, B.-L.Yu, S. Kurian, et al., "Transparent Backsheets for Bifacial Photovoltaic Modules,". Proceedings *47th IEEE Photovoltaic Specialists Conference (PVSC),* (2020) 1651. https://doi.org/10.1109/PVSC45281.2020.9300924.
[6] K. Arihara, et al., " A. Reliability and Long Term Durability of Bifacial Photovoltaic Modules Using Transparent Backsheet,". *Jpn. J. Appl. Phys. 57* (2018) 08RG15. https://doi.org/10.7567/JJAP.57.08RG15.
[7] M. D. Kempe, D. C. Miller, A. Zielnik, D. Montiel-Chicharro, et al., "Survey of Mechanical Durability of PV Backsheets," Proceedings, 44th IEEE Photovoltaic Specialist Conference, (2017), 3208.
[8] M. R. VanLandingham, "Review of Instrumented Indentation," J. Res. Natl. Inst. Stand. Technol., 108 (2003) 249.

Studying time-series of wet leakage resistances for modules with various backsheet types

Claudia Buerhop, Oleksandr Stroyuk, Julian Zöcklein, Jens Hauch, Ian Marius Peters

High Throughput Methods in Photovoltaics, Forschungszentrum Jülich GmbH, Helmholtz Institute Erlangen-Nürnberg for Renewable Energy (HI ERN), Immerwahrstraße 2, 91058 Erlangen, Germany

c.buerhop-lutz@fz-juelich.de, Tel.: 0049-9131-9398-177, FAX: 0049-9131-9398-199

ABSTRACT During the last years PV-modules fail very often due to their backsheet material, and cause inverter issues. IV-curves and power measurements are not affected and therefore other methods are necessary to disclose the errors. We carried out time-series of wet leakage current measurements and imaging methods on field-aged PV-modules with various backsheets. The backsheet material was identified with field-suitable near infrared absorption spectroscopy. We found three main behaviours: 1) high leakage resistances (LR) and LR decrease for modules with double fluoropolymers, 2) medium LR and moderate LR decrease for modules with polyamide backsheets and 3) medium LR and strong LR drop for those with single-fluoropolymer backsheets, especially with thin fluorinated coatings (FC). We show that with water ingress into the backsheet and encapsulant the leakage current increases. Furthermore, UV-fluorescence is able to differentiate different backsheet materials, no differences were observed for electro- or photoluminescence. The fact that modules with FC backsheets increasingly show electrical insulation issues during operation makes the relevance for field operation clear.
Keywords: PV-modules, degradation, polymers, backsheet

INTRODUCTION

Degradation of polymer components of crystalline silicon solar modules is currently recognized, along with Si wafer failures, as one of the most important factors affecting performance and lifetime of both ground- and roof-based solar installations [1]. Resulting insulation issues can have severe impact on the yield and energy production because of potential inverter shut-offs. We have reported earlier that field-aged modules show a weather-sensitive behavior caused by the degradation of backsheets (BS) and depending on the BS composition and structure [2]. While the BS composition can be identified in the field by spectroscopic methods, such as NIRA [3], almost no visual or spectral signs of early BS degradation resulting in the weather-dependent behavior can be distinguished by visual inspection or NIRA measurements. For this reason, additional tests of the BS insulation state, such as wet leakage resistance (LR) measurements, are required for a reliable analysis of the performance of solar parks. However, LR techniques, though being popular, can be deployed only in the lab and therefore combination of LR tests with other field-compatible tests such as fluorescence spectroscopy/imaging, should be developed.

EXPERIMENTAL PROCEDURE

The insulation state of solar modules was studied by performing wet leakage resistance measurements. Therefore, field-aged PV-modules were immersed in an aqueous electrolyte bath (filled with DI water containing a background salt electrolyte and a surfactant to provide conductivity and alleviate surface tension effects, respectively) for several days. Regularly, time-series of electric and spectroscopic characterizations were carried out: J-V curves under different irradiation levels, characterization of BS and encapsulant by NIRA, photoluminescence (PL) and electroluminescence (EL) imaging of silicon wafers as well as UV-excited fluorescence (UVF) imaging of the encapsulant.

RESULTS AND DISCUSSION

The present work was motivated by a recent field study of a multi-MW solar power station located in Germany where we observed a number of failures, including potential-induced degradation (PID), insulation issues, BS cracking, discoloration, and delamination, etc. By using NIRA spectroscopy we were able to identify the BS structure for more than 500 modules and tentatively associate certain module failures with specific BS compositions. In particular, modules with polyamide (PA) based BS showed a tendency for PID events and extensive BS cracking/chalking. Modules with thin air-side layers of fluorinated coatings (FC), polyethylene terephtalate (PET) cores and inner polypropylene (PP) layers showed numerous cases of insulation issues (resulting in modules being electrically "bridged", that means shortening the string by the number of bridged modules), inner and air-side cracking, discoloration, and BS delamination due to destruction of PP layer.

The insulation issues led to inverter shut-offs [4] and had a negative impact on the yield. The shut-offs showed a tendency to occur more often with each year of exploitation [2] as illustrated by annual changes of recorded threshold value of inverters connected to modules of different power classes (Figure 1).

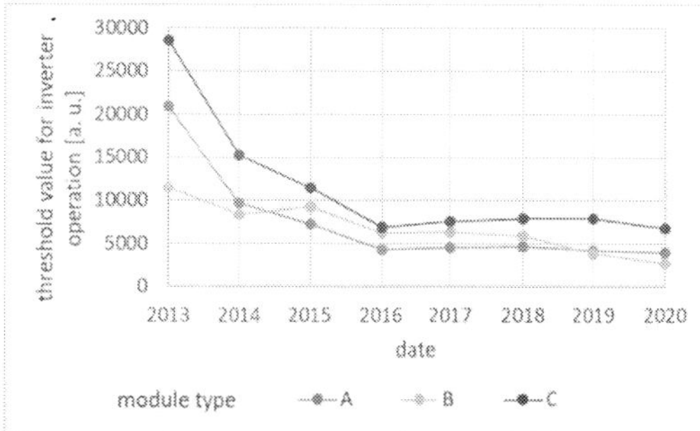

Figure 1: averaged threshold value as an indicator for the starting value of inverters with different types of modules observed through 2013-2020.

We found that disconnected ("bridged") modules with FC/PET/PP showed no obvious spectral signs of chemical BS transformations, indicating that additional analysis methods are required to identify such modules among non-affected ones. Extensive cracking of PA-based BS were also not accompanied by changes in the NIRA spectra or drastic drops in the module power, additionally indicating that potential insulation problems cannot be addressed solely by conventional J-V and EL measurements, as well as NIRA tests of the BS materials.

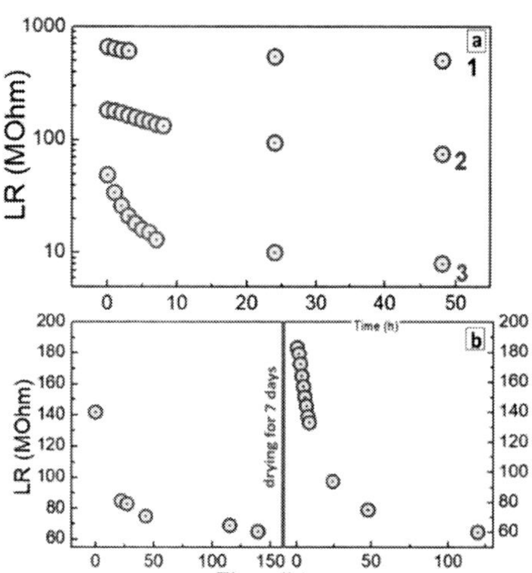

Figure 2: leakage resistances with time, a) modules with differing backsheets (1: double-fluoropolymer, 2: non-fluoropolymer, 3: single-fluoropolymer), b) LR-evolution of single-fluoropolymer with drying interval after 160 h immersion time

In view of this, we subjected a series of selected modules to a complex study using lab-based LR measurements combined with spectral observations with NIRA, and module characterization with EL/PL and UV-excited fluorescence imaging. As a reference to the modules with FC/PET/PP and PA BSs tested in this work, we used modules of the same power class with robust PVF/PET/PVF BSs (Tedlar like BS, PVF - polyvinyl fluoride) that typically show a high insulation stability.

We found that both the starting LR_0 (at $t = 0$) and the rate of LR reduction during the immersion depend on the BS type. In particular, the modules with Tedlar-type BSs showed the highest LR_0's of 600-700 MΩ and the slowest LR decrease when immersed (Figure 2a). The PA-based modules showed intermediate LR_0's (100-200 MΩ) and a faster LR decay, while the modules with FC/PET/PP BSs revealed the lowest LR_0's (40-50 MΩ) and the fastest reduction in insulation upon immersion. These observations are confirmed by our field findings, indicating that the fraction of modules with FC/PET/PP BS showing insulation issues is much higher than for modules with other BS types.

When extracted from the bath and left for drying the modules of all tested types restore their insulation level and lose it again at new immersion (Figure 2b) showing a reversible character of the observed LR changes.

In contrast to changes of LR with immersion time, the measured module power remained for all modules and all backsheet types unchanged.

PL/EL images of the modules taken during their contact with water show no particular changes in integral intensity.

Air-side NIRA spectra of modules with PA-based BSs showed no appreciable changes after water immersion of the modules for up to 72 h. In contrast, for the modules with FC/PET-based BSs distinct evolutions in NIRA spectra can be observed upon immersion in the spectral range of 1300-1500 nm and particularly at 1910 nm (Figure 3a).

Both ranges are characteristic for water vibrations indicating that PET core layer starts to adsorb water upon the module immersion. A ratio of intensities of water-related 1910 nm band and characteristic PET-related band at 1660 nm ($A_{1910\ nm}/A_{1660\ nm}$), which can be used as a measure of water content in the BS increases with the immersion time (Figure 3b, squares) reaching a plateau at longer times.

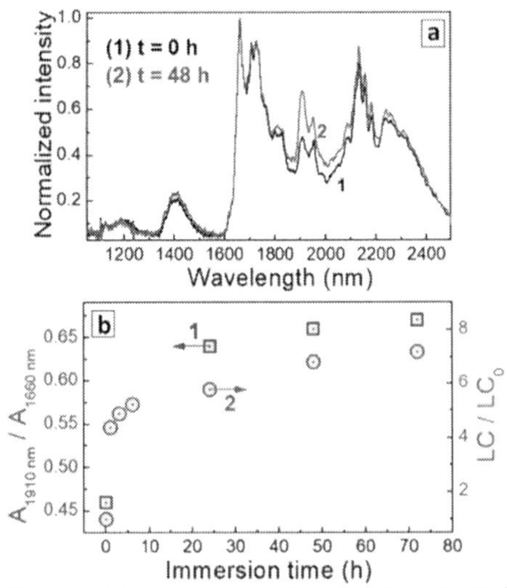

Figure 3: (a) NIRA spectra of air side of a FC/PET-based BS prior to (curve 1) and after immersion to water for 72 h (2); (b) Intensity ratio of NIRA peaks at 1920 nm and 1660 nm (squares 1) and relative increase of the leakage current (LC/LC_0, circles 2) as functions of the immersion time for a module with FC/PET-based BS.

When modules are extracted from the bath, the NIRA shows reversible spectral events indicating gradual water evaporation and return close to the original state within 24 h.

Under UV light excitation the tested modules showed encapsulant fluorescence patterns particular for each BS type (Figure 4a-c) depending on the BS composition.

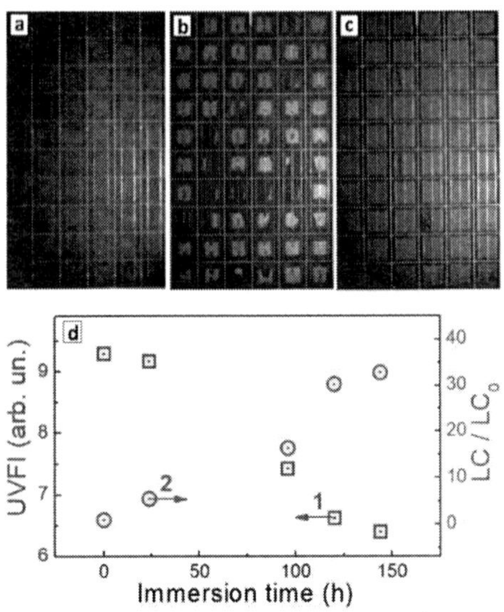

Figure 4: (a-c) UV-excited fluorescence (UVF) images of modules with PVF/PET/PVF BS (a), PA BS (b), and FC/PET/PP BS (c); (d) integral UVF intensity (UVFI, squares 1) and relative increase of LC (circles 2) of a FC/PET/PP-based module as function of the immersion time.

The relative leakage current (LC, normalized to LC_0 at t = 0 h) increases with the immersion time, the shape of both dependences being very similar indicating the same origin of both changes in spectra and in LC.

We found that the integral UVF intensity decreases for longer immersion periods (>48 h). This is most likely due to gradual water penetration into EVA layers and fluorescence quenching by dissolved oxygen. This conclusion is supported by a leakage current increase observed simultaneously with the UVF quenching (Figure 4d). This observation indicates that UVF measurements can potentially be used for the field evaluation of the insulation state of solar modules.

Changes in EL- and PL-images with respect to longer immersion periods were not observed.

SUMMARY

A combined approach is proposed to assess the insulation state of crystalline solar modules depending on their BS composition which unites electric measurements (J-V, LR), NIRA spectroscopy of BS materials, and module PL/EL/UVF imaging. The LR was found to depend on the BS type decreasing from robust double-fluoropolymer BS to polyamide BS to single-fluoropolymer BSs. This trend mirrors the climate-dependent changes in the field performance of corresponding module types. We found clear spectral signs of water ingress both to BS and encapsulant layers of single-fluoropolymer BS-based modules which are directly proportional to leakage current. The presented results are relevant for the development of new diagnostic tools for field assessments of the module insulation state.

ACKNOWLEDGEMENT

HI ERN gratefully thanks the German Federal Ministry for Economic Affairs and Energy (BMWi) for financial funding of the project COSIMA (FKZ: 0324291A). The work was supported by ZIM Project "PolymerCHECK" (No. 16KN083038).

REFERENCES

[1] A. Fairbrother, et al.; 7 - Degradation Processes and Mechanisms of Backsheets; Durability and Reliability of Polymers and Other Materials in Photovoltaic Modules; H. E. Yang, R. H. French and L. S. Bruckman; 2019.

[2] C. Buerhop, et al., 47th IEEE PVSC (2020) 0320 .

[3] O. Stroyuk, et al.; Solar Energy Materials and Solar Cells 216 (2020) 110702.
[4] C. Buerhop, et al.; SOLMAT 231 (2021).

PERFORMANCE ANALYSIS OF SHADED PV MODULE POWER ELECTRONIC SYSTEMS

Franz P. Baumgartner*, Roman Vogt, Cyril Allenspach, Fabian Carigiet
[1] ZHAW, Zurich University of Applied Sciences, School of Engineering, Institute of Energy Systems and Fluid Engineering, Technikumstrasse 9, 8401 Winterthur, Switzerland
*phone: +41 (0) 58 934 7232; e-mail: bauf@zhaw.ch; www.zhaw.ch/=bauf

ABSTRACT: In the last decade decentralized Modul Level Power Electronics (MLPE) equipment has gained tremendous market share due to the potential to operate each Photovoltaic (PV) module in their optimum power point even in partial shading condition. The total losses of the group of decentralised DC/DC converter combined with the coupled centralised DC/AC inverter not always offer an advantage to the standard String Inverter System (SINV), like in the unshaded moments of high-power operation of a PV system at noon. The customer expects a clear answer about the quantified gain in annual power of a roof top system either operated by MLPE or SINV. Today, even the experienced planner is not able to elaborate these numbers in an economic efficient way. This is the case due to a lack of complex geometrical data of the shading obstacles and absence of software tools which are able to simulate the MLPE and SINV by calculating the shade of each solar cell in all PV modules together with an appropriate loss model of all used power electronic components. Up to know no standards exist to measure the set of MLPEs in the lab and the manufactures have not proposed detailed loss models up to now, whereas only max efficiency number of the MLPE are stated in their data sheets. This paper shows that detailed loss measurements performed in the lab, provided up to three percent higher losses of the MLPEs in the relevant operation area commonly used through a year of operation. It is recommended to use a very narrow range of numbers of MLPE in the string for high efficiency power conversion, due to the fact, that losses increase by 1.5% if the input/out voltage ratio of MLPE differ 5% from unity. A concept is presented to estimate the final so-called shading adaption efficiency which is based on the efficiency measurement of the MLPE in the indoor lab at a few operation points and by using weighting factors. Thus, the comparison of the shading adaption efficiency is given, either for different MLPEs or SINVs power electronic systems for a typical PV system with shading, relative to the same aggregated sum of maximum decentralised DC power at the PV Modules. Finally, one example of a tilt PV roof top system with partial shading of a chimney is given, were the standard SINV shows 1.2% percent higher losses estimated for a whole year of operation compared to a MLPE system. This value will change if the number of MLPE in the string is modified.
Keywords: inverter, optimizer, power electronic, MLPE, performance, measurement, shading losses, SINV

1 INTRODUCTION

Today, several millions of MLPE components are in operation worldwide, with still very high growth rates in the markets of PV systems on rooftops. Their unique ability to operate each single PV module in the individual absolute maximum power point, under different shading conditions of each module, is the most important door opener in the market. In literature, we are not able to find clear evidence to prove typical numbers of annual performance increase of such PV rooftop systems equipped with MLPE in the double-digit range relative to standard SINV. Surveys have been published on the topic of testing of typical row shading in outdoor conditions for the purpose of large-scale greenfield installations.[1,2] Outdoor comparison of smaller roof top PV systems is critical, because the expected yield difference is a low one-digit number, which is close to the outdoor measurement uncertainties. Additionally, having an equal diffuse light condition on both installations MLPE and SINV is challenging.

Back in 2007, the relevance of about 3% difference in annual inverter efficiency on the market was demonstrated due to the dependency on the chosen DC generator voltage of a single inverter. Due to the cost share of the inverter of about 10% of total system costs this was leveraged to about 30% of specific inverter prices.[5] As a follow up, the market player promoted development of standards [6] and today in the inverter data sheets we will find the difference of annual inverter efficiency at different DC voltage levels without any simulation needed for the planers/customers decision. Today each qualified PV planer must also quantify the benefit in yield of MLPE versus SINV, conventional serial connected PV modules to feed standard string inverter for each PV roof top system. Two types of MLPE are available on the market, the buck-boost converter, and the buck converter. The latter easily may be added to very few numbers of PV modules in a conventional SINV string, beneficial to get out most of the power. This works, even at the high string currents of the inverter, in some cases of partial shading as it is shown in the right part of Fig. 1 without activating the PV module by-pass diodes. That simple concept without any further control and communication within the DC string needed will fail for higher numbers of partial shaded modules, which may lead to a final DC string voltage which is below the limit of the input stage of the DC/AC inverter itself.

However, commercial PV planning tools using efficient annual yield simulation are facing two relevant challenges. First, the measurement of all the shading obstacles on the roof or nearby objects must be transferred into the 3-dimensonal model which takes typically several hours. Second, commercial PV planning tools apply annual yield simulation generally in a one-hour simulation time steps and are typically not able to analyse partial shading of all solar cells individually on each PV module and do not provide an appropriate loss model of the different types of MLPEs. Finally, on the planer side, this effort to perform this PV fine planning in manpower and equipment if e.g., drones are used, will exceed the value of the today's typical margin of a single-family solar roof installers.

Thus, other practical solutions will be developed to take the limited budget of the final customer for this MLPE versus SINV decision into account as it will be presented in this paper.

2 APPROACH - COMPARISON SHADING LOSSES

In this paper we are proposing to quantify the annual

improvement in performance of different power electronic components on some representative typical shading cases on single family PV rooftop systems with defined obstacles. The applied method is based on indoor measurement of the power optimizers and string inverter using solar array simulator for the partial shading emulation of each PV module. In detail, the analyses need the entanglement of indoor efficiency measurement of all the MLPE components of the system that are individually powered by several DC power supplies emulating the shaded set of PV modules in a minute time interval. Thus, the parameter settings of this power supplies are calculated from ray simulations as shown in Fig. 1 with a high-precision surface resolution of the shadow on top of each solar cell of all modules. This leads to the individual PV module current voltage characteristics.[3]

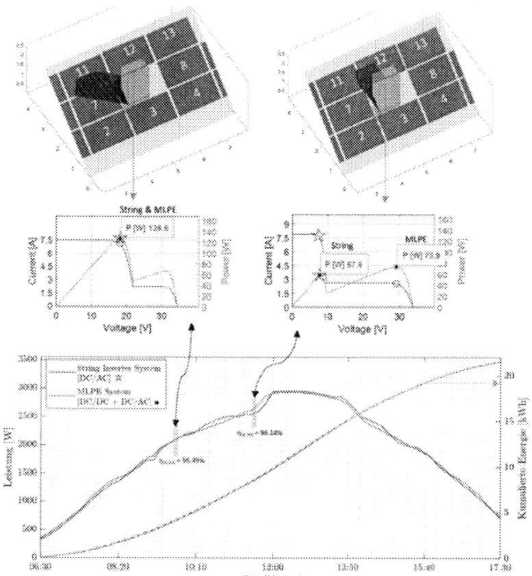

Figure 1: Two examples of the IV characteristics of shaded module no. 12 with a beneficial gain of MLPE on the right side relative to the operation point of the conv. SINV.

3 MEASUREMENT SETUP INDOOR LAB

In the subsequent laboratory performance measurement of the string inverter systems and MLPE systems, both systems are powered by the same current voltage characteristics settings of the power supplies, representing the individual shaded PV modules on a roof top installation. To test and develop these procedures, the ZHAW IEFE laboratory equipment consists of 10 Keysight E4360 SAS and several Newtons4th precision power analyzers of type PPA1500 and PPA5500. Notably, with this setup and the direct power measurement capability of the PPAs, the efficiency of the Device under Test (DUT) can generally be tested and measured with a GUM uncertainty of below ±0.5% (k=2) at standard operating conditions of most MLPE (50W<P<400W, 20V<U<100V, 1A<I<10A). Accordingly, this approach offers better reproducibility and lower uncertainties compared to outdoor measurements.

As an example, for such a laboratory analysis, the MLPE could be tested at different operation condition not only partial load but also ratio of DC input versus DC output voltage. The results offer much lower efficiency values around 97% than the given values of 99% in the manufactures data sheet, although operated in the permitted working range of the number of optimizers.[4] A MLPE loss model will be presented based on the input current and the voltage ratio of the DUT.

The most challenging part is the selection of typical shading cases and typical partial shaded PV roof top systems. In practice, these approaches are not independent of the type of PV power electronics chosen. Installers will typically increase the number of PV modules on the roof top by using MLPE but are not able to guarantee a certain yield number. Using string inverter during the design stage the number of PV modules is expected to be smaller, and thus the shading rate will be lower, in most of the cases. The millions of MLPE customers are facing this problem today. It is not very likely that these detailed analyses will be performed by the installer business because of the excessive measurement effort of all the relevant shading objects and correct orientation of each single PV module in the simulation. The process of selecting the relevant typical shading cases must consider the experience of several groups in different countries. Our concept of performance comparison is intended to be further developed within international collaborations like the IEA and others.

Finally, these collaborations should lead to typical shading cases, with a few selected solar array simulator DC settings of involved PV module, respectively. Thus, PV power electronic components either MLPE or string inverter will be tested on the same settings and the results as number of annual performance shading losses, could be compared by the customers. Nevertheless, some individual shading situation of a specific customer with the most matching typical cases will be analysed in this work and therefore, the customers will receive useful practical information.

4 RESULTS

First results from typical moderate shading conditions, such as that of a chimney on a single-family solar roof, show a shading performance gain of about two percent relative to the string inverter. But, for low shading conditions the annual performance of conventional high-efficient string inverter systems is higher, if the number of chosen MLPE DC/DC converter is not optimized for an input to output voltage ratio of 1. This mismatch configuration leads to MLPE efficiency values nearly 2% lower compared to the given efficiency number in the data sheet.

Our finding should be a starting point for comparing these kinds of results throughout independent research and test labs using the same methodology. This should lead to more reliable performance data of MLPE, as well as to a fairer and more understandable comparison to string inverter under partial shading condition.

4.1 MLPE single component characteristics

The losses in the power electronic converters arises in the individual components such as

- passive components in the inductors and capacitors by the ohmic losses
- semiconductor switches by ohmic and switching losses
- socket losses due to power supply of all other components of drives and control circuits

In Fig 2 the characteristics of the losses are shown as a function of input current of the DC/DC converter with socket losses of below 2 W for the buck mode. Fig. 3 shows that additionally the ratio of input to output voltage is needed to optimize the simple loss model as given in equation (1) where the coefficients are polynomials of U_{in}.

$$P_{loss} = c_0 + c_1 I + c_2 I^2 \qquad (1)$$

Generally, it is well known that a set of small power electronic components will lead to higher total losses compared to a single device e.g. DC/DC converter as it is implemented in a SINV. Improvements of MLPE efficiency with high bandgap semiconductors can reduce the gap.

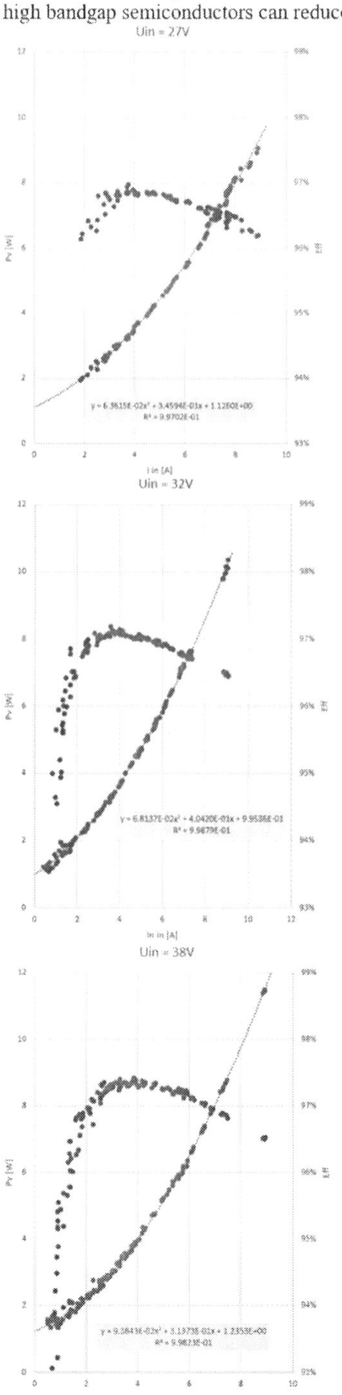

Figure 2: The measured losses at the ZHAW IEFE lab are strongly correlated to the input current of the DC/DC optimizers for the buck converter mode. [9]

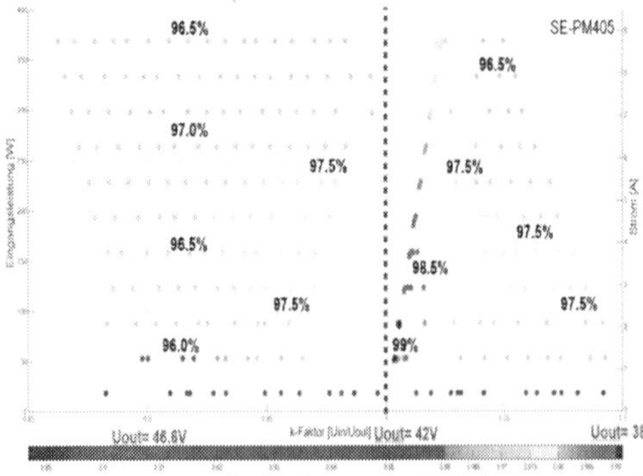

Figure 3: Measured efficiency of MLPE is between 96% and 97.5% at constant input voltage of 42V at an input power level between nominal 400W and 10% of nominal power and a DC/DC voltage ratio is about +/-10% around unity for most of the typical operating points (P405, Solar Edge data sheet efficiency 99%).[7]

5 QUANTIFY PV EFFICENCY AT SHADING

First results from typical moderate shading conditions, such as that of a chimney on a single-family solar roof, show a practical shading performance gain of MLPE of about two percent relative to the SINV systems. Here an approach is presented to quantify the electrical output of the system relative to highest possible DC input power, which is reached if all PV modules individually are operated at their MPP each.

5.1 SHADING ADAPTION EFFICIENCY

By using the equation (1) the annual DC/AC efficiency is calculated for PV power electronic system a consisting of a set of shading objects. (Fig. 3) The so-called shading adaption efficiency gives the ratio of output AC power versus the maximum available sum of aggregated DC power from each of the k PV modules in the string if all of them are operated in their individual absolute MPP. This hypothetical DC power is generated by 100% efficient DC/DC converters and will not be reach in real systems. In laboratory setups, as described in chapter 3, this $\eta_{shad,a}$ will be calculated by the measured P_{ac} power at a characteristically shading situation a which leads to a specific IV characteristic of each PV module. If the real MLPE is able to track the PV module in this absolute MPP this DC input power, according to equation (2), will be measured in the lab and summed up to the total DC input power of all i PV modules, used in the denominator of that equation. Testing SINV systems the measured DC input power on the plugs of the inverter could be smaller than the term $\sum_{i=0}^{k} P_{mod,i}$ for at situations, where the string DC current will not lead to operate each PV module in their absolute MPP. In that case the value of the denominator is modified.

$$\eta_{shad,a} = \frac{P_{ac}}{\sum_{i=0}^{k} P_{mod,i}} \qquad (2)$$

The annual average $\eta_{shad,a}$ will be found by applying the concept of weighting factors for a few typical shading moments in a year. Thus, the weighting factors $a_{shad,a,n}$ are multiplied by the measured efficiency $\eta_{shad,a,n}$ of MLPs or SINV inverter systems in the laboratory, at that limited set of operating points n, at a characteristically shading situation a as given in equation (3).

$$\eta_{shad,a} = \sum_{n=1}^{N} a_{shad,a,n} \cdot \eta_{shad,a,n} \qquad (3)$$

The binning of this characteristically shading situations a will be found at several moments n in the annual mapping as it is shown in Fig. 4 and illustrated by the numbers therein. The weighting factors are representing the energy amount during the year at similar shading adaption efficiencies which are different for MLPE and SINV.

Here the mapping shows the difference of efficiency of MLPE versus SINV relative to the annual energy DC output of MLPE. The value of the individual weighting factors $a_{shad,a,1}$ and $a_{shad,a,2}$ to $a_{shad,a,n}$ are found by the aggregation over the whole year at this typical shading situation a including several PV modules and MLPEs or SINV in the plant. A similar approach is found by the established calculation of the average string inverter efficiency according to the standards, based on the irradiance binning at different locations, known as the EUR efficiency or California inverter efficiency.[6]

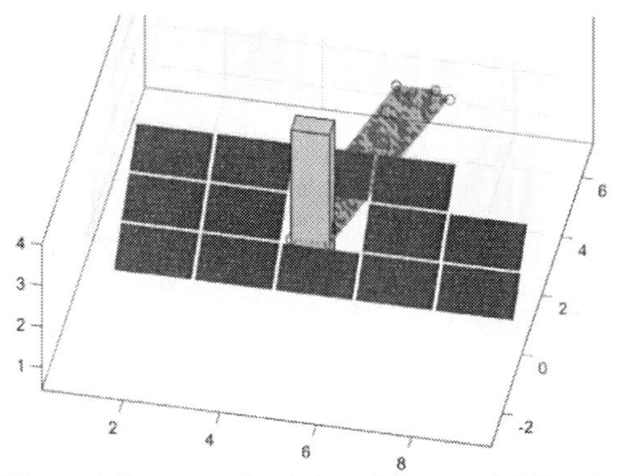

Figure 4: Ray trace simulation of direct and diffuse irradiation on each part of a single solar cell on all PV modules in the south oriented roof top PV systems a consisting of chimneys shading obstacle. [4, 7]

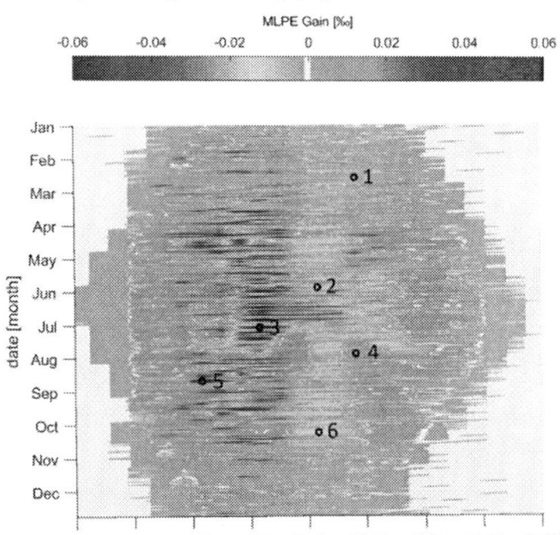

Figure 5: Mapping of power gain of MLPE versus SINV DC/AC Systems for hours and das over a year for the roof top PV systems system a consisting of the shading object chimneys as shown in Fig. 3. Each marked point 1 to 6 could by one of n characteristic test set ups for each module for the efficiency test in the indoor lab – equation (2). [4, 7]

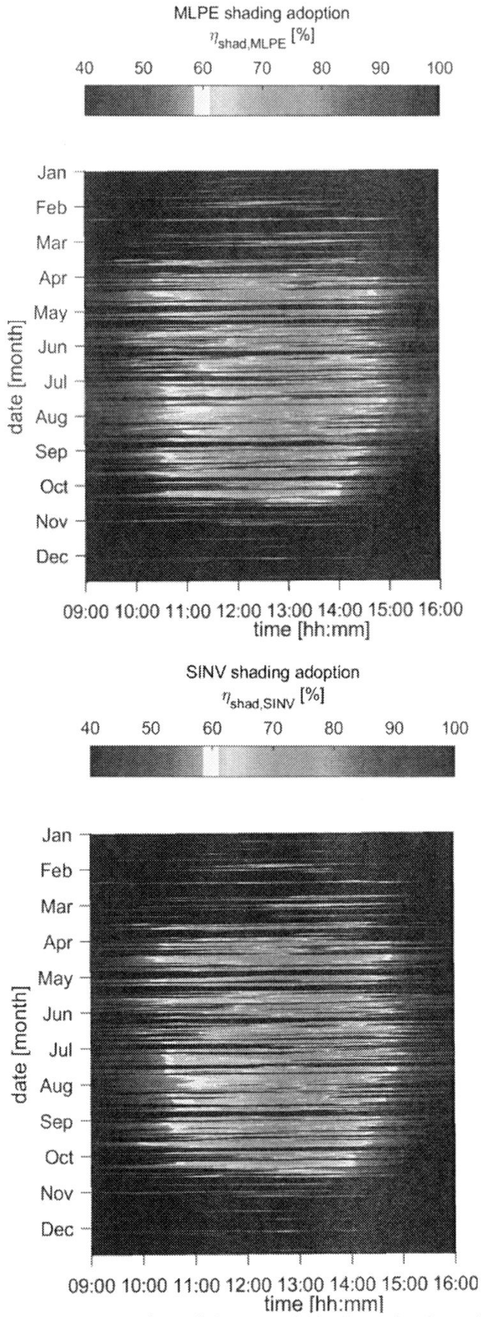

Figure 6: Mapping of the annual shading adaption efficiency, relative to the nominal PV generator power for the MLPE and below for the string inverter system SINV for the shading system given in Fig. 3 using equation (1).

Each PV system consists of number i of PV modules. The concept is looking forward using several limited measurement moments in the lab e.g. 10 to by performed with the power electronic equipment to by applied by the different weighting factors for several other shading types a.

Other laboratories are invited to perform their test on other MLPE equipment or extend the concept to other shading systems. Table I should be extended in sizes of the PV systems, like different number of PV module on single-family houses and other shading objects like, dormer, other chimney, venting pipes, trees, neighbouring buildings.

Table I: Annual average shading adaption efficiency for a few typical shading objects *a* and commercial MLPE or SINV products used in the single family roof top system according to equation (2) and (3) with the defined weighting factors based on the laboratory tests of the commercial equipment. The used string inverter for II shows an average euro efficiency of 97.5%.

Systems of PV and shading cases	Shading objects *a*	I MLPE	II SINV
Single roof tilt chimney	1	96.6%	95.8%
Single roof tilt dormer	2	x	x
Single roof tilt tree	3	x	x
Flat roof Ventilation pipe	4	x	x

6 CONCLUSIONS AND OUTLOOK

Indoor laboratory measurements of the efficiency of a commercial DC/DC MLPE shows more than 1.5% higher losses than the data sheet value if the voltage ratio input to output differ more than 5% from unity. This leads to the recommendation to the PV planers that only a certain number of optimizers, plus minus one will lead to low losses compared to SINV systems in the unshaded case e.g. at noon.

First results from typical moderate shading conditions, such as that of a chimney on a single-family solar roof, show a shading performance gain of MLPE of 1.2% to the SINV. But, for low shading conditions the annual performance of conventional high-efficient string inverter systems is higher, if the number of chosen MLPE DC/DC converter is not optimized for an input to output voltage ratio of close to unity. This mismatch configuration leads to MLPE efficiency values up to 3% lower compared to the given efficiency number in the data sheet.

This research work will by continued for the next two years at ZHAW IEFE within the national research project EFFPVSHAD and within the follow up for the new IEA PVPS Task 13 international collaboration on MLPE activities is requested. [9, 4] Our findings could be a starting point for a systematic comparison of the performance of MLPE versus SINV to engage other independent research and test labs to compare results by similar methodology. This should lead to more reliable performance data of MLPE versus SINV on their manufactures data sheets, as well as to a fairer and more understandable and thus, economical comparison of PV output under partial shading condition.

6 REFERENCES

[1] C. Deline et. al., "Photovoltaic Shading Testbed for Module-Level Power Electronics: 2016 Performance Data Update," NREL, Golden, Colorado, United States of America, Rep. TP-5J00-62471, Sep. 2016

[2] W. T. Franke, "The Impact of Optimizers for PV-Modules", SDU ELE. ENG., Syddansk Universitet; sdu.dk; May 2019

[3] C. Allenspach, V. Echavarri Castro, S. Richter, C. Meier, F. Carigiet, F. P. Baumgartner; 'MODULE-LEVEL POWER ELECTRONICS UNDER INDOOR PERFORMANCE TESTS', in Proceedings of the 37th EUPVSEC, Lisbon, 2020

[4] Report IEA-PVPS T13-15:2021 Performance of New Photovoltaic System Designs; ISBN 978-3-907281-04-8; chapter 3.1.4; https://iea-pvps.org/research-tasks/performance-operation-and-reliability-of-photo-voltaic-systems/

[5] F.P. Baumgartner, H. Schmidt, B. Burger, R. Bründlinger, H. Häberlin, M. Zehner, Status and rele-vance of the DC voltage dependency of the inverter efficiency, in: EUPVSEC,2007. Milano

[6] R. Bründlinger, N. Henze, H. Häberlin, B. Burger, A. Bergmann, F. Baumgartner, prEN 50530 – the New European standard for performance characterisation of PV Inverters, in: EUPVSEC, 2009

[7] R. Vogt, Bachelor Thesis, ZHAW SoE IEFE, 2021

[8] Swiss national research project EFFPVSHADE funded by the Swiss Federal Buero of Energy, by the contract number, SI_502247-01, started in 2021

[9] F. Baumgartner, electrosuisse bulletin, 2021 May, *ISSN* 2673-6322 (Online) https://www.bulletin.ch/de/news-detail/optimizer-nur-ein-hype-oder-die-zukunft.html

DEVELOPMENT OF MISSION PROFILES FOR HUMIDITY MODELS IN THE RELIABILITY TESTING OF PV INVERTERS

Ramanathan Thiagarajan[1], Peter Hacke[1], and Jack Flicker[2]
[1] National Renewable Energy Laboratory, Golden, CO, 80401 USA
[2] Sandia National Laboratories, Albuquerque, NM 87123, USA

ABSTRACT: To understand the impacts of humidity on photovoltaic (PV) inverters, mission profiles were developed to accurately describe the different processes and rates based on the environmental factors of temperature, relative humidity, and irradiance. The operating environment of the device introduces stress factors dependent on component temperature, relative humidity, and component voltage. The development of profiles started from looking at historical climate conditions at Cocoa, Florida, a hot and humid location. The profile developed from the field location was compared with existing humidity test profiles in industrial and military standards to determine acceleration factors. Prior to the evaluation of the mission profiles, critical components in the PV inverter were identified based on vendor discussions along with failure modes of interest. The profiles considered in this study were demonstrated using a 4-kVA string inverter. The measurements from the critical components including film capacitor were substituted in the device manufacturer-specified equation to determine the severity of each test. The severity of each test was compared against the severity of the 24-hour result to determine the acceleration factor of each test. The acceleration factors were determined to be within a range of 1.2-6 for the existing humidity tests from standards when compared against the mission profile developed from field data. This optimal humidity/thermal profile will be passed to the International Electrotechnical Commission (IEC) standards development group for inverter qualification testing.

Keywords: Inverter, Reliability, Qualification and Testing, and System Performance.

1 INTRODUCTION

Photovoltaic (PV) inverters installed in the field are subjected to multiple stresses throughout their life. Among the various stresses, temperature and relative humidity play a very important role during the life of the PV inverter. High ambient temperature in addition to the temperature increase contributed by powering of the device causes degradation to temperature-sensitive devices, such as electrolytic capacitors used in DC-link applications and transistor modules [1]. The effects of temperature can be easily visualized through a physics-based electrothermal model [2]. On the other hand, humidity produces long term effects on the electronic components inside the inverter. Humidity leads to corrosion, which is accelerated by residual halogens in various devices. Humidity diffuses through potting or conformal coatings of the inverter, and under condensing conditions, it precipitates, causing corrosion. Powering the inverter simultaneously drives out humidity, but it also drives failures associated with ion migration.

The current testing methods for electronic components including capacitors involves accelerated testing using constant temperature, relative humidity, and voltage bias. The Hallberg-Peck paper [3] provides acceleration factors for total time in testing and it treats the effect of moisture ingress into the plastics. In [4], the remaining lifetime estimate of the metallized capacitor was estimated against standard operating conditions accounting for temperature variation, relative humidity variation, and voltage variation. Capacitor manufacturers also use the temperature–humidity–bias test to calculate acceleration factors for capacitors operating under high temperature and high humidity variation [5]. The design qualification testing for the capacitors is performed at a uniform temperature, voltage, and humidity setpoint throughout the test duration. In the field operation of a PV inverter throughout 24 hours of the day, the inverter, and the critical components inside the inverter, including the film capacitors, are subjected to varying irradiances that drive inverter power, varying temperature, and varying humidity which can introduce different failure mechanisms due to these transient processes compared to standard qualification testing.

To understand these transient processes, a representative 24-hour profile was chosen from the field data of a hot and humid location in Florida. Section II deals with the development of the mission profiles. Section III discusses the validation of the profiles along with the selection of the critical inverter components sensitive to humidity effects. Section IV covers the results of the tests along with the acceleration factors. Section V discusses the conclusions from the tests.

2 DEVELOPMENT OF MISSION PROFILES

Metrological year data of Cocoa, Florida, a location known for its humidity, was chosen for the representative profile. The data consisted of 8 months of ambient temperature, dew point temperature, relative humidity, and global horizontal irradiance sampled at a rate of 0.05 Hz / 1 sample per 20 seconds.

For the representative model for humidity, a day with warm rainy conditions coupled with low irradiance data was chosen. Warm, rainy conditions were chosen to replicate the hot and humid conditions that lead to condensation, and low-irradiance conditions were chosen because high irradiance leads to greater output power heating up the power conversion equipment and driving away the moisture.

The high-resolution metrological data were narrowed down to the day (24-hour profile) with warm, rainy conditions with low irradiance using the following methodology:

1. A daily average of the metrological data was created. Along with the raw data, a difference in ambient temperature and dew point temperature was included in the daily average data. These data were plotted against time to select the days of interest. The days of interest included the days with least difference of ambient and dew point temperature, implying very high relative humidity, and low irradiance. Figure 1 shows the daily average of the metrological year with separate subplots for temperature, relative humidity, and solar irradiance.

2. From Figure 1, using visual inspection, 10 days of interest were chosen to narrow down the dataset. For each of these days, the maximum and minimum relative humidity was obtained and used to compute the difference in relative humidity. The days with lowest difference in relative humidity were further selected.
3. From the days with the lowest difference in relative humidity, the maximum irradiance of the days was listed. Then, the day with lowest maximum irradiance was chosen.
4. In summary, the day with a combination of the lowest difference in relative humidity and the lowest maximum irradiance was chosen as the representative 24-hour profile for the PCE device under test. Figure 2 shows the 24-hour representative profile selected from the high-resolution metrological data.

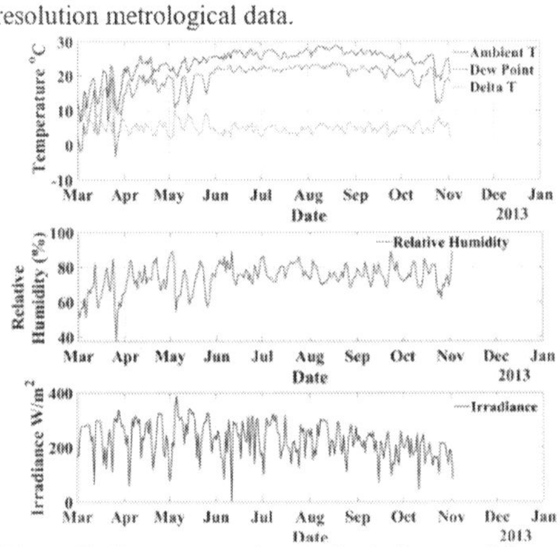

Fig. 1: Daily average of metrological year of Cocoa

This mission profile presents conditions where the inverter experiences high-humidity, zero-power conditions in the initial hours of the profile, leading to condensation. Then, as the temperature increases through powering of the profile, it leads to further retention of moisture. At the end of the profile, the temperature decreases as the device is powered off. These temperature and humidity conditions coupled with low irradiance throughout the day makes this profile a good candidate for testing humid conditions.

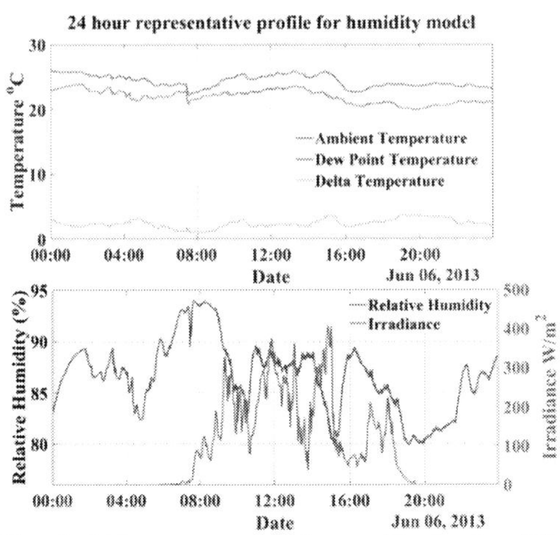

Fig. 2: 24-hour representative profile for humidity model

To demonstrate the mission profile, a U.S. market, 120/240-V, split-phase, 4-kVA string inverter with two independent DC maximum power point tracker inputs for differing PV module strings was characterized and tested. DC power was supplied to the inputs with Keysight N8937APV solar array simulators. The power at the inverter inputs and output was monitored with Yokogawa WT 330 power analyzers. A diagram of the apparatus is shown in Figure 3.

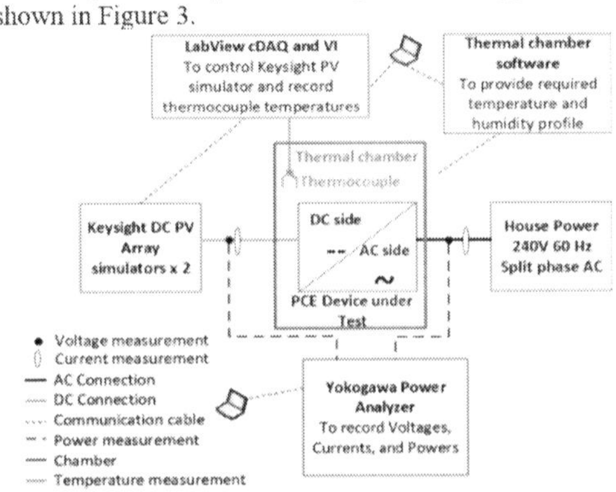

Fig. 3: Experimental setup for validation of humidity mission profile

The critical components of the inverter were identified based on the status review of the PV equipment reliability [6], the PVQAT Failure Modes and Effects Analysis sheet [7] on inverters, and on further discussion with vendors. Based on these discussions, the components under interest were narrowed down to film capacitors, solid-state relays, and electrolytic caps. With this assumption, thermocouples were placed on these components to measure the temperature and to calculate the relative humidity on their surfaces.

Considering the time, it takes to soak the chamber to the given relative humidity setpoint, and with very little variation of temperature throughout the given day, the 1-minute, 24-hour profile for the given day was reduced to a 2-hour step, 24-hour profile, as shown in Figure 3. The experiment was performed inside the chamber with a tolerance of ±1°C and ±3% relative humidity.

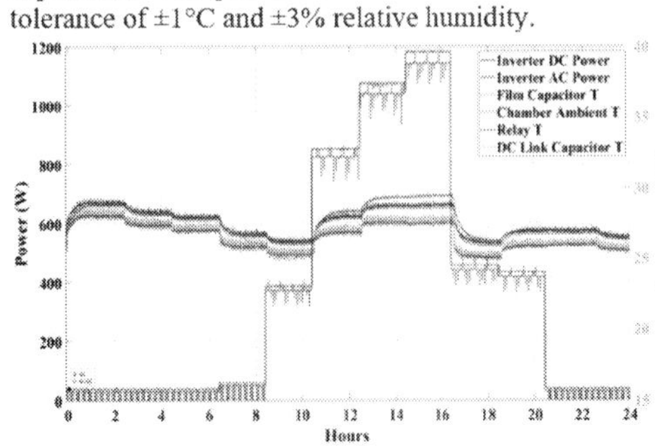

Fig. 4: DC and AC power measurements - 24-hour rainy-day profile

The power and temperature values measured from this experiment were used as input, along with the relative humidity data, to determine the severity of the humidity test. The ambient temperature, T, and relative humidity, RH were used to determine the dew point temperature, TD, of the setup [14]. The dew point temperature and measured capacitor temperature were used to determine the relative humidity on the capacitor surface. The dew point and ambient temperature are expressed in °C, and relative humidity is expressed in percentage, %.

$$TD = 243.04 * \frac{\ln\left(\frac{RH}{100}\right) + \left(\frac{17.625 * T}{243.04 + T}\right)}{17.625 - \ln\left(\frac{RH}{100}\right) - \left(\frac{17.625 * T}{243.04 + T}\right)} \quad (1)$$

$$RH = 100 * \frac{exp\left(\frac{17.625 * TD}{243.04 + TD}\right)}{exp\left(\frac{17.625 * T}{243.04 + T}\right)} \quad (2)$$

Based on this formulae, the calculations were repeated for 40°C and 93% RH damp heat test, as recommended in method 103B of MIL-STD-202G, shown in Figure 5. With ambient conditions of 40°C and 93% RH, the dew point temperature, TD, is estimated to be 38.65°C. Based on the dew point temperature, TD, and the capacitor temperature of 42.2°C, the relative humidity on the surface of the capacitor is estimated to be 83.19%, shown in Figure 5.

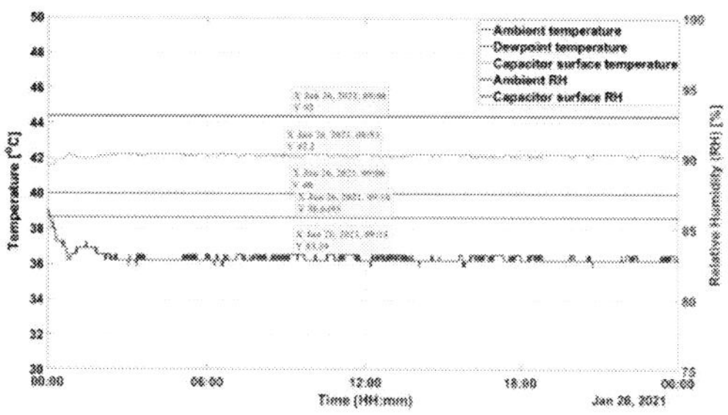

Fig. 5: Temperature and relative humidity data from MIL-STD-202G standard method 103B

The same procedures were repeated for MIL-STD-750D (moisture resistance test at 10% of rated power) and MIL-STD-810G (natural cycle humidity test at 10% of rated power). The capacitor temperatures measured from the experiments are plotted in Figure 6.

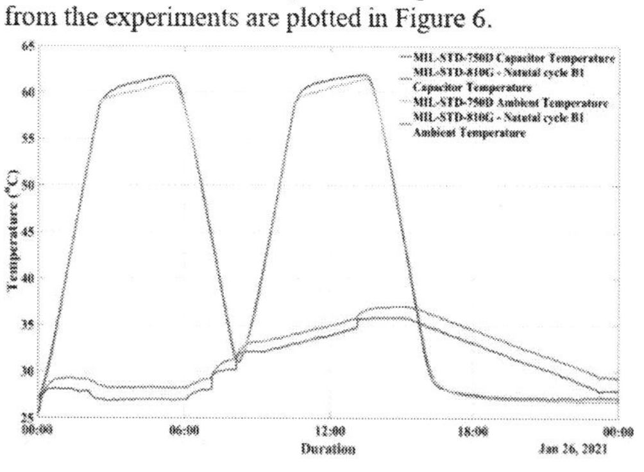

Fig. 6: Capacitor temperature measurements from standard tests

The severity of any humidity test is estimated based on the manufacturer-specified equation for humidity testing [9]. The manufacturer-specified equation is based on the severity test mentioned in the International Electrotechnical Commission (IEC) 60068-2-78 [8].
The severity, S, of any humidity test can be defined as the product of two factors:

$$S = \rho \times F_D \quad (3)$$

Where ρ is the density of water vapor in air, which is a function of both relative humidity and temperature, expressed in g/m³. Psychrometric charts were used to evaluate the density of water vapor in air when the temperature and relative humidity are known.
F_D is the thermal acceleration factor for diffusion.

$$F_D = \exp\left(\frac{-E}{RT}\right) \quad (4)$$

In this equation, E is the activation energy for diffusion (1.0E4 cal/mol), R is the universal molar constant for gases = 1.987 cal/mol K, and T is the Temperature in kelvins = T (in °C) + 273.

4 RESULTS

The 24-hour experiments performed for the representative rainy day in Florida were completed, along with the humidity tests from the industrial and military standards mentioned, in addition to recording of AC and DC power data along with temperature measurements. The measurements from this test were substituted in the manufacturer-specified equation to determine the severity of the test condition and compare it to the baseline data of the Florida rainy-day profile, as shown in Table 1.

Table I: Severity and acceleration factors for the humidity profiles under consideration

Name of the profile	Severity	Acceleration factor
Florida 24-hour rainy-day profile	6.98E-05 (Baseline)	1
MIL-STD-202G (40°C 90% RH steady-state at 10% of rated power)	0.000129	1.84
MIL-STD-750D (moisture resistance test at 10% of rated power)	0.000135	1.9326
60°C 85% RH (modified IEC 62093 Damp heat test at 10% of rated power)	0.000474	6.756
MIL-STD-810G (natural cycle humidity test at 10% of rated power)	7.81E-05	1.118

Severity at Florida 24 hour profile, $S = \Sigma (\rho \times F_D) = 6.98 \times 10^{-5}$

Severity at 40°C 93% RH (MIL-STD-202G), $S = \Sigma (\rho \times F_D) = 1.088 \times 1.181 \times 10^{-5} = 0.000129$

Comparing the MIL-STD-202G humidity test with that of the Florida rainy-day profile showed a severity ratio of 1.8, implying that the MIL-STD-202G humidity test stresses the device 1.8 times faster than the baseline calculation.

5 CONCLUSION

The mission profiles for the humidity modeling were developed using field data from Florida based on the conditions of a warm, rainy day with low irradiance. This profile was compared against the humidity test profiles from existing industrial and military standards for testing the humidity of the electronic components. These tests were validated using a 4-kVA inverter inside a thermal chamber. The severity of the tests was estimated using the manufacturer-specified equations, and acceleration factors were determined.

6 ACKNOWLEDGEMENTS

This work was authored in part by the National Renewable Energy Laboratory, operated by Alliance for Sustainable Energy, LLC, for the U.S. Department of Energy (DOE) under Contract No. DE-AC36-08GO28308. Funding provided by U.S. Department of Energy Office of Energy Efficiency and Renewable Energy Solar Energy Technologies Office. The views expressed in the article do not necessarily represent the views of the DOE or the U.S Government. The U.S. Government retains and the publisher, by accepting the article for publication, acknowledges that the U.S. Government retains a nonexclusive, paid-up, irrevocable, worldwide license to publish or reproduce the published form of this work, or allow others to do so, for U.S. Government purposes.

7 REFERENCES

[1] Hacke, Peter, Daniel Clemens, Ramanathan Thiagarajan, Jack Flicker, and Hironobu Igarashi. "Evaluation of the DC bus link capacitors and power transistor modules in the qualification testing of PV inverters." Progress in Photovoltaics: Research and Applications 29, no. 7 (2021): 675-683.
[2] Nagarajan, Adarsh, Ramanathan Thiagarajan, Ingrid L. Repins, and Peter L. Hacke. Photovoltaic inverter reliability assessment. No. NREL/TP-5D00-74462. National Renewable Energy Lab.(NREL), Golden, CO (United States), 2019.
[3] Hallberg, Örjan, and D. Stewart Peck. "Recent humidity accelerations, a base for testing standards." Quality and Reliability Engineering International 7, no. 3 (1991): 169-180.
[4] Gallay, Roland. "Metallized film capacitor lifetime evaluation and failure mode analysis." arXiv preprint arXiv:1607.01540 (2016).
[5] Michelazzi, Marco, Evangelista Boni, Davide Montanari, Luigi Barbieri, T. Chow, Esa Matero, Stefano Sartini, and Francesco Bergamaschi. "RFI X2 capacitors for high humidity environment." CARTS International (2014).
[6] Hacke, Peter, Sumanth Lokanath, Paul Williams, Arvind Vasan, Paul Sochor, GovindaSamy TamizhMani, Hirofumi Shinohara, and Sarah Kurtz. "A status review of photovoltaic power conversion equipment reliability, safety, and quality assurance protocols." Renewable and Sustainable Energy Reviews 82 (2018): 1097-1112.
[7] PVQAT. Task Group 11 failure modes, mechanisms, and effects analysis for inverter components https://pvqataskforceqarating.pbworks.com/f/FMEA%20 Jan%2027%5B1%5D.xlsm accessed 30 June 2020 (use https://pvqataskforceqarating.pbworks.com for request of password
[8] IEC 60068-2-78 Environmental testing – Part 2-78: Tests – Test Cab: Damp heat, steady state
[9] Sochor, P., and H. Igarashi. "IEC 62093 Ed. 2 BOS Components for PV Systems-Design Qualification PV Inverter Reliability." In IEC-TC82 WG6 Meeting Tokyo. 2015.
[10] Standard, Military. "Mil-std-750D, test methods for semiconductor devices." vol 7 (1995): 47-49.
[11] Specification, Military. "MIL-STD-202G-Method 204D." Test method standard: electronic and electrical component parts. Vibration, high frequency. Department of Defense (2002): 6.
[12] MIL-STD-810G. "Environmental Engineering Considerations and Laboratory Tests." (2008).
[13] Film capacitors, General technical information, available at: https://www.tdk-electronics.tdk.com/download/187688/480aeb04c789e45ef5bb9681513474ba/pdf-generaltechnicalinformation.pdf
[14] Temperature, Relative Humidity, and Dewpoint temperature conversions, available at: https://bmcnoldy.rsmas.miami.edu/humidity_conversions.pdf

MODULE-INVERTERS (MICROINVERTERS): INFLUENCE OF MODULE SIZE ON CONVERSION EFFICIENCIES AND ENERGY YIELDS

Stefan Krauter, Jörg Bendfeld
Paderborn University, EET-NEK
Pohlweg 55, D-33098 Paderborn, Germany
Phone: +49 5251 60 2301, Fax: +49 5251 60 3235, E-mail: Stefan.Krauter@upb.de

ABSTRACT: Efficiencies and energy yields of microinverters available on the market during 2014–2021 have been measured, compared, and ranked. Conversion efficiencies as a function of load have been measured indoors with high accuracy and ranked according to Euro- and CEC weightings. Energy yields have been measured outdoors via identical and calibrated crystalline silicon PV modules of 215 W_P (until 2020) and 360 W_P (starting 2021). Inverters with two inputs have been fed by two of those modules. DC input, AC power output and energy yield of each micro-inverter have been recorded by individual calibrated electricity meters. CEC and EU efficiency rankings have been computed and compared. To assess the influence of PV module size, two extremes have been investigated: A rather small module with 215 W_P - as it has been used 10 years ago, and a brand-new module (2021) offering 360 W_P. Both types of modules contain 60 solar cells in series connection. Appling the low-power modules, the challenge for the different micro-inverters has been during weak-light conditions, using the high-power modules, some inverters temporarily reach their power limits and yield is reduced. A method using a reference configuration of inverter & module and a linear equation $y = ax + b$ resulting in the actual yield, any module & inverter configuration can be characterized by just the coefficients a and b.

Keywords: Microinverter, Conversion Efficiency, Energy rating, Weak-Light Performance, Overload

1 INTRODUCTION

Microinverters are inverters that are connected primarily to a single PV module (occasionally to two modules, as indicated in the tables; very few are available for four modules, but these are not considered here), so each module–inverter combination acts as an independent power plant. The microinverter consists of a maximum power point tracker (MPPT), the DC-AC inverter, and an islanding protection unit (see e.g., [1]). For higher power requirements, several module-inverter combinations are interconnected in parallel on the AC output side. This configuration offers various advantages: Easier planning and easier installation, easy up- and downscaling of a power plant, including extensions or repair that could be carried out even during power plant operation. Logistics is simplified. Effect of shadowing is very limited, and due to low system voltages, potential induced degradation (PID) does not occur. An excellent overview of the development and the advantages of microinverters has been compiled by H. Oldenkamp [2]. However, costs of power plants based on micro inverters are about 10–20% higher. Some of the inverters cannot be operated by themselves and require a control unit (often combined with a remote shutdown option and a monitoring system), or a protective device for grid interfacing (depending on national regulations), thus adding extra costs. Also, conversion efficiency may not be as high as for central inverters. Due to smart master–slave concepts centralized solutions with multiple but relatively large inverters may offer higher yields under weak light conditions. [3] is giving a performance comparison of a microinverter, a power-optimizer, and a central inverter.

2 INDOOR MEASUREMENTS

Due to the reproducible test conditions in the indoor lab, the inverters have been examined individually with predefined and controlled input data. Input has been a PV module simulator with data being set corresponding to the modules used in the outdoor test. The main output data being recorded is the delivered AC power of the inverters. Besides input power, output is also a function of input voltage. If input voltage is getting too low, the inverters even stop operating. The following examinations are based on the possible range of input data (including voltage) given the specific PV module also used for the outdoor investigation

2.1 Peak–, EU– and CEC– conversion efficiency

Peak efficiency is often reached close to the maximum load of the inverter. Peak efficiency in data sheets is not a helpful value since most of the time the inverters operate in the range of 20% to 40% of their rated power – at least under non-arid conditions. Consequently, an adequately weighted efficiency is a more adequate measure to rate conversion devices. One type of weighted efficiency is the so-called "European Efficiency" η_{Euro} – it is calculated according to:

$$\eta_{Euro} = 0.03 \cdot \eta_{5\%} + 0.06 \cdot \eta_{10\%} + 0.13 \cdot \eta_{20\%} + 0.1 \cdot \eta_{30\%} + 0.48 \cdot \eta_{50\%} + 0.2 \cdot \eta_{100\%} \quad (1)$$

The other is the "CEC efficiency" by the California Energy Commission (CEC). CEC efficiency is computed as an average value of DC-AC conversion efficiencies at six pre-defined outputs between 10% and 100% of the rated power (with an emphasis on higher irradiance levels) is determined by:

$$\eta_{CEC} = 0.04 \cdot \eta_{10\%} + 0.05 \cdot \eta_{20\%} + 0.12 \cdot \eta_{30\%} + 0.21 \cdot \eta_{50\%} + 0.53 \cdot \eta_{75\%} + 0.05 \cdot \eta_{100\%} \quad (2)$$

For the "European Efficiency", weighting factors for high relative power values are lower.

2.2 Test arrangements and results

Ten different output power values from zero up to the maximum operational power of the PV modules have been considered (they are roughly identical for all seven micro inverters with 240 to 260 Watt rated power). The values used for the single input inverters (adjusted by controlling the current) are: 10 W, 25 W, 50 W, 75 W, 100 W, 125 W, 150 W, 175 W, 200 W, and 225–260 W. The averaging time for the high precision power meter used (Zimmer LMG 670) was set to 500 ms. More than one micro inverter showed an unstable AC power output for values below 100 W. Since 2019, the test setup has been changed, the power values used for the inverters (adjusted by controlling the current) are continuously increased in 1024 steps from 0 to maximum. Each step takes eight seconds while the measurement duration is still 500 ms. Figure 1 shows an example for the new measuring procedure.

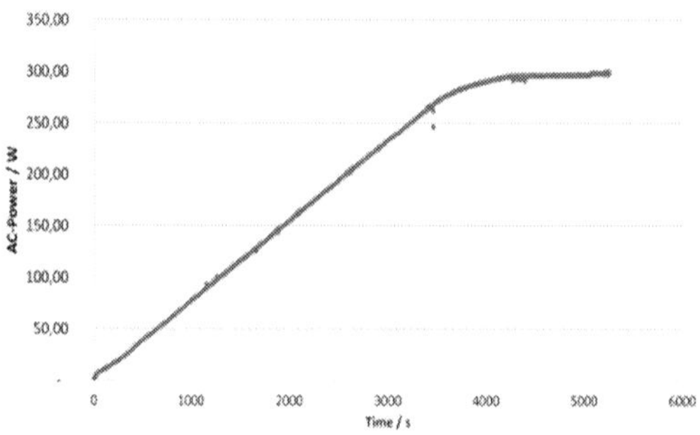

Figure 1: Measured AC power output (in Watt) as a function of measurement duration (in seconds)

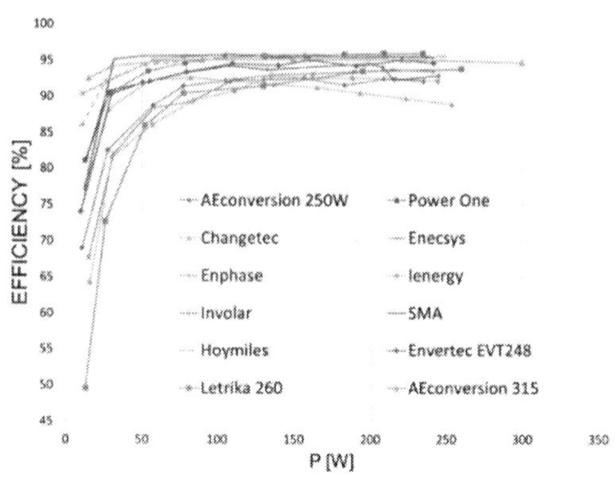

Figure 2: Measured DC-AC conversion efficiencies as a function of power output, for microinverters with single PV module inputs

The measured DC-AC conversion efficiencies of all inverters are shown in Fig. 2 and Fig. 3. Based on those measurements, the European (EU) efficiencies and the CEC efficiencies for the micro inverters have been calculated according to (1) and (2), eleven micro-inverters are designed for single modules, five inverters have inputs for 2 modules: Involar MAC 500, the APS Y 500, the Envertech EVT 560, the WVC 600 and the WVC 700. The WVC 600* inverter stopped working at a measured power of 250 W. After a test run on higher temperatures the inverter failed constantly. Since the documentation of the WVC 600 and WVC 700 inverter has been extremely poor, its rated power has been assumed. For this reason, the WVC 700 is shown twice. First, with the assumed rated power of 600 W, then with the assumed rated power of 700 W. The maximum measured power of the WVC 700 inverter was 600 W only.

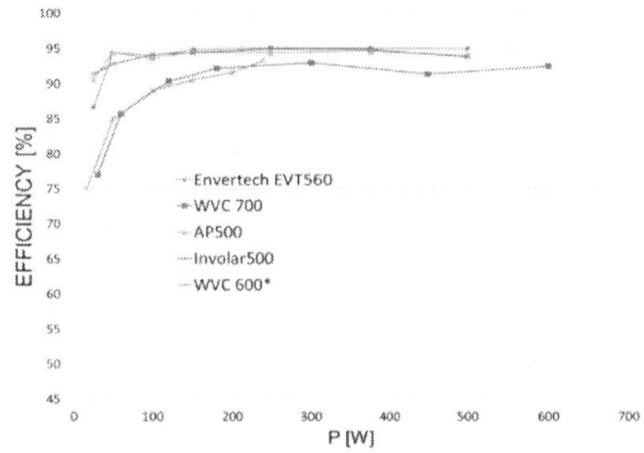

Figure 3: Measured DC-AC conversion efficiencies as a function of power output, for five inverters that have inputs for two PV modules

The ranking considering the European conversion efficiency is shown in Table I: AEconversion/Aptronic 250 W and Enecsys have the same EU-efficiency, thus sharing rank number 13.

TABLE I: Ranking of all tested microinverters by "European Conversion Efficiency", according to (1), same types as listed in Table III

Rank	Manufacturer	European-Efficiency	Relative eff. vs #1
1	SMA	95.4 %	100.0 %
2	Enphase	95.2 %	99.8 %
3	Hoymiles	95.0 %	99.5 %
4	AEconversion 315	94.9 %	99.5 %
5	Envertech560	94.61 %	99.2 %
6	Power One	94.6 %	99.2 %
7	Involar 500	94.3 %	98.8 %
8	APSystems 500	94.1 %	98.6 %
9	Envertech	93.2 %	97.7 %
10	Involar	92.7 %	97.2 %
11	WVC 700 (600W)	91.6 %	96.0 %
12	Changetech	90.9 %	95.3 %
13	Aptronic	90.4 %	94.7 %
13	Enecsys	90.4 %	94.7 %
14	Ienergy	89.9 %	94.3 %
15	Letrika 260	88.7 %	93.0 %
16	WVC 700 (700W)	73.3 %	76.8 %
17	WVC 600*		0.0 %

*failed.

TABLE II: Ranking of all tested microinverters by "CEC Efficiency", according to (2), same types as listed in Table III

Rank	Manufacturer	CEC-Efficiency	Relative eff. vs #1
1	Enphase	95.55 %	100.0%
2	Power one	95.46 %	99.9%
3	Hoymiles	95.40 %	99.8 %
4	SMA	95.06 %	99.5 %
5	AEconversion 315	94.98 %	99.4 %
6	Envertech 560	94.64 %	99.0 %
7	Involar 500	94.60 %	99.0 %
8	APSystems 500	94.50 %	98.9 %
9	Envertech	94.05 %	98.4 %
10	Involar	93.86 %	98.2 %
11	Enecsys	92.00 %	96.3 %
12	WVC 700 (600W)	91.60 %	95.9 %
13	Letrika 260	91.50 %	95.8 %
14	Ienergy	91.45 %	95.7 %
15	AEconversion 250	91.21 %	95.5 %
16	Changetech	90.90 %	95.1 %
17	WVC 700 (700W)	87.50 %	91.6 %
18	WVC 600*	*	0.0 %
	* failed		

3 OUTDOOR MEASUREMENTS

3.1 Test arrangement and measurements

For the PV outdoor test lab (see Fig. 4) installed on the roof of the University of Paderborn (51.707° N, 8.771° E), a specific test system has been employed. Each PV module (QC-CO2 by Hanwa Q-cells) consists of 60 solar cells. In stated plant, these equal and calibrated modules are the input for each micro inverter. The goal of this investigation is to analyze the performance of the inverters in the lab and under real operating conditions comparing their energy yield simultaneously at the same climatic conditions and solar irradiance. The modules have been measured and calibrated first in 2011 and then again in 2021: The degradation during these 10 years has been within the measurement tolerance (-2 W_p, at an accuracy of ±3%), and is therefore neglectable. The climatic conditions have been monitored during the whole test period. The meteorological monitoring equipment consists of two calibrated pyranometers in the plane of the module (CMP 21 and SP 2 lite by Kipp & Zonen), a 3-D ultrasonic anemometer (by Thies), a thermo-hygro sensor (by Thies) and a thermo-hygro meter with a wind sensor WXT520 (by Vaisala).

Each microinverter has been directly connected to a calibrated electrical energy meter with a S_0-interface. To secure an accurate yield measurement, the calibrated electrical energy meters are replaced on a regular base with new freshly calibrated ones. All S_0-interfaces have been connected to a server-based data acquisition system.

Figure 4: PV outdoor laboratory at Paderborn University for energy yield comparison of microinverters during 2014-2020 using eight equal, calibrated PV modules of 215 W_p each as inputs.

TABLE III: Ranking of microinverters by energy yield, using the 215 W_p modules [4].

Rank	Manufacturer	Type	Relative yield vs. ABB
1	Involar	MAC 500	100.7 %
2	Power One / Aurora / ABB	Micro-0.25-i	100.0 %
3	APSystems	YC 500	99.3 %
4*	Hoymiles	MI 600	97.4%*
5	SMA	Sunny Boy 240	95.2 %
6	Enphase	M 215	95.2 %
7	Involar	MAC 250	94.2 %
8*	Envertech	EVT 300	94.0%*
9*	WVC 700	700 (600 W)	91.7%*
10	AEconversion / Aptronic	INV 250-45	92.5 %
11	Envertech	EVT 248	92.1 %
12	IEnergy	GT 260	91.5 %
13	Enecsys	SMI-S-240W	88.7 %
14	Hoymiles	MI 250	78.4 %
15	Changetec	ELV 300-25	75.6 %

Measurements indicated by * are preliminary only.

WVC 600 failed, insufficient data has been recorded.

AEconversion 315 is still under investigation.

4 NEW CONFIGRATION

The new configuration for the tests, using ten 360 W_p modules (lower row, from left) is shown in Figure 5. Modules are made by Solarwatt (Eco 120M Style 360 W_p), power of each module has been measured in the factory in Dresden (Germany). Additionally, one module has been sent for a precision measurement to ISFH in Hameln (Germany). It turned out that the factory measurements have been very accurate (362 W_p vs. 359.34 W_p ±3% at ISFH in July 2021).

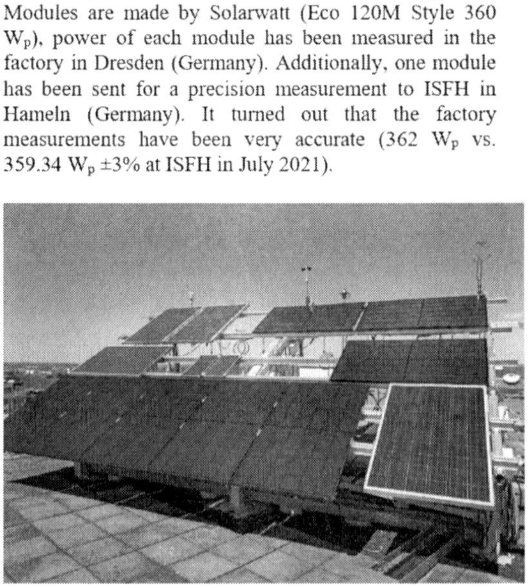

Figure 5: New configuration of PV outdoor laboratory for energy yield comparison of microinverters using eight equal, calibrated PV modules of 360 W_p each as inputs.

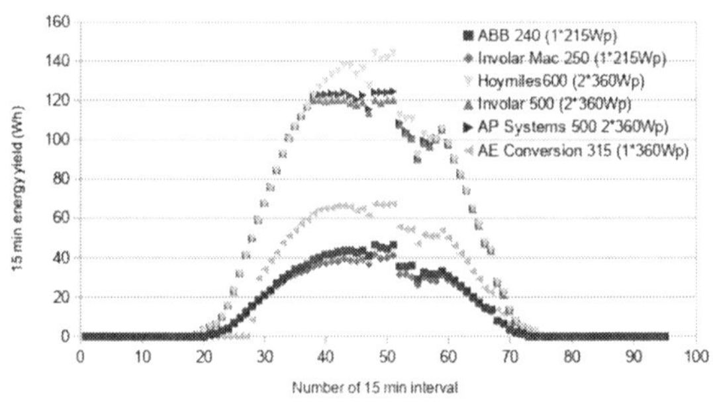

Figure 6: Electrical energy yields of different inverters and different PV modules during a mostly clear day (some clouds in the afternoon).

Besides the effects already observed with the 215 W_p modules, such as distinct conversion efficiencies at different irradiance levels, speed of MPPT algorithms, minimum thresholds for initiating operation; additionally, temporal saturation effects are occurring at some inverters with the new 360 W_p modules. The resulting electrical energy yields during the course of a day for the different configurations of microinverters and modules are shown in Figure 6. To some extent, the above-mentioned effects can be observed. While the different types of effects make it quite cumbersome to predict an energy yield for a certain configuration at a certain location, a more consumer-friendly yield-predicting method has been elaborated.

5 METHOD TO DESCIBE ENERGY YIELD CHARACTERISTICS OF MICROINVERTER & MODULE COMBINATIONS

To ease the characterization of a specific combination of PV module & microinverter, a linear equation has been applied to a well investigated reference characteristics of a very good inverter without issues for low irradiance, MPPT, and saturation. The inverter chosen as a reference has been the Enphase M 215, which ranked #1 at the CEC-efficiency rankings, see Table II. At the single module inverters, the Power One / ABB Micro-0.25-i and the SMA Sonny Boy 240 ranked better at the yield comparison (see Table III), but the manufacturing of those has been discontinued.

Plotting a function of the actual yield (y) over the reference yield (x), that function would be $y = a\,x + b$ with the trivial coefficients $a = 1$ and $b = 0$ for the reference configuration (Enphase M 215 with the Q-cell 215 W_p module). Figure 7 shows the original configuration with the inverters for single modules and the 215 W_p modules attached.

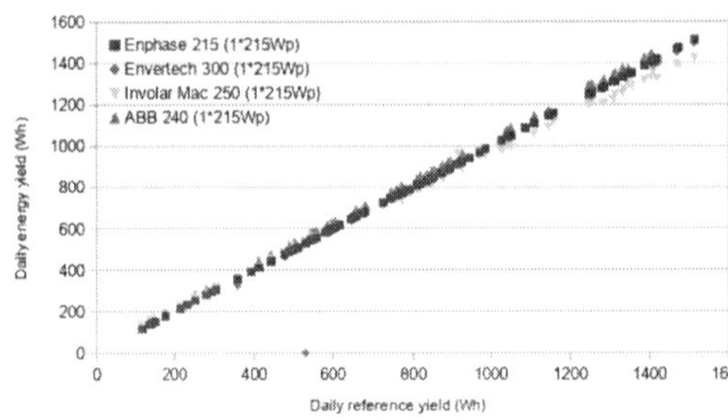

Figure 7: Electrical energy yields of different inverters for single modules with a 215 W_p module attached. Daily reference yield (x-axis) is the yield achieved by the Enphase 215 with a single 215 W_p module applied.

The coefficients of the different inverters for the rel. yield equation $y = ax + b$ are elaborated in Table IV: It can be observed, that for low daily yields, the Involar MAC 250 is performing a little bit better than the reference, so b is above 0, for high yields it performance is decreasing (relative to the reference), so a is above 1. For the Envertech EVT 300 the characteristics is vice versa: Performance at low yields is worse than the reference, so b is negative; relative performance is increasing towards high reference yields, so steepness of curve is higher, resulting in an $a > 1$.

TABLE IV: Coefficients for rel. daily yield $y = \boldsymbol{a}\,x + \boldsymbol{b}$ (referenced to Enphase 215, all with a **single 215 W_p module**)

Manu-facturer	Type	*a*	*b* (Wh)
Involar	MAC 500	0.923	+43.35
Power One / Aurora / ABB	Micro-0.25-i	1.011	+25.90
Envertech	EVT 300	1.020	-33.45
Enphase	M 215	1.000	±0.00

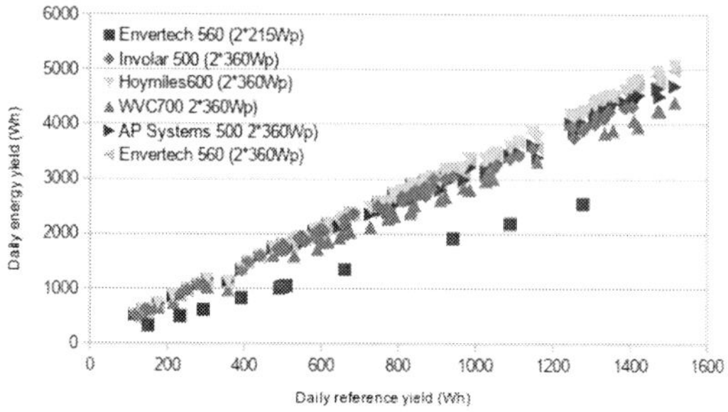

Figure 8: Electrical energy yields of different inverters for two modules with two 215 W_p or 360 W_p modules attached. Daily reference yield (x-axis) is the yield achieved by the Enphase 215 with a single 215 W_p module applied.

Figure 8 shows the characteristics of different microinverters that are able to serve two modules, either with two 215 W_P (older measurements) or two 360 W_P module (latest measurements). Table V shows the corresponding coefficients a for "steepness" and b for "offset" of the relative daily yield curve.

TABLE V: Coefficients for rel. daily yield $y = \boldsymbol{a}\, x + \boldsymbol{b}$ for microinverters serving **two modules**, either 215 W_p or 360 W_p types (refferred to Enphase 215 with a 215 W_p module)

Manu-facturer	Type (module power)	a	b (Wh)
Envertech	EVT 560 (2 x 215 W_P)	1.983	+37.80
Envertech	EVT 560 (2 x 360 W_P)	3.227	+109.97
Hoymiles	600 (2 x 360 W_P)	3.217	+152.23
Involar	MAC 500 (2 x 360 W_P)	2.889	+180.70
AP System	YC 500 (2 x 360 W_P)	2.953	+254.77
WVC	WVC 700 (2 x 360 W_P)	2.750	+172.39

The coefficients of determination R^2 for all regressions of the measurement values to determine the coefficients a and b have been in the vicinity of 0.99 or above.

6 CONCLUSION

The use of a reference configuration together with the two coefficients of a linear equation enables a simple method to describe quite accurately the daily yield performance of any microinverter in combination with any PV module, even with under- or oversized ones.

7 OUTLOOK

While prices of PV modules are decreasing at a higher pace than prices for microinverters, we will see configurations with oversized modules and more saturated microinverters more often in the future. Also, there is a legal limit in some countries for so called "balcony PV" of "plug-in PV" for the maximum inverter power (e.g., in Germany: 600 W) which makes PV overpowering an attractive option. This underlines the necessity of a method (e.g., as described) to extrapolate energy yield.

8 REFERENCES

[1] W-F. Lai, S-M. Chen, T-J. Liang, K-W. Lee, A. Ioinovici: "Design and Implementation of Grid Connection Photovoltaic Inverter", Proceedings IEEE Energy Conversion Congress and Exposition (2012) 2426.

[2] H. Oldenkamp, I.J. de Jong: "The return of the AC-module inverter", Proceedings 24th European Photovoltaic Solar Energy Conference (2009) 3101.

[3] D. Stellbogen, P. Lechner, M. Senger: "Field and laboratory performance characterisation of microinverter and Power optimizer systems", Proceedings 32nd European Photovoltaic Solar Energy Conference (2016) 1654.

[4] S. Krauter, J. Bendfeld: "Micro-Inverters: An Update and Comparison of Conversion Efficiencies and Energy Yields". Proceedings of 37th European Photovoltaic Solar Energy Conference (2020).

This is a short version of the paper submitted to the journal Solar RRL.
It has been selected by the Executive Committee of the EU PVSEC 2021 for submission to Solar RRL.

38th European Photovoltaic Solar Energy Conference and Exhibition

ANALYSIS AND DEVELOPMENT OF A FAULT-TOLERANT POWER CONVERTER FOR SOLAR PV APPLICATIONS

Filba-Martinez, Alber; Cabre-Piqueras, Claudia; Trilla, Lluis; Paradell, Pol; Dominguez-Garcia, Jose Luis
Catalonia Institute for Energy Research (IREC)
Jardins de les Dones de Negre 1 2nd floor, 08930, Sant Adria de Besòs, Barcelona (Spain)

ABSTRACT: Due to the ongoing growth of Solar PV generation on the energy share, it is of major interest to achieve cost reductions for this technology. When considering PV plant lifetime and reliability, the power converter element is responsible for approximately 50% of total failures on a solar PV plant. Thus, this paper proposes a fault-tolerant multistring converter to address this challenge. This converter is able to survive a first fault achieving fault tolerance in both DC-DC and DC-AC stages. In order to validate this design, a 5 kW lab-scale prototype has been tested under emulated faults successfully. Also, a quantitative evaluation of the proposed fault-tolerant converter has concluded that by a total cost increase of only 4% and an efficiency reduction of a 0.5%, this topology almost doubles the converter lifetime expectancy in comparison to the conventional setup.
Keywords: fault-tolerant; photovoltaic; reliability

1 AIM AND APPROACH USED

In the last years, Solar PV generation is continuously growing on the energy share and it is expected to be up to 8500 GW worldwide according to IRENA [1]. In order to achieve this, further cost/LCOE reductions are required. Such reduction may be achieved through several drivers as reduction of investment costs, reduced operation and maintenance costs, reduction of capital costs, and by increasing the energy generated during the system lifetime by means of increased efficiency or lifetime extension.

When considering PV plant lifetime and reliability, the power converter element is responsible for approximately 50% of total failures on a solar PV plant [2]. In addition, the typical expected lifetime of PV plant inverters (from string-based to central configurations) is about 10 to 15 years, thus requiring at least one full replacement when considering a typical PV-plant expected lifetime of 25 years. Thus, it is clear that there is the need to reduce the impact of failures on the inverter as well as increase its lifetime to be on par with the other components.

In this regard, key innovations to handle such challenges are required. This paper presents a fault-tolerant multistring inverter, showing the global benefits of the proposed technology in terms of enhanced system lifetime and cost, as well as a prototype of the proposed converter under lab testing.

2 SCIENTIFIC INNOVATION AND RELEVANCE

The proposed fault-tolerant multistring inverter is shown in Figure 1. It comprises a conventional two-stage inverter, identified in Figure 1 by the elements marked in black, and additional elements to implement the fault-tolerance capability, identified in Figure 1 by the elements marked in red. In the proposed solution, fault tolerance is achieved by means of adding redundancy with the redundant switching leg comprised of switches S_{RH} and S_{RL}. Employing leg-level redundancy allows an optimum compromise between system cost, performance, and reliability [3].

The proposed fault-tolerant converter design has been developed as a trade-off of two-fold conditions: to be able to survive a first fault and continue operation without downgrading the inverter nominal power, and achieve

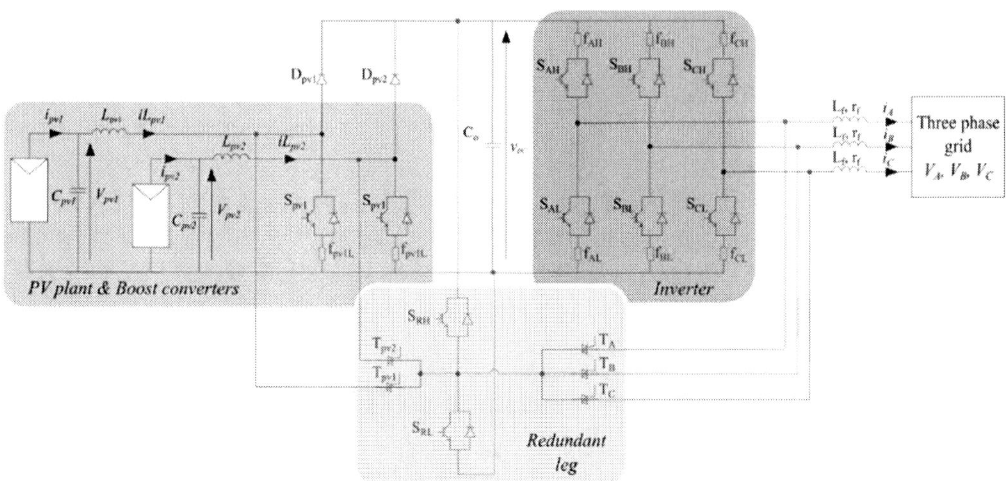

Figure 1: General schematic of proposed fault-tolerant multistring inverter topology. The converter first stage comprises two interleaved dc-dc boost converters. The second stage comprises a two-level three-phase voltage-sourced dc-ac converter.

fault-tolerance on both stages of the inverter with a reduced increase on the converter cost.

The principle of operation of the fault-tolerant converter is the following: In pre-fault conditions the redundant leg and TRIACS T_A, T_B, T_C, T_{pv1}, and T_{pv2} are disabled. When a fault occurs in one of the operating switches, the fault is first isolated by the fuses in the faulty leg (in case a short-circuit fault occurs), the faulty leg is identified, and the redundant leg is connected in place of the faulty leg by enabling the corresponding TRIAC. For instance, if a short-circuit fault occurs in S_{AH}, fuses f_{AH} and f_{AL} would be immediately blown, isolating the faulty leg, the faulty leg would be identified with the proper fault diagnosis algorithm, TRIAC T_A would be enabled, and S_{AH} and S_{AL} command signals would be redirected to S_{RH} and S_{RL}, respectively. When an open-circuit fault occurs, fuse action is not required, since this type of fault inherently isolates the faulty leg.

3 RESULTS

In order to prove the feasibility of the proposed fault-tolerant inverter, a 5 kW lab-scale prototype has been developed and experimentally testes. The values of the components in the converter are shown in Table I. The inverter control, fault-diagnosis algorithm, and system monitoring have been implemented employing the fast-prototyping platform OP4510 from OPAL-RT. Figure 2 shows the inverter prototype and Figure 3 shows the system console. The experimental tests were run under the conditions listed in Table II.

Figure 2: Fault-tolerant inverter prototype.

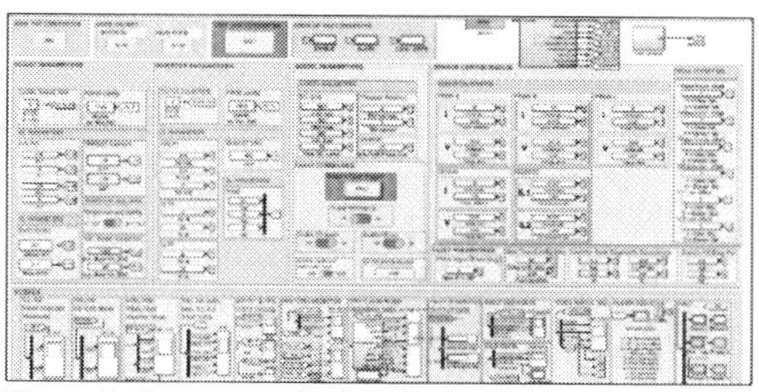

Figure 3: System monitoring platform.

In addition, a quantitative evaluation of the proposed fault-tolerant inverter main benefits has been performed, including the potential cost, overall efficiency and converter expected lifetime. Such analysis has been carried out in [4] and it is based on probabilistic analysis for failures and percentage impact approximations for cost.

Table I: Converter prototype component parameters

Component		
C_{pv1}, C_{pv2}	700	μF
L_{pv1}, L_{pv2}	2.5	mH
C_o	1400	μF
L_f	4.6	mH
R_f	0.315	Ω

Table II: Test conditions

Boost converter		
V_{pv1}, V_{pv2}	140	V
i_{Lpv1}, i_{Lpv2}	7.2	A
Inverter		
V_{DC}	175	V
Grid		
$V_{LL,rms}$	70	V
$V_{LN,rms}$	40	V

3.1 Prototype testing

To test the capabilities of the prototype, faults on both converter stages have been emulated. These faults are forced by leaving a switch in a branch open, acting as an open-circuit fault. The faulty switches that the tests have been run on are: S_{AH} and S_{AL} in the DC-AC stage, and S_{pv1} in the DC-DC stage.

Figure 4 and Figure 5 show the behavior of the fault-tolerant converter when exposed to a fault on S_{AH} and S_{AL}, respectively. When a fault is emulated on S_{AH}, phase A current has no positive values. Similarly, the same happens with negative values when a fault is emulated on S_{AL}. As it can be seen on both figures, once the fault-diagnosis algorithm detects and locates the fault, it reconfigures the converter such that redundant leg works in place of the failed leg, allowing to recover the converter normal operation. In both cases, the converter is able to detect and handle the fault in less than two grid periods. During the time period between the fault and the reconfiguration, the converter control is able to control the dc-link voltage (v_{DC}) close to its reference value (V_{DC}).

Figure 6 shows the response for a fault on the DC-DC stage. When a fault is emulated on one of the boost converters, the current passing through the inductance is zero. In this case, the system is also reconfigured in less than two grid periods.

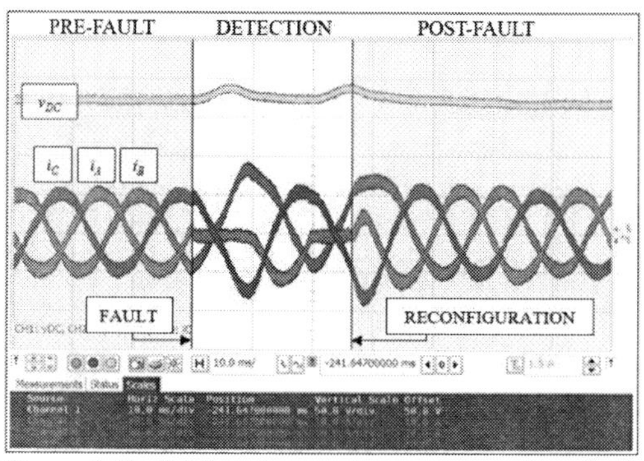

Figure 4: Fault-tolerant converter response under a fault on DC-AC stage switch S_{AH}.

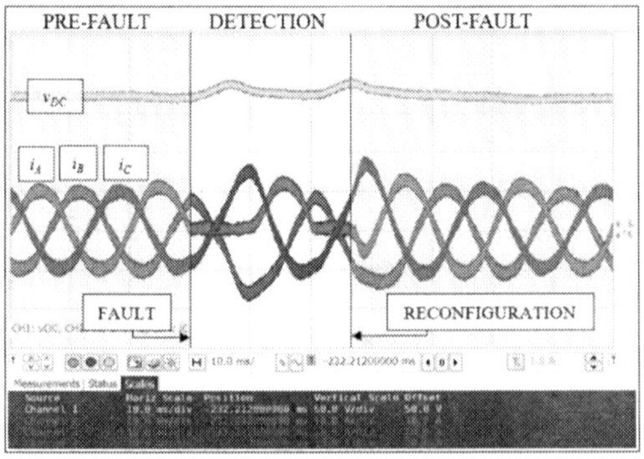

Figure 5: Fault-tolerant converter response under a fault on DC-AC stage switch S_{AL}.

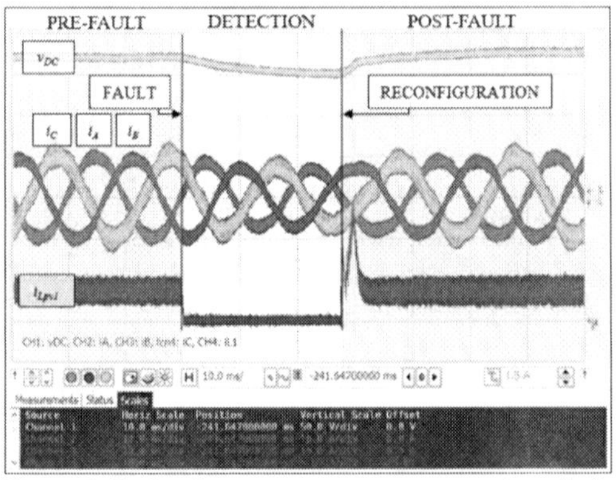

Figure 6: Fault-tolerant converter response under a fault on DC-DC stage switch S_{pv1}.

3.2 Cost analysis

The cost analysis has been performed based on the number of components required and their impact on the total cost of the power converter. In this regard, the switching components and ancillary circuits represent less than 20% of the total cost of the converter. Considering an increase of about 20% of components due to the redundant structure, the total cost increases by 4%.

3.3 Efficiency analysis

During pre-fault operation, the proposed converter will feature power losses similar to the conventional structure, with a slight increase due to the losses in the fuses in series with each switch. Nonetheless, these losses can be generally neglected due to the low-ohmic value of the fuses. In post-fault operation, conduction losses will also occur at the TRIAC enabled for the reconfiguration. The losses in the TRIAC will be approximately equal to the conduction losses of the redundant switching leg once it is enabled. Considering a pre-fault converter efficiency of 95% and that conduction losses account for half of the semiconductor losses, the converter efficiency would only be reduced a 0.5%.

3.4 Component lifetime analysis

For the evaluation of the component lifetime, a Markov Chain-based reliability model was applied, following failure probabilities and component failure rates. This evaluation leads to a converter lifetime increase of approximately 95%; i.e., the lifetime of the conventional converter can be doubled when employing the proposed fault-tolerant structure.

4 CONCLUSIONS

In this paper, an innovative and cost-effective fault-tolerant converter for solar PV applications has been introduced. The concept design, prototype development and lab-test have been presented. In addition, the proposed design is targeting the best trade-off of cost increase vs performance improvement. As shown the reliability of the system is greatly increased. This is accomplished with a minor economic impact, which is easily counteracted by the increase of energy yielded by the system during its extended lifetime. This solution is especially interesting for floating PV systems or PV plants located in remote areas which are harder to reach and show higher operating and maintenance costs with a large mean time to repair.

5 REFERENCES

[1] IRENA. Global Energy Transformation: A Roadmap to 2050; International Renewable Energy Agency: Abu Dhabi, UAE, 2019, p. 52. ISBN 978-92-9260-121-8.

[2] Formica, T.J.; Khan, H.A.; Pecht, M.G. The Effect of Inverter Failures on the Return on Investment of Solar Photovoltaic Systems. IEEE Access, 2017, 5, 21336–21343.

[3] Zhang, W.; Xu, D.; Enjeti, P. N.; Li, H.; Hawke, J. T.; Krishnamoorthy, H. S. Survey on fault-tolerant techniques for power electronic converters, IEEE Trans. Power Electron., 2014, vol. 29, no. 12, 6319–6331.

[4] Renaudineau, H.; Paradell-Solà, P.; Trilla, L.; Filba-Martinez, A.; Cardoner, D.; Domínguez-García, J.L. Reliability Assessment of a Fault-Tolerant PV Multistring Inverter. Energies, 2020, 13, 6525.

This project has received funding from the European Union's Horizon 2020 research and innovation programme under SUPER PV project (grant agreement No 792245) and form AGAUR under the project CORV (with reference 2019 LLAV 00045). The authors acknowledge the support provided by AC3E (ANID/Basal/FB0008).

CONTEXT-SENSITIVE PV PLANT COMPONENTS BENCHMARKING BASED ON MONITORING DATA

W. Vanheusden[1], J. Ascencio-Vásquez[1], K. De Brabandere[1], S. Lindig[2], D. Moser[2], M. Richter[1]
[1]3E sa, Quai à la Chaux 6, 1000 Brussels, Belgium
[2]Institute for Renewable Energy, EURAC Research, Viale Druso 1, 39100 Bolzano, Italy
E-mail: wannes.vanheusden@3e.eu

ABSTRACT: In the framework of the new H2020 project TRUST PV, a benchmarking methodology is being developed. The objective is analyzing how different contextual parameters – such as micro-climate and environment, different assets composed of different components from different manufactures in different configurations – influence the distribution of performance Key Performance Indicators (KPIs). This is achieved by applying big data analytics to over 6.5 GW of solar monitoring data coming from more than 10.000 PV sites monitored through 3E's asset management platform SynaptiQ. In this paper, the data requirements, and data architecture for this type of benchmarking are presented as well as several preliminary case studies. However, the work is still ongoing within the TRUST PV project and needs to be further extended and validated. The end goal of this component-benchmarking tool is to support technical decisions of different stakeholders (manufacturers, designers, operators, financial entities, etc.), to improve the assumptions taken in PV yield and degradation modelling.
Keywords: PV Inverter, PV Module, Monitoring, Benchmarking

1 INTRODUCTION

The deployment of PV systems continues to show amazing growth worldwide. The use of smart monitoring platforms together with sophisticated communication devices helps maximizing the energy yield in real-time, trigger alarms in case of failures, and track the performance evolution of PV components (mainly PV inverters and PV modules). In the framework of the H2020 TRUST-PV project [1], a context-sensitive PV plant component benchmarking tool is under development based on monitoring data from over 6.5 GW of PV plants under operation and Big Data analytics, which will be directly linked to the Cost Priority Number (CPN) methodology [2], which is further developed within the TRUST-PV project. To achieve those ambitious goals, 3E's SynaptiQ monitoring platform [3] is being exploited. This platform has over 10.000 PV plants connected to it, working with over 60 PV inverter manufacturers, and over 200 PV module manufacturers (up to August 2021). The comprehensive detail of information of each component, and monitoring data at different electrical levels (sub string, string, inverter, PV plant), together with an organized data architecture, enables the exploitation of the data for a component-benchmarking of PV modules and PV inverters. The goal of this component-benchmarking tool is to support technical decisions of different stakeholders (manufacturers, designers, operators, financial entities, etc.), improve the assumptions taken in PV yield and degradation modelling, inverter reliability, and the maintenance of a large database with operational Key Performance Indicators (KPIs) coming from real-world PV plants in different climate zones globally.

The use of smart information flow schemes, and statistical approaches (e.g., machine learning and big data algorithms) are presented as basis for a commercial component-benchmarking of operational PV plants. The conceptual methodologies to process a very large dataset and their expected results are given. This work is relevant for the whole PV community because it is setting the baseline for component-benchmarking of PV modules and PV inverters, and with slight adaptations, it could be extended to any PV system component (transformers, cables, etc.) or other energy sectors (Wind, Grids, etc.). In addition, the methods and outcomes from the component-benchmarking can be linked to the CPN methodology towards the geographical identification of costs and benefits of O&M practices

In this work, we first explore one of the largest databases with real PV plants connected globally (3E SynaptiQ Solar). Then, the main components, technical strategies and data architecture are discussed towards a commercial component-benchmarking tool for different components of PV plants. Lastly, several preliminary case studies are presented to better illustrate how the tool could be used. However, the work is still ongoing and needs to be further extended and validated within the TRUST PV project the coming year. This will happen in collaboration with the consortium partners – EURAC, Baywa r.e., Enel Green Power, Innosea and Huawei – leading to interesting functionalities such as the Cost Priority Number (CPN) and optimal replacement cycles for predictive maintenance.

2 THE 3E SYNAPTIQ DATABASE

The exploration of the 3E Solar SynaptiQ monitoring platform shows very attractive numbers to run meaningful statistics on the database. The diversity of data can be observed in Figure 1. The high number of different inverter and module manufacturers can support a comprehensive component benchmarking tool.

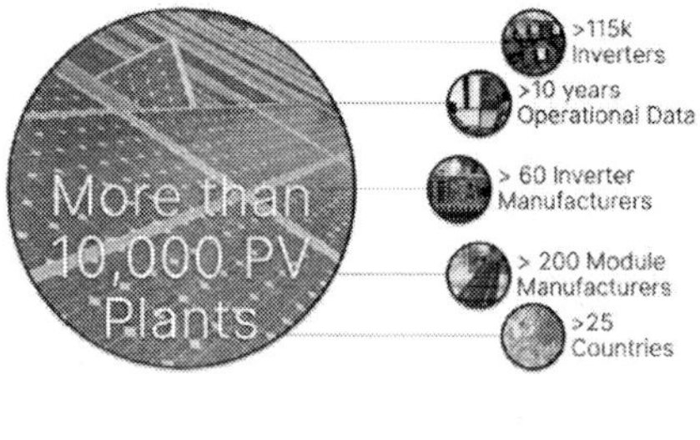

Figure 1: Number of components connected to 3E Solar SynaptiQ Platform

3 METHODOLOGY

3.1 Technical strategies

Smart monitoring platforms allow the recording of variables at many levels of PV systems. Thus, in the best case, data availability can be very large, and consequently the computational resources to run such component-benchmarking exercise can grow accordingly. Considering such limitations, two strategies are addressed:

1. Component-benchmarking of PV Plant (Modules and Inverters combined)

This is a limited but straightforward method to get firsthand insights to compare the performance of PV assets as overall systems. Here, the operational KPIs used are the ones at the AC output of the system, and the features are a combination of the PV modules and inverters.

2. Component-benchmarking of PV Modules and PV Inverters separately

This is a more comprehensive holistic approach where the tool would:

Process the data around the PV modules (DC variables of the PV system), where relevant timeseries are irradiance, ambient temperature, DC power, DC voltage, DC current, thermal losses, angle of incidence (AOI) losses or spectral losses (see Figure 2).

Process the data around the PV inverters (DC input and AC output), where the relevant indicators can be the irradiance, temperatures, DC power, DC voltage, DC current, AC power, AC voltage, AC current, clipping losses, thermal losses (see Figure 3).

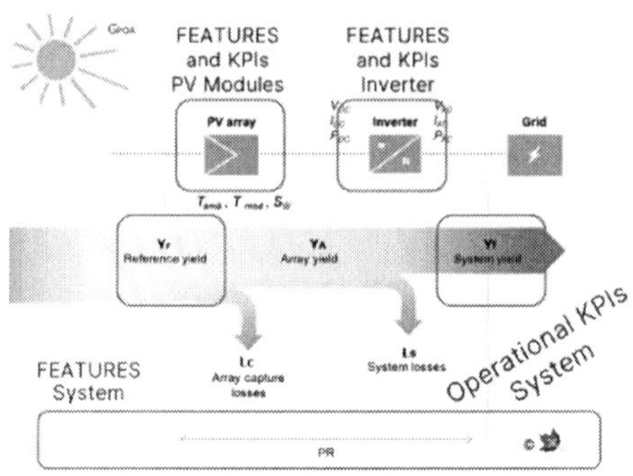

Figure 2: Schematic of PV-plant benchmarking

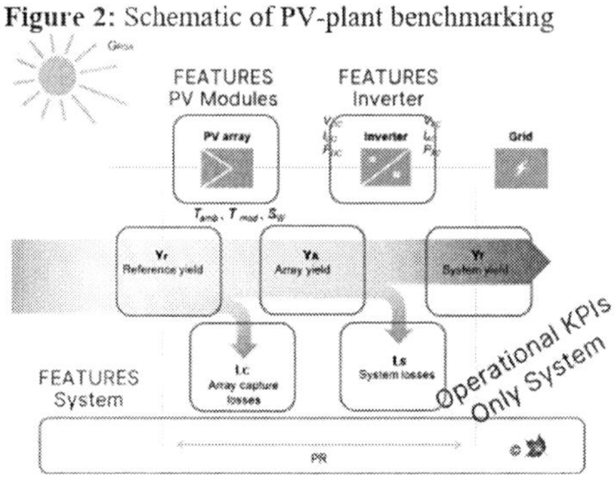

Figure 3: Schematic of PV Module and PV Inverter benchmarking separately

3.2 Main components

Both inverters and modules are represented by Operational KPIs and Features. The features can be divided in type-related or geographically related (countries, regions, KGPV zones [4], etc.), while the operational KPIs will depend on the PV plant data availability, including also the output of the CPN methodology (see Figure 4). The approach is visually explained in Figure 5, where the combination of different Operational KPIs and Features will help to get relevant statements and to support decision-makers in PV. As example, we compare one Operational KPI (Performance Ratio) versus Feature 1 (PV Module Type), for Feature 2 (Rooftop mounting) and Feature 3 (Temperate Climate).

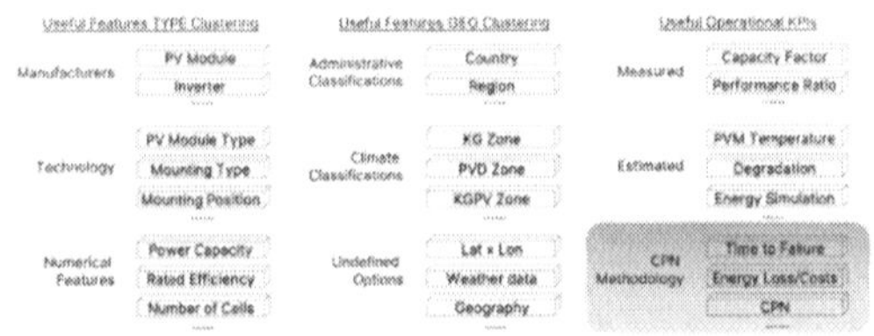

Figure 4: List of useful features (type and geo related) and operational KPIs (including CPN methodology)

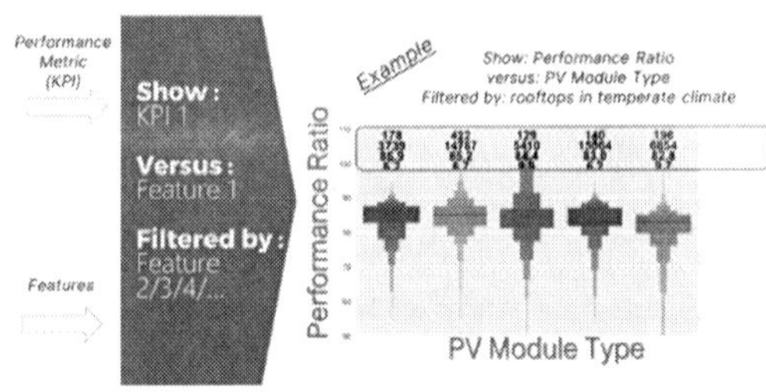

Figure 5: Benchmark KPI with several feature filters

3.3 Data architecture

The data architecture of the proof of concept component-benchmarking tool based on monitoring data (in this case, coming from the 3E Solar SynaptiQ monitoring platform) is presented in Figure 6. The database is divided into Features and Operational KPIs per plant and then unified after different data processing blocks for cleaning and extracting of statistics. Finally, linked by the same ID number, the All-in Features, All-in Timeseries, and All-in Statistics will provide the requested insights on-demand.

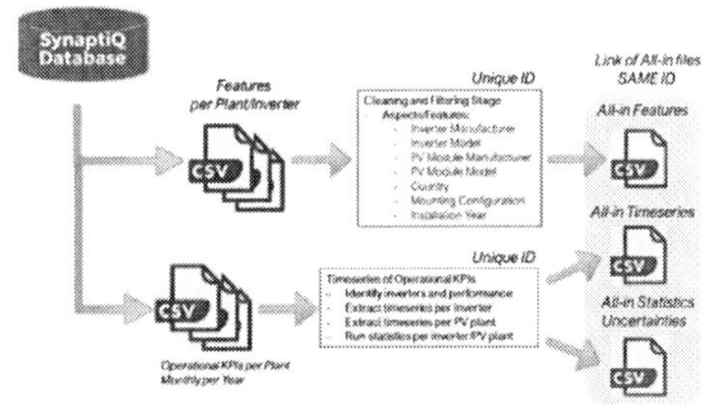

Figure 6: Simple visualization of the data architecture behind the component-benchmarking approach

4.1 Case study 1

In Figure 7, the performance of a case study of rooftop PV systems in a specific environment using operational data of 2019 are benchmarked. A portfolio of 7 PV plants is benchmarked to 65 PV plants using inverters of brand "C", and PV modules of brand "X". The performance ratio of this portfolio is under the average, which can be a sign of alert for the operators. The best performing PV system (composed of Inverter "D" and PV Module "Y") has shown 86.78% as performance ratio. Similar insights can be achieved by analyzing the Energy Performance Index (EPI) and Availability. The number of systems, inverters and devices connected can help to evaluate the uncertainties of the statements.

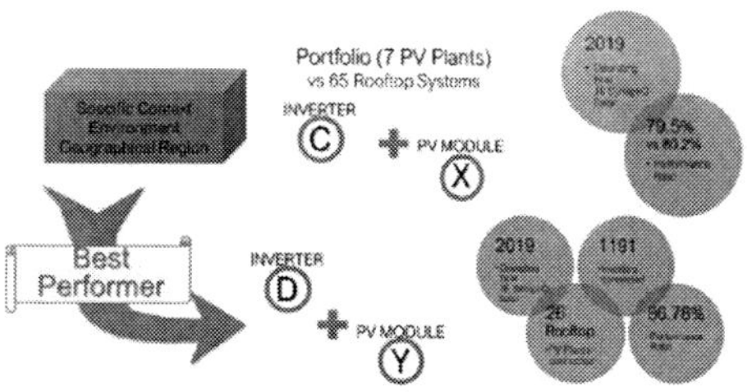

Figure 7: Example of the outcomes from a component-benchmarking of PV modules and PV inverters

4.2 Case study 2

The benchmarking tool can also allow to follow certain trends within the PV industry. For example, the DC/AC ratio of inverters installed globally each year in Figure 8 clearly shows a trend towards oversizing, a result of a shifting financial optimum due to declining PV module costs in combination with reduced solar energy tariffs. The benchmarking allows to further refine these results by filtering on features such as irradiation, climate zone, plant size, etc.

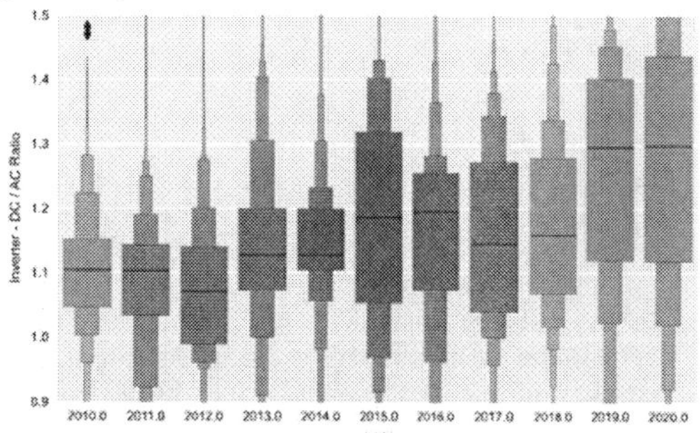

Figure 8: Example of the outcomes from component-benchmarking for PV market trends

5 CONCLUSION AND OUTLOOK

In this work, the data requirements for a first proof of concept of benchmarking of PV module manufacturers, PV inverters manufacturers and a combination of specific PV module' and inverter manufacturers were presented. Furthermore, the data structure needed for such benchmarking tool was illustrated, which comprises of three main blocks mentioned as All-in Features, All-in Timeseries and All-in Statistics. Finally, these methods were applied to benchmark PV modules and inverters in different countries using the database of the 3E SynaptiQ monitoring platform, consisting of over 6.5 GW and 10.000 connected PV plants.

In doing so, this work aims to be the cornerstone for solar PV component-benchmarking, but the discussed strategies could be easily extended to different industries (for example application to wind energy or storage components). However, the work is still ongoing and needs to be further extended and validated within the TRUST PV project the coming years. This will happen in collaboration with the consortium partners – EURAC, Baywa r.e., Enel Green Power, Innosea and Huawei – leading to interesting functionalities such as the Cost Priority Number (CPN) and optimal replacement cycles for predictive maintenance.

ACKNOWLEDGMENTS

This project has received funding from the European Union's Horizon 2020 research and innovation programme under GA. No. 952957.

REFERENCES

[1] "H2020 TRUST-PV project." https://trust-pv.eu/ (accessed Aug. 24, 2021).

[2] D. Moser, M. Del Buono, U. Jahn, M. Herz, M. Richter, and K. De Brabandere, "Identification of technical risks in the photovoltaic value chain and quantification of the economic impact: Economic impact of technical risks in photovoltaic plants," Prog. Photovolt: Res. Appl., vol. 25, no. 7, pp. 592–604, Jul. 2017, doi: 10.1002/pip.2857.

[3] "3E SynaptiQ Solar." https://www.3e.eu/synaptiq/ (accessed Aug. 24, 2021).

[4] J. Ascencio-Vásquez, K. Brecl, and M. Topič, "Methodology of Köppen-Geiger-Photovoltaic climate classification and implications to worldwide mapping of PV system performance," Solar Energy, vol. 191, pp. 672–685, Oct. 2019, doi: 10.1016/j.solener.2019.08.072.

This is a short version of the paper submitted to the journal Progress in Photovoltaics.
It has been selected by the Executive Committee of the EU PVSEC 2021 for submission to Progress in Photovoltaics.

38th European Photovoltaic Solar Energy Conference and Exhibition

RE-USE OF DECOMMISSIONED PV MODULES: OPPORTUNITIES AND TECHNICAL GUIDELINES

Arvid van der Heide[1], Loic Tous[1], Karsten Wambach[4], Jef Poortmans[1,2,3], Jan Clyncke[5], Eszter Voroshazi[1]
[1]: imec (Energyville campus), Thor Park 8320, 35Genk, Belgium
[2]: KU Leuven, department ESAT – ELECTA, Kasteelpark Arenberg 10, 3001 Leuven, Belgium
[3]: UHasselt, Martelarenlaan 42, 3500 Hasselt, Belgium
[4]: bifa Umweltinstitut, Am Mittleren Moos 46, 86167 Augsburg, Germany
[5]: PV CYCLE, Brand Whitlocklaan 114/5, 1200 Brussels, Belgium
arvid.vanderheide@imec.be

ABSTRACT: Since huge numbers of PV modules are expected to be discarded over the next decades, it is important to think about end-of-life management for those modules, like recycling and re-use. However, the re-use of decommissioned modules is a quite complex subject since there are concerns from technical, economic and legislative point of view. An evaluation of possible applications for second-hand modules shows that currently, the use of these modules in high-income countries is only interesting to repair PV systems (usually still receiving feed-in tariff) and to replace old modules to extend PV system lifetime. For developing regions in Africa and Asia, second-hand modules are interesting to build new small to medium size PV systems (often off-grid). The typical decommissioned module is a crystalline silicon glass-backsheet module from a utility power plant. Most modules originate from plants that have been partly damaged by severe weather or from repowered plants. Currently, technical requirements to qualify PV modules for re-use are lacking. In the legislation also a clear criterion is needed for a module to be considered as a functional product instead of waste, since it is not an easy yes/no situation like for a typical electronic device. In this paper, it is proposed to set a clear performance threshold at 70% of the original power. Also guidelines for a low-cost quality inspection and cost-effective module repair are given. With this, we aim to open the dialogue on a commonly accepted protocol and criteria. Currently, the worldwide re-use market size is estimated to be around 1 GWp/year, of which 0.3 GWp/year originating from Europe (mainly Germany, with Italy rapidly coming up). Many second-hand modules are shipped to developing countries without recycling facilities which might be a risk for disposal on the longer term. To create a healthy and sustainable market for re-use of modules it will be important that standards for re-use module evaluation become available and that the existing electronic waste legislation will be adapted for energy producing products like PV modules.

Keywords: Sustainability, Qualification and Testing, Decommissioned PV modules

1 INTRODUCTION

The cumulative installed PV capacity reached 623 GW at the end of 2019 and is expected to reach over 1 TW by 2025 according to the IEA PVPS [1]. The rapid steady increase of global PV power will lead to large amounts of PV module waste in the future. IRENA published an estimated cumulative waste of 1.7-8 million tons by 2030 [2], making clear that adequate solutions for end-of-life management of PV modules will be needed. There are several options to treat waste, from preparing for re-use to disposal as schematically shown in the waste hierarchy used in the European Waste Framework Directive, see Figure 1. Moving down in the triangle means moving to a less favourable option from an environmental perspective. By preparing PV modules for re-use, PV module waste can be turned into a product again, moving up through the red line that separates product and waste.

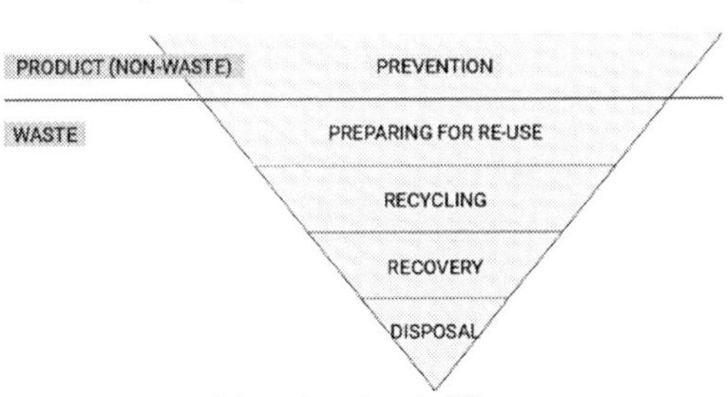

Figure 1: Waste hierarchy triangle [3].

In the European waste legislation, there is a strong focus on avoiding waste disposal as much as possible, and to stimulate the re-use of discarded products. However, from a technical, economic, and legislative point of view, the re-use of these decommissioned modules is not straightforward at all.

From a technical point of view, there can be concerns whether the module still meets the requirements for performance, safety and reliability, especially if some repair had to be done. An important problem concerning performance is that the functionality cannot easily be tested like for a radio or other electronic device, but involves equipment like a solar simulator. An even larger problem is that it is not obvious which is the lower limit of remaining power performance of a module to be acceptable for re-use.

From economic point of view, there are also important prerequisites for re-use of modules to be successful. Most importantly, there should be interesting applications where the re-use of modules should be preferable above using new ones. Since new modules are cheap and have higher efficiency than old modules, the choice to install second-hand modules instead of new ones is not obvious at all.

This paper addresses important technical issues of re-using PV modules, but since it is impossible to exclude the economic issues in this complex subject of module re-use, they will also be addressed where necessary. To put things in perspective, the current practices in the world of second-hand modules will also be discussed briefly, based on interviews conducted with major actors. Finally, an outlook for module re-use will be given.

2 THE PATH TO RE-USE OF PV MODULES

2.1 Re-use applications for decommissioned PV modules

For a successful re-use of decommissioned PV modules, the first thing is to have relevant applications. For several possible applications, the advantages and disadvantages are listed in Table I.

Table I: List of possible applications for re-use of PV modules with main advantages and disadvantages.

Applications for re-use of PV modules	Main advantages (+) and disadvantages (-) for the potential customer
Replacing damaged modules, usually due to severe weather. In case of subsidised PV plants, operators are often required to replace the damaged modules by (nearly) identical ones to avoid losing subsidies.	+ No loss of (high) subsidies, possibility to complete the installation again in an easy way. - Finding modules (nearly) identical to the damaged ones can be very difficult leading to high module costs.
Replacing all modules to extend PV plant operation beyond the initial design life.	+ Low module costs + Module dimensions adapted to existing infrastructure. - Low module efficiency and remaining lifetime, no/little warranty.
Using modules to create a new PV plant of commercial- or utility-scale.	+ Low module costs. - Low efficiency and lifetime, no/little warranty. - Limited savings: modules costs are only ~1/3 of total PV system costs.
Installing modules in new residential PV systems.	+ Low module costs. - Low efficiency, and lifetime, worse aesthetics. - Limited savings: modules costs are only ~1/4 of total PV system costs.
Export of modules to developing regions in Africa or Asia for installation in new small/medium sized PV systems (often off-grid).	+ Low module costs. - Low efficiency and lifetime, no/little warranty.

The disadvantages of installing second-hand modules in new residential, commercial/industrial or utility-scale PV plants in Europe are considered to outweigh any potential benefit. So the main applications for re-use of standard PV modules are: (1) replacing damaged modules in an existing PV system to ensure (nearly) identical replacement, (2) replacing modules of a plant to extend PV plant lifetime, (3) exporting modules to developing regions for installation in new small to medium size PV systems (often off-grid).

2.2 Available decommissioned PV modules

The next important issue is to know where decommissioned modules can be found, and which module properties can be expected.

In the figure below, it is shown how much power per module technology has been produced in the past:

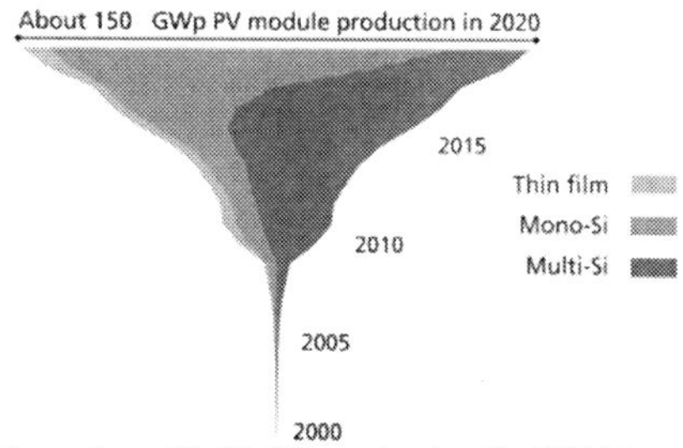

Figure 2: Annual worldwide PV production (in GWp) by technology [4].

It is clear that almost all currently installed modules are of crystalline silicon (c-Si) technology, thin film contributes only ~5%. The graph also shows that module production became much larger after 2010, which means huge amounts of discarded panels can be expected from around 2030 considering an average lifetime of 20 years.

Concerning the module construction, it is important to note that the decommissioned c-Si modules currently available have been made with a back sheet as rear cover, since glass-glass standard modules became more popular only in recent years. BIPV modules will be neglected in the following since their installation share is only ~1%, while they are often custom made and installed in small- to medium-sized systems which makes re-use inefficient.

In practice, almost all decommissioned PV modules have been removed from utility-scale power plants (> 1 MWp) since removal from smaller systems is inefficient. A first reason for removal can be a severe weather event (e.g. hailstorm, lightning strikes, etc.) destroying only part of the modules while other modules are still fine for re-use. A second reason can be the occurrence of repairable defects: the PV modules are removed under warranty and replaced by identical new ones. A third reason can be "re-powering" meaning that relatively old PV modules are all removed and replaced by latest generation PV modules with higher performance.

It should be mentioned that the 20 year lasting feed-in tariff subsidies still have a huge impact on the re-use of modules in Europe for now and the near future. On the one hand, repowering will be delayed until the feed-in tariff expires, on the other hand some rare module types can be sold for prices 10 times higher than new modules since people need them to replace damaged modules in PV systems with feed-in tariff to avoid losing subsidies.

2.3 Preparing decommissioned modules for re-use

The legislative definition of preparing for re-use is "checking, cleaning or repairing operations, by which products or components of products that have become waste are prepared so that they can be re-used".

The quality inspection and possible repair can be done either at the decommissioning site or at a treatment location with more verification and repairing facilities. Before removing and verifying the individual modules of a PV plant, the general input data should be collected (PV module types, dimensions, nominal electrical data, serial numbers, number of modules per string, etc.). The owner/operator of the decommissioning site may have additional information (current age, predominant defect types, current average module performance, etc.). In combination with monitoring results a good view on the general condition of the modules can be obtained.

During the quality inspection, the modules can be sorted into 3 different classes:

- Class 1: Re-use without further handling is possible
- Class 2: Further measurement and/or repair needed
- Class 3: Recycling

The process of inspection and sorting of modules for both location types is shown below:

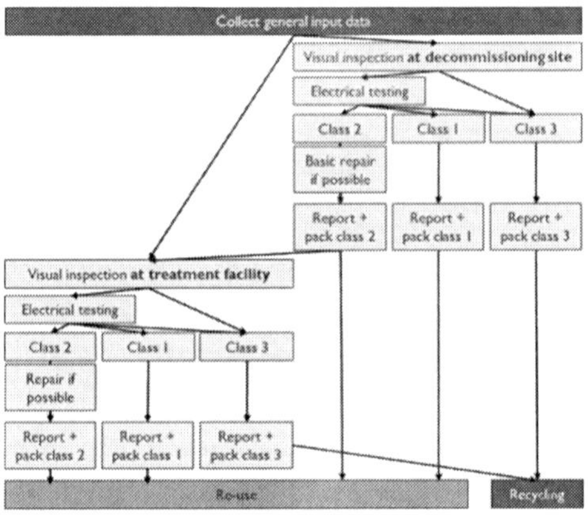

Figure 3: Quality inspection sequence to prepare PV modules for re-use performed at decommissioning site (light blue) or at a treatment facility (pink).

As mentioned in figure 3, any electrical testing should be done only after passing a quick visual inspection, since it would be a waste of time and money to do this on modules with severe visible defects. Table II shows an example of a visual inspection checklist.

Table II: Example of checklist for visual inspection

Visual inspection:

Glass broken	>10% of cells broken	Moisture or corrosion in module	Broken ribbons	Burn marks on cells or ribbons (hot spots)
□	□	□	□	□
Backsheet damaged (cracked, peels-off)	Delamination at edges	Snail trails	Air bubbles at module edges, on or between cells	
□	□	□	□	
Frame warped or broken	Junction box damaged or loose	Cables damaged or loose	Connector(s) damaged	Optional: check robustness of junction box, cables, connectors according to IEC61730, MST42
□	□	□	□	□

Only modules with either snail trails, damaged junction box, cables, local back sheet damage or connectors should be submitted to electrical testing while the presence of any other visual defect (broken glass, >10% broken cells, broken ribbon, burn marks, large area edge delamination, broken frame, etc.) should lead to modules being put directly to waste.

The most basic electrical testing of module performance is the measurement of short circuit current (I_{sc}) and open circuit voltage (V_{oc}) in daylight using a hand-held multimeter. These I_{sc} and V_{oc} can then be compared to the original module datasheet values. Since differences in I_{sc} and V_{oc} will only show very severe defects, it is strongly recommended to measure a full *I-V* curve for each module. Nowadays, portable *I-V* curve recorders are available that include irradiation and temperature sensors to correct to standard test conditions.

A difficult point here is to decide when the module is not (sufficiently) functional anymore. For usual electronic equipment it is easy to test on functionality, it either works or not. However, for a PV module the only way to judge its functionality is to set a lower limit for the remaining power, and this value will always remain a bit arbitrary. A proposal could be to take 70% of the original power for this lower limit. This unclarity in functionality up to now has also a very important consequences for the waste legislation, since non-functional products are considered to be waste.

After the electrical power of the PV module has been measured, the following minimum electrical safety tests should be performed using an isolation test recorder:

- Ground continuity tests with all frame parts
- Isolation resistance

The recommended tests have been described in IEC61730-2, MST13 (ground continuity, to check if all frame parts are electrically connected) and MST16 (isolation resistance, determination if the module is sufficiently well-insulated between current-carrying parts and the frame or the outside world). Modules failing these safety tests should be put in the recycling class, or alternatively to a class for lower system voltages (<60 V).

Modules that were sent to class 2 need to be repaired before re-use. A list of defects with repair solutions and repair costs is presented in table III.

Table III: Module defects repairable in a reasonable way, with solutions and costs (for Western Europe) estimated based on experience of field operators.

Defect	Repair solution & costs/module
Failed bypass diode(s) in junction box	Replace bypass diode(s) by comparable ones. Not possible for junction boxes filled with potting material. Costs 20 euro.
Failed bypass diode(s) with potting in junction box	Remove junction box and replace by new one (including diodes) not requiring potting. Costs 60 euro.
Junction box with internal or external damage	Remove junction box and replace it by a comparable new one. Costs 60 euro.
Damaged cables	Replace cables (including connectors). Costs 20 euro.
Damaged or missing cable connector(s)	Mount new connector(s) on cables. Costs 5-10 euro.
Cracked backsheet over entire surface	Coating application on top of the original backsheet: Still under development.

The listed repair costs would be too high if every module would need to be repaired, given the low prices of brand new modules. However, usually only a few percent of the modules need to be repaired, except for the cracked backsheet defect. Since these repair costs can be distributed among all the modules, repair is still cost-effective. Full area backsheet cracking is related to material composition, so it occurs on every module. Several companies and institutes are working on promising repair solutions, but it is not yet clear if these will be widely applied in practice .

3 CURRENT RE-USE MARKET OVERVIEW

The PV re-use market is growing with around 15 companies worldwide, trading an estimated 1 GWp/year, of which around 0.3 GWp/year originates from Europe. These estimates are based on discussions with actors in the field of second-hand modules, since neither site decommissioning nor PV reselling is being registered. In Europe, 5 German companies are active in this sector, which is logical since Germany was the earliest adopter of PV, and has the largest installed power of Europe as well. Most of the other big players are registered in China. Decommissioned PV modules with lower performance and lifetime expectations are mostly in demand in developing regions. These include African countries but also certain regions of the Middle East and South East Asia which seek extreme low-cost PV systems with less quality/aesthetic requirements. A rather small (around 10 MWp/year) and temporary market still exists in (Western) Europe where feed-in tariff regulations often require "similar" replacement modules in case of damage. Overall, re-used or repaired PV modules are not competitive for new residential, commercial and utility-scale PV installations in high-income countries. PV module re-use in these countries is today only viable for niche applications and its uptake will depend on the issue if/how CO_2 footprint will be integrated in electronic product cost.

The trend to ship second-hand modules to countries without PV recycling facilities and adequate waste regulations increases the risks on disposal instead of recycling. However, the same is valid there for new modules.

4 CONCLUSIONS AND OUTLOOK

The evaluation of possible applications for second-hand modules showed that the use of these modules in high-income countries is only interesting to repair PV systems that still receive feed-in tariff, and maybe to replace old modules to extend PV system lifetime. For developing regions, second-hand modules are interesting to build new small to medium size PV systems that are often off-grid. The typical decommissioned module is a crystalline silicon glass-backsheet module from a utility power plant. Most modules originate from plants that have been partly damaged by severe weather or from repowered plants. Guidelines for a low-cost quality inspection and cost-effective module repair have been given. To start with, a visual inspection excludes modules with shattered glass, bended frames, hot spot damage, etc. for which it does not make sense to do any further inspection. Since it is not cost-effective to spend much time and money on checking the modules intended for re-use, only an outdoor/indoor *I-V* measurement and basic safety tests can be performed.

An important difficulty is the criterion for a module to be still functional, since it is not a clear easy yes/no criterion like for a typical electronic device. In this contribution, a value of 70% of the original power is proposed to open the discussion with the different actors. We believe that a common set of testing requirements and performance thresholds in a standard is critical to ensure quality, that can trigger acceptance from users and create a clear boundary with modules that should be considered as waste. Currently, the worldwide market size for second-hand modules is estimated to be ~1 GWp/year, of which 0.3 GWp/year originating from Europe (mainly Germany). Many second-hand modules are shipped to developing regions without recycling facilities, which might finally result in module disposal.

It is important to create international standards for the quality inspection of second-hand modules. Developments in this direction are made within the European project CIRCUSOL, within the Photovoltaics Quality Assurance Task Force (PVQAT) task group 15 and by Sustainable Electronics Recycling International (SERI). Recently the point has also been raised within IEC TC82, the first goal is to write a Technical Report on the subject.

Concerning waste legislation, the EU is currently the only region that has included PV modules. In the EU, PV modules are considered to be electronic waste, but even this legislation should be more adapted to PV modules (originally written for usual electronics instead of energy-producing equipment). Of course, it is important to create also adequate PV waste legislation in non-EU countries.

Looking 10-20 years ahead, there will be important positive changes for the potential of module re-use, since decommissioned modules:

- Will be mostly of glass-glass type: less risks (fewer cell cracks, lower degradation, no backsheet issues).
- Will be mostly of multi-wire type: power less sensitive to cell cracks.
- Will probably not be much less efficient than new ones.
- Will come from systems without feed-in tariff: no need to wait 20 years before repowering.
- Will be probably cheaper compared to new ones than now, due to more availability and limited possibility for future price lowering for new modules.

While the better quality and reliability of decommissioned modules mentioned in the first 3 points are positive for "re-users", they are likely to induce a delay in repowering. However, assuming that repowering will finally take place anyway, it is expected that the potential for re-use of modules will still increase on the longer term.

5 ACKNOWLEDGEMENTS

The authors thank Mónica Alemán, Ivan Gordon, Bart Vermang, Sebastien Lizin, Boris Mertvoy and Kerstin Baumann for their valuable contributions to this work.
This work has been funded in part by the CIRCUSOL project, that has received funding from the European Union's Horizon 2020 research and innovation program under grant agreement 776680.

6 REFERENCES

[1] G. Masson and I. Kaizuka "Trends in Photovoltaics Applications 2020", Report IEA PVPS Task 1 (2020)
[2] IRENA report "End-of-Life Management Solar Photovoltaic Panels" (2016)
[3] https://ec.europa.eu/environment/waste/framework/ about the Directive 2008/98/EC on waste (Waste Framework Directive)
[4] S. Phillips and W. Warmuth, "Photovoltaics Report", Fraunhofer ISE/PSE Projects, (2021)

FOR A MORE SUSTAINABLE FUTURE – MYLAR® UVHPET™ BACKSHEETS CONTAINING POLYMERIC POST-CONSUMER RECYCLED CONTENT

Emily Parnham[1], Steven Davies[1], Dmytro Stratiychuk[1], Sven Kaeselau[2]

DuPont Teijin Films

The Wilton Centre, Redcar, TS10 4RF, UK[1]

BP 1681, Rue General Patton, Contern, Lux[2]

ABSTRACT: Over the last 10 years there has been an exponential increase in photovoltaic (PV) installations [1]. During this period of rapid growth there has been limited focus on the end of life strategies for these modules, but now with waste from the PV industry forecast to grow to 1.7 million tons by 2030 and approximately 60 million tons of cumulative PV panel waste by 2050 [2], there is an urgent need to address both the production materials used and their eventual disposal options, to ensure the industry is thoroughly applying the principles of eco design and the circular economy. This paper will focus on the development using post-consumer recycled polyethylene terephthalate (rPET) as a feedstock in the DuPont Teijin Films LuxCR™ process where the rPET is converted back by glycolysis [3] into the monomer unit bis(2-hydroxyethylterephthalate) or BHET which is chemically indistinguishable from virgin monomer. This base monomer is then repolymerised and is subsequently converted into a wide range of biaxially-oriented polyethylene terephthalate (BOPET) films maintaining the same mechanical, physical and optical properties as virgin material. The use of glycolysis as a relatively low carbon form of chemically recycling offers the PV industry not only a viable end of life option for the backsheet material, but also the ability to incorporate recycled material back into PV modules, boosting the circularity of the photovoltaic market and contributing to the EU's plastic strategy goal of using 10 million tonnes of recycled plastic by 2025 [4].

Keywords: Backsheet, Recycling, PET

1 INTRODUCTION

The huge increase in waste demands means that all parts of the PV industry must assess the products they manufacture and drive innovation that reduces carbon foot print and increases circularity at the end of module service life. This need for the industry to take more ownership of the end of life of its products is increasingly being underpinned by legislation with the European WEEE Directive (Directive 2002/96/ EC) and the revised WEEE Directive of 2014 establishing a legal framework for the collection and treatment of PV waste based on the principle of extended-producer-responsibility.

PV modules typically have between 5 and 10% polymeric material by weight [2] comprising encapsulate and backsheet material. For the backsheet, Mylar® UVHPET™ already provides a halogen free, lower carbon alternative to fluorinated products and does not represent an issue in terms of hazardous materials in high temperature waste processing such as incineration and pyrolysis. PET is also the most widely recycled plastic in Europe with over 1.5 million tonnes of recycled material being processed through best in class mechanical and chemical recycling processes. DuPont Teijin Films have consistently driven innovation in the polyester industry from the initial invention and development of the polymer in the 1940s through to its current innovative sustainability strategy. The development of the LuxCR™ process utilises the available post-consumer recycled polyethylene terephthalate (rPET) and incorporates it into high quality biaxially-oriented polyethylene terephthalate (BOPET) finished films maintaining optical, physical and mechanical properties.

The three main objectives when developing the LuxCR™ process were:

1. Enabling the use of rPET in high temperature packaging applications
2. Ensuring the mechanical properties were comparable to virgin material to allow the use in demanding, long life non-packaging applications eg photovoltaic backsheets
3. To open the possibilities of using lower quality recyclate which does not currently have an economically viable use

The use of post-consumer recycled content in PET films is not new with well-established markets for mechanically recycled bottle polymer being used in thermoformed trays, fibres and to a lesser extent bottles. Although mechanical recycling will continue to play an important role in the circular economy, it does have limitations with regards to the degradation of optical, physical and mechanical properties. This limits the amount of times an item can be recycled and ultimately excludes its use in highly demanding applications such as photovoltaics.

The LuxCR™ process addresses this issue by depolymerising mechanically recovered PET via the process of glycolysis (Figure 1), back into the monomer unit bis(2-hydroxyethylterephthalate) or BHET, which is chemically indistinguishable from virgin monomer. This base monomer is then repolymerised and is subsequently converted into a wide range of BOPET films indistinguishable from virgin material. Contamination is removed during the process through a combination of monomer and polymer filtration units and by vacuum extraction which runs for several hours at temperatures between 270-300°C. The resulting range of products have the same mechanical, physical and optical properties as virgin material.

Perhaps the biggest benefit of the LuxCR™ process is the potential to open up a wider range of used plastics which previously had been deemed to be unrecyclable due to their low molecular weight (M_w). The molecular weight of a plastic determines its mechanical and physical properties to a large extent. For PET particularly, lower M_w corresponds to increasing brittleness of the polymer. Up until now, for PET, this has necessitated using reclaimed high M_w bottle polymer to give sufficient mechanical properties of the recycled material. The LuxCR™ process would open the way to reclaim lower quality PET from sources such as fibre and food trays.

The reduction in M_w is one of the rate limiting factors within the mechanical recycling infrastructures and a key advantage of the depolymerisation processes is that low M_w material can be transformed into a high M_w material in significantly less time using less energy and costing less money than other options such as solid-state polymerisation.

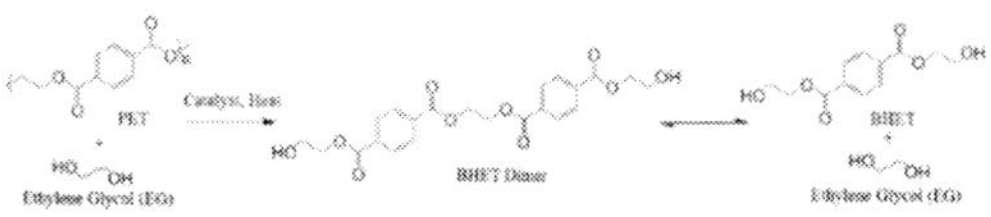

Figure 1: Schematic of glycolysis of PET to form BHET and EG

2 RESULTS AND CONCLUSION

A study was carried out to investigate the impact on M_w and colour of mechanically recycling PET polymer. Both properties are important for PET used in PV backsheet applications. Customers expect to purchase a clear or white backsheet with a consistent colour ie $\Delta E \leq 1$. The M_w must remain high to ensure the film is not brittle and can be processed in the BOPET filming process as well as ensuring a hydrolysis resistant [5] PET end product.

PET polymer chip was processed through an extruder 5 times to simulate the process of mechanically recycling PET. Figure 2 shows graphically the yellowing of the film by plotting the Yellowness Index (YI D1925) values against number of extrusion cycles. Even after only one extrusion cycle there is a visible change in colour. Figure 3 shows a photo of the polymer after 1 to 5 extrusion passes. This visually shows the yellowing effect that multiple cycles of mechanical recycling has on PET.

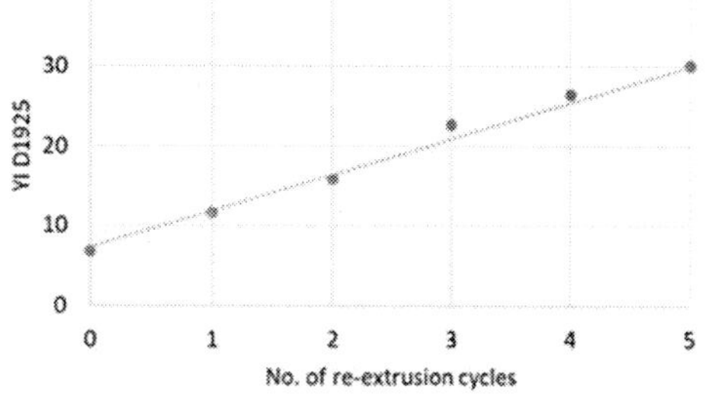

Figure 2: Graph of number of extrusion cycles against YI D1925

Figure 3: Polymer re-extruded 1 to 5 times increasing in extrusion cycles from left to right

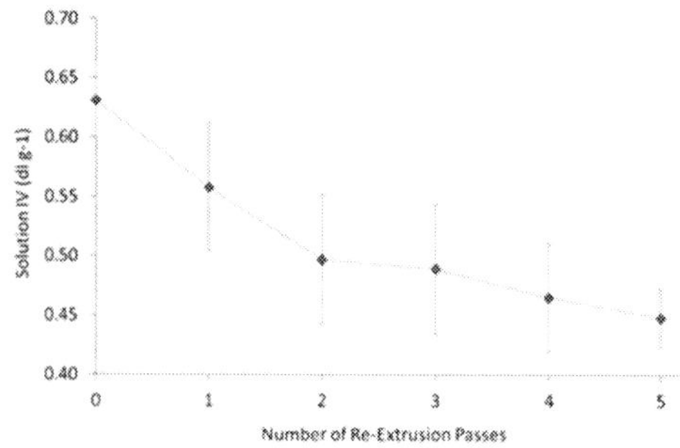

Figure 4: Graph of number of extrusion cycles against IV

Figure 4 shows the drop in M_w (measured as a solution Intrinsic Viscosity) as number of extruder cycles increases. If this material were to be used in a PV backsheet, we would expect a significant reduction in mechanical properties and hydrolysis resistance leading to a material not fit for purpose.

In comparison, a 19 micron clear film was synthesised with and without rPET through the LuxCR™ process. The rPET was depolymerised through the process of glycolysis then combined with virgin monomer and EG. The BHET and virgin monomer were added in a 1:1 ratio by weight. Esterification was followed by polycondensation to produce PET polymer which was then made into BOPET film. Mechanical properties of these films were measured and the results are graphed in Figure 5 below. The data confirms that there is no significant difference between the PET film made with 100% virgin starting material verses that made with 50% rPET via the LuxCR™ process.

L, a*, b* were measured for the two films and ΔE calculated giving a value of 0.84. A value of $\Delta E \leq 1$ is classed as not being a perceivable colour difference by the eye. As the value of ΔE is < 1 we can conclude that the two films are indistinguishable with regards to colour.

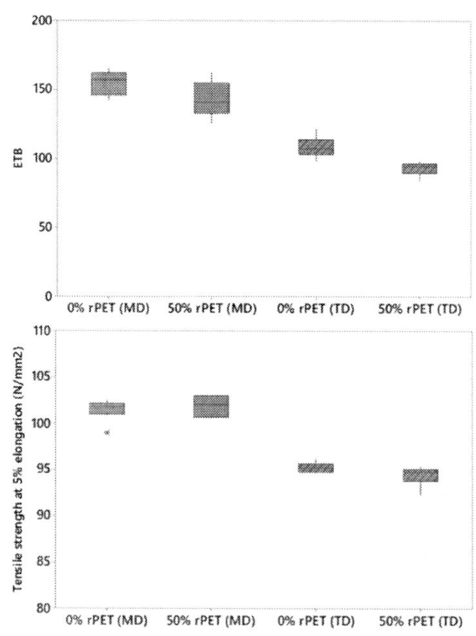

Figure 5: Graphs to compare the mechanical properties of PET films without and with 50% rPET manufactured using the LuxCR™ process

A 33% rPET Mylar® UVHPET™ SPV9R PV backsheet material was synthesised via the LuxCR™ process. This film was tested against a reference sample and molecular weights compared and found to be comparable. Hydrolysis was tested via Damp Heat Test (DHT) 85°C/85% humidity. As would be expected with two materials with comparable molecular weights, the hydrolysis performance was the same as illustrated in Figure 6.

These films have been exposed to accelerated UV aging accordance to IEC 62788-7-2 A3 conditions for 4000hrs. Figure 7 shows that whilst thet REF sample started more yellow – the final b* value after accelerated weathering of the REF and the 33% rPET Mylar® UVHPET™ SPV9R PV backsheet material were the same. Figure 8 shows the elongation to break data for the two materials and again there is no difference between the ref and the rPET material after 4000hrs accelerated weathering.

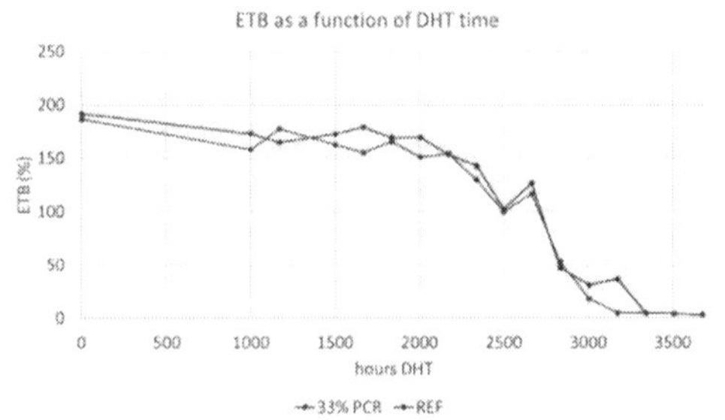

Figure 6: Graph to compare the mechanical properties of Mylar® UVHPET™ films without and with 33% rPET manufactured using the LuxCR™ process

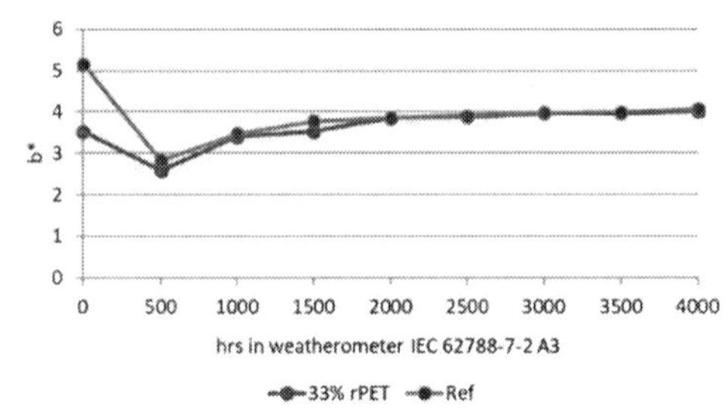

Figure 7: Graph to compare colour change of Mylar® UVHPET™ films without and with 33% rPET manufactured using the LuxCR™ process after 4000hrs of accelerated weathering according to IEC 62788-7-2 A3

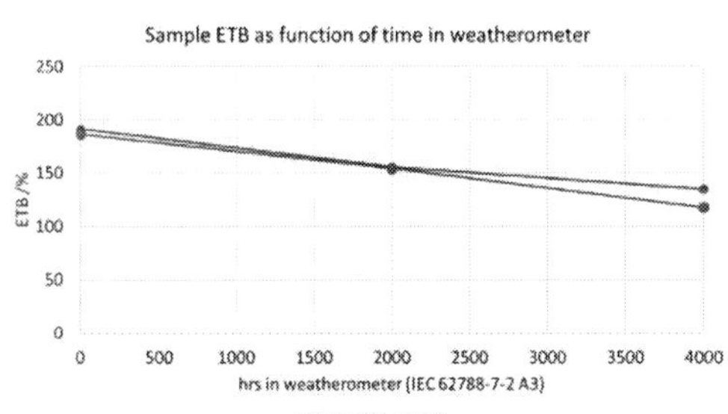

Figure 8: Graph to compare mechanical properties of Mylar® UVHPET™ films without and with 33% rPET manufactured using the LuxCR™ process after 4000hrs of accelerated weathering according to IEC 62788-7-2 A3

3 CONCLUSIONS

In conclusion this paper has covered the advantages of chemical recycling vs mechanical recycling with regards to maintaining a high M_w and hence excellent mechanical properties which are indistinguishable from film made using 100% virgin starting materials.

The use of glycolysis potentially offers the PV industry not only a viable end of life option for the backsheet material, but also the ability to incorporate recycled PET backsheet material back into PV modules, boosting the circularity of the photovoltaic market and contributing to the EU's plastic strategy goal of using 10 million tonnes of recycled plastic by 2025 [4]. The carbon foot print is dramatically reduced by the use of recycled content with an estimated carbon foot print saving of 1.4Kgs for every 1Kg of virgin material replaced [6].

An additional benefit of the use of recycled content in PET backsheet applications is the positive message given to consumers that plastic packaging can be diverted either from landfill or marine litter and can be recycled back into technically demanding long life applications.

This innovative work has been awarded the Inter Solar Award 2021.

4 REFERENCES

[1] End-of-Life pathways for photovoltaic backsheets, Fraunhofer Institute for Environmental, Safety and Energy Technology UMSICHT, Oberhausen, 2017.

[2] IRENA/IEA PVPS Task 12, End-of-life Management: Solar Photovoltaic Panels, 2016.

[3] A.Sheel, D. Pant; Recycling of Polyethylene Terephthalate Bottles, Plastics Design Library, 2019, 61-84.

[4] European strategy for plastics – voluntary pledges, published: 04/03/2019.

[5] K. Looney, B. Brennan; Modelling the correlation between DHT and True field lifetimes for PET based backsheet, PVSEC 2014.

[6] Results stem from the "Carbon Footprint Calculator" of the IFEU – Institute for Energy and Environmental Research Heidelberg.

ENERGY PAYBACK TIME OF PHOTOVOLTAIC ELECTRICITY GENERATED BY PASSIVATED EMITTER AND REAR CELL (PERC) SOLAR MODULES: A NOVEL METHODOLOGY PROPOSAL

Marc Salibi[12], Frederik Schönberger[12], Qendresa Makolli[12], Erion Bousi[12], Saker Almajali[12], Lorenz Friedrich[2]
[1]Albert-Ludwigs-Universität Freiburg, Fahnenbergplatz, 79085 Freiburg im Breisgau
[2]Fraunhofer-Institut für Solare Energiesysteme ISE, Heidenhofstraße 2, 79110 Freiburg im Breisgau

ABSTRACT: Renewable energy (RE) capacity is projected to surge to an 85% share of global electricity generation by 2050, the photovoltaic (PV) share specifically is expected to increase from 1% to 22%. Increasing shares of RE in the grid mix will influence energy performance indicators, such as energy payback time (EPBT). To accurately calculate and interpret energy performance indicators, these influences must be captured. In this paper, the IEA PVPS Task 12 methodology to calculate EPBT and non-renewable EPBT (NR-EPBT) is applied. The method is evaluated quantitatively, based on the implications introduced by the grid efficiency parameter. A modified methodology for calculating EPBT (M-EPBT is proposed and applied within a range of three global average grid efficiency scenarios. M-EPBTs are found to be significantly lower for all scenarios. In the mid scenario (ηaGlobal = 33%), M-EPBT values are compared to the respective EPBTs for three installation locations, which are lowered by following magnitudes: -29% for Rio De Janeiro (Brazil), -21% for Ottawa (Canada), and -23% for Catania (Italy). Future EPBT projections until 2050 show that the M-EPBT can represent the expected energy intensity improvements of the PV system and balance of system (BOS) technologies without an altering effect of local grid efficiency at the PV system installation location more effectively.
Keywords: Energy payback time; PERC; grid efficiency; energy performance

1 INTRODUCTION

Today, more than 90% of the photovoltaic (PV) market share is dominated by crystalline-silicon wafers, with aluminum back surface field (Al-BSF) identified as the most common industrial process. However, passivated emitter and rear cell (PERC) technology is gaining more attention and is predicted to hit a market share of almost 60% by 2027 [1].

To compare the energy performance of different PV module technologies and system setups, but also to compare different energy generation technologies, the energy payback time (EPBT) is a widely applied indicator to illustrate energy performance. EPBT of mono-crystalline PV systems has decreased by 12% over the last 24 years as the cumulative production doubled, indicating a powerful learning rate for PV systems, and providing a positive energy performance assessment in comparison to other energy sources [2]. EPBT of PV systems is expected to keep on decreasing as the technology itself enhances.

Mainly, three parameters affect the EPBT values: cumulative energy demand (CED), the installation location's grid efficiency, and the PV system's specific yield. In this paper, the currently followed methodology of the International Energy Agency (IEA) PVPS Task 12 for calculating PV system's EPBT and non-renewable-EPBT (NR-EPBT) is evaluated quantitatively, based on the implications introduced by the considered grid efficiency [3], and thereafter, a new methodology for calculating the EPBT is proposed. It should be noted that the proposed methodology is not limited to PERC solar modules specifically, but can be utilized for all PV module technologies, through altering the related parameters accordingly.

2 METHODOLOGY

2.1 EPBT and NR-EPBT methodology

The IEA PVPS Task 12 defines EPBT as the period, captured in years, required for a renewable energy system (PV system in our case) to generate the same amount of energy (primary energy equivalent) to what was used to produce the system itself [3]. Equation (1) illustrates the integral elements of the EBPT computation methodology [3].

$$EPBT=\frac{(CED_{mat}+CED_{manuf}+CED_{trans}+CED_{inst}+CED_{EOL})}{\left(\left(\frac{E_{agen}}{\eta_G}\right)-CED_{O\&M}\right)} \quad (1)$$

CED_{mat}: CED (in MJ) to produce materials comprising PV system
CED_{manuf}: CED (in MJ) to manufacture PV system
CED_{trans}: CED (in MJ) to transport materials during lifecycle
CED_{inst}: CED (in MJ) to install the system
CED_{EOL}: CED (in MJ) for end-of-life management
E_{agen}: Mean annual electricity generation (in $kWh_{electric}$)
$CED_{O\&M}$: CED (in MJ) for operation and maintenance
η_G: Grid efficiency, primary energy to electricity conversion at the demand side ($\frac{kWh_{electric}}{MJ}$)

As can be drawn from equation (1), the EPBT accounts in the numerator for the energy that is embedded in the PV system, including the CED for material, manufacturing, transport, installation as well as for end-of-life energy expenses. In the denominator, the mean annual energy generation is converted by the installation location local grid efficiency to primary energy equivalents (PE-eq.) with the energy demands for operation and maintenance ($CED_{O\&M}$) being subtracted.

Besides the EPBT, the IEA PVPS Task 12 proposed the NR-EPBT as an indicator to capture a PV system's energy performance. The NR-EPBT is calculated analogously to the EPBT methodology, except replacing primary energy (PE) with non-renewable primary energy (NR PE) for both the manufacturing and generation side. NR-EPBT quantifies the time needed to compensate for the employed NR PE in a PV system's life cycle. The calculation of EPBT and NR-EPBT do not differ greatly in the case of country or region dominated by non-renewable energy sources. However, in the case of high renewable energy (RE) share in the grid mix, these values differ significantly, and it seems impossible to distinguish in different EPBT analyses between technological progress in efficient PV production technology and share of RE (or NR) in the production and installation locations' grid mix. In this context, due to this methodological implication, the NR-EPBT is not seen as a relevant indicator for energy performance within this work and subsequently not further applied in the following analyses.

2.2 Grid efficiency

The electrical yield of a PV system is dependent on its installation location with respect to local irradiation and climate patterns [3]. Apart from that, EBPT is additionally biased by the average grid efficiency at each individual installation site. The grid efficiency serves as a conversion rate to translate generated PV electricity to PE-eq. The grid efficiency is calculated with the following equation as taken from [4]:

$$\eta_{G,I} = \frac{E_{grid,I}}{CED_{E,I}} \quad (2)$$

$CED_{E,I}$: PE-eq. (in MJ) for average electricity mix of the grid at installation location I to generate $E_{grid,I}$
$E_{grid,I}$: Electric energy supplied by the grid at installation location I

In that context, $CED_{E,I}$ is the PE-eq. for the average electricity mix at the installation location. It is calculated as the sum of the CEDs of all energy sources in the grid mix, as given by the Ecoinvent CED method's subcategories. This approach of summing the CEDs is indeed problematic, since Ecoinvent states that: 'We refrain from giving an aggregated total of the CED-indicators' [5]. Due to the different approaches followed at each subcategory, summing the CEDs can lead to energetic inconsistencies. An increased share of RE in the grid mix can lead to higher grid efficiencies, because of an intrinsic considered 100% conversion efficiency for RE. Conventional electricity generation technologies are accounted with fundamental different conversion efficiencies. In the case of coal power, the upper heating value of the feed stock is accounted as PE-eq, for nuclear power the PE-eq. in the isotopes. This inconsistent approach results in much lower conversion efficiencies compared to the assumed 100% efficiency for RE generation in the Ecoinvent database. Hence, when PV systems are manufactured at one location and installed in different locations, electricity grids with high RE deployment lengthen EPBT values [2,4].

At the manufacturing location, grid efficiency influences embedded CED of the PV products. As PV and balance of system (BOS) manufacturing lines are fed with electrical and thermal energy, power plants convert PE upstream the production line to secondary energy carriers, such as electricity and heat. Subsequently, assuming a steady manufacturing process at a particular production location and increasing average grid efficiency, lowers the CED of PV and BOS manufacturing and therewith shortens PV systems' EPBT [6].

Consequently, EPBT implies two contrary effects with respect to increasing RE penetration in future power grids. First, increasing the grid efficiency reduces module and BOS' CED in PV manufacturing. Second, increasing grid efficiency alters the translation of output electric energy to PE-eq on the generation side, resulting in longer EPBT. Considering the stated opposing trends, EPBT might not be able to illustrate improving energy performance caused by future module efficiency and manufacturing CED improvements adequately [2,7].

2.3 EPBT dataset

All data used for this work was taken from an attributional life cycle assessment (LCA) study of Friedrich et. al [4], including the life cycle inventory (LCI) of a 15 kWp Czochralski Passivated Emitter Rear Cell (Cz PERC) system and respective CEDs for production and operation, as well as location specific irradiance and grid efficiency parameters necessary for yield assessment.

In order to compare the EPBT values, China as a PV system production location is considered. China has the highest manufacturing share of PV modules and BOS components [2].

3 RESULTS

3.1 Novel approach for EPBT methodology

As stressed by Friedrich et al. [4] and Raugei [7], grid efficiency tends to have a significant influence on the amount of generated PE-eq. on the generation side of PV systems. To contradict the distorting effect of increasing RE share on the EPBT indicator, the M-EPBT is proposed. In this light, it is the novel approach's goal to maintain the energy saving implication of high grid efficiency in manufacturing, in other words lowering the CEDs in the production of PV systems, and to exclude its payback time lengthening effect on the generation side. To fulfill the latter outlined requirements, the location dependent grid efficiency (η_G) in the denominator (equation (1)) is substituted by the average global grid efficiency ($\eta_{gGlobal}$) in equation (3). The concept of average global grid efficiency is further explained later in this section.

$$\text{M-EPBT}= \frac{(CED_{mat}+CED_{manuf}+CED_{trans}+CED_{inst}+CED_{EOL})}{\left(\left(\frac{E_{agen}}{\eta_{aGlobal}}\right) - CED_{O\&M}\right)} \quad (3)$$

CED_{mat}: CED (in MJ) to produce materials comprising PV system
CED_{manuf}: CED (in MJ) to manufacture PV system
CED_{trans}: CED (in MJ) to transport materials during lifecycle
CED_{inst}: CED (in MJ) to install the system
CED_{EOL}: CED (in MJ) for end-of-life management
E_{agen}: Mean annual electricity generation (in $kWh_{electric}$)
$CED_{O\&M}$: CED (in MJ) for operation and maintenance
$\eta_{aGlobal}$: Average global conventional grid efficiency $(\frac{kWh_{electric}}{MJ})$

The proposed methodology alteration shall implement following principles in the M-EBPT indicator:

a. M-EPBT magnitudes shall be significantly less influenced by the electricity grid mix at the installation location, especially by the EPBT lengthening impact of RE sources.
b. Energy savings in the CED for manufacturing will be considered. Improvements in grid efficiency at the production location, due to increasing RE penetration, will lower CED for manufacturing processes.

c. In contrast to NR-EPBT, the proposed method considers RE that is consumed in the production phase.
d. Module efficiency improvements will be depicted in the indicator. Alteration of generated PE-eq by increasing grid efficiency no longer compensates module and energy efficiency improvements in manufacturing.
e. To develop an indicator which allows facile intercountry comparison, $\eta_{aGlobal}$ is assumed constant for all installation locations.

In other words, the proposed indicator illustrates how many years it will take for a PV system to produce as much electricity as could be produced by the average global electricity generation sources, using the same amount of PE.

With respect to the fact that global grid mixes will develop toward 100% RE in the future, it may seem contradicting that the generated energy is converted partly to NR energy-eq in the proposed method. However, given that from a theoretical perspective, RE (e.g., PV systems) still replaces formerly installed NR energy in future highly RE grids, as similar to the argumentation for NR-EPBT in the IEA Task 12 methodology guidelines [3]. Hence, the M-EPBT represents a viable method to explicitly illustrate the strict substitution of NR electricity generation by RE sources.

To incorporate the above explained concept numerically, $\eta_{aGlobal}$ is defined within the following sensitivity range:

i) low scenario $\eta_{aGlobal} = 26\%$
ii) mid scenario $\eta_{aGlobal} = 33\%$
iii) high scenario $\eta_{aGlobal} = 40\%$

$\eta_{aGlobal}$ is implemented within a sensitivity span to account for the fact that a reliable value of the average conventional global grid efficiency of conventional generation was not found in literature. As mentioned in [4], the global average may be approximated with the average efficiency of thermal (conventional) power plants lowered by transmission losses. Having this in mind, we defined the mid scenario with $\eta_{aGlobal} = 33\%$ and the respective low and high scenarios with a deviation of +/- 7%, which is aligned to the conventional grid efficiency deviation span stated by Ritchie et al. [8].

Following the Ecoinvent LCA data and CED methodology, it is known that certain locations, due to low efficient conventional power generation, are characterized by significantly lower grid efficiencies, such as Jaipur, India (Indian grid) with 17.5% and Cape Town, South Africa with 21.9 % [4]. Nevertheless, these locations represent a minority and are expected to increase efficiency in the future due to technological improvement. Moreover, there exists locations with high grid efficiencies, such as Kinshasa, Republic of Congo, where the grid efficiency is 73.5%. EPBT values of such high grid efficiency locations are set to be lowered with the new methodology. Within this approach, the resulting M-EPBT allocates comparable energy performance values for PV systems installed at various locations, with varying PV electric yield (due to solar irradiation), and varying local grid efficiencies. An overview of the input data for the EPBT and M-EPBT is presented in table I.

Table I: Parameters to assess energy payback indicators (EPBT and M-EPBT) for the installation locations Ottawa (Canada), Rio de Janeiro (Brazil), and Catania (Italy)

		Dependency: Installation location		
Installation Location of the PV System	**Energy Performance Indicators**	**Grid efficiency** *(kWh/ kWh)*	**aGlobal-grid efficiency** *(kWh/ kWh)*	**PV yield** *((kWh/kW)/a)*
Ottawa, Canada (Canadian grid)	EPBT	42.8%		1,260
	M-EPBT upper		40%	
	M-EPBT mid		33%	
	M-EPBT lower		26%	
Rio de Janeiro, Brazil (Brazilian grid)	EPBT	46.6%		1,438
	M-EPBT upper		40%	
	M-EPBT mid		33%	
	M-EPBT lower		26%	
Catania, Italy	EPBT	40.4%		1,586
	M-EPBT upper		40%	
	M-EPBT mid		33%	
	M-EPBT lower		26%	

3.2 M-EPBT results

As can be extracted from figure 1, the M-EPBT magnitudes range from 0.67 years to 1.02 years in Catania, Italy, 0.72 years to 1.12 years in Rio de Janeiro, Brazil and between 0.83 years to 1.29 years in Ottawa, Canada. Besides the M-EPBT high scenario for Catania, Italy, where $\eta_{aGlobal} = \eta_G$, the M-EPBTs display distinctly lower values compared to the respective EPBTs that are plotted on the left side of the x-axis in figure 1 for the respective three installation locations.

The broadest difference between EPBT and M-EPBT with respect to the mid grid efficiency scenario ($\eta_{aGlobal} = 33\%$) is found in Rio De Janeiro with a M-EPBT decrease of 29%, whereas Catania and Ottawa range from -21% to -23% compared to EPBT. Energy performance influencing factors, such as location specific yield and total CED for system manufacturing are considered equally for EPBT and M-EPBT within one location. Thus, the divergence between EPBT and M-EPBT is to be explained with the modified grid efficiency at each installation site. As outlined from a theoretical perspective in section 2.2, it may be observed that the share of RE sources in a particular installation location's grid mix is directly correlated to the extension of EPBT. As the proposed M-EPBT method implies an equal $\eta_{aGlobal}$ for the three installation locations, the proportional share between M-EPBT and EPBT illustrates the lengthening effect of high RE penetration in installation location's grid mixes. Hence, the higher the RE share in the respective installation grid mix, the larger the difference between EPBT and M-EPBT. Since Rio de Janeiro displays concurrently the highest share of RE in the grid (grid efficiency = 46.6%) and the largest difference between EPBT and M-EPBT, the presented results support this hypothesis. Subsequently, the proposed M-EPBT methodology serves as an explicit indication for i) the specific conditions at each installation location, in particular for the magnitude of solar radiation, the minor impacts of operation and maintenance energy expenses, and CED for transport from manufacturing to installation location, and ii) for grid efficiency and energy efficiency improvements at the manufacturing location.

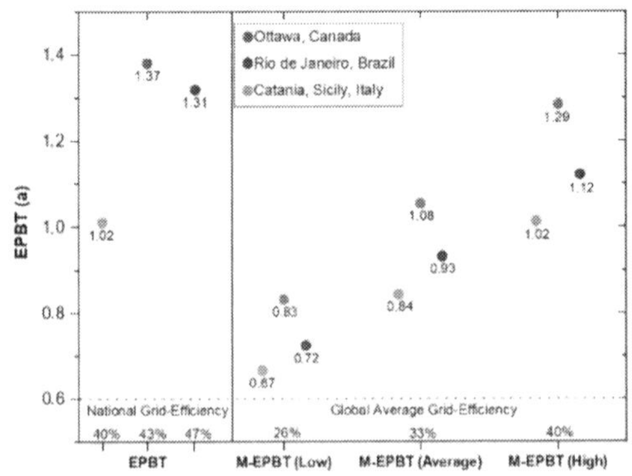

Figure 1: EPBT and M-EPBT of a 15 kWp slanted rooftop Cz PERC PV system produced in China for the installation locations Ottawa (Canada), Rio de Janeiro (Brazil), and Catania (Italy). Depicted from LCA and EPBT dataset of Friedrich et al. [4]

3.3 Application of EPBT and M-EPBT in future scenario

PV systems are expected to be more efficient and less energy demanding in the future, due to technological developments [7]. According to the technology roadmap published in 2018 by the International Renewable Energy Agency (IRENA) [9], by 2050, RE sources are expected to reach an 85% share in the worldwide electricity generation mix, with solar PV's share increasing from 1% to 22% [9]. The EU has already pledged for zero carbon emissions by 2050 [9].

Moreover, in the last decade, the average silicon module efficiency has increased from 12% to 17% today, with the number being even higher for PERC modules (21%) [12]. This trend is expected to continue in the future, driven by solar cell efficiency increase. An efficiency of 24% is projected by 2050 for one junction crystalline-Si (c-Si) modules [10], limited mainly by the maximum theoretical efficiency of 29.4% of c-Si cells [11]. In addition, the production chain of c-Si modules and BOS components have improved in the last years, leading to less energy intensive production methods. In 2006, the first LCA reports showed a lifecycle energy demand for mono-crystalline modules produced in Europe of 36 MJ-eq/Wp [12], while today the CED of PERC solar modules produced in Europe is 8.6 MJ-eq/Wp [9]. Although these values cannot be directly compared due to differences in electricity supply, they clearly show a trend of future reduced CED for module production.

In order to demonstrate the above-mentioned trends and their effect on the future EPBT and grid efficiency we follow the method proposed by Raugei [7] and apply the mid scenario value for $\eta_{aGlobal} = 33\%$. The main target of the following methodology is to demonstrate future trends and effects that will arise with the current EPBT framework, rather than predicting exactly the EPBT in the future.

According to [7], the grid efficiency is the result of the weighted sum of the life cycle efficiencies of each energy source in the grid:

$$\eta_G = \sum_i w_i * \eta_{i,LC} \quad (4)$$

i: Energy source
w_i: Share of the energy source *i* in the grid (in %)
$\eta_{i,LC}$: Lifecycle primary energy-to-electricity conversion efficiency of the corresponding energy source (in %)

The lifecycle efficiency is calculated differently for thermal power plants and RE sources, because of the different approaches discussed above in the CED calculations by the Ecoinvent database [5]. For thermal power plants (fossil, nuclear, biomass), the grid efficiency equation is as follows, according to [7]:

$$\eta_{th,LC} = \frac{E_{out}}{CED_{pp} + CED_p} = \frac{E_{out}}{CED_{pp} + \frac{E_{out}}{\eta_{th}}} \quad (5)$$

E_{out}: Total produced electricity (in MJ)
CED_p: PE in the feedstock (in MJ)
CED_{pp}: Additional (non-feedstock) PE required over the system's lifetime (in MJ)
η_{th}: Lifecycle primary energy to electricity conversion efficiency of the thermal electricity production system (in %)

As for PV solar energy, the grid efficiency equation is as follows, according to [7]. The same formulation as the PV energy can also be applied for hydro and wind, but with different conversion efficiencies [5,7]:

$$\eta_{PV,LC} = \frac{E_{out}}{CED_{pp} + CED_p} = \frac{E_{out}}{CED_{pp} + \frac{E_{out}}{\eta_{inv}}} \quad (5)$$

E_{out}: Total produced AC electricity over the whole PV lifetime (in MJ)
CED_p: PE in the captured solar radiation (in MJ). This represents the DC electricity that the solar module produces
CED_{pp}: Additional (non-solar) PE required over the system's lifetime (in MJ)
η_{inv}: Primary energy-to-electricity conversion efficiency of the PV system (taken as the average inverter efficiency = 93.5% according to [5] (in %)

Two different scenarios are considered:

- *Improved technology scenario:* Improved PV system technology, where the PV specific yield is expected to increase due to an increase in module and BOS efficiencies until 2050 [10]. The CED of PERC systems is decreasing linearly by 1% per year until 2050 [7]
- *Stagnant technology scenario:* CED, module efficiencies, and PV yield are kept constant as in the reference year 2020

In both scenarios, the share of solar energy increases from 1% in 2020 to 22% in 2050 [9]. Moreover, the installation location is chosen to be the average European grid and solar irradiation, with PERC modules produced in China [4].

Based on equation (6), the lifecycle efficiency of PV solar is found to be 83.8% in 2020 and 2050 in the stagnant technology scenario, and 87.5% in 2050 in the improved technology scenario. By assuming that the lifecycle efficiency of all the other technologies remain stable, the projected grid efficiency is calculated for both scenarios based on equation (4).

The grid efficiency increases steeply, reaching 43.8% in the stagnant scenario and 44.6% in the improved scenario. The difference between the two scenarios is rather small, indicating that the increase is mainly driven by the increased share of PV, rather than the technological development. Table II summarizes the given input parameters, and the grid efficiency results.

Table II: Main input parameters for current and future projections of M-EPBT in 2050, for two different scenarios, as well as the projected grid efficiency changes [1-4]

PV system parameters	2020	2050	
		Stagnant Technology	Improved Technology
Production Location	China		
Installation Location, Capacity	Average EU, 15 kWp rooftop		
Average Solar Irradiation [(MWh/m²)/a]	1.33		
PV Yield (GJ/a)	58.9		70.2
CED PV System (GJ)	219		153
Solar Share in Grid Mix (%)	1	22	
Calculated Grid Efficiency (%)	33	43.8	44.6

Figure 2 shows that the EPBT in the technology improvement scenario is decreased steeper for the modified methodology (-41%) compared to the standard methodology (-21%).

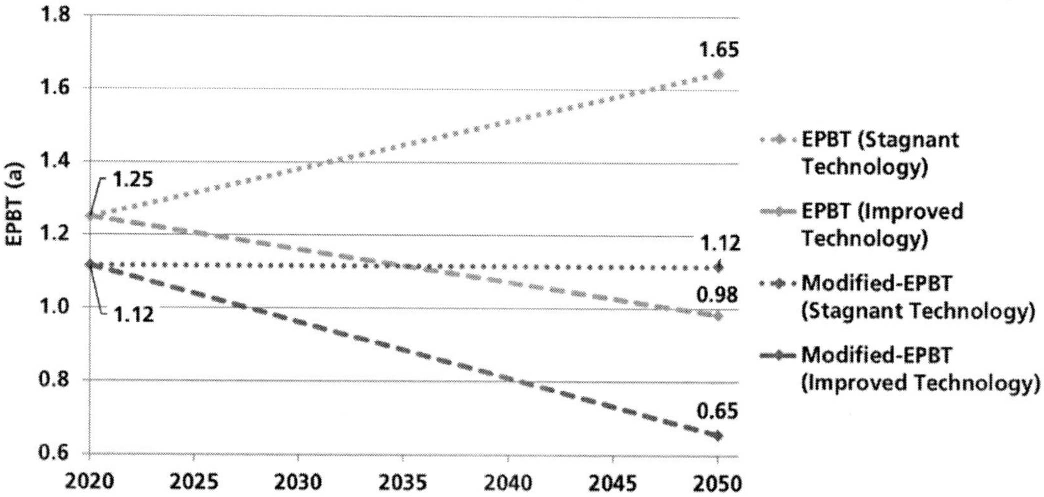

Figure 2: EPBT evolution until 2050, as calculated by the current IEA methodology (EPBT) and the proposed M-EPBT, for two different scenarios

In the standard EPBT case, the counteracting effects of increased grid efficiency and reduced CED_{PP} lead to a limited EPBT reduction. Moreover, in the stagnant technology scenario, the EPBT calculated with the standard methodology, has increased more than 30%, reaching 1.65 years in 2050, dictated mainly by the installation location's grid efficiency rise. M-EPBT remains stable in the stagnant technology scenario throughout the years.

For both scenarios, it is demonstrated that the standard EPBT methodology cannot fully express future technological improvements of the PV system technology, due to the imminent EPBT value rise. The newly proposed methodology provides a solution to translate the technology development into reduced EPBT. In this way, the improvements in the system yield and production technology can be depicted in the M-EPBT, despite the increase of the share of RE sources.

4. CONCLUSION

This work focused on the EPBT analysis of PERC modules and the proposal of a new methodology (M-EPBT) for the calculation of the EPBT, in order to overcome some identified weaknesses of the current IEA Task 12 approach.

The conducted analysis demonstrates selected shortcomings of the EPBT. Due to its methodological dependency on grid efficiency, particularly at the PV system's installation location, the EPBT is not seen as a valid energy performance indicator for grid mixes characterized by high RE generation. As emphasized, increasing grid efficiency influences the conversion of output PV electric energy to PE-eq on the generation side, resulting in longer EPBTs. Hence, technological improvement in module technology and energy saving in production will be (partly) compensated by this effect. Regarding future grids being dominated by RE, the EPBT could no longer deliver sound indication for PV system's energy performance.

Also, the second IEA methodology approach NR-EPBT, is seen as insufficient to illustrate energy performance. In detail, the NR-EPBT lacks by only considering NR PE-eq in manufacturing. Especially considering worldwide increasing RE penetration, it is seen as inadequate from a methodological perspective to exclude highly efficient RE manufacturing in the PV system's CED. Second, as the IEA NR approach assumes, the same amount of electric output (E_{agen}) is solely produced by the PV system, it is divided by the NR grid efficiency, which itself may reach higher values than the total grid efficiency by method. Again, under consideration of rising RE deployment, this efficiency increasing effect can dramatically amplified the (E_{agen}/η_G) term in the NR approach, resulting in delusively longer

NR-EPBT values for locations with high RE power generation.

To address the stated distorting effects, the M-EPBT was introduced. The novel approach maintains the primary energy saving implication of high grid efficiency in manufacturing (by including the non-renewable and renewable PV system CEDs) and excludes its payback time lengthening effect on the generation side. To fulfill the latter outlined requirement, the installation location grid efficiency (η_G) was substituted by the average global conventional lifecycle grid efficiency ($\eta_{aGlobal}$).

The EPBT and M-EPBT were projected for a 2050 scenario, including module efficiency and manufacturing CED improvements. As expected, the M-EPBT appears to decrease more steeply (-41%) from 1.12 years to 0.65 years, in comparison to the standard EPBT approach (-21% from 1.25 years to 0.98 years), simply because the previously mentioned counteracting effects are eliminated. Thus, the proposed methodology provides a solution to illustrate the technological development undistorted in the M-EPBT energy performance indicator.

All in all, many qualitative and quantitative parameters shall be considered in the evaluation and assessment of EPBT as an indicator. Above all, clear and unambiguous definitions of underlying assumptions and system boundaries, from the researchers, are cornerstones of LCA in general, and EPBT in particular.

5. REFERENCES

[1] Fischer, M., Woodhouse, M., Herritsch, S., Trube, J. (2021). International Technology Roadmap for Photovoltaic (ITRPV): 2020 Results. VDMA. Retrieved 15 August 2021, from https://itrpv.vdma.org/en/

[2] Fraunhofer ISE. (2021). Photovoltaics Report. Fraunhofer Institute for Solar Energy ISE. Retrieved 20 August 2021, from https://www.ise.fraunhofer.de/de/veroeffentlichungen/studien/photovoltaics-report.html

[3] Frischknecht, R., Stolz, P., Heath, G., Raugei, M., Sinha, P., de Wild-Scholten, M. (2020). Methodology Guidelines on Life Cycle Assessment of Photovoltaics 2020, IEA-PVPS Task 12, Report T12-18:2020, 4th edition. Retrieved 15 January 2021, from https://iea-pvps.org/key-topics/methodology-guidelines-on-life-cycle-assessment-of-photovoltaic-2020/

[4] Friedrich, L., Nold, S., Müller, A., Rentsch, J., Preu, R. (2021). Global Warming Potential and Energy Payback Time Analysis of Photovoltaic Electricity by Passivated Emitter and Rear Cell (PERC) Solar Modules. IEEE Journal Of Photovoltaics, submitted.

[5] Hischier R., Weidema B., Althaus H.-J., Bauer C., Doka G., Dones R., Frischknecht R., Hellweg S., Humbert S., Jungbluth N., Köllner T., Loerincik Y., Margni M., and Nemecek T. (2010). Implementation of Life Cycle Impact Assessment Methods, ecoinvent report No. 3, v2.2. Swiss Centre for Life Cycle Inventories, Dübendorf. Retrieved 15 January 2021, from https://www.ecoinvent.org/database/older-versions/ecoinvent-version-2/reports-on-ecoinvent-2/reports-on-ecoinvent-2.html

[6] Palmer, G., Floyd, J. (2017). An Exploration of Divergence in EPBT and EROI for Solar Photovoltaics. Biophysical Economics And Resource Quality, 2(4). doi: 10.1007/s41247-017-0033-0

[7] Raugei, M. (2012). Energy pay-back time: methodological caveats and future scenarios. Progress In Photovoltaics: Research And Applications, 21(4), 797-801. doi: 10.1002/pip.1249

[8] Ritchie, H., Roser, M. (2020). Energy mix. Our World in Data (2021): Energy mix. Our world in data. Retrieved 29 January 2021, from https://ourworldindata.org/energy-mix

[9] IRENA (2018). Global Energy Transformation, a Roadmap to 2050. Retrieved 29 January 2021, from https://www.irena.org/-/media/Files/IRENA/Agency/Publication/2018/Apr/IRENA_Report_GET_2018.pdf

[10] Fraunhofer ISE (2015). Current and Future Cost of Photovoltaics. Long-term Scenarios for Market Development, System Prices and LCOE of Utility-Scale PV Systems. Study on behalf of Agora Energiewende. Retrieved 29, January 2021, from https://www.ise.fraunhofer.de/en/publications/studies/studie-current-and-future-cost-of-photovoltaics-long-term-scenarios-for-market-development-system-prices-and-lcoe-of-utility-scale-pv-systems.html

[11] Richter, A., Hermle, M., Glunz, S. (2013). Reassessment of the Limiting Efficiency for Crystalline Silicon Solar Cells. IEEE Journal Of Photovoltaics, 3(4), 1184-1191. doi: 10.1109/jphotov.2013.2270351

[12] Fthenakis, V., Alsema, E. (2006). Photovoltaics energy payback times, greenhouse gas emissions and external costs: 2004–early 2005 status. Progress In Photovoltaics: Research And Applications, 14(3), 275-280. doi: 10.1002/pip.706

TERAWATT SCALE PV BY 2050 AND COMPETITION FOR MINERALS: THE CASE OF SILVER AND COPPER

Philippe Macé[1], Elina Bosch[1], Monica Aleman[1]
[1]Becquerel Institute
146 Rue Royale, 1000 Brussels, Belgium
Phone: +32 2 514 78 04

ABSTRACT:
The role of the deployment of solar PV systems in possibly triggering competition for silver and copper supply, in the frame of a terawatt scale market, is presented in this paper, including the methodology implemented to determine it. Possible strategies to reduce the risk of competition, such as increasing the contribution of secondary production in raw material supply, are analysed. The opportunity of shifting from Ag to Cu in PV cells' metallization is measured. The interest of efficiency increases is discussed, as well as the attractiveness of shifting from single-junction c-Si to perovskites, in single-junction configuration or in tandem with c-Si. The contribution of the various components of PV systems is also highlighted. Eventually this research permits to evaluate to which extent the objective to limit global warming to 1,5°C by 2100 is realistic, considering the constraints on silver and copper potentially caused by solar PV.
Keywords: Sustainable, Environmental Effect, Minerals, Silver, Copper, Resources Scarcity

1 INTRODUCTION

1.1 Context

All sectors of the economy have a role to play in order to maintain the global average increase of temperature below 2,0°C, as ambitioned by Paris Climate Agreement [1]. In this regard, the energy transition, which aims at reducing the environmental footprint of the energy sector, is crucial. The work to be accomplished is enormous and this transition, or revolution, will itself have impacts, as it requires to use technologies that consume a wide range of minerals as raw materials, directly and indirectly. Among these technologies, solar PV systems occupy a central role. Indeed, as studied by various researchers, in order to have a chance to achieve the Paris Climate Agreement target, it will be necessary to reach a cumulative installed PV capacity in 2050 ranging from 20 to 60 TWp [2] [3].

1.2 Problem definition & objectives

The ambition of this paper is to evaluate to which extent a trajectory towards a terawatt-scale global PV market by 2050 will put two crucial raw materials in solar PV systems, namely silver and copper, under pressure and if this could trigger competition with other sectors using these resources.

The question of resource consumption has already been partially investigated in the literature, although the issue of competition has not been investigated yet, and an « absolute » evaluation of the demand for the raw materials is often considered, without putting it in regard of other usages [4] [5] [6] [7] [8] [9]. In addition, considering the extremely rapid evolution of the sector, both in terms of technological and market development, the data and assumptions used in the estimations can rapidly become obsolete, thus often making past conclusions irrelevant. This has for example been witnessed for various recent studies, relying on what can be considered as outdated data [10] [11] [12].

In this paper, four mature PV module technologies (respectively based on p-type monocrystalline silicon PERC cells, n-type monocrystalline silicon HJT, CdTe and CIGS cells) as well as emerging ones (perovskites and perovskites/c-Si tandem) are considered. Moreover, the demand due to other components of PV systems is also taken into account, including inverters, mounting structures, cabling and electrical components, among others.

The analysis focuses on silver and copper minerals, and among the technological pathways studied, one will envisage the substitution of silver by copper (together with nickel) for the metallization of solar PV cells. This permits to assess the interest of such strategy from a resource depletion point of view. This constitutes a new argument since it has mostly been discussed in the industry due to cost reasons [13].

2 METHODOLOGICAL APPROACH

The different steps of the modelling can be divided in five meta-categories, which are presented below.

2.1 Perimeter of the analysis

As explained in the introduction, the analysis has been conducted for multiple key components of PV systems, to obtain a vision as exhaustive as possible.

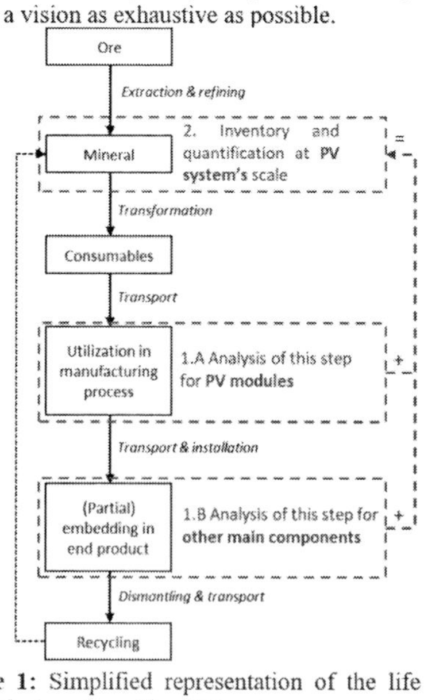

Figure 1: Simplified representation of the life cycle's steps of mineral used in a PV system.

Two different strategies have been adopted to quantify

the amount of silver and copper used to produce these components. Firstly, a detailed review of state-of-the-art manufacturing processes has been conducted for PV modules. It includes estimations of processes' efficiencies, quantities of materials and consumables used, and the performances of the end products, among others. Secondly, a simplified analysis has been conducted for inverters, mounting structures, cabling and junction boxes, as well as transformers, due to the lack of available data. For these components, the existing literature was screened to estimate the quantity of embedded materials. These quantities were then derived to assess the consumption of the two target minerals.

The overall strategy is summarized on the diagram shown in Figure 1. Note that only "direct" consumption is considered. For example, transportation steps are excluded, and the materials' consumption of machinery and production tools necessary to produce and install the components has not been taken into account.

2.2 Future PV market developments

The first question to answer is what the exact level of the terawatt scale PV deployment will be by 2050. Thus, PV market development scenarios until 2050 were defined.

Multiple scenarios exist in the literature. These have been developed by researchers, private entities such as energy majors (Total and Shell, among others) or advisory companies, as well as international institutions such as IRENA or the IEA. In our case, we selected three scenarios developed by various research teams, from DNV GL, Stanford University and Lappeenranta University of Technology [1] [2] [3].

These scenarios were selected because they allow to depict three possible pathways, each of different scale, which permits to take into account the uncertainty associated with such prospective exercise. The first scenario, called "Minimal transition" and based on the projections of advisory firm DNV GL, is not sufficient to respect Paris Climate Agreement goal of limiting global warming to 2,0°C, as admitted by the analysts themselves [1]. This could be seen as a "business-as-usual" pathway. On the other hand, the "Ambitious transition" scenario, based on Stanford's researchers' projections, should permit to do limit temperature increase to 2,0°C [2]. Finally, the third scenario "Total transition" should permit to limit global warming to 1,5°C by fully decarbonizing the world's energy demand [3].

Table I: Selected PV market development scenarios

Scenario's name	Publishing organisation	Source	Cumulative PV (TWp)	
			2030	*2050*
Minimal transition	DNV GL	[1]	4.1	12.4
Ambitious transition	Stanford	[2]	7.4	21.6
Total transition	LUT	[3]	10.2	63.4

The next sub-step consists in projecting what will be the shares, in the defined market development scenarios until 2050, of the analysed PV module technologies, in order to compute the entire demand for silver and copper due to the deployment of PV systems. This an extremely difficult exercise, and the source of multiple uncertainties. Indeed, as recent history has shown, it is extremely difficult to anticipate, even in the short term, how PV cells and modules technologies will evolve and what products and features will be commercially viable and become mainstream.

The ITRPV was used as a main source to define the PV market shares based on the cell technology [15]. Nonetheless, it only provides an overview until approximately 2030. Thus, additional assumptions have been made on technological developments in the PV sector until 2050. Note that this prospective exercise was conducted for each PV cell technology but also other parameters that can potentially impact the consumption of copper and/or silver or the performances of systems. This encompasses, among others, the size of modules, the size of wafers used, the penetration of bifacial modules or the type of installation (rooftop or ground-mounted). For all these parameters, assumptions were made regarding their future penetration on the PV market.

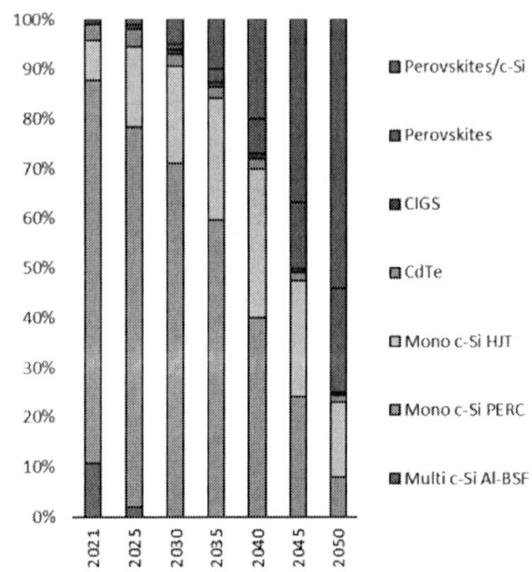

Figure 2: Projected PV market share per cell technology used in the calculations.

2.3 Material intensity of PV systems

Then, the material intensity has been evaluated and modelled, in terms of silver and copper consumption, for the selected PV cells and module technologies, both commercial and emerging ones, as well as for other key PV system components.

Focusing first on PV cells and modules, the entire manufacturing workflow has been analysed, for each device technology. This has been done via an extensive literature review and exchange with specialists of the sector. In the case of crystalline silicon technologies, for various parameters influencing average consumption of silver and copper, assumptions have been made based on the latest edition of ITRPV [15]. This encompasses for instance average silver consumption for cell metallization, average cell efficiencies, processes efficiencies or cell-to-module ratios. But, again, estimated values are only provided until 2030. Hence, values until 2050 have been projected, assuming that the trend between 2020 and 2030 can be replicated between 2031 and 2050. This is based on the logic that the industry will keep on gaining experience and developing innovations, but at half the pace it used to. Ranges are provided, as various c-Si technologies are studied.

For CdTe, the most impactful evolving parameter is the estimated efficiency. It is based on datasheet of FirstSolar modules as well as various publications [16] [17]. For CIGS, typical efficiencies reached by current commercial products (Avancis, NICE Solar Energy, Solibro) are averaged to obtain today's efficiency, while the literature is used to estimate future efficiencies [16] [18]. Then, concerning perovskites, the situation is even more challenging, as there is no commercial product available in the market. All assumptions are based on the scientific literature [15] [19] [20] [21]. In the case of perovskites on c-Si tandem, cell efficiencies are partially estimated on the basis of latest ITRPV, using the cell-to-module ratio to obtain module's nominal power [15].

Some of the key parameters evolving across the studied period are summarized in the tables below. This includes parameters impacting performances of cells and modules as well as those directly impacting the final demand for the two studied minerals.

Table II: Key evolving parameters for c-Si technologies (Al-BSF, PERC and HJT).

Parameter	Unit	Values		
		2021	*2030*	*2050*
Cell efficiency	[%]	19,4%–24,0%	20%–25,5%	26,5%–28%
Ag paste consumption	[mg/cm²]	0,29–0,69	0,18–0,35	0,11–0,17
CTM ratio	[%]	98,1%–101,0%	99,3%–103,3%	100,5%–105,7%

Table III: Key evolving parameters for CdTe and CIGS

Parameter	Unit	Values		
		2021	*2030*	*2050*
Module efficiency CdTe	[%]	18,0%	21,0%	24,0%
Module efficiency CIGS	[%]	15,0%	18,0%	22,0%

Table IV: Key evolving parameters for perovskites technologies (single-junction and tandem).

Parameter	Unit	Values		
		2021	*2030*	*2050*
Module efficiency pks	[%]	/	23,0%	26,0%
Cell efficiency pks/c-Si	[%]	/	28,0%	35,0%

Concerning other crucial components of PV systems, namely cabling and connectors, inverters, junction boxes and transformers as well as mounting structures, the assumptions are also based on the literature [22] [23] [24] [25] [26] [27]. Note that in the case of inverters, a range is provided as the relationship of weight composition to power is not linear.

Table V: Key evolving parameters for other PV system components.

Component	Unit	Ag	Cu
Cable & connectors	[kg/m]	/	0,26
Inverters	[kg/kVA]	/	3,1 – 180,1
Junction boxes	[kg/MVA]	/	20
Transformers	[kg/MVA]	/	80

Note that for all these components, it is assumed that the values are constant, due to a lack of information. Also, mounting structures do not have any influence, as they do not contain any silver or copper.

2.4 Minerals demand & production

Focusing on the two minerals analysed, i.e. silver and copper, their global demand on a yearly basis has been estimated until 2050, in addition to the demand due to PV deployment, in order to estimate the risk of competition. Multiple sources have been used, such as the French « Bureau de Recherches Géologiques et Minières » and the US Geological Survey [31] . In addition, annual reports published by trade associations representing the companies active in the production or transformation of these metals have been used [30] [32].

Regarding demand growth due to other sectors of the economy until 2050, a simplified assumption is made. It is assumed that the Compound Annual Growth Rate (CAGR) on the period 2021-2050 will be similar to what was witnessed on the period 1989-2019. A shorter, more recent period could have been considered, but as shown on the table below, it would have resulted in a pretty similar CAGR hypothesis.

Table VI: Average past CAGR of demand.

Period	Ag	Cu
1989-2019	1,7%	2,7%
1999-2019	2,2%	2,3%
2009-2019	1,9%	2,2%

Then, the level of yearly primary production must be projected. It is done based on the Hubbert peak theory [27]. This model provides an approximation of the production rate of a resource over a time period, recognizing the finite nature of materials. Future primary production is thereby modelled using the following equations:

$$P(t) = \frac{2 * P(t_{MAX})}{1 + \cosh(b * (t - t_{MAX}))}$$

and

$$UR = 4 * \frac{P(t_{MAX})}{b}$$

Where P(t) is the global primary production in year t, $P(t_{MAX})$ is the peak global primary production, in year t_{MAX}, b is the curve coefficient and UR is the level of ultimate resources[1]. The main results of the application of these formulas are provided in the table below. This will permit a comparison with the demand, helping to characterize the risk of seeing possible competitions arise.

Table VII: Main results of the modelling of future production.

Mineral	Primary production (kt/year)		Peak production year
	2030	*2050*	
Ag	28	25	2033
Cu	26.393	37.203	2077

[1] Ultimate resources of a mineral give an estimation of the total available quantities of this mineral in Earth's crust, including non-identified quantities, i.e. inferred ones, as well as those that are not yet exploitable, from a technical or economic point of view.

Finally, the level of reserve[2] and ultimate resource of silver and copper are also obtained from the literature [29]. They are indeed necessary for the simulation of production and allow to feed the analysis of the level of pressure and competition that could potentially arise for the supply of the two studied minerals.

Table VIII: Recapitulative table of key data for the two target minerals. All data are the based on the most recent information available, mainly 2019.

Parameter	Ag	Cu
Demand (kt/year)	31	28.000
Share due to PV	5%	2%
Primary production (kt/year)	27	20.000
Reserves (kt)	560	870.000
Ultimate resources (kt)	2.900	5.600.000

2.5 Prospective comparison

Finally, the cumulated level of demand for the various market and technological scenarios are put in regards of the level of production, reserves and ultimate resource of silver and copper, to estimate the risk of competition. A comparison of annual levels is done as well. Altogether, these elements permit to characterize the risk of competition.

3 RESULTS

3.1 Cumulated demand

Here below are presented the relative shares of the global cumulative demand on the period 2021-2050. In other words, it shows the split between the demand due to PV systems' deployment and the demand due to other sectors. As a reminder, on this graph and on the others, "TM" refers to the "Minimal transition" scenario, "TA" refers to the "Ambitious transition" scenario and "TT" refers to the "Total transition".

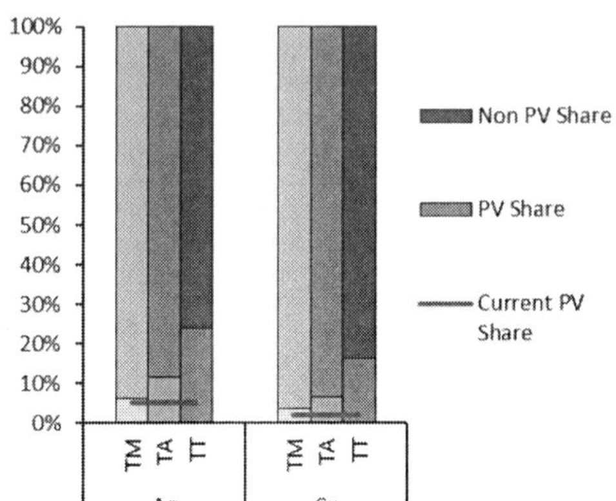

Figure 3: Relative shares of the global cumulative demand, for the three analysed PV market scenarios.

In both cases, the depicted evolution on the 2021-2050 requires attention. Indeed, for the three scenarios, the share of PV in the global cumulated demand for the two analysed minerals is foreseen to significantly increase compared to current levels. In particular, in the "TT" scenario, the only one that should allow humanity to maintain average temperature increase at 1,5°C, demand due to PV systems' deployment should reach approximately 25% of the global cumulated demand for Ag and approximately 16% for Cu, on the 2021-2050 period. Only in the case of the "TM" scenario, which is insufficient to respect Paris Climate Agreement, should this share of PV in global demand remain around today's level. Although, it would still be exceeded.

3.2 Annual demand and production

Following the study of the structure of global cumulated demand, the resulting data can be further analysed by looking at annual levels of demand and production.

On the two following figures, the stacked bars show the annual demand for the investigated metal, both due to the deployment of PV systems (in orange) and other usages (in blue). These usages are for example the electronic and electrical devices or heating systems in the case of copper, or jewellery and photography in the case of silver. The yellow curve gives the estimated level of primary production of the studied metal, per year, calculated thanks to the Hubbert peak theory, as explained in section 2.

In both cases, one can notice that the annual primary production is always below the annual demand, for all PV market development scenarios. It means that secondary production, i.e. recycling, is crucial to close the gap and satisfy global demand. In the case of silver (Figure 4) the gap between annual demand and annual primary production is first relatively limited but keeps on increasing as primary production decreases. Indeed, as shown on Table VII in section 2, it is estimated that peak annual production will be reached around 2033.

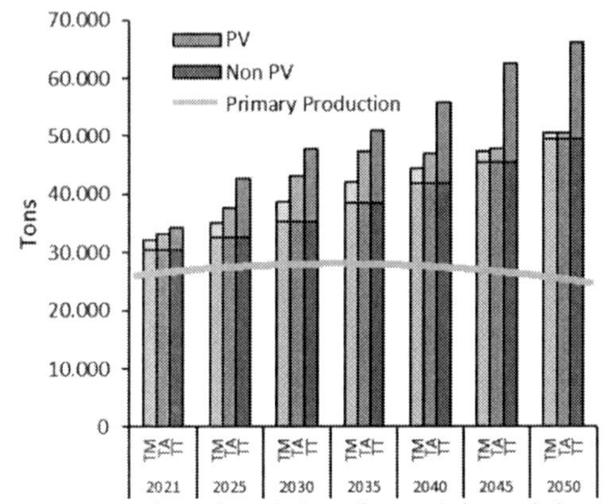

Figure 4: Estimated evolution of annual demand (bars) and annual primary production (curve) of Ag.

Concerning copper, primary production is expected to keep on rising on the 2021-2050 period, on a yearly basis, as the peak production's year is projected to be 2077. Although, demand is foreseen to increase at a faster pace, which will lead to a worsening of the gap between annual demand and annual production.

[2] Reserves quantities are quantities of a mineral that are demonstrated, i.e. their existence in Earth's crust is proven, and whose exploitation is technically possible and economically viable.

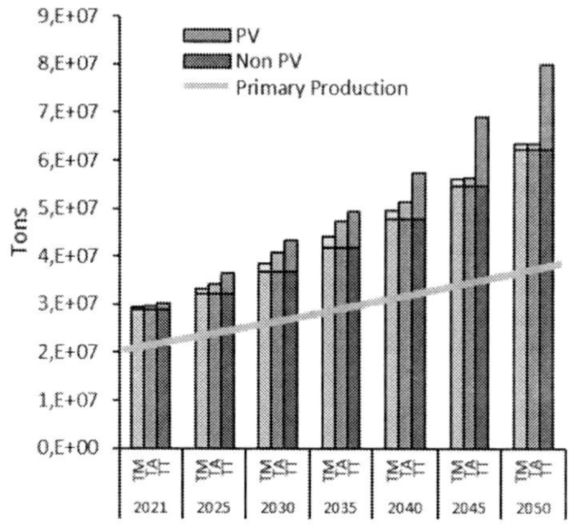

Figure 5: Estimated evolution of annual demand (bars) and annual primary production (curve) of Cu.

On both graphs, the demand linked to the PV sector keeps on increasing for the three PV market scenarios until approximately 2035. Then, on the 2036-2050 period, in the "Minimal transition" and "Ambitious transition" scenarios, the pressure due to PV systems' deployment decreases, while it keeps on increasing in the "Total transition" scenario. Indeed, in this scenario it is assumed that annual PV capacity installed keeps on increasing until 2050, in order to be able to limit global temperature rise to 1,5°C. This significantly contributes to increase the gap between annual demand and annual production, which can trigger competitions if recycling does not increase as well.

3.3 The contribution of secondary production

After having mentioned the gap between annual demand and annual primary production in the previous sub-section, the secondary production levels required to compensate for this gap can also be computed, to estimate the necessary contribution of recycling. This allows to evaluate if recycling can realistically help to ease the potential risk of competition.

On the figure below, the current recycling rate (on an annual basis) of the two studied metals is shown by the red bar. For each scenario, the white dots with a blue circle represent the computed annual recycling rates required to cover the gap between annual primary production and annual demand, for each year of the 2021 to 2050 period.

One can first notice that in the case of silver, on all years and for the three scenarios, the annual recycling rate would have to be higher than its current level. As we advance on the studied period, this necessary recycling rate increases, with a maximum of 62% reached for the "Total transition" scenario in 2050. The median annual recycling rate on the period would have to range from 33% in the case of the "Minimal transition" scenario to 45% in the "Total transition" scenario, which is significantly higher than current annual recycling rate reached for silver, standing slightly below 20%.

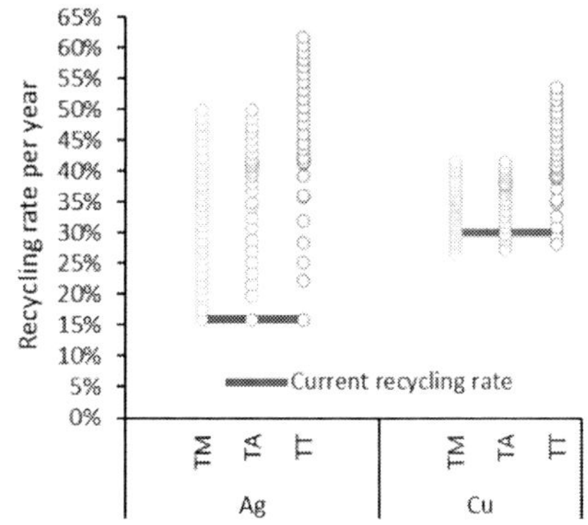

Figure 6: Annual recycling rate required for secondary production to cover the gap between annual demand and annual primary production, on the 2021-2050 period.

The case of copper is less extreme. Indeed, until approximately 2025, the required annual recycling rate is foreseen to stand below or around current annual recycling rate, which equals 30%. After 2025, in the "Minimal transition" and the "Ambitious transition" scenarios, this required recycling rate would steadily grow, reaching approximately 40% per year. In the "Total transition" scenario, on the contrary, the growth would be much steeper, and the annual recycling rate required to compensate for the gap between annual demand and annual primary production would have to reach almost 55% by 2050, almost twice as high as current rate.

For both studied metals, yearly secondary production would have to be substantially improved in the coming decades. In the case of silver, depending on the scenario, this rate would have to increase every year by 1,1 to 1,5 in percentage points, on average, to compensate for the rising gap between annual demand and annual primary production. In the case of copper, depending on the scenario, this yearly increase would have to stand between 0,4 and 0,8 in percentage points, on average. A tremendous effort considering that global demand will grow as well, in absolute terms.

3.4 Pressure on reserves and ultimate resources

The last element to investigate in order to identify if competition could appear for the studied metals is the level of reserves and ultimate resources compared to global cumulated demand.

On the following figure, one can see that both for silver and for copper, the global demand on the 2021-2050 period do not exceed the level of ultimate resources. In the case of silver, 44% to 55% of this ultimate resources' level is reached, depending on the scenario, while in the case of copper, 25% to 28% is reached.

The picture is much gloomier when looking at reserves' levels, as estimated in 2019. In both cases, they should be exceeded by global cumulated demand on the 2021-2050 period. This 2019 level of reserves is surpassed approximately in 2035 in the case of silver, for all three scenarios, and around 2040 in the case of copper. Depending on the considered PV market scenario, global cumulated silver demand could reach 229% to 283% of estimated silver reserves in 2019. In the case of copper, the global cumulated demand on 2021-2050 period could reach 159% to 183% of estimated copper reserves in 2019.

Nonetheless, one should keep in mind that the notion of reserves varies in time, as it depends on the average price of the studied metal. Indeed, it indicates the quantities of minerals that can be extracted from Earth's crust, without losing money. Hence, it could vary upwards as time goes by and the balance between supply and demand varies.

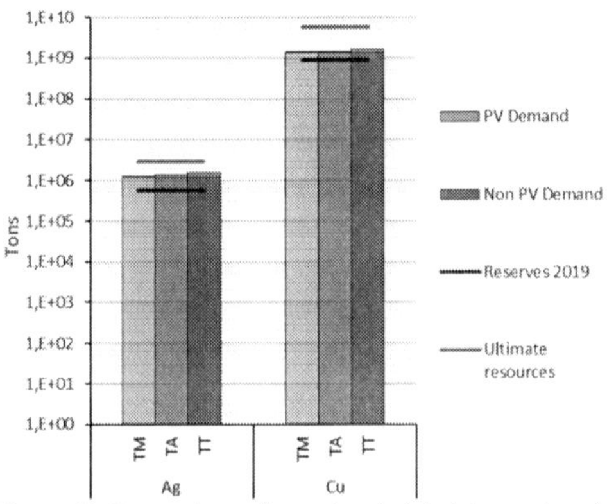

Figure 7: Comparison of cumulated global demand on the 2021-2050 period with reserves and ultimate resources (logarithmic scale).

4 DISCUSSION

4.1 Risk of competition

Based on the results presented in the previous section, it is possible to characterize the risk of competition for the two studied metals. As the level of ultimate resources is far from being reached on the analysed period, but pressure on reserves is high and annual demand is consistently estimated to be higher than annual primary production, the risk of competition can be considered as serious, for both metals.

Then, the analysis can be further developed by looking at secondary production. In the case of silver, the yearly primary production levels are anticipated to decrease, and the gap with annual demand to rise. As a consequence, yearly recycling rates would have to be significantly and steadily increased compared to current levels, which would be extremely challenging. On the other hand, in the case of copper, primary production would keep on rising per year, on the 2021-2050 period. Even if the average recycling rate per year would have to increase compared to today's level, it seems achievable, except in the case of "Total transition" scenario.

Considering this, the risk of competition for silver supply is clearly higher than the risk of competition for copper supply. Nevertheless, one should bear in mind that copper is at least as critical as silver for PV, as it is widely used in cabling and key power electronics components, such as inverters, junction boxes and transformers.

4.2 Role of PV

It is worth highlighting that although the role of PV is substantial, it remains limited with regards to global demand. As calculated in this paper, the deployment of PV systems on the 2021-2050 period should at worse be responsible of 16% of global cumulated demand of copper, a figure that would rise to 25% for silver. Moreover, the growth of PV systems' deployment does not seem to trigger competition by itself. It is only one contributing sector among others, in a general trend of growing material demand. Indeed, if when comparing the annual growth in demand due to PV in the coming decades to the historical annual growth rates of global demand, it can be seen that PV would not put on metals' production a higher pressure than what has already been witnessed in the past decades.

4.3 Substitution of Ag by Cu

Among the possibilities to reduce the cost of solar PV, the replacement of silver by nickel and copper in cell metallization has been a hot topic in the industry for years. While the direct economic advantage seems straightforward considering the prices differences between the two metals, it could also generate additional benefits in the longer term.

Assuming that the quantities consumed for cell metallization are comparable whether it is done with silver or copper, as their respective conductivities are extremely similar. Assuming also that cell efficiency would not be impacted, such substitution would have no influence on the level of risk of competition for copper supply due to PV. This can be explained by the difference of scale in demand. Indeed, material intensities of PV systems for these two metals are highly different, standing on average for the three market scenarios at <10t/GWc for silver and >4.000t/GWc for copper, on the studied period. Thus, the additional few tons per GWc in copper consumption caused by this substitution at cell level would not weight much in comparison with the thousands of tons per GWc in copper consumption coming from cabling, connectors and power electronics.

5 CONCLUSIONS

As the level of ultimate resources is far from being reached on the analysed period, but pressure on reserves is high and annual demand is consistently estimated to be higher than annual primary production, the risk of competition can be considered as serious, for both metals. Then, when looking at secondary production levels, it is clear that the risk of competition is even more serious for silver compared to copper, as recycling will have to tremendously increase.

It can also be concluded that the growth of PV systems' deployment does not seem to trigger competition by itself. It is only one contributing sector among others, in a general trend of growing material demand. Indeed, demand due to PV in the coming decades would not put on metals' production a higher pressure than what has already been witnessed in the past decades.

Moreover, the analysis presented in this paper shows that a terawatt scale PV market aiming at achieving Paris Climate Agreement, i.e. staying well below 2°C of global warming, would be extremely challenging to achieve, although not impossible in theory. Indeed, compared to the "Minimal transition" and "Ambitious transition" scenarios, the "Total transition" scenario would highly increase the risk of competition for the supply of the two analysed metals. This is true even for copper, in spite of a rising primary production and high level of ultimate resources.

To reduce this pressure on supply and the risk of competition, as well as to maximize the chances of success towards decarbonization of energy supply, cell and module efficiency increases are crucial. They are indeed the most effective way to reduce material intensity of the sector.

It was also shown that another potential mean of reducing the risk of competition for silver, a rare metal, is the substitution of Ag by Cu for cell metallization. Such shift would eliminate the risk of competition for silver supply, while having no impact on the risk of competition for copper supply, due to the major difference of scale in material intensity.

Finally, many questions remain with regards to future supply of raw materials. For instance, what about other strategic metals for PV, such as aluminum or nickel? What would be the indirect environmental impacts of material supply (extraction, refining, ...)? In addition, what about the demand due to other devices needed to succeed on transitioning towards a decarbonized economy, such as wind turbines, batteries or electrolyzers? Further research has to be conducted, to ensure the most sustainable pathways are chosen. But one thing is already certain: a global shift towards more "material sobriety" is imperative.

6 REFERENCES

[1] REN21, "Renewables 2020 Global Status Report," Paris, June 2020.

[2] M. Z. Jacobson, A. M. Delucchi, M. A. Cameron, I. P. Manogaran, Y. Shu and A.-K. von Krauland, "Impacts of Green New Deal Energy Plans on Grid Stability, Costs, Jobs, Health, and Climate in 143 Countries," *One Earth*, vol. 1, no. 4, pp. 449-463, December 2019.

[3] M. Ram., D. Bogdanov, A. Aghahosseini, A. Gulagi, A. Oyewo, M. Child, U. Caldera, K. Sadovskaia, J. Farfan, L. Barbosa, M. Fasihi, S. Khalili, B. Dalheimer, G. Gruber, T. Traber, F. De Caluwe, H.-.. Fell and C. Breyer, "Global Energy System based on 100% Renewable Energy – Power, Heat, Transport and Desalination," Lappeenranta University of Technology and Energy Watch Group, Lappeenranta (Finland) and Berlin (Germany), March 2019.

[4] J. Harmsen, A. Roes and M. Patel, "The impact of copper scarcity on the efficiency of 2050 global renewable energy scenarios," *Energy*, vol. 50, pp. 62-73, January 2013.

[5] A. Stamp, P. Wäger and S. Hellweg, "Linking energy scenarios with metal demand modeling – The case of indium in CIGS solar cells," *Resources, Conservation and Recycling*, vol. 93, pp. 156-167, 2014.

[6] S. Davidsson and M. Höök, "Material requirements and availability for multi-terawatt deployment of photovoltaics," *Energy Policy*, vol. 108, pp. 574-582, 2017.

[7] J. McNerney, G. Kavlak, R. Jaffe and J. Trancik, "Metals Production Requirements for Rapid Photovoltaics Deployment," *Energy & Environmental Science*, 2015.

[8] J. Jean, P. Brown, R. Jaffe, T. Buonassisi and V. Bulovic, "Pathways for solar photovoltaics†," *Energy & Environmental Science*, vol. 8, pp. 1200-1219, 2015.

[9] P. J. Verlinden, "Future challenges for photovoltaic manufacturing at the terawatt level," *Journal of Renewable and Sustainable Energy*, no. 12, 2020.

[10] S. Teske and et al., Achieving the Paris Climate Agreement Goals - Global and Regional 100% Renewable Energy Scenarios with Non-energy GHG Pathways for +1.5°C and +2°C, Springer, 2019.

[11] S. Carrara, P. Alves Dias, B. Plazzotta and C. Pavel, "Raw materials demand for wind and solar PV technologies in the transition towards a decarbonised energy system," Publication Office of the European Union, Luxembourg, 2020.

[12] F.-W. Wellmer, P. Buchholz, J. Gutzmer and et al., Raw Materials for Future Energy Supply, Springer, 2019.

[13] K. Gawlinska-Necek and et al, "The Use of Copper in Solar Cells and Modules," in *37th EU PVSEC*, Lisbon, Portugal, 2020.

[14] DNV GL, "Energy Transition Outlook 2021: A global and regional forecast to 2050," DNV GL AS, Hovik (Norway), 2021.

[15] VDMA, "International Technological Roadmap for Photovoltaics - 11th Edition," 2021.

[16] A. Kuypers and et al., "Thin Film PV Roadmap," Solliance, 2018.

[17] Y. Chen and et al., "From Laboratory to Production: Learning Models of Efficiency and Manufacturing Cost of Industrial Crystalline Silicon and Thin-Film Photovoltaic Technologies," *IEEE Journal of Photovoltaics*, vol. 8, no. 6, pp. 1532-1538, 2018.

[18] S. Paetel, "Roadmap CIGS towards 25% Efficiency," in *7th International Workshop on CIGS Solar Cell Technology*, 2016.

[19] N.-G. Park and a. et, "Scalable fabrication and coating methods for perovskite solar cells and solar modules," *Nature Reviews Materials*, vol. 5, pp. 333-350, 2020.

[20] S. Albrecht and et al., "On top of commercial photovoltaics," *Nature Energy*, vol. 2, 2017.

[21] EPKI, "Perovskite-based photovoltaics: A unique chance for European PV-industry," European Perovskite Initiative, 2019.

[22] R. Frischknecht, R. Itten and et al., "Life Cycle Inventories and Life Cycle Assessment of Photovoltaic Systems," International Energy Agency (IEA) PVPS Task 12, 2015.

[23] Z. Yu, W. Ma, K. Xie, G. Lv, Z. Chen, J. Wu and J. Yu, "Life cycle assessment of grid-connected power generation from metallurgical route multi-crystalline silicon photovoltaic system in China," *Applied Energy*, vol. 185, pp. 68-81, 2017.

[24] M. Hegedic and et al., "Life Cycle Assessment of Power Transformer - Case Study," in *MOTSP 2016 International Conference Management of Technology – Step to Sustainable Production*, Porec, Croatia, 2016.

[25] P. Razdan and P. Garrett, "Life Cycle Assessment of Electricity Production from an onshore V112-3.45 MW Wind Plant - Version 1.1," Vestas Wind Systems A/S, Aarhus, Denmark, July 2017.

[26] L. Tschümperlin, P. Stolz, F. Wyss and R. Frischknecht, "Life cycle assessment of low power solar inverters (2.5 to 20 kW)," Swiss Federal Office of Energy SFOE, 2016.

[27] M. Caduff and et al., "Power-Law Relationships for Estimating Mass, Fuel Con-sumption and Costs of Energy Conversion Equipments," *Environmental Science & Technology*, vol. 45, no. 2, pp. 751-754, 2011.

[28] H. Sverdrup and K. V. Ragnarsdóttir, "Natural Resources in a Planetary Perspective," *Geochemical Perspectives*, vol. 3, no. 2, 2014.

[29] H. Sverdrup and al., "Investigating the sustainability of the global silver supply, reserves, stocks in society and market price using different approaches," *Resources, Conservation and Recycling*, vol. 83, pp. 121-140, 2014.

[30] The Silver Institute, "World Silver Survey 2020," The Silver Institute, 2020.

[31] U.S. Geological Survey, "Mineral Commodity Survey," 2020.

[32] Copper Alliance, "Copper Demand & Long-Term Availability," [Online]. Available: https://copperalliance.org/about-copper/long-term-availability/. [Accessed 30 June 2020].

ECO-DESIGN AND ENERGY LABELING FOR PHOTOVOLTAIC MODULES, INVERTERS AND SYSTEMS – ENABLING A SUSTAINABLE VALUE CHAIN IN THE EU?

Andreas Wade, International Thin-Film Solar Industry Association (PVthin), Brussels, Belgium
Thomas Sauer, President & CEO, EXXERGY GmbH, IECRE PV SWG Convenor, Gräeflfing, Germany
Holger Neuhaus, Fraunhofer Institute for Solar Energy Systems ISE, Freiburg, Germany
Leonhard Probst, Fraunhofer Institute for Solar Energy Systems ISE, Freiburg, Germany
Roger Taylor, Director Business Development, EXXERGY GmbH, Frankfurt, Germany
David Moser, Institute for Renewable Energy, Eurac Research, Bolzano, Italy
Carsten Rohr, Vice-Chair, European Solar Manufacturing Council (ESMC), Brussels, Belgium
Raffaele Rossi, Senior Policy and Market Analyst, SolarPower Europe, Brussels, Belgium

ABSTRACT: As the PV sector continues to grow, becoming an increasingly important contributor to the electricity supply mix, proving true sustainability is becoming increasingly important to support the promise of a viable, long-lasting, and eco-friendly electricity source. Various data suggest room for improvement when it comes to reliability and sustainability, and therefore, several initiatives from the private sector and from governments – including the EU Commission – have been launched.
Following the conclusion of the European Commission's preparatory study for Eco-Design, Energy Labeling, Green Public Procurement and Ecolabeling for PV modules, inverters and systems, a Joint Mission Group (JMG) of international PV experts evaluated the proposals and provided recommendations on the parameters, structure and benchmarks for the implementation of sustainable product policy instruments for PV in the European Union. The closely interlinked topics of product lifetime, quality and reliability, energy yield and sustainability performance require a holistic approach, to ensure comparability and transparency for electricity prosumers and a level playing field for the value chain actors – the combination of mandatory product requirements, backed by quality, reliability & sustainability standards, information requirements on performance through prosumer labels and incentivization of more sustainable products can become the backbone of a reinstated European PV value chain along all life cycle stages. Included in the recommendations of the JMG is the concept of an "Environmental Impact Index".
Keywords: Sustainability, Photovoltaic, PV, module, inverter, amount of damage, internal defects, lifetime, PV power plant, insurance claim case, EU Commission, Energy Label, Eco-Design, Ecodesign, Eco-Label, Ecolabel, Green Public Procurement, GPP, Joint Mission Group, JMG, ETIP PV, SolarPower Europe, PVThin, ESMC, IECRE, World Energy Outlook, Product Environmental Footprint, PEF, Product Environmental Footprint Category Rules, PEFCR, Corporate Social Responsibility, CSR, Environmental, Social & Government, ESG, Environmental Impact Index, EII

1 INTRODUCTION

The number of renewable energy power plant installations worldwide has increased exponentially over the last 20+ years. The PV market has evolved from one of relatively small-scale applications to a mainstream electricity source with a trend towards large utility scale PV power plants, while residential, commercial and industrial applications continue to represent a major share of installed capacity as well. While having experienced this incredibly dynamic growth over the last 20 years, the solar sector currently represents only approx. 3-4% of worldwide annual electricity production (depending on the source of the data). As the PV sector becomes increasingly visible as a viable electricity generation source aiming for the next order of magnitude - representing 20% in the electricity mix and beyond at some point in the future - it is now a good time to discuss the reliability, predictability, and performance of PV power plants and as a consequence, the improvement potential for sustainability and government initiatives related thereto.

Building on this train of thought, this article will provide a short introduction of concerns related to the sustainability of PV power plant performance, and hence, sustainability; predominantly related to technical performance. Following, the EU Commissions' sustainability initiatives as well as the status of the consultation input by industry will be discussed.

2 CONCERNS RELATED TO SUSTAINABILITY IN THE PV SECTOR

There are various technical reasons why a PV power plant can underperform or completely fail. A review of the downwards cost trend of turnkey PV power plants over the last 20 years is very impressive. However, whether some stakeholders involved in the PV power plant value chain exercise poor quality practices because of distressed financials is another question. Regardless of the motivation, within the realm of the energy transition, there is an increasing concern about whether the PV "value proposition" lives up to the promise of providing a sustainable means of generating electricity. This means that the reputation of PV may eventually be at stake.

Some evidence on the viability of this concern is given by an insurance claim case study in which the root causes for more than 3.600 insurance claim cases has been conducted, all of which occurred in the Northern hemisphere at latitudes North of 35°. The following outlines an excerpt of a significantly more detailed study[1].

The insurance claim records date from the time between January 2012 and June 2017 so that a history of 5.5 years of insurance claim cases was covered in the study. The spread of damages where compensation to asset owners was paid is significant with a peak value of more than 110% relative to investment (>1,100 EUR per 1,000 EUR investment). Looking at the relationship of damage to nameplate power, the maximum damage was 3,250 EUR/kWp installed capacity. Note that both maximum numbers originate from different claim cases. Not surprisingly, no linear correlation could be drawn for either relationship, neither for the damage amount versus investment nor for the damage amount versus the installed capacity. The decline in frequency of claimed damages over the amount of loss follows a nearly logarithmic relationship: Damage cases with high amounts of loss are rare whereas minor damage cases occur more frequently. Figure 1 shows the frequency of damage claim cases on a logarithmic scale as a distribution of the amount of loss data in €/kWp for those claim cases for which the amount of loss has been covered.

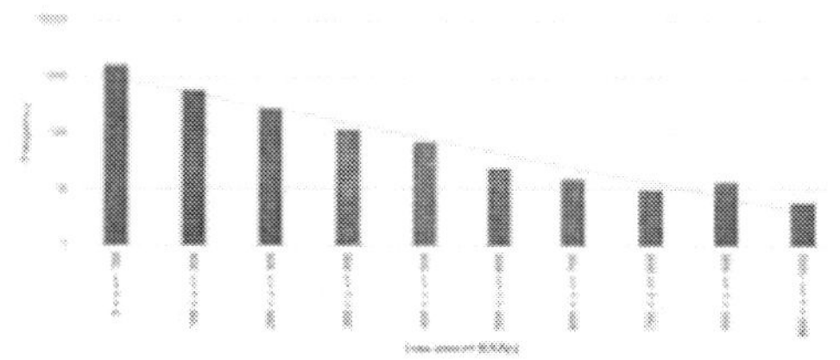

Figure 1: Distribution of amount of loss classification

One of the findings of a "deep-dive" analysis reveals that the average amounts of loss incidents over the service life of a PV power plant increases significantly for internal defects, see

Figure 2. shows the average amount of damage resulting from so called "internal defects". Internal defects are defects resulting from bad manufacturing or installation practices, hence, result from the internal integrity – or health – of the PV power plant. Besides the insight of a "to be expected" increase in amounts of damage over lifetime, another important finding is that average amounts of damage can be quite significant and higher than expected prior to the study. The standard deviation of the average amount of damage varies between 14.8% and 32.5%, the average amount of loss significantly increases with service lifetime of a PV power plant. It is important to note that many internal defects subject to the assessment have not been insured.

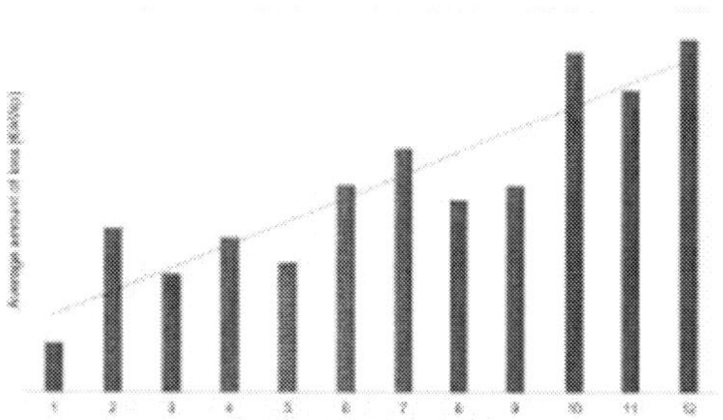

Figure 2: Average amount of loss as a function of service life (internal defects)

More insights of the insurance claim case study and questions related to investment risk mitigation have been discussed in the past[2].

The results naturally trigger an additional aspect of how to look at sustainability. Besides using "eco-friendly" (ideally recycled) materials and components, producing good quality products and systems throughout the value chain that are designed and constructed for longevity is very important aspect of sustainability. Further important aspects relate to the "carbon footprint" producing and decommissioning a PV power plant versus the actual energy yield of a PV power plant. Reparability and end-of-life recyclability of components are also important aspects to be factored into the sustainability "equation".

3 EU AND INDUSTRY SUSTAINABILITY INITIATIVES

The factors discussed above are only a small excerpt of many – some obvious – triggers of why sustainability has eventually become a paramount goal for governments as well as for investors. As stated in the EU Green Deal Communication[3], the just transition to climate neutrality and a sustainable future is the overarching objective of all EU policies and actions. The decarbonisation of the energy sector, investments into environmentally friendly technologies and helping industry innovate are essential elements of this transition. The proposed European Climate Law[4] and the Green Deal Key Actions[5] build on the Clean Energy Package's Renewable Energy Directive (RED II)[6] targets for the mass deployment of solar, wind, hydro and other forms of sustainable energy conversion technologies. Solar photovoltaics could play the leading role in this transition: it is quickly becoming the most economic form of electricity generation worldwide, and consequently the IEA has declared PV to be the "new king of electricity" in its World Energy Outlook 2020[7]. The pathway to terawatts of solar being deployed required a number of dedicated policy and standardisation activities to ensure that the photovoltaic (PV) value chain grows in a manner that is resource efficient, environmentally safe and protective of human health. The PV industry today is working effectively towards decoupling economic growth from negative environmental impacts and resource consumption.

Over the last five years, the PV industry has pro-actively engaged with regulators, policy makers and wider stakeholder groups, to quantify the environmental performance of PV technologies and demonstrate the tangible advantages of the different PV technologies available. In the EU, the PV industry participated in the Product Environmental Footprint (PEF) Pilot Phase[8],and developed sectoral Product Environmental Footprint Category Rules (PEFCR) for Photovoltaic Modules used in photovoltaic power systems for electricity generation[9]. This validated the environmental performance of PV technologies in the EU, and helped better inform decisions on what EU sustainable product policies would be most appropriate for this category of products. The PEF pilot phase, the development of the PEFCR as well as the related pilot and screening studies, clearly identified the environmental hotspots in the life cycle of PV systems, aiding in the development of voluntary industry standards to address these hotspots at a global level[10], [11].

Building on the results of the PEF pilot phase, the European Commission added photovoltaic panels and inverters to the work program for Eco-Design in 2016 and extended the Preparatory Study[13] carried out from 2017 to 2019 to also assess whether sustainable product policy instruments such as the EU Energy Label, Ecolabel and Green Public Procurement would be appropriate for the PV industry, see Figure 3.

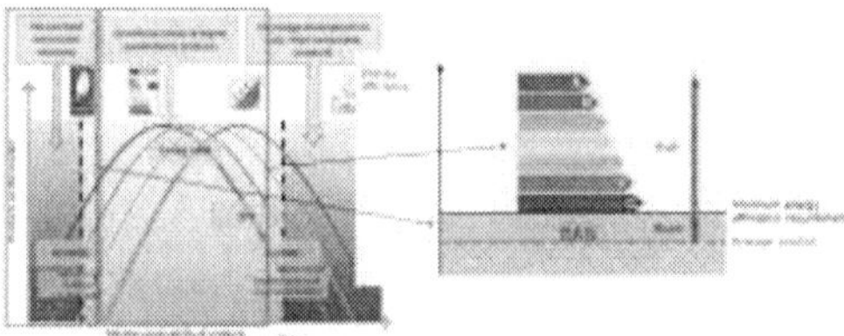

Figure 3: Overview of policy instruments considered by the EU Commission

The Commission's policy scenario evaluation concluded that the best way to further regulate PV modules was via a combination of mandatory and voluntary policy instruments. This scenario evaluation considered mandatory instruments such as Eco-Design measures for photovoltaic modules and inverters, augmented by the use of the Energy Label for residential PV systems, and voluntary instruments such as Green Public Procurement Criteria. The latter aspects will be developed between 2020 and 2023 to focus on a number of sustainability, quality, durability, circularity and performance criteria[14].

4 THE JOINT MISSION GROUP: INCLUDING THE INDUSTRY PERSPECTIVE

In order to inform the development of these criteria, representatives of the photovoltaic value chain came together to set up a Joint Mission Group under the umbrella of the European Technology Innovation Platform Photovoltaics (ETIP PV) - in cooperation with SolarPower Europe, PVThin, the European Solar Manufacturing Council, and IECRE - to review the results of the preparatory study and provide recommendations for the next regulatory discussions.

As outlined below (Figure 4), between June 2020 and January 2021, the Joint Mission Group (JMG) followed a structured approach to provide a comprehensive review and set of recommendations to support the discussions on the compulsory policy instruments of Eco-Design and the Energy Label – complemented by a dedicated investigation of the cross-cutting and more general sustainability criteria.

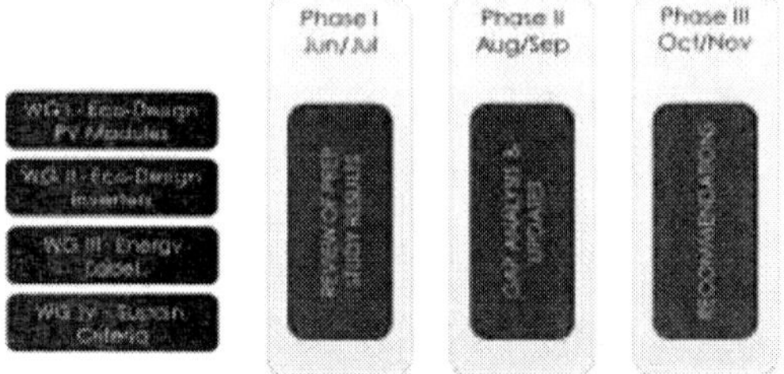

Figure 4: Joint Mission Group approach (2019)

The Expert Input Paper comments on the policy recommendations set out in the final report[14] and extends those to reflect on identified gaps and required updates from the perspective of the solar industry & solar research community. The first part of that report is structured as direct feedback to the policy recommendations. The second part of the report provides additional recommendations from the expert group to achieve the policy objectives.

A final comment in the introduction is to recommend periodic reviews of the directives every 2 years.

5 SUMMARY OF RECOMMENDATIONS FOR POLICY MAKERS

5.1 General remarks

Photovoltaics is a proven technology capable of making a substantial contribution to a sustainable global energy system. Its widespread use in all geographic regions, versatility in application and modularity in scale enables a socially acceptable energy transition by offering distributed electricity generation, employment and new business opportunities. [15]

The Joint Mission Group welcomes the policy recommendation on the introduction of eco-design requirements for photovoltaic modules and inverters in the EU. These future requirements should be based on standards, which determine the service life, energy yield and degradation – which are the most important parameters influencing the sustainable performance of these components. Given the longevity of the components and the fast evolution of new products & technology concepts, reference values established through accelerated life cycle testing and lifetime yield prediction should provide minimum requirements for performance guarantees, replacement and reparability within the eco-design regulations.

An Energy Label recommendation for residential scale photovoltaic systems is more challenging. Such a system runs a risk of being complex because of the number of important variables and consequently, of confusing and disincentivising the electricity prosumer – when every kilowatt peak of solar electricity generation capacity installed provides significant environmental and societal benefits in achieving the green transition of the EU Economy. An Energy Label should not disincentivise development of specific applications, but rather ensure transparency of environmental and quality performance of the system components deployed to allow conscious and informed choices.

The Joint Mission Group has provided detailed feedback on the issues it has identified and provided recommendations for addressing these issues. In many cases, these recommendations have been adopted.

5.2 Additional recommendations

As well as commenting on the specific proposals, additional recommendations pertaining Eco-Design requirements for PV modules and inverters, as well as considerations and calculations on a potential Energy Label for residential photovoltaic systems were provided.

In addition, a holistic approach for sustainability evaluation in the form of an Environmental Impact Index (EII) is proposed and referenced to applicable standards and methodologies. This Environmental Impact Index that is further elaborated in section 5.6 would satisfy the information requirements established in the policy recommendations and would help to trigger consumer interest as well as market pull for more sustainable products. Ultimately, further discussions of the EEI framework - also possibly including quality and reliability requirements - could lead to the definition of multi-dimensional green public procurement requirements for PV electricity (and components) in the EU, satisfying the objectives of the EU Green Deal & Green Recovery ambitions.

5.3 Eco-Design for Photovoltaic Modules

The key variables in determining the life-time total energy yield of a PV module are its service life and its degradation rate. The eco-design proposal assumes a 30 year lifetime for all module technologies and a degradation rate determined by a procedure and common for all modules regardless of the technology. This is considered by industry experts to be unreliable and time-consuming. The JMG assessed 3 alternatives to this approach and recommends a performance guarantee for module degradation and service life will be complied by the module manufacturers. If the modules fail to achieve this performance the manufacturers have to replace the modules in the event of reduced performance. The manufacturer will therefore carefully set the guaranteed degradation and lifetime and carry out appropriate tests. An important addition to the alternative approach is the establishment of minimum guarantee requirements that are comprehensively defined in the eco-design directive.

5.4 Eco-Design for Photovoltaic Inverters

The environmental impact of photovoltaic inverters is largely determined by its expected lifetime. Significant improvement of this factor could be achieved by focussing on the inverter's reparability. This should not be met by defining minimum warranty requirements, as this could lead to negative incentives such as replacing the inverter during the warranty time, which could increase the overall footprint.

In order to promote inverter reparability, the Eco-Design measures should ensure the availability of spare parts and that the inverter is constructed to allow for access to and replacement of identified parts, and that spare parts are available over a long period of time after purchase. Spare parts should be replaceable with the use of commonly available tools and without permanent damage to the inverter. In order to enhance the repair market, manufacturers would have to ensure the availability of repair and maintenance information for professional repairers.

5.5 Energy Label for residential-scale systems

The policy recommendation on the introduction of an energy label suggests a label for the entire solar photovoltaic system deployed on residential rooftops. Here, a small number of system performance factors such as the energy yield of the module, the efficiency of the inverter, the orientation of the module and the location are taken into account. Given the overarching policy objective to improve the sustainability performance of the different system components, the proposed methodology for the determination of the energy performance falls short on its ability to provide component level differentiation.

An informed electricity prosumer in the European Union should have comprehensive and holistic sustainability performance data available for the photovoltaic modules and inverters upon purchasing those components. The lifecycle environmental impact of these system components is well understood, and lifecycle hotspots have been identified and can be effectively addressed by creating market pull for more sustainable products. The introduction of an independently validated combined Energy & Environmental Impact Index, embedded in a quality conformity assurance framework is seen as the most effective means to enable this transparency and to induce the continuous improvement in sustainability performance of these product groups.

It is important that unbundled renewable energy certificates are excluded in the supply chain of electricity for module production and rather the country's electricity mix has to be used to prevent double counting or misallocation of lifecycle impacts. In addition, consideration should be given to whether ESG (environment, social and governance) criteria should be taken into account for the assessment of manufacturing and the manufacturer in order to promote an ecological, social and transparent way of doing business in the EU.

To inform the future discussions on the development of a calculation methodology for residential scale photovoltaic systems, the Joint Mission Group performed and presented a number of benchmark calculations for reference systems, using the specified PV System Calculator as well as commercial energy yield prediction and assessment tools.

5.6 Holistic evaluation of sustainability performance - Environmental Impact Index (EII)

The environmental impact of a PV power plant requires a system perspective. Naturally, the environmental impact of each component is an indispensable compliance criterion; however, it is ultimately the entire system that needs to comply just as well with sustainability criteria.

Assessing the environmental impact of any product or system is a complex task. An environmental impact index is a possible means to translate the complexity of an environmental assessment into an abbreviated form. The effect of such an index would be to condense the various influencing factors of industrial or other human activity on the environment in a summarizing, comprehensive index in a way so that impacts from alternative actions can be compared and add to product or system evaluation. Figure 5 illustrates the elements of the proposed EII concept.

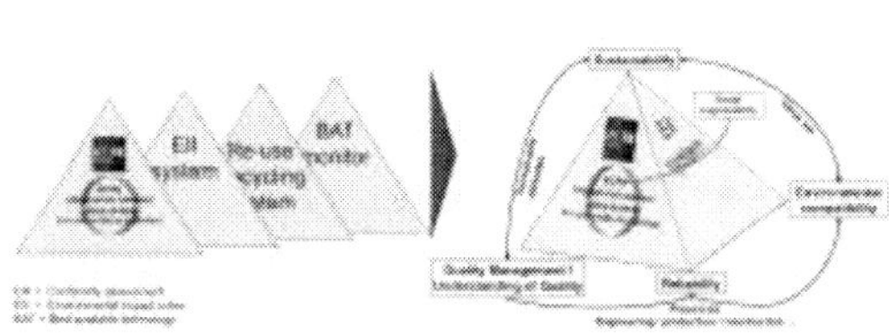

Figure 5: Elements of the Environmental Impact Index (EII)

Yet subject to further development, Figure 6 and Figure 7 outline the state of discussion on details related to the proposed concept, scale, and evaluation criteria of an EII scheme for modules, while Figure 8 outlines the corresponding details for inverters. The potential evaluation criteria are still subject to further discussion. The right hand side of Figure 6 and Figure 8 depict a possible visualization of the resulting EII.

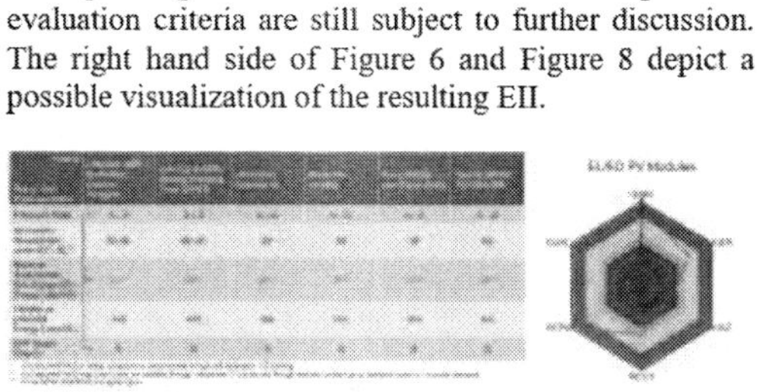

Figure 6: Details of the EII concept and visualization for modules

Figure 7: Possible evaluation criteria for the EII for modules

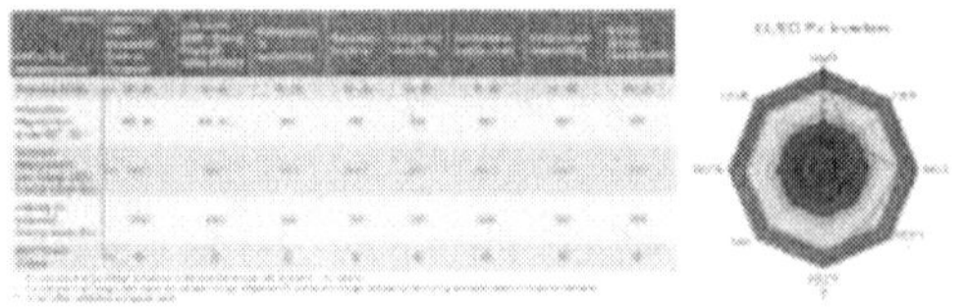

Figure 8: Details of the EII concept and visualization for inverters

The development of such an index could be undertaken in close cooperation between the standardization body called for by the EU Commission and an industrial expert group that can be managed by a neutral (non-profit) entity and the results incorporated into assessments for GPP and Eco Label.

5.7 Recommendation to include CSR (Corporate Social Responsibility) criteria

Considering the EU strategy on promoting CSR and encouraging businesses to adhere to international guidelines[16], several initiatives have been launched starting from the EC communication 2011[17], to the directive 2014/95/EU[18] and the study commissioned by EU parliament 2020[19] . CSR and RBC (Responsible Business Conduct) have been clearly identified as the core concepts to manage the negative impacts on society and the environment by preventing and mitigating them including the global supply chain. The last study published in November 2020, covers the international law instruments on CSR from the UNGPs[20] (United Nations Guiding Principles on business & human rights), the ILO - MNE Declaration [21] (International Labour Organization – Multinational Enterprises) and the OECD guidelines and guidance[22] (Organization for Economic Co-operation and Development). The international instruments are considered as "minimum thresholds the companies can take into account in the context of CSR debate". As a strong support of the United Nations 2030 Agenda for sustainable Development and associated Sustainable Development Goals (SDGs), the European Commission "has taken a very active approach to CSR at EU level". The EU analysis in the context of the COVID crisis has raised that the implementation of CSR within companies implies extra costs, however, creates in parallel significant value creation in terms of innovation and benefits. Moreover, the European Commission has raised the clear demand of the citizens for sustainable business initiatives. Accelerating the incentives on CSR legislative approach, the European Commission has announced the revision process 2014/95/EU in 2020 to "strengthen the foundations for sustainable investment".

The development of sustainable finance, so called green financing models based on ESG (Environmental, Social & Governance) is expanding very rapidly and will necessarily impact the energy sector investment framework. Indeed, on the front line, and to answer the carbon neutrality target, renewable energy investments are expected to be one of the most important release of funds in the coming years. Many European oil companies such as BP (UK), Shell (UK/NL), Eni (Italy), Total (France), Respol (Spain), Equinor (Norway) have announced in 2020 that their objectives to significantly reduce their oil and gas activities. In the meanwhile announcing their intentions to invest massively in low-emission activities, including mostly renewable energies. This type of investment will necessarily be deployed in the context of green financing. The photovoltaic sector will have to meet sustainability criteria including social criteria that are required to access financing from the "sustainable" taxonomy and integrate the portfolios of major financial institutions.

6. CONCLUSIONS

The European Commission is progressing the development of three new directives to improve the energy transition, enabling energy security and resilience for critical energy technology in the European Union and to ensure the sustainability of PV components and systems as the alternative.

Two of these directives, Eco-design for residential PV power systems and Eco Label for PV modules and inverters will be mandatory requirements for products sold within the EU. The Green Public Procurement (GPP) requirements will be voluntary.

The industry based Joint Mission Group has been actively involved in considering and providing feedback and recommendations on the EC proposals for Eco-design and Eco Labels.

A number of recommendations that have so far not been included within the regulatory design and the reason why they are considered relevant are highlighted in this paper.

The industry and scientific expert community has demonstrated a common approach to this topic, which has paramount importance, and will continue to participate in the discussions.

REFERENCES

[1] Consortium project PVScan – Evaluation research of quality assurance and assessments of PV modules in solar power plants – work package 5 titled "current insurance products and detection of relevant data" (original language: German)

[2] EPJ Photovoltaics 10, 7 (2019): Solar investment risk mitigation – are we all on the same page?

[3] Communication from the Commission: The European Green Deal; COM(2019) 640 final

[4] Proposal for a REGULATION OF THE EUROPEAN PARLIAMENT AND OF THE COUNCIL establishing the framework for achieving climate neutrality and amending Regulation (EU) 2018/1999 (European Climate Law); COM(2020) 80 final

[5] Annex to the Communication on the European Green Deal Roadmap - Key actions; COM(2019) 640 final

[6] DIRECTIVE (EU) 2018/2001 OF THE EUROPEAN PARLIAMENT AND OF THE COUNCIL of 11 December 2018 on the promotion of the use of energy from renewable sources (recast)

[7] IEA World Energy Outlook 2020, https://www.iea.org/reports/world-energy-outlook-2020

[8] Product Environmental Footprint (PEF) Pilot Phase, https://ec.europa.eu/environment/eussd/smgp/ef_pilots.htm

[9] PEFCR PV electricity v. 1.1; https://ec.europa.eu/environment/eussd/smgp/pdf/PEFCR_PV_electricity_v1.1.pdf

[10] Wade, Andreas, Philippe Stolz, Rolf Frischknecht, Garvin Heath, und Parikhit Sinha. „The Product Environmental Footprint (PEF) of photovoltaic modules-Lessons learned from the environmental footprint pilot phase on the way to a single market for green products in the European Union“. *Progress in Photovoltaics: Research and Applications*, 2017. https://doi.org/10.1002/pip.2956.

[11] Wade, Andreas, Parikhit Sinha, Karen Drozdiak, Dustin Mulvaney, und Jessica Slomka. „Ecodesign, Ecolabelling and Green Procurement Policies – enabling more Sustainable Photovoltaics?“ In *WCPEC-7, Proceedings of the 7th World Conference on Photovoltaic Energy Conversion*. Hawaii, USA: IEEE, 2018.

[12] Ecodesign Working Plan 2016 – 2019, COM(2016) 773 final

[13] JRC, Solar Photovoltaic modules, inverters and systems preparatory study; https://susproc.jrc.ec.europa.eu/solar_photovoltaics/

[14] Preparatory study for solar photovoltaic modules, inverters and systems – Final report; Dodd, N., Espinosa, N., Van Tichelen, P., Soares, A., JRC B.5 unit, VITO, 2020

[15] See also: Photovoltaic Solar Energy: Big and Beyond – Sustainable Energy to Limit Global Warming to 1.5 Degrees – Vision and Claims of the European Technology and Innovation Platform for Photovoltaics (ETIP PV), 2020

[16] Corporate Social Responsibility – Recommendations to the European Commission By the subgroup on « CSR » of the Multi-Stakeholders Platform on the Implementation of the SDGs in the EU https://ec.europa.eu/info/sites/info/files/recommendations-subgroup-corporate-social-responsibility_en.pdf

[17] Directive 2014/95/EU amending directive 2013/34/EU as regards disclosure of non-financial and diversity information by certain large undertakings and groups https://eur-lex.europa.eu/LexUriServ/LexUriServ.do?uri=COM:2011:0681:FIN:EN:PDF

[18] https://eur-lex.europa.eu/legal-content/EN/TXT/PDF/?uri=CELEX:32014L0095&from=EN

[19] Corporate social responsibility (CSR) and its implementation into EU Company law https://www.europarl.europa.eu/RegData/etudes/STUD/2020/658541/IPOL_STU(2020)658541_EN.pdf

[20] United Nations Guiding Principles on business & human rights https://www.ohchr.org/Documents/Publications/GuidingPrinciplesBusinessHR_EN.pdf

[21] Tripartite Declaration of Principles concerning Multinational Enterprises and Social Policy (MNE Declaration) - 5th Edition (2017) https://www.ilo.org/empent/areas/mne-declaration/lang--en/index.htm

[22] OECD – Guidelines for Multinational Entreprises http://www.oecd.org/daf/inv/mne/48004323.pdf & OECD - Due Diligence Guidance for RBC http://mneguidelines.oecd.org/OECD-Due-Diligence-Guidance-for-Responsible-Business-Conduct.pdf

A COMPARATIVE LIFE CYCLE ASSESSMENT OF PV MODULES – INFLUENCE OF DATABASE AND BACKGROUND SYSTEM

Sina Herceg[1*], Ann-Kathrin Briem[2], Marie Fischer[1], Peter Brailovsky[1], Tobias Dannenberg[1], Michael Held[2]
1 Fraunhofer-Institute for Solar Energy Systems ISE, Freiburg, Germany
2 Fraunhofer-Institute for Building Physics IBP, Stuttgart, Germany
* sina.herceg@ise.fraunhofer.de

ABSTRACT: The installed capacity of photovoltaic modules is increasing worldwide. Simultaneously, sustainable production is becoming more important in the PV sector. To ensure sustainable growth, politics and industry are acting in bringing forward environmental labels and certificates. Oftentimes, values for standards and thresholds are based on life cycle assessment (LCA). Despite all efforts taken towards the comparability of the ecological performance of products within one product category, there are still uncertainties about the reproducibility of results calculated with different LCA software tools and databases. Therefore, we compare the life cycle impact of a state-of-the-art Passivated Emitter and Rear Contact (PERC) solar cell assessed with two of the most commonly used software/database combinations (SimaPro/Ecoinvent and GaBi/GaBi database). We found conformity between the top processes of our hot spot analysis and good reproducibility for categories such as *Climate change* and *Resource use, fossils*. For categories sensitive to the underlying regionalization and for processes that are strongly affected by the underlying production mix, we found large deviations. We therefore conclude that results from different databases/software combinations should not be mixed in comparative LCA. Instead, a harmonized database should be used, as it is already done in certain cases.

Keywords: CO_2-footprint, Sustainable, Module Manufacturing, PERC, Product Environmental Footprint

1 INTRODUCTION

With a steadily growing market [1], the environmental impact of the PV value chain is increasingly becoming the focus of legal regulations and standards, such as the EU energy label or the EU eco-design directive. The Product Environmental Footprint Category Rules (PEFCR) recommend Life Cycle Assessment (LCA) based methods to quantify the environmental impacts of products and thereby aim to harmonize the modelling approach within one product category [2,3]. Although the carbon footprint of PV electricity generation is low compared to conventional fossil-based electricity, various other environmental impacts are found along the PV value chain [4,5]. Next to a high energy demand, local emissions originating from exhaust air and wastewater streams can be associated with PV production facilities. As production energy consumption is gradually decreasing and the share of clean energy is constantly increasing in most countries, it is becoming more relevant to quantify the locally caused emissions with more accuracy. Strict regulations and innovative concepts, such as zero liquid discharge factories, also support the efforts of reducing factory emissions in the PV supply chain.

Numerous LCA studies on PV technologies and PV systems can be found in the literature. In 2012, a harmonization study of the National Renewable Energy Laboratory (NREL) screened almost 400 studies related to LCAs of PV and found only 13 studies regarding c-Si technologies that met the criteria to be harmonized, mostly due to missing transparency [6]. Since then, many more studies have been published, however, the call for more transparency, methodological consistency and the use of up-to-date inventories remains [7,8]. Furthermore, most studies focus only on greenhouse gas emissions in their analysis and exclusively evaluate the global warming potential. Additional impact categories are often neglected, although they address important environmental issues in the context of the PV value chain [9,10,11]. Nevertheless, with the PEFCR for PV in place, more studies start to evaluate the more diverse EF 3.0 indicators as a set [12].

Different software/database combinations can be used to perform an LCA study. The most widely used software packages are SimaPro, GaBi, Umberto and openLCA, either in combination with the Ecoinvent or the GaBi LCI-database. Since the implementation of Life Cycle Impact Assessment (LCIA) methods and background data differ between all these software/database combinations, differences in LCIA results can occur. For example, the underlying methodological approaches for the modelling of background data can vary from database to database. Besides the possibility of differences in the boundary conditions of available LCI inventories and included processes (e.g. reference years, technology references, differences in production mixes), the LCA databases might also follow different methodological approaches (e.g. allocation method, system boundaries) for the conception of life cycle inventories, which then affect the calculated LCIA results. Several studies have been published that report major differences for certain indicators using calculations from different LCA software and databases [13,14,15]. Some of these studies focus on the different implementation of LCIA methods and characterization factors [16,17,18], whereas others also focus on the differences in the implemented LCI database [13,14]. These studies also find that the use of the Ecoinvent database often leads to higher environmental impacts in comparison to the GaBi database. Furthermore, these studies identify the energy use as one source of error, due to differences in reference years between databases. Since the energy use has a high influence in the background systems of production processes of materials and intermediates, these differences usually strongly affect the overall environmental burden.

Since commercial LCA databases contain a multitude of datasets, the differences that other studies find for a specific product system, e.g. for packaging [14], cannot easily be generalized to other product systems, e.g. PV systems. Furthermore, characterization factors vary between different impact assessment methods. Therefore, in this study we specifically address the topic of comparability of PV LCA results calculated with the EU recommended EF 3.0 impact assessment method.

2 METHODS

All the efforts taken towards a comparability of the ecological performance of products within one product category raise the question of reproducibility of results between different stakeholders. Therefore, we perform an LCA of a PV module equipped with multi-busbar interconnected Passivated Emitter and Rear Contact (PERC) solar cells processed on a recharged Czochralski (Cz) grown mono-Si wafer. The investigated module is a c-Si bifacial Selective Emitter (SE) 60-cell p-type Cz PERC module. The functional unit is chosen as 1 m^2 PERC module with the described specifications. The focus lies on the manufacturing phase, excluding the usage (electricity generation) and end-of-life. Therefore, the system boundary is chosen as cradle-to-gate. The chosen focus is based on the aim to raise potentials for green manufacturing pathways in the PV value chain. The assessment includes a Chinese (CN) and a German (DE) based production scenario, for which the datasets are chosen according to the location whenever possible. For the impact assessment, the EF 3.0 methodology is chosen. Furthermore, we use two different LCA software tools with different life cycle inventory (LCI) databases for the modelling of the background systems respectively. The analysis is carried out using the GaBi software and the GaBi LCI-database version CUP 2021.1 and, simultaneously, the SimaPro software and the LCI database Ecoinvent 3.6. To ensure the comparability of the two models, the same up to date LCI is used to model the foreground system. The defined system boundary, functional unit (1m^2 of PV module produced) and life cycle impact assessment method (EF 3.0) are implemented consistently in both models and compared in detail.

3 RESULTS

By comparing the results generated with two of the most commonly used LCA-software tools and LCI-databases, we are able to show how the implementation of certain process flows and allocation choices in the background system influence the LCA results even when exactly the same data is used to model the foreground system. On the one hand, a hot spot analysis of 16 indicators of the EF 3.0 LCIA method shows comparable results for those categories that are expected to be most sensitive for impacts from energy consumption, namely *Climate Change* and *Resource use, fossils* (see Table 1). Overall, the main impacts are caused by electricity use in almost all categories, but also the main materials on the module level, such as glass, aluminum for the frame and copper, as well as the silver content in the metallization pastes have a significant impact. On the other hand, a number of impact categories, namely *Eutrophication, freshwater* and *Resource use, minerals and metals*, show significant differences between the two database systems. Regarding *Ozone depletion*, the comparison is not possible, due to the fact that with the Montreal Protocol in place no relevant emissions are released in the analyzed foreground system. However, some background system datasets still contain some flows that influence the results in this category, which do not reflect the real situation. The indicator *Ozone depletion* yields results close to zero in GaBi and similarly low in Ecoinvent and is, therefore, neglected in our in-depth analysis. In total, impacts are higher for the calculations with Ecoinvent. Table I provides an overview of the results for the 16 analyzed EF 3.0 indicators, presented as relative difference of the Ecoinvent results compared to the GaBi results.

Table I: Overview of the differences in mid-point results for all 16 recommended EF 3.0 indicators. Comparison between Ecoinvent and GaBi results given as percentage of GaBi result.

EF 3.0 indicator	Unit	Ecoinvent / GaBi	
		DE	CN
Acidification	[H+ eq.]	207%	148%
Climate Change	[kg CO_2 eq.]	115%	105%
Ecotoxicity, freshwater	[CTUe]	564%	788%
Eutrophication, freshwater	[kg P eq.]	27605%	28000%
Eutrophication, marine	[kg N eq.]	197%	160%
Eutrophication, terrestrial	[N eq.]	180%	151%
Human toxicity, cancer	[CTUh]	269%	227%
Human toxicity, non-cancer	[CTUh]	200%	202%
Ionizing radiation, human health	[kBq U235 eq.]	115%	153%
Land Use	[Pt]	110%	173%
Ozone depletion	[CFC-11 eq.]	-	-
Particulate matter	[Disease inc.]	173%	94%
Photochemical ozone formation	[NMVOC eq.]	179%	147%
Resource use, fossils	[MJ]	142%	124%
Resource use, mineral and metals	[kg Sb eq.]	1660%	1711%
Water use	[m^3 world eq.]	243%	382%

Figure 1 presents the weighted and normalized results of the PERC module production, divided into the most relevant process steps. Using Ecoinvent, module production has the highest impact, followed by cell production, polysilicon processing and wafer production, whereas in GaBi, module production is followed by polysilicon processing, wafer production and cell production. This difference can be traced back to the higher values in the category *Resource use, minerals and metals* in the Ecoinvent scenarios (see Figure 2), which are especially significant for the metallization paste in cell production. Furthermore, Ecoinvent includes background datasets for PV infrastructure (namely Panel Factory, Cell Factory and Wafer Factory) that are not included in the GaBi database. Those infrastructure processes cause noticeable impacts, especially in the categories *Land use* and *Resource use, minerals and metals*.

The influence of each category to the overall weighted and normalized single-score EF 3.0 is illustrated in Figure 2. The SimaPro/Ecoinvent results are dominated by the category *Resource use, minerals and metals*, which contributes to over half of the weighted score (EF 3.0). For the GaBi results, the categories *Climate change, Resource use, fossils* and *Resource use, minerals and metals* are the most relevant. The German scenario (DE) has a lower impact compared to the Chinese scenario (CN) for both software/database systems, especially in the categories *Acidification, Climate change, Eutrophication, marine* and *terrestrial, Particulate matter, Photochemical ozone formation* and *Water use*. Simultaneously, the categories

Ionizing radiation and *Eutrophication, freshwater* are significantly higher in the DE scenario in the results of both software/database systems. These differences can be explained by the differences in the regional electricity grid mix. In the case of China, this is dominated by coal and hydro power, leading to higher impacts in the above-mentioned categories, whereas the German grid mix has a higher share of biomass and biogas as well as nuclear power, which are the source of emissions the categories *Eutrophication, freshwater* and *Ionizing radiation*. *Resource use, minerals and metals* is in a comparable range for the DE and CN scenarios.

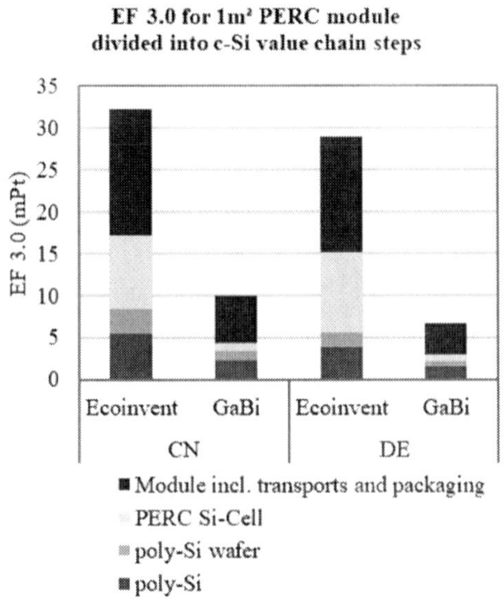

Figure 1: Overall EF 3.0 results (weighted and normalized) for 1 m² PERC module divided into the main process steps.

4 DISCUSSION

4.1 Differences in electricity mixes

The main differences in several impact categories can be traced back to the underlying electricity generation mix for China and Germany. First, the Chinese grid mix has a higher share of coal and, therefore, a larger impact in the *Climate Change* category leading to higher results for the Chinese scenario compared to the German scenario in both databases. Second, the GaBi electricity mix results in a slightly lower impact per kWh compared to the Ecoinvent electricity mix for both the Chinese and the German grid mix. This is mainly due to different reference years (GaBi: reference year 2017; Ecoinvent: reference year 2014/15). The differences in the electricity mixes regarding the impact category *Climate Change* is shown in Figure 1 (top). The deviation in the results for 1 m² PERC module between the databases of 15 % for the German scenario and 5 % for the Chinese scenario can be almost completely traced back to the differences in the electricity mixes, which are 15 % and 27 %, respectively. The smaller deviations in the total *Climate Change* results of the CN production scenarios of GaBi and Ecoinvent of 5 % compared to the differences of the results of electricity mixes can be explained by the use of the differences in the aluminum datasets, which are used for the modelling of the CN aluminum frames (GaBi: CN; Ecoinvent: GLO).

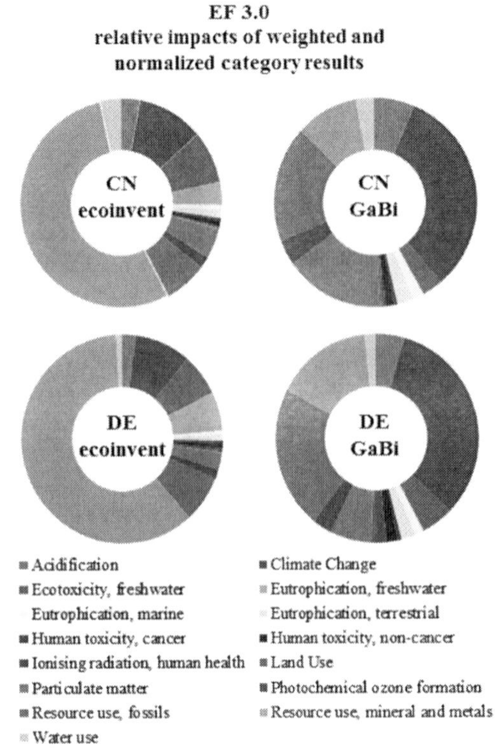

Figure 2: Relative share of each category result to the weighted and normalized endpoint EF 3.0 for each production location (DE/CN) and each software/database system (Ecoinvent/GaBi), respectively.

If the weighted and normalized EF 3.0 is taken into account, it can be seen in Figure 1 (bottom), that the differences between the databases are higher compared to the more reliable/stable *Climate Change* category (Figure 1, top). These differences are, however, even larger in the results for 1 m² PERC module, consequently the influence of the different electricity mixes does not explain the entire deviation. Following the electricity mix, the influence of metals was observed to be high and a major source of differing results between the databases (see section 4.2).

Since the main impact in almost all categories originates from energy consumption, the respective country specific electricity grid mix has the highest influence. Depending on the exact production location, the country specific mix, however, might not represent the local conditions accordingly and regional mixes might paint a clearer picture. Furthermore, a number of PV companies have joined the "RE100" initiative and aim to use 100 % renewable energy by 2050 the latest [19]. Ambitious companies want to reach this goal much earlier. Therefore, currently predominant emissions from electricity generation will be reduced in the future, emphasizing the need to improve manufacturing techniques to unlock further sustainability potentials.

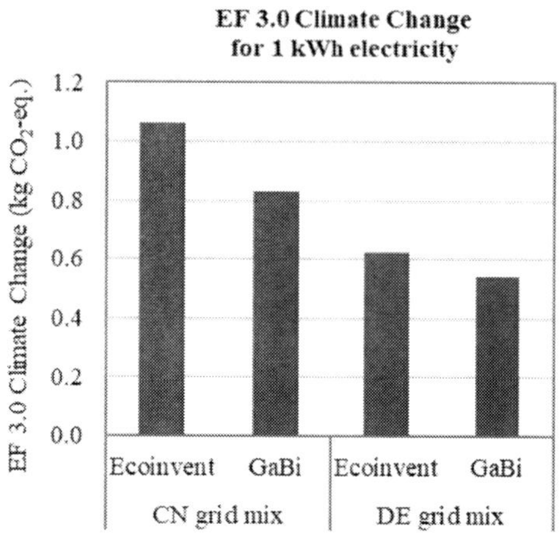

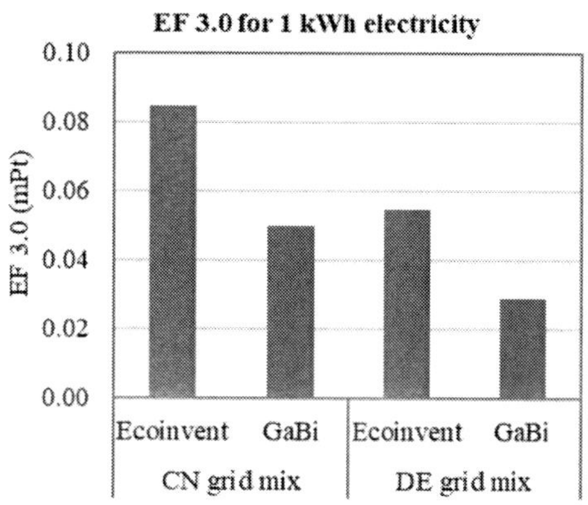

Figure 3: Comparison of the impacts of the electricity grid mix for China (CN) and Germany (DE) in the Ecoinvent and GaBi databases. The upper graph shows the mid-point results for the impact category EF 3.0 *Climate Change* (in CO_2-equivalents) of the analyzed grid mixes, whereas the lower graph shows the normalized and weighted EF 3.0 single-score results, respectively.

4.2 Differences in metals (Ag, Cu, Al)

The use of certain metals in PV module production is another cause for differences between the analyzed results, especially in the category *Resource use, minerals and metals*. The impact of the silver content in the metallization pastes is almost 45 times higher in Ecoinvent/SimaPro compared to GaBi. Also, the aluminum frame of the PV module has a higher impact in the Ecoinvent/SimaPro results. The large difference in EF 3.0 results can be traced back to the category *Resource use, minerals and metals* for the most part in addition to the category *Eutrophication, freshwater*. The different production mix in the background datasets for global production is the major cause for those deviations. Furthermore, differences in the allocation of burdens from multi-output processes, such as metal mining occur between the two software/database combinations.

A significantly higher impact of silver and aluminum in Ecoinvent/SimaPro compared to GaBi can also be observed in the categories *Eutrophication, terrestrial, Particulate matter* and *Photochemical Ozone Formation*. Furthermore, aluminum as well as copper have a larger impact the category *Human toxicity, non-cancer*.

4.3 Other relevant observed differences

The impact in the category *Eutrophication, freshwater* is substantially higher in Ecoinvent/SimaPro compared to GaBi. Those impacts are mainly caused by polysilicon production, electricity use as well as by the metals silver, aluminum and copper. Characterization in EF 3.0 for *Eutrophication, freshwater* only considers phosphate compounds. In Ecoinvent, the emissions can be traced back to the usage of coal in the energy mix since coal mining releases phosphates into the freshwater bodies. In GaBi, relevant contributions to *Eutrophication, freshwater* are mainly caused by waste water processing, agricultural processes and electricity generation from biomass and biogas in the background systems.

The category *Water use* is strongly influenced by regionalized water flows. Small differences in the background models can cause strong deviations. To get meaningful results in this category, regionalization must be implemented consistently and with a level of great detail. For a dedicated water footprint, all flows should be meticulously checked. For example, if a water flow in a standard dataset from the database is not regionalized, it is taken into account using average impact factors for Water scarcity. Depending on the region, this average factor can be many times higher compared to a correctly regionalized flow and have a potentially large influence on the category result. Similarly, the category *Land use* is also strongly dependent on specific modelling and regionalization of land use activities related to the product system, and, consequently, sensitive to flows in background datasets that do not reflect the specific conditions of the foreground system.

5 CONCLUSIONS

Our research has shown that LCA with different database/software combination produces comparable results for the categories that are considered reliable, such as *Climate change* or *Resource use, fossils*. Furthermore, a hot spot analysis revealed the same environmental hot spots of the PV module under consideration in both analyses. Still, categories that are sensitive to small changes, such as *Water use* or *Ecotoxicity*, differ greatly in our analysis regarding absolute numbers and, in some instances, the identified hot spots. Furthermore, Ecoinvent results were higher in all categories, which is consistent with the observations by other authors.

These differences are to be expected for the absolute results when using different database and software combinations, as the models are based on varying methodological approaches and implementation of background datasets. As LCA is a complex approach, which considers thousands of different energy and material flows in each analysis, none of the alternatives we compared can be considered as advantageous. All models have different advantages and disadvantages, and it is important to acknowledge these uncertainties. The analysis demonstrated in this work identifies current sensitivities of PV LCAs in Ecoinvent and GaBi regarding the PEF indicators and, therefore, provides valuable information for decision-makers and other LCA practitioners.

The interpretation and communication of LCA results requires a strong background knowledge as well as a detailed knowledge of the product system under study and the chosen LCIA method. Different databases and software tools should not be mixed for comparative LCA. This must also be taken into account when LCA result from different studies are compared. Instead, a harmonized database should be used as it is already done with the Environmental Footprint database on EU level or on national level.

Meanwhile, efforts to achieve comparability in LCA studies are taken by regulators and LCA practitioners. This work contributes to this by providing a detailed analysis using the two most commonly applied LCA software and database systems and recommended impact indicators.

6 ACKNOWLEDGEMENTS

This work is conducted within the project "Green Manufacturing" funded by the German federal Ministry for Economic Affairs and Energy (BMWi, funding code: 03EE1048A). In the project, experts from leading research institutions and industrial companies are working over a period of three years to develop and evaluate ecologically and economically efficient PV production processes.

7 REFERENCES

[1] IRENA (2019) Global energy transformation: A roadmap to 2050 (2019 edition). International Renewable Energy Agency. Abu Dhabi. Available online at https://www.irena.org/-/media/Files/IRENA/Agency/Publication/2019/Apr/IRENA_Global_Energy_Transformation_2019.pdf.

[2] European Commission (2020) Product Environmental Footprint Category Rules (PEFCR). Photovoltaic modules used in photovoltaic power systems for electricity generation. Available online at https://ec.europa.eu/environment/eussd/smgp/pdf/PEFCR_PV_electricity_feb2020_2.pdf, checked on 2/1/2021.

[3] Wade, A.; Stolz, P.; Frischknecht, R.; Heath, G.; Sinha, P. (2018) The Product Environmental Footprint (PEF) of photovoltaic modules-Lessons learned from the environmental footprint pilot phase on the way to a single market for green products in the European Union. In Prog. Photovolt: Res. Appl. 26 (8), pp. 553–564. DOI: 10.1002/pip.2956.

[4] Stamford, L.; Azapagic, A. (2018) Environmental Impacts of Photovoltaics: The Effects of Technological Improvements and Transfer of Manufacturing from Europe to China. In Energy Technology 6, pp. 1148–1160.

[5] Müller, A.; Friedrich, L.; Reichel, C.; Herceg, S.; Mittag, M.; Neuhaus, D.H. (2021) A Comparative Life Cycle Assessment of Silicon PV Modules: Impact of Module Design, Manufacturing Location and Inventory. Solar Energy Materials and Solar Cells, 230,111277. doi.org/10.1016/j.solmat.2021.111277

[6] Hsu, D.; O'Donoughue, P.; Fthenakis, V.; Heath, G.; Kim, H.; Sawyer, P.; Choi, J.; Turney, D. (2012) Life Cycle Greenhouse Gas Emissions of Crystalline Silicon Photovoltaic Electricity Generation. Journal of Industrial Ecology, Volume 16, Issue s1 p. S122-S135 https://doi.org/10.1111/j.1530-9290.2011.00439.x

[7] Bracquene, E.; Peeters, J.R.; Dewulf, W.; Duflou, J.R. (2018) Taking Evolution into Account in a Parametric LCA Model for PV Panels, Procedia CIRP 69 389–394.

[8] Koppelaar, R.H.E.M. (2016) Solar-PV energy payback and net energy: Meta-assessment of study quality, reproducibility, and results harmonization, Renewable and Sustainable Energy Reviews (2016)

[9] Gerbinet, S.; Belboom, S.; Léonard, A. (2014) Life Cycle Analysis (LCA) of photovoltaic panels: A review. Renewable and Sustainable Energy Reviews 38. 747–753.

[10] Bhandari, K.; Collier, J.; Ellingson, R.; Apul, D. (2015) Energy payback time (EPBT) and energy return on energy invested (EROI) of solar photovoltaic systems: A systematic review and meta-analysis, Renewable and Sustainable Energy Reviews 47. 133–141.

[11] Peng, J.; Lu, L.; Yang, H. (2013) Review on life cycle assessment of energy payback and greenhouse gas emission of solar photovoltaic systems, Renewable and Sustainable Energy Reviews 19. 255–274.

[12] Blanco, C.F.; Cucurachi, S.; Peijnenburg, W.J.G.M.; Beames, A.; Vijver, M.G. (2020) Are Technological Developments Improving the Environmental Sustainability of Photovoltaic Electricity? Energy Technol. 2020, 8. doi: 10.1002/ente.201901064

[13] Kalverkamp, M., & Karbe, N. (2019) Comparability of life cycle assessments: modelling and analyzing LCA using different databases. In Cascade use in technologies 2018 (pp. 51-63). Springer Vieweg, Berlin, Heidelberg. doi: 10.1007/978-3-662-57886-5_8

[14] Pauer, E., Wohner, B., & Tacker, M. (2020) The Influence of Database Selection on Environmental Impact Results. Life Cycle Assessment of Packaging Using GaBi, Ecoinvent 3.6, and the Environmental Footprint Database. Sustainability, 12(23), 9948. https://doi.org/10.3390/su12239948

[15] Silva, D., Nunes, A. O., da Silva Moris, A., Moro, C., & Piekarski, T. O. R. (2017) How important is the LCA software tool you choose Comparative results from GaBi, openLCA, SimaPro and Umberto. In Proceedings of the VII Conferencia Internacional de Análisis de Ciclo de Vida en Latinoamérica, Medellín, Colombia (pp. 10-15).

[16] Herrmann, I. T., & Moltesen, A. (2015) Does it matter which Life Cycle Assessment (LCA) tool you choose? –a comparative assessment of SimaPro and GaBi. Journal of Cleaner Production, 86, 163-169. https://doi.org/10.1016/j.jclepro.2014.08.004

[17] Speck, R. L. (2014) A comparative analysis of commercially available life cycle assessment software. Michigan State University. Packaging.

[18] Silva, D. A. L., Nunes, A. O., Piekarski, C. M., da Silva Moris, V. A., de Souza, L. S. M., & Rodrigues, T. O. (2019) Why using different Life Cycle Assessment software tools can generate different results for the same product system? A cause–effect analysis of the problem. Sustainable Production and Consumption, 20, 304-315. https://doi.org/10.1016/j.spc.2019.07.005

[19] RE100 climate group: List of members. Available online at https://www.there100.org/re100-members, checked on 1/30/2021.

This is a short version of the paper submitted to the journal EPJ Photovoltaics.
It has been selected by the Executive Committee of the EU PVSEC 2021 for submission to EPJ Photovoltaics.

SUSTAINABLE MATERIAL FLOWS IN THE PV SECTOR: WHAT WORK REMAINS TO BE DONE TO ACHIEVE SDG 12?

Estelle Gervais, Sina Herceg, Sebastian Nold, Karl-Anders Weiß
Fraunhofer Institute for Solar Energy Systems ISE, Heidenhofstr. 2, 79110 Freiburg, Germany
estelle.gervais@ise.fraunhofer.de

ABSTRACT: The large-scale deployment of photovoltaics (PV) is a central pillar in decarbonizing energy systems and reaching climate goals. Although PV is inherently associated to environmental awareness, it is not immune to reputational risks nor exempt of a responsibility for transparency and sustainability leadership. Over the last years, advances in the PV industry have mainly been shaped by cost-reduction targets. We identified in previous works 16 topics where the PV sector comes short in addressing the United Nations Sustainable Development Goal 12 (SDG 12) "Ensure sustainable consumption and production patterns". In this paper, approaches to address these sustainability gaps are proposed. The best-practices identified for the PV sector cover all aspects of sustainability as defined by SDG 12 – from resource use through corporate reporting and risk assessment to due diligence and waste management. Insights on methodological needs to improve sustainability accounting in PV are also provided. The compiled list of actions needed, although not intended to be exhaustive, constitutes a starting point for stakeholders to achieve more sustainability in PV value chains.
Keywords: Sustainable, Environmental Effect, Manufacturing and Processing, Recycling

1 INTRODUCTION

For the past 15 years, sustainable practices associated with the production and disposal of photovoltaic (PV) modules have been a growing subject of concern. The use of critical materials [1], incidents linking PV manufacturing to chemical pollution [2] and limited waste regulation outside the EU [3] have been cited as some hotspots in the sector. More recently, the allegations of forced labor in the Xinjiang Province of China, upon which the global solar industry is dependent for polysilicon, brought to public light the lack of traceability in supply chains [4]. The terawatt-scale deployment of solar energy, necessary to achieve long-term climate goals [5,6], have the potential to aggravate the current gaps existing in the sector. Reputational risks are also not to be overlooked. Initiatives from policymakers, industrials and researchers have multiplied to advance sustainability in PV. A snapshot of this sustainability landscape is provided in Fig. 1.

Despite a wave of activities, the PV value-chain appears relatively fragmented. Sustainable practices linked to the use phase and raw material supply are limited. Trends such as the switch from slurry-based wafering to diamond wire sawing, the continuous reduction of silver paste use per cell and the production of thinner wafers [7] have admittedly shaped a more efficient use of resources. They are however primarily driven by cost reduction: Their potential to sufficiently contribute to a sustainable manufacturing from an environmental and social point of view remains uncertain. While Life Cycle Assessment (LCA) is a well-established method for stakeholders to quantify their environmental impacts, there is no consensus on how to measure circularity [8] and criticality [9] and integrate them into product development. Scientific debate on how to perform social LCA is also still ongoing. Sustainable practices as reported by PV manufacturers can be characterized by a lack of transparency and common assessment frameworks [10].

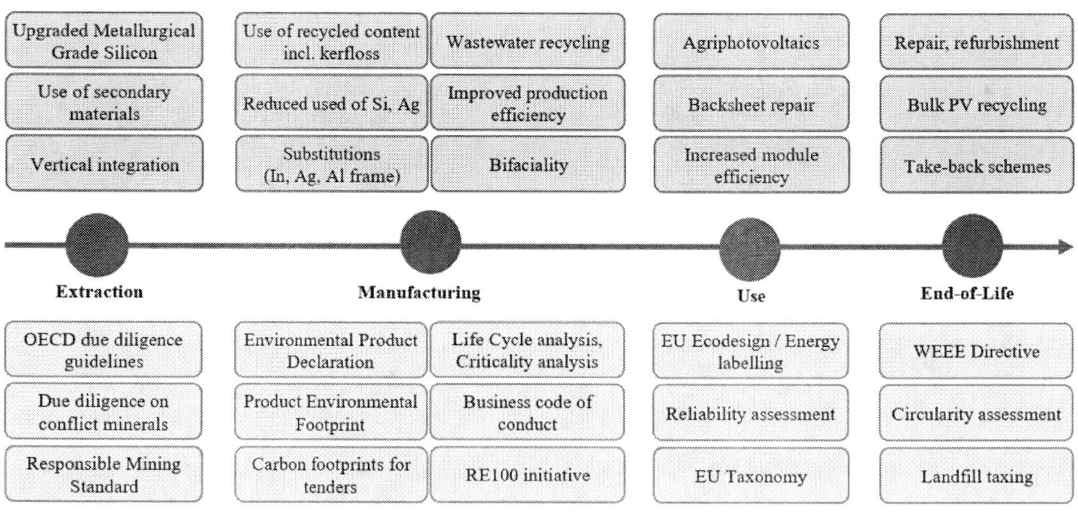

Figure 1: Exemplary snapshot of the sustainability landscape over the PV value chain. Practices implemented or in development for industrial applications are in green; policies, research instruments and standards are in grey.

In this paper, we review the state-of-the art sustainability challenges faced by the PV sector and provide insights on how to improve existing shortcomings as well as sustainability assessment. The concept of sustainability as proposed by the United Nations Sustainable Development Goal 12 (SDG 12) "responsible consumption and production" is hereby used. Its framework is well adapted for translating global needs into business practices over the entire value-chain, from raw material extraction to End-of-Life (EoL), and unites different sustainability topics, from environmental aspects to social responsibility. This paper aims at paving the way for a structured course of action in the PV sector towards sustainability in its broad sense.

2 APPROACH

SDG 12 calls for the implementation of sustainable consumption and production patterns. According to the UN, it is about "doing more and better with less. It is also about decoupling economic growth from environmental degradation, increasing resource efficiency and promoting sustainable lifestyles" [11]. SDG 12 is a holistic approach, broken down into 8 targets, requiring collaboration among stakeholders to achieve transformative changes.

Based on the scope of SDG 12 originally created for nations, sustainability targets for the PV sector were defined by Gervais et al. [10]. A gap analysis highlighted a need for progress in 16 areas or "sustainability strategies" displayed in Table I. The challenges associated with the operation of a PV power plant are not listed but could also be integrated in the SDG 12 framework (e.g. by considering land as a natural resource).

This paper builds upon these previous results and specifies what work remains to be done to implement each sustainability strategy.

3 RESULTS

The reader is referred to the peer-reviewed version of this paper for the comprehensive results. As an example, approaches to advance the sustainability strategies "reduce material intensity" and "ensure the social and environmental performances of suppliers" are briefly discussed.

3.1 Reduce material intensity

A lifecycle inventory is necessary to determine the total material demand from raw material extraction to module use, repair and recovery and must include components of the PV system, as well as production losses, auxiliary materials and utilities. The unused extraction, i.e. the materials that have to be moved to gain access to the coveted materials and tailings should further be taken into account [12].

Instead of targeting overall resource efficiency through increased module performance, individual raw materials should be prioritized for the development of reduction measures. This selection needs to be based on environmental, social and economic considerations combined. For instance, around 10% of primary silver production is estimated to be linked to artisanal and small-scale mining which relies on a mostly unskilled workforce and is associated to poor health and safety practices. Silver mining shows further high environmental hazard potential as it might be linked to acid mine drainage, heavy metal concentrations in deposits and toxic chemicals use for extraction and processing [13]. Silver also accounts for 10% of PV module costs [14].

Cell design adaptations such as increasing the number of busbars and thinner wafers are already paving the way towards a reduced use of selected materials [7]. More ambitious approaches such as kerfless wafering or frameless modules are further required. Short-term targets for material intensity reduction beyond considerations on material cost savings need to be defined and their avoided environmental and social impacts assessed.

Table I: Sustainability targets and strategies for the PV sector under the lens of SDG 12 [10]

# SDG	Topic	Target for the PV sector	Sustainability strategy for the PV sector
12.1	Instruments	Collectively support coherent sustainable practices in the PV sector	Define common sustainability goals with all stakeholders
			Implement methodology to assess progress towards sustainability
12.2	Natural resources	Achieve the sustainable management of natural resources	Reduce material intensity
			Reduce water use
			Reduce (fossil-fuel based) energy consumption
12.4	Hazardous waste & Chemicals	Achieve the environmentally sound management of chemicals and wastes	Optimize the treatment of production line outputs
			Avoid the release of harmful substances during extraction, use and EoL
12.5	Waste generation	Substantially reduce PV material waste generation	Increase the collection rate of EoL modules
			Extend the use phase of products
			Increase the efficiency of EoL recycling
			Increase the use of recycled materials in production
12.6	CSR & reporting	Ensure stakeholder well-being and demonstrate sustainability commitment	Standardize sustainability reporting
			Standardize corporate social responsibility (CSR) practices
12.7	Sustainable procurement	Adopt sustainable procurement practices	Improve the supply security of products
			Ensure the social and environmental performances of suppliers
12.8	Public awareness	Inform consumers on the sustainability of product/company	Democratize sustainability labelling for PV modules

Recommended actions:

- Prioritize raw materials for efficiency measures based on multidimensional sustainability assessment (environmental, social, economic)
- Report at corporate level total primary material intensity in kg/MWp
- Accelerate the switch to far-less material intensive production techniques

3.2 Ensure the social and environmental performances of suppliers

Corporates are expected to take full responsibility for the supply chains associated with their goods and services. This is however mostly done on a voluntary basis, as due diligence regulations only concern a few countries or conflict minerals such as tin. The adoption of mandatory due diligence legislation for instance at EU scale could foster progress in human rights and environmental standards and create a level playing field. As already done by some top tier PV manufacturers, suppliers and their environmental and social performances can also be fully disclosed in sustainability reports [15].

In particular, raw material procurement for the PV industry needs to gain in transparency and traceability via third-party verification. The Certification of Raw Materials (CERA) [16] is an example of a standardized certification framework ensuring environmental, social and economic sustainability in extraction, processing and trading. Schemes similar to the Battery Passport introduced by the EU could be used in PV supply chains. Distributed ledgers such as blockchain associated to an evolution towards smart factories are a promising tool to support supply chain due diligence and should be tested for PV. Challenges regarding data-quality and implementation costs need to be addressed in the process.

Recommended actions:

- Adopt sustainability certification schemes for mineral value chains
- Disclose sustainability performances of suppliers in corporate reporting
- Advance applied research on product tracking related challenges as closing data gaps, ensuring data quality and data exchange processes

ACKNOWLEDGEMENT

This work was supported by the Fraunhofer Innovation Program within the research project “Souveräne Wertschöpfungszyklen SVC”. The authors would like to thank the project partners for the constructive collaboration and input.

REFERENCES

[1] E. Gervais, S. Shammugam, L. Friedrich, T. Schlegl, Renewable and Sustainable Energy Reviews **137**, 110589 (2021)

[2] D. Mulvaney, IEEE Spectr. **51**, 30 (2014)

[3] R. Deng, N. L. Chang, Z. Ouyang, C. M. Chong, Renewable and Sustainable Energy Reviews **109**, 532 (2019)

[4] L. Murphy, N. Elimä, In Broad Daylight: Uyghur Forced Labour and Global Solar Supply Chains., 2021

[5] IRENA, World Energy Transitions Outlook: 1.5°C Pathway, 2021

[6] N. M. Haegel, H. Atwater, T. Barnes, C. Breyer, A. Burrell, Y.-M. Chiang, S. de Wolf, B. Dimmler, D. Feldman, S. Glunz, J. C. Goldschmidt, D. Hochschild, R. Inzunza, I. Kaizuka, B. Kroposki, S. Kurtz, S. Leu, R. Margolis, K. Matsubara, A. Metz, W. K. Metzger, M. Morjaria, S. Niki, S. Nowak, I. M. Peters, S. Philipps, T. Reindl, A. Richter, D. Rose, K. Sakurai, R. Schlatmann, M. Shikano, W. Sinke, R. Sinton, B. J. Stanbery, M. Topic, W. Tumas, Y. Ueda, J. van de Lagemaat, P. Verlinden, M. Vetter, E. Warren, M. Werner, M. Yamaguchi, A. W. Bett, Science (New York, N.Y.) **364**, 836 (2019)

[7] VDMA, International Technology Roadmap for Photovoltaic (ITRPV): 2020 Results. 2020 Results, No. 12. Edition, 2021

[8] S. Herceg, M. Dick, E. Gervais, K.-A. Weiß, Conceptualized Data Structure for Sustainability Assessment of Energy and Material Flows: Example of a PV Life Cycle, 2021

[9] D. Schrijvers, A. Hool, G. A. Blengini, W.-Q. Chen, J. Dewulf, R. Eggert, L. van Ellen, R. Gauss, J. Goddin, K. Habib, C. Hagelüken, A. Hirohata, M. Hofmann-Amtenbrink, J. Kosmol, M. Le Gleuher, M. Grohol, A. Ku, M.-H. Lee, G. Liu, K. Nansai, P. Nuss, D. Peck, A. Reller, G. Sonnemann, L. Tercero, A. Thorenz, P. A. Wäger, Resources, Conservation and Recycling **155**, 104617 (2020)

[10] E. Gervais, S. Herceg, S. Nold, K. A. Weiss, IEEE Journal of Photovoltaics (2021)

[11] United Nations, *Goal 12: Ensure sustainable consumption and production patterns*, https://www.un.org/sustainabledevelopment/sustainable-consumption-production/, accessed Jun 29, 2021

[12] VDI, VDI 4800 Part 2: Resource efficiency: Evaluation of raw material demand. Evaluation of raw material demand, 2018

[13] G. Dehoust, A. Manhart, P. Dolega, R. Vogt, C. Kemper, A. Auberger, F. Becker, C. Scholl, A. Rechlin, M. Priester, Environmental Criticality of Raw Materials: An assessment of environmental hazard potentials of raw materials from mining and recommendations for an ecological raw materials policy. An assessment of environmental hazard potentials of raw materials from mining and recommendations for an ecological raw materials policy, 2020

[14] E. Bellini, *Silver accounts for 10% of PV module costs*, https://www.pv-magazine.com/2021/03/04/silver-currently-accounts-for-10-of-pv-module-costs/, accessed Jun 30, 2021 (2021)

[15] Canadian Solar, *Sustainability Report 2019* (2021)

[16] CERA, *The Certification of raw materials*, https://www.cera-standard.org/, accessed Oct 29, 2020

ASSESSING PLAUSIBLE ENVIRONMENTAL IMPLICATIONS OF A GROUND-MOUNTED PEROVSKITE/SI TANDEM PV SYSTEM

Caterin Salas-Redondo[1*], Carlos-Felipe Blanco[2], Jean-Baptiste Puel[1,3], Arthur Julien[1], Faïza Soumane[1,5], Stefano Cucurachi[2], Willie J.G.M. Peijnenburg[2,4], Lars Oberbeck[1,5]
[1]Institut Photovoltaïque d'Ile-de-France, 18 Boulevard Thomas Gobert, 91120 Palaiseau, France
[2]Leiden University, 2300 RA Leiden, The Netherlands
[3]Électricité de France, R&D center, Boulevard Gaspard Monge, 91120 Palaiseau, France
[4]Center for Safety of Substances and Products, National Institute of Public Health, and the Environment, 3720 BA Bilthoven, The Netherlands
[5]TotalEnergies, 2 Place Jean Millier, 92078 Paris La Défense Cedex, France
*caterin.salas-redondo@ipvf.fr

ABSTRACT: We have previously reported a comprehensive life cycle assessment (LCA) for large area (> 2 m^2) perovskite/Si tandem solar modules with 30% power conversion efficiency (PCE), 30 years lifetime and an annual degradation of 0.5%, manufactured in a gigawatt-sized industrial scale production plant. We did an extensive compilation of the life cycle inventory (LCI) thanks to direct contacts with equipment manufacturers, who provided the required data to deliver a perspective to real-life conditions. In the present study, we prepared an updated LCA that includes an upgraded LCI as well as the addition of the end-of-life (EoL) stage and the comparison of electricity mix scenarios at the production stage. These scenarios are essential in a near future where renewable energies are expected to have a larger contribution in the global energy mix. The key result found in this work is the relevance of the performance and reliability of the photovoltaic (PV) system to achieve lower environmental impacts. We tested different assumptions using a detailed yield model for the PV system which considered optimistic, realistic, and pessimistic degradation rates, lifetime and PCE, among other parameters to investigate this matter. In a realistic scenario, we observed that for all the impacts assessed, the perovskite/Si tandem based system shows a decrease of between 30-45% in the contributions to each impact category, compared to the Si-module based reference PV system. Moreover, contemplating different EoL scenarios permits to address strategic methodologies to implement a circular economy in the life cycle of the tandem system. Finally, the analysis anticipating future electricity mix scenarios allows to define conditions to achieve further decreases of the environmental impacts.

Keywords: c-Si, CO_2 footprint, Environmental Effect, Heterojunction, Large Grid-connected PV systems, LCA, Perovskite, Photovoltaic, Recycling, Tandem.

1 AIM AND APPROACH USED

The goal of this study is to evaluate the life-cycle environmental impacts of a large area (> 2 m^2) four-terminal two-wires (4T2) perovskite/Si tandem module, mass-produced in a 1 GW production plant. The environmental impact associated with the direct electricity generation from such module in a ground-mounted utility-scale photovoltaic (PV) system over its lifetime is also studied. The PV system is assumed to be installed in Marseille, France because it is in the region with the highest global horizontal irradiation of the country [1].

The silicon heterojunction (SHJ) technology bottom cell is used in the tandem, and a SHJ-based PV system as a comparative reference in the analysis. We consider SHJ solar cells for different reasons: (i) SHJ is expected to gain a market share of about 10% in 2024 and 17% by 2030 [2], (ii) it has been shown to be a very good silicon bottom cell to yield the highest perovskite/Si tandem efficiency [3], and (iii) in SHJ cells, the pn-junction is created through a low-temperature deposition of doped amorphous Si, instead of using diffused pn-junction formed at high temperatures in conventional c-Si solar cells. This decreases the thermal budget and therefore the energy required for cell production, resulting in a potential reduction in environmental impacts [4].

The analyzed impact categories are chosen according to the recommendations given in the International Reference Life Cycle Data System handbook of the European Commission's Joint Research Centre [5]. The life cycle assessment (LCA) system boundaries include the impact of mining and processing raw materials necessary to produce cells, modules, and the balance-of-system (BOS) components, operation, maintenance, and different scenarios at the end-of-life (EoL) phase.

As a basis for the environmental impact assessment, we use a functional unit of 1 kWh of generated energy from a ground-mounted PV system over its lifetime. We tested different assumptions to estimate the energy output of the PV tandem module, considering optimistic, realistic, and pessimistic degradation rates, lifetime, and cell efficiencies, among other parameters. The various scenarios were examined in a self-developed model to calculate the energy yield for tandems and silicon single junction under realistic conditions [6]. Finally, we apply the future electricity mix in the foreground system to anticipate possible conditions that allow for further decrease of environmental impacts.

2 SCIENTIFIC INNOVATION AND RELEVANCE

The rapid progress of perovskite thin-film technology and its potential market entry as single junction or in tandem configuration [7-8], justify the need to conduct an LCA, to guide efforts towards product designs and developments with minimum environmental impacts. LCAs have already been published for single junction perovskite solar cells [9], and in tandem configurations with silicon at lab, pilot

[10-11] or pre-industrial scale [12]. However, to this date and to the extent of our knowledge, our previous report [13], with a comprehensive LCA for large area (> 2 m^2) perovskite/Si tandem solar modules at gigawatt-sized industrial scale, was the first one of its kind. We did an extensive compilation of the life cycle inventory (LCI). This was possible thanks to direct input from equipment manufacturers who provided the required data to deliver a perspective of future production conditions.

In the present study, we report an LCA of a utility scale PV system, that includes an upgraded LCI of the perovskite/Si tandem as well as the addition of the EoL stage, and the comparison of different electricity mix scenarios at the production stage. This consideration is essential in a near future, where renewable energies are expected to have a larger share in the global energy mix.

Utility-scale power plants will continue to dominate the PV market accounting for about 60% of all cumulative installations by 2030 [1], with a further increase by 2050 to meet future low-carbon electricity needs. In that future, we expect that highly efficient tandem PV systems will play a key role in the decarbonization of the energy mix. In a nutshell, we provide for the first-time relevant insights in the environmental aspects of a large-scale deployment of perovskite/Si tandems modules.

3 RESULTS

3.1 Environmental profile comparison between perovskite/Si tandem and Si-based PV systems

The environmental impacts of generating 1 kWh of electricity with a perovskite/Si tandem vs a SHJ module-based ground-mounted PV system is shown in Figure 1. In a realistic scenario (see scenario RE, Table I), a decrease of between 30-45% was found for the perovskite/Si module-based system, across all impact categories. The better performance achieved with the tandem system, due to the higher power conversion efficiency and resulting lower equivalent area to produce 1 kWh, enables the lower environmental impact. This is an indication of the environmental advantages and benefits of choosing a perovskite/Si tandem module vs. a single-junction c-Si module in future PV systems, under the assumptions made in this study (see Table I).

Table I: Most relevant parameters of the PV tandem module in the variation of the optimistic (OP), realistic (RE) and pessimistic (PE) scenarios, as considered in this study.

	OP	RE	PE
Initial efficiency	30%	28%	26%
Annual degradation rate	0.3%	0.5%	0.7%
Lifetime (years)	30	25	20
Average energy yield, Marseille [a), b)] (kWh/m²/year)	431.0	398.8	372.8

[a)] Obtained with an in-house model based on optical, electrical, and thermal modeling of PV tandem modules [5]. Variations of perceived spectral irradiance, site temperature, geometrical aspects (albedo, utility scale PV system configuration) are considered. Source for weather data is EnergyPlus Weather Data

[b)] For comparison, the electricity yield of a c-Si PV power plant in Spain is about 200 kWh/m²/year [14] (meaning, about 180 kWh/m²/year if installed in Marseille)

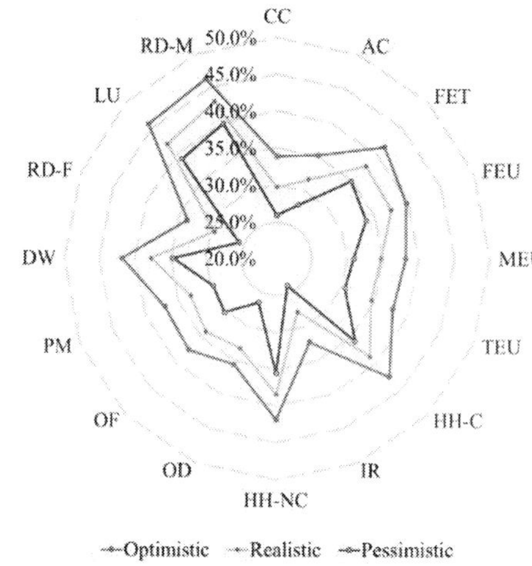

Figure 1: Relative comparison between a perovskite/Si tandem vs a SHJ module-based ground-mounted PV system for various impact categories. Optimistic (green), realistic (yellow) and pessimistic (red) scenarios. CC: climate change; AC: acidification; FET: freshwater ecotoxicity; FEU: freshwater eutrophication; MEU: marine eutrophication; TEU: terrestrial eutrophication; HH-C: human health, carcin. effects; IR: ionising radiation; HH-NC: human health, non-carcin. effects; OD: ozone layer depletion; OF: photochemical ozone creation; PM: particulate matter; DW: depletion of water; RD-F: resource depletion, fossils; LU: land use; RD-M: resource depletion, minerals, and metals.

3.2 Future electricity mix scenarios in sensitivity analysis

A projected electricity mix for the years 2030 and 2050 was created in a foreground system and applied to the fabrication of perovskite and SHJ solar cells (including silicon and wafer production), as well as module assembly. The future electricity mix of the years 2030 and 2050 is based on the roadmap to net zero emissions given by the IEA, where solar and wind power race ahead, raising the share of renewables in total generation from 29% in 2020 to nearly 90% in 2050, complemented mainly by nuclear and hydropower [15].

Lower environmental impacts could be obtained with an electricity mix dominated by low-carbon energy sources in which the greenhouse gas (GHG) intensity of electricity generation falls from 255 gCO_2e/kWh in 2019 to 86 ± 11 gCO_2e/kWh in 2030[1] [15] and further down in 2050. Figure 2 shows the potential reduction of using the forecasted electricity mixes IEA (2030) and IEA (2050) compared to that of ecoinvent ei3.6 (RER).

There are small negative environmental credits accounting for about 0.2% in the impact category of resource depletion (RD-M) of minerals and metals, mostly because of the material demand of solar and wind technologies and their large share in the electricity mix. This is also observed in the impact category of land use (LU) for the 2050 scenario, where renewable energies account for a share of 90% of the mix, from which solar and wind represent about 70% of the share.

Minimizing the carbon footprint of the electricity generation is also reflected in a potential decrease of in between 5 – 10 % in the climate change (CC) impact

[1] These values represent indicative intensity levels that would allow the EU to achieve a net 55% reduction in greenhouse gases compared with 1990 by 2030 and carbon neutrality by 2050.

category. Furthermore, the reduction in the ionizing radiation category comes from the decrease of 0.5 and 25% share of nuclear power in the years 2030 and 2050, respectively.

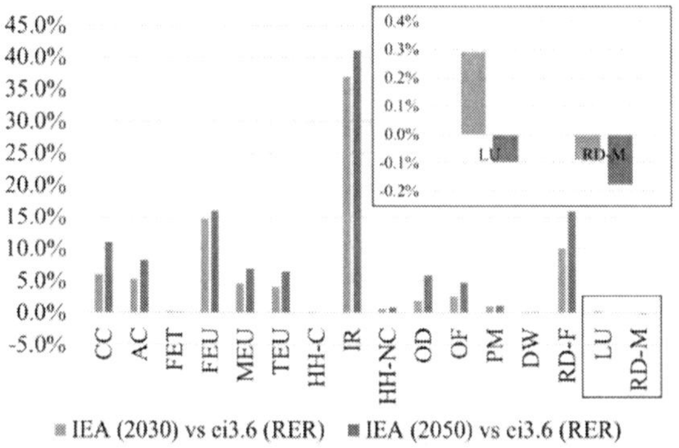

Figure 2: Environmental benefits in different impact scores of the perovskite/Si PV system on ecoinvent3.6 background database (ei3.6 (RER)) and foreground electricity mix projections to the years 2030 (IEA (2030), light green) and 2050 (IEA (2050), dark green). Future electricity mixes are based on the roadmap to net zero emissions given by the IEA [16]. Inset: zoom of results for LU and RD-M impact categories. CC: climate change; AC: acidification; FET: freshwater ecotoxicity; FEU: freshwater eutrophication; MEU: marine eutrophication; TEU: terrestrial eutrophication; HH-C: human health, carcin. effects; IR: ionising radiation; HH-NC: human health, non-carcin. effects; OD: ozone layer depletion; OF: photochemical ozone creation; PM: particulate matter; DW: depletion of water; RD-F: resource depletion, fossils; LU: land use; RD-M: resource depletion, minerals, and metals.

3.3 End-of-Life scenarios in sensitivity analysis

As the PV installations reach the final decommissioning stage, recycling is mandatory in Europe. Nowadays the volume of the recycled PV panels is still negligible, as most of these devices are still in operation. This situation will change in 10-30 years from now, as the volume of EoL PV modules is expected to multiply by a factor of 10-100, reaching millions of tons of PV waste [17]. Therefore, EoL management for PV technology is crucial as it will spawn new industries, support economic value creation, and is consistent with a global shift to sustainable long-term development [18].

This study considers a circular approach to manage the PV waste generated from a ground-mounted perovskite/Si tandem system, beyond the current recycling of PV modules i.e., shredding/crushing, and further sorting. This generally leads to downcycling, assuming (due to the lack of data reported) a very low quality of regenerated materials produced and partially landfilled. We consider scenarios such as the recovery of silver from silicon cell recycling, as well as flat glass recycling with corresponding treatment of the perovskite layer. Some studies review disposal scenarios of perovskite [19] or silicon solar cells [20] separately, but no dedicated EoL approaches have been pursued for tandem structures, as we do here.

Two scenarios are considered as detailed in Table II: Case 1 covers the traditional recycling schemes of used crystalline silicon PV modules in a first-generation recycling process, where (crushed) glass, aluminum and copper are recovered, and the rest is incinerated [21]. Case 2 follows the treatments proposed in reference [22] with further adjustments to use pyrolysis instead of incineration during delamination [23]. The LCI of both cases is adapted to the PV module tandem configuration.

The following conditions are attributed to the EoL methods: (i) treatment efforts and emissions are fully attributed to the treatment service, and (ii) potential environmental benefits are included due to avoiding primary material supply.

The possible avoided burdens of Case 2 EoL recycling are reflected throughout all impact categories, mostly with 35% reduction of resource depletion and 7% less contribution to climate change, as shown in Figure 3. The impact category of metal and mineral resource depletion (RD-M) shows the highest benefits when aiming at the recovery of metals included in the solar cells. Such benefits are not observed when applying a traditional recycling scheme (Case 1). This reduction becomes relevant in the situation where more metal resources are required in a terawatt-scale deployment of PV, as partially illustrated through the negative credits of RD-M in Figure 2.

Table II: Description of the end-of-life scenarios considered in this study. Case 1 is the standard first-generation recycling of silicon modules. Case 2 applies further treatments to recover metals from the silicon solar cells. Both cases are adapted to PV tandem modules.

Recycling stage	Case 1	Case 2	Targeted material
Disassembly	Mechanical disassembly (Aluminum frame and junction box)		Al, Cu
Delamination	Shredding and sorting Incineration	High frequency heat knife Pyrolysis	Crushed glass, EVA
Metal recovery	NA	Acid leaching and electrolysis	Si, Ag, In, Cu

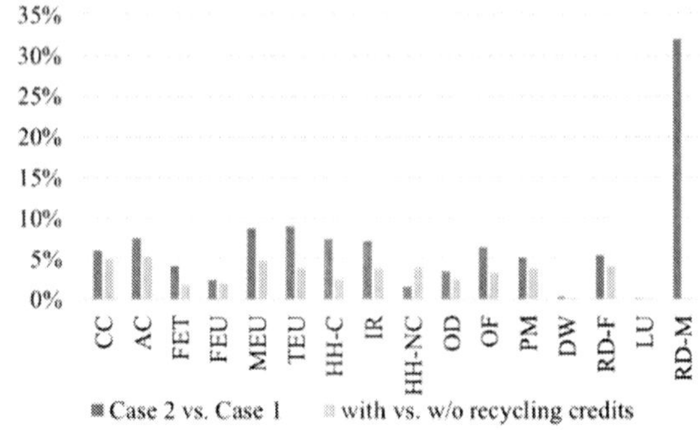

Figure 3: Potential reduction of environmental impacts through end-of-life recycling. Case 1 refers to the 1st generation recycling. Case 2 relates to standard treatments to recover metals in the silicon cells (acid leaching and electrolysis). Both cases are described in the text. CC: climate change; AC: acidification; FET: freshwater ecotoxicity; FEU: freshwater eutrophication; MEU: marine eutrophication; TEU: terrestrial eutrophication; HH-C: human health, carcin. effects; IR: ionizing radiation; HH-NC: human health, non-carcin. effects; OD: ozone layer depletion; OF: photochemical ozone creation; PM: particulate matter; DW: depletion of water: water use; RD-F: resource depletion, fossils; LU: land use; RD-M: resource depletion, minerals and metals.

4 CONCLUSION AND FUTURE WORK

We report that for all environmental impact categories assessed, the perovskite/Si tandem based PV system performs better compared to the single-junction SHJ-module based reference PV system. The key result of this work is the relevance of the performance and reliability of the PV system to achieve lower environmental impacts. A comparison between optimistic, realistic, and pessimistic scenarios was performed to investigate this matter. Moreover, investigating different EoL scenarios permits to address strategic methodologies to implement a circular economy in the life cycle of the tandem system. Implementation of mid- and long-term recycling treatments could provide an overall perspective to efficiently treat the PV waste. This could go in hand with a techno-economic assessment to evaluate the economic feasibility behind the technological solutions which could be proposed.

Furthermore, we demonstrated that lower environmental impacts could be obtained with an electricity mix dominated by low-carbon energy sources. This was possible thanks to the application of future electricity mixes in the foreground processes of silicon cell production.

Finally, as more PV installations will be combined with storage systems with a trend of 4% in 2020 to 50% in 2030 [2] and expected to increase further towards full electrification scenarios in 2050, steps forward are needed to include the storage solutions in the LCA system.

5 ACKNOWLEDGEMENTS

The authors thank all co-workers involved in the development of perovskite/Si tandem modules at IPVF. The authors also thank all the equipment manufacturers who have provided detailed technical information about their production tools, as well as experts in the field of PV waste management for the discussions. This project is supported at IPVF by the French Government in the frame of the program of investment for the future (Programme d'Investissement d'Avenir - ANR-IEED-002-01). Researchers at Leiden University were funded by the European Union's Horizon 2020 research and innovation program within the project SiTaSol under grant agreement No 727497.

6 REFERENCES

[1] The World Bank. 2019. Global solar atlas 2.0

[2] VDMA Photovoltaic equipment. 2021. International Technology Roadmap for photovoltaic (ITRPV), 12th edition.

[3] Messmer C, et al. 2020. The race for the best silicon bottom cell: Efficiency and cost evaluation of perovskite–silicon tandem solar cells. Prog Photovolt Res Appl. 1–16.

[4] van Sark W, Korte L, Roca F. 2012. Physics, and technology of amorphous-crystalline heterostructure silicon solar cells. Springer.

[5] European Commission-Joint Research Centre - Institute for Environment and Sustainability: International Reference Life Cycle Data System (ILCD) Handbook-Recommendations for Life Cycle Impact Assessment in the European context. EUR 24571 EN. Luxemburg. Publications Office of the European Union; 2011.

[6] Julien A et al. 2020. Backside light management of 4-terminal bifacial perovskite/silicon tandem PV modules evaluated under realistic conditions. Optics Express 28(25): 37487-37504.

[7] Perovskites take steps to industrialization. 2020. Nat Energy 5:1.

[8] Verma A, et al. 2020. Towards industrialization of perovskite solar cells using slot die coating. J Mater Chem C 8:6124-6135.

[9] Maranghi et al. 2019. Environmental Profile of the Manufacturing Process of Perovskite Photovoltaics: Harmonization of Life Cycle Assessment Studies. Energies 12: 3746.

[10] Celik I, et al. 2016. Life Cycle Assessment of perovskite PV cells projected from lab to fab. Sol. Energy Mater. Sol. Cells 156: 157-169.

[11] M. de Wild-Scholten, 2017. Deliverable 3.1 - Life Cycle Analysis of CHEOPS technologies and benchmarking: Screening, SmartGreenScans, Groet, NL.

[12] Alberola-Borras JA et al. 2018. Perovskite Photovoltaic Modules: Life Cycle Assessment of Pre-industrial Production Process. iScience 9: 542-551.

[13] Salas-Redondo C, et al. 2020. LCA of Perovskite on Si Tandem PV Modules at Industrial Scale. 37th EU PVSEC proceedings: 754 – 764.

[14] Frischknecht R, et al, 2020, Life Cycle Inventories and Life Cycle Assessment of Photovoltaic Systems, International Energy Agency (IEA) PVPS Task 12, Report T12-19:2020.

[15] Greenhouse gas emission intensity of electricity generation in Europe. Online. Last visited: August 13, 2021. https://www.eea.europa.eu/data-and-maps/indicators/overview-of-the-electricity-production-3/assessment-1

[16] IEA. 2021, Net Zero by 2050, IEA, Paris

[17] IRENA and IEA-PVPS, 2016. End-of-Life Management: Solar Photovoltaic Panels. International Renewable Energy Agency and International Energy Agency Photovoltaic Power Systems.

[18] Weckend S, Wade A, and Heath G. 2016. End of Life Management: Solar Photovoltaic Panels. United States: N. p.

[19] Serrano-Lujan, L., et al. 2015. Tin- and Lead-Based Perovskite Solar Cells under Scrutiny: An Environmental Perspective. Adv. Energy Mater., 5: 1501119.

[20] Lunardi, M.M. et al. 2018. Comparative Life Cycle Assessment of End-of-Life Silicon Solar Photovoltaic Modules. Appl. Sci., 8: 1396.

[21] Wambach K, Heath G, and Libby C. 2017. Life Cycle Inventory of Current Photovoltaic Module Recycling Processes in Europe, IEA-PVPS Task 12, Report T12-12:2017.

[22] Latunussa, C.E.L. et al. 2016. Life Cycle Assessment of an innovative recycling process for crystalline silicon photovoltaic panels. Sol. Energy Mater. Sol. Cells., 56: 101 – 111.

[23] Pyrolysis treatment data were obtained from an environmental equipment manufacturer.

The Impact of the Lamination Process on the Adhesion Properties at the Glass-Encapsulant Interface and Damp Heat Stability of PV Modules

Aksel Kaan Öz[1], Christoph Herzog[1], Christine Wellens[1], Djamel Eddine Mansour[1], Martin Heinrich[1], Achim Kraft[1]

[1]Fraunhofer Institute for Solar Energy Systems ISE, Heidenhofstraße 2, 79110 Freiburg, Germany

Corresponding Author: Aksel Kaan Öz | +49 (0) 761 4588 5025 | e-mail: aksel.kaan.oez@ise.fraunhofer.de

ABSTRACT: This paper presents an extensive study of the influences of the lamination process on the adhesion properties of the glass-encapsulant interface of different laminates. The investigations were conducted on small-scale laminates which were laminated with different material combinations at peak lamination temperatures from 140°C to 180°C. As a measure of the long-term durability, the laminates were exposed to damp heat (DH) up to 3000 hours. We investigated the development of the adhesion between encapsulant and glass by peel tests, the measurement of the gel content and the colour change of the backsheet by measurements of the yellowness index. The peel test results show that the relative maintained peel force (RMPF) varies in the range from 15.7-73.0% depending on the material and the lamination process. It is obtained that the EVA laminates showed less degradation in peel force after DH than POE samples and a higher lamination temperature was beneficial in peel force and gel content before and after ageing for cross-linking encapsulants such as EVA and POE. For the used TPO the peel force did not change for different lamination temperatures. We showed that with lamination temperatures up to 180 °C EVA, POE and TPO modules can be manufactured and withstand DH ageing but they may show a slightly increased yellowing of the rear side.

Keywords: PV modules, peel test, damp heat, backsheet, degradation

1.INTRODUCTION

A current topic in the Photovoltaic (PV) module industry is the process time reduction to achieve higher throughput for module manufacturing [1]. However, these modules also need to operate for long lifetimes of more than 25 or even 30 years [1,2]. A good adhesion between the module layers may prevent delamination, which is a defect that occurs in the field and repeatedly leads to lifetime issues [3]. Another frequently observed indication of PV module ageing is the yellowing of the materials [5].
The adhesion strength at the different interlayers of the laminate and the yellowing after ageing are influenced by the choice and interaction of the encapsulant materials and the process parameters during lamination. Therefore, to ensure good adhesion, it is essential to choose the lamination parameters and materials carefully [6]. Module manufacturers aim to increase their lamination temperatures from 150°C to higher temperatures [7]. The goal is to speed up the lamination process and realize a higher throughput of module production. But so far, there are not many studies about the influences of high lamination temperatures on the long-term adhesion properties of the material compound in the module.
Damp heat ageing is a common way to accelerate the ageing of PV modules by driving humidity into the module package which leads to a decrease in adhesion strength of the interfaces [8,9]. To evaluate the stability of the multilayer composite of a PV-module, peel tests after damp heat (DH) tests with long durations can be conducted.
This paper investigates long-term effects of lamination temperatures > 150°C in terms of peel force measurements and evaluation of the discolouration in terms of the yellowness index before and after DH 1500 and DH 3000 exposure. Additionally, the different failure types that occurred during the peel tests and the gel contents of the EVA encapsulant were analysed in detail.

2.MATERIAL AND METHODS

2.1 Sample Preparation

Small scale laminates (200 mm x 200 mm) with polypropylene (PP) and polyethylene terephthalate (PET) based backsheets were laminated with ethylene-co-vinyl acetate (EVA), polyolefin elastomer (POE), and thermoplastic polyolefin (TPO) based encapsulants at peak laminations temperatures of 140 °C, 160 °C, and 180 °C. The lamination time varied between 4-10 minutes as shown in Table 1.

Table 1: Parameters of the Lamination Process

Temperature [°C]	Curing time [min.]
140	10
160	10
180	4

The specifications and properties of the different encapsulation materials that were used in this study are shown in Table 2. The material combinations represent common, commercially available products from the PV market.

2.2 Damp Heat Test

The influences of the lamination process on the adhesion properties under water vapour ingress were investigated by exposing the modules to the damp heat test (85 °*C*, 85 % r.*h*.) for 1500 h and 3000 h. The modules were characterized before and after the DH-test regarding their peel force, gel content and yellowness index.

2.3 Peel Tests

The peel force at the glass-encapsulant interface is measured with the PSE peel-off instrument at an angle of 90° (Figure 1). The peel tests were conducted at the centre of the small-scale laminates according to the BS EN ISO 8510 standards [9] by cutting strips with the width of 10 mm and length of 200 mm[10]. These strips were peeled off with a speed of 50 mm/min at room temperature. The relative maintained peel force (RMPF) of the laminates after been exposed to the DH 3000 test is calculated with the formula:

$$RMPF = \frac{Peel\ force\ after\ DH\ 3000}{Initial\ peel\ force}\ x\ 100\%$$

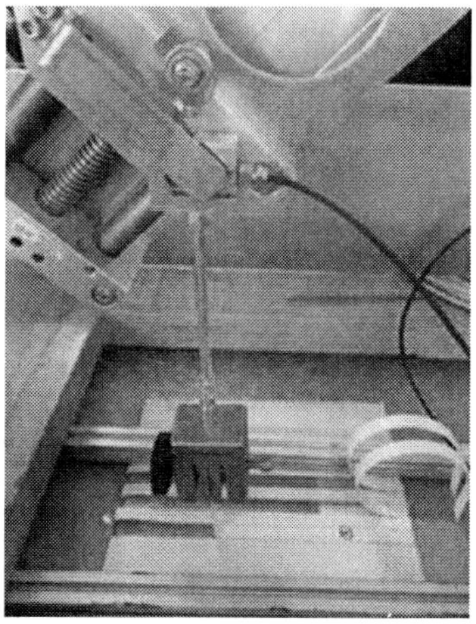

Figure 1: 90° Peel test at the glass-encapsulant interface

2.4 Gel Content Measurements

To investigate the correlation of the peel force at the glass encapsulant interface and gel content of the encapsulant, the Soxhlet extraction method was used to measure the gel content. In this method, small pieces of the encapsulant, extracted from the laminate samples are dipped into a solvent for a certain period [11]. The solvent and the period changes according to the type of encapsulant that is measured. The gel content was calculated with the following formula [2];

$$Gel\ content\ (\%) = \frac{M_2}{M_1}\ x\ 100$$

where M_1 represents the initial weight of the specimen and M_2 its weight after the extraction and drying process.

2.5 Colour Measurements

The degradation of encapsulants in laminates can also be investigated by using colour measurements. The tests, on both sides of the laminate, were performed according to the EN ISE 11664-4 standards by characterising the laminates with respect to their yellowness index (YI) [12]. To do this, a portable ColorLite sph900 spectrophotometer with an observer angle of 10° was used.

Table 2: Specifications and properties of the encapsulation materials used in this work

Layer	Material	Thickness [mm]	Water Vapour Transmission Rate [at 38 °C, 90 % r.h.]
front glass	safety glass	3	0
encapsulant	EVA	0.5	n.a
encapsulant	POE	0.5	n.a
encapsulant	TPO	0.4	n.a
backsheet	PET/PET/Primer	0.295	2.4
backsheet	PET/Aluminium/PET/Olefin	0.33	n.a
backsheet	PP	0.36	0.6

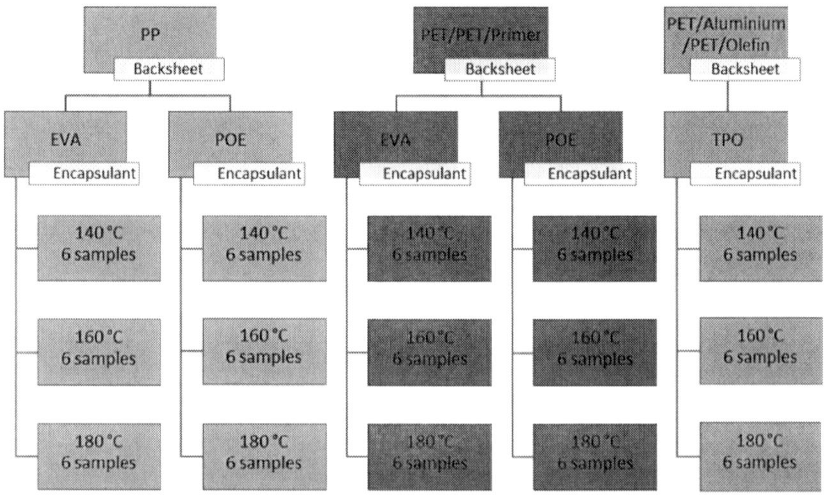

Figure 2: Material combinations and number of laminates built for the study

3.RESULTS

In this section, the different lamination processes are evaluated according to the gel content and peel tests and considering the influences of the water vapour ingress after DH characterised according to the trend in gel content, peel force and yellowness index.

3.1 Gel Content

The change in gel content of the EVA based laminates according to their peak lamination temperature and lamination time is analyzed directly after lamination and after 1500 and 3000 h DH.
For each experimental point, two laminates were used, and measurements were taken from two random locations of these laminates. Each measurement point shown in Figure 3, corresponds to the average value of these four measurements with the error bars indicating the standard deviation.

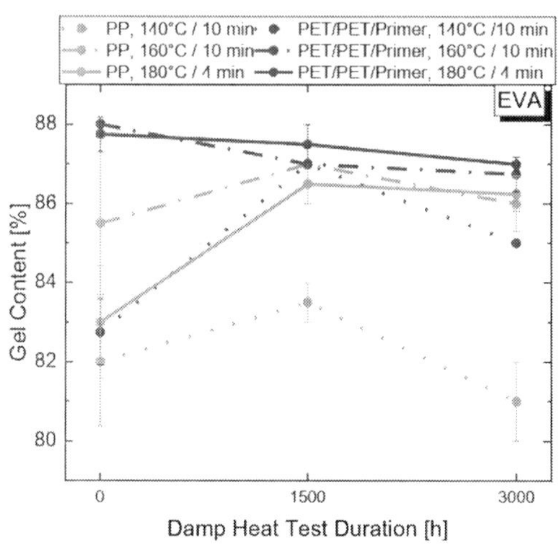

Figure 3: Gel-content of the EVA laminates with PP or with PET backsheet directly after lamination, after 1500 h and 3000 h damp heat test.

For the un-aged laminates, it was found that the gel content of the 160 °C laminates was slightly higher than the ones laminated at 180 °C. This happened due to the lower lamination time at 180 °C. The 160 °C laminates had a curing time of 10 minutes, while the 180 °C laminates had a curing time of 4 minutes.
After being exposed to DH 1500, the laminates with the PP backsheet showed an increased gel content, while the laminates with the PET backsheet show a slight decrease in gel content. Exposing further with DH 3000, the gel content of all samples decreased slightly.
As reason for this behaviour, we suspect that the exposure to an elevated temperature during the DH-test may lead to a post-crosslinking in the encapsulation material. Another effect is a degradation of the material due to the humid environment which slowly leads to a decrease in gel content. For samples with initially comparable low gel content the post-crosslinking dominates while further exposing to humidity or with a high gel content from the start the degradation effect due to humidity dominates.
Overall, all samples show an acceptable gel content before and after each DH step.

3.2 Peel Test Failure Type Analysis

In this section and the next section peel forces at the glass encapsulant interface before and after DH for the different lamination temperatures and lamination setups are analyzed. Firstly, an investigation of the different failure modes is performed and in the second part the peel force is investigated.
Each peel force measurement was done on two laminates with each three peel tests, respectively. This section analyses the type and number of failures that were observed during the peel tests in correlation to the used materials and the peak lamination temperature of the laminates. Some of the laminates showed multiple types of failures, and in that case, the failure type that occurred first was taken into consideration. We observed three failure types: a) delamination of backsheet in sub-layers, b) delamination at the backsheet-encapsulant interface and c) rupture of the backsheet. Representative examples are shown in Figure 4.

Figure 4: Photograph of the back sheet failure mode occurred during the peel test: Delamination of Backsheet in Sub-Layers (a), Delamination at the backsheet-encapsulant interface (b), and rupture of the Backsheet (c)

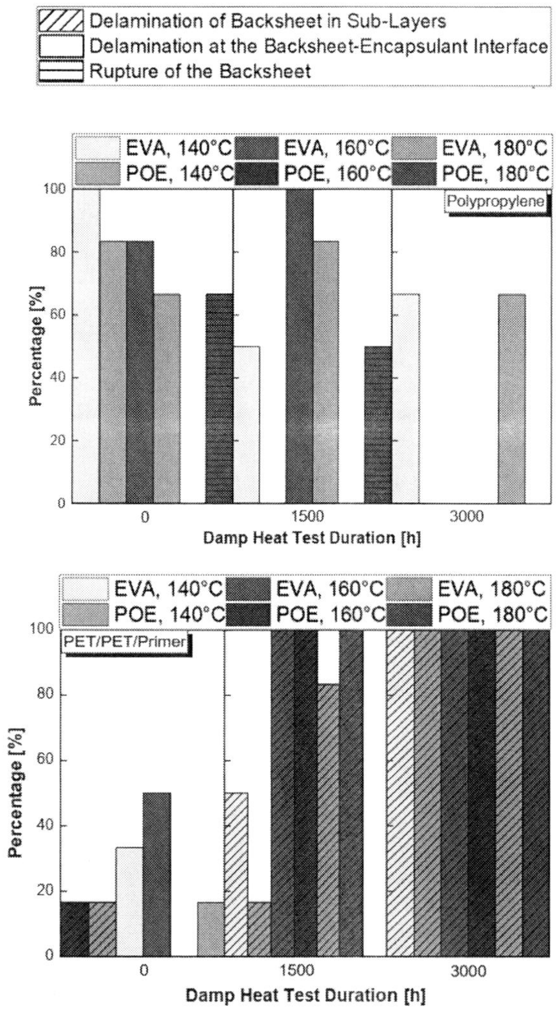

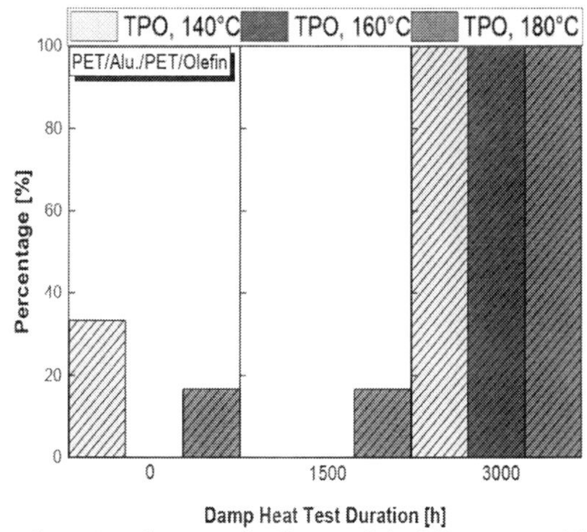

Figure 5: Failure type distribution of the different material combinations

Figure 5, shows the different failure modes depending on the materials and the degradation times [13].

For the samples with PET/Aluminium/PET/Olefin backsheet, the only failure mode was delamination of backsheet in sub-layers. These occurred for 40.7 % of the samples and indicates a potential issue of the backsheet itself under long DH exposure. A similar trend was observed for the laminates with the PET/PET/Primer backsheet. Here, the major failure type before ageing was delamination at the backsheet-encapsulant interface, while it changed to delamination of the back sheet in sub-layers after damp heat exposure. Again, it suggests an issue of the backsheet under long DH exposure.

For PP backsheets the most observed failure was a delamination at the the backsheet-encapsulant interface however for high DH exposure the encapsulant-glass interface could be measured. A rupture of the backsheet was observed in cases for POE encapsulant with an overall high peel force (see next section).

3.3 Peel Test

Each measurement point in this section corresponds to the average peel force of the six peel tests. The standard deviation is indicated with the error bars. Depending on the failure mode (shown in the previous section), the sample size for individual measurements points may be smaller, since only the peel force of the glass- encapsulant interface is evaluated here. , shows the measured peel-force over the lamination peak temperature before DH-test. The colours represent the different backsheets, PET/Aluminium/PET/Olefin in orange, PET/PET/Primer in blue and PP in green, while the symbols show the used encapsulant.

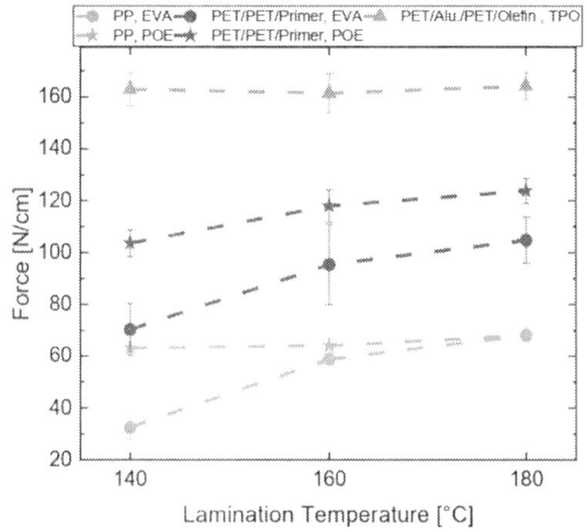

Figure 6: Initial peel forces obtained from the glass-encapsulant interface

The laminates with the TPO encapsulant have a consistent peel force of around 160 N/cm independent from the peak lamination temperature. The reason is that TPO does not require any cross-linking it just needs to exceed its melting point during lamination to form a good contact. EVA and POE require a certain time-temperature budget for the cross-linking reaction; therefore, we observe differences for the different lamination temperatures [2,14]. The peel forces for lamination temperatures of 180 °C are the highest even with shorter lamination times. For the PP backsheet, the peel forces are consistently lower than for the PET backsheet since it does not facilitate a primer layer. However, it is already a strong indication that the lamination setup was not sufficient for this parameter setup. This sample also showed the lowest gel content as seen in Figure 3.

The influence of the damp heat stability on the peel force for the different material combinations is presented in Figure 7 while the RMPFs calculated from these graphs are shown in Table 3.

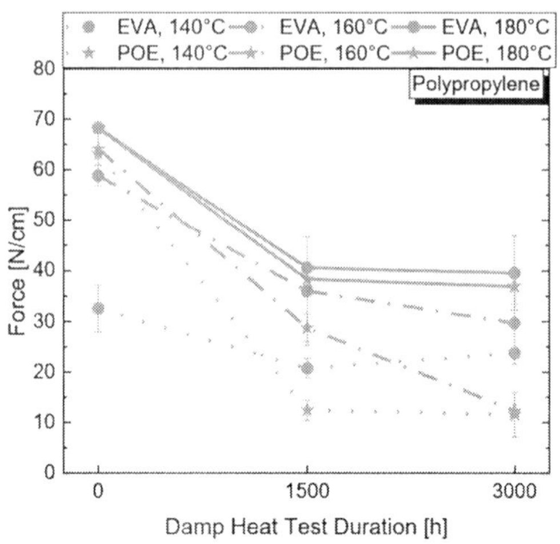

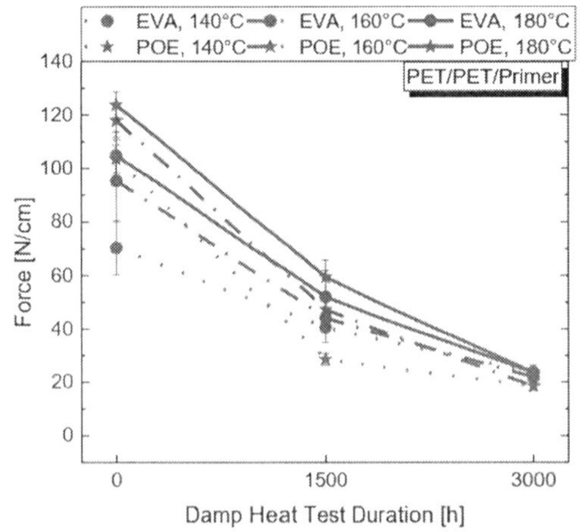

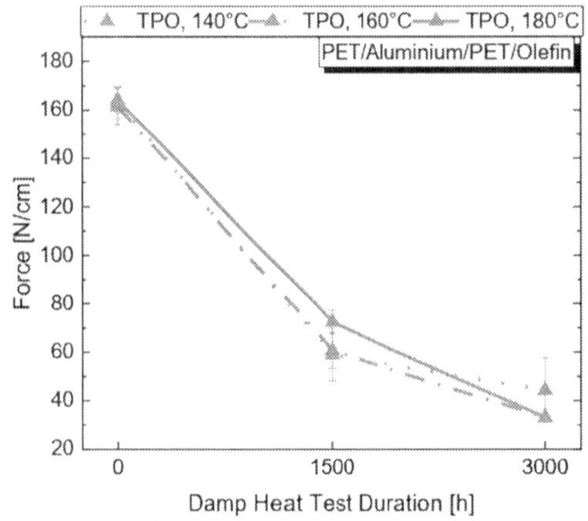

Figure 7: Peel force at the glass-encapsulant interface after 1500 h and 3000 h DH for the laminates with polypropylene, PET/PET/Primer, and PET/Aluminium/PET/Olefin backsheets respectively.

The graphs show that the highest peel force degradation occurred at the laminates which used the PET/Aluminium/PET/Olefin backsheet with a TPO encapsulant. While these laminates had peel forces of around 160 N/cm before the damp heat exposure, they dropped to 33-45 N/cm after exposing them to DH for 3000 hours. However, the absolute values of peel force are still amongst the highest overall and at DH 3000 all samples failed because of a delamination within the backsheet.

A strong drop in peel force is also observed for the laminates with the PET/PET/Primer backsheet. These laminates dropped from around 120 N/cm to 18-25 N/cm after DH 3000.

The laminates with polypropylene backsheet on the other hand had much lower peel forces (~65 N/cm) before DH testing compared to the other two backsheets, but after the damp heat exposure they degraded less than the others and their peel forces dropped to ~30 N/cm. At most of the polypropylene laminates, it is observed that the degradation rate lowered a lot by becoming around 15 % or even less, after DH 1500.

Comparing the results to the gel content measurements in Figure 3 it can be deduced that the samples with low gel content (before ageing) also expose a lower peel force. After ageing the peel force is reduced for all sample but still a lower gel content also means a lower peel force. This correlation between peel force and gel content is expected, since a more cross-linked material should show a stronger adhesion force. However, a quantification was not possible in this work.

Comparing the results to the moisture ingress as it can be seen in Table 2, that the PP backsheet has a comparably low WVTR [6,15]. Therefore, a lower degradation is expected. The decrease in peel force of the PET/Aluminium/PET/Olefin laminates on the other hand is not yet fully understood, because it is expected that the aluminium layer functions as a moisture barrier. However, since delamination of the backsheet in sub-layers occurred in all cases after DH 3000, further investigations should consider whether the backsheet is inherently degraded.

Comparing the different lamination conditions, it is observed that independent of the material combination, a low initial peel force is also followed by a low peel force after DH 1500 or DH 3000. Therefore, a higher process temperature is beneficial for higher peel forces initially and after DH exposure.

It is also observed, that independent of the lamination conditions, POE has a lower RMPF and therefore shows a higher degradation rate than EVA, which indicates that POE may be more susceptible to degradation by water vapor in terms of peel force.

Table 3: List of the used materials and the related average peel force values before ageing, after DH 3000 and the relative change of the peel force

Backsheet	Encapsulant	Lamination Temperature [°C]	Initial Peel Force [N/cm]	Peel Force after DH 3000 [N/cm]	Relative Maintained Peel Force [%]
polypropylene	EVA	180	68.23	36.88	54.1
PET/PET/Primer	EVA	180	104.81	23.54	22.4
polypropylene	EVA	160	58.80	29.66	50.4
PET/PET/Primer	EVA	160	95.36	21.80	22.9
polypropylene	EVA	140	32. 52	23.75	**73.0**
PET/PET/Primer	EVA	140	70.33	23.43	33.3
polypropylene	POE	180	68.23	37.24	54.6
PET/PET/Primer	POE	180	123.78	23.62	19.1
polypropylene	POE	160	64.09	12.27	19.2
PET/PET/Primer	POE	160	117.93	18.48	15.7
polypropylene	POE	140	63.12	11.55	18.3
PET/PET/Primer	POE	140	103.59	18.53	17.9
PET/Aluminium/PET/Olefin	TPO	180	**164.17**	33.12	20.2
PET/Aluminium/PET/Olefin	TPO	160	161.44	33.12	20.5
PET/Aluminium/PET/Olefin	TPO	140	162.91	**44.29**	27.2

3.4 Colour Measurements

To study the optical degradation, colour measurements were conducted. The results were evaluated at the front side and the rear side of the laminates and are shown in Figure 8.

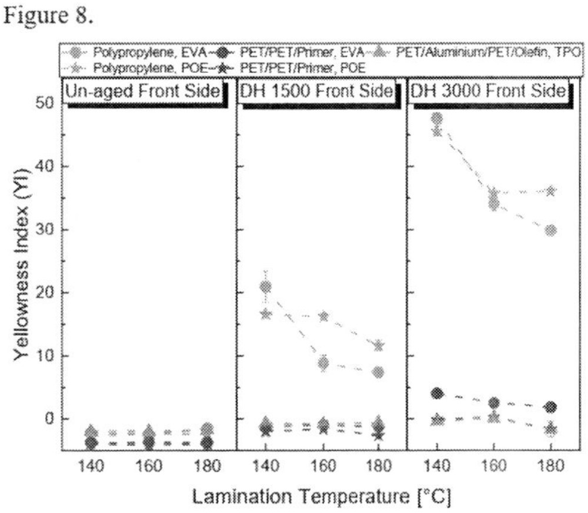

(a)

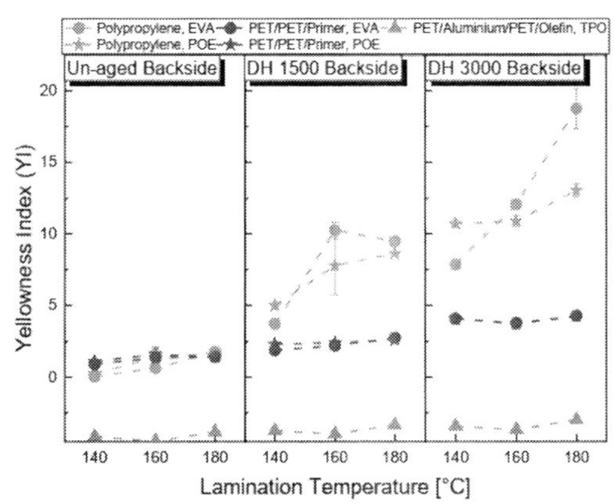

(b)

Figure 8: Yellowness index of the laminates with different material combinations (blue, green, orange) at different peak lamination temperatures on the (a) front side and (b) rear side of the laminate after lamination (left), after DH 1500 (middle), and after DH 3000

Before ageing the YI is quite similar for all lamination conditions when measured from the front side. After DH exposure (DH 1500 and DH 3000) the 180°C laminates, show on the front side a lower YI in contrast to the 140°C laminates for the PP backsheet and for a slightly lesser

degree for the other backsheets. However, when considering measurements on the rear side, the YI of the backsheets increases for higher lamination temperatures after ageing. A possible interpretation is that the backsheet is yellowing because of chemical reactions within the backsheet and within the encapsulant. At the rear, the process within the backsheet is dominating and an initial "burning" of the backsheet for higher lamination temperatures may increase during ageing.
For the front side the dominating factor might be the amount of the uncross-linked encapsulants. The polypropylene laminates had already a lower degree of cross-linking to begin, and the material might be more sensitive to chemical reactions caused by this uncross-linked particles. However, with a higher crosslinking degree (linked to higher lamination temperatures) these effects may be reduced.
Overall, PP backsheet shows the highest YI after degradation. The melting point of PP is much lower compared to the other backsheets that are used in this study. This might have caused chemical changes in the backsheet material during the lamination process which in turn led to the high YI values.

4. CONCLUSION

An extensive study to gain a deeper understanding of the influences of the lamination conditions and the DH stability of small-scale laminates featuring different backsheets and encapsulation materials was conducted. We used three different peak lamination temperatures (140°C, 160°C, and 180°C) with different curing times, and exposed the laminates to DH for 1500h and 3000 h.
The results indicated that laminates with POE and the EVA encapsulant show higher peel forces when laminated at higher temperatures, while the TPO-based laminates showed consistent peel forces independent of the peak lamination temperature. This also generally applies for the peel force after DH exposure. POE samples showed overall a higher peel force initially, while EVA did generally degrade less than POE during DH exposure. The degradation rate with PET based backsheets was higher than with the PP backsheets but the peel force after ageing was still higher than with PP backsheets. The results also show a correlation between the gel content and the measured peel forces. Laminates with a low degree of cross linking had a low peel force. Our interpretation would be that the peel force is increasing parallel with higher gel contents until a certain degree of cross linking where the influence of the gel content becomes negligible.
The results of the failure type analysis indicate that the failure type that occurred during the peel test is mainly affected by the backsheet structure. The PP backsheet was the only one at which rupture of backsheet occurred while both PET based backsheets showed delamination of the backsheet in sub-layers after the DH 3000 for all test samples.
The colour measurements showed an increase in yellowing index from the rear side for increased lamination temperature but a decrease in yellowing index from the front side, which may hint to different yellowing reactions inside and outside of the module.
Overall, it can be concluded that module laminated with elevated temperatures of up to 180 °C can be manufactured and they withstand DH3000 tests in terms of peel force and yellowing. This may allow for faster lamination processes.

5. ACKNOWLEDGEMENT

This work was supported by the German Federal Ministry for Economic Affairs and Energy (BMWi) under the contract number 0324287C, acronym GEPARD.

References

1. VDMA. Interenational Technology Roadmap for Photovoltaic (ITRPV) 2020 Results (2021).
2. Hirschl, C. *et al.* Determining the degree of crosslinking of ethylene vinyl acetate photovoltaic module encapsulants—A comparative study. *Solar Energy Materials and Solar Cells* **116,** (2013).
3. Wu, D., Zhu, J., Betts, T. R. & Gottschalg, R. Degradation of interfacial adhesion strength within photovoltaic mini-modules during damp-heat exposure. *Prog. Photovolt: Res. Appl.* **22,** 796–809 (2014).
4. J. Zhu, D. Wu, D. Montiel-Chicharro, T.R. Betts, R. Gottschalg. Realistic Adhesion Test for Photovoltaic Modules Qualification, 218–223 (2018).
5. Peike, C. Inauguraldissertation. Albert Ludwigs Universtiy, 2014.
6. Zhu, J. *et al.* Influences of Lamination Condition on Device Durability for EVA-Encapsulated PV Modules. 5 pages / 31st European Photovoltaic Solar Energy Conference and Exhibition; 2465-2469; 10.4229/EUPVSEC20152015-5CV.2.2 (2015).
7. Sraisth, Achieving Faster Lamination Process for Crystalline Photovoltaic Modules by using Latest Lamination Technologies (2017).
8. Mansour, D. E. *et al.* Effect of Backsheet Properties on PV Encapsulant Degradation during Combined Accelerated Aging Tests. *Sustainability* **12,** 5208 (2020).
9. Peike, C. *et al.* PV module degradation in the field and in the lab - how does it fit together? 5 pages / 29th European Photovoltaic Solar Energy Conference and Exhibition (2014).
10. BS EN 28510–1:1993, ISO8510-1:1990. *Adhesives— peel test for a flexible-bonded-to-rigid test specimen assembly-Part 1: 90° peel* (1993).
11. Jonai, S., Hara, K., Tsutsui, Y., Nakahama, H. & Masuda, A. Relationship between cross-linking conditions of ethylene vinyl acetate and potential induced degradation for crystalline silicon photovoltaic modules. *Jpn. J. Appl. Phys.* **54,** 08KG01 (2015).
12. *DIN EN ISO/CIE 11664-4:2020-03, Farbmetrik_-Teil_4:CIE_1976 L*a*b* Farbraum (ISO/CIE_11664-4:2019); Deutsche Fassung EN_ISO/CIE_11664-4:2019* (Beuth Verlag GmbH).
13. Wu, D. *et al.* Effects of different lamination conditions on the reliability of encapsulation materials of pv modules: adhesion strength (2014).
14. Agroui, K., Jaunich, M. & Arab, A. H. Analysis Techniques of Polymeric Encapsulant Materials for Photovoltaic Modules: Situation and Perspectives. *Energy Procedia* **93,** 203–210 (2016).
15. Krauter, S., Pénidon, R., Hanusch, M., Lippke, B. & Grunow, P. in *Proceedings of the ISES Solar World Congress 2011,* edited by K. Vajen (International Solar Energy Society, Sunday, August 28, 2011 - Friday, September 2, 2011), pp. 1–9.
16. Oreski, G. *et al.* Performance of PV modules using co-extruded backsheets based on polypropylene. *Solar Energy Materials and Solar Cells* **223,** 110976 (2021).

PERFORMANCE ANALYSIS AND CTM SIMULATIONS OF PV MODULES WITH VARIOUS ENCAPSULANT MATERIALS

Meriç Çalışkan[1], Fırat Es[1]
[1] KalyonPV Research and Development Center, Kalyon Güneş Teknolojileri Üretim A.Ş., 06909 Ankara, Turkey
*Corresponding author: Phone: +905374108157, Email: mcaliskan@kalyonpv.com

ABSTRACT: Photovoltaic (PV) modules are multilayer systems that are bonded adhesively to each other to create interfaces of different components including glass/encapsulant, interconnections/encapsulant, encapsulant/cell. One of the most important steps in the formation of mentioned interfaces during module fabrication is encapsulation of the solar cells. The main purpose of this step is to provide electrical isolation, mechanical support, thermal conduction and protect the cells from the exterior and interior impacts. The encapsulation materials are highly prone to aging and it directly influences the long term characteristics of PV modules. Therefore, the proper selection of encapsulation layers is very critical for the module design. To the prevent degradation of modules, polymer-based materials such as polyolefin elastomers (POEs), ethylene vinyl acetate (EVAs) and thermoplastic polyolefins (TPOs) are commonly utilized as encapsulation layers. In this study, the primary objective is observing effects of encapsulant films on the electrical performance of glass-glass modules. After that, paper will focus on the investigation of the cell to module (CTM) losses for different encapsulant based modules.
Keywords: encapsulant, EVA, POE, TPO, encapsulant, CTM, PID

1 INTRODUCTION

In photovoltaic module fabrication industry, EVA is predominating material of choice for encapsulation of modules due to its low cost, transparency, easy accessibility and know-how on processability [1]. Nowadays, different types of encapsulant materials attract the attention of researchers both for academic and industrial applications. The polyolefin based next generation encapsulants so called thermoplastic polyolefin and polyolefin elastomer have several advantages compared to ethylene vinyl acetate based encapsulants. Polyolefin based films possess higher volume resistivity, improved photoinduced degradation (PID) resistance and enhanced thermo-mechanical properties [2], [3].

During module lamiation process, encapsulant materials undergo crosslinking reaction which leads to chemical modification of the polymers. In this step weak thermoplastic polymer converts into to thermoset structure by creating chemical bonds between polymer chains [4]. In this way, mechanical property of the films enhances. Generally, encapsulant films require high level crosslinking degree above 70-80 %, in order to support PV module over a long period in the field [5], [6]. Main drawback of the EVA films is the acetic acid formation after the crosslinking reaction which accelerates the yellowing process of modules in the power plant [7]. Moreover, the usage of polyolefin containing films prevents the side product formation since these materials have non-polar groups on their polyethylene backbone which protects module from the aging process [8], [9]. Despite its disadvantages, EVA is widely used for the encapsulation of solar cells in the industry. This study shows the comparison of EVA, POE and TPO films and deals with the proper encapsulant material selection on the module manufacturing with respect to their effects on the electrical performance of the solar modules.

For the photovoltaic module manufacturing, the efficiency and output power of the module excessively depends on components used in the module design technologies. The fabrication of module from cells causes different physical power loss and gain mechanisms. The understanding and analysis of cell-to-module gains and losses mechanisms is vital to enhance photovoltaic modules perfomances, features of cells and as well as the materials [10], [11]. Generally, the total power sum of individual cells is higher than the assembled module power and this difference is called CTM losses. While predicting the performance of module, it is necessary to demonstrate several factors including optical, geometrical as well as electrical that impact the CTM ratio [12].

2 EXPERIMENTAL

2.1 Qualification of encapsulants

For the qualification of encapsulant materials gel content analysis and peeling strength tests were carried out. The crosslinking degree of the materials were investigated with the help of Soxhlet extraction method. In order to examine encapsulant features with the Soxhlet extraction, samples must be laminated under the same conditions as a normal solar module.

Two layers of 20cm*20cm encapsulants namely EVA, POE and TPO in an envelope of a Teflon sheet were putted between two pieces of small size solar glasses (50cm*50cm) and lamination process were performed with the proper recipe. After completion of the lamination process, encapsulants were taken out of the Teflon envelope and cut into 4x4 mm pieces. Small pieces of each encapsulant were desiccated and weighted (M1), after that the cured encapsulants were placed to the Soxhlet thimble and they were dipped into toluene solvent. Extraction process was performed with the suction effect. Then uncured part of the sample inside the Soxhlet thimble was removed, dried (80 °C), desiccated and weighted again (M2). The weight fraction (M2/M1) provides the gel content. Additionally, with the help of horizontal strip peeling tension testing machine, peeling strength of the three encapsulant sample with glass layer were detected.

2.2 Module production process

A standard module production process contains several steps including front glass loading, encapsulant cutting and placement, stringing, module lay-up, autobussing, back glass loading, edge sealing, lamination, attachment of the junction box, curing and a power measurement. In the KalyonPV module line, identical 144 half-cut bifacial monocrystalline PERC full square solar cell based glass/glass (2.5 mm thickness) solar modules with three different encapsulant layers were fabricated with the proper lamination parameters.

In order to investigate the behaviours of encapsulant layers and performance of modules in dependence of the encapsulation sheets, lamination parameters should be optimized on the module fabrication process. The proper lamination recipes for the module production were obtained after the several optimization studies and Table I represents the selected recipe.

Table I: Optimal lamination recipes for different encapsulant samples

	TPO	POE	EVA
Lamination temperature (C)	145	146	138
Lamination pressure (mbar)	500	800	800
Vacuum time (min)	6	7	6
Pressing time (min)	3	13	12

2.3 Electrical characterization of the modules

The electrical properties of the produced modules before and after PID aging were investigated by means of current-voltage (I-V) characteristics measurements. The power at maximum point (Pmax), voltage at maximum point (Vmax), current at maximum point (Imax), open circuit voltage (Voc), short circuit current (Isc), fill factor (FF) and the efficiency of the modules were defined as electrical parameters and mentioned parameters of the test modules were evaluated under illumination at standard test conditions. Standard test condition represents AM1.5 G spectral distribution at 25°C and 1000 W/m^2 illumination. In order to determine electrical performances PASAN solar simulator which is calibrated by a golden module per shift was utilized.

Besides from the electrical characterization, uniformity and electrical homogeneity of the produced module were also analysed. These analyses were performed by the means of SC Solar EL testing imaging equipment.

Electrical inhomogeneities occurs due to extrinsic and intrinsic defects. Intrinsic defects arise from the dislocations, shunts or process failures. On the other hand, extrinsic defects can be caused by microcracks, cracks, broken fingers or corrosions that observed on the PV cells. Furthermore, EL images reveal defects and microcracks in the inner structures of the PERC solar cells that make up a solar module. In this analysis, an external power supply was connected to the module and current passed into the positive terminal and out the negative terminal. That current flow cause solar cell to emit light in non-visible range and emitted light detected with a sensitive CCD camera. In the EL images, damaged areas or defects of the modules appear dark as compared to undamaged areas.

2.4 Aging test

Degradation behavior of the solar modules with respect to electrical performances were detected by means of PID aging.

PID effect can occur when the module's voltage potential and leakage current drive ion mobility within the module between the other components such as glass or frame. This effect reduces power output as well as the Isc and the Voc of the module. Thus, it directly affects the electrical performance of the solar modules.

In this study, three identical modules with different type of encapsulant (EVA, POE and TPO) were subjected to PID test under the 85 % RH and maximum system voltage at 60 °C for 96 hours.

2.5 CTM analysis

Necessary geometrical, electrical and optical information of module materials are collected and classified according to product data sheets and laboratory measurements. All related data was processed in order to perform CTM analyses of three different encapsulant layer containing modules via software namely SmartCalc.CTM developed by Fraunhofer ISE.

3 RESULTS AND DISCUSSION

Table II and Table III represent the I-V parameters of the three different encapsulant based identical modules before and after PID effect tests respectively.

Table II: I-V parameters of different encapsulant based modules before PID test

Product type	Voc (V)	Isc (A)	Pmax (W)	Vmax(V)	Imax(A)	FF (%)	Eff (%)
POE	48.9	10.2	401.2	41.3	9.7	80.33	19.73
TPO	48.8	10.2	399.4	41.2	9.7	80.19	19.58
EVA	49.1	10	401.1	41.3	9.7	79.73	19.74

Table III: I-V parameters of different encapsulant based modules after PID test

Product type	Voc (V)	Isc (A)	Pmax (W)	Vmax(V)	Imax(A)	FF (%)	Eff (%)
POE	48.6	10.1	393.8	41.1	9.5	80.61	19.38
TPO	48.8	10.1	394.8	41.2	9.5	80.24	19.43
EVA	48.9	10.1	394.4	41.2	9.1	79.97	19.41

As can be seen from the I-V characteristics of the various encapsulant based solar test modules, electrical performances revealed minor differences. Graphical represantation of the power changes of samples before and after PID aging can be seen in Figure 1.

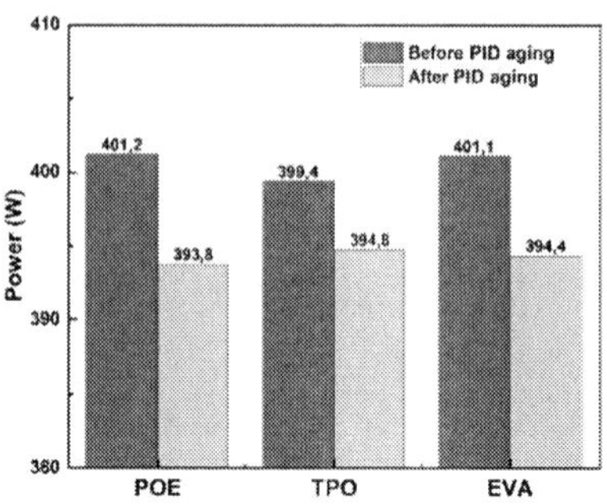

Figure 1: Power changes of samples after PID aging

Upon accelerated PID aging test, no significant power loss was observed for the fabricated modules with different encapsulants. According to test results, standard deviation on module power of EVA containing module is higher as compared to POE and TPO based glass-glass modules. Standard deviation of module powers after PID test were calculated as 2.84 %, 1.75 % and 1.43 % for EVA, TPO and POE films respectively.

Gel content analysis results reveal that POE and EVA

have 86.3 % and 82.32 % crosslinking degree respectively. By comparison, the TPO film shows no crosslinking reaction. During lamination process of the TPO sample, curing and crosslinking reaction not occurred since interfacial adhesion takes place with grafted functional group on polymer chain or with comonomers that already present in the TPO film. After lamination, chemical composition of TPO does not change owing to absence of crosslinking reaction [13]. Conversely, as can be seen from the crosslinking result POE and EVA samples show curing process during lamination. This reaction brings undesired changes especially for the EVA films. Due to presence of by-products such as gaseous and some volatiles after crosslinking reaction of EVA samples, EVA based modules tends to delamination, corrosion and discoloration [14].

Peeling strength with the glass material of the corresponding POE, TPO and EVA films were found as 87.4 N/cm, 90.7 N/cm and 98.9 N/cm respectively (see Table IV)

Table IV: Crosslinking degree and peeling strength with glass values of different encapsulants

Sample	Crosslinking degree (%)	Peeling strength with glasss (N/cm)
POE	86.3	87.4
TPO	non-crosslinking	90.7
EVA	82.32	98.9

Figure 2 represents the electroluminescence images of the three differing film based fabricated solar modules before and after PID conditioning test.

a)
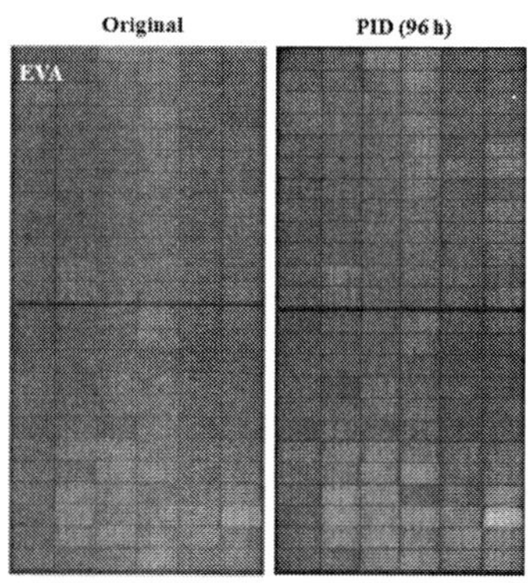

b)
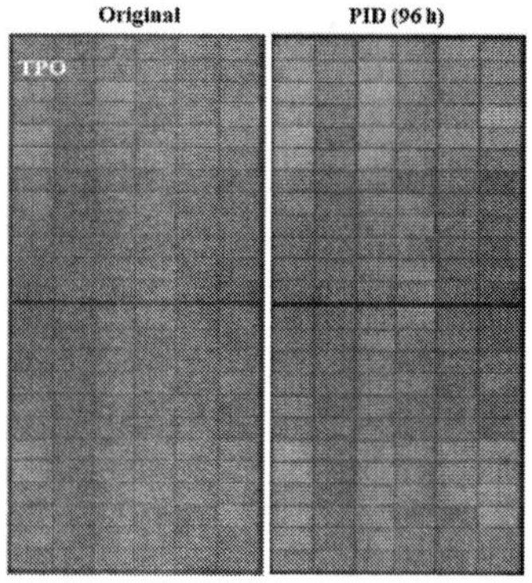

c)
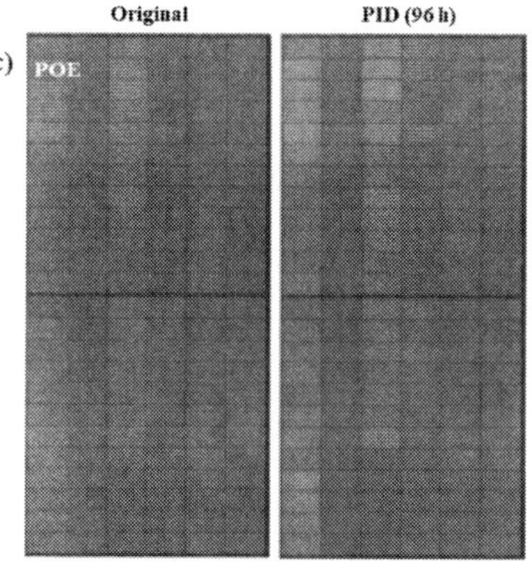

Figure 2: EL images of the modules before and after PID test for a) EVA, b) TPO, c) POE samples

It can be stated that there is no yellowing or ribbon corrosion effect after the accelerated PID test for POE and TPO based module. On the other hand, EVA based module shows darker areas especially on the upper left corner of the solar panel as can be seen in Figure 2a. After 96 hours system voltage application , EVA encapsulant based module showed signs of silver grid corrosion on the surface of cells and that situation was not detected in TPO or POE containing samples. Polyolefin based materials show higher chemical inertness than EVA and less prone to develop PID. The related PID phenomena that observed on the EVA samples can be explained from the acetic acid formation during hydrolysis process of crosslinking reaction of the EVA encapsulation [14].

To obtain detailed overview of the loss and gain mechanism of the module systems, CTM simulations was performed for all the corresponding module structures. Using SmartCalc. CTM software, the influence of materials that are used in the construction of module such as encapsulants, glass, cell connectors can be asssessed and analyzed. Figure 3 illustrates the waterfall power diagram of the 144-half cut mono PERC bifacial glass-glass different encapsulant based modules.

a)
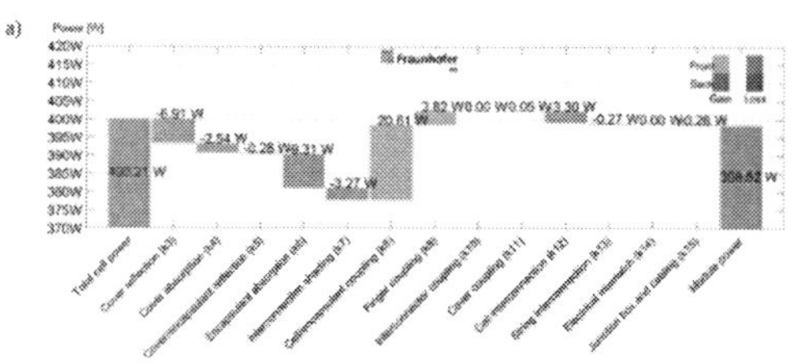

b)
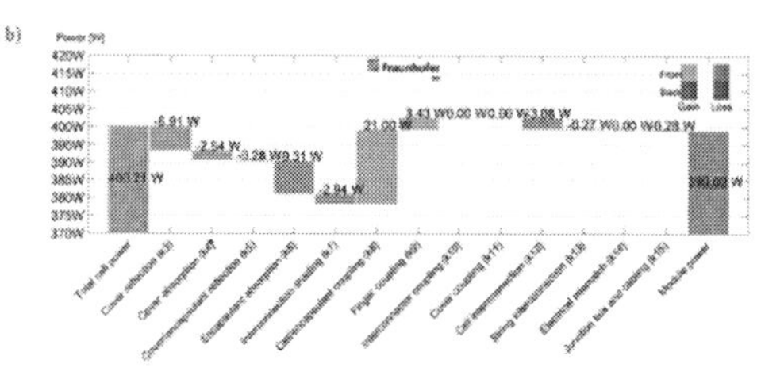

c)

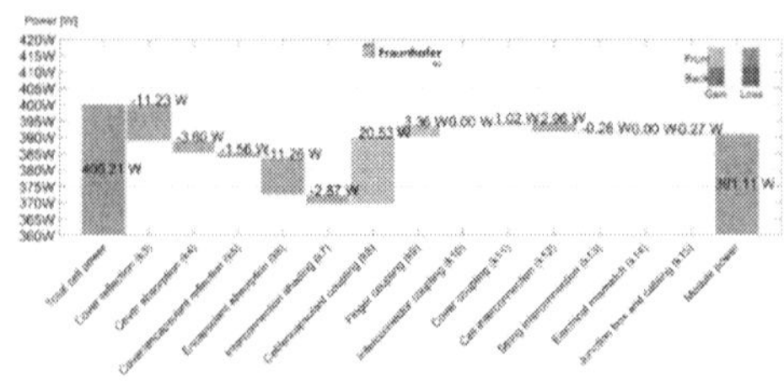

Figure 3: Waterfall power diagram of the 144 half cut mono PERC bifacial glass-glass module based on a)POE b) TPO c) EVA samples

Checking the waterfall power diagram of the corresponding modules, it can be stated that EVA sample containing module reveal more power loss in accordance with cell to module construction. As compared to polyolefin containing modules, EVA modules shows higher encapsulant absortion loss and cover reflection losses. Geometrical and electrical losses and gains values seems similar for all type of modules.

CTM ratios were calculated by dividing total cell power to the module power that are obtained from the software. CTM ratios are as following; 99.57 % for POE, 99.71 % for TPO and 97.73 % for EVA based samples (see Table V).

Table V: CTM ratios of the corresponding modules

Sample	Total cell power (W)	Module power (W)	CTM ratio (%)
POE	400.21	398.52	99.57
TPO	400.21	399.02	99.71
EVA	400.21	391.11	97.73

As can be seen the CTM diagrams of the samples, cell/encapsulant coupling gain is predominant for all of the module desings. The cell encapsulant coupling gain was found for 20.61 W for POE, 21 W for TPO and 20.53 W for EVA based solar modules. Encapsulant materials provides optical coupling gains due to their ability to reduce reflection from the front surface of the solar cells. Total internal reflection redirects the scattared light that falls into gaps between spaces. Then redirected lights are easily absorbed by solar cells. In this way module can provide more power by reducing optical CTM losses. Therefore, the CTM loss can be decreased by selecting the proper module materials especially encapsulant layers.

4 CONCLUSIONS

In accordance with these results, the strategy of replacement of EVA films with PO based encapsulants will pave a way in order to enhance the module performance and increase the solar panel stability in the power plants. Since polyolefin based encapsulant less prone to PID effect,they are very promising materials to provide structural and mechanical strength for solar modules. The results of CTM power ratio simulations have a good agreement with the actual I-V parameters. According the simulation results, it can be suggested that in order to observe higher power gain from the cell-encapsulant coupling, as an alternative of EVA film utilization on the glass-glass modules, PO based film can be also selected as an encapsulation layer.

References

[1] S. Jiang, K. Wang, H. Zhang, Y. Ding, and Q. Yu, "Encapsulation of PV Modules Using Ethylene Vinyl Acetate Copolymer as the Encapsulant," *Macromol. React. Eng.*, 2015, doi: 10.1002/mren.201400065.

[2] B. M. Chen, C. Y. Peng, J. L. Cho, and G. A. Porter, "Optimization of solar module encapsulant lamination by optical constant determination of ethylene-vinyl acetate," *Int. J. Photoenergy*, vol. 2015, 2015.

[3] M. C. López-Escalante, L. J. Caballero, F. Martín, M. Gabás, A. Cuevas, and J. R. Ramos-Barrado, "Polyolefin as PID-resistant encapsulant material in PV modules," *Sol. Energy Mater. Sol. Cells*, 2016, doi: 10.1016/j.solmat.2015.10.009.

[4] B. Adothu *et al.*, "Newly developed thermoplastic polyolefin encapsulant–A potential candidate for crystalline silicon photovoltaic modules encapsulation," *Sol. Energy*, 2019, doi: 10.1016/j.solener.2019.11.018.

[5] G. Oreski *et al.*, "Properties and degradation behaviour of polyolefin encapsulants for photovoltaic modules," *Prog. Photovoltaics Res. Appl.*, vol. 28, no. 12, pp. 1277–1288, 2020, doi: 10.1002/pip.3323.

[6] C. Hirschl *et al.*, "Determining the degree of crosslinking of ethylene vinyl acetate photovoltaic module encapsulants - A comparative study," *Sol. Energy Mater. Sol. Cells*, 2013, doi: 10.1016/j.solmat.2013.04.022.

[7] A. Morlier *et al.*, "Influence of the curing state of ethylene-vinyl acetate on photovoltaic modules aging," 2013.

[8] B. Camino, *Designing New Materials for Photocathodes*. 2017.

[9] G. Oreski *et al.*, "Properties and degradation behaviour of polyolefin encapsulants for photovoltaic modules," *Prog. Photovoltaics Res. Appl.*, vol. 28, no. 12, pp. 1277–1288, 2020, doi: 10.1002/pip.3323.

[10] I. Haedrich, U. Eitner, M. Wiese, and H. Wirth, "Unified methodology for determining CTM ratios: Systematic prediction of module power," *Sol. Energy Mater. Sol. Cells*, vol. 131, no. 7, pp. 14–23, 2014, doi: 10.1016/j.solmat.2014.06.025.

[11] J. Eymard *et al.*, "Performance Simulations Of A 72-Cell , A-Si Het Module With Different Tab-Interconnection Geometries

[12] M. Mittag, T. Zech, M. Wiese, D. Blasi, M. Ebert, and H. Wirth, "Cell-to-Module (CTM) Analysis for Photovoltaic Modules with Shingled Solar Cells," pp. 1531–1536, 2018, doi: 10.1109/pvsc.2017.8366260.

[13] B. Adothu, P. Bhatt, S. Zele, J. Oderkerk, F. R. Costa, and S. Mallick, "Investigation of newly developed thermoplastic polyolefin encapsulant principle properties for the c-Si PV module application," *Mater. Chem. Phys.*, 2020, doi: 10.1016/j.matchemphys.2020.122660.

[14] B. Sultan and E. Sörvik, "Thermal degradation of EVA and EBA—A comparison. I. Volatile decomposition products," *J. Appl. Polym. Sci.*, 1991, doi: 10.1002/app.1991.070430917.

Methodology for Evaluating Solar Module Encapsulant Materials

Hsin-Hsin Hsieh[1], Syh-Homg Chen[1], Meng-Tzu Lai[2], Chin-Ping Huang[2], Cheng-Wen Kuo[3], Ta-Ming Kuan[3], Cheng-Yeh Yu[3]
[1] Center for Measurement Standards, Industrial Technology Research Institute, Hsinchu, 310, Taiwan ROC.
[2] Material and Chemical Research Laboratories, Industrial Technology Research Institute, Hsinchu, 310, Taiwan ROC.
[3] TSEC Corporation, Hsin-Chu 30351, Taiwan, R.O.C.

Polymeric encapsulant material is the key to high performance and reliability of solar module. It is known that the most common encapsulant Ethylene Vinyl Acetate (EVA) has been widely used in c-Si solar module for past decades. According the literature and reports, it is revealed that most failure modules were caused by the degradation of encapsulant though the solar modules were qualified and certificated with IEC standards, in which only one parameter is controlled during the test, but it is performed in sequence. In fact, a real environment is a complex mixture of UV exposure, temperature variation, and humidity change. And the key factor degrading the EVA performance is still unclear. Therefore, by studying the key factors that cause failure, we can establish accelerated aging test to fast screen the applicability of encapsulant materials. In this research, a methodology for evaluating solar module encapsulant materials is constructed to characterize the reliability and performance.
keywords : Polymeric encapsulant; UV exposure; temperature variation; humidity change

1. Introduction

With the diverse development of technologies, the requirements of electricity is increasing. And photovoltaic module technology is one of the commonest renewable energies for its stability, high performance, and easy-installation. In general, PV manufacturers can provide a minimum warranty of > 20 years in power generation. For these reasons, PV modules are available to install on the rooftop or large-scale field site.

Despite the fact that corrosion/winds issues are highly valued for harsh environment, the failures in module are frequently discussed in PV systems. According to IEV-PVPS report, three typical failure scenarios for c-Si PV modules are light-induced degradation (LID), potential induced degradation (PID), and EVA degradation [1]. It is noted that EVA degradation is a failure mode that occurs constantly from the beginning through the wear-out state. Although the EVA discoloration is not the determinant factor for failure, the degradation of these polymeric materials will lead to metallization corroded due to the formation of acetic acid. In fact, the report further indicates that the EVA degradation may not only considered as an aesthetic issue, but also decrease the current output due to the lower transmittance.

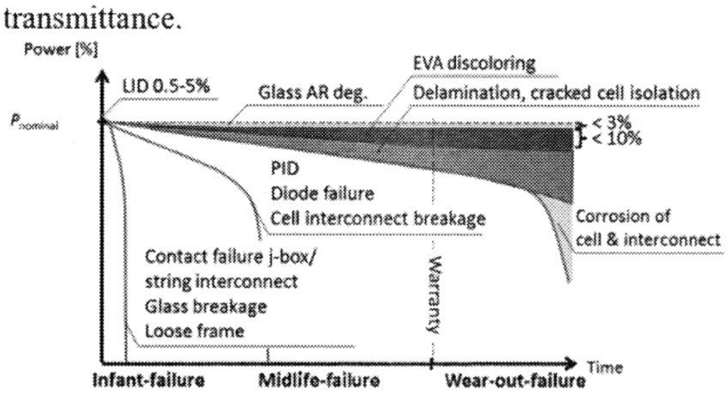

Fig. 1 Failure modes of c-Si PV modules [1].

Apart from the EVA discoloration, the occurrence of PID is also corresponding to the insufficient insulation of the EVA material. It is reported that polymeric materials of high volume resistivity significantly reduce the occurrence of PID [2-4]. The literature shows that the encapsulant is the key material to maintain the performance and extend the lifetime of PV modules.

With the increasing demanding on PV performance and reliability, more and more encapsulant materials are developed to apply in lamination technology. According to the analyses, the market share of EVA will decrease and materials like polyolefin, thermoplastic polyurethane, and PVB, will increase the market share [5~6]. These materials are laminated in PV modules for various filed application. For example, white EVA is a good option for large-scale power plant for its enhancement in optical reflection; polyolefin is preferred in glass-glass module for its high reliability and performance [7]. However, the certifications for PV products generally take few month and expenses. To shorten the testing duration, key tests referred from testing standards and experiences applied to evaluate the availability and qualification in PV module.

2. Testing methods

Considering the effectivity and hierarchy, PID, silver discoloration, weathering series test of UV-TC-HF, and adhesion test are proposed to fast evaluate the performance and reliability of encapsulant.

(1) PID test: for encapsulant material, the anti-PID ability is mainly determined by the volume resistivity. Thus, the PID test setup is mainly composed of two electrodes and a power supplier. The testing sample is positioned between the electrodes and then introduce an applied voltage (shown in Fig. 2). On the other hand, a chamber test based on IEC 61215:2021 is also carried out in solar modules with EVA based encapsulants [9].

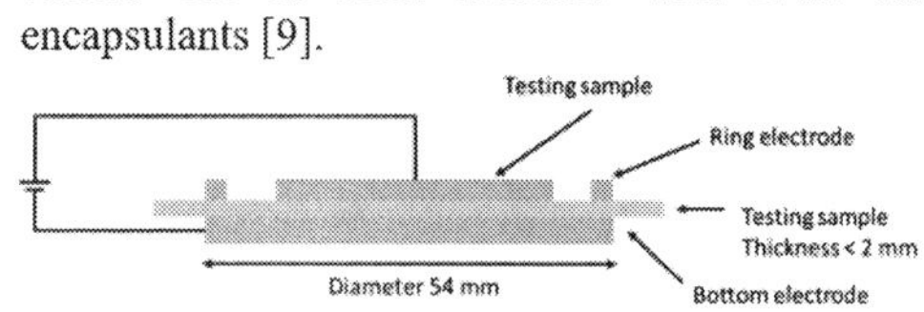

Fig. 2 PID test setup for encapsulant material.

(2) Reliability series test: (UV-TC-HF) series test referred from testing standard IEC 61215 is to simulate the module when exposed under actual

weathering condition [9]. To differentiate the reliability and performance of encapsulants, testing cycle are repeated and the degrading rate in each aging step is carefully recorded.

(3) Ag discoloration: This test simulates the cause of the color change of the silver wire. The encapsulant material is immersed in a chemical solution for 168 hours. After analyzing the light penetration characteristics, it can be seen that the immersed sample is passed through a visible ultraviolet light machine to perform the packaging material. The color change behavior is calibrated, and the spectral absorption measurement is performed at a specific wavelength (424 ~ 425) nm as a criterion for judgment.

(4) Adhesion test: By evaluating the bonding strength between the packaging material and each interface, the bonding ability between the packaging material and the backsheet, glass, solar cell and other layers can be initially evaluated. In order to comply with the 90-degree bonding strength test system specified by ASTM and ISO standards, the test system is shown in the figure below. After the test of this system, a special fixture was designed to improve the problem that the test angle is less than 90 degrees caused by the extension of the sample, so that the sample can be maintained at an angle of 90 degrees during the test. In addition, this test system can also be used with accelerated aging test, series test of interface reliability analysis after exposure, high temperature and high humidity test.

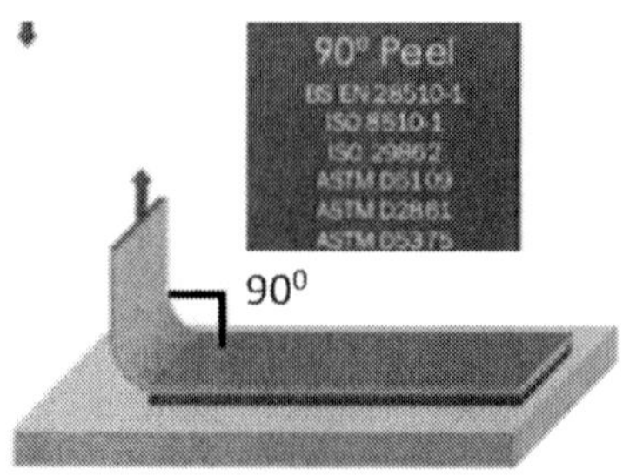

Fig. 3. Adhesion test setup

3. Results

PID test

The volume resistivity of encapsulant materials are shown in Fig. 4. With a common lamination temperature at 150 °C, the encapsulants can reach the high resistivity > 10^{14} Ω-cm. It is noted that the POE type performs highest volume resistivity of 2.315×10^{16} Ω-cm, while the EVA based encapsulants are around (0.21~1.75) $\times 10^{15}$ Ω-cm. Furthermore, it is observed that the volume resistivity is linear to the lamination temperature at (110~150) °C. Among the testing results, the volume resistivity of POE type encapsulant is significantly temperature dependent, of 6.00×10^{14} Ω-cm/ °C, while other EVA based are (1.96~4.00)$\times 10^{13}$ Ω-cm/°C.

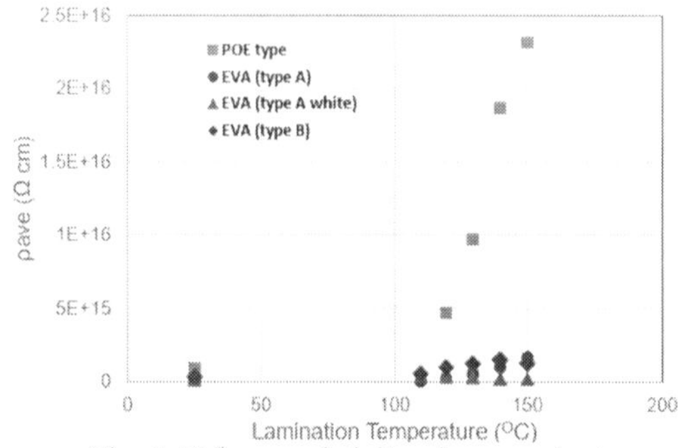

Fig. 4. Volume resistivity of encapsulant

It is well-know that IEC standard has a testing duration of 96 hr for PID module test. However, the qualified modules are still observed PID in field applications. Thus, considering the long-term performance of PID resistance of solar modules, the power degradation of PID module test with prolonged testing duration is conducted and shown in Fig. 5. It is observed that the all the samples are stable in the of PV frame materials in the initial condition are revealed in Fig 2. It is observed that the power degrades generally with the PID duration. However, a dramatic power decrease in a commercial encapsulant (sample A) at 384 hr, and a recovery is detected at 480 hr. It should be due to the residual moisture resulting in decreasing shunting resistance of solar module. On the other hand, it is

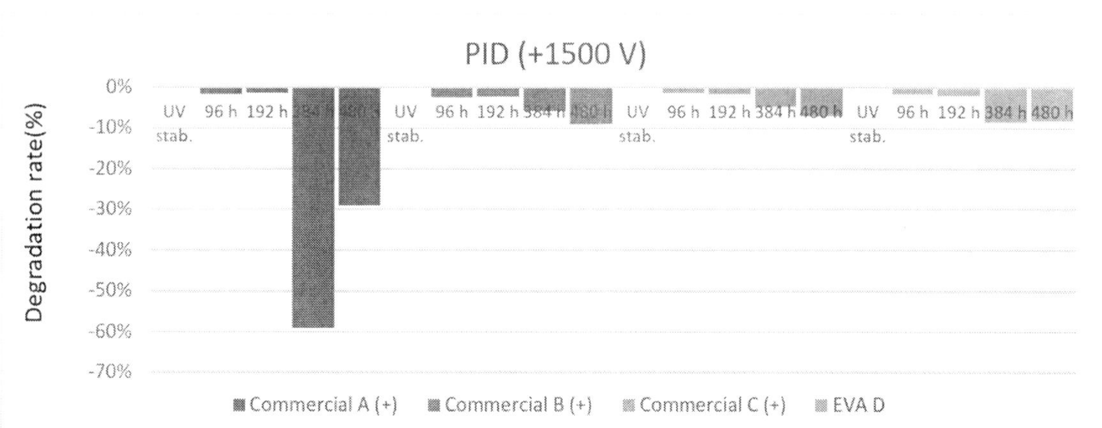

Fig. 5 Power degradation of PID with various encapsulant materials.

detected that all the encapsulant materials pass the criteria of power drop (< 5 %) for the requirements of IEC 61215:2021. However, the power decay caused by PID becomes evident when perform untill 384 hr. It is suggested that extend the testing duration

Reliability series test

The power degradation of solar module undergoing UV-TC-HF series test repeated 3 cycles is shown in Fig. 6. According to the testing results, it is seen that the power degradations for four kinds of encapsulants are generally less than 5 % after 3 cycles. In the beginning, the power is observed degrades < 3 % in UV test, and the following thermal cycling (TC) and humidity freeze (HF) tests continuously degrade power. However, it can be seen that the modules has a large degradation after humidity freeze (HF) test (shown as bars in purple and dark navy) and a slight recovery after UV exposure test (shown as light blue and dark red).

The IEC standards shows a temperature change of (-40~85) °C for both TC and HF tests. Besides, HF test accompanies a high humidity of 85 % RH for 10 days. The temperature difference (ΔT=125 °C) and high humidity should deteriorate the encapsulants and result in a power degradation. And UV exposure is followed with a dry condition at 60 °C for 7 days, which helps to decrease the humidity of solar modules/encapuslants and then recover the power. Therefore, it is assumed that for encapsulant materials, the influence of temperature changes and humidity is greater than UV exposure test.

Ag discoloration

The spectrum analysis results of the silver metallization color change test are shown in the figure 7. According to the testing results, the absorption peak of the encapsulant material after the accelerated catalysis of the chemical solution can be seen obvious changes. Except for the self-developed material (new material), it shows a low absorbance of 0.05 at wavelength 450 nm, and the rest of the samples all show an absorbance change of greater than 0.05 at wavelength 450 nm, showing that the current material research and development test results have better chemical stability than the other two commercial competing products.

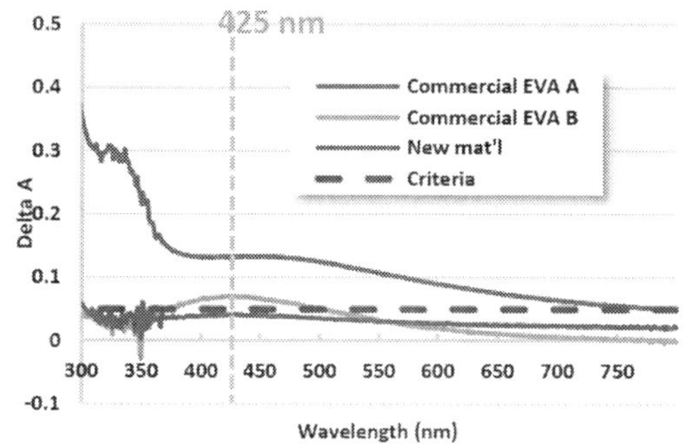

Fig. 7 Ag discoloration test.

Adhesion test

The figure 7 is the tensile stress curve of the glass/encapsulant material interface. By evaluating the peeling type and quantitative value of the tensile curve, the uniformity of the additive distribution and the interface strength can be observed. During the tensile test, the improved tensile test system can maintain the interface between the encapsulant material and the glass at 90 degrees to avoid plastic deformation of the encapsulant material and cause misjudgment. On the other hand, observing the tensile curve shows that the initial stage of the test is elastic deformation. It can be seen from the test results that the tensile test curves of the three materials show continuous and uniform stripping characteristic. And the interface strength of the three encapsulant materials is quite high (higher than the recommended value for packaging materials/glass: 50 N/cm). Among them, EVA materials A and B have higher tensile strength

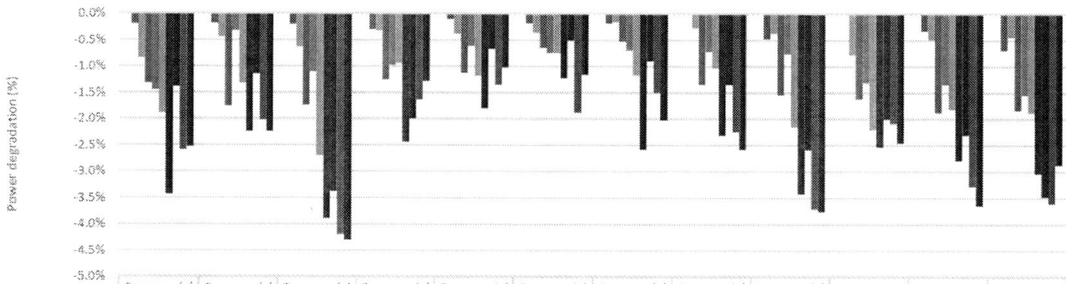

	Commercial material A	Commercial material A	Commercial material A	Commercial material B	Commercial material B	Commercial material B	Commercial material C	Commercial material C	Commercial material C	EVA material D	EVA material D	EVA material D
Initial	0%	0%	0%	0%	0%	0%	0%	0%	0%	0%	0%	0%
Loss after UV1	-0.20%	-0.18%	-0.20%	-0.31%	-0.10%	-0.18%	-0.18%	0.12%	-0.47%	1.62%	-0.33%	-0.69%
Loss after TC1	-0.84%	-0.45%	-0.63%	-0.33%	-0.39%	-0.37%	-0.16%	-0.27%	-0.37%	-0.78%	-0.49%	-0.45%
Loss after HF1	-1.33%	-1.75%	-1.74%	-1.26%	-1.12%	-0.65%	-0.53%	-1.34%	-1.54%	-1.62%	-1.87%	-1.83%
Loss after UV2	-1.45%	-0.33%	-1.11%	-0.98%	-0.61%	-0.75%	-0.70%	-0.72%	-0.76%	-1.31%	-1.34%	-1.55%
Loss after TC2	-1.89%	-1.33%	-2.70%	-0.94%	-1.18%	-0.75%	-1.17%	-1.03%	-2.16%	-2.21%	-1.81%	-1.87%
Loss after HF2	-3.44%	-2.24%	-3.89%	-2.44%	-1.79%	-1.22%	-2.58%	-2.31%	-3.43%	-2.54%	-2.79%	-3.03%
Loss after UV3	-1.39%	-1.14%	-3.38%	-1.99%	-0.67%	-0.51%	-0.90%	-1.34%	-2.59%	-2.01%	-2.30%	-3.48%
Loss after TC3	-2.59%	-2.02%	-4.20%	-1.63%	-1.35%	-1.87%	-1.49%	-2.25%	-3.72%	-2.09%	-3.28%	-3.60%
Loss after HF3	-2.53%	-2.24%	-4.30%	-1.28%	-1.02%	-1.16%	-2.03%	-2.58%	-3.76%	-2.46%	-3.65%	-2.87%

Fig. 6 Power degradation of solar modules with various encapsulant materials.

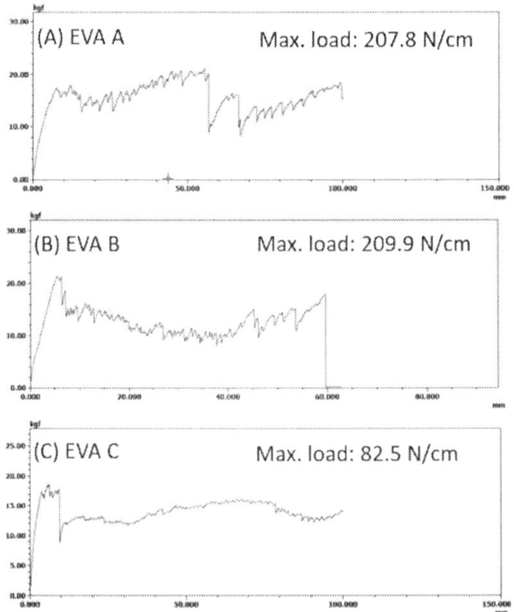

Fig. 7 Adhesion strength of encapsulant and glass.

and higher interface strength.

Conclusion

This research illustrates the methodology evaluating encapsulant materials for solar module lamination. Four important characteristics considering interfacial mechanical property, long-term reliability, chemical stability, and insulation safety are widely observed for evaluation. Some key conclusion can be addressed below:

(1) Volume resistivity of encapsulants is corresponding to PID resistance. And the PID module test should be conducted > 192 hr for a comprehensive evaluation.
(2) UV-TC-HF series test shows that temperature change and humidity are main degradation factors for encapsulants.
(3) Chemical stability help inhibit the Ag discoloration of encapsulant. The accelerated aging test shows that some commercial EVA is highly sensitive to chemicals, while a developed material represents a good stability.
(4) Mechanical property: an improved tensile test system can keep the peeling angle at 90 ° between the encapsulant material and the glass to avoid plastic deformation of the encapsulant material.

References

[1] Review of failures of photovoltaic modules, IEA-PVPS T13-01(2014).

[2] Juliane Berghold, Simon Koch, Benny Frohmann, Peter Hacke, and Paul Grunow, Properties of Encapsulation Materials and Their Relevance for Recent Field Failures, Proc. 40th IEEE Specialist conference, Denver, pp. 1987-1992, 2014.

[3] B. M. Habersberger, P. Hacke, L. Madenjian, IEEE Specialist conference, IEEE 7th World Conference on Photovoltaic Energy Conversion, Waikoloa, pp. 2018.

[4] R. Tanaka, and H. Zenkoh, Encapsulant effect on PID durability of various crystalline PV cells, NREL PV reliability Workshop, 2015.

[5] Cornelia Peike, Ingrid Hädrich, Karl-Anders Weiß, and Ines Dürr, Overview of PV module encapsulation materials, Photovoltaic International, pp.85~92. 2013.

[6] International Technology Roadmap for Photovoltaic, ITRPV, 2021.

[7] G. Oreski, A. Omazic, G. C. Eder, Y. Voronko, L. Neumaier, W. Muhleisen, C. Hirschl, G. Ujvari, R. Ebner, and M. Edler, Properties and degradation behavior of polyolefin encapsulants for photovoltaic modules, Prog. Photovolt Res Appl. 2020; **28**:1277~1288.

[8] M. Kohl, Accelerated service life testing of photovoltaic modules, in Service life prediction of exterior plastics, Springer, pp. 199~213, 2014. DOI: 10.1007/978-3-319-06034-7 13.

[9] IEC 61215:2021, Terrestrial photovoltaic (PV) modules - Design qualification and type approval - Part 1: Test requirements.

STABILITY OF INKS USED FOR MASKING METALLIC INTERCONNECTS IN BIPV MODULES

Alejandro Borja Block[1*], Alessandro Virtuani[1], Christophe Ballif[1,2]
[1] École Polytechnique Fédérale de Lausanne (EPFL), Institute of Electrical and Micro Engineering (IEM), Photovoltaics and Thin Film Electronics Laboratory (PVLAB), 2000 Neuchâtel, Switzerland
[2] CSEM, PV-center, 2000 Neuchâtel, Switzerland

ABSTRACT: The necessity to reduce carbon emissions has pushed European institutions to foster the adoption - for new buildings - of Nearly Zero Energy Buildings, which need to minimize energy consumption with energy generated on site. This opens up an enormous potential for the integration of renewable energy sources in the built environment. However, a massive introduction of PV into buildings should be accompanied by an increased sensitivity in terms of the aesthetic appeal of the modules. Inks masking the metallic interconnects in building integrated photovoltaic (BIPV) elements might be an attractive option to produce more aesthetically pleasant modules. In this work, we investigate the stability of three commercial inks (1 solvent based and 2 UV curable) using two common PV configurations (glass-EVA-glass and glass-EVA-backsheet). We lay down a testing protocol to screen the stability of inks supplied by different manufacturers using damp heat (DH 85°C, 85%RH) and ultraviolet exposure conditions (UV 65°C, 60 W/m^2). Preliminary results show a halo in the surroundings of the coated metallic interconnects for UV curable inks when exposing these samples to direct UV, while no halo is observed when damp heat is performed first and for the solvent-based ink indicating that it is more stable.

Keywords: Building Integrated PV (BIPV), aesthetics, reliability, metallic interconnects, ink

1 INTRODUCTION

There is still huge area of opportunity to implement renewables in buildings. Worldwide, buildings account for 33% of the final energy demand, with 86.4% of it coming from non-renewable energy and traditional biomass [1]. Thanks to the great technological advances of the recent years, solar energy is nowadays competitive and one of the cheapest sources for generating electricity, compared to more conventional power generation technologies [2].

International governments target to reduce greenhouse gases emissions by lowering pollution from buildings. Some of the European Union (EU) initiatives are the Building Energy Performance Directive [3] and the Directive on Renewable Energy [4]. Such actions aim ambitious objectives and renewable energy, principally solar, are necessary to achieve them. Incorporating solar energy into constructions is an attractive way to generate self-sustaining buildings and use on a wise manner the space to generate electricity cleanly.

Unfortunately, as of today several factors affect the integration of BIPV into the built-environment such as the skepticism about the PV technology from architects and other stakeholders, the lack of tools to help the implementation (i.e. Building Information Modeling), and the high manufacturing costs compared to mainstream modules, [5], [6]. Additionally, when integrating PV in the built environment, aesthetics may play a major role, as particularly architects may be extremely sensitive to this aspect. To boost the adoption of BIPV, it is therefore necessary to address these different issues simultaneously. As an example, when choosing modules for façades (or for other installations that may be visible by a broader public) the visual appearance of the PV module is often judged as a critical element by architects. Multiple solutions have being used to camouflage the presence of PV modules into the built-environment, these include, colorizing foils, ceramic printed glass covers and spectral-selective photonic morpho structures. These solutions, which allow in some cases to completely mask the presence of the PV modules, lead however frequently to a considerable decrease of the module power output [7].

In this work, we focus on masking the metallic interconnects (also called metallic ribbons) used to contact the different solar cells and strings in a PV module. In the past solar cells made use of three to four ribbons to contact adjacent cells. In recent years, this number is increasing considerably from five metallic ribbons per solar cell to nine or even more.

Ribbons have a metallic and reflective appearance that sometimes is not judged as attractive. For this reason, we try to mask their presence by using an ink, which should increase the uniform visual appearance of the modules.

In this work, we lay down a protocol to screen the stability of the inks used for masking metallic ribbons interconnects in BIPV modules.

The long-term goal of this research project is to manufacture an ink-dispensing tool to mask ribbons in a cost and time efficient way.

2 EXPERIMENTAL

We prepare samples with a usual PV configuration (glass-EVA-glass and glass-EVA-backsheet) with encapsulated ink coated metallic ribbons and we perform a protocol to assess the stability of the inks. Reflectance measurements on the masked metallic ribbons and visual inspection are performed to assess the appearance of the samples.

2.1 Sample preparation

Two usual PV configurations are prepared with commercial materials (see figure 1).

Squared sized pieces of 5 cm x 5 cm glass, EVA and backsheet (PPE) are cut and 3 cm x 0.5 cm metallic interconnects are coated with three different black colored commercial inks (A, B and C).

The stack of materials are formed, positioning the coated metallic ribbons in a centered place, and laminated with a standard recipe for the used EVA which involved maximum temperatures of 150°C.

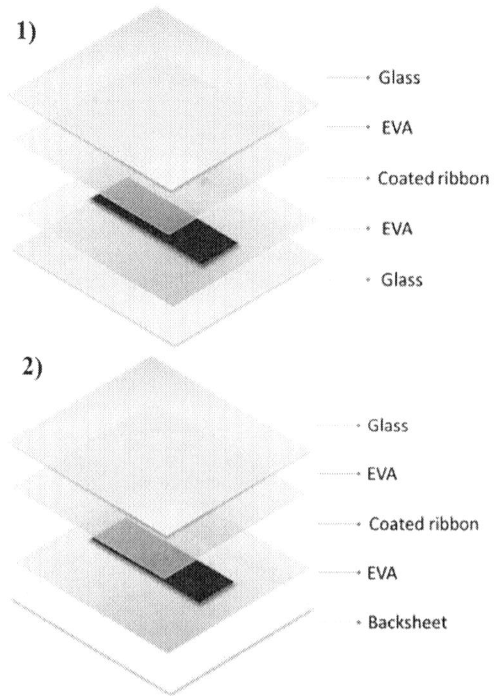

Figure 1: sketch of the samples structures configurations used to test the effect of the ink after accelerated aging tests. (1) glass-glass (G/G) and (2) glass-backsheet (G/Bs) configurations.

2.2 Testing sequences

The samples are aged following two main sequences shown in figure 2.

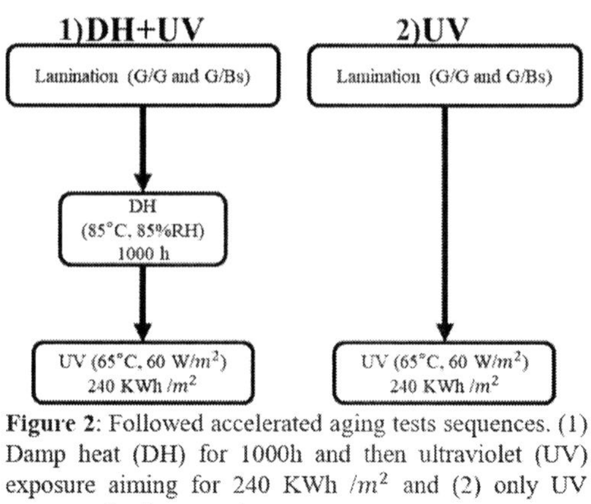

Figure 2: Followed accelerated aging tests sequences. (1) Damp heat (DH) for 1000h and then ultraviolet (UV) exposure aiming for 240 KWh $/m^2$ and (2) only UV exposure for the same dose.

All sample configurations were subjected to sequence DH+UV or UV, test are ongoing for ink C (see table 1). The idea behind these two sequences is to identify if humidity in G/G and G/Bs configurations produces a different effect in the samples with masked metallic ribbons after UV conditions.

Table I: status of the followed sequences and type of ink

Ink	Status Sequence DH+UV	Status Sequence UV	Type of ink
A	Performed	Performed	Solvent based
B	Performed	Performed	UV curable
C	Ongoing	Ongoing	UV curable

The IEC 62788-7-2 [8] standard is followed for all the samples aiming for a total duration of at least 4000 hours. A Q-Sun Xenon chamber (Q-Lab) is used with the following exposure conditions:

- UV dose: 60 W/m^2
- Chamber air temperature: 65°C
- Black panel temperature: 90°C
- Irradiance: 0.8 W/m^2at 340 nm
- Relative humidity: 20%

After 4000 hours of aging test, a total UV dose of 240 kWh/m^2 will be reached. This corresponds roughly to an outdoor exposure of ~4 years in a Central European temperate climate. This is the UV reached on the front glass of the samples, not the UV reaching the ink inside the samples. The glass absorbs a small fraction of around 10% of the total UV dose.

Damp Heat (DH, at 85°C, 85% of relative humidity) is performed according to the IEC 61215-2 standard [9], using a Weiss-Voetsch C 180, -40 chamber.

2.3 Sample characterization techniques

The goal of using ink in PV is to change the aesthetics of the modules. The measurements involved in this work are aiming to examine if the ink is preserving its appearance and if a visual change is occurring over time in the samples.

UV-Vis-NIR spectroscopy

Ultraviolet(UV)/visual (Vis)/near IR (NIR) spectroscopy is used to study the optical properties of the metallic ribbons, but only the reflectance from 300nm to 900nm is measured. A UV-Vis-near-infrared spectrophotometer Lambda 950 from Perkin Elmer (Waltham, USA) is used to perform the measurements. Spectra are recorded from 300 to 900 nm with a measurement increment of 10 nm using an integrating sphere from Labsphere (North Sutton, USA) in order to measure reflectance. As the measuring aperture of this equipment has a diameter of 3cm and the metallic ribbons have a rectangular shape of 3cm x 0.5cm a mask was manufactured to be positioned on the equipment with the same dimensions of the metallic ribbons. As a mask is used post data treatment is done to obtain the corrected data using the following formula:

$$R_{corrected} = \frac{(R_{sample+mask} - R_{mask}) * R_{ref}}{R_{mask+ref} - R_{mask}} \quad (1)$$

Where:

-$R_{sample+mask}$: reflectance of the coated metallic ribbon with the mask

-R_{mask}: reflectance of only the mask

-R_{ref}: reflectance of a high reflective surface (reference spectralon)

-$R_{mask+ref}$: reflectance of the mask with the reference spectralon

-$R_{corrected}$: corrected reflectance of the coated metallic ribbon

Visual inspection
Periodic visual inspections were performed at high and constant illumination to minimize artefacts.

3 RESULTS

All coated metallic ribbons, irrespective of the ink used, endured a lamination process used for PV modules without any modification to their visual appearance, showing that all inks are compatible with conventional lamination processes. Due to the mechanical stresses caused by the thickness of the metallic ribbons, which is of 0.4mm, some of the G/G configuration samples were cracked or after lamination presented some bubbles, this was not the case for G/Bs configuration.
Performing reflectance measurements before and after stress testing the samples (sequence UV and DH+UV), showed no change in the reflectance (between 300nm and 900nm). This is true for all inks and both G/G and G/Bs samples, as can be seen in figures 3 to 5. The masked metallic ribbons have preserved up to now their black colored appearance.
As the non-coated metallic ribbons are highly reflective if the ink covering them was disappearing, an increase in reflectance would be noticed in the measurements, which we did not observe.
Presently the full testing sequence has been applied to ink A and B. For ink C, measurements are still ongoing.
Figures 3 to 5 contain as well the results of the visual characterization performed on the different samples before and after the stress tests sequences.

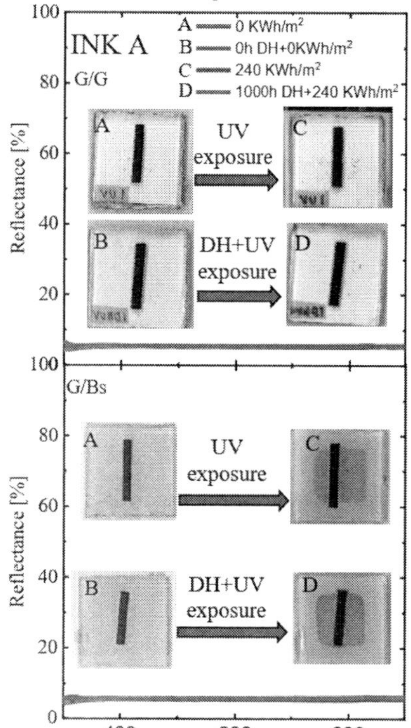

Figure 3: Reflectance and visual inspection of ink A, before and after performing the different stress test sequences (UV and DH+UV). The reflectance does not change , G/G do not present any significant change besides a light yellowing of the EVA and G/Bs generate a square shaped yellow effect in the inner position of the samples after exposing the samples to UV.

All samples with G/Bs configuration show a yellowing effect after UV exposure following both sequences, this effect is also shown in the work of J. H. Kim et al. [10], DH does not seem to contribute to this effect as it only appears on UV exposure conditions. Furthermore, none of the G/G configuration samples presents it.
A halo is observed in the surroundings of the ribbon in all configurations of ink B and C, which are UV curable, being more noticeable for ink C, following the sequence that only involves UV exposure (see figures 4 and 5). The halo has not appeared up to these exposure conditions following the sequence of DH+UV exposure, tests are ongoing. It is believed that ink B and C are causing this effect because the halo is observed in all configurations when these inks are present after the sequence with only UV exposure. G/Bs configuration samples were manufactured with and without unmasked metallic interconnects to compare effectively the different effects (see figure 6).
Ink C produces a larger halo than ink B in the surroundings of the metallic ribbon, while it is inexistent with ink A indicating that the solvent based ink is more stable.
The change in appearance might be caused by ink migration to the encapsulant, further investigation will be performed in order to understand this mechanism and the reason why the halo is produced with UV exposure but not with DH and then UV exposure.

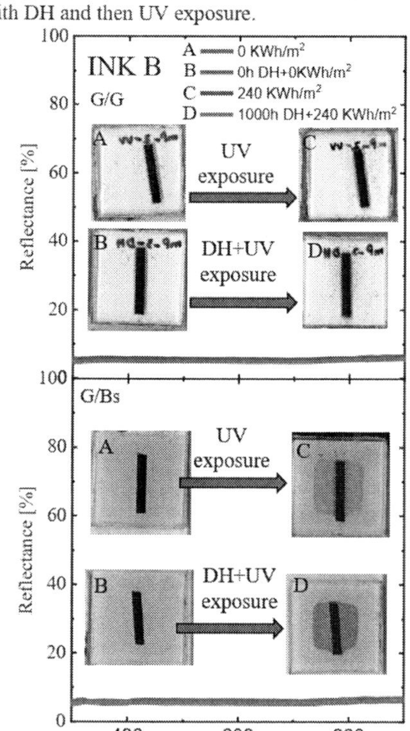

Figure 4: Reflectance and visual inspection of ink B, before and after performing the different stress test sequences (UV and DH+UV). The reflectance does not change and G/Bs generate a square shaped yellow effect

in the inner position of the samples after exposing them to UV. All samples following the UV sequence present a slight halo in the surroundings of the metallic ribbon, while samples following the DH+UV sequence do not present it.

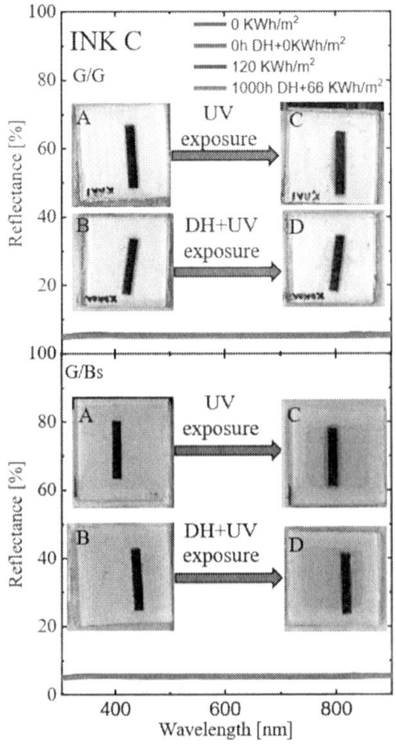

Figure 5: Reflectance and visual inspection of ink C, before and after performing the different stress test sequences (UV and DH+UV). The reflectance does not change and G/Bs generate a square shaped yellow effect in the inner position of the samples after exposing them to UV. All samples following the UV sequence present a halo in the surroundings of the metallic ribbon, while samples following the DH+UV sequence do not present it. Tests are ongoing.

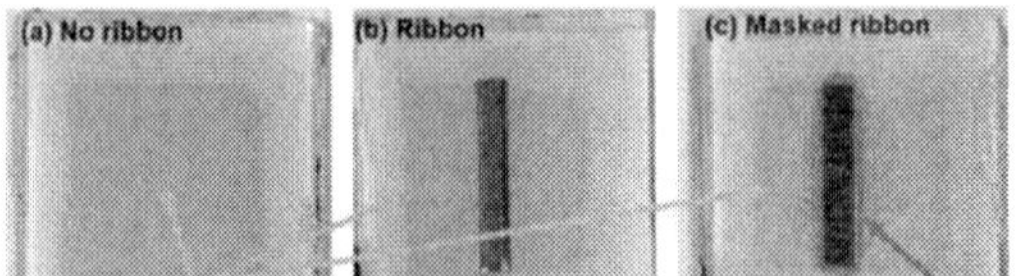

Figure 6: Samples in a G/Bs configuration exposed to a UV dose of 120 kWh/m^2 (a) without metallic interconnect (b) with an uncoated metallic interconnect and (c) with masked metallic interconnect using ink C. The yellowing effect shown in (a), (b) and (c) does not depend on the ribbon nor the ink and the halo visible only in (c) is related to the ink.

There are two main effects which can be observed in figure 6: (1) the yellowing effect for the G/Bs configurations; and (2) the appearance of the halo surrounding the metallic ribbon. The halo appears for both in the G/G and G/Bs configurations, being slightly milder for the G/G configuration. Both effects appear after UV exposure conditions. We observe that following the DH+UV exposure sequence the halo is not observed up to now. The exact reason for this is not clear and further investigation is ongoing to explain this behavior.

4 CONCLUSION

The purpose of this work is to investigate the stability of three commercially available black inks used to mask the metallic ribbons employed to connect the solar cells and strings in a PV module. All inks were compatible with the conventional lamination process. The samples were exposed to 1000h of DH (85°C, 85%RH) and then to 240 kWh/m^2 (corresponding to approximately 4 years of operation in a Central European temperate climate) in DH+UV sequence and only to 240 kWh/m^2 in UV sequence, tests are still ongoing for ink C.
Up to this exposure conditions two main effects are seen. One is the change in color in the surroundings of the coated metallic ribbons, this halo occurs on ink B and C, which are both UV curable inks, being more present on ink C, while it is not present on ink A, a solvent-based ink, indicating that it is more stable. The halo is observed following the sequence that only involves UV exposure. Up to these exposure conditions, in the samples following the sequence DH+UV the halo is not observed, this might be because the ink could increase its adhesion with time. Ongoing experiments will provide a deeper understanding on the underlying degradation mechanism for the halo effect that might be caused by ink migration to the encapsulant. The other effect is a square shaped yellowing that appears after UV exposure conditions in all G/Bs samples. DH and the inks do not seem to contribute to this yellowing effect in the G/Bs samples.
The reflectance is remaining constant indicating that the metallic ribbons have preserved their aesthetics.
These preliminary results suggest that the ink choice to modify the visual appearance of BIPV modules need to be considerate and it is important to have a protocol to assess the stability of the inks and screen them. The testing protocol used in this work shows the importance of long-term UV exposure to screen inks.
We are continuing this research activity by exposing the samples to further accelerated aging conditions and testing other material configurations. The long-term goal of this research project is to manufacture an ink-dispensing tool to mask ribbons in a cost and time efficient way.

5 REFERENCES

[1] REN21, *Renewables 2020 Global Status Report*. 2020.

[2] ITRPV, "International Technology Roadmap for Photovoltaic," 2019. [Online]. Available: https://itrpv.vdma.org/en/ueber-uns.

[3] European Commission, "DIRECTIVE (EU) 2018/844 OF THE EUROPEAN PARLIAMENT AND OF THE COUNCIL of 30 May 2018 amending Directive 2010/31/EU on the energy performance of buildings and Directive 2012/27/EU on energy efficiency (Text with EEA relevance)," *Off. J. Eur. Union*, vol. 2018, no. October 2012, pp. 75–91, 2018, [Online]. Available: https://eur-lex.europa.eu/eli/dir/2018/844/oj.

[4] European Commission, "Directive (EU) 2018/2001 of the European Parliament and of the Council of 11 December 2018 on the promotion of the use of energy from renewable sources (Text with EEA relevance.)," *Off. J. Eur. Union*, vol. 2018, no. October 2003, pp. 1–25, 2018, [Online]. Available: https://eur-lex.europa.eu/eli/dir/2018/2001/oj.

[5] PVsites, "BIPV market and stakeholder analysis and needs Project report," 2016.

[6] C. Ballif, L.-E. Perret-Aebi, S. Lufkin, and E. Rey, "Integrated thinking for photovoltaics in buildings," *Nat. Energy*, vol. 3, no. 6, pp. 438–442, Jun. 2018, doi: 10.1038/s41560-018-0176-2.

[7] C. Kutter *et al.*, "DECORATED BUILDING-INTEGRATED PHOTOVOLTAIC MODULES: POWER LOSS, COLOR APPEARANCE AND COST ANALYSIS," in *35th European Photovoltaic Solar Energy Conference and Exhibition*, 2018, pp. 1488–1492, doi: 10.4229/35thEUPVSEC2018-6AO.8.6.

[8] IEC, "IEC TS 62788-7-2: Measurement procedures for materials used in photovoltaic modules - Part 7-2: Environmental exposures - Accelerated weathering tests of polymeric materials," 2017.

[9] IEC, "IEC 61215-2: Terrestrial photovoltaic (PV) modules – Design qualification and type approval – Part 2: Test procedures," 2016.

[10] J. H. Kim *et al.*, "Nanomechanical and Fluorescence Characterizations of Weathered PV Module Encapsulation," *IEEE J. Photovoltaics*, vol. 11, no. 3, pp. 725–730, May 2021, doi: 10.1109/JPHOTOV.2021.3053657.

6 ACKNOWLEDMENTS

This project has received funding from the European Union's Horizon 2020 research and innovation programme under the Marie Skłodowska-Curie grant agreement No 754354. This work has been funded in part by the European Commission (EC) under the H2020 Be-SMART (#818009). We gratefully acknowledge support from all PV-Lab team members.

Three-Dimensional Multi-Ribbon Back-Contact Interconnection: Latest Results on Reliability Testing

Rik Van Dyck[1,2*], Tom Borgers[2], Jonathan Govaerts[2], Arvid van der Heide[2], Loic Tous[2], Jef Poortmans[1,2,3], Aart Willem van Vuure[1]

[1]KU Leuven, Oude Markt 13, 3000 Leuven, Belgium

[2]imec (EnergyVille campus), Thor Park 8320, 3600 Genk, Belgium

[3]University of Hasselt, Martelarenlaan 42, 3500 Hasselt, Belgium

*Corresponding author, work phone: +32 16 28 19 41, e-mail: rik.vandyck@imec.be

Sub-topic 4.1: PV Module Design, Manufacture, Performance and Reliability

ABSTRACT
Back-contact solar cells are gaining terrain in the PV industry, although the lack of a standardized and low-cost interconnection method inhibits their market growth. To overcome this challenge, the three-dimensional multi-ribbon interconnection is proposed by imec as a novel interconnection concept for highly efficient back-contact cells. It is based on an interconnection fabric of encapsulant with incorporated metal ribbons that form a three-dimensional network. Module fabrication consists of a layup phase and a lamination step. In the layup step, a stack is established with the frontsheet, a front encapsulant, the cells, the interconnection fabric a glass or polymer backsheet. During the lamination step, soldering and encapsulation happen simultaneously. In previous research, the concept was tested in two four-cell modules with interdigitated back-contact (IBC) cells, both passing three times the IEC 61215 standard for thermal cycling reliability testing. In this work, these modules were tested using damp-heat and humidity-freeze tests. No significant sign of degradation was observed based on electrical characterization and electroluminescence imaging after consecutive 1000 hours of damp-heat and 50 cycles of humidity-freeze exposure.

Keywords: back contact, PV module, interconnection, multi-ribbon

1 INTRODUCTION

Back-contact solar cell technology is an interesting concept because of its potential to enable highly efficient modules with increased aesthetical appearance compared to conventional, two-side contacted cells. These assets are especially important for applications such as vehicle and building integrated PV. Also, the 2020 ITRPV roadmap predicts an increase in back-contact solar cell technology employment to 5% by 2030 [1]. One of the main hurdles in increasing manufacturing volumes of modules using these cells is the lack of a standardized and low-cost interconnection method. Multiple institutes have proposed various types of interconnection approaches, all in different stages of their development [2]–[6]. A more detailed review of these approaches can be found in the literature and is beyond the scope of this work [7], [8].

This work presents the recent developments of imec's three-dimensional (3D) multi-ribbon interconnection. The technology is based on an interconnection fabric, consisting of an encapsulant sheet with incorporated metal ribbons. The metal ribbons are incorporated using a hybrid stitch-weaving method, which involves making parallel cuts and passing the metal ribbons through this. In this way, a three-dimensional network of metal ribbons is established. An image of the interconnection foil for a four-cell module is shown in fig. 1. The horizontal en vertical ribbons are separated by the encapsulant. At specified locations, both ribbons cross at the same side of the encapsulant, forming a connection point. The module fabrication happens in two steps: a layup step and a lamination step. During the layup step, a stack is established consisting of a front glass sheet, a front encapsulant sheet, back-contact solar cells, the interconnection fabric, and a glass or polymer back sheet. Until now, only glass is used as back sheet material to benefit from the back-contact cells being bifacial. An exploded view of the stack is shown in fig. 1. During the lamination step, the stack is heated in a vacuum chamber while a membrane applies pressure. The interconnection ribbons of the fabric are coated with a low-melting temperature solder. During the heating process, the solder alloy melts to form solder joints between the crossing ribbons in the fabric on the one hand and between the ribbons and the cell metallization on the other hand. Simultaneously, the encapsulant melts and embeds the cells and interconnection between the front- and backsheet. The advantages of this 3D multi-ribbon interconnection technology are its flexibility in design and potential cost savings on cell metallization, copper consumption, and reduced number of processing and alignment steps. More details on the interconnection concept can be found in [9].

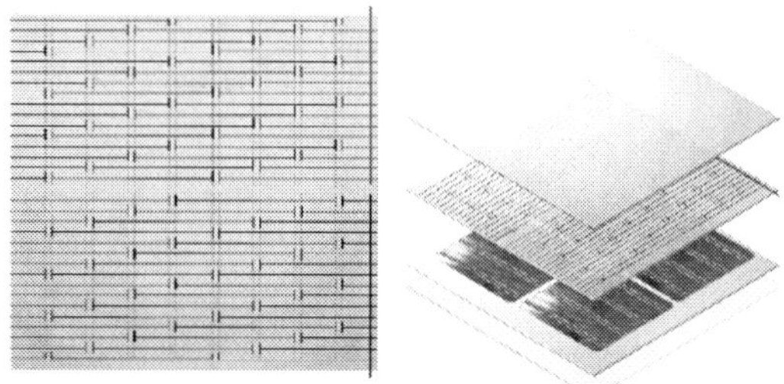

Figure 1: An illustration showing the interconnection fabric consisting of an encapsulant foil with incorporated metal ribbons (left) and an exploded view of the module layup stack (right). From bottom to top the stack consists of a front sheet, encapsulant foil, solar cells, aligned interconnection fabric, and backsheet. The red and black colour indicates a positive and negative electrical polarity, respectively.

The technology was first introduced at the Silicon PV conference in 2019 where it was implemented in 1-cell modules using lower efficiency metallization wrap-through (MWT) cells [10]. The capability of the encapsulant to function as an insulator between ribbons of opposite polarity was shown. In a recent publication, the technology was demonstrated in more detail and a tailored encapsulant was implemented [9]. The tailored encapsulant was a glass fibre reinforced thermoplastic polyolefin (TPO), where characterization on thermal expansion and rheology prove that the addition of the glass fibre reinforcement was beneficial. Two modules with four interdigitated back-contact (IBC) cells were fabricated for this, and reliability testing using thermal cycling from -40°C to 85°C according to the IEC61215 standard proves their reliability, passing three times the prescribed test of 200 cycles.

In this work, the aim is to use different reliability testing approaches in the form of damp-heat and humidity-freeze testing. Since there was only limited degradation on the four-cell modules from the previous study, it was opted to reuse them here. The metallization design of the cells is shown in fig. 2. Although the technology is in development for direct finger soldering to eliminate the need for busbar printing and potentially save material cell metallization costs, the modules for these studies use cells where these busbars are still present.

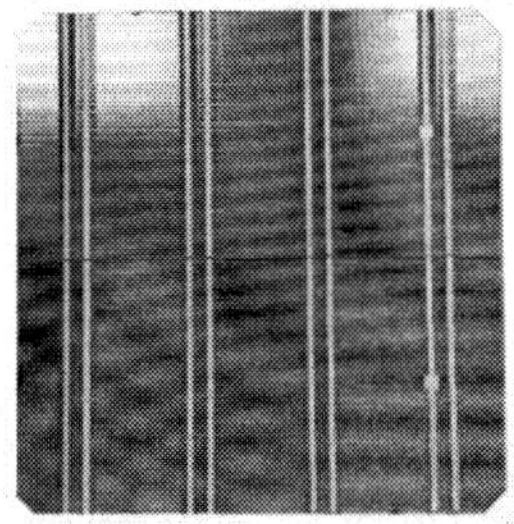

Figure 2: An image of the backside of the interdigitated back-contact cells. The metallization consists of horizontal, continuous fingers and 4 sets of 2 vertical busbars.

2 EXPERIMENTAL SETUP

2.1 Module preparation

The modules were fabricated using a hand-made interconnection fabric with a 10% glass fibre reinforced TPO foil as encapsulant material. The metal ribbons have a copper core of 800x70 μm^2 and are coated in a 12 µm thick $Sn_{57}Bi_{42}Ag_1$ solder alloy (T_{melt} = 139°C). A stack (as shown in fig. 1) is established consisting of a glass front sheet, encapsulant, four IBC cells, an interconnection fabric aligned to the cell metallization, and a glass backsheet. The front- and backsheet are both low Fe tempered 35x35 cm^2 glass plates with a thickness of 3 mm and without anti-reflection coating. During a lamination step, the stack is preheated to 165 °C in a vacuum for 10 minutes before 700 mbar of membrane pressure is applied for 17 minutes. Fig. 3 shows the front- and backside of a module next to an electroluminescence image, showing that the interconnection was functional. The two modules had a fill factor (FF) of 74.9% and 71.3%. The modules then underwent 600 thermal cycles from -40°C to +85°C before they were used for this work, after which the FF dropped slightly to 74.5% (-0.4%) and 70.3% (-1.0%). These are considered as the initial FF's for this work. The initial power at the maximum power point (P_{mpp}) was 20.18 W and 19.45 W for modules 1 and 2 respectively. The lower FF and P_{mpp} of the second module are attributed to a short circuit in one of the cells, causing a part of the cell not being connected. This is further discussed in section 3.

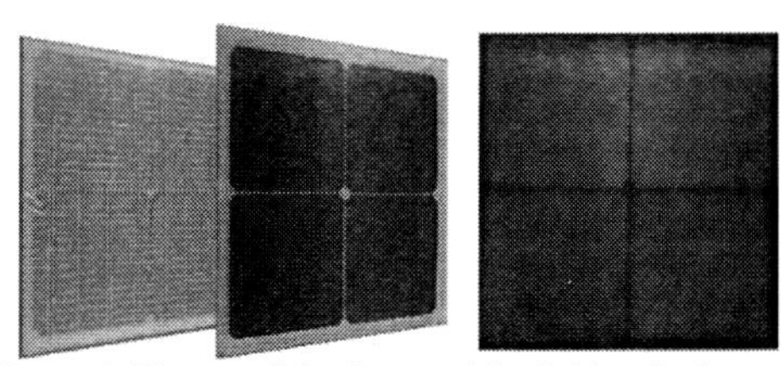

Figure 3: Picture of the front and backside of a four-cell module using the multi-ribbon interconnection (left) and an electroluminescence image (right)

2.2 Reliability testing

Two types of reliability tests were performed in a climate-controlled chamber. The first test is damp-heat, during which the temperature is kept constant at 85°C and the relative humidity at 85%. Electrical characterization is done after 150, 500, and 1000 hours of exposure. After reaching 1000 hours, the humidity-freeze cycling test was started. One humidity-freeze cycle consists of 20 hours of exposure to 85°C and 85% relative humidity, ramping to -40°C in 1.5 hours and holding at -40°C for 1.5 hours. Then ramping to 25°C before measuring the electrical performance or executing another cycle. Electrical characterization is done after each +/- 3 cycles.

3 RESULTS AND DISCUSSION

3.1 Damp-heat testing

The measured FF as a function of the hours in damp-heat testing is shown in fig. 3. Apart from a slight FF increase for module 2 after 150 cycles, the FF for both modules remains constant indicating no degradation of the interconnection or cell metallization. Also, the power output did not change significantly, with a P_{mpp} after 1000 hours of 20.30W (+0.12W) and 19.49W (+0.04W) for modules 1 and 2, respectively.

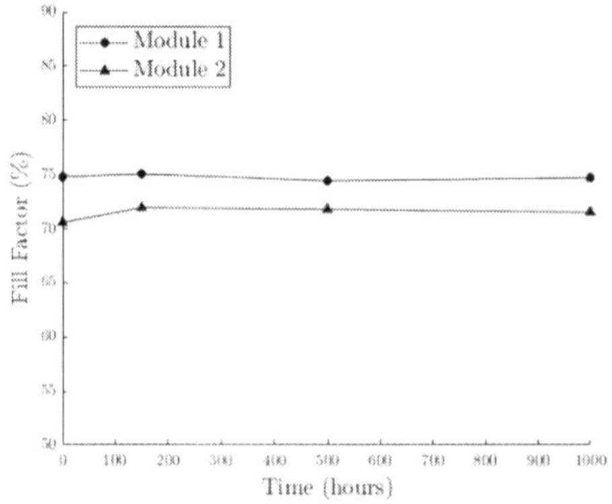

Figure 4: The fill factor of two four-cell modules during damp-heat testing. No significant degradation of the modules is observed after 1000 hours.

3.2 Humidity freeze testing

The graph in fig. 5 shows the FF for both modules during the humidity-freeze testing. During humidity-freeze cycling, the FF of module 1 increased from 74.7% to 77.1% (this is further discussed in the next section). After this sudden increase, the electrical performance for this module remained relatively constant. For module 2, the FF remained constant until the 8th cycle. Then it increased by 1.6 %.

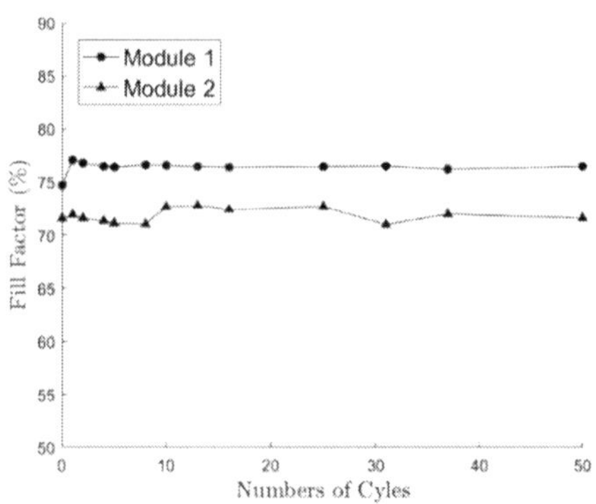

Figure 5: The fill factor of two four-cell modules during humidity freeze testing. No significant degradation of the modules is observed after 50 cycles, passing the IEC standard for humidity-freeze testing.

3.3 Discussion

The slight performance increase for module 2 after the first damp-heat testing and the last two humidity-freeze cycles might be attributed to an interconnection problem disappearing. Analysing the current-voltage curves indicated a short circuit path in one of the cells and/or the associated interconnect ribbons on top of that cell, which was then reduced during damp-heat testing and almost completely removed during humidity freeze testing. Indeed, the electroluminescence images as given in fig. 6 show initially a darker area at the right edge of cell (2,2), which is disappearing after 150 hours of damp-heat testing and is almost completely gone after the last humidity-freeze cycle.

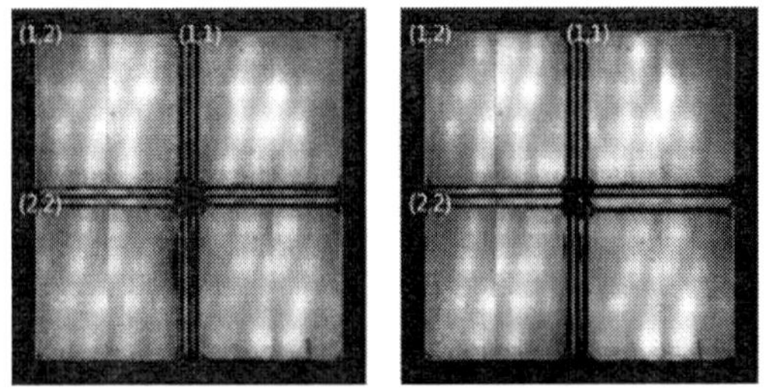

Figure 6: Electroluminescence image of module 2 before (left) and after (right) consecutive 1000 hours of damp-heat and 50 humidity-freeze cycles. No sign of large degradation is observed.

Comparing the current-voltage curves of module 1 after 0 and 1 humidity-freeze cycles suggests there is a small short circuit that disappears. In the electroluminescence images, of which the first and last one is given in fig. 7, there is no clear indication of where this could be. The only observed difference is the disconnected finger on the bottom left of cell (2,2), which is connected after 150 hours of damp-testing. This might either be a problem on cell metallization or module interconnection level.

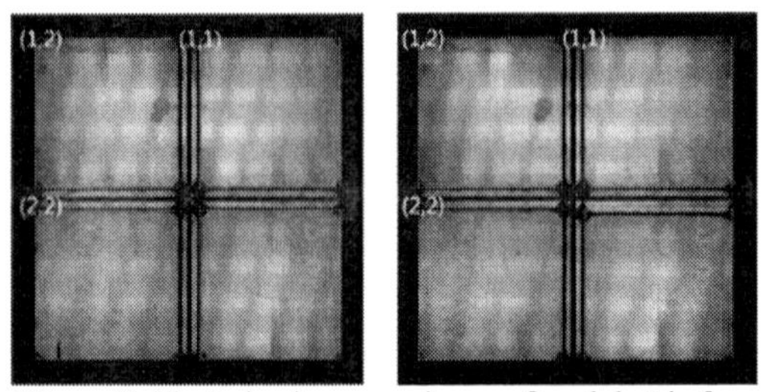

Figure 7: Electroluminescence image of module 1 before (left) and after (right) consecutive 1000 hours of damp-heat and 50 humidity-freeze cycles. No large differences are observed.

As a final overview, table 1 shows the FF of both modules after each of the consecutively executed reliability tests. This shows that the technology has great potential to be a reliable option for back-contacted interconnection modules. However, some variability in the FF due to short-circuiting is observed and this emphasizes the importance of the alignment requirement. Margins and their related tolerances for the alignment should be considered carefully when designing the interconnection fabric and cell metallization.

Table I: The fill factor of the modules after each test.

	FF 1 (%)	FF 2 (%)
Initial	74.9	71.3
600 thermal cycles [9]	74.5	70.3
1000 hours damp-heat	74.7	71.6
50 cycles humidity freeze	76.0	72.3

Also, the effect of removing the busbars on the cell and soldering directly onto the fingers should be studied in more detail. The ribbon-finger connection point is much smaller compared to a ribbon-busbar connection, meaning these latter ones will most likely be more reliable.

4 CONCLUSION

Four-cell modules with IBC cells connected using imec's three-dimensional multi-ribbon interconnection passed 600 thermal cycles from -40°C to +85°C according to the IEC61215 standard in a previous study. They were used again in this work and underwent 1000 hours of damp-heat and 50 cycles of humidity-freeze reliability testing. Both modules passed the tests, and no clear sign of degradation was observed from the electrical performance and electroluminescence imaging. A small short-circuit was observed at the start of the experiment, which was partially removed due to the testing conditions. The short-circuits were likely caused by uncontrolled solder-flow or misalignment of the interconnection fabric. The results prove the potential of the technology as an interconnection for highly efficient back-contact cells, however, some challenges like alignment requirements and controlling the solder flow need to be addressed in future development.

5 ACKNOWLEDGEMENTS

The authors gratefully acknowledge imec's SiPV industrial affiliation programme and its partners, Fonds Wetenschappelijk Onderzoek through the SB Ph.D. Fellowship under grant number 1S14120N, the support of the European Union through the funded H2020 project HighLite under Grant Agreement no. 857793, the partial funding by the Kuwait Foundation for the advancement of Sciences under project number P115-15EE-01 and ISC Konstanz for providing IBC cells.

6 REFERENCES

[1] VDMA, "International Technology Roadmap for Photovoltaic (ITRPV), Eleventh Edition," 2020.

[2] D. Rose, S. Daroczi, P. Helm, W. Palz, and H. A. Ossenbrink, "Development and Manufacture of Reliable PV Modules With 17% Efficiency," in *Proceedings of the 20th European Photovoltaic Solar Energy Conference and Exhibition*, 2005.

[3] A. Halm *et al.*, "The Zebra Cell Concept - Large Area N-Type Interdigitated Back Contact Solar Cells And One-Cell Modules Fabricated Using Standard Industrial Processing Equipment," in *Proceedings of the 27th European Photovoltaic Solar Energy Conference and Exhibition*, 2012, pp. 4–7.

[4] A. Drews *et al.*, "HIP-MWT Solar Cells-Pilot-Line Cell Processing and Module Integration," in *Proceedings of the 27th European Photovoltaic Solar Energy Conference and Exhibition*, 2012.

[5] U. Eitner, D. Eberlein, and M. Tranitz, "Interconnector-based module technology for thin MWT cells," *Proc. 27th Eur. Photovolt. Sol. Energy Conf. Exhib.*, pp. 3461–3464, 2012.

[6] I. J. Bennett, M. J. A. A. Goris, and W. Eerenstein, "Reducing the Cost of MWT Module Technology Based on Conductive Back-Sheet Foils," in *Proceedings of the 28th European Photovoltaic Solar Energy Conference and Exhibition*, 2013, pp. 3111–3113.

[7] I. Bennett, E. Bende, M. Goris, and W. Eerenstein, *An overview of developments in foil-based back-contact modules*. 2014.

[8] T. Borgers, E. Voroshazi, J. Govaerts, J. Szlufcik, and J. Poortmans, "Multi-wire interconnection technologies weaving the way for back contact and bifacial PV modules," in *Proceedings of the 44th Photovoltaic Specialist Conference*, 2017, pp. 1–4.

[9] R. Van Dyck, T. Borgers, J. Govaerts, J. Poortmans, and A. W. van Vuure, "Three-dimensional multi-ribbon interconnection for back-contact solar cells," *Prog. Photovoltaics Res. Appl.*, vol. n/a, no. n/a.

[10] R. Van Dyck, T. Borgers, J. Govaerts, E. Voroshazi, A. W. Van Vuure, and J. Poortmans, "Woven multi-ribbon interconnection for back-contact cells: Extending the functionality of the encapsulant," in *AIP Conference Proceedings*, 2019, vol. 2147.

IMPROVED EDDY-CURRENT PROBE FOR NON-DESTRUCTIVE CHARACTERIZATION OF ELECTRICAL CONTACTS IN PV MODULES

L. Neumaier[1*], M. Lenzhofer[1], Ch. Hirschl[1], J. Kosel[1]

[1] Silicon Austria Labs GmbH, Europastr. 12, 9524 Villach, Austria
Corresponding Author: E-Mail: lukas.neumaier@silicon-austria.com, T: +43 664 8820014
E-Mail: martin.lenzhofer@silicon-austria.com, christina.hirschl@silicon-austria.com, juergen.kosel@silicon-austria.com

ABSTRACT: Within this work we are presenting an improved eddy-current based analysis method for the inspection of electrical contacts and solder joints in PV modules. The basic concept as well as first results were already shown at last year's EU PVSEC 2020. The concept proofed that a fast, non-destructive (contactless) and portable characterization without powering a PV module is feasible. Hence, it has the potential to be used as handheld, contact free inspection system for O&M applications on site, as well as an industrial tool for quality control and process monitoring directly inside of the PV production line. Within the Austrian flagship project "Sustainable Photovoltaics" further research is carried out aiming to improve this eddy-current analysis method. In a first step towards an improved system, an enhancement version of the sensor probe was developed by changing the single coil configuration to a multi coil setup. Due to the specific design, the system is eliminating negatively influencing effects and because of the differential measurement approach high gain factors are possible to increase the systems sensitivity.

Keywords: characterization, solder joint issues, eddy currents, measurement tool

1 INTRODUCTION

At the last year's EU PVSEC 2020 we presented an eddy-current based analysis method for the inspection of electrical contacts and solder joints in PV modules [1]. It was shown that the method can be applied to analyze the quality of solder joints, i.e., to distinguish between sufficient and insufficient soldering joint quality on customized sample laminates. The concept already proofed that a fast, non-destructive (contactless) and portable characterization without powering a PV module is feasible. Hence, it has the potential to be used as handheld, contact free inspection system for O&M applications on site, as well as an industrial tool for quality control and process monitoring directly inside of the PV production line. Within the Austrian flagship project "Sustainable Photovoltaics" further research is carried out aiming to improve this eddy-current analysis method. Here, the major objectives are i) increasing the sensitivity, ii) integrating a reference/calibration directly in the probe and iii) verifying the method for solder joint characterization under load conditions on-site. In a first step towards an improved system, an enhancement version of the sensor probe was developed by changing the single coil configuration to a multi coil setup.

Analysis methods based on eddy currents are already established for non-destructive testing in aerospace and metal industries to determine failures and flaws in conductive materials. Since electrical issues of PV modules are one of the most prominent PV module failures [2], which typically lead to power degradation, eddy current analysis is perfectly suited for a non-destructive, fast and accurate acquisition of electrical information about material and structural properties. The method is based on the physical phenomenon of electromagnetic induction. If a coil that is supplied by an alternating current, thus generating an oscillating magnetic primary field, is brought close to a conductive material, a circular flow of electrons (eddy current) will appear in the conductor. The eddy current will in turn generate its own secondary magnetic field, which interacts with the coil described through mutual inductance. Any changes in the thickness of the conductor or conductivity (defects, flaws) alter the amplitude and pattern of the eddy current and its resulting magnetic field (which can be measured e.g., via a voltage measurement). Hence, electrical defects can be analyzed with an appropriate sensor probe.

In addition to the concept presented at last year's PVSEC conference (2020), which is based on a single coil probe, significant improvements, due to the utilization of a multicoil system to gain higher sensitivity, are introduced within this work.

2 APPROACH

For evaluating the proposed method, a specially prepared single cell mini module, comprising two defective solder joints, was utilized. After artificially aging in a climatic chamber (100 thermal cycles; -40°C to +80°C, according to IEC 61215) the degradation effects were analyzed by electroluminescence (EL) imaging and quantified by greyscale analyses (see Figure 1).

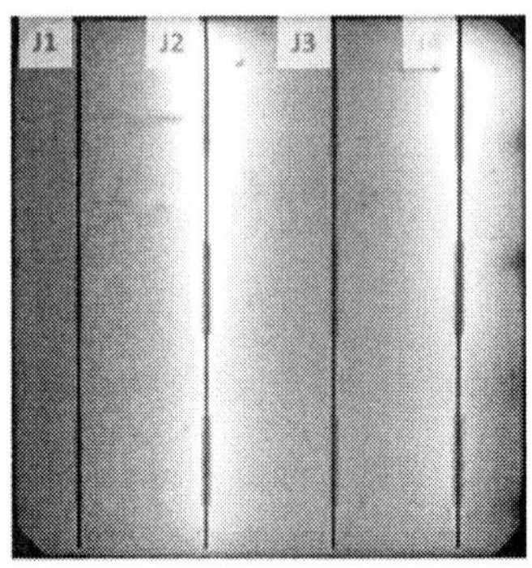

Figure 1: Electroluminescence image of a single cell module comprising 2 defective solder joints (J1 & J3); grey value analysis was used to determine degradation

For the characterization of the solder joints, eddy current measurements are performed [3]. In general, they are always compared to an appropriate reference standard like air, or a sample with well-known properties, also considering and compensating temperature influences. The proposed differential sensor probe design eliminates extensive post processing and compensation procedures because they are implicitly performed by the system design. The coil dimensions are chosen according to the dimensions of commonly used interconnections within PV modules. For field generation a long inner coil is used (excitation coil), while the outer two are the measurement and reference coils, respectively. These coils are connected in an anti-series way, leading to a voltage compensation in air. In addition, they are separated by a defined space (distance of one coil length), for ensuring that their fields do not influence each other. One coil is used as reference to air inducing U_1, while the other one is in contact to the sample inducing U_2. Due to the occurring eddy currents in the solder joint U_2 is changed, leading to a measurable voltage $U_{diff} = U_2 - U_1$ (see Figure 2). This setup removes any common mode effects by differential measurement and higher gain factors can be applied to increase the systems sensitivity.

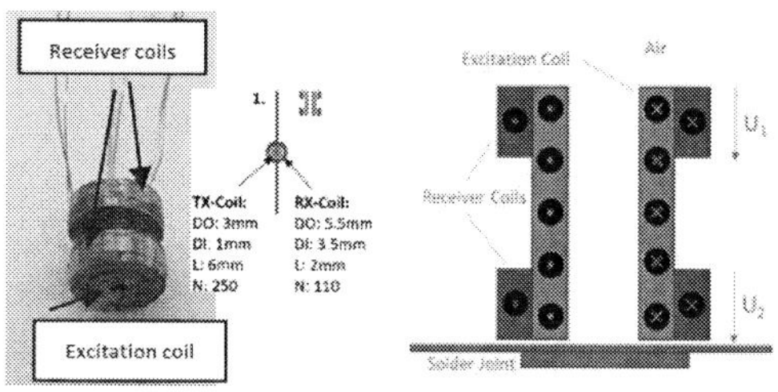

Figure 2: Multi-coil configuration and setup of the improved differential eddy current sensor probe

To ensure proper handling, the coils were attached to a customized probe holder (see Figure 3).

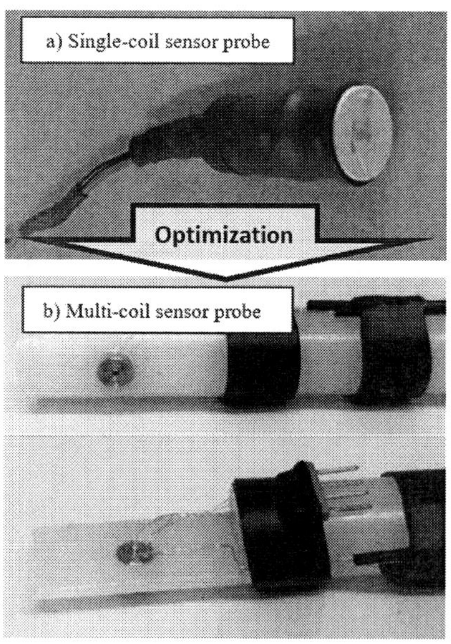

Figure 3: Measurement setups of the (a) single coil (previous version) and (b) the multi-coil sensor probe

For conducting the measurements, the sensor probe is connected to a SR830 lock-in amplifier unit, which is configurated to supply the excitation coil and the receiver coil is fed to the input channel of the measurement unit. The phase is accordingly adjusted to the internally set and locked frequency, to ensure maximal sensitivity.

3 RESULTS

Figure 4 & Figure 5 show the output voltage (induced voltage) of the multi-coil probe as function of the frequency, whereby consistency with the EL image analysis could be proofed. The difference between "good" and "bad" solder joint quality is reflected in a voltage change, whereas sensitivity is higher at higher frequencies.

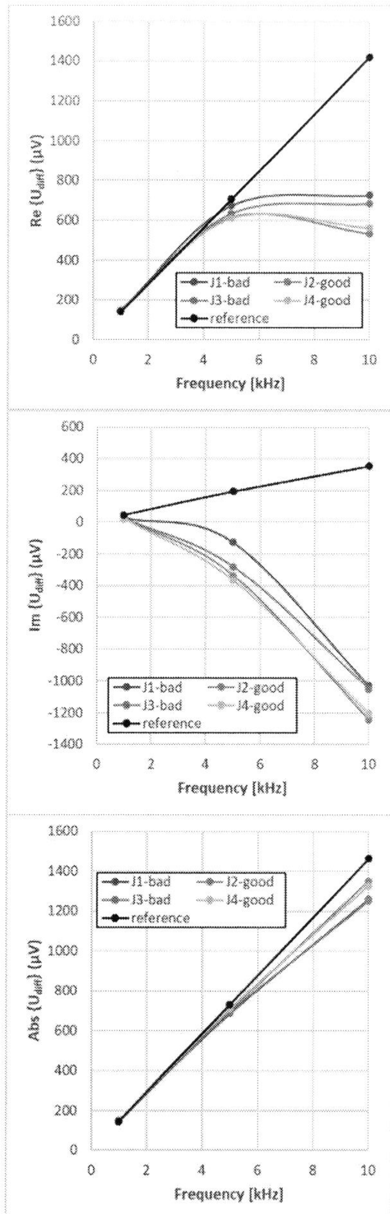

Figure 4: Real part, imaginary part and absolute value of the induced voltage at 4 solder joints (J1-J4) and the reference (air). The different color plots represent different solder joint qualities ("good" was soldered according to target quality, "bad" was not soldered (electrical contact just by lamination pressure))

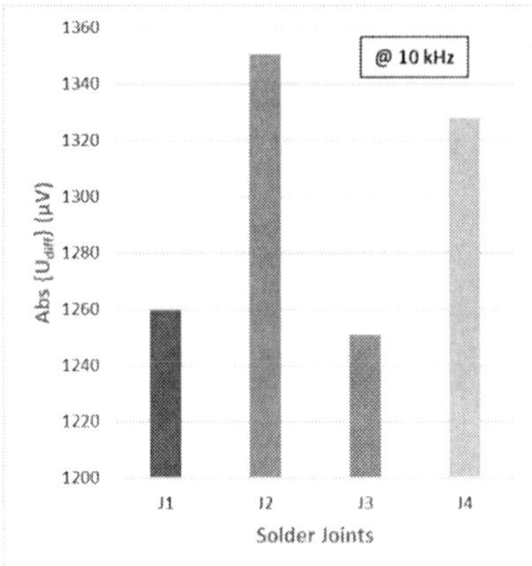

Figure 5: Absolute value results @ 10 kHz

4 CONCLUSION & OUTLOOK

The new improved multi-coil probe is a significant step towards a portable system that can be used for sensitive measurements to characterize solder joints in PV modules within a realistic scenario. Due to the specific design, the system is eliminating negatively influencing effects and because of the differential measurement approach high gain factors are possible to increase the systems sensitivity. Nevertheless, the system is still too sensitive to positioning and handling. Minor changes in the sample alignment are resulting in deviating results. Further design optimizations (bigger coil dimensions, excitation coil change to outer position, etc.) are planned to overcome these issues. It is expected to get more reliable and reproducible results and minimized negative effects due to misalignment and handling.

5 ACKKNOWLEDGEMENT

This work was conducted as part of the Austrian "Energy Research Program" project PVRe2 - Sustainable Photovoltaics, funded by the Austrian Climate and Energy Fund and the Austrian Research Promotion Agency (FFG), both are gratefully acknowledged.

6 REFERENCES

[1] L. Neumaier, W. Mühleisen, M. Lenzhofer, P. Malago, Ch. Hirschl. Eddy-Current Analysis Method for Non-Destructive Characterization of Electrical Contacts and Solder Joints in PV Modules. 37th European Photovoltaic Solar Energy Conference and Exhibition 2020. 10.4229/EUPVSEC2020-4CO.4.1

[2] M. Halwachs, et.al., Error analysis of aged modules with cracked polyamide backsheets, Renewable Energy Volume 139, August 2019, Pages 1040-1060. https://doi.org/10.1016/j.renene.2019.02.135

[3] D. L. Atherton, Remote field eddy current inspection, IEEE Transactions on Magnetics, Volume 31, Number 6, Nov. 1995, Pages 4142-4147. https://doi.org/10.1109/20.489888

ELECTRICALLY CONDUCTIVE ADHESIVE INTERCONNECTS: HOW LOW CAN YOU GO?

Daniel Tune,* Karl Wienands, Ingo Ullmann, Maria Ignacia Devoto, Tudor Timofte, Andreas Halm
International Solar Energy Research Center (ISC) Konstanz, Rudolf-Diesel-Str. 15, 78467 Konstanz, Germany
*Corresponding author email: daniel.tune@isc-konstanz.de

ABSTRACT: One of the ways to reduce the material cost of shingled modules is to reduce the mass of the relatively expensive electrically conductive adhesive (ECA) that is required per shingled joint. This work explores some factors limiting such reductions by producing shingled cells and minimodules that are then characterised in terms of their electrical and mechanical performance. It is found that the mechanical properties depend heavily on the ECA mass. However, the electrical properties show no such dependence, even for joints in which the ECA mass and thus mechanical strength is so low that the joints cannot be handled without breakage.

Keywords: Electrically Conductive Adhesive; Shingling; Electrical Properties; Mechanical Properties

1 INTRODUCTION

Several commercially available alternative module concepts use electrically conductive adhesive (ECA) interconnects rather than traditional soldering approaches. Examples include shingled modules and modules based on conductive backsheets, where traditional soldering is not applicable, and modules based on heterojunction cells, where the lower thermal budget of the passivation layers precludes high temperature soldering. As an alternative to soldering, ECA has several advantages including the lack of lead, tailorable elasticity, higher resistance against stresses during thermal cycling or in high vibration environments such as vehicle integration or space solar, and the possibility of precise and fine pitch application via screen/stencil printing or pressure-time/jet dispensing. The main disadvantages of ECA compared to soldering are the higher per-module cost of the material, and initial retooling expenses due to the significantly different module assembly processes required. The minimum amount of ECA required is determined relative to the initial mechanical strength and electrical properties (bulk and contact resistivity) of the interconnects, as well as those same properties after extended aging of the modules (through damp heat and thermal cycling tests). The minimum amount of ECA per module can be reduced by varying the filler type and %content in parallel to adjusting the polymer adhesive matrix, as well as by using different print layouts (lines, dots, others) of both the ECA and the metallisation structures (busbars, fingers) to which the ECA must bond.

Here we add to the already significant body of work done on ECAs over the last decade or so (e.g. refs [1–9]) by determining and then exploring strategies to reduce further the minimum amount of ECA required using common commercially available products for the specific case of jet-dispensed ECA used in shingled joints.

The approach used is to dispense ECA in either solid lines or lines of dots onto busbars and then use these to produce shingled joints. The shingled joints are assessed in terms of their mechanical performance (in-plane shear strength) as well as their electrical performance (from IV characterisation).

2 SAMPLES AND DEPOSITION

Samples were manufactured using either pressure-time (needle) dispensing or jet-dispensing. The substrates were 6BB G1 mono PERC cells and three different ECAs (one epoxide and two acrylates) were selected as providing a broad snapshot of the currently available commercial products. For pressure-time dispensing, the smallest reliable and reproducible deposition amount possible was used, depending on cartridge pressure, needle size, and line speed. For jet dispensing, the deposition amount was primarily controlled by varying the line speed, and thus dot separation, as well as the open and close times of the jet cycle. As shown in **Figure 1**, the deposition profile varied from full lines to a series of dots with a separation of approximately 4 mm. For each ECA, a total of 15 samples were produced for mechanical measurements and 9 samples for electrical characterization. Of note is that the line process speed was much faster for jet dispensing (100–400 mm/s) than for pressure-time dispensing (10–30 mm/s).

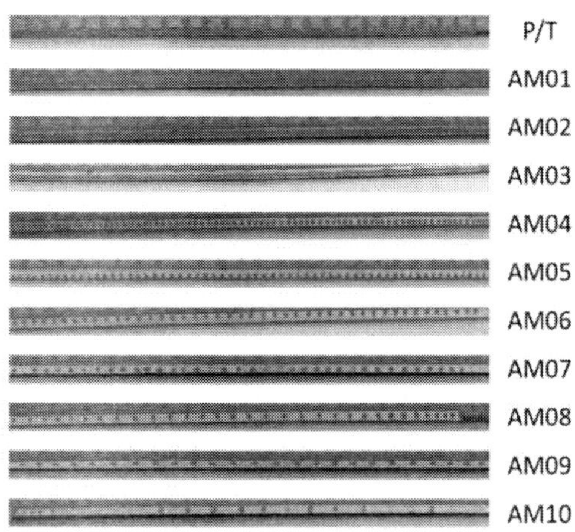

Figure 1. Photographs of the 11 different deposition profiles used in the study (taken after mechanical testing).

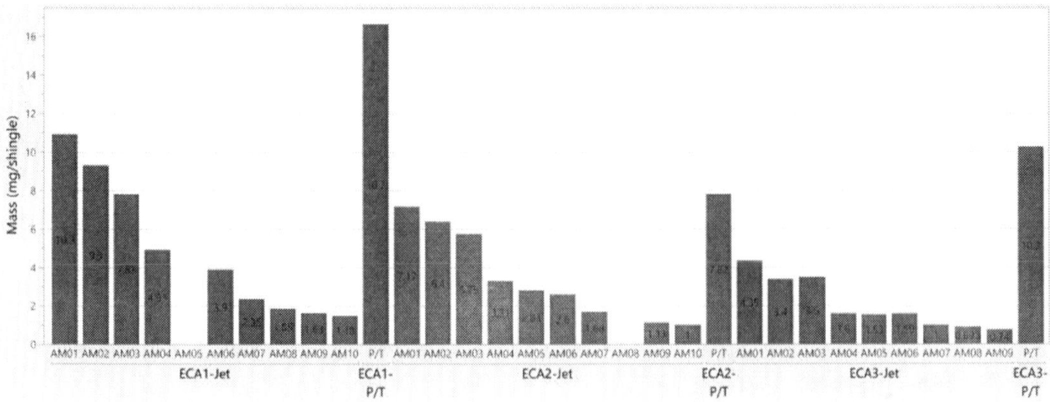

Figure 2. Laydown masses of each of the three ECAs investigated in the study. The masses of lines deposited using pressure-time (needle) dispensing are indicated as 'P/T', as well as the masses of 10 different amounts of ECA deposited by jet-dispensing with variations in line speed and open close time of the jetting cycle.

3 MECHANICAL PERFORMANCE

The tensile strength of ECA joints is a critical parameter for the manufacturability, performance, and long-term stability of PV products that use ECA. The tensile strengths of ECA joints manufactured in this study were measured using a custom-built measurement fixture from Zwick Roell GmbH as shown in **Figure 3**.

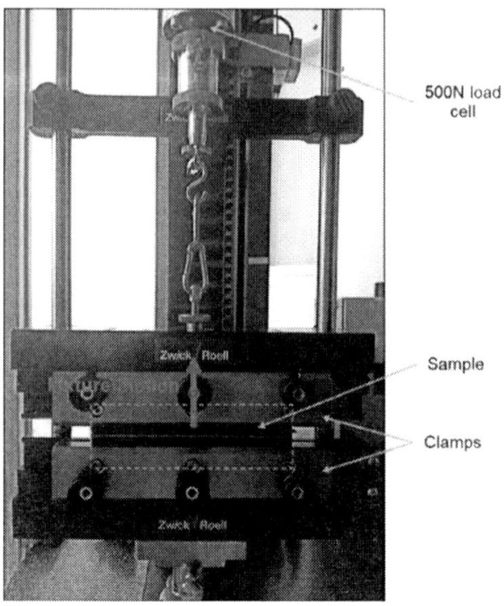

Figure 3. Photograph showing the setup of the custom shear force measurement tool used in the study.

In the fixture, the two sides of a two-piece shingle cell were held securely in place by metal clamps coated with a soft but non-yielding polymer and then the upper part of the fixture was drawn upwards to apply lateral shear to the adhesive joint. The tensile force to break was measured with a 500 N load cell and the travel distance and type of break (silicon cracking or joint separation) were recorded. The data are summarized in **Figure 4** and reveal the following key points:

- For all **pressure-time samples**, the tensile strength of the joints was greater than the material strength of silicon → **break in silicon**, not in the ECA joint
- For **jet-dispensed samples**, the tensile strength of the joints depended on both the **brand/product** and **deposition amount** of ECA
- For all breaks in joints, the type of break was **cohesive failure** of the ECA or **adhesive failure** of the ECA-metal interface, not cohesive or adhesive of the cell metallisation
- For ECA2 and small deposition amounts of ECA3, the mechanical strength of the joints was **too low to handle** the samples without joint failure

4 ELECTRICAL PERFORMANCE

Two-piece shingle cells (6BB mono PERC) were cut into thirds and assembled into glass-backsheet minimodules with soldered ribbons and EVA encapsulant as shown in **Figure 5**. The IV characteristics of these cells were obtained with a standard flasher and the electrical performance was assessed as shown in **Error! Reference source not found.**. A figure of merit consisting of the difference between the pseudo fill factor extracted from the Suns-Voc curve and the actual fill factor calculated from the I–V curve was used in addition to the usual short circuit current density, open circuit voltage, fill factor, and efficiency. For all cells, the fill factor was reduced by edge recombination due to the small cell area relative to the four cut edges. Some samples with low ECA mass were not viable due to fragility during handling or to joint failure as a result of soldering-induced bowing. The following observations can made regarding the IV data:

- In this study, the ECA1 (epoxide) data exhibited a larger spread than ECA2 or ECA3 (both acrylates)
- **No significant difference** was observed between the pressure-time and jet-dispensed samples
- **No significant dependence** of the IV characteristics on the mass of ECA in the joints was observed, even for the lowest ECA mass

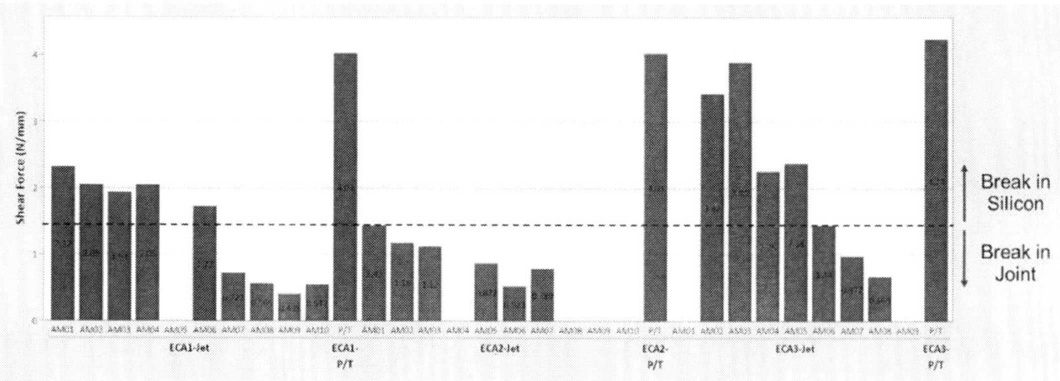

Figure 4. Tensile strengths of the ECA joints tested. The line at ~1.5 N/mm is set based on observations of the failure type.

Figure 5. 20 x 20 cm² minimodule with 6 samples, each consisting of two pieces of 6BB G1 mono PERC shingles bonded with ECA and contacted by soldered ribbons for electrical measurement

5 SUMMARY AND OUTLOOK

The mechanical and electrical performance of shingle cells made using pressure-time and jet-dispensed ECA was assessed as a function of the mass of ECA used in the joints. ECA masses as low as 4.7 µg/mm (0.74 mg per G1 shingle) were investigated and potentially viable (mechanically strong enough for handling) shingle cells with ECA mass as low as 6.3 µg/mm (1 mg per G1 shingle, equivalent to 384 mg per 64-cell-equivalent 6BB shingle module) were realised. Line process speeds as high as 400 mm/s were investigated, with 200 mm/s being the fastest speed to produce mechanically viable shingle joints. At this speed, a dispensing system consisting of 6 parallel jet-dispensing heads could produce 6BB G1 shingle strings with a cycle time of 1 s, yielding 3600 wafers per hour.

The mechanical strength of the shingle joints was observed to depend strongly on both the ECA mass and the ECA brand/product. In contrast, the electrical performance of the joints was independent of the ECA mass within the range of masses and brands/products in this study. Thus, mechanical strength is the key determining factor limiting reductions in the ECA mass per shingle joint. This means that a key focus of efforts to reduce the cost of shingled modules should be on methods to improve the mechanical strength of the joints since there is clearly more room at the bottom in regards to electrical performance. That said, degradation tests (damp heat and temperature cycling) of all samples are ongoing to determine if there is a dependence on ECA mass, and this will likely add a further parameter that must be considered when reducing the mass. Additional work involves similar investigations of cell-to-ribbon joints using ECA and variations to the cell metallisation geometry beyond simple lines are also ongoing.

6 ACKNOWLEDGEMENTS

This work was supported by the German Federal Ministry for Economic Affairs and Energy (BMWi) as part of the HoSSa project with reference number 03EE1014D.

7 REFERENCES

[1] U. Eitner, T. Geipel, S.-N. Holtschke, M. Tranitz. Energy Procedia 27 (2012) 676. https://doi.org/10.1016/j.egypro.2012.07.128

[2] A. Schneider, R. Harney, S. Aulehla, F. Demiralp, S. Koch, J. Scheurer, C. Seeger. Energy Procedia 55 (2014) 509. https://doi.org/10.1016/j.egypro.2014.08.017

[3] G. Beaucarne. Energy Procedia 98 (2016) 115. https://doi.org/10.1016/j.egypro.2016.10.087

[4] T. Geipel, L. C. Rendler, M. Stompe, U. Eitner, L. Rissing. Energy Procedia 77 (2015) 346. https://doi.org/10.1016/j.egypro.2015.07.049

[5] T. Geipel, M. Meinert, A. Kraft, U. Eitner. IEEE J. Photovoltaics 8(4) (2018) 1074. https://doi.org/10.1109/JPHOTOV.2018.2828829

[6] M. Estruga, L. Theunissen, A. Ardizzone, B. Willems, A. Henckens, H. B. Nv. 35th European PV Solar Energy Conference and Exhibition (2018). https://doi.org/10.4229/35thEUPVSEC20182018-1CO.1.2

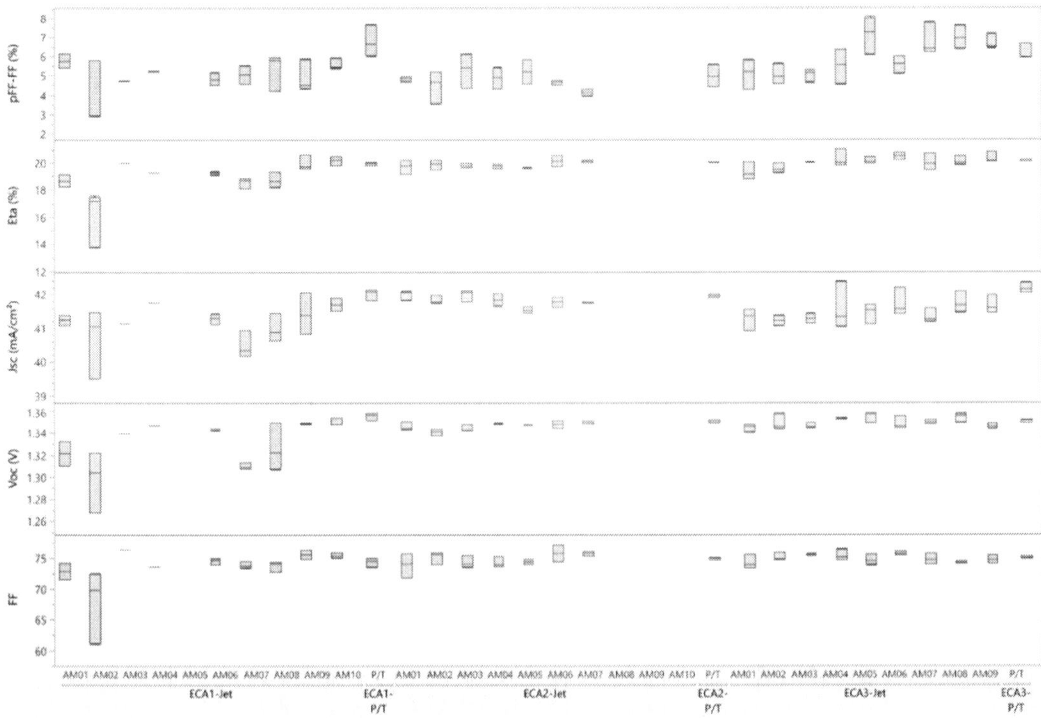

Figure 6. IV characteristics of the two-piece shingled cells used in the study.

[7] T. Geipel, V. Nikitina, L. P. Bauermann, E. Fokuhl, E. Schnabel, D. Erath, A. Krieg, A. Kraft, T. Fischer, R. Lorenz, D. Breitenbücher. 36th European PV Solar Energy Conference and Exhibition (2019). https://doi.org/10.4229/EUPVSEC20192019-4CO.3.3

[8] C. Kaiser. 37th European PV Solar Energy Conference and Exhibition (2020). https://doi.org/10.4229/EUPVSEC20202020-4AV.2.6

[9] M. I. Devoto, T. Timofte, A. Halm, D. Tune. 37th European PV Solar Energy Conference and Exhibition (2020). https://doi.org/10.4229/EUPVSEC20202020-4AV.1.26

PRODUCTION PROCESS FMEA: EFFECTIVE PROCEDURE TO DETECT MAJOR PROCESS RELATED RELIABILITY RISKS AND BETTER SPECIFY RELIABLE PRODUCTIONS WINDOWS

Bengt Jaeckel[1,2], Matthias Pander[1,2]*, Daniel Philipp[3], Karl-Anders Weiß[3]
[1]Fraunhofer Center for Silicon Photovoltaics, Otto-Eissfeldt-Str. 12, 06120 Halle (Saale), Germany
[2] Fraunhofer Institute for Microstructure of Materials and Systems IMWS, Halle, Germany
[3] Fraunhofer Institute for Solare Energy Systems ISE, Freiburg, Germany
phone: +49-345-5589-5135, fax: +49-345-5589-5999, bengt.jaeckel@csp.fraunhofer.de

ABSTRACT: PV module failures can be a result of a variety of causes. They can lead to catastrophic failures, like glass breakage, or are observable as a loss in output power over time. Glass breakage of crystalline Silicon PV modules for instance is typically either a design flaw (too thin, not strong enough, incorrect specifications), a result of misuse (incorrect installations e.g. in a region of high snow loads or hail impact) or dramatic non-predictable weather events (flooding, extreme hail or storm events). Those are generally not related to the PV module production or design (for thin-film technologies this statement is not correct!). In this contribution we evaluate field observations and try to point them to production (process) related influencing factors. Based on a detailed FMEA of failures and possible production factors we prioritized the influence and weight "risk" of detection and failure. The FMEA is based on contributions from two research institutions, two module manufacturers and one manufacturer of production equipment. This interdisciplinary approach allowed extensive experience along module production and failure analyses could be considered. The processes with the highest priority numbers are followed up in terms of process variations and a detailed indoor and outdoor test study.
Keywords: PV Module, Production, FMEA, Defects, Reliability

1 INTRODUCTION

Photovoltaic (PV) industry is always pressured to enhance efficiency of cells and modules [1], optimize production processes to reach higher yields while at the same time further improve the longevity of the product and reduce costs. Most modules come with a performance warranty of 25 years and some manufacturers state even 30 or more years [2].

Most failures known today are more infant mortality issues, typically occurring within the first 5 to 10 years of operation [3]-[11]. Solar cell and module development phases are in the range of 6 to 12 months. However, accelerated testing in climate chambers should be as short as possible and is in the range of some weeks to a few months. Therefore, highly accelerated stress tests would be needed to really speed processes up by a factor of around 100. In durability and reliability testing reasonable aging acceleration can mostly only be achieved with a factor of about 10.

Therefore, it must be well understood which effects the test protocol should address and which possible root cause is behind the degradation or failure mode.

Within this contribution all relevant production processes of a standard PV c-Si modules were evaluated with a Failure Mode and Effects Analysis (FMEA) protocol. The adapted FMEA procedure is set to focus on a more in-depth understanding of the severances of production process variations and material variations within a given type of material. The results of the FMEA should lead to the installation of a standardized procedure for future material and machinery changes within the manufacturing process. The protocol should therefore help to avoid production caused defects and so should enhance overall quality and the risk on field returns can be lowered and better estimated.

Optimization and enhancement of production processes is essential for reliable and cost-competing products. The aim of the production process FMEA approach is to basically do a Design of Experiment (DoE) to optimize allowed production windows to either achieve reliable products or to improve production yield (more products per hour or better utilization of materials). The designed experiments should help to better address the specific related degradation and failure-causing effects in the products. Especially the interdisciplinary contributions by PV module manufactures, research institutes and equipment producers are hereby new and unique to our state of knowledge.

2 FMEA APPROACH

2.1 General aspects

A *Failure Mode and Effects Analysis (FMEA)* study was undertaken to identify, prioritize and rate field defects with the aim to achieve a deeper understanding to overcome in the long term such defects. Following the methodology of the FMEA approach the Risk Priority Number RPN is the multiplication of the rating values for **S**everity S, probability of **O**ccurrence O and probability of **D**etection D (in the factory). Latter is important as we want to focus on variations of module production processes and not on design or specific climatic / weather events. Severity, Occurrence and Detection can have numbers from 1 (no impact, 100% detection probability) to 10 (severe impact, very low probability of detection, safety hazard) resulting in Priority numbers (Risk Priority Numbers RPN) from 1 (best product, no problems → not likely) to 1000 (very high risk of catastrophic failure → can happen for new, fast and not in detail developed products or newly implemented processes). Table 1 gives the details for the evaluation of the FMEA criteria.

2.2 Focus points for PV module production FMEA

For the PV module manufacturing FMEA the following steps were investigated:

1. Source material verification and incoming inspections (e.g. quality of cells, homogeneity of e.g. backsheets and Encapsulation material)
2. Storage of incoming and outgoing goods
3. Soldering of solar cells (stringing)
4. Layup of cells and strings

5. Contacting of cell strings and contact formation to junction box
6. Lamination
7. Framing

Depending on the type of manufacture the emphasis/priority may lay on different topics and depends on process and machinery understanding. In this contribution the focus is on the general aspects for PV module manufacturing that can be used for any factory. Special topics e.g. for Sondermodule [13] [14] production, as used for vehicle (ViPV) or building integration (BiPV), will be discussed elsewhere.

The first and second point refer to verifying specifications (based on product design, improper material selection in the design phase is out of the scope of this FMEA) and ensure that only materials within specifications are used for production. Obviously, this is essential to produce reliable PV modules. However, the measurable material variations compared to specifications are typically small but there is a certain hazard not to detect deviations which might become critical during later operation.

Similarly, the layup of cell strings and the framing process only have a minor impact on product reliability. Therefore, those were not focused on in detail. It must be pointed out that misaligned strings or not as designed spacings between cells and strings have an esthetic drawback and might be serious "defects" for Sondermodule.

Highest emphasize was given to the electrical contact formation (e.g. soldering parameters) and lamination process. Soldering is a complex process of electrical contact formation and the reliability strongly depends on materials (solder, solar cell) and process (temperature, time for heating and cooling). Lamination is typically a heat and pressure-controlled process to achieve cross-linking of the materials (in case of e.g. EVA). Several steps, including liquefaction of the encapsulant to achieve a homogeneous film of encapsulation material around all components, a vacuum process to remove any air and moisture bubbles and cross-linkage byproducts. Dependent on material quality (mixture of the encapsulant) and the lamination process various things can go wrong and do not lead to the defined specifications like transparency and degree of crosslinking. Besides the crosslinking the process steps are similar also for other encapsulation materials like POs and quite identical for POEs. However also alternatives exist like using silicone-based materials where such multi-step lamination processes are not needed. Other obstacles like correct and homogenous mixture of used two component materials must be overcome.

Based on the FMEA results different parameters are specified that will serve as the base for the specification of tests and to optimize module production.

3 RESULTS AND DISCUSSION

As described in the FMEA approach the focus was set on experiments only for two major concerns namely electrical contact formation (soldering) and the lamination process. Both are essential for the long-term reliability of the PV modules and therefore uncontrolled and incorrect assessments of process variations can lead to

Table 1: FMEA evaluation criteria

Probability of **Occurrence O**	Impact / **Severity S**	Probability of **Detection D**	Points
unlikely	hardly noticeable	high The error is detected in any case by common measures, so that a very prompt correction is possible	1
Very low	rather insignificant Aesthetic, very little impact on performance, no impact on safety, does not fall under warranty	moderate Procedures are in place to enable early detection. However, measurement and test results can be misinterpreted, for example, or incorrect assessments can lead to the defect being overlooked / disregarded.	2 to 3
low	moderately serious error measurable and clear influence on performance, increased clarification effort with customers to be expected, no influence on safety. Possibilities for compensation or repair are possible at reasonable expense	low Only detectable with accurate and comprehensive testing which is rarely performed on small samples - time of detection likely to be after module delivery	4 to 6
Moderate	serious error Measurable and clear influence on performance, influence on electrical safety or fire. High liability risk and high follow-up costs to be expected (e.g. module replacement).	very low not detectable with common QA measures. Process monitoring does not provide any clues. Detectable with very elaborate tests, but these are not routinely performed. Time of detection for sure only after module delivery.	7 to 8
High	extremely serious error Can lead to massive injuries or death	unlikely There are practically no effective early detection methods. Defects only become evident in operation after delivery.	9 to 10

significant losses in long term durability and so represent an increased quality issue.

A selected part of the FMEA with the highest PN numbers is summarized in Table 2 and will be introduced and discussed in the next section separately.

3.1 Electrical contact formation

The electrical contacts are essential for the output power of the PV modules. Each cell must be interconnected in series to achieve a reasonable voltage and a certain current the power inverters can work with effectively. Therefore typically 60 to 78 cells are interconnected in a PV module where in a system strings of 15 to 30 modules are connected in series. This results in a few thousand cells connected to an inverter input. This underlines the importance of well electrically interconnected solar cells, even if there are some backups like bypass diodes installed in the PV modules.

Four selected results are presented in Table 2 for soldering that go from an RPN number of 48 (acceptable) to 168 (can have significant impact). The severity of the selected processes is quite similar (S = 7 - 8) and can be of significant concern. The occurrence for each is not very high, but still must be considered (O = 3 - 5). However, the largest difference is within the detection probability (D = 2 or 8). With a D = 2 the process deviation is quite easy to detect, where a D = 8 is a more difficult one and may require in-depth and costly analytics or detection equipment.

The first two items "Soldering" #1 and #2 deal with soldering flux application. Two cases were discussed where the first one discussed the problem of contaminated flux and the second with no flux application. As for both the occurrence was very low and the more obvious case of no flux application can be detected very well (no soldering) no follow up actions were discussed.

The third item is about ribbon pre-treatment within the stringer machine. The quality of the ribbon crimp is quite difficult to assess (D=8). The occurrence in the used machinery of the participating manufacturers was discussed to be very low no further follow up was done here.

The last item (#4) deals with process time and temperature profile during the soldering process. Such variations occur more frequently (e.g., due to productions stops due to cell breakage) and therefore have a more important impact on overall production yield. Therefore, the focus for follow-up tests was set to that observation even that the RPN-number was not very high. To investigate the issue on long term stability of the modules special module were built with the particular defect and long-term test-to-fail experiments, including extended Thermal cycling tests per IEC 61215-2 MQT 11, were started. Those tests are still ongoing. Preliminary results are given in Figure 1 and show that those overheated cell strings have still good long-term stability and behave very similar to normal processed strings.

Table 2: Selected part (highest PN-number) of process FMEA: Severity S: 1 [very low risk]-10 [deadly]; Occurrence O: 1[very low] – 10 [very high, certain]; probability of Detection D: 1 [certain – fault will be caught on test] – 10 [fault is undetectable].

Process filter	Fault / Question	Severity S	Occurrence O	PN (S x O)	Detection D	RPN (S x P x D)
Soldering #1	• Flux is contaminated • Occurs rarely as typical flux nozzles will be clogged. What is the impact on solder joints if impurities are integrated in the joint?	7	3	**21**	8	**168**
Soldering #2	• Flux is not applied • Occurs if flux nozzles are clogged, maintenance intervals not well chosen What happens to the contacts?	8	3	**24**	2	**48**
Soldering #3	• Ribbon crimp: material damage of the ribbon • Rare occurrence, typically only if broken tooling used What is the impact of pre-damaged ribbons on long term stability of the PV module?	7	2	**14**	8	**112**
Soldering #4	• Hot cell string: Over-soldered cells due to machine stop • Occurs several times per day It is unclear what long-term effect hot cell strings have. Currently such strings are sorted out. But is that necessary?	7	5	**35**	2	**70**
Lamination #1	• Low or too high degree of cross linking of EVA encapsulant • Occurs if process control for heating phase faulty. Heating phase 20-30% too long or too short • Observation: Modules look normal. No drift of cells within the laminate How is module durability effected with over cured EVA?	5	5	25	7	175
Lamination #2	• Low degree of cross linking of EVA encapsulant • Occurs if fault in Laminator, bad material or to fast lamination process What happens for gel contents in the range of 65-80% or lower?	5	5	25	7	175

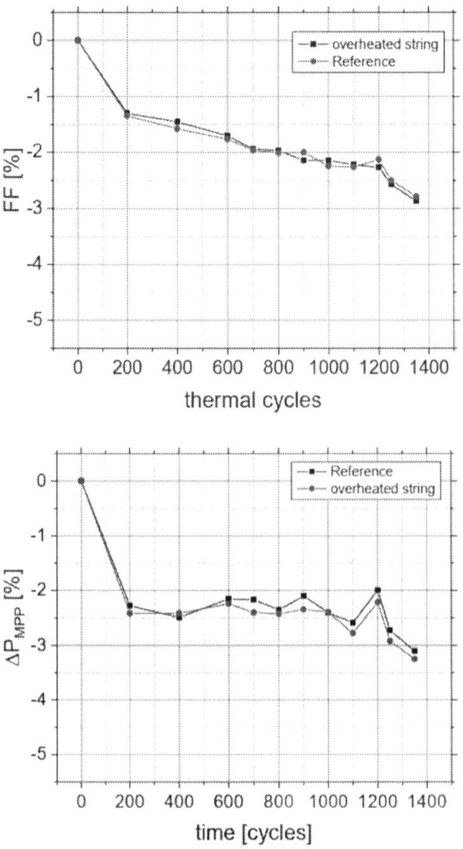

Figure 1: Preliminary results of extended Thermal cycling (TC) testing of Modules containing "overheated cell" string: Top: Trend in fill factor, bottom: trend in loss of power.

3.2 Lamination

Besides soldering the lamination process is the other important step in module manufacturing. As lamination is often the bottleneck process optimization is often done here with the focus on reduction of process time. Lamination is a complex process with several parameters to balance out and further significantly depends on material properties but is also influenced by environmental parameters like temperature of the manufacturing facility. Smaller variations in material properties and environmental parameters are summarized to impact the degree of crosslinking only a bit rising the question on the influence if the degree of cross linkage varies within +/- 10% ("Lamination #1 in Table 2). Typically, such modules show no obvious optical findings directly after manufacturing and detection is quite challenging (D = 7).

Modules with a very low degree of crosslinking are either made on purpose by significantly accelerated lamination processes or are caused by a defect of the laminator or by very poor quality of the material (Lamination #2). The latter one should be detected earlier within the incoming goods inspection. A defect on a heater in a laminator can occur and may not be detected immediately and depends on laminator type and sensors in the laminator. Therefor it can happen that several batches are processed prior the problem is detected (D = 7). Depending on module tracking in the factory such modules can be traced back, but also depends on sensors and tracking of modules throughout the module production process. Based on long term experience low cross-linked modules show a higher probability of delamination. An example of a 6-year-old module is shown in Figure 3 on the left side. Based on the RPN-number and the importance for the production quality and yield this process flaw was further evaluated and samples were built to better judge the direct influence comparing good and low degree of cross linking. Preliminary results of extended damp heat testing are given in Figure 2 for module output power and first delamination effects after ~1000 h of Damp-heat in Figure 3 on the right side. Standard modules with high degree of cross linkage do not show such delamination. However, the power loss difference after 1000 h is not huge (2 vs 3 %), but the trend of the low degree cross linkage sample (red – low GC) is expected to be stronger than for a normally processed module where loss of about 3 - 5 % are expected after 2000 h.

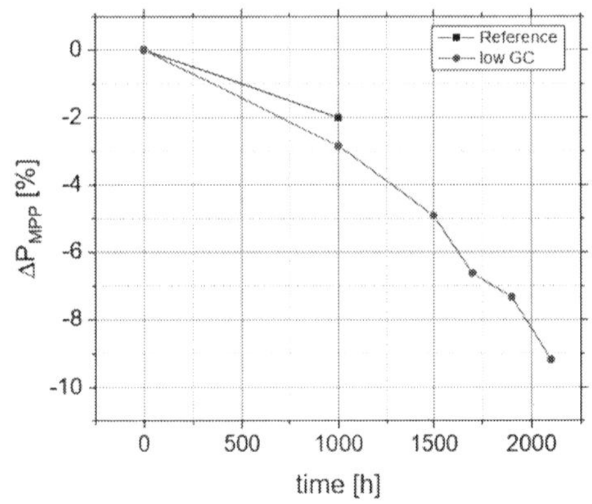

Figure 2: Preliminary results of extended Damp-heat (DH) testing of samples containing normally cured EVA modules (degree of cross linking ~85%) and module with low degree of cross linking of less than 50%.

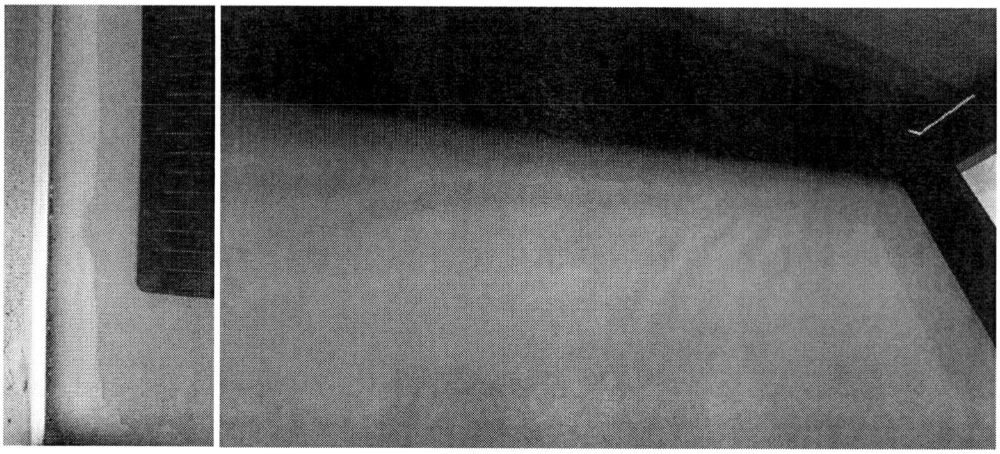

Figure 3: left: delamination and bubble formation of field module from central Europe after approx. 6 years, right: large area delamination and bubble formation of specially prepared low-gel-content PV module after damp-heat exposure (prior damp heat back-sheet was optically OK), and similar to field observations e.g. in [12].

4 SUMMARY AND CONCLUSION

The interdisciplinary FMEA by manufacturers, equipment developers and research institutes lead to a long list of processes and a short list of important parameters to prioritize on. The developed list can now be used to assess new processes or materials. The FMEA should help to better understand risky processes and to avoid the implementation of materials and processes that will lead to non-reliable products.

The given results are not new to insiders [12][16]-[18], but the methodology to achieve the understanding is far clearer and more reproducible if each step is well documented.

The results of the initial FMEA will be used to assess other encapsulation materials combined with different processes and module structures in future work. Material wise the FMEA structure should help to figure out critical processes to e.g. adapt the current lamination process for PO or POE materials [19]. Further it should help to understand the impact of influence while using either a polymer backsheet or a Glass as back-side insolation.

5 ACKNOWLEDGEMENTS

The authors gratefully acknowledge the financial support by the German Federal Ministry for Economic Affairs and Energy for financial support of the project "ProModul - Simulationsbasierte Prozessoptimierung für neue Zell- und Modulkonzepte sowie Prüfprozedurentwicklung für beschleunigte Materialqualifizierung und Gebrauchsdauerabschätzung" under grant no. 0324354.

Furthermore, a special thank you to the industry partners from machinery, module and metrology manufacturing that supported the discussions within the FMEA process.

6 REFERENCES

[1] VDMA, "International Technology Roadmap for Photovoltaic," 12th edition Results 2020, 2021. https://itrpv.vdma.org/en/ (accessed Jun. 07, 2021).

[2] Solar Power Europe, "Global Market Outlook 2020-2024",

[3] J. C. Coello et al., "Analysis of the Degradation of 735 Commercial Crystalline Silicon Modules after First Operation Year", 25th European Photovoltaic Solar Energy Conference - Valencia, 2010

[4] D. Jordan et al, "Photovoltaic Degradation Rates - an Analytical Review", Progress in Photovoltaics: Research and Applications, 2011, 18

[5] L. G. Garreau-Illes et al., "Backsheet and Module Durability and Performance and Comparison of Accelerated Testing to Long Term Fielded Modules", 28th European Photovoltaic Solar Energy Conference - Paris, 2013

[6] D. Jordan et al., "Compendium of photovoltaic degradation rates", Progress in Photovoltaics: Research and Applications, 2016, 24, 978-989

[7] J. L. Garcia et al., "Degradation analysis of PV module technologies in a moderate subtropical climate", 36th European Photovoltaic Solar Energy Conference - Marseille, 2019

[8] D. Jordan et al., "PV field reliability status - Analysis of 100 000 solar systems", Progress in Photovoltaics: Research and Applications, 2019

[9] K. Kiefer et al., "Degradation in PV Power Plants: Theory and Practice", 36th European Photovoltaic Solar Energy Conference - Marseille, 2019

[10] P. Lechner et al., "Bewertung von Auffälligkeiten und Schäden an der Rückseitenfolie von PV-Modulen", PV Symposium Bad Staffelstein, 2019

[11] M. Bolinger et al, "System-level performance and degradation of 21 GWDC of utility-scale PV plants

in the United States", Journal of Renewable and Sustainable Energy, 2020, 12, 043501

[12] N. Bosco et al., "Defining Threshold Values of Encapsulant and Backsheet Adhesion for PV Module Reliability", 45th IEEE Photovoltaic Specialists Conference, 2017

[13] Fraunhofer CSP leading theme 'Sondermodules', website link: https://www.csp.fraunhofer.de/de/leitthema-sondermodule.html

[14] B. Jaeckel, "Lightweight PV modules: challenges and possible solutions for reliable designs", Sophia Workshop 2021

[15] K. Agroui et al, "Effect of crosslinking on EVA-based encapsulant properties during photovoltaic module fabrication process", *Revue des Energies Renouvelables*, 2015, 18, 303-314

[16] J. Berghold, "Properties of Encapsulation Materials and Their Relevance for Recent Field Failures", 40th IEEE Photovoltaic Specialists Conference Denver, Colorado, 2014

[17] N. Bosco et al., "Environmental Influence on Module Delamination Rate", 46th IEEE Photovoltaic Specialists Conference, 2018

[18] D. E. Mansour et al., "Correlation between EVA Degree of Crosslinking and Moisture Ingress into PV Laminates", 37th European Photovoltaic Solar Energy Conference and Exhibition, 2020

[19] G. Oreski et al., "Properties and degradation behaviour of polyolefinen capsulants for photovoltaic modules", Progress in Photovoltaics: Research and Applications, 2020

ADVANCED DEGRADATION MODELLING OF PHOTOVOLTAIC MODULES AND MATERIALS

Karl A. Berger[1*]; Karl Knöbl[2]; Barbara Brune[3], Irene Ortner[3], Sarka Grubner[3]; Gernot Oreski[4], Chiara Barretta[4]; Bernhard Kubicek[1]; Lukas Neumaier[5], Markus Feichtner[6], Yuliya Voronko[7], Gabriele C. Eder[7]

[1] *AIT, Austrian Institute of Technology GmbH, Austria, karl.berger@ait.ac.at*
[2] *University of Applied Sciences Technikum Vienna, Austria, karl.knoebl@technikum-wien.at*
[3] *Applied Statistics GmbH, Vienna, Austria, barbara.brune@applied-statistics.at*
[4] *PCCL, Polymer Competence Center Leoben, Austria, Gernot.Oreski@pccl.at*
[5] *SAL Silicon Austria Labs GmbH, lukas.neumaier@silicon-austria.com*
[6] *KIOTO Photovoltaics GmbH, Markus.Feichtner@kioto.com*
[7] *OFI, Austrian Research Institute for Chemistry and Technology, Austria, Gabriele.Eder@ofi.at*

ABSTRACT: The recently started Austrian R&D project "ADVANCE! - Advanced Degradation Modelling of Photovoltaic Modules and Materials!", a collaboration of Austrian applied science research and a module manufacturer, addresses innovative and complex statistical and data processing methods to enhance analysis and modelling of climate related degradation processes of photovoltaic modules. As a starting point, data from a previous project, that dealt with module design adaption to different outdoor climatic conditions, will be used, and enhanced by additional data derived from measurements and observations. In the first step the focus is on applying data science approaches on PV module stress testing results to achieve insight in underlying PV module degradation processes and to support manufacturing. An example for applying mixed effects modelling on power degradation data from 6-cell mini modules by means of pairwise comparison of aging actions using a base action, and one with an additional stress, is given, resulting in relative degradation values. Beside investigations on mini modules with identical BOM and different aging actions applied, variations in modules BOM are also investigated. In the next steps feature extraction from data will be continued and network structural equation models implemented. Of course, outdoor data analysis is necessary too, to understand module and system performance.
Keywords: PV-modules, stress testing and ageing, degradation modes, modelling.

1 AIM AND APPROACH

Due to the large effort in terms of time and equipment necessary for reliability testing of PV modules, the PV community has always endeavoured to obtain service life estimates, based on an extrapolation of measurement and characterization data from accelerated aging tests or modelling.

Thus, a recently started Austrian R&D activity ("Advance! - Advanced Degradation Modelling of Photovoltaic Modules and Materials!") addresses innovative and complex statistical and machine learning data processing methods for digital analysis and improved modelling of the time and stress-dependent performance (degradation & reliability) of PV modules.

2 PROJECT OBJECTIVES

A wide variety of modelling approaches (statistical, chemical-physical-electrical) are developed and applied in order to identify interrelations between

(i) the performance decline of operating PV modules
(ii) the specific degradation behaviour of the materials and material composites used
(iii) the stress conditions applied.

These interrelations will be used for innovative material and component development and the definition of predictive maintenance specifications. Figure 1 gives a schematic overview.

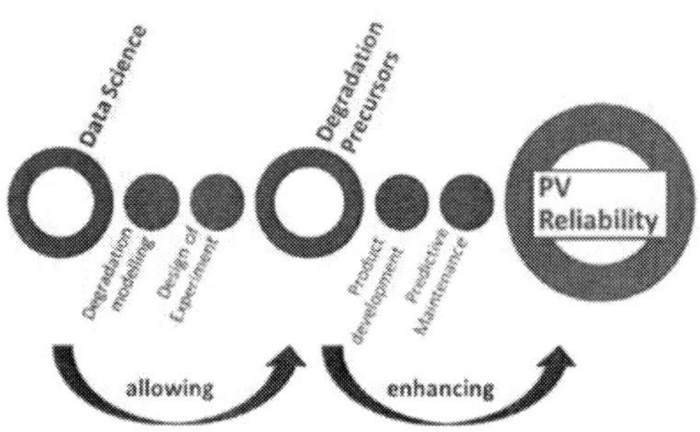

Figure 1: Schematics of the ADVANCE! project objectives.

This highly interdisciplinary research project is intended to open up new paths in the digital analysis of the long-term and degradation behaviour of PV modules and to lay the foundations for future highly efficient material developments for PV and predictive maintenance requirements for PV systems. Scientific innovation and relevance.

3 DATA BASE STRUCTURE, DATA TREATMENT

3.1 Test specimen and aging actions test schemes

Within the former Austrian flagship project INFINITY (2015-2018), [1] hundreds of framed mini modules with front glass and backsheet, including 6 monocrystalline cells each, were produced. One batch of such modules was with identical bill of materials (BOM), using an EVA encapsulant and a PET based backsheet, the other with varying BOM, with two backsheet types (CPO and PPF) and three encapsulants (EVA, POE, TPO), respectively.

The "Aging Actions" and names are inspired by the Koeppen-Geiger climate zones [2], see Table I, and [1], now extended by calculated effective module surface humidity values.

Table I: Combined and sequential aging actions applied.

Aging action	Duration		Temperature T Module	Temperature T Chamber	Rel. Humidity H Module	Rel. Humidity H Chamber	Irradiance I	DML	Salt	TC	Intervalls
Reference	1000h	1		85°C		85%	-	-	-	-	constant
Moderate1	1000h	1	113°C	85°C	30,4%	85%	1000 W/m²	-	-	-	constant
Moderate2	1000h × 7 cycles	1	78 °C	60°C	18,1%	40%	1000 W/m²	-	-	-	48h
		2		85°C		85%	-	-	-	-	96h
Moderate3	1000h × 7 cycles	1	78°C	60°C	18,1%	40%	1000 W/m²	-	-	-	48h
		2		85°C		85%	-	-	-	-	96h
		3		-	-	-	-	1000 c	-	-	24h
Moderate4	1000h × 7 cycles	1	78°C	60°C	18,1%	40%	1000 W/m²	-	-	-	48h
		2		85°C		85%	-	-	-	-	96h
		3		60°C	-	-	-	-	Salt mist	-	24h
Moderate5	3000h × 7 cycles	1	78°C	60°C	18,1%	40%	1000 W/m²	-	-	-	48h
		2		85°C		85%	-	-	-	-	96h
		3		-40/+85°	-	-	-	-	-	50 c	300h
Alpine	2000h × 4 cycles	1		85°C		85%	-	-	-	-	250h
		2	119°C	85°C	24,9%	85%	1200 W/m²	-	-	-	250h
		3		-	-	-	-	1000 c	-	-	24h
Tropical1	3000h	1		85°C		85%	-		-	-	constant
Tropical2	3000h	1		90°C		90%	-		-	-	constant
Arid	1000h	1	129°C	95°C	15,7%	30%	1200 W/m²		-	-	constant

These values in column six were calculated for the modules surface under irradiation during damp-heat chamber conditions by means of measured surface temperature, water-saturation pressure and dew point [3].

Combined and cyclic stresses were applied to 6-cell mini modules, with three of them, in parallel, a "module group", always to increase statistical reliability. In between tests, a wide variety of measurements was used: electrical I-V-curve measurement at STC and parameter extraction, (polymere) molecule spectroscopic and color index measurements as well as different imaging techniques like UV fluorescence (UVF) and electroluminescence (EL).

3.2 Data base

All measurement results were organized in an optimized database, which forms the information base for setting up models for climate sensitive aging and degradation processes. The database, using SQLite in conjunction with peewee [4, 5], was structured around the modules (DUT), which were logically connected via their specific aging module group with strictly associated aging actions and instances, Figure 2. Data triples for the three modules treated in parallel were logically grouped in the database with the corresponding measurement results.

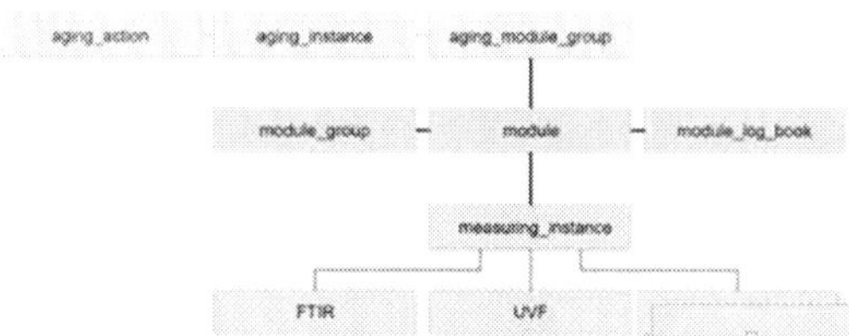

Figure 2: Example for the database structure

Cumulated exposure-hours were calculated as time in climate chamber (h) × rel. humidity H (%), irradiance I (W/m²) and temperature T (°C), respectively. The three graphs in Figure 3 give the cumulated stresses applied during the different ageing actions, summarized in Table I.

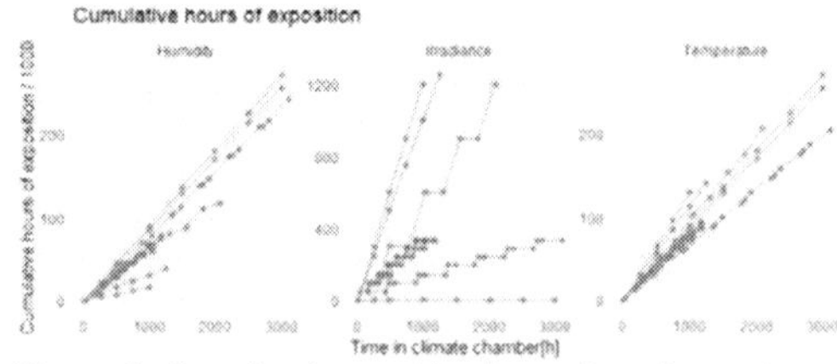

Figure 3: Cumulated exposition hours for aging actions.

4 FIRST RESULTS

4.1 Modelling power degradation under varied stress input, identical module BOM

For having a basic stress applied (Aging action 1) and an additional stress in another aging action (Aging action 2) with the other parameters unchanged, and identical module BOM, pairwise comparisons, classical statistical mixed effects modelling [7] can be applied to the artificial ageing of modules with identical BOM, here an EVA encapsulant and a PET based backsheet. Changes in normalized MPP Power at STC (P_{MPP}) were fitted by means of model equation (1) using a ramp for the initial power gain, and an overlay of quadratic terms, one for the ageing action without, the other for the one introducing an additional stress, and an additional random term accounting for variation between similarly built and treated modules:

$$P_{MPP,\,norm}(t) = 1 + \beta_1\, t\, 1\left\{t < {}^{500h}/_{3000h}\right\} + \\ + \beta_2\, t^2 + \beta_3\, t^2\, 1\{feature\} + b_{module}\, t^2 + e(t) \quad (1)$$

The fitted coefficients for the different aging action pairs (cf. Table I), are depicted in Table II.

Table II: Fitted mixed effects modelling coefficients for the modules P_{MPP} - pairwise comparisons.

Aging action 1	Aging action 2	Additional stress impact / feature	Coefficients β_1	Coefficients β_2	Coefficients β_3	Coefficient ratio $(\beta_2+\beta_3)/\beta_2$
Tropical 1	**Moderate 1**	**irradiation**	0.0257	-0.0431	-0.2588	**7,0***
Moderate 1	Moderate 2	constant vs. sequential I	0.0292	-0.3320	0.1174	0.64*
Moderate 2	Moderate 3	DML	0.0278	-0.1955	-0.2754	2.41*
Moderate 2	Moderate 4	Salt mist	0.0249	-0.1566	0.0069	0.96
Moderate 2	Moderate 5	Temperature cycles	0.0207	-0.0103	-0.0315	4.06
Tropical 1	**Tropical 2**	**Increase in T and H**	0.0294	-0.0552	-0.0767	**2.39***

The rightmost column gives a relative measure how much the additional stress forces degradation. The * indicates significant results applying a 95% confidence interval. In the following figures the time series of measured (left) and fitted (right hand side) P_{MPP} are compared:

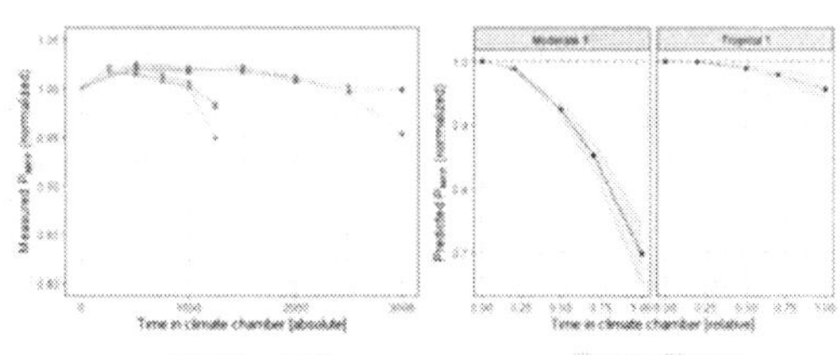

Figure 4: "Tropical 1" (dashed lines) and "Moderate 1" (solid lines) P_{MPP} comparison. Left side measured, right side modeled, i.e. predicted power degradation.

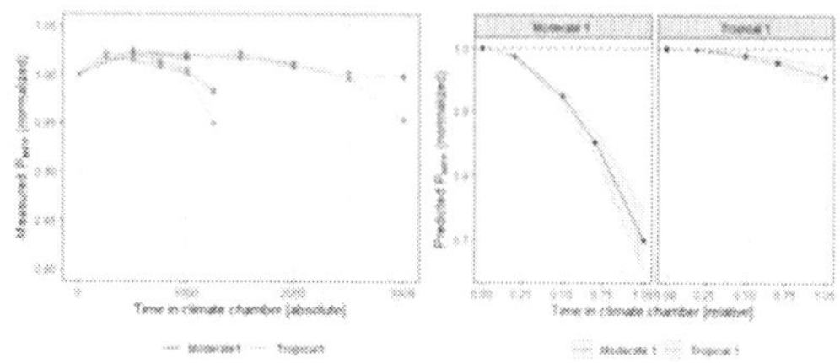

Figure 5: "Tropical 1" (dashed lines) and "Tropical 2" (solid lines) P_{MPP} comparison. Left side measured, right side modeled, i.e. predicted power degradation.

In Figure 4 for the "Tropical 1" vs. "Moderate 1" (DH 85/85, additional feature 1000 W/m² irradiation), and in Figure 5 "Tropical 1" vs. "Tropical 2" (DH 85/85 → DH 90/90) are compared.

4.2 Power degradation under identical stress, module BOM varied

Power degradation under DH stress was investigated for six different module types, having two backsheet variants (CPO and PPF) and three different encapsulants (EVA, POE, TPO). The measured (absolute) P_{MPP} for these six module types, represented by groups of three modules with identical BOM is compared in Figure 6, with a measurement interval of 1000 h.

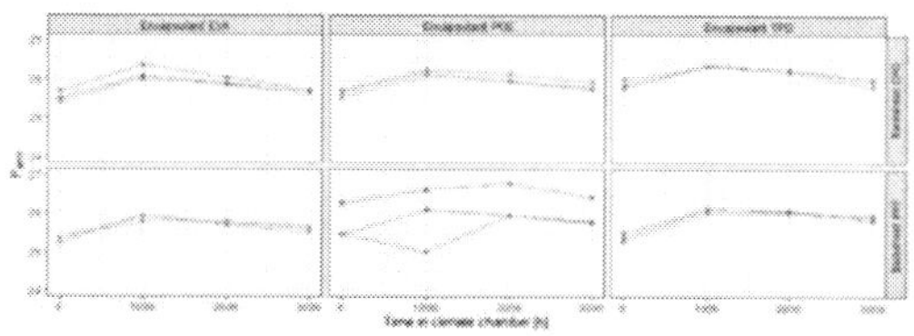

Figure 6: Comparison of P_{MPP} measured, upper row CPO, lower PPF backsheets. Left side EVA, middle POE, right hand side TPO encapsulants. Aging action: up to 3000 h DH 85/85.

Changes in P_{MPP} and EL images are correlated to material degradation behaviour of the polymers.

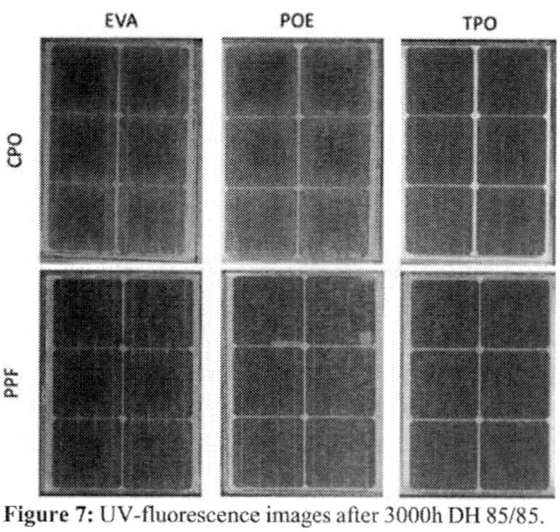

Figure 7: UV-fluorescence images after 3000h DH 85/85. Upper row CPO, lower PPF backsheets, left side EVA, middle POE, right hand side TPO encapsulants.

Such changes are e.g. visible in the UV-fluorescence images of the modules / encapsulants and UV-VIS-spectra [7, 8], given as an example in Figure 7 for the state after 3000 h DH 85/85.

5 SUMMARY AND OUTLOOK

Examples for investigations on mini modules tested by means of artificial aging actions, using combined and cyclic stresses and first evaluation and modelling steps for power degradation are explained. Data fits and modelling steps, applied to the existing measurement results, e.g. like the ones given as an example in chapter 4.1, and extracted values, e.g. shifts and intensity changes in molecule spectroscopy data will extend our data base.

For feature extraction from EL and UVF images (as depicted in chapter 4.2) deep learning-based image classification algorithms will be used, see [9], 5CV.2.11 in this EU PVSEC proceedings as well.

These data will then be used to generate mechanistic/inferential pathway models [10], by Network Structural Equation Models implemented in the statistical language R [11], netSEM package [12].

In addition to the data derived from artificial aging procedures, data from fielded PV systems, failure protocols, PV plant monitoring data and degradation rates will also be used [13,14] to compare with results from artificial aging, and therefore improve degradation modelling and forecasting.

ACKNOWLEDGEMENT

ADVANCE! receives funding by the Austrian government, by means of the Austrian Federal Ministry for Climate Action, Environment, Energy, Mobility, Innovation and Technology (bmk.gv.at), represented by the Austrian Research Promotion Agency (FFG, https://www.ffg.at/), under contract No. 881133, within the 6. e!MISSION call.

REFERENCES

[1] Gabriele C. Eder, Yuliya Voronko, Simon Dimitriadis, Karl Knöbl, Gusztáv Újvári, Karl A. Berger, Martin Halwachs, Lukas Neumaier and Christina Hirschl; Climate specific accelerated ageing tests & evaluation of ageing induced electrical, physical and chemical changes; Prog Photovolt Res Appl. 2018; 1–16; DOI: 10.1002/pip.3090

[2] Kottek et al. (2006), Rubel and Kottek (2010) and Rubel et al. (2017). Updated Köppen-Geiger climate classification web site, http://koeppen-geiger.vu-wien.ac.at/

[3] The Engineering Toolbox, https://www.engineeringtoolbox.com/water-vapor-saturation-pressure-d_599.html

[4] SQLite database engine, https://sqlite.org

[5] Object-relational mapping tool peewee, http://docs.peewee-orm.com/en/latest/

[6] Pinheiro, Jose C., and Douglas M. Bates. Mixed-Effects Models in S and S-PLUS. Statistics and Computing. New York: Springer, 2000

[7] M. Köntges, A. Morlier, G. Eder, E. Fleis, B. Kubicek, and J. Lin; Review: Ultraviolet

Fluorescence as Assessment Tool for Photovoltaic Modules, IEEE J. Photovoltaics, vol. 10, no. 2, pp. 1–18, 2020.

[8] G. C. Eder, Y. Voronko, P. Grillberger, B. Kubicek, and K. Knöbl, UV-Fluorescence measurements as tool for the detection of degradation effects in PV-modules, Vienna, Austria: Proceedings of the European Weathering Symposium EWS, pp. 205-214, 2017.

[9] L. Neumaier, J. Scherer, Ch. Schwarzlmüller, B. Kubicek, F. Mödritscher, Ch. Hirschl; Deep learning-based image feature extraction for predicting climate related degradation of PV modules. 5CV.2.11, EU-PVSEC 2021

[10] R.H. French, R. Podgornik, T.J. Peshek, L.S. Bruckman, Y. Xu, N.R. Wheeler, A. Gok, Y. Hu, M.A. Hossain, D.A. Gordon, P. Zhao, J. Sun, G. Zhang, Degradation science: Mesoscopic evolution and temporal analytics of photovoltaic energy materials, Current Opinion in Solid State and Materials Science, Volume 19, Issue 4, 2015

[11] Statistical computing language R, www.r-project.org

[12] Wei-Heng Huang, Nicholas Wheeler, Addison Klinke, Yifan Xu, Wenyu Du, Amit K.Verma, Abdulkerim Gok, Laura Bruckman and Roger French; netSEM: Network Structural Equation Modeling. R package version 0.5.1., (2018). https://CRAN.R-project.org/package=netSEM

[13] M. Köntges, G. Oreski, U. Jahn, M. Herz, P. Hacke, K. A. Weiss, G. Razongles, M. Paggi, D. Parlevliet, T. Tanahashi and R. H. French; Assessment of Photovoltaic Module Failures in the Field. Report IEA-PVPS T13-09:2017, ISBN 978-3-906042-54-1, 2017.

[14] M. Halwachs, L. Neumaier, N. Vollert, L. Maul, S. Dimitriadis, Y. Voronko, G.C. Eder, A. Omazic, W. Mühleisen, Ch. Hirschl, M. Schwark, K.A. Berger, R. Ebner; Statistical evaluation of PV system performance and failure data among different climate zones; Renewable Energy 139 (2019), p. 1040 - 1060; https://doi.org/10.1016/j.renene.2019.02.135

ACCELERATED ACID CORROSION TESTING OF SOLAR CELLS: TEST TO FAILURE

Andrew Fairbrother[1], Luca Gnocchi[1], Baptiste Bergerot[1], Nicolas Rospars[1], Alessandro Virtuani[1], Christophe Ballif[1,2]

1 – École Polytechnique Fédérale de Lausanne (EPFL), Institute of Electrical and Micro Engineering (IEM), Photovoltaics and Thin-Film Electronics Laboratory, Rue de la Maladière 71b, 2002 Neuchâtel (Switzerland)
2 – CSEM PV-Center, Rue de Jaquet-Droz 1, 2002 Neuchâtel (Switzerland)
Email: andrew.fairbrother@epfl.ch

ABSTRACT: Corrosion is one main lifetime limiting degradation modes of photovoltaic (PV) modules. The process is usually gradual and it can take decades to become a factor that significantly affects performance because of the slow accumulation of moisture and acetic acid (from encapsulant degradation). In this work accelerated corrosion tests were performed on modules with Al-back surface field (BSF) and silicon heterojunction (SHJ) solar cells by immersion in water or acetic acid baths. After lamination the top encapsulant/glass were removed so that only the front size of the cells was exposed. During the tests to failure three factors were varied: acid concentration, temperature, and electrical bias. SHJ solar cells were found to be more resistant to degradation than Al-BSF cells. For all factors studied the presence of acid accelerated degradation compared to the control tests with water. Elevated temperatures accelerated the rate of degradation by several times. Application of electrical bias led to an initial drop in current, but these modules eventually outperformed the non-biased modules. Lead oxides were the primary degradation products detected, found to accumulate mostly along the printed metallization in proximity to the busbar. The accelerated corrosion test methodology is being optimized to match characteristics observed in field modules with corrosion, and can readily be applied to other solar cell, metallization, and interconnection technologies to assess their long-term durability.
Keywords: Degradation, Experimental Methods, Module, Qualification and Testing, Reliability

1 INTRODUCTION

Photovoltaic (PV) module lifetime is influenced by a variety of degradation and failure phenomena. While there are many tests to assess design quality and early- or mid-life failure modes, there are few to determine the mechanisms and impacts of end-of-life degradation modes such as corrosion. The damp heat test is the main accelerated test for corrosion in PV modules.[1,2] However, the conditions are very aggressive – 85°C and 85% relative humidity – and may induce degradation in modules that is not observed in field operation. Moreover, the test only involves water as a chemical stressor, even though it is known that acetic acid is present in field modules with concentrations up to 0.13% v/v (volume/volume) after several years.[3,4] Thus a more accurate test for corrosion in PV modules should include acetic acid as a stressor and be performed at lower temperatures.[1,2]

Acetic acid is generated by the degradation of ethylene vinyl acetate (EVA) encapsulants, and it can take several years to accumulate to appreciable levels. This is because the degradation of EVA is an autocatalytic process, so that the rate of generation and accumulation of degradation products such as acetic acid is not linear with time. At first it will increase slowly, and then rapidly, with the module performance decreasing correspondingly.[1,2,5] Kyranaki et al. and Tanahashi et al. [1,2] have developed acetic acid-based corrosion tests for solar cells, the former by immersion and the latter by vapor exposure. They both observe mostly current and fill factor losses, but propose two different mechanisms, either solder or finger corrosion. Additionally, the variety of test variables means that methods must be optimized to produce meaningful results.

In this work an accelerated corrosion test method is developed based on the immersion of modules into acetic acid baths. Tests were performed on modules with aluminum-back surface field (Al-BSF) and silicon heterojunction (SHJ) cells. Modules had the top EVA/glass removed after lamination so that only the front-side of the cells were exposed. Three factors were varied, including acetic acid concentration, temperature, and electrical bias. The presence of acid accelerated degradation compare to the water exposure tests, but very high concentrations gave results that might not be representative of field performance. Power loss in the acid-exposed cells was mostly driven by reductions in fill factor. Increased temperature also accelerated corrosion, as it is a thermally activated chemical process. The application of bias led to an initial reduction in current, but over time biased cells outperformed unbiased cells. Finally, SHJ cells were found to be more resistant to acid-induced degradation than Al-BSF cells. The test to failure methodology developed in this work can be used on a variety of cell, metallization, and interconnection technologies to assess and compare their degradation and failure behavior.

2 EXPERIMENTAL

2.1 Module fabrication

Monocrystalline silicon modules with Al-BSF or SHJ solar cells were prepared with a half-laminated structure as illustrated in Figure 1a, using glass front and back covers and an ethylene vinyl acetate (EVA) encapsulant. An ethylene tetrafluoroethylene (ETFE) release layer was inserted between the solar cell and top layer EVA so that it formed a weakly adhered interface. After lamination, the top layer ETFE/EVA/glass was pried off of the solar cell, resulting in the half-laminated construction that leaves the cell, metallization, and interconnects exposed for the corrosion tests. This structure isolated the bottom of the cell so the acid baths only affected the front of the cells. The structure also ensured the cells were flat and able to be handled without breakage during the tests. The solar cell top metallization was based on screen printed silver with five (Al-BSF) or four (SHJ) busbars, and cell interconnects were 1.5 mm wide flat copper wires coated with a lead-tin based solder. For Al-BSF cells these were soldered onto the busbars, and for SHJ cells they were attached using a silver-based electrically conductive adhesive (ECA).

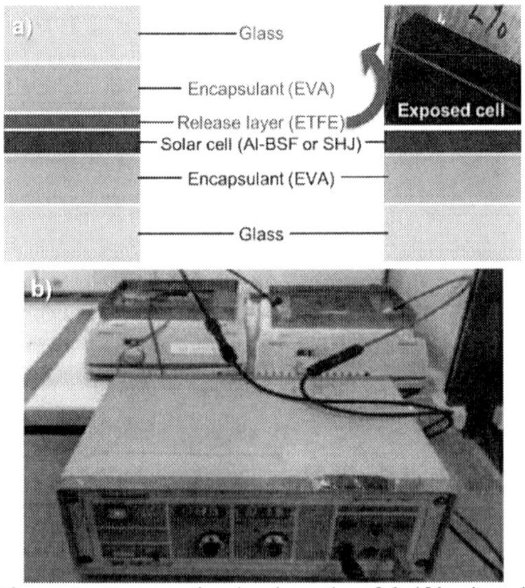

Figure 1. Cross-sectional schematic of half-laminated module construction after lamination (a, left) and after top layers removed (a, right), and photograph of the accelerated corrosion test setup for high temperature and electrical bias (b).

2.2 Corrosion testing

After measuring the initial module characteristics (described in 2.3), they were immersed in acetic acid baths using conditions shown in the test matrix of Table 1. The three test variables included acid concentration (0% to 10% v/v), temperature (20°C or 60°C), and cell electrical bias (0 A or 8.5 A). These stress levels were selected to mimic or intensify field conditions. Older field modules have been measured containing up to 0.13% v/v acetic acid,[3,4] while in this work stronger concentrations were used to accelerate corrosion processes. For simplification, in the rest of this work acid concentrations are referred to as simple percentages, instead of % v/v. 60°C is a typical field module operating temperature in temperate climates.[6] Finally, 8.5 A corresponds approximately to the current during max power point operation of the solar cells under AM1.5 illumination. During the development of the methodology not all possible combinations of conditions were explored, and some levels remain to be tested (e.g. lower acid concentrations (<1%) and higher temperatures (≥60°C)). The baths and modules were placed inside of metal trays and covered during the tests to limit evaporation. For the elevated temperatures hot plates were used to heat the solutions, and for biasing the modules were connected to a power supply, as seen in Figure 1b, and left in the dark.

Table 1. Accelerated corrosion test variables and levels. Note that not all combinations of levels were used.

Acid conc. [% v/v]	Temperature [°C]	Electrical bias [A]
0	20	0
1	60	8.5
2		
5		
10		

2.3 Module and material characterization

Periodically during the corrosion tests the modules were removed from their solutions, rinsed, and dried before being characterized. Modules were visually inspected, and then light current-voltage (I-V) characteristics were measured under AM1.5 conditions. Electroluminescence was measured with the modules under 8.5 A bias. After these tests, modules were reintroduced to their acid baths. Corrosion tests were carried out for 300 to 600 hours of total exposure.

For some tests the specimens were prepared for optical and scanning electron microscopy (SEM) to examine changes to cell and metallization microstructure. Energy dispersive x-ray (EDX) and Raman spectroscopies were used to identify degradation products.

3 RESULTS AND DISCUSSION

3.1 Al-back surface field modules

Most of the tests in this work utilized Al-BSF cells. Many advanced solar cell technologies (e.g. passivated emitter and rear cell (PERC)) utilize similar front metallization and interconnect materials, based on silver printed pastes and soldered interconnects. Thus Al-BSF cells were used to establish a baseline for the corrosion tests. Additional tests were performed on SHJ solar cells (in section 3.2), and are planned for PERC cells.

3.1.1 Visual inspection

The visual aspect of the modules changed over the course of the tests. Most notable was a discoloration of the acid exposed modules (Figure 2a). This likely came from accumulation of corrosion products, for example, lead oxides are yellow, brown, or red in color depending on their oxidation state. In the high acid concentration tests (>1%) ribbon detachment was fairly common (Figure 2b), a behavior also observed in [1]. This detachment failure tended to occur before significant degradation was observed (i.e. >5% change in optoelectronic properties), so the remainder of the corrosion tests used either water or 1% acid. After prolonged exposure the solder on the cell and string interconnect ribbons began corroding away, even to the point of leaving the copper core exposed. By visual inspection it appeared that the solder joint to the busbar was the cell component most likely to fail, a finding consistent with the lower oxidation potentials of lead-tin solder compared to copper and silver.[7]

Figure 2. Visual inspection of modules during the corrosion tests, including discoloration (a) and cell interconnect ribbon detachment (b).

3.1.2 Module performance

The main optoelectronic properties that were monitored included the max power point (P_{MPP}), short circuit current (I_{SC}), open circuit voltage (V_{OC}), and fill factor (FF). Fill factor was the main parameter that changed during the tests, and led to corresponding reductions in the max cell power. Short circuit current generally did not change much either, except in some cases noted below. Open circuit voltage was relatively stable for all tests, not decreasing more than 3% of their initial values. Figure 3 shows a summary of the power loss for the Al-BSF module tests, including water and acid (1%).

Acid concentration

In all tests the presence of acid accelerated module degradation compared to the water tests. In fact, power loss was never greater than 5% for the water tests in any test conditions. The initial rate of power loss was similar for all acid concentrations (not shown), indicating that acetic acid was not a rate-limiting reactant for concentrations above 1%. Further tests will include lower acid concentrations, including levels measured in field modules (i.e. <0.13%). Ribbon detachment eventually occurred in all tests with acid concentration above 1%. When detachment occurred cell power would drop by tens of percent, and even to zero in the most severe cases. Because of this, it was not possible to decouple gradual corrosion effects with this sudden detachment.

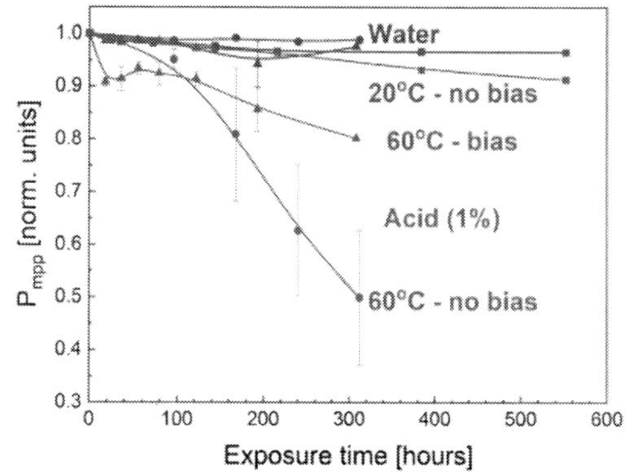

Figure 3. Normalized cell power loss as a function of exposure time for water and acid (1%) exposed samples at 20°C and 60°C, with and without electrical bias.

Temperature

Increased temperature (60°C versus 20°C) resulted in a much faster rate of power loss. After 200 hours the low temperature modules (without bias) had undergone about 4% power loss, compared to nearly 30% for the high temperature modules. Even after 550 hours the low temperature modules had only lost 9% of their initial power. Indeed, because corrosion is a chemical process it is thermally activated, thus corrosion rates would be accelerated during the daytime when modules are warmest.

Electrical bias

The application of bias to the cells resulted in a rapid reduction in short current (about 8%) in the first few hours of exposure. After this, current remained stable, and fill factor losses became the main contributor to power loss. These modules outperformed the unbiased modules up to 300 hours, with power loss of only 20%, compared to 50%. Because corrosion is an electrochemical process, it was expected that an electrical bias would alter the mechanism and rate of power loss. Electroplating of copper was noticed on the biased cells, and this process may have contributed to the improved performance over the unbiased cells.

3.1.3 Material-level characterization

SEM was used to examine the cell microstructure during the tests. The most notable feature included accumulations along the printed metallization, a feature also observed in field modules.[7,8] In water these had a flake-like morphology (Figure 4a), while in acid they exhibited a fine, cluster-like morphology (Figure 4b). EDX showed these features to be rich in lead and oxygen and Raman spectra (Figure 4c) showed characteristic peaks of PbO_2.[9] Peak shape and position sometimes varied by point, which may be a laser-induced heating effect, similar to behavior observed in [9]. Metal acetates and other oxides were not detected at any points.

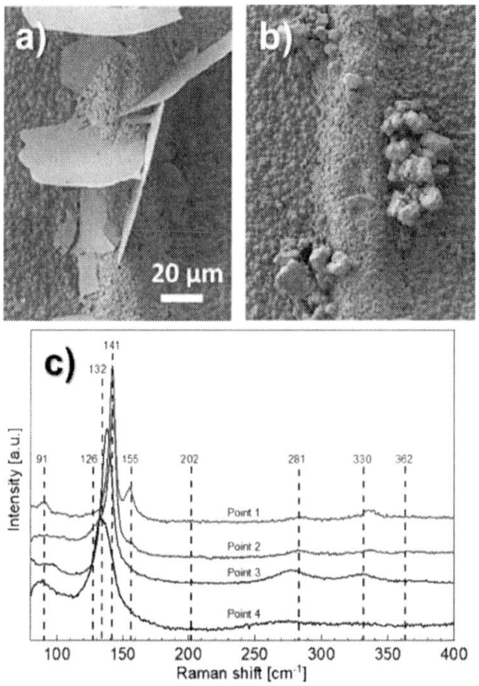

Figure 4. SEM along the finger of solar cell exposed at 20 °C in water (a) and 1% acid (b), and Raman spectra of corrosion products in acid exposed cells indicating the presence of lead oxides (c).

3.2 Silicon heterojunction modules

SHJ solar cells are a high efficiency technology based on crystalline silicon coated on both sides with amorphous silicon layers. This passivates the cell surface, giving rise to higher cell voltages than for Al-BSF or PERC cells. Typically cell interconnect ribbons are attached with an ECA instead of solder, because the thin amorphous silicon layers are sensitive to high process temperatures.

3.2.1 Visual appearance

The visual appearance during testing was similar to the Al-BSF modules (section 3.1.1), with discoloration of the acid exposed modules. Interconnect ribbon detachment was not observed in any of the tests.

3.2.2 Module performance

Figure 5 shows the power of SHJ and Al-BSF cells in water and 1% acid at 60°C. The initial rate of power loss is similar up to 150 hours (about 10%), but after this the Al-BSF cells degraded much more quickly. After 300 hours the Al-BSF cells had lost 50% of their initial power, compared to only 20% for the SHJ cells. The cell interconnect ribbons are attached by a silver-based ECA, which is clearly more resistant to acid-induced corrosion than the lead-tin solder. This is also evident from the limited corrosion observed along silver fingers for both types of cells.

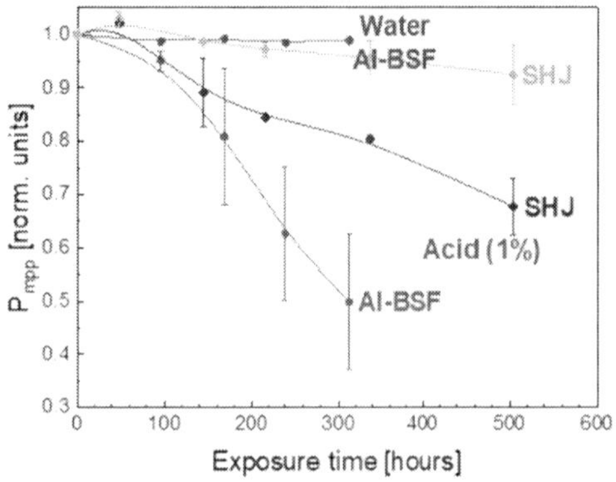

Figure 5. Normalized cell power loss as a function of exposure time for SHJ and Al-BSF cells in water and acid (1%) at 60°C and without electrical bias.

4 CONCLUSIONS

In this work an accelerated corrosion test method was developed. The tests to failure included immersion of half-laminated solar cells (front-side exposed) in acetic acid baths of varying concentration, temperature, and cell bias. High acid concentrations, above 1%, resulted in rapid degradation due to ribbon detachment. Higher temperatures accelerated module power loss by several times, mostly by fill factor reduction, and sometimes current reduction, and features match those seen in field modules with corrosion-related degradation. The application of electrical bias initially led to a drop in current, but eventually these cells outperformed non-biased cells. Finally, SHJ cells were more resistant to corrosion effects than Al-BSF, a behavior likely related to the use of a silver-based ECA for cell interconnection. An optimized test methodology can be used to improve end-of-life assessment for a variety of solar cell, metallization, and interconnect technologies.

5 ACKNOWLEDGEMENTS

This work has been funded in part by the European Commission (EC) under the H2020 project Be-SMART (#818009), and the Swiss National Science Foundation (SNF) under the project REBI-PV (COST IZCOZ0_182967).

6 REFERENCES

[1] N. Kyranaki, J. Zhu, R. Gottschalg, T. Betts, Investigating the degradation of front and rear sides of c-Si PV cells exposed to acetic acid, in: Loughborough University, 2018. https://doi.org/[u'https://doi.org/10.4229/35thEUPVSEC20182018-5CV.3.47'].

[2] T. Tanahashi, N. Sakamoto, H. Shibata, A. Masuda, Electrical detection of gap formation underneath finger electrodes on c-Si PV cells exposed to acetic acid vapor under hygrothermal conditions, in: 2016 IEEE 43rd Photovoltaic Specialists Conference (PVSC), IEEE, Portland, OR, USA, 2016: pp. 1075–1079. https://doi.org/10.1109/PVSC.2016.7749778.

[3] T.J. Trout, W. Gambogi, T. Felder, K.R. Choudhury, L. Garreau-Iles, Y. Heta, K. Stika, PV Module Durability –connecting field results, accelerated testing, and materials, in: 44th IEEE Photovoltaic Specialist Conference (PVSC), IEEE, Washington, DC, USA, 2017: pp. 2312–2317.

[4] T. Doi, S. Sakamoto, Y. Hara, S. Suzuki, A. Masuda, Possible Measure of Reliability for Crystalline-Si Photovoltaic Modules, 29th European Photovoltaic Solar Energy Conference and Exhibition. (2014) 2566–2569. https://doi.org/10.4229/EUPVSEC20142014-5DO.16.5.

[5] M. Köntges, S. Kurtz, C. Packard, U. Jahn, K.A. Berger, K. Kato, Performance and reliability of photovoltaic systems: subtask 3.2: Review of failures of photovoltaic modules: IEA PVPS task 13: external final report IEA-PVPS, International Energy Agency, Photovoltaic Power Systems Programme, Sankt Ursen, 2014.

[6] International Electrotechnical Commision, IEC TS 63126:2020 - Guidelines for qualifying PV modules, components and materials for operation at high temperatures, (2020).

[7] W. Oh, S. Kim, S. Bae, N. Park, S.-I. Chan, Y. Kang, H.-S. Lee, D. Kim, Migration of Sn and Pb from Solder Ribbon onto Ag Fingers in Field-Aged Silicon Photovoltaic Modules, International Journal of Photoenergy. 2015 (2015) 1–7. https://doi.org/10.1155/2015/257343.

[8] N.C. Park, J.S. Jeong, B.J. Kang, D.H. Kim, The effect of encapsulant discoloration and delamination on the electrical characteristics of photovoltaic module, Microelectronics Reliability. 53 (2013) 1818–1822. https://doi.org/10.1016/j.microrel.2013.07.062.

[9] L. Burgio, R.J.H. Clark, S. Firth, Raman spectroscopy as a means for the identification of plattnerite (PbO2), of lead pigments and of their degradation products, Analyst. 126 (2001) 222–227. https://doi.org/10.1039/b008302j.

COMPATIBILITY OF CROSSLINKING ENCAPSULANTS WITH SMART WIRE CONNECTION TECHNOLOGY: WAYS TO PREVENT BUBBLES FORMATION

I. Dmitriev[1], S. Yakovlev[1], N. Glebova[2], A. Nechitailov[2], E.Terukova[1], D. Andronikov[1], D. Orekhov[1], I. Shakhray[3]
[1]R&D Center of Thin Film Technologies in Energetics, 194064, St.-Petersburg, Russia
[2]Ioffe Physical-Technical Institute of the Russian Academy of Sciences, 194021, St.-Petersburg, Russia
[3]Avelar Solar Technology, Russia

ABSTRACT: The possibility of application of crosslinking encapsulants POE and EVA with SmartWire Connection Technology in glass/glass, glass/backsheet and frontsheet/backsheet module structures was studied. The main obstacle of the modules lamination with EVA and POE encapsulants is an intensive formation of bubbles under the SmartWire electrode connecting polymer film. The reason of this is volatile organic compounds release from the encapsulants because of the chemical reactions during their crosslinking. Several approaches have been proposed to resolve this problem: 1) to find the technological window for producing modules with the POE and EVA encapsulants without bubbles; 2) to use connecting electrode films which is less inclined to accumulate the bubbles 3) to ensure smoother heating of the module stack using pin-equipped design of laminator.

Keywords: SmartWire Connection Technology, lamination, crosslinking, encapsulants, EVA, POE.

1 INTRODUCTION

SmartWire Connection Technology (SWCT) is a photovoltaic module concept, developed by Meyer Burger, which is an effective alternative to standard bus-bar technology. SWCT implies an interconnection design of the multiple copper wires laminated in connecting film connected to the PV cells through the connecting film. This design brings new requirements to the selection of an appropriate encapsulant material. Generally, non-crosslinked encapsulants like thermoplastic polyolefin (TPO) are used for SWCT modules lamination. However, TPO encapsulants have their own disadvantages as compared with crosslinking ones, like ethylene vinyl acetate (EVA) and polyolefin elastomer (POE), especially when the increased creep stability is required (e.g., for frameless modules).

It was noticed that the main obstacle of the SWCT modules lamination with EVA and POE encapsulants is an intensive formation of bubbles under the connecting film (Figure 1). In this work, the origin of the bubbles embedded inside the SW film in the PV modules is proposed to be most likely volatile organic compounds (VOCs) release from the encapsulants because of the chemical reactions during their crosslinking [1]. Bubbles appear and collect because of the connecting electrode film acts as a trap for these gas compounds.

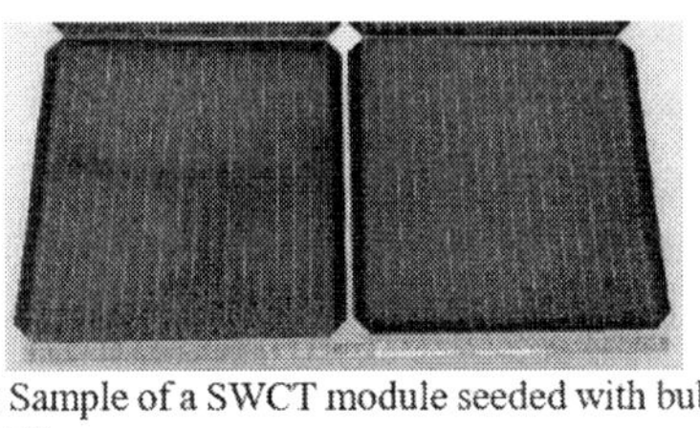

Figure 1: Sample of a SWCT module seeded with bubbles under SW film.

The following factors that affect this problem can be identified:
1) The volume of VOCs which the encapsulant is inclined to emit during crosslinking;
2) The properties of the connecting electrode films which could more or less inclined to accumulate the bubbles,
3) The design of the laminator, which could provide more or less controllable heating of PV module stack.

The work outlined the following ways to solve the problem of bubbles:
1) to find the technological window for lamination of modules with the POE and EVA encapsulants without bubbles;
2) to use connecting electrode films which is less inclined to accumulate the bubbles;
3) to use laminator equipped with preheating on pins system.

The aim of this work is to study possibility of using EVA and POE for SWCT modules production and to identify the problem areas and solutions.

2 EXPERIMENTAL DETAILS

Solar panels for investigation were produced from the HJT solar cells with thickness of 130 μm connected with SWCT.

The next designs of the modules were prepared: Glass/glass (G/G), Glass/backsheet (G/BS), frontsheet/backsheet (FS/BS) (Figure 2).

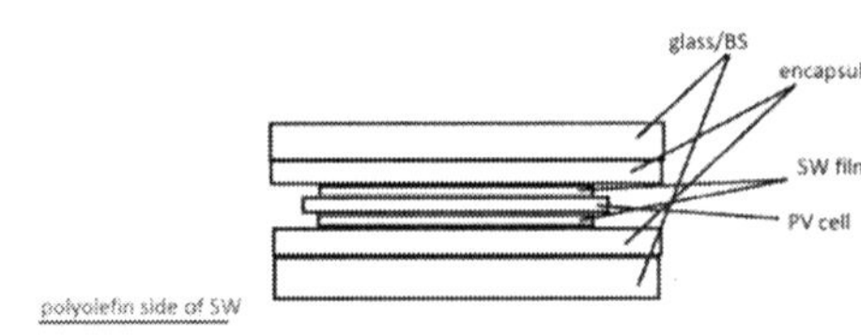

Figure 2: The scheme of the design of the module stacks before lamination.

The following encapsulanting films are used for lamination: POE 400 g/m^2, EVA 400 g/m^2 and three-layer films EVA/POE/EVA 520 g/m^2. Smart Wire connecting films are PET/PO and POE/PO where is PET (polyethylene terephthalate) and POE (crosslinked polyolefin) is a currier substrate for PO which is a polyolefin thermoplastic adhesive layer.

The wires are InSn alloy coated copper wires of 250 μm diameter.

Single chamber membrane laminators have been used

for laminations.

Differential thermal analysis techniques have been used to understand behaviour of crosslinking encapsulants (Simultaneous Thermal Analysator (TGA/DSC) Mettler-Toledo TGA/DSC 1).

The appearance of bubbles after the lamination was evaluated visually.

3 RESULTS AND DISCUSSION

3.1 DSC and DTA investigation of POE and EVA encapsulants

On the DSC and TGA curves weight decrease is observed at the stage of crosslinking (>150°C) (Figure 3). The weight loss percentage is determined to be about 0.02 wt% for EVA and 0,2 wt% for POE at the temperature of 200°C. The weight loss has been identified to be due to loss of volatile organic compounds (VOCs) produced in the EVA or POE curing and thermal degradation process. It can be noted that, during crosslinking reaction EVA releases less VOCs, than POE.

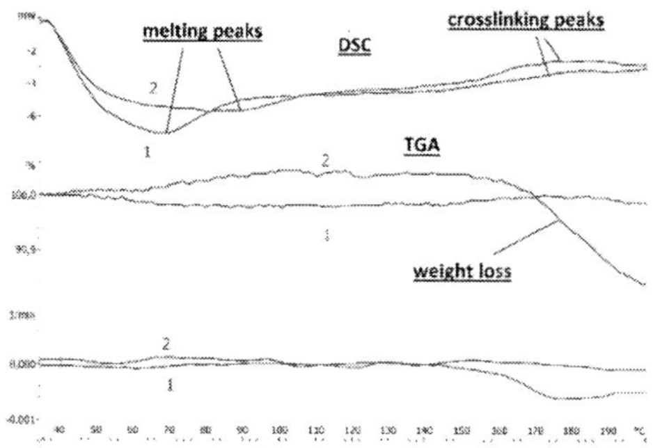

Figure 3: DTA thermogramms of EVA (1) and POE (2) encapsulants.

3.2 Optimization of lamination of the modules

Glass/glass SW modules based on POE encapsulants were fabricated. It was noticed that the appearance of bubbles strongly depends on the technological regime of lamination. The peroxide crosslinking reaction proceeds through the mechanism of thermal activation, so the number of bubbles under the film increases with an increase in the time or temperature of lamination. An interesting question is whether it is possible to find conditions that allow laminating the module without the formation of bubbles, but with a sufficiently high value of the gel content. The results of such an experiment are presented in Figure 4. Lamination regime with parameters T=140°C, t_{vac}=8 min, t_{ramp}=1 min, t_{press}=7 min at 700 mbar made it possible to produce glass-glass modules without bubbles. The POE gel content was 65%.

A similar experiment was repeated for laminating of the modules with three-layer encapsulating films EVA/POE/EVA (Table I). However, both for G/G modules and for G/BS modules it is not possible to determine the temperature-time optimum of lamination without the appearance of bubbles.

The noted difference in the behavior of POE and EVA/POE/EVA laminating materials is because these films are from different manufacturers, which use different chemical crosslinking systems, differing for amount of VOCs emitted.

Unfinished crosslinking	Optimum	Overexposing/ overheating

GC 0 ÷ 50% No bubbles	GC ~ 65% No bubbles	GC ~ 70-85% A lot of bubbles

Figure 4: Results of technological optimization of G/G modules to avoid bubbles and get maximal crosslinking.

Table I: Results of technological optimization of G/G and G/BS modules to avoid bubbles and get maximal crosslinking.

Lamination parameters (Tlam (tvac; tramp; tpress) pressure)	Module G/G	Module G/BS
135°C (5; 1; 15) 800 kPa	EVA is crosslinked, POE is not crosslinked No bubbles	POE and EVA are crosslinked Gel content 88% Bubbles
138°C (5; 1; 15) 800 kPa	Amount of bubbles are less	Bubbles
140°C (7; 1; 10) 800 kPa	POE and EVA are crosslinked Gel content 90% POE and EVA are crosslinked	Bubbles

3.3 Type of connecting electrode film

The connecting electrode film is a two-layer material including a thermally stabilized substrate and a thermoplastic adhesive layer for fixing the contact wires. The substrate is usually made of PET film or cross-linked polyolefin film (POE), and the adhesive layer is thermoplastic polyolefin (PO).

The connecting electrode films from three different manufacturers were investigated. Two of them had a PET/PO structure (PET/PO-1 and PET/PO-2), and the third had a POE/PO structure (Figure 5). The experiment was provided for one-cell FS/BS modules with POE encapsulant. Lamination regime was as follows: T=150°C, t_{vac}=5 min, t_{ramp}=1 min, t_{press}=20 min at 700 mbar. The aim was to carry out visual inspection of bubbles appearance after lamination.

Encapsulant type	Connecting film type		
	PET/PO-1	PET/PO-2	POE/PO
POE	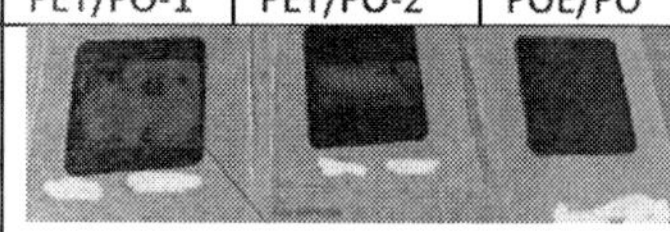		
EVA/POE/EVA			

Figure 5: Results of experiments of FS/BS modules lamination with the different connecting and

encapsulating films.

The experimental results showed that the tendency of bubble formation is more pronounced for PET/PO films. POE/PO films showed a clean surface with almost no bubbles or voids. The type of encapsulant used did not have a pronounced effect. It is likely that the POE substrate of the connecting electrode film eliminates the accumulation of VOCs under the SW. Thus, films of the POE/PO type can be considered as a promising solution to this problem.

3.4 Design of the laminator

It was assumed that the preheating mode can affect the dynamics of VOCs accumulation under the SW film. A smoother preheating of the module can be carried out using the support pins system on the heating plate of the membrane laminator. As a result of the experiments carried out using a pin-equipped laminator, it was possible to select the lamination mode of the G/G module with POE encapsulant in which no bubbles were formed. Lamination regime was as follows: T=150°C, t_{vac}=5 min, t_{ramp}=1 min, t_{press}=30 min at 700 mbar. The POE gel content was 75%. The use the laminator not equipped with pins with the same lamination mode did not allow to achieve the absence of bubbles.

Table II: Results of lamination of G/G modules.

Laminator	Photo
Membrane laminator without pin system	
Pin-equipped membrane laminator	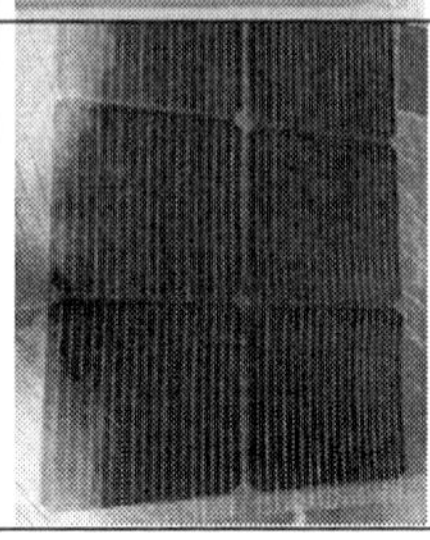

4 CONCLUSIONS

Crosslinkable POE and EVA encapsulants can be used for SWCT. To avoid the bubbles, it is necessary to use the lowest possible temperature and lamination time. However, this route may not be effective in the case of the release of a large amount of VOCs products. Encapsulants from different suppliers differ in the volume of VOCs. Structure of Smart Wire film makes a difference – POE based connecting film looks as a good solution. Different module structures (G/G or G/BS or FS/BS) requires different approach to lamination conditions to eliminate bubbles. Pin-equipped design of membrane laminator allows for a smoother heating of the module stack, which significantly lowers the accumulation of the bubbles.

REFERENCES

1] C. Peike, I. Hädrich, K.-A. Weiß, I. Dürr. "Overview of PV module encapsulation materials", Photovoltaics International, 19, p. 85-92. (2013).

HOT CELLS IN HIGH-POWER PHOTOVOLTAIC MODULES WITH SOLAR CELLS FROM LARGER SILICON WAFER FORMATS

Robert Witteck, Michael Siebert, Iris Kunze, and Marc Köntges
Institute for Solar Energy Research Hamelin (ISFH)
Am Ohrberg 1, 31860 Emmerthal, Germany

ABSTRACT: We demonstrate that partial shading of a single c-Si solar cell in a state-of-the-art high-power photovoltaic (PV) module with three bypass diodes can result in hot cells with critical module peak temperatures of 196 °C. The PV module under test features 144 bi-facial PERC+ half-cells of M6 silicon wafer format with a cell area of (166 x 83) mm^2. The module's measured maximum power output is 437 W_P. We test three cells of the PV module in an extended IEC 61215-2 MQT09 hot-spot endurance test for 5 h in a steady-state sun simulator. All shaded cells reach steady-state temperatures above 175 °C. For the hottest cell we measure peak temperatures of 196 °C at the module surface. After the test the PV modules shows a discoloration of the white backsheet at the position of the shaded solar cells. This is likely to result in a module power loss or module failure due to encapsulation discolorations, de-soldering of the cell interconnection, backsheet cracking of even glass breakage. Our results are essential for improving the long-term stability of new high-power modules with M10- or M12-sized cells and power outputs over 500 W_P against partial shading. New high-power modules require hot cell mitigation techniques that go beyond three bypass diodes per module or heat-resistant module materials.
Keywords: PV Module, Reliability, Shading

1 INTRODUCTION

Since the beginning of 2020, the major Asian manufacturers are pushing with new high-power solar modules onto the market, which have outputs of 500 W_P to 700 W_P [1]. Furthermore, modules with an output power of 800 W_P have been announced [2]. Partial shading of a solar cell in such a high-power solar module can force the shaded cell into reverse bias operation. The shaded solar cell then consumes electrical power generated by the other unshaded cells in the same cell string and heats up. This heating is called the hot cell effect [3], [4]. It can significantly affect the long-term reliability of solar modules with high power output.

In an unshaded solar module, the individual solar cell voltages add up and each solar cell is generating the same current. All solar cells are working at the same operating point and generate power. If a single solar cell is partially shaded, the current of this cell is reduced proportionally to the degree of shading. Due to the series connection this cell has to operate at the same current as the unshaded cells in the string. In case the short circuit current of the shaded cell is reduced below the operating current of the remaining cell, the operating point of the shaded solar cell is shifted from the positive to the negative voltage range. Instead of generating electrical power, this cell consumes power. In the past, this was especially a problem for solar modules with multicrystalline Si (mc-Si) solar cells [5], [6]. A negative cell voltage can lead to a local junction breakdown in mc-Si solar cells, where the power is converted to heat at a small area of the solar cell [7]. As a result, the solar cell becomes hot in a small spot. Therefore, this effect is called hot-spot effect.

However, today the majority of state-of-the-art solar modules usually use monocrystalline silicon solar cells [8], which feature homogeneous properties and are typically free of hot-spot effects. In the case of shading, the dissipated power is distributed homogeneously over the cell area, resulting in a homogenously heating of the cell. Bypass diodes usually limit the dissipated power in a shaded cell and therefore this effect was not critical for solar modules with a power below 450 W_P and smaller cell areas in the past.

In fact we predicted in a recent study with simulations that for new high-power modules the hot cell effect can result in reliability problems [4]. The higher power output of new high-power modules is mainly achieved by larger silicon wafer formats, i.e. M6, M10 or M12 wafers resulting in solar cells with edge lengths of 166 mm, 182 mm or 210 mm. Due to the larger solar cell area the generated current of a solar cell increases, which eventually also increases the cell and module power. A partially shaded solar cell therefore also dissipates more power. Furthermore, the area to perimeter ratio increases with increasing cell size and the cooling of the cell over the cell edge is less efficient. In [4] we showed, that shading neighboring cells from different strings in a half cell module (e.g. cell 1L and 1R in Fig. 1) increases the module surface temperature compared to a single shaded cell. These effects can result in critical high cell temperatures and associated reliability problems for the solar module.

In this work we investigate a high-power half-cell module with a nameplate rated peak power of 440 W_P in the IEC61215 MQT09 hot-spot endurance test to experimentally demonstrate the relevance of the hot cell effect. We determine the worst-case shading fractions of three solar cells and apply these shading fractions to the individual cells within a steady-state irradiance experiment. During the illumination test we monitor the module surface temperature in the vicinity of the shaded cells. We measure a maximum module surface peak temperature of 196 °C, which may reduce module performance or cause a module failure in the long run.

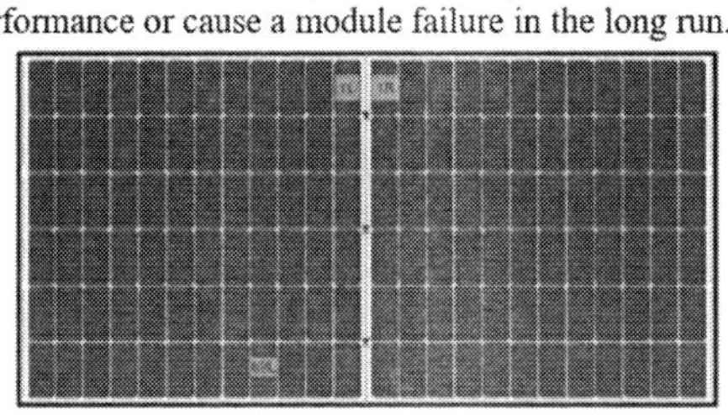

Figure 1: Schematic representation of the solar module under test. The red labels indicate the cells that are shaded in the test according to IEC 61215:2016 MQT09 standard.

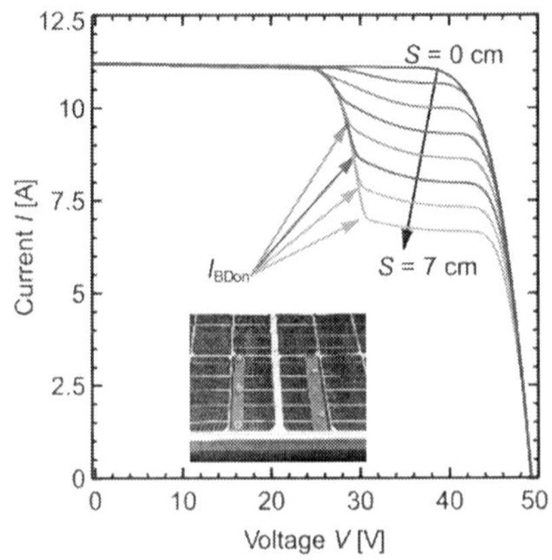

Figure 2: Measured current voltage $I(V)$ data of the tested module when cell 1L is shaded by a strip of varying width S from 0 cm to 7 cm. I_{BDon} indicates the point where the bypass diode starts conducting. The inset indicates the orientation of the shading strips.

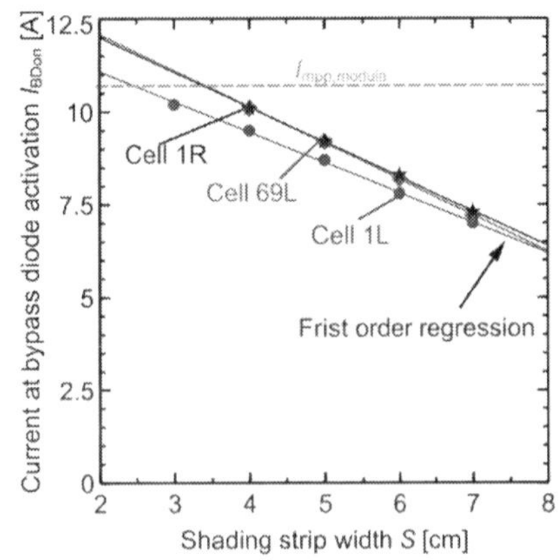

Figure 3: Module current I_{BDon} when the bypass diode starts conducting as function of the shading strip width applied to the cells 1R, 1L, and 69L. The dotted lines show the results of the first order regression modeling.

2 METHODOLOGY

Figure 1 shows a schema of the PV module we test in our steady-state sun simulator. The module features 144 PERC+ half cells each with an area of (166 x 83) mm² (M6 format). The glass-backsheet module consists of two submodules denoted as submodule L and submodule R. Both submodules are connected in parallel and in each submodule 72 cells are connected in series. Submodule L consists of cells 1L through 72L and submodule R consists of cells 1R through 72R. In each submodule there are 6 parallel strings. The module features three bypass diodes, each in a separate junction box in the center of the module and connected anti-parallel to two double strings. In our experiment we shade three solar cells denoted cell 1L, 1R and 69L as indicated in Figure 1. We determine the worst-case shading for each solar cell by measuring the current-voltage $I(V)$ characteristics of the module with our class AAA module flasher according to the IEC 61215:2016 MQT09.1 - f) - 2) procedure. We illuminate the solar module in a continuous light sun simulator for 5 h, with the worst-case shading condition applied to each of the three solar cells simultaneously. The irradiance in the module plane is set to the maximum value of 1100 W/m² according to the IEC61215:2016 MQT09 standard. During the test we continuously measure the irradiance with two thermopile pyranometers, the ambient temperature, and the average PV module temperature by applying four thermocouples to the rear side of the module. The temperature of the shaded cells is measured with a calibrated infra-red thermometer. During the test an air condition maintains an average PV module temperature of 50 °C. After the test we remeasure the $I(V)$ characteristics of the module, document the optical changes and make an ultra-violet fluorescence (UVFL) image [9] of the module's front and rear side.

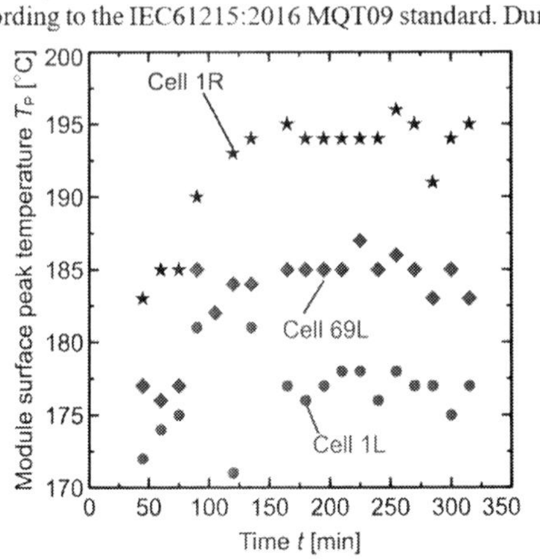

Figure 4: Measured module surface peak temperature during the IEC61215 MQT09 hot-spot endurance test at the position of the three shaded solar cells 1R, 1L, and 69L (cf. Fig 1).

3 EXPERIMENTAL RESULTS

Figure 2 shows the $I(V)$ measurement results of the PV module when cell 1L is shaded with a 1 cm to 7 cm wide cover strip. The inset in Fig. 2 shows a photo indicating how we apply the strips to partially shade the cells of the PV module during the test. The shading strip fully covers the cell in the direction parallel to the long cell edge. From the $I(V)$ measurements we determine the current where the bypass diode is activated I_{BDon}. The point when the bypass diode is activated can be recognized by the kink in the curve. We denote the current at this point as I_{BDon}.

Figure 3 shows I_{BDon} as a function of the shading strip width S for the three examined cells. Symbols indicate the derived values for I_{BDon} and the dashed lines the first order regression to the data. The worst-case shading condition is achieved when I_{BDon} is just below module's current at the maximum power point $I_{mpp,module}$. In this case the current of the shaded cell is highest and thus the dissipated power, before it is limited by the bypass diode. We infer the worst-case shading by linear extrapolation, since for small S the determination of the kink, and hence I_{BDon} is ambiguous.

We apply each shading fraction to the three cells at the same time and illuminate the module in a steady-state sun simulator. Figure 4 shows the module's peak surface temperatures we measure in the vicinity of the shaded cell as a function of the test time. All cells reach a temperature above 175 °C with a maximum peak surface temperature of 196 °C for cell 1R. Table 1 summarizes the shading strip width and the maximum measured module surface peak temperature.

Figure 5 shows the UVFL images from the front and rear side of the module after the hot-spot endurance test.

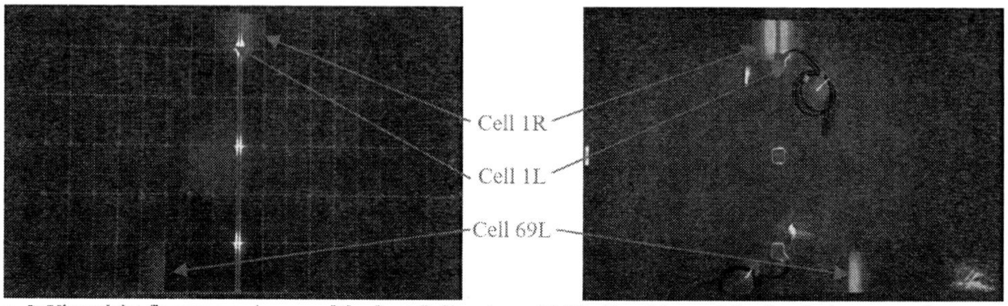

Figure 5: Ultra-violet fluorescence images of the front (left) and rear (right) side of the module after the hot-spot endurance test. The higher fluorescence in the regions of the shaded cells indicates that the encapsulation polymer and backsheet are affected from the hot cells.

The images reveal an increased fluorescence signal at the position of the of the shaded cells, which is most prominent in the image of the module's rear side.

Table 1: Shading strip width and maximum module surface peak temperature

Cell	Shading strip width S [cm]	Maximum surface peak temperature T_P [°C]
1L	2,7	181
1R	3,7	196
69L	3,7	187

Figure 6 shows a photo of the module's rear side after the hot-spot endurance test. In the photo, strong brownish discolorations of the backsheet are visible, which correlate with the positions of the shaded cells.

4 DISCUSSION

During our illumination experiment we measure a module peak surface temperature well above the typical lamination temperature for the PV module manufacturing process of 150 °C to 160 °C. The temperature in the vicinity of all shaded cells exceeds 150 °C after 20 min of constant illumination. After 120 min the temperature of all cells saturates, and the cells are in thermal equilibrium with a maximum temperature of 196 °C measured for cell 1R. This rapid heating of the cells makes the hot cell effect relevant for any kind of permanent shading in the field, such as plant growth near the illuminated front side of the module, soiling by moss, bird droppings or similar kinds of small dirt features on the PV module.

Considering the thermal conductivity of the encapsulation polymer and the backsheet we infer that the temperature at the cell surface is significantly hotter than the measured temperature at the module's surface [4]. The high cell temperature particularly affects the encapsulation polymer, which is in direct contact with the cell surface. If the cell exceeds a critical temperature, this can lead to a decomposition of the polymer. This is visible in the UVFL images in Fig. 5 showing an increased fluorescence signal in the vicinity of the shaded cells. Morlier *et al.* showed that this increased fluorescence correlates with the decomposition of the encapsulation polymers [9]. This decomposition manifests in a yellowing of the polymer, which potentially reduces the module's current and eventually the module output power [9]. Additionally, we could smell the reaction gases of melting polymer after 20 min of testing indicating the beginning of decomposition of the encapsulation and backsheet materials. A reaction of the backsheet is also evident when considering Fig. 6, which shows a brownish discoloration of the backsheet at the positions of the shaded cells. Thus, hot cells may cause cracking of the backsheet, which eventually results in a module failure [10].

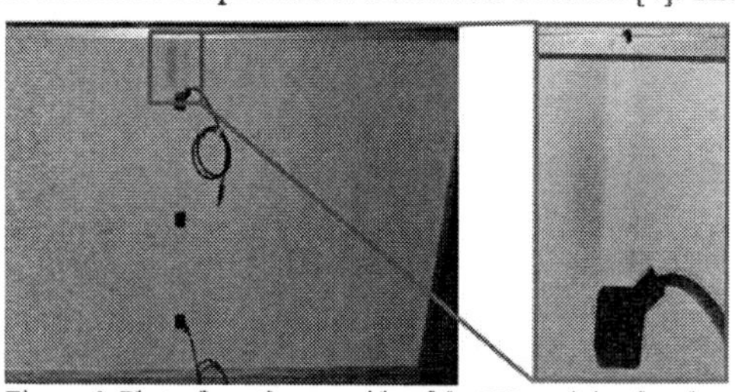

Figure 6: Photo from the rear side of the PV module after the hot-spot endurance test. The hot cells damage the module's backsheet which is visible by the discoloration of the white backsheet in the vicinity the shaded cells. Note, that large fractions of the backsheet areas are affected due to areal heating.

In our previous work we also showed within thermal simulations that for solar cells from larger wafer formats like M10 or M12 can reach even higher module temperatures [4]. This further accelerates the decomposition of the polymers since higher temperatures enhance the reaction kinetics. Furthermore, the critical temperature for polymer decomposition is already reached at lower irradiances, which occur more frequently for modules operating outdoors. Higher temperatures may also result in solder joint failures. The liquidus temperature of Sn63Pb37 solder is at 183 °C and for lead-free solders at around 220 °C. Above this temperature, leaching of the silver contact may occur [11], resulting in solder joint failures. Hot cells may also increase the risk of glass breakage due to temperature gradients across the glass above 150 °C [12].

5 SUMMARY

Shading individual solar cells in a high-power module with M6 wafer based half cells results in hot cells with a maximum peak temperature at the module surface of 196 °C. We measure module surface temperatures of over 181°C in the vicinity of all shaded cell, which is well above the typical PV module lamination temperature of max. 160 °C. Visual inspections and UVFL images indicate that the hot cells damage the module encapsulation materials. This may result in a yellowing of the encapsulation polymer, which eventually reduces the module power output. The strong discoloration of the backsheet indicates that it is also damaged by the hot cells. Considering field operations for multiple years this may result a cracking or delamination of the backsheet [13], [14], which will in the long run end in a module failure.

Our results suggest that the hot cell effect will have even more severe consequences on the long-term stability of modules with solar cells made of M10 or M12 wafers with a module power of 500 W_P to 600 W_P. As cell area and module power increase, the power dissipation in a shaded cell increases, leading to even higher module temperatures.

A solution to address the hot cell issue is to employ fewer cells per bypass diode, e.g., using six bypass diodes as in scenario or employing module level power electronics to operate each substring at the individual MPP. Using module materials with an enhanced heat resistivity might be another solution.

6 ACKKNOWLEDGEMENTS

The results were generated in the SUPERPV project (grant agreement no. 792245) funded by the European Union's Horizon 2020 Research and Innovation Program.

6 REFERENCES

[1] S. K. Chunduri and M. Schmela, 'Advanced Solar Module Technology 2021 Edition', TaiyangNews, 2021.

[2] 'SNEC 2020: How modules topping 800W and new BIPV innovations stole the show', *PV Tech*. https://www.pv-tech.org/editors-blog/snec-2020-how-modules-topping-800w-and-new-bipv-innovations-stole-show (accessed Jan. 12, 2021).

[3] C. Comes, 'Mitigate Hot-Spot Risks in PV Modules of Very High Output Power', presented at the PV Module Forum 2020, Cologne, Germany, Feb. 18, 2020. [Online]. Available: https://www.tuv.com/landingpage/en/pv-module-forum/

[4] R. Witteck, M. Siebert, S. Blankemeyer, H. Schulte-Huxel, and M. Kontges, 'Three Bypass Diodes Architecture at the Limit', *IEEE J. Photovolt.*, vol. 10, no. 6, pp. 1828–1838, Nov. 2020, doi: 10.1109/jphotov.2020.3021348.

[5] K. Bothe *et al.*, 'Luminescence emission from forward- and reverse-biased multicrystalline silicon solar cells', *J. Appl. Phys.*, vol. 106, no. 10, p. 104510, 2009, doi: 10.1063/1.3256199.

[6] O. Breitenstein *et al.*, 'Understanding junction breakdown in multicrystalline solar cells', *J. Appl. Phys.*, vol. 109, no. 7, p. 071101, 2011, doi: 10.1063/1.3562200.

[7] F. Fertig, S. Rein, M. Schubert, and W. Warta, 'Impact Of Junction Breakdown In Multi-crystalline Silicon Solar Cells On Hot Spot Formation And Module Performance', 2011.

[8] Mechanical Engineering Industry Association (VDMA), *International Technology Roadmap for Photovoltaic*. 2020.

[9] A. Morlier, M. Siebert, I. Kunze, G. Mathiak, and M. Kontges, 'Detecting Photovoltaic Module Failures in the Field During Daytime With Ultraviolet Fluorescence Module Inspection', *IEEE J. Photovolt.*, vol. 7, no. 6, pp. 1710–1716, 2017, doi: 10.1109/jphotov.2017.2756452.

[10] Y. Lyu *et al.*, 'Drivers for the cracking of multilayer polyamide-based backsheets in field photovoltaic modules: In-depth degradation mapping analysis', *Prog. Photovolt. Res. Appl.*, vol. 28, no. 7, pp. 704–716, 2020, doi: 10.1002/pip.3260.

[11] D. Shangguan, *Lead-free solder interconnect reliability*. ASM international, 2005. doi: 10.1108/ssmt.2005.21917bae.005.

[12] Y. Yang, L. Miao, and C. L. Chow, 'Relative significance of temperature gradient components on cracking behavior in glass panes under thermal radiation', *Appl. Therm. Eng.*, vol. 131, pp. 837–848, Feb. 2018, doi: 10.1016/j.applthermaleng.2017.12.060.

[13] O. Stroyuk, C. Buerhop-Lutz, A. Vetter, J. Hauch, and C. J. Brabec, 'Nondestructive characterization of polymeric components of silicon solar modules by near-infrared absorption spectroscopy (NIRA)', *Sol. Energy Mater. Sol. Cells*, vol. 216, p. 110702, Oct. 2020, doi: 10.1016/j.solmat.2020.110702.

[14] D. E. Mansour *et al.*, 'Effect of Backsheet Properties on PV Encapsulant Degradation during Combined Accelerated Aging Tests', *Sustainability*, vol. 12, no. 12, Art. no. 12, Jan. 2020, doi: 10.3390/su12125208.

APPROACHES FOR A LIGHTWEIGHT MODULE WITH LAMINATED MATERIALS

Martin Heinrich[1], Andreas J. Beinert[1], Pascal Romer[1], Li Carlos Rendler[1], Felix Basler[1], Stefan Wendlandt[2], Dirk Holger Neuhaus[1]

[1] Fraunhofer Institute for Solar Energy Systems ISE, Heidenhofstrasse 2, 79110 Freiburg, Germany
[2]PI Photovoltaik-Institut Berlin AG, Wrangelstr. 100, 10997 Berlin, Germany
*corresponding author, e-mail to: martin.heinrich@ise.fraunhofer.de

ABSTRACT: Lightweight photovoltaic (PV) modules are the key technology to access new markets such as industrial buildings and mobile applications. In this work we investigate methods to reduce the weight of a module while still being able to withstand mechanical loads according to IEC 61215. Finite element modelling (FEM) shows that increasing stiffness of the backsheet has more impact on the stress of the solar cells and displacement of the module under mechanical load than increasing the encapsulants stiffness. Therefore, two approaches are evaluated to increase the Young's modulus and hence stiffness: Firstly, laminating glass fiber layer onto the rear side of a module and secondly laminating honeycomb structures into the module. The results show that an increase in Young's modulus by glass fiber layers or honeycomb structures are sufficient to provide a significant rigidity to show a similar displacement during mechanical load tests as conventional module setups. Therefore, embedded glass fibers or honeycomb structures in the backsheet can be used to achieve lightweight modules with thinner or even without front glass. Additionally, backsheets with embedded glass fibers may be a viable option for modules with thinner front glass which currently require a rear glass.

Keywords: Lightweight Modules, Integrated PV, Glasfibre, composite materials, mechanical load, digital prototyping, mechanical FEM

1 INTRODUCTION

Lightweight modules are crucial for a wide variety of potential applications ranging from industrial buildings with a limited roof load to applications on commercial vehicles where the potential payload needs to be maximized [1–6]. Further applications include aircrafts or temporary buildings where the load-bearing capacity is optimized for an overall lightweight structure [7]. Additionally, glass thickness of the front glass by standard modules is aimed to be reduced to (e.g.) reduce optical losses [8]. This can lead to a lighter module but usually a rear glass instead of a backsheet is used in those cases to provide sufficient mechanical stability.

In this work we investigate the influence of the encapsulant and rear of the module on the mechanical stability. The aim is to reduce the weight of the PV module by decreasing the thickness of the front glass in a glass-foil PV module and avoiding the need of a frame for stability or even replacing the front glass with polymeric materials. Therefore, we perform FEM simulations to investigate potential improvements of the mechanical setup of the module. We identify the Young's modulus, i.e. the stiffness of the rear sheet as one major influencing factor as opposed to e.g. the Young's modulus of the encapsulant. To increase the effective Young's modulus of the rear sheet we laminate different types of glass fiber layers onto the rear side of a laminate and perform mechanical load tests. To even further increase the stiffness, we investigate the behavior of different honeycomb structures during lamination. Suitable honeycomb structures are identified and used for module manufacturing and mechanical load testing.

2 FEM Simulations

We performed FEM simulations of a glass foil laminate with a 2 mm glass and 350 μm backsheet to analyze the effect of different Young's moduli for the encapsulant material and backsheet material.

The FEM model used is based on previous studies considering large deflections [9, 10]. For simplicity, the metallization and ribbons are not considered in the FEM model. Figure 1 shows a schematic cross section of the simulated material stack.

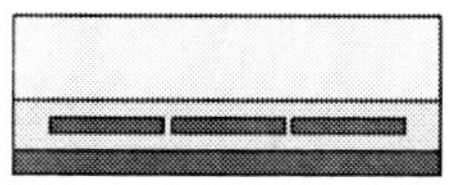

Figure 1 Cross section of the model for FEM simulations.

The mechanical load simulations consider clamping of the module at fixed locations at 20% from the edge of the long module side (see Fig. 2). Since the module is symmetrical only a quarter of the module is simulated as shown in Fig. 2.

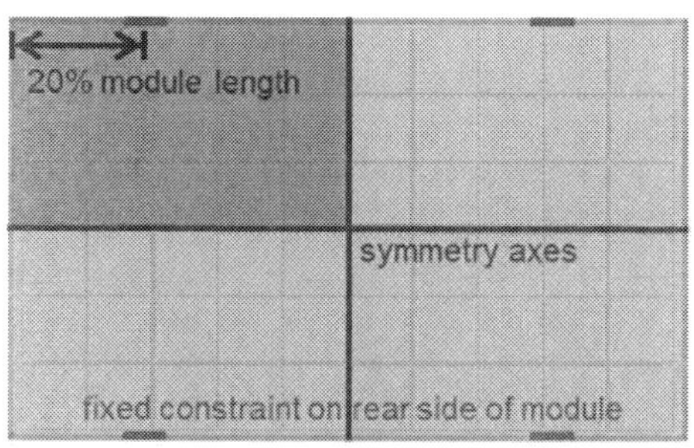

Figure 2 Simulation setup of module fixture and module area (marked with green mask), adapted from [9].

The simulations are than performed with different equivalent loads of 2400 Pa and the stresses for each load are provided as the max. first principle stress σ_1 in MPa, which is the tensile stress equivalent.

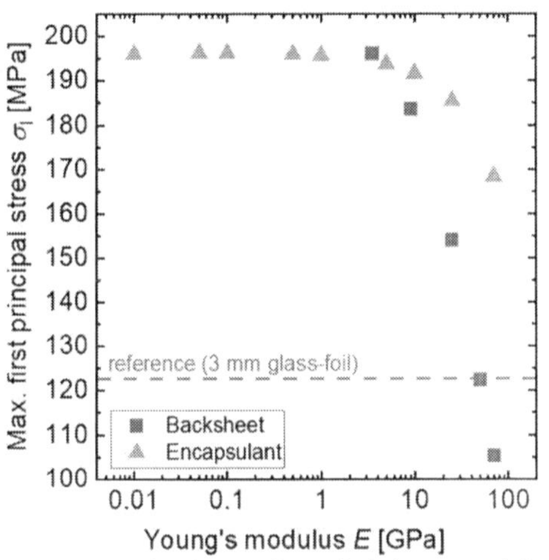

Figure 3 Simulated max. first principle stress of the solar cell depending on the Young's modulus for backsheet and encapsulant at 2400 Pa. The first principle stress of a 3 mm glass-foil module is referenced as grey line.

Figure 3 shows simulations of a setup with a 2.1 mm glass as front sheet and different Young's modulus of the backsheet and encapsulant material. Only for very high values of the encapsulants Young's modulus the max. first principal stress of the solar cells is decreasing. For backsheets the stress of the solar cell decreases almost linearly with increasing the Young's modulus of the backsheet. For comparison the first principle stress of a standard module with 3 mm glass is shown as well. Therefore, when the backsheet Young's modulus is above 50 GPa, a 3 mm glass could be replaced by a 2.1 mm glass.

3 EXPERIMENTAL REALIZATION

3.1 Approaches

To achieve a high Young's modulus of the backsheet two approaches were investigated: Firstly, glass fiber layers were laminated onto the rear of the backsheet and secondly honeycomb structures were laminated onto the rear of the backsheets. Figure 4 shows the two approaches.

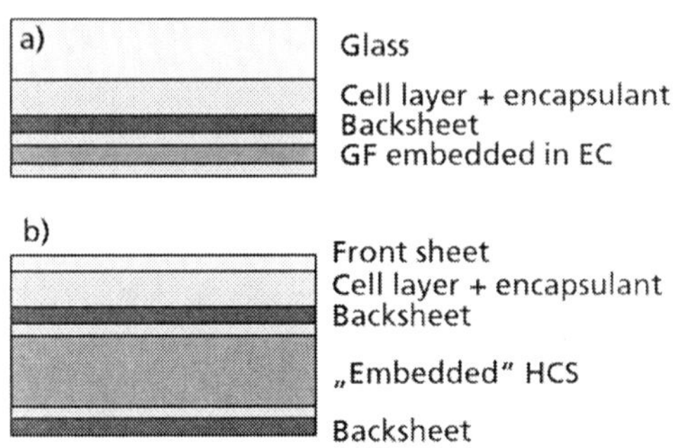

Figure 4 Lamination setups of the two approaches: a) glass fiber layer (GF) embedded into encapsulation material (EC) laminated onto the rear of the backsheet and b) honeycomb structure (HCS) embedded between backsheets (BS) laminated onto the rear of a backsheet.

For approach a) we investigated woven or stacked glass fiber materials with a glass fiber density between 163 g/m² and 320 g/m². For approach b) we compare different materials of honeycomb structures in particular Aluminum, aramid, and polypropylene with a height of 10 mm.

3.2 Methods

Lamination of the modules was done with a Ypsator by Bürkle, Germany with lamination temperature of 155 °C for 12 min at a pressure of 900 mbar. We used a one-process approach in which the glass fibers and honeycomb structures were attached to the module during the lamination process. For the glass, solar grade glass with a thickness of 2.1 mm or 3.2 mm was used, as encapsulant we used a 450 µm EVA and as backsheet we used a 300 µm thick, PET-based backsheet and as polymer front sheet we used a 300 µm thick also PET-based transparent front sheet.

Every setup was evaluated optically after lamination and peel tests were performed on the honeycomb structures to ensure a good adhesion. The adhesion force was tested with a setup manufactured by PSE AG, Germany, which pulls the sample at a 90° angle from each other. The samples for the pull tests were prepared such as to measure the interface between the encapsulant and the honeycomb structure with a width of 10 mm over a length of at least 100 mm.

The measurements of the displacement under mechanical load were done directly on the unframed laminates with three clamps on the long module side with 20% off of each edge and one in the center (see Fig. 5). The cups enforce a local pressure onto the module which evens out to a homogeneous pressure load. The own weight of the cups translates to a load of 500 Pa but they can be lifted to apply also a lower pressure. The tests have been done until a slipping of the module occurs, i.e. the module slips out of the clamps during the load test, or until 2400 Pa as specified in the mechanical load test of IEC 61215.

Figure 5 Measurement setup for mechanical load test with the clamped laminates.

4 RESULTS

4.1 Pretests of glass fiber layers

Figure 6 Stacked glass fiber material (left) and woven glass fiber material (right).

For manufacturing of PV modules with glass fibers, different materials have been tested: 1. Stacked glass fibers, where several fiber layers are stacked on top of each other, 2. Woven glass fibers, here the glass fibers are woven into each other and 3. A stack of several stacked glass fibers with EVA between each glass fiber material. (Fig. 6 shows images of the individual glass fiber materials). The initial tests showed that lamination was possible and a good adhesion between the glass fibers and the encapsulation material was formed.

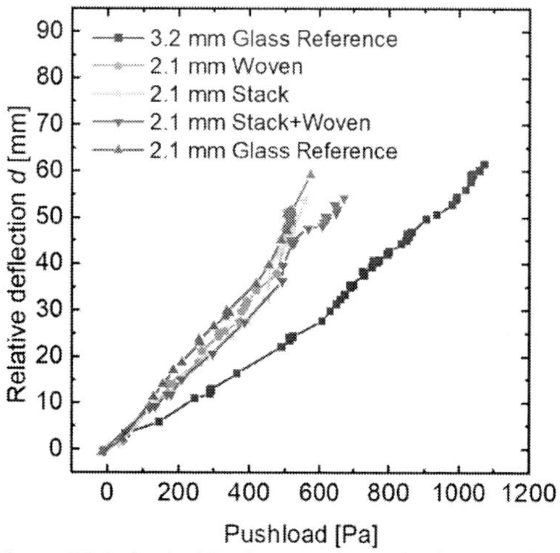

Figure 7 Mechanical load test results on laminates with different glass fiber layers in comparison with a 3.2 mm reference glass and a 2.1 mm reference glass.

The mechanical load tests on the glass fiber layer laminates showed that only a slight improvement regarding the 2.1 mm reference was possible and the overall displacement was very similar to the displacement of the 2.1 mm reference sample. The 3.2 mm reference samples showed a significantly lower displacement. Comparing the different glass fibers with each other, the stacking of a woven and stacked glass fiber layer shows the lowest deflection. Additionally, the difference between woven and stacked fiber layer was only very little. Therefore, further tests were only performed with a stacked fiber layer and a stack of several stacked fiber layers.

4.1 Pretests of honeycomb structures

For the pretests we investigated a metallic honeycomb structure made of aluminum, and two organic structures made of aramid and polypropylene, respectively. Aramid is a synthetic fiber material used in aerospace applications for its lightweight but also rigid structure. Lamination tests on polypropylene resulted in some slight deforming of the structure since the material was only on the edge of withstanding the temperatures present during lamination. For Aluminum and Aramid no visual damage of the structures after lamination was observed.

Table 1: Peel test results of the different materials.

Material	Peel force [N/cm]	Standard deviation [N/cm]
Aluminum	32	±2
Aramid	11	±10
Polypropylene	16	±10

The peel tests in Tab. 2 show that only with Aluminum a sufficient and reproduceable adhesion with our encapsulation material was possible. Therefore, further tests have been performed with the Aluminum honeycomb structure.

4.3 Mechanical load tests

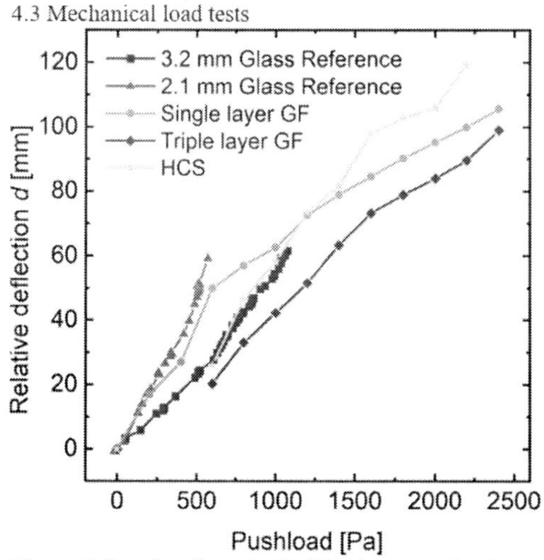

Figure 8 Results of mechanical load tests on laminates with single layer stacked glass fiber (GF), triple layer stacked GF, and Aluminum honeycomb structure (HCS) as well as reference laminates with a standard 2.1 mm and 3.2 mm front glass.

The results of the mechanical load tests of the different samples with the references are shown in Fig. 8. As discussed in Sec. 4.1 the single glass fiber layer does not show a significant improvement in the displacement compared to the reference without glass fiber layer. However, the results of three glass fiber layers stacked at the backsheet show an even lower deflection as the 3.2 mm reference sample. Therefore, a sufficient glass fiber layer setup embedded into the backsheet may be able to replace a rear glass when using thinner front glass in terms of mechanical stability.

Table 2: Weight of the different samples.

Material	Weight in [kg]
3.2 mm glass reference	15
2.1 mm glass reference	11
Single layer glass fiber	12
Triple layer glass fiber	15
Honeycomb structure	5

The honeycomb structure shows a similar deflection under mechanical load as the 3.2 mm reference sample. Comparing the weights of the sample as shown in Tab. 2, it is obvious that a lightweight setup can only be achieve, when moving to very thin front sheets, such as polymeric foils. Since the honeycomb structure is able to maintain a similar deflection under mechanical load, it can be deduced that even an extremely reduced stiffness of the front sheet can be compensated by an increased stiffness of the backsheet. On the other hand, a significant weight reduction may not be achieved when just using glass fiber layers. However, it should be noted, that if the glass fibers would be integrated into the backsheet directly, rather than laminating it on its rear side, a much lower weight of the modules could be achieved than shown in Tab. 2.

5 CONCLUSIONS

In this work we investigated the alteration of the stiffness (i.e. the Young's modulus) of the encapsulant and backsheet of a solar module to achieve a light weight setup in FEM simulations and experiments. The FEM simulations revealed that an alteration of the encapsulant stiffness may not be sufficient to increase the mechanical stability of a module. However, when using a backsheet with a Young's modulus of above 50 GPa it is possible to achieve similar stresses in the solar cell at 2400 Pa mechanical load when using a 2.1 mm front glass compared to a module with a 3.2 mm front glass and standard backsheet.

The increased stiffness of the backsheet was achieved in experiments applying glass fiber layers or honeycomb structures onto the backsheet. The mechanical load tests show that with a honeycomb structure a similar deflection under mechanical load can be achieved when using a 1 mm thick Aluminum honeycomb structure with a 300 µm thick polymeric front sheet compared to a reference laminate with 3.2 mm front glass and standard backsheet. The weight of the honeycomb laminate is significantly lower (3 times lower) than the weight of the 3.2 mm reference laminate.

A stack of three glass fiber layers embedded onto the rear of a laminate with 2.1 mm front glass shows even a lower deflection and therefore a higher mechanical stability than the 3.2 mm reference laminate. Therefore, embedded glass fiber layers into the backsheet may be a viable alternative to the rear glass which is usually used when using 2.1 mm front glass in terms of mechanical stability. However, a significant weight reduction may not be achieved in comparison to the 3.2 mm reference laminate.

6 ACKNOWLEDGEMENTS

The authors thank all colleagues from Fraunhofer ISE who contributed to this work. This work was supported by the German Federal Ministry for Economic Affairs and Energy (BMWi) under the contract number 03SBE0005B.

6 REFERENCES

[1] T. E. Kuhn, C. Erban, M. Heinrich, J. Eisenlohr, F. Ensslen, and D. H. Neuhaus, "Review of Technological Design Options for Building Integrated Photovoltaics (BIPV)," *accepted for publication in the Energy and Buildings special issue review articles from the editors*, 2020.

[2] A. C. Martins, V. Chapuis, A. Virtuani, and C. Ballif, "Robust Glass-Free Lightweight Photovoltaic Modules With Improved Resistance to Mechanical Loads and Impact," *IEEE J. Photovoltaics*, pp. 1–7, 2018, doi: 10.1109/JPHOTOV.2018.2876934.

[3] A. C. Oliveira Martins, V. Chapuis, A. Virtuani, L.-E. Perret-Aebi, and C. Ballif, "Hail Resistance of Composite-Based Glass-Free Lightweight Modules for Building Integrated Photovoltaics Applications," 2017, doi: 10.4229/EUPVSEC20172017-6BV.3.62.

[4] C. Ballif, A. Virtuani, and F. Lisco, "Optimisation of the Frontsheet Encaspulant for Increased Resistance of Lightweight Glass-Free Solar PV Modules," *37th European Photovoltaic Solar Energy Conference and Exhibition*, pp. 777–783, 2020, doi: 10.4229/EUPVSEC20202020-4BO.11.5.

[5] Jonathan Govaerts, Tom Borgers, Arvid van der Heide, "ON THE ROAD TOWARDS VEHICLE INTEGRATION: GLASS-FIBRE REINFORCED ENCAPSULATION ENABLING LIGHT-WEIGHT AND CURVED MODULES," *37th EUPVSEC European PV Solar Energy Conference and Exhibition, 7-11 September 2020, Lisbon, Portugal.*

[6] C. Kutter, F. Basler, D. H. Neuhaus, M. Heinrich, J. Markert, and L. E. Alanis, "Integrated Lightweight, Glass-Free PV Module Technology for Box Bodies of Commercial Trucks," *37th European Photovoltaic Solar Energy Conference and Exhibition*, pp. 1711–1718, 2020, doi: 10.4229/EUPVSEC20202020-6DO.11.6.

[7] J. M. Vega de Seoane *et al.*, "Versatile & Lightweight Transparent Composite Technology for Advanced BIPV Architectural Solutions," in *Proc. of 35th European Photovoltaic Solar Energy Conference and Exhibition*, Brussels, Belgium, 2018. Accessed: Jan. 14 2020.

[8] ITRPV, "International Technology Roadmap for Photovoltaic (ITRPV): 2020 Results, April 2021," VDMA Photovoltaic Equipment, Frankfurt 12, Apr. 2021. Accessed: May 10 2021.

[9] A. J. Beinert, P. Romer, M. Heinrich, M. Mittag, J. Aktaa, and H. Neuhaus, "The Effect of Cell and Module Dimensions on Thermomechanical Stress in PV Modules," *IEEE J. Photovoltaics*, vol. 10, no. 1, pp. 70–77, 2020, doi: 10.1109/JPHOTOV.2019.2949875.

[10] P. Romer, G. Oreski, A. J. Beinert, D. H. Neuhaus, and M. Mittag, "More Realistic Consideration of Backsheets Coefficient of Thermal Expansion on Thermomechanics of PV Modules," in *Proceedings of the 37th European Photovoltaic Solar Energy Conference and Exhibition (EU PVSEC)*, online, 2020, pp. 772–776.

SHINGLED SOLAR MODULE FOR BIPV APPLICATION

Junjie Zhu*, Nathan Roosloot, Gaute Otnes, Sean Erik Foss
Institute for Energy Technology
Instituttveien 18, 2007 Kjeller, Norway

ABSTRACT: Building Integrated Photovoltaics (BIPV) is a potential solution for zero emission buildings. Different aspects ranging from aesthetics, yield, safety, cost, etc. need to be further developed to meet the requirement from the market for further growth. Shingle module technology is a very good approach for BIPV application due to the advantage on both aesthetics and energy output. However, the reliability of the shingle modules is an issue making the PV industry hesitant. Here, the shingle modules fabricated with two different kinds of adhesives were investigated. The electroluminescence image and current-voltage measurement shows that all the modules passed 1000 thermal cycles from -40°C to+85°C without failure.
Keywords: Building Integrated PV (BIPV), Shingle, Stability

1 INTRODUCTION

Solar energy is one of the most promising forms of renewable energy, which has developed greatly in the recent decade. Building Integrated Photovoltaics (BIPV), in which the solar module is integrated into the building envelope to replace conventional building materials, is a trend for zero emission buildings. Thanks to the technology development and cost reduction, the interest in BIPV products installation is increasing.

Shingled module technology is using a concept similar as shingle tiles on the rooftop, where a sliced cell is placed on top of the next one, with an overlap at the busbar area and connected by an electrically conductive adhesive [1]. It therefore removes the need for interconnection ribbons and eliminates the spacing between the cells. It can not only improve the aesthetic aspect but also power output for more area usage. In addition, the sliced cell will generate less current compared to the traditional module, as a result, the resistive power loss can be also reduced and the risk of hot spot due to shading can also be minimized. Therefore, the shingle module technology can be the most promising choice for BIPV application. However, the shingle module development is not as fast as expected. The earliest shingle module concept was published in 1950s [2], it started to be commercialized for mass production just few years ago. Even though, the concern about shingle module stability is still there in the PV industry since the adhesive used for interconnection is mostly based on polymer material. In addition, BIPV modules will experience higher operating temperatures compared to open-rack mounted modules due to the lack of sufficient ventilation [3], which will place higher stresses on the interconnects.

In the present work, BIPV modules with shingled interconnection were fabricated with two different conductive adhesives. The performance of these modules was investigated especially by extended thermal cycle testing in a climate chamber, with intermediate characterization steps.

2 EXPERIMENTAL

The passivated emitter and rear cell (PERC) architecture was used for the shingled modules. The 6' PERC solar cell (P_{max}=5.2W) was cut into 5 strips. Screen printable conductive adhesives from two

* Email: junjie.zhu@ife.no, Tel: +47-9947-4138

different suppliers were applied for the interconnection. The following modules with those two adhesives are marked as 'series A' and 'series B'. After stacking and curing, two strings with 10 strips each was interconnected by a special designed ribbon to avoid cell cracking due to thermal stresses introduced during soldering. Polyolefin elastomer (POE) was used as the encapsulant material. A sandwich structure of Glass(4mm)/POE/Cell/POE/Glass(4mm) was used for the module fabrication. Five modules of each conductive adhesive were fabricated.

Electroluminescence (EL) and Current-Voltage (IV) under 1000 W/m^2 illumination were used as the main tools to characterize the electrical performance of the modules. In addition, the adhesion of the interconnection between the cells with different adhesives was measured using a peel off machine. Based on the IEC 61215, the thermal stability of the modules was evaluated in a climate chamber with thermal cycling (TC) temperatures ranging from -40 °C to +85 °C. 1000 thermal cycles was carried out. More details on the extended thermal cycling experiments can be found in ref [4]. During the positive temperature slope, a current equal to the I_{sc} of the modules was applied, while at all other moments, a current equal to 1% of the I_{sc} was applied to monitor circuit integrity. Before putting the modules into the climate chamber for TC testing, they were stabilized by current soaking in the dark for 40 h at I_{sc}. One of the B series modules was kept stored in the dark outside the climate chamber, as a reference for IV measurement.

3 RESULTS

Fig.1 show a picture of one of the shingle modules. The lack of busbars and interconnect ribbons between the cells gives a homogeneous appearance. Although the ribbon for string interconnection can still be seen here, it can be easily hidden or covered with coloured tape. Therefore, compared to a traditional module, it can be better coordinated with different colour elements as Poly Vinyl Butyral (PVB) or glass. Due to the uniformity of its colour, the module can thus be an excellent choice for BIPV applications, meeting aesthetic requests from architects.

Fig. 1 Picture of a shingle module which is suitable for BIPV visual aspect

The average STC maximum power (P_{max}) with standard deviation for Module A and B series is 17.91 ± 0,06 W and 18.06 ± 0,04 W, respectively. This corresponds to a cell-to-module loss of 13-14%, which is higher than expected. We attribute this at least partly to areas of sub-optimal contacting by the conductive adhesive (as can be observed in the EL-image in Fig.3), most likely due to processing issues. These are defects that should be avoidable in an optimized factory environment. In addition, there is some uncertainty in the absolute power value given by our solar simulator, as we do not have a certified reference module which is matched in size and bill of material to the tested devices. Comparing the two module types the average power of module A series is ~0.15 W lower than that of Module B. Although both V_{oc}, I_{sc} is slightly higher for Modules B, the contribution from FF due to less series resistance (R_s) is significant. The R_s in

Module B on average is ~3.2% less than that of Module A.

Although the R_s in Module B is less than Module A leading to higher conductivity, the adhesion between the strips is lower. The peel off tests shows that an average of 9.9±0.7 N/mm^2 and 7.1±0.5N/mm^2 is obtained for module A and B, respectively.

Fig.2 show the plot of electrical performance before TC, after TC_{200}, TC_{500}, TC_{800} and TC_{1000}. It can be seen that most of the modules start to degrade after TC_{500} and some modules from series B start to degrade after TC_{800}.However, it is clear that even after TC_{1000}, these modules have no problem meeting the requirements in IEC61215 of less than 5% reduction in P_{max}. Actually, all modules show a small increase in P_{max} after TCs. This is also reported by others with conductive adhesive used for shingled modules [5,6].

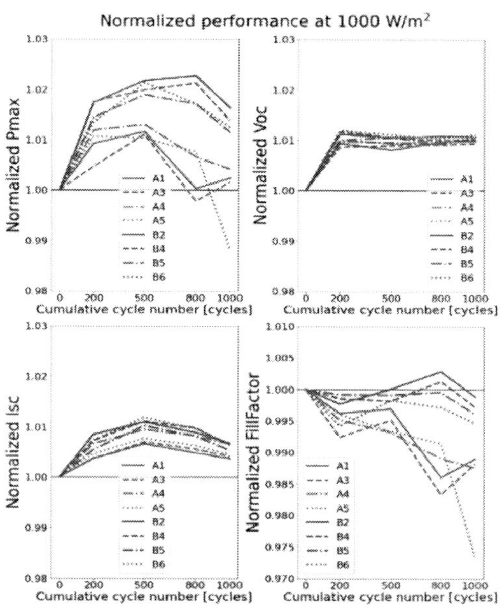

Fig. 2 Normalized IV results for modules before and after TC_{1000}

Fig.3 shows the typical EL image for the modules from module A and B series before and after TC_{1000}. The difference between the area with and without conductive adhesive connects the busbar pads of adjacent cells is obvious. The interconnection area where the conductive

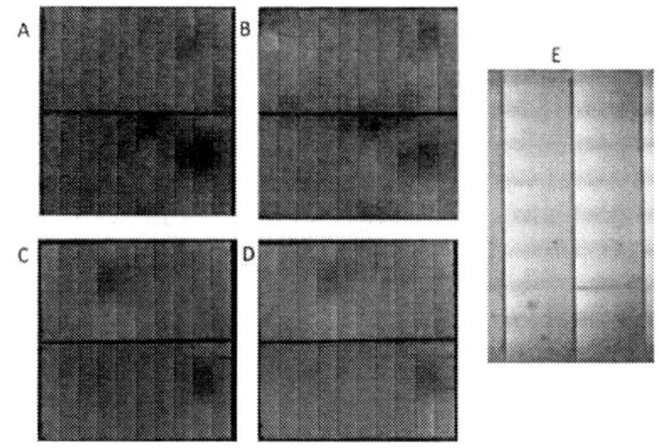

Fig. 3 Typical EL images for module-type A before TC (A) and after TC_{1000} (B); module-type B before TC (C) and after TC_{1000} (D); Magnification of interconnection area (E)

adhesive is applied shows more bright in the EL (showed more clearly in Fig. 3E). It may give an indication that the electrical performance can be improved if the conductive adhesive between strips can be distributed continusely along the busbar. In addition, some local defects are detected before TC, which may be attributed to bad interconnections in some strips due to the lab process. Compared to the initial state of the modules, the EL show no indications of degradation after TC_{1000}, which is in accordance with the IV results.

4 SUMMARY

Shingled modules with two different electrically conductive adhesives have been fabricated. The color of the cell areas in the module is homogeneous, which makes it a promising interconnection technology for BIPV applications, where aesthetics is important. Avoiding space between cells should also allow for higher power output per area. The two kinds of shingled modules show stable electrical performance without significant degradation after 1000 thermal cycles. These are promising results for BIPV applications.

5 REFERENCE

[1] G. Beaucarne, Materials challenge for shingled cells interconnection, *Energy Procedia* 98 (2016) 115.

[2] D.M. Chapin, C.S.Fuller, , G.L. Pearson, A new silicon p-n junction photocell for converting solar radiation into electrical power, *J. Appl. Phys.* 25 (1954), 676.

[3] M.Koehl, S. Hamperl, M. Heck, Effect of thermal insulation of the back side of PV modules on the module temperature. *Prog. Photovolt: Res. Appl, 24 (2016), 1194.*

[4] N Roosloot, J Zhu, SE Foss, G Otnes, Extended Thermal Cycling of Shingled Cell Interconnects. *IEEE 48th Photovoltaic Specialists Conference (PVSC)*, (2021), 2328.

[5] L. Theunissen, et al., Electrically Conductive Adhesives as Cell Interconnection Material in Shingled Module Technology, *AIP Conference Proceedings* 1999 (2018), 080003.

[6] D. Tonini et al., Shingling technology for cell interconnection: Technology aspects and process integration, *Energy Procedia*, 150 (2018) 36.

ACKNOWLEDGMENT

This project has received funding from the European Union's Horizon 2020 Research and Innovation Programme under grant agreement 818009. The information reflects only the author's view and the Commission is not responsible for any use that may be made of the information it contains.

SIMULATION TOOL FOR THE PERFORMANCE OPTIMIZATION OF COLORED PV MODULES

Charlotte Pfau[1], Abbas Raza Bangash[1,2], Christian Hagendorf[1], Marko Turek[1]
[1] Fraunhofer Center for Silicon Photovoltaics, Otto-Eißfeldt-Straße 12, 06120 Halle (Saale), Germany
[2] Anhalt University of Applied Sciences, D-06366 Köthen (Anhalt), Germany

ABSTRACT: Building integrated module technologies are expected to meet a growing market in the next years. In contrast to large-scale power plants, the optical appearance of the modules is a major criterion in addition to energy output for their success as building components. In particular, PV modules for architecture, display facades or urban noise protection may be individually designed by colored patterns based on ink-jet printing technologies. In our work, we set up an optical simulation tool that calculates energy output of colored PV modules from the optical properties of the bare printing inks. The model is experimentally tested using common inkjet inks for CMYK+W digital printing on glass substrates. Optical spectroscopy on specially prepared ink samples are implemented in order to provide the required transmission and reflection coefficients as input parameters. Simulated efficiency predictions can be performed for colored PV modules with various ink multilayer stacks and thus for different possible color shades obtained by subtractive color mixing. In case of a double-print-layer including white ink, we can improve the power loss estimation from about 20% obtained by a simple approach to about 5% accuracy by considering individual optical losses and gains in the layer stack.
Keywords: Optical Losses, Optical Properties, Simulation, Colored PV modules, Building Integrated PV(BIPV)

1 INTRODUCTION: MOTIVATION AND RELEVANCE

Improved aesthetics through flexibility in color design is one of the key criteria to increase the value and acceptance of BIPV and is thus an important contribution to its wide-spread distribution [1]. One common way of producing colored solar modules is digital printing on module front glasses [2]. From an architectural point of view, a vast range of color shades of BIPV modules is required. The respective color shade is created by printing ink colors on top of each other in printing grids. However, color-printing PV modules also strongly impacts the module performance by modifying its quantum efficiency. Thus, a reliable and quantitative prediction of the interplay between complex color shades based on multi-ink layers and module power has to be established. However, due to the vast number of possible color shades obtained by combination of the basic colors, it is practically unfeasible and uneconomical to determine the module efficiency experimentally in advance for each single-color shade. Therefore, we have set up a new experimental approach combined with a calculation method which enables quantitative efficiency predictions for various color shades. In this way, the optical properties of the basic inks as input parameters need to be determined only once. In this paper, we consider translucent inks typically used in digital printing for subtractive CMYK color mixing, supplemented with a white ink, which is necessary to make the translucent print colors visible on the dark cell background. In a first evaluation step of the approach, three types of single-cell solar modules with front glasses printed with different fully covering ink layers were fabricated and electrically characterized regarding their performance and quantum efficiency. We show for a module with single and double print layers including white, that we can achieve a prediction of the module's current losses with a stepwise improved accuracy.

2 APPROACH AND INNOVATION

Our approach takes the spectral dependent optical losses and gains in the module layer stacks well as the

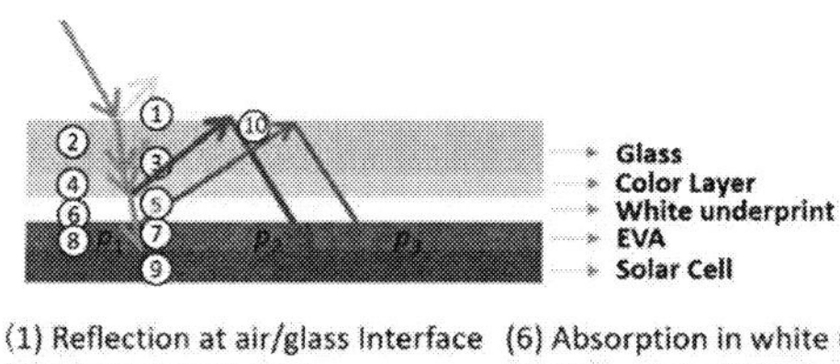

(1) Reflection at air/glass Interface
(2) Absorption in glass
(3) Reflection at glass/color Interface
(4) Absorption in color layer
(5) Reflection at color/white underprint interface
(6) Absorption in white underprint
(7) Reflection at EVA/white underprint interface
(8) Absorption in EVA
(9) Reflection at cell/EVA interface
(10) Internal total reflection and subsequent light redirection gains

Figure 1: Individual optical losses and gains in the material stack of a color printed solar module

cell's internal quantum efficiency IQE into account. The J_{sc} of the mini-module under AM1.5G illumination is calculated in a geometrical optics approach via

$$J_{sc,module} = q * \int_{1200}^{280} AM1.5 * IQE * \sum_n p_n * d\lambda \quad (1),$$

where q is elementary charge and $AM1.5$ the photon flux of the AM1.5G sun spectrum [3]. The p_ns are the light intensity fractions of the different possible direct and indirect light paths into the cell. The p_ns are calculated from the reflection and transmission coefficients R_i and T_l, respectively, determined for the individual layers l & interfaces i of the stack from optical spectroscopy on module components.

Fig. 1 schematically illustrates the different relevant light paths with optical losses and gains in a module stack with two ink layers. The fraction of light reaching the solar cell is basically determined by the optical losses along the possible direct light paths into the cell, which is indicated as p_1 in Fig. 1. Internal reflections can guide the light indirectly to the cell leading to further light paths, exemplary indicated as p_2 and p_3 in Fig. 1. It should be noted that in the case of halftone printing, several layer stacks are present next to each other, containing different and sometimes no ink layers (unprinted areas). In this case, the number of p_ns in Eq. (1) increases enormously, all the

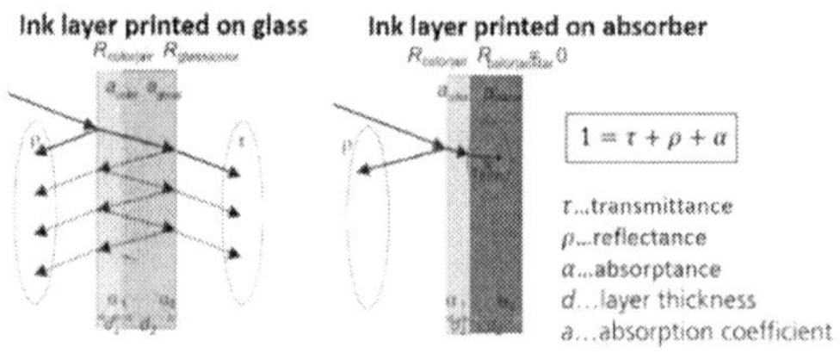

Figure 2: Optical conditions of ink on glass and on an absorber yielding reflection suppression on the backside.

light fractions p_ns must then be calculated individually from the respective area ratios of the printed and unprinted areas by considering the whole system of different layer stacks.

The required reflection and transmission coefficients for glass, EVA and the cell are determined with the modified Fresnel equations for vertical light incidence in analogy to the method described in [4] using hemispherical reflectance and transmittance spectra of the individual module components.

In order to deduce the required optical coefficients of the ink layers we have developed a novel method using special samples with a single ink layer on a spectrally broadband backside absorber for reflection spectroscopy. The special black broadband absorber foil [5] effectively suppresses all reflections from the ink layers backside in the whole relevant spectral range. Thus, the reflectance spectrum of the color layer printed on a black absorber directly corresponds to the reflection coefficient $R_{\text{color-air}}$ of the air-color interface, as illustrated in Fig. 2, enabling for the subsequent determination of T_{color} from the transmittance spectrum of printed glass.

However, the efficiency calculation approach applied here is well suited for transparent materials with negligible scattering [6]. Since this is not true for the white ink, in contrast to all other materials of the module front stack, the white layer is considered as a system whose front and back side interface is an a infinitely thin Lambertian scatterer with a scattered light fraction given by the reflection coefficient determined with the absorber method and whose bulk volume has an absorption coefficient and is scatter-free. Large scattering angles result in total internal reflections at the glass/air interface and thus additional relevant light paths, shown exemplarily in Fig. 1 as p_2 and p_3. These optical gains are considered in the model by adding these light fractions to Eq. (1).

3 EXPERIMENTAL METHODS

The spectral characterization of the inks and module components (optical spectra and IQE) was done only once before one-cell mini modules with different prints were simulated and prepared for electrical characterization. The transmittance and reflectance spectra $\tau(\lambda)$ and $\rho(\lambda)$, respectively, were measured with a *Perkin Elmer Lambda 1050* double beam spectrometer equipped with an integrating sphere. Subsequently, single-cell solar modules with front glasses printed with different ink layers were prepared. Four different types of modules were chosen: one with only a yellow ink layer (Y), one with only a white ink layer (W), one with a yellow plus a white ink layer (Y+W) as well as an unprinted reference (ref). As module materials *Sunpower* solar cells, black

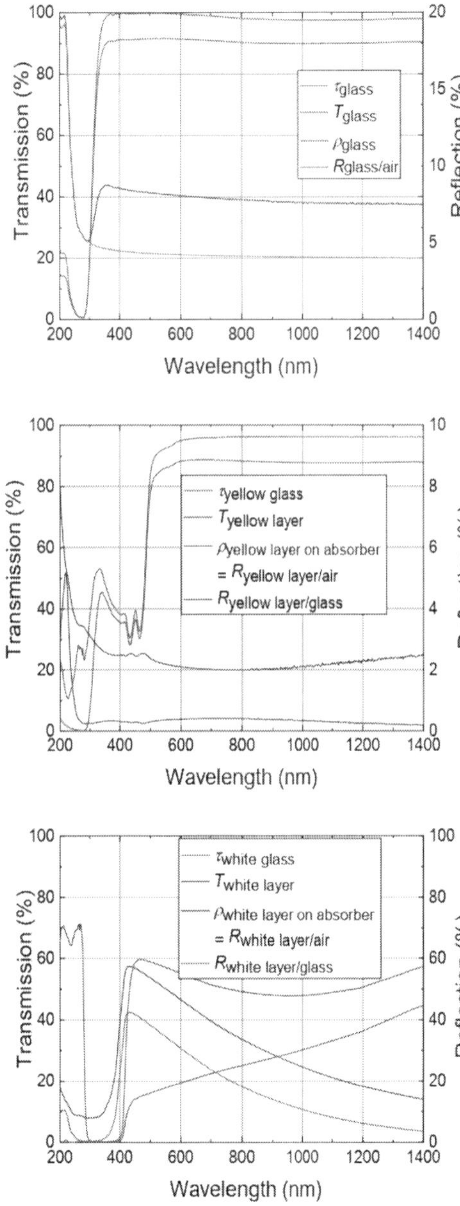

Figure 3: Measured transmittance (τ) and reflectance (ρ) spectra of glass (top) and yellow (middle) and white (bottom) print layer samples and deduced transmission (T) and reflection (R) coefficients.

backsheet, EVA encapsulation foil and 4 mm thick clear float glass were laminated. For printing the backside of the module glasses, commercially available UV curing ink jet inks and a *Handtop HT3200* digital printer with a resolution of 1200 x 600 dpi and a ink drop size of 14 pl were used, whereas the layer thickness was set to 100% (one layer of full coverage). The electrical characterization of the cells and mini-modules (EQE and J_{sc}) was performed with a *LOANA* tool.

4 RESULTS AND CONCLUSIONS

The required coefficients R and T of the single layer

components glass and EVA are determined from their transmittance and reflectance spectra by solving the simplified Fresnel equations [4]. For the cell-EVA interface an effective reflection coefficient is used and estimated in analogy to method described in [4]. Fig. 3a shows the measured hemispherical reflectance and transmittance and the deduced reflection and transmission coefficients, exemplary for an unprinted module glass.

In contrast to the glass and EVA, separate samples, i.e. ink layers, cannot be produced from the ink, as it has to be printed on a substrate. If only printed on a glass, the exact determination of the inks´ reflection coefficient is not possible due to several unknown coefficients of this stack and to multiple reflections as illustrated in Fig. 2 left. Thus, a new ink characterization method is introduced here, using a backside absorber: For the determination of the reflection and transmission coefficients of a ink layer, the reflectance of the ink layer printed on a black broadband absorber foil [5] and the transmittance of the ink printed on glass needs to be measured. The reflectance spectrum of the ink layer printed on a black absorber directly corresponds to the reflection coefficient $R_{ink\text{-}air}$. With known $R_{air\text{-}ink}$, $R_{air\text{-}glass}$ and T_{glass} the reflection coefficient $R_{glass\text{-}ink}$ and subsequently the transmission coefficient T_{ink} can be calculated from the transmittance spectrum of the printed glass [7]. In Fig. 3b and c the corresponding spectra and deduced ink layer coefficients are shown for yellow and white, respectively.

With the derived coefficients of all interfaces and layers in the stack, the fraction p_n of light (or photon flux) reaching can be calculated by multiplying the individual fractional optical losses ((1-R) and T) along the corresponding light passes to the cell. For the considered module layer stack with a yellow and a white print layer p_1 is calculated via:

$$P_1 = \left(1 - R_{glass|air}(\lambda)\right) * \left(T_{glass}(\lambda)\right) * \left(1 - R_{glass|color}(\lambda)\right) * \left(T_{color}(\lambda)\right) * \left(1 - R_{color|whiteunderprint}(\lambda)\right) * \left(T_{white}(\lambda)\right) * \left(1 - R_{white|EVA}(\lambda)\right) * \left(T_{EVA}(\lambda)\right)\left(1 - R_{EVA|cell}(\lambda)\right) \quad (2)$$

Considering only the optical losses along the direct light path to the cell, i.e. by using p_1 in Eq. (1) and ignoring the optical gains p_2 and p_3, yields the stepwise J_{sc} loss under AM1.5G illumination, which is shown in Fig. 3 for the yellow and white printed solar module. The maximum short-circuit current without the optical losses in the module front (i.e. cell-IQE-level J_{sc}) is calculated to be 40.9 mA/cm^2. This current is reduced to 10.1 mA/cm^2 after inclusion of the module components including the two ink layers (module-level J_{sc}) into the model. For this extreme case, Fig. 4 illustrates where the screws need to be tightened to optimize the colored module´s quantum efficiency. In particular, the reflection losses at the white layer should be emphasized. However, the white ink and the associated reflections are necessary to make the translucent CMY inks visible on the dark solar cell background. Note that usually printing grids are used and for the most colors only partially covering print layers are necessary, which increase the efficiency. Furthermore, almost half of the losses could be avoided with white inks that are transparent in the IR. In this work, however, it is of interest how good the prediction accuracy is, which will be discussed in the following.

An interesting quantity for colored modules is the relative short circuit loss compared to an unprinted reference mini module, i.e. $\Delta J_{sc} = 1\text{-}(J_{sc,colored}/J_{sc,\,unprinted})$.

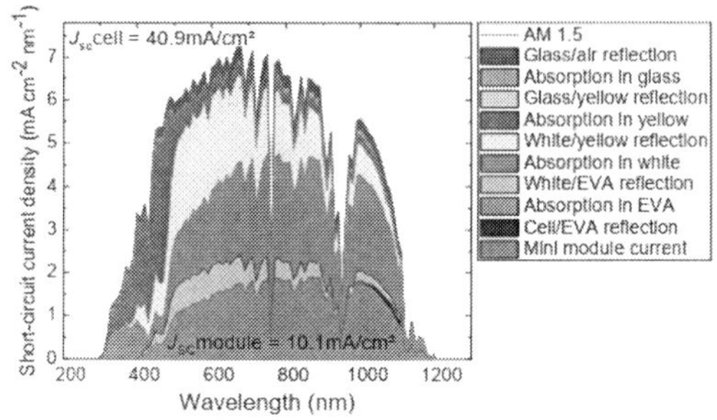

Figure 4: Calculated, spectrally resolved short-circuit-current density from cell-IQE-level to module-stack-level.

With the calculated and measured reference module J_{sc} of 36.3 mA/cm^2 and 36.1 mA/cm^2, respectively, the predicted ΔJ_{sc} of the Y+W printed module is 72%. The measured ΔJ_{sc} for Y+W module, on the other hand, is 64%, which fits to the fact that the optical gains are not considered for the Y+W module.

Due to scattering of the white layer and subsequent total internal reflection at the glass-air interface two other relevant light paths are possible having the fraction p_2 and p_3, being

$$P_2 = \left(1 - R_{glass|air}(\lambda)\right) * \left(T_{glass}(\lambda)\right) * \left(R_{glass|whiteunderprint}(\lambda)\right) * \left(T_{glass}(\lambda)\right) * 0.33 * \left(T_{glass}(\lambda)\right) * \left(1 - R_{glass|whiteunderprint}(\lambda)\right) * \left(T_{white}(\lambda)\right) * \left(1 - R_{white|EVA}(\lambda)\right) * \left(T_{EVE}(\lambda)\right)\left(1 - R_{EVA|cell}(\lambda)\right) \quad (3)$$

$$P_3 = \left(1 - R_{glass|air}(\lambda)\right) * \left(T_{glass}(\lambda)\right) * \left(1 - R_{glass|whiteunderprint}(\lambda)\right) * \left(T_{white}(\lambda)\right) * \left(R_{white|EVA}(\lambda)\right) * \left(T_{white}(\lambda)\right) * \left(1 - R_{glass|whiteunderprint}(\lambda)\right) * \left(T_{glass}(\lambda)\right) * 0.33 * \left(T_{glass}(\lambda)\right) * \left(1 - R_{glass|whiteunderprint}(\lambda)\right) * \left(T_{white}(\lambda)\right) * \left(1 - R_{white|EVA}(\lambda)\right) * \left(T_{EVE}(\lambda)\right)\left(1 - R_{EVA|cell}(\lambda)\right) \quad (4)$$

These equations are valid under the simplified assumption that in each case 33% of the light which reaches the glass-air interface is above its critical angle of total internal reflection (33% due to assumed Cosin-Lambert-Scattering considered in the calculation). In an extended modeling step, the optical gains are taken into account in the calculation of the current, i.e. p_2 and p_3 are no longer neglected in Equ. 1.

In Fig. 4 the ΔJ_{sc}s calculated with three different methods are compared to the measured value: In method 1 the widely used procedure is applied, in which the quantum efficiency is only multiplied with the transmittance of the printed glass to calculate the J_{sc}. In method 2 all the individual optical losses are considered in the J_{sc} calculation (only direct light paths (p_1) in Eq. (1)). In method 2 all the individual optical losses and gains are considered in the J_{sc} calculation (direct and indirect light paths (p_1, p_2 and p_3) in Eq. 1). Compared to the simple conventional procedure, the model which includes the individual losses at the stacks´ layers and interfaces already represents a significant improvement. It can be clearly seen that, we could improve the J_{sc} loss estimation accuracy from about 20% obtained by a typically used simple approach to about 5% by considering individual losses and gains in the layer stack.

Fig. 6 gives an overview of the calculated (method 3) and measured ΔJ_{sc} of all analyzed printed mini modules, whereas in the calculation of the yellow module p_2 and p_3

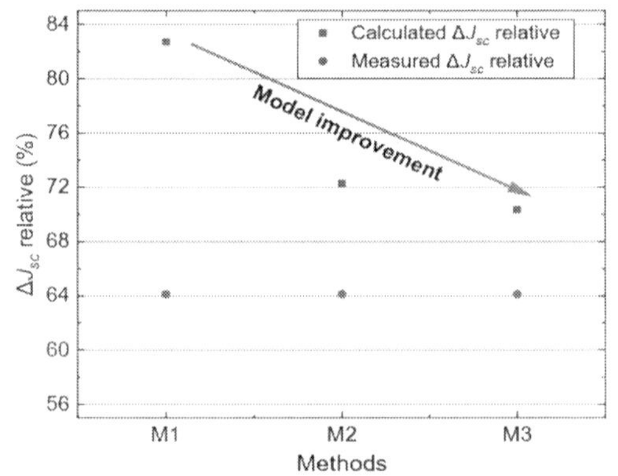

Figure 5: Relative short circuit current losses calculated with 3 different methods and measured values

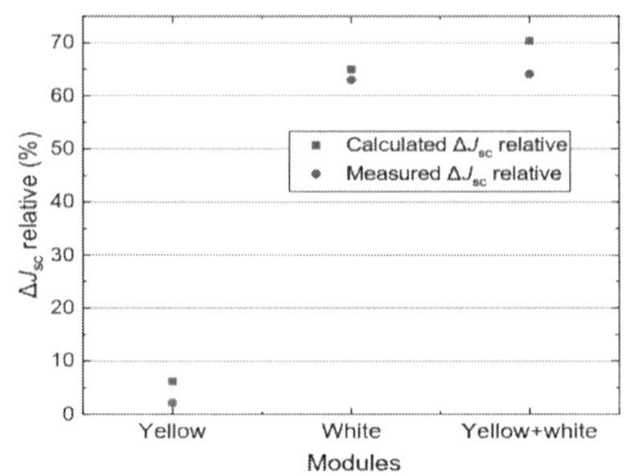

Figure 6: Relative short circuit current losses calculated for different printed modules (yellow module, white module and yellow + white underprint module) and measured values with method 3 including losses and gains.

was zero due to the absence of white ink. The loss estimation accuracy (5% or better) was achieved also for the other modules. The comparison of the modules shows that the white ink basically determines the module's efficiency whereas the translucent yellow ink results only in minor efficiency losses, which can be related to the inks' absorption at about 420 nm where the IQE and sun intensity are relatively low.

5 SUMMARY AND OUTLOOK

We have set up a new experimental approach combined with a calculation model that allows the prediction of a colored module's performance with multiple print layers. It is intended to facilitate module and building designers selecting inks, print patterns and color tones. The longterm goal is to develop a tool which allows for efficiency prediction for modules with arbitrary color shades, created by printing ink colors on top of each other in printing grids, on the basis of optical measurements on single bare ink samples only. For this purpose, a special optical characterization method was developed in which the respective inks are characterized on a spectral broad band absorber foil. In a first model validiation step carried out in this work, different solar modules with up to two fully covering ink layers where fabricated and measured. We could show for the printed modules, that we achieve a prediction of the module's current losses with an accuracy from about 20% obtained by a simple approach to about 5% by considering individual optical losses and gains in the layer stack. Due to the generalized nature of our approach, the model can can be transferred printing grids.

References

[1] G. Eder, G. Peharz, R. Trattnig, P. Bonomo, E. Saretta, F. Frontini, C. S. Polo López, H. R. Wilson, J. Eisenlohr, N. M. Chivelet and S. Karlsson, "COLOURED BIPV Market, Research and Development" IEA International Energy Agency, 2019.

[2] M. H. Saw, J. P. Singh, Y. Wang, K. E. Birgersson and Y. S. Khoo, "Electrical performance study of colored C-Si building-integrated PV Modules" IEEE, vol. 10, no. 4, pp. 1027 - 1034, 2020IEA International Energy Agency, 2019.

[3][Online].Available:https://www.pveducation.org/pvcdrom/appendices/standard-solar-spectra [Accessed 07 09 2021]

[4] H. Hanifi, C. Pfau, M. Turek, J. Schneider, "A practical optical and electrical model to estimate the power losses and quantification of different heat sources in silicon based PV modules" Elsevier, vol. 127, pp. 602-612, 2018.

[5] Klemens Ilse, Charlotte Pfau, Paul Miclea, Stephan Krause and Christian Hagendorf, "Quantification of abrasion-induced ARC transmission losses from reflection spectroscopy," IEEE 46th Photovoltaic Specialists Conference (PVSC), pp. 2883-2888, 2019.

[6] Seifert, Gerhard, Schwedler, Isolde, Schneider, Jens and Wehrspohn, Ralf B., Light Management in Solar Modules, Wiley-VCH Verlag GmbH & Co. KGaA., 2015.

[7] Abbas Raza Bangash, "Colored glass for photovoltaics modules: Investigation of the influence of different printing inks, printing grades on the efficiency and color reproduction in a solar module." [Master's thesis, HS Anhalt, Köthen, Germany], 2021.

LIGHTWEIGHT GLASS-FREE SOLAR MODULES BASED ON POLYCARBONATE AND FIBERGLASS PROTECTIVE SHEETS

S.Yakovlev[1], I.Dmitriev[1], E.Schebet[1], K.Emtsev[1], D.Andronikov[1], D.Orekhov[1], I.Shakhray[2]
[1]R&D Center of Thin Film Technologies in Energetics (TFTE), 194064, St.-Petersburg, Russia
[2]Avelar Solar Technology, 117342, Moscow, Russia

ABSTRACT: The possibility of using polycarbonate and fiberglass sheets for the manufacture of a lightweight module was studied. The stability of the lightweight solar modules assembled from Si-heterojunction solar cells (HJT) and polycarbonate and fiberglass protective sheets in a damp heat test and thermocycling test as per IEC 61215 standard was investigated. Resistance to hail impact of lightweight module with polycarbonate and fiberglass protective frontsheets was compared. The UV stability of polycarbonate and fiberglass sheets was investigated.

Keywords: lightweight module, heterojunction, thermocycling test, damp heat test, UV test, hail test.

1 INTRODUCTION

The main disadvantage of conventional solar modules with glass is their high weight per square. Traditional glass/backsheet modules with a glass thickness of 3.2 mm weigh about 11,5 kg/m^2 and glass/glass with a glass thickness of 2.5 mm – 16 kg/m^2. Such modules have good environmental and mechanical stability, but at the same time, they are not always applicable for BIPV applications due to their large weight. Therefore, there is a need to search for new materials and to develop a lightweight glass-free solar modules. In previous works, the environmental and mechanical stability of semi-flexible modules based on PET protective sheets has already been considered [1-3].

The aim of this work is to study the possibility of using polycarbonate and fiberglass sheets for the manufacture of a lightweight module. The stability of the lightweight solar modules assembled from Si-heterojunction solar cells (HJT) and polycarbonate and fiberglass protective sheets in a damp heat, thermocycling and UV tests as per IEC 61215 standard was investigated. Also the resistance to hail impact of lightweight module with polycarbonate and fiberglass protective frontsheets was compared.

2 EXPERIMENTAL DETAILS

Solar panels for investigation were produced from the HJT solar cells with thickness of 130 mkm connected with Busbar ECA technology. The design of the modules was follow as:

1) Polycarbonate Novattro clear frontsheet and backsheet with thickness 1 mm and two layers of EVA encapsulant with thickness 450 mkm (see Fig.1a);

2) Fiberglass Fibroser STD-FX clear frontsheet with thickness 1 mm, PET-based backsheet with thickness 400 mkm and two layers of EVA encapsulant with thickness 450 mkm (see Fig.1b).

For damp heat, thermocycling and hail test panel with 2x3 cells was fabricated and for UV test – panel with 1x1 cells.

The compatibility of these materials with backsheets based on PET was analyzed. It was found that modules with polycarbonate frontsheet and PET backsheet bend strongly after lamination due to the very different coefficients of thermal expansion, while modules with fiberglass frontsheet did not bend and remain flat on the surface. In addition, it turned out that these materials have good adhesion only to EVA encapsulants, and not TPO/POE.

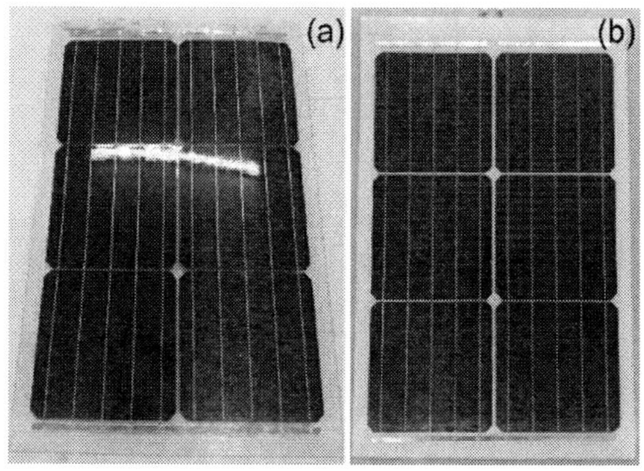

Figure 1: Modules with 2x3 HJT cells for climatic tests: (a) based on polycarbonate Novattro; (b) based on fiberglass Fibroser STD-FX.

Damp heat, thermocycling, UV and hail tests were conducted according to IEC 61215-2:2016. Samples were hung during test. The edges of the modules were not sealed. Samples were characterized by visual inspection, IV testing and EL imaging prior, during and after test.

3 RESULTS AND DISCUSSION

3.1 Hail test

Hail impact test was conducted using in-house developed test system providing shots with 25 mm ice balls at 23 m/sec velocity according to IEC 61215-2:2016. Samples were mounted on a rigid plate using thin double-sided tape. Each sample was hit with a three hail stone. Samples were characterized by IV testing and EL imaging prior and after the hail testing. Results of hail test effect are presented in Table 1 and Figure 2.

Table I: Degradation of electrical characteristics of lightweight panels with 2x3 cells after hail test.

Frontsheet	**Polycarbonate 1 mm**		**Fiberglass 1 mm**	
Test conditions	Before	After	Before	After
Isc, A	8,68	8,68	8,89	8,87
Pmpp, W	29,95	29,48	30,76	30,10
FF, %	78,34	77,21	78,70	77,29
Power degradation, $\Delta P/P_{before}$*100%	~1,6 %		~2,1 %	

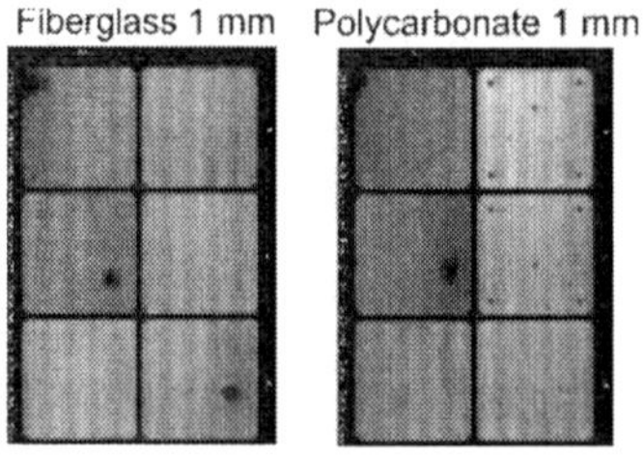

Figure 2: EL images of lightweight modules after hail test.

3.2 Thermocycling test

Lightweight modules 2x3 cells based on polycarbonate Novattro and fiberglass STD-FX sheets were fabricated for thermocycling test. Results of test are presented in Table 2. Measurement of IV curves and EL showed that both design of modules have high power degradation after 200 cycles (>20 %). As follow from EL images the main reason for the degradation of IV curves is a decrease in the contact between the busbars and the contact grid of the cells (see Fig.3). This can be explained by the movement of the tires during the test due to the different thermal expansion of the module materials. This is confirmed by visual inspection, which shows a strong bending of the tires in the space between the cells in the string (see Fig.4)..

Table II: Degradation of electrical characteristics of lightweight panels with 2x3 cells after thermocycling test.

Characteristics	Sample with polycarbonate 1 mm		Sample with fiberglass 1 mm	
Test conditions	Initial	After 200 cycles	Initial	After 200 cycles
Uoc, V	4,39	4,39	4,40	4,39
Isc, A	8,88	8,83	8,65	8,54
Umpp, V	3,64	3,17	3,66	2,92
Impp, A	8,36	7,66	8,17	7,34
Pmpp, W	30,44	24,27	29,94	21,43
FF, %	78,03	62,61	78,68	57,18
Power degradation, $\Delta P/P_{initial}*100\%$	~20,3 %		~28,4 %	

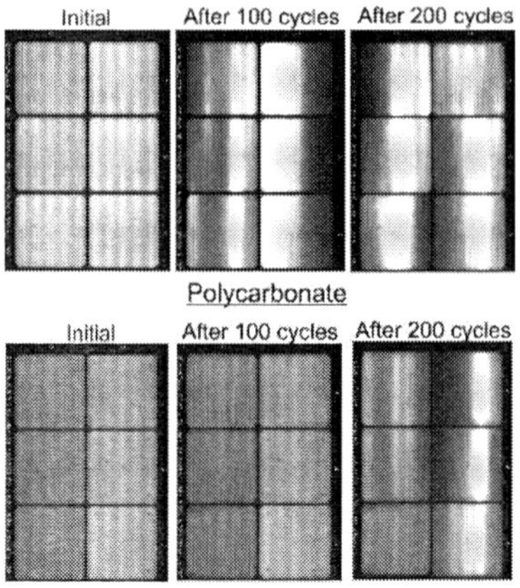

Figure 3: EL images of lightweight modules before and after 200 cycles of thermocycling test.

Figure 4: Bending of the tires during thermocycling test.

3.3 Damp heat test

Lightweight modules 2x3 cells for damp heat test were made from polycarbonate and fiberglass sheets. The results of damp heat effect are presented in Table 3.

Table III: Degradation of electrical characteristics of lightweight panels with 2x3 cells after damp heat test.

Characteristics	Sample with polycarbonate 1 mm		Sample with fiberglass 1 mm	
Test conditions	Initial	After 1000 hours	Initial	After 1000 hours
Uoc, V	4,39	4,38	4,39	4,35
Isc, A	8,75	8,72	8,64	7,99
Umpp, V	3,63	3,57	3,65	2,72
Impp, A	8,33	8,24	8,16	6,49
Pmpp, W	30,29	29,38	29,76	17,69
FF, %	78,9	76,93	78,43	50,84
Power degradation, $\Delta P/P_{initial}*100\%$	~3,0 %		~40,6 %	

The power degradation of module with fiberglass Fibroser is very large (>40 %). The El images show that the decrease in power is associated with cells degradation due to the high permeability water vapor of fiberglass sheet (see Fig.5). Visual inspection also showed a strong yellowing fiberglass and delamination along the busbars.

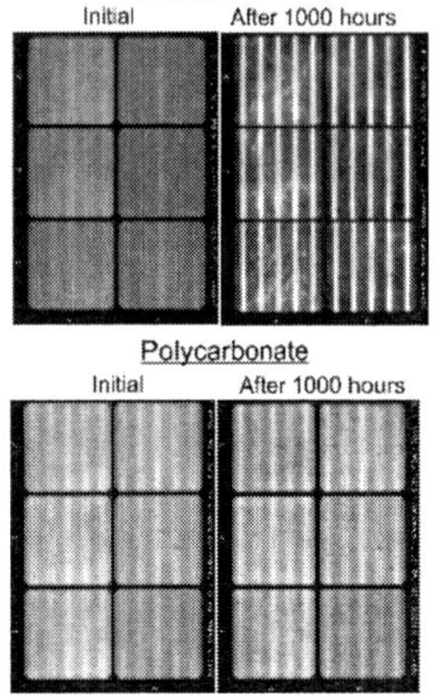

Figure 5: EL images of lightweight modules before and after 1000 hours of damp heat test.

The power degradation of module with polycarbonate Novattro is significant less (~3%). But cracks appeared on the surface of polycarbonate after 600 hours of damp heat testing. Yellowing of the busbars was also observed.

3.4 UV test

Samples 1x1 cell were made for research UV stability of polycarbonate and fiberglass sheets. Both of samples was irradiated 60 kW*hour/m². The module based on polycarbonate didn't show any power degradation and visual changes after UV test, while the fiberglass sheet yellowed strong that led to tangible power degradation of based-on module (see table 4).

Table IV: UV test results for module with frontsheet based on fiberglass Fibroser.

Test conditions	Initial	After irradiation 60 kW*hour/m²
Uoc, V	0,73	0,73
Isc, A	8,78	8,36
Umpp, V	0,60	0,59
Impp, A	8,21	7,82
Pmpp, W	4,89	4,58
FF, %	75,96	75,14
Power degradation, $\Delta P/P_{initial}*100\%$	~6,3 %	
Yellowing of frontsheet based on fiberglass after UV test		

4 CONCLUSIONS

Damp heat, thermocycling, UV and hail tests for two designs of lightweight module were conducted according to IEC 61215-2:2016. The module with fiberglass Fibroser frontsheet and PET-based backsheet showed high power degradation after all tests. Main disadvantages of this sheet are strong yellowing during UV test and high permeability water vapor during damp heat. The module with polycarbonate Novattro frontsheet and backsheet passed tests better. It has minimal degradation after 150 cycles of thermocycling, 1000 hours of damp heat and 60 kW*h/m² of UV irradiation, but after damp heat was observed appearance of cracks on the surface and yellowing of the busbars inside the module.

REFERENCES

1] A. Abramov, D. Andronikov, K. Emtsev, D. Orekhov, I. Shakhray, E. Terukov, E. Terukova, S. Yakovlev. "Super Lightweight Flexible HJT Solar Panel", 35th EU PVSEC, p.1227-1229 (2018).

[2] S. Yakovlev, E. Schebet, K. Emtsev, D. Andronikov, A. Abramov, D. Orekhov, I. Shakhray. "Mechanical Stability of Semi-Flexible Solar Panels", 36th EU PVSEC, p.1040-1041 (2019).

[3] S. Yakovlev, E. Schebet, K. Emtsev, D. Andronikov, A. Abramov, D. Orekhov, I. Shakhray. "Environmental Stability of Semi-Flexible HJT Solar Panels", 37th EU PVSEC, p.1117-1119 (2020).

412 W SOLAR MODULE USING SILICON HETEROJUNCTION CELLS AND SHINGLE INTERCONNECTION

Marina Foti(1)*, Marco Galiazzo(2), Lorenzo Cerasti(2), Enrico Sovernigo(2), Paolo Fugolo(2), Cosimo Gerardi(1), Alfredo Guglielmino(1), Grazia Litrico(1), Marcello Sciuto(1), Antonio Spampinato(1), Antonino Ragonesi(1), Francesco Rametta(1), Andrea Canino(3), Agata Carbonaro(3), Fabrizio Coco(3), Agnese Di Stefano(3), Fabrizio Bizzarri(4)

1. Enel Green Power S.p.A, 3SUN, Contrada Blocco Torrazze Zona Industriale 95121, Catania, Italy
2. Applied Materials Italia Srl, Via Postumia Ovest 244, Olmi di San Biagio di Callalta (TV) Italy
3. Enel Green Power S.p.A, Innovation Lab Contrada Passo Martino Zona Industriale 95121, Catania, Italy
4. Enel Green Power S.p.A, Solar Innovation, V.le Regina Margherita 125, Roma, Italy

*Corresponding author: marina.foti@enel.com

ABSTRACT: We present results of full size bifacial PV module combining Silicon Heterojunction (SHJ) solar cells and Shingle interconnection. SHJ solar cells, based on crystalline silicon (c-Si) absorber and thin hydrogenated amorphous silicon (a-Si:H) passivation and selective contacts, have demonstrated to reach efficiencies higher than 25% at industrial scale. On the other hand, shingling technology applied to solar cell allows the elimination of the space between the cells; increases the active surface per module; removes exposed busbars, reducing the overall resistive losses and promoting a highly aesthetic appearance. Moreover, shingling, due to low-temperature process, is completely suitable for HJT cells and is currently the only industrial solution compatible with thinner wafers (<120 um) and larger wafer formats.

At present time, there are not commercial modules using both technology and there are very few data in literature. Previous studies on minimodules based on SHJ cells showed promising results in terms of both power output as well as reliability.

In this work, we show the results of a full size bifacial PV module (72 cells equivalent), glass-to-glass assembled. We obtained 9% increment of power output on shingled module with respect to the reference module based on ECA- busbar interconnections. Shingling technology allows to reach a Cell-To-Module (CTM) factor equal to 1 and to maintain high bifaciality factor (>90%).

Keywords: Silicon Heterojunction Solar Cell; Shingling; Bifacial

1 INTRODUCTION

Shingled interconnection concept applied to c-Si solar cells has been discovered in 1960 and it has been extensively investigated for more than 26 years [1].

In shingled modules solar cells are cut into strips and overlapped to form the so called "strings", with the characteristic roof-tiles shape. Shingling allows the elimination of the space between the cells; increases the active surface per module; removes exposed busbars, reducing the overall resistive losses and promoting a highly aesthetic appearance.

The rise of high-efficiency and highly bifacial solar cell technologies has brought shingle technology developments back to the forefront, with ITRPV forecasting a market share around 10% by 2030 [2]. Originally applied to residential and commercial applications, recently Sunpower and Tongwei deployed this technology in the the utility-scale market [3].

It is known that standard (i.e. soldering based) module processing are not completely suitable for SHJ cells [4] due to the high soldering temperatures (>280°C) and thermo-mechanical stress present in the interconnection process. Shingling instead provides industrial feasibility thanks to the use of Electrically Conductive Adhesive (ECA) to join the portions of cell, establishing a permanent bond after curing at less than 200 °C. Due to this low-temperature interconnection, shingling is currently the only industrial solution compatible with thinner wafers (<120 um) and larger wafer formats.

Previous studies on minimodules based on SHJ cells showed promising results in terms of both power output as well as reliability [5], here we present the first data about full size shingled modules using SHJ solar cell precursors.

These precursors, combining crystalline silicon (c-Si) absorber and thin hydrogenated amorphous silicon (a-Si:H) passiving allow to reach efficiencies higher than 25% [6].

2 EXPERIMENTAL SET-UP AND RESULTS

Reference modules based on ECA- busbar interconnections have been processed in EGP 3SUN factory in Catania. The factory consists in fully automated production line for cells and modules manufacturing with annual capacity of 200MWp/year. Modules are bifacial glass-glass with 72 cells monocrystalline n-type SHJ cells [7, 8].

Most of the process steps to fabricate shingled modules (cells precursor up to ITO deposition and strings assembly in modules) have been carried out in 3SUN fab for a better result comparison.

A. Cell precursors

The solar cell precursors were realized using n-type M2 Silicon substrates, which were inspected in terms of contamination, mechanical defects and electrical characteristics. They were processed in wet bench where a sequence of chemical wet processes carrying out surface texturing and cleaning. The a-Si:H intrinsic and doped layers were deposited on both side of the textured wafer using a Plasma Enhanced Chemical Vapor Deposition reactor (PECVD), to form the emitter (on the rear side) and the collector (on the front side) of the solar cell. To complete the electrical contact Transparent Conductive Oxide (TCO) ITO was deposited through a sputtering method on both wafer sides to enhance light transmission and collect the electrical charges produced inside the cell.

B. Shingling cells and strings

For this work we processed 655 precursors with frontside Double Printing and the following parameters:

- Screen mesh and opening: 325 wires per inch, 16 um wire diameter, 42 um opening front and 46 um rear
- Paste: low-temperature Ag paste, 93-95% solid content

The frontside linewidth is on average 68 um after curing. The line height average is 24 um, so the overall finger aspect ratio is 0.35. Paste laydown is 103 mg for the cell frontside and 184 mg for the rearside (Fig.1).

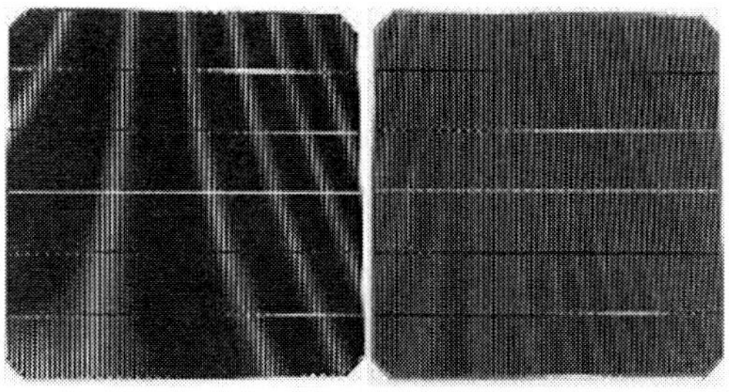

Figure 1: Pictures of front (left) and rear (right) of HJT cell with shingling layout. In this configuration the pattern is asymmetrical

Double Printing metallization process has been selected because charge carriers have to travel through all the finger length to be collected by the busbars, as presented in [9]. In addition, keeping front to back busbar alignment as accurate and precise as possible is important to consistently control the cell overlapping during the shingling process.

In order to perform the electrical testing for the cells with shingling layout a dedicated contacting unit has been used, with the aim of directly contacting fingers instead of front and back busbars.

Efficiency distribution of the processed cells is substantially equivalent to the one achieved in production, ranging from 22.2% to 22.9% absolute (Fig.2).

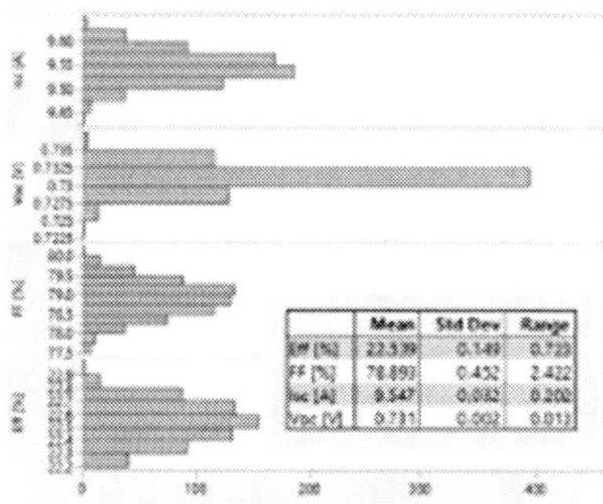

Figure 2: Key electrical parameters for the processed batch have been measured as reported in the histogram. Average efficiency was 22.54%.

The main steps of the stringing process in shingling are: 1) cell scribing and mechanical cleaving, 2) deposition of the ECA on the Ag pads present on the busbars, 3) alignment of the cut cells, or shingles, to form the string, 4) attachment of the end ribbons to the string via ECA and 5) final curing process.

The fully automated equipment adopted for this test is based on a laser platform with galvanometer scanner, a screen printing system to deposit the ECA with a typical printing speed of 200-300 mm/s, linear motion units to handle precisely the shingles, an integrated device to deposit and align the end ribbons to the string and a continuous oven able to perform the curing with a maximum process window of 200°C and 90 seconds. The equipment has been tested at 4,000 wph in a dual lane configuration (Fig.3).

Figure 3: Overview of the Applied Materials Sonetto tool used for the string assembly process.

C. Module layout and manufacturing

Shingled module layout, including string size and connection, has been studied to have module output electrical parameters in line with the standard ones.

Results are compared with standard EGP production module, obtained by cells of similar efficiency. We used same module size and materials type. Standard modules have been processed in a full automated way, instead the shingled module were processed manually in more steps, including shipment of strings to the different sites.

Comparison of electrical parameters and Current-Voltage, Power-Voltage curves at Standard Test Condition (STC) is reported in Fig. 4.

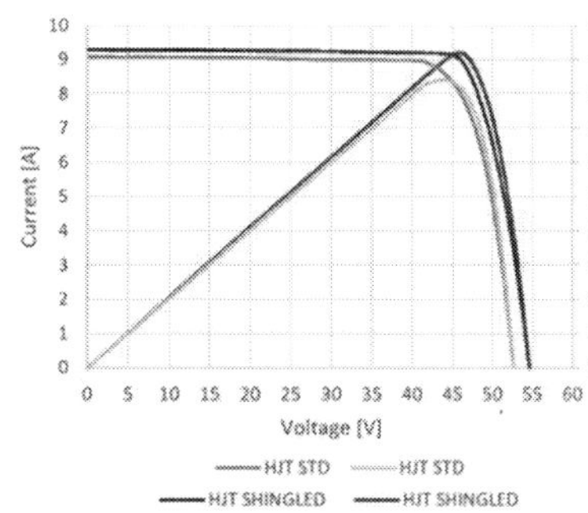

	Isc [A]	Voc [V]	FF [%]	Pmax [W]	Eff [%]
Shingled	9.28	54.73	81.06	411.91	20.81
Standard	9.08	52.74	79.08	378.78	19.14
difference	0.20	1.99	1.98	33.13	1.67
% difference	2.20	3.77	2.50	8.75	8.73

Figure 4: Comparison of electrical performance of standard and shingled module at STC.

We observe that module output power of shingled module is 35 W higher with respect to the standard, corresponding to a 9% increase. The result is in line with the increased active area (4%) and reduced resistive loss. Shingling technology allows to reach a Cell-To-Module (CTM) factor equal to 1 (CTM is 0.96 for the standard module). Shingled module exhibits a bifaciality factor value comparable with the standard (>90%).

Picture of shingled module, together with Electroluminescence (EL) images are showed in Fig. 5. Despite the manual process, we did not observe significant defectivity in shingled modules.

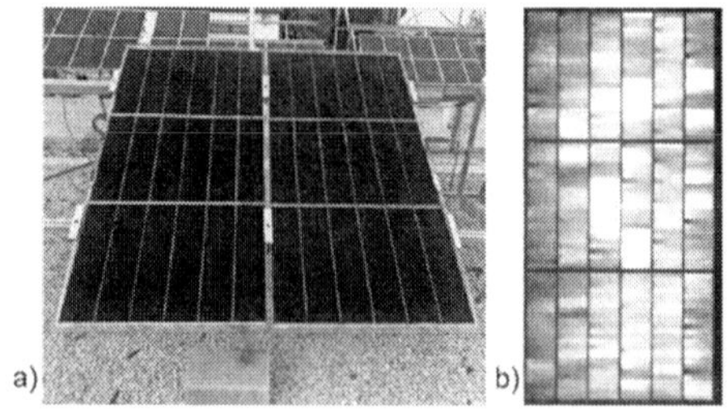

Figure 5: a) Two shingled module in outdoor testing. b) Electroluminescence (EL) image of shingled module showing no defectivity.

3 CONCLUSIONS

In this work, we showed 21% efficiency on full size modules (72 cells equivalent), obtained by the combination of two highly efficient PV technologies: silicon heterojunction cells (SHJ) and shingling interconnection.

Thanks to the increased active area and reduced resistive loss, we obtained 9% increment of power output on shingled module with respect to the reference module based on ECA- busbar interconnections.

4.1 Acknowledgment

The authors want to acknowledge Applied Materials laboratory for processing the cells and CEA-INES (Armand Bettinelli and Samuel Harrison) for providing the metallization layout.

4.2 References

[1] D. M. Chapin, C. S. Fuller and G. L. Pearson, "A New Silicon p-n Junction Photocell for Converting Solar Radiation into Electrical Power," Journal of Applied Physics 25, p. 676, 1954.

[2] International Technology Roadmap for Photovoltaic (ITRPV), "Results 2019 including maturity report 2020," Eleventh Edision, Oct 2020.

[3] PV, "SNEC 2020 Wrap-Up II: Large formats, high module power output," in SNEC 2020, 2020.

[4] S. K. Chunduri and M. Schmela, "Heterojunction solar Technology," TaiyangNews, 2020.

[5] C. Carrière, V. Barth, A. Bettinelli, S. Harrison, A. Derrier, L. Cerasti and M. Galiazzo, "Toward shingling interconnection with SHJ solar cells," in EU PVSEC, 11 September 2020.

[6] W. Favre et al, "25% efficient large area silicon solar cell: paving the way for premium PV manufacturing in Europe" in EU PVSEC, 11 September 2020

[7] G. Condorelli et al., "High Efficiency Hetero-Junction: From Pilot Line To Industrial Production," 2018 IEEE 7th World Conference on Photovoltaic Energy Conversion (WCPEC) (A Joint Conference of 45th IEEE PVSC, 28th PVSEC & 34th EU PVSEC), Waikoloa Village, HI, 2018, pp. 1970-1973, doi: 10.1109/PVSC.2018.8548197.

[8] G. Condorelli et al., "Initial Results of Enel Green Power Silicon Heterojunction Factory and Strategies for Improvements," 2020 47th IEEE Photovoltaic Specialists Conference (PVSC), Calgary, OR, 2020, pp. 1702-1705, doi: 10.1109/PVSC45281.2020.9300806

[9] M. Galiazzo et al, 'Status and Perspectives of Metallization Technologies for High Volume Manufacturing' available at http://www.metallizationworkshop.info

[10] Y. Zhang et al, 'Knotless Screen Printing for c-Si Solar Cells' available at http://www.metallizationworkshop.info

CROSSLINKING KINETICS OF PHOTOVOLTAIC MODULE ENCAPSULANTS - INVESTIGATION OF SELECTED EVA AND POE GRADES

Baloji Adothu[1], Robert Pugstaller[1], Martin Tiefenthaler[1], Francis Reny Costa[2], Sudhanshu Mallick[3], and Gernot M. Wallner[1]

[1]Johannes Kepler University Linz – Institute of Polymeric Materials and Testing & Christian Doppler Laboratory for Superimposed Mechanical-Environmental Ageing of Polymeric Hybrid Laminates.
[2]Borealis Polyolefine GmbH, St.-Peterstraße 25, A-4021, Linz, Austria
[3]Metallurgical Engineering and Materials Science, Indian Institute of Technology Bombay, Mumbai, Maharashtra, India 400076. The National Centre for Photovoltaic Research and Education (NCPRE), Indian Institute of Technology Bombay, Mumbai, Maharashtra, India-400076

ABSTRACT: Solar photovoltaic (PV) module encapsulation technology has grown rapidly worldwide in the last few years. Various novel encapsulation materials have been introduced in the PV market. Thermoplastic polyolefin elastomer (POE) encapsulant is a highly relevant alternative to the most commonly used crosslinking ethylene-vinyl acetate copolymer (EVA). During lamination, encapsulation materials are frequently crosslinked or cured, which allows for reducing creep at elevated temperatures. Crosslinking agents based on organic peroxides are well established. The main objective of this paper was to implement an isothermal, dynamic-mechanical analysis (DMA) method to investigate and compare the crosslinking kinetics of EVA and POE encapsulants. Isothermal experiments were carried out in torsional mode at different temperatures. The gelation point was deduced from the crossover-point of the real and imaginary part of the modulus. Comparable activation energies (E_a) of 149 and 152 kJ/moles were obtained for EVA and POE, respectively. By assessment of complex viscosity, the degree of crosslinking was deduced. Interestingly, the degree of crosslinking at gelation point was a factor of about three higher for POE, which revealed a significantly lower viscosity in the fully cured state. This paper will help in understanding the crosslinking kinetics, which is very beneficial for the PV industry and encapsulant manufacturers for quickly assessing the modification of any encapsulant crosslinking process.
Keywords: Activation energy, Crosslinking kinetics, Dynamic mechanical analysis, Encapsulants, Photovoltaics,

1 INTRODUCTION

Solar photovoltaic (PV) technology has now become the fastest-growing sector in renewable energy. The PV industry is experiencing an intense rivalry between the various PV module technologies, components, and PV encapsulant material manufacturing industries [1]. A PV module comprises different features such as solar glass, a front encapsulant, solar cells, a rear encapsulant, and a backsheet. The encapsulant is crucial for current PV technology providing special mechanical, optical, and electrical property profiles, including long-term stability under service-relevant stress conditions. EVA is the most commonly used crosslinking encapsulant in the PV industry. Although significant improvements in material quality and overall performance have been achieved, EVA is still critical due to the uptake of water, formation of acetic acid degradation products, yellowing, and delamination, or bubble formation problems [2,3]. Polyurethane (PU), ionomer, silicone elastomer, or polyvinyl butyral (PVB) are alternative encapsulants to EVA [4]. PV researchers continuously focus on the development of new encapsulation materials with an improved price/performance ratio [5].

Currently, in the PV industry, polyolefin encapsulants are gaining the highest potential for substituting EVA [6]. Thermoplastic polyolefin (TPO) is a non-crosslinking material [3], whereas thermoplastic polyolefin elastomer (POE) is a crosslinking encapsulant like EVA. During the PV module lamination process, the crosslinking encapsulants undergo significant chemical modification from the weaker thermoplastic state into a robust wide-meshed, cross-linked state [6]. Proper and uniform crosslinking is important to ensure reliable and durable PV modules. An improperly cross-linked encapsulant shows strong adverse effects like discoloration/formation of unsaturated compounds or delamination of PV modules [4]. A fully crosslinking encapsulant exhibits better adhesion strength, higher optical transmission, less mechanical creep, better thermal and UV radiation stability. To understand the crosslinking process behavior, it is necessary to study the kinetics of different crosslinking encapsulants and to adjust lamination and processing conditions accordingly.

Several investigators have used dynamic mechanical analysis (DMA) to study the viscoelastic behavior of polymeric encapsulation materials [7–11]. DMA provides isothermal, non-isothermal, time & frequency sweep test methods to characterize the viscoelastic behavior. The temperature sweep test is the most commonly used for EVA allowing for characterization of viscosity, degree of crosslinking, melting, and glass transition temperature. Isothermal-time sweep test is another simple method to adjust lamination process conditions for crosslinking encapsulant materials. Very limited studies are available for isothermal curing inestigation [12].

The main objective of this work was to implement a dynamics mechanics analysis-based testing method to assess crosslinking kinetics of commercially available EVA and POE encapsulants. By systematic comparison, potential differences in their activation energy for curing reactions and their overall properties (modulus, viscosity, degree of curing) should be established. The isothermal crosslinking kinetics was studied through dynamic mechanical analysis (DMA) in a torsional mode.

2 EXPERIMENTAL

2.1 Materials

Commercially available EVA and POE encapsulant films were used in this isothermal crosslinking kinetic study. The thickness was 0.45 and 0.54 mm, respectively. The encapsulant films were packed in envelopes and kept in a vacuum box. This confirms the original encapsulant film has no direct interaction with the surrounding environment. This can minimize the loss of additives from the encapsulant films.

2.2 Dynamic Mechanical Analysis (DMA)

DMA is a material characterization technique for assessment of the viscoelastic properties at low strains either as a function of a linear heating rate or time at a given temperature. Isothermal DMA experiments at 130, 140, and 150 °C were carried out in torsional mode at a frequency of 1Hz, The diameter of EVA and POE disc-like samples was 25mm. Complex viscosity, as well as storage and loss modulus, were evaluated. Gelation time t_{gel} was defined by the cross-over point of real and imaginary parts of the modulus.

2.3 Modeling

Arrhenius modeling is a well established method to describe the temperature-dependent kinetics of any chemical reaction. Arrhenius model for a reaction rate (k) can be written as:

$$k(T) = A\exp\left(\frac{-E_a}{RT}\right) \qquad (1)$$

Where T is the temperature, A is a material constant, R is the gas constant, and E_a is the activation energy. The activation energy for crosslinking reaction of EVA and POE encapsulants is calculated by using equation (1).

3 RESULTS AND DISCUSSION

The changes in viscoelastic properties of encapsulants during the crosslinking process can be directly observed through the DMA materials characterization technique. This section describes the DMA isothermal results of crosslinking kinetics of encapsulant grades based on EVA and POE.

3.1 Shear Modulus

The crosslinking reaction is initiated by the decomposition of the peroxide crosslinking agent leading to higher values of viscosity and shear modulus. Figure 1 depicts the temperature-dependent storage and loss modulus variation with respect to the crosslinking time. At a certain threshold, the storage modulus and loss modulus reached equal values. This crossover time is called "gelation point/time" (t_{gel}) [11-12]. Below the gelation point, an encapsulant is in a molten state, whereas an encapsulant is in a solid state above the gelation point. Gelation time was quite different for the investigated encapsulants. The t_{gel} value shifted towards lower crosslinking time as the temperature increased.

For EVA, a factor of about three lower gelation times were obtained. Accordingly, the maximum value of storage modulus in the fully cured state was significantly higher for EVA. Presumably, the type or concentration of peroxide was differing. Moreover, the chemical interaction between peroxides and EVA or POE could be different due to varying chemical structure of CHO-polymers such as EVA or CH-polymers such as POE. By investigating the crosslinking kinetics at different isothermal crosslinking temperatures (130, 140, and 150 °C), the temperature dependency of gelation time was evaluated from Fig.1.

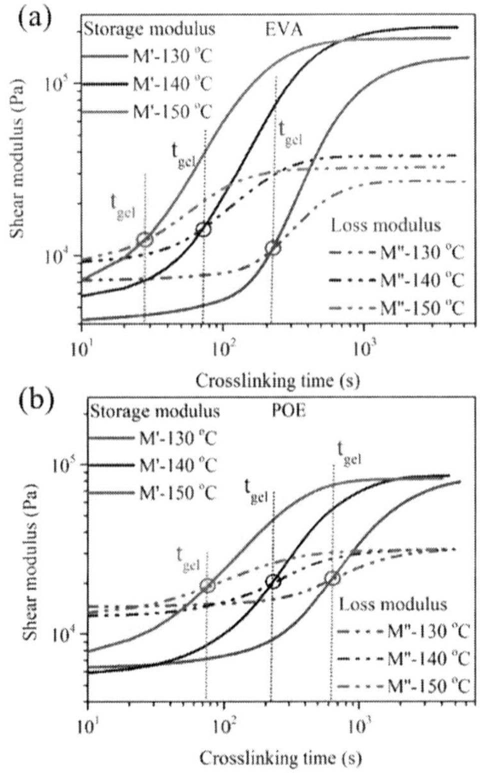

Figure 1: Temperature-dependent shear modulus of (a) EVA and (b) POE as a function of crosslinking time.

In equation (1), the reaction rate k is insufficient to predict the activation energy. In the crosslinking reaction approach, crosslinking time (t) is the most crucial parameter to reach a specific threshold value [12]. So, the Arrhenius equation can be re-written as

$$\ln(t_{gel}) = C - \left(\frac{-E_a}{RT}\right) \qquad (2)$$

Where C is a specific material constant. The Arrhenius variables alter when there is a change in reaction mechanisms or even degradation of the polymeric network. It should be noted that the estimated reaction threshold time t_{gel} is just valid in a given specific temperature range.

As shown in Fig. 2, the obtained gelation time values could be well described by an Arrhenius fit. Activation energy (E_a) values of 148.8 and 152.2 kJ/moles were obtained for EVA and POE, respectively. These values are found at 0.99 goodness of the linear fitting coefficient (R^2). Due to similar values of activation energy, most likely, the same type of peroxide was used, however, at a higher concentration for EVA.

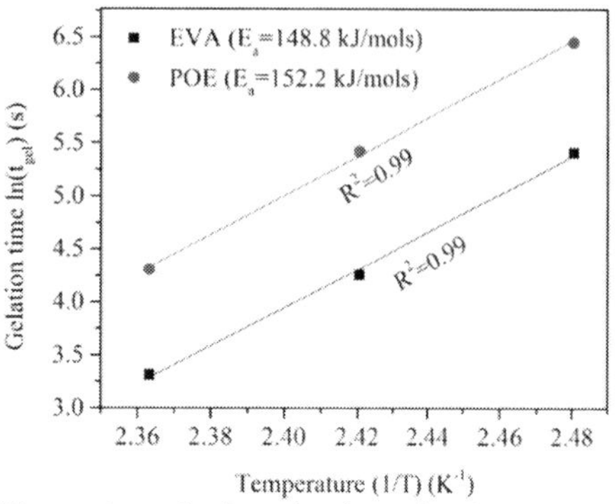

Figure 2: Arrhenius plot for estimating the activation energy for isothermal crosslinking of EVA and POE.

As discussed earlier, the crosslinking reaction increases the viscosity and stiffness of the encapsulant material. The variation of complex viscosity of encapsulants with crosslinking time at different isothermal temperature values is shown in Fig. 3. The saturation of complex viscosity indicates completion of the crosslinking reaction. The degree of crosslinking was calculated from the recorded complex viscosity by using equation (3):

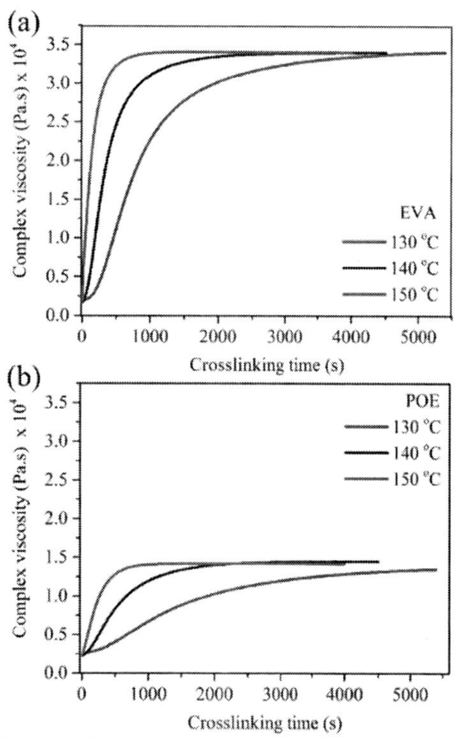

Figure 3: Temperature-dependent complex viscosity of (a) EVA and (b) POE.

$$\chi = \frac{\eta^*(t)-\eta^*(t_o)}{\eta^*(t_f)-\eta^*(t_o)} \qquad (3)$$

Where η^* (t) is the complex viscosity at time t, η^* (t_o) is the complex viscosity at the beginning of experiment time t_o, η^* (t_f) is the final complex viscosity after reaching the saturation or level off state at the end of the crosslinking process.

Fig. 4 shows the calculated degree of crosslinking variation at a different isothermal temperature. For EVA, a factor of about three higher complex viscosity was obtained in the cured state. Interestingly, differences in the degree of crosslinking at the gelation point were ascertained. While it was just 5 % for EVA, gelation of POE occurred at a degree of crosslinking of 18 %. This is a further indicator for a lower amount of crosslinking agents in POE.

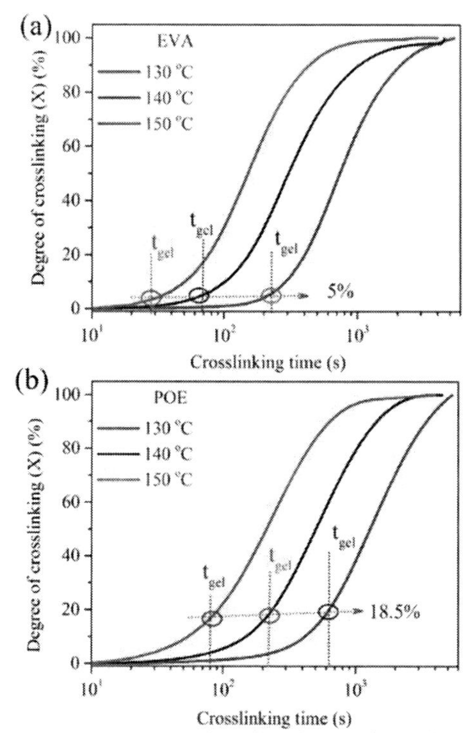

Figure 4: Temperature-dependent degree of crosslinking of (a) EVA, and (b) POE as a function of curing time.

4 SUMMARY AND CONCLUSIONS

The investigation clearly confirmed the high sensitivity and capability of isothermal dynamic mechanical analysis for characterization of the curing kinetics of peroxide-modified encapsulants.

According to theory, the peroxide crosslinking reaction was faster at higher temperatures. As a consequence, storage modulus and complex viscosity curves were shifted at elevated temperatures towards lower crosslinking times. It was ascertained that the maximum storage modulus and complex viscosity values were significantly higher for EVA as compared to POE in the fully cured state. This could presumably be attributed to variations in the formulation of the curing additive package or to differences in the chemical interactions of curing agents with CHO-polymers such as EVA and CH-polymers such as POE. Quite a different gelation time values were obtained for EVA and POE. While the gelation was faster for EVA, the time to full-cure was comparable for both encapsulant grades. By Arrhenius modeling of the temperature-dependent t_{gel} values, activation energy values of 149 and 152 kJ/moles were obtained for EVA and POE, respectively. Interestingly, the comparable activation energy values did not reflect the differences in crosslinking kinetics of the investigated materials.

A full understanding of curing reactions is of high relevance for adequate processing and lamination of PV modules. Moreover, pre-testing on the encapsulant level would allow for fast and cost-efficient development and screening of novel encapsulant materials for PV module lamination.

ACKNOWLEDGEMENTS

The financial support by the Austrian Federal Ministry for Digital and Economic Affairs, the National Foundation for Research, Technology, and Development, and the Christian Doppler Research Association are gratefully acknowledged.

REFERENCES

[1] B. Adothu, S. Chattopadhyay, P. Bhatt, P. Hui, F.R. Costa, S. Mallick, Early-stage identification of encapsulants photobleaching and discoloration in crystalline silicon photovoltaic module laminates, Prog. Photovoltaics Res. Appl. (2020). doi:10.1002/pip.3269.

[2] A.W. Czanderna, F.J. Pern, Encapsulation of PV modules using ethylene vinyl acetate copolymer as a pottant: A critical review, Sol. Energy Mater. Sol. Cells. 43 (1996) 101–181. doi:10.1016/0927-0248(95)00150-6.

[3] B. Adothu, P. Bhatt, S. Chattopadhyay, S. Zele, J. Oderkerk, H.P. Sagar, F.R. Costa, S. Mallick, Newly developed thermoplastic polyolefin encapsulant–A potential candidate for crystalline silicon photovoltaic modules encapsulation, Sol. Energy. 194 (2019) 581–588. doi:10.1016/J.SOLENER.2019.11.018.

[4] A. Omazic, G. Oreski, M. Halwachs, G.C. Eder, C. Hirschl, L. Neumaier, G. Pinter, M. Erceg, Relation between degradation of polymeric components in crystalline silicon PV module and climatic conditions: A literature review, Sol. Energy Mater. Sol. Cells. 192 (2019) 123–133. doi:10.1016/j.solmat.2018.12.027.

[5] O. Hasan, A.F.M. Arif, Performance and life prediction model for photovoltaic modules: Effect of encapsulant constitutive behavior, Sol. Energy Mater. Sol. Cells. 122 (2014) 75–87. doi:10.1016/j.solmat.2013.11.016.

[6] B. Adothu, F.R. Costa, S. Mallick, Damp heat resilient thermoplastic polyolefin encapsulant for photovoltaic module encapsulation, Sol. Energy Mater. Sol. Cells. 224 (2021) 111024. doi:10.1016/j.solmat.2021.111024.

[7] M. Jaunich, M. B??hning, U. Braun, G. Teteris, W. Stark, Investigation of the curing state of ethylene/vinyl acetate copolymer (EVA) for photovoltaic applications by gel content determination, rheology, DSC and FTIR, Polym. Test. 52 (2016) 133–140. doi:10.1016/j.polymertesting.2016.03.013.

[8] G.C.E. Ch. Hirschl, L. Neumaier, S. Puchberger, W. Mühleisen, G. Oreski, M.K. R. Frank, M. Tranitz, M. Schoppa, M. Wendt, N. Bogdanski, A. Plösch, Determination of the degree of ethylene vinyl acetate crosslinking via Soxhlet extraction: Gold standard or pitfall?, Sol. Energy Mater. Sol. Cells. 143 (2015) 494–502. doi:10.1016/j.solmat.2015.07.043.

[9] W. Stark, M. Jaunich, Investigation of Ethylene/Vinyl Acetate Copolymer (EVA) by thermal analysis DSC and DMA, Polym. Test. 30 (2011) 236–242. doi:https://doi.org/10.1016/j.polymertesting.2010.12.003.

[10] K. Agroui, G. Collins, G. Oreski, M. Boehning, A.H. Arab, D. Ouadjaout, Effect of crosslinking on EVA-based encapsulant properties during photovoltaic module fabrication process, Rev. Des Energies Renouvelables. 18 (2015) 2–303. http://www.cder.dz/download/Art18-2_13.pdf.

[11] G.M. Wallner K. Grabmayer and G. Oreski, Physical characterization of eva embedding materials for solar cells, in: 25th Eur. Photovolt. Sol. Energy Conf. Exhib. 6-10 Sept. 2010, Val. Spain, 2010: pp. 4033–4035. doi:10.4229/25thEUPVSEC2010-4AV.3.22.

[12] S.H. Schulze, A. Apel, D. Daßler, C. Ehrich, Cure state assessment of EVA-copolymers for PV-applications comparing dynamic-mechanical, dielectric and calorimetric properties, Sol. Energy Mater. Sol. Cells. 143 (2015) 411–417. doi:10.1016/j.solmat.2015.07.024.

[13] W. Du, L. Tan, Y. Zhang, H. Yang, H. Chen, Dynamic rheological investigation during curing of a thermoset polythiourethane system, Int. J. Polym. Sci. 2019 (2019). doi:10.1155/2019/8452793.

PV MODULE TRANSPORTATION IN TRUCKS WITH TWO DIFFERENT FLOOR DESIGNS

Devan P. Vasudevan and Anil Kottantharayil
National Centre for Photovoltaic Research and Education, and Department of Electrical Engineering,
Indian Institute of Technology Bombay, Mumbai, Maharashtra, 400076, India.

ABSTRACT: This work aims to understand the nature of the vibration experienced by the PV modules during transportation over 800 kms on two different types of containers. The methodology used for the analysis of transportation vibration data generated during PV module transportation is reviewed. Transportation vibration data was collected from 5 PV module pallets placed in two containers, one with a metal floor and another one with a wooden floor. It is shown that signal-triggered data sampling result in less loss of information compared to time-triggered data sampling. It was found that the PV modules placed in the container with wooden floor showed lower amount of damage as compared to the PV modules in the container with metal floor. The vibration along the axis perpendicular to the PV module plane was severe in the 10-20 Hz range.
Keywords: PV Module, Transportation, Vibration, Micro-cracks

1 INTRODUCTION

PV industry uses different types of trucks or containers for transporting PV modules to power plants in large volume. Previous studies analyzed the transportation vibration data from PV modules inside standard pallet of 20 modules and cassette type packing of 2 modules [1]. The work reported that the vertical vibration profile obtained for the PV module transportation on Indian roads were not fitting with the ASTM D4169 Assurance Level II [2], the main reference of IEC 62759-1:2015 [3]. Another study focused on the lateral vibrations experienced by the PV modules placed vertically in a 25 PV module pallet [4]. Lateral vibrations generated during the transportation of vertically placed PV module pallets were immoderate. The study also recommended to incorporate the laboratory simulation of the lateral vibration for the PV module pallet in the IEC62759 standard. The analysis methodology used in previous studies [1],[4] were adopted from Singh et al. [5].

This study attempts to review the analysis methodology and update it with the required modifications. IEC 62759-1: 2015 standard recommends to conduct same series of tests for the PV module transportation for all type of trucks [2]. This study also focuses on finding the dependence of the truck type on the vibration experienced by the PV modules over a longer distance. PV module transportation vibration data collected from containers with 2 different floor types were analyzed. One container had wooden floor and the other container had metal floor. The experiment was conducted on two fully packed containers which covered around 800 kms.

2 EXPERIMENTAL DETAILS

2.1 PV Module, Pallet and Container

Crystalline silicon PV Modules of size 1.95 m X 0.99 m with 72 cells and a maximum power of 320 W were used in this study. Two containers were used for transporting 35 PV module pallets. One container had wooden floor and the other container had metal floor. The transportation started from the PV module manufacturing unit and ended at the installation site covering around 800 kms. Fig. 1 shows the arrangement of PV module pallets inside the two containers. The circle at the top right corner of the figures of both containers indicates the driver's side. Each box represents the top view of the pallets. 5 PV module pallets which were under observation are marked in red. The pallets were handled using forklift, electric pallet truck or hand pallet truck.

25 PV modules were kept under observation. These 25 PV modules were distributed in 5 pallets, P1, P2, P3, P4 and P5. 5 PV modules were distributed in each pallet, as two PV modules at both the ends and one PV module at the middle. Fig. 2 shows the arrangement of PV modules inside the pallet. One PV module in each pallet was attached with an accelerometer to collect vibration data. The PV modules which were under observation are marked in yellow.

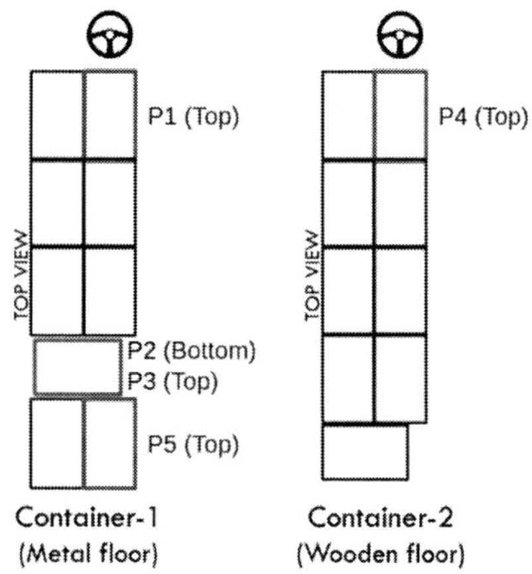

Figure 1: Top view of the pallets arranged over 2 containers; red boxes indicates the PV module pallets under observation.

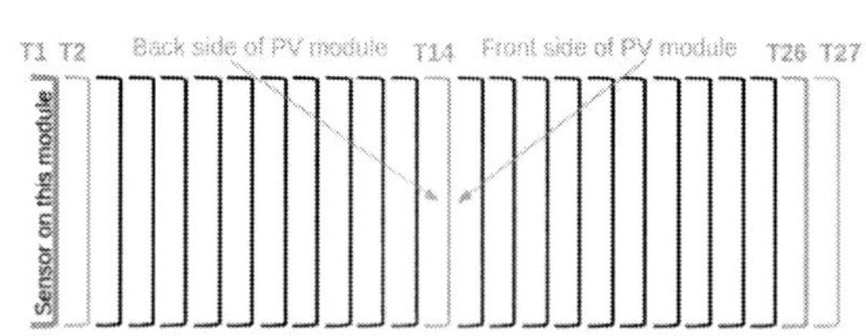

Figure 2: Arrangement of PV modules inside the pallets; yellow color indicates the PV modules kept under observation.

2.2 Instrumentation

Five GCDC X16-4 accelerometers [6] with extended battery backup were used for this study. The accelerometers were attached to the center of the backsheet with 3M 950 double sided adhesive tape. The sensors were set to record continuous 3-axes vibration data at 800 Hz sampling rate. From this data, vibration data events were sampled in two ways: (i) time-triggered events where a vibration data event of 2 seconds duration was sampled once in every 10 minutes, and (ii) signal-triggered events where a vibration data sample of 2 seconds was sampled when the magnitude of the vibration data exceeds 1.5 g (15 m/s2). A pre-sampling of 1 sec was used. Hence, the signal-triggered samples would consist of 1 second vibration data before triggering and 1 second after triggering. This would make sure that the threshold crossing point of the peak would be in the middle of the data sample.

3 RESULTS AND DISCUSSIONS

Transportation vibration data were analyzed using Power Spectral Density (PSD) profiles. Electroluminescence (EL) images taken before and after transportation were compared to detect micro-crack generation or propagation due to transportation and handling.

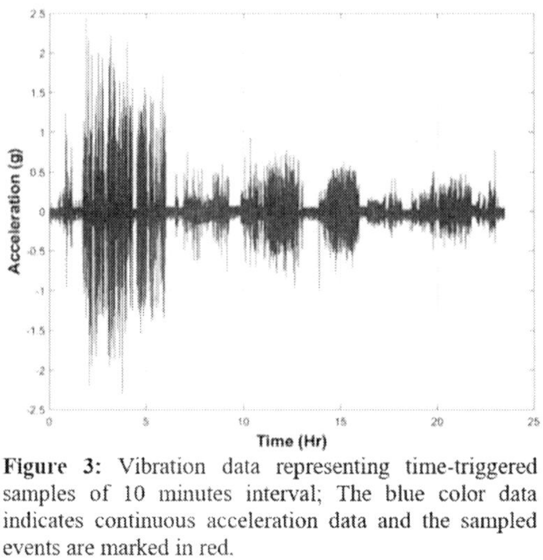

Figure 3: Vibration data representing time-triggered samples of 10 minutes interval; The blue color data indicates continuous acceleration data and the sampled events are marked in red.

PSD profiles were generated for the individual signal-triggered and time-triggered events. Linear averages of these PSD profiles were taken to find the final PSD profile. It was then benchmarked with ASTM D4169 Assurance Level II, which is the main reference of IEC 62759-1:2015 standard.

PSD profiles generated using this methodology resulted in PSD profiles which were not severe as compared to the PSD profiles obtained from previous studies [1],[4]. This led to a detailed analysis of the time-triggered and signal-triggered data. It was carried out by comparing PSD profile generated from the signal-triggered and time-triggered data separately. Fig. 3 shows vibration data representing time-triggered samples of 10 minutes interval. Fig. 4 shows vibration data representing signal-triggered events at 1.5 g threshold. In both the figures, the blue color data indicates continuous acceleration data recorded by the accelerometers and the sampled events are marked in red.

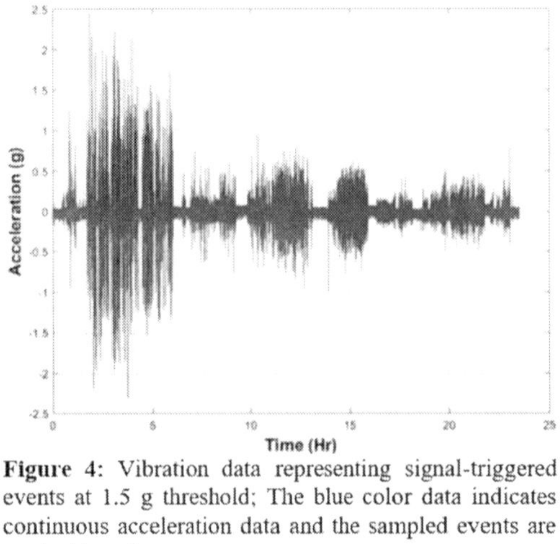

Figure 4: Vibration data representing signal-triggered events at 1.5 g threshold; The blue color data indicates continuous acceleration data and the sampled events are marked in red.

From Figures 3 and 4, it may be observed that there were more number of time-triggered events sampled than signal-triggered events. The severity of time-triggered events was less as compared to the signal-triggered events. Time-triggered events were sampled even when the container was stationery. In order to understand the impacts of this analysis scheme, the PSD profiles of these data were plotted separately. Figures 5 and 6 shows the PSD profiles generated from the time-triggered events alone and signal-triggered events alone, respectively. From these figures, it may be observed that the severity of the PSD profile generated from time-triggered samples was less as compared to the PSD profile generated from signal-triggered events. Hence, signal-triggered samples should be used for the PSD profile analysis. In the remainder of this work, PSD profiles were generated from signal-triggered events alone.

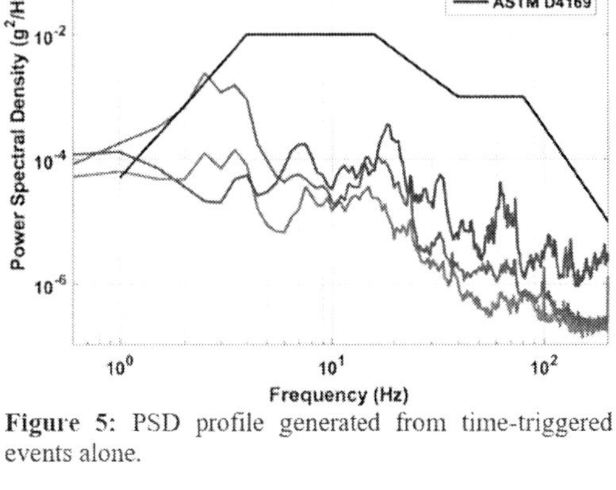

Figure 5: PSD profile generated from time-triggered events alone.

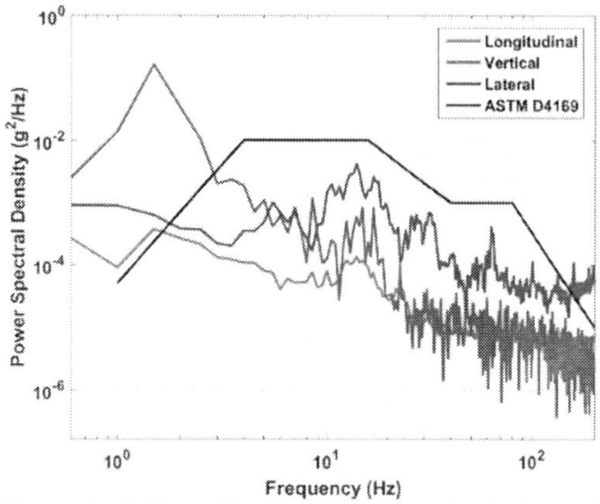

Figure 6: PSD profile generated from signal-triggered events alone.

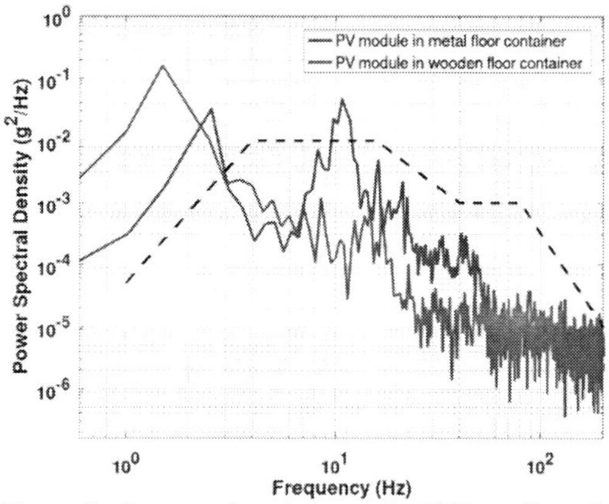

Figure 7: A comparison between the PSD profiles of vertical vibration data from PV modules placed inside the pallet P4 (placed in container with wooden floor) and pallet P1 (placed in container with metal floor).

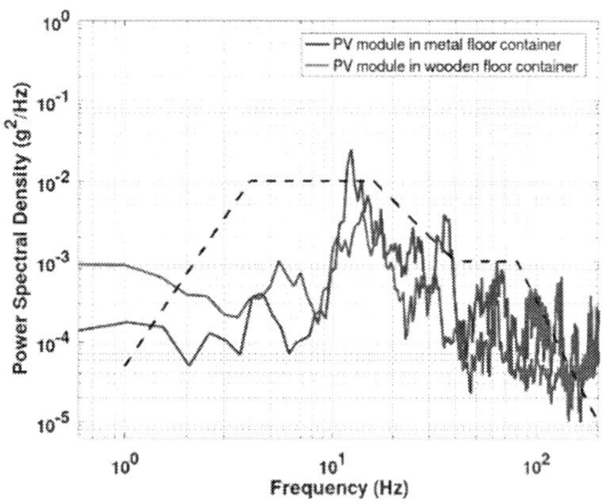

Figure 8: A comparison between the PSD profiles of lateral vibration data collected from PV modules placed inside the pallet P4 (placed in container with wooden floor) and pallet P1 (placed in container with metal floor).

Fig. 7 shows the comparison between PSD profiles of the vertical vibration data collected from the PV modules placed inside pallet P1 (placed in container with metal floor) and inside pallet P4 (placed in container with wooden floor). At lower frequency (1-3 Hz), vertical vibration magnitude was severe for the PV module placed in container with wooden floor. But a peak in the vertical vibration data was observed at 11 Hz for the vibration data from the PV module placed in container with metal floor. But it was absent in case of the PV module placed in container with wooden floor. Similar peak was observed from the lateral axis of the truck (along perpendicular to the PV module plane), in the previous studies [1],[4].

Fig. 8 shows the comparison between PSD profiles of the lateral vibration data from the PV modules placed inside pallet P1 (placed in container with metal floor) and inside pallet P4 (placed in container with wooden floor). From 10 Hz onwards, lateral vibrations from the PV module placed in container with metal floor were more severe than the vibrations from the PV module placed in container with wooden floor.

Table I: Crack statistics of 5 PV modules each from pallets P1 (placed in container with metal floor) and P4 (placed in container with wooden floor)

Type of micro-cracks generated or propagated	Number of cracks from PV modules in P1 (metal floor)	Number of cracks from PV modules in P4 (wooden floor)
New mode A cracks generated	19	0
New mode B cracks generated	0	0
New mode C cracks generated	1	0
Mode A transformed to mode B	1	5
Mode A transformed to mode C	1	0
Mode B transformed to mode C	0	1
Number of existing damaged cells before transportation (out of 5 PV modules)	53	151
Number of damaged cells during transportation (out of 5 PV modules)	22	6

Table I shows the crack statistics of 5 PV modules each from pallets P1 (placed in container with metal floor) and P4 (placed in container with wooden floor). Among the 5 PV modules placed in the container with wooden floor only 6 cells were damaged out of 5 modules during transportation and handling. In case of container with metal floor, 22 cells out of 5 modules were damaged during transportation and handling. Hence, it may be concluded that the wooden floor dampened the vibrations more effectively compared to metal floor. Influence of other parameters like the condition of the containers, skills of the drivers, etc. different between two containers could have also played a role in influencing this observation.

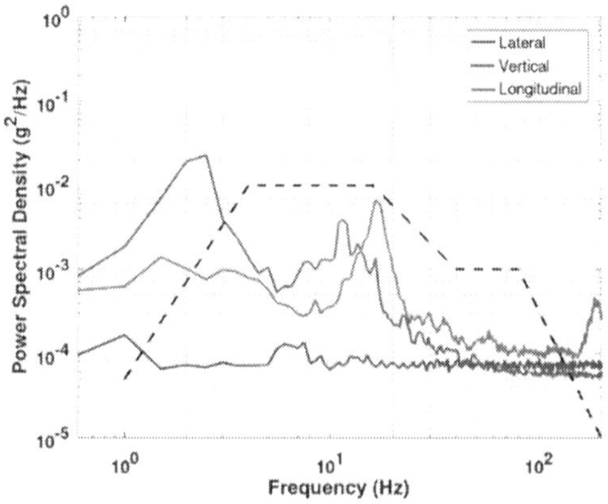

Figure 9: PSD profiles of 3-axes vibration data collected from the PV module inside pallet P3. The axis perpendicular to the PV module plane was oriented towards longitudinal direction of the container.

Fig. 9 shows the 3-axes PSD profiles of the vibration data collected from the PV module inside pallet P3. The axis perpendicular to the PV module plane was oriented towards longitudinal direction of the container. In the PSD profile, a peak in the longitudinal vibration data was observed in 10-20 Hz range. Previous study reported similar peak in the lateral direction when the axis perpendicular to the PV module plane was along lateral direction of the truck [4]. Hence, it may be concluded that, irrespective of the relative orientation of the module surface vis-à-vis the direction of travel of the truck, vibrations along the axis perpendicular to the module plane is quite significant and should be used for laboratory simulations.

4 CONCLUSIONS

This work investigated the vibration generated on PV modules during transportation in two types of trucks with different floor designs. This work also reviewed the analysis methodology used for the PSD analysis. For the PSD profile analysis of the vibration data generated during PV module transportation, only signal-triggered events should be used. From the comparison between PV modules placed in two type of containers, the PV modules placed in the container with wooden floor showed lower amount of damage as compared to the PV modules placed in the container with metal floor. A peak in the vertical vibration data was observed at 11 Hz for the data from the PV module placed in container with metal floor. But it was absent in case of the PV module placed in container with wooden floor. Irrespective of the direction of travel of the truck, vertically placed PV module experiences severe vibrations in 10-20 Hz range along the axis perpendicular to the module plane. Hence, the vibration along the axis perpendicular to the PV module plane aligned over lateral/longitudinal direction of the truck may be included for the laboratory simulation of the transportation of vertically placed PV module pallets.

ACKNOWLEDGEMENT

Ministry of New and Renewable Energy (MNRE), Government of India for funding the National Centre for Photovoltaic Research and Education (NCPRE). PV module manufacturer, who supported this study. Mr. Ajeesh Alath for the help in data collection.

REFERENCES

[1] D. P. Vasudevan, P. Bhatt, and A. Kottantharayil. "Impact of Transportation on Indian Roads, on PV Modules." In 2019 IEEE 46th Photovoltaic Specialists Conference (PVSC), pp. 1529-1532. IEEE, 2019.

[2] ASTM D4169-16: Standard Practice for Performance Testing of Shipping Containers and Systems, 2016.

[3] IEC 62759-1:2015: Photovoltaic (PV) modules-Transportation testing- Part 1: Transportation and shipping of module package units, 2015.

[4] D. P. Vasudevan, P. Bhatt, and A. Kottantharayil. "Lateral vibrations experienced by vertically placed PV modules in the pallet during transportation." In 2020 47th IEEE Photovoltaic Specialists Conference (PVSC), pp. 1323-1325. IEEE, 2020.

[5] S. P. Singh *et al.*, "Measurement and Analysis of Truck and Rail Shipping Environment in India," in Packaging Technology and Science, vol. 20, pp. 381-392, 2007.

[6] GCDC X16-4 USB Accelerometer Data Logger User Manual, www.gcdataconcepts.com.

INVESTIGATIONS ON THE TEMPERATURE DEPENDENCY OF THE SHUNT RESISTANCE IN CIGS THIN-FILM SOLAR MODULES USING DARK IV-CURVE MEASUREMENTS

Lukas Gerstenberg, Pavan Kumar Panda, Viktor Wesselak
Institut für Regenerative Energietechnik, University of Applied Sciences Nordhausen
Weinberghof 4, 99734 Nordhausen, Germany
E-Mail: lukas.gerstenberg@hs-nordhausen.de

ABSTRACT: For the application of Potential-Induced Degradation detection via dark current-voltage curve measurement, the temperature dependency of electrical module parameters plays a vital role. For the detection of shunting-type PID, the shunt resistance can serve as an indicator. The experiments show that the shunt resistance of Copper-Indium-Gallium-Selenide Thin-Film modules is strongly temperature dependent. In a 90 K range between -10 °C and 85 °C, the shunt resistances of two modules from two different manufacturers both show a negative temperature coefficient over the whole temperature interval. For module A, the shunt resistance changes by a factor of around five and for module B, the factor is around three. The decrease of the shunt resistance with increasing temperature is monotonically decreasing and non-linear for both modules. The correlation between temperature and shunt resistance does not show any hysteresis effect.
Keywords: CIGS, degradation, shunt resistance

1 INTRODUCTION

The shunt resistance R_{Sh} of solar modules is a parameter that can be strongly affected by potential-induced Degradation (PID). For crystalline Silicon (c-Si) photovoltaic (PV) modules there are promising studies that allow early detection of PID based on the measurement of the dark current-voltage (IV) curve with focus on the shunt resistance [1, 2]. This method is applicable to modules where the essential PID mechanism is the so-called shunting-type PID (PID-s). Negative potential of the cell against ground causes positive sodium ions (Na^+) to migrate into the active cell area and accumulate in stacking faults. The accumulation of charges causes increased recombination and lowers the shunt resistance R_{Sh} of the cell or module [3].

For Thin-Film Solar Modules (TFSM) the range of possible PID mechanisms is large and versatile. Superstrate TFSM (amorphous Silicon (a-Si) and Cadmium Telluride (CdTe)) show susceptibility to corrosion of the transparent conductive oxide (TCO) layer. Here an electrochemical process including water ingress leads to corrosion of the TCO layer and thereupon delamination of the TCO from the front glass [4, 5]. Furthermore, CdTe TFSM also show shunting-type PID in an early stage of accelerated PID testing. In an earlier study, it was shown that the investigated CdTe TFSM showed continuous decrease of the shunt resistance and power over the complete testing time of the accelerated PID test. TCO delamination occurred rapidly in the middle of the test [4]. The continuous decrease of the shunt resistance during PID progress allows the application of shunt resistance monitoring for CdTe TFSM for the early detection of PID.

Substrate Copper-Indium-Gallium-Selenide (CIGS) TFSM show a number of changes in module parameters undergoing PID stress. Besides a decrease of I_{SC} and V_{OC} [6], shunting was observed in cells [7] and modules [6]. The detected change in R_{Sh} was shown as an outcome of PID stress. In this work, we show results of preliminary experiments that are necessary to conduct before applying the in-situ dark IV-curve method to CIGS TFSM. As reported in [8] for c-Si modules, the temperature dependency of R_{Sh} is a key influence on the accuracy of this method. For the future application of the method on CIGS TFSM, the temperature dependency on the R_{Sh} of modules from two manufacturers A and B was studied in detail.

2 LITERATURE REVIEW: TEMPERATURE DEPENDENCE OF SHUNT RESISTANCE

In the electrical One-Diode Model of a PV module, the shunt resistance represents the ability of the cell or module to withstand leakage of charges through the cell and along the edges of the cell, see Fig. 1. Shunt pathways depend on the cell technology and can differ with manufacturing processes and material composition. For CIGS TFSM the main shunt pathways are ohmic paths caused by pinholes in the absorber layer and other defects usually located at grain boundaries. Furthermore, the diffusion of metal into the absorber can also increase the shunt conductance [9].

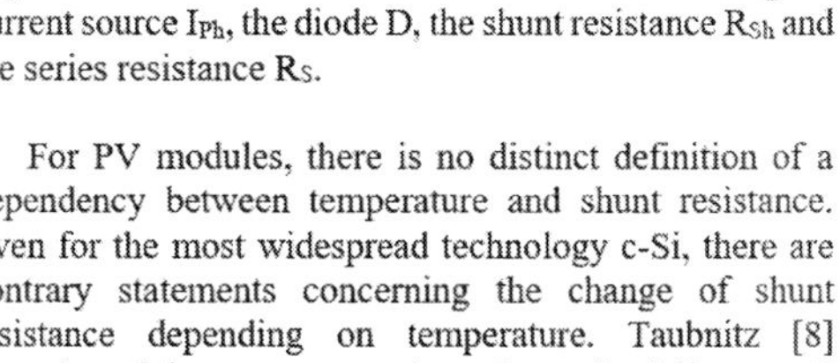

Figure 1: One-Diode Model of a PV module including the current source I_{Ph}, the diode D, the shunt resistance R_{Sh} and the series resistance R_S.

For PV modules, there is no distinct definition of a dependency between temperature and shunt resistance. Even for the most widespread technology c-Si, there are contrary statements concerning the change of shunt resistance depending on temperature. Taubnitz [8] investigated the temperature dependency for different c-Si modules and found that the module of one manufacturer showed a positive temperature coefficient (TC) for the shunt resistance, while a module of a different manufacturer showed a negative TC. One module even showed a temperature dependent TC, thus changed the sign for temperatures above 49 °C. Kronthaler [10] and Ruschel [11] both report that no clear correlation between temperature and shunt resistance could be found for their experimentations on c-Si modules. A simulation by Diatta [12] showed a non-linear, negative TC for silicon cells.

For CIGS, there are only a small number of studies

available reporting on the temperature dependency of the shunt resistance. Williams [9] mentions a linear relationship between shunt resistance and temperature for CIGS cells. There is no further explanation on how this assumption originates. Theelen [13] describes a detailed investigation on the temperature dependency of all CIGS cell parameters and reports a non-linear negative TC for the shunt resistance between 20 °C and 105 °C. The cause for a decrease in shunt resistance with higher temperature may arise from the decline of the depletion zone and an increase in the saturation current of the diode.

From the variety of dependencies shown, it becomes apparent that the temperature behaviour needs to be determined for each individual module.

3 MEASUREMENTS

Two CIGS modules experienced a temperature range from -10 °C to 85 °C to determine the temperature dependency of their shunt resistances. Each temperature was set and then held for half an hour. During the whole experiment, the dark IV curve was measured to later extract the shunt resistance from the data using equation (eq.) (1). Here V_{SC} describes the voltage near the short circuit point on the IV curve and I_{SC} is the short circuit current. The delta describes a small interval between two points close to the short circuit point.

$$R_{Sh} = \frac{\Delta V_{SC}}{\Delta I_{SC}} \qquad (1)$$

The temperature range is -10 °C to 85 °C since it includes both the actual temperature range in the field in central Europe (-10 °C to around 70 °C) and the typical temperature for accelerated PID tests in climatic chambers which is 85 °C. Higher temperature were avoided to ensure that the encapsulation endures the measurements without permanent damages. In the chosen interval, the set temperatures are -10 °C, 10 °C, 30 °C, 50 °C, 70 °C and 85 °C.

Each module, one of manufacturer A and one of manufacturer B were set on a rack inside the chamber and four Pt100 temperature sensors were attached – two on the front glass and two on the back glass. In a first run, the temperature was increased from -10 °C to 85 °C and in a second run, the temperature was decreased from 85 °C to -10 °C to detect possible differences between temperature rise and fall. The dark IV curve was measured in an interval of on minute.

4 RESULTS AND DISCUSSION

Figure 2 shows the shunt resistance and mean module temperature (mean of the four Pt-100 readings) over the duration of the temperature increase for the CIGS module from manufacturer A. At -10 °C, the module shunt resistance is 3.2 kΩm² and there is a significant decrease with every rise in temperature. At 85 °C, the module shunt resistance reached 0.65 kΩm², thereby decreased by a factor of around five. The higher mean variation of the R_{Sh} values at lower temperatures arises from the limited absolute resolution of the current measurement of the dark IV curve measurement equipment of 6 µA. At low temperatures, the IV curves show small currents in the I_{SC} region and thereby the absolute resolution of the current measurement affects the calculation of the shunt resistance. Overall Fig. 2 shows that the change of R_{Sh} gets smaller with higher temperatures. Between -10 °C and 10 °C, the change of R_{Sh} is 0.787 kΩm² and between 50 °C and 70 °C, the change is 0.42 kΩm². The change in shunt resistance is non-linear.

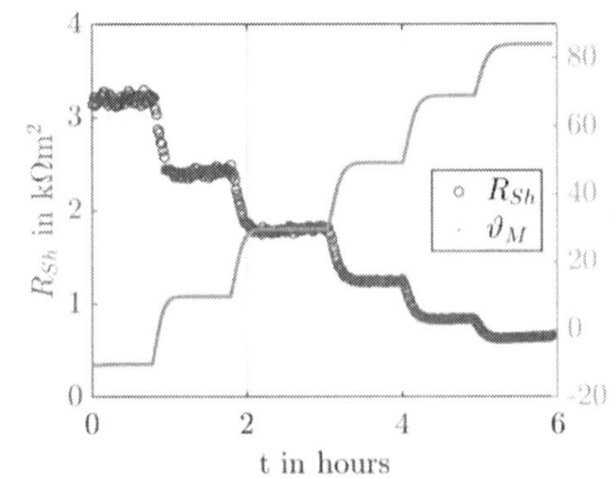

Figure 2: Shunt resistance R_{Sh} (blue circles, left axis) and mean module temperature (green dots, right axis) over time for a CIGS module from manufacturer A.

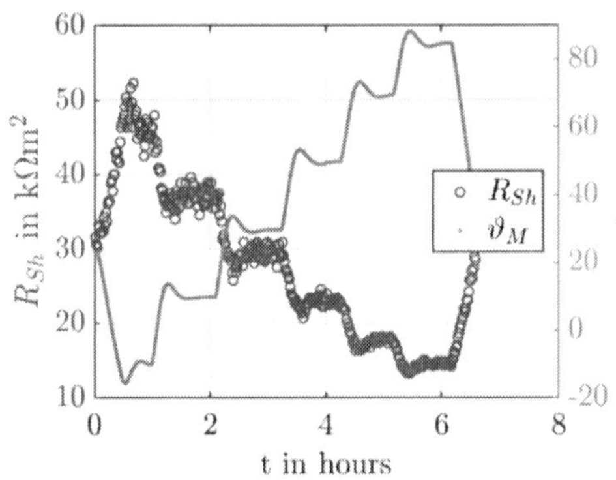

Figure 3: Shunt resistance R_{Sh} (blue circles, left axis) and mean module temperature (green dots, right axis) over time for a CIGS module from manufacturer B.

The module from manufacturer B experienced the same test sequence. Fig. 3 shows the modules shunt resistance and mean temperature over the testing time. The shunt resistance shows values one order of magnitude higher than the module of manufacturer A. Additionally, there is a transient temperature stabilization visible after each temperature increase. This might be caused by the larger module size, which leads to a higher position of the temperature sensors in the climatic chamber. In this position, the airflow is higher, causing the sensors to register the temperature overshoot. The thermal capacity of the module however, causes the increase of the CIGS junction temperature to occur slowly, which becomes apparent through the almost stepwise change of the shunt resistance values. Similar to the measurements of module A, also the module of manufacturer B shows higher variance in the R_{Sh} values at lower temperatures. Altogether, due to lower slope at the I_{SC} point for this module, the mean variance of the shunt resistance values is higher for module B. Also similar to module A, module B shows less change in R_{Sh} for higher temperatures, which shows a non-linear relationship between the cell temperature and the shunt resistance.

In a second run, the temperature decreased from 85 °C to -10 °C. Figure 4 shows the shunt resistance and mean temperature for module A during the temperature decrease. During the beginning temperature rise from 20 °C to 85 °C, the decrease in the shunt resistance becomes apparent. The following stepwise decrease of the temperature caused the shunt resistance to rise. Similar to the run with temperature increase (Fig. 2), the change of R_{Sh} is non-linear. The settling values of R_{Sh} are similar to the values during the heating-up, which shows that there is no sign of hysteresis effect caused by the temperature increase or decrease in module A.

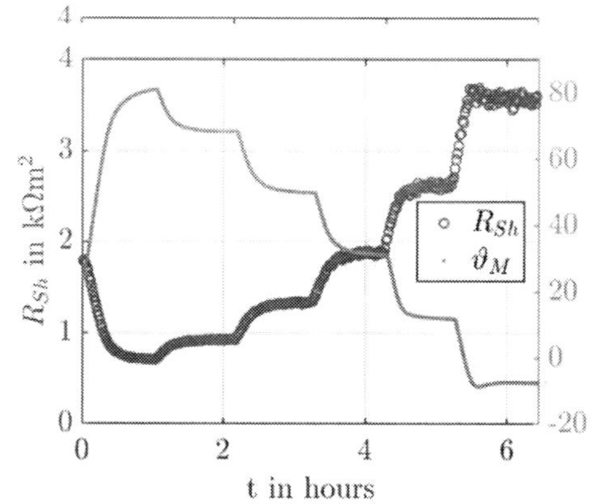

Figure 4: Shunt resistance R_{Sh} (blue circles, left axis) and mean module temperature (green dots, right axis) over time for a CIGS module from manufacturer A during temperature decrease.

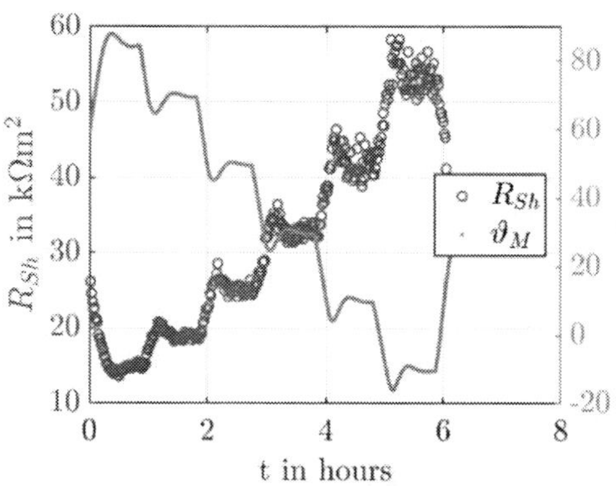

Figure 5: Shunt resistance R_{Sh} (blue circles, left axis) and mean module temperature (green dots, right axis) over time for a CIGS module from manufacturer B during temperature decrease.

Figure 5 shows the measurements for the module of manufacturer B. Here, similar to the heating-up run, the temperature shows overshoots caused by the regulation of the chamber temperature. This time, the shunt resistance also shows some overshoots. For low temperatures, again the IV curve slope becomes quite low which causes the accuracy of the R_{Sh} calculation (eq. (1)) to decrease and the values show a higher variance. The settling values are comparable to the heating-up run. For temperatures above 30 °C, the R_{Sh} values are slightly higher during the cool down run compared to the heating-up run. A possible cause for the difference is the impact of the absolute resolution of the current measurement, which becomes more apparent at low current values, i.e. at high R_{Sh} values.

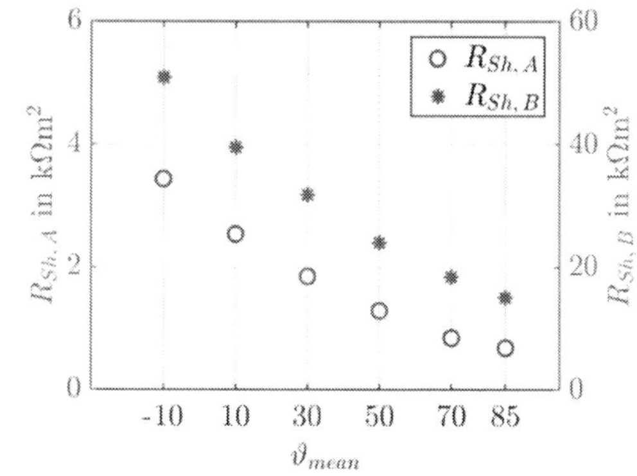

Figure 6: Shunt resistance R_{Sh} of CIGS modules from manufacturer A (blue circles, left axis) and B (blue stars, right axis) over module temperature. The values of R_{Sh} are the mean value from the heating-up and cool down runs from Fig. 2 to Fig. 5.

For a better visibility of the correlation between temperature and shunt resistance, Figure 6 shows the module shunt resistances of manufacturer A and B over the module temperature. The figure shows the mean values of R_{Sh} of the heating-up and cool down run after thermal stabilization. In this notation, the non-linear relationship between R_{Sh} and temperature becomes even more apparent. For module A, there is a nearly linear relationship between R_{Sh} and temperature between -10 °C and 30 °C, after which the rate of R_{Sh} decrease reduces. The module of manufacturer B shows a similar behaviour. Even though these two CIGS modules from different manufacturers show a similar trend of R_{Sh} change depending on temperature, the literature review illustrates that the temperature dependency must be determined for each individual module.

5 SUMMARY AND OUTLOOK

The determined temperature dependency of two CIGS modules from two manufacturers shows a non-linear correlation. The absolute R_{Sh} values for Module B are around one order of magnitude greater than the values of module A. Both modules show a non-linear, negative temperature coefficient. In the 95 K range from -10 °C to 85 °C, module A shows a change of R_{Sh} by a factor of around five. In module B, R_{Sh} changes by a factor of around three. This significant decrease with higher temperature does not show any hysteresis effect in both modules; it is thereby immediately reversible.

For the identification of PID-s in CIGS modules via measurements of the shunt resistance, it is essential to consider this temperature dependency. Thereby it is possible to distinguish between a change in R_{Sh} that is introduced by temperature and a change in R_{Sh} that is introduced by PID-s progress.

ACKNOWLEDGEMENT

The authors would like to thank the German Federal Ministry of Education and Research (BMBF) for funding the project "PV-FeldLab" within the funding program Forschung an Fachhochschulen under the support code FKZ 13FH600IB6.

REFERENCES

[1] Florides, M. et al.: Early Potential Induced Degradation (PID) Detection in the Field: Solar Cell Shunt Resistance Measurement at different Bias Conditions. In. 35th European Solar Energy Conference and Exhibition, 2018

[2] Hacke, P. et al.: In-Situ Measurement of Crystalline Silicon Modules Undergoing Potential-Induced Degradation in Damp heat Stress testing for Estimation of Low-Light Power Performance. In.: 23rd Workshop on Crystalline Silicon Solar Cells & Modules: Materials and Processes, 2013

[3] Hacke, P. et al.: Elucidating PID Degradation Mechanisms and In Situ Dark I V Monitoring for Modeling Degradation Rate in CdTe Thin Film Modules. In: IEEE Journal of Photovoltaics, Vol. 6 (2019), pp. 1635-1640

[4] Gerstenberg, L. et al.: CdTe Module Degradation during Dry and Damp Accelerated Aging Tests in Respect to the Leakage Current Pathway. In: 37th European Photovoltaic Solar Energy Conference and Exhibition, 2020

[5] Weber, T. et al.: Electroluminescence on the TCO corrosion of thin film modules. In: 25th European Photovoltaic Solar Energy Conference and Exhibition, 2010

[6] Boulhidja, S. et al: Experimental Evidence of PID Effect on CIGS Photovoltaic Modules. In: Energies 2020

[7] Muzzillo, C. P. et al.: Potential Induced Degradation of Cu(In,Ga)Se2 Solar Cells: Alkali Metal Drift and Diffusion Effects. In: IEEE Journal of Photovoltaics, Vol. 8 (2018), pp. 1337-1342

[8] Taubnitz, Ch. et al.: Potential-Induced Degradation: An Improved Unterstanding of Mechanisms and Influence Factors. In: 32nd European Photovoltaic Solar Energy Conference and Exhibition, 2016

[9] Williams, B. L. et al.: Identifying parasitic current pathways in CIGS solar cells by modelling dark J-V response. Progress in Photovoltaics, Research and Publications, Vol. 23 (2015), pp. 1516-1525.

[10] Kronthaler, L. et al.: Temperature Dependence of Shunt and Series Resistance of Different PV Module Technologies. In: 28th European Photovoltaic Solar Energy Conference and Exhibition, 2014

[11] Ruschel, C. S. et al: Experimental analysis of the single diode model parameters dependence on irradiance and temperature. In: Solar Energy, Vol. 217 (2021), pp. 134-144,

[12] Diatta, I. et al.: Temperature effect on the shunt resistance of a white biased silicon solar cell. In: Journal of Scientific and Engineering Research Vol 4(8), pp. 11-19

[13] Theelen, M. et al: Determination of the temperature dependency of the electrical parameters of CIGS cells. In. Journal of Renewable and Sustainable Energy, Vol. 9 (2017), pp. 021205-1- 021205-16

COMPLETE RECOVERY OF CRYSTALLINE SILICON PHOTOVOLTAIC MODULES BY THE EARLY DETECTION OF POTENTIAL INDUCED DEGRADATION

Michalis Florides, George Makrides and George E. Georghiou
Photovoltaic Technology Laboratory, FOSS Research Centre for Sustainable Energy,
Department of Electrical and Computer Engineering, University of Cyprus
Panepistimiou 1 Avenue, 1678, Nicosia, Cyprus
Phone: +357 22 894397, Fax: +357 22 895370
florides.michalis@ucy.ac.cy, makrides.georgios@ucy.ac.cy, geg@ucy.ac.cy

ABSTRACT: High voltage (HV) photovoltaic (PV) systems are affected by potential induced degradation (PID). PID decreases the output power of PV modules, however, it is a reversible process and could be recovered. Researchers demonstrated that the earlier PID is detected, the higher and faster the recovery. This work presents the experimental results comparing the PID recovery of PV modules under early and late PID detection. It was shown that complete recovery is possible if PID is detected at an early stage (< 1% power loss).

Keywords: photovoltaic (PV), potential induced degradation (PID), detection, forward DC resistance (FDCR)

1 INTRODUCTION

Potential induced degradation (PID) is a major performance loss issue in photovoltaic (PV) systems [1]–[11]. Laboratory tests on PV modules from 13 manufacturers have revealed that more than 50% of them are prone to PID under various test conditions [12], [13] affecting the performance and reliability of high voltage PV systems due to their increased degradation rate.

The power generation of a PV system could be reduced by more than 25% due to PID [3]–[5], [7], [8], [10]. The extend to which PID will progress, depends on the PV system operating voltage and the environmental conditions [14]–[16]. It was demonstrated that PID is reversible and recovery is possible if recovery measures are applied at the early stages of PID development [17]–[19].

We have demonstrated [20]–[26] that the measurement of the forward DC resistance (FDCR) of PV cells at low forward bias conditions could be used for the early detection (< 1% power loss) of PID of the shunting type (PID-s) that affects p-type crystalline PV cells [27] which dominate the market.

This work presents the experimental results comparing the PID recovery of PV modules under early and late PID detection. PID was monitored by using a prototype PID monitoring circuit (PMC) with the FDCR method embedded.

2 METHODOLOGY

2.1 Experimental Setup

Multi-cell PV modules were used as test samples. The PV samples were commercial 60-cell PV modules (1640x992x40 mm) made of 60 p-doped base and n-doped emitter (emitter pointing towards the front glass) polycrystalline cells, high transmission low iron tempered glass (AR coated), EVA encapsulant and composite film back-sheet. The modules were rated for a system voltage of 1000 V DC and were protected by 3 diodes in the junction box.

An indoor test setup was implemented that allowed the application of a high voltage (HV) bias under dark conditions by using a power supply unit (PSU). The samples were placed in an environmental chamber to be able to control their temperature (Fig. 1). The aluminium frame of the sample was connected to the negative (-ve) terminal of the HV PSU. The positive (+ve) terminal of the sample was connected to the positive (+ve) terminal of the HV PSU. Both electrical terminals of the sample were connected to the PID monitoring circuit (PMC). Aluminium foil was used to accelerate PID progression. The foil was placed on top of the front glass of the sample and was electrically connected to the aluminium frame of the module.

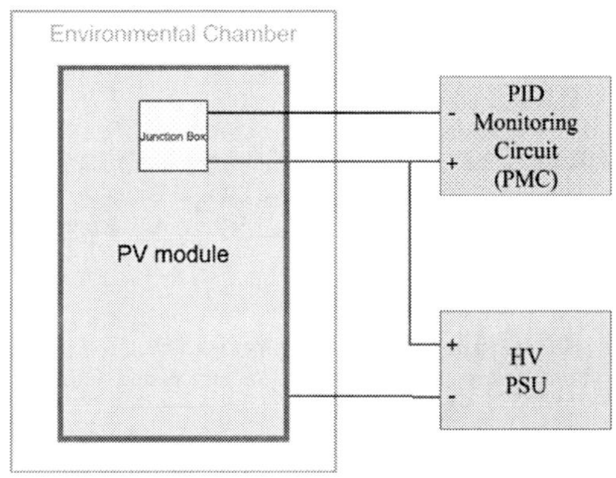

Figure 1: Indoor test setup for inducing and monitoring PID in multi-cell PV modules. The PMC electrical connection to the PV module was 2-wire.

2.2 PID Degradation

PID was induced by applying a voltage of -1000 V DC under dark conditions (environmental chamber). The temperature of the PV samples (T_s) was set to 70 °C (±0.5 °C) by using the environmental chamber to accelerate PID degradation. The chamber and sample temperature were measured by using RTD sensors.

2.3 PID Recovery

PID was recovered by applying a voltage of +1000 V DC (voltage recovery) or a temperature of

70 ° (thermal recovery) or a combination of both under dark conditions (environmental chamber). The temperature of the PV samples (T_s) was controlled by using the environmental chamber. The chamber and sample temperature were measured by using RTD sensors.

2.4 PID Progression

The PID progression (degradation or recovery) was monitored by measuring the FDCR at low forward bias conditions (< 10 mA) by using the PMC. An electrical current was applied to the electrical terminals of the PV sample and the FDCR was calculated based on the measured voltage. The PMC was measuring the FDCR continuously with a 2-wire measurement (4-wire was not necessary due to the low bias currents during the measurement). The measurement was conducted with a sampling rate of 10 seconds for a short period of time, i.e. every 10 seconds a measurement was acquired for a duration lower than 350 ms. This was performed in order not to bias the sample continuously and possibly affect PID progression.

The PMC was able to measure the FDCR while the HV bias was being applied in order to avoid any capacitive currents, which could affect PID progression, if the HV was switched on/off every 10 seconds.

2.5 Output Power Measurement

The output power (P_{out}) of the PV samples was measured at standard test conditions (STC). The output power was measured by using light current-voltage (LIV) curves (solar simulator). The maximum power measurement was performed in less than 30 minutes after the HV bias was removed in order to minimise PID recovery effects.

2.6 Electroluminescence Imaging

Electroluminescence (EL) images were captured for the PV samples after each output power (P_{out}) measurement (see before). The images were captured at a current of 1 A (more sensitive to PID). The measurement was conducted in less than 30 minutes after the output power measurement in order to minimise PID recovery effects.

2.7 Equipment and Measurement Accuracy

The equipment used during the experiments and its measuring accuracy are shown in Table I.

Table I: Equipment used and measurement accuracy.

Equipment	Measurement Accuracy
HV PSU (Keithley 2410)	< 0.1%
Source-Meter device (Keithley 2430)	< 0.1%
Solar simulator (Pasan A+A+A+)	< 1.0%*
PID monitoring circuit (PMC)	< 0.5%
Temperature sensors (RTD)	< 0.5%

* The combined accuracy was < 1% (temporal instability < 0.15%, irradiance uniformity < 0.51% and maximum power measurement < 0.23%).

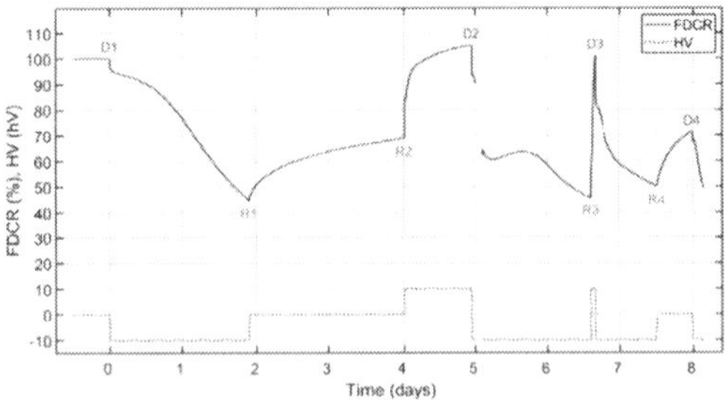

Figure 2: Forward DC resistance (FDCR) and high voltage (HV) against time (PID progression) for multiple PID degradation/recovery cycles in a 60-cell PV sample. All of the PV cells of the sample were subjected to degradation/recovery. A 'D' and 'R' label indicates degradation and recovery respectively. The HV value indicates whether the sample was subject to degradation (-1000 V), voltage recovery (+1000 V) or thermal recovery (0 V). The sample temperature (T_s) was 70 °C ±0.5 °C and the relative humidity (RH) < 10%. The initial (100%, before PID) FDCR was 8 kΩ. The 148 minutes interruption between day 5 and 6 is due to missing data.

3 RESULTS

3.1 Early Recovery

PID decreases the output power (P_{out}) of PV modules. The power could be recovered if PID is detected at an early stage. It has been demonstrated that the output power of PV modules is not fully recoverable, at least within the time duration of the tests, if the power loss due to PID is > 10% [17]–[19]. Therefore, the early PID detection may allow for the full recovery of the output power of PV modules in a short time and preserve their lifetime.

To demonstrate the advantage of early detection on the recovery, various degradation (D)-recovery (R) cycles were applied to a 60-cell PV sample as shown in Fig. 2. The PID progression was monitored by the PMC with the FDCR method embedded which is able to detect PID before 1% power loss (P_{loss}). Every time the FDCR value was reduced by approximately 50% by a degradation (D) event, a recovery (R) event was initiated. A more detailed description is given below.

1. Before day 0: reference measurement.
2. D1: Degradation started.
3. R1: The FDCR reached 44.6%, the sample was thermally recovered at 70 °C.
4. R2: The sample recovered slowly to 69% and a voltage recovery was applied. The FDCR fully recovered within a few hours and exceeded its initial value (105.3%). This is probably due to residual sodium ions in the PV cells before the PID test had started.
5. D2: The FDCR reduced quickly to 64.7% and after several hours to 45.5% in an unexpected way. The fast reduction could be because some sodium ions did not cross into the glass and returned quickly into the PV cells p-n junction. The unexpected increase-decrease of the FDCR could be because some ions crossed the

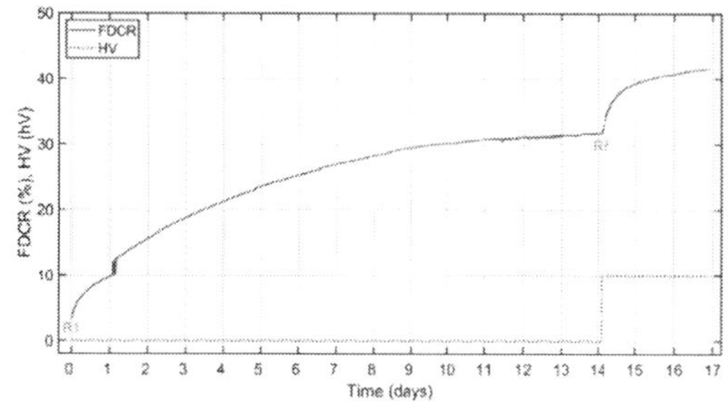

Figure 3: Forward DC resistance (FDCR) and high voltage (HV) against time (PID progression) for PID recovery in a 60-cell PV sample. All of the PV cells of the sample were subjected to recovery. The 'R' labels indicate different recovery stages. The HV value indicates whether the sample was subject to voltage recovery (+1000 V) or thermal recovery (0 V). The sample temperature (T_s) was 70 °C ± 0.5 °C and the relative humidity (RH) was < 10%. The initial (100%, before PID) FDCR was 8 kΩ.

p-n junction into the p-region reducing the shunting of the PV cells until more ions moved from the glass into the junction.

6. R3: The sample was voltage recovered and the FDCR reached 101.4%.
7. D3: The FDCR reduced to 50% within 1 day.
8. R4: Thermal recovery, the FDCR increased to 71.6% within 0.5 days.
9. D4: The FDCR reduced to 49.5% after a few hours.

From the results, it seems that the degradation/recovery time depends the PID history of the PV sample. Also, it appears that after a voltage recovery, the sample degrades faster than after a thermal recovery.

Furthermore, an important result stems by measuring the output power (P_{out}) of the sample before and after the test. The output power after the test was 0.43% higher than the initial power (within the 0.6% measurement accuracy of the solar simulator). Thus, no power was lost during the degradation/recovery cycles. This is because the PV sample was recovered early, every time the FDCR decreased approximately 50% from its initial value. Therefore, detecting PID at an early stage allows for the full recovery of the PV sample.

3.2 Late Recovery

In addition to the early PID recovery, late PID detection recovery (late recovery) was examined. A 60-cell PV sample was degraded and then was recovered. The results are shown in Fig. 3. The output power (P_{out}) of the PV sample was measured as well at various stages during the recovery (Table II). Initially, the PV sample was subjected to thermal recovery for 14 days (R1). It can be seen that the FDCR increased by 28.3% while the output power increased by 5.7%. A similar FDCR change (24.4%) occurred in the early PID thermal recovery in just 2.1 days while the output power change was < 1% (R1 event in Fig. 2). This indicates that if PID is detected at an early stage the recovery of the PV sample is faster. After the thermal recovery, voltage and thermal recovery was applied for 2.8 days (R2). The FDCR increased by 9.8%, however, the output power increase was < 1% (considering the 1% measurement accuracy of the solar simulator). This suggests that the output power of a PV sample recovers with a high rate at the beginning which decreases as the output power reaches the initial (before PID) power of the sample. This is supported by the electroluminescence images in Fig. 4 that show a significant increase in the luminosity of the PV cells from L1 to L2, whereas the luminosity change from L2 to L5 is insignificant. The image at L0, before PID degradation, reveals that the PV sample had not fully recovered yet.

4 CONCLUSION

Potential induced degradation (PID) is a major performance loss issue in photovoltaic (PV) systems. Researchers demonstrated that PID is reversible and recovery is possible if recovery measures are applied at the early stages of PID development. This work presented the experimental results comparing the PID recovery of PV modules under early and late PID detection. The PV modules were subjected to a series of degradation and early recovery events under different conditions (thermal recovery and/or voltage recovery). The output power (P_{out}) was measured before and after the events and did not change by more than 1% (within the measurement uncertainty of the solar simulator). This suggests that the early PID detection allows the complete power recovery of PV modules from PID in a short time. This is a promising outcome since it could preserve the lifetime of PV modules in the field. From the late PID detection, the output power of the PV module recovered with a high rate at the beginning and decreased as the output power reached the initial (before PID) power of the module. The overall rate of recovery was slower compared to early recovery.

Table II: Forward DC resistance (FDCR) and output power (P_{out}) change (Δ) with PID recovery at different PID levels (L) in a 60-cell PV sample.

PID level (L)	Time (Days)	FDCR (kΩ)	FDCR (%)*	ΔP_{out} (%)**
L1	0	0.28	3.48	-8.53
L2	4.8	1.86	23.1	-2.45
L3	8.4	2.32	28.8	-2.35
L4	14.1	2.56	31.8	-2.81
L5	16.9	3.35	41.6	-2.03

The PID recovery was progressing from L1 to L5.
The FDCR was measured at 70 °C and P_{out} at STC.
The initial, before PID (L0), FDCR was 8 kΩ and P_{out} was 260.1 W.
$* \% = \frac{V_{Lx}}{V_{L0}} * 100$, value at each level (Lx) and value at L0.
$** \% = \frac{V_{Lx}-V_{L0}}{V_{L0}} * 100$, value at each level (Lx) and value at L0.

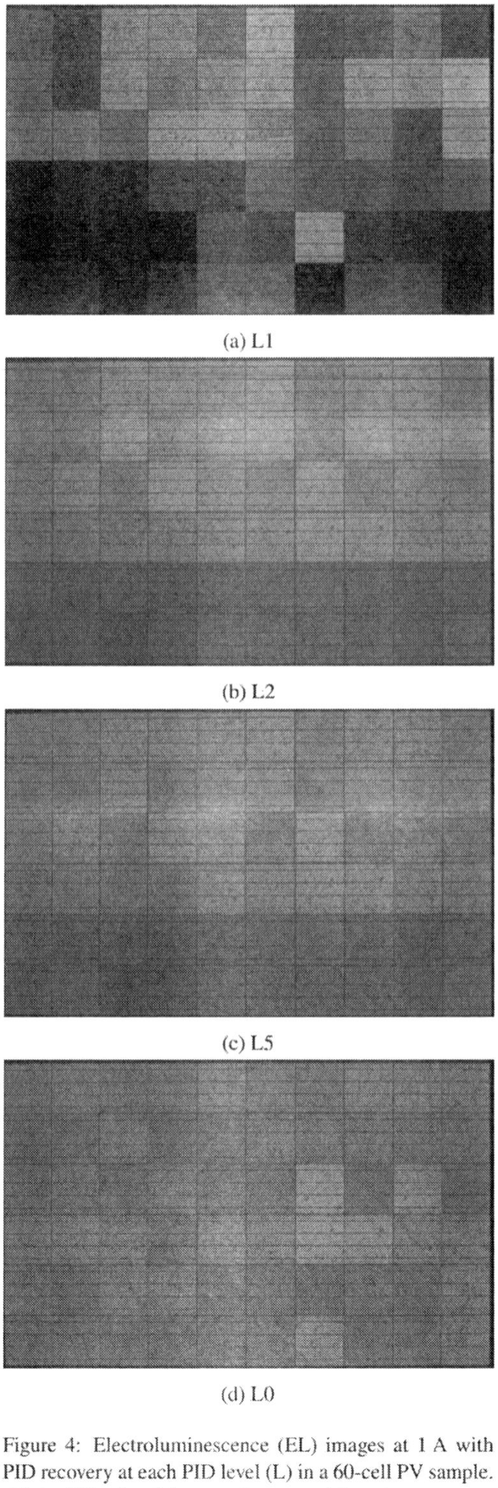

(a) L1

(b) L2

(c) L5

(d) L0

Figure 4: Electroluminescence (EL) images at 1 A with PID recovery at each PID level (L) in a 60-cell PV sample. All the PV cells of the sample were subjected to recovery. For the PID levels, see Table II. The EL image before PID degradation (L0), is shown as well for comparison. The lower the luminosity of the PV cells, the higher the PID degradation.

5 REFERENCES

[1] H. Yang, F. Wang, H. Wang, J. Chang, D. Song, and C. Su, "Performance deterioration of p-type single crystalline silicon solar modules affected by potential induced degradation in photovoltaic power plant," *Microelectronics Reliability*, vol. 72, pp. 18–23, 2017. DOI: 10.1016/j.microrel.2017.03.017.

[2] A. Bouaichi, A. A. Merrouni, C. Hajjaj, *et al.*, "Experimental investigation of potential induced degradation (PID) impact and recovery on crystalline photovoltaic systems," in *International Conference on Electronic and Renewable Energy*, Saidi, 2018, pp. 623–629. DOI: 10.1007/978-981-13-1405-6_71.

[3] J. Huang, H. Li, Y. Sun, H. Wang, and H. Yang, "Investigation on potential-induced degradation in a 50 MWp crystalline silicon photovoltaic power plant," *International Journal of Photoenergy*, vol. 2018, pp. 1–7, 2018. DOI: 10.1155/2018/3286124.

[4] J. Hylsky, D. Strachala, P. Vyroubal, P. Cudek, J. Vanek, and P. Vanysek, "Effect of negative potential on the extend of PID degradation in photovoltaic power plant in a real operation mode," *Microelectronics Reliability*, vol. 85, pp. 12–18, 2018. DOI: 10.1016/j.microrel.2018.04.003.

[5] M. A. Islam, M. Hasanuzzaman, and N. A. Rahim, "Investigation of the potential induced degradation of on-site aged polycrystalline PV modules operating in Malaysia," *Measurement*, vol. 119, pp. 283–294, 2018. DOI: 10.1016/j.measurement.2018.01.061.

[6] F. Martinez-Moreno, G. Figueiredo, and E. Lorenzo, "In-the-field PID related experiences," *Solar Energy Materials and Solar Cells*, vol. 174, pp. 485–493, 2018. DOI: 10.1016/j.solmat.2017.09.037.

[7] H. Wang, P. Zhao, J. Chang, D. Song, and S. Sang, "Performance variation of dark current density-voltage characteristics for PID-affected monocrystalline silicon solar modules from the field," *Microelectronics Reliability*, vol. 81, pp. 320–327, 2018. DOI: 10.1016/j.microrel.2017.11.003.

[8] J. Huang, Y. Sun, H. Wang, and J. Zhang, "Regular and irregular performance variation of module string and occurred conditions for potential induced degradation-affected crystalline silicon photovoltaic power plants," *Energies*, vol. 12, no. 22, p. 4230, 2019. DOI: 10.3390/en12224230.

[9] G. Sun, X. Tu, and R. Wang, "Research on the potential-induced degradation (PID) of PV modules running in two typical climate regions," *Clean Energy*, vol. 3, no. 3, pp. 222–226, 2019. DOI: 10.1093/ce/zkz007.

[10] H. Wang, X. Cheng, H. Yang, W. He, Z. Chen, and L. Xu, "Potential-induced degradation: Recombination behavior, temperature coefficients and mismatch losses in crystalline silicon photovoltaic power plant," *Solar Energy*, vol. 188, pp. 258–264, 2019. DOI: 10.1016/j.solener.2019.06.023.

[11] Y. R. Golive, S. Zachariah, R. Dubey, *et al.*, "Analysis of field degradation rates observed in All-India survey of photovoltaic module reliability 2018," *IEEE Journal of Photovoltaics*, vol. 10, no. 2, pp. 560–567, 2020. DOI: 10 . 1109 / JPHOTOV . 2019.2954777.

[12] M. Ebert, ""Reliability of solar modules and systems" - R&D competencies," Fraunhofer, Tech. Rep., 2015. DOI: 10.13140/RG.2.1.1328.3280.

[13] J. Carolus and M. Daenen, "Potential induced degradation (PID): a test campaign at module level," Tech. Rep., 2016.

[14] S. Hoffmann and M. Koehl, "Effect of humidity and temperature on the potential-induced degradation," *Progress in Photovoltaics: Research and Applications*, vol. 22, no. 2, pp. 173–179, 2014. DOI: 10.1002/pip.2238.

[15] P. Hacke, S. Spataru, K. Terwilliger, *et al.*, "Accelerated testing and modeling of potential-induced degradation as a function of temperature and relative humidity," *IEEE Journal of Photovoltaics*, vol. 5, no. 6, pp. 1549–1553, 2015. DOI: 10.1109/JPHOTOV.2015.2466463.

[16] M. Koehl and S. Hoffmann, "Impact of rain and soiling on potential induced degradation," *Progress in Photovoltaics: Research and Applications*, vol. 24, no. 10, pp. 1304–1309, 2016. DOI: 10.1002/pip.2786.

[17] B. Jaeckel, M. Cosic, and J. Arp, "PID effect of c-Si modules: Study of degradation and recovery to more closely mimic field behavior," *Israel Journal of Chemistry*, vol. 55, no. 10, pp. 1091–1097, 2015. DOI: 10.1002/ijch.201400201.

[18] J. Oh, S. Bowden, and G. TamizhMani, "Potential-induced degradation (PID): Incomplete recovery of shunt resistance and quantum efficiency losses," *IEEE Journal of Photovoltaics*, vol. 5, no. 6, pp. 1540–1548, 2015. DOI: 10.1109/JPHOTOV.2015.2459919.

[19] J. Carolus, W. D. Ceuninck, and M. Daenen, "Irreversible damage at high levels of potential-induced degradation on photovoltaic modules," in *IEEE Photovoltaic Specialits Conference*, 2017.

[20] M. Florides, G. Makrides, and G. E. Georghiou, "Early potential induced degradation (PID) detection in the field: Voltage measurement methods," in *European Photovoltaic Solar Energy Conference and Exhibition*, Amsterdam, 2017, pp. 1677–1681. DOI: 10 . 4229 / EUPVSEC20172017-5BV.4.61.

[21] M. Florides, G. Makrides, and G. E. Georghiou, "Characterisation of the shunt resistance due to potential induced degradation (PID) in crystalline solar cells," in *IEEE World Conference on Photovoltaic Energy Conversion*, Waikoloa, Hawaii, 2018, pp. 0695–0699. DOI: 10 . 1109 / PVSC.2018.8547500.

[22] M. Florides, G. Makrides, and G. E. Georghiou, "Early detection of potential induced degradation by measurement of the forward DC resistance in crystalline PV cells," *IEEE Journal of Photovoltaics*, vol. 9, no. 4, pp. 942–950, 2019. DOI: 10.1109/JPHOTOV.2019.2910235.

[23] M. Florides, A. Livera, G. Makrides, and G. E. Georghiou, "Shunt resistance relation to power loss due to potential induced degradation in crystalline photovoltaic cells," in *IEEE Photovoltaic Specialists Conference*, Chicago, 2019, pp. 1950–1954. DOI: 10.1109/PVSC40753.2019.8980722.

[24] M. Florides, G. Makrides, and G. E. Georghiou, "Early detection of potential induced degradation in the field: Testing a new methodology on silicon PV modules," in *European Photovoltaic Solar Energy Conference and Exhibition*, Lisbon, 2020, pp. 1392–1395. DOI: 10 . 4229 / EUPVSEC20202020-5DO.4.4.

[25] M. Florides, G. Makrides, and G. E. Georghiou, "Electrical and temperature behavior of the forward DC resistance with potential induced degradation of the shunting type in crystalline silicon photovoltaic cells and modules," *IEEE Journal of Photovoltaics*, vol. 11, no. 1, pp. 16–25, 2021. DOI: 10 . 1109/ JPHOTOV.2020.3030191.

[26] M. Florides, G. Makrides, and G. E. Georghiou, "Early detection of potential induced degradation in the field: Testing a new method for silicon PV modules," in *IEEE Photovoltaic Specialists Conference*, Fort Lauderdale, 2021, pp. 0950–0953. DOI: 10.1109/PVSC43889.2021.9518482.

[27] W. Luo, Y. S. Khoo, P. Hacke, *et al.*, "Potential-induced degradation in photovoltaic modules: a critical review," *Energy & Environmental Science*, vol. 10, pp. 43–68, 2017. DOI: 10.1039/c6ee02271e.

LETID: ELECTRICAL PARAMETERS TREND OF BIFACIAL SILICON MODULES USING ACCELERATED AGING TESTS AND STATISTICAL MODELLING

G. Plessis*[1], J. Dupuis[1], O. L. Rhazi[1], E. Sandré[2], K. Radouane[2]

[1]EDF R&D, EDF Lab Les Renardières, Avenue des Renardières, 77250 Moret Loing et Orvanne, France
[2]EDF Renouvelables, 100 Esplanade du Général de Gaulle, 92932 Paris La Défense Cedex, France
*Corresponding Author: gilles.plessis@edf.fr

ABSTRACT: This work aims at supporting a better understanding of the LeTID phenomenon. It relies on accelerated stress test under current injection to emulate light exposure for three batches of bifacial PV modules (mono and cast mono silicon modules). Indoor characterization is carried out regularly at different irradiance levels and electrical parameters are estimated from it. The paper goes through a qualitative analysis describing the evolution of electrical characteristics and a quantitative one relying on descriptive statistics to identify the drivers behind the LeTID degradation, their role during degradation and regeneration phase and the unbalanced response to different irradiance levels. The analysis reports that a convergence of the long-term stabilization power regardless of the injected current and highlights that the performance losses are driven globally by Isc, then Voc and FF. Isc has a prominent role in the degradation phase and FF drives the power recovery for the regeneration phase. In addition, LeTID phenomenon appears significantly stronger at low irradiance levels for the rear face. The maximal degradation is reduced by 0.2%/(100W/m^2) between 200 W/m^2 and 1200 W/m^2.
Keywords: PV module; Bifacial; LeTID; Simulation

1 INTRODUCTION

Light and elevated Temperature-Induced Degradation (LeTID) has become a main concern for new photovoltaic projects, especially when it comes to technologies such as Passivated Emitter and Rear Cell modules (PERC modules) with performance losses ranging from 0% to more than 9% in laboratory tests [1-6]. This phenomenon is still not yet fully understood. A recent overview on the last researches on the LeTID phenomenon can be found in [7].

In [2-3], we have highlighted the different behavior observed on bifacial modules between the front and rear side, its impact on bifaciality coefficient and consequence on PV plants yield. In this paper we report a convergence for long term stabilization regardless of the injected current and performance losses driven globally by Isc, then Voc and FF on the same batches of modules.

2 EXPERIMENTAL DETAILS

2.1 Modules types and carrier induced degradation (CID) parameters

The table 1 details the module specifications used for these tests. The module batch comes from a tier one manufacturer. 4 modules using monocrystalline p-PERC cells were tested.

These modules were also used in previous papers for initial LeTID analysis with other type of modules. Results can be found in [2-3].

Table I: PERC module specifications. SP is related to the presence of a white screen-print between the solar cells to boost front side power.

	PERC 1
Cell number	60
Silicon type	monocrystalline
Average bifaciality factor	65.5%
Rear side material	Glass + SP
Power	300 Wp

CID testing was used to activate LeTID phenomenon. The protocol for LeTID evaluation in silicon modules is still under debate and institutes have tested currents ranging from Isc-Impp (0.5 to 1 A usually) to Isc (8 to 10A) and temperatures ranging from 75 to 85°C. This yielded different LeTID impacts as the kinetic is modified. The detection of the maximal degradation caused by the phenomenon can vary by several percent or even be undetectable [1-5]. Recent results suggest that using current injection of Isc-Impp gives the best representability for LeTID degradation and regeneration as carrier density is quite similar to 1 sun illumination in MPP mode and because indoor results are in agreement with some real power plants [6]. The drawback of using low current values is a longer testing time to render the LeTID behavior. Even if high temperatures can accelerate the phenomenon, the modules should not be operated at temperatures above 85°C as that is the maximum working temperature allowed for a classical solar module. Carrying out the experiment at high temperatures entails the risk of activating other unwanted failure modes.

As no information on the aspects of the rear side of bifacial modules was available in the literature, we chose to evaluate LeTID using CID and a large current injection range: 0.5 A (≈ Isc-Impp), 5A (≈ Isc/2) and 10A (≈ Isc). One module was tested for each current. PERC 3 modules had been tested only at 0.5A and 5A current injection because of limited number of modules available. The temperature was set to 75°C to avoid working at too high temperature with the risk of activating failure modes unrelated to LeTID. No light soaking at low temperature was performed priori to LeTID testing. Modules were stored in the dark before the test.

2.2 Current voltage and electroluminescence characterization

Electrical characterization of the photovoltaic modules was carried out using a PASAN flash tester AAA SunSim 3b. The complete measurement system is located in an air-conditioned room with an ambient temperature of 25±1°C and a module temperature control under standard testing conditions (STC) of 25±0.5°C.

Calibration of the flash tester is performed using PERC gold modules calibrated at Fraunhofer ISE with ±2.5% power uncertainty. The module temperature is measured by Pt100 temperature sensors attached to the back of the module. The reproducibility was estimated at ±0.5% based on the reference modules used during the experiment.

Electroluminescence (EL) measurements are done on a Greateye Lumi Solar Professional system. The imaging system resolution is 250 µm/pixel. The applied voltage range for this setup ranges from 0 to 150 V. The system is equipped with two CCD Matrix cameras characterized by a sensor format of 2048 x 512 pixels, a pixel size of 13.5 µm x 13.5 µm and high sensitivity in the near-infrared region, which is suitable for our silicon PV modules EL images. The spatial resolution of the system is estimated at 0.5 mm, for a 50 mm focal length, a numerical aperture of 1.4 and 0.95 m of working distance. Applied voltage was adjusted to get EL images of PV modules in short circuit conditions.

3 CARRIER INDUCED DEGRADATION RESULTS

As shown is Figure 1 and Figure 2 for monocrystalline modules, it is well known that the change of the excess carrier density by current injection strongly modify the impact of LeTID effect. We also demonstrate that for monocrystalline modules tested, main electrical parameters are all stabilizing at similar values whatever the tested current injection. This means that stabilized value can be obtained without using a slow current injection and can help to forecast module behavior in a PV field.

Moreover, if we focus on the lowest current injection (0.5 A is representative of a 1 sun exposure) we can see that the maximal degradation point happens at the same time between all key electrical parameters of the modules (V_{OC}, Isc and power) except for FF. The regeneration trend is not the same between all of them. This is also true between the different face as compared between Figure 1 and Figure 2 but also between different technologies (not shown here) as the three batches tested are all showing specific regeneration profile.

This mean that we can rely on LeTID testing with low injected current to determine the maximum degradation point. The testing at low current need to be extended to forecast how the module will behave during the regeneration phase to better anticipate the yield of future PV plants but not till a stabilization is observed. A test at higher current can be used in a reliable way to estimate long-term stabilization.

4 STATISTICAL MODELLING

4.1 Drivers of P_{MPP} variations

Computing correlations between electrical characteristics and P_{MPP} variations give insight on the parameters that must be chosen with care for modelling and informs on the underlying phenomena.

Table II presents Kendall and Spearman correlations for monocrystalline modules (PERC 1) for all indoor measurements between P_{MPP} variations and the three main components, i.e. Isc, Voc and FF. Even though FF contribution is low, it as to be noted than P_{MPP} is also highly correlated; 0.86 and 0.96 for Kendall and Spearman correlations respectively [8-9]. On the overall the main driver for P_{MPP} variation is Isc.

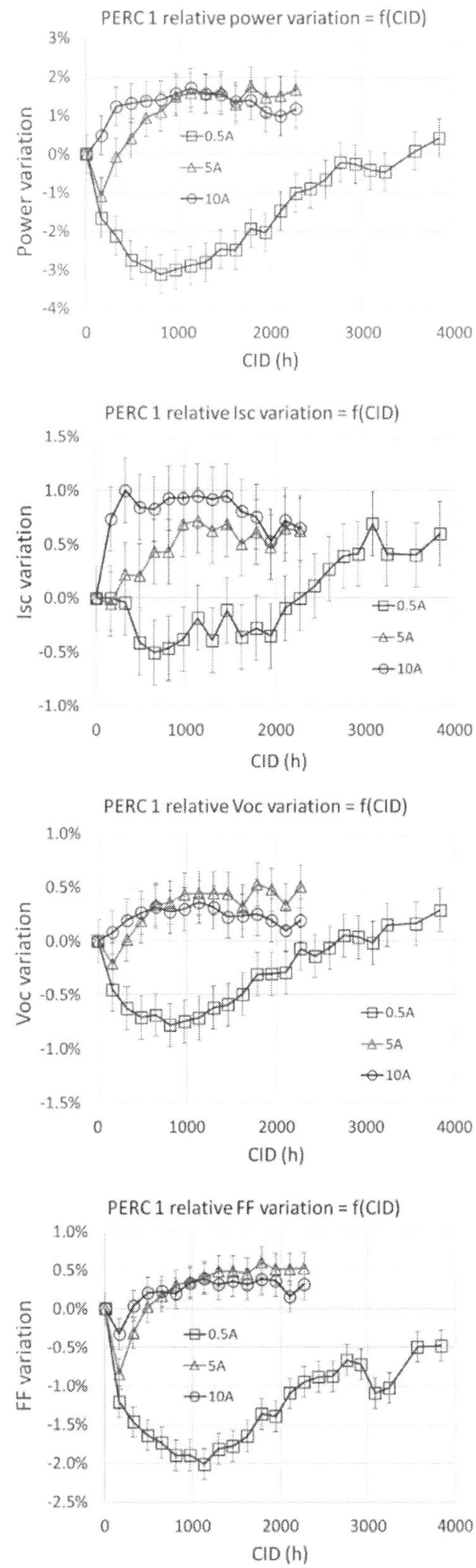

Figure 1: Bifacial monocrystalline modules evolution for the front side with CID testing

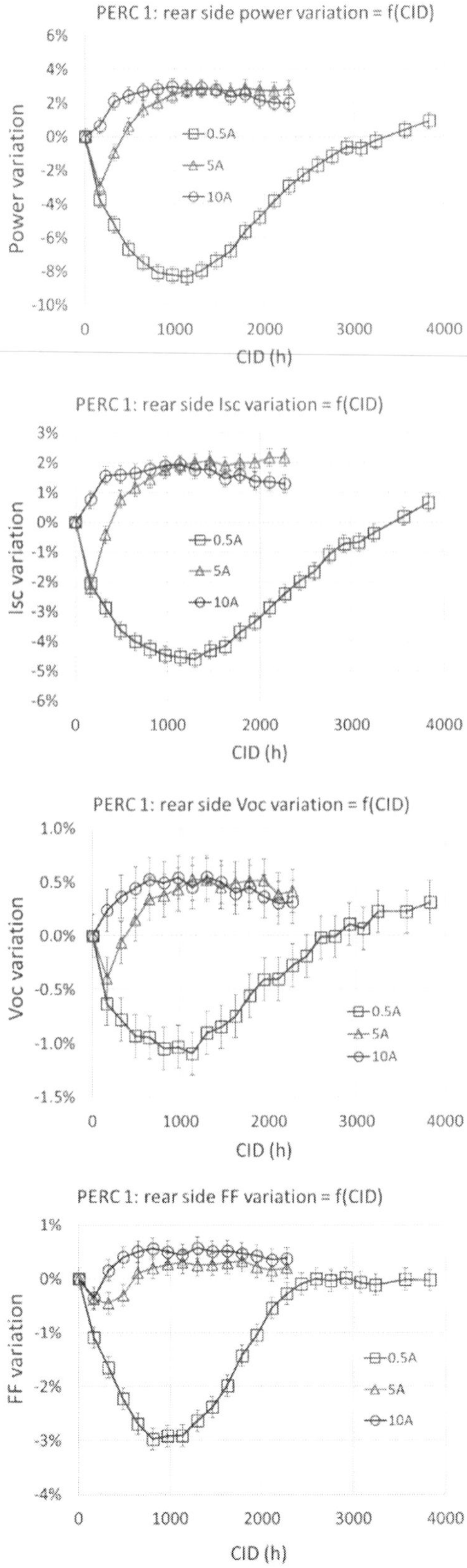

Figure 2: Bifacial monocrystalline modules evolution for the rear side with CID testing

Table II: Correlation between electrical parameters

	Kendall	Spearman
Isc	0.77	0.91
Voc	0.71	0.88
FF	0.63	0.80

4.2 Electrical parameters evolution

Comparing evolutions of main components, i.e. Isc, Voc and FF against P_{MPP} informs on the different dynamics at place and the underlying phases. On the Figure 3, electrical characteristics are plotted on the y-axis and P_{MPP} variations on the x-axis for the monocrystalline module at low current injection. Progress in hour through the accelerated stress test is color coded for each measurement point @ STC and markers are used to describe electrical characteristic (I for Isc, V for Voc…).

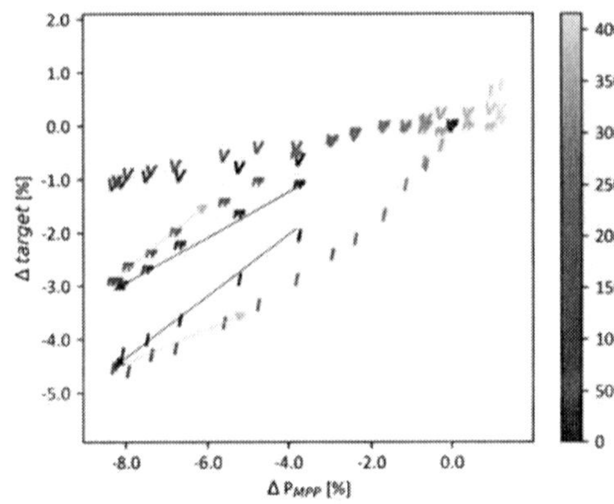

Figure 3: Evolutions of Isc, Voc and FF against P_{MPP}

A hysteresis pattern can be observed describing different dynamics for the Isc and FF compared to the P_{MPP} and a change after the maximal degradation point.

Isc demonstrates a more important slope during the degradation phase (arrow in dark blue) compared to the regeneration phase (arrow in yellow), therefore relatively to P_{MPP}, Isc tends to slow down during the regeneration phase. For FF, the opposite phenomenon can be observed, the slope is more important for the regeneration phase (arrow in yellow), indicating this time that the FF is regenerating faster than P_{MPP}. For Voc, the 2 phases are almost overlying, indicating a similar rate of change compare to P_{MPP}. In short Isc is driving the degradation and FF is driving the regeneration.

4.3 Irradiance dependency on maximal degradation

Experimental results also indicate a non-homogeneous degradation response to the irradiance levels for the rear face. The lower the irradiance level, the higher the maximal degradation is. For the monocrystalline modules, maximal P_{MPP} degradation goes from -10.0% at 200 W/m² to -7.9% at 1200 W/m² on the rear face. A linear dependence is found with a coefficient of 0.2% / (100 W/m²) with statistical significance (p-value below 1e-4) as described in Table III.

Table III: Linear relationship between maximal degradation and irradiance levels

	Rear side
Irradiance dependence	0.21% / (100 W/m²)
95% confidence interval (low)	0.18% / (100 W/m²)
95% confidence interval (high)	0.24% / (100 W/m²)

5 CONCLUSIONS

The analysis reports that a convergence of the long-term stabilization power regardless of the injected current. However, to capture the maximal degradation a low injected current should be used.

The study also highlights that the performance losses are driven globally by Isc, then Voc and FF. Isc has a prominent role in the degradation phase and FF drives the power recovery for the regeneration phase. In addition, LeTID phenomenon appears significantly stronger at low irradiance levels for the rear face. The maximal degradation is reduced by 0.2%/(100W/m²) between 200 W/m² and 1200 W/m².

6 ACKNOWLEDGMENTS

The authors are very grateful to the EDF R&D laboratory team for the help provided in performing module testing.

7 REFERENCES

[1] E. Fokuhl, T. Naeem, A. Schmid, P. Gebhardt, T. Geipel and D. Philipp, Proceedings of the 36th European Photovoltaic Solar Energy Conference, Marseille, France (2019) 816-821.
[2] J. Dupuis, G. Plessis, G. El Hajje, E. Lajoie-Mazenc, E. Sandré, K. Radouane and P. Dupeyrat, Proceedings of the 37th European Photovoltaic Solar Energy Conference, Marseille, France (2020) 802-810.
[3] J. Dupuis, G. Plessis, G. El Hajje, E. Lajoie-Mazenc, E. Sandré, K. Radouane and P. Dupeyrat, Prog Photovolt Res Appl. 29 (2021), 694-704
[4] M. Pander, M. Turek, J. Bauer, T. Luka and C. Hagenforf, Proceedings of the 35th Euro-pean Photovoltaic Solar Energy Conference, Brussels, Belgium (2018) 1265-1268.
[5] F. Kersten, P. Engelhart, H-C. Ploigt, A. Stekolnikov, T. Lindner, F. Stenzel, M. Bartzsch, A. Szpeth, K. Petter, J. Heitmann and J. W. Müller, Sol. Ener. Mat. & Solar Cells, 142 (2015) 83-86.
[6] F. Kersten, F. Fertif, K. Petter, B. Klöter, E. Herzog, M. B. Strobel, J. Heitmann and J. W. Müller, Energy Procedia, 124 (2017) 540-546.
[7] D. Chen, M. Vaqueiro Contreras, A. Ciesla, P. Hamer, B. Hallam, M. Abbott and C. Chan, Prog Photovolt Res Appl. (2020) 1–22
[8] Kendall, M. G. "A new measure of rank correlation." Biometrika 30, no. 1–2 (June 1, 1938): 81–93. https://doi.org/10.1093/biomet/30.1-2.81.
[9] Hollander, Myles, and Douglas A. Wolfe. Nonparametric Statistical Methods. Wiley Series in Probability and Mathematical Statistics. New York: Wiley, 1973.

POTENTIAL INDUCED DEGRADATION (PID) FREE MODULE DESIGN VIA INTERRUPTION OF THE ELECTRIC FIELD

Kai Sporleder[1,2,*], Bengt Jäckel[1], Sebastian Schindler[1], Sebastian Dittmann[1,2], Ralph Gottschalg[1,2]
[1] Fraunhofer Institute for Microstructure of Materials and Systems IMWS, Otto-Eißfeldt-Straße 12, 06120 Halle (Saale), Germany
[2]Anhalt University of Applied Sciences, Bernburger Straße 55, 06366 Köthen, Germany

*Corresponding author: Tel. +49 (0) 345/5589-5070, kai.sporleder@hs-anhalt.de

ABSTRACT: Potential-Induced Degradation (PID) is a major reliability problem and can lead to significant power and energy yield losses and even total loss of the affected systems. If a PV module is operated under PID conditions, in short, if there is a high voltage drop between the solar cells installed in the module and the front glass, which is in contact with the grounded frame, PID may occur. Electric fields in the magnitude of MV/cm through the glass, encapsulation and the dielectric passivation layers of the solar cell cause a leakage current. Ions or electrons flow to the cell and can lead to severe degradation. PID is almost independent of the cell technology and affects different generations of solar cells. The approach presented here shields the dielectrics of the Si solar cell from electric fields and interrupts the leakage current. The proposed concept can be applied to different cell technologies.
Keywords: PV Module, PID, degradation, reliability

1 INTRODUCTION: RELEVANCE AND AIM AND DELIVERY

PID of solar modules is known as a critical degradation mechanism. Technologically, the cause of PID lies in the electrical connection of PV modules in the PV system. Due to the series connection of solar modules in a string, the module voltages add up. Whereas in the past most systems were designed for a system voltage of 1.000V, modern systems are built with system voltages up to 1.500V. For efficiency gains most systems have a so called floating potential and the system voltage is almost symmetric around ground. However, some inverter types have a fixed negative potential of around 400V. Therefore, smaller systems and operation in warmer climates lead to a significant number of modules operated at negative potentials versus ground promoting PID-s of standard p-type c-Si cells.

At the ends of a string, voltages of several hundred of volts can be reached with respect to the grounded frame [2].

Degradation of PV modules can range from a reversible minor loss in module power to irreversible total power loss. Moisture or a combination of impurities and moisture cause conductive films to form on the module glass and the backside. Latter will be important for bifacial cells and the severances will depend on module construction and the resistivity of the front and the back cover. A voltage drop occurs between the surface of the module and the solar cells inside. Under these conditions, PID occurs over a wide range of module technologies: n-PERT [3], IBC [4], p-PERC [5] perovskite tandem materials [6, 7], bifacial solar cells can degrade on both sides, front and back [8]. After damp-heat pre-conditioning Si-HJT was demonstrated to suffer from PID, too [9].

The underlying physical mechanisms are different. Currently, PID by shunting (PID-s) [10, 11], PID by polarization (PID-p) [3, 12, 13], PID by corrosion (PID-c) [5, 14, 15] and PID by delamination (PID-d) [16] are known. All of them take place because an electric field is generated by the cover material (glass and polymers) and the functional coatings of the solar cell (Si_3N_4, Al_2O_3, SiO_2). Mitigation strategies can be applied at solar cell, PV-module, or PV-system level [17] Our approach is to interrupt the PV module internal electric field, independent of the underlying solar cell type. For this purpose, we use a transparent (no light should be absorbed) and conductive (shielding) layer between glass and EVA that interrupts the electrical field.

2 EXPERIMENTAL

Industrial PERC solar cells from p-type Cz wafers are laminated into mini modules, see Figure 1. The cells are known to be highly sensitive to PID-s.

While standard module, Figure 1a), uses a 3 mm soda lime glass, vinyl acetate (EVA) as an encapsulant and a backsheet, the shielded module, Figure 1b), is realized by using a glass with transparent conductive oxide, with sheet resistance of $R_\square = 10\ \Omega$ at the glass/EVA layer. The TCO layer is connected to the solar cell grid.

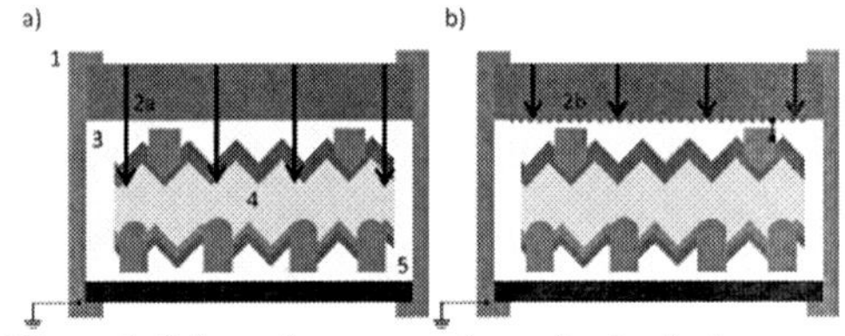

Figure 1: Schematic cross sections of a standard minimodule (a), and the shielded mini module during PID-Test. 1-grounded frame, 2a: standard glass, 2b: glass with TCO coating, 3: encapsulant, 4: solar cell, 5: back-sheet. Solar cell re-drawn from [1]

Four mini modules are built each, for the standard and the shielded module design. Initial power of the standard and the shielded mini module is first compared by means of current-voltage (i-v) characteristics. Then, a PID test is performed to demonstrate the shielding effect of the TCO layer. PID test was performed at 85 °C for 24 h. A voltage of $U_{PID} = 1000$ V is applied, with solar being at negative potential compared to the grounded Al foil at the front surface. To assess impact of PID stress, the mini modules undergo electroluminescence (EL) imaging and IV-measurements before (initial) and after (final) the PID test.

3 RESULTS AND DISCUSSION

3.1 Performance of the standard and the shielded mini module

Exemplarily, one IV-curve of each module design is shown in Figure 2. The average results for the four standard and the four shielded modules are given in Table 1. In the initial state, the maximum power, P_{max}, of the shielded mini modules is roughly 16 % lower. This is due to a decreased current caused by the absorption of the incident light in the used glass and the TCO layer itself. The charge carrier generation is reduced by the missing photons. This is accompanied by a slightly reduced V_{OC}, which depends logarithmically on the charge carrier density. However, the increase of the fill factor, FF, is addressed by the reduced current, since lower currents, reflect in lower power losses at the series resistance at the maximum power point.

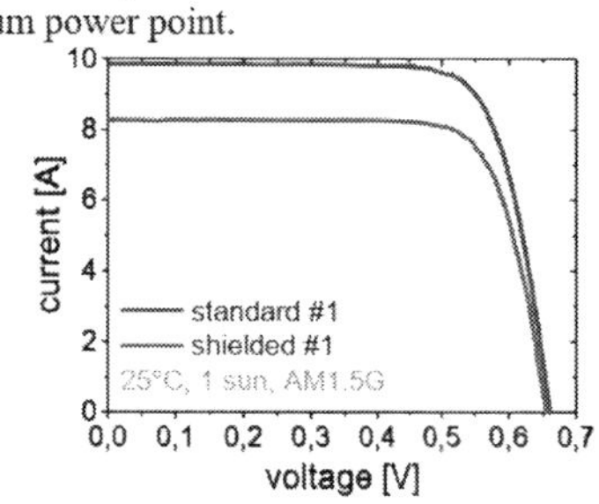

Figure 2: Initial state i-v curves, exemplarily given for one of the four mini modules. Standard module, red line; shielded module blue line. Measurements are performed under standard testing conditions.

3.2 PID related power losses

All eight mini modules were subjected to a PID for 24 h, at 85 °C and -1000 V (PID-s configuration). After the PID test the standard modules show a homogeneous decrease in the EL signal, see Figure 3 a). Whereas the shielded minimodule, Figure 3 b), with the additional TCO layer shows no visible change in the EL signal. The result is underlined by the measured IV characteristics.

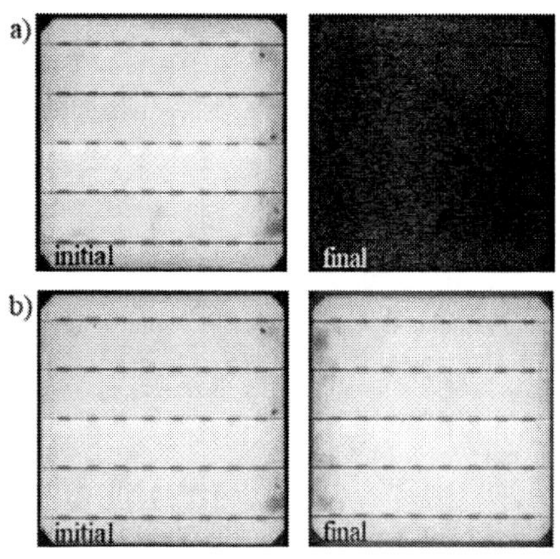

Figure 3: EL-images at the initial and final state for the standard (left) and the shielded (right) mini module. Images at forward bias U = 0.8 V, integration time, t = 10 s

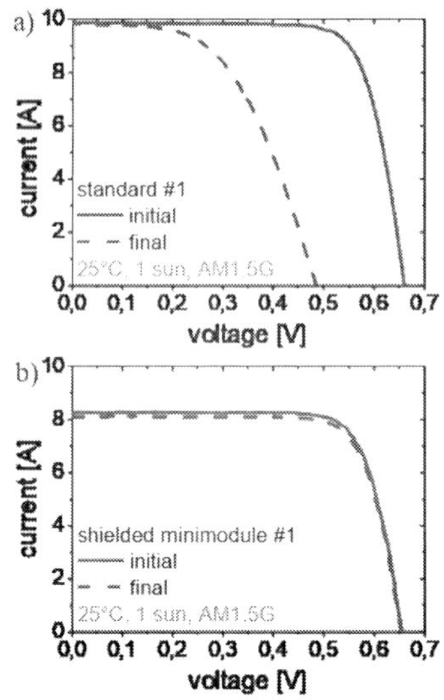

Figure 4: I-v curves, exemplarily given for one of the four mini modules of the standard and the shielded design each. Solid line refers to initial state, dashed line to final state measurement.

After the PID test, the standard module shows a change in maximum power, P_{max}, of ΔP_{max} = -50.9 % ± 2 %. The mini modules with the additional TCO layer between glass and encapsulant show a change, ΔP_{max} = -4.4 % ± 1 %. The degradation deviations given in percent were determined from the four samples, showing nice reproducibility.

4 SUMMARY

A new and innovative approach for the protection of solar cells against PID related power losses is presented. As the external potential due to the PV system configuration is nicely shielded away from the solar cell this concept can be utilized for any kind of PID type. Furthermore, it can be applied on both sides, if needed, to protect bifacial cells. Further proof-of concept measurements are ongoing to verify and proof certain applications.

However, gain of reliability is achieved at a cost of reduced power gain, as conductive layer reduced transmittance. The presented 16% loss can most likely be easily reduced as the used TCO-glasses were standard conductive glasses, not optimized for PV application and therefore also had additional losses in the glass itself.

The addition of a conductive layer as a PID barrier addresses the module design and hence is easy to implement. Here, the concept was demonstrated on bifacial PERC solar cells, but in general it is independent of the underlying cell technology. Future experiments on other cell types are planned.

ACKNOWLEDGEMENT

This work was funded within the project CTS1000+ (03EE1028A). The authors would like to thank Uwe Müller and Florian Kirchberg for sample preparation and support with the measurements.

REFERENCES

[1] T. Dullweber, C. Kranz, R. Peibst *et al.*, "PERC+: industrial PERC solar cells with rear Al grid enabling bifaciality and reduced Al paste consumption," *Progress in Photovoltaics: Research and Applications,* vol. 24, no. 12, pp. 1487-1498, 2016.

[2] S. Pingel, O. Frank, M. Winkler *et al.*, "Potential induced degradation of solar cells and panels." pp. 002817-002822.

[3] W. Luo, P. Hacke, K. Terwilliger *et al.*, "Elucidating potential-induced degradation in bifacial PERC silicon photovoltaic modules," *Progress in Photovoltaics: Research and Applications*, vol. 26, no. 10, pp. 859-867, 2018.

[4] V. Naumann, T. Geppert, S. Großer *et al.*, "Potential-induced Degradation at Interdigitated Back Contact Solar Cells," *Energy Procedia,* vol. 55, pp. 498-503, 2014.

[5] K. Sporleder, V. Naumann, J. Bauer *et al.*, "Root cause analysis on corrosive potential-induced degradation effects at the rear side of bifacial silicon PERC solar cells," *Solar Energy Materials and Solar Cells,* vol. 201, 2019.

[6] Z. Purohit, W. Song, J. Carolus *et al.*, "Impact of Potential-Induced Degradation on Different Architecture-Based Perovskite Solar Cells," *Solar RRL*, pp. 2100349, 2021.

[7] J. Carolus, T. Merckx, Z. Purohit *et al.*, "Potential-Induced Degradation and Recovery of Perovskite Solar Cells," *Solar RRL,* vol. 3, no. 10, pp. 1900226, 2019.

[8] J. Carolus, J. A. Tsanakas, A. van der Heide *et al.*, "Physics of potential-induced degradation in bifacial p-PERC solar cells," *Solar Energy Materials and Solar Cells,* vol. 200, 2019.

[9] A. Masuda, C. Yamamoto, Y. Hara *et al.*, "Influence of hygrothermal stress on potential-induced degradation for homojunction and heterojunction crystalline Si photovoltaic modules," *Japanese Journal of Applied Physics,* vol. 59, no. 7, 2020.

[10] V. Naumann, D. Lausch, A. Hähnel *et al.*, "Explanation of potential-induced degradation of the shunting type by Na decoration of stacking faults in Si solar cells," *Solar Energy Materials and Solar Cells,* vol. 120, pp. 383-389, 2014.

[11] V. L. Naumann, D. Hagendorf, Chr., "Sodium Decoration of PID-s Crystal Defects after Corona Induced Degradation of Bare Silicon Solar Cells," *Energy Procedia,* vol. 77, pp. 397–401, 2015.

[12] R. Swanson, M. Cudzinovic, D. DeCeuster *et al.*, "The surface polarization effect in high-efficiency silicon solar cells," in 15th PVSEC Shanghai, China, 2005.

[13] K. Sporleder, V. Naumann, J. Bauer *et al.*, "Time-Resolved Investigation of Transient Field Effect Passivation States during Potential-Induced Degradation and Recovery of Bifacial Silicon Solar Cells," *Solar RRL*, pp. 2100140, 2021.

[14] K. Sporleder, V. Naumann, J. Bauer *et al.*, "Microstructural Analysis of Local Silicon Corrosion of Bifacial Solar Cells as Root Cause of Potential-Induced Degradation at the Rear Side," *physica status solidi (a),* vol. 216, no. 17, 2019.

[15] K. Sporleder, V. Naumann, J. Bauer *et al.*, "Local Corrosion of Silicon as Root Cause for Potential-Induced Degradation at the Rear Side of Bifacial PERC Solar Cells," *physica status solidi (RRL) – Rapid Research Letters,* vol. 13, no. 9, 2019.

[16] P. K. Hacke, M. Wohlgemuth, J. Li, J. Shen, Y., "Potential-Induced Degradation-Delamination Mode in Crystalline Silicon Modules," *Workshop on Crystalline Silicon Solar Cells and Modules: Materials and Processes*, 2016.

[17] V. I. Naumann, K. Hagendorf, Chr., "On the Discrepancy between Leakage Currents and Potential-induced Degradation of Crystalline Silicon Modules," *28th European Photovoltaic Solar Energy Conference - Paris*, pp. 4, 2013.

CALCULATION OF THE SHORT-CIRCUIT CURRENT OF COLORED BIPV MODULES UNDER FIELD CONDITIONS BY APPLICATION OF SPECTRALLY AND ANGLE RESOLVED MEASUREMENT DATA

Lionel Clasing[1], Simon Schaaf[1], Ulf Blieske[1], Nicholas Riedel-Lyngskær[2], Adrián A. Santamaría Lancia[2], Nils Reiners[3]

[1]Cologne University of Applied Sciences, Faculty for Process Engineering, Energy and Mechanical Systems, Cologne Institute for Renewable Energy, Betzdorfer Str. 2, 50679 Cologne, Germany
[2]Technical University of Denmark, Department of Photonics Engineering, Roskilde, Denmark
[3]Fraunhofer Institute for Solar Energy Systems ISE, Electrical Energy Storage Department, Freiburg, Germany

ABSTRACT: The aim of this work is to determine the short-circuit current of colored solar glasses under field conditions, which includes varying solar spectral irradiance, oblique irradiation angles and fractions of diffuse light. Therefore, the field performance of colored building-integrated photovoltaic modules is calculated with spectral solar resource data and angular-dependent spectral responsivity measurements from laboratory testing. This method allows to examine the behavior of modules under test by application of physical laboratory measurements and spectral solar resource data. In order to validate the short-circuit current calculations, monitoring data from field tests are used. In order to assess the results, the short-circuit current is compared with a model using the incidence angle modifier. The results show that the application of angular-dependent spectral responsivity measurements allows to reduce relative mean bias error from 2.26% to 0.25% and relative root mean square error from 2.72% to 1.57% for the blue colored building-integrated photovoltaic module compared to the incidence angle modifier model, respectively.
Keywords: BIPV, colored solar glass, short-circuit current, angular-dependent spectral responsivity, IAM

1 INTRODUCTION

Colored glasses are of high interest for building-integrated PV (BIPV) applications such as the integration of modules in façades. Using colored glass allows a wide variety of designs for architects and planners. However, to introduce modules with colored glasses to the market, it is crucial to give realistic yield predictions to the investors. As for BIPV modules, the orientation is generally not ideal and thus the share of suboptimal angles of incidence (AOI) on the module surfaces is much higher than at standard photovoltaic (PV) systems. Therefore, it is important to have accurate models for the non-perpendicular incidence cases. The incidence angle modifier (IAM) is normally used to account for reflection losses at the glass-air interface. According to the IEC standard 61853-2:2016, the IAM is derived from the measured change in short-circuit current I_{SC} with changing AOI [1]. A few PV system simulation tools use additional spectral correction factors for specific cell and module types. Such spectral corrections are generally given as simple correlations with air mass [2, 3]. These standard procedures do typically not include spectral measurements and do not consider that the IAM is not only a function of the AOI, but is also varying with different spectra [4]. Therefore, the angular-dependent spectral responsivity (ASR) is applied for the first time to calculate the I_{SC} in a physical measurement setup under field conditions. Prior works about ASR data have focused on laboratory measurements [5, 6] or theory [7, 8].

2 CALCULATION AND MEASUREMENT OF THE SHORT-CIRCUIT CURRENT

2.1 PV field measurements and data pre-processing

In order to investigate the spectral and angular-dependent influence of colored glass for application in BIPV modules in detail, single-cell laminates (200 mm x 200 mm) with blue and grey colored 3.2 mm glass samples, monocrystalline silicon solar cells (156 mm x 156 mm) equipped with two busbars, black back sheet and transparent ethylene-vinyl acetate encapsulant were examined in this study. The color of the glass is obtained by a thin film multi-layer deposition on the inner glass surface. The outer glass surface is treated, which results in diffuse reflection to achieve improvements in aesthetics and avoid glare effects. In a measurement campaign the data of BIPV modules with different colored glasses and a control reference sample with standard ultra-white solar glass have been recorded under field conditions at the DTU campus in Risø, Denmark. Short-circuit current I_{SC} and back sheet temperature T_M of the test samples were measured in 10-second intervals each for horizontal orientation in July and August 2018. The samples are shown in Figure 1. It can be expected that the backsheet temperature T_M under field conditions differs from the temperatures in the laboratory measurements, therefore the measured I_{SC} values are temperature corrected. Furthermore, a high-resolved climate dataset with spectral and broadband irradiation, ambient temperature, relative humidity, precipitation and windspeed data for the period under consideration has been recorded at the same location.

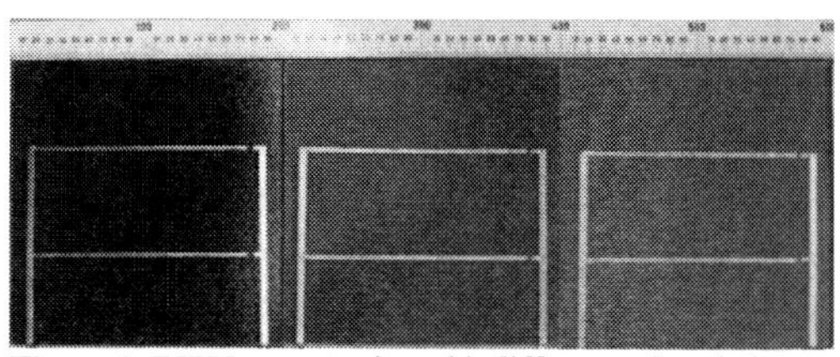

Figure 1: BIPV test samples with different colored glasses used for measurements and calculations

The irradiation data includes broadband global horizontal irradiance GHI and diffuse horizontal irradiance DHI measured with two class A pyranometers in 10-second intervals. Direct normal irradiance DNI is measured with a class A pyrheliometer in 10-second intervals, as well. Additionally, spectrally resolved DHI and DNI are measured by two diffraction grating spectroradiometers. The spectrometers are specified for a wavelength range between 300 nm and 1100 nm with a full width half

maximum (FWHM) spectral resolution of 7 nm in 1 nm steps. The spectral measurement data is recorded in 5-minute intervals [9].
Since crystalline silicon solar cells can convert wavelengths up to 1.2 µm, it is necessary to extend the missing wavelength range between 1.1 µm up to 1.2 µm of the measured spectral irradiance dataset for calculation of the I_{SC}. This is done by means of the "Simple Model of the Atmospheric Radiative Transfer of Sunshine" (SMARTS) by Gueymard [10, 11]. The SMARTS model provides spectral GHI, DNI and DHI, but is limited to clear-sky conditions. For that reason, clouds are taken into account in the extended data by application of a cloud scaling factor according to Ernst et al. [12]. However, the spectrally resolved irradiance data is measured by two spectrometers in series, which results in a time offset of about 15 seconds between the start of the first measurement and end of the second measurement. In case of high cloud variability, the irradiance can change significantly during this period. To counteract the resulting uncertainties, the dataset is filtered using a variability index VI according to Stein et al. [13]. For this purpose, the broadband GHI values are used in a 2-minute interval around the respective measured value. If the VI is greater than 1.1, the corresponding measured value is removed. This filters only those values where the irradiance remains approximately identical between the measurements of the two spectrometers in the dataset.

2.2 Laboratory measurements and data pre-processing

The ASR measurements in this work were performed by using a modified experimental setup in a laboratory at the TH Köln campus Deutz, Germany [14]. The setup is based on a solar simulator with filter wheel monochromator for PV solar cells (FiMo SR EQE, Aescusoft). It consists of a solar simulator, an upstream optical chopper, two filter-monochromators, an automated rotational sample stage for PV single cell laminates and a data acquisition system. The setup is divided into two rooms, which are only connected by an aperture for the light beam. The illumination source of the solar simulator is a xenon short-arc lamp (1000 W Osram XBO) connected to a stabilized DC power supply. The double filter wheel monochromator contains 40 different bandpass filters and can hence provide a wavelength range from 300 to 1188 nm with a FWHM spectral resolution of 25 nm. The wavelength steps of the measurement are specified by the bandpass filters and are equally distributed. The optical chopper is coupled with a lock-in amplifier to achieve high-precision measurements of the short-circuit current I_{SC} at a specific wavelength.
According to the IEC 60904-8 standard, the short-circuit current conditions must be ensured during spectral responsivity (SR) measurements, which is achieved by a 4-quadrant chopper. Furthermore, an appropriate bias illumination is used, as specified in the standard. However, the standard stated that it is sufficient to perform the SR measurements only for one irradiance intensity and one module temperature [15]. Subsequently, the SR measurement results are linearly interpolated and thus allow calculations with 1 nm spectral resolution.
The ASR measurements were performed using a fully automated rotational sample stage. A second aperture is used to reduce the size of the monochromatic light spot to a small homogenous area. This light spot is projected completely onto the solar cell, even at high irradiation angles. The angular-dependent measurements were performed based on the IEC 61853-2 standard, i.e. from -60° up to +60° AOI in 10° steps and down to -80° respectively up to +80° AOI in 5° steps [1]. As significantly increasing uncertainties were observed in an interlaboratory round robin test by Riedel-Lyngskær et al. [16] for AOI > 80°, the I_{SC} calculations in this study are performed only up to AOI ≤ |±80°|.
The ASR measurements are linearly interpolated to 1° AOI resolution to obtain an appropriate level of detail in I_{SC} calculations. In the AOI range between the defined 10° and 5° AOI steps, measurements with 1° AOI steps have been performed to determine the deviations between measured and linearly interpolated values. The measured and interpolated values show a good agreement, hence the AOI measurement steps recommended by the IEC 61853-2 standard are used in this study.
The ASR plot shows the spectral and angular-dependent effects of the blue glass on the BIPV modules characteristics. However, increasing losses at higher AOI and wavelengths > 500 nm can be observed.

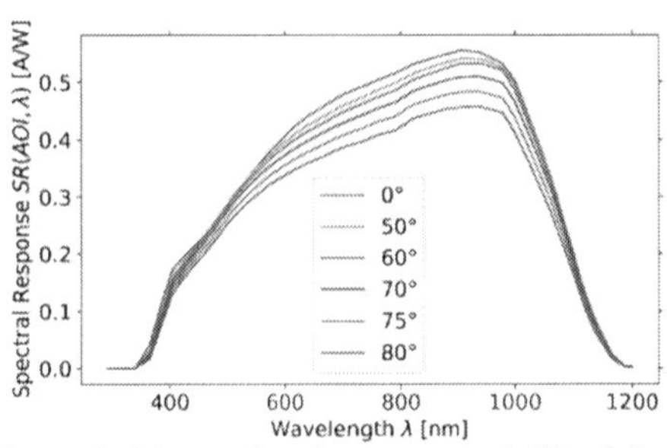

Figure 2: Measured and pre-processed SR of the blue colored PV test sample for different angle of incidence AOI

2.3 IAM measurement and simulation for direct and diffuse beam irradiance

According to the IEC 61853-2 standard the IAM for direct beam irradiance can be measured and determined per equation (1) [1].

$$\mathrm{IAM} = \frac{I_{SC}(\theta)}{I_{SC}(0°) * \cos(\theta)} \qquad (1)$$

Another particular interest is given on modelling the influence of the IAM in diffuse beam irradiance in order to calculate losses by oblique light in a high-level of detail. For that reason, the IAM is applied to the diffuse beam irradiance for calculation of the short-circuit current I_{SC}. In case of horizontal module orientation, only the diffuse sky irradiance must be considered. Hence, the direct and diffuse beam irradiance of the ground reflection can be neglected and a diffuse IAM for ground reflection is not required. However, Martin and Ruiz [17] presented an analytical model to simulate the IAM for standard solar glass based on only one shape parameter, the a_r-value. Furthermore, they presented a method to apply the IAM to the sky diffuse beam irradiance, which is used in this work to calculate the I_{SC} by application of ASR laboratory measurements. The experimental setup described in section 2.2 is used with broadband illumination to measure the IAM of each sample. The shape parameter a_r is calculated by using the conducted IAM laboratory measurements and iterative fitting to the simulated IAM calculated by equation (2) [17]:

$$\text{IAM} = \frac{1 - \exp(^{-\cos(\theta)}/_{a_r})}{1 - \exp(^{-1}/_{a_r})} \tag{2}$$

In Figure 3 IAM measurements and simulations with Martín and Ruiz model [17] are shown. It can be observed that the simulation is not exactly fitting the measurements. In order to find the best fit, a comparison between different IAM simulation models has been done. The results show that the best agreement is shown by the model of Martín and Ruiz, followed by the ASHREA model [18]. Based on these findings, the model of Martín and Ruiz [17] is used for the simulation of the IAM for I_{SC} calculations with ASR model.

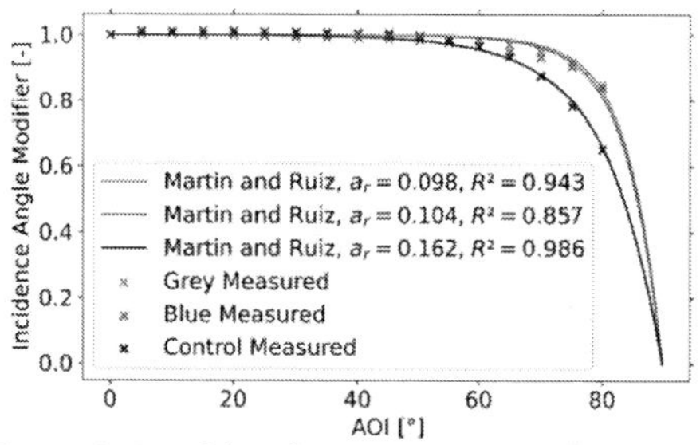

Figure 3: Resulting shape parameters a_r for Martín and Ruiz simulation model *[17]* determined by iterative fitting to IAM laboratory measurements for different colored BIPV modules

Additionally, the IAM model is used for direct beam irradiance in order to calculate the I_{SC} with perpendicular SR_0 to compare the results.

2.4 Mathematical expression of short-circuit current

Generally, the short-circuit current I_{SC} of a solar PV module can be calculated by the simplified equation (3).

$$I_{sc} = A^* \int E(\lambda)^* \, SR(\lambda, \theta) \, d\lambda \tag{3}$$

The surface area of the module A is multiplied by the integral over the global in-plane spectral irradiance on the PV solar module at a specific wavelength λ and ASR at a specific wavelength λ and AOI θ.

The global in-plane spectral irradiance must be specified more in detail to calculate the I_{SC} with an appropriate accuracy. Therefore, the measured DNI is converted to direct in-plane irradiance by multiplying it to the cosine of the AOI. Reflection losses caused by varying AOI are taken into account in the angular-dependent SR(λ,θ). Consequently, an IAM model is not applied for direct beam irradiance in the ASR approach.

A diffuse IAM model, as described in section 2.3, is applied to the measured DHI. $IAM_{diff,sky}$ accounts for the reflection losses of the diffuse sky irradiance, caused by the AOI. The direction of light rays of the diffuse irradiation is assumed isotropic. As a result, the $I_{SC,ASR}$ of a BIPV module is calculated for horizontal orientation by application of ASR by the following equation (4).

$$I_{SC,ASR} = A^*\left(\int DNI(\lambda)^* \cos(\theta)^* SR(\lambda, \theta) \, d\lambda + \int DHI(\lambda)^* IAM_{diff,sky}{}^* SR_0(\lambda) \, d\lambda\right) \tag{4}$$

Nevertheless, the IAM_{dir} model proposed by Martín and Ruiz [17] is used by multiplying with SR_0 for direct beam irradiance to determine the $I_{SC,IAM}$. The $I_{SC,IAM}$ is utilized to compare the ASR model with the model proposed by Martín and Ruiz for standard solar glass and is calculated by equation (5).

$$I_{SC,IAM} = A^*\left(\int DNI(\lambda)^* \cos(\theta)^* IAM_{dir}{}^* SR_0(\lambda) \, d\lambda + \int DHI(\lambda)^* IAM_{diff,sky}{}^* SR_0(\lambda) \, d\lambda\right) \tag{5}$$

2.5 Model analysis and comparison

In order to analyze and compare the calculated short-circuit current I_{SC} for the respective colored module under test, a residual error RE is calculated per equation (6).

$$RE = I_{SC,Measured} - I_{SC,Calculated} \tag{6}$$

In this equation the residual error is calculated by the difference of the measured and calculated I_{SC}. Thus, the residual error can be used to analyze and validate the calculations by application of the ASR in comparison with IAM_{dir} model for DNI.

A mean bias error MBE is calculated to summarize the error for the dataset under consideration according to equation(7).

$$MBE = \frac{1}{n}\sum_{i=1}^{n} I_{SC,Measured} - I_{SC,Calculated} \tag{7}$$

Furthermore, a root mean square error RMSE is calculated by equation (8).

$$RMSE = \sqrt{\frac{1}{n}\sum_{i=1}^{n}\left(I_{SC,Measured} - I_{SC,Calculated}\right)^2} \tag{8}$$

3 RESULTS

Figure 4 shows the results of the measured and calculated short-circuit current I_{SC} of the blue colored BIPV module on 3rd of August 2018. The I_{SC} is calculated according to equation (4) by the ASR model with horizontal module orientation. In addition, the residual error for each timestamp calculated by equation (6) is shown.

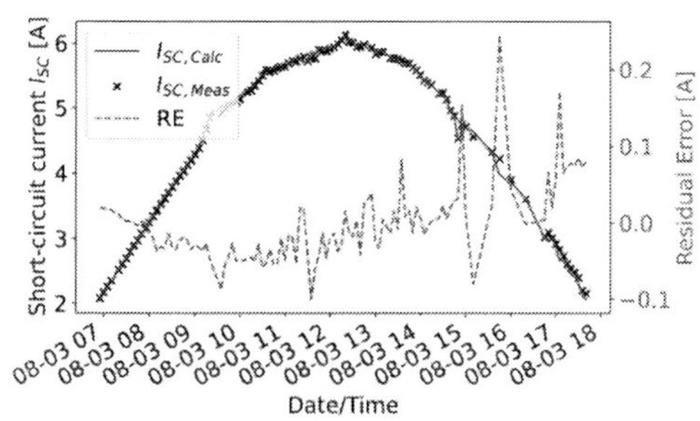

Figure 4: Measured and calculated short-circuit current I_{SC} (ASR model) for a blue colored PV module with horizontal orientation on 3rd August 2018

The 3rd of August has mostly clear-sky conditions, especially in the morning. In the afternoon its partly cloudy and cloud variability increases. The application of a VI model by Stein et al. [13], as described in Section 2.1, filters a few timestamps at which the cloud variability is above the specified threshold. A good agreement between measurement and calculation can be observed for clear-sky conditions. Between 7 am and approximately 2.30 pm the residual error shows a maximum absolute error of 100.85 mA and a MBE of 19.94 mA. From 2.30 pm on, the cloud variability is partly increased and a few filtered timestamps can be observed. The error increases up to a maximum of 245.31 mA at 4.40 pm. The largest errors are either before or after VI filtered timestamps.

3.1 Model validation and comparison

A model validation is done to assess the short-circuit current I_{SC} calculations for two different colored BIPV modules and a control reference with standard solar glass. Furthermore, a comparison is done with the IAM model proposed by Martín and Ruiz **[17]**. Figure 5 shows the residual error of $I_{SC,ASR}$ model in terms of direct beam irradiance as a function of the AOI. Additionally, the residual error of the $I_{SC,IAM}$ is shown. $I_{SC,IAM}$ is calculated by applying the IAM model to account for angular optical losses. Therefore, the SR_0 with perpendicular AOI is used. Finally, a linear regression model is calculated, which shows a slightly increasing trend with increasing AOI.

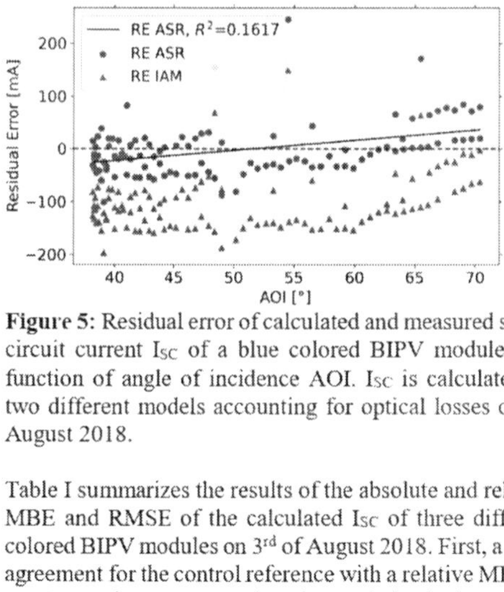

Figure 5: Residual error of calculated and measured short-circuit current I_{SC} of a blue colored BIPV module as a function of angle of incidence AOI. I_{SC} is calculated by two different models accounting for optical losses on 3rd August 2018.

Table I summarizes the results of the absolute and relative MBE and RMSE of the calculated I_{SC} of three different colored BIPV modules on 3rd of August 2018. First, a good agreement for the control reference with a relative MBE of -0.29% and -0.41% can be observed for both models. Second, the results of the ASR model show a decrease of the relative MBE from 2.26% to 0.05% for the blue colored test sample compared to the IAM model. Third, the comparison for the grey colored test samples shows that both models show nearly identical results.

Table I: Summarized residual error results of short-circuit circuit I_{SC} calculation of three different BIPV modules with ASR and IAM model on 3rd August 2018

ASR model	Control	Blue	Grey
MBE [mA]	-14.69	2.41	-27.11
MBE_{rel} [%]	-0.29	0.05	-0.55
RMSE [mA]	54.22	50.29	60.1
$RMSE_{rel}$ [%]	1.06	1.1	1.21
IAM_{dir} model			
MBE [mA]	-20.86	103.8	-33.28
MBE_{rel} [%]	-0.41	2.26	-0.67
RMSE [mA]	52.22	116.31	63.69
$RMSE_{rel}$ [%]	1.02	2.53	1.28

A residual error analysis is done for a time period of 11 days in July and August. The dataset contains 519 timestamps and is filtered by a VI > 1.1, as described in section 2.1. The results for the blue colored BIPV module are presented in Figure 6.

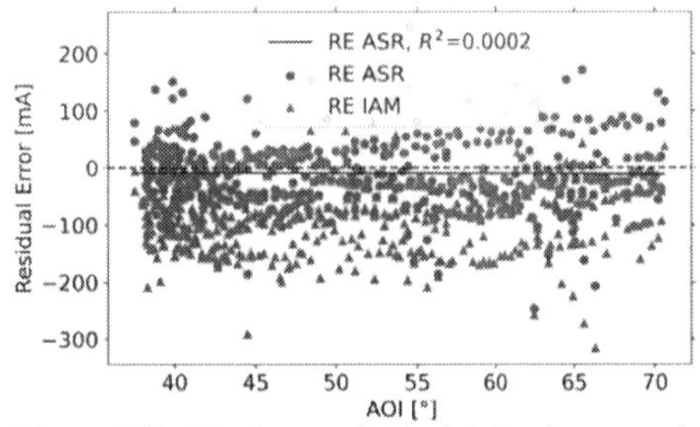

Figure 6: Residual error of calculated and measured short-circuit I_{SC} current of a blue colored BIPV as a function of angle of incidence AOI. I_{SC} is calculated by two different models accounting for optical losses with from 31st July until 10th August 2018.

Table II summarizes the MBE and RMSE of the performed I_{SC} calculations for 11 days in July and August. In comparison with the analysis of one day, the ASR model shows slightly larger MBE and RMSE of 0.25% and 1.57% for the blue colored BIPV module. The residual error for the control reference and grey BIPV module is nearly equal for both calculation models, although the ASR model consistently shows slightly lower errors.

Table II: Summarized statistical metrics for residual error of short-circuit current I_{SC} calculation of three different BIPV modules with ASR and IAM model from 31rd July until 10th August 2018

ASR model	Control	Blue	Grey
MBE [mA]	-9.1	9.74	-21.85
MBE_{rel} [%]	-0.21	0.25	-0.54
RMSE [mA]	56.47	59.85	71.5
$RMSE_{rel}$ [%]	1.32	1.57	1.76
IAM_{dir} model			
MBE [mA]	-16.13	86.29	-23.89
MBE_{rel} [%]	-0.37	2.26	-0.59
RMSE [mA]	58.72	103.56	75.82
$RMSE_{rel}$ [%]	1.37	2.72	1.87

5 SUMMARY

The calculated short-circuit I_{SC} of BIPV modules with different colored glasses is presented in this work. Therefore, the ASR is applied for the first time in a calculation model for direct beam irradiance in order to calculate the I_{SC} of colored BIPV modules under field conditions. The calculations are validated with field measurement data of 11 days in July and August for horizontal module orientation. The results of the calculations show a good agreement with the conducted field measurements with a MBE of 9.74 mA and -21.85 mA for the blue and grey colored module, respectively. Thus, the ASR model allows to reduce relative MBE from 2.26% to 0.25% and relative RMSE from 2.72% to 1.57% for the blue colored BIPV module compared to the IAM_{dir} model, respectively.

In future works the I_{SC} is calculated and validated for BIPV modules with a tilt angle by application of a transposition model for diffuse sky and ground reflected irradiance. Furthermore, the ASR should be additionally applied for diffuse sky irradiance instead of a diffuse IAM model.

6 ACKNOWLEGDEMENTS

This project has been financed by the European Fonds for Regional Development North Rhine-Westphalia (EFRE.NRW)

7 REFERENCES

[1] *IEC 61853-2:2016 Photovoltaic module performance testing and energy rating - part 2: spectral responsivity, incidence angle and module operating temperature measurements,* International Electrotechnical Commission, 2016.

[2] M. Lee and A. Panchula, "Spectral correction for photovoltaic module performance based on air mass and precipitable water," in *43rd Photovoltaic Specialists Conference (PVSC)*, Portland, OR, USA, 2016.

[3] B. H. King, C. W. Hansen, D. Riley, C. D. Robinson and L. Pratt, Procedure to Determine Coefficients for the Sandia Array Performance Model (SAPM), Sandia National Laboratories, 2016.

[4] F. Plag, I. Kröger, T. Fey, F. Witt and S. Winter, "Angular-dependent spectral responsivity—Traceable mea- surements on optical losses in PV devices," *Progress in Photovoltaics: Research and Applications, Vol. 26,* pp. 565-578, 2017.

[5] I. Geisemeyer, N. Tucher, B. Müller, H. Steinkemper, J. Hohl-Ebinger, M. C. Schubert and W. Warta, "Angle Dependence of Solar Cells and Modules: The Role of Cell Texturization," *IEEE Journal of Photovoltaics Vol. 7,* pp. 19-24, Jan. 2017.

[6] N. Reiners, U. Blieske and S. Siebentritt, "Investigation on the Angle and Spectral Dependence of the Internal and the External Quantum Efficiency of Crystalline Silicon Solar Cells and Modules," *IEEE Journal of Photovoltaics Vol. 8,* pp. 1738-1747, Nov. 2018.

[7] M. R. Lewis, E. M. Tonita, C. E. Valdivia, R.-J. K. L. J. Obhi, M. I. Bertoni and K. Hinzer, "Angular Dependence of Textured Bifacial Silicon Heterojunction Solar Cells for High Latitudes," in *IEEE 46th Photovoltaic Specialists Conference (PVSC)*, Chicago, IL, USA, 2019.

[8] S. Pal, A. Reinders and R. Saive, "Simulation of Bifacial and Monofacial Silicon Solar Cell Short-Circuit Current Density Under Measured Spectro-Angular Solar Irradiance," *IEEE Journal of Photovoltaics, Vol. 10, No. 6,* pp. 1803-1815, 2020.

[9] N. Riedel, A. Thorseth, A. A. Santamaria Lancia, S. Thorsteinsson, P. Poulsen, B. Iandolo, R. S. Davidsen and G. Benatto, "Direct Beam and Diffuse Spectral Irradiance Measurements in a Nordic Country Analyzed With the Average Photon Energy Parameter," in *IEEE 7th World Conference on Photovoltaic Energy Conversion (WCPEC) (A Joint Conference of 45th IEEE PVSC, 28th PVSEC & 34th EU PVSEC)*, 2018.

[10] C. Gueymard, SMARTS, A Simple Model of the Atmospheric Radiative Transfer of Sunshine: Algorithms and Performance Assessment, 1679 Clearlake Rd., Cocoa, FL 32922: Professional Paper FSEC-PF-270-95, Florida Solar Energy Center, 1995.

[11] C. Gueymard, "Parameterized Transmittance Model for Direct Beam and Circumsolar Spectral Irradiance," *Solar Energy (71:5),* pp. 325-346, 2001.

[12] M. Ernst, H. Holst, M. Winter and P. P. Altermatt, "SunCalculator: A program to calculate the angular and spectral distribution of direct and diffuse solar radiation," *Solar Energy Materials and Solar Cells, Vol. 157,* pp. 913-922, Dec. 2016.

[13] J. Stein, C. Hansen and M. Reno, "The Variability Index: A New and Novel Metric for Quantifying Irradiance and PV Output Variability," in *World Renewable Energy Forum*, USA, 2012.

[14] K. Meisenzahl, N. Schneble, M. Volk, U. Blieske, L. Clasing, J. Müller-Ost, J. Münzberg and P. Hakenberg, "Measuring the Incidence Angle Modifier of Optically Uncoupled Glass for PV Application," in *36th European Photovoltaic Solar Energy Conference and Exhibition*, Marseille, France, 2019.

[15] *IEC 60904-8:2014 Photovoltaic devices - Part 8: Measurement of spectral responsivity of a photovoltaic (PV) device,* International Electrotechnical Commission, 2014.

[16] N. Riedel-Lyngskær, A. A. P. F. Santamaría Lancia, I. Kröger, M. R. Vogt, C. Schinke, R. S. Davidsen, M. Amdemeskel, M. J. Jansen and P. S. L. H. Manshanden, "Interlaboratory comparison of angular-dependent photovoltaic device measurements: Results and impact on energy rating," *Progress in Photovoltaics: Research and Applications,* December 2020.

[17] N. Martín and J. M. Ruiz, "A New Model for PV Modules Angular Losses under Field Conditions," *International Journal of Solar Energy,* pp. 19-31, Vol. 22, No. 1 2002.

[18] A. Souka and H. Safwat, "Determination of the optimum orientations for the double exposure flat-plate collector and its reflections," *Solar Energy Vol .10,* pp. 170-174, 1966.

THE IMPACT OF REAL ALBEDO VALUES ON ENERGY ESTIMATION FOR BIFACIAL MODULES

Hugo Sánchez[1,2], Sebastian Dittmann[1,3], Carlos Meza[1,2], Ralph Gottschalg[1,3]

[1]Anhalt University of Applied Sciences, Bernburger Str. 55, 06366 Köthen, Germany
E-mail: husanchez@itcr.ac.cr, Tel.: +49 3496 67 2354

[2] Electronics Engineering School, Costa Rica Institute of Technology, Costa Rica
[3]Fraunhofer-Center for Silicon Photovoltaics CSP, Halle, Germany;

ABSTRACT Albedo is one of the crucial variables in the energy estimation for bifacial photovoltaic modules due to the modules' capability to collect electricity from both front- and rear-sides. There are several modeling tools for energy estimation of these modules. However, all of them use a constant albedo value for all period. Our paper reports on the impact of using real measured values of albedo in the modeling process. The results show that there is a difference of up to 7% when the value of the albedo set constant instead using a real measured value. This result translates into an inaccurate prediction affecting the making-decision process for stakeholders in the adoption of this technology.

Keywords: Albedo, bifacial PV modules, photovoltaic, modeling

1 INTRODUCTION

Bifacial technology becomes a popular technology with a marked share of 35% worldwide [6]. Bifacial modules can collect electricity from the front incident and ground reflected irradiation. This is possible because it exposes both the front and back solar cells, changing the common backsheet with glass or transparent backsheet. According to the literature [1]-[5], this change improves the electrical generation up to 30 %.

However, albedo, the diffuse reflectivity of a surface, has a significant contribution to the estimation of energy for bifacial modules [7]. It has almost a proportional linear relation compare to bifacial gain in the solar modules. However, there is evidence that albedo values can dramatically shift, based on sun angle, seasonal irradiance, type, and seasonal variation (due to the vegetative ground cover, degree of backside shading, presence and degradation of snow, and prevalence of airborne particles). Interesting experimental results about these changes, in several locations around the world, can be found in [8] and [9].

Simulation tools or energy yield prediction models use constant value of albedo for all year or simulation period. This might lead to inaccuracy if the simulated energy yield of a given installation. It is mandatory to improve the simulations tools for bifacial modules with realistic data, to get more accurate predictions. With that, empowering stakeholders on the investment and implementation of bifacial systems [10]-[11].

The present work presents the impact of using real measured data of albedo in comparison to a fixed albedo value and its influence on the energy prediction of bifacial module.

2 EXPERIMENTAL METHODOLOGY

To analyze the variation of albedo an experiment was designed to measure the rear- and front-side irradiance. For this purpose, an albedometer was installed at the outdoor test field of the Anhalt Photovoltaic Performance and Lifetime Laboratory (APOLLO) at Anhalt University of Applied Sciences in Bernburg, Germany. The albedometer consists of a combination of two pyranometers (both Kipp&Zonen SMP10). One is mounted facing the sky the other one inverted to measure the ground-reflected irradiance. Figure 1 shows the measurement set-up with the horizontal measurement and the measurement of the rear-side plan-of-array (POA) at 35°.

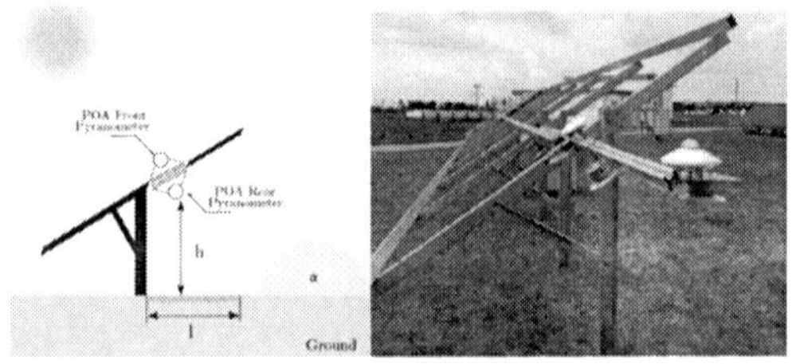

Figure 1: Albedometer installed at APOLLO, Bernburg Germany.

The sky facing pyranometer is equipped with the standard sunscreen meanwhile the inverted pyranometer is equipped with a glare screen as requested by Kipp&Zonen. According to the manual, the glare screen has an angle of 5° and is fitted to the pyranometer to prevent direct irradiance at high sun azimuth angels. Irradiance data are measured in 10 seconds intervals. The measurement period started in October 2019 and finished in September 2020. During the measurement period the substrate changed from green grass to yellow due to a very dry summer.

Two different energy yield prediction models have been investigated, one proposed by Purdue University (PUB) [12] and one by Solar World [13]. Simulation for a single bifacial module were carried out using datasheet values for module's electrical characterization. Table 1 shows the main parameters used for both tools for a single bifacial module in a landscape orientation.

Table 1: Main simulation parameters used in the simulation tools

Parameter	*Value*	*Unit*
Latitude	51.773	°
Longitude	11.763	°
Tilt Angle	35	°
Module height	1	m
Cleareance height	1.5	m
Azimuth Angle [a]	180	°
Front side efficiency	18	%
Bifaciality	90	%
Temperature Coefficient	-0.4319	%/K
U_0 Constant heat transfer	22.7	$W/m^2/K$
U_1 Convective heat transfer	6.84	$W.s/m^3/K$

[a] Facing south

By definition albedo is measured horizontal. Since in this study we consider rear-side plan-of-array irradiance at 35° measurements we define the rear-front side ratio as α [8].

$$\propto = \frac{G_{rear}}{G_{front}} \qquad \text{(Eq. 1)}$$

3 RESULTS

Figure 2 shows the monthly values of α and its distribution. The seasonal variation of α is clearly visible ranking from 0.08 to 0.25 with a mean value of 0.15.

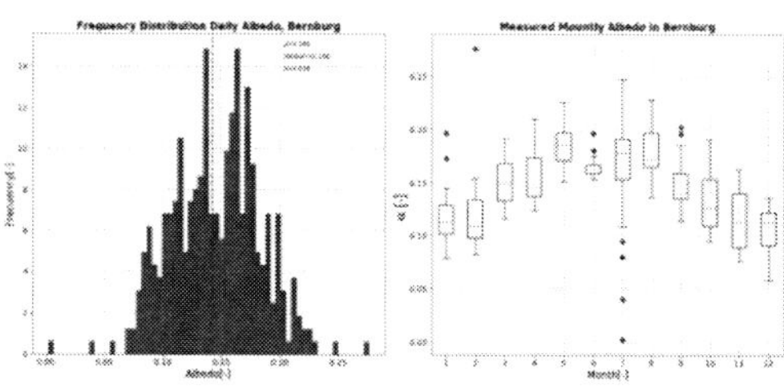

Figure 2: Histogram and monthly distribution of α measured at the test side in Bernburg.

3.1. Energy estimation using constant value over the year and the median of real measured α values per month

Figure 3 shows the results for the PUB (Figure 3a) and the Solar World (Figure 3b) model using a constant value of α in comparison with the real measured median values of each month as shown in Figure 2. As it can be observed there is an over estimation if a constant value of α is used. In the case of the PUB model it represents only minor difference of about 1 kWh/kWp/a. In the case of the model from Solar World the difference corresponds to 3.7 kWh/kWp/a. However, single month show higher deviation such as the winter month due to the seasonal changes of α.

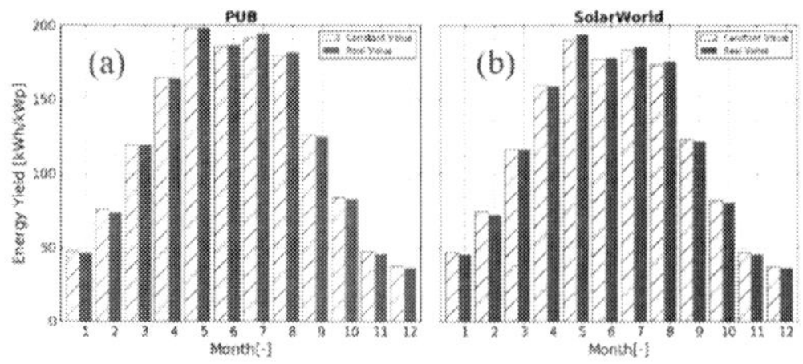

Figure 3: Comparison of the simulated energy yield using constant α value versus real measured values.

As a summary, Table 2 shows the annual mean values for RMSE and MBE for both models, PUB and Solar World.

Table 2: Annual mean values for RMSE and MBE.

Model	*RMSE [%]*	*MBE [%]*
PUB	6.92	0.386
Solar World	7.61	0.390

3.2. Energy estimation using constant value over the year and real measured α values

Figure 3 only considers the monthly median value. To get a deep analysis about the seasonal variation of α. Figure 4 shows a comparison between the RMSE and the MBE using the constant value compare to the minimum, the maximum and the mean measured α value as shown in Figure 2.

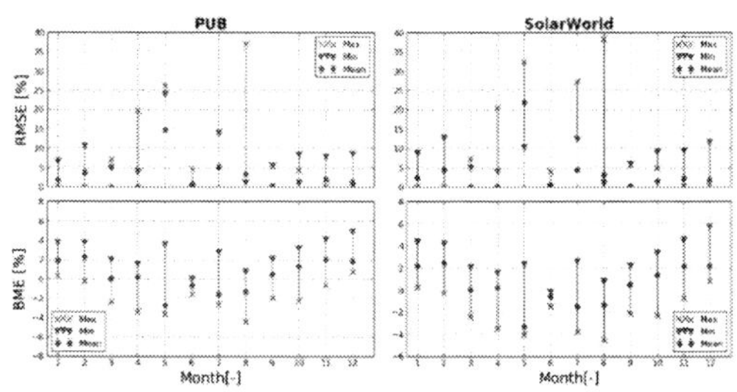

Figure 4: RMSE and MBE for the real measured α value.

As it can be seen, there is seasonal variability in the estimation due to the ratio between the rear- and front-side irradiation. This variability also affects the energy estimation. Most of the software tools take a constant albedo value as a default parameter. This situation generates a less accurate result. In this study, the variance becomes higher in the winter period (November to February). During the summertime, the MBE is lower (up to 2.5%). Meanwhile, during the wintertime, the MBE reaches up to 6%. In both cases, refers to if the minimum α value is used. On the other hand, the RMSE shows the difference between the two values through the square of the mean. In August, the maximum value reaches a deviation up to 40%, due to the variability of the maximum value of α.

For the cases if the median value is used the maximum deviation reaches 2%. And it keeps the same behavior as when the maximum value is used. The winter months correspond to the period with the higher deviation. The errors that come from the use of the minimum and maximum values of the albedo also reflect the sensitivity of the simulation tools.

From Table 2, the PUB model presents less variability of the two methods using a real measured α values. Both methods are acceptable, due to the values of the MBE (<0.5%), in comparison to the real values during the year. However, the comparison reflects the importance of using the real α value. The variability of the RMSE corresponds to the critical values when the evaluation of the maximum albedo is considered.

5 SUMMARY AND OUTLOOK

We presented the effect of the rear-front-side ratio (α) is crucial in the process of modeling for bifacial modules. Most of the simulation tools use a constant value of α. However, this situation is far from reality. It was demonstrated that α shows seasonal variation and effects the simulated energy yield of a single bifacial module. The deviation can reach up to 7%. Our future research will include the analysis with other simulation tools including the analysis of other variables such as non-uniformity and seasonal variation of the spectral of the rear-side irradiance.

4 REFERENCES

[1] R. Guerrero-Lemus, R. Vega, T. Kim, A. Kimm, and L. Shephard, "Bifacial solar photovoltaics–a technology review",Renewable and sustainable energy reviews,vol. 60, pp. 1533–1549, 2016.

[2] A. Cuevas, A. Luque, J. Eguren, and J. del Alamo, "50 per cent more outputpower from an albedo-collecting flat panel using bifacial solar cells",Solar Energy, vol. 29, no. 5, pp. 419–420, 1982.

[3] U. A. Yusufoglu, T. H. Lee, T. M. Pletzer, A. Halm, L. J. Koduvelikulathu, C.Comparotto, R. Kopecek, and H. Kurz, "Simulation of energy production bybifacial modules with revision of ground reflection",Energy Procedia, vol. 55,pp. 389–395, 2014.

[4] C. Deline, S. MacAlpine, B. Marion, F. Toor, A. Asgharzadeh, and J. S. Stein,"Evaluation and field assessment of bifacial photovoltaic module power rating methodologies", in2016 IEEE 43rd Photovoltaic Specialists Conference (PVSC),IEEE, 2016, pp. 3698–3703.

[5] J. Appelbaum, "Bifacial photovoltaic panels field",Renewable Energy, vol. 85,pp. 338–343, 2016.

[6] ITRPV,International technology roadmap for photovoltaic (itrpv)-2019 results, 2020.

[7] H. Sánchez, C. Meza, S. Dittmann, and R. Gottschalg, "The effect of clearance height, albedo, tilt and azimuth angle in bifacial pv energy estimation using different existing algorithms", in III Ibero-American Congress of Smart Cities(ICSC-CITIES 2020), 2020.

[8] S. Dittmann, H. Sanchez, L. Burnham, R. Gottschalg, S.-Y. Oh, A. Benlarabi,B. Figgis, A. Abdallah, C. Rodriguez, R. Rüther,et al., "Comparative analysis of albedo measurements (plan-of-array and horizontal at multiple sites world-wide", in 36th European Photovoltaic Solar Energy Conference (EU PVSEC), 2019,pp. 1388–1393.

[9] Giuliano L. Martins, Sebastian Dittmann, Carlos D. Rodríguez-Gallegos, Hyeonwook Park, Soo-Young Oh, Woo Kyoung Kim, Laurie Burnham, Ben Figgis, Ricardo Rüther, Ralph Gottschalg, "Energy contribution of rear-side irradiance for bifacial Photovoltaic modules", in 36th European Photovoltaic Solar Energy Conference (EU PVSEC), 2019

[10] T. S. Liang, D. Poh, and M. Pravettoni, "Challenges in the pre-normative char-acterization of bifacial photovoltaic modules",Energy Procedia, vol. 150, pp. 66–73, 2018

[11] R. Kopecek and J. Libal, "Towards large-scale deployment of bifacial photo-voltaics",Nature Energy, vol. 3, no. 6, p. 443, 2018

[12] G. Razongles, L. Sicot, M. Joanny, E. Gerritsen, P. Lefillastre, S. Schroder, andP. Lay, "Bifacial photovoltaic modules: Measurement challenges",Energy Pro-cedia, vol. 92, pp. 188–198, 2016.

[13] X. Sun, M. R. Khan, C. Deline, and M. A. Alam, "Optimization and perfor-mance of bifacial solar modules: A global perspective",Applied energy, vol. 212,pp. 1601–1610, 2018.

[14] M. Kutzer, A. Fülle, A. Jahnke, J. Hahn, S. Wendt, D. Neuhaus, A. Witzig, andK. Kutzer, "Ertragssteigerung durch bifaciale modultechnologie", inProc. 31stSymp. Photovolt. Sol. Energy, 2016, pp. 1–10.

EFFICIENCY LOSS IN COLOURED PHOTOVOLTAICS: ESTIMATING THE CONTRIBUTION FROM REFLECTION LOSS AND ABSORPTION LOSS

Arne Røyset[1], Tore Kolås[1], Matas Rudzikas[2], Alexander Ulyashin[3],
[1] SINTEF Industry, Høgskoleringen 5, 7460 Trondheim Norway
[2] Center for Physical Sciences and Technology, Sauletekio av 3, LT-10257 Vilnius, Lithuania
[3] SINTEF Industry, Forskningsveien 1, 0314 Oslo, Norway

ABSTRACT: Building integrated photovoltaics (BIPV) offers efficient utilisation of land area and local renewable energy generation. Attractive aesthetic aspects and efficient energy generation are important drivers for market acceptance of solar panels as building elements. Coloured modules improve aesthetic aspects at the expense of efficiency loss and increased cost. For a given lightness and colour, a minimum energy loss due to reflected photons is unavoidable. Additional efficiency loss may come from parasitic absorption in the colour providing layers and from unwanted reflection and absorption in the nonvisible near infrared wavelength range. In this paper we present a novel method to identify the different loss contributions: visible reflection loss, near infrared reflection loss, and parasitic absorption loss. The methodology is applied to silicon solar cells where colouration is introduced by interference from a thin indium tin oxide layer on top of the SiN_x:H antireflection coating. The method presented here will be a useful tool to optimize efficient colouring of solar cells.
Keywords: Building Integrated PV (BIPV), Energy Performance, Module

1 INTRODUCTION

In recent years we have seen an increased scientific literature on coloured solar modules for building integrated photovoltaics (BIPV) with a range of different colours and different techniques for introducing the colour to the solar module. Less attention has been paid to the energy loss caused the colouring of the solar module. Market acceptance of BIPV requires attractive appearance with a minimal energy loss [1]. Traditional PV modules have a dark colour, but coloured modules with increased lightness leads to an increased energy loss due to the added reflection that is needed for the increased lightness. There may also be an additional parasitic absorption from the materials that is applied to the solar module to create the colour.

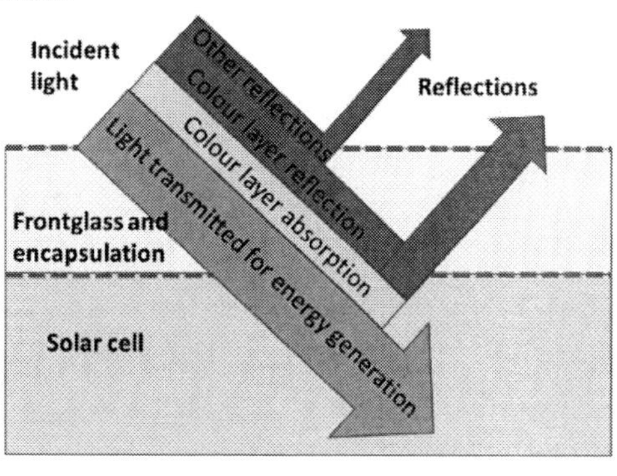

Figure 1: Schematic of light that is reflected, absorbed and transmitted in the solar module.

In a previous paper [2], we introduced a method to quantify the reflection loss caused by colouring of solar modules, and the impact of lightness, hue and saturation on the energy efficiency loss. There we presented reflection measurements for a range of commercial solar modules and estimated the energy loss this reflection introduces. We also introduced a novel colour performance index (CPI) that is a figure of merit of how energy efficient the module lightness is introduced.

In this paper we develop the method further and include absorption loss in the analysis, and a method to characterise the contributions from reflectance loss and absorption loss to the reduced power conversion efficiency.

Based on measurements of module efficiency and reflectance, combined with numerical calculation of reflection induced efficiency loss, we can estimate the parasitic absorption introduced by the colouring elements in the module. Efficient colouring requires that the parasitic absorption and NIR reflection are kept at a minimum. This estimation procedure is a useful tool to optimise the performance of BIPV modules and to achieve attractive appearance at a minimal efficiency loss.

2 MATERIALS AND METHODS

The coloured solar cells were made from commercial multicrystalline silicon solar cells with a single layer antireflection coating (SLARC). The single layer of SiNx:H with a thickness of about 80 nm resulted in a reflectance minimum at 600 nm giving a dark blue colour.

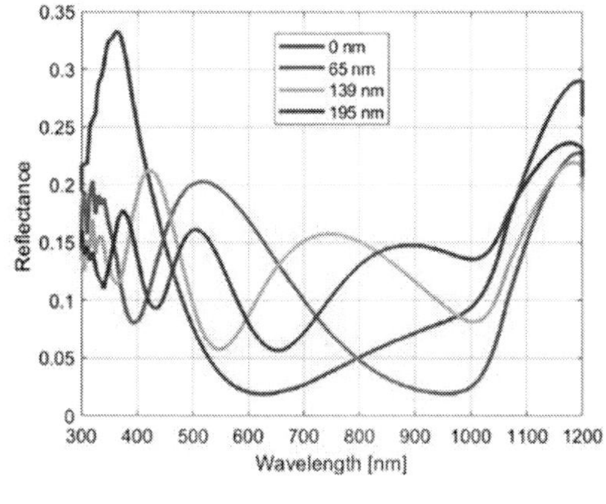

Figure 2: Measured solar cell reflectance spectra for 0, 65, 139, and 195 nm ITO thickness

Double layer antireflection coating (DLARC) cells were formed by adding a layer of magnetron sputtered transparent indium tin oxide (ITO). The ITO thickness was varied from 20 to 240 nm by adjusting the sputter durationThereby a total of 20 different colours were created due to the combined interference of the SiNx:H and and ITO layers. The variation in ITO thickness resulted in a variation of the lightness, hue and saturation of the colour. Example spectra for 0, 65, 139, and 195 ITO thickness are displayed in figure 2.

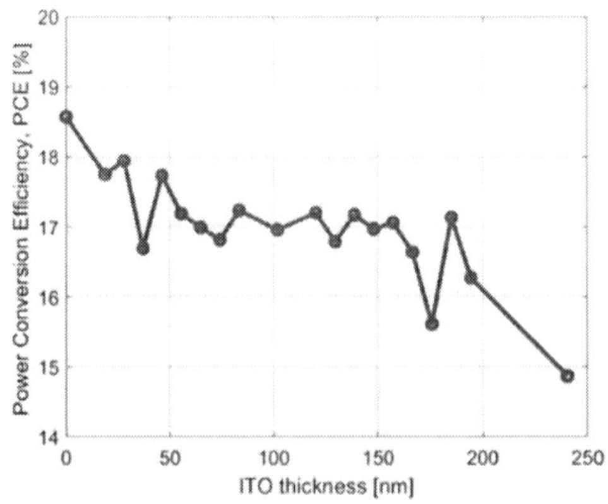

Figure 3: Measured solar cell power conversion efficiency versus ITO thickness.

Power conversion efficiency (PCE) were measured, and the results are given in figure 3. More details on the fabrication and measurements are given in [3].

The model we used for the estimation of efficiency loss is based on our previous results which investigated the relation between colour, reflection loss, and efficiency penalty. We define a colour induced relative efficiency E(t) and relative efficiency loss P(t) as

$$E(t) = \frac{PCE_M(t)}{PCE_{00}} = 1 - P(t)$$

where $PCE_M(t)$ is the measured power conversion efficiency, and t is the ITO thickness. The total relative efficiency loss P(t) is the sum of the contribution from reflection and absorption losses, $P(t)=P_A(t)+P_R(t)$. $P_A(t)$ is the relative efficiency loss contribution from absorption increase due to colouration layer while $P_R(t)$ is the relative efficiency loss caused by reflection. PCE_{00} is a model reference efficiency representing the theoretical case with zero cell reflectance and zero absorption in the colouration layer. In our experimental case, this corresponds to no ITO layer, t=0.

$$PCE_{00} = \frac{PCE_M(t=0)}{1 - P_R(t=0)}$$

The reference efficiency PCE_{00} is the efficiency that would result in the measured $PCE_M(t{=}0)$ if there were no parasitic absorption in the t=0 reference cell. The relative loss $P_R(t)$ caused by reflection are estimated by a model calculation of the change in short circuit current as

$$P_R(t) = \frac{\Delta PCE_R(t)}{PCE_{00}} = \frac{\int_{300}^{1200} \lambda R(t,\lambda) I(\lambda) IQE(\lambda) d\lambda}{\int_{300}^{1200} \lambda I(\lambda) IQE(\lambda) d\lambda}$$

Here $PCE_R(t)$ is the reflection contribution to reduced efficiency, λ is the wavelength, I(λ) is the incident radiation (AM 1.5G), IQE(λ) is the internal quantum efficiency, and R(t,λ) is the measured spectral reflectance of the solar cell. More details on the model are given in [2] The contribution from absorption $P_A(t)$ can then be determined from the relation $P_A(t)=P(t)-P_R(t)$.

3 RESULTS AND DISCUSSION

By applying this method to the measured spectra and power conversion efficiencies of the DLARC cells we have calculated the contribution from reflection loss and absorption loss as displayed in Figure 4, while the contribution from reflection and absorption to the reduction in measured PCE is displayed in figure 5.

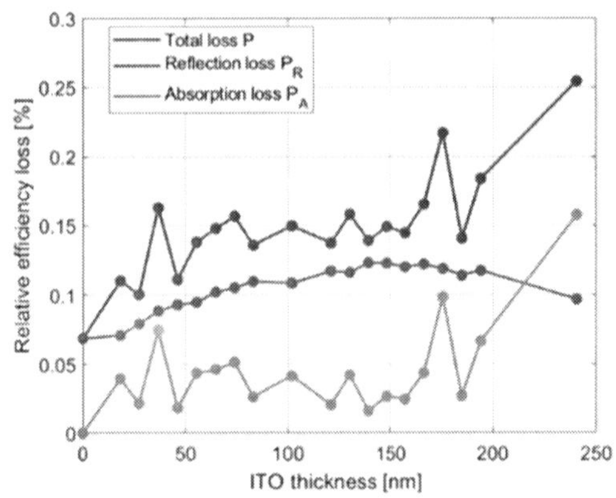

Figure 4: Total relative efficiency loss and contribution from reflection loss and absorption loss

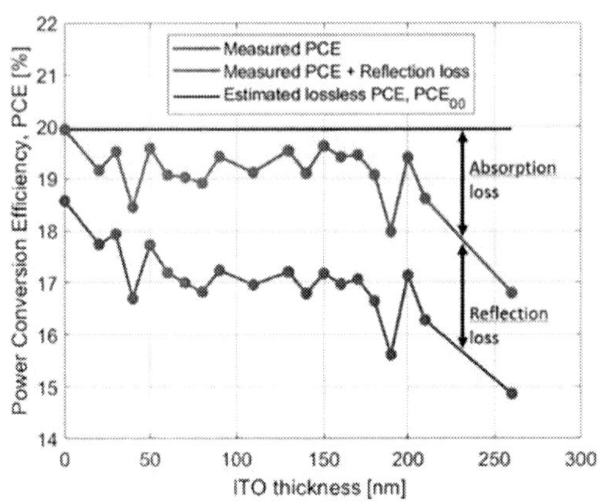

Figure 5: Measured power conversion efficiency PCE, estimated lossless power conversion efficiency PCE00, and the contribution from absorption and reflection losses to colourisation induced PCE reduction.

The results in figure 5 demonstrates a case where the reflection loss is greater than the parasitic absorption loss. This is expected from interference colours but may not be the case for pigment coloured PV.

An important appearance characteristic of coloured solar cells is the lightness, which can be quantified by the CIE Y value. The CIE Y value represents the reflected light weighted by the visual eye sensitivity. Increasing the lightness requires more of the incident light to be reflected and causes an increased efficiency loss. In a previous paper [2] we introduced a Colour Performance Index (CPI) as a Figure of Merit to quantify how efficiently the addition of colouring layer can increase the lightness at a minimal efficiency penalty as CPI(t)=Y(t)/P(t).

How Y, P, and CPI varies with ITO thickness in our experimental case is displayed in figure 6. The Y and CPI has a peak near 70 nm thickness, while the variation in efficiency has a smaller variation. The reason why there are only a small covariation between Y and P is caused by the relative contribution to efficiency loss in the three spectral ranges UV (300-400 nm), VIS (400-700 nm), and NIR (700-1200 nm): $P(t)=P_{UV}(t)+P_{VIS}(t)+P_{NIR}(t)$.

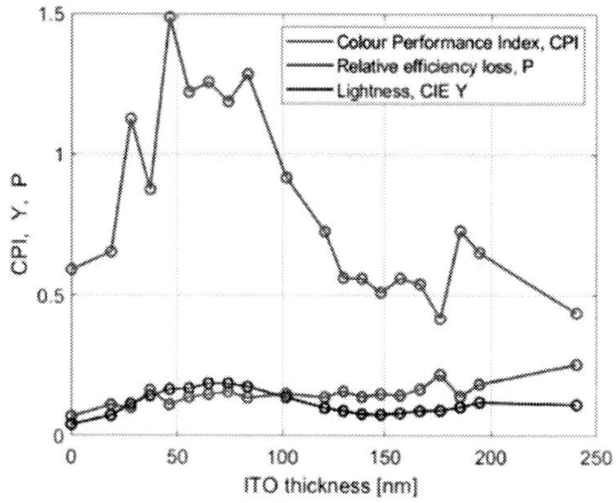

Figure 6: Colour Performance Index CPI, Lightness CIE Y, and relative efficiency loss versus ITO thickness

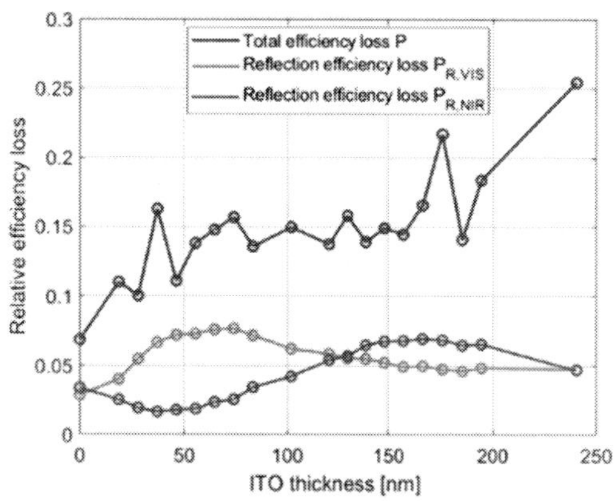

Figure 7: Total relative efficiency loss and the contribution from visible and near infrared spectral range to reflection loss.

The contribution from P_{VIS} and P_{NIR} versus thickness is displayed in figure 7, illustrating that VIS reflectance contributes to lightness and efficiency loss, but also the contribution from NIR reflectance that only contributes to efficiency loss and not to lightness and colour.

This illustrates that efficient colouring of solar cells requires the NIR reflectance to be kept at a minimum. It is observed that ITO thickness of 45 nm results in a green colour with high lightness and low NIR reflectance resulting in an energy efficient colouring and high CPI. The cell with 195 nm ITO thickness also has a green colour, but in this case the interference also results in a high NIR reflectance, a higher total reflectance loss, and a much lower CPI. It also illustrates that green-yellow colours (peak at 500-550 nm) gives the highest CPI because this is the wavelength range with the highest eye sensitivity. One way to improve the CPI is to optimize the reflectance spectrum further by adding more interference layers, but that may be at the expense of added fabrication cost.

4 CONCLUSIONS

In conclusion we have proposed and demonstrated a methodology for estimating the reflectance and absorption contribution to efficiency loss in coloured solar modules.

The method presented here will be a useful tool to optimize energy efficient colouring of solar cells.

REFERENCES

[1] C. Ballif, L.E. Perret-Aebi, S. Lufkin, E. Rey, Integrated thinking for photovoltaics in buildings, Nat. Energy, 3 (2018), pp. 438-442

[2] A Røyset, T Kolås, BP Jelle, Coloured building integrated photovoltaics: Influence on energy efficiency, Energy and Buildings, 208, 109623, 2020

[3] M Rudzikas, A Setkus, M Stange, J Ulbikas, A Ulyashin, Simple interference based colorization of Si based solar cells and panels with ITO/SiNx:H double layer antireflective coatings, Solar Energy, 207, 218-227, 2020

ANGLE OF INCIDENCE STUDY AT PHOTOVOLTAIC MODULES WITH POLYMER FRONT SHEET

Stefan Wendlandt[1], Sophie Heller[1], Jonathan Govaerts[2], Arvid van der Heide[2], Jarno Kaakkunen[3], Tuukka Savisalo[3], Arnaud Morlier[4], Daria Raine[5] and Dominic Röder[5]

[1] PI Photovoltaik-Institut Berlin AG, Wrangelstr. 100, D-10997 Berlin, Germany
Phone: +49 (30) 814 52 64-0, e-mail: wendlandt@pi-berlin.com
[2] imec-EnergyVille, Thorpark 8320, 3600 Genk, Belgium
[3] Valoe Oyj, Insinöörinkatu 5, FI-50150 Mikkeli, Finland
[4] ISFH, Am Ohrberg 1, 31860 Emmerthal, Germany
[5] Fraunhofer ISE, Heidenhofstraße 2, 79110 Freiburg im Breisgau, Germany

ABSTRACT:

For years, photovoltaic systems have been successfully installed on a large scale on open spaces as well as on (BAPV) and on buildings (BIPV). For the applications, PV modules in the classic glass-backsheet or glass-glass design are mostly used because their properties are well known. For BIPV and further on for automotive applications (VIPV) it makes sense to reduce the weight of the modules in order not to overload the building's loadbearing capacity statics as well as for vehicles to minimize their energy consumption. One approach for weight reduction is by replacing glass with polymer materials. The light coupling on typical PV modules is largely optimized by means of an anti-reflective coating (ARC) on the front glass. This has not yet been done for polymer front sheets, but this paper will take the first step and characterize different lightweight approaches based on polymer front sheets with different solar cell types (SHJ, IBC) for their skewing behavior and then give an evaluation of the electrical yield for angle-dependent installation as well as for different locations.

Keywords: light weight, energy yield, angle of incidence

1 INTRODUCTION

A typical pv-module has glass as front sheet. In order to decrease the optical losses and increase energy yield the glass is anti-reflective coated (ARC). Optimised parameters of those layers are a thickness of $d = 120$nm and a refraction index $n = 1.23$ [1]. Another approach to couple more sun rays into the module is to increase the glass area by structuring the glass on the air/glass interface. To modell the AOI effect an empirical model has been developed Martin *et al.* [2]. In 2006 Grunow *et la.* showed data for outdoor AOI measurements [3].

Within is study we will show the optical properties of polymer front layers as well as to show the optical differences between samples with glass and polymer front sheets.

2 EXPERIMENTAL

2.1 Test Samples

For our investigations we designed and laminated four different samples with polymer front sheets and different refractive index. The designs and sample properties are shown in Table 1.

Table I: Module Design

Sample number	A Front sheet: polycarbonate	B Front sheet: Polycarbonate
Design	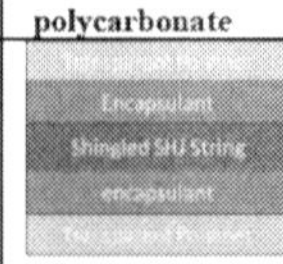 	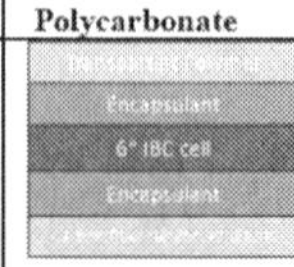 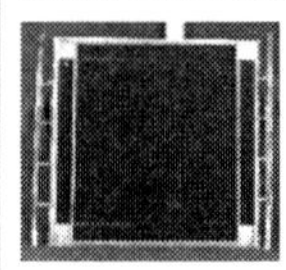
Shape	Flat	Flat
Sample number	C The design is presented in [4]	D (1/2 cell)
Design	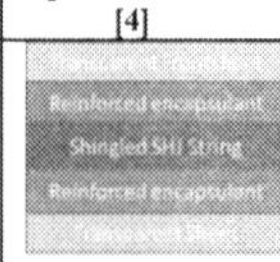 	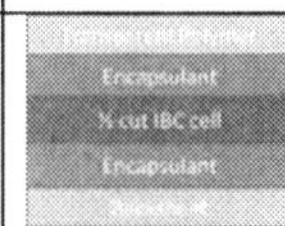 
Shape	Flat	Flat

At the test cell in the centre of the sample the IV-Curves were measured. The surrounding cells are dummy ones. Those cells will simulate the internal reflection as it exists in full size modules.

The determination of the optical losses will be done by compare the results of the polymer frontsheet samples with identical samples with ARC glass (thickness 3.2 mm) as front sheet.

2.2 Testing

In a first step the spectral transmission of the polymer front sheets and encapsulation materials like was measured. The data were recorded with a Lambda 1050 UV/VIS/NIR spectrophotometer device. The measurement interval of the wavelength was $d\lambda = 1$ nm. Additional to the spectrometer analysis the surfaces were investigated with a profilometer to receive information on the roughness of the outside of the polymer frontsheets.

We also determined the angle of incidence (AOI) behaviour of all samples with polymer frontsheet as well as of the reference samples with ARC glass as frontend. The IV Curves were measured in 10° steps and at >60° every 5°. In order to eliminate the light reflection inside of the flasher – a second measurement with a mask 40 cm in front of the module were done.

Figure 1: Sample A mounted in the AOI set-up.

The following equation has been applied to evaluate the AOI measurements.

$$rel\ I_{sc} = \frac{I_{sc,total} - I_{sc,diff}}{\cos(\alpha)} \quad (1)$$

The next chapter presents the results.

3 RESULTS

The following figures display the results of the spectral transmission of the encapsulation materials and polymer frontsheets.

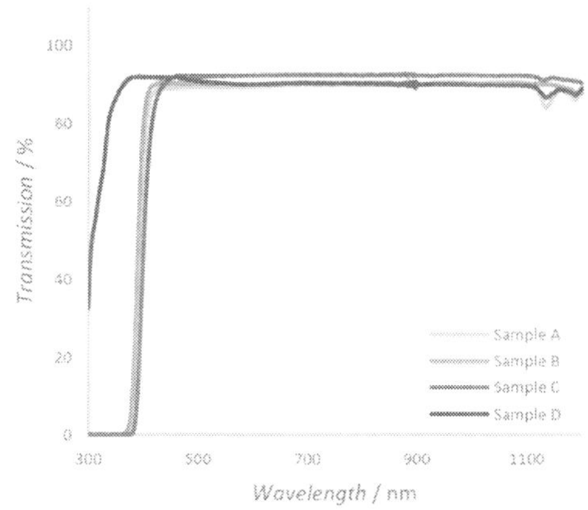

Figure 2: Spectral transmission of the different frontsheets.

Nearly all samples a show similar transmission except sample D. That sample is additional UV open, while other blocking it. In the range of 400…1150 nm the transmission of all samples is with around 90% and slightly lower than glass.

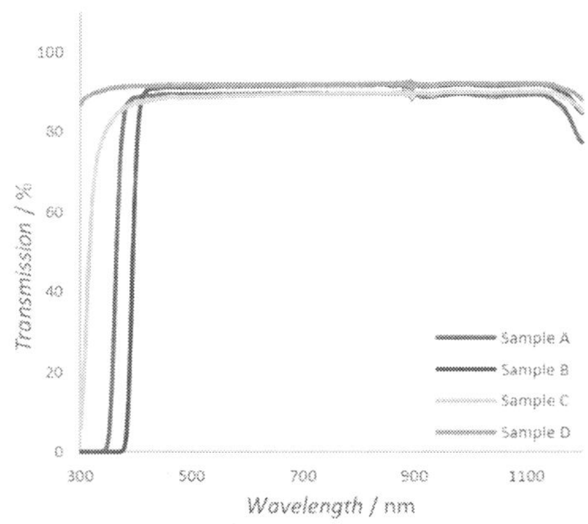

Figure 3: Spectral transmission of the different encapsulation materials.

Samples C & D are UV open compared to the other samples. The transmission rate of around 90% in the range of 400…1150 nm is quite similar. At wavelength > 1150 nm a reduction in transmission is visible. It is assumed that the reduction acts like a heat shield for the cell.

The roughness of the samples was investigated since (micro)structures on the outside of the frontglass can lead to a higher performance at higher incidence angle. The results of the profilometer analysis are shown next.

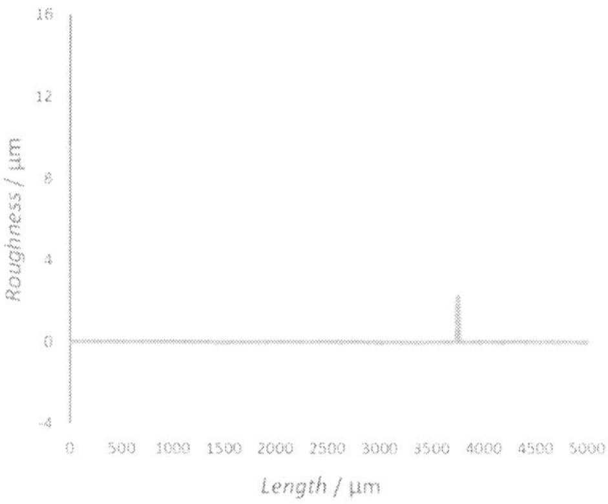

Figure 4: Roughness of samples A.

The figure makes clear that sample A has a smooth surface. The peak at length 3,750 µm seems to be an aberration.

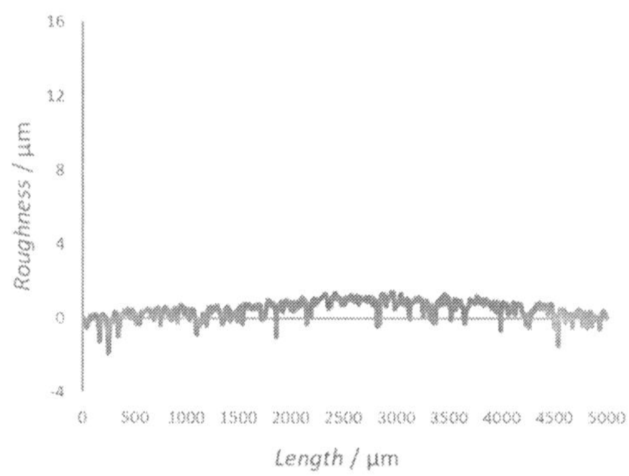

Figure 5: Roughness of samples B.

Sample B's surface is rougher than the one form Sample A. However, the surface sample C rougher.

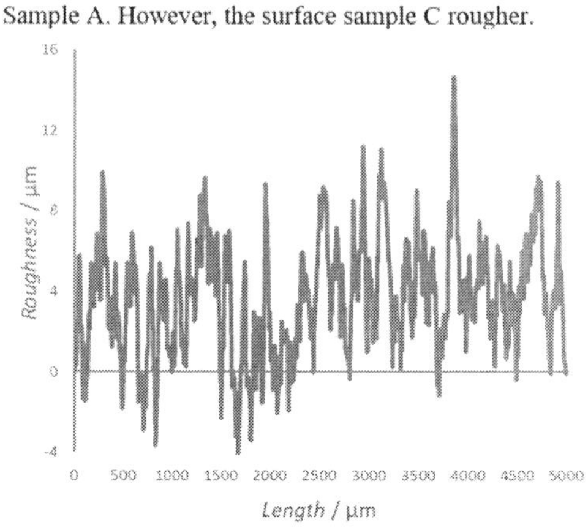

Figure 6: Roughness of sample C.

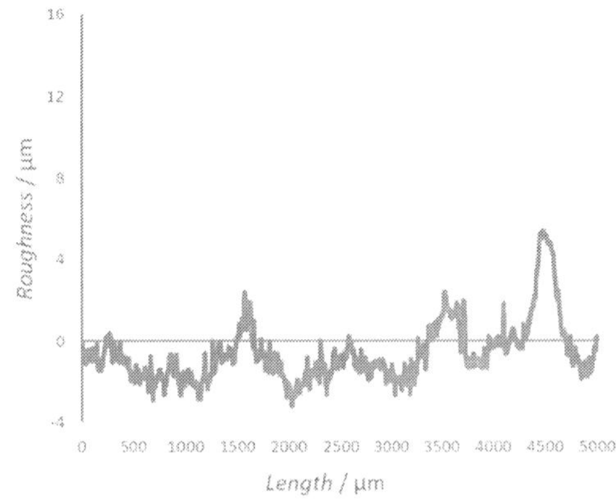

Figure 7: Roughness of samples D.

The surface of sample D has some peaks, but it seems smoother compared to sample C. The roughness results were evaluated based on the R_a arithmetic centerline roughness value what is defined as:

$$R_a = \frac{1}{l}\int_0^l |z(x)|\, dx \qquad (2)$$

and the R_q square center roughness value

$$R_q = \sqrt{\frac{1}{l}\int_0^l z^2(x)\, dx} \qquad (3)$$

Next to them also the Peak and Valley were considered for analysis the data. Table 2 presents the average results of four roughness measurements per sample.

Table 2: Evaluation of the surface investigations

	Sample A	Sample B	Sample C	Sample D
R_a / nm	35.6	822.7	2066.8	785.6
R_q / nm	67.2	1051.4	2988.0	1022.0
Peak / nm	755.2	1544.8	13292.8	3964.6
Valley / nm	-115.6	-4236.4	-6994.9	-2648.2

Table 2 displays the different surface properties. Summarised sample A has flattest surface while sample C is mostly structured. It can be assumed that the structure is a copy of the membrane from the laminator. If the power output benefits from the structure at high incidence angle will be shown next.

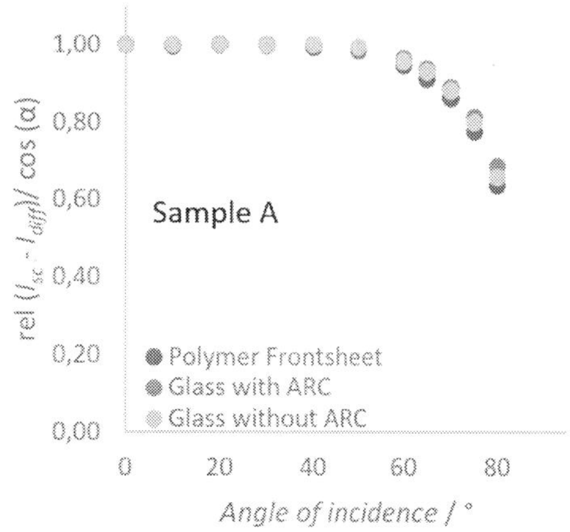

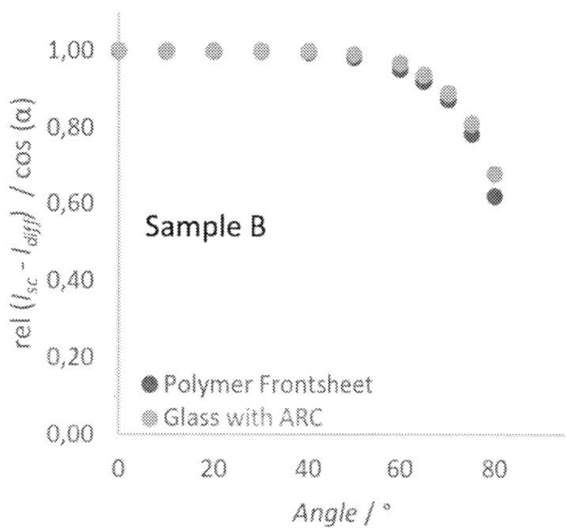

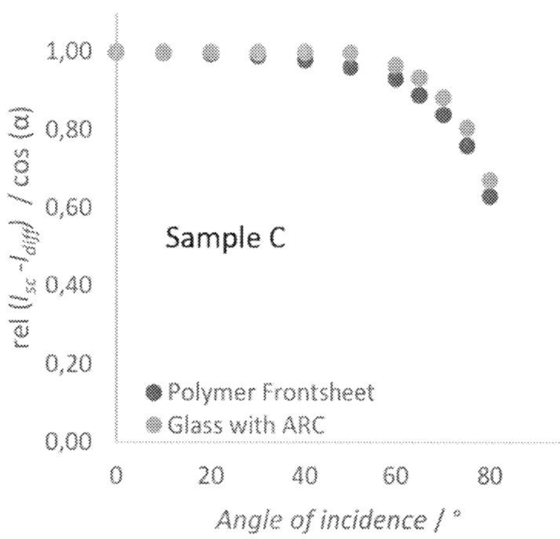

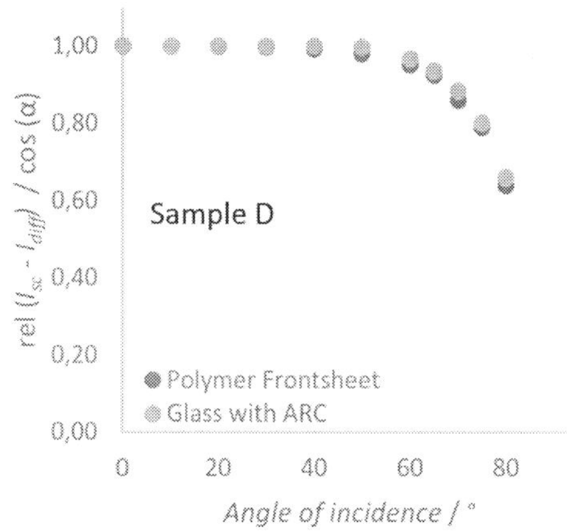

Figure 7: AOI results all the samples with polymer frontsheet and ARC glass as well as with glass without ARC for sample A.

All shapes following the Fresnel equation $(\frac{|| + =}{2})$. At all samples it is visible that for angles >40° there is no major difference in the AOI between glass and polymer frontsheet samples. At angles >40° the glass samples show a higher current.

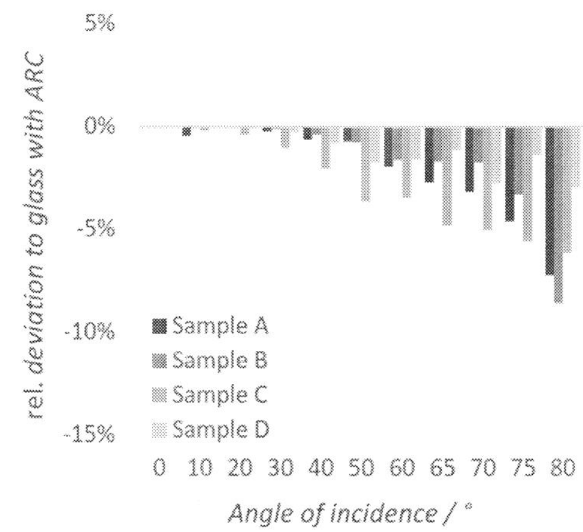

Figure 8: Relative deviation from polymer frontsheet sample to reference sample ARC glass.

In this plot, the angle-dependent optical losses of the polymer frontsheet samples are compared to the identical references with ARC glass. Sample C shows the largest deviation over all angles. In a final step we simulated the energy yield with the different AOI measured data.

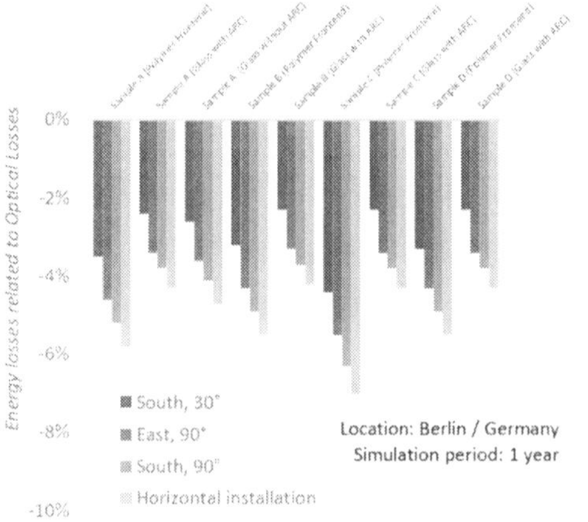

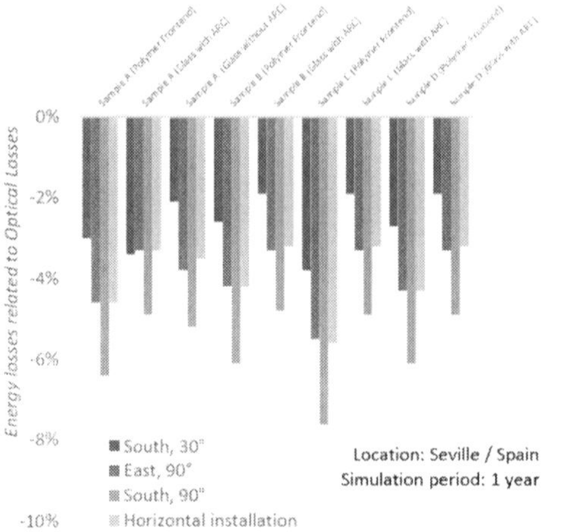

Figure 9: Energy losses related to optical losses due to different frontsheets.

At both sites, optical losses are lowest at 30° south orientation (Berlin: < -4.6%; Seville: < -3.0%). In Germany the max optical losses for a polymer frontsheet sample is -7.0% and for Spain -7.6%.

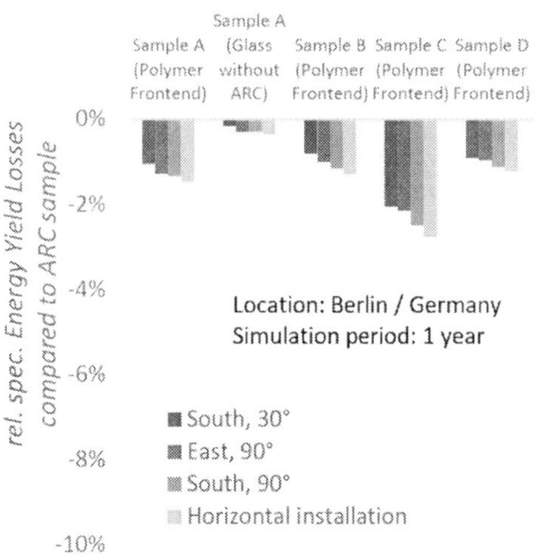

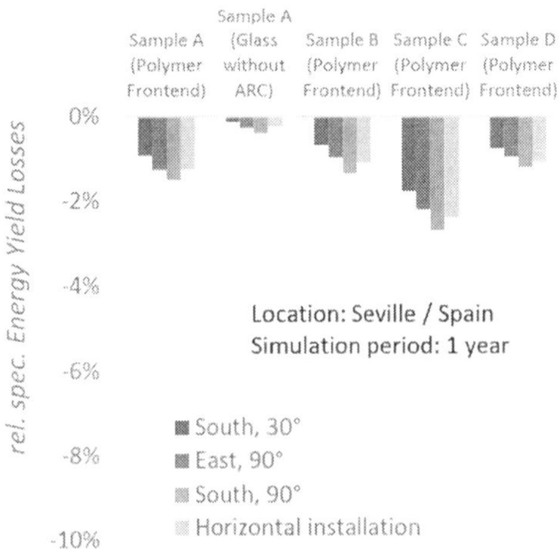

Figure 10: Energy yield losses due to different frontsheets relative to ARC glass.

The simulated energy yield considering the different optical properties leads to a reduction of ≤-2.8% (worst case: Horizontal installation) for Berlin / Germany. For Seville / Spain we simulated a worst-case reduction of -2.7 % (vertical south installation). However, the losses depend strongly on the installation orientation. The simulations were done with PVsyst for period of one year.

4 SUMMARY

In our investigations we designed and laminated four different samples with polymer front sheets and identical samples with ARC-glass. We did analysis of the optical properties from the polymer front sheets and encapsulation materials like spectral transmission and did investigations of the surface roughness from the polymer frontsheet. We determined the angle of incidence (AOI) on the samples, determined the angle-dependent optical losses compared to the ones with ARC-Glass and finally we simulated the energy yield losses due to the different frontsheets. We can summarize:

- All polymer frontsheets show a transmission of ~90% at λ > 400nm. Sample D's frontsheet is UV open, while the other samples UV light blocking.
- The polymer frontsheet samples have a different roughness. The range covers from nearly flat (Sample A) until 12μm (Sample C).
- The ARC-Glass references show a better AOI compared to the polymer frontsheet samples. Also, the glass without ARC of sample A is slightly better than the ones with polymer frontend.
- Sample C shows the largest and sample D has the smallest deviation over all angles compared to the ARC-Glass.
- No clear correlation between the surface roughness and the optical losses could be seen.
- In Germany the max optical losses for a polymer frontsheet sample is -7.0% and for Spain -7.6%.
- The relative energy yield losses due to different optical properties compared to samples with ARC glass are ≤ -2.0% for Berlin/Germany and < -2.7% for Seville/Spain.

5 ACKNOWLEDGEMENTS

This study was supported by the European Union's Horizon 2020 Programme for research, technological development and demonstration under Grant Agreement number 857793 (Project name: HighLite).

6 REFERENCES

[1] B. Litzenburger, J. Dittrich: *„PV Modules with Anti-Reflective Coated Glass: Performance Simulation and Outdoor Measurements of Specific Energy Yield Gain"*; 27th European Photovoltaic Solar Energy Conference and Exhibition

[2] N. Martin, J.M. Ruiz: *„A New Model for PV Modules Angular Losses Under Field Conditions"*, J. Sol. Energy 22 (2002) 19–31.

[3] P. Grunow, S. Krauter: *„Modelling of the Encapsulation Factors for Photovoltaic Modules"*, Proc. IEEE 4th World Conference on Photovoltaic Energy Conversion, Hawaii, 2006, p. 2152.

[4] J. Govaerts *et al.*: *"The Potential of Glass-Fibre-Reinforcement: (Thermo-)Mechanical Testing of Light-Weight PV Modules, this conference" this conference"*

BEYOND-STANDARD EQUIVALENT CELL TEMPERATURE (ECT) EVALUATION

Georgi H. Yordanov[1,2*], Mohammed Gofran Chowdhury[1,2,3], Jens D. Moschner[1,2], and Arvid van der Heide[2,3]
[1] KU Leuven, ESAT-ELECTA, Leuven, Belgium; [2] EnergyVille, Genk, Belgium; [3] *imec*, Leuven, Belgium.
* Corresponding author; E-mail: georgi.yordanov@energyville.be

ABSTRACT: The IEC Standard (60)904-5:1993/2011 limited the calculation of Equivalent Cell Temperature (**ECT**) of PV devices from the V_{OC} to irradiance above 0.2 Sun and a constant ideality factor. Increased error at low irradiance was attributed to a varying temperature coefficient (TC) of the V_{OC}. In previous works, we showed that this was in fact due to rise of the ideality factor (near Open Circuit, OC) at low irradiance, and developed individual corrections for different monofacial c-Si PV modules allowing for accurate **ECT** evaluation down to ~0.03 Sun. In the present work, we compare the **ECT** to module back temperatures (T_{back}) measured with up to 16 sensors in two outdoor testing campaigns involving 6 monofacial, glass-backsheet and 5 bifacial, glass-glass c-Si modules made with PERC cells. We present a method to derive the irradiance-dependent ideality factor, n_{OC}, allowing for accurate **ECT** evaluation and outdoor performance characterization also at the very low irradiance. Correction methods for the rated V_{OC} and its TC are presented. We show the possible uses and challenges of the **ECT**–T_{back} plots over in-plane irradiance (ΔT plots) and propose a simple effective irradiance in the bifacial PV case.
Keywords: PV Module, Outdoor characterisation, Thermal Performance, c-Si, Monofacial, Bifacial

1 INTRODUCTION

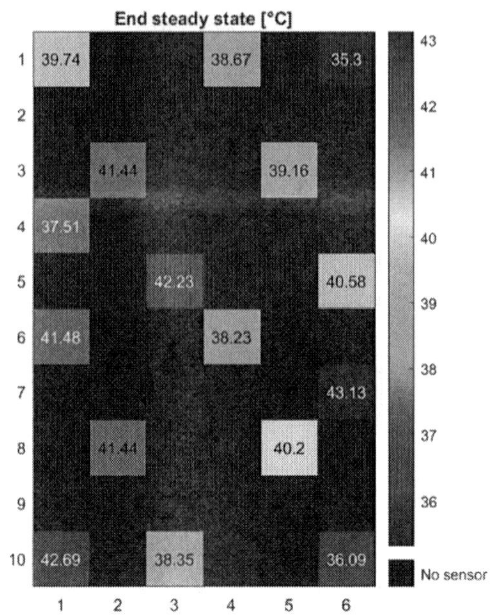

Figure 1: Sixteen back-of-module temperature readings of a monofacial, glass-plastic, framed c-Si PV module under steady-state, full-sun conditions in November, 2019 in Genk, Belgium. View is from the back.

Outdoor characterization and rating of PV modules requires the measurement of, besides in-plane irradiance, also an effective operating temperature. While bifacial modules also require the rear-side in-plane irradiance, those devices most often have a glass-glass design making the temperature sensing at the back very inaccurate. Even for glass-plastic PV modules, the back temperature can differ from the effective one by several degrees, and given that the typical temperature coefficients (TCs) of power of e.g. c-Si PV are in the order of 0.4%/°C, this can lead in an error of a few %. Another issue with real-life testing of PV modules is the temperature's spatial non-uniformity; see Fig. 1. Temperature sensor accuracy within ±1°C is usually required in the characterization of PV modules [1, 2]. We are evaluating the accuracy of external temperature sensing compared to the determination of the Equivalent Cell Temperature (**ECT**) from electrical measurements.

2 MEASUREMENTS

2.1 Temperature sensing

From Fig. 1 it is clearly seen that, even at steady state, the back temperatures of the two central cells differ by 4°C while other cells are even hotter or cooler. At high irradiance, a temperature gradient is present across the encapsulation of the module, and its surfaces are cooler than the cells. Some authors using film thermocouples (slim sensors recommended for flat surfaces) have adopted an average cell-to-back temperature difference of 2.5±1°C under solar irradiance of 1 Sun (1.0 kW/m^2) [3]. Common practice is to use one or two sensors per module placed behind central cells [4, 5] (also in our previous, unpublished works for customers). Foiled sensors were shown to measure more accurately than ones placed in a ceramic case [6], whereas tubed Pt100 RTDs have resulted in a severe underestimation [7]. We found the flat, wire-wound Pt100 RTDs of type NWT-F by Ephy-Mess to be particularly suitable for PV module back sensing. The total uncertainty in the module's effective cell temperature obtained with back sensing is 3-5°C at Standard Testing Conditions (STC) that is equivalent to rated power uncertainty of 1.5-2.5% for c-Si modules [8].

2.2 Temperature derivation from electrical parameters

To solve the thermal non-uniformity and transient issues, one can use the 'instantaneous' open-circuit voltage, V_{OC}, of the PV module to evaluate its 'real-time' effective, or equivalent cell temperature (**ECT**) as defined in the IEC Standard (60)904-5:1993 (1st Edition, withdrawn) and 60904-5:2011 (2nd Edition) [1, 7, 9]. (One should bear in mind that, during I-V curve measurement, the OC and the maximum-power point (MPP) are recorded at somewhat different times, depending on the particular setup and PV device's capacitance, ideally within a few tens to a few hundred milliseconds.) The V_{OC} reacts instantaneously to changes in irradiance and cell temperature, whereas the back-of-module temperature is affected by thermal transients (delays). The equivalent cell temperature (ECT) was defined as the uniform p-n junction temperature at which the measured electrical output of the PV device would be reproduced [1, 9]. In [5], the formulation given in [9] was used:

$$ECT \equiv T_2 = T_1 + \frac{1}{\beta}\left[V_{OC2} - V_{OC1} + DN_S \, ln\left(\frac{G_1}{G_2}\right)\right], \quad (1)$$

where $\mathbf{D} = \mathbf{n}k(\mathbf{T_2}+273.15\ \mathrm{K})/q$ is the "diode thermal voltage" at $\mathbf{T_2}$ (including the ideality factor at OC, **n**, sometimes denoted $\mathbf{n_{OC}}$); **k** is Bolzmann's constant; **q** is the elementary charge; and $\mathbf{N_S}$ is the number of serially-connected cells in a module. Prior knowledge is required of the open-circuit voltage $\mathbf{V_{OC1}}$ at reference testing conditions (a set of irradiance $\mathbf{G_1}$ and cell temperature $\mathbf{T_1}$ which may or may not equal the ones defined at STC). In addition, the temperature coefficient (TC) of $\mathbf{V_{OC}}$, β (in units V/K), and the thermal voltage **D** must be known in advance. Since **D** is linearly related to $\mathbf{T_2}$, Eq. (1) is easily solved for $\mathbf{T_2}$. Furthermore, the ratio of the corresponding short-circuit currents $\mathbf{I_{SC1}/I_{SC2}}$ can be used instead of $\mathbf{G_1/G_2}$, if the short-circuit current of the PV device scales linearly with irradiance. Since the TC of $\mathbf{I_{SC}}$ is very small (in the order of +0.05%/°C for c-Si PV), and since the ratio is in a logarithm, the error due to the different temperatures corresponding to $\mathbf{I_{SC1}}$ and $\mathbf{I_{SC2}}$ will be negligibly small.

The IEC procedure assumes: 1) a linear dependency of the $\mathbf{V_{OC}}$ on temperature; 2) that the device's I-V curve is governed by the one-exponential equation; 3) a constant **n**; 4) a constant **β**; 5) an Ohmic series resistance; and 6) a reasonably high shunt resistance. The last assumption is considered valid for most of the commercially available PV devices [1]. The procedure is however not recommended for irradiance below 0.2 Sun (0.2 kW/m^2), where the temperature coefficient **β** is said to drop rapidly. In general, the involved assumptions need verification for each device type, given the short innovation cycles of some PV technologies [10]. The method is only applicable to *non-shaded* PV devices.

It was shown in [5] that in fact the ideality factor at OC rises at low irradiance, whereas the **β** is used in Eq. (1) at the reference irradiance $\mathbf{G_1}$, after the translation of the $\mathbf{V_{OC2}}$ from $\mathbf{G_2}$ to $\mathbf{G_1}$. Individual corrections to the ideality factor **n** at OC at low irradiance were implemented in [5] for some of the studied monofacial c-Si PV modules, allowing for accurate ECT evaluation down to ~0.03 Sun. In [11], the ECT procedure was extended to PV devices with bilinear variation of the ideality factor **n** with irradiance and temperature.

The IEC standard's 2nd Edition uses a dimensionless "diode thermal voltage", **a**, and the β in relative units of K^{-1} [1]. In fact, in Eq. (5) therein, the "factor **a**" is wrongly replaced with the thermal voltage **D** times $\mathbf{N_S}$, thus mixing dimensionless and voltage units in one expression. A correct form of that equation would divide the thermal voltage **D** by the reference open-circuit voltage, $\mathbf{V_{OC1}}$. Furthermore, either the sign of that term should be opposite or the irradiance ratio in the logarithm should be reciprocal ($\mathbf{G_1/G_2}$ as in the Standard's withdrawn 1st Edition). To our knowledge, the Standard's 2nd edition is currently being reviewed [12].

The present paper builds on the **ECT** study in [5] but extends its evaluation below 0.2 Suns using a more general and straight-forward method applicable to arbitrary variation (rising) of the $\mathbf{n_{OC}}$. A hypothesis about an alternative method to evaluate the β expressed in [5] is refined in the present paper with detailed math and shows very promising.

3 ECT SENSITIVITY & PARAMETERS FIXING

Once Eq. (1) is solved for the ECT, by partial differentiation it is possible to derive the ECT errors due to (small) errors in each of the involved variables [5]. This is best visualized when plotting the difference of **ECT** – $\mathbf{T_{back}}$ over irradiance (ΔT plots). Fig. 2 shows the effect of varying the rated $\mathbf{V_{OC}}$ ($\equiv \mathbf{V_{OC1}}$), while similar errors in the measured $\mathbf{V_{OC2}}$ would result in similar **ECT** errors but of an opposite sign. A 'well-behaving' ΔT plot is linear with irradiance (with some spread) and should have a 0°C limit in the absence of solar forcing (at zero irradiance). Provided that the other parameters, **n** and β, are correct, the ΔT plot can be used to derive a correction $-\beta\Delta\mathbf{ECT}$ of the rated $\mathbf{V_{OC}}$. Interestingly, the plot in Fig. 2 is from a multicrystalline Si PV module with cells made of an earlier-generation solar-grade Si (Elkem Solar Silicon™, vintage 2008), and still maintains a remarkable overall linearity down to very low irradiance (0.03 Suns).

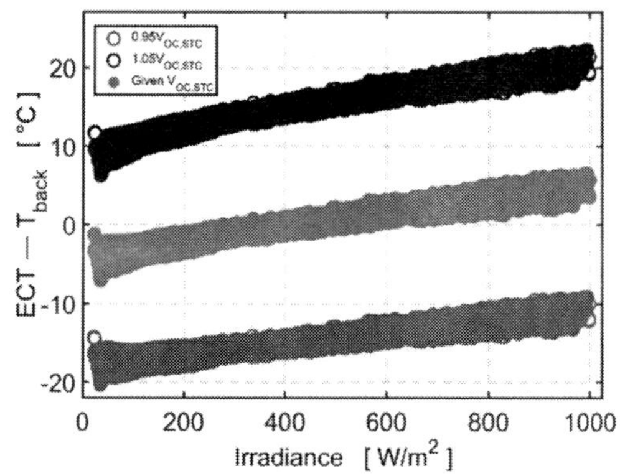

Figure 2: ΔT plots with a given value and ±5% of the STC-rated $\mathbf{V_{OC}} \equiv \mathbf{V_{OC1}}$ [5].

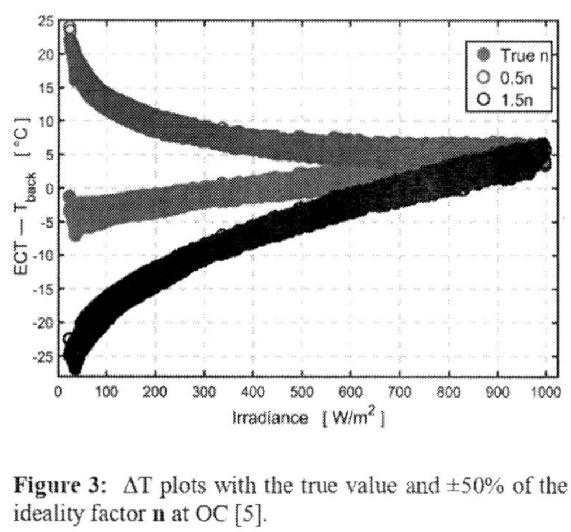

Figure 3: ΔT plots with the true value and ±50% of the ideality factor **n** at OC [5].

Similarly, the ΔT plots in Figs. 3 and 4 show the effect of an erroneous ideality factor **n** at OC and an erroneous TC of the $\mathbf{V_{OC}}$, β. The corresponding (linear-term) error expressions are given in Eqs. (2) and (3). (Note that the error in the given value of the rated V_{OC} had not been corrected for (yet) in either of the Figs. 2-4, resulting in a negative shift of the 'linear' ΔT plots.)

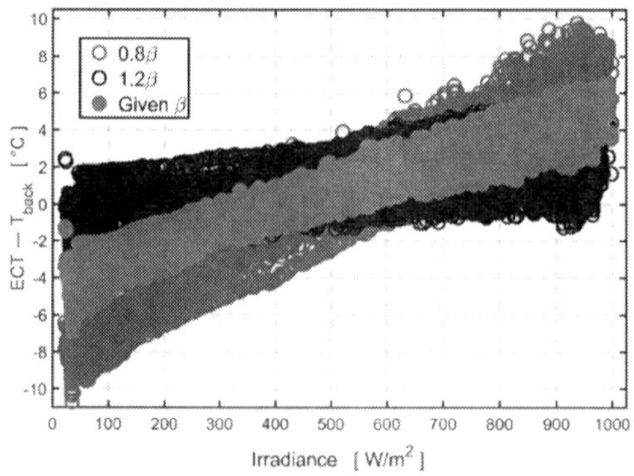

Figure 4: ΔT plots with the given value and ±20% of the TC β [5].

$$\Delta ECT_n = -\Delta n\left[\frac{N_S k(ECT+273.15\ K)\ln G}{q\beta+nkN_S\ln G}\right] \quad (2)$$

$$\Delta ECT_\beta = -\frac{\Delta\beta}{\beta}(ECT - 25°C) \quad (3)$$

Note that errors in the **n** have no effect at the reference irradiance ($G_2 = G_1$), since the ratio $G = G_2/G_1$ in the logarithm in Eq. (2) (the relative irradiance) becomes unity. On the other hand, the lower (or higher) the irradiance G_2 is compared to the reference one G_1, the larger is the effect of the same error in **n** on the **ECT**. We will use the STC irradiance as reference G_1 by default (as well as the STC reference temperature, $T_1 = 25°C$), and hence **G** is the measured irradiance G_2 expressed in relative units or in Suns.

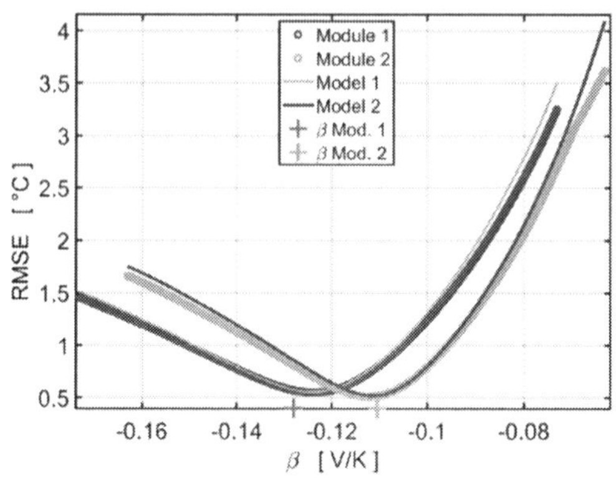

Figure 5: RMSE of linear fits to multiple ΔT plots made with trial values of β from [5] for two monofacial mc-Si PV modules. The solid lines are models based on Eq. (4) and the plus symbols denote the β values obtained from fitting outdoor V_{OC} data over T_{back} at 1 Sun.

Assuming that the given value of β (from an indoor test at an independent, certified laboratory) was very accurate [5], the effect on **ECT** for different values is proportional to the relative error of the TC, times the difference of the **ECT** from the STC temperature; see Eq. (3). The spread in the plots with erroneous $\beta \equiv \beta'$ becomes generally larger, while they remain linear but with a different slope; see Fig. 4. However, the ΔT plot with the least spread (i.e., with the least RMSE of the linear fit) does not necessarily correspond to the true value of β, yet the two can be very close. An approximate RMSE(β') model is given in Eq. (4) and the minimum's location is given in Eq. (5); see Fig. 5.

$$RMSE(\beta') \approx \sqrt{\frac{\sum\left[r-(r+R)\frac{\Delta\beta}{\beta'}\right]^2}{N}} \quad (4)$$

$$\beta_{RMSE} \approx \beta\left[1+\frac{\sum r(r+R)}{\sum R(r+R)}\right] \quad (5)$$

The summation in the above two equations runs over the number of data points, N; β' is the trial value of the TC and $\Delta\beta$ is its deviation from the empirical value β; **r** are the residuals from the linear fit to the ΔT plot corresponding to β; and **R** are the residuals from the linear fit of T_{back} over the irradiance **G**, which are independent on the trial value β' and are typically of a much higher magnitude than the **r**. The 2nd term in the square brackets in Eq. (5) ranged within ±3% for most of the c-Si PV modules tested in [5] and equaled −2.0% for a CIGS module. In fact, Eq. (5) gives us yet another estimate for β (besides the one used so far, obtained from linear fitting of the V_{OC} over T_{back} at irradiance of about 1 Sun), where **r** is calculated with the classic β value:

$$\beta \approx \frac{\beta_{RMSE}}{\left[1+\frac{\sum r(r+R)}{\sum R(r+R)}\right]} \quad (6)$$

Surprisingly, for most but one of the PV modules tested in [5], the two estimates of β differ by ±1-2% (median −0.62%, sigma 1.63%). This is much lower than the uncertainty typical for the classic method using outdoor data. Unfortunately, not all the back-temperature measurements yield such a 'well-behaving' ΔT plot (subject to linearization and of a small spread) as that presented in Fig. 2.

4 ECT EVALUATION BELOW 0.2 SUNS

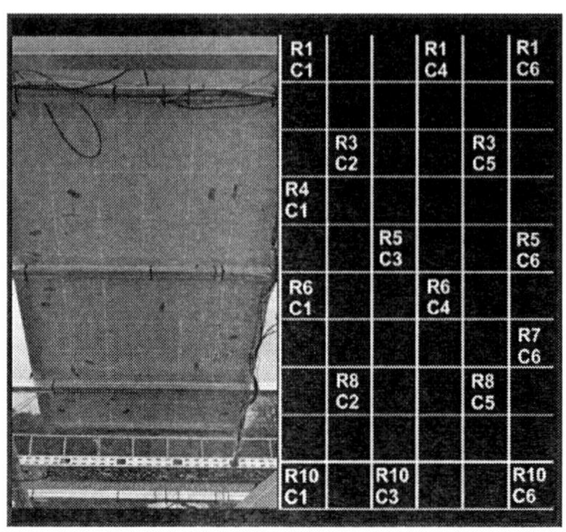

Figure 6: Back view of the PV module represented in Fig. 1 with locations of the 16 back T sensors denoted according to each cell's row and column numbers.

In outdoor PV testing, just a few back T sensors are usually attached to each PV module. In our campaigns, we mostly use two redundant Pt100s behind two central cells. Unfortunately, this does not always yield benign ΔT plots (linear with irradiance above 0.2 – 0.4 suns). We therefore ran a dedicated campaign with a single 60-cell, glass-plastic monofacial PV module with monocrystalline PERC cells and equipped it with 16 thin Pt100 back T sensors as depicted in Fig. 6. The module's I-V curves were measured several times every minute during daytime between November 2019 and June 2020. In between, the module was operated at the current level of the last-known MPP. The module was facing due South, at a tilt angle of 35° from horizontal. The setup was located at the roof of the EnergyVille 1 building in Genk, Belgium.

We analyzed the ΔT plots from each of the back T sensors from the period between 18 May and 22 June 2020. Even for such a short period, only 4 or 5 of the plots were benign, while the others were either bifurcated or had a very large spread especially at high irradiance, with the corresponding cells seemingly jumping between two or more different thermal regimes. We continued the analysis with back T data sets from the sensors denoted R3C5, R5C3, R6C4, and R8C2.

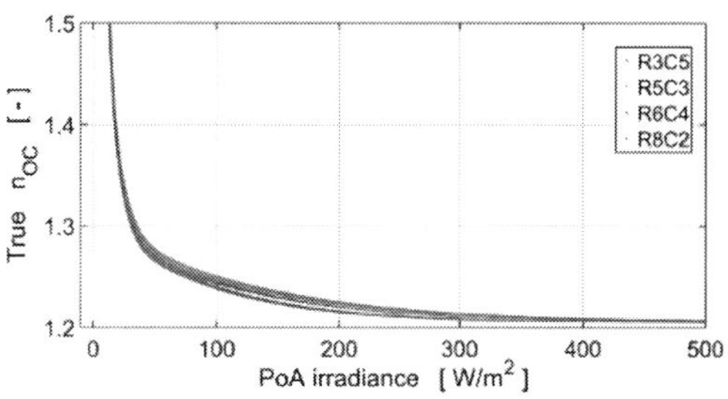

Figure 7: Estimates of the low-irradiance $\mathbf{n_{OC}}$ behavior of the PV module represented in Fig. 6 based on four (out of 16) Pt100 back T sensors giving benign ΔT plots.

In Fig. 7, we present four estimates of the low-irradiance behavior of the ideality factor $\mathbf{n_{OC}}$ of the module in Fig. 1 which were obtained by linearization of benign ΔT plots from four back T sensors using Eq. (2). The linearized ΔT plots were used to correct the corresponding **ECT** estimates at low irradiance. The corrected STC values of the $\mathbf{V_{OC}}$ are: 39.724 V; 39.756 V; 39.738 V; and 39.661 V (median 39.731 V, very close to the indoor rated value of 39.74 V). With the median value, the four **ECT** estimates are practically identical over most of the irradiance range and within ±0.4°C from their median below 400 W/m^2, down to about 10 W/m^2; see Fig. 8. This suggests a very accurate **ECT** estimation after linearization of the (benign) ΔT plots, which should allow for reliable outdoor performance characterization over the entire practical ranges of irradiance and operating temperature.

We used the methodology described above in the outdoor characterization of 5 more monofacial modules of the same type and 5 bifacial, glass-glass modules of the same form factor made with PERC cells. Those were tested outdoors between September 2018 and September 2019 with the same tilt and azimuth as the module studied above but were equipped with just two Pt100 back T sensors each, placed behind the central cells R5C3 and R6C4 as in Fig. 6. In the case of the bifacial modules, we found that an effective in-plane irradiance equaling the sum of the medians from 4 front and 4 rear irradiance sensors (c-Si reference cells with thermal correction) was good enough to use in our ΔT plots. The ideality factors of the modules at high irradiance were determined from indoor-measured I-V curves with the method of the **m-I** plots described in [13]. Unfortunately, not all the back T sensors of every module yielded benign ΔT plots. From those who did, we estimated the true underlying OC ideality factors (see Figs. 9 and 10).

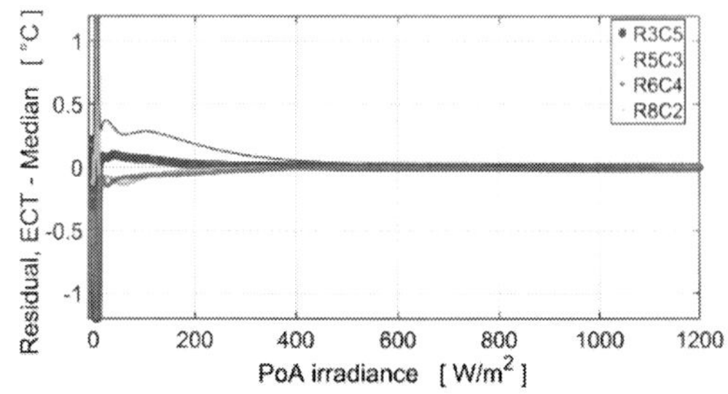

Figure 8: Residuals of the four **ECT** estimates vs. their median.

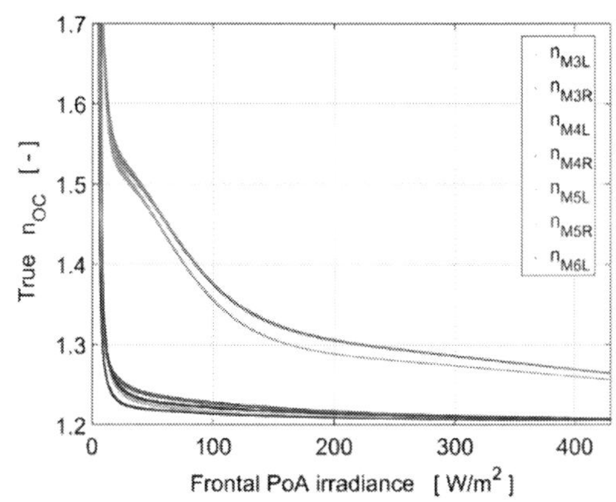

Figure 9: Estimates of the low-irradiance $\mathbf{n_{OC}}$ behavior of the other 5 monofacial c-Si PV modules based on their left and/or right back T sensors giving benign ΔT plots.

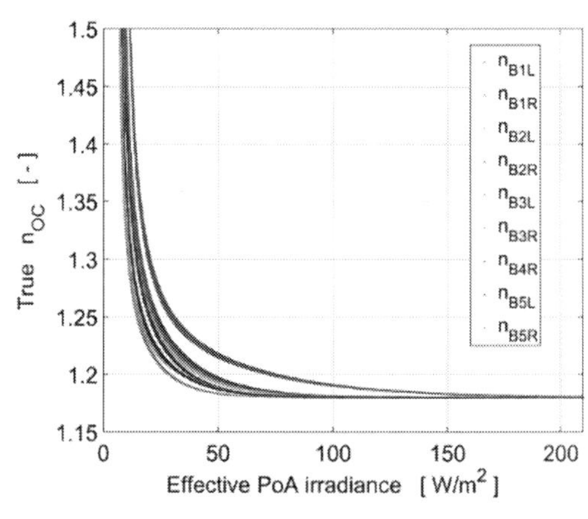

Figure 10: Same as in Fig. 9 but for the 5 bifacial c-Si PV modules.

While the left and right back T sensors of each module yielded very agreeable estimates of its ideality factor behavior at low irradiance, one extreme case was identified among the monofacial devices, as seen in Fig. 9. After the campaign's end, electroluminescence imaging of that module revealed that it had developed cracks in some cells, and it is well known that unpassivated crystal surfaces act as recombination loci and shunt the p-n junction. The remaining monofacial modules and practically all the bifacial modules showed a rather constant ideality factor down to about 0.1 Sun. The monofacial module represented in Fig. 7 was an intermediate case, with a rising ideality factor below 0.2 Sun. Practically all devices show a drastic increase in the ideality factor below 50 W/m^2 but most PV performance models are not interested in such very low irradiance conditions. Overall, the **ECT** evaluation down to irradiance as low as ~50 W/m^2 should be possible with our methodology.

As for the alternative method for evaluating the true β from the linear part of benign ΔT plots that was presented here, it needs further refinement, as the right-hand side of Eq. (6) contains a variable that is dependent on β. If the initial guess (classically determined value) for the β contains a significant error, the method may not be applicable, and one may not know a priori if this is the case.

5 CONCLUSIONS AND FUTURE WORK

Our work empowers the outdoor PV module testing by improving and extending the **ECT** evaluation from $\mathbf{V_{OC}}$ beyond the limitations of the IEC Standard 60904-5. This should allow for much more accurate outdoor performance and energy rating of c-Si and potentially other types of PV. We show the possible rising of the ideality factor at low irradiance beginning below or above the Standard's limit of 0.2 Sun, depending on the PV device. We mention a few equation errors in the current 2nd Edition of the Standard from year 2011 that is currently being reviewed. Also, the TC of the $\mathbf{V_{OC}}$, β, is only used at the reference irradiance (typically 1.0 Sun or another high enough value), and the possible challenge at low irradiance mentioned by the Standard is due to an eventual ideality factor rising there.

REFERENCES

[1] IEC Standard 60904-5:2011 (2nd Ed.), "Photovoltaic devices - Part 5: Determination of the equivalent cell temperature (ECT) of photovoltaic (PV) devices by the open-circuit voltage method."

[2] A. Jagomägi et al., "European network of PV outdoor testing – steps towards harmonized procedures," Proc. 24th EU PVSEC, Hamburg, pp. 3432–3438, 2009.

[3] K. Whitfield and C.R. Osterwald, "Procedure for determining the uncertainty of photovoltaic module outdoor electrical performance," *Progress in Photovoltaics: Research and Applications* **9**, pp. 87–102, 2001.

[4] H. Müllejans, W. Zaaiman, and R. Galleano, "Analysis and mitigation of measurement uncertainties in the traceability chain for the calibration of photovoltaic devices," *Measurement Science and Technology* **20**, Article ID: 075101, DOI: 10.1088/0957-0233/20/7/075101, 12 pp, 2009.

[5] G.H. Yordanov, "Characterization and analysis of photovoltaic modules and the solar resource based on in-situ measurements in southern Norway," Ph.D. thesis, Trondheim, 2012.

[6] M. Prorok *et al.*, "Reducing uncertainty of PV module temperature determination based on analysis using data gained during outdoor monitoring," Proc. 23rd EU PVSEC, Valencia, pp. 2865–2871, 2008.

[7] S. Krauter and A. Preiss, "Comparison of module temperature measurement methods," Proc. 34th IEEE PVSC, Philadelphia, pp. 000333–000338, 2009.

[8] R. Gottschalg, "Performance characterization of photovoltaic modules," Proc. 35th IEEE PVSC, Honolulu, pp. 001265–001270, 2010.

[9] IEC Standard (60)904-5:1993 (1st Ed., Withdrawn), "Photovoltaic devices - Part 5: Determination of the equivalent cell temperature (ECT) of photovoltaic (PV) devices by the open-circuit voltage method."

[10] T. Huld et al., "Mapping the performance of PV modules, effects of module type and data averaging," *Solar Energy* **84**, pp. 324–338, 2010.

[11] G.H. Yordanov, O.-M. Midtgård, and T.O. Saetre, "Equivalent cell temperature calculation for PV modules with variable ideality factors," Proc. 38th IEEE PVSC, Austin, TX, pp. 000505–000508, 2012.

[12] Tony Sample, European Commission JRC, personal communication dated Sep 6, 2021.

[13] G.H. Yordanov, T.O. Saetre, and O.-M. Midtgård, "Differential analysis of PV module I-V curves," Proc. 28th EU PVSEC, Paris, 2013, pp. 3207–3211.

DESIGN AND DEVELOPMENT OF SOLAR CELL INTEGRATED MOISTURE AND TEMPERATURE SENSORS FOR PHOTOVOLTAIC MODULES

Jaimin Navin Bhai Patel[1], Esther Fokuhl[1], Karthika Sheeja Prakash[1], Andreas Beinert[1], Viktor Wesselak[2], Paul Gebhardt[1], Daniel Philipp[1]
[1] Fraunhofer Institute for Solar Energy Systems Heidenhofstr. 2, 79110 Freiburg, Germany
[2] Institute for Renewable Energy Technologies (in.RET), University of Applied Sciences Nordhausen, Weinberghof 4, D-99734 Nordhausen, Germany

ABSTRACT: We are presenting the design and development of a capacitive moisture sensor and a resistive temperature sensor integrated into solar cells. The sensors are produced by screen printing on solar cell wafers using a silver paste commonly used in solar cell production. The sensor cells are designed to be integrated into PV modules. For evaluation of the sensors, measurements of imprinted wafers laminated between two layers of encapsulant were performed. Impedance measurements on the moisture sensor indicate an exponential dependency of the capacitance on the moisture content in the encapsulant at constant temperatures. The temperature sensors show a high sensitivity and reproducibility.

Keywords: Capacitance, Impedance, Resistance, Moisture, Temperature, Screen Printing, Sensor

1 INTRODUCTION

Degradation of PV modules under long-term operational conditions can be attributed to different climatic conditions like humidity, high temperature, Ultraviolet (UV) irradiation, temperature changes, and mechanical stress. Out of these, moisture ingress and high temperature are important environmental factors that are influenced by the climatic conditions on-site and the PV module design. The moisture typically penetrates from the back sheet and from the module edges through the encapsulation polymers. It can lead to electrochemical corrosion of cells and metallization as well as yellowing of the encapsulant [1]. High and low temperature cycles can cause thermomechanical fatigue and cracks in the cell, and broken interconnectors [2]. High cell temperature reduces the efficiency of the module [3].

Huyberechts et al. enabled humidity monitoring inside of PV modules using alumina substrates containing screen-printed interdigitated electrodes [4]. The permittivity of the polymer was monitored by electrochemical impedance spectroscopy. Carlsson et al. monitored moisture ingress into a polymer by using a conductive coating and a fork-type geometry on a glass substrate acting as the moisture-sensitive component [5]. Jankovec et al. developed an in-situ measurement technique based on commercial digital humidity sensors laminated in PV-module-like structures and mini-modules [6]. For the latter approach, the module EVA thickness had to be adapted to the commercial humidity sensors, because of their size.

As an advantage over these sensor concepts, the solar cell integrated sensor presented in this work can be produced with standard screen-printing processes by using materials common in solar cell production and are an extension to the previously presented solar cell integrated stress sensors [7]. Therefore, the sensor cell can be easily integrated with other solar cells into the module. PV module integrated moisture and temperature sensors can help to understand the root causes of degradation in the field and indoor tests by monitoring stress due to moisture and temperature.

2 APPROACH

2.1 Humidity sensor design

An interdigitated capacitor with comb-shaped electrodes is used as moisture sensor. Ethylene-vinyl acetate (EVA), which is used as encapsulant in our experiments, functions as dielectric of the capacitor. A schematic diagram of the humidity sensor is presented in Figure 1. The design parameters were chosen based on requirements of the printing process and calculations on the expected capacitance values as listed in Table 1.

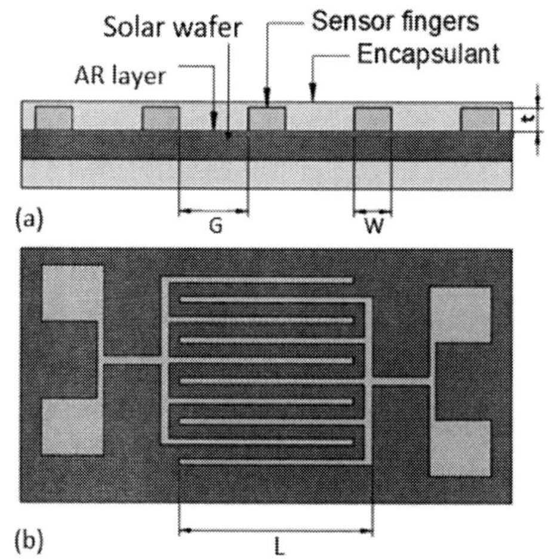

Figure 1: Schematic drawing of the moisture sensor (a) Cross-sectional view of sensor electrodes on a solar wafer laminated between two sheets of encapsulant (b) Top view of the sensor. L, W, and t are the length, width, and thickness/height, where G is the spacing between the fingers.

Table 1: Design parameters of the humidity sensor.

Parameters	
Number of fingers (N)	10
Length of electrode (L)	5 mm
Gap between fingers (G)	200 µm
Width of the finger (W)	60 µm
Thickness of the finger (t)	20 µm

2.2 Temperature sensor design

The design of the temperature sensor is based on the principle of resistance temperature sensors. A similar approach with a meander-shaped structure and a lower printing height, which led to PT-100-like characteristics, has been evaluated in previous work [7]. In this work, we are using the same paste and printing process as used for the humidity sensors and therefore chose a simplified design for the temperature sensor as shown in the schematic diagram in Figure 2.

The parametric equation for resistance depends on the length (L), cross-section (A), and resistivity (ρ) of the material:

$$R = \rho \frac{L}{A} \qquad (1)$$

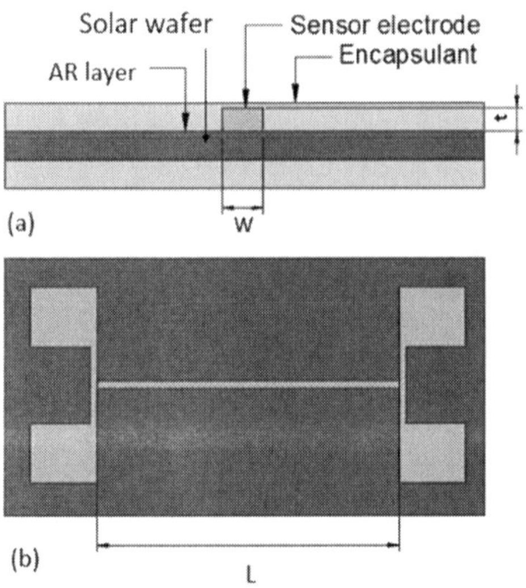

Figure 2: Schematic drawing of the temperature sensor (a) Cross-sectional view of the sensor on a solar wafer laminated between two sheets of encapsulant (b) Top view of the sensor. L, W, and t are the length, width, and thickness/height of the electrode.

The design dimensions for the temperature sensor are listed in table 2.

Table 2: Design parameters of temperature sensor

Parameters	
Width of the finger (W)	60 µm
Length of the electrode (L)	13.75 mm
Thickness of the electrode (t)	20 µm

Resistance values of the conductor at any temperature are determined by the below equation (2),

$$R = R_{ref}\left(1 + \alpha\left(T - T_{ref}\right)\right) \qquad (2)$$

Where R is the resistance at temperature T in Ω, R_{ref} is the resistance at reference temperature T_{ref} in Ω, T is the temperature in °C, T_{ref} is the reference temperature in °C (usually 20 °C), and α is the temperature coefficient of resistance for the material. For calibration purposes, a quadratic polynomial equation is used and it is defined for PT-100 temperature sensors in IEC 60751.

3 EXPERIMENT SETUP

3.1 Sample preparation

The sensors are printed on the front side of solar wafers by using a screen-printing process. In this experiment, low-temperature and high-temperature silver paste are used for the metallization of the sensor. After the screen printing of the sensors, the wafers are cut horizontally into 4 strips with a laser. The sensing part is connected by soldering high-temperature resistive wires to the contact pads.

To enable a fast moisture ingress and egress into the samples and therefore faster moisture content changes in the polymeric material at the sensor locations, the sensor stripes were laminated between two layers of EVA without front glass and polymeric back sheet. To monitor reference values for the moisture inside the laminates during the climate chamber investigations, a commercial humidity sensor (SHT31D) was also placed next to the wafer strips between the EVA sheets (see Figure 3 (c)). The lamination of the samples was carried out in a standard lamination process for PV modules with a lamination time of 25 minutes and a lamination temperature of 160 °C.

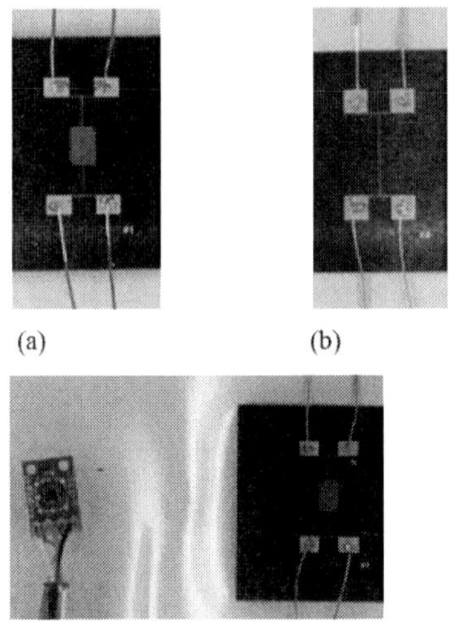

Figure 3: (a) laminated screen-printed humidity sensor (b) laminated screen-printed temperature sensor (c) commercial humidity sensor laminated between the EVA sheets with a sensor strip.

3.2 Measurement setup

For the measurement of the capacitive values of the humidity sensor, an LCR meter was used, which measures the real and imaginary components of impedance. The effective test voltage used for this experiment is 1 V. For the capacitance analysis, 100 Hz, 120 Hz, 1 kHz, 10 kHz, and 100 kHz frequencies were applied. An automated measurement setup [8] was used as shown in Figure 4. A data logger (Agilent 34970A) was connected to the LCR meter. The data logger had three measurement cards, which were used as switch units in this setup. This way, numerous test samples can be monitored simultaneously.

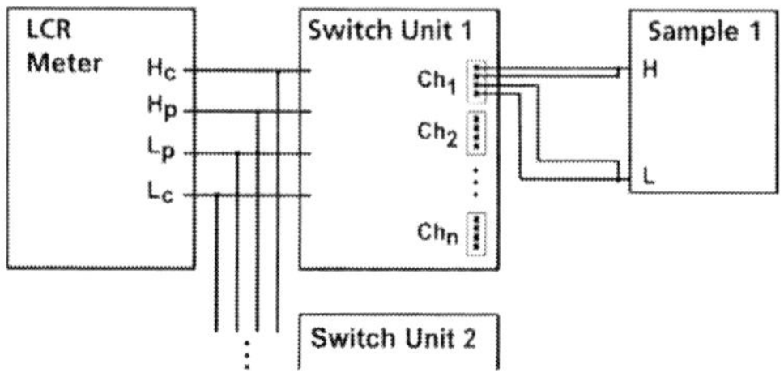

Figure 4: Schematic diagram of the LCR meter setup. Where Ch is demonstrated as Channels; Hc – High current; Hp – High potential; Lp – Low potential; Lc – Low current [8].

The resistance of the temperature sensors was measured by a multifunction device (Agilent 34980A), which contains a multimeter and a datalogger.

3.3 Test conditions

In order to evaluate the sensors, a range of environmental conditions were defined and controlled in a climate chamber.

The temperature range was set from -40 °C to 85 °C for the evaluation of the temperature sensor. The temperature inside the chamber was changed in steps of 20 °C for every step between -40 °C and 80 °C and a step of 5 °C between 80 °C and 85 °C. The holding time at constant temperature was approximately 2 h in this experiment. To verify the reproducibility of the sensor performance, the test between 20 °C to 85 °C was repeated once.

For the humidity sensors, the relative humidity was controlled in steps between 20 %RH and 85 %RH at constant temperatures of 85 °C and 60 °C. The humidity inside the chamber was changed in steps of 10% RH for every step between 20 %RH and 80 %RH and a step of 5 %RH between 80 %RH and 85 %RH. The holding time at constant relative humidity was approximately 1 h and 30 min at 85 °C and 2 h and 30 min at 60 °C. To investigate the reproducibility of the sensor response, the test at 85 °C was repeated 5 times ("five-cycle test").

4 TEST RESULTS

4.1 Setup capacitance

The sensors are not directly connected to the measurement device. The electrical properties of wires and datalogger cards (see Figure 4) contribute to the measured capacitance [8]. The measurement setup showed capacitive behaviour with a phase shift between voltage and current of $\varphi \approx 90°$ [8]. Thus, these parts of the measurement are considered as setup capacitance C_S, which is parallel to the sample. From this assumption, we can correct the capacitance of the sensor by subtracting C_S from the measured capacitance values C_{measured}. For a cable length of 2 m, C_S amounts approximately 355 pF ± 5 pF with neglectable humidity and temperature dependence.

4.2 Capacitance measurement of the humidity sensor

The test conditions used for analysing the humidity sensor are described in section 3.3. Only the sensors printed with low-temperature paste (LTP) are evaluated since the characterization of the high-temperature paste (HTP) sensors showed a low insulation resistance between the electrodes. This was most likely caused by the firing step, which was performed at a temperature between 800 °C and 850 °C. Therefore, the insulation effect of the AR layer is reduced because the AR layer deposition process (PECVD) was held at temperature between 165 °C to 400 °C [9].

Figure 5 shows the corrected capacitance values (after subtracting C_S) of a sensor printed with low-temperature paste. The measurements were taken with different frequencies at 85 °C and relative humidity setpoints between 20 % and 85 %RH. The humidity values on the x-axis correspond to the values measured by the commercial humidity sensor inside the sample after reaching a steady state.

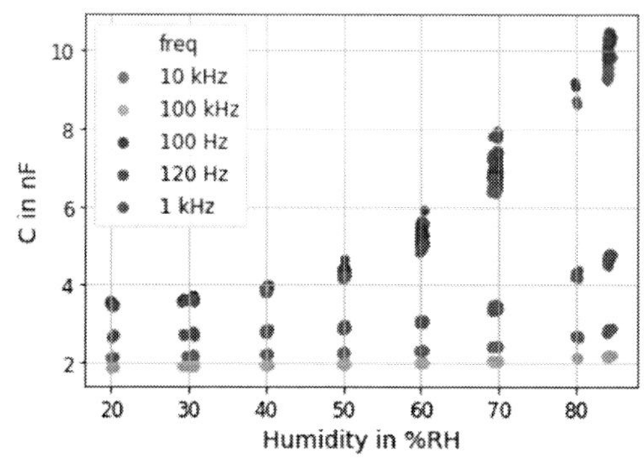

Figure 5: Sensor response at different frequency levels.

The capacitance values of a sensor was measured at different frequencies with increasing humidity. At 10 kHz and 100 kHz, the sensitivity is significantly lower than at lower frequency levels. At 100 Hz and 120 Hz, comparably high fluctuations of the measurement values were observed at humidity values above 50 %RH. The measurement values at 1 kHz represent a compromise with a moderate sensitivity and a comparably low fluctuation and were therefore used for further evaluation.

Figure 6 illustrates the capacitance values of a humidity sensor determined at 1 kHz and different humidity levels during the five-cycle test at 85 °C. The sensor response at constant temperature can be described by an exponential fit:

$$C = m + Ae^{B\%RH} \qquad (3)$$

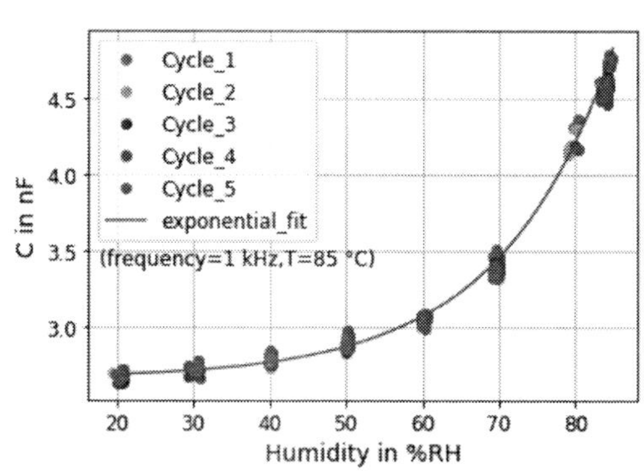

Figure 6: Measured capacitance of humidity sensor at 1 kHz frequency with exponential curve fit of the five-cycle test at 85 °C.

Additional measurements were performed at 60 °C and different steps of RH. Figure 7 shows the sensor measurements at 85 °C (first cycle measurement data from 5-cycle tests) and 60 °C. The sensor response shows similar behaviour at 60 °C as at 85 °C but with lower sensitivity to changes in relative humidity and with lower values at dry conditions. The lower sensitivity at 60 °C can be explained by the absolute moisture content, which is lower at the same RH values.

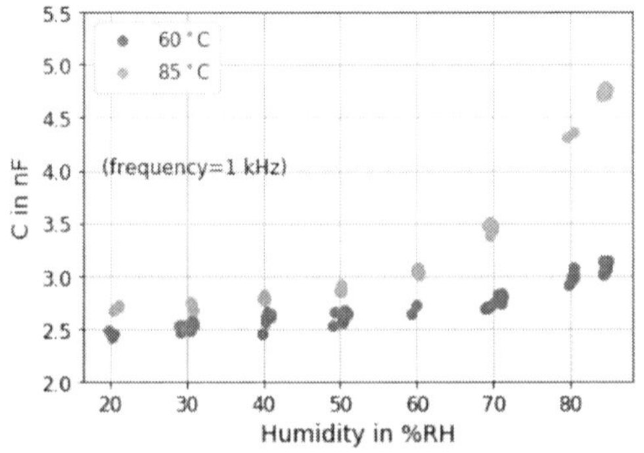

Figure 7: Capacitance measurement at different temperatures (e.g., 60 °C and 85 °C) at 1 kHz.

4.3 Resistance measurement of the temperature sensor

In order to characterize the temperature sensor, the resistance of the sensor was measured at different temperature ranges. Figure 8 and Figure 9 represent readings of the high-temperature paste and low-temperature paste printed sensors. In contrast to the humidity sensor, the firing step used for the high-temperature paste printed sensors had no negative effect because a high insulation resistance between paste material and wafer is not necessary for this measurement.

The sensor response of both pastes is different because of their different temperature coefficient of resistance α. The measured resistance changes within 20 °C steps are at least 4 % for the low-temperature paste and 5 % for the high-temperature paste printed sensor.

The sensors were calibrated using a method for PT-100 temperature sensors, according to IEC 60751. The resistance values at the temperature setpoints were fitted to the following polynomial equation:

$$R_{calculated} = R_0(1 + At + Bt^2) \qquad (4)$$

Where, R_0 is the reference resistance at 0 °C, A and B are polynomial constants. Table 3 shows the constant values of high- and low- temperature paste.

Table 3: Constant values of polynomial fitted curve

	LTP	HTP
R_0	0.926	0.489
A	3.23e-3	3.86e-3
B	3.82e-7	-3.58e-7

The resistance value at reference temperature R_0 varies because of different temperature co-efficients for both pastes (see equation (2)). The constants A and B show similarities in values but the polarity is different for constant B.

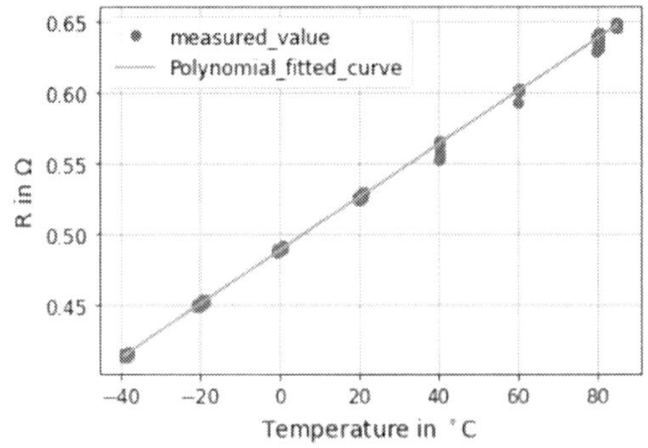

Figure 8: Temperature dependent resistance of a sensor, which was printed with high-temperature paste

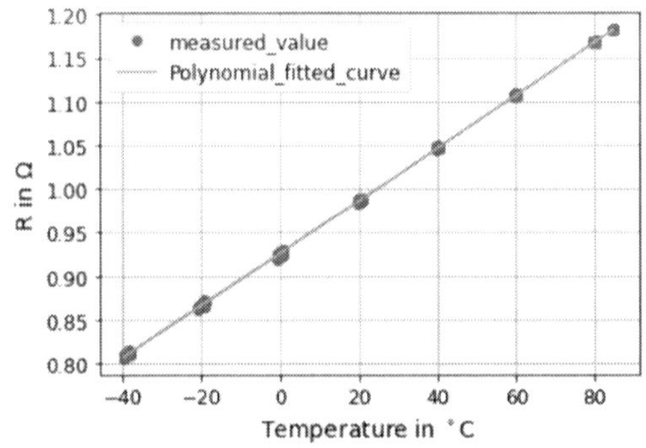

Figure 9: Temperature dependent resistance of a sensor, which was printed with low-temperature paste

The experiment with low-temperature paste was repeated to investigate the reproducibility of the temperature sensor (see Figure 10). The measured values are reproducible at the tested temperature range.

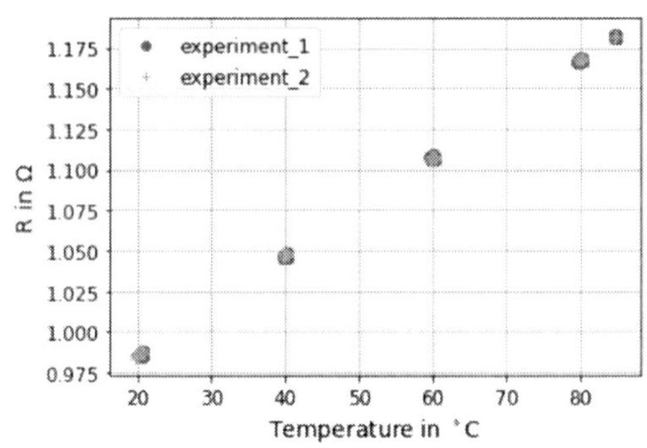

Figure 10: Temperature dependent resistance values from two experiments (low-temperature paste sensor); where experiment_1 is at temperatures between -40 °C and 85 °C test data points and experiment_2 is at temperatures between 20 °C to 85 °C test data points

5 SUMMARY AND OUTLOOK

We have presented the design and proof of concept of screen-printed, solar cell integrated humidity and optimized temperature sensors.

The humidity sensor was analysed under different relative humidity conditions between 20 %RH and 85 %RH at constant temperatures of 85 °C and 60 °C. The sensor gives an exponential response to increasing moisture in the EVA and shows high reproducibility. The temperature sensor was evaluated at temperatures between -40 °C and 85 °C. The sensor gave a polynomial response, following the calibration function according to IEC 60751, and showed high reproducibility.

As a further step, a full calibration of the humidity sensors is planned. Since the measurement values of the humidity sensor are not solely humidity dependent but also change with temperature, calibration steps at various temperature levels are required. In the future, both the temperature and the humidity sensor are planned to be integrated into PV modules for experimental purposes and on-field measurements. For the application on-field, a new measurement setup, ideally with wireless data transmission, will be needed.

6 ACKNOWLEDGEMENTS

The authors would like to thank Denis Erath, Michael Linse, Timo Wentzel, Diana Witt, Georg Mülhöfer, Heinrich Berg, Sandor Stecklum and Felix Basler for sample preparation, technical help and insightful discussions during experiment works.

7 REFERENCES

[1] Bouaichi A, Merrouni AA, El Hassani A, Naimi Z, Ikken B, Ghennioui A, Benazzouz A, El Amrani A, Messaoudi C. Experimental evaluation of the discoloration effect on PV-modules performance drop. *Energy Procedia* 2017;119:818–27, doi:10.1016/j.egypro.2017.07.107.

[2] Quansah DA, Adaramola MS, Takyi G. Degradation and longevity of solar photovoltaic modules—An analysis of recent field studies in Ghana. *Energy Sci Eng* 2020;8:2116–28, doi:10.1002/ese3.651.

[3] Razak A, Irwan Y, Leow WZ, Irwanto M, Safwati I, Zhafarina M. Investigation of the Effect Temperature on Photovoltaic (PV) Panel Output Performance. *International Journal on Advanced Science, Engineering and Information Technology* 2016;6:682, doi:10.18517/ijaseit.6.5.938.

[4] Guido Huyberechts LF. In situ formation of humidity-sensitive devices for the evaluation of solar panel encapsulations.

[5] Carlsson T, Halme J, Lund P, Konttinen P. Moisture sensor at glass/polymer interface for monitoring of photovoltaic module encapsulants. *Sensors and Actuators A: Physical* 2006;125:281–7, doi:10.1016/j.sna.2005.07.022.

[6] Marko Jankovec, Eleonora Annigono, Christophe Ballif, Marko Topič. In-situ Determination of Moisture Diffusion Properties of PV Module Encapsulants Using Digital Humidity Sensors. *IEEE J. Photovoltaics* 2018.

[7] Beinert AJ, Imm M, Benick J, Becker F, Seitz S, Heinrich M, Paul O, Glunz SW, Aktaa J, Eitner U, Neuhaus H. Silicon solar cell–integrated stress and temperature sensors for photovoltaic modules. *Prog Photovolt Res Appl* 2020;28:717–24, doi:10.1002/pip.3263.

[8] Esther Fokuhl, Heinrich Berg, Sandor Stecklum, Djamel Eddine Mansour, Daniel Philipp, Paul Gebhardt. MONITORING THE MOISTURE INGRESS INTO PV MODULES BY MEASURING CAPACITIVE CHARACTERISTICS. *European Photovoltaic Solar Energy Conference and Exhibition (EU PVSEC)* 2020.

[9] Yimao Wan, Keith R. McIntosh, and Andrew F. Thomson: Characterisation and optimisation of PECVD SiNx as an antireflection coating and passivation layer for silicon solar cells (*AIP Advances* 3, 032113 (2013)).

CLASSIFICATION OF UNCERTAIN I-V CURVES IN PV MODULES BASED ON CURRENT AND VOLTAGE EVALUATION

Li Feng[1], Nowshad Amin[3], Jingwei Zhang[2], Kun Ding[2], Frank.U. Hamelmann[1]
[1]Solar Computing Lab –Bielefeld University of Applied Science, Artilleriestraße 9, 32427, Minden, Germany
[2] College of Mechanical and Electrical Engineering, Hohai University, 213022, Changzhou, China
[3]Institute of Sustainable Energy, Universiti Tenaga Nasional, Kajang, Malaysia

ABSTRACT: A current–voltage (I-V) curve is a graphical representation of the relationship between the voltage applied across a solar cell and the current flowing through the solar cell. Key properties of solar cells can also be extracted from the shape and details of the curve, enabling greater insight into its solar energy conversion ability and efficiency. The work focuses on the uncertain I-V curves classification from a big online database based on the current and voltage evaluation. The open-circuit voltage, short-circuit current, current and voltage at MPP on the I-V curves are derived to define the ratios according to the characteristic information of curve shapes. Then, the real I-V curves were acquired from the measurement system and used to verify the proposed method. The results show the proposed method achieved good effects of identifying the curves types and it can be used to meet data requirements of further scientific research.

Keywords: PV modules; I-V curves; curves classification; current and voltage ratios

1 INTRODUCTION

With environmental issues being paid attention, the global goals of 100% renewable energy are carried out to change energy supply from conventional fossil fuel sources to renewable energy in recent years [1],[2]. Germany try to achieve the transition to 100 % renewable energies in all energy sectors (electricity, heat, mobility, industry) by 2030, especially PV growth reaching 86 GW per year [3].

The current – voltage (I-V) characteristics of photovoltaic (PV) modules represent the detailed description of its solar energy conversion ability and efficiency. Knowledge of the electrical I-V characteristics of PV modules is critical for determination of the module ' s output performance. With the technology development, the I-V curve measurement devices are no longer limited to mobile equipment, more and more online uninterrupted devices are used in the PV field and store the data into online or local databases. However, the measurement procedures will be affected by many uncertain factors, such as module faults, meteorological conditions and ambient environment changes, the accuracy of sensors, response speed of devices, equipment failures and staff mistakes, which will result in complicated data types. Therefore, when the I-V curves are exported from databases to be used for scientific research, data selection is necessary to extract uncertainties on performance and measurement procedures, so that the selected data meet research requirements.

In view of this, a classification method based on voltage and current evaluation is presented in this paper to classify common I-V curves of PV modules. Firstly, I-V curve quantities (e.g., the open-circuit voltage, short-circuit current, voltage and current at the maximum power point (MPP) of the PV module) are identified to define the current and voltage ratios. Then, the curve classification strategy is developed by utilizing current and voltage ratios according to the characteristic information of curve shapes. Finally, current and voltage ratios are evaluated at real-time voltage and current data are compared with the corresponding thresholds to divide curve types.

2 METHOD OF CURVES CLASSIFICATION BASED ON CURRENT AND VOLTAGE EVALUATION

2.1 Ratios definition of current and voltage

The I-V curve represents the photovoltaic (PV) quality of solar modules. And it is described by following I-V equation [4].

$$I = I_{ph} - I_0\left(e^{\frac{q(V+IR_s)}{nKT}} - 1\right) - \frac{V + IR_s}{R_{sh}} \tag{1}$$

Where, I_{ph} is the photocurrent, I_0 is the saturation current of diode, R_s is series resistance, R_{sh} is shunt resistance, q is the electronic charge ($q \approx 1.6\times10^{-19}$ C), n is the ideality factor of diode, K is the Boltzmann constant ($k \approx 1.38\times 10^{-23}$ m^2 kg s^{-2} K^{-1}), T is the temperature of solar cell. I, V are the current and voltage of a solar cell at maximum power point.

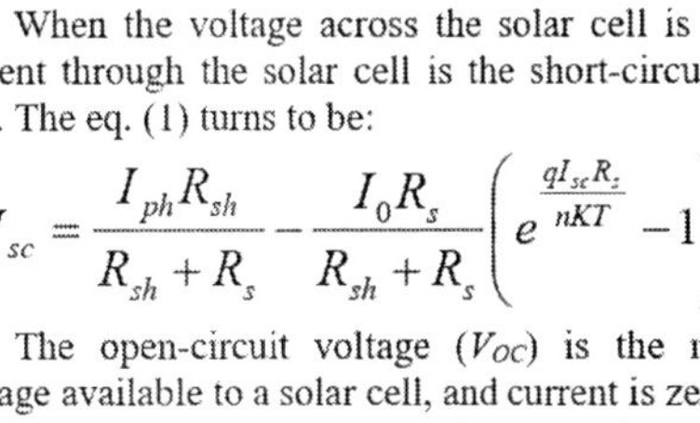

Figure 1: Equivalent circuit of Single-diode model

When the voltage across the solar cell is zero, the current through the solar cell is the short-circuit current (I_{sc}). The eq. (1) turns to be:

$$I_{sc} = \frac{I_{ph}R_{sh}}{R_{sh} + R_s} - \frac{I_0 R_s}{R_{sh} + R_s}\left(e^{\frac{qI_{sc}R_s}{nKT}} - 1\right) \tag{2}$$

The open-circuit voltage (V_{OC}) is the maximum voltage available to a solar cell, and current is zero.

$$V_{oc} = I_{ph}R_{sh} - I_0 R_{sh}\left(e^{\frac{qV_{oc}}{nKT}} - 1\right) \tag{3}$$

A special point on a power – voltage (P-V) curve that corresponds to the maximum value of the product of voltage and current, or the maximum output power. According to the differentiation of the power of the solar cell to the voltage, it can be found that it is equal to zero [5]. Hence:

$$\frac{dP}{dV} = \frac{d(IV)}{dV} = 0 \tag{4}$$

$$\frac{dI}{dV} = -\frac{I_m}{V_m} \tag{5}$$

Then, the open-circuit voltage (V_{oc}), short-circuit current (I_{sc}), current (I_m) and voltage (V_m) at MPP on the I-V curves are applied in expressing the ratios of current ($R_{\Delta I}$) and voltage ($R_{\Delta V}$), as following. The detailed description is shown in Fig.2.

$$\mathrm{R}_{\Delta \mathrm{I}} = \frac{(\mathrm{I}_{sc} - \mathrm{I}_{m})}{\mathrm{I}_{sc}} \tag{6}$$

$$\mathrm{R}_{\Delta \mathrm{V}} = \frac{(\mathrm{V}_{oc} - \mathrm{V}_{m})}{\mathrm{V}_{oc}} \tag{7}$$

Actually, output characteristics of PV modules are decided by modules materials and efficiency. The ratios for normal curves should be equal to the values which are calculated by the open-circuit voltage ($V_{oc,stc}$), short-circuit current ($I_{sc,stc}$), current ($I_{m,stc}$) and voltage ($V_{m,stc}$) at MPP from datasheet under standard condition. Besides, the degradation caused by natural aging appears with time. So the degradation parameters of current (β_1) and voltage (β_2) will add to the ratio expressions under STC.

$$\mathrm{R}_{\Delta \mathrm{I,stc}} = \frac{\beta_1 * (\mathrm{I}_{sc,stc} - \mathrm{I}_{m,stc})}{\mathrm{I}_{sc,stc}} \tag{8}$$

$$\mathrm{R}_{\Delta \mathrm{V,stc}} = \frac{\beta_2 * (\mathrm{V}_{oc,stc} - \mathrm{V}_{m,stc})}{\mathrm{V}_{oc,stc}} \tag{9}$$

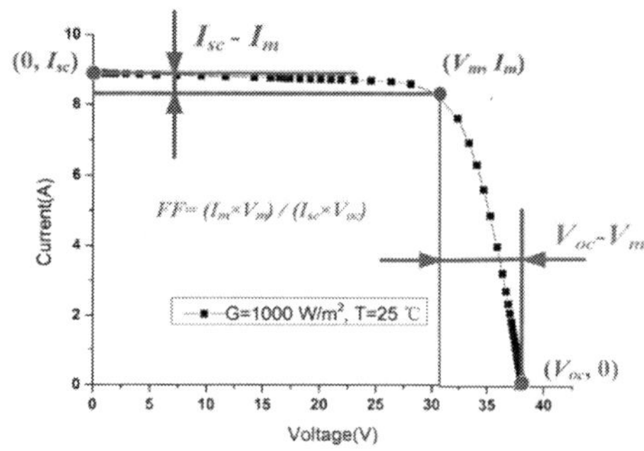

Figure 2: The ratio expressions of current and voltage

2.2 Strategy of curves classification

In this section, the eq. (6)-(9) are used to develop the classification strategy of uncertain I-V curves in a big online database. The detailed procedure of curves classification is implemented in Fig.3 and described as below:

Step 1: The I-V curves of the PV modules, irradiance (S) and temperature(T) are obtained from database.

Step 2: The I_{sc}, V_{oc}, I_m, V_m, S, T are extracted from I-V curves.

Step 3: The $R_{\Delta I}$, $R_{\Delta V}$, $R_{\Delta I,stc}$, $R_{\Delta V,stc}$ are calculated based on the parameters in step 2.

Step 4: The normal and abnormal curves are determined by the values of $R_{\Delta I,stc}$ and $R_{\Delta V,stc}$.

Step 5: The second-order partial derivative of the current relative to the voltage ($\partial^2 I$ / $\partial^2 V$) based on basic I-V equation of single – diode PV module is used to capture the number of inflection points and knee points on I-V curves [6]. Then, the steps of I-V curves caused by partial shadows is detected through the number of inflection points from the abnormal I-V curves in step 4.

Step 6: The remaining curves are classified according to different ratio ranges of different curve types. Ratios ranges are obtained through a large number of statistical analysis of curve types in the test PV system.

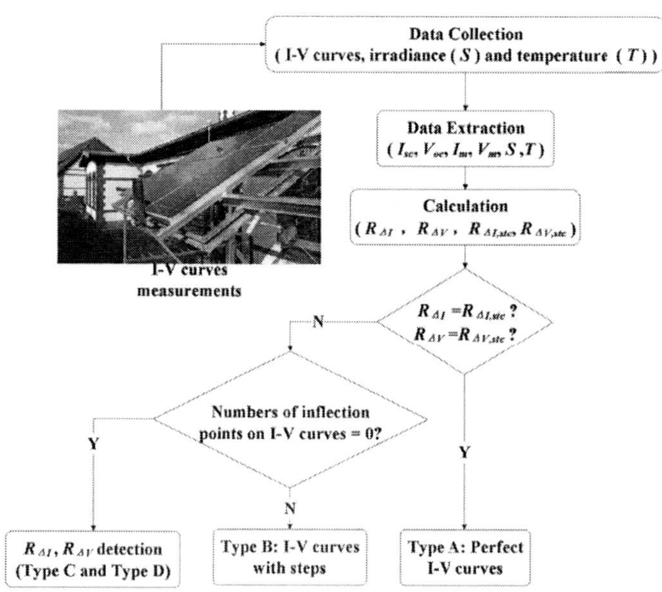

Figure 3: Flowchart of the proposed classification strategy

3 DATA COLLECTION

One I-V curves measurement system was installed at campus Minden in Bielefeld University of Applied Sciences. This performance test system integrates various types of photovoltaic modules (such as thin-film photovoltaic modules, monocrystalline silicon modules, polycrystalline silicon modules, CdTe modules), I-V curve test equipment, maximum power point (MPP) tracking equipment, weather station, spectrometer, database, data acquisition and storage system. It can achieve the on-line accurate measurement of I-V curves with various modules types and performance monitoring of PV modules. At present, the system has collected monitoring data from PV modules for 7 years, which include electrical data of PV module and meteorological data. These data will lay a solid data foundation for the goals of this paper. Besides, the poly-Si module was used in section 4. The detail specification of the used module is as shown in Table I.

Figure 4: Performance test system of PV modules

Table I: The specification of poly-Si module under standard condition (STC)

Parameters	Variable	Values
Maximum Power	P_{ref}	180 W
Voltage at MPP	V_{ref}	35.3V
Current at MPP	I_{ref}	5.1A
Open Circuit Voltage	$V_{oc,ref}$	44.8V
Short Circuit Current	$I_{sc,ref}$	5.45A
Temperature coefficient of I_{sc}	α	0.003%/ ℃
Temperature coefficient of V_{oc}	β	-0.35%/ ℃

4 RESULTS AND DISCUSSION

The data from 6:30 to 18:30 in an I-V measurements system are selected to show the application of the proposed approach in curves classification of PV modules. The ratios for a whole day are from 0.1 to 0.7, as shown in Fig.4. The results of classification are expressed by 4 curve types in Fig.5. The perfect I-V curves in Fig.5(a) are easily found with the ratios being approximately equal to the values of 0.1 ($R_{\Delta I,stc}$) and 0.2 ($R_{\Delta V,stc}$) for the day at 12:00 to 16:00.

The I-V curves with steps caused by partial shadows from the trees around modules. When the substrings in the PV module are short circuited through bypass diode because of severe partial shadow, the slope of the whole I-V curve will drop sharply, as shown in Fig.5(b). Partial shadows have a big effect on the current, so the ratio of current increases from 0.2 to 0.6, ratio of voltage is reduced to 0.12.

Actually, when irradiance and temperature are low, even the solar cells in the PV module will still work normally, but the slope of the curves between the maximum power point and the short-circuit current point is greater. So that the ratios increased slightly to the range of 0.2-0.3. Curve type D shows that the curves are not smooth because of zigzagged points on the whole I-V curves. The possible reason is that solar cells are working with few lights in the dawn and dusk, the curves have an open circuit voltage and very small short circuit current. And the efficiency of the modules is very low too. So the accuracy of the device and the sensor cause more measurement error between points, which leads to the zigzagged shape. In this case, the ratios of current and voltage are very similar and more than 0.5.

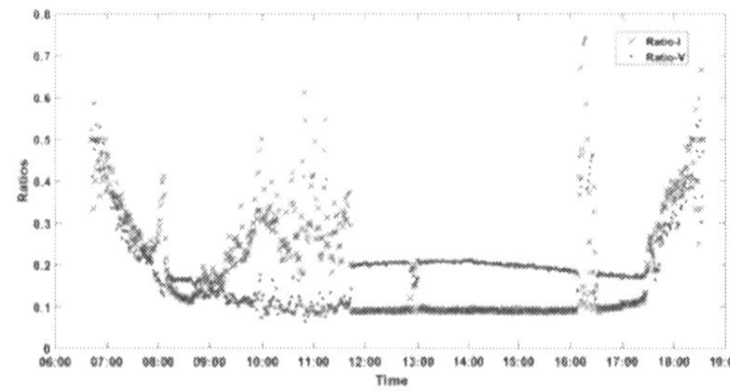

Figure 4: The ratios of current and voltage within one day

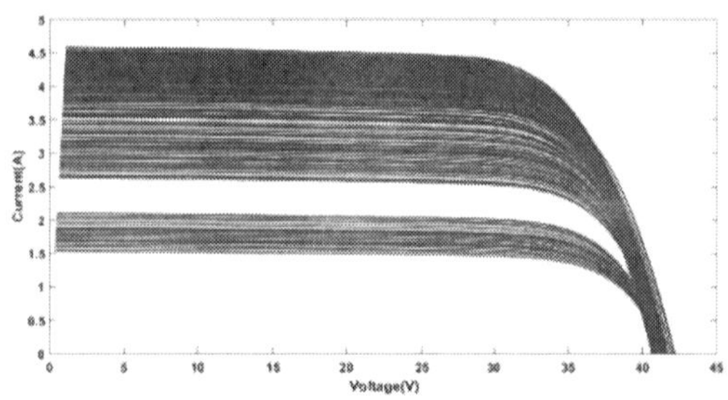

(a) Type A: Perfect I-V curves

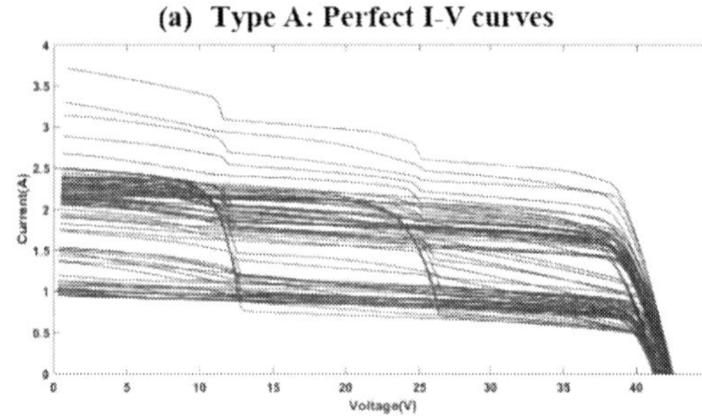

(b) Type B: I-V curves with steps

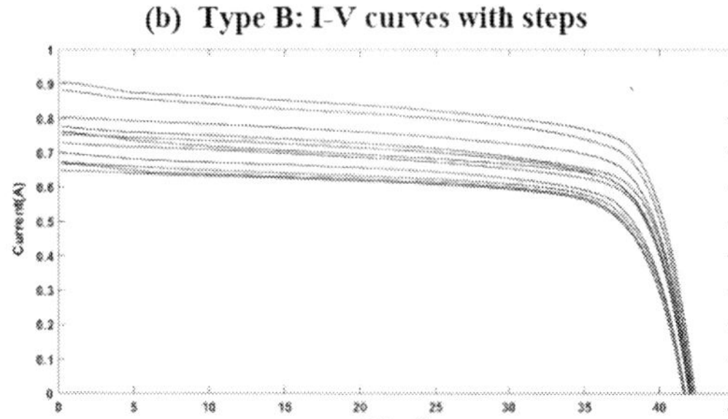

(c) Type C1: I-V curves with low irradiance

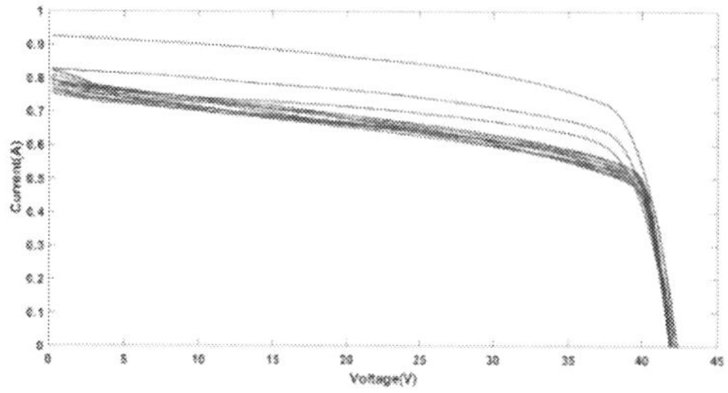

(c) Type C2: I-V curves with low irradiance

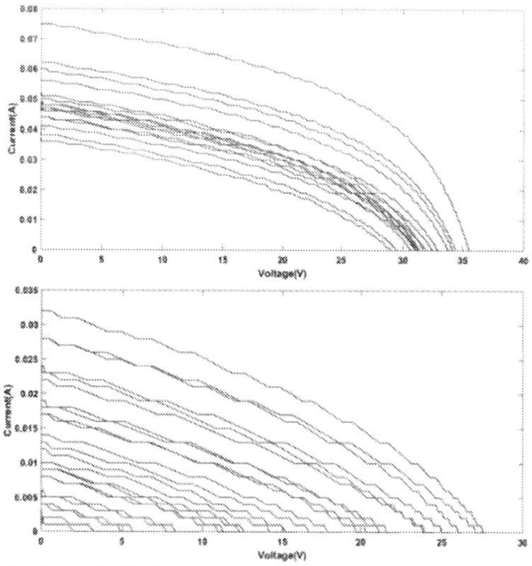

(d) Type D: I-V curves in the dawn and dusk
Figure 5: Results of curves classification

5 CONCLUSION

The classification of uncertain I-V curves in PV modules was developed based on current and voltage evaluation. The open-circuit voltage, short-circuit current, current and voltage at MPP were used to describe the ratios to identify the I-V curve from an online database. The measured I-V curves for one day was applied in verifying the proposed method. The results show the good effect of the defined ratios for classifying the uncertain I-V curves, when it included partial shadows and low irradiance. Besides it can be expanded to select normal I-V curves for many years to investigate natural aging of modules (power losses), material characteristics and uncertainty effects on performance and measurements. Also, it could be applied in testing the error tolerance of measurement devices, such as failures and accuracy of devices and sensors.

6 ACKNOWLEDGMENT

This work was supported by the project of Adaptive Ertragsprognose mit Data-Mining im PV-Feld auf Grundlage einer digitalen Signatur der PV-Module und der Systemkomponenten (PV Digital 4.0, No. 13FH020PX6), Germany. It was also supported by the National Natural Science Foundation of China (Grant No. 51777059) and Changzhou Sci & Tech Program (Grant No. CJ20200074), China.

7 REFERENCE

[1] https://www.oneearth.org/100-renewable-energy/.
[2] Manish Ram, Dmitrii Bogdanov,et al..Lappeenranta University of Technology Research Reports 'Global Energy System based on 100% Renewable Energy', 2019.
[3] Thure Traber, Hans-Josef Fell,Franziska Simone Hegner. 100 % renewable energies for Germany by 2030, 2021.
[4] Manan Mittala, Birinchi Borab, Sahaj Saxenaa, Anshu Mli Gaur.Performance prediction of PV module using electrical equivalent model and artificial neural network.Solar Energy, 176:104 – 117,2018.
[5] Bilge Kaan ATAY, Ulaş EMİNOĞLU.A new approach for parameter estimation of the single-diode model for photovoltaic cells/modules.Turkish Journal of Electrical Engineering & Computer Sciences, 27: 3026 – 3039,2019.
[6] Yuanliang Li, Kun Ding,et.al.. A fault diagnosis method for photovoltaic arrays based on fault parameters identification. Renewable Energy 143:52-63,2019.

INFLUENCE OF LIGHT, TEMPERATURE AND CURRENT ON STABILIZED OUTPUT POWER AND ENERGY YIELD OF CDTE PV MODULES

Matthias Pander[1,2]*, Bengt Jaeckel[1,2], Matthias Ebert[1,2]
[1]Fraunhofer Center for Silicon Photovoltaics, Otto-Eissfeldt-Str. 12, 06120 Halle (Saale), Germany
[2] Fraunhofer Institute for Microstructure of Materials and Systems IMWS, Halle, Germany
phone: +49-345-5589-5215, fax: +49-345-5589-5999, matthias.pander@csp.fraunhofer.de

ABSTRACT: It is well known that thin film modules like CdTe have certain power meta instabilities depending on their short- and long-term history. For CdTe extensive dark storage (e.g. due to transportation or modules from warehouse) typically lead to a lower output power and requires extensive light stabilization prior indoor STC characterization. IEC 61215-1-2 states requirements for standard stabilization for new, out of the box CdTe modules. For single modules that might be feasible, but to verify larger batches extensive indoor light soaking is cost and time prohibitive and therefor alternative procedures to reliable determine the output power of CdTe modules is needed. However, current standards do not cover any particular stabilization for field aged modules.

Pre-treatment and storage of the module play an important role in the quickly to determine output power. Short term illumination by flash-testing/measurement increases measured output power and basically compensate for short term meta stabilities. However, the final power depends on pre-history and on illumination duration and intermediate short-term dark storage. Different modes of operation (Voc, Isc and Pmpp) and different temperatures are investigated to accelerate stabilization utilizing artificial light. In general, higher temperature lead to an accelerated change of characteristic parameters of the PV module. The operation mode of the module during light soaking (Isc, Voc or Pmpp) does also influence the change behavior of the output power and to an immediate response of the module. In the last experiment "stability" was reached after approx. 30 hours of light soaking at 1.000 W/m².

Therefor at least three different effects have to be considered, namely dark-storage, fast to compensate light effects (ms-range) and long-term saturation effects in the range of days.

Keywords: PV Module, Degradation, Reliability, CdTe, Light soaking, stabilization

1 INTRODUCTION

It is well known that thin film modules utilizing a-Si, CIGS or CdTe as light absorbing materials have certain power meta instabilities depending on their short- and long-term history [1]. For CdTe extensive dark storage (e.g. due transportation or modules from storage) typically lead to a lower output power [2]-[8] and requires extensive light stabilization prior a reliable indoor STC characterization. IEC 61215-1-2 [9] states requirements for standard stabilization for new modules. However, this procedure is typically used to stabilize and verify output power of fielded CdTe modules with unknown long- and short-term history. Furthermore, the technology of CdTe modules changed over the years and may make stabilization unique for a particular production lot.

The purpose of this work is to investigate different indoor treatments to better understand the light- and dark storage behavior of CdTe modules. The aim is to understand the effects of short- and long-term light exposure to determine a reliable / reproducible stabilization procedure for CdTe modules that can be applied to evaluate the output power of new and field returned CdTe modules. This understanding is further helpful also for better evaluate results during type approval as e.g. during Damp-heat test or PID test meta stabilities due to prolonged dark storage that could affect pass or fail decision [10].

It was found in [11] that operation in open circuit mode leads to increased degradation up to 12 %. In this work we analyze the impact of the different operation modes (short circuit Isc, open circuit Voc and maximum power point Pmpp) and module temperatures during light exposure of CdTe modules.

As of today, no standardized and reliable method exist to determine output power of field aged and long term dark stored CdTe modules. Based on the age and the need to re-evaluate output power of module in the field, e.g. to verify warranty expectations after 10 years [12][13][14], such procedures are needed by test labs to serve O&M, inspectors and owner needs.

2 EXPERIMENTAL SETUP

2.1 Sun Simulators

2.1.1 Flash tester

For STC standard performance measurements a AAA class solar simulator from Berger Lichttechnik was used. The pulse length is 10 ms. The long-term reproducibility of the power is 0.3 % relative based on data of stable crystalline silicon reference modules.

2.1.2 Light Soaker

A constant sun simulator is used to study the long-term light soaking behavior. The light source has a constant irradiance and at least a Class B-spectrum per IEC 60904-9 [15]. The homogeneity is class C. The samples were measured continuously during the light soaking experiments with in-situ monitoring. Besides logging of module temperature all relevant PV module parameters were measured utilizing IV-curve sweeps (Isc, Voc and Pmpp).

2.2 Experimental procedure

2.2.1 Short-term light pulse treatment

A very simple procedure is to repeat the flash test at 1000 W/m² and 25°C several times to see if meta stabilities in the CdTe device cause power drops or gains. With that procedure light exposure in the ms-range can be achieved while each light pulse is separated from the next by about a minute. After the initial activation the intervals are extended to 15 min.

2.2.2 Temperature dependence

The power change under constant irradiation at different temperatures is investigated. For reference the settings defined by IEC 61215-1-2 are a module temperature of 50°C and maximum power point as point of operation. STC peak power is determined after light soaking intervals of 10 kWh/m² that is typically done in 8-12 hours. Stabilization is reached if module output power is stabile within 2 % in two subsequent intervals.

To investigate the influence of temperature under constant illumination Mini-Modules were used. The experimental setup is shown in Figure 1. The temperature was logged with a temperature sensor on the top of the module. As soon as the thermal equilibrium is reached, the modules were irradiated with a light source consisting of 9 LEDs. An intensity of one sun (1000 W/m²) was set. The illumination time was chosen depending on the expected degradation time. The mini-modules were connected to a current-voltage meter and the IV curve was automatically recorded every 5 minutes. The Voc is used as most reliable parameter to record the regeneration/degradation process.

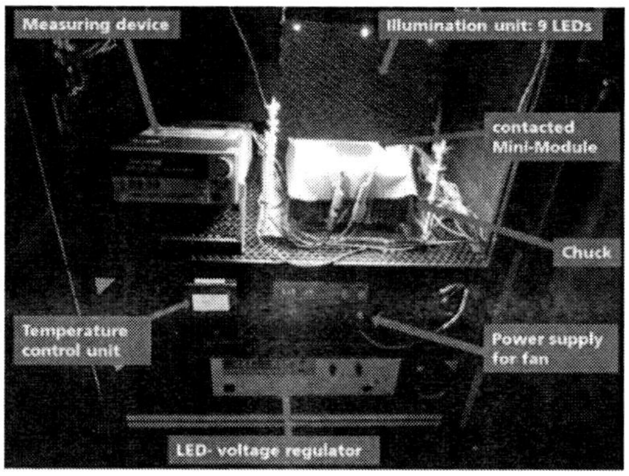

Figure 1: Test setup for Light soaking experiment of Mini-modules

2.2.3 Mode of operation

After the analysis of the temperature dependence, we analyze the impact of the different operation modes (short circuit Isc, open circuit Voc and maximum power point Pmpp) during light exposure of CdTe modules. The open circuit mode is important, because between PV park set up a significant time interval can elapse until grid connection while the modules are already being irradiated.

2.3 Samples

The samples originate from a commercial CdTe module line. Depending on the experiment either full-size modules or mini-module samples prepared out of standard modules (see e.g. Figure 2) are used. The mini-modules consist of six cells connected in series and allow for more test variations while keeping the absorber properties more constant. Additionally, the environmental conditions can be better controlled on a sample size of 10cm x 10cm vs 120cm x 60cm. All samples initially come directly from the factory. They were stored in the dark under laboratory conditions and conditioned by a dark anneal at 85°C for minimum 72 h before the experiments if not stated otherwise.

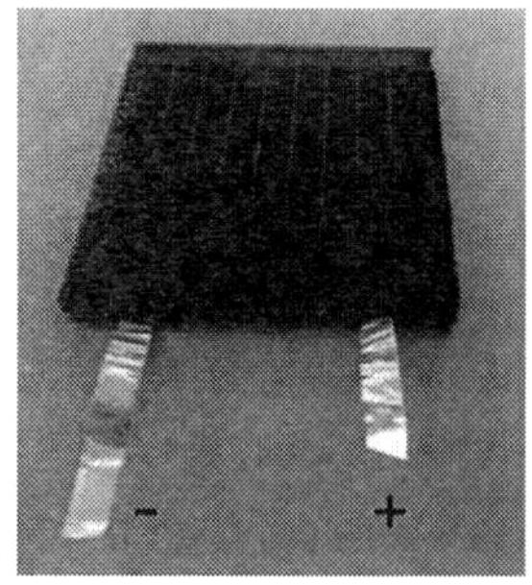

Figure 2: CdTe Mini-Module (10 cm x 10 cm)

3 EXPERIMENTAL RESULTS

This section highlights the experimental results of different conditioning procedures for CdTe modules.

3.1 Short-term light pulse treatment results

Figure 3 shows the effect of short-term light pulse treatment for a full-size module with respect to Pmpp and Voc change vs time. The initial Pmpp/Voc value is marked with a larger square as reference point for the fresh module.

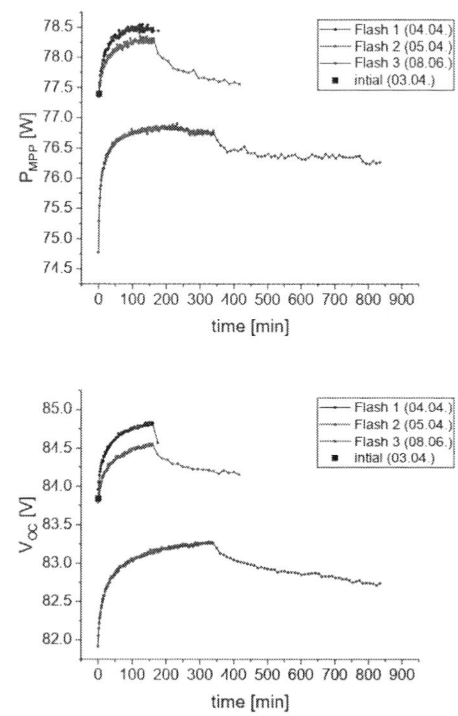

Figure 3: Short term illumination (~10 ms) using a flash tester. Every 80 s a flash at 1000 W/m² was performed. After 160 min (Flash 1 and Flash 2) the duration between flashes was extend to 15 min. Flash 3 series was performed after ~2 months of dark storage with 80s intervals first and after 5 hours extended to 15 min. An obvious lower power is measured (>3%).

The first series of flash tests was done in intervals of 80 s with a 10 ms light exposure. The module improved by about 1 W in performance. After one day the experiment was repeated and a very similar behavior was obtained. However, even after 1 day of dark storage, the initial (red curve) started below the black and did not reach the same end value. To investigate the effect on time between the light pulses the delay time was increased to 15 min. The module shows immediately a slow decrease in output power and Voc underlining the short-term meta stabilities of the CdTe samples with main impact in Voc. The power variation is mainly determined by increase and decrease in Voc respectively, the fill factor (FF) and the short circuit current (Isc) remain largely stable. This indicates that defects are formed in the semiconductor material, which reduce the effective lifetime.

The procedure of short-term pulses and different delay times was repeated after the module was stored in the dark for 2 months (blue curves) without pre-treatment. Obviously, the module improved in Pmpp (about 1.5-2 W) and Voc but started at a significant lower level. Even an extended repeat of the 10 ms flashs did not lead to a full recovery to the initial parameters of Voc and Pmpp. After increasing the duration between flashes both parameters drop again. However, on a little different time constant compared to the new module.

3.2 Temperature dependence

The tests on Mini-Modules were carried out at different temperatures between 25 °C and 85 °C. Results are shown in Figure 4. Mini-module MiMo 2 and 29 were measured at 25 °C and show initially the short-term regeneration as found during short-term light pulse treatment. After that the power slowly decreases. The degradation of the Mini-Modules was found to be clearly temperature dependent. All samples lose Voc voltage after a maximum is reached in a time scale of minutes to some hours. This is quite important to note as the time is short and temperatures are not extremely high compared to what modules see in the field. At 50 °C there is still an initial voltage gain visible. At higher temperatures, the degradation is significantly faster. Above 75 °C the degradation is so fast that the performance gain at the beginning is no longer present or could not be evaluated. As it can be seen even similar prepared samples show variations of initial voltage but the qualitative degradation behavior is comparable.

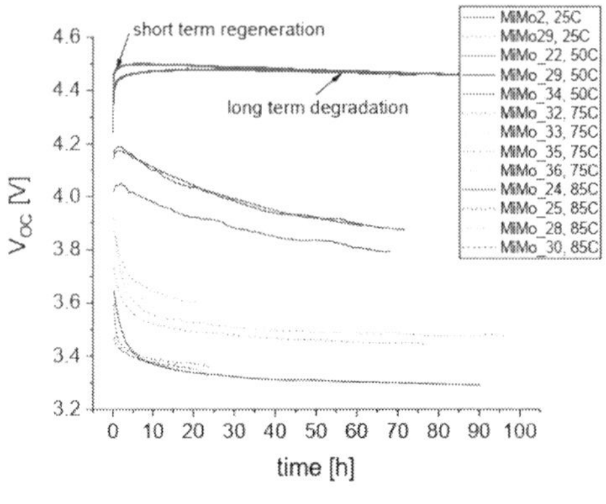

Figure 4: Voc-change of CdTe Mini-Modules under constant illumination at different cell temperatures

Assuming that Voc degradation is described by an exponential decay (equation (1)), the time constants of the degradation were determined. These time constants are shown in the Arrhenius plot in Figure 5. Data at 25°C is omitted since it was not possible to achieve a reasonable fit. The degradation rates vary significantly for the different samples. Nevertheless, an Arrhenius relationship between degradation rate and inverse temperature can be assumed. Based on this correlation, an activation energy of the defects of (0.558 ± 0.065) eV was determined.

$$V_{OC}(t) = \Delta V_{OC} \cdot e^{-\frac{t}{\tau(T)}} + V_{eq} \qquad (1)$$

V_{OC} – open circuit voltage
ΔV_{OC} – change of open circuit voltage
V_{eq} – open circuit voltage in equilibrium
t – time
$\tau(T)$ – time constant
E_A – activation energy
R – universal gas constant

$$\tau(T) = \tau_0 \cdot e^{-\frac{T}{E_A R}} \qquad (2)$$

E_A – activation energy
R – universal gas constant
T – absolute temperature

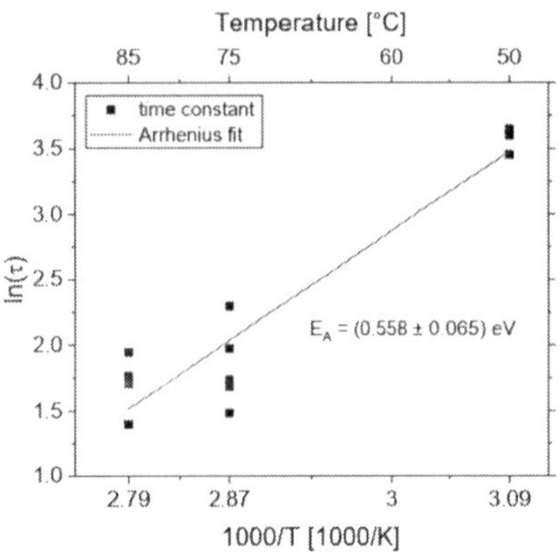

Figure 5: Arrhenius plot for Mini-Module degradation

For selected samples, PL images were taken before and after treatment. Three different effects of photoluminescence migration could be observed. All samples show a reduction of PL intensity for the whole area by an average of 10 – 20 %. This general reduction in charge carrier lifetime suggests large area degradation and correlates with Voc loss. However, in areas with locally increased recombination rates, marked by dark areas in the PL image, a local regeneration, i.e. healing of defects, can also be detected. The PL intensity even increases relative to the value of the initial image. This can be seen in Figure 6 in area R1.

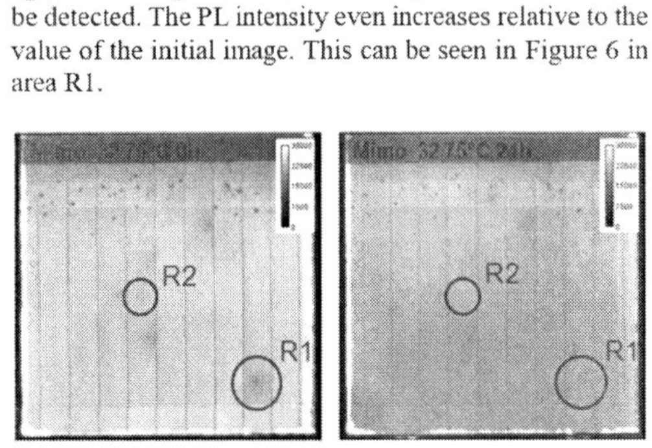

Figure 6: PL images before and after treatment. R1: local regeneration.

In addition to the disappearance of such recombination centers, local degradation was observed. Areas with locally increased recombination obviously spread to neighboring areas as a result of the treatment. This can be seen in Figure 7 in area R4. On the same figure, local regeneration can also be seen in area R3. Thus, it appears that certain degradation and regeneration processes are taking place simultaneously.

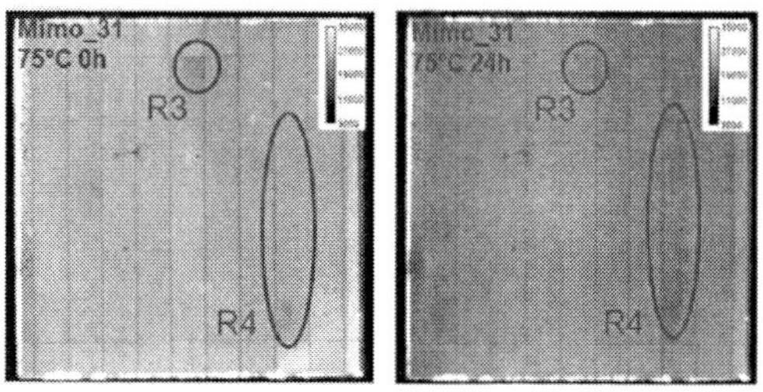

Figure 7: PL images before and after treatment. R3: local regeneration, R4: local degradation

3.3 Mode of operation

In a third set of experiments the mode of operation of the module during illumination was investigated. Most easy to setup are Isc and Voc module, while Pmpp requires a Pmpp-Tracker but is most relevant for field applications. The general understanding between these modes is also important to understand module behavior in PV system where several strings are connected in parallel while not all strings are having the same voltage behavior e.g., by partial shading or inhomogeneous degradation.

Figure 8 shows a long-term light exposure experiment where the module was set to different modes of operation. The module temperature during the exposure time was quite constant at around 75 °C. The change from Pmpp to Isc and from Isc to Voc always resulted in an upwards jump in open circuit voltage. It was found that the rate of decay increases when switching to open circuit. Over the whole period of light exposure, the voltage dropped by approx. 7 V, ~10 %.

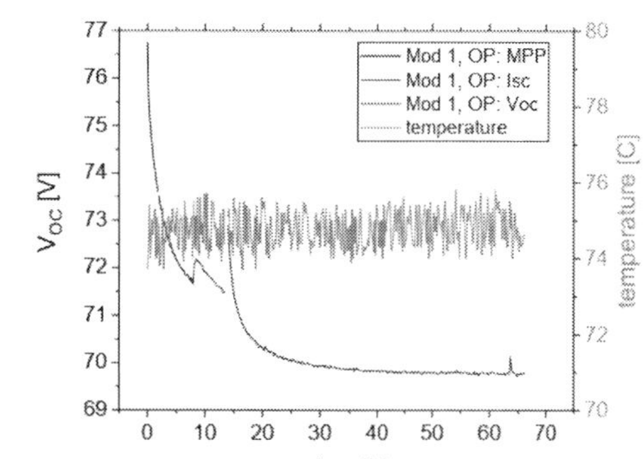

Figure 8: Influence analysis of operation mode during light soaking: Voc plot for Module #1 including monitored temperature (constant) starting on the left in Pmpp (black), Isc (red) and more right in Voc-mode (blue)

The reproducibility was investigated with three modules. The results are displayed in Figure 9. The voltage dependence (Figure 9 a)) is for all three modules similar except the magnitude of absolute drop in voltage and the jump during switch between the modes of operation.

Figure 9 b) shows the trend of the three modules for Pmpp. The jumps are not as pronounced as for Voc, but the points of change in operation mode are still visible. Overall, the module power dropped during that time in the order of 10 %. Figure 9 c) shows the trend for Isc. Not much is visible here. The change of the open circuit voltage is again the main parameter describing the module degradation.

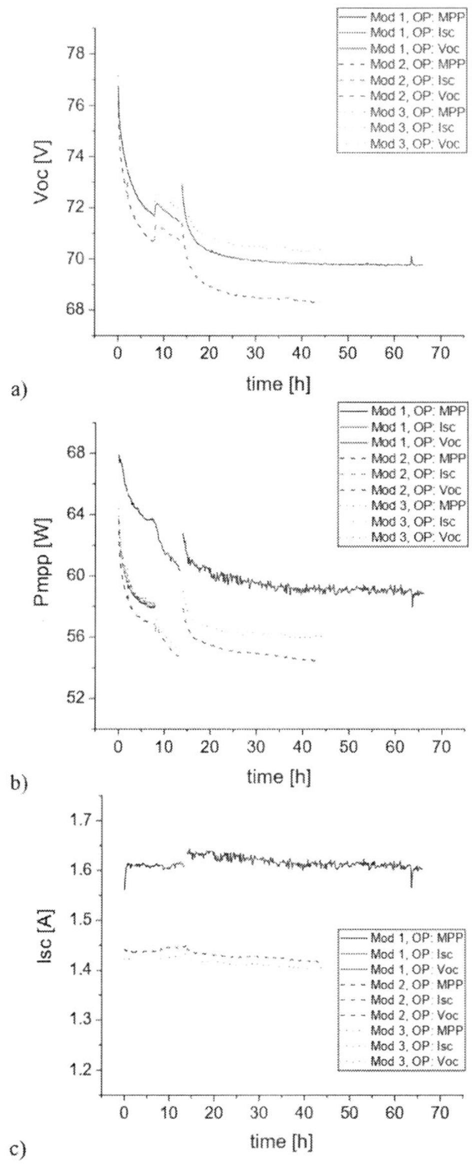

Figure 9: a), b) and c) comparison of three different modules to study reproducibility for the different parameters Voc, Pmpp and Isc, respectively (module temperature: 75°C)

4 DISCUSSION

The results of the tested modules in the study show a short-term increase in output power while exposed to a quite small dose of light. Longer exposure showed a significant (~10%) decrease in open circuit voltage and in a similar magnitude for Pmpp. The temperature dependence found can be used to shorten the duration of the stabilization experiments. A temperature of 50 °C corresponds to the temperature range of 40-60 °C required in IEC 61215-2 [9]. Under these conditions, the expected duration to complete stabilization is more than 80 hours. This is within expectations for stabilization of (new) CdTe modules from this production period and supports the requirements for stabilization of "new" CdTe modules according to IEC 61215-1-2, which requires three measurements with two consecutive intervals of 20 kWh/m² each to show stability within 2 %, typically resulting in 4-6 intervals and thus a range of 80-120 kWh/m². By increasing the test temperature to 75 °C and 85 °C, most modules reached a stable operating point after about 30 hours and 20 hours, respectively. This leads to an acceleration by a factor of 4-8.

If other degradation mechanism within the initial stabilization can be excluded, higher module temperature can be used to accelerate the test, and therefore safe time and cost. The mode of operation plays also a role within that setup. Voc operation seems to further accelerate degradation. The preferred, while most realistic, is using Pmpp mode.

5 SUMMARY AND CONCLUSION

To summarize the results at least three different effects have to be considered. Namely, dark-storage (influencing the initial power and the stabilized power), fast to compensate light effects (minutes range) and long-term saturation effects in the range of days. The extent of the effects depends on the type of module (microstructure of the CdTe absorber stack [16]), the age in the field, and the duration of dark storage. However, since only one particular CdTe absorber was studied in this publication, the magnitude of the effects cannot be generalized. The test results show why stabilization of field-aged modules and prolonged dark storage do not always yield conclusive results. It is therefore considered necessary to re-evaluate and, if necessary, adjust the stabilization procedures in case of a significant change of the absorber stack. The documentation of the test temperature and irradiation time are the most important parameters. Further work is needed to better understand the stabilization of CdTe modules during light exposure and prolonged dark storage.

6 ACKNOWLEDGEMENTS

The authors gratefully acknowledge the financial support of the project "ProMo" (grant no. 1604/00103) by the State of Saxony-Anhalt and the European Regional Development Fund.

The authors would like to thank Felix Neduck for performing the degradation experiments and Tabea Luka and Rico Meier for supervising and planning the project work.

7 REFERENCES

[1] Cueto J. A. del von Roedern, B., "Long-term Transient and Metastable Effects in Cadmium Telluride Photovoltaic Modules", Progress in Photovoltaics: Research and Applications, 2006, 14, 615-628

[2] Dobson K., "Stability of CdTe/CdS thin-film solar cells", Solar Energy Materials & Solar Cells, 2000, 62, 295-325

[3] Erra, S., "An effective method of Cu incorporation in CdTe solar cells for improved stability", Thin Solid Films, 2007, 515, 5833-5836

[4] Mermoud A., "Performance Assessment of a Simulation Model for PV Modules of any Available technology", 25th European Photovoltaic Solar Energy Conference - Valencia, 2010, 6

[5] Rimmaudo, I., "Ageing of Cdte Devices by Copper Diffusion", 27th European Photovoltaic Solar Energy Conference - Frankfurt, 2012, 5

[6] Deceglie, M. G. Silverman, T. J. Marion, B. Kurtz, S. R., "Temperature-dependent light-stabilized states in thin-film PV modules", IEEE PVSC - Conference Proceeding, 2015

[7] Deline, Chr. A., "Metastable Electrical Characteristics of Polycrystalline Thin-Film Photovoltaic Modules upon Exposure and Stabilization", Journal of Photonics for Energy, 2012, 2, 13

[8] Silverman, T. Deceglie, M. Marion, B. Kurtz, S., "Performance stabilization of CdTe PV modules using bias and light", 41st IEEE Photovoltaic Specialists Conference, 2014

[9] IEC 61215-1-2, "Terrestrial Photovoltaic (PV) Modules - Design Qualification and Type Approval - Part 1-2: Special requirements for testing of thin-film Cadmium Telluride (CdTe) based Photovoltaic (PV) modules, 2016

[10] Hacke, P., "Correction for Metastability in the Quantification of PID in Thin-Film Module Testing", 44th IEEE Photovoltaic Specialists Conference, Washington, 2017

[11] Sinapis, K., "Outdoor degradation of CdTe modules at open circuit and at maximum power point", 42nd IEEE Photovoltaic Specialists Conference, New Orleance, USA, 2015

[12] Jordan D. C. Kurtz, S. R., "Photovoltaic Degradation Rates - an Analytical Review", Progress in Photovoltaics: Research and Applications, 2011, 18

[13] Jordan, D. C. Kurtz, S. R. VanSant, K. Newmiller, J., "Compendium of photovoltaic degradation rates", Progress in Photovoltaics: Research and Applications, 2016, 24, 978-989

[14] Sastry, O., "Degradation Analysis of Different PV Modules After Prolonged Field Operation", 26th European Photovoltaic Solar Energy Conference - Hamburg, 2011, 5

[15] IEC 60904-9, "Photovoltaic devices –Part 9: Solar simulator performance requirements"

[16] Strevel, N., "Improvements in CdTe module reliability and long-term degradation through advances in construction and device innovation", Photovoltaic international, 2014

DOUBLE-SIDED CHARACTERIZATION OF FULL-SIZE BIFACIAL PV MODULES BASED ON LOW-COST LED BIAS LIGHT

Teodora S. Lyubenova, Robert P. Kenny, David Shaw, Diego Pavanello, Juan Lopez-Garcia
European Commission, DG Joint Research Centre, European Solar Test Installation
v. E. Fermi 2749, I-21027 Ispra (VA), Italy
*Tel: +39 0332786574, Fax: +39 0332 789 992, e-mail: Teodora.LYUBENOVA@ec.europa.eu

ABSTRACT: The present study is focused on the design and application of simultaneous double-side illumination approach for the characterization of full- size commercial silicon bifacial module in order to reproduce the real conditions with both front and rear illumination. The testing procedure is performed by using a prototype system based on Light Emitting Diode (LED) rear bias light coupled to a large-area long-pulse solar simulator that illuminates the front side of the device. The viability of the method to fulfill the IEC TS 60904-1-2 requirements for characterization of bifacial devices is tested. "Self-reference" methodology is applied using an accredited reference values (ISO/IEC 17025) obtained by single-side illumination of the same test device. Both, the proposed approach and prototype testing system for large-scale modules have demonstrated good performance in comparison with the single-side method. The current-voltage (I-V) curves and key electrical parameters are very similar. Slightly higher P_{max} (0.3-0.9%), BiFi (0.5%) and FF (0.3-0.7%) are obtained. The rear-side non-uniformity in the double-side illumination system varies with applied voltage and distances between the LED and the module, classifying the LED in Class B (>200 W/m^2) or Class C (<100 W/m^2). The proposed testing system configuration would be convenient and economically viable for industrial applications of bifacial modules.

Keywords: Bifacial PV modules, Double-side illumination, LED, solar simulator, characterization

1 INTRODUCTION

Bifacial crystalline Silicon (Si) photovoltaic (PV) devices are competitive and promising technology because they can produce additional output energy in comparison with the conventional monofacial modules. Depending on module design and mounting, climate zone and albedo characteristics, the bifacial devices could produce from 5% to 30% more electrical power that their monofacial variants. This PV technology is receiving growing interest on the market and it is expected to increase rapidly in near future. The International Technology Roadmap for Photovoltaic (ITRPV) 2019 published that in 2020 bifacial cells account for about 20% of the total world PV cell market and predicted that by 2030, this share will increase to 70% [1]. For bifacial PV modules, the market share for 2020 stands at about 12% and is predicted to increase to about 30% by 2030 [2]. The reliable electrical characterization of the performance of the bifacial modules still present lot of challenges, but is key to the successful market penetration.

The publication of the technical specification IEC TS 60904-1-2 (TS) for the measurement of current-voltage characteristics of bifacial PV devices [3] standardised various methods used to measure and label a bifacial PV device. Three different approaches are considered for the electrical performance of bifacial PV modules: natural sunlight; a solar simulator with adjustable irradiance level for single-side illumination using the equivalent irradiance method; and a solar simulator with double-sided illumination. The single-side illumination is the most used approach for bifacial PV module testing, but arguably, the method that reproduces the real conditions best with both front and rear illumination is the double-side method.

At a module scale, this approach represents technical and logistical challenges due to the associated cost of using two controlled light sources instead of one [4] and it is considered that the current equipment is not suitable for this setup and requires significant and potentially expensive modifications [5]. Thus, developing and validating a cost-effective approach that could be used in the industry is an important step for the transition toward bifacial modules dominating the global market. Recent advances in high-power Light Emitting Diode (LED) technology have resulted in a number of LED-based sun simulators coming to market. Based on a previous work on small-scaled prototype [6], the aim of this work is to extent the study towards the development and validation of a large area double-sided solar simulator based on an existing large-area long pulse flash solar simulator and a low-cost LED-based rear bias light. This system configuration would be convenient and economically viable for industrial testing applications of bifacial modules. The system parameters such as the distance to the module and LED power settings on the rear irradiance non-uniformity are studied. The results are validated using equivalent measurements on an accredited single-side pulsed solar simulator setup.

2 METHODOLOGY

2.1 Sampling

The device under test (DUT) is a commercially available n-type bifacial c-Si glass/glass frameless module of 60 cells (6x10, 156.75 × 156.75mm) without any partial "self-shading" on the rear side due to the junction box, lips of the frame or label. The total module area is 1.67 m^2. The module bifaciality is φ=0.67.

2.2 Experimental set up

In the applied methodology the DUT is used as "module self-reference" (MSR), using an analogous approach to the one published in [7, 8] and in Standard IEC 60904-7 (Section 3) [9]. The DUT has been previously measured by an accredited single-side illumination approach on a pulse solar simulator PASAN IIIB (PIIIB) (details in section 2.2.1). The summarized procedure is as follows:

- First, current-voltage (I-V) curves are measured using

the single-side illumination when the module reaches 25°C±0.1°C, which provides a measure of short-circuit current (Isc) with minimal uncertainty (±0.86%, k=2) due to the highly uniform irradiance distribution of the solar simulator PIIIB (NU is ±0.75%). The corresponding uncertainty values are: V_{oc} (±0.22%, k=2) and for P_{max} (±1.1%, k=2). The I-V curves were obtained at 1000W/m^2 (STC, front and rear side) and at equivalent irradiances 1067W/m^2 and 1134W/m^2 (front side) corresponding to rear side irradiance levels of 100W/m^2 and 200W/m^2 in accordance with the TS. The power gain yield (BiFi) is calculated.

- Second, the measurements from the single-side illumination are used as a reference for the double-side illumination method. The long-pulse simulator irradiance is adjusted to obtain the same I_{sc} at STC with single-side illumination (i.e. with the LEDs turned off). In the case of double-side illumination, by adjusting the applied voltage, and hence intensity control of the LEDs, an effective irradiance is generated such that the short circuit current I_{sc} matches that obtained using the PIIIB at the desired equivalent irradiances.

- Third, validation of the results in terms of I-V curves similarity measured at both set ups is performed. Rear irradiance distribution is analysed.

2.2.1 Single-side illumination approach

The single-side illumination method is performed using a pulsed solar simulator (PASAN IIIB, 3x3m^2, flash 10ms) at the European Solar Test Installation (ESTI), an ISO/IEC 17025 accredited calibration laboratory for PV device testing and calibration. ESTI has been recently accredited, for the first time worldwide, for I-V measurements of bifacial PV modules applying this approach according to IEC TS 60904-1-2. The method provides an accredited reference value and allow for testing the other approaches proposed in the IEC TS 60904-1-2 as in the present study.

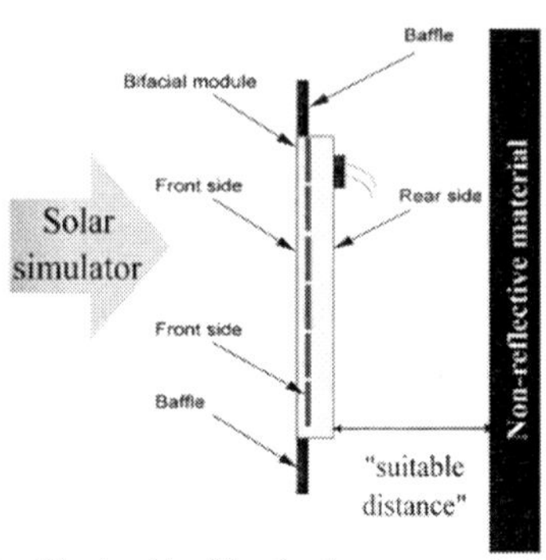

Figure 1: Single-side illumination approach based on equivalent irradiance method

The single-side approach follows the definition of equivalence irradiance (G_{Ei}, G_{Ei}= 1000 W/m^2+φ*G_{Ri}; G_{Ri} rear-side irradiance levels, i.e. 100 W/m^2 and 200 W/m^2) and thus involves separate measurements of each side of the bifacial module, using a single light source and non-reflective background with irradiance < 3 W/m^2, at any point (covering the non-illuminated side) as it establishes in the IEC TS 60904-1-2 [2]. The experimental set up parameters are listed in Table I and visual representation is shown in Fig.1. The I-V curve measurements were performed in a forward sweep mode (with a nominal sweep time of 10 ms) at 1000 W/m^2 and temperature 25.0 ± 0.1°C and corrected to STC (1000 W/m^2, 25°C and AM1.5G). For all measurements involving the PIIIB simulator, the so-called multi-flash (MF) method [10] was employed as the 10 ms pulse is too short for accurate measurement of the high capacitance module under test (DUT). In this method, the final I-V curve is constructed from typically 30-40 individual flashes. Each point of the I-V curve was obtained by applying a constant voltage to the module, which was then subjected to a flash from the solar simulator at the desired irradiance level. The resultant module I-V curve was then obtained by performing an interpolation of the measured points [11, 12].

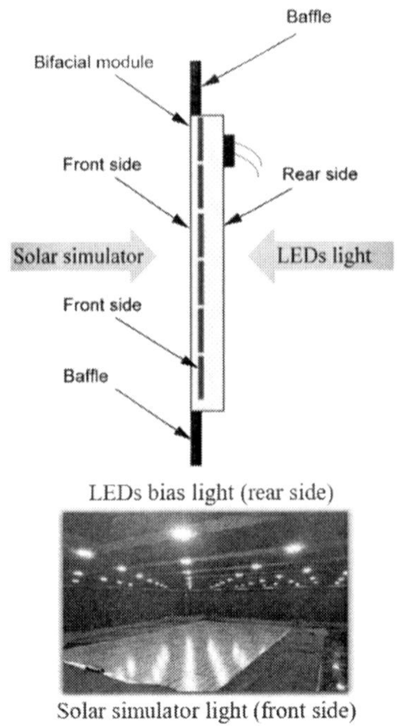

Figure 2: Representation of a double-sided illumination method using two light sources (top); Picture of applied illumination system composed by LED bias light on the rear side and long-pulsed solar simulator on the front side (bottom)

2.2.2 Double-side illumination approach

The experimental setup consists of two separate light sources which can be either used simultaneously or as single light sources (Fig. 2). Here, we are using the simultaneous configuration that enables illumination of the rear side by LED bias light and the front side by the long-pulse simulator (Fig. 2 bottom). For the front-side illumination, a large-area long pulsed flash solar simulator (Spire SLP4600 with two Xenon linear lamps and pulse duration up to 100ms at 1000W/m^2, class AAA spectrum classification with spatial non-uniformity ~1.45% and ambient temperature control) was used. For the rear-side illumination, a low-cost LED-based bias light on the top consisting of 55 Chip on Board (CoB) LEDs at different distances was applied. Pulsed operation mode ensures a very long lifetime of the LEDs and programmed for a steady state output over a 4s time period facilitates synchronization with the pulsed simulator it is paired with. The irradiance may be controlled by adjusting the drive voltage and/or the distance to the test device. By adjusting the distance between 25 cm and 60 cm combined with acceptable drive voltages, the irradiance in the test plane may be controlled in a range between ~60 W/m^2 and ~350 W/m^2.

The technology chosen was an LED integrated array COB (circuit on board) module, where each 30 mm diameter LED light emitting surface has between one to two hundred individual series and parallel connected blue emitters below the orange-red local phosphor covering. A matrix of 55 of these low-cost COBs was then envisaged to cover the test area of the pulsed solar simulator with sufficient uniformity. The LED production "bin" was chosen to be nearly the highest power and photon-flux available on the market, however for higher reliability and repeatability the 'second-bin' flux level was chosen with at least 13000 lumens at 4000K at 25°C. We supply all COB LEDs in parallel with a single programmable voltage from a commercial 1 kW laboratory PSU, at a variable voltage between 65V and 78V. An Arduino control system was used to adjust the PSU output into 2000 ms pulses for testing purposes, interface to the ESTI control systems and to raise the array power above 500W/m^2, or to reduce the measured irradiance to well below 100 W/m^2 (nominal 200W/m^2).

The spatial rear irradiance non-uniformity of the LED bias light was obtained by moving across the test plane (divided in sections with area of 20 cm×20 cm each) a large reference cell. The cell is a crystalline-silicon with an active area of 15 cm×15 cm. It is masked so that only the active area is exposed to light. The resulting measurements are represented as spatial maps.

The non-uniformity of the irradiance on the test plane is calculated according to the standard IEC 60904-9 [13] with the following equation:

NU = (maxIRR - minIRR)/(maxIRR + minIRR) (Eq.1)

where maxIRR and minIRR are, respectively, the maximum and the minimum irradiance values measured on the test plane. NU is given as percentage (%).

The non-uniformity of the LED bias light is predicted to be less than 5 % over an illumination area of 130cm × 200cm in order to fulfil the requirements defined in the IEC 60904-9 [13].

3 RESULTS AND DISCUSSION

3.1 I-V measurements

Figure 3a shows the reference I-V curves of the single-side approach (PIIIB) at 1000W/m^2 (STC, front and rear) and at equivalent irradiance illuminations i.e. 1067W/m^2 and 1134Wm^2 according the IEC TS 60904-1-2. The corresponding electrical parameters are listed in Table II.

In Fig. 3b the I-V curves obtained by the simultaneous bifacial illumination measured at effective irradiances are shown. Here, an extra measurement was made in reverse sweep mode at STC conditions to confirm the shape of the curve. In addition, testing at STC with and without black cover were performed to analysis the difference between both set ups (data not shown).

The obtained double-side I-V curves are not perfectly smooth as in the single-set up, but their shapes are very similar. The reverse sweep overlaps with the foreword mode, thus excluding any abnormal electrical behaviour of the module. The P_{max} value obtained from the measurement with black cover is 0.27% higher than without it.

In Fig. 3c, both approaches are compared. The two I-V curves measured by PIIIB and Spire-LED at effective irradiances are very close match. It can be observed that the curves are very similar in shape and irregularities and this reveals a good agreement of the proposed methodology. In both cases, there is an absence of measurement artefacts due to internal capacitance or other resistance effects.

The differences between I_{sc} values are very small ~0.05% as expected. The V_{oc} values exhibit higher deviation ~0.1% (Table II), but well within the measurement uncertainties. Higher output power (P_{max}) is generated by the double-side approach having the lowest value at G_{front}=1000W/m^2 and G_{R2}=200W/m^2 with deviation ±0.3%. The overall deviation from the single illumination is less than ±1%. The power gain yield (BiFi) is linear with applied irradiance (Fig. 3d), being slightly reduced relative to the single-method due to increasing P_{max} values. The BiFi deviation is ±0.51% (Table II).

Table III shows Fill factor (FF) measured at both illumination conditions. The $FF_{STC\ FRONT}$ refers to single-sided irradiance 1000W/m^2 for both approaches without rear side light source. The FF for the double-side approach is noted to be about ~ 1% higher (Table III). These slightly increasing results could be mainly associated with inherent characteristics of the pulse solar simulator related with non-homogenous irradiance distribution. It has been published that the FF under bifacial irradiance is typically found to be higher [14] that coincides with our findings. The influence of the use of black cover in the FF is noted to be insignificant 0.026% higher FF measured at STC with black cover in comparison with the one without it (data not shown). The mean deviation in FF between single and double illumination was found to be ±0.59%, ranging from a minimum deviation of ±0.33% to a maximum of ±0.74%. Lower deviation is observed for higher irradiances and could be associated with better rear-side uniformity distribution.

When combined with a commercial Class AAA pulsed solar simulator, double-side illumination produces results very close to those obtained with the Class AAA solar simulator using equivalent irradiance and single-side illumination. Overall, the deviation in the key parameters is less than ±1%.

3.2 Rear-Side Irradiance Measurements

The prototype system based on LED bias light has demonstrated good performance, enabling illumination of the rear side of a bifacial module at variable light levels up to ~350 W/m^2 by adjusting of supplied voltages from 65V to 78V. Linear behaviour of irradiance with applied voltage is investigated for values higher than 65V, being non-linear bellow this ranges (Fig. 4 bottom). The NU distribution changes with the distance between the module and the LEDs and with the supplied voltage. It can be observe from the Fig. 4 (top) that at lower irradiances and for all distances, the NU increases ranging from ±6% to ±11%, classifying the simulator in Class C, while at higher irradiances the NU decreases significantly for almost all distances. The optimal distance selected for the study is ~51cm, displaying the lowest NU values (Table IV) and classifying the solar simulator in Class B according the IEC 60904-9. The non-uniformity below ±5% means it meets the requirements of the IEC TS 60904-1-2 for use with bifacial modules.

Figure 3: Electrical parameters of a bifacial PV module: (a) I-V curve of single-side illumination; (b) I-V curve of double-side illumination; (c) Comparison of I-V curves of single and double-side illumination approaches; (d) Linear fitting of BiFi power gain yield for both approaches

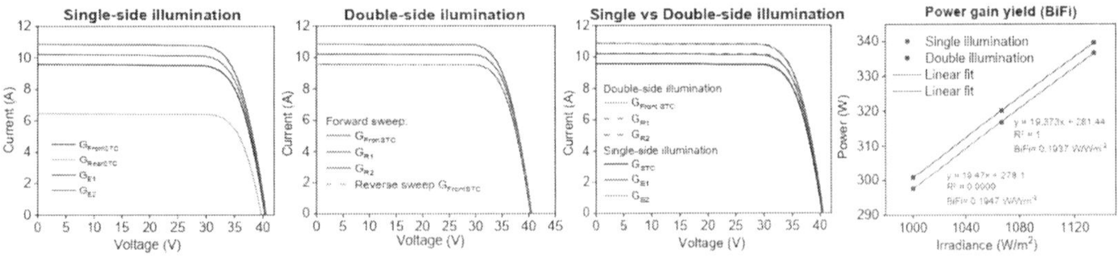

(a) (b) (c) (d)

Table I: Experimental set up parameters of single and double-side approaches

	Irradiance (W/m^2)	G_{Front}	G_{Rear}	Isc (A)
Single (PIIIB)	$G_{Front\ STC}$	1000	-	
	$G_{Rear\ STC}$	-	1000	$Isc_{\_STC}$= 9.58
	$G_{E1\ 100}$	1067	-	$Isc_{E1/R1}$=10.19
	$G_{E2\ 200}$	1134	-	
Double (SPIRE+LED)	$G_{Front\ STC}$	1000	-	$Isc_{E2/R2}$=10.85
	G_{R1}	1000	100	
	G_{R2}	1000	200	

Table II: Electrical parameters measured by single and double -side illumination approaches

	Isc_{STC} (A) Voc_{STC} (V)		$Isc_{GE1/GR1}$ (A) $Voc_{GE1/GR1}$ (V)		$Isc_{GE2/GR2}$ (A) $Voc_{GE2/GR2}$ (V)		$P_{max\ Front\_STC}$ (W)	$P_{max\ GE1/GR1}$ (W)	$P_{max\ GE2/GR2}$ (W)	$P_{maxBiFi100}$ (W)	$P_{maxBiFi200}$ (W)	BiFi (W/Wm^{-2})
Single	9.585	40.351	10.195	40.470	10.851	40.568	297.2 (±1.1%)	316.3 (±1.1%)	336.1 (±1.1%)	316.67 (±1.37%)	336.14 (±2.13%)	0.1947 (±0.024%)
Double	9.588	40.304	10.190	40.446	10.866	40.519	299.54*	319.32	337.25	318.91	338.28	0.1937
Δ (%)	0.031	-0.116	-0.049	-0.059	0.138	-0.121	0.79	0.95	0.34	0.71	0.64	-0.51

Table III: Fill factor (FF) measured by single and double -side illumination approaches

Fill factor (%)	$FF_{STC\ FRONT}$	$FF_{GE1/GR1}$	$FF_{GE2/GR2}$
Single-side	76.97	76.65	76.35
Double-side	77.51	77.22	76.6
Δ (%)	0.70	0.74	0.33

As it can be observed from the data in Table IV, at 65V (~100W/m²) the NU value is higher i.e. ±7.7% and do not achieve the requirements of the TS. At higher voltages (irradiance superior than ~200W/m²), the values decrease significantly achieving the lowest NU of ±3.4% and classifying the prototype system in Class B, that state to be suitable for reliable characterization of bifacial devices. In Table IV, column "Class B", shows calculated NU data that the LED bias light has to reach in order to be classified in the Class B range.

Spatial distribution maps of the rear irradiance made by the LED on the test plane at 51cm distance and for different voltages are depicted in Fig. 5. The square on each image represents the module size and position on the test plane. The axis indicates the dimensions of the long-pulsed solar-simulator illuminated surface (x=200cm and y=130cm). It has to be mentioned that non-reflective baffles are used for covering the illuminated area to avoid parasite reflections, as the DUT is smaller than the simulator's area. The colour map corresponds to the measured irradiance, with blue being the lowest and red the highest. The measured irradiance values are shown on the *y axis* of each image expressed in W/m².

In Fig.5 it can be analyse that the irradiance distribution is not homogeneous along the testing area, displaying areas with lower and higher illumination. Higher NU is observed for the map at 65V, where large part of the area is under lower illumination (62-68W/m²) than the expected 100W/m². This correspond to ±7.7% (Table IV). Correction is applied for the G_{R1} set up i.e. 66.6V according to Fig. 4 bottom to achieve the ~ 100W/m². Improvement is observed on the map corresponding to 70V expressed in larger red area. The irradiance values are close to ~200W/m² and practically corresponds to the 71V applied for the approach according to Fig. 4 bottom. The best performance of this study correspond to the 75V and 77V with the largest red area. At 77V the irradiance achieves ~380W/m² and it is also related with the lowest NU values measured (3.4%) in this work.

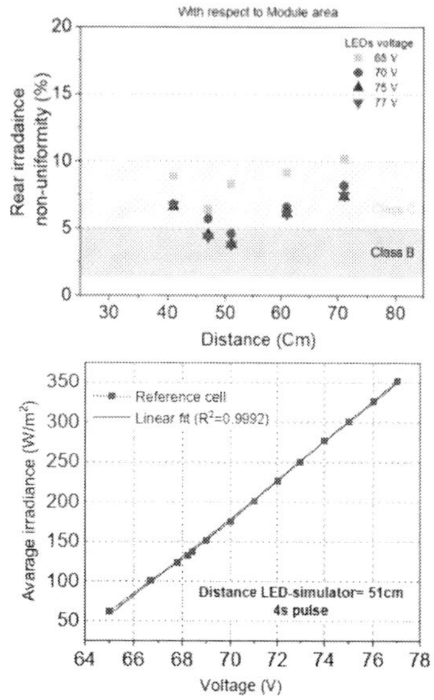

Figure 4: Rear irradiance non-uniformity (%) of a LED as a function of the distance for several applied LED voltages (top). Grey zones represents the Class B and C solar simulator according to the IEC 60904-9 (top). LED irradiance levels measured with reference cell on different voltage (bottom)

Table IV: Non-uniformity parameters measured using the LED simulator at ~51cm distance from the module

Voltage (V)	G_{min} (W/m^2)	G_{max} (W/m^2)	NU (%)	G_{avg} (W/m^2)	Class B (%)	STD (%)
65	61.5	71.9	7.7	67.7	65.0	2.7
70	176.3	194.5	4.9	187.0	175.9	5.1
75	303	324.6	3.4	315.1	293.7	7.6
77	353	378.1	3.4	366.2	342.1	8.7

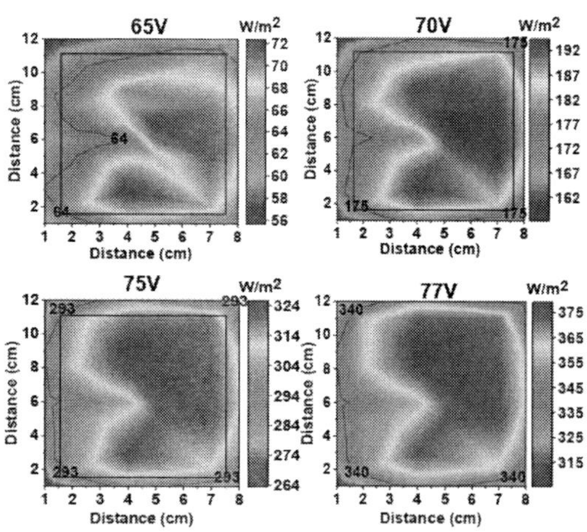

Figure 5: Spatial irradiance distribution maps of the rear side of the module, placed on the test plane at ~51cm of the LED and for different LED power supply voltages. The square represents the standard 60-cells module size (axis x=200cm; y=130cm)

4 CONCLUSION

The paper presents a double-side illumination approach for characterization of large-sale commercial bifacial devices based on a prototype system composed of a long-pulse solar simulator and a LED rear bias light. A module self-reference procedure was proposed and outlined. The obtained results have demonstrated good performance in comparison with the single-side measurements based on a commercial Class AAA simulator using validated standard methodology. The electrical performance demonstrates practically identical current-voltage (I-V) curves and similar key parameters (deviation less than ±1%) which validates the method for characterization of bifacial devices according to IEC TS 60904-1-2. Slightly higher P_{max}, BiFi and FF are measured and are mostly associated with inherent characteristics of the long-pulse solar simulator. The rear-side non-uniformity (NU) in the double-side illumination system varies with applied voltage and distance between LEDs and the device, which would result in a classification of the system in Class B (>200 W/m^2) or Class C (<100 W/m^2) according to the IEC 60904-9. The LED bias light instrument demonstrates good performance, enabling illumination levels higher than 350 W/m^2 at 77V applied LED voltage. Overall, the suitability of the proposed method in the IEC TS 60904-1-2 for the characterisation of bifacial PV devices was experimentally tested and the developed system configuration would be a convenient and low cost approach for further industrial implementation.

5 ACKNOWLEDGEMENTS

The authors would like to acknowledge Francesco Noseda for the technical support in the construction of the LED mounting structure. The research was carried out within the project (PV-ENERATE) leading to this application has received funding from the EMPIR programme co-financed by the Participating States and from the European Union's Horizon 2020 research and innovation programme.

6 REFERENCES

[1] ITRPV-2019 International Technology Roadmap for Potovoltaic (ITRPV) 10th edition (2019)

[2] Report IEA-PVPS T13-14:2021, "Bifacial Photovoltaic Modules and Systems: Experience and Results from International Research and Pilot Applications" (2021)

[3] IEC TS 60904-1-2 Ed. 1 Measurement of current-voltage characteristics of bifacial photovoltaic (PV) devices (2019)

[4] S. Roest, W. Nawara, B.B. Van Aken, E. Garcia-Goma, "Single side versus double side illumination method IV measurements for several types of bifacial PV modules", 33rd European Photovoltaic Solar Energy Conference and Exhibition (2017) 1427-1431.

[5] B. Newman, A. Carr, K. Groot, N.J.J. Dekker, B.B. Van Aken, A. Vlooswijk, A. Van de Loo, "Comparison of bifacial module laboratory testing methods", 33rd European Photovoltaic Solar Energy Conference and Exhibition, (2017) 1632-1635.

[6] D. Shaw, J. Lopez-Garcia, R. Kenny, L. Pinero-Prieto,

E. Ozkalay,"Design study of a double-side illumination solar simulator for bifacial silicon PV modules characterisation based on low-cost LED bias light", 35th European Photovoltaic Solar Energy Conference and Exhibition, (2018) 1001 - 1005.
[7] P. F. Ndione et al., "Combining Indoor and Outdoor Measurements to Lower Uncertainty in PV Modules Performance," 2020 47th IEEE Photovoltaic Specialists Conference (PVSC), 2020, pp. 2185-2187, doi: 10.1109/PVSC45281.2020.9300723
[8] D. H. Levi, C. R. Osterwald, S. Rummel, L. Ottoson and A. Anderberg, "Self-reference procedure to reduce uncertainty in module calibration," 2017 IEEE 44th Photovoltaic Specialist Conference (PVSC), 2017, pp. 467-471, doi: 10.1109/PVSC.2017.8366752
[9] IEC 60904-7 Ed.4 Photovoltaic devices - Part 7: Computation of the spectral mismatch correction for measurements of photovoltaic devices (2019)
[10] J. Lopez-Garcia, B. Haile, D. Pavanello, A. Pozza and T. Sample, "Characterisation of n-type bifacial silicon PV modules" Proc. 32rd European Photovoltaic Solar Energy Conference and Exhibition, (2017) 1724-1729.
[11] N. Ferretti, Y. Pelet, J. Berghold, V. Fakhfouri and P. Grunow, "Performance testing of high-efficient PV modules using single 10ms flash pulses" Proc. 28th European Photovoltaic Solar Energy Conference and Exhibition, (2013) 3184-3187.
[12] A. Virtuani, G. Rigamonti, P. Beljean, G. Friesen, M. Pravettoni and D. Chianese, "A fast and accurate method for the performance testing of high-efficiency C-Si photovoltaic modules using a 10Ms single-pulse solar simulator", Proc. 38th IEEE Photovoltaic Specialists Conference, (2012) 496-500.
[13] IEC 60904-9 Photovoltaic devices - Part 9: Solar simulator performance requirements (2007)
[14] C. Reise, M. Rauer, M. Mittag, A. Schmid, Photovoltaics International, "Bifacial characterization and simulation, From bifacial PV cells to bifacial PV power plants – the chain of characterization and performance prediction", (2018) 1-10

Development of mobile laboratory for photovoltaic testing service

Si-Xian Li *, Chia-Jen Lin, Wei-Yu Lin, Chao-Fan Hsieh, Teng-Chun Wu, Shu-Tsung Hsu

Photovoltaics Technology Center, Industrial Technology Research Institute, Hsinchu 30010, Taiwan

(*E-mail: sven@itri.org.tw)

Abstract — **This paper demonstrates the development of mobile photovoltaic test laboratory. The fully enclosed cabinet design can not only shield the ambient light but also control the temperature conditions. On-site test deployment can also be faster and more stable. The innovative design based on an expandable mechanism and oil-hydraulic stretch platform has allowed to implement a solar simulator of tunnel type in a vehicle that weighs less than 3,500 kg and to have the stable optical performance. The Spectral match, non-uniformity of irradiance and temporal instability of irradiance of the simulator remain the parameters defined as class A+A+A+, the characteristic of spectrum also with high SPC and low SPD accredited IEC 60904-9:2020.**

I. Introduction

Photovoltaic modules account for about 30 % of the cost of photovoltaic power plants. To reduce investment risks and improve the reliability of power plant operations, the inspection and testing of PV modules before installation is necessary. At present, the module inspection during the construction of the power plant must be transported from the site to the test laboratory for power measurement and EL testing. It is not easy to carry out the transportation back and forth, which also increases the risk of breakage of the cell due to vibration. According to research statistics, at least 5% of the customer complaints of PV modules damages are directly related to the logistics process of the modules [1]. On the other hand, considering the cost and time required to travel between the testing laboratories and power plants, the number of sampling tests is also very limited. The current laboratory testing method is difficult to perform a large number of tests. Not only the cost of testing has exceeded the price of the module itself, but the lengthy delivery and testing time is often prone to delays in the construction of the power plant.

With the development of the Mobile PV Test Lab, on-site tests in PV modules with the same reliability of an accredited conventional laboratory can be performed directly in the PV plants. The main advantage is that the installation duration could be reduced of the modules be sent to and back from the stationary laboratory, production down-time due to modules disconnection is minimized, aside from eliminating any breakage risks due to transportation of the modules to the stationary laboratory.

II. DEVELPOPMENT OF THE Mobile PV Test Lab

The Mobile PV Test Lab as shown in Fig. 1 is designed to perform the following tests: visual inspection, peak power determination in STC, electrical insulation, IR thermography and electroluminescence imaging. The design of the mobile laboratory ensures the reliability of the results as if conducted in a conventional laboratory. Its innovative design based on an expandable mechanism and oil-hydraulic stretch platform has allowed us to implement a solar simulator of tunnel type manufactured by PASAN in a vehicle that weighs less than 3,500 kg and to have the stable optical performance. The Spectral match, non-uniformity of irradiance and temporal instability of irradiance of the simulator remain within the parameters that define it as a class A+A+A+ simulator, due to its light source type of xenon lamp, the characteristic of spectrum also with high AM1.5 spectral coverage (SPC) and low AM1.5 spectral deviation (SPD) accredited IEC 60904-9:2020. The environmental conditions during the performance of the tests are controlled.

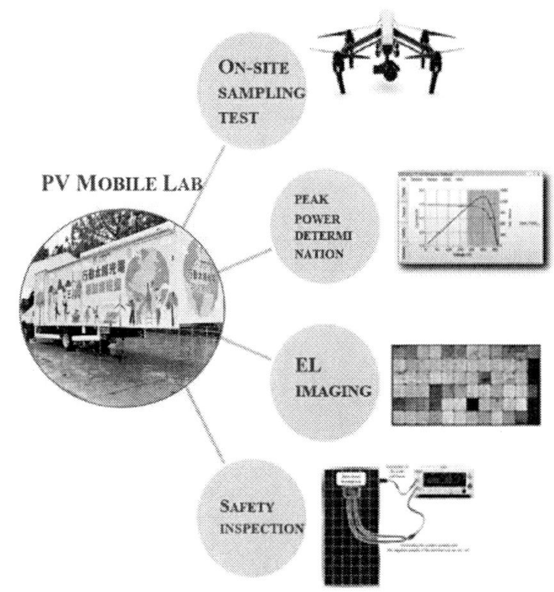

Fig 1. Test services of Mobile PV Test Lab

The innovative mechanical power contraction and expansion cabinet can meet the needs of large space, darkroom, and constant temperature room required for solar simulator testing and EL testing. The length of the cabinet can be expanded from 4750 mm to 7650 mm. It can also have the mobility and maintainability of a minivan as shown in Fig. 2. On-site test deployment can also be faster and more stable.

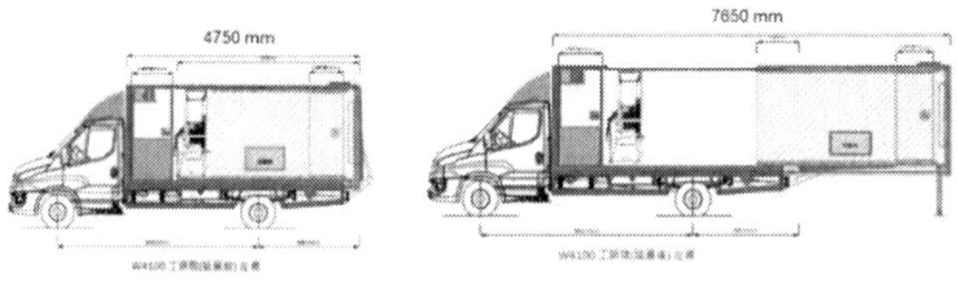

Fig 2. Innovative mechanical power contraction and expansion cabinet

The architecture of the previous study [2, 3] uses the conventional sliding cabin technology of the camper. The smaller inner cabinet is used as the sliding and extended part, and the larger outer cabinet is used as the main body fixed to the vehicle platform. This method limits the light source of the solar simulator, which is arranged on the sliding cabinet. The drawback of this design is that the unstable extended cabinet may cause the illumination optical path misaligned. In this work, the light source is placed in the fixed cabinet in the front of the trunk. The irradiation surface (module) is placed in the extendable rear cabinet, which is supported by a hydraulic supporting platform to improve the problem that the alignment of the light path may be changed unexpectedly.

Due to the limited space in the chosen IVECO New Daily truck, especially at the carriage, the standard height and width of PASAN SunSim 3C tunnel are 245 cm x 245 cm must be zoo out to the 180 cm x 215 cm and it is a challenge to make the non-uniformity of irradiance in the test plan could be kept under 1 %, special secondary optical designed light path use one screen walls form the light tunnel of the Mobile PV Test Lab. However, other light rays than direct light from the lamp to the DUT will have at least 1 reflection within the light tunnel, the stray light could be reduced by the design of mat surface of the wall that simulation by the optical model ray tracing, shown as Fig 4.

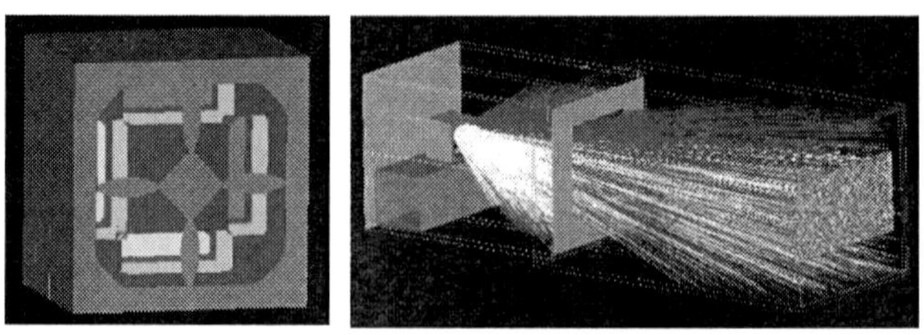

Fig 4. the stray light analysis that by the optical model ray tracing

Fig 5. The view of Mobile PV Test Lab operating

III. RESULT AND DISCUSSION

It is demonstrated that the solar simulator light source set at conventional sliding cabinet may cause unexpected optical path misalignment as shown in Fig. 6(a). If the optical path is 550 cm, the light spot displacement will be 9.6 cm per degree misalignment. Fig. 6(b) is shown that the sliding cabinet will be more stable with hydraulic supporting platform.

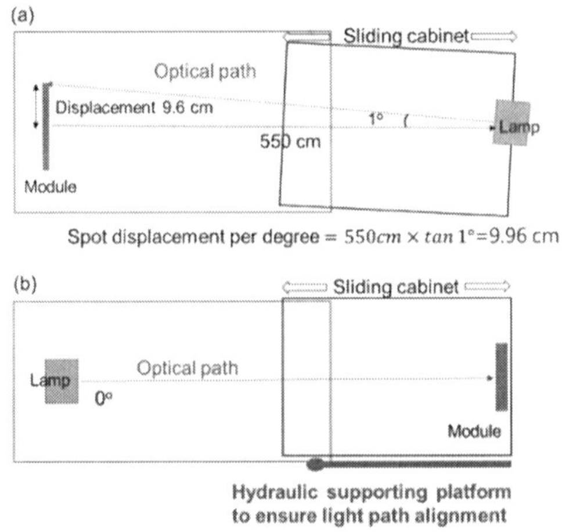

Fig 6. PV Mobile Lab (a) Conventional sliding cabinet design. (b) Sliding cabinet with hydraulic supporting platform.

The performance of the solar simulator inside the PV Mobile Test Lab can reach Class A+A+A+, the characteristic of spectrum also with high SPC and low SPD shown in table 1. The five properties include:

1. Spectral match
2. Non-uniformity of irradiance
3. Temporal instability of irradiance
4. AM1.5 spectral coverage (SPC)
5. AM1.5 spectral deviation (SPD)

The measurement results of Spectral match to the AM1.5 reference spectrum is in the range of 0.875~1.125, which can meet the Class A+ level in the wavelength range 300 nm to 1 200 nm required by the IEC 60904-9:2020 standard [4], the Spectral irradiance curves of PV Mobile Test Lab compare with the AM 1.5 reference spectral irradiance shown as Fig 7. The Non-uniformity of irradiance is 0.93 % after calibration, which is less than 1 % required by Class A+ in the range of 200 cm × 100 cm, the irradiance pattern shown as Fig 8. The measurement result of temporal instability of irradiance is 0.37 %, which is also less than 1 % required by Class A+ during the 10 ms data acquisition period, shown as Fig 9. In addition, the parameters measurement result of SPC is 100 % which a high value is more desirable and the measurement result of SPD is 14.7 % which a low value is more desirable, the value of SPC higher and value of SPD lower mean that the spectral irradiance curves closer to the AM 1.5 reference spectral irradiance and the measurement uncertainty due to the spectral mismatch will be reduce. On the other hand, if the value of SPC is lower and the value of SPD is higher, the measurement uncertainty may still high, even though the spectral match is class A+. In addition, Heating and cooling air conditioners are installed, and the temperature measured by the module can be controlled within (25±2) ℃ under testing required by the IEC 61215-2:2021 standard [5].

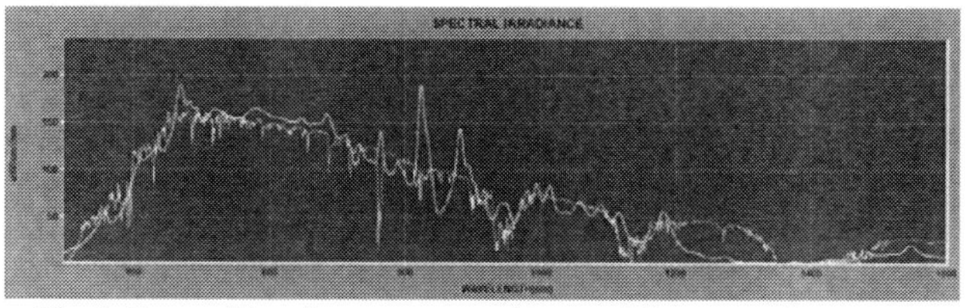

Fig 7. Spectral irradiance curves of PV Mobile Test Lab compare with the AM 1.5 reference spectral irradiance

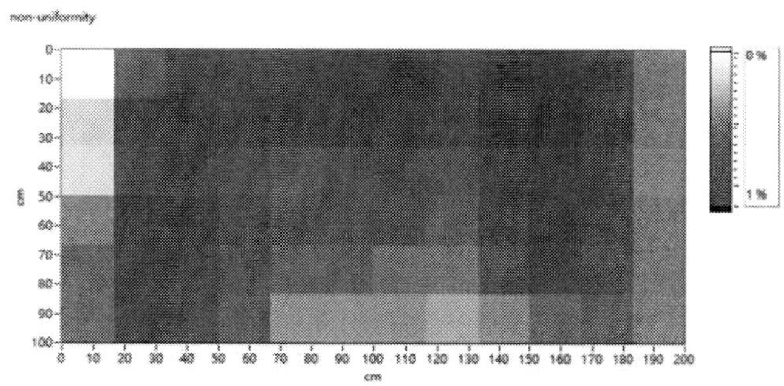

Fig 8. Irradiance pattern in the area of 200 cm × 100 cm

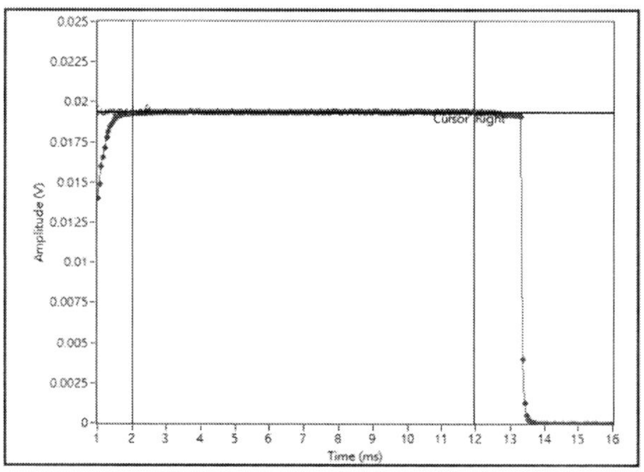

Fig 9. Data acquisition period

Table 1. Performance of the solar simulator of the Mobile PV Test Lab

Spectral match					
Wavelength (nm)	Irradiance (W/m^2) Solar Simulator	Energy Distribution (%) AM1.5G	Energy Distribution (%) Solar Simulator	Coincidence	Judgment
300-470	125.17	14.55	16.61	0.88	A+
470-561	141.19	16.41	16.74	0.98	A+
561-657	144.75	16.82	16.67	1.01	A+
657-772	149.46	17.37	16.63	1.04	A+
772-919	143.89	16.72	16.66	1.00	A+
919-1200	155.88	18.12	16.69	1.09	A+
Non-uniformity of irradiance					
Test area (cm^2)			200 cm × 100 cm		
Numbers of sampling			72		
Non-uniformity of spatial irradiance (%)			0.93		
Judgment			A+		
Temporal instability of irradiance					
Test condition			Long term (LTI)		
Data acquisition period (ms)			10		
Temporal instability (%)			0.37		
Judgment			A+		
AM1.5 spectral coverage (SPC)					
100 %					
AM1.5 spectral deviation (SPD)					
14.7 %					

IV. CONCLUSIONS

This paper demonstrates the development of mobile photovoltaic Test laboratory. In this work, the light source is placed in the fixed cabinet in the front of the trunk. The irradiation surface (module) is placed in the extendable rear cabinet, which is supported by a hydraulic supporting platform to improve the problem that the alignment of the light path may be changed unexpectedly. Moreover, the performance of the solar simulator inside the Mobile PV Test Lab can reach Class A+A+A+, the characteristic of spectrum also with high SPC and low SPD accredited IEC 60904-9:2020. On-site test deployment can also be faster and more stable.

REFERENCES

[1] Marc Köntges, Ulrike Jahn, Karl A. Berger, Kazuhiko Kato, Thomas Friesen, Haitao Liu, and Mike Van Iseghem. "Review of Failures of Photovoltaic Modules" *Report IEA-PVPS* T13-01:2014.

[2] Jorge Coelloa, Leonardo Péreza, "On-site quality control of photovoltaic modules with the PV MOBILE LAB", *2013 ISES Solar World Congress*.

[3] F. P. Baumgartner, T. Achtnichl, "SWISS MOBILE FLASHER BUS", *Proceedings of the 24th European Photovoltaic Solar Energy Conference*, Hamburg 21st Sept. 2009.

[4] International Electrotechnical Commission. "*Standard IEC 60904-9: 2020, 3rd edition: Photovoltaic devices – Part 9: Solar simulator performance requirements*".

[5] International Electrotechnical Commission. "*Standard IEC 61215-2: 2021, 2nd edition: Terrestrial photovoltaic (PV) modules – Design qualification and type approval –Part 2: Test procedures*".

ANALYSIS OF PERFORMANCE AND DEFORMATION FOR PHOTOVOLTAIC MODULE UNDER DIFFERENT WIND SPEEDS AND MULTI-TYPE MECHANICAL LOADINGS

Shu-Tsung Hsu*, Wei-You Lin, San-Yu Ting
Center for Measurement Standards, Industrial Technology Research Institute
Rm. 407, Bldg. 78, No. 195, Sec. 4, Chung Hsing Rd., Hsin-Chu 31057, Taiwan
*E-mail: andersonhsu@itri.org.tw

ABSTRACT: For years, photovoltaic systems were damaged often by typhoons, hurricanes and cyclones. Therefore, considerable research efforts in PV field have been investing in the development of new mechanical loads technologies. In particular, strong gusts always generated the non-uniform wind loads, and induced the extra dynamic force such as vibrate, torque and moment on PV system. To date, the effects due to wind loads on PV modules have typically been verified by using different mechanical loads test (uniform load) described in IEC 61215 MQT 16 (Static Mechanical Loads, SML) and IEC TS 62782 (Dynamic Mechanical Loads, DML). Nevertheless, these documents are still not sufficient to fulfil those uneven loads are due to various wind velocity, wind direction angle and module tilt. Therefore, this paper aims to provide an overview of current best practices by measuring the characteristics of twelve PV mono modules (M1-M12) subject to three wind speeds (41.4 m/s, 50.9 m/s, 61.2 m/s). It seeks to identify these samples' data like performance and deformation by different test procedure (DML, SML), but with two types of designed loads (uniform, non-uniform). Results will be helpful to determine the required standards like non-uniform wind loads test for PV module.
Keywords: Deformation, Non-uniform, Photovoltaic, Module, Wind

1 INTRODUCTION

Typhoon, hurricane and cyclone are the world's most frequently occurring natural disasters that easily cause photovoltaic (PV) systems installed outside to be damaged seriously. For example, the breakages from glass covers and cells are mainly due to these strong gust to induce their deformation experienced the different change or higher change rate. Furthermore, the installation of PV system need to concern with uplift and moment produced by uneven loads due to wind. Currently, two analytical models like Mean Extended Wind Load (MEWD) [1] and Mean Surface Pressure Pattern (MSPP) [2], and both have been designed to simulate the non-uniform wind loads under the specific environmental conditions include wind speed (V), module tilt (α) and wind direction angle (β).

For years, considerable research efforts in PV field have been invested in the development of wind loads test technologies (e.g., wind tunnel test, CFD, vibration, dynamic mechanical loads, etc.) that will offer new applications to simulate the various wind loads test, or develop new characteristic methods (e.g., mechanical stress, material analysis, defects inspection, etc.) or both. In particular, wind gust generates the non-uniform pressure load or moment effect on PV system [1, 6, 7].

IEC 61215-2 MQT 16 [3] is designed for the type approval of PV module, and provides the static mechanical load (SML) test. To pass the tests, the modules need to withstand a uniform test stress of 2,400 Pa (BS10 [2] and safety factor 3), and even 5,400 Pa in cases of higher snow load classes. IEC TS 62782 [4] provides a test method (± 1,000 Pa, 1,000 cycles, 3 to 7 cycles per minute) for performing a dynamic mechanical loading (DML) test in which the module is supported at the design support points and a uniform load normal at the module surface is cycled in alternating negative and positive directions. In IEC 62938 [5], a test apparatus was developed to simulate inhomogeneous snow loads based on standardized load calculations from the Euro code. Field failures could be well reproduced by applying loads of the relevant snow load zone to identical modules in the laboratory. The test helps to design modules for high snow load classes. Nevertheless, these three IEC standards still lack of the non-uniform wind loads test under high wind speed classes.

Therefore, this paper aims to provide an overview of current best practices for measuring the PV modules (α= 25°) subject to three wind speeds (V = 41.4 m/s, 50.9 m/s and 61.2 m/s). It seeks to identify both the deformation characteristics (e.g., max. displacement) and performance data (e.g., I-V, EL) of twelve test modules (Mono, 72 cell), by two cyclic test procedures like DML test (1000 cycles, six cycles per minute) and SML test (One hour to front and back surfaces in turn, total test time is six hours). In addition, both ML tests use two types of designed loads (uniform and non-uniform). Upon the non-uniform loads, the pressure loads refer to front wind scenario (β=180°), and the suction loads refer to back wind scenario (β=0°). In this work, there are six modules evaluated by designed wind loads refer to non-uniform MSPP (V, α = 25°, β) [2], and the other six modules test by the according average loads of each MSPP. Results will be helpful to measure and determine the content of wind loads test for PV module.

2 TEST EQUIPMENT

Below are four major test capacities.

- Non-uniform Dynamic Mechanical Loads (NUDML) system (see Fig. 1a): owns 18 (3x6) independent loads (max. = ±12,000 Pa).
- EL system (see Fig. 1b): Z4-PLUS contains 4 high-resolution CCD cameras in NIR technique, biased current 8A.
- I-V system (see Fig. 1c): BERGER Pulsed Solar Simulator, STC: 1000 W/m^2, 25 °C, AM 1.5G.
- Deformation Measurement System (DMS, see Fig. 2): There are 8 units of displacement sensors allocated under the back side (middle line, X = W/2) for test module installed on NUDML system. The dada acquisition triggered by Keithley 6510 with two units of model 7710 multiplexer module (20 channels per unit) with scan rate 0.00125 sec per channel. The max sampling rate for each displacement sensor is 25 pc per second.

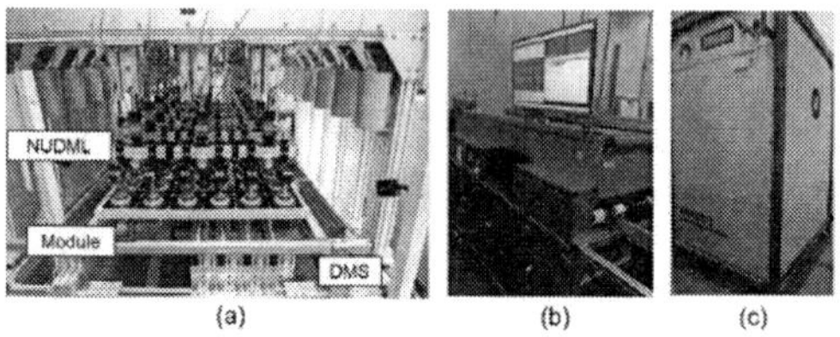

Figure 1: (a) NUDML system (eighteen pistons), and DMS (32 units displacement sensors) (b) EL system (c) Solar simulator

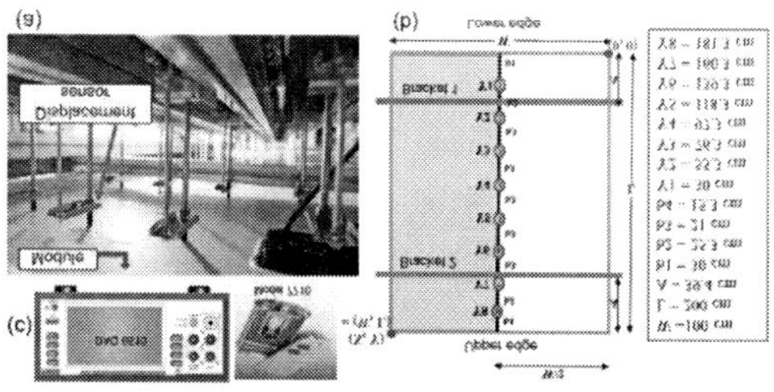

Figure 2: (a) Displacement sensors under the back of module (b) Diagram of allocation at X=W/2 for eight displacement sensors (L: length of module, W: width of module) (c) Dada acquisition of Keithley 6510 and model 7710 multiplexer module

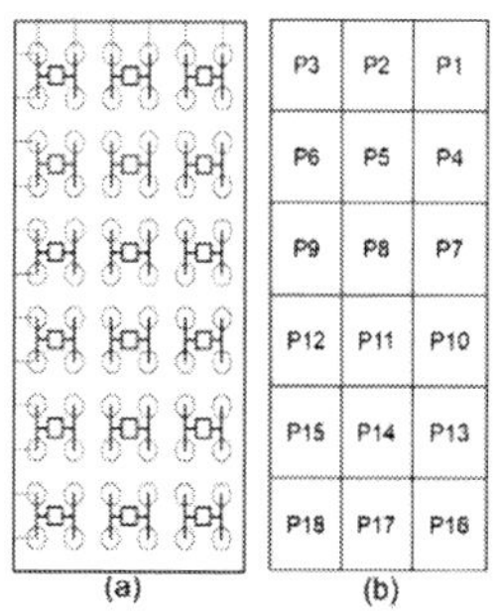

Figure 3: (a) Diagram of NUDML system with eighteen (3x6) independent loads, each piston with 4 suction cups (b) eighteen mean pressure loads of MSPP@p_1~p_{18}

3 TEST PLAN AND TEST ITEMS

3.1 Test samples and fixtures

- Test samples: twelve mono 72 cells modules (1968 x 992 x 40 mm) named M1~M12, and qualified by type approval testing [3].
- Racking: two brackets with 110 x 4 x 4 cm, the distance between each bracket to nearby edge of module is 39.4 cm (see A value in Fig. 2(b)).
- Screw: M8 bolts and mounting torque is 20 N·m.

3.2 Pre-test and post-test

- Measure the performance of the module in accordance with IEC 61215-2 Clause 4.2.
- Perform an insulation test in accordance with IEC 61215-2 Clause 4.3.
- Perform a wet leakage current test in accordance with IEC 61215-2 Clause 4.15.
- Perform an EL [8] scan by using a forward bias current 1A to identify the changes in test module.

3.3 Mechanical loads test

This work designed a series tests and referred to both DML and SML procedures (cyclic test), under two different designed loads (see Table I and Table II), e.g., uniform and non-uniform. In order to simulate the influence due to three wind speeds (41.4 m/s, 50.9 m/s, 61.2 m/s) by using NUDML system, the first designed load is non-uniform type and refer to Mean Surface Pressure Pattern (MSPP) [2]. MSPP@p_1~ p_{18} can fulfil the requirement of eighteen independent loads of NUDML system (see Fig. 1a and Fig. 3), by separating the wind pressure deviation distribution on the module surface into 18 zones and averaging them. The second designed load is uniform type, and this load equals to the average of each required MSPP.

Table I: Test condition (MSPPs see Table II) and twelve test samples (M1-M12)

	severity 1 V=41.4 m/s BS13	severity 2 V=50.9 m/s BS15	severity 3 V=61.2 m/s BS17
Non-uniform loads	MSPP1 MSPP2	MSPP3 MSPP4	MSPP5 MSPP6
DML	M3	M7	M11
SML	M4	M8	M12
Uniform loads	AVE(MSPP1) = +2565 Pa AVE(MSPP2) = -2448 Pa	AVE(MSPP3) = +3876 Pa AVE(MSPP4) = -3702 Pa	AVE(MSPP5) =+5601 Pa AVE(MSPP6) = -5352 Pa
DML	M1	M5	M9
SML	M2	M6	M10

Definition: AVE: average; BS: Beaufort scale (see Table 1 in [2]).

Table II: Six cases of MSPP@p_1~p_{18}, unit = Pa, safety factor = 3

Pressure load β=180°, α = 25°			Suction load β=0°, α = 25°		
MSPP1@p_1~p_{18}(V=41.4 m/s)			MSPP2@ p_1~p_{18}(V=41.4 m/s)		
1193	677	1193	-3686	-4579	-3686
1972	1076	1972	-3334	-3587	-3334
2568	1051	2568	-2411	-2192	-2411
3570	3045	3570	-2275	-1586	-2275
3447	4416	3447	-1982	-1608	-1982
3131	4119	3131	-1115	-925	-1115
MSPP3@ p_1~p_{18}(V=50.9 m/s)			MSPP4@ p_1~p_{18}(V=50.9 m/s)		
1803	1023	1803	-5571	-6922	-5571
2981	1627	2981	-5039	-5422	-5039
3881	1589	3881	-3645	-3314	-3645
5397	4603	5397	-3439	-2397	-3439
5211	6674	5211	-2997	-2431	-2997
4733	6226	4733	-1685	-1398	-1685
MSPP5@ p_1~p_{18}(V=61.2 m/s)			MSPP6@ p_1~p_{18}(V=61.2 m/s)		
2607	1479	2607	-8054	-10007	-8054
4309	2352	4309	-7285	-7838	-7285
5611	2297	5611	-5269	-4791	-5269
7802	6655	7802	-4972	-3465	-4972
7533	9649	7533	-4332	-3514	-4332
6842	9001	6842	-2436	-2021	-2436

Note: Wind direction angle β=180° for front wind and β=0° for back wind; "+" pressure force, "-"suction force; MSPPs data are developed by Bernoulli principle (see Eq. 1 [2]). If some of the eighteen designed loads (p_1~p_{18}) less than the minimum piston load (e.g., 20 kgf) of ML system, then using 20 kgf instead of these designed load.

$$C_p = (p_a - p_\infty) / (0.5\rho V_a^2) = (p_b - p_\infty) / (0.5\rho V_b^2) \quad (1)$$

$$\Delta P_{max} = (P_{max}(1) - P_{max}(2)) / P_{max}(1) \quad (2)$$

4 DISCUSSION

4.1 Performance data (see Table III) and safety data (see Table IV)

In Table III, the power loss over 5% are only happened on the modules (M9, M11) of severity 3 for DML case, and the causes could be due to the visible cracks in their EL images (see Fig. 4). In Table IV, all the insulation resistance times the area of module (1.95 m^2) are not less than 40 MΩ·m^2.

4.2 EL data (see Fig. 4)

The higher severity level got the greater wind speed and wind pressure. Three modules (M7, M11, M12) due to non-uniform force have visible cracks, and two modules (M7, M11) of them due to DML procedure own the maximum pressure happened at the lower edge then cause larger cracks and damage, but the maximum suction force happened at the upper edge is less damage. In addition, two modules (M9, M10) due to uniform force also have visible cracks, and module (M9) due to DML procedure owns the greater damage happened near center of module.

Table III: Performance data (ΔP_{max}, see Eq. (2))

Test procedure	DML	SML	DML	SML
Loading type	Uniform		Non-uniform	
Severity 1, V=41.4 m/s				
Module	M1	M2	M3	M4
P_{max} (1)	366.118	365.607	364.977	369.347
P_{max} (2)	362.231	361.682	361.555	369.928
ΔP_{max}(%)	1.06	1.07	0.94	-0.16
Severity 2, V=50.9 m/s				
Module	M5	M6	M7	M8
P_{max} (1)	367.664	367.083	366.155	366.306
P_{max} (2)	367.815	367.793	361.357	366.798
ΔP_{max}(%)	-0.04	-0.19	1.31	-0.13
Severity 3, V=61.2 m/s				
Module	M9	M10	M11	M12
P_{max} (1)	368.390	367.406	369.751	370.659
P_{max} (2)	337.872	361.829	350.802	361.709
ΔP_{max}(%)	8.28	1.52	5.12	2.41

Note: (1) pre-test, (2) post-test (for Table III, Table IV and Fig. 4)

Table IV: Safety data

Module		Insulation test		Wet leakage test
		Leakage current (μA)	Resistance (MΩ)	Resistance (MΩ)
M1	(1)	0.3	46975	3335
	(2)	0.4	34970	1976
M2	(1)	0.3	>50000	3525
	(2)	0.8	15571	2120
M3	(1)	0.4	>50000	1743
	(2)	0.6	35284	1171
M4	(1)	0.5	33759	1561
	(2)	0.7	29512	1195
M5	(1)	0.6	32294	1986
	(2)	0.8	26118	1212
M6	(1)	0.6	39628	2026
	(2)	0.7	28021	1768
M7	(1)	0.3	>50000	1949
	(2)	0.4	43767	1544
M8	(1)	0.3	>50000	2868
	(2)	0.6	41352	2162
M9	(1)	0.4	42685	2594
	(2)	0.6	30816	2196
M10	(1)	0.5	37285	1921
	(2)	0.6	22549	1525
M11	(1)	0.4	48942	1976
	(2)	0.6	37452	1495
M12	(1)	0.4	>50000	1879
	(2)	0.5	42628	1687

4.3 Deformation data (see Fig. 5)

In Fig. 5, the higher severity level got the greater deformation. The fixtures of two brackets will restrain the deformation type of test modules. For example, the middle line on test module (at X=W/2), the deformation due to pressure force is smaller than the influence of suction force, and the maximum deformation of the uniform force case occur near the center of the module. But the largest deformation of the non-uniform force case occurs near lower edge (at Y=0) and upper edge (at Y=L). Furthermore, three modules (M7/ DML, M8/ SML and M12/ SML) in the case like pressure force of non-uniform loads, the similar warpage deformation happened near the upper edge caused by support of the bracket 2.

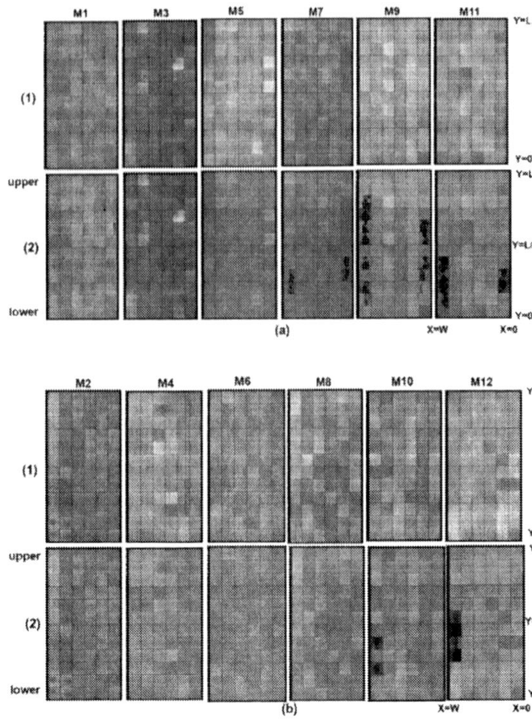

Figure 4: EL images for (a) DML procedure (b) SML procedure. Note: uniform loads cases (M1, M2, M5, M6, M9, M10); non- uniform loads cases (M3, M4, M7, M8, M11, M12)

5 CONCLUSION

Non-uniform wind loads test for PV module is necessary, and severity test for different wind speed is important, too. In this work, two types of designed loads (uniform and non-uniform) and two test procedures (DML and SML) have been evaluated successfully on twelve PV modules. The required environmental factors include three wind velocity (41.4 m/s, 50.9 m/s, 61.2 m/s), two wind direction angle (0°, 180°) and one module tilt (25°). Both NUDML system and MSPPs data can fulfil these specific test requirement, especially focus on the higher wind velocity from some severe weather like typhoon or hurricane. This research provides an overview of current best practices for measuring the performance of PV modules, and seeks to identify what additional test condition can do determination of the correct wind loads test on PV modules even their fixtures, and how these measures might be standardized in future [9]. This work contributes to the ongoing discussions in IEC TC82 [10] or PVQAT [11] about test condition of different wind actions on PV systems, and helps PV industry to have a common testing method to refer to when desired. Furthermore, results can also apply to the wind requirements for PV system on land and FPV system on water [12].

6 ACKNOWLEDGEMENT

The authors appreciate the technical support from Dr. Che-Wei Hsu at TSEC in Taiwan.

7 REFERENCES

[1] S. T. Hsu, T. C. Wu, Simulated Wind Action on Photovoltaic Module by Non-uniform Dynamic Mechanical Load and Mean Extended Wind Load, Energy Procedia (2017) pp. 94-101, DOI: 10.1016/j.egypro.2017.09.401.
[2] S. T. Hsu, W. Y. Lin, S. J. Wu, Environmental Factors for Non-uniform Dynamic Mechanical Load Test due to Wind Actions on Photovoltaic Module, Energy Procedia (2018) pp. 50-57, DOI: 10.1016/j.egypro.2018.09.008.
[3] IEC 61215-2:2020, Crystalline Silicon Terrestrial Photovoltaic (PV) Modules - Design Qualification and Type Approval.
[4] IEC TS 62782:2016, Photovoltaic (PV) Modules – Cyclic (Dynamic) Mechanical Load Testing.
[5] IEC 62938:2020, Photovoltaic (PV) Modules, Non-uniform Snow Load Testing.
[6] M. Assmus, M. Koehl, Experimental Investigation of the Mechanical Behavior of Photovoltaic Modules at Defined Inflow Conditions, Journal of Photonics for Energy (2012), vol. 2, pp. 022002-1 to 022002-11.
[7] K. C. Su, K. M. Chung, S. T. Hsu, Numerical Simulation of Wind Loads on Solar Panels, Modern Physics Letters B, Vol. 32, Nos. 12 & 13 (2018) 1840009 (5 pages), DOI: 10.1142/S0217984918400092.
[8] IEC TS 60904-13, Photovoltaic Devices – Part 13: Electroluminescence of Photovoltaic Modules.
[9] S. T. Hsu, Standardization Work of Non-uniform Wind Loads Test on PV Module, 37th European Photovoltaic Solar Energy Conference and Exhibition (2020), pp. 894-899, DOI: 10.4229/EUPVSEC20202020-4CO.3.2.
[10] https://www.iec.ch/dyn/www/f?p=103:14:10044622981200::::FSP_ORG_ID:2753.
[11] https://www.pvqat.org/.
[12] S. T. Hsu, K. C. Su, Wind and Wave Effect on Floating Solar Panel, 37th European Photovoltaic Solar Energy Conference and Exhibition (2020), pp. 1887-1890, DOI: 10.4229/EUPVSEC20202020-6CV.2.3.

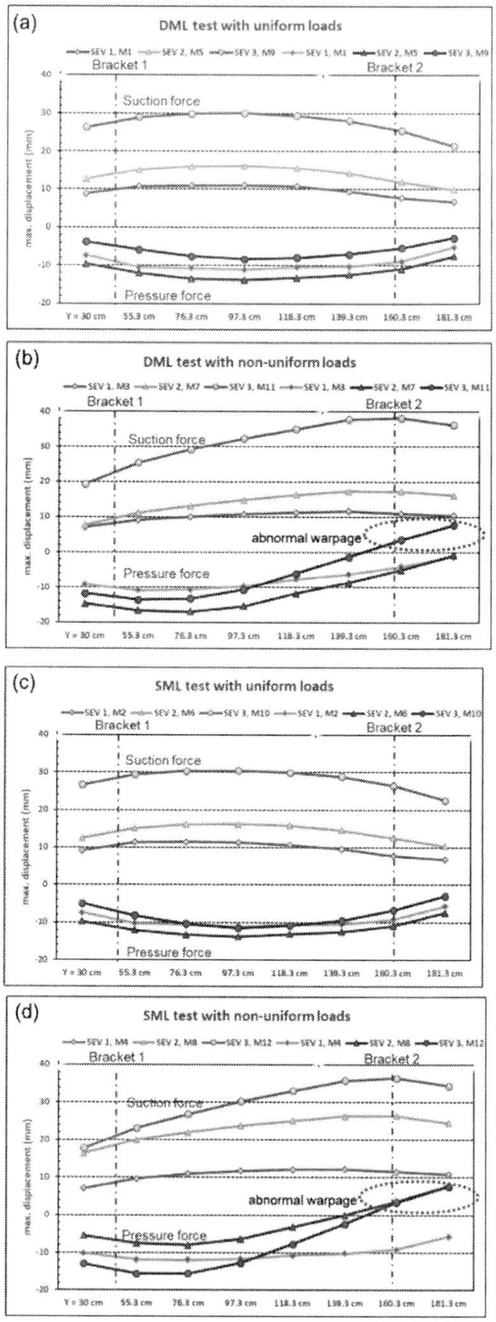

Figure 5: Four cases of deformation data (X=W/2, Y=Y1~Y8) for both pressure and suction force. DML test procedure (a) uniform load (b) non-uniform load, and SML test procedure (c) uniform load (d) non-uniform load. Positive data means "pull up", negative data means "push down". Hollow symbol means suction force, filled symbol means pressure force

MULTI-CRITERIA ANALYSIS METHOD TO EVALUATE DIFFERENT ENCAPSULATION MATERIALS FOR PV MODULES AND PROPOSING A SUITABLE CANDIDATE

Hesam-Edin Hayati Soloot[1], Amir Hayati Soloot[1], Shahab Moghadam[1]
1- SolarEdition, Oslo (Norway)

ABSTRACT: The power loss in photovoltaic (PV) modules due to degradations in the encapsulant layers accounts for more than 10% of the power loss within the module. Choosing a suitable material for the encapsulant layer has a significant impact on the price and performance of the PV module. The performance analysis of PV modules in the near- and long-term are usually referred to as reliability and durability in the literature. While the reliability issue to a large extent and acceptable limits is covered by the different international standard tests, the main challenges are related to the durability of the PV modules in the different field installations. Durability issues couldn't be solved without putting forward a general rule of thumb for choosing a suitable encapsulation material. Then, this paper proposes a general multi-criteria analysis method for doing that. After that, this analysis is applied to compare three conventional encapsulation materials including EVA, PVB, and TPO. Finally, this study proposes a suitable candidate with long-lasting durability and providing better performance for the PV module encapsulation based-on this multi-criteria analysis.
Keywords: Encapsulation, Durability, Reliability, PV module, EVA

1 INTRODUCTION

Renewable energy will outstrip their traditional and dirty fossil fuel-based rivals in power generation by 2025 [1-3]. Among all renewable energy technologies, the photovoltaic (PV) module has been dominating technology for many years [1]. As a matter of fact, the new annual added solar capacity accounted for more than 48% of the total installed renewable energy capacity in 2020 [4]. This indicates the PV module has become a reliable power generation technology. Thus, this industrial dominance brings reliability and durability issues of PV modules into the central focus [5, 6].

PV modules consist of several layers with different materials in which each layer fulfills specific duties. As a dominant type of PV modules, crystalline silicon (c-Si) modules with more than 90% market share are embedded in polymeric encapsulant layers [7-10].

The performance analysis of PV modules in the near- and long-term is usually referred to as reliability and durability in the literature [5, 6, 11, 12]. While the reliability issue to a large extent and acceptable limits is covered by the different international standard tests with conducted accelerated aging and stress testing [13, 14], the main challenges are related to the durability of the PV modules in different field installations. Because the standard reliability tests could not ensure the long-lasting service lifetime of PV modules [6, 12].

On the other hand, the power loss in PV modules caused by the encapsulation material accounts for more than 10% of the power loss distribution [15]. Thus, both the reliability of PV modules for mass production and the durability of them for long service lifetime up to 25-30 years are depending on the careful choice of materials for the encapsulant layers [16].

There is a famous Golden triangle in the PV module production that scientific researchers and PV module producers consider to assess the technical feasibility and commercialization capability of that product in the destination market [10, 17]. This Golden triangle, as depicted in Fig. 1, is tied to the three points: Price, Performance, and Efficiency.

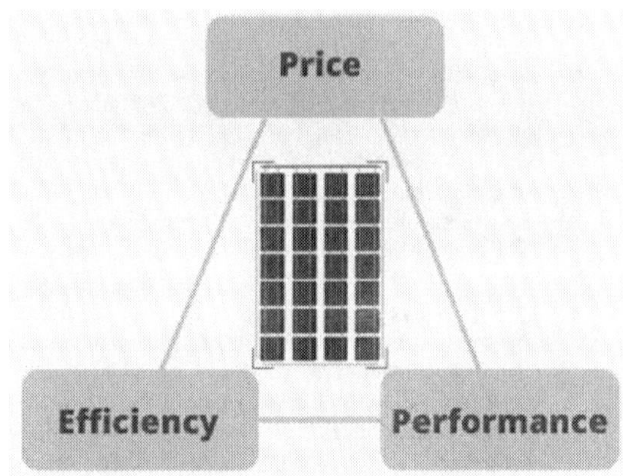

Figure 1: Golden Triangle for PV Module Production

There have been the most important criteria for selecting a suitable encapsulation material that can be categorized under these three pillars. These criteria include low price, high UV durability, good light transmittance, good electrical insulation, long and sustainable structural life, and low moisture ingression [5, 18-20].

They are necessary to meet for any encapsulation material as the desirable requirements for utilizing in the PV module manufacturing but are not sufficient for the optimal performance of PV modules. Besides the performance, price is another pillar for choosing a suitable encapsulant [21]. In order to distinguish suitable materials, we need to emphasize the importance of all three points equally.

There are several failure modes that lead to degradation of the third pillar, the efficiency, in addition to the other two ones. In other words, these failure modes have impact on the time-dependent performance or durability of the PV modules [22].

The encapsulation material is the most influencing component of the PV module for time-dependent failures including discoloration, delamination, potential-induced degradation (PID), corrosion, snail trails and hotspots [23, 24].

Therefore, it is crucial to propose necessary and sufficient criteria for selecting a suitable encapsulation material to retain the optimal performance of PV modules during their standard qualification tests and their 25-30 years of service as well as its cost.

The low cost, lightweight, flexibility, and easier assembly of polymeric materials have made them the most widely used material in the PV module encapsulation. First of all, Polyethylene (PE) was initially chosen as a polymeric encapsulant because of its simple structure and low cost. However, the main drawback was its opaque and translucent appearance [25, 26]. After that, PE-based materials in which PE was the backbone along with different side groups were chosen to solve transparency challenges and keep PE advantages [25].

One of these PE-based polymeric materials is Ethylene-vinyl acetate (EVA) in which vinyl acetate (VA) was added to PE to form a copolymer. EVA has become the mainstream material for commercial use due to its low cost, adequate transparency, and flexibility since 1981[5, 25, 27]. Today, EVA is the most popular encapsulation material with more than 80% market share [28]. The specific advantages of EVA encompass the ease of processability, low cost, low weight, good optical and mechanical properties, and more than 35 years of established history for field installation [5, 29].

The factors which cause deterioration of the functionality of EVA include acetic acid formation (deacetylation), moisture ingression, temperature, and ultraviolet (UV) radiation [29-31]. In the late 1990s, it was reported that EVA turned yellow/brown color (discoloration) due to UV radiation that led to reduction of its light transmittance, and eventually the performance loss in the PV module [18, 22, 30]. Discoloration of EVA appears in several degrees and extends from yellow to dark brown. Depending on the severity of discoloration and environmental conditions of installation location, the rising of temperature at the discolored area may cause hotspots or large hot areas in the PV module [33]. It is worth mentioning that VA content of EVA is the main driver of deacetylation which has a direct effect on the degradations of PV modules such as discoloration.

Moreover, it has also been reported that EVA was prone to moisture ingress which caused speed up several time-dependent failures such as discoloration, corrosion of metallization on top of the solar cells [18, 31, 32]. Delamination and bubble formation are the other failures due to moisture ingress in the PV modules which were encapsulated by EVA [31, 32]. Finally, potential-induced degradation (PID) can become a significant failure when both of the high temperature and high value of water vapor or humidity are frequently detected in the installation location of PV modules [34]. As EVA is prone to moisture ingress and in presence of high UV radiation as well as high temperature, PV modules which encapsulated with EVA have enormous potential to degrade with this phenomenon. These deteriorating factors, sometimes lonely and usually along with each other during the period of field operation, cause the aforementioned time-dependent failures.

A number of studies have suggested special additives to avoid regarding failures of EVA. A part of them proposed peroxide as the cross-linking agent to ensure thermal stability of EVA for final lamination in the PV module fabrication process and provide good adhesion to the glass and backsheet as well [22, 27, 30, 35, 36]. Good adhesion is partly prohibited from delamination. Some others have put forward the UV stabilizer, for instance Hindered Amine Light Stabilizers, and UV absorber like Cyasorb UV 531 that mitigates somewhat the discoloration [36-38]. But, in parallel to their positive effects, they also have negative effects that can change chemical and physical properties of EVA during the lamination process and under the environmental conditions of installation location [30, 38-41].

Therefore, the introduction of a new non-EVA type encapsulant with low-cost, high-performance, and high durability could provide a solution to discoloration, delamination, corrosion, and PID problems [21, 22, 30, 38].

Non-cross-linking thermoplastic polyolefin, hereafter TPO, have four distinct advantages over EVA. Two of them are associated with the needless to the cross-linking additive. First, this feature in the PV module encapsulation leads to eliminating not only all the chemicals needed for cross-linking or curing but also the cross-linking reaction byproduct. The second advantage is the reduction of the lamination cycle time due to omitting the cross-linking process in the PV module fabrication. The other two advantages are related to the intrinsic nature of TPO. There is no need for a UV stabilizer and UV absorber as well as it will not result in the acetic acid formation [30, 31].

These features of TPO are crossing the barriers in front of utilizing EVA as the encapsulation material and keep PV modules away from time-dependent failures which are observed for EVA.

Hence, to accurately assess and analyze the potential of TPO as a suitable alternative for EVA, the paper is conducted a multi-criteria analysis method which requires for the comparison of TPO with the other encapsulation materials such as EVA and polyvinyl butyral (PVB). This is the key objective of this paper. In addition, it indicates implicitly the strengths and weaknesses of these three encapsulation materials.

2 MATERIAL & METHOD

There are two encapsulant layers in a conventional c-Si PV module. The first and front encapsulant layer is located between the glass and solar cells, covering surface number 2, while the second and the rear encapsulant layer is embedded between solar cells and the backsheet, covering surface number 4, as shown in Fig. 2. These two layers have six main functionalities.

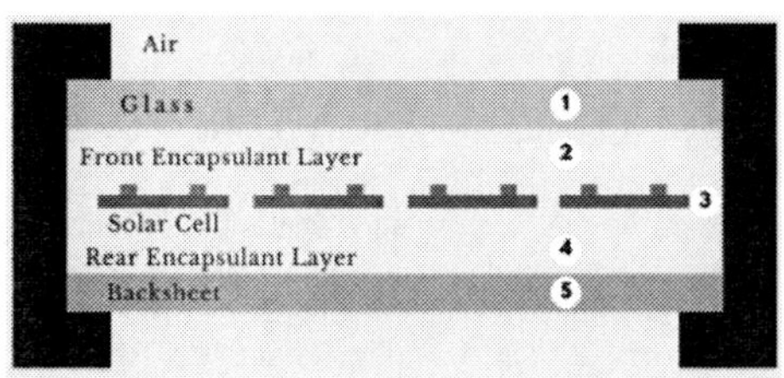

Figure 2: A conventional c-Si PV module with five layers

First, an encapsulant should provide structural and mechanical supports for solar cells and their circuit components. Second, it must have low light absorption, high optical transmittance, and an adapted refractive index to maximize the optical performance of a module. Third, it has to guarantee the electrical safety and provide the electrical insulation. Fourth, it should have well-thermal stability and conductivity to dissipate heat. Fifth, an encapsulant must have a reliable structural life during its lifetime operation, especially under UV radiation. Finally, it prevents the PV module's components from permeating exterior or environmental agents such as water and oxygen [5, 18, 22, 23, 30, 36, 41]. As shown in Fig. 3, all of these functionalities must be provided by an encapsulant at possibly low material and production price to become commercially viable.

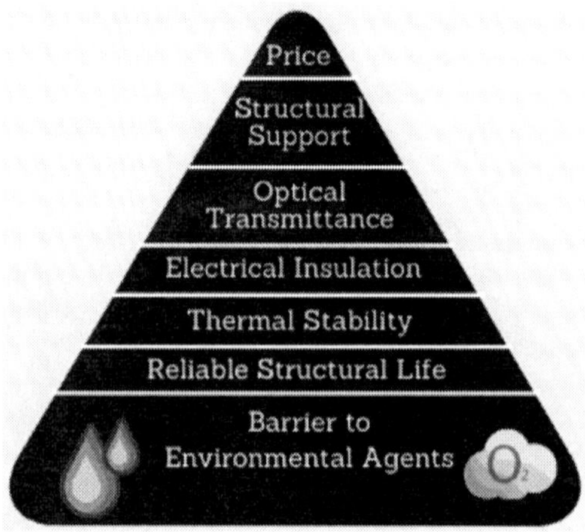

Figure 3: Most important functionalities of an encapsulant

Based on these functionalities different authors and researchers have put forward the minimum and desirable requirements which should be met by a material as an encapsulant of PV modules. This was done to gather all the required items in one place to become a qualification and quantitative checklist for experts.

Czanderna and et al. [22] have prepared one of the old and highly reputable versions of this checklist. They gathered together a list of general criteria for encapsulation materials and their minimum requirements, as shown in Table. 1. But they have not specified which criteria are more important than others for taking into consideration and simplicity of decision making to choose a better encapsulant among a group of materials. Hence, this study takes one step forward and indicates not only which criteria are vital for consideration but also explains the rationale behind their selection.

Table I: Categorizing General and Desirable Requirements for Encapsulation Materials

Characteristic	Minimum Requirements
Optical:	Total hemispherical transmission above 90% over the wavelength range from 300nm to 1100nm, Hazing or clouding (None at 80°C, 100% RH), UV absorption degradation (None at wavelength >350nm).
Chemical:	Hydrolysis (none at 80°C, 100% RH), Chemical inertness (no reaction with embedded Cu coupons at 90°C).
Mechanical:	Mechanical creep (None at 90°C), Tensile Modulus (⩽20.7MPa at 25°C).
Thermal:	Glass Transition Temperature(T_g), T_g⩽-40°C, Resistance to Thermal oxidation (Stable up to 85°C).
Fabrication:	Fabrication temperature (≤170°C), Fabrication pressure (≤1atm), Minimum fabrication thickness (≥0.150 mm).
Environmental:	Human hazards (non-toxicity & non-odor), water absorption (⩽0.5% wt% at 20°C/ 100% RH).

An encapsulant must have the total hemispherical transmission, or in short the optical transmittance, above 90% over the wavelength range from 300nm to 1100nm and show no sign of hazing or clouding at 80 degrees Celsius and 100% relative humidity (RH) [22, 38]. This results in higher optical performance and an increase in the cell-to-module (CTM) ratio and higher power output of the PV module.

Some authors categorized the other part of electromagnetic radiation of sunlight, meaning the UV radiation, under the optical properties of the encapsulation material [38, 42]. Due to the detrimental effect of the UV part of solar radiation on the polymer chain of encapsulants such as EVA, PV module producers are unavoidable to add an UV stabilizer to their encapsulation material. This additive increases the UV cut-off value of the encapsulant. The UV absorption has not to be seen at wavelengths above 350nm. The inevitable consequence of not adding the UV stabilizer is quicker discoloration. It is important to note that discoloration is not a single-agent failure mode [29].

Sometimes, PV module producers use encapsulation materials with different optical reflectivity, optical transmittance and UV cut-off for the front and rear encapsulant layers. The motive behind it is to enhance the entrance light and to boost the light-recycling effect inside the PV module, which results in gaining higher CTM ratio as well as more output power [43-46]. To this end, they utilize encapsulation materials with high UV transmission for the front encapsulant layer and conventional UV cut-off for the rear encapsulant layer.

This configuration not only helps to improve the module performance but also promotes its reliability, thanks to reducing yellowing at the interface between the rear encapsulant layer and backsheet [44-46]. The UV cut-off range for PV modules is greater than 280 nm and below 350 nm [38].

As a chemical reaction, hydrolysis is related to the chemical disintegration of an encapsulation material due to the interaction with water molecules [29]. EVA is prone to hydrolysis which also helps to form acetic acid [23]. The possibility of viscoelastic or Mechanical creep of the encapsulant is associated with the crystallinity and melting transition temperature of encapsulation material [35, 41]. Low Young's modulus (< 50 MPa) is required for encapsulant layers to dissipate the heat and to relieve thermal stresses in the PV module [22, 30, 47].

To achieve a better PV module performance and thermal stability is essential an encapsulation material to have low glass transition temperature below (-40) degrees Celsius [22, 31]. As an environmental concern, an encapsulation material should not have health and safety hazards for humanity and the environment. Some encapsulation materials have not any environmental hazard by itself but produce hazardous byproduct during lamination process or in the PV module lifetime because of environmental agents such as water or oxygen. As shown in Table. 1, Czanderna and et al have pointed to one of them [22].

This checklist has advantages and disadvantages in order to be reliable for decision-makers. As advantages of the checklist, it covered the structural support and optical transmittance items of the most important criteria for selecting encapsulation material. However, it covered the thermal stability and barrier to environmental agents partly and not sufficiently entered items for making sure the electrical insulation and reliable structural life of the encapsulation material.

These criteria in the best situation are necessary for choosing the PV module's encapsulation material and help to pass the reliability tests while there is a demand for the multi-criteria decision making list which makes sure both the reliability and durability of the PV modules. Therefore, in the next section, this paper will propose a needed multi-criteria analysis method which helps decision-makers to quickly and accurately take into account its desirable material for encapsulating PV modules.

2.1 Multi-criteria Analysis Method

In this section, we explain the necessary and sufficient criteria which need to be considered when decision-makers want to choose their best option for the encapsulating of PV modules. In the previous section, we explained some of them but to fill the gaps and present a universal and unique method to assess different encapsulation materials, we explain the remaining ones. It is worth mentioning that this multi-criteria list provides a basic framework for comparison of different encapsulants.

From the optical point of view, the refractive index (n) of the encapsulation material should also be added to the optical criteria. This is because of the importance of the index matching at the front encapsulant - glass and the rear encapsulant - backsheet interfaces [48]. The index matching between the front encapsulant – glass and rear encapsulant - backsheet interfaces leads to reduction in the reflective loss at the former interface and increases the reflective gain at the latter one. Both the reduction in the optical loss and increase in the optical gain, as factors of the CTM ratio, enhance the power output of the module. The latter accounts part of the indirect optical gains within the PV module which is called the light recycling effect [43-46]. This parameter is less pronounced in the literature as one of the most important criteria for selecting suitable encapsulation material.

O. Dupré and et al. [49] have shown the importance of this light recollection mechanism in the CTM ratio of PV modules with reflective (e.g. white) backsheets. They have found that about 45% of the light incident in these areas can be recollected for standard cell spacing (2-3 mm) and a backsheet reflectance of 87%. Thus, to mitigate the reflective loss and boost the reflective gain the refractive index of the encapsulant material is suitable to be between 1.48 and 1.5 at λ= 680nm, near the refractive index of glass [48].

A material with a higher volume resistivity is desirable for encapsulation. Since it reduces the ion mass transfer from the glass to the solar cells of a PV module. Ion mass transferability is one of the degradation mechanisms which reduces the electrical insulation of encapsulant and increases leakage current and leads to PID [19, 23, 34, 50].

The moisture or humidity ingress in PV modules is measured by the water vapour transmission rate (WVTR). It is necessary to define a criterion to assess the oxygen transmission rate (OTR) as well. Czanderna and et al. have specified 0.5% weight percent (wt) at 20 degrees Celsius. It means 5gr of water penetration when there is 1000gr of water. Encapsulation materials with higher WVTR are very prone to delamination and bubble formation.

Glass transition temperature, melting point temperature, degree of crystallinity, and thermal stability can generally be regarded as criteria associated with the thermal stability of the encapsulation material. But each one has its role to guarantee the reliability and durability of the encapsulant.

Glass transition temperature is a temperature that the encapsulant is in the brittle/glassy and rubbery/flexible states below and above that temperature. The required amount for this criterion is lower or equal to (-40) degrees Celsius [31, 51].

There are two melting peaks which are observed in the endothermic reaction curve in the differential scanning calorimetry (DSC) technique. The highest peak is the actual melting point temperature. The transition between these two melting peaks is called the melt transition. The melt transition shows the thermal behavior of the encapsulant. Most of the authors have suggested 85 degrees Celsius as the minimum requirement for the melting point temperature based on the IEC 61215 standard. But considering melting point temperature above 100 degrees Celsius ensures the preventing of the mechanical creep at temperature above 85 degrees Celsius [35, 38]. It is one of the parameters which was conducted in the international reliability tests while the amount of it is not sufficient to guarantee the durability of the encapsulation material during PV module's lifetime.

The encapsulation material with low degree of crystallinity is preferred. Since low degree of crystallinity provides greater flexibility and high mechanical resistance for the encapsulant. The degree of crystallinity depends on the area under two melting peaks. The minimum requirement for the degree of crystallinity of the encapsulant should be lower than 20% [30].

The requirement of thermal stability is being stable under temperature equal or above 300 degrees Celsius. Defects in solar cells is the main reason for hot spot formation in a PV module. The temperature in the hotspot areas reaches up to 300 degrees Celsius [52]. The PV module should have its conventional performance at this temperature or above.

A higher adhesion strength has a two-sided effect on the PV module performance. First of all, it is a barrier to water/oxygen permeability and second it prevents delamination and bubble formation. It is also one of the established criterion for the durability of the PV module for long service lifetime up to 25-30. The average peel force (N) has been taken from 10 mm to 100 mm and divided with 1 cm peel sample width to obtain the final adhesion strength. The minimum required amount for the adhesion strength is 75 Newton per centimeter [30, 38].

As the last but not least requirement, it is relevant to the health, safety and environment (HSE) issues. Encapsulation materials should not have made any HSE issue during the fabrication process such as harmful chemical reaction or byproduct and throughout their longevity. As a transcend and futuristic criterion, it can be imposed recyclability on the encapsulation material which is the circular economic indicator for production companies or PV module producers. This criterion is in line with the carbon-neutral scenarios and meeting climate change initiatives [53, 54].

All these most important criteria and their minimum requirements have been tabulated in Table. 2. This multi-criteria decision making list makes sure both the reliability and durability of the PV modules. In order to implement this multi-criteria analysis method, we have evaluated it to compare three encapsulation materials including EVA, PVB, and TPO.

Table II: Most Important Criteria and Their Minimum Requirements (Multi-criteria Analysis Method)

Criteria	Minimum Requirement
Optical Transmittance (OT)	≥ 90%
Refractive Index (RI)	1.48 - 1.50
UV	280nm<UV $_{cut\text{-}off}$ ≤350nm
Volume Resistivity (VR)	$\geq 10^{12}$ Ω cm
WVTR	5gr/m^2-day
Melting Point Temperature (MPT)	≥ 85 °C
Glass Transition Temperature (GTT)	≤ -40°C
Modulus of Elasticity (ME)	≤50 MPa
Degree of Crystallinity (DC)	≤20%
Thermal Stability (TS)	≥300°C
Adhesion Strength (AS)	≥75 N/cm
HSE hazard	None

2.2 Multi-criteria analysis for comparison among three encapsulation materials

As the key objective of this paper, the paper tries to implement this multi-criteria analysis method. This method makes decision makers capable of comparing different encapsulation materials with each other to choose a suitable one for the PV module's encapsulant, especially a new non-EVA type encapsulant. Here, the paper is made a comparison among EVA, PVB, and TPO with the help of this method for assessing reliability and durability of them.

The first criterion that has demonstrated to have great differences among these three encapsulation materials is the volume resistivity. As mentioned in the 2-1 section, an encapsulation material with a higher volume resistivity is a suitable candidate for encapsulant layers. TPO has the greatest volume resistivity among the three encapsulation materials. TPO's volume resistivity is greater by two orders of magnitude from EVA and four orders of magnitude from PVB. Due to this factor along with no acid acetic formation, TPO is resistant to PID as one of the time-dependent failure modes [23, 34, 50].

EVA has relative high volume resistivity, while it decreases when temperature raises up. The other factors which have impacts on the PID failure are the acetic acid formation, the high temperature and high value of water vapor or humidity [23, 34]. The latter factor is compared by WVTR of each encapsulant. Then, only TPO has WVTR below 5gr/m^2-day. Therefore, EVA is also the wrong choice for encapsulating.

The low glass transition temperature resolves the brittlement and delamination failure of the encapsulant when PV modules experience an extremely low temperature in the installation location [30]. TPO has the lowest glass transition temperature among them [36]. The high melting transition temperature determines an encapsulation material undergoes no significant change during lamination in the fabrication process [38].

The melting point temperature of TPO is closer to the minimum requirement and its two melting peaks or its area under the melting transition is greater than EVA. Then, it is obvious that the degree of crystallinity of TPO is higher than EVA, as shown in Table. 3. It is noted that both the degree of crystallinity of TPO and EVA are lower than the minimum requirement. Although EVA has a lower degree of crystallinity compared to TPO, its crystallinity increases when exposed to the UV radiation [55]. It is worth mentioning that the low crystallinity of EVA provides greater flexibility at moderate temperatures.

TPO's lower Young's modulus is another reason in favor of having better performance and thermal behavior than EVA. While PVB has the lowest among them [36]. The thermal stability of EVA is below the requirement while TPO 's thermal stability is well above it [30]. It is important to note that one of the major roles of adding a cross-linking agent to EVA is going up the thermal stability of it. In spite of this fact, its thermal stability is lower than the required amount. This results in giving opportunity to hotspot formation in PV modules in hot climate zones.

A standard EVA has shown 80-100 Newton per centimeter adhesion strength while TPO has indicated close to 200 Newton per centimeter at the front encapsulant-glass and rear encapsulant-backsheet interfaces [30]. Hence, it puts up a protective barrier to mass transports (such as water vapor or oxygen) and delamination.

TPO without need to cross-linking processes eliminates all the cross-linking reactions and its harmful byproduct as well as reduces the lamination cycle time. As a matter of fact, TPO's lamination cycle time is significantly shorter than EVA. This feature gives three distinct advantages including more stable lamination, higher adhesion strength, and higher total fabrication in the production line [30, 38].

Removing these time and energy-consuming processes which have an impact on the final price and performance of the PV modules and reduce price-level or the Levelized Cost of Electricity (LCOE) of PV module technology. As the last but definitely not the least point, TPO is compatible with most modern recycling processes [56,57]. Hence, it is in well agreement with the circular economy initiative to tackle the increasing PV module waste generation issue in the forthcoming years.

Table 3: Multi-criteria Analysis of Different Encapsulation Materials

Criteria	EVA	PVB	TPO
OT	91%	91%	92%
RI (n)	1.48-1.49	1.48	1.48-1.49
UV*	350 nm	-	240
VR	1.0×10^{14} Ω cm	4.8×10^{12}	2.0×10^{16}
WVTR	34 gr/m^2-day	40.5	3.3
MPT	47 - 68°C	-	49 - 94
GTT	(- 40) - (-34) °C	12 - 20	(- 60) - (- 40)
ME	≤68 MPa	≤12	≤32
DC	4.6%	-	7.8%
TS	Stable up to 260°C	-	Stable up to 400
AS	80-100 N/cm	-	200
FT**	140 - 160°C	140 - 160	140 - 160
LCT***	13 -15 min	-	9 min
HSE	Acetic Acid	None	None

* UV absorption degradation has been obtained at 10% of the representative solar-weighted transmittance.

** FT: Fabrication Temperature
*** LCT: Lamination Cycle Time

TPO is a suitable candidate and acceptable alternative to EVA due to its having overwhelming superiority in the majority of aforementioned important criteria, Table. 3, which not only ensures the reliability and durability of the PV modules but also removes the barriers of EVA and operates at the optimal level when undergoing different reliability and field tests. Moreover, it meets the requirements of the Golden triangle. It is estimated that TPO will reduce the dominant share of EVA in the market and grab close to 20% share within in the next 10 years, however, other materials will have kept their low market shares for niche applications [9].

3 CONCLUSION

The PV module has been dominating technology for many years among all renewable energy technologies and extending its market share year on year in the energy sector. Then, the durability issue of PV modules is playing a crucial role to maintain this momentum.

More than 10% of the power loss in the PV modules is directly relevant to the failures in their encapsulants during their lifetimes. EVA as the dominant encapsulation material in the PV module segment with over 80% market share has suffered from several time-dependent failures such as discoloration, delamination, PID, and in some worse cases hotspots. These time-dependent failures are related to the intrinsic nature of EVA, its chemical composition, the environmental and extrinsic motivators. These failures seriously jeopardize the durability of the PV module.

To be needless to additives like as a cross-linking agent, UV absorber, and UV stabilizer in addition to no acetic acid formation makes TPO unique for being an encapsulation material. TPO not only has no harmful byproduct but also can remove the barriers of EVA and operate at the optimal level when undergoing different reliability and field tests.

This paper proposed a general multi-criteria analysis method for assessing the durability of three encapsulation materials including EVA, PVB, and TPO. This multi-criteria analysis method to a great extent ensures the reliability and durability issues associated with the encapsulation material of PV modules. It is also noted that this method has pointed out how PV module producers can switch from the linear economy to the circular economy. Based on the method, this study concluded that TPO is a suitable candidate and acceptable alternative to EVA. Most alternative encapsulation materials need chemical additives such as peroxide cross-linking agent and UV absorber to function properly and to avoid related time-dependent failures. While TPO without need for these time and energy-consuming processes which have impact on the final price and the performance of the PV modules during their service life, to high-level ensures the durability of PV modules. Therefore, TPO meets all three pillars of the Golden triangle for commercializing in the PV module market.

4 REFERENCES

[1] Future of Solar Photovoltaic, IRENA report, 2019.
[2] Soloot, H. E. H., Agheb, E., Soloot, A. H., & Moghadam, S. (2020, December). A SWOT Analysis of Two Protection Strategies Due to the Expansion of Renewable Distributed Generation on Distribution Network. In 2020 15th International Conference on Protection and Automation of Power Systems (IPAPS) (pp. 49-52). IEEE.
[3] Soloot, H. E. H., & Beyrami, H. The Floating Energy Hub Based on Multi-renewable Energy Systems Meets the Energy Trilemma Objectives.
[4] Renewable capacity statistics 2021 International Renewable Energy Agency, IRENA report, Abu Dhabi, 2021.
[5] Oreski, G., Omazic, A., Eder, G. C., Voronko, Y., Neumaier, L., Mühleisen, W., ... & Edler, M. (2020). Properties and degradation behaviour of polyolefin encapsulants for photovoltaic modules. Progress in Photovoltaics: Research and Applications, 28(12), 1277-1288.
[6] Ishii, T., & Masuda, A. (2017). Annual degradation rates of recent crystalline silicon photovoltaic modules. Progress in Photovoltaics: Research and Applications, 25(12), 953-967.
[7] Polman, A., Knight, M., Garnett, E. C., Ehrler, B., & Sinke, W. C. (2016). Photovoltaic materials: Present efficiencies and future challenges. Science, 352(6283).
[8] Fraunhofer Institute for Solar Energy Systems, "Photovoltaic report," 2020, available: https://www.ise.fraunhofer.de/content/dam/ise/de/documents/publications/studies/Photovoltaics-Report.pdf
[9] International Technology Roadmap for Photovoltaic (ITRPV), 2020 Results, 2021.
[10] Meng, L., You, J., & Yang, Y. (2018). Addressing the stability issue of perovskite solar cells for commercial applications. Nature communications, 9(1), 1-4.
[11] Kurtz, S., Sample, T., Wohlgemuth, J., Zhou, W., Bosco, N., Althaus, J., ... & Kondo, M. (2015, June). Moving toward quantifying reliability-the next step in a rapidly maturing PV industry. In 2015 IEEE 42nd Photovoltaic Specialist Conference (PVSC) (pp. 1-8). IEEE.
[12] Osterwald, C. R., & McMahon, T. J. (2009). History of accelerated and qualification testing of terrestrial photovoltaic modules: A literature review. Progress in Photovoltaics: Research and Applications, 17(1), 11-33.
[13] International Electrotechnical Commission, Terrestrial photovoltaic (PV) modules—design qualification and type approval, IEC 61215 series Ed. 1.0, 2016.
[14] International Electrotechnical Commission, Photovoltaic (PV) module safety qualification, IEC 61730 series Ed. 2.0, 2016.
[15] Yang, H., Wang, H., Cao, D., Sun, D., & Ju, X. (2015). Analysis of power loss for crystalline silicon solar module during the course of encapsulation. International Journal of Photoenergy, 2015.
[16] Mansour, D. E., Barretta, C., Pitta Bauermann, L., Oreski, G., Schueler, A., Philipp, D., & Gebhardt, P. (2020). Effect of Backsheet Properties on PV Encapsulant Degradation during Combined Accelerated Aging Tests. Sustainability, 12(12), 5208.
[17] Yeo, J. S., & Jeong, Y. (2020). Pathway toward market entry of perovskite solar cells: A detailed study on the research trends and collaboration networks through bibliometrics. Energy Reports, 6, 2075-2085.
[18] Hasan, O., & Arif, A. F. M. (2014). Performance

and life prediction model for photovoltaic modules: Effect of encapsulant constitutive behavior. Solar Energy Materials and Solar Cells, 122, 75-87.
[19] Luo, L., Liu, P., Zhang, K., Tang, G., Hou, H., Li, B. G., & Wang, W. J. (2020). Vinyl-Functionalized Polyolefins for Fast Photovoltaic Cell Encapsulation. ACS Applied Polymer Materials, 2(7), 2571-2577.
[20] Yang, H. E., French, R., & Bruckman, L. (Eds.). (2019). Durability and Reliability of Polymers and Other Materials in Photovoltaic Modules. William Andrew.
[21] Agroui, K., Jaunich, M., & Arab, A. H. (2016). Analysis techniques of polymeric encapsulant materials for photovoltaic modules: Situation and perspectives. Energy Procedia, 93, 203-210.
[22] Czanderna, A. W., & Pern, F. J. (1996). Encapsulation of PV modules using ethylene vinyl acetate copolymer as a pottant: A critical review. Solar energy materials and solar cells, 43(2), 101-181.
[23] Omazic, A., Oreski, G., Halwachs, M., Eder, G. C., Hirschl, C., Neumaier, L., ... & Erceg, M. (2019). Relation between degradation of polymeric components in crystalline silicon PV module and climatic conditions: A literature review. Solar energy materials and solar cells, 192, 123-133.
[24] Köntges, M., Kurtz, S., Packard, C. E., Jahn, U., Berger, K. A., Kato, K., ... & Friesen, G. (2014). Review of failures of photovoltaic modules.
[25] Hsu, H. Y., Hsieh, H. H., Tuan, H. Y., & Hwang, J. L. (2010). Oxidized low density polyethylene: A potential cost-effective, stable, and recyclable polymeric encapsulant for photovoltaic modules. Solar energy materials and solar cells, 94(6), 955-959.
[26] Kempe, M. D., Jorgensen, G. J., Terwilliger, K. M., McMahon, T. J., Kennedy, C. E., & Borek, T. T. (2007). Acetic acid production and glass transition concerns with ethylene-vinyl acetate used in photovoltaic devices. Solar energy materials and solar cells, 91(4), 315-329.
[27] Klemchuk, P., Ezrin, M., Lavigne, G., Holley, W., Galica, J., & Agro, S. (1997). Investigation of the degradation and stabilization of EVA-based encapsulant in field-aged solar energy modules. Polymer degradation and stability, 55(3), 347-365.
[28] Cattaneo, G., Faes, A., Li, H. Y., Galliano, F., Gragert, M., Yao, Y., ... & Perret-Aebi, L. E. (2015). Lamination process and encapsulation materials for glass-glass PV module design. Photovoltaics International, 27, 1-8.
[29] de Oliveira, M. C. C., Cardoso, A. S. A. D., Viana, M. M., & Lins, V. D. F. C. (2018). The causes and effects of degradation of encapsulant ethylene vinyl acetate copolymer (EVA) in crystalline silicon photovoltaic modules: A review. Renewable and Sustainable Energy Reviews, 81, 2299-2317.
[30] Adothu, B., Bhatt, P., Chattopadhyay, S., Zele, S., Oderkerk, J., Sagar, H. P., ... & Mallick, S. (2019). Newly developed thermoplastic polyolefin encapsulant–A potential candidate for crystalline silicon photovoltaic modules encapsulation. Solar Energy, 194, 581-588.
[31] Adothu, B., Mallick, S., & Kartikay, P. (2019, June). Determination of Crystallinity, Composition, and Thermal stability of Ethylene Vinyl Acetate Encapsulant used for PV Module Lamination. In 2019 IEEE 46th Photovoltaic Specialists Conference (PVSC) (pp. 0491-0494). IEEE.
[32] Swonke, T., & Auer, R. (2009, August). Impact of moisture on PV module encapsulants. In Reliability of Photovoltaic Cells, Modules, Components, and Systems II (Vol. 7412, p. 74120A). International Society for Optics and Photonics.
[33] Kaplani, E. (2012). Detection of degradation effects in field-aged c-Si solar cells through IR thermography and digital image processing. International Journal of Photoenergy, 2012.
[34] López-Escalante, M. C., Caballero, L. J., Martín, F., Gabás, M., Cuevas, A., & Ramos-Barrado, J. R. (2016). Polyolefin as PID-resistant encapsulant material in PV modules. Solar Energy Materials and Solar Cells, 144, 691-699.
[35] Kempe, M. D., Miller, D. C., Wohlgemuth, J. H., Kurtz, S. R., Moseley, J. M., Shah, Q. A., ... & Vanderpan, C. E. (2015). Field testing of thermoplastic encapsulants in high-temperature installations. Energy Science & Engineering, 3(6), 565-580.
[36] Peike, C., Hädrich, I., Weiß, K. A., Dürr, I., & Ise, F. (2013). Overview of PV module encapsulation materials. Photovoltaics International, 19, 85-92.
[37] Pern, F. J., & Czanderna, A. W. (1992, December). EVA degradation mechanisms simulating those in PV modules. In AIP Conference Proceedings (Vol. 268, No. 1, pp. 445-452). American Institute of Physics.
[38] Adothu, B., Bhatt, P., Zele, S., Oderkerk, J., Costa, F. R., & Mallick, S. (2020). Investigation of newly developed thermoplastic polyolefin encapsulant principle properties for the c-Si PV module application. Materials Chemistry and Physics, 243, 122660.
[39] Jaunich, M., Böhning, M., Braun, U., Teteris, G., & Stark, W. (2016). Investigation of the curing state of ethylene/vinyl acetate copolymer (EVA) for photovoltaic applications by gel content determination, rheology, DSC and FTIR. Polymer Testing, 52, 133-140.
[40] Jentsch, A., Eichhorn, K. J., & Voit, B. (2015). Influence of typical stabilizers on the aging behavior of EVA foils for photovoltaic applications during artificial UV-weathering. Polymer Testing, 44, 242-247.
[41] Adothu, B., Bhatt, P., Kartikay, P., Zele, S., Costa, F. R., Oderkerk, J., & Mallick, S. (2019, June). Determination of Crystallinity and Thermal Stability of Newly Developed Thermoplastic Polyolefin Encapsulant for the c-Si PV Module Application. In 2019 IEEE 46th Photovoltaic Specialists Conference (PVSC) (pp. 0487-0490). IEEE.
[42] McIntosh, K. R., Cotsell, J. N., Cumpston, J. S., Norris, A. W., Powell, N. E., & Ketola, B. M. (2009, June). An optical comparison of silicone and EVA encapsulants for conventional silicon PV modules: A ray-tracing study. In 2009 34th IEEE Photovoltaic Specialists Conference (PVSC) (pp. 000544-000549). IEEE.
[43] Saw, M. H., Khoo, Y. S., Singh, J. P., & Wang, Y. (2017). Cell-to-module optical loss/gain analysis for various photovoltaic module materials through systematic characterization. Japanese Journal of Applied Physics, 56(8S2), 08MD03.
[44] Saw, M. H., Dong, P., Singh, J. P., Liang, T. S., Chai, J., Wei, Y., ... & Khoo, Y. S. (2018, June). Potential of white EVA as PV Encapsulant: Performance and reliability Study. In 2018 IEEE 7th World Conference on Photovoltaic Energy Conversion (WCPEC)(A Joint Conference of 45th IEEE PVSC, 28th PVSEC & 34th EU PVSEC) (pp. 2832-2837). IEEE.
[45] López-Escalante, M. C., Fernández-Rodríguez, M., Caballero, L. J., Martín, F., Gabás, M., & Ramos-Barrado, J. R. (2018). Novel encapsulant architecture on the road to photovoltaic module power output increase. Applied Energy, 228, 1901-1910.

[46] Gong, H., Wang, G., Zheng, L., & Gao, M. (2015). Reliability and durability impact of high UV transmission EVA for PV modules. In 31st European Photovoltaic Solar Energy Conference and Exhibition (EUPVSEC) (pp. 2061-2065).
[47] Agroui, K., & Collins, G. (2014). Determination of thermal properties of crosslinked EVA encapsulant material in outdoor exposure by TSC and DSC methods. Renewable energy, 63, 741-746.
[48] Hanifi, H., Pfau, C., Dassler, D., Schneider, J., Schindler, S., Turek, M., & Bagdahn, J. (2016). Investigation of cell-to-module (CTM) ratios of PV modules by analysis of loss and gain mechanisms. Photovolt. Int, 32, 89-99.
[49] Dupré, O., Levrat, J., Champliaud, J., Despeisse, M., Boccard, M., & Ballif, C. (2018, August). Reassessment of cell to module gains and losses: Accounting for the current boost specific to cells located on the edges. In AIP Conference Proceedings (Vol. 1999, No. 1, p. 090001). AIP Publishing LLC.
[50] Köntges, M., Oreski, G., Jahn, U., Herz, M., Hacke, P., Weiss, K. A., ... & French, R. H. (2017). Assessment of photovoltaic module failures in the field. IEA PVPS Task, 13.
[51] Agroui, K., Collins, G., & Farenc, J. (2012). Measurement of glass transition temperature of crosslinked EVA encapsulant by thermal analysis for photovoltaic application. Renewable Energy, 43, 218-223.
[52] Deng, S., Zhang, Z., Ju, C., Dong, J., Xia, Z., Yan, X., ... & Xing, G. (2017). Research on hot spot risk for high-efficiency solar module. Energy Procedia, 130, 77-86.
[53] Sica, D., Malandrino, O., Supino, S., Testa, M., & Lucchetti, M. C. (2018). Management of end-of-life photovoltaic panels as a step towards a circular economy. Renewable and Sustainable Energy Reviews, 82, 2934-2945.
[54] Farrell, C. C., Osman, A. I., Doherty, R., Saad, M., Zhang, X., Murphy, A., ... & Rooney, D. W. (2020). Technical challenges and opportunities in realising a circular economy for waste photovoltaic modules. Renewable and Sustainable Energy Reviews, 128, 109911.
[55] Ottersböck, B., Oreski, G., & Pinter, G. (2017). Comparison of different microclimate effects on the aging behavior of encapsulation materials used in photovoltaic modules. Polymer Degradation and Stability, 138, 182-191.
[56] Mülhaupt, R. (1999). Ziegler-Natta catalysis and propylene polymerization. In Polypropylene (pp. 896-920). Springer, Dordrecht.
[57] Omazic, A., Oreski, G., Edler, M., Eder, G. C., Hirschl, C., Pinter, G., & Erceg, M. (2020). Increased reliability of modified polyolefin backsheet over commonly used polyester backsheets for crystalline PV modules. Journal of Applied Polymer Science, 137(30), 48899.

THERMAL CYCLE ANALYSIS ON SHINGLED GLASS – GLASS SAMPLES WITH SHJ CELLS

Stefan Wendlandt[1], Carolyn Carriere[2], Vincent Barth[2], Samuel Harrison[2] and Armand Bettinelli[2]

[1] PI Photovoltaik-Institut Berlin AG, Wrangelstr. 100, D-10997 Berlin, Germany
Phone: +49 (30) 814 52 64-0, e-mail: wendlandt@pi-berlin.com
[2] CEA-LITEN, INES, 50 avenue du Lac Léman, F-73375 Le Bourget-du-Lac, France

ABSTRACT:

For years photovoltaic (PV) modules were manufactured with 60 or 72 cells in glass / back sheet designs. Since R&D teams around the world simulated the potential to increase the module efficiency by modifying the designs, the industry develops new production machines to realize those. One such novel approach is the shingling – here the wafer is cut into small stripes. In a module, the stripes are overlapping each other at the ends and an ECA interconnects two cells together. To study the thermal long-term stability, we build mini-modules in various designs with shingled SHJ cells and applied an extended thermal stress sequence (3x IEC) to those samples. We analysed the ageing process with image-based and electrical setups.

Keywords: stability, shingled, heterojunction

1 INTRODUCTION

Since wafer-based solar cells become less expensive it makes sense to cover the pv-module surface full of those and reach high cell-to-module (CTM) [1] values. For that singled approach the wafers have to cut into smaller piece. Typical cut shingled pieces are 1/5 or 1/6. Shingled cells were not soldered – this are connected with an Electrically Conductive Adhesives (ECA) [2]. It has to be conductive and mechanical flexible at the same time. It can be assumed that such a connection is susceptible to thermo-mechanical stresses, therefor we decided to laminate our samples in a symmetric glass-glass design in order to protect the solar cell in the neutral axis from tensile and compressive stresses. [3]

2 EXPERIMENTAL

In this paper we studied the thermal stability of shingled single heterojunction samples laminated in glass – glass design. The layout of such a sample is shown in Figure 1 & 2.

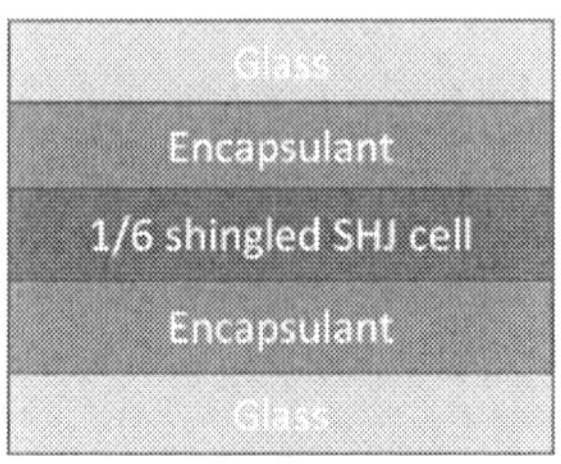

Figure 1: Design of the shingled glass – glass SHJ mini modules.

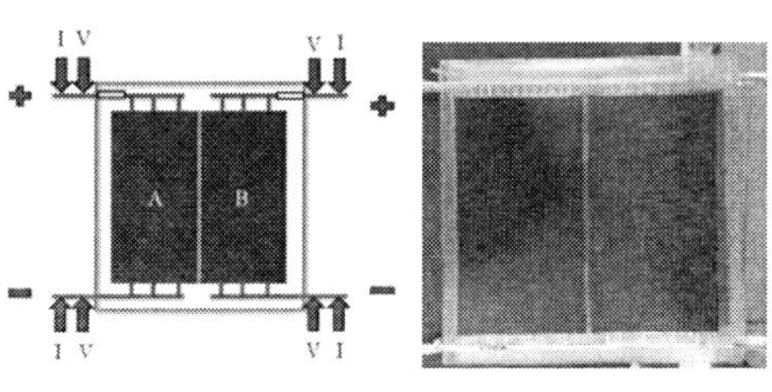

Figure 2: Layout of a shingled glass – glass SHJ mini modules (left: electrical layout; right: realized sample).

Each mini module is configured into two shingle strings independent gathered in one mini module. Per string there are twelve 1/6 cutted shingled SHJ cells in series connected. The samples vary into the cell thickness and the cell geometry (full and pseudo square (PSQ)). A detailed sample overview is displayed in table 1.

Table 1: Sample overview

Cell thickness µm	Module Nb	String name	Type	Batch
120	037 039	**Highlite 037BH*** Highlite 039AH	Full PSQ	H
120	040 041	Highlite 040A ; Highlite 040B Highlite 041A ; Highlite 041B	Full PSQ	I
120	042 044	Highlite 042A ; **Highlite 042B** Highlite 044A ; Highlite 044B	Full PSQ	J
120	046 047	Highlite 046A; **Highlite 046B** Highlite 047A ; Highlite 047B	Full PSQ	K
120	048 037 039	Highlite 048A ; Highlite 048B; Highlite 037AL Highlite 039BL	Full Full PSQ	L
120	032	Highlite 032A ; Highlite 032B	Full	F
TOTAL :	10			
Cell thickness µm	**Module Nb**	**String name**	**Type**	**Batch**
160	013	Highlite 013A ; Highlite 013B	Full	A
160	018	Highlite 018A ; Highlite 018B	Full	B
160	024	Highlite 024A ; Highlite 024B	Full	C
160	028	Highlite 028A; Highlite 028B	PSQ	D
160	029	Highlite 029A ; Highlite 029B;	Full	E
160	035 036	Highlite 035A ; Highlite 035B Highlite 036A ; **Highlite 036B**	Full	G
TOTAL :	7			

In total 17 mini modules were manufactured to which we applied the test sequence shown in Figure 3.

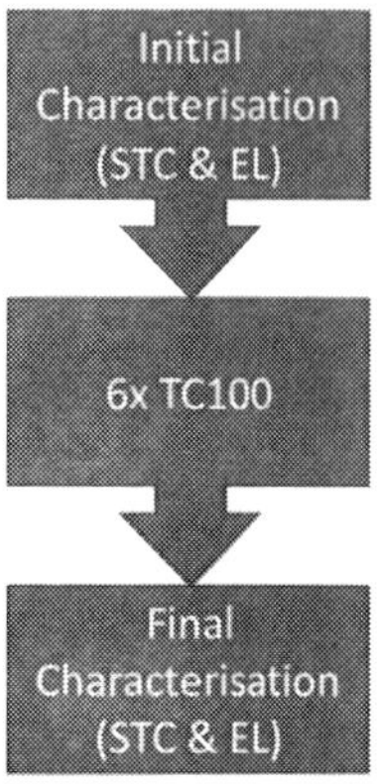

Figure 3: Applied test sequence.

Initially we carried out IV characterization at STC and electroluminescence (EL) imaging, followed by 6x TC 100 (according to the IEC 61215). After each step of TC 100, intermediate IV characteristics and EL images were taken. The full test sequence ends with the final IV and EL characterization. Upon the EL imaging, due to the strong emitting NIR signal of the SHJ cells, the injected current was set to 0.07x Isc.

3 RESULTS

To have more statistics on that box plots diagrams were plotted for the 120 µm and 160 µm solar cells. Figure 3 shows the box plot diagram of the relative to initial Pmax degradation.

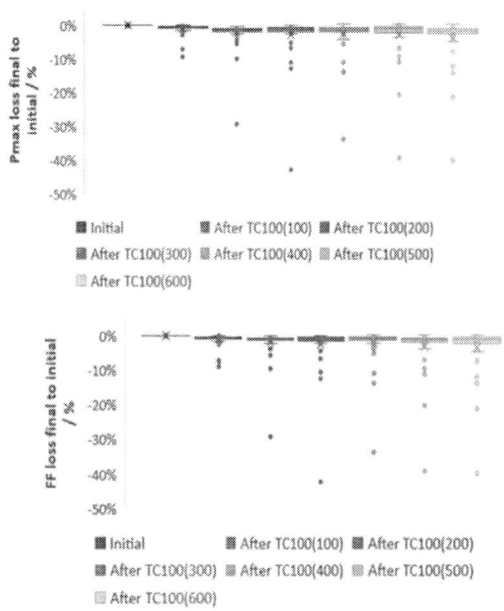

Figure 3: Degradation in Pmax (top) and FF (bottom) of all samples.

Some samples show a strong initial degradation (Ploss >5% after 1x TC100). While Isc and Voc are stable (not shown here), the FF degrades as well. Therefore, it seems those samples have an interconnecting problem. A deeper analysis shows there is a interconnecting problem between the first/last cell in a string and the ribbon. To have more statistics on that box plots diagrams were plotted for the 120 µm and 160 µm solar cells.

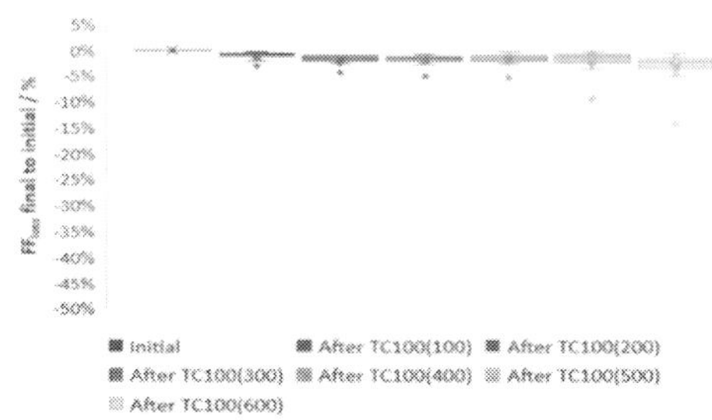

Figure 4: Pmax Degradation of the samples with 160 µm.

The Samples (n = 16) with wafer thickness of 160µm showed 6x TC100 an average Ploss relativ to initial of -3.2% (-2.4% without the samples with initial degradation).

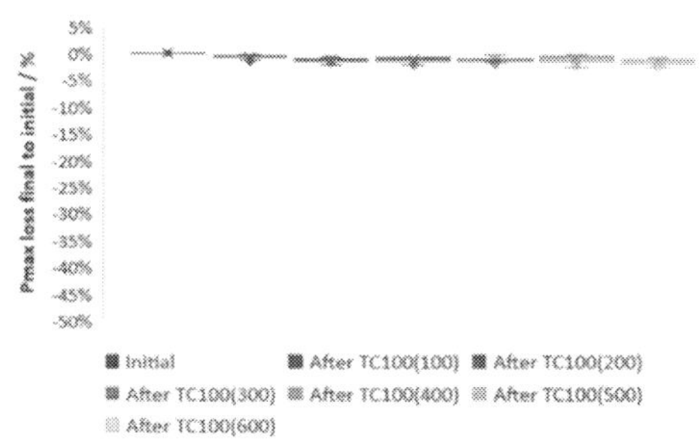

Figure 5: Pmax Degradation of the samples with 120 µm.

The Samples (n = 16) with a wafer thickness of 120µm showed after 6x TC100 average Ploss relativ to initial of -1.5% (without the strong initial degraded samples).

The electroluminescence and electrical data for four samples at different TC stages are presented here.

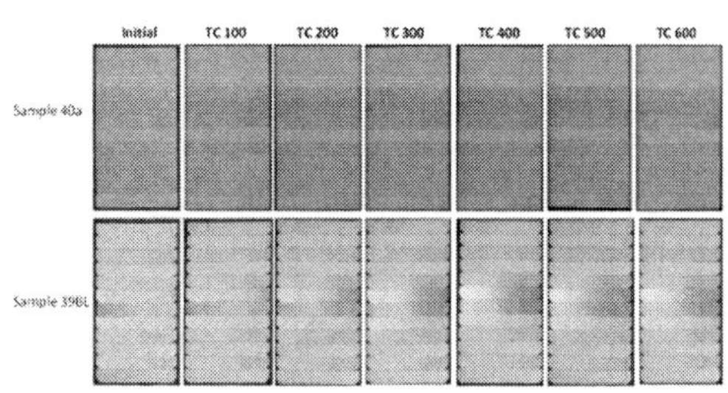

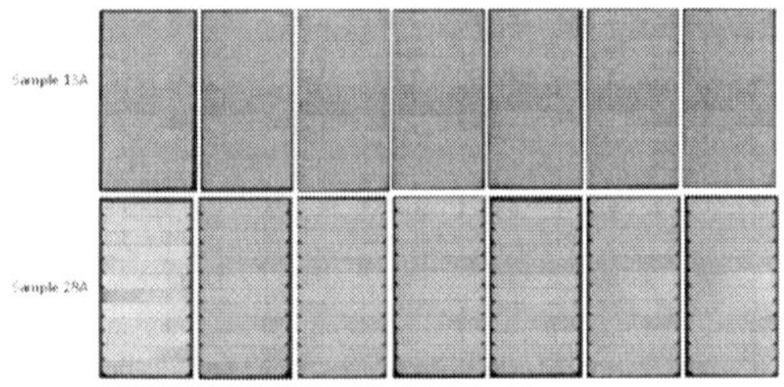

Figure 6: Electroluminescence images of the different samples at different test levels.

Except sample 39BL no major additional changes to the initial state of the modules are to report. On 39BL an interconnection problem becomes visible.

The relative changes of the electrical data from the same samples are shown next.

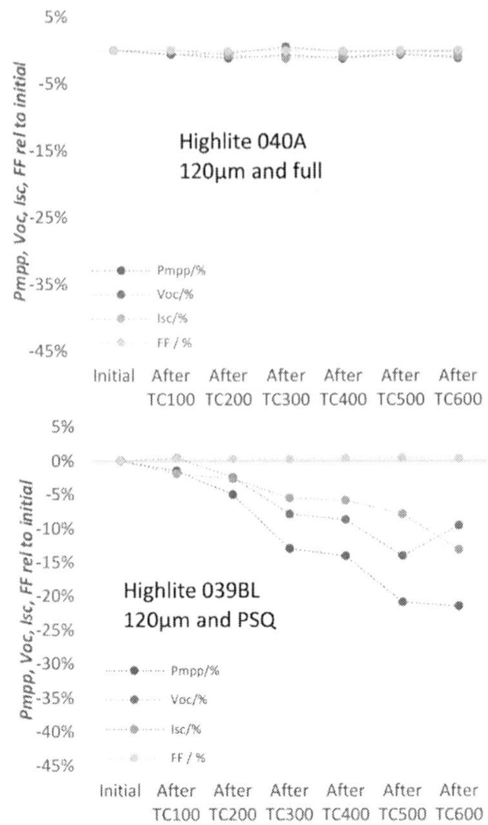

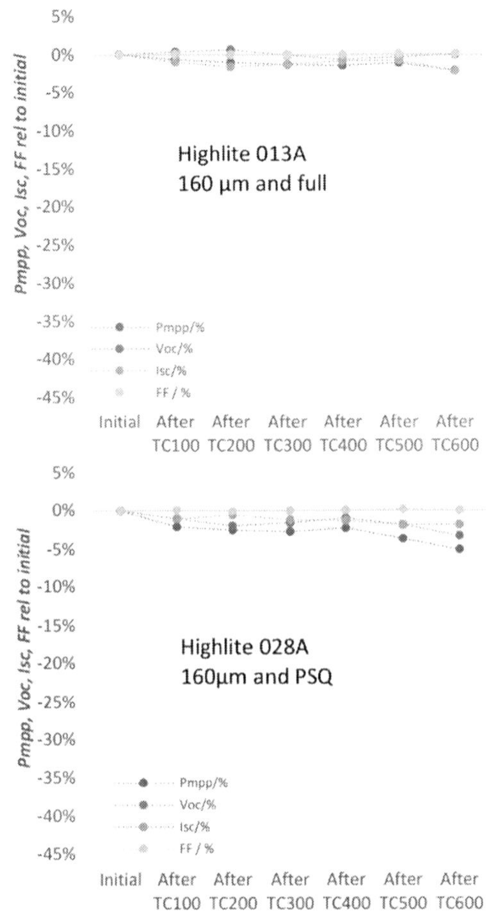

Figure 7: Degradation of the electrical data due to temperature cycle stress of the different samples.

Sample 39BL shows the largest degradation, followed by sample 28A. Both samples are PSQ but with different cell thicknesses. Sample 39BL cells are 120 µm thick and sample 28A cells are 160 µm.

4 SUMMARY

We have designed and thermal cycle tested glass - glass mini modules with shingled SHJ solar cells. The main variation of the solar cells was the wafer thickness. Batch 1 had 160µm and batch 2 had 120µm. We applied six cycles of TC-Test (-40°C ... +85°C) without current injection to those. That activity was assisted by initial, intermediate, and final STC and EL analyses.

We noticed at same samples an initial degradation in Pmax and FF. A deeper investigation made sure that at those was an interconnection issue at the first/last cell and the copper ribbon, while the shingled cell to cell connection was properly functioning.

After completing 6x TC100 we noticed that both thicknesses (160µm and 120µm) showed a comparable degradation without taking the samples with the initial degradation into account. In detail the 160µm samples showed an average in Ploss of -2.4% (without the samples with initial degradation) and the 120µm samples showed an average in Ploss of of -1.5% (without the strong initial degraded samples). At that point it can summarized that no higher risk on thinner solar cells for application at shingled modules in glass – glass design could be determined. Furthermore, we also observed at samples with PSQ a stronger degradation which isn't fully understood yet.

5 ACKNOWLEDGEMENTS

This study was supported by the European Union's Horizon 2020 Programme for research, technological development and demonstration under Grant Agreement number 857793 (Project name: HighLite).

6 REFERENCES

[1] A. Mondon, N. Klasen, E. Fokuhl, M. Mittag, M. Heinrich, H. Wirth: „*Comparison of Layouts for Shingled Bifacial PV Modules in Terms of Power Output, Cell-to-Module Ratio and Bifaciality*", 35th European Photovoltaic Solar Energy Conference and Exhibition

[2] J. W. Baik *et al.*:" Reliability of the industrial shingling module by ECA characteristics", 37th European Photovoltaic Solar Energy Conference and Exhibition

[3] S. Wendlandt, M. Gebrelul, S. Heller, P. Dörder, L. Podlowski: „*Characterization and Long Term Stability Analysis at Photovoltaic Modules with Shingled Cell Strings*", 37th European Photovoltaic Solar Energy Conference and Exhibition

ENABLING MEASUREMENT OF PV MODULE CURVATURE: TOWARDS CHARACTERIZATION OF THERMOMECHANICAL RESIDUAL STRESSES

Ichrak Rahmoun[1,2], Tristan Le Carre[1], Bertrand Chambion[1], Eeva Mofakhami[1], Aude Derrier[1], Jean-Luc Bouvard[2] and Pierre-Olivier Bouchard[2]
[1] Univ. Grenoble Alpes, CEA, Liten, INES, ITE INES.2S, 73375 Le Bourget du Lac, France.
[2] CEMEF Mines ParisTech, PSL Research University, CNRS UMR 7635, CS 10207, 06904 Sophia-Antipolis Cedex, France.

ABSTRACT: Assessment of the effect of the lamination process parameters on the reliability of PV modules is a key step to extend their lifetime. In this work, a new method of measurement of the deflection PV modules is presented. This method consists in mapping the surface of both sides of a PV module positioned on 4-point supports. The measurement repetition on both sides aims at compensating for the gravity effect induced by the self-weight of the PV module. The position of the supports is analytically optimized. The error induced by this protocol of measurement is numerically estimated.
An experimental measurement campaign is carried out to characterize the impact of lamination process parameters on the residual deflection of the PV modules. First, twelve glass plates are tested. The results showed that their initial shape is not flat with asymmetric deflection. Then, the PV modules are manufactured using the same glass plates while varying architecture, lamination temperature, and cooling type. For glass-backsheet architecture, the deflection decreases when the lamination temperature increases, which is not observed in glass-glass architecture. The type of cooling has also an effect, as the deflection of ambient air-cooled PV modules is higher than that of cooling press-cooled PV modules.

Keywords: Curvature measurement, Thermomechanical stress, PV module architecture, PV module manufacturing, FEM.

1 INTRODUCTION

Thermomechanical residual stresses have a degrading effect on the reliability and lifetime of photovoltaic (PV) modules. Measuring these residual stresses is a key step in understanding the contribution of certain process parameters in their forming stage. However, residual stress measurement methods are limited. Tippabhotla et al. suggested a synchrotron X-ray micro-diffraction method to evaluate residual stresses in solar cells surface in PV module after interconnection and lamination processes [1]. Beinert et al. provided a measurement of thermomechanical stresses in solar cells and PV module by confocal micro-Raman spectroscopy [2]. These methods do not allow a global measurement of the deflection of the PV module.

The post-lamination thermomechanical residual stresses are translated into a residual curvature of the PV module, which is linked to deflection. In this work, we present a new method of measuring the global deflection of PV module. Then, this method is validated numerically and experimentally.

Indeed, the deflection (ω_x,ω_y) of PV module can be measured experimentally using distance sensors (laser or mechanical sensor). Moreover, local orientations (φ_x, φ_y) can be calculated directly using the following Eq. (1) as shown in Fig. 1:

$$\varphi_x = \frac{\partial \omega}{\partial y}; \quad \varphi_y = \frac{\partial \omega}{\partial x} \quad (1)$$

Then, the variation of local orientation along principal directions is used to calculate local curvature (κ_{xx},κ_{yy}) in the associated coordinate system as follow:

$$\kappa_{xx} = \frac{\partial \varphi_x}{\partial y}; \quad \kappa_{yy} = \frac{\partial \varphi_y}{\partial x} \quad (2)$$

Several assumptions, similar to Kirchhoff-Love [3] thin plate model can be used to estimate bending strains at the surface of a PV module of thickness t from local curvature:

$$\varepsilon_{xx} = \pm\frac{t}{2}\kappa_{xx}; \quad \varepsilon_{yy} = \pm\frac{t}{2}\kappa_{yy} \quad (3)$$

In this document, we describe the experimental protocol developed for the estimation of PV module residual deflection. The measurement deviation induced by this protocol has been estimated numerically. Afterward, a first experimental validation of the protocol was carried out on a glass plate and several PV module architectures. Then, an experimental measurement campaign was realized using the developed deflection measurement protocol to study the influence of certain parameters of the lamination process such as temperature and cooling type.

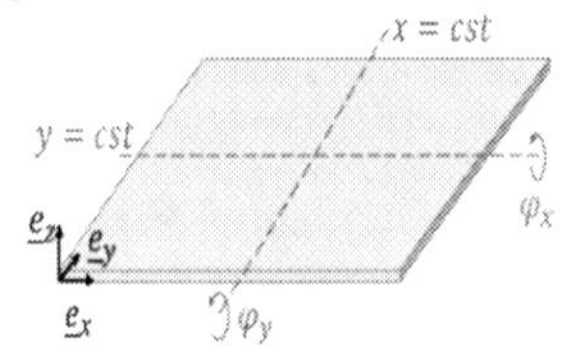

(a) Representation of coordinate system and local rotations

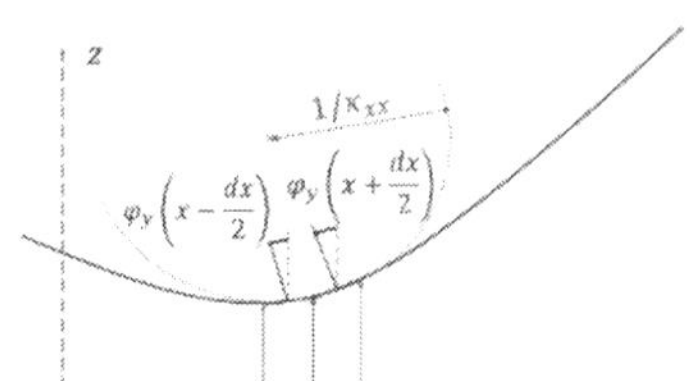

(b) Calculation of plate curvature from deflection measure, considering cross-section in (Oxz) plane

Figure 1: Geometric notation of the problem

2 DEFINITION OF THE EXPERIMENTAL PROTOCOL

2.1 Compensation of gravity effect

The main issue is to find a method to measure the

deflection without constraining the PV module shape. Indeed, placing it directly on a plane support/table might impose an artificial flatness, while putting it on the edge would constrain the edge shape.
The solution we suggest is to put the PV module on four one-point supports, distributed according to the laminate symmetries. This distribution of supports shall not constrain the measures, assuming the residual deflection to be symmetric. However, gravity effects occur due to self-weight of PV module. The additional deflection induced by gravity might even be higher than the residual deflection we are trying to measure, if the protocol is not carefully thought out.

This method is inspired from monocrystalline silicon wafers measurement to eliminate the gravity issue. Indeed, SEMI standards [4] suggest repeating the measure on both sides of the wafer in order to compensate its self-weight (see Fig. 2).

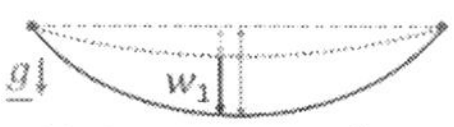

(a) Measurement on side 1

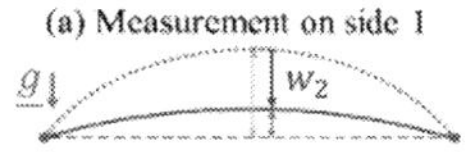

(b) Measurement on side 2

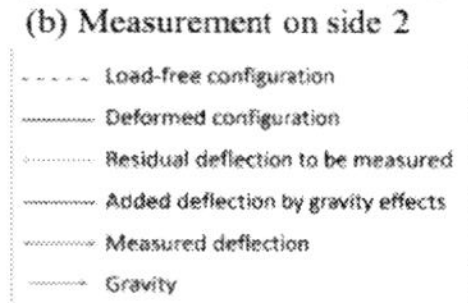

Figure 2: Schematic explanation of the experimental protocol used to compensate gravity effects

2.2 Supports position optimization

Gravitational induced deflection strongly depends on the supports position. The analytic study is reduced to a one-dimension optimization of the support position, along a beam. This simplification offers a right estimation with much easier calculations than a two-dimension optimization on a plate. The deformation in the transverse direction can be neglected, as the PV module is a thin plate. Thus, in the 1D study the one-point support is a simple support.

We consider a beam of length $2l$, placed on two supports at an equal distance a from the center of the beam (Fig. 3a). The beam is assumed homogeneous, linearly elastic with a Young modulus E, a moment of inertia I and a linear density λ. The structure undergoes its self-weight, equivalent to a homogeneous linear loading $-\lambda g e_y$, where g is the gravitational acceleration. The vector e_x denotes the direction of the beam and the origin $x = 0$ is set in its center.

The application of Euler-Bernoulli theory allows neglecting the shear deformation. Normal deformation of the beam is also neglected in front of bending. Furthermore, we apply small strains assumption. The boundary condition along the symmetry plane is a sliding support with imposed orthogonality.

In the deformed configuration (Fig. 3b), the deflection at x-axis is denoted by $\omega(x)$. Particularly, the deflection of interest will be the one in the center of the beam $\omega_C = \omega(x = 0)$ and at the end $\omega_E = \omega(x = l)$.

Castigliano's second theorem is used to calculate directly the deflection at the center and at the end of the beam:

$$\omega_C = \frac{\lambda g a^2}{4EI}\left[\frac{5a^2}{6} - (l-a)^2\right] \quad (4)$$

$$\omega_E = \frac{\lambda g (l-a)}{2EI}\left[\frac{(l-a)^3}{4} + a(l-a)^2 - \frac{2a^3}{3}\right] \quad (5)$$

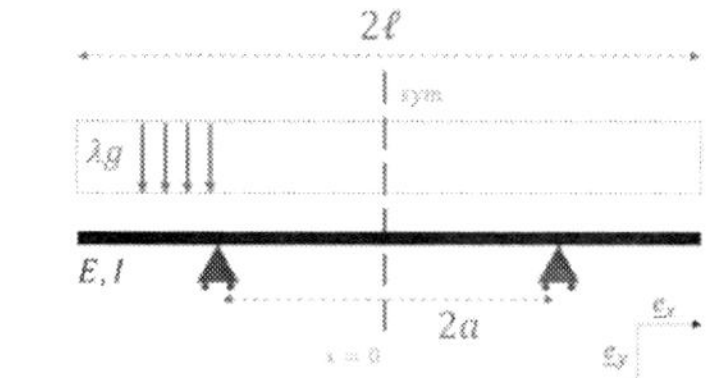

(a) The beam (in black) is placed on two sliding supports (in blue), submitted to self-weight loading (in orange)

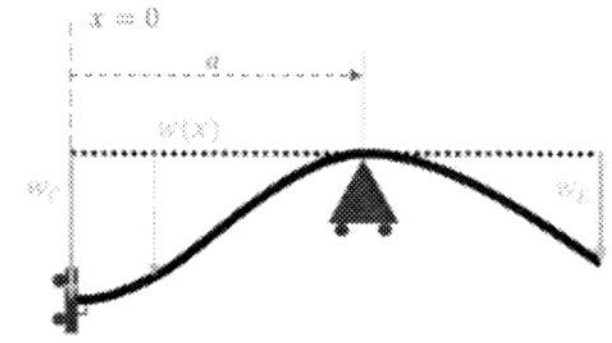

(b) Deflection notations (in yellow), describing displacement from initial to deformed configuration (respectively dashed and continuous line)

Figure 3: Schematic representation of the problem: (a) loading case and (b) notations in deformed configuration

Several cases have been considered for the support position, corresponding to different optimization criteria (Fig. 4):

- **Case 0:** This reference case corresponds to the choice of putting the support at the end of the beam ($a = l$). This choice maximizes the deflection in the center ω_C, which could be minimized.
- **Case 1:** The second option would be to minimize the deflection in the center with respect to a, where $\partial \omega_C / \partial a = 0$. However, this minimum corresponds to a negative value of deflection in the center, while the deflection at the end of the beam ω_E increases. It occurs with a support around the third of the length: $a = 0.347l$.
- **Case 2:** In order to avoid the reverse deflection in the center, another criterion would be to minimize the absolute value of ω_C, resulting in $a = 0.523l$. However, if this choice ensures a zero deflection at the center of the beam, the deflection at the end of the beam is still not taken into consideration.
- **Case 3:** Finally, the most general criteria would be to minimize the maximum deflection along the beam. This case can be expressed mathematically, by defining the optimal position as follows:
$$a_{opt} = \underset{a \in [0,l]}{\operatorname{argmin}}\left(\max_{x \in [0,l]} \omega(x)\right) \quad (6)$$
The extreme deflection of the beam is reached either in its center or at its end, depending on the support position. Thus, the optimal value according to this criterion will be respected when $\omega_C = \omega_E$, giving $a = 0.554l$.

Case 3 was selected, as it is the criteria that minimize the maximum value of deflection along the whole beam.

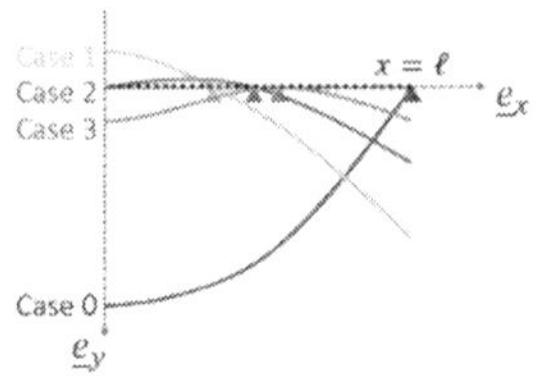

Figure 4: Undeformed (in black) and deformed shape of the beam for different cases of support position, corresponding to a choice of optimization criteria

3 NUMERICAL VALIDATION OF EXPERIMENTAL PROTOCOL

Theoretically, it is assumed that the effect of gravity will be identical on both sides. Nevertheless, the validity of this assumption depends on several parameters including the effect of the structure especially when the PV module has an asymmetrical architecture.

A 2D Finite Element (FE) model is defined using Abaqus Software in order to validate the experimental protocol. We assimilate the PV module to a $1m \times 2m$ glass shell plate, as it is the thickest material of the laminate. A constant curvature κ_{xx} is imposed along direction e_x. The curvature is null along the transverse direction e_y (see Fig. 5). This arbitrary curvature is meant to represent the residual deformation of PV modules after lamination. We chose a thickness of $3mm$ for a glass-backsheet architecture and $6mm$ for a glass-glass architecture. The problem has two geometrical symmetries and loading with respect to the planes (Oxz) and (Oyz), at the center of the PV module. Thus, only a quarter of the structure is considered. The support is modeled by a 3D discrete rigid half sphere. A contact interaction is defined between the support and the bottom face of the laminate, with normal hard contact and frictionless sliding. Linear quadrilateral shell elements with reduced integration were used to mesh the structure. The mesh is refined near to the contact zone with the support. A mesh analysis was performed to assess the accuracy of the numerical results.

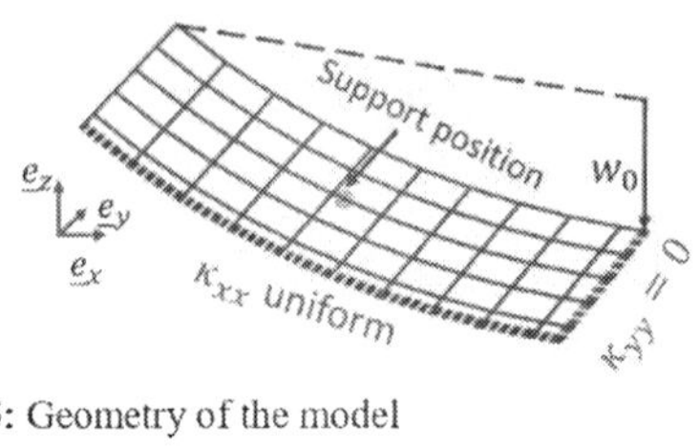

Figure 5: Geometry of the model

In order to assess the validity of the experimental protocol, we compare the initial deflection imposed to the model with the deflection estimated using the approach described above. Considering the notations from Fig. 5, the real deflection we are trying to measure is ω_0. Thus, the error criteria for the numerical validation is:

$$\epsilon = \frac{\omega_1 - \omega_2}{\omega_0} \quad (7)$$

The results are represented in Table I, for a real deflection from $0.1mm$ to $10mm$. The error induced by the protocol is inferior to 0.75%. It can also be noticed that the protocol induces an under-estimation of the deflection measurement, as the error is negative. Finally, the results shows also that glass-glass architecture is less affected by structural effects compared to glass-backsheet architecture, as the error is at least 20 times lower.

Table I: Error induced by the experimental protocol for different value of residual deflection, computed with ABAQUS

Residual deflection ω_0 (mm)	Measurement error ϵ (%)	
	Glass-Backsheet	Glass-Glass
0,1	-0,7200	0,0126
1	-0,7391	-0,0270
10	-0,5817	-0,0297

4 EXPERIMENTAL DEFLECTION PROBING

A preliminary study has been led, in order to prove the relevance of the experimental approach before a larger experimental campaign. The experiments have been made using a three dimension coordinate measuring machine developed by a partner Mitutoyo. This machine offers a high precision probing of the measured PV module (up to $7\mu m$). The coordinate measurement is performed by a contact inductive distance sensor, moving in the horizontal plane using motorized axes. The machine offers several options for the number and density of measured points. The program chosen is a mapping of the sample surface, with one measurement point every $5cm$.

The samples chosen for this preliminary study are:

1) A glass plate, presumed to be perfectly flat.
2) A laminated glass-backsheet PV module.
3) A glass-free symmetric PV module with polymeric front and back layers, which is particularly bowed.

The maximum size of the sample was limited by the measuring area of the machine used. Thus, we chose for the glass-backsheet sample a 4×4 cells PV module and the associated glass size for samples 1 and 2. The third sample is a mini-module with only four cells, which is smaller.

The experimental protocol plans to put the module on four supports, instead of placing it on the machine plate. A rigid frame linking the four supports was made. Stops are added to ensure that the PV module is always placed at the same position. The experimental set up is shown on Fig. 6.

In order to calculate the real deflection, a Python code was developed to post process data files of the coordinate measurement machine. On each side, a point was measured every $5cm$. By flipping the PV module, an offset was observed between the measurements on each side. Therefore, a linear interpolation was established to compare the deflection at the same point on both sides.

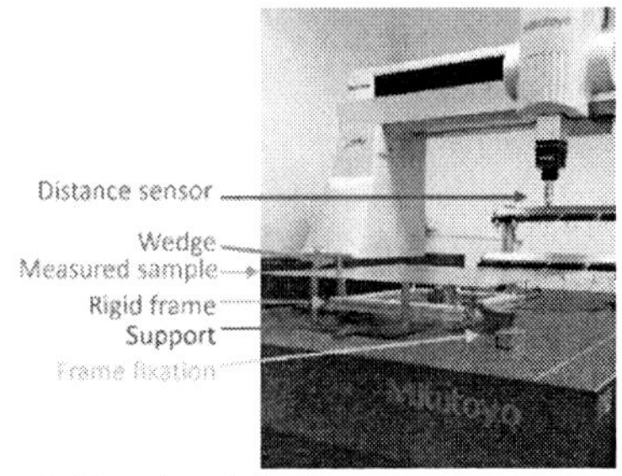

Figure 6: Experimental set up

For each sample, the real deflection ω_0 free of gravity effect is calculated. The effect of gravity ω_G is also

computed.
Fig. 7 displays the experimental results of the glass-backsheet PV module. The planarity error is defined as the distance between maximum and minimum deflection. Several conclusions can be drawn from the experimental results:

- The glass plate is not as flat as expected, with a planarity error of $1.35mm$.
- The deflection of the glass-backsheet sample is smaller than the glass plate (planarity error of $1{,}15mm$. During lamination, the PV module is heated and pressed. Thus, it might cause stress relaxation in the glass plate.
- The gravity on the 16 cells PV module samples has a bigger effect in the direction x than y, as it is the longer side ($4cm$ longer, ie 5% of the length).
- The residual deflection of the PV module without glass appears to be much larger than for other samples ($21.3mm$). The gravity-induced deflection seems to be inclined in the y direction, due to a mis-centering of solar cells with respect to the polymer front and rear faces.

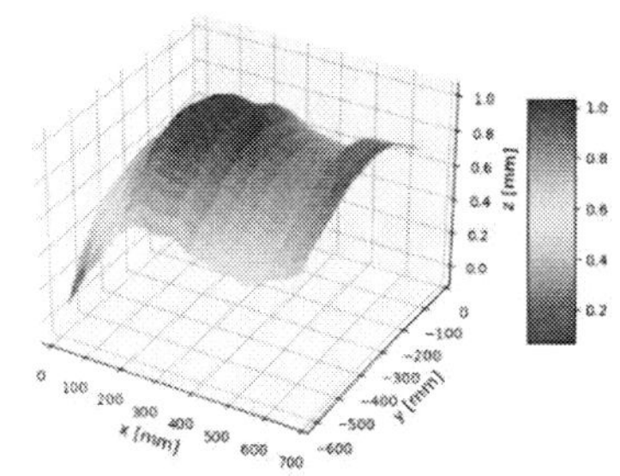

(a) Real deflection ω_0

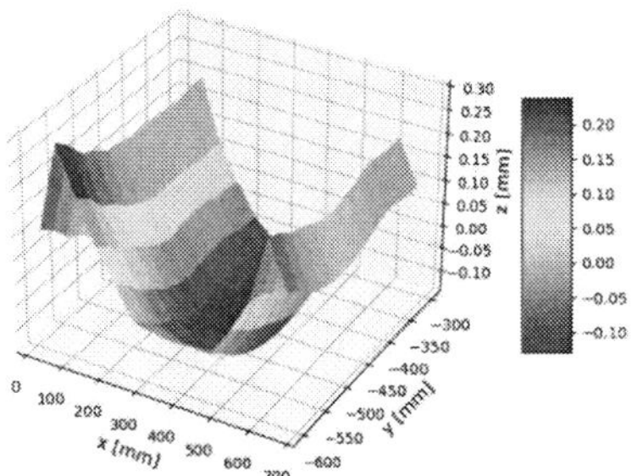

(b) Effect of gravity ω_G

Figure 7: Experimental results for the glass-backsheet PV module (sample 2)

Several improvements should be brought to this measurement protocol, in particular the flatness of the supports, as well as the measurement positions at each side in order to avoid a possible error linked to the linear interpolation between the measurement points.

5 PARAMETRIC EXPERIMENTAL STUDY OF THE LAMINATION PROCESS AND ITS IMPACT ON THE RESIDUAL DEFLECTION

5.1 Preparation and manufacturing of PV modules

As discussed in the previous section, the glass plates are not initially flat. Therefore, a measurement of the initial deflection of the glass plates is necessary before integrating them into the PV module. Using the same protocol presented previously, the initial shapes of twelve glass plates were measured. The quality of the supports was improved (see Fig. 8) and the density of the measurement points was increased by keeping the same position on each side to avoid interpolation errors between the measurement points.

Figure 8: The improved experimental set up

Fig. 9 presents the real deflection ω_0 of a glass plate. It can be noticed that the initial shape of the glass plate is not flat and its curvature is not symmetrical. Fig. 10 shows the planarity error measured for twelve glass plates. The median planarity error in the glass plates is equal to $1.433mm \pm 0.034$. This residual curvature present in a sampling of glass plates can eventually be caused by the nature of the cooling after their manufacture.

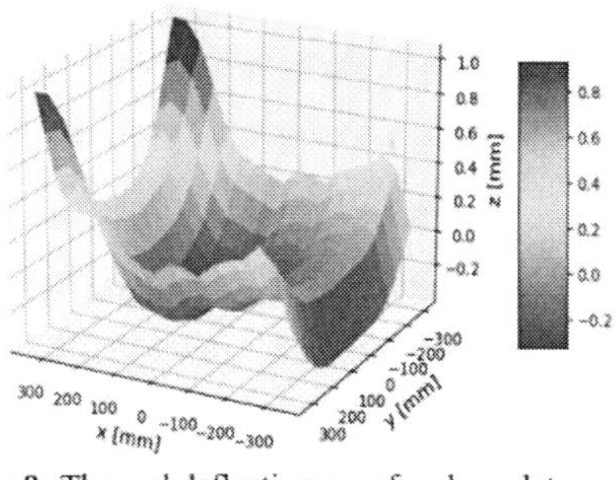

Figure 9: The real deflection ω_0 of a glass plate

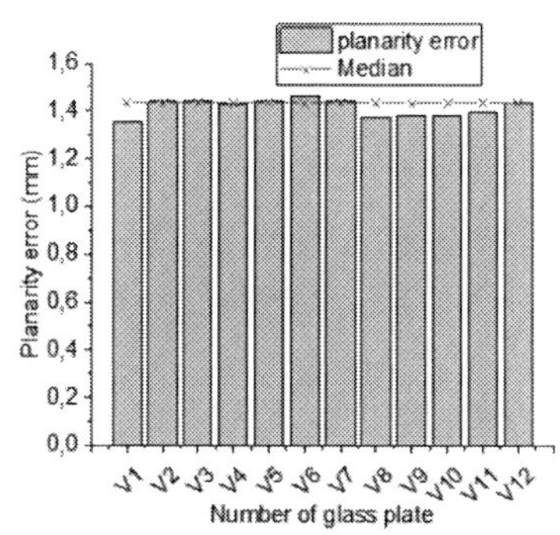

Figure 10: Planarity error of the twelve glass plates

These glass plates were used to manufacture the PV modules for this study. For glass-glass modules, the combined plates have approximately the same planarity error in order to distinguish the effect of the lamination process parameters on their initial deflection.

Eight 4×4 solar cells PV modules were manufactured by varying three process parameters as shown in Table II. The heterojunction (HJT) solar cells were not interconnected. A thermoplastic polyolefin

Table II: Parameters study of the lamination process

Architecture	Temperature of lamination (°C)	Type of cooling
Glass-Glass	100	Cooling press
	156	Cooling press
		Ambient air
	170	Cooling press
Glass-Backsheet	100	Cooling press
	156	Cooling press
		Ambient air
	170	Cooling press

(TPO) encapsulant was used to encapsulate the PV module layers. For glass-backsheet PV modules, a polypropylene (PP) backsheet was used. The layers were stacked considering the initial concave shape of the glass plates as shown in Fig. 11. The glass-glass PV modules have undergone an additional step in the double-sided heating plate compared to the glass-backsheet modules. This is an optional stage, developed more especially for glass-glass PV modules, to avoid edge pinching caused by the membrane and to ensure symmetrical heating of the PV module. During this step, the glass-glass PV modules were pressed with a pressure three times higher than the pressure applied in the membrane.

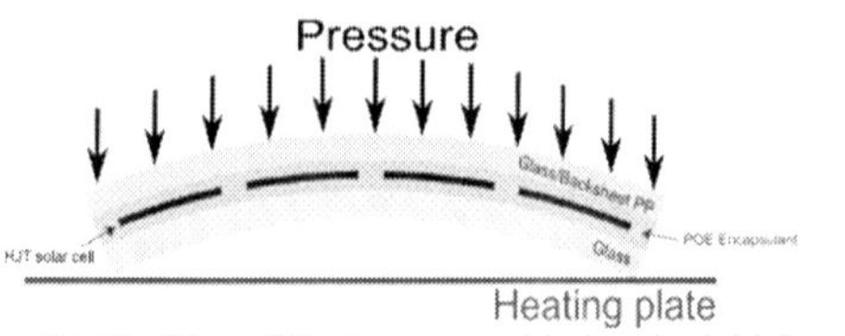

Figure 11: Stacking of the layers considering the initial concave shape of the glass plates

5.2 Results and discussion

Fig. 12 displays the planarity error of PV modules regarding the architecture, temperature of lamination and type of cooling. For the glass-backsheet architecture, by comparing the planarity error of PV modules cooled in a cooling press and encapsulated at different temperatures, we notice that the higher the lamination temperature is, the less the residual deflection of the PV module will be. However, for the glass-glass architecture the lowest planarity error is observed in the PV module laminated at $100°C$. The temperature effect is not proportional to planarity error in glass-glass architecture since two glass plates with different residual curvature are combined.

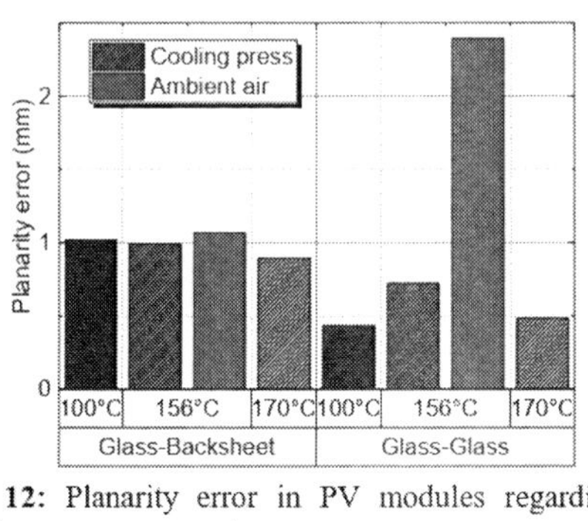

Figure 12: Planarity error in PV modules regarding lamination process parameters

On the other hand, for both architectures, the highest residual deflection is found in the case of ambient air cooling as in the cooling press stage, pressure is applied to maintain the flatness of the PV module during cooling. The effect of the type of cooling is higher on the glass-glass architecture where the planarity error increased by 230% compared to the PV module cooled with the cooling press. This difference is probably due to the supplementary stage of the double-sided heating plate, which applies a high pressure compared to the membrane stage.

Since the glass-glass PV modules have undergone an additional double-sided heating plate step with a higher pressure than that applied to the glass-backsheet modules in the membrane, it is uncertain to judge whether the lower planarity error found in the glass-glass modules is mainly related to an architecture effect or to the applied pressure.

In general, PV modules cooled in a cooling press have a lower deflection than the glass plates. Therefore, the lamination process with controlled cooling can attenuate the residual curvature of the glass plates through temperature or pressure or the creation of the laminate.

6 CONCLUSION AND FUTURE WORK

This work presents a new experimental method to measure the global deflection of a PV module using a 3D coordinate machine. An optimization of the positions of the supports used in the experimental protocol was studied analytically and numerically. Then, a numerical study allowed to estimate the error of measurement of the deflection related to the effects of gravity. Finally, experimental measurements were carried out to validate the experimental methodology.

An experimental measurement campaign using this protocol was carried out in order to study the influence of lamination process parameters on the deflection of the PV modules. A 3D numerical model will be used to valid the numerical/experimental correlation between the levels of residual curvature.

ACKNOWLEDGMENT

The authors would like to thank the technical team of Mitutoyo France for providing the experimental measurement on the three dimension coordinate machine.

This work was supported by the French National Program "Programme d'Investissements d'Avenir - INES.2S" under Grant ANR-10-IEED-0014-01.

REFERENCES

[1] Tippabhotla, Sasi Kumar et al., "From Cells to Laminate: Probing and Modeling Residual Stress Evolution in Thin Silicon Photovoltaic Modules Using Synchrotron X-Ray Micro-Diffraction Experiments and Finite Element Simulations," Progress in Photovoltaics: Research and Applications, vol. 25, pp.791--809, 2017.

[2] Beinert, Andreas J. et al., "Enabling the Measurement of Thermomechanical Stress in Solar Cells and PV Modules by Confocal Micro-Raman Spectroscopy," Solar Energy Materials and Solar Cells, vol. 139, pp.351--360, 2019.

[3] Timošenko, Stepan P. and Woinowsky-Krieger, Sergius, "Theory of Plates and Shells," 2. ed. reissued,

Engineering Societies Monographs, New York: McGraw-Hill, 1987.
[4] SEMI M1-0302, "Specifications for Polished Monocrystalline Silicon Wafers," Semi, 2001.

Analysis of the thermomechanical behaviour of concrete facade elements with integrated photovoltaic modules

Paul Schenk*, Sebastian Schindler, Matthias Pander, Ulli Zeller, Bengt Jaeckel, Matthias Ebert
Fraunhofer Center for Silicon Photovoltaics CSP, Otto-Eissfeldt-Str. 12, 06120 Halle (Saale), Germany
Fraunhofer Institute for Microstructure of Materials and Systems IMWS, Walter-Huelse-Str. 1, 06120 Halle, Germany
*=corresponding author: phone: +49-345-5589 5220, fax: +49-345-5589 5999, paul.schenk@csp.fraunhofer.de

ABSTRACT: Solar integration in the building envelope still represents a niche market, despite legal requirements and regulation, e.g. nearly zero-energy buildings (NZEB), which, in addition to a high overall energy efficiency, requires alternative energy generation in or near the building. A recent study entitles that building facades offering a so far untapped potential for the integration of photovoltaics. This results in great potential for building-integrated photovoltaics (BIPV), where high-quality design solutions integrated into facades and building components will be increasingly in demand.

In the project SOLARcon, exposed concrete facade elements are designed and build in which small PV modules are integrated. Due to the special design and the connection between the concrete element and the PV module, new questions arise in regard to mechanical stresses and temperature interactions on both the concrete building unit and the PV module compared to conventional PV modules with aluminium frames. Within this paper, first results of finite element analysis (FEA) are presented. We investigated how temperature distribution between photovoltaic module and the surrounding concrete facade element takes place and how resulting thermally induced mechanical stresses influence the service life of the photovoltaics.

Keywords: Simulation, Building Integrated PV (BIPV), Thermal Performance

1 INTRODUCTION

Solar integration in the building envelope still represents a niche market, despite legal requirements and regulation, e.g. nearly zero-energy buildings (NZEB) [1], which, in addition to a high overall energy efficiency, requires alternative energy generation in or near the building. A recent study entitles that building facades offering a so far untapped potential for the integration of photovoltaics [2]. This results in great potential for building-integrated photovoltaics (BIPV), where high-quality design solutions integrated into facades and building components will be increasingly in demand.

In the SOLARcon project, concrete elements with integrated PV are designed and analyzed with regard to thermo-mechanical load, among other things. A proven method for these simulations is the finite element analysis (FEA). It has already been successfully applied to standard modules [3]-[6].

Temperature changes usually lead to the expansion or shrinkage of materials and components. If these changes in shape are restricted by geometric properties or if different materials are used to cause different expansions, this usually leads to thermally induced mechanical stresses in the components. In the worst case, this leads to the failure/destruction of individual components. The PV modules integrated into the concrete facade heat up to 50°C and more as a result of solar radiation (800 W/m² irradiation, 25°C ambient temperature, free-standing module) [6]. The concrete element, on the other hand, heats up more slowly, which means that each PV module must be considered a temperature hotspot in the concrete façade.

In order to classify the feasibility of integrating PV modules in concrete elements, FEA will be used to simulate the heat transfer and the resulting temperature input from the PV modules into the concrete element. In addition, the thermal induced stresses of the components are to be determined in order to define critical geometric areas.

2 MOTIVATION

Due to the geometry/design of the SOLARcon facade elements (Figure 1), a new installation situation arises for the PV module. The PV module is no longer fixed to a base frame with an aluminum frame or as a frameless variant with clamps, but is embedded and glued into the concrete body. This means that the rear and side surfaces of the module are completely insulated by adhesive and concrete, whereas with conventional modules the rear surfaces remain free and serve for heat dissipation. Thus, encasing the module in concrete is not only a new installation variant, but also a new thermal load case for the material.

This potentially higher load on the PV module can lead to mechanical problems in addition to less electrical efficiency [7][8]. Possible failure scenarios can be glass breakage in the modules, cracks in the concrete edging, spalling of concrete pieces or detachment of the PV modules.

The detachment problem can be considered as already solved by using an approved adhesive and is therefore no part of the simulations. In order to better predict the mechanical and thermal problems, the thermo-mechanical behavior of the concrete facade elements is simulated.

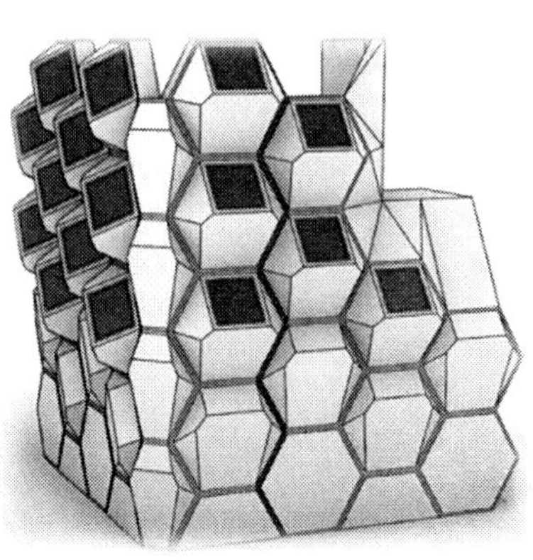

Figure 1: Design study of SOLARcon concrete facade element with integrated PV modules (dark blue), here using the example of a building corner [9]

3 MODELING AND SIMULATION

The simulations are carried out with ANSYS Workbench. The basic simulation model consists of two simulation modules, a steady-state thermal simulation and a static mechanical analysis. The results from the temperature simulation are imported into the mechanical simulation as a load case. The stresses in the components are then generated by the temperature expansion and the resulting deformations.

In the simulation, a concrete facade element with seven integrated PV modules is considered. The geometry used is simplified as much as possible so that it consists of only three different components. It consists of the concrete element, the adhesive and the PV module (Figure 2). The PV module is also simplified and shown as a pure glass plate. The classical PV components such as cell, busbars and EVA can be omitted for these considerations, since it is to be examined specifically if glass breakage is to be expected with the modules. The PV modules used have an edge dimension of 186x186 mm² and are 8 mm thick. This corresponds to the thickness of a glass-glass module as manufactured and used in the project. The module cut-out in the concrete has an edge dimension of 190x190 mm² and a depth of 13 mm. These dimensions result in an expansion joint width for the adhesive of 2 mm all around and a thickness of 5 mm between the concrete and the back of the module. Three different commercially available adhesives are considered in the simulation, including a silicone-based adhesive (adhesive 1) and two PU-based adhesives (adhesive 2 and 3).

The geometry is meshed with tetrahedral and hexahedral elements. Tetrahedral elements are used for the concrete body and hexahedral elements for the adhesive and the PV module (Figure 2). The complete meshing is done with an element size of 5 mm. This results in a still reasonable number of elements and acceptable calculation time.

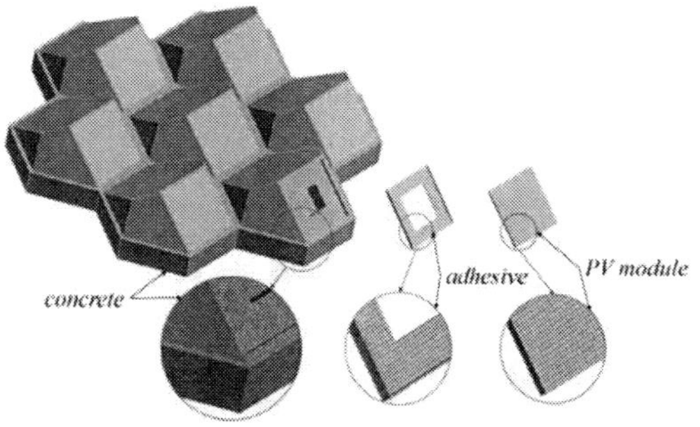

Figure 2: Geometry and mesh setup for basic simulation starting from the concrete façade over the used adhesive to attach the PV module

The temperature simulations are based on the physics of the solar cell. The cells convert the incoming light energy into electrical energy and heat. Modern cells in solar modules have an efficiency η of about 20 %, which means in first order approximation that 80 % of the incoming light irradiation is converted into heat [10]. Since these simulations cannot include irradiation, the heating must be simulated by suitable boundary conditions. The simulation component for the temperature simulation contains the boundary condition "internal heat generation", which requires the power density φ of a body as an input component and then calculates the resulting temperature. The power density for the solar cell can be calculated analog to equation 1.

$$\varphi_{Si,wafer} = (1-\eta) * \frac{\text{irradiation}}{\text{material thickness}} \quad (1)$$

Since the entire PV module is represented in the simulation model as a single body, the power density of the cell must be scaled to the volume of the module, as in equation 2.

$$\varphi_{modul} = \varphi_{Si,wafer} * \frac{t_{Si,wafer} * A_{Si,wafer}}{t_{modul} * A_{modul}} \quad (2)$$

In order to establish a temperature balance, a heat release to the environment must be defined in addition to the energy supply. There are various factors playing a role such as ambient temperature and wind speed. In the simulation, the temperature boundary condition convection is used for this purpose. This specifies how much power per area and Kelvin is emitted to the environment. To calculate the convection of a concrete body there are different complex models depending on the application. [11] In addition to these models, there are also various guide values for generalized applications. This is where the value of 5W/m²*K for free convection of gases at surfaces comes from, which is used in this simulation. [12] The model with the defined temperature boundary conditions is shown in Figure 3.

In addition to the boundary conditions for the temperature simulation, the model for the static-mechanical simulation also needs boundary conditions that describe the bearing of the geometry. In reality, the concrete facade elements would be supported and fixed by inserted building anchors. These anchors are usually connected to the reinforcement of the concrete. Since no reinforcement is considered in the simulation, the building anchor cannot be inserted as a stand-alone geometry component. Within the simulation the support is simplified by two surfaces imprinted on the back side of the concrete body. To avoid static overdetermination, both surfaces are considered together as a single surface and the boundary conditions in the three space directions are assigned as follows (Figure 3):

- Displacement in X-direction = 0, for the lower left point of the left surface
- Displacement in Y-direction = 0, for both surfaces
- Displacement in Z-direction = 0, for the bottom edge of both surfaces

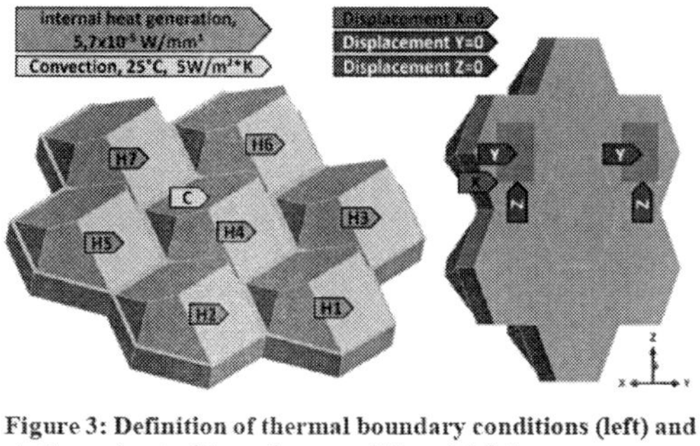

Figure 3: Definition of thermal boundary conditions (left) and static mechanical boundary conditions (right)

The first simulations showed that the selected meshing was not fine enough in certain areas to represent certain narrower areas of the concrete body, such as the module edging. This resulted in very high stresses in individual element nodes, which did not match the rest of the stress profile. For this reason, a stress convergence calculation was performed. For this purpose, the element size of the concrete body was gradually reduced from 10 mm to 3.5 mm and increased from approx. 1.4 million elements to approx. 23.6 million elements.

In Figure 4 the results for the stress convergence (red) and temperature convergence (black) for all three adhesives are shown. It can be clearly seen that the refinement of the mesh has a smaller influence on the calculated temperature than on the calculated stress. It can also be seen that the selected element size of 5 mm is not sufficiently accurate, since the calculated stresses increase significantly with further refinement (evaluation criterion: change of less than 5 % for two successive refinement steps).

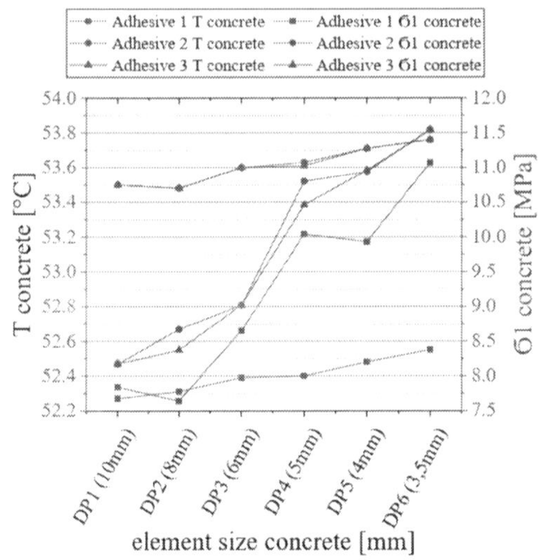

Figure 4: Calculation convergence for temperature (black) and stress (red) of the concrete body via mesh refinement from design point 1 (DP1) = 10 mm to DP6 = 3.5 mm for all three adhesives

To avoid increasing calculation times by further refinement and because the first results showed that the model behaves symmetrically, the element size remains at 5 mm. For further mesh refinement a sub-model is used.

For the submodeling, the middle element was "cut out" of the concrete body and considered in a separate calculation step. This modeling technique enables the use of smaller bodies with higher mesh resolution and still use a significantly smaller number of elements than in the general model. Figure 5 shows the geometry and meshing of the sub-model that was used. The initial results showed critical stress areas in the corners of the module enclosure. So, the area where the PV module is enclosed was separated from the rest of the concrete body for modeling and meshed with a different element type even finer than the rest of the concrete body. This separation is not a physical separation, the two halves of the concrete still form one part for the calculation. The adhesive and the PV module are also more finely meshed than in the overall model.

To ensure a more realistic load distribution in the sub-model, the results from the simulations of the global model were imported into the sub-model as temperature load and displacement boundary condition (Figure 6). In order to calculate the stresses in the sub-model, the global model must be solved first.

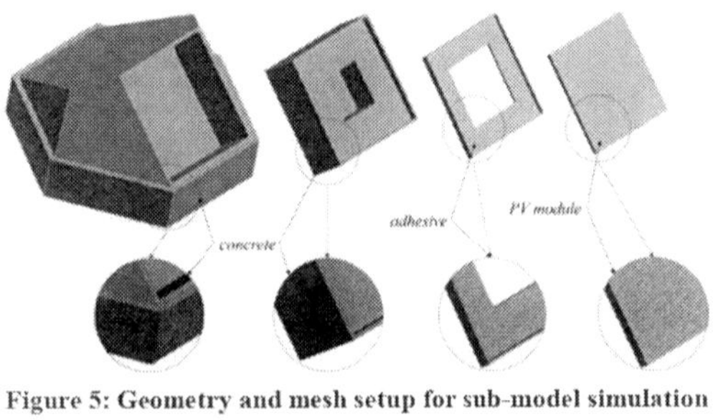

Figure 5: Geometry and mesh setup for sub-model simulation starting from the splitted concrete element over the used adhesive to attach the PV module

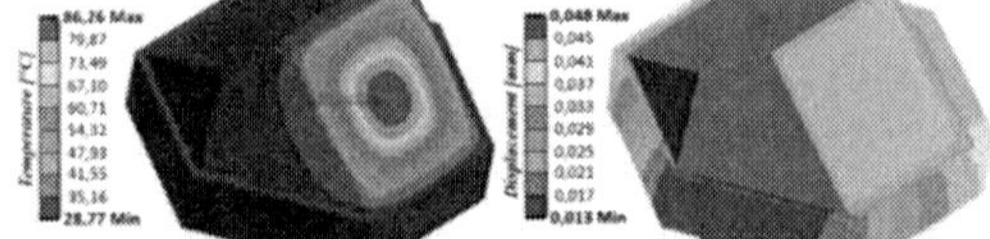

Figure 6: Imported body temperature from the global model to the sub-model (left) and imported cut boundary constraint from the global model to the sub-model (right)

4 RESULTS

The simulated thermo-mechanical loads and stresses in the model for Adhesive 1 at 25°C ambient temperature are shown graphically in Figure 7 and Figure 8. The results for all three adhesives at 25°C ambient temperature (in the following load case 1) are compared in Table 1. Table 2 compares results for all three adhesives at -5°C ambient temperature (in the following load case 2).

Figure 7 shows the simulated temperatures (top) and stresses (middle=general model, bottom=sub-model) in the PV module. Figure 8 behaves analogously only for the concrete. For both the PV module and the concrete, the 1st principal stress is evaluated because brittle materials such as glass and concrete fail under the maximum tensile stress [13].

For both, load case 1 and load case 2, the calculated temperatures are very close for all adhesives in the PV module and concrete. For the stresses, there are larger deviations between the three adhesives in terms of relative differences to the absolute value (e.g. temperature is only 2 % higher but stress in PV module is 10 % higher). Here, the PU-based adhesives induce higher stress than the silicone-based adhesives, which is due to the stiffer material behavior. In the range of the analyzed loads, the two PU-based adhesives behave almost identically. The calculated stresses in the sub-model are about 50 % lower than in the global model. The differences in the maximum stress between the global model and the sub-model may be due to the mesh adjustments and the avoidance of mesh singularities. Another reason may be the influence of the selected support and the resulting deformations (see section 4.2).

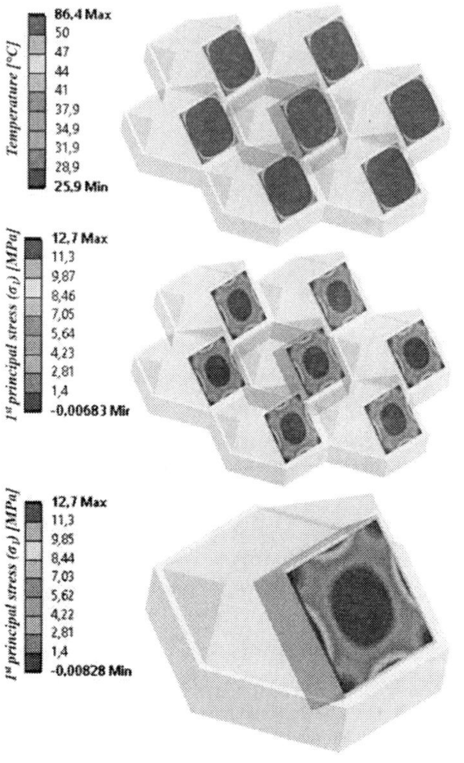

Figure 7: Simulation results for the PV module (top temperatures, middle stress in the global model, bottom stress in the sub-model)

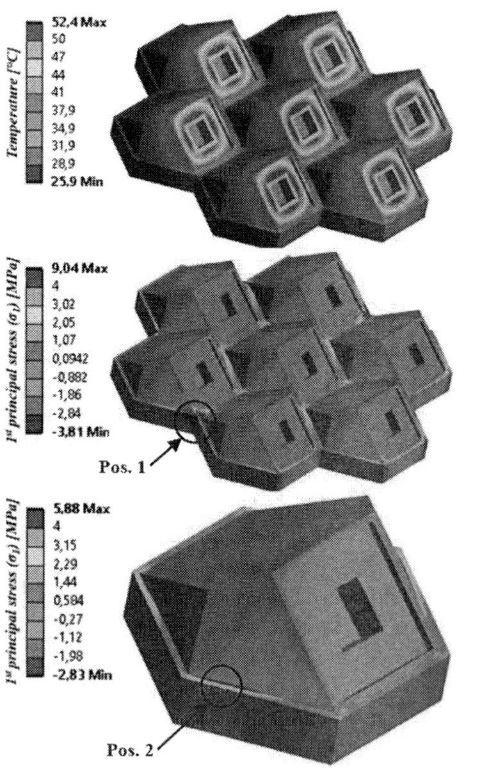

Figure 8: Simulation results for the concrete (top temperatures, middle stress in the global model, bottom stress in the sub-model)

Table 1: Simulation results for 25°C ambient temperature

Result parameter	adhesive 1	adhesive 2	adhesive 3
T_{PV} [°C]	86.38	85.18	85.18
$T_{concrete}$ [°C]	52.41	53.62	53.62
$\sigma_{1,PV}$ [MPa]	12.69	14.01	14.01
$\sigma_{1,concrete}$ [MPa]	9.04	9.45	9.45
$\sigma_{1,PV,sub\text{-}model}$ [MPa]	12.67	14.08	14.08
$\sigma_{1,concrete,sub\text{-}model}$ [MPa]	5.88	6.8	6.74

Table 2: Simulation results for -5°C ambient temperature

Result parameter	adhesive 1	adhesive 2	adhesive 3
T_{PV} [°C]	56.38	55.18	55.18
$T_{concrete}$ [°C]	22.41	23.62	23.62
$\sigma_{1,PV}$ [MPa]	12.5	14.78	14.78
$\sigma_{1,concrete}$ [MPa]	9.02	9.27	9.27
$\sigma_{1,PV,sub\text{-}model}$ [MPa]	12.47	14.83	14.83
$\sigma_{1,concrete,sub\text{-}model}$ [MPa]	5.87	6.02	6.02

4.1 Geometry study

In addition to the calculated load cases, simulations were also carried out to compare different geometric influences. Among other things, the influence of the gap width between the PV module and the concrete was considered, but also the use of thinner glass-foil modules, which results in a reduction of the cut-out depth into the concrete, was considered. The geometry variations are shown graphically in Figure 9. In total, 4 different combinations were calculated for load case 1 and adhesive 1. The results of the geometry study are compared in Table 3.

The results show that the adjustments made have little or no influence on the calculated stresses. Especially the stresses in the concrete nearly did not change in the global model as well as in the sub-model. Only the stresses in the PV module increase noticeably in the variants with reduced module thickness. This is due to the higher surface pressure at the rim and the lower moment of resistance of the thinner panes. The higher temperatures may be due to the adjustment of power density (reduced volume of PV modules) and reduced cooling because of a reduction in surface area.

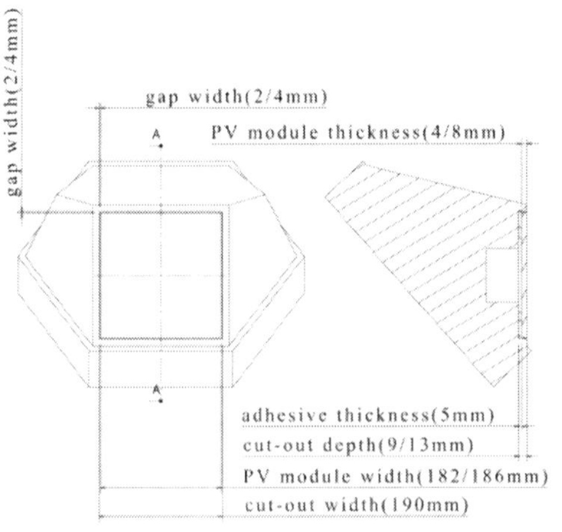

Figure 9: Variations of certain geometry parameters for the geometry study

Table 3: Simulation results for geometry study

Result parameter	2mm gap width, 8mm module thickness	2mm gap width, 4mm module thickness	4mm gap width, 8mm module thickness	4mm gap width, 4mm module thickness
T_{PV} [°C]	86.38	97.43	88.23	100.37
$T_{concrete}$ [°C]	52.41	52.95	52.75	53.48
$\sigma_{1,PV}$ [MPa]	12.69	15.82	12.71	16.36
$\sigma_{1,concrete}$ [MPa]	9.04	8.66	8.52	9.28
$\sigma_{1,PV,sub\text{-}model}$ [MPa]	12.67	15.85	12.72	16.41
$\sigma_{1,concrete,sub\text{-}model}$ [MPa]	5.88	5.83	5.9	5.77

4.2 Mechanical support considerations

The maximum stress in the global model is at the edge between the outer two concrete elements (Pos. 1, Figure 8). In the sub-model, the maximum stress is located at the transition between the base and the bevel (Pos. 2, Figure 8).

The calculated stresses in the global model and their position in the geometry would argue against the middle concrete element as a sub-model, but these stresses only occur due to the calculated deformation of the global model. And the calculated deformation, in turn, is related to the selected support of the geometry (Figure 3). Due to the chosen support, the concrete element, when viewed horizontally, can bend downwards. In Figure 10 the deformation is exaggerated. Since only the area where the facade anchors are located is supported and a displacement in the direction of the building wall is prevented here, the rest of the concrete element bends freely "downwards". In reality, the element is also not restricted in this direction by a mechanical mounting, but if it were to deform in this direction, it would press against the building wall and support itself there. However, this support effect is not recognized in the simulation. Thus, the maximum stress at this point is caused by a deformation that cannot occur in reality.

Figure 10: Exaggerated deformation result of the concrete element for support variant 1

For this reason, different support concepts were evaluated in a further model study. As in the geometry study from Chapter 4.1, only load case 1 and adhesive 1 were considered here. The different bearing variants are compared in Figure 11. The calculated stresses for each bearing variant are listed in Table 4.

In the case of bearing variant 2, the surrounding lower edge of the global model was supported in addition to the two surfaces in the Y-direction (house wall). This prevents the deflection shown in Figure 10, but the supported edge is loaded so strongly that the calculated stresses (also due to mesh singularities in this area) in the global model are even higher than in variant 1. In support variant 3, not only the edge, but also a surrounding surface was additionally supported in the Y-direction. This leads to the fact that the stresses in the global model decrease in comparison to variant 1. However, the maximum is now also in one of the supported surfaces due to the artificially generated restraint.

Looking at the stresses in the sub-models of the respective bearing variants, it is noticeable that there is nearly no difference between variants 2 and 3 and that both are lower than the stress in the sub-model of variant 1. This is due to the fact that the displacements at the cutting edges are taken over from the global model as a load in the sub-model. By reducing the deformation in variants 2 and 3, the load for the sub-model is also reduced.

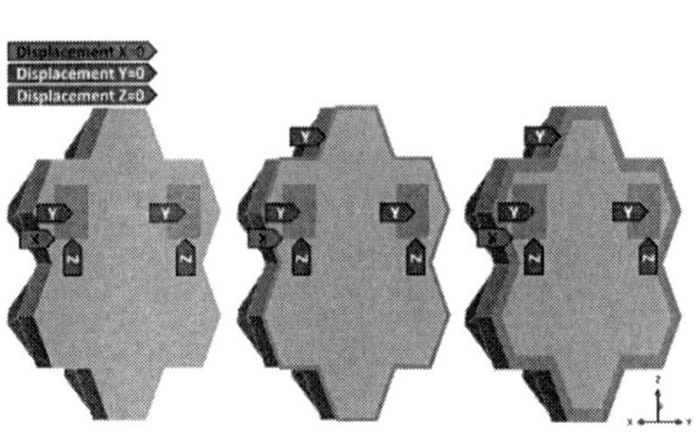

Figure 11: Variations of mechanical support (variant 1 (original one) left, variant 2 middle, variant 3 right)

Table 4: Simulation results for mechanical support variants

Result parameter	variant 1	variant 2	variant 3
$\sigma_{1,concrete}$ [MPa]	9.04	17.08	6.73
$\sigma_{1,concrete,sub\text{-}model}$ [MPa]	5.88	4.47	4.43

5 CONCLUSIONS AND OUTLOOK

The results of the simulations do not indicate any critical stresses (the critical material parameters have been agreed with the project partners, but cannot yet be specified during the development process) in the PV module. The stresses here are far below the permissible stresses for safety glass. Therefore, glass breakage is unlikely due to the thermo mechanical stress. The electrical performance of the used modules must be considered separately due to the high temperatures and evaluated more precisely with suitable experiments.

Despite the use of a sub-model and the additional considerations regarding the support variants, there are still areas in the concrete element with unacceptable tensile stresses. Here, further considerations are needed regarding the effects on the facade element. If and to what extent cracks will grow in the concrete here cannot be predicted at this point in time. In the real component, the concrete body has an additional steel reinforcement which, among other things, should also absorb tensile forces/stresses. Similar to the electrical performance of the modules, suitable experiments are required here that can represent this load case.

Furthermore, in addition to the steady-state thermal simulation, a transient temperature simulation is also planned in the course of the project. This will allow the different rates of heating and different rates of expansion of the individual parts to be simulated more accurately than with the steady-state variant. Compared to static temperature simulation, the transient temperature simulation requires detailed temperature and irradiation profiles over daily or weekly periods. These are currently prepared in order to use them for the further simulations.

6 ACKNOWLEDGEMENTS

The authors thank the Federal Ministry of Economics and Energy for the financial support of the work in the project "SOLARcon" (grant no. 03EE1016B).

7 REFERENCES

[1] European Parliament. (2010). Directive 2010/31/EU of the European Parliament and of the Council on the energy performance of buildings (recast). Brussels

[2] Behnisch et al. „Anwendungsszenarien von Geomassendaten zur Modellierung von Grünvolumen und Solarflächenpotenzial“, 2020

[3] S. Dietrich et al., Introducing a reliability concept based on probabilistic material data of glass for PV modules, 26th EU-PVSEC, 2011

[4] M. Pander et al., Lifetime estimation for solar cell interconnectors, 28th EU-PVSEC, 2013

[5] M. Pander et al., Digital Prototyping – application of numerical methods in module development, 36th EU-PVSEC, 2019

[6] Ulli Zeller, The influence of air flow on the temperature distribution within crystalline solar modules, 28th PVSEC, 2013

[7] Hanifi H., Dassler D., Evaluation and Comparison of PV Modules with Different Designs of Partial Cells in Desert and Moderate Climates, 45th IEEE Photovoltaic Specialists Conference, 2018, 8, 1266-1273

[8] Hanifi H., Pander M., Loss analysis and optimization of PV module components and design to achieve higher energy yield and longer service life in desert regions, Applied Energy, 2020, 280, 116028

[9] Yield-Optimized Small Photovoltaic Module Solutions in Combination with Concrete Facades Elements; Schindler, S., Schenk, P., Ebert, M., Huth S., Heller A., Hülsmeier F., Reise J., Wilhelm, K., Butler, M., Grebe, R., Erban, C., Hirsch, D.; 38. European Photovoltaic Solar Energy Conference and Exhibition; 2021; Lisboa

[10] Peter Würfel, Physik der Solarzelle, Spektrum Akademischer Verlag / Springer Spektrum, 2000

[11] VDI Gesellschaft: VDI-Wärmeatlas. Wiesbaden: Springer Berlin Heidelberg, 2005.

[12] Cerbe/Hoffmann: Einführung in die Thermodynamik: Von den Grundlagen zur technischen Anwendung, Hanser Fachbuchverlag Leipzig, 2002

[13] Gross, D.; Hauger W.; Technische Mechanik 2 – Elastostatik; Springer Verlag, 8. Auflage 2005, S. 67-68

SPATIALLY RESOLVED LEAKAGE CURRENT DENSITY IN PHOTOVOLTAIC MODULES

H. Nagel, M. Glatthaar, D. Philipp, H. D. Neuhaus and S. W. Glunz

Fraunhofer Institute for Solar Energy Systems (ISE), Heidenhofstraße 2, 79110 Freiburg, Germany
Email: henning.nagel@ise.fraunhofer.de

ABSTRACT: Due to the non-negligible electrical conductivity of module building materials, small leakage currents flow between the grounded module frame and the active cell matrix in photovoltaic (PV) modules under normal operation conditions, which can lead to significant degradation, such as e. g. potential-induced degradation of the shunting type (PID-s). In general, the degradation rate is strongly correlated with the leakage current density, which is highly dependent on the position in large-area modules due to the voltage drop between the frame and the cells. In this work, we present easy-to-use mathematical equations for rapid analytical calculation of the spatial leakage current density distribution in PV modules as a function of distance from the grounded module frame. The validity of the equations is confirmed by numerical simulation of a resistor network. As an application, the leakage current density distribution in a standard c-Si PV module was analytically calculated as a function of temperature and relative humidity, and the width of the circumferential module area threatened by PID-s was predicted under accelerated test conditions.
Keywords: degradation, modelling, module

1 INTRODUCTION

In general, photovoltaic modules consist of the active solar cell matrix and a two-layer system on the front as well as the back side. For crystalline silicon cell arrays, for example, the front side package often consists of a polymer encapsulant (usually ethylene vinyl acetate, EVA) and a glass cover. The back side is made of a polymer encapsulant and a polymer backsheet or glass cover. Due to the high system voltage and the non-negligible conductivity of the materials used and the interfaces formed, leakage currents occur which are the driving force for degradation, e. g. potential-induced degradation of the shunting type [1], potential-induced degradation of the surface passivation [2] or electrochemical corrosion [3].

The total leakage current can be calculated using a circuit model in which the resistances of the individual current paths are combined into a single resistance value [4]. However, this method does not provide information about the lateral distribution of the leakage current density. Therefore, it is not possible to predict a higher degradation rate localised at the perimeter of the module. A systematic approach to predict and understand the main factors for local degradation is to measure the conductivity of the materials and interfaces as a function of temperature and humidity, and then calculate the spatial leakage current density distribution $J_l(x)$. It has been shown before that J_l can be effectively calculated by i) setting up a resistor network and ii) subsequent numerical simulation [5,6]. This method involves some effort, since the resistor network must be built up from several thousand individual resistors to achieve good spatial resolution, and the subsequent numerical simulation can be time consuming. In this work, we present fast and easy-to-use analytical calculations of the leakage current density in large-area PV modules as a function of distance from the grounded module frame.

2 MODEL FOR RESISTOR NETWORK OF PV MODULES

The module package on the front side is electrically described by a two-dimensional resistor network representing a grounded (metal) frame, a laterally conductive glass/air interface, a laterally and vertically conductive glass cover, a laterally conductive glass/encapsulant interface (featuring infinite vertical conductivity because it is very thin), a laterally and vertically conductive encapsulant and a solar cell array to which a uniform module voltage is applied, see schematic drawing in Fig. 1. The formula symbols for the parameters are listed in Table 1 together with a short description of their meaning. For the resistor network, we set up and analytically solved the coupled partial differential equations arising from Kirchhoff's and Ohm's laws. The mathematical derivation of the equations is extensive, but straight forward. A detailed

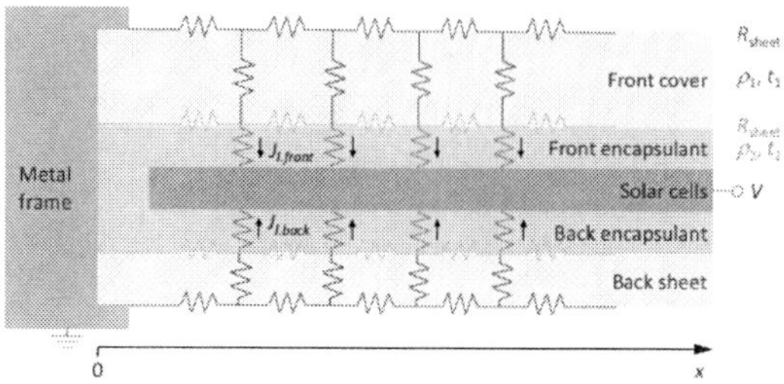

Fig. 1: Schematic drawing of a c-Si PV module cross section showing leakage current paths and electrical material and interface parameters.

Table 1: Parameters for simulation of the resistor network representing the module's front side.

Parameter	Unit	Description
R_{sheet1}	Ω	Sheet resistance of the surface of the front cover
R_{sheet2}	Ω	Lumped sheet resistance of the front cover, front cover/encapsulant interface and encapsulant
ρ_1	Ω cm	Resistivity of the front cover
ρ_2	Ω cm	Resistivity of the front encapsulant
t_1	cm	Thickness of the front cover
t_2	cm	Thickness of the front encapsulant
V	V	Voltage between solar cell and grounded module frame
x	cm	Distance from the module edge
$J_{l,front}$	A/cm²	Leakage current density on the front of the solar cells

description is beyond the scope here and will be published elsewhere. Table 2 shows the result. The obtained equations allow a quick calculation of the leakage current density in PV modules as a function of the distance to the module frame, e. g. using a spreadsheet program such as Microsoft Excel. In the Results section, a verification of an analytically calculated local leakage current density profile is presented by comparison with the profile numerically calculated using the electronic circuit simulator software LTspice XVII [7] and the same device parameters.

Unless otherwise stated, in this work the outer solar cells of the cell matrix are assumed to be close to (without direct electrical contact with) the grounded metal frame. This model can be extended to include an additional distance L between the cells and the module frame by introducing a voltage U_{offset} that drops at the circumferential edge of the module across encapsulants, covers and interfaces. In [5], it was shown by two-dimensional finite element based numerical solution of Maxwell's equations that in the case of high sheet resistance of the module surface and for standard sodalime cover glass, EVA encapsulant, and Tedlar-Polyester-Tedlar backsheet, the electrical field in the region between the cell and the frame is directed horizontally towards the frame. Then, a simple one-dimensional model can account for a voltage offset

$$U_{offset} = R_{sheet3} L \frac{I}{w}$$

which reduces the voltage between the cell matrix and the grounded module frame. I and w are the total leakage current through the perimeter of the module and the length of the perimeter, respectively. R_{sheet3} is a lumped sheet resistance given by

$$R_{sheet3} = \left(\frac{1}{R_{sheet.front\_surface}} + \frac{1}{R_{sheet.front\_cover}} + \frac{1}{R_{sheet.interface.front\_cover/front\_encapsulant}} + \frac{1}{R_{sheet.front\_encapsulant}} + \frac{1}{R_{sheet.back\_surface}} + \frac{1}{R_{sheet.back\_cover}} + \frac{1}{R_{sheet.interface.back\_cover/back\_encapsulant}} + \frac{1}{R_{sheet.back\_encapsulant}} + \frac{1}{R_{sheet.interface.front\_encapsulant/back\_encapsulant}}\right)^{-1}.$$

Table 2: Equations for analytical calculation of the local leakage current density.

$$J_{l,front} = \frac{V}{\rho_2 t_2}\left(\left(\frac{R_{sheet1} c_1}{p_1} - c_1 p_1 \rho_1 t_1\right) exp^{-p_1 x} + \left(\frac{R_{sheet1} c_2}{p_2} - c_2 p_2 \rho_1 t_1\right) exp^{-p_2 x}\right) \quad (1)$$

$$c_1 = -\frac{p_1 p_2^2}{(p_1^2 - p_2^2) R_{sheet1}} \quad (2)$$

$$c_2 = \frac{p_1^2 p_2}{(p_1^2 - p_2^2) R_{sheet1}} \quad (3)$$

$$p_{1,2} = \sqrt{\frac{R_{sheet1} + R_{sheet2}}{2\rho_1 t_1} + \frac{R_{sheet2}}{2\rho_2 t_2} \pm \sqrt{\left(\frac{R_{sheet1} + R_{sheet2}}{2\rho_1 t_1} + \frac{R_{sheet2}}{2\rho_2 t_2}\right)^2 - \frac{R_{sheet1} R_{sheet2}}{\rho_1 t_1 \rho_2 t_2}}},$$

'+' applies to p_1, '-' to p_2 (4)

$$R_{sheet2} = \left(\frac{1}{R_{sheet.front\_cover}} + \frac{1}{R_{sheet.interface.front\_cover/front\_encapsulant}} + \frac{1}{R_{sheet.front\_encapsulant}}\right)^{-1} \quad (5)$$

Outside the distance L at the module edge, the front and back sides of the module are electrically decoupled by the intermediate solar cell matrix, so the local leakage current density on each side can be calculated separately using the equations in Table 2 but different electrical material und interface parameters. In this work, we focus on the calculation of the local leakage current density on the front side of the solar cells to obtain information about the width of the module edge affected by PID-s. PID-s takes place in the emitter of the cells, which is generally located on the front side of the PV module.

It should be noted that it is not directly the leakage current density, but the area-related cumulated charge density that is decisive for the local degradation. It is simply the local leakage current density on the solar cell times the duration of the current flow. In the case of fluctuating leakage current densities, e. g. due to changes in electrical resistivity caused by temperature fluctuations, the leakage current density is integrated over time.

3 RESULTS

3.1 Validation of the analytical calculation

Fig. 2 shows a comparison of analytically and numerically calculated leakage current density as a function of the distance to the module frame. 1000 V were applied between the cell matrix and the frame. The simulated cross-sectional module layout was 3.2 mm sodalime glass/0.42 mm EVA/cell matrix. The electrical material parameters were taken from [6] for a module temperature of 60 °C and a relative humidity (RH) of 45 % (which affects the conductivity of the front surface). The obtained sheet resistances are listed in Table 3.

In the numerical simulation, we varied the node spacing of the resistor network between 1 and 0.1 cm. It can be seen from Fig. 2 that the resulting leakage current curves converge with that determined analytically and show very good agreement at a node spacing of 0.1 mm. Since the spatial resolution of the numerical simulation increases with finer node spacing, we conclude that the analytical calculation is well confirmed.

Table 3: Sheet resistances of the layers in the simulated module at 60 °C and 45 % RH.

Layer	R_{sheet} (Ω/sq)
Glass surface	$6.93 \cdot 10^{11}$
Glass	$3.09 \cdot 10^{11}$
Interface between glass and EVA	$1.06 \cdot 10^{11}$
EVA	$2.33 \cdot 10^{13}$

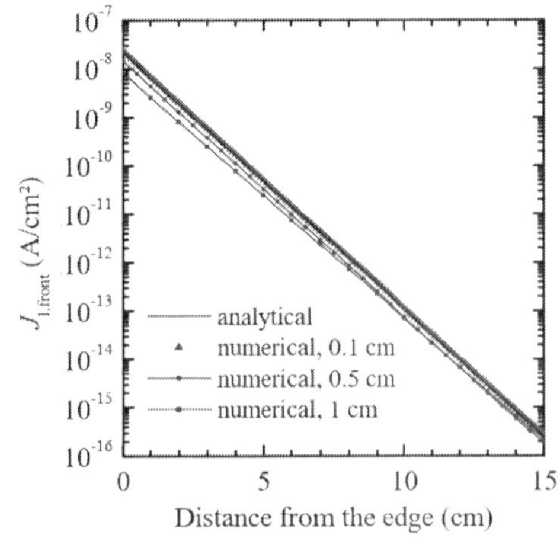

Fig. 2: Comparison of analytically and numerically calculated leakage current density through the front side of a PV module. The numbers in the legend indicate the node distances in the numerical simulation.

3.2 Application of the analytical calculation: variation of temperature and relative humidity

The equations presented in Table 2, together with the electrical parameters of materials used for module fabrication, provide the means to calculate the effects of environmental conditions on the local leakage current density. We again assumed the cross-sectional module layout 3.2 mm sodalime glass/0.42 mm EVA/cell matrix, a voltage of 1000 V applied between the cell matrix and module frame, and used the previously measured electrical parameters as a function of temperature and RH [6]. Fig. 3 shows the result for two different relative humidities of 45 and 85 % and a module temperature ranging from 40 to 70 °C. From the figure, it can be seen that the influence of RH on the leakage current density at a certain distance from the module edge is even greater than the influence of the increased conductivity of materials and interfaces due to the increase in module temperature. Obviously, the increase of the conductivity of the surface of the front glass with higher RH has a great effect.

The grey line in Fig. 3 marks a current density of $1x10^{-10}$ A/cm², which results in a charge density of $6x10^{-5}$ C/cm² deposited on the front of the encapsulated solar cells after 6.9 days. In case of PID-s prone cells, this charge density leads to a 90 % power drop [8]. It can be seen from the figure that at 70 °C and 85 % RH, a complete 6inch solar cell at the module edge is at risk by this degree of degradation.

It is interesting to note that the analytical solution for the local leakage current density described in Section 2 is a linear superposition of two exponential functions, see Equation (1). From Fig. 3 it can be seen that it can often be approximated by a single

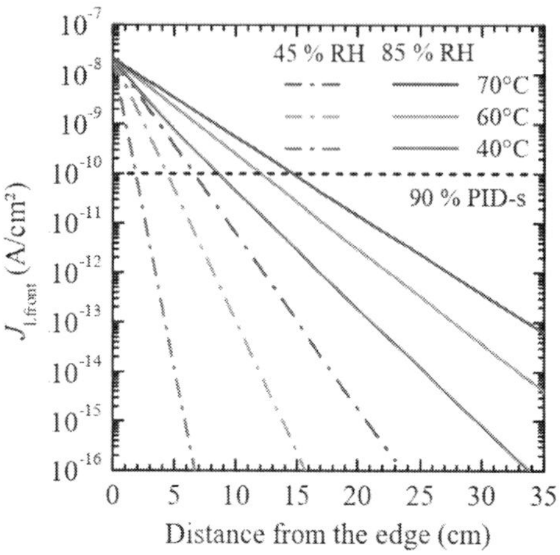

Fig. 3: Analytically calculated local leakage current density through the front side of a PV module with the cross-sectional layout 3.2 mm glass/0.42 mm EVA/cell array. The bias voltage between cell array and grounded module is 1000 V. The dashed black line indicates the expected local 90 % power degradation of PID-s prone c-Si solar cells after 6.9 days.

exponential function. Whether this is a good approximation depends on the ratio of the sheet resistance of the glass surface R_{sheet1} and the lumped sheet resistance R_{sheet2} of the front cover, front cover/encapsulant interface and encapsulant.

4 SUMMARY

Easy-to-use equations for analytical calculation of leakage current density in PV modules as a function of distance from the grounded module frame were presented. The result was compared with that obtained from numerical simulations of a resistor network. By varying the node spacing in the resistor network, it was found that the numerical solution approaches the analytical solution for a small node spacing of 0.1 mm, showing i) the validity of the analytical solution and ii) the need for a very fine resistor network for accurate numerical simulation, which can thus become quite time consuming. As an application, the effect of temperature and relative humidity on local leakage current density in a standard c-Si PV module was calculated analytically using the equations in a spreadsheet software along with previously measured material parameters. In combination with climate data and degradation rates as a function of area-related charge density, it may be possible to rapidly predict local degradation of PV modules due to leakage current as a function of exposure time in different climates.

ACKNOWLEDGEMENT

Part of this work was funded by the German Federal Ministry for Economic Affairs and Energy within the project "PID-s" under contract number 0325748A.

REFERENCES

[1] J. Berghold, O. Frank, H. Hoehne, S. Pingel, B. Richardson and M. Winkler, "Potential Induced Degradation of Solar Cells and Panels", *Proceedings of the 25th European Photovoltaic Solar Energy Conference* (2010) 3753.

[2] R. Swanson, M. Cudzinovic, D. DeCeuster, V. Desai, J. Jürgens, N. Kaminar, W. Mulligan, L. Rodrigues-Barbosa, D. Rose, D. Smith, A. Terao, and K. Wilson, "The Surface Polarization Effect in High-Efficiency Silicon Solar Cells", *Proceedings of the 15th International Photovoltaic Science & Engineering Conference* (2005) 410.

[3] G. R. Mon and R. G. Ross, "Electrochemical Degradation of Amorphous-Silicon Photovoltaic Modules", *Proceedings of the 18th IEEE PV Specialists Conference* (1985) 1142.

[4] J. A. del Cueto and T. J. McMahon, "Analysis of Leakage Currents in Photovoltaic Modules Under High-Voltage Bias in the Field", Progress in Photovoltaics: Research and Applications 10 (2002) 15.

[5] N. Shiradkar, E. Schneller and N. G. Dhere, "Finite Element Analysis Based Model to Study the Electric Field Distribution and Leakage Current in PV Modules Under High Voltage Bias", *Proceedings of SPIE 8825* (2013) 88250G.

[6] H. Nagel, M. Glatthaar, and S. W. Glunz, "Quantitative Assessment of the Local Leakage Current in PV Modules for Degradation Prediction", *Proceedings of the 31st European Photovoltaic Solar Energy Conference* (2015) 1817.

[7] https://www.analog.com/en/design-center/design-tools-and-calculators.html#LTspice

[8] A. Raykov, H. Nagel, D.-J. Amankwah and W. Bergholz, "Climate Model for Potential-Induced Degradation of Crystalline Silicon Photovoltaic Modules", *Proceedings of the 27th European Photovoltaic Solar Energy Conference* (2012) 3399.

Electrical bias stabilisation of power of the photovoltaic technologies

Ankit Mittal[1], Nicole Zechner[1], Marcus Rennhofer[1], Gusztáv Újvári[1]
[1]Austrian Institute of Technology, Giefinggasse 4, 1210 Vienna, Austria
* email: ankit.mittal@ait.ac.at

ABSTRACT: In scope of stabilization of PV modules technologies, it has been a challenge for many measurement labs to stabilize the power in a shorter interval of time. Since, the usual standard procedure as stated in IEC 61215-1-x (x: technology type) is to keep to modules under illumination for a certain number of hours at its MPP load. An effort is being made to make the stabilization of modules faster and more reproducible before the actual performance determination. Since 2011, however, methods have also been proposed that work purely electrically, i.e. without illumination [1-3]. Therefore, this study aims to establish dark bias methods in comparison to conventional, exposure methods. Our goal is to establish a shorter, purely electrical stabilization procedure to mitigate the metastable effects in thin-film and faster measurements (thin-film and crystalline). The measurement results and reliability in regard to initial power stabilization for different technologies were compared.
Keywords: Stabilization, Dark biasing, thin-films, crystalline silicon

1 INTRODUCTION

To accurately model outdoor performance of photovoltaic devices, the state reached in outdoor deployment must be well represented in indoor power measurements with simulated sunlight [4].

While thin film PV devices have been in production for decades, their performance characterization is still a topic of ongoing research. This results partly from the fact that while all thin film cells exhibit some sort of metastable behavior, the technologies used are diverse and each is physically very different from the well established crystalline silicon cell [5]. Precise measurements are hindered by reversible metastable changes. These changes range in time scale from a few seconds to a few days. Another problem is the possiblity that metastable effects are confused with permanent degradations [6]. Understanding of metastabilities will lead to more accurate characterization of PV devices [7].

To generate this outdoor exposed state, prolonged light exposure by natural or artificial light is the standard procedure according to the international standard IEC 61215/1-1-x [8]. But the standard also allows for the use of other stabilizing procedures, if the given standardized specifications are met [8]. One approach to alternative methods of stabilization for thin film photovoltaic devices is the application of a forward bias without light exposure ("dark bias") as proposed by [9], [10] and [6]. Even though this purely electrical stabilization procedure yielded favorable results throughout literature ([4,6,9,8] little was published on the topic in the last five years.

2 MOTIVATION

The most important parameters for sizing a photovoltaic power plant are the exposure to sunlight in the location of the power plant and the power output of the solar modules used [13,14].

Nearly all photovoltaic devices exhibit performance changes when exposed to light, but the scale of these changes varies widely between different technologies [15]. While crystalline silicon devices are relatively stable on medium time intervals (minutes to days) and mostly suffer from long term degradation [14], thin film photovoltaic devices experience metastable effects (timed from seconds to days) that interfere with performance measurements [6]. The procedure was introduced for amorphous silicon, historically the first thin film photovoltaic technology [15]. Other thin film technologies are currently subjected to the same stabilization procedure before power measurements, even though the method is not optimized for them [11].

In [10] electrical forward bias was proposed as an alternative to light soaking for stabilization of thin film photovoltaic devices.

One of the preconditioning methods was dark bias with a comparatively low biasing current of 1/3 of the short circuit current. Somewhat surprisingly, this method yielded good results that had not been found in any previous experiments where much higher currents had been used. Since all four different modules types (CdTe, CIGS, CZTS) tested gave good results for dark bias, it was decided to look into the method further.

3 METHODOLOGY

Stabilization of modules power through biasing in dark have been proposed and published in the1990s [11] and a number of publications came from it in the time from 2010 to 2015, mainly from the NREL [11] in 2010, [9] in 2012, [6] in 2014 and [4] in 2015.

3.1 Stabilization Criterion

According to the IEC 61215-2 the stabilization step was continued until the stabilized electrical power (below 2% overall change) was reached according to the mentioned formula (1) for three consecutive steps of stabilization in a way that

$$\Delta P = (P_{max} - P_{min}) / P_{average} \leq 0.02 \qquad (1)$$

with Pmax the largest value of P over the three cycles, Pmin the smallest value of P over the three cycles and Paverage the averaged power of three consecutive cycles. When this stabilization criteria was met, the power output of the module was considered stabilized according to IEC 61215. Measurements were stopped either after the condition $\Delta P \leq 0.02$ was met, or it appeared that no stabilization was possible in a reasonable amount of time.

3.2 Measurement procedure

The full scale CIGS and cSi modules first underwent an initial measurement in the IV flasher before starting with the biasing experiments. This measurement was used to determine the bias current, taking the average of the two I_{SC} values of each technology. The biasing current for the thin film modules (CIGS) was set at 1/3 of the averaged I_{SC} of the two modules as:

I_{bias}, CIGS $=1/3\ (I_{SC1} + I_{SC2})/2$

I_{bias}, c-Si $=2/3\ (I_{SC1} + I_{SC2})/2$

For the mono-c-Si modules, a biasing current of 2/3 of the initial Isc was used. The higher current used for the crystalline silicon modules was applied because a higher current would be needed to find a measurable change in the generally stable c-Si technology.

Table I: Current biasing parameters for the full scale modules (CIGS and c-Si), according to change in temperature. The current was chosen higher for crystalline silicon, but for CIGS a lesser biasing current was chosen to avoid any degradation.

Measurement step	Temperature	Current Bias	Module Type
1	50 °C	1/3 I_{SC}	CIGS
1	50 °C	2/3 I_{SC}	c-Si
2	65 °C	1/3 I_{SC}	CIGS
2	65 °C	2/3 I_{SC}	c-Si
3	75 °C	1/3 I_{SC}	CIGS
3	75 °C	2/3 I_{SC}	c-Si

4 RESULTS

All the modules (c-Si and CIGS) reached stabilization during biasing at each chosen temperature. The difference have been observed in the amplitude of the electrical parameters in comparison to their initial starting point at each temperature.

CIGS

In figure 2, electrical paramteres Pmax, fill factor there is a pronounced difference between modules CIGS 1 and CIGS 2, where CIGS 1 has larger changes in all these parameters. Changes in Isc and Voc are on a very small scale, below the resolution of measuring instruments featured here.

The overall improvement in fill factors leads to a higher Pmax. A higher temperature biasing noteworthy tendency for higher temperatures to also bring larger deviation (not strictly, but on average). This is the most pronounced in the first step of biasing and less pronounced in the second and third steps. In figure 4. It is an observable improvement in FF in the CIGS modules, whereas no significant change in the c-Si modules.

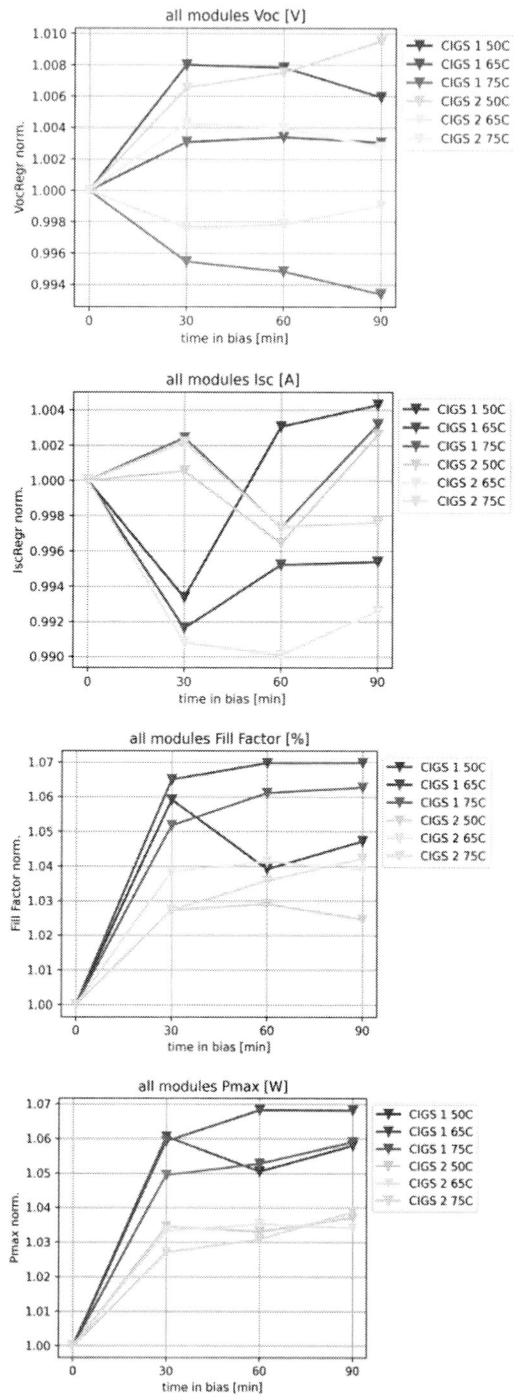

Figure 2: Stabilization of CIGS modules using dark bias procedure. It shows the overview of electric parameters during biasing at different temperatures.

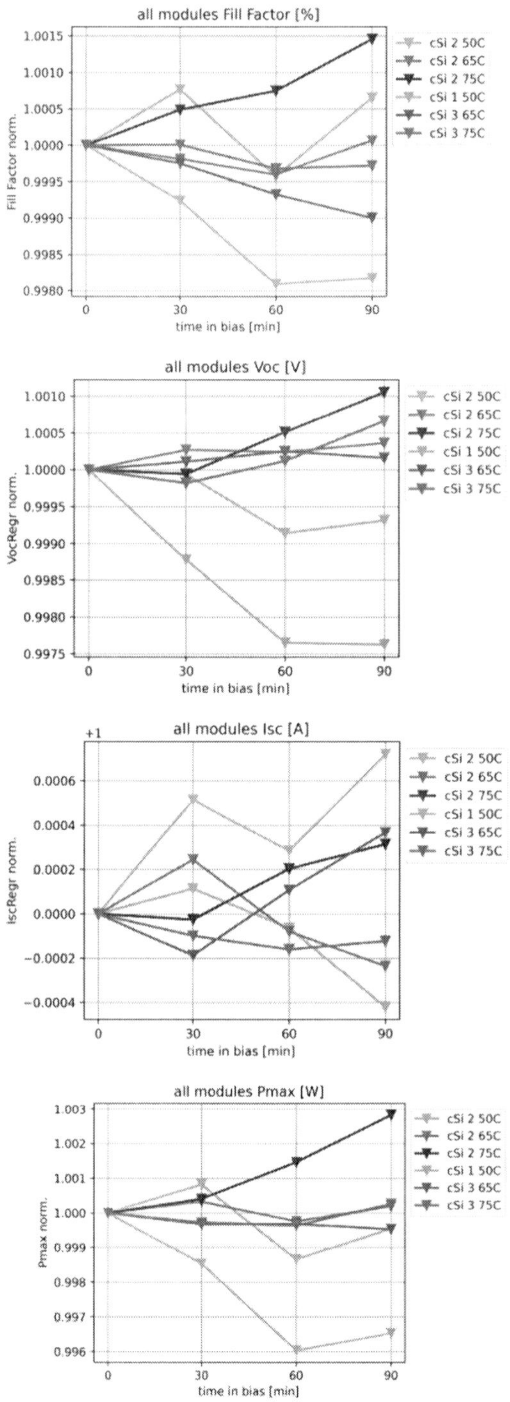

Figure 3: Stabilization of Crystalline silicon modules using dark bias procedure. It shows the overview of electric parameters during biasing at different temperatures.

Crystalline silicon

As observed in figure 3, all changes happened on extremely small scales (< 0.5%). Bias appears to cause both degrading, rearranging mechanisms at 50°C as well as ordering mechanisms at 75°C. Voc and Vmpp appear to be sorted by temperature (the higher the temperature, the larger these parameters). This indicates that 50°C are not enough to trigger annealing.

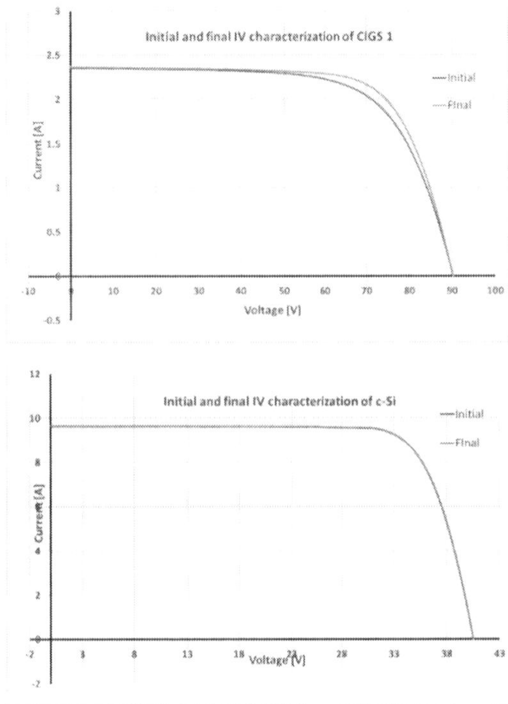

Figure 4: Variation of temperature in combination of dark biasing in the CIGS and crystalline silicon technology. The initial curve (blue) is without any dark biasing and the final curve is the one after the finishing of the dark biasing at 75°C.

Comparing the IV characteristic from initial unstabilized power to the stabilization after 75°C dark bias measurement shown in figure 4 suggest a clear change in performance of the CIGS modules. For the CIGS technology a substantial change in power was observed during each step of stabilization, but on the contrary crystalline silicon a bare minimum change was observed.

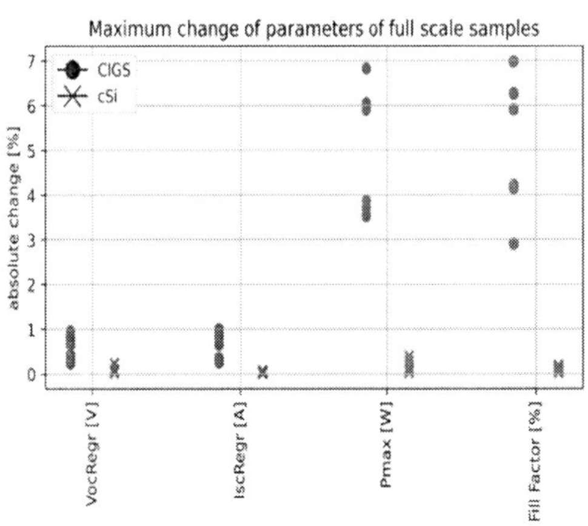

Figure 5 : Absolute change of the most important

electrical parameters of CIGS (dots) and c-Si (crosses) modules.

In table 2, the average change in the power from its initial measurement to its final stabilized power at the particular temperature is shown. An absolute changes in the electrical parameters can be seen in figure 5. It clearly shows that due to the metastable behavior of CIGS technology, a higher amplitude of change was observed of approximately 5% during each stabilization procedure. Whereas the c-Si technology showed less than 0.5% change in the power.

The overall change in the module power according to the formula described in section 3.1, was well below the requirement, as shown in figure 6. The module passed the dark biasing procedure at each temperature step.

Table 2: Average change of maximum power output of full scale modules, averaged by temperature.

Temperature [°C]	CIGS: average change [%]	c-Si: average change [%]
50	4.88	0.27
65	5.17	0.04
75	4.89	0.16

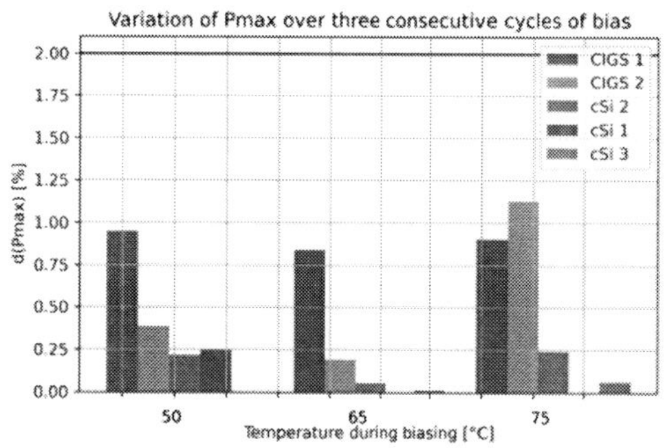

Figure 6: Overview of changes in parameters for all full scale modules between dark biasing procedures, as well as the 2%-marker indicating stabilization.

5 DISCUSSION

It can be inferred from figure 6, that all full scale modules achieved stabilization during dark biasing at different temperatures. The deviation in power are well below the 2%-threshold for both technologies. Crystalline silicon technology showed a deviation in power of less than 0.25%, whereas the CIGS technology showed less than 1% of deviation in power.

Since the power changes for c-Si modules were much smaller than for the thin film CIGS modules, it can be result of using lower current or the shorter duration of stabilization cycle. Also, c-Si is known for forming boron and oxygen metastable complexes under illumination, which can account for one of the reasons why the modules didn't show degradation in power under dark bias.

CIGS modules were able to avoid the short-term metastablities and also long-term metastabilities, since the former is susceptible to the light and the stabilization in dark helps eliminating it. The long-term metastabilities can be stabilized by the application of elevated temperatures and biasing, which was exactly carried out during the dark biasing.

The modules showed stabilization faster than under illumination, as stated in the standard IEC 61215-1-x procedure. Further experiments are needed to achieve comparable results like during outdoors exposures or under simulated sunlight.

6 CONCLUSION

All the module achieved stabilization using the proposed methods. The effort of this study was to reduce time of power stabilization procedure. Since It can ease up the procedure for the labs without requiring a light induced stabilization. Crystalline silicon modules didn't seem to show detrimental effects due to this stabilization procedure.

Module power rating and stabilization deal with standardized reproducibility in the lab, which can be resolved by using dark biasing procedure.

There the best way for uniform measurement still seems to measure dark stored, mainly for the thin-film technology.

7 REFERENCES

1] T.J. Silverman, M.G. Deceglie, B. Marion, S.R. Kurtz „Performance Stabilization of CdTe PV modules using bias and light", in IEEE Journal of Photovoltaics, vol. 5, no. 1, pp. 344-349, Jan. 2015.

[2] C. Deline et al. "Electrical Bias as an Alternate Method for Reproducible Measurement of Copper Indium Gallium Diselenide (CIGS) Photovoltaic Modules", conference paper (NREL/CP-5200-56078), SPIE Optics + Photonics 2012, San Diego, California, Aug. 2012

[3] S. Dittmann, A. Virtuani, G. Friesen and F. Serrano, "Current-Soaking and Dark Storage Effects of Polycrystalline Thin Film Solar Modules", 2014 IEEE 40th Photovoltaic Specialist Conference (PVSEC), Denver, CO, 2014.

[4] M. G. Deceglie, T. J. Silverman, K. Emery, D. Dirnberger, A. Schmid, S. Barkaszi, N. Riedel, L. Pratt, S. Doshi, G. Tamizhmani, et al., "Validated method for repeatable power measurement of cigs modules exhibiting light-induced metastabilities," IEEE Journal of Photovoltaics, vol. 5, no. 2, pp. 607–612, 2014.

[5] T. J. Silverman, U. Jahn, G. Friesen, M. Pravettoni, M. Apolloni, A. Louwen, W. van Sark, M. Schweiger, G. Belluardo, J. Wagner, et al., "Characterisation of performance of thin-film photovoltaic technologies," 2014.

[6] T. J. Silverman, M. G. Deceglie, B. Marion, and S. R. Kurtz, "Performance stabilization of CdTe PV modules using bias and light," in 2014 IEEE 40th Photovoltaic Specialist Conference (PVSC), pp. 3676–3681, 2014.

[7] S. Novalin, M. Rennhofer, and J. Summhammer, "Electrical metastabilities in chalcogenide

photovoltaic devices," Thin solid films, vol. 535, pp. 261–264, 2013.

[8] IEC 61215/1-1-x, "Terrestrial photovoltic (PV) modules - Design qualification and type approval," standard, International Electrotechnical Commission, 2016.

[9] C. Deline, A. Stokes, T. J. Silverman, S. Rummel, D. Jordan, and S. Kurtz, "Electrical bias as an alternate method for reproducible measurement of copper indium gallium diselenide (CIGS) photovoltaic modules," in Reliability of Photovoltaic Cells, Modules, Components, and Systems V, vol. 8472, p. 84720G, International Society for Optics and Photonics, 2012.

[10] R. Sasala and J. Sites, "Time dependent voltage in CuInSe2 and CdTe solar cells," in Conference Record of the Twenty Third IEEE Photovoltaic Specialists Conference-1993 (Cat. No. 93CH3283-9), pp. 543–548, IEEE, 1993.

[11] J. A. del Cueto, C. A. Deline, S. R. Rummel, and A. Anderberg, "Progress toward a stabilization and preconditioning protocol for polycrystalline thin-film photovoltaic modules," in 2010 35th IEEE Photovoltaic Specialists Conference, pp. 002423–002428, IEEE, 2010.

[13] H. H ̈aberlin, "Photovoltaik," AZ-Verlag/VDE-Verlag, Aarau, 2007.

[14] A. Wagner, Photovoltaik Engineering. Springer, 5 ed., 2019.

[14] M. Gostein and L. Dunn, "Light soaking effects on photovoltaic modules: Overview and literature review," in 2011 37th IEEE Photovoltaic Specialists Conference, pp. 003126–003131, IEEE, 2011.

New developments in accelerated weathering tests for Back-Contact modules

Guido J.W. Meijers[1], Leonid Pastukhov[2], Robert H.C Janssen[2], Frank van Duijnhoven[1]

[1]Endurans™ Solar, Urmonderbaan 22, 6167RD Sittard-Geleen, The Netherlands

[2]DSM Applied Science Centre, Urmonderbaan 22, 6167RD Sittard-Geleen, The Netherlands

Abstract— **Back-contact solar modules made with crystalline silicon cells are, to-date, amongst the highest peak power density devices on the market and are predicted to retain this status also in the coming year. One of the challenges for this technology was for some time the availability of reliable supply of conductive backsheet, back-contact solar cells and back-encapsulants which has been addressed for some years now. Still the industry is continuously searching for better and cheaper bill-of-materials, however the introduction of these materials is very time-consuming as quite some accelerated ageing tests are needed to certify the solar modules, typically over a period of 5-6 months of time. In this study, three different methods have been investigated, showing a similar ageing profile achieved in a shorter timeframe. The first method is the use of a single-contact solar device, which is much more sensitive during ageing and thus gives a quicker indication of changes in material properties. The second method is the use of a thermal shock chamber instead of thermal cycling. Third method makes use of damp-heat testing at elevated temperatures.**

Keywords—Back-contact modules, conductive backsheets, series resistance, accelerated testing, thermal shock, damp-heat.

I. Background and Introduction

Although the development of solar materials and processes is a very active field, many of the innovations do not reach commercialization, take too long to reach the market, or the full value of the technology is not well captured. This is due to several reasons. Many material and process innovations in solar module manufacturing are hampered by the complexity of the different process steps and materials, the lack of harmonization between the different materials and processes as well as the unavailability of turnkey solutions over the whole value chain. High switching costs and low degree of automation (high labor cost) may disqualify some of the newer technologies. In addition, generating sufficient reliability data for material screening and development purposes is a time intensive process, of minimum three to six months. In addition, before a bill-of-materials is accepted by modules manufactures, components needs to be certified according to various solar standards to ensure a constant electricity gain for at least 20 years. One of the key standards to fulfill is the IEC 61215 (approval on the design). Module manufacturers often require to exceed 2-3 times the official norm, which then easily takes 5-6 months in total.

Back-contact solar modules made with crystalline silicon cells are facing similar challenges, although they are amongst the highest peak power density devices on the market and are predicted to retain this status also in the coming years [1]. Especially modules built with interdigitated back contact (IBC) cells are leading in this respect (see the crystalline silicon cell record tables [2]). Apart from utilizing such high efficiency cells, back-contact solar module efficiencies also benefit from innovative cell interconnection techniques that minimize cell to module losses (CTM), the most promising of which is to use a conductive (metallic) back-sheet that is connected to the rear of the cells with a conductive paste or solder [3, 4].

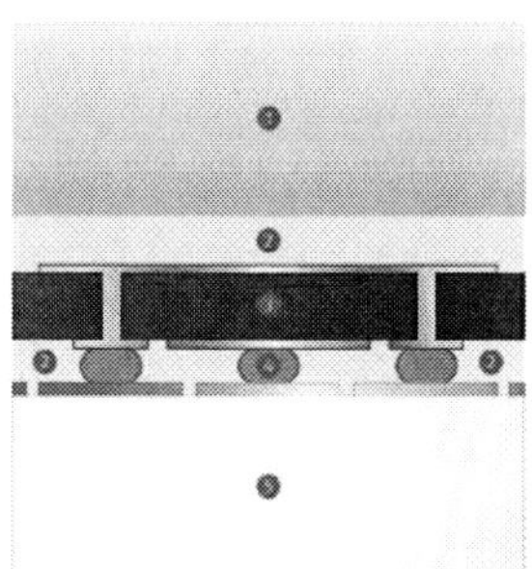

Figure I-1 Schematics of the built-up of a back contact module: 1-glass; 2- encapsulant; 3- solar cell; 4-conductive adhesive; 5-conductive backsheet; 6-copper; 7- conductive foil.

Another great advantage of such back-contact solar modules is that these modules can be assembled automatically which increases their reliability considerably. The back-contact technology is thus very promising, however still only a small portion of the solar market is utilized with these modules. Tailored and faster accelerated ageing tests, simulating degradation of back-contact PV module technologies under given stress conditions, but still comparable to those observable in real-life, are key for efficient and fast product improvement and acceptance of back-contact PV technologies.
This work focuses on techniques for measuring single-contact resistance in emerging conductive backsheet technology upon ageing and on the importance of this approach on enabling detection of power loss upon faster testing cycles and without

the need of fabricating full-size modules. The overarching goals are to develop faster and reliable ageing protocols accessible for implementation in testing facilities across the world and to enable adoption of conductive backsheet at faster pace, and yet reliable.

II. EXPERIMENTAL

A. *Single-point resistance measurement*

The electrical resistance on individual conductive-contacts sandwiched between cell and conductive backsheet was measured via a 4-point probe approach. A schematic of the chosen sample geometry is shown is Figure II-1.

Single cell devices were built according to standard lamination procedures with a conductive back-sheet structured according to Fig. 1 and electro-conductive adhesive stencil printed onto locations 1-16. These sites 1-16 connect to the front side of a standard JA Solar MWT cell through the cell vias. Cell vias 1-5-9-13, 2-6-10-14, 3-7-11-15 and 4-8-12-16 were looped through at the front side of the cells via cross connectors soldered onto the cells (see Fig. 2). Finally, this allows the resistance of individual electro-conductive contacts to be assessed with a 4-point probe measurement.

Figure II-2 Sample overview for single-contact resistance measurement on conductive backsheets.

B. *Thermal shock and thermal cycling testing*

In this study, 2x2 mini-modules have been prepared for thermal cycling and for thermal shock testing and characterized by IV power measurements before and after 100, 200, 300 , 400 and 600 (only TS) cycles. Thermal shock was performed using a two chamber system in which mini-modules are rapidly moved from hot and cold temperatures (Figure II-3). In both tests, the cold temperature is set on -40 °C and the hot temperature is set on 85°C. Another large difference between the tests is that the dwell time in the TS chamber was much shorter compared to the dwell time in the TC chamber.

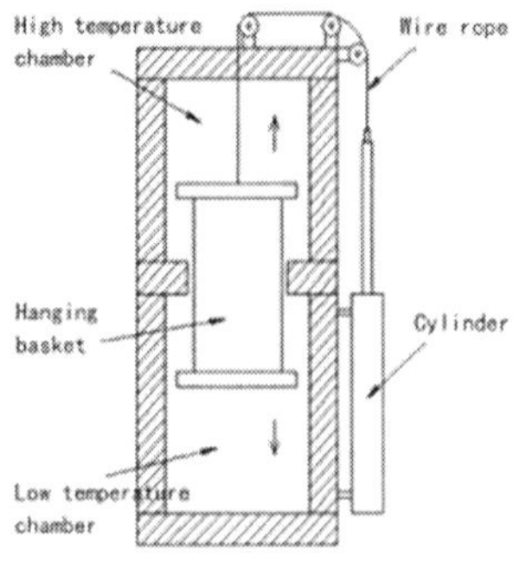

Figure II-4 Schematics of the thermal shock chamber

Four-cell back-contact mini-modules have now been aged in both a TS- as a TC-chamber. The ageing of the module was measured by performing a flash-test every 100 cycles, after which the change in power output was calculated.

C. *Damp-heat testing at elevanted temperatures*

The last ageing method which has shown to give a much earlier indication of the ageing behavior of new materials is the use of a damp heat chamber at higher temperature. As described in the Arrhenius law, the rate of chemical reactions scale like $e^{-E_{act}/RT}$, i.e. most chemical reactions double their speed with every 10 °C temperature increase. Now this phenomenon is used to receive results quicker by performing the DH test at temperatures higher than 85 °C (comparing the data to results generated at 85 °C). We now investigated the ageing of 4-cell mini-modules at 75, 85 (reference) and 95 °C to be able to make a solid evaluation on the influence of the temperature during this test. Again 2x2 back-contact mini-modules were produced and aged in a damp heat chamber which were all set on a relative humidity of 85 %, but with a different temperature (75, 85, 95 °C).

III. RESULTS AND DICUSSION

A. *Single point resistance measurement upon ageing*

Results now show that only about 100 thermal cycles are needed to receive sufficient change in the measured contact resistance (Figure III-1). Normally minimum 400 thermal cycles are needed to measure a significant change in the power output of mini- modules. The ageing time needed to receive a first adequate conclusion is therefore already done after 1 month, and thus not after 4 months. More details will be given during the IEEE event.

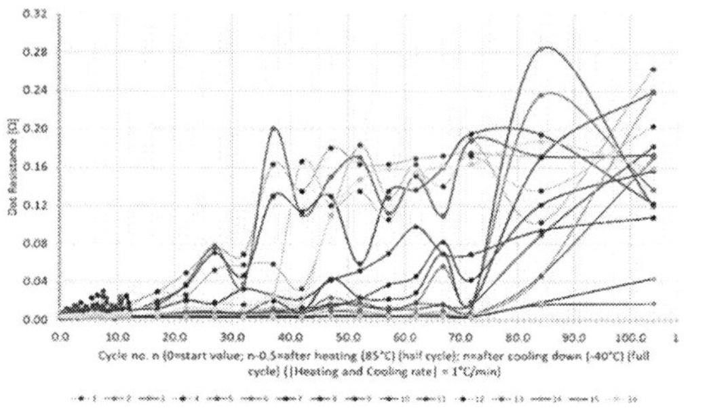

Figure III-2. Contact resistance measurement on different conductive adhesive dots during thermal cycling.

B. Thermal shock testing

The results show that 400 thermal cycles are needed to obtain about 2 % of power decay, and also 400 cycles of thermal shock are needed to obtain about the same 2 % power decrease. However, the TC-results took about 4 months including measurement and analysis time to complete, whereas with thermal shock only 3-4 weeks were needed to obtain the same results and conclusions.

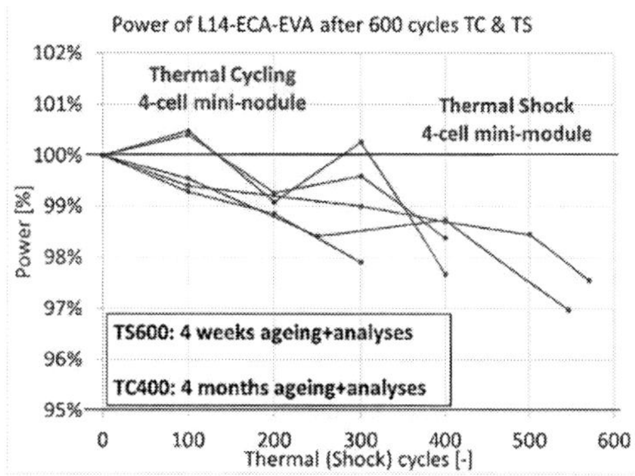

Figure III-3Field factor decay upon thermal cycling versus thermal shock test.

C. Damp-heat testing at elevanted temperatures

Results show that a clear power decrease is noticed after about 2500 hours of ageing at 95 °C, and a similar power decrease is measured after 4000 hours at 85 °C. So again, results are obtained significantly faster using this different approach of damp heat ageing.

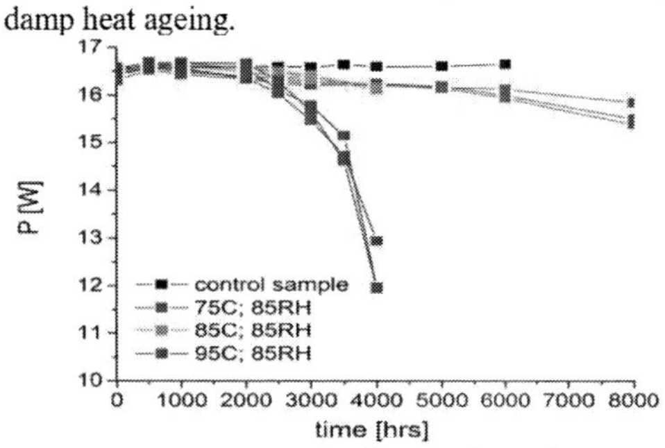

Figure III-4Power output during damp heat using different temperatures

IV. SUMMARY

Back-contact solar modules made with crystalline silicon cells are, to-date, amongst the highest peak power density devices on the market and are predicted to retain this status also in the coming year. However, the technology is still only a small portion of the current solar market but is growing rapidly. One of the challenges for this technology was for some time the availability of reliable supply of conductive backsheet, back-contact solar cells and back-encapsulants which has been addressed for some years now. Still the industry is continuously searching for better and cheaper bill-of-materials, however the introduction of these materials is very time-consuming as quite some accelerated ageing tests are needed to certify the solar modules. The challenge is that about 5-6 months of time is needed before the customer receives sufficient ageing data, and in this article a clearly faster evaluation has been shown of two of the most significant ageing tests, thermal cycling and damp heat. Three different methods have been investigated which show that a similar ageing profile can be received, but in a shorter timeframe. The first method is the use of a single cell solar device, which is much more sensitive during ageing and thus gives a quicker change in properties. The second method is the use of a thermal shock chamber instead of thermal cycling, because this chamber can be set on the same temperatures as TC but the cycle time is much shorter. Thus, a similar ageing profile can be seen in only 3-4 weeks, where 4 months is needed in a thermal cycling-chamber. The third method which shows a great benefit, is ageing in a standard damp heat chamber, but at a higher temperature. Because ageing at a higher temperature will be faster (Arrhenius), significant ageing is received already after 2500 hours and not after the normal 4000 hours. .

REFERENCES

[1] ITRPV Eighth Edition, "International Technology Roadmap for Photovoltaic Results 2016", 2017.

[2] M. A. Green, Y. Hishikawa, E. D. Dunlop, D. H. Levi, J. Hohl-Ebinger, and A. W. Ho-Baillie, "Solar Cell Efficiency Tables (version 51)", Progress in Photovoltaics, 26(1), 3, 2018.

[3] I. J. Bennett, E. Bende, M. J. A. A. Goris and W. Eerenstein, "An Overview of Developments in Foil-Based Back-Contact Modules" presented at 29th European Photovoltaic Solar Energy Conference and Exhibition, Amsterdam, The Netherlands, 22-26 September 2014.

[4] G.J.W. Meijers; I. J. Bennett, P. Di Lorenzo, P. Pasmans. I. Suez, "Real benefits from conductive backsheets from back-contacted solar modules" presented at 37th European Photovoltaic Solar Energy Conference and Exhibition, Virtual, 7-11 September 2020.

[5] R.H.C Janssen; F. van Duijnhoven; I.J. Bennett; J.J. Xu; "Development of an Electrically-Conductive Backsheet for Back-contact based PV-Modules" presented at 34th European Photovoltaic Solar Energy Conference and Exhibition, Virtual, 7-11 September 202

END-OF-LIFE MANAGEMENT OF SOLAR PHOTOVOLTAIC PANELS IN INDIA- IDENTIFICATION OF NEEDS, STAKEHOLDERS & CHALLENGES

Kishore Ganesan[1], Dr. César Valderrama[2]
1 Sustainable Energy Consultant, No.11/12, Kalaivani Street, Rajakilpakkam, Chennai-600073, India, email address: kishoreganesh18@gmail.com, Mobile: +919962720398
2 Associate Professor, Chemical Engineering Department, Universitat Politècnica de Catalunya, Barcelona TECH and Programme Director MSc SELECT, email address: cesar.alberto.valderrama@upc.edu

ABSTRACT: Solar Photovoltaic systems create a positive impact by bringing energy equity, security, and environmental sustainability to societies across the World. Given the current volume of installations, especially in India, and those planned in the future, the need to sustainably manage Solar PV panels at the end of their useful life is undeniable. This study outlines the need for the sustainable end of life management of crystalline silicon solar panels in India. Apart from avoiding environmental impact due to unsafe disposal, sustainable end of life management can relieve raw material supply constraints. A material criticality analysis is conducted on forecasted PV panel production in India until 2050. The results highlight the various raw material supply constraints and the volume of materials embedded within PV panels until 2050 with respect to the current production volumes. This study also provides an interactive stakeholder map displaying the location and the related parameters of potential PV waste generation sites and downstream recyclers in India for end-of-life management of PV panels. A preliminary zone-wise analysis has been carried out to identify a geographical mismatch between stakeholder locations that could affect the transportation of waste in the future and tentatively quantity the number of PV takeback and recycling centres needed in India in the near future. This map can serve as a reference for future studies to be carried out for location optimization of PV takeback and recycling centres in India.

Keywords: c-Si, Recycling, PV Material, PV Module, Environmental Effect

1 INTRODUCTION AND MOTIVATION

According to IRENA, India is expected to be the fourth-largest PV waste producer in the World, with 4.5 to 7.5 million tonnes of PV waste by 2050 [1]. Currently, India has no regulations in place to tackle solar waste specifically. However, India's solar PV installations will increase exponentially, making the country a growing market for PV waste without regulations.

In relation to Solar PV waste, the Ministry of New and Renewable Energy had developed a Concept Note/ Blueprint on Management of Antimony Containing Glass from End-of-Life of the Solar PV Panels [2]. The ministry of environment, forests, and climate change has drafted a national resource efficiency policy mandating establishment of recycling centers for Solar PV panels. It is specified in the policy that four authorized PV panel dismantling facilities would be established by 2025 and 8 such facilities by 2030. A target of 85 % recovery rate of materials by weight from discarded PVs is also set in the draft policy [3]. Ministry of New and Renewable Energy (MNRE) has also initiated a feasibility study to identify best practices/policies needed for the end-of-life management of solar PV panels in India [4]. However, there are no reported studies in the literature on various aspects of end-of-life management of PV panels in India. The country is currently devoid of PV recycling infrastructure on a commercial scale [5].

Challenges in end-of-life management of PV panels with respect to transportation and downstream processing for secondary raw material production is also highlighted. India is the fifth largest country in the world with respect to PV installation capacity. Around 90% of annual PV module installations are currently supplied through imports and in the wake of the COVID- 19 pandemic, the Government of India has laid out a vision for "Self-Reliant India", which is bound to increase domestic production of products in the solar PV value chain in the country. On the other hand, the recycling sector is not well organized in India [6]. In India, 95% of the approximately 2 million tons of e-waste generated per year is collected and processed by informal micro-entrepreneurs under unsafe work conditions and with unsound recycling practices [7]. Therefore, the conducted study on end-of-life management of solar PV in India can help proactively plan for circular PV life cycle and formalize the recycling infrastructure for solar waste in the country.

This study outlines the need for the sustainable end of life management of crystalline silicon solar panels in India. Apart from avoiding environmental impact due to unsafe disposal, sustainable end of life management can relieve raw material supply constraints. A material criticality analysis is conducted on forecasted PV panel production in India until 2050.

This study also provides an interactive stakeholder map displaying the location and the related parameters of potential PV waste generation sites and downstream recyclers in India for end-of-life management of PV panels.

A preliminary zone-wise analysis has been carried out to identify a geographical mismatch between stakeholder locations that could affect the transportation of waste in the future and tentatively quantity the number of PV takeback and recycling centers needed in India in the near future. This map can serve as a reference for future studies to be carried out for location optimization of PV takeback and recycling centers in India.

2 METHODOLOGY

2.1 Material criticality analysis for establishing the need for End-of-Life Management of PV panels in India

Raw material criticality analysis for PV module production has been carried out in the Indian context in this study. The aim is to emphasize the need for urban mining of waste PV panels to lessen the burden of raw material supply constraints in the future. The material criticality is identified by comparing the forecasted total raw material demand till 2050 as opposed to the current production, import, export scenario for materials involved in PV module production in India. Crystalline silicon PV technology is primarily focused on this study owing to its enormous market share at present and as predicted in the future. The methodology used in this study is depicted in Figure 1.

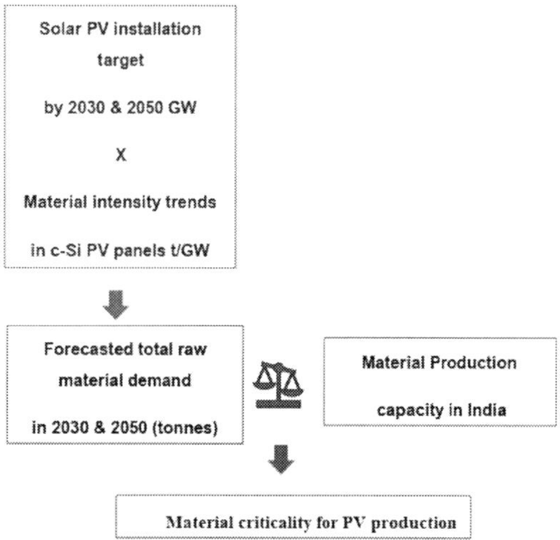

Figure 1. Methodology for material criticality analysis on PV panel production in India

2.2 Stakeholder Map for End-of-Life Photovoltaic Panel management in India

The data sources and assumptions used to construct the map are illustrated in Table I.

Table I Assumptions and Data sources for reverse logistics map

Data Source & Assumptions	
Solar PV state-wise installed capacity	Ministry of New and Renewable Energy (MNRE)
Utility-scale power plant location, Year of commissioning	Department of Economic Affairs (Cross checked with State renewable agencies & Satellite view of Google maps) >= 100 MW plants
PV waste projection (Assumption)	Panel: CS6X-320P Canadian Solar (Conservative estimate)
Glass plants (Float, fibre glass, container)	All India Glass Manufacturers' Federation (AIGMF)
Aluminium & metal recyclers	ENF Recycling Directory
E-waste recyclers	Ministry of Electronics & Information Technology (MeitY) >= 5000 MTA

The stakeholder map points to geographical mismatches between power plants and downstream recyclers in India. This map could serve as a base for future reverse logistics study to optimise transportation distances. It can also help interested businesses tentatively decide processing capacity and locate take-back and recycling centres relative to the displayed waste generation sites and downstream treatment sites.

A preliminary analysis to assess the capacity required and location for PV recycling centres across India was performed based on the stakeholder map. The analysis was carried out for five geographical zones similar to the electricity grid demarcation, namely northern, southern, western, eastern, and northeastern, as shown in Figure 2

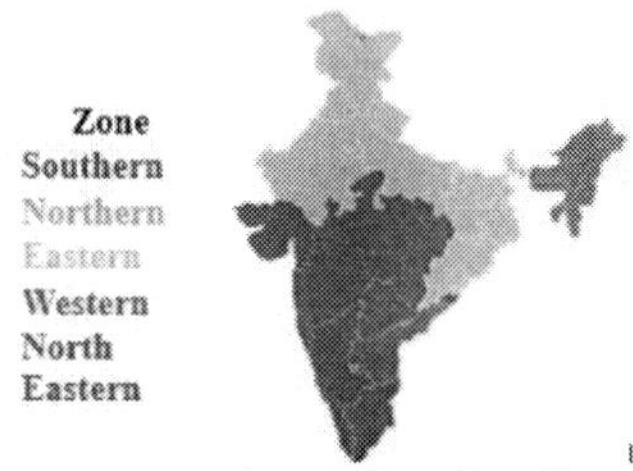

Figure 2 Zone wise split of India for recycling infrastructure criticality

Solar PV waste expected by 2030 from PV installations was calculated for each of the zones based on the assumption of module type, power capacity, and weight used in the stakeholder map with the assumption of a 25-year lifetime. In addition, an analysis of off-grid solar waste potential in India was also carried out. The solar panel was assumed to have the material composition, Assuming 100% recovery of materials from PV panels, the recyclable individual material capacity was calculated for each of the zones and compared with the existing recycling capacity to analyse the criticality of the zones for self-processing of waste and the need for excessive transportation.

3 RESULTS

The results have been presented with the needs for End-of-Life Management of PV panels in India emphasized through material criticality analysis and the identification of stakeholders and challenges via an interactive stakeholder map.

3.1 Need for End-of-Life Management for Solar PV panels in India – Material Criticality Analysis

Considering a logistic growth of Solar PV installations in India in realization of the 2030 target and onwards to 764 GW Solar PV capacity by 2050, the projection in Figure 3 has been modelled. In India, on assumption of 25 years lifetime for PV panels, significant End of Life PV systems arise after 2040. In the analysis, it is assumed that all the retired PV capacity would be repowered. Logistic function is used to model the growth of Solar PV as it is considered more realistic compared to exponential growth. The model based on [8] is calculated as follows,

$$Cumulative\ PV\ installation(Year) = \frac{L}{1 + e^{k(x_0 - Year)}}$$

Table II Variables used in forecasting trends in future PV installations in India

L	Maximum projected value of Cumulative PV capacity	600 GW
k	Expected growth rate of PV installations	15%
X_o	Year of maximum growth (mid-point of curve)	2030

The values of L, k and xo shown in Table II assume that India will achieve the set targets of 60 GW, 300 GW and 764GW of PV Capacity in 2022, 2030 and 2050, respectively. The years involving re-addition of capacity are indicated in green in Figure 3.

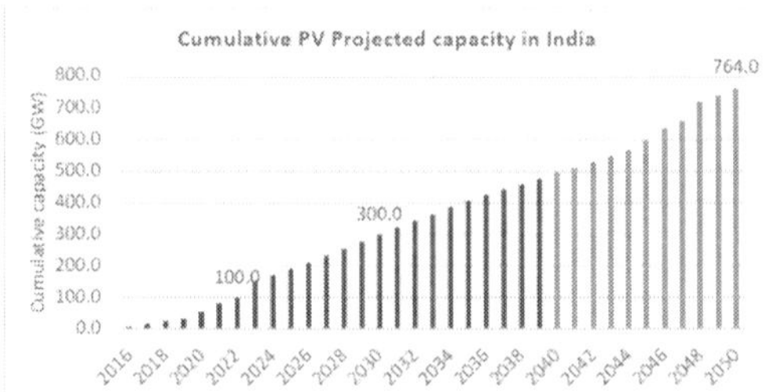

Figure 3 Predicted Logistic growth of cumulative Solar PV capacity in India (2020-2050)

Material usage trends needs to be studied to forecast the raw material demand in the future owing to the impending large-scale installations of Solar PV across the world. Country wise it is essential to identify vulnerabilities in the supply chain and to assess whether future raw material demand can be satisfied with scalable production capacities. The major elements that make up a c-Si PV system are Concrete, Steel, Glass, Aluminium, Copper, Plastics, Silicon & Silver. Demand for Concrete & Steel which are used to install PV system support structures seem to be non-critical as they would represent just 1% and 3% of their current world production capacity in 1 TW PV installation respectively [9]. Therefore, the rest of the materials used in PV module production are considered for this analysis.

The following analyses assumes that the current market share of 95% for c-Si PV technologies would remain until 2050 with modules having lifetime of 25 years. The material intensities (t/MWp) are varied under three scenarios as shown in Table III and depicted in Figure 4.

Table III Scenario description for c-Si PV module material demand [10]

Scenario	Glass, Al, Cu, Plastics	Silver (Ag)	Silicon (Si)
	% reduction by 2050 from current demand in 2019		
Low Demand Scenario (LDS)	20%	95%	75%
Medium Demand Scenario (MDS)	10%	90%	50%
High Demand Scenario (HDS)	0%	75%	25%

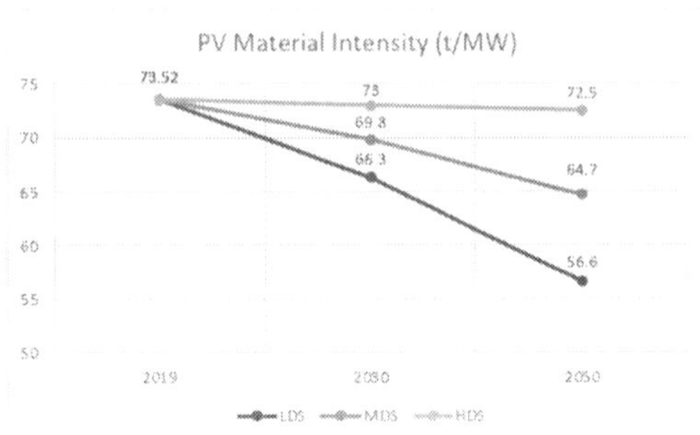

Figure 4 PV Material Intensity variation across scenarios

While the material intensities of PV module materials are expected to gradually reduce by 2050, the materials which are constituents of PV cell like Silicon and Silver are forecasted to have drastic reductions in their material intensities owing to their high contribution to overall module price.

2019 (India) at 8.8GW/year		
Material intensity (t/MW)		**Annual Material Demand (000' t)**
Glass	46.4	408.32
Aluminium	12.5	110
Copper	4.6	40.48
Silver	0.02	0.176
Silicon	4	35.2
Plastics	6	52.8
Total	**73.52**	**650**

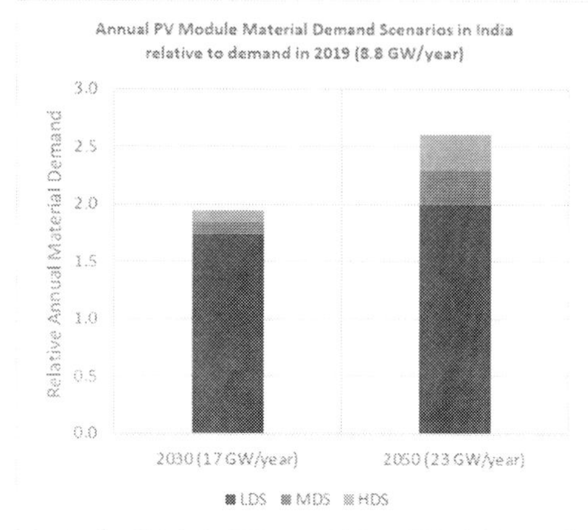

Figure 5 Annual PV module Material Demand in India relative to demand in 2019

It can be seen from Figure 5 that annual material demand for c-Si PV installations in India is expected to almost double across scenarios in 2030 and is expected to at least double (LDS) or increase at most by 2.6 times (HDS) in 2050 from the current demand of 650 thousand tonnes in 2019. Alternatively, in 2030 about 11.5 to 12.4 million tonnes of materials would be embedded exclusively in PV panels and the number further reaches 34 to 42 million tonnes by 2050 in India based on the targets as specified in Figure 3. These values also include the materials that would be present in the End of Life panels after 2040 (assuming lifetime of 25 years).

India is the fifth largest country in the world with respect to PV installation capacity. Around 90% of annual PV module installations are supplied through imports as shown in Figure 6. Domestic PV production capacity in India includes 3.1 GW of PV cell and 11 GW of PV module manufacturing. Moreover only 50% and 28% of the cell and module domestic capacity respectively is utilised for annual demand. In addition, there are no existing poly silicon and ingot/wafer production facilities in India. Energy security demands that at least 60-70% of manufacturing capacity in India to satisfy the annual demand [11]. In the wake of COVID-19 pandemic the Government of India has laid out a vision for Atmanirbhar Bharat (Self Reliant India) which is bound to increase domestic production of products in solar PV value chain in the country.

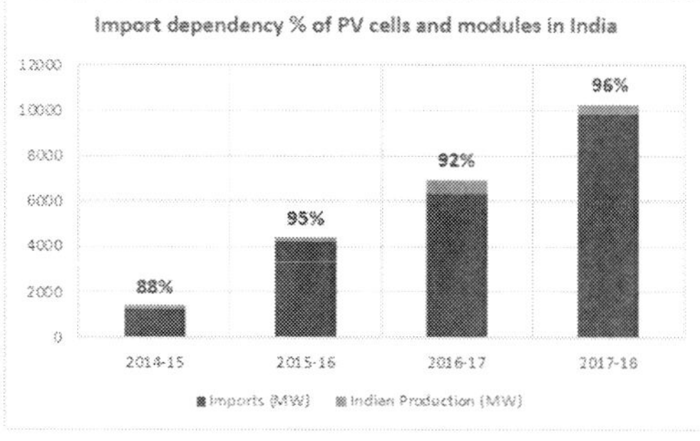

Figure 6 Import dependency of PV cells and modules in India [6]

Assuming India's domestic manufacturing capabilities in PV cells and modules increase significantly to satisfy future installation demand, the following material criticality analysis has been carried out based on Medium demand scenario as specified in Table III. Materials required upstream and generated downstream in PV module life cycle are analysed in Indian domestic production & trade context as show in Figure 7 and Table IV.

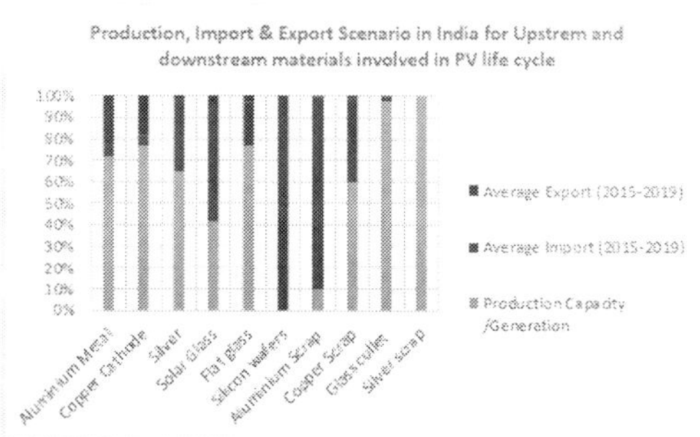

Figure 7 Production, Import & Export scenario in India for Upstream and downstream materials involved in PV module life cycle [12, 13]

Aluminium metal and Copper cathode are the only materials that are net exported in India. Although copper cathode is exported, copper ore is imported in India. In addition, aluminium scrap demand for secondary aluminium production is satisfied 90% through imports and likewise around 40% copper scrap for annual usage is imported [14]. This proves to be a significant driver for material recovery from Solar PV panels to satisfy Al, Cu scrap demand in future. Figure 8 displays the annual material demand for Aluminium ,Copper and Silver upto 2050 as a percentage of current annual production capacity in India.

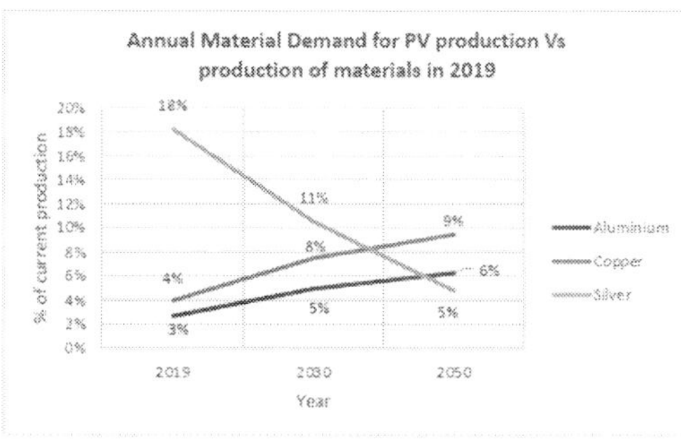

Figure 8 Annual material demand projected in India Vs current production of materials in 2019

Silver, Poly Silicon, and low iron tempered glass for solar panels (solar glass) are found to be critical materials in India with regards to increasing PV installations through domestic manufacturing in the future.

Table IV PV related materials- Scenario in India Material Production Capacity /Generation

Material	Production Capacity /Generation (Thousand tonnes)	Average Import (2015-2019) (Thousand tonnes)	Average Export (2015-2019) (Thousand tonnes)	HS Code
Aluminium Metal	4.126	0.35	1.298	7601
Copper Cathode	1.018	0.07	0.245	7403
Silver	0.00096	0.0004	2.837E-05	7106
Solar Glass	0.132	0.1704	0.0152	70071900
Flat glass	1.32	0.344	0.0584	7005
Silicon wafers	0	0.0049	0.000051	381800
Aluminium Scrap	0.12	1.0986	0.0051	7602
Copper Scrap	0.3	0.1946	0.006	7404
Glass cullet	1.97	0.047	0.00583	7001

Silver is mined and produced by Hindustan Zinc Limited in India at 966 tonnes capacity per year [14]. The yearly industrial demand for silver (1142 tonnes) is based on electronics, photovoltaics, Ethylene oxide catalysts, light duty vehicles industries and might include hybrid, BEVs in the future. As shown in Figure 8, the industrial demand for silver is satisfied through recycling of silver (25%) and imports (44%) apart from the production share for industrial usage (31%) [14]. Currently the PV cell manufacturers import silver paste for front and back metallization of PV cells. Assuming the future installations are carried out with domestically manufactured PV cells, with current industrial demand already exceeding India's production capacity, the additional demand needs to be satisfied through recycling or imports. In 2030, the Annual PV installation demand would necessitate an increase in silver import dependency by 7% from the current annual import volume of 501 tonnes. Therefore, there is need to shift the future silver supply in India for PV cells from imports to recycling or reduce the silver demand in future domestic PV manufacturing by substitution.

Figure 8 Industrial demand for Silver in India supplied through production, imports and recycling (2010-2019) [15]

Poly Silicon/ Silicon Ingots /Silicon wafers have currently no manufacturing facility in India and the current demand is satisfied completely through imports. The future demand for Poly Silicon in domestic PV industry necessitates a production capacity of 34000-60000 tonnes per annum by 2030. Additionally, production of Poly Silicon is challenging with only countries like South Korea, US, Germany and Japan supplying it commercially. This would serve as an opportunity for India to indigenize production of such materials with high supply chain risks [16].

Low iron tempered solar glass is currently produced by only Borosil Renewables in India with production capacity of 0.132 million tonnes per year against an annual demand of 0.217 million tonnes. Solar glass becomes highly critical due to future PV installations in India that potentially demand 0.8 Million tonnes in 2030 to roughly 1 Million tonnes per annum in 2050 as per medium demand scenario. This necessitates a six-fold increase from current production volume if the domestic production is preferred by 2030. In terms of the flat glass market in India, solar glass would constitute 60% of current production capacity in 2030 and 75% in 2050.

Therefore, this analysis clearly emphasises the need for resource efficiency in PV manufacturing Industry and sustainable material recovery through proper end-of-life management of solar PV panels to reduce the raw material burden in India and the world. Figure 10 provides a clear picture on the amount of materials locked up in installed PV panels as a percentage of India's production capacity for those respective materials.

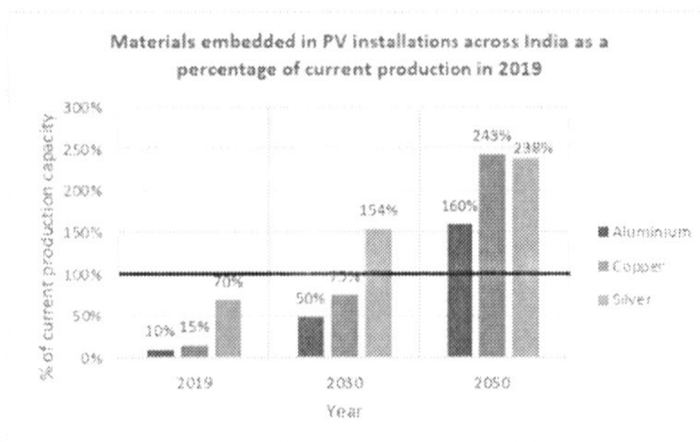

Figure 10 Materials embedded in PV installations across India as a percentage of current production in 2019

3.2 Stakeholder Map and Zone wise analysis

The stakeholder map designed using Tableau Software shown in Figure 11 can be accessed here.

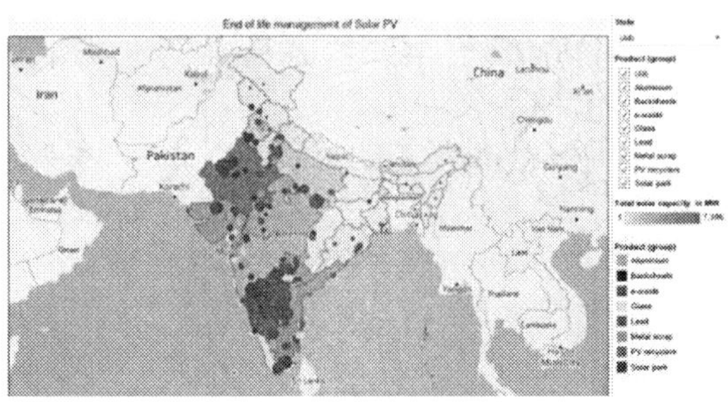

Figure 11 Interactive Stakeholder Map for EoL management of Solar PV panels in India (created using Tableau software)

This dedicated map, focusing on stakeholders of End of Life Solar PV, provides an initial picture of the current infrastructure of downstream recyclers and future potential of PV waste state wise in India. For a selected state, based on the balance between PV waste volume to be generated and downstream processing capacities, the need for transport or decision on extra capacity addition for downstream recycling can be realized.

Industrial areas are dedicated spaces allocated by the respective Indian state governments for entrepreneurs at economical prices, ensuring infrastructure availability in the form of roads, electricity, water, telecommunications, and postal facilities. Future Industrial areas planned in the major hotspot states could be the potential locations for future PV recycling centres and downstream capacity expansion.

3.2.1 PV waste hotspots in India

India has 35.74 GW of Solar Photovoltaic installations as of August 2020. It is clear that the states with most PV installations would be the hotspots for future waste generation.

Utility-scale power plants comprise more than 90% of the total PV installations in the country. The state of Karnataka holds 21.6% of the total Utility-scale PV installations in the country, followed by Rajasthan at 15.2% and Tamil Nadu at 12.1%, as shown in Figure 12. The advantage of such large installations of Utility-scale PV plants is that the bulk of future waste generation is centralised with respect to these plants' locations compared to the possibility of extremely distributed sites due to rooftop PV and off-grid solar.

With respect to rooftop installations, Gujarat leads with 23.8% of the total 3.088 GW installations in India as of August 2020. It is followed by Rajasthan at 11% and Maharashtra at 8.3%, as seen in Figure 12. Therefore, emphasis must be given to these states to track rooftop installations and update registries so that future PV takeback programs can benefit from it.

On comparison of India's leading states concerning Utility & rooftop-type Solar PV installations, southern states of Karnataka, Tamil Nadu are among the top 5 in both cases apart from the northern state of Rajasthan and western state of Gujarat.

On the contrary, focus on off-grid solar installations suggests a different set of leading states in India. Off-grid lighting products usually consist of Solar lanterns, street lighting, and home lighting systems. Uttar Pradesh holds 29% of the total off-grid lighting products distributed across the country. It is followed by the eastern states of Bihar and Jharkhand at 18% and 8%, respectively. The northeastern state of Assam holds 7% of the total distributed solar lighting products. Apart from lighting applications, off-grid solar products also cater to water pumping for irrigation and standalone electricity generation. The eastern state of Chhattisgarh leads both in standalone PV installations with 16.3% of the total installation of 192 MW and in Solar photovoltaic pump installations with 25% of total solar pump installations.

With respect to the end of life management, off-grid solar products are more critical as they are prone to disposal due to their size and shorter lifetime compared to utility and rooftop installations and difficult to takeback owing to their wide distribution in remote places. It is also to be noted that there is no reported research specifically on the topic of end of life management of off-grid solar products in the literature.

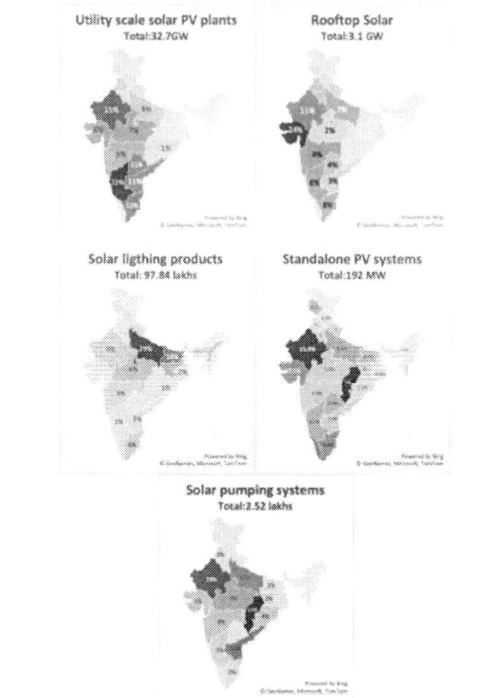

Figure 12 Hotspot Indian States with respect to Solar PV installations as of August 2020

3.2.2 Downstream recycling scenario

The prominent downstream recyclers associated with crystalline silicon PV waste products are glass manufacturers (flat glass, foam, fibre glass, container glass industries), Aluminium refiners or remitters, e-waste recycler (cable, junction box), copper refiners. Further, the services of material recyclers dedicated to processing silicon, silver, plastics, and lead would be needed based on the severity of high-value material recovery by the PV recycler.

In order to assess the criticality associated with the introduction of PV waste stream on the capacity of downstream recyclers, zone wise analysis was carried out.

In this analysis, an assumption is made on the cause of waste generation as only through the end of life without considering the premature failures and quality of the PV panels installed. The installed PV panels are assumed to have a lifetime of 25 years. India had a surge in annual installation around the scale of 8 GW from 2015. Therefore, significant volumes of PV waste stream can be expected around 2040. The zone-wise waste projection is based on the same proportion as the installed capacities of PV plants in the respective zones. According to their market share, the project waste panels are assumed to be 95% of c-Si PV panel type. According to IRENA [1], the material intensity of the installed PV panels around 2015 is assumed to bound between 70-80 t/MW for c-Si PV panels. As the analysis determines the capacity needed for PV and downstream recycling centres, an estimate of 100% material recovery from PV panels is assumed. For estimating the number of PV recycling centres needed, the capacity is assumed to be 5000 metric tonnes per annum based on the current reference of a commercial-scale recycling facility in France.

The results of the analysis are detailed in Figure 13.

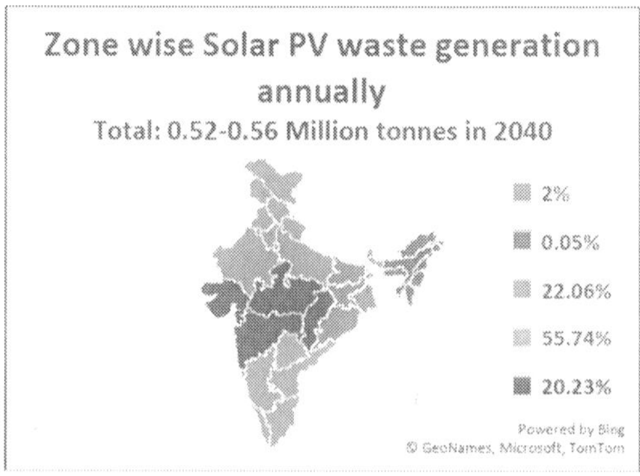

Figure 13 Zonal analysis of Solar PV waste generation annually

3.2.3 Glass cullets

Table V Zone wise analysis of glass cullets generated from PV recycling

Zone	Cullet Capacity (kT per annum)	Projected PV waste by varying material intensities annually in 2040 (kilo ton per annum)		(% of waste projected to recycling capacity)
		70 t/MW scenario	80 t/MW scenario	
Southern	624	205	234	35
Northern	539	81	93	16
Eastern	75.9	7	8	10
Western	731	74	85	11
North Eastern	No major facility	0.19	0.22	-

The glass cullets generated from the PV recycling facility can directly be used in glass manufacturing. Usage of cullets in glass furnace reduces the energy demand and displaces the usage of equivalent primary raw materials. The various type of glass manufacturers like solar glass, flat glass, container glass, fibre glass, and foam glass can utilize the cullets based on descending order of quality.

The glass manufacturing capacity usually denoted in tons per day is converted to annual capacity assuming 80% operation time through the year. Cullet usage in manufacturing varies according to the type of glass manufacturers, such as 40% for flat glass and 60% for container, fibre, and solar glass. Accordingly, all manufacturers' annual glass manufacturing capacity was translated to cullet processing capacity, as shown in Table V.

It can be seen that PV recycling can supply 35% of cullets to various glass manufacturers depending upon quality in the southern region, followed by a 16% supply in northern and 10 & 11% supply in eastern and western regions, respectively. There are no major glass manufacturers in the northeastern region; either the local informal sector can utilise the cullets or significant transportation is needed to supply the cullets to glass manufacturers in the eastern region.

It is to be noted that the capacity presented in this analysis has been derived from significant assumptions owing to the lack of publicly available capacity or cullet usage data from major glass manufacturers in India. The results, however, establish a significant percentage of cullet contribution from PV recycling in the future for glass manufacturers. Assuming an expansion of glass manufacturing facilities and increased cullet usage in future production, it is possible to accommodate the cullet supply from PV recycling.

As seen from the material criticality analysis in section 3.1, India must accelerate the establishment of solar glass manufacturing facilities given the drive towards self-reliant PV manufacturing to realise the ambitious solar PV installation targets and the presence of only one major solar glass manufacturer currently. High-quality glass cullets from PV recycling can help solar glass manufacturing close the circular chain and improve resource efficiency.

3.2.4 Aluminium frame

Table VI Zonal analysis of Al frame scrap from PV recycling

Zone	Capacity (kilo tonne per annum)	Projected PV waste by varying material intensities annually in 2040 (kilo ton per annum)		(% of waste projected to recycling capacity)
		70 t/MW scenario	80 t/MW scenario	
Southern	81.7	53	60	69
Northern	392	21	24	5.7
Eastern	0.65	1.8	2.1	300
Western	235	19	22	8.7
North Eastern	No major reported facility	0.049	0.057	-

c-Si Solar PV panels usually contain AlMg3 anodized aluminium alloy frame for structural stability. Based on the PV recycling techniques, it is clear that aluminium frames can be removed without contamination of other materials from PV panels. Hence, they can directly be remelted to form secondary aluminium ingots.

In the Indian context, currently, 90% of aluminium scrap is imported. Thus, Al frame scrap from PV recycling becomes a valuable waste stream with respect to the trade scenario and for energy demand reduction by displacing primary materials.

From the results showcased in Table VI, PV recycling would contribute volume of Al frame scrap that represents 69% of total processing capacity in the southern region. Given that there is already net import of scrap and assuming the re-melters run close to full capacity, there is a significant need for capacity expansion to accommodate scrap from this new flow of waste stream (PV panels). The analysis also points to processing capacity shortage in the eastern region and no processing capacity in the northeast with significant capacity in the northern region, thus confirming a geographical mismatch. This would lead to excessive transportation of Al scrap from eastern and northeastern regions to the northern region creating economic and environmental impact.

3.2.5 e-waste from PV recycling

Table VII Zonal analysis of e-waste from PV recycling

Zone	Capacity (kilo tonne per annum)	Projected PV waste by varying material intensities annually in 2040 (kilo ton per annum)		(% of waste projected to recycling capacity)
		70 t/MW scenario	80 t/MW scenario	
Southern	192	7	8	4
Northern	447	2.6	3	1
Eastern	5.5	0.23	0.26	4
Western	138	2	3	2
North Eastern	Major facility not reported	0.006	0.007	-

c-Si PV recycling produces cable and junction box scraps that need to process downstream by e-waste recyclers. Copper can be retrieved from both these e-waste products.

The analysis suggests an accommodative volume of e-waste stream from PV recycling by e-waste recyclers event with the current capacity as shown in Table VII.

3.2.6 Other recoverable material streams

High-value PV recycling of c-Si panels could yield silver, copper, aluminium, silicon, tin, lead, plastics (EVA, PVF composites).

As discussed in 3.1, silver is a highly valuable material stream that is net imported in India. Hence recovery of quality silver from PV recycling can significantly contribute to silver production in the country. Copper and Aluminium from solar cells can add to the already recoverable streams from cable and frame of PV panels. Lead present in solder needs to be recovered in order to avoid disposal hazards. Although it has been indicated that exposure risk is under limits for the lead content present in a single PV panel [17], the disposal risk of a large volume of PV waste in a single location still needs to be studied. Also, the recovery of material from a resource efficiency perspective could be beneficial.

The plastics waste from PV panel that comprises ethylene-vinyl acetate, polyvinyl fluoride, polyethylene terephthalate needs to be investigated for secondary usage in other applications. Energy (thermal and electricity) can be recovered from burning the waste in a Waste to Energy (WtE) plant under a controlled environment. Exhaust gas treatment is necessary due to the presence of fluorine in the waste.

Silicon recovery is significant, with the current emphasis on circulating high-quality silicon back to solar PV production [18]. Given India's push to establish polysilicon and silicon wafer production facilities, the recovery from the end of life panels would add significant energy demand reduction compared to the production with complete primary materials.

3.2.7 Future PV takeback and recycling centres

Assuming 5000 MTA PV recycling capacity, the number of facilities needed to process future PV waste across zones is indicated in Table VIII.

Table VIII Zonal distribution of PV recycling centres

Zone	PV waste projected (kilo tonne per annum)		PV recycling centres (5000 MTA)
	70 t/MW scenario	80 t/MW scenario	
Southern	297	339	59-68
Northern	117	134	23-27
Eastern	10	12	3
Western	108	123	22-25
North Eastern	0.28	0.32	1

The above preliminary analysis can be built upon to optimise PV recycling centres' placement with respect to economic and environmental considerations.

3.3 Challenges

- Different types of Solar PV systems (Utility, rooftop, Offgrid) are concentrated on different parts of the country and each needs a separate strategy to sustainably manage End of life management
- There is definite presence of downstream recycling capabilities in India for recovered materials like aluminium, copper, glass ,e-waste from End of life Silicon PV panels. However currently there is a geographical mismatch between the solar waste producer and recycler which emphasizes the impending transportation impact
- There is a need to develop PV recycling infrastructure and strategically place them near the solar PV installation clusters in order to minimize transportation impact.
- For high value recycling (recovery of silicon, silver and encapsulants) there is a need to develop downstream recycling technologies and on a policy side material specific quotas would allow recovery of these materials.
- Standards for secondary raw materials needs to be developed in order to smoothly promote circular use of material without downcycling

4 CONCLUSION

This is a first of its kind study on identification of need for sustainable end of life management of PV panels in India from the perspective of material criticality for PV panel production.

The work also provides an initial mapping and identification of potential stakeholders for end-of-life management of PV panels in India and visually points to the impending challenges in takeback, recycling and downstream processing of end-of-life PV panels owing to the geographical mismatch among stakeholder locations.

5 REFERENCES

[1] Weckend, S., A. Wade, and G.A. Heath, End of Life Management: Solar Photovoltaic Panels. 2016, National Renewable Energy Lab.(NREL), Golden, CO (United States) (IRENA).
[2] Ministry of New and Renewable Energy (MNRE), G.o.I., Concept Note on Antimony content in Solar PV Glass. 2017.
[3] Ministry of Environment, Forest, Climate Change G.o.I., Draft policy on resource efficiency 2019.
[4] Feasibility study commissioned by MNRE with NSEFI, Solar Power Europe, Saur Energy, 2020.
[5] PV waste management in India, EU TCP, Report, 2020
[6] Ministry of New and Renewable Energy, G.o.I., Import of Solar panels. Lok Sabha Questions AU397, Indian Parliament 2018.
[7] Informal sector at Work in Selampur, New delhi, E[co]work 2020.
[8] W. Hoffmann, A.M. Modules at a Price of 10 $ct/Wp - Dream or Reality? New Approach for Future Cost and Price Predictions. in 36th European Photovoltaic Solar Energy Conference and Exhibition. 2019.
[9] Pierre Verlinden, A.P. Future Sustainability Challenges for PV Manufacturing at the Multi Terawatt Level. in PV Magazine Roundtable 2020. PV Magazine.
[10] Carrara, S., Alves Dias, P., Plazzotta, B. and Pavel, C., Raw materials demand for wind and solar PV technologies in the transition towards a decarbonised energy system. Publications Office of the European Union, 2020.
[11] Ministry of New and Renewable Energy (MNRE), G.o.I., Concept Note on Solar PV Manufacturing Scheme. 2017.
[12] Statistics, U.N.T., UN Comtrade Database.
[13] Ministry of Commerce & Industry, G.o.I., Export Import Data Bank. (2015-2019).
[14] Ministry of Mines, G.o.I., Monthly summary of Non ferrous minerals & metals production. 2020.
[15] Institute, T.S., World Silver Survey. 2020.
[16] Centre for Energy Finance , C.f.E., Environment and Water, Solar PV Manufacturing: Reassessing Priorities in the Post-COVID-19 Era. 2020.
[17] P. Sinha, G.H., A. Wade, K. Komoto, Human health risk assessment methods for PV, Part 3: Module disposal risks. International Energy Agency (IEA) PVPS Task 12, Report T12-16:2020., 2019.
[18] Heath, G.A., et al., Research and development priorities for silicon photovoltaic module recycling to support a circular economy. Nature Energy, 2020.

GREENHOUSE GAS EMISSIONS AVOIDANCE BY PHOTOVOLTAIC PLANTS ON THE ROAD TO CARBON NEUTRALITY

José Távora [a], Maria João Cortinhal [b, c], Mónica Meireles [d]
[a] Iscte – Instituto Universitário de Lisboa, Business Research Unit (bru iscte), jose.tavora@gmail.com
[b] Iscte – Instituto Universitário de Lisboa, Iscte Business School – Dep. Métodos quantitativos para a Gestão e Economia
[c] CMAF-CIO, Faculdade de Ciências da Universidade de Lisboa
[d] Iscte – Instituto Universitário de Lisboa, Iscte Business School – Dep. Economia, Business Research Unit (bru iscte)
[a, b, d] Av. das Forças Armadas, P 1649-026, Lisboa, Portugal, [c] Lisbon P 1749-016, Campo Grande, Portugal

ABSTRACT: Roadmaps to carbon neutrality challenge GHG emissions savings by solar PV energy: current systems are still largely produced using fossil fuels and PV plants occupy for decades land areas that might capture and store atmospheric carbon. We take primary emissions from three PV technologies and secondary emissions from vegetation clearing in PV power plant sites and compare them to lifecycle emissions savings from PV energy over continental Portugal assuming that the carbon intensity of the electrical grid is decreased linearly to near zero in thirty years. The balance is revealed by the emissions payback time (CPBT) metric. In our approach, time affects emissions over a PV plant's lifetime in three ways: overall decreasing grid carbon intensity, yearly avoided grid emissions, and yearly cancelled carbon capture due to land cover clearing - a forest of planted eucalyptus trees. Results for mono-Si reveal mean emissions payback times of 14 years for the lowest irradiation places, and 10.5 to 12.5 years for 93% of the territory. Secondary emissions increase the mean value to 19, with a maximum of 24 years. These figures strongly recommend that national regulators enforce checks on primary and secondary emissions by PV systems sited on forested land.
Keywords: CO2 footprint, Environmental effect, Sustainable, Carbon payback time, Secondary emissions.

1 INTRODUCTION

Massive deployment of solar photovoltaic energy is currently a main strategy to decarbonize electricity supply in roadmaps to carbon neutrality being deployed in many countries, including Portugal [2]. Solar PV plants contribute to reduce GHG emissions by replacing fossil fuel power generators. But their emissions avoidance role will be challenged by the steep decarbonizing process because PV modules are still largely produced using fossil fuels and PV power plants occupy for decades large areas of land that might capture and store atmospheric carbon.

Our research addresses the interplay among the declining carbon intensity of the electrical grid, the GHG emissions of the materials and manufacture of PV systems, and the emissions caused by cutting down vegetation when installing PV power plants, and how these factors modify emissions avoidance. We assess the balance between the emissions avoiding effects of PV power plants and their own primary and secondary emissions in continental Portugal. The GHG Emissions Payback Time also referred as Carbon Payback Time (CPBT) is used to appraise the result. Primary emissions will be CO_2 equivalent releases from materials and manufacture of PV components. And secondary emissions will be CO_2 releases from land transformation: cutting down the vegetation cover and cancelling its constant carbon-capture activity. Time is a key factor in our analysis: the emissions from the electrical grid will steadily decline due to the roadmap to carbon neutrality [2] but at the same their release occurs *continuously* and therefore their effect on the atmosphere declines as time passes. The same happens with the CO_2 not captured from the atmosphere due to vegetation clearing – equivalent to a CO_2 emission. On the contrary, the GHG emissions from PV materials and manufacture are conventionally released as an *impulse* just prior to the installation and are not affected by the passage of time. In our assessment, secondary emissions will be caused by cutting down planted eucalyptus forests, one of largest biomes in continental Portugal [9].

Our work is close to [1][4][6][10][14] but based on a detailed geospatial approach that includes nineteen plant biomes and considers the effect of time on emissions. [1] and [10] study secondary emissions from cut down forests in the US. [6] and [14] present PV metrics across Europe and the world, respectively, but do not include secondary emissions nor the evolution of emissions with time. [14] report regions in Europe with emissions payback times over 10 years, namely in France, Sweden, and Norway. [4] report carbon payback times for PV plants in Brazil showing how a low carbon intensity electrical grid may lead to carbon payback times near or over PV plant lifetimes. They have also point out that emissions from PV systems are expressed mostly before operation while grid emissions will be spread over PV plant lifetimes, although their work does not develop the subject.

2 MATERIALS & METHODS

To obtain primary GHG emissions we combined monocrystalline silicon (m-Si), cadmium telluride (CdTe), and copper indium selenide (CIS) product data from, respectively, Jinko™, First Solar ™, and Solar Frontier ™ with LCA results from [12] as displayed in Table I. The 'product stage' of PV components was chosen as system boundary, consistently with the sources. The 'functional unit' is the alternating current output of the inverters. Solar yields and losses for continental Portugal were obtained by several processes. For average yearly irradiation and inclination, spectrum, and temperature and low irradiation losses on south-oriented, optimally inclined panels we queried PVGIS ™ at 3751 locations, forming a 5.5 km resolution grid, through a custom developed program. Lifetime average degradation losses for PV modules were taken from product datasheets (Cf. Table I), except degradation for CIS, taken from [7]. System losses were fixed at 4%.

One-off CO_2 emissions from clearing eucalyptus forests were obtained from tree carbon content in [5] and maps in [9]. We assumed that trees cut down would be converted at once into CO2 emissions, a reasonable simplification since eucalyptus trees are used mostly for wood fuel and wood pulp. Measurements of yearly carbon capture by eucalyptus plantations were taken from [15]. Total PV plant areas were assumed to be four times PV panel areas - a conservative, revised figure by the authors, lower than the value estimated in [17]. We also assumed that the whole PV station area was initially covered by trees. Values for the current and planned future GHG grid emissions were taken from [2] and [8].

The timing of GHG emissions by a source influences its global warming effect [3][11][13]. Compared to an emission occurring at t_0 an emission happening t years in the future will stay in the atmosphere t years less at the end of an analytical time horizon. The method of Time Adjusted Warming Potentials (TAWP) [11] integrates emissions timing in global warming calculations through TAWP coefficients that adjust later emissions so that they can be compared at some initial time $\mathbf{t_0}$. Table 2, below, illustrates the method applied to electrical grid emissions in the next 30 years assuming the Portuguese roadmap to carbon neutrality [2], and to secondary emissions from cancelled carbon capture by a eucalyptus forest. The analytical time horizon was chosen to be 100 years – a conservative approach that enables comparing TAWP results with PAS 2050 results shown in the last column. The TAWP coefficients were calculated by a 3rd degree polynomial regression provided by [11] as supplementary information.

Emissions balancing are illustrated by the GHG Emissions Payback Time metric, also known as Carbon Payback Time (CPBT), calculated from Equation 1.

$$CPBT = \frac{CC_{mat} + CC_{manuf}}{\frac{CC_{avoided\ LT}}{LT}} \quad (1)$$

CC_{mat} and CC_{manuf} are the climate change impacts of the materials and manufacturing of the PV system components, in gCO_2e. $CC_{avoided\ LT}$ represents emissions avoided during the PV plant's lifetime, LT, also in gCO_2e.

Table I: Technical and environmental characteristics of PV components.

Tech.	Origin	Capacity @STC [W_P]	Panel area [m^2]	Eff. [%]	LT average Degradation [%]	Emissions [$kgCO_2e/kW_P$]	Emissions [$kgCO_2e/m^2$]	Energy mat & fab. [MJ/kW_P]	Energy mat & fab [MJ/m^2]
m-Si	China	345	1.687	20.45	9.89	1980	398.96	26500	5339.7
CdTe	USA	450	2.475	18.18	8.20	650	118.18	10500	1909.0
CIS	Japan	144	1.228	11.73	0.30	1050	123.12	17600	2063.7

Table II: Calculations of the effect of time on emissions (T.A. = Time-adjusted).

Time [years]	TAWP AT = 100 years	Grid carbon intensity [gCO_2/kWh)]	Grid carbon int. (T.A.) [gCO_2/kWh]	Secondary emissions [$kgCO_2/m^2$]	Secondary emissions (T.A.) [$kgCO_2/m^2$]	PAS 2050 general case N= 30 years
0	1	254.0	254.0	35.23	35.23	
1	0.9918043	245.6	243.6	9.24	9.16	0.0330
2	0.9836389	237.2	233.3	9.24	9.09	0.0327
3	0.9755018	228.8	223.2	9.24	9.01	0.0323
(...)	(...)	(...)	(...)	(...)	(...)	(...)
28	0.7752386	18.5	14.4	9.24	7.16	0.0240
29	0.7671133	10.1	7.8	9.24	7.09	0.0237
30	0.7589607	1.7	1.3	9.24	7.01	0.0233
	TAWP avg. weighting factor	Lifetime average	Time-adjusted lifetime average	Lifetime total	Time-adjusted lifetime total	PAS 2050 weighting factor
	0.8794	123.6	113.3	312.4	277.9	0.8450

Table III: Mean and limit values for carbon payback time distributions.

Metric	m-Si			CdTe			CIS		
	Min.	**Mean**	Max.	Min.	**Mean**	Max	Min.	**Mean**	Max.
CPBT **Carbon Payback Time** [years]	9.8	**11.2**	14.4	3.1	**3.6**	4.6	4.9	**5.5**	7.1
CPBT **with secondary emissions** [years]	16.7	**19.3**	24.3	10.6	**12.2**	15.5	16.0	**18.3**	23.0

3 RESULTS

Table III presents the mean value and limits of CPBT distributions of the three PV technologies, for primary and total (primary and secondary) emissions. Regarding primary emissions, CdTe and CIS have much lower carbon payback times than monocrystalline silicon, as expected. However, when adding secondary emissions, their CPBTs increase dramatically due to the additional weight in the numerator of Equation 1. CdTe, for instance, features 118.2 $kgCO_2e/m^2$ of primary and 277.9 $kgCO_2e/m^2$ of secondary emissions, so the latter are 2.4 times higher than the former. Thin-film advantages over m-Si are thus strongly reduced as can be seen from Table III. Figure 1 shows the spatial distribution of CPBT for monocrystalline silicon over the eucalyptus biome, for primary emissions. Figure 2 shows a similar result for total emissions. Analysis of the histogram of primary emissions of monocrystalline silicon over continental Portugal reveals that 93% of the territory has carbon payback times of between 10.5 to 12.5 years.

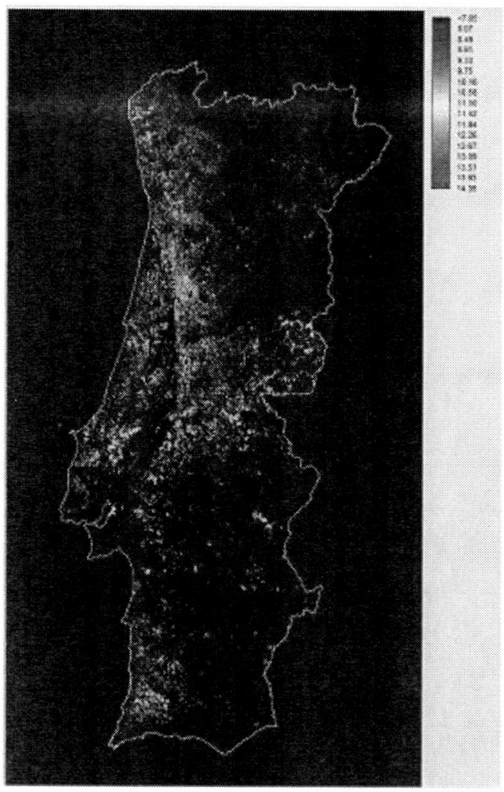

Figure 1: CPBT on eucalyptus areas – primary emissions

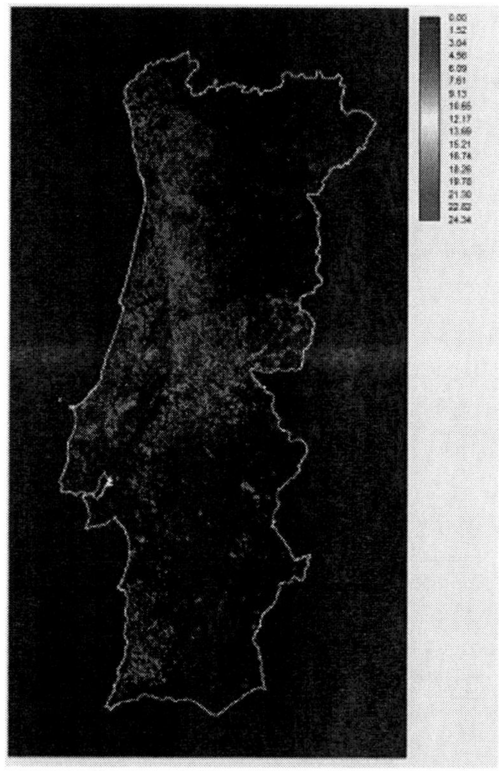

Figure 2: CPBT on eucalyptus areas – total emissions

4 CONCLUSIONS

Our results clearly demonstrate that primary emissions alone are responsible for large values of carbon payback times for the dominant PV technology – crystalline silicon. This happens in a country with one of the best solar resources in Europe but with a national electricity grid with just an intermediate level of carbon intensity. The reason is the introduction in our analysis of an inescapable reality: roadmaps to carbon neutrality will steeply reduce the carbon intensity of the grid in the next decades leaving PV power plants installed *now* (on in the next few years) with reduced emissions saving capabilities. Another reason is at play, stressed for instance in [4]: importing PV technology with high or moderate carbon footprint and implementing it in countries with a low grid carbon intensity (Brazil or France, for instance) will weaken or even cancel any emissions avoidance advantage.

Secondary emissions have been mostly neglected in the appraisal of emissions avoidance by PV systems, except for [1][10]. Our example shows, however, that in the case of forested land being cleared for PV power plants their contribution can be determinant. In fact, as PV technology reduces its inherent carbon footprint (e.g., through "solar for solar" approaches), secondary emissions and other impacts in ecosystem services will come to the foreground. We believe that soon, for some technologies and implementation regions GHG avoidance by PV power plants will cease to be a relevant justification to change land use: PV projects will instead be evaluated by their ability to not degrade ecosystems or by introducing new ecosystem services improving overall benefits - including climate positive outcomes as it may happen with agrivoltaics.

A main recommendation from our research is that national regulating authorities should enforce checks on carbon footprint credentials of PV technologies, as in already happens in one European country [16]. Secondary emissions from PV implementation should also in environmental impact assessments, together with primary emissions.

Our work resorts to simplifications, which do not affect its general conclusions: the vegetation to cleared occurs in the full area of the power station and is fully cut down; soil carbon remains undisturbed by vegetation removal; initial vegetation clearing results in products with a very short lifecycle compared to the lifetime of the plant; and PV recycling does contribute to negative emissions. The assumptions are all plausible, depending on each case. Much research work remains to be done, namely about carbon capture measurements for the main ecosystems in the country, considering local climate conditions [15].

5 REFERENCES

[1] Anctil, A., & Fthenakis, V. (2012). Greenhouse gases emissions and energy payback of large photovoltaic power plants in the northeast united states. In Conference Record of the IEEE Photovoltaic Specialists Conference (pp. 753–756).

[2] APA. (2019). Roteiro para a neutralidade carbónica 2050 - Anexo Técnico Energia Indústria - Versão 07.01.2019. http://apambiente.pt/_zdata/CONSULTA_PUBLICA/2019/RoteiroNeutralidadeCarbonica/Anexo_tecnico_Energia_Industria.pdf.

[3] CARBON TRUST (2008*). Guide to PAS 2050 – How to assess the carbon footprint of goods and services*. ISBN 978-0-580-64636-2.

[4] Constantino, G., Freitas, M., Fidelis, N., & Pereira, M. G. (2018). Adoption of photovoltaic systems along a sure path: A life-cycle assessment (LCA) study applied to the analysis of GHG emission impacts. Energies, 11(10).

[5] Costa-Pereira, T., Amaro, A., Borges, M., Silva, R., Pina, A., & Canaveira, P. (2019). Portuguese national inventory report on greenhouse gases, 1990-2017. Amadora, Portugal. https://doi.org/10.1016/j.solmat.2013.08.037

[6] De Wild-Scholten, M., Cassagne, V., & Huld, T. (2014). Solar Resources and Carbon Footprint of Photovoltaic Power in Different Regions in Europe. Proceedings of the 29th EU PVSEC, 3421–3430.

[7] De Wild-Scholten, M. J. (2013). Energy payback time and carbon footprint of commercial photovoltaic systems. Solar Energy Materials and Solar Cells, 119, 296–305.

[8] DGEG. (2020). DGEG - Balanços e indicadores energéticos. Retrieved March 12, 2020, from http://www.dgeg.gov.pt/

[9] DGT. (2019). Carta de Ocupação do Solo de Portugal continental (COS). http://mapas.dgterritorio.pt/ DGT-ATOM-download/ COS_Final /COS2018_v1/ COS2018_v1.zip

[10] Fthenakis, V., & Kim, H. C. (2009). Land use and electricity generation: A life-cycle analysis. Renewable and Sustainable Energy Reviews, 13(6–7), 1465–1474.

[11] Kendall, A. (2012). Time-adjusted global warming potentials for LCA and carbon footprints. International Journal of Life Cycle Assessment, 17(8), 1042–1049.

[12] Leccisi, E., Raugei, M., & Fthenakis, V. (2016). The energy and environmental performance of ground-mounted photovoltaic systems - A timely update. Energies, 9(8).

[13] Levasseur, A., Lesage, P., Margni, M., Deschěnes, L., & Samson, R. (2010). Considering time in LCA: Dynamic LCA and its application to global warming impact assessments. *Environmental Science and Technology*, *44*(8), 3169–3174.

[14] Louwen, A., Schropp, R. E. I., van Sark, W. G. J. H. M., & Faaij, A. P. C. (2017). Geospatial analysis of the energy yield and environmental footprint of different photovoltaic module technologies. Solar Energy, 155, 1339–1353.

[15] Pereira, J. S., Mateus, J. A., Aires, L. M., Pita, G., Pio, C., David, J. S., … Rodrigues, A. (2007). Net ecosystem carbon exchange in three contrasting Mediterranean ecosystems - The effect of drought. Biogeosciences, 4(5), 791–802.

[16] République Française, 2020. Cahier des charges de l'appel d'offres portant sur la réalisation et l'exploitation d'Installations de production d'électricité à partir de l'énergie solaire 'Centrales au sol'. Annexe 2: Méthodologie de l'évaluation carbone simplifiée.

[17] Távora, J., Cortinhal, M. J., & Meireles, M. (2020). Land use intensity and land occupation of utility-scale photovoltaic power plants in continental Portugal. In 37th European Photovoltaic Solar Energy Conference and Exhibition (pp. 1975–1978).

CARBON FOOTPRINT ANALYSIS OF CIGS THIN-FILM PV MODULES WITH FOCUS ON BUILDING-INTEGRATED APPLICATIONS

Peter Borowski[1], Franz Grömmer[2], Franz Karg[1]
[1] AVANCIS GmbH, Otto-Hahn-Ring 6, 81739 München, Germany
[2] Technical University of Dresden, Institute of building construction, August-Bebel-Straße 30, 01219 Dresden, Germany
phone: +49 89 219620-464, email: peter.borowski@avancis.de

ABSTRACT: Carbon footprint (CF), being the most important parameter of a life-cycle assessment (LCA) of a product, is becoming a vital quantity both in the field of energy production and in the sector of building materials. Focusing on the three production stages A1 to A3 ("cradle-to-gate", DIN EN 15804), the carbon footprint of a Copper-Indium-Gallium-Selenide (CIGS) thin-film photovoltaic (PV) module of the manufacturer AVANCIS is computed for the production in Germany. From the main result of 59 kg CO_2eq per module, deduced quantities are reported: carbon footprint per area, per kW_p, per kWh electricity generated, energy payback time as well as the carbon payback period. For two types of PV systems in Munich, Germany, values for the latter in the range of 1 to 1.6 years are obtained. The paper briefly discusses the relevance and sources of input data and compares the quality of simple estimates with the result obtained with a commercially available software for LCA. It is found that energy consumption during production and the module's content of glass and steel account for approximately 80% of the total carbon footprint of a glass-glass CIGS thin-film PV module. Different to crystalline silicon PV modules, the contribution of the material forming the semiconductor is small and below 10%.
Keywords: CO_2 footprint, Building Integrated PV (BIPV), CIGS

1 INTRODUCTION

The manufacturing of materials for building construction accounts for an estimated 10% of global carbon dioxide emissions [1] and is therefore a significant contributor to global warming. Photovoltaics (PV) and particularly building-integrated photovoltaics (BIPV) can be a decisive factor for a reduction of this share. To quantify the impact of PV on the carbon footprint (CF) of a building, it is necessary to know the carbon footprint of the PV system and therefore its PV modules. The CF of a PV module can then be compared to conventional, passive facade elements that are being replaced by BIPV modules. Also, an analysis of the contributors to the CF shows the most promising ways towards ultra-low carbon production. The carbon footprint of a product or an entity is measured in grams of carbon dioxide equivalents (CO_2eq) and represents the contribution to global warming by the production or use of the product related to the damaging potential of one gram of pure carbon dioxide emitted into the atmosphere. Changing a building's envelope from passive to active using (BI)PV, a building has the potential to turn its net lifetime carbon emission from positive to negative numbers by replacing fossil-based electricity that it would otherwise consume from the grid.

The impact on global warming is one of several impact categories of a standardized life-cycle assessment (LCA) or environmental product declaration (EPD) [2], with other impact categories being, e.g., eutrophication or human toxicity potential. With a certified LCA, an EPD can be obtained, a standardized and publicly available document summarizing the environmental impacts of the specific product. For electricity production, CF analysis is a well-established tool for comparing various technologies (e.g., [3]) and literature studying the LCA or the carbon footprint of PV modules is available in larger quantities, with a few focusing on the same thin-film technology as in the present paper ([4] and references therein). Comprehensive EPDs, however, are still rare for PV modules[1]. Within the construction sector, EPDs are trending to become a demanded and necessary declaration for different types of products [5]. Once established, EPDs can serve as a basis for decision-making for an architect or property developer motivated to build carbon-neutral buildings or needing to act within governmental or municipal regulations regarding the climate effect of buildings.

In its *Roadmap to a Resource Efficient Europe* from 2011 [6], the European Commission has identified the building sector as a key contributor. Better construction and use of buildings in the EU would influence 42% of the final energy consumption, about 35% of greenhouse gas emissions and more than 50% of all extracted materials. The application of active components like PV instead of passive building materials promises to strongly support those ambitions. But among the milestones set by the EU for 2020 was also the consistent application of an LCA for all building components, and a PV building component irrespective of its active nature would not make an exception.

The present paper, focusing on the carbon footprint of a PV module as a building component, is organized as follows. In the next section, we briefly state the aim of the work and describe the specific PV module considered, followed by a more detailed description of the approaches applied for calculating the carbon footprint. In section 3, the results of the analysis are presented, relating to different aspects of the 'hybrid' product that a BIPV module is. The main contributors to the CF are identified and the results are related to alternative construction materials. We conclude with a broader discussion and offer recommendations in the last section.

---
[1] The authors are aware of publicly available EPDs of only four PV module manufacturers.

2 AIM AND APPROACH

2.1 Aim

The aim of the present study is to quantify the carbon footprint (CF) of a PV module manufactured by AVANCIS and identify the main contributors and therefore the leverages for a further reduction of the CF. We comment on the approach and compare the CF of a PV module for building-integrated photovoltaics (BIPV) to alternative, electrically passive facade materials.

2.2 Details of the product and its production

We consider the production of a copper indium gallium selenide (CIGS) based thin-film PV module in the factory Fab2 of AVANCIS located in Torgau, Germany. The PV module considered is a glass-glass module with dimensions of 1587 × 664 mm^2 (area 1.054 m^2) weighting approximately 17 kg, whose photoelectric activity is realized by a thin (~2 μm) stack of metallic and semiconductor layers. **Figure** *1* shows an AVANCIS SKALA module as well as its application in a vertical PV system on the facade of Fab2. Mounting of the frameless module is accomplished by two steel backrails glued to the backside of the glass-glass laminate. A pair of junction boxes including a bypass diode and connected by a bypass cable facilitates the electrical connection.

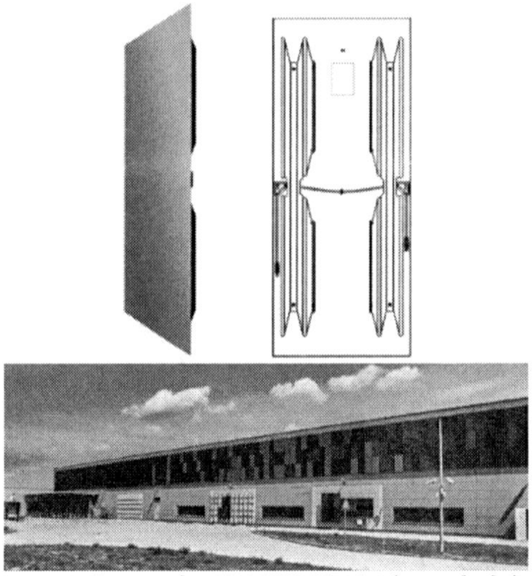

Figure 1: Top panel: AVANCIS SKALA in standard size (dimensions 1587 × 664 × 38 mm^3) in its variety 7003 (blue) and a technical drawing of its rear side showing the steel backrails for mounting; bottom panel: A facade PV system consisting of AVANCIS SKALA modules of different colours on the south-west wall of Fab2 in Torgau, Germany.

Two product families are being produced in Fab2 of AVANCIS: PowerMax® for free-field PV systems and SKALA for building-integrated PV systems, particularly for facade applications [7]. Whereas PowerMax® comes in a fixed size and colour (black), SKALA is available in various lengths and colours of the sunny side and is produced with a stronger focus on the visual integration of the PV module into the building. The differences in composition and processing of these two product families has only a negligible impact on the carbon footprint of the final product. The presented analysis is therefore valid for modules of both product families. Differences exist when relating the CF to the electrical power or yield of the modules, since the coloured SKALA modules optimized for quality of visual impression provide a slightly lower electrical power than the conversion efficiency-optimized PowerMax® modules.

The cadmium- and lead-free module is produced in an integrated process of several consecutive coating and structuring steps without any wet-chemical processes (no special waste water treatment needed) and is summarized in **Figure** *2* schematically together with an extract of the bill of materials (BOM).

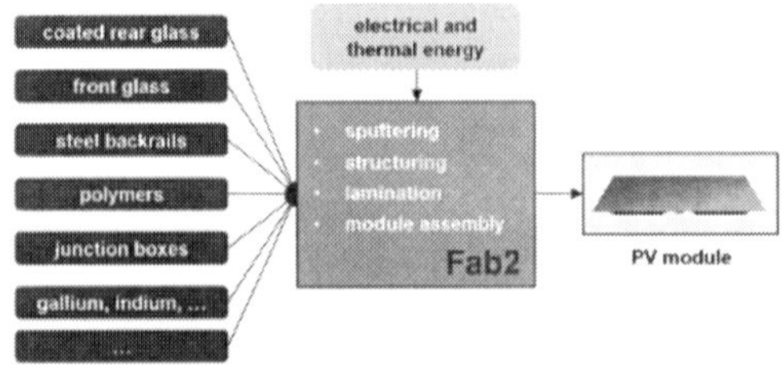

Figure 2: Input and output of the production process in Fab2 including a shortened list of the bill of materials. Input: coated rear glass, front glass, steel backrails, polymers, cables, junction boxes, connectors, sputter targets, process gases, electricity, natural gas, cooling water, ...; output: PV module.

2.3 Approach

We started our analysis with a simple compilation of the main materials used for the production of one module and a rough number for the electricity consumption of Fab2. Data of the involved material's CF was obtained from various publicly available databases (e.g., [8]), with some gaps due to lacking CF data for some of the more specific materials (e.g., molybdenum or indium). For those, very rough and conservative estimates were used. The CF of the module is then computed by simply adding up the individual contributions of its components.

In a subsequent and independent approach, a detailed model was set up in the commercial software GaBi [9], now considering detailed material uses, scrap, use factors and estimates for the transport of the materials from the component suppliers to Fab2 in Germany. Whereas the major components of the PV module are easily implemented, the unknowns in the simple approach mentioned above – the carbon footprint of some of the minor contributors – did not disappear due to a limited material database included in the standard version of the software. It proved to be difficult to obtain reliable input data for some of the minor components of a thin-film PV module even in the commercial software GaBi. Thus, in the final LCA, we still needed to rely on assorted sources for the input data ([10], among others), making the final result to some extent dependent on the specifics of the sources chosen.

For the more detailed approach, the energy consumption of Fab2 was analyzed on a production process level rather than globally as in the rough approach. The electricity consumption of all processing steps was deduced per processed module during phases near full production capacity. The electricity consumption of the Fab facilities (cooling, supply of clean dry air, pressured air, lighting, ...) strongly depends on the environmental conditions. As a way of averaging, we looked at the ratio of electricity consumption for facilities and for the production machinery over a full year. We find that the electricity consumption for facilities leads to an increase of electricity consumption per module produced of about 40%. Fab2 currently consumes electricity of the standard German electricity mix with a CF of 366 g CO_2eq/kWh in 2020 [11] [12]. Own-consumption of electricity produced in R&D PV systems on the Fab's premises is currently negligible (<1%). Natural gas and water are consumed by facilities and are included in the analysis, however, with minor contributions to the CF. As will be shown in section 3, the CF of an AVANCIS PV module is dominated by only three contributors with well-known component CFs: electricity, glass and steel. Main conclusions about the CF can therefore be drawn already from a coarse and simple calculation, rather than a detailed model, which is a first important result of the present study.

Scrap occurring during production (e.g., due to glass breakage) is considered in the present study. For the input materials (**Figure** *2*), we include use factors such that not only the amount of material built into one module is accounted for, but rather the total consumption of the material divided by the number of modules produced, which is relevant particularly for the processes including sputtering targets.

In the present analysis, we restrict ourselves to the limited life cycle cradle-to-gate (phases A1 to A3 [2]), leaving out a detailed analysis of the use phase, deconstruction, recycling or disposal. A brief discussion on these phases is presented in section 3.

As system boundaries we chose an individual module including its share of standard packaging (box of 20 modules). The carbon footprint of the substructure holding the modules, inverter, cables, thermal insulation or the building wall or roof are not included. As this specific choice can serve for reasonable comparability between PV modules, is it less likely to lead to good comparability between a (BI)PV module and an alternative realization of a building's envelope.

3 RESULTS AND DISCUSSION

3.1 Carbon footprint and its contributors

As a result of the analysis described in section 2, we compute a carbon footprint of approximately 59 kg CO_2eq for one AVANCIS PV module. **Figure** *3* shows the main contributors, with electricity consumption of Fab2 constituting the largest share.

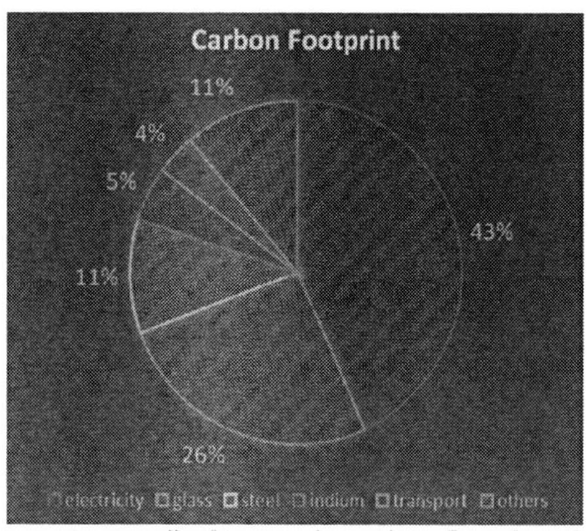

Figure 3: Contributions to the carbon footprint of an AVANCIS SKALA PV module.

The three components with the biggest mass (front glass, rear glass, steel backrails) dominate the CF caused by the materials. Further contributions to the CF summarized as "other" in **Figure** *3* with contributions larger than 1% are: silicone glue, polymers (lamination foil, edge sealant, junction boxes, cable insulation) and packaging (cardboard and EPS). The remaining components and facilities (natural gas, cables, Al, Mo, Ga, Se, ZnO, …) contribute less than 5% of the total CF of a module. The polymers, although having a significant share of the mass of a module, contribute only little to the CF, with the notable exception of the silicone-based polymer employed as backrail glue. With the exception of indium, all the ingredients for the semiconductor layers only contribute marginally to the CF, mainly due to their small amounts needed for the production of one PV module (typically less than 10g). Indium sticks out, since in the database used, its production-related CF is assumed to be much higher than those of the other ingredients (e.g., 600 times the CF of copper and about 100 times that of Ga or Mo). This particular assumption is contradicted by data from alternative databases for CF and we are convinced that the share stated in **Figure** *3* is an upper bound.

It is important to note that CF calculations for a product as complex as a PV module come with a relatively high uncertainty, to which the biggest contributor in the presented case is the assessment of energy consumption for production. The main materials (glass and steel) might have a comparably precise CF, partially declared in EPDs, but small uncertainties as well as consideration of scrap during production can have an impact on the reliability of the product's CF due to their large contribution. For many of the minor contributors, the choice of representation in the LCA or even the choice of database used can affect the final result (as discussed earlier for the case of indium). These uncertainties in combination with differences in system boundaries have to be kept in mind when basing decisions between competing products on CF data. Our estimate for the uncertainty of the presented CF analysis due to the mentioned factors is at least +/- 10% of the final figure.

3.2 Carbon footprint for varying functional units and derivatives

Depending on the mode of comparison, a functional unit for the CF other than "per module" is preferable. A useful functional unit for facade materials is one square meter. **Table** *I* lists the carbon footprint of an AVANCIS PV module with respect to several functional units, as well as the carbon footprint of a kWh of electricity produced by the PV module under certain external conditions (**Table** *II*).

Table I: CF of an AVANCIS PV module with regard to various functional units. (*)The CF of one kWh electricity produced assumes a lifetime of 25 years and a rough estimate for system losses. For simplicity, yield data estimates were obtained using PVGIS [13]. Only the CF of the module is considered, neglecting the CF of the inverter, the substructure and of the other material needed for an operating PV system. See **Table** *II* for the two cases.

functional unit	carbon footprint in kg CO_2eq
per module	59
per m^2	56
per kg	3.5

per kW$_p$ (PowerMax® 160W)	370
per kW$_p$ (SKALA 140W)	420
per kWh$^{(*)}$ – case 1	0.014
per kWh$^{(*)}$ – case 2	0.023

Table II: Cases considered.

case	system specifics in Munich, Germany
case 1	free-field, south-faced, 30°, PowerMax® 160W
case 2	facade, south-faced, 90°, SKALA 140W

With an estimate of the electrical yield during operation of a PV module, an average electrical energy payback time can be computed as the time during which the module has produced as much electrical energy as has been needed for its production. Within many current feed-in schemes that prioritize electricity from renewable sources against fossil-based energy, every kWh of energy produced by a PV module replaces the CF of the average country-wide electricity mix. The same holds for own-consumption in case the consumer would use the country-specific average electricity mix. Under these conditions, a country-specific carbon payback time for a PV module can be given, quantifying the time a PV module must have operated to replace the CO_2 equivalents that have been emitted for its production. **Table *III*** states both of these quantities for an AVANCIS PV module for the two cases considered, where for reasons of simplicity we have based the calculation on the CF of the German electricity mix of 2020.

Table III: Energy payback time and time to compensate its CF of production of an AVANCIS PV module for two scenarios.

	time in years
case 1	
energy payback time	0.4
time to compensate its CF of production	1.0
case 2	
energy payback time	0.7
time to compensate its CF of production	1.6

3.3 Discussion of the results

The obtained result of approximately 59 kg CO_2eq per module and roughly 56 kg CO_2eq per square meter fits well within earlier studies for CIGS thin-film modules ([4] and references therein) although direct comparison is difficult with the strong influence of the changing CF of electricity over the years. Thin-film modules typically achieve a lower CF per module and per module area compared to PV modules based on crystalline silicon (cSi) wafers, mainly due to a greatly reduced mass of the photovoltaically active material (e.g., [14] based on older data). The typically higher photovoltaic conversion efficiency of cSi modules reduces the gap when considering CF per electrical power (kW$_p$) or electricity generated (kWh). For both cSi and thin-film PV module technologies, the CF of the generated electrical energy lies far below that of electricity generated in power plants burning coal or gas (500 to 800 g CO_2eq/kWh) and within the range of other significant renewable sources (11 to 38 g CO_2eq/kWh) [3].

For building-integrated PV (BIPV), a comparison to construction material that might be replaced by BIPV modules is most relevant. **Table *IV*** lists a few of those in addition to typical glass-based facades. The area-based CF of AVANCIS SKALA lies in between those of passive facade materials that might directly be replaced by the PV module and glass facades that are more comparable in terms of its composition. For a fair comparison, however, only full facade systems should be considered, i.e., for a ventilated curtain wall system (a typical application of current BIPV modules), the building's wall, thermal insulation and mounting structure would need to be included. The main point, however, not considered in simple comparisons of EPDs or CF is that a BIPV facade is an active facade, generating 'green' energy during operation with the potential of compensating and over-compensating the carbon footprint caused by its production. A passive construction element as, e.g., an aluminum polymer compound, will always have a positive CF with the exemption of wood-based products that might facilitate a carbon sink (until it is disposed of by incineration). For a BIPV module, over-compensation of the production-caused CF is reached after the times given in **Table *III***. For case 2 considered, an AVANCIS module installed in a south-faced facade in Munich, Germany replaces about 2600 kWh of 'grey' electricity within 25 years, worth about 950 kg CO_2eq, assuming a constant CF of the German electricity mix for simplicity. Subtracting the CF of its production, one square meter of AVANCIS SKALA contributes a strongly negative CF to the carbon balance of the building (last row in **Table *IV***). An increasing share of renewable energy sources within the electricity mix will both lower the production-related CF of a PV module and its impact by replacing 'grey' electricity during its use phase.

Table IV: Carbon footprint per square meter area for various facade materials as obtained from publicly available EPDs. The values shown are for the stages A1 to A3 (cradle-to-gate), except for the last row. Note, that the numbers given cannot directly be compared among systems that facilitate a full facade setup and others that require an additional wall and mounting structure in a building's envelope. The last row states the value for an AVANCIS SKALA module in a facade (case 2) including the replacement of 'grey' electricity during 25 years.

material	CF in kg CO_2eq/m^2
architectural glass	260
glass curtain wall	120, 160
AVANCIS SKALA	56
aluminium facade system (thickness 1mm)	28
aluminum polymer compound	27
steel facade system (thickness ~1mm)	16
natural stone (thickness 30mm)	4
wood-plastic composite (WPC)	-0.6
AVANCIS SKALA incl. use phase	-900

3.4 Potential for further decrease in carbon footprint for CIGS modules

Although the considered PV modules by AVANCIS come with a relatively low production-related carbon footprint, further reduction is a worthwhile goal. Our analysis clearly identifies the electricity consumption of the module-producing factory as the main leverage for further reduction. This can be achieved both by decreasing the electricity consumption and by using electricity with a lower

CF. Reduction of electricity demand is subject of continuous optimization of the production processes, reduction of scrap and increase of material usage. In addition, reduction of energy demand (and therefore also opex costs) is of course a main goal in the development of future production equipment with significant steps already achieved during past and current R&D activities at AVANCIS (e.g., [15]). Reducing the CF input of the electricity consumed can be achieved by switching from 'grey' to 'green' energy suppliers or by the generation of electricity by own PV plants on the premises of the factory. AVANCIS is currently following the latter with an example shown in **Figure 1**.

Further reductions of CF could be achieved by sourcing at suppliers who guarantee a lower-than-standard CF for their products. Significant effects could be achieved for the suppliers of glass and steel, whereas switching suppliers for the other materials would only have minor effects if the main contributors remain unchanged. The same holds true for transport-related carbon emissions. Both the production of glass and steel require high amounts of energy (primarily for high process temperatures) and major reductions in the CF of the respective production have been made [16] [17], making further strong reductions in the near future unlikely [18].

Recycling (both of the scrap material incurring during the production of a PV module as well as modules after end-of-life) has the potential for reducing the CF of a PV module further in a cradle-to-cradle analysis. Whereas recycling of scrap during production is well established (but not considered in the present cradle-to-gate analysis), due to the long lifetime of AVANCIS PV modules, it is yet difficult to assess the precise impact of the dismantling and recycling of modules after end-of-life. For this, detailed comparisons of various scenarios for recycling are necessary. From the literature, a reduction of the cradle-to-cradle CF is expected to lie in the range of 5% [19] [20]. Separating steel, cables and junction boxes from the glass-glass laminate is an established process as well as their recycling. Glass from laminates is typically downcycled (e.g., to foam glass), although processes exist for the separation of the less abundant materials forming the semiconductor [21].

4 CONCLUSION

In the present study, we have calculated the carbon footprint of a commercially available CIGS thin-film PV module that is particularly well-suited for building-integrated PV. The results obtained lie within typical values known from literature and we identify the main leverages for a further decrease of the carbon footprint. Switching the energy supply of the production plant to renewable sources – own-consumption of PV-produced energy on-premise being the most attractive way – and lowering the energy consumption both of the production machinery and the facilities can lead to a carbon footprint as low as 32 kg CO_2eq/m^2, which translates into 8 g CO_2eq/kWh for a free-field PV system and 13 g CO_2eq/kWh for a south-facing facade-installed module in Munich. Also by continuing with 'grey' electricity, the currently decreasing CF of the German electricity mix [11] will further reduce the CF of a PV module being manufactured in Germany.

For the PV module studied, we do not see an advantage of using a commercial software against a simple calculation using publicly available data for the CF of materials. Differences between inventory databases lead to uncertainties in the computed CF that are larger than most of the smaller contributions that one might miss in a simplified approach. In the present case with low amounts of rare materials, the impact of strongly differing databases does not alter the main conclusions of the CF analysis. For products with a higher share of those materials, however, choice of database can be decisive and hinders comparability of EPDs when differing databases have been used. We therefore see a need for harmonized CF databases that ideally would be in the public domain.

Comparing the results of LCAs, EPDs or CF calculations for assumed similar products without a detailed analysis of the respective system boundaries and methodologies needs to be handled with great care. There is a need for establishing a commonly accepted methodology, which, for the case of BIPV, is currently being worked on [22]. Guidance might be found in other areas of the construction sector, where, e.g., regionally differing standards and requirements (insulation, fire protection, ...) make direct comparison of individual construction products difficult, as well.

As for other construction materials, recycling or disposal are ideally included in a comprehensive LCA, although these topics are difficult to quantify for products whose expected lifetime is longer than the time its technology has been around. It needs to be discussed, whether production of 'green' electricity during the use phase of a (BI)PV module (and therefore the displacement of fossil-based 'grey' electricity) shall be included in an LCA/EPD. This, however, will not only be strongly location- and system-specific but also depend on the development of the CF of the to-be-replaced 'grey' electricity – a number very difficult to predict for the typically long lifetimes of PV modules.

For BIPV modules, we recommend to state not only one but two figures for the carbon footprint, to account for the hybrid nature of BIPV (both building material and electricity generator). The carbon footprint per square meter is necessary for an architect to compute the contribution of, e.g., a PV facade to the total CF of the building. Providing the CF per produced kilowatt-hour for a specific installation site then serves as a measure for CO_2 savings during the operational phase of the BIPV system. Providing only the area-related number does not account for the module being able to offset its carbon footprint during use. The energy-related CF (per kWh), however, is system- and location-dependent and cannot be generalized for building planners. We also note that any comparison of an area-specific production-related CF of construction materials might need corrections if expected lifetimes differ greatly among considered products.

Summarizing the aforementioned items, we see a strong need to standardize methods leading to comparable CFs and EPDs between, e.g., competing construction products. In view of the formal characteristics of an EPD combined with still insufficient precision of the input data, we recommend to include a careful discussion of the uncertainties when comparing different LCAs in the present stage.

5 REFERENCES

[1] United Nations Environment Programme, „2020 Global Status Report for Buildings and Construction," 2020.

[2] „DIN EN 15804: Sustainability of construction works - Environmental product declarations - Core rules for the product category of construction products," 2019.

[3] S. Schlömer, T. Bruckner, L. Fulton, E. Hertwich, A. McKinnon, D. Perczyk, J. Roy, R. Schaeffer, R. Sims, P. Smith und R. Wiser, „Annex III: Technology-specific cost and performance parameters. In: Climate Change 2014: Mitigation of Climate Change Contribution of Working Group III to the Fifth Assessment Report of the Intergovernmental Panel on Climate Change," Cambridge University Press, 2014.

[4] A. Briem, „Life cycle assessment of CIGS PV modules: update of current production conditions in Germany and investigation of a planned factory in China," *Proc. of the 35th EUPVSEC*, 2018.

[5] J. Anderson, „ConstructionLCA's 2021 Guide to Environmental Product Declarations (EPD)," 2021. [Online]. Available: https://infogram.com/constructionlcas-2021-guide-to-epd-1h1749vwlxy7l6z.

[6] Euopean Commission, „Roadmap to a Resource Efficient Europe," 2011.

[7] *www.skalafacade.com.*

[8] VTT Technology, „Carbon footprint for building products," 2013.

[9] *GaBi: Software and database contents for Life Cycle Engineering.*

[10] *openLCA (openlca.org).*

[11] P. Icha, T. Lauf und G. Kuhs, „Entwicklung der spezifischen Kohlendioxid-Emissionen des deutschen Strommix in den Jahren 1990 - 2020," Umweltbundesamt, 2021.

[12] *www.carbonfootprint.com.*

[13] „PVGIS: Photovoltaic geographical information system (https://ec.europa.eu/jrc/en/pvgis)".

[14] F. Liu und J. van den Bergh, „Differences in CO2 emissions of solar PV production among technologies and regions: Application to China, EU and USA," *Energy Policy*, p. 111234, 2020.

[15] B. Cord, A. Amrhein, M. Fürfanger, O. Weiß und A. Schmalstieg, „Advanced R&D platform for highest efficient CIGS-Modules," *Proc. of the 31st EUPVSEC*, pp. 1266-1270, 2015.

[16] Glass Alliance Europe, „The European glass sector contribution to a climate neutral economy," 2021.

[17] World Steel Association, „Steels's contribution to a low carbon future and climate resilient societies," 2020.

[18] U.S. Department of Energy, „Bandwidth Study on Energy Use and Potential Energy Savings Opportunities in U.S. Glass Manufacturing," 2017.

[19] P. Stolz und R. Frischknecht, „Life Cycle Assessment of Photovoltaic Module Recycling," Swiss Federal Office of Energy SFOE, 2016.

[20] IEA PVPS Task 12, „Life Cycle Assessment of Current Photovoltaic Module Recycling," 2018.

[21] M. Marwede, W. Berger, M. Schlummer, A. Mäurer und A. Reller, „Recycling paths for thin-film chalcogenide photovoltaic waste – Current feasible processes," *Renewable Energy*, pp. 220-229, 2013.

[22] IEA PVPS Task 15, *Enabling Framework for the Development of BIPV.*

ADAPTING PVGIS TO TRENDS IN CLIMATE, TECHNOLOGY AND USER NEEDS

Ana Gracia Amillo*[1], Nigel Taylor[1], Ana M. Martinez[1], Ewan D. Dunlop[1], Panagiotis Mavrogiorgios[2], Fernando Fahl[2], Gianpiero Arcaro[3], and Irene Pinedo[1]
[1] European Commission, Joint Research Centre, Viale Fermi, 2749, Ispra, Italy
[2] GFT Italia S.r.l., Milan, Italy
[3] PIKSEL S.r.l. - Via Ernesto Breda, 176 – Milan, Italy
ana.gracia-amillo@ec.europa.eu

ABSTRACT: The Photovoltaic Geographical Information System tool (PVGIS) of the European Commission is a very popular online tool which provides open access high-quality and high-resolution data at continental scale of solar radiation and other climatological variables, like ambient temperature and wind speed. PVGIS also provides estimates of the power output of different photovoltaic (PV) technologies and system configurations. These datasets and estimates are based on data from 2005 to 2016. The present study describes the processes adopted to update the various datasets with improved solar radiation and climatological data and an extended time frame up to 2020. A preliminary analysis on the long-term trends of those variables will be presented along with the implications of considering the updated datasets in the estimated PV potential at continental scale. Preliminary results show that the addition of data from 2017 to 2020 leads to increased levels of irradiation and PV production in most European locations. The analysis presented here is part of an ongoing project to analyze in further detail the long-term trends of the various variables involved in the PV output estimation and the implications on the PV performance at continental scale.
Keywords: solar radiation, PV performance, long-term trends, modelling

1 INTRODUCTION

The PVGIS (Photovoltaic Geographical Information System) tool was developed 20 years ago by the Joint Research Centre of the European Commission with the aim of supporting the deployment of solar energy systems, especially in developing countries [1, 2]. This online tool (https://ec.europa.eu/jrc/en/pvgis) provides high-quality and high-spatial and temporal resolution data of solar radiation as well as other climatological variables like ambient temperature and wind speed at continental scale. PVGIS also provides estimates of the PV energy yield from different PV technologies and PV systems configurations. These data can be obtained either as complete time series of hourly values or as average values at different time scales like yearly or monthly. Besides, PVGIS also produces Typical Meteorological Year (TMY) datasets which are commonly used in building energy efficiency analysis to comply with the requirement of the Energy Performance of Buildings Directive [3]. PVGIS does not require registration or installation of any software and it applies no restrictions to the uses given to the data. Additionally, it is very easy and intuitive to use, making it a very user-friendly online application. This, in addition to the quality of the data provided, makes PVGIS a very popular tool. For example, in 2020 the tool was visited by almost 2.5 million users who on average sent over 30 million data request per month to PVGIS.

PVGIS is under continuous development and at present version 5.1 is available. Both implemented models and available datasets have been improved and updated over the years, and new functionalities have been added as well. For example, the API (Application Programming Interface) which offers a robust and fast access to the data by means of a harmonized nomenclature enabling users to automatize calls with scripts written in various languages. Version 5.2 is expected to be released at the end of 2021 including substantial improvements. One of the main changes regards the update of the underlying datasets both in terms of quality and resolution, and by extending the time period used. While PVGIS 5.1 datasets cover the time period from 2005 to 2016, the updated datasets will include data up to 2020.

Pfeifroth et al's analysis [4] of ground measurements and satellite retrieved (SARAH-2) solar radiation data in Europe shows an overall brightening from the 1980s up to 2015, amounting to between 1.9 and 2.4 W/m^2/decade, although significant geographic variations were observed. The current satellite-based dataset available in PVGIS 5.1 based on SARAH-1 covering the time period between 2005 and 2016 does not appear to show systematic variations. However, there is evidence that more recently climate change effects may be becoming more evident. In fact, the European Environmental Agency notes that the decade 2010-2019 was the warmest on record for European land temperatures [5].

The update and additional years included in PVGIS datasets will enable taking these climate change effects into consideration in the estimated PV performance provided by PVGIS.

The present paper first analyses the validity of the new and updated datasets against ground measurements, to then focus on the preliminary analysis of possible long-term trends of the extended datasets. Secondly, the impact of the new datasets on the estimated PV performance at continental scale are analysed as well.

2 PVGIS DATASETS AND PROPOSED UPDATE

2.1 PVGIS 5.1

The main datasets available in PVGIS 5.1 providing data for the area of Europe and Africa cover the time period between 2005 and 2016. In terms of solar radiation data two main products are available. The first one contains geostationary satellite-retrieved irradiance values based on the CMSAF's SARAH-1 product [6], which has been obtained using a modified Heliosat method and the SPECMAGIC algorithm [7]. The data available in PVGIS has hourly values estimated from the analysis of one single image per hour, while the original CMSAF hourly values are the weighted average of half-hourly values obtained processing various images per hour. The spatial resolution of SARAH-1 data is about 5 km at nadir.

In order to provide data for areas not covered by the geostationary satellite images, PVGIS 5.1 includes solar radiation data from ERA5 reanalysis product developed by the ECMWF [8]. This dataset contains hourly values with a spatial resolution of 31 km.

PVGIS also has data of other variables like ambient temperature or wind speed, which are input to the PVGIS model used to estimate the performance of PV systems. Data of other climatological variables like relative humidity included in the Typical Meteorological Year tool of PVGIS are also available. These data come from the ECMWF reanalysis product ERA-Interim [9], which contains data at 3 hours intervals. However, for PVGIS 5.1 these values have been interpolated in order to provide hourly values. The spatial resolution of this dataset is 81 km.

PVGIS also contains spectral correction coefficients obtained from the analysis of the spectrally resolved irradiance of SARAH-1 product over 2011, which are considered in the PV output estimation. PVGIS also includes a 90 m resolution Digital Elevation Model which has been used to define the surrounding horizon at every location in the geographical area covered by PVGIS. This last information is used in the estimation of the available in-plane irradiance for every simulated plane.

2.2 Proposed datasets update

One of the main updates of the various datasets available in PVGIS regards the time period used, in particular to include data from 2017 up to 2020.

The SARAH-1 solar radiation data will be replaced by new CMSAF solar radiation product SARAH-2.1 [10]. As with the SARAH-1 original datasets, only one image per hour will be used from the original 30-minute data available in CMSAF's SARAH 2.1 product.

Regarding the solar radiation data from ERA5 reanalysis product, the updated dataset will include the most recent data between 2017 and 2020.

Another significant upgrade affects the climatological data, since ERA-Interim data will be replaced by ERA5-Land, which has improved temporal and spatial resolution of hourly data and 9 km respectively [11].

The other datasets available in PVGIS regarding spectral effects and terrain information will remain unchanged.

A complete preliminary version of the updated datasets has been prepared, and the data transformed into the time series used by PVGIS.

3 VALIDATION AGAINST GROUND STATION MEASUREMENTS

The accuracy of the updated and extended datasets, which now cover the time period between 2005 to 2020, has been analyzed comparing these datasets to ground measurements.

3.1 BSRN stations

A series of 15 stations from the BSRN network have been used for the validation study. The one minute resolution of the measurements allows an accurate comparison with the satellite-based data, as they have been retrieved from the analysis of one single image and can therefore been considered as "instantaneous" data. Figure 1 shows the spatial distribution of the stations considered. It should be noted that not all of these stations have measurements over the complete time period of 16 years.

Figure 1: Distribution of the considered BSRN stations.

3.2 Hourly global horizontal irradiance

The hourly values of global horizontal and direct normal irradiance (W/m^2) from SARAH 1, SARAH 2.1 and ERA5 data between 2005 and 2020 (2016 for SARAH 1) have been validated against ground measurements from the BSRN network. The BSRN one minute data have been preprocessed before being used in the validation. For the comparison against the "instantaneous" satellite-based data (derived from single image processing), the measurements taken during the ten-minute interval centered at the time of the satellite image have been averaged to obtain the measured value to be used for the satellite-based data validation. Since the reanalysis data itself is hourly, these have been compared to the hourly average values calculated from the minute data. As an example of the various validation comparisons performed in the present study, Figure 2 shows the results of the validation of the global horizontal irradiance values of satellite-based products SARAH 1 and SARAH 2.1. Figure 2a) contains the absolute and relative values of the mean bias difference MBD (W/m^2) and rMBD (%), respectively. Figure 2b) shows the root mean square difference (RMSD, W/m^2) in absolute and relative (rRMSD, %) terms, while figure 2c) shows the number of hours used in each validation study.

We should note that data from SARAH 1 does not cover the time period from 2017 to 2020, and thus the number of data points considered is different for SARAH 1 and SARAH 2.1. Therefore, the values shown of the metrics of the validation are meant to be an indication of how each derived product compares to ground station data maximizing data availability, but should not directly be compared for the two products.

While SARAH 1 values tend to underestimate ground measurements, in most locations SARAH 2.1 shows positive MBD. Regarding the RMSD values, the SARAH 2.1 results are lower than the values obtained from the previous CM SAF product.

The results from the BSRN station of Izaña should not be taken into consideration in the overall validation. Due to its specific conditions and location, data from this ground station is not representative of the pixel area in the satellite image. The satellite-based irradiance highly underestimates the available solar resource since this station is often above the cloud cover seen in the satellite image, as noted in [12]

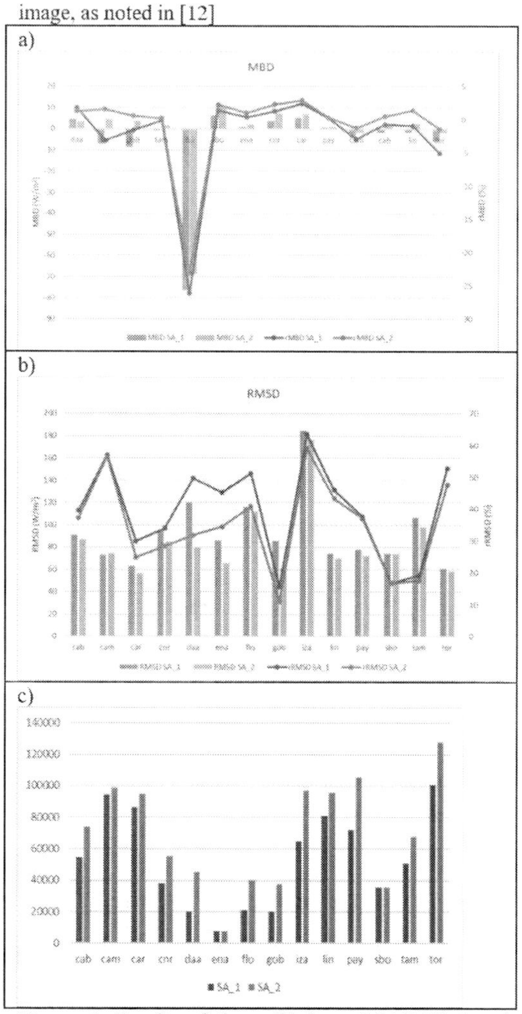

Figure 2. Results of the validation of SARAH 1 and SARAH 2.1 hourly values of global horizontal irradiance against BSRN measurements, along with the number of data points considered in every station and database.

3.3 Hourly ambient temperature

The hourly values of ambient temperature from ERA-Interim and ERA5-Land have also been compared to the measurements of the BSRN stations. Table I shows the overall results when all data from all stations is considered as a whole. The time period considered for ERA-Interim validation goes from 2005 to 2016, while ERA5-Land data extends to 2020.

Table I: Results of the validation of the ERA-Interim and ERA5-Land ambient temperature values

	ERA-Interim	ERA5-Land
MBD (°C)	0.9	0.3
rMBD (%)	6.9	2.5
RMSD (°C)	4.6	2.8
rRMSD (%)	35.64	20.9

The higher resolution, both temporal and spatial, of ERA5-Land with regard to ERA-Interim improves the quality of the estimates according to first analysis.

4 OBSERVED TRENDS

4.1 Implications at European locations

The implications of adding data from the period 2017 to 2020 in the long-term averages have been analysed for Europe. Two planes have been considered, the horizontal plane and an optimally inclined plane at every location. As an example of the results, Figure 3 shows the trends observed for Lisbon (Portugal), as host city of the 38th EU PVSEC. In Figure 3a the yearly global horizontal irradiation (kWh/m^2) from SARAH 1 between 2005 and 2016 is compared to the values of SARAH 2.1 since 2005 up to 2020. The discontinuous lines show the trends in the observed yearly values, whose slope is indicated in the legend. Figure 3b shows the yearly PV output of an optimally inclined c-Si system (kWh/kWp) calculated on the one hand considering SARAH 1 solar radiation data and the climatological values from ERA-Interim, against the values calculated with the same PVGIS' PV model using SARAH 2.1 solar radiation data and the wind speed and ambient temperature from ERA5-Land.

It can be observed in Figure 3 how the trend of both SARAH 1 and SARAH 2.1 is similar up to 2016, but when more recent years are considered, the trend becomes more flat or slightly positive. While the trends observed for SARAH 1 data between 2005 and 2016 are slightly negative in 55% of the locations considered, only 3% of them show negative trends when SARAH 2.1 data from 2005 to 2020 are analysed.

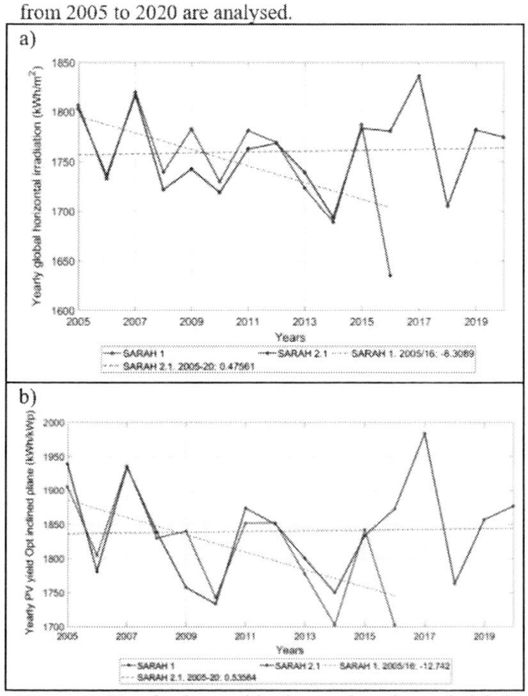

Figure 3. Trends observed in the yearly values of global horizontal irradiation and PV yield from SARAH 1 and SARAH 2.1 for Lisbon.

4.2 Implications on the available irradiation at continental scale

The solar radiation database used in PVGIS 5.1 based

on the SARAH 1 values of CMSAF contains data between 2005 and 2016. However, the updated dataset created from CMSAF's SARAH 2.1 product contains data from 2005 to 2020. The difference on the yearly average irradiation (kWh/m^2) values from SARAH 1 dataset and the values calculated based on SARAH 2.1 dataset have been analysed at continental scale. Figure 4a shows the relative difference between the yearly average values calculated from the two datasets over the same time period: 2005 to 2016. The histogram of the observed values is also shown on the right-hand side. Figure 4b shows the relative difference between SARAH 1 2005-2016 yearly averages and the values obtained from SARAH 2.1 2005-2020 period. The third comparison (Figure 4c) is between the yearly average values obtained from the twelve-year period of SARAH 1 data and the most recent 12 year period available from SARAH 2.1 (2009-2020). From the first comparison it is observed that SARAH 2.1 values in most locations are between 1 to 3% higher than the average values from SARAH 1 in the same time period between 2005 and 2016. However, when most recent years are considered in SARAH 2.1 dataset (Figure 4b), the average yearly values of global horizontal irradiation increase, resulting in most regions in an overestimation compared to SARAH 1 values of 3 to 4%, although there are regions where the difference is up to 10%. In general differences are higher in mountainous regions and they tend to increase with latitude. However, locations in southern Europe show lower values from SARAH 2.1 than SARAH 1 even when most recent years are considered in SARAH 2.1. This could derive from the characteristics of the algorithms and input data used in both CMSAF products and a decrease in irradiance levels in those latitudes. Further analysis including, if available, high time resolution ground data is needed.

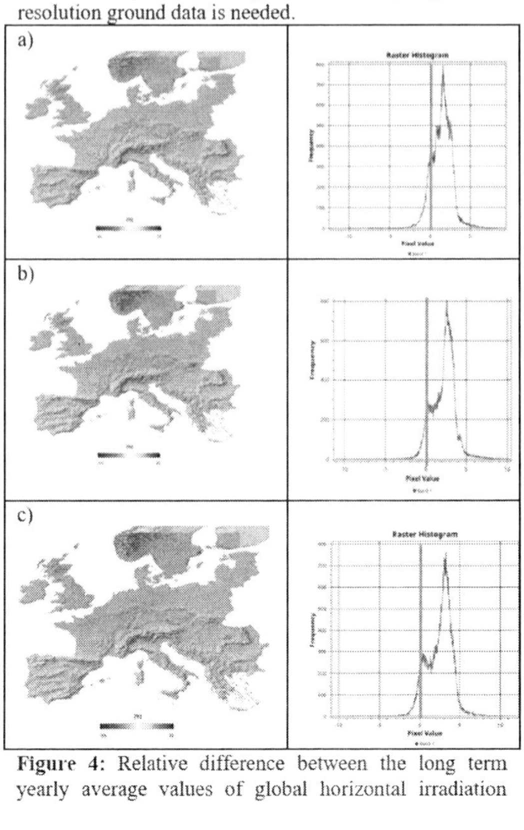

Figure 4: Relative difference between the long term yearly average values of global horizontal irradiation calculated with SARAH 1 2005-2016 data, compared to the values obtained from SARAH 2.1 datasets considering the time periods of 2005-2016 (a), 2005-2020 (b) and 2009-2020 (c).

4.2 Implications on the PV production at continental scale

The PV energy yield has been calculated using the model implemented in PVGIS which uses the in-plane irradiance, the ambient temperature and the wind speed as input data. The model has been applied to the two different sets of input data as described at the beginning of Section 4. SARAH 1 and ERA-Interim on one hand, and SARAH 2.1 and ERA5-Land on the other hand. Similarly to the comparisons shown in Figure 4, Figure 5 shows the difference in percentage between the PV energy yield estimated with these two sets of input data for a 1 kWp crystalline silicon system installed at optimal inclination angle in every location.

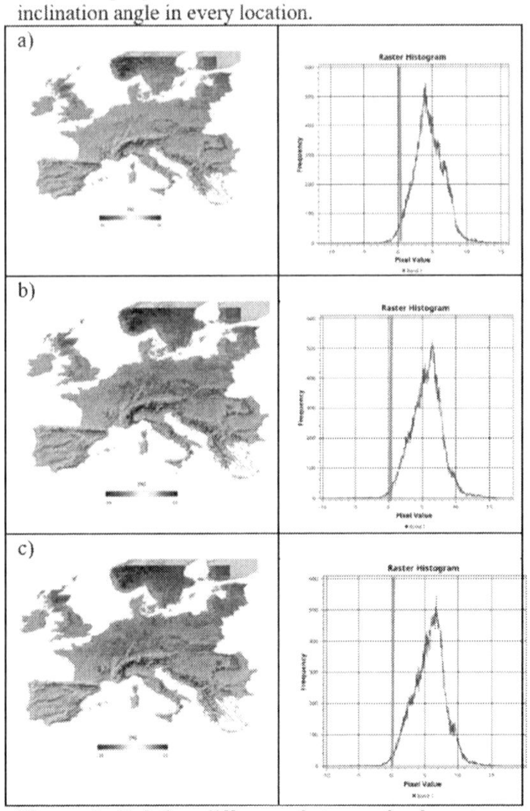

Figure 5: Relative difference between the long term average values of annual PV energy yield calculated for 1 kWp c-Si optimally inclined PV system, obtained from SARAH 2.1 and ERA5-Land datasets considering the time periods of 2005-2016 (a), 2005-2020 (b) and 2009-2020 (c). compared to that calculated with SARAH 1 and ERA-Interim 2005-2016 data.

The observed relative differences are higher than those observed in the case of the horizontal irradiation. This may be due to the increased SARAH 2.1 irradiance values, but also due to the use of the higher resolution ERA5-Land data. The differences observed in mountainous regions and northern latitudes in Fig 4 are increased in the PV yield values. Almost all regions in Europe obtain higher values from SARAH 2.1 and ERA5-Land than with SARAH 1 and ERA-Interim, regardless of the years considered from the SARAH 2.1 datasets. Although, considering most recent years results in higher differences as well.

5 CONCLUSIONS AND FUTURE WORK

The most recent available products of CMSAF and ECMWF have been analysed in order to update the solar radiation and climatological datasets available in PVGIS. The solar radiation data SARAH 2.1 from CMSAF, along with the ERA5-Land climatological data from ECMWF have been analysed, validated and prepared for integration in PVGIS. The validation against ground measurements of the BSRN network has demonstrated the accuracy of the updated datasets. Preliminary results show that including data from 2017 to 2020 increases both the yearly average irradiation and the PV output from PV systems in most locations in Europe by a few percentage points, in comparison to the values obtained from the PVGIS 5.1 datasets, which cover the time period between 2005 and 2016.
A new version of PVGIS is expected to be released by the end of 2021 with the new and extended datasets of SARAH 2.1 and ERA5-Land. The new version will also include an updated PV performance model which is currently under study and therefore has not been applied in the present study.
Further analysis will focus on quantifying the effect of the combination of the updated datasets when used as input data of the PV model on the resulting PV performance.

6 ACKNOWLEDGEMENTS
Authors would like to thank the Satellite Application Facility on Climate Monitoring (CMSAF), European Centre for Medium-Range Weather Forecast (ECMWF) and Baseline Surface Radiation Network (BSRN) for making all their datasets publicly available, and Deutscher Wetterdienst (DWD) for their technical support in the development of the new updated datasets which will be implemented in the new version of PVGIS, PVGIS 5.2.

7 REFERENCES

[1] Huld T., Müller R., Gambardella A. A new solar radiation database for estimating PV performance in Europe and Africa. *Solar Energy*, 86, 1803-1815 (2012).
[2] Huld T., Pinedo Pascua I., Gracia Amillo A., Urraca R., Dunlop, E. PVGIS version 5: improvements to models and features. In: 33rd European Photovoltaic Solar Energy Conference and Exhibition, 25-29 September 2017. EU PVSEC conference proceedings, 2017, ISBN 3-936338-47-7, ISSN 2196-100X, p. 2069-2074.
[3] Directive 2010/31/EU of the European Parliament and of the Council of 19 May 2010 on the energy performance of buildings (recast) 2021. (http://data.europa.eu/eli/dir/2010/31/2021-01-01)
[4] Pfeifroth U., Sanchez-Lorenzo A., Manara V., Trentmann J., Hollmann R. Trends and Variability of Surface Solar Radiation in Europe Based On Surface- and Satellite-Based Data Records. *Journal of Geophysical Research: Atmospheres*, 123, 1735–1754 (2018), doi: 10.1002/2017JD027418.
[5] https://www.eea.europa.eu/data-and-maps/indicators/global-and-european-temperature-10/assessment
[6] Muller, R., Uwe, P., Trager-Chatterjee, C., Cremer, R., Trentmann, J., Hollmann, R. Surface Solar Radiation Data Set - Heliosat (SARAH) - Edition 1.. Satellite Application Facility on Climate Monitoring (CM SAF), 2015.
http://dx.doi.org/10.5676/EUM_SAF_CM/SARAH/V001
[7] Müller, R., Behrendt, T., Hammer, A., Kemper, A. A new algorithm for the satellite-based retrieval of solar surface irradiance in spectral bands. *Remote Sens.* 4, 622–647 (2012). http://www.mdpi.com/2072-4292/4/3/622.
[8] Atmospheric Reanalysis ERA5 products. Copernicus Climate Change Service (C3S) at the European Centre for Medium-Range Weather Forecast.
https://confluence.ecmwf.int/display/CKB/ERA5%3A+data+documentation
[9] Dee, D.P., Uppala, S.M., Simmons, A.J., Berrisford, P., Poli, P., Kobayashi, S., Andrae, U., Balmaseda, M.A., Balsamo, G., Bauer, P., Bechtold, P., Beljaars, A.C.M., van de Berg, L., Bidlot, J., Bormann, N., Delsol, C., Dragani, R., Fuentes, M., Geer, A.J., Haimberger, L., Healy, S.B., Hersbach, H., Hólm, E.V., Isaksen, L., Kållberg, P., Köhler, M., Matricardi, M., McNally, A.P., Monge-Sanz, B.M., Morcrette, J.-J., Park, B.-K., Peubey, C., de Rosnay, P., Tavolato, C., Thépaut, J.-N., Vitart, F. The ERA-Interim reanalysis: configuration and performance of the data assimilation system. *Quart. J. Roy. Meteorol. Soc.* 137 (656), 553–597 (2011). http://dx.doi.org/10.1002/qj.828.
[10] Pfeifroth U., Trentmann J., Kothe S. Validation report Meteosat Solar Surface Radiation and Effective Cloud Albedo Climate Data Record. SARAH-2.1 climate data records (2019).
DOI: 10.5676/EUM_SAF_CM/SARAH/V002_01.
[11] Atmospheric Reanalysis ERA5-Land product. Copernicus Climate Change Service (C3S) at the European Centre for Medium-Range Weather Forecast. https://www.ecmwf.int/en/forecasts/dataset/ecmwf-reanalysis-v5-land.
[12] Urraca R., Huld T., Gracia-Amillo A., Martinez-de-Pisona F.J., Kasparc F., Sanz-Garcia A. Evaluation of global horizontal irradiance estimates from ERA5 and COSMO-REA6 reanalyses using ground and satellite-based data. Solar Energy 164, 339-354 (2018) https://doi.org/10.1016/j.solener.2018.02.059

IMPROVING THE PREDICTION OF DNI VIA PHYSICS-BASED SIMULATION OF ALL-SKY CIRCUMSOLAR RADIATION

Yu Xie[1], Jaemo Yang[1], Manajit Sengupta[1], and Yangang Liu[2]
[1]National Renewable Energy Laboratory, 15013 Denver W Pkwy, Golden, CO 80401
[2]Brookhaven National Laboratory, 99 Rochester St., Upton, NY 11973

ABSTRACT: Direct Normal Irradiance (DNI) is one of the most used quantities to quantify the magnitude of solar energy resource. The concept of DNI is often interpreted differently for ground measurements and solar forecasting by numerical weather prediction (NWP) models, leading to substantial bias during evaluation of DNI forecasts especially under cloudy-sky conditions. To eliminate the bias, we use the Fast All-sky Radiation Model for Solar applications with DNI (FARMS-DNI) to provide a physics-based solution of solar radiation in the circumsolar region. The FARMS-DNI is implemented in the Weather Research and Forecasting model with solar extensions (WRF-Solar) to forecast day-ahead DNI in the north America. By comparing with conventional predictions from WRF-Solar and satellite observations from the National Solar Radiation Data Base (NSRDB), we found significant improvements in our prediction of DNI.
Keywords: solar radiation, DNI, solar forecasting

1 INTRODUCTION

Following the definition of direct normal irradiance (DNI) in the ISO 9488 standard (https://www.iso.org), DNI represents "the quotient of the the radiant flux on a given plane receiver surface received from a small solid angle centerred on the sun's disk to the area of that surface". However, most radiative transfer models used in the NWP models compute direct solar radiation on the basis of the Lambert law assuming an infinitely narrow beam along the solar incident angle [1, 2]. This assumption ignores the scattered solar radiation in the circumsolar region, and thus often leads to a significant underestimation of DNI in the cloudy atmosphere. On the other hand, models empirically estimate DNI based on long-term observation of all-sky global horizontal irradiance (GHI) and DNI and often overestimate DNI in the cloudy atmosphere [3]. To precisely compute all-sky circumsolar radiation, Xie et al. [4] developed FARMS-DNI that is a physical solution of scattered solar radiation that falls in the circumsolar region. This model uses a precomputed lookup table of cloud transmittance and reflectance to compute solar radiation for differential solid angles. It further uses a finite-surface integration algorithm to efficiently infer the radiation received by a surface perpendicular to the direction of the sun.

To increase the computational efficiency, we parameterized the cloud transmittance and reflectance using functions of solar zenith angle, cloud type, cloud optical thickness, and cloud effective particle size. FARMS-DNI with the parameterization has been implemented in the Weather Research and Forecasting model with solar extensions (WRF-Solar) [5, 6] to improve the prediction of DNI. The performance of the DNI forecast is evaluated using satellite observations in North America.

2 METHODOLOGY

To investigate the impact of the new DNI approach on solar forecasting, WRF-Solar with FARMS-DNI is configured to provide day-ahead prediction over the Contiguous U.S. using 9 km grid spacing. The day-ahead forecast spanning April 2014 is produced for every 15 minutes by WRF-Solar. Each simulation is initialized at 6 UTC with a forecast length of 48 hours. The National Centers for Environmental Prediction Global Forecast System (0.5°×0.5°; 3-hour intervals) forecast data are used for the initial and boundary conditions provide to the WRF-Solar model. Following the NCEP operational forecasting of the atmosphere using the High-Resolution Rapid Refresh (HRRR) model, we consider an extended domain covering the contiguous United States (the purple shadow in Fig. 1). The spatial resolution of the solar forecasting is 9 km, with a domain consisting of 211,447 grid boxes.

The forecast of DNI is evaluated using satellite data from the National Solar Radiation Data Base (NSRDB) [7-11]. The native NSRDB data (4km) are aggregated to the WRF-Solar grids (9km). In addition, the NSRDB data are produced over the ocean to evaluate the WRF-Solar forecast for its full extent in the North America. In this study, the 30-minute resolution satellite-based data are compared with the DNI forecasts in the matching time steps.

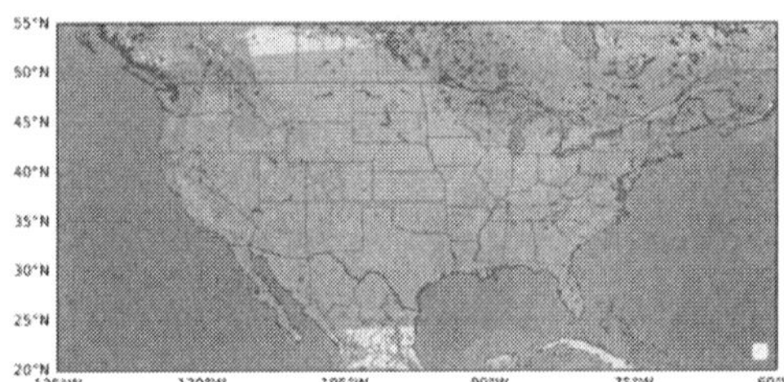

Fig. 1 The geographic domain of the solar forecasting by WRF-Solar.

3 FORECAST EVALUATION

The comparison of the new model against the Lambert law using global statistical metrics (computed with all pairs of observations and predictions for April 2014) shows that FARMS-DNI decreases root mean square error (RMSE), bias and mean absolute error (MAE) (Figure 1). There is an improvement from using FARMS-DNI with a reduction of 10% in RMSE, a reduction of 69% in bias, and a reduction of 9% in MAE compared to the Lambert law. The temporal variation of RMSE, Bias, and MAE for the DNI forecasts calculated with all available data at a given forecast lead time are presented in Fig. 2.

The FARMS-DNI shows RMSE and MAE reductions for DNI forecasts in the range of 1-16% and -3-13%, respectively, compared to the Lambert law across all forecast lead times. Considering bias, the models behave differently with positive (FARMS-DNI) and negative (Lambert law) values. This is attributed to FARMS-DNI that produces larger direct irradiances under cloudy-sky conditions. The performance of the two models has lower differences under clear-sky conditions.

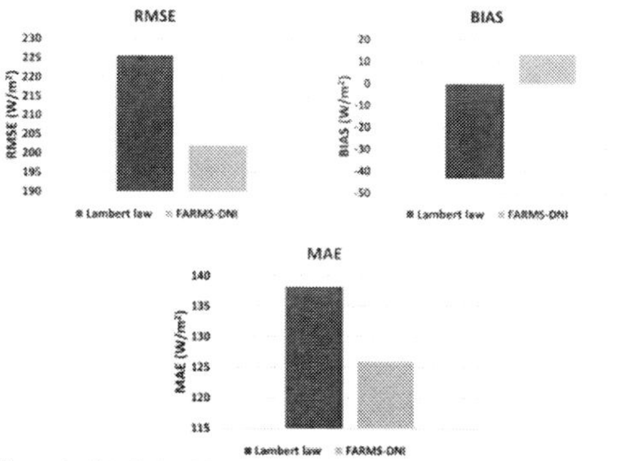

Fig. 2 RMSE, bias, and MAE of direct irradiance computed using the Lambert law and FARMS-DNI.

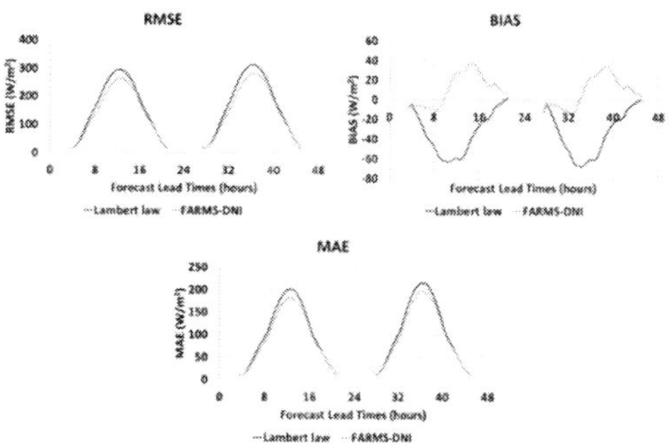

Fig. 3 Temporal variation of RMSE, bias, and MAE of direct irradiance computed using the Lambert law and FARMS-DNI.

Figure 4 illustrates the RMSE, bias, and MAE of direct irradiance in the North America. WRF-Solar with the Lambert law substantially underestimates the DNI in virtually the entire domain, whereas FARMS-DNI primarily matches the spatial allocation of DNI obtained by the satellite data. The results demonstrate negative correlations between the magnitude of DNI and the relative bias in the forecast using both the Lambert law and FARMS-DNI. For example, the DNI forecasts in the western United States show moderate biases compared to the oceanic area and the eastern area, particularly around New England and the Great Lakes region. This is attributed likely to the fewer clouds on average in the western United States and consequently the reduced uncertainty in cloud forecasting with smaller fluctuations in solar radiation. The negative correlations become less apparent in the forecasts by FARMS-DNI owing to the improvements in the cloudy-sky DNI simulations.

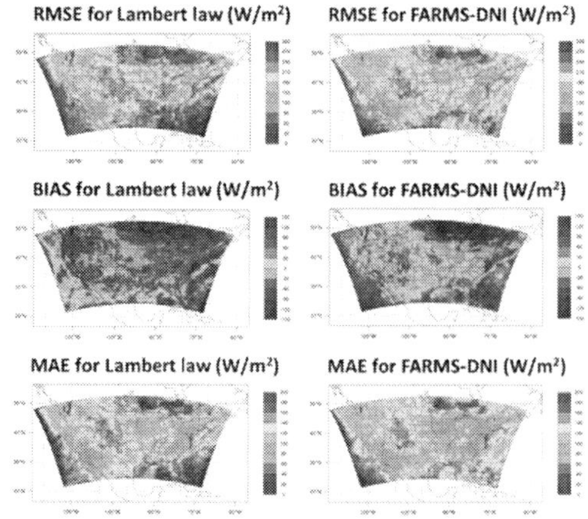

Fig. 4 RMSE, bias, and MAE of direct irradiance in the North America.

4 CONCLUSIONS

This study employs FARMS-DNI in the state-of-the-art solar forecasting model, WRF-Solar, to make physics-based predictions of all-sky radiation in the circumsolar region. The performance of FARMS-DNI is evaluated against satellite-based data in an extended domain covering the contiguous United States. The forecasts based on FARMS-DNI are compared with the Lambert law which is used by the latest version of WRF-Solar, as well as a variety of other NWP models to predict direct solar radiation. It is revealed from the satellite-based data that the DNI forecasts by the Lambert law are systematically underestimated. These biases are significantly reduced by implementing FARMS-DNI to account for the scattered radiation in the circumsolar region. Although the forecast errors for some individual scenes are reduced by FARMS-DNI, they are nonignorable. The negative correlations between the DNIs and their relative biases indicates that the accuracy of DNI forecasting is severely influenced by the long-term cloud coverage in the area.

ACKNOWLEDGMENTS

This work is supported by U.S. Department of Energy Office of Energy Efficiency and Renewable Energy Solar Energy Technologies Office. The National Renewable Energy Laboratory, operated by Alliance for Sustainable Energy, LLC, for the U.S. Department of Energy (DOE) under Contract No. DE-AC36-08GO28308. The Brookhaven National Laboratory is operated by the Brookhaven Science Associates, LLC (BSA), for the U.S. Department of Energy under Contract No. DE-SC0012704.

The views expressed in the article do not necessarily represent the views of the Department of Energy or the U.S. Government. The U.S. Government retains and the publisher, by accepting the article for publication, acknowledges that the U.S. Government retains a nonexclusive, paid-up, irrevocable, worldwide license to publish or reproduce the published form of this work, or allow others to do so, for U.S. Government purposes.

This research was performed using computational resources sponsored by the Department of Energy's Office of Energy Efficiency and Renewable Energy and located at the National Renewable Energy Laboratory.

REFERENCES

[1] E. J. Mlawer, S. J. Taubman, P. D. Brown, M. J. Iacono, and S. A. Clough, "RRTM, a validated correlated-k model for the longwave," *J. Geophys. Res.*, vol. 102, pp. 16663-16682, 1997.

[2] K. N. Liou, *An introduction to atmospheric radiation*, 2nd ed. Amsterdam; Boston: Academic Press, 2002, p. 583.

[3] E. Maxwell, "A quasi-physical model for converting hourly global horizontal to direct normal insolation," presented at the Solar Energy Research Inst., Golden, CO, 1987.

[4] Y. Xie *et al.*, "A physics-based DNI model assessing all-sky circumsolar radiation," *iScience*, vol. 22, p. doi.org/10.1016/j.isci.2020.100893, 2020.

[5] P. Jimenez *et al.*, "WRF-Solar: Description and clear-sky assessment of an augmented NWP model for solar power prediction," *Bull. Amer. Meteor. Soc.*, vol. 97, pp. 1249-1264, 2016.

[6] S. Haupt *et al.*, "Building the Sun4Cast system: Improvements in solar power forecasting," *Bull. Amer. Meteor. Soc.*, pp. 121-135, 2018.

[7] M. Sengupta, Y. Xie, A. Lopez, A. Habte, G. Maclaurin, and J. Shelby, "The National Solar Radiation Data Base (NSRDB)," *Renew. Sustain. Energy Rev.*, vol. 89, pp. 51-60, 2018.

[8] Y. Xie, M. Sengupta, and J. Dudhia, "A Fast All-sky Radiation Model for Solar applications (FARMS): Algorithm and performance evaluation," *Sol. Energy*, vol. 135, pp. 435-445, 2016.

[9] Y. Xie and M. Sengupta, "A Fast All-sky Radiation Model for Solar applications with Narrowband Irradiances on Tilted surfaces (FARMS-NIT): Part I. The clear-sky model," *Sol. Energy*, vol. 174, pp. 691-702, 2018.

[10] Y. Xie, M. Sengupta, and C. Wang, "A Fast All-sky Radiation Model for Solar applications with Narrowband Irradiances on Tilted surfaces (FARMS-NIT): Part II. The cloudy-sky model," *Sol. Energy*, vol. 188, pp. 799-812, 2019.

[11] A. Habte, M. Sengupta, C. Gueymar, A. Golnas, and Y. Xie, "Long-term spatial and temporal solar resource variability over America using the NSRDB version 3 (1998–2017)," *Renew. Sustain. Energy Rev.*, vol. 134, p. 110285, 2020.

CHARACTERIZING THE CONVERGENCE AND ROBUSTNESS OF THE KERNEL DENSITY MAPPING METHOD FOR SITE-ADAPTATION OF GLOBAL HORIZONTAL IRRADIATION IN WESTERN EUROPE

Loïc Yezeguelian[a,b], Christophe Vernay[a], Dr. Thomas Carrière[a] , Prof. Philippe Blanc[b]

[a]SOLAÏS, 55 allée Pierre Ziller, 06560 Sophia Antipolis, France
[b]Centre Observation, Impacts, Energy, ARMINES / MINES ParisTech, CS 10207, 06904 Sophia Antipolis, France

ABSTRACT: The estimation of the annual solar resource at a given location is a crucial part of the financing of new photovoltaic power plants. The standard practice is to use satellite-derived databases of ground irradiance estimations. Although these databases have a long-term history that properly captures the climatological variations of a given site, they are not calibrated with ground measurements and therefore may suffer from significant biases. In this paper, we focus on Western Europe and propose to use ground measurements of global horizontal irradiation (GHI) combined with a Kernel Density Mapping algorithm to calibrate several satellite databases (Solargis, HelioClim-3 and CAMS Radiation); we then show that their biases can be reduced from roughly ±3 % down to 0.5 %. More importantly, we perform a sensitivity analysis on the minimal amount of ground measurements required to perform a proper calibration and show that 12 months is the perfect duration to achieve optimal calibration performance. Additionally, we compare the calibration performance on several different satellite databases and show that a combination of several databases can lead to lower overall uncertainty.
Keywords: GHI calibration, site-adaptation, Kernel Density Mapping, Quantile Mapping, Solargis, HelioClim-3, CAMS Radiation, bias removal, convergence, campaign duration

Table I: Notations

CDF	:	Cumulative Distribution Function
CS	:	when indexed, refers to clear-sky
GHI	:	Global Horizontal Irradiation
I	:	interpolation interval of the CDF
KDE	:	Kernel Density Estimator
KDM	:	Kernel Density Mapping
KSI	:	Kolmogorov-Smirnov Integral
LT	:	when indexed, refers to the long-term
$meas$	:	when indexed, refers to the ground measurements
MBE	:	Mean Bias Error
PDF	:	Probability Density Function
$RMSE$	:	Root Mean Square Error
sat	:	when indexed, refers to the satellite database
ST	:	when indexed, refers to the short-term measurement campaign
$\tilde{x}$	:	uncalibrated satellite point of data
x	:	calibrated satellite point of data

1 INTRODUCTION

The profitability of photovoltaic (PV) projects as well as the financial structuring strongly rely on the global performance of the PV plant which is commonly estimated with percentiles 50 % (also known as P50 in the industry) and 10 % (P90) of the yearly energy yields during its future multi-decadal lifetime. The uncertainty over such decisive assessments mainly comes from the imperfect knowledge of the local long-term irradiation (typically the Global Horizontal Irradiation – GHI). [1] The best practice for deriving typical (P50) or conservative (P90) meteorological year at a given location now consists in using satellite-based databases because of their long-term history (typically 15+ years). However, the uncertainty in the satellite GHI estimation still jeopardizes the development of large-scale PV projects for which feed-in tariffs get closer to network parity.

As for wind industry in the last decades, site-adaptation techniques are now emerging in the PV industry for correcting the GHI assessment and more precisely for removing the Mean Bias Error (MBE) of the satellite-based estimation, which is experienced whatever the modeling time step. Such a local calibration process requires local measurements of the GHI over a short-term period of several months, which in turn requires the setup of a measurement campaign at the early phase of the project. The collected data are then coupled to calibration algorithms that compare the satellite data with the measurements to model the errors. The resulting correction, in the form of a filter function, is propagated to the long-term satellite derived GHI time series.

Numerous types of calibration filters have been explored in the scientific literature. [2] Almost all these methods are mainly grouped around two mathematical procedures, namely multi-parameter regression and Quantile Mapping. The former consists in modeling the measured time-series by a combination of several relevant parameters from the satellite dataset or solar-position retrieval algorithms. Such regression can be linear or non-linear and makes use of every available data representing the instantaneous atmospheric and solar configuration of the measurement point, such as solar elevation, clearness index, etc. However, no consensus has been reached over the best combination of variables. Regression also proves to be geographically inconsistent and to strongly rely on local fluctuations of environmental parameters with post-calibration biases drastically fluctuating around zero. As for Quantile Mapping, it is rather based on a blind matching of probabilistic distributions of both satellite and measured time-series. The Cumulative Distribution Function (CDF) of the satellite irradiance data is transformed into that of the measurements covering the same time period. The mathematical transformation, taking the form of the filter, is then applied to the long-term. The benefit of this approach is that it does not depend on the geography or meteorology of the studied site. It is mathematically constructed to remove any distribution errors in the satellite data and should cancel the MBE between the long-term and the short-term without making use of any spatial or temporal parameter other than the one it intends to correct.

While comparative studies confronting different algorithms are already available, this study intends to stress and assess the behavior of Quantile Mapping-based processes from a practical perspective. It is notably in line with the related activity of the Task 16 of the International Energy Agency program PVPS and the corresponding papers from Polo et al. [3] The main objective is to characterize the performances of the Kernel Density Mapping (KDM) method, derived from Quantile Mapping and commonly used for site-adaptation, with an innovative focus on both convergence (campaign duration) and robustness (choice of the database and spatial homogeneity). Developing an operational calibration solution raises indeed practical issues. Notably, addressing the problem of measurement campaign duration as well as selecting the best suited satellite database is mandatory prior to any site-adaptation procedure. The variability in performance caused by these choices is addressed in this paper.

To identify the performance indicators on the long-term calibration and draw reliable conclusions, experiments were conducted on sites where historical irradiance measurement data are available and with good quality. Their goal is to evidence convergence visually and numerically through a statistical approach for a number of test conditions.

2 SITE-ADAPTATION USING KDM METHODOLOGY

Providing an unbiased estimation of the GHI is the major challenge for the "solar resource" community as the size of the PV projects drastically increases (100+, even 1000 MWp) while the corresponding business plans are getting tighter. Addressing the problematic of site-adaptation is therefore mandatory to reduce uncertainty of the planned energy yield and therefore its financial risk.

The depicted calibration method consists in correcting long-term satellite-database GHI time-series using short-term horizontal pyranometer measurements. For each of the reference sites, a large number of "virtual" measurement campaigns of variable duration (ranging from one to 24 months) are simulated. Each measurement campaign leads to a specific performance (*i.e.*, long-term GHI measurement vs. calibrated GHI time-series) which is then used to study the sensitivity of the calibration process.

A calibration method highly adapted to the elimination of the MBE for a given site is KDM. Quantile Mapping removes systematic errors between two datasets by matching the CDF of the modeled time-series on its corresponding truth data counterpart. For every estimated point of data $\tilde{x}$ (in this case the GHI estimation from the satellite model to be calibrated), the associated calibrated value x is the one that should have the same image when applying the measured dataset CDF function:

$$\forall \tilde{x} \in I_{ST_{meas}} \cap I_{ST_{sat}}, x = {CDF_{ST_{meas}}}^{-1} \circ CDF_{ST_{sat}}(\tilde{x})$$

Where $I_{ST_{meas}}$ and $I_{ST_{sat}}$ are the respective intervals modeling the observed ranges of $CDF_{ST_{meas}}$ and $CDF_{ST_{sat}}$. Index ST refers to the short term, i.e., the measurement period.

The construction of the CDF is proving to be a significant issue. As the support interval of possible GHI values is indeed continuous, that the associated CDF also is. Time-series being discrete, the easiest way to interpolate a CDF using a finite ensemble of draws is to empirically build a step function. The dataset is sorted, and the function is incremented with a constant step at each new value of GHI. Such piecewise constant function can raise inversion issues when applying ${CDF_{ST_{meas}}}^{-1}$ to subintervals where GHI points of data are scarce. In particular, the extreme high values of irradiance may thus be subject to "vertical bar" phenomena due to the non-strict monotonicity of ${CDF_{ST_{meas}}}^{-1}$. KDM methodology aims at countering these purely mathematical issues. It uses a Kernel Density Estimator (KDE) to propagate the distribution of the finite dataset of discrete GHI values in a continuous space (over the ensemble of real numbers). Instead of being considered as a probabilistic Dirac, the presence density, i.e., the Probability Density Function (PDF) of each new irradiance occurrence is expanded over $\mathbb{R}$. In the case under study, the shape of the selected kernel is a centered reduced Gaussian normalized by the number of samples N.

$$\forall s \in \mathbb{R}, \quad \begin{cases} PDF(s) = \dfrac{1}{Nh}\sum_{k=1}^{N} \dfrac{1}{\sqrt{2\pi}} e^{-\frac{(s-x_k)^2}{2h^2}} \\ CDF(s) = \displaystyle\int_0^s PDF(t)dt \end{cases}$$

Where h is the bandwidth of the gaussian kernel. The higher h, the smoother the PDF. Yet the bandwidth needs to be low enough to maintain sufficient curve details. In these works, $h = 3$ proved to be the best balance between smoothness and a good level of details.

It is then possible to fall back into a finite interval by truncating the CDF while removing its intercept and dividing it by the difference between the maximum and minimum value. The most relevant interval can be $I' = [0, \max(GHI_{LT_{sat}})]$, where LT refers to the long term satellite period, but it still leads to vertical bar phenomena if the maxima of the long-term satellite time-series and the measured maxima are too far apart ($CDF_{LT_{sat}}$ or its real counterpart that would have originated from the long-term pyranometer measurement campaign will reach the horizontal asymptote 1 too quickly before the other). $CDF_{ST_{meas}}$ and $CDF_{ST_{sat}}$ are therefore interpolated on interval I:

$$I = [0, \min\{argmin(|CDF_{ST_{meas}} - 0.99|), argmin(|CDF_{ST_{sat}} - 0.99|)\}].$$

It must be noted that the value 0.99 has been proposed as it generates the expected effect.

The KDM methodology can be visualized as the function modifying every date point of a time-series so that the curve of its KDE estimated CDF matches the wanted one:

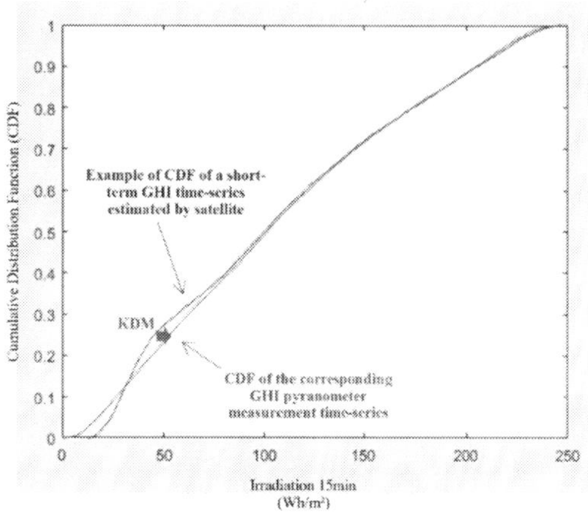

Figure 1: Visualization of the Kernel Density Mapping of a GHI time-series

It should be noted that KDM can be applied whatever the duration of the short-term or long-term time-series and whatever the database used.

3 EXPERIMENTAL PROCESS

3.1 Ground data

These works make use of seven ground-based stations located in Western Europe and belonging to the BSRN network (Baseline Surface Radiation Network) which added value mainly consists in the high-quality of the available data. [4] The one-minute GHI measurements from the ground stations over a ~12-year period (2007+) are considered as the long-term reference to be reached.

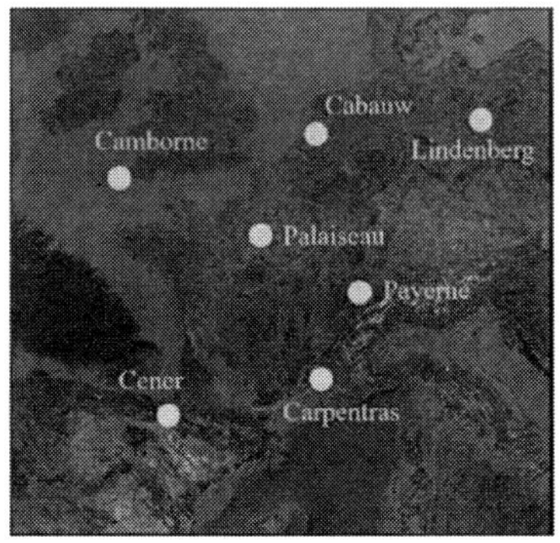

Figure 2: Location & names of the seven BSRN ground-based stations in Western Europe

Table II: Useful details for every station used. $GHI_{LT\ meas}$ is the average perceived 15-min GHI observed over the entire time series during periods when the sun is above the horizon

Name	Country	Lat(°)	Lon(°)	Elev(m)	Long-term period used	Climate	$GHI_{LT\ meas}$ (Wh/m²)
Cabauw	Netherlands	51.9711	4.9267	0.0	2007-2018	Oceanic	72.1
Camborne	UK	50.2167	-5.3167	88.0	2007-2017	Oceanic	77.3
Carpentras	France	44.0830	5.0590	100.0	2007-2018	Mediterranean	109.3
Cener	Spain	42.8160	-1.6010	471.0	2009-2019	Oceanic	97.7
Lindenberg	Germany	52.2100	14.1220	125.0	2007-2017	Continental	76.1
Palaiseau	France	48.7130	2.2080	156.0	2007-2018	Continental	85.5
Payerne	Switzerland	46.8150	6.9440	491.0	2007-2013	Continental	84.7

The heterogeneity in the seven ground-based stations, in term of climate and latitude, allows gaining confidence in the results and conclusions when addressing projects in Western Europe. Though, pyranometers are operated independently from one site to another. To equalize the data and remove any inconsistent measurement point, a quality check was applied [5]. This filtering has the effect of discarding corrupted, unmeasured, or unrepresentative pieces of data.

3.2 Satellite databases

Three satellite-based databases were used as the long-term GHI estimation to be calibrated: Solargis [6], HelioClim-3 (version 5 – HC3v5) [7] and CAMS Radiation (version 3.2 – Copernicus Atmosphere Monitoring Service) [8]. The two first ones are commercial products commonly used by the PV actors while the third one is freely provided by the EU-funded Copernicus Service.

As mentioned later, the errors made post-calibration for a given site seem to be relatively variable depending on the year of measurement, even for an identical database. Thus, it appeared to be coherent to cross-calibrate time series from the same campaign but from different databases to compensate for some of the respective biases. For that purpose, an experimental Hybrid database was also built using a regression of Solargis and HC3v5. It aims to find the best combination of several databases to enhance correlation with the short-term measurement campaign when the calibration function is mapped.

3.3 Monte-Carlo based assessment

The BSRN stations provide enough historical data to allow for experimentation based on statistical draws. To characterize the robustness of the described methodology on plausible data, it has been proposed a Monte-Carlo based variability assessment of the short-term measurement campaigns. For each of the seven sites, one iteration of the process can be summed up as follows:

- A starting date is randomly chosen.
- The corresponding short-term campaign with a specified duration in months is extracted from the

measured dataset; quality check and shadow removal are applied.
- All CDF needed are computed using the described KDM methodology.
- The Quantile Mapping function is interpolated on *I* and applied to the satellite long-term.
- Statistical errors are retrieved by comparing with the measurement long-term. MBE, Root Mean Square Error (RMSE) and Kolmogorov-Smirnov Integral (KSI) were chosen to assess the performance of the calibration process with respect to the target which is the actual long-term BSRN data.

This algorithm was run for *n* starting dates. For each of them, campaign durations were selected between one and 24 months. The statistical analysis of errors provides information on the random distribution as well as the observed median value of indicators with respect to the number of months used.
$n = 100$ was found to be enough iterations. Going over this number was not considered relevant as additional data would have great chance to overlap already existing results, bringing no further details. The experiment was conducted on the three chosen satellite databases as well as the Hybrid and respective performances were compared. Visual convergence was demonstrated using boxplot (i.e., statistical box) representation. This representation allows to easily see the distribution of the results of the n experiments for each campaign duration in months.

4 RESULTS

When experimenting the Monte-Carlo process, boxplots perfectly illustrate the expected behavior of increased confidence when working with longer measurement campaigns. Several statistical errors were monitored, the most essential being MBE followed by KSI. Since the aim of the calibration is to estimate accurately the yearly sum of irradiation, it is essential than the calibrated data is unbiased. Measures of the accuracy of the calibrated time series compared to the raw measurements (such as RMSE or MAE) are less important. These indicators provide an exhaustive view of the representativeness of the calibrated time-series obtained by demonstrating the systematic deviation and the discrepancy in terms of CDF post-calibration. RMSE is also shown in order to control that the dispersion of random errors does not increase substantially while calibrating.

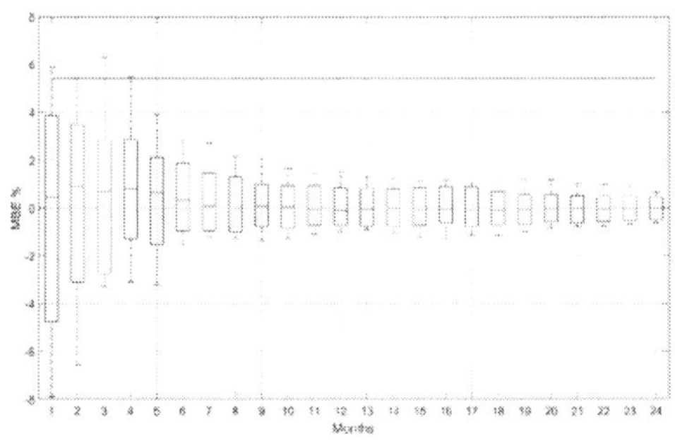

Figure 3: MBE convergence (in percent with respect to the long-term reference, Y-axis) after calibration of HC3v5 at Cener, Spain. Duration of the measurement campaign is expressed in months (X-axis). Horizontal blue line is the raw error of the satellite-based database (i.e. without calibration) while the boxplots depict the MBE spread after the calibration using campaigns with same duration (median in red, percentiles P10 and P90 – thus 80 % of the samples – as blue rectangle).

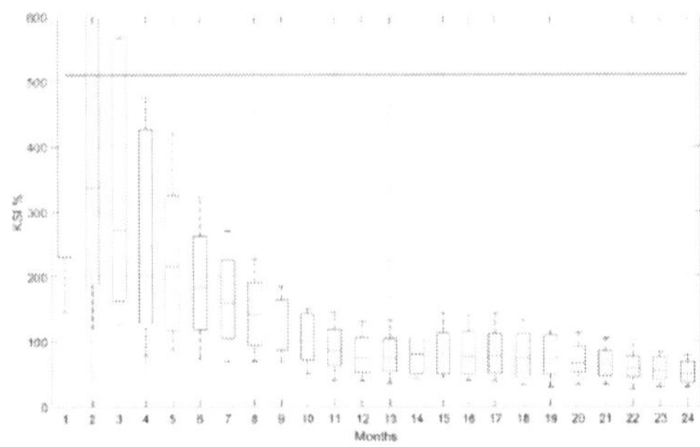

Figure 4: KSI convergence after calibration of HC3v5 at Cener, Spain. This KSI test is used to compare two statistical samples, i.e., to measure the distance between two CDF. It is possible to normalize this parameter to obtain a relative value of KSI; a value greater than 100 % indicates a significant difference between the two CDFs with a confidence level of 95 %.

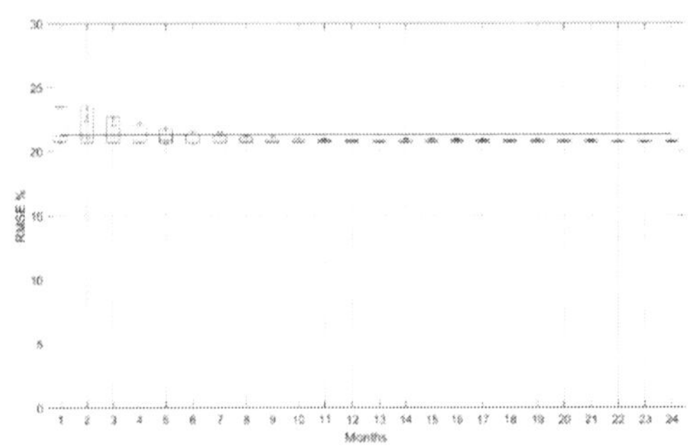

Figure 5: RMSE convergence (for 15-min time-step) after calibration of HC3v5 at Cener, Spain.

Results show that running a 12-month measurement campaign allows capturing the whole seasonal information for an efficient long-term calibration. Using more than twelve months does not bring much added value as depicted in Figures 2 and 3 (one specific site, one specific database, 100 campaigns) where MBE and KSI drastically decrease along with the campaign duration up to 12 months, then reach a plateau. A bounce-back effect can even be observed for campaigns lasting between one and two years, reflecting the fact that some seasonal effects are doubled while others are not. To reach a performance at least better than for 12 months, measuring for twice as long is a minimum, which is incompatible with the time constraints of most PV project developments.

Table 2 depicts for each of the seven locations the raw performances of the satellite-based databases along with the median of the MBE, KSI and RMSE (bold) resulting from 12-month measurement campaigns with random starting dates (100 draws). P25 and P75 are also shown and were preferred to P10 and P90 because they are less impacted by exceptional results. Such a convergence can be observed whatever the site and the database, with different rate and convergence point. However, when focusing on the MBE, running the adapted KDM with a Hybrid database systematically improves the performances as depicted in Figure 6 where the performances for 12-month measurement campaigns are compared depending on the database.

Table III: Raw MBE, KSI and RMSE (no calibration, in bold) and median (P50) MBE, KSI and RMSE (one hundred measurement campaigns of twelve months) for the seven BSRN stations and for the three satellite-based databases (plus Hybrid). P25 and P75 are also shown to illustrate confidence.

BSRN	Database	MBE [%]				KSI [%]				RMSE [%]			
		Raw	Calib.			**Raw**	Calib.			**Raw**	Calib.		
			P25	P50	*P75*		*P25*	P50	*P75*		*P25*	P50	*P75*
CAB	Sol.	**-2.5**	*-0.1*	0.4	*0.9*	**355**	*96*	125	*160*	**25.1**	*27.9*	28.0	*28.1*
	HC3v5	**-0.9**	*-1.3*	0.5	*2.2*	**316**	*122*	160	*201*	**27.8**	*27.9*	28.1	*28.3*
	CAMS	**6.1**	*-1.4*	0.2	*1.4*	**665**	*112*	143	*189*	**31.1**	*31.2*	31.3	*31.4*
	Hybr.		*-0.8*	-0.3	*0.2*		*241*	255	*270*		*24.2*	24.3	*24.3*
CAM	Sol.	**-2.6**	*0.2*	0.4	*1.0*	**244**	*73*	80	*99*	**26.6**	*26.9*	27.0	*27.0*
	HC3v5	**3.9**	*-0.8*	-0.6	*0.3*	**291**	*56*	63	*69*	**26.8**	*28.2*	28.3	*28.5*
	CAMS	**4.2**	*-1.1*	-0.1	*1.5*	**365**	*56*	100	*131*	**37.6**	*35.1*	35.2	*35.3*
	Hybr.		*-0.8*	-0.4	*0.0*		*169*	191	*212*		*25.6*	25.7	*25.7*
CAR	Sol.	**0.4**	*-0.2*	0.3	*0.9*	**89**	*71*	88	*123*	**15.0**	*15.0*	15.0	*15.1*
	HC3v5	**3.4**	*0.2*	0.7	*1.2*	**392**	*84*	121	*153*	**15.5**	*15.2*	15.3	*15.3*
	CAMS	**1.5**	*-0.4*	0.9	*1.5*	**324**	*111*	132	*199*	**19.8**	*17.9*	18.0	*18.1*
	Hybr.		*-0.2*	0.0	*0.4*		*122*	140	*181*		*14.6*	14.6	*14.6*
CEN	Sol.	**0.8**	*-0.3*	0.1	*0.4*	**275**	*75*	94	*118*	**20.7**	*20.8*	20.9	*21*
	HC3v5	**5.4**	*-0.4*	-0.1	*0.4*	**506**	*62*	75	*94*	**21.5**	*20.7*	20.8	*20.8*
	CAMS	**2.3**	*-1.5*	0.0	*1.3*	**395**	*78*	143	*189*	**26.1**	*26.1*	26.2	*26.2*
	Hybr.		*-0.5*	-0.3	*0.0*		*220*	242	*275*		*19.9*	19.9	*20.0*
LIN	Sol.	**-2.8**	*-0.5*	0.0	*0.5*	302	*79*	92	*110*	**25.3**	*25.3*	25.4	*25.5*
	HC3v5	**-0.6**	*-0.8*	0.0	*0.9*	**259**	*88*	97	*124*	**26.3**	*26.5*	26.5	*26.7*
	CAMS	**2.1**	*-0.4*	0.9	*1.9*	**359**	*96*	136	*179*	**34.7**	*32.4*	32.5	*32.6*
	Hybr.		*-0.8*	-0.4	*-0.1*		*214*	235	*282*		*24.2*	24.3	*24.3*
PAL	Sol.	**-2.9**	*-0.3*	0.3	*0.8*	**324**	*80*	115	*131*	**23.5**	*23.5*	23.6	*23.6*
	HC3v5	**3.6**	*-0.4*	0.5	*0.9*	**449**	*76*	87	*103*	**23.3**	*23.3*	23.3	*23.3*
	CAMS	**4.3**	*-0.9*	0.6	*1.6*	**547**	*113*	146	*185*	**29.2**	*29.5*	29.6	*29.6*
	Hybr.		*-0.3*	0.0	*0.3*		*223*	250	*282*		*22.1*	22.1	*22.2*
PAY	Sol.	**1.1**	*-0.7*	-0.4	*0.3*	**139**	*65*	82	*96*	**22.2**	*23.3*	23.4	*23.5*
	HC3v5	**-0.4**	*-0.8*	-0.3	*0.2*	**201**	*64*	76	*94*	**25.4**	*26.4*	26.5	*26.6*
	CAMS	**5.5**	*-0.6*	-0.1	*0.3*	**384**	*54*	73	*104*	**29.8**	*28.1*	28.3	*28.4*
	Hybr.		*-0.6*	-0.4	*-0.0*		*163*	211	*249*		*22.5*	22.5	*22.6*

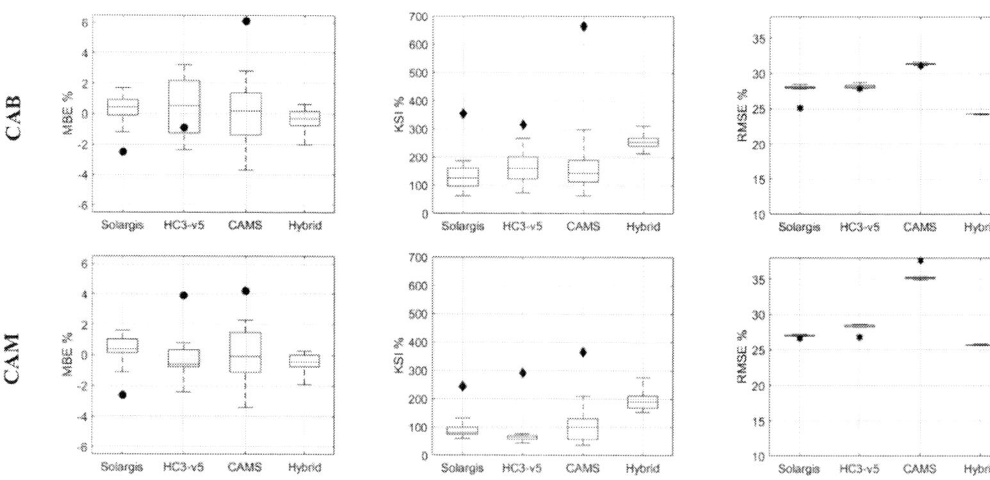

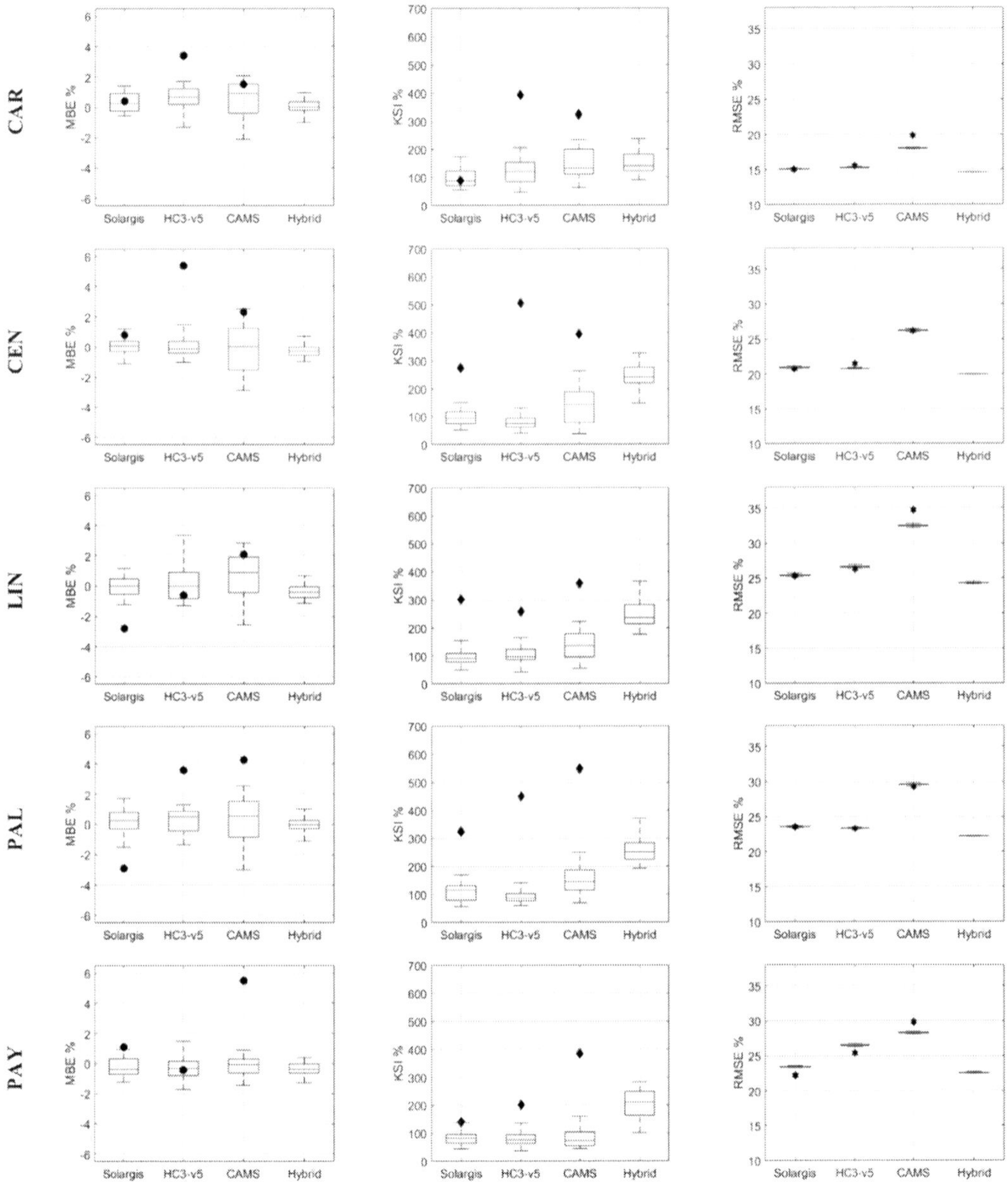

Figure 6: Comparative performances for 12-month measurement campaigns at seven BSRN locations. The black point is raw error of the satellite-based database (i.e., without calibration) while the boxplots depict indicator spread after calibration (P25 & P75).

It is then possible to see the overall behavior of each database by aggregating the data from the seven sites.

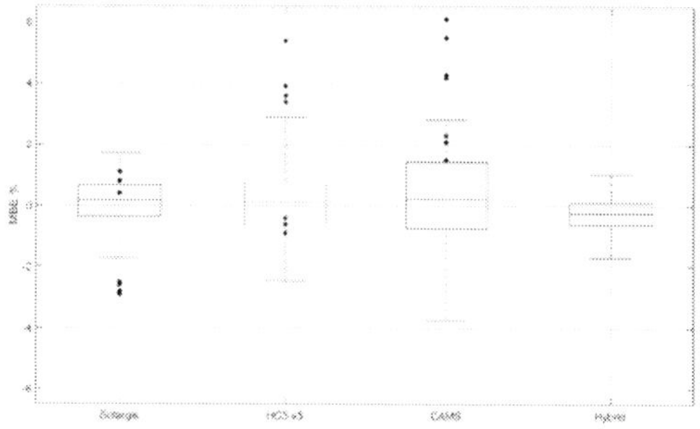

Figure 7: Overall performances for 12-month measurement campaigns at seven BSRN locations. The black points are the seven raw error of the satellite-based database (i.e., without calibration) while the boxplots depict MBE spread after calibration for the 7*100 compiled calibration processes.

The median results for the seven locations show that a 12-month measurement campaign systematically reduces the MBE when compared to the initial performance (*i.e.,* without calibration) when the initial bias is not already close to zero. Moreover, the median of the MBE systematically remains between -0.6 % and 0.9 % for all databases. Finally, the Hybrid solution systematically shows better performances with 50 % of the MBE (P25-P75 interval) remaining between -0.6 % and 0.1 % and 80% (P10-P90 interval) between -1% and 0.4%. It is however slightly negatively biased by -0.2% but this behavior can either be considered negligible or be corrected as almost constant on every given site.

When looking at Solargis results, the satellite time-series already has extremely low bias before calibration for two instances (Carpentras and Cener). Quantile Mapping displays a bias distribution which might look then worst. This cannot be considered as an issue as Solargis or Hybrid time-series systematically output very low residual post-calibration bias. In a real situation where nor raw nor calibrated bias can be computed, the use of site-adaptation guarantees to obtain reasonable remanent bias value while the uncorrected can typically range anywhere between -3.5% and 3.5% in Europe (as mentioned in both Solargis and HC3 validation reports). It is fair to state that most of calibration processes conducted in Western Europe using Solargis on any measurement campaign will lead to a global systematic error inferior to 1%. This is less true when making use of HC3v5 which can be calibrated extremely well in some cases (Cener) but poorly in other instances (Cabauw). CAMS is for its part not suitable for KDM calibration, showing too much fluctuation. No investigation has been conducted here to understand these differences between the databases. The respective models used are not visible in the resulting time-series which thus take the form of black box dataframes.

Other indicators demonstrate notable behaviors post-calibration. KSI is systematically reduced to reach values around 100% when using Solargis, HC3v5 and CAMS. This level of performance indicates that both satellite and measured CDFs are close together throughout all the interval: the two time-series are drawn from comparable random variables. Hybrid displays higher KSI on every BSRN station and is the least performing database for this one statistical error. Regarding the RMSE, it is however the one that shows the best results with levels consistently below the other three.

It is worth noticing that KDM does not have drastic impact on RMSE. This indicator indeed only highlights random errors committed by the satellite database, totally undetectable for Quantile mapping. Its monitoring still gives evidence that the method does not disperse the data, but only distorts its overall distribution keeping its original precision. Mathematically, Quantile Mapping only distorts the point cloud in the satellite vs true data scatterplot so that there is a symmetrical distribution of points around the bisector line $y = x$.

5 CONCLUSIONS

Raw bias analysis on high quality historical BSRN data proves that top-tier satellite-based databases may have systematic errors which significantly jeopardize the reliability of a yield assessment. Assuming that the systematic bias is temporally stationary, long-term satellite time-series can theoretically be perfectly compensated using calibration on short-term in-situ measurements. Yet experiments show that the performance fluctuates depending on several input features.

An effective benchmark of possible configurations allows the characterization of the calibration parameters to be favored by the surveyor in order to maximize the representativeness of the corrected irradiance data. Once the method has been set, the three main parameters that can affect site adaptation are the start date of the campaign, its duration, and the choice of the satellite database. While the first is rarely adjustable in a professional context, the other two must be considered to effectively lower the uncertainty compared to raw data. The study carried out makes it possible to rule statistically on the latter. The three major conclusions are:

- 12 months of measurement is the optimal duration. This is the minimum period to capture all seasonal effects. It is also found that an extension between 12 and 24 months does not bring any improvement and may even degrade the level of performance.
- Satellite databases are not equally suitable for calibration. Among the three tested, Solargis seems to be the most consistent due to systematically less distributed post-calibration biases on various locations. Using HelioClim-3v5 or CAMS seem to lead to higher uncertainty.
- Satellite databases can be combined after being independently calibrated in order to benefit from fluctuating performances, thus leading to better performances.

As previously pictured, a good choice of parameters can drastically decrease uncertainty on remanent bias and lead to a much higher level of confidence: calibrated time-series biases typically range between -0.5% to 0.5% while raw values range between -4% and 4%. Consequences are double: P50 yields are more representative and prone to better bankability; gap between P90 and P50 can be drastically reduced since uncertainty on the satellite database is dropped from 3% (as suggested by validation reports) to somewhere around 0.5%. PV developers could then benefit from a slightly higher debt leverage due to a P90 being closer to P50.

As a perspective, it would be interesting to extend this analysis to other calibration methods and to have a better understanding of why some campaigns work better than others.

6 REFERENCES

[1] Suri, M., Huld, T., Dunlop, E. D., Albuisson, M., Lefèvre, M., & Wald, L. (2007). Uncertainties in photovoltaic electricity yield prediction from fluctuation of solar radiation. In *22nd European Photovoltaic Solar Energy Conference*.

[2] Polo, J., Fernández-Peruchena, C., Salamalikis, V., Mazorra-Aguiar, L., Turpin, M., Martín-Pomares, L., et al. (2020). Benchmarking on improvement and site-adaptation techniques for modeled solar radiation datasets. *Solar Energy*, *201*, 469-479. https://doi.org/10.1016/j.solener.2020.03.040

[3] Polo, J., Wilbert, S., Ruiz-Arias, J. A., Meyer, R., Gueymard, C., Súri, M., Cebecauer, T. (2016). Preliminary survey on site-adaptation techniques for satellite-derived and reanalysis solar radiation datasets. *Solar Energy*, *132*, 25–37. https://doi.org/10.1016/j.solener.2016.03.001

[4] Ohmura, A., Dutton, E. G., Forgan, B., Fröhlich, C., Gilgen, H., Hegner, H., ... & Wild, M. (1998). Baseline Surface Radiation Network (BSRN/WCRP): New precision radiometry for climate research. *Bulletin of the American Meteorological Society*, *79*(10), 2115-2136.

[5] Espinar, B., Blanc, P., Wald, L., Hoyer-Klick, C., Homscheidt, M. S., & Wanderer, T. (2012). On quality control procedures for solar radiation and meteorological measures, from subhourly to monthly average time periods. *EGU General Assembly, Vienna, Austria*, 22-27.

[6] Šúri, M., Cebecauer, T., & Skoczek, A. (2011, September). SolarGIS: Solar data and online applications for PV planning and performance assessment. In *26th European photovoltaics solar energy conference*.

[7] Blanc, P., Gschwind, B., Lefèvre, M., & Wald, L. (2011). The HelioClim project: Surface solar irradiance data for climate applications. *Remote Sensing*, *3*(2), 343-361.

[8] Schroedter-Homscheidt, M., Arola, A., Killius, N., Lefèvre, M., Saboret, L., Wandji, W., et al. (2016). The Copernicus atmosphere monitoring service (CAMS) radiation service in a nutshell. *Proc. SolarPACES16*, 11-14.

OPTIMIZING METHODOLOGY FOR ESTIMATING GLOBAL HORIZONTAL IRRADIANCE (GHI) USING SOLAR PHOTOVOLTAICS' OUTPUT AC POWER MEASUREMENTS

Muhammad Ammar Khan[1], Dan-Eric Archer[2], Nelson Sommerfeldt[1]
[1] KTH Royal Institute of Technology, Brinellvägen 68, SE-10044 Stockholm
[2] CheckWatt AB, Svärdvägen 3A, 182 33 Danderyd, Sweden
Corresponding author: Muhammad Ammar Khan, E-mail: makhan7@kth.se

ABSTRACT: In this paper, optimization of a model for estimation of global horizontal irradiance (GHI) using PV's AC power output is proposed such that errors in estimated GHI are minimized. Modification of the developed model is carried out to adopt the methodology into different programming environment and to make it consistent with the available datasets. Results show that the developed model on Python language has Mean Absolute Percentage Error (MAPE) of 5.77% compared to GHI measurements using pyranometers for estimation of one month's data. Modifications are done in detection of clear sky periods and minimization of loss functions for estimation of orientation in contrast with other methods. During model optimization it was also found that Ineichen/Perez's Clear Sky GHI model, Direct Insolation Simulation Code (DISC) Model for Direct Normal Irradiance (DNI) estimation and Perez Model for sky diffuse irradiance calculations had minimum MAPE. Moreover, modelling parameters in the model were also adjusted to find best estimates of GHI. It was observed that timestep of 10 mins and past data of 120 days for estimation of orientation weights and error function were suitable. Number of iterations were also reduced to 10 from a starting number of 30 for estimation of GHI considering MAPE and computation times. Therefore, the main purpose of this research is to develop an optimized model that could be commercialized aiming to convert a PV system into virtual GHI sensor providing; information of solar potential and serve as an input data for solar forecasting or performance monitoring.
Keywords: Global Horizontal Irradiance (GHI), Photovoltaic, Optimization, GHI Estimation, Orientation Assessment, Mean Absolute Percentage Error (MAPE)

1 INTRODUCTION

During the past few years, advancement in PV technologies, has substantially increased utilization of solar PV for electricity generation on all scales. As per the IEA's Renewables 2019 report, solar PV is the most adopted form of renewable energy in the market and is also the leading contributor in renewable energy transitions. Moreover, it also predicted that by 2024 renewables will increase by 50% as compared to 2019, with solar PV being 60% of the total capacity enhancement [1].

Market acceptance and growth of such magnitudes has encouraged solar community to develop ingenious techniques to further improve efficiency of solar PV cells and many efforts are being made across the globe to formulate new methods for enhancing performance monitoring of PV systems. One such part of PV system monitoring is utilization of solar irradiance data to carry out performance calculations or to forecast system outputs. However, to do so, accurate solar data is required which currently usually involves expensive instrumentation and control systems. These systems when combined with large scale installations are usually cost effective but small-scale or residential installations usually lack these sensory equipment as they do not have such economic leverage [2].

Second option, alongside instrumentation is to use satellite based solar resource data for carrying out performance monitoring or solar forecasting. However, the data provided by the satellites does not have good spatial-temporal accuracies. The data provided by the satellites is based on meteorological images taken by satellites having limited spatial resolution of about 3 km and temporal resolution of 15 minutes. Moreover, these satellites mostly orbit in equatorial region and results provided for Northern and Southern Hemisphere contain considerable variations compared to on-ground measurements [3]. On the contrary, a lot of work is also being done on satellite data and results obtained from these efforts are very encouraging [4].

Importance of accurate solar resource data serves as a key factor for business development. Therefore, in collaboration with CheckWatt AB, a solar company based in Sweden, research work was carried out to develop a model that estimates Global Horizontal Irradiance (GHI) using solar PV's AC power output, location coordinates and ambient temperature conditions [5]. The model developed is inspired by the work carried out by Nespoli et al. [6]. The developed model aims to become a commercial product that can be integrated with CheckWatt AB's performance monitoring system and to serve as a source of solar resource data especially in areas of Northern Hemisphere. GHI is referred as the component of extraterrestrial solar irradiance observed on a surface, placed horizontal on the earth [7].

This paper hence contributes to studies carried out in [5] and [6] by highlighting the modifications adopted to develop a model on different programming environment, and data that was in line with organizational utilization, having better computational performance. Moreover, this paper discusses a study conducted on the optimization of developed model over a range of parameters such that MAPE compared with on-ground sensors are minimal and results obtained are better than the satellite data. Lastly, with the aid of this study runtimes of the developed model can be improved playing an integral role in commercial applicability of this model.

2 GHI ESTIMATION MODEL

Utilization of solar PV power output measurements for estimation of GHI is not a new concept and different techniques have been devised based on the available datasets with PV power output being of the dependent variables. However, in most of the literature details pertaining to modules are required. Especially information regarding modules' orientation and type. Since the aim was to develop a model that could serve for both small-scale/residential installations and large-scale plants, it was very difficult to get estimates of the orientations of smaller systems. Hence, the model proposed in [6] was taken as a reference to develop a model in Python programming with modifications in accordance with available programming language packages and data requirements. The model detailed in [5] contains two parts, same as [6], with the first part dealing with module orientation estimation based on clear sky time periods and second part dealing with GHI estimation using gradient descent technique with reference GHI coming from Clear Sky GHI Estimation Models. Overall, the main equation followed in GHI estimation method can be summarized using (1)

$$GHI = min||P_{mea} - P_{est}|| \quad (1)$$

2.1 Orientation Assessment

Without orientation, it is difficult to compute plane of array (POA) irradiance components for solar PV systems. PV power outputs are specific to one orientation and with any change in either the tilt or azimuth of the PV system, the amount of solar insolation received by the modules changes, thus causing a change in the output. Therefore, information regarding orientation is of prime importance and a methodology is adopted where clear sky periods are identified first. Several different techniques have been adopted for determining clear sky periods, summary of which is provided in [5]. These clear sky periods are then combined with GHI values proposed by clear sky models in [8-11] and a range of proxy orientations to compute estimated PV power outputs. Errors between estimated PV power outputs and measured PV power outputs for clear sky periods only are computed and a loss function is defined against the computed error. Error minimization techniques are used to obtain weights associated with each proxy orientation and weights with non-zero values specify probable orientation of the system. With weights determined, errors for clear sky periods are plotted against sun positions to identify any probable shading patterns to ensure that such instances do not interfere with GHI Estimation in the next step. Details of step-by-step methodology adopted for orientation assessment could be found in [5]. Figure 1 provides an illustration of orientation assessment.

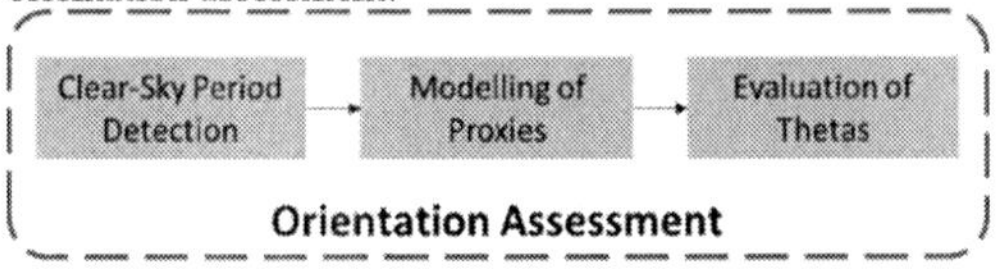

Figure 1: Orientation Assessment sub-parts [5]

2.2 GHI Estimator

After obtaining estimates of orientation and error function against sun positions, model moves onto the second part. In this part, error function is firstly utilized to compute trust function having values in the range of 0 to 1 with values close to 0 highlighting time of the day where shading is observed. Second step within this part is to detect outlier values using Tukey's outlier detection method and interquartile range of errors between measured and estimated PV power values. Default values, as proposed in [6], are used in [5] also for both these steps. Third and the last step in GHI estimation methodology is application of GHI Estimator sub-part whereby gradient-descent technique is used to estimate GHI. Clear sky GHI values are utilized to develop initial guess GHI values for each timestep. This initial guess GHI is used to compute new estimated PV power outputs. Then based on the error between estimated and measured power values, GHI is adjusted in an iterative loop until loop termination conditions are met. Technical details associated with iterative loops are present in [5]. At the end of this sub-part final values of GHI estimates are available. Figure 2 provides an outline of GHI Estimator sub-parts.

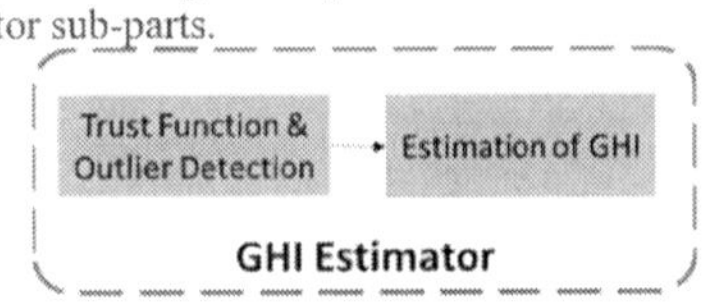

Figure 2: GHI Estimator sub-parts [5]

To further illustrate, the inputs and outputs of the GHI Estimation Model, Figure 3 depicts a flow diagram incorporating above mentioned Orientation Assessment and GHI Estimator parts.

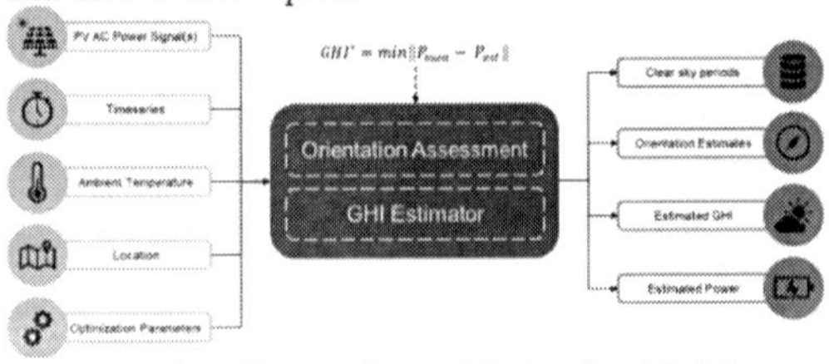

Figure 3: Flow diagram for GHI Estimation Model

3 ADOPTION OF GHI ESTIMATION MODEL

Implementation of GHI Estimation model into a new programming environment and on different datasets required modifications into the model proposed in [6]. Moreover, the aim was to develop a commercial product that could be integrated into the performance monitoring system of CheckWatt AB easily. Python programming was used for development of the model in [5] and open-source python packages like NumPy [12], Pandas [13], SciPy [14], scikit-learn [15] and pvlib [16] were used. Major modifications during the implementation of GHI Estimation model includes different methodology for determination of clear sky periods and use of Python's built-in functions for minimization of loss functions for estimation of proxy orientations.

3.1 Clear Sky Period Detection

In [5], different clear sky detection techniques were discussed that have been used to identify clear sky periods using PV power measurements and several different variables. Hence, a methodology is proposed in [5] where bimodality test is removed, and a gaussian probability distribution curve is fitted on the power distributions. Gaussian distribution curves are fitted such that for every power distribution there are two curves only. Curve with highest mean is selected and using its standard deviation upper and lower bounds for clear sky periods for those sun positions are set. All the values lying within these ranges for defined sun positions are selected as clear sky periods, whereas other values are disregarded. This modification was introduced to decrease computation time with a slight reduction of quality of estimation. It was observed that computation times were reduced by $1/10^{th}$ of the time taken with bimodality test and MAPE only increased by 1 percentage point for one-month GHI Estimations. Considering commerciality of the product this change was accepted for CheckWatt AB's system.

3.2 Minimization of loss function

As proposed both in [5] and [6], errors between measured and estimated power values are computed and then Huber's and Tukey's loss functions are defined for those errors. Using minimization techniques, values of these loss functions are reduced to obtain coefficients referred as thetas in [5] defining probable orientation of the system. In [6], MATLAB's built-in function of fmincon () was used for minimization whereas no function with same performance is available in Python. Closet option to fmincon () is scipy.optimize.minimize () function that was used in [5] along with least squares methodology. For same datasets, the results obtained from [5] were close to [6] but not similar. This resulted in some difference in estimated GHI values which were adjusted using optimization techniques.

3.3 Other Modifications

Comparing methodology in [5] with [6], some other modifications were also observed. Computation of proxy orientations is carried out using pre-defined Python package to speed up the process whereas authors of [6] developed their own code for determining proxy orientations. Both the methodologies offer flexibility in entering a range of proxy orientations dependent on the user requirement. Moreover, computation of trust function required mapping of errors over sun positions and aggregation of these errors for 2° discretized sun zenith and azimuth angles. This aggregation is done differently in [5] using Python specific packages but the results obtained are quite close to the results in [6].

4 OPTIMIZATIONS OF GHI ESTIMATION MODEL

Models developed in [5] and [6] did differ in the way they were adopted to meet the development constraints, but both [5] and [6] lacked proper optimization of modelling parameters and different irradiance models used in the algorithm development. Hence, this study is conducted on top of work carried out in [5] to optimize inputs and modelling parameters such that better estimation of GHI could be carried out. These optimizations were carried considering the available functions present in pvlib [16] python package.

4.1 Irradiance Models

Incident solar irradiance on a PV system is defined by plane of array components which are computed using modules' orientation, GHI, Direct Normal Irradiance (DNI) and Diffused Irradiance. GHI serves as the major dependent variable for computation of all the components and hence through this research work, target was to obtain good estimates of it [5].

However, to start the estimation process reference values were required that could be adjusted to fit real values. To do so, GHI values defined by clear sky models were selected to serve as a reference. Diving into the clear sky models; Haurwitz Model, Simplified_solis Model and Ineichen/Perez Model were selected as probable models to check effect of clear sky models on results [8]. GHI estimates using each model was carried out and results analyzed to depict which model is best suited.

DNI serves as a major contributing component of solar irradiance utilized for solar PV output. Since data from no other source was considered in the model, transposition models estimating DNI using GHI, and orientation details were utilized. DISC Model [17], DIRINT Model [18] and Erbs Model [19] were implemented in developed model and results observed to highlight model with minimal error.

With models for GHI and DNI discussed, third step was to study which methodology should be used for estimation of sky diffused irradiance. After going through literature and considering the functions provided in pvlib python package, seven sky diffused irradiance modelling techniques were studied. These models include Isotropic, Klucher, Haydavies, Reindl, King and Perez Models [20]. For Perez Model, several different selector options were also provided in pvlib [16] and hence, an optimization of 5 different selectors were also carried for this model only.

4.2 Modelling Parameters

Following irradiance models' optimization, study of modelling parameters was also carried. Three modelling parameters; time resolution of input data, number of days for input of past data and number of iterations for gradient descent based GHI estimation were analyzed.

Number of iterations required to tune initial guess GHI values to actual GHI values were varied with an interval of 5. Therefore, model performance was studied from minimum of 5 iterations to maximum 30 iterations. Iterations more than 30 were not selected considering increased processing times.

For accurate orientation assessment, large datasets were required to have enough clear sky periods forming the base on which whole model stands. However, considering processing times and probable constraints related to available datasets both variations in timestep and number of days for input of past data were studied. Data of time resolution 1 min, 5 mins, 10 mins, 15 mins, 30 mins and 60 mins was studied. Number of days for input data were varied with an interval of 15 previous days, starting from 15 days up to 180 days. Moreover, combined effect of variation in numbers of days and time resolution was also considered.

4.3 Methodology and Test Sites

With optimization strategies defined in the previous sub-sections; next step was to define how these optimizations were to be carried. Data used in [6] was obtained and the results were used a benchmark to compare results obtained using model proposed in [5] and after optimizations. Moreover, data from Öckerö Ishall, a site located in Sweden with CheckWatt AB's installations was also studied. Data from this site was studied because the data obtained from [6] had 10 mins time resolution and had data of past 5 months only. On contrary, data for Öckerö Ishall had a time resolution of 1 min with data for multiple years available. 1-min resolution input data was taken as the reference and was resampled to 5-mins, 10-mins, 15-mins, 30-mins, and 60-mins using NumPy functions in python. Average power and solar irradiance measurements were obtained when data was resampled. Moreover, since this site was already located in Sweden, open source meteorological and solar resource data from SMHI STRÅNG was also used [21].

To start off the optimization process, data for past 5 months from [6] at 10 mins resolution was used to compute orientation assessment. Once orientation assessment was conducted, GHI estimation for the month of August 2016 was carried out. Following was the order selected for study of irradiance models:

a) Clear Sky Models
b) DNI Estimation Model
c) Sky Diffuse Irradiance Estimation Models,

for estimation of GHI. When one irradiance model was optimized, it was incorporated into the model before optimizing the next model.

With irradiance models optimized, number of iterations were varied on the same dataset. For this study, processing times were also analysed in conjunction with MAPE, to properly weigh out the decision for optimal number of iterations considering commercial aspect of the model also.

Lastly, time resolution variations and number of days for input of past data were studied on Öckerö Ishall's datasets. Results of the previous optimizations carried out on data from [6] were incorporated when carrying out this optimization study.

4 RESULTS

Results of the model proposed in [5] were already benchmarked with results from [6]. Like the study in [5], MAPE was selected as the indicator to analyze optimizations.

4.1 Irradiance Models

In this sub-section results obtained from the optimization of irradiance models are discussed in detail.

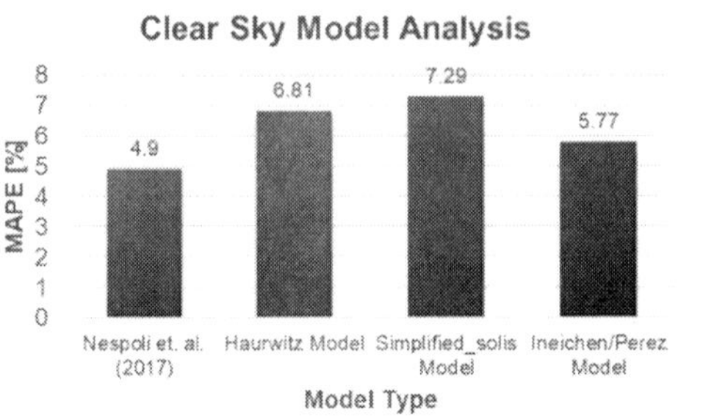

Figure 4: MAPE with different clear sky GHI Models for data from [6]

Figure 4 outlines the results obtained when different clear sky GHI models were used. From the results use of Ineichen/Perez Model, resulted in minimum MAPE, and should be considered in estimation GHI. Haurwitz Model was used in [6] and in contrast, for model in [5], use of this model doesn't give the best results.

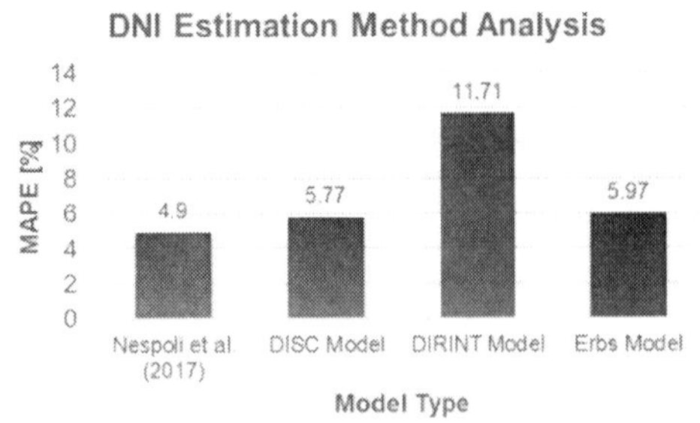

Figure 5: MAPE with different DNI Estimation Models for data from [6]

Looking into DNI estimation models' result from Figure 5, it is observed that DISC Model provides the best results. Campbell_norman model [22] is also present in pvlib and can be utilized for DNI estimation. It is observed that due to limited information regarding solar transmittance factors this model was not used but could be tested in future with the values of transmittance factors available. Erbs Model provided close estimates to DISC Model but can have limitations at the time of dusk and dawn [23].

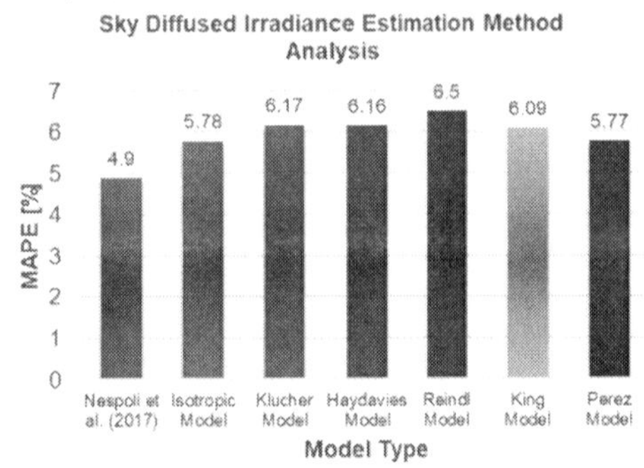

Figure 6: MAPE with different Sky Diffused Irradiance Estimation Models for data from [6]

Results obtained for estimation of GHI using different sky diffused irradiance models showed much less variation in MAPE. From Figure 6, it is observed that Perez model and Isotropic model are the best options when used with Ineichen/Perez Clear Sky GHI model and DISC model for DNI Estimation.

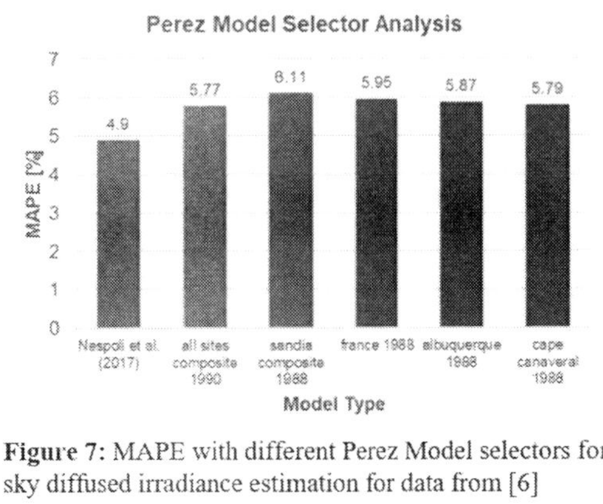

Figure 7: MAPE with different Perez Model selectors for sky diffused irradiance estimation for data from [6]

Similar to the results of different sky diffused irradiance models, Figure 7 shows that different Perez

Model selectors have very small variations in MAPE. All sites composite 1990 selector provided the best results and was also the one used in [6] also.

4.2 Modelling Parameters

In this sub-section, effect of varying different modelling parameters on MAPE of estimated GHI will be detailed.

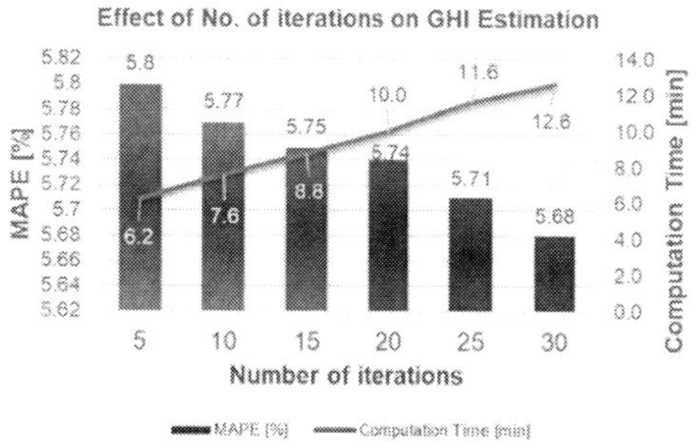

Figure 8: Effect of changing number of iterations on MAPE and computation time for GHI Estimation.

Figure 8 provides very useful information for a relationship between computation time and MAPE with increasing number of iterations. The results also show, the ease of flexibility that this GHI estimation model offers. Overall with increase of five iterations in gradient-descent loop of GHI esitmation [5], difference in MAPE is only about 0.2-0.3% for 30 days GHI Estimation. In parallel to MAPE, processing or computation time significantly increases with about 1.5 min on average for every additional five iterations. Therefore, from the results 10 iterations per timestep appears to be a good option, specifically considering the commercial applications.

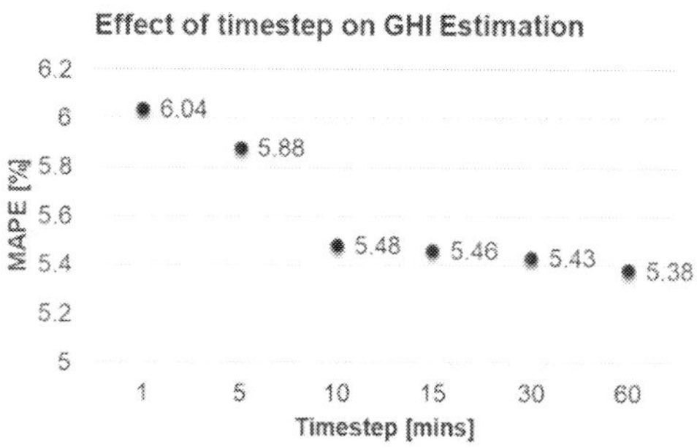

Figure 9: MAPE with different time resolution of input data for Öckerö Ishall

Data from Öckerö Ishall was utilized for carrying out further optimization studies as explained in sub-section 4.3. Figure 9 depicts that the data with time resolution of 10-mins and above provided lower values of MAPE. This was observed because number of data points decreased when down sampling the data. Considering the results obtained while conducting these studies, it was observed that 10-mins data provided good estimates of GHI values and can be considered as good time resolution for GHI as weather variations in 10-mins resolutions are not much. Moreover, 10-mins data provided enough data points to obtain good trust function results as compared to 30-mins or 60-mins data which had a lot of gaps for different sun positions. Figure 10 (a) and (b) provides an insight of 10-mins and 60-mins trust functions to confirm this result.

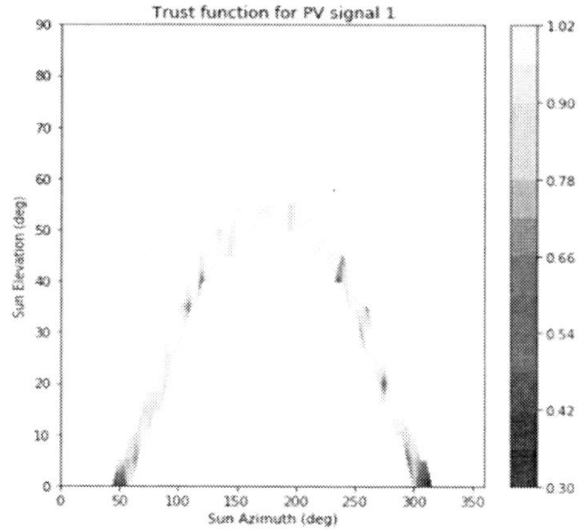

(a)

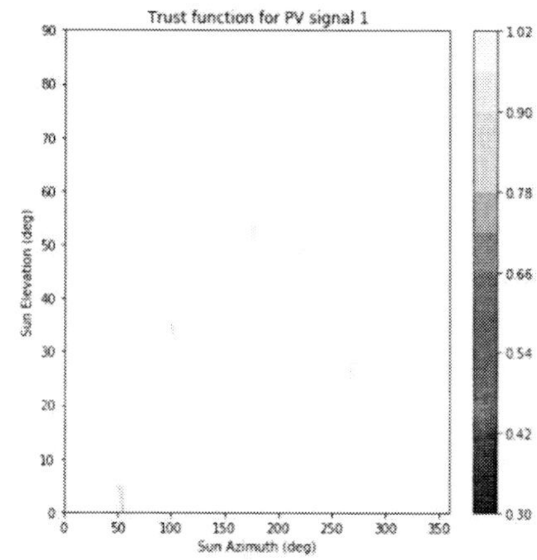

(b)

Figure 10: Trust Functions developed during GHI Estimation process to identify shading patters for input data of time resolution, (a) 10-mins, (b) 60-mins

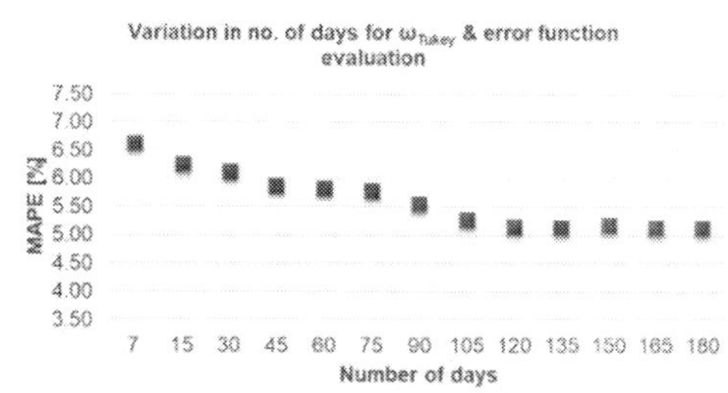

Figure 11: MAPE with different number of days for input of past data for Öckerö Ishall

Based on the results depicted in Figure 11, 120 number of days for input of past data provided best results to estimate weights for orientation assessment and error functions. Increasing the number of days for input of past data did not have much effect on MAPE but 120 days or more do cater enough seasonal variations that aid in GHI Estimation.

With time resolution and number of days for input of past data being studied separately, another study of combined effect of both modelling parameters was also carried out and the results are shown in Figure 12.

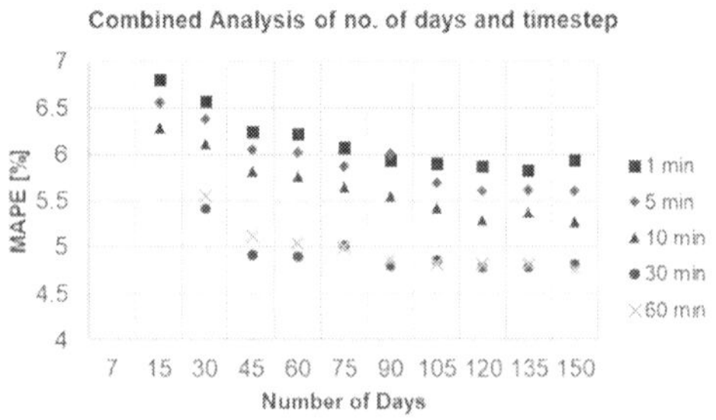

Figure 12: Combined effect of time resolution and number of days for input of past data on MAPE of GHI Estimation for data from Öckerö Ishall

From Figure 12, 10-mins time resolution and 120 number of days for input of past data again provided good results. Time resolutions of 30-mins and 60-mins did provide significantly lower MAPE values but were ignored as trust functions were not very accurate and weather variations were also not catered for.

5 DISCUSSION & CONCLUSION

Results from optimization of model developed in [5], inspired by the work carried out in [6], show that small optimizations can result in reduction of MAPE. Based on these studies, it is concluded that Ineichen/Perez Model is best suited for obtaining values of clear sky GHI that are then utilized for GHI estimates. Secondly, DISC Model for DNI Estimation and Perez Model for estimation of sky diffused irradiance provided least MAPE values when used in GHI Estimation model. Optimization of modelling parameters when carried out showed that using 10 iterations in gradient-descent loop for determination of GHI values resulted in lower processing times and adequate MAPE values. Moreover, time resolution of 10-mins for input data and 120 number of days for input of past data used in estimation of weights for orientation assessment and error functions are the values that results in minimal MAPE and accurate GHI estimates, close to the values obtained using model in [6]. Work carried out in this study could be further extended to cater modifications proposed in irradiance models. Moreover, Campbell Norman and Bird method could also be studied for DNI Estimation. Lastly, flexibility of this model could be further studied using combined effect of parameters mapping errors in matrices, achieving better results considering processing times also.

6 REFERENCES

[1] Renewable capacity growth between 2019 and 2024 by technology - IEA, Paris, November 15, 2019. - https://www.iea.org/data-and-statistics/charts/renewable-capacity-growth-between-2019-and-2024-by-technology.

[2] Iyengar Srinivasan [et al.] SolarClique: Detecting Anomalies in Residential Solar Arrays [Conference] ACM SIGCAS COMPASS. - Menlo Park and San Jose, CA, USA: ACM, 2018.

[3] Mecikalski J. R. [et al.] Convective Initiation and Lightning Prediction: The potential of the MTG-FC imagery mission [Conference] EUMETSAT 5th MTG MTM. - 2007.

[4] Bright Jamie M. [et al.] Improved satellite-derived PV power nowcasting using real-time power data from reference PV systems. Solar Energy. -Australia: Elsevier, 2017. - Vol. 168.

[5] M. A. KHAN and F. NAVEEN, Performance Evaluation of distributed Solar PV Installations, Dissertation, 2020.

[6] Nespoli Lorenzo and Medici Vasco: An Unsupervised Method for Estimating the Global Horizontal Irradiance from Photovoltaic Power Measurements [Journal]. Switzerland: Elsevier, 2017. – Solar Energy Vol. 158

[7] Solar Resource Glossary. [Online]. Available: https://www.nrel.gov/grid/solar-resource/solar-glossary.html#g

[8] M. Reno, C. Hansen, and J. Stein, "Global Horizontal Irradiance Clear Sky Models: Implementation and Analysis", Sandia National Laboratories, SAND2012-2389, 2012.

[9] P. Ineichen and R. Perez, "A New airmass independent formulation for the Linke turbidity coefficient", Solar Energy, vol 73, pp. 151-157, 2002.

[10] R. Perez et. al., "A New Operational Model for Satellite-Derived Irradiances: Description and Validation", Solar Energy, vol 73, pp. 307-317, 2002.

[11] P. Ineichen, "A broadband simplified version of the Solis clear sky model," Solar Energy, 82, 758-762 (2008).

[12] NumPy Reference. [Online]. Available: https://numpy.org/doc/stable/reference/

[13] API reference – Pandas. [Online]. Available: https://pandas.pydata.org/docs/reference/index.html

[14] SciPy documentation. [Online]. Available: https://docs.scipy.org/doc/scipy/reference/

[15] API Reference – Scikit-lean. [Online]. Available: https://scikit-learn.org/stable/modules/classes.html

[16] Pvlib – python. [Online]. Available: https://pvlib-python.readthedocs.io/en/stable/index.html

[17] Maxwell, E. L., "A Quasi-Physical Model for Converting Hourly Global Horizontal to Direct Normal Insolation", Technical Report No. SERI/TR-215-3087, Golden, CO: Solar Energy Research Institute, 1987

[18] Perez, R., P. Ineichen, E. Maxwell, R. Seals and A. Zelenka, (1992). "Dynamic Global-to-Direct Irradiance Conversion Models". ASHRAE Transactions-Research Series, pp. 354-369

[19] D. G. Erbs, S. A. Klein and J. A. Duffie, Estimation of the diffuse radiation fraction for hourly, daily, and monthly-average global radiation. Solar Energy 28(4), pp 293-302, 1982. Eq. 1

[20] Loutzenhiser P.G. et. al. "Empirical validation of models to compute solar irradiance on inclined surfaces for building energy simulation" 2007, Solar Energy vol. 81. pp. 254-267

[21] SMHI STRÅNG. [Online]. Available: http://strang.smhi.se/

[22] Campbell, G. S., J. M. Norman (1998) An Introduction to Environmental Biophysics. 2nd Ed. New York: Springer.

[23] pvlib.irradiance.erbs. [Online]. Available: https://pvlib-python.readthedocs.io/en/stable/generated/pvlib.irradiance.erbs.html

UNCERTAINTY OF TILTED IRRADIANCE MEASUREMENTS USING PHOTODIODES AND REFERENCE CELLS

Anton Driesse[1], Stefan Wilbert[2], Anne Forstinger[3]

[1] PV Performance Labs, Emmy-Noether-Str. 2, Freiburg, Germany,
anton.driesse@pvperformancelabs.com

[2]German Aerospace Center (DLR), Almeria, Spain
[3]CSP Services GmbH, Cologne, Germany

Irradiance is one of the most important parameters for the solar industry. However, it is very challenging to measure accurately, therefore it is all the more important to quantify the uncertainty of irradiance measurements. Current best practices tend to use worst-case assumptions and apply them more-or-less uniformly, possibly leading to overly pessimistic uncertainty evaluations. In this work we evaluate the uncertainty of individual tilted irradiance measurements of time series data using two different approaches: first, we compare the measurements from five different devices of each type operating in parallel for a GUM type A evaluation; and second, we calculate the uncertainty associated with several key properties of the instruments (spectral response, directional response, temperature response) and combine them in a GUM type B evaluation for comparison. Correlations seen between the A and B results in one year of data at two very different sites indicate that the second method should be suitable for uncertainty evaluations in other locations.

Keywords: Solar radiation, plane-of-array, uncertainty, pyranometer, reference cell, photodiode, spectral response, directional response, monitoring, system performance

1 INTRODUCTION

Solar irradiance measurements provide essential information at all stages of the PV system life cycle. Site selection, system design, certification, operation, maintenance, trouble-shooting, upgrading, expansion, and even decommissioning—all business decisions rely on accurate solar irradiance data. Most modern sensors used to measure global (or hemispherical) solar irradiance fall into three broad categories: thermopile pyranometers, photodiode pyranometers, and photovoltaic reference cells, and within these categories are instruments in various price and accuracy classes. While irradiance measured with pyranometers and reference cells are two fundamentally different quantities, the users of both instrument categories share the same need to understand the uncertainty of their measurements.

In this work we evaluate the uncertainty of individual tilted irradiance measurements in time series data using two different approaches based on [1] and [2]: first, we compare the measurements from five different devices of each type operating in parallel for a GUM type A evaluation; and second, we calculate the uncertainty associated with several key properties of the instruments (spectral response, directional response, temperature response) and combine them in a GUM type B evaluation. In this manner we obtain two estimates for the uncertainty of each measurement in each time series. The aim is to validate the type B approach at two locations for which such data is available, and thereby demonstrate that the type B method is suitable for uncertainty evaluations in other locations.

Realistic assessments of uncertainty substantially increase the usefulness and value of irradiance measurements. Not only can they be helpful in quality control procedures, for example to identify outliers, but they are essential inputs to determine uncertainties of key performance indicators (KPI), or uncertainties of aggregate values.

2 METHOD

As mentioned in the introduction, this work evaluates measurement uncertainties for two categories of sensors using two methods and data from two locations. The two categories, photodiode pyranometers and PV reference cells, are each represented by five products as listed in **Table 1**. Outdoor operating data were collected in Freiburg, Germany and Albuquerque, USA, and relevant instrument properties were measured by JRC Ispra and PV Performance Labs, Freiburg. The latter are described in more detail in [3] and [4].

Table 1 Sensor identification

Manufacturer	Model	Category
Apogee Instruments	SP-110	Photodiode pyranometer
Eko Instruments	ML-01	
Kipp & Zonen	SP-Lite2	
LI-COR	LI-200	
Skye Instruments	SKS-1110	
EETS	RC01	Reference cell
Fraunhofer ISE	11311102	
Ingenieurbuero Mencke & Tegtmeyer (IMT)	SiS-02-Pt100	
	Si-02-Pt100	
NES - Mess- und Meldesysteme	SOZ-03	

There are multiple differences between these product categories, but one is particularly important for the uncertainty analysis: the measurand for the photodiode pyranometer is clearly defined as the broadband hemispherical irradiance incident on the measurement plane, whereas the measurand for the reference cells lacks such a precise definition. Consequently, for the photodiodes the "true" value of the measurand can be estimated with superior thermal instruments (in our case even multiple instruments), whereas for the reference cells a substitute must be found. In this work we use the most practical (although somewhat optimistic) estimate that the true value is the mean of the five instruments. These estimates of the true value are called *reference values* in the following, and deviations from the reference values are called *measurement errors*.

2.1 Type A evaluation based on multiple observations under the same conditions

Type A uncertainty evaluations use the observed dispersion of multiple measurements to infer a level of uncertainty. In this analysis the dispersion between simultaneous readings (one-minute average values) from five similar instruments is used to infer the level of uncertainty of measurements using an *arbitrary instrument* in the category. This is a suitable objective as for average users the instruments in a certain category or class are more or less equivalent.

Measurement errors are calculated as the difference from the reference values (as discussed above), and the sample standard deviation (std) of these five errors is normally adopted as the standard uncertainty (k=1). Since the number of readings is small and the normality of their distribution unproven, alternate approaches could be considered.

In the case of the photodiodes the mean of the five readings usually differs from the thermopile reference value, which represents a bias that is not reflected in the standard deviation alone. Therefore, the bias is treated like an additional source of uncertainty and combined with the sample standard deviation as the root of the sum of squares (RSS):

$$u = \sqrt{std^2 + bias^2} \quad (1)$$

Since the reference value for the reference cells is chosen to be their mean value, the bias is always zero and the uncertainty is estimated only by the sample standard deviation.

2.2 Type B evaluation based on a priori knowledge about input quantities

The aim of the second evaluation is to quantify the separate sources of uncertainty that contribute in the overall uncertainty individually, and combine them using the standard rules of the GUM and the measurement equation. The measurement equation describes how the measurand is calculated from the instrument's output signal, which is a small voltage for these 10 irradiance sensors.

Table 2 (located near the end of this article) lists and describes the main characteristics that influence the instruments. The table also makes clear that some additional information is needed to complete the uncertainty estimates, such as sun angles and air temperature, but by far the most complex of these is spectral irradiance. We will discuss this point next since it is central to the measurement equation that will follow.

2.3 Uncertainty of spectral mismatch

Differences between device spectral responses and the ideal or reference spectral response lead to measurement errors, called spectral mismatch errors. These errors (and likewise the uncertainty associated with them) depend on the spectrum of the incident irradiance. Therefore, spectral irradiance is a required input for the type B method.

Since spectral irradiance measurements are usually unavailable, the spectral irradiance must be simulated for all possible sky conditions. The SEDES2 algorithm is one of the few options available for this task, and is used for example in [5]. We choose a different approach here motivated by the observation that the simulated spectra do not need to be completely accurate: only the correct spectral mismatch factors are needed. The chosen solution uses a clear-sky spectrum simulated using SMARTS2 [6] to estimate the SMM for all times when beam irradiance is "strong", which means clear-sky index for beam irradiance is greater than 0.5. We call these conditions "open-sky". At other times when the sun is obscured by clouds SMM is not calculated at all, but rather a simple empirical relationship is used to estimate its uncertainty.

The clear-sky assumption is tested using tilted spectral irradiance measurements and simulations in Freiburg, Germany (30° tilt) and Golden, Colorado (40° tilt) [7]. **Figure 1** and **Figure 2** show the SMM calculated using the average photodiode spectral response at the Colorado location, highlighting the values for summer (red), winter (blue), and behind plane of array (black).

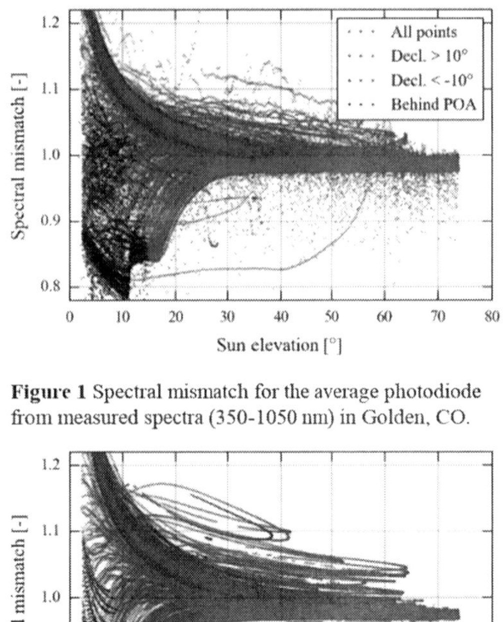

Figure 1 Spectral mismatch for the average photodiode from measured spectra (350-1050 nm) in Golden, CO.

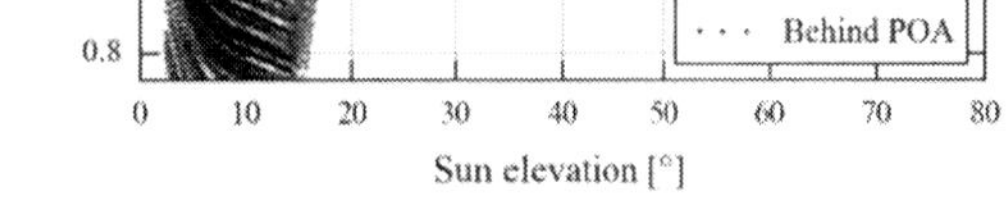

Figure 2 Spectral mismatch for the average photodiode from clear-sky simulated spectra (350-1050 nm) in Golden, CO.

The similarity between the SMM from measured and simulated spectral data is visually striking and indeed the histogram in **Figure 3** shows that the differences are most often quite small. Nevertheless, at higher sun elevations there is a distinct negative bias of a 1-2% on the SMM from simulations which is not seen in the second location, Freiburg. This could be due to the source atmospheric data (CAMS global reanalysis [8]), incorrect assumptions about the local or zonal albedo, or limitations of the SMARTS model itself. Similar analysis for the photodiodes in Freiburg, as well as for the reference cells in both locations showed SMM values generally somewhat closer to unity, and consequently also slightly smaller deviations than shown in **Figure 3**.

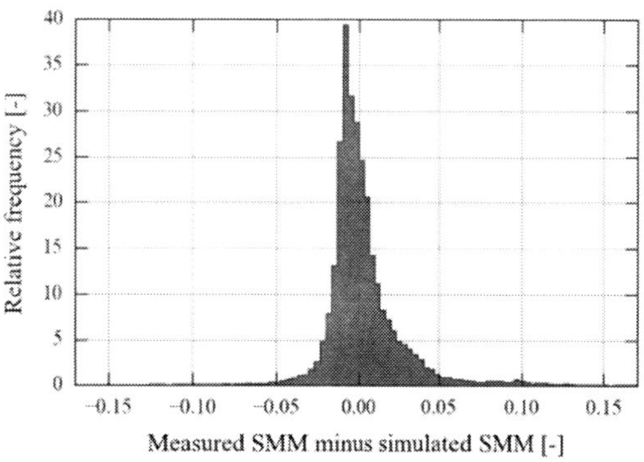

Figure 3 Differences between the SMM from measured and simulated spectra for the average photodiode in Golden, CO.

The identification of open-sky conditions is generally easy because intermediate DNI conditions occur much less frequently than low or high. Furthermore, high DNI conditions usually account for a large portion of the total annual energy (> 75% in Freiburg during the period of the data), thus being able to predict the spectral mismatch factor for 75% of the available energy is a valuable capability on its own. Further validation for additional climate zones should be carried out for general application of this mismatch prediction method, but since the two validation locations are similar to our two irradiance sensor locations we proceed to use it here for the uncertainty analysis.

To estimate uncertainty due to SMM, we calculate the SMM for each of the five sensors in each group, determine the SMM error by subtracting the reference value (which is 1.0 in the case of the photodiodes, and the mean of five for the reference cells), and finally combine the mean and standard deviation of the errors using (1).

2.4 The measurement equation

The required measurement equation is developed beginning with the most basic equation possible, which defines a linear relationship between the irradiance G, the measured signal V via the calibration factor or responsivity R_0:

$$G_{m,basic} = \frac{V}{R_0} \tag{2}$$

This basic equation is standard for photodiode measurements. However photovoltaic devices usually exhibit a linear temperature dependency that increases the apparent responsivity at higher temperatures. Thus a temperature correction can be incorporated into the measurement equation as follows:

$$G_{m,tc} = \frac{V}{R_0(1 + \alpha(T - 25))} \tag{3}$$

where α is the unit-less linear temperature coefficient and T is the instrument temperature in °C. This equation is usually recommended for use with reference cells and is frequently implemented in the electronics of powered reference cell products, where a typical value for α is 500 ppm. If the value of α for a photodiode is known, this equation can also be used for temperature correction of photodiode measurements.

Additional characteristics of sensors can be expressed in general as a function f_R that modifies the original responsivity R_0:

$$G_m = \frac{V}{R_o \cdot f_R(\vec{C}, \vec{P}) \cdot (1 + \alpha(T - 25))} \tag{4}$$

where C and P are vectors of measurement conditions and sensor properties (characteristics) respectively. When included in the measurement equation this responsivity modification serves as a correction and in general improves the accuracy of the measured irradiance. (It should be noted here that the temperature correction could also be placed within the modifier function but is retained in its usual place for familiarity.)

To incorporate directional response characteristics or incidence angle modifier (IAM) into the measurement equation a separation of beam and diffuse irradiance is needed, which is done via the beam fraction, f_{beam}:

$$f_R = \begin{bmatrix} f_{beam} \cdot IAM_{beam}(\theta) + \\ (1 - f_{beam}) \cdot IAM_{diff}(\beta) \end{bmatrix} \tag{5}$$

For the open sky case the spectral mismatch is likewise calculated separately for beam and diffuse irradiance, but the two must be immediately combined into a single global SMM_{clear}. This is because the SMM for beam and diffuse vary much more strongly than the combined SMM for global irradiance, and this individual variability should not feed into the uncertainty assessment for global irradiance. The new modifier function is thus:

$$f_R = \begin{bmatrix} f_{beam} \cdot IAM_{beam}(\theta) + \\ (1 - f_{beam}) \cdot IAM_{diff}(\beta) \end{bmatrix} \cdot SMM_{clear} \tag{6}$$

which can be written more compactly as:

$$f_R = IAM_{clear}(f_{beam}, \theta, \beta) \cdot SMM_{clear} \tag{7}$$

Similarly for sky conditions without beam irradiance the responsivity modifier would be:

$$f_R = IAM_{diff}(\beta) \cdot SMM_{variable} \tag{8}$$

These last two equations are blended by using a new factor f_{open} which varies between zero and one, just like f_{beam}:

$$f_R = \begin{bmatrix} f_{open} \cdot IAM_{clear}(f_{beam}, \theta, \beta) \cdot SMM_{clear} \\ +(1 - f_{open}) \cdot IAM_{diff}(\beta) \cdot SMM_{variable} \end{bmatrix} \tag{9}$$

The value of f_{open} is obtained by applying a simple clipping or sigmoid function to the clearness index for beam irradiance. And finally the non-linearity (NLIN) is appended as a function of the measured voltage:

$$f_R = \begin{bmatrix} f_{open} \cdot IAM_{clear}(f_{beam}, \theta, \beta) \cdot SMM_{clear} \\ +(1 - f_{open}) \cdot IAM_{diff}(\beta) \cdot SMM_{variable} \end{bmatrix} \cdot NLIN(V) \tag{10}$$

With this responsivity modifier substituted into equation (4) it is possible to implement correction functions. More importantly for this work, however, the equation defines the propagation path from uncertainties in the inputs (both operating conditions and sensor properties) and uncertainty in the output: the measured irradiance. Even if measured irradiance is calculated without corrections using the most basic equation (2), the combination of equations (4) and (10) is needed to calculate the type B uncertainty of that measured irradiance.

It should be stressed that not all sources of uncertainty are included in these calculations. Possible bias due to incorrect calibration factors was removed by applying a scaling factor to the measurements of each instrument, which is essentially a post-calibration. This was done using stable irradiance values in the range 800–1100 W/m^2 at times and angle of incidence less than 5°. Furthermore, the instruments were aligned with great care and cleaned frequently, usually twice per week. Finally, the comparisons between sensors are relative measurements, therefore logger calibration errors cancel out. A complete uncertainty analysis for field measurements with these instruments should include additional allowance for uncertainty due to calibration, alignment, soiling, aging, data logger and perhaps other installation-specific factors, and the combined uncertainty would be larger than those shown in this paper.

The propagation of uncertainties through this equation is not simple and is therefore not done analytically as shown in [2]. Instead, the open-source software package *uncertainties* [9] is used, which calculates all the required partial derivatives internally. Hence we can proceed directly from the descriptions of methods and inputs to results.

3 RESULTS AND DISCUSSION

After calculating the uncertainties using the two methods we need to evaluate their similarity. A key aspect of the measurement uncertainty is its time-varying nature due to fluctuating operating conditions. For solar energy applications such fluctuations are often visualized in the form of sun-path diagrams, where each pixel represents the mean value of a parameter for all times where the sun occupied that position in the sky. We will therefore show a small selection of sun-path diagrams depicting some general observations, followed by more direct comparisons.

Non-linear linearity is by far the smallest of the effect considered, and as shown in **Figure 4** it manifests itself primarily at low irradiance. The uncertainty due to temperature response (**Figure 5**) is somewhat higher, and is greatest in winter, with a slight lean towards morning for photodiodes. For reference cells this seasonal trend is inverted since it is in the hot summer that operating temperatures are furthest from the reference temperatures, but overall the uncertainty level is lower because a temperature correction is always done. Directional response uncertainty in **Figure 6** overshadows the previous two, and as the angle of incident approaches 90°, the uncertainty of the angle itself further accentuates the uncertainty of the IAM. Over most of the angular range reference cells have lower uncertainty, primarily because they are compared to their mean value, and not to the Lambertian ideal. The final example in **Figure 7** shows the uncertainty due to spectral mismatch obtained using the clear-sky spectral simulations. Once again reference cells have the benefit of being compared to each other, and their uncertainty due to SMM using clear-sky spectra lies below 1% almost always.

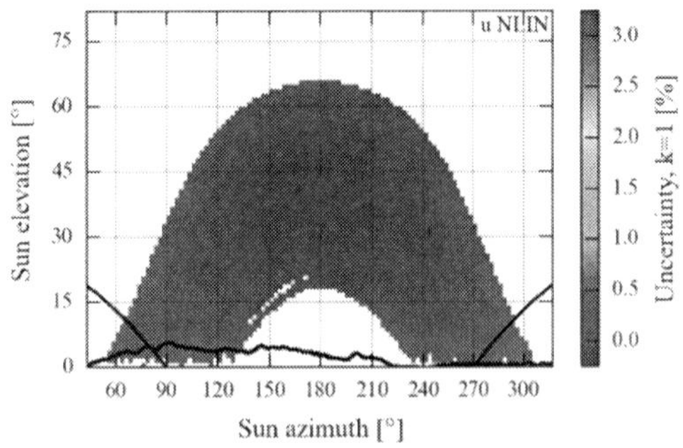

Figure 4 The uncertainty contribution of non-linearity using the example of photodiodes in Freiburg.

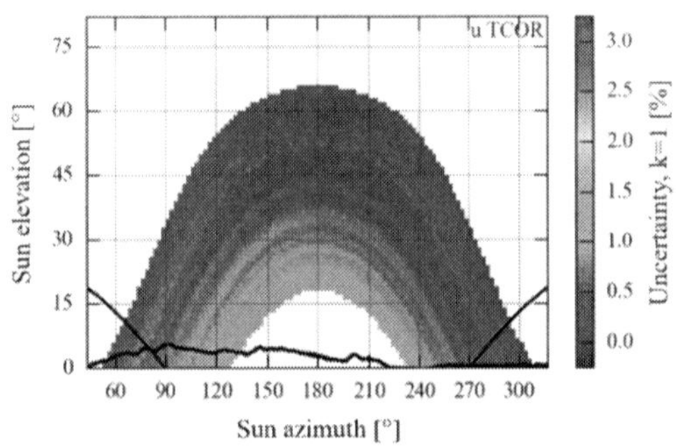

Figure 5 The uncertainty contribution of temperature response using the example of photodiodes in Freiburg.

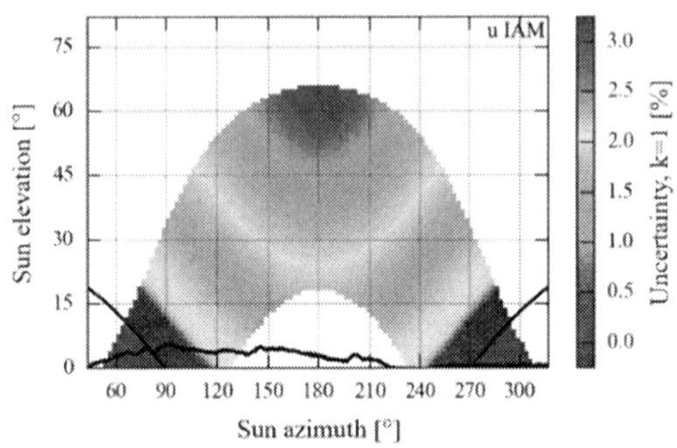

Figure 6 The uncertainty contribution of directional response using the example of photodiodes in Freiburg.

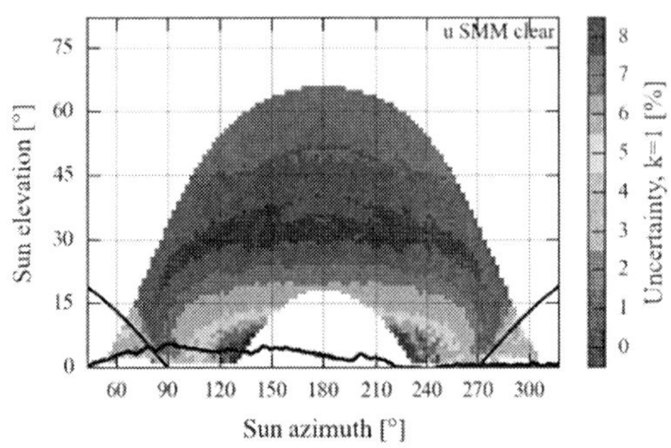

Figure 7 The uncertainty contribution of spectral response for open-sky conditions using the example of photodiodes in Freiburg.

The final two sun path diagrams (**Figure 8** and **Figure 9**) show the full type A and type B uncertainties for photodiodes under all sky conditions in Freiburg. Many matching trends and patterns can be seen in these diagrams, which is a first indication that the type B procedure described in this paper is indeed capable of producing a very plausible and realistic uncertainty profile. However, this is the best of four cases considered. Some challenges of the other three cases will be seen more clearly in other graphical representations.

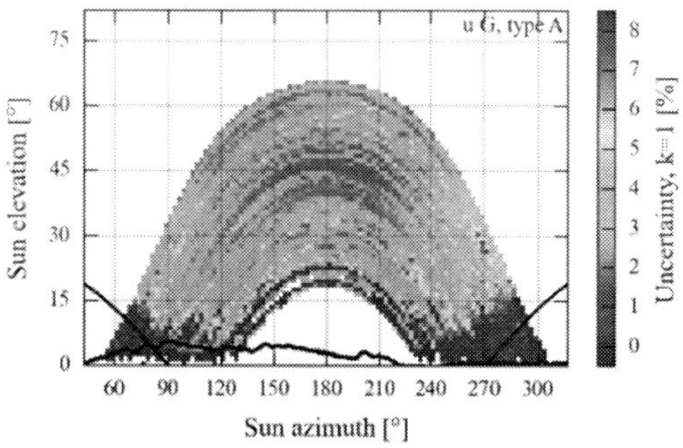

Figure 8 The time-dependent uncertainty of photodiode irradiance measurements in Freiburg estimated from measurement errors w.r.t a broadband reference.

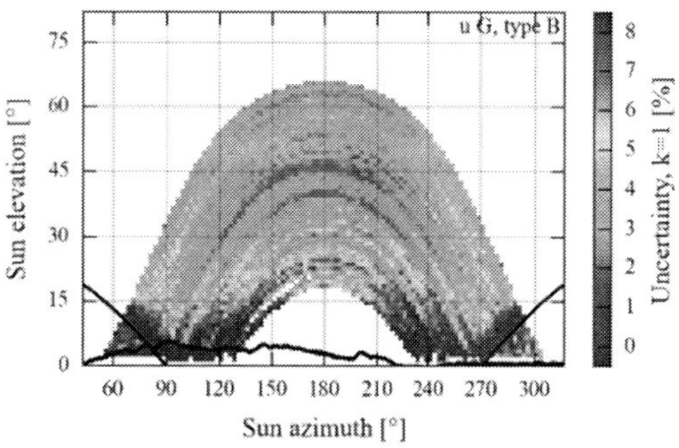

Figure 9 The time-dependent uncertainty of photodiode irradiance measurements in Freiburg estimated by combining multiple component sources of uncertainties.

While these sun-path diagrams provide useful insights, a more compact and direct comparison is the traditional scatter plot of Type B vs. Type A. These are shown in **Figure 10** (located near the end of this article). For photodiodes, (a) and (b), there is a clear correlation between the two uncertainty estimates albeit with some overestimation in the type B. Uncertainties reach substantially higher values in Albuquerque (b) which is due at least in part to the higher mismatch factors at high irradiance at that location. This is a group bias that propagates into the overall type B uncertainty. The type A estimate picks up the same effect through higher deviations from the broadband reference irradiance.

For the reference cells, (c) and (d), the correlation during non-open-sky conditions is more clearly visible than for the open-sky conditions. In fact, for open-sky conditions the clearest feature is the substantial overestimation of type B. The lack of correlation here is rooted in the non-normal distribution of the characteristics of the five products, in particular the directional responses.

Finally, we go back to the measurements of each individual sensor to evaluate whether they reflect the expectations implied by the calculated uncertainty. For example, when an uncertainty u (k=1) is given, the expectation is that roughly 68% of the measurements will have an error less than u, and 95% will have an error of less than $2u$ (k=2). The final four charts in **Figure 11** plot the CDF of the measurement errors for each individual instrument divided by the estimated uncertainty in order to assess this aspect. For the photodiodes, (a) and (b), the shapes and position of the CDF are not far from the Gaussian dotted line, indicating that the measurement errors are in line with expectations: i.e. for most products more than 68% of the errors were smaller than the k=1 uncertainty level, and similarly for most products more than 95% of the errors were smaller than the expanded k=2 uncertainty. Overall the five photodiodes do not differ substantially and it appears appropriate to use the same type B uncertainty estimate for all.

For the reference cells, (c) and (d), the picture differs substantially. The uncertainty estimates for the group are much too high for three of the cells, and too low for the other two. Reference cell errors are calculated with respect to the mean of the group and the separation in this diagram highlights the fact that deviations from the mean are more systematic than random. Consequently, the type B uncertainty estimate calculated for the group is not appropriate for any of the members of the group.

4 CONCLUSIONS

The dual uncertainty analysis carried out in this work produced consistent uncertainty estimates for the group of five photodiodes in two very different locations. This paves the way for using the more flexible type B method to estimate tilted measurement uncertainty for arbitrary photodiode pyranometers at arbitrary locations and in arbitrary orientations. Nevertheless, some caution—and preferably additional validation—would be recommended for use in more extreme climates and measurement situations.

For the group of five reference cells the same consistency could not be observed because the group contained two units having unique designs and operating characteristics. Thus, uncertainty estimates suitable for all five members of the group could not be determined. This is not a failure of the method as such, but rather a consequence of a problem with the input.

A more fundamental obstacle to estimating the uncertainty for reference cells is the lack of a precise definition of the measurand, or alternatively of the required instrument characteristics. While there are legitimate reasons for having reference cells with different characteristics for different purposes, each needs a standardized definition against which measurement errors and uncertainties can be gauged.

Although not apparent from the results shown in this paper, the strongest contributor to the uncertainty of photodiode measurements was by far the spectral mismatch. By contrast the reference cells demonstrated stronger variability in directional response. Another observation related to both of these aspects is that to calculate spectral mismatch on a tilted plane it is essential to consider the beam and diffuse spectral irradiance separately. They always differ strongly and must be mixed in the correct proportions.

The calculation steps that lead to the uncertainty estimates are complex, but much can be automated. In the context of low-cost sensors such calculations could be offered as a service to achieve economies of scale, or alternatively typical year uncertainty profiles could be generated in the form of look-up tables (LUT) as a one-time effort. The method is of course not inherently limited to use with low-cost sensors, but for thermopile pyranometers the spectral mismatch effect is generally so small that the large effort of simulating spectral irradiance can perhaps be spared.

Looking beyond uncertainty estimates, the approach of using clear-sky spectra to estimate spectral mismatch for a broader range of sky conditions holds promise for use in correction procedures for photodiodes pyranometers. But perhaps even more importantly, this approach could be used to estimate energy gains and losses in PV power plants due to spectral mismatch.

5 ACKNOWLEDGEMENTS

The authors thank the German Ministry for Economic Affairs and Energy for funding this study within the SOLREV project (contract number 03EE1010). We would also like to acknowledge the supporters and partners of the PVSENSOR project, which provided core data for this study—especially JRC Ispra and Sandia National Labs. We are also grateful to various organizations that made supporting data available, such as Fraunhofer ISE and the National Renewable Energy Lab.

6 AUTHOR CONTRIBUTIONS

Anton Driesse: Conceptualization, Data curation, Formal analysis, Methodology, Software, Validation, Visualization, Writing – original draft, Writing – review & editing. **Stefan Wilbert**: Conceptualization, Writing – review & editing. **Anne Forstinger**: Writing – review & editing.

7 REFERENCES

[1] Joint Committee for Guides in Metrology Working Group 1. "Guide to the Expression of Uncertainty in Measurement," 2008.

[2] "ASTM G213-17 Standard Guide for Evaluating Uncertainty in Calibration and Field Measurements of Broadband Irradiance with Pyranometers and Pyrheliometers," ASTM International, 2017. doi: 10.1520/G0213-17.

[3] A. Driesse and W. Zaaiman, "Characterization of Global Irradiance Sensors for use with PV Systems," Proc. 42nd IEEE PVSC, 2015.

[4] A. Driesse et al., "Indoor and Outdoor Evaluation of Global Irradiance Sensors," 30th EU-PVSEC, Hamburg, 2015.

[5] A. Forstinger et al "Uncertainty calculation method for photodiode pyranometers," accepted for publication in Solar RRL, 2021.

[6] Gueymard, C.A., "SMARTS2, Simple Model of the Atmospheric Radiative Transfer of Sunshine: Algorithms and performance assessment," Florida Solar Energy Center, 1995.

[7] A. Andreas and T. Stoffel, "NREL Solar Radiation Research Laboratory (SRRL): Baseline Measurement System (BMS)," Golden, Colorado, 1981, NREL Report No. DA-5500-56488, doi: 10.5439/1052221.

[8] A. Inness et al., "The CAMS reanalysis of atmospheric composition," Atmos. Chem. Phys., vol. 19, no. 6, pp. 3515–3556, Mar. 2019, doi: 10.5194/acp-19-3515-2019.

[9] Eric O. Lebigot, "Uncertainties: a Python package for calculations with uncertainties", version 3.1.6 https://pypi.org/ project/uncertainties/.

Table 2 Summary of the uncertainty components evaluated in the type B analysis.

Source of uncertainty	Details for photodiodes	Details for reference cells	Additional notes
Directional response or incidence angle modifier (IAM)	The ideal directional response is Lambertian. Uncertainty varies by angle and is estimated by the RSS of the standard deviation and bias at that angle. The IAM uncertainty for diffuse irradiance is 0.2%.	The ideal directional response is defined as the average directional response of the 5 devices at a given angle. Uncertainty varies by angle, and is estimated as the standard deviation of the responses at that angle. The IAM uncertainty for diffuse irradiance is 0.2%.	The uncertainty on the angle of incidence is 0.5°.
Spectral response, which leads to spectral mismatch (SMM)	The ideal mismatch factor is 1.0. The uncertainty varies by spectrum, and is estimated by the RSS of the standard deviation and bias under that spectrum for open-sky conditions.	The ideal mismatch is the average mismatch of the 5 devices under a given spectrum. The uncertainty varies by spectrum, and is estimated by the standard deviation of the 5 mismatch values under that spectrum for open-sky conditions.	For non-open-sky conditions the average mismatch is estimated at 1.0 and the uncertainty decreases as total POA irradiance increases.
Temperature response or correction (TCOR)	Signal is not adjusted for temperature. Assume a fictional α=0 for the measurement equation and use the measured ambient temperature. u_α = 500 ppm u_T= 4°C	Signal is already adjusted using the internal measured temperature T and factory-specified temperature coefficient α. The uncertainty is added with a second fictional adjustment with α=0. u_α = 80 ppm u_T= 1°C	Through the measurement equation this uncertainty grows as the temperature deviates further from 25 °C
Linearity or non-linearity (NLIN)	Some photodiodes show substantially higher non-linearity than reference cells. $u_{lin,max}$=0.6%	This is a negligible factor for most reference cells. $u_{lin,max}$=0.2%	This uncertainty is zero at 1000 W/m² and increases linearly away from this value to the maximum at 0 W/m²
Calibration or normalization (CAL)	All measurements were scaled to produce the same reading at the standard reference conditions AM1.5G and 25 °C. u_{cal}=0.5%	All measurements were scaled to produce the same reading at the standard reference conditions AM1.5G and 25 °C. u_{cal}=0.5%	Since the reference conditions do typically not occur naturally, high irradiance conditions were used together with corrections for spectrum and temperature.
Datalogger (LOG)	The logger uncertainty is considered to be covered in the uncertainty of calibration/normalization.	The logger uncertainty is considered to be covered in the uncertainty of calibration/normalization.	The uncertainty of the logger used here is very small compared to other factors.

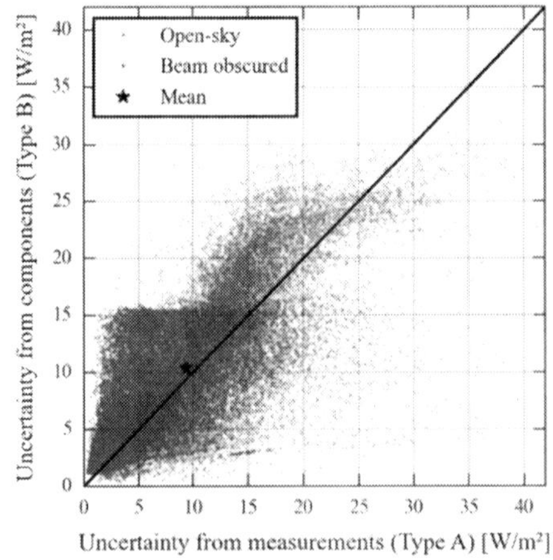

(a) Photodiodes in Freiburg

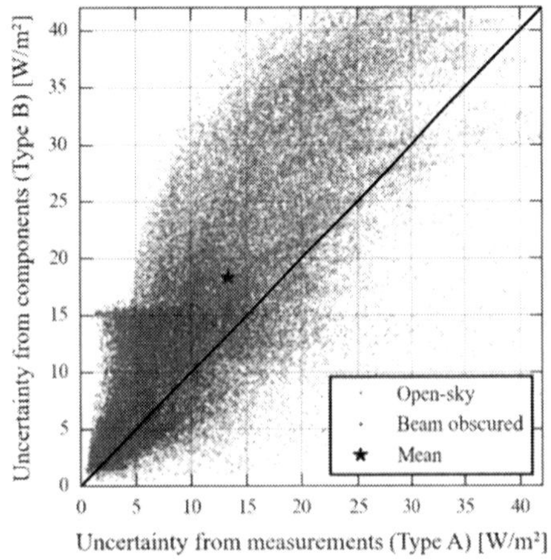

(b) Photodiodes in Albuquerque

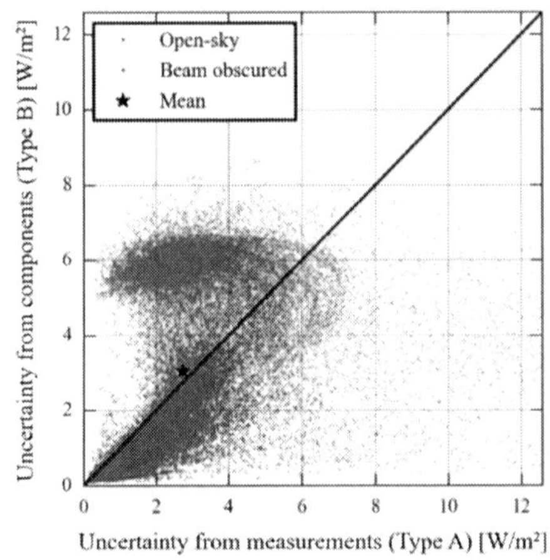

(c) Reference cells in Freiburg

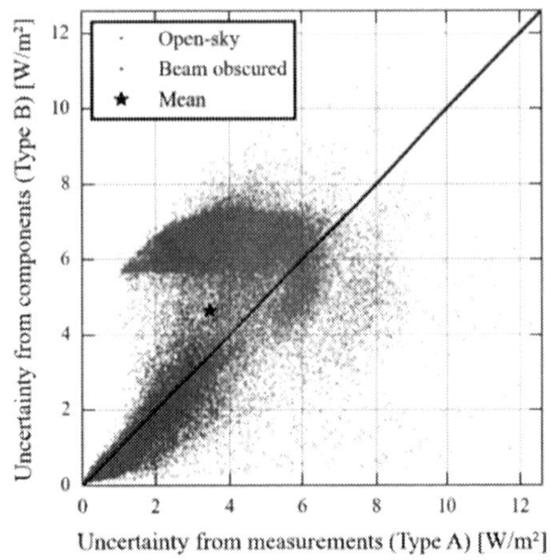

(d) Reference cells in Albuquerque

Figure 10 Scatter charts plotting the type B uncertainty estimates against the type A estimates. The uncertainties during open-sky conditions are shown in red. Note that the scale of the lower two graphs is only 30% of the upper graphs, reflecting the fact that the photodiodes are compared against a broadband reference whereas the reference cells are compared against the mean of the group.

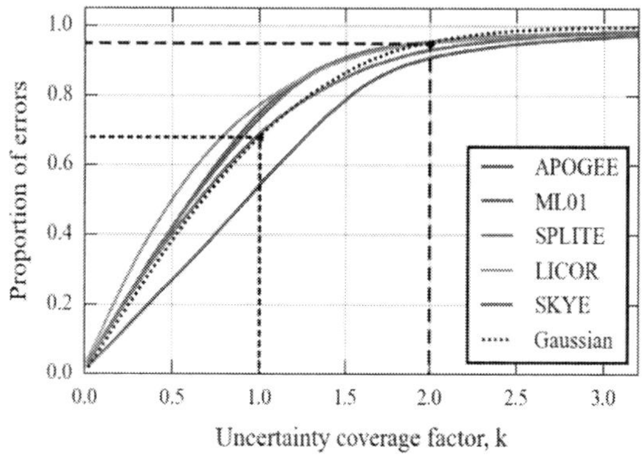

(a) Photodiodes in Freiburg

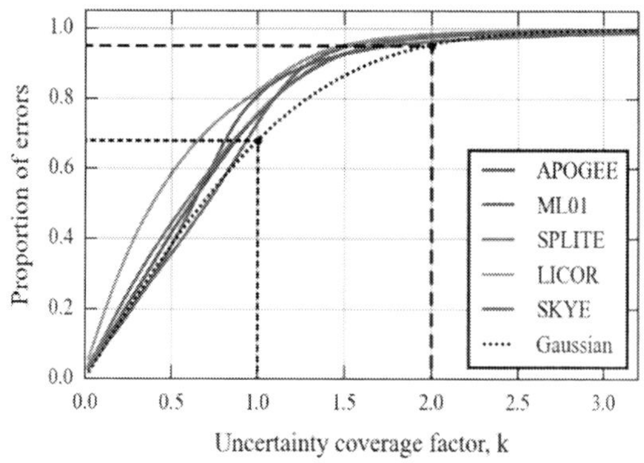

(b) Photodiodes in Albuquerque

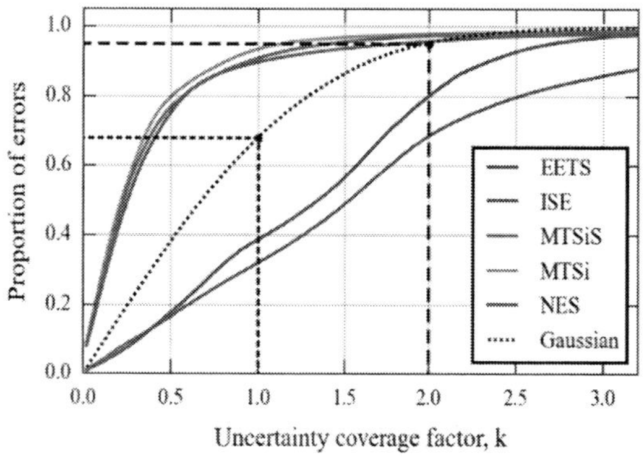

(c) Reference cells in Freiburg

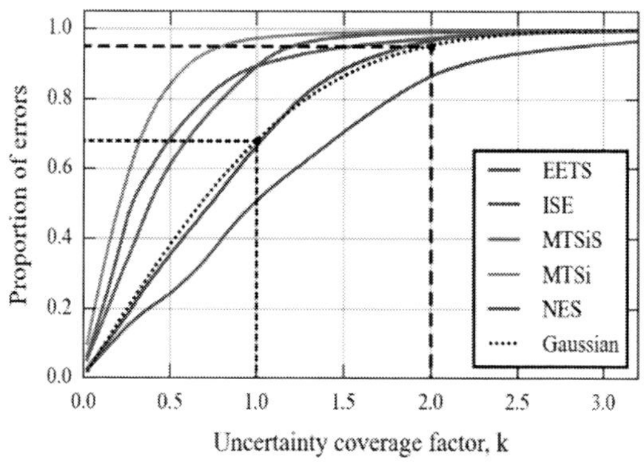

(d) Reference cells in Albuquerque

Figure 11 Cumulative distributions functions (CDF) of measurement errors normalized by the type B uncertainty estimates. When a curve lies above the dotted Gaussian line, the errors are smaller than expected based on the type B uncertainty estimates; when it lies below, the errors were larger than expected.

A COMPARATIVE ANALYSIS OF FORECASTING METHODS FOR PHOTOVOLTAIC POWER AND ENERGY GENERATION WITH AND WITHOUT EXOGENOUS INPUTS

Anna Starosta, Komal Kaushik, Purav Jhaveri, Nina Munzke, Marc Hiller
Karlsruhe Institute of Technology,
Hermann-von-Helmholtz-Platz 1,
76344 Eggenstein-Leopoldshafen, Germany

ABSTRACT: With drive towards a green energy sector and significant cost reduction of photovoltaic (PV) technology, global PV capacity has increased multi folds in the last decade. PV power increases the variability of power supply due to the sun's changing position and weather volatility. Battery systems tackle these problems. For an optimized battery charging management, accurate PV energy forecasts are vital. The present work compares five 1-day-ahead PV power prediction models for a PV array of 8.64 kWp at KIT with 30° tilt and a 15° eastward orientation. The power prediction is used for intelligent battery charging management. The models are the offline persistence forecast (PF), an online forecast based on numerical weather prediction (NWP) and the machine-learning based offline Facebook prophet (FBP), support vector regression (SVR) and multilayer perceptron (MLP). According to the needs of intelligent battery charging management, the hourly PV energy forecast for the rest of the day is evaluated and compared for the different methods. The prediction methods are configurable for arbitrary inclinations and orientations. To the knowledge of the authors, FBP is compared for the first time to other models in the context of PV prediction. The results are evaluated for one year of data between March 2020 and February 2021. When comparing the performance of the models for different times of the day, SVR and MLP outperform the other models around noon, while the PF and most of the time NWP and FBP outperform the SVR and MLP in the morning and evening. When evaluating the models over one year, SVR outperforms the other model's power and energy prediction. At the same time, the other models have similar power prediction performance, but varying energy prediction performance. When evaluating the models over the meteorological seasons, there are striking differences of the models' performance. The SVR performs mostly best but is outperformed by the NWP in spring. At the same time, the NWP performs worst in winter. This is associated to reasons of spatial resolution of the NWP data. With the two best models SVR and MLP, it is shown that endogenous values generally suffice, while the partial outperformance of the NWP still motivates for a further investigation into the use of exogenous input.

Keywords: Battery storage and control, Evaluation, Photovoltaic, Software, PV Array, Solar Radiation

1 INTRODUCTION AND LITERATURE REVIEW

Photovoltaic (PV) power increases globally and has environmental and economic benefits. Its variable availability can be tackled by battery systems. Optimal battery charging management requires accurate energy forecasts [1]. The PV field of KIT Campus North with 102 PV arrays provides an attractive data basis for the research on power prediction. The power prediction is transformed into an energy forecast that serves the battery load management of a Li-ion battery system coupled to a PV array. The battery is charged so that it is full before sunset to prevent aging. Furthermore, the energy forecast is used to estimate the state of charge and state of health for batteries in advance [2]. Solar irradiation and resulting PV power can be predicted in several ways. While some methods use exogenous inputs such as the ambient temperature from a global forecasting system (GFS [3]) ("online" forecasts) [4, 5] others only use data collected on site ("offline" forecasts) [6, 7]. While the latter often use statistical methods in form of different time series analyses methods (persistence forecast (PF)), the former typically use numerical weather prediction (NWP) models. A combination of methods in order to merge advantages is common [8] but it has also been shown that univariate forecasting can achieve similar or even better accuracy compared to multivariate models [9, 10]. The PF as well as the NWP are based on physical equations that calculate the PV power output through considering the sun's position and radiation onto the tilted and specifically oriented PV array [11]. The difference between PF and NWP is the calculation of the global horizontal irradiance and using historical data in case of the PF [11] while using irradiance, albedo and ambient temperature from the GFS in case of the NWP [11, 12, 13].

In recent years, research has increasingly dealt with the application of machine learning (ML) in order to more accurately forecast the PV power and take advantage of patterns in historical data [14]. Popular and common approaches include support vector regression (SVR) and artificial neural networks (ANN), specifically deep learning methods [15]. SVR is the most extensively used ML algorithm in the field of PV power forecasting [9] because of its ability to find non-linear relationships and the small number of parameters to tune which is less likely to overfit the data [10]. Another approach is the time series model called Facebook prophet (FBP) which is faster than other statistical methods [16] and has shown promising results for PV forecasting [16]. In search of an optimal application of ML to PV forecasts, comparisons are regularly made [17, 18, 19] to find the best model, whereas the findings differ. To the knowledge of the authors, FBP is compared for the first time to other models in the context of PV prediction.

The present work introduces and compares five 1-day-ahead PV power prediction models for an example PV array, namely an online forecast based on NWP which uses exogenous and multivariate input and the offline, univariate PF, ML based FBP, SVR and multilayer perceptron (MLP) that only use endogenous input. The models' advantages and disadvantages are examined as well as their results compared and evaluated. The results are evaluated regarding the error metrics *MAE* and *RMSE* for specific times of the day, over the course of one year as well as for the meteorological seasons.

2 DATA INPUT

The 1 MWp PV field of KIT Campus North located at 49.1°N and 8.44°E provides an extensive dataset of several years of measured PV data at 1 second resolution. Here, a 8.64 kWp PV array with 30° tilt and a 15° eastward orientation has been chosen to train and test the different models. The considered meteorological seasons are divided into spring (March to May), summer (June to August), autumn (September to November) and winter (December to February). The time range of spring 2020 till winter 2020/2021 is considered.

3 FUNDAMENTALS

3.1 Physical Principles

The PV power is calculated using the direct irradiance I_{dir} excluding the incident angle losses contained in the incident angle modifier IAM_b [20], the diffuse irradiance I_{dif} [21, 22] and the ground diffuse (reflected) irradiance I_{ref} [13] in the plane of the PV array, as in

$$P_{PV} = \eta_{PV} \cdot A_{PV} \cdot (IAM_b \cdot I_{dir} + I_{dif} + I_{ref}) \qquad (1)$$

where η_{PV} denotes the efficiency of the PV array and A_{PV} denotes the PV panel size. A more detailed overview can be found in [23]. The global irradiance from which the direct and diffuse irradiance is derived as well as other parameters are obtained depending on the offline or online approach and are therefore explained in subsections 4.1 and 4.2 respectively.

3.2 Machine Learning and Time-Series Forecasting

In Machine-learning (ML), PV power time series forecasting can be viewed as a supervised regression problem [24]. In this work, only the past values of the time series as endogenous values are used. So called direct multi-step ahead methods are applied.

Deep Learning is a subfield of ML and uses artificial neural networks (ANNs) as its basic structure. The ANN comprises an input and output layer as well as a variable number of hidden layers. The layers consist of neurons, also called nodes or units. The most fundamental form of an ANN is a feed-forward network where the information flow happens in a forward direction [25].

An essential part of building an ML model is the training of the model minimizing a loss function while using training data and afterwards testing the model on unseen testing data.

3.3 Energy Forecast

The energy forecast uses the forecasted power values *P(t)*. The expected energy $E(t_i)$ for the rest of the day at time point t_i corresponds with the integration of the power *P(t)* from t_i until the end of the day *T*, as in:

$$E(t_i) = \int_{t_i}^{T} P(t)\, dt \qquad (2)$$

4 FORECASTING MODELS

4.1 Persistence Forecast

The Persistence Forecast (PF) model is an offline model that only uses endogenous parameters. It is based on the physical equations explained in section 3.1 and assumes persisting weather conditions for a specific time horizon [1, 4]. In this work, a day-ahead prediction P_{PF} is made by scaling the clear-sky power curve of the predicted day $P_{CS,T}$, as described in Equation (3)

$$P_{PF} = k_{PF} \cdot P_{CS,T} \qquad (3)$$

with the scaling factor k_{PF} based on the generated PV energy of the previous day $E_{Real,T-1}$ and the PV energy $E_{CS,T-1}$ that would have been generated on a clear-sky day, as described in Equation (4)

$$k_{PF} = \frac{E_{Real,T-1}}{E_{CS,T-1}} \qquad (4)$$

The clear-sky power curve $P_{CS,T}$ is calculated using Equation (1). As the PF model only has access to offline parameters, the global horizontal irradiance is calculated using the approaches in [26, 27] as further explained in [11]. The albedo is calculated using the equations in [13, 21].

Advantages of the PF model are its independency on exogenous variables or larger datasets, its simplicity, robustness and short computing times of less than one minute. Disadvantage is its reduced accuracy for the 1-day-ahead prediction.

4.2 Numerical Weather Prediction

The Numerical Weather Prediction (NWP) model is an online forecasting model that uses the exogenous inputs ambient temperature, global horizontal irradiance and albedo coefficient. The NWP is based on the Global Forecasting System (GFS) provided open-source by [3]. Similar to the PF, the calculation of PV power is based on Equation (1), while the PV efficiency η_{PV} is calculated depending on the GFS's predicted ambient temperature [28]. The global horizontal irradiance and the albedo coefficient are taken from the GFS's data and are handled using the equations in [21, 26]. The diffuse irradiance comprises the differentiation between a twilight zone before sunrise and after sunset using the equations in [21, 22] and times of direct sunlight using the equations in [21, 29].

The NWP data is updated every 6 h and its data is available in a 1-hour temporal resolution and a 0.25° spatial resolution. In this work, the midnight update is used for the 24 h day-ahead prediction.

Advantages of the NWP are the possibility to use a vast range of predicted parameters. Disadvantages are its need for an online connection, the large GFS files that need to be downloaded and processed and therefore increased computing times of one to two hours for this work's NWP model.

4.3 Facebook Prophet Algorithm

The Facebook Prophet Algorithm (FBP) is a univariate time series forecasting tool and uses only offline time series PV power data as an input. FBP is based on an additive regression model stated as described in [16]

$$y(t) = g(t) + s(t) + h(t) + e(t) \qquad (5)$$

Table I: Choice of FBP parameters

Parameter	**Value**
growth	linear
seasonality mode	multiplicative
change point prior scale	10
fourier order	15

with the time-dependent functions for trends *g(t)*, periodic changes *s(t)*, irregular predictable occasions *h(t)* and independent identically distributed noise *e(t)* (see Equation (5)). Advantages of the FBP are, that the statistical method is relatively simple and robust to outliers, missing values and sudden changes. Disadvantages are its dependency on larger datasets [16]. In this work, the parameters were tuned using manual tuning, also known as exhaustive search, on training data. An input of the prior 3 months' data showed to be optimal. This work's FBP model needs 2.5 minutes to load the input data and about 40 seconds to calculate. The optimized parameters are as shown in Table I.

4.4 Support Vector Regression

Support Vector Regression (SVR) is a supervised ML technique for regression analysis. Here, only offline PV power data is used as input. SVR is able to handle non-linear data using a kernel function, a set of mathematical functions, which model the input data into a higher dimensional space. Goal is to find a hyperplane (decision boundary) in an n-dimensional space that classifies the data points. SVRs are based on so called maximum-margin classifiers as the distance of the data points closest to the hyperplanes (support vector) is maximized and therefore the optimal orientation and location of the hyperplane is achieved. At the core of the technique, the distance between the data points outside a permitted marginal error ε, so called slack variables ξ and ξ*, are minimized, as in [30]

$$\min\left(\frac{1}{2}\|w\|^2 + C\sum_{i=1}^{n}|\xi_i|\right) \tag{6}$$
$$s.t.\, |y_i - w_i x_i| \leq \varepsilon + |\xi_i|$$

with the target y, the features x, the weight w and the tunable penalty of error term C (see Equation (6)). The parameters were tuned using random search optimization with the search space 'rbf'-kernel, $C = [0.1, 0.5, 1, 10, 16]$, the tube-defining $\varepsilon = [0.01, 0.1, 1]$, the kernel coefficient $\gamma = [0.1, 0.01, 1, 10, 100, 1000]$. The parameters were tuned on the basis of 1.3 years and 11 days input data predicting according days in winter and summer. The optimized parameters are shown in Table II. Furthermore, a training data set of the prior 6 months of PV power data of the predicted day showed to be optimal. SVR is the most extensively used ML algorithm in the field of PV power forecasting [9] because of its ability to find non-linear relationships and the small number of parameters that have to be tuned. Therefore, it is less likely to overfit the data [10]. Furthermore, this work's SVR needs about 5 minutes

Table II: Choice of SVR parameters

Parameter	**Value**
kernel	*rbf*
C	16
γ	0.01
ε	0.1

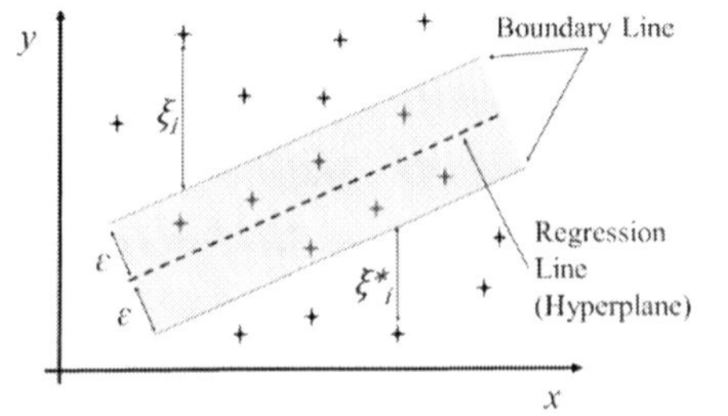

Figure 1: Basic principle of the SVR model. All errors outside an error tolerance ε are optimized towards the regression line.

for processing the input data while the algorithm itself needs about 12 seconds. A disadvantage is, that it depends on larger datasets.

4.5 Multilayer Perceptron

Multilayer Perceptrons (MLPs) are feed-forward ANNs with one or more hidden layers of non-linearly-activating nodes. The hidden layers consist of so called perceptrons which are multiple input single output units processing binary data z with a chosen non-linear activation function f. The unit input x is multiplied by weights w which are adjusted during the training phase in order to minimize the output error. For each unit, the basic principle looks as follows, where a bias is added for a non-zero output as described in Equation (7) [31]

$$z = \begin{cases} 0, & \sum_{i=1}^{m} w_i x_i + bias < 0 \\ 1, & \sum_{i=1}^{m} w_i x_i + bias \geq 0 \end{cases} \tag{7}$$

$$\hat{y} = f(z)$$

with $\hat{y}$ being the output of the unit. MLPs learn via backpropagation in order to adjust the unit weights for minimizing the error of the overall output calculated as described in Equation (8)

$$\mathcal{E}(n) = \frac{1}{2}\sum_j e_j^2(n) \tag{8}$$

The number of hidden units and layers has to be optimized and tuned [32]. Figure 2 shows the basic operating principle of the MLP model for the 1-day-ahead PV power prediction. The input data has an hourly resolution.

ANNs are popular deep learning algorithms for PV power forecasting that are currently further investigated in literature [6, 32]. At the same time it is difficult to determine the high number of hyper-parameters [33] and the MLP depends on larger datasets. The MLP used in this work needs about 5 minutes to load the input data and about 4 minutes to apply the MLP algorithm. After a first manual tuning, the hyper-parameters were further optimized using random search optimization. The parameters were tuned on the basis of 1.3 years and 11 days input data predicting according days in winter and summer. The search space includes 3 input neurons, 2 hidden layers, 2 to 40 neurons in the hidden layers, "adam" optimizer and the "relu" and "softplus" activation functions. The optimized hyper-parameters are found in Table III. Furthermore, a training data set of the prior 6 months of PV power data of the predicted day showed to be optimal.

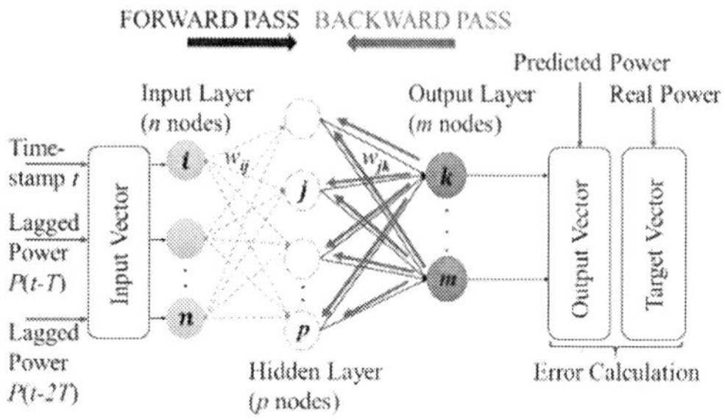

Figure 2: Basic operating principle of the MLP model. The timestamp *t* and lagged power from 1 and 2 days prior (*P(t-T)* and *P(t-2T)*) are used as inputs for the input layer. The values are processed via the hidden layers and the output error is calculated. The weights are tuned via backpropagation [31].

Table III: Choice of MLP parameters

Parameter	Value
optimizer	*adam*
neurons in input layer	3
neurons in hidden layer 1	18
neurons in hidden layer 2	26
activation functions	*relu, softplus*
neurons in output layer	1
epochs	11688
batch size	10000

5 DATA PRE- AND POST-PROCESSING

FBP processes the data automatically while SVR and MLP need feature engineering [24]. Date time features including hours of the day as well as the power value of a previous time step, called lag feature, are chosen. Furthermore, the data is transformed via standardization and normalization. After prediction, the results are inversely transformed again. For SVR, the lag combination of power *P(t-T)* is chosen, where *t-T* denotes the value one-day prior. For MLP, the combination of *P(t-T)* and *P(t-2T)* performed slightly better. The input data has a 1-hour resolution.

Prediction data outside the sunrise and sunset times (at nighttime) were set to zero so that the models are only compared for relevant times of PV power generation.

6 EVALUATION METRICS

6.1 Mean Absolute Error

The mean absolute error *MAE* shows the mean absolute deviation of the prediction values *y* from the actual values *x* for each time point *t* and the time range *n*. The *MAE* is less sensitive towards outliers.

$$MAE = \frac{\sum_{t=1}^{n} |y(t) - x(t)|}{n} \quad (9)$$

6.2 Root Mean Square Error

The root mean square error *RMSE* is more sensitive towards outliers as it regards the square of the deviations, as in:

$$RMSE = \sqrt{\frac{\sum_{t=1}^{n}(y(t) - x(t))^2}{n}} \quad (10)$$

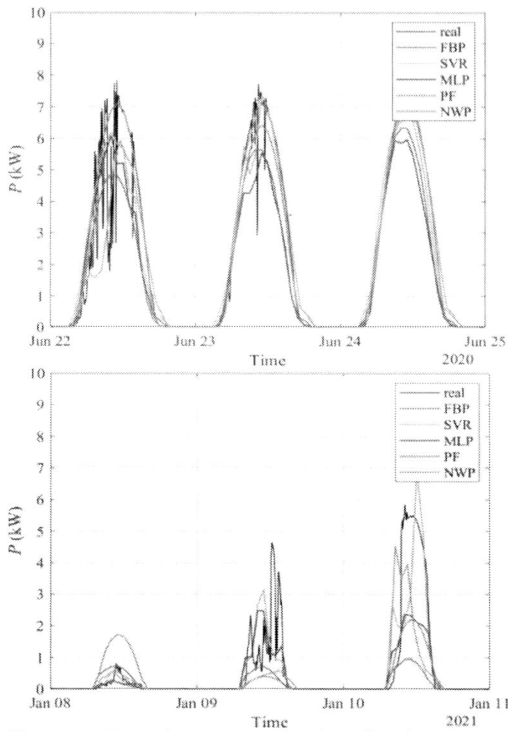

Figure 3: Exemplary measured and predicted power on three days in summer 2020 (upper panel) and winter 2021 (lower panel) comparing the FBP, SVR, MLP, PF and NWP model.

7 RESULTS

7.1 Real and Predicted Power in Winter and Summer

Before comparing the models via error metrics, example days in summer and winter are shown to give a first impression on the different models and their differing ability to react to weather patterns.

Figure 3 (upper panel) shows three exemplary days in summer 2020 with the real power in kW as the black curve and the different models in colors. The fluctuating real power curves on June 22nd and June 23rd indicate moving clouds, otherwise there are long time periods of clear-sky conditions with a peak power of 7 kW due to the PV array's tilt and orientation. Comparing the daily changing performance of models, the FBP model adapts too slowly to the almost clear-sky conditions, ranging between 4.7 kW on June 22nd and 6.5 kW on June 24th. The SVR model performs well and only underestimates the power of June 22nd and June 24th by about 0.5 kW to temporarily 1.5 kW. The MLP model underestimates all three days by 1.5 to 2 kW. The PF model predicts the power well as there are little weather changes and nearly persisting weather conditions. The NWP model also performs well and only underestimates the power of June 22nd by 1.2 kW.

Figure 3 (lower panel) shows three exemplary days in winter 2020/2021. Here, the real power fluctuates highly. The first exemplary day, January 8th, shows the typical power pattern for a day with a thick cloud layer with a maximum PV power of 0.8 kW. January 9th shows changing clouds with an occasional increase in solar irradiance with a PV power up to 4.5 kW. January 10th shows even more solar irradiance with an initial lack of light in the morning and early noon and a peak power of 6 kW. Here, the reduced peak power of about 6 kW is obvious due to winter time. For these three days, the models' adaptability to weather changes is required. The FBP model again adapts too slowly and underestimates the power of January 9th and January 10th by up to 4.5 kW. The SVR model predicts the power well. The MLP model predicts the power well, too, but underestimates the power on January 10th by far with a difference of 4.5 kW. The PF model does not cope with the weather changes well as the assumption of persisting weather conditions of the previous day do not hold true. Therefore, the PF model overestimates the power by 0.8 kW (January 8th) and underestimates the power by 4 kW (January 9th) and by 3.4 kW (January 10th). The NWP model also mostly underestimates the power by an average of 2.4 kW but performs better than the PF model.

7.2 Real and Predicted Energy in Winter and Summer

According to Equation (2), the energy is calculated for each power curve. As a result, Figure 4 (upper panel) shows the energy prediction on three exemplary days in summer. Again, there are significant differences between the models. The SVR and PF predict the energy similarly well under- or overestimating the energy at maximum by 5 kWh. The MLP and FBP perform continuously bad, underestimating the generated energy by up to 15 kWh.

Figure 4 (lower panel) shows the energy prediction on three exemplary days in winter. The most accurate energy prediction is performed by the SVR model under- or overestimating the energy at maximum by 1 kWh. The FBP and PF perform worst, under- or overestimating the energy up to 10 kWh. Most notably, the PF does not adapt due to the high differences in weather conditions between each day.

Both the power and energy forecasts on the exemplary days give the motivation to find which model works best under what circumstances.

7.3 Comparison of PF, NWP, FBP, SVR and MLP

The model comparison comprises two approaches. First, the errors of the power prediction is calculated from spring 2020 till winter 2020/2021 over one year for the time of the day (see Equations (9) and (10)).

Figure 5 shows the corresponding model errors over the day. Due to the lack of normalization, it is not possible to compare a model with itself for different times of the day. The focus lies on the comparison of the different models with each other. It stands out that there are differences between the morning, noon and evening times.

Figure 5 (upper panel) shows the *MAE* of the power predictions' error over the day. During noon time between 8 am and 3 pm (UTC), SVR performs best with a *MAE* of around 1 kW, followed by the MLP, NWP and lastly the FBP and PF models with a MAE of up to 1.75 kW. Strikingly, the performance of the models turns around at morning and evening times between about 4 am to 8 am (UTC) and 3 pm to 8 pm (UTC). Here, the PF and NWP model work best in the morning and the PF model in the evening, which indicates that a prediction based on physical equations has an advantage calculating sunrise and sunset times with more accuracy than machine-learning models can, although the FBP model is similarly good at detecting the regularities. The worst performance is observed for the MLP model with a *MAE* of up to 0.5 kW.

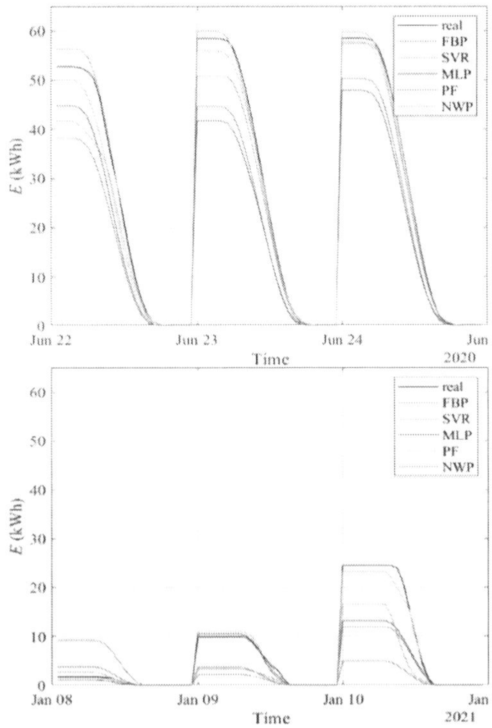

Figure 4: Exemplary measured and predicted energy for the remaining day on three days in summer 2020 (upper panel) and winter 2021 (lower panel) comparing the FBP, SVR, MLP, PF and NWP.

Table IV: Errors over one year for each model

	PF	NWP	FBP	SVR	MLP
MAE (kW)	0.51	0.47	0.55	0.42	0.51
RMSE (kW)	1.15	1.05	1.12	0.88	1.01
MAE (kWh)	4.83	3.65	4.89	1.81	3.21
RMSE (kWh)	9.24	7	8.49	3.8	5.98

Figure 5 (lower panel) shows the *RMSE* of the power predictions' error over the day. The course is similar to that of the *MAE* with the SVR model performing best during the day with a *RMSE* up to 1.5 kW. During noon time, the PF model shows to be more prone to outliers and performs slightly worse than the FBP. Similarly, the NWP is more prone to outliers than the MLP. The difference between the *RMSE* of the MLP and NWP is about 0.1 kW higher than that of the *MAE*. Still, there are high differences in the models at mornings and evenings, where the PF and NWP models based on physical equations and the FBP model with its ability to detect patterns perform better than the SVR and MLP models. Here, the difference in *RMSE* is more noticeable in the morning than the evening. The bad performance of the MLP increases from a *MAE* of about 0.5 kW to a *RMSE* of about 0.8 kW and shows to be derived especially from outliers.

In a second approach, the errors (see Equations (9) and (10)) of the power and energy are calculated over time ranges, namely the different meteorological seasons and one year.

Table IV shows the errors of power and energy forecasts for the entire year under consideration. It is obvious that the SVR model outperforms all other models irrespective of the power or energy prediction with a *MAE*

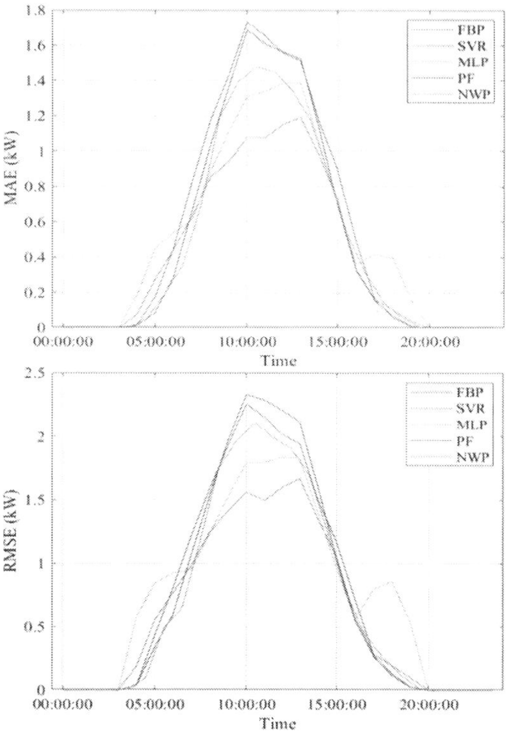

Figure 5: Error of the power prediction calculated over one year (spring 2020 – winter 2020/2021) for specific times of the day.

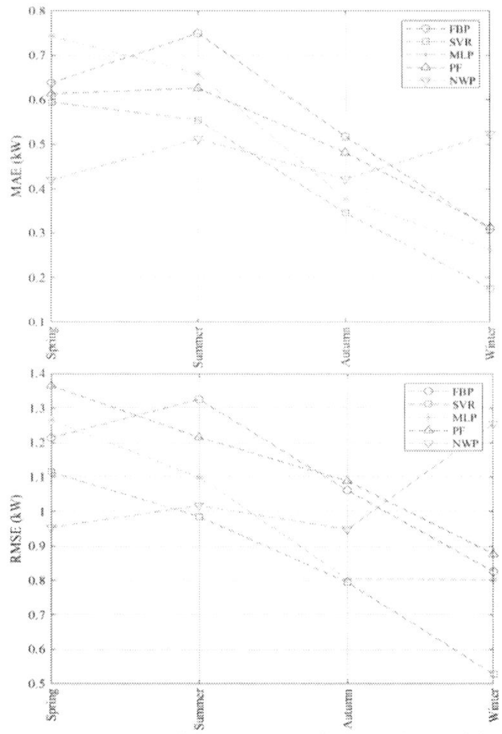

Figure 6: Error of the power prediction calculated for different seasons, namely spring, summer, autumn and winter 2020/2021.

of 0.42 kW and 3.8 kWh and a *RMSE* of 0.88 kW and 3.8 kWh. Concerning the power prediction, PF, NWP, FBP and MLP perform in a similar range with a *MAE* of about 0.5 kW. The energy prediction shows higher differences in performance, where the MLP model performs second best, followed by the NWP, FBP and lastly PF model with a *MAE* of 4.83 kWh and *RMSE* of 9.24 kWh.

In order to evaluate the accuracy of the models in more detail, the *MAE* and *RMSE* are also calculated for the different meteorological seasons.

Figure 6 (upper panel) shows the *MAE* of the power prediction over the different seasons. It is striking that the performance of the models can differ a lot depending on the time of the year. While the NWP model performs best in spring and summer with a *MAE* of 0.42 kW and 0.5 kW respectively, it performs worst in winter with about 0.52 kW. A similar contrast is observed for the MLP model which performs worst and second worst in spring and summer with 0.74 kW and 0.66 kW respectively, while it performs second best in autumn and winter with 0.4 and 0.28 kW respectively. The bad performance of the MLP in spring and summer indicates that the high errors observed in the morning and evening in Figure 5 are mostly to be related to the spring and summer time. In general, the PF and FBP model perform rather bad, while the PF's *MAE* is slightly better than that of the FBP. The SVR model performs overall well and best for autumn and winter with 0.35 and 0.18 kW respectively.

Figure 6 (lower panel) shows the *RMSE* of the power prediction over the different seasons. The difference of the course of the *RMSE* to the one of the *MAE* are to be found in a slightly better performance of the MLP and SVR with a *RMSE* between 0.1 kW and 0.35 kW lower than the respectively worst performing model, while the PF model performs slightly worse now representing the worst performing model in spring, autumn and winter with 1.36, 1.12 and 0.89 kW respectively. This indicates, that MLP and SVR are less prone to outliers, whereas PF is. The NWP model shows a decline in performance in winter from a difference of 0.2 kW *MAE* to 0.35 kW *RMSE* to the next best model which indicates that the NWP is more prone to outliers especially in winter. It is to be assumed that the partly bad performance of the NWP is related to spatial differences in the accuracy of the GFS data. Overall, the SVR model generally performs best again with a *RMSE* of 1.11, 1.0, 0.8 and 0.53 kW in spring, summer, autumn and winter respectively, with only the NWP model performing better in spring with 0.96 kW.

Figure 7 (upper panel) shows the *MAE* of the energy prediction over the different seasons. Similar to the error values over one year, the seasonal evaluation of the energy prediction shows stronger differences between the models. Especially the PF model worsens compared to the FBP model with a *MAE* of 5.8 kWh, 4.7 kWh and 3.5 kWh in spring, autumn and winter respectively and only outperforms the FBP model in summer with a *MAE* of 5.6 kWh compared to 6.4 kWh for the FBP. The MLP and SVR model clearly improve and the SVR model strengthens its superiority with a *MAE* of 1.6 kWh, 1.1 kWh and 0.85 kWh in summer, autumn and winter respectively, except for spring time, where the NWP still performs best with 2.85 kWh compared to 3.6 kWh for the SVR. The SVR's improvement is most apparent in summer and autumn.

Figure 7 (lower panel) shows the *RMSE* of the energy prediction over the different seasons. The development and stronger clarity of model differences is similar to that of the *MAE* where again the PF model shows to deal worse

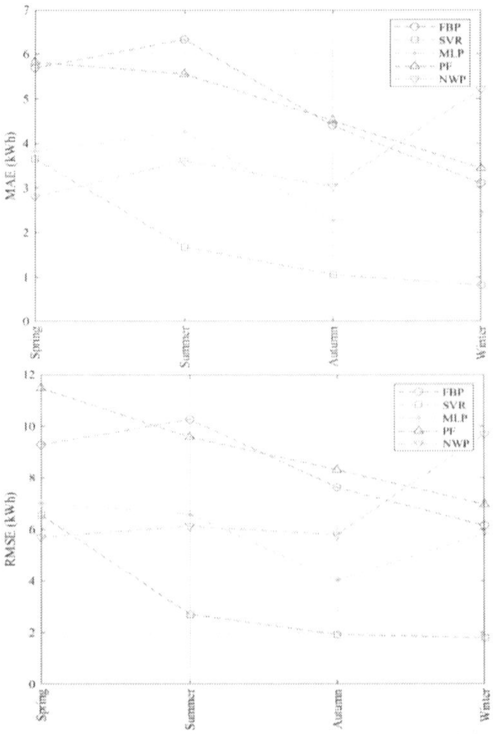

Figure 7: Error of the energy prediction calculated for different seasons, namely spring, summer, autumn and winter 2020/2021.

with outliers with a *RMSE* of 11.5 kWh, 9 kWh and 7 kWh in spring, autumn and winter respectively, where only the FBP model is worse in summer with 10.2 kWh, while the SVR does not seem to be much affected by outliers with a *RMSE* of 6.5 kWh, 2.8 kWh, 2 kWh and 1.95 kWh in spring, summer, autumn and winter respectively, only outperformed by the NWP model in spring with a *RMSE* of 5.8 kWh.

8 CONCLUSION AND OUTLOOK

The present work introduces different 1-day-ahead PV power prediction models for a PV array of 8.64 kWp at KIT with a tilt of 30° and an eastward orientation of 15°. The considered models namely the persistence forecast (PF), numerical weather prediction (NWP), Facebook prophet algorithm (FBP), support vector regression (SVR) and multilayer perceptron (MLP) are compared.

There is a striking difference of performance offered by the above-mentioned models during noon, morning and evening. The SVR and MLP models outperform the other models at noon, while the PF, (most of the time) NWP and FBP outperform the SVR and MLP at morning and evening times.

An evaluation over one year shows the SVR to be outstanding, while the model could not outperform the NWP model in spring. Other than the SVR, all other models have similar power prediction performance and only the energy prediction shows differences. The MLP performs second best, followed by NWP, FBP and lastly PF.

It is shown that an evaluation of the models' accuracy for the meteorological seasons throughout a considered year from spring 2020 till winter 2020/2021 shows significant differences of the models' performance depending on the time of the year. The strong differences in the NWP model's accuracy indicate a spatial influence of the GFS data and motivate for a differentiated use of the model depending on the location.

It is expected that with further tuning of the MLP's hyper-parameters, it's performance during morning and evening could be increased.

With the two overall most accurate models SVR and MLP, it is shown that endogenous values generally suffice, while the partial outperformance of the NWP still motivates for a further investigation into the use of exogenous input.

Concluding, it makes sense to look at the differences in times of the day and seasons, as different models perform differently well for different times of the day and year. The advantages of the different models can therefore temporally be combined into according hybrid models.

As the PV predictions are used for battery charging management, further investigations into intraday-forecasts are necessary. An according PF and NWP model is already available at the institute and will be further evaluated.

This will be part of future work.

9 REFERENCES

[1] J. Barry und J. Thomas, „Online and Offline PV Power Forecasts for Optimal Control of Storage Systems,“ *33rd European Photovoltaic Solar Energy Conference Classhighlight SpanExhibitions*, pp. 2729-2732, November 2017.

[2] B. Verma und N. Munzke, „Performance of In-house Li-ion Battery Storage Systems Based on Various Strategies,“ *EU PVSec Proceedings*, 2018.

[3] NOAA, „NCEP Products Inventory: Global Products,“ National Weather Service, [Online]. Available: https://www.nco.ncep.noaa.gov/pmb/products/gfs/. [Zugriff am 09 09 2021].

[4] R. H. Inman, H. T. C. Pedro und C. F. M. Coimbra, „Solar Forecasting Methods for Renewable Energy Integration,“ *Progress in Energy and Combustion Science*, pp. 535-576, 2013.

[5] E. Lorenz, J. Kühnert, D. Heinemann, K. P. Nielsen, J. Remund und S. C. Müller, „Comparison of Global Horizontal Irradiance Forecasts Based on Numerical Weather Prediction Models with Different Spatio-temporal Resolutions,“ *Progress in Photovoltaics: Research and Applications*, pp. 1626-1640, 2016.

[6] H. T. C. Pedro und C. F. M. Coimbra, „Assessment of Forecasting Techniques for Solar Power Production Without Exogenous Inputs,“ *Solar Energy*, pp. 2017-2028, 2012.

[7] V. Badescu, „3D Isotropic Approximation for Solar Diffuse Irradiance on Tilted Surfaces,“ *Renewable Energy*, pp. 221-233, 2002.

[8] J. Antonanzas, N. Osorio, R. Escobar, R. Urraca, F. J. Martinez-de-Pison und F. Antonanzas-Torres, „Review of Photovoltaic Power Forecasting,“ *Solar Energy*, pp. 78-111, 2016.

[9] M. Rana, I. Koprinska und V. Agelidis, „2D-interval Forecasts for Solar Power Production,“ *Solar Energy,* pp. 191-203, 2015.

[10] M. Rana und A. Rahman, „Multiple Steps Ahead Solar Photovoltaic Power Forecasting Based on Univariate Machine Learning Models and Data Re-sampling,“ *Sustainable Energy, Grids and Networks,* p. 100286, March 2020.

[11] J. A. Thomas, „Optimisation Method for the Clear Sky PV Forecast Using Power Records from Arbitrarily Oriented Panels,“ in *7th International Conference on Renewable Energy Research and Applications (ICRERA)*, Paris, 2018.

[12] D. T. Reindl, W. A. Beckman und J. A. Duffie, „Diffuse Fraction Correlations,“ *Solar Energy,* pp. 1-7, 1990.

[13] P. G. Loutzenhiser, H. Manz, C. Felsmann, P. A. Strachan, T. Frank und G. M. Maxwell, „Empirical Validation of Models to Compute Solar Irradiance on Inclined Surfaces for Building Energy Simulation,“ *Solar Energy,* pp. 254-267, 2007.

[14] A. Mellit, A. Massi Pavan, E. Ogliari, S. Leva und V. Lughi, „Advanced Methods for Photovoltaic Output Power Forecasting: A Review,“ *Applied Sciences,* p. 487, 2020.

[15] Y. LeCun, Y. Bengio und G. Hinton, „Deep Learning,“ *Nature,* pp. 436-444, 2015.

[16] M. M. Hasan Shawon, S. Akter, M. K. Islam, S. Ahmed und M. M. Rahman, „Forecasting PV Panel Output Using Prophet Time Series Machine Learning Model,“ *2020 IEEE Region 10 Conference (Tencon),* 2020.

[17] H. Long, Z. Zhang und Y. Su, „Analysis of Daily Solar Power Prediction with Data-driven Approaches,“ *Applied Energy,* pp. 29-37, 2014.

[18] M. Abuella und B. Chowdhury, „Solar Power Forecasting Using Support Vector Regression,“ in *arXiv preprint,* 2017.

[19] S. Theocharides, G. Makrides, G. E. Georghiou und A. Kyprianou, „Machine Learning Algorithms for Photovoltaic System Power Output Prediction,“ in *2018 IEEE International Energy Conference (ENERGYCON)*, 2018.

[20] W. De Soto, S. A. Klein und W. A. Beckman, „Improvement and Validation of a Model for Photovoltaic Array Performance,“ *Solar Energy,* pp. 78-88, 2006.

[21] J. E. Hay, „Calculating Solar Radiation for Inclined Surfaces: Practical Approaches,“ *Renewable Energy,* pp. 373-380, 1993.

[22] B. Y. H. Liu und R. C. Jordan, „The Interrelationship and Characteristic Distribution of Direct, Diffuse and Total Solar Radiation,“ *Solar Energy,* pp. 1-9, 1960.

[23] W. Holmgren, C. Hansen und M. Mikofski, „pvlib Python: A Python Package for Modeling Solar Energy Systems,“ *Journal of Open Source Software,* p. 884, 2018.

[24] J. Brownlee, Introduction to Time Series Forecasting with Python: How to Prepare Data and Develop Models to Predict the Future, Machine Learning Mastery, 2017.

[25] I. Goodfellow, Y. Bengio und A. Courville, Deep Learning, MIT Press, 2016.

[26] R. Perez, P. Ineichen, K. Moore, M. Kmiecik, C. Chain, R. George und F. Vignola, „A New Operational Model for Satellite-derived Irradiances: Description and Validation,“ *Solar Energy,* pp. 307-317, 2002.

[27] P. Ineichen und R. Perez, „A New Airmass Independent Formulation for the Linke Turbidity Coefficient,“ *Solar Energy,* pp. 151-157, 2002.

[28] E. Skoplaki und J. A. Palyvos, „On the Temperature Dependence of Photovoltaic Module Electric Performance: A Review of Efficiency/Power Correlations,“ *Solar Energy,* pp. 614-624, 2009.

[29] D. T. Reindl, W. A. Beckman und J. A. Duffie, „Diffuse Fraction Correlations,“ *Solar Energy,* pp. 1-7, 1990.

[30] M. Awad und R. Khanna, „Efficient Learning Machines: Theories, Concepts and Applications for Engineers and System Designers,“ *Springer nature,* p. 268, 2015.

[31] K.-L. Du und M. N. S. Swamy, „Multilayer Perceptrons: Architecture and Error Backpropagation,“ *Neural Networks and Statistical Learning,* pp. 97-141, 2019.

[32] D. L. Chester, „Why Two Hidden Layers are Better Than One,“ *Proceedings IJCNN,* pp. 265-268, 1990.

[33] F. Hutter, H. Hoos und K. Leyton-Brown, „An Efficient Approach for Assessing Hyperparameter Importance,“ in *International Conference on Machine Learning*, 2014.

[34] M. Rana, I. Koprinska und V. Agelidis, „Univariate and Multivariate Methods for Very Short-term Solar Photovoltaic Power Forecasting,“ *Energy Conversion and Management,* pp. 380-390, 1 August 2016.

[35] A. Mellit und A. M. Pavan, „A 24-hour Forecast of Solar Irradiance Using Artificial Neural Network: Application for Performance Prediction of a Grid-connected PV plant at Trieste, Italy,“ *Solar Energy,* pp. 807-821, 2010.

[36] S. J. Taylor und B. Letham, „Forecasting at Scale,“ *Peer Preprints,* 2017.

SPATIO-TEMPORAL MACHINE LEARNING METHODS FOR MULTI-SITE PV POWER FORECASTING

Rafael E. Carrillo, Baptiste Schubnel, Jelena Simeunovic, Renaud Langou and Pierre-Jean Alet
CSEM PV-Center
Rue Jacquet-Droz 1, 2002 Neuchâtel, Switzerland
Emails: *rafael.carrillo@csem.ch, baptiste.schubnel@csem.ch, jelena.simeunovic@csem.ch, renaud.langou@csem.ch, pierre-jean.alet@csem.ch*

ABSTRACT: In this paper we present recent advances at CSEM in multi-site PV forecasting that rely entirely on past production data. The presented methods take a graph signal processing perspective and exploit the intuition that a dense network of PV stations provides enough information for short-term prediction horizons (up to six hours ahead) with a high temporal and spatial resolution. The proposed forecasting models are based on recent advances in graph machine learning to capture the spatio-temporal dependencies of the PV production data. The proposed methods were evaluated in two data sets for an entire year: 1) production data from 304 real PV systems, and 2) simulated production of 1000 PV systems, in both cases distributed over Switzerland. The best of the proposed methods achieves a daytime NRMSE (over all sites) of 15.5% for the real data set and 12% for the synthetic data set for 6 hours ahead prediction and 7.5% for the real data set and 3.5% for the synthetic data set for 15 minutes ahead prediction. Furthermore, the proposed non-linear methods outperform state-of-the-art methods with NWP as inputs on time horizons between 0 h and 5 h.
Keywords: Solar radiation, Photovoltaic, Forecasting, Machine Learning.

1 INTRODUCTION

Power production forecasting is a critical technology for enabling large-scale penetration of photovoltaic (PV) generation into the power grid. Improved forecasting leads to lower net generation costs in the power system and to reduced curtailment of PV production. Forecasting accuracy is limited both by the chaotic nature of the weather and by the uncertainty on the physical response of PV systems to given weather conditions due to variable temperature coefficients, non-linear behavior at low irradiance, or complex local shading. Therefore, state-of-the-art PV forecasting approaches combine machine learning models with numerical weather predictions (NWP) that in general have a limited spatial resolution, which poses a challenge for accurate production forecasting [1].

Many approaches have been proposed to improve PV forecasting accuracy with higher spatial resolution. Since the high variability of solar irradiance mainly comes from cloud movement and its stochastic blocking of sunlight, state-of-the-art methods have focused on predicting the cloud dynamics using ground-based cameras, satellite images, and NWP [2], [3]. While ground-based cameras provide precise local information, they are expensive to deploy and maintain over a large number of PV stations. On the other hand, satellite based-images and NWP provide wide-area observational data and excel at long-term forecasts, but perform rather poorly for short-term and high spatial resolution forecasts. Precise local numerical weather forecasts may be accessible by dedicated meteorological providers, but are often very costly to acquire or require heavy processing. Therefore, recent studies have proposed data-driven methods that use multi-site spatio-temporal past data for PV forecasting [4]–[7].

In this paper we present graph-based machine learning methods developed at CSEM for multi-site PV forecasting which rely entirely on production data and exploit the intuition that PV systems provide a dense network of simple weather stations. Graph signal processing (GSP) is an emerging field that allows the processing of signals defined in irregular domains by using graphs to capture their interdependence. Recently, graph neural networks (GNN) have attracted a lot of attention due their expressive power and ability to infer information from complex data [8]. Hence, we use graphs to model the spatio-temporal dependencies among PV systems. The underlying idea in our methods is that temporal correlations between the outputs of PV systems are linked with their spatial relationship, and thus PV systems can effectively be used as ground-level weather stations to forecast the production of their neighbors.

The present work focuses on the short-term prediction horizon (six hours ahead, 15 minutes resolution). We present one linear spatio-temporal autoregressive (STAR) model and two GNNs, the graph-convolutional long short term memory (GCLSTM) and the graph-convolutional transformer (GCTrafo) models. The proposed methods were evaluated in two data sets for an entire year: 1) production data from 304 real PV systems, and 2) simulated production of 1000 PV systems, in both cases distributed over Switzerland. The proposed methods were benchmarked against state-of-the-art methods for multi-site PV forecasting and single-site methods that use NWP as inputs.

2 METHODOLOGY

2.1 Forecasting models

The spatio-temporal autoregressive (STAR) model was introduced in [6]. It consists of an auto-regressive linear model that uses a group Lasso estimator to select the relevant predictor plants for each individual production node. Model coefficients are fitted over two months of data and are used to predict power production for the next two weeks period. After two weeks, the training data are shifted, the models are refitted and the process continues. The STAR model uses power production data from the past 3 hours as input to the model. A key point to achieve good results with this method is the use of a tailor made data-driven normalization for each plant signal, that is able to extract the clouds mask from the original power signals. Further details of this model can be found in [6].

The GNN models were introduced in [7]. The Graph-

Convolutional Long Short Term Memory (GCLSTM) model is based on an encoder-decoder LSTM architecture. The encoder LSTM is used to encode the past data information (similar to a Kalman filter) and the decoder uses this encoded information to make the prediction over the entire horizon. LSTM standard matrix multiplications are replaced by graph convolutions to model spatial information spreading between the PV systems. In our GCLSTM implementation, the Laplacian is learnt in the course of training, starting from a fixed but large nearest neighbor graph. The Graph-Convolutional Transformer (GCTrafo) model is inspired by attention-based models used in natural language processing [9]. It uses a multi-head convolutional mechanism with a learnt graph Laplacian operator at each head level. Using this method, the different heads are able to focus on different dynamic scales and increase prediction power. As opposed to the STAR model, GCLSTM and GCTrafo don't use a tailor made normalization scheme, but just apply standard normalization procedures and add astronomical clear-sky data over the prediction horizon to the PV plant production signals as an additional source of information. Both models use production data from the past 4 hours. For further details on the GCLSTM and the GCTrafo models we refer the reader to [7].

2.2 Datasets and training

Two datasets are used in our study. The first dataset, dubbed the real dataset, consists of 304 PV plants across Switzerland, recorded by Solar-Log devices from BKW's subsidiary Solare Datensysteme GmbH (SDS). The PV plants are spread inhomogeneously over the entire country, with a density reflecting the population density. The second dataset, dubbed the synthetic dataset, has 1000 PV plants and has been generated with statistical models that match the statistics of the real dataset in terms of location density, size, orientation and pitch angles. Production time series were simulated using historical weather data (with high temporal and spatial resolution) and the generated plant characteristics using PVlib package. Both datasets have a 15 minutes resolution for the years 2016-2017.

Both GCLSTM and GCTrafo architectures are trained using the entire 2016 year and evaluated on the entire 2017. The GNN models were trained on a single GPU using stochastic gradient descent. The STAR model is evaluated in the entire 2017 year and its coefficients are updated every two weeks using the prior two months for fitting.

2.3 Evaluation and metrics

The proposed models are compared over the year 2017 on the two datasets. Both GCLSTM and GCTrafo architectures are trained on the first year of available data (2016). The STAR model is trained with the two months prior to the evaluation period. We use as evaluation metric the normalized root mean-squared error (NRMSE), normalized by the peak production over the evaluation period. Night times are excluded from the error computations.

3 RESULTS

3.1 Multi-site comparison

We start by comparing the proposed models (STAR, GCLSTM, GCTrafo) for multi-site PV power forecast. Figure 1 and Figure 2 show the forecast error evolution over the prediction horizon (in discrete time steps of 15 minutes) for the real dataset and the synthetic dataset, respectively. We use as benchmark the spatio-temporal convolutional neural network (STCNN) model from [5]. For each prediction step the median value (solid line) and the interquartile distance (shadow bounds) over all sites are shown. Results show that in the real dataset GCLSTM outperforms all other methods for the entire prediction horizon. However, the GCTrafo model outperforms GCLSTM and STAR models for middle range predictions (4 to 6 hours ahead) in the synthetic dataset.

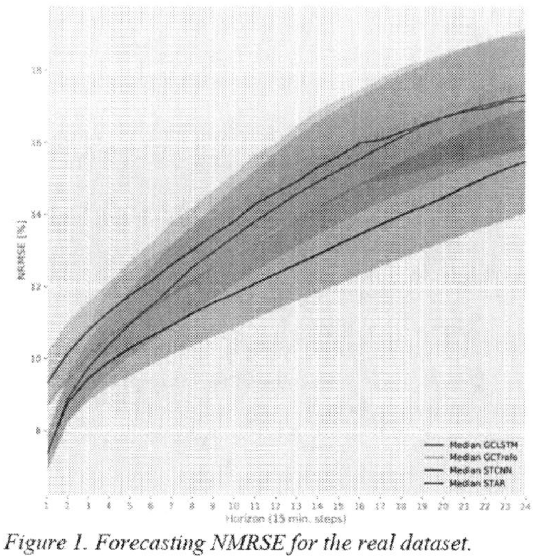

Figure 1. Forecasting NMRSE for the real dataset.

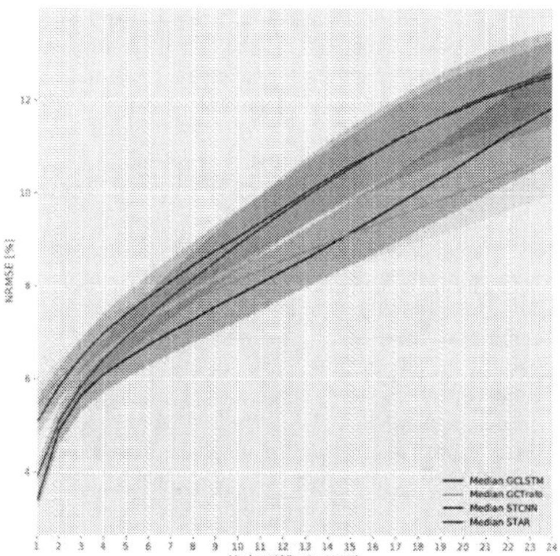

Figure 2. Forecasting NMRSE for the synthetic dataset.

3.2 Single-site comparison

For the real dataset, we have also included the forecasts for single sites and benchmarked the performance of the proposed methods against two state-of-the-art method that uses NWP for two sites in the central part of Switzerland: Bern and Bätterkinden. As comparison we use forecasts based on support vector regressors (SVR) with NWP as inputs (global irradiance and temperature) [10] and an encoder-decoder LSTM (EDLSTM) that uses both past production data from the site and NWP (global irradiance and temperature) for the prediction horizon [7]. The forecasts for Bern were computed using historical NWP from the global forecast system (GFS) that have a temporal resolution of 3 hours. The forecasts for Bätterkinden were computed using historical NWP from Meteotest with a temporal resolution of 1 hour. Figure 3 and Figure 4 show the results of the comparison for Bätterkinden and Bern, respectively. We observe that in both cases, the GNN models (GCLSTM and GCTrafo) outperform the single site predictions on time horizons between 0 and 5 hours and are better at capturing local phenomenon, whereas the linear auto-regressive model degrades faster and is outperformed by the single site methods after 2 hours ahead.

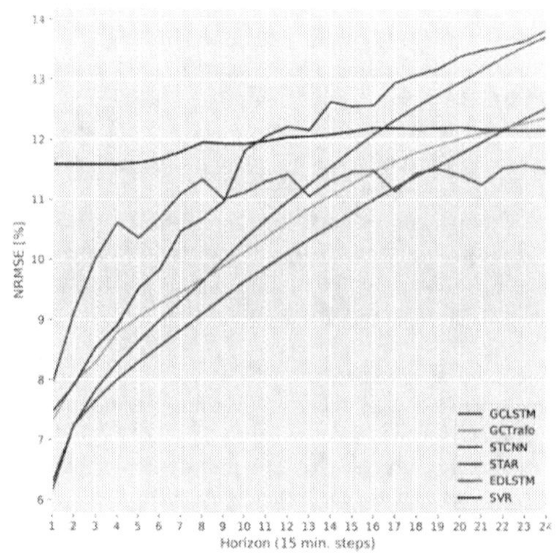

Figure 3. Forecast NMRSE for Bätterkinden. Single-site methods use NWP with 1 hour resolution.

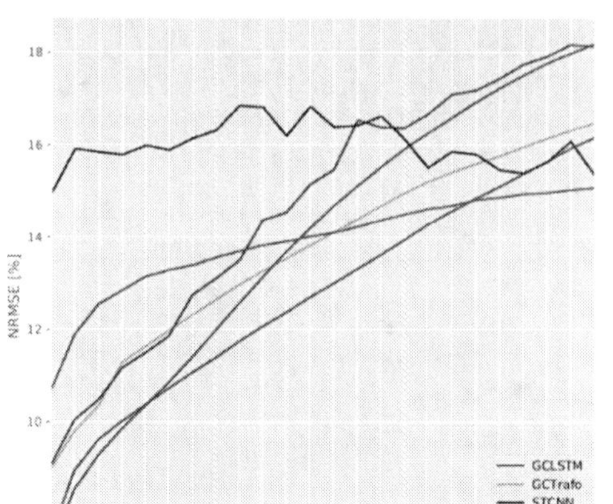

Figure 4. Forecasting NMRSE for Bern. Single-site methods use NWP with 3 hours resolution.

4 CONCLUSIONS

In this paper, we presented and compared novel graph-based machine learning methods for short-term (six hours ahead, 15 minutes resolution) multi-site PV production forecast that exploit the spatio-temporal correlation between different systems. The best of the proposed methods achieved a daytime NRMSE (over all sites) of 15.5% for the real data set and 12% for the synthetic data set for 6 hours ahead prediction and 7.5% for the real data set and 3.5% for the synthetic data set for 15 minutes ahead prediction. We showed that nonlinear methods using learnable graph convolutions outperform linear methods for intra-day forecasts. Moreover, the non-linear methods outperform state-of-the-art methods with NWP as inputs on time horizons between 0 and 5 hours, confirming the intuition that PV systems can effectively be used as distributed weather stations for forecasting purposes. Both GNN models were trained on a single GPU for the 1000 nodes case and can be scaled to a higher number of nodes using multiGPU computing, making them appealing for grid management applications with large number of nodes. A live PV forecasting demonstrator with 650 nodes in the Netherlands was implemented using the GNN models (http://portal.csem.ch:9107/).

5 REFERENCES

[1] J. Antonanzas, N. Osorio, R. Escobar, R. Urraca, F. J. Martinez-de-Pison, and F. Antonanzas-Torres, 'Review of photovoltaic power forecasting', *Solar Energy*, vol. 136, pp. 78–111, Oct. 2016.

[2] R. Marquez and C. F. M. Coimbra, 'Intra-hour DNI forecasting based on cloud tracking image analysis', *Solar Energy*, vol. 91, pp. 327–336, May 2013.

[3] H. S. Jang, K. Y. Bae, H.-S. Park, and D. K. Sung, 'Solar Power Prediction Based on Satellite Images and Support Vector Machine', *IEEE Transactions on Sustainable Energy*, vol. 7, no. 3, pp. 1255–1263, Jul. 2016.

[4] X. G. Agoua, R. Girard, and G. Kariniotakis, 'Short-Term Spatio-Temporal Forecasting of Photovoltaic Power Production', *IEEE Transactions on Sustainable Energy*, vol. 9, no. 2, pp. 538–546, Apr. 2018.

[5] J. Jeong and H. Kim, 'Multi-Site Photovoltaic Forecasting Exploiting Space-Time Convolutional Neural Network', *Energies*, vol. 12, no. 23, Art. no. 23, Jan. 2019.

[6] R. E. Carrillo, M. Leblanc, B. Schubnel, R. Langou, C. Topfel, and P.-J. Alet, 'High-Resolution PV Forecasting from Imperfect Data: A Graph-Based Solution', *Energies*, vol. 13, no. 21, Art. no. 21, Jan. 2020.

[7] J. Simeunović, B. Schubnel, P.-J. Alet, and R. E. Carrillo, 'Spatio-temporal graph neural networks for multi-site PV power forecasting', *arXiv:2107.13875 [cs, eess]*, Jul. 2021, Accessed: Jul. 30, 2021. [Online]. Available: http://arxiv.org/abs/2107.13875

[8] A. Ortega, P. Frossard, J. Kovačević, J. M. F. Moura, and P. Vandergheynst, 'Graph Signal Processing: Overview, Challenges, and Applications', *Proceedings of the IEEE*, vol. 106, no. 5, pp. 808–828, May 2018.

[9] A. Vaswani *et al.*, 'Attention is all you need', in *Proceedings of the 31st International Conference on Neural Information Processing Systems*, Long Beach, California, USA, Dec. 2017, pp. 6000–6010.

[10] M. Boegli, M. Pierro, D. Moser, and P.-J. Alet, 'Machine learning techniques for forecasting single-site PV production', Brussels, Sep. 2018.

ASSESSMENT OF CLOUD MASK FORECASTS FROM THE WRF-SOLAR ENSEMBLE PREDICTION SYSTEM

Jaemo Yang[1], Ju-Hye Kim[2], Manajit Sengupta[1], Pedro A. Jimenez[2], Yu Xie[1]

[1]National Renewable Energy Laboratory, Golden, CO, USA
[2]National Center for Atmospheric Research, Boulder, CO, USA

ABSTRACT:
Numerical weather prediction (NWP) models are important tools used by government agencies and the renewable energy enterprise to forecast solar radiation. Cloud prediction, a key process of NWP models, has the highest impact on the accuracy of solar forecasting. This study uses satellite observations from the National Solar Radiation Database (NSRDB) to evaluate the cloud mask forecast by the Weather Research and Forecasting-Solar Ensemble Prediction System (WRF-Solar EPS). Preliminary analysis of the data in 2018 demonstrates the need for further improvement in predicting thin and low-level clouds. The information obtained from this work will be used to enhance WRF-Solar EPS in reproducing the cloud field over the contiguous U.S. and reducing solar forecasting errors.

1 Aim and Approach

The main purpose of this study is to develop a method to evaluate cloud forecasts from numerical weather prediction (NWP) models. The limitation in the current verification of gridded solar forecasts, which focus only on the evaluation of NWP outputs for the prediction of solar irradiance using simple statistical metrics, has motivated a new approach in this study to assess the prediction accuracy of solar irradiance. Forecasting of clouds in NWP is the key factor in predicting solar irradiance because that directly impacts the solar irradiance under overcast or partial cloudy-sky conditions (i.e., solar irradiance forecasts are influenced by errors stemming from the cloud forecasts); thus, it is essential to implement an evaluation of cloud forecasts from NWP models.

In addition, forecast developers and users need information for model performance at arbitrary locations on the NWP model grid. If we use only ground observations for the model evaluations, it is not possible to acquire the spatial distribution of forecast errors on the model grid. In this case, satellite-derived solar radiation data offer the opportunity for an in-depth analysis to assess gridded forecasts over a wide range of regions.

In this study, we suggest a method for validating cloud products from the Weather Research and Forecasting-Solar Ensemble Prediction System (WRF-Solar EPS), which is an enhanced version of WRF-Solar [1], to provide grid operators with high-quality probabilistic solar forecasts. This work includes three objectives:

- Assessment of cloud mask forecasts from WRF-Solar against the National Solar Radiation Database (NSRDB)
- Analysis of cloud detection metrics (e.g., hit rate, false alarm rate, Kuiper's skill score)
- Development of evaluation techniques to verify WRF-Solar EPS using the cloud detection metrics.

2 Scientific Innovation and Relevance

2.1 WRF-Solar EPS

WRF-Solar (https://ral.ucar.edu/projects/wrf-solar) [1] is the state-of-the-art NWP model that provides prediction of solar irradiance with global horizontal irradiance (GHI), direct normal irradiance, and diffuse horizontal irradiance specialized for solar power applications. Recently, WRF-Solar EPS) [2] was developed through a collaboration between the National Renewable Energy Laboratory and the National Center for Atmospheric Research under a U.S. Department of Energy project to provide probabilistic forecast information on solar irradiance. WRF-Solar EPS generates ensemble members by using a technique to stochastically perturb key forecasting variables identified through the tangent linear sensitivity analysis [3] for six WRF-Solar modules, controlling the solar irradiance and the formation and dissipation of clouds.

2.2 Satellite Observations

The NSRDB [4] is used for the evaluation of cloud mask forecasts from WRF-Solar EPS. The NSRDB is a publicly available data set of satellite-derived observations that provides meteorological data in North and Central America (https://nsrdb.nrel.gov/). Because the domain of WRF-Solar EPS covers the contiguous United States (CONUS) as well portions of the Pacific and Atlantic oceans, we reprocessed the currently available NSRDB to produce additional data over the ocean. This work aims to evaluate the forecasts of WRF-Solar EPS for its full extent, including the ocean and CONUS. In addition, the NSRDB has been aggregated from the native NSRDB grid (2-km or 4-km resolution based on year) to the WRF-Solar grid (9 km) to efficiently handle the two big data sets (Fig. 1).

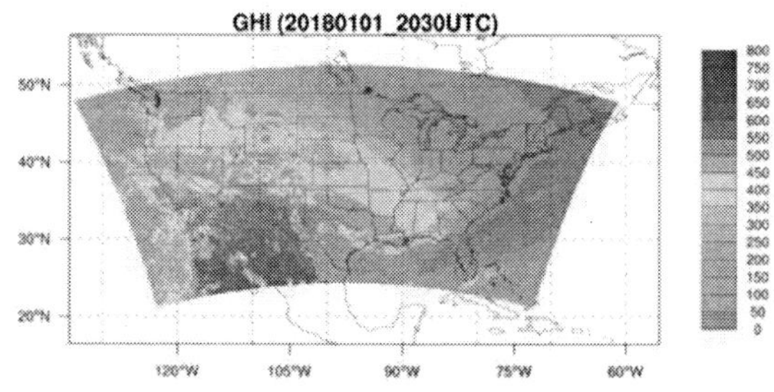

Figure 1. GHI from 9-km NSRDB aggregated to WRF-Solar EPS grid (20:30 UTC 01 January 2018).

2.3 Data Processing for NSRDB and WRF-Solar EPS

To analyze the capability of WRF-Solar EPS in predicting the cloud mask, we need to filter the NSRDB and the forecasts of WRF-Solar EPS for clear- and cloudy-sky conditions. The following are the main steps of the filtering algorithm:

- Step 1: An absolute difference between clear-sky GHI and all-sky GHI is calculated for each individual pixel and time step:

$$DIFF = |GHI_{Clear} - GHI_{All\ Sky}| \quad (1)$$

- Step 2: Each pixel is filtered by the conditions in Table 1. If DIFF from Eq. (1) is smaller than 1 W/m^2, the pixel is treated as clear sky; otherwise, the pixel is treated as cloudy sky. For ensemble forecasts simulated from WRF-Solar EPS, additional criteria are employed. For example, if more than 50% of the ensemble members satisfy the condition of DIFF $\geq$ 1.0 W/m^2, that pixel is treated as cloudy sky. Note that nighttime data are excluded by filtering with a threshold for inclusion (0° < solar zenith angle < 85°) in this analysis.

Table 1. Criteria for data-processing of NSRDB and WRF-Solar EPS for clear-/cloudy-sky conditions

	NSRDB	**WRF-Solar EPS**
Clear sky	DIFF < 1.0 W/m^2	> 50% of ensemble members are: DIFF < 1.0 W/m^2
Cloudy sky	DIFF $\geq$ 1.0 W/m^2	$\geq$ 50% of ensemble members are: DIFF $\geq$ 1.0 W/m^2

After filtering is completed, the cloud detection metrics are calculated based on a contingency table (Table 2). The four categories in Table 2 indicate binary representations for the cloud occurrences for the NSRDB and WRF-Solar EPS. We compute the total number of frequencies that corresponded to each category (i.e., a, b, c, and d) and then calculate the following metrics [5] to quantify the results:

$$POD_{clear} = \frac{a}{a+b} \times 100\% \quad (2)$$

$$POD_{cloudy} = \frac{d}{c+d} \times 100\% \quad (3)$$

$$FAR_{clear} = \frac{c}{a+c} \times 100\% \quad (4)$$

$$FAR_{cloudy} = \frac{b}{b+d} \times 100\% \quad (5)$$

$$HR = \frac{a+d}{a+b+c+d} \times 100\% \quad (where\ 0 \leq HR \leq 100\%) \quad (6)$$

$$KSS = \frac{a \cdot d - c \cdot b}{(a+b) \cdot (c+d)} \times 100\% \quad (where\ -100\% \leq KSS \leq 100\%) \quad (7)$$

Table 2. Contingency matrix for WRF-Solar EPS and NSRDB.

		WRF-Solar EPS	
NSRDB	Scenario	Clear	Cloudy
	Clear	a	b
	Cloudy	c	d

3 Results

We analyze the hit rate (HR) to quantify the performance of the cloud mask forecasts from 10 stochastic-based ensemble members simulated from WRF-Solar EPS. The monthly variation of hit rate computed with all pairs of observations and predictions for 2018 is presented in Fig. 1. Figure 1a shows that the hit rates are smaller in summer than in the other seasons. Two examples of spatial distribution of hit rate in February and July are shown in Figs. 1b and 1c. In February 2018, some regions show high hit rates (>80%), which can be attributed to winter storms on the east part of CONUS (Fig. 1b). Apparently, WRF-Solar EPS accurately represents nonconvective events, as shown from the high hit rates in February 2018, compared to the low hit rates over the central part of the U.S. unrepresented events in July 2018 (Fig. 1c). The unrepresented events are apparently related to the higher uncertainty in predicting the location and timing of the occurrence of convection in summer.

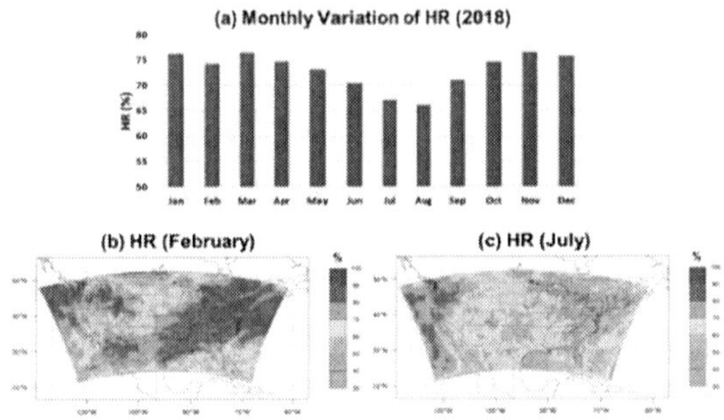

Figure 2. (a) Monthly variation of hit rate for 2018 and spatial distribution of hit rate in (b) February and (c) July.

Because the hit rate represents an overall efficiency of the model performance for both clear- and cloudy-sky conditions, the probability of detection (POD) is analyzed separately for the two sky conditions. Figure 3 exhibits an example of POD_{clear} and POD_{cloudy} analyzed for July 2018. As expected, WRF-Solar EPS provides high POD_{clear} scores (>80%) for CONUS in July (Fig. 3a). On the other hand, there are POD_{cloudy} values that are lower than 40% in the central parts of the United States (Fig. 3b). As stated previously, we found that a large portion of mismatched clouds under cloudy-sky conditions, indicated by the low POD_{cloudy}, resulted in the the regions of low hit rates in Figure 2c.

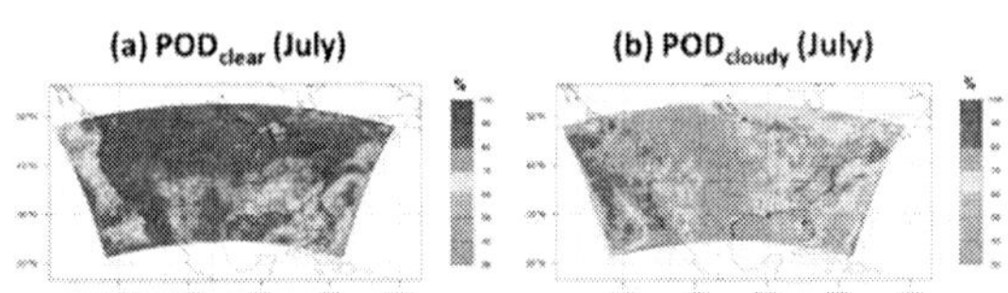

Figure 3. Spatial distribution of POD in (a) clear-sky and (b) cloudy-sky conditions in July 2018.

As shown in Fig. 3, WRF-Solar EPS produces more mismatched pixels for cloud mask forecasts in cloudy-sky than clear-sky conditions; therefore, we need to investigate which types of clouds cause difficulties in predicting cloud masks from WRF-Solar EPS. To further assess the cloud mask forecasts of WRF-Solar EPS, we analyzed POD_{cloudy} classified into various cloud optical depth ranges with data obtained from the NSRDB (Fig. 4a). In terms of the capability of WRF-Solar EPS in predicting clouds, POD_{cloudy} increases as optical depth increases, and this pattern occurs for all seasons. This indicates that we will need to focus on further improvements in the prediction of thin clouds in WRF-Solar EPS. Because the cloud types can be grouped by altitude of occurrence, we additionally analyzed the performance of WRF-Solar separately for low-, middle- and high-level clouds. For this analysis, we calculated false alarm rate (FAR) in cloudy conditions as a metric to evaluate cloud mask forecasts for the three levels of cloud heights. The cloud top height (CTH) extracted from the WRF-Solar runs was used to classify the clouds into three groups: CTH less than 2 km are classified as low-level clouds, CTH between 2 km and 6 km are classified as middle-level clouds, and the rest are classified as high-level clouds. WRF-Solar produces higher FAR_{cloudy} when predicting low-level clouds compared to FAR_{cloudy} for high-level clouds (Fig. 4b). Consistent with the results of POD_{cloudy} shown in Fig. 4a, patterns of the classified FAR_{cloudy} for different CTHs are similar throughout the seasons, while noting that WRF-Solar produces lower POD_{cloudy} and higher FAR_{cloudy} in summer than winter.

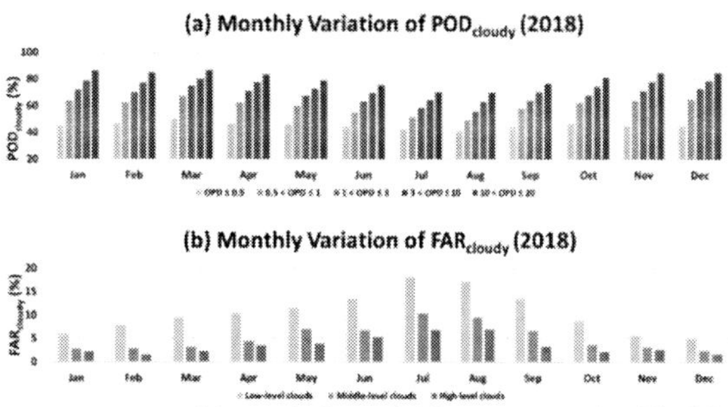

Figure 4. Monthly variation of (a) POD_{cloudy} classified in different cloud optical depths and (b) FAR_{cloudy} classified in three cloud levels for 2018.

4 Conclusions

In this research, we present a new approach for evaluating cloud mask forecasts simulated from WRF-Solar EPS. The newly processed NSRDB data sets, which were aggregated to the 9-km WRF-Solar grid, are used to assess the model performance in predicting clouds over the ocean as well as for CONUS. Preliminary results indicate that we will need further development in WRF-Solar to improve the prediction skills for low-level and thin clouds. Future studies will be needed to consider the lack of skill in detecting thin and low-level clouds from satellites. The proposed evaluation method is proven to be useful when assessing cloud mask forecasts from WRF-Solar EPS and when quantifying its performance by using satellite-derived observations, and it can be extended to forecast evaluations in general.

Acknowledgments

This work was authored by the National Renewable Energy Laboratory, operated by Alliance for Sustainable Energy, LLC, for the U.S. Department of Energy (DOE) under Contract No. DE-AC36-08GO28308. Funding provided by the U.S. Department of Energy Office of Energy Efficiency and Renewable Energy Solar Energy Technologies Office. The views expressed in the article do not necessarily represent the views of the DOE or the U.S. Government. The U.S. Government retains and the publisher, by accepting the article for publication, acknowledges that the U.S. Government retains a nonexclusive, paid-up, irrevocable, worldwide license to publish or reproduce the published form of this work, or allow others to do so, for U.S. Government purposes.

This material is also based upon work supported by the National Center for Atmospheric Research, which is a major facility supported by the National Science Foundation under Cooperative Agreement No. 1852977.

References

[1] Jimenez, P.A., J.P. Hacker, J. Dudhia, S.E. Haupt, J.A. Ruiz-Arias, C.A. Gueymard, G. Thompson, T. Eidhammer and A. Deng, 2016: WRF-Solar: Description and Clear-Sky Assessment of an Augmented NWP Model for Solar Power Prediction. Bulletin of the American Meteorological Society, 97, 1249-1264.

[2] Kim, J.H., P.A. Jimenez, J. Dudhia, J. Yang, M. Sengupta, and Y. Xie, 2020: Probabilistic forecast of all-sky solar radiation using enhanced WRF-Solar: Preprint. Golden, CO: National Renewable Energy Laboratory. NREL/CP-5D00-77693. https://www.nrel.gov/docs/fy20osti/77693.pdf.

[3] Yang, J., J.H. Kim, P.A. Jimenez, M. Sengupta, J. Dudhia, Y. Xie, A. Golnas, and R. Giering, 2020: An efficient method to identify uncertainties of WRF-Solar variables in forecasting solar irradiance using a tangent linear sensitivity analysis. Solar Energy, 220, 509-522.

[4] Sengupta, M., Y. Xie, A. Lopez, A. Habte, G. Maclaurin, and J. Shelby, 2018: The national solar radiation data base (NSRDB). Renewable and Sustainable Energy Reviews, 89, 51-60.

[5] Karlsson, K.G. and E. Johansson, 2013: On the optimal method for evaluating cloud products from passive satellite imagery using CALIPSO-CALIOP data: example investigating the CM SAF CLARA-A1 dataset. Atmospheric Measurement Techniques, 6(5), 1271-1286.

A HYBRID SOLAR IRRADIANCE NOWCASTING APPROACH: COMBINING ALL SKY IMAGER SYSTEMS AND PERSISTENCE IRRADIANCE MODELS FOR INCREASED ACCURACY

Bijan Nouri[1*], Niklas Blum[1], Stefan Wilbert[1], Luis F. Zarzalejo[2]
[1] DLR, Institute of Solar Research, Paseo de Almería 73-2, 04001 Almería, Spain
[2] CIEMAT Energy Department – Renewable Energy Division, Av. Complutense 40, 28040 Madrid, Spain

ABSTRACT: New technical challenges arise due to the growing share of distributed and intermittent solar power generation. Intra-minute and intra-hour variabilities, when insufficiently accounted for, could lead to grid instabilities. Shortest-term forecasts, often referred as nowcast, can anticipate such variabilities with high resolutions in space and time. Therefore, nowcasts are suited for fine-grained control applications in solar power plants and electrical grids. Highly resolve nowcasts can be obtained by all sky imager (ASI) systems. Sky conditions are analyzed and irradiance nowcasts a derived from the hemispherical sky images. In this work, ASI based nowcasts and persistence based nowcasts are combined to a hybrid nowcasts approach. This hybrid approach archived a reduction of the root mean square deviation of 12% over a 62-day lasting validation data set.

Keywords: Grid Integration, Grid Stability, Pyranometer, Solar Irradiance Nowcasts, All Sky Imager

1 AIM AND APPROACH USED

Reliable nowcasts of the solar irradiance become increasingly important, considering the rapid growth of solar power penetration. Applications include energy grid control [1] as well as ramp rate control within utility scale PV power plants [2]. Intra-hour and intra-minute variability of the incoming solar irradiance is mainly caused by clouds. Highly resolved observations of the sky can be obtained by all sky imagers (ASI), consisting of upward facing cameras with fisheye lenses. In recent years, methods have been developed to automatically detect clouds [3], evaluate the cloud radiative effect [4] and identify cloud motion vectors from such sky images [5]. If multiple cameras with overlapping viewing angles are available, cloud geolocalization can be performed using stereo photography [6]. A combination of such processing steps can be applied to derive irradiance nowcast [7]. Further methods of obtaining irradiance nowcasts are persistence approaches, which also serve regularly as control reference forecasting models. The performance of persistence approaches is often competitive in intra-hour nowcasting [8], considering the challenges of ASI based nowcasting systems [9]. This work presents a real time capable hybrid ASI and persistence nowcasting approach, which combines the strengths while reducing the respective weaknesses of both approaches.

2 SCIENTIFIC INNOVATION AND RELEVANCE

A nowcasting system based on two off-the-shelf Mobotix Q25 surveillance cameras has been developed at CIEMAT's Plataforma Solar de Almería (PSA). Hemispherical sky images with a resolution of 4.3 MP and an exposure time of 160 µs are taken every 30 s. The ASI nowcasting system also uses ground based irradiance measurements, taken directly next to one of the cameras. A detailed description of the used ASI system, with the already mentioned processing steps of cloud detection, evaluation of the radiative effect, tracking and geo-localization, is presented in [7]. The final output consists of global horizontal irradiance (GHI) maps, covering an area of up to 64 km² with a spatial resolution of 20 m and lead times up to 20 min ahead.

The utilized persistence nowcast is based on the work from Ineichen & Perez [10]. The prevailing Linke turbidity factor (TL) is calculated using the GHI measured by a pyranometer, h the height of the site, I0 the solar constant, r0 the average distance sun earth, r the current distance sun earth, α the solar altitude angle and AM the air mass.

$$TL = \left(\left(-1 \cdot \log\left(GHI/\left((5.09 \cdot 10^{-5} \cdot h + 0.868) \cdot I_0 \cdot (r_0/r)^2 \cdot \sin(\alpha)\right)\right) \cdot (1/(3.92 \cdot 10^{-5} \cdot h + 0.0387) \cdot AM)\right) - \exp\left(-\frac{h}{8000}\right)\right) \cdot 1/\exp\left(-\frac{h}{1250}\right) + 1 \quad \text{Equation 1}$$

For the persistence nowcast, TL is kept constant while α and AM are calculated for the lead times > 0 min. Subsequently, the GHI persistence nowcast is calculated with the corresponding rearranged Equation 1.

A sliding validation of recent historical nowcast is performed between both nowcasting approaches and the ground based irradiance measurement. For this purpose, RMSD is calculated in real time according to Equation 2 using the pyranometer's GHI (GHI_{pyr}) of the past 5 minutes and the corresponding GHI nowcasts $GHI_{LT,j}$ for each delivered lead time. The index j indicates the two distinct nowcast approaches and the index LT the lead time. One RMSD is calculated for the ASI nowcast, another RMSD for the persistence nowcast. t_i describes the corresponding time stamp and n the selected time interval of 5 minutes.

$$RMSD_{LT,j} = \left[\frac{1}{n}\sum_{i=1}^{n}\left(GHI_{pyr}(t_i) - GHI_{LT,j}(t_i)\right)^2\right]^{0.5} \quad \text{Equation 2}$$

Subsequently, current nowcasts $GHI_{LT,j}$ of both approaches for a given lead time are combined to the hybrid nowcast using the accuracy weighting approach described by [11] according to Equation 3.

$$GHI_{LT} = \frac{1}{\sum_{j=1}^{2}\frac{1}{RMSD_{LT,j}}} \cdot \sum_{j=1}^{2}\frac{GHI_{LT,j}}{RMSD_{LT,j}}$$

Equation 3

The ASI based, persistence based as well as the hybrid nowcasting approach are validated at PSA with 8 reference ground based stations distributed over an area of 1 km². The validation data set contains 62 variable days between 01.09.2019 and 30.11.2019. Figure 1 illustrates the GHI conditions within the validation data set.

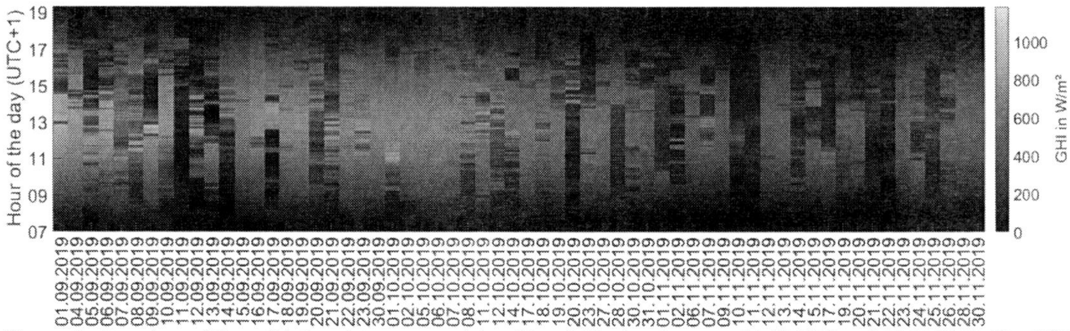

Figure 1: Overview of the validation data set containing 62 variable days between 01.09.2019 and 30.11.2019. The GHI corresponds to the arithmetic average of 8 reference pyranometers distributed over an area of 1 km².

The nowcasting accuracy strongly depends on the prevailing conditions. Comparatively high accuracy is reached during rather low variable conditions consisting of clear sky or overcast conditions. In contrast, lower nowcasting accuracies are reached during highly variable conditions, especially under conditions with fast moving low layer clouds or under complex multi-layer conditions. Overall, validation results for a given time interval without consideration of the prevailing conditions can lead to false impressions of the accuracy. Therefore, the validation data is discretized into eight distinct variability classes from clear sky (class 1) to overcast (class 8) according to [12]. Table I lists a short description of the variability classes.

Table I: Overview on used variability classes

Class	Sky conditions	Clear sky index	variability
1	Mostly clear sky	Very high clear sky index	Low variability
2	Almost clear sky	High clear sky index	Low variability
3	Almost clear sky	High/intermediate clear sky index	Intermediate variability
4	Partly cloudy	Intermediate clear sky index	High variability
5	Partly cloudy	Intermediate clear sky index	Intermediate variability
6	Partly cloudy	Intermediate/low clear sky index	High variability
7	Almost overcast	Low clear sky index	Intermediate variability
8	Mostly overcast	Very low clear sky index	Low variability

Figure 2 illustrates the distribution of the variability classes as well as the average GHI for each class within the validation data set. Around 57% of the data is comprised of the low variability conditions classes 1, 2 and 8. Intermediate and highly variable conditions comprise 34% and 9% of the data set, respectively. Most clear sky conditions at low solar altitude angles are classified in class 2, which results in comparably low GHI values within this class. The average GHI over the entire data set is 384 W/m². For the validation, RMSD is used as error metric, utilizing the entire data set discretized over the lead times. Furthermore, the validation is performed discretized over the lead times as well as the variability classes.

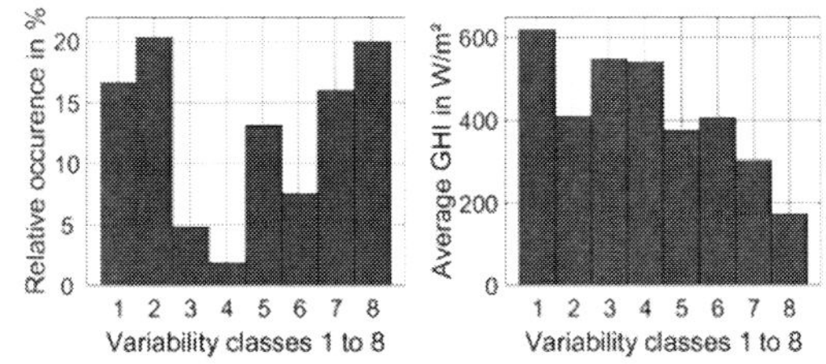

Figure 2: Distribution of the occurring variability classes (left) and average GHI within the classes (right) of the validation data sets.

3 RESULTS AND CONCLUSION

The overall validation results are shown in Figure 3. The ASI and persistence based nowcasts achieve similar RMSDs, with a slight advantage of the ASI based approach for the lead times 6 to 15 minutes ahead. In contrast, the hybrid nowcasting approach shows a significant performance improvement. An average reduction of around 12.6% in RMSD is visible over all lead times compared to the ASI or persistence based approach. Figure 4 shows the validation results discretized over the variability classes. Unsurprisingly, the lowest RMSD values are observed for the clear sky conditions class 1 and 2. Furthermore, as expected, the persistence approach shows an especially low RMSD for these low variable conditions. The third low variability condition class 8 (overcast) shows also low RMSDs. However, especially for the high lead times, the RMSDs are by a factor of up to three higher than for class 1 conditions. All three nowcasting approaches show similar RMSDs for class 8. For the intermediate variability condition class 3, the ASI based approach shows a significant advantage compared to the persistence approach. The remaining intermediate variability classes show similar performance of the ASI and persistence based nowcast, with a slight advantage for the persistence in class 5 and a slight advantage for the ASI in class 7. Both approaches are outperformed by the hybrid approach in all intermediate variability conditions. As expected, the persistence approach yields high RMSDs for the highly variable conditions classes 4 and 6. The ASI approach outperforms the persistence approach clearly in these cases, especially for class 4, whereas the ASI and hybrid approach show similar performances for both highly variable classes.

It was shown that ASI and persistence based

nowcasting approaches show similar overall nowcast performances. However, clear differences are visible when discretizing the validation data in distinct variability conditions. As expected, persistence shows the best performance for low variability conditions, whereas it is outperformed by the ASIs during highly variable conditions. The clear division of strengths within distinct conditions is exploited by the hybrid approach, which clearly outperforms the ASI and persistence approach. The presented real time combination method is a promising approach to improve ASI based irradiance nowcasts.

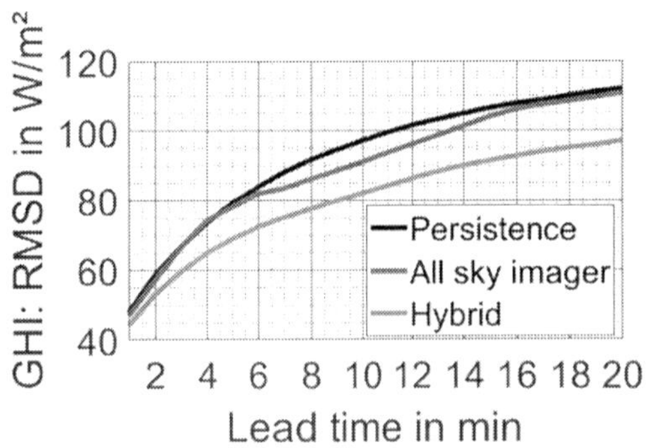

Figure 3: Overall RMSD over lead times of all three approaches.

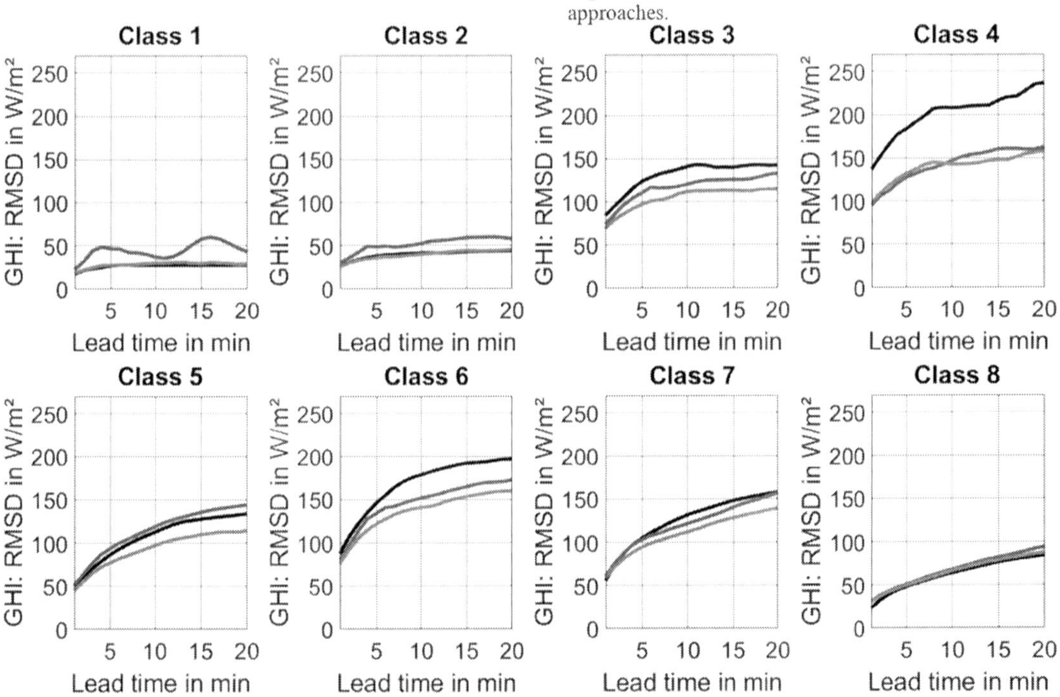

Figure 4: RMSD discretized over variability classes and lead times of all three approaches

ACKNOWLEDGEMENTS

This research received funding by the German Federal Ministry for Economic Affairs and Energy funded within the WobaS-A project (GrantAgreement no. 0324307A). Additional funding was received by the European Union within the H2020 program under the Grant Agreement no.864337 (Smart4RES).

REFERENCES

[1] Wan, C., Zhao, J., Song, Y., Xu, Z., Lin, J., & Hu, Z. (2015). Photovoltaic and solar power forecasting for smart grid energy management. *CSEE Journal of Power and Energy Systems*, *1*(4), 38-46.

[2] Cirés, E., Marcos, J., de la Parra, I., García, M., & Marroyo, L. (2019). The potential of forecasting in reducing the LCOE in PV plants under ramp-rate restrictions. *Energy*, *188*, 116053.

[3] Hasenbalg, M., Kuhn, P., Wilbert, S., Nouri, B., & Kazantzidis, A. (2020). Benchmarking of six cloud segmentation algorithms for ground-based all-sky imagers. Solar Energy, 201, 596-614.

[4] Tzoumanikas, P., Nikitidou, E., Bais, A.F., Kazantzidis, A., 2016. The effect of clouds on surface solar irradiance, based on data from an all-sky imaging system, Renew. Energy 95, 314–322.

[5] Peng, Z., Yu, D., Huang, D., Heiser, J., Yoo, S., Kalb, P., 2015. 3D cloud detection and tracking system for solar forecast using multiple sky imagers Sol. Energy 118, 496–519.

[6] Nguyen, D.A., Kleissl, J., 2014. Stereographic methods for cloud base height determination using two sky imagers. Solar Energy 107, 495–509.

[7] Nouri, B., Wilbert, S., Blum, N., Kuhn, P., Schmidt, T., Yasser, Z., ... & Pitz-Paal, R. (2020, December). Evaluation of an all sky imager

based nowcasting system for distinct conditions and five sites. In AIP Conference Proceedings (Vol. 2303, No. 1, p. 180006). AIP Publishing LLC.

[8] Kumler, A., Xie, Y., & Zhang, Y. (2018). A New Approach for Short-Term Solar Radiation Forecasting Using the Estimation of Cloud Fraction and Cloud Albedo (No. NREL/TP-5D00-72290). National Renewable Energy Lab.(NREL), Golden, CO (United States).

[9] Kuhn, P., Nouri, B., Wilbert, S., Prahl, C., Kozonek, N., Schmidt, T., Yasser, Z., Ramirez, L., Zarzalejo, L., Meyer, A., Vuilleumier, L., Heinemann, D., Blanc, P. and Pitz-Paal, R. Validation of an all-sky imager based nowcasting system for industrial PV plants, Prog Photovolt Res Appl., 1-14 (2017).

[10] Ineichen, P., Perez, R., 2002. A new airmass independent formulation for the Linke turbidity coefficient. Solar Energy 73, 151–157.

[11] Meyer, R., Geuder, N., Lorenz, E., Hammer, A., Ossietzky, C. Von, Oldenburg, U., Beyer, H.G., 2008. Combining solar irradiance measurements and various satellite-derived products to a site-specific best estimate, in: SolarPACES Conference. pp. 1–8

[12] Nouri, B., Wilbert, S., Kuhn, P., Hanrieder, N., Schroedter-Homscheidt, M., Kazantzidis, A., Zarzalejo, L., Blanc, P., Kumar, S., Goswami, N., Shankar, R., Affolter, R. and Pitz-Paal, R. Real-Time Uncertainty Specification of all Sky Imager Derived Irradiance Nowcasts. Remote Sens. 11(9), 1059 (2019).

CLOUD SEGMENTATION AND CLASSIFICATION TO IMPROVE THE SHORT-TERM PV FORECASTING USING ALL SKY IMAGER

Dhia Eddine Melliti, Duy Long HA*,Philippe Marechal, Sylvie Ghalila-Sevestre
Univ Grenoble Alpes, CEA, LITEN, DTS, SMSP, INES, F-38000 Grenoble, France
E-mail : Duy-Long.Ha@cea.fr

Jerome Lehaire, Thomas Capelle, Guillaume Tremoy
Steadysun, 17 Avenue du Lac Léman, 73370 Le Bourget-du-Lac

ABSTRACT: Accurate prediction of photovoltaic energy remains a challenge, as PV production is dependent on fluctuating weather conditions, such as solar irradiance which relies on cloudy conditions. In this context, our work is based on a specific "All Sky Imager" (ASI) , integrating "fish-eye" concave lenses with a 180° field of view.
This paper presents a novel approach of cloud segmentation using ASI to improve - PV production forecasting. First, we improve the identification of various components in the images (clouds, sun, noise, etc.). Then, we study the very short-term impact of clouds through a sky images segmentation and tracking processing. Finally, we correlate the sky-level segmented image to fluctuations in the actual on-site solar irradiance measurements.
The results show that the segmentation is efficient in clear and overcast sky conditions. However, high precision irradiance accuracy in partially cloudy sky conditions is difficult to obtain due to the chaotic impact of the circumsolar region and different cloud opacities.
Keywords: short-term forecasting; all sky imager; image processing; cloud segmentation

1 INTRODUCTION

Steadysun [1] develops and operates an advanced solar power production forecasting technology at intra-hour time horizon named Steady-Eye. This unique application delivers one of the most accurate forecasting solutions available on the market to anticipate solar energy variability. This work comes from a R&D cooperation within CEA-INES [2] to meet the needs of Steadysun to improve very short-term forecasting of PV production technology.

In this paper, we propose a segmentation algorithm to detect clouds on the images produced by Steadysun's ASI. The algorithm is able to filter noise, flares around the sun. It also classifies clouds in two types: opaque and thin. These results will be used later to predict solar irradiance over a ten-minute to half-hour horizon by Steadysun.

2 METHODOLOGY

2.1 Camera calibration

The camera model is obtained beforehand with standard checkerboard based calibration method. This step is crucial as the camera model parameters are very important to understand the space as it is perceived by the camera then, transpose it (un-distortion) into the real world.

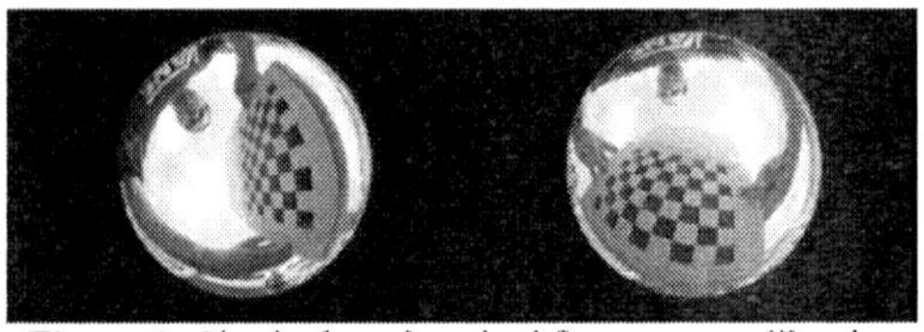

Figure 1: Checkerboard method for camera calibration

Scaramuzza et al. [3] toolbox is well known as a camera calibration tool. This toolbox uses black and white equal-square checkerboard images placed at different positions and tilt angles around the camera (figure 1). Using the known dimensions of the black and white square checkerboard pattern, the model estimates the desired parameters of the camera model.

In this paper, we propose a different approach than the one mentioned previously avoiding extra image acquisition. Indeed, this approach is based on the sun positon to estimate the camera model parameters as mentioned in [4]. As a matter of fact, the sun position $Xs=(xs,ys)$ (in pixels) is tracked on each image. In addition, the angular solar position $\Phi s=(\theta s,\phi s)$ is estimated using the NREL solar position algorithm [5]. The NREL algorithm takes the terrestrial coordinates (latitude, longitude, altitude) and time of the observation location as inputs, and outputs the solar zenith angle θs and the solar azimuth angle ϕs. Then, a linear transfer function between the solar coordinates $Xs=(xs, ys)$ and the angular solar coordinates $\Phi s=(\theta s, \phi s)$ is calculated. In general, these parameters must be calculated for the camera at the calibration step and stored as initialization to the model to use them for later calculations.

2.2 Fish-eye lens model

Sky images in our model are acquired by a whole-sky imager installed locally at the research sites considered. The camera is located 200m from the solar irradiation measuring equipment. The imaging system consists of a high-resolution camera (3-megapixel RGB CMOS sensor) with an equisolid angle projection fish-eye lens (1.27mm focal length) attached to it, which provides a 187° Horizontal Field Of View (HFOV) and 168 ° Vertical Field Of View (VFOV). The optical system is enclosed in a waterproof box with a plastic dome above the lens. The box protects the camera from temperature, humidity, variations in wind and other environmental factors. Images are acquired at a resolution of (2048 x 1536) pixels and are saved at a sampling rate of one frame per minute.

Optical systems introduce some unwanted effects on the image, rendering the perception of the environment inaccurate. Generally, in most cameras, straight lines in the real world are mapped as straight lines in the image captured. However, the use of the fisheye lens introduces a radial distortion effect. This phenomenon is noticeable in the border region of the image, where the levels distortion level is extremely noticeable. This radial distortion causes pixels on the image plane to move from their ideal position in the standard camera model, along a radial axis in the fisheye image plane.

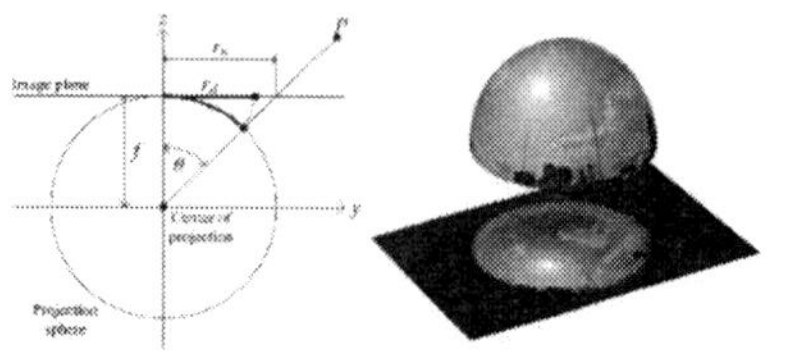

Figure 2: Geometric representation of the fish-eye model used and projection of the dome image on the plan

The model we used to simulate the behavior of the fisheye lens as a function of the focal length f of the camera and the angle θ, inspired by [10], is given by the following equation:

$$r_d = k_1 * \sin(\theta/2) + k_2 \quad (1)$$

The relationship between the radial distance r_d from the center of the image and the angle θ made by a pixel on the dome with respect to the vertical axes allows us to project a captured image onto the dome [6]. The parameters k_1 and k_2 of equation (1) were estimated using a linear regression in the practical case by calculating the apparent position of the sun on the test images using the polar coordinates $X_{s,p} = (r_s, \phi_s)$.

The last step in the calibration is to find the orientation of the camera relative to the north-south axis of the Earth. The parameters we considered are:

α : camera offset from the North-South axis.

γ : azimuth angle between a point on the image and the North-South axis.

By calculating the coordinates (ϕ_s, γ_s) of the sun, we will use these to find the offset of the North-South axis, through another linear regression of equation (2).

$$\gamma = \alpha - \phi \quad (2)$$

2.3 Clear Sky Library (CSL)

The binary segmentation is based on Clear Sky Library (CSL) similar to the approach described in [7, 8]. The CSL is a database that provides a reference with respect to the RBR (Red to Blue ratio R / B) values of the RGB image for each pixel according to the zenith and azimuth angle of the image calculated from historical images in clear sky conditions.

The RBR level is higher in cloudy regions in the image, as well as in the solar and circumsolar regions. The RBR is the highest near the sun and decreases further away from the center of the sun. The RBR also increases near the horizon for large values of the zenith angle due to optical distortion and higher aerosol concentrations towards the horizon. The segmentation results based on the CSL and Otsu thresholding ,which will describe in details in the next section, are presented in figure 3.

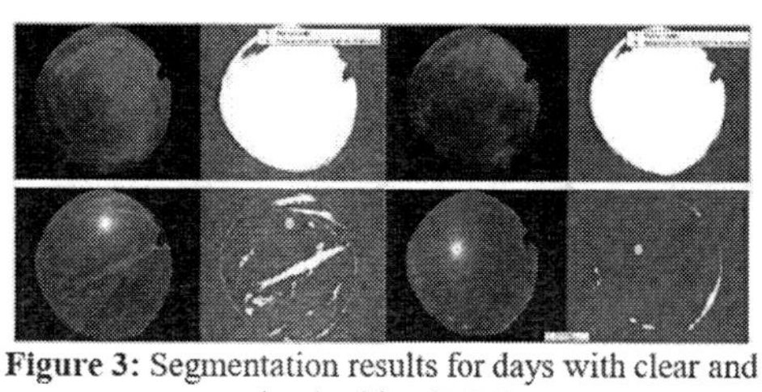

Figure 3: Segmentation results for days with clear and cloudy skies (INES)

For the construction of the CSL, a polynomial model using the RBR values of the clear sky images is estimated. The second degree polynomial model (9 coefficients) is calculated through the RBR image according to the zenith angle (in pixels) and the azimuth angle of the sun (in pixels) according to the following model:

$$RBR(\theta_p, \psi_{ps}) = \sum_{i=0}^{2}\sum_{j=0}^{2} p_{ij} * \theta_p^{\,i} * \psi_{ps}^{\,j} \quad (3)$$

For each clear sky image, the coefficients of the polynomial model (equation 3) are calculated and stored in the CSL with the corresponding zenith, azimuth angles and date/time of day. The representation of the angles θ_p, ψ_{ps} and θ_z is described in figure 4 with θ_z the solar zenith angle, fixed for each image interpolation by the polynomial model.

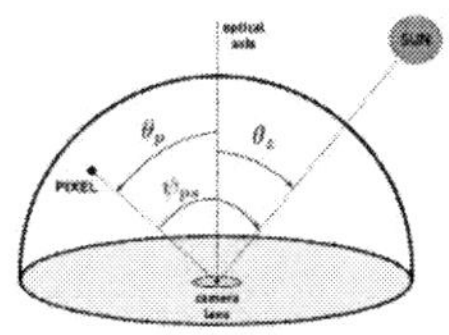

Figure 4: θ_p, ψ_{ps} and θ_z angles for the CSL model

3 IMAGE SEGMENTATION

This section will be reserved for the describing of the segmentation procedure for cloud detecting and the different techniques implemented ensure a good segmentation quality.

3.1 Image filtering

Before we start the segmentation process, we filter the dataset to remove outliers. Some examples of outliers are presented in figure 5.

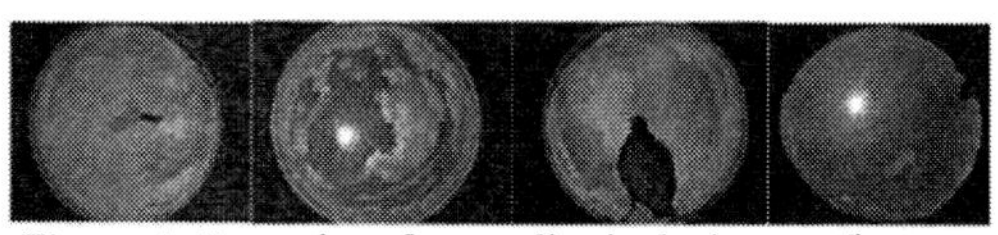

Figure 5: Examples of anomalies in the images (insects, scratches, birds, flare)

The images with abnormalities are removed from the dataset through visual inspection or treated with specific algorithms, which will be described later on. Otherwise, they will affect the final cloud-detection decision image.

3.2 Image cleaning

The surface of the dome during real operational conditions is exposed to dirt or scratches which will result in additional scattering of light with specific patterns that change with the position of the sun and the time of day. Also, the reflections from the sun hitting the image sensor which become visible at high solar altitudes (for low values the zenith angle or the direct beam is almost orthogonal to the image plane) result in a red circular spot in the camera dome (flare). The impacts of this stray light were qualitatively assessed by visually inspecting clear sky images such as those in Figure 6.

Figure 6: Raw images artefacts and flares

These artefacts have consequences later on in the CSL construction and image segmentation. In fact, in the case of CSL, these spots present false high RBR values which biases the estimation of the polynomial model of the CSL. In addition, if left untreated, these spots will result in false detections later in the segmented image or be interpreted as false clouds.

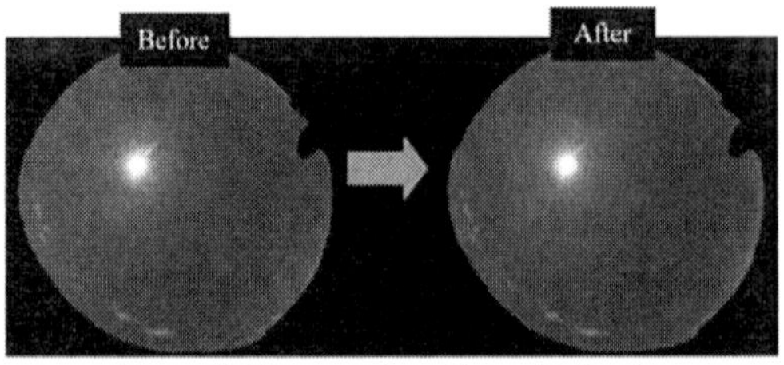

Figure 7: Flares and artefacts elimination

To solve this problem, we apply a Gaussian filter with parameters μ and σ in the specific regions where these artefacts and flares exist. An illustration of the resulting image after treatment is presented in figure 7.

3.3 Segmentation parameters

As previously exposed, the algorithm to detect clouds is based on the RBR. There are 4 cases where the detection procedure applied is different. Each of these cases is determined from the state of the sun (hidden, clear) and the circumsolar region (hidden, clear). Two different indicators are calculated, the Sun Disk Factor SDF and the Circumsolar Factor CF.

The circumsolar factor is defined as the ratio of clouds detected in the circumsolar zone compared to the total surface of the considered circumsolar region. The method used for detecting clouds in this region is based on the RBR of the considered image. According to the values of CF, we distinguish two cases:

$$if\ CF < threshold_{CF} :\ circumsolar\ clear$$

$$else\quad :\ cicumsolar\ occulted$$

For the SDF, we calculate the area of the saturated region of the image (Mean(RGB) > 240) in the solar perimeter (a region of radius 300 pixels for example). Then, we divide this area by a disk surface of radius 50 pixels (fixed sun radius). This will give us a variable solar ratio factor SDF that we will later use to determine if the sun is occulted.

$$if\ SDF < seuil_{SDF} :\ solar\ disk\ clear$$

$$else\quad :\ solar\ disk\ occulted$$

Considering a fixed sun radius simplifies the problem as the real sun radius in the image does not vary much in the considered region of interest of zenith angle [-85°,85°]. A similar approach was proposed in [8] were this hypothesis was tested and validated.

3.4 Segmentation steps

At this stage, we can now start the process of cloud detection. At first, we consider the RGB image in question (Figure 8-a). We start by eliminating the artefacts and flares from the image as mentioned in (4.1) and (4.2). After this step, considering the position of the sun in the image, we calculate the corresponding zenith and azimuth angles. Then, we look in the CSL data base (which we prepared beforehand) for a similar day (or closest) in terms of zenith and azimuth angles. The RBR image from the database for the corresponding values of zenith and azimuth is then generated using formula (3) and the polynomial coefficients found in the CSL (Figure 8-c). After subtracting the RBR of the CSL image (Figure 8-c) from the RBR of the raw image (Figure 8-b), we obtain a difference image of the RBR values (figure 8-d).

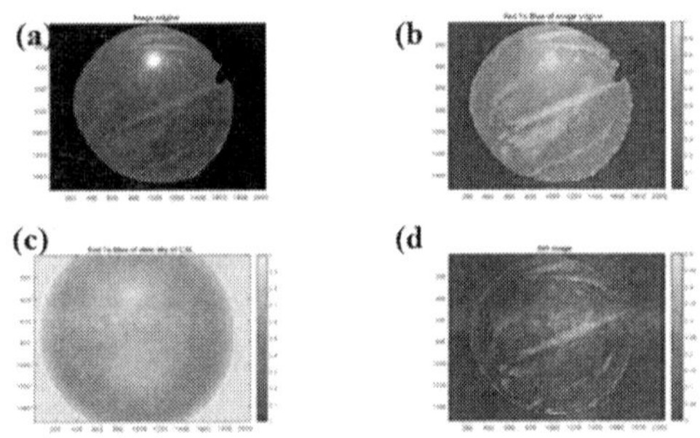

Figure 8: Cloud segmentation steps with CSL

By comparing the pixels of the difference image (figure 8-d) to an Otsu threshold calculated through the whole difference image, the final resulting image which indicates the presence of clouds or not, is obtained. If the pixel values in the difference image is above the Otsu threshold, the method classifies these pixels as cloudy ones which indicates the presence of clouds in that region. Otherwise, the pixels are bright (no clouds).

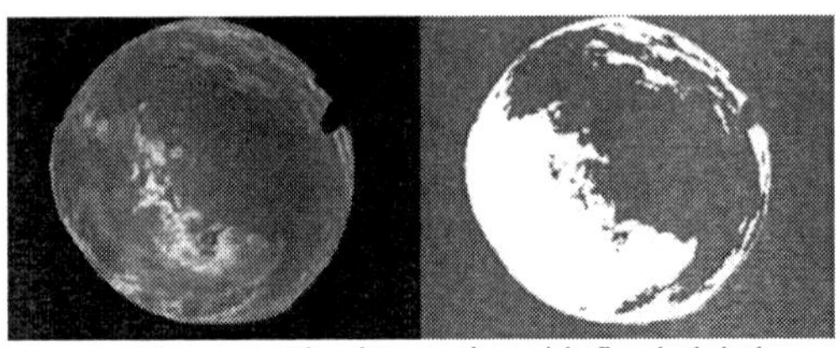

Figure 9: Cloud Detection with fixed global threshold

However, inspired by [8], we introduce a subtle correction in the form of an exponential Otsu threshold to improve detection and avoid misclassification especially in the closest vicinity of the sun (circumsolar region), which affects the very short-term solar forecasting in general and specifically in our model. The exponential Otsu threshold proposed in [8] is in the form:

$$\delta_{diff}(\psi_{ps}) = \delta_{diff,0}\left(1 + \exp\left(\frac{-\psi_{ps}}{200}\right)\right) \quad (4)$$

With ψ_{ps} in degrees, δ_{diff0} is the Otsu threshold of the difference image and 200° is a scaling factor, which has been fixed through manually testing a set of batch images. Using the same logic as before, we then compare the calculated δ_{diff} with the the difference image by classifying the pixel as cloudy if the pixel value is above δ_{diff} and clear otherwise. However, we have exploited this result but with some modifications and improvements described in the following table:

Table 1: Different threshold settings for cloud detection

Case	Sun	Circumsolar	CSL	Threshold	Expression
1	Occluded	Occluded	No	Fixed threshold	Threshold = 0.7
2	Clear	Occluded	yes	δ_{diff0} = $Thresh_{Otsu}$	$\delta_{diff}(\psi_{ps}) = \delta_{diff,0}(1 - \delta_{diff,0} * \exp\left(\frac{-\psi_{ps}}{200}\right))$
3	Occluded	Clear	yes	δ_{diff0} = $Thresh_{Otsu}$	$\delta_{diff}(\psi_{ps}) = \delta_{diff,0}(1 + \exp\left(\frac{-\psi_{ps}}{200}\right))$
4	Clear	Clear	yes	$\delta_{diff0} = 0.17$	$\delta_{diff}(\psi_{ps}) = \delta_{diff,0}(1 + 0.5 * \exp\left(\frac{-\psi_{ps}}{200}\right))$

These settings are site independent and are the same for all studied sites.

4 RESULTS AND VALIDATION WITH FIELDS TESTS

4.1 Cloud detection validation

The first validation step is visual. In fact, we inspect and compare the segmented images with the real images of the camera in order to check the quality of cloud detection especially in the circumsolar region. This first validation is carried out on the entire database available on the various studied sites.

Second, once the clouds are detected, we verify whether the sky situation described by the mask generated by our model is in accordance with the irradiance measured on the ground. In other to confirm the matching of generated image and the actual GHI measurements of the site, we apply a certain logic described by the following table:

Table 2. Comparison logic of the GHI measurements to the model

Case	Sun	Circumsolar	Condition	Solar irradiance
1	Occulted	Occulted	$G_m \leq G_{cs}$	Occulted
2	Clear	Occulted	$G_m > G_{cs}$	Diffused
3	Occulted	Clear	$G_m < G_{cs}$	Occulted
4	Clear	Clear	$G_m = G_{cs}$	Transmitted

G_m: *Irradiance measured* – G_{cs} : *Irradiance in clear sky condition*

In the case of occulted sun, the measured value should be lower than the theoretical value of the clear sky model. Also, if the sun is clear, the location of the circumsolar zone determines whether the irradiance reaching the sensor is transmitted or diffused. In the case of transmission, the measured irradiance is equal to that of clear sky conditions, otherwise in the case of scattering, the measured irradiance is much higher.

To validate these hypotheses, we compare in another step the values of irradiance measured on site to those of clear sky irradiance from a model, which generates these values taking into account the longitude, altitude and latitude of the site [9]. The comparison is made according to the formula:

$$Diff_{irradiance}(\%) = \frac{\text{Irradiance}_{\text{measured}} - \text{Irradiance}_{\text{clear sky}}}{\text{Irradiance}_{\text{clear sky}}} \quad (5)$$

After that:

If $|Diff_{irradiance}| < 12\,\%$: *sun and circumsolar clear*

If $Diff_{irradiance} < 12\%$: *sun occulted*

else : *sun and circumsolar occulted*

This logic will be used subsequently for the validation of the segmentation results.

4.2 Results for cloudy and clear sky days

The validation of the segmentation quality was first done visually (Figure 12) and on the entire database of test sites.

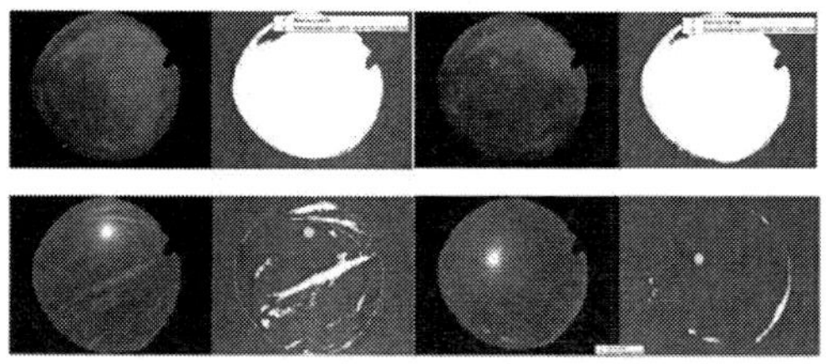

Figure 12: Segmentation results for overcast and clear sky days.

The model shows good performance for days with overcast and clear sky. First, by eliminating flares and artefacts around the sun thoroughly, secondly in terms of the deployment of the CSL and its usefulness for days with clear sky conditions and thirdly, the effectiveness of the fixed threshold for overcast skies (Table 1 - Case 1) and the adaptive threshold (Table 1 - Case 4) especially in the circumsolar region which mainly impacts the GHI forecast in the very short term.

For the GHI validation of our cloud detection method, we only have GHI measurements for the INES site. The GHI validation results are summarized in the following table:

Table 3: GHI validation results for clear and overcast days

Site	Data available	Clear sky	Overcast	Precision [%]	
INES	2018 (20 days over 12 months)	2 days	13 days	Clear sky	87 - 98
				Overcast	98 - 100

4.3 Results for days with partly cloudy skies

The proposed model is more beneficial to be applied in partially cloudy conditions. In fact, the variations of measured GHI are very fluctuating and thus intriguing in this case. This underlines the importance following the evolution of clouds especially in the circumsolar region, to better understand and correlate the impact of clouds on very short-term GHI fluctuations. A better understanding of this phenomenon will subsequently make it possible to provide a better, reliable forecast of solar irradiance.

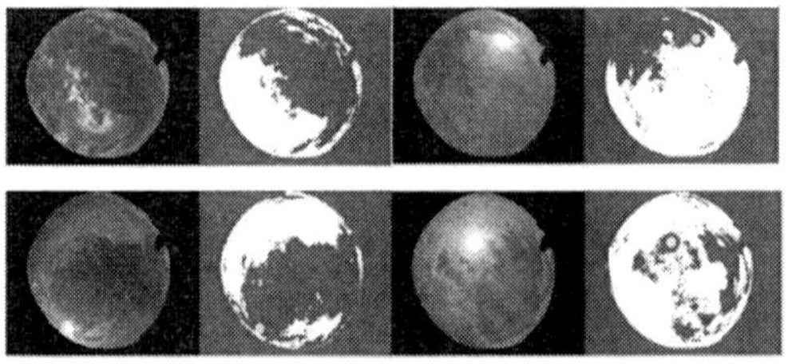

Figure 13: Segmentation results for a partly cloudy sky (INES)

The results for the partly cloudy data set are displayed in table 4, as well as for the visual validation part in Figure 13.

Table 4. GHI validation results for partially cloudy days

Site	Data available	Partly cloudy	Precision [%]
INES	2018 (20 days over 12 months)	5 days	51 - 71

The model accuracy in partially cloudy days (51% - 71%) versus clear and overcast ones ((87% - 98%) and (98% -100%) respectively) illustrates the variability of cloud impact on GHI measurements and the difficulty to pinpoint with precision it's impact on all the data used, in the case of partially cloudy days. This large variability leads to greater uncertainties later on, over the assessments made on the future value of GHI. This is due to the different types of clouds (opaque, thin, permeable ... etc.) as we have concluded.

The validation using in-site irradiance measurements shows expectable results for clear sky and cloudy conditions. However, the segmentation should be improved specially for partially cloudy days, which will be the focus of our future work.

4 CONCLUSION

In this paper, we have presented method for cloud detection and segmentation using preexisting techniques, which we have adapted locally for our study site and improved, based on data driven approaches.

The validation of our method shows good results for clear sky and overcast days. For partially cloudy sky conditions, there is still some improvements to work on to distinguish between different types and shapes of clouds. Indeed, based on the results obtained on partially cloudy days, we noticed that the opacity of the clouds approaching the circumsolar region and the sun in particular, influence the transmissibility of solar radiation to different degrees. The impact is measurable as soon as the clouds cross the circumsolar region, where measurements start to become chaotic. This is why a classification of the clouds according to the opacity level is essential to more precisely distinguish the transmissivity of clouds and subsequently, to increase the precision of the model.

One approach is a cloud classification procedure based on sky images histogram analysis. The impact of cloud opacity on irradiance transmissivity will be developed in future works. Another perspective is to distinguish between different types of clouds by a classification procedure based on color (RGB) and texture (gradient for example). This step should help identifying the impact of clouds on solar irradiance and subsequently, improve the forecast.

Finally, the main finding is that there is no "one best model" for predicting solar irradiance, but some models give better results than others do in specific sky and meteorological conditions. It would be interesting to find a way to hybridize the different models we have according to the forecast horizon, different site conditions and optical systems in order to try to keep the forecast error as low as possible.

6 REFERENCES

[1] https://www.steady-sun.com/

[2] https://www.ines-solaire.org/

[3] D. Scaramuzza, A. Martinelli, and R. Siegwart, 'A Toolbox for Easily Calibrating Omnidirectional Cameras', in 2006 IEEE/RSJ International Conference on Intelligent Robots and Systems, Beijing, China, Oct. 2006, pp. 5695–5701, doi: 10.1109/IROS.2006.282372.

[4] B. Urquhart, B. Kurtz, and J. Kleissl, 'Sky camera geometric calibration using solar observations', Atmospheric Measurement Techniques (Online), vol. 9, no. 9, Sep. 2016, doi: 10.5194/amt-9-4279-2016.

[5] I. Reda and A. Andreas, 'Solar Position Algorithm for Solar Radiation Applications (Revised)', NREL/TP-560-34302, 15003974, Jan. 2008. doi: 10.2172/15003974.

[6] C. Hughes, P. Denny, E. Jones, and M. Glavin, 'Accuracy of fish-eye lens models', Applied optics, vol. 49, pp. 3338–47, Jun. 2010, doi: 10.1364/AO.49.003338.

[7] J. Du, Q. Min, P. Zhang, J. Guo, J. Yang, and B. Yin, 'Short-Term Solar Irradiance Forecasts Using Sky Images and Radiative Transfer Model', Energies, vol. 11, p. 1107, May 2018, doi: 10.3390/en11051107.

[8] M. Caldas and R. Alonso-Suárez, 'Very short-term solar irradiance forecast using all-sky imaging and real-time irradiance measurements', Renewable Energy, vol. 143, pp. 1643–1658, Dec. 2019, doi: 10.1016/j.renene.2019.05.069

ANALYSIS OF AUTOMATIC FAULT DETECTION METHODS FOR COMMERCIALLY OPERATED PV POWER PLANTS

E.A. Sarquis Filho*, Klaus Kiefer**, Nicolas Holland**, Bernd Kollosch***, Björn Müller****, P.J. Costa Branco*
* IDMEC, Instituto Superior Técnico, Universidade de Lisboa, Av. Rovisco Pais, 1049-001 Lisboa, Portugal,
eduardosarquis@tecnico.ulisboa.pt, pbranco@tecnico.ulisboa.pt
** Fraunhofer Institute for Solar Energy Systems ISE, Heidenhofstrasse 2, 79110 Freiburg, Germany,
klaus.kiefer@ise.fraunhofer.de
*** Pohlen Solar GmbH, Am Pannhaus 2-10, 52511 Geilenkirchen, Germany
**** Enmova GmbH, Basler Str. 115, 79115 Freiburg, Germany

ABSTRACT: Automatic fault detection (AFD) is a key technology to optimize the Operation and Maintenance of photovoltaic (PV) systems portfolios. Although several recent studies suggest methods and procedures for automatic fault detection, they lack a common basis for testing their effectiveness under real operating conditions. In this study, we have selected and implemented a list of methods suitable for the data available in the standard monitoring of commercial systems. The performance of these methods was tested using measurements collected over 58 months on 85 rooftop-type PV systems installed in Germany. The maintenance records from these systems were used to measure the accuracy of the fault detection alerts and to analyze the sensitivity and specificity of each method. The best performing was able to identify almost one third (29.6%) of the faults observed in PV systems with an average of 15.8% false alerts. The method with the least false alerts (11.7%) could identify 19% of the faults. The AFD methods tested show different sensitivity depending on the type of fault, which suggests that a combination of methods can increase the overall sensitivity.
Keywords: PV System, System Performance, Operation & Maintenance, Defects, Automatic Detection.

1 INTRODUCTION

Automatic fault detection is a key technology to optimize the Operation and Maintenance (O&M) of photovoltaic (PV) systems portfolios. In the literature, several recent studies suggest methods and procedures for automatic fault detection [1-4].

The diversity of methods is due to the different sensors and signals, pre-processing techniques, comparison methods, and decision criteria. These characteristics influence the accuracy of detection and the type of faults that can be detected. Attempts to classify and categorize these methods lack a common basis for testing their effectiveness under real operating conditions. This is critical because several of these studies have not been experimentally validated with data from actual PV systems. They are based only on simulations [4-10] or small PV systems built in laboratories and tested under a few examples of failure conditions [11-15]. Some methods described in the literature require special sensors for collecting data at high sampling frequency [9, 13, 16] or at unusual points in the circuit [5, 11], which make them unsuitable for the cost-effective monitoring of commercially operated PV power plants.

In this paper, several studies have been reviewed to gather information about the different automatic fault detection (AFD) methods published, e.g., the features used for fault detection and the comparison/decision methods used to detect the abnormal behavior. We have selected only methods suitable for the data available in the standard monitoring of commercial systems. The performance of these methods is tested under real operating conditions, using the monitoring database of a large portfolio of PV systems with reference to actual maintenance logs to validate the alerts.

The relevance of this study is the use of actual monitoring data from 85 commercially operated PV power plants to quantify the performance of the AFD methods by measuring the percentage of faults detected and the number of false alerts generated. This study also addresses sensitivity according to fault type, revealing which methods can be combined to achieve higher sensitivity.

This work is part of the research project OptOM (see acknowledgements), within which an integrated, cost-optimized operational management concept for the economic lifetime of PV power plants will be developed and implemented. The study of fault detection complements previous studies on long-term performance loss [17] in pursuing a cost-optimized maintenance plan to correct faults and reduce power losses.

2 FIELD DATA AVAILABLE

All field data used in this study comes from 85 rooftop PV systems installed in Germany. One such system can be seen in Figure 1. The size of these systems ranges from 55 to 1550 kWp, 66% have less than 300 kWp. All systems combined have a total of 944 inverters and 137169 PV modules. The large majority (94%) of the inverters are string inverters with rated power up to 25 kWp; and 33 inverters have rated power between 400 and 600 kWp and are central inverters. The nominal power of PV modules ranges from 145 to 280 W, with the majority being 220 W.

The data used in this study was collected from January 2016 to October 2020 (58 months). The monitoring data available comes mostly from the sensors available in the inverters, which measure the current, voltage, and power signals at the input and output of the device. Additionally, each system contains one or more irradiance sensors (reference cells) installed in the same plane as the PV panels. The data is acquired every minute or faster and averaged values are stored every 5 minutes. This monitoring setup resembles the common practices adopted in commercial medium size PV systems. As a premise of the project, only data from standard monitoring equipment will be used, so that the results can be applied to any system without incurring additional costs for sensor installation.

Figure 1: Example of a PV system (87 kWp) from the portfolio.

2.1 Fault historic

During the period of 58 months, 1843 unique monitoring and maintenance tickets were created for all 85 PV systems combined, summing 20428 affected days. Tickets are created on the day the fault is identified and are repeated daily until the problem is corrected. As the original ticket data does not follow a strict categorization scheme, an automatic classification of the failures into categories was done by searching for keywords in the ticket description. Figure 2 depicts the distribution of the tickets according to their categories. Table 1 contains a description of the faults included in each category.

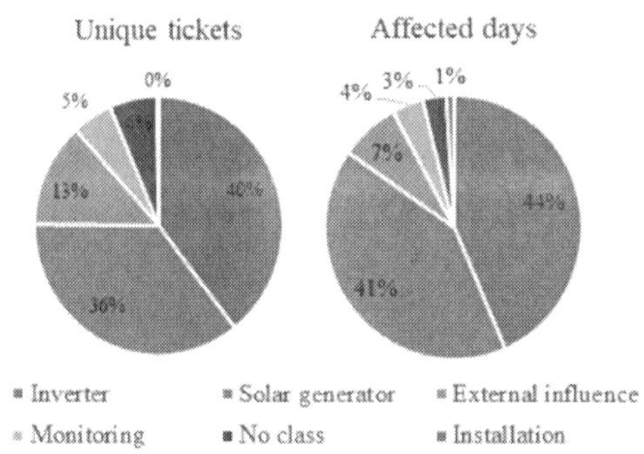

Figure 2: Percentage distribution of maintenance tickets according to their fault category.

Table I: Fault categories and description.

Fault category	Description
Solar generator	Defective module
	Insulation fault
	Hotspots
	String failure
Inverter	Inverter breakdown
	Inverter disc. from the busbar
	Inverter switched off
	Inverter switches on delayed
	Inverter failed
	Insulation fault
	No inverter connected to the busbar
External influence	Snow
Monitoring	Problem at the monitoring cabinet
	Router problem
	Irradiance sensor defective
	Irradiance sensor w/ broken glass
	Irradiance sensor shaded
	Sensor shows wrong values
No class	General system fault
Installation	Irradiance sensor misaligned

3 FAULT DETECTION METHODS TESTED

Automatic fault detection methods follow a general flow that starts with processing the measurements to extract relevant information (also known as features), followed by a comparison/decision step. In this study, several publications have been reviewed to map the methods applied for the feature extraction and the comparison/decision steps. The review also included different methods used for creating a reference model (a digital twin), since it is frequently used within the feature extraction step in the automatic fault detection process.

3.1 Feature extraction methods

The feature extraction step aims to obtain information from the raw measurements that can be correlated to the occurrence of faults. Some faults may show a greater correlation with a specific feature, and lower correlation with other features. Hence, it is important to understand which faults are most critical and to map which features are suitable for their detection. Some features are useful for detecting faults of any type, while others show high sensitivity to a specific type of fault. For example, the Performance Ratio (PR) can be affected by any fault that reduces the overall efficiency of the PV system. In comparison, high-frequency harmonics in DC voltage measurements can reveal the occurrence of short circuits in PV strings.

In the literature, there are different features used for fault detection in PV systems. However, not all of them can be calculated based on the data available. A method that requires special or unusual measurements cannot be applied without installing additional sensors. In this study, we have selected only the feature extraction methods that can be applied to the data already available in the standard monitoring of commercial systems (as described in Section 2). The tested feature extraction methods are summarized in Table II.

Table II: Feature extraction methods tested.

Feature Extraction Method	Ref.
Raw measurements	[4]
Median of DC currents	[14]
Performance Ratio (PR)	[24]
Ratio between measured and expected power (P_{ratio})	[7, 18]
Ratio between measured and expected current and voltage	[20, 19, 10]
Difference between expected and measured power (P_{diff})	[6, 19, 8, 10, 9]
Difference between expected and measured capture losses (CL_{diff})	[20]

The calculation of some features requires the definition of expected values, which can be based on data from the system itself or even from neighboring systems. For example, in [14], the median value of the inverters' DC current input is used as a reference for identifying inverters operating in an abnormal condition. Nonetheless, in general the definition of an expected value requires the creation of a reference model. In the next section, an overview of the methods adopted in the reviewed literature is given.

3.2 Methods for creating a reference model

Reference models for PV systems can be created using an analytical approach, i.e., calculating the output power based on equations describing the system's physical characteristics. For example, the single or double diode model has been widely adopted to estimate the output characteristics of PV arrays. In [6, 9, 10], the parameters for these models were calculated based on the information available in the PV module datasheet. In contrast, in [19, 20], the parameters were estimated directly from the characteristic curves measured on the installed panels.

As pointed out by [20], the simulation of a PV system using PV module parameters given by the manufacturer does not consider the mismatch effects in the PV array configuration, resulting in an over-estimation of the output power in the simulation results. On the other hand, it is rare to have IV curve data available for the installed PV array.

A different approach, commonly used to create reference models, adopts statistical rather than analytical models. Generically, these models learn the behavior of the system based on its measurements. The minimum information needed is irradiance and the targeted output variable, typically AC power. PV module temperature measurements are useful for refining the quality of the model, but these are not always available. Within the reviewed literature, this approach has been implemented using different methods. In [12], a linear auto-regressive model with exogenous inputs is used to estimate the AC power output using the latest AC power and irradiance measurements. In [8], an artificial neural network has been used for estimating the output AC power using the irradiance and module temperature as inputs. In the present study, the ANN implemented is of the feed-forward type, with two hidden layers (12 neurons each) using multilayer perceptrons (MLP) with hyperbolic tangent activation function. In addition to these models, the authors also tested a second-order polynomial regression model, a simpler statistical model to implement and define coefficients.

A promising approach to creating reference models combines analytical equations from the PV domain with learning methods to adjust their coefficients according to measurements. In [21], the authors propose two equations to calculate the efficiency of the PV modules according to irradiance and temperature, and then estimate the power output according to the module area installed. These equations are adaptable to any system by adjusting three parameters according to the measurements. In [22], a chain of physical models is used to estimate several energy conversion steps – from irradiance to output AC power. The models' parameters are identified based on the measurements using an optimization procedure applying a genetic algorithm (GA). These methods require temperature measurements, which are not available in this study and therefore could not be tested.

Table III: Methods for building a reference model.

Methods	Ref.
Single diode model	[6, 19, 20]
Two diodes model	[9, 10]
Artificial Neural Network (ANN)	[8]
Auto-regressive model with exogenous inputs (ARX)	[12]
Polynomial regression	Suggested
Empirical PV module model	[21]
Chain of models + GA	[22]

3.3 Comparison and decision methods

The second part of the fault detection process is where the extracted feature values are analyzed to assess how far from the expected behavior the system is operating. The variable nature of the weather is reflected in the PV systems' behavior and the features extracted from it. To define the decision boundaries and avoid false alerts, elaborated ways to assess the features have been developed in the literature.

To reduce the influence of these oscillations, the authors in [7] calculate the exponentially weighted moving average (EWMA) of the target signal before comparing it to the predefined control limits, which are based on the statistical analysis (standard deviation) of fault-free measurements. The result is a method affected by seasonalities since it is not sensible to gradual changes in the system performance. The control limits are to be carefully defined to avoid the occurrence of false alerts.

An adaptive threshold (AT) comparison method described in [12] uses EWMA to adjust the control boundaries gradually. This allows automatic calibration to seasonalities and long-term changes in the system performance. One drawback, however, is that the control limits are updated even if a failure occurs, which might lead to clear the fault alert before it has been corrected. The careful selection of the parameters (forgetting factor and tolerance window) can reduce this effect but hardly make it immune to the problem.

The comparison method implemented by [9] uses the confidence interval obtained from a standard normal distribution. The confidence interval is very sensitive and can easily achieve 100%; thus, the effectiveness of this method depends heavily on the accuracy of the reference.

Based on the observations of the AT method, the authors suggest checking if the signal is within the control limits before adjusting the threshold. This conditional adaptive threshold (CAT) avoids adjusting the reference to a fault condition while maintaining seasonal adaptability, making the method more robust.

The identification of abnormal behavior can also be done using an outlier detection algorithm. In [14], the authors apply the Hampel outlier identifier (HOI) method to the DC current measurements from different inverters in the same system. A fundamental limitation of this method lies in its need for at least three input signals, what makes it not suitable for systems with only two inverters.

Table IV shows a summary of the comparison and decision methods tested.

Table IV: Comparison and decision methods tested.

Methods	Ref.
Exp. Weighted Moving Average (EWMA)	[7]
Adaptive threshold (AT)	[12]
Confidence interval (CI)	[9]
Conditional Adaptive Threshold (CAT)	Suggested
Hampel Outlier Identifier (HOI)	[14]

3.4 Methods tested

The different methods for each step in the automatic fault detection (AFD) process were combined to create a list of possible solutions as described in Table V.

Table V: List of Automatic Fault Detection methods tested.

AFD	Feature	Ref. Model	Comp./Dec.
1	DC Current	-	HOI
2	Daily PR	-	EWMA
3	Daily PR	-	AT
4	Daily PR	-	CAT
5	P_{diff}	ANN	EWMA
6	P_{diff}	ARX	EWMA
7	P_{diff}	Poly Reg	EWMA
8	P_{diff}	ARX	CI
9	P_{ratio}	ARX	EWMA
10	CL_{diff}	ARX	EWMA

4 TESTING METHODOLOGY

Using the available data described in Section 2, each AFD method listed in Table V was implemented and tested. The reference models were created using the first year (365 days) of measurements as training data when required.

The output of the comparison step is a continuous variable that is translated into a fault alert by applying a threshold. The receiver-operator characteristic (ROC) curve illustrates the fault detection capability for all possible thresholds. This method was originally developed for operators of military radar receivers to measure the performance of binary classifiers. It offers a comprehensive way to compare the effectiveness of the tested methods [23].

The fault alerts are calculated on a daily basis to be compared with the daily maintenance tickets. Each daily fault alert will be classified as True Positive (TP), False Positive (FP), True Negative (TN), or False Negative (FN). The TP are interpreted as the detected faults, and the TN are the normal operation. The FN are the undetected faults, and the FP are the false alarms. Considering the monitoring and maintenance tickets' imperfection (or late reaction) due to technically or economically limited inspection routines, some fault alerts may appear days before they are detected and recorded in the maintenance log. Thus, a FP is reclassified as a Hidden Fault (HF) if it comes right before a TP. The count of occurrence in those classes is used for calculating the sensitivity and specificity using the following equation.

$$\text{Sensitivity} = (TP + HF) / (TP + HF + FN) \quad (1)$$
$$\text{Specificity} = TN / (TN + FP - HF) \quad (2)$$

The sensitivity measures how many of the existing faults were correctly detected. The specificity measures how many of the negatives are truly negative. In addition, the percentage of false alerts in relation to the average alerts per year per system is calculated for a more practical understanding of the specificity of the methods.

5 RESULTS AND DISCUSSION

The ROC curves obtained for each of the AFD processes tested are depicted in Figure 3. High sensitivity levels could not be achieved without decreasing the specificity, i.e., increasing the occurrence of false alerts. In the period analyzed, the maintenance tickets of those 85 systems contain a record of almost 52 faults per year per system. An AFD method with 83% specificity would generate an equivalent number of false alerts in this scenario. Such large number of false alerts is not tolerable in large portfolios, thus achieving high specificity is critical for an AFD. For example, if false alerts should be limited to 16% (10 days per year per system), the minimum acceptable specificity would be around 97%. Under these conditions, the maximum achievable sensitivity of the individual methods evaluated is about 29.6%. Table VI presents the overall sensitivity, specificity, and the total percentage of false alerts for each AFD.

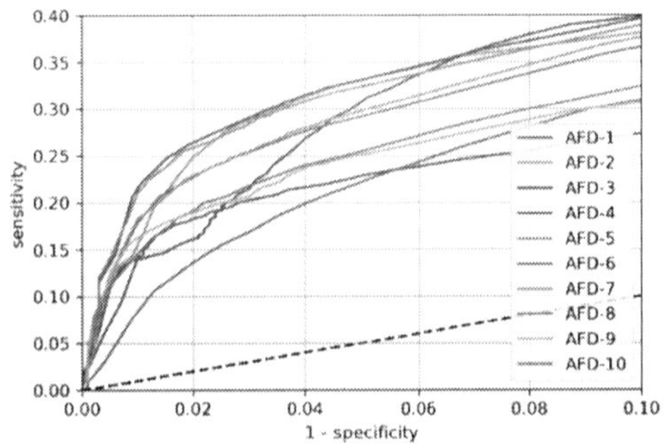

Figure 3: ROC curves obtained for each of the AFD process tested.

Table VI: Result of the Automatic Fault Detection methods test.

AFD	Sensitivity	Specificity	False alerts
01	19.0%	**97.7%**	**11.7%**
02	29.1%	96.9%	15.7%
03	16.9%	97.0%	15.2%
04	22.4%	96.9%	15.7%
05	25.6%	96.9%	15.3%
06	**29.6%**	96.8%	15.8%
07	26.1%	96.8%	16.1%
08	21.6%	97.1%	14.8%
09	21.6%	96.8%	16.1%
10	29.1%	96.8%	16.0%

The results show that none of the tested methods was capable of detecting all faults. The method proposed in [14], which compares the DC current measurements with their instant median values, was able to detect 19% of the faults with the lowest number of false alerts. This method does not require any external reference information, such as irradiance measurements, thus, it might be simple to implement.

Comparing AFD 2, 3 and 4, all of which use the Daily PR, one can observe that the best comparison method was the EWMA (see Figure 4). The EWMA proved to be a good comparison method when the tolerance is set according to one year's standard deviation, i.e., a large tolerance window. This reduces false alerts while detecting the larger deviations. The CAT method suggested by the authors achieved a similar sensitivity with a smaller tolerance window, which indicates good functioning of the adaptive threshold. However, the higher number of false alerts indicates that it may not be suitable for dealing with data with a high level of oscillation or noise.

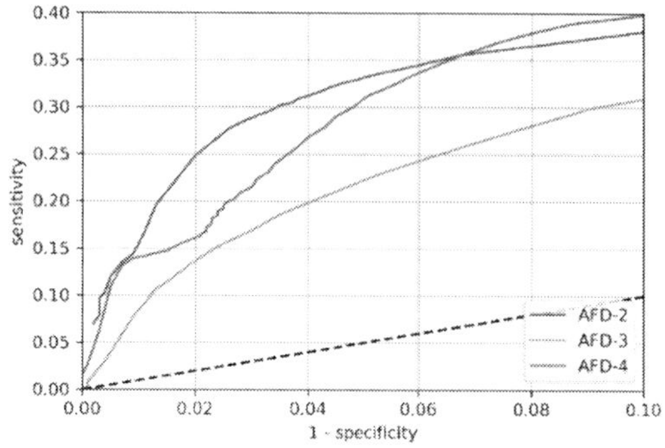

Figure 4: ROC curves obtained for AFD with the same feature (Daily PR) but different comparison methods.

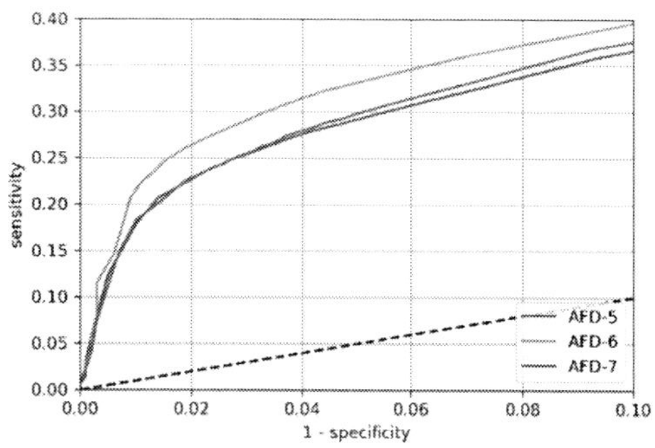

Figure 5: ROC curves obtained for AFD with the same feature (P_{diff}) and comparison method (EWMA), but different reference model.

Comparing AFD 5, 6 and 7 in Figure 5, we can see that the reference built with ARX achieved better results. Polynomial regression showed similar results to the simple ANN implemented here, with the advantage of being much simpler to implement and train.

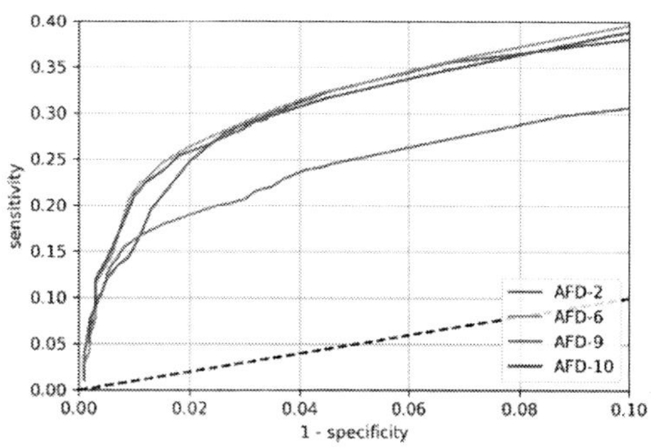

Figure 6: ROC curves obtained for AFD with the same reference model (ARX) and comparison method (EWMA), but different features.

Figure 6 shows a comparison of AFD 2, 6, 9 and 10. We can see a similar performance for the use of daily P_{diff}, and daily CL_{diff}, which could be related to the absence of temperature measurements to better refine the model dependent features. The results with daily PR were almost as good as the best, revealing that this simple feature can also be a good source of information for fault detection.

The AFD methods tested show different sensitivity depending on the fault category (see Figure 7). Despite being the most sensitive method overall, AFD-6 showed relatively low sensitivity to external influence (snow) and installation (misaligned irradiance sensor) faults. These fault categories were better detected by AFD-4 and AFD-7, respectively. Therefore, it is possible to increase sensitivity by using different methods simultaneously.

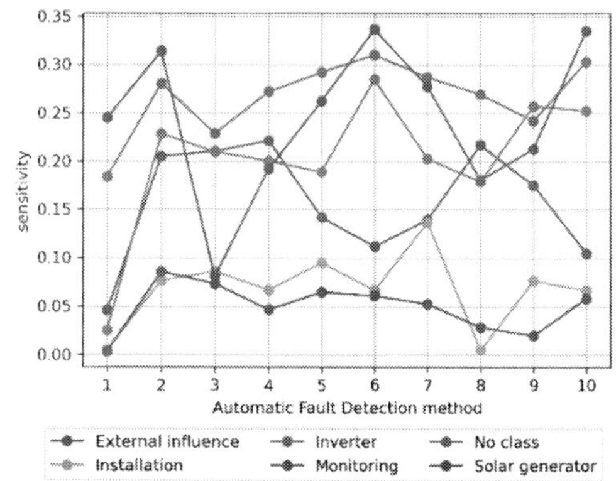

Figure 7: Sensitivity of the AFD according to the fault category.

It is expected that the increase in the specificity leads to the decrease in the sensitivity. However, none of the AFDs reached sensitivity higher than 40% while also maintaining high specificity (above 90%). It is important to note that maintenance records are not a perfect reference, they reflect the actions of the inspection and maintenance staff, but there may be undetected faults. Thus, part of the false alerts may actually be undetected failures. On the other hand, maintenance records are not necessarily associated with problems that significantly affect system performance. Some records are repeated daily even if there is no power loss at the time, e.g., insulation faults in the solar generator may not cause problems every day, only on rainy days. This could explain the low maximum sensitivity observed in all ROC curves. Nonetheless, maintenance records remain the best record of the plant's history and the best available reference for measuring the quality of AFD methods.

The difficulty in achieving high specificity may also be due to the natural presence of noise in the measurements. This reinforces the importance of using a noise-resistant comparison method. Alternatively, an additional filtering step could be used to identify persistent alerts before triggering a failure alert.

6 FINAL CONSIDERATIONS AND OUTLOOK

A list of methods for automatic failure detection available was tested using field data from a portfolio of 85 commercially operated PV power plants located in Germany. These methods are inexpensive to implement. The best performing one was able to identify almost one third of the faults observed in PV systems. The results suggest that a combination of methods can increase the overall sensitivity.

The information required for adapting any of these methods for different power plants can be obtained from the system planning (e.g. equipment's rated power) or extracted directly from the measurements (e.g. the average and standard deviation). At least one year of measurements might be needed for tuning the methods' parameters.

The continuation of this research will investigate the power loss associated with faults and the sensitivity of the methods with respect to power loss. Increasing the sensitivity by combining AFD methods and adapting them for different scenarios will also be studied.

7 ACKNOWLEDGEMENTS

The authors would like to thank all co-workers at Fraunhofer ISE, Pohlen Solar, Enmova and at IDMEC for supporting this work. This work was partly supported by FCT, through IDMEC, under LAETA, project UID/EMS/50022/2019. This work was partly financed by the Federal Ministry for Economic Affairs and Energy of Germany (BMWi) within the framework of the research program "Innovations for the energy transition" funded under the contract number 03EE1058.

8 REFERENCES

[1] S. R. Madeti and S. N. Singh, "A comprehensive study on different types of faults and detection techniques for solar photovoltaic system," *Solar Energy*, vol. 158, pp. 161–185, Dec. 2017.

[2] A. Mellit, G. M. Tina, and S. A. Kalogirou, "Fault detection and diagnosis methods for photovoltaic systems: A review," Renewable and Sustainable Energy Reviews, vol. 91, pp. 1–17, Aug. 2018.

[3] D. S. Pillai, F. Blaabjerg, and N. Rajasekar, "A Comparative Evaluation of Advanced Fault Detection Approaches for PV Systems," IEEE Journal of Photovoltaics, vol. 9, no. 2, pp. 513–527, Mar. 2019.

[4] M. N. Akram and S. Lotfifard, "Modeling and Health Monitoring of DC Side of Photovoltaic Array," IEEE Transactions on Sustainable Energy, vol. 6, no. 4, pp. 1245–1253, Oct. 2015.

[5] M. Alajmi and I. Abdel-Qader, "Fault detection and localization in solar photovoltaic arrays using the current-voltage sensing framework," in 2016 IEEE International Conference on Electro Information Technology (EIT), May 2016, pp. 0307–0312.

[6] M. Davarifar, A. Rabhi, A. E. Hajjaji, and M. Dahmane, "New method for fault detection of PV panels in domestic applications," in 3rd International Conference on Systems *and Control*, Oct. 2013, pp. 727–732.

[7] E. Garoudja, K. Kara, A. Chouder, S. Silvestre, and S. Kichou, "Efficient fault detection and diagnosis procedure for photovoltaic systems," in *2016 8th International Conference on Modelling, Identification and Control (ICMIC)*, Nov. 2016, pp. 851–856.

[8] L. L. Jiang and D. L. Maskell, "Automatic fault detection and diagnosis for photovoltaic systems using combined artificial neural network and analytical based methods," in *2015 International Joint Conference on Neural Networks (IJCNN)*, Jul. 2015, pp. 1–8.

[9] A. Umana and A. P. S. Meliopoulos, "Detection of cell-level fault conditions within a photovoltaic array system," in *2016 IEEE/PES Transmission and Distribution Conference and Exposition (T D)*, May 2016, pp. 1–5.

[10] M. Tadj, K. Benmouiza, A. Cheknane, and S. Silvestre, "Improving the performance of PV systems by faults detection using GISTEL approach," *Energy Conversion and Management*, vol. 80, pp. 298–304, Apr. 2014.

[11] Y. Hu *et al.*, "Online Two-Section PV Array Fault Diagnosis with Optimized Voltage Sensor Locations," *IEEE Transactions on Industrial Electronics*, vol. 62, no. 11, pp. 7237–7246, Nov. 2015.

[12] A. E. Lazzaretti *et al.*, "A Monitoring System for Online Fault Detection and Classification in Photovoltaic Plants," *Sensors*, vol. 20, no. 17, Art. no. 17, Jan. 2020.

[13] Z. Yi and A. H. Etemadi, "Fault Detection for Photovoltaic Systems Based on Multi-Resolution Signal Decomposition and Fuzzy Inference Systems," *IEEE Transactions on Smart Grid*, vol. 8, no. 3, pp. 1274–1283, May 2017.

[14] Y. Zhao, B. Lehman, R. Ball, J. Mosesian, and J.-F. de Palma, "Outlier detection rules for fault detection in solar photovoltaic arrays," in *2013 Twenty-Eighth Annual IEEE Applied Power Electronics Conference and Exposition (APEC)*, Mar. 2013, pp. 2913–2920.

[15] E. A. Sarquis Filho, F. C. Santos, and P. J. Costa Branco, "Development of Predictive Maintenance Algorithms for Photovoltaic Systems Using Synthetic Datasets," in *37th European Photovoltaic Solar Energy Conference and Exhibition*, Oct. 2020, pp. 1584–1589.

[16] G. Wang, C. C. Youn, and A. M. Stankovic, "DC-side high impedance ground fault detection for transformerless single-phase PV systems," in *2015 North American Power Symposium (NAPS)*, Oct. 2015, pp. 1–6.

[17] K. Kiefer, B. Farnung, B. Müller, K. Reinartz, I. Rauschen, and C. Klünter, "Degradation in PV Power Plants: Theory and Practice," in *36th European Photovoltaic Solar Energy Conference and Exhibition*, Oct. 2019, pp. 1331–1335.

[18] N. Gokmen, E. Karatepe, B. Celik, and S. Silvestre, "Simple diagnostic approach for determining of faulted PV modules in string based PV arrays," *Solar Energy*, vol. 86, no. 11, pp. 3364–3377, Nov. 2012.

[19] E. Garoudja, F. Harrou, Y. Sun, K. Kara, A. Chouder, and S. Silvestre, "Statistical fault detection in photovoltaic systems," *Solar Energy*, vol. 150, pp. 485–499, Jul. 2017.

[20] S. Silvestre, A. Chouder, and E. Karatepe, "Automatic fault detection in grid connected PV systems," *Solar Energy*, vol. 94, pp. 119–127, 2013.

[21] W. Heydenreich, B. Müller, and C. Reise, "Describing the World With Three Parameters: A New Approach to PV Module Power Modelling," in *23rd European Photovoltaic Solar Energy Conference and Exhibition*, 1-5 September 2008, Valencia, Spain, Nov. 2008, pp. 2786–2789.

[22] D. E. Guzman Razo, B. Müller, H. Madsen, and C. Wittwer, "A Genetic Algorithm Approach as a Self-Learning and Optimization Tool for PV Power Simulation and Digital Twinning," *Energies*, vol. 13, no. 24, Art. no. 24, Jan. 2020.

[23] T. Fawcett, "An introduction to ROC analysis," *Pattern Recognition Letters*, vol. 27, no. 8, pp. 861–874, Jun. 2006.

[24] T. Shimakage, K. Nishioka, H. Yamane, M. Nagura, and M. Kudo, "Development of fault detection system in PV system," in *2011 IEEE 33rd International Telecommunications Energy Conference (INTELEC)*, Oct. 2011, pp. 1–5.

FAULT DETECTION IN OPERATION AND MAINTENANCE OF PV SYSTEMS

Atse Louwen[1*], Daniele Miorandi[2], Christian Torrero[2], Federico Venturini[2], David Moser[1]
[1]EURAC Research, Bolzano, Italy - atse.louwen@eurac.edu; david.moser@eurac.edu
[2]U-Hopper, Trento, Italy - daniele.miorandi@u-hopper.com, christian.torrero@u-hopper.com, federico.venturini@u-hopper.com
*Corresponding author: Viale Druso 1, 39100 Bolzano, Italy. Phone: +39 0471 055 736

ABSTRACT: Concurrently with the strong decline in installed PV system prices, the margin for operation and maintenance (O&M) operators has decreased, as O&M costs are commonly budgeted as a fixed percentage of installed system costs. Hence, there is a need to improve the efficiency of O&M operations. In the PV4.0 project, we are aiming to improve O&M efficiency across the value chain, by implementing Industry 4.0 concepts. One key part of these concepts is to implement an automatic fault detection system, that is used to quickly detect faults in PV plant operation, minimizing electricity generation losses. Furthermore, this fault detection system is to be used to automatically generate maintenance tickets, so corrective maintenance can be performed as soon as possible. In this paper, we present the development of two neural network (NN) based machine learning models. Model A is a four-layer NN, trained to classify individual datapoints as either normal or faulty. Detected faults are subsequently classified as either a hard (zero power output) or soft (power output higher than zero but lower than nominal) failures. Model B is a model that attempts to predict whether the number of failures in the next days exceed a certain threshold. Our results indicate that Model A has a fault detection rate of 79% in our testing dataset, while the performance of Model B is lower, at a fault detection rate of 49% for 1-day-ahead predictions, and only 30% for 2-day-ahead predictions. Further work will focus on improving the failure detection rate for both models. For Model B, we will additionally focus on improving the input dataset, so that better predictions might become possible. Model A will be implemented in a real-time O&M environment to test its performance in real operating conditions.
Keywords: PV system monitoring; fault detection; machine learning; neural networks; classification; prediction

1 INTRODUCTION

In the past decades, the global installed capacity of PV systems has dramatically grown. While incentive schemes have incentivised PV system installations especially in early years, this strong growth can mainly be attributed to a very strong decrease in installed PV system costs. As a result of these strong cost reductions, and of the fact that Operation and Maintenance (O&M) costs are commonly budgeted as a fixed percentage of installed system costs, financial margins for O&M contractors are decreasing. In order to make O&M more cost-effective, and in order to (further) improve the performance of PV systems, the efficiency of O&M operations should increase. By applying digitalization along the O&M value chain, operations can become more cost-effective, and one key issue here is the timely and automatic detection of faults in PV plant operation that leads to a loss in electricity generation and thus revenue. Automated fault detection could decrease at the same time the cost of O&M by reducing labour intensive tasks and requirements, as well as potentially decreasing the plants levelized cost of electricity when faults are detected more quickly. Further gains can be achieved if faults can be predicted from a comprehensive set of monitoring data, where the aim is to prevent electricity generation losses from happening by signalling a possible upcoming fault and sending a field operator to inspect and fix any found issues. Fault detection is a very active research field in PV, with a variety of approaches being studied.

Within the PV4.0 project [1], we are developing a software platform aimed at improving the efficiency of operation and maintenance (O&M), by integrating and fully digitalizing O&M workflows. In this project, we have already developed and implemented an asset management system to keep track of all PV plant components, including a geo-localisation of these assets in the field. A ticketing system to track PV plant failures and planned and corrective maintenance activities has been linked to this asset management system, and a smartphone application has been developed, in order to allow for efficient and standardised data exchange with these systems, as well as to allow operators in the field to create and process maintenance tickets.

Still under development is a decision support system linked to the ticketing system, that leverages all plant information and the CPN methodology [2, 3] to prioritise tickets and suggest optimal corrective maintenance procedures. A final and key part of this platform will be a subsystem that analyses plant monitoring data in real-time, to automatically detect faults, triggering automated generation of maintenance tickets and the associated decision support system. Here, we present the development of two models that aim to detect faults in real-time, and to leverage the results of this first model to attempt to predict failures before they cause losses in electricity generation.

2 METHODS

The fault detection algorithm will consist of two different models, to 1) detect the presence of failures resulting in loss of energy generation, and 2) to attempt to predict failures before they cause a loss in energy generation. Several subroutines are preliminarily used to label the records in the dataset as either "Normal" or "Failure", i.e., they serve the purpose of identifying historical failures in the training data. Subsequently, using said labels, the first part of the algorithm, from here on named **Model A**, implements a classification by means of a Neural Network (NN) [4] that will be used to label new data points as either normal or faulty operation. Failures are then given a preliminary classification as being either a hard failure (zero power output or failure affecting more than half of inverters), soft failure (failure with power output higher than zero).

The second part of the algorithm, from here on named **Model B**, investigates the application of Long Short-Term Memory (LSTM) NNs [5], to predict the overall behaviour of an inverter in the next days to come, i.e., whether the inverter will display a (relatively) large number of daily failures or not.

2.1 Input Data

For our research we used a dataset containing 4 years of operation (between 2014 and 2018) of a 7MW PV plant located in Southern Europe. Due to confidentiality, the exact location of the PV plant is not specified.

The PV plant consists of 6 subsystems, each with two inverters, for a total of 12 inverters. The data were recorded at 10-min resolution and include measurements of DC and AC voltage of each inverter (three phases), active and reactive power, power delivered to the grid, grid frequency, power factor and temperature of the MPPT tracker. Global horizontal (GHI) and plane-of-array (GPOA) irradiance, weather parameters (temperature, wind speed, wind direction, relative humidity, and precipitation) are also available.

Figure 1 shows a simple overview of our input dataset. As can be seen in this figure, inverter 12 shows a low power output during the whole 4-year period of measurement, while the nominal power and connected PV module capacity of this inverter is similar to the others. Hence, this inverter was completely excluded from this analysis.

2.2 Data pre-treatment

Both manually and subsequently using several heuristics subroutines, the raw input data were pre-treated, to label obviously faulty data. Also, night-time hours, as well as low irradiance measurements (using a POA irradiance threshold of 25 W/m^2), were labelled and excluded from training and testing datasets Moreover, as one of the inverters seems to be operating in a soft failure mode during the whole 4-year period of measurement, this inverter was excluded from further analysis.

After excluding records as explained here above, approximately 70% of the remaining ones were used for training, 20% for validation and 10% for testing purposes. The dataset was split in chronological order, the training set being composed of the oldest records, with the validation and test sets containing the most recent 30% of the data.

2.3 Technical description of Model A

The first part of our analysis focuses on the development of a classification model that labels individual datapoints as being representative for either "normal" or "faulty" operation.

The model is a feed-forward NN made up of four layers, each one displaying standard Deep learning tools like *Batch Normalization* and *Dropout*. The former is a technique that smoothens the profile of the loss function (i.e., the cross entropy) and that, as such, simplifies and speeds up the search for a suitable minimum. The latter is a tool that prevents the model from overfitting, i.e., allows the model not to worsen its performances when applied to brand new data. The non-linearity interspersed in the layers is the *Scaled Exponential Linear Unit* (SELU), though tests showed that - essentially - any other function would yield comparable results.

Prior to being inputted into the model, data underwent standard pre-processing operations: missing values were replaced with values obtained from time interpolation, categorical variables were one-hot encoded in order to be converted into numbers, and each numerical attribute underwent a quantile transformation to map its values into a tight range centred at 0 (empirically, said transformation – and alternative/similar ones – proves to be highly beneficial during model training). Moreover, "interaction" features were added to the initial set of attributes by multiplying every couple of numerical variables displaying a poor correlation (i.e., lower than 0.1 - 0.15).

Software-wise, the NN and said pre-processing transformations were implemented within Jupyter notebooks making use of standard Python libraries like - among others - PyTorch (for initializing and training the model), NumPy and pandas (for pre-processing).

2.4 Technical description of Model B

Using the results from Model A described in the section above, a subsequent model was developed and trained, attempting to predict the occurrence of failures in the next two days.

This model, Model B, is a 1-layer LSTM network taking as input, for each individual inverter, a sequence of integers corresponding to the number of failures displayed, day by day, by the inverter in the time interval between the

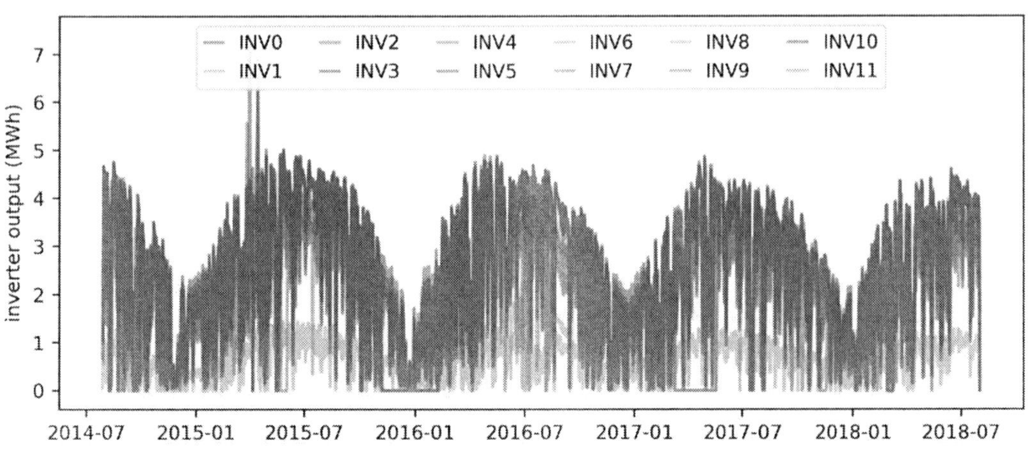

Figure 1: Overview of the power output of all 12 inverters over the whole time period of the input dataset. The graphs shows the daily electricity generation for all inverters. Here it can be noted that inverter 12 has performance issues over the whole measurement period.

last maintenance and the current date. Said sequence, which is obviously arranged so to respect the chronological order, is used by the model to classify the two days to come as "red" or "green", depending on whether the number of daily failures will be larger or smaller than 10 (representing a failure time of 1.67 hours). In order to have a final output with the desired shape, the output of the LSTM is fed into a fully-connected layer, the latter being a component of the model that gets trained together with the LSTM itself.

The whole model, trained by minimizing the binary cross entropy, is provided with *Dropout* but not with *Batch Normalization* because of the sequential handling of input data.

It is worth stressing that Model B works as a classifier and not as a regressor, i.e., it returns the "red"/"green" labels directly, without outputting any integer corresponding to an estimated number of failures for the next days.

Model B was implemented and trained using the same software tools as Model A, i.e., Python libraries like PyTorch, NumPy and pandas within a Jupyter notebook.

3 RESULTS

The results of the model training and testing are shown below. After pre-treatment of the data the NN was trained using the training dataset consisting of the first 70% of the whole dataset. Our testing results use the last 10% of the dataset. Our results indicate that when excluding night-hour records, 3.7% of the data was found to suffer from one of the two mentioned failure modes.

3.1 Model Training

The data - labelled with the heuristic subroutines - were used to train the neural network to solve a standard classification problem. The neural network takes all available (i.e., weather and electrical) features as input, in order to try to predict in a binary way which label ("failure" or "normal") should be assigned.

Shown in Figure 2, the prediction loss in the testing dataset stabilizes after around 150-200 epochs of model training. In this graph it can be clearly observed that the loss for the training dataset (red line) is always below that of the testing dataset (blue line) which is the desired/typical behaviour. The relatively large distance

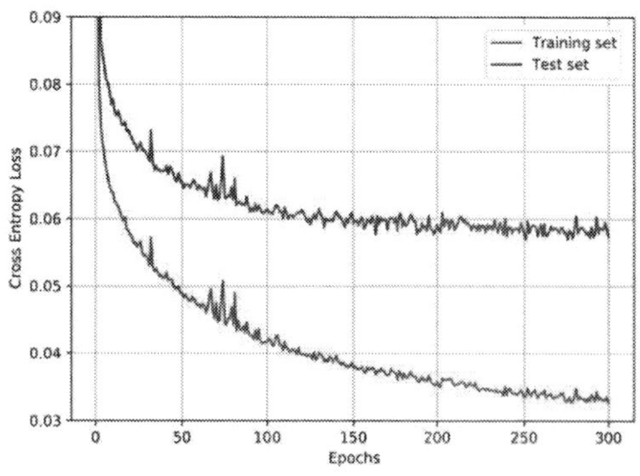

Figure 2: Behaviour of the cross entropy loss for the training and test set, as a function of the training progress measured in epochs, for the classification model described in Section 2.3.

between the two lines might reek of overfitting on the training dataset (thus leading to a less generalizable model), as - ideally - these two lines should lie very close to each other. However, the gap between training and test loss is not known a priori and can very well be non-negligible. Furthermore, several model setups resulting into almost overlapping training and test losses were examined, but all of them showed much lower performances for both the training and the test set. Finally, since the dataset was split into training, validation and test set, we were able to double-check the stability of the model performances on unseen data and to exclude any overfitting more confidently.

In Figure 3, the behaviour of the loss function for the prediction model is shown. Even though, after sufficient training epochs, the loss scores for both training and test set reach a plateau where the loss values of both datasets are relatively close, nevertheless the test loss lies *below* the training loss and not (slightly) above as it should be: this, together with the fact that this mutual position holds throughout the whole model training, is a sign that Model B does not perform as hoped. In principle, the fact that the test loss is below the training loss could be explained saying that the test set is a particularly favourable partition of the original dataset where the model is able to perform better than the training set. However, this could happen when the test set is very small compared to the training set, but this does not hold in the current case because, in terms of number of samples, the test set is big enough to avoid this phenomenon.

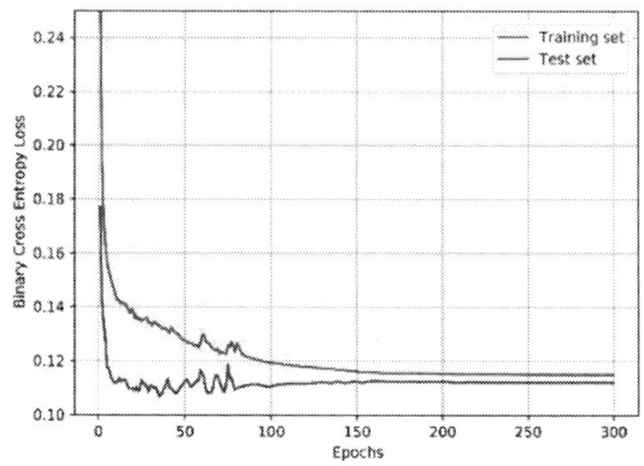

Figure 3: Behaviour of the binary cross entropy loss for the training and the test set, as a function of the training progress measured in epochs, for the prediction model described in Section 2.4.

3.2 Validation of the classification model results

In this section, we show and discuss the results of the classification model (Model A) by displaying the so-called confusion matrices for the training and test set. A confusion matrix compares the number of datapoints with actual normal or faulty behaviour, with those labelled normal or faulty by the two models.

Altogether, Model A yields quite accurate results. As shown in Table I and Table II, only a limited number of datapoints is erroneously labelled. This NN results into an F-score - computed only on the "Failure" class - of 0.899 for the training set, and of 0.843 for the test set. This corresponds to 1.1% and 1.9% of the datasets being labelled erroneously, respectively. In the training and test sets, roughly 13% and 21% of the failures are incorrectly labelled as being normal operativity. This represents a failure detection rate of just under 80% for the latter.

Further work will mainly aim to improve this particular score. On the other hand, only 0.4% and 0.6% of normal operativity datapoints are incorrectly labelled as faulty operativity for these two sets respectively.

Table I: Confusion matrix for the classification model as computed on the training set

		Actual Class	
		Normal	Failure
Predicted	Normal	828495	6990
Class	Failure	3136	45309

Table II: Confusion matrix for the classification model as computed on the test set

		Actual Class	
		Normal	Failure
Predicted	Normal	92534	1348
Class	Failure	561	5133

Figure 4 shows the number of faults that our algorithm found for all inverters, while Figure 5 displays the occurrence of selected fault labels over time, as an indication of the high number of failures in this dataset. In the latter figure, we only show the faults that occur at least for 2 hours during a day. Figure 6 shows an example of some of the detected faults for a selected set of inverters and period of time.

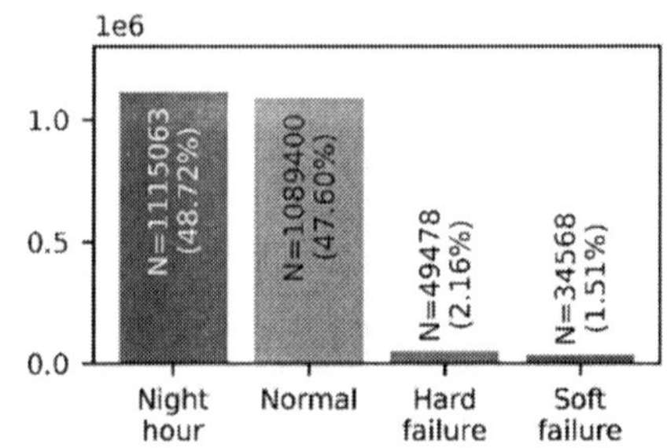

Figure 4: Overview of absolute and relative label counts in each category

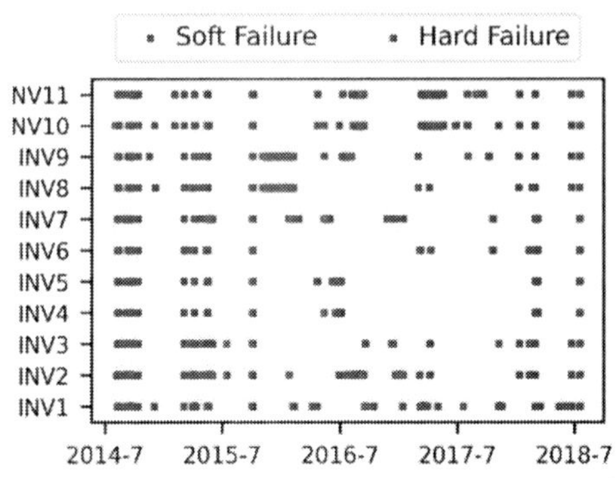

Figure 5: Overview of errors identified for the 11 inverters being studied. Blue-green markers indicate "soft-failures": failures where power was less than expected but larger than zero. Dark red areas indicate "hard failures" (power is zero while irradiance is greater than zero). Only days with failures that are persisting for 4 hours or more are shown here.

Out of roughly 2.5 million entries, expectedly almost half are labelled as night-time hours. Roughly 3.7% of all records fall into one of the two fault categories. Excluding night-time hours, this figure increases to 7.2% of the remaining records.

3.3 Validation and analysis of the prediction model results

The results for the prediction model (Model B) are shown below. This model was used to predict whether a day in the future should be classified as "green" (less than 10 faulty operation points) or "red" (at least 10 faulty operation points). These predictions were carried for the next two days to come.

First, we show the confusion matrices for the model when used to predict one day ahead, for the training and the test set in Tables III and IV respectively. In Tables V and VI, we show the same results when the model is used to make predictions two days ahead.

From the results in Tables III and IV, we can observe that the prediction model yields much less accurate results as the classification model. Almost half of the red days are correctly predicted as such in the test set and no days are incorrectly labelled as red days (Table IV).

Table III: Confusion matrix for the prediction model as computed on the training set (1-day-ahead predictions).

		Actual Class	
		Green	Red
Predicted	Green	10282	157
Class	Red	177	654

Table IV: Confusion matrix for the prediction model as computed on the test set (1-day-ahead predictions).

		Actual Class	
		Green	Red
Predicted	Green	1595	19
Class	Red	0	18

For the two-day ahead predictions, shown in Table V and VI the results were found to be substantially less accurate. In the test set, again no days are incorrectly labelled as red days, but the number of days correctly predicted to be red days drops to 30%.

Table V: Confusion matrix for the prediction model as computed on the training set (2-day-ahead predictions).

		Actual Class	
		Green	Red
Predicted	Green	10249	231
Class	Red	155	635

Table VI: Confusion matrix for the prediction model as computed on the test set (2-day-ahead predictions).

		Actual Class	
		Green	Red
Predicted	Green	1578	38
Class	Red	0	16

4 CONCLUSIONS

The aim of this work was to develop two machine learning based models for 1) fault detection in individual datapoints (Model A) and 2) fault prediction using the results from the fault detection model (Model B). To this

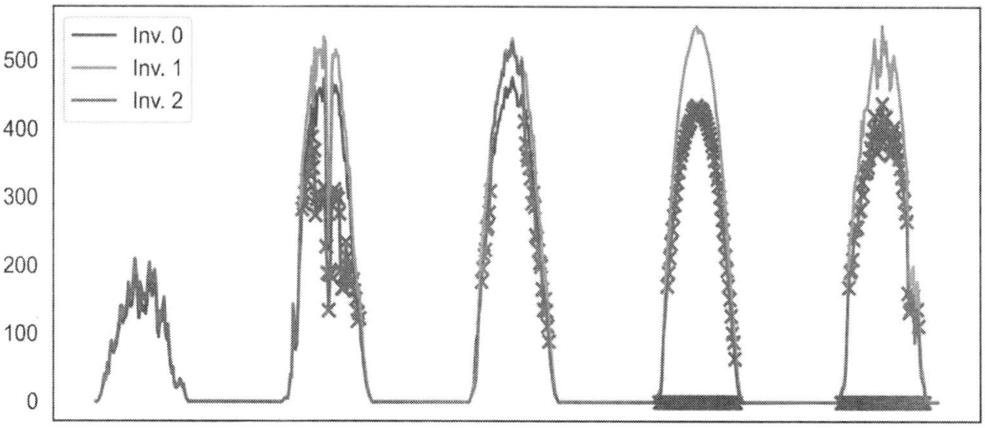

Figure 6: Example of detected faults in both categories. Inv. 1 (orange) shows normal operativity in the whole period while Inv. 2 shows a "soft failure" (indicated by green crosses) and Inv. 0 shows a "hard failure" for the last 2 days of this plot.

end, we developed and analysed two neural network-based models.

Model A is a four-layer NN model that was developed to classify individual datapoints as either faulty or normal. This model was trained on a dataset with fault labels that were determined using manual approach This model was able to classify faults with an F-Score of 0.843 in our test set, corresponding to a failure-detection rate of approximately 79%, and a false positive rate of only 0.6%.

Model B is also an NN based model, which was developed to predict the occurrence of failures one or two days ahead. This model classifies the upcoming days as either "green" (with less than 10 failures) or "red" (with at least 10 failures). Here, we found that this model produces much less accurate results compared to Model A, which was in line with our expectations. The one-day ahead predictions correctly labelled roughly 50% of red days, while in the two-day ahead predictions, only about 30% of red days were correctly predicted. In both one-day- and two-day-ahead results, no days were incorrectly labelled as red days (e.g., the false positive rate was zero), though tis might likely be a coincidence related to the test set used in the current study.

Further work on Model A will mainly focus on reducing the number of false negatives and on investigating the robustness of the algorithm when using lower-quality input data. Furthermore, the model will be deployed in an O&M setting to analyse its performance under real operating conditions and to automatically detect faults within real-time data. In this context, the model will also be used for automatic generation of maintenance tickets. Additionally, we will investigate the expected improvements when leveraging a peer-to-peer comparison of inverters in this fault detection model, which was not yet included in the current implementations. Finally, we will analyse how the model should be optimised for real-time detection, considering for instance a moving window for the training dataset to ensure the model can correct for gradual performance degradation.

As for the fault prediction model (Model B), further work will focus on improving the failure-detection rate, although we consider the potential for improvement of this model to be limited. In fact, contrary to mechanical systems that usually undergo a slow deterioration (often preceded by some symptoms - think, for example, to the sort of "whistle" heard when brake pads start being worn out), malfunctions in electrical systems typically occur suddenly with scarce (if any) signals. Hence, there is limited information in the currently included input variables that would allow to detect any sudden failure. Although PV systems exhibit gradual performance losses, these are not the types of failures we are trying to identify here. Besides studying alternative means of prediction (using, for example, approaches based on clustering of the available input parameters), we speculate the current setup of Model B might nevertheless yield better results when inputted a dataset cleaner than the one currently at our disposal. In fact, in this dataset, data were not annotated manually to indicate past failures, moreover, no information about corrective or scheduled maintenances (instrumental for the workflow of Model B) was made explicit but, rather, had to be indirectly inferred from the data.

Among the algorithms that were developed, especially Model A offers the potential to make an important step towards fully digitalized O&M workflows, by automatically labelling faulty data from real PV plants (thus allowing for quicker field-intervention). The first step will be to deploy and test the model in real-time operational conditions. However, we aim to improve the algorithms up to a fault detection accuracy of at least 95% before using it to create annotated datasets to train other (types of) models.

5 ACKNOWLEDGEMENTS

This work was by the European Regional Development Fund (ERDF) Provincia autonoma di Bolzano-Alto Adige 2014–2020, under Project number FESR1128 "Use of Industry 4.0 and Internet of Things logics in the photovoltaic sector—PV 4.0".

We would like to acknowledge the support of Saidea S.r.l., our key partner in the project PV4.0. Saidea is providing the IT backbone of the PV4.0 platform, and as such is responsible for the implementation of the different components of PV4.0. Among other activities, they have implemented the operation and maintenance ticketing system and the PV plant component management system, and are currently in the process of implementing the fault detection Model A as described in this paper, and will support us in real-world pilot testing of this model.

6 REFERENCES

[1] A. Louwen, L. Koester, S. Lindig, A. Astigarraga, and D. Moser, "PV4.0: Combining Asset Management, PV Measurement Data and the Cost Priority Number Method in a Digital Infrastructure for PV Plant O&M," in *37th European Photovoltaic Solar Energy Conference and Exhibition*, 2020, pp. 1378–1380.

[2] D. Moser, M. Del Buono, U. Jahn, M. Herz, M. Richter, and K. De Brabandere, "Identification of technical risks in the photovoltaic value chain and quantification of the economic impact," *Prog. Photovolt. Res. Appl.*, vol. 25, no. 7, pp. 592–604, Jul. 2017, doi: 10.1002/pip.2857.

[3] G. O. Hernández, S. Lindig, D. Moser, and P. Chiantore, "Optimization of the Cost Priority Number (CPN) Methodology to the Needs of a Large O&M Operator," in *Proceedings of the 36th European Photovoltaic Solar Energy Conference and Exhibition*, 2019, pp. 1613–1617. doi: 10.4229/EUPVSEC20192019-5CV.4.19.

[4] I. Goodfellow, Y. Bengio, and A. Courville, *Deep Learning*. The MIT Press, 2016.

[5] S. Hochreiter and J. Schmidhuber, "Long Short-Term Memory," *Neural Comput.*, vol. 9, no. 8, pp. 1735–1780, Nov. 1997, doi: 10.1162/neco.1997.9.8.1735.

FAILURE DIAGNOSIS AND TREND-BASED PERFORMANCE LOSSES ROUTINES FOR THE DETECTION AND CLASSIFICATION OF INCIDENTS IN LARGE-SCALE PHOTOVOLTAIC SYSTEMS

Andreas Livera[1], Marios Theristis[2], Joshua S. Stein[2] and George E. Georghiou[1]
[1]PV Technology Laboratory, FOSS Research Centre for Sustainable Energy,
Department of Electrical and Computer Engineering, University of Cyprus, Nicosia, 1678, Cyprus
[2]Sandia National Laboratories, Albuquerque, New Mexico, 87185, USA

ABSTRACT: Failure diagnosis (detection and classification) in photovoltaic (PV) systems through data diagnostic algorithms is a fundamental task that ensures quality of operation and significantly improves the performance and reliability of operating PV systems. The scope of this work is to present the development of Failure Diagnosis Routines (FDRs) and Trend-based performance Losses Routines (TLRs) for diagnosing PV underperformance issues due to failure occurrences and performance loss events. The proposed routines complement the developed Data Quality Routines (DQRs) and operate exclusively on recorded electrical and meteorological measurements. The proposed routines were experimentally validated on a large-scale PV system installed in Larissa, Greece. The results demonstrated the effectiveness of the routines for detecting system underperformance issues and accurately classifying the detected issues into zero power production incidents, degradation, soiling and snow losses. The failure detection stage of the FDRs achieved a detection accuracy of 97.3% for zero power production incidents during daylight hours. A precision accuracy of 96.32% was obtained by the FDRs when classifying zero power production due to fault incidents, while the TLRs achieved 91.66% classification accuracy.
Keywords: change-point techniques, data quality, failure diagnosis, performance loss factors, photovoltaics.

1 INTRODUCTION

As reported recently by SolarPower Europe (SPE), solar photovoltaic (PV) is set to become the dominant source of electricity generation in Europe [1]. While the PV market has grown rapidly, the solar industry is continuously struggling to minimize downtimes and loss of energy generation [2]. In this domain, a main challenge surrounds enhancing the lifetime field performance of PV systems through online performance monitoring and optimized operation and maintenance (O&M) strategies.

Remote and real-time fault diagnosis (detection and classification) in PV systems is a fundamental task that ensures quality of operation and significantly improves the energy generation [3]. Different approaches have already been proposed in the literature for remote and automated fault detection [4–6]. However, distinguishing root causes and accurate classification of PV underperformance issues remains a research challenge.

Since PV power plants continuously generate large amounts of data, data analytic applications are becoming more valuable. Interpreting efficiently the mere data can provide meaningful information about different failures/losses and their causes. Following data-driven insights, PV plant owners can perform maintenance actions to maximize the plant output power and minimize loss of energy generation. However, best practices lack of a methodology for the accurate differentiation of failure types and trend-based performance loss mechanisms (reversible and irreversible).

To bridge the gap that exists in the literature, Failure Diagnosis Routines (FDRs) and Trend-based performance Losses Routines (TLRs) were developed for classifying underperformance issues in large-scale PV systems. The proposed routines operate exclusively on recorded electrical and meteorological measurements and complement the developed Data Quality Routines (DQRs). The FDRs and TLRs incorporate statistical, comparative and change-point (CP) algorithms and focus mainly on differentiating common PV failures and performance losses such as zero and reduced power production incidents, degradation, soiling and snow losses. This is the first attempt in differentiating faults from reversible and irreversible performance losses for optimizing O&M activities (i.e., informing O&M teams about the underperformance issue to act accordingly and recover some of the performance and financial losses). Historical field data from a PV power plant installed in Larissa, Greece were used to benchmark the proposed diagnostic procedure.

2 METHODOLOGY

The proposed methodology for the detection and classification of incidents in PV systems consists of the DQRs for data validity and sanity, the FDRs and TLRs for detecting PV underperformance incidents (due to failures and/or performance losses) and accurately classifying the different fault causes and performance loss factors. The developed routines operate on the readily available recorded measurements (meteorological and electrical data) obtained from the monitoring system.

2.1 Experimental setup

The developed FDRs and TLRs were benchmarked experimentally using a PV dataset containing historical measurements obtained from a PV power plant installed in Larissa, Greece. The PV plant (with an installed capacity of 1.8 MWp) comprises of 7824 poly-crystalline-Silicon (poly c-Si) PV modules of 230 W_p. It is installed in ground mounted configuration, at a fixed tilt angle of 25° due to South. The PV modules are connected in series to form 326 strings at the inputs of 4 grid-connected inverters (82 strings are connected to Inverter 1 and 3, while 81 strings are connected to Inverter 2 and 4) and there are 4 transformers.

The PV system performance and the prevailing meteorological conditions were recorded according to the requirements set by the IEC 61724-1 [7] and stored with the use of a monitoring platform. The monitoring platform comprises of solar irradiance, wind, temperature and electrical operation sensors connected to a central data acquisition (DAQ) system that stores measurement data at a resolution of 1 second. The recorded meteorological measurements include the in-plane irradiance (G_I) measured with a pyranometer, ambient temperature (T_{amb}), module back-surface temperature (T_{mod}), inverter temperature (T_{inv}), relative humidity (RH), wind speed (W_s) and direction (W_a). The electrical data include the string DC current (I_{DC}), voltage (V_{DC}) and power (P_{DC}) and AC output power (P_{out}). Additional yields and performance parameters such as the PV array energy yield (Y_A), the final PV system yield (Y_f), the reference yield (Y_r) and the monthly DC performance ratio (PR) were also calculated [8]. Lastly, weather data (e.g., snowfall and rainfall measurements) were sourced from Modern-Era Retrospective analysis for Research and Applications, Version 2 (MERRA-2) [9]. These were downloaded at hourly intervals through the soda-pro web interface.

2.2 PV dataset of observed measurements

The recorded outdoor field measurements as well as the calculated performance parameters and yields, were used to create the test PV dataset. The dataset contains 15-minute average measurements obtained from the 4 grid-connected inverters. Over the evaluation period (April 01, 2013 to December 31, 2018), different types of underperformance incidents (i.e., plant was offline, low DC power production, soiling, degradation, etc.) occurred during the operation of the PV plant. Information about the outage's period, failure types and loss mechanisms were recorded in a maintenance (incidents) log.

It is worth noting here that the issues reported in the log and their classification are highly dependent on the monitoring system level (i.e., module, string and/or inverter level). It should also be noted that the test PV plant is currently monitored by an asset solar company; thus, the detected failures were resolved (i.e., repaired) within the agreed response time (as stated in the contract made between the plant owner and the O&M company) [10].

2.3 Data Quality Routines

Initially, the DQRs were applied to the constructed PV dataset to ensure data validity. The DQRs methodology was used to detect invalid measurements (that may indicate equipment malfunctions and/or faulty PV operation) and to provide insights about possible failures and their root causes [2]. The DQRs process comprises of 8 steps, including initial data statistics (Step 1), consistency examination (Step 2a), identification of monitoring availability (Step 2b - additional checks added to Step 2 for failure detection), data filtering (Step 3), detection of invalid data (Step 4), determination of missing data mechanism and rate (Step 5), treatment of invalid values and dataset reconstruction (Step 6), data aggregation (Step 7) and final data statistics (Step 8) [11].

2.4 Power simulation model

For the predicted DC power, the simplified version of the Sandia PV Array Performance Model (SAPM) [12] proposed by Huld et al. [13] was used. The power model was selected due to the smaller number of input parameters (only the in-plane irradiance and module temperature measurements are required) and its high prediction accuracy for a range of PV module technologies (especially for c-Si PV modules) under different climatic conditions (e.g., clear-sky, cloudy and partly cloudy conditions) [14].

The model was trained using in-plane irradiance and module temperature measurements based on a 10:90% train and test set approach. The train set (i.e., covering approximately a period of 6 months) contained fault free data that were used for the model's training process (deriving the model's coefficients), while 54 months of data (containing normal and fault conditions) were used for the test process. The goodness of the model's fit was evaluated using the correlation coefficient (R), the coefficient of determination (R^2), the mean absolute percentage error (MAPE) and the root mean square error (RMSE) [15]. The predictive model was then used as reference models in the failure diagnostic procedure by performing comparisons between the actual (measured) and predicted (simulated) power.

2.5 Failure Diagnosis Routines

The FDRs comprise of a failure detection and a classification stage. The detection stage (depicted in Fig. 1) was based on a comparative assessment between the measured and simulated energy yield against fixed and/or dynamic threshold levels (TLs). More specifically, the measured DC power production of the PV plant was compared to the predicted power production for each data sample. A failure was detected when the absolute error (AE), defined as the absolute difference between the predicted and measured DC power, exceeded the TL.

In this work, a fixed TL was used. The set TL was calculated by multiplying the power of the array at Standard Test Conditions (STC) with the combined yield uncertainty of the model. The combined uncertainty was calculated by deriving the partial derivatives of the model's inputs [16]. Upper and lower control limits (UCL and LCL) of the daily power production were also calculated for determining the acceptable power deviation range [17]. Finally, outlier detection algorithms were also employed to detect data anomalies in electrical measurement data and in PV performance time series and profiles [17].

Once fault conditions (power reduction from 20-100%) were identified, a classification procedure was followed to classify the incidents into different categories; zero power production incidents (power reduction from 85-100%), reduced current, voltage, or both (power) production incidents (power reduction from 20-85%).

The classification stage of the FDRs was based on data-driven algorithms, which were used to postprocess the pattern of the affected electrical measurements through the application of logic trees. For the classification procedure, environmental conditions (i.e., snow, humidity, rainfall measurements) were also utilized to correlate PV system underperformance with the site weather conditions. To assess the performance of the classification algorithms, the precision accuracy metric was used [18].

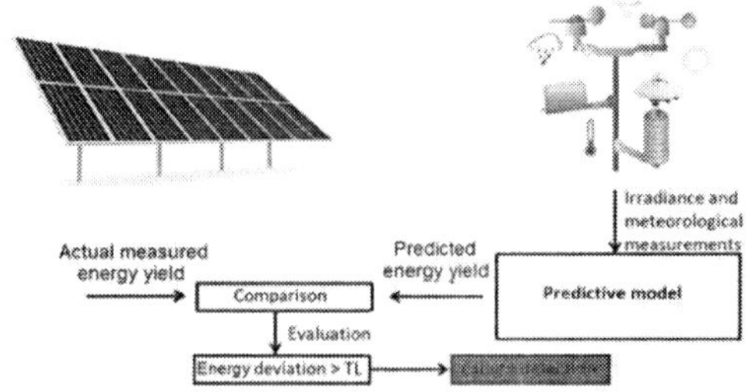

Figure 1: Schematic diagram of the failure detection stage of the FDRs.

2.6 Trend-based performance Losses Routines

Trend-based performance losses refer to linear and nonlinear drops in performance time series and profiles that may affect the produced power of a PV system by up to 20% [19]. In this work degradation (irreversible mechanism), soiling (reversible mechanism) and snow (reversible mechanism) losses were investigated.

The TLRs for degradation and soiling loss were extracted by utilizing the *RdTools* open-source python library [20]. In particular, the Year-on-Year (YoY) method was utilized for estimating the linear degradation rate (R_D). In parallel, the stochastic rate and recovery (SRR) method [21] was used for estimating the soiling profiles from the daily DC performance data. The Combined Degradation and Soiling (CODS) method was then used for distinguishing degradation from soiling loss [22].

Finally, snow losses were detected by correlating low PV performance due to loss incidents with the site's weather conditions (e.g., snowfall and ambient temperature measurements). The low PV performance was detected by leveraging an anomaly detection algorithm (i.e., the Seasonal Hybrid Extreme Studentized Deviates [S-H-ESD] algorithm) [2].

2.7 Failures and losses differentiation

A CP algorithm was used for differentiating fault occurrences from loss events as well as for distinguishing reversible from irreversible performance losses. In particular, the Facebook prophet (FBP) algorithm was used [23,24]. FBP detects the number and location(s) of CP(s) by capturing statistical changes in the slopes of defined segments. It distributes "potential" CPs uniformly along the selected range of the time series' trend and then compares the slopes to extract the significant CPs (by performing comparisons against a set TL). The FBP algorithm calibration process was performed as reported by Theristis et al. [23] (the number of changepoints to be distributed was set to 100, the *changepoint_range* argument was set to 1, and the flexibility, *changepoint_prior_scale* argument, was varied from 0.05 to 2.5 for the PR time series).

The flexibility of the algorithm was initially adjusted (set to 2.5) to capture changes due to performance loss factors and then re-adjusted (set to 0.05) to capture degradation rate changes only by avoiding the influence of outliers (e.g., faults) and temporary effects (e.g., soiling and snow losses).

For validation purposes, the exact linear time (PELT) algorithm was used [25]. The PELT technique can detect both changes in the mean and variance. Changes in the mean indicate performance loss events, while changes in the variance indicate a fault occurrence.

3 RESULTS

3.1 Data quality routines

The DQRs were initially applied to the test dataset to include daylight measurements only (irradiance values between 20 W/m^2 and 1300 W/m^2). The process did not include data imputation, neither correction to fully capture the exhibited patterns during fault/loss conditions. The DQRs methodology was then used for providing insights about possible failures and their root causes. This was achieved by applying physical limits and performing statistical/comparative tests.

Visual inspection of diagnostic plots (such as the power-irradiance scatter plot, power production and performance series time plots) was also deemed sufficient for detecting PV operational problems (such as zero or reduced power production, inverter clipping) and data issues (such as gaps/holes due to communication loss, storage faults, outages, missing measurements). PV operational and data quality issues that can be detected by visual inspection of the DC power time series plot are depicted in Fig. 2.

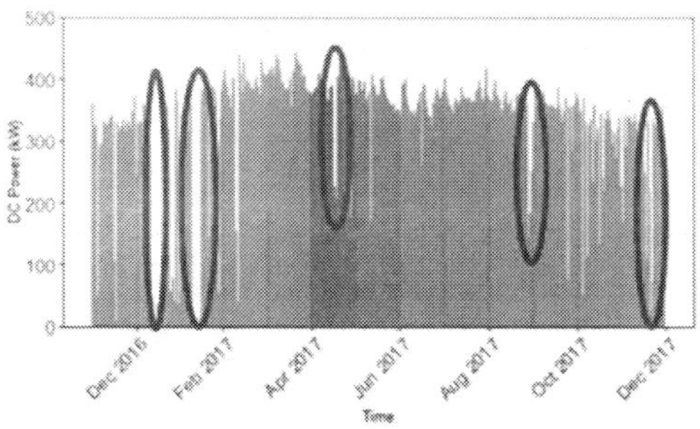

Figure 2: Measured DC power of inverter 1 over a 1-year time period. Examples of detected PV operational and data quality issues are circled in red.

3.2 Normal operation

Normal operation describes the fault-free operation of the PV system. During normal operating conditions, the PV system is affected only by environmental factors. In such cases, the measured DC power should be in good agreement with the predicted power (see Fig. 3). Over the test period, the Huld model predictions exhibited an R^2 of 0.99, a MAPE of 8.6% and a RMSE of 0.13.

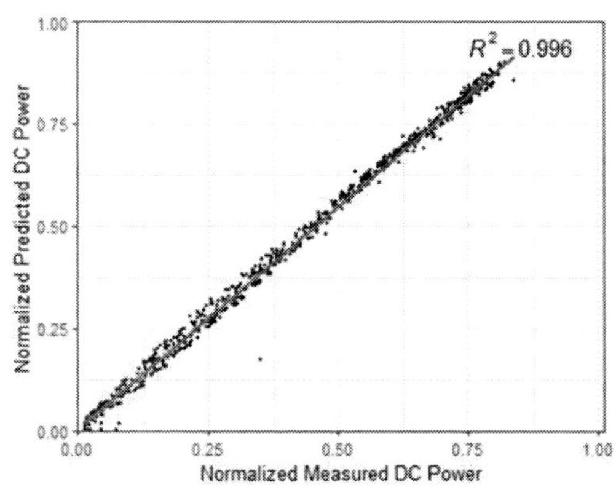

Figure 3: Normalized predicted and measured DC power of inverter 1 for a 1-week time period in August.

During normal operation, the AE was below the set TL and the measured DC power exhibited similar behavior to the predicted DC power (see Fig. 4).

3.3 Fault/Loss PV operation

Under fault/loss conditions, the PV system is underperforming leading to zero or reduced power production. PV underperformance issues may be attributed to fault occurrences and/or loss incidents. In this work, zero power production due to fault events and reduced (or nonzero) power production due to loss mechanisms were studied.

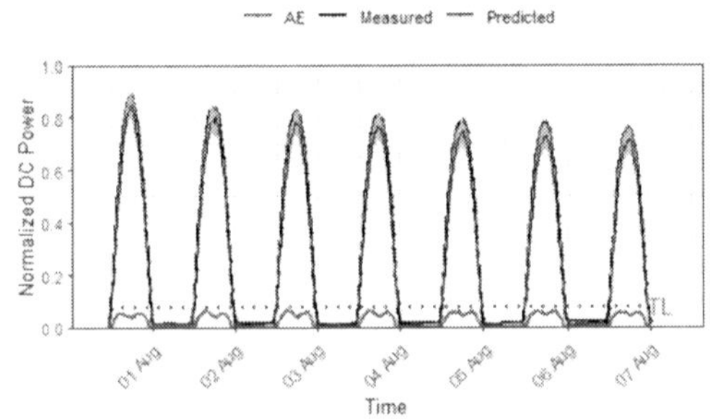

Figure 4: Normalized predicted and measured DC power of inverter 1 measurement data for a 1-week time period in August under normal operation. The control limits are grey shaded.

3.3.1 Zero power production incidents due to fault incidents

During zero power production, the measured DC power was nearly zero (power reduction from 85% to 100%) and significantly lower than the predicted one. The comparative algorithm detected power discrepancies as the AE exceeded the set TL (see Fig. 5, 24th of October). During the inverter failure occurrence (on the 24th of October), a CP in the variance was detected by the PELT algorithm. Information extracted from the maintenance log of the test PV plant indicated a ground fault error on the 24th of October.

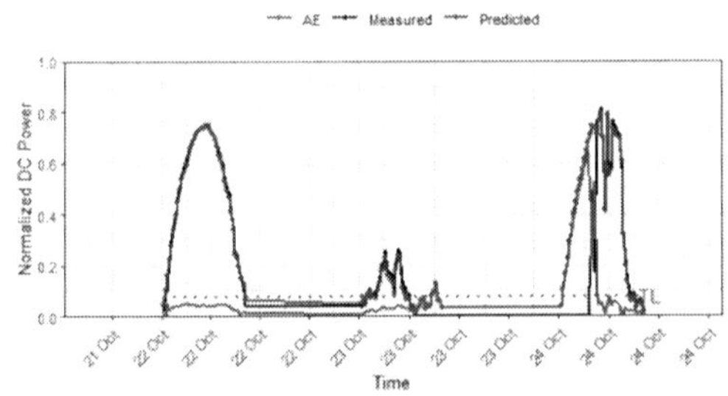

Figure 5: Normalized predicted and measured DC power of inverter 1 during a zero power production incident fault incident.

During the test period, the failure detection stage of the FDRs detected 2274 incidents of zero DC power production during daylight hours, achieving a detection accuracy of 97.3% (1602 True Positive, 567 True Negative, 45 False negative and 15 False Positive values, respectively). The detected fault data points may be attributed to inverter shutdown failures, grid failures, ground faults, maintenance events, etc.

For the classification stage of the fault, a logic tree was constructed. The algorithm considers the amount of power reduction (85-100%), the affected electrical parameters (e.g., reduced current and/or voltage) and the detected CP type (variance or mean CP). The algorithm then post-processes these features with the site's weather conditions (low or zero snowfall and rainfall) to classify the underperformance incident. The classification algorithm for zero power production due to fault incidents achieved a precision accuracy of 96.32% (since 7233 data points out of 7509 were correctly classified over the test period).

3.3.2 Nonzero power production due to loss factors

Over the evaluation period, the R_D was -0.90%/year (for inverter 1) when utilizing the comparative YoY method from the *RdTools* library (see Fig. 6).

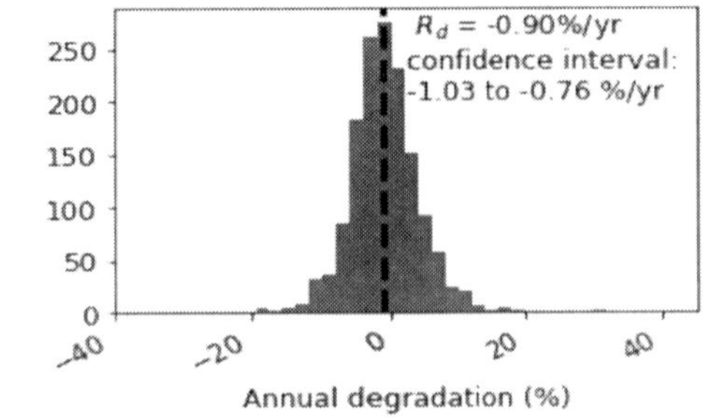

Figure 6: Export from the *RdTools* depicting the annual degradation distribution of inverter 1 measurement data over the evaluation period.

The SRR method was then used for estimating the soiling loss (the P50 insolation-weighted soiling ratio for inverter 1 was 0.954, while the 68.2 confidence interval was 0.946-0.960). The results for inverter 1 soiling analysis are shown in Fig. 7 (along with the soiling profiles), demonstrating soiling rates ranging from 0 to -0.6%/day.

To distinguish the detected irreversible (degradation) from reversible (soiling) losses, the CODS for combined degradation and soiling analysis was used resulting to P50 average soiling ratio and degradation rate of 0.983 and -0.72%/year, respectively, for inverter 1 measurement data.

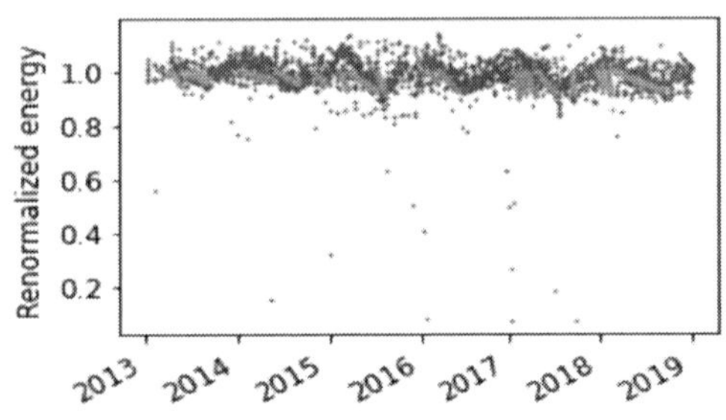

Figure 7: Exports from the *RdTools* depicting the renormalized energy and soiling profiles (coloured in orange) of inverter 1 over the evaluation period.

For snow loss detection, the *PR* time series was examined for "low" *PR* values (reduction up to 20%) and data anomalies during months with low ambient temperatures (an example is shown in Fig. 8).

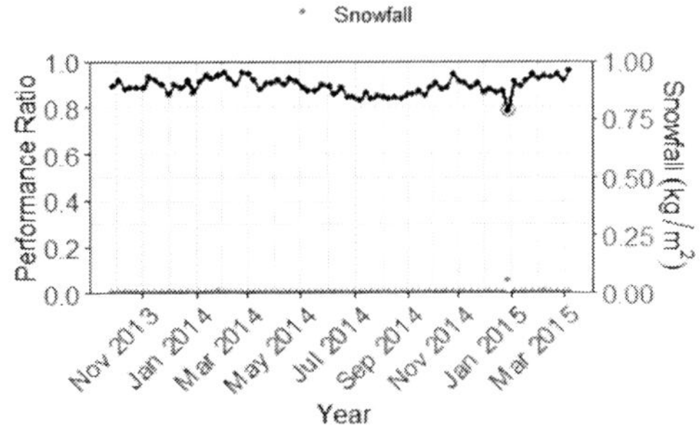

Figure 8: Weekly *PR* time series of the PV system

over the period from September 2013 to March 2015. Snowfall measurements are indicated with green points, while the detected data anomaly due to snow coverage is circled in red.

The PV underperformance was then correlated with the site's environmental conditions (i.e., snowfall and temperature measurements). In cases of PV underperformance during winter months with "low" ambient temperature ($T_{amb} < 5°C$ – site dependent value), the incident was attributed to extreme weather conditions (i.e., snow coverage). During the test period, low PR values during December/January 2015 and 2017 were detected, which were attributed to snowfall.

3.4 Fault and performance loss differentiation

To distinguish degradation from other fault types and loss factors, the FBP flexibility was initially adjusted to 2.5 for the *PR* time series. The performance loss rate (*PLR*) extracted by applying the ordinary least squares (OLS) method on the FBP trend was -0.99%/year (see Fig. 9).

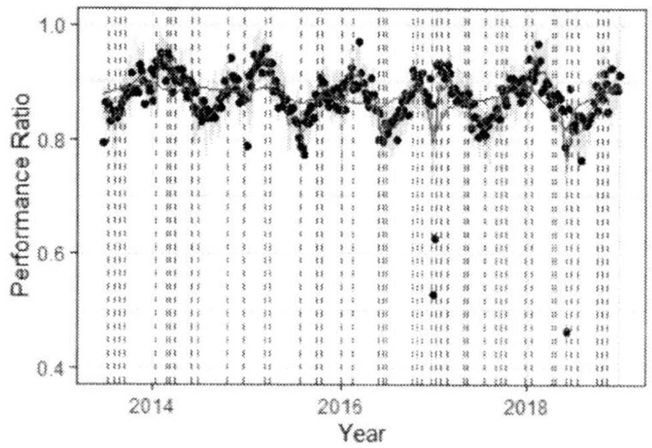

Figure 9: Weekly *PR* time series (black dots) of the test PV system along with the potential CPs coloured in red dashed lines. The red solid line indicates the extracted trend, the blue solid line indicates the FBP fit, while the blue shaded area is the uncertainty.

Different trend-based performance losses and faults will exhibit different type (i.e., variance or mean CP) and sequence of CPs (e.g., continuous or random CPs) as well as different rate of change (e.g., positive or negative).

For the investigated *PR* time series, 14 significant CPs (12 due to loss mechanisms and 2 due to fault occurrences) were detected. The TLRs achieved a precision accuracy of 91.66% (1 CP was misclassified) when classifying the detected performance loss incidents. The incidents' classification was verified against the maintenance log of the PV system and the site's environmental conditions.

More details on how to distinguish faults from reversible and irreversible performance losses are provided in the paper submitted to the Progress in Photovoltaics Special Issue (Journal Peer Review Process).

The FBP flexibility was then re-adjusted (and set to 0.05) to avoid the influence of outliers and temporary effects (see Fig. 10). In this case, a R_D of -0.49%/year was obtained. Even though, degradation contributes to the *PLR*, most of the exhibited performance losses were due to reversible phenomena and fault incidents.

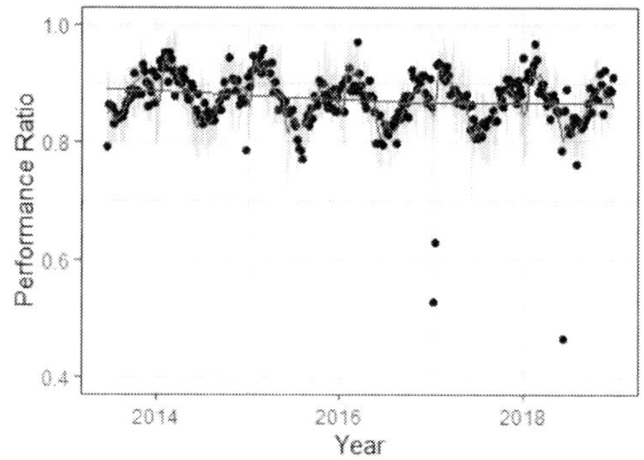

Figure 10: Weekly *PR* time series (black dots) of the test PV system along with the extracted trend coloured in red. The blue solid line is the FBP fit, while the blue shaded area indicates the uncertainty.

4 CONCLUSIONS

In this work, FDRs and TLRs were developed for classifying the fault and loss incidents using a data-driven approach. The routines were benchmarked experimentally using historical inverter level measurements from a PV power plant installed in Larissa, Greece.

The results demonstrated that the developed routines could diagnose specific operational problems occurring in PV systems, exhibiting high detection and precision accuracies for diagnosing failure occurrences and performance loss mechanisms. The failure detection stage of the FDRs achieved a detection accuracy of 97.3%. By post processing the detected failure pattern and the site weather conditions, a precision accuracy of 96.32% was obtained when classifying zero power production incidents due to fault incidents, while the TLRs achieved 91.66% classification accuracy.

An important remark is that the proposed routines use readily available field measurements obtained from monitored PV systems. The developed routines can be applied for real-time monitoring and/or for mass time series analysis of PV plants given their main technical specifications (e.g., system characteristics and drawings). This is the first attempt towards the differentiation of faults from reversible and irreversible performance losses for optimizing O&M activities (i.e., informing O&M teams about the incident to act accordingly and recover lost energy).

Finally, only inverter measurements were available in this study and differentiation of faults from loss factors was performed at the inverter level. Future work will focus on benchmarking the FDRs and TLRs at string and/or module level, extending the PV underperformance categories and including more fault types such as partial shading, vegetation and PV module level failures.

5 ACKNOWLEDGEMENTS

The work of Andreas Livera and George E. Georghiou was funded through the ROM-PV project, which is supported under the umbrella of SOLAR-ERA.NET Cofund. The UCY authors would like to acknowledge the financial support from the Research and Innovation Foundation (RIF) of Cyprus. SOLAR-ERA.NET is supported by the European Commission within the EU

Framework Programme for Research and Innovation HORIZON 2020 (Cofund ERA-NET Action, N° 691664).

The work of Marios Theristis and Joshua S. Stein was supported by the U.S. Department of Energy's Office of Energy Efficiency and Renewable Energy (EERE) under the Solar Energy Technologies Office Award Number 34366. Sandia National Laboratories is a multimission laboratory managed and operated by National Technology & Engineering Solutions of Sandia, LLC, a wholly owned subsidiary of Honeywell International Inc., for the U.S. Department of Energy's National Nuclear Security Administration under contract DE-NA0003525. This article describes objective technical results and analysis. Any subjective views or opinions that might be expressed in this article do not necessarily represent the views of the U.S. Department of Energy or the Unites States Government.

Alectris Hellas IKE is kindly acknowledged for providing the data of the test PV plant.

6 REFERENCES

[1] SolarPower Europe [SPE], 100% Renewable Europe How To Make Europe's Energy System Climate-Neutral Before 2050, 2020.

[2] A. Livera, M. Theristis, A. Charalambous, J.S. Stein, G.E. Georghiou, Decision support system for corrective maintenance in large-scale photovoltaic systems, in: 48th IEEE Photovolt. Spec. Conf., 2021: pp. 1–6.

[3] European Technology and Innovation Platform for Photovoltaics (ETIP PV), Photovoltaic Solar Energy: Big and Beyond Sustainable energy to limit global warming to 1.5 degrees, 2019.

[4] M. Mansouri, M. Hajji, M. Trabelsi, M.F. Harkat, A. Al-khazraji, A. Livera, H. Nounou, M. Nounou, An effective statistical fault detection technique for grid connected photovoltaic systems based on an improved generalized likelihood ratio test, Energy. 159 (2018). https://doi.org/10.1016/j.energy.2018.06.194.

[5] A. Livera, M. Florides, M. Theristis, G. Makrides, G.E. Georghiou, Failure diagnosis of short- and open-circuit fault conditions in PV systems, in: 45th IEEE Photovolt. Spec. Conf. PVSC 2018, 2018.

[6] A. Livera, G. Makrides, J. Sutterlueti, G.E. Georghiou, Advanced Failure Detection Algorithms and Performance Decision Classification for Grid-connected PV Systems, in: 33rd Eur. Photovolt. Sol. Energy Conf. Exhib. (EU PVSEC), 2017.

[7] IEC 61724-1:2017, Photovoltaic system performance - Part 1: Monitoring, (2017).

[8] M. Theristis, V. Venizelou, G. Makrides, G.E. Georghiou, Chapter II-1-B – Energy yield in photovoltaic systems, in: Kalogirou, S.A. (Ed.), McEvoy's Handb. Photovoltaics, Third Ed. Acad. Press, 2018: pp. 671–713.

[9] Global Modeling and Assimilation Office (GMAO) (2015), MERRA-2 tavg1_2d_slv_Nx: 2d,1-Hourly,Time-Averaged,Single-Level,Assimilation,Single-Level Diagnostics V5.12.4, Greenbelt, MD, USA, Goddard Earth Sciences Data and Information Services Center (GES DISC, (n.d.). https://doi.org/10.5067/VJAFPLI1CSIV.

[10] SolarPower Europe [SPE], Operation & Maintenance Best Practice Guidelines / Version 4.0, 2019.

[11] A. Livera, M. Theristis, E. Koumpli, S. Theocharides, G. Makrides, J. Sutterlueti, J.S. Stein, G.E. Georghiou, Data processing and quality verification for improved photovoltaic performance and reliability analytics, Prog. Photovoltaics Res. Appl. (2020). https://doi.org/10.1002/pip.3349.

[12] D.L. King, W.E. Boyson, J.A. Kratochvil, Photovoltaic array performance model, SANDIA Report SAND2004-3535, 2004. https://doi.org/10.2172/919131.

[13] T. Huld, G. Friesen, A. Skoczek, R.P. Kenny, T. Sample, M. Field, E.D. Dunlop, A power-rating model for crystalline silicon PV modules, Sol. Energy Mater. Sol. Cells. 95 (2011) 3359–3369. https://doi.org/10.1016/j.solmat.2011.07.026.

[14] E. Koumpli, Impact of Data Quality on Photovoltaic (PV) Performance Assessment, Doctoral Thesis University of Loughborough, Doctoral Thesis, Loughborough University, UK, 2017.

[15] A. Livera, M. Theristis, G. Makrides, J. Sutterlueti, S. Ransome, G.E. Georghiou, Performance analysis of mechanistic and machine learning models for photovoltaic energy yield prediction, in: 36th Eur. Photovolt. Sol. Energy Conf. (EU PVSEC), 2019: pp. 1272–1277.

[16] G. Makrides, B. Zinsser, M. Schubert, G.E. Georghiou, Energy yield prediction errors and uncertainties of different photovoltaic models, Prog. Photovoltaics Res. Appl. 21 (2013) 500–516. https://doi.org/10.1002/pip.1218.

[17] Y. Zhao, B. Lehman, R. Ball, J. Mosesian, J.-F. De Palma, Outlier Detection Rules for Fault Detection in Solar Photovoltaic Arrays, 28th IEEE Annu. Appl. Power Electron. Conf. Expo. (2013).

[18] B. Lantz, Machine Learning with R, Packt Publishing Ltd, 2013. https://doi.org/10.1007/978-981-10-6808-9.

[19] D.C. Jordan, S.R. Kurtz, Photovoltaic degradation rates - An analytical review, Prog. Photovoltaics Res. Appl. 21 (2013) 12–29. https://doi.org/10.1002/pip.1182.

[20] M.G. Deceglie, D. Jordan, A. Shinn, C. Deline, RdTools : An Open Source Python Library for PV Degradation Analysis degradation rate, (2018) 1–15.

[21] M.G. Deceglie, L. Micheli, M. Muller, Quantifying Soiling Loss Directly from PV Yield, IEEE J. Photovoltaics. 8 (2018) 547–551. https://doi.org/10.1109/JPHOTOV.2017.2784682.

[22] A. Skomedal, M.G. Deceglie, Combined Estimation of Degradation and Soiling Losses in Photovoltaic Systems, IEEE J. Photovoltaics. 10 (2020) 1788–1796. https://doi.org/10.1109/JPHOTOV.2020.3018219.

[23] M. Theristis, A. Livera, C.B. Jones, G. Makrides, G.E. Georghiou, J.S. Stein, Nonlinear Photovoltaic Degradation Rates: Modeling and Comparison Against Conventional Methods, IEEE J. Photovoltaics. 10 (2020) 1112–1118. https://doi.org/10.1109/JPHOTOV.2020.2992432.

[24] M. Theristis, A. Livera, L. Micheli, B. Jones, G. Makrides, G.E. Georghiou, J. Stein, Modeling nonlinear photovoltaic degradation rates, in: 47th IEEE Photovolt. Spec. Conf., 2020.

[25] R. Killick, I. Eckley, changepoint: An R Package for changepoint analysis, Lancaster Univ. 58 (2013) 1–15. https://doi.org/10.1359/JBMR.0301229.

TOPIC 5: PV Systems – Modelling, Design, Operation and Performance; 5.3 Operation, Performance and Maintenance of PV Systems

PERFORMANCE IMPUTATION TECHNIQUES FOR ASSESSING COSTS OF TECHNICAL FAILURES IN PV SYSTEMS

S. Lindig[1,2*], A. Louwen[1], M. Herz[3], J. Ascencio-Vásquez[2,4], D. Moser[1], and M. Topič[2]

[1]Institute for Renewable Energy, EURAC Research, Viale Druso 1, 39100 Bolzano, Italy
[2]University of Ljubljana, Faculty of Engineering, Tržaška cesta 25, 1000 Ljubljana, Slovenia
[3]TÜV Rheinland Energy GmbH, Am Grauen Stein, 51105 Cologne, Germany
[4]3E sa, Quai à la Chaux 6, 1000 Brussels, Belgium
*Phone: +39 0471 055 715; E-mail: sascha.lindig@eurac.edu

ABSTRACT: In the framework of the new H2020 project TRUST PV, the Cost Priority Number methodology, a cost-based failure modes & effect analysis method, is tailored for its application onto real-case studies during the operational phase of a PV plant. Thereby, it is possible to calculate the energetic and economic losses for each individual failure by combining actual technical characteristics of the failure tickets, the operational costs of fixing those failures, and the power loss due to the failure occurrence. In this work, different empirical, machine learning and univariate models have been tested to find the optimal model to impute power data for all possible scenarios of missing performance data. The parameter space comprises of the amount and type of missing data, the ratio between training and test set and the predictor availability. Overall, the most desirable setting for highly accurate data imputation is the availability of a neighbouring string/plant or measured climate data as predictor with a high training/test set ratio for small test sets and ratios between 80/20 to 50/50 for bigger test sets (longer than 12 hours). In general, machine learning imputation models and empirical power models perform similar for smaller test set sizes and machine learning models are preferrable for bigger data gaps. Univariate methods should be avoided if possible.

Keywords: PV systems, PV system performance, Data imputation, Data quality

1 INTRODUCTION

In recent years, the photovoltaic (PV) industry experiences considerable advancements in the operation and maintenance (O&M) of PV power plants. The aim is to increase plant performance and availability while decreasing costs. During the operational phase of a PV system, O&M concepts are key tools to achieve these goals through a decrease in the number and severity of performance impairing events. Advanced O&M routines consist of smart and site-specific preventive maintenance solutions and improved corrective maintenance, which is characterized by a reduced detection time of an appearing failure and consequently a reduced downtime of PV plant components in which the failure is remedied [1].

Possible risks, leading to a decrease in PV performance, are faster detected and counteracted against through a better understanding of typical PV system behaviour and advanced detection and prevention techniques. A method to map and understand the likeliness and severity technical failures may have on PV plant performance, together with the associated costs, is the Cost Priority Number (CPN) method. The initial CPN, a cost-based failure modes & effect analysis (FMEA) method, was developed in the H2020 project Solar Bankability [2] with the aim to assess the economic impact of failures in PV projects and is calculated in €/kWp/year. In the framework of the new H2020 project TRUST PV [3], the CPN is improved by adapting parameters for its application onto real-case studies. The first version of the CPN methodology was based on the development of failure scenarios. Instead, within TRUST PV, tickets of O&M activities coming from SCADA systems of operating PV plants are categorized and evaluated. The CPN is used to accurately calculate costs/benefits for each individual ticket in order to understand and further improve the effectiveness of operators and maintainers. To do so, the accurate estimation of energy losses due to recorded failures is essential. However, several data issues such as gaps, communication errors and high measurement uncertainties make this task quite challenging.

In this work, we test widely used data imputation models considering all possible scenarios of missing power/energy data, which include the following parameter:

- Type of missing data
- Amount of missing data
- Ratio training set / test set
- Predictor availability

For each scenario, the tested models have been benchmarked. Optimized imputation approaches for all considered cases were identified by minimizing the following statistical metrics: relative root-mean-square-error (rRMSE), relative mean bias error (rMBE), relative difference (rD) and the absolute difference (aD). The overall aim is to find the optimal imputation solution for each possible scenario in terms of available predictor data and missing data rate to automatize the imputation process to a high degree.

2 COST PRIORITY NUMBER METHODOLOGY

In the H2020 project Solar Bankability [2], the CPN methodology has been developed to quantify appearing degradation modes and other performance impairing effects in PV plants. The focus was on the assessment of risks connected to investments in PV projects [4]. Thereby, the methodology assesses the economic impact based on factors such as performance reduction and downtime. Within the project, the CPN was applied to theoretical scenarios to calculate extreme values for the CPN metric, expressed in €/kWp/year. All phases of a PV plant life cycle (from product testing to decommissioning) have been included and a nomenclature for possible failures has been created in form of a technical risk matrix [5]. Later, the CPN has been adapted towards the needs of a large O&M operator by assessing real tickets from the ticketing system of an operating PV plant manually [6]. In the H2020 project TRUST PV [3], the assessment of real tickets is automatized to not only study the economic impact of individual technical failures, but also to draw useful statistics in terms of failure appearance likeliness and severity. Therefore, the technical risk matrix has been updated. The TRUST PV risk matrix only includes failures from within the operational phase of a PV plant divided by component, subcomponent, and individual failure. Overall, 344 failures have been identified. Additionally, the following optional information on individual failures can be supplied: cause (what is the failure cause), accountability (which party is accountable), detection (how was the failure detected), origin (can the failure be attributed to a root-cause stemming from a specific phase of the PV plants life-cycle) and solution (how was the failure fixed). The workflow of assigning a cost to a failure and thereby calculating the CPN is the following:

1. Failure appearance in PV plant
2. Failure detection
3. Creation of ticket in SCADA system
4. Classification of failure according to TRUST PV's Risk Matrix
5. Resolution of failure
6. Statistical analysis of failure using the CPN methodology

The CPN is calculated by the sum of costs due to downtime and fixing costs due the failure appearance:

$$CPN\ [€/kWp/year] = C_{down} + C_{fix}$$

The two categories of costs are defined as:

a) Economic impact due to downtime (C_{down}):
   - failures that cause downtime or power loss
   - considers time from failure to repair/substitution (thereby including detection time, response time, repair time, shutdown time)
   - missing income related to the sale of electricity, defined by the feed in tariff, power-purchasing agreement or missing savings generated by PV plants installed on roofs/facades defined as retail cost of electricity
   - important to pay attention to component level of failure as other components might be affected

b) Economic impact due to repair (C_{fix}):
   - cost of detection
   - cost of labour
   - cost of transportation
   - cost of repair/substitution

The scope of this work is to reliably calculate the energy loss caused by an appearing failure to determine the economic impact due to downtime. A technical failure leads to a partial or complete reduction of the energy produced by a PV plant. With the help of this study, we determine how much energy a plant would produce without the appearance of a failure and the difference between the estimated energy production and the reduced actual production is the amount lost through the appearance of the failure.

3 PV PLANT

For this work, performance timeseries data of an experimental PV plant installed and operated at the Airport Bolzano Dolomiti (South Tyrol/Italy) have been used. The data are recorded and stored by EURAC Research. Longitude and latitude of the system are 46.4625°N and 11.3299°E respectively, and the plant is situated 240 m above sea-level. The system was installed in 2010 with a fixed tilt of 30° and an orientation of 8.5° west of south. The plant consists of 20 poly-crystalline silicon PV modules with an overall nominal power of 4.2 kWp. Additionally, climate data are recorded at a nearby meteorological station including plane-of-array irradiance and ambient temperature.

4 PARAMETER

Incoming performance data of PV systems can have different measured signals available. In order to cover all possible scenarios of missing data for calculating the energy loss, we created a parameter space including four different variables, namely the type and amount of missing data, the ratio between training and test set as well as the predictor availability.

4.1. Type of missing data

Missing data can usually be categorized as being randomly or continuously distributed. In the scope of this work, we only work with continuous missing data as this is the predominant type in case of failure appearance. Furthermore, data can be categorized based on pairwise or listwise missingness. In case of listwise missing data, the complete column (power/energy and all available measured predictors) is missing, whereas in case of pairwise missing data only the power/energy measurement is affected.

4.2. Amount of missing data

This parameter is expressed through downtime of the plant or plant component. It depends on the failure type in combination with the effectiveness of the O&M activities, driven by detection and response time on one hand and repair time paired with spare part management on the other hand. In order to simulate different grades of missingness, the tests set size (data to impute) was varied from 4 hours to 10 days in five steps.

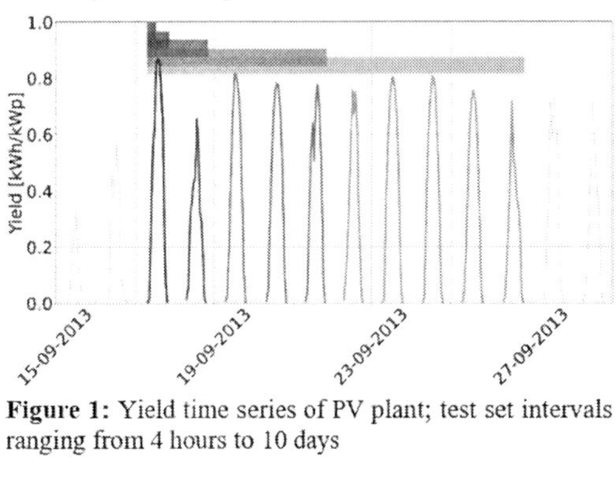

Figure 1: Yield time series of PV plant; test set intervals ranging from 4 hours to 10 days

In Figure 1, the intervals of the test sets are shown exemplarily. Thereby, the test set position is drawn randomly from the time series, which ranges from 2012 until 2019 spanning seven years of data.

4.3. Ratio training set / test set

After assigning the test set size and position, the training set is selected as the data preceding the test set in seven different training/test set splits. Ultimately, 35 different datasets are rendered for the imputation. The training set is thereby the dataset used to train a data imputation model and the test set the missing energy data to impute. Across all different settings the missing data rate ranges from 4 hours to 10 days. Different training/test set ratios are tested to find the amount of required training data to fit the imputation models sufficiently. The amount of training data is a trade-off between providing enough data for model fitting while capturing the instantaneous performance of the PV plant.

4.4. Predictor availability

It can be expected that different PV plants have different predictors available through varying data acquisition and monitoring concepts. That is why different predictors and predictor sources have been selected within this study:

Table 1: Predictor groups and predictors used for data imputation.

MEASURED PARAMETER	SATELLITE DATA	NO PREDICTOR AVAILABLE
Climate data	**ERA5 [7] + tilt & azimuth**	
$G_{POA_{meas}}$ $T_{amb_{meas}}$	$G_{POA_{ERA5}}$ $T_{amb_{ERA5}}$	
Neighbouring string/plant	**CAMS [8] + tilt & azimuth**	
Power	$G_{POA_{Cams}}$	

Table 1 shows the different groups of predictors together with the specific parameters. The first group are measured climate data from a meteorological station on-site. If a neighbouring plant or a neighbouring string/inverter (in case of higher resolution monitoring) is available, the power signal of this neighbour is used. In this study, the neighbour is a poly-crystalline silicon PV plant installed on the same site as the test system having the same tilt and orientation while using modules from a different manufacturer. Both systems have the same configuration of front-glass/cell/backsheet while being installed on an open rack. This is important for the temperature evolution across the systems and thereby for the performance/efficiency evolution. For reanalysis/ satellite data, ERA5 data [7] as well as CAMS data [8] were selected as they are freely available in high resolution. In order to compute plane-of-array data [9, 10], tilt and azimuth of the PV plant in question must be known. Figure 2 shows the correlation [11] between the target variable (power of test system) and the tested predictors in hourly resolution. It is visible that the power signal of the neighbouring plant has the highest correlation with the target variable, rooting from the same installation conditions and the same PV technology. As expected, the measured plane-of-array irradiance (G_{POA}) also correlates highly, followed by irradiance values retrieved through CAMS and ERA5 data. Ambient temperature correlates to a lesser degree with the produced power.

The last predictor option is that no data are available for imputation. In this case, only univariate imputation methods can be applied and tested.

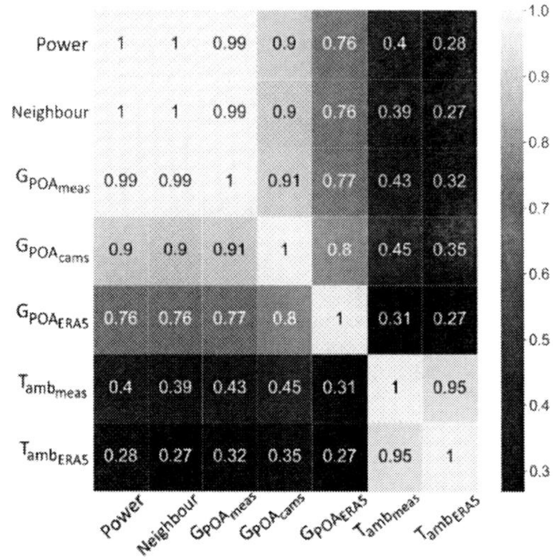

Figure 2: Heatmap showing the Pearson correlation **[11]** between the target variable (Power) and the tested predictors in hourly resolution

5 IMPUTATION MODELS

Overall, 14 different models have been tested. The applicability of the models always depends on the availability of predictors. The models are coming from three different families: empirical, regression & machine learning and univariate models.

5.1. Empirical power prediction models

Empirical models predict the power of a PV system as a function of climate inputs, usually irradiance and temperature.

For the application of most models the module temperature is required. The Nominal Operating Cell Temperature Model [12] has been used to calculate the module temperature. In this study, three different empirical models have been tested.

a) PVWatts model [13]

The PVWatts model relates the incoming irradiance and cell temperature to the PV system power with:

$$P = P_{nom} \frac{G_{POA}}{1000W/m^2}(1 + \gamma(T_{mod} - 25°C))$$

Here, γ is the temperature coefficient given by the module manufacturer. The model is provided by NREL as an online PV performance modelling application [14].

b) PVGIS model [15]

The PVGIS model combines regression of normalized irradiance and PV module temperature with six empirical coefficients and is a variant of King's model [16] (used in the PVGIS online tool [17]). In this work, modelled module temperature and measured/modelled plane-of-array irradiance is used without normalization.

$$P = G_{POA}(P_{nom} + k_1 \ln(G_{POA}) + k_2 \ln(G_{POA})^2 + T_{mod}(k_3 + k_4 \ln(G_{POA}) + k_5 \ln(G_{POA})^2) + k_6 {T_{mod}}^2)$$

k_1 to k_6 are the empirical coefficients which are fitted with non-linear least squares.

c) Three parameter model

This model was benchmarked among different heuristic models by Ding et al. [18] and proved to provide accurate results. It is described by:

$$P = \left(aG_{POA} + bG_{POA}^2 + cG_{POA} ln\left(\frac{G_{POA}}{1000\frac{W}{m^2}}\right)\right)(1 + \gamma(T_{mod} - 25°C))$$

Here, a, b and c are fitting coefficients, again determined with non-linear least squares.

5.2. Regression & Machine Learning models

Several widely used regression and machine learning models have been tested as alternatives to more classical approaches of predicting power by using empirical models. All models have been implemented with the Python library scikit-learn [19]. For the models *b)* to *g)* hyperparameter tuning has been carried out with a pre-defined parameter space using an exhaustive searching approach.

a) Multivariate linear regression

The first model is an ordinary least-squares linear regression of the predictor/s versus power.

b) K-nearest neighbours

K-nearest neighbour regression creates data clusters based on training data. The predicted value will be approximated based on the predictors in the cluster space through weighted averaging. The size of the clusters is defined by the parameter K.

c) Decision tree

Decision trees design sets of logical decisions to divide data into nodes by minimizing the residual sum of squares within a node for a given tree depth.

d) Random forest

A random forest regressor constructs a number of decision trees and returns the average prediction of the individual trees to improve accuracy [20]. It fits the trees using bootstrapping and selecting an optimum split.

e) Extra tree

The extra tree regressor is very similar to random forest but uses the whole training sample for fitting individual decision trees and splits the tress randomly [21].

f) Gradient boosting regressor

When using gradient boosting regressors, trees are grown sequentially using information from previously grown trees. Thereby, additionally to the depth of the tree and the number of nodes the learning rate is variable [22].

g) Histogram based gradient boosting regressor

This model is implemented within scikit-learn based on the LightGBM framework [23]. It provides a novel gradient boosting regressor including new techniques to find the optimal split point of trees by reducing the number of data instances and reducing the feature space.

5.3. Univariate models

As mentioned before, the situation might arise where no training data are available to predict the lost energy. In this case, univariate imputation methods are the only option. Here, only energy data before and after the outage instance are used to train the models and to perform the imputation. The models have been implemented using the R packages imputeTS [24] as well as forecast [25, 26].

a) Average value

This method simply averages hourly values in the training set by the hour of the day. The imputed value is the average for the specific hour.

b) Random

Missing values are replaced by drawing a random value from the training dataset.

c) Kalman

This is a two step-process. First, the time series (training data) is transformed into a state space model using ARIMA via seasonal decomposition. Data imputation is performed via Kalman smoothing of the created model. The Kalman filter joins the probability distributions of the univariate data and estimates of unknown variables at each timeframe, allowing the prediction when no recorded data is available [27].

d) Interpolation

This model imputes data via linear interpolation. For seasonal data, such as performance time series, seasonal decomposition is first carried out.

5.4. Model evaluation metrics

The presented models have been evaluated based on the parameter introduced in Section 3 using four different statistical metrics. In this study, a full outage of the whole plant has been assumed. The following parameter were used to test the best imputation scenarios and to evaluate the accuracy of the different models:

$$rRMSE = \frac{\sqrt{\frac{\sum_{i=1}^{N}(\tilde{y}_i - y_i)^2}{N}}}{\frac{1}{N}\sum_{i=1}^{N} y_i} * 100\%$$

$$rMBE = \frac{\sum_{i=1}^{N}(\tilde{y}_i - y_i)}{\sum_{i=1}^{n} y_i} * 100\%$$

$$aD = \left|\sum_{i=1}^{N} y_i - \sum_{i=1}^{N} \tilde{y}_i\right|$$

$$rD = \frac{\left|\sum_{i=1}^{N} y_i - \sum_{i=1}^{n} \tilde{y}_i\right|}{\sum_{i=1}^{n} y_i}$$

The relative root mean square error (rRMSE) helps to identify imputation scenarios with large individual errors between predicted value and target value. The relative mean bias error (rMBE) shows if a set of predictors (or a model) tends to over- or underestimate the target value. The absolute difference (aD) tells us by how many kWh our prediction deviates from the target. This parameter is very interesting as the outcome can be directly used to assess in economic impact of the imputation inaccuracy. Finally, the relative difference (rD) sets this parameter into relation to the overall lost energy to better intercompare the results on varying timescales.

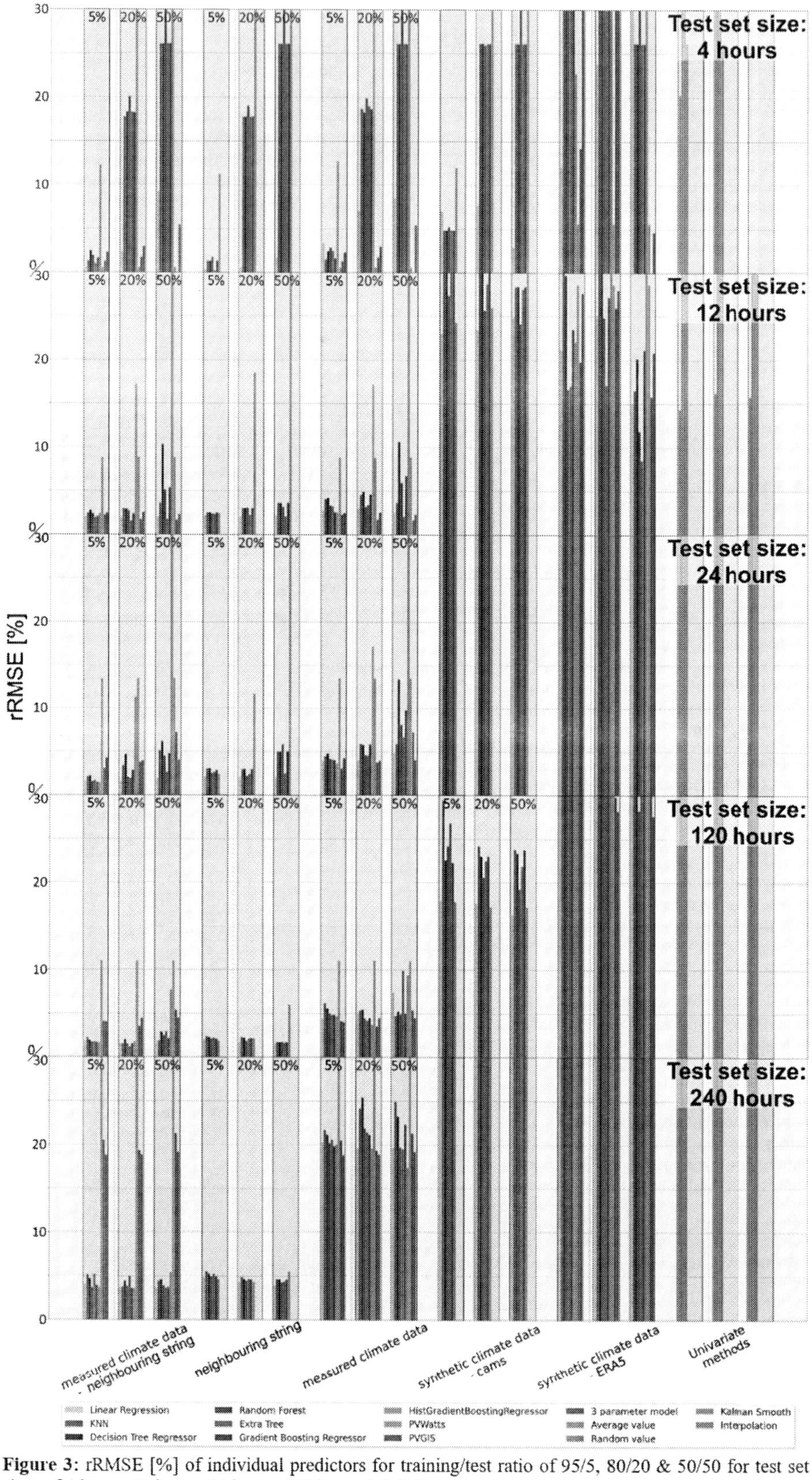

Figure 3: rRMSE [%] of individual predictors for training/test ratio of 95/5, 80/20 & 50/50 for test set sizes of 4 hours, 12 hours, 24 hours, 120 hours & 240 hours

6 RESULTS

The results of this study are shown in Figure 3 as well as the following figures in the Appendix. In the figures, the statistical metrics introduced in Section 5.4. are depicted. In the figures, the test set size increases from top to bottom and from left to right the different predictors are shown. For each predictor, the training/test set ratios of 95/5, 80/20 and 50/50 are included. For example, a 95/5 ratio for a 12-hour test set equals to 228 hours of training data versus 12 hours of test data, an 80/20 ratio equals 48 hours vs. 12 hours and a 50/50 ratio equals 12 hours vs. 12 hours.

Looking at the overall results, it is visible that a neighbouring string/plant is the most desirable predictor followed by measured climate data, indicated by the lowest prediction errors across all error metrics. Except for very short time series imputation (4 hours), lower training/test ratios (80/20 to even 50/50) perform as good and partially even better compared to very high training/test ratios of 95/5. The reason for the bad performance for a 80/20 or 50/50 split for a test set size of 4 hours is because the amount of training data is simply too low. It is assumed that 95/5 splits for longer imputation periods do not perform too well because a very long training dataset contains data from very different seasons of the year and temperature dependent performance deviations throughout the seasons might introduce a bias.

Looking at the predictors neighbouring string and/or measured climate data, it is visible that the rRMSE (Figure 3) is slightly increasing with a increasing test set size while the absolute and relative difference (Figure 4 & Figure 5) are almost constant for the individual best performing models. The bias (Figure 6) is decreasing with an increasing test size. In terms of imputation methods, both machine learning and empirical models with fitting coefficients perfom similarly well for small test set sizes while machine learning models are superior when being applied on larger test sets, especially if measured climate data are used.

PVWatts is the worst performing empirical model among the tested ones, probably due its nature of being generally applicable without any training of fitting coefficients. The three parameter model and the PVGIS model show a comparable accuracy. For measured climate data, impuation using these two models has a similar, partially even lower, bias compared to machine learning approaches.

For small test set sizes, multivariate regression shows also reliable results. In contrast, histogram based gradient boosting regression performs poor for absolute training set sizes below 60 hours, especially when looking at the rRMSE. Across the tested scenarios and machine learning models, the decision tree regressor and the extra tree regressor perform well.

If no measured predictors are available, two imputation options have been identified, either the usage of satellite climate data or univariate imputation. Univariate methods are not performing very good, they really should be a last resort if no other option is available. The average value method with a medium/low training/test ratio shows the best performance among the univariate methods. For this particular location, the accuracy using both satellite data sources is low compared to measured predictors. Looking at the rRMSE in Figure 3 as well as the absolute difference in Figure 4, CAMS seems to produce slightly better results compared to ERA5 reanalyis data. For CAMS data, the bias tends to be lower for a high training/test ratio for small test set sizes and for a lower training/test ratio for bigger data holes to impute. As CAMS does not provide ambient temperature data, only machine learning imputaton methods were tested. The best performing model changes depending on the test set size and the training/test ratio, but in general multivariate regression seems to perform well in comparison for small test set sizes, and histogram based gradient boosting regression for bigger test sets.

We expect to see lower imputation errors for satellite and reanalysis data for PV systems installed in areas with less complex climate conditions and less mountainous terrain. Bolzano is located at multiple climate borders. It is situated in a valley amid mountains at the foot of the Dolomites. Satellite data are by experience highly unreliable in this location. System performance of PV plants installed in more stable climatic conditions with a simpler terrain can usually be predicted with higher accuracy using satellite reanalysis data.

7 CONCLUSIONS & OUTLOOK

In this work, different imputation models for energy data of a PV plant, installed in Bolzano/Italy, have been studied. Thereby, empirical models, machine learning approaches as well as univariate methods have been tested. The imputation accuracy has been evaluated in dependency on the available predictors, the amount of data to be imputed and the ratio between training and test set size. This work is part of a larger study to estimate the costs and impact of an appearing failure in the field for operating PV systems (cost priority number – CPN). One of the primary costs is the energy lost due to the appearance of a failure. Reliable methods to accurately calculate the energy loss covering all possible scenarios in terms of predictor data availability and training/test set characteristics are required to estimate the economic impact of technical failures.

This study has shown that a neighbouring plant/string is the most desirable predictor to impute performance data, followed by measured climate data. In most parameter settings, the application of machine learning models delivers higher imputation accuracies. Short time series can be imputed with high accuracy using simple multivariate regression or empirical models including fitting coefficients. Results with higher accuracy for longer time series are calculated using decision tree or extra tree regression. Also, histogram based gradient boosting regression performed very well if the training set has more than 60 entries (tested on data with hourly resolution). Data holes exceeding 12 hours can reliably be imputed with lower training/test set ratios of 80/20 up to 50/50 splits. Lower training/test set ratios have two advantages:

- computation time is lower for model optimization & application
- long training sets tend to introduce a seasonal performance related bias in the models, which might also introduce a bias in the results

If no measured predictors are available, freely obtainable reanalysis/satellite data such as CAMS or ERA5 reanalysis data are a viable choice and preferred to univariate data imputation. The imputation using satellite data was subject to high uncertainties for the PV plant in question. That is because the plant is installed in complex terrain and at the border of several climate zones. It is expected to see more reliable imputation results for systems in flat terrain and clearly defined climate zones.

In the future, this study will be extended to several PV systems, installed in different climate and terrain conditions, to define general best practices as well as to prepare automatized scripts for the imputation of missing performance data. Furthermore, an adapted version of cross validation for time series will be used to generalize the results for individual plants by removing data artefacts. Finally, it is foreseen to carry out an economic sensitivity analysis across the variability of the results to understand the impact of imputation uncertainties on the calculated CPN value.

ACKNOWLEDGMENTS

This project has received funding from the Horizon 2020 programme of the European Union under GA. No. 952957 – Project TRUST PV.

[7] was downloaded from the Copernicus Climate Change Service (C3S) Climate Data Store. The results contain modified Copernicus Climate Change Service information 2020. Neither the European Commission nor ECMWF is responsible for any use that may be made of the Copernicus information or data it contains.

REFERENCES

[1] National Renewable Energy Laboratory, Sandia National Laboratory, SunSpec Alliance, and the SunShot National Laboratory Multiyear Partnership (SuNLaMP) PV O&M Best Practices Working Group, "Best Practices for Operation and Maintenance of Photovoltaic and Energy Storage Systems; 3rd Editions," Golden, CO: National Renewable Energy Laboratory, 2018.

[2] "H2020 Solar Bankability Project," [Online]. Available: http://www.solarbankability.org/home.html. [Accessed 20 07 2021].

[3] "H2020 TRUST PV Project," [Online]. Available: https://trust-pv.eu/. [Accessed 11 01 2021].

[4] D. Moser et al., "Identification of Technical Risks in the PV Value Chain and Quantification of the Economic Impact on the Business Model," *Progress in Photovoltaics: Research and Applications,* vol. 25, no. 7, pp. 592-604, 2017.

[5] H2020 Project Solar Bankability, "Technical Risk Matrix," 2017.

[6] G. Oviedo-Hernández et al., "Optimization of the Cost Priority Number (CPN) Methodology to the Needs of a Large O&M Operator," in *36th EU PVSEC*, Marseille, 2019.

[7] Copernicus-Climate-Change-Service-ERA5, "Fifth generation of ECMWF atmospheric reanalyses of the global climate. Copernicus Climate Change Service Climate Data Store (CDS)," 2020. [Online]. Available: https://cds.climate.copernicus.eu/cdsapp#!/home . [Accessed 15 July 2020].

[8] "CAMS: Surface solar radiation data," [Online]. Available: https://atmosphere.copernicus.eu. [Accessed 12 01 2021].

[9] W. Holmgren, C. Hansen and M. Mikofski, "Pvlib python: A python package for modeling solar energy systems," *J. Open Source Softw.*, vol. 3, no. 29, p. 884, 2018.

[10] H. Hottel and B. Woertz, "The Performance of Flat Plate Solar-Heat Collectors," *Trans. ASME,* vol. 64, pp. 64-91, 1942.

[11] J. Benesty et al., "Pearson correlation coefficient," in *Noise reduction in speech processing*, Springer, 2009, pp. 37-40.

[12] R. G. Ross, "Flat-Plate Photovoltaic Array Design Optimization," in *14th IEEE Photovoltaic Specialists Conference*, San Diego, CA, USA, 1980.

[13] NREL, PVWatts Version 5 Manual, National Renewable Energy Laboratory, 2014.

[14] NREL, "PVWatts Calculator," 2019. [Online]. Available: https://pvwatts.nrel.gov/index.php. [Accessed 20 07 2021].

[15] T.A. Huld et al., "A power-rating model for crystalline silicon PV modules.," *Solar Energy Mater. Solar Cells,* vol. 95, pp. 3359-3369, 2011.

[16] D. L. King, W. Boyson and J. Kratochvill, "Photovoltaic Array Performance Model," Sandia National Laboratories, Alberquerque, 2004.

[17] European Commission, "Joint Research Centre Photovoltaic Geographical Information System (PVGIS)," [Online]. Available: https://ec.europa.eu/jrc/en/pvgis. [Accessed 20 07 2021].

[18] K. Ding, Z. Ye and T. Reindl, "Comparison of Parameterisation Models for the Estimation of the Maximum Power Output of PV Modules," *Energy Procedia,* vol. 25, p. 101–107, 2012.

[19] F. Pedregosa et al., "Scikit-learn: Machine Learning in Python," *J. Mach. Learn. Res.,* vol. 12, p. 2825.2830, 2011.

[20] L. Breiman, "Random Forests," *Machine Learning,* vol. 45, pp. 5-32, 2001.

[21] P. Geurts, D. Ernst and L. Wehenkel, "Extremely Randomized Trees," *Machine Learning,* vol. 63, pp. 3-42, 2006.

[22] J. Friedman, "Greedy Function Approximation: A Gradient Boosting Machine," *Ann. Stat.,* vol. 29, pp. 1189-1232, 2001.

[23] G. Ke et al., "LightGBM: A Highly Efficient," *Adv. Neural Inf. Process. Syst.,* vol. 2017, pp. 3149-3157, 2017.

[24] S. Moritz and T. Bartz-Beielstein, "imputeTS: Time Series Missing Value Imputation in R," *The R Journal,* vol. 9, no. 1, pp. 207-218, 2017.

[25] R. Hyndman, Y. Khandakar, "Automatic time series forecasting: the forecast package for R," *Journal of Statistical,* vol. 26, no. 3, pp. 1-22, 2008.

[26] R. Hyndman et al., "forecast: Forecasting functions for time series and linear models. R package version 8.11," 2020. [Online]. Available: http://pkg.robjhyndman.com/forecast>.

[27] R. E. Kalman, "A New Approach to Linear Filtering and Prediction Problems," *Journal of Basic Engineering,* pp. 35-46, 1960.

APPENDIX

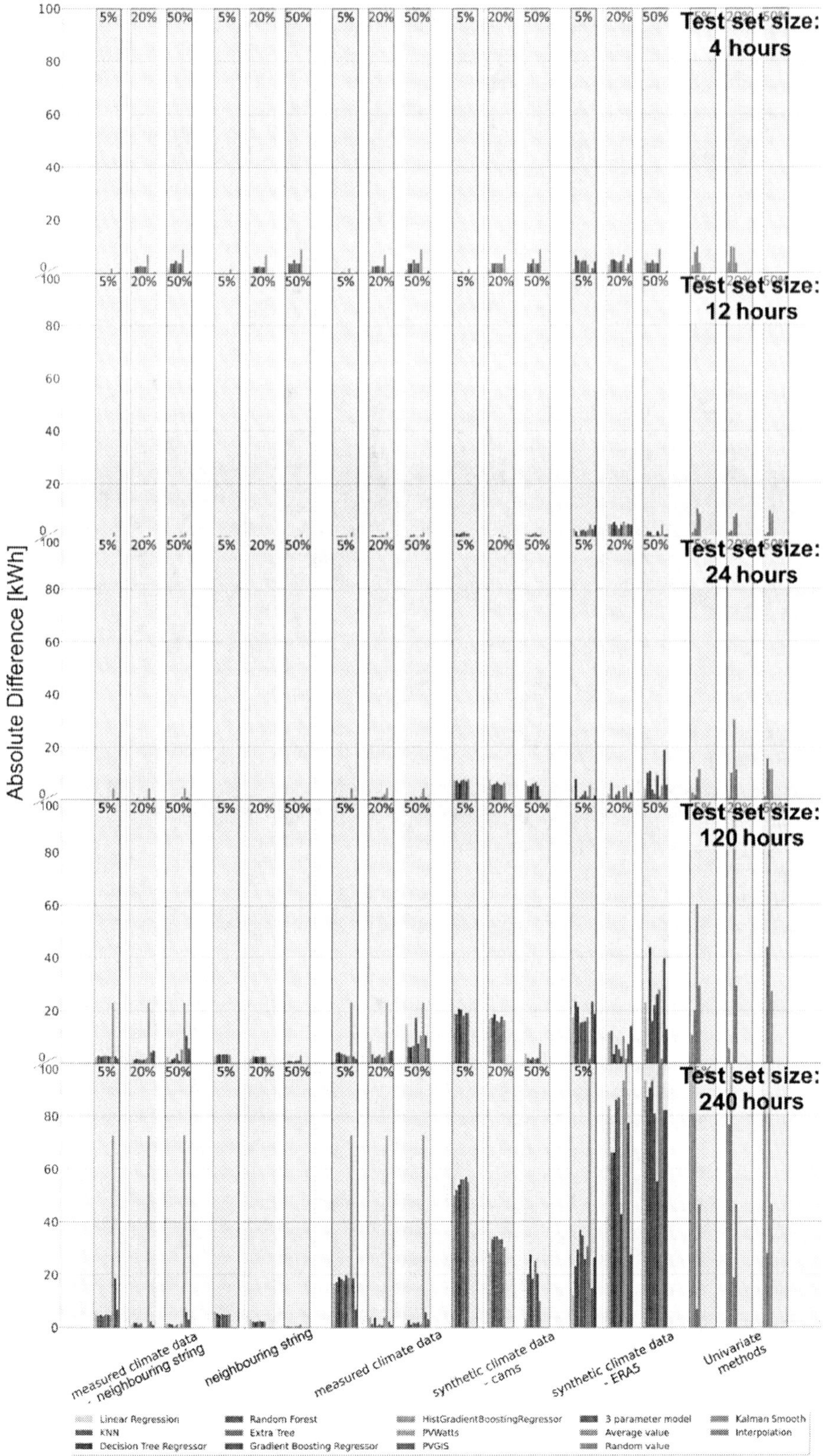

Figure 4: Absolute difference [kWh] of individual predictors for training/test ratio of 95/5, 80/20 & 50/50 for test set sizes of 4 hours, 12 hours, 24 hours, 120 hours & 240 hours

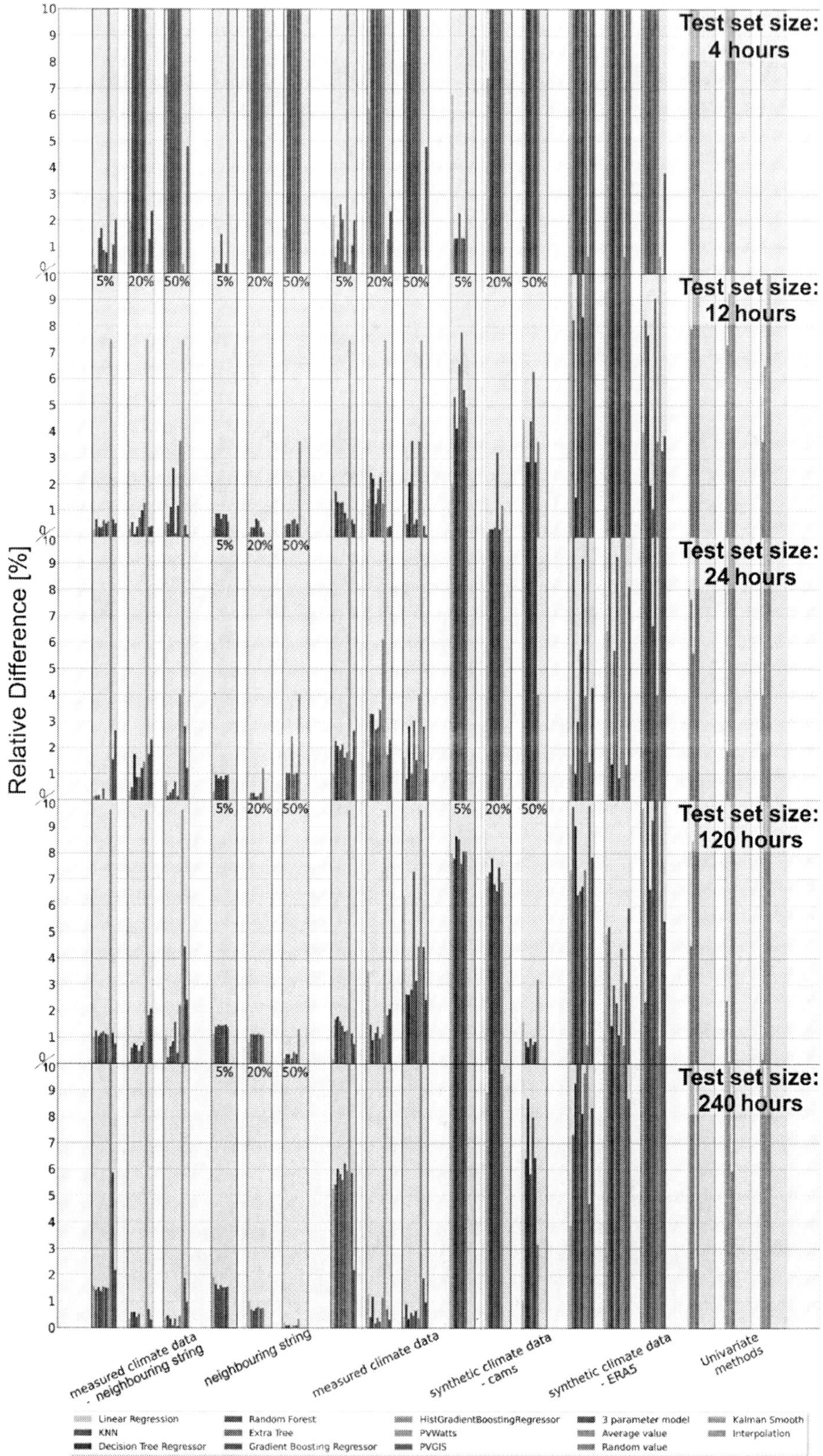

Figure 5: Relative difference [%] of individual predictors for training/test ratio of 95/5, 80/20 & 50/50 for test set sizes of 4 hours, 12 hours, 24 hours, 120 hours & 240 hours

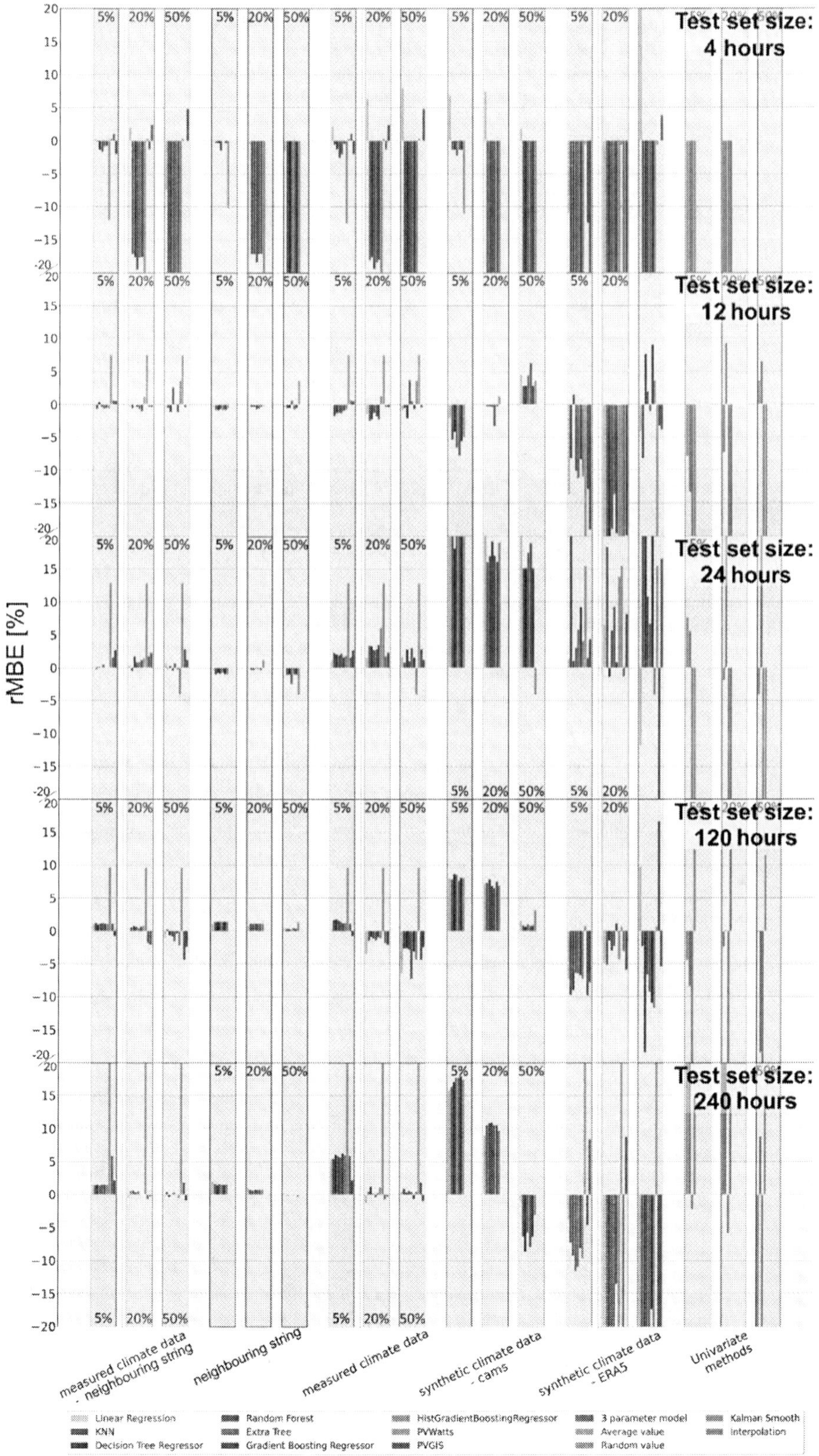

Figure 6: rMBE [%] of individual predictors for training/test ratio of 95/5, 80/20 & 50/50 for test set sizes of 4 hours, 12 hours, 24 hours, 120 hours & 240 hours

LONG-TERM DEGRADATION ASSESSMENT OF FIVE DIFFERENT PHOTOVOLTAIC TECHNOLOGIES IN AUSTRIA

Robert Höller and Kadrije Cërriku
University of Applied Science Upper Austria, School of Engineering
Stelzhamerstrasse 23, 4600 Wels, Austria
Email: robert.hoeller@fh-wels.at; Phone: +43 50805 44250

ABSTRACT: A PV rooftop test stand with five different PV module technologies is in operation at the University of Applied Sciences Upper Austria at Wels Campus since June 2007. This study aims to evaluate the long-term outdoor performance of each photovoltaic array utilizing measurement data of the ten-year period from 2009 to 2019, identify possible causes that led to underperformance, and determine the degradation rate of modules. The degradation rate of each PV array was calculated using two different meteorological datasets. The first dataset is based on the meteorological measurement station on the roof of the university building at Wels Campus, the second dataset is from SolarGIS. For the determination of the degradation rates, two different methods were applied: the trend of normalized energy and the classical seasonal decomposition statistical (CSD) method.
Keywords: PV System, degradation, performance

1 INTRODUCTION

Five different PV technologies (CIS, CdTe, a-Si, monocrystalline (c-Si) and polycrystalline (p-Si)) were deployed outdoors on the rooftop of the University of Applied Sciences Upper Austria in Wels in June 2007 [1]. Factors that influence degradation of PV modules in field operation include meteorological parameters such as temperature, humidity, precipitation, dust, snow and solar irradiation. Additionally, at the array level, shading and module mismatch contribute to degradation [2]. These factors impose significant stress over the lifetime of a PV system, resulting in the reduction of durability, which must be quantified through the measurement of the rate of performance degradation.

The long-term performance and degradation assessment of these PV technologies in the continental climate of Austria can be of particular interest to intended PV projects in the area [3]. In particular, the degradation rate is a relevant parameter for energy yield simulations and PV business models. The PV market is dominated by crystalline silicon technologies and this fact can be attributed to their outstanding long-term performance, comparatively high efficiencies and the reduced end costs. Thin-film PV modules have more or less disappeared on the European PV market in recent years.

The international standard on PV performance monitoring (IEC 61724-1:2017) recommends the usage of final energy yield, temperature-corrected performance ratio and energy efficiencies as best metrics for the assessment of PV systems performance [4]. Therefore, many researchers base their study of PV system performance on the performance ratio and the energy yield evaluation [5]. According to Fraunhofer Institute for Solar Energy Systems (ISE), the performance ratio of current PV systems shows considerable improvement compared to the years before 2000, when it was around 70%. Nowadays, the PV systems performance ratio ranges between 80 % and 90 % [6].

2 SYSTEM DESCRIPTION

The rooftop PV system of University of Applied Sciences Upper Austria (Wels, Austria, latitude 48.17°N, longitude 14.03°E, elevation 317 m) is composed of five types of PV technologies: copper indium selenide (CIS), amorphous silicon (a-Si), cadmium telluride (CdTe), polycrystalline silicon (p-Si) and monocrystalline silicon (c-Si), which were deployed outdoors in June 2007. The 176 modules are south oriented and have a fixed tilt angle of 32°. The PV system comprises of two Würth-Solergy Solar Star 1500 string inverters, each connected to 13 CIS PV modules, each rated at 75 Wp. The nominal power output of these 26 CIS PV modules at STC is 1950 Wp. One SMA Sunny Boy 2100TL inverter is connected to a string of 12 Sharp's c-Si PV modules, each rated at 185 Wp. The nominal power output of these 12 c-Si PV modules at STC is 2200 Wp. One SMA Sunny Boy 1700 inverter connected to 9 parallel-connected strings of 4 Kaneka a-Si PV modules, each rated at 60 Wp. The nominal power output of these 36 a-Si PV modules at STC is 2160 Wp. One SolarMax 3000C inverter connected to 12 parallel-connected strings of 4 Antec CdTe PV modules, each rated at 43 Wp. Out of 48 installed CdTe modules, 20 broke after installation and were not substituted. The nominal power output of the remaining 28 CdTe PV modules at STC is 1204 Wp. Three Fronius IG-30 inverters, each connected to 2 parallel-connected strings of 9 series-connected Schott p-Si PV modules, each rated at 165 Wp. The nominal power output of these 54 p-Si PV modules at STC is 8910 Wp. Figure 1 shows several images of the rooftop PV system as well as the solar radiation measurement station.

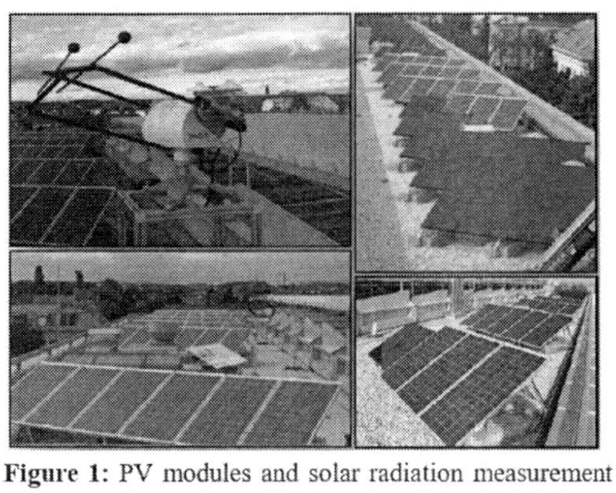

Figure 1: PV modules and solar radiation measurement station on the rooftop of University of Applied Science Upper Austria at Wels Campus. Two pyranometers and a pyrheliometer are mounted on a tracker for the measurement of global, diffuse and direct irradiance.

In Figure 2 an overview of the system, including the PV modules, inverters and string connections is shown.

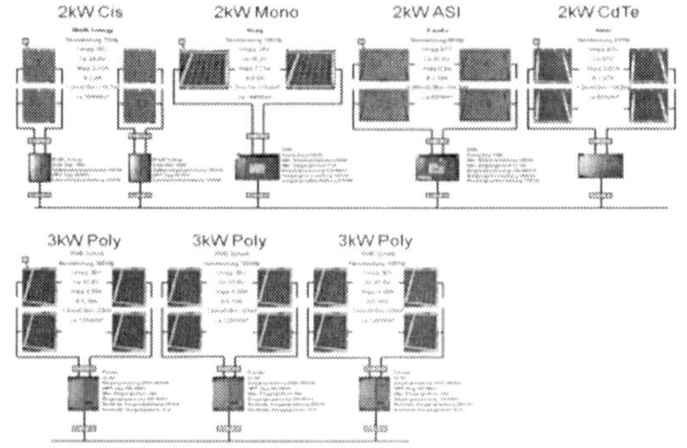

Figure 2: PV System description

3 DATA SOURCE AND QUALITY CONTROL

The analysed meteorological dataset contains meteorological (GHI, temperature, module temperature, wind speed, wind direction, humidity) and electrical parameters of each string, measured every ten minutes for a period of eleven years from 2008 to 2020. In addition, a dataset from SolarGIS that contains meteorological parameters for every fifteen minutes for a period of twenty years from 2000 to 2020 is applied. The first step was to resample both datasets at an interval of fifteen minutes for the period from 2009 to 2019. The year 2008 was discarded due to sensor problems. After data resampling, both datasets were synchronised due to summer time shifts twice per year. Plane-of-array (POA) irradiance was calculated from the measured GHI using the Reindl method.

The meteorological and electrical parameters underwent several Quality Control (QC) checks, applying both thresholds and statistical checks of the dataset, both for the meteorological and the inverter data. Details can be found in Cerriku [7]. Table I shows the data availability of each year after QC, and Figure 3 shows the overall data availability for the whole period after QC.

Table I: Data availability of each year

	2009	2010	2011	2012	2013	2014	2015	2016	2017	2018	2019
DA	87.9	93.4	93.2	94.0	72.5	80.8	91.9	81.9	81.6	61.2	85.4

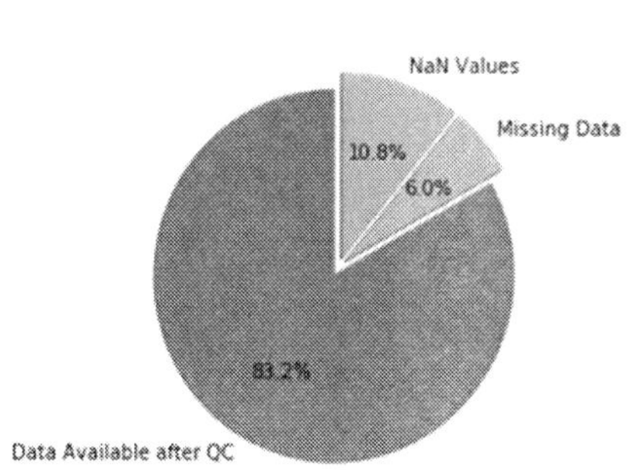

Figure 3: Overall data availability after data quality control

As a result of the QC checks 0.5% of GHI, 1.1% of DHI and 0.5% of DNI data were rejected. The electrical data where checked if they fit within the lower and higher MPP range of the corresponding inverter. Data availability (DA) for each year of the considered 11-year period was calculated and it varies in the range of 80.8 – 94 %. Only the years 2013 and 2018 have a lowest DA of 72.5 % and 61.2 %, respectively, due to maintenance and calibration of the sensors.

4 METHODOLOGY

A number of methodologies for measuring and calculating degradation rate of PV) technologies is reported in the literature [2]. Several statistical analysis methods are existing for calculating degradation rates: linear regression, classical seasonal decomposition (CSD), autoregressive integrated moving average, locally weighted scatterplot smoothing, and year-on-year (YOY) analysis using RdTools.

RdTools is an open-source Python library created from NREL and evaluation of the degradation rate of photovoltaic energy systems is its main functionality [8]. RdTools can analyse different data frequencies (hourly, daily, weekly etc.), but the best results are obtained with frequencies lower than one hour. Normalization of the dataset is done with the PVWatt modelled power. As input, the energy yield, in-plane irradiance, and the cell temperature time series are used. Along with the time series, some metadata need to be provided: latitude, longitude, time zone, azimuth, tilt angle, temperature coefficient of power and the rated power of the PV array.

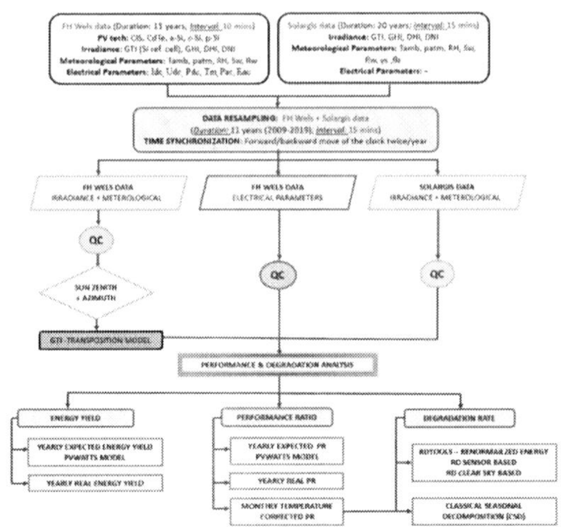

Figure 4: Flowchart of the applied methodology

RdTools can also calculate the clear-sky-based degradation rate. The clear-sky RD aims to be a solution for the bias that sensor-based data creates, because of the existing sensor drifts or problems that occur during recalibrations. Normalization is followed by data filtering which aims to remove all non-zero yield data from outages and data that create bias and noise in the data analysis. Data points, where the normalized energy is lower than 1 %, are removed. After the filtering process data is aggregated (into daily, weekly or monthly data) in order to avoid the impact of erroneous data points usually occurring in the morning and near evening. Finally, the degradation rate is estimated from the data included in the analysis. By default, the value of the confidence interval is 68.2 %, which belongs to 1 σ standard deviation. The confidence interval can be changed to match the intended percentage of data that an analysis requires.

In addition, the classical time series decomposition (CSD) was applied. It reduces the seasonality of the times series through mathematical smoothing of the trend, which is calculated using a symmetrical moving average over 12 months.

Figure 4 shows a flowchart of the applied methodology in this study to determine energy yield, performance ratio and degradation rates.

Soiling losses are modelled using the Kimber soiling model, which assumes that soiling builds up at a daily rate unless the rain threshold (selected as 10 mm), the soiling rate is 0.0015 corresponding to urban areas, the grace period is set as 5 days and max soiling value is set to 0.3. A manual cleaning performed on June 20-21, 2014 was also considered. Light-induced-degradation (LID) was assumed at 1.5%, for all other losses the PVWatts default values have been applied.

5 RESULTS AND DISCUSSION

Figure 5 shows the real and simulated annual performance ratio and energy yield for one exemplary string (c-Si). The average predicted PR of the c-Si array for the considered period of 11 years from 2009 till 2019 array is estimated to be 0.90 while the average real PR is 0.87. The figure shows the trendline of the predicted and the measured PR and energy yield using the measured meteorological, irradiance and electrical data. The relative difference between the expected and the real energy generated from c-Si array when considering FH Wels data is 10 %, as shown in Table II.

The same analysis was performed using the meteorological and irradiance data from SolarGIS and the measured electrical data. The in-plane irradiance measured from SolarGIS is higher than the calculated in-plane irradiance using the Reindl model. The predicted performance ratio remains the same (0.90) since the predicted AC power of the PVWatts model also increases at higher irradiance values. The average real PR is 0.82. The predicted energy yield using SolarGIS data is higher compare to FH Wels because of the increased values of SolarGIS in-plane irradiance.

The Year-on-Year (YOY) degradation analysis using RdTools for both the measured and SolarGIS data is illustrated in Figure 6. The difference between the clear-sky-based and sensor based is higher for measured than for SolarGIS data. It confirms that the irradiance sensor drift has an influence and the clear-sky-based results are more reliable.

The CSD method was applied on the temperature corrected monthly PR of c-Si array for both datasets. Linear regression of the smoothened trendline, after the seasonality and the residuals component has been removed, gives an annual RD value of 0.53 % (0.01 % higher than the clear-sky-based RD of RdTools) for the measured data and 0.41 % (0.1 % lower than the sensor-based RD of RdTools) for SolarGIS data. In Figure 7 the degradation rate determined by the CSD method based on measured data and SolarGIS data for the c-Si string is shown.

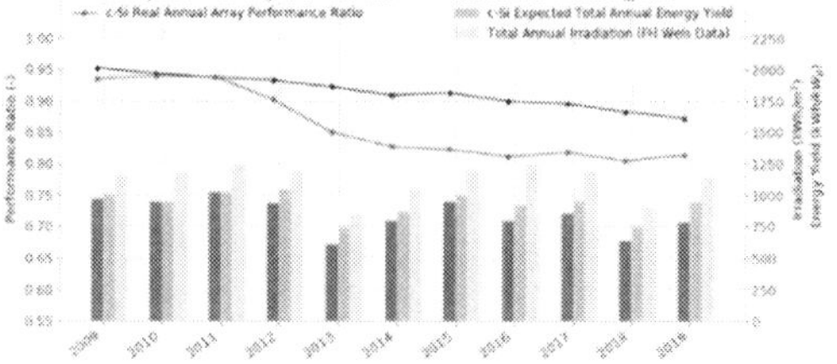

Figure 5: Evolution of real and expected performance ratio and energy yield of the c-Si-2 string based on measured irradiation datat at Wels.

Table II: Expected versus real energy of c-Si array

c-Si Array's Expected vs Real Energy (kWh) & the Relative Difference (%)											
	2009	**2010**	**2011**	**2012**	**2013**	**2014**	**2015**	**2016**	**2017**	**2018**	**2019**
Exp.	2220	2103	2275	2331	1676	1931	2220	2042	2109	1665	2097
Real	2147	2101	2258	2087	1353	1756	2102	1768	1900	1396	1750
Rel. Diff.	3	0	1	12	24	10	6	15	11	19	20
Relative Difference = 10 %											

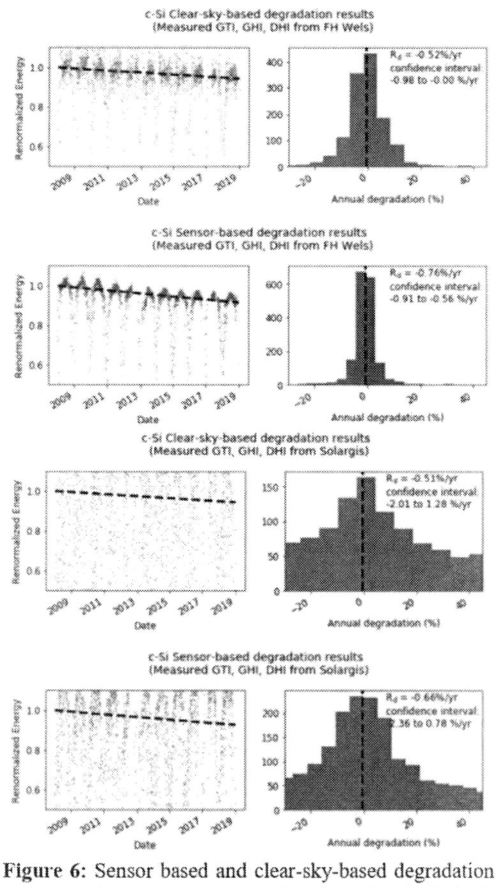

Figure 6: Sensor based and clear-sky-based degradation rate using the year-on-year technique of RdTools for both measured and SolarGIS data for c-Si array.

The average degradation rate for the c-Si array is 0.6 %/ year, as shown in Table IV.

Table IV: Degradation rates of the c-Si array for measured and SolarGIS datasets.

c-Si Array's Degradation Rate (% / year)		
Method	**FH Wels**	**SolarGIS**
RdTools (Clear-sky-based)	0.52	0.51
RdTools (Sensor-based)	0.76	0.66
CSD	0.53	0.41
Avg. RD = 0.6 % / year		

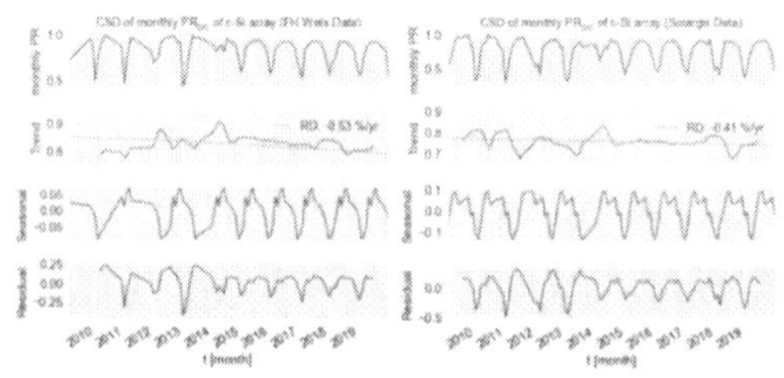

Figure 7: Degradation rate using the CSD method based on measured data and SolarGIS data for the c-Si string.

An overview of the main results of the yield and PR data analysis is shown in Table V. The PR and the specific yield of one the CIS-2 string is very low, because several of the modules were getting defect during the analysed period. Both in terms of PR and specific yield, the monocrystalline modules (c-Si) show the best performance.

Table V: Yearly average values of both FH Wels and SolarGIS datasets, for the degradation rate (RD), expected and real performance ratio, expected and real final yield, and their respective relative differences.

Degradation Rate of each PV Array[%/yr] (measured met. data)								
	CIS-1	**CIS-2**	**CdTe**	**a-Si**	**c-Si**	**p-Si-1**	**p-Si-2**	**p-Si-3**
RdTools (Clear-sky-based)	0.96	0.55	0.77	1.05	0.52	0.55	0.66	1.23
RdTools (Sensor-based)	1.52	0.91	1.28	1.71	0.76	0.87	1.23	0.99
CSD	1.81	1.45	0.76	1.56	0.53	1.09	1.34	5.43
Avg. RD (%/yr)	1.43	0.97	0.94	1.44	0.6	0.84	1.08	2.55

The summarized results of the average values of the degradation ratios from both datasets (measured met. data and SolarGIS) are shown in Tables IV and V. The average degradation rate of the PV technologies ranges between 0.6 and 2.55 % for the measured dataset.

Table VI: Degradation rates (%/year) of all PV technologies applying the measured met. dataset

Degradation Rate of each PV Array[%/yr] (SolarGIS data)								
	CIS-1	**CIS-2**	**CdTe**	**a-Si**	**c-Si**	**p-Si-1**	**p-Si-2**	**p-Si-3**
RdTools (Clear-sky-based)	1.45	0.41	0.46	1.6	0.53	0.48	1.12	3.03
RdTools (Sensor-based)	1.39	0.68	0.21	0.79	0.66	1.05	1.69	2.48
CSD	1.29	0.78	0.56	1.36	0.41	1.09	1.29	5.18
Avg. RD (%/yr)	1.38	0.62	0.41	1.25	0.53	0.87	1.37	3.56

The PV technology that performs best in the climate of Upper Austria) is c-Si with a degradation rate of 0.6 %/year and an average performance ratio of 0.85.

Table VII: Degradation rates (%/year) of all PV technologies applying the SolarGIS dataset

Degradation Rate of each PV Array[%/yr] (SolarGIS Data)								
	CIS-1	**CIS-2**	**CdTe**	**a-Si**	**c-Si**	**p-Si-1**	**p-Si-2**	**p-Si-3**
RdTools (Clear-sky-based)	1.45	0.41	0.46	1.6	0.51	0.48	1.12	3.03
RdTools (Sensor-based)	1.39	0.68	0.21	0.79	0.66	1.05	1.69	2.48
CSD	1.29	0.78	0.56	1.36	0.41	1.09	1.29	5.18
Avg. RD (%/yr)	1.43	0.97	0.94	1.44	0.60	0.84	1.08	2.55

6 REFERENCES

[1] https://goo.gl/maps/2DDFJCjrp8FgXnXB9

[2] A. Phinikarides, N. Kindyni, G. Makrides, G. E. Georghiou, Review of photovoltaic degradation rate methodologies, Renewable and Sustainable Energy Review, 40, 143–152 (2014).

[3] A. Högl, R. Höller, and Y. Ueda, Performance- und Degradationsanalyse von PV Modulen und Systemen, 15. Symposium Energieinnovation, Graz (2018).

[4] IEC 61724-1:2017, Photovoltaic system performance - Part 1: Monitoring.

[5] W.G.J.H.M. van Sark, N.H. Reich, B. Müller, A. Armbruster, K. Kiefer, Ch. Reise, Review of PV performance ratio development, World Renewable Energy Forum 2012, Annual Conference (2012). DOI: 10.13140/2.1.2138.7204

[6] Fraunhofer ISE. Photovoltaics Report (2020).

[7] K. Cërriku, Long-term Outdoor Performance Evaluation and Degradation Assessment of Different Photovoltaic Technologies, Master Thesis, University of Applied Sciences Upper Austria, Wels (2020).

[8] M.G. Deceglie, D., Jordan, A., Shinn, C., Deline, RdTools : An Open Source Python Library for PV Degradation Analysis degradation rate. National Renewable Energy Laboratory, 1-15 (2018).

PREDICTING PV SELF-CONSUMPTION IN VILLAS WITH MACHINE LEARNING

Fabian Galli and Nelson Sommerfeldt*
KTH Royal Institute of Technology, Dept. of Energy Technology
Brinellvägen 68, 10044 Stockholm, Sweden
*nelson.sommerfeldt@energy.kth.se, Tel: (+46) 8 790 81 54

ABSTRACT: The price for solar photovoltaic (PV) generation fed into the network from a prosumer system is often lower than the price earned with self-consumption, thereby making it a critical input to the economic analysis. This study aims to create a self-consumption model that can be applied during the planning phase of a residential PV system in Sweden, when detailed knowledge of the system and/or building load are often limited. Eight regression models are trained using 1.08 million unique data points built from measured villa loads and simulated PV generation to calculate self-consumption with an hourly time step. Five of the trained models are found to deliver acceptable performance with an R^2 greater than 0.9. Random Forest is the best performer with R^2=0.985 and a mean absolute error of 1.5 percentage points while only requiring annual building load, PV generation, and latitude as inputs. This is noteworthy given that SC can vary by +/- 20 percentage points for a given solar fraction and is a marked improvement over previous estimation techniques. The resulting models are limited to Swedish villas and future studies using this technique in other regions and building types with local building loads are recommended.
Keywords: Prosumer, Modelling, Grid-Connected, Economic Analysis, PV Systems

1 INTRODUCTION

Self-consumption is an increasingly important issue with distributed photovoltaic (PV) systems in buildings [1]. In Europe, most countries have shifted away from the feed-in tariff, and are using a self-consumption model where PV generation is used to displace loads in the building, valued at the retail price, and overproduction is sold to the network for a lower price, often times at wholesale [2]. This makes self-consumption a critical assumption in the planning of PV systems in buildings, where economic optimization of PV capacity is a trade-off between installed prices and self-consumption [3].

Ideally, the historical hourly or sub-hourly electric load and insolation data at the specific building would be used in modelling. However, in practice this can be hard to acquire, requiring considerable time and effort for such an early analysis phase. One study by McKenna et al. [4] in the UK was able to use one-minute empirical data from 218 homes with PV to derive a linear regression model with three terms – annual building demand, annual PV generation, and the percentage of electricity demand during the daytime. This is an improvement on the method most analysts and policy makers use today, which are educated estimates of self-consumption based on limited public statistics with high variance [3, 5]. This study builds on the work of McKenna et al. and contributes to the knowledge of PV systems through a novel investigation of machine learning methods of PV self-consumption prediction. Successful models will further reduce techno-economic uncertainties in the PV planning process for installers and owners.

2 OBJECTIVE AND SCOPE

This study aims to create an improved self-consumption prediction model that can be applied in the planning phase for prosumer PV systems. The model should be capable of using readily available data for any given location of interest within a region and be applied in a techno-economic optimization. Given the diversity of building design, energy supply systems, and climate, this study is limited to Swedish villas due to data availability.

3 METHODOLOGY

The self-consumption model is generated using regression based, machine learning methods from the open-source, SciKit-Learn libraries written in the Python programming language. Seven commonly used algorithms are tested, including; k nearest neighbors (k-NN), random forest, multi-layer perception (MLP), linear regression, polynomial regression, ridge regression, and lasso regression. Each algorithm is optimized considering specific model parameters and input selection to produce the best possible outcome. Model performance is measured using mean absolute error (MAE), mean bias error (MBE), the coefficient of determination (R^2) and R^2-adjusted. Additionally, the linear regression model presented by McKenna et al. [4], which is the only paper known to the authors that has presented a regression model to predict self-consumption, will be recreated here using Swedish data.

The models are trained to accept up to five input parameters; annual building load (kWh/yr), annual PV generation (kWh/yr), PV array azimuth (Degrees, north as zero), PV array tilt angle (Degrees, horizontal as zero), and latitude (Degrees north). The single output is annual self-consumption as a ratio with the range 0-1 (i.e. 0-100%). Training data is generated using a combination of measured building electricity loads and modeled PV generation using an hourly time step to calculate self-consumption. The combination of all PV systems and building loads create 1.08 million unique data points. Model training is performed using the k-fold cross validation method with k=10 folds and feature scaling to give all inputs equal standing in the regression.

3.1 Building Loads

The measured hourly load data comes from 108 villas in Karlstad, Sweden between the years 2015 to 2019, resulting in 540 villa-years of load profiles. Figure 1 shows the annual electricity consumption of each villa and year in the data set, ranging from 1.5 MWh/year up to 44 MWh/year. The diversity in the load data is suitable for creating a model that can make accurate estimations for a variety of villas with different load patterns and total loads. The raw load data had several missing days, with the largest being the month of November 2016. In most cases, the gaps are small enough to copy a similar period using a visual inspection. For the month long gap, the Finkelstein-Schafer method [6] is used to identify which of the other year's November data is most typical for each villa, which is then used to fill the gap. This is the same approach used to create meteorological year (TMY) climate data [7, 8].

Figure 1: Distribution of annual loads from villas

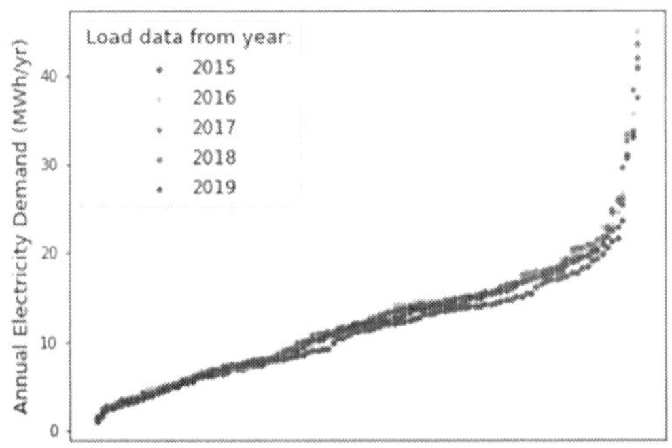

The load profiles for each building has a distinct distribution over the year due to number of occupants, size, envelope efficiency, and heating equipment. Homes with electric heating see greater proportions of load in the winter due to the northern climate. Homes without electric heating see a small difference due to lighting and a general trend to spend more time indoors during winter. This is visualized in Figure 2 where the monthly energy consumption for five sample villas are presented. The 34 MWh villa in purple stands out with dramatically higher summertime loads, suggesting that it uses direct electric water and space heating, whereas the 12.3 and 20.5 MWh villas are most likely relying on heat pumps. The 2.5 and 6.5 MWh/yr villas are most likely to use non-electric heat sources, leaving their loads for appliances only, and therefore more consistent throughout the year.

Figure 2: Monthly load pattern samples

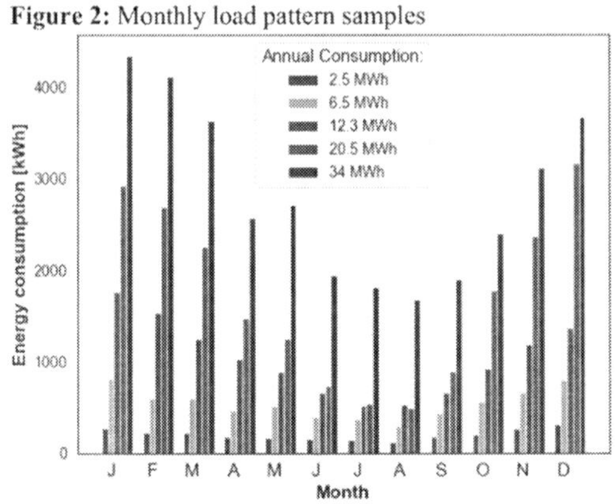

3.2 PV Simulation

PV power production is simulated using the PVWatts model in System Advisor Model [9] with TMY climate data from the PVGIS-SARAH database [10]. This means the hourly building load patterns do not correspond exactly to the climate data, however since TMY climate data is used in long-term PV performance prediction, this arrangement most represents the intended application for the models. Due to the change of sun paths caused by geographical location, different latitudes are used to create PV generation profiles for five locations in Sweden. Locations with approximately the same annual global horizontal irradiation (+\- 0.4%) are chosen to limit the differences to irradiance patterns. Latitudes in the southern part of Sweden are selected due to 88% of Sweden's population living south of 62°N and is where most of future PV installations can be expected. The extreme weather in the north is also likely to create more extreme outliers that could affect a general model applicable for the majority of Sweden.

The PVWatts simulations are based on a 1 kW_p system with a typical performance ratio of 85% [11, 12] achieved with a fixed inverter efficiency of 98% and remaining losses of 12%. To cover the full range of possible orientations, profiles for azimuths ranging from 0° to 350° in steps of 50°, and tilts of 0° to 90° in steps of 22.5° are generated, resulting in 40 unique generation profiles for each location. An example of PV yields (in kWh/kW_p) at the 55.4°N location are shown in Figure 3 and the monthly distributions of hourly generation for each location with a 22.5° tilt, 150° azimuth orientation are given in Figure 4. These results demonstrate the diversity of the generation profiles applied in the training data.

Figure 3: PV yield by orientation

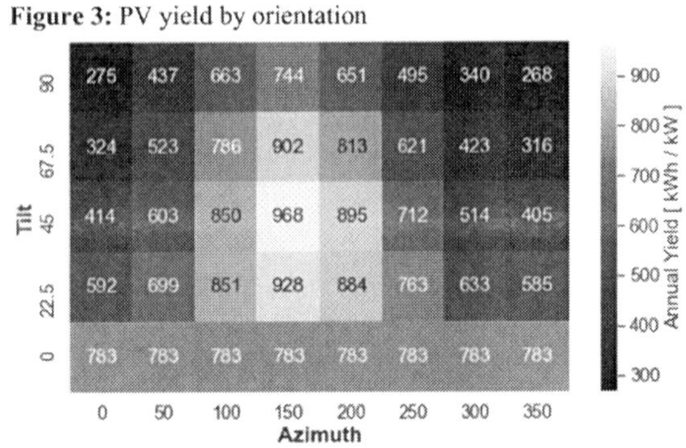

Figure 4: Box plots of hourly generation by month and latitude for a representative 1 kW_p system

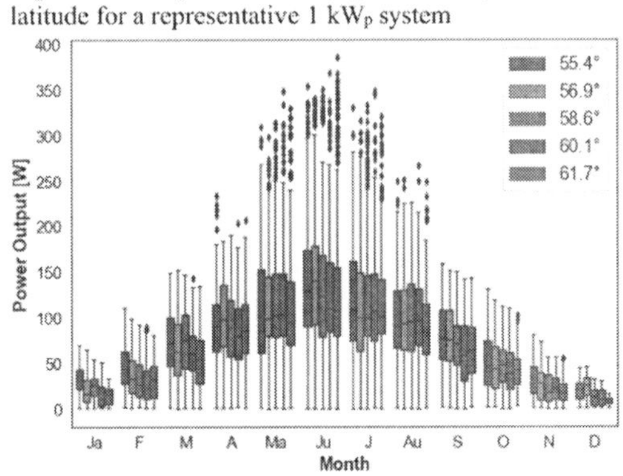

To train the models for a range of PV capacities, the generation profiles are scaled linearly to represent smaller and larger systems. With each orientation's yield, a maximum capacity is determined with each villa and year such that the PV system generates 100% of the annual demand. This capacity is then split in to 10 possible system sizes, i.e scaling factors. A summary of the PV system parameters and their ranges are presented in Table I. In total, 2000 different PV generation profiles are simulated.

Table I: PV system parameters

	Unit	Min	Max	Step
PV Capacity	% of Load	10	100	10
Azimuth	Degrees	0	360	50
Tilt Angle	Degrees	0	90	22.5
Latitude	Degrees	55.4	61.8	~1

3.3 Self-Consumption Data

Each of the 2000 simulated PV generation profiles are combined with each of the 540 villa loads to result in 1.08 million unique self-consumption observations. Self-consumption is defined as the ratio of PV generation used directly in the villa over the total generation. It is calculated hourly and summed for the year to create a single self-consumption data point for training.

To qualify the suitability of the dataset, average, maximum and minimum self-consumption values are compared with measured values from Stridh [5] in Figure 5. The x-axis is presented as solar fraction, which is the ratio of annual PV generation to annual building load. Individual measurements are represented as blue dots and the blue line is a fitted trend line, both from [5]. The orange line is the averaged curve for all 540 building load profiles with one PV system orientation (22.5° tilt, 150° azimuth).

Figure 5: Training data results compared to measured self-consumption values from [5]

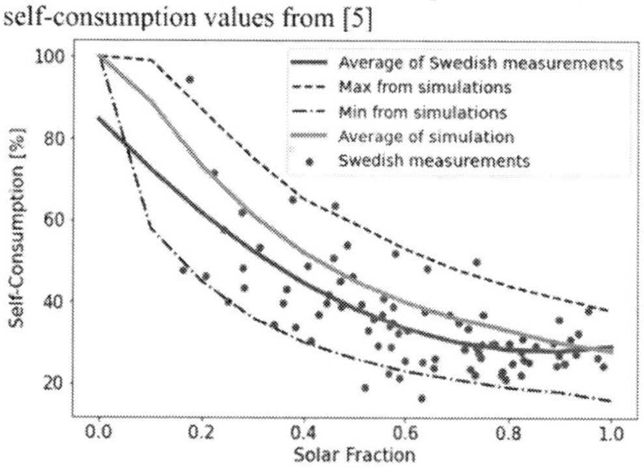

The average curve from the simulated dataset is consistently higher than the measured by about five percentage points, until around an 80% solar fraction where the two curves begin to converge. However, it should be noted that the measured data is sourced from a few hundred villas during 2018, making it a relatively limited dataset for validation. The orientation of each measured system is also unknown, and the comparison of one simulated PV orientation to a diverse range of orientations will be a source of error. However, this is the only publically available documentation of self-consumption the authors are aware of Sweden, and this comparison suggests the simulated dataset is a reasonable representation of Swedish villas with PV systems.

4 RESULTS

Model development, and thus results generation, is performed in two steps; 1) create the best possible version of each model, and 2) rank/compare the models. Several of the models have auxiliary parameters, all of which are optimized parametrically. A summary of the final parameters are listed in Table II and the full parametric analysis can be found in [13].

Table II: Parameters and final values for each model

Model	Final Parameter Values
Linear	N/A
Polynomial	Degree: 6
Ridge	Degree: 6, Alpha: 0.01
Lasso	Degree: 5, Alpha: 0.01
k-NN	Neighbors: 26
Random Forest	Trees: 130, Max depth: 25
MLP	Hidden layers: 2, Neurons: 8 & 6

4.1 Input Variable Selection

To assess the impact of each input feature on the model performance, a sensitivity analysis is conducted where the variables are removed one at a time. The minimum viable inputs are assumed to be building load, and PV, negating the need to test each input individually. Each combination is represented in Table III with an assigned test number.

Table III: Input variables for each selection test

Test	Load	PV	Tilt	Azimuth	Latitude
1	•	•	•	•	•
2		•	•	•	•
3	•		•	•	•
4	•	•		•	•
5	•	•	•		•
6	•	•	•	•	
7	•	•			•
8	•	•		•	
9	•	•	•		
10	•	•			

The result of the analysis is presented in Figure 6 where the performance of each model is shown with R^2. The patterns are similar with all models, where the worst performance is attained when either building load or PV generation is removed. Removing one or more of the other inputs, tilt, azimuth and latitude, had almost no effect on the performance. It even somewhat improved the performance of the Random Forest and k-NN models.

Figure 6: R^2 for each test and model

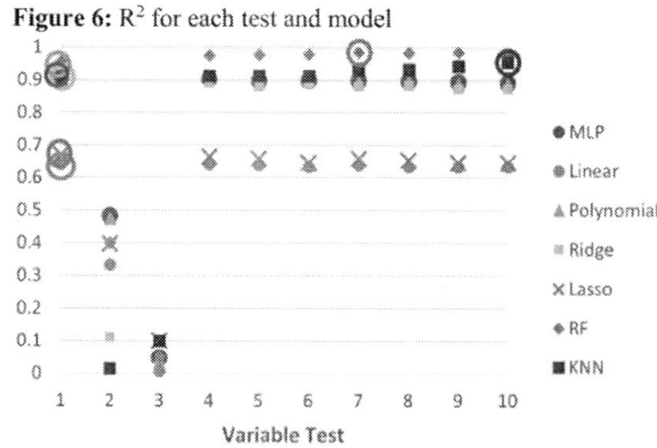

4.2 Model Comparison

The final results from each optimized model is shown in ranked order in Table IV. R^2-adjusted is omitted since the values are functionally identical to R^2, suggesting no occurring overfitting of the models. Random Forest proved to have the lowest MAE and highest R^2, with a mean absolute error of only 1.5 percentage points. As shown in Figure 5, this is noteworthy given that self-consumption can vary by +/- 20 percentage points for a given solar fraction [5]. It also noteworthy that the top two performing models, Random Forest and k-NN, performed better with fewer variables, similar to the McKenna model which only uses three variables [4]. In the case of k-NN, only building load and PV generation are needed for predictions, which is similar to what most analysts would use today. Linear regression performed the worst, with an R^2 of 0.641, followed closely by the McKenna model at 0.646. In their paper, McKenna et al. [4] found an R^2 of 0.754. An R^2 value of 0.9 is a common threshold for model validation, and the results here suggest that five of the seven (assuming 0.897 can be rounded to 0.9) could be considered suitable predictors.

Table IV: Final model performance values

	MAE	R^2	MBE
Random Forest	0.015	0.985	1.00e-04
k-NN	0.025	0.956	-1.70e-04
MLP	0.038	0.907	7.35e-06
Polynomial	0.038	0.907	1.33e-05
Ridge	0.038	0.897	-3.73e-07
Lasso	0.058	0.670	1.59e-07
Linear	0.074	0.641	2.12e-07
McKenna et al.	0.073	0.646	-1.02e-08

As with most of the models, Random Forest also has an error bias of nearly zero, which is also shown in Figure 7 by the residual distribution. The high kurtosis of the residual distribution ($\sigma = 0.035$) suggests most predictions will have a low error, but that a small number of outliers (tails) can result in errors up to 15 percentage points.

Figure 7: Residual distribution for Random Forest

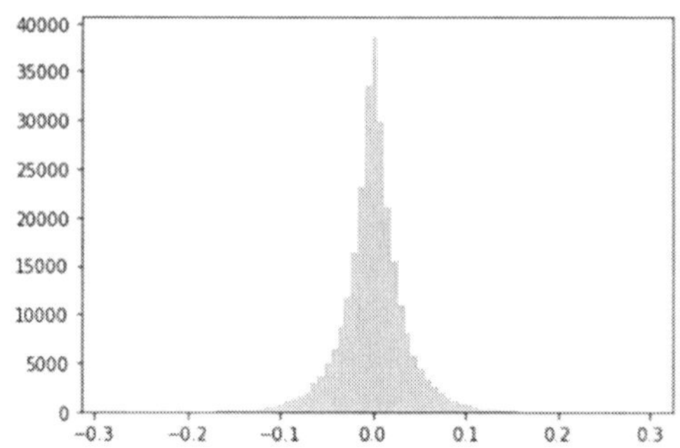

5 DISCUSSION

Even though most models performed best with all input features included, removing tilt, azimuth and latitude had almost no effect on the model's prediction ability. This suggests that using only annual PV yield and load as inputs could be a good trade-off between model simplicity and performance. This is also the most commonly available data in real projects during the planning phase, further promoting their removal for practical use.

In the paper by McKenna et al. [4], the reported model performance in terms of R^2 is better than the results found here using the same methodology but different dataset. The exact cause of the difference is unknown and merits further investigation. One speculative possibility is the relative homogeneity of the data used by McKenna et al., which is sourced from approximately 300 PV systems over two years in northeast Britain, in contrast to the more diverse load and PV profiles used here. At the same time, it is possible that the homogeneity of southern Sweden is similar to that of northeast Britain and studies from more regions are needed to help build further understanding of these models on self-consumption prediction.

A key feature of this study is the use of simulated PV data, enabling a large dataset to be generated. Likewise, this feature is also a limitation in that is decouples the load profiles from the climate data used in PVWatts, potentially leading to incorrect self-consumption values. The impact of this potential error is mitigated to some degree by the testing of multiple locations within Sweden and multiple years of load data.

The original motivation for testing multiple locations was to alter the sun path, however each location also includes random variations in irradiance due to other factors, e.g. clouds. Since all of this information is included in the dataset, it is possible the errors due to climate mismatch between the loads and generation are minimized, but does require further analysis to quantify. In the thesis on which this paper is based [13], a limited analysis on how self-consumption varies over the years shows annual changes of up to 14 percentage points. Future work that uses spatially and temporally coordinated weather data will quantify the error, if any, within this dataset and model. It should be noted, however, that using a single location could potentially limit the applicability of the resulting models for other locations in Sweden.

This study aimed to create a generic self-consumption model to cover most potential PV systems in Swedish villas across most habituated regions of the country. While the approach used is believed to have done so, an empirical validation is still missing. The nature of simulation will always make it possible to generate larger datasets as compared to measured systems. It is unclear at this point if a dataset of a few hundred measured systems is superior or inferior to a one million simulations. Measured systems will likely include their own artefacts, such as local shading, that could influence their applicability in a general purpose model. Regardless, future work should aim to validate the machine learning models trained with simulated data with measured systems, and even other models trained with measured data.

At the time of writing, there is relatively little availability of PV self-consumption data. For example, the only study found by the authors for Sweden published self-consumption for approximately 400 villas over one year [5]. Given the rapid growth of PV and data collection, it is likely that many future studies can be generated on this topic that will improve the self-consumption models.

6 CONCLUSIONS

This study aims to improve the prediction of PV self-consumption for use in techno-economic optimization during the planning phase of PV systems in Swedish villas. The results have shown that the Random Forest machine learning technique produces the best performance with an R^2 of 0.985 and an MAE of 0.015. This is a marked improvement over the only other similar study found from McKenna et al. [4] where a linear regression model generated an R^2 of 0.754. Moreover, many of the models can be used for satisfactory predictions of self-consumption, where all models except the linear regression and Lasso regression, have an R^2 at or above 0.9 and a low average error between 1.5 to 4 percentage points of annual self-consumption. This result demonstrates the superior performance of statistical machine learning models over simpler regression methods.

It is also concluded that while optimal model prediction is found with most of the tested inputs, nearly identical performance can be had with only PV generation and building load. This makes the models highly practical given that limited information is usually available during the planning phase. Indeed, incorrect self-consumption estimations can significantly harm the value of a PV system, particularly if it is oversized, and the models trained in this study are a marked improvement over existing techniques in industry. Future work is needed to further define and validate the models, however statistical machine learning methods are poised to provide considerable value to PV system design.

7 ACKNOWLEDGEMENTS

The authors are grateful to the Swedish Energy Agency for funding this research via the Design for Everyday Energy Efficiency program, project number 48103-1. Thanks also goes to David Stoltz and Fredrik Balderud at Karlstad Energi for supplying the load data.

8 REFERENCES

[1] R. Luthander, J. Widén, D. Nilsson, and J. Palm, "Photovoltaic self-consumption in buildings: A review," Applied Energy, vol. 142, pp. 80–94, 2015, doi: 10.1016/j.apenergy.2014.12.028.

[2] I. Masson, G; Kaizuka, IEA PVPS report - Trends in Photovoltaic Applications 2020. 2020.

[3] N. Sommerfeldt, "Solar PV in prosumer energy systems," KTH Royal Insitute of Technology. Phd. Thesis, 2019.

[4] E. McKenna, J. Pless, and S. J. Darby, "Solar photovoltaic self-consumption in the UK residential sector: New estimates from a smart grid demonstration project," Energy Policy, vol. 118, no. April, pp. 482–491, 2018, doi: 10.1016/j.enpol.2018.04.006.

[5] B. Stridh, "Utvardering av egenanvandning av solel i Sverige," 2020.

[6] J. M. Finkelstein and R. E. Schafer, "Improved Goodness-Of-Fit Tests," Biometrika, vol. 58, no. 3, pp. 641–645, 1971.

[7] O. S. Ohunakin, M. S. Adaramola, O. M. Oyewola, and R. O. Fagbenle, "Generation of a typical meteorological year for north-east, Nigeria," Applied Energy, vol. 112, pp. 152–159, 2013, doi: 10.1016/j.apenergy.2013.05.072.

[8] S. Pusat, I. Ekmekçi, and M. T. Akkoyunlu, "Generation of typical meteorological year for different climates of Turkey," Renewable Energy, vol. 75, pp. 144–151, 2015, doi: 10.1016/j.renene.2014.09.039.

[9] National Renewable Energy Laboratory, "System Advisor Model." 2020.

[10] A. G. Amillo, T. Huld, and R. Müller, "Eastern Meteosat Satellite , Models and Validation," pp. 8165–8189, 2014, doi: 10.3390/rs6098165.

[11] M. Dhimish, "Performance Ratio and Degradation Rate Analysis of 10-Year Field Exposed Residential Photovoltaic Installations in the UK and Ireland," Clean Technologies, vol. 2, no. 2, pp. 170–183, 2020, doi: 10.3390/cleantechnol2020012.

[12] B. ; van Sark, W.G.J.H.M.; Reich, N.H.; Müller and Ch. Armbruster, A.; Kiefer, K.; Reise, "Review of PV Performance Ratio Development," World Renewable Energy Forum, WREF, vol. 6, 2012.

[13] F. Galli, "Predicting PV Self-consumption in Villas with Machine Learning," KTH Royal Insitute of Technology. M.Sc. Thesis, 2019.

THE ROLE OF COLLECTIVE SELF-CONSUMPTION IN THE PROMOTION OF PV IN MULTI-APARTMENT BUILDING FACADES

R. Amaro e Silva[1], C. Gerçek[2], M. C. Brito[3], G. Pontes Luz[3]

[1] MINES ParisTech, PSL Research University, Centre Observation, Impacts, Energy (O.I.E.), 06904 Sophia Antipolis, France

[2] Saxion University of Applied Sciences, Academy of Life Science, Engineering & Design (LED), Sustainable Energy Supply research group (DEV), 7513 AB Enschede, Netherlands

[3] Instituto Dom Luiz, Faculdade de Ciências da Universidade de Lisboa, Campo Grande, 1749-016 Lisbon, Portugal

ABSTRACT: Even though buildings in urban environments constitute large consumption clusters, the exploitation of building facades for PV generation to meet this demand through self-consumption is yet largely residual. This work aims to assess the potential of collective self-consumption to unlock higher PV penetration levels in multi-apartment buildings, saturating the available rooftop area. Using real 15-minute electricity demand data, weather data and a typical building shape for the city of Lisbon, Portugal, the area availability and PV potential are assessed using Sketchup and BIMsolar. Then, Calliope estimates the optimal PV deployment while considering a range of techno-economical constraints. Results corroborate, on the one hand, the limitations of individual self-consumption. On the other hand, they point out that collective setups can increase the capacity of an optimal system by a factor of 2, boosting the deployment of vertical PV (which can represent up to 50% of the capacity of the PV system). Collectives, as aggregators of diverse loads, also led to an increase of both the self-sufficiency and self-consumption rates of 6 % and 7 %, respectively, in median.
Keywords: Self-consumption, Building Integrated PV (BIPV), Grid-connected

1 INTRODUCTION

Despite the fact that buildings in urban environments constitute dense clusters of electricity demand, individual self-consumption promotes considerably small photovoltaic (PV) systems for residential dwellings (e.g., 70% of self-consumption installations in Portugal have no more than 1 kWp [1]). The resulting small surface area needed for PV deployment mostly leads to rooftop-only installations (where the PV yield, i.e. the annual generation per unit of installed capacity, is higher), even in multi-apartment buildings. This has hindered the exploitation of building facades, even though research has vastly shown that such structures are essential to achieve high penetration levels of PV electricity in city centers [2], as well as contribute to a more manageable generation profile [3].

At the European Union level, the recent introduction of collective frameworks such as collective self-consumption and energy communities [4] raises new opportunities. The aggregation of prosumers under a single collective PV system has the potential to minimize the mismatch between PV generation and load [5] and, thus, generation surpluses.

By improving self-consumption, collective frameworks can be expected to drive the deployment of larger PV systems and, potentially, the exploitation of building facades. The present work aims to quantify these potential impacts in multi-apartment buildings, in terms of self-consumption performance and added PV capacity. To do so, a 5-story building is assumed; load, weather and building shape data for the city of Lisbon, Portugal, is considered.

2 METHODS

2.1 Simulating multi-apartment buildings

Following the approach proposed in [6], virtual buildings were simulated by considering random subsets from a larger data set of apartment-level electricity load profiles. This allows to adopt a Monte Carlo approach and generate a number of samples to be analyzed, even when the available data is limited. This results in a distribution of outputs which can be interpreted as a proxy to the building-to-building variability.

2.2 Estimating PV potential on building surfaces

Sketchup was used to design a multi-apartment building, while BIMSolar [7], [8] was used to estimate the PV potential of each building surface.

BIMsolar is a PV modelling tool focused on 3D environments that accounts for building geometry, shading losses, and PV system losses. From it, the PV potential of each building surface and the number of modules it can serve as a platform are estimated. This tool demands a variety of weather inputs: air temperature at surface level; dew-point temperature; relative humidity; air pressure at surface level; infrared, global, direct, and diffuse irradiance at the surface level; and wind speed and direction.

For the present work, a single building was considered. This simplistic setting disregards the potential facade unavailability and shadowing due to adjacent and surrounding buildings. However, it allows inferring the maximum potential for vertical PV deployment due to the shift from individual to collective self-consumption.

2.3 Estimating optimal PV deployment

The optimal PV capacity allocated to a given grid-connect multi-apartment building is obtained using Calliope, an open-source multi-energy system modelling and optimization framework [9], [10]. This is built on top of three structural elements: locations (or nodes); links which interconnect locations; and technologies, allocated in locations and/or links, which are responsible for the generation, consumption, storage or conversion of energy.

One of the main particularities of Calliope is its flexibility, enabling users to custom design energy system layouts, as well as a wide range of technical and economic parameters (e.g., conversion efficiency, various kinds of costs and area occupancy for a given technology, the available area in a given location; or the direction energy flow is allowed in a given link).

Once an energy system is designed, Calliope builds a linear optimization model that, when solved, outputs the optimal energy system configuration and operation according to a given cost function (by default, the total system cost). Specifically, the model returns the optimal capacity for each technology allocated in each location and link, but also a set of time series with, e.g., energy fluxes, resource usage, or dispatch decisions.

Thus, for the assumed multi-apartment building two energy system layouts were designed in Calliope: i) the individual self-consumption, where PV capacity is allocated at the apartment level, with surplus generation being exported directly to the external grid; ii) the collective counterpart, where a single, centralized, PV system can be deployed and the prosumers share its generation.

2.4 Self-consumption metrics

Two common performance metrics are considered when evaluating self-consumption systems: the rates of self-consumption (SCR) and self-sufficiency ratios (SSR). The first corresponds to the fraction of PV generation that is consumed on-site (i.e., the complement of the PV surplus fraction).

$$SCR = \frac{\int PV_{generation}(t)\,dt - \int PV_{surplus}(t)\,dt}{\int PV_{generation}(t)\,dt}$$

The second, on the other hand, quantifies the fraction of electricity needs that the PV generation is able to meet on-site.

$$SSR = \frac{\int PV_{generation}(t)\,dt - \int PV_{surplus}(t)\,dt}{\int Load(t)\,dt}$$

3 DATA

This work is done using 15-minute electricity consumption profiles from 18 residential apartments located in Lisbon, Portugal, for 2013 [5]. The contracted power associated to the various apartments ranges from 3.45 to 10.35 kVA. Due to data privacy issues, no geographical information regarding the apartments' location was provided.

Based on the available data, 25 five-story multi-apartment VBs were designed and simulated for each considered scenario. Regarding building layout, this work assumes a typical building height of 16.7 m and rooftop area of 144 m2 (10 apartments; 14.4 m2 per apartment) [11]. As only rooftop area and building height were available the rooftop was assumed to be flat and square-shaped (i.e., 12 m each side).

To estimate the potential in each building surface, a small 2 kWp PV system (eight 250 Wp mc-Si modules) was designed and replicated to fill each surface. Row spacing was accounted to avoid considerable shadowing losses based on the work of Brecl et al. [12], as it considers a similar latitude and module tilt to this work. The authors suggest a spacing factor (the ratio between the distance between two consecutive rows and the width of the considered module) of 2.46 to achieve annual energy losses below 2.5%. Additionally, only 23% of the rooftop area was considered technically viable, as suggested by Izquierdo et al. for buildings in very dense and highly populated areas [13]. This information was incorporated by only considering 55% and 42% of the N-S and E-W dimensions, respectively. Thus, based on the module size and the considered constraints, it was possible to infer that the rooftop has enough space for 8 PV modules, whereas each facade 56. Figure 2 illustrates the resulting BIMSolar designs.

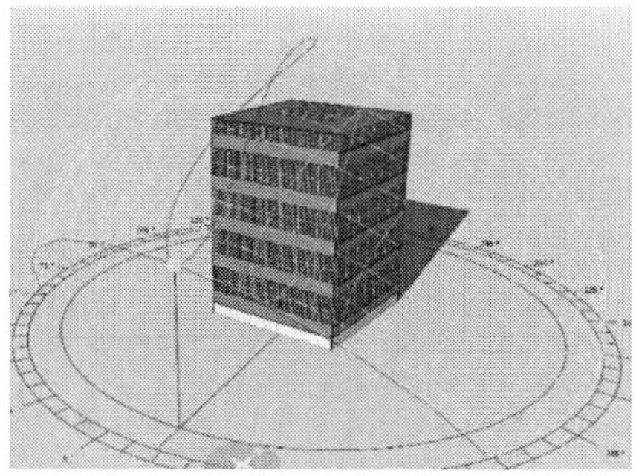

Figure 1: Illustration of the designed 5-storey building layout assuming a south-facing orientation, with the solar trajectory throughout the year for Lisbon.

Most of the weather variables required by the BIMSolar software (listed before) were provided by the SoDa – Solar radiation data company [14], [15] with 15-minute resolution records. Only the longwave irradiance was obtained from the ERA5-reanalysis [16], with 1 hour and 30 km temporal and spatial resolution.

BIMSolar cannot model with more detail than hourly resolution. Thus, the 15-minute weather data was converted to hourly averages and the BIMSolar outputs were then downscaled again to 15 minutes using a linear interpolation. While it is undeniable that this processing steps may impact the variability of the PV timeseries, two factors downplay this issue: i) the present work is focused in vertical PV which, due to its higher dependence on diffuse irradiance, is less variable than an optimal-tilted rooftop profile; ii) it has been shown that the higher variability of electricity demand (which was kept at the original resolution) dominates the generation demand mismatch [17].

Lastly, Table 1 summarizes the considered Calliope techno-economic parameters.

Table I: Parameters assumed in Calliope.

Parameter	Assumed value
Electricity cost	15.57 c€/kWh [18]
PV unitary cost	2 €/Wp [19]
PV surplus remuneration	0 c€/kWh
PV lifetime	25 years

4 RESULTS

4.1 Virtual Buildings characterization

As mentioned in the previous section, 25 VBs were designed for this study. Figure 2 illustrates the distribution of VB annual demand and the corresponding daytime fraction, peak load, and standard deviation ($\sigma_{VB\ load}$). The latter can be interpreted as a proxy to load variability.

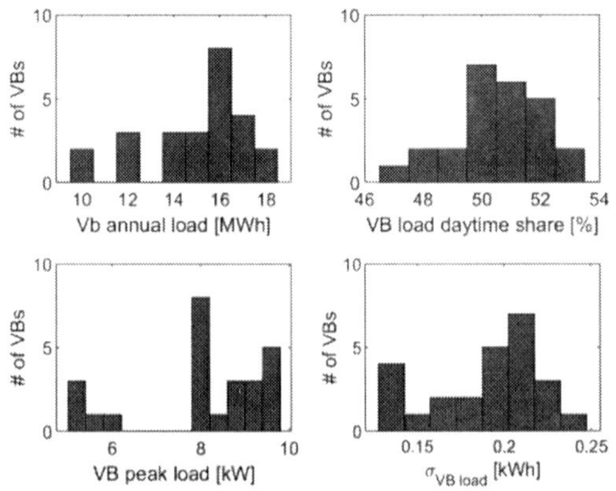

Figure 2: Distribution of VB characteristics due to Monte Carlo approach.

4.2 Individual self-consumption and the shift to a collective framework

In an individual self-consumption setting, the optimal configuration for the various VBs assumed at least an installed capacity of 1.75 kWp. Only 20% of the cases had vertical PV installed, always in the S-facing façade, which never surpassed 0.75 kWp (equivalent to 37.5% of a fully exploited rooftop). The sharing of a single PV system by a prosumer collective was shown to promote a larger deployment of PV capacity in 94% of the VBs. In this context, the installed capacity increased between 14.3% and 100% (median 75%), and at least 88% of the cases deploy PV in the facades (corresponding to a median share of 43% of the total installed PV capacity).

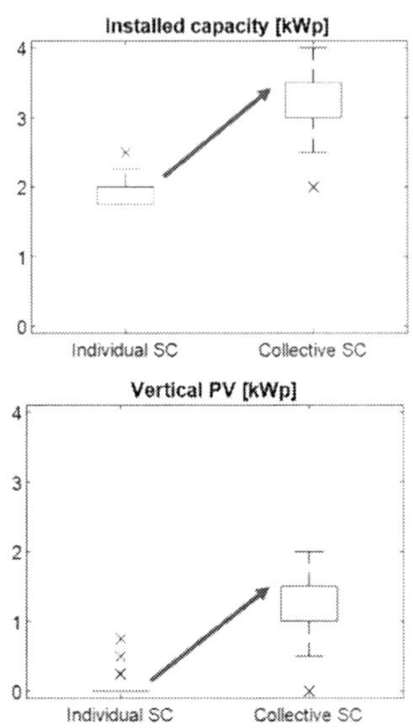

Figure 3: Distribution of the installed PV capacity and corresponding vertical PV share, considering two self-consumption settings. Shifting from an individual to a collective framework almost doubles the installed capacity and, as the rooftop saturates, the deployment of PV on the S-facing facade.

With the deployment of small PV systems, it is logical that the achieved SSR is low. Similarly, the achieved SCR is high, as surpluses are not economically valued (and, thus, represent a small share of the overall PV generation). Figure 4 illustrates this.

Shifting to a collective framework led to 6% and 7% increases in the median SSR and SCR, respectively, as also shown in Figure 4. On one hand, the deployment of larger systems increases the penetration of PV electricity. On the other, it is the aggregation of prosumers with different generation profiles which improves the load-PV generation match and leads to a more efficient PV integration.

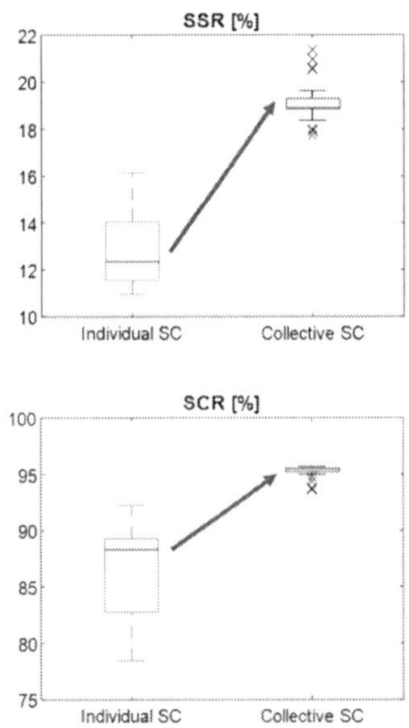

Figure 4: Distribution of the SSR and SCR metrics, considering two self-consumption settings. Shifting from an individual to a collective framework improves the SSR, as more PV is installed, and the SCR, as the PV generation is shared between various prosumers.

5 CONCLUSIONS

The present work highlights the limitations of individual self-consumption in a multi-apartment building, the benefits of shifting to a collective framework and the ability of the Calliope software to model both settings.

For simulated multi-apartment buildings with 5 storys and 10 apartments located in the city of Lisbon, individual self-consumption promotes the installation of PV systems smaller than 2.5 kWp, with low SSR values and a scarce usage of building facades. However, collective setups promote higher penetration levels of PV electricity, while improving the self-consumption performance (i.e., minimizing surplus exportation).

It is important to note that such positive impacts could be further boosted by shifting more load to the daytime through storage or demand side management and, thus, poses an interesting line of research.

6 ACKNOWLEDGEMENTS

This research has received funding from the European Union's Horizon 2020 research and innovation program under grant agreement no 957682 and presented at the 38th EUPVSEC thanks to an ITC Conference Grant from PEARL-PV COST Action CA16235. The authors also gratefully acknowledge the support of FCT- project UIDB/50019/2020 - Instituto Dom Luiz.

There were also contributions which are important to highlight as they made this work possible: the Solar radiation Data (SoDa) company, providing most of the used weather data and the Lisboa E-Nova energy and environment agency of Lisbon which provided the typical dimensions for the assumed building configuration.

7 REFERENCES

[1] "Public records from Direção Geral da Energia e Geologia (DGEG), accessed through the Observatório Fotovoltaico platform." .

[2] S. Freitas and M. C. Brito, "Solar façades for future cities," *Renew. Energy Focus*, vol. 31, pp. 73–79, 2019, doi: 10.1016/j.ref.2019.09.002.

[3] S. Freitas and M. C. Brito, "Non-cumulative only solar photovoltaics for electricity load-matching," *Renew. Sustain. Energy Rev.*, vol. 109, pp. 271–283, 2019, doi: 10.1016/j.rser.2019.04.038.

[4] D. Frieden, A. Tuerk, J. Roberts, S. D'Herbemont, and A. Gubina, "Collective self-consumption and energy communities: Overview of emerging regulatory approaches in Europe," 2019.

[5] V. Reis, R. H. Almeida, J. A. Silva, and M. C. Brito, "Demand aggregation for photovoltaic self-consumption," *Energy Reports*, vol. 5, pp. 54–61, 2019, doi: 10.1016/j.egyr.2018.11.002.

[6] M. B. Roberts, A. Bruce, and I. MacGill, "A comparison of arrangements for increasing self-consumption and maximising the value of distributed photovoltaics on apartment buildings," *Sol. Energy*, vol. 193, pp. 372–386, Nov. 2019, doi: 10.1016/j.solener.2019.09.067.

[7] "PVSITES BIMsolar & PVSITES Plugins." .

[8] S. Quintana, P. Huang, P. Saini, and X. Zhang, "A preliminary techno-economic study of a building integrated photovoltaic (BIPV) system for a residential building cluster in Sweden by the integrated toolkit of BIM and PVSITES," *Intell. Build. Int.*, vol. 13, no. 1, pp. 51–69, 2021, doi: 10.1080/17508975.2020.1765134.

[9] S. Pfenninger and B. Pickering, "Calliope: a multi-scale energy systems modelling framework," *J. Open Source Softw.*, vol. 3, no. 29, p. 825, Sep. 2018, doi: 10.21105/joss.00825.

[10] "Calliope website." https://www.callio.pe/ (accessed Mar. 01, 2021).

[11] "Lisbon municipal buliding stock database (source: Lisboa E-Nova, energy and environment agency of Lisbon)." .

[12] K. Brecl and M. Topič, "Self-shading losses of fixed free-standing PV arrays," *Renew. Energy*, vol. 36, no. 11, pp. 3211–3216, 2011, doi: 10.1016/j.renene.2011.03.011.

[13] S. Izquierdo, M. Rodrigues, and N. Fueyo, "A method for estimating the geographical distribution of the available roof surface area for large-scale photovoltaic energy-potential evaluations," *Sol. Energy*, vol. 82, no. 10, pp. 929–939, Oct. 2008, doi: 10.1016/j.solener.2008.03.007.

[14] L. Wald *et al.*, "SoDa: a project for the integration and exploitation of networked solar radiation databases," *Environ. Commun. Inf. Soc.*, no. 2, p. 9, 2002, [Online]. Available: http://www.soda-is.org/doc/enviroinfo2002.pdf.

[15] "Solar Radiation Data (SoDa)." http://www.soda-is.com/eng/index.html (accessed Mar. 01, 2021).

[16] H. Hersbach *et al.*, "The ERA5 global reanalysis," *Q. J. R. Meteorol. Soc.*, vol. 146, no. 730, pp. 1999–2049, Jul. 2020, doi: 10.1002/qj.3803.

[17] T. Beck, H. Kondziella, G. Huard, and T. Bruckner, "Assessing the influence of the temporal resolution of electrical load and PV generation profiles on self-consumption and sizing of PV-battery systems," *Appl. Energy*, vol. 173, pp. 331–342, Jul. 2016, doi: 10.1016/j.apenergy.2016.04.050.

[18] ERSE, "Relatório anual sobre os mercados de eletricidade e de gás natural em 2019," 2020. [Online]. Available: https://www.erse.pt/media/en0plsfa/relatório_ce-2019.pdf.

[19] F. M. Camilo, R. Castro, M. E. Almeida, and V. F. Pires, "Economic assessment of residential PV systems with self-consumption and storage in Portugal," *Sol. Energy*, vol. 150, pp. 353–362, Jul. 2017, doi: 10.1016/j.solener.2017.04.062.

HOW DIGITALIZATION CAN BE A DRIVER FOR SUPPORTING MORE AND MORE THE SOLAR DEPLOYMENT?

E. Saretta[1]*, P. Bonomo[1], F. Frontini[1], V.K. Nguyen[2], W. Maeder[2]

[1] SUPSI-ISAAC, CH 6850 Mendrisio, Switzerland
[2] CADCAMATION, CH 1213, Onex, Switzerland
* mail to: erika.saretta@supsi.ch

ABSTRACT: Digitalization is a non-reversible process that is affecting all sectors because it allows enhancing efficiency, increasing performances and optimizing costs. Considering the need of reducing the reliance on fossil fuels through renewables, the digital transition in the photovoltaic (PV) sector can be a key to increase efficiency, productivity and competitiveness of the European PV sector. Indeed, the current value chain of the European PV sector is often characterized by analogue and fragmented processes that should be overcame to support greater solar deployment and PV market growth in Europe. The adoption of a more open and collaborative digital-based approach characterized by data-sharing among different stakeholders and more integrated information thread from the design till O&M can provide direct benefits in the optimization of the PV process, increased performances, and reduction of costs. To support such digitalization, a novel PV Information Management (PIM) approach has been envisioned based on the BIM process typical for the AEC sector. This novel BIM-based approach tailored for supporting PV cost-reduction has been implemented within the European H2020 project, SUPERPV(www.superpv.eu). In accordance with the PIM objectives, new digital features have been developed to address a more collaborative approach in the PV sector: i) advanced functionalities introduced in the existing BIMSolar® software for improving the simultaneous design, performance simulation and cost assessment of medium and large PV systems, ii) a proof-of-concept for aggregating all relevant information into a Digital Twin platform for post-construction management and lifecycle assessment of the whole PV system toward circular economy whereas sustaining reductions in greenhouse gas emissions.

Keywords: PV plant design, PV plant simulation, cost-assessment, PIM, digital twin

1 INTRODUCTION

Even though photovoltaic (PV) costs are reducing and the cost of PV electricity is decreasing globally, European PV manufacturers are still facing a drop in the global competitiveness. Among the several reasons of this issue, one is represented by the high fragmentation of the European value chain that affects the final cost of PV components and plants.

To improve PV competitiveness, the SuperPV project has been proposed to achieve significant LCOE reduction by adopting hybrid combination of technological innovations and data management methods along the PV value chain, with the final aim to accelerate large scale deployment in Europe [1]. Indeed, the SuperPV innovations regards:

1. PV Module innovations
2. Module Level Power Electronics
3. PV system integration and process innovation based on digitalization and data management solutions.

Each one of these innovations aims at reducing LCOE costs as shown in Figure 1.

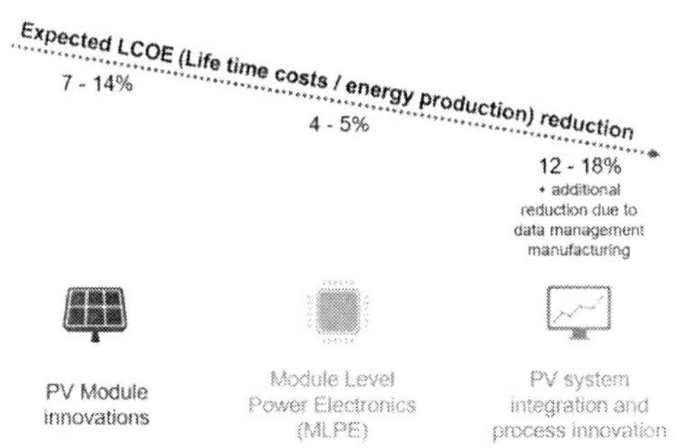

Figure 1: SuperPV project planned impact on LCOE

Regarding the PV system integration and process innovation, the main challenge of the European PV landscape is today characterized by overcoming the current fragmentation of the value chain thanks to the opportunities provided by digital approaches and data management solutions which are successfully applied in other sectors to enhance efficiency, increase quality and optimize costs, without the need for a vertical integration typical of other major solar actors.

2 THE PIM APPROACH

In the framework of the SuperPV project, the solution proposed to deal with the PV sector fragmentation is represented by the development and adoption of a digital approach that is aimed at bridging the informative gaps among the main key-stages of the PV process, namely the manufacturing, Engineering, Procurement and Construction (EPC), and Operation and Maintenance (O&M) stages. Such a digital approach has been named integrated **PV information management (PIM)**.

PIM consists into a digital and collaborative approach based on data-sharing among different stakeholders and the main stages of the PV process. As it has been demonstrated in the Building Information Modeling/Management (BIM) approach, data-sharing and collaboration among different stages and stakeholders allows a dramatic reduction of time and efforts, mistakes and extra-costs that affects the final cost [2].

In the case of PIM and SuperPV project, the opportunity to share data of PV products from the manufacturing to the EPC (Engineering, Procurement, and Construction) stage allows transferring relevant information till the operation of the real PV plant. Moreover, by monitoring these data on-site and analyzing them, new informative patterns can arise not only to improve O&M and reduce related costs but also to enhance other stages. For instance, this information from O&M can be used to provide manufacturers with feedback on PV product quality in a PLM (Product Lifecycle Management) perspective or to give feedback to the EPC stage about real performances of a specific project, as shown in Figure 2.

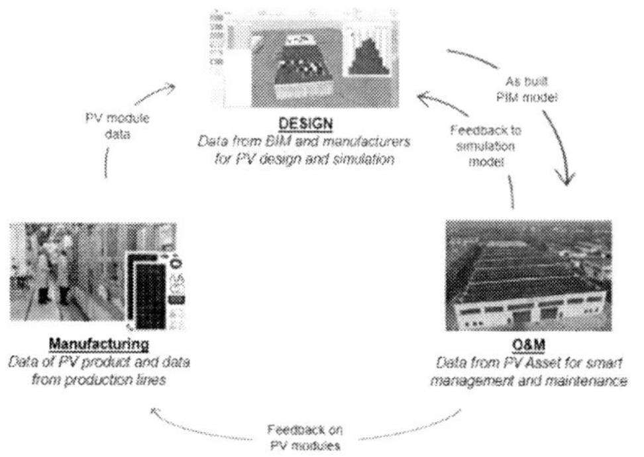

Figure 2: Overview of the PIM approach

3 RESULTS AND DISCUSSION

Even though the majority of PV stakeholders are fully aware of the potential benefits of digitalization, "the prevailing approach to digital innovation in energy is to create "point solutions," which apply technology to narrow uses without fundamentally changing how people work" [3].

Therefore, in the SuperPV project, the PIM approach has been implemented into a prototype digital platform that enable an informative thread from the design till O&M by supporting the control of PV cost-effectiveness. In this context, two solutions will be described:

- the prototype SuperPV **digital platform for design and simulation** to support the cost reduction of PV systems;
- the proof-of-concept of the tool for extending the PIM-based platform functionalities for O&M operations through a **Digital Twin** Model with a specific focus on Product Lifecycle Management **(PLM) of PV modules.**

3.1 Prototype SuperPV digital platform for design and simulation

Such a digital platform has been developed upgrading the current BIMSolar® software [4], [5] – mainly oriented to BIPV – to include new tools for the cost-effective design of medium and large PV plants such as Capital Expenses (CAPEX) and Levelized Cost of Electricity (LCOE) calculators and simultaneous variants comparison. Indeed, the peculiarities of this platform are represented by:

- the possibility to design medium and large PV plants starting from a IFC (Industry Foundation Classes) file that is an open international standard (ISO 16739-1:2018). In this way, EPC companies can share data with other stakeholders involved in the design stage (e.g. civil engineers, architects and infrastructure works stakeholders which are adopting more and more BIM and IFC),
- the chance to manage data and parameters associated to PV components in a PLM (Product Lifecycle Management) perspective. In fact, the PLM approach enable to define data and properties of the components (as designed and as built) which are then stored within the file and are "transferred" from one stage to another in order to be accessible for updates or other purposes (e.g. maintenance) along the PV process and lifetime,
- the inclusion of decision-supporting tools to assess and evaluate the PV CAPEX and LCOE of different PV design scenarios and components directly into the platform since the early-design stages in fast and easy way, thus affecting the final cost of PV due to the reduction of extra-costs,, mistakes, information losses, etc. in the EPC stage.

3.2 Proof-of-concept of the SuperPV Digital Twin for PLM of PV modules

Once the PV plant model is obtained, and the PV system is realized on site a 3D model with embedded information of all energy, economic and material data can be validated. Hence, such an informative 3D model can be directly used in a seamless workflow also for the O&M stage without losing data and thus reducing the number of request for information (RFI) during the setup of O&M models and PV Digital Twins. In the O&M stage, indeed, the deep knowledge of the "as-built" PV system is extremely important to set appropriate and cost-effective maintenance operations by means of Digital Twins.

Considering the whole PV plant lifecycle, the possibility to continually update data and information about the PV components thanks to Digital Twins, IoT sensors and monitoring systems allows to track all the events and changes sustained while in use. In this way, the complete lifecycle of components installed in the PV plant can be accessed in a unique digital "source-of-truth" towards a PLM perspective.

To evaluate the impacts of the proposed innovations on the cost reduction targets, a small group of real PV stakeholders has been invited to provide preliminary estimation within the project activities. In particular, PV stakeholders have been asked to indicate the impacts of the proposed new functionalities introduced in the SuperPV platform, such as:

- Impact on Design (e.g. import of 3D CAD/BIM model of the context, PV module creation, geometrical design of the PV plant)
- Impact on energy yield estimation (e.g. electrical design and simulation)
- Impact on economic estimation and cost-control (e.g. CAPEX and LCOE calculation)
- Impact on rework and iterative tasks (e.g. ease to change PV components, variants comparison)
- Impact on O&M (seamless workflow).

As a result, the total impact of the proposed collaborative platform on the EPC and O&M stages has been estimated in the range from 14% to 55%, when O&M is considered.

4 CONCLUSIONS

Considering the need for increasing competitiveness of the European PV value-chain to support the energy transition, the proposed approach towards a digitized process is expected to streamline information exchange among the stakeholders. In this paper, the collaborative digital PIM approach developed in H2020 SuperPV project, has been presented and, beyond that, the digital solutions capable to enable a digital informative thread thus avoiding point solutions have been described. Specifically, the impact of these digital solutions on cost-reduction on the EPC and O&M stages has been estimated in a preliminary assessment in the range of 14% to 55%. However, even though the PV stakeholders group considered for this estimation was limited, the obtained results suggested that there is a quite large potential provided by PV digitalization in terms of cost-reduction along the PV value chain, that could foster PV deployment in Europe. The development of a PIM approach and the digital solutions of the SuperPV prototype digital platform are aimed not only at reducing risks and unforeseen extra-costs, but also at impacting the PV process with a higher quality, by supporting increasing reliability and performances of PV systems.

5 ACKNOWLEDGMENTS

This paper has been developed in the framework of the SuperPV project. The SuperPV project has received funding from the European Union's Horizon 2020 Research and Innovation Programme under Grant Agreement No. 792245. Authors would like to thank also IREC (specifically, Dr. J. L. Dominguez and Mr. P. Paradell Solà) for the support in the economic assessment, as well as BNW-ENERGY (Dr. T. Haarberg) and Becquerel Institute (specifically, Mr. P. Macè and Ms. E. Bosch) for their supervision in economic assessment.

6 REFERENCES

[1] SuperPV Project. https://www.superpv.eu/

[2] Harvey, Bernstein e Stephen, Jones. McGraw Hill Construction SmartMarket Report: The Businees Value of BIM for Construction. 2014. SmartMarket Brief: BIM Advancements No. 01 - 2015 (https://www.smartmarketbrief.com/reports/SMBrief-BIM-Advancements-01.pdf)

[3] A. Booth, N. Patel, M. Smith, Digital transformation in energy: Achieving escape velocity. McKinsey & Company (2020)

[4] H. Ossenbrink, A. J. Waldau, N. Taylor, I. P. Pascua, S. Szabó, Perspectives on future large-scale manufacturing of PV in Europe. JRC Report (2015)

[5] International Finance Corporation – World Bank Group. Utility-Scale Solar Photovoltaic Power Plants: A project developer's guide (2015)

This is a short version of the paper submitted to the journal Progress in Photovoltaics.
It has been selected by the Executive Committee of the EU PVSEC 2021 for submission to Progress in Photovoltaics.

ADVANCED ANALYTICS ON IV CURVES AND ELECTROLUMINESCENCE IMAGES OF PHOTOVOLTAIC MODULES USING MACHINE LEARNING ALGORITHMS

Authors: Vedant Kumar, Pranav Maheshwari
PV Diagnostics, c/o Fototentia Diagnostics Private Limited
Address: B/702, Satellite Garden 1, Gen AKV Road, Goregaon east, Mumbai-400063, MH, India
Correspondence: pranav@pv-diagnostics.com; Phone: +91-9820721990

ABSTRACT: Advanced analysis and monitoring of photovoltaic solar modules is required to maintain the reliable operations of photovoltaic plants. Hence it requires diagnostics through current-voltage (IV) curves, electroluminescence (EL) imaging and other measurement techniques. The analysis through IV characterization provides the discerning insight about the quantitative measure of solar module performance, while the image characterization methods on EL images can capture spatial defects with microscopic resolution such as micro cracks, broken cells interconnections, shunts, among many other defect types. The fusion of these two methods with supervised and unsupervised machine learning can generate unique insight with classification, regression, and dimension reductions on IV-EL data. In this study, we have performed the IV-EL correlation by classifying the IV data based on EL image annotation (where the class information is coming from EL image). The feature vectors consist of IV curve parameters as well as statistical features. We have done the binary classification for identifying defect vs. no defect. Furthermore, we had applied feature selection algorithms before applying the classification algorithms. We have performed the classification of various defect types by applying the Random Forests (RF) and XGBoost algorithm while identifying the top features. The accuracy was achieved greater than 91% and 94% respectively for supervised methods on the top 5 features. This correlation of IV- EL measurement could benefit in quick identification of various defect types in PV modules with only IV curve parameters, given the classification models are modeled using large scale data sets and tuned optimally.

Keywords: IV Curve, Electroluminescence Imaging, Modelling, Classification, Machine Learning, Random Forests, XGBoost

1 INTRODUCTION

The global installation of large PV plants has grown significantly in the last decade. The global installed solar photovoltaic (PV) capacity exceeded 500 GW at the end of 2018, and an estimated additional 500 GW of PV capacity is projected to be installed by 2022–2023, bringing us into the era of TW-scale PV [1]. It is proven that PV technology can only be economical if PV modules achieve reliability and high performance for 25 -30 years under field conditions [2]. The installation and operations of PV modules undergo complex processes which often lead to degradation due to various exposure to environmental conditions as well as mechanical stress during transportation [3]. The most common stressor from environmental conditions are temperature, humidity, wind, soiling and UV, which leads to degradation modes that include cell cracking, series resistance increases, discoloration, hot spots, encapsulant chemical changes, and many others [4] [5].

To maintain the PV plant operation and power generation at the maximum capacity, it is obligatory to perform the diagnostic of PV modules on recurrent basis. Identifying the defects and faults in photovoltaic (PV) modules rapidly is a challenging task yet it can benefit the solar power generation industry exceedingly. There are numerous measurement techniques available for PV modules test and diagnostics. One of the most effective diagnostics techniques to assess the module degradation and performance are maximum power point tracking and current-voltage (IV) curve tracing. The IV curve tracing provide point-in-time data regarding the electrical state of the test module [4]. Of these, IV tracing performs a complete electrical sweep from short-circuit current (ISC) to open-circuit voltage (V_{OC}) of an illuminated PV module. The decreasing short-circuits current (I_{SC}) and shunt resistance (R_{SH}) are sometimes related to cracked cells [6], and increasing series resistance (R_S) can be related to corrosion of cell metallization or interconnects. But IV curve features alone cannot provide the insights regarding degradation mechanisms leading to power loss.

There are range of different other imaging techniques such as electroluminescence (EL), photoluminescence, ultraviolet (UV) fluorescence, and IR thermography, which can be utilized to collect high density quantitative data that provide information on cell variability and other module defects [7]. The EL imaging is one such technique, which helps in identifying PV module defects such as cell cracks, potential induced degradation (PID), crystal dislocations in a multi crystalline wafer, finger interruptions, humidity corrosion, shunted bypass diode (sub-string failure) and many others. This is possible because EL images are rich in spatial information [4].

There is a scarcity of research of correlation studies of IV data and electroluminescence (EL) imaging of PV modules. Many insights can be drawn from IV curve and EL image correlation which can lead to identification of various defect types. There are few research articles in similar domain, but experiments in those studies have been performed in controlled environment [8]. In this research, we have devised a unique machine learning approach for identifying various defect types through correlation of IV curve data and EL images, where we have trained our classification model on features generated from IV curve data and class information from EL image analysis. The idea is, once the machine learning model learns on a large-scale data set of IV curves and counterpart EL images, it will predict the various defect types of PV modules on the IV curve feature vectors itself with high accuracy.

Our goal is to predict module-level defects from IV curve features derived with algorithmic and machine learning techniques. Prediction of defect types at module level without an imaging technique such as EL would enable high-speed characterization of modules defects in outdoor field conditions. We provide an overview of the experimental study, data collection, algorithms and computational methods like image processing, feature extraction, feature engineering, feature selection, and machine learning algorithms in section 2 and section 3. In section 4, we present the results and discussion of various clustering and classification algorithms. The final section 5 contains conclusions of the research

2 DATA PROCESSING

The data set consist of 6051 IV curve and corresponding EL images of field installed 60/72 cell modules of different manufactures at multiple locations across India from 27 projects with age ranging from 1 to 8 years. We collect IV data through an IV curve tracer device along with solar irradiance and module temperature as per methodology following IEC 61829 [9] [10]. We have translated our data set to standard test conditions (STC) (1000 W/m2, 25°C) for obtaining standardized IV data as per translation procedure mentioned in IEC 60891 [11].

We have extracted the following features from IV Curve: maximum power (Pmax), fill factor (FF), current at maximum power (I_{MP}), voltage at maximum power (V_{MP}), current at short circuit (I_{SC}), voltage at open circuit (V_{OC}), radius of curvature (ROC) near maximum power point (MPP), slope at short circuit (R_{SH}) and slope at open circuit (R_S). We have built linear regression model to estimate the values of R_{SH} and R_S. We have added another feature which is area under the IV curve (AUC_{IV}) which has been calculated with trapezoidal rule. Apart from the mentioned features, few statistical features like skewness, kurtosis and entropy are also computed based on the IV data distribution.

Time difference between measurement of IV curve & Electroluminescence imaging is only 6-36 hours for all the projects. This time duration is very small (~0.01% of module lifetime), which is unlikely to cause any significant pressure impact to reflect changes in PV modules. Identification of defect types have been done through analysis of EL images. The EL images were analyzed as per the IEC 60904-13 standard [12]. The process by which we capture EL images add the module orientation variability in the image. Annotation of various defect types (as mentioned in Table I and Table II) have been performed by two experts from PV Diagnostics.

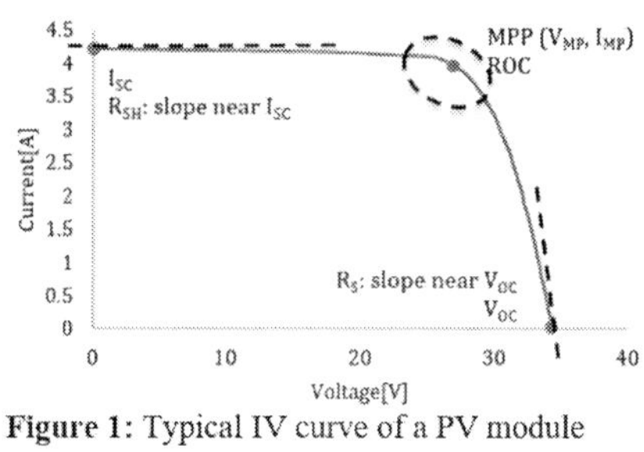

Figure 1: Typical IV curve of a PV module

Table I: Defect annotations on EL images

Sr. No.	Defect Type
1	Cracks (micro, partial, full, compound/spider)
2	Solder failure (partial, full)
3	Potential induced degradation (PID)
4	Back sheet scratch, corrosion
5	Diode fault

After annotation, we have classified the EL Images in to following classes: Cracks (C), Solder failures (SF), PID and Corrosion (PNC). As shunted diode (diode faults) can be easily identified through VOC of module, it is not included in classification. Since PV modules can develop multiple defects, so we have created total 8 possible combinations shown in Table II

Table II: Defect class label information of EL images

Defect Class	Class Definition
ALL	All defects class (C, SF, PNC)
False	No defects
C	Cracks
SF	Solder Failure
PNC	PID and Corrosion
C_SF	Cracks and Solder Failure
C_PNC	Cracks and PID
SF_PNC	Solder Failure and PID

After performing the feature engineering step and annotations for all the defect types, we have performed the data pre-processing where we have dealt with missing values, outliers and data normalization and standardization using the open-source pandas library to prepare the dataset for classification. The process is shown in Figure 2.

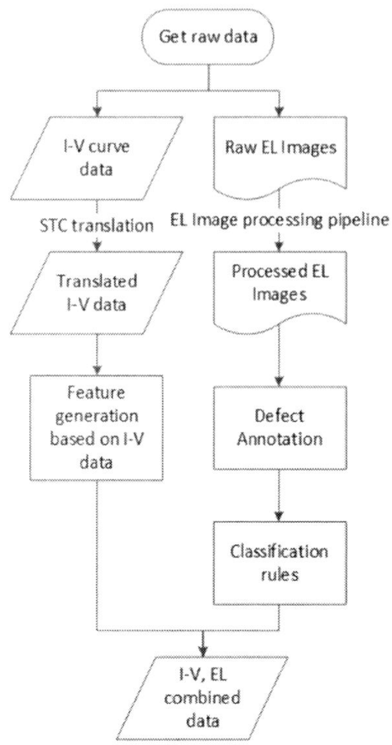

Figure 2: Process flow for IV and EL data preparation

3 MODELLING AND ANALYSIS

The feature vectors that we have extracted from IV data and class information of EL data have been combined to form a new dataset for machine learning driven analysis as show in final step of Figure 2. Then pre-processing steps have been performed including data cleaning, outliers handling, normalization, and standardization. Further, while performing the exploratory data analysis, we have identified a class imbalance. Therefore up-sampling method with Synthetic Minority Oversampling Technique (SMOTE) has been implemented to deal with class imbalance problem.

We have first compared the pool of supervised machine learning algorithms for binary classification task and usages for defect prediction. This step aims to identify the best performance classifiers and the key trends among different types of supervised machine learning algorithms. We have explored Support Vector Machine (SVM), Logistic Regression (LR), K Nearest Neighbor (KNN), Navie Bayes (NB), Decision Trees (DT), Random Forest (RF), and Extreme Gradient Boosting (XGB). We have identified that the Random Forest and XGBoost outperformed the other algorithms in terms of accuracy.

We have also looked at other performance metrics such as precision, recall and F1 score and finalized the RF and XGBoost for further ML analysis tasks. The decision is also based on ease of use, scalability of the algorithms. The various defect types as mentioned in the Table II were clubbed to create binary class with any defects (TURE) and no defect (FALSE). While performing the binary and multi class classification, we have found the class 'C', 'SF_C' and 'ALL' were reducing the accuracy. We have created two distribution dataset, where in one distribution we have kept all the defect classes and in other distribution we have dropped the C and SF_C classes.

The feature importance and feature selection approach have been implemented to improve the model's accuracy and reduce the complexity with RF and XGBoost to find out the top features. The RF and XGBoost have been implemented with hyperparameter of large parameter space and selected the top model with respect to accuracy performance metric (refer Figure 3 for complete process flow and implementation steps)

3.1 Machine Learning classifiers

Random Forests is an ensemble of ML techniques that builds multiple decision tree classifiers on random sub-samples of the training dataset. Each decision tree predicts the response by following the tree's decisions from the root to the leaf. The output of each decision tree is then averaged to determine the prediction [13]. RF's main advantage is leveraging the power of many randomly selected trees to represent the solution. Thus, instead of using one decision tree, RF uses all the decision trees to determine the classification; this procedure reduces errors and uncertainties [14]. In this study, the number of trees selected were ranging from 50 to 500 in step size of 25. The minimum number of samples that are required to split an internal node is set to 25. The maximum depth of the tree is set from 20 to 200 [15]. The training algorithm for RF applies the general technique of bootstrap aggregating or bagging to tree learners.

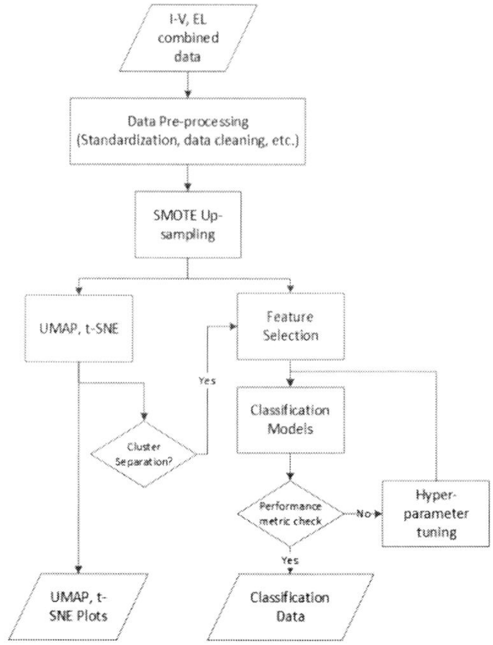

Figure 3: Flow diagram showing various steps of data modelling and machine learning development

The other classification ML algorithm we have used is XGBoost. XGBoost was developed by Chen, T. & Guestrin, C. [16], a scalable tree boosting system that is widely used by data scientists, providing state-of-the-art results on many problems. XGBoost is an optimized distributed gradient boosting library designed to be highly efficient, flexible and portable. It implements machine learning algorithms under the Gradient Boosting framework. It provides a parallel tree boosting (also known as GBDT, GBM that solve many data science problems in a fast and accurate way.

Open-source Python was utilized in XGBoost implementation for classification of defect types with feature vectors. Various metrics were considered to determine the goodness of fit of these models. These metrics include accuracy, F1 Score, precision-recall curve, number of false positives (FP), true positives (TP), false negatives (FN) and true negatives (TN) at various decision thresholds, and calibration curves of the estimated probabilities.

3.2 Feature Selection and Hyper-Parameter Tuning

All the initially computed features as mentioned in Table 1 have been used. The dataset is divided into training and validation in a ratio of 80:20. The initial model has been built using this split of data and we have used 10-fold cross validation to confirm the stability of algorithm. The initial training accuracy was ~75% while validation accuracy of ~70%. Based on the confusion matrix of initial results, we dropped class C, class C_SF and class ALL as the accuracy values for these defects were significantly lower than other classes.

Afterwards, we built the random forests and XGBoost classification models using only selected classes and computed the feature importance for both models. Then, we have arranged the features in decreasing order of importance as show in Figure 4 and Figure 5. Next, we have selected the top 5 features to further enhance the model performance. To study the influence of hyperparameters on the RF classifiers' performances, combinations of hyperparameters are used, with each combination leading to one classifier (refer Table III). Similarly, for XGBoost grid search was performed by selecting the hyperparameters sets mentioned in Table III below to further optimize the model.

Table III: Hyperparameters for RF and XGBoost models

RF Parameters	XGBoost Parameters
maximum depth ∈ {20, 40,..., 200}	maximum depth ∈ {5, 10, 15}
number of trees ∈ {50, 75, ..., 500}	number of trees ∈ {500, 1000, 1500, 2000}
split criterion ∈ {Gini, entropy}	mini split loss ∈ {0, 0.5, 2, 10}
max feature ∈ {sqrt, log2}	learning rate ∈ {0.01, 0.1, 0.3, 0.5}

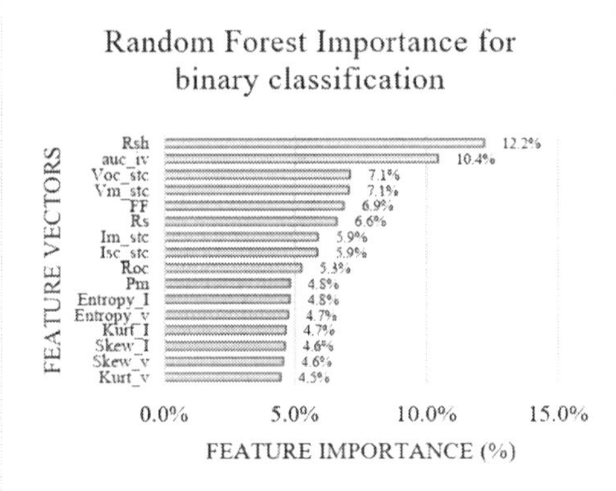

Figure 4: Feature Vector Importance ranked from highest to lowest for RF classifiers

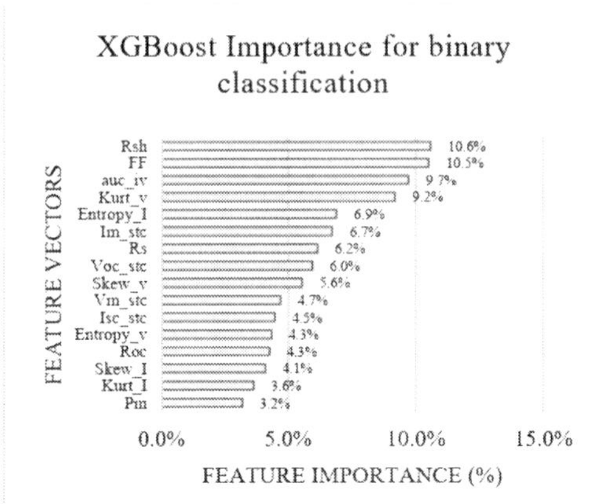

Figure 5: Feature Vector Importance ranked from highest to lowest for XGBoost classifiers

4 RESULTS AND DISCUSSION

The compassion among various supervised machine learning were performed. It is evident from the box-plots of classifier's accuracy for 10 fold cross validation that XGBoost and RF have better inter quartile range and good average accuracy compare to SVM with kernel trick, Navie Bayes, K Nearest Neighbors and Decision Trees. Both RF and XGBoost have average accuracy of ~84% when all class considered and ~94% when C and SF_C class dropped as depicted in Figure 6.

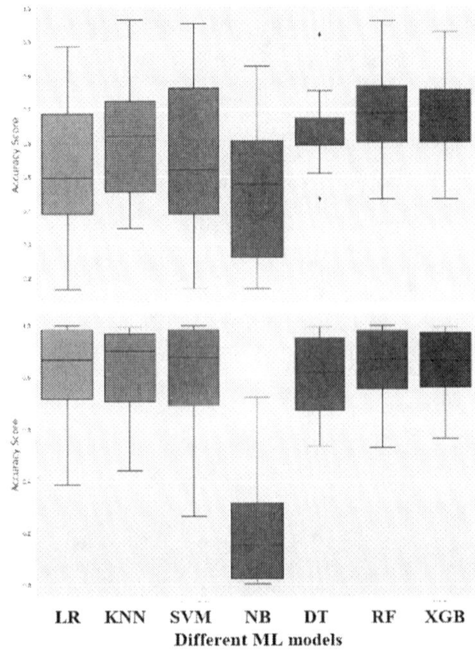

Figure 6. Box plots distribution of different ML models' accuracy for (top) all classes and (bottom) selected classes

The feature set V1 comprises all feature vectors. While feature vector set V2 comprises of top 5 selected feature vectors as per feature importance scores calculated for both classification models, respectively. We have found that the V2 set of RF model includes R_{SH}, AUC_{IV}, V_{OC}, V_{MP} and FF feature vectors as shown in Figure 4. Similarly, for XGBoost model, V2 set includes R_{SH}, FF, AUC_{IV}, Kurt_v, Entropy_I feature vectors as shown in Figure 5.

We have compared the performance scores on validation set using both feature vector sets V1 and V2, when they are used as inputs for both ML classifiers RF and XGBoost as shown in Table IV. For both cases, vector V2 outperforms the vector V1 in all performance measures, achieving higher accuracy and F1 score. A slight difference can be observed between RF and XGBoost classifiers. It has also been observed through a comparison between the training and validation sets that the data has not been over-fitted. We have created the receiver operating characteristic (ROC) curve by plotting true positive rate against false positive rate to measure the performance of the classifiers. The area under the ROC curve (AUC) represents results with almost perfect discrimination with AUC score of 0.95 for RF and 0.91 for XGBoost model, as shown in Figure 7 and 8 respectively.

Table IV: Binary classification results

Feature Vectors	**Feature Vectors V1**		**Feature Vectors V2**	
ML Classifies	RF	XGBoost	RF	XGBoost
F1 Score	90.5	91.3	92.5	95.7
Accuracy	89.36	93.3	90.6	94.3
Recall	90.0	93.9	92.2	97.0
Precision	91.0	91.6	92.7	94.0

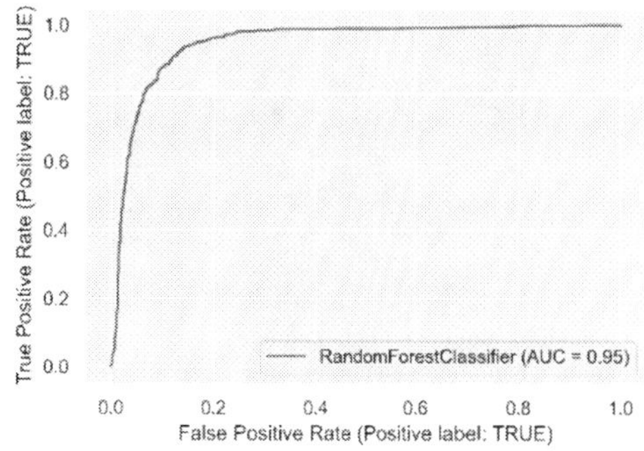

Figure 7: ROC curve for RF classification showing 0.95 AUC score depicting excellent discrimination

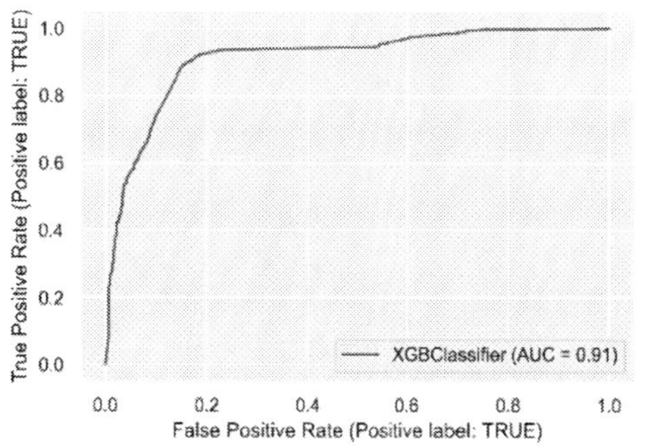

Figure 8: ROC curve for XGBoost classification showing 0.91 AUC score depicting excellent discrimination

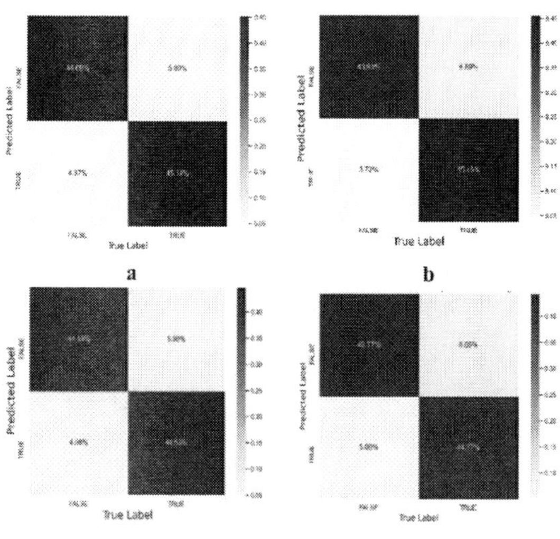

Figure 9: Confusion Matrix for (a) RF with all feature vectors set V1, (b) RF with top 5 feature vectors set V2, (c) XGBoost with all feature vectors set V1 (d) XGBoost with top 5 feature vectors set V2.

5 CONCLUSION

The early phase detection of defects as cracks, solder failure, potential induced degradation and finger interruption in solar cells is important to estimate power generation and life of PV modules. Analyzing EL images to locate and identify these failures is typically a time-consuming manual process and requires expert knowledge. In this paper, a machine learning-based defect identification method was presented for IV-EL Correlation. The technique uses EL images for the class information to classify three major classes of defects using feature vectors based on IV curve parameters and statistical features. It has been shown that defects in EL images correlate well with trends seen in IV curve data. The developed feature vector achieves an accuracy and F1-score which are quite promising for defects identification in PV Industry. While performing the model training-testing, we observed that the overall accuracy and other performance metrics decreased with inclusion of crack defect class. This is because, the impact of cracks on I-V data is very complex to analyze and it also depends on severity levels of different types of cracks. Thus in future, with increased dataset, the accuracy can be further increased for the dropped classes too.

As the proposed method requires less computation power and time, it will be valuable for outdoor IV inspection including handled IV inspection, Internet of things (IOT) based cloud inspection through module monitoring. The learned models can be easily integrated on a software as a service (SaaS) platform and deployed on cloud for processing large scale datasets as XGBoost provides a parallel tree boosting and its codes can run on distributed environment (Hadoop, SGE, MPI).

6 ACKNOWLEDGEMENT

The authors thank PV Diagnostics for providing all the R&D and logistical support. PV Diagnostics has provided a constant support throughout this process. We would also like to express our sincere thanks to the operations team, backend analysis team and the EL image analysis experts who were instrumental in collecting, pre-processing, annotating, and reviewing the dataset. This helped the authors to build optimal ML models for defect identification using IV-EL data. We would also express our gratitude to Mr. Ankur Gahlot who helped us in the inception of this analysis.

6 REFERENCES

[1] N. Haegel, H. Atwater, C. Barnes, A. Burrell, Y. Chiang, S. De Wolf, B. Dimmler, D. Feldman and S. Glunz, "Terawatt-scale photovoltaics:Tranform global energy," Science, no. 364, pp. 836-838, 2019.

[2] V. Sharma and S. Chandel, "Performance and degradation analysis for long term reliability of solar photovoltaic systems: A review," Renewable and Sustainable Energy Reviews, vol. 27, no. Nov, pp. 753-767, 2013.

[3] U. Eitner, S. Kajari-Schrder, M. Kntges and H. Altenbach, "Thermal Stress and Strain of Solar Cells in Photovoltaic Modules," in Shell-like Structures: Non-classical Theories and Applications," Springer Berlin Heidelberg, pp. 453-468, 2011.

[4] M. Kongtes, S. Kurtz, U. Jahn, K. A. Berger, K. Kato, T. Friesen, H. Liu and M. V. Iseghem, "IEA-PVPS T13-01 Performance and Reliability of Photovoltaic Systems: Review of Failures of PV Modules," International Energy Agency, 2014.

[5] D. Jordan, J. Silverman, J. Wohlgemuth, S. Kurtz and

K. VanSant, "Photovoltaic failure and degradation modes," Progress in Photovoltaics: Research and Applications, vol. 25, no. 4, pp. 318-326, Apr, 2017.
[6] T. Pletzer, J. V. Molken, S. Ribi, O. Breitenstein and J. Knoch, "Influence of cracks on the local current–voltage parameters of silicon solar cells," Progress in Photovoltaics, vol. 23, no. 4, pp. 428-436, Apr, 2015.
[7] T. Trupke, J. Nyhus and J. Haunschild, "Luminescence imaging for inline characterisation in silicon photovoltaics," Physica Stuatus Solidi (RRL), vol. 5, no. 4, pp. 131-137, 2011.
[8] J. Fada, "Correlation of I–V curve parameters with module-level electroluminescent image data over 3000 hours damp-heat exposure," in IEEE Photovoltaic Specific Conference, Washington D.C., Jun, 2017.
[9] International Electrotechnical Commission, "IEC 60904-1 Photovoltaic devices - Part 1: Measurement of photovoltaic current-voltage characteristics," 2020.
[10] International Electrotechnical Commission, "IEC 61829 Photovoltaic (PV) array - On-site measurement of current-voltage characteristics," 2015.
[11] International Electrotechnical Commission, "IEC 60891 Photovoltaic devices - Procedures for temperature and irradiance corrections to measured I-V characteristics," 2009.
[12] International Electrotechnical Commission, "IEC TS 60904-13 Photovoltaic devices - Part 13: Electroluminescence of photovoltaic modules," 2018.
[13] L. Breiman, Random forests Machine Learning, 2001.
[14] A. Ziegler, R. James, D. Witten, T. Hastie and R. Tibshirani, An introduction to statistical learning: With applications in R, 2016.
[15] L. Breiman, J. Freidman, C. Stone and R. Olshen, Classification and Regression Trees, CRC Press, Florida, USA, 1984.
[16] T. Chen and C. Guestrin, "XGBoost: A scalable tree boosting system," in 22nd ACM SIGKDD International Conference on Knowledge Discovery and Data Mining, San Francisco, CA, USA, Aug, 2016.

AUTOMATIC FAULT DETECTION AND CLASSIFICATION IN PV SYTEMS BY THE APPLICATION OF MACHINE LEARNING ALGORITHMS

Ganesh Deepak Rupakula[1], David Dassler[1,2], Stephanie Malik[1], Matthias Ebert[1], Ralf Schmidt[3]
[1]Fraunhofer Center for Silicon Photovoltaics CSP
Otto-Eissfeldt-Strasse 12, 06120 Halle (Saale), Germany
[2]Anhalt University of Applied Sciences, Faculty EMW
Bernburger Strasse 55, 06366 Koethen, Germany
[3]GETEC green energy GmbH, Magdeburg, Germany

ABSTRACT: For an effective functioning of photovoltaic (PV) plants, it is important to identify the potential faults (technical defects or yield loss due to harsh environmental conditions) that can occur during the operation of the plant and thereby derive possible maintenance work. This ensures a high energy yield from the PV plant throughout the operating time. This research work explains the approach to establish a full detection and classification model of a large-scale grid-connected PV power plant at the inverter level by the application of different machine learning (ML) algorithms. This work initially explains a statistical method to identify different faults at the inverter level based on the necessary model parameters which include module temperature, voltage and current at the inverter; solar irradiation and ambient temperature from a sensor. A machine algorithm and another statistical method are applied to this data for identification of normal functioning, fault detection and fault classification. Three important faults are detected in the present research namely, snow cover, inverter outage and degradation. An Artificial Neural Network (ANN) and a novel statistical method are proposed to detect these defects. The accuracy of the algorithms and the optimized parameters have been described and the possibility of integrating the algorithms into the monitoring platform of a PV power plant has been discussed.

Keywords: defects, PV system, monitoring, inverter, performance, Machine Learning, classification

1 INTRODUCTION

Photovoltaic (PV) installations are increasing drastically worldwide every year. Only in Germany, the total installed capacity rose from 2 MW in 1990 to 53.6 GW in 2020. [1]. The capacity of PV systems is expected to rise in the future as well. As installations rise, monitoring of these plants becomes a key importance in order to ensure smooth uninterrupted green energy generation. Several interruptions can occur during the running phase of the plants due to various defects or other operational problems. Some maintenance problems include improper cleaning of PV panels, defective strings/inverters and vegetation growth near the PV panels. Weather-related problems include snow event covering the PV panels and dust accumulation leading to soiling losses [2].

Such defects can affect the current, voltage and power and thus, lead to decrease in the energy yield of the PV pant. This in turn results in higher operation and maintenance of the plants and thereby higher costs which affects the return on investment. Hence, advanced methods are needed to detect such abnormalities in advance for a more efficient and cost-effective maintenance leading to smooth and constant energy yield of the plants.

There are different components in the PV power plant whose data can be used for the analysis and developing the machine learning algorithms. Components whose data can be considered include PV array, string, combiner box and inverter through which the performance of the PV systems can be assessed. However, previous research mainly focused on module and string level [3] [4][5]. This work considered inverter data as inverter is one of the crucial components of the plant and a small abnormality in the data implies a huge energy yield loss. Moreover, inverter data is easier to use and interpret when compared to a module level or string-level data. Hence, for the defect detection and algorithm development, inverter data is considered.

Additionally, several defects can be detected and interpreted based on the data from the inverter. Defects like snow cover, vegetation, degradation in PV panels, partial shading and inverter outage can be interpreted from the data [2]. This research work however does not focus on all defects but only on inverter outage, snow cover and degradation. The reason is because the considered PV power plant had robust monitoring checks due to which other defects never happened in the considered time period. Therefore, the algorithm developed can predict the above-mentioned defects; other defects like partial shading and vegetation growth is out of the scope of the research.

This research work focusses on detecting and interpreting such defects that occurred in the past using the monitoring data of the PV plant. Machine learning models have been used to detect such defects and predict the defect pattern based on basic input data from the PV power plant. For this work, five years of data from 2014 to 2018 has been considered for analysis.

This paper is intended to help PV power plant owners reduce the maintenance costs and effort with the advanced algorithms for early detection of the possible faults. This in turn ensures an undisturbed energy yield and cash inflows throughout the operation of the power plant.

2 METHODOLOGY

2.1 Power plant data

The data considered for the research is taken from a multiple MW photovoltaic power plant containing numerous inverters near Halle, Germany. Each inverter's nominal power ranges from 500 kW - 720 kW. Five years of irradiation data, module and ambient temperature from the plant's sensors along with the inverter data (current, voltage, module temperature, energy yield and power) have been taken for the analysis. The data is obtained from the plant's corresponding monitoring platform and is pre-analyzed to ensure sufficient data quality.

2.2 Previous work

Research was already performed previously with the same data to predict different parameters of the PV systems using Artificial Neural Networks (ANN). The optimized ANN model took input data such as irradiation, temperature, sun position and predicted the current, power, voltage and energy yield of the PV system with decent accuracy [6].

This work will consider the predicted values i.e. current, voltage and power of the ANN model along with the actual values for the prediction of defects. Additionally, irradiation, module temperature and ambient temperature will also be considered for the ML model development. Two different methodologies are applied here to identify different defects based on time scales: Hourly Fault Detection (HFD) and Monthly Fault Detection (MFD). Further explanation will be provided in the next sections.

2.3 Method HFD:

This method is used to identify the snow cover and inverter outage events from normal functioning based on hourly-resolution data. The other method MFD used is to identify degradation in the PV panels from the inverter data based on daily-resolution data and observing the behavior for a month. This method will be explained in the next section. A single methodology is not possible because in order to identify degradation, hourly resolution data will not be clear and cause errors in misinterpretation of data. Hence, two different methodologies are needed.

As mentioned before, ANN was used for the prediction of current, voltage, power and energy yield. In this work, ANN algorithm will be employed as well to analyze and accurately predict the state of the PV system-i.e. normal functioning, snow cover or inverter outage.

ANN consists of nodes, also called neurons, in different layers. Each node is completely connected to the nodes of the adjacent layer imitating the neural network of the brain. These connections have an activation function which act like signal from the incoming node layer to the final output layer. Moreover, the strength of the signal can be varied by the weight of the neuron. The weights for the neuron are initially pre-defined. The neurons automatically adjust the weights as the algorithm is running and this adjustment is based on a parameter called learning rate. As in any other machine learning algorithm, ANN splits the input data into training and testing data set. 70- 75 % of the data would be used for the training of the ANN and the remaining for testing the data set. During this training process, the weights of the neurons are adjusted. Further discussion regarding different parameters in ANN are out of scope of this paper and the reader is suggested to read [7] for an in-detail explanation.

For this methodology, ANN was utilized via Keras Library in Python 3.8 [8]. For this case, multi-class classification algorithm is employed for the detection of the defects. There are numerous parameters in the algorithm that need to be optimized for an accurate prediction. Different values for each parameter have been trained and tested to ultimately obtain an accuracy of at least 95 %. Parameters like activation function in other layers, optimizers, number of layers and learning rate are varied to achieve the accuracy.The parameters and the values are shown in Table 1. However, some parameters are considered constant for the algorithm which include the loss function-categorical cross entropy and the activation function at the output layer- SoftMax function [7]

Table 1: Different hyper parameters of ANN and the values that were tested to achieve the best prediction accuracy.

Hyper parameter	Values tested
Units per layer	20, 50
Learning rate	0.1, 0.01, 0.001
Activation function	ReLU, tanh, sigmoid
Optimiser	Adam, SGD
Number of layers	1, 2,3
Epochs	100,150, 200

For an accurate determination of algorithm's performance, confusion matrix is employed. A confusion matrix is used in classification algorithm to differentiate the predicted classes to the actual classes. This differentiation helps identify the performance of the algorithm. Rows of the confusion matrix represents the instances in an actual class while the columns represent instances in a predicted class or vice-versa. The correct instances of predictions are located in the matrix's diagonal while the wrong instances are located in the adjacent part [9]. Therefore, this metric is used to differentiate correctly predicted instances of snow cover and inverter outage events from the falsely predicted instances. This helps to further analyze the algorithm's performance.

The input data for the algorithm was also experimented to find the best possible parameters that needs to be given for the prediction. The parameters from [6] have also been considered for an effective prediction. A total of 10 parameters have been initially taken and experimented until the desired accuracy is achieved. The parameters included the irradiation, ambient temperature, module temperature, actual current, voltage, power and energy yield. Additional parameters from [6] are the predicted current, voltage, power and energy yield. All parameters were put together and fed to the algorithm. Five years of data from 2014 to 2018 was available. One-year data-2017 was taken for validation at a later stage and hence the initial training-testing data set consisted of the data from the years 2014, 2015, 2016 and 2018.

One of the goals is also to find out the minimum size of the input data for training and testing the algorithm. If such a tool should be used by a power plant operator, the person needs to initially train the algorithm to employ for the plant. However, if the algorithm needs a huge amount of data-set only for training, that could be a disadvantage while an operator might not have enough data for the training. Hence by optimizing the size of the input data, it would be relatively easy for the algorithm to be employed for a PV power plant.

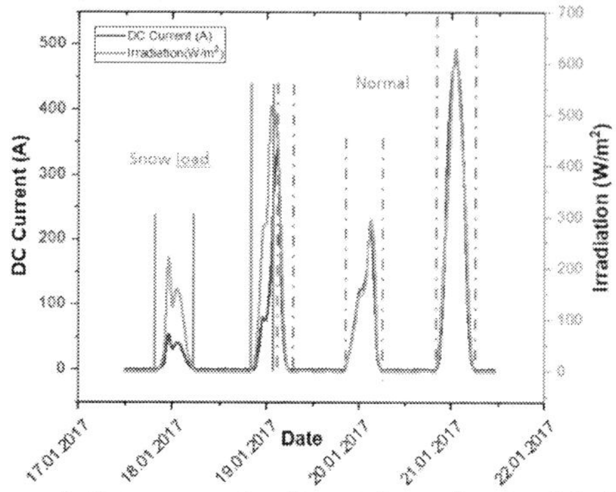

Figure 1: Data preparation for training and testing it in the algorithm.

All the instances of defects are known based on the data observation and through the reports from the power plant operator from 2014 to 2018. These reports mainly consisted of the date and type of events of disturbance during the power plant operation. Based on this, a defect number was manually assigned to every hour of the data of the inverter. A normal functioning event was assigned 0, snow cover as 1 and inverter outage as 2. For this work, the data only during the sun-hours was taken and the data during the night was ignored for accurate prediction of the algorithm. **Figure** *1* shows an instance of the defect numbering based on the monitoring data of the inverter. On 18.01.2017, there has been a snow event, hence all the sun-hours from 9:00 am until 17:00 pm are assigned a defect number 1. From 20.01.2017-22.01.2017, the PV systems functioned normally, hence all the sun hours are assigned a defect number 0. On 19.01.2017, the snow cover was until afternoon 14:00 pm and by 15:00 pm, all snow melted. Hence, a defect number 1 was assigned from 9:00 am until 14:00 pm and number 0 was assigned from 15:00 pm until 17:00 pm.

The ANN algorithm would be trained and tested to predict the defect numbers based on the input data for every hour. Therefore, the training and testing process was done to find out the best hyper parameters, input parameters and the least possible input data for effective training purposes.

The complete algorithm has been mentioned in **Figure 3**. In the initial step, only one inverter data was chosen for initial training and testing purpose. The main reason is due to the fact that many inverter outage events and snow events were recorded in this data. In the train-test phase, along with snow and inverter outage event dates, only two normal days of every month were chosen and fed into ANN. This was done as the number of occurred defects were very low and if too many normal events are fed, the algorithm will not be able to identify the defect accurately.

After this phase, the algorithm is given a data-set consisting 80 random days in 2017. Based on the result, the parameters are adjusted until an accuracy of 95 % is achieved. The algorithm is always validated with a foreign data set to see the performance and improve the accuracy. The aim of the work is to develop a tool containing the algorithms and the statistical methods for effective monitoring of the PV systems and defect detection. It is important to know how the algorithm performs on a completely new data-set. Therefore, the developed algorithm is always tested on a new foreign data-set to see the performance, identify the deficiencies, optimize the parameters and make it more robust.

Once the desired accuracy is reached, the algorithm with the optimum parameters is trained and tested on the data of three other inverters. Here the inverters are chosen based on their capacity. As mentioned before, inverters range from 500-720 kW. By training and testing on these three inverters, the complete range of the inverters in the power plant will be covered. The ANN algorithm is trained and tested and later, validated by feeding foreign data of four other random inverters of the plant. The validation data is selected in such a way that it consisted of significant number of defect events along with normal functioning events. The accuracy limit has been again set to 95 % and additional optimization is done accordingly. The goal of the algorithm development is reached once it performs with an accuracy of atleast 95 %.

2.4 Method MFD:

Degradation of the PV modules is hard to detect and slowly reduces the energy yield of the modules over time. One of the reasons is that the power and the yield produced by degraded modules is slightly lower than the actual value, hence a significant difference would not be observed at the first glance. Additionally, this defect can be misunderstood by other anomalies. Its effect becomes significant only after it shows a notable difference in the yield when compared to the original yield six months or one year ago. For this power plant, degradation of many PV modules occurred at one region and the data was recorded by one of the inverters. A traditional method to identify the degradation is to compare the performance ratio of this inverter data to another inverter data with the same rated capacity i.e inverter covering the same number of PV arrays. However, for this method, another inverter data is needed. **Figure** *2* shows the traditional method for detection of degradation at the Inverter number 2.1.

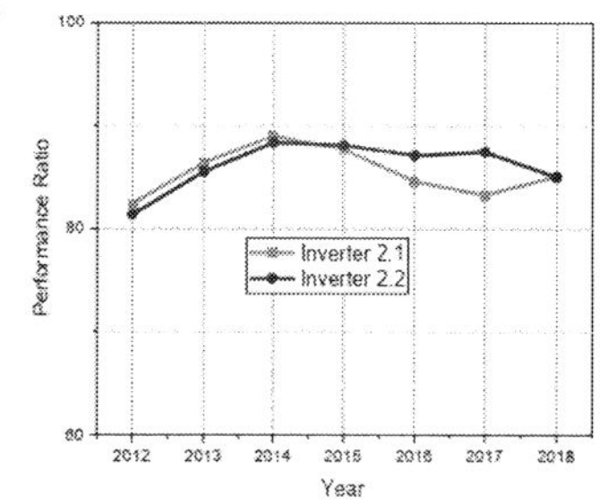

Figure *2:* Degradation detection at Inverter 2.1 by comparing with Inverter 2.2 data.

Using the monitoring data, one can observe that the power plant detected the degradation using another inverter data as reference. Reports from the power plant operator and the data observed in **Figure 2**, it can be concluded that degradation occurred from November 2015 and lasted until October 2017. Due to difficulty in detection, the degradation of the modules was identified very late and the replacement of the PV modules was done after almost two years. A disadvantage of this method is one needs to see the plot in a yearly time-scale to find a significant difference in the performance. One of the goals of this work is to develop a method for early detection of such degradation based on daily-resolution data.

The data from [6] is again used for the detection. After a series of trials, it has been found that the actual voltage and predicted voltage values can be used to identify the anomaly. As mentioned in [6], the predicted voltage is evaluated using ANN solely based on the irradiation, module temperature, ambient temperature and the sun's position. It can be assumed as the ideal voltage and this can be compared with actual voltage to identify the defect. Since the degradation event did not occur during 2014, the inverter data of the year 2014 was considered healthy and taken as reference to compare with that of the data during which the degradation occurred. Hence, for this method, one year of data is required as well. For ease of working and to reduce noise, mean daily-resolution data is evaluated from the available hourly-resolution data. A scattered plot and regression lines are drawn for one month for the reference data 2014 and the other test data.

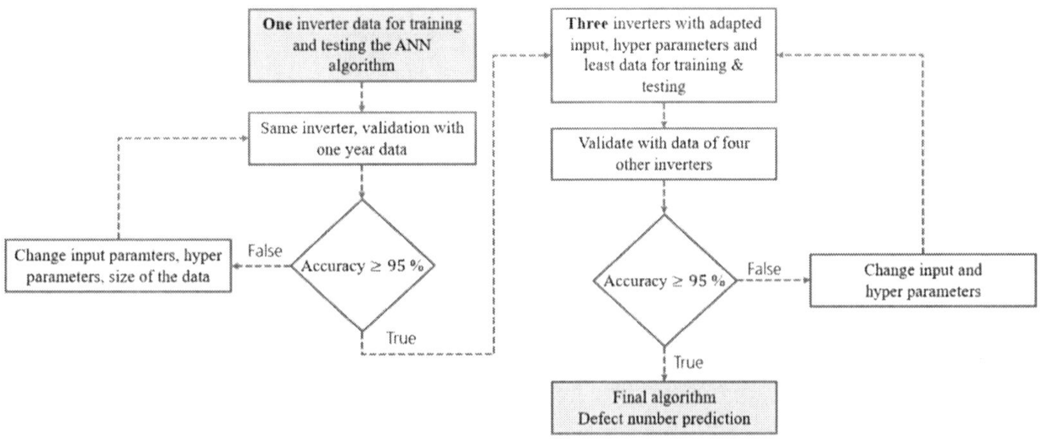

Figure 3: Flowchart of method HFD.

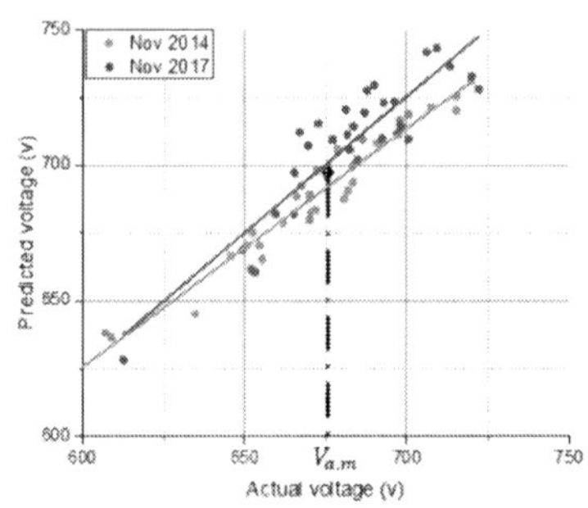

Figure 4: Linear regression lines plotted based on the data of November 2014- reference and November 2017. The two lines are in coherence.

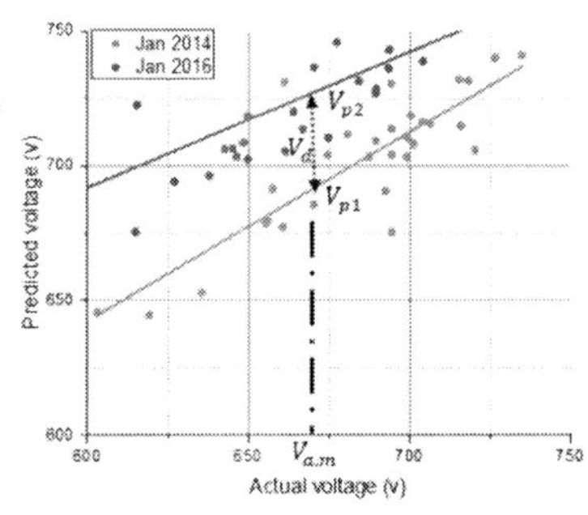

Figure 5: Linear regression lines plotted based on the data of January 2014- reference and January 2016. The two lines are not in coherence.

Figure 4 shows the regression line plots for the reference month November 2014 and another month November 2017. The two lines are in coherence. However, in **Figure 5**, the regression line plots for the reference month January 2014 and another month January 2016 are far apart and not at all in coherence. To detect the degradation, a statistical method can be used here by firstly identifying the mid-point at X-axis - $V_{a,m}$. Based on the mid-point, the corresponding Y-points V_{p1} and V_{p2} are evaluated. The difference between the two- V_d is calculated and compared to a pre-defined threshold value V_t. If V_d is greater than V_t, it implies there is degradation in PV panels during the selected month. The statistical method in the form of flowchart is shown in **Figure 6**. The threshold voltage V_t is evaluated based on a series of trials and its validity will be explained in the results.

3 RESULTS

3.1 Results HFD:

As explained in **Figure 3**, the algorithm is firstly trained and tested on one inverter data. Here the algorithm's accuracy or confusion matrix on the testing data-set is not discussed and is primarily focused on validation data-set.

Table *2* shows the optimal input parameters and input data necessary for the algorithm to reach a 95 % accuracy. Many trials were run by changing the train-test data set size from 2014- 2018, 2015-2018, 2014-2016 to only 2018. After the trails, it has been observed that even with only 1-year data- 2018, the algorithm has performed well. Therefore, one-year data is sufficient to train-test the algorithm for fault detection. The optimized hyper parameters of ANN have been shown in

Table *3*.

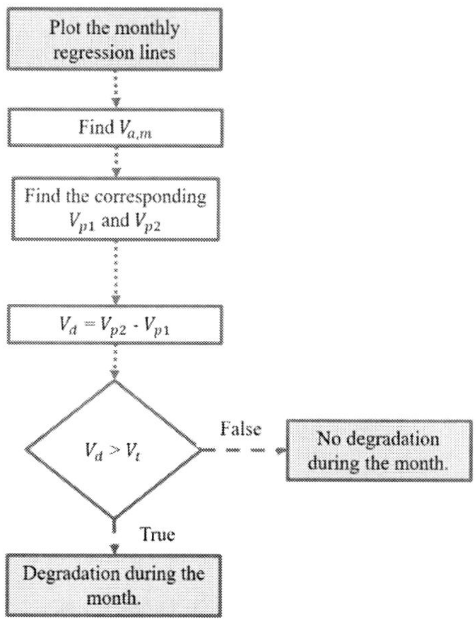

Figure 6: Flowchart of the method MFD

With these optimal parameters and data, validation was performed on 80 random days of 2017 data. The confusion matrix is shown in **Figure 7**. There were totally 934 data-points present in the data set consisting of 80 days. The accuracy of prediction here is indeed 99.3 %, however most of the time the plant was functioning normally and snow occurred only on 5 days and no inverter outage events were reported. Hence, the algorithm has identified normal functioning of the data set for the rest of the days with high precision. There are two deviations to notice- three events where the actual state was normal while the algorithm predicted it as snow cover and another four events where the actual state was snow cover but the prediction was normal. False negative is more important than false positive i.e the case with four wrong predictions is more important to focus. Since, each event represents an hour in a day, four wrong predicted hours among a data-set of 934 hours is tolerable, hence the algorithm has performed well with the detection of the fault in the inverter data.

Table 2: The optimal input parameters and input data needed for ANN for the fault prediction.

Input parameters
Irradiation
Module temperature
Actual current
Predicted current
Actual Power
Predicted Power
Actual voltage
Actual energy yield

Input data
one year- 2018

Table 3: Optimal hyper parameters for ANN to predict the faults with the best accuracy.

Hyper parameter	**Values**
Units per layer	50
Learning rate	0.01
Activation function	ReLU
Optimiser	Adam
Number of layers	3
Epochs	150

With these parameters, the algorithm is then trained and tested with the data of three other inverters each covering different number of PV panels in the plant, hence inverters with different capacities. The main purpose, as explained before, is to make the algorithm more universal for the entire power plant by training it with different inverter data-sets. After training and testing, the algorithm is then validated with a data-set of four other inverters. This time, the data-set for validation is selected in such a way that many snow cover events are present along with normal functioning events. Once again, the train-test accuracy will not be discussed here and only, the accuracy with foreign data will be explained.

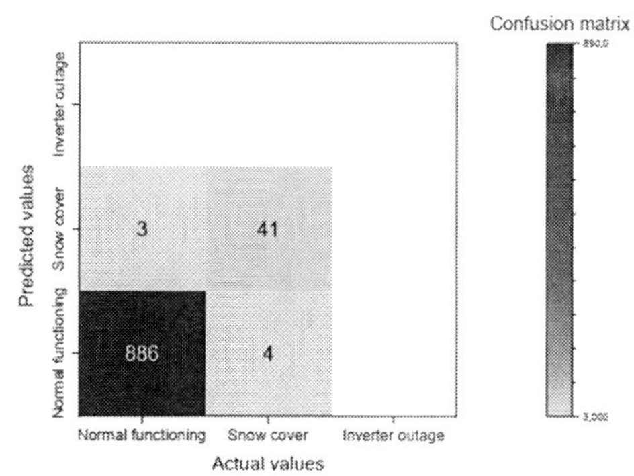

Figure 7: Confusion matrix for validation of the algorithm initially developed with the data containing 80 days of the year 2017.

As shown in **Figure 8**

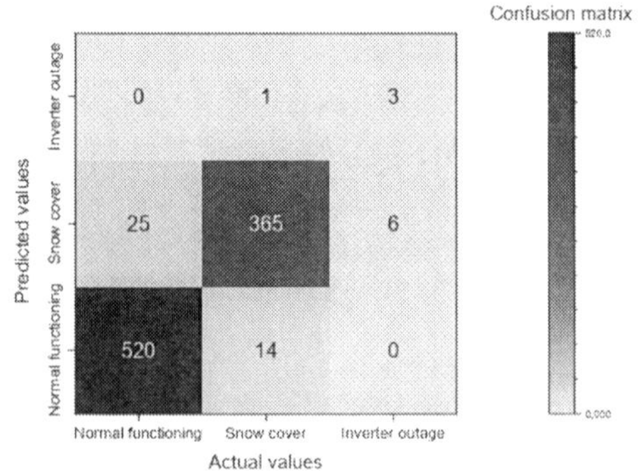

Figure 8: Confusion matrix for validation of fault prediction using selected data from four different inverters.

, the algorithm has detected majority of snow cover and normal functioning events and few inverter outage events. The outage event has actually occurred only for one day only at one inverter. Hence, enough data was not available as in snow cover for 1:1 performance testing. Once again, the false negative deviation is considered in more detail i.e the case where 14 events were actually snow cover while the prediction was normal. When compared to a total number of 934 events where snow fall occurred, the algorithm predicted 365 events correctly with an error in prediction of 3.69 %. Moreover, the overall accuracy of the prediction was 95.07 %. Therefore, it can be concluded that with the input data of one-year, defined input parameters and the hyper parameters, the algorithm can detect the two faults in the PV power plant using any inverter data.

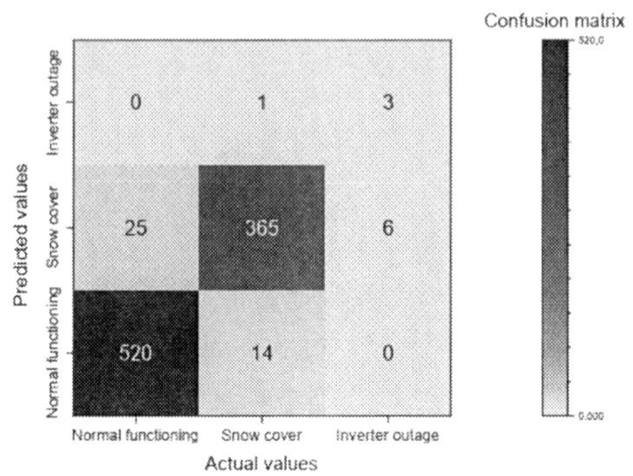
Confusion matrix
Predicted values
Inverter outage
Snow cover
Normal functioning
0
1
3
25
365
6
520
14
0
Normal functioning
Snow cover
Inverter outage
Actual values

Figure 8: Confusion matrix for validation of fault prediction using selected data from four different inverters.

3.2 Results MFD:

One-year data- 2014 for one inverter is taken as reference and the algorithm shown in **Figure *6*** is applied for each month from January 2015 to December 2018. The actual degradation status and the prediction is shown in **Figure 9**. Among 48 months, the algorithm has performed accurately by predicting the months with degradation events and the ones with no degradation. Only two events were observed when the algorithm failed to predict the status of degradation accurately. However, the overall accuracy is 95.83 %. Hence, this algorithm is deemed suitable to predict the status of degradation of the PV system using the inverter data for any given month.

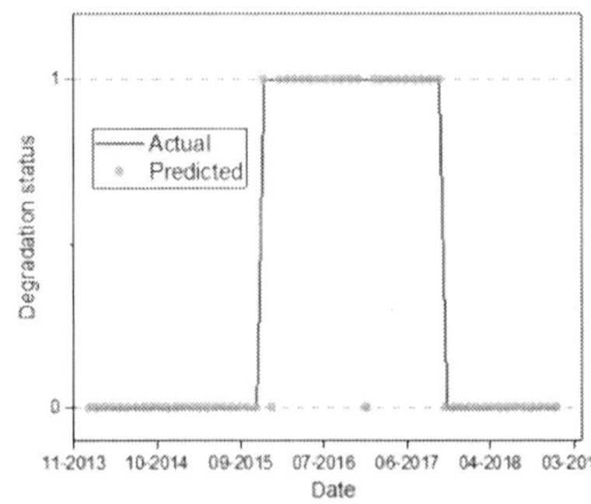

Figure 9: Actual degradation status of the PV systems and predictions using the statistical method and inverter data.

In order to make sure that the algorithm works accurately for other inverters as well, seven random inverters at the power plant were chosen and the algorithm was applied for 48 months for each inverter. All the PV modules connected to these inverters were healthy and no degradation was reported. The results are mentioned in **Figure *10***. Here, the accuracy states for number of correct predictions out of the 48 months tested. It can be seen that the algorithm performed with an average accuracy of 97 %, i.e. among 48 months tested, the algorithm predicted the status of degradation for 47 months correctly. Therefore, this algorithm can be employed for the entire PV power plant. In order to employ it for other PV power plants, minor changes might be needed based on inverter capacities and the size of the plants for this algorithm to predict the status of degradation of the PV arrays.

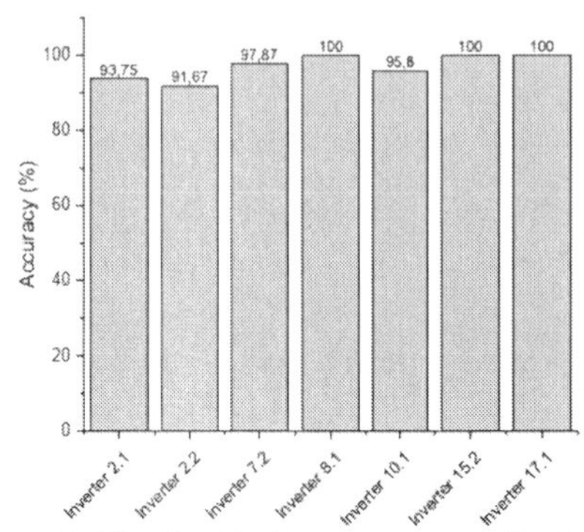

Figure 10: The algorithm's accuracy to predict the status of degradation using the data from other inverters.

4 CONCLUSION

The purpose of this research was to identify different possible faults in PV systems using machine learning algorithm with the inverter data. Three important faults were identified in the given PV power plant- namely snow event, inverter outage and degradation of PV arrays. To develop the algorithms for the fault detection, the model and the values from [6] were also considered. For the first two faults, an ANN algorithm has been developed to detect them and differentiate them from the normal functioning based on hourly-resolution data. The algorithm has been initially developed and validated on one inverter data. The input parameters, hyper parameters and input data have been optimized to achieve an accuracy of 95 %. This algorithm has been eventually validated on other inverters. The main purpose is to make the algorithm detect and classify the faults for an inverter of any capacity. For degradation, it was not possible to detect based on the hourly values of voltage or current or energy yield produced. Hence, with daily resolution data, status of degradation was predicted with high accuracy by plotting regression lines for one month. The main objective of detecting and classifying the three faults along with normal functioning has been achieved. There is a possibility that additional defects can be detected, for instance, partial shading or vegetation and so, the two algorithms can further be modified to detect these faults. Therefore, a robust predictive maintenance tool employing these algorithms can be developed for a PV power plant based on the inverter data for early automatic detection of the faults. This results in effective maintenance of the systems, decrease in O&M costs and finally, ensures a constant energy yield from the power plant throughout its operation.

ACKNOLEDGEMENT

This work was funded by the German Federal Bank Investitionsbank Sachsen-Anhalt, the German Federal State Saxony-Anhalt and the European Regional Development Fund (EFRE) in the project Mon-KI (No. 2004/00009).

5 REFERENCES

[1] Photovoltaics Report by Fraunhofer Institute for Solar Energy Systems, ISE with support of PSE Projects GmbH,27 July 2021, Freiburg.

[2] IEA PVPS Task 13, Subtask 2, Analytical Monitoring of Grid-connected Photovoltaic Systems - Good Practices for Monitoring and Performance Analysis, Report IEA-PVPS T13-03:2014, ISBN 978-3-906042-18-3, March 2014.

[3] S. R. Madeti und S. Singh, „Modeling of PV system based on experimental data for fault detection using KNN method," Solar Energy, volume 173, pg 139-151, October 2018.

[4] M. Dhimish, V. Holmes, B. Mehrdadi, M. Dales und P. Mather, „Photovoltaic fault detection algorithm based on theoretical curves modelling and fuzzy classification system," Energy, pp. 276-290, 2017.

[5] M.Ejgar, B. Momin, T. Ganu, Intelligent monitoring and maintainence of solar plants using real-time data analysis, 2017 ICCE-Asia.

[6] David Dassler, Shreyas Bangalore Kuppanna, Stephanie Malik, Ralf Schmidt, Matthias Ebert, Training and evaluation for yield-driven detection of losses in PV systems utilizing artificial neural networks, 47th IEEE Photovoltaic Specialists Conference, Calgary, OR, pp. 1903-1909, 2020.

[7] Kevin Gurney, An introduction to neural networks, 1997.

[8] Keras Documentation, Keras: The Python Deep Learning library, https://keras.io/.

[9] Confusion matrix theory, https://en.wikipedia.org/wiki/Confusion_matrix

RESEARCH ON DIGITAL TWIN SYSTEM OF PHOTOVOLTAIC ARRAY

*Kun Ding[1], Xiang Chen[1], Jingwei Zhang[1], Shuai Weng[2], Li Feng[3], Frank U. Hamelmann[3]
[1]College of Mechanical and Electrical Engineering, Hohai University, Changzhou, China
[2]Changzhou Key Laboratory of Photovoltaic System Integration and Production Equipment Technology, Changzhou, China
[3] Solar Computing Lab, University of Applied Science Bielefeld, Minden, Germany
*Corresponding author(dingk@hhu.edu.cn)

ABSTRACT: To ensure the smooth operation of photovoltaic (PV) power plants and provide stable power output for the power system, a reliability evaluation of the power output of the PV power plant is required. PV array is a significant part of the PV power plant. The modelling technique of the digital twin is used to build a digital twin model of the PV array that maintains a simultaneous mapping to the output characteristics of the physical PV array and can therefore be used as the basis for PV array reliability assessment. Based on the existing PV array modeling research, the dynamic bayesian network (DBN) is used to realize the adaptive update of PV array model parameters and complete the construction of the digital twin model. The digital twin model enables real-time derivation of model parameters, prediction of output power, and reliability assessment of PV array. The research conducted in this paper is a preliminary attempt at digital twin technology in the PV field, hoping to promote digital twin application in the PV field.
Keywords: PV array, modelling, digital twin

1 INTRODUCTION

PV arrays work under complex and variable environmental conditions for a long time, so the output power of PV arrays is in an unstable state, and the degradation process of PV arrays is also dynamic and non-linear [1]. To enable PV power to serve the power system better, the PV array should be effectively monitored to adjust the operating status of the PV power plant in time and eliminate the faults in the PV array in time to ensure the smooth operation of the PV power plant. Therefore, the introduction of digital twin technology to realize real-time situational awareness of the output characteristics of PV arrays is conducive to regulate and control the operating status of PV power plants. The digital twin is a simulation process that integrates multidisciplinary, multi-physics, multi-scale, and multi-probabilities. The real-time situational awareness of physical entities is realized by establishing digital twins with real-time mapping relationships with physical entities [2]. The digital twin concept was first proposed by Michael Grieves and has been applied in many fields such as aerospace, industrial manufacturing, smart workshops, and power systems [3].

The modelling ideas of digital twin model of PV array in this paper can be divided into three steps: the establishment of the PV array mathematical model; the construction of the digital twin data; real-time derivation of model parameters. The establishment of the mathematical model of the PV array is based on the research of the modelling methods of PV module and PV array. The characterization of the output characteristics of PV arrays under normal and fault conditions is achieved by improving the existing PV array model. The digital twin data includes the measured data of the PV array, the feature values after processing the measured data, and the mathematical model parameters of the PV array. The degradation rate of the parameters of the mathematical model of PV array is analyzed by the measured data, and the model parameters are deduced by combining with the dynamic bayesian network (DBN).

2 DIGITAL TWIN MODEL OF PV ARRAY

First, the digital twin model framework for PV arrays is introduced; then, the specific methods for establishing the digital twin model are described.

2.1 Framework of digital twin

In this paper, based on the five-dimensional digital twin model [4], combined with scientific issues in the PV, a framework of the digital twin model for PV arrays is constructed, as shown in Figure 1. The physical model expresses the working mechanism of the PV array from a mathematical point of view and explores the factors that affect the output of the PV array. Digital twin data includes physical data and virtual data. The physical data have the measured ambient, electrical, and equipment information in the actual PV array operation process. The virtual data is the characteristic parameters after processing the measured data through a mathematical model and data-driven algorithm. The digital twin data provides a reliable data source for the adaptive deduction of the digital twin model of the PV array and provides a guarantee for the performance evaluation and power prediction of the PV array. The adaptive deduction of the digital twin model can be realized by the deduction of DBN [5] based on the degradation rate of characteristic parameters of the PV array. Based on the study of the degradation rate of model parameters, the mapping relationship between model parameters and the actual operating state of PV arrays is established to further realize the characterization of the degradation process of PV array.

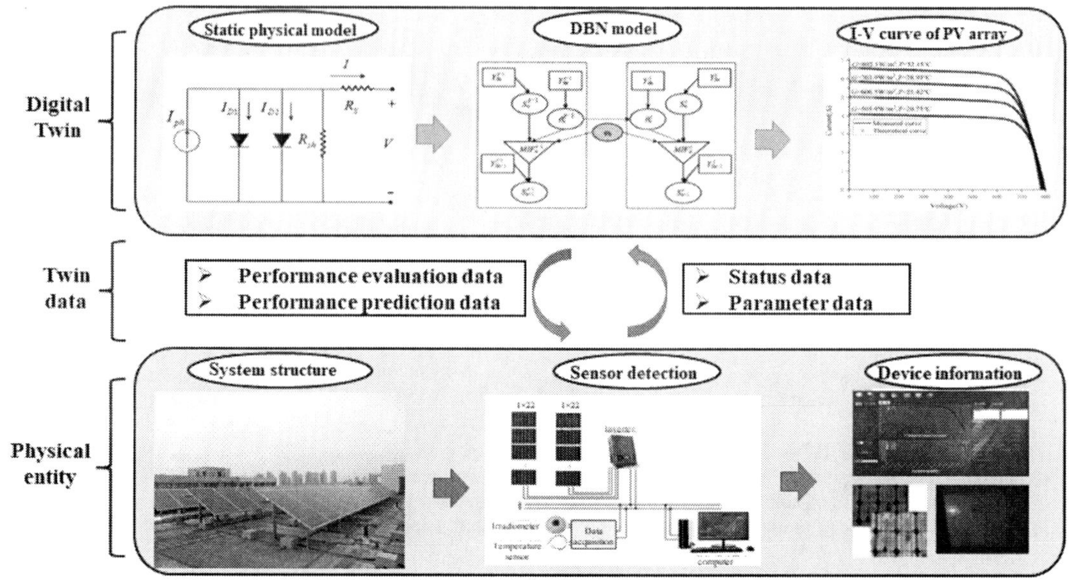

Figure 1 The framework of the digital twin model of the PV array

2.2 Construction methods of digital twin model for PV array

A. PV array mathematical modeling

The mathematical model of the PV module is analyzed first in this paper, and then a mathematical model of the PV array is established based on the principle of superposition, which provides a theoretical basis for the establishment of the digital twin model. Since the dual-diode model in the solar cell model has higher accuracy than other mathematical models, especially funder low irradiance levels. The simulation results are still consistent with the measured IV curve of the PV module, so the dual-diode model is used as the structural unit of the static model of the PV array. The equivalent circuit diagram of the dual-diode model is shown in Figure 2 [6].

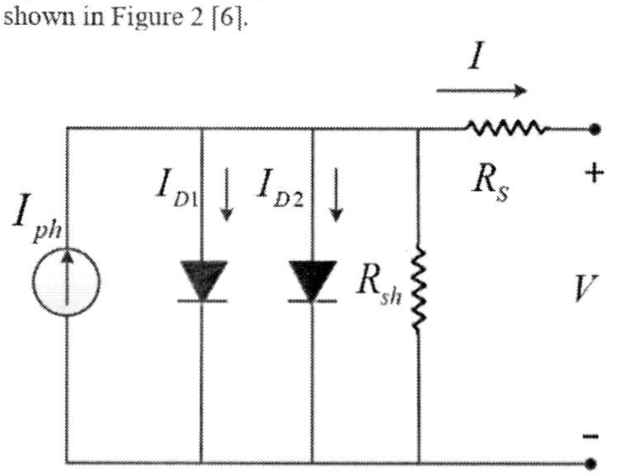

Figure 2 Equivalent circuit diagram of the mathematical model of the double-diode PV cell

According to the mathematical equations of the dual-diode model, it can be known that the model has seven unknown parameters, I_{ph}, I_{01}, I_{02}, n_1, n_2, R_s, and R_{sh} [6]. Due to the excessive number of unknown parameters in the model, the solution process is difficult, so the model parameters need to be simplified. The simplified model parameters include short-circuit current, open-circuit voltage, series resistance and parallel resistance, which can directly reflect the output characteristics of PV module. The parameters such as ideal factor and reverse saturation current, which are not easy to be measured, are simplified or replaced by other parameters. The simplified model parameters are changed from seven to four [6], and can be solved by analytical method or intelligent optimization algorithm.

Figure 3 is a method flow chart for rapidly modeling of PV arrays. The mathematical model of the PV array can be established based on the improved double diode model of the PV cell and the PV array modelling method described earlier.

B. Methods for processing digital twin data

The digital twin data is the core element of digital twins model. The digital twin data includes the measured data of the actual PV array, theoretical data of the digital twin model and application information. At the same time, After processing and fusion of data from different levels, they together form the digital twin data. The role of these data is used to facilitate the better functioning of the digital twin model, which is also known as the blood of the digital twin. [4]. The measured data of the PV array includes the ambient, electrical, equipment, and characteristic parameters, which are processed by mathematical model and data-driven algorithm. Because the data in the actual measurement process have the error data, faulty data, etc, the incomplete data is reconstructed by interpolation, and the faulty data is analyzed and restored to normal data under the corresponding environmental parameters. After pre-processing the measured data, the analytical method and intelligent optimization algorithm can be combined to extract the model parameters of the measured data, and the Variational Modal Decomposition (VMD) algorithm is used to extract the trend item of the data as a feature vector [7].

The PV array is in a dynamic degraded state during actual operation. The factors related to the degradation of the PV array and the degradation rate of model parameters need to be researched, which is the basis for realizing the adaptive update of the digital twin model parameters. Studies have shown that changes in ultraviolet radiation, temperature and humidity in the outdoor environment are the main environmental factors that degrade the performance of PV arrays. As a result, the changes of environmental factors are reflected in different phenomena, such as ethylenevinyl acetate (EVA) yellowing, line aging and material corrosion in PV modules. The degradation of PV array model parameters thus proceeds slowly, the most intuitive result is the reduction of PV array output power.

DBN is a probabilistic graph model, which transforms the uncertain parameters in the model into random nodes, and deduces the model backward by observing the data of the nodes [8]. The posterior probability distribution of the updated state node is used to estimate the parameters of the PV array model, so it can be used as a theoretical method for adaptively updating the parameters of the PV array model. Combining DBN to deduce and update the parameters of the digital twin model, the output power of the PV array can be estimated, and its performance can be evaluated.

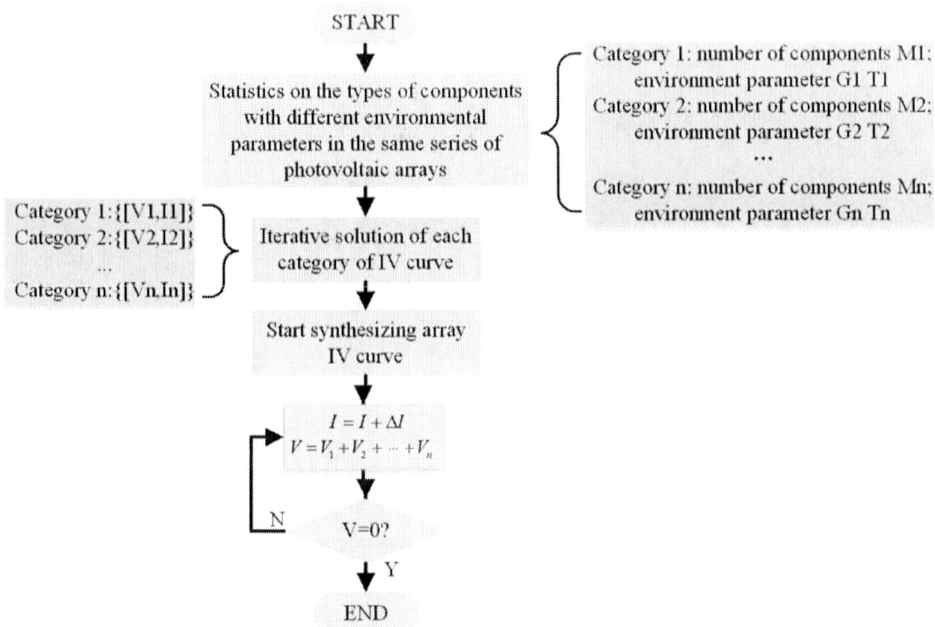

Figure 3 Rapid calculation process of IV curve for PV array

3 APPLICATION OF DIGITAL TWIN

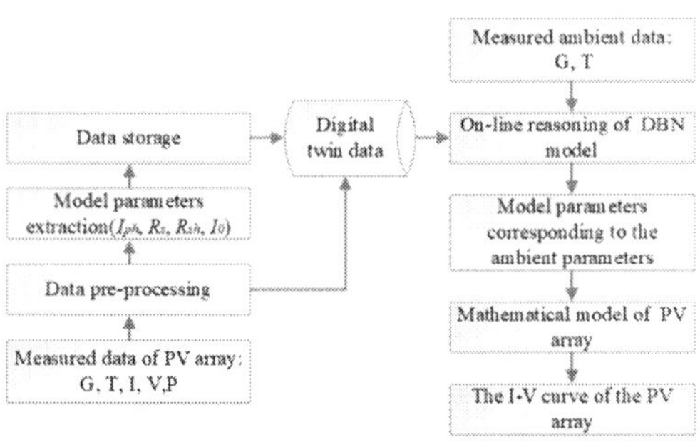

Figure 4 The modeling process of the digital twin model for PV array

The modeling process of the digital twin model for the PV array is shown in figure 4. The digital twin model of the PV array can be used to simulate and analyze the output characteristics of the PV array in the normal state, short-circuit state, and shading state to verify the effectiveness of the model proposed in this paper. Figure 5 is the result of the experiment. The short-circuit fault configuration is to short-circuit one PV module in the array. The partial shading failure is achieved by shading two PV modules in the PV array, where the shading is a plastic film with a transmittance of 0.45. The output characteristic of the array in the normal state is the I-V curve under different irradiance and temperature conditions. From the experimental results, the PV array digital twin model can accurately characterize the output characteristics of the physical PV array, so it has the function of real-time situational awareness of the digital twin model. And the simulation results of the PV array as shown in Figure 6, the left part shows the measured I-V curve and the theoretical I-V curve under the partial shading condition of the PV array, and the right is the three-dimensional simulation for the PV array.

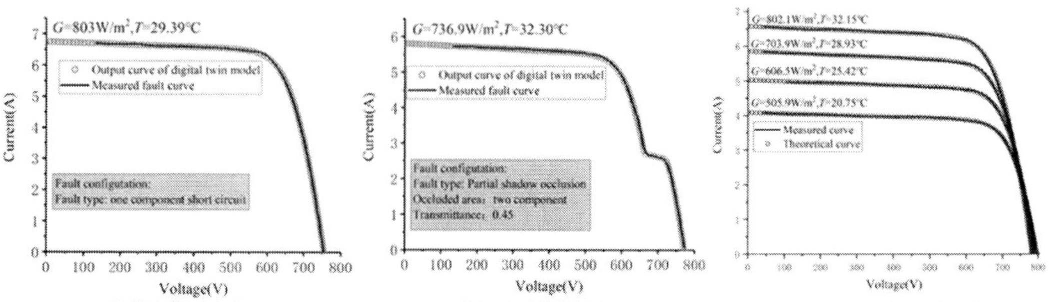

(a) Short circuit fault (b) Partial shading fault (c) I-V curve under normal conditions

Figure 5 I-V curves in different states

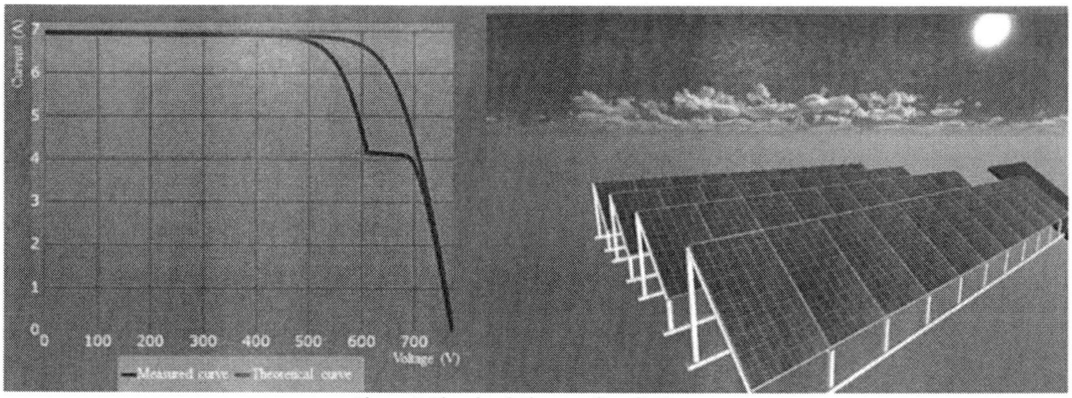

Figure 6 The simulation results of PV array

4 CONCLUSIONS

In this paper, we introduce the theory of digital twin modeling and the related theory of the PV array model to realize the establishment of the PV array digital twin model. The PV array digital twin model includes the PV array mathematical model and the real-time derivation of the model parameters to achieve the real-time characterization and prediction of the PV array output characteristics. The constructed PV array digital twin framework includes the physical PV array, the digital twin model, and the digital twin data.

The digital twin model of the PV array can accurately simulate the output characteristics of the PV array, including the I-V characteristic curve of the PV array under normal and fault conditions. Due to the limited ability of the digital twin model to characterize the real-time state of the PV array, the real-time derivation and updating method of DBN-based PV array model parameters still needs further improvement. This is one of the subsequent research directions regarding the application of digital twin in PV array modeling.

5 ACKNOWLEDGEMENTS

This work was supported by the National Natural Science Fundation of China (Grant No. 51777059), Qing Lan Project of Jiangsu Province and Changzhou Sci&Tech Program (Grant No. CJ20200074). It was also supported by Adaptive Ertragsprognose mit Data-Mining im PV-Feld auf Grundlage einer digitalen Signataur der PV-Module und der Systemkomponenten (PV Digital 4.0,No. 13FH020PX6).

6 REFERENCE

[1] Bruckman L S, Wheeler N R, Ma J , et al. Statistical and Domain Analytics Applied to PV Module Lifetime and Degradation Science. IEEE Access, 2013, 1:384-403.

[2] Korth B, Schwede C, Zajac M. Simulation-ready digital twin for realtime management of logistics systems. In: Proceedings of the 2018 IEEE International Conference on Big Data (Big Data).Seattle, USA: IEEE, 2018. 4194-4201

[3] Grieves M W. Product lifecycle management: the new paradigm for enterprises. International Journal of Product Development, 2005, 2(1-2): 71-84

[4] TAO F, ZHANG M, LIU Y, et al. Digital twin driven prognos-tics and health management for complex equipment. CIRP Am-nals-manufacturing Technology, 2018,67(1): 169-172

[5] LI C, MAHADEVAN S, et al. Dynamic Bayesian network for aircraft wing health monitorin digital twin. AIAA Joumal, 2017, 55(3): 930-941

[6] PETRONE G RAMOS-PAJA C Modeling of photovoltaic fields in mismatched conditions for energy yield evaluations. Electric Power Systems Research, 2011, 81(4): 1003-1013

[7] DRAGOMIRETSKIY K, Z0SS0 D. Variational mode decomposition. IEEE Transactions on Signal Processing, 2014, 62(3): 531-544

[8] XIE H, SHI J, LU W. et al. Dynamic Bayesian networks in electronic equipment health diagnosis. Proceedings of Prognostics and System Health Management Conference (phm-chengdu) Washington, D. C, USA: IEEE, 2016: 1-6

AUTO-PARAMETRIZING THE DIGITAL TWIN OF PHOTOVOLTAIC POWER SYSTEMS

Arttu Tuomiranta[1], Imre T. Horváth[1], Arnaud Schils[1], Karel De Brabandere[2], Eszter Voroshazi[1], Emmanuelle Bertrand[3], Bader S. Aldalali[4], Ivan Gordon[1], Andreas Wabbes[3], Stijn Scheerlinck[3]
[1]imec (partner in EnergyVille), Thor Park 8320, 3600 Genk, Belgium
[2]3E, Quai à la Chaux 6, Brussels, Belgium
[3]ENGIE Laborelec, Rodestraat 125, 1630 Linkebeek, Belgium
[4] Electrical Engineering Department, Kuwait University, Kuwait

ABSTRACT: Physics-based digital twins of photovoltaic (PV) systems involve many uncertain parameters whose assumed values can be erroneous or unknown. In this paper, we propose an auto-identification algorithm that can be used to verify or approximate such uncertain parameters. The algorithm automatically parametrizes a PV digital twin based on operational data i.e., inverter-measured system performance and weather data. The performance of the initial parametrization step of the algorithm was evaluated in three use cases where plane-of-array (POA) orientation, basic electrical layout, or all system design factors are unknown. The evaluation was performed based on real operational data collected at five utility-scale PV systems considering different training data periods. The auto-identification of POA orientation performs well. With a yearly dataset, the annual POA irradiation estimates resulting from the auto-identified POA angles stayed within ±5% of the real value at all the sites. The auto-identification of electrical layout resulted in a higher uncertainty due to the larger number of unknown parameters involved. Even so, the electrical layout approximated solely based on one year of operational data resulted in a total number of modules within ±10% of the real value at all the sites.
Keywords: Calibration, Monitoring, Simulation, System Identification, System Performance

1 INTRODUCTION

The estimation of technical performance indicators is the key functionality of the digital twin of a solar photovoltaic (PV) power system. PV system performance can be predicted at various levels of model greyness. Black-box models are built on the empirically derived correlations between the historical time series of performance measurements and the temporally variable ambient predictors. They are ideal for forecasting the performance of an individual PV system with long time series of performance data available. A pure black-box model can be greyed by integrating known physical dynamics within the box and between the box and its environment. The more physics is integrated, the fewer historical data are required and the more general performance estimates can be obtained. The maintained links to physics also allow for a more straightforward analysis of system failures. That is because unlike in black-box modelling, the impact of various system design factors can be analyzed without training data that range across the relevant factor domains such as various array orientations.

In contrast to the black-box approach, the construction of a PV digital twin as a grey box requires the parameters of the incorporated physical relationships. By varying these parameters, the performance model can be applied to different kinds of systems. The present study addresses use cases where some of these parameters are unknown or potentially erroneous. The considered three use cases include

1) plane-of-array (POA) orientation i.e., the tilt angle (β_{poa}) and the azimuthal orientation angle (γ_{poa}) of the modules,

2) basic electrical layout parameters i.e., the number of parallel strings ($N_{str,mpp}$) connected to each monitoring unit and the number of modules per string ($N_{mod,str}$), and

3) all PV system design parameters.

In all the three use cases, the degraded module current-voltage (*I-V*) characteristics under standard test conditions (STC) are unknown, which is common with operational PV systems.

Hence, the objective was to develop a methodology for auto-parametrizing PV digital twins solely based on operational data i.e., inverter-monitored system performance time series and weather data.

In the prior art, PV system identification is typically performed using a black-box approach – Hammerstein-Wiener models (e.g., [1]–[3]) in particular. No published work was found to apply a grey-box approach to the auto-parametrization of PV systems. In the PV literature, the use of the grey-box models seems to be limited to the extraction of the parameters of the equivalent circuit models for solar cells. A system-level grey-box approach is worth trying because the related physical phenomena are well-known and not highly complex. Its training requires less data and computation capacity than that of a black-box model.

The integration of physical dynamics in the auto-parametrization algorithm makes the procedure more general. It allows to estimate the effect of changes in design factor levels, for which there are no measurements available. This enables the sensitivity analysis of the interactions between different parameters, the estimation of degraded *I-V* characteristics, and a further diagnosis of the state of the system comparing different time domains, ambient conditions, and systems. This is useful, for example, if the design of a PV system is retrofitted several years after the start of the operation when datasheet information is no longer valid. In general, the grey-box approach is useful when it is important not only to estimate energy yield accurately but also to give possible reasons for the evolution of system performance. The automatic identification itself is useful whenever parameter values are missing or suspected to be erroneous. It also has the potential to significantly reduce the time required for manual parametrization especially if there is a fleet of PV systems to be considered instead of a single system.

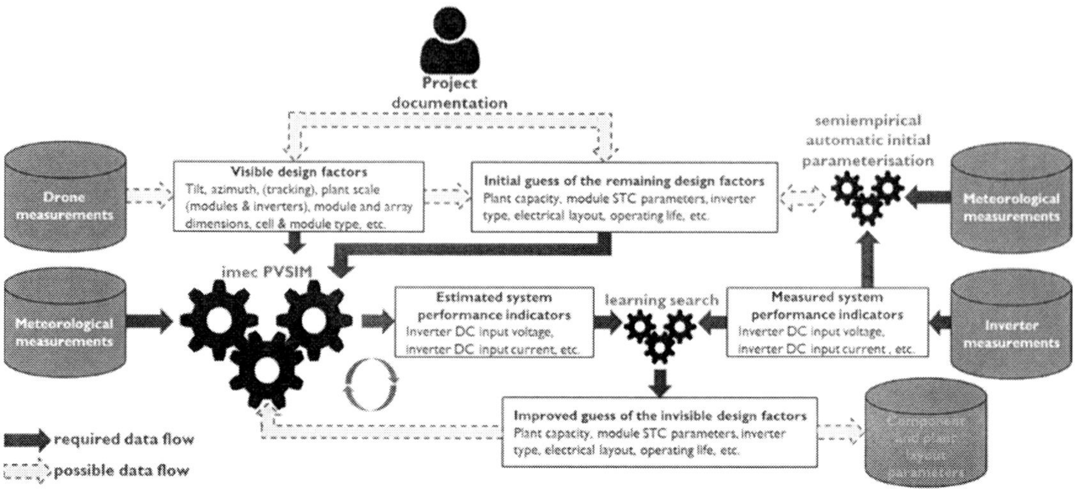

Figure 1: Process flow of the proposed auto-parametrisation algorithm for PV digital twins

2 ALGORITHM

The process flow of the proposed auto-parametrization algorithm is presented in Fig.1. In addition to the required operational data, the parametrization accuracy of the algorithm can be improved by inputting any known parameters from project documentation (e.g., module and inverter datasheets) or visible parameters derived from drone images. These data are optional, however. All the remaining unknown parameters are estimated at the novel process of semiempirical automatic initial parametrization, which outputs the estimates as an initial guess of all the parameters required by the grey-box-like system performance model: imec's PV system simulation framework (imec PVSIM in Fig. 1) ([4], [5]). The system performance indicators output by the simulation framework are then compared with the inverter measurements to evaluate the fit. The initial fit is improved by changing the parameter vector and rerunning the simulation. This learning cycle is repeated until 1) the mean bias error (MBE) of the power estimates over the entire timeframe of the input data becomes zero and 2) the root mean square error (RMSE) at zero-MBE is not greater than any of the RMSE levels obtained for previous iterations. The cycle is adapted from the improved learning search algorithm of [6] by adding physical constraints to its exploration steps.

The initial parametrization uses simple correlations and well-known physical links between solar geometric, solar meteorological, and PV system performance parameters. Shading occurrence is first roughly approximated by comparing global horizontal irradiance (GHI) data with current measurements. The solar positions at which the mean current-irradiance ratio is less than half of the 90th percentile level of the ratio are determined as shaded. The approximation of β_{poa} and γ_{poa} can be done independently of the other system design parameters making it particularly suitable for the initial parametrization. In the present method, it is done by comparing the unshaded current measurements under clear-sky conditions with time series of solar beam angle of incidence calculated for all possible plane orientations i.e., β_{poa}-γ_{poa} combinations. After determining the first guess of the POA orientation, the shadow map is recreated based on POA irradiance instead of GHI. The second guess of the POA orientation is then generated using the corrected shadow map.

The electrical layout is also approximated at the initial parametrization. Unlike the POA orientation, $N_{str,mpp}$ and $N_{mod,str}$ cannot be estimated independently of the modules' I-V parameters. If the module properties and the total number ($N_{mod,mpp}$) of the modules connected to the monitoring unit are unknown, the present method approximates modules' operating current and voltage from polynomial fit equations derived from the public module database of the California Energy Commission (CEC) [7]. The fitted curves and the used data are shown in Fig. 2. Based on the correlation of Fig. 2a, non-degraded STC cell efficiency (η_{stc}) is first estimated as a function of the inverter-measured ratio of maximum power point (MPP) voltage (V_{mpp}) to open-circuit voltage (V_{oc}). V_{oc} and the

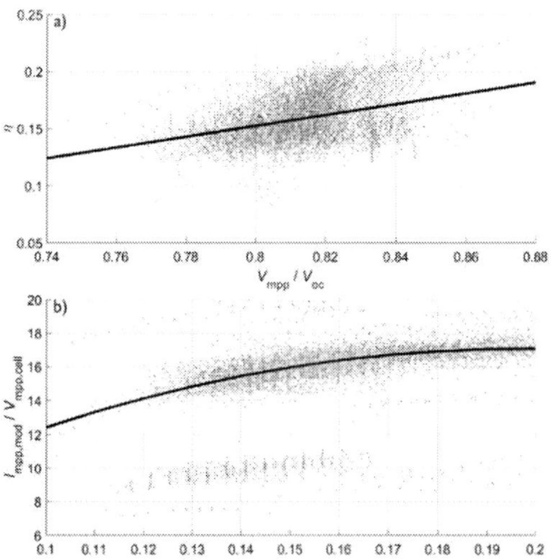

Figure 2: Polynomial fits derived from the module data published by the California Energy Commission (CEC) [7] for monofacial modules equipped with 60 or 72 cells based on crystalline silicon wafers showing the correlation between a) the voltage ratio of maximum power point (MPP) to open circuit (OC) and the cell conversion efficiency (η_{stc}) and b) η_{stc} and the cell current-voltage ratio at MPP

corresponding V_{mpp} are approximated as the highest recorded voltage level and the mean voltage under similar ambient conditions over the entire timeframe, respectively. Since the predictor is a ratio, it is not a problem that the voltage is measured at the inverter level but the fit functions are derived from module data. Using the correlation of Fig. 2b, the cells' I-V ratio at the MPP can be estimated using the approximated η_{stc}. The temperature dependence of η is inferred from the temperature dependence of the inverter-measured power. $N_{mod,mpp}$ is estimated by using the temperature-corrected η together with the operational data. A set of all $N_{str,mpp}$-$N_{mod,str}$ combinations roughly amounting to $N_{mod,mpp}$ is created. First, the physically unfeasible combinations are omitted and, finally, the combination resulting in the closest fit with the estimated I-V ratio is determined as the most likely layout.

Unlike the correlations used in the initial parametrization, the imec PVSIM is a detailed model, which captures, for instance, the heat transfer between module layers and the diode-like nature of solar cells. Hence, the learning search cycle is used to finally optimize the values of the coupled, degraded I-V parameters of the modules and the system. Fig. 3 shows how through the learning search, the estimated set of I-V parameters converges from the initial guess into the optimal set that minimises the model bias at one of the monitored systems (S1 in Table I).

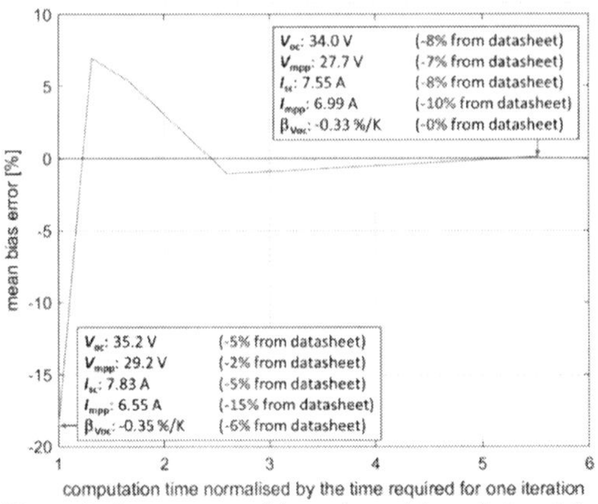

Figure 3: Convergence of the estimated degraded module current-voltage characteristics into the optimal parametrisation at one of the test sites (S1 in Table I)

3 VALIDATION

This section first describes the methodology for validating the auto-parametrization algorithm in the three use cases specified in section 1. Subsections 3.2-4 then present the results of the validation for each use case.

3.1 Methodology

The accuracy of the proposed methodology for the initial parametrization of β_{poa}, γ_{poa}, $N_{str,mpp}$, and $N_{mod,str}$ was evaluated based on inverter measurements and satellite-based ambient data for the five PV systems specified in Table I and denoted as S1–S5. The data have a resolution of 15 minutes and cover the year 2019. In addition to the full-year-based parametrization, six shorter data subsets were tested: three monthly and three 3-daily periods centered around June 21, September 23, and December 22.

The operational data from the five systems were collected by the MPP trackers of their inverters. The number of monitoring MPP trackers (N_{mpp} in Table I) ranges from one at S1 to 85 at S3. Each monitoring MPP tracker supplies its own system performance time series and therefore, the amount of available system performance data varies across the five systems.

Table I: PV systems used for evaluating the proposed auto-parametrisation algorithm for PV digital twins

	S1	S2	S3	S4	S5
$N_{mod,mpp}$	2838	66	44	36	21
$N_{str,mpp}$	129	3	2	2	1
$N_{mod,str}$	22	22	22	18	21
β_{poa}	10	15	30	15	15
γ_{poa}	142	142	180	180	142
N_{mpp}	1	6	85	56	6

The performance of the algorithm is evaluated by first calculating the deviations of the auto-identified parameter estimates from the real values. The errors are then reported as the MBE across the systems' all MPP trackers for each combination of a system and a training period. Additionally, the total impact of the auto-identification errors is quantified by POA irradiance (estimated as per [8]) with the POA orientation parameters and by $N_{mod,mpp}$ with the electrical layout parameters.

3.2 Initial parametrization of plane-of-array orientation

Table II and Table III show that with a full year of available operational data, the MBE of the auto-identified POA parameters stays within ±10 degrees. As shown in Table IV, these deviations result in a MBE within ±2% in the annual POA irradiation. With its range-normalized and averaged mean absolute error (MAE) of 5.4%, the initial parametrization of γ_{poa} is slightly more accurate than that of β_{poa}, whose averaged and normalized MAE is 6.9%.

Table II: Mean bias error [°] of auto-identified array tilt angle for each combination of a site and a training data period

	S1	S2	S3	S4	S5
full year	0.0	1.5	3.9	1.5	0.5
summer month	10.0	4.0	-18.8	7.9	3.3
autumn month	2.0	2.0	23.4	5.9	1.5
winter month	8.0	5.2	9.0	8.3	6.0
3 Jun days	4.0	-1.0	-3.7	6.3	-2.0
3 Sep days	-6.0	-2.7	-21.3	-7.4	-4.3
3 Dec days	2.0	-6.3	-14.9	-6.1	-5.8

Table III: Mean bias error [°] of auto-identified array azimuth angle for each combination of a site and a training data period

	S1	S2	S3	S4	S5
full year	2.0	9.2	-2.2	3.4	8.2
summer month	20.0	11.0	0.1	8.9	9.7
autumn month	8.0	10.3	-1.0	-26.1	14.5
winter month	31.0	32.2	-15.0	10.3	33.5
3 Jun days	16.0	-3.5	-0.3	-24.4	-6.5
3 Sep days	-18.0	17.8	-90.2	-49.2	19.8
3 Dec days	32.0	42.3	-56.4	-1.9	39.5

It can be concluded based on Table IV that a full year of operational data is required to identify the POA parameters at a sufficient accuracy for keeping the annual POA irradiation estimates within ±5% of the real irradiation. If an accuracy of ±10% is acceptable, also monthly training periods in summer and equinoctial seasons seem to provide sufficiently accurate parametrizations.

Table IV: Mean bias error [%] of annual global in-plane irradiation estimated based on auto-identified array orientation parameters for each combination of a site and a training data period

	S1	S2	S3	S4	S5
full year	0.1	1.4	-2.0	0.4	0.8
summer month	6.5	2.7	-7.9	2.2	2.3
autumn month	1.7	1.7	-2.7	1.2	1.6
winter month	5.4	3.3	-11.2	2.3	3.5
3 Jun days	3.2	-0.9	-10.1	1.4	-1.7
3 Sep days	-4.8	-0.9	-14.5	-4.9	-2.1
3 Dec days	1.9	-3.5	-9.0	-4.2	-3.1

3.3 Initial parametrization of basic electrical layout

If $N_{mod,mpp}$ is correctly fixed and the module properties known, the algorithm determines $N_{str,mpp}$ and $N_{mod,str}$ without any error in all the considered datasets. Table V and Table VI show the performance of the algorithm when the input $N_{mod,mpp}$ is unknown. In this case, the variability in the performance of the algorithm between the sites is wider with electrical layout than with POA orientation. At all the sites, the resulting MBE in $N_{mod,mpp}$ (Table VII) stays within ±15% with the yearly training data and with the monthly summer and autumn datasets. With the monthly winter and 3-daily datasets, the estimates become highly uncertain with a maximum $N_{mod,mpp}$ underestimation of 49% with the monthly winter dataset for S3.

Table V: Mean bias error [%] of auto-identified number of parallel strings for each combination of a site and a training data period when the total number of modules is unknown

	S1	S2	S3	S4	S5
full year	4.7	0.0	17.1	3.6	0.0
summer month	8.5	0.0	21.2	3.6	0.0
autumn month	8.5	0.0	0.0	3.6	0.0
winter month	-1.6	0.0	-38.8	-43.8	0.0
3 Jun days	12.4	0.0	11.0	53.6	0.0
3 Sep days	-1.6	0.0	0.0	3.6	0.0
3 Dec days	-12.4	-22.2	-27.1	-33.0	0.0

Table VI: Mean bias error [%] of auto-identified number of modules per string for each combination of a site and a training data period when the total number of modules is unknown

	S1	S2	S3	S4	S5
full year	0.0	-0.8	-4.9	8.5	-0.8
summer month	4.5	-0.8	-4.6	-2.0	-0.8
autumn month	0.0	6.8	-7.6	2.5	-0.8
winter month	0.0	9.1	-18.5	-1.0	5.6
3 Jun days	-9.1	4.5	-9.7	-12.6	11.1
3 Sep days	0.0	1.5	-4.1	5.5	-0.8
3 Dec days	-4.5	9.1	-22.4	-3.3	-10.3

Table IV and Table VII show that the impact of the initial parametrization of electrical layout on the accuracy of energy yield prediction is more severe than that of POA orientation with most site-period combinations. This is because the initial parametrization of electrical layout involves a larger number of unknown parameters that need to be approximated – the degraded *I-V* characteristics for example. The impact of these initial parameterization steps of both POA orientation and electrical layout is, however, reduced by the final learning search cycle of the algorithm that re-estimates parameters to minimize the RMSE and the MBE of the system performance estimates (Fig. 3).

Table VII: Mean bias error [%] of the total number of modules estimated based on auto-identified electrical layout parameters for each combination of a site and a training data period

	S1	S2	S3	S4	S5
full year	4.7	-0.8	3.3	12.5	-0.8
summer month	13.5	-0.8	2.1	3.9	-0.8
autumn month	8.5	6.8	-7.6	6.2	-0.8
winter month	-1.6	9.1	-48.9	-43.5	5.6
3 Jun days	2.2	4.5	-0.3	36.2	11.1
3 Sep days	-1.6	1.5	-4.1	9.0	-0.8
3 Dec days	-12.4	-15.4	-36.7	-35.1	-10.3

3.4 Initial parametrization solely based on operational data

The initial parametrization of POA orientation does not require any information about the remaining system design factors other than the site coordinates. If the site location is unknown, the algorithm first approximates the pair of site coordinates by searching for the pair at which the calculated solar geometry best matches with the inverter-measured current or power data. This added uncertainty does not notably worsen the estimation accuracy at any of the system sites. With some site-period combinations, β_{poa} and γ_{poa} are actually estimated more accurately after approximating the site coordinates. The reason for this surprising finding is probably the spatial error of the satellite-based irradiance data.

Subsection 3.3 presents the accuracy of the estimation of $N_{str,mpp}$ and $N_{mod,str}$ when $N_{mod,mpp}$ is unknown but the module type known. If also the module type is unknown, the algorithm estimates the required module properties based on the inverter measurements and the fitting equations derived from a public module library as described in section 2.

Table VIII-Table X present the evaluation results for this parametrization of electrical layout solely based on operational data. The comparison of these results with the corresponding results in subsection 3.3 shows that the chosen method to approximate module properties works well. With some site-period combinations, the auto-identification of the module properties even results in a more accurate identification of electrical layout than the identification with an available module datasheet. The available module data are the least useful with the yearly dataset, with which the auto-identification of module properties performs especially well. However, the average MAE of the case where also module properties are auto-identified (17.5%) is 7.2 percentage points higher than the case with available module properties (10.3%).

Table X shows that if there is only one year of operational data available for the initial parametrization of electrical layout, the estimated total number of modules is likely to stay within ±10% of the real value.

Table VIII: Mean bias error [%] of auto-identified number of parallel strings for each combination of a site and a training data period when only operational data are available

	S1	S2	S3	S4	S5
full year	3.1	0.0	11.8	3.6	0.0
summer month	-7.0	0.0	14.7	-6.3	0.0
autumn month	-3.9	0.0	0.0	3.6	0.0
winter month	-45.0	-33.3	-41.8	-41.1	0.0
3 Jun days	-12.4	0.0	0.0	3.6	0.0
3 Sep days	-22.5	0.0	0.0	3.6	0.0
3 Dec days	-31.8	-33.3	-35.9	-39.3	0.0

Table IX: Mean bias error [%] of auto-identified number of modules per string for each combination of a site and a training data period when only operational data are available

	S1	S2	S3	S4	S5
full year	4.5	-0.8	-5.1	3.2	0.8
summer month	4.5	-12.1	-11.9	11.0	-17.5
autumn month	-4.5	-5.3	-10.3	-7.5	-4.8
winter month	9.1	13.6	-19.5	-1.7	-10.3
3 Jun days	-27.3	3.0	35.5	-21.3	-5.6
3 Sep days	-18.2	-15.2	-8.6	-8.6	-20.6
3 Dec days	4.5	12.1	-24.1	-15.7	-21.4

Table X: Mean bias error [%] of the total number of modules estimated based on auto-identified electrical layout parameters for each combination of a site and a training data period when only operational data are available

	S1	S2	S3	S4	S5
full year	7.8	-0.8	0.1	7.1	0.8
summer month	-2.7	-12.1	-7.2	2.3	-17.5
autumn month	-8.2	-5.3	-10.3	-3.7	-4.8
winter month	-40.0	-24.2	-52.3	-41.6	-10.3
3 Jun days	-36.3	3.0	35.5	-18.6	-5.6
3 Sep days	-36.6	-15.2	-8.5	-4.7	-20.6
3 Dec days	-28.7	-25.3	-47.7	-46.2	-21.4

4 CONCLUSION

This paper presented a novel auto-parametrization methodology for digital twins of PV systems. The proposed algorithm automatically approximates system design factors based on operational data of weather and system performance. The performance of the initial parametrization step of the algorithm was evaluated in three use cases where 1) plane-of-array orientation, 2) basic electrical layout, or 3) all system design factors are unknown. The evaluation was performed based on real operational data collected at five utility-scale PV systems. The data were additionally divided into seven different subsets to analyze the impact of training period on the performance of the algorithm.

The auto-identification of plane-of-array (POA) orientation angles (tilt and azimuth) performs well with yearly datasets. The annual POA irradiation estimates resulting from the auto-parametrized angles stayed within ±5% of the real value at all the sites. Even with the monthly summer and autumn datasets, an accuracy of ±10% was reached. The basic electrical layout (numbers of parallel strings and of modules per string) could be derived without any error with all the datasets if the total number of modules ($N_{mod,mpp}$) was known. With an unknown $N_{mod,mpp}$, the performance of the algorithm clearly deteriorated. The $N_{mod,mpp}$ estimated based on the auto-identified layout parameters stayed within ±15% with the yearly training data and with the monthly summer and autumn datasets. The auto-parametrization solely based on operational data did not notably reduce accuracy when identifying POA orientation. The overall accuracy of the auto-identification of electrical layout decreased due to the unavailable module datasheet information. However, with a full year of operational data, this reduction was minor or non-existent implying that the chosen method to approximate module properties based on operational data performs well. The $N_{mod,mpp}$ estimated solely based on a full year of operational data stayed within ±10% of the real value at all the sites.

Through the given examples, it was shown that the automatic identification of PV design parameters is possible solely based on operational data. The results are promising but especially with electrical layout, this automatic parametrization still seems too uncertain to entirely replace manual verification. The accuracy of the auto-identification of electrical layout is particularly sensitive to the assumed evolution of the time-dependent degradation of module current-voltage characteristics. Thus, the future work to improve the accuracy should concentrate on more accurate degradation models to be integrated in the algorithm.

ACKNOWLEDGEMENTS

This work was executed within the imec.icon project ANALYST PV, a research project bringing together academic researchers (imec-PV and imec-IPI-Ghent) and industry partners (3E, Sitemark, AllThingsTalk, and Engie Laborelec). The ANALYST PV project was co-financed by imec and received project support from Flanders Innovation & Entrepreneurship (project nr. HBC.2019.0050) and Innoviris. The work in this paper was also partially funded by the Kuwait Foundation for the Advancement of Sciences under project number CN18-15EE-01.

REFERENCES

[1] N. Patcharaprakiti *et al.*, "Modeling of single phase inverter of photovoltaic system using Hammerstein–Wiener nonlinear system identification," *Curr. Appl. Phys.*, vol. 10, no. 3, Supplement, pp. S532–S536, May 2010, doi: 10.1016/j.cap.2010.02.025.

[2] A. Alqahtani, M. Alsaffar, M. El-Sayed, and B. Alajmi, "Data-Driven Photovoltaic System Modeling Based on Nonlinear System Identification," *Int. J. Photoenergy*, vol. 2016, p. e2923731, Jun. 2016, doi: 10.1155/2016/2923731.

[3] R. Celikel and A. Gundogdu, "System identification-based MPPT algorithm for PV systems under variable atmosphere conditions using current sensorless approach," *Int. Trans. Electr.*

Energy Syst., vol. 30, no. 8, p. e12433, 2020, doi: 10.1002/2050-7038.12433.

[4] H. Goverde *et al.*, "Energy yield prediction model for PV modules including spatial and temporal effects," in *29th European Photovoltaic Solar Energy Conference and Exhibition*, Amsterdam, Netherlands, Nov. 2014, pp. 3292–3296. doi: 10.4229/EUPVSEC20142014-5CV.2.28.

[5] I. T. Horváth *et al.*, "Photovoltaic energy yield modelling under desert and moderate climates: What-if exploration of different cell technologies," *Sol. Energy*, vol. 173, pp. 728–739, Oct. 2018, doi: 10.1016/j.solener.2018.07.079.

[6] T. Huang, C. Zhang, H. Ouyang, G. Luo, S. Li, and D. Zou, "Parameter Identification for Photovoltaic Models Using an Improved Learning Search Algorithm," *IEEE Access*, vol. 8, pp. 116292–116309, 2020, doi: 10.1109/ACCESS.2020.3003814.

[7] California Energy Commission, "PV_Module_List_Full_Data_ADA.xlsx." 2021. Accessed: Feb. 05, 2021. [Online]. Available: https://www.energy.ca.gov/sites/default/files/2020-08/PV_Module_List_Full_Data_ADA.xlsx

[8] R. R. Perez, P. Ineichen, R. D. Seals, J. Michalsky, and R. Stewart, "Modeling daylight availability and irradiance components from direct and global irradiance," *Sol. Energy*, vol. 44, no. 5, pp. 271–289, 1990, doi: 10.1016/0038-092X(90)90055-H.

APPLICATION OF MACHINE LEARNING TO ASSESS THE THERMAL BEHAVIOUR OF PV MODULES IN DIFFERENT CLIMATE ZONES

Julián Ascencio-Vásquez[1,2*], Ismail Kaaya[3], Karl-Anders Weiss[3], Marko Topič[2]

1 Envision Digital, Redwood City, California, United States
2 University of Ljubljana, Faculty of Electrical Engineering, Ljubljana, Slovenia
3 Fraunhofer Institute for Solar Energy Systems ISE, Freiburg, Germany

*Phone: +34 623 44 83 01; Email: julian.ascencio@envision-digital.com

ABSTRACT: Identifying the main factors for PV module performance losses, degradation and failures are still under research. Researchers have developed physical models to characterize the weather and the local environment where PV modules operate. , Indeed, published physical models can help to estimate PV long-term yield and degradation rates. These models highly depend on the operating PV module temperature, UV irradiance, and relative humidity as main inputs. In most cases, these input variables need to be modelled. For example, several models to estimate PV module temperature are available. However, improvements in their accuracy are still needed. Therefore, in this paper, we apply Machine Learning approaches to identify the combination of climate variables needed to model the PV module temperature. We then propose a modifiedtFaiman model that includes the effect of UV irradiance. Lastly, the modified model is benchmarked with other physical temperature models and shows an improvement over other models.
Keywords: Photovoltaics; PV Module; Thermal model; Machine learning

1 INTRODUCTION

PV module temperature has been identified as the second most crucial weather factor for the performance and reliability of PV systems. To model this parameter, researchers have developed physical models to represent the heating and cooling of modules from weather variables such as irradiance, ambient temperature and wind speed, which characterize the local weather where PV modules are operating.

However, the accuracy of the available PV module temperature models is still another challenge that needs improvement. The available temperature models have shown different accuracy levels in different climate zones [1]–[3]. This implies that these models might be sensitive to the local weather variables which need to be considered during the modelling.

Therefore, in this paper, we use Machine Learning (ML) approaches (e.g., Artificial Neural Network - ANN) to evaluate the influencing weather variables by assessing the predicted PV module temperature's accuracy using different combinations of input weather variables. After assessing the possible combinations and how they affect the predictions in different locations, we propose an improved and generalized PV module temperature model based on the most influencing weather variables.

The potential of ML approaches is being exploited in many scientific fields (Photovoltaics is no exception). These models use advanced algorithms supported by code libraries freely available, for example, in R and Python. They show better estimations than physical models. However, the physical behaviour is lost in the modelling (seen as a black box). Here we show the benefits of using ML learning to understand and improve physical models. The approach is applied to PV module temperature. This work's relevance is reflected by the PV industry's need to predict degradation rates (based on physical models), which require highly accurate estimations of PV module temperature as the most influencing factor of degradation [4], [5]. We apply machine learning approaches to assess input variables' relationship with the estimated PV module temperature accuracy. The aim is to develop more accurate PV module temperature models that consider all identified influencing variables.

We present the benchmarking of physical models and ML models for PV module temperature. The ML approaches: artificial neural network (ANN), linear regression (LR), and Light Gradient Boosting Machine (LightGBM) are used in this study. The different feature sets are determined with ambient temperature, global irradiance, wind speed, relative humidity and UV irradiance. The results are proved in three different locations (Zugspitze – Germany, Gran Canaria – Spain, and Negev – Israel).

1.1. Importance of PV module temperature in the lifetime yield of PV modules

To analyse performance and degradation behaviours in PV modules and systems, researchers use mathematical methods. In most cases, the performance metrics used to analyse PV performance, such as the performance ratio (PR), require temperature corrections (e.g. temperature corrected PR has shown better performance than traditional PR [6]). However, because the module temperature is required instead of the ambient temperature, it must be calculated using different temperature models [7], [8].

It has been shown that the performance of these temperature models depends on the operating weather conditions. Since the accuracy of the temperature models influences the PR calculations [9], it is crucial to improve the accuracy of the temperature models to have reliable and realistic performance assessments. Additionally, physical models used to calculate the degradation rates of PV modules and systems are based on major influencing factors (i.e. module temperature, UV and relative humidity). Our recent study [10] shows that the uncertainty in degradation rates can be as high as 15% due to module temperature estimations.

In summary, Figure 1 shows the expected impact of PV module temperature on the lifetime yield. At higher PV module temperatures, the PV module efficiency will decrease and the PV energy yield too [1], [11]. At the same time, increasing the temperature will increase the climate stresses perceived by the modules, increasing the degradation rates [4], [10]. Those two components: lower yield and higher degradation rates, will decrease the expected lifetime yield (sum of the production over the lifespan). Those facts highlight the importance of the PV module temperature.

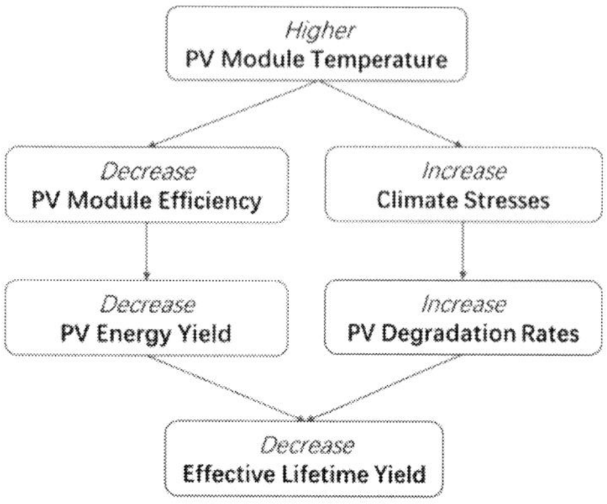

Figure 1. A flowchart about the impact of PV module temperature on the effective lifetime yield.

2 A BRIEF OVERVIEW OF PHYSICAL MODELS AND MACHINE LEARNING APPROACHES

2.1 Physical models for PV module temperature

In general, any physical model is supposed to empirically and explicitly show a particular component's thermal, mechanical or electrical behaviour. An example of this, and the core of this work, is the use of physical models to estimate the PV module temperature (T_{mod}).

The Ross Model [12] needs ambient temperature and global irradiance to predict the Tmod based on the parameter (K_{ross}) which is related to a physical process, as indicated in their metrical units. The Ross model has been widely used in literature and even served as the PV industry standard.

A considerable improvement in thermal modelling was adding the wind speed (WS) to the equations with the Faiman model [13] and the King model [14]. By adding the WS, the cooling effects are now taken into account. The Faiman model is being used due to its high accuracy and low sensitivity in different climates [3], [15]. Many other models are being proposed in the literature, but Ross and Faiman remain simple, meaningful and accurate models.

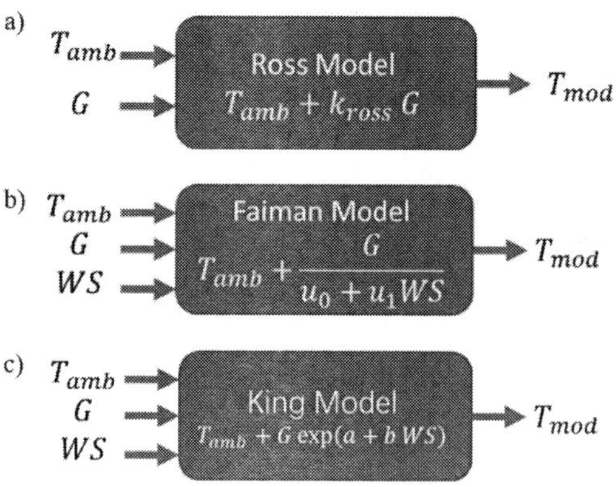

Figure 2. Representation of physical models for the estimation of PV module temperature. a) Ross model, b) Faiman model, and c) King model.

2.2 Machine learning approaches used for PV module temperature estimations

Machine Learning approaches are becoming more and more popular and robust thanks to the rapid advance of computational resources. They are also becoming more popular in the Photovoltaics (PV) industry, showing excellent results [16]. However, the physical processes of expected output are somewhat "hidden" in this link of different mathematical layers. See Figure 3 for the case of Artificial Neural Network (ANN) design that could be applied to estimate the PV module temperature.

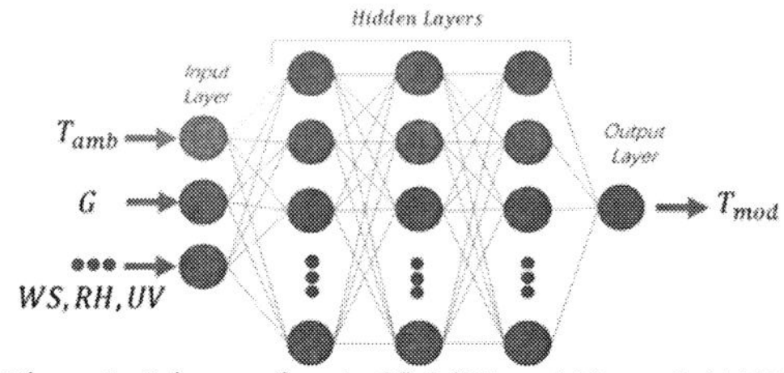

Figure 3. Scheme of an Artificial Neural Network (ANN) model applied to PV module temperature.

A highlight of this work is the use of the Light Gradient Boosting Model (LightGBM) [17] for the predictions of PV module temperature. In Figure 4, a schematic of this algorithm is presented. As stated in [16], LightGBM is a novel GBDT model that involves two novel techniques to deal with the problems of finding the optimal split point: gradient-based one-side sampling (GOSS) and exclusive feature bundling (EFB). The GOSS is applied to reduce the number of data instances, and EFB is used to reduce the feature space.

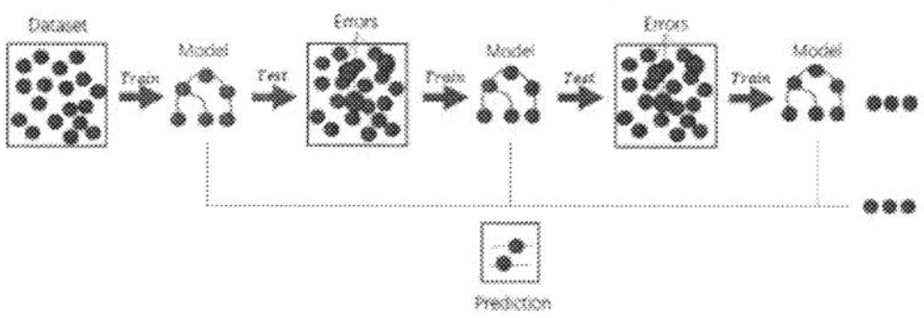

Figure 4. Scheme of the LighGBM model applied to PV module temperature.

In this work, by using ML approaches, we aim to identify new relevant weather variables to improve thermal models' accuracy. We look to a comprehensive set of weather variables: ambient temperature, relative humidity, ultraviolet irradiance, global irradiance, and wind speed. Those variables are combined in different Feature Set (FS) to predict the PV module temperature using Linear Regression (LR), ANN and LightGBM. The approach is repeated in three different locations: Gran Canaria - Spain, Negev - Israel, and Zugspitze – Germany, to check any climate correlationtion effects .

3 METHODOLOGY

3.1. Creation of feature sets and prediction errors using different models in different climates

Empirical models are calibrated using defined input variables (irradiance, temperature and wind speed), while machine learning approaches can be trained using any set of input variables, which in this paper are called Feature Sets (FS).

The combination of different weather parameters is combined in the FS shown in Table 1.

Table 1. Feature Sets (FS) used to train the machine learning approaches, where G is the plane-of-array irradiance, T_{amb} is the ambient temperature, WS is the wind speed, RH is the relative humidity, and UV is the ultraviolet irradiance.

	G	T_{amb}	WS	RH	UV
FS1	X	X			
FS2	X	X	X		
FS3	X	X		X	
FS4	X	X	X	X	
FS5	X	X	X		X
FS6		X	X		X
FS7		X			X
FS8		X		X	X

3.2. Numerical pipeline to estimate PV module temperature with physical and empirical models.

Figure 5 shows the pipeline from input data to the upgrade of physical models. The data collected in all three sites are read and parsed, followed by clean and filter stages.

Once the datasets are clean, considering all feature variables and target variables, the sets are split into training, testing and validation. The training sets will help (1) calibrate the physical models and (2) train the machine learning models.

Then, testing and validation datasets will be used to calculate the accuracies of the predictions, which will support the visualization and pitfall identification.

The outcome is expected to end up in the design or upgrade of physical models.

Finally, the new physical model is benchmarked against the previous physical and ML models.

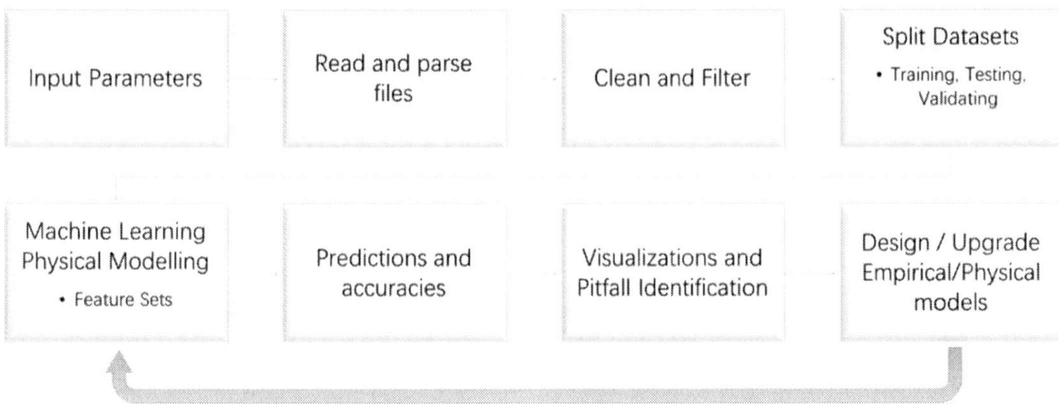

Figure 5. Flowchart of Thermal Modelling using Ross, Faiman, and King models.

4 RESULTS

4.1. Impact of different weather conditions on the PV module temperature prediction

In Figure 6, the physical models and machine learning models were benchmarked for predicting the PV module temperature under different weather conditions (i.e., climates):

- The system in Negev, Israel, operates in a Desert climate, showing more stable weather, with high temperatures and irradiances. The MBE shows that physical models get better results than machine learning.
- In Gran Canaria, Spain, the climate can be defined as Tropical, described as stable weather over the year but with sudden changes over days, including rapid wind speed changes. This location shows better results with physical and machine learning, including wind speed in their input feature sets.
- For Zugspitze, Germany, the climate is volatile due to the high altitude. In this location, the physical models cannot achieve the accuracies of the machine learning approaches.

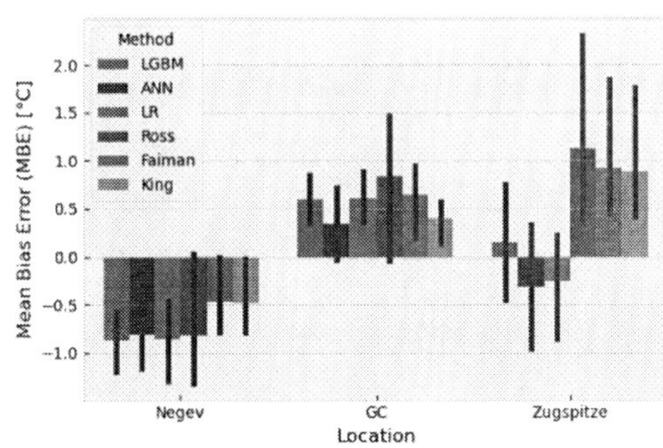

Figure 6. Mean-Bias-Error (MBE) of PV module temperature estimations using different physical and ML approaches in three locations.

Figure 7 shows the MBE versus the type of Feature Set used during the training stage. The outcomes of this study are listed here:

- The use of wind speed in physical models result in a great accuracy improvement.

- The lack of irradiance in the feature sets for machine learning shows unexpected, very high uncertainties.
- Including relative humidity or ultraviolet irradiance to WS, T_{amb} and G to the feature sets of machine learning show the best results of this study.

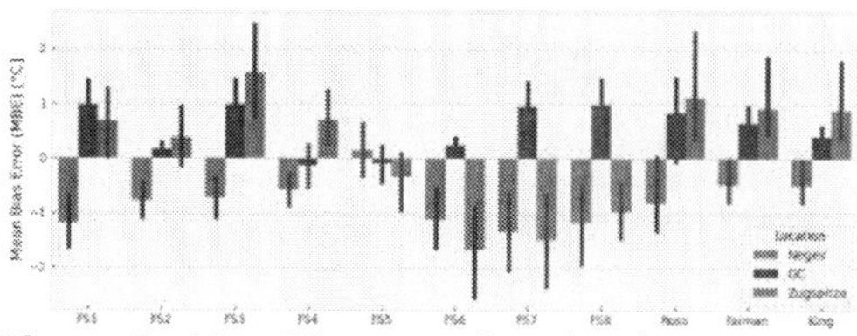

Figure 7. Mean-Bias-Error (MBE) of PV module temperature estimations using different feature sets in three locations.

4.2. Modified PV module temperature model

The outcomes of the benchmark shown above were sufficient to guide the upgrade of a physical model. The Faiman model has been considered with a new term related to ultraviolet irradiance. This parameter improves the predictions at lower temperature and irradiance values and includes a modelled UV irradiance as 5% of the PoA irradiance. The units of the term U_0 U_1 and U_2 are [W/°Cm²], [Ws/°Cm²] and [W.°C/m2] respectively. The proposed method is shown in Eq.1.

$$T_{mod} = T_{amb} + \frac{H_{poa}}{U_o + U_1 * Ws} - \frac{U_2}{1[W/m^2] + 0.05 * H_{poa}} \quad \text{Eq. 1}$$

4.3. Accuracy of physical models

The modified Faiman model is benchmarked to the Ross, Faiman and King models. Figure 8 shows the MBE of the models for the three different locations using different physical models. The proposed model gets better predictions in Zugspitze and comparable predictions in Negev and Gran Canaria. In Figure 9, the overall result for each location is shown, highlighting that the proposed model shows a MBE below 0.2 °C.

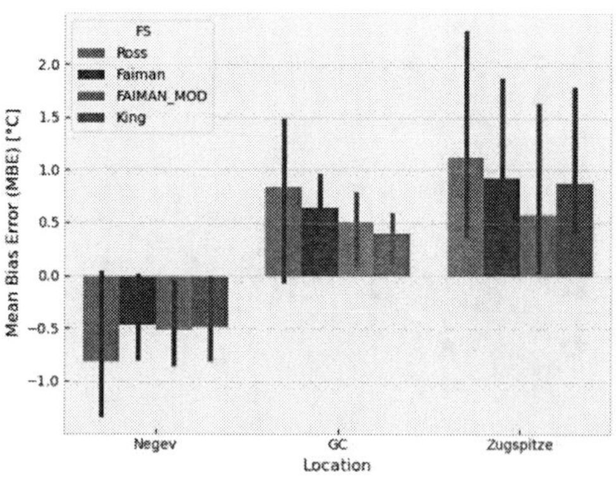

Figure 8. Mean-Bias-Error (MBE) of PV module temperature estimations using different physical models in three different locations.

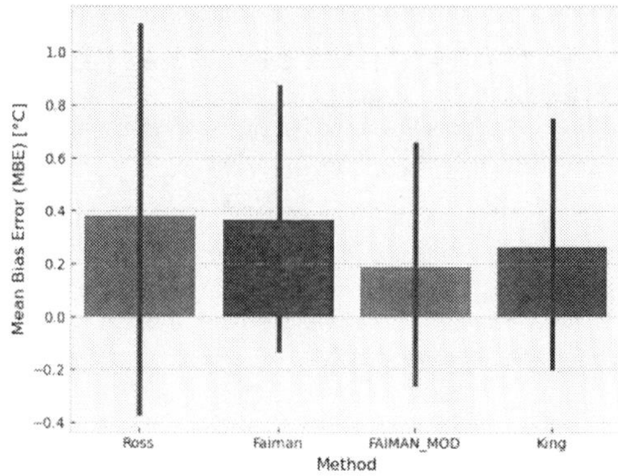

Figure 9. Mean-Bias-Error (MBE) of PV module temperature estimations using different physical models.

5 CONCLUSIONS

In this research, the benchmarking of physical models and machine learning approaches has been shown. The designed pipeline helped to understand and identify the potential improvements of physical models to predict PV module temperature.

The machine learning approaches were seen as a useful black box that provides accurate and fast results. However, using only ML, it is hard to understand the physical processes behind the numerical results. That is why a large number of feature sets, including different weather conditions, has been considered.

A new physical model has been proposed by extending the Faiman model and adding a new term as the modelled ultraviolet irradiance. On average, the proposed model gets better accuracies than Ross, Faiman and King models in all three climates.

REFERENCES

[1] J. Ascencio-Vásquez, K. Brecl, and M. Topič, "Methodology of Köppen-Geiger-Photovoltaic climate classification and implications to worldwide mapping of PV system performance," *Sol. Energy*, vol. 191, pp. 672–685, 2019, doi: https://doi.org/10.1016/j.solener.2019.08.072.

[2] J. Ascencio-Vásquez, I. Kaaya, K. Brecl, K.-A. Weiss, and M. Topič, "Global Climate Data Processing and Mapping of Degradation Mechanisms and Degradation Rates of PV Modules," *Energies*, vol. 12, no. 24, 2019, doi: 10.3390/en12244749.

[3] C. Schwingshackl *et al.*, "Wind Effect on PV Module Temperature: Analysis of Different Techniques for an Accurate Estimation," *Energy Procedia*, vol. 40, pp. 77–86, 2013, doi: https://doi.org/10.1016/j.egypro.2013.08.010.

[4] I. Kaaya, M. Koehl, A. P. Mehilli, S. de Cardona Mariano, and K. A. Weiss, "Modeling Outdoor Service Lifetime Prediction of PV Modules: Effects of Combined Climatic Stressors on PV Module Power Degradation," *IEEE J. Photovolt.*, vol. 9, no. 4, pp. 1105–1112, 2019, doi: 10.1109/JPHOTOV.2019.2916197.

[5] A. Bala Subramaniyan, R. Pan, J. Kuitche, and G. TamizhMani, "Quantification of Environmental Effects on PV Module Degradation: A Physics-Based Data-Driven Modeling Method," *IEEE J.*

Photovolt., vol. 8, no. 5, pp. 1289–1296, 2018, doi: 10.1109/JPHOTOV.2018.2850527.

[6] T. Dierauf, A. Growitz, S. Kurtz, J. L. B. Cruz, E. Riley, and C. Hansen, "Weather-Corrected Performance Ratio," Apr. 2013, doi: 10.2172/1078057.

[7] P. Mora Segado, J. Carretero, and M. Sidrach-de-Cardona, "Models to predict the operating temperature of different photovoltaic modules in outdoor conditions," *Prog. Photovolt. Res. Appl.*, vol. 23, no. 10, pp. 1267–1282, 2015, doi: https://doi.org/10.1002/pip.2549.

[8] M. Koehl, M. Heck, S. Wiesmeier, and J. Wirth, "Modeling of the nominal operating cell temperature based on outdoor weathering," *Sol. Energy Mater. Sol. Cells*, vol. 95, no. 7, pp. 1638–1646, 2011, doi: https://doi.org/10.1016/j.solmat.2011.01.020.

[9] R. A. Fajarin and E. A. Setiawan, "Analysis Corrected Performance Ratio on Photovoltaic Through Four Temperature Cell Model," *E3S Web Conf.*, vol. 67, p. 01026, 2018, doi: 10.1051/e3sconf/20186701026.

[10] I. Kaaya, J. Ascencio-Vásquez, K.-A. Weiss, and M. Topič, "Assessment of uncertainties and variations in PV modules degradation rates and lifetime predictions using physical models," *Sol. Energy*, vol. 218, pp. 354–367, 2021, doi: https://doi.org/10.1016/j.solener.2021.01.071.

[11] S. Lindig, J. Ascencio-Vásquez, J. Leloux, D. Moser, and A. Reinders, "Performance Analysis and Degradation of a Large Fleet of PV Systems," *IEEE J. Photovolt.*, vol. 11, no. 5, pp. 1312–1318, 2021, doi: 10.1109/JPHOTOV.2021.3093049.

[12] Ross, R.G., Jr, "Design Techniques for Flat-Plate Photovoltaic Arrays," in *Proceedings of the 15th IEEE Photovoltaic Specialists Conference*, Orlando, Florida, 1981, pp. 811–817.

[13] D. Faiman, "Assessing the outdoor operating temperature of photovoltaic modules," *Prog. Photovolt. Res. Appl.*, vol. 16, no. 4, pp. 307–315, 2008, doi: https://doi.org/10.1002/pip.813.

[14] J. A. Kratochvil, W. E. Boyson, and D. L. King, "Photovoltaic array performance model.," Aug. 2004, doi: 10.2172/919131.

[15] W. F. Holmgren, C. W. Hansen, and M. A. Mikofski, "pvlib python: a python package for modeling solar energy systems," *J. Open Source Softw.*, vol. 3, no. 29, p. 884, 2018, doi: 10.21105/joss.00884.

[16] J. Ascencio-Vásquez, J. Bevc, K. Reba, K. Brecl, M. Jankovec, and M. Topič, "Advanced PV Performance Modelling Based on Different Levels of Irradiance Data Accuracy," *Energies*, vol. 13, no. 9, 2020, doi: 10.3390/en13092166.

[17] G. Ke *et al.*, "LightGBM: A Highly Efficient Gradient Boosting Decision Tree," in *Proceedings of the 31st International Conference on Neural Information Processing Systems*, Red Hook, NY, USA, 2017, pp. 3149–3157.

ESTIMATING DAY AHEAD PHOTOVOLTAIC PRODUCTION DISTRIBUTION FUNCTION FOR THE RISK ASSESSMENT OF CONTROL ENERGY PROVISION.

Matthias Steinbrecher, Bernhard Kubicek, Marcus Rennhofer
AIT Austrian Institute of Technology
Vienna, Austria

ABSTRACT: To enable the potential business case of advanced energy market participation of photovoltaic generator systems, not only the day ahead forecast of PV production needs to be known, but also the likelihood of strong deviations from the predicted production. By using an analytic predictor to model PV production, that is then used as an additional input feature for black box machine learning, a generalized predictor can be created. By clustering forecastable ambient conditions, the typical deviations of the generalized predictor can be estimated in historic data, to estimate the likelihood distribution of forecasted future production.
Keywords: expert-system, grid stability, monitoring, large grid-connected PV systems, PV Market, Shading

1 INTRODUCTION

At the time of writing, large scale photovoltaic generator systems are established in most European countries. While the market is becoming more saturated, the question arises if grid support functions could be used as a secondary business case to further decrease amortization times.

While the phase shifting behavior of inverters can react to conditions with too much local production towards the grid (i.e. overvoltage) by phase shifts, on can also decrease grid feedin, if the non-local situation demands it. Furthermore, also a power reserve can be established, that would only be provided in the case of grid power demand. For this business case, the energy markets demand a day ahead promise of operating reserves, and fees need to be paid in case a promised reserve cannot be delivered.

This rises the demand for a time resolved day ahead production forecast, that also is able to estimate the likelihood of deviations.

2 DATA PREPROCESSING

In typical use cases, historical data is available of a PV system that needs to be forecast. The older the system is, the better its degradation can be integrated, however a minimum of one year is commonly mandatory so that the four seasons are covered. This monitoring data often contains problems that need to be understood prior to using that data for further evaluations, see Fig. 1.

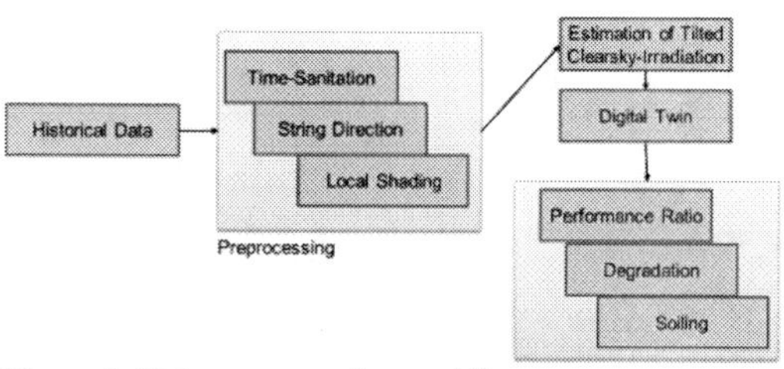

Figure 1: Data preprocessing workflow

It was found that the timestamps of a PV system often contain unreliable or incorrect time zone information, as well as problems with daylight-saving times, either through human failure or firmware problems. If no internet time server synchronization is used, gradual drifts of up to 30 minutes per year were observed, due to temperature dependence of clock quartzes resonant frequency.

This effects can be analyzed by comparing GPS-based sunrise and sunset times with data-derived counterparts, see Fig. 2. Additional problems arise as some monitored values might be averaged within a time interval (e.g. production power), while typically sensor values (e.g. irradiation) might be momentary values, causing an artificial shift of half the sampling period. Commonly, weather stations use a different time source than the ensemble of inverters.

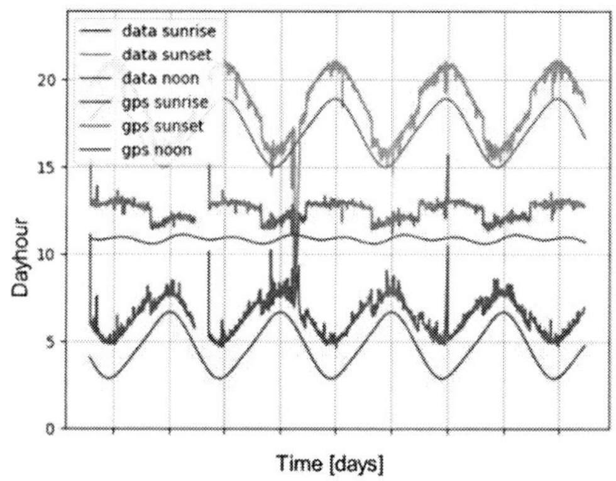

Figure 2: The detection of time zone, daylight-saving time shifting, or of a general drift of machine time is based on location dependent calculated astronomical sunrise/sunset times, as well as data derived equivalents.

If the individual strings' orientations are not know or unreliable, the orientation can be found using numerical optimization. This already requires corrected or reliable time stamps, as a half hour timeshift corresponds to 7.5° in the azimuth. Using a customized clearsky-condition detection, a data based map of the suns position can be created, that for discretized directions contains the temperature-corrected production power, see Fig. 3. The same can be calculated for arbitrary hypothetical module orientations, and the correlation accuracy can be used as optimization goal.

Furthermore, the comparison of observed power to calculated clearsky directional power defines a map that shows local shading, as well as regions where the possibly imperfect estimated orientation causes deviations, see Fig 4. This shading map can be used to include local and seasonal shading effects in the predictor.

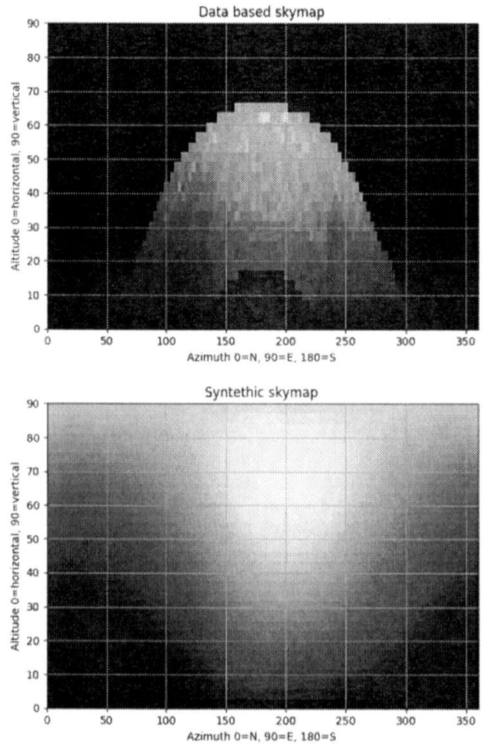

Figure 3: The data based clearsky irradiation map (top) is compared to maps based on various orientations (bottom). The orientation of the strings' modules is numerically obtain by optimization.

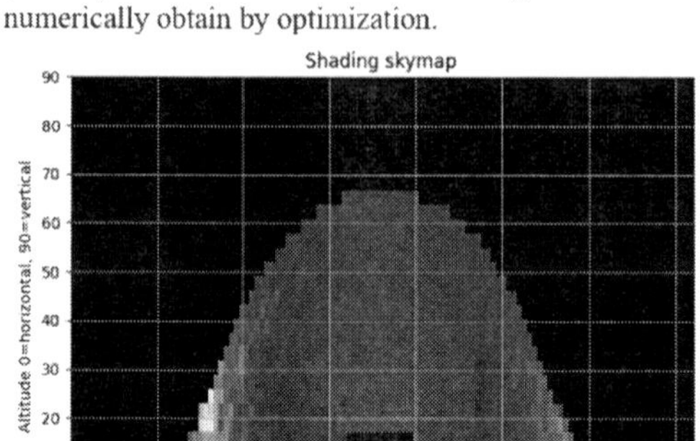

Figure 4: The ratio between synthetic clear sky and observed clearsky can be used to find local shading sources, in this case a wind turbine west (240°) of the PV system.

3 ANALYTIC PREDICTOR

As soon as the orientation of the system is known, on can use again numerical optimization to fit a one dimensional model of the systems AC production. This modle includes the load dependent inverter loss as well as the decreased low-light output of the PV modules. In Fig. 5 additional fit parameters for the maximum production power, the turn on power, the effective watt peak, and the nonlinearity are indicated. For this optimization, it is useful to give data points varying importance, so that volatile conditions as well as cloudy conditions are decreased in relevance for the numerical fit.

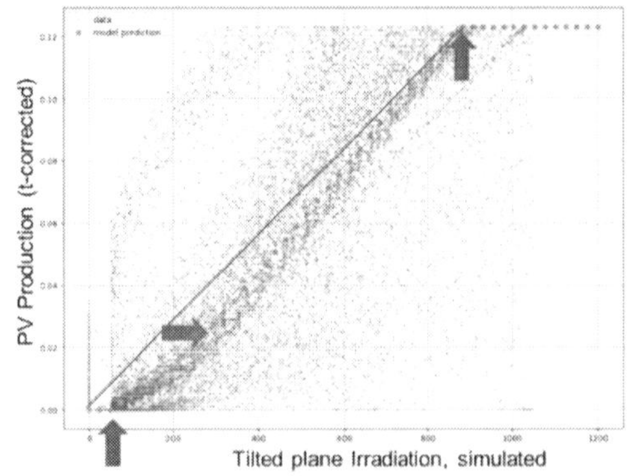

Figure 5: Using numerical optimization, the strings parameters (turn on point, low light decrease, Wp, clipping power) can be obtained. Thereby the deviation from clearsky and the intensity of second order change rate is used to decrease individual points optimization importance.

To check the analytic predictors accuracy, the historic data (or parts of a control sample) can be used as input. In Fig. 6, this exhibits very little deviations. This is explained thereby that the largest deviations would be caused by discrepancies in irradiance, while the local sensor obviously describes a perfect irradiation "nowcast".

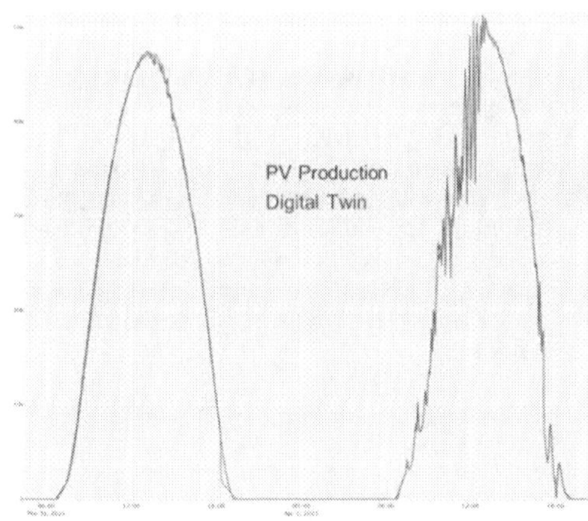

Figure 6: While performing a now-casting using historically collected data (including a local irradiation sensor), the accuracy of the numerical model is accurate.

4 MACHINE LEARNING

While the analytic predictor uses only few input parameters its prediction, many more parameters ("features") can be used to improve predictions. For forecasting purposes, features are only useful if the they can be predicted sufficiently. Such features include the phase within the seasonal cycle (seasonal soiling), the phase within the day (thermal mass behavior of the mounting location), the sun's sky altitude and azimuth, or the impact angle of the sun rays onto the module (i.e. angular dependent reflection and its complex soiling interaction).

These features can be used as input variables to create a machine learning system to do e.g. a black box predictor alternative to the analytic predictor. Commonly, more capacity for "intelligence" of such a system requires more internal unknown parameters that need to be fitted with an increasing amount of data and corresponding numerical complexity.

Hence it is beneficial to include the prediction of the analytic predictor as additional input feature of the machine learning system: A lot of the "heavy lifting" is already performed thereby, and the machine learning only needs to evaluate the residual deviations between reality and analytic predictor, that can be attributed to systematic changes (opposed to random effects). On one hand, this approach exhibits physical properties of the system useful for error detection. This enables small complexity machine learning, requiring less learning data and less computation time.
Multiple ML systems were compared. It was found that multidimensional linear regression is similarly outperformed by both shallow neural networks and linear regression trees. The latter two showed nearly identical prediction accuracy. The mean of the absolute error in the case of irradiation-perfect nowcasting was found to be 1.7% of the peak production power for the linear regression, while the other methods enabled 1.2-1.3% accuracy. These values do not translate directly into forecasting accuracy, as explained in the following. Detailed comparisons of the machine learning will be published soon [1].

5 SITUATION-AWARE ERROR FORECASTING
One can categorize operating conditions into arbitrary groups of tilted irradiations. For all conditions within an irradiation interval, the relative deviation of prediction and reality can be evaluated for all historic times. The probability density function (PDF) of these deviations can approximated by a histogram, see Fig 7. In this histogram, the uncertainty of the forecast in a given situation bin is shown by a typically asymmetric non-trivial distribution with "fat tails". In financial mathematics, this fat tails are the relevant parts to evaluate risk, as they describe the likelihood of strong deviations. Hence, also for the risk assessment of day ahead energy reserve provision, they need to be known and evaluated.

The situation bins can not only be defined by a one-dimensional clustering, but also by adding a second discrete dimension, e.g. using the module or ambient temperature. So far unevaluated, but also a metric describing the cloud-clustering might be useful, as the impact on the prediction risk is high [2].

An analytic predictor was used to facilitate black box machine learning. The total deviation of the combined predictor is analyzed for clustered ambient conditions. This allows to not only perform a prediction of the most likely system reaction to forecasted ambient conditions, but also to predict the PDF of the likely deviations. In Fig. 8 and 9, the PDF is visualized by the plus/minus one standard deviation.
In case the PDFs for the day ahead forecast of irradiation and temperature are known, by numerically convolution or by monte carlo sampling, the joint PDF of both digital twin and forecasting can be calculated.

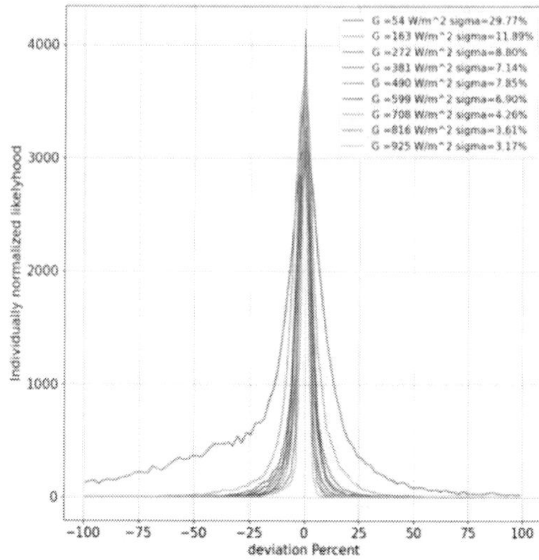

Figure 7: Differences between predictor and real data can be categorized by ambient situations, i.e. into irradiance groups. Within each group, a histogram exposes a systematic offset and the expected deviation distribution of the predictor.

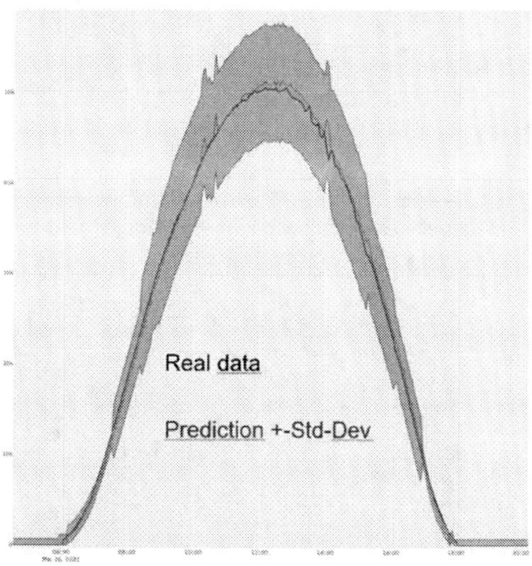

Figure 8: Comparison of the predictor and the real data on a sunny day: The one-standard deviation band is indicated in brown, although for each instance the full expected probability density (PDF) function is known.

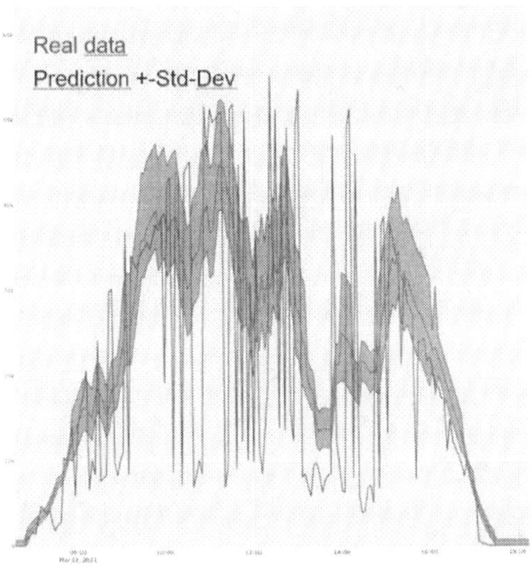

Figure 9: Comparison of the predictor and the real data on a volatile cloudy day: The one-standard deviation band is exceeded, as the events are rare and only explained by the tails of the PDF.

6 OUTLOOK and CONCLUSIONS

It was demonstrated that machine learning can be applied hand in hand with expert based models. This results in both physically interpretable, inter-compare-able system parameters expressed in analytic functions, and the ability to cover more complex systematic interactions using machine learning.

Aspects of this work were performed within the European horizon 2020 project under grant agreements 646039 and 775970.

While the project is still in early stages at the time of writing, system monitoring data, weather data and weather prediction data is currently accumulated, to hopefully soon estimate the accuracy of the total prediction, as well as the accuracy of the prediction of the deviation probability density function. Furthermore, the market models are currently modeled and will be published in due time.

[1] M. Steinbrecher, diploma thesis, technical university of Vienna, to be published, (2021).
[2] S. Stoyanov et al, "Fat-tailed models for risk estimation", Working Paper Series in Economics, Karlsruhe institute of technology, 30, (2011).

QUANTIFYING PERFORMANCE LOSS RATES OF PHOTOVOLTAIC MODULES USING GROUND-BASED VS SATELLITE-BASED METEOROLOGICAL DATA

Ebrar Özkalay[1,2*], Alessandro Virtuani[1], Andrew Fairbrother[1], Artur Skoczek[3], Gabi Friesen[2], Christophe Ballif[1,4]
1 – École Polytechnique Fédérale de Lausanne (EPFL), Institute of Electrical and Micro Engineering (IEM), Photovoltaics and Thin-Film Electronics Laboratory, Rue de la Maladière 71b, 2002 Neuchâtel (Switzerland)
2 – University of Applied Sciences and Arts of Southern Switzerland (SUPSI-PVLab), Via Flora Ruchat-Roncati 15, 6850 Mendrisio, Switzerland
3 -Solargis s.r.o., Mytna 48, 811 07 Bratislava (Slovakia)
4 – CSEM PV-Center, Rue de Jaquet-Droz 1, 2002 Neuchâtel (Switzerland)
Email: ebrar.ozkalay@supsi.ch

ABSTRACT: Accurate assessment of the long-term performance of photovoltaic (PV) systems is critical for manufacturers, investors, plant owners and O&M companies. Many PV systems, especially residential and some commercial/industrial systems, are not equipped with meteorological monitoring systems such as irradiance and temperature sensors. Whereas to calculate the performance ratio (PR), a common performance measure, the plane of array irradiance (G_{POA}) is an input parameter. The irradiance data is the largest contribution to uncertainty in PR. Therefore, in this study, we first evaluate satellite-based irradiance and compare to ground-based irradiance data. Next, we investigate the accuracy of using the satellite-based meteorological data in long-term performance analysis of three test modules in the absence of ground-based data. We use (1) ground-based and (2) satellite-based data separately as input irradiance. We calculate 60 PR time series and performance loss rates (PLR) for each irradiance data using different filtering methods and performance metrics. As a representative PLR value, we used the mean value (excluding outliers) of the various PLR values obtained as suggested by IEA PVPS Task 13. For the first module, this method gave almost identical PLR values (-0.5% error) when ground- and satellite-based data are used. For the other two modules, the values are not identical, but standard deviations largely overlap. We show that when the mean of PLR values is used, it is possible to accurately evaluate the long-term performance using satellite-based meteorological data.
Keywords: performance loss rate (PLR), ground-based meteorological data, satellite-based meteorological data, reliability, data filtering

1 INTRODUCTION

The long-term performance of photovoltaic (PV) systems is usually determined by assessing the system's performance ratio (PR), which corresponds to the measured energy yield divided by the reference yield, the latter depending on the plane of array irradiance (G_{POA}) and nameplate capacity. Temperature effects, soiling, cable, inverter, and shadowing losses are major parameters affecting the PR.

PV installations for research usually make use of ground-based monitoring systems, which include PV yield and meteorological monitoring (properly maintained irradiance and module temperature sensors) systems with high-frequency sampling. The meteorological parameters are significant in calculating PR and assessing the long-term performance of a PV system. In the past, residential PV systems (or even larger commercial systems) were not equipped with monitoring systems because of the added cost, whereas nowadays they increasingly make use of less sophisticated yield monitoring systems, but without any meteorological measurements such as irradiance and temperature. Utility-scale systems increasingly make use of well-equipped monitoring systems. Still, in several cases due to absence of the sensors, costs, poorly maintained or calibrated sensors, satellite-derived meteorological data or data form nearby meteorological ground stations prove to be a valuable alternative in assessing a PV plant's performance.

Data from ground weather stations can be accurate, but stations are sparsely located and thus not viable data sources for all PV systems. In such cases, satellite-derived meteorological data can be a valuable alternative in assessing a PV system's performance. However, the uncertainty of such data is generally much larger – depending on the temporal resolution (e.g. hour, day, month, and year) – than ground on-site data measured with properly maintained ground-based sensors [1]. Hence, the performance metric, PR, calculated using satellite-derived data and the assessment of the PV system's long-term performance tends to be less accurate. It is however known that when longer integration times are used to determine insolation (kWh/m^2) (e.g. hour, day, month, and year) the discrepancy between ground-based and satellite-based insolation data is considerably reduced, leading to a reduction in uncertainty.

In this work, the satellite-based G_{POA} at 20° tilt was evaluated and compared to ground-based measurements. Then, the long-term performance of three modules, monitored for 3.5 years, were analyzed using ground-based and satellite-based monitoring systems. Computed PLR values using different sources of meteorological data (ground- and satellite-based) were compared.

2 EXPERIMENTAL

2.1 Ground- and Satellite-based Meteorological Data

Table I provides a definition of the ground-based and satellite-based meteorological monitoring systems that have been used in this work. Ground G_{POA} (20° tilt and -4° azimuth from South) was monitored every minute between December 2015 and May 2019 by well-calibrated (and maintained) thermopile pyranometer installed on the outdoor test site of SUPSI, in Canobbio (Switzerland, 46.02°N and 8.91°E). The temperature of the modules were monitored using a Pt100 sensor attached on the rear side of the modules at 1-minute intervals during the outdoor deployment. Satellite-derived G_{POA} (20° tilt and -4° azimuth from South) and ambient temperature (at 2 meters height) data were retrieved with 5-minute frequency for the same duration of time.

The Solargis satellite-based solar radiation is calculated by numerical models, which are parameterized by a set of inputs characterizing the cloud transmittance, state of the atmosphere and terrain conditions. In the Solargis approach, the clear-sky irradiance is calculated by the simplified SOLIS model [2]. This model allows fast calculation of clear-sky irradiance from the set of input parameters. The clouds are the most influential factor, modulating clear-sky irradiance. The effect of clouds is calculated from the satellite data in the form of a cloud index (cloud transmittance). All-sky irradiance in each time step is calculated by coupling the clear-sky global horizontal irradiance with the cloud index. More information regarding the Solargis satellite-based data is presented in the report [3].

Table I. Definition of ground-based and satellite-based input meteorological data.

	Ground-based Monitoring System	**Satellite-based Monitoring System**
G_{POA}	On-site pyranometer (properly maintained and calibrated)	Satellite-derived
Module Temperature	Temperature sensor (back of module) (properly maintained and calibrated)	Calculated from estimated ambient temperature using Ross' approximation [4]

2.2 Performance Loss Rate Analysis

In this work the PLR of three crystalline silicon (c-Si) PV modules were calculated. The DC yields were monitored for almost four years between December 2015 and May 2019 at SUPSI, in Canobbio, Switzerland. The electrical performance parameters of the modules were acquired using maximum power point (MPP) trackers at 1-minute intervals. In this work, not a single PLR analysis method was applied, but various PLR analysis methods applied to each long-term performance of PV modules. Each PLR analysis method is a combination of data filtering method, performance metric and statistical analysis tool. We created 60 PLR analysis methods using 30 different data filtering methods and two different performance metrics.

2.2.1 Data Filtering

The data filtering methods were created using irradiance, power, instantaneous PR, power-irradiance relation and clear-sky filters with different thresholds (see Table II).

The irradiance threshold filter is one of the most used filters to remove anomalies due to measurement equipment, low irradiance conditions (e.g. cloud coverage or nighttime), and high irradiance conditions (e.g. extreme irradiance due to cloud reflections). Four different low irradiance thresholds (50 W/m^2, 100 W/m^2, 200 W/m^2 and 500 W/m^2) were used here. The high irradiance threshold was kept constant at 1250 W/m^2.

The power threshold filter is another commonly adopted filter similar to the irradiance threshold filter. Data removal is done based on the nominal power (P_{nom}) at standard test condition (STC). To remove failures in power measurement (P_{meas}), the low power threshold, which can be 1% of the nominal power, is commonly used. Conversely, 102% or 120% of the nominal capacity were used as the high power thresholds.

The next filter removes data depending on the instantaneous PR (PR_{ins}) which is 5-minute in this study. The standard deviation of PR_{ins} is calculated and the less likely PR values are removed using one and two standard deviations ($\pm1\sigma$ and $\pm2\sigma$) as thresholds. In addition, another threshold method was used to remove very low and high PR_{ins} outside the range 0.3 and 1.2 [5].

The power-irradiance filter removes unlikely measurement points using the non-linear relationship between the power and the irradiance [6]. In addition, a clear sky filter, RdTools' Clear Sky Index (CSI) filter was applied [7]. The filtered data is here referred to as "CSI". The corresponding unfiltered dataset is referred to as "All Days".

Table II. Summary of the data filtering techniques used in this work. The parameters changed are written in bold.

Irradiance Threshold	**ClearSky**
• **50** W/m^2 < G_{POA} < 1250 W/m^2 • **100** W/m^2 < G_{POA} < 1250 W/m^2 • **200** W/m^2 < G_{POA} < 1250 W/m^2 • **500** W/m^2 < G_{POA} < 1250 W/m^2	• RdTools' CSI [7]
Power Threshold	**Power - Irradiance** [6]
• 0.01 x P_{nom} < P_{meas} < **1.2** x P_{nom} • 0.01 x P_{nom} < P_{meas} < **1.02** x P_{nom}	• LQ: **20th**, UQ: **80th** • LQ: **10th**, UQ: **90th** • LQ: **5th**, UQ: **95th**
Instantaneous PR (5-minute)	
• $\pm$**1**σ • $\pm$**2**σ • **0.3** < PR_{ins} < **1.2**	
G_{POA}: Plane of array irradiance, P_{nom}: Nominal power, P_{meas}: measured power, PR_{ins}: instantaneous PR, LQ: Lower quantile, and UQ: Upper quantile.	

We initially created 15 data filtering methods by combining four of the above-mentioned data filtering approaches (irradiance threshold, power threshold, power-irradiance and instantaneous PR filters). Then, each data filtering method was made with and without CSI clear sky filtered input data for G_{POA}. Hence, in total, 30 different filtering methods were created for the PLR analysis.

2.2.2 Performance Metrics and Statistical Analysis Method

The monthly aggregated PR and temperature corrected PR (PR_{cor}) were used as the main performance metrics in the long-term performance analysis of the three PV modules installed at SUPSI [8]. PR is calculated from the ratio between the energy yield of an array (Y_f) and reference yield calculated over a given temporal interval (Y_r) [9].

The module temperature is not available when satellite-derived meteorological inputs are used. For this reason, the Ross model [4] was utilized to estimate the temperature of the modules starting from the available satellite-based ambient temperature data. The following approximate equation is used to model the cell temperature (T_{cell}):

$$T_{cell} = T_{amb} + \frac{NOCT_{mpp} - 20^{\circ}C}{800\ W/m^2} * G_{POA} \quad (1)$$

where T_{amb} is the satellite-based ambient temperature (at 2 meters height) and $NOCT_{mpp}$ is the nominal operating cell temperature under load conditions. More advanced temperature models, which use ambient temperature and wind speed together to improve the accuracy, were not used since satellite-based wind data were not available.

A statistical method, the seasonal and trend decomposition using LOESS (STL), was used to compute PLR since it is insensitive to outliers and seasonality [10, 11].

3 RESULTS AND DISCUSSION

3.1 Evaluation of Satellite-based Irradiance Data

We evaluated the satellite-based irradiance data with respect to ground-based data. The G_{POA} values below 50 W/m^2 and above 1250 W/m^2 were filtered out to exclude anomalies. The time-series of normalized bias of satellite-derived G_{POA} ($G_{POA,satellite}$) for monthly temporal resolution was calculated using the following formulas with respect to ground-based G_{POA} ($G_{POA,ground}$):

$$\text{Norm.Bias}_{G_{POA}}[\%] = \frac{G_{POA,satellite} - G_{POA,ground}}{G_{POA,ground}} \times 100 \qquad (2)$$

Fig. 1 shows monthly normalized bias series of $G_{POA,satellite}$. The monthly $G_{POA,satellite}$ is overestimated in summers and underestimated in winters (sinusoidal-like trend). CSI clear sky filtering did not have significant impact on the accuracy of monthly $G_{POA,satellite}$.

3.2 Performance Loss Rate Analysis

Fig. 2 shows a conceptual flowchart of our PLR analysis approach, including data filtering, performance metrics and statistical method. Initially, long-term PV performance data was merged with satellite-based monitoring data. Then, 30 different data filtering methods were applied, and 30 time-series were created. For each time-series, monthly aggregated PR and PR_{cor} metrics were calculated. This results in 60 monthly performance metric time-series in total (30 for PR and 30 for PR_{cor} series). Then, the STL statistical method was applied on PR and PR_{cor} time-series to compute PLR. Therefore, 60 PLR using satellite-based monitoring ($PLR_{satellite}$) were calculated for each PV performance data. The same approach was applied using ground-based meteorological data as well. Hence, 60 PLR using ground-based monitoring (PLR_{ground}) were created for each PV performance data in addition to 60 $PLR_{satellite}$.

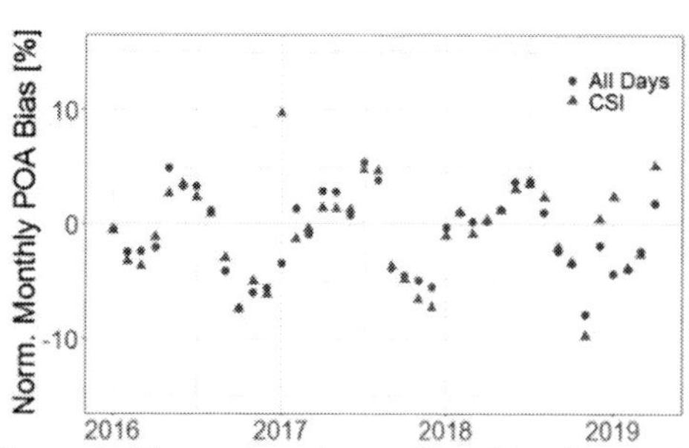

Figure 1. Time-series of normalized bias for monthly aggregated satellite-derived G_{POA} with respect to ground-measured insolation ("All days": no clear sky filter and "CSI": RdTools' clear sky filter).

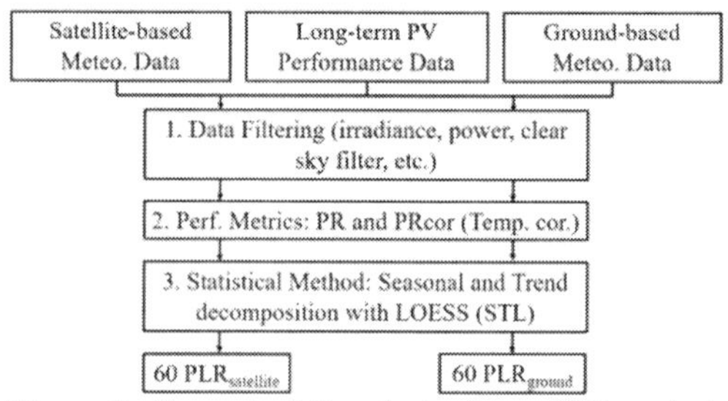

Figure 2. Conceptual flowchart of our PLR analysis approach including data filtering, performance metric computation and statistical method steps.

As previously mentioned, three PV modules were monitored for 3.5 years at SUPSI. For each PV module, 60 $PLR_{satellite}$ and 60 PLR_{ground} values were computed by following the methods summarized in Fig. 2. Fig. 3a shows the 60 PLR_{ground} values of one of the monitored modules (module-3). The PLR_{ground} values are mostly similar, but we observe some outliers. PLR_{ground} values outside a band of $\pm 2\sigma$ (standard deviation) were excluded to determine $PLR_{ref,ground}$ which is the mean of the remaining PLR_{ground} values (Fig. 3a). Using the mean of the different PLR_{ground} values as a reference ($PLR_{ref,ground}$) is an approach proposed in the work of IEA PVPS Task 13 [12].

Fig. 3b shows the 60 $PLR_{satellite}$ values with the $PLR_{ref,ground}$ for the same module, module-3. The $PLR_{satellite}$ values are more scattered than the PLR_{ground} values, which may be due to uncertainty of satellite data and temperature modelling. The process was repeated for the other two modules as well (not shown). The PLR methods leading to outliers were not the same for the all the three modules.

It is difficult to establish a uniform way of calculating PLR using satellite-based data, which is suitable to evaluate long-term performance of all the modules. In this work, we could not find a single uniform way to compute PLR for all the three modules when satellite-based meteorological data is used. Therefore, the method explained previously to determine the $PLR_{ref,ground}$ using PLR_{ground} values was utilized to determine a single representative $PLR_{satellite}$ ($PLR_{ref,satellite}$) from the computed $PLR_{satellite}$ values using various data filtering methods and performance metrics. The $PLR_{satellite}$ values outside a band of $\pm 2\sigma$ were removed and the mean of the remaining $PLR_{satellite}$ values represents the $PLR_{ref,satellite}$.

Fig. 4 shows the $PLR_{ref,ground}$ and $PLR_{ref,satellite}$ of each module. For module-3, the $PLR_{ref,ground}$ and $PLR_{ref,satellite}$ are almost identical whereas for module-1 and module-2, $PLR_{ref,satellite}$ values fall slightly outside of the 1σ band surrounding $PLR_{ref,ground}$. However, there are overlaps between the standard deviations computed for $PLR_{ref,ground}$ and $PLR_{ref,satellite}$ for the both modules. Considering the relatively large uncertainty associated to outdoor measurements, satellite-derived insolation data and long-term performance analysis, we do not spot major differences between the $PLR_{ref,ground}$ and $PLR_{ref,satellite}$ calculated for each module. This demonstrates that an accurate long-term performance assessment using satellite-based monitored data instead of ground-based monitored data is in principle possible by computing the mean of various PLR values obtained from different data filtering methods and performance metrics.

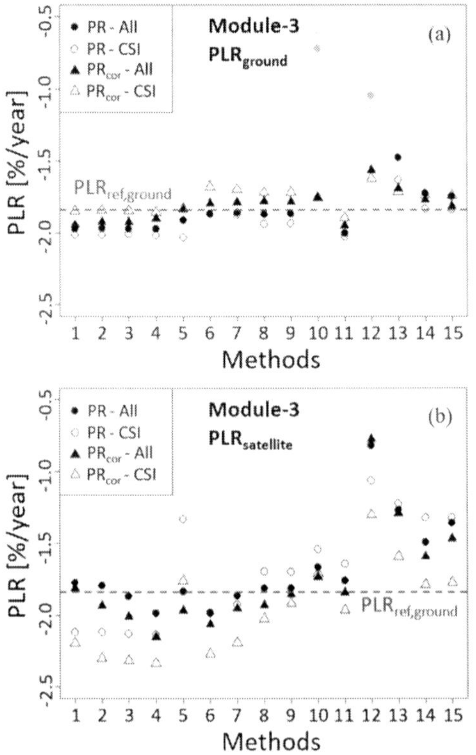

Figure 3. (a) PLR_{ground} values with $PLR_{ref,ground}$ (red dashed line), the mean of the remaining (without outliers) PLR_{ground} values (dark) for module-3. Gray PLR_{ground} values are outliers. (b) 60 $PLR_{satellite}$ values with $PLR_{ref,ground}$ for module-3.

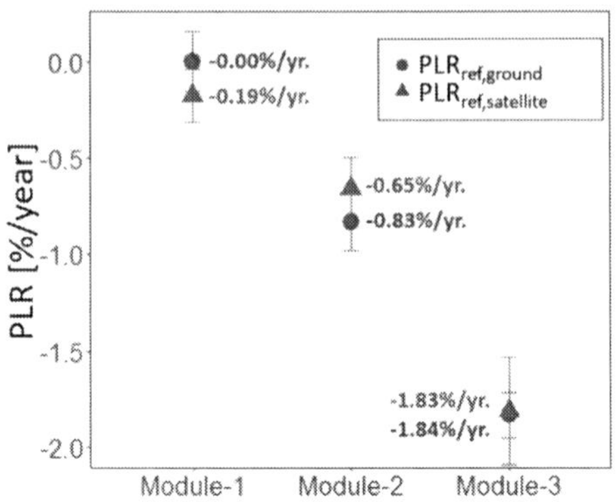

Figure 4. Reference performance loss rates using ground- and satellite-based monitoring, $PLR_{ref,ground}$ and $PLR_{ref,satellite}$, of the three modules with their standard deviation.

4 CONCLUSIONS

In this work we found that satellite-derived insolation data have seasonal bias compared to the reference measurements performed with ground-based irradiance sensors, leading to underestimation in winters and overestimation in summers. Consequently, the monthly performance ratio using satellite-based monitoring, $PR_{satellite}$, is underestimated in summers and overestimated in winters with respect to the monthly performance ratio using ground-based monitoring, PR_{ground}.

The long-term performances of three modules monitored for 3.5 years were analyzed using the set of different PLR analysis methods, which have different data filtering methods (combination of various filtering approaches such as irradiance, power thresholds, clear sky filters, etc.) and use temperature corrected and not corrected PR as the performance metrics. Averaging PLR values computed using various methods is an approach proposed in the work of IEA PVPS Task 13. The mean of a set of PLR values computed using satellite-based monitoring gave an accurate PLR with respect to the ground-based mean PLR. For one of the modules, this method gave almost identical mean PLR values (-0.5% error) when ground- and satellite-based data are used. For the other two modules, the satellite-based mean PLRs were slightly outside of the one standard deviation band of the corresponding ground-based mean PLRs. However, the standard deviations estimated for ground- and satellite-based mean PLRs for the two modules overlap. Considering the various uncertainty sources such as outdoor measurement and satellite-based meteorological monitoring, there is no major difference between the mean PLRs computed using different meteorological monitoring systems. This work shows that assessing long-term performance using satellite-based meteorological data can be possible if the mean of PLR values from various methods is used instead of using a single PLR analysis method.

5 ACKNOWLEDGEMENTS

This work was supported in part by the Swiss National Science Foundation under COST IZCOZ0_182967. The authors would like to thank SUPSI PVLab (University of Applied Sciences and Arts of Southern Switzerland), the researchers of the BIPV-Innovative Building Skin Team, and all the staff members who contributed in supplying the relevant data for this work. Additional (collaborative) support was provided by the COST Action PEARL PV (CA16235).

6 REFERENCES

[1] P. Ineichen, "Long term satellite global, beam and diffuse irradiance validation," in *International Conference on Solar Heating and Cooling for Buildings and Industry*, 2013.

[2] P. Ineichen, "A broadband simplified version of the Solis clear sky model," *Solar Energy*, vol. 82, no. 8, pp. 758-762, 2008.

[3] T. Cebecauer, B. Schnierer, A. Skoczek, N. Suriova, J. Betak, V. Madlenakova and M. Suri, "GIS data layers of Global Horizontal Irradiation, Air Temperature and Photovoltaic Electricity Potential," Solargis, 2017.

[4] M. Mattei, G. Notton, C. Crisfari, M. Muselli and P. Poggi, "Calculation of the polycrystalline PV module temperature using a simple method of energy alance," *Renewable Energy*, pp. 553-567, 2006.

[5] S. Lindig, D. Moser, B. Muller, K. Kiefer and M. Topic, "Application of dynamic multi-step performance loss algorithm," in *IEEE Photovoltaic Specialists Conference*, 2020.

[6] J. Ascencio-Vásquez, J. Bevc, K. Reba, K. Brecl, M. Jankovec and M. Topič, "Advanced PV performance modelling based on different levels of irradiance data accuracy," *Energies*, 2020.

[7] D. C. Jordan, C. Deline, S. R. Kurtz, G. M. Kimball and M. Anderson, "Robust PV degradation methodology and application," *IEEE Journal of Photovoltaics*, vol. 8, pp. 523-531, 2018.

[8] IEA-PVPS Task13, "Photovoltaic Module Energy Yield Measurements: Existing Approaches and Best Practice," International Energy Agency, 2018.

[9] IEC, "IEC61724:1998 Photovoltaic System Performance Monitoring—Guidelines for Measurement, Data Exchange and Analysis," International Electrotechnical Comission, Geneva, 1998.

[10] R. Cleveland, W. Cleveland, J. McRae and I. Terpenning, "STL: A Seasonal-Trend Decomposition Procedure Based on Loess," *Journal of Official Statistics*, 1990.

[11] S. Lindig, I. Kaaya, K. A. Weis, D. Moser and M. Topic, "Review of Statistical and Analytical Degradation Models for Photovoltaic Modules and Systems as Well as Related Improvements," *IEEE Journal of Photovoltaics*, vol. 8, pp. 1773-1786, 2018.

[12] S. Lindig, D. Moser, A. J. Curran, K. Rath, A. Khalilnejad, R. H. French, M. Herz, B. Müller, G. Makrides, G. Georghiou, A. Livera, M. Richter, J. Ascencio-Vásquez, M. van Iseghem, M. Meftah, D. Jordan, C. Deline, W. van Sark, J. S. Stein, M. Theristis, F. Baumgartner and W. Luo, "International collaboration framework for the calculation of performance loss rates: Data quality, benchmarks, and trends (towards a uniform methodology)," *Progress in Photovoltaics: Research and Applications*, vol. 29, no. 6, p. 573–602, 2021.

QUALITY ASSURANCE OF THE PHOTOVOLTAIC POWER PLANTS INSTALLATION STAGE – A COMPLEMENTARY STRATEGY BASED OF PHOTOLUMINESCENCE AND STEADY-STATE THERMOGRAPHY

Lukas Koester[1,2,*], Alexander Astigarraga[1], Sascha Lindig[1], Atse Louwen[1], Marco Antinori[3], David Moser[1] and Giampaolo Manzolini[2]
[1]Institute for Renewable Energies - EURAC Research, Viale Druso 1, 39100 Bolzano, BZ, Italy
[2]Dipartimento di Energia, Politecnico di Milano, Via Lambruschini 4, 20156 Milano, MI, Italy
[3]Viridis Energia Srl, Via Galvani 24, 20124 Milano, MI, Italy
*Phone: +390471055715 ; E-mail: lukas.koester@eurac.edu, alexander.astigarraga@eurac.edu, sascha.lindig@eurac.edu, atse.louwen@eurac.edu, marco.antinori@viridisenergia.com, david.moser@eurac.edu, giampaolo.manzolini@polimi.it

ABSTRACT: Procurement, installation and operation are different phases of a photovoltaic (PV) plant life cycle. Each of these phases come with different risks. These risks and their possible effects are shown in this paper with particular focus on PV modules, explained along the real construction of a PV plant in northern Italy. The quality is tested during all phases; before procurement to ensure module quality, after transport and installation to detect the impact of wrong handling, and during the operational phase. Electroluminescence, current-voltage curve measurements, infrared thermography and daylight photoluminescence were performed. Electroluminescence and current-voltage curve measurements were carried out during procurement and after transport and installation to evaluate the impact of module mishandling. The damage wrong handling of a PV module can cause is shown. During the first year of operation, two infrared thermography inspections were performed to assess the health status of the PV plant and to detect possible further degradation modes, which have not been noticed before. The development of hot spots over half a year is analyzed and severe cell cracks were found as root causes, detected through daylight photoluminescence imagery. Further, the applicability of daylight photoluminescence and the detectability of degradation modes in images of two varying distances is investigated.
Keywords: Field inspection, Thermography, Photoluminescence, Electroluminescence

1 INTRODUCTION

Along the life cycle of a photovoltaic (PV) power plant different phases can be defined. Development, design, installation, operation, and decommissioning. The solar bankability project [1] defined technical risks which can occur along these phases. Risks during the installation phase are, for example, bad planning and procurement of components and wrong handling during installation or operation. This can lead to degradation or failures. The goal is to identify all initial risks, prevent and reduce them to a minimum number.

Among the 20 most common risks identified by solar bankability project [2] two are: inadequate component testing during the procurement phase and incorrect degradation estimation over time. While the first one can be avoided, the latter might be considered correctly but can change due to mishandling in the transport and installation phase, resulting in yields differing from planned. Further, transportation and handling, especially during construction of the components, are identified as common risks.

The technical risks the PV module experiences during the different phases of its value chain might result in precursors or root causes of failures becoming evident only in a later stage. Typically, the focus on PV module defect detection is on the operational phase. There, degradation modes (DM) or faults may appear due to several years of environmental and mechanical stressors between others. However, the effects of transport and installation, where mishandling can lead to direct failures or increased degradation over time [3], are often underestimated.

Several inspection methods are available for detection of PV module DM and failures. While current-voltage (I-V) curves directly show the electrical behavior of a PV panel and quantify resulting power losses, it is not feasible to perform on every module in a PV plant. Regarding visual inspection (VI), it could be done much faster, but only allows the detection of visible DM in an advanced stage. Therefore, the common practice for O&M in PV plants results in a yearly infrared thermography (IR-T) inspection [4] as the main quality assurance method allowing the detection of several DM [5]. Conversely, as it only shows thermal anomalies in the modules, it often does not allow the accurate identification of the root causes, which are necessary for warranty claims. Electroluminescence (EL) is another characterization method allowing the detection of several DM in detail [6] and is usually performed on conspicuous modules identified during IR-T inspection. Depending on the number of modules and related costs, it could be done either by sending the modules to a laboratory or by EL night [7], daylight [8] or mobile laboratory [9] inspection in the field.

This paper describes the analysis of a 1 MWp PV plant in central north Italy during its procurement, installation and first year of operational phase, focused on the PV modules. A detailed verification of module quality was done before procurement using I-V curve measurements and EL imaging. Mishandling during transport and installation was observed and resulted in several issues. Modules affected during these steps were identified and got analyzed separately by I-V and EL four months after installation. The first goal was to detect other mishandled modules and later to get the health status of the whole PV plant. In total, three IR inspections were performed in the first year. In the present case, instead of EL, daylight photoluminescence (dPL) [10]–[12] was applied on the field. This novel method is based on a similar principle as EL. dPL is performed on detected hot spots (HS) during IR-T inspection to get more detailed information. Further, a share of arbitrarily chosen modules is inspected by dPL to get a general idea of the PV plant status. All inspections performed are listed in **Table 1** which gives further information about time, particular reasons and number of modules investigated during each inspection. Finally, the performance ratio (PR) is calculated on a monthly average for the 1.5 years of inspections.

The paper is structured in six sections. Section 2 provides PV plant information and detailed description of performed characterization techniques with special attention on the dPL methodology. Afterwards, in section 3, the inspection results are summarized followed by the conclusions obtained from these investigations in section 4. Section 5 analyzes the usability of dPL especially in terms of image quality for different distances. The paper is finalized with a discussion in section 6, which also includes an outlook on the future of dPL, EL and IR-T as inspection methods for quality assurance.

Table 1: Timetable of inspections performed. Specified is time in months relatively to installation of the PV modules, the reason for the respective inspection and the number of modules under inspection. Abbreviations used: I-V: current-voltage curve, EL: electroluminescence, IR-T: infrared thermography, UAV: unmanned aerial vehicle, dPL: daylight Photoluminescence

Time [months]	Inspection	Reason	Modules
-1	I-V + EL	Before procurement	30
0	Visual	After transport	All
1	IR-T (ground)	After installation	All
4	I-V + EL	Transportation Hot spots Visual issues	20 10 9
5	IR-T (ground)	Check by owner	All
11	IR-T (UAV)	Annual check by O&M	All
16	dPL	Details of HS + arbitrary share of plant	214

2 PV PLANT AND INSPECTION METHODS

The conducted inspections were performed by different suppliers and in different conditions. Each process is explained below.

2.1 PV Plant

The PV plant under investigation is in central north Italy, in a humid subtropical climate. It consists of around 3000 c-Si modules with passivated emitter and rear cell (PERC) technology. Each module is rated by the producer with 310 W nominal power, resulting in a total PV plant power of 0.96 MWp. The modules are installed in double rows and portrait orientation at 25° tilt. Installation date was June 2019, and the plant is equipped with a monitoring system measuring plane of array irradiance, ambient and module temperature.

2.2 Current-Voltage Curve

The I-V curves were recorded in laboratory conditions with a solar simulator rated class A (IEC 60904-9 [13]). The flash testing was performed following IEC 60904-1 [14]. Due to high capacity of the PERC modules, the I-V curve of each module was conducted by combination of multi-flash measurements consisting of 20 flashes. The solar simulator is evaluated with an uncertainty of 2.8% on maximum power $P_{\max}$. All power values given in the paper are $P_{\max}$.

2.3 Infrared Thermography

IR-T describes the acquisition of a spatially resolved thermal image of a PV module or plant under steady-state operating conditions with high irradiances [6]. Several DM result in temperature increase of different patterns like spots in a cell (*type-C* cell cracks, i.e. broken cell), whole cells (shadowing, delaminated cell), or even whole module parts (e.g. short-circuited bypass diode) [6].

In total, of three IR-T measurements were taken. Two were performed from ground using hand-held IR cameras: the first immediately after installation with irradiances around 800 W/m² and the second after 5 months of operation with irradiances around 500 W/m². The third IR-T was performed as an aerial proximity survey 11 months after installation using an unmanned aerial vehicle (UAV) at approximately 600 W/m². All IR-T measurements were performed by third parties. The results are taken from the compiled reports showing the temperature difference ΔT of the HS compared to the rest of the module. It is assumed that the emissivity was measured and HS calculated correctly in contrast to the average module temperature.

2.4 Electroluminescence

The principle of EL is to use a PV module as a light emitting diode by applying a forward current to it. This results in radiative recombination of carriers in the Si, thus photon emission in the short-wave infrared regime. This luminescence signal can be captured and allows statements about the series resistance of the solar cells and the detection of degradation modes like cell cracks, connection problems, shunts, and potential induced degradation [15].

The EL images were recorded in laboratory following IEC 60904-13 [15] using a 640x512 pixel InGaAs camera in dark conditions. For the first measurement cycle, only EL images were recorded. For the second cycle (inspection of modules after transport/installation) background subtraction was performed and in addition to the short circuit current (I_{sc}) image, a $0.1 \times I_{SC}$ image was recorded.

Cell cracks detected in the EL images are valued following the standardized cell characterization [15]: Micro cracks not influencing the signal intensity of neighboring cell areas are rated as *type A* cracks, while cracks reducing the intensity of a cell part to zero in the I_{SC} as well as in the $0.1 \times I_{SC}$ image are considered as *type C*. Finally, if the area shows luminescence signal in the low-current image, it is a *type B* crack.

2.5 Daylight Photoluminescence

Like EL, dPL works by capturing radiative emission from the solar cell Si. Instead of back powering the module, excitation is triggered by the exposure to sunlight. Compared to the strong sun irradiance, the share of luminescence signal is not visible without subtraction of a background image free of luminescence radiation. For acceptable image quality and signal-to-noise ratio (SNR), several images of each state are recorded and averaged [10]. The emission of luminescence signal can be altered by changing the operation point of the PV module between open circuit (OC) and one with high current flow:

maximum power (MP) or short circuit (SC), for example. The high currents in the latter prevent carriers from participating in radiative recombination, while OC has no current flow and forces the carrier to recombine.

Altering the operating points can be done by different methods [11], [12], [16]. As the whole process of dPL is novel, no standard exists describing the best practice. The method used here was done by manually switching off the whole PV string at combiner box (CB) level and recording 50 images of each operating point batched, as described in [12]. Therefore, the complete process is done manually and by 2 persons: one controlling and moving the camera setup, the other switching the CB. To compare DM detectability and imaging speed, a combination of overview (four modules per image) and close-up (one module per image) images are performed and will be evaluated in this work. The spatial resolution plays an important role as InGaAs sensor cameras have relatively low resolutions.

The image quality is evaluated by calculating the SNR. The high number of images needs adjustment of the SNR_{50} calculation suggested for EL images in [15], resulting in SNR_{AVG} derived by Mantel et al. [17] for daylight EL. This equation is here used for dPL:

$$SNR_{AVG} = \frac{\sum_k \left(avg_{i \in N}(OC_i(k) - MP_i(k))\right)}{\sum_k \left(\left|avg_{i \in N_1}(OC_i(k)) - avg_{j \in N_2}\left(OC_j(k)\right)\right| \cdot K\right)} \quad (1)$$

With $K = \sqrt{0.5} \cdot \left(\frac{2}{\pi}\right)^{-0.5}$, k the pixel index, N the numbers of pictures taken for each working point, N_1 the odd numbers and N_2 the even numbers of pictures taken at OC, with pixel intensities $OC_i(k)$ and $MP_i(k)$ for each pixel k.

For dPL images a distinction between *type B* and *type C* cell cracks is not possible by recording a low-current image. Therefore, the analysis treats black cell areas as *type C* cracks, while areas with only slightly reduced luminescence intensity next to cell cracks are defined as *type B* cracks.

2.6 Performance Ratio

The performance ratio (PR) sets the final yield (Y_f) of a PV plant in relation to its reference yield (Y_r). It is thereby an indicator of the PV plant performance quality [18]:

$$PR = \frac{Y_f}{Y_r} = \frac{E_{\text{ac}} \cdot G_{\text{STC}}}{G_{\text{POA}} \cdot P_{\text{N}} \cdot uptime} \quad (2)$$

Here, E_{ac} is the output energy on AC side, G_{STC} the irradiance under standard test conditions (i.e. 1000 W/m²), G_{POA} the plane-of-array irradiance and P_{N} is the nominal power of the PV plant. Additionally to [18], $uptime$ is included as availability factor of the PV plant. This is usually covered by filtering irradiance data for instances where no power measurements were available. In the present case, all values were given on a monthly average, resulting in a monthly calculated PR. G_{POA} was measured by the pyranometer installed on site.

2.7 Self-Shading Simulation

To estimate the effect of self-shading on the plant energy production, a digital version of the PV plant is simulated in Python. Module specifications taken from the data sheet and the irradiance data taken from ERA5 reanalysis climate data [19] feed this simulation. The lost irradiance is calculated, which is directly relatable to losses in generated electricity. As the modules are installed in portrait, it is assumed that the captured direct irradiance is reduced by the proportion the first solar cell row in a module is shaded. As soon as the first cell row is completely shaded, the whole module is assumed to lose all direct irradiance for electricity generation, and thus only diffuse irradiance is considered. In the whole PV plant, the upper and lower modules in one row belong to different electrical strings. Therefore, the captured irradiance in the upper row is reduced only if reached by shadows. For better comparability, the share of captured irradiance is calculated as 100 % of irradiance minus the lost one in 15 minutes resolution.

3 INSPECTION RESULTS

The results are given in chronological order following the inspections stated in **Table 1** but summarized for different stages.

3.1 I-V and EL Inspection - Pre-Procurement

30 PV modules chosen by the producer were analyzed by terms of I-V curves and EL images one month before installation. One module showed *type A* and *B* cracks in several cells, another one strong mismatch in two of its cells. All other modules looked overall healthy but with minor anomalies like the example module shown in **Figure 1**. DM like finger failures (**Figure 1**, frame 1), *type A* micro cracks or shunts due to inhomogeneous resistance in the Si (**Figure 1**, frame 2) were found. The I-V curves showed no striking features. The strongly cracked module was found to have the lowest power with 303.2 W, while all other modules range between 305.8 and 308.6 W with an average of 307.2 W.

3.2 I-V and EL Inspection - Post Transport and Installation

Although the analysis of all 39 modules was done at the same time, this part is separated into three causes: problems during transport which led to visible defects (20 modules), visible defects after observed mishandling during installation (9 modules) and HS detected during IR-T inspection after installation (10 modules). In total, seven modules were in a status which didn't allow to mount them in the test structure. Thus, they are not considered in the analysis.

Post transport: Two of the 20 modules were excluded from inspection due to heavy damage. The other 18 modules showed an average power of 302.2 W, therefore 2.6% below rated. The variation reached from 293.1 (5.5%) to 305.2 W (1.6% below rated). Regarding EL, the

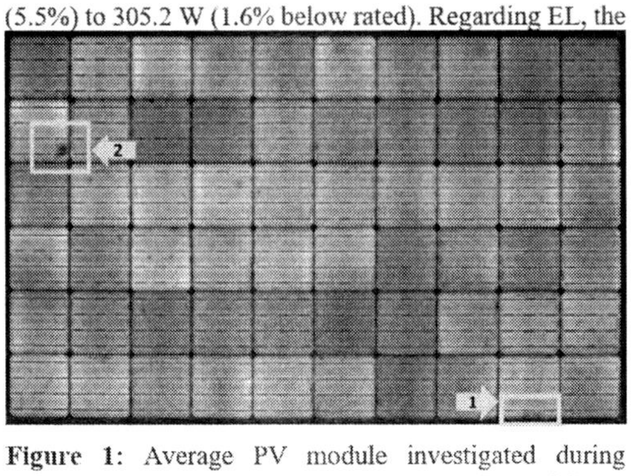

Figure 1: Average PV module investigated during procurement phase with minor DM such as finger failures (1) and anomalies in the Si series resistance (2).

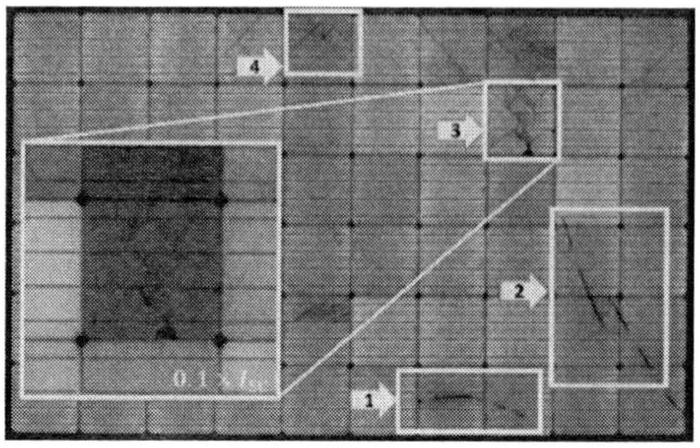

Figure 2: EL image of PV module after transport, not installed. Highlighted DM are: backsheet scratches (1), backsheet scratches with growing out micro cracks (2), *type A, B* and *C* cracks (3) and little black dot with micro cracks growing out of it (4). Cell in 3 is shown as well in 0.1×I_SC image to verify *type C* crack.

inspection revealed that none of the modules was damage free: 17 had *type A* cracks, 8 showed *type B* cracks and 4 *type C* cracks, mostly in multiple cells. Significant mismatch could be found as well in three modules. Moving on to the backsheets, very striking were scratches visible by the naked eye. In the EL images they were visible as bold black stripes going over several cells, easily distinguishable from cell cracks. Often, these scratches had micro cracks growing out of them. Scratches were found in 16 of the 18 modules, of which 13 showed proper scratches of different lengths (few centimeters up to a meter) and 12 showed little punctures from which micro cracks grew. An example module showing several of the detected DM is shown in **Figure 2**.

Visible defects after installation: After installation, another 9 modules were found to have visible defects: seven of these with scratched backsheets, one with scratched frame and one with burn marks on the frame. Only five of these modules (4 with backsheet scratches and the one with burn marks on the frame) were able to get investigated in the laboratory. They showed an average power of 298.9 W (range: 297.9 to 300.6 W), thus 3.6% below rated power. The EL images show cracks, *type A* in all 5 modules, *type B* in 3 and *type C* in 2 modules. Additionally, all modules were affected with backsheet injuries, from small punctures to severe scratches with multiple cracks growing out.

Hot spots after installation: The ten modules found to have HS are divided into two categories. Four modules with ΔT between 5 and 20 °C, six below 5°C. One of the latter was broken and not able to be inspected further. The other eight modules had an average power of 299.1 W. Separated by HS intensities: 299.7 W average for HS below 5°C, 298.4 W average for the hotter ones. The EL inspection showed that two highly cracked modules were the ones with lowest power. Three modules showed micro cracks growing out of backsheet punctures. One had strong mismatch of 5 cells and two modules showed minor mismatches. Except two modules all showed micro cracks and 4 even *type B* cracks.

3.3 Ground IR-T - Five Months Post Installation

Five months after installation, the whole PV plant was inspected with a handheld IR camera from ground. A total of 37 hot spots were reported. 18 modules with HS ΔT < 5 °C (HS3), 5 with ΔT > 20 °C (HS1) and 14 between 5 and 20 °C (HS2). Three modules with HS1 were exchanged after the measurement: one because the glass was broken and another as the HS was affecting multiple cells. Most modules showed only one cell with hot spot, while only five showed multiple cells hot. Four of these had HS3 and one ΔT = 6.2 °C as maximum temperature

Figure 3: IR-T image taken from ground after 5 months of installation. One cell creating a HS with ΔT=29.7 °C and adjacent cell showing a fraction of the cell heated can be seen.

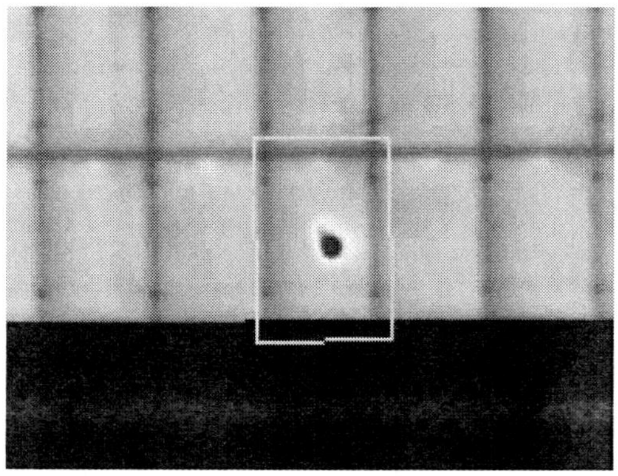

Figure 4: IR-T image taken from UAV 11 months after installation. Visible hot spot with ΔT=35.2 °C. Same HS as in **Figure 3**.

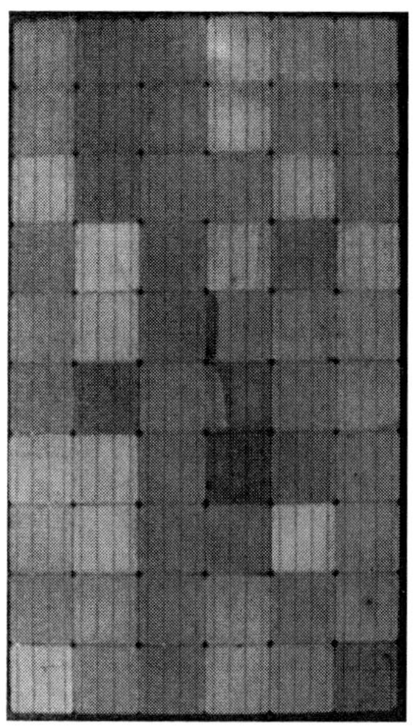

Figure 5: dPL close-up image of the same module shown in IR-T images in **Figure 3** and **Figure 4**. Cell cracks *type B* are visible in two cells in the center. Also visible other DM such as *type A* cracks and shunts in the Si.

difference of the hotter cells. **Figure 3** is an example IR-T image taken during this inspection. The module had one whole cell and a fraction of another cell showing a HS1.

3.4 Aerial IR-T - Eleven Months Post Installation

Almost one year after installation, an aerial IR-T inspection was performed by the O&M operator, as it was contractually constituted. The report showed only HS with $\Delta T > 5$ °C. 15 were detected in total: three of them HS1 and the rest HS2. The hottest HS presented $\Delta T = 55.6$ °C in a cell of a module not affected before. Instead, 10 of the detected HS were already known from the previous IR-T inspection. The development of these 10 HS is given in **Table 2**. The new IR-T image of the module presented in **Figure 3** is shown in **Figure 4**.

3.5 daylight Photoluminescence inspection

The dPL inspection was performed on 214 modules in overview mode (4 modules per image, one time only 2 modules per image). Further, 24 of these modules were recorded in a close-up image (one module fitting the sensor): 21 of these planned, as they contained a known HS, and 3 more spontaneously chosen, as they showed an image artefact in the overview image indicating severe damage.

Close-up images: Out of 24 modules, 11 were with *type A* cracks, 15 with *type B* cracks, 13 with *type C* cracks. Further, 11 showed cell mismatches, 8 presented defects in the Si and one a backsheet scratch. Additionally, other DM such as repetitive patterns in the Si due to cell welding processes were detected in 7 of the inspected modules. None of these modules were DM free. As an example, **Figure 5** shows the dPL image of the module with the HS presented in both previous figures.

Overview images: in the other 190 modules, 39 with minor cracks (*type A* and/or *B*, not clearly identifiable) and 3 with *type C* cracks were detected. Further, 92 modules showed mismatched cells, 19 Si defects, 4 scratches on the backsheet and another 4 other DM or anomalies. In total, 121 of the 190 modules were affected by at least one DM.

3.6 Performance Ratio and Captured Irradiance

The monthly PR is calculated following eq. 2 and it is shown with a red solid line in **Figure 6**. While the initial PR was 80% in June 2019, it increased to 86% in October and then decreased to a minimum of 71% in December. An increase to 86% can be observed in February the following year which is followed by a slow decrease to 77% in August.

Between February and October, the system operates under site-specific optimal conditions. In Winter, considerable self-shading occurs which peaks in December, resulting in captured irradiance of only 84%. **Figure 6** shows this captured irradiance (green dot-dash line) as well as ambient temperature (blue dashed line). It can be observed that the irradiance captured by the plant drops during winter months, resulting in the mentioned performance drop.

Table 2: Ten modules showing HS in both IR-T measurements and their ΔT development. Additionally, the main DM causing the HS is presented.

Module	5 Months IR-T ΔT in °C	11 Months IR-T ΔT in °C	Root cause (DM) detected in dPL image
1	3.1	6.5	*Type C* crack
2	3.2	12.3	*Type C* crack
3	3.6	16.6	*Type C* crack
4	4.9	6.5	*Type C* crack
5	7.2	5.5	*Type C* crack
6	8.3	9.2	*Type C* crack
7	9.2	5.2	*Type C* crack
8	9.7	7.4	*Type A* crack / not detectable
9	21.4	32	*Type C* crack
10	29.7	35.2	*Type B* crack

4 DISCUSSION

4.1 The Impact of Transport and Installation

The inspection during procurement shows an overall healthy status for the PV modules. Although the average power is 0.9% below rated power, this lays in the margin of uncertainty of the measurement process. The average EL image looks overall DM free, although numerous minor anomalies like inhomogeneous signal intensities,

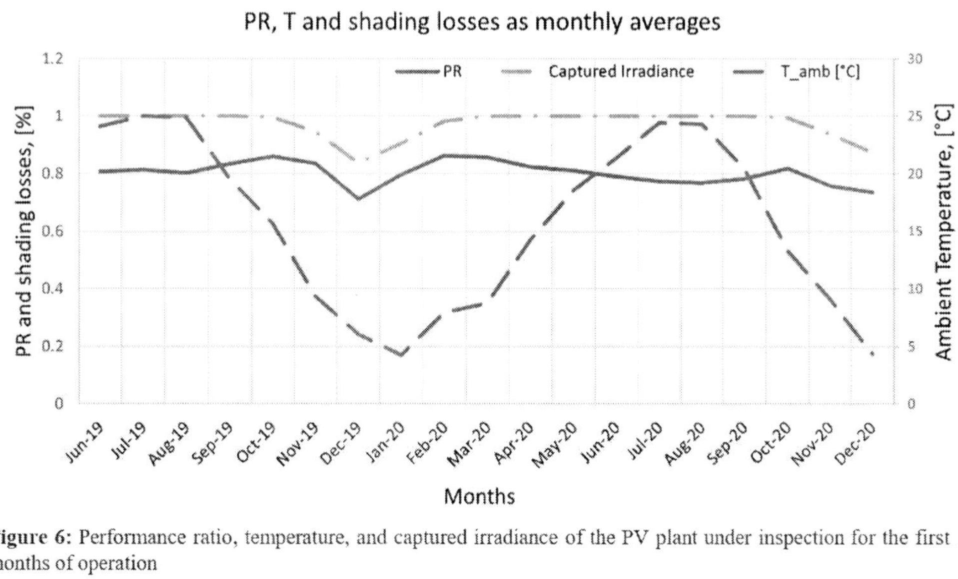

Figure 6: Performance ratio, temperature, and captured irradiance of the PV plant under inspection for the first 19 months of operation

minor finger interruptions and shunts in the Si (cf. **Figure 1**) are visible and might be a sign for lower production quality. Two modules with severe DM, i.e. 6.6% of the modules, encourage a general inspection (at least with IR-T) of all modules before installation. Further, it shall be emphasized that not a single module reached the power it was rated.

The impact that wrong transport had on the inspected modules is alarming. While the power dropped only to 2.6% below rating, the EL images show the real impacts. None of the investigated modules was free of DM. The main issue detected were backsheet scratches in almost all modules. While some scratches were up to a meter long, small dots with growing out micro cracks were probably caused by punctuated hits against the backsheet (e.g. by other modules corners during transport). The scratches may cause immediate safety issues if the isolation is compromised and lead to enhanced degradation of the backsheet. Almost half of the modules showed cell cracks of different severity. Although only severe *type C* cracks come with an observable reduced power measured. Nevertheless, the high number of cracks are expected to cause faster degradation and resulting power losses [20].

Mishandling of modules during installation can result in further initial damage. The damage visible in EL images of modules identified only after the installation was less severe than the one observed in the modules damaged during transport. Nevertheless, the average of measured nominal power in the installed modules was another percentage point lower, thus 3.6% below rated power. It must be considered, that even less damaged, the exposure to sunlight might have caused light induced degradation (LID) [6]. This theory is supported by the inspection of nine modules, selected based on detected HS, whose measured power was 3.5% below data sheet values. The lowest measured power was found in two heavily cracked modules, resulting in a performance of 5.5% below the rated for the combination of *type C* crack with assumable LID. The percentage of modules which got affected by mishandling during installation cannot be evaluated, as it is not known at which time of installation and for how long the mishandling occurred.

4.2 Development of Hot Spots and dPL Analysis

Between the two IR-T measurements lays a time of six months. While a few of the modules showing HS in the first measurements have been exchanged, the main share of 34 detected HS kept installed. As both IR-T inspections have been done by different parties, divergent categorizations have been chosen for HS classification. Thus, 18 of the 34 HS detected with $\Delta T < 5$ °C from the first report would not have been shown as a HS in the

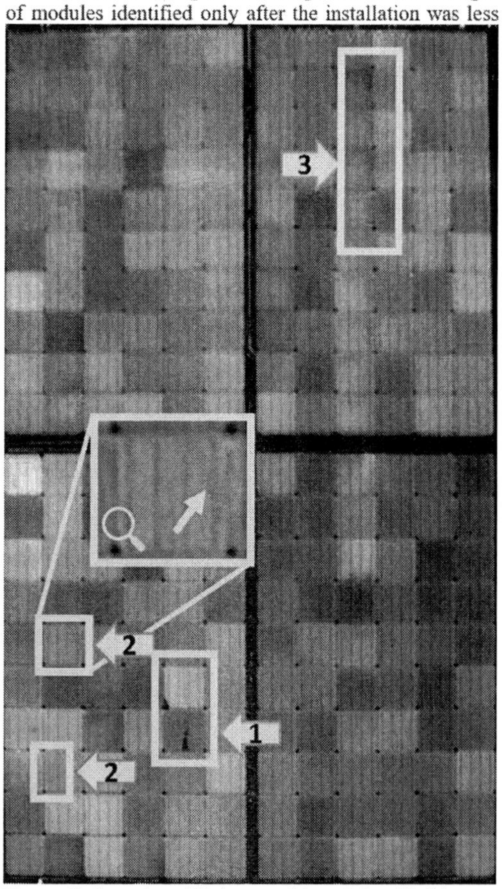

Figure 7: Overview dPL image of 4 modules. Visible *type C* crack (1), almost not visible *type A* crack (2) (including zoom for better visibility), and image artefact indicating a scratch in the backsheet (3). Overall cells with varying signal intensity (chess pattern).

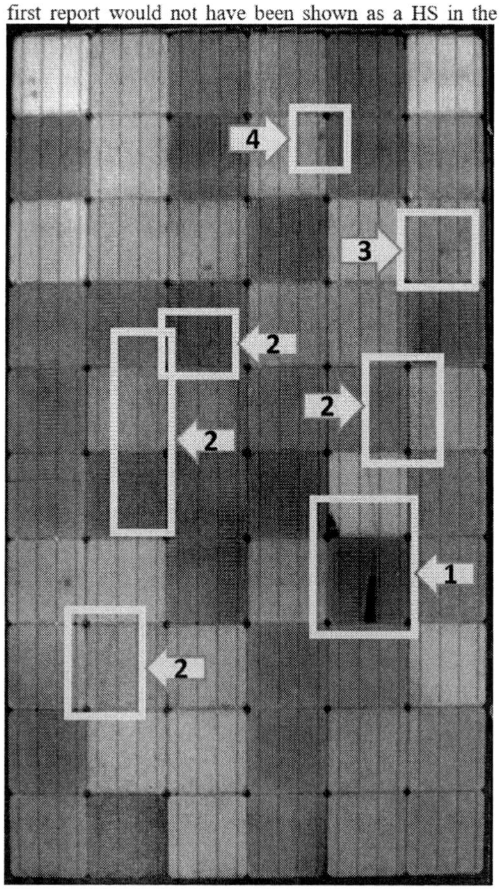

Figure 8: Close-up dPL image of the module in the bottom left in **Figure 7**. Clearly visible *type C* cracks (1), visible *type A* cracks (2), punctuated impact with growing out micro cracks (3), and shunted Si material (4). Cells in mismatch (chess pattern).

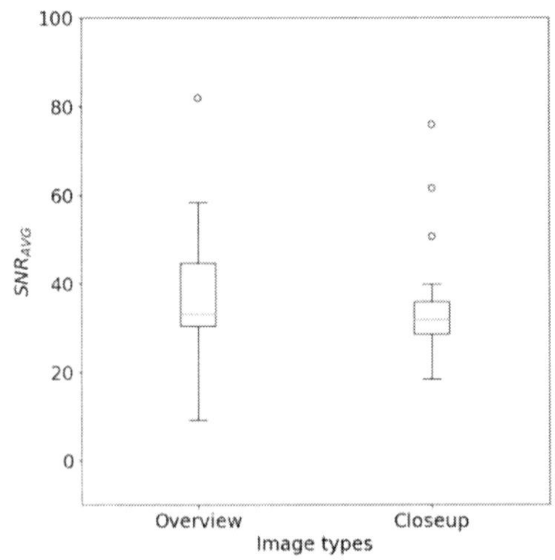

Figure 9: SNR_{AVG} values calculated for the different image types acquired by dPL technique: overview and close-up images. The boxplots represent a total of 54 overview and 24 close-up images.

second report. Overall, the second measurement shows 15 HS of which 10 can be connected to the previous measurement. **Table 2** shows the temperatures of HS detected in both measurements and the root causes detected during dPL inspection. They were in all cases severely broken cells. For three HS2 modules (modules 5, 7 and 8 in **Table 2**) a decrease in HS temperature from one measurement to the other is shown. For ten more modules the HS (all HS2) must have dropped below 5 °C for the second measurement resulting in no detection. This might be due to differing imaging conditions (ground vs. aerial, different irradiance during measurements) and its impact on image acquisition and PV module working conditions. Another explanation can be dirt or shadowing as a cause of HS in the first measurement, which was not present anymore during the second. All other modules with HS in both measurements show increase in ΔT. Four of these in a low range of around 3 °C. Instead, two hardly perceptible HS in the first measurement (modules 2 and 3) quadrupled their temperature difference. The two hottest HS in the first measurement increased by 50% (module 9) and 19% (module 10), respectively.

As this analysis indicates, for most HS severely cracked cells were found as root cause in the dPL inspection. In total, 17 of the 24 close-up images investigated with dPL showed *type C* cracks in the hot cell.

The overview dPL images offer less spatial resolution and the detectability is influenced as discussed in the following section. In this study not all DM in the overview images could be identified clearly. If that was the case, the most likely looking DM was chosen. Although only in 11 of the 190 modules *type A* cracks could be found, 31 modules with *type B* and 3 further with *type C* let assume that there are much more *type A* cracked cells, which are not visible in the overview images. Considering detected cracks, a not negligible share of about 20% of the modules experienced cell cracks of different severity. Attention should also be drawn on about 50% of the modules showing mismatches between the cells. The irregular distribution of these cells over a module in a chess-pattern like behavior is probably a symptom for light and elevated temperature induced degradation (LeTID) [21], resulting in high power losses of around 6% in the first months after installation.

4.3 Performance Ratio Analysis

As expected, due to increased performance at lower module temperatures, the PR shows an increase during autumn as shown in **Figure 6**. The strong decrease in performance between October and February is expected to be caused by self-shading losses. The simulation of the plant performance in respect to the layout supports this assumption. It shows a decrease in captured irradiance to 84% in December 2019. The PR evolution follows this trend closely.

Comparing July and August of 2019 and 2020, a loss in PR of 3-4% can be observed; in combination with the observed chess-board patterns in the dPL images, presumably caused by LeTID. Although, the average temperature for these two months is comparable, the performance loss remains an approximation as long as no temperature corrected PR is calculated. Due to data constraints, the temperature correction could not be performed at this point.

5 USABILITY OF DAYLIGHT PHOTOLUMINESCENCE

Because of the limited resolution of InGaAs cameras, an overview image comes with a significant decrease in spatial resolution compared to close-up images. In fact, the spatial resolution in the close-up images recorded accounts for 0.21 – 0.31 cm/px (bottom to top, due to module tilt) and in the overview images 0.39 – 0.70 cm/px. Thus, severe cracks (*type B* or *C*) are usually visible in the overview images, while micro cracks (*type A*) not. This is shown in comparison between **Figure 7** and **Figure 8**. The first is an overview image containing a module with known HS in the bottom left. It clearly shows the *type C* crack **Figure 7** (1) as can be verified in **Figure 8** (1). The *type A* cracks clearly visible in **Figure 8** are difficult to detect in the overview image, if at all. Further DM are marked in the images. Nevertheless, electrical mismatches between cells and abnormalities in the Si of a solar cell are easily detectable in both images.

A second impact on image quality is due to the weather conditions and the imaging procedure. As the imaging is done batched (first 50 OC images followed by 50 MP images), changes in irradiance bias the background intensities for the different WPs. Moreover, as the luminescence signal is just a few percent of the total image signal, changes in irradiance can cover it completely. This effect is enhanced due to the manual handle switching in the CB, which results in longer and less stable imaging times. Irradiance correction can be performed to reduce the effect [10] but is not done for the images presented here. In either case, a clear sky and stable irradiances are needed for reliable results.

Due to the manual imaging process, the quality of different images varied, as can be seen by the SNR_{AVG} values calculated for each image and separated by image types (**Figure 9**). The spread in the overview images, 10 to 60, is relatively high. Nevertheless, all images fulfil the requirements for outdoor imaging. The same is the case for close-up images where the spread, 15 to 40, is more focused.

6 CONCLUSIONS

This paper shows several inspection steps in the planning, installation, and operational phase of a PV plant. PV modules were analyzed by I-V and EL in the procurement phase, as well as after transport and installation. Further, after 5 and 11 months of operation respectively, an IR-T inspection was performed on the whole plant. After 16 months of installation, a share of 7% of the PV modules were inspected by dPL.

It is shown, even if the quality of PV modules is ensured during procurement, that transport and wrong handling during installation can have a great impact on the quality of modules installed in the PV plant. The inspected modules showed several damages from broken glass (immediately excluded for reinstallation) to visible backsheet scratches (which might result in safety issues) and increased number of cracked solar cells. Even if the latter did not noticeably impair the initial performance, they are prone to enhanced degradation processes. Therefore, if problems during transport or wrong installation are noted, it is highly recommended to inspect the modules in question precisely. Otherwise, they can lead to performance loss, enhanced degradation and safety issues during operation.

IR-T is a fast inspection method with the possibility to show the presence of many DM. Unfortunately, root cause specification needs further inspection by means of EL. Nowadays, new inspection methods like dPL are available allowing similar results as EL with the positive side of not interfering in the connection at the PV plant, neither dismounting the module for laboratory check. Being a new imaging method, only non-standardized procedures, equipment or software are available, summarized in the latest PVPS report by the International Energy Agency on mobile test equipment for PV (Werner et al. [9]). Therefore, repeatability and quality depend on the method used and user skills. In terms of speed, dPL is to the current day far behind IR-T. On the other hand, in terms of DM detection, dPL is competitive due to the detectability of root causes behind HS. The same considerations can be done for EL, which experienced strong development in the last years in terms of aerial EL inspections, easier and safer powering tools and daylight applications [9].

Once a DM is present, its degradation intensifies. In consequence, nevertheless the inspection method used, regular inspections are recommended to avoid safety issues and ensure the plant operation.

6 ACKNOWLEDGEMENTS

This project has received funding from the Horizon 2020 programme of the European Union under GA. No. 952957 – Project TRUST PV.

[19] was downloaded from the Copernicus Climate Change Service (C3S) Climate Data Store.

The results contain modified Copernicus Climate Change Service information 2020. Neither the European Commission nor ECMWF is responsible for any use that may be made of the Copernicus information or data it contains.

7 REFERENCES

[1] "Solar Bankability project website." Accessed: Aug. 18, 2021. [Online]. Available: http://www.solarbankability.org/home.html

[2] C. Tjengdrawira, M. Richter, and I.-T. Theologitis, "Best Practice Guidelines for PV Cost Calculation: Accounting for Technical Risks and Assumptions in PV LCOE," Solar Bankability Project, Dec. 2016.

[3] C. Olschok, M. Schmid, R. Haas, and G. Becker, "Inappropriate Exposure To Pv Modules: Description And Effects Of Handling Defaults," in *28th EU-PVSEC*, Frankreich, Oct. 2013, p. 4. doi: 10.4229/28thEUPVSEC2013-4AV.4.41.

[4] M. Heisz and M. G. Paredes, "Operation & Maintenance - Best Practices Guidelines," Solar Power Europe, Best practice guidelines Version 3, Dec. 2018.

[5] G. Cipriani *et al.*, "Application of Thermographic Techniques for the Detection of Failures on Photovoltaic Modules," in *2019 IEEE International Conference on Environment and Electrical Engineering and 2019 IEEE Industrial and Commercial Power Systems Europe (EEEIC / I&CPS Europe)*, Genova, Italy, Jun. 2019, pp. 1–5. doi: 10.1109/EEEIC.2019.8783525.

[6] M. Köntges, S. Kurtz, C. Packard, U. Jahn, K. A. Berger, and K. Kato, *Performance and reliability of photovoltaic systems: subtask 3.2: Review of failures of photovoltaic modules: IEA PVPS task 13: external final report IEA-PVPS.* Sankt Ursen: International Energy Agency, Photovoltaic Power Systems Programme, 2014.

[7] S. Koch, T. Weber, C. Sobottka, A. Fladung, P. Clemens, and J. Berghold, "Outdoor Electroluminescence Imaging Of Crystalline Photovoltaic Modules: Comparative Study Between Manual Ground-Level Inspections And Drone-Based Aerial Surveys," p. 5.

[8] G. Alves dos Reis Benatto *et al.*, "Drone-Based Daylight Electroluminescence Imaging of PV Modules," *IEEE J. Photovoltaics*, vol. 10, no. 3, pp. 872–877, May 2020, doi: 10.1109/JPHOTOV.2020.2978068.

[9] W. Herrmann *et al.*, *Qualification of Photovoltaic (PV) Power Plants using Mobile Test Equipment: International Energy Agency Photovoltaic Power Systems Programme: IEA PVPS Task 13: Report IEA-PVPS T13-24:2021.* International Energy Agency, Photovoltaic Power Systems Programme, 2021.

[10] R. Bhoopathy, O. Kunz, M. Juhl, T. Trupke, and Z. Hameiri, "Outdoor photoluminescence imaging of solar panels by contactless switching: Technical considerations and applications," *Prog Photovolt Res Appl*, p. pip.3216, Dec. 2019, doi: 10.1002/pip.3216.

[11] M. Guada *et al.*, "Daylight luminescence system for silicon solar panels based on a bias switching method," *Energy Sci Eng*, pp. 1–15, Jul. 2020, doi: 10.1002/ese3.781.

[12] L. Koester, A. Astigarraga, S. Lindig, and D. Moser, "Development Of Daylight Photoluminescence Technique For Photovoltaic Modules And Investigation Of Temperature Dependency," in *37th*

EU-PVSEC, Online, Sep. 2020, p. 6. doi: 10.4229/EUPVSEC20202020-4CO.4.2.

[13] *IEC 60904-9:2007 Photovoltaic devices – Part 9: Classification of solar simulator characteristics*, 2.0. Geneva, CH: International Electrotechnical Commission (IEC), 2007.

[14] *IEC 60904-1:2020 Photovoltaic devices - Part 1: Measurement of photovoltaic current-voltage characteristics*, 3.0. Geneva, CH: International Electrotechnical Commission (IEC), 2020.

[15] *IEC TS 60904-13 - Photovoltaic Devices - Part 13: Electroluminescence of Photovoltaic Modules*. Geneva, CH: International Electrotechnical Commission (IEC), 2018.

[16] R. Bhoopathy, O. Kunz, M. Juhl, T. Trupke, and Z. Hameiri, "Outdoor photoluminescence imaging of photovoltaic modules with sunlight excitation," *Prog Photovolt Res Appl*, vol. 26, no. 1, pp. 69–73, Jan. 2018, doi: 10.1002/pip.2946.

[17] C. Mantel *et al.*, "SNR Study of Outdoor Electroluminescence Images under High Sun Irradiation," in *2018 IEEE 7th World Conference on Photovoltaic Energy Conversion (WCPEC) (A Joint Conference of 45th IEEE PVSC, 28th PVSEC & 34th EU PVSEC)*, Waikoloa Village, HI, Jun. 2018, pp. 3285–3289. doi: 10.1109/PVSC.2018.8548264.

[18] *IEC 61724-1:2017 - Photovoltaic system performance, part 1: Monitoring*. Geneva, CH: International Electrotechnical Commission (IEC), 2017.

[19] H. Hersbach *et al.*, "ERA5 hourly data on single levels from 1979 to present," *Copernicus Climate Change Service (C3S) Climate Data Store (CDS)*, 2018, doi: 10.24381/cds.adbb2d47.

[20] M. Köntges, I. Kunze, S. Kajari-Schröder, X. Breitenmoser, and B. Bjørneklett, "The risk of power loss in crystalline silicon based photovoltaic modules due to micro-cracks," *Solar Energy Materials and Solar Cells*, vol. 95, no. 4, pp. 1131–1137, Apr. 2011, doi: 10.1016/j.solmat.2010.10.034.

[21] M. Pander *et al.*, "Benchmarking Light and Elevated Temperature Induced Degradation (LeTID)," *35th EU-PVSEC*, vol. 35th European Photovoltaic Solar Energy Conference and Exhibition; 1265–1268, p. 4, 2018, doi: 10.4229/35THEUPVSEC20182018-5CV.1.46.

REPAIR AND PREVENTIVE MAINTENANCE OF PV MODULES WITH DEGRADING BACKSHEETS USING FLOWABLE SILICONE SEALANT

Guy Beaucarne[1], Gabriele Eder[2], Emmanuel Jadot[1], Yuliya Voronko[2], Wolfgang Mühleisen[3]
[1]Dow Silicones SRL, Parc Industriel zone C, 7180 Seneffe, Belgium
[2]OFI, Austria
[3]Silicon Austria Labs, Austria

ABSTRACT: PV modules with a degrading backsheet are a problem for the complete PV value chain. Some types of backsheets are known to develop cracks because of an aging-induced change in the mechanical characteristics of the material during operation in the field, which results in a loss of insulation resistance. In this work, we present a solution for repair and preventive maintenance based on a single component flowable silicone sealant. The method fills the cracks present in the backsheet with an insulating material, restores the insulation resistance, and provides a protective layer to avoid subsequent degradation. The solution was successfully implemented on the back of PV modules with co-extruded polyamide backsheet ('AAA') which showed deep cracks following degradation 5-7 years of operation in a solar park. The repaired modules had a restored insulation resistance of several hundred MΩ as observed with wet leakage testing, and maintained a high resistance after accelerating aging. This repair technology can be done in the field and is an alternative solution to module replacement.

Keywords: backsheet repair, cracks, degradation, preventive maintenance; flowable silicone,

1 INTRODUCTION

Over the last few years, several solar park operators have observed a premature degradation of some PV modules. These modules were manufactured and installed between 2010 and 2015 using co-extruded polyamide AAA backsheets, which turned out to degrade in the field although they had passed accelerated tests at the time of manufacturing.

Several repair solutions have been proposed and are being studied[1, 2]. In this conference contribution we present a method involving coating the degraded backsheet with a flowable silicone sealant.

2 EXPERIMENTAL DETAILS

2.1 Flowable silicone sealant

The material that was used in this study is a flowable silicone sealant called DOWSIL™ 7094 Flowable Sealant. It is a single component material that consists of a mixture of a filler (calcium carbonate particles) and a PDMS polymer that will cross-link ('cure') upon exposure to ambient moisture. The sealant is formulated to be flowable rather than a paste. The moisture cure system results in an inward directed curing process starting from the surface in contact with air, with the cure rate slowing down as the barrier for moisture diffusion becomes wider. After cure, the material is a solid elastomer with a high elongation at break (400%), which develops good adhesion to backsheet materials. A layer of 0.20 mm cures in 45 min. This flowable sealant has a tensile strength of 1.2 MPa.

2.2 Tests at backsheet level

In order to study the adhesion of the material on backsheet materials, T-peel test samples were prepared. Rectangular strips of different backsheet materials (AAA, TPT, KPK and PET) were glued together with about 1 mm of the flowable sealant. The peel force was measured as a function of displacement. Some samples were subjected to 1000 h of damp heat treatment prior to peel testing.

2.3 Tests at module level

The flowable sealant was applied on the back of PV modules with AAA backsheets that had previously been operating in a PV-plants for 5 to 7 years and had developed deep longitudinal cracks, leading to leakage currents under wet conditions. Some modules were repaired in the lab (horizontal, backsheet facing up), while others were repaired in the field. The modules were first cleaned by wiping the backsheet surface with a damp cloth and after that dried immediately with a dry cloth. Then, the flowable sealant was dispensed on the areas of the backsheet with the cracks. With a spatula, then material was pushed to fill the cracks and then spread and smoothened into a uniform flat layer over the whole surface as shown in Figure 1. The modules repaired in the field were left to operate for about two months before being dismounted and brought to the lab for further study.

a) b)

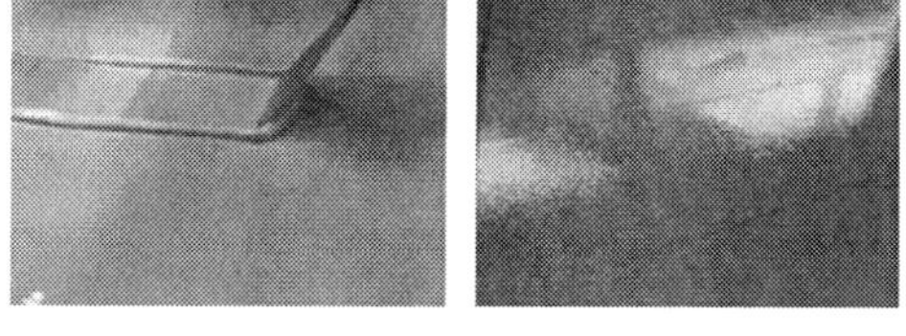

Figure 1: Application of the silicone sealant on a PV module. a) spreading of sealant with spatula, b) back surface of the module right after application

The repaired modules were observed with in-plane and cross-sectional optical microscopy. Some modules were exposed to damp heat conditions (85 °C, 85% relative humidity). Insulation resistance under wet leakage test conditions was measured before and after damp heat.

3 RESULTS AND DISCUSSION

3.1 Results of tests at backsheet level

Fig. 2 shows a typical peel test profile. The peel process reaches a steady state, but there is some local variation, leading to an irregular plateau.

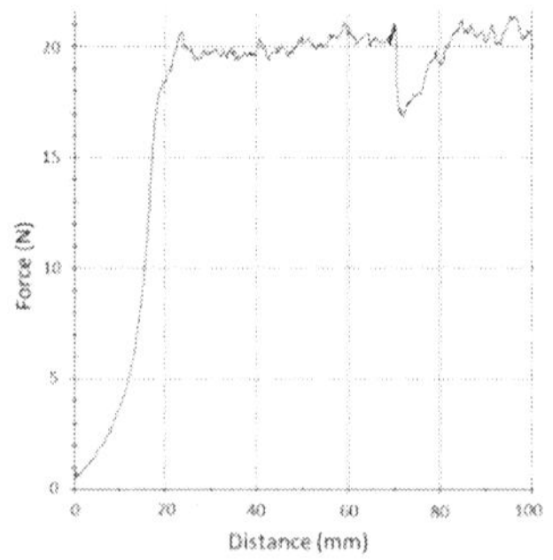

Figure 2: Typical peel test curve obtained during this study

A compilation of the results of the adhesion study on backsheets can be found in Table 1. As can be seen, the material provides good adhesion on all the different types of backsheets, including AAA backsheets, with 100% cohesive failure in all cases. Exposure to damp heat did not change the situation fundamentally.

Table I: Results of the adhesion study

Backsheet type	$F_{peel,max}$ (N/mm) after cure	$F_{peel,max}$ (N/mm) after cure and 500 h at 85°C, 85 RH	$F_{peel,max}$ (N/mm) after cure and 1000 h at 85°C, 85 RH
AAA	0.60	0.74	0.80
TPT	0.74	0.71	0.59
KPK	0.99	0.75	0.63
PET	0.87	0.68	0.49

Failure mode: 100 % cohesive for all samples in all tested conditions

3.2 Results of tests at module level in the laboratory

After coating, the surface is flat and only the deep cracks are still visible as somewhat darker areas. The small cracks are no longer visible. Figure 3 shows optical microscope pictures of the cross-section of a cracked and coated backsheet sample. It can be seen that the cracks in the backsheet have been filled with the flowable silicone sealant and in addition a protection layer of ~100μm of the silicone has been formed on the whole backsheet outer surface.

The PV module that underwent the repair procedure was electrically measured showing a restored insulation resistance giving values above the resistance requirement for safe operation. After damp heat treatment, the module maintained a high insulation and again passed the wet leakage test (see Table II).

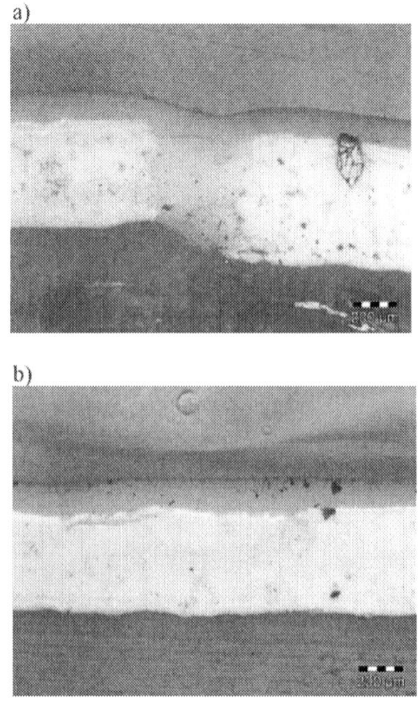

Figure 3: Cross-sectional optical microscope photos of repaired backsheet. a) area with crack, b) area without crack

Table II: Electrical measurements of module under study before and after repair

Before repair		**After repair**	
'Dry' $R_{insulation}$: measurement following MQT 15 but without water immersion	Wet leakage test result	Wet leakage test result	Wet leakage test result after 1000h at 85°C, 85 RH
	Failed	Passed	Passed
> 1000 MΩ	$R_{insulation}$= 0 MΩ	$R_{insulation}$ >1000 MΩ	$R_{insulation}$= 254 MΩ

3.3 Results of tests at module level in the field

The protective layer on the back of the modules that had been repaired in the field showed no sign of cracking or coming off. The microscopic images showed similar 'filled cracks' features (Figure 4) as the modules that had been repaired in the laboratory.

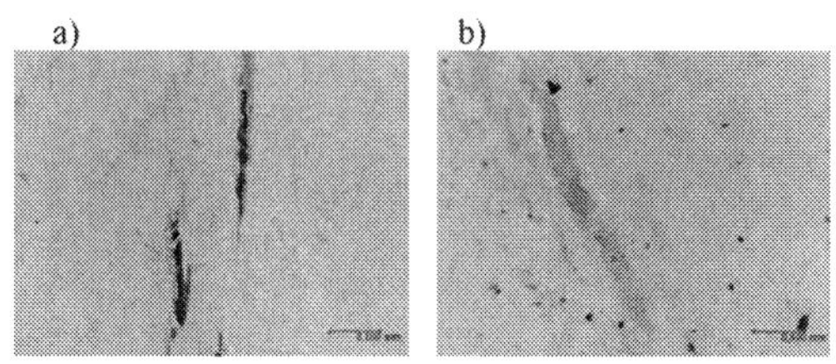

Figure 4: Planar optical microscope photos of repaired backsheet. a) before repair, b) after repair

4 DISCUSSION

The data obtained so far on the repair solution with the flowable silicone sealant indicate that it is effective in restoring electrical insulation and enabling continued operation. Early results also indicate that the module status is stabilized, but further long-term testing is needed.

It is not clear yet which application approach is the best. The easiest and fastest way to coat the module is to place the module horizontally, pour some flowable silicone sealant at one end of the module, and then spread it with a squeegee. A module can be coated this way in less than a minute. However, this requires temporarily dismounting the module and mounting it again, which also takes time. Another approach is to apply the material on mounted modules. This can be done with a spray process but smoothing with spatula is needed to ensure a consistent thickness and ensure that the cracks are filled. This approach saves the dismounting and mounting time but requires more time for the coating itself.

The best method will likely be the result of economic optimization, considering labor cost, allowable down time and material consumption.

5 CONCLUSIONS

In this work, we presented a solution for repair and preventive maintenance based on a flowable silicone sealant. The method fills the cracks present in the backsheet and provides a top layer with an insulating material, restoring insulation resistance, and providing a protective layer to avoid subsequent degradation. This repair solution can be done in the field and can lead to a much lower cost than module replacement or off-site repair. The repaired modules maintain their insulation resistance after accelerated aging and during post-repair outdoor exposure so far.

As there is an estimated 12 GW of installed modules with AAA backsheets, a large part of which will develop issues in the coming years, this solution, both as module repair and preventive maintenance, could lead to massive savings for various players in the value chain.

6 ACKNOWLEDGMENTS

The work at OFI and SAL was supported by the Austrian “Energy Research Program” project “PV Re2 - Sustainable Photovoltaics” (FFG No. 867267) funded by the Austrian Climate and Energy Fund and the Austrian Research Promotion Agency (FFG).

7 REFERENCES

[1] Y. Voronko, G. Eder, C. Breitwieser, W. Mühleisen, L. Neumaier, S. Feldbacher, G. Oreski and N. Lenck, *Repair options for PV modules with cracked backsheets*, Energy Science & Engineering (00), 1-13 (2021).

[2] Y. Voronko, G. Eder, C. Breitwieser, W. Mühleisen, L. Neumaier, F. Sonja and G. Oreski, *Repair options for PV modules with cracked polyamide backsheets*, in *37th European Photovoltaics Solar Energy Conference* (online, 2020).

ADVANCED METHOD TO DETERMINE THE GAIN OF BIFACIAL PV SYSTEMS

Dirk Stellbogen, Peter Lechner, Jonathan Kriening, Olaf Schanz, Dieter Geyer
Zentrum für Sonnenenergie- und Wasserstoff-Forschung Baden-Württemberg (ZSW)
Meitnerstrasse 1, 70563 Stuttgart, Germany
Tel.: +49-711-7870221, Fax: +49-711-7870230, dirk.stellbogen@zsw-bw.de,

ABSTRACT: An experimental bifacial PV test installation has been setup to evaluate a novel approach for determining the bifacial gain, which applies an internal monofacial reference established by covering bifacial PV modules on the back. The direct comparison of bifacial and 'monofacial' operation allows monitoring the bifacial gain of a PV plant in operation and solves the issue of finding an appropriate reference. The bifacial gain observed in the test system was about 7 %, proving the usefulness of bifacial PV modules also on mounting racks not optimized for bifacial PV. Measurements of albedo and backside irradiance enable the detailed evaluation of the bifacial PV performance. However, measurement of the backside irradiation is complex and over- or underestimation has to be carefully avoided. The method of internal monofacial reference shall be developed further for supervision and fault detection in Bifacial PV plants.

Keywords: Bifacial PV, Performance, Quality and Testing, Irradiation Measurement, Albedo

1 INTRODUCTION

PV modules with bifacial technology are increasingly being applied in PV plants, particularly in the power market but also in other applications [1]. However, methods for projecting the energy output of such systems and the gain derived from the bifacial sensitivity are presently associated with complex calculations, large computing efforts and significant uncertainties [2].

In the same way, the examination of those projections is hard for owners and investors as is the checking of the system functioning for the operators due to the complex interaction of multiple direct and reflected irradiance components, bifacial PV module properties and impacts from system components.

For assessing the actual gain from the bifacial PV modules (and check it against the forecast) an appropriate reference is required. However, the comparison to monofacial PV installations with modules of unequal geometry, optical or thermal properties in a different system set-up imposes margins of uncertainty, which may make the judgement vague.

Further, the evaluation of operational monitoring data of a bifacial PV installation and the distinction between module and system performance requires the acquisition of irradiance data, for which the choice of sensors and their positions is not trivial.

To address these issues, ZSW is conducting a comprehensive field study with a bifacial PV system applying an ideal monofacial reference at its outdoor test site Widderstall (South Germany).

2 APPROACH

2.1 The concept of internal monofacial reference

In the test installation of ZSW, a monofacial reference to the PV modules operating bifacially is established by using a subset of the bifacial modules and inhibiting their use of backside irradiation by a foil cover. In this way, the geometry as well as the physical composition of the reference modules is equal to the bifacial PV devices under test and they are operating in the same system environment. Effects from the foil applied at the back surface to the utilization of front side irradiance due to absorption as well as to the heat transfer can be determined by laboratory characterisation and operating temperature measurements, respectively.

Besides the PV modules artificially made sensitive to irradiance only at the front side, the experimental setup also comprises some modules with the front irradiance blocked off to extract just the backside contribution to the bifacial PV module performance.

The PV modules with a cover at the back surface will be called 'monofacial' modules in this report, while the modules with cover on the front will be called 'mono backside' modules.

2.2 Measurements and evaluation

The measurement scheme emphasises the monitoring of the multiple components of the solar irradiation as well as of the performance of the bifacial PV and the monofacial reference at the module and system level:

- Irradiation: measurement of albedo, front and backside plane-of-array irradiance
- PV module: monitoring of individual bifacial modules with equal properties, some of them covered on the back or front surface
- PV system: DC performance of two parallel module strings, with one covered at the back surface

The equivalence of the two strings was proven in an initial verification phase during which the PV modules in both strings operated in the original bifacial mode before the foil cover was applied to the back of the modules of the 'monofacial' reference string.

2.3 Evaluation

The combination of measurements taken at the PV system and PV module level with dedicated irradiance measurements allows for a detailed assessment of the bifaciality effects.

The primary performance figure is the bifacial gain, which is the relative yield increase of the bifacial PV related to the yield of the monofacial reference system:

$$BG = \frac{E_{bif} - E_{monof}}{E_{monof}}$$

The bifacial gain BG depends on the amount of irradiation falling on the backside of the bifacial PV array, the relative conversion efficiency for the back irradiance, and on the non-ideal system efficiency due to the inhomogeneity of the irradiance reflected to the back of the array, shading effects from the mounting structure, and module mismatch. It can thus be expressed as

$$BG = \frac{G_{back}}{G_{front}} * BF * \left(1 - \beta_{Loss,back}\right)$$

where

$\frac{G_{back}}{G_{front}}$ = relative irradiation on the back side

BF = relative conversion efficiency for back irradiation (Bifaciality Factor)

$\beta_{Loss,back}$ = system losses for utilization of backside irradiance

The bifacial PV experiment at ZSW has been designed to enable an accurate determination of the Bifacial Gain as well of the decomposition into the irradiance contribution, conversion efficiency and loss factor.

Further, a bifacial backside performance ratio PR_{back} can be defined as the ratio of the bifacial gain to the additional irradiance proportion received at the back:

$$PR_{back} = BG / \frac{G_{back}}{G_{front}}$$

$$PR_{back} = BF * \left(1 - \beta_{Loss,back}\right).$$

3 SETUP AND INSTRUMENTATION

3.1 Layout of the test system

The bifacial PV test installation is shown in Figure 1. It comprises 27 bifacial PV modules with PERC technology, which divide into a PV generator of 22 PV modules and five individually measured modules placed in between on the same rack.

The STC power of the bifacial modules has been determined with the laboratory flash sun simulator separately for irradiation on the front and the backside. It showed as average for the ensemble

- STC power measured from front: 318 Wp
- STC power measured from the back: 221 Wp

which gives a bifaciality factor of 69.5%.

The laboratory characterisation also showed that for the PV modules covered at the back with a black foil the STC power for front irradiation was reduced by 2.5%.

Figure 2 gives a schematic of the PV module layout on the rack. The PV generator consists of two strings with 11 modules each. The bifacial PV modules and the reference 'monofacial' modules are distributed in such way that there is an equal number of modules for both strings in each of the array rows and that positioning effects regarding ground reflected irradiance and operating temperature are balanced.

Also shown in the figure are the modules monitored individually with one of them in full bifacial operation, one 'monofacial' module covered with a foil at the backside and three modules with front cover.

3.2 Irradiation measurement

Measurements of the incoming irradiance are taken by ventilated pyranometers in the plane of the array from the front as well as from the back, where the redundant instruments are mounted at the top edge of the array (Figure 3). A measurement of the ground reflected irradiance by a down facing pyranometer to determine the albedo is set up close by (Figure 4).

Further, three bifacial PV modules are placed in a column in the three rows of the array with the module fronts covered so that they can only receive irradiance from the back (schematic in Figure 5). Thus, they operate as sensors for the backside irradiation in the respective module position by measuring their short circuit currents.

Figure 1: Bifacial PV module installation at the ZSW test site: PV generator with two module strings and embedded individual modules, with some covered on the front or rear surface; the structure is not optimized for bifacial PV.

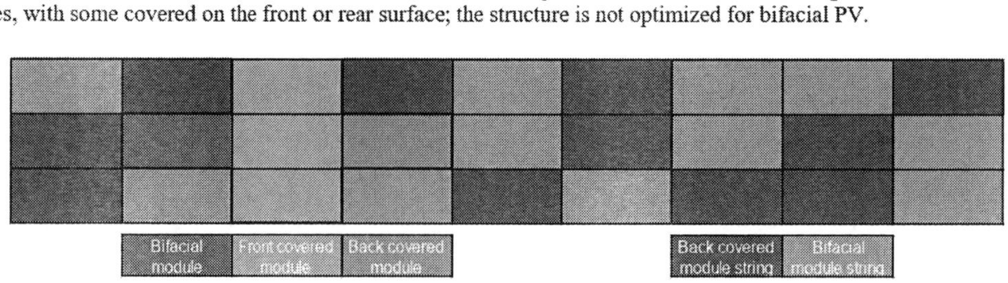

Figure 2: Layout of the PV modules on the rack: the light and dark blue modules constitute the bifacial PV module string and the reference 'monofacial' module string, respectively, of the generator; the other modules are monitored individually.

Figure 3: Backside irradiance measurement with ventilated pyranometers mounted at the top of the array.

Figure 4: Measurement of ground reflected irradiance with a ventilated pyranometer to determine the albedo.

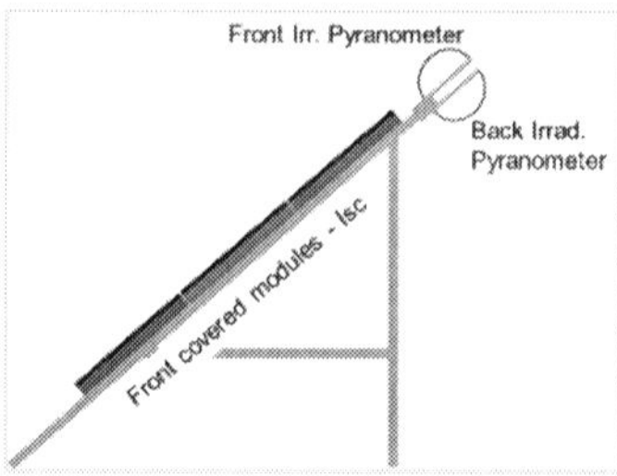

Figure 5: Schematic of three bifacial PV modules mounted in the different rows of the array and covered at their front to monitor the available backside irradiance.

3.3 PV performance monitoring

The two PV module strings of the generator are connected to an inverter with separate MPP tracking inputs. For each of the strings, the actual DC power, voltage and current are measured with high precision transducers.

The individually monitored PV modules are connected to electronic measuring loads, which continuously perform synchronous IV scans of the PV devices to deliver the values of the MPP, short circuit and open circuit points as well as the full characteristic curves.

The operating temperature of the PV modules monitored individually as well as of selected PV modules in the generator strings are measured with resistive sensors attached to their back surfaces.

4 MEASUREMENT RESULTS

4.1 Irradiation

The development of the observed albedo over the monitoring period of one year up to now is displayed in (Figure 6). The albedo parameter is mostly in the range of 0.20 to 0.25 besides snowing events in winter pushing the ground reflection to the ideal case with albedo values of 0.8 and above.

The albedo features some seasonal variation with lower values in summer according to changes in the grass vegetation. Its considerable short time spread is reflecting the daily variation of the incidence angle caused by the course of the sun.

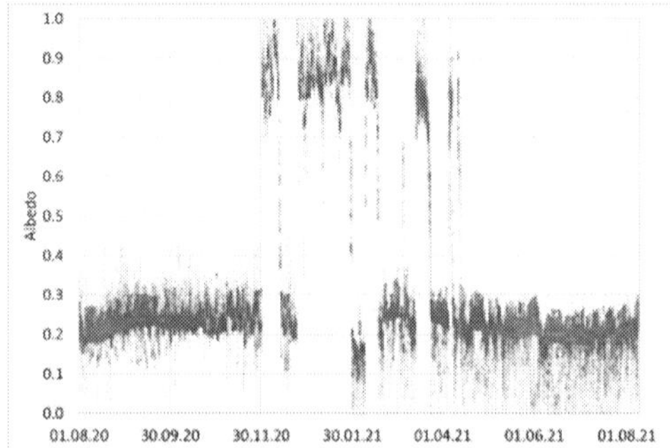

Figure 6: Development of the albedo measured at the test installation; high values in the winter season are due to snow on the ground.

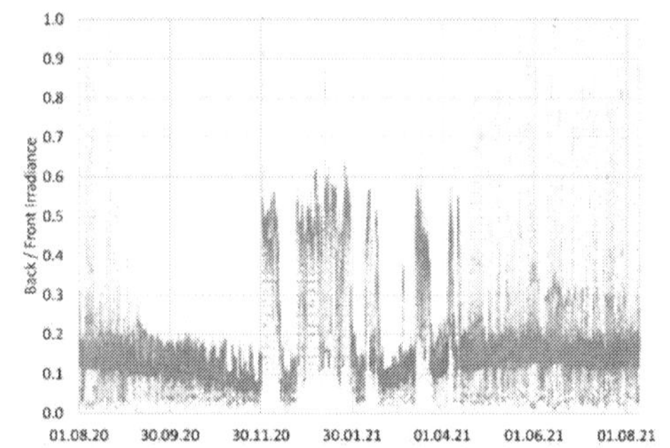

Figure 7: Development of the back to front irradiation ratio measured at the test rack; higher values in the winter period are associated with snow events.

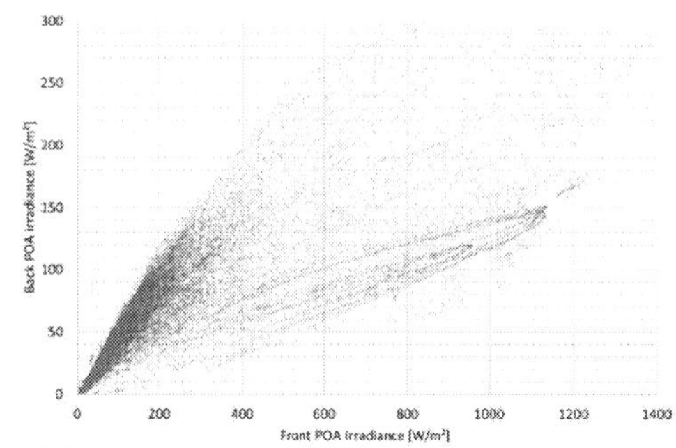

Figure 8: Back side irradiance in the plane-of-array versus the front irradiance, for snow events only (albedo >0.4)

Figure 7 shows the ratio of the irradiance observed in the plane-of-array at the rear to the front irradiance along the monitoring period. The values are mostly in the range of 0.10 to 0.20.

A seasonal reduction from summer to winter is observed, which is attributed to the combined effect of the lower sun path and the high measurement position chosen at the top of the array. With snow on the ground, the backside irradiance increases to about half of the front side insolation. However, this is generally associated with low front POA irradiance, i.e. is occurring at cloudy or overcast skies (see Figure 8).

The three individually monitored PV modules, which are covered at the front and positioned in the different rows of the array, exhibit a considerable inequality in the amount of available irradiation at the backside and its seasonal course. Figure 9 displays for each of these PV modules the ratio of the effective back irradiance expressed by its short circuit current to the average of the short circuit currents.

The module in the middle row has usually below average back irradiance with a slight seasonal variation. For the modules in the top and bottom positions, a significant dependence on the course of the sun is observed. In summer with high sun declination, the top module always has the most back irradiance whereas the bottom module has the least. However, from autumn to early spring with lower sun elevation, the pattern changes. At bright skies with much direct light, the bottom module receives the most irradiation reflected from the ground and the module at the top has the least back irradiance. At cloudy or overcast situations with mostly diffuse light, the top positioned module retains the advantage obtained from its larger viewing field.

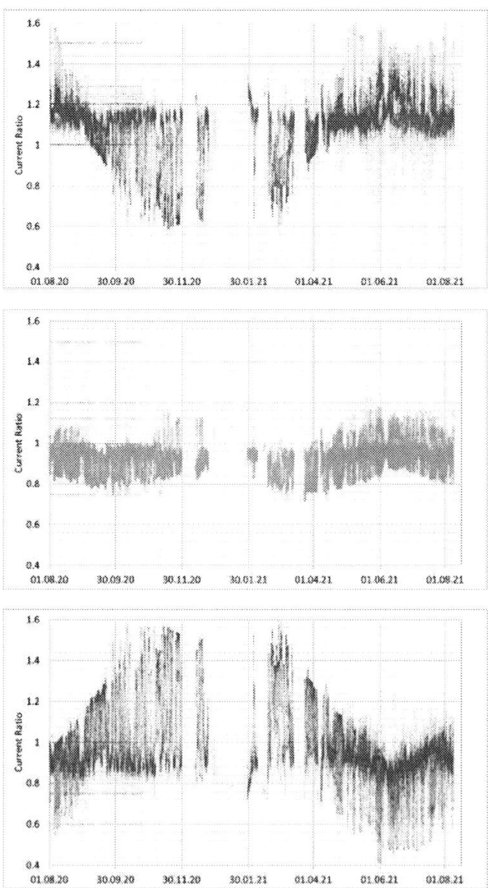

Figure 9: Development of the relative effective backside irradiance for the three modules covered at the front as expressed by their individual short circuit current value related to the mean of the short circuit currents (snow events omitted); red: module in top row, green: middle row, blue: bottom row.

An example of the daily course of the monitored irradiance components is given in Figure 10 for a bright day in spring. The diagram shows the front and back irradiance measured in plane-of-array as well as the ground reflected irradiance that gives the albedo and the effective backside irradiance for the three front covered modules.

From the figure, the proportion of the irradiation received at the front and at the back can be perceived. It also displays the considerable difference in the measurements of rear irradiance by the pyranometer at the array top and by the sensing modules in the different rows.

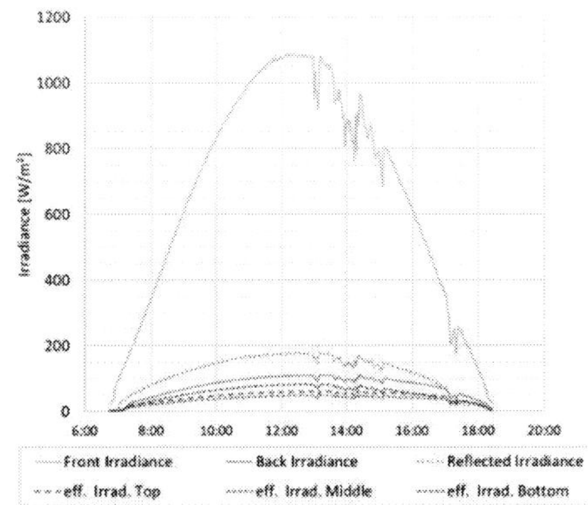

Figure 10: Course of the measurements of front and back irradiance, ground reflected irradiance and effective rear irradiance derived for the front covered sensing modules in the three rows of the array; bright day in spring.

4.2 PV performance

The power curves of the bifacial PV module string and the monofacial reference string with modules covered at the back is displayed in Figure 11 for the same bright day in spring together with the respective curves for the bifacial and reference modules in individual monitoring. From the diagram, the surplus achieved by the bifacial PV modules can be perceived over the course of the day.

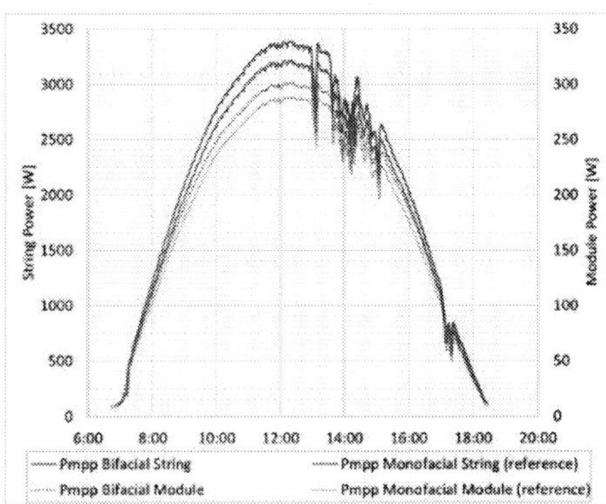

Figure 11: Course of the electrical power delivered by the bifacial PV modules and the 'monofacial' reference modules in the PV generator strings and in individual module monitoring, respectively; bright day in spring.

In Figure 12, the development of the instantaneous power ratio of the two PV module strings is shown. In an extended initial verification phase, the PV modules in both strings were operated in their original bifacial mode without covers at the module back surfaces. In this way, it was proven that no systematic bias of power exists between the strings. After fitting the modules in the reference string with the back covers, the relative gain of the bifacial string over the 'monofacial' reference shows in each one-minute data point.

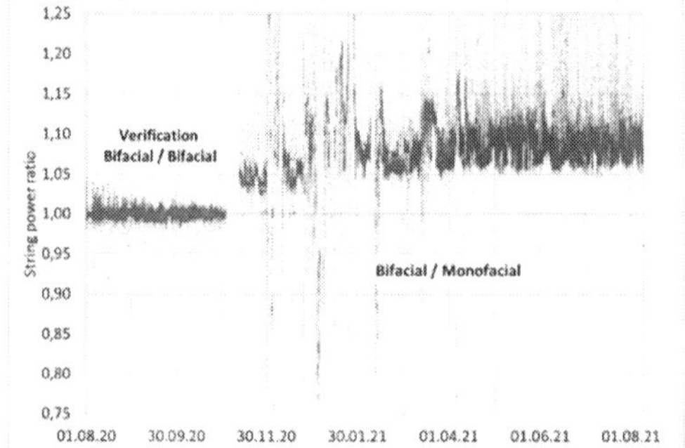

Figure 12: Ratio of the delivered electrical power of the two generator strings; blue: verification phase, red: evaluation phase with the modules of the 'monofacial' reference string covered with a foil at the back.

The surplus of the bifacial PV string is mostly in the range of 5 % to 10 %. At snow events with high ground albedo, the gain reaches higher values. However, operating points with lower power from the bifacial string are also observed, which is probably due to dispersed snow cover on the module rack. For this reason, all snow events with albedo > 0.40 are excluded from the integral performance assessment.

5 EVALUATION

From the measurement data, the bifacial gain BG at the PV module string level as well as the relative irradiance on the backside G_{back}/G_{front} as defined in section 2.3 can be deduced.

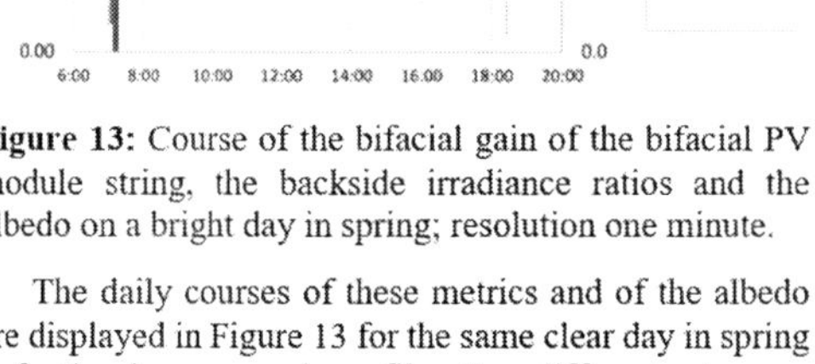

Figure 13: Course of the bifacial gain of the bifacial PV module string, the backside irradiance ratios and the albedo on a bright day in spring; resolution one minute.

The daily courses of these metrics and of the albedo are displayed in Figure 13 for the same clear day in spring as for the above example profiles. Two different values are assessed for the backside irradiance ratio, namely the back irradiance measured with a pyranometer at the top of the array and the mean effective irradiance derived from the three individually monitored modules with front cover. As can be seen, the metrics relating to these two figures differ a lot. The measured back irradiance appears quite high whereas the derived effective back irradiance is striking low, indicating that the latter value is underestimating the irradiation available at the array back whereas the former one is overestimating it.

Some spikes on the profile of the bifacial gain in Figure 13 are induced by the inverter operation. The u-shape of the curves reflects the impact of the sun position changing over the day. The rear irradiance contribution and the power gain are somewhat larger in the afternoon when the sun is in the west of the installation.

The relation of the bifacial gain on the string level to the front irradiance is illustrated in Figure 14 displaying the individual data points of the one-minute monitoring intervals (snow events filtered out). It shows that for high irradiance conditions occurring at clear skies the bifacial gain is rather well defined in the range of about 5 % to 8 %, whereas at lower irradiances, the span widens and the majority of data points goes to somewhat higher values.

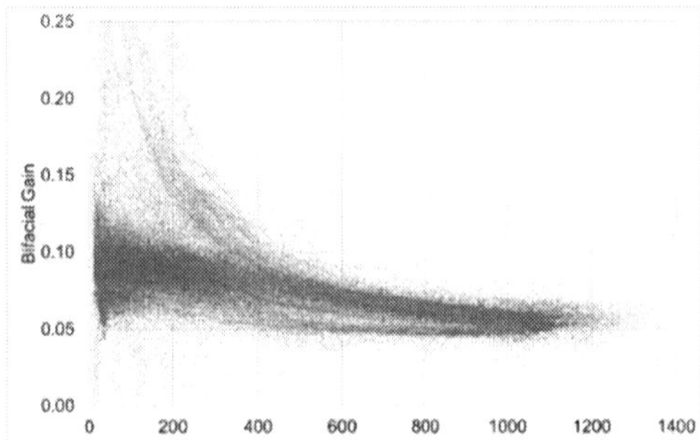

Figure 14: Instantaneous one-minute values of gain of the bifacial PV string in relation to the irradiance measured at the front side.

The development of the daily integral gain of the bifacial PV module string versus the 'monofacial' reference PV module string is presented in Figure 15 together with the respective sum of front irradiation (monitoring data associated with snow events have been excluded). In spring and summer, the daily bifacial gains range from about 7 % to 10 % with some spread according to the variation of the irradiation, while in winter the figures are a bit lower.

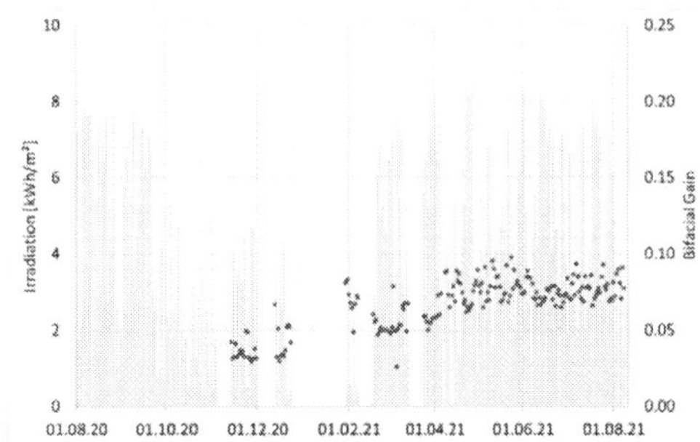

Figure 15: Daily total values of the bifacial gain of the bifacial PV module string with the daily sum of front irradiation given for reference; snow events omitted.

In Figure 16, the daily total bifacial gain is displayed together with the daily integral backside irradiance ratios computed in two different ways as discussed above. The values of the metric related to measurement of the back facing pyranometer at the top of the array are rather high, about two times the final bifacial energy gain. They also show a marked seasonal variation with a substantial decrease in winter.

When the metric is calculated with the mean effective back irradiance derived from the three front covered modules, the seasonal differences are less pronounced. However, these values are much lower and only slightly higher than the bifacial gain suggesting underestimation of the available irradiance at the rear of the PV array.

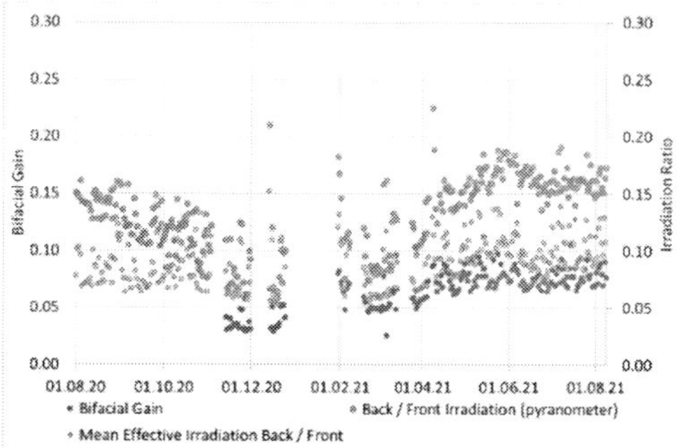

Figure 16: Daily total values of the gain of the bifacial PV module string together with the daily backside irradiance fractions related to the rear pyranometer measurements and to the effective back irradiance received by the front covered sensing modules, respectively.

A summary of the integral bifacial performance over the evaluation period is given in Table I including a discrimination of the winter and spring to summer periods (all metrics are determined with data from snow events excluded).

The total bifacial gain over the evaluation period from November 2020 to July 2021 was determined to 6.6 % with the winter period from November to March showing a bifacial gain of 4.8% and the spring-summer period from April to July delivering a bifacial gain of 7.4%.

Table I: Bifacial performance summary

Period	*BG*	G_{back}/G_{front}	
		Pyrano-meter	Effective Irradiation
November 2020 – March 2021	4.8%	8.9%	6.6%
April 2021 – July 2021	7.4%	15.8%	9.5%
November 2020 – July 2021	**6.6%**	13.8%	8.6%

Note: data from snow events not included

The table also states the relative contribution of backside irradiation for the particular periods applying the two different metrics related to the measurement of the pyranometer facing back and the mean effective backside irradiation derived from the front covered PV modules. For the whole evaluation period, they give the highly dissenting figures of 13.8 % and 8.6 %, respectively, for the backside irradiation ratio.

In terms of the bifacial backside performance ratio $PR_{back}=BG/(G_{back}/G_{front})$, this would give the numbers of 48 % and 77 %. With $PR_{back}=BF^*(1-\beta_{Loss,\ back})$ and BF=0.69 determined for the bifacial PV modules in test, this would give a dubiously high loss factor of almost one third applying the irradiation metric based on the back pyranometer at the top and an even negative loss factor when using the other metrics based on the module backside effective irradiation. This underlines the indication of an overestimation of the available backside irradiation by the former metric and an underestimation by the latter. Hence, a better metric has to be addressed by the further investigations of the experimental installation.

6 SUMMARY AND CONCLUSIONS

An experimental bifacial PV test installation has been setup to exercise and evaluate a novel approach for determining the bifacial gain. It applies an internal monofacial reference established by covering a subgroup of the bifacial PV modules on the back thus inhibiting them from utilising irradiance falling on the rear side.

The direct contrasting of bifacial and 'monofacial' operation allows monitoring the bifacial gain of a PV plant in operation and solves the issue of finding an appropriate reference and avoids complex correction calculations associated with substantial uncertainties when relating to different PV module types, installation set-ups or sites.

The gain of the bifacially operating module string in the investigated PV system was about 7 % for the evaluation period. This surplus was reached despite the high non-uniformity of the irradiation reaching the modules from the back due to the mounting rack not being optimized for bifacial PV. This proves that substantial bifacial gain can be achieved as well when utilizing structures not originally intended for bifacial applications, e.g. in system retrofitting.

Additional measurements of albedo and backside irradiance enable the analysis of the bifacial PV performance and its decomposition into the contribution of rear irradiation, bifacial conversion and system efficiency. However, the measurement of the backside irradiation is complex due to its inhomogeneity and over- or underestimation has to be carefully avoided.

The method of internal monofacial reference cannot only be applied for determining the bifacial gain and verifying the yield projections. It can also be utilised for the operational monitoring in a bifacial PV power plant.

For example, the initial value of the derived bifacial gain represents a realistic performance benchmark of the particular installation determined by the PV module product, the mounting setup and system properties. Thus, changes of the observed bifacial performance may indicate changes in operating conditions, system failures or degradation processes.

The method presented shall therefore be developed further for supervision and fault detection in bifacial PV plants.

REFERENCES

[1] Kopecek, R.; Libal, J.: Bifacial Photovoltaics 2021: Status, Opportunities and Challenges. Energies 2021, 14, 2076

[2] IEA Report: IEA PVPS Task 13 Performance, Operation and Reliability of Photovoltaic Systems - Bifacial Photovoltaic Modules and Systems: Experience and Results from International Research and Pilot Applications (2021)

TERRAIN-FOLLOWING SINGLE-AXIS TRACKING PV SYSTEMS: ADVANTAGES AND PERFORMANCE ANALYSIS

[1]Amir Asgharzadeh Shishavan, [1]Venkata Rahul Abbaraju, [1]Aron Dobos, [2]Francesco Borrelli
[1]Nextracker, Inc., Fremont, CA USA
[2] Department of Mechanical Engineering, University of California, Berkeley, CA USA

ABSTRACT: In this work, we evaluated the performance of the north-south terrain-following single-axis tracking photovoltaic systems on multiple ground undulation types. Such tracker systems are gaining interest as a viable solution to significantly reduce the costly grading operations and qualify more lands for the deployment of photovoltaic systems. However, the performance of such systems may decrease because of the resulting less optimal incidence angle of direct light, the mismatch between modules, and potential row-to-row shading. Our sensitivity study suggests that while terrain-following trackers can have a reasonable performance loss (or even performance gain) on a certain range of north-south slopes, their performance may drop significantly at high north-south slopes. Furthermore, high angles between adjacent torque tubes (tubes) within a tracker can increase the mismatch between the modules and consequently reduce the energy production of the system. Also, the angle between the tubes can cause additional row-to-row shading on the modules if standard tracking algorithms are used. Therefore, in designing terrain-following tracker solutions, special consideration needs to be given to undulation levels and ground slopes to avoid high performance losses.
Keywords: Tracking, Energy Performance, Modeling

1 INTRODUCTION

Single-axis tracking (SAT) is the most common and cost-effective strategy for the deployment of utility-scale photovoltaic (PV) systems. [1] SAT systems were originally designed for flat and semi-flat sites. However, the availability of lands with flat terrain is increasingly becoming limited. Additionally, preparing the site for installation demands a large amount of civil work which is time-consuming and expensive (the grading cost in the site preparation can be as high as 15% of the project budget). Moreover, the land disruption can have a negative environmental impact causing erosion and flooding. [2] Therefore, the need for installing SAT systems on undulating lands is becoming more prominent. One solution to install trackers on the north-south (N-S) undulating land is to modulate post height along the length of the tracker keeping the torque tube straight along the average ground slope. However, this method is more limited to low undulating grounds. Using this installation method in high undulating sites is expensive because it requires taller posts (especially for high-wind sites). To overcome the mentioned challenges, terrain-following tracker solutions have been introduced which allows the trackers to follow the terrain in the N-S orientation while keeping the heights of the posts the same. An example of a terrain-following tracking solution is shown in Fig. 1. However, such solutions may hurt the performance of the systems by causing modules to have a higher angle of incidence (AOI) and introducing additional mismatch because of different irradiance levels on the modules in the same electrical string. Moreover, using standard tracking algorithms on these trackers will result in additional row-to-row shading on the modules due to its non-linear axis (torque tube). In this work, we evaluated the impact of such terrain-following solutions on the performance of SAT systems on various ground undulation types. The results from this study can assist with designing the terrain-following tracker solutions based on the ground undulation types, and the allowable angle between tracker tubes to avoid large performance losses. It is worthwhile to mention that the focus of this study was only on the energy performance aspect of these tracker systems. Determining the economic viability of such solutions requires a project-based techno-economical analysis on the levelized cost of energy (LCOE).

Figure 1: An example of N-S terrain-following tracker system.

2 GROUND UNDULATION MODELING

To study the impact of terrain-following tracker solutions on the performance of SAT systems, we modeled an electrical string consisting of 32 modules on a tracker system. There are five piles underneath this string which divides it into four sub-strings with eight modules each (four tubes per string). Additional modules on the north and south ends of the string as well as adjacent rows were considered to avoid the edge effects. We evaluated the performance of such string on three terrain types: (a) convex terrain, (b) concave terrain, and (c) rolling terrain. These scenarios are shown in Fig. 2 and include most of the terrain possibilities below a tracker in a real site. Blue rectangles in Fig. 2 represent the tracker tubes. Each tube is installed under eight PV modules. We defined two angle parameters to perform the sensitivity study for different terrain categories. Angle α is the angle between adjacent tubes on each tracker. In this work, we assumed the angle between all the adjacent tubes is the same. We also defined angle β as the average N-S axis tilt of the tracker (average N-S ground slope) measured from the south. A positive β shows a south-facing slope while a negative β represents a north-facing slope.

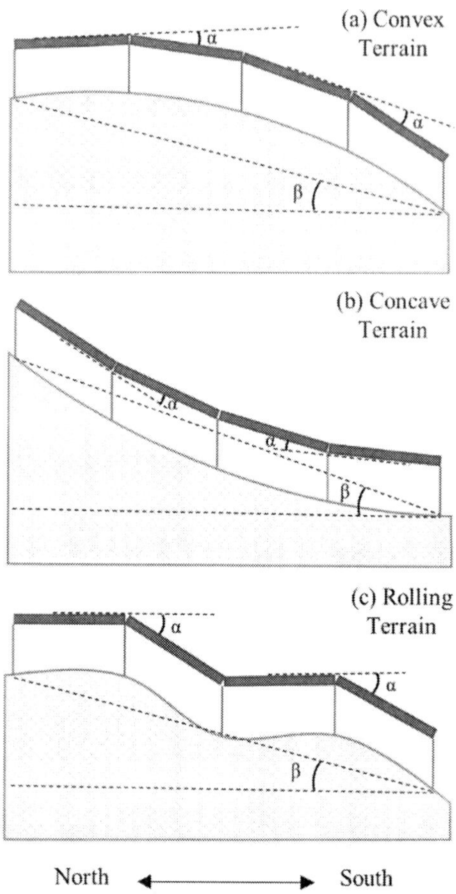

Figure 2: Multiple terrain possibilities were considered in this study: (a) convex terrain, (b) concave terrain, and (c) rolling terrain. In each terrain category, α shows the angle between adjacent tubes (blue rectangles), and β shows the average axis tilt of the tracker on the undulating ground measured from the south. Each blue rectangle represents eight PV modules in a 32-module electrical string.

3 IRRADIANCE AND ELECTRICAL MODELING

We implemented a model based on open-source pvlib [3], RADIANCE [4], and PVMismatch [5] tools to study the performance impact of N-S terrain-following trackers. This model creates a 3D scene of three one-in-portrait tracker rows on different undulation types discussed in the previous section using the RADIANCE ray-tracing tool. Next, using the tool, we created a sky scene based on the Perez model [6] and calculated the plane of array (POA) (direct, diffuse, and ground-reflected irradiance) profile on each module of the string in the middle tracker row. We calculated irraidance on the 12 equially-spaced points along the panel height. By capturing the irradiance on multiple points on the panel we are able to account for the impact of potential irregular shade patterns on the modules. Examples of rendered scenes from RADIANCE are shown in Fig. 3 for convex, concave and rolling terrains.

The irradiance data is then fed into the PVMismatch model which is based on a two-diode electrical model to calculate the string-level I-V curve and determine the maximum power point tracking (MPPT) DC power of the string while accounting for the mismatch between modules. In this study, we modeled split-cell modules which are more shade tolerant compared to the traditional full-cell modules and are expected to dominate the PV module market in the future.

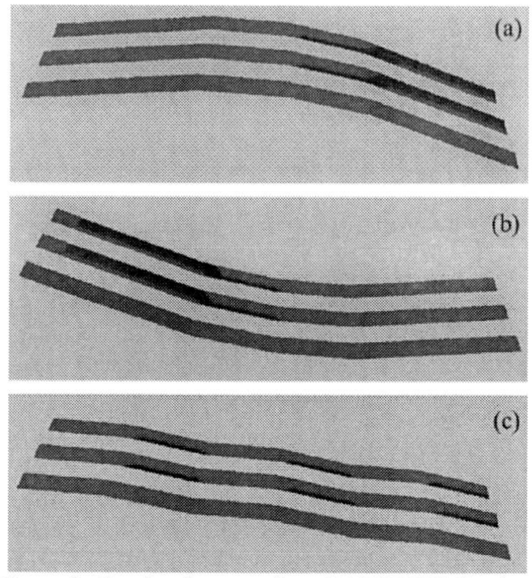

Figure 3: Rendered scenes from RADIANCE for three trackers on (a) convex, (b) concave, and (c) rolling terrains at 7:40 PM, on June 21st for a system in Sacramento, CA, USA. α and β angles in these examples are 9°, and 10°, respectively.

In this study, we assumed the backtracking algorithm (60° range of motion) with the tilted axis option available in pvlib. This is a modified version of the standard backtracking algorithm for the trackers on the N-S slope. The axis tilt used in this algorithm is the average N-S slope of the tracker or β. As shown previously in Fig. 3, row-to-row shading can occur on terrain-following trackers when the sun is low at the sky even with using the N-S slope (axis tilt) modified version of the standard backtracking algorithm. The N-S slope modification to the tracking algorithm ensures no shading when the tracker tube is straight ($\alpha = 0°$). However, when there is an angle difference between the tracker tubes ($\alpha \neq 0°$), this algorithm can cause shading on the panels. It is worthwhile to mention that using the standard backtracking algorithm without N-S slope modification would result in larger shades on the panels of terrain-following trackers and therefore lead to higher performance loss. Examples of tracker angles for the standard backtracking algorithm (without axis tilt option) and the axis tilt modified version of the algorithm are shown in Fig. 4. At these simulations, a ground coverage ratio (GCR) of 0.4 and an axis tilt of 10° are assumed.

4 RESULTS AND DISCUSSION

4.1 Impact on clear days

We performed a sensitivity study and calculated the daily DC energy yield of a system in Sacramento, CA, USA location (latitude: 38.58°, longitude: -121.60°) for various α and β angles on the summer solstice (June 21st), the fall equinox (September 21st), and the winter solstice (December 21st) to capture the seasonality effect. We assumed a clear sky weather model in these simulations.

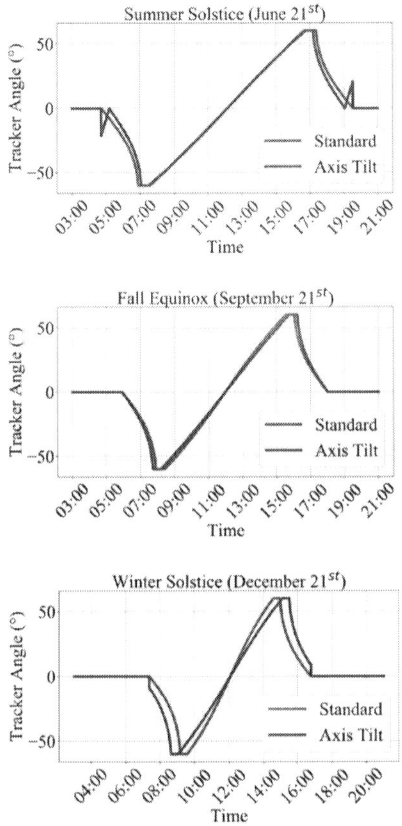

Figure 4: Comparison of the tracker angles computed with the standard backtracking algorithm versus the backtracking algorithm with the axis tilt option for a system with β = 10°.

To understand the impact of the average N-S ground slope (β), we compared the performance of a tracker system with a straight torque tube along the average ground slope (α = 0°) for different β angles. Fig. 5 shows the performance impact of the system for different N-S slopes. As shown, slopes facing north reduce the performance of the system and the percentage reduction can be as high as 3.6%, 22.9%, and 64.5% for extreme ground slopes (20° slopes) on clear days on the summer solstice, fall equinox, and winter solstice, respectively. The reason relies on the fact that by shifting the module plane from flat toward the north, AOI increases, and modules receive less effective irradiance (in the northern hemisphere). The performance loss is however less significant at lower north-facing slopes. For the 5° north-facing slope, the performance loss is 0.2%, 4.5%, and 16.1% on the summer solstice, fall equinox, and winter solstice, respectively. On the other hand, south-facing slopes can increase the production by reducing the AOI (in the northern hemisphere) for fall equinox and winter solstice. However, in the summer solstice, the system's performance reduces on high south-facing slopes. The reason is that by using the backtracking algorithm with the axis tilt option, unlike fall and winter, the AOI of the system on the south-facing slope is higher than the tracker system on flat land for the large periods of morning and evening in summer. Therefore, the effective irradiance received by the modules is lower in these timeframes and as a result, the daily energy yield of the system on the south-facing slope becomes lower than the system installed on flat ground in summer.

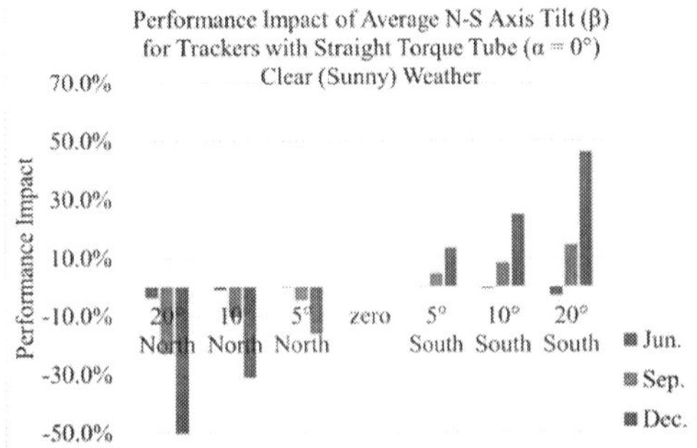

Figure 5: Performance impact of an SAT system in Sacramento, CA with a straight torque tube along the various average N-S ground slopes.

Next, we calculated the performance loss associated with using the terrain-following tracker solutions for different terrain types. Fig. 6 shows the performance loss of the system for different α and β angles and terrain types. The baseline for this study was the non-terrain-following tracker with a straight torque tube along the average ground slope (α = 0°) at each angle β. Therefore, the performance losses shown in Fig. 6 are due to only the row-to-row shading and the electrical mismatch between the modules (the impact of AOI loss was separately discussed and shown in Fig. 5 and it is not included in the Fig. 6 results).

Fig. 6 shows that the results for the convex and concave terrains are similar (because of the same AOI for the modules in the concave and convex scenarios, and similar shade patterns on the modules during back-tracking hours). Our results suggest that large angles between adjacent tubes (α) can reduce the performance of the system significantly (especially at high N-S slopes) by increasing the mismatch between the modules (caused by different AOI and hence different irradiance received by the modules) as well as the resulting row-to-row shading. On the other hand, having α angles of 1° or less minimizes the mismatch loss and the row-to-row shading loss.

Moreover, the mismatch and shading losses for the convex and concave terrains are more than the losses in their rolling terrain counterpart scenarios. The underlying reason is that the modules of a string on a rolling terrain have less variation in the N-S slope and therefore the received effective irradiance does not vary much between the modules and consequently mismatch loss is lower. Also, as shown in the example scenario of Fig. 3, row-to-row shading resulting from the rolling terrain is less than the convex or concave terrains.

The results from this study indicate that consideration of a proper limit for the allowable angle between tubes and the average N-S ground slope is crucial when designing and installing terrain-following trackers on N-S undulating lands to avoid high performance loss.

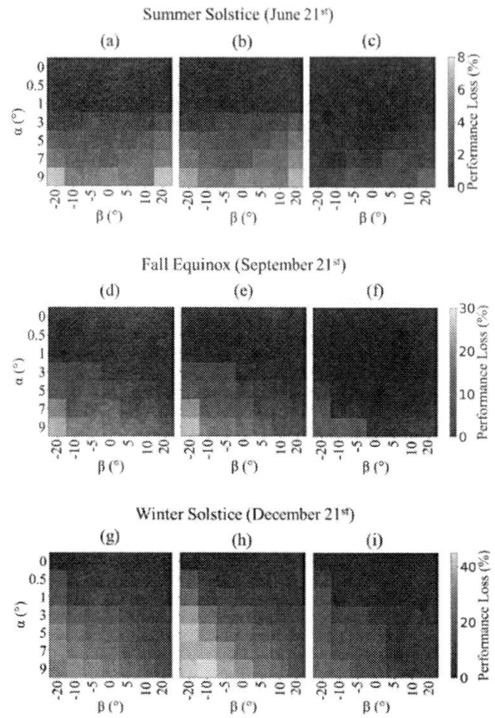

Figure 6: Performance loss (combined mismatch and shading losses) of the terrain-following tracker system for different angles between tubes (α) with respect to a linear tube (α = 0°) at each average N-S ground slopes (β) on convex (a, d, g), concave (b, e, h), and rolling (c, f, i) terrains for clear days on the summer solstice, fall equinox, and winter solstice.

4.2 Impact of cloudy weather condition

So far we have evaluated the performance of the terrain-following tracker solutions under clear (sunny) weather conditions. Under cloudy conditions, the loss associated with the terrain-following systems is lower. The impact of AOI loss under the diffuse conditions is lower, and therefore the electrical mismatch between modules is lower. Furthermore, the row-to-row shading is less impactful on the performance of the terrain-following systems under cloudy weather conditions. Fig.7 shows the performance impact of various N-S slopes (β) for an SAT system with a straight torque tube (α=0°) on a clear versus cloudy day on the fall-equinox day. As shown, the performance loss (or gain) is lower on the cloudy condition by up to ~4%. In this analysis, we assumed an 80% diffuse index (diffuse fraction of the global horizontal irradiance).

Fig. 8 compares the performance loss of the system for various α and β angles under clear and cloudy conditions on the fall equinox. Similar to Fig. 6, the baseline for this analysis was the tracker with a straight torque tube (α = 0°) for each β slope. This comparison is performed on only convex terrains. As shown in the figure, the resulting mismatch and row-to-row shading losses are less under the cloudy condition compared to the clear sky conditions.

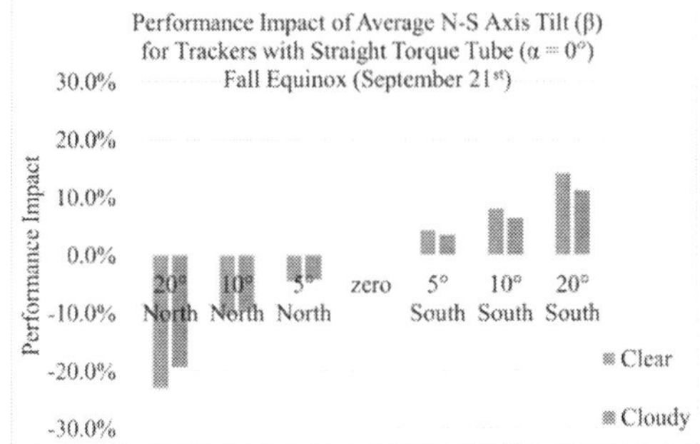

Figure 7: Performance impact of an SAT system in Sacramento, CA with a straight torque tube along the various N-S slopes on a clear vs cloudy day on the fall equinox.

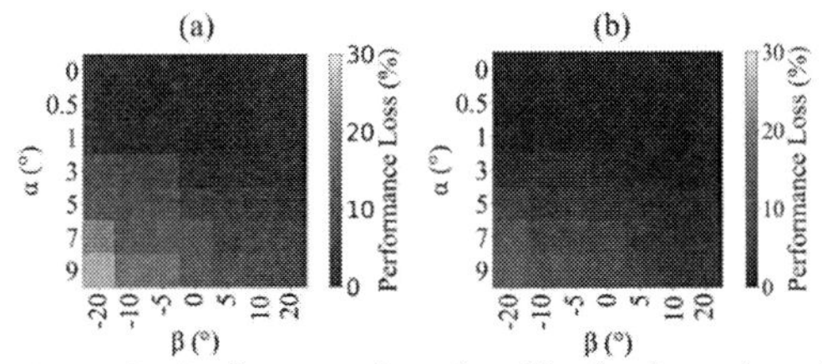

Figure 8: Performance loss (combined mismatch and shading losses) of the terrain-following tracker system for different angles between tubes (α) with respect to a linear tube (α = 0°) at each average N-S ground slopes (β) on convex terrains on a clear day (a) and a cloudy day (b) on the fall equinox.

5 CONCLUSIONS

Terrain-following solutions for single-axis tracker systems can reduce cost and installation time for solar projects. However, the performance modeling of such trackers is more challenging due to different angle-of-incidence and hence different irradiance received by the modules in the electrical strings. Additionally, using standard tracking algorithms result in irregular shade patterns on the module, and 3D modeling is required to accurately account for this impact. Therefore, a careful modeling approach should be utilized to account for the mismatch between modules and the row-to-row shading impact. In this work, we developed a model using open-source software to calculate the performance of terrain-following single-axis trackers on N-S undulating lands.

Furthermore, we performed a sensitivity study to understand the impact of N-S ground slope and the angle between adjacent tubes in a terrain-following solution on the performance of the system. Our results suggest that accurate engineering considerations should be taken into account in the design and installation of such systems to minimize performance loss. Steep North-facing ground slope can significantly reduce the performance of the systems and therefore a limit on the allowable average N-S slope should be considered. Additionally, high allowable angles between the adjacent sub-trackers (tubes) in one string can increase the mismatch and row-to-row shading losses of the system. Therefore, a limit on this parameter should be considered as well to avoid a large performance drop in the system. Our results suggest that using an angle of less than 1° between adjacent tubes can significantly minimize the mismatch and row-to-row shading losses of the system.

6 REFERENCES

[1] M. Bolinger, J. Seel and D. Robson, "Utility-Scale Solar: Empirical Trends in Project Technology, Cost, Performance, and PPA Pricing in the United States – 2019 Edition," 2019.

[2] R. R. Hernandez, S. B. Easter, M. L. Murphy-Mariscal, F. T. Maestre, M. Tavassoli, E. B. Allen, C. W. Barrows, J. Belnap, R. Ochoa-Hueso, S. Ravi and M. F. Allen, "Environmental impacts of utility-scale solar energy," *Renewable and Sustainable Energy Reviews*, vol. 29, pp. 766-779, 2014.

[3] W. F. Holmgren, C. W. Hansen and M. A. Mikofski, "pvlib python: a python package for modeling solar energy systems," *Journal of Open Source Software*, vol. 3, p. 884, 2018.

[4] G. J. Ward, "The RADIANCE Lighting Simulation and Rendering System," in *Proceedings of the 21st Annual Conference on Computer Graphics and Interactive Techniques*, New York, NY, USA, 1994.

[5] M. Mikofski, B. Meyers and C. Chaudhari, *PVMismatch Project: https://github.com/SunPower/PVMismatch*, Richmond, CA: SunPower Corporation, 2018.

[6] R. Perez, R. Seals and J. Michalsky, "All-weather model for sky luminance distribution—Preliminary configuration and validation," *Solar Energy*, vol. 50, pp. 235-245, 1993.

ANALYSIS OF THE IRRADIANCE NON-UNIFORMITY ON THE PERFORMANCE OF VERTICAL AND TILTED BIFACIAL PV MODULE ARRAYS

Marina Calcagnotto Mascarello, Robert. P. Kenny[1], Teodora S. Lyubenova,
Ana Gracia-Amillo, Juan Lopez-Garcia
European Commission, Joint Research Centre (JRC), Ispra (VA), Italy
[1]Tel: +39 0332789287, e-mail: robert.kenny@ec.europa.eu; marinacmascarello@gmail.com;
teodora.lyubenova@ec.europa.eu; ana.gracia-amillo@ec.europa.eu, juanlopezgarcia1981@gmail.com.

ABSTRACT: The bifacial crystalline silicon (Si) photovoltaic (PV) module market has grown significantly in recent years thanks to their increased performance in comparison to conventional monofacial PV modules. Nonetheless, due to the multiple variables that influence the performance of such devices, their characterization lacks consistency. In particular, when it comes to bifacial systems operating under real operating conditions there are gaps to be filled in order to guarantee meaningful characterization for the long-term energy yield assessment. This can be achieved by the analysis of field data for different technologies, rack configurations and climates. Based on previous work on single bifacial modules, this work extends the analysis of the behaviour of the rear-side irradiance of outdoor bifacial test arrays and their energy yield during one year. The rear irradiance non-uniformity (NU) analysis is crucial for long-term degradation rate and bifacial gain predictions (and hence bankability) in medium and large utility-scale solar systems with high-efficiency bifacial modules.
Keywords: bifacial, PV array, performance, non-uniformity, rear-side irradiance.

1 INTRODUCTION

According to the International Technology Roadmap for Photovoltaic (ITRPV), bifacial module technology is expected to increase from a current 10% market share to a 35% share in 2030 [1]. A key factor enabling the development of bifacial technology is the standardisation that will ensure the rapid growth of the bifacial PV market. When it comes to energy rating, an extension of the IEC 61853 standard series is proposed for bifacial modules, aiming to help end-users evaluate and select appropriate bifacial modules with the relevant mounting configuration and climatic profile appropriate to their intended use [2]. It is interesting to include in this proposal some variables which have an influence in the performance of a module installed in a bifacial system, operating under real outdoor conditions [3,4]. This specific work is intended to provide some insights which can contribute for this objective.

2 EXPERIMENTAL SETUP

2.1 Modules mounting and configuration

Two test arrays have been installed at the European Solar Test Installation (ESTI), in Ispra (northern Italy, 45°48'43.4" N and 8° 37' 37.4"E, at 220 m above sea level) considered to be a moderate subtropical climate (-10 °C to +35 °C, with less than 90% RH). Each array is composed of ten bifacial modules with a total nominal power of 2.94 kW, configured as two rows of five modules in landscape orientation, lower edge 1 m from the ground. Both arrays are shown in Figure 1, the first installed on a fixed open rack South oriented with a tilt angle of 30° while the other is vertically mounted in an East/West-facing vertical configuration. The ground underneath the system is composed of aged concrete foundation surrounded by gravel, with average albedo of 0.17. The two arrays were connected to respective SMA inverters. Additionally, a 3kWp equator oriented fixed rack with monofacial c-Si modules close to the tilted bifacial PV module array was used as reference for the energy yield analysis. The modules installed are commercially available p-PERC type 60 cells frameless glass-glass type without any partial self-shading in the rear due to the junction box or label or mounting structure. The indoor characterisation and installation have been previously described in [5]. As defined by IEC TS 60904-1-2, bifaciality is the ratio between rear and front side of a bifacial device at STC, regarding short-circuit (φ_{Isc}) current, open-circuit voltage (φ_{Voc}) and output power (φ_{Pmax}). The equivalent irradiance levels G_{Ei} are determined as follow:

$$G_{Ei} = 1000 + \varphi \cdot G_{Ri} \quad (1)$$

$$\varphi = \min(\varphi_{Isc}, \varphi_{Pmax}) \quad (2)$$

where G_{Ri} is the rear side irradiance levels and φ is the module bifaciality described as the minimum value between the φ_{Isc} and φ_{Pmax}. For the devices installed, the average value of the bifaciality coefficients are $\varphi_{Isc} = 66.3 \pm 0.7$ % and $\varphi_{Pmax} = 67.3 \pm 0.9$ %.

Figure 1 - South-oriented tilted (top) and vertical East-West oriented (bottom) bifacial PV modules test arrays.

One set of I-V curves for one selected bifacial module are plotted in Figure 2, which were measured following the accredited procedure for single-side illumination method according to IEC TS 60904-1-2 with a pulsed solar simulator at different irradiances and G_{ei} for the front and rear side. It is important to highlight that no discrepancy among the modules was identified and the behaviour of I-V curves is similar for all of them.

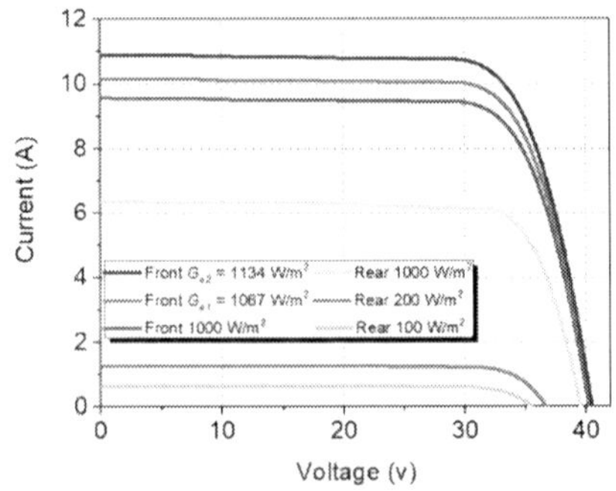

Figure 2 - Indoor I-V curves at different front and rear irradiances of a bifacial PV module installed in the test arrays measured according to IEC TS 60904-1-2.

The single-side illumination method was used first by illuminating the front side only with three intensity levels: 1000 W/m², G_{e1} equal to 1067 W/m² and G_{e2} equal to 1134 W/m², which are the effective irradiance calculated based on rear irradiances of 100 W/m² and 200 W/m², respectively. The resultant curves are the three higher ones from Figure 2, whereas the three lower curves are obtained by illuminating the rear side only, with levels of 1000, 200 and 100 W/m².

2.2 Long-term monitoring
The in-plane irradiance on both sides is measured with a series of calibrated c-Si reference cells placed around the perimeter of the array of modules in order to avoid shading the rear side of the modules. Figure 3 shows a representation of the sensors' arrangement: on the tilted array there are 12 on the rear side and one on the front side, in the central position of the system (dark blue); on the vertical rack there are 10 on each side, mounted back-to-back. The placement choice took into consideration having a representative value for each module of the system, especially for the tilted array, in which larger non-uniformities were expected. On the rear-side of the tilted array additional sensors of the same type attached to a movable structure (represented in green colour, letters a-e) were used for shorter term measurements, which are described in the next section.

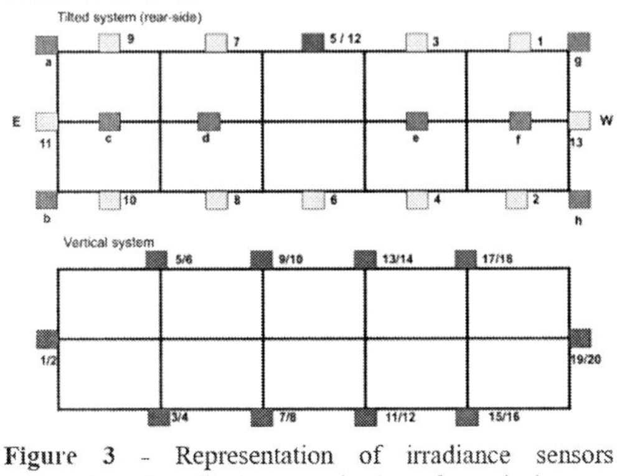

Figure 3 - Representation of irradiance sensors positioning for tilted array (top) and vertical array (bottom).

The module temperature is measured with a surface mounted Pt100 sensor placed on the rear side of a selected module. Measurements of irradiance, DC voltage, DC current, temperature and DC energy were taken over a full year at 1-2 minute intervals.

2.3 Short-term rear-side irradiance measurements
As described, the calibrated reference cells are placed around the outer edges of the arrays, thus the irradiance measured and consequently the non-uniformity could not be representative for one entire module, especially for the ones placed more to the centre of the rack, where shading effects are more pronounced. Aiming to characterize the distribution of the rear-side irradiance and assess possible variations between positions, sensors were placed in the central positions and on both corners of the outer edge of the rack during some days in winter, first on the East side and then on the West side. Some variation of aging and consequently of albedo of the concrete base under the tilted rack, and which influences the reflected component of G_{rear} at the East-most side is observed. For this reason, one experiment was conducted in order to achieve uniform albedo of this surface by covering it with a white plastic material. The measured albedo using an albedometer is approximately 0.5. Another experiment was carried out covering the rear-side of the tilted system with a dark opaque material in order to verify the coefficients of the linear equation which correlates irradiance and maximum power current for the front side only. Effective irradiance G_{eff} was calculated considering the small amount of rear-side light leakage measured - of the order of 20 W/m² - and the bifaciality φ_{Isc} = 0.66. Doing so, the equation obtained is a good match to the one obtained from indoor measurements, having the same slope and slightly higher intercept. Further discussion on this follows in session 3.3.

2.4 Simple irradiance model evaluation
The proposed method for the energy rating of bifacial modules described in [2] uses the three-component anisotropic model of Perez [6] for the calculation of hourly rear side irradiance values. This validity of this approach is studied by comparing estimated values of G_{rear} calculated by applying the irradiance model with measured data, thus estimating the differences due to factors not taken into consideration by the model. Measured direct and global irradiance and albedo are used as input data and applied for the different planes of inclination used in our test arrays. Given that this model does not include any effects of shading, which are particularly significant in the tilted system, an alternative set-up was placed, aiming to evaluate G_{rear} on a non-shaded surrounding. The two sensors were mounted on a structure composed of horizontal bars (approximately 2 m long) fixed on the outer edge of a tilted rack, with the same tilt angle as the main tilted rack, at heights of 1 and 2 m above ground level. The configuration can be seen in Figure 4. The experiment was conducted in winter time with gravel ground cover (as seen in the image of Figure 4) and with the increased albedo of 0.5 described above.

3 RESULTS AND DISCUSSION
3.1 Rear side irradiance non-uniformity
The non-uniformity of the irradiance, *G non-uniformity*, is defined as by equation (3):

$$G\ Non-uniformity = \frac{G_{Max}-G_{Min}}{G_{Max}+G_{Min}} \quad (3)$$

Where G_{Max} and G_{Min} are the maximum and the minimum irradiance values measured.

Figure 4 - Assembly of sensors for short term rear-side irradiance measurements.

For the NU analysis one representative clear sunny day for each season was selected. Clear sunny days are selected as being most relevant to this study, since it was observed that for overcast and cloudy days the influence of NU is not very significant.

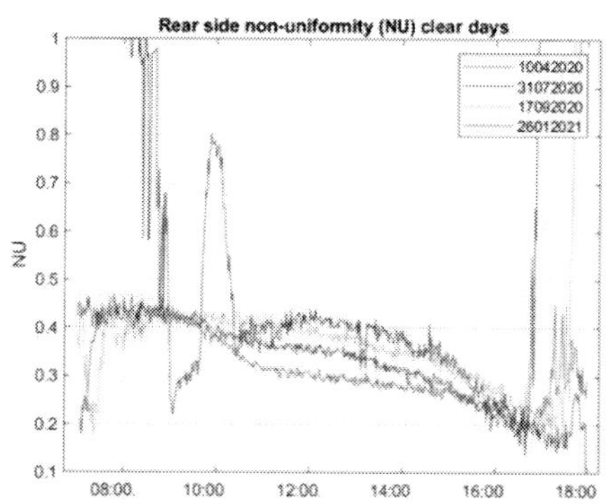

Figure 5 Non-uniformity of the rear-side irradiance for the tilted system on clear days in different seasons.

For the tilted system during the central hours of the day NU varies between 0.2 and 0.45 throughout the year, as can be seen in the samples from Figure 5. Additionally, there is a trend of higher NU during the morning, which is caused by the non-uniform albedo of the ground below the East corner, as already mentioned. For summer the daily average values are the lowest, whereas winter has average values that are higher than the other seasons. In addition, in wintertime the high uniformities from early morning occur at a later time, therefore they are the only one visible in the graph, even though they are always present. For example, for the summer day represented in Figure 5 the highest values of NU are observed in the first hours of the morning (before 7 a.m.) and later hours in the afternoon (after 7 p.m.). However, the absolute values of irradiance at these moments of the day are very low, consequently the total energy yield of these periods represent a low fraction of the daily yield. Therefore an approach was taken to estimate a more representative average value which accounts for the fact that the NU is more significant in moments of high energy production. In this approach the NU of each time step is weighted according to the fraction of energy produced with respect to the total daily energy yield and then averaging all resultant values. For the representative summer day shown in Figure 6 the calculated average of NU is 0.36 whereas the weighted value after is somewhat lower at 0.31.

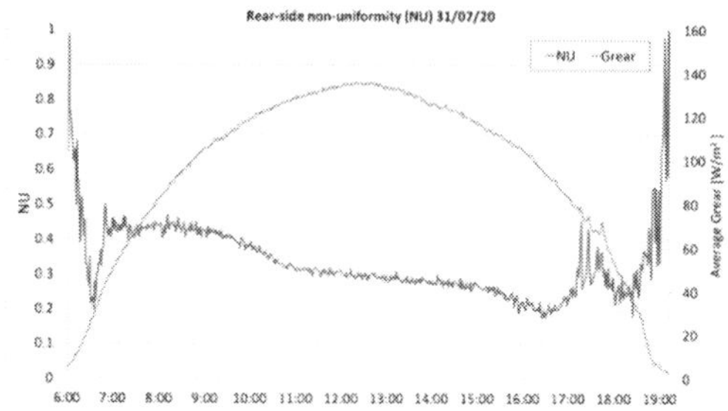

Figure 6 Average rear-side irradiance and non-uniformity for the tilted system on a representative clear summer day.

Following the same approach, a weighted average NU for the whole year was calculated for the tilted system, in this case weighing the daily average NU according to the ratio between the daily yield and the total yearly yield. The weighted average of NU results in values 58% lower for winter and 24% lower for fall, but summer and spring, due to the higher energy production, see an increase of 14% and 12% respectively in comparison to the average calculated with absolute values. The yearly average weighted NU for the tilted system is equal to 0.35, which is 15% lower than the average calculated without weighting.

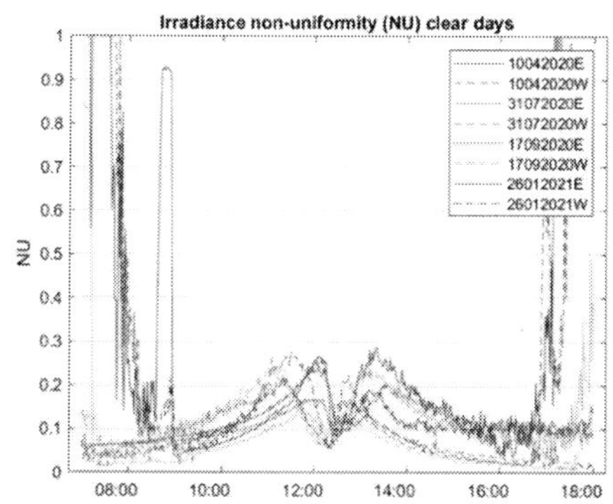

Figure 7 Non-uniformity of irradiance on both sides of the E-W vertical system on selected clear days in different seasons.

Regarding the vertical system, the NU was calculated for both sides (East and West), with "rear-side" here to be understood as the side which is not receiving direct irradiance (West side before solar noon and East side after solar noon), or else referred as non-sun-facing side. As can be observed in Figure 7, the behaviour of NU is seasonal: for the East side the average values of NU are lower for fall (0.11) and higher for winter (0.14). The maximum values happen around noon (1-2pm solar time), reaching a maximum of 0.28 in the winter day. The West side presents average values a bit lower and quite similar for all seasons, of around 0.1. The same summer day was analysed separately for the vertical system as shown in Figure 8. It can be seen that the non-sun-facing side has a higher NU for most of the day and that the peaks also happen on hours of very low irradiance. Considering the NU values for this configuration are relatively low in comparison to the tilted system, the weighted NU was not calculated.

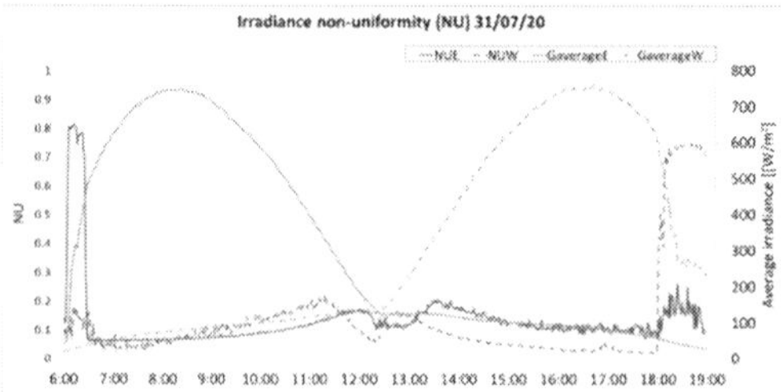

Figure 8 Average irradiance and non-uniformity for both sides of the vertical system on a representative clear summer day.

3.2 Rear-side irradiance spatial distribution

The measurements from the sensors placed on the edge of the rack showed less variation among them, which is expected given that this section is less affected by shading. When analysing the central positions, the one closer to the ground (referred as bottom) presents higher values, while the measurement from the middle position is a lot closer to the one from the top sensor. The biggest difference was identified at the East side of the rack, with an absolute difference of 33% in average in the morning hours (9am to 1pm) and 22% in the afternoon (1pm to 5 pm). This is likely due to the higher reflected component from the lighter concrete base, as already mentioned. To avoid any biased caused by the albedo non-uniformity, the West side was also analysed and the results are shown in Figure 9. The difference still occurs but is lower, with an absolute difference of 19% in average during the morning and 8% during the afternoon. The middle and top position agree within 10% during the whole day (9 am to 5 pm) in both sides. For the next set of sensors placed even more to the centre of the rack, the absolute G_{rear} values are found to be lower, since the centre of the rack is more affected by ground-shading. In this case, as it can be seen in (b) the top and middle sensor show an even closer match. This is an indication that the NU is lower for the modules in the top row of the system.

3 Output current versus irradiance analysis

When measuring G_{rear} in different positions for a whole rack, the question of representativeness arises. Since measurements of G_{rear} were taken in 12 different positions in the tilted rack, the aim was to identify one single value of G_{rear} representative of the whole system. Values of maximum power point current measured (I_{mp}) were analysed against values of G_{eff} calculated with different values of G_{rear}. Current was chosen as the variable to be analysed because it is much less affected by temperature then power and the variables analysed show very little variation, given that G_{rear} has a small weight in G_{eff}. Since G_{eff} considers the bifaciality, it could be expected that the same linear equation that relates I_{mp} and G_{front} would apply to G_{eff} calculated with a representative G_{rear}. The equation applied is the one obtained as explained in section 2.3. For one representative clear day in summer, different values of G_{rear} were applied in the calculation of G_{eff}, as represented in **Error! Reference source not found.**.

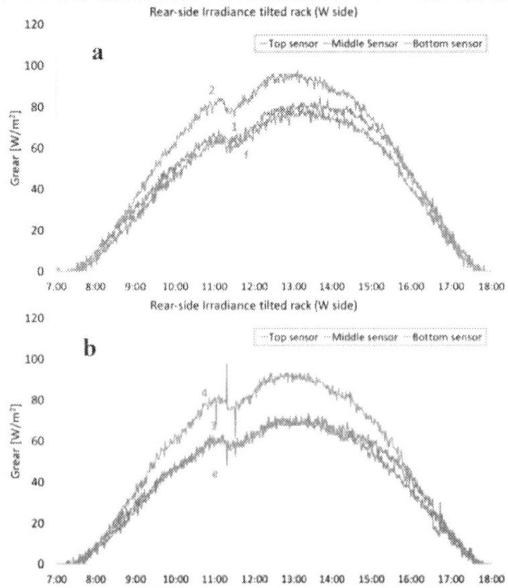

Figure 9 Rear side irradiance on the West side of the tilted rack measured on positions **a.** 1(top), 2(bottom) and f (middle) **b.** 3(top), 4(bottom), e(middle)

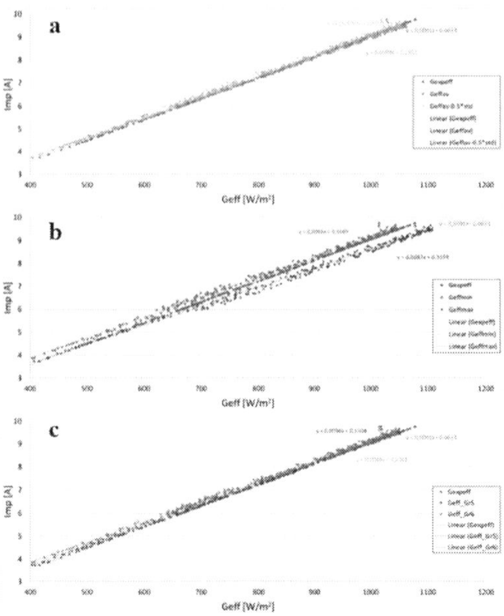

Figure 10 Series of G_{eff} versus I_{mp} in comparison to the reference equation using different values of G_{rear} for a representative summer day - data and trend lines of G_{rear} being (a) average value of all sensors and average minus 0.5 standard deviation; (b) maximum and minimum values; (c) values from central positions of the rack (sensors 5 and 6).

The average and average -0.5 standard deviation (a) give a good match to the reference, with variation of 0,82% and 1,99% in average respectively. Nonetheless, multiple sensors are needed to be able to have reliable average and standard deviation values, so another approach would be to analyse specific sensors. In case the sensor chosen has extreme values the equation tends to deviate from the reference, as seen for maximum and minimum rear-side irradiance values (b). Whereas when using data from the sensors in the central position of the rack (5 on the top and 6 on the bottom) the result is much more close to the reference equation (c), with average variation of 1,56% and 2,46% respectively, which are comparable to the sets of (a) calculated with measurements of multiple sensors.

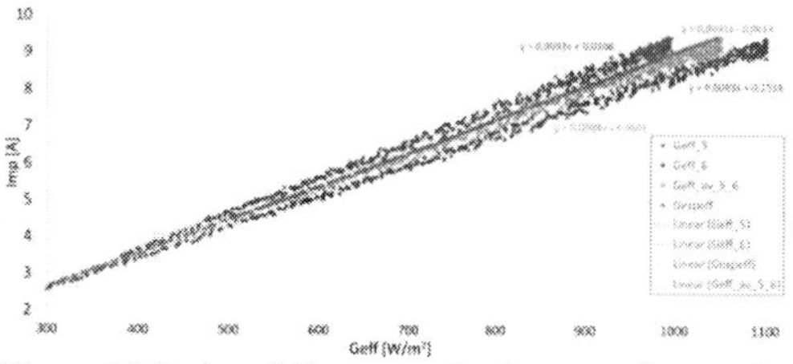

Figure 11 Series of G_{eff} versus I_{mp} in comparison to the reference equation for a winter day with increased albedo using positions 5,6 and average of 5 and 6 as rear side irradiance.

The same calculation was performed for a clear day in winter, with the increased albedo (white material covering the ground under the tilted rack), as shown in Figure 11. In this case, given that it is winter time it is expected that measurements at top and bottom have a higher deviation, so the average of both of them was also plotted. Although this is a simple analysis, the results are consistent and the average value of both sensors in the most central position of the rack turns out to be the best approximation, with average variation of -1,35% in respect to the reference.

3.4 Evaluation of rear-side irradiance model

For the vertical system, in general the model gives a good approximation for both East and West sides especially to the average value of irradiance. One representative clear day in summer is shown in Figure 12, for the West side and the East side. For this day the average error is 4% (East) and -6% (West) over the whole day, which indicates that applying this simple model can give a good prediction of G_{front} and G_{rear} for the vertical configuration during the day (excluding early morning and late afternoon).

For the tilted system, the comparison is not so straightforward given that the impact of the reflected component is much higher, thus shading of the ground and consequently NU play an important role. The comparison for one representative clear day is shown in Figure 13, in which it is possible to see that the model gives an overestimation both in comparison to the average and the minimum value, which was expected since the model does not consider any shading effect of the rack on the ground and uses a constant value of albedo.

The alternative configuration with two sensors has also shown results which are consistent. From the plot of Figure 14 it is possible to see that due to shading of the nearby rack starting in the late morning and increasing in the afternoon as the sun moves to the West, there is a bigger divergence from the model to the measured data. This confirms that the divergence of the real data in respect to the model is due to shading. It is also important to highlight that the top and bottom sensor have shown very similar results in this case, as expected since this configuration is less impacted by shading as there are no modules providing shading. When evaluating the interval from 9 to 1 pm the average error values are 8,5% (bottom sensor) and 10,5% (top sensor).

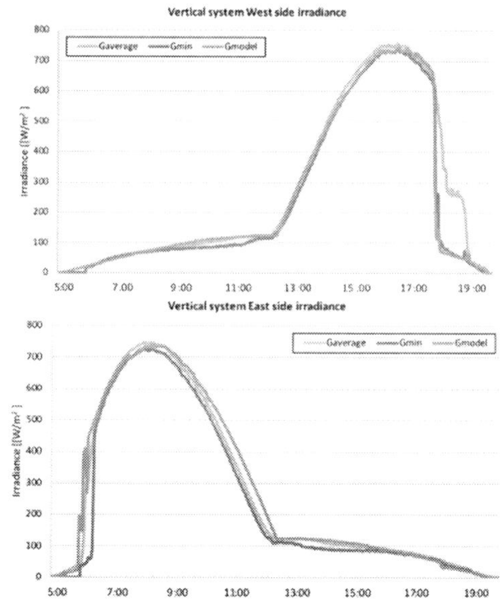

Figure 12 Comparison of irradiance modelled and measured (average and minimum value) for West and East side of vertical array on a representative clear day in summer.

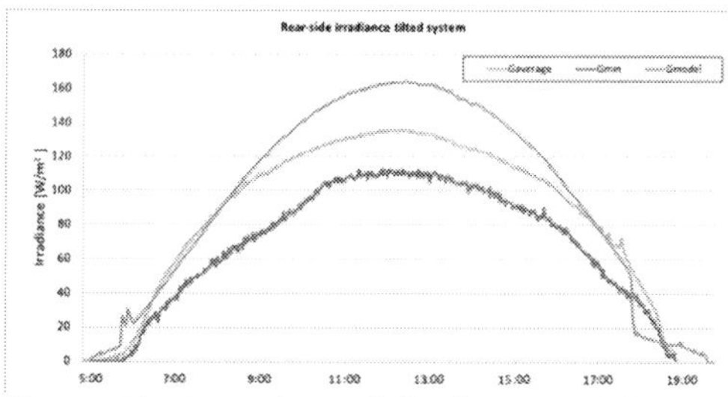

Figure 13 Comparison of irradiance modelled and measured (average and minimum value) for the tilted array on a representative clear day in summer.

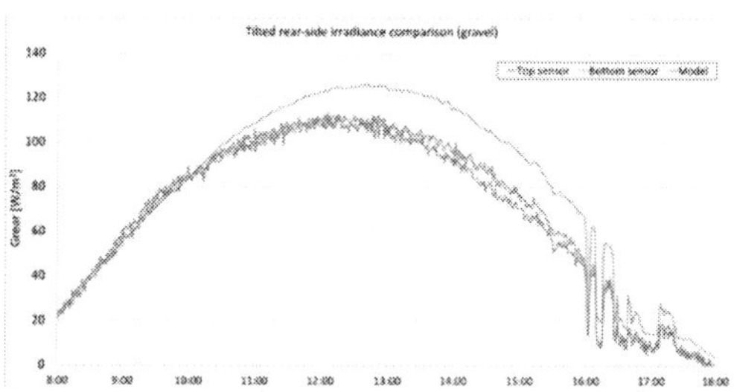

Figure 14 Comparison of irradiance modelled and measured (top and bottom sensors) for the alternative configuration on a winter day.

For the case of increased albedo, as shown in Figure 15 there is a more pronounced difference between top and bottom sensor results, as expected. During the morning hours the top sensor gives a better match to the model. This difference depending on height suggests that an improved model may also need to take into account the height from the ground.

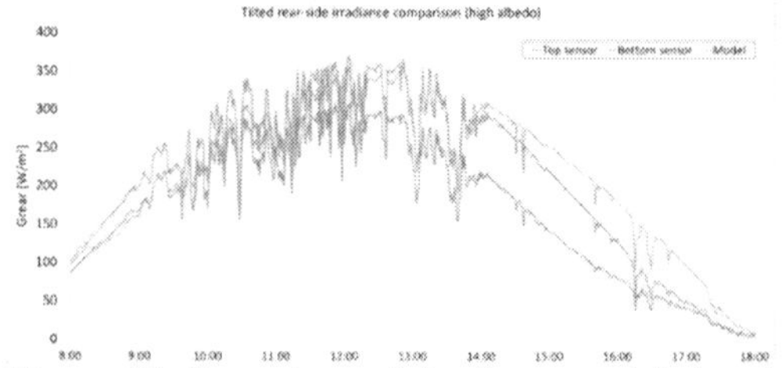

Figure 15 Comparison of irradiance modelled and measured (top and bottom sensors) for the alternative configuration on a winter day with increased albedo.

3.5 Energy yield analysis

The DC energy yield for each month of both bifacial arrays and a closely located and similarly mounted monofacial c-Si array is represented on Figure 16. The tilted bifacial array output exceeds the others all throughout the year, with an average gain in comparison to the monofacial array of 15.8%. The vertical array has a lower energy yield during most of the year, except for the months of June and December, which is also the month of lowest yield for all systems. This configuration has as a disadvantage the low bifaciality coefficient of the modules, which causes the performance of the system to be poor during the afternoon hours. In average the yield of this system is 5% lower than the reference around the year and this difference is less pronounced during the central months of the year, which are responsible for a great share of the annual energy production. Previous work in the same location with a single module of the same bifaciality as the ones used in this work has shown a bifacial gain of around 25% [7]. This demonstrates that systems composed of multiple modules have additional effects – such as the rear-side non-uniformity- which cause a significant reduction on the effective bifacial gain. This dataset will be used in future work for comparison of the calculated estimations.

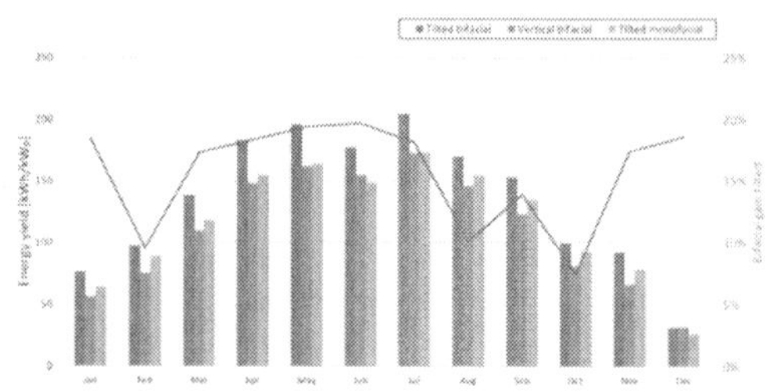

Figure 16 Yearly energy yield for tilted and vertical bifacial systems in comparison to a tilted monofacial and bifacial gain of tilted system (blue line).

4 CONCLUSIONS

For the vertical configuration the impact of non-uniformity is not so relevant as for the tilted system. Additionally, the simple irradiance model applied gives a good approximation to the average irradiance measured.

When it comes to the tilted system, the analysis of I_{mp} and G_{eff} resulted in a behaviour of modules in a system subject to real outdoor conditions comparable to that expected from the indoor characterization. A simple model applied to calculate rear-side irradiance needs to be adjusted to include ground shading in order to avoid over estimation of energy production. The albedo is one important parameter both on the analysis of real outdoor data of bifacial tilted systems and for model application; further work on the influence of this parameter is needed. A suitable de-rating of albedo may give sufficiently good values of rear side irradiance for energy rating purposes.

Regarding a representative value of rear-side irradiance for the tilted system, the sensors placed in the central positions (i.e. those most consistently impacted by shading) have shown a satisfactory result in comparison to values based on measurements of multiple sensors. This may be an advantage when monitoring the performance of larger-scale bifacial systems, since only a small number of sensors may be sufficient.

The results of this work will contribute to on-going work focusing on the extension of IEC61853 energy rating standard to bifacial modules.

5 ACKNOWLEDGEMENTS

Authors would like to thank Mr Francesco Noseda for the technical support, Flavio and Nico Ambrogio for the electronic implementation, Wim Zaaiman for the help and Gambale s.r.l for the arrays set-up.

6 REFERENCES

[1] ITRPV-2020 International Technology Roadmap for Photovoltaic (ITRPV) 11th edition, 2020.

[2] Gracia-Amillo, A.M., Kenny, R. P., Friesen, G., Reise, C., Lopez-Garcia, J. Extension of energy rating to bifacial modules – proposals from the PV-enerate project - JRC Technical Report, 2020.

[3] J. Bonilla-Castro, M. Herz, C. Monokroussos, M. Schweiger, Energy yield comparison between biacial and monofacial PV modules: real world measurements and validation with bifacial simulations, 35th EUPVSEC, 2018, pp. 1236.

[4] C. Deline, S. Macalpine, B. Marion, F. Toor, A. Asgharzadeh, J.S. Stein, Assessment of Bifacial Photovoltaic Module Power Rating Methodologies-Inside and Out, J. Photovoltaics, 7 (2017) 575.

[5] Lopez-Garcia, J., Urraca, R., Kenny, R., Sample, T., Comparison of the energy yield of vertical (E-W orientation) and tilted (equator facing) bifacial module arrays, 37th EU PVSEC, pp. 883–889,2020.

[6] Perez, R., Ineichen, P., Seals, R., Michalsky, J., Stewart, R. Modelling daylight availability and irradiance components from direct and global irradiance, Solar Energy (1990), 44(5), 271-289.

[7] Gracia-Amillo, A., Lopez-Garcia, J., Kenny, R.P., Sample, T., Energy yield analysis of bifacial PV modules: different technologies and configurations, 36th EU PVSEC, pp. 1165–1170, 2019.

TECHNICAL PERFORMANCE EVALUATION OF BIPV AND BAPV SYSTEMS

Patrik Ollas
Department of Energy and Resources
RISE Research Institutes of Sweden
Industrigatan 4, 501 15 Borås, Sweden
patrik.ollas@ri.se

Jon Persson
Department of Energy and Resources
RISE Research Institutes of Sweden
Industrigatan 4, 501 15 Borås, Sweden
jon.persson@ri.se

Peter Kovacs
Department of Energy and Resources
RISE Research Institutes of Sweden
Industrigatan 4, 501 15 Borås, Sweden
peter.kovacs@ri.se

ABSTRACT: In this work, seven residential solar photovoltaic systems are evaluated for their technical performance under a full year's operation in Sweden, at latitude 57.72. The system technologies are selected to reflect available products on the Swedish market and include three building attached (BAPV) and five building-integrated (BIPV) systems. The evaluation is done for the performance ratio (PR) and temperature corrected PR, and the results highlight the seasonal variation and effect from the modules' temperature dependency. The annual losses due to operation at elevated temperature are also quantified for each system, using a temperature-compensated power output, and the results show that the losses are 1-3.2% on the measured yield, depending on the PV technology and extent of roof integration (BIPV or BAPV).
Keywords: Building Integrated PV (BIPV), Building Attached PV (BAPV), System Performance, Performance Ratio, Temperature Dependency

1 INTRODUCTION

Installations of photovoltaic systems on buildings can in principle be done in two different ways: Building Attached PV (BAPV) and Building Integrated PV (BIPV), where the former refers to installations of PV modules that are not an active component of the building's climate screen, while the latter typically replaces existing building elements with PV modules, e.g., roof tiles or facade materials [1]. The market segment for both types has significantly grown in the last years, where BAPV accounts for most installations, reaching more than 600 GW globally. While the size of the BIPV segment is more difficult to determine, it is estimated to be around 1 GW globally [2].

Typically, solar photovoltaic (PV) systems are evaluated using the performance ratio (PR) which describes the specific energy output over a normalized irradiance [3]. However, this definition does not account for the module's temperature dependency which is an important aspect when evaluating the performance [4, 5]. In [6], two methods are introduced to account for the temperature dependency. One using the reference temperature (25°C) from the standard test conditions (STC) and the other the irradiance-weighted average annual module temperature. Both methods adjust the PV output using the instantaneous temperature difference and the temperature coefficient. A simulation study was done in [7] to quantify the resulting seasonal variation in energy yield for four geographical locations and two arrays with different temperature coefficients. Results show that the annual temperature losses were around 1% for the high-latitude locations (Ottawa and Tampere). The effect from varying the module temperature was quantified to the range 1—2.2% from a study of 23 different modules with varying PV technologies [8]. The referred study introduces a way to quantify the losses for elevated temperature operation by adjusting the measured output using the module's temperature coefficient.

The purpose of this work is to evaluate the annual and weekly performance of seven residential photovoltaic systems from operation in Sweden (latitude 57.72). The metric used for the evaluation is the performance ratios—as is and temperature corrected, where the temperature corrections are done with reference to the standardised test conditions (STC) temperature and the irradiance-weighted average annual module temperature, as defined in [6]. All data is based on Class B quality measurements [6]. Quantification is also made of the annual losses from evaluated temperature operation using the standardised reference temperature (25°C) and the irradiance-weighted average module temperature to adjust for the power output.

2 METHODS

In this section the system selection process is motivated, the test sites introduced, and the evaluation methods explained.

2.1 Selection of Systems and Test Site Layout

The selection of systems was made to reflect the residential market in Sweden and to cover its variance in buyers' choices. Seven PV systems were selected with a peak power ranging between 3–6 kWp, and with five systems being Building Integrated (BIPV) and the other three Building Attached photovoltaic (BAPV). The systems also differ in terms of PV technology, i.e., thin film/crystalline, and the type of optimization, i.e., module or string. All systems were oriented due south (180 azimuth) during the test and with a 45° tilt angle to the horizontal. The latter represents a suitable angle at the site's latitude to maximise the annual output [9]. Table I shows the specifications of each of the systems. The system designs and installations were set and performed by the respective system's supplier and all systems were positioned to avoid shading and reflection from surrounding objects. The temperature coefficient in Table I refers to the effect on power output and is taken from the datasheet of the respective PV modules.

Table I Technical specifications of the seven systems used for the performance evaluation. The temperature coefficient refers to the effect on the power output.

	System Identification						
	A	B	C	D	E	F	G
Peak Power [W]	3080	2975	3900	4080	3245	4800	3289
BIPV/BAPV	BIPV/BAPV	BIPV	BAPV	BIPV	BIPV	BAPV	BIPV
Cell Technology[1]	CIGS	CdTe	c-Si	c-Si	c-Si	c-Si	CIGS
Maximum Power Temperature Coefficient [%/K]	-0.33	-0.21	-0.37	-0.41	-0.42	-0.37	-0.33

Table II Summary of measured quantities.

Data	Measurement	Unit	Comment
e^i	Generated energy for system's A–G	kWh	Measured AC quantity after the PV inverter.
t^i_m	Module temperatures	°C	Each system array is equipped with 1-3 temperature sensors attached to the back side of the modules with an adhesive.
POA	Global solar in-plane irradiance	W/m^2	Measured in the array plane at both test sites in Figure 1. Systems A-E are measured with one sensor, and F-G with a separate sensor.

2.2 Data Acquisition

Measurements of energy output and irradiance were made with a 30-second temporal resolution and ambient and module backside temperatures with a 60-second resolution. This was made using two separate data acquisition systems from the two test sites. In the evaluation, the data was averaged to 10-minute values to match the requirement of Class B data acquisition [6]. Figure 1 shows the two test sites, having separate ambient temperature sensors and in-plane pyranometers.

Figure 1 Overview of the two test site installations, where **(a)** shows systems A—E and **(b)** systems F—G[2].

A summary of the measured quantities is given in Table II. The data from the two acquisition systems were then merged to a common time vector and missing data points in the resampled data set were acquired using linear interpolation. Obvious measurement errors, system down-periods, and other abnormalities were removed before the performance evaluation. Energy measurements were made using energy meters with 1000 pulses per kWh and a measurement uncertainty of approx. 1%. Temperature readings were made with PT100 sensors with an absolute error of <±0.5K, and the irradiance measurements were made using both pyranometers and silicon sensors with an uncertainty of 2%. The system's module temperature was measured via sensors on the backside of the panels using 1-3 sensors per system.

2.3 Performance Evaluation

The Performance Ratio (PR^i) is defined in [3] as

$$PR^i(t) = \frac{y_f(t)}{y_r(t)} = \frac{\frac{e^i_{out}(t)}{P^i_r}}{\frac{H(t)}{G_{ref}}} \qquad (1)$$

where $e^i_{out}(t)$ is the measured energy output, P^i_r the array's rated power, $H(t)$ the measured solar irradiance in the module's plane, and G_{ref} the reference value for the irradiance. In this study, the latter was set to 1000 W/m^2 in the evaluation to represent the irradiance level at standard test conditions (STC) defined in IEC 61215 [10]. Using (1) gives the relation between the measured and theoretical output for each time step, "t", and the metric can be evaluated for a defined period by calculating the ratio of the sum in the numerator and denominator. However, this definition does not account for the module's temperature dependency. As noted in [6], a deficit in (1) is the lack of compensation for the elevated temperature and its effect on the PV system's performance. To overcome this, a compensation term is introduced in [6] to account for the difference between module and STC reference temperature (25°C). This term, PR_{STC}, is defined in [6] as

$$PR^i_{STC}(t) = \frac{\frac{e^i_{out}(t)}{P^i_r}}{\frac{H(t)}{G_{ref}}} \frac{1}{c_k^{STC}(t)} \qquad (2)$$

where the compensation term, $c_k^{STC}(t)$, is given as

$$c_k^{STC}(t) = \left(1 + \delta_i\left(t^i_m(t) - 25\right)\right) \qquad (3)$$

with δ_i as the temperature coefficient for the PV module (see Table I) and $t^i_m(t)$ the measured module temperature. Another alternative from [6] is the weather-corrected PR defined as

[1] c-Si = monocrystalline silicon.

[2] The system in the bottom right corner was equipped with a battery storage and excluded from this analysis.

$$PR^i_{corr}(t) = \frac{\frac{e^i_{out}(t)}{P^i_r}}{\frac{H(t)}{G_{ref}}} \frac{1}{c^{corr}_k(t)} \quad (4)$$

where $c^{corr}_k(t)$ is given as

$$c^{corr}_k(t) = \left(1 + \frac{\delta_i}{100}\left(t^i_m(t) - T^{avg,i}_m\right)\right) \quad (5)$$

where $T^{avg,i}_m$ is the irradiance-weighted average annual module temperature and calculated as

$$T^{avg,i}_m = \frac{\sum H(t) t^i_m(t)}{\sum H(t)} \quad (6)$$

Since the module temperature is weighted against the irradiance, night-time values (with zero irradiance) are discarded from the evaluation. The reason for using (2) and (4) is to reduce the seasonal variations that are otherwise observed when the PR is calculated from (1).

Over time, the performance ratio can be expressed using a linear regression [11] as

$$y_f = \overline{pr} \cdot y_r \quad (7)$$

where $\overline{pr}$ is the average performance ratio over a defined time and with y_f and y_r from (1).

Using the energy output and the system's rated power from Table I enables an evaluation of the specific energy output per installed peak power [12] as

$$E^i_S = \frac{\int_{t_1}^{t_2} p_i(t)dt}{P^i_r} \quad (8)$$

where p_i is the measured power output during t_1–t_2.

To quantify the effect of the elevated temperature on the annual energy output, the temperature correction terms from (3) and (5) are used to compensate for the measured power output [8] as

$$p^i_{STC}(t) = p^i(t) \frac{1}{\left(1+\delta_i\left(t^i_m(t)-25\right)\right)} \quad (9)$$

$$p^i_{corr}(t) = p^i(t) \frac{1}{\left(1+\frac{\delta_i}{100}\left(t^i_m(t)-T^{avg,i}_m\right)\right)} \quad (10)$$

and these compensated power outputs are then used together with the numerator in (8) to calculate the annual temperature corrected energy.

3 RESULTS

The seven systems are evaluated for their performance ratio (PR) using the three alternative ways presented in Section 2.3. The measured power outputs are compensated for the elevated temperature operation and the annual yields are compared to the measurements to quantify the effect from the module's temperature dependency.

In Figure 2, the weekly variation in PR for System A is shown for the three alternatives introduced in (1), (2), and (4). Despite the effort of reducing the seasonal variations with (2) and (4), there is still a significant variation observed during the winter period. This is caused by periods of low radiation which fails to initiate any power generation from the arrays. Figure 3 shows an example of two consecutive weeks in December where System A does not generate any output due to the low irradiance. It can also be observed from Figure 2 that without the temperature compensation, PR using (1) is lower than the others for the periods with high irradiance and higher in the winter. However, the absolute differences between the three PR representations are within ±0.05 annually.

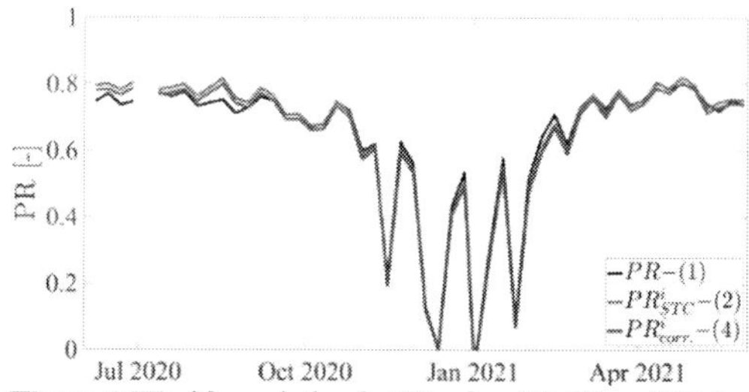

Figure 2 Weekly variation in PR using (1), (2) and (4) for System A.

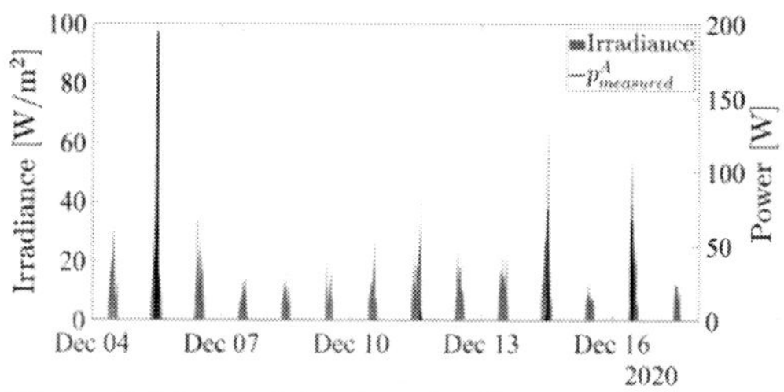

Figure 3 Measured in plane irradiance and System A power output for two consecutive weeks in December.

Comparing the two temperature-compensated PRs using (2) and (4), the results from System A's annual performance show that the STC representation is always higher than the method using the irradiance-averaged module temperature. The reason for this is that the weighted average annual module temperature for System A, $T^{avg,i}_m$ used in (5), is 30.8°C which can be compared to 25°C used in (3). Table III summarises the annual irradiance-average module temperatures from (6) for all seven systems where it can be noted that all of them exceed the STC reference temperature of 25°C. With reference to the extent of roof-integration outlined in Table I, it is evident that the BIPV systems (B, D, E, and G) have a higher temperature on average than the BAPV systems (C and F), and that the hybrid (System A) is somewhere in between these two.

Table III Calculated irradiance-weighted average annual module temperature.

	System Identification						
	A	B	C	D	E	F	G
$T^{avg,i}_m$	30.6	34.5	26.9	31.0	30.3	26.6	38.3

The weekly variation in PR using (1), (2), and (4) for systems B-G are shown in Figure 5 where the seasonal variations from Figure 2 are again noted. Overall, the relative comparison between the PRs follows the trends of System A, with (1) being lower during the warmer periods and higher during the winter. A few exceptions are noted though, one is during December for System B, where the temperature corrected PRs are well above the non-corrected. This is caused by missing module temperature data for a few days with power output, resulting in an overestimation of the temperature corrected values. Noticeable is also the results seen for System's C and F (Figure 5b and Figure 5e) where the temperature-corrected PRs' remain lower than (1) after the winter period. To explain this, Figure 4a and Figure 4b show the measured module temperature as a function of irradiance[3] for System C and F respectively. Each recorded data point is color-coded and the colorbar refers to the daily sequence number (day 1-365) from the start of the measurements. In both figures, the temperature records from the early days (blue) are higher for the same irradiance than in the final stage (yellow), thus affecting the PRs correction terms in (3) and (5). For System G in Figure 5, the temperature corrected PRs significantly overestimates the performance during the warmer periods. This is caused by the installation of the system, located on top of an attic, see the top system in Figure 1. For this system the measured temperature at the back of the modules is affected by the attic temperature, resulting in both an overestimation of the module temperature and also a time-shift in the temperature curve due to the building's inertia.

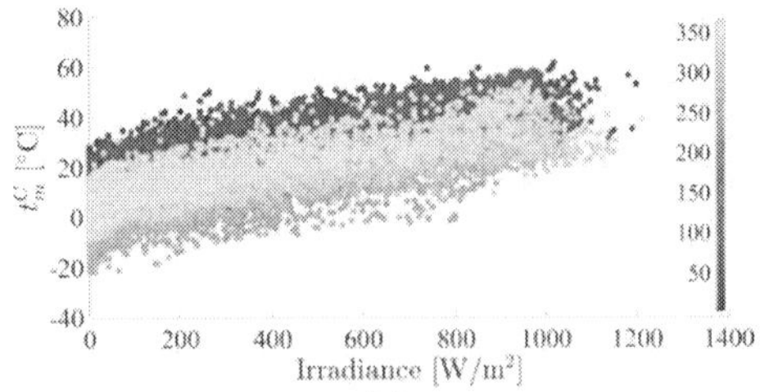

(a)

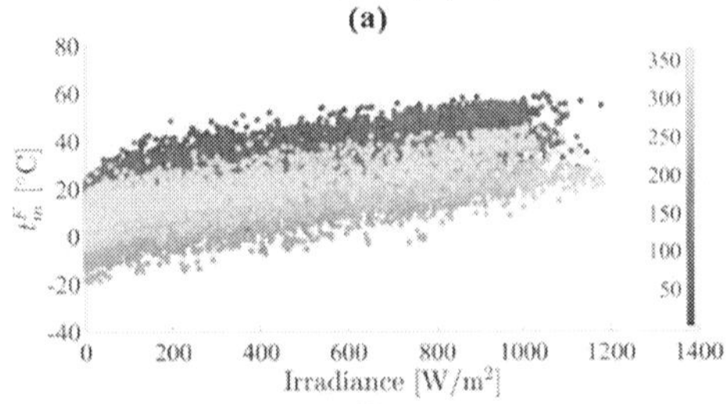

(b)

Figure 4 Continuous record of measured module temperatures as a function of irradiance for System C **(a)** and F **(b)**.

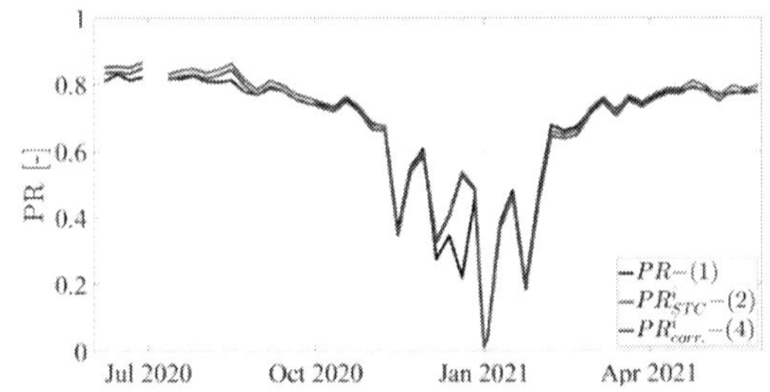

(a) – System B

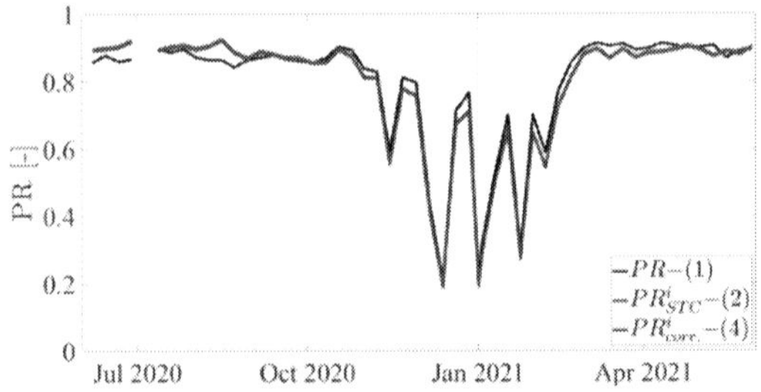

(b) – System C

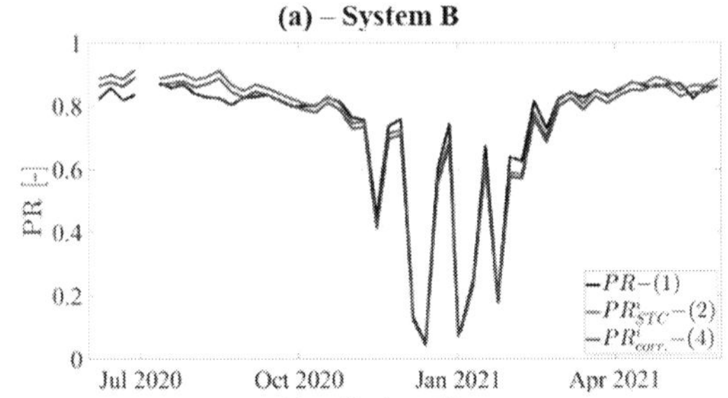

(c) – System D

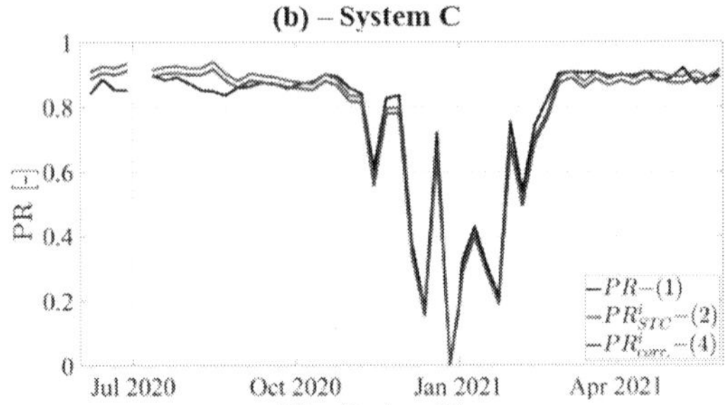

(d) – System E

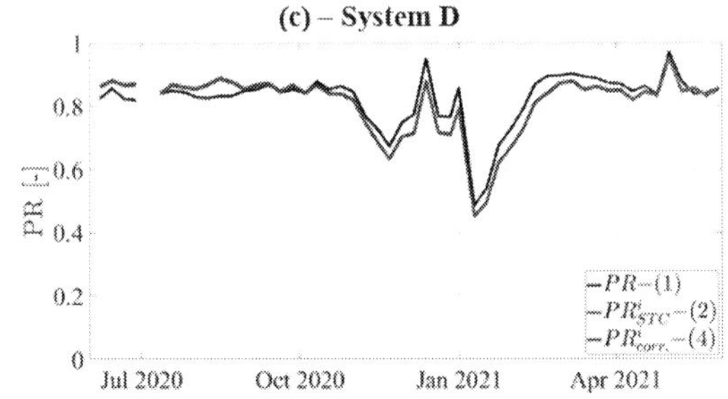

(e) – System F

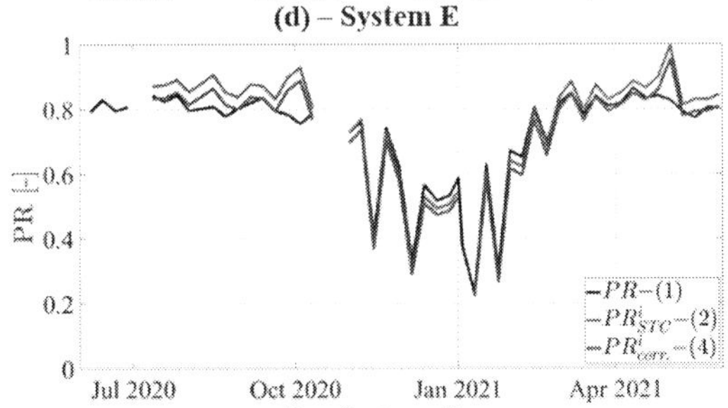

(f) – System G

Figure 5 Weekly variation in PR for systems B—G using (1), (2) and (4).

[3] The authors acknowledge that the module temperatures are also affected by the ambient temperature, but here the analysis is done only on the irradiance level since it is the factor used in the PR equations.

A comparison of the measured and temperature corrected power outputs—using (9) and (10)—is shown in Figure 6 for an arbitrary day's operation with clear-sky. The measured module temperature, t_m^A, is also shown together with the STC reference (25°C) and the annual irradiance-average temperature, T_m^{avg}. As seen from the comparison, the measured power is higher in the morning when the module temperature is lower than the reference temperatures used for the compensations in (3) and (5). Around 9 a.m., the module temperature exceeds the STC and the corresponding temperature corrected power, p_A^{STC}, becomes higher than the measured output. Similar, when the annual irradiance-average temperature is reached, $p_A^{corr.}$ is higher than the measured values. Since T_m^{avg}>25 for System A, the STC compensated power using (9) will always be higher than the corresponding value from (10). With the temperature corrected power outputs, the annual energy for each system is calculated using (8) and the resulting quantities are shown in Figure 7a for all systems together with their extent of roof-integration (BIPV or BAPV) and temperature coefficient from Table I. Comparing the temperature corrected outputs, the STC is the highest due to the previous reasoning of 25<T_m^{avg}. The STC compensated yields are within 3.2% (<1% for the annual irradiance-average temperature corrected) compared to the measured quantities for all but System G, suggesting that the measured module temperature is an overestimation of the true experienced value. The differences in temperature compensated outputs from the two methods are shown in Figure 7b with reference to the measured values. Noticeable is that $E_{corr.}^i$ differs within 1% for all systems except G, and this is as expected from [6]. As for E_{STC}^i, where the power is corrected to the STC temperature, the temperature compensated output is 1.0—3.2% higher, meaning that the losses due to the elevated temperature are in the same magnitude. Comparing these findings with the related study in [8] and the results from the BAPV mounted c-Si's show that the results align well in terms of annual losses from elevated temperature operation.

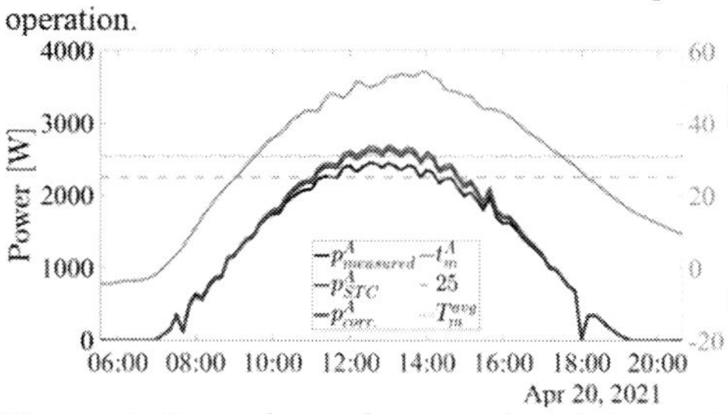

Figure 6 Comparison of measured and temperature corrected power output for System A.

The temperature corrected PR using (4) is described in [6] to give accurate results from the yearly PR, compared to the PR using (1), and that the correction using the STC temperature gives an overestimation of the performance. In Figure 8, the yearly PRs using (7) for the three representations are shown as a function of specific energy output, calculated from (8) for all systems except G. From this, the claim made in [6] is confirmed, where the difference in yearly PR using (1) and (4) are insignificant, while the results using (2) shows an overestimation of the PR in the range 0.005—0.02.

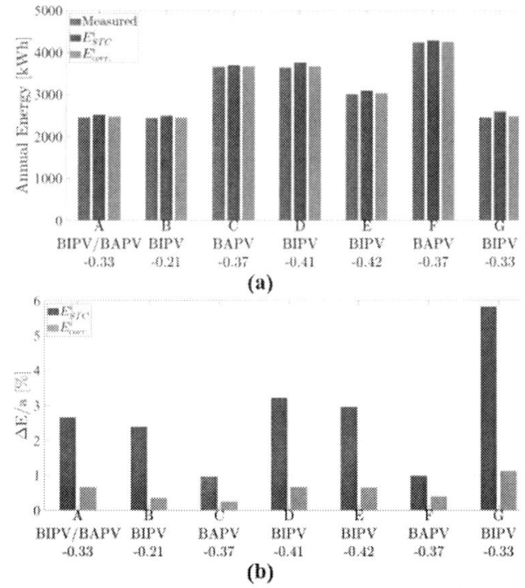

Figure 7 Annual energy from measurements and the two temperature-corrected power outputs **(a)**, and the difference between the temperature corrected values and the measured output **(b)**.

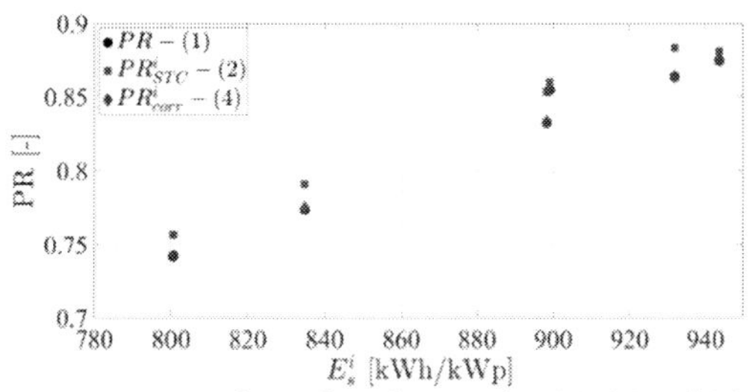

Figure 8 Comparison of yearly PR using (1), (2) and (4) as a function of specific energy output for each system.

4. DISCUSSION & CONCLUSION

This study has compared the annual performance of seven residential PV systems during operation in Borås, Sweden. The systems have been evaluated for the performance ratio (PR) using three different representations. Results show that regardless of the method to calculate PR, a significant seasonal variation is noted. Concluding that the low radiation at this high latitude during the winter period makes PR estimations highly uncertain.

An estimation of the annual effect from elevated module temperature operation on the energy yield shows that for the BAPV mounted c-Si technology the losses are 1%. And that the BIPV mounting of the same technology—with the modules also having a higher temperature dependency—adds 2 percentage points to these losses.

Results show that the temperature correction using the annual irradiance-average module temperature shows good coherence with the measurements, both in terms of yearly PR and annual energy yield.

The results also highlight the difficulty in acquiring accurate measurements of the module temperatures and the effect it has on the temperature corrected powers and performance ratios. As many temperature sensors are mounted out of reach, taped at the back of the modules, it is difficult to validate their attachment to the surface over time. Hence, the first months' data is considered more reliable than temperature data from later in the measurement period.

ACKNOWLEDGEMENT

The work is financially supported by the Swedish Energy Agency Test Lab. The authors would also like to acknowledge the thesis work from Gulala Safari and Awaz Rashid at Mälardalen University titled "Technical Performance Evaluation of Solar Photovoltaic Systems".

REFERENCES

[1] K. Berger, A. Belén, Cueli, S. Boddaert, M. Del Buono, V. Delisle, A. Fedorova, F. Frontini, P. Hendrick, H. Ishii, C. Kapsis, J.-T. Kim, P. Kovacs, N. Martin, L. Chivelet, M. Maturi, A. Machado, A. Schneider and H. Rose Wilson, "International definitions of "bipv"," International Energy Agency. Photovoltaic Power Systems Programme, 2018.

[2] G. Masson and I. Kaizuka, "Trends in photovoltaic applications 2020," International Energy Agency. Photovoltaic Power Systems Programme, 2020.

[3] International Electrotechnical Commission, "Photovoltaic system performance monitoring-guidlines for measurement, data exchange and analysis," International Electrotechnical Commission (IEC), 1998.

[4] S. Dubey, N. Sarvaiya and B. Seshadri, "Temperature dependent photovoltaic (pv) efficiency and its effect on pv production in the world–a review," *Energy Procedia*, vol. 33, pp. 311-321, 2013.

[5] M. Theristis, V. Venizelou, G. Makrides and E. Georghiou, "Energy yield in photovoltaic systems, in: McEvoy's handbook of photovoltaics," *Elsevier*, pp. 671-713, 2018.

[6] European Committee for Electrotechnical Standardization, "Photovoltaic system performance - Part 1: Monitoring," European Committee for Electrotechnical Standardization, 2017.

[7] P. Manganiello, J. Govaerts, I. T. Horvarth, M. G. Chowdhury, G. H. Yordanov, H. Goverde, B. Aldalali, I. Beausoleil-Morrison, S. Valkealahti, K. Lappalainen and J. Poortmans, "Tuning electricity generation throughout the year with PV module technology," *Renewable Energy*, vol. 160, pp. 418-427, 2020.

[8] A. Louwen, A. C. de Waal, R. E. I. Schropp, A. P. C. Faaij and W. G. J. H. M. van Sark, "Comprehensive characterisation and analysis of PV module performance under real operating conditions," *Progress in Photovoltaics*, vol. 25, pp. 218-232, 2016.

[9] M. Z. Jacobson and V. Jadhav, "World estimates of pv optimal tilt angles and ratios of sunlight incident upon tilted and tracked pv panels relative to horizontal panels," *Solar Energy*, vol. 169, pp. 55-66, 2018.

[10] International Electrotechnical Commission, "Terrestrial photovoltaic (PV) modules - Design qualification and type approval - Part 2: Test procedures," European Committee for Electrotechnical Standardization, 2017.

[11] A. Woyte, M. Richter, D. Moser, N. Reich, M. Green, S. Mau and H. G. Beyer, "Analytical monitoring of grid-connected photovoltaic systems. good practices for monitoring and performance analysis," IEA PVPS Task 13, 2014.

[12] C. P. Chioncel, D. Kohake, L. Augustinov, P. Chioncel and G. O. Tirian, "Yield factors of a photovoltaic plant," *Acta Technica Corviniensis-Bulletin of Engineering*, vol. 2, pp. 63-66, 2010.

[13] T. Dierauf, A. Growitz, S. Kurtz, J. L. B. Cruz, E. Riley and C. Hansen, "Weather-corrected performance ratio," National Renewable Energy Lab (NREL), 2013.

[14] K. Keeratimahat, A. Bruce and I. MacGill, "The impact of dc/ac ratio on short-term variability of utility-scale pv plants," in *IEEE Power & Energy Society General Meeting (PESGM)*, 2019.

[15] R. S. Faranda, H. Hafezi, S. Leva, M. Mussetta and E. Ogliari, "The optimum pv plant for a given solar dc/ac converter," *Energies*, vol. 8, no. 6, pp. 4853-4870, 2015.

[16] N. Müller, S. Z. P. Kouro, P. Wheeler, G. Bittner and F. Girardi, "Energy storage sizing strategy for grid-tied pv plants under power clipping limitations," *Energies*, vol. 12, no. 9, 2019.

DATA-DRIVEN ASSESSMENT OF SOILING LOSS IN PHOTOVOLTAIC PLANTS

Kaan Yurtseven[1,2], Engin Karatepe[1,*], Esref Deniz[2]
[1]Department of Electrical - Electronics Engineering, Dokuz Eylul University, Izmir, Turkey
[2]Entegro Enerji Sistemleri, Izmir, Turkey
*Email: enginkaratepe@gmail.com

ABSTRACT: Utilizing reliable monitoring and available data in order to ensure that the PV plants become best-return-on investments is possible thanks to ever going developments in data-driven algorithms. One of the major challenges in PV power generation is the soiling events caused by the dusty environments and climatic conditions resulting in the reduction of the performance of the PV plants. Since the soiling is a complex event whose characteristics are forged by a number of external stochastic phenomena, a number of studies are conducted by researchers. Soiling in PV plants is a continuous subject that emerges over time as an important drawback in the performance of the PV plants. In that point, it is crucial to automatically determine appropriate cleaning schedules in order to achieve a high-profit PV plant operation. In this study, a method is presented which only uses the DC current inputs of the inverters to determine the soiling loss without using any environmental sensors. The proposed method is tested by conducting experimental works on three different PV plants which are connected to the power grid in Turkey. The results show that the proposed method is effective for analyzing the performance reductions due to soiling in the PV plants without using any environmental sensors.
Keywords: PV System, Operation & Maintenance, System Performance, Soiling Loss, Shading

1 INTRODUCTION

There is a rapid growth in the investments in photovoltaic (PV) power generation, and it is considered as the favorite among other renewable energy technologies [1]. Unceasing advances in technology made PV plants to be considered as secure investments. In that point, it is crucial to utilize reliable monitoring and available data in order to ensure the maximum energy yield. By doing so, it is possible to maximize the efficiency and profit through appropriate operation and maintenance (O&M) schedules.

One of the major challenges in PV power generation is the soiling events caused by the dusty environment and climatic conditions [2]. Moreover, PV soiling is considered a complex phenomenon since it is strongly dependent on geographical features of the PV site, seasonal variations, environmental and weather conditions [3]. Since the soiling is a complex event whose characteristics are forged by a number of external stochastic phenomena, it drew attention in the literature. In [4], the performance degradation caused by dust deposition on PV plants considering the type of solar collectors, geographical location, local climate, and exposure period of the collectors absent any manual cleaning is presented. In addition, it is observed in [5] that the daily change in a PV module's cleanness is inversely proportional to the daily average ambient dust concentration and the relative humidity, directly proportional to the wind speed. Furthermore, in [6], the correlations are significantly weaker where seasonal variations in relative humidity are weak. In that point, it is obvious that there is a challenge which is to determine the loss due to soiling in the PV plant site in order to ensure efficient operation of the PV plant. The solution that first comes to mind in this situation is to install a number of sensors to detect these several parameters and conduct studies on the efficiency of the PV plant. However, a huge number of sensors are needed for an accurate representation of the PV plant. Even though a number of sensors are installed in the PV plant, it is not possible to fully specify the nature of the PV plant [7]. Overcoming the demanding nature of sensor installation is possible by developing data-driven methods thanks to the fact that today's inverters allow users to gather several types of electrical data.

In this paper, a novel data-driven method that only uses the data available from the inverters in the PV plant is presented for the assessment of soiling loss. The proposed method consists of five stages. Firstly, the days with clear-sky conditions are selected by observing the DC current values. Secondly, the selected DC current values are filtered in order to obtain only the values of a narrow interval during the noon period. Thirdly, the maximum amount of irradiance is calculated by using the well-known mathematical model as a function of time taking into account the Sun's path and the properties of relevant PV plants, latitude, longitude, and tilt angles. Fourthly, the theoretical DC current values are calculated by scaling the mathematically obtained irradiance with the maximum power point currents of the PV modules. Fifthly, the mathematically calculated DC current values are compared with the current values of the DC side which receives the maximum irradiance on the site in terms of the values transformed into moving averages for a more revealing understanding. This method eliminates the requirement for environmental sensors in analyzing the soiling events at the PV plant sites.

2 METHODOLOGY

The proposed method is based on only the DC current values of the inverter which are easily available from either the built-in monitoring system of the inverters or a custom monitoring system. In PV plants distinguishing between soiling and abnormal conditions is a challenge because they all result in similar electrical outputs. In order to overcome this problem, the gathered data needs to be undergone preprocessing. The preprocessing is done by filtering out the data that belong to cloudy-sky and abnormal conditions [7]. Herein, daily DC current values are checked for clear-sky and cloudy-sky conditions as shown in Figure 1. It is worthy to note that instead of dealing with the data gathered throughout the day, only a narrow interval during noon, when the irradiance is received fairly by all the dc sides of inverters, is used in the proposed method as depicted in Figure 2. Each narrower cycle in the box here corresponds to the samples between 12:00 and 14:00 at noon. After the preprocessing, the DC current data is compared to the corresponding theoretical clear-sky DC current values. In order to theoretically calculate the DC current values, first the well-known mathematical model is used to characterize the irradiance received by the PV plant sites, see the book [8] for further information. In that point, the maximum power point current (Impp) value of the PV modules used in the PV plant and the topology information of the PV plant is utilized.

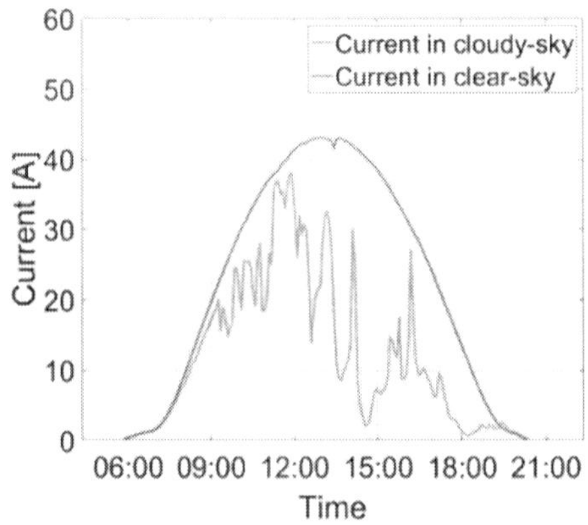

Figure 1: DC current in cloudy-sky and clear-sky conditions.

By scaling the ideal Impp values obtained from the manufacturer with the mathematically generated irradiance values, the theoretical DC current values are calculated. In the proposed method, moving averages of the theoretical and actual DC current values are calculated before the comparison in order to obtain more explicit results. Moreover, the comparison between the moving averages of the actual and theoretical DC current values of the DC side, which received the most irradiance during the noon period, is done in terms of the percentage of the difference due to soiling loss.

3 EXPERIMENTAL RESULTS

This section is dedicated to the experimental results which are obtained from three different PV plants connected to the power grid in Turkey. The detailed information about those PV plants is given in Table I.

The mathematically modeled and actual DC currents of the DC side which receives the maximum irradiance on the site in the first PV plant are given in Figure 3. The daily DC current values are 5-min samples between 12:00 and 14:00. The DC current values are transformed into moving averages for a more revealing understanding as shown in Figure 4 along with their percentage differences. In the PV plant there are 2850 PV modules whose Impp value is 9.39 A at STC. It can clearly be seen from Figure 4a that actual DC current values follow a lower path than the mathematical DC currents resulting in an increase in the difference throughout the consecutive clear-day conditions, as shown in Figure 4b. It is worthy to note that consecutive clear-days do not have to be sequential in terms of the days of the year. The increase in the difference is caused by soiling in the PV plant.

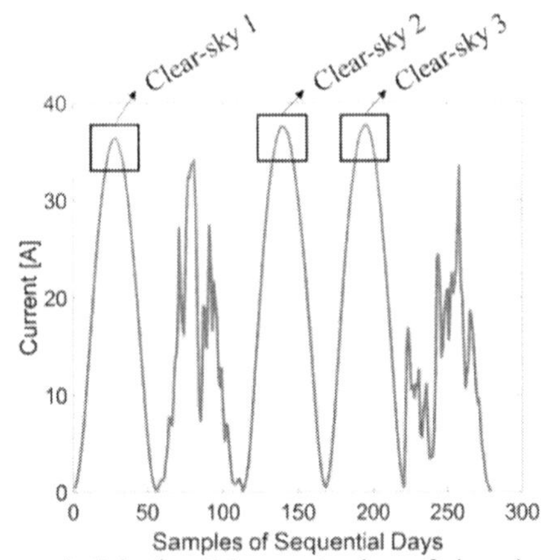

Figure 2: Selecting DC current values of clear-sky days during the noon period.

It is known that in the early days of September in order to overcome significant soiling loss, the PV plant is cleaned manually which results in a near-zero difference between the mathematical and actual DC current values. It is worthy to note that other than the performance increases due to manual cleaning there are other increases due to natural cleaning such as rainfalls during the autumn season.

Likewise mathematical and actual DC currents of a DC side in the second PV plant are given in Figure 5. The daily DC current values are 15-min samples between 12:00 and 14:00. In the PV plant there are 1924 PV modules whose Impp value is 8.48 A at STC. The moving averages of mathematical and actual DC current values with their differences are given in Figure 6. It is clear from Figure 6a that there are two time periods when the actual DC current values are lower than the mathematical DC currents due to soiling in the PV plant. Moreover, it is known that in the last days of August and mid-October the PV plant is cleaned manually. Therefore, the difference curve tends to have near-zero values after cleanings, as shown in Figure 6b.

Table I: PV plant specifications

PV Plant	Peak Power	Number of PV Modules	Inverters	*PV Module I_{mpp}
1	1083 kWp	2850	18 x 50 kW	9.39 A
2	500 kWp	1924	8 x 50 kW + 1 x 20 kW + 1 x 15 kW	8.48 A
3	356 kWp	1320	6 x 50 kW	8.71 A

**Standard test conditions (STC)*

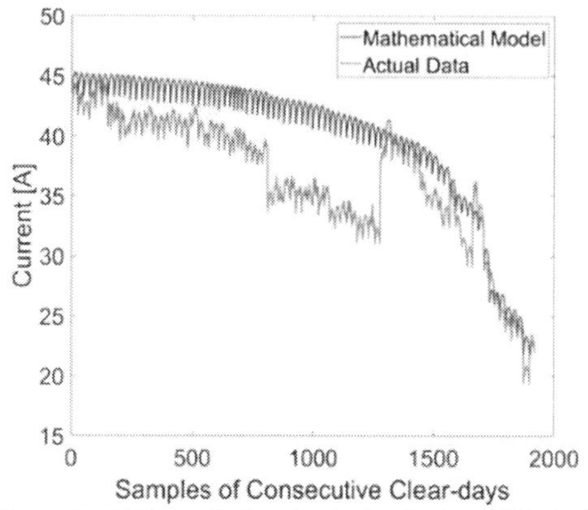

Figure 3: Mathematical and actual outputs of PV plant 1 without moving average (5 min sampling interval)

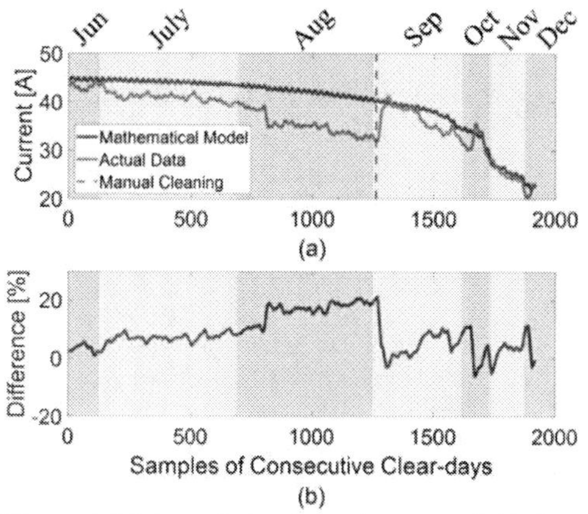

Figure 4: Mathematical and actual outputs of PV plant 1.

The mathematical and actual DC currents of a DC side in the third PV plant are given in Figure 7. The daily DC current values are 5-min samples between 12:00 and 14:00. In the PV plant there are 1320 PV modules whose Impp value is 8.71 A at STC. It can be seen from Figure 8a that during the early samples there is a huge difference between the mathematical and actual DC current values due to soiling in the PV plant. It is known that the PV plant is manually cleaned in the early days of October which results in near-zero differences as shown in Figure 8b.

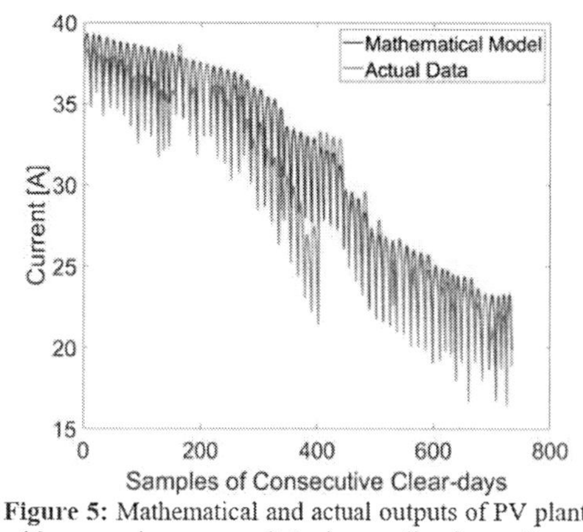

Figure 5: Mathematical and actual outputs of PV plant 2 without moving average (15 min sampling interval)

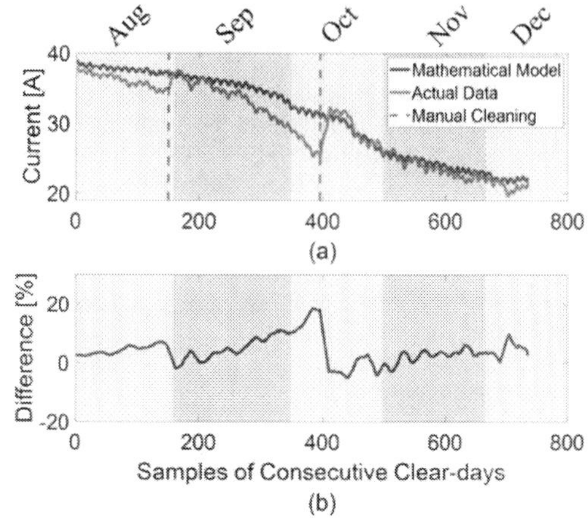

Figure 6: Mathematical and actual outputs of PV plant 2.

In the light of the above result, it is obvious that the cleaning schedules mostly depend on the tolerance of the PV plant owner to the power loss amount. With the proposed method, a desired difference threshold value can be set, and the cleaning is scheduled accordingly.

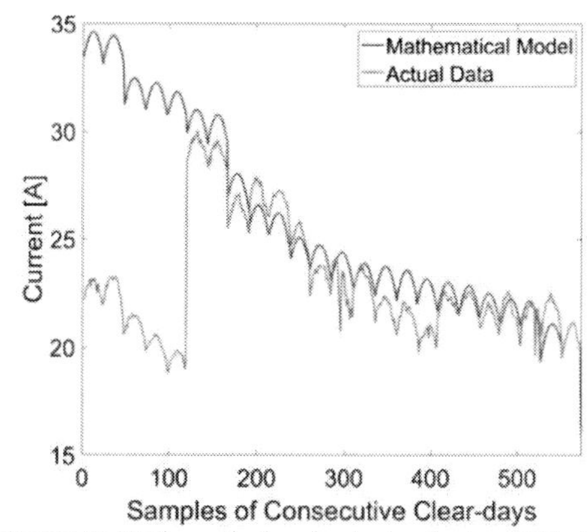

Figure 7: Mathematical and actual outputs of PV plant 3 without moving average (5 min sampling interval)

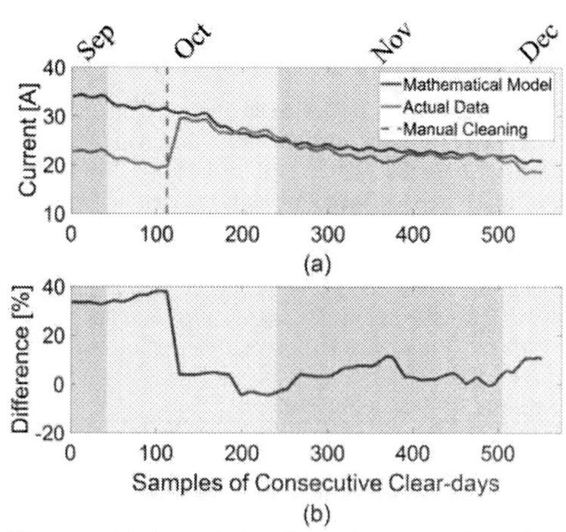

Figure 8: Mathematical and actual outputs of PV plant 3.

4 CONCLUSION

This paper presents a method that can quantify the soiling in a PV plant without using any additional environmental sensors. In this study, it is observed that the soiling results in a difference between actual and mathematically modeled DC currents in clear-sky conditions. The rate of this difference is the parameter that helps O&M crews to schedule the cleaning of the PV plant in order to achieve an efficient and high-profit PV plant operation.

5 ACKNOWLEDGEMENT

This work was supported by the Scientific and Technological Research Council of Turkey (TUBITAK) with the project number 7190391 under TUBITAK-TEYDEB 1507 support program.

5 REFERENCES

[1] A. Jäger-Waldau, "Snapshot of Photovoltaics—February 2020," *Energies*, vol. 13, no. 4, p. 930, Feb. 2020, doi: 10.3390/en13040930.

[2] W. Javed, B. Guo, B. Figgis, L. Martin Pomares, and B. Aïssa, "Multi-year field assessment of seasonal variability of photovoltaic soiling and environmental factors in a desert environment," *Solar Energy*, vol. 211, pp. 1392–1402, Nov. 2020, doi: 10.1016/j.solener.2020.10.076.

[3] L. Micheli, E. F. Fernández, M. Muller, and F. Almonacid, "Extracting and Generating PV Soiling Profiles for Analysis, Forecasting, and Cleaning Optimization," *IEEE Journal of Photovoltaics*, vol. 10, no. 1, pp. 197–205, Jan. 2020, doi: 10.1109/JPHOTOV.2019.2943706.

[4] A. Sayyah, M. N. Horenstein, and M. K. Mazumder, "Energy yield loss caused by dust deposition on photovoltaic panels," *Solar Energy*, vol. 107, pp. 576–604, Sep. 2014, doi: 10.1016/j.solener.2014.05.030.

[5] B. Guo, W. Javed, B. W. Figgis, and T. Mirza, "Effect of dust and weather conditions on photovoltaic performance in Doha, Qatar," in *2015 First Workshop on Smart Grid and Renewable Energy (SGRE)*, Mar. 2015, pp. 1–6, doi: 10.1109/SGRE.2015.7208718.

[6] R. R. Cordero *et al.*, "Effects of soiling on photovoltaic (PV) modules in the Atacama Desert," *Scientific Reports*, vol. 8, no. 1, Art. no. 1, Sep. 2018, doi: 10.1038/s41598-018-32291-8.

[7] K. Yurtseven, E. Karatepe, and E. Deniz, "Sensorless fault detection method for photovoltaic systems through mapping the inherent characteristics of PV plant site: Simple and practical," *Solar Energy*, vol. 216, pp. 96–110, Mar. 2021, doi: 10.1016/j.solener.2021.01.011.

[8] J. A. Duffie and W. A. Beckman, *Solar Engineering of Thermal Processes*. John Wiley & Sons, 2013.

ESTIMATION OF SNOW LOSS FOR PHOTOVOLTAIC PLANTS IN NORWAY

Mari B. Øgaard[1,2*], Heine N. Riise[2], and Josefine H. Selj[1,2]
[1]Department of Technology Systems, University of Oslo, Kjeller, Norway
[2]Department of Renewable Energy Systems, Institute for Energy Technology, Kjeller, Norway
*Corresponding author. E-mail: mari.ogaard@ife.no, Tel.: +47 976 356 08,
Adr.: Instituttveien 18, 2007 Kjeller, Norway

ABSTRACT: Large PV plants are increasingly common in locations with colder climates where snow can lead to significant PV power loss. For these locations, estimates of snow loss is necessary for accurate PV yield modeling. Robust estimation of snow loss is, however, challenging. Snow-induced loss is expected to vary with climate, weather, and PV plant design. In this work, we estimate snow loss from historical data for a set of PV plants in Norway. To extend the snow loss dataset, 12 years of weather data and a modified adaption of the Marion snow loss model are used to simulate snow loss for the analyzed PV plants over time. For the historical data, we observe variations in annual losses for the same system of more than 10 percentage points. For some of the systems, we find losses in a range from 0 to 100 % for the same month. As expected, systems with colder climates have higher loss than systems in warmer climates, and systems with higher tilt has lower loss than systems with lower tilt. With snow loss modeling we get improved understanding of typical and extreme values, and the potential inter-annual variation in monthly and annual snow loss.

Keywords: PV System, System Performance, Modeling, Soiling, Snow

1 INTRODUCTION

As cost reductions have made photovoltaics (PV) a favorable choice also in colder climates, deployment rates in regions with snow falls are rapidly increasing [1–3]. Snow on PV modules may lead to significant power loss. For certain locations snow fall can result in zero electricity production in the winter season and more than 30 % annual loss [4]. Consequently, it is an important loss mechanism to consider in PV system models to get accurate assessments of the expected energy generation from PV plants in snow-affected locations. Snow-induced PV power loss is expected to vary from year to year, between different system configurations and between different locations. To get accurate snow losses for a specific system, a model taking into account the different influential parameters is therefore necessary. Recent research has demonstrated that for snow-affected locations the uncertainty in yield estimations [5–8] and forecasting [9] can be reduced if snow loss models are included. Despite this, snow loss models are often not implemented in PV simulation software. The System Advisor Model (SAM) has implemented the model suggested by Marion et al. [5,6], but in other software, snow is either not considered [10] or estimated by constant soiling values [11] with little guidance on how these constant values should be obtained.

Accurate snow loss modeling is, however, challenging, because the parameters influencing the snow cover and resulting PV system loss are manifold. The influential parameters range from weather conditions (precipitation, temperature, irradiance, wind, etc.), to installation and technology specific configurations (tilt, module technology/orientation, objects obstructing snow sliding etc.) [1,12] and type of snow [4]. Multiple snow loss models have been suggested [4], but validation is typically lacking [6]. To include all the parameters influencing snow cover and resulting loss in a physical model is challenging, and most suggested models for PV snow loss are based on empirical correlations [4].

In our previous work [13], we show that the snow loss model suggested by Marion et al. [3], where empirical correlations are used to model natural snow clearing, performs better than models where snow loss is directly estimated based on empirical correlations between power loss and system and weather data. Ryberg et al. [6] and van Noord et al. [14] also find acceptable correlation between estimated and modeled snow loss using the Marion model.

To estimate the snow coverage on PV modules, the Marion model aims to predict: 1) presence of snow cover on PV modules, 2) when snow is cleared off the modules, and 3) the snow clearing rate. The separation of these three processes in the model, enables improvement of the model by developing the modeling of each process by either using additional physical modeling or collecting more empirical data. In the model, the snow clearing rate is estimated with an empirical snow clearing coefficient. Many different parameters related to system design and weather/snow conditions are assumed to impact how fast the snow is cleared [13]. Frameless modules [15], empty space below modules [12] will promote sliding, for instance. With more data from different system configurations in different climates, we would get improved understanding of which parameters that impact the snow clearing rate the most, and consequently also get better values for the snow clearing coefficient and its potential variation.

In our evaluation of the model [13], we estimate the snow clearing coefficient from the snow loss data for the analyzed system, and we observe that for thin snow covers, the natural snow clearing rate is faster than the clearing rate of thicker covers [13]. By introducing separate snow clearing coefficients for thin and thick snow covers, reduced error in modeled snow loss is achieved. This also seems to make the model more general: when using snow depth dependent snow clearing coefficients we get better results when we model losses for systems with similar technical configurations with the same coefficients compared to when we use one single coefficient [16]. For transferability, it is important that we can use the same empirical coefficients for systems with similar technical configurations.

In addition to the challenge of accurate snow loss modeling, there is a lack of established guidelines on how to take snow losses into account when used in e.g. PV yield modeling or PV system dimensioning. Input data for the snow loss estimation, temporal resolution of the loss parameter, inter-annual variations and the impact of climate change need to be discussed. As pointed out by Marion et al. [5], typical meteorological year (TMY) values are not sufficient to use as input in snow loss modeling for PV yield assessments. Because snow is not one of the parameters considered in the derivation of TMY, TMY data does not necessarily represent a typical snow year. Using a long time series of meteorological data, enabling quantification of typical values and the inter-annual variability is suggested instead [5]. It is, however, important to use recent data. Because of climate change, historical snow data might not be representative for future snow conditions. In Norway, it is estimated that climate change will lead to reductions in snow depth and length of snow season, and an increase in snowline elevation [17]. Temporal resolution of the snow loss parameter is to our knowledge not much discussed in the literature. In the simulation tool PVsyst, monthly constant snow losses are used for PV simulations [11]. While this can be sufficient in assessments of total yield, this will not sufficiently describe the potential inter- and intraday variation. This variation can be relevant in system dimensioning, in particular for hybrid/battery systems.

In this work, we estimate the snow loss for a set of PV plants in Norway. Two different system designs are evaluated: commercial systems with modules installed with low tilt angles on flat roofs, and residential systems on tilted roofs. The aim of this analysis is to describe the variations in both monthly and annual snow losses, with respect to both time, location and system configuration, and to discuss how this could be included in e.g. PV yield modeling. The losses are estimated using both historical data and simulations based on longer time series of weather data and a modified adaption of the Marion snow loss model.

2 METHODOLOGY

2.1 PV system data

Seven PV installations in Norway with a total installed capacity of 1.6 MWp are analyzed. The evaluated dataset is the same as the dataset used to validate the modified snow loss model in [16], but some of the data series are extended in time. Two different system types are evaluated: residential systems on tilted roofs, and commercial large-scale systems on flat roofed buildings. The commercial systems have modules installed with low tilt and east/west orientation. This configuration is not optimal for total annual production in Norway, but is commonly used on flat roofed buildings to increase the packing density and reduce the seasonality of the production profile. The modules are installed in portrait orientation at the residential systems, and landscape orientation at the commercial system. All the PV modules are crystalline silicon. Apart from some variations in exact orientation, and tilt for the residential systems, the installations of the same type are assumed technically identical. Tilt and length of analysis period for the systems are given in Table I.

Table I: Module tilt and length of analysis period for analyzed systems

System ID	Tilt	Analysis period
Residential systems		
R1	26	Jan 2019 – June 2021
R2	40	Jan 2018 – June 2021
R3	24	Jan 2019 – June 2021
Commercial systems		
C1	10	Jan 2015 – June 2021
C2	10	Jan 2017 – June 2021
C3	10	Jan 2018 – June 2021
C4	10	Jan 2018 – June 2021

The measured energy of the PV systems is collected from the inverters. For the commercial systems, the effective in plane irradiance and the module temperature is measured by reference cells. The residential systems have no on-site sensors. For all the locations, snow depth and snow fall data are collected from seNorge.no [18] and temperature and global horizontal irradiation (GHI) data are collected from nearby weather stations [19].

As illustrated in Figure 1, the analyzed systems are situated in three different geographic regions in Norway (East, West and Central), and in three different Köppen-Geiger (KG) [20] climate zones (Humid continental climate (Dfb), subarctic climate (Dfc) and oceanic climate (Cfb)). This gives variation in snow and weather conditions between the locations. Figure 2 shows 16 years of snow depth data for the four different combinations of geographic region and climate zone.

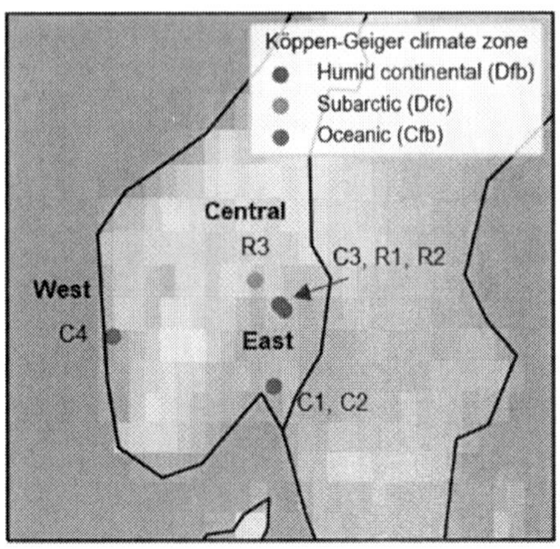

Figure 1: Location on the map for the analyzed systems. The locations are labeled with geographic region and climate zone is given by the marker color.

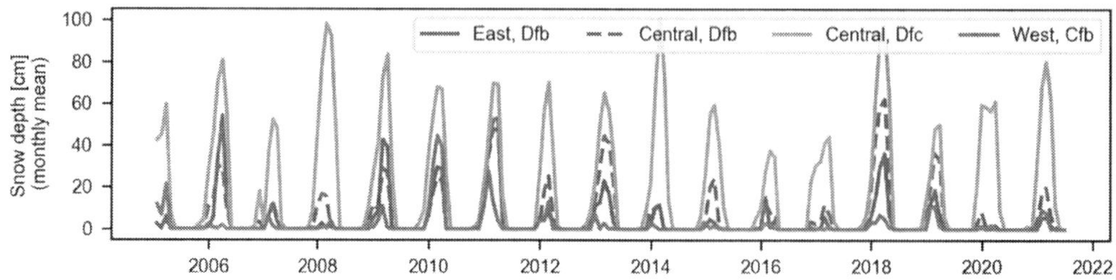

Figure 2: Sixteen years of snow depth data for the four combinations of geographic region and KG climate zone in the analyzed dataset.

2.2 Estimation of snow loss from historical PV data

To estimate historical snow losses from PV monitoring data, it is necessary to get an accurate estimate of what the energy production could have been if there was no snow. This requires an accurate model that considers all other losses of the PV system, and an efficient method to separate snow losses from other losses. To correctly estimate snow losses, it is especially important to take into account other wintertime losses such as losses caused by low irradiance, and high angles of incidence. These types of losses are typical for high latitude locations in the wintertime [21], and can introduce increased uncertainty in PV system modeling if not properly accounted for.

To estimate expected PV module power output for the commercial systems, the effective irradiance measured by the reference cells and the measured module temperature are used as input to a single diode model in pvlib python [22] to model PV module power output, using the procedure described in [13]. For the residential systems, detailed module data and onsite measurements are not available. Effective irradiance and module temperature are modeled in pvlib from measurements of GHI and ambient temperature from nearby weather stations. The GHI measurement is decomposed using the Erbs model [23] to estimate diffuse irradiance, and the Disc [24] model for direct irradiance. When modeling the in plane irradiance for the systems, the Hay and Davies' 1980 model [25] is used to determine the in plane diffuse irradiance from the sky. From the modeled in plane global irradiance, the effective irradiance is calculated by adding reflection losses using an incident angle modifier based on the physical model described in [26]. The module temperature is modeled using the PVsyst temperature model [27]. The expected power output from the modules is modeled using PVWatts [28].

The described PV module power output models do not take into account all the relevant losses (all other losses than snow-induced losses) of the systems. From the energy performance index (*EPI*) of the system, the ratio between measured and modeled energy, we observe that the calculated value is below 1. Additionally, the *EPI* has a systematic seasonal component suggesting higher losses in the winter months, also in periods without snow. We assume that the significant losses not accounted for in the model, can be estimated with a constant and a seasonal component. To accurately find the seasonal components for the analyzed systems, seasonal trend decomposition is performed on the daily *EPI*, after filtering out time periods with snow on the ground (which introduces a non-systematic seasonal component). Seasonal trend decomposition is suggested by [29] as a method to find and correct the seasonal component in PV performance metrics. The deviation between 1 and the median of the seasonally corrected *EPI* is used as an estimate of the constant system losses. These two components are then used to correct the modeled PV module output to find the expected system output. By this way aiming to take all other significant losses into account, the snow loss is then estimated to be the difference in expected system output and measured system output in periods where the snow data suggests snow on the ground.

An additional uncertainty in this methodology is that snow cover on the irradiance sensors can lead to underestimation of snow losses. To reduce this uncertainty, the reference cell measurements from the commercial systems were controlled and corrected by the external GHI data. Pyranometers is expected to have lower risk for full snow cover than reference cells, because of the shape and elevation of the sensor, and better ventilation and maintenance.

2.3 Modeling snow loss with the modified Marion model

In the Marion snow loss model [5] the presence of a new snow cover is assumed to happen after snow fall. The model further assumes that natural snow clearance will happen during melting. Melting is predicted to happen during the following conditions:

$$T_{amb} > G_{POA}/m. \quad (1)$$

T_{amb} is the ambient temperature, G_{POA} is the in plane irradiance and m is an empirically defined value of -80 W/(m^2 °C). During melting, the snow will be cleared by sliding or direct melting on the modules [4]. To estimate the reduction in snow coverage in the melting period, measured in fractions of the system height, the tilt of the modules and an empirical snow clearing coefficient (sc) is used:

$$Snow\ slide\ amount = sc * sin\ (tilt). \quad (2)$$

Based on these assumptions, the snow coverage on the modules is estimated, and the corresponding power loss calculated. If a module substring is partially covered by snow, the power output is assumed to be zero. This way, it is taken into account whether the modules are installed in portrait or landscape orientation. The pvlib python [22] implementation of the Marion model is used in this work to model the relative snow loss. To estimate the absolute energy loss, the modeled relative snow loss is multiplied with the modeled energy output of the system, modeled using the procedure described in Section 2.2.

In the development of the snow loss model, Marion et al. found *sc* to be 0.20 [5] for roof mounted systems. This value is the default *sc* in the implementation of the model in pvlib python [22] and the PV modeling software SAM [6]. The snow clearing coefficient is, as previously discussed, expected to depend on different system and module designs [13], because technical aspects can either promote or obstruct natural snow clearing [1]. In our evaluation of the model, we found that snow clearing is slower for the systems we have analyzed [13,16] compared to the validation systems the Marion model is based on. A possible explanation for the difference is higher roof interference for the systems that we have evaluated. In our evaluation of the model we also find that the rate of snow clearing is influenced by the thickness of the snow cover [13]. We therefore add a small modification to the Marion snow loss model by introducing a snow depth dependent *sc*. Because the dataset in this work is the same as in [16], we use the snow depth dependent snow clearing coefficients from [16] that gave the best modeling results. As also described in [16], we use snow depth data from the ground to separate between thin and thick snow covers for the commercial system where there is little sliding. For the residential systems where there is more sliding and where snow depth data from the ground are less representative, we use cumulative snow fall data as an indicator for snow cover thickness.

2.4. Simulation of snow losses for longer time series

To simulate losses for the analyzed systems over time, to get improved understanding of typical losses, we use long time series of weather and snow data to model snow losses, as proposed by [5]. GHI and ambient temperature data for all the locations from the time period 2005-2016 and the ERA5 database is collected from PVGIS [30]. The expected module power output for all the systems is modeled as described for the residential systems in section 2.2. System loss of 7 % is added using the PVWatts system loss function with default loss values for mismatch, wiring, LID, connections and name plate rating [28]. Snow losses are then modeled using the same procedure as described in 2.3.

3 RESULTS

3.1 Snow loss estimated from historical PV data

Figure 3 shows the annual historical snow loss for the analyzed systems (both system configurations) estimated from historical PV data. The loss is given relative to the mean expected annual yield. The mean value is chosen to avoid variations in the loss caused by variation in the total annual irradiation. We observe large variations in snow losses from year to year, and between different systems.

As expected, we observe that weather, system design and climate on snow losses seem to impact the snow losses. The inter-annual variation in snow losses for the systems, as well as the variation in losses between systems located in the same climate zone, but in different locations (C1 and C3), can be explained by typical variations in weather between different locations and different years. C3, R1 and R2 are located in the same area, but R2 has lower loss than R1 every year, and C3 typically has higher loss than both. This could be explained by the impact of tilt on the snow clearing, as snow clearing is inversely proportional with tilt. C1 and C2 have the same technical configuration and experiences the same weather as they are co-located, and their estimated losses are very similar. C4 located in oceanic climate typically has lower losses than the identical systems (C1-C3) located in humid continental climate. R3, located in a subarctic climate, typically has higher losses than R1, which has almost the same tilt but is located in a humid continental climate.

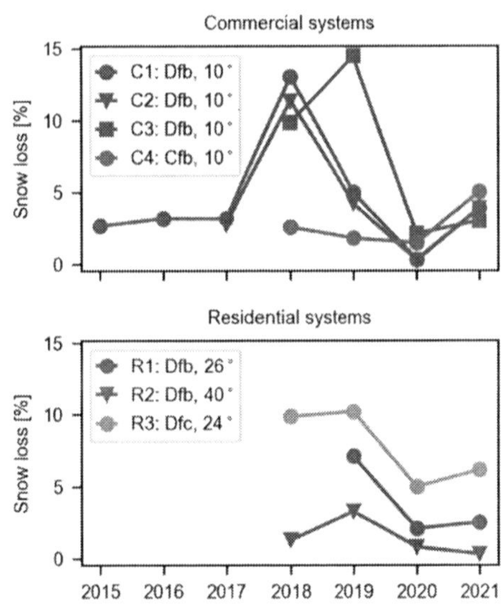

Figure 3: Annual snow loss for the analyzed systems, estimated from PV data. The losses are given relative to mean expected annual yield for the analysis period. The systems in humid continental climate is plotted in blue, green represents oceanic climate and orange represents subarctic climate.

Figure 4 shows the monthly losses for all the full years in the analysis period. Large variations in the monthly loss value are observed for several of the months. For most of the datasets the loss is typically increasing during late autumn, reaching its highest peak in midwinter, before it decreases in the spring. The snow data do, however, not follow the same trend. Typically, the locations have the most snow in the late winter months, but this also corresponds with higher temperatures.

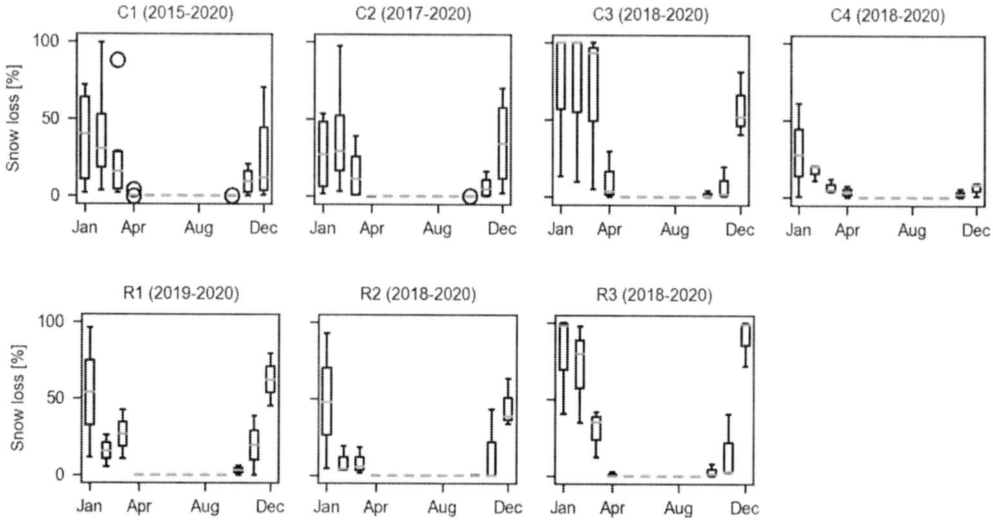

Figure 4: Monthly snow loss for the analyzed systems in the analysis period (given in the subfigure title), estimated from historical PV data. The estimated losses for each month is plotted using a boxplot to show the interannual variation. The box extends from the first to the third quartile values of the monthly loss data, with a line on the median. The whiskers extend to maximum 1.5 multiplied the interquartile range. Outliers are given as circles.

3.2 Simulated snow loss

Based on the results presented in Figure 3 and Figure 4, it is not always clear what would be the best estimate for typical annual and monthly snow losses for the analyzed systems. Especially for the locations with large snow losses, there can be large variations for the same month between different years. With potentially large variations from year to year, estimating typical snow loss for short time series might give an output that is not necessarily representative for the system configurations and the location. Based on this, selecting a representative snow loss value for e.g. a PVSyst simulation seems challenging, as long time series for different system designs and locations would be needed.

Figure 5 and Figure 6 show the correlation between snow loss estimated from PV data and modeled snow loss using the modified Marion snow loss model for respectively annual and monthly losses. Both on the monthly and annual time scales we observe a linear relationship between modeled losses and losses estimated from historical PV data, indicating that the model can be used to predict the losses on both time scales. Some uncertainty in the prediction can, however, be expected. As seen in the figures, there are some deviations between modeled loss and loss estimated from PV data.

In Figure 7 and Figure 8 the simulated monthly and annual snow losses for the analyzed systems using 12 years of irradiation and temperature data from PVGIS is presented. With the longer time series, we get a better understanding of what is typical losses, and what the potential variation and the extreme values could be. With the longer time series, we now see for all of the systems that the losses are highest during mid-winter. Some of the systems get higher monthly median losses than what we observed in Figure 4. This suggests that the years in the analysis period used to estimate losses from historical data are not necessarily years that represent the long-term trend.

Using longer time series and modeling could also enable estimation of snow losses for locations where PV data is lacking. Additionally, future snow losses could be estimated using output data from climate models giving data for the future. To avoid the impact of extreme values, we propose to utilize the median value of the modeled losses as an estimate of the monthly/annual snow losses in yield simulations.

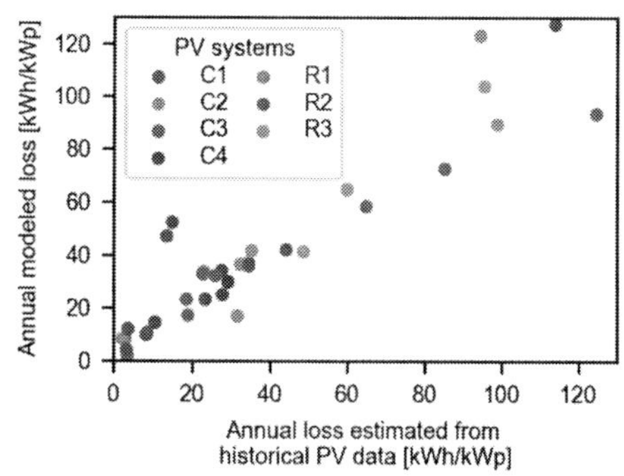

Figure 5: Annual modeled absolute loss compared to loss values estimated from historical PV data.

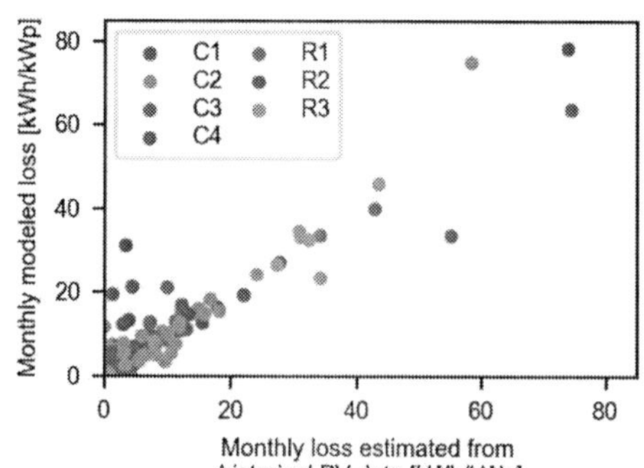

Figure 6: Monthly modeled absolute loss compared to loss values estimated from historical PV data.

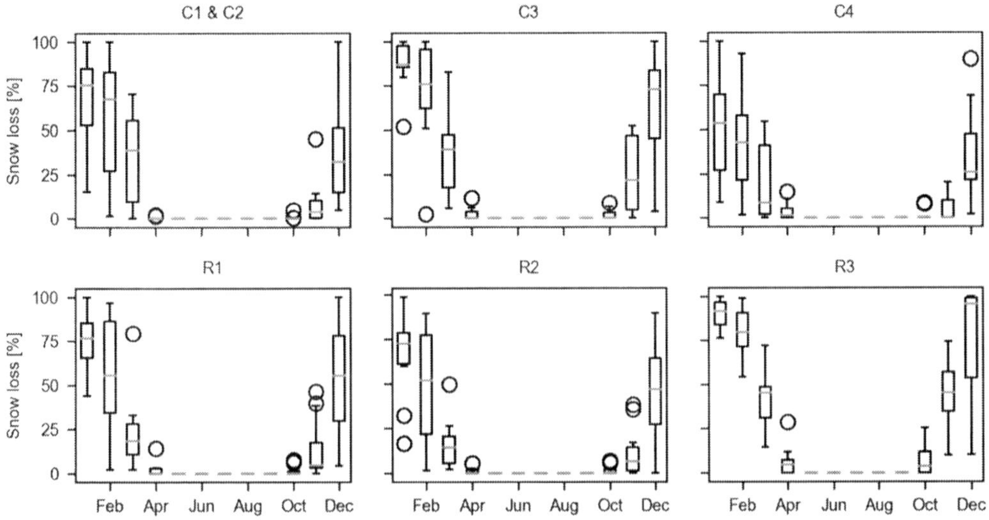

Figure 7: Monthly simulated losses for the analyzed systems, based on 12 years of PVGIS data and the modified Marion snow loss model. The simulated losses for each month is plotted using a boxplot to show the interannual variation. The box extends from the first to the third quartile values of the monthly loss data, with a line on the median. The whiskers extend to maximum 1.5 multiplied the interquartile range. Outliers are given as circles.

Using simulations to estimate PV systems snow loss could in addition to the loss value and estimation on interannual variability, also give realistic production profiles on daily and hourly timescale, which is useful in system size optimization and when building synthetic data series or adding synthetic performance loss for testing of e.g. fault detection algorithms [29]. The uncertainty in the modeling on high time resolutions is likely too high for e.g. monitoring purposes where the modeled PV output should match measured data, but to describe how snow losses vary within a day and from day to day, the modeling is useful.

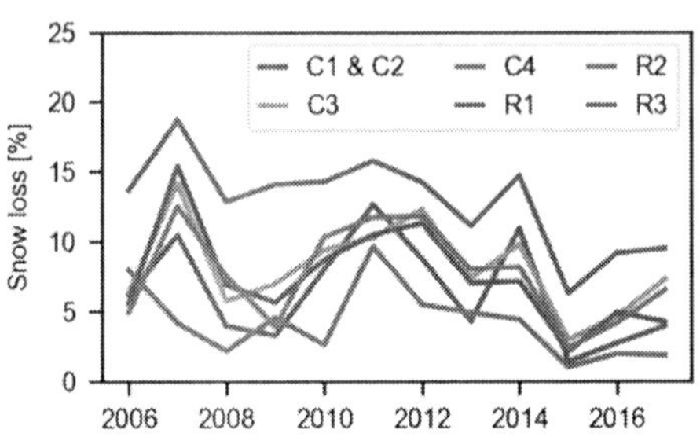

Figure 8: Simulated annual snow loss for the analyzed systems, based on 12 years of PVGIS data and the modified Marion snow loss model. The losses are given relative to mean expected annual yield for the analysis period.

4 CONCLUSIONS

In this work, we estimate annual and monthly snow loss for a set of PV plants in Norway. In both annual and monthly losses, we observe large interannual variations, and we see that systems in colder climates typically have higher losses than systems in warmer climates. We also observe that higher tilt gives reduced losses, confirming previous studies. A modified adaption of the Marion snow loss model where snow depth is considered in the snow clearing modeling is used with 12 years of weather data to simulate losses for a longer time series, to get improved understanding on the potential interannual variation in snow losses. We find that snow loss modeling is a useful tool for estimating monthly or annual snow losses for use in yield modeling when long time series of snow loss data for a given type of system in a given location is not available.

5 REFERENCES

[1] L. Burnham, D. Riley, J. Braid, Design considerations for photovoltaic systems deployed in snowy climates, 37th Eur. Photovolt. Sol. Energy Conf. Exhib. (2020) 1626–1631.
[2] B. Hashemi, A.M. Cretu, S. Taheri, Snow loss prediction for photovoltaic farms using computational intelligence techniques, IEEE J. Photovoltaics. 10 (2020) 1044–1052.
[3] IEA, PVPS Trends in photovoltaic applications 2020, 2020.
[4] R.E. Pawluk, Y. Chen, Y. She, Photovoltaic electricity generation loss due to snow – A literature review on influence factors, estimation, and mitigation, Renew. Sustain. Energy Rev. 107 (2019) 171–182.
[5] B. Marion, R. Schaefer, H. Caine, G. Sanchez, Measured and modeled photovoltaic system energy losses from snow for Colorado and Wisconsin locations, Sol. Energy. 97 (2013) 112–121.
[6] D. Ryberg, J. Freeman, Integration, validation, and application of a PV snow coverage model in SAM, Golden, 2017.
[7] T. Townsend, L. Powers, Photovoltaics and snow: An update from two winters of measurements in the Sierra,

in: 2011 37th IEEE Photovolt. Spec. Conf., IEEE, 2011: pp. 3231–3236.
[8] L.B. Bosman, S.B. Darling, Performance modeling and valuation of snow-covered PV systems: examination of a simplified approach to decrease forecasting error, Environ. Sci. Pollut. Res. 25 (2018) 15484–15491.
[9] E. Lorenz, D. Heinemann, C. Kurz, Local and regional photovoltaic power prediction for large scale grid integration: Assessment of a new algorithm for snow detection, Prog. Photovoltaics Res. Appl. 20 (2011) 760–769.
[10] PVGIS, Data sources and calculation methods, (2020). https://ec.europa.eu/jrc/en/PVGIS/docs/methods (accessed December 22, 2020).
[11] PVsyst, Soiling loss, (2020). https://www.pvsyst.com/help/index.html.
[12] N. Heidari, J. Gwamuri, T. Townsend, J.M. Pearce, Impact of snow and ground interference on photovoltaic electric system performance, IEEE J. Photovoltaics. 5 (2015) 1680–1685.
[13] M.B. Øgaard, B.L. Aarseth, Å.F. Skomedal, H.N. Riise, S. Sartori, J.H. Selj, Identifying snow in photovoltaic monitoring data for improved snow loss modeling and snow detection, Sol. Energy. 223 (2021) 238–247.
[14] M. van Noord, T. Landelius, S. Andersson, Snow-Induced PV Loss Modeling Using Production-Data Inferred PV System Models, Energies. 14 (2021) 1574.
[15] D. Riley, L. Burnham, B. Walker, J.M. Pearce, Differences in snow shedding in photovoltaic systems with framed and frameless modules, in: 2019 46th IEEE Photovolt. Spec. Conf., IEEE, 2019: pp. 558–561.
[16] M.B. Øgaard, H.N. Riise, J.H. Selj, Modeling Snow Losses in Photovoltaic Systems, in: 2021 48th IEEE Photovolt. Spec. Conf., 2021.
[17] T. Saloranta, J. Andersen, Simulations of snow depth in Norway in a projected future climate (2071-2100), 2018.
[18] NVE, seNorge, (2019). www.senorge.no.
[19] Norsk Klimaservicesenter, Observasjoner og værstatistikk, (2020). https://seklima.met.no/observations/.
[20] H.E. Beck, N.E. Zimmermann, T.R. McVicar, N. Vergopolan, A. Berg, E.F. Wood, Present and future köppen-geiger climate classification maps at 1-km resolution, Sci. Data. 5 (2018).
[21] M.B. Øgaard, H.N. Riise, H. Haug, S. Sartori, H. Selj, Photovoltaic system monitoring for high latitude locations, Sol. Energy. 207 (2020) 1045–1054.
[22] W.F. Holmgren, C.W. Hansen, M.A. Mikofski, pvlib python: a python package for modeling solar energy systems, J. Open Source Softw. 3 (2018) 884.
[23] D.G. Erbs, S.A. Klein, J.A. Duffie, Estimation of the diffuse radiation fraction for hourly, daily and monthly-average global radiation, Sol. Energy. 28 (1982).
[24] E.L. Maxwell, A quasi-physical model for converting hourly Global Horizontal to Direct Normal Insolation, Sol. Energy Res. Inst. (1987).
[25] J.E. Hay, J.A. Davies, Calculation of the solar radiation incident on an inclined surface, in: Proc. First Can. Sol. Radiat. Data Work., 1980.
[26] W. De Soto, S.A. Klein, W.A. Beckman, Improvement and validation of a model for photovoltaic array performance, Sol. Energy. 80 (2006) 78–88.
[27] PVsyst, Array Thermal losses, (2021). https://www.pvsyst.com/help/index.html.
[28] A.P. Dobos, PVWatts Version 5 Manual, Golden, CO (United States), 2014.
[29] Å.F. Skomedal, M.B. Øgaard, H. Haug, E.S. Marstein, Robust and fast detection of small power losses in large-scale PV systems, IEEE J. Photovoltaics. 11 (2021) 819–826.
[30] T. Huld, R. Müller, A. Gambardella, A new solar radiation database for estimating PV performance in Europe and Africa, Sol. Energy. 86 (2012) 1803–1815.

MODELLING AND IMPACT OF SOLAR ECLIPSES ON PV ENERGY PRODUCTION

Julián Ascencio-Vásquez[1,2*], Rodrigo Amaro e Silva[3], Elias Urrejola[4], Kristijan Brecl[2], Marko Topič[2]

1 Envision Digital, Redwood City, California, United States
2 University of Ljubljana, Faculty of Electrical Engineering, Ljubljana, Slovenia
3 Centre O.I.E, MINES ParisTech, PSL, France
4 ATAMOSTEC, Antofagasta, Chile

*Phone: +34 623 44 83 01; Email: julian.ascencio@envision-digital.com

ABSTRACT: Total solar eclipses are events that rarely occur but can highly impact the operation of electrical grids. During a few minutes, the Sun can be fully covered by the moon, blocking the sun rays towards the Earth surface. This phenomenon has a considerable effect on the electrical generation of PV systems and, depending on the level of deployment of this technology, can compromise the stability of electrical grids, causing perturbations and making the grid prone to failures. For this reason, the accurate yield forecasting of PV plants during a solar eclipse can help grid planners to handle this event in the most cost-efficient way. In this work, we present preliminary results of the modelling and quantification of PV plants energy production and losses during total solar eclipses. The TDP model is used to model the PV performance under clear-sky conditions. The results show around 10% of daily PV energy yield and 13% of daily irradiation during the total solar eclipse seen in Chile on the 14th December 2020.

Keywords: Photovoltaic; Eclipse; System Performance; TDP; Forecasting

1 INTRODUCTION

Despite relatively short and infrequent events, the abnormal solar irradiance ramps introduced by solar eclipses raises challenges to the operation of power systems that rely on PV generation [1]. Total solar eclipses are events that rarely occur but can highly impact the operation of electrical grids. During a few minutes, the Sun can be fully covered by the moon, blocking the sun rays towards the Earth surface. This phenomenon has a considerable effect on the electrical generation of PV systems and, depending on the level of deployment of this technology, can compromise the stability of electrical grids, causing perturbations and making the grid prone to failures. For this reason, the accurate yield forecasting of PV plants during a solar eclipse can help grid planners to handle this event in the most cost-efficient way.

Figure 1 illustrates the moon's position between the Earth and the Sun, blocking the sun rays reach the surface. Some of the latest total solar eclipses were seen in Europe on the 20th March 2015 at around 10 am with a magnitude of 1.045. In North America on the 26th February 2017 at around 6 pm (M1.031), and two in South America on the 2nd July 2019 and 14th December 2020 with magnitudes of 1.046 and 1.025, respectively.

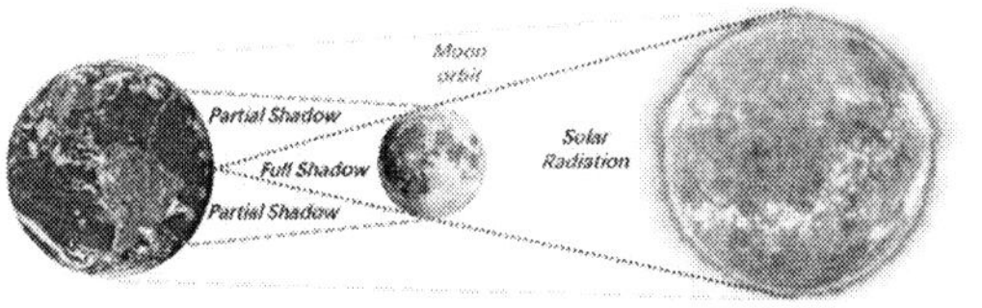

Figure 1. Illustration of solar eclipses.

These rare events led to the integration of an eclipse-focused package in traditional numerical weather prediction models [2] and have piqued the interest of researchers in various countries such as Germany [3], [4], the Czech Republic [5], the United States of America [6], [7], Australia [8], or India and southeast Asia [9].

The solar eclipse event that occurred on the 20th March 2015 in Europe has been well-documented. Libra et al. assessed the eclipse impacts for a 510 W_p small-scale system installed in Prague [5], measuring a cumulative -0.42 W/ W_p ramp down followed by a cumulative 0.57 W/ W_p ramp up (i.e. a -63% and +335% relative variations). Additionally, a 13.5% energy loss for that day was estimated. Kohler et al. forecasted the impacts of this same eclipse for the solar resource and PV generation in Germany. Using the COSMO-DE numerical weather prediction model, the authors estimated a maximum irradiance ramp down of -390 W/m^2 (a 66% relative reduction) for the worst-case scenario (i.e. assuming clear sky conditions). Their results were then extended to PV power gradients of 6.20 GW and 19.9 GW across the country, assuming 37 GW_p of PV. On the same date, a partial solar eclipse was visible in Ljubljana, Slovenia. It covered around 60% of the Sun. The solar irradiance was reduced to around 1/3 of the expected value. Unfortunately, the sky was partially covered with clouds; therefore, an exact energy loss was not measured.

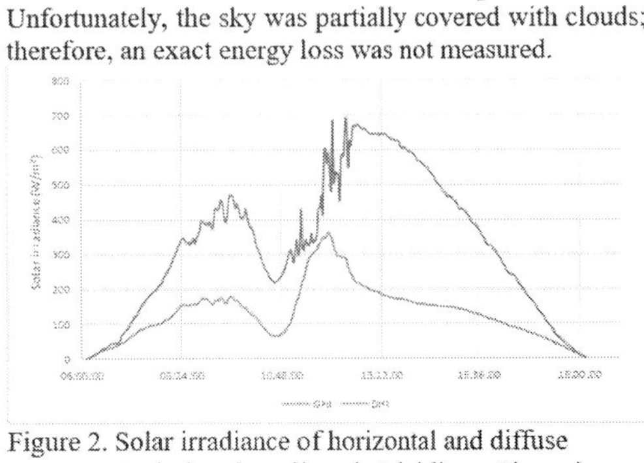

Figure 2. Solar irradiance of horizontal and diffuse components during the eclipse in Ljubljana, Slovenia (15th March 2015).

Zack et al. forecasted the impacts of the 21st August 2017 eclipse on the power system operated by the California Independent System Operator (CAISO) in the United States of America [7]. Using the Weather Research and Forecasting (WRF) numerical weather prediction model, the authors estimated a maximum deficit of 0.55 W/Wp and solar generation losses of 8577 MWh.

For the solar eclipse in 2019 in Chile, in [10], the energy

losses due to obscuration were estimated between 20–40% for global horizontal irradiance and 25–50% for direct normal irradiance over the Chilean territory.

In this work, we focus on the solar eclipse impacting Chile on the 14th December 2020. We combine several state-of-the-art models to accurately calculate the impact of total solar eclipses on PV systems' energy production. Using historical data and the Typical Daily Profiles (TDP) model [11], a clear-sky model of the real PV plant operation is created. In further work, the LightGBM [12] will predict the losses during future solar eclipses.

2 DATA AVAILABILITY AND APPROACH

2.1. PV production data

Chile is a very interesting country for research in photovoltaics (PV), as it is characterized by a remarkable solar resource and has been identified as the best country to invest in renewables according to the Climatescope 2020 [13]. Additionally, Chile's national open data policies allow researchers to retrieve hourly data of power plants from the Coordinador Electrico and, thus, work with real operational data that can be retrieved. As this work addresses the impact of the 2020 total solar eclipse in Chile, we focus on the operational data of PV power plants.

In previous work [11], a deep investigation about the PV performance of the systems in Chile has been presented, highlighting not only the highest irradiation levels worldwide but also the very high PV yields (+3000 kWh/kW_p in best cases). Over 3GW of installed PV capacity is already operating in the country, represented by roughly half of tracking systems and half of fixed-mounted systems.

2.2. Weather data

Historical weather data, mainly global horizontal irradiance, has been retrieved from the public ERA5 climate reanalysis dataset [14].

2.3. Modelling and Quantification approach

The combination of several physical and statistical models and freely available data sources contribute to accurately estimate PV systems' energy losses due to total solar eclipses at any geographical location (see the scheme in Figure 3).

First, the irradiance losses due to the solar eclipse are estimated taking into account three different elements: i) the modelled irradiance, which was retrieved from the ERA5 dataset and interpolated using numerical techniques (linear and spline); ii) the clear-sky PV performance was extracted from the operational data using the TDP model; and iii) the eclipse motion (Sun, Moon and Earth), which is modelled using the PyEphem Python package [15]. A PV performance model based on Machine Learning approaches (LightGBM) will be used to forecast the losses during future solar eclipses.

The outcome of the scheme allows getting the eclipse percentage at any date, the irradiance losses and PV production losses due to total solar eclipses.

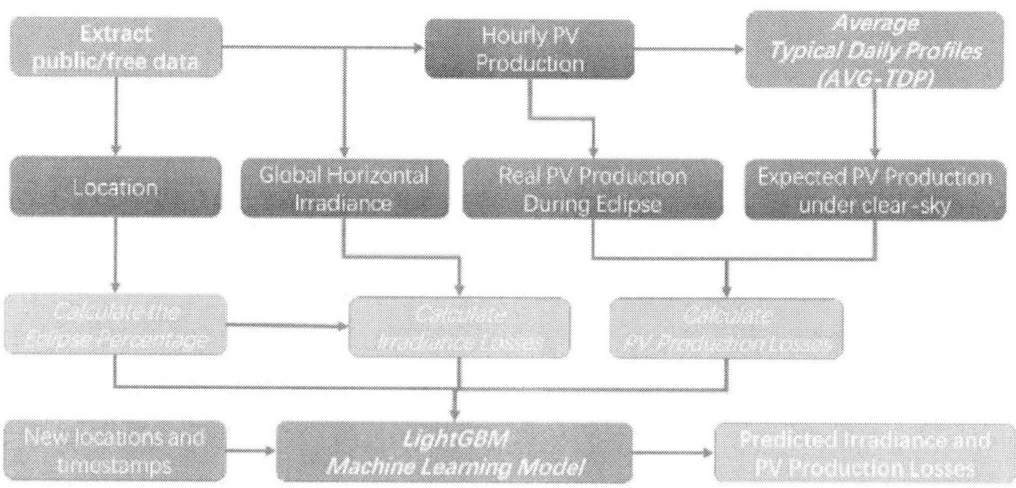

Figure 3. Flowchart to estimate the PV performance losses due to solar eclipses.

3 RESULTS

3.1. Application of the TDP model

The TDP model is used on hourly long timeseries to create daily profiles of PV systems. The Average TDP (AVG-TDP) shows the expected PV power per month under clear sky conditions using a percentile 90th.

Figure 4 shows the steps to build an AVG-TDP. The long time series of raw PV production data are extracted and grouped month by month. Then, grouping by the hour, the percentile 90th will approximate the production at the specific hour and month under clear sky conditions.

This procedure is applied to each PV system with sufficient data before the eclipse of the 14th December 2020.

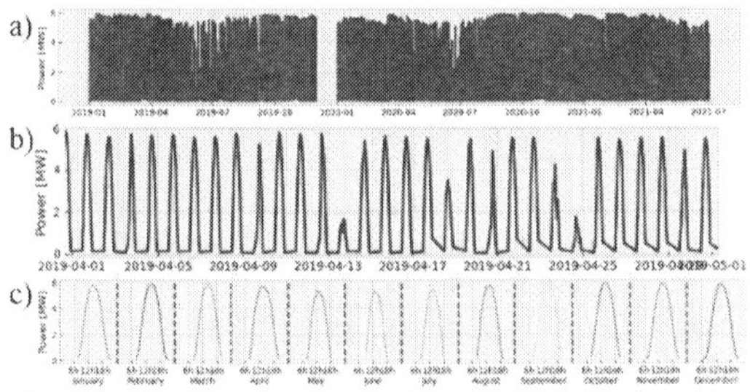

Figure 4. a) Long-term raw hourly PV production data. b) Extracted data for April. c) Average TDP for the specific PV system.

3.2. Calculation of irradiance losses

The irradiance losses are calculated for all locations where PV systems are deployed using the approach described above, and an example is shown in Figure 5, where the irradiance extracted from ERA5 is modified by applying the percentage of eclipse during the 14th December 2020.

The irradiance losses will be quantified by subtracting the irradiance data without eclipse losses versus the irradiance with the eclipse losses.

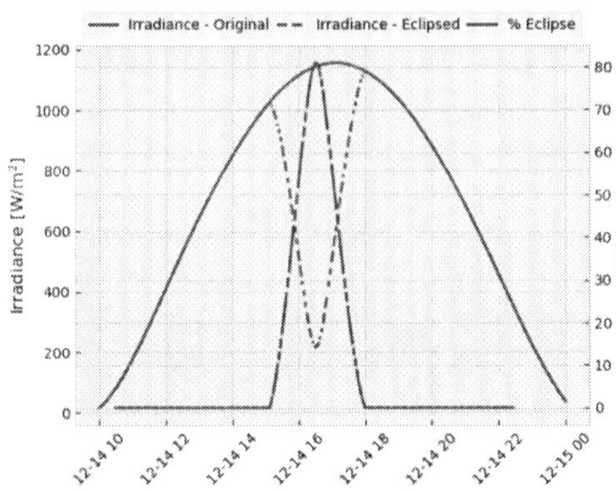

Figure 5. Example of irradiance timeseries and eclipse percentage during the eclipse 14th December 2020.

3.3. Calculation of PV production losses

To quantify the PV production losses, it is important to identify the operation of the PV system under normal conditions (non-eclipsed). For this, we use the AVG-TDP to identify the operation under clear skies. In Figure 6, the PV production of one system is shown, where the AVG-TDP for December is compared to the PV production experienced on 14th December 2020. The spline interpolation from hourly to minutely resolution is applied to the PV production data to get more realistic results. The PV production losses due to the eclipse will be calculated by subtracting the AVG-TDP and the interpolated PV production data on that specific date.

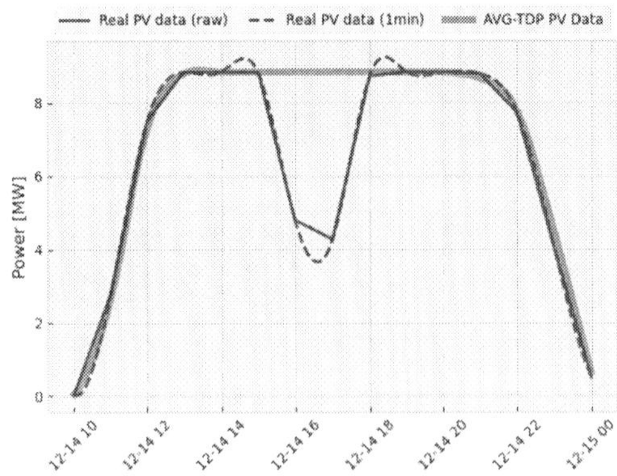

Figure 6. Example of PV production timeseries and impacted by the solar eclipse on14th December 2020.

3.4. Irradiance and PV Fleet losses

The procedures mentioned above for irradiance and PV production are applied to all the systems operating in Chile. In Figure 7, the normalized daily losses are presented. For the global horizontal irradiance, the losses range from 11.0% to 21%, with an average of 13% daily losses. For PV production losses, the normalized daily losses range between 2.5% to 20%, with an average of 10%.

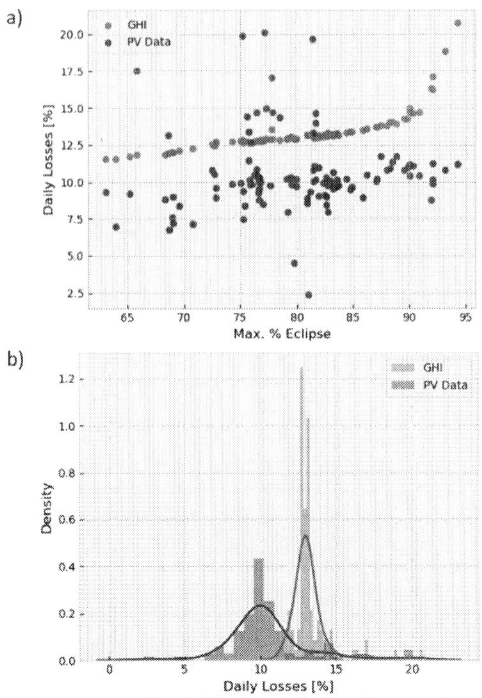

Figure 7. Normalized daily losses for irradiance and PV production data. a) Daily losses versus the maximum percentage eclipse observed. b) Histogram and KDE approximation of the normalized daily losses.

4 CONCLUSIONS

The quantification of irradiance and PV losses can help the TSOs and DSOs, to anticipate losses during solar eclipses. In this research work, the impact of the solar eclipses is highlighted with confirmed cases in Europe (2015), North America (2017) and Chile (2019). Furthermore, a comprehensive methodology for the quantification of past and future events is proposed. The methodology uses public input data to be processed with the TDP model to extract the PV performance of systems under clear sky conditions.

The irradiance losses in Chile on the 14th December 2020 were, on average a 13% daily, while PV production was on average 10%. This approach will be in the future applied to new events in different geographical regions. In this study, the TDP proves to be a valuable tool for modelling any PV system.

REFERENCES

[1] K. Burges, M. Döring, R. Kuwahata, U. Bachmann, and R. Sikora, "Operating Power Systems with high Photovoltaic Generation during a Solar Eclipse Challenges and Operational Measures," 2014.

[2] A. Montornès, B. Codina, J. W. Zack, and Y. Sola, "Implementation of Bessel's method for solar eclipses prediction in the WRF-ARW model," *Atmospheric Chem. Phys.*, vol. 16, no. 9, pp. 5949–5967, May 2016, doi: 10.5194/acp-16-5949-2016.

[3] S. Killinger, N. Kreifels, B. Burger, B. Müller, G. Stiff, and C. Wittwer, "Impact of the Solar Eclipse from 20 th March 2015 on the German Electrical Supply-Simulation and Analysis," *Energy Technol.*, vol. 4, no. 2, pp. 288–297, Feb. 2016, doi: 10.1002/ente.201500228.

[4] C. Kohler *et al.*, "Assessing the impact of a solar eclipse on weather and photovoltaic production," *Meteorol. Z.*, vol. 25, no. 1, pp. 55–69, 2016, doi: 10.1127/metz/2015/0672.

[5] M. Libra, P. Kouřím, and V. Poulek, "Behavior of Photovoltaic System during Solar Eclipse in Prague," *Int. J. Photoenergy*, vol. 2016, 2016, doi: 10.1155/2016/2653560.

[6] A. Sundararajan, T. O. Olowu, L. Wei, S. Rahman, and A. I. Sarwat, "Case study on the effects of partial solar eclipse on distributed PV systems and management areas," *IET Smart Grid*, vol. 2, no. 3, pp. 477–490, 2019, doi: 10.1049/iet-stg.2019.0002.

[7] J. W. Zack and A. W. S. Truepower, "Pre-event Estimation , Real-time Forecasting and Post-event Analysis of the Impact of the Aug 21 , 2017 Solar Eclipse on Renewable Generation in California," *7th Sol. Integr. Workshop*, 2017.

[8] A. Millican and D. Spoor, "Power System Operation during Solar Eclipses: International Lessons Applied in Australia," in *7th International Conference on Smart Grid, icSmartGrid 2019*, 2019, pp. 58–63. doi: 10.1109/icSmartGrid48354.2019.8990673.

[9] A. M. Nobre *et al.*, "On the Impact of a Solar Eclipse across 100 PV Systems in India and Southeast Asia," in *37th European Photovoltaic Solar Energy Conference and Exhibition*, 2020, pp. 1413–1417. doi: 0.4229/EUPVSEC20202020-5DO.5.5.

[10] A. Castillejo-Cuberos, J. M. Cardemil, and R. Escobar, "Analyzing Regional and Local Changes in Irradiance during the 2019 Total Solar Eclipse in Chile, Using Field Observations and Analytical Modeling," *Energies*, vol. 14, no. 17, 2021, doi: 10.3390/en14175352.

[11] J. Ascencio-Vásquez, J. C. Osorio-Aravena, K. Brecl, E. Muñoz-Cerón, and M. Topič, "Typical Daily Profiles, a novel approach for photovoltaics performance assessment: Case study on large-scale systems in Chile," *Sol. Energy*, vol. 225, pp. 357–374, 2021, doi: https://doi.org/10.1016/j.solener.2021.07.007.

[12] J. Ascencio-Vásquez, J. Bevc, K. Reba, K. Brecl, M. Jankovec, and M. Topič, "Advanced PV Performance Modelling Based on Different Levels of Irradiance Data Accuracy," *Energies*, vol. 13, no. 9, 2020, doi: 10.3390/en13092166.

[13] BloombergNEF, "Climatescope 2020."

[14] H. Hersbach *et al.*, "The ERA5 global reanalysis," *Q. J. R. Meteorol. Soc.*, vol. 146, no. 730, pp. 1999–2049, 2020, doi: https://doi.org/10.1002/qj.3803.

[15] B. C. Rhodes, *PyEphem: Astronomical Ephemeris for Python*. 2011.

SUITABLE SILICON SOLAR CELL TECHNOLOGIES FOR USE IN SPACE APPLICATIONS

Valentin D. Mihailetchi[1], Christoph Peter[1], Lejo Koduvelikulathu[1], Razvan Roescu[1], Radovan Kopecek[1], Kaveh Rouhani[2], Sergiu Pop[2], Murat Okandan[2]
[1]International Solar Energy Research Center (ISC) Konstanz, Rudolf-Diesel-Str. 15, D-78467 Konstanz, Germany
[2]mPower Technology, 5901 Indian School Road, NE Albuquerque, NM 87110, USA
*Corresponding author: valentin.mihailetchi@isc-konstanz.de

ABSTRACT: In this paper, we present an overview of various commercially or R&D available c-Si solar cell concepts and find the most suitable technologies for space applications. First, we use a three-dimensional device simulation tool and fit a SRH recombination model that best predicts the relative degradation of a known *p*-type and *n*-type solar cells upon proton irradiation with various energies and fluences. Then, the resulted SRH recombination parameters are used to predict the degradation rate and end-of-lifetime (EOL) performance of various cell technologies for any given space mission providing as input the equivalent proton and electron fluences. Based on this study, we are able to indentify the properties and key design features of c-Si cells most suited for use in space applications.
Keywords: silicon solar cell, space applications, proton irradiation

1 INTRODUCTION

Until early 1990 crystalline Si (c-Si) solar cells were the main technology for space satellite applications. Since then the decrease in fabrication costs of GaAs-based (III/V) cells made these the predominant technology due to higher efficiency and lower degradation as compared to c-Si single junction cells. However, today's large increase in space missions and constellations require a PV capacity in the order of tens of MW/year, with the potential to further increase in the coming years with the planned launch of thousands of more satellites in low-earth-orbit (LEO) for broadband internet connection. The present III/V solar cells production capacity worldwide is less than 2 MW/year, with significant capital investment required to increase capacity.

The recent efficiency improvements of c-Si solar cells combined with the GW/year production capacity for terrestrial applications and the substantial decrease in cost, make the c-Si technologies ideal candidates for space applications. Thus, silicon solar cells are currently attracting attention again and they are expected to return into the space market especially for short-term missions.

Many different c-Si solar cell concepts are currently in mass production or in R&D pilot lines with efficiencies exceeding 25% under AM1.5g spectrum.

2 METHODS

2.1 Terrestrial Si cell concepts candidates for space applications

High efficiency c-Si solar cells have been developed recently and implemented in mass-production for terrestrial use [1-3]. Figure 1 shows a schematic cross-section view of various c-Si cell concepts based on *p*-type or *n*-type doped wafers that are currently in mass production (a-d) or in R&D pilot lines (e-h). Among them, the Al-BSF – a *p*-type cell concept – has been the longest in production for terrestrial PV power generation and it is the most extensively used concept for space applications. This is because of its simple fabrication process with low-cost and relatively good performance in the harsh outer space conditions. However, due to lack of rear side passivation, and with the improvements in wafer quality, diffusion, passivation, and metallization, this cell concept reaches its efficiency limit for today's process capabilities. Therefore a higher efficiency PERC concept is dominating more than 50% of current PV production. The highest efficiency cell concepts in mass-production today are those based on n-type wafers, such as passivated emitter rear totally diffused (n-PERT), TOPCon, silicon heterojunction (SHJ), and interdigitated back contact (n-IBC). The n-IBC cell concept is reaching efficiency >25% in mass-production. A low cost n-IBC ZEBRA® cell concept developed by ISC Konstanz is also currently in mass-production with average efficiency of 23.5% [3].

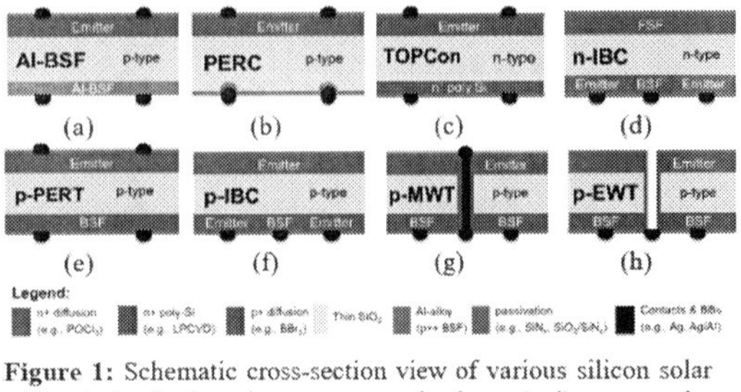

Figure 1: Schematic cross-section view of various silicon solar cells technologies in mass production (a-d) or under development in R&D labs (e-h) chosen for this study. The legend shows the properties of main features of the cell.

In this study we aim to investigate suitability for space applications of these commonly available concepts in mass-production as well as other *p*-type based cell concepts, such as those shown in Figure 1(e-h), that are under development and which can easily be implemented in the production line. For example; the *p*-IBC, metal-wrap-through (*p*-MWT) and emitter-wrap-through (*p*-EWT) concepts can all be fabricated in the same line with *n*-IBC ZEBRA by changing the wafer type and wafer drilling and metallization. One should note that a PERC-MWT concept and the *n*-type silicon heterojunction (SHJ) are already in mass-production, but for simplicity we did not consider them in this study.

2.2 Modelling of proton/electron irradiated cells

The main cause of solar cell degradation in space is due to bombardment with protons and electrons either trapped in the earth's Van Allen radiation belts or ejected in solar events. One of the most commonly used analytical model for the degradation of the solar cells upon proton or electron irradiation is given by the basic relationship for minority-carrier diffusion length (L_n)

reduction upon particle irradiation:

$$1/L_n^2=1/L_0^2+K_L*\Phi \qquad (1)$$

where L_0 is the initial diffusion length before irradiation, K_L is the diffusion length damage coefficient, and Φ is the proton fluence. The K_L depends on the energy spectrum of the particles, the properties of the Si cell such as doping type and resistivity, temperature and light intensity. The $K_L*\Phi$ product is proportional to the number of recombination centers (N_t) introduced by irradiation and their effective capture cross-section for electrons (σ_n) and holes (σ_p) which is thus affected by the minority charge carrier density generated by light.

There are basically two main methods being used to model solar cell degradation in space. The first method was developed by NASA Jet Propulsion Laboratory (JPL) [4,5] and uses an extensive set of experimental data for relative K_L determination. The second method, which needs less experimental data, was developed by US Naval Reseach Laboratory (NRL) and it is based on calculation of the displacement damage dose for a given mission using the energy dependence of the K_L. The relative damage coefficients are calculated from the nonionizing energy loss (NIEL) for protons and electrons. A good overview of the two methods is given by Messenger *et al.* [6].

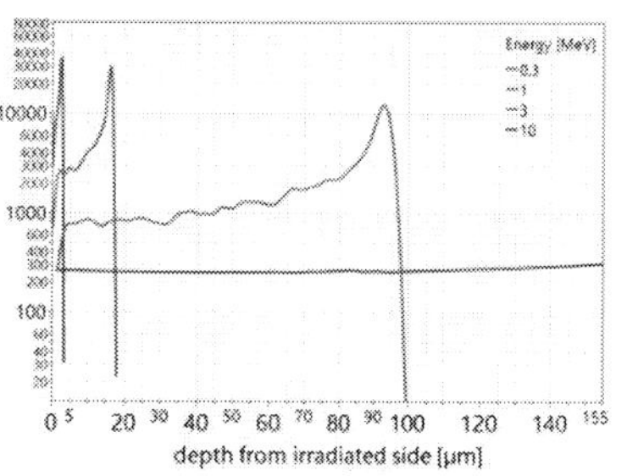

Figure 1: Typical vacancies produced and the penetration depth by proton collision with Si at various energies (see legend), as calculated by SRIM [8].

Both JPL and NRL methods give similar good results and are relatively straightforward for electrons and protons that traverse the solar cell without significant change of their energy. This is the case for proton stopping range larger than the cell thickness, corresponding to energies above ~3.6 MeV for a typical 155 µm bare Si cell, resulting in a relatively uniform distribution of defects throughout Si bulk [7]. In this case the simplified model of the diffusion length L_n reduction throughout the bulk given by Equation 1 is sufficient to characterize the degradation of irradiated solar cells. However, for protons that stop inside the semiconductor, the calculation has to take into account the fact that the relative damage increases rapidly at the end of proton path with virtually no damage caused to the Si bulk beyond their stopping range, as shown by the calculated SRIM [8] damage distribution profiles in Figure 2. This leads to a significant variation in the L_n with the location in the Si bulk. Consequently, approximations have to be used to account for the variation in damage along the proton track. Thus, from Equation 1, an effective diffusion length L over the entire bulk is assumed such that the degradation of the cell can be predicted. Hence, an empirical K value has to be found by fitting the experimental data for any given irradiation conditions and cell parameters, limiting the prediction capability only to those cell concepts and irradiation conditions for which the experimental data exists. In 2002 Messenger *et al.* updated the NRL modelling approach in order to quantitativelly model the response of the silicon solar cell to protons with stopping range within Si cell [9].

In this work, we used a 3-dimensional device modeling approach that takes into account the actual damage profile N_t caused by the proton irradiation with a known energy and fluence, and performs a fit of the Shockley-Read-Hall (SRH) recombination parameters (i.e., E_t, σ_n and σ_p) for the Si bulk that yield the performance of an experimental irradiated p- and n-type Si cell. This device simulation model takes into account local variation of the diffusion length within the Si crystal upon any given irradiation conditions. The N_t profile for a given proton energy and fluence was taken from the calculated damage distribution using SRIM [8], and used as an input SRH trap density in the device simulation tool Quokka3 [10] in order to compute the cell performance. It was found that three SRH recombination levels with distinct E_t, σ_n and σ_p parameters [11] are sufficient to adequately predict the degradation of a Si cell given as input parameter the calculated damage profile N_t for a given energy and fluence. For fitting these paramters, the experimental data for only one proton energy with at least two fluence levels are needed for a p-type or n-type cell with at least two based resistivities.

Figure 3 shows an example of the resulted fit to the experimental data for the remaining factor in Jsc×Voc (top row) and cell efficiency (bottom row) as a function of fluence for a PERC concept with two different bulk resistivities. The cells were irradiated from the front side with a 5 MeV proton energy at three different fluence levels. Similar good aggrement is obtained for an experimental n-type cell (data shown in Figure 5) using the same SRH fit paramters. No physical significance should be attributed to these fitted SRH parameters. Nevertheless, once the E_t, σ_n and σ_p parameters are known, the EOL performance of any other Si cell technology could be predicted from the device simulation given the equivalent proton energy and fluence for the orbital mission.

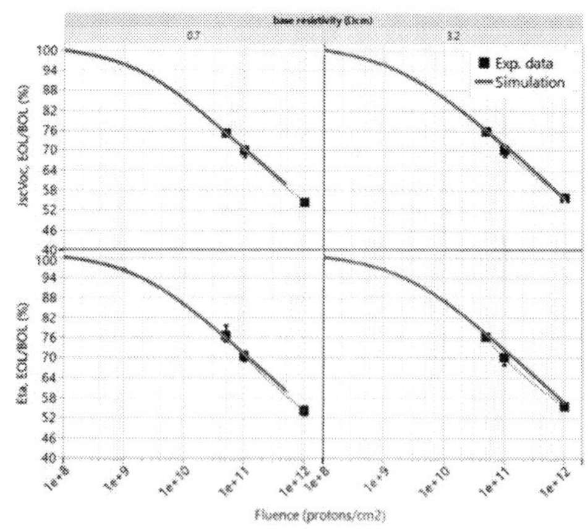

Figure 3: Experimental data and device simulation (see legend) of the remaining factor (EOL/BOL) in JscVoc product and cell efficiency as a function fluence for a PERC cell concept with two different bulk resistivities irradiated with 5 MeV protons. The IV measurements are perfoemed under AM1.5g and 25°C.

This modelling methodology uses significantly less experimental data and does not require prior determination of K_L. Furthermore, since a full device simulation can be performed for any EOL conditions, the calculated bulk lifetime can be used to determine K_L. The resulted K_L for 5 MeV protons with a fluence of 1e11 cm^{-2} is show in Figure 4 as a function of base resistivity. These simulated K_L values are in good agreement with previously reported experimental data for similar irradiation conditions [11, 12].

The modelling methodology presented here, by fitting a SRH model for the Si bulk, can be applied for electron irradiated cells in a similar way.

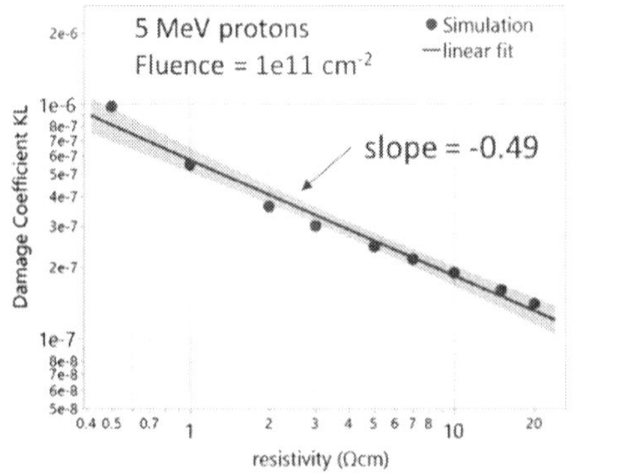

Figure 4: Diffusion length damage coefficient K_L as a function of bulk resistivity extracted from the device modelling for the PERC cell concept.

3 RESULTS

3.1 Modelling versus experimental data

Once the SRH recombination model for the Si bulk is found from fitting the experimental data, the device simulation can then be used to predict the degradation of other cell concepts or for other irradiation conditions. One should mention that this modelling approach would apply for those experimental irradiation conditions that produce predominatly point defects in the bulk, which is the case for Earth's orbits.

Figure 5 shows the simulation and experimental data results of the remaining factor in JscVoc product as a function of proton energy and fluence for PERC (0.9 Ωcm), p-EWT (0.9 Ωcm), and a TOPCon (1 Ωcm) cell concepts. The experimental samples used in this study were fabricated at ISC Konstanz R&D lab using all industrial processing tools. The cells were fabricated on M2 size wafers and subsequently cut to 20×20 mm^2 for proton irradiation. The irradiation runs were performed at Rubitec GmbH in Germany (lab 1) with some conditions repeated at Texas A&M (lab 2) in USA. Each experimental point comprises of 5-15 mini-cells. All measurements (BOL and EOL) were done under STC conditions (AM1.5g, 25°C) and no defect annealing was performed after irradiation.

The simulations for the beginning-of-lifetime (BOL) cell performance were performed using the best-know input parameters that were known to us from the experimental cell, such as wafer resistivity, n^+ and p^+ doping profiles, surface and bulk recombination parameters, cell geometry and metallization layout. A good quantitative agreement to the experimental data was obtained at least for the proton energies up to 10 MeV and fluences up to 1×10^{12} p/cm^2. Furthermore, from the results shown in Figure 5 one can calculate a deviation factor (Δ) to the experimental data, which is given by:

$$\Delta=[\mathrm{EOL/BOL(exp)} - \mathrm{EOL/BOL(sim)}] / [\mathrm{EOL/BOL(exp)}]$$

We obain Δ values (data not shown) well below ±5% for proton energies between 1-10 MeV whereas for the 0.3 MeV the Δ values are slightly larger (especially for the highest fluence). For 0.3 MeV the damage profile is very shallow and close to the front surface, with no damage caused on most of the Si bulk, as shown in Figure 2. This makes the device numerical simulation sensitive to the simplified assumptions of the input cell parameters (e.g., junction depth and Rsh) and to numerical convergence.

Using the same SRH recombination model for the Si bulk found by fitting the experimental data for one energy level in Figure 4, the device modelling predicts very well the degradation of the cells for other tested conditions. This good aggrement with the experimental data demonstrate that this modelling methodology can be applied to predict EOL cell performance as well as to optimize device fabrication process for space applications. Hence, the EOL of other c-Si cell technologies and designs, for which there are no experimental data available, can readily be simulated and optimized using this device modelling methodology.

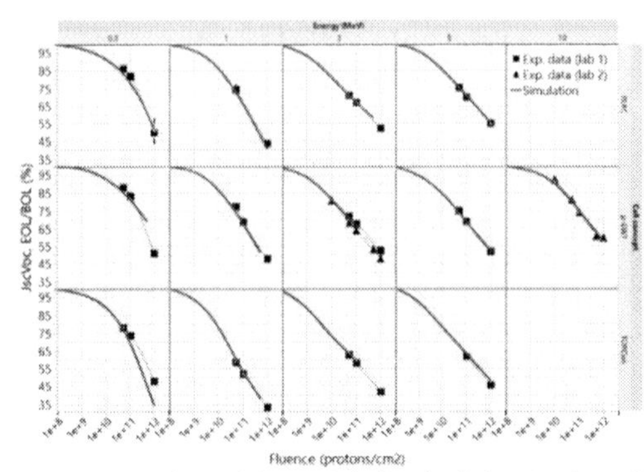

Figure 5: Experimental data and device simulation (see legend) of the remaining factor (EOL/BOL) in JscVoc product as a function of proton energy and fluence for PERC, p-EWT, and TOPCon cell concepts.

3.2 Characteristics of Si cell concepts for space applications

In order to compare the EOL expected performance of various Si cell technologies considered in this study (see Figure 1), we have taken as an example a Low Earth Orbit (LEO) satellite mission with a 560 km circular orbit and a 53° inclination. Using SPENVIS [13] one can estimate the 10 MeV equivalent proton fluence 1 MeV equivalent electron fluence as a function of mission duration. Assuming NASA AP-9 and AE-9 radiation belt models and a 100 μm cover glass protecting the solar cells with an infinity back shielding, we have used the device simulation model to predict the remaining factor and absolute power density P_{max} under AM0 and 30°C for various cell concepts as a function of mission time. The results are plotted in Figure 6 for all cell concepts considered in this study. The cell thickness was assumed to be 155 μm for all concepts. The bulk resistivity was assumed to be 3 Ωcm for the *p*- and *n*-IBC concepts whereas for all other concepts was assumed to be 1.5 Ωcm. A simulation study was conducted to find the optimum restivity range for each cell concept using as criteria the maximum remaining factor (data not shown). This yielded an optimum resitivity range between 1-5 Ωcm for all cell concepts.

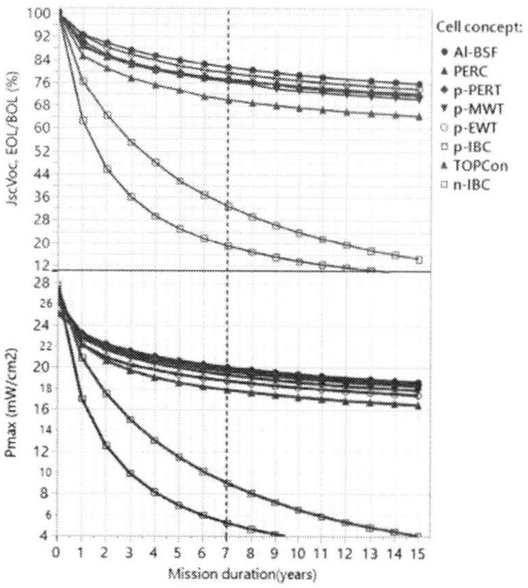

Figure 6: Simulated remaining factor in JscVoc product and absolute power density as function of mission duration for an LEO (560 km) circular orbit with a 53° inclination. A 100 µm cover glass and an infinite back shielding was assumed. For all cell concepts it was assumed a cell thickness of 155 µm and the AM0 spectrum and 30°C.

As it can be observed from the simulated data in Figure 6, the IBC cell concepts show substantial higher degradation as compared to the other cell technologies and are not suited for space mission although they potentially provide higher BOL power. Another important observation from these results is that the *n*-type based cell concepts degrade more (typically by 7-10%) than the same concepts on *p*-type wafers, eg., TOPCon versus *p*-PERT, PERC, Al-BSF, *p*-EWT, or *p*-MWT.

In order to maximize the EOL performance, one must chose a cell concept fabricated on *p*-type doped wafers with front collecting p-n junction (emitter) and front metal contact (e.g., Al-BSF, PERC, PERT, MWT), or front collecting junction and rear contact (such as EWT). From this, the most attractive cell concepts are the *p*-MWT or p-EWT, which combine the features of a front collecting junction but with back contact interconnection technology. The effect of other cell parameters for space application such as thickness, passivation and ARC layers, doping regions and cell geometry can easily be estimated using the device modelling and will be subject of our future study.

4 SUMMARY AND OUTLOOK

We developed a methodology to model the degradation of c-Si solar cells upon proton (and electron) irradiation using a full device electrical simulation with a fitted multi-level Shockley-Read-Hall (SRH) recombination model for the Si bulk. This modelling show a good agreement to the experimental data for proton energies at least up to 10 MeV and fluences up to 1×10^{12} cm^{-2}. We used this device modelling to evaluated EOL performance under given space conditions (e.g., LEO mission) of many terrestrial high efficiency low-cost c-Si solar cell concepts. Based on this study, the properties and key design features of Si cells for use in space applications (with the highest radiation hardness) are:

- Based on p-type doped wafers (1-5 Ωcm),
- Front collecting junction and front/back contact (e.g., BSF, PERC, PERT, MWT),
- Front collecting junction and all back contact (e.g., EWT).

Further studies are in progress evaluating annealing behavior of the experimental silicon cells.

REFERENCES

[1] A. Wang, "High efficiency PERC and PERL silicon solar cells", PhD. Thesis, (UNSW, Sydney, 1992).
[2] K. Yoshikawa *et al.*, *Nature Energy* 21, 17032-1 – 17032-8 (2017).
[3] R. Kopecek et al., "*ZEBRA technology: Low cost bifacial IBC solar cells in mass production with efficiency exceeding 23.5%,*" Conf. Rec. IEEE Photovolt. Spec. Conf., vol. 2020-June, pp. 1008–1012 (2020).
[4] B.E. Anspaugh, "*The Solar Cell Radiation Handbook Addendum 1: 1982±1988*", JPL Publication, Addendum 1, p69-82 (1989).
[5] B.E. Anspaugh, "*GaAs Solar Cell Radiation Handbook*", JPL Publication, p96-99 (1996)
[6] S .R. Messenger *et al.*, Prog. Photovolt: Res. Appl. 9, p103-121 (2001).
[7] Y. Morita *et al.*, "*Anomalous degradation in silicon solar cells subjected to high-fluence proton and electron irradiations,*" J. Appl. Phys., vol. 81, no. 9, pp. 6491–6493 (1997).
[8] J. F. Ziegler, J. P. Biersack, and M. D. Ziegler, "*SRIM, Stopping and Range of Ions in Matter*", www.srim.org.
[9] S. R. Messenger, E. A. Burke, G. P. Summers and R. J. Walters, "*Application of displacement damage dose analysis to low-energy protons on silicon devices,*" in IEEE Transactions on Nuclear Science, vol. 49, no. 6, p2690-2694 (2002)
[10] A. Fell, http://quokka3.com.
[11] J. R. Srour, S. Othmer and K. Y. Chiu, "*Electron and Proton Damage Coefficients in Low-Resistivity Silicon,*" in IEEE Transactions on Nuclear Science, vol. 22, no. 6, p2656-2662 (1975).
[12] M. Yamaguchi et al., IEEE Trans. Electron Devices, p2133 (1999).
[13] D. Heynderickx, B. Quaghebeur, and J. Wera, "*New radiation environment and effects models in the European Space Agency's Space Environment Information System (SPENVIS)*", SPACE WEATHER, vol. 2, S10S03, (2004)

LIGHT WEIGHT INTERCONNECTION WEAVE FOR SPACE PV

Tom Borgers[1], Jozef Szlufcik, Manuel Van den Storme[3], Guy Van den Storme[3], Christian Brandt[2], Donato Girolamo[2], Shumit Das[2], Julo Verdonck[5], Philippe Nivelle[3], Jonathan Govaerts[1], Geert Doumen[1], Luc Vastmans[1], Sander De Vrieze[6], Eszter Voroshazi[1], Jef Poortmans[1]

1: imec (Partner in EnergyVille)/Thor Park 8320 B-3600 Genk, Belgium 2: ESA/Keplerlaan 1, 2201 AZ Noordwijk, The Netherlands 3: VdS Weaving/De Bruwaan 16 B-9700 Oudenaarde 4: UHasselt (Partner in EnergyVille)/Thor Park 8320 B-3600 Genk, Belgium 5: Thales Alenia Space Belgium/Gaston Geenslaan 9 B-3001 Heverlee, Belgium 6: Centexbel/Technologiepark 70 B-9052 Zwijnaarde, Belgium

ABSTRACT: This work reports on the first results of a new concept for a flexible blanket based on weaving technology, which could potentially be employed as a carrier structure for the photovoltaic assembly (PVA) within a flexible solar array (SA) architecture. Though further away from current proposed solutions, the promising potential of such a weaving approach is describing the creation of an electrically functional and integrated flexible carrier with minimal weight and stowage volume.
Keywords: Solar Array, GEO, Solar Cell Assembly, weave, conductor integration, weaving, flexible blanket

1 BACKGROUND

The ever-existing need for developing high reliability low-cost products for space applications could pave the way for developing concepts of modular PVA systems. This technology can be applied to e.g., communication satellites or other electrically powered Geostationary Earth Orbit (GEO) applications.

The trend towards higher power demands, e.g., due to electrical propulsion, implies increasing the power demanded from the solar array through increasing solar array size and/or efficiency. With state-of-the art solar array (SA) concepts based on rigid panel substrates, only a limited size increase is possible due to mass and mechanical limitations and especially due to stowage volume constraints. To tackle this, several flexible deployable solar arrays are introduced or in development. [1] [2] [3]

Important parameters for Photovoltaic powered generators are specific power (W/m²), stowage volume efficiency (W/m³) and the structural mass of the system. Moreover, material compatibility to GEO in-orbit conditions, such as radiation, temperature etc., is crucial to guarantee long-term operation and reliability. Important aspects for cost reduction are materials and manufacturing cost and the level of automation during production and assembly; the increasing demand for powerful solar arrays is translated in an increasing number of proposals and ideas for low-cost solar array development and innovations.

Figure 1: Example of a rollable SA (ROSA) [3]

The main goal of the activity is to demonstrate the use of technical textiles as a cost-efficient, rollable, integrated, functional and light-weight carrier substrate for flexible PVAs and explore its electrical and mechanical functionalization. Through material characterization, design, and weave prototyping, the integration of the electrical cable network and support features in a flexible weave will be investigated. Besides the flexibility and adaptability of textiles to high performance and low-weight requirements of the PVA, they potentially come with a cost advantage due to the limited material diversity use, and integrated fabrication method.

2 MODULAR SOLAR ARRAY APPROACH

The proposed woven weave, with a width of minimum that of an SCA string, up to that of the solar array, a maximum length defined by the solar array, functions simultaneously as carrier substrate for Solar Cell Assemblies (SCA), the electrical circuit structure, and the mechanical support solar array blanket. This results in an integrated PVA weave that can be assembled directly to a mechanical solar array carrying structure, while directly integrating the wiring into the weave. Multiple blankets could be connected mechanically and electrically to form a large solar array integrated weave to create the total desired size (and power output) (Figure 2). Such integrated weave modules could be employed within rollable or foldable flexible solar arrays.

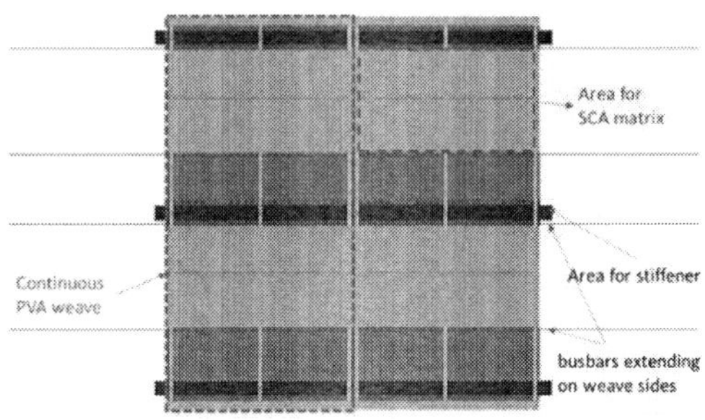

Figure 2: Modular integrated PVA weave

The goal of this structure is to improve structural weight

and stowage volume of the solar array, while simultaneously improving reconfigurability, potentially reducing total cost, and still allowing automated SA manufacturing processes. Aim of the activity is to optimize the material selection for GEO environment compatibility and aim to reduce the number of materials in order to increase reliability.

3 MATERIAL SELECTION AND EVALUATION

A basic evaluation of materials used for this weaving concept is required to ensure compatibility with the operational environment. Therefore, a test group of yarns was pre-selected based on their material properties and potential resistance to GEO radiations.

Pre-screening tests of the yarn materials (8 materials in total) under reduced test regime included (half fluence) electron- and proton-, and (100 ESH) UV- irradiation. The evaluation was done through visual inspection and tensile testing based on the EN ISO 2062 (2009) standard. Additionally, static shrinkage tests were performed on non-aged yarns; limited shrinkage might be of importance to reduce stress on the SCA's after gluing, or sagging of the weave in the SA structure.

A short overview of the visual and tensile results of the tested yarn materials is listed in Figure 3 and Table 1.

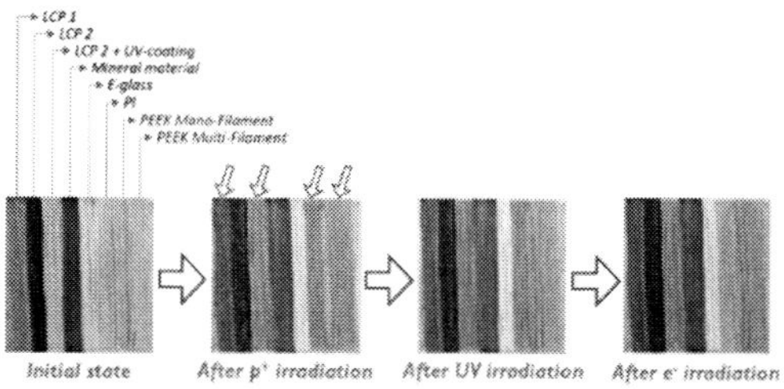

Figure 3: Visual evolution of test yarns during reduced regime irradiation test

Color darkening was observed for the Aramid, LCP (Liquid Crystal Polymer) and multifilament PEEK yarn after proton irradiation (indicated by the red arrows, Figure 3). No additional changes were noticed after further UV and then electron irradiation.

Table 1: Yarn tensile- and shrinkage test overview

	Aramid	LCP	LCP + Coating	PEEK multi	PEEK mono	PI	Mineral material	E-glass
Initial avg. Max. Load	29986	38690	38506	45359	598	2801	7604	6220
Initial avg. tenacity at Max. Load [cN/Tex]	186.97	231.68	230.58	276.58	46.33	26.35	50.69	44.97
Avg. tenacity at Max. Load after aging (% of initial)	56.83	13.51	8.88	7.07	57.25	64.89	71.28	249.41
Shrinkage (non-aged yarns) at 180 °C (%)	-0.12	0.06	0.02	-0.08	4.42	0.17	0.31	-1.47

Both LCP yarns and the PEEK yarn experienced considerable mechanical degradation; however, the PEEK monofilament yarn performed considerably better than its multifilament counterpart. The Aramid-based yarn performs better keeping 57% of the initial tenacity at maximum load after ageing. PI yarn appeared to be the best performing polymer-based material based on remaining tenacity (cN/Tex) at maximum load after aging (65% relative to initial value).

The ageing tests had less influence on Glass fibre and the mineral fibre material. However, in terms of initial tenacity at maximum load, this last material outperformed the E-glass fibre. The mineral material still retains 71% of its tenacity at maximum load value after ageing, with no degradation in Tensile strain at maximum load. A possible hypothesis of the maximum load degradation might be related to the sample preparation (fibre damage of on the sample holder). The estimated tenacity values of E-glass are very unlikely; seen the initially better performance of the mineral material, E-glass is not considered in the further development.

Static shrinkage tests on non-aged yarn samples at 180 °C revealed no large variations between the different materials except for monofilament PEEK showing much more shrinkage. This might be related to the production method (extrusion) and relaxation of the polymer at high temperature.

A basic representative test weave of the best performing polymer-based yarn (PI) and mineral-based yarn was prepared and tested according to following test cycle:

- Electron and proton irradiation at 2 energy levels, total fluence
- UV irradiation 1500 ESH

Intermediate visual inspections, and thermo-optical measurements were done before and after irradiation tests (Figure 4).

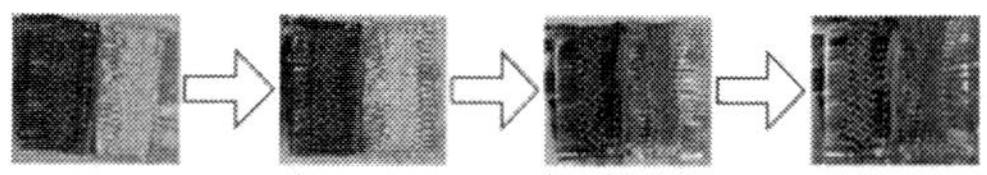

Figure 4: Visual evolution of test weaves during irradiation tests

Although no visual effects were noticed after electron irradiation on both materials, the PI sample showed waving, discoloration and became fragile and hard to manipulate after proton- and UV irradiation, in contrast to the mineral material sample.

Table 2: Thermo-optical measurements

Sample	Absorbance		Emissivity	
	Initial	After irradiation	Initial	After irradiation
PI	0.38	0.59	0.92	0.92
Mineral material	0.83	0.81	0.95	0.94

Thermo-optical measurements revealed no change in emissivity for both samples, however absorbance changed for the PI sample (Table 2).

Table 3: Tensile test results on aged weaves

		Mineral material	PI
Yarn max. load [cN]	*Non-woven, non-aged yarn*	7604	2801
	yarn from non-aged weave	7468	2469
	yarn from aged weave	4741	978
Weave Max. load [N]	*non-aged*	507	181
	aged (calculated)	310-316	63-72

A tensile test comparison with non-aged weave samples was foreseen on the aged weave, but unfortunately, the sample turned out to be damaged and frayed by handling and mounting to be employed in further mechanical testing. Therefore, tensile tests were performed on individual warp yarns of the aged weaves. Based on tensile measurements and a comparison with tensile tests on non-aged weaves of both materials with the tensile test results of non-woven and non-aged yarns and non-aged warp yarns from a weave (calculated to a similar number of warp yarns as used in the tested weaves), an estimated maximum load for both aged weaves were calculated (Table 3). These estimated weave residual strengths after irradiation revealed that the mineral material-based weave preserved 62 % to 65% of the non-irradiated weave strength (507 N). The aged PI-based weave only retained 35 % to 40 % of the non-irradiated weave strength (181.0 N). From the tests it was observed a change in mechanical behavior of the material after irradiation suggesting a material embrittlement after irradiation.

4 PVA WEAVE TESTS

From these aging test results, a passive (electrically non-functional) PVA weave design was created with respect to the different requirements and restrictions of the SA design and environment, based on the mineral yarn material. A symmetric weave style was selected, suitable for adhesion of the SCA strings. UD-tapes (unidirectional tapes), working as load carrying elements and attachment points for the PVA, are embedded in the two outer sides and in the middle of the warp (Figure 5, Left). These tapes combine continuous carbon fibres in a matrix of polymer resin resulting in a combined composite CTE resembling that of typical carbon fibre structures avoiding waving and sagging of the PVA structure. The resin allows lamination during post-weave processing. The tapes are combined to weft-oriented UD tapes to create an interwoven reinforcement harness: after lamination wherein the yarns crossing the tapes are embedded in the tape resin, the resulting weave features a UD-tape "harness"; it could be used to suspend the weave between interfaces of solar array structures.

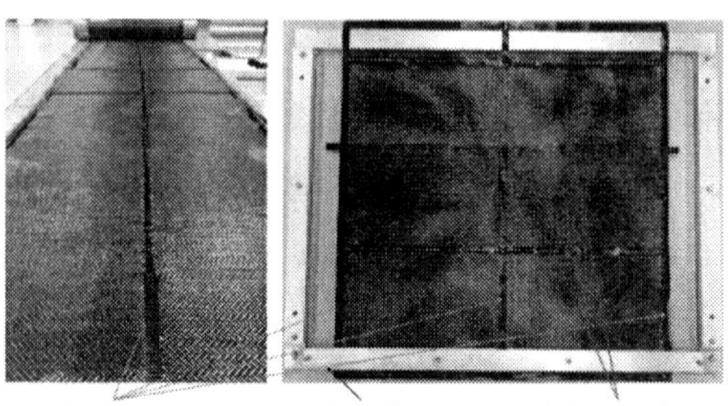

Figure 5: Left: continuous weave after weaving **Right:** Passive PVA weave mounted in test frame

5 ELECTRICALLY FUNCTIONAL PVA WEAVE

Following up on the experience from the passive weave tests, an active weave is developed, adding electrical functionality. All production methods developed for the active and passive weave are developed with automated PVA fabrication in mind, together with an industrial weaving company.

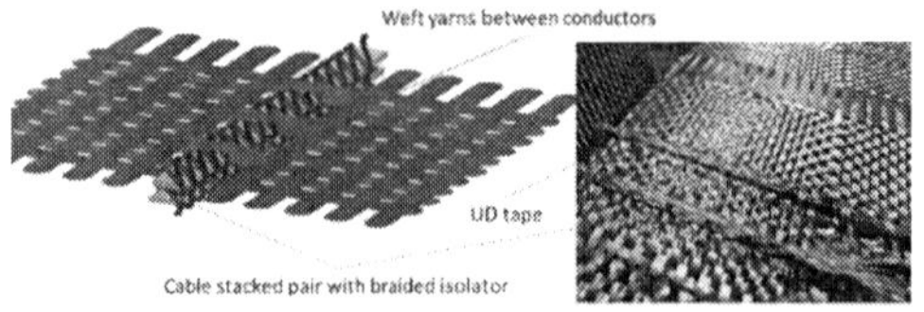

Figure 6: Detail of braided cable pair integrated in weave

The smart 3D-woven weave interweaves conductive cables and insulative yarns. The yarns provide the mechanical integrity and a carrying base for the SCA strings, simultaneously ensure electrical insulation where needed, and alignment of the cables. As in the passive weave, additional weft-oriented conductors are foreseen for string busbar interconnection. The conductors in the warp direction have an electrically insulating braided yarn jacket with a metal flat-braided ribbon core. These cables allow electrical interconnection of the PVA busbars to the solar array power management and convertors at the base of the solar array.

When a specific braiding density for the cable jacketing is used, the inner conductive braided core is still accessible by solder that is liquefied during the solder reflow process; hence the braided jacket allows selective interconnection of busbars and cables: when the weft PVA busbar ribbon is crossing over the warp cables, no electrical contact is created unless a solder joint is created through the pores of the braided cable jacketing (Figure 7); This potentially allows a soldered interconnection without locally having to remove the isolative jacketing from the cable.

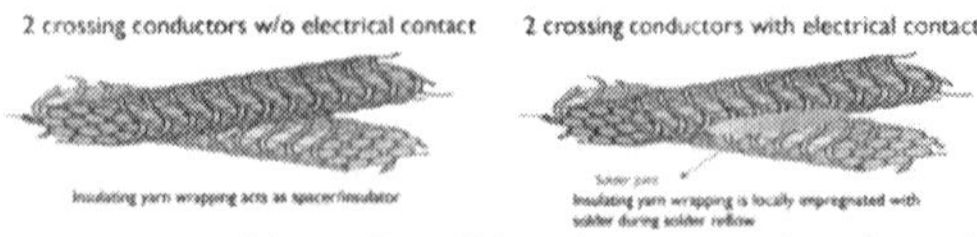

Figure 7: Cable and/or ribbon interconnection through braided isolator

A potential additional advantage of using a braided cable structure might be its ability to compensate excessive material CTE differences between different materials and components; the 3D-oriented strands and fibres allow a high degree of flexibility and stretch allowing the cable to accommodate dimensional changes during thermal cycles. Limiting material diversity by using the same material for cable jacketing and the weave reduces the risk of material incompatibility (degradation, CTE differences) potentially increasing reliability and reducing cost.

As an example, the weave can be sized to accommodate a matrix of 4 strings of 4 SCA's of standard 6x12 cm^2 size. The design is adaptable to larger PVA sizes and different PVA configurations (Figure 8).

In the warp of the weave, between the UD tapes, warp yarns are distributed along with conductive cable stack pairs. As can be seen in figure 6, a cable stack is a combination of 2 flat braided silver-plated copper filament ribbons, covered in a braided isolative jacket from the mineral yarn (Figure 6).

Yarns are introduced in the weft during weaving, mainly in the area where SCA strings will be attached/glued on the weave in later steps, and in locations where the warp cable stacks must remain well-aligned. The weft yarns are crossing below, in between and above the cables of the warp stacked cable pairs to secure their position in the weave and acting as an extra isolative layer with the environment.

At the locations where potential SA carbon interface points are located, the warp UD tapes are not interwoven with the weft yarns, thereby selectively interweaving the warp cables and warp yarns.
As weaving is a continuous operation, this pattern is repeated. As such, the resulting weave is a continuous weave that can have a total length in accordance with the total length of the SA.

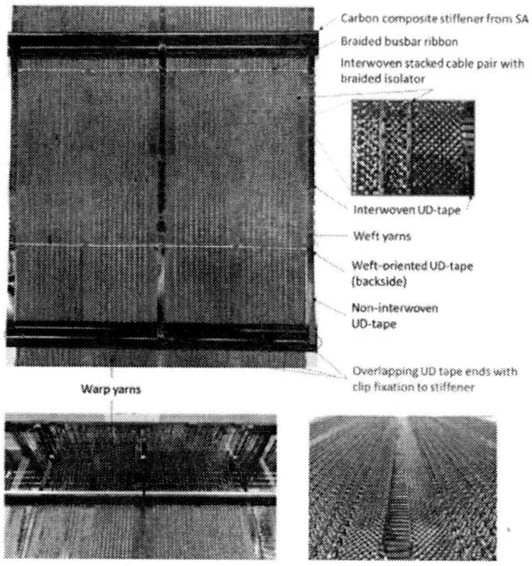

Figure 8: First version of a functional PVA weave, pictures, and details of a functional weave version

After weaving, UD tapes are placed in weft orientation, together with flat-braided conductive ribbons aligned on top of these tapes. The weave area between these two busbar ribbons is foreseen for gluing of SCA strings. During a lamination step, the warp (interwoven) UD tapes will embed all weft yarns and UD tapes that are crossing, in the resin of the UD tapes. The resulting structure creates a frame (harness) of UD tapes in which the weave is suspended.
The outer electrical contact pads of the SCA string can be welded or soldered on the braided busbar; the braided busbar allows a thermally mechanical stress-free connection. Local cutting of the busbar will create the desired SCA string interconnection scheme. The busbar ends can be soldered or welded with the PVA weave-integrated braided cable pairs.

6 PVA WEAVE CLIP ATTACHMENT TO SA STRUCTURE

To attach the weave to a load carrying SA structure, a preliminary clip design was made. The advantage of this attachment system is the limited weight (0.178 g), combined with the potential to use resistant metal coatings for GEO environment and low topography. At this time, the attachment is assumed to connect to CFRP based tape springs such as described in Ref [1 to 3]. At the areas where the UD tapes are not interwoven with weft yarns (Figure 5), the UD tapes are attached to the carbon interface points of the SA structure: an opening in the tape is made through which a metal clip is pushed.

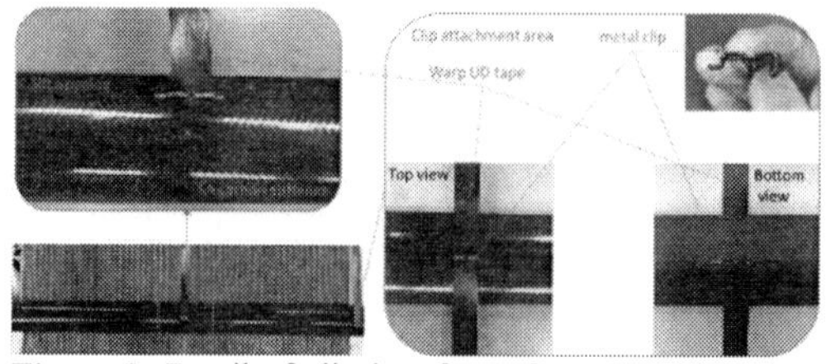

Figure 9: Detail of clip interface of weave with carbon interface point.

The clip is on its turn pushed through a slit opening in the carbon interface point of the SA (Figure 9). First tensile tests were performed on this interface by fixing the ends of 2 UD tapes, attached to a carbon interface point with a clip, in the gauges of the pull tester; a promising maximum tensile force of 200 N was reached before one of the UD-tapes was torn out at the clip area. Additional assessment of this attachment system is ongoing.

7 CONCLUSION AND OUTLOOK

In this work we report on new textile fabrics based on weaving technology that could be used as flexible carrier substrates for PVA modules that could be employed within flexible solar array architectures. We discussed first test results of prototype textiles, different functionalities and their benefits that are combined into this concept, potentially allowing for efficient, lightweight, and flexible interconnection of SCA's in different configuration architectures.

8 ACKNOWLEDGMENTS

The development is in the framework of next-generation flexible solar arrays within the ESA call "GT17-137TI: Preparation of enabling space technologies and building blocks". The authors gratefully acknowledge The European Space Agency ESA, Thales Alenia Space Belgium, VdS Weaving and Centexbel. This project has received funding from the European Space Agency for research, technological development, and demonstration in the framework of ESA General Support Technology Programme under ESA Contract Agreement No. 4000129045/19/NL/FE/vr (WARP project).

9 REFERENCES

[1] M. Straubel et.al, "Evaluation of Different Architectural Concepts for Huge Deployable Solar Arrays for Electric Propelled Space Crafts," in 14th European Conference on Spacecraft Structures, Materials and Environmental Testing (ECSSMET), Toulouse, France, 2016.
[2] "Solar arrays can now be SWIRLed up," The European Space Agency, 10 February 2020. [Online]. Available: https://www.esa.int/Enabling_Support/Space_Engineering_Technology/Shaping_the_Future/Solar_arrays_can_now_be_SWIRLed_up.
[3] "DSS to provide ROSA solar array for NASA's Double Asteroid Redirection Test (DART) Mission," DSS Deployable Space Systems, 21 April 2018. [Online]. Available: https://www.dss-space.com/post/dss-to-provide-rosa-solar-array-for-nasas-double-asteroid-

redirection-test-dart-mission.

FIRST MECHANICAL STUDY ON LIGHTWEIGHT MICROCONCENTRATORS SYSTEMS FOR SPACE APPLICATIONS

Victor VAREILLES *
Univ. Grenoble Alpes
CEA, LITEN, DTS, SMSP, LAM, INES
F-38000 Grenoble, France
victor.vareilles@cea.fr

Anderson BERMUDEZ-GARCIA
Univ. Grenoble Alpes
CEA, LITEN, DTS, SMSP, LAM, INES
F-38000 Grenoble, France
anderson.bermudezgarcia@cea.fr

Jérôme FRANCOIS
Univ. Grenoble Alpes
CEA, LITEN, DTS, SMSP, LPM, INES
F-38000 Grenoble, France
jerome.francois@cea.fr

Yannick VESCHETTI
Univ. Grenoble Alpes
CEA, LITEN, DTS, INES
F-38000 Grenoble, France
yannick.veschetti@cea.fr

Mohamed AMARA
Univ Lyon, CNRS, INSA Lyon, ECL, UCBL, CPE Lyon, INL, UMR5270
F-69621 Villeurbanne CEDEX, France
mohamed.amara@insa-lyon.fr

Philippe VOARINO
Univ. Grenoble Alpes
CEA, LITEN, DTS, SMSP, LAM, INES
F-38000 Grenoble, France
philippe.voarino@cea.fr

Fabien CHABUEL
Univ. Grenoble Alpes
CEA, LITEN, DTS, SMSP, LAM, INES
F-38000 Grenoble, France
fabien.chabuel@cea.fr

*Corresponding author: victor.vareilles@cea.fr

Abstract— The French Commission for Atomic and Alternative Energies (CEA) has been investigating a lightweight honeycomb core composite structure for micro-concentrators for space applications. This new design requires a double stage honeycomb sandwich, a front layer to support the optical system with a glass front sheet and a second honeycomb to provide stiffness and mechanical strength to the module. Mechanical studies were conducted to design a lightweight structure suitable for the space environment, meeting the requirements for stiffness, adhesion between the components and mechanical strength.
Keywords— concentrators, mechanical structures, micro-concentration, optics

1 INTRODUCTION

1.1 CPV for space

Future spacecraft will require higher-powered photovoltaic subsystems in order to achieve missions that are more ambitious. For space applications, constraints on solar array size, mass, and storage volume have encouraged the development of efficient Multi-Junction Solar Cells (MJSC). While their higher power levels have helped to increase spacecraft payload capabilities, MJSC for space are composed of expensive III-V semiconductor materials (e.g., Ge, Ga). New generation solar array architectures are being proposed and studied. One of these approaches is Concentrator PhotoVoltaics (CPV) arrays, which demonstrated the highest efficiency ever achieved by any solar technology at the cell, module, and system levels [1]. Most CPVs used in spacecraft utilize either reflective or refractive elements to focus the sunlight onto a much smaller solar cell area, typically MJSC.

The key interest of concentrators is to rely on the usage of these high-efficiency MJSC while reducing the cells' area by orders of magnitude, then reducing the cost. In addition, the use of CPV also enables significantly higher conversion efficiency than cells operating under AM0 conditions. In CPVs, the supplementary mass related with the optical system and mechanical support structures could be seen as a drawback, however, it comes with a major advantage: improved radiation hardness, for instance, in Earth orbits and around the Jovian system. CPVs can also be used as a complementary energy source to the Radioisotope Thermoelectric Generator (RTG) for deep space missions, by operating the PV cells with an irradiance close to one sun.

Several CPV systems have been developed over several decades. The PASP+ (Photovoltaic Array Space Power Diagnostics Plus) program was the first satellite with on-board CPV arrays. *In-situ* measurements demonstrated that CPV presented the smallest degradations induced by space radiation, when compared to the other 12 array technologies [2]. Then, in 1998, the first mission to rely upon a solar concentrator was successfully launched. The *Deep Space 1* spacecraft using SCARLET (Solar Concentrator Array with Refractive Linear Element Technology) has proven the viability of CPVs in space; during its 38-month mission, it powered the instruments and the ion propulsion engine, which propelled the spacecraft throughout its entire journey [3].

Furthermore, advanced approaches designs including microscale (< 900 µm^2) photovoltaics have the potential to extremely reduce the optical profile, thus lowering mass, while having moderate concentration factors (< 50X). Moreover, other advantages of using micro-cells include passive thermal dissipation, which avoids heat sinks and reduce mass through the mitigation of CTE mismatch effects between cell and substrate, due to small dimensions [4],[5].

At the CEA, we have worked on a low concentration system that respects the angular tolerance specifications but its specific power (W/kg) was too low [6],[7]. The aim of this work is to study the mechanical behaviour of different honeycomb structures in order to choose the most suitable configuration for a new lightweight proposal. This innovative design integrates a network of optics inserted in a honeycomb structure [8], which suggest that high power levels per unit mass (W/Kg) can be achieved. Taking into account the mass of PVA mechanical support, it could realistically exceed 300W/kg (at incident AM0 spectrum) and serve as a drop-in replacement for existing CIC technology at a substantially lower cost.

1.2 Standard solar array panel constituents

Current standard PVA for space applications is an assembly of cells protected by cover glass, so-called CIC (Cover Glass Interconnected Cell) or SCA (Solar Cell Assembly), which is attached using durable adhesives and bonding agents to a rigid and lightweight structural substrate (aluminium honeycomb core) and carbon-reinforced composites, as shown in Figure 1.

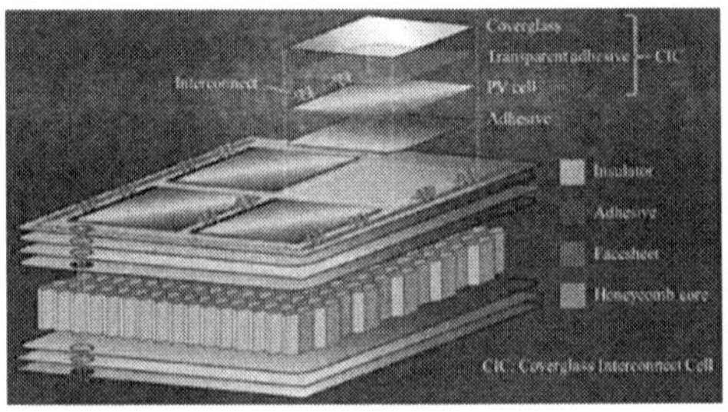

Figure 1: Solar array panel constituents (exploded view), image adapted from Francis *et al.* (2005) [9] and taken from Bermudez-Garcia *et al* (2020)[10]

1.3 Studied structure

The studied structure is as presented on the scheme below (Figure 2). A first honeycomb core structure on the bottom is made of two face sheets bonded to both faces of the honeycomb, using epoxy-based adhesive. A second honeycomb (top) is then bonded to the bottom one using the same adhesive. The cells of the top honeycomb are partially filled with silicone. The front glass plate is ultimately bonded to the top honeycomb.

Integrating the optics into the top honeycomb removes the need for an additional support for the optics, which would increase the mass of the overall system while maintaining high mechanical properties.

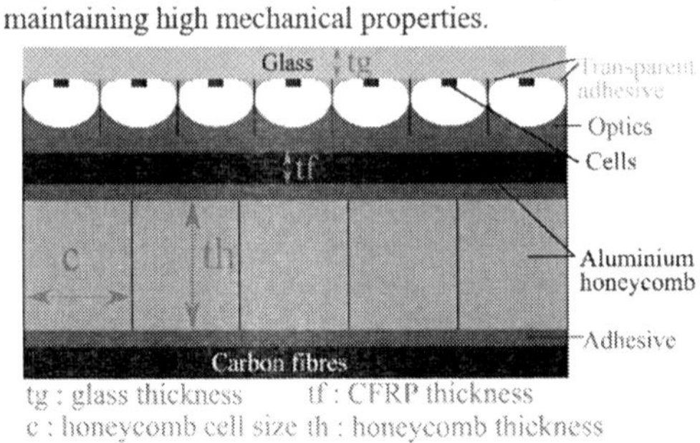

Figure 2: Scheme of the complete structure

1.4 Materials

Materials used for this structure are listed below:

- Aluminium honeycomb is used as a lightweight core for the sandwich and provides high shear and compressive strength.
- Bottom and middle face sheets are made or Carbon Fibres Reinforced Plastic (CFRP). Carbon fibres are commonly used for space applications because of their high strength-to-weight and stiffness-to-weight ratio.
- The front face sheet is made of borosilicate glass. This material is chosen for its excellent optical properties, as sunlight will go through this sheet before reaching the cells. It also helps protecting the optics and the solar cells from energetic particles radiation in the space environment. However, its low mechanical strength makes it unsuitable as the main structural part of the structure.
- Optics are made of silicone, using an injection moulding process. The typical elastic modulus of silicone rubber is around 1 MPa, whereas the elastic modulus of glass is around 70,000 MPa. Thus, we decided not to add the silicone optics for this particular study, as it will not influence the stiffness of the structure.

2 MECHANICAL STUDY

2.1 Experiments preparation

In our design, a large glass plate will form the top of the structure. Because of glass brittleness, which may lead to crack propagation, the bottom honeycomb structure must support the majority of the stress so that the top one is relieved. Furthermore, it should be as light as possible to reduce launching costs. In order to design a light and stiff-enough structure suited for space environment, the influence of various parameters on the stiffness, the stiffness-to-weight ratio and the modes of failure are studied. A summary can be found in Table I. We chose to study three parameters, which gave us 12 configurations. We used the JMP® software (website: https://www.jmp.com/en_us/home.html) to build an experiment plan to minimize the number of experiments. Ultimately, we tested the configurations in bold, summarized in Table II.

Table I: Parameters experimental values

Parameter	Level 1	Level 2	Level 3
Honeycomb	3/16 - 10mm	1/4 - 20mm	/
CFRP plies	2	4	/
Glass thickness	300 µm	400 µm	500µm

2.2 Sample preparation and test procedure

As the mechanical strength of glass is higher under compressive stress and less subject to local defects, the structure was tested as shown in Figure 4 (the glass on the top) to avoid premature failure of the structure. Some samples were also tested with the glass facing downwards. The orientation of the honeycomb also influences the stiffness of the structure; therefore, all samples were made following the L direction (Figure 3). Five test samples were made for each structure to ensure the reproducibility of the tests.

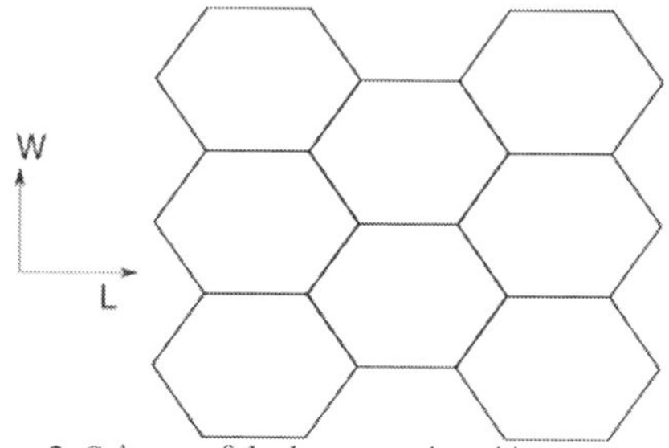

Figure 3: Scheme of the honeycomb and its two directions

The structures were processed at the CEA in two steps: first the CFRP sheets and the two honeycombs were laminated in a vacuum membrane laminator. The glass sheets were then bonded on the top honeycomb with a transparent and rigid epoxy adhesive.

2.3 Experimental setup

To determine the stiffness of the aforementioned structures, 4-point flexural testing was used. Sample geometry was chosen according to NF T 54-606 standards [11]. This test also allows evaluating the adhesion force between the components by examining the fracture surfaces and the area between the loading points. With sufficient adhesion, we should see no sign of delamination.

Tests were carried out thanks to an INSTRON 3365 device with a 5 kN load cell and a fixture compatible with ASTM D7250 standards [12] (Figure 4). We used quarter point loading with uniformly distributed load. The deflection of the sandwiches was measured using an INSTRON SVE2 video extensometer with ±1% accuracy. Figure 5 shows the load-displacement curves obtained for structure #4 as an example.

Figure 4: Experimental setup comprising the INSTRON 3365 device with a 5 kN load cell and 4-point flexural testing with a sample under test

2.4 Bending stiffness

The bending stiffness of the structures under 4-point bending with punctual loading is given by the following equation:

$$D = \frac{11 \times F \times L^3}{768 \times w}$$

- F is the load
- L is the support span
- w is the maximum deflection

To compare samples with different geometries, we present D_n, the bending stiffness divided by the width (b) of the samples in the rest of this paper.

Experimentally, the slope of the load-displacement curve (S) is used:

$$D_n = \frac{11 \times S \times L^3}{768 \times b}$$

In practice, there is little difference with uniformly distributed load so we chose to use this equation for punctual load. We also present the specific bending stiffness D_ρ, which is D_n divided by the area density of the samples. Both D_ρ and D_n are presented in Figure 6. We observed a good reproducibility of the experiments. The highest standard deviation was obtained for structure #2 at 7.7% of D_n. Standard deviation for the others was below 5%.

Table II: Summary of the configurations. The ones in bold were tested experimentally

#	Honeycomb	Glass thickness	CFRP
1	3/16 - 10 mm	300 µm	2 plies
2	**3/16 - 10 mm**	**400 µm**	**2 plies**
3	**3/16 - 10 mm**	**500 µm**	**2 plies**
4	**3/16 - 10 mm**	**300 µm**	**4 plies**
5	3/16 - 10 mm	400 µm	4 plies
6	3/16 - 10 mm	500 µm	4 plies
7	**1/4 - 20 mm**	**300 µm**	**2 plies**
8	1/4 - 20 mm	400 µm	2 plies
9	1/4 - 20 mm	500 µm	2 plies
10	1/4 - 20 mm	300 µm	4 plies
11	**1/4 - 20 mm**	**400 µm**	**4 plies**
12	**1/4 - 20 mm**	**500 µm**	**4 plies**

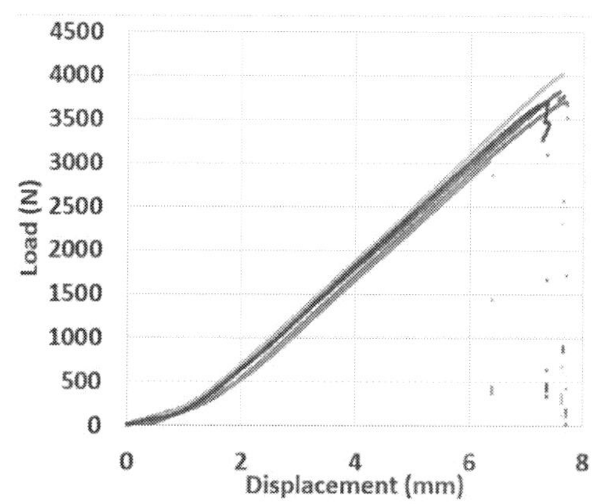

Figure 5: Load-deflection curves for configuration #4. Each colour represents a test sample.

For structure #6 one curve was not exploitable because delamination between the bottom CFRP sheet and the honeycomb happened. As only one sample had this problem, we believe that this is due to a manufacturing defect created during sample preparation.

Structure #3 could not be brought to failure because of our load cell limitation. Its bending stiffness can nonetheless be determined.

We used the JMP® software to analyse our experimental data, which gave us the influence of the parameters (LogWorth) and the bending stiffness for the other configurations (lighter colours in Figure 6). The p-value indicates the level of certainty of the influence, a lower p-value meaning a more precise LogWorth value. Please note that the bending stiffness computed by JMP should be treated as an order of magnitude rather than a definitive value.

Table III: Effect of the parameters on the bending stiffness (computed by JMP®)

Parameter	Effect (LogWorth)	p-value
Honeycomb	2.27	0.005
CFRP thickness	2.21	0.008
Glass thickness	1.68	0.021

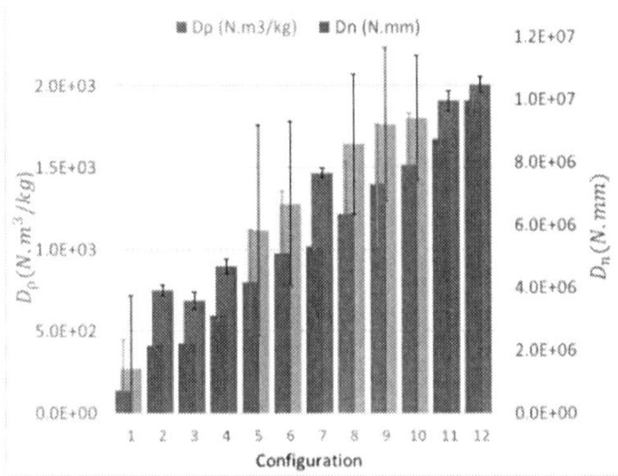

Figure 6: Experimental (darker) and computed by JMP® (lighter) bending stiffness D_n and specific bending stiffness D_ρ

2.5 Mass

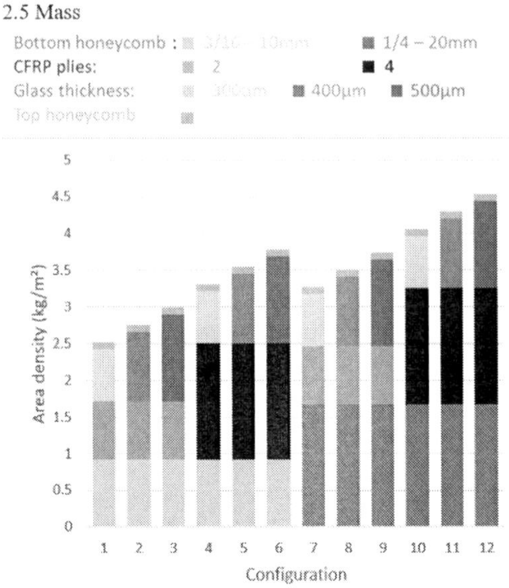

Figure 7: Area density of the different components

Figure 7 shows the area density percentage of the components for each configuration. Combined with Table III, we can deduce which parameters should be changed to decrease the mass without sacrificing too much stiffness. A thicker honeycomb and a thinner glass plate should be used to reach this goal.

2.6 Failure modes

We observed different failure modes:

- Glass failure
  - Failure under the loading areas under compressive stress (e.g. Figure 10)
    - Happened mostly with the thinnest glass (300 µm thick)
  - Failure on the edges of the loading areas under shear stress
    - Happened mostly with thicker glass
- Failure of the bottom CFRP under tensile stress
- Crushing of the top honeycomb under compressive stress (e.g. Figure 9)
- Failure of the top fibres under compressive stress (e.g. Figure 9)

Figure 8: Wrinkling of the top honeycomb due to shear stress

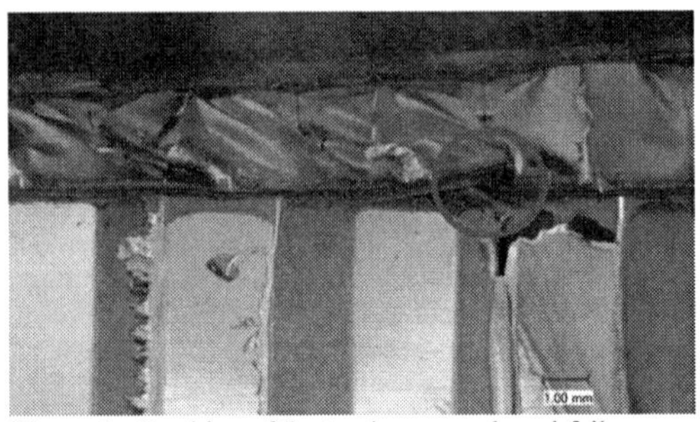

Figure 9: Crushing of the top honeycomb and failure of the top CFRP

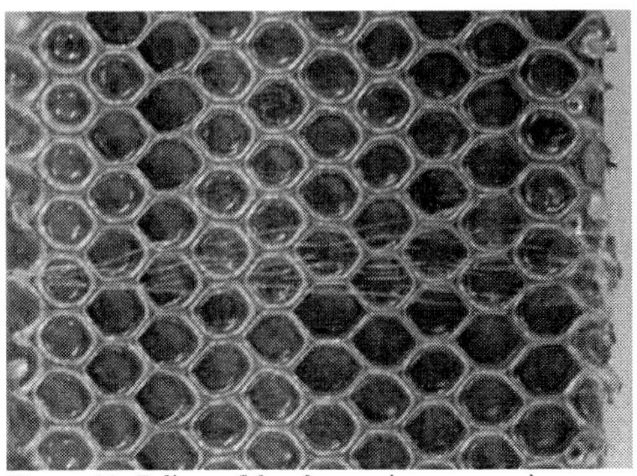

Figure 10: Failure of the glass under compression

Some structures exhibit multiple modes of failure and some are most likely induced by others. As an example, the top CFRP never failed without the glass. We can assume that the failure of the glass leads to an increase of the stress of the top CFRP, causing its failure.

On the other hand, it is not straightforward to determine if the failure of the glass leads to the crushing of the top honeycomb or if the crushing induces high shear stress in the glass, causing its failure.

In every case, the top honeycomb was damaged after the test. We can notice wrinkles (e.g. Figure 8) due to shear stress between the loading and support points on the samples and some due to compression under the loading areas. A top honeycomb with a higher shear strength could be used to prevent its deformation but would also increase the shading of the cells, so the choice will depend on the significance of shear stress for our application.

The configuration that proved to be the most effective at protecting the cover glass was #2, for which the glass did not break during tests for three samples.

2.7 Adhesion

4-point bending also allow checking if delamination occurs between the loading areas. As mentioned earlier, delamination of the bottom CFRP happened for one test sample (#11) and we believe this is due to manufacturing defects. We also noticed that no visible delamination of the glass occurred. Even after glass cracking in the case of compression failure, the glue film still held the glass in the place. However, in case of shear failure of thinner glass, it shattered and the bond with the honeycomb was broken.

2.8 Correlation between mechanical tests and their results

Based on our experiments, some configurations better suited for our needs, namely a high stiffness, a low mass and protecting the glass will be studied further. Configurations with a thicker bottom honeycomb and glass and thinner CFRP like 8 and 9 seem to be the most promising ones.

Regarding the mass of the system, a less dense bottom honeycomb could be used to significantly reduce the mass while maintaining a high stiffness.

3 CONCLUSION AND PERSPECTIVES

This study is the preliminary step to manufacture a lightweight micro-concentrator module for space applications. The bending stiffness and failure of 12 different structures were studied using 4-point bending tests. The mass of the different components was determined to identify ways to make the structure lighter without decreasing its mechanical properties. Configurations with a thicker glass (400, 500 µm) and thinner CFRP (2 plies) proved to be more efficient at protecting the cover glass and distributing the stress (e.g. structure 2). However, decreasing the thickness of the glass and increasing the one of the CFRP is a good way to increase the specific stiffness, thus a compromise needs to be made.

The next step of this study is to correlate the experimental results with analytical and numerical simulations. This study will then be carried out further with the optimisation of the structure, mainly increasing its specific bending stiffness. Simulations will be carried out to study the influence of additional parameters and reach our goal. This study will also continue with the investigation of the thermal effects due to space environment on the focal point of the optic system.

4 ACKNOWLEDGEMENT

The authors gratefully thank "Institut Carnot Energie du Futur" for funding this project.

This work is part of INES.2S which has received support from the French government under the « Programme d'Investissements d'Avenir" (ANR-10-IEED-0014-01).

REFERENCES

[1] M. Wiesenfarth, S. P. Philipps, A. W. Bett, Fraunhofer ISE / K. Horowitz, S. Kurtz, National Renewable Energy Laboratory NREL, USA | Version 1.3, April 2017

[2] H. Curtis, M. Piszczor, P. Severance, D. Guidice and D. Olson, "Early results from the PASP Plus flight experiment," in *Proceedings of 1994 IEEE 1st World Conference on Photovoltaic Energy Conversion - WCPEC (A Joint Conference of PVSC, PVSEC and PSEC)*, 1994

[3] M. I. Eskenazi, D. M. Murphy, E. L. Ralph and H. I. Yoo, "Testing of dual-junction SCARLET modules and cells plus lessons learned," in *Conference Record of the Twenty Sixth IEEE Photovoltaic Specialists Conference - 1997*, 1997.

[4] Domínguez, N. Jost, S. Askins, M. Victoria, and I. Antón, A review of the promises and challenges of micro-concentrator photovoltaics, AIP Conference Proceedings 1881, 080003 (2017).

[5] Ritou, P. Voarino, O. Raccurt, Does micro-scaling of CPV modules improve efficiency? A cell-to-module performance analysis, Solar Energy 173, 789-803 2018.

[6] P. Voarino, A. Ritou, C. Seraine, M. Roux, R. Couderc, Microconcentration for space applications: A 7.6 X TIR design, AIP Conference Proceedings 2149 (1), 050008, 2019

[7] P. Voarino, A. Ritou, B. Boulanger, C. Seraine, R. Couderc, H. Boirard, E. Rapp, "A space self-integrated micro-concentrator design: the SLIT," 2019 European Space Power Conference (ESPC), Juan-les-Pins, France, 2019, pp. 1-6, doi: 10.1109/ESPC.2019.8932023.

[8] A. Bermudez-Garcia, P. Voarino, O. Raccurt, patent pending, Concentrateur optique à structure alvéolaire, FR2013856, 21/12/2020.

[9] Robert W. Francis, Charles Sve, and Timothy S. Wall, "Thermal Cycling Techniques for Solar Panels", Crosslink Fall, Vol. 6, No. 3, 2005

[10] A. Bermudez-Garcia, P. Voarino, O. Raccurt, "Environments, needs and opportunities for future space photovoltaic power generation: A review", Applied Energy, Volume 290, 2021.

[11] NF T 54-606 standards: *Structures sandwichs à base de plastique. Essais de flexion*, 1987

[12] ASTM D7250/D7250M standards: *Standard Practice for Determining Sandwich Beam Flexural and Shear Stiffness*, 2020

WHAT IS THE CORRECT EFFICIENCY FOR TERRESTRIAL CONCENTRATOR PV DEVICES?

Harald Müllejans[1], Stefan Winter[2], Martin A. Green[3] and Ewan D. Dunlop[1]
[1]European Commission, Joint Research Centre, Ispra, Italy
[2]Physikalisch-Technische Bundesanstalt (PTB), Bundesallee 100, 38116 Braunschweig, Germany
[3]Australian Centre for Advanced Photovoltaics, University of New South Wales, Sydney, Australia

ABSTRACT: In this paper we assess the calculation of efficiency for concentrator PV (CPV) devices critically. Based on the general definition and conventional assumptions we argue that the efficiency for CPV devices should not be referenced only to the direct incident irradiance but to the total irradiance which includes also the diffuse and albedo components. The total (spectral) irradiance which would be incident onto a CPV device when the direct irradiance is exactly as already defined in the direct reference spectrum AM1.5D is calculated here and denominated as AM1.5GN. Based on this we derive a correction factor to obtain the thermodynamic efficiencies of CPV from the conventionally quoted efficiency with respect to the direct irradiance only. We illustrate the concept by applying it to the latest published world record efficiencies for CPV.
Keywords: Concentrators, Concentrator cells, solar cell efficiencies, solar radiation, calibration

1 INTRODUCTION

The aim of this work is to assess whether the efficiencies determined for different terrestrial PV technologies are derived in a consistent manner, in particular in the case of terrestrial CPV technology, which uses the direct reference spectrum (AM1.5D listed in IEC 60904-3 [1] and ASTM G173-03 [2]). The question we address is whether this is the most appropriate spectrum to use. Or should rather the full spectrum incident onto a CPV device be used? And if, what would this spectrum actually be? Is the global reference spectrum AM1.5G (AM1.5 listed in IEC 60904-3 and ASTM G173-03) directly applicable? Or do we require a new definition?

While the two spectra AM1.5D and AM1.5G are derived from the same calculation, the former is incident perpendicular onto the PV device (corresponding to a device tilt of about 48°), whereas for the latter the direct component is incident at an angle of about 11° against the normal as the device tilt is assumed to be 37°. Considering this, what would be the global normal spectrum AM1.5GN incident onto a PV device mounted at 48° tilt?

This paper provides answers to these questions. We will also address the equivalence of the reference spectra defined by IEC and ASTM, as they have adopted slightly different approaches in their definition.

Traditionally efficiency for CPV has been referenced to AM1.5D neglecting the diffuse and albedo components of the global irradiance incident onto the device. Here we argue to adopt the real thermodynamic efficiency based on the global irradiance onto the CPV device.

The additional reference spectrum AM1.5GN calculated here is relevant in the evaluation of terrestrial CPV devices. The irradiance incident onto the receiving plane of these devices is AM1.5GN, consisting of the already known direct beam irradiance (including circumsolar) AM1.5D, and additionally contributions from diffuse sky irradiance and albedo. CPV devices usually then only use the direct beam irradiance due to their construction and optical arrangement, thereby not utilising the irradiance components from diffuse sky and albedo. However, when evaluating the efficiency of CPV technology the full incident irradiance should be used rather than just the direct component, which is allowed to "enter" the device. It is a characteristic of the CPV technology to only use the direct component of irradiance and neglect the others. Not taking the global contribution into account in the calculation of the efficiency gives an apparent advantage to CPV technology suggesting efficiencies higher than they are in reality, in particular when comparing to non-concentrating terrestrial PV technologies.

For example, for thin film technologies such as amorphous silicon the irradiance available in the spectral band of their sensitivity is 500 W/m^2 to 600 W/m^2; nevertheless the efficiency is calculated using the full 1000 W/m^2 irradiance available in AM1.5G. The limitation of a technology to only use a part of the available wavelength range of the overall incident irradiance should not be a reason to ignore the full quantity of available irradiance. CPV technologies partly overcome this limitation by using multi-junction devices, thereby harvesting a large range of the available wavelengths (but excluding the diffuse and albedo irradiance components), leading to higher calculated output and efficiencies. The higher efficiency due to the extra junctions is rightly assigned to the technology, but should not be inflated artificially further by referencing to the direct irradiance rather than the global irradiance.

This issue becomes particularly relevant as it is possible to design CPV systems that use both the direct and diffuse components of sunlight as pointed out already 35 years ago [3] and recently demonstrated by a high efficiency "hybrid" CPV/PV module [4].

The efficiency of a PV device is defined as the ratio of electric power generated by it to the irradiance incident on it. This should be applied consistently to all PV technologies. For non-concentrating terrestrial PV devices calibrated under standard test conditions the incident irradiance is 1000 W/m^2 and the efficiency calculation uses this value independent on how much of this irradiance is actually used by the PV technology. Accordingly, for terrestrial CPV technology which effectively uses the direct irradiance normally incident, for the calculation of the conversion efficiency the incident irradiance should be considered which is the total global irradiance incident onto these devices at AM1.5. Here we will derive this spectrum AM1.5GN and apply it to the calculation of CPV efficiencies.

2 METHODS

2.1 Definition of efficiency

According to IEC TS 61836 *Solar photovoltaic energy systems - Terms, definitions and symbols* [5] clause 3.1.17, the *Conversion efficiency* is defined as the *Ratio of electric power generated by a PV device per unit area to its incident irradiance*.

Here we argue that this definition should be strictly followed. In the following we will call this the thermodynamic efficiency, in order to distinguish it from the efficiencies quoted conventionally, which for CPV devices use only the direct component of the incident irradiance.

2.2 History of reference spectrum definition: IEC and ASTM

IEC 60904-3 from ed. 2 onwards and ASTM G173-03 have both used SMARTS2 2.9.2 [6] for calculating the global and direct reference spectra. A general description of SMARTS can be found in [7] and the references therein. IEC chose to extend the spectral irradiance delivered by SMARTS (in the wavelength range up to 4000 nm) by a component beyond that wavelength up to infinity, and then scale (factor 0.9971 quoted in IEC 60904-3) the total such that the total irradiance for the global reference spectrum up to infinite wavelengths is equal to 1000 W/m^2.

For the IEC an estimate of the irradiance beyond the wavelength of 4000 nm was required. This value was determined to be 2.555669 W/m^2 from 4000 nm to infinity (C. Gueymard and F. Fabero, private communication) from an irradiance model different than SMARTS, which, however, showed good agreement with SMARTS in the wavelength range 3000 nm to 4000 nm.

ASTM on the other hand chose to use the SMARTS output directly without any scaling and extension beyond 4000 nm as the reference spectra. The result of this is that the spectral shape of the IEC and ASTM reference spectra is identical, while the integrated total irradiance is not and differs by 0.29%, with ASTM being the higher values.

2.3 Calculation of reference spectrum AM1.5GN in accordance with IEC 60904-3

The reference spectra in both IEC 60904-3 and ASTM G173-03 are calculated for air mass (AM) 1.5, i.e. at a solar zenith angle of 48.236°. The direct reference spectrum AM1.5D is perpendicularly incident onto a tilted plane of 48.236°, whereas the global reference spectrum AM1.5G is then calculated for a tilt angle of 37°, i.e. the direct beam is incident under angle of 11.236° against the normal.

The first step was to calculate the global reference spectrum AM1.5GN incident onto a PV device mounted with tilt angle 48.236°. This angle is already provided by SMARTS as it is used for the calculation of the direct beam at AM1.5.

We used the same software (SMARTS2 2.9.2) with the identical input parameters as listed in IEC 60904-3 except for the tilt angle which was set to 48.236°. This provided a spectral irradiance output from 280 nm to 4000 nm wavelengths.

The SMARTS output was scaled by the factor 0.9971 (same as used in IEC 60904-3 for AM1.5G and AM1.5D spectra) and then all three spectra compared (Figure 1).

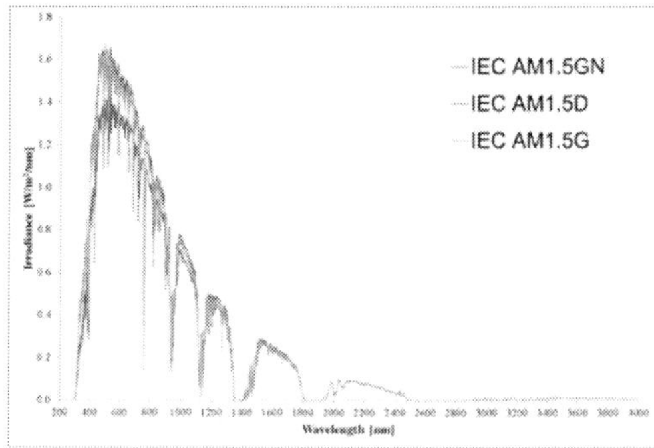

Figure 1: Reference spectra AM1.5G and AM1.5D from IEC 60904-3 compared to the newly calculated AM1.5GN spectrum.

2.4 Calculation of reference spectrum AM1.5GN in accordance with ASTM G173-03

ASTM uses the calculation output from SMARTS2 2.9.2 directly as the reference spectrum, so no further calculations or scaling are required. ASTM G173-03 only defines the total irradiance up to 4000 nm with no estimation of the irradiance beyond that wavelength.

2.5 Estimation of far infrared (FIR) integral (only required for IEC)

As IEC 60904-3 defines the total irradiance up to infinite wavelength, the missing irradiance (above 4000 nm) was estimated based on the information available from the other two reference spectra (AM1.5G and AM1.5D) and some simple physical considerations. This provided the total irradiance for the global normal incidence spectrum AM1.5GN.

Various approaches were tested to estimate the irradiance contained in the wavelength regime from 4000 nm to infinity, which we designate FIR integral. For more details, see results section.

i) the FIR integral from AM1.5D provides a lower bound for the global spectrum AM1.5GN under exactly the same angle;
ii) in addition to i): estimate of additional albedo contributions based on AM1.5G;
iii) in addition to i): estimate of additional albedo contributions based on difference between AM1.5G and AM1.5D;
iv) simple scaling of the FIR integral for AM1.5G by the total irradiance up to 4000 nm;
v) appropriate rounding.

2.6 Calculation of thermodynamic efficiency of terrestrial CPV devices

Based on the above considerations we derived a scaling factor (determined on the considerations above) to convert conventionally reported efficiencies of CPV devices according to AM1.5D (ASTM or IEC) to thermodynamic efficiencies (based on AM1.5GN) and recalculated the world record efficiencies extracted from table 5 of "Solar cell efficiency tables (version 58)" [8]. These recalculated thermodynamic efficiencies for CPV devices can be directly compared to the efficiencies of non-concentrating PV devices in the other tables of [8].

3 RESULTS

3.1 AM1.5GN

The three reference spectra (AM1.5G, AM1.5D and AM1.5GN) as calculated by SMARTS2 2.9.2 are plotted in Figure 1.

3.2 Integrated irradiance

The total irradiance for the raw SMARTS output up to 4000 nm is 1029.32 W/m^2 which gives 1026.33 W/m^2 when scaled with 0.9971 (Table I). Table I provides also the relevant components of the SMARTS output and the integrated irradiances as discussed below.

The additional irradiance in the FIR integral was estimated as follows.

i) From AM1.5D we already know that the FIR integral is 2.47 W/m^2, so for the global spectrum AM1.5GN under exactly the same angle this gives 1028.80 W/m^2.

This is a lower bound giving a minimum irradiance of 1028.80 W/m^2 in AM1.5GN. However, there will be also contributions from the diffuse sky irradiance and the albedo. For the diffuse sky irradiance we estimate a negligible amount based on the proportionality to Rayleigh scattering with 1/(wavelength)^4 (Figure 2).

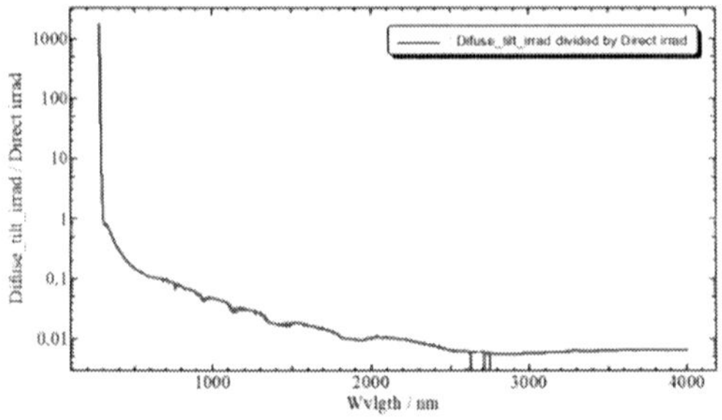

Figure 2: Diffuse irradiance divided by direct irradiances function of wavelength. The decline of the curve is mainly caused by the Rayleigh scattering.

The albedo, however, is estimated to contribute in particular as the irradiance due to albedo increases from 23.87 W/m^2 for 37° tilt to 39.58 W/m^2 for 48.236° tilt (Table 1). The following estimates were made:

ii) In addition to the baseline from i): For the spectrum AM1.5G we have 2.4% irradiance due to the albedo and a FIR integral of 2.53 W/m^2. So we estimate that this portion will increase proportional to the increase in albedo irradiance when tilted:

39.58 / 23.87 * 2.4% * 2.53 W/m^2 = 0.10 W/m^2

giving a total irradiance of 1028.90 W/m^2.

iii) In addition to the baseline from i): For the spectra AM1.5G and AM1.5D we have a FIR integral of respectively 2.53 W/m^2 and 2.47 W/m^2. The latter, however, is incident under an angle of 11.236°. Therefore onto the 37° tilted plane we only have

2.47 * (cos(11.236°) = 2.42 W/m^2.

The difference is then (2.53 – 2.42) W/m^2 = 0.11 W/m^2 due to albedo. This will increase with the ratio of albedo:

(2.53 - 2.42) * 39.58 / 23.87 W/m^2 = 0.18 W/m^2

giving a total irradiance of 1028.98 W/m^2.

iv) The FIR integral for AM1.5G of 2.53 W/m^2 is simply scaled by the total irradiance up to 4000 nm

1026.33 / 997.47 * 2.53 W/m^2 = 2.60 W/m^2

giving a total 1028.93 W/m^2.

v) In practice the spectra in IEC have always been rounded to give integer total irradiance up to infinite wavelength according to the raw SMARTS output up to 4000 nm. Therefore, we just round to 1029 W/m^2.

It is not clear which of the approximations is more correct. However, this is more of an academic question, as the final values are very close to each other (Table I). Therefore, we propose the simple approach v) which is fully justified by the other estimates made independently and consistent with the other reference spectra in IEC 60904-3 and also consistent with ASTM (see below).

Table I: Irradiances (in W/m^2) as discussed in the text and respective scaling factors.

wave-lengths [nm]			Spectrum	AM1.5G	AM1.5D	AM1.5GN	Ratio	
	SMARTS2 2.9.2		zenith angle [°]	48.236	48.236	48.236	AM1.5D / AM1.5GN	
			tilt angle [°]	37	48.236	48.236		
280 to 4000		Integrated irradiance [W/m²]	Total	1000.37	900.14	1029.32	0.8745	
			direct	880.7	897.91	897.91		
			diff+circ	95.8	2.23	91.83		
			albedo	23.87	NA	39.58		
	ASTM G173-		Total	1000.4	900.1	1029.3	0.8745	
			Nominal	1000	900	1029	0.8746	
	IEC 60904-3 2019 ed.4 scaling 0.9971		Total	997.47	897.53	1026.33	0.8745	
280 to ∞				1000	900	1028.80	0.8748	i
						1028.90	0.8747	ii
						1028.98	0.8747	iii
						1028.93	0.8747	iv
						1029	**0.8746**	v

The ASTM adopts the calculated SMARTS output integrated to 4000 nm rounded to one decimal therefore this gives 1029.3 W/m^2. Similarly, ASTM gives 1000.4 W/m^2 and 900.1 W/m^2 for AM1.5G and AM1.5D respectively. However, in the text of ASTM G173-03 these irradiances are explicitly stated as being sufficiently close to just use them as 1000 W/m^2 and 900 W/m^2. Therefore, for AM1.5GN the total irradiance according to ASTM would be 1029 W/m^2.

3.3 Scaling factor for conversion of conventional to thermodynamic efficiencies.

The total irradiance derived for AM1.5GN according to the principles of IEC and ASTM is in both cases 1029 W/m^2. A CPV device installed at a tilt angle of 48.236° receives a total global irradiance of 1029 W/m^2, out of which 900 W/m^2 are from the direct irradiance component, both in the IEC and ASTM standard spectral definitions. Typical non-concentrating solar cells installed in this plane would utilize this irradiance fully, with a minor reduction due to angular effects in the range of only 1% in comparison to an ideal Lambert detector [9].

However, CPV devices cannot convert the additional diffuse irradiance into electricity, due to the optical arrangement. Therefore, the conventionally reported efficiencies in [8] for CPV devices as calculated based on AM1.5D with a total irradiance of 900 W/m^2, are overestimated by a factor of 1029 W/m^2 / 900 W/m^2 = 1.1433. The thermodynamic efficiencies based on AM1.5GN are obtained by multiplying with the ratio of the integrated intensities, i.e. a factor of 900 W/m^2 / 1029 W/m^2 = 0.8746 (Table ITable I: Irradiances (in W/m^2) as discussed in the text and respective scaling factors.).

3.4 New thermodynamic efficiency for CPV

Table II shows the world record efficiencies for CPV devices, as original quoted efficiencies [8] and the thermodynamic efficiencies calculated based on the work here. Most entries in the table are CPV devices referenced to AM1.5D, all of which have been multiplied by the correction factor of 0.8746.

There are, however, two exceptions. Firstly, the luminescent mini-module had been referenced already to AM1.5G, therefore the efficiency quoted is already the thermodynamic efficiency. Secondly, the hybrid module CPV/PV [4] has to be assessed. Originally, its efficiency was based on AM1.5D for the CPV part and on AM1.5G for the PV part. However, to be consistent with the approach taken here, the efficiency was recalculated for the AM1.5GN spectrum. Under the latter the contribution of the CPV part will be identical (as AM1.5D is the same subset and assuming no influence of the slightly higher overall global irradiance), whereas the PV part will produce differently. According to [4] the PV part produced power partly due to scatter from the direct beam at the optics and partly due to the diffuse component in the global spectrum. Based on the information contained in [4] the former will be invariant (as the direct part of the beam is identical) whereas the latter will scale with the larger diffuse part in AM1.5GN with respect to AM1.5G. The thermodynamic efficiency is then 33.5% (Table II).

Table II: Thermodynamic efficiencies for terrestrial CPV devices as recalculated according to the proposal here. Note that one entry (coloured in light green) was referenced already to the global spectrum AM1.5G and therefore need not be changed. For the hybrid module (coloured in light blue) the recalculation is more complex (see text for details).

TABLE 5 NEW	Version 58	0.8746
Classification	Effic. (%)	corr. Effic. (%)
Single-junction cells		
GaAs	30.5	26.7
Si	27.6	24.1
CIGS (thin film)	23.3	20.4
Multi-junction cells		
AlGaInP/AlGaAs/GaAs/GaInAs(3)	47.1	41.2
GaInP/GaAs; GaInAsP/GaInAs	46.0	40.2
GaInP/GaAs/GaInAs/GaInAs	45.7	40.0
InGaP/GaAs/InGaAs	44.4	38.8
GaInAsP/GaInAs	35.5	31.0
Minimodules		
GaInP/GaAs; GaInAsP/GaInAs	43.4	38.0
GaInP/GaInAs/Ge; Si	40.6	35.5
Modules		
Si	20.5	17.9
Three-junction (3J)	35.9	31.4
Four-junction (4J)	38.9	34.0
Hybrid module		
Four-junction (4J)/bifacial c-Si	34.2	**33.5**
'Notable exceptions'		
Si (large area)	21.7	19.0
Luminescent minimodule	7.1	7.1
4J minimodule	41.4	36.2

3.4 Impact of calibration using reference spectra from IEC and ASTM

As already noted the ASTM reference spectra have the same shape as those in IEC but an intensity which is 0.29% higher. Therefore, here we examine how this would affect the calibration of PV devices in general.

The most common calibration of PV devices is a secondary calibration against another PV device including spectral mismatch correction according to IEC 60904-7 [10]. As the equation for the latter contains the reference spectrum in numerator and denominator, any absolute scaling of them is irrelevant and the calculated spectral mismatch will be identical.

For primary calibrations of PV reference devices according to IEC 60904-4 [11] the situation is more complex because there are several methods listed in annex A of that standard.

i) for the solar simulator method (SSM) against a spectroradiometer (eq. A.15 in [11]) the calculated calibration values would differ depending on the reference spectrum used as the absolute reference spectral irradiance is used in the numerator (and only there);

ii) for the SSM against a cavity radiometer as well as the global and direct sunlight method (GSM and DSM) the spectral mismatch correction is calculated according to eq. A.3 in [11] which requires to take all integrals up to infinite wavelengths. As in practise, this is not possible, a suitable truncation of the integrals or extension of the spectral data is required. The effect on the final calibration value will depend on the details of this procedure. In general, if done consistently the choice of reference spectrum will not matter, as similar to the equation in IEC 60904-7, the spectral irradiance of the reference spectrum is in numerator and denominator and any scaling will cancel. However, it might be tempting to only perform the integral over the product of spectral irradiance and spectral responsivity explicitly, but for the integral of the reference spectral irradiance take the known values (e.g. 1000 W/m^2) without explicitly performing the integration. In this case, the choice of reference spectrum will affect the final result.

iii) for the differential spectral responsivity calibration (DSR) eq. A.11 in [11] contains the reference spectral irradiance in the numerator and the integrated irradiance in the denominator. The latter is then stated in eq. A.12 in [11] as 1000 W/m^2; in this case, the choice of reference spectrum is important. In order to resolve this the integrated irradiance in eq. A.11 in [11] should be calculated explicitly, and then the choice of reference spectrum would be irrelevant.

4 DISCUSSION

Which efficiencies should be quoted for CPV devices? For standard flat plate terrestrial PV the efficiencies traditionally quoted based on AM1.5G are already the thermodynamic efficiencies, as the AM1.5G contains all incident irradiance. Using AM1.5GN instead of AM1.5G would not change these numbers, as the higher irradiance in the latter would be accompanied by a proportionate increase in the power output of the PV device. This actually assumes that the power output of the device is linear over the irradiance range from 1000 W/m^2 to 1029 W/m^2 and that there are no spectral effects due to potential small differences between AM1.5G and AM1.5GN in spectral shape. Both these assumptions are reasonable.

But what about CPV efficiencies? Traditionally these efficiencies have been reported with respect to AM1.5D, i.e. the direct beam which is effectively used by the devices, but neglecting the diffuse component of the spectrum as included in AM1.5G or AM1.5GN. Thereby the quoted efficiencies are higher (as in the calculation the power output of the device is divided by a lower irradiance). However, these conventionally reported efficiencies are not the thermodynamic efficiencies as defined in IEC TS 61836. The latter we have calculated and presented in this paper.

Obviously, it is possible to make any convention on how to define and calculate efficiencies of PV and in particular CPV devices. For the latter up to now this has been using AM1.5D. However, with the introduction of hybrid CPV/PV devices this concept is not practical any more. In addition, the question remains, whether the comparison of efficiencies between flat plate non-concentrating PV and CPV is appropriate in general. Based on the current conventions it is not, as we have demonstrated here. The efficiencies of CPV routinely quoted are inflated due to the selection of a reduced irradiance, which makes them appear significantly higher than those of non-concentrating PV. In order to report thermodynamic efficiencies, which are comparable between the different technological approaches, the CPV efficiencies should at least be corrected downward by a factor of 0.9 = 900 W/m² / 1000 W/m², i.e. the ratio of integrated intensities between AM1.5D and AM1.5G. However, this would still not be fully correct, as the irradiance contained in AM1.5D is accompanied by diffuse component such that the total is 1029 W/m² as in AM1.5GN. Therefore, the correction factor as derived here is 0.8746.

As already mentioned any convention can be made. Efficiencies could also be calculated and quoted for totally different spectra (both in shape and integrated irradiance). But does this make sense in the PV field, where the definition of the reference spectra is derived from calculations based on reasonable assumptions to obtain solar irradiances resembling those of natural sunlight? The existing reference spectra ASM1.5G and AM1.5D as well as the newly calculated AM1.5GN represent realistic scenarios, and we advocate that device efficiencies should be calculated based on these scenarios, thereby providing the thermodynamic efficiencies as presented here.

We are well aware that the efficiency is only part of the story. In reality, the energy production of PV devices is relevant, which has to take into consideration the energy rating (IEC 61853 [12]). However, this is a complex topic far beyond the scope of this paper.

If we wish to proceed such that efficiencies quoted for PV resemble in some way real life, then we should extend this approach also to CPV; in this case, we propose that it is more appropriate and correct to use henceforth thermodynamic efficiencies.

5 CONCLUSION

The IEC 60904-3 equivalent reference spectrum AM1.5GN incident on a PV device tilted at 48.236° was calculated as a function of wavelength up to 4000 nm. The integrated irradiance up to this wavelength was determined to be 1026.33 W/m² and the total irradiance up to infinite wavelength 1029 W/m². Similarly, AM1.5GN according to ASTM G173-03 has an integrated irradiance up to 4000 nm of 1029.32 W/m², and ASTM would define the total irradiance only up to this 4000 nm range as 1029.3 W/m2 and suggest to ignore the decimal and round to 1029 W/m².

We have calculated the global normal incident reference spectrum AM1.5GN for CPV. Based on this we have calculated the thermodynamic efficiencies of CPV devices. The latter are traditionally referenced to 900 W/m² whereas the calculation should use 1029 W/m² (as shown here), i.e. CPV efficiencies are inflated and should be corrected by a factor of 0.8746. Then the efficiency for CPV devices would be directly comparable to the efficiency for non-concentrating terrestrial PV devices and would also be consistent with the definition of PV device efficiency as provided in IEC TS 61836.

6 ACKNOWLEDGEMENTS

The authors gratefully acknowledge private communications from F. Fabero and C. Gueymard concerning the history of calculation of the global reference spectrum in IEC 60904-3.

7 REFERENCES

[1] IEC 60904-3:2019 *Photovoltaic devices - Part 3: Measurement principles for terrestrial photovoltaic (PV) solar devices with reference spectral irradiance data* https://webstore.iec.ch/publication/61084

[2] ASTM G173-03(2020) *Standard Tables for Reference Solar Spectral Irradiances: Direct Normal and Hemispherical on 37° Tilted Surface* https://www.astm.org/Standards/G173.htm

[3] A Goetzberger, W.Stahl, B.Voss *Comparison of yearly efficiency and cost of energy for stationary, tracking and concentrating PV systems* Proc 7th EUPVSEC, Sevilla (1986) 250-255

[4] JF Martínez, M Steiner, M Wiesenfarth, G Siefer, SW Glunz, F Dimroth *Power rating procedure of hybrid concentrator/flat-plate photovoltaic bifacial modules* Prog Photovolt Res Appl. **29** (2021) 614–629

[5] IEC TS 61836:2016 *Solar photovoltaic energy systems - Terms, definitions and symbols* https://webstore.iec.ch/publication/59630

[6] SMARTS Version 2.9.2 spectral model code is available, free of charge, subject to the author's license agreement, at https://www.nrel.gov/grid/solar-resource/smarts.html

[7] CA Gueymard, C Myers, K Emery *Proposed Reference Irradiance Spectra for Solar Energy Systems Testing* Solar Energy **73** (2002) 443-467

[8] MA Green, ED Dunlop, J Hohl-Ebinger, M Yoshita, N Kopidakis, X Hao *Solar cell efficiency tables (version 58)* Prog Photovolt Res Appl. **29** (2021) 657– 667

[9] S Winter, D Friedrich, T Gerloff *Effect of the angle dependence of solar cells on the results of indoor and outdoor calibrations* Proc 25th EUPVSEC, Valencia (2010) 4304-4306

[10] IEC 60904-7:2019 *Photovoltaic devices - Part 7: Computation of the spectral mismatch correction for measurements of photovoltaic devices* https://webstore.iec.ch/publication/65654

[11] IEC 60904-4:2019 *Photovoltaic devices - Part 4: Photovoltaic reference devices - Procedures for establishing calibration traceability* https://webstore.iec.ch/publication/66079

[12] IEC TS 61853 series *Photovoltaic (PV) module performance testing and energy rating - Parts 1-4*

INTRODUCING THE THIRD EDITION OF THE BEST PRACTICES HANDBOOK FOR THE COLLECTION AND USE OF SOLAR RESOURCE DATA FOR SOLAR ENERGY APPLICATIONS

Jan Remund[1], Aron Habte[2], Manajit Sengupta[2], Christian Gueymard[3] and Stefan Wilbert[4]
[1] Meteotest AG, Fabrikstrasse 14, CH-3012 Bern, Switzerland, jan.remund@meteotest.ch
Tel. 0041 31 307 26 26, Fax 0041 31 307 26 10
[2] NREL, Golden CO, USA
[3] Solar Consulting Services, Colebrook NH, USA
[4] DLR, Almeria, Spain

ABSTRACT: Sunlight is the fuel for all solar energy generation technologies. For any generation source and its users, knowledge of the quality and future reliability of the fuel is essential for accurate control and analysis of system performance and to determine the financial viability of projects. Information about the resources and forecasts are concluded in the Solar Resource Handbook [1]. In spring 2021 the National Renewable Energy Laboratory (NREL) published the third edition of the widely used Solar Resource Handbook. This handbook was developed by participants of International Energy Agency (IEA) Photovoltaic Power Systems (PVPS) Task 16 and is the main and final outcome of the first phase of IEA PVPS Task 16 (2017–2020). This update includes contributions from 41 authors. Compared to the 2nd edition, it has been significantly enhanced (from 225 to 335 pages), with new chapter on relevant meteorological parameters. The chapters on measurement and modeled data uncertainty, solar forecasting, and applications of solar resource data to various solar energy technologies and products have also been heavily updated. This paper includes the major updates as well as two examples of updated descriptions.
Keywords: Solar radiation, grid integration, modelling, solar resources, solar forecasts

1 THE 3rd EDITION

The solar PV industry has expanded rapidly in the last few years based on ongoing technical evolution and shrinking prices. The size of installations as well as PV penetration have grown – both enhancing the need for accurate solar data for planning and operation. Induced by this need there have been significant enhancements in the body of knowledge in the areas of solar resource assessment and forecasting. This third edition of the handbook [1] updates and enhances the initial versions and presents the state of the art in a condensed form for all of its users.

The structure of the handbook has been slightly updated from the previous version. Chapter 1 lays out the need for high-quality and reliable solar resource data to support the rapidly growing industry, and Chapter 2, as with the previous editions, provides a basic tutorial on solar resources. Chapter 3 presents a comprehensive overview of best practices for measuring solar resource, including information gained under collaborative work completed under Task 16, and Chapter 4 summarizes techniques used to develop estimates of solar resource from weather satellite data and numerical model predictions. Chapter 5 is a new chapter describing additional meteorological parameters beyond solar radiation that are required for accurate modeling and performance analysis of solar systems. Chapter 6 lists a number of commonly used data sets available both in the public and private sectors (updated from the previous editions), and Chapter 7 provides important information on both measured and modeled solar data uncertainty. Chapter 8 provides an update on recent developments in the ability to forecast the solar irradiance components over time horizons spanning nowcasting to hours-ahead and days-ahead. All of this information leads up to Chapter 9, which provides data application techniques for the various stages of project development, from pre-feasibility to routine operations, as shown in Table 1. The outlook for future work is summarized in Chapter 10.

System Size →

Phase ↓	Small	Medium	Large
1. Pre-feasibility & Planning	• Long-term averages • Monthly data • Solar cadastres / maps • Simple shading analysis	• TMY • Hourly data • Shading analysis	• Long-term satellite data • Hourly data
2. Feasibility			• Satellite data • Time series (>10 y) • Ground meas. (> 1 year) • Shading analysis • Further site and technology- specific meteo. parameters (e.g. albedo, soiling)
3. Due diligence & Finance		• Satellite data • Time series (>10 y) • Minute data • Shading • Further site and technology- specific meteo. parameters (e.g. albedo, soiling)	• Satellite data • Time series (>10 y) • Ground meas. (> 1 year) • Minute data • Shading analysis • Further site and technology- specific meteo. parameters (e.g. albedo, soiling)
4. Operation & Maintenance	• Simple monitoring	• Local measurements • Forecasts	• Local measurements • Forecasts

Table I: Data application techniques for the various stages of project development

Project developers, engineering procurement construction firms, utility companies, system operators, energy suppliers, financial investors, organizations involved in planning and managing solar research programs, and others involved in solar energy systems planning and development will find this handbook to be a valuable reference for the collection and interpretation of solar resource data. Readers are encouraged to provide feedback to the authors for future revisions and an expansion of the handbook's scope and content.

2 TWO HIGHLIGHTS

Two examples of strongly updated chapters are described here. The first is about the so called "Solar Constant" and the other about Probabilistic forecasts.

2.1 Solar Constant

The total radiant power from the Sun is nearly constant. The solar output is called the total solar irradiance (TSI). TSI was commonly called the solar constant (SC) until slight temporal variations were discovered. The solar constant is now defined as the long-term mean TSI.

Both TSI and solar constant are made independent from the actual Sun-Earth distance (a deterministic quantity) by evaluating them at 1 astronomical unit (AU).

Small but measurable changes in the Sun's output and TSI are related to its magnetic activity. There are cycles of approximately 11 years in solar activity, which are accompanied by a varying number of sunspots (cool, dark areas on the Sun) and faculae (hot, bright spots). TSI increases during high-activity periods because the numerous faculae more than counterbalance the effect of sunspots. From an engineering perspective, these TSI variations are relatively small, so the solar constant concept is still useful and perfectly appropriate in solar applications.

Figure 1 shows the latest version of a composite TSI time series based on multiple spaceborne instruments that have monitored TSI since 1978 using a variety of instruments and absolute scales [2].
The solar constant (always evaluated at 1 AU) is shown as a horizontal solid line at 1361.1 W/m^2 – a value that is interestingly intermediate between what NASA promulgated in the early 1970s (1353 W/m^2) and those that followed and were used in the solar literature until recently, e.g., 1367 or 1366.1 W/m^2.

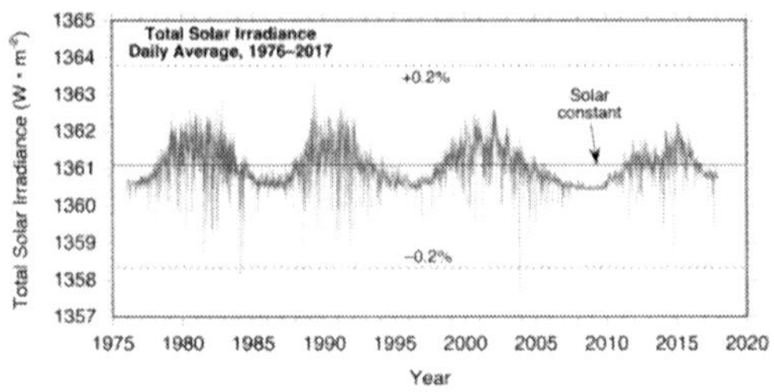

Figure 1: Four solar cycles show the temporal variations of daily TSI in a composite reconstruction of the 1976–2017 time series based on observations from spaceborne radiometers after corrections and gap-filling. Image based on data published in Gueymard [2].

2.2 Probabilistic forecasts

A forecast is inherently uncertain and a proper assessment of its associated uncertainty offers the grid operator a more informed decision-making framework. For example, a deterministic forecast that includes predictions intervals is of genuine added value and, if appropriately incorporated in grid operations, might permit an increase in the value of solar power generation [3].

This section is restricted to the univariate context that corresponds to probabilistic forecasts that do not consider the spatiotemporal dependencies generated by stochastic processes such as solar power generation. Two types of solar[1] probabilistic forecasts are considered in the handbook: *quantile forecasts* and *ensemble forecasts* (i.e., those using the Ensemble Prediction System, or EPS). Quantile forecasts are quite versatile probabilistic models and as such might address different forecasting time horizons, whereas EPS forecasts generally provide day-ahead probabilistic forecasts. Further, a verification framework is considered for evaluating the quality of solar probabilistic forecasts. The evaluation framework is based on visual diagnostic tools and a set of scores mostly originating from the weather forecast verification community [4]. What follows provides an overview of the basic concepts related to solar probabilistic forecasting methods with an emphasis on the specific associated verification metrics. Comprehensive overviews regarding forecasting methods and the verification of solar probabilistic forecasts metrics can be found in [5–7].

In contrast to deterministic forecasts, probabilistic forecasts provide additional information about the inherent uncertainty embodied in numerical weather prediction models (NWP). The probabilistic forecast of a continuous variable, such as solar power generation or solar irradiance, takes the form of either a cumulative distribution function (CDF), $F(Y)$, or a probability density function (PDF).

The predictive distribution can be summarized by a set of quantiles. A quantile corresponds to the threshold value below which an event, y, materializes with a probability level, τ. Prediction intervals (also called interval forecasts) can be inferred from the set of quantiles. Prediction intervals define the range of values within which the observation is expected to be with a certain probability (i.e., its nominal coverage rate) [8].

For example, a central prediction interval with a coverage rate of 95% is estimated by using the quantile $q_{\tau=0.025}$ as the lower bound and $q_{\tau=0.975}$ as the upper bound. Figure 2 shows an example of the probabilistic forecasts of solar irradiance where prediction intervals have been computed for nominal coverage rates ranging from 20% to 80%.

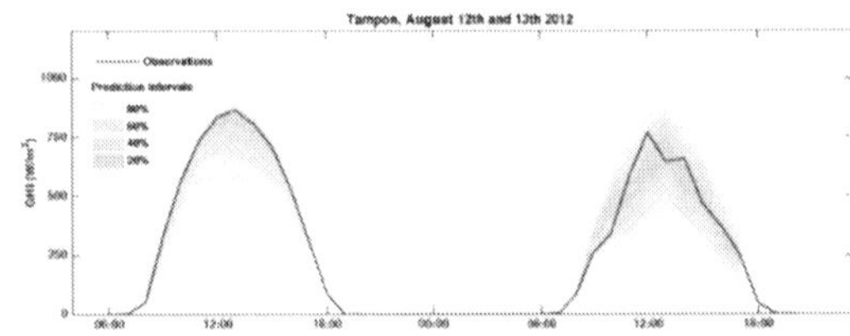

Figure 2. Example of probabilistic solar irradiance forecasts: two days of measured GHI at Le Tampon, France, and associated 1-hour ahead forecasts with prediction intervals (yellow) generated with the quantile regression forest model. The yellow shade indicates the range of probability (20–80%) in comparison with the actual observations (black line).

3 REFERENCES

[1] Sengupta, M., Habte, A., Wilbert, S., Gueymard, C. & Remund, J. *Best Practices Handbook for the Collection and Use of Solar Resource Data for Solar Energy Applications*. (2021). https://iea-pvps.org/key-topics/best-practices-handbook-for-the-collection-and-use-of-solar-resource-data-for-solar-energy-applications-third-edition/ https://www.nrel.gov/docs/fy21osti/77635.pdf

[2] Gueymard, C. A. "A Reevaluation of the Solar Constant Based on a 42-Year Total Solar Irradiance Time Series and a Reconciliation of Spaceborne Observations." *Solar Energy* 168: 2–9 (2018).

[1] The term *solar forecast* encompasses solar irradiance forecasts and PV power forecasts.

[3] Morales, J. M., A. J. Conejo, H. Madsen, P. Pinson, and M. Zugno. 2014. *Integrating Renewables in Electricity Markets, International Series in Operations Research and Management Science*. Springer, Boston, MA.

[4] Wilks, D. S. 2014. *Statistical Methods in the Atmospheric Sciences: An Introduction*. Elsevier.

[5] Van der Meer, D. W., J. Widén, and J. Munkhammar. 2018. "Review on Probabilistic Forecasting of Photovoltaic Power Production and Electricity Consumption." *Renewable and Sustainable Energy Reviews* 81: 1484–512.

[6] Antonanzas, J., N. Osorio, R. Escobar, R. Urraca, F.-J. Martinez-de-Pison, and F. Antonanzas-Torres. 2016. "Review of Photovoltaic Power Forecasting." *Solar Energy* 136: 78–111. http://dx.doi.org/10.1016/j.solener.2016.06.069.

[7] Lauret, P., M. David, and P. Pinson. 2019. "Verification of Solar Irradiance Probabilistic Forecasts." *Solar Energy* 194: 254–71. https://doi.org/10.1016/j.solener.2019.10.041.

[8] Pinson, P., H. A.Nielsen, J. K. Møller, H. Madsen, and G. N. Kariniotakis. 2007. "Non-Parametric Probabilistic Forecasts of Wind Power: Required Properties and Evaluation." *Wind Energy* 10: 497–516. https://doi.org/10.1002/we.230.

COMBINATION OF PHYSICS BASED SIMULATION AND MACHINE LEARNING FOR PV POWER FORECASTING OF LARGE POWER PLANTS

Nicolas Holland[1], Xiulan Pang[2], Wiebke Herzberg[1], Jefferson Bor[1] and Elke Lorenz[1]
Fraunhofer Institute for Solar Energy Systems ISE, Heidenhofstr. 2, 79100 Freiburg, Germany[1]
SPIC Huanghe Hydropower Development Co., LTD. HHDC[2]

ABSTRACT: Forecasting of PV power essentially relies on forecasting of the solar irradiance. There are different sources of irradiance forecasts like numerical weather predictions as well as satellite and measurement-based methods. Those different sources are typically combined to derive optimized forecasts for different forecast horizons. To derive PV power forecasts on this basis, new machine learning methods are a valuable addition to physics-based simulations. In our approach we combine not only different sources of irradiance forecasts but also the physics-based simulations with machine learning methods in order to produce optimized PV forecasts for large scale power plants and apply the method on a power plant (>800MW) in the Qinghai province of China. This PV power plant is frequently subject to curtailment by the grid operator for large power values, which presents a challenge for model training that we address by applying a mix of active power control thresholding and a machine learning model trained to detect curtailment in PV power measurements.

Keywords: PV Forecast, Satellite, Machine learning

1 INTRODUCTION

Forecasting the power output of large PV power plants is often attached with significant challenges ranging from the presence of different technologies within the power plant to external power regulations that effect the power production and can have little to do with weather situations. We developed a PV power forecasting method applied and tested for a power plant of over 800MW installed capacity in the Qinghai province of China. Due to grid regulations, power plant operators are obligated to provide power production forecasts to the grid operator and are only allowed to feed in power according to their predictions [1]. In cases where the predicted power is below what could actually be produced, grid operators may apply curtailment, which leads to missed opportunities for selling power and, in cases in where the produced power is significantly below the predicted power, there can be penalty fines that have to be paid to the grid operator. This situation creates a demand for reliable forecasting. Due to the nature of the regulation the 2-hour ahead forecast plays an important role in the evaluation of the forecasts.

Our first approach was a model, that relied on numerical weather predictions (NWP) and measurement based global horizontal irradiance (GHI) forecasts (persistence) which were combined using a regression model to create an optimized GHI forecast. This GHI forecast was then used as input for a PV power simulation alongside ambient temperature data resulting in a power forecast. The method is largely based on previous work by Kühnert et. al [2] and yielded promising results, we first described the approach at the IEEE PVSEC-46 [3].

Since then we improved the forecast performance by introducing satellite based GHI forecasts and by replacing the regression model with a neural network. Instead of combining GHI forecasts, we apply PV power simulations on the GHI forecasts first and use the neural network to combine the power predictions into an optimized PV power prediction.

One of the major challenges to model training is the fact that many of the power measurements are affected by curtailment. When training a model with respect to power measurements, curtailed power creates a bias to underestimate the true power output potential, because the power plant output is artificially lower than what weather and GHI inputs would suggest. We address this challenge by both applying curtailment detection methods on the training data and therefore removing parts of the dataset that would hinder model training, and by adjusting the model training loss function such that curtailment artefacts that remain undetected within the training data have a less severe impact on the resulting model. Both the introduction of satellite data, the neural network and the curtailment detection improve the forecasting system to the point where it can achieve the level of accuracy required by the grid regulation.

2 DATA

2.1 On-site measurements

The Qinghai based power plant has an over 800MW installed capacity and contains different PV technologies including mono- and bifacial modules as well as tracked and fixed orientations.

GHI measurements were taken with a horizontal standard pyranometer and power measurements were obtained for the overall PV power output of the entire plant. Both GHI and power measurements were in 1-minute resolution, which were averaged to 15-minute time series. The grid regulation requires forecasts to be in 15-minute resolution. Next to the measurements, the on-site available data includes Active Power Control (AGC) data, which is a signal sent by the grid operator to regulate and curtail the power plant. Our curtailment detection methods rely both either directly or implicitly on this AGC data.

The data which our analysis is based on ranges from March 2020 to April 2021 and has a few weeks of missing data in August 2020. AGC data was available from March to Mai of 2020.

2.2 Numerical weather predictions

In addition to on-site measurements the forecasting method relies also on numerical weather predictions. We use forecasts by the European Centre for Medium-Range Weather Forecasts (ECMWF), which are updated four times per day and have a forecast horizon of up to several days. To resample the hourly ECMWF values to a 15-minute resolution, the values are divided by modelled clearsky irradiance to compute clearsky indices, which then are interpolated to produce 15-minute values. We use the clearsky model by Dumortier [4] to model clearsky irradiance. Temperature forecasts by ECMWF were used as input to the PV simulation method.

3 METHOD

The entire PV power forecast procedure consists of several methods that are combined and adjusted to form one forecasting system. First, we use methods for predicting GHI irradiance like measurement-based persistence forecasts and satellite-based forecasts. Next, these forecasts together with external NWP data are input to a physics-based PV power simulation, which are than the bases for a neural network combi model. The neural network was trained on a dataset that was subject to our curtailment detection, which itself consists of an AGC-based thresholding method or a machine learning model for cases in which AGC data is not available.

3.1 Persistence forecasts

Persistence GHI forecasts (G_{pfc}) are based on both on-site GHI measurements and a clearsky irradiance model. Given a GHI measurement at time t and assuming the current weather situation does not change, a persistence forecast is computed by multiplying the current clearsky index with a clearsky model,

$$G_{pfc,t+1} = \frac{G_{meas,t}}{G_{cs,t}} G_{cs,t+1}$$

with $G_{meas,t}$ and $G_{cs,t}$ being the GHI measurement and clearsky model prediction at time t. Both the irradiance measurements and modelled clearsky irradiance are averaged to 15-minute resolution time series.

3.2 Satellite based forecasts

Satellite based GHI forecasts (G_{sat}) are derived from satellite images using the heliosat method [5] and cloud motion vectors [6]. We use satellite data in the visible spectrum from the geostationary Himawari satellite operated by the Japan Meteorological Agency. The satellite makes new measurements every 10 minutes, which canonically leads to GHI forecasts with a 10-minute resolution. The core aspect of the forecasting algorithm is the calculation of cloud index information from satellite images which in combination with clearsky models produces GHI estimations. Using two consecutive cloudindex images allows to compute a vector field describing the cloud motion between the two images. This vector field is referred to as either optical flow or cloud motion vectors. Assuming that the cloud movement is persistent the vector field is then used to extrapolate cloud movement into the future thereby creating cloudindex forecasts and ultimately GHI forecasts.
The cloud motion vectors are computed using the deepflow method [7] implemented in the OpenCV library [8]. The entire satellite based GHI forecast process can be summarized as this,

$$\text{heliosat}(CI_t) \rightarrow G_{sat,t}$$

$$\text{deepflow}(CI_{t-1}, CI_t) \rightarrow \binom{u}{v}_t$$

$$\text{map}\left(CI_t, \binom{u}{v}_t\right) \rightarrow CI_{t+1}$$

$$\text{heliosat}(CI_{t+1}) \rightarrow G_{sat,t+1}$$

With heliosat and deepflow being the in [6] and [7] described methods, CI_t the derived cloudindex at time t, $\binom{u}{v}_t$ the cloud motion vector and CI_{t+1} or $G_{sat,t+1}$ the forecasted cloudindex or irradiance. The forward extrapolation in time is done via a simple inverse mapping,

$$CI_{t+1}(x,y) = CI_t(-u_t(x,y) + x, -v(x,y) + y)$$

The resulting 10-minute resolution GHI forecasts are resampled to match the 15-minute resolution of the other irradiance and power time series. Figure 2 shows an example of a satellite image as it is used in the forecasting process, the Qinghai region is in the upper left area of the image.
Another approach that we observed with great interest is the satellite-based forecasting method by Harty et. al. [9]. While our cloud motion vectors used for cloudindex forecasting are solely derived from optical flow computed from satellite images, they combine optical flow with wind vector fields derived from numerical weather predictions and compute cloud motion vectors that appear to be outperforming the optical flow only forecasts and could potentially improve our system as well in future works.

Figure 2: Himawari satellite image in visible spectrum, image taken on 2021-08-04 00:10 UTC. Satellite images like this one are the bases for cloud motion vector and satellite irradiance forecasts. Image by Japan Meteorological Agency [10].

3.3 PV simulation

The first step in creating PV power forecasts is to simulate the power plant with different GHI forecasts as input. The splitting of GHI into its direct and diffuse components is done using the DIRINT model [11] and the conversion from horizontal irradiance to plane of array irradiance is done using the model by Perez et al [12]. We use the implementations of those models from the PVLIB library [13]. The models used to derive PV power from plane of array irradiance are implemented in-house and are described in [14].
Since the power plant consists of different technologies and configurations the simulation is helped by organizing the power plant into suitable sub-plants which are simulated independently. The results of those simulations

are then summed up to describe the entire power plant output. We apply this simulation approach on every GHI forecast individually so that, for all $s \in \{nwp, pfc, sat\}$

$$\text{simulate}(G_s, T, \theta) \rightarrow P_s$$

with G_s being the irradiance forecast generated through method s and T, θ being the ambient temperature and sun angles ie. zenith and elevation. Figure 3 shows the result of the PV simulation applied on satellite GHI forecasts with a 2-hour forecast horizon.

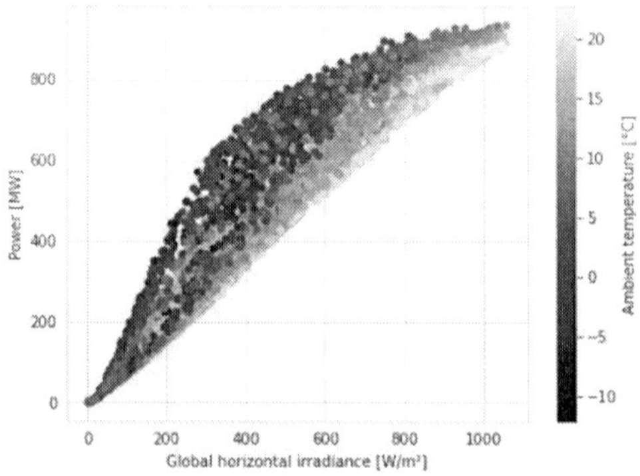

Figure 3: Scatter plot of PV simulation and GHI satellite data with ambient temperature from ECMWF forecasts. We see a spread of power values for similar GHI values, this can be explained by them occurring at times with different solar angles, which can have a large effect especially on parts of the power plant with tracked modules. We also see a temperature dependency, as illustrated here with different colors.

3.4 Curtailment detection

Figure 4 shows an example of a day where power measurements were likely curtailed, and the curtailment was detected through one of our models. Our methods rely on either the availability of AGC data to make a direct prediction over whether a power measurement was affected by curtailment or a prediction is made by a machine learning model trained on curtailment information, which in turn can have been produced using AGC. The actual truth of when the power plants were curtailed is held by the grid operator and was in this case not known to us. Our curtailment detection method was first described in a talk given at PVSEC-30 2020 [15].

3.4.1 Curtailment detection using AGC

The AGC approach is basically a thresholding approach where the difference between AGC and power measurement is compared with a fixed percentage of the overall installed capacity,

$$C_{AGC}(P_t) = \begin{cases} 1, & |AGC_t - P_t| < 1\% \, inst.kWp \\ 0, & otherwise \end{cases}$$

with $C_{AGC}(P_t)$ being a truth value over whether the power measurement at time t was curtailed. The percentage threshold was adjusted through manual inspection.

3.4.2 Curtailment detection using RDF

In the absence of AGC information the curtailment detection relies on a random decision forest (RDF) which was trained on labels that were produced by the AGC method. Besides both power and irradiance measurements P_{meas}, G_{meas} the input features used to make RDF predictions include simulated power P_{sim} and clearsky index $\frac{G_{meas,}}{G_{cs,}}$ and a set of time series features that are derived from those four. We use the Scalable Hypothesis method [16] to derive those features, which is a two-step process. First, we compute a large set of time series features that could potentially be relevant, before rating them individually as to how well every feature can predict the ground truth curtailment labels. Individual features are rated using the Benjamini-Yekutieli procedure [17] and because this process produces relevant time series features without manual feature engineering, it is often referred to as automatic feature extraction. For the feature computation we use the implementation from the tsfresh library [18] and for the RDF we use the implementation in the scikit-learn library [19]. The training set for the RDF model was the time range from March 2020 to Mai 2020 in which AGC data was available.

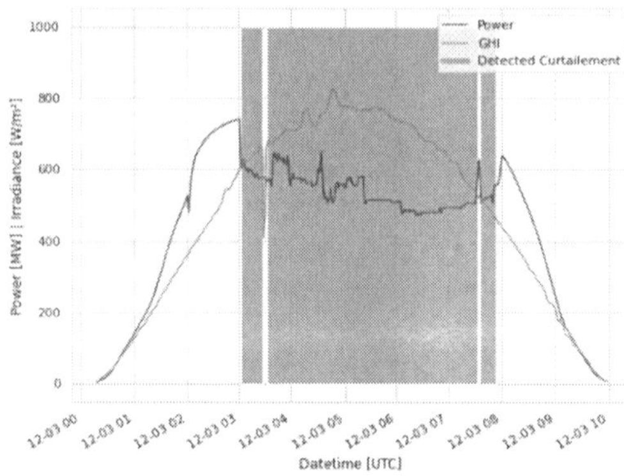

Figure 4: Power and GHI measurements for single day in December. There are visible curtailment effects in the power measurement which were detected by the RDF model. Note that the model was trained on and is therefore meant to be used on minute resolution data.

3.5 Model training

To combine the different power forecasts into a single best power prediction we use a densely connected neural network consisting of two hidden layers which take as input both the three different power forecasts as well as the original GHI forecasts and solar zenith angles. The neural network is trained to produce complete forecasts, ie all the forecast steps are produced by the same network at once. The differentiation between forecast steps is achieved by adding the forecast horizon as an additional input feature. This enables the model to learn weighting satellite, persistence, and NWP inputs dependent on the forecast horizon. The training is done using Adam optimization [20] and we use the network implementation from the keras library [21]. Model training was done on the curtailment filtered dataset.

3.5.1 Train/test split

The entire dataset we used for the analysis ranges from March 2020 to April 2021, the data used for forecast model training ranges from March 2020 to September 2020 with everything starting from October 2020 being solely used for testing.

3.5.2 Loss function

We started our forecast model training minimizing the root-mean-square-error (RMSE) as a loss function. In the regular RMSE, given by

$$RMSE = \sqrt{\frac{1}{N}\left(\sum_t (P_{p,t} - P_{m,t})^2\right)}$$

with $P_{p,t}, P_{m,t}$ being the predicted and measured power at time t, over- and underestimations are treated equally. However, even with curtailment detection applied on training data, there are still issues which can have effects on model training. Both undetected curtailment as well as simple lack of data in situations with high GHI values, and a correlation clearly smaller than one introduces incentives for the model training to produce models which tend to underestimate the true output potential of the power plant for larger GHI values. To mitigate the effect of this tainted training data we make an adjustment to the loss function of the model training. We introduce a skewed version of the RMSE where over- and underestimations are weighted differently by preset parameters α, β

$$\sqrt{\frac{1}{N}\left(\alpha \sum_{t\in\Omega_\alpha} (P_{p,t} - P_{m,t})^2 + \beta \sum_{t\in\Omega_\alpha} (P_{p,t} - P_{m,t})^2\right)}$$

with

$$\Omega_\alpha = \{t: P_{p,t} < P_{m,t}\}$$
$$\Omega_\beta = \{t: P_{p,t} > P_{m,t}\}$$

and $\alpha > \beta$, the exact values for α, β are defined through manual inspection. This way the underestimations are weighted higher, leading to more impact on the model training, which is desired assuming that some of the overestimations are caused by curtailment and can therefore be accepted.

3.6 Metrics

In addition to the RMSE as a performance metric we also consider the accuracy metric issued by Chinese grid operators, which is used to assess and regulate power plants in the Qinghai province as well as other provinces [1],

$$1 - 2\sum_{t=1}^{N}\left(\left|\frac{P_{p,t}}{P_{p,t} + P_{m,t}} - \frac{1}{2}\right| \frac{|P_{p,t} - P_{m,t}|}{\sum_{i=1}^{N}|P_{p,i} - P_{m,i}|}\right)$$

with N the number of values per day and $P_{p,t}, P_{m,t}$ being the predicted and measured power at time t. The metric is used to assess the forecast accuracy of entire days and is typically applied on the 2 hours ahead forecast. Only value pairs with either the predicted or measured power above 3% of the installed capacity are considered in the assessment.

4 RESULTS AND DISCUSSION

4.1 Curtailment

The detection of curtailment is an important part of both model training and evaluation, as curtailed power measurements can introduce strong biases into models trained on them. While in the presence of AGC information, we prefer using the thresholding method to detect curtailed events, our results show that the machine learning based RDF method can be a feasible alternative for situations without AGC information. Figure 5 shows the curtailment detection results for the two methods.

The dataset is split into two parts, with the first part being used as training set for the RDF model and the second as test set. Without curtailment detection there is a bias in the training set of 9.2% with respect to the installed capacity, and 6.2% in the test set. With curtailed datapoints filtered out, there remain biases of 1.9% on each of the two parts of the dataset, which is a big improvement but also suggests that there might be undetected curtailment.

When comparing the AGC method with the RDF method on the part of the dataset where both methods are applicable, we find that the AGC method flags 810 of the 5855 datapoints while the RDF method flags 882 datapoints, with 633 being flagged by both methods in agreement. In total, the curtailment detection process removes 15.2% of datapoints in the entire dataset. The figure also shows that there is an underrepresentation of situations with large power outputs in the dataset.

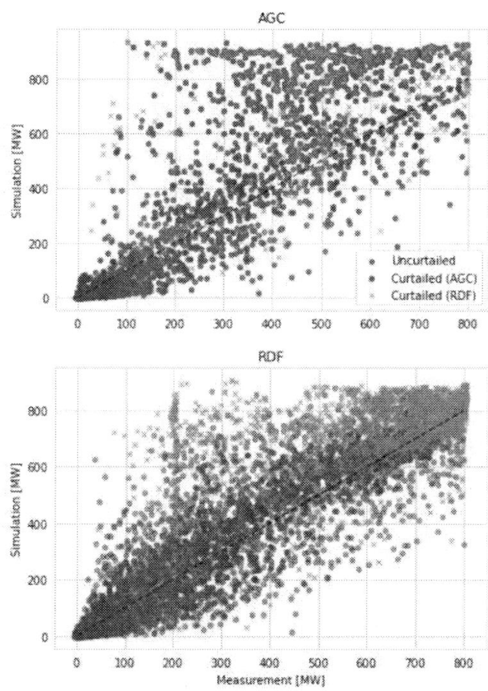

Figure 5: Scatter plots of satellite-based over measured power with curtailment detected using the AGC and RDF methods. Top: Training set (March to Mai 2020) with AGC available. Bottom: Test set (June 2020 to April 2021) without AGC available. There appear to be common events where power measurements are curtailed to 200MW, many of which were identified by the curtailment detection.

4.2 Forecast accuracy

The forecasts are evaluated using both the RMSE metric, which is widely used, especially in PV power forecasting and the daily accuracy metric, which is used in the Qinghai province where the power plant is located. Figure 6 compares power forecasts made from satellite or NWP data as well as combi models trained with regular RMSE and our skewed RMSE as loss functions for the curtailment filtered part of the dataset. Both the skewed and regular combi models have learned to reduce the scattering while simultaneously introducing a bias towards underestimating the power for situations with above 700MW power output. This can be caused by undetected curtailment still being present in the training data and because in cases with a correlation < 1.0 the RMSE is being reduced by reducing the amount of extreme values. The skewed model has learned to be less biased towards underestimation for high power values in comparison, but also has more overestimations in situations with small values.

Since the daily accuracy metric is of importance for the Qinghai region, figure 7 shows the distribution of daily accuracies from the skewed combi model. We find that the combi model achieves to meet the 75% accuracy threshold requirement in 50.8% of days. It is important to note that datapoints that were affected by the grid operator through curtailment are to be excluded from this evaluation metric. Since it is only the grid operator who knows the exact ground truth over curtailment status, their forecast evaluation may differ, and our evaluation is meant to generate an expectation of what their results could be.

Detailed evaluations for the different forecast models and datasets with respect to both RMSE and daily accuracy are shown in table 1 and 2 respectively. Figure 8 shows an example of a single forecast for a day with mostly clearsky conditions which illustrates the effect of either undetected curtailment or underrepresented high irradiance situations in the training set on both combi models and how the skewed model has learned to mitigate the effect.

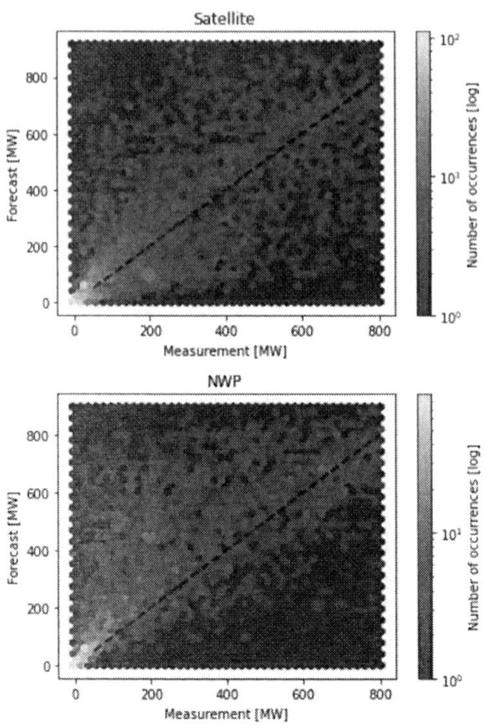

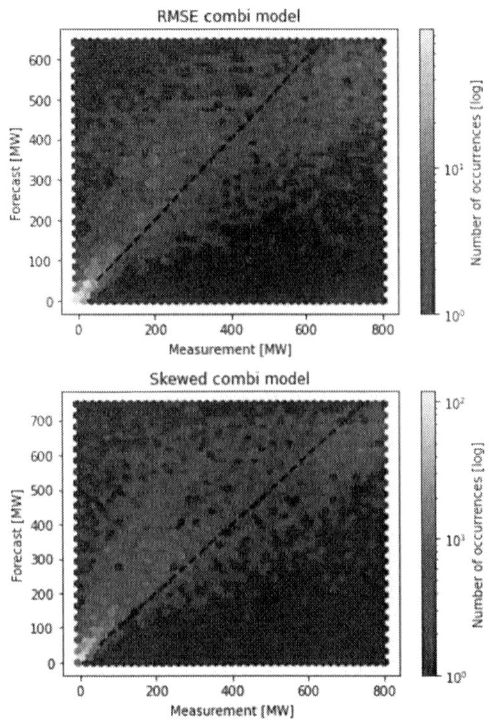

Figure 6: Scatter plots of forecasted power over measured power for the 2-hour ahead horizon comparing satellite and NWP based forecasts with the combi models. Note that the Forecast axis do not have equal range on all the plots.

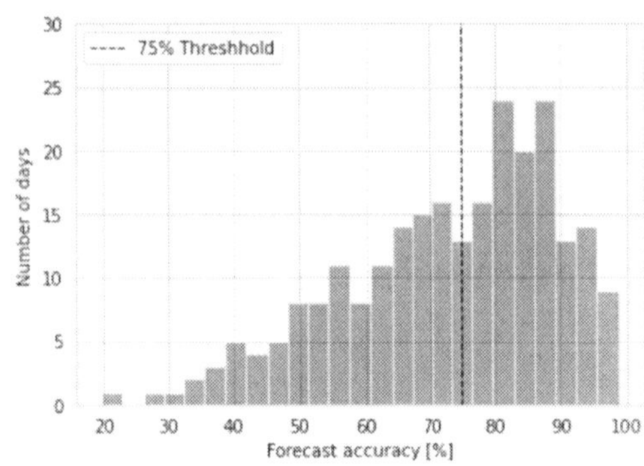

Figure 7: Distribution of daily accuracy values for the skewed combi model evaluated on the entire dataset with curtailment filtered out. With 75% accuracy being the threshold set by the energy regulation which must be met as a minimum requirement in order to avoid being penalized for providing insufficient forecasts.

When comparing the model performances on the different datasets in table 1 and 2 we find that

- All the models, as expected, have higher RMSE values on the dataset that is not curtailment filtered.
- The satellite-based forecast outperforms the NWP forecast on the 2-hour ahead horizon.
- Since the combi models have reduced scattering, their RMSE values are lower than the satellite and NWP forecasts.

- The RMSE of the model that was trained on RMSE is evidently lowest.
- Some of the performance differences found with respect to RMSE were also found with respect to the daily accuracy metric, e.g. both the combi models were outperforming the satellite and NWP forecasts and that the combi models appear to perform better on the evaluation set, which indicates fewer curtailment situations being present there.

Dataset	Satellite	NWP	CombiS	CombiR
T_A	22.7%	23.4%	19.1%	17.0%
E_A	18.5%	18.9%	14.6%	14.7%
F_A	21.9%	22.5%	18.3%	16.5%
T_C	19.3%	20.5%	17.9%	15.0%
E_C	17.7%	19.3%	15.4%	13.8%
F_C	19.0%	20.3%	17.5%	14.7%

Table 1: Results for RMSE metric computed on the training dataset (T), the test dataset (E) and the full dataset (F) for both the curtailment filtered part of the dataset (C) and all the data (A). The RMSE values are normed with respect to the installed capacity. CombiS is the combi model trained with the skewed RMSE and CombiR the model with the regular RMSE.

Dataset	Satellite	NWP	CombiS	CombiR
T_A	65.7%	68.9%	71.0%	73.6%
E_A	76.7%	77.2%	79.3%	80.5%
F_A	71.1%	73.0%	75.1%	77.0%
T_C	64.2%	68.2%	69.0%	72.2%
E_C	74.3%	74.7%	76.5%	78.1%
F_C	68.9%	71.3%	72.5%	75.0%

Table 2: Mean of daily accuracy metric evaluated on the same data as in table 1. Both the RMSE and skewed combi models outperform the forecasts based solely on satellite or NWP data having similar performance improvements when compared to the RMSE metric.

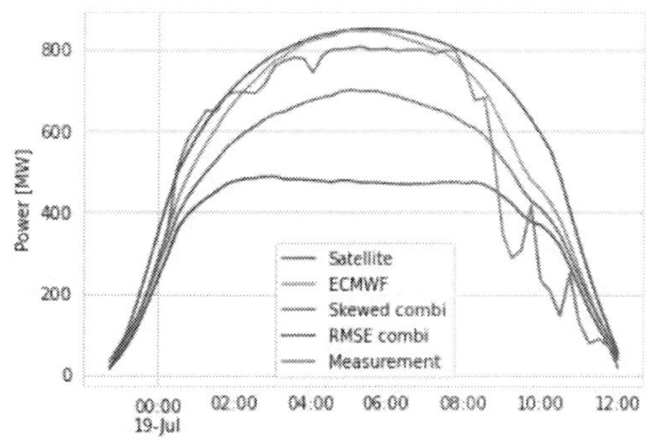

Figure 8: PV power forecasts computed on a day in July 2020 compared with the power measurement of that day. At the time at which the forecasts were computed both the satellite and NWP based methods predicted mostly clearsky conditions which for the majority of the day turned out to be the case. Despite the optimistic input predictions both the skewed and RMSE combi models appear to have bias towards underestimating the power production with the RMSE combi model being significantly more pessimistic. It is precisely this pessimistic behavior for otherwise optimistic situations, which lead us to introduce the skewing of the loss function into the model training.

5 SUMMARY AND OUTLOOK

With machine learning in general relying heavily on clean and rich datasets, the absence of reliable clean ground truth data makes model training under this condition a challenge. Our results show that model training for power forecast enhancement is feasible even with curtailment issues tainting power measurements and introducing biases. Our curtailment detection method was able to find many of the curtailed measurements, but the training of the model remains to be challenging, especially because the dataset gets reduced and not all the curtailment events are detected. Through the introduction of a skewed loss function that treats over- and underestimations differently, the aim of avoiding large underestimations for situations with high power values is achieved. However, in this first approach it is also related to reduced forecast quality for low power values. Further investigations on metrics that can reflect model performances under assumed curtailment and model training specialized for curtailment situations will follow, as will further curtailment detection methods, for example, through the usage of uncurtailed reference arrays.

6 REFERENCES

[1] Northwest China Energy Regulatory Bureau of National Energy Administration of the People's Republic of China, http://xbj.nea.gov.cn/website/Aastatic/news-196276.html

[2] Wolff, B., J. Kühnert, E. Lorenz, O. Kramer, D. Heinemann: Comparing support vector regression for PV power forecasting to a physical modeling approach using measurement, numerical weather prediction, and cloud motion data", Solar Energy, Volume 135, October 2016, Pages 197-208, ISSN 0038-092X, doi:10.1016/j.solener.2016.05.051.

[3] Holland, Nicolas, et al." Solar and PV forecasting for large PV power plants using numerical weather models, satellite data and ground measurements." 2019 IEEE 46th Photovoltaic Specialists Conference.

[4] Dumortier, D., 1995. Modelling global and diffuse horizontal irradiances under cloudless skies with different turbidities. Daylight II, jou2-ct92- 0144, final report vol. 2. Tech. rep., CNRSENTPE.

[5] Hammer A, Heinemann D, Hoyer C, Kuhlemann R, Lorenz E, Muller R, Beyer HG. Solar energy assessment using remote ¨ sensing technologies. Remote Sensing of Environment 2003; 86(3): 423–32.

[6] J. Kuhnert, E. Lorenz, D. Heinemann: 'Satellite-Based Irradiance and Power Forecasting for the German Energy Market' in Solar Energy Forecasting and Resource Assessment, Editor: Jan Kleissl (Elsevier 2013).

[7] Weinzaepfel, P., J. Revaud, Z. Harchaoui, C. Schmid, 2013: Deepflow: Large displacement optical flow with deep matching. – ICCV – IEEE International Conference on Computer Vision 1385–1392.

[8] Bradski, G., 2000: The opencv library. – Dr. Dobb's Journal of Software Tools.

[9] Harty, Travis M., William F. Holmgren, Antonio T. Lorenzo, and Matthias Morzfeld. "Intra-hour cloud index forecasting with data assimilation." *Solar Energy* 185 (2019): 270-282.

[10] Himawari satellite image, Full Disk, Meteorological

Satellite Center of Japanese Meteorological Agency, http://www.data.jma.go.jp/mscweb/data/himawari/, visited 2021-08-04.

[11] Perez, R., P. Ineichen, E. Maxwell, R. Seals and A. Zelenka, (1992). "Dynamic Global-to-Direct Irradiance Conversion Models". ASHRAE Transactions-Research Series, pp. 354-369

[12] Perez, R., Ineichen, P., Seals, R., Michalsky, J., Stewart, R., 1990. Modeling daylight availability and irradiance components from direct and global irradiance. Solar Energy 44 (5), 271-289.

[13] William F. Holmgren, Clifford W. Hansen, and Mark A. Mikofski. "pvlib python: a python package for modeling solar energy systems." Journal of Open Source Software, 3(29), 884, (2018). https://doi.org/10.21105/joss.00884

[14] Muller, B., L. Hardt, et al. (2015)." Yield predictions for photovoltaic power plants: empirical validation, recent advances and remaining uncertainties." Progress in Photovoltaics: Research and Applications: published online

[15] Holland, Nicolas, et al." Trainable curtailment detection for the enhancement of PV power forecasting." 2020 PVSEC-30 & GPVC 2020, November **8** - **13**, 2020 ICC JEJU, Jeju, Republic of Korea

[16] Christ, M., Kempa-Liehr, A.W. and Feindt, M. (2016). Distributed and parallel time series feature extraction for industrial big data applications. ArXiv e-prints: 1610.07717 URL: http://adsabs.harvard.edu/abs/2016arXiv161007717C

[17] Benjamini, Y. and Yekutieli, D. (2001). The control of the false discovery rate in multiple testing under dependency. Annals of statistics, 1165–1188

[18] tsfresh, (2020). Retrieved from https://tsfresh.readthedocs.io/en/latest/index.html

[19] Pedregosa et al., Scikit-learn: Machine Learning in Python, JMLR 12, pp. 2825-2830, 2011.

[20] Kingma, Diederik P., and Jimmy Ba. "Adam: A method for stochastic optimization." *arXiv preprint arXiv:1412.6980* (2014).

[21] François Chollet et. al., Keras, (2020), GitHub repository, https://github.com/keras-team/keras

AUTHOR INDEX

WIP – Renewable Energies
Sylvensteinstr. 2
81369 Munchen
Germany

ISBN 978-1-7138-5213-1

38th European Photovoltaic Solar Energy Conference and Exhibition (EU PVSEC 2021)

Online
6-10 September 2021

Volume 3 of 3

WIP RENEWABLE ENERGIES

38th European Photovoltaic Solar Energy Conference and Exhibition (EU PVSEC 2021)

Online
6-10 September 2021

Volume 3 of 3

Editors:

J.M. Almeida Serra
R. Kenny
P. Helm

ISBN: 978-1-7138-5213-1

Printed from e-media with permission by:

Curran Associates, Inc.
57 Morehouse Lane
Red Hook, NY 12571

Some format issues inherent in the e-media version may also appear in this print version.

WIP – Renewable Energies
Sylvensteinstr. 2
81369 Munchen
Germany

Phone: +49 89 72012735
Fax: +49 89 72012791

wip@wip-munich.de

Additional copies of this publication are available from:

Curran Associates, Inc.
57 Morehouse Lane
Red Hook, NY 12571 USA
Phone: 845-758-0400
Fax: 845-758-2633
Email: curran@proceedings.com
Web: www.proceedings.com

38th European Photovoltaic Solar Energy Conference and Exhibition

Proceedings of the International Conference

06 September – 10 September 2021

Edited by:

J.M. ALMEIDA SERRA
University of Lisbon - FCiencias.ID
Lisbon, Portugal

R. KENNY
European Commission Joint Research Centre
Ispra, Italy

P. HELM
WIP Renewable Energies
Munich, Germany

Coordination of the Technical Programme:

European Commission Joint Research Centre
Via E. Fermi 1
21020 Ispra (VA)
Italy

Institutional Support:

European Commission

UNESCO – United Nations Educational, Scientific and Cultural Organization, Natural Sciences Sector

WCRE – World Council for Renewable Energy

Supporting Organisations:

APVI – Australian PV Institute

EREF – European Renewable Energies Federation

ESA – European Space Agency

ESMC – European Solar Manufacturing Council

EUFORES –The European Forum for Renewable Energy Sources

EUREC – The Association of European Renewable Energy Research Centres

ETIP PV – European Technology & Innovation Platform PV

GÜNDER – The Turkish Solar Energy Society

IEA PVPS – International Energy Agency, Photovoltaic Power Systems Programme

REN21 – Renewable Energy Policy Network for the 21st Century

Institutional PV Industry Cooperation:

SolarPower Europe

SOLARUNITED

Supporting Associations:

AIE – European Association of Electrical Contractors

AREA – African Renewable Energy Alliance

CABA – Continental Automated Buildings Association

EU PVSEC realised by:

WIP Renewable Energies
Sylvensteinstr. 2, 81369 München, Germany
Tel: +49 89 720 12 735, Fax: +49 89 720 12 791
Email: pv.conference@wip-munich.de
www.photovoltaic-conference.com
www.wip-munich.de

Proceedings produced and published by:

WIP
Sylvensteinstr. 2, 81369 München, Germany
Tel: +49 89 720 12 735, Fax: +49 89 720 12 791
Email: pv.conference@wip-munich.de
www.photovoltaic-conference.com
www.wip-munich.de

38th EUROPEAN PHOTOVOLTAIC SOLAR ENERGY CONFERENCE AND EXHIBITION
06 SEPTEMBER – 10 SEPTEMBER 2021

EU PVSEC COMMITTEES

INTERNATIONAL SCIENTIFIC ADVISORY COMMITTEE (ISAC)

Chair

P. Szymanski, European Commission Joint Research Centre, Director of Energy, Transport and Climate, Petten, The Netherlands

Committee Members

P. Frankl, Head of the Renewable Energy Division, International Energy Agency, France
M. Getsiou, European Commission, DG RTD, Brussels, Belgium
S.W. Glunz, Head of Division "Solar Cells – Development and Characterization", Fraunhofer ISE, Freiburg, Germany
R. Kenny, European Commission Joint Research Centre, Directorate for Energy and Transport and Climate, Ispra, Italy
P. Malbranche, General Director, CEA INES, France
P. Menna, ret. European Commission, DG ENER, Brussels, Belgium
S. Nowak, Managing Director of NET Nowak Energy & Technology, St. Ursen, Switzerland
W.C. Sinke, Principal Scientist Solar Energy, TNO Energy Transition, The Netherlands
M. Topič, Chairman ETIP PV, Head of Laboratory of Photovoltaics and Optoelectronics of the University of Ljubljana, Slovenia
P. Verlinden, Director at Amrock Pty Ltd, Visiting Professor at Sun Yat-Sen University, Guangzhou, China
A. Grassi, Managing Director, ETA - Renewable Energies, Florence, Italy
P. Helm, Scientific & Managing Director, WIP Renewable Energies, Munich, Germany

CONFERENCE EXECUTIVE COMMITTEE

Conference General Chair

J.M. Almeida Serra, University of Lisbon- FCiencias.ID, Portugal

Technical Programme Chair

R. Kenny, European Commission Joint Research Centre, Directorate for Energy and Transport and Climate, Ispra, Italy

Committee Members

H. Ossenbrink, ret. European Commission Joint Research Centre, Italy
P. Helm, Scientific & Managing Director, WIP Renewable Energies, Munich, Germany
W.C. Sinke, Principal Scientist Solar Energy, TNO Energy Transition, The Netherlands
S. Nowak, Managing Director of NET Nowak Energy & Technology, St. Ursen, Switzerland
A. Grassi, Managing Director, ETA - Renewable Energies, Florence, Italy
J. Bergmiller, Managing Director Events, WIP Renewable Energies, Munich, Germany
J. de Gregorio, Head of Unit, Scientific Services and Cooperation, WIP Renewable Energies, Munich, Germany

2021 SCIENTIFIC COMMITTEE
Topic Organisers and Paper Review Experts

Topic 1 – New Materials and Concepts for Photovoltaic Devices

A. Martí Vega, UPM, Madrid, Spain
P. Pearce, University of Cambridge, United Kingdom
J. Poortmans, imec, Leuven, Belgium
B. Rech, HZB, Berlin, Germany

J.P. Connolly, GeePs, Gif sur Yvette, France
N.J. Ekins-Daukes, UNSW Australia, Sydney, Australia
J. Govaerts, imec, Genk, Belgium
M.A. Hernandez-Fenollosa, UPV, Valencia, Spain
J.C. Hummelen, University of Groningen, The Netherlands
J.-P. Kleider, CNRS/GeePs, Gif sur Yvette, France
D. König, UNSW Australia, Sydney, Australia
I. Konovalov, University of Applied Sciences Jena, Germany
C. Lévy-Clément, CNRS, Thiais, France
Y. Okada, University of Tokyo, Japan
I. Ramiro, NOVA University Lisbon, Caparica, Portugal
F. Roca, ENEA, Portici, Italy
M. Rusu, HZB, Berlin, Germany
R.E.I. Schropp, Driebergen, The Netherlands
T. Tayagaki, AIST, Tsukuba, Japan
W. Warta, Fraunhofer ISE, Freiburg, Germany

Topic 2 – Silicon Materials and Cells

K. Bothe, ISFH, Emmerthal, Germany
D. Bredemeier, ISFH, Emmerthal, Germany
J. Cárabe, CIEMAT, Madrid, Spain
M. Despeisse, CSEM, Neuchâtel, Switzerland
F. Ferrazza, ENI, San Donato Milanese, Italy
S.W. Glunz, Fraunhofer ISE, Freiburg, Germany
G. Hahn, University of Konstanz, Germany
M. Hermle, Fraunhofer ISE, Freiburg, Germany
S. Peters, Hanwha Q CELLS, Bitterfeld-Wolfen, Germany
S. Riepe, Fraunhofer ISE, Freiburg, Germany
P. Roca i Cabarrocas, LPICM-CNRS, Palaiseau, France
M. Topic, University of Ljubljana, Slovenia
A.W. Weeber, TNO Energy Transition, Petten, The Netherlands
P. Wohlfart, Singulus Technologies, Kahl am Main, Germany

O. Anspach, PV Crystalox Solar, Erfurt, Germany
A. Augusto, Arizona State University, Tempe, USA
N. Azkona, UPV/EHU, Bilbao, Spain
D.L. Bätzner, Meyer Burger Research, Hauterive, Switzerland
M.P. Bellmann, SINTEF, Trondheim, Norway
M. Benmansour, CEA, Le Bourget-du-Lac, France
M. Boccard, EPFL, Neuchâtel, Switzerland
R. Brendel, ISFH, Emmerthal, Germany

T.M. Bruton, Progress in Photovoltaics, Woking, United Kingdom
A. Ciesla, UNSW Australia, Sydney, Australia
A. Danel, CEA, Le Bourget-du-Lac, France
W. Dauksher, Arizona State University, Tempe, USA
P. Delli Veneri, ENEA, Portici, Italy
A. Descoeudres, CSEM, Neuchâtel, Switzerland
S. Dubois, CEA, Le Bourget-du-Lac, France
T. Dullweber, ISFH, Emmerthal, Germany
M. Ernst, ANU, Canberra, Australia
P. Fath, RCT-Solutions, Konstanz, Germany
W. Favre, CEA, Grenoble, France
S. Gall, HZB, Berlin, Germany
C. Hagendorf, Fraunhofer CSP, Halle (Saale), Germany
K. Hesse, Wacker Chemie, Burghausen, Germany
J. Horzel, Fraunhofer ISE, Freiburg, Germany
J. John, imec, Leuven, Belgium
A. Meeder, Geo-En Energy Technologies, Berlin, Germany
J. Meier, Meier Technologies, Berlingen, Switzerland
A. Metz, Alzenau, Germany
D. Muñoz, CEA, Le Bourget-du-Lac, France
R.C.G. Naber, Hanwha Q CELLS, Vaassen, The Netherlands
H.-J. Nonnenmacher, Meyer Burger, Hohenstein-Ernstthal, Germany
Y. Ohshita, Toyota Technological Institute, Nagoya, Japan
A. Otaegi, UPV/EHU, Bilbao, Spain
K. Petter, Hanwha Q CELLS, Bitterfeld-Wolfen, Germany
R. Preu, Fraunhofer ISE, Freiburg, Germany
J. Rentsch, Fraunhofer ISE, Freiburg, Germany
J. Schmidt, ISFH, Emmerthal, Germany
M.C. Schubert, Fraunhofer ISE, Freiburg, Germany
L. Serenelli, ENEA, S. Maria di Galeria - Rome, Italy
K. Sporleder, Fraunhofer CSP, Halle, Germany
B. Terheiden, University of Konstanz, Germany
R. Turan, METU, Ankara, Turkey
P.J. Verlinden, AMROCK, McLaren Vale, Australia
M. Vetter, IPHT, Jena, Germany
Q. Wang, Jinko Solar, Haining, China
W. Wu, Tongwei Solar, Chengdu, China
A. Yamada, Tokyo Institute of Technology, Japan
G. Yang, Delft University of Technology, The Netherlands
D.L. Young, NREL, Golden, USA
J. Zhao, CSEM, Neuchâtel, Switzerland

Topic 3 – Perovskites and Other Non-Silicon Materials and Devices, Multijunctions/Tandems

G. Bardizza, European Commission JRC, Ispra, Italy
G. Flamand, imec, Leuven, Belgium
J.C. Goldschmidt, Fraunhofer ISE, Freiburg, Germany
S. Nicolay, CSEM, Neuchâtel, Switzerland
S. Paetel, ZSW, Stuttgart, Germany
R. Po, eni, Novara, Italy
A. Redinger, University of Luxembourg, Luxembourg
G. Siefer, Fraunhofer ISE, Freiburg, Germany
A.N. Tiwari, EMPA, Dübendorf, Switzerland
W. Tress, Zurich University of Applied Sciences, Winterthur, Switzerland
S.C. Veenstra, TNO Energy Transition, Eindhoven, The Netherlands

S. Albrecht, HZB, Berlin, Germany
N. Barreau, University of Nantes, France
V. Bermudez Benito, QEERI, Doha, Qatar
M.J. Brites, LNEG, Lisboa, Portugal
R. Campesato, CESI, Milan, Italy
G. Coletti, TNO Energy Transition, Petten, The Netherlands
S. Cros, CEA, Le Bourget-du-Lac, France
T. Dalibor, Avancis, Munich, Germany
B. Dimmler, NICE Solar Energy, Schwäbisch Hall, Germany
D. Dimova-Malinovska, Bulgarian Academy of Sciences, Sofia, Bulgaria
C.J. Fell, CSIRO, Mayfield West, Australia
I. Gordon, imec, Leuven, Belgium
S. Huang, Macquarie University, Sydney, Australia
S.J.C. Irvine, Swansea University, St. Asaph, United Kingdom
S. Ishizuka, AIST, Tsukuba, Japan
R.R. King, Arizona State University, Tempe, USA
P. Kratzert, SOLARWATT, Dresden, Germany
C.-F. Lin, National Taiwan University, Taipei, Taiwan
D. Lincot, CNRS, Palaiseau, France
M. Lira-Cantú, CIN2, Barcelona, Spain
M.Ch. Lux-Steiner, HZB, Berlin, Germany
M. Meuris, imec, Diepenbeek, Belgium
U.W. Paetzold, Karlsruhe Institute of Technology, Germany
A. Romeo, University of Verona, Italy
V. Sittinger, Fraunhofer IST, Braunschweig, Germany
M. Theelen, TNO/Solliance, Eindhoven, The Netherlands
G. Timò, RSE, Piacenza, Italy
E. Unger, HZB, Berlin, Germany
A. Virtuani, EPFL, Neuchâtel, Switzerland
A. Walter, CSEM, Neuchâtel, Switzerland
S. Yerci, METU, Ankara, Turkey

Topic 4 – Photovoltaic Modules and BoS Components

C. Buerhop-Lutz, HI ERN, Erlangen, Germany
R. Gottschalg, Fraunhofer CSP, Halle (Saale), Germany
W. Herrmann, TÜV Rheinland Energy, Cologne, Germany
N.M. Pearsall, Northumbria University, Newcastle upon Tyne, United Kingdom
A. Wade, Viessmann, Berlin, Germany
K. Wambach, Wambach-Consulting, Petersdorf, Germany

S. Arancón, WIP Renewable Energies, Munich, Germany
P. Bauwens, Ghent University, Belgium
G. Beaucarne, Dow Silicones Belgium, Seneffe, Belgium
N. Bosco, NREL, Golden, USA
C. Camus, LayTec, Berlin, Germany
S. Capaccioli, ETA - Florence Renewable Energies, Italy
R. Einhaus, Apollon Solar, Saint Priest, France
G. Friesen, SUPSI, Canobbio, Switzerland
V. Fthenakis, Columbia University, New York, USA
W. Gambogi, DuPont, Wilmington, USA
G. Graditi, ENEA, Portici, Italy
Y. Hishikawa, AIST, Tsukuba, Japan
U. Jahn, VDE Renewables, Alzenau, Germany
A.R. Lagunas, CENER, Sarriguren-Navarra, Spain
J. Lopez-Garcia, European Commission JRC, Ispra, Italy
M. Mittag, Fraunhofer ISE, Freiburg, Germany
C. Monokroussos, TÜV Rheinland, Shanghai, China
J. Moschner, KU Leuven, Belgium
M.A. Muñoz-García, UPM, Madrid, Spain
H. Nagel, Fraunhofer ISE, Freiburg, Germany
H. Nussbaumer, ZHAW, Winterthur, Switzerland
S. Oktik, Kadir Has University, Istanbul, Turkey
G. Oreski, PCCL, Leoben, Austria
W. Palitzsch, LuxChemtech, Freiberg, Germany
M. Pander, Fraunhofer CSP, Halle (Saale), Germany
J.N. Roy, IIT Kharagpur, India
T. Sample, European Commission JRC, Ispra, Italy
R. Sinton, Sinton Instruments, Boulder, USA

Topic 5 – PV Systems – Modelling, Design, Operation and Performance

J.M. Almeida Serra, University of Lisbon - FCiencias.ID, Portugal
F. Al-Shakarchi, Entech Smart Energies, Quimper, France
I. Antón Hernández, UPM, Madrid, Spain
A.M. Gracia Amillo, European Commission JRC, Ispra, Italy
P. Lechner, ZSW, Stuttgart, Germany
G. Mütter, Enery, Vienna, Austria
H. Ossenbrink, ret. European Commission Joint Research Centre, Italy
S. Taylor, European Space Agency, Noordwijk, The Netherlands
W.G.J.H.M. van Sark, Utrecht University, The Netherlands

C. Alonso-Tristán, UBU, Burgos, Spain
P. Blanc, MINES ParisTech, Paris, France
M. Grottke, WIP Renewable Energies, Munich, Germany
D. Guida, ENEL Green Power, Roma, Italy
P. Hacke, NREL, Golden, USA
V. Khorenko, Azur Space, Heilbronn, Germany
K. Kiefer, Fraunhofer ISE, Freiburg, Germany
K. Lappalainen, Tampere University, Finland
J. Lin, PV Guider, Taipei, Taiwan
A. Louwen, Eurac Research, Bolzano, Italy
O. Mayer, Bayern Innovativ, Nürnberg, Germany
A. Migan-Dubois, GeePs, Gif-sur-Yvette, France
D. Moser, Eurac Research, Bolzano, Italy
K. Nishioka, University of Miyazaki, Japan
C. Nyman, Soleco, Porvoo, Finland
C. Protogeropoulos, EEPS, Athens, Greece
A.H.M.E. Reinders, University of Twente, Enschede, The Netherlands
J. Remund, Meteotest, Bern, Switzerland
M. Richter, 3E, Brussels, Belgium
S. Rodrigues, Forschungszentrum Jülich, Germany
J.A. Ruiz-Arias, University of Malaga, Spain
M. Sengupta, NREL, Golden, USA
J.S. Stein, Sandia National Laboratories, Albuquerque, USA
D. Stellbogen, ZSW, Stuttgart, Germany
J. Tavares Pinho, UFPA, Belém, Brazil
M. Theristis, Sandia National Laboratories, Albuquerque, USA
M. Victoria, Aarhus University, Denmark

Topic 6 – PV Applications, Integration and Storage

G. Adinolfi, ENEA, Portici, Italy
F.P. Baumgartner, Zurich University of Applied Sciences, Winterthur, Switzerland
F. Dolci, European Commission JRC, Petten, The Netherlands
F. Frontini, SUPSI, Canobbio, Switzerland
K.-P. Kairies, ACCURE Battery Intelligence, Aachen, Germany
U. Muntwyler, Bern University of Applied Sciences, Burgdorf, Switzerland
R. Pestana, R&D Nester, Sacavém, Portugal
R. Schlatmann, HZB, Berlin, Germany
A. Scognamiglio, ENEA, Portici, Italy
E. Voroshazi, CEA, Le Bourget-du-Lac, France
I. Weiss, WIP Renewable Energies, Munich, Germany

P.-J. Alet, CSEM, Neuchâtel, Switzerland
P. Alonso, BWP Baumanagement, Munich, Germany
K. Baert, KU Leuven, Heverlee, Belgium
G. Barchi, Eurac Research, Bolzano, Italy
P. Bonomo, SUPSI, Mendrisio, Switzerland
R. Bründlinger, AIT, Vienna, Austria
F. Carigiet, ZHAW, Winterthur, Switzerland
M Centeno Brito, University of Lisbon, Portugal
J. da Costa Fernandes, University of Applied Sciences Offenburg, Germany
K. Ding, Forschungszentrum Jülich, Germany
M. Dörenkämper, TNO Energy Transition, Eindhoven, The Netherlands
G.C. Eder, OFI, Vienna, Austria
V. Efthymiou, University of Cyprus, Nicosia, Cyprus
W. Folkerts, TNO, Eindhoven, The Netherlands
B. Gaiddon, HESPUL, Lyon, France
I.B. Hagemann, Architekturbüro Hagemann, Aachen, Germany
C. Helmke, Clake Energy, Berlin, Germany
B. Herteleer, KU Leuven, Gent, Belgium
B. Kämpfen, kämpfen zinke + partner, Zürich, Switzerland
J.-M. Klein, CEA, Le Bourget-du-Lac, France
P. Lalanne, HydroClapeyron, Dinard, France
B. Lim, ISFH, Emmerthal, Germany
A.C. Neves Foles, University of Évora, Portugal
T. Reijenga, BEAR-iD, Gouda, The Netherlands
E. Román Medina, Tecnalia, Derio, Spain
E. Schüpbach, BFH, Bern, Switzerland
K. Sinapis, Trespa International, Weert, The Netherlands
L.H. Slooff-Hoek, TNO Energy Transition, Petten, Netherlands
L. Tautenhahn, Avancis, Torgau, Germany
R.M.E. Valckenborg, TNO, Eindhoven, The Netherlands

Topic 7 – Finance, Markets and Policies

C. Breyer, Lappeenranta University of Technology, Finland
M. Getsiou, European Commission DG RTD, Brussels, Belgium
P. Malbranche, CEA, Le Bourget-du-Lac, France
G. Masson, Becquerel Institute, Brussels, Belgium
T. Nordmann, TNC Consulting, Feldmeilen, Switzerland
S. Nowak, NET Nowak Energy & Technology, St. Ursen, Switzerland

G. Arrowsmith, EUREC, Brussels, Belgium
S. Caneva, WIP Renewable Energies, Munich, Germany
J. Donoso Alonso, UNEF, Madrid, Spain
I. Kaizuka, RTS Corporation, Chuo-ku, Japan
D. Polverini, European Commission DG GROW, Brussels, Belgium
N. Taylor, European Commission JRC, Ispra, Italy
E. Vartiainen, Fortum Growth, Finland

Awards Coordinators

Student Awards Coordinator
A.H.M. Smets, Delft University of Technology, The Netherlands

Poster Awards Coordinator
J. Cárabe, CIEMAT, Madrid, Spain

SUBJECT INDEX

Topic Code

1 New Materials and Concepts for Photovoltaic Devices
2 Silicon Materials and Cells
3 Perovskites and Other Non-Silicon Materials and Devices, Multijunctions/Tandems
4 Photovoltaic Modules and BoS Components
5 PV Systems – Modelling, Design, Operation and Performance
6 PV Applications, Integration and Storage
7 Finance, Markets and Policies

Session Type

P = Plenary Session
O = Oral Session
V = Visual Session

Day Codes

A = Monday, 06 September 2021
B = Tuesday, 07 September 2021
C = Wednesday, 08 September 2021
D = Thursday, 09 September 2021
E = Friday, 10 September 2021

e.g. 1AO.1 ⇒ 1= New Materials and Concepts for Photovoltaic Devices, A=Monday, O=Oral session, 1=Session 1

TABLE OF CONTENTS
of EU PVSEC 2021 Proceedings Papers

Oral PRESENTATIONS 4BO.5 Interconnects and Soldering

Oral PRESENTATIONS 4CO.1 Polymers and Encapsulation of PV Modules

Oral PRESENTATIONS 4CO.2 PV Module Backsheets

Oral PRESENTATIONS 4CO.3 Performance of BOS Components

Oral PRESENTATIONS 4CO.4 Recycling, Repair and Reuse of PV Modules

Oral PRESENTATIONS 4DO.11 Sustainability Aspects of PV

FOREWORD

The EU PVSEC 2021, as the second online edition of the European PV Solar Energy Conference and Exhibition was a great success and has shown once again that a sustainable energy supply is becoming ever more important and only together, through scientific endeavor and cooperation and political commitments and investments that this goal can be achieved. The Conference provided once again a stage for the dissemination of work that will help ensure ever increasing quantities of PV energy and its integration into the current and future energy network. New discussion sessions focused highly relevant topical questions, provided key lessons for upcoming developments. This year 1,300 participants from 62 countries took part in this most inspiring platform. In the exhibition 16 companies from all parts of the world welcomed online visitors and presented their products and services.

The EU PVSEC 2021 Proceedings give a comprehensive overview of the photovoltaic solar sector, its current status and future prospects in science, research, innovation, development and deployment on almost 3,800 pages. Selection for inclusion in the conference was made by paper review experts and topic organisers (see the listing on pages III-V), to whom we express our sincere gratitude for their comprehensive review work and overall contribution to the success of the conference. In addition to the 364 submitted papers, the proceedings include 103 presentations (slides) shown during the plenary and oral presentations as well as 195 poster files of the visual presentations. In total this amounts to 662 publications.

The 2021 EU PVSEC Proceedings maintain our commitment to providing quick and open access to high quality scientific results. They constitute a powerful tool for targeted and quick information search and retrieval, enabling you to search by keyword, paper title and authors. We are sure that these features will help to simplify the use and exploit the full potential of this extensive source of information.

The Conference Proceedings are published as downloadable DVD file and are also fully accessible online. A DOI code (Digital Object Identifier) has been assigned to each paper. This ensures unequivocal and permanent identification and full citability. The papers can be viewed and downloaded in a full free open access from the EU PVSEC Proceedings website www.eupvsec-proceedings.com.

We are confident that these Proceedings will play an important role in providing a comprehensive overview of the current actors and activities in the global PV sector and that they will disseminate information on the state-of-the-art of technologies and applications. This can generate further research, add momentum to innovation and promote interest in PV worldwide.

We would like to cordially thank all authors and participants of the EU PVSEC 2021 for their contributions and look forward to welcoming you in 2022 at the WCPEC-8, the 8th World Conference on Photovoltaic Energy Conversion in Milan, Italy.

The Editors

INTRADAY POST PROCESSING OF SOLAR IRRADIANCE FORECASTS FROM WRF MODEL USING SATELLITE-DERIVED DATA IN FRENCH GUIANA.

Maha SALLOUM, Jeremy MACAIRE, Sara ZERMANI, Jessica BECHET, Antoine PRIMEROSE, Laurent LINGUET
Company / Institute(s)
University of French Guiana
UMR ESPACE-DEV, BP 792, 97 337 Cayenne, France
maha.salloum@ird.fr

ABSTRACT :
The objective of this study is to improve day ahead Weather Research and Forecasting Model (WRF) forecasts in any geographical location in the Inter Tropical Zone (ITZ). The WRF model is a physical one with high biased outputs. A post processing step is required to reduce the WRF model output's bias. This step is obtained by post processing WRF output with a Kalman filter bias correction method (hybrid physical statistical model). We proposed to use Satellite images as observation instead of ground measurements for the post processing of WRF model. Since Satellite images have a broader coverage than ground measurements. We validated our methodology over three months of Global Horizontal Irradiance (GHI) data from six stations located in French Guiana, a South America territory with a tropical climate regulated by the Intertropical-Convergence Zone. To compute the model accuracy, we used three metrics; the mean bias error, mean absolute error and root mean square error. For all sky conditions and all stations merged, the post-processing method using satellite-derived data has a similar performance to the post-processing method using ground-based irradiation measurements. Satellite-derived data perform better than ground one for clear sky conditions with low variability. Moreover, we note a better performance with satellite-derived data for high variability and independent of sky condition.

Keywords: Intraday solar irradiance forecast, Satellite derived data, Post processing WRF, Kalman filter.

1 INTRODUCTION

Accurate forecast of solar irradiance is crucial for photovoltaic PV central management to optimize electrical network management [1], [2]. Several authors proposed and validated several methods for GHI forecasts in various territories. Nevertheless, only a few studies focused on GHI forecasts in the inter tropical zone (ITZ). The ITZ is a zone with high potential but also with high variability due to the rapid cloud transformations caused by the intertropical convergence zone [3]. Due to this variability, it is difficult to obtain accurate GHI forecasts. French Guiana is a territory located in the inter tropical zone (ITZ).

For a large area geographic data collection, there is no better observational tool than an NWP [4]. A NWP provides more information up to several days ahead. Weather Research and Forecasting model (WRF) a numerical model can compute a forecast up to 48 H with high spatial resolution. However, it is well-known that this model's output is highly biased. Therefore, a post processing step is required to reduce model uncertainties [5], [6]. Kalman filter have been used extensively in a number of areas, including post-processing of numerical weather prediction model outputs [7], [6].

Recently, [7] applied Kalman filter with historical measurements of solar irradiance to correct the prediction of the Canadian numerical model GEM Global Environmental Multi Scale. [8] proposed a Model Output Statistics (MOS) method for improving the day ahead solar GHI forecasts from Weather Research and Forecasting (WRF) model in Reunion Island. Yet, results found for territories located in the ITZ were consistently higher than those found for similar climate and models in extra tropical areas. Moreover, this improvement is limited on the ground stations.

In our study, we propose to improve hourly global horizontal irradiance (GHI) forecasts from Weather Research and Forecasting model (WRF) [9] using a Kalman filter post-processing method based on GHI satellite estimates. GHI satellite estimates are obtained following the method developed in [10]. GOES satellite observations are converted to GHI estimates using an optimized Heliosat-2 method [11]. To our knowledge this is the first attempt of using the kalman filter post processing method with GHI estimates obtained from satellite observations which has border coverage compared to ground measurements and good spatial resolution.

2 DATA

We tested and validated our methodology over six stations in French Guiana. French Guiana is a French territory located in the inter-tropical zone.

It has overall a tropical rainforest climate. French Guiana experiences two seasons a rainy season and a dry season [12].

2.1 GROUND MEASUREMENTS

We exploited hourly GHI measurement from six meteorological stations of the French national weather service (Meteo-France) equipped with pyranometers of CMP6 Kipp & Zonen (World Meteorological Organization 2008), The six stations are shown in Figure 1.

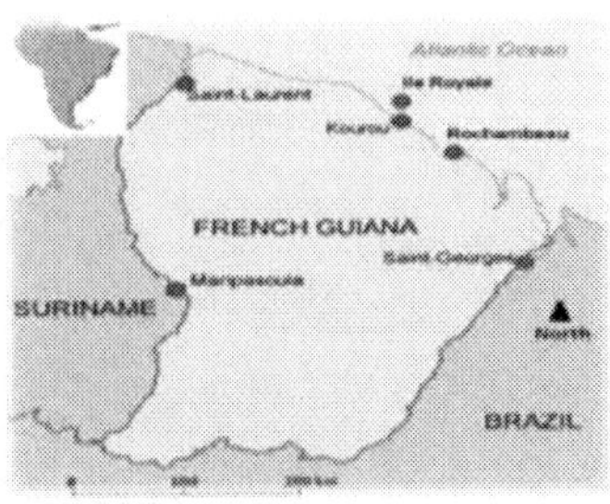

Figure 1 Meteorological stations in French Guiana.

We chose data from three representative months of 2016. September, May, represent the dry and rainy seasons, respectively. The third month March describes the latency period between the ITCZ southward to the northward motion over French Guiana. Depending on the selected year March either behaves as a rainy or a dry season month. Stations are characterized in Tab 1.

2.2 SATELLITE DATA

GHI estimates were derived from satellite images. We downloaded two images (slots) by hour, one at -20 minutes and the other at -50 minutes from time T. So to get the estimation for time T we calculate the mean value, Tab 1.

Station	*Latitud*	*Longitud*	*Altitude*	*Nb_Ground*	*Nb_Satellite*
RO (Rochambeau)	4.81	-52.37	4	18270	36540
SG (Saint-Georges)	3.88	-51.80	6	19046	38092
MP (Maripasoula)	3.63	-54.03	104	17373	34746
SL (Saint-Laurent)	5.48	-53.90	4	18652	37304
KR (Kourou)	5.12	-52.44	12	16563	33126
IR (Ile Royale)	5.28	-52.58	48	18153	36306

Tab 1 .Latitude, longitude and altitude of the studied ground meteorological stations in French Guiana.

A comparison between these estimations and ground measurement sets was necessary before using satellite estimates as observation. The rRMSE and rMBE of the satellite estimates compared to the ground measurement are shown in Figure 2 (for more details on the metrics see VALIDATION METRICS section).

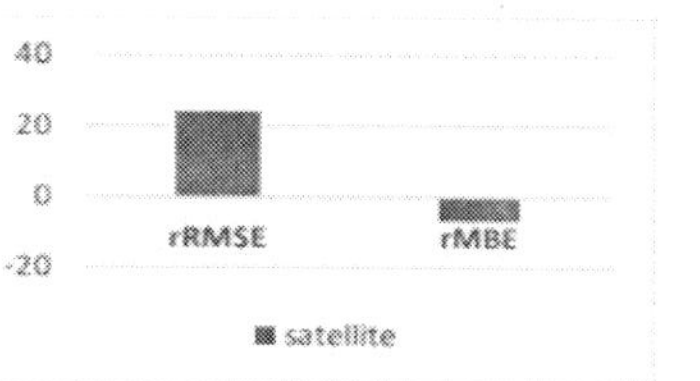

Figure 2 rMBE and rRMSE for satellite estimates data.

These differences can be explained by the distance between the geographic location of the station and its nearest pixel from image satellite. Figure 3, shows the distribution of ground measurements depending on the month for Royale island station in 2016. It can be observed that both ground and satellite data have a similar tendency throughout the year that is consistent with the three seasons mentioned above. The distribution of the data during the day is also similar. Both sets behave alike, showing its maximum values at midday and following the expected pattern, increasing in the mornings and decreasing in the afternoons.

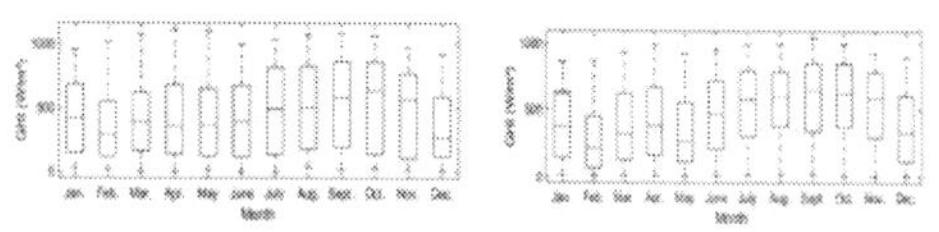

a. GHI ground b. GHI Satellite

Figure 3 Hourly Data distribution by month for IR in 2016.

2.3 WEATHER FORECASTS

We exploited hourly solar irradiance values from Weather Research and Forecast model in Mars, May and September of 2016. WRF forecasts have a spin up period of 12h and a forecast horizon ranging between +12 hours and +48 hours. This data were produced by [13], they have a 1 hour output frequency.

3 METHOD AND VALIDATION METRIC

3.1 METHOD

The different steps of the method used to improve the GHI forecasts in the ITZ are shown Figure 4.

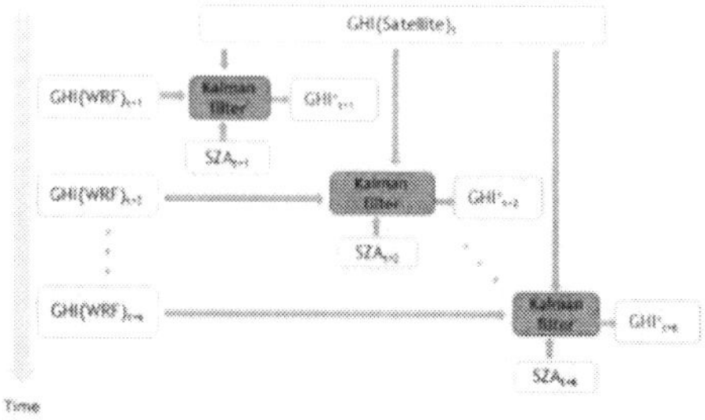

Figure 4 procedure flow for horizons h ranging from 1 to 6 hours to update WRF GHI forecasts.

The GHI_GOES estimates at a given time are used to update GHI_WRF forecasts for horizons ranging from 1 to 6 hours. Figure 4 illustrates the procedure of the proposed method where the intraday forecasting method proposed in this work is a hybrid model. Indeed, it combines in a statistical model, namely the linear Kalman filter, the outputs of a numerical forecasting model, the satellite estimation measurements made at the experimental site and the cosine of the zenith solar angle. We post processed these WRF outputs using a Kalman Filter bias correction method [7] to refine the forecast of WRF. We defined Kalman filter's covariance matrix (P, W, V) as in Figure 5. In this study, the Kalman Filter bias correction method uses past hour observations to correct WRF outputs. The Kalman filter was chosen as the post-processing algorithm because it removes the bias in GHI forecasts without the need for a long archive of historical data.

At the current time T, satellite data estimate or ground measurement is used as observation. Kalman filtering, a post-processing method, is applied on GHI_WRF of T+1 to T+6 hours ahead using in order to improve GHI 1 to 6 hours ahead forecasts. This hybrid method (WRF+KALMAN) allows us to broadly update the multi hours a head forecast of GHI from WRF forecasts done the day before. Our objective is to increase the accuracy of the solar irradiance forecast in French Guiana for use in an operational setting. Figure 4, illustrates a global view of the proposed method.

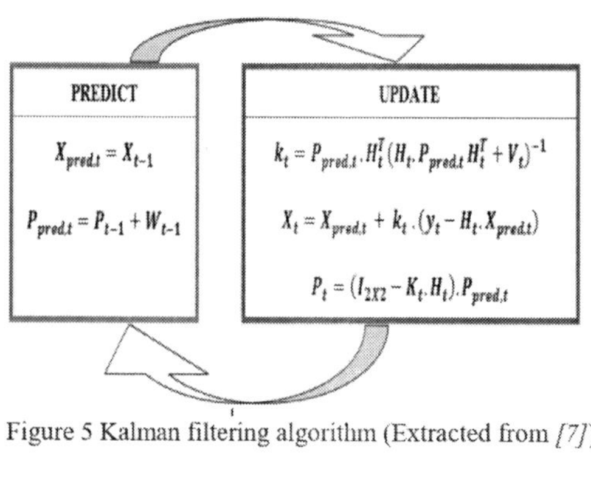

Figure 5 Kalman filtering algorithm (Extracted from *[7]*).

We compared the post processing method using satellite observations against ground measurements over six meteorological stations.

3.2 VALIDATION METRICS

we also compared the post processing method using satellite observation against ground measurements according to [14] to assess the accuracy of using satellite derived GHI. The Root Mean Square Error (RMSE) is used when significant errors are undesirable (eq 1), Mean Bias Error (MBE) to indicate if the model overestimates or underestimates the actual value (eq 2). And Mean Absolute Error (MAE) measures the average of the absolute errors between the forecasted and the measured values (eq 3). These three performance criterions used to evaluate each model's accuracy are defined as:

$$RMSE = \frac{1}{N} \cdot \sum_{i} (y_i - x_i)^2 \qquad 1$$

$$MBE = \frac{1}{N} \cdot \sum_{i} (y_i - x_i) \qquad 2$$

$$MAE = \frac{1}{N} \cdot \sum_{i} |y_i - x_i| \qquad 3$$

Where, $\mathbf{y_i}$ represents the ith forecasted GHI value and $\mathbf{x_i}$ represents the ith measured GHI value. relative values of these metrics were obtained by a normalization step to the mean value of the ground measured GHI for the considered period.

In order to quantify the GHI variability, we use two indexes based on clear sky Index K_C (eq 4):

$$K_c = \frac{GHI_{obs}}{GHI_{ESRA}} \qquad 4$$

Where, GHI_{obs} is the observed GHI and GHI_{ESRA} the European solar radiation atlas (ESRA) clear sky model [14]. The sky conditions were considered as clear when $Kc > 0.65$, cloudy when $0.4 < Kc < 0.65$ and overcast when $Kc < 0.4$ [6].

4 RESULTS AND DISCUSSION

We compared the post processing method using satellite estimates as observations against ground measurements. As we mentioned before, three months of 2016 were selected in French Guiana. WRF forecast errors before and after processing using ground-measured data and satellite-derived data from all stations are discussed.

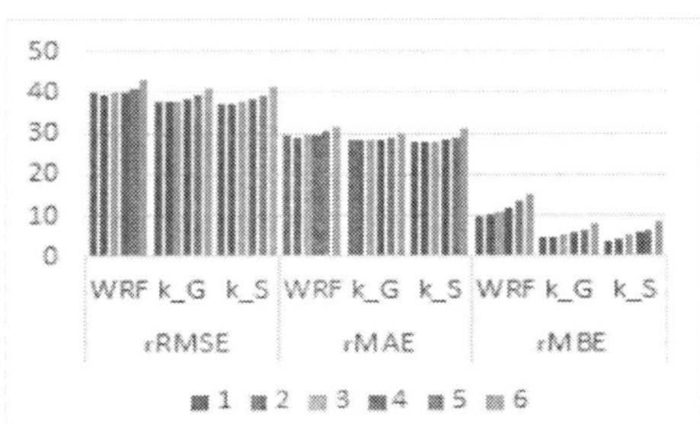

Figure 6 Forecast performance post-processing WRF with ground data k_G and satellite-derived data k_S for horizons of 1, 2, 3, 4, 5, 6 for three months of 2016.

For all sky conditions of horizon from 1 to 6, hourly GHI forecasts from the WRF model were improved using satellite estimations at six ground stations in French Guiana. For all stations merged, we found that results using satellite estimates as observation in Kalman model are similar to these when ground measurements are used to improve WRF forecasting performance for time horizon ranging from 1 to 6 hours, Figure 6.

Hourly post-processing of GHI forecasted with the Kalman filtering using satellite gave better results than direct day ahead WRF output forecasts as in Figure 6. We also found that the post processing methods using satellite derived data perform similarly as the post processing methods using ground measurements for all time horizon. Using satellite-derived data as observation decreases the rMBE of WRF forecasts to more than the half, from 9.93% 15.4% to 3.7% 8,5% for 1H to 6H ahead, respectively. For all stations, we found that using satellite-derived data as observation decreases rRMSE, rMAE and rMBE of WRF from 1 to 6H ahead.

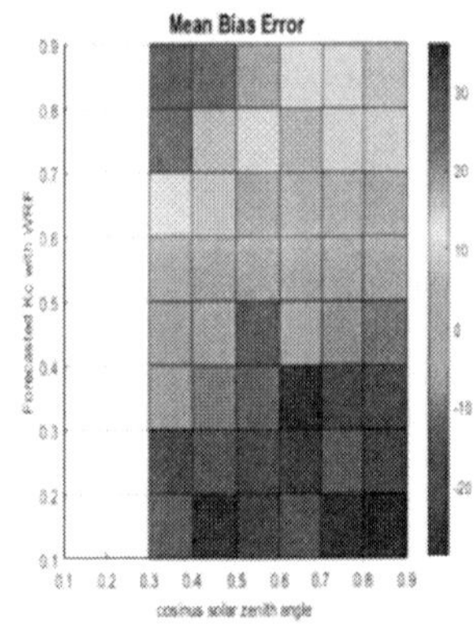

Figure 7 Variation of the bias with kc forecasted by WRF and SZA.

Figure 7 is a graphical representation of the MBE depending on the forecasted clear sky Index kc^* and the cosines of Solar Zenith Angle SZA. Figure 7 allowed us to conclude that when a clear sky is forecasted then the model overestimates GHI. However, if a cloudy sky is forecasted, then the model underestimates GHI. WRF forecast seems reliable for clear days and inaccurate for cloudy days. Accuracy evaluation of WRF gives a rRMSE of 51,39% and a rMBE of 9,67% all stations merged in 2016.

The graphical representation of the MBE depending on the forecasted kc^* and the SZA allowed us to conclude that the Kalman filtering algorithm corrects high underestimation and overestimation for all time horizon. Results in Figure 8 show that the application of the Kalman hybrid method with satellite data offers a good performance for forecasting cloudy skies when the solar angle is high.

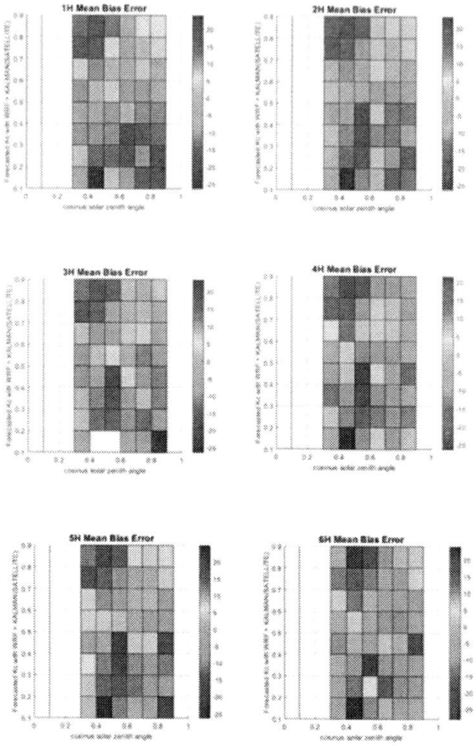

Figure 8 Bias corrected after Kalman filtering using satellite data.

For all stations, hourly GHI forecasts were tested against ground-based irradiation measurements. WRF post-processing method via kalman filter improves WRF forecasts. For all-sky conditions, the post-processing method using satellite-derived data performs close to the post-processing method using ground-based irradiation measurements. Satellite-derived data perform better than ground one for clear sky conditions with low variability. Moreover, we note a better performance with satellite-derived data for high variability and independent of sky condition.

5 CONCLUSIONS

The purpose of the current paper is to study a globally applicable approach to improve day ahead GHI forecasts from WRF model in French Guiana a territory of the ITZ. These forecasts were tested in terms of the rMBE, rMAE and rRMSE, against ground measurements along three representative months of 2016. We apply the post-processing method the Kalman filtering to the WRF forecasts using satellite data as observations instead of ground measurements. Our work focused on rRMSE and rMBE reduction of GHI forecasts hour by hour for the day ahead. Accuracy evaluation showed that satellite data improve WRF forecasts and the post processing using satellite data and measured data had a very close performance by a rRMSE difference 0.04% to 0.06% according to the time horizon. For time horizon of 4h, post processing method with satellite data has slightly better rMBE than post processing using ground measurements for all stations.

As perspective, results could be improved by applying more elaborated pre-processing combined with our method and it will be a good perspective of future work that can improve substantially the forecasting performance. The methodology presented here may be extended to longer time horizons. It would also be interesting to test This method to other sites, climates or to other similar climate regions in the world, at least, accounting for the GOES satellite coverage.

References

[1] B. Elliston and I. MacGill, 'The potential role of forecasting for integrating solar generation into the Australian National Electricity Market', p. 12, 2010.

[2] R. Ahmed, V. Sreeram, Y. Mishra, and M. D. Arif, 'A review and evaluation of the state-of-the-art in PV solar power forecasting: Techniques and optimization', *Renewable and Sustainable Energy Reviews*, vol. 124, p. 109792, May 2020, doi: 10.1016/j.rser.2020.109792.

[3] 'guyane.pdf'.

[4] P. Bauer, A. Thorpe, and G. Brunet, 'The quiet revolution of numerical weather prediction', *Nature*, vol. 525, no. 7567, pp. 47–55, Sep. 2015, doi: 10.1038/nature14956.

[5] P. Lauret, M. Diagne, and M. David, 'A Neural Network Post-processing Approach to Improving NWP Solar Radiation Forecasts', *Energy Procedia*, vol. 57, pp. 1044–1052, 2014, doi: 10.1016/j.egypro.2014.10.089.

[6] M. Diagne, M. David, J. Boland, N. Schmutz, and P. Lauret, 'Post-processing of solar irradiance forecasts from WRF model at Reunion Island', *Solar Energy*, vol. 105, pp. 99–108, Jul. 2014, doi: 10.1016/j.solener.2014.03.016.

[7] S. Pelland, G. Galanis, and G. Kallos, 'Solar and photovoltaic forecasting through post-processing of the Global Environmental Multiscale numerical weather prediction model: Solar and photovoltaic forecasting', *Progress in Photovoltaics: Research and Applications*, vol. 21, no. 3, pp. 284–296, May 2013, doi: 10.1002/pip.1180.

[8] T. Soubdhan, J. Ndong, H. Ould-Baba, and M.-T. Do, 'A Robust Forecasting Framework based on The Kalman Filtering Approach with a Twofold Parameter Tuning Procedure: Application to Solar and Photovoltaic Prediction', *Solar Energy*, p. 18, 2016.

[9] M. Diallo, 'Solar irradiance forecast and assesment in the intertropical zone', p. 161, 2018.

[10] T. Albarelo, 'Estimation de l'Irradiation Solaire sur le Plateau des Guyanes: apport de la Télédétection Satellite', p. 108, 2017.

[11] D. Mouhamet, A. Tommy, A. Primerose, and L. Laurent, 'Improving the Heliosat-2 method for surface solar irradiation estimation under cloudy sky areas', *Solar Energy*, vol. 169, pp. 565–576, Jul. 2018, doi: 10.1016/j.solener.2018.05.032.

[12] ph Héritier, *Le climat guyanais ; petit atlas climatique de la Guyane française*. 2011.

[13] M. Diallo, 'Estimation et prévision de l'éclairement solaire au sol en zone intertropicale', 2019.

[14] M. David, H. M. Diagne, and P. Lauret, 'outputs and error indicators for solar forecasting models', p. 6, 2012.

RECURRENT NEURAL NETWORK FOR SHORT-TIME POWER FORECAST FOR A 540 KWP GRID-CONNECTED PV PLANT INSTALLED AT THE UNIVERSITY OF SÃO PAULO, BRAZIL

Wellington Winícius Ferreira Fonseca[1]
Fernando Ramos Martins[2]
Roberto Zilles[1]

[1]Universidade de São Paulo, Instituto de Energia e Ambiente
[2]Universidade Federal de São Paulo, Departamento de Ciências do Mar

wellingtonwfonseca@usp.br
fernando.martins@unifesp.br
zilles@usp.br

ABSTRACT: A RNN model was developed to predict up to three hours ahead of the production of a 540 kWp grid-connected PV plant installed at the Energy and Environment Institute in São Paulo, Brazil. The RNN's predictors are meteorological and environmental data and a set of power output data of two subset PV plants of 156 kWp each. Two RNN models were set concerning the two typical climate seasons - dry and rainy months. The aerosol optic depth data provided by MERRA-2 reanalysis model was included as a predictor. A brute force algorithm determined the architecture with three hidden layers and thirty neurons in each layer. We evaluated the accuracy of the PV power forecast through Mean Squared Error (MSE), Nominal Root Mean Squared Error (NRMSE), and Skill Score (SS) comparing to a persistence model. Both forecast models resulted in a good fit comparing observed and forecasted data, with NRMSE ranging from 8.01 to 13.88 kWh. The method applied in this work consistently outperformed the persistence model for all evaluated scenarios, especially during rainy months. SS values observed range from 0.72 to 0.91. Finally, the inclusion of aerosol data has slightly improved the forecast model performance during dry months.
Keywords: PV power forecasting; machine learning; artificial neural networks;

1 INTRODUCTION

Energy transition is currently a topic of utmost importance to society given its intrinsic relationship with the quality of life, welfare, the required infrastructure to support socioeconomic development and, finally, with the environmental issues associated with climate change and emission of atmospheric pollutants [1]. During the last few decades, the scientific community and society have been extensively discussing over climate change associated with anthropic action, especially in terms of the increasing emissions of greenhouse gases. The most recent global commitment to the climate seeks to keep the average temperature increase on the planet below 1.5 ° C [2]. In this context, there is an ongoing worldwide effort to increase the participation of renewable energy sources, mostly from intermittent resources, such as solar PV.

Brazil has a great solar PV potential over the entire territory, with low variability throughout the year [3]. Over the last years, grid-connected distributed PV power generation has been growing rapidly in Brazil due to a considerable decrease in implementation costs and the population's overall interest in renewable energy sources [4]. The large scale and rapid implementation of intermittent energy sources, however, poses challenges for power supply operators and integrators. Along with power storage systems, power forecasting models are among the main strategies to support the rapid growth of solar PV power generation, mitigating the management challenges of grid-connected systems. Recently, forecasting techniques, especially those using AI methods, e.g. Artificial Neural Networks (ANNs), have been gaining attention due to high accuracy and rapid development of new methods [5 – 12]. The development of forecast power models that consider local atmospheric and operational conditions is of utmost relevance, especially in dense urban areas. The city of São Paulo is the most populous city in Brazil, presenting a demographic density over 8,100 habitants/km^2 [13]. Thus, São Paulo presents a strong anthropic activity, especially due to high concentration of industries and circulation of vehicles. One particular feature of local atmospheric condition in São Paulo is a high concentration of aerosols during dry months [14].

This paper proposes a machine learning based technique for a short-term power forecast model for a grid-connected 540 kWp PV plant installed at the University of São Paulo in Brazil. We chose a multilayered Recurrent Neural Network (RNN) as the method to forecast the hourly solar PV power up to three hours ahead. The RNN models were developed using data under different atmospheric conditions: dry climate months (from April to September) and for rainy months (from October to March). We also evaluated the influence of atmospheric aerosol as an RNN predictor. Therefore, the approach used in this paper consists of developing two different forecast models. We employed the same RNN topology for both cases.

The remainder of this paper is organized as follows: Section 2 describe the PV system and available data. In Section 3 we present the machine learning method and statistical tools applied in this paper. In section 4 we present and discuss the training procedure results as well as the comparison between forecasted and measured for unprecedented PV output data. Finally, the conclusions of this paper are presented in section 5.

2 PV SYSTEMS AND DATA DESCRIPTION

The 540 kWp PV plant referenced in this work is installed at the University of São Paulo, in Brazil, with coordinates of 23°33'32" S e 46°44'06" W. The PV plant consists of four subset systems. CTPV system with a capacity of 156 kWp and installed over a ground support structure. BAPV system with a capacity of 156 kWp installed on the rooftop of the Brasiliana Library. CRPV system with a capacity of 78 kWp installed on the rooftop of IEE-USP administration building. Finally, BIPV system with a capacity of 150 kWp installed on the IEE-USP parking lot. All subsystems consist of polycrystalline silicon modules of 260 Wp. Fig. 1 and Fig. 2 display the studied systems as well as measurement station (S) installed at the Solar Photovoltaic Laboratory (LSF).

Figure 1: PV systems installed at the Energy and Environment Institute of the University of São Paulo

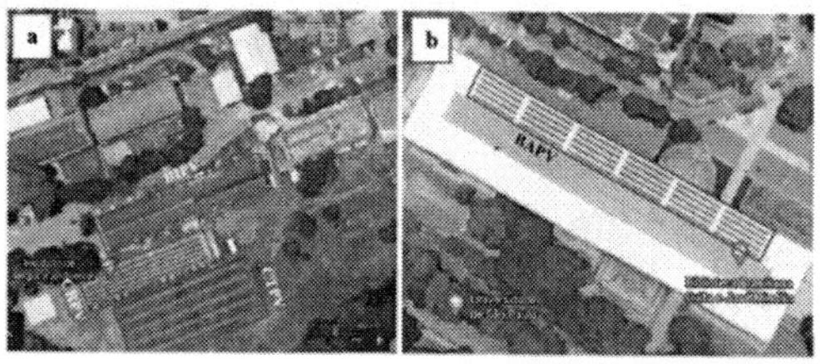

Figure 2: (a) Energy and Environment Institute. Yellow square 'S' is the automatic weather station installed at Solar PV Laboratory (LSF). Red rectangles are PV systems BIPV, CRPV and CTPV. (b) PV system BAPV installed in the rooftop of Brasiliana Library. Source: Adapted Google Earth

The dataset used to develop the RNN forecasting model consists of a time series from January/2017 to November/2020 containing hourly data of Power generation (kWh), Global Horizontal Irradiation (kWh/m^2), relative humidity (%), average environmental temperature (C°), wind speed (m/s) and Aerosol Optical Depth (AOD) at the wavelength of 550 nm provided by MERRA-2 reanalysis database. We split the dataset into dry months (April to September) and rainy months (October to March) to train the forecast model.

3 METHODOLOGY

3.1 Artificial Neural Networks (ANNs)

Artificial Neural Networks are a powerful statistical tool composed of simple processing units, commonly called neurons, which compute mathematical functions from a provided time series [6]. The topology of an ANN consists of an input layer, one or more hidden layers and one output layer. Each neuron receive information from the previous layer and compute the activity level (v_j) (1) by summing weighted individual inputs ($w_{ji}x_i$) along with their biases (b_j) and transfer this processed information to neurons on the next layer through an activation function ($\varphi(v_j)$), e.g. hyperbolic tangent (2). Since ANNs are inspired in the biological brain, the process of transferring information through neurons is commonly called synapse.

$$v_j = \sum_{i=1}^{n} w_{ji}x_i + b_j \quad (1)$$

$$\varphi(v_j) = \frac{1 - e^{-v_j}}{1 + e^{-v_j}} \quad (2)$$

A conventional Feedforward ANN can solve complex non-linear problems [15], but cannot retain information from previous iterations. In this sense, Recurrent Neural Networks (RNNs) have been gaining attention in solving problems where previous states of a time series influence the current state [7 – 12]. Therefore, we chose a multilayered RNN as the machine learning approach to short-term forecast the hourly solar PV power up to three hours ahead.

3.2 RNN structure

A brute force algorithm determined the best topology being a multilayered RNN with three hidden layers and thirty neurons each (Fig. 3). The study also evaluated the influence of atmospheric aerosol as a predictor for the forecast model. Therefore, the approach used in this work consists in developing two different forecast models. The first including AOD as a predictor and the latter without AOD data. We applied the same method and topology for both models. The forecasting RNN models were trained using a combination of power generation time series from CTPV and BAPV and validated using power generation data from BIPV.

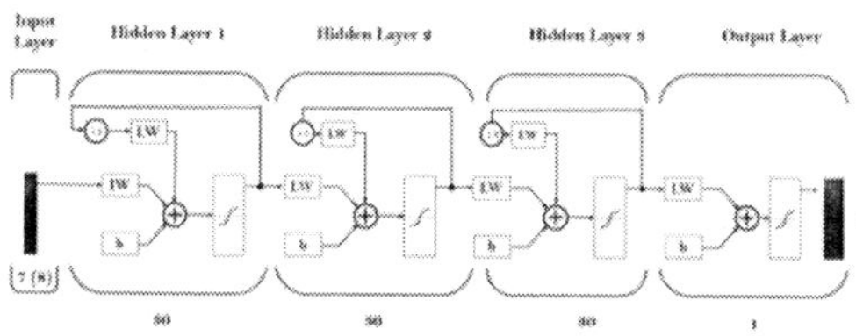

Figure 3: Recurrent Neural Network Topology. The RNN has three hidden layers with thirty neurons each. The Input vector has seven or eight predictors depending on whether aerosol is disregarded or not, respectively.

3.3 Training procedure

The forecasting model was developed using the software MATLAB. We used the Resilient Backpropagation (RP) algorithm to train the RNN, being well suited for time series pattern recognition and lighter computational requirements. A delay of up to three iterations was considered in each hidden layer of the RNN. The data records were divided into training, validating, and testing data sets in a proportion of 60%, 20% and 20%, respectively. The Training set adjusts weights and biases in each iteration. The Validation set is used as a stopping criterion when the model fails to improve performance after a given number of consecutive iterations. Lastly, the Test set is used to measure the model's performance for unprecedented data. We used the hyperbolic tangent (2) as the activation function. Mean squared error (3) is the performance indicator used during the training procedure.

3.4 Statistical validation of RNNs

In order to evaluate the performance of the proposed RNN models, we compared the forecasted PV power versus the observed power production of the PV system. The statistical indexes used in this study are Mean Squared Error (MSE) (3), Nominal Root Mean Squared Error (NRMSE) (4), Skill Score (SS) (5) and Pearson's Correlation (R).

$$MSE = \frac{1}{N}\sum_{i=1}^{N}(P_{out} - P_{forecast})_i^2 \quad (3)$$

Where P_{out} is the observed PV generation and $P_{forecast}$ is the forecasted power output from the RNN model, and N is the number of available data records.

$$NRMSE = P_{kWp} \cdot 1h \cdot \sqrt{\frac{1}{N}\sum_{i=1}^{N}(P_{out} - P_{forecast})_i^2} \quad (4)$$

Where P_{kWp} is the maximum power capacity of the referenced PV system, given in kWh. The NRMSE allows to evaluate the uncertainty of the RNN model in terms of electric generation's unit.

$$SS_{MSE} = 1 - \frac{MSE_{forecast}}{MSE_{persistence}} \quad (5)$$

Where $MSE_{forecast}$ is the MSE for the forecasted PV output provided by the RNN model compared to the observed PV power generation, and $MSE_{persistence}$ is the MSE for the forecasted PV output provided by a persistence model. We developed a smart persistence model [16] as displayed in (6).

$$Persistence = Kt_i \cdot TOA_{i+h} \cdot \eta \cdot A_{mod} \cdot n_{mod} \quad (6)$$

Where Kt_i is the Clearness Index at the time i, TOA_{i+h} is the Irradiance at the top of atmosphere calculated h hours ahead. η is the module efficiency (15.92%). A_{mod} is the area of an individual module (1.63 m^2) and n_{mod} is the number of modules installed for the referred PV system.

4 RESULTS AND DISCUSSION

4.1 RNN training results

We trained both forecasting models under different atmospheric conditions over varying time horizons using CTPV and BAPV datasets. We observed a MSE of 2.5 – 6.9 x 10^{-3} for dry months and 3.5 – 8.5 x 10^{-3} for rainy months.

The training NRMSEs were calculated in reference to a 156 kWp PV system. We observed NRMSEs of 7.86 – 12.33 kWh in the dry months if AOD was included as RNN predictor, and 8.4 – 12.99 kWh otherwise. In the rainy months, we achieved NRMSE values between 9.54 – 14.43 kWh if AOD data was included and 9.27 – 13.64 kWh otherwise. The models performed considerably well, with an overall error of 5.04% – 9.25% compared to installed capacity.

The R index measures the linear correlation between forecasted value and measured PV output. R values equal to positive one means a perfect positive correlation between variables. We observed R of 0.862 – 0.953 in the dry season, and R of 0.803 – 0.927 in the rainy season.

As expected, the errors observed during rainy months were higher than those observed during dry months due to more cloud conditions variability associated with the rainy season. However, the RNN models still performed considerably well during the rainy season, with correlation factor above 0.8 for scenarios and error below 10%, indicating a good performance in operational situations when forecasting is more challenging.

In general, based on the SS, the inclusion of aerosol data has improved the precision of the RNN model for dry months, while decreasing precision during wet months.

4.2 Performance Analysis of RNN forecasting models

In order to evaluate the performance of the proposed RNN forecasting models, we prepared an experimental setup using unprecedented data from the BIPV subsystem. As described in section II, BIPV has an installed capacity of 150 kWp located close to the automatic weather station S. We compared forecasted versus observed BIPV PV power output, in kWh, for different time horizons as well different atmospheric conditions. In Fig. 4 and Fig. 5, we present a scatter plot colored with Kernel's Density analysis. Each line displays the comparison of the forecasting outputs for one, two, and three hours ahead, respectively. The left column presents a comparison when AOD is used to feed the RNN models. The right column presents the comparison results for the case without AOD data.

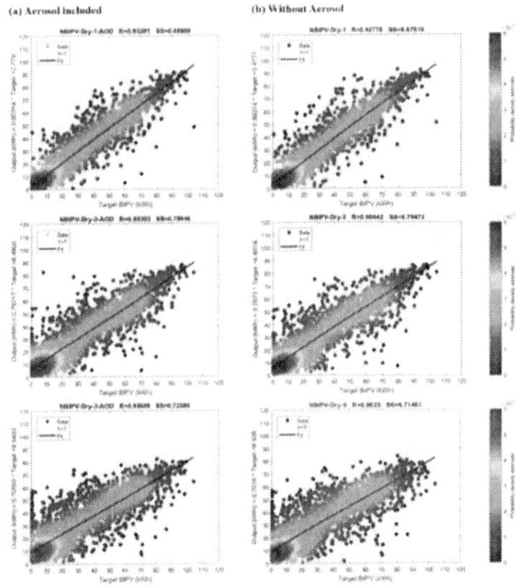

Figure 4: Measured vs. Forecasted BIPV power in kWh for 1h (first row), 2h (second row) and 3 hours (third row) ahead during dry months using AOD data (left column) and without AOD data (right column).

The SS index measures the forecasting ability of a model in comparison to a reference model. Positive SS values mean a gaining in performance, while negative values mean the opposite. In this paper, we used a smart persistence model (6) as the reference model. The proposed RNN models outperformed persistence for all atmospheric and climate conditions.

We observed higher density along the main diagonal for both atmospheric conditions, indicating the RNN models output is similar to measured data (Target). During dry months, the RNN models using AOD data achieved 8.01, 10.87, 13.16 kWh for NRMSE deviations at one, two, and three hours ahead, and SS values were 0.888, 0.799, and 0.726, respectively. In cases of AOD was not used to feed the RNN models, the NRMSE values were from 8.42 to 13.43 kWh, and SS from 0.876 to 0.715 for the three forecast horizons.

The deviations increased during rainy months, as shown in Fig.5. Using AOD as an RNN predictor, the NRMSE deviations were 10.07, 12.32, 13.88 kWh for one, two, and three hours ahead, respectively. The SS values achieved 0.89, 0.848, and 0.81 for the same time horizons.

As observed in the previous section, the performance improved in rainy months without the AOD input. In this case, the NRMSE values were 9.93, 12.31, 13.63 kWh, and the SS achieved 0.9, 0.848, 0.82 for one, two, and three hours ahead.

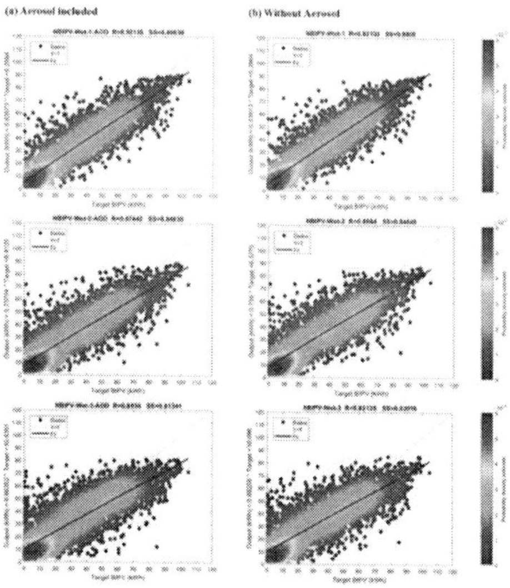

Figure 5: Measured vs. Forecasted BIPV power in kWh for 1h (first row), 2h (second row) and 3 hours (third row) ahead during wet months using AOD data (left column) and without AOD data (right column).

The results observed for the BIPV subsystem behaved in compliance with what was observed during training, validation, and testing, even though BIPV production data was not considered during the training procedure. These results indicate that the proposed machine learning method demonstrated to be fairly robust and performed well, presenting a potential to be fit to forecast different PV systems in proximity area, provided under similar environmental and operational conditions. The inclusion of reanalysis modeled AOD resulted in a slight enhancement of the RNN model performance during the dry season for all evaluated scenarios using the BIPV dataset. Considering aerosol is the most significant solar resource modulator in clear sky conditions [14], the results presented in this paper highlight the potential importance of the inclusion of this atmospheric parameter, especially in dense urban areas. However, it is important to reinforce that the observations are bound to modeled AOD values and further assessment is welcomed and encouraged.

5 CONCLUSIONS

This paper has proposed a forecasting model for a grid-connected 540 kWp PV system installed at the University of São Paulo, in Brazil. The proposed RNN models were trained using local power output data, meteorological, atmospheric and environmental data collected nearby the PV systems. A further assessment using MERRA-2 reanalysis modeled aerosol data was carried out. As expected, the RNN model provided PV power forecasts presenting higher NRMSEs during rainy months. However, the proposed model has presented an SSMSE for 3 hours ahead higher than presented in dry months, indicating better performance than persistence even on conditions where forecasting is usually more complex. Overall, the aerosol data has improved the model performance during dry months by 0.9% for validating data. This result raises attention to the importance of this atmospheric variable, especially in highly polluted urban areas with high potential for PV power. The authors welcome and recommend further studies in different areas and use surface measurements for further assessment of the influence of aerosol on power forecast accuracy.

ACKNOWLEDGEMENT

The authors thank the Energy and Environment Institute (IEE-USP) and the Solar Photovoltaic Laboratory (LSF) for providing the data used in this research. Also, thanks are due to Coordination for the Improvement of Higher Education Personnel (CAPES) for financial support of this research. Thanks are due to CNPq for research fellowship for Fernando Martins and Roberto Zilles.

REFERENCES

[1] J. Goldemberg,; O. Lucon. "Energia e meio ambiente no Brasil". Estudos Avançados, 21(59), 2007, pp. 7–20. https://doi.org/10.1590/S0103-40142007000100003.

[2] UNFCCC. Conference Of The Parties (COP). Paris Climate Change Conferece-November 2015, COP 21. Adoption of the Paris Agreement. Proposal by the President., v. 21932, n. December, p. 32, 2015.

[3] E.B. Pereira; F.R. Martins; A.R. Gonçalves; R.S. Costa; F.L. Lima; R. Rüther; S.L. Abreu; G.M. Tiepolo; S.V. Pereira; J.G. Souza. "Atlas Brasileiro de Energia Solar", second ed. Brazilian Institute for Space Research, 2017, São José dos Campos, Brazil. DOI: 10.34024/978851700089.

[4] P.V. Gomes, N.K. Neto, L. Carvalho, J. Sumaili, J.T. Saraiva, B.H. Dias, V. Miranda, S.M. Souza. "Technical-economic analysis for the integration of PV systems in Brazil considering policy and regulatory issues", Energy Policy, vol. 115, pp. 199-206. 2018. https://doi.org/10.1016/j.enpol.2018.01.014.

[5] F.J.L. Lima, F. R. Martins, E. B. Pereira, E. Lorenz, D. Heinemann. "Forecast for surface solar irradiance at the Brazilian Northeastern region using NWP model and artificial neural networks", Renewable Energy, Volume 87, Part 1, 2016, pp. 807-818, https://doi.org/10.1016/j.renene.2015.11.005.

[6] R.A. Guarnieri. "Emprego de redes neurais artificiais e regressão linear múltipla no refinamento das previsões de radiação solar do modelo ETA". Dissertation (MSc.) – Meteorology Graduate Program of National Institute of Space Research (INPE). São José dos Campos, 2006.

[7] M. Malvoni; M.G. De Giorgi; P.M. Congedo. "Photovoltaic power forecasting using statistical methods: impact of weather data". IET Science, Measurement & Technology, 8(3), 90–97. (2014). doi:10.1049/iet-smt.2013.0135.

[8] A. Mellit; A. M. Pavan. "A 24-h forecast of solar irradiance using artificial neural network: Application for performance prediction of a grid-connected PV plant at Trieste, Italy". Solar Energy. Vol. 84 Issue 5, pp. 807–821, 2010, doi:10.1016/j.solener.2010.02.006.

[9] E. G. Kardakos, M. C. Alexiadis, S. I. Vagropoulos, C. K. Simoglou, P. N. Biskas and A. G. Bakirtzis, "Application of time series and artificial neural network models in short-term forecasting of PV power generation," 2013 48th International Universities' Power Engineering

Conference (UPEC), Dublin, Ireland, 2013, pp. 1-6, doi: 10.1109/UPEC.2013.6714975.

[10] F. Almonacid, P.J. Pérez-Higueras, E. F. Fernández, L. Hontoria, "A methodology based on dynamic artificial neural network for short-term forecasting of the power output of a PV generator, Energy Conversion and Management, Vol. 85, 2014, pp. 389-398, https://doi.org/10.1016/j.enconman.2014.05.090.

[11] K. Dahmani, R. Dizene, G. Notton, C. Paoli, C. Voyant, M.L. Nivet. "Estimation of 5-min time-step data of tilted solar global irradiation using ANN (Artificial Neural Network) model", Energy, Vol. 70, 2014, pp. 374-381, https://doi.org/10.1016/j.energy.2014.04.011.

[12] A. Yona, T. Senjyu, T. Funabashi and C. Kim, "Determination Method of Insolation Prediction With Fuzzy and Applying Neural Network for Long-Term Ahead PV Power Output Correction," in IEEE Transactions on Sustainable Energy, vol. 4, no. 2, pp. 527-533, April 2013, doi: 10.1109/TSTE.2013.2246591.

[13] IBGE – Instituto Brasileiro de Geografia e Estatística. Available at < https://www.ibge.gov.br/cidades-e-estados/sp/sao-paulo.html>. Date of access 02/03/2021.

[14] T.T. de Almeida Albuquerque, M. de Fátima Andrade, R.Y. Ynoue. "Characterization of atmospheric aerosols in the city of São Paulo, Brazil: comparisons between polluted and unpolluted periods". Environ Monit Assess 184, 969–984. (2012). https://doi.org/10.1007/s10661-011-2013-y

[15] S. Sobri; S. Koohi-Kamali; N.A. Rahim. "Solar photovoltaic generation forecasting methods: A review". Energy Conversion and Management, 156(), 459–497. (2018). doi:10.1016/j.enconman.2017.11.019

[16] J. Antonanzas; N. Osorio; R. Escobar; R. Urraca; F.J. Martinez-de-Pison; F. Antonanzas-Torres. "Review of photovoltaic power forecasting". Solar Energy, [S.L.], v. 136, p. 78-111, out. 2016. Elsevier BV. http://dx.doi.org/10.1016/j.solener.2016.06.069.

POST PROCESSING OF DAY-AHEAD SOLAR IRRADIANCE FORECAST USING SATELLITE DERIVED DATA IN FRENCH GUIANA

Jérémy Macaire[1], Maha Salloum[1], Jessica Bechet[1], Sara Zermani[1], Laurent Linguet[1]
[1]Université de la Guyane – UMR Espace-Dev
jeremy.macaire@gmail.com ; laurent.linguet@univ-guyane.fr ; sara.zermani@univ-guyane.fr ; dbedbh@yahoo.com ; jessica.bechet@univ-guyane.fr

ABSTRACT: The aim of this work is to improve day-ahead hourly Global horizontal irradiance (GHI) forecast from Weather Research and Forecasting (WRF) model using post-processing techniques trained with satellite estimates of hourly irradiance. Generally, post-processing techniques are trained with ground measures of the irradiance, however, in French Guiana, only 6 ground stations are available. To face this issue, we investigate the use of satellite-derived GHI data, which are available over the whole territory, instead of ground data to fit different post-processing techniques to correct the bias of WRF GHI forecasts. We compared two post-processing techniques, Artificial Neural Network and Support Vector Machine, both trained with ground-based solar irradiance data first, and then with satellite-derived solar irradiance data. We showed that post-processing techniques trained with satellite-derived data improved the accuracy of the model. The relative Root Mean Squared Error (rRMSE) and the relative Mean Bias Error (rMBE) decrease, respectively, from 48.69 % to between 40.85 and 41.81 % and from 19.14 % to between -1.52 and -8.76 %. We also showed that using satellite-derived data degrades only slightly the forecast accuracy compared to the techniques using ground data (around 1 % of difference for the rRMSE). Moreover, Support Vector Machine post-processing model overperformed the Artificial Neural Network model.

Keywords: Solar forecasting, post-processing, satellite-derived estimation, ground solar irradiance measures, machine learning

1 INTRODUCTION

Energetic transition leads to an increase of the integration of the photovoltaic power in the energy mix. This also increases the need of accurate day-ahead solar irradiance forecast.

Physical forecasting models are far more performing than statistical models or machine learning models for day-ahead solar irradiance forecasts. Then, to obtain them, Numerical Weather Prediction (NWP) models are used.

However, NWP outputs are still inaccurate because they cannot exactly describe all the physical processes in the atmosphere (Rincón et al. 2018 [1]). So, they need to be improved with post-processing techniques. Post-processing methods estimate the bias of NWP and then remove it to correct the forecasts.

Several papers already showed post processing techniques efficiency to correct day ahead solar irradiance forecasts from WRF. Lauret et al. [2] proposed an Artificial Neural Network (ANN) as post-processing technique in Reunion Island which improved accuracy of the forecast and gave a bias of -0,25 %. Rincon et al. [3] combined a Kalman filter and a MOS to post-process WRF in Paraguay which decreased the relative mean bias error from 21 % to -0,7 %. Diagne et al. [4] proposed a Kalman filter post-processing technique which greatly improves forecast accuracy. However, according to the knowledge of the authors, most of the post-processing techniques use measured ground data because they are the most accurate solar data. Nevertheless, they are not available on any points in the world.

Nevertheless, in French Guiana, only 6 ground stations are available (**Fig. 1**).

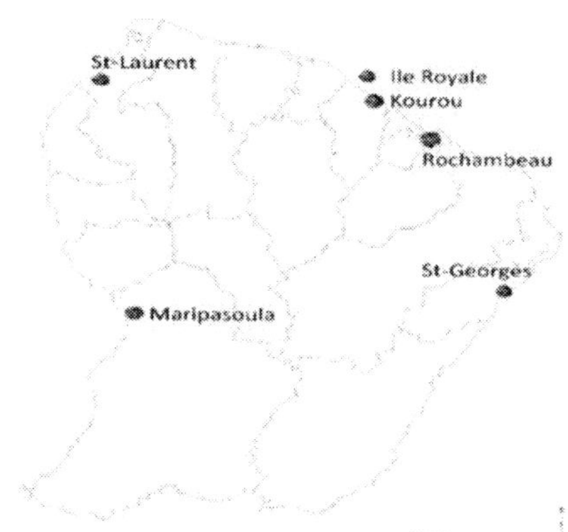

Figure 1: Ground measurements stations of solar irradiance in French Guiana

The objective is to show that the use of satellite data in a MOS could improve accuracy of the WRF day-ahead forecasts.

As first post-processing method we use the model proposed by Lauret et al. 2014. It takes two features: clear sky index forecasted by WRF (kc*) and cosinus of solar zenithal angle (SZA). The second post-processing method is an Optimizable Support Vector Machine (SVM). Both were trained with ground data irradiance as targets at first, then with satellite-derived data.

Firstly, we will present the data used in this paper. Then, we will explain the methods used. Next, we will discuss the results. We will finish with a brief conclusion.

2 DATA

2.1 WRF Outputs

We chose WRF as NWP because it is an open-access model with a huge community. Over 30 000 researchers use WRF around the world. To feed our local Numerical Weather Prediction model, the forecasts of a global model are needed to be used as boundary conditions. We chose the Global Forecasting System (GFS) model as input. GFS has the advantage to be in open access and to be more efficient than the Integrated Forecasting System (IFS). We collected the data over 2016 on the University Corporation for Atmospheric Research (UCAR) site.

The physical parameterization of WRF was taken from Diallo et al. 2018 [5], which adapted the model WRF to be more accurate for solar forecasting in French Guiana. We only changed simulation parameters (timestep, spatial resolution, ...) to be able to run the model on local servers in reasonable time limit.

However, running WRF remains time consuming. So, at this time we only have WRF outputs for 6 months from 2016 (February, March, April, May, June, September).

2.2 Satellite-derived and ground-based data

Concerning satellite data, GHI estimates were derived from GOES 13 satellite images according to the developed method in Albarelo et al. 2017 [6]. We downloaded two images (slots) by hour from GOES 13, one at -20 minutes and the other at - 50 minutes from hour h. So, to get the estimation for hour h we calculate the mean value.

Ground measures were extracted from Meteo France.

3 METHODS

3.1 Post-processing

Post-processing is a technique used to estimate the bias of the GHI forecasts. We used 2 features to estimate this bias: the Clear Sky index, Kc, and the cosinus of the Solar Zenithal Angle, cos SZA.

The **Fig. 2** shows a schema of the post-processing method.

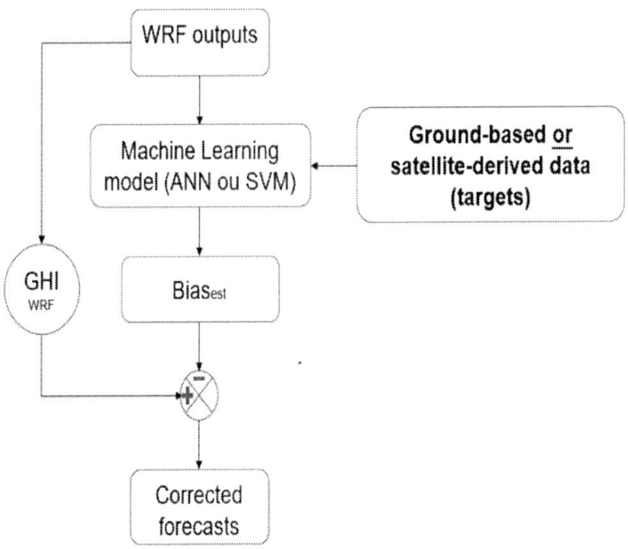

Figure 2: Schema of post-processing method

The principle is to train a machine learning to estimate the bias according the two features. Then, we correct the forecasts by removing the bias estimated from the WRF outputs (equation 1).

$$GHI_C = GHI_{WRF} - Bias_{est} \quad (1)$$

With GHIc, solar irradiance corrected values (W/m²), GHI_{WRF}, WRF solar irradiance forecast (W/m²), and $Bias_{est}$, the estimated bias.

3.2 Artificial Neural Network

ANN are non-linear fitting models. They are able to recognize patterns in data and are now popular in solar forecasting (Diagne et al. 2013 [7]). ANNs have a neuronal structure, with an input layer, which only collects features, one or some hidden layers, which processes the features, and the output layer, which gives the result.

The ANN we used is the same as the model proposed by Lauret et al. 2014. It is an ANN with 4 hidden neurons, with hyperbolic tangent as activation function and Bayesian regularization as training function. The structure of this ANN is showed in **Fig. 3**.

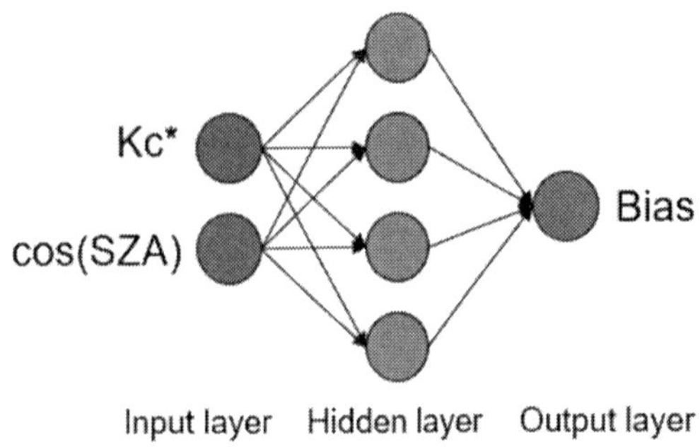

Figure 3: Structure of the ANN

3.3 Support Vector Machine

We also used an Optimizable Support Vector Machine (SVM). SVM are supervised machine learning models which are efficient for both classification problems and non-linear regression problems. The optimizable SVM finds itself, during the training phase, its best hyperparameters (MATLAB Documentation [8]). The best kernel function is a gaussian, when trained with ground data, and a polynomial function when trained with satellite data.

3.4 Validation

In this case study, six months of GHI forecasts obtained using WRF are corrected using both ground-measurement and satellite-derived data. The two post-processing methods are tested across six stations (**Fig. 1**) for 6 months of 2016 (February, March, April, May, June, September). The validation was made over 30 days randomly chosen among the 6 months for the six stations. Finally, 83 % of the data were used for the training and 17 % for the validation.

We use three metrics to compare the accuracy of the models. The relative root mean square error (rRMSE), which penalizes great errors, the relative mean bias error (rMBE), and the relative mean absolute error (rMAE).

$$rRMSE = \frac{\sqrt{\frac{1}{N} * \sum_i (y_i - x_i)^2}}{\sum_i x_i} * N * 100$$

$$rMBE = \frac{\sum_i (y_i - x_i)}{\sum_i x_i} * 100$$

$$rMAE = \frac{\sum_i |y_i - x_i|}{\sum_i x_i} * 100$$

With y, solar irradiance forecasted values (W/m²), x, measured values of solar irradiance (W/m²), and N, amount of data.

4 RESULTS

Table I: Metrics results of the different models

	rRMSE%	rMBE%	rMAE%
Satellite estimations	22.34	-3.24	16.53
WRF	48.69	19.14	35.96
WRF+Lauret PP	40.23	-4.72	31.75
WRF+Lauret PP (sat)	41.81	-8.76	33.35
WRF+SVM	40.04	-1.52	30.97
WRF+SVM (sat)	40.85	-4.86	31.94

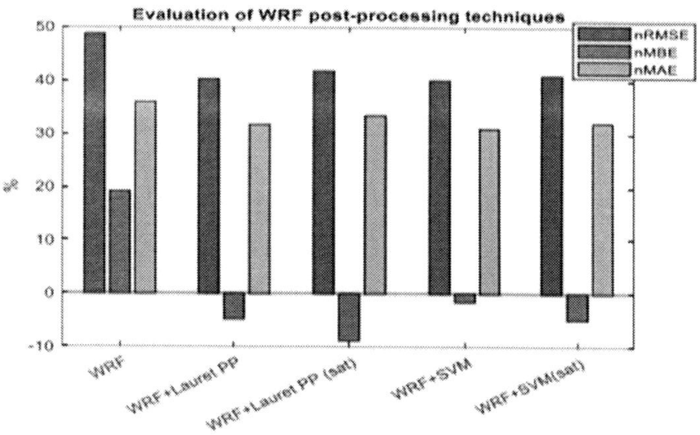

Figure 4: Comparison between the different models

First, we will discuss about the metrics results of the different models tested presented in tab 1 and fig. 2.

The Lauret et al. post-processing decreases rRMSE (48,69 % to 40,23 %) and rMAE (35,96 % to 31,75 %) compared to WRF alone. The rMBE is greatly improved (19,14% to -4,72 %) however, the bias is overestimated by the model which leads to an underestimation of the forecast. When the model is trained with satellite data, the rRMSE and the rMAE compared to WRF alone decreases (48,69 % to 41,82 % and 35,96 % to 33,35 % respectively). Moreover, trained with the data satellite, the model also overestimates the bias which leads to a rMBE of -8,76 %. The underestimation is greater in this case than the case with ground data because of the own bias of satellite estimates (-3,24 %).

The SVM post-processing greatly decreases rRMSE (48,69 % to 40,04 %) and rMAE (35,96 % to 30,97 %) compared to WRF alone. This is the most efficient post-processing technique. The SVM trained with satellite data improves the forecasts accuracy compared to WRF alone and only slightly degrades performance compared to the SVM trained with ground data. It shows we can use satellite data as target to train WRF post-processing models.

Moreover, we can notice that regardless the model is trained with ground measures or satellite-derived data, the SVM is more efficient than the ANN. The SVM shows its efficiency particularly for the rMBE (Between -4.72 and -8.76% for the ANN against between -1.52 and -4.86% for the SVM). It is certainly due to the fact that contrary to the SVM, the structure of the ANN is not optimal.

Now, we will discuss about the dispersion of the forecasts against the observations.

The **Fig. 5** shows the comparison between forecasts and observations for WRF forecast and for WRF forecast with postprocessing. We notice that for all post-processing techniques, the highest solar irradiances are underestimated. Moreover, the lowest solar irradiances are overestimated and forecasts are very scattered. It is due to the variability of cloudy hours.

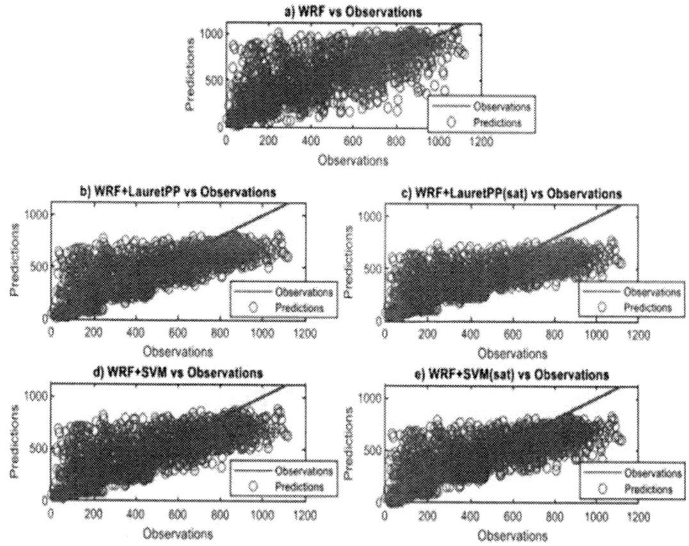

Figure 5: Scatter plot between observations and day-ahead forecasts of hourly irradiance from a) WRF, Lauret post-processing trained with b) ground data, c) satellite data, SVM post-processing trained with d) ground data, e) satellite data.

The **Fig. 6** shows the mean bias error according to kc* and cos (SZA) for each model. We notice that for all post-processing techniques, when kc* is near to cloudy hours and cos (SZA) is near to 1 (which means hours in middle of the day), the models failed to fit correctly the bias.

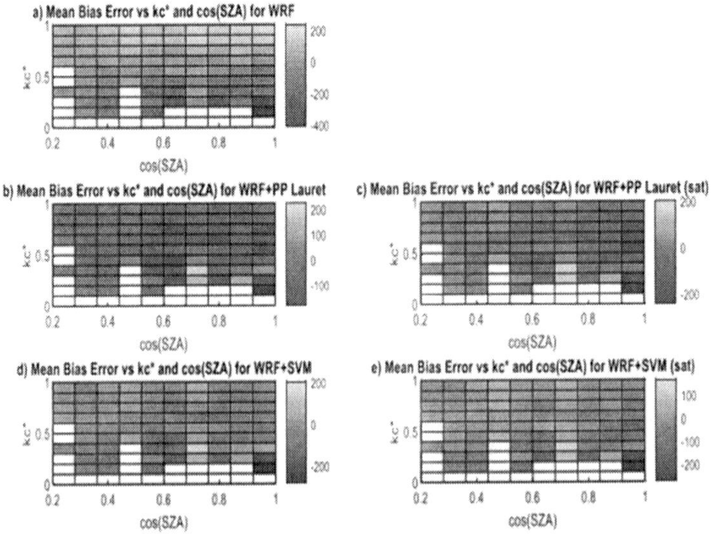

Figure 6: Mean bias error vs kc* and cos (SZA) from a) WRF, Lauret post-processing trained with b) ground data, c) satellite data, SVM post-processing trained with d) ground data, e) satellite data.

5 CONCLUSION

In conclusion, we showed that two post-processing techniques trained with satellite data have proven their efficiency greatly improving accuracy of WRF forecast, although slightly degrading accuracy compared to the post-processing techniques trained with ground data. Moreover, SVM overperformed the ANN proposed by Lauret et al., as post-processing technique. It is certainly due to the fact that contrary to the SVM, the structure of the ANN is not optimal. However, all methods used overestimate the bias of the highest irradiance values, which lead to an underestimation of the irradiance. The underestimation is greater with models trained with satellite-derived data, as their own bias is added. We also notice, for cloudy hours, a strong scatter of the forecasts, which means post-processing struggles to model the bias. So, some improvements could be added to this work. In a future work, we will study the contribution of a clustering method and the use of more features in order to improve post-processing techniques for cloudy hours and avoid underestimation for the highest irradiation values. The use of machine learning with more complex structure will also be considered.

6 REFERENCES

1. Rincón, A., Jorba, O., Frutos, M., Alvarez, L., P. Barrios, F., A. González, A.: Bias correction of global irradiance modelled with weather and research forecasting model over Paraguay. Solar Energy 170, 201-211(2018).
2. Lauret, P., Diagne, M., David, M.: A Neural Network Post-Processing Approach to Improving NWP solar Radiation Forecasts. Energy Procedia 57, 1044-1052 (2014)
3. Rincon, A., Jorba, O., Frutos, M., Alvarez, L., Barrios, F., Gonzalez, J.: Bias correction of global irradiance modelled with weather and research forecasting model over Paraguay. Solar Energy 170 (2018), 201-211
4. Diagne, M., David, M., Boland, J., Schmutz, N., Lauret, P.: Post-processing of solar irradiance forecasts from WRF model at Reunion Island. Solar Energy 105, 99-108(2014).
5. Diallo, M. 2018 : Estimation et prédiction de l'ensoleillement en zone intertropicale.
6. Albarelo, T. 2017 : Estimation de l'irradiation solaire sur le plateau des Guyanes : apport de la Télédétection satellite, 108.
7. Diagne, M., David, M., Lauret, P., Boland, J., Schmutz, N.: Review of solar irradiance forecasting methods and proposition for small-scale insular grids. Renewable and Sustainable Energy Reviews 27, 65-76(2013).
8. MATLAB Documentation on Optimizable SVM : https://fr.mathworks.com/help/stats/choose-a-classifier.html#bunt0oj

NOVEL INTRADAY PHOTOVOLTAIC PRODUCTION FORECASTING ALGORITHM USING DEEP LEARNING ENSEMBLE MODELS

Spyros Theocharides[1, *], George Makrides[1], Marios Theristis[2], George E. Georghiou[1]
[1]PV Technology Laboratory, FOSS Research Centre for Sustainable Energy,
Department of Electrical and Computer Engineering, University of Cyprus, Nicosia, 1678, Cyprus
[2]Sandia National Laboratories Albuquerque, NM 87185, USA
*corresponding author email: theocharidis.spyros@ucy.ac.cy

ABSTRACT: The high penetration level of photovoltaic (PV) systems to the utility grid and the intermittent nature of the generated power, introduces new challenges for the stability of electricity grids. The scope of this study is to present a novel intraday (up to 5-hours ahead forecasting) PV power production forecasting algorithm that is fully data-driven and based on machine learning ensemble principles. The methodology followed to develop the forecasting ensemble comprised of a cluster of Bayesian neural networks (BNN) that were trained with different exogenous variables, lag structures and estimation window duration. Preliminary obtained results demonstrated that the resulting hour-ahead forecasting ensemble model showed accuracies beyond the state-of-the-art (<3.3%). Finally, the implemented ensemble was able to forecast 5 points in time (5-hours ahead) with an accuracy <8%.
Keywords: artificial neural networks, forecasting, performance, photovoltaic, power.

1 INTRODUCTION

During the last decades, electricity system operation has been upgraded involving PV production forecasts and most commonly adopted PV production forecasting approaches are based on time series analysis techniques [1].

In addition, physical models for PV production forecasting have been introduced [2], [3] but their ability to forecast the power output of PV systems is limited since data related to system characteristics and behaviour should be provided. Therefore, a share of research is devoted to the development of more sophisticated and flexible prediction techniques using data driven models based on machine learning algorithms [4]–[9]. To develop a PV power forecasting model with identical data, weather classification and machine learning techniques may be performed [10], [11]. Moreover, ensembled models comprised of combination of features of a physical model combined with artificial neural networks (ANNs) were presented elsewhere [7], [12].

Furthermore, the system's behaviour can be estimated by processing recent PV operational data-sets. using the classical approach of the feedforward neural network (FFNN). This method was inspired by the method of biological neurons and is the simplest type of ANNs. The classical approach of the FFNN with an input, a hidden and an output layer of linear and non-linear activation functions can be viewed as a convenient way to forecast the power output of PV systems.

In addition, ANNs are efficient for online training due to their capability of reflecting the information of new instances on a model by changing the weight values only.

Another approach is to utilise Numerical Weather Prediction (NWP) models to forecast weather variables [3], [13]–[15]. NWP models deliberates the atmosphere as an fluid thus the concept of weather forecasts is to parametrise the state of the fluid at a certain time (t) by utilising the fluid- and thermo-dynamic equations to forecast the state of the fluid at time (t + n) [16]–[18]. The utilisation of mathematical equations and models to simulate the atmosphere are not accurate, thus the NWP methods are associated with forecasting errors and biases [19], [20].

Previous work in the area of short-term (30 minutes to 6 hours ahead) solar irradiance forecasting utilising satellite data and sky images demonstrated errors of 20% to 25% [21]. Another study that focused on the development of a mid-term solar forecasting model using meteorological parameters as inputs to an ANN achieved root mean square error (RMSE) values in the range of 10 to 15% [22].

In addition, parametric models for forecasting PV production have already been developed [2], [3] but their capacity to forecast the power output of PV systems is a critical task since accurate knowledge of system characteristics and behaviour should be provided. Therefore, a huge share of research is devoted for the development of more sophisticated and flexible prediction techniques using non-parametric models based on machine learning algorithms [4], [6], [8], [10]. Moreover, numerous studies have used weather classification combined with machine learning techniques in order to feed the models with identical data at the training stage [11].

In this work, an intraday (1 to 5 hours ahead) PV power production forecasting utilising an ensemble of data-driven machine learning models was introduced. Specifically, 1-year data from a reference poly-crystalline silicon (poly-c-Si) PV system in Cyprus were used for training and testing the model and the forecasting performance results showed that the application of this method improves the intraday PV production forecasting accuracy.

2 EXPERIMENTAL SETUP

The DER-Grid smart facility of the FOSS Research Centre for Sustainable Energy, at the UCY in Nicosia, Cyprus is a flexible and scalable testing and demonstration facility for smart grid and other advanced energy technologies. The infrastructure includes a test-bench grid-connected PV system used for the forecasting analysis of this study. The test-bench PV system comprises of 5 poly-crystalline Silicon (poly-c-Si) PV modules of rated power 235 Wp that are installed in an open-field arrangement at the optimal yearly energy yield inclination angle of 27.5°.

Furthermore, the PV system was connected to a state-of-the-art data acquisition (DAQ) system, used for the monitoring and storage of meteorological and PV operational data. The DAQ system is connected to high accuracy meteorological and PV operational measuring sensors. The performance of the system and the prevailing weather conditions were recorded according to the requirements set by IEC 61724 [23]. Specifically, the meteorological measurements include the in-plane global irradiance (G_I), relative humidity (RH), wind direction (W_a) and speed (W_s), as well as ambient temperature (T_{amb}). The PV system operational measurements include maximum power current (I_{mp}), voltage (V_{mp}) and power (P_{mp}), as measured on the DC side of the system [24].

3 METHODOLOGY

The methodology followed to develop the intraday PV production forecasting ensemble included the construction of 3 BNN models, each successively trained at different regimes (exogenous data, lag structures and window duration). The yearly performance dataset of the test-bench PV system was divided into a training and testing set for the purpose of this evaluation. More specifically, the training set was used to construct the best-performing model by obtaining the optimal hyperparameters for the resulting BNN model. The devised BNN models consisted of four input parameters (G_I, T_{amb}, elevation angle of the sun (α) and azimuth angle of the sun (φ_s)), those parameters and training range was investigated in a previous study [25]. Each model was trained at different historical data lags by shifting backwards the produced hourly average power output to map prior G_I, T_{amb}, α and φ_s data points to form a cluster of 3 BNN models. The first BNN cluster was trained with 3 backward shifted successive datasets that mapped the input features to the immediate power data-points. Similarly, the second and third BNN cluster were trained with backward shifted data starting from the second and third immediate data-point, respectively. The results of all models were merged to an output stage regression voting ensemble (RVE), which provided the power output forecasts at time horizons from 1- to 5-hours ahead.

The results of all BNN clusters were merged to an output stage regression voting ensemble (RVE), which provides the output forecast. Figure 1 depicts the lag structure of historical constructed datasets employed to train each BNN cluster. At the output stage, each cluster provided the forecasts to an output stage RVE, which provided the power output forecast by averaging the values.

The forecasting performance accuracy was evaluated using the test set and based on commonly employed statistical performance metrics such as the mean absolute percentage error (MAPE), which calculates the prediction accuracy of a forecasting method, the RMSE that describes the standard deviation of the forecasting errors and the normalised root mean square error that is the RMSE (nRMSE) relative to the nominal capacity of the PV system.

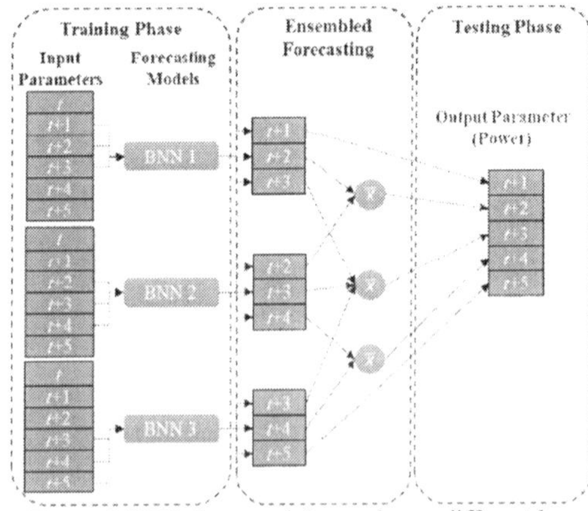

Figure 1 BNN cluster training regime at different lag structures.

3.1 Artificial Neural Networks

The strength of artificial neural networks (ANNs) relies on their ability to approximate non-linear functions through a supervised learning process (a set of input and output data is available). Training ANN models is performed by applying a learning algorithm to a cost function. More specifically, the training step in neural networks is formulated in terms of the minimisation of a loss function, which in this case comprises of data and regularisation terms. In this study, Bayesian inference was applied to the training algorithm of a deep learning neural network. A BNN is a type of ANN that is constructed by introducing stochastic components into the network architecture (activation and weights) to simulate multiple possible models with an associated probability distribution. To this end, BNNs are considered a special case of ensemble learning, where instead of training one single model, a set of models is trained and their predictions are aggregated [26]. The Bayesian paradigm is based on the principle that probability is a measure of belief in the occurrence of events and that prior beliefs influence posterior beliefs. The marginal probability distribution of the predicted output was computed based on the Bayesian posterior.

3.2 Model performance assessment

The forecasting performance accuracy was assessed based on several predefined metrics when the test set was applied to the developed algorithms. The metrics commonly used in PV production forecasting applications include the MAPE, RMSE and nRMSE to the nominal PV system peak power:

$$MAPE = \frac{100}{n} \times \sum_{i=1}^{n} \left| \frac{y_{actual,i} - y_{predicted,i}}{y_{actual,i}} \right| \quad (1)$$

$$nRMSE = \frac{100}{P_{nominal}} \times \sqrt{\frac{1}{n} \times \sum_{i=1}^{n} (y_{actual,i} - y_{predicted,i})^2} \quad (2)$$

where $y_{actual,i}$ and $y_{forecasted,}$ is the actual and predicted irradiance and power respectively, $Y_{nominal}$ is the nominal peak power of the PV system (1365 W).

4 RESULTS

The model comprised of 4 input parameters and

optimised based on [25] (forecasted *GHI*, forecasted T_{amb}, α and φ_s).

The performance evaluation results provided evidence that high accuracies beyond the state-of-the-art can be achieved by employing ensemble deep learning approaches. Table I summarises the performance accuracy of the ensemble model. The results showed that the 1-hour ahead model demonstrated the lowest errors (nRMSE of 3.49% and MAPE of 3.24%).

Table I: Statistical analysis of the derived results

Forecasts	Metrics	
	MAPE (%)	nRMSE (%)
1-hour	3.24	3.49
2-hour	4.69	5.14
3-hour	4.84	6.06
4-hour	5.71	6.19
5-hour	7.47	7.92

The nRMSE values at each forecasting horizon (1- to 5-hours ahead) over the test set period, are depicted by the box-plot in Figure 2. The 1-hour ahead forecasting demonstrated the lowest nRMSE, while the highest nRMSE was exhibited at a horizon of 5-hours ahead. This provides evidence that the forecasting error progressively increases and is highly dependable to the forecasting horizon.

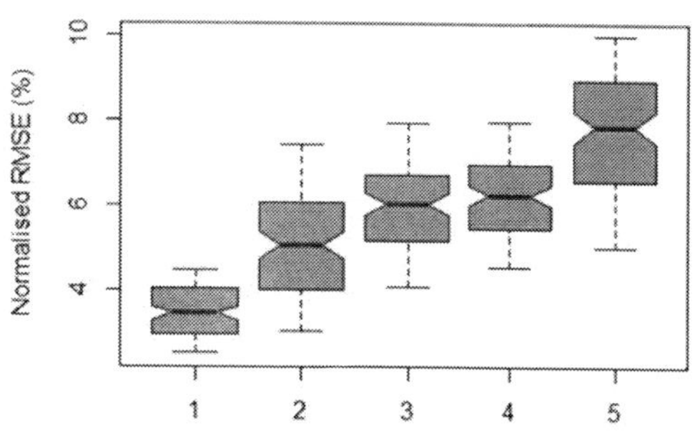

Figure 2: Boxplot of the nRMSE distribution at the different forecasting horizons, over the test set evaluation period for the ensembled methodology

Additionally, the impact of the RVE on the forecasting accuracy at horizons of 2- to 4-hours ahead was investigated by comparing the forecasts of the first BNN cluster and the ensemble. The comparative results, presented in Figure 3, showed that absolute nRMSE improvements in the range of 2%-4% can be achieved by employing the RVE. The ability of a BOC to improve further the forecasting accuracy at forecasting horizons of 5 hours will be investigated in the full manuscript.

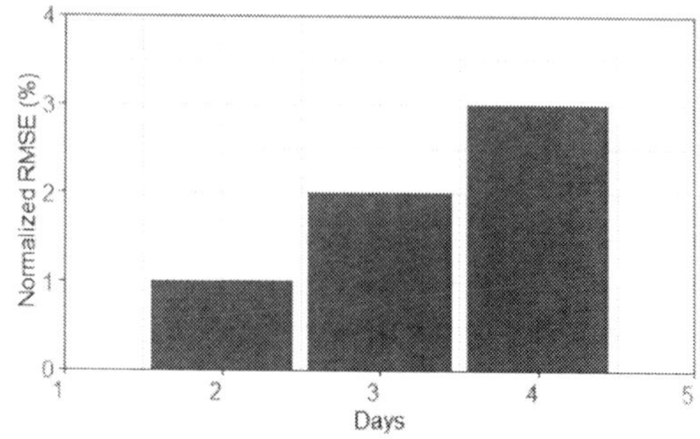

Figure 3: Absolute nRMSE difference between cluster BNN 1 and the ensemble at time horizons of 2- to 4-hours ahead.

5 CONCLUSIONS

The scope of this study is to propose an accurate intraday (up to 5-hours ahead forecasting) PV production forecasting algorithm. To this end, an intraday forecasting ensemble that comprised of a cluster of BNN models was implemented by utilizing only on-site solar irradiance and temperature measurements. The approach followed to develop the models was to construct 3 BNNs, each trained under different historical lags and historical window durations.

The derived results demonstrated that the highest performing accuracy was acquired from the 1-hour ahead forecasting model with nRMSE and MAPE of 3.60% and 3.28% respectively, while the lowest accuracy was demonstrated by the 5-hours ahead model with nRMSE and MAPE of 8.63% and 7.82% respectively. Absolute nRMSE improvements in the range of 2%-4% can further be achieved by employing the RVE.

6 ACKNOWLEDGEMENTS

This work was funded through the ELECTRA project, which is co-financed by the European Regional Development Fund of the European Union under the grant agreement number INTEGRATED/0918/0071. The authors from the University of Cyprus would also like to acknowledge the financial support from the Research & Innovation Foundation of Cyprus. Sandia National Laboratories is a multimission laboratory managed and operated by National Technology & Engineering Solutions of Sandia, LLC, a wholly owned subsidiary of Honeywell International Inc., for the U.S. Department of Energy's National Nuclear Security Administration under contract DE-NA0003525. This paper describes objective technical results and analysis. Any subjective views or opinions that might be expressed in the paper do not necessarily represent the views of the U.S. Department of Energy or the United States Government.

7 REFERENCES

[1] S. Pelland, J. Remund, J. Kleissl, T. Oozeki, and K. De Brabandere, "Photovoltaic and Solar Forecasting: State of the Art," *Int. Energy Agency Photovolt. Power Syst. Program. Rep. IEA PVPS T14*, pp. 1–40, 2013, doi: 978-3-906042-13-8.

[2] E. Lorenz, T. Scheidsteger, J. Hurka, D. Heinemann, and C. Kurz, "Regional PV power prediction for improved grid integration," *Prog. Photovoltaics*

Res. Appl., vol. 19, no. 7, pp. 757–771, 2011, doi: 10.1002/pip.1033.

[3] S. Pelland, G. Galanis, and G. Kallos, "Solar and photovoltaic forecasting through post-processing of the Global Environmental Multiscale numerical weather prediction model," *Prog. Photovoltaics Res. Appl.*, vol. 21, no. 3, pp. 284–296, May 2013, doi: 10.1002/pip.1180.

[4] I. Ashraf and a Chandra, "Artificial neural network based models for forecasting electricity generation of grid connected solar PV power plant," *Int. J. Glob. Energy Issues*, vol. 21, no. 1, pp. 119–130, 2004, [Online]. Available: http://inderscience.metapress.com/content/N0R76HDTKC19GWVH.

[5] D. Caputo, F. Grimaccia, M. Mussetta, and R. E. Zich, "Photovoltaic plants predictive model by means of ANN trained by a hybrid evolutionary algorithm," 2010, doi: 10.1109/IJCNN.2010.5596782.

[6] S. A. Kalogirou, "Applications of artificial neural-networks for energy systems," *Appl. Energy*, vol. 67, no. 1–2, pp. 17–35, 2000, doi: 10.1016/S0306-2619(00)00005-2.

[7] A. Gandelli, F. Grimaccia, S. Leva, M. Mussetta, and E. Ogliari, "Hybrid model analysis and validation for PV energy production forecasting," in *Proceedings of the International Joint Conference on Neural Networks*, 2014, pp. 1957–1962, doi: 10.1109/IJCNN.2014.6889786.

[8] A. Mellit and A. M. Pavan, "A 24-h forecast of solar irradiance using artificial neural network: Application for performance prediction of a grid-connected PV plant at Trieste, Italy," *Sol. Energy*, vol. 84, no. 5, pp. 807–821, 2010, doi: 10.1016/j.solener.2010.02.006.

[9] F. Almonacid, P. J. Pérez-Higueras, E. F. Fernández, and L. Hontoria, "A methodology based on dynamic artificial neural network for short-term forecasting of the power output of a PV generator," *Energy Convers. Manag.*, vol. 85, pp. 389–398, 2014, doi: 10.1016/j.enconman.2014.05.090.

[10] C. Chen, S. Duan, T. Cai, and B. Liu, "Online 24-h solar power forecasting based on weather type classification using artificial neural network," *Sol. Energy*, vol. 85, no. 11, pp. 2856–2870, 2011, doi: 10.1016/j.solener.2011.08.027.

[11] J. Shi, W. J. Lee, Y. Liu, Y. Yang, and P. Wang, "Forecasting power output of photovoltaic system based on weather classification and support vector machine," 2011, doi: 10.1109/IAS.2011.6074294.

[12] A. Dolara, F. Grimaccia, S. Leva, M. Mussetta, and E. Ogliari, "A physical hybrid artificial neural network for short term forecasting of PV plant power output," *Energies*, vol. 8, no. 2, pp. 1138–1153, 2015, doi: 10.3390/en8021138.

[13] S. E. Haupt *et al.*, "Building the Sun4Cast System: Improvements in Solar Power Forecasting," *Bull. Am. Meteorol. Soc.*, vol. 99, no. 1, pp. 121–136, 2018, doi: 10.1175/BAMS-D-16-0221.1.

[14] D. P. Larson, L. Nonnenmacher, and C. F. M. Coimbra, "Day-ahead forecasting of solar power output from photovoltaic plants in the American Southwest," *Renew. Energy*, vol. 91, pp. 11–20, 2016, doi: 10.1016/j.renene.2016.01.039.

[15] X. Sun and T. Zhang, "Solar Power Prediction in Smart Grid Based on NWP Data and an Improved Boosting Method," in *Proceedings - 1st IEEE International Conference on Energy Internet, ICEI 2017*, 2017, pp. 89–94, doi: 10.1109/ICEI.2017.23.

[16] P. Lynch, "The origins of computer weather prediction and climate modeling," *J. Comput. Phys.*, vol. 227, no. 7, pp. 3431–3444, 2008, doi: 10.1016/j.jcp.2007.02.034.

[17] L. F. Richardson, *Weather prediction by numerical process, second edition*. 2007.

[18] R. Kimura, "Numerical weather prediction," *J. Wind Eng. Ind. Aerodyn.*, vol. 90, no. 12–15, pp. 1403–1414, 2002, doi: 10.1016/S0167-6105(02)00261-1.

[19] H.-L. Xue, X.-S. Shen, and J.-F. Chou, "A forecast error correction method in numerical weather prediction by using recent multiple-time evolution data," *Adv. Atmos. Sci.*, vol. 30, no. 5, pp. 1249–1259, 2013, doi: 10.1007/s00376-013-2274-1.

[20] L. Gao, H.-L. Ren, J.-P. Li, and J.-F. Chou, "Analogue correction method of errors and its application to numerical weather prediction," *Chinese Phys.*, vol. 15, no. 4, pp. 882–889, 2006, doi: 10.1088/1009-1963/15/4/038.

[21] E. Lorenz, A. Hammer, and D. Heinemann, "Short term forecasting of solar radiation based on satellite data," 2004, doi: 20.06.-23.06.2004.

[22] R. Marquez and C. F. M. Coimbra, "Forecasting of global and direct solar irradiance using stochastic learning methods, ground experiments and the NWS database," *Sol. Energy*, 2011, doi: 10.1016/j.solener.2011.01.007.

[23] A. Drews, H. G. Beyer, and U. Rindelhardt, "Quality of performance assessment of PV plants based on irradiation maps," *Sol. Energy*, vol. 82, no. 11, pp. 1067–1075, Nov. 2008, doi: 10.1016/j.solener.2008.04.009.

[24] G. Makrides, B. Zinsser, M. Norton, G. E. Georghiou, M. Schubert, and J. H. Werner, "Potential of photovoltaic systems in countries with high solar irradiation," *Renewable and Sustainable Energy Reviews*, vol. 14, no. 2. pp. 754–762, 2010, doi: 10.1016/j.rser.2009.07.021.

[25] S. Theocharides, G. Makrides, A. Livera, M. Theristis, P. Kaimakis, and G. E. Georghiou, "Day-ahead photovoltaic power production forecasting methodology based on machine learning and statistical post-processing," *Appl. Energy*, vol. 268, no. December 2019, p. 115023, 2020, doi: 10.1016/j.apenergy.2020.115023.

[26] Z. H. Zhou, *Ensemble methods: Foundations and algorithms*. 2012.

SOLAR IRRADIANCE FORECASTING USING NUMERICAL WEATHER PREDICTION MODEL AND INSAT-3D BASED SATELLITE MODEL

Akriti Masoom and Ankit Bansal
Department of Mechanical and Industrial Engineering
Indian Institute of Technology Roorkee, Uttarakhand, India- 247667

ABSTRACT: The solar radiation received by India, 5000 trillion kWh in a year, is much greater than the total annual consumption of energy. This amount of available energy can be used to generate power through photovoltaic (PV) and concentrated solar power (CSP) applications. Solar radiation is dependent on the geography of the region along with azimuth and inclination angles, Sun-Earth position hour and atmospheric attenuations. Atmospheric barriers like clouds and aerosols can affect the radiation up to 70-80%. Solar radiation is subjected to reflection scattering and absorption by air molecules, clouds, and aerosols in the atmosphere. Clouds can block most of the direct radiation. Modern solar energy forecasting systems rely on real-time Earth observation from the satellite for detecting clouds and aerosols. Moreover, a complete utilization of the solar energy depends on its proper integration with power grids. However, the incorporation of solar based energy generation into electricity grids suffers technical challenges due to its variable nature. Hence, accurate and independent (from ground data) solar irradiance forecasting plays a crucial role in the grid integration of solar based generations. The poor resolution of the ground data of surface solar radiation presents the need to look into the satellite-based solar irradiance modeling. In the present study, a hybrid model based on numerical weather prediction and satellite model is presented for day-ahead irradiance forecasting.

1 INTRODUCTION

The operational day-ahead solar irradiance forecasts are needed to know the electric output from photovoltaic (PV) systems in advance for day-to-day operation and scheduling. The regional NWP models (e.g., WRF) are majorly used for forecast horizons beyond a few hours due to higher forecast quality as compared to the other methods. However, these models suffer from inherent biases due to the coarser global model inputs (e.g., GFS) used to run these models. Hence, forecast quality improvements are required for regional models. The NWP forecasts can be improved by incorporating the satellite and/or ground based data into the model's initialization or can be used in the post-processing of the forecasts. In this work, the ground-based data is used only to evaluate the forecast quality of the NWP model. Ground-based observations are a major source of observation data, however, their spatial distribution is generally coarse that limits their usage in the post-processing of the NWP data.

In this work, we focused on the integration of the satellite based model with the Weather Research and Forecasting (WRF) model to better forecast the solar irradiance with an aim to make an operational day-ahead forecasting model.

2 METHODOLOGY

2.1 Operational day-ahead forecasting system

A simple scenario for daily execution of day-ahead forecasting is presented in Figure 1. The time of correction is taken from 01 UTC for an NWP forecast initialized at 18 UTC as the GFS data initialized at 18 UTC is available at 00 UTC. The corresponding satellite correction can be performed at 01 UTC to produce a corrected forecast of 01 UTC of next day [1]. The Forecast time horizon is from 18 UTC on first day to 12 UTC of third day. For example, if the forecast is initialized at 18 UTC on May 1, then first valid forecast will be made for 03 UTC of May 3 which has a forecast horizon of 33 h and so on.

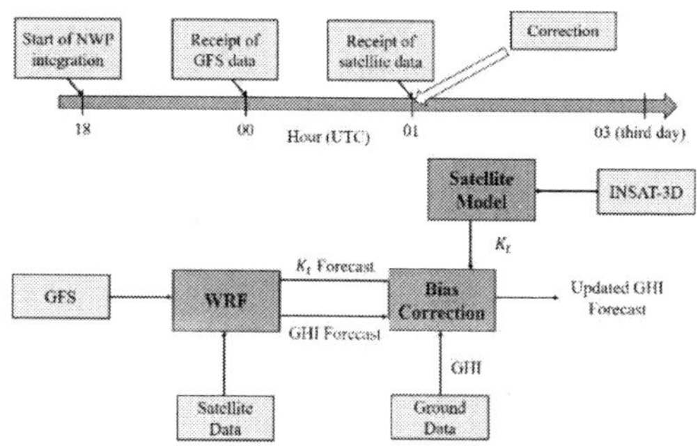

Figure 1: WRF forecast improvement with satellite correction

2.2 Numerical weather prediction model

WRF domain The WRF forecasts are generated at a spatial resolution of 1 km and temporal resolution of 30 min. Two-way nesting was carried out with the outer domain at 4 km resolution and the inner domain at 1 km resolution as shown in Figure 1. The physics schemes used for WRF simulations are shown in Table 1. WRF simulations were run with the default clouds and aerosols schemes to keep the complexity of the model to minimum as well as to save the computational cost involved.

Table I: WRF model configuration

Physics	Scheme
Microphysics	New Thompson
Cumulus	Tiedtke
Shortwave radiation	RRTMG
Longwave radiation	RRTMG
Planetary boundary layer	Mellor-Yamada-Janjic

2.3 Satellite model

The satellite based ADI model uses cloud indices calculated from different satellite channels to obtain an averaged cloud index which is termed as all day index as presented by authors in [2]. The ADI is calculated as follows

$$\bar{n} = \frac{1}{N}\sum_{i=1}^{N} n_i \qquad \text{(i)}$$

where n_i represents the cloud indices calculated from different satellite bands. The cloud index is calculated as

$$n = \frac{BTD - BTD_{min}}{BTD_{max} - BTD_{min}} \qquad \text{(ii)}$$

where BTD represent brightness temperature difference. BTD_{min} and BTD_{max} represent the minmum and the maximum BTD values for a 15-day period.

$$k = \begin{cases} 1.2 & if \quad n \leq -0.2 \\ 1 - n & if \quad -0.2 < n \leq 0.8 \\ 1.661 - 1.781n + 0.725n^2 & if \quad 0.8 < n \leq 1.05 \\ 0.09 & if \quad 1.05 < n \end{cases} \qquad \text{(iii)}$$

$$GHI = k.GHI_{CS} \qquad \text{(iv)}$$

3 RESULTS

3.1 Satellite image processing

BTD images contain multiple geographical features. Hence, features like clouds and dust need to be extracted from the combined image so as to estimate the irradiance on the surface. Figure 2 shows a detailed feature extraction and generation of cloud indices from satellite images. BTD12 and IDDI images are segmented using k-mean clustering. Then, the image segments with the largest area are removed from the segmented images to filter out the features like landmass, ocean, grassland, etc. that are not desired in solar irradiance estimation. The remaining segments represent the presence of dust, fog, and clouds. Hence, the final cloud indices are generated for May 02 2018 at 01 UTC.

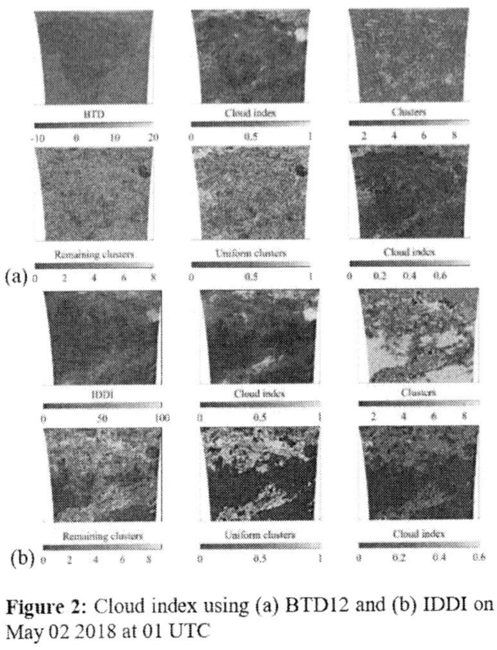

Figure 2: Cloud index using (a) BTD12 and (b) IDDI on May 02 2018 at 01 UTC

3.2 WRF forecast bias correction

The radiation based on the satellite images is estimated using the methodology presented in Section 2.3. The cloud indices are combined to generate ADI using Equation (i) which is then converted into clear-sky index k using Equation (iii). Figure 3 shows the GHI estimated from the satellite model at 07 UTC on May 02 which is compared with the GHI forecasted by WRF model which was initialized at 18 UTC on May 01. Based upon the comparison the bias produced by the WRF model is corrected by the WRF forecast for 07 UTC on May 03.

(a) (b) (c)

Figure 3: GHI maps from satellite model, WRF forecast and the corresponding corrected day ahead WRF forecast of next day

4 DISCUSSIONS

4.1 Validation with ground measurements

The results generated in Section 3.2 are copared with the BSRN ground measurements of GHI for May 03 at 07 UTC as shown in Figure 4. Day 1 represent the day of WRF initiation which is May 1 in this case. Day 2 represent the day of satellite correction and Day 3 deals with the corrected WRF day-ahead forecasts. The comparison with the BSRN GHI showed an error of 11% from satellite GHI and 13% from WRF forecast for Day 2. The WRF forecast for Day 3 showed an error of 14% that improved to 10% with the satellite correction.

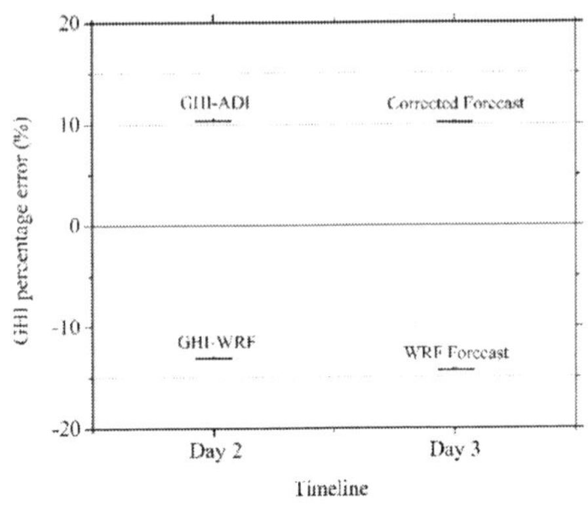

Figure 4: Comparison of WRF GHI with BSRN ground based GHI.

5 CONCLUSIONS

The methodology presented in this work deals with the daily operation of mesoscale NWP model with satellite based bias correction. The WRF model was found to overestimate GHI as compared to the satellite model that underestimated the GHI but with lesser error. The satellite based correction was performed to obtain day-ahead corrected WRF forecast that was improved by 30%.

References

[1] T. Watanabe, H. Takenaka and D. Nohara, Post processing correction method for surface solar rradiance forecast data from the numerical weather model using geostationary satellite observation data Solar Energy 223 (2021), 202-216, 2021.

[2] Y. Kashyap, A. Bansal, A. K. Sao and A. Hammer, Model for Estimation of Global Horizontal Irradiance in the Presence of Dust, Fog, and Clouds, IEEE Transactions on Geoscience and Remote Sensing 56 (12), 7030-7037, 2018.

COMPARISON AND VALIDATION OF IRRADIANCE DATA: SATELLITE METEOROLOGICAL DATASET MERRA-2 VS. METEONORM AND GERMAN WEATHER SERVICE (DWD)

Arash Khatibi, Stefan Krauter
Paderborn University, Faculty of Computer Science, Electrical Engineering and Mathematics
Electrical Energy Technology – Sustainable Energy Concepts (EET-NEK)
Pohlweg 51, 33098 Paderborn, Germany
Email: arash.khatibi@uni-paderborn.de

ABSTRACT: Access to precise meteorological data is crucial to be able to plan and install renewable energy systems such as solar power plants and wind farms. In case of solar energy, knowledge of local irradiance and air temperature values is very important. For this, various methods can be used such as installing local weather stations or using meteorological data from different organizations such as Meteonorm or official Deutscher Wetterdienst (DWD). An alternative is to use satellite reanalysis datasets provided by organizations like the National Aeronautics and Space Administration (NASA) and European Centre for Medium-Range Weather Forecasts (ECMWF). In this paper the "Modern-Era Retrospective analysis for Research and Applications" dataset version 2 (MERRA-2) will be presented, and its performance will be evaluated by comparing it to locally measured datasets provided by Meteonorm and DWD. The analysis shows very high correlation between MERRA-2 and local measurements (correlation coefficients of 0.99) for monthly global irradiance and air temperature values. The results prove the suitability of MERRA-2 data for applications requiring long historical data. Moreover, availability of MERRA-2 for the whole world with an acceptable resolution makes it a very valuable dataset.
Keywords: Power potential estimation; Photovoltaic; Solar radiation; Temperature measurement; Satellite data; Meteonorm; MERRA-2; DWD

1 INTRODUCTION

When designing a PV or wind power system, it is essential to choose system components in a way to be able to fulfill the demand at all times with minimum over- or underestimation. Not meeting these conditions leads to electricity shortage, possibly a grid outage and higher costs. For the correct sizing of the system, long-time historical local meteorological data is required. Various research and articles have already shown the interest and importance of data analysis for renewable energy systems including topics like analysis of historical data, solar plant and wind farms production and prediction, forecasting model of irradiance, temperature and wind speed [1] [2] [3] [4] [5] [6] [7]. In this paper, the focus is on analysis and validation of historical data (MERRA-2) which is provided by NASA. In [8], MERRA-2 solar and wind parameters are evaluated for some stations which were tried to be as near as possible and as equally high as possible (in altitude) to a local weather station. In this study, the stations are chosen freely and more stations from Germany are selected to verify MERRA-2 quality against DWD dataset as well.

Firstly MERRA-2 dataset will be shortly introduced. Later Meteonorm and DWD data will be discussed. Later on, MERRA-2 measurements of global horizontal irradiance and air temperature will be compared to Meteonorm and DWD datasets for various locations. For this, Meteonorm and DWD are considered as the reference since they contain locally measured data and they are standard and widely used by renewable systems designers.

2 MERRA-2

Modern-Era Retrospective analysis for Research and Applications dataset version 2 (MERRA-2) offers a major variety of meteorological parameters for each of its locations. The data is available for over 41 years and is available for all the geographical coordinates with 0.5° in latitudinal and 0.625° in longitudinal direction [9]. It covers the earth with 576 equally distributed latitude values and 361 equally distributed longitude values. The distance resolution of two consecutive measurement points is 50 kilometers in latitudinal direction [9]. In case of solar energy, the RAD dataset in MERRA-2 is freely available since January 1st 1980. RAD dataset contains the reanalysis satellite values such as surface incoming shortwave flux (global horizontal irradiance), surface net downward shortwave flux, cloud are fraction for low, medium and high clouds, surface albedo for different wavelengths, surface skin temperature and other parameters. The MERRA-2 single-level diagnostics (SLV) dataset offers single-level diagnostics parameters, such as specific humidity, surface pressure, 2-, 10- and 50-m temperature, and eastward/northward wind velocity since 1980. The full list of parameters for all different datasets in MERRA-2 can be found in [9]. Some of the parameters important for designing solar energy systems can be found in Table I.

Table I: Some sample parameters in the RAD & SLV datasets of MERRA-2 [9].

Parameter Name	Description	Units
ALBEDO	"Surface albedo"	1
CLDTOT	"Total cloud area fraction"	1
SWGDN	"Surface incoming shortwave flux"	W/m²
SWGNT	"Surface net downward shortwave flux"	W/m²
SWGDNCLR	"Surface incoming shortwave flux assuming clear sky"	W/m²
SWGNTCLR	"Surface net downward shortwave flux assuming clear sky"	W/m²
TS	"Surface skin temperature"	K
T2M	"2- meter air temperature"	K

In this study, MERRA-2 data for global irradiance and air temperature is studied and verified for various locations to see whether it can be considered as a countable option for a site's solar energy potential evaluation. For this, version 8.0 of Meteonorm dataset and software and Germany's National Meteorological Service (DWD) is used as local measurements and reference for comparison.

3 METEONORM

Meteonorm is a product of Meteotest that is "a leading company in weather, climate, environment and computer science" [10]. Meteonorm is the name of a software/dataset product that gathers worldwide weather data from 8325 weather stations around the world and five stationary satellites [11]. Meteonorm is "a global standard and a powerful tool for solar energy applications, building design, heating and cooling systems, education, agriculture, forestry and many more" [12]. For this paper Meteonorm software version 8.0.4 is used for finding the suitable locations and obtaining the required data. Although Meteonorm allows the user to choose any location on the globe freely, not all of the locations have local measurements. In this paper, since the validity of MERRA-2 is the question, it must be validated in comparison with real local measurements. Therefore, only the stations with fully locally measured data are found and extracted in Meteonorm.

The datasets in Meteonorm are available with hourly, daily, monthly, and yearly resolutions. The hourly dataset in Meteonorm is logged in local time of the site and the user may need to shift the timestamps when comparing with other datasets [13]. Another point about Meteonorm hourly datasets is that it only offers the data for a single *typical year*. This hourly data is not a real measured year, but it is mathematically calculated based on the historical measurements of the location. As Meteonorm says, this data is different from an average year. For instance, in an average year in a location in northern hemisphere, a summer day usually shows high values of irradiance and temperature because it is averaged through many years. In Meteonorm and in a typical year distribution, a summer day for the same location can still show low very irradiance levels because it represents what really can happen in real situation. Due to this point, it is not possible to validate MERRA-2 hourly values with Meteonorm hourly values. However, still if the hourly data for typical year is averaged through months and years, it can correspond to the long-term mean to compare with MERRA-2. Meteonorm is able to export the averaged values through months and year in a customized format for using in this study.

4 DWD

"The Deutscher Wetterdienst is a public institution with partial legal capacity under the Federal Ministry of Transport and Digital Infrastructure. Germany's National Meteorological Service, the Deutscher Wetterdienst (DWD), is responsible for meeting meteorological requirements arising from all areas of economy and society in Germany" [14].

DWD offers various meteorological services, such as "short- and long-term recording, monitoring, and evaluation of meteorological processes in the atmosphere, its structure, and composition" [15]. The forecasting of meteorological processes, the operation of the necessary measuring and observation systems, and also the provision, storage, and documentation of meteorological data and products are other tasks of this organization [15].

DWD offers various datasets, such as air temperature, pressure, cloud type, cloudiness, precipitation, pressure, soil temperature, solar irradiance and duration, visibility, wind speed, wind direction, etc. for 1 min, 10 min, daily, hourly, monthly, and multi-annual intervals [16] [17]. Since these data are measured locally and by installation of various measuring devices, they offer a high precision. DWD has about 56 stations/locations for solar energy data and 606 stations/locations for temperature and humidity around Germany [16] [17]. DWD provides Meteonorm with its meteorological measurements for the locations in Germany [13]. This can also be verified by comparing the geographical coordinates of weather stations in Meteonorm and DWD in Germany. Therefore, for the comparison of MERRA-2 with DWD average measurements, the locations in Germany are chosen in Meteonorm as well.

5 METHODOLOGY

For the comparison Meteonorm and DWD datasets are considered as the reference data and MERRA-2 will be validated in comparison to them. Since MERRA-2 is available for the whole world, choosing of locations were started in Meteonorm. The locations are chosen in a way that they only contain 100% locally measured data and the share of satellite or interpolated measurements are 0%. Moreover, Meteonorm itself uses DWD data for locations in Germany meaning that the weather stations available in Germany are identical to the data that DWD offers. 4 sites are chosen inside Germany to include DWD datasets and 4 other locations are chosen around the globe in a way to include all solar irradiance levels like very sunny, moderate, and low irradiance locations.

Next, the closest datapoints to the Meteonorm/DWD locations in MERRA-2 are found to reach the highest possible precision in comparison. Beside haversine distance of nearest points, altitude is another important factor that must be considered when comparing meteorological parameters like irradiance and temperature.

Table II: Chosen Meteonorm/DWD stations vs. their corresponding MERRA-2 datapoints.

Station	Meteonorm/ DWD Coordinates	MERRA-2 Coordinates	Distance (km) ≈	Altitude difference (m) ≈
Alice Springs (Australia)	23.800° S, 133.900° E	24° S, 133.75° E	26.96	57
Belmullet (Ireland)	54.233° N, 10° W	54° N, 10° W	25.91	37
Calama/El Loa (Chile)	22.5° S, 68.900° W	22.5° S, 68.75° W	15.41	162
El Kharga (Egypt)	24.450° N, 30.533° E	25.5° N, 30.625° E	10.78	29
Erfurt/Bindersleben (Germany)	50.98° N, 10.97° E	51° N, 11.25° E	20.33	20
Fürstenzell (Germany)	48.550° N, 13.350° E	49° N, 13.125° E	17.5	130
Innsbruck Univ. (Austria)	47.267° N, 11.400° E	47.5° N, 11.25° E	28.26	303
Schleswig (Germany)	54.533° N, 9.55° E	54.5° N, 9.375° E	11.79	36
Würzburg (Germany)	49.767° N, 9.967° E	50° N, 10° E	25.69	33

Table II details the chosen locations and their corresponding geographical coordinates in both Meteonorm/DWD and MERRA-2 datasets. Distance and altitude difference of compared points are shown as well.

Table III contains the mutual parameters and their corresponding names in Meteonorm and MERRA-2 that are compared in this paper.

Table III: Chosen parameters and their dedicated names in Meteonorm and MERRA-2.

Parameter	Name in Meteonorm	Name in MERRA-2
Global radiation horizontal	Gh	SWGDN
Air temperature	Ta	T2M

In Meteonorm, Gh is the parameter name of mean daily global horizontal irradiation in Wh/m^{2}d for each month of the year. This parameter and its values can be extracted for any location ready for comparison. In case of MERRA-2, the global irradiance values (SWGDN) are continuously measured and averaged over an hour (hourly resolution) with unit W/m^2 [9]. For comparison of these two values, sum of SWGDN hourly values in MERRA-2 has been calculated for every day (in Wh/m^{2}d). After that, these values of daily irradiated energy are averaged for each month through the year through the whole period.

Time period for each station and parameter could be different in Meteonorm. Because of this, extracting of MERRA-2 data for each parameter and location must be done for multiple times. Table IV shows the uncertainty and available period of the measurements in Meteonorm that are considered when extracting MERRA-2 data. Unfortunately, duration limit of Meteonorm does not allow us to use longer period of data which is already available in MERRA-2.

Table IV: Time periods and uncertainty of available data for Meteonorm/DWD stations.

Station	Radiation Period	Gh Uncertainty	Wind and Temperature Period	Ta Uncertainty
Alice Springs (Australia)	1996–2015	2%	2000–2019	0.3 °C
Belmullet (Ireland)	1996–2015	3%	2000–2019	0.3 °C
Calama/El Loa (Chile)	1996–2015	7%	2000–2019	0.3 °C
El Kharga (Egypt)	1996–2015	4%	2000–2019	0.3 °C
Erfurt/Bindersleben (Germany)	2004–2013	3%	2000–2019	0.3 °C
Fürstenzell (Germany)	1997–2015	2%	2000–2019	0.3 °C
Innsbruck Univ. (Austria)	1996–2015	3%	2000–2019	0.3 °C
Schleswig (Germany)	1996–2015	4%	2000–2019	0.3 °C
Würzburg (Germany)	1996–2015	3%	2000–2019	0.3 °C

For MERRA-2 analysis, a program in Python is developed to access and preprocess the data, simplify and accelerate obtaining the required data for the wanted location, timeframe and also for modifying the data to achieve the corresponding parameter in Meteonorm.

6 RESULTS AND DISCUSSION

Figure 1 shows comparison diagrams of the values of average daily irradiance between MERRA-2 and Meteonorm for every month of the year. It can be seen that the highest difference happens in summer months of the year that have higher irradiance levels for all locations.

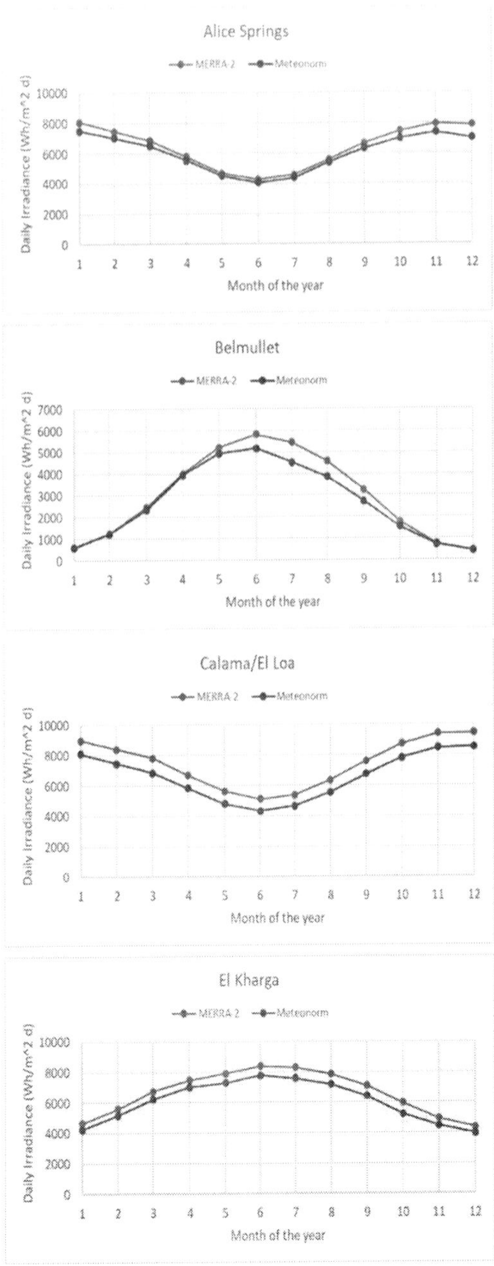

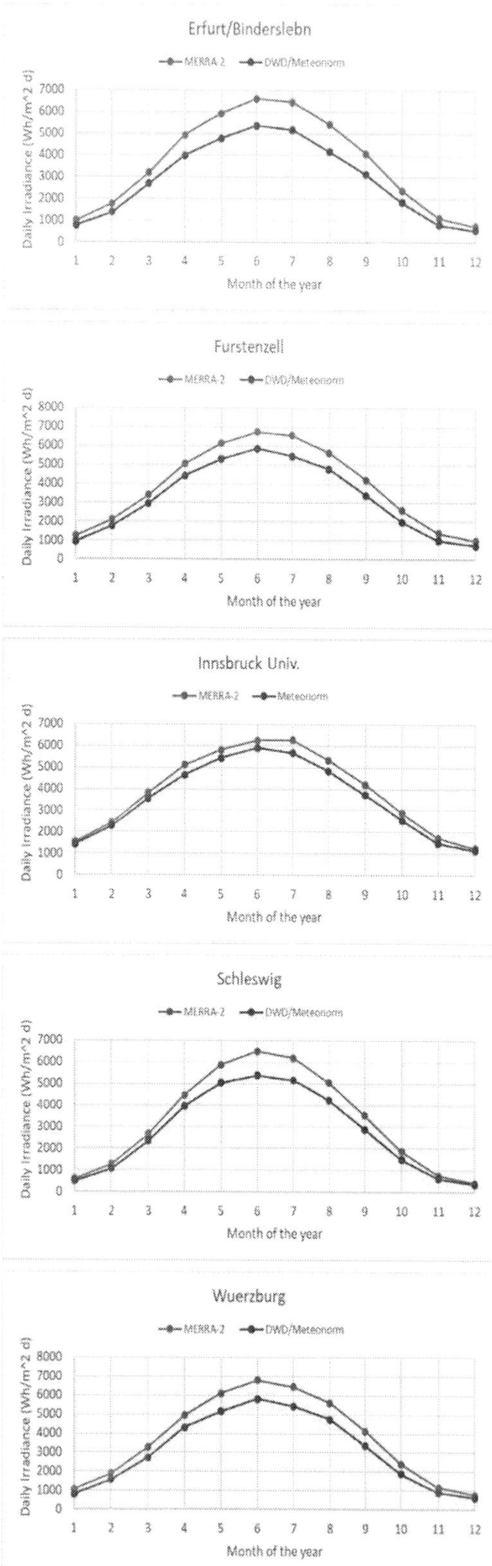

Figure 1: MERRA-2 and Meteonorm comparison for monthly average values of daily irradiance for all locations (through the years from Table IV, column "Radiation period").

In the locations of Alice Springs, Belmullet, El Kharga and Innsbruck the differences are minimal considering Meteonorm typical year data as reference. Especially for Alice Springs, Belmullet and Innsbruck, this result is pretty outstanding considering that the datapoints in Meteonorm and MERRA-2 are more than 25 kilometers away from each other. In case of Innsbruck the datapoints are even 303 meters different in altitude. For all other sites, the differences are higher however, it is interesting to see that diagrams follow each other's distribution very precisely. Another point is that MERRA-2 seems to overestimate the irradiance values in all cases. This overestimation does not happen to an equal extent for all stations. For instance for Alice Springs and Belmullet, it only happens in few months of the year and in the rest of the year the values are almost equal. At this point no pattern can be concluded and no correction may be implemented.

Table V contains the calculated correlation coefficients and other errors for average daily global horizontal irradiance levels for all the locations. All correlation coefficients are higher than 0.99. This confirms that MERRA-2 follows the local measured values almost perfectly. The lowest value is for Belmullet (0.9936) and the highest correlation is for Calama/El Loa that has an altitude difference of 162 meters.

Mean absolute error and root mean square error values indicate a better understanding of the quality of MERRA-2. MAE values are between 297.42 and 873.42 Wh/m^{2}d. The highest MAE is still an acceptable error considering that the average daily irradiance levels in Calama/El Loa reaches 8500 Wh/m^{2}d in December and datapoints are 162 meters different in altitude. On the other hand, the second lowest MAE is for Innsbruck (330.98 Wh/m^{2}d) which is very interesting for a location with 28.26 km distance and 303 m difference in altitude between its datapoints.

Table V: Errors and correlation coefficients of MERRA-2 data vs. Meteonorm/DWD data of monthly average values of daily global irradiance (through the years from Table IV, column "Radiation period").

Global Irradiance	***R***	***MAE*** (Wh/m^{2}d)	***RMSE*** (Wh/m^{2}d)	***NRMSE***
Alice Springs	0.9963	381.55	431.57	0.127
Belmullet	0.9936	297.42	428.74	0.090
Calama/El Loa	0.9994	873.42	876.62	0.208
El Kharga	0.9977	560.02	572.41	0.151
Erfurt/Bindersleben	0.9988	750.49	856.85	0.178
Fürstenzell	0.9985	625.29	676.19	0.132
Innsbruck Univ.	0.9986	330.98	365.01	0.077
Schleswig	0.9991	511.88	621.36	0.123
Würzburg	0.9994	605.61	676.58	0.131

Root mean square error (RMSE) magnifies the extreme differences in values of two datasets when compared to MAE. That is the reason for RMSE to be always higher than MAE. In other words, RMSE and MAE are close to each other for the locations that do not have extremely different values in Meteonorm and MERRA-2. For this, normalized root mean square error (NRMSE) could possibly give a better view of the quality of MERRA-2, because the RMSE is normalized by the full range of values meaning maximum measurement minus minimum measurement. For Belmullet, Erfurt and Schleswig the value of RMSE is moderately higher than MAE and this means that the differences in sunnier months of the year are high. In case of Belmullet despite having a relatively higher RMSE to MAE, looking at NRMSE (0.090) will help to verify that the results are still good. Additionally, NRMSE for Innsbruck (0.077) confirms the low error of MERRA-2 considering highest distance and altitude difference between its datapoints.

Figure 2 shows the average air temperature distribution for every month of the year for all the considered sites. The diagrams are almost perfectly the same except for Calama/El Loa and Innsbruck stations. By looking at Table II, it can be verified that these two locations have very high difference in their altitudes between their Meteornom and MERRA-2 points. For both these locations, MERRA-2 points stand in higher altitude and that explains why MERRA-2 air temperature values are lower. The results for El Kharga too seems to differ slightly from Meteonorm although the datapoints for this station have a relatively short distance and altitude difference.

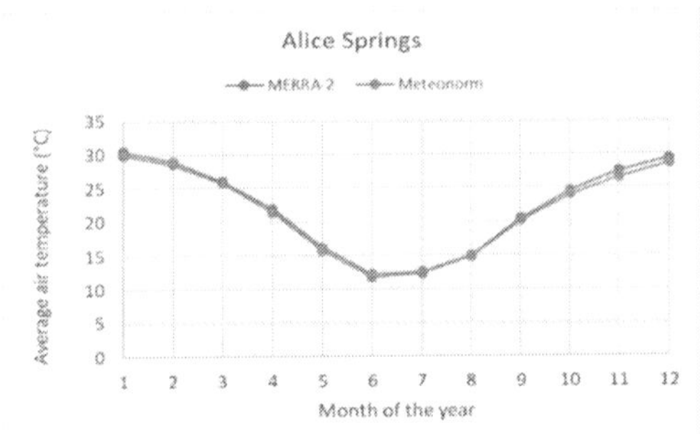

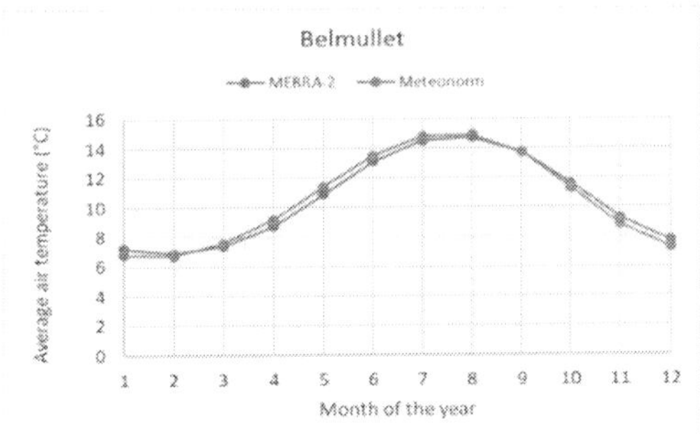

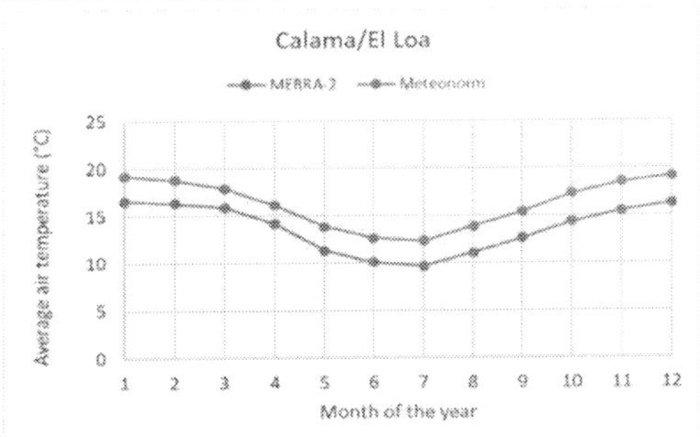

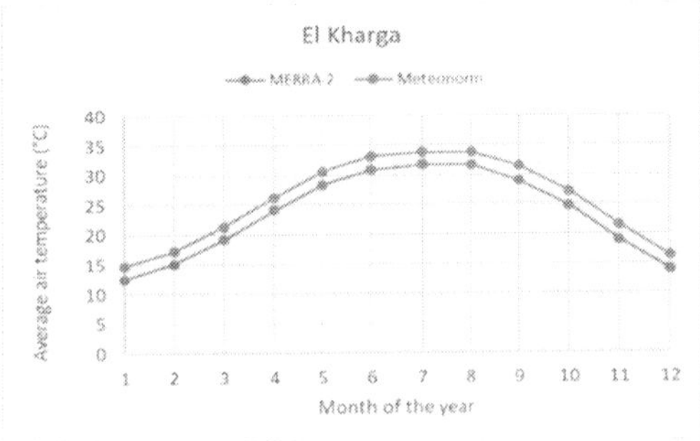

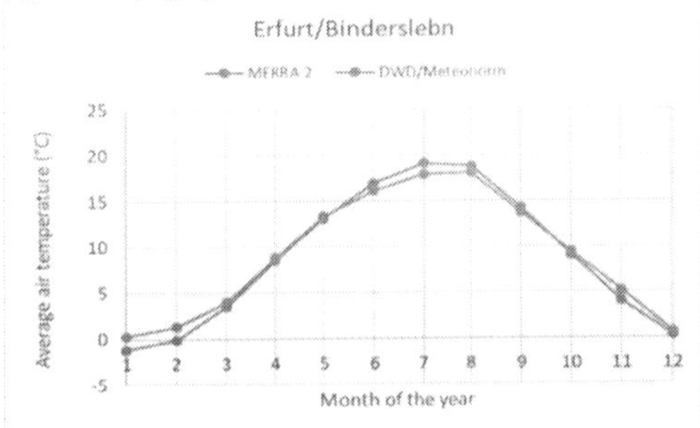

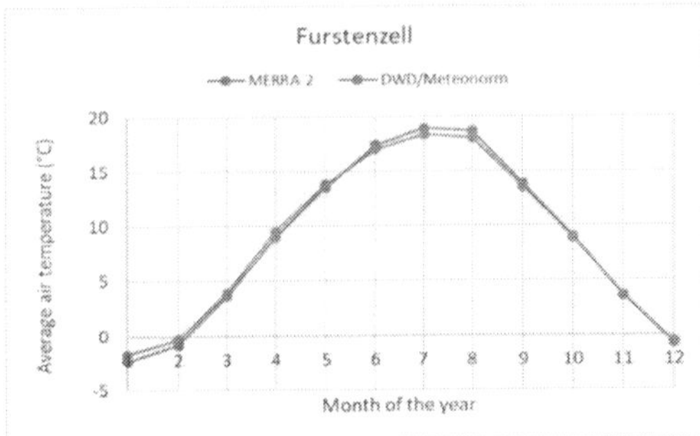

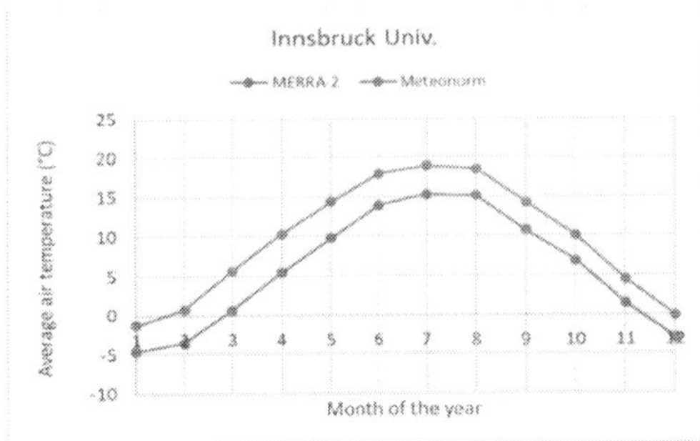

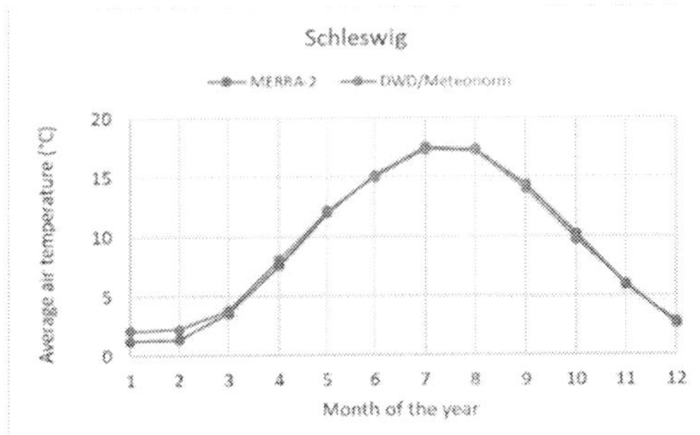

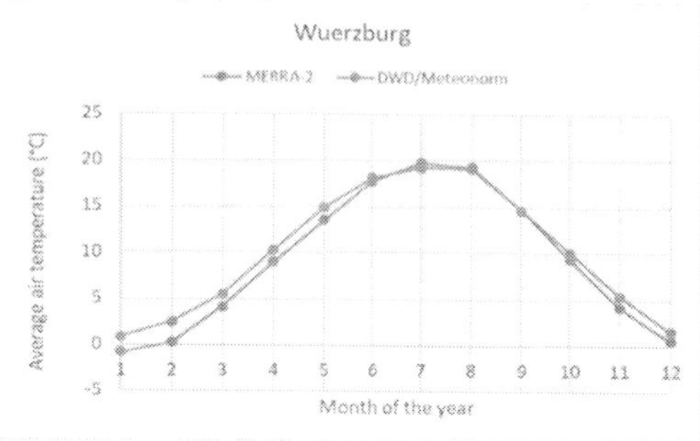

Figure 2: MERRA-2 and Meteonorm/DWD comparison for monthly average values of air temperature for all locations through years 2000–2019.

Table VI shows the calculated values for correlation coefficients, MAE, RMSE and NRMSE errors for monthly average air temperature. The correlation coefficients are all higher than 0.99 that confirms a very precise correlation for MERRA-2 to local measurements. MAE values are very outstanding except Calama/El Loa, El Kharga and Innsbruck as expected from the diagrams in Figure 2. Again for Calama/El Loa and Innsbruck, the altitude difference might be the reason. Looking at RMSE column, it is verified that the values are minimally different from the corresponding MAE values, and this concludes that the differences in corresponding air temperature between MERRA-2 and Meteonorm values are not extreme. NRMSE values are also very good except those three stations mentioned above. As expected, Calama/El Loa has the highest NRMSE still due to very high altitude difference and distance.

Table VI: Errors and correlation coefficients of MERRA-2 monthly average values of air temperature in comparison to Meteonorm/DWD through years 2000–2019.

Air Temperature	***R***	***MAE*** (°C)	***RMSE*** (°C)	***NRMSE***
Alice Springs	0.9995	0.50	0.55	0.030
Belmullet	0.9945	0.32	0.35	0.043
Calama/El Loa	0.9917	2.61	2.63	0.381
El Kharga	0.9998	2.27	2.28	0.120
Erfurt/Bindersleben	0.9987	0.75	0.886	0.049
Fürstenzell	0.9992	0.37	0.42	0.021
Innsbruck Univ.	0.9951	3.80	3.87	0.192
Schleswig	0.9984	0.37	0.45	0.029
Würzburg	0.9979	1.01	1.19	0.065

7 CONCLUSIONS

Rural areas in underdeveloped countries very often do not have access to the electricity network – even at growing needs for energy. Planning and building renewable energy systems is vital for these locations and their inhabitants. Since local weather stations are not available everywhere, MERRA-2 and its comparable reanalysis datasets become very attractive alternatives since they are freely available for the whole world and cover long periods of time. In this paper, the validity of MERRA-2 is examined for eight various locations around the world. The locations cover a wide range of distances (from 10 km to 28 km away) and altitude differences (from 20 m to 303 m) to the nearest local weather stations. Local measurements are taken from Meteonorm and DWD sites that have zero share of satellite or interpolated data.

After several comparisons and calculations of errors (Table V and Table VI), it can be verified that MERRA-2 is a valuable and stable dataset considering its availability since 1980 and coverage for the whole world. Comparing monthly average values, MERRA-2 shows a good correlation with local ground measurements. MAE values of 297 to 873 Wh/m^2d are calculated for monthly averaged values of daily global horizontal irradiated energy. In case of monthly averaged values of air temperature, MAE of 0.32 to 3.80 °C are calculated in comparison to Meteonorm and DWD stations. Highest error values happened in the locations that are very different in altitude or distance with their corresponding local weather stations however, the errors might not be due to low quality of MERRA-2. The errors are mainly happening since no MERRA-2 measurement point is exactly placed on a local weather station and therefore cannot be compared with the local measurements of its exact position. Considering hourly measurements, MERRA-2 values differ from the values of their nearby weather stations. However, it still can be said that it is countable enough considering its availability and high correlation to ground measurements especially for the applications requiring long historical average measurements and also for any location around the world with no access to reliable local ground measurements.

8 REFERENCES

[1] J. Santos, Y. Sakagami, R. Haas, J. C. Passos, M. N. Machuca, W. C. Radünz, E. Dias and M. M. Lima, "Wind Speed Evaluation of MERRA-2, ERA-Interim and ERA-5 Reanalysis Data at a Wind Farm Located in Brazil," in *ISES Solar World Congress 2019/IEA SHC International Conference on Solar Heating and Cooling for Buildings and Industry 2019*, 2019.

[2] D. Yang and J. M. Bright, "Worldwide validation of 8 satellite-derived and reanalysis solar radiation products: A preliminary evaluation and overall metrics for hourly data over 27 years," *Solar Energy*, pp. 3-19, 01 11 2020.

[3] T. Huld, R. Urraca, A. Gracia Amillo and J. Trentmann, "A global hourly solar radiation data set using satellite and reanalysis data," in *33rd European Photovoltaic Solar Energy Conference and Exhibition*, Amsterdam, 2017.

[4] P. Juruš, K. Eben, J. Resler, P. Krč, I. Kasanický, E. Pelikán, M. Brabec and J. Hošek, "Estimating climatological variability of solar energy production," *Solar Energy*, pp. 255-264, 12 2013.

[5] S. Pfenninger and I. Staffell, "Long-term patterns of European PV output using 30 years of validated hourly reanalysis and satellite data," *Energy*, pp. 1251-1265, 01 11 2016.

[6] Y. Sianturi, Marjuki and K. Sartika, "Evaluation of ERA5 and MERRA2 reanalyses to estimate solar irradiance using ground observations over Indonesia region," in *AIP Conference*, 2020.

[7] S. P. Durrani, S. Balluff, L. Wurzer and S. Krauter, "Photovoltaic yield prediction using an irradiance forecast model based on multiple neural networks," *Journal of Modern Power Systems and Clean Energy*, pp. 255-267, 05 03 2018.

[8] A. Khatibi and S. Krauter, "Validation and Performance of Satellite Meteorological Dataset MERRA-2 for Solar and Wind Applications," *Energies*, vol. 14, no. 4, 02 09 2021.

[9] Bosilovich, M. G., R. Lucchesi and M. Suarez, "MERRA-2: File Specification. GMAO Office Note No. 9 (Version 1.1)," 2016. [Online]. Available: https://gmao.gsfc.nasa.gov/pubs/docs/Bosilovich785.pdf. [Accessed 02 09 2021].

[10] "About Us," Meteotest, [Online]. Available: https://meteotest.ch/en/company/about-us. [Accessed 02 09 2021].

[11] Meteonorm, "Meteonorm documents," [Online]. Available: https://meteonorm.com/assets/downloads/broschuere-mn-7.1.pdf. [Accessed 02 09 2021].

[12] Meteonorm, "Meteonorm introduction," Meteonorm, [Online]. Available: https://meteonorm.com/en/. [Accessed 02 09 2021].

[13] Meteotest AG, J. Remund, S. Müller, M. Schmutz, D. Barsotti, P. Graf and R. Cattin, "Meteonorm Documents," 10 2020. [Online]. Available: https://meteonorm.com/en/meteonorm-documents. [Accessed 02 09 2021].

[14] "Deutscher Wetterdienst (DWD)," [Online]. Available: https://www.dwd.de/EN/aboutus/aboutus_node.html. [Accessed 09 2021].

[15] "Sphere of tasks," Deutscher Wetterdienst, [Online]. Available: https://www.dwd.de/EN/aboutus/tasks/task_node.html. [Accessed 02 09 2021].

[16] DWD, [Online]. Available: ftp://ftp-cdc.dwd.de/pub/CDC/observations_germany/climate/hourly/solar/ST_Stundenwerte_Beschreibung_Stationen.txt. [Accessed 09 2021].

[17] DWD, [Online]. Available: ftp://ftp-cdc.dwd.de/pub/CDC/observations_germany/climate/hourly/air_temperature/historical/TU_Stundenwerte_Beschreibung_Stationen.txt. [Accessed 09 2021].

THE LONG-TERM OF THE ALBEDO STABILITY UNDER DIFFERENT WEATHER CONDITIONS

*Sergio Suarez[1], Eneko Ortega[2], Juan Carlos Jimeno[2], Jose Rubén Gutierrez[2], Vanesa Fano[2], Jose Manuel Rivas[1], Gustavo Navas[1], Ignacio Fernández[1], Sofia Rodríguez-Conde[1]
[1]Enertis Applus, Av. Bruselas 31, 28108 Madrid, Spain; www.enertis.es
[2]EHU/UPV, Escuela de Ingeniería de Bilbao, C/ Plaza Ingeniero Torres Quevedo,1, 48013 Bilbao

Phone: +34 91 651 70 21, *sergio.suarez@enertis.es

ABSTRACT: Fluctuations at surface albedo measurements are widely known because of seasonality, weather, and heterogeneity on the ground, among others. Additionally, even some studies provide an analysis of cloud enhancement (CE) and albedo enhancement (AE) due to sky cloudy conditions. In this paper, we report about surface albedo long-term changes under different weather conditions compared to a clear sky reference dataset built up using two different measurement systems and two different soils. A good statistical relationship between cloudiness and surface albedo long term stability has been found. The final results exposed in this paper will help to optimize the duration and characteristics of the albedo measurement campaigns.
Keywords: Monitoring, Photovoltaic, Pyranometer, Stability, Qualification and Testing, Surface albedo

1 INTRODUCTION

Bifacial technology has experienced a strong impact on the photovoltaic sector in recent years, manifesting an important growth of this technology in the market. The key factor on bifacial technology comes from his bifacial gain thanks to the irradiance reflected by the ground.

This reflective property calls albedo tells you how much irradiance can be reflected by a surface. The determination of the surface albedo has consequently been a theme that has developed strongly, due to its close relationship with this technology.

The measurement of surface albedo is usually carried out using two sensors like c-Si cells or pyranometers (one oriented to the ground and the other towards the sky) where the division between the two signals gives the albedo value.

There are different approaches to the albedo measurement methodology. The experiment shown in this paper aims to help find the most cost-effective solution analyzing the long-term stability of the albedo as a function of different weather conditions.

2 ALBEDO VS SKY DATASET

2.1 Experimental set-up

The experimental set-up is made up of two types of albedometers: one consisting of dual pyranometers and another one consisting of two c-Si cells. A sample of 56 days has been fully tested over six months under different soil combinations, as figure 1 shows, using natural gravel (A) and artificial white soil (B).

Albedo measurements have been collected using the same datalogger at 1-s time step resolution to compare both configurations. The system has been placed at 1 meter height between the sensor and ground during the whole experiment whereas the representative shadow effects[1] are avoided. In this presentation, the results found when comparing the albedo measurement obtained by two different measurement systems under various soils and weather conditions will be presented.

It should be noted that the environment of the measurements throughout the experiment for the cell and pyranometer systems has remained constant.

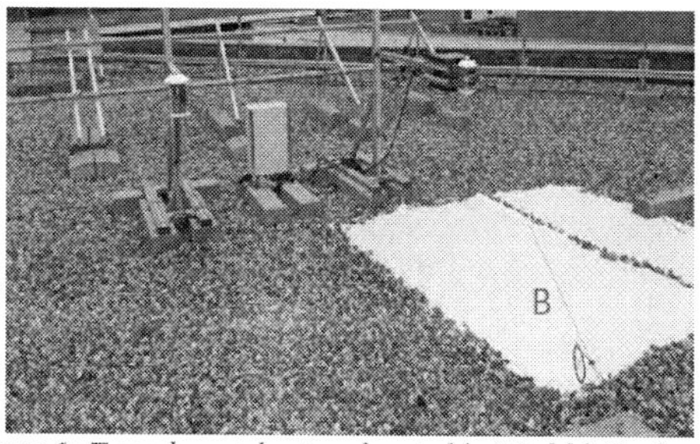

Figure 1: Experimental set-up located in Madrid, Spain.

2.2 Parametrization

The availability of daylight during the day was studied using two different evaluation methods. Both methods were applied to find the cloudiness characteristic parameters of each day of the dataset to compare them with albedo stability.

The first method was carried out using the K_t index clearness to find different cloudiness degrees and compare with the albedo fluctuations. On the other hand, a second method was used based on the half-day sunshine duration SSN[2], parameter used to show the influence of cloudiness and turbidity due to many possible cloudiness types and cloud covers[3].

The cloudiness and sunshine parameterization of characteristic daily illuminance conditions can be done using both indexes as shown Figure 2. In this work, this classification was compared with the mean albedo (MA) and standard deviation albedo (SDA) values.

$$K_t = \frac{G}{G_{ex}}$$

$$SSN = \begin{cases} 1 & if \ \frac{(G_h - D_h)}{Zenith\ distance} > 120\ W/m^2 \\ 0 & \end{cases}$$

Figure 2: Clearness index K_t and Sunshine duration SSN

Where G_{ex} is the extraterrestrial irradiance (1326 W/m^2), G_h and D_h are hourly Global and diffuse irradiance respectively and Zenith distance is 90° minus the body's altitude above the horizon.

2.3 Clear sky reference dataset

Using both parameters commented previously we have differentiated three different day types according to the clearness index mean (MK_t), its standard deviation (SDK_t) and the absolute sunshine duration value (SSN).

For instance, a K_t boxplot is shown in figure 3 where clear days correspond to high MKt and low SDK_t values, partially cloudy days refer to slight low MKt and moderate SDK_t levels and finally, cloudy days correspond to relatively low MKt with regard to very large SDK_t signals.

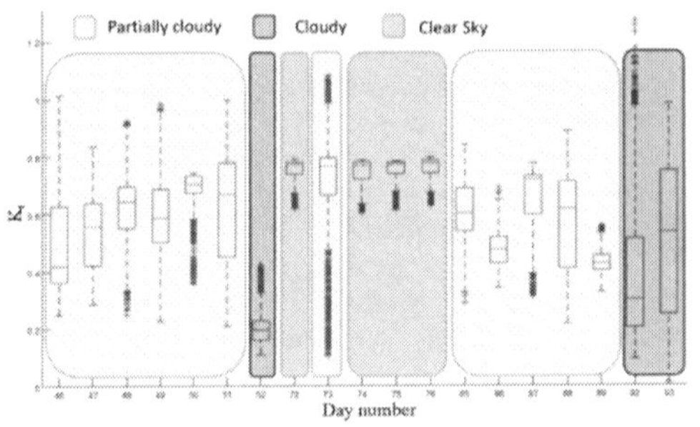

Figure 3: K_t Box plot diagram used for day classification

Additionally, SSN is used to analyse this dataset because of the existence of cloud enhancement (CE) and albedo enhancement (AE) phenomena caused by transient effects produced by different cloud types.
The implications of these transient effects in PV systems are being studied because of their negative potential impact on safety[4].

Even though SSN index shows days #52 and #92 are the same type of day, different cloudiness types are observed (see Figure 3). This is because day #92 with a $K_t \geq 1$ shows CE phenomenon caused by the passage of clouds[5], showing a different behavior than day #52 which has $K_t \leq 0.2$, meaning a very dense cloudy day.

If the albedo values found for the cells system and artificial white ground for day #19 are plotted (Figure 4), some effects are observed. On the one hand, clear sky albedo values show very good stability with respect to its mean value (with an average albedo of 54%). One the other hand, partially cloudy and cloudy days boxplots show much more dispersion in the albedo values due to a larger K_t dispersion.

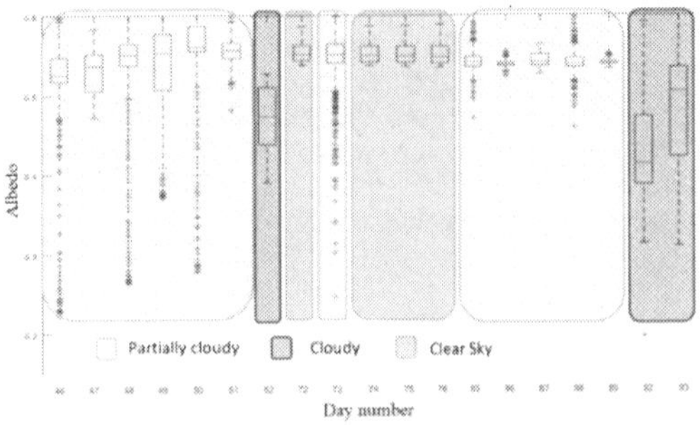

Figure 4: Albedo boxplot diagram based on the day classification.

Additionally, cloudy days not only show higher SDA values, but lower MA values with respect to partially cloudy days.

The dataset shows strong correlation between MA and SDA values and K_t and SSN indexes.

3 EXPERIMENTAL RESULTS

The results of both measurement systems and both soils are showed in this work. At this point, the aim is trying to find the quantitative correlation between K_t and MA.

A statistical linear fit between albedo and K_t index was done for each ground type and measurement system, where Pearson coefficient (R^2), p-value and root mean square error (RMSE) were calculated to understand how well all groups fit.

For example, Figure 5 shows the linear fit using data from the albedometer composed by c-Si cells and the artificial white ground, where the red line is the linear fit and green dots are the errors between estimated and measured data.

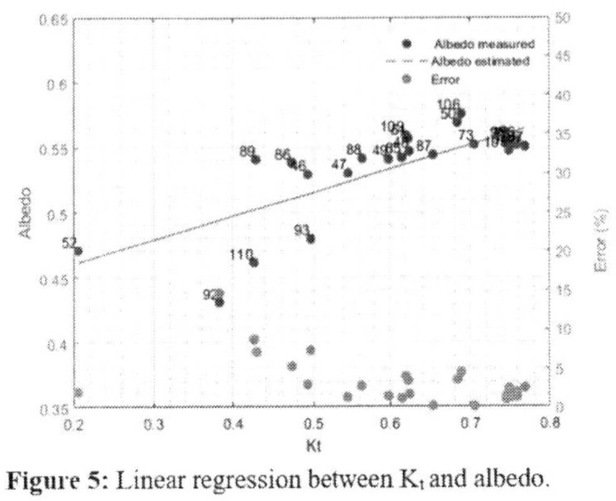

Figure 5: Linear regression between K_t and albedo.

There is a moderate direct correlation between K_t and albedo (R^2=0.59) and good statistically significance ($p<0.01$), however the errors statistically increase as cloudy days accumulate. For example, if all the dataset points of this experiment are included, the RMSE value would be 4.46% (which corresponds with an uncertainty of ± 8.92% for a coverage factor K=2).

Figure 6 shows a clear difference between partially

cloudy data, cloudy and clear sky data, regardless of their deviations. We consider a climatic effect such as rain or another related to high cloudiness can significantly affect the average value of the sample.

As shown below, we split data into two linear regression models, in which we have separated rainy days (correlated with high cloudiness) and dry days with or without partial cloudiness.

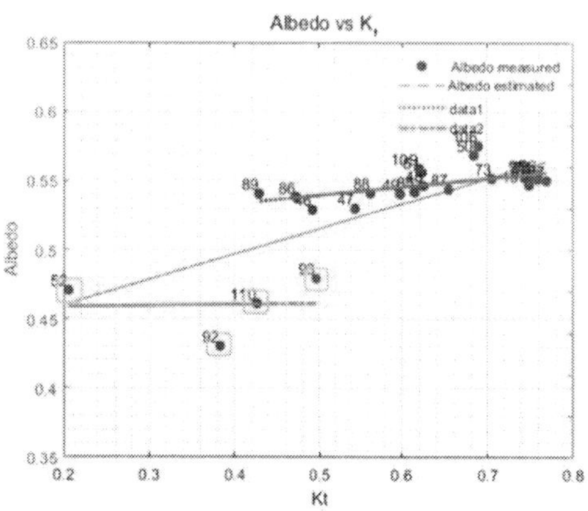

Figure 6: Separate linear regressions after splitting in the rainy and cloudy days.

Yellow dots correspond to cloudy data from sky dataset, split in another linear regression. Similar slope could show a similar albedo vs k_t behavior with some reduction-bias of albedo average because of some climatic uncontrolled variable like rain or another weather element correlated with high cloudiness and not tested in this work.

Table 1 collects all the linear regression parameters by every group tested, skipping the rainy days. The linear regression parameters show below are the slope and the interception (B) of a linear fit.

Table I: Linear regression parameters summary (skipping the rainy days)

Sensor	Soil	Slope	B	R^2	RMSE (%)
Pyrs	White	0.061	0.48	0.72	1.27
Pyrs	Gravel	0.065	0.21	0.89	2.29
Cells	White	0.061	0.51	0.77	1.51
Cells	Gravel	0.066	0.19	0.70	3.81

During this analysis, a systematic RMSE reduction for all groups has been found. For instance, the RMSE value decreases from 4.46% to 1.51% for the case of the white soil and cell system previously mentioned.

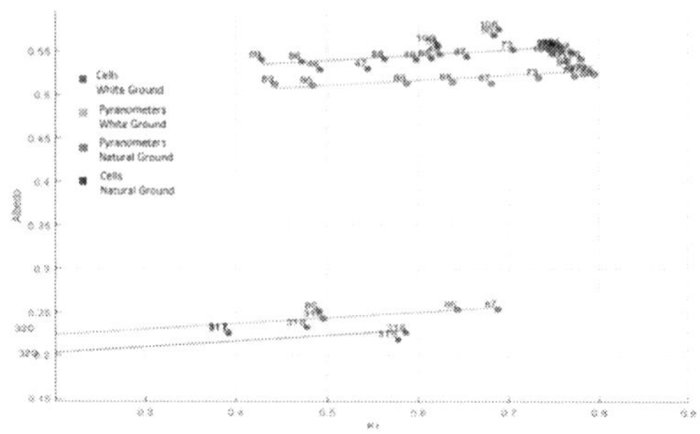

Figure 7: K_t vs albedo of all measurement systems are graphically compared excluding all rainy days.

Similar tendencies can be observed in Figure 7 comparing albedo values for a certain soil type (white and natural ground) regardless the type of albedometer used. Slight differences are related to the different spectral response of pyranometer and cell albedometers. The spectral effects in albedo have been widely studied [6].

To conclude, a good statistical relationship between surface albedo and K_t and SSN indexes was found using different experimental set-ups under constant soils. This experiment has allowed checking the long-term stability of the albedo measurements.

From this work periodic measurement campaigns under clear sky conditions are highly recommended because of their very low RSME, uncertainty and high representativeness, whereas long-term measurements could be affected by inherent cloudy conditions.

4 CONCLUSIONS

- Direct correlation between SSN and K_t and albedo measurements was found.
- Very strong decrease in albedo values has occurred during rainy episodes.
- Using non-rainfall days long-term albedo measurements shows high stability and similarity with short term measurements

5 REFERENCES

[1] Y. Qin and H. He, "A new simplified method for measuring the albedo of limited extent targets," *Sol. Energy*, vol. 157, no. September, pp. 1047–1055, 2017, doi: 10.1016/j.solener.2017.09.027.

[2] D. Stanislav and K. Richard, "Classification of daylight conditions in cloud cover situations," *Light Eng.*, vol. 23, no. 1, pp. 4–14, 2015.

[3] R. Kittler, R. Perez, and S. Darula, "A set of standard skies. Characterizing daylight conditions for computer and energy conscious design. NASA LaRC Global Solar Mapping View project Calibration of an artificial sky with sun after refurbishment View project," no. July 2015, 1998, doi: 10.13140/RG.2.1.4798.7048.

[4] M. Lave, J. Kleissl, and E. Arias-Castro, "High-frequency irradiance fluctuations and geographic smoothing," *Sol. Energy*, vol. 86, no. 8, pp. 2190–2199, 2012, doi: 10.1016/j.solener.2011.06.031.

[5] C. A. Gueymard, "Cloud and albedo enhancement impacts on solar irradiance using high-frequency measurements from thermopile and photodiode radiometers. Part 1: Impacts on global horizontal irradiance," *Sol. Energy*, vol. 153, pp. 755–765, 2017, doi: 10.1016/j.solener.2017.05.004.

[6] M. Gostein, B. Marion, and B. Stueve, "Spectral Effects in Albedo and Rearside Irradiance Measurement for Bifacial Performance Estimation," *Conf. Rec. IEEE Photovolt. Spec. Conf.*, vol. 2020-June, pp. 0515–0519, 2020, doi: 10.1109/PVSC45281.2020.9300518.

OPEN-SOURCE GEOSPATIAL DATA SERVICE WITH AN APPLICATION IN IRRADIANCE MODELLING FOR VIPV

Evgenii Sovetkin*, Neel Patel, Andreas Gerber, and Bart E. Pieters

IEK5-Photovoltaics, Forschungszentrum Jülich
52425 Jülich, Germany

ABSTRACT: We present an open-source service, PV-GRIP (Photovoltaics Geographic Raster Image Processor), that provides unified access to multiple types of geospatial data and performs a series of computations relevant to photovoltaics (PV). PV-GRIP features distributed computation design and provides decentralised data storage facilitating sharing data and PV computations across the Internet.

The necessity of such a tool is motivated by the growing range of PV applications where the local environment has a substantial influence on the performance of PV modules (e.g., building-integrated (BIPV) and vehicle-integrated photovoltaics PV (VIPV)). Hence, PV-GRIP collects various geospatial data; combining multiple geospatial data sources can improve yield simulations for those PV applications.

Additionally, we introduce a Simple Sky Dome Projector (SSDP) C-library for modelling irradiance. It computes sky domes according to the Perez all-weather sky model and projects this sky on a tilted surface. Unlike similar libraries, in SSDP, sky parameters are easy to change, which allows to leverage accuracy and computation time.

Paper demonstrates the utility of PV-GRIP service with an application in VIPV. The data provided by our service is used for irradiance modelling for a vehicle travelling in various areas in Germany.

Keywords: geospatial data, BIPV, VIPV, distributed computation, irradiance modelling

1 INTRODUCTION

Recently an increasing number of photovoltaics (PV) applications appeared, where the local environment surrounding PV modules plays a significant role in their performance. In such applications, PV modules no longer find themselves unobstructed from the sky dome. The local environment starts playing a substantial role in the way modules are exposed to the sun, which eventually affects the modules' performance.

Among those PV applications are building-integrated PV (BIPV), vehicle-integrated PV (VIPV), road-integrated PV, and agricultural PV (Agri-PV). In the VIPV and BIPV, modules are affected by shading and reflections from neighbourhood buildings and vegetation. For the VIPV, when a vehicle is moving, the environment and shading conditions are highly dynamic. For the Agri-PV, plants' seasonal cycles affect albedo and, hence can influence the performance of bifacial modules.

In order to study the environmental impact, one requires to describe it. Usually, this means that one associates a set of values with a particular location on the Earth. Hence, we refer to such data as *geospatial data*.

As we shall see later, different types of geospatial data are available that are relevant to PV. Moreover, access to such data is spread among many data providers. Thus, the first motivation behind PV-GRIP is to have **unified access to various geospatial data sources**.

One of the characteristics of geospatial data is that it occupies large storage volumes. For example, high-resolution aerial imaging can occupy several terabytes of compressed image data for a single region in Germany. Hence, when one works with multiple data sources, it poses a challenge that data does not fit on a single machine, and the storage speed restricts access to data.

The property of geospatial data to be big motivates the design of the PV-GRIP tool. Therefore, we choose a **distributed computing/storage design**. Such design is easily scalable, allowing PV-GRIP to be used by many users simultaneously and perform computations over large geographical areas.

In addition to serving the data, PV-GRIP uses several libraries to perform various PV computations. Such computations include **irradiance and shading simulations**. We demonstrate the utility of those computations in PV-GRIP with irradiance modelling for a VIPV application. There we compute a plane-of-array irradiance for a module installed on a vehicle that moves along a given route. Such computation requires data from multiple sources: weather data (global horizontal and diffused horizontal irradiance) and topographical data (describing buildings and vegetation).

The irradiance computations are facilitated by several third-party libraries, as well as libraries we develop. In this paper, we introduce our open-source

*Corresponding author: e-mail: e.sovetkin@fz-juelich.de

Simple Sky Dome Projector (SSDP) library. The SSDP computes sky domes according to the Perez all-weather sky model [26] and can project this sky on a tilted surface. In addition, it can process topographical data and compute a horizon and thus take shading into account.

Originally, PV-GRIP is based on the "open-elevation" project, [2]. PV-GRIP extends this project to fully distributed and easily scalable, accommodating different data types and providing a series of computations relevant to PV. The source code of the PV-GRIP is available at [25][1], and we provide reproducible Docker images available from [11].

Lastly, we remark that the current target audience for the PV-GRIP tool are researchers in PV. Therefore, we welcome contributions and suggestions to the project, believing that this can establish better simulation models for several PV applications.

The paper is organised as follows. Section 2 elaborates on currently included supported geospatial data. Section 3 discusses current simulations tools included in PV-GRIP. There we also introduce our SSDP library and demonstrate a small example of an application in VIPV simulations. Section 4 elaborates on the distributed design of PV-GRIP. Lastly, Section 5 summarises the paper and provides a view of the potential future of PV-GRIP.

2 GEOSPATIAL DATA

In this section, we provide examples of geospatial data currently supported by PV-GRIP and provide the data sources. For every type of geospatial data, we argue its relevance to the PV module performance.

As mentioned previously, geospatial data describes the environment on the Earth, i.e., every geographical location is associated with a set of numerical values. This section discusses several such data types that are integrated into PV-GRIP and relevant to PV in general.

PV-GRIP provides such data in a uniform way across different geospatial data sources. For example, a user request defines a rectangular area with the desired resolution and a data source, and PV-GRIP outputs a geospatial raster image. In addition, a user can select the output from various data types, such as GeoTIFF, pickle, or png.

PV-GRIP relies on many libraries. The main libraries for processing geospatial data are the GDAL [16] (for processing geospatial raster images) and PDAL [8] (for processing 3d point cloud). Other libraries are more specific to a data source and cited in the following subsections.

[1] `https://github.com/esovetkin/pvgrip`

2.1 Topographical data

Topographical data describes the surface elevation of the Earth. Different data sources provide terrain data with different resolutions. For example, the shuttle radar topography mission (SRTM, [32]) provides coarse data identifying general terrain outlook. In contrast, areal LIDAR measurements create more accurate 3d terrain models, where objects like buildings, cars, and vegetation are visible.

The topographical data is relevant for PV as it allows computation of a horizon, and for instance, enables calculation of shadings on any day of the year.

PV-GRIP combines multiple data sources and a wide range of data types. For our application, we use the following openly accessible data:

- a satellite-based elevation measurement (SRTM) data [32, 13],
- light detection and ranging (LIDAR) points 3d data, and aerial imaging in North Rhine-Westphalia (NRW), [1, 10, 6],
- OpenStreetMap data, [24],
- and the National Solar Radiation Database (NSRDB) weather data, [30, 23].

An example of the SRTM data is demonstrated in Figure 6. Figure 6 depicts a region sampled at $600\text{m} \times 600\text{m/pixel}$ covering the Italian peninsula, Alps, eastern Balkans, Sicily, and Corsica. Note, the maximum resolution of the SRTM data is $90 \times 90\text{m}^2$ per pixel.

This base topography provided by SRTM data is augmented with higher resolution LIDAR data, where such data exists. We sample the LIDAR scan points with a resolution of $0.3 \times 0.3\text{m}^2$ per pixel, computing various statistics on the points occurring in each $0.3 \times 0.3\text{m}^2$ region. Figures 1 and 2 demonstrate the so-called "hillshade" images generated from the LIDAR point clouds, [39]. Figure 1 is generated using the maximum value statistic that captures vegetation and buildings. Figure 2 is generated with the minimum value statistic, and it gives a view without the influence of trees.

2.2 Aerial imaging

Another type of raster data is aerial imaging that provides orthographic images of the Earth's surface. Such imaging is commonly performed in visible and near-infrared (NIR) light spectra. The NIR light indicates the presence of vegetation, as vegetation reflects much light in this spectrum. For the PV, such data is relevant, as one can use it to study the albedo of surfaces and vegetation at different seasons.

In the NRW data we use in our application, [10], these aerial orthophotos are given with a resolution of $0.1 \times 0.1\text{m}^2$ per pixel. Figure 3 demonstrates an

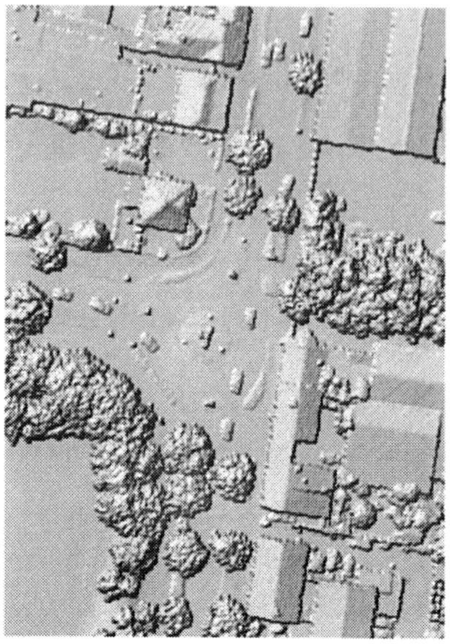

Figure 1: "Hillshade" visualisation of terrain sampled from LIDAR with the **maximum** statistic (sampled at 50cm/pixel). Maximum resolution: ≈ 30cm/pixel

Figure 2: "Hillshade" visualisation of terrain sampled from LIDAR with the **minimum** statistic (sampled at 50cm/pixel). Maximum resolution: ≈ 30cm/pixel

example of such images (in RGB) and Figure 4 (in NIR).

2.3 Vector maps data

Commonly used maps like OpenStreetMaps provide geospatial vector data. Such data describes streets, buildings, and other geographical objects as polygons with vertices in a longitude/latitude coordinate system. The OpenStreetMap data can serve as an alternative to the topographical data, providing locations of objects on the ground. Currently, PV-GRIP includes a basic render engine, [33], for the OpenStreetMap data, allowing to query positions of different objects, such as buildings, forests, water, and roads. Figure 5 demonstrates the same region as in Figures 1 and 3 identifying locations of buildings, roads, and a green area.

3 SIMULATIONS METHODS

In addition to providing unified access to various geospatial data, PV-GRIP focuses on delivering a set of simulation tools allowing different computations with geospatial data relevant to PV. In this, we discuss a set of tools currently included in PV-GRIP, and we demonstrate an example of irradiance computations for a VIPV application.

3.1 GRASS

The GRASS GIS, [17], ecosystem includes a variety of tools to work with geospatial data. For GRASS, in PV-GRIP, we utilise a single function, "r.sun".

The "r.sun" function implements a solar irradiance model that computes direct, diffuse and reflected solar irradiation for given atmospheric conditions and time. The atmospheric conditions are described by the Linke atmospheric turbidity coefficient, real-sky beam and diffuse radiation coefficients (to model cloudiness).

Further, "r.sun" can be used to obtain binary maps of shadows. For that, one computes solar incidence angle raster map, where "nan" values correspond to areas where the sun is not directly visible. Figure 7 shows an example of such a binary map of shadows.

3.2 SSDP

The SSDP (Simple Sky Dome Projector) is a library for modelling irradiance. It computes sky domes according to the Perez all-weather sky model, [26], and projects this sky on a tilted surface. Unlike similar libraries, in SSDP, sky parameters are easy to change, which allows to leverage accuracy and computation time. Furthermore, SSDP allows more efficient computations for routes.

The SSDP implements a sky dome for diffuse light modeling. The sky dome consists of a variable sized hexagonal mesh. A user can specify the number of zenith discretizations. In our examples below, we use the value of 10, which corresponds to 269 sky patches. Higher values lead to more accurate shadow shapes.

The intensity of each sky patch may be computed using either a uniform sky approximation or the

Figure 3: Aerial RGB orthographic image (sampled at 50cm/pixel). Maximum resolution: 10cm/pixel

Figure 4: Aerial NIR orthographic image highlighting vegetation (sampled at 50cm/pixel)

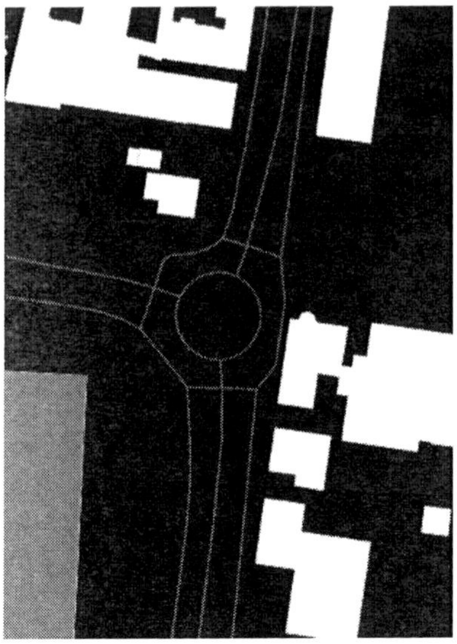

Figure 5: OSM sampled data indicating of streets (red lines), buildings (white) and a playground (green)

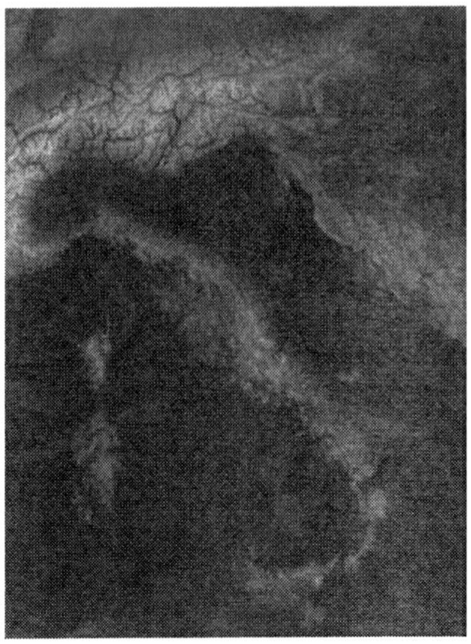

Figure 6: A region from SRTM data covering the Italian peninsula, Alps, eastern Balkans, Sicily, and Corsica (600m/pixel)

Figure 7: Shadow cast on the 1st July 08:00

previously mentioned Perez all-weather sky model, which is commonly used for PV applications (e.g., in PVlib [19]). Note, that the hexagonal mesh is only used for diffuse light and thus does not include the sun itself. The sun is treated separately as a point source. The solar position is computed using the PSA algorithm [4].

In addition to modelling the sky dome, SSDP implements various tools to handle topographies. Most importantly, SSDP can compute a horizon for a specific location in a topography. The final component in SSDP is a transposition model to project the irradiance on a tilted surface, including the shading by the horizon. The surface tilt and height are specified independently of the topography. However, SSDP also implements tools to offset a surface from the topography and optionally also rotates the tilted surface by the surface normal (e.g., it can adapt the tilt of a vehicle-mounted PV panel to a tilted road).

The SSDP supports several simulation modes:

- Static: For one or more tilted surfaces at specified locations, the irradiance is computed for all provided time, GHI, and DHI values.
- Integral Static: The irradiance is integrated over the provided time, GHI, and DHI values for one or more tilted surfaces at specified locations.
- Route: The irradiance is computed moving through a list of provided locations. This mode is specifically included for VIPV applications.

Figure 8 shows the effective irradiance computed for the same region as in Figure 1. The computations are performed in the normal direction to the surface, with GHI $= 500$ W/m^2 and DHI $= 200$ W/m^2.

Figure 9 depicts integrated effective irradiance over a week from the 3rd until the 10th of June 2015. The GHI and DHI values are the measurements at the location from the same region in Germany.

Table I: Comparison of integrated irradiance along a route with and without the influence of local environment for a horizontal surface. Atmospheric irradiance values (GHI, DHI) are taken from historical observation.

LIDAR simulations	SRTM simulations
191 Wh/m^2	237 Wh/m^2

3.3 Application in VIPV simulations

In this section, we demonstrate a simple example of PV-GRIP usage for an application in VIPV simulations. We utilise data from two sources: topographical data and the irradiance weather data (currently not included by default in PV-GRIP).

We compute the integrated plane-of-array irradiance of a horizontal module on top of a car that moves along a given route. For such computations in SSDP, we use the recorded time and position of the vehicle, the global horizontal irradiance (GHI), and the diffused horizontal irradiance (DHI) atmospheric values at the travel time provided by NSRDB, [30, 23]. For the horizon computations, we use SRTM and LIDAR data sources.

Figure 10 depicts a route along which we made our computations. Most of the distance is covered on a highway. Table I compares integrated irradiance along the chosen way, computed using coarse SRTM data (without building and vegetation), and more accurate LIDAR data (sampled with the maximum statistic). Table I registries a 24% decrease in integrated irradiance. Such decline can be attributed to the shading cast from the non-urban highway's environment.

4 DISTRIBUTED ARCHITECTURE

In this section, we elaborate on the distributed architecture design of PV-GRIP. We motivate such design since geospatial data is big, and a successful tool operating with such data must be easily scalable in storage and computation power.

Additionally, the service has to run on a range of heterogeneous hardware, from a commodity desktop to a cloud platform nodes. Hence, all services in PV-GRIP are built around the easily deployable Docker containers [12]. Docker allows easy scalability to several hundreds of physical nodes, with fault-tolerance being achieved by replicating services.

Figure 8: Irradiance at GHI $= 500$ W/m^2, DHI $= 200$ W/m^2

Figure 9: Integrated irradiance over a week from the 3rd until the 10th of June 2015

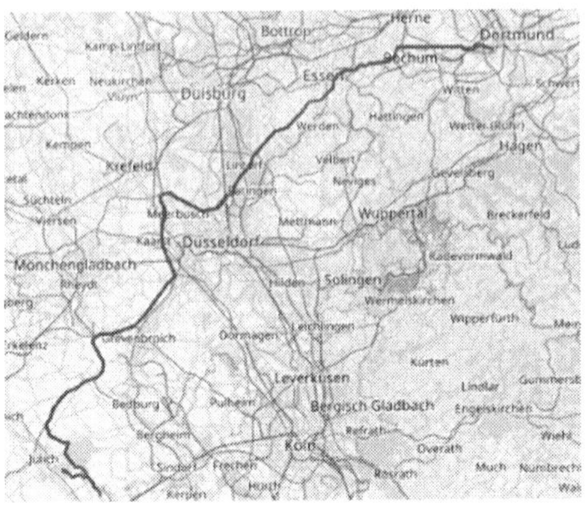

Figure 10: An example of a route travelled by car (mostly on a highway)

Moreover, our design is optimised when different services *may be* located far away from each other. The physical distance between the nodes introduces latency in communications, which is usually not a problem in HPC systems with high-throughput network systems.

Figure 11 depicts a schematic representation of the distributed network architecture of PV-GRIP. It consists of a series of services that communicate with each other inside a network.

For the network architecture, we utilise a virtual private network system (e.g., Innernet [22] or Tailscale [34]) that works on top of Wireguard [38]. Such a system establishes a secure peer-to-peer network communication between the nodes. The Wireguard provides a security layer with minimal overhead. We remark that precise network implementation is irrelevant for PV-GRIP; the only requirement is that services can communicate with each other.

In Figure 11, different services are indicated with an ellipse, whereas black rectangles indicate physical machines. Note, each physical computer may accommodate one or several services at once.

PV-GRIP consists of four service types. We elaborate on each of these services in the following subsections. Those services are:

- *webserver* service (Section 4.1): listens to user requests and sends back requested data or a result of some computation.
- *broker* service (Section 4.2): manages the exchange of short messages between nodes. For example, when a task is completed, the result is announced via a message.
- *compute* service: performs various computational tasks. The tasks queue is managed by a tasks management system (Section 4.3).
- *storage* service (4.4): manages data, its access and replication among multiple physical machines.

The following shows a typical flow of a data query process.

1. A user makes a query to a webserver.
2. The webserver generates a queue of tasks and sends it to a message broker.

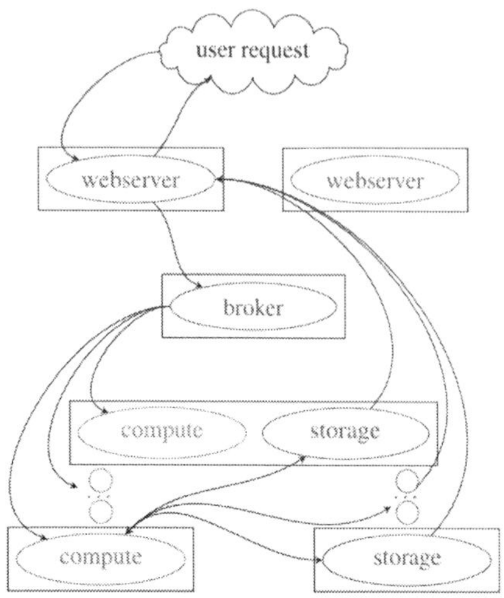

Figure 11: The distributed network architecture of PV-GRIP. Coloured ellipses represent different services. Black rectangles represent different physical machines. Arrows represent the flow of data and messages. A user sends a request to a webserver. The webserver generates a dependency graph of tasks and sends it to the broker. The broker instructs free compute nodes to start tasks. All required data needed to perform tasks is queried from the storage, and the results of all tasks are saved on the storage. The final result is queried by a webserver on its completion and is sent back to the user.

3. The message broker distributes queue tasks among the compute nodes.

4. All data required for a task is retrieved from the distributed storage, and the task results are saved in the distributed storage.

4.1 Webserver

The webserver is the single service that is open to the outside world. Users send queries to the webserver to obtain specific data or to run simulations. Multiple webserver services can run in parallel on multiple nodes. A round-robin DNS [37] has to be configured in the network to balance the webserver requests.

Currently, we use Bottle [5] for the webserver, which provides a simple API to the PV-GRIP services. However, in the future, we can add a more sophisticated web interface.

4.2 Broker

The message broker is a service that facilitates the exchange of short messages between services, provides a distributed in-memory hashtable, as well as a locking mechanism. All such requirements are satisfied with the Redis [28] message broker.

The locking mechanism guarantees no duplicated tasks, and the distributed hashtable is used to track user requests and corresponding tasks queues.

Currently, the Redis broker is the single service that is not duplicated in the network, and hence, it is a single point of failure. In the future, the Redis broker could be replaced by a more convenient service.

4.3 Task management system

For the task management system, we use Celery [7]. Tasks execution parameters and task results (usually a path to a file in a distributed storage) are distributed using messages via the broker.

Celery allows allocating different tasks to different queues, allowing certain tasks to be executed by specific nodes in the network. For example, this allows running computations that require GPU on nodes that have such devices.

4.4 Storage

The storage of PV-GRIP consists of two parts: a distributed storage and a local cache. The local cache contains data required for a task. The size of the local cache is bounded in size and has the recently used replacement policy. The distributed storage stores geospatial data, *geospatial index* as well as results of different tasks.

The storage is fault-tolerant by data being replicated among different physical nodes. Data replication also assists with parallel access to data from multiple computation nodes avoiding local storage speed bottlenecks.

PV-GRIP permits multiple types of distributed storage. Currently, we operate with Apache Cassandra [3, 14] and InterPlanetary File System (IPFS, [21, 20, 15]).

Apache Cassandra is a NoSQL distributed database that can be spawned easily within the Docker environment. We utilise Cassandra as a distributed storage and as a geospatial index.

Files in Cassandra are split into smaller chunks, with the chunk's order stored in a table indexed by the hash of the corresponding chunks.

The spatial index is a vital component of a tool that operates with geospatial data. It allows an efficient search of data that represents a specific region on the Earth. PV-GRIP uses our implementation of the spatial index that is hosted in the Cassandra and is based on a combination of Geohash [40, 35], Rtree [29, 36] and polygon geometry operations [31]. Our spatial index works similarly to the index in the geodatabase systems like PostGIS [27]. The choice of the spatial index implementation is motivated by its distributed design, opposing the centralised architecture of a PostGIS database. The source code for the spatial index and file storage can be found in [14].

Alternative to our implementation of the file storage in Cassandra, we utilise IPFS in conjunction with the IPFS-cluster. The IPFS is a protocol and peer-to-peer network for storing and sharing data. IPFS-cluster implements data replication among different nodes in the cluster network.

In our experience, Apache Cassandra is more expensive in terms of RAM than IPFS. IPFS can be run on heterogeneous machines and utilise under 500MB RAM. IPFS-Cluster in the CRDT mode provides more automatic safeguards against node failures than Apache Cassandra. In addition, IPFS permits lazy pinning of data that accelerates tasks' execution and data exchange.

Lastly, IPFS storage operated in a public network allows an easy way to share data and computation results between independent parties that use PV-GRIP. Hence this brings PV-GRIP a step closer to the decentralised design. The source code of the scripts for the IPFS based storage can be found in [15].

5 FUTURE OF PV-GRIP

In this paper, we presented a PV-GRIP service that allows gathering a range of geospatial data and provides a series of irradiance computations. We also introduced an open-source SSDP library to model irradiance.

An example with VIPV irradiance simulation shows the utility of PV-GRIP. We demonstrated that even an open highway environment provides shading that can cause a 24% drop in irradiance.

PV-GRIP is an ongoing project, and we would like to extend it further in the following ways. Firstly, PV-GRIP data sources collection can be appended by **other types of geospatial and remote sensing data**. So, for example, one can add soil-quality data for plant growth models assisting in bifacial Agri-PV-related applications. Further, in the subsequent releases of PV-GRIP, the NSRDB, [30, 23], and Sentinel Data, [9], will be added.

Currently, we focus on open-source data. However, the design of PV-GRIP is not restricted to free data providers, and one can include proprietary data providers. For example, Google Street View data and traffic congestion data can be beneficial in VIPV related computations.

Secondly, PV-GRIP can be extended by more **advanced optical models**. For some applications, irradiance modelling with the GRASS and SSDP libraries can be less accurate than more computationally intensive ray-tracing models.

Generative models can also be used to generate artificial LIDAR-like data for the regions where such data is not available by leveraging public data available in other sources.

PV-GRIP aims to be a **decentralised service** in the future, where data and computations results are shared among anyone across the Internet. IPFS distributed storage can facilitate such a goal.

Lastly, we remark that the current target audience for the PV-GRIP tool are researchers in PV. In the future, PV-GRIP can be made with a more user-friendly interface allowing a broader audience. We welcome contributions and suggestions to the project, believing that this can contribute to establishing better simulation models for several integrated PV applications.

ACKNOWLEDGEMENTS

The authors gratefully acknowledge the Helmholtz Data Federation (HDF) for funding this work by providing services and computing time on the HDF Cloud cluster at the Jülich Supercomputing Centre (JSC), [18].

This work is supported by the HGF project "Living Lab Energy Campus (LLEC)", as well as the "Street" project (Street — Einsatz von hocheffizienten Solarzellen in elektrisch betriebenen Nutzfahrzeugen, Förderkennzeichen: 0324275A).

References

[1] *3d-measurement data in North Rhine-Westphalia*. URL: https://www.bezreg-koeln.nrw.de/brk_internet/geobasis/hoehenmodelle/3d-messdaten/index.html.

[2] *A free and open-source elevation API*. URL: https://github.com/Jorl17/open-elevation.git.

[3] *Apache Cassandra: an open source NoSQL distributed database*. URL: https://cassandra.apache.org/.

[4] M. Blanco-Muriel, D. C. Alarcón-Padilla, T. López-Moratalla, and M. Lara-Coira. "Computing the solar vector". In: *Solar Energy* 70.5 (2001), pp. 431–441.

[5] *Bottle is a fast, simple and lightweight WSGI micro web-framework for Python*. URL: https://bottlepy.org/docs/dev/.

[6] H. Brüggemann. "Geo data infrastructure for North Rhine-Westphalia". In: 126 (Jan. 2001), pp. 181–186.

[7] *Celery: task queue management system*. URL: https://docs.celeryproject.org/en/stable/getting-started/introduction.html.

[8] P. Contributors. *PDAL Point Data Abstraction Library*. Nov. 2018.

[9] *Copernicus Open Access Hub (previously known as Sentinels Scientific Data Hub)*. URL: https://scihub.copernicus.eu/.

[10] *Digital orthophotos in North Rhine-Westphalia*. URL: https://www.bezreg-koeln.nrw.de/brk_internet/geobasis/luftbildinformationen/aktuell/digitale_orthophotos/index.html.

[11] *Docker containers for PV-GRIP*. URL: https://hub.docker.com/r/esovetkin/pvgrip.

[12] *Docker containers*. URL: https://www.docker.com/resources/what-container.

[13] T. G. Farr, P. A. Rosen, E. Caro, R. Crippen, R. Duren, S. Hensley, M. Kobrick, M. Paller, E. Rodriguez, L. Roth, et al. "The shuttle radar topography mission". In: *Reviews of geophysics* 45.2 (2007).

[14] *File storage and spatial index with Apache Cassandra*. URL: https://github.com/esovetkin/cassandra_io.

[15] *File storage with IPFS*. URL: https://github.com/esovetkin/ipfs_io.

[16] GDAL/OGR contributors. *GDAL/OGR Geospatial Data Abstraction software Library*. Open Source Geospatial Foundation. 2021.

[17] *Geographic Resources Analysis Support System*. URL: https://grass.osgeo.org/.

[18] B. Hagemeier. "HDF Cloud–Helmholtz Data Federation Cloud Resources at the Jülich Supercomputing Centre". In: *Journal of large-scale research facilities* 5.FZJ-2020-00569 (2019), A137.

[19] W. F. Holmgren, C. W. Hansen, and M. A. Mikofski. "pvlib python: a python package for modeling solar energy systems". In: *The Journal of Open Source Software* 3.29 (2018), p. 884.

[20] *IPFS Cluster: pinset orchestration*. URL: https://cluster.ipfs.io/.

[21] *IPFS: a peer-to-peer hypermedia protocol*. URL: https://ipfs.io/.

[22] *Innernet: a private network system that uses WireGuard under the hood*. URL: https://github.com/tonarino/innernet.

[23] *National Solar Radiation Database*. URL: https://nsrdb.nrel.gov.

[24] OpenStreetMap contributors. *Planet dump retrieved from https://planet.osm.org*. https://www.openstreetmap.org. 2017.

[25] *PV-GRIP: Photovoltaics Geographic Raster Image Processor*. URL: https://github.com/esovetkin/pvgrip.git.

[26] R. Perez, R. Seals, and J. Michalsky. "All-weather model for sky luminance distribution—preliminary configuration and validation". In: *Solar energy* 50.3 (1993), pp. 235–245.

[27] *PostGIS: a spatial database extender for PostgreSQL database*. URL: https://postgis.net/.

[28] *Redis: in-memory data structure store*. URL: https://redis.io/.

[29] *Rtree: a ctypes Python wrapper of libspatialindex*. URL: https://pypi.org/project/Rtree/.

[30] M. Sengupta, Y. Xie, A. Lopez, A. Habte, G. Maclaurin, and J. Shelby. "The national solar radiation data base (NSRDB)". In: *Renewable and Sustainable Energy Reviews* 89 (2018), pp. 51–60.

[31] *Shapely: Manipulation and analysis of geometric objects in the Cartesian plane*. URL: https://pypi.org/project/Shapely/.

[32] *Shuttle radar topography data*. URL: https://dds.cr.usgs.gov/srtm/version2_1/.

[33] *Smrender: a rule-based renderer for OSM data*. URL: https://github.com/rahra/smrender.

[34] *Tailscale: a secure network that just works*. URL: https://tailscale.com/blog/how-nat-traversal-works/.

[35] *Wikipedia: Geohash*. URL: https://en.wikipedia.org/wiki/Geohash.

[36] *Wikipedia: R-tree*. URL: https://en.wikipedia.org/wiki/R-tree.

[37] *Wikipedia: Round-robin DNS*. URL: https://en.wikipedia.org/wiki/Round-robin_DNS.

[38] *Wireguard: fast, modern, secure vpn tunnel*. URL: https://www.wireguard.com/.

[39] *gdaldem: tools to analyze and visualize DEMs*. URL: https://gdal.org/programs/gdaldem.html.

[40] *python-geohash: a fast, accurate python geohashing library*. URL: https://pypi.org/project/python-geohash/.

PV 2-STATE: A SIMPLE BUT ACCURATE SHORT-TERM PV POWER FORECASTING TOOL

Marius Paulescu[1], Nicoleta Stefu[1], Andreea Sabadus[1], Ciprian Dughir[2], Sorin Bojin[1]
[1]Faculty of Physics, West University of Timisoara, V Parvan 4, 300223 Timisoara, Romania
[2]Electronics and Telecommunications Faculty, Politehnica University Timisoara, V Parvan 2, 300223 Timisoara, Romania

ABSTRACT: This paper presents an innovative two-state model for short-term PV power forecasting (further identified as PV 2-state). The two states are differentiated by sunshine number, a binary quantity stating whether the Sun is shining or not. The model operates as follows: if the Sun is shining, the PV power is estimated using a real-time adjusted PV power clear-sky model; otherwise, the estimated clear-sky PV power is adjusted with the cloud transmittance. Sunshine number is the only physical quantity directly forecasted. Results of testing the PV two-state model highlight its ability to issue high-accurate forecasts. Each time the sunshine number is correctly predicted, the forecasts trace with accuracy the measurements. Due to the binary character of the sunshine number, its prediction is always a challenge. Even if a great effort was paid for increasing the accuracy in predicting sunshine number, further work is required.
Keywords: PV power; forecasting; sunshine number.

1 INTRODUCTION

The power-grid equilibrium is continuously changing with the load fluctuation. This equilibrium is furthermore threatened by the current trend towards a higher renewable energy penetration in the power grid. The inherent variability of wind and solar sources may induce significant energy fluctuation into the power-grid. As a response, the smart-grid concept is more and more present in the grid management. Thus, forecasting the output power of wind and solar plants become key tasks to provide smartness to the grid. Accurate forecasts will enable the computers to take near real-time control actions for balancing the grid. Basically, the smart-grid management requires intra-hour forecasts of the renewable output power, the technical constraints being related to the start-up time of conventional power sources connected as balancing instruments.

A huge effort is paid all over the world for increasing the accuracy of forecasting the energy production of PV plants [1]. Typically, a procedure for forecasting the output power of a PV plant involves two stages: (1) a realistic modeling of the PV plant operation and (2) forecasting the solar resource availability. Differently, this paper introduces a model which condenses the two stages in single stage. The accuracy of forecasting the PV power is strongly conditioned by the accuracy of forecasting the solar resource. Intra-hour forecasts of PV power are usually generated based on two types of models: (1) statistical extrapolation of the in-plane solar irradiance time series measured in-situ [2] and (2) all-sky imager-based algorithms [3].

This paper presents an innovative two-state model for intra-hour PV power forecasting (further identified as PV 2-state). The two states are differentiated by the sunshine number (SSN), a quantity stating whether the Sun is shining or not (see Sec. 2.2). The model operates as follows: if the Sun is shining, the PV power is estimated using a real-time adjusted PV power clear-sky model; otherwise, the estimated clear-sky PV power is adjusted with the cloud transmittance. Therefore, SSN is the sole quantity directly forecasted. The clear-sky model for PV power forecasting can be a simple empirical equation fitted on the data measured in-situ. The PV 2-state tool emerges naturally as an upgrade of our two-state model for short-term forecasting of solar irradiance [4, 5].

The accuracy of the PV 2-state model is assessed in comparison with the standard persistence. The performances and limitations of the model are studied in relation to the forecasting horizon and the variability in the state-of-the-sky. Generally, PV 2-state demonstrates notable performance in all performed tests.

2 DATA

2.1 Experimental setup

Data recorded on the Solar Platform of the West University of Timisoara, Romania (http://solar.physics.uvt.ro/srms) are used in this study. The town of Timisoara (45°46'N; 21°25'E and 85 m asl) has a temperate climate, (Köppen climate classification Cfb [6]). The Solar Platform is equipped with various facilities for conducting studies on energy – meteorology field. This study was conducted with data recorded on FORPV – an experimental setup designed for testing the PV power forecasting models (Fig. 1). The experimental setup consists of a PV system (540 Wp) with the PV modules connected to a pure resistive constant load. The horizontal global and diffuse solar irradiances and the main atmospheric parameters are measured with an automatic Delta-T Devices WS-GP2 weather station equipped with a SPN1 Pyranometer. The in-plane solar irradiance is measured with a Kipp & Zonen SMP10 pyranometer (class A, ISO9060:2018). All the parameters are recorded 24/7 at equal time intervals of 4 seconds.

2.2 Post-processed data

The performance of a PV plant depends essentially on whether the direct solar radiation is incident or not on the PV modules surface. The rapid variation of solar irradiance causes the so-called solar ramp problem, the term referring to the grid management when solar irradiance changes rapidly, causing a massive change in the output power. An appropriate parameter describing the relative position of the Sun and clouds has been introduced in [7]. This is the sunshine number, defined as a time dependent random binary variable:

$$SSN_t = \begin{cases} 0 & \text{if the sun is covered by clouds at time } t \\ 1 & \text{otherwise} \end{cases} \quad (1)$$

Figure 1: The experimental setup FORPV. Photo taken on May 2021.

Series of *SSN* values have been derived using the WMO sunshine criterion [8]: the Sun is shining at time t if the direct-normal solar irradiance exceeds 120 W/m²:

$$SSN_t = \begin{cases} 1 & \text{IF } \left(G_t - G_{d,t}\right)/\sin h_t > 120\ \text{W/m}^2 \\ 0 & \text{OTHERWISE} \end{cases} \quad (2)$$

where G_t and $G_{d,t}$ denote the measured global and diffuse solar irradiance at the time t and h_t is the Sun elevation. The average value of *SSN* over a given time interval Δt is equal to the relative sunshine.

3 PV 2-STATE MODEL

Basically, the PV 2-state model connects in an innovative manner an empirical model for estimating the clear-sky PV power with a statistical model for forecasting SSN. The model equation is (the forecast is issued at time $t-\Delta t$):

$$\hat{P}_t = \begin{cases} \alpha_{cs} P_{0,t} & \text{IF } SSN_t = 1 \\ \gamma_c P_{0,t} & \text{IF } SSN_t = 0 \end{cases} \quad (3)$$

where $\hat{P}_t$ is the forecasted value of PV power at time t. $P_{0,t}$ is the estimated PV power under clear sky at time t. α_{cs} is a dynamic correction applied to the atmospheric transmittance encapsulated into the $P_{0,t}$ equation. γ_c represents the attenuation applied to $P_{0,t}$ according to the actual cloud transmittance. Both parameters α_{cs} and γ_c are estimated simultaneously by a linear regression applied to data measured in a period Δt prior to the forecasting moment. An important finding was reported in [9]: a minimum error is reached if Δt equals the forecasting horizon. Each day, the procedure of computing α_{cs} and γ_c runs as follows. At the time when the Sun elevation reaches a predefined value, α_{cs} and γ_c are initialized: $\alpha_{cs}=1$, which means that $P_{0,t}$ is assumed to be an accurate estimator for the measured clear-sky PV power and $\gamma_c = 0.29$, as the multiannual average of cloud transmittance. Further, α_{cs} and γ_c are systematically updated before generating a new forecast.

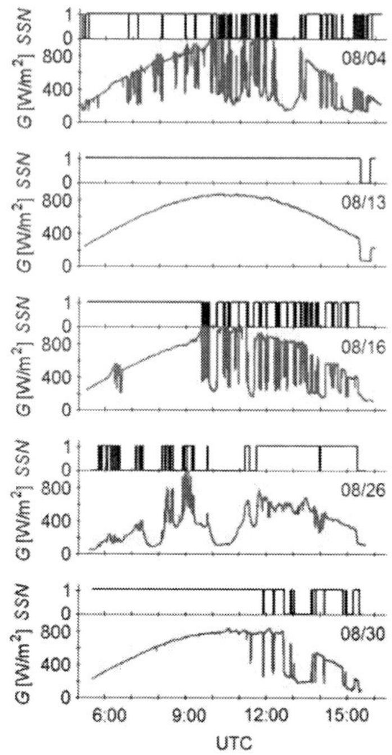

Figure 2: Measured global solar irradiance G and sunshine number *SSN* in five days of August 2019.

In order to do that, the solar irradiance measurements recorded during a period Δt before the moment of generating the forecast are separated into two bins with respect to SSN. The measurements from the bin SSN = 1 are used for estimating the actual value of α_{cs} while the measurements from the bin *SSN* = 0 are used for estimating the actual value of γ_c. If a bin contains less than three measurements, the corresponding parameter, α_{cs} or γ_c, keeps the previous estimated value. Hence, the atmospheric transmittance and the cloud transmittance in Eq. (3) are always related to the most recent measurements.

4 PV 2-STATE ASSESSMENT

4.1 Model's implementation

PV power under clear sky at time t was estimated using an empirical equation fitted on data collected in-situ in 12 clear-sky days spread from January to July 2019:

$$P_{0,t}\left(G_t, T\right) = -50.4 - 1.33T + 1.256G_t - 0.00678G_t^2 + \\ +1.85\cdot 10^{-5} G_t^3 - 1.9265\cdot 10^{-8} G_t^4 + 6.8\cdot 10^{-12} G_t^5 \quad (4)$$

G_t represents the in-plane solar irradiance under clear-sky. It was evaluated in two steps. Firstly, direct-normal and diffuse solar irradiances on the horizontal plane were estimated with a clear-sky solar irradiance model [10]. Secondly, G_t was calculated by transposing the horizontal components on the PV modules plane (facing South, tilt 30°) using the very simple isotropic transposition model.

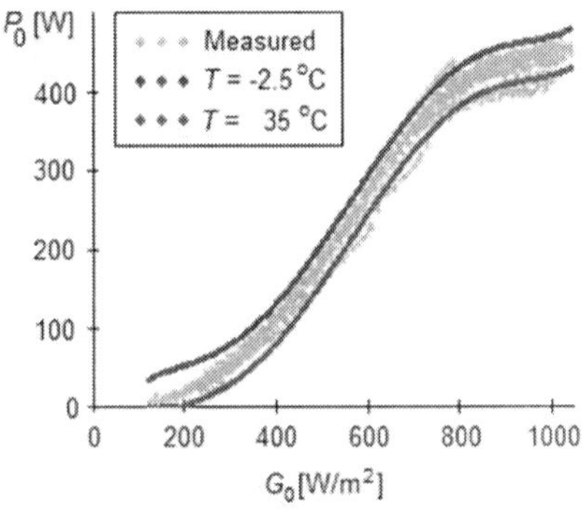

Figure 3. Estimated output power of FORPV under clear-sky conditions (Eq. 4) at two different values of the atmospheric temperature.

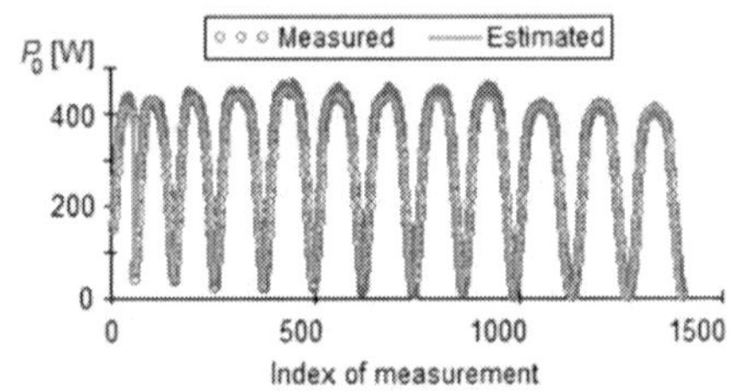

Figure 4: Estimated (Eq. 4) vs. measured output power of FORPV during the clear-sky days from fit period.

The variable T from Eq. (4) represents the air temperature. Its effect on the FORPV output power is well illustrated in Fig 3. In our implementation of the PV 2-state model, we assumed that during forecast horizons shorter than 20 minutes, T does not change significantly. Thus, the measured value of T at the time when a forecast was issued, was also used in Eq. (4). Figure 4 highlights the high-quality of Eq. (4) in replicating the FORPV output power under clear sky.

The ARIMA(0,1,0) model was considered for nowcasting SSN regardless of the forecast horizon. This choice is motivated by the conclusions of Refs. [4, 11] which, using different approaches, demonstrated that ARIMA(1,0,0) and ARIMA(0,1,0) are adequate solutions for nowcasting SSN.

4.2 PV 2-state model accuracy

Figure 5 displays the percentage of forecasts PF accurate to within a given tolerance interval TOL around the measurement. The PV 2-state model performs considerably better than the persistence model. During the day 08/26 (mostly cloudy), for a tolerance interval TOL =10% the precision of the PV 2-state model is double to the precision of the persistence model. Precision is a general characteristic of the PV 2-state model. In general, for $TOL < 20\%$, PF in case of the 2-state model is far greater than PF in case of persistence. Roughly, at TOL = 50%, the success rates of the two models become comparable, but the PV 2-state model still preserves an advance. This is a consequence of the fact that the PV 2-state model equation is dynamically adapted to the actual measurements of the output power, ascribing it a high-precision when SSN is correctly predicted (Fig 6).

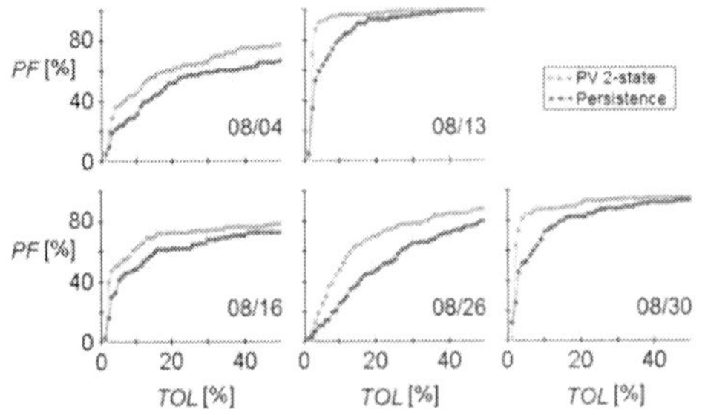

Figure 5: Percentage of forecasts PF accurate to within a given tolerance interval TOL centered on the measurement. Forecast horizon = 5 min.

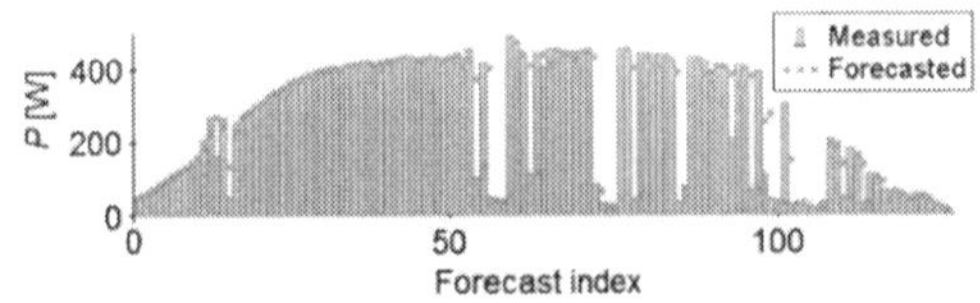

Figure 6: Measured and forecasted output power of FORPV in 08/16. Forecast horizon = 5 min.

Table I: Indicators of the PV two-state model accuracy in nowcasting the output power of FORPV. Forecast horizon = 5 min.

Date	08/04	08/13	08/16	08/26	08/30
State-of-the-sky	High variability	Clear sky	Variable afternoon	Overcast	Low variability
nMBE	-0.022	0.001	0.006	-0.030	0.007
nRMSE	0.566	0.008	0.472	0.399	0.123
SS	0.111	0.717	0.132	0.247	0.192
PF [%]	47.7	96.8	64.2	52.0	87.3

We can deduct from here that the more stable the solar radiative regime, the better Eq. (3) will work.

Table I presents the model's performance for a forecast horizon of 5 minutes. Skill score is evaluated in terms of $RMSE$ over persistence. The values of the statistical indicators emphasize a better performance of the two-state model, both in terms of accuracy ($nRMSE$, $nMBE$) and precision (PF). A significant improvement over persistence in stable sunny sky (SS = 71.7%) and overcast sky (SS = 24.7%) can be noted. The stability of the state-of-the-sky influences dramatically the forecasting accuracy. As example, at a forecast horizon of 5 minutes, in 08/13 (a stable clear sky day) the value of the relative root mean square is $nRMSE$ = 0.8%, while in 08/04 (a day with high variability in the state-of-the-skt) $nRMSE$ = 56.6%.

Figure 7 presents the accuracy of nowcasting the FORPV output power by the PV 2-state model in 08/30 with respect to the forecast horizon. As the forecast horizon increases, a slight decrease in the PV 2-state model accuracy is observed. Skill score decreases almost linearly with the increasing of the forecast horizon, indicating a higher rate of decrease in the accuracy of the PV 2-state model than that of the persistence model.

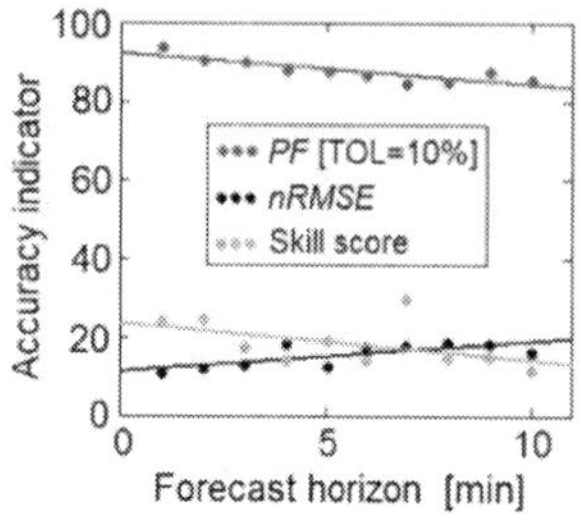

Figure 7: Accuracy of nowcasting the FORPV output power by the PV 2-state model in 08/30 with respect to the forecast horizon. All indicators are in percent.

Based on the PV 2-state model, a simple and reliable procedure for forecasting the energy delivered by a PV system in a given period can be applied. On short, the energy $\hat{E}$ delivered in Δt is forecasted by summing-up the output power values $\hat{P}(FH_i)$ predicted by the PV 2-state model at one minute sampling:

$$\hat{E} = \frac{1}{60}\sum_{i=1}^{N}\hat{P}(FH_i) \quad [\text{Wh/m}^2] \qquad (5)$$

where FH_i is the forecast horizon i. For example, if Eq. (3) is applied for nowcasting the energy delivered in the next 5 minutes, then $N = 5$ and $\hat{P}(FH_1)$ represents the forecasted value of PV power at one minute ahead, $\hat{P}(FH_2)$ at two minutes ahead, and so on.

5 CONCLUSIONS

A new model for intra-hour PV power forecasting is introduced. The PV 2-state model demonstrates the ability to issue high-accurate forecasts. When SSN is correctly predicted, the forecasts trace with accuracy the measurements. A notable advantage of the PV 2-state model is its modularity, which allows for increasing the forecasting accuracy by fine tuning its components: the PV power clear-sky model, the dynamic adjustment or the model for nowcasting SSN. Although many efforts were devoted to increasing the accuracy of nowcasting SSN [12], further research is required. Certainly, the accuracy of the PV 2-state model depends straightforwardly on the ability to make a right prediction of the state-of-the-sky, which, considering the binary nature of SSN, is always a big challenge. The major limitation standing against an increase in forecasting accuracy of SSN is also associated with persistence. Forecasting SSN based on sky-imagery is expected to generate a strong increase in the overall performance of the PV 2-state model, aiming beyond the state-of-the-art.

6 ACKNOWLEDGMENTS

This work was supported by a grant of the Romanian Ministry of Education and Research, CCCDI-UEFISCDI, Project No. PN-III-P2-2.1-PED-2019-3942, within PNCDI III.

7 REFERENCES

[1] U. Das U, K.S. Tey, M. Seyedmahmoudian, S. Mekhilef, M. Idris, W. Deventer, B. Horan, A. Stojcevski. Renewable and Sustainable Energy Reviews 81 (2018) 912.
[2] R. Blaga, A. Sabadus, N. Stefu, C. Dughir, M. Paulescu, V. Badescu. Progress in Energy and Combustion Science 70 (2019) 119.
[3] Q. Paletta, G. Arbod, J. Lasenby. Solar Energy 224 (2021) 855.
[4] M. Paulescu, O. Mares, E. Paulescu, N. Stefu, A. Pacurar, D. Calinoiu, P. Gravila, N. Pop, R. Boata. Energy Conversion and Management 79 (2014) 690
[5] E. Paulescu, M. Paulescu. Renewable Energy 143 (2019) 985.
[6] M. Kottek, J. Grieser, C. Beck, B. Rudolf, F. Rubel. Meteorologische Zeitschrift 15 (2006) 259.
[7] V. Badescu, M. Paulescu. Atmospheric Research 101 (2011) 194.
[8] World Meteorological Organization. Guide to Meteorological Instruments and Methods of Observation. WMO-No.8/2008
[9] D. Costa, O. Mares. AIP Conference Proceedings 1634 (2014) 151.
[10] Biga AJ, Rosa R. Solar Energy 23 (1979) 61–67.
[11] V. Badescu, M. Paulescu. Meteorology and Atmospheric Physics 112 (2011) 139.
[12] M. Brabec, M. Paulescu, V. Badescu. Solar Energy 102 (2014) 272.

CLOUD HEIGHT ESTIMATION USING ALL SKY IMAGERS

João Esteves[1], Rui Pestana[1,2], Yang Cao[1], Nuno Pinho da Silva[1]

[1]Centro de Investigação em Energia REN-State Grid, S.A, Rua Cidade de Goa, nº4-B, 2685-038 Sacavém, Portugal, +351 210011387, joao.esteves@rdnester.com
[2]REN – Rede Eléctrica Nacional, S.A., Rua Cidade de Goa, nº4-B, 2685-038 Sacavém, Portugal

ABSTRACT: The solar forecast research continues to increase in the academic and industry world to provide tools and methodologies with increasingly performances. The levels of penetration of this renewable energy source is growing and more optimized tools are needed in order manage the electrical grid in a proper and safe way. With the new electricity market designs reaching shorter gate closers, emphasizing the importance of forecast tools for solar energy. The use of all-sky imager (ASI) applied to the solar forecast is one approach being researched in the industry. The ASI images provide a bottom-up approach and in the case of renewable solar energy, it can assist the decision-making by allowing an almost-real time knowledge of the sky and cloudiness behavior. In this work, it is presented a methodology that allows to estimate the cloud height using two ASI installed in two neighborhood photovoltaic power plants.
Keywords: solar forecast, all-sky imager

1 INTRODUCTION

In recent years, the solar power installed capacity increased significantly all over the world. The total installed annual PV capacity for PV reached at least 627 GW in 2019. In these statistics China is leading the market with a total capacity of 204.7 GW cumulative capacity installed [1]. In Fig. 1 it is shown the evolution of the annual PV installed installations.

With the need of these solar power plants to participate in the electricity markets, forecasting tools for different time horizons plays an important role.

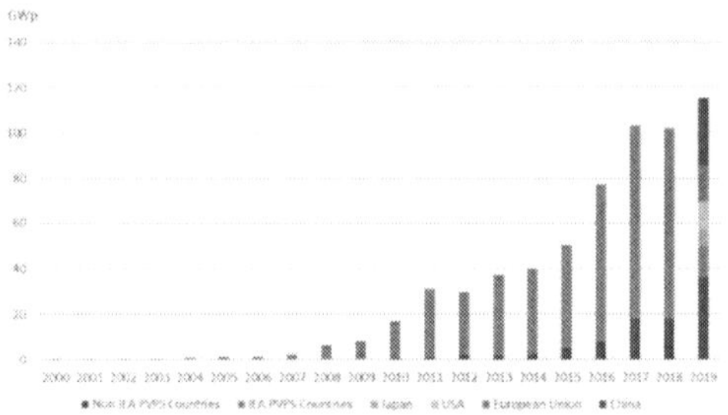

Figure 1: Evolution of annual PV installations [1]

In the subject of short-time forecasting, the use of ASI is under the spotlight in the scientific and industry community. Several methodologies are used to tackle this problem, from purely mathematic to sophisticated methods using artificial intelligence. For instance, in [2] the author developed an automatic system of cloud observation (SONA) that provides cloud cover percentage using neural networks in all-sky images. The ASI images are stored in time periods from 1 to 5 minutes and real time processed to obtain cloud cover percentage during daytime. In [3], the authors proposes that there are differences in optical properties of atmospheric aerosols and cloud particles introduce the visual distinctions between clear and cloudy regions of the sky. By identifying the red-blue-ratio value of each pixel, the cloud segment can be easily distinguished. Cloud Motion Vectors (CMV) are then used to estimate cloud height. In a different study [4], a study using five all-sky imagers and seven camera distances is performed. This study used off-the-shelf surveillance cameras. Pairs of two cameras are used to calculate the cloud height. In [5], a method is proposed for estimating the height of clouds. This method uses a triangulation procedure.

The work presented in this paper aims to present a methodology developed to estimate the cloud height using two ASI installed in two neighboring photovoltaic power plants.

2 METHODOLOGY

The proposed methodology uses two ASI installed in in two neighborhood photovoltaic power plants. By using images of both ASI for the same time point, it is intended to detect the same cloud point and estimate the cloud height. This estimation is made by applying a trigonometric reason to triangulate the same cloud in both sites. The general overview of the developed methodology can be seen in Fig. 2 and each module is explained in each sub-chapter.

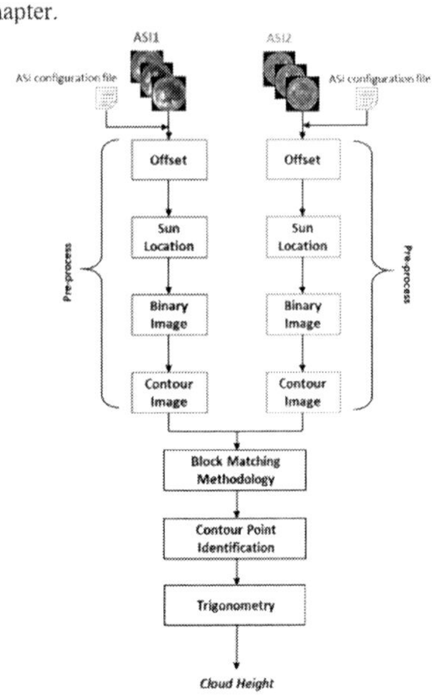

Figure 2: Developed methodology overview

2.1 Pre-process: Offset

In order to have both ASI perfectly aligned, to avoid relative orientation errors in the cloud matching, to each one of the ASI are applied an offset angle.

2.2 Pre-process: Sun Location

The computational knowledge of the Sun location in the ASI image is an important factor that contributes to analyze and determine clouds in the sky. This is important due to the fact that the circumsolar region, area near the Sun, have higher brightness and must be taken into consideration. The Sun Location in the sky can be determined by zenith and azimuth angles. If one considers zenith angle and azimuth angle as θ_Z and γ_S, the latitude of camera site as φ, equator angle as δ, and hour angle as ω, the following equations are given:

$$\theta_Z = \cos^{-1}(\sin\varphi \cdot \sin\delta + \cos\varphi \cdot \cos\delta \cdot \cos\omega) \quad (1)$$

$$\gamma_S = \sin^{-1}\left(\frac{\cos\delta \cdot \sin\omega}{\sin\theta_Z}\right) \quad (2)$$

$$\delta = 23.45° \cdot \sin\left[2\pi \cdot \frac{284 + dayth}{365}\right] \quad (3)$$

$$\omega = (dayhour - 12) \cdot 15° \quad (4)$$

In equation (3), $dayth$ represents the day number of the year, ranging from 1 (January 1st) to 365 (December 31st). In equation (4), $dayhour$ is a representation of the time affected by the longitude zone of the ASI site:

$$dayhour = hour + \frac{minutes}{60} + \frac{Longitude * 4}{60} \quad (5)$$

Lastly, the Sun location in terms of image pixels is calculated using equation (6) and equation (7). An example of the Sun location applied to a set of images is shown in Fig. 3.

$$sunloc_x = imCenter_x - \theta_Z \cdot \sin(\gamma_S) \quad (6)$$

$$sunloc_y = imCenter_y + \theta_Z \cdot \cos(\gamma_S) \quad (7)$$

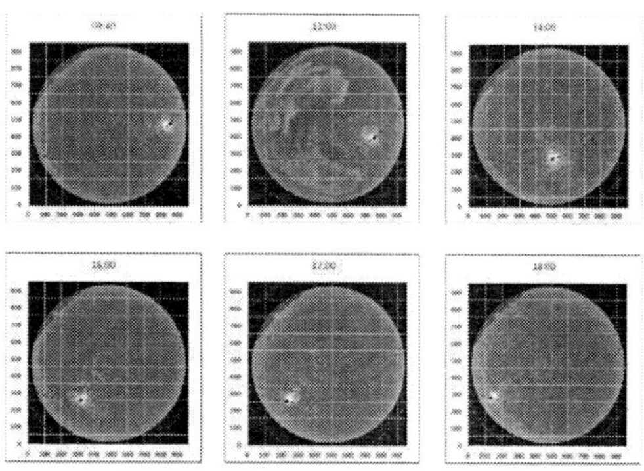

Figure 3: Setting the Sun trajectory in the ASI images

2.3 Pre-process: Binary Image

The original image of the ASI is a colorized image, red-blue-green (RGB), and this feature brings computational problems when identifying clouds and sky pixels. The red-blue-ratio (RBR) methodology was used. This ratio is used given the fact that when compared with the clean and cloudless atmosphere, the clouds and the aerosols enhance red versus blue intensity. Thus, increasing the RBR value. The RBR can be defined as [6]:

$$RBR = \frac{R}{B} = 1 + \frac{R - B}{B} \quad (8)$$

R and B represent the red and blue channel magnitude for each pixel in the image. This equation is applied to each pixel in the image, transforming it into a RBR image. The binary image is obtained by using a threshold value. Such transformation can be seen in the left side of Fig. 4.

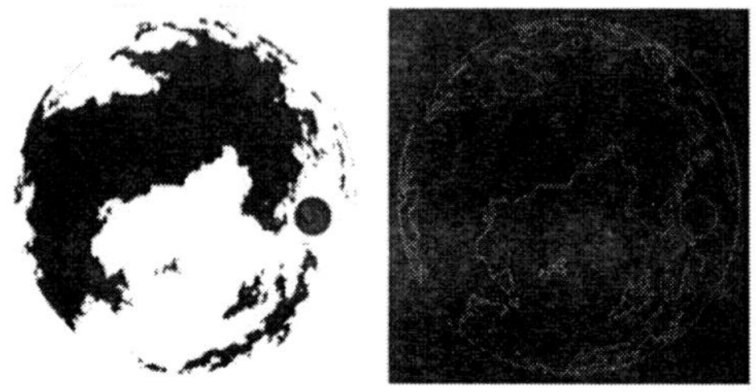

Figure 4: Binary image and contour image

2.4 Pre-process: Contour Image

The Binary image as an important role to clearly distinguish the clouds from the sky. This is especially important when trying to find the same cloud in both ASI's images. Nevertheless, the contour of the cloud also plays an important role in the developed methodology. After the cloud has been identified in both ASI's images, a single point in the contour must be selected to perform the cloud height calculations. This image transformation can be seen in the right side of Fig. 4.

2.5 Block Matching Methodology

In order to estimate the cloud height, it is necessary to be able to detect the same cloud in both ASI images. This is achievable, by defining a grid of searching zones, also known as blocks. The block size, in number of pixels, is of great importance and affects the performance of the search for a matching cloud. Blocks with higher size leads to wide areas of clouds and sky whereas very small blocks contain little information to perform a satisfactory comparison. For the scope of this work, the grid of searching zones is made by considering blocks that have ten percent of the binary image size. One of the ASI is considered as the base camera, in this case ASI1. In practice, this means that for each one of the ASI blocks, a search is going to be performed, comparing with all blocks from the second ASI, ASI2. This can be seen in Fig. 5.

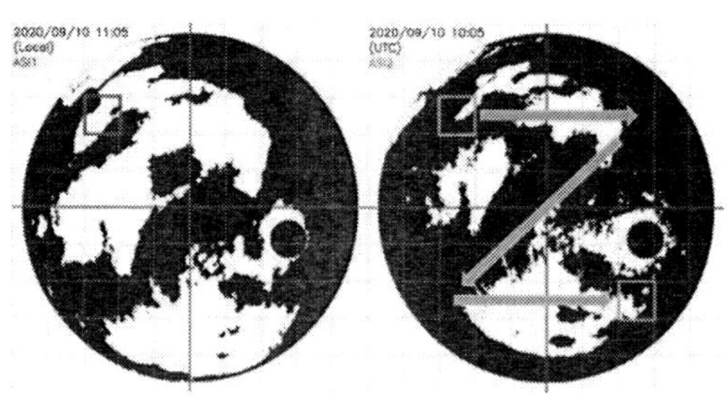

Figure 5: Block search methodology

The block matching methodology is based on the comparison between blocks of two different images. These blocks can be seen as matrices of the same size that contains binary values. The comparison is computed using the Structural Similarity Index (SSIM) which is highly detailed in [7]. This type of methodology is suitable when comparing small parts of a picture at the time, as the developed block matching methodology employs. The base block with the higher similarity rating is retained, as well as the correspondent block in the second ASI.

2.6 Contour Point Identification

The next step in the methodology, once two matching blocks are found, is to search for a similar point in the cloud contour to proceed with the trigonometry calculations. In Fig. 6 one ca see the results of: i) successfully finding matching blocks (left side) and, 2) successfully finding matching cloud contour points inside the matching blocks (right side).

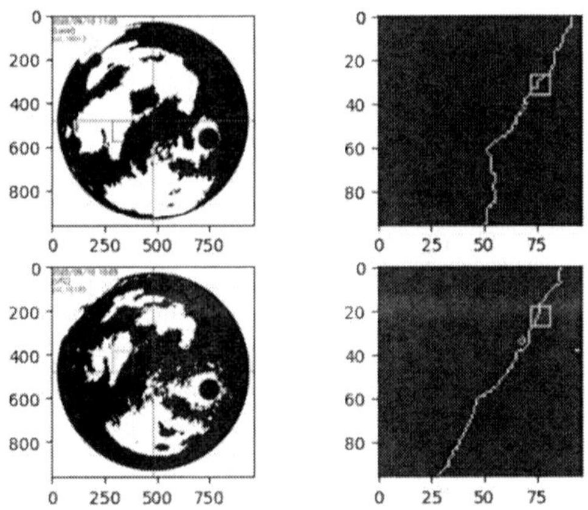

Figure 6: Contour point methodology

2.7 Trigonometry

The last step to estimate the cloud height consist of solving a set of Cartesian equations that allow to triangulate both cloud points. The equation of a line can be written in the Cartesian form as follows:

$$\begin{gathered}\vec{a} = a_x\vec{\imath} + a_y\vec{\jmath} + a_z\vec{k}\\ \vec{r} = \vec{a} + \lambda\vec{b} \rightarrow \vec{b} = b_x\vec{\imath} + b_y\vec{\jmath} + b_z\vec{k}\\ \vec{r} = x\vec{\imath} + y\vec{\jmath} + z\vec{k}\end{gathered} \quad (9)$$

Given the fact that the center of the ASI image is also the center of the three dimensional axis, the set of equations can also be rearranged in the following manner:

$$\frac{x}{a_x} = \frac{y}{a_y} = \frac{z}{a_Z} = k \quad (10)$$

In the scope of this work, the goal is to intercept two points on two ASI's images. These two points, in theory, must represent the same spatial point but seen from different places. A set of Cartesian equations are defined for each ASI:

$$\begin{cases}\dfrac{X_1}{ax_1} = k_1\\ \dfrac{Y_1}{ay_1} = k_1\\ \dfrac{Z_1}{az_1} = k_1\end{cases} \qquad \begin{cases}\dfrac{X_2}{ax_2} = k_2\\ \dfrac{Y_2}{ay_2} = k_2\\ \dfrac{Z_2}{az_2} = k_2\end{cases} \quad (11)$$

In equation (11), index 1 represents ASI1 and index 2 represents ASI2. In addition, the ASI2 set of equations can be rewritten using the relative distance of the cameras in the three-dimensional axis, $(\Delta d_x, \Delta d_y, \Delta d_z)$:

$$\begin{cases}X_2 = Y_1 + \Delta d_x\\ Y_2 = Y_1 + \Delta d_y\\ Z_2 = Z_1 + \Delta d_z\end{cases} \quad (12)$$

Using equation (11) and (12), one can formulate the amplitudes of the lines, k_1 and k_2:

$$k_1 = \frac{\Delta d_x \dfrac{ay_2}{ax_2} - \Delta d_y}{ay_1 - \dfrac{ax_1 ay_2}{ax_2}} \quad (13)$$

$$k_2 = \frac{k_1 az_1 + \Delta d_z}{az_2} \xrightarrow{Z_2 = Z_1} k_2 = \frac{k_1 az_1}{az_2} \quad (14)$$

Reaching this stage, to solve the Cartesian equations and calculate the cloud height (Z), the necessary angles to solve equation (13) and (14) must be retrieved from the ASI images, as shown in Fig. 7.

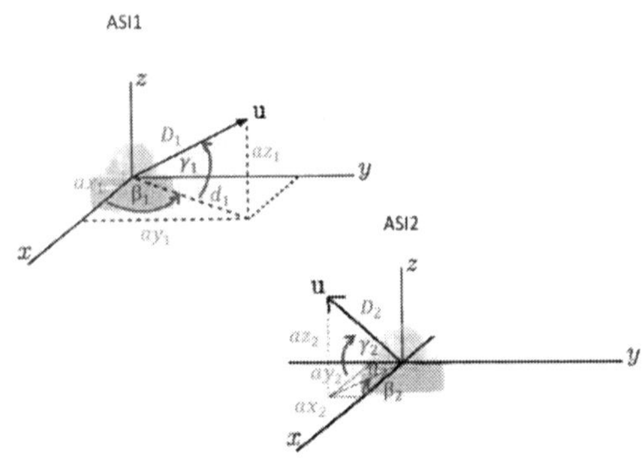

Figure 6: Two referecials (ASI's) pointing to the same point.

By analyzing Fig. 6, and can see that (ax_1, ay_1, az_1) and (ax_2, ay_2, az_2) can be defined in terms of two angles, (β_1, γ_1) and (β_2, γ_2).

$$\begin{cases}ax_1 = d_1\cos(\beta_1)\\ ay_1 = d_1\sin(\beta_1)\\ az_1 = D_1\sin(\gamma_1)\end{cases} \qquad \begin{cases}ax_2 = d_2\cos(\beta_2)\\ ay_2 = d_2\sin(\beta_2)\\ az_2 = D_2\sin(\gamma_2)\end{cases} \quad (15)$$

$$\begin{cases}d_1 = \dfrac{ax_1}{\cos(\beta_1)}\\ D_1 = \dfrac{d_1}{\cos(\gamma_1)}\end{cases} \qquad \begin{cases}d_2 = \dfrac{ax_2}{\cos(\beta_2)}\\ D_2 = \dfrac{d_2}{\cos(\gamma_2)}\end{cases} \quad (16)$$

With this set of equations the cloud height can be calculated. The angles that are retrieved from the ASI images are illustrated in Fig. 7.

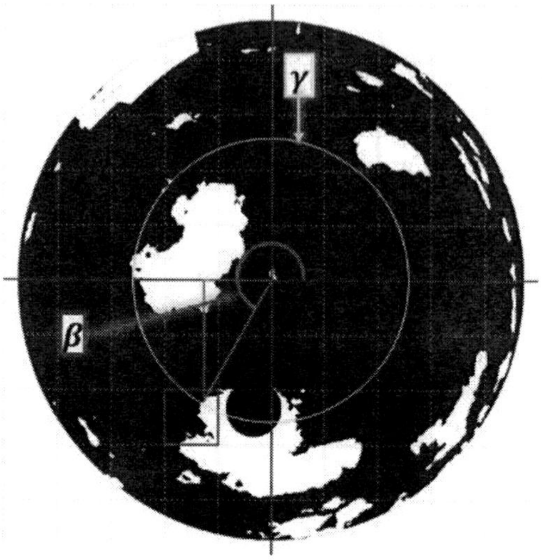

Figure 7: ASI angles

3 CASE STUDY

To test the performance of the methodology a case study was defined in a sequence of four images, as can be see Table I. The selected case study to be presented in this paper represents a sequence of four images with a time step of five minutes. This is important to verify the variance of the cloud height estimation. The selected time points represent a day with high clouds and low wind speeds, which makes the clouds behavior less volatile.

By analyzing the results obtained, and depicted in Table I, one can observe some notes:

- The clouds are being mapped at a similar height, representing a good indication of the performance of the methodology.
- Taking advantage of this particular set of images, in which the clouds are high enough to be easily identifiable (by one's eyes) in both ASI's, the block matching methodology is able select correct matches.
- The changes in position of the clouds from one image to another (5 minutes apart) is not so significant, indicating lower values for wind speed for that height.

To illustrate the results, one can observe Fig. 8 for the first two time points.

Table I: Case study definition

#	DATE (UTC)	β_{ASI1}	γ_{ASI1}	β_{ASI2}	γ_{ASI2}	Height (km)
1	10/09/2020 10:05	195.5	65.6	147.1	62.7	5.78
2	10/09/2020 10:10	261.6	64.2	155.9	81.8	4.69
3	10/09/2020 10:15	249.4	59.1	192.1	52.2	4.97
4	10/09/2020 10:20	283.8	41.9	288.1	60,0	6.49

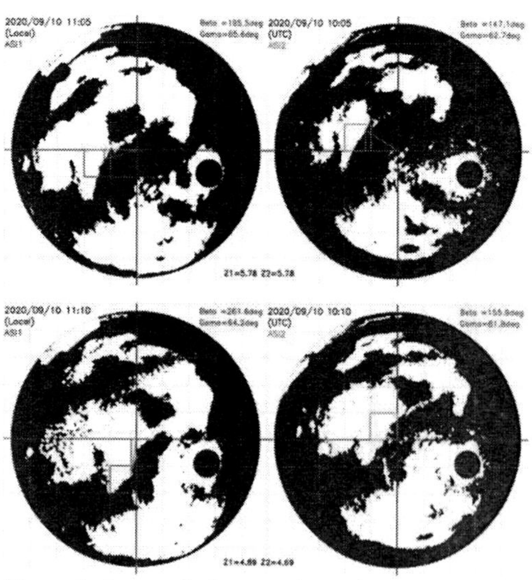

Figure 8: Case study first two time points

4 CONCLUSIONS

The presented methodology relies on the use of two ASI located in neighborhood photovoltaics power plants. The ASI images are subjected to a set of image processing processes in order to have the necessary features. After the pre-process, an algorithm analyzes the ASI images to find the same cloud in both sites and to find a cloud match point. With a matching cloud point, a set of trigonometric equations are solved, retrieving information from the ASI images, and the cloud height is estimated.

6 ACKNOWLEDGEMENT

This work was supported by CEPRI (Chinese Electric Power Research Institute), in the scope of the R&D NESTER (Centro de Investigação em Energia REN-SGCC, S.A.) project 'Renewable Energy Power Forecasting Model Research and Prediction Results Analysis'.

7 REFERENCES

[1] IEA International Energy Agency, Photovoltaic Power Systems Programme, Task 1 Strategic PV Analysis and Outreach, 2020 – Snapshot of Global PV Markets, Report IEA-PVPS T1-37: 2020.

[2] Y. Gonzalez, C. López and E. Cuevas, "AUTOMATIC OBSERVATION OF CLOUDINESS: ANALYSIS OF ALL-SKY IMAGES," in TECO-2012, WMO

Technical Conference on Meteorological and Environmental Instruments and Methods of Observation, Brussels, Belgium, October 16–18., Santa Cruz de Tenerife, Canary Islands, Spain, 2012
[3] C. Zhang, Y. Du, X. Chen and E. G. Lim, "Cloud Motion Forecasting and Cloud Base Height Estimation Using Two Low-Cost Sky Cameras," in IEEE Conference on Energy Internet and Energy System Integration, 2018.
[4] P. Kuhn, B. Nouri, S. Wilbert, N. Hanrieder, C. Prahl, L. Ramirez, L. Zarzalejo, T. Schmidt, T. Schmidt, Z. Yasser, D. Heinemann, P. Tzoumanikas, A. Kazantzidis, J. Kleissl, P. Blanc and R. Pitz-Paal, "Determination of the optimal camera distance for cloud height measurements with two all-sky imagers," Solar Energy, vol. 179, pp. 74-88, February 2019.
[5] F. M. Janeiro, P. M. Ramos, F. Wagner and A. M. Silva, "Developments of low-cost procedure to estimate cloud base height based on a digital camera," Measurment, vol. 43, pp. 684-689, 2010.
[6] M. S. Ghonima, B. Urquhart, C. W. Chow, J. E. Shields, A. Cazorla and J. Kleissl, "A method for cloud detection and opacity classification based on ground based sky imagery," Atmospheric Measurment Techniques, vol. 5, p. 2881–2892, 2012.
[7] Z. Wang, A. Conrad Bovik, H. Rahim Sheikh and E. P. Simoncelli, "Image Quality Assessment: From Error Visibility to Structural Similarity," IEEE Transactions on Image Processing, vol. 13, no. 4, pp. 600-612, April 2004.

DAY-AHEAD PV GENERATION FORECASTING BASED ON DEEP LEARNING APPROACH

Despoina Kothona[1], Afroditi Zamanidou[1], Ioannis P. Panapakidis[2], Georgios C. Christoforidis[1]
[1]University of Western Macedonia, [2]University of Thessaly
d.kothona@uowm.gr, a.zamanidou@uowm.gr, panapakidis@uth.gr, gchristoforidis@uowm.gr

ABSTRACT: The development of robust photovoltaic (PV) power forecasting models is essential for the secure and economic integration of PV power into the intra-day and day-ahead energy markets. Since the day-ahead solar power forecasting is a more challenging procedure, at the present paper we employ a Deep Learning approach and specifically the Long-Short Term Memory (LSTM) model, due to its ability to capture long-term dependencies between the timeseries. The performance of the model is examined based on a selection of different inputs: a) historical data, b) weather predictions and c) historical data alongside with weather predictions. The results point out that the utilization of meteorological forecasts can increase the forecasting accuracy of the model from 5.3% to 29%.
Keywords: PV power forecasting, Long Short-Term Memory Unit, Energy Market

1 INTRODUCTION

The extensive integration of large-scale PV plants into the power system introduces new challenges to intra-day and day-ahead electricity markets. According to the present situation, the PV owners do not participate in the energy trading processes and the produced PV power is consumed in priority as clean energy. However, as the energy markets of European Union's countries are being harmonized with the requirements of the Target Model, the PV owners will be able to participate into the energy trading [1]. Considering this, the development of accurate forecasting models is vital in order to prevent the penalties of mismatching between the forecasted and the delivered power [2].

Several PV power forecasting methods have been proposed in the literature, referring to different forecasting horizons [3], [4]. The models referring to predictions from several minutes to several hours ahead aim to ensure the integration of PV power into the intra-day energy market. On the other hand, the day-ahead forecasting models are essential for the smooth participation of PV power into the day-ahead energy markets [5].

Solar power generation highly depends on climate conditions and even in nowcasting models, i.e. models referring to several minutes ahead, there is a deviation between the real and forecasted power [6]. Therefore, the day-ahead forecasting is a challenging task. In the literature, several day-ahead models have been proposed, based on Artificial Intelligence (AI) techniques, i.e., Machine Learning (ML) and Deep Learning (DL) methods. Specifically, in [7] the authors have examined the performance of well-established ML and DL techniques, such as Linear Regression Interacts (LRI), Support Vector Regression (SVR) with different kernel functions, Feed Forward Neural Network (FFNN) and LSTM. The results highlight that SVR and LSTM provide more accurate forecasts. Moreover, according to the results in [8], the DL models and more precisely the LSTM model is a more appropriate forecaster as the forecasting horizon increases. This is also concluded in [9] and [10].

Since PV power production mainly depends on the time of the day, we deal with a timeseries problem. In fact, we attempt to predict the PV power of the next day and so we need a predictor able to capture long-term dependencies between timeseries. Accordingly, in the present paper, we select the LSTM model, which is a type of Recurrent Neural Network (RNN) that lies in the field of deep learning. In contrast with the previous researches, at the present paper we also develop the solar irradiation forecasts timeseries. Additionally, we formulate five different scenarios in order to examine how the forecasting accuracy is affected when different variables are considered as inputs. The results of the present paper highlight that the forecasting error is reduced when the meteorological forecasts are considered as input, while the most accurate predictions are achieved when both historical data and meteorological forecasts are utilized.

2 METHODOLOGY

2.1 Long-Short Term Memory model

At the present research, the vanilla LSTM model is selected for the forecasting process. LSTM is a recurrent neural network and has been widely used in timeseries forecasting applications due to its ability to detect dependencies between different time series and to provide forecasts based on the information of previous time steps. This ability derives from the structure of the unit, consisting of four layers, which is able to control the information flows from the current unit to the next ones [11].

2.2 Climate Forecasting

2.2.1 Solar irradiation

A database of temperature and cloud cover climate prediction variables is developed for the purpose of this study [12], [13]. Estimating climate predicting variables of solar radiation is based on total monthly solar radiation energy density distribution (kWh/m^2 and month) among the month days, using as a parameter the average solar altitude per day [14]. Based on the Sachsamanoglou – Makrogianni [14] method, the total solar radiation is given by the (1):

$$I_{glo} = A \cdot \sin(\beta) + B \qquad (1)$$

where sin(β) is the solar height at a certain geographical point and A and B are parameters of solar irradiation, as presented in Table I [14].

It is assumed that: (a) when cloud cover is up to 30%, A_{high} and B_{high} are used in (1) for high solar irradiation conditions, (b) when 30%<cloud cover≤60%, A_{mean} and B_{mean} are used in (1) for mean solar irradiation conditions, and (c) when cloud cover is greater than 60%, A_{low} and B_{low}

are used in (1) for low solar irradiation conditions. The values of parameters A and B are reproduced for each time step using linear interpolation. Calculation of components and total solar irradiation on an inclined surface are done as presented in [15].

Table I: Parameters of solar irradiation [14].

Month	A_{low}	B_{low}	A_{mean}	B_{mean}	A_{high}	B_{high}
January	995	-48	1102	-47	1156	-12
February	913	-37	1061	-43	1177	-15
March	880	-41	1050	-39	1184	-11
April	872	-41	1045	-37	1158	-10
May	875	-43	1051	-35	1170	-15
June	878	-42	1056	-34	1171	-13
July	878	-42	1062	-36	1144	-12
August	871	-38	1064	-35	1180	-15
September	906	-45	1083	-42	1194	-19
October	983	-44	1106	-40	1256	-14
November	932	-47	1163	-42	1244	-15
December	886	-50	1094	-46	1253	-16

The deviation between the predicted and the actual power is estimated as:

$$Error = I_f(i) - I_a(i) \quad (2)$$

where I_a is the actual solar irradiation and I_f is the forecasted one. Additionally, in Figure 1 the hourly error of the predictions is presented. The figure refers to the period from 08/25/2019 – 08/19/2020.

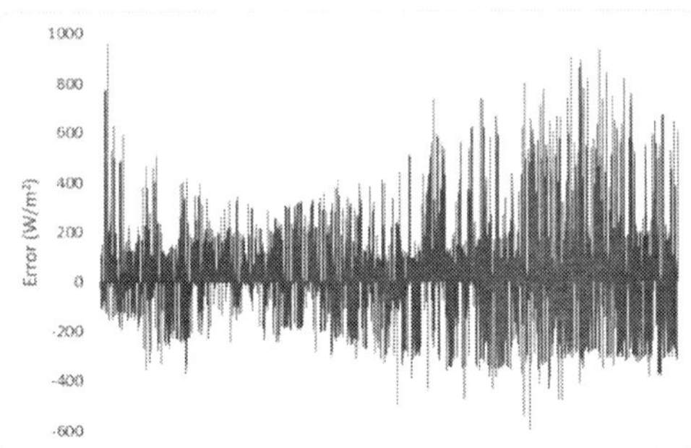

Figure 1: Deviation between real and forecasted solar irradiation

2.2.2 Temperature

The database of temperature climate prediction variable is developed, gathering data from two climate prediction websites: [12] for the period 08/25/2019-10/01/2019, and [13] for the period 10/02/2019-08/19/2020. Hourly timeseries of temperature climate prediction variable were gathered at 6am of every day for the next 24 hours.

2.2.3 Cloud Cover

The database of cloud cover climate prediction variable is developed, gathering data from two climate prediction websites: [12] for the period 08/25/2019-10/01/2019, and [13] for the period 10/02/2019-08/19/2020. Hourly timeseries of cloud cover climate prediction variable were gathered at 6am of every day for the next 24 hours.

2.3 Evaluation metrics

The model's evaluation is based on a set of six well-known performance indices [16], [17] and [18]:

- Absolute error (AE)
- Mean Absolute Error (MAE)
- Normalized Root Mean Square Error (NRMSE)
- Mean Absolute Range Normalized Error (MARNE)
- R-squared (R^2)
- Skill Score (SS)

Equations (3) – (8) present these metrics.

$$AE = \left|P_a(i) - P_f(i)\right| \quad (3)$$

$$MAE = \frac{1}{N}\sum_{i=1}^{N}\left|P_a(i) - P_f(i)\right| \quad (4)$$

$$NRMSE = \frac{1}{N}\sqrt{\sum_{i=1}^{N}\left(P_a(i) - P_f(i)\right)^2} \quad (5)$$

$$MARNE = \frac{1}{N}\sum_{i=1}^{N}\frac{\left|P_a(i) - P_f(i)\right|}{\max(P_a)} \times 100 \quad (6)$$

$$R^2 = 1 - \frac{\sum_{i=1}^{N}\left(P_a(i) - P_f(i)\right)^2}{\sum_{i=1}^{N}\left(P_a(i) - \overline{P_a}\right)^2} \quad (7)$$

$$SS_k = 1 - \frac{k_m}{k_{ref}} \quad (8)$$

where, P_a indicates the real produced power and P_f is the predicted one, N is the total number of the observations of the test set, $\overline{P_a}$ is the average actual PV power production of the system, k denotes the indicator, i.e., AE, MAE, NRMSE and MARNE, m indicates the case and *ref* is the reference model.

3 HISTORICAL DATA DESCRIPTION

The historical PV power data that we utilize in our research are collected from a PV plant located in northern Greece. The system's nominal capacity is 10 kWp and the recording frequency is 15 min. Moreover, the historical meteorological data are obtained from the nearest meteorological station. Specifically, the solar irradiation and the ambient temperature are available with a recording frequency of 10 min. Finally, the data referring to the meteorological forecasts are per hour.

A data preprocessing of the PV power and climate historical data is conducted in order to perform hourly day-ahead solar power forecasting. Since the LSTM deals with timeseries forecasting, an autocorrelation analysis is performed on the PV power data, in order to define the optimal number of previous time-steps to be used as inputs to the model. In Figure 2, the autocorrelation curve is illustrated for six sequential days. It is clear that when we want to predict the power generation at time t of the next day, the most correlated values are the t-24 and t-25, i.e. the PV power generation at time t and t-1 of the current day.

However, in order to participate in the day-ahead energy market, the predictions of the next day's power production should be provided until 12:00 of the current day. This influences the time step's selection for the forecasting process. For power forecasts until 12:00 of the next day (d+1) we can use the recorded data of the current day. On the other hand, forecasts referring to time after 12:00 derives from the data of the previous day (d-1). The proper selection of the lags considering the forecasting hour is illustrated in Figure 3.

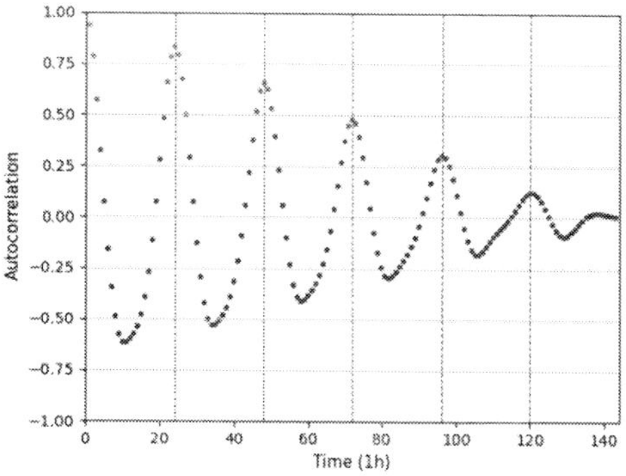

Figure 2: Pearson's autocorrelation curve

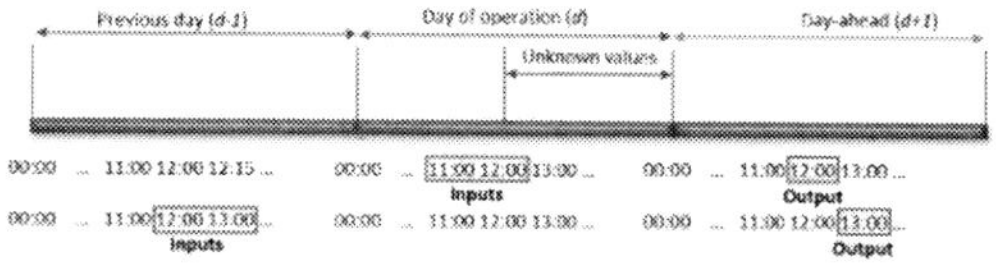

Figure 3: Day-ahead forecasting process

4 EXAMINED SCENARIOS

In order to examine the contribution of the climate predictions in PV power forecasting applications, five different cases have been formulated. Let D and t denote the current day and the hour we attempt to predict, respectively. Also, variables V and F refer to the historical variables (power, solar irradiation, temperature) and the climate predictions (solar irradiation, temperature, cloud cover), respectively.

In Case#Base the training of the forecasting model is based only on historical values of PV power production and meteorological data (solar irradiation and temperature). Specifically, in order to predict the power generation at time t of the next day (D+1), we use as inputs the PV power, the solar irradiation and the temperature of times t and t-1 of the current day, i.e. V_{t-1}^{D}, V_t^{D}. However, since the available data of the current day covers a period from 00:00 – 12:00, for the predictions referring to hours after 12:00, we use as inputs the available data of the previous day D-1, i.e. V_{t-1}^{D}, V_t^{D-1}.

In Case#1a and Case#1b on the other hand, we utilize only climate forecasts for the training. The main difference between them is that at the former solar irradiation, temperature and cloud cover data are utilized, while on the latter, the cloud cover timeseries is excluded from the training process. In both cases, in order to predict the PV power at time t of the next day (D+1), we use as input the climate predictions referring to time t of the next day, i.e. F_t^{D+1}.

Finally, Case#2a and Case#2b are a combination of Case#Base with Case#1a and Case#1b respectively. The training process is based on both historical values (power, solar irradiation and modules temperature) and meteorological predictions (solar irradiation, temperature and cloud cover). The variables of the climate forecasts used in these cases, are the same as in Case#1a and Case#1b, respectively. Table II illustrates the inputs for each case.

The results of the prementioned models are compared with the persistence model, i.e. the naive method. The persistence model assumes that the PV generated power at time t of the next day (D+1) remains the same as time t of the current day (D). However, if we want to forecast the generated power at time t of D+1, and t refers to time after 12:00, we utilize the historical data from the previous day (D-1) [18].

Table II: Utilized inputs per case.

Cases	Historical Data	Weather Forecasts		
		Solar Irradiation	Cloud Index	Temperature
Case# Base	✓	×	×	×
Case#1a	×	✓	✓	✓
Case#1b	×	✓	×	✓
Case#2a	✓	✓	✓	✓
Case#2b	✓	✓	×	✓

5 RESULTS

The forecasting results are presented in Table III. It is clear from Case#1a and Case#2a that when the cloud index of the weather forecasting is included in the forecasting process, we can achieve more accurate predictions. Moreover, in Case#2a and Case#2b, where the model is trained with historical values and weather forecasts, the forecasting accuracy is significantly improved. Additionally, based on the SS_{MAE} index, we observe that only the utilization of historical data is insufficient for day-ahead PV power forecasting applications, since the indicator has negative value and the model has lower performance than the persistence model.

Table III: Performance metrics of cases.

Cases	MAE (kW)	NRMSE (%)	MARNE (%)	SS_{MAE} (%)
Persistence Model	0.511	14.933	6.256	-
Case#Base	0.530	12.775	6.484	-3.6
Case#1a	0.434	10.376	5.311	15.1
Case#1b	0.484	11.227	5.925	5.3
Case#2a	0.363	9.704	4.441	29.0
Case#2b	0.407	10.984	4.977	20.4

Moreover, an analysis is conducted based on the MAE. Specifically, the MAE index has been assessed per day for each case. Afterwards, we create eight bins with a step of 0.1 kW and we calculate the number of days that lies in each bin. The results are presented in Table IV. It is obvious that in Case#Base the majority of days has a MAE error within a range of 0.4 – 0.6 kW. However, when only weather forecasts are used as input, i.e. Case#1a and Case#1b, the error lies between 0.2 kW and 0.6 kW.

Besides, when historical timeseries are used alongside with weather forecasts, i.e. Case#2a and Case#2b, the MAE index is even smaller. Precisely, in Case#2a 46 out of 96 days has a MAE within a range 0 - 0.2 kW.

Based on the above analysis, the visualization of the real and forecasted PV power curves is complicated, due to the high number of bins and scenarios. In order to reduce the number of clusters and present the actual and predicted curves, hierarchical clustering is utilized and the ward method is employed in order to separate the days according to the MAE. The maximum number of clusters is set to three. Case#1b and Case#2b are excluded from the analysis, since Case#1a and Case#2a provide more accurate predictions. Considering these, in Figure 4 the real and the predicted curves of each cluster are presented for Case#Base, Case#1a and Case#2a. Additionally, the range of each cluster is presented in Table V. Based on Figure 4 and Table V, Cluster#1 of each case includes the forecasted days with the minimum daily MAE, while Cluster#3 includes the days with the highest daily MAE. Based on Figure 4 and Table V it is clear that the days belonging to Cluster#1 of Case#2a have the smallest forecasting error compared to Cluster#1 of the other cases. Furthermore, although the upper limit of Cluster#1 is lower in Case#2a than the upper limit of Case#1a, the number of days in the cluster is higher. Finally, although Cluster#3 of Case#2a contains the highest number of days compared to Cluster#3 of Case#Base and Case#1a, the lower limit is significantly reduced (Table V).

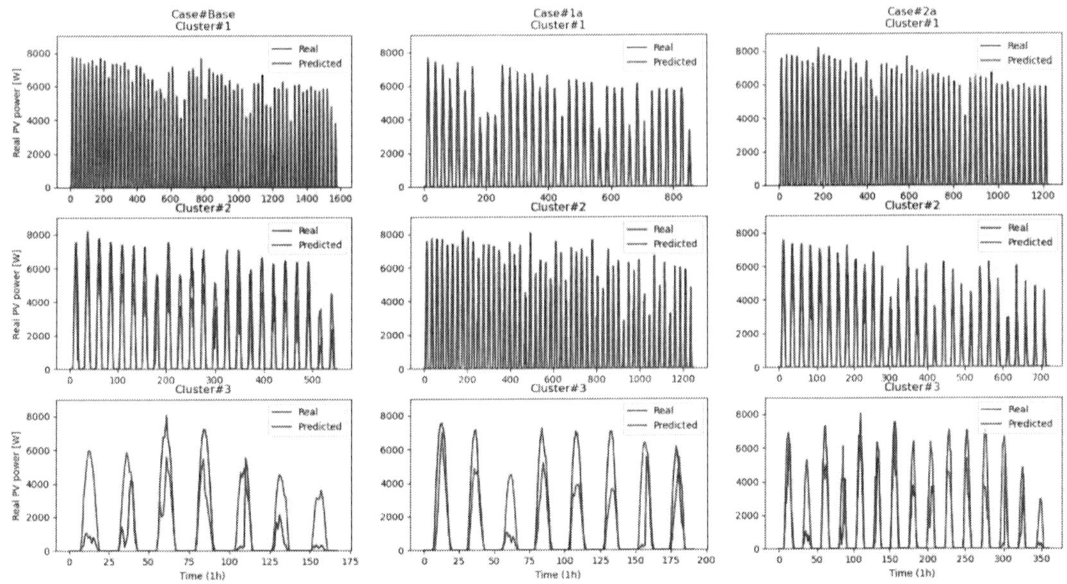

Figure 4: Actual and predicted PV power curves per cluster

Table IV: Number of days per case within a range of MAE.

CASE / MAE (kW)	Case# Base	Case# 1a	Case# 1b	Case# 2a	Case# 2b
(0, 0.1)	0	0	0	8	3
(0.1, 0.2)	0	2	1	8	17
(0.2, 0.3)	3	27	4	30	15
(0.3, 0.4)	18	20	23	16	18
(0.4, 0.5)	30	19	35	15	15
(0.5, 0.6)	20	10	19	9	12
(0.6, 0.7)	13	10	8	5	8
(0.7, 0.8)	5	2	1	1	4

Table V: Lower and upper limit of MAE per case for each cluster.

Cases	Limits	Cluster#1	Cluster#2	Cluster#3
Base	Lower	236.224	567.143	862.341
	Upper	557.394	754.2	1485.44
1a	Lower	176.672	339.638	732.708
	Upper	327.827	671.925	1096.533
2a	Lower	73.554	336.663	553.986
	Upper	313.11	521.911	1143.054

In Figure 4, the PV curves, included in each cluster, are not sequential, even if they are presented continuously. Hence in Figure 5 the real and the forecasted power curves of 19 sequential days are illustrated. The presented days are randomly selected and cover a period from 12/11/2020 to 30/11/2020. If a more detailed examination of the curves is made, we can observe that Case#2a, at the majority of days, provides more accurate predictions.

The scatter plots of real and forecasted power for the cases are presented in Figure 6. From these plots we can get additional information about the AE between the actual and the predicted power, since the plots share a common colormap representing the AE. Additionally, for each case a linear regression analysis is conducted and the R^2 indicator is shown at each graph. From Figure 6 we can conclude that the AEs of Case#2a are minimized. Also, the R^2 metric has the greatest value, meaning that the relationship between the real and the forecasted power is stronger.

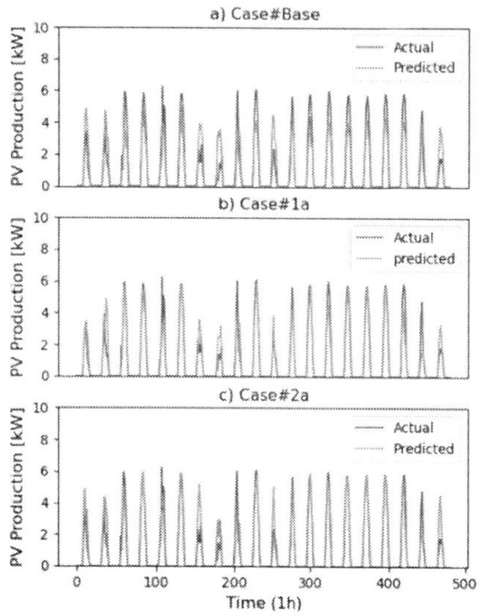

Figure 5: Actual and predicted PV power curves

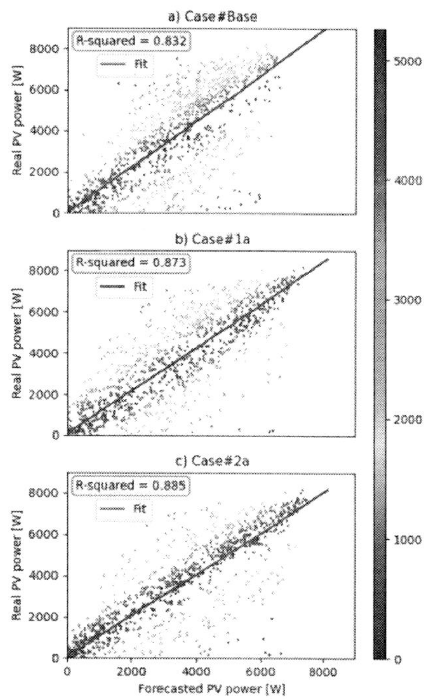

Figure 5: Scatter plots of real and forecasted PV power per case

6 CONCLUSIONS

At the present paper the LSTM model is employed and the contribution of weather forecasts in day-ahead PV power forecasting is examined. Five different cases have been developed, in terms of input data, and their performance has been compared to the persistence model.

The results highlight that the utilization of historical data, without having any information about the weather conditions of the next day, is inadequate in day-ahead forecasting. On the other hand, the utilization of both historical and weather forecasts can significantly improve the predictions' accuracy. Finally, we conclude that the cloud index prediction of the next day is a significant factor that should be considered as input in the forecasting process.

ACKNOWLEDGEMENTS

This work has been supported by Project [PANAMA: Prescriptive analytics and advanced workforce management for optimized O&M of solar power plants], which is supported under the umbrella of SOLAR ERA.NET Cofund 2 by General Secretariat For Research and Technology. SOLAR-ERA.NET is supported by the European Commission within the EU Framework Programme for Research and Innovation HORIZON 2020 (Cofund ERA-NET Action, N° 786483).

REFERENCES

[1] J. Kong and J. Jung, "Development of Incentive Model for Photovoltaic Generators to Participate in a Day-ahead Electricity Market in South Korea," presented at the Asia (ISGT Asia), Chengdu, China, China, 2020.

[2] D. Yang *et al.*, "Operational solar forecasting for grid integration: Standards, challenges, and outlook," *Solar Energy,* 224 (2021), pp. 390-937.

[3] R. Ahmed *et al.*, "A review and evaluation of the state-of-the-art in PV solar power forecasting: Techniques and optimization," *Renew. Sust. Energy Rev.,* 124 (2020), p. 109792.

[4] O. Gandhi *et al.*, "Review of power system impacts at high PV penetration Part I: Factors limiting PV penetration," *Solar Energy,* 210 (2020), pp. 181-201.

[5] M. Pierro *et al.*, "Machine learning-based PV power forecasting methods for electrical grid management and energy trading," *Machine Learning and Data Science in the Power Generation Industry,* (2021), pp. 165-194.

[6] A. Catalina *et al.*, "Combining Numerical Weather Predictions and Satellite Data for PV Energy Nowcasting," *IEEE Transactions on Sustainable Energy,* 11 (3) (2020), pp. 1930-1937.

[7] C. L. Dewangan *et al.*, "Combining forecasts of day-ahead solar power," *Energy,* 202 (2020), p. 117743.

[8] G. Li *et al.*, "Recurrent Neural Networks Based Photovoltaic Power Forecasting Approach," *Energies,* 12 (13) (2019), p. 2538.

[9] U. K. Das *et al.*, "Forecasting of photovoltaic power generation and model optimization: A review," *Renew. Sust. Energy Rev.,* 81 (2017), pp. 912-928.

[10] H. Zhou *et al.*, "Short-Term Photovoltaic Power Forecasting Based on Long Short Term Memory Neural Network and Attention Mechanism," *IEEE Access,* 7 (2019), pp. 78063 - 78074.

[11] K. Greff *et al.*, "LSTM: A Search Space Odyssey," *IEEE Trans Neural Netw Learn Syst,* 28 (10) (2017).

[12] *Weather Underground (2019).* Available: https://www.wunderground.com/ [Accessed:

[13] *YR.* Available: https://www.yr.no/en. [Accessed:

[14] C. Sachsamanoglou and T. Makrogiannis, "General Meteorology Courses," *Athens: Art of Text Publications,* (2004).

[15] S. A. K. D.G. Erbs, J.A. Duffie "Estimation of the diffuse radiation fraction for hourly, daily and

monthly-average global radiation," *Solar Energy,* 28 (4) (1982), pp. 293-302.
[16] H. T.C.Pedro and C. F.M.Coimbra, "Assessment of forecasting techniques for solar power production with no exogenous inputs," *Solar Energy,* 86 (7) (2012), pp. 2017-2028.
[17] J. Zhang *et al.*, "A suite of metrics for assessing the performance of solar power forecasting," *Solar Energy,* 111 (2015), pp. 157-175.
[18] S. Pelland *et al.*, "Photovoltaic and Solar Forecasting: State of the Art," *IEA PVPS Task 14, Subtask 3.1,* (2013).

Using on-site Measurement Data and Laboratory Test Data of PV Modules for Evaluating the Performance Degradation

Ming-Wang Chen [1,*], Chun-Hung Lin[1], Cheng-I Chen[2]

[1]Taiwan electric research & testing center, Taoyuan, Taiwan
[2]National Central University, Taoyuan, Taiwan
*Corresponding author. Tel.: +886 3 4839090; Fax: +886 3 4838119
E-mail: vrains6246@ms.tertec.org.tw

ABSTRACT

Photovoltaic (PV) modules are exposed to a variety of climatic loads during outdoor operation. As time goes by, these lead a number of electric output power performance loss. Many researches show that the module degradation has been evaluated with the test result in laboratory. However, it will inevitably lead to damage for the module during transportation or disassembling/installation. In order to reduce the risk, we compared the electric characteristic of PV system from on-site measurement data and laboratory test data. Then, the related trend and correlation of performance degradation has been analyzed. Applying the result of on-site measurement data we can reduce the relevant risk for testing in laboratory. The I-V curve characteristics from on-site measurement data of more than three years have been collected and corrected to standard test condition (STC) according to IEC 60891 for analyzing their performance degradation of various type modules. The impacting factors for the performance degradation of various type modules have been also researched and the result can help the manufacturers improve their product designs and quality controls. For enhancing the analysis accuracy, before correcting the characteristic we filtered the error information by irradiance and measurement time, and also utilized the density based clustering algorithm (DBSCAN) to robust the correctness. Besides, for verifying the modules damage condition, the I-V curve and electroluminescence (EL) have been checked in the laboratory. The results show that the maximum output power of some modules is reduced in range from 0.7%/year to 2.9%/year. The year average output power of module for on-site measurement is different from the test result in the laboratory about 5~20W under STC. However, through our statistical analysis and appropriate correcting, the attenuation trend of the on-site measurement and laboratory test data is nearly similar. Following the trend, we can evaluate the performance degradation form on-site measurement data, and reduce the risk for testing in the laboratory for specific climate conditions. And the electric output of the PV field can be estimated in the meanwhile and also applied for the power prediction for following years. This paper can be a reference guideline for the industry to improve products, which helps to evaluate the performance degradation of module products and upgrade their product quality.

Keywords: Photovoltaic, Power Generation Performance, Performance Degradation, Reliability Research.

I. INTRODUCTION

The demand for environmental protection and the progressive exhaustion of traditional fossil energy sources have increased the interest in integrating renewable energy sources into the existing power systems. Harnessing solar energy directly as electricity is already seeing massive worldwide installation and commissioning of PV systems, from small power of street lighting to large-scale MW power generation plant. Consider the environmental and economic issues, the government has conducted a policy to subsidize the individuals on installing the photovoltaic system from both financial assistance and buy-back assurance such as Feed-in Tariff (FIT) scheme. And the power plant uses the high-efficiency modules with Voluntary Product Certification (VPC) can increase the FIT purchase rate by 6%.

Application process of VPC in Taiwan includes the laboratory testing, carbon footprint, and factory examination. The standards of laboratory testing has IEC 61215: 2005 (CNS 15114), IEC 61646: 2008 (CNS 15115), and CNS 15118. Above of these standard are only focus indoor testing data. In order to understand the performance degradation of modules, using the outdoor measurement unit to fit the on-site data to standard test condition (STC). And using the indoor testing data and outdoor fitting data to observe the trend. Institutions are working on research topics related to accelerate aging or power performance.

K. K Khanum et al. [1] have studied the remainder of the spectra is absorbed by the PV system, and could influence its output. It is possible that dust actually absorbs the unwanted IR that would otherwise increase cell temperature, degrading its output. N. Santhosh and B. Prasad [2] has proposed a method to improve the efficiency by spectral sharing. Approximately 25-55% of efficiency can be possible with proposed method. Y. Chiba et al [3] have obtained result that estimated power generation from 2013 to 2016 was calculated using two kinds of temperature coefficient (γ). For both two kinds of γ, it is found that the difference between the estimated and the actual power generation amount is gradually close to zero with the exposure period. M. U. Saleh et al. [4] have proposed the novel use of spread-spectrum time-domain reflectometry (SSTDR) for detecting and locating disconnection faults in photovoltaic (PV) power plants. Disconnection faults are manually introduced and correctly identified and localized applying the proposed technique.

II. ANALYZING TOOL

2.1 DBSCAN Algorithm

Density Based Spatial Clustering of Application with Noise (DBSCAN) [5] is a density clustering algorithm. Its advantage is that can automatically cluster data points without specifying the number of clusters in advance. The cluster shape can be of any size scale and can be used for any data point distribution. The result has the ability to not change the clustering by the order of data reading. The DBSCAN algorithm has only two input parameters, namely the neighbor radius *Eps* (ε-Neighborhood) and the minimum number of clusters (*MinPts*), which can be adjusted according to their analysis requirements, which can be expressed as equations (1) and (2). The DBSCAN grouping concept is shown in Figure 1.

$$N_{Eps}(p) = \{q \in D \mid dist(p,q) \le Eps\} \quad (1)$$

$$|N_{Eps}(p)| \ge MinPts \quad (2)$$

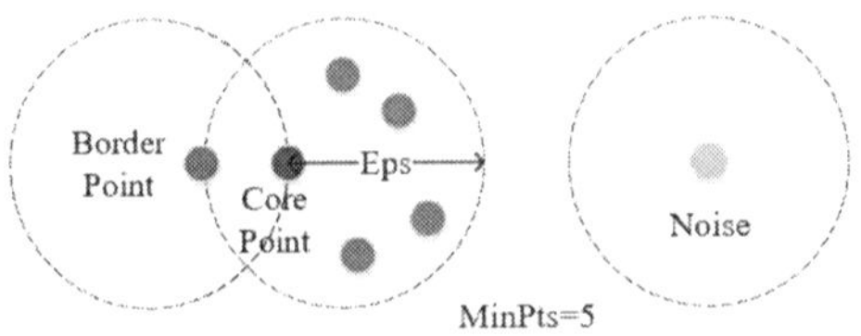

Figure 1 DBSCAN grouping concept

The step of DBSCAN algorithm is following:

step1. Find the data point p and hasn't been clustered. If data point satisfies equation (1) and (2) at the same time, *Eps*(p) is regarded as the new cluster *CDi* and p is the core point of *CDi*, and put in For the candidate set seeds, if equation (1) and (2) cannot be satisfied, the data point p is marked as noise (Noise), find the next ungrouped data point and repeat **step 1**.

step2. If the data point p is the core point of *CDi* and the data point, take q from the candidate set seeds. If equation (1) and (2) are satisfied, the data point p is regarded as the core point of *CDi* and *NEps*(q) All unclassified data points are stored in the candidate set, and data point q is deleted from the candidate set seeds.

step3. If the data point cannot satisfy equation (1) and (2) at the same time, the data point q is regarded as a border point (Border Point), although *NEps*(q) is not regarded as the same cluster as *CDi* and within *NEps*(q) All ungrouped elements are not stored in the candidate set seeds.

step4. Take an element from the candidate set seeds and repeat **step2** and **step3** until there is no element in the candidate set seeds.

step5. Repeat **step1** until all elements in the data point are grouped.

2.2 I-V characteristic correction of PV modules

Because cells are made of semiconductor materials, the principle of power generation is to illuminate the cells through total irradiance, and the semiconductor materials absorb light energy and convert them into electrical energy. Each cell module has a set of voltage-to-current (I-V) characteristic curves and I-V correction parameters, which can be used to make the module reach the maximum photoelectric conversion efficiency.

In order to provide a basis for the performance evaluation of PV modules, Bureau of Standards, Metrology and Inspection (BSMI) of Economic Affairs published in November 101 the application of the CNS 15600 national standard (IEC 60891) "Photovoltaic devices - Procedures for temperature and irradiance corrections to measured I-V characteristics". When measuring the I-V characteristics of the semiconductor of cells, the standardized procedures to be followed when measuring the I-V correction parameters are specified and the reference basis for the calculation of the correction factors [6], [7].

There are three correction procedures applicable to other temperature and irradiance conditions, which can correct the measured I-V characteristics under Standard Test Condition (STC). The first is through the CNS 15600 first edition standard formula; the second is the algebraic correction replacement method, the correction result is better when the irradiance is corrected (>20%); the two of these require a known PV and modified parameters. The third is the interpolation method, which does not require known correction parameters as input items, and can cover at least 3 I-V curves of the temperature and irradiance range of the applied correction method.

Generally, the irradiance G should be calculated based on the short-circuit current (I_{RC}) measured by the PV reference device defined in CNS 13059-2 and the correction value ($I_{RC,STC}$) in the STC. The correction reference should be 25°C and 1,000W/m^2 of the specific temperature coefficient (α_{RC}, the unit is 1/°C) and reference device temperature (T_{RC}).

$$G = \frac{1000\,\mathrm{Wm}^{-2} I_{RC}}{I_{RC,STC}}\left[1 - \alpha_{RC}(T_{RC} - 25^\circ\mathrm{C})\right] \quad (3)$$

If the PV reference device and the test sample have a spectral match, otherwise the spectral can be corrected by CNS 13059-7.

$$I_2 = I_1 + \left(\frac{G_2}{G_1} - 1\right) + \alpha(T_2 - T_1) \quad (4)$$

$$V_2 = I_1 - R_S(I_1 - I_2) - KI_2(T_2 - T_1) + \beta(T_2 - T_1) \quad (5)$$

where, I_1 and V_1 are measurement characteristic point coordinates; I_2 and V_2 are corresponding point coordinates of the modified characteristic; G_1 is irradiance measured from reference device; G_2 is Irradiance measured in STC ; T_1 is the module temperature measured from reference device; T_2 is the module temperature measured in STC; I_{SC} is the test sample measures the short-circuit current at G_1 and T_1; α and β are temperature coefficient of current and voltage; Rs is the series internal resistance of test sample; K is curve correction function.

Equation (4) is only suitable at the irradiance of the I-V curve is a fixed value, and it is not suitable at the irradiance will attenuate, so it can be converted to equation (6).

$$I_2 = I_1 + \frac{G_1'}{G_{SC}} I_{SC}\left(\frac{G_2}{G_1'} - 1\right) + \alpha(T_2 - T_1) \quad (6)$$

where, G_{SC} is the irradiation measured at Isc; G_1' is irradiation measured values obtained from each I-V data。

2.3 Power generation correction of PV modules (system)

Power generation amount of PV modules (system) is calculated by the following formula [8]:

$$P_{\max} = P_{\max\_stc} \times \frac{G}{G_{stc}} \times \{1 + \gamma(T - T_{stc})\} \tag{7}$$

where $P_{\max\_stc}$ is measured initial power output of each module under STC; G is the global solar irradiance measured using the pyrometer (CMP10 and CMP11); G_{stc} is the solar irradiance of 1000 W/m^2 defined by STC; γ is the temperature coefficient of maximum output power; T is the measured temperature of the back sheet surface of each PV module, and T_{stc} is 25 °C. G and T are measured by every 5 min.

The degradation rate of PV modules can be determined by accelerated ageing test methods, on-site comparison verification methods or other effective methods. By using on-site comparison verification methods, after testing its initial maximum output power, install it in the same environment as other components produced in the same batch and run it to generate electricity normally, and measure its maximum output power again after one year from the date of operation. The formula is calculated by the following:

$$P_{\max\_degradation} = \frac{P_{\max\_start} - P_{\max\_now}}{P_{\max\_start}} \times 100\% \tag{8}$$

where $P_{\max\_start}$ is initial value; $P_{\max\_now}$ is present value.

III. RESULT

3.1 System structure

The PV testing field architecture diagram analyzed in this paper is shown in from Figure 2. The power system in Taipei is that each PV modules of equipment number 1-CH1 to 9-CH4 are connected to the micro-inverter, and equipment numbers 10-CH1 to 12-CH4 are connected to an inverter.

These modules are all integrated to the internal electricity of the testing field. The PV testing field in Taipei installs 19 types of photovoltaic modules with a total device capacity of 14.2kW. All of the data collection part measures the I-V Curve every 5 minutes. The measurement time is from 6 am to 7 pm daily, and the data is measured. The modes include irradiation, module temperature, open-circuit voltage, short-circuit current, maximum output voltage, current, power, and fill factor, etc.

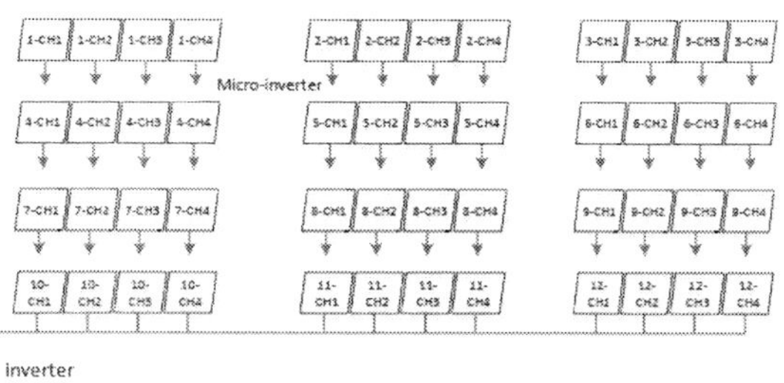

Figure 2 The PV testing field architecture diagram in Taipei

Due to different manufacturing processes, PV modules have different performance characteristics and parameters related to the modules. This paper analyzes two adjacent modules for correcting the power generation characteristics of the modules measured under normal conditions. The test conditions are discussed with the two types for the laboratory measured under standard test conditions. The measured power generation characteristics of the modules measured, and the possible influences on the power generation performance are analyzed. Based on the long-term collected power generation performance Data analysis and verification whether the attenuation situation of the module is similar to the difference measured by the laboratory comparison method to assist in the subsequent evaluation of the reference basis for the use of the module. The related module parameters can be queried in the module specifications of each manufacturer. Among them, the temperature coefficient of short-circuit current is α, the temperature coefficient of open-circuit voltage is β, and the temperature coefficient of maximum output power is γ. The parameter values of the modules analyzed in this article as shown in Table 1.

Table 1 Module parameter table

	Type 1	Type 2
Maximum output power(W)	300	295
Maximum output voltage (V)	34.08	31.69
Maximum output current (A)	8.82	9.31
Open-circuit voltage(V)	40.02	38.87
Short-circuit current(A)	9.55	9.92
γ (1/°C)	-0.45%	-0.44%
β (V/°C)	-7.43%	-0.33%
α (A/°C)	0.47%	0.07%

3.2 STC Comparison and Long-term Observe

This paper analyzes the measurement data of module model from 2018 to 2020. The data screening part only considers the data under the relatively complete illumination on that day. The data for that day needs to cover up to 1,000W/m^2 data, and in order to eliminate the data error of the measurement, by using the DBSCAN algorithm. Use the CNS 15600 (IEC 60891) standard to correct the screened data to standard test conditions, and use the linear regression tool to find the relationship between the maximum output power and the amount of irradiation. The amount of data collected and analyzed in each empirical field is shown in Table 2.

Table 2 Analysis data statistics from 2018 to 2020

Year / Month	2018		2019		2020	
	Original data	Filtered data	Original data	Filtered data	Original data	Filtered data
Jan.	4,868	1,257	4,868	988	4,868	1,322
Feb.	4,552	953	4,552	1,153	4,552	1,574
Mar.	4,846	680	4,846	1,320	4,846	1,118
Apr.	4,726	735	4,726	952	4,726	1,453
May	4,846	895	4,654	653	4,846	1,512
Jun.	4,726	953	4,726	950	4,726	1,460
Jul.	4,846	985	4,846	1,032	4,846	1,440
Aug.	4,846	1,013	4,846	980	4,846	1,520
Sep.	4,726	895	4,726	960	4,726	1,320
Oct.	4,746	934	4,756	1,100	4,656	1,250
Nov.	4,846	680	4,846	1,320	4,846	1,118
Dec.	4,726	735	4,726	952	4,726	1,453

The Taipei testing field which the observed modules is installed type 1 at 11-CH1 and 11-CH2, and type 2 at from 10-CH1 to 10-CH4. By testing at laboratory, the power generation characteristics of the PV module can be adjusted to STC to facilitate observation of whether the module is attenuated. If the last data before disassembly and assembly of the module is used for preliminary evaluation of its power generation performance relationship, it can be summarized as shown in Table 3. P_{max_year} is yearly of on-site measurement statistics data, P_{max_stc} is yearly of laboratory data. The measured maximum output power after correction refers to the national standard CNS 15600 for outdoor data module temperature and radiation. The maximum output power of the module under STC is the maximum output power data of the module measured at 1,000W/m^2 and 25°C when the module by laboratory.

Table 3 Yearly of average maximum output power

Year / P_{max}	First data		2018		2019		2020	
	Type1	Type2	Type1	Type2	Type1	Type2	Type1	Type2
P_{max_year}	-	-	282.15	275.12	281.32	270.54	280.12	267.13
P_{max_stc}	299.1	295.3	-	-	297.5	292.9	297	291.6

-: data missing or unavailable.

From Table 3 the average degradation rate can be calculated at Table 4. The beginning of data collection is not complete, it recuses the data missing or unavailable. Now the rule has been formulated, the trend of outdoor measurement can be complete.

Table 4 Yearly of average degradation rate

	1st year		2nd year		3rd year	
	Type1	Type2	Type1	Type2	Type1	Type2
P_{max_year}	-	-	0.29%	1.66%	0.43%	1.26%
P_{max_stc}	-	-	-	-	0.17%	0.44%

-: data missing or unavailable.

The average value of the maximum output power calculated by linear regression is used to bring in the data after the linear regression equation with the solar illuminance at 1,000W/m^2 to obtain the statistical value of the maximum output power of the module for that month. This method can be used to research and analyze the power generation performance of the module, explore the use of different models of modules under the same operating environment, analyze the power generation performance of the module, and compare it with the maximum output power of the module under standard test conditions. The average maximum output power of the used module is compared with the maximum output power under STC to evaluate the possible power distribution of the module. The yearly statistics of the modules in this demonstration field are shown in Figure 3. The average degradation rate is type2 more than type1. Further using electroluminescence (EL), it can been obtained to Figure 4 and Figure 5. The type 2 encapsulation material has some cracked, it may be indirect the power generation.

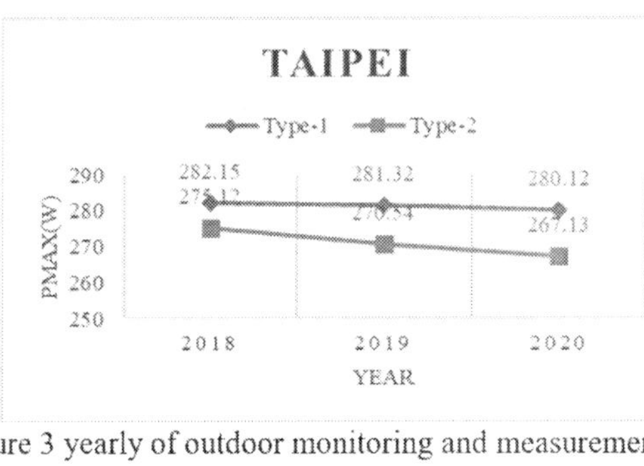

Figure 3 yearly of outdoor monitoring and measurement data

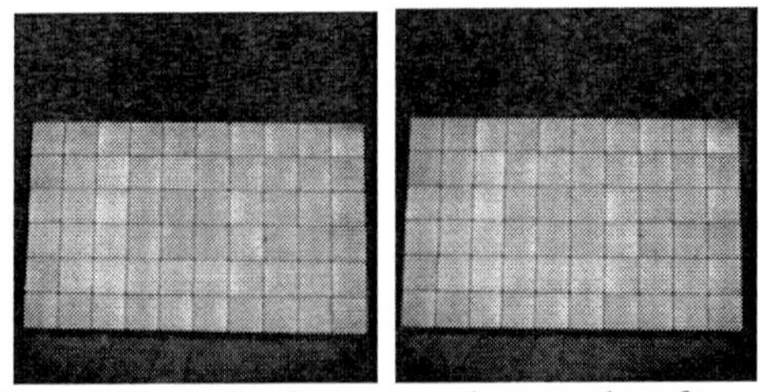

Figure 4 Electroluminescence photography of type-1 (left: 2019; right: 2020)

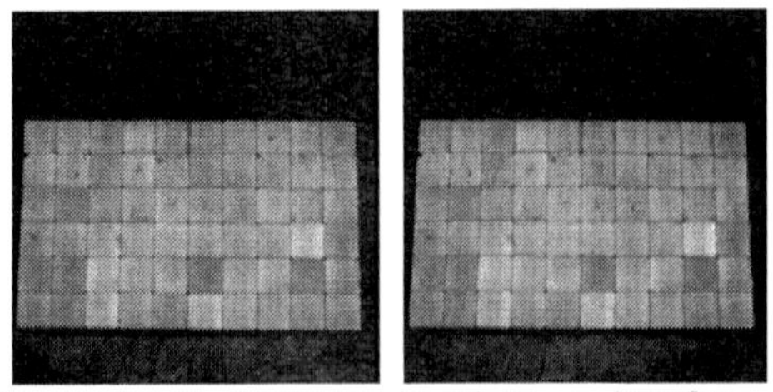

Figure 5 Electroluminescence photography of type-2 (left: 2019; right: 2020)

IV. CONCLUSION

In this paper, we observed the module data of PV modules installed in different climatic empirical fields from 2018 to 2020. The power generation performance of modules using the same model under different conditions. The experimental shows that the PV module under the same operator condition might its material and usage are different. The attenuation situation is different under different models. Reasonable evaluation. If the module used in the PV system is evaluated later, the model may have a lower power generation. Follow-up will continue to observe the changes in its long-term power generation performance to obtain more accurate results.

ACKNOWLEDGEMENT

This study was supported by the Bureau of Standards, Metrology and Inspection (BSMI) under the Ministry of Economic Affairs (MEA).

REFERENCES

[1] K. K Khanum, A. Rao, N. C. Balaji, M. Mani, and P. C Ramamurthy, "Performance evaluation for PV systems to synergistic influences of dust, wind and panel temperatures: spectral insight," in *Proceeding of 2016 IEEE 43rd Photovoltaic Specialists Conference* (PVSC), 2016.

[2] N. Santhosh and B. Prasad, "Efficiency improvement of a solar PV-panel through spectral sharing by combination of different panels," in *Proceeding of 2016 IEEE Students' Conference on Electrical, Electronics and Computer Science* (SCEECS), 2016.

[3] Y. Chiba, T. Ishii, R. Sato, S. Choi, and A. Masuda, "Comparison between estimated and actual power generation amounts of photovoltaic modules at Tosu City in Japan," in *Proceeding of 2018 IEEE 7th World Conference on Photovoltaic Energy Conversion* (WCPEC), 2018.

[4] M. U. Saleh, C. Deline, S. Kingston, N. K. T. Jayakumar, E. Benoit, J. B. Harley, C. Furse, and M. Scarpulla, "Detection and localization of disconnections in PV strings using spread-spectrum time-domain reflectometry," *IEEE Journal of Photovoltaics*, Vol. 10, No. 1, 2020.

[5] Z. Zhu, X. F. Bai1, W. Zhang, R. Liu, and Y. Xu, "Identifying equivalent load model of power systems with wind power integration," *2016 UKACC 11th International Conference on Control* (CONTROL), 2016.

[6] CNS 15600:2012, Photovoltaic devices-Procedures for temperature and irradiance corrections to measured I-V characteristics, 2012.

[7] IEC 60891:2009, Photovoltaic devices - Procedures for temperature and irradiance corrections to measured I-V characteristics, 2009.

[8] T. Ishii, R. Sato, S. Choi, Y. Chiba, and A. Masuda, "Development of a practical method of estimating electric power from various photovoltaic technologies with high precision," *Japanese Journal of Applied Physics*, Vol. 56, No. 8S2, pp. 08MD05, 2017.

TEMPERATURE COEFFICIENTS OF PHOTOVOLTAIC MODULES UNDER PARTIAL SHADING CONDITIONS

Oscar Kwame Segbefia[1], Basant Raj Paudyal[1], Ingunn Burud[2], Tor Oskar Sætre[1]
[1]Department of Engineering Sciences, University of Agder, 4879 Grimstad, Norway
[2]Department of Physics, Norwegian University of Life Sciences, 1433 Ås, Norway

ABSTRACT: Partial shading of photovoltaic (PV) modules leads to mismatch losses which cause local hotspots. Hotspots can lead to various failure mechanisms, hence, power degradation in PV plants. This work investigates the temperature sensitivity of field-aged multicrystalline silicon PV modules under uniform partial shading conditions. Thermal images suggest that hotspots are localized to busbars at the shaded area. The results estimate that, the temperature coefficient of V_{mpp} (β_{Vmp}) of the field-aged PV modules under STC ranges from -0.41 %/°C to -0.48 %/°C. However, under partial shading conditions, the absolute variation in the β_{Vmp} is >2.00 %/°C for the 3 field-aged PV modules studied. The results for the temperature coefficient of I_{mpp} (β_{Jmp}) ranges from ca. -0.08 %/°C to 0.01 %/°C and ca. -1.00 %/°C to -0.90 %/°C under non-shading and partial shading conditions, respectively. However, the severity of the temperature sensitivity under partial shading conditions also depends on the degradation state of the PV modules.
Keywords: partial shading, hotspot, temperature coefficient, efficiency

1 INTRODUCTION

Solar irradiance conditions influence the overall efficiency, reliability, and lifespan of photovoltaic (PV) devices [1-3]. Partial shading has been identified as a denominator for the reduction in power and degradation in PV plants since the 1990s [4]. Shading of PV plants is an issue in all parts of the world, depending on the local surroundings/obstacles near the installation, even under optimum solar irradiation conditions [5]. However, shading can be more difficult to avoid and have a greater impact on yield at high latitudes, due to low solar incidence angles in the winter [4]. Partial shading, one of the main factors of mismatch losses, induces several reliability issues in PV devices [6]. These reliability issues can lead to significant power losses and accelerated degradation in PV plants, which in turn can lead to higher levelized cost of energy (LCOE) of PV power [6, 7].

Usually, mismatched cells due to shading may become reverse biased which leads to local hot spots or joule heating [1]. Hence, PV modules are equipped with bypass diodes to mitigate the effect of reverse biasing [7]. Yet, it has been discovered that bypass diodes are limited in preventing hot spots entirely [7, 8]. The problem of local hot spots in PV plants is yet to be fully resolved [1, 6].

Algorithms based on the Shockley diode equation have been propounded for optimizing the performance of bypass diodes via maximum power point tracking (MPPT) mechanisms [5, 6]. When a bypass diode is activated, a whole string of solar cells is typically bypassed. The I-V curve will also change in a characteristic way, giving a substantial loss of power for the panel.

A good understanding of the influence of temperature variations under different climatic (irradiance) conditions is vital for developing mitigation techniques for monitoring PV module performance and reliability [2, 3, 5, 7, 9, 10]. According to Sun et al. (2018, 2019), the Suns-V_{mpp} technique could be used to monitor and diagnose PV array failure mechanisms at maximum power point (MPP) [11]. However, the method depends on data sheet information (which is sometimes unavailable in the case of old PV plants) and numerical fitting. Recently, a technique was proposed for detecting partial shading effects on a PV module using the temperature corrected maximum power point current - voltage (I_{mpp} - V_{mpp}) curves [8].

Temperature profiles of PV devices can be an indicator of the degradation state of a PV module during field operation [10, 12-14]. Temperature effects on PV devices are accurately accounted for by their temperature coefficients: change in device parameter per unit change in temperature [15, 16]. Hence, incorporating the temperature coefficients of PV modules under partial shading conditions in mitigation algorithms for diagnosing and resolving mismatch losses, due to hot spots has the potential of improving diagnostic methods for solar power plants [15, 16].

In the present work, the temperature sensitivity of field-aged multicrystalline silicon PV modules operating under custom-made uniform partial shading conditions is investigated. The temperature sensitivity of the modules is investigated by evaluating the temperature coefficients of maximum power, efficiency, open circuit voltage, short circuit current and maximum power point voltage (V_{mpp}) and current (I_{mpp}) of the field-aged PV modules under both non-shading and partial shading conditions.

2 METHODS

In the year 2000, 50 multicrystalline silicon PV modules were installed in Dømmesmoen, Grimstad (58.3447° N, 8.5949° E) in Norway. In 2011, the PV modules were decommissioned and kept securely for research purposes. At the time of decommissioning, the PV modules' average power was 90.16 W: about 10 % decrease in nominal power of 100 W [17]. In the present work, the temperature profiles of 3 of these field-aged PV modules have been investigated under a custom-made uniform partial shading conditions in real field operating conditions. In the entire investigation, the shade is placed at the same position of the PV module, shading ca. 3.40 % of the PV module area. With this shading procedure, 6 cells out of the 72 cells of the PV module are partially shaded simultaneously during each measurement. Each PV module consists of (12 x 2) series connected cells and 3 substrings with 2 bypass diodes.

The 3 sampled PV modules were taken through infrared (IR) thermal imaging using the Fluke Ti400 Infrared Camera by following the IEA prescribed procedure [18]. Measurements were done in real field outdoor conditions for both non-shading and shading

conditions. IR thermal images were taken after the PV modules were sun soaked for at least 30 minutes.
Performance data for the analysis was acquired by way of current-voltage (I-V) curve electrical measurements on the 3 selected field-aged PV modules using a handheld I-V 500w I-V Curve Tracer, following the procedure according to the IEC 60904- 1. These measurements provided information on the maximum power (P_{max}), open circuit and maximum power point voltage (V_{oc}, V_{mpp}), short circuit and maximum power point current (I_{sc}, I_{mpp}), fill factor (*FF*), irradiance, and module temperature characteristics of each module at Standard Test Conditions (STC). STC specifies cell temperature of 25 °C, an irradiance of 1000 W/m^2 and air mass 1.5 (AM 1. 5) spectrum for commercial PV modules. Measurements were done at varying irradiance conditions (700 - 1030 W/m^2), and the I-V tracer used converts all measurements to STC automatically. This means the operating conditions were optimally resolved by the device to minimize errors in measuring and recording data. However, the guidelines of IEC 61853-1 were observed to ensure that the wind speed and in-plane irradiance (G_I) during the measurements were within the recommended limits to ensure the accuracy of the results [19]. The distribution of the G_I and PV module operating temperature (T_m) during the measurement period is illustrated in Figure 1. The in-plane irradiance during the studies ranges from 700 - 1030 W/m^2, which falls within the ± 30 % limit specified in the IEC 61853-1 protocol [19].

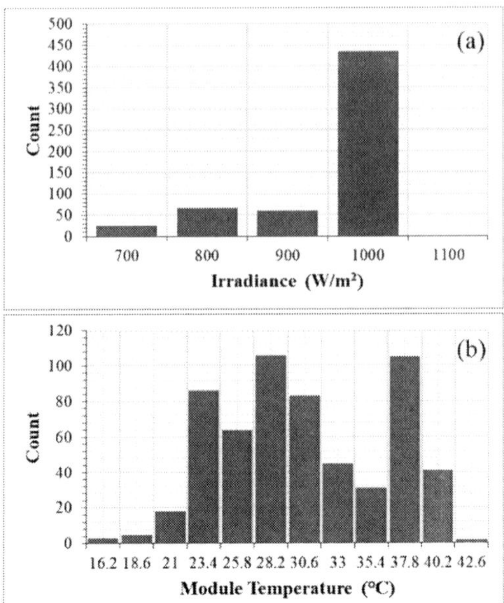

Figure 1: Distribution of (a) in-plane irradiance and (b) the corresponding PV module operating temperatures during the outdoor studies.

The model for terrestrial PV module assumes that, under MPP conditions, the PV module efficiency (η_m) assumes a linear function of the PV module temperature (T_m) and in-plane solar irradiance (G_I) [13, 20, 21].

$$\eta_m = \eta_r[1 - \beta_r(T_m - T_r) + \gamma log_{10}G_I] \quad (1)$$

Where η_r and β_r are the module reference efficiency and temperature coefficient, respectively at reference temperature, T_r. γ is the solar radiation absorption coefficient: the ratio of transmitted to incident solar radiation. However, high solar absorptivity is desired of PV devices [22]. So, most often, the γ is considered negligible and taken as zero [13, 20, 21]. Hence, (1) approximates to

$$\eta_m = \eta_r[1 - \beta_r(T_m - T_r)] \quad (2)$$

The slope of the linear regression plot of η_m versus (T_m - T_r) in (2) gives $\eta_r\beta_r$. That is, a graph of a parameter x which obeys (2), versus T_m fits the linear equation $y = mx + c$, with a slope, m related to β_x and the intercept, $c = \eta_r$ as

$$\beta_x = {}^{m}/_{c} \quad (3)$$

The value from (3) gives the temperature coefficient of x (β_x) in °C^{-1} [23]. The relative temperature coefficients are determined using regression plots of the electrical parameters (P_{max}, V_{oc}, I_{sc}, FF, V_{mpp}, I_{mpp}, and η_m) versus T_m. The temperature coefficients of the PV modules under the partial shading and non-shading conditions are then compared.

3 RESULTS AND DISCUSSION

Table I summarizes the results from the current-voltage (*I-V*) measurement of one of the field-aged PV modules in its present state and indicates that over the 20 years period, the efficiency of the module has reduced significantly due to degradation in P_{max}, V_{oc}, I_{sc}, FF, V_{mpp}, I_{mpp}, and η_m. However, under partial shading conditions for the given case shown in Figure 2a, the power reduction was more than 60 % for the PV module.

Generally, P_{max} degradation for all the 3 PV modules investigated under partial shading (PS) conditions was more than 50 %. From Table I and Figure 2b, it seems partial shading conditions affect I_{sc} more than the V_{oc}. Even so, the degradation in P_{max} is directly related to the value of current and voltage at maximum power point, which are I_{mpp} and V_{mpp}, respectively. That is

$$P_{max} = I_{mpp} \cdot V_{mpp} \quad (4)$$

Under partial shading conditions, see Figure 2b, both I_{mpp} and V_{mpp} are far lower. According to Xiao et al. [24], 2020,

Table I: Average electrical parameters of field-aged PV module C in the year 2021 at STC. Measurements were done at 991 - 1019 W/m^2: 22.4 - 37.6 °C and 995 - 1027 W/m^2: 22.5 - 39.1 °C for non-shading and shading conditions, respectively.

Parameter	P_{max} (W)	V_{oc} (V)	I_{sc} (A)	V_{mpp} (V)	I_{mpp} (A)	FF (%)	η (%)
Data sheet	100	21.60	6.70	16.70	6.00	70	13.00
No shading	77.16	19.96	5.97	15.09	5.11	65	10.03
Partial shading	37.45	19.68	5.31	13.16	2.86	36	4.87

degraded Schottky contacts produce localized hotspots, which leads to bypass diode failure and hence, power degradation and possible permanent damage to panel. Degradation in P_{max} due to localized hotspots was also observed elsewhere [5, 9]. That is, the value of P_{max} reduces according to (4) when I_{mpp} and V_{mpp} decrease.

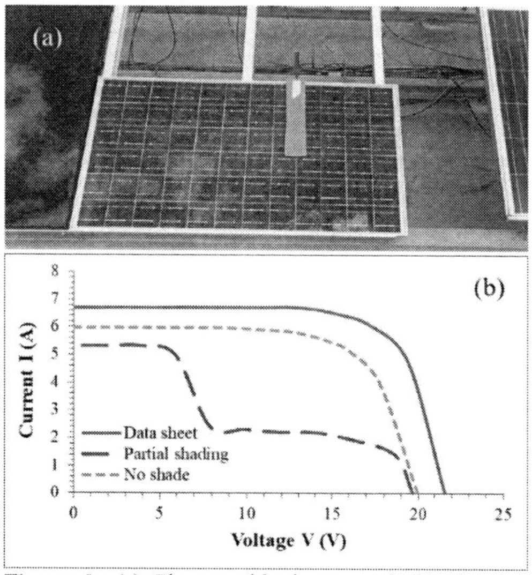

Figure 2: (a) Photographic image and (b) electrical profiles of a PV module C partially shaded intentionally during the investigation.

The P_{max} degradation as a result of lower I_{mpp} and V_{mpp} is known to be due to the temperature profiles of the PV modules under different shading conditions; which induces hotspots [3, 5]. In addition, the asymmetrical temperature distribution in PV modules can be a precursor for other degradation mechanisms which can affect the long-term performance reliability of PV modules [5, 7]. This is evident in the values reported in Table I. The values for both I_{mpp} and V_{mpp} are far lower under partial shading conditions.

The IR thermal images of the 3 field-aged PV modules are shown in Figure 3. The thermal profile of the modules under non-shading conditions varies for each PV module due to different defects and fault modes affecting each of the modules, see Figures 3a, 3b, and 3c. The defect cells show darker red colours in the thermal image due to localized hotspots [10]. Figures 3d, 3e, and 3f are the corresponding thermal images of PV modules A, B, and C, respectively under partial shading conditions. Figure 3 suggests that under partial shading conditions, hotspots are formed around the shaded area, but along the affected busbars. That is, in Figures 3d, 3e, and 3f, the hotspots are restricted to the upper left side of the PV module where the shade covers. The lower portion of the modules appears to be less affected due to the partial shading at the upper part of the module. This is because the busbars at the lower parts of the modules are not covered by the shade.

Another observation from Figure 3 is that, under partial shading conditions, formation of hotspots due to defects and fault modes decreases. This is because reversed biased current at the shaded area, hence, less or no current flows through the rest of the modules, including defect affected areas. This means that a combination of thermal imaging and shading could be used to isolate PV module material defects and fault modes from other defects such as soiling. This is because material defects will not show significant thermal signatures under partial shading conditions. In addition, the presence of defect and fault mode types can also influence the geometry or distribution of hotspots around the shade affected area of the PV module. The shape of the hotspots formed under partial shading conditions for the 3 modules is different and appears to depend on the type of defect at the shaded area, see Figure 3.

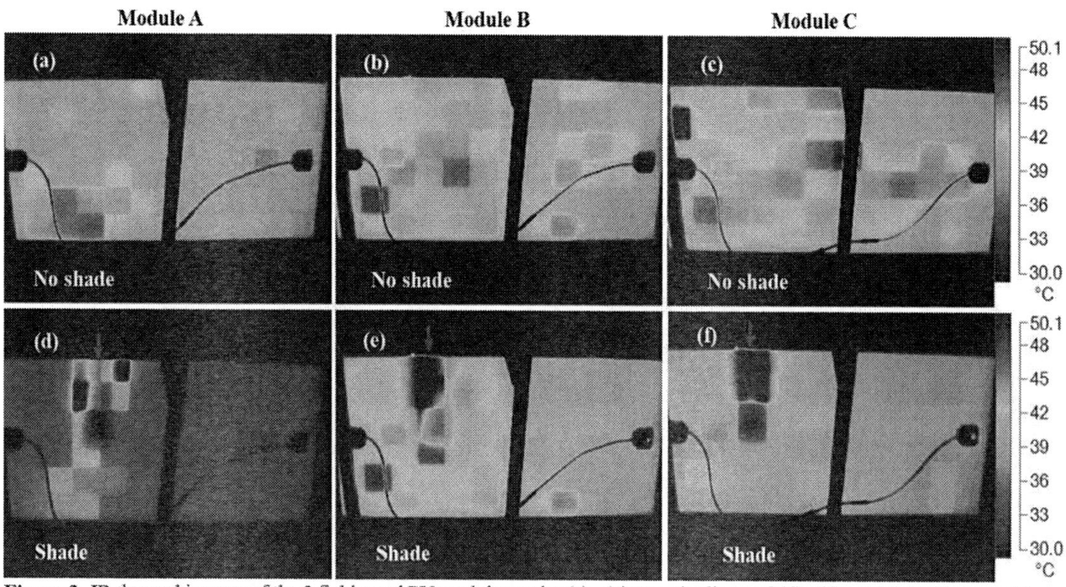

Figure 3: IR thermal images of the 3 field-aged PV modules under (a) - (c) non-shading and (d) - (f) partial shading conditions, respectively. The red arrow in (d) - (f) shows the position of the customized shade (as shown in Figure 2a) during the measurements. The measurements were done in real field outdoor conditions. The thick vertical blue line in the middle of each image is due to the metal frame of the rack (behind the module) used for holding the PV modules in place during measurements.

The dependence of I_{mpp} and V_{mpp} on the PV module temperature and irradiance conditions is well known [3, 8, 25]. Figure 4 illustrates the influence of varying irradiance conditions on the PV module operating temperature (T_m), P_{max}, V_{mpp}, and I_{mpp} under nominal operating conditions in non-shading and partial shading conditions. The values of ΔP_{max}, ΔV_{mpp}, and ΔI_{mpp} are the differences between the respective measured values under each condition (as in Figure 4) and the average value under nominal conditions in Figure 4a. Figure 4 suggests that T_m increases with increasing G_I, even under partial shading conditions, but reveals opposite effects on P_{max}, V_{mpp}, and I_{mpp}.

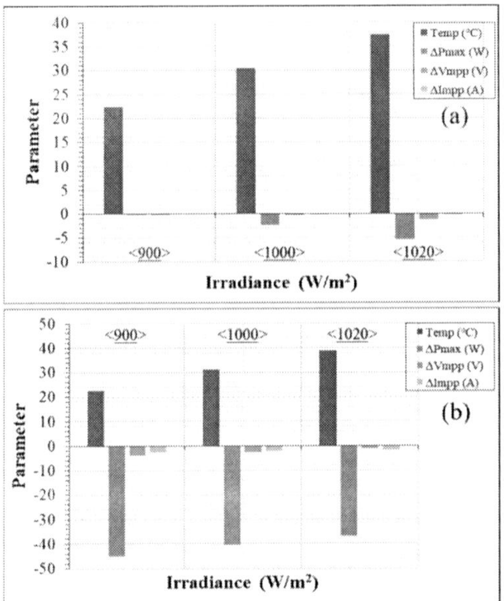

Figure 4: Evolution of PV module operating temperature, P_{max}, V_{mpp}, and I_{mpp} under varying irradiance conditions under (a) non-shading and (b) partial shading conditions. Δ is the difference between a parameter x and the average x-value in Figure 4a.

It appears the drop in P_{max}, V_{mpp}, and I_{mpp} increases with increasing temperature and irradiance in the absence of shading effects, see Figure 4a. However, under partial shading conditions, the drop in P_{max}, V_{mpp}, and I_{mpp} decreases when T_m and in-plane irradiance (G_I) increases. Another observation is that, under partial shading conditions, the percentage degradation in I_{mpp} becomes more critical to the degradation in P_{max} and worsen under higher G_I and T_m.

The dependence of P_{max} and FF on T_m of the field-aged PV modules is shown in Figure 5. In Figure 5a, the P_{max} and T_m profiles of the field-aged PV modules show relative dependence on each other in the absence of any shading condition. The FF also does not show any significant variation as T_m increases. However, under partial shading conditions, the trend is different, see Figure 5b.

Under partial shading conditions, the P_{max} and FF curves are not as smooth as compared to how they appear when the module is not affected by any shading condition. It is observed that the unsteadiness of current flow under partial shading conditions makes the P_{max} and FF curves are less smooth. Figure 5b suggests that the effect of T_m on P_{max} and FF degradation is more critical under partial shading conditions. As much as the P_{max} and FF depend on T_m, they are influenced by the in-plane irradiance (G_I) as well [2, 3]. The influence of T_m appears to be more pronounced under partial shading conditions which resulted in the surge in the P_{max} and FF profiles at T_m > 30 °C, see Figure 5b.

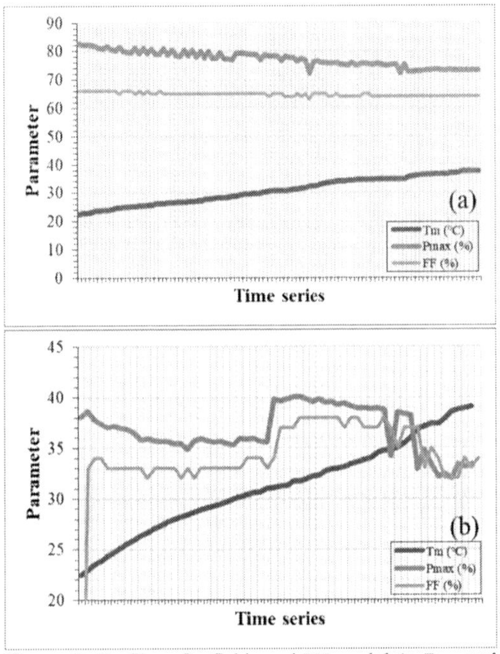

Figure 5: Profiles of a field-aged PV module's P_{max} and fill factor (FF) at varying T_m under (a) non-shading and (b) partial shading conditions in time series.

Hot spots from mismatch losses arise from uneven temperature distributions over the PV cell/module due to reverse biasing [6]. Figure 6 is Box and Whisker plots of the evolution of the temperature sensitivity of a field-aged PV module in the absence of shading and in the presence of partial shading conditions. Figure 6a is the absolute variation in the temperature coefficients of P_{max} (β_{Pmax}), V_{oc} (β_{Voc}), I_{sc} (β_{Jsc}), and FF (β_{FF}) of the field-aged PV module at STC in the absence of any shading condition. The absolute variation in the temperature coefficients under partial shading and low irradiance conditions are represented in Figure 6b and Figure 6c, respectively.

According to Figure 6, the absolute variation in the β_{Pmax}, β_{Voc}, β_{Jsc}, and β_{FF} increases under partial shading conditions. However, the variation in β_{FF} dominates the cause for the variation, hence, the higher variation in β_{Pmax} [3, 7, 8]. Figure 6 also suggests that at STC, the influence of β_{Voc} in the degradation in β_{Pmax} is fairly constant even under partial shading conditions. In contrast, under partial shading conditions, β_{Jsc} is more affected than β_{Voc}, see Figure 6c. Figure 6 also suggests that the β_{Jsc} of the partially shaded modules under both STC and lower irradiance conditions remains fairly constant.

Partial shading effects could be better diagnosed from the I_{mpp} and V_{mpp} characteristics of a PV module [8]. Previous investigations reported a minimal influence of temperature on I_{mpp}, ca. <0.04 %/K [3, 9, 14, 23, 25].

However, the temperature sensitivity of a PV module could be influenced by different degradation modes [26]. Therefore, the temperature sensitivity of P_{max}, hence, I_{mpp} and V_{mpp} of a PV module could be influenced by different degradation modes such as cracks and moisture induced degradations [16, 27]. The temperature coefficients of V_{mpp} (β_{Vmp}) and I_{mpp} (β_{Jmp}) of one of the field-aged PV modules is presented in Figure 7. The experimental investigation on the field-aged PV modules were done under the conditions ca. STC (980 - 1030 W/m^2).

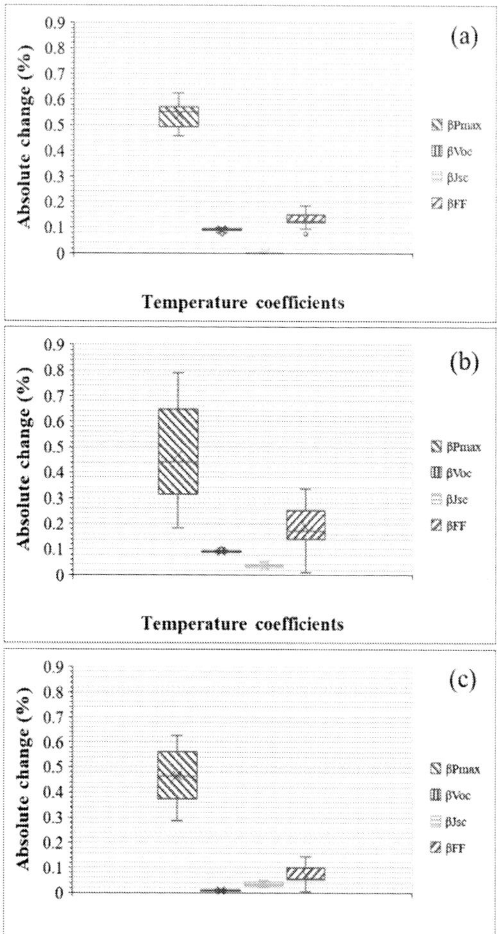

Figure 6: Evolution of temperature coefficients of a field-aged PV module under (a) STC conditions (in the absence of shading), (b) STC and (c) 700 - 800 W/m^2 conditions, but under partial shading conditions. The ends of the boxes are the lower and upper quartiles (interquartile range), the internal lines and x-marks indicate the median and mean, respectively.

Under non-shading conditions (in the absence of shading), the outdoor experiments estimate the β_{Vmp} and β_{Jmp} to be -0.0681 V/°C (-0.41 %/°C) and -0.0041 A/°C (-0.08 %/°C), respectively, see Figure 7a. The outdoor experimental studies on the same module under partial shading conditions give the results in Figure 7b. This estimates the β_{Vmp} and β_{Jmp} to be -0.0875 V/°C (-0.44 %/°C) and -0.0194 A/°C (-0.66 %/°C), respectively. Table II summarizes the results from the temperature sensitivity profiling of the 3 selected field-aged PV modules under non-shading and partial shading conditions.

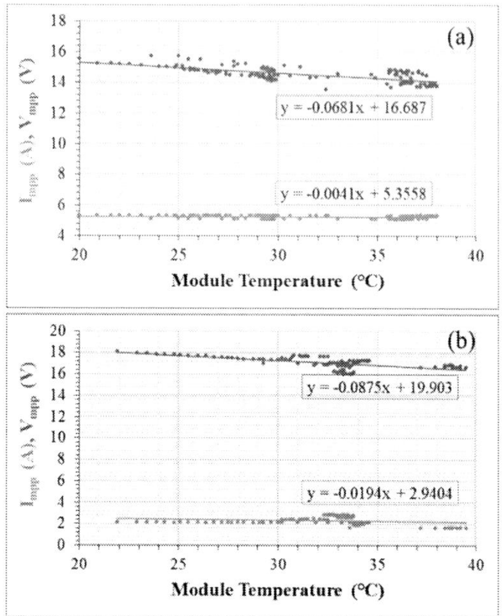

Figure 7: Evolution of temperature coefficients of I_{mpp} (β_{Jmp}) and V_{mpp} (β_{Vmp}) of a field-aged PV module under (a) non-shading and (b) partial shading conditions.

The results estimate the β_{Vmp} to be between ca. -0.41 %/°C and -0.48 %/°C under nominal conditions, which agrees with previous findings [3, 14, 25]. These results indicate that over the 20 years, the β_{Vmp} remains largely constant, and is not affected by the different defects that affected the PV modules. Yet, under partial shading conditions, the estimate for the PV module β_{Vmp} ranges from ca. -0.44 %/°C to 2.66 %/°C, see Table II. Increase in V_{mpp} under partial shading conditions was also observed elsewhere [8]. The disparity in β_{Vmp} is due to different defect mechanisms which trigger different characteristic current flux in the field-aged PV modules [8, 28]. This was observed during outdoor experiments on the field-aged PV modules when the in-plane irradiance is >1020 W/m^2, irrespective of the T_m. Strikingly, the β_{Vmp} for PV module C is positive under partial shading conditions, probably due to specific defect(s), see Table II. Comprehensive investigation on the typical effect of different defects mechanisms on the temperature profiles of the field-aged PV modules is underway.

In contrast, the results for the β_{Jmp} ranges from ca. -0.08 %/°C to 0.01 %/°C under non-shading conditions. These results differ from the 0.04 %/°C reported previously for multicrystalline silicon PV modules [3, 14,

Table II: Temperature coefficients of I_{mpp} (β_{Jmp}) and V_{mpp} (β_{Vmp}) of the field-aged PV modules investigated.

PV module	A		B		C	
Parameter (%/°C)	β_{Jmp}	β_{Vmp}	β_{Jmp}	β_{Vmp}	β_{Jmp}	β_{Vmp}
Partial shading*	-1.00	-0.36	-0.66	-0.44	-0.90	2.66
No shade	0.01	-0.41	-0.08	-0.41	0.01	-0.48

25]. Variation in the β_{Jmp} with irradiance is reported in other studies [12, 14]. Surprisingly, PV module B shows a negative β_{Jmp} under nominal conditions, most likely due to degradation issue(s).

Also, under partial shading conditions, the estimate for the β_{Jmp} for the modules ranges from ca. -1.00 %/°C to -0.90 %/°C, see Table II. The β_{Jmp} appears to be negative for all the modules measured under partial shading conditions. Obviously, this could be caused by the defects (e.g., cracks, moisture ingress, optical degradation, potential induced degradation, etc.) that has affected the field-aged PV modules. More importantly, the results show that the β_{Vmp} and β_{Jmp} increase under partial shading conditions, hence, affect P_{max}. β_{Jmp} is negative due to its dependence on *FF* which has a negative temperature dependence [25]. It is assumed that the degradation in both β_{Vmp} and β_{Jmp} stems from parasitic series and shunt resistance losses emanating from different moisture induced failure mechanisms affecting the field-aged PV modules as reported elsewhere [25, 29].

The influence of the temperature coefficients of efficiency ($\beta_{\eta m}$) on one of the field-aged PV modules under non-shading and shading conditions is shown in Figure 8.

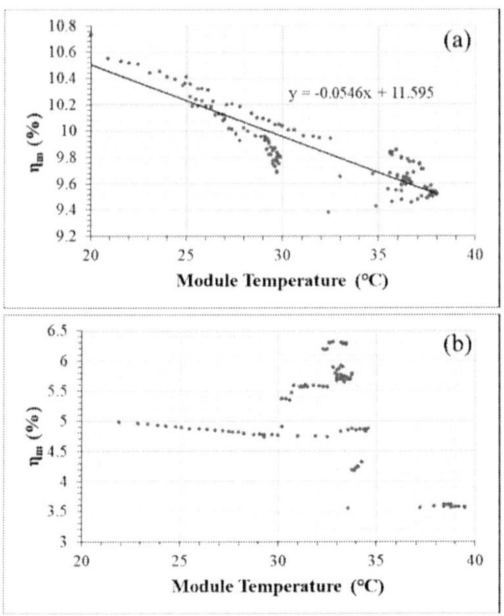

Figure 8: The temperature coefficients of efficiency of a field-aged PV module at ca. STC under (a) non-shading and (b) partial shading conditions.

The figure suggests that PV modules are more sensitive to temperature effects under partial shading conditions. That is, under STC and in the absence of any form of shading, the $\beta_{\eta m}$ is ca. -0.47 %/°C, refer to Figure 8a and (3). However, under partial shading conditions, the $\beta_{\eta m}$ is estimated to be ca. -0.92 %/°C, refer to Figure 8b and (3). The estimated $\beta_{\eta m}$ for the 3 field-aged PV modules is recorded in Table III.

The $\beta_{\eta m}$ under non-shading and shading conditions ranges from -0.41 to -0.48 %/°C and -0.59 to -1.18 %/°C, respectively for the 3 field-aged PV modules. In addition, Figure 8 estimates that at STC (i.e., at 25 °C), the PV module efficiency (η_m) reduces to 10.23 % and 5.62 % for non-shading and partial shading conditions, respectively.

Table III: Temperature coefficients of efficiency, η_m ($\beta_{\eta m}$) of the 3 field-aged PV modules investigated.

PV module $\beta_{\eta m}$ (%/°C)	A	B	C
Shade	-1.18	-0.92	-0.59
No shade	-0.41	-0.47	-0.48

4 CONCLUSIONS

The temperature coefficients of 20 years field-aged PV modules have been investigated under real field operation conditions in Grimstad, southern Norway. Thermal images show that hotspots are mostly localized to the busbars that are closer to the shade affected area of the module. The results show that, under partial shading conditions, drop in the temperature coefficient of fill factor (β_{FF}) dominates for power degradation, especially under partial shading conditions. Furthermore, it is estimated that under partial shading conditions, degradation in P_{max}, V_{mpp}, and I_{mpp} decreases when the PV module operating temperature (T_m) and in-plane irradiance increases, due to the dependence of V_{mpp}, and I_{mpp} on fill factor. That is, under partial shading conditions, the dependence of P_{max} on the T_m could be different from what is known.

Also, the results estimate that, the temperature coefficient of V_{mpp} (β_{Vmp}) of the field-aged PV modules under STC is ca. -0.41 %/°C to -0.48 %/°C, which agrees with previous findings [3, 25]. The results suggest that, over the 20 years, the β_{Vmp} of the modules remains largely constant and is not affected by the degradation state of the PV modules. However, under partial shading conditions, the absolute variation in the β_{Vmp} is >2.00 %/°C for the 3 field-aged PV modules studied. The results for the temperature coefficient of I_{mpp} (β_{Jmp}) ranges from ca. -0.08 %/°C to 0.01 %/°C and ca. -1.00 %/°C to -0.90 %/°C under non-shading and partial shading conditions, respectively. The disparity in the β_{Jmp} values from the 0.04 %/°C reported previously for multicrystalline silicon PV modules could be due to the degradation states of the field-aged PV modules.

The results suggest that the PV modules have undergone some form of degradation over the 20 years. More importantly, the studies demonstrate that the temperature coefficients of field-aged PV modules increase under partial shading conditions. Yet, the severity of the temperature sensitivity under partial shading conditions also depends on the degradation state of the PV modules. It appears that thermal images under partial shading conditions could be useful for identifying PV modules affected by material defects from those affected by other defects such as dirt. However, more investigations are underway to explore the full scope of these observations. This may have the potential in the future to be used as a diagnostic tool in field installations.

REFERENCES

[1] A. Dolara, G. C. Lazaroiu, S. Leva, and G. Manzolini, "Experimental investigation of partial shading

scenarios on PV (photovoltaic) modules," *Energy*, vol. 55, pp. 466-475, 2013.

[2] B. R. Paudyal and A. G. Imenes, "Investigation of temperature coefficients of PV modules through field measured data," *Solar Energy*, vol. 224, pp. 425-439, 2021.

[3] M. Seapan, Y. Hishikawa, M. Yoshita, and K. Okajima, "Temperature and irradiance dependences of the current and voltage at maximum power of crystalline silicon PV devices," *Solar Energy*, vol. 204, pp. 459-465, 2020.

[4] A. Woyte, J. Nijs, and R. Belmans, "Partial shadowing of photovoltaic arrays with different system configurations: literature review and field test results," *Solar energy*, vol. 74, no. 3, pp. 217-233, 2003.

[5] H. Mohammed, M. Kumar, and R. Gupta, "Bypass diode effect on temperature distribution in crystalline silicon photovoltaic module under partial shading," *Solar Energy*, vol. 208, pp. 182-194, 2020.

[6] O. Kunz, R. J. Evans, M. K. Juhl, and T. Trupke, "Understanding partial shading effects in shingled PV modules," *Solar Energy*, vol. 202, pp. 420-428, 2020.

[7] K. A. Kim and P. T. Krein, "Reexamination of photovoltaic hot spotting to show inadequacy of the bypass diode," *IEEE Journal of Photovoltaics*, vol. 5, no. 5, pp. 1435-1441, 2015.

[8] M. Seapan, Y. Hishikawa, M. Yoshita, and K. Okajima, "Detection of shading effect by using the current and voltage at maximum power point of crystalline silicon PV modules," *Solar Energy*, vol. 211, pp. 1365-1372, 2020.

[9] M. Zaimi, H. El Achouby, A. Ibral, and E. Assaid, "Determining combined effects of solar radiation and panel junction temperature on all model-parameters to forecast peak power and photovoltaic yield of solar panel under non-standard conditions," *Solar Energy*, vol. 191, pp. 341-359, 2019.

[10] O. K. Segbefia, A. G. Imenes, I. Burud, and T. O. Sætre, "Temperature profiles of field-aged multicrystalline silicon photovoltaic modules affected by microcracks," in *2021 IEEE 48th Photovoltaic Specialists Conference (PVSC)*, 2021: IEEE, pp. 0001-0006.

[11] X. Sun, R. V. K. Chavali, and M. A. Alam, "Real-time monitoring and diagnosis of photovoltaic system degradation only using maximum power point—the Suns-Vmp method," *Progress in Photovoltaics: Research and Applications*, vol. 27, no. 1, pp. 55-66, 2019.

[12] R. Dubey *et al.*, "Measurement of temperature coefficient of photovoltaic modules in field and comparison with laboratory measurements," in *2015 IEEE 42nd Photovoltaic Specialist Conference (PVSC)*, 2015: IEEE, pp. 1-5.

[13] E. Skoplaki and J. A. Palyvos, "On the temperature dependence of photovoltaic module electrical performance: A review of efficiency/power correlations," *Solar energy*, vol. 83, no. 5, pp. 614-624, 2009.

[14] D. L. King, J. A. Kratochvil, and W. E. Boyson, "Temperature coefficients for PV modules and arrays: measurement methods, difficulties, and results," in *Conference record of the twenty sixth IEEE Photovoltaic Specialists Conference-1997*, 1997: IEEE, pp. 1183-1186.

[15] M. A. Green, "General temperature dependence of solar cell performance and implications for device modelling," *Progress in Photovoltaics: Research and Applications*, vol. 11, no. 5, pp. 333-340, 2003.

[16] O. Dupré, R. Vaillon, and M. A. Green, "Physics of the temperature coefficients of solar cells," *Solar energy materials and solar cells*, vol. 140, pp. 92-100, 2015.

[17] D. Verma, M. Tayyib, T. O. Saetre, and O.-M. Midtgård, "Outdoor Performance of 10 year Old a-Si and Poly-Si Modules in southern Norway conditions," in *2012 38th IEEE Photovoltaic Specialists Conference*, 2012: IEEE, pp. 002368-002371.

[18] U. Jahn, M. Herz, M. Köntges, D. Parlevliet, M. Paggi, and I. Tsanakas, *Review on infrared and electroluminescence imaging for PV field applications: International Energy Agency Photovoltaic Power Systems Programme: IEA PVPS Task 13, Subtask 3.3: report IEA-PVPS T13-12: 2018*. International Energy Agency, 2018.

[19] I. 61853-1, "Photovoltaic (PV) module performance testing and energy rating–Part 1: Irradiance and temperature performance measurements and power rating," 2011.

[20] M. Siegel, S. Klein, and W. Beckman, "A simplified method for estimating the monthly-average performance of photovoltaic systems," *Solar Energy*, vol. 26, no. 5, pp. 413-418, 1981.

[21] D. Evans, "Simplified method for predicting photovoltaic array output," *Solar energy*, vol. 27, no. 6, pp. 555-560, 1981.

[22] M. Keevers and M. Green, "Extended infrared response of silicon solar cells and the impurity photovoltaic effect," *Solar Energy Materials and Solar Cells*, vol. 41, pp. 195-204, 1996.

[23] B. H. King, C. W. Hansen, D. Riley, C. D. Robinson, and L. Pratt, "Procedure to determine coefficients for the Sandia Array Performance Model (SAPM)," *SAND2016-5284*, vol. 1256510, 2016.

[24] C. Xiao, P. Hacke, S. Johnston, D. B. Sulas-Kern, C. S. Jiang, and M. Al-Jassim, "Failure analysis of field-failed bypass diodes," *Progress in Photovoltaics: Research and Applications*, vol. 28, no. 9, pp. 909-918, 2020.

[25] M. Kasu, J. Abdu, S. Hara, S. Choi, Y. Chiba, and A. Masuda, "Temperature dependence measurements and performance analyses of high-efficiency interdigitated back-contact, passivated emitter and rear cell, and silicon heterojunction photovoltaic modules," *Japanese Journal of Applied Physics*, vol. 57, no. 8S3, p. 08RG18, 2018.

[26] E. L. Meyer and E. E. Van Dyk, "Assessing the reliability and degradation of photovoltaic module performance parameters," *IEEE Transactions on reliability*, vol. 53, no. 1, pp. 83-92, 2004.

[27] C. Buerhop, D. Schlegel, M. Niess, C. Vodermayer, R. Weißmann, and C. Brabec, "Reliability of IR-imaging of PV-plants under operating conditions," *Solar Energy Materials and Solar Cells*, vol. 107, pp. 154-164, 2012.

[28] E. Molenbroek, D. Waddington, and K. Emery, "Hot spot susceptibility and testing of PV modules," in *Photovoltaic Specialists Conference*, 1991, vol. 1: Las Vegas, pp. 547-552.

[29] O. K. Segbefia, A. G. Imenes, and T. O. Sætre, "Moisture ingress in photovoltaic modules: A review," *Solar Energy*, vol. 224, pp. 889-906, 2021.

PERFORMANCE ANALYSIS OF PV MODULES INSTALLED IN THE ALPINE REGION

Fabian Carigiet[1]*, Danilo Grunauer[2], Franz P. Baumgartner[1]
[1] ZHAW, Zurich University of Applied Sciences, School of Engineering, Institute of Energy Systems and Fluid Engineering, Technikumstrasse 9, 8401 Winterthur, Switzerland
*phone: +41 (0) 58 934 7292; e-mail: fabian.carigiet@zhaw.ch; www.zhaw.ch/=cari
[2] EKZ, Electricity Utility of the Canton of Zurich, Ueberlandstrasse 2, 8953 Dietikon, Switzerland

ABSTRACT: Several mono- and bifacial mc-Si PV modules were installed at a location 2500 m above sea level in the Alps and grouped in six segments with different inclinations 30°, 70° and 90°. The PV modules and the meteorological conditions are monitored minutely and compared to a 30° tilted PV module installed in the urban region of Zurich. During the analysis period between October 2018 and September 2020, the yield loss due to snow coverage was minimal. The highest loss was 2.3 % (2018/19) and 3.6 % (2019/20) for the 30° inclined PV modules evaluated by the introduced snow coverage model based on electrical and weather data. The two segments with a 30° inclination showed a 20.9 % to 27.2 % higher yield than the PV module installed in the urban region, mostly produced in the winter season. The bifacial alpine energy yield is about twice as high as that of the urban PV module from November to May. The highest yields of 1800 Wh/Wp in 2018/19 and 1696 Wh/Wp in 2019/20 was measured at the 70° tilted bifacial PV module without losses due to row shading as it is expected in PV plants.
Keywords: alpine region, bifacial, monitoring, performance, photovoltaic

1 INTRODUCTION

The Alps are known for having more sunny days in winter than in the Swiss midlands because there are less foggy days at higher altitudes. Photovoltaic installations in the Alps could therefore be a part of the future electricity generation based entirely on renewables and produce more electricity during wintertime feeding the higher electricity demand. Additionally, the generated PV electricity could be used locally such as in the ski areas and in the numerous existing mountain villages.

The average annual irradiance is up to 1.5 times higher in the Swiss Alps than in the Swiss midlands [1]. The average snow cover duration was 190 days for locations 2000 m above sea level in the years 2011 to 2016 [1]. This increases the electricity production of PV further due to the high albedo effect and thus, steeper tilt angles of PV installations can be considered without having to accept major energy yield losses as it is the case in urban region. In a case study in Austria, the performance of a PV installation on the Pitztal glacier (2900 m.a.s.l.) was compared to a reference PV installation in the valley (625 m.a.s.l.). The high alpine PV plant showed an additionally yield of 25 % in the year 2016 and 2017. The production profile was more balanced throughout the year, with the winter electricity accounting for 40 to 50 % of the total production [2].

In 2017, the electricity utility of the canton of Zurich (EKZ) installed several mono- and bifacial mc-Si PV modules in the alpine region (figure 1). The goal is to analyse the impact of the climatic conditions of the location at 2500 m above sea level on the PV module performance. The most interested climatic parameters are the module temperatures, the snow coverage of the PV modules and the effect of the high albedo during wintertime. Furthermore, the PV modules in the test setup have different inclinations for determining the individual performances during high albedo conditions.

In this work, the low light and temperature behaviours are modelled and the performance factors [3] are extracted and compared for each installed segment. The module temperature of the bifacial PV modules needs to be modelled because the temperature sensor could not be mounted directly behind the solar cells. The results of the 30° tilted segment are compared to a PV module installed with the same inclination in the urban region Dietikon, Switzerland, which is used for a long-term performance study and therefore well-known and analysed [4]. Additionally, the annual irradiation differences as well as the yield differences between the alpine and the urban region in Switzerland are elaborated. Finally, the influence of the weather-based parameters (irradiance, albedo and snow coverage) and the influence of the PV model-based parameters (inclination, bifaciality, module temperature, low-light behaviour and an aggregated factor accounting for spectral and degradation behaviour) on the PV module performance are determined.

2 APPROACH AND MEASUREMENT SETUP

The PV test facility is installed at the Totalp near the town Davos, Switzerland. There are 6 different PV segments indicated with letter from A to F with three to four PV modules each as shown in figure 1. The tilt angle can be individually chosen for each segment. The segments A and B have a tilt angle of 30°, C and D are tilted by 70° and the last segments E and F are installed vertically. The segment B is shown in figure 1 with a 50° tilt angle. For the analysis period, the inclination of this segment was also 30°. All PV modules are south oriented without nearby shading objects in front. The PV modules on the segments D and E are bifacial mc-Si modules from PVP (PVP-GE285M). The other segments have monofacial PV modules PVP-GE280M either with or without frame (segment A and C with frame). Figure 2 shows the detailed PV modules layout in the PV plant. Each module is connected to a commercial power optimizer (SolarEdge OPA300 for monofacial and OPA400 for bifacial PV modules), which ensures the MPPT. Different electrical, thermal and meteorological quantities such as voltage, current, module and ambient temperatures, wind speed, irradiances in every plane of array (POA) as well as the reflected irradiance due to the albedo for the segments C to F are measured every minute. The small weather station and the 6 PV segments are additional monitored by several webcams to analyse snow coverage.

Figure 1: Setup of the PV test facility at the Totalp, Davos, Switzerland (Source: EKZ [5]). Pyranometers were used as irradiance sensors.

The Totalp test power plant is one of the highest PV plants in Switzerland, located 2500 m above sea level and it is equipped with a high-quality measurement system. The analysis period ranging from 1st of October 2018 to 30th of September 2020 showed a very high measurement uptime of 94.4 % and 95.3 %. Unless otherwise mentioned in this work, minutely averaged data were used for the analyses.

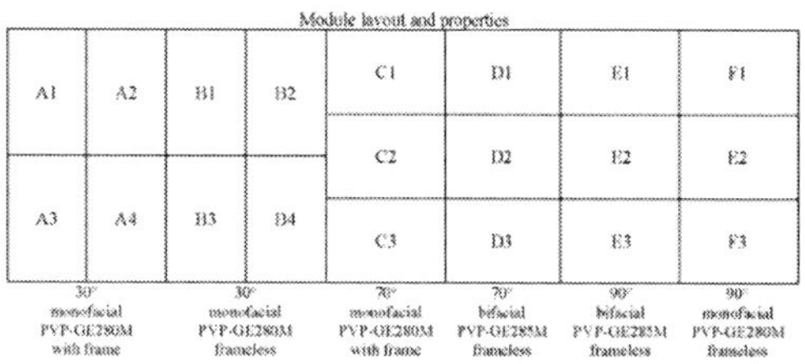

Figure 2: PV module layout and properties at the test facility.

3 MODELLING

For the analyses performed in this work, four models are used and described. First, the snow covering need to be detected to address the yield losses in terms of snow. Then, the low-light and thermal modelling is required as an input to the PV performance model. The performance model is used to quantify the influences of weather-based or PV model-based parameters as well as to model the PV power output without snow covering when there is snow on the PV modules.

3.1 Snow cover detection

To identify the snow coverage, an algorithm using only measured electrical and weather parameters is developed (no webcam image processing used). Therefore, the time has to be known when a snow cover was present. For this purpose, different days were selected with and without snow coverage according to the webcam images. For these days, daily patterns, MPP voltage vs currents diagrams and various parameter combinations, such as MPP current per irradiance, were analysed and statistically evaluated. Based on that, the snow cover detection algorithm was elaborated and visualised in the flow chart in figure 3a. This algorithm was then tested on different days and compared with webcam images [6]. This algorithm is not a general solution for snow detection on the PV module surface. It was developed just for the purposes of successfully quantifying the yield losses due to snow coverage within this work. Figure 3b shows the resulting plot when the algorithm is applied on the measurements of segment B on 29th of December 2019.

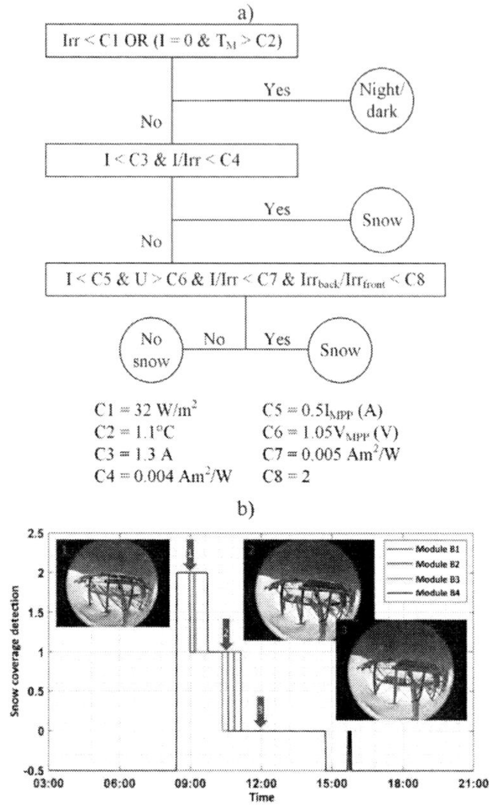

Figure 3: The flow chart a) shows the algorithm that is used to detect the snow coverage of the PV module installed in the Alps. "Irr" stands for measured irradiance and "I" for the measured current at MPP. The plot b) shows the snow coverage detection for the case on 29th of December 2019.

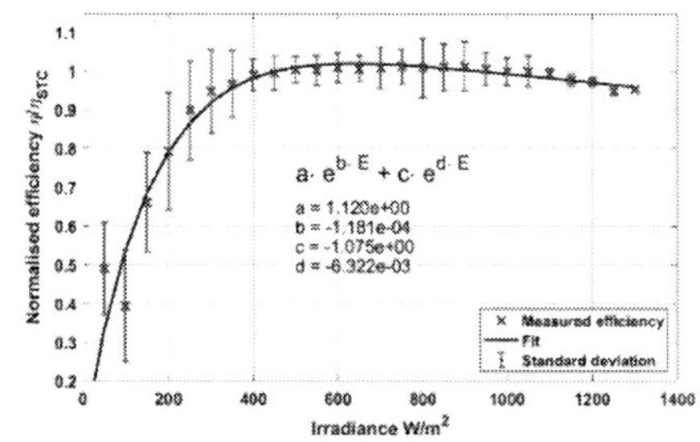

Figure 4: Normalised efficiency of the PV module C3 for the 26 irradiance intervals between 50 and 1300 W/m^2. The double exponential fit is used in the PV performance model.

3.1 Low-light behaviour of the PV modules

For the analysis of the low-light behaviour, the measured module outputs were sorted into bins with ±25 W/m^2 intervals at irradiance levels between 50 – 1300 W/m^2. This high upper limit of irradiance is due to the high albedo in the Alps. Additionally, the data were selected only when clear sky conditions were met according to an algorithm from the SANDIA labs [7]. For each interval, the efficiency at 25 °C was calculated using the linear regression between the measure module power and the measured module temperature. A more detailed description of this method can be found in the papers [3, 4]. Figure 4 shows the resulting low-light behaviour evaluated for the PV module C1.

3.2 Thermal model

The module temperature modelling uses irradiance, ambient temperature and wind measurements as model inputs. The model is based on the thermodynamic law of conservation of energy, which considers the solar radiation Q_S, the generated electrical energy W_{PV}, the convection Q_C and the heat radiation of the modules to the sky $Q_{m\rightarrow s}$ and the ground $Q_{m\rightarrow g}$. The time derivative of the energy balance is described in the differential equation (1) by using the equation (2) to (5).

$$C_m \frac{dT_m}{dt} = \dot{Q}_S - \dot{Q}_C - \dot{Q}_{m\rightarrow s} - \dot{Q}_{m\rightarrow g} - P_{PV} \quad (1)$$

$$\dot{Q}_S = E_{POA}\, A_m \quad (2)$$

$$\dot{Q}_C = A_m\, U(v_w)\, (T_m - T_a) \quad (3)$$

$$\begin{aligned} &\dot{Q}_{m\rightarrow s} = \varepsilon_m\, \varepsilon_s\, F_v\, \sigma\, A_m\, (T_m^4 - T_s^4) \\ &with\, T_s = 0.0552\, T_a^{1.5} \\ &and\, \varepsilon_s = 0.72 + 0.005\, \vartheta_a^{1.5} \end{aligned} \quad (4)\ [8]$$

$$\dot{Q}_{m\rightarrow g} = \varepsilon_m\, \varepsilon_g\, F_v\, \sigma\, A_m\, (T_m^4 - T_g^4) \quad (5)$$

The following simplification or assumptions were made for the modelling of the PV module temperature:

- The temperature of the ground T_g is equal to the ambient temperature T_a.
- The emissivity ε_s and the temperature T_s of the sky are modelled according to Mittag et al. [8].
- The emissivity ε_m of the PV module and ε_g of the ground are assumed to be 0.9.
- The view factors F_v are set equal to 1.

The empirical model [9] was used to determine the convective heat transfer coefficients, modelling constant free convection with U_0 and forced convection, which increases linearly with wind speed, with U_1. The clear-sky data were averaged over 10 min to consider equation (1) as quasi-stationary. The linear regression was used to determine the convective heat transfer coefficients from equation (6). Only irradiance levels between 990 and 1010 W/m^2 were considered in this analysis.

$$\frac{\dot{Q}_S - \dot{Q}_{m\rightarrow s} - \dot{Q}_{m\rightarrow g} - P_{PV}}{A_m\, (T_m - T_a)} = U_0 + U_1\, v_w \quad (6)$$

The average heat transfer coefficient U_0 was found to be 8.8 W/m^{2}K. The coefficient U_1 modelling the forces convection depends on the tilt angle of the PV resulting to average heat transfer coefficients of 4.7 Ws/m^{3}K (30°), 3.4 Ws/m^{3}K (70°) and 2.6 Ws/m^{3}K (90°).

The PV module heat capacity C_m could be determined by minimising the root mean square error (RMSE) between the measured module temperature and the simulated module temperature by solving the differential equation (1). The average heat capacity was found to be 32.7 kJ/K.

This analysis was performed only with the monofacial PV modules. The temperature sensor of the bifacial PV module could not be mounted directly behind the solar cells which made this analysis impossible for this type of module. Thus, the convective heat transfer coefficients and module heat capacity had to be assumed to be equal to those values of the monofacial PV modules with the corresponding tilt angles. This assumption can be made because the monofacial modules are also glass-glass modules.

3.3 PV performance model

The power output of the PV modules installed at the Totalp, Davos, are modelled by equation (7). The following parameters serve as input:

- Irradiance at the module level E_{POA} including the reflected irradiance on the rear side for the bifacial modules.
- PV module area A_m.
- Efficiencies of the modules (16.9 % for monofacial and 17.2 % for bifacial PV modules at STC) including their low-light behaviour $\eta(E_{POA})$.
- Temperature coefficient δ according to the manufacturer's datasheet (-0.42 %/K for the mono- and -0.4 %/K for the bifacial PV modules).
- The performance factor k_P for modelling the spectral influences, degradation, setup errors and irradiance inhomogeneities on the frontside and backside of the bifacial PV modules

$$P_{PV} = k_P\, E_{POA}\, A_m\, \eta(E_{POA})\, (1 + \delta\, \Delta T) \quad (7)$$

The performance factor k_P was determined after all other parameters and models had been carried out. For this purpose, the RMSE between the measured power and the simulated power has been minimised.

The individual performance factors that influence the performance of the PV modules on the Totalp was determined for the period from 1st of October 2019 to 30th of September 2020. The other two individual weighted and averaged performance factors k_{LL} and k_T were calculated. The product of the factors k_P (spectral influences, degradation, measurement setup), k_{LL} (low light behaviour) and k_T (temperature behaviour) represents the performance ratio PR (8). The resulting performance factors and ratios are given and discussed in the section 4.

$$PR_{DC} = \frac{W_{PV}\, 1\frac{kW}{m^2}}{G_{POA}\, P_{STC}} = k_P\, k_{LL}\, k_T \quad (8)$$

4 RESULTS

First, the measured yields of the alpine installation are presented. Then, the modelled yields are compared to the measured yields and the performance factors are determined for the different tilted PV modules in the Alps and compared to the reference PV module in the urban area (230 Wp, Sunways). Finally, the yield losses due to the snow coverage of the PV modules are elaborated.

4.1 Measured yield and performance ratio

The PV module yield was analysed for the period ranging from 1st of October 2018 to 30th of September 2020. The measurement setup had a very high uptime of 94.4 % and 95.3 % for the two years. For the comparison to an urban PV module, the data sets from both sites were matched for this yield analysis reducing the uptime slightly to 92.3 % and 91.2 %, respectively.

Table I: Analysed irradiations, PV module yields and PR for the individual PV segment on the Totalp and the reference module (230 Wp, Sunways) in the urban region during the periods 2018/19 and 2019/20. For the bifacial PV modules, a distinction was made between the PR with respect to only the irradiation on the frontside and a second PR that considers the sum of the front and backside irradiation.

Segments	A	B	C	D	E	F	Dietikon	Units
	1st of October 2018 to 30th of September 2019 (Uptime 92.3 %)							
G_{POA}	1644.2	1658.8	1539.9	1539.9	1530.3	1530.3	1294.3	kWh
G_{Albedo}	-	-	-	701.9	733.9	-	-	kWh
Yield	1502.4	1428.2	1526.4	1799.7	1662.5	1373.9	1181.4	kWh/kWp
PR_{DC}	0.914	0.861	0.991	1.169	1.086	0.898	0.913	-
PR_{DC} **(incl. rear side irr.)**	-	-	-	0.803	0.734	-	-	-
	1st of October 2019 to 30th of September 2020 (Uptime 91.2 %)							
G_{POA}	1533.4	1541.5	1373.3	1373.3	1365.9	1365.9	1353.1	kWh
G_{Albedo}	-	-	-	618.7	640.9	-	-	kWh
Yield	1372.5	1315.9	1383.8	1696.1	1563.9	1221.4	1234.2	kWh/kWp
PR_{DC}	0.895	0.854	1.008	1.235	1.145	0.894	0.912	-
PR_{DC} **(incl. rear side irr.)**	-	-	-	0.851	0.779	-	-	-

Table I includes the annual measurement results for the alpine installation as well as the reference PV module installed in Dietikon, the urban area of the canton of Zurich. Table II shows the monthly yield ratios of the bifacial and the monofacial PV modules meaning the monthly yield of the bifacial setups are divided by the corresponding yield of the monofacial setup with same inclination at Totalp test setup.

Table II: Ratio of the monthly energy yield of the bifacial module segments to the monofacial module segments in the Totalp installation for the time period 2019/20.

Month	Oct	Nov	Dec	Jan	Feb	Mar
Seg D/C	1.15	1.24	1.21	1.16	1.26	1.28
Seg E/F	1.20	1.35	1.27	1.20	1.27	1.26
Month	**Apr**	**May**	**Jun**	**Jul**	**Aug**	**Sep**
Seg D/C	1.27	1.31	1.35	1.12	1.13	1.13
Seg E/F	1.26	1.43	1.55	1.23	1.20	1.18

The annual PR of the segment A in 2018/19 was 0.914 and thus equal to the PR of the reference module in Dietikon with the same 30° south inclination. In the second year, the PR of segment A was 1.7% % higher than that of the PV module in the urban area. Segment B had PRs that were 5.2% and 5.8%, respectively, lower than the urban reference module. The monofacial PV modules of segment C (70°) had a very high PR between 0.99 and 1.01. Segment F (90°), on the other hand, had a lower PR of 0.898 and 0.894, although the irradiation yield was almost identical to that of segment C. The reason can be seen in the performance factor k_P of the segment C in the subsection 4.3, which is much higher than the according factor of segment F. With a very high probability, there must be an error in the setup of the measurement system.

The PR of the bifacial PV modules of segment D (70°), considering the albedo radiation on the rear side, was 0.803 and 0.851. For the bifacial PV modules of segment E (90°), a PR of 0.734 and 0.779 was observed. One reason for the low PR compared to the monofacial PV modules is the overestimation of the influence of the irradiation on the rear side in the performance ratio calculation. This effect was also observed in the evaluations of the simulations in the subsection 4.2 where the simulations showed an overestimation of the energy yield between 3.4 % and 12.8 %. A second point is a module failure of segment D (70°) and E (90°) from 17 October to 5 November 2018 as described in the subsection 4.2.

The monthly energy yields ratios between bifacial PV modules of segments D and E and the corresponding segments C and F with the same inclination show that the energy yields of the bifacial PV modules are between 12 % and 55 % higher than the monofacial PV modules. Especially in the months of May, June and July, the additional yield for the analysed period can be between 42 % and 55 %. Over the entire period from October 2019 to September 2020, an additional yield of 22.6 % and 28.8 % could be measured for segment D (70°) and segment E (90°), respectively.

The additional yield of the PV modules installed in the alpine region compared to the urban PV module in Dietikon depends strongly on the respective year. An additional yield of 27.2 % (segment A) and 20.9 % (segment B) could be measured in 2018/19. The following year showed a significantly lower additional yield of 11.2 % (segment A) and 6.6 % (segment B). Overall, a yield difference between the two years of 14.0 % to 17.3 % was observed on the Totalp in Davos.

For the year 2019/20, segment A (30°) showed an increase in yield of 4.1 % compared to the segment B (30°) and 11.2 % compared to Dietikon (30°). The PV modules of segment F (90°) showed a yield reduction of 12.2 %. compared to segment A (30°) and thus produced about the same as the PV module in Dietikon. The use of the bifacial PV modules at the given tilt angles resulted in an additional yield of 26.2 % (segment D 70°) and 15.5 % (segment E 90°) compared to segment A (30°). Compared to the reference module in Dietikon, these were 37.4 % and 26.7 % respectively.

Table III: Modelled and measured energy yield of the analysed PV segments in the Alps for period from 1st of October 2018 to 30th of September 2019.

Segment	A	B	C	Units
Modelled	1541.3	1521.9	1593.8	kWh/kWp
Measured	1559.8	1482.5	1577.4	kWh/kWp
Difference	-1.19	2.66	1.04	%
Segment	D	E	F	Unit
Modelled	2054.4	1947.8	1392.0	kWh/kWp
Measured	1870.2	1727.0	1420.6	kWh/kWp
Difference	9.85	12.79	-2.02	%

4.2 Model verification

The modelled energy yields for every PV segment are compared to the corresponding measurement results for the two years. Table III and figure 5 show the results for the time period from 1st of October 2018 to 30th of September 2019. Table IV and figure 6 includes the data for the following year (1st of October 2019 to 30th of September 2020)

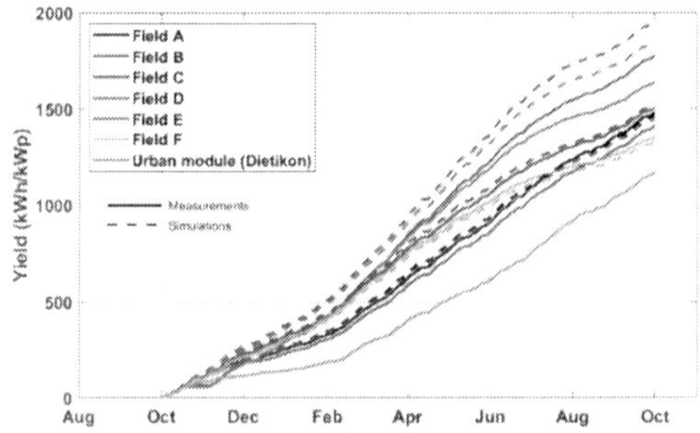

Figure 5: Measured and simulated cumulated yield of the six different segments installed in the Alps in 2018/2019. The measured yield of the reference PV module in the urban region is included for the same time period.

Table IV: Modelled and measured energy yield of the analysed PV segments in the Alps for period from 1st of October 2019 to 30th of September 2020.

Segment	**A**	**B**	**C**	**Units**
Modelled	1445.3	1424.9	1433.4	kWh/kWp
Measured	1449.1	1390.0	1455.3	kWh/kWp
Difference	-0.26	2.51	-1.51	%
Segment	**D**	**E**	**F**	**Unit**
Modelled	1844.3	1741.6	1247.2	kWh/kWp
Measured	1783.3	1641.6	1282.4	kWh/kWp
Difference	3.42	6.09	-2.75	%

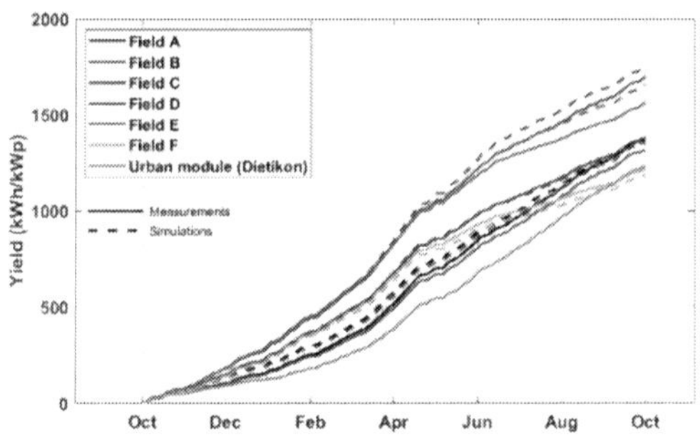

Figure 6: Measured and simulated yield of the six different segments installed in the Alps in 2019/2020. The measured yield of the reference PV module in the urban region is included for the same time period.

The modelled yields for the monofacial modules for 2018/19 and 2019/20 are within 3 % of the measured yields. In the last year, the modelled yields of the bifacial module deviate 3.42 % (segment D) and 6.09 % (segment E) from the measured yields. The overperformances of the simulation of the bifacial PV modules occurs mainly in the summer months as shown in figure 5. In the first year 2018/19, the model of the bifacial PV modules had a significant overperformance of 9.85 % (segment D) and 12.79 % (segment E). The reason for this deviation between the two consecutive years is the power outage of the two segments (reason unknown). However, the measurement system worked, which is why these measurements were included in the evaluation. According to the simulation, the yield losses were 88.4 kWh/kWp for segment D and 82.2 kWh/kWp for segment E, which corresponds to a share of 4.3 % and 4.2 % of the simulated annual energy yield, respectively. Taking this into account, the model overperformance is 5.55 % and 8.59 % for that year and it is again evident that the rest of the overperformance occurs in the summer months (see figure 6). For both years, this overperformance is driven by the overestimation of the albedo radiation on the backside of the PV modules.

4.3 Performance comparison alpine vs urban

The performance model of equation (8) serves as the basis. From this model, the individual average power-weighted performance factors could be determined. The product of the factors k_P (spectral influences, degradation, measurement setup), k_{LL} (low-light behaviour) and k_T (temperature behaviour) represents the DC performance ratio PR_{DC}. Table V lists the individual performance factors, the PRs calculated from these values and the modelled PR. The performance factors of the reference module in the urban environment in Dietikon serve as a comparison. These reference values were analysed for the year 2012 [3], two years after commissioning.

The performance factors k_P for alpine installations are lower than in Dietikon except for the segment C. This factor includes spectral influences, degradation and uncertainties in the measurement setup, because the alpine measurement setup does not have four-wire measurement technology. An additional uncertainty are the power optimisers that are used for MPPT. Segment C has a performance factor k_P of 1.055, which indicates are failure in the setup.

The low-light behaviour of the PV reference polycrystalline silicon module from Sunways in the urban area leads to an increase in performance. This is no longer the case with the PV modules on the Totalp. The determined low-light characteristics of the PV modules installed alpine region show a lower increase in performance with irradiance levels between 500 W/m^2 and 800 W/m^2 compared to the PV module in Dietikon. Furthermore, the maximum irradiance levels on the Totalp are significantly higher than in the urban area.

The weaker performance due to the low-light behaviour is partly compensated by the better temperature behaviour. The bifacial PV modules have the lowest k_T values because more solar radiation is converted into heat due to the bifaciality. The results of the temperature influence of the bifacial PV modules should be treated with caution, as the temperature simulation could not be verified with temperature measurements so far.

4.4 Yield loss due to snow coverage

The yield losses due to snow coverage were close to inexistent for the vertical installations (segment E and F) and the 70° bifacial installation (D). The losses were between 0.12 % and 0.31 %. The 70° monofacial installation in segment C had snow coverage losses of 1.5 % for both years. The losses of the 30° tilted PV segments were between 2.3 % (2018/19) to 3.6 % (2019/20) of the corresponding annual yield.

Table V: The performance ratios according to equation (8) were determined by analysing the data in the year 2019/20 which is two years after commissioning. The PR was calculated based on these performance factors as well as based on the simulated power. The performance factors of the reference module in the urban region in Dietikon, Switzerland, were calculated for the year 2012, which corresponds also to two years after commissioning [3].

Segments	A	B	C	D	E	F	Urban [3]
k_P (spectral, degradation, setup)	0.942	0.906	1.055	0.938	0.886	0.925	0.963
k_{LL} (low-light)	0.965	0.968	0.961	0.961	0.960	0.957	1.007
k_T (temperature)	1.002	1.003	1.003	0.976	0.969	0.992	0.960
PR_{DC} (performance factors)	0.912	0.879	1.017	0.880	0.825	0.878	0.930
PR_{DC} (simulation)	0.898	0.880	0.996	0.881	0.826	0.873	-
Difference (simulated - measured)	-1.53 %	0.09 %	-2.03 %	0.17 %	0.17 %	-0.61 %	-

5 CONCLUSIONS AND OUTLOOK

The yield losses due to snow coverage were surprisingly lower than expected. The vertical installation had hardly any losses. The highest yield loss of 3.6 % was observed at the segment B in the year 2019/20.

The advantages of lower temperatures in the alpine area resulted in a performance increase of 0.9 % to 4.3 % compared to the urban area. The low-light performance loss in the alpine area relative to STC was 3.2 % to 4.3 % compared to the urban area.

The additional yield in the alpine region differed heavily between the two analysed years. In the period 2018/19, the yield was between 14.0 % to 17.3 % higher compared to the period 2019/20. The segments A and B with a 30° inclination had a 27.2 % and 20.9 %, respectively, higher yield than the PV module installed in the urban region with the same tilt angle during the period 2018/19. In the second analysed year, the additional yield was lower and resulted in a factor of 11.2 % and 6.6 %, respectively. The vertically installed monofacial PV modules (segment F) had a significantly lower yield, which was 11.0 % lower than that of the 30° tilted PV modules (segment A). Thus, it produced about the same amount of electricity as the PV module in the urban region, but mostly during the winter period. The use of the bifacial PV modules under the given tilt angles 70° and 90° resulted in an additional yield of 23.5 % and 13.9 %, respectively with respect to the modules in segment A. The additional yields of the alpine-installed PV modules for 2019 occurred mainly in the months February up to May. While the PV module in the urban region produced around 100 Wh/Wp electricity per month during this period, the monofacial PV modules in the Alps generated 1.5 times as much electricity. The measured yield of the bifacial PV modules was twice as high during these months.

The measurement data from this test facility is continuously evaluated for the most recent year. In addition, the albedo radiation should be better reflected in the modelling. For this purpose, new diffuse and direct irradiance measurements are being analysed and evaluated. A further point will be the adaptation to the size of PV power plants, in which row shading must be considered and albedo effects need to be adjusted.

6 REFERENCES

[1] A. Kahl, J. Dujardin and M. Lehning, 'The bright side of PV production in snow-covered mountains', Proceedings of the National Academy of Sciences, 116, 2019, DOI: 10.1073/pnas.1720808116

[2] P. Egger, H. Skopetz and W. Woyke, 'Ertragsanalyse einer alpinen Photovoltaik-Anlage', Proceedings of 11. Internationale Energiewirtschaftstagung an der TU Wien, 2019

[3] F. Carigiet et al., 'Energy Rating Based on Thermal Modelling of Five Different PV Technologies', in Proceedings of the 29th European Photovoltaic Solar Energy Conference and Exhibition, Amsterdam, 2014 DOI: 10.4229/EUPVSEC20142014-5CV.2.34

[4] F. Carigiet, F. Baumgartner and C. Brabec, 'Long-term power degradation analysis of crystalline silicon PV modules using indoor and outdoor measurement techniques', Renewable and Sustainable Energy Reviews, 144, 2021, DOI: 10.1016/j.rser.2021.111005

[5] www.ekz.ch/blue/de/innovation/2020/Totalp.html

[6] T. Läuchli, 'Alpine Solaranlage Totalp', Bachelor thesis, ZHAW, School of Engineering, Winterthur, 2021.

[7] J. S. Stein, C. W. Hansen and M. J. Reno, 'Global horizontal irradiance clear sky models: implementation and analysis.', SAND2012-2389, 2012. DOI: 10.2172/1039404.

[8] M. Mittag et al., 'Thermal Modelling of Photovoltaic Modules in Operation and Production', in Proceedings of the 36th European Photovoltaic Solar Energy Conference and Exhibition, Marseille, 2019, DOI: 10.4229/EUPVSEC20192019-4CO.2.4.

[9] D. Faiman, 'Assessing the outdoor operating temperature of photovoltaic modules', Prog. Photovolt. Res. Appl., 16, 2008, DOI: 10.1002/pip.813.

PV PLANT MONITORING NEEDS BOTH – DATA ANALYSIS AND ON-SITE INSPECTION

W. Mühleisen[1], L. Neumaier[1], J. Kosel[1], C. Hirschl[1], R. Rattenberger[2], P. Prasser[3]
[1] SAL Silicon Austria Labs GmbH, Europastr.12, 9524 Villach, Austria, (lukas.neumaier@silicon-austria.com, juergen.kosel@silicon-austria.com, christina.hirschl@silicon-austria.com); Corresponding author: E-Mail: wolfgang.muehleisen@silicon-austria.com, T: +43 66488200144
[2] University of Applied Sciences Burgenland, Steinamangerstraße 21, 7423 Pinkafeld, Austria, (ruediger.rattenberger@holzdiesonne.net)
[3] KIOTO Photovoltaics GmbH, Solarstr.1, 9300 St.Veit, Austria, (peter.prasser@kioto.com)

ABSTRACT: Photovoltaic systems can be controlled with several approaches. By means of monitoring or with sample measurements on site. On a 9-year-old photovoltaic system with 970 kWp in the south of Austria, both approaches were pursued to investigate whether a cost-effective monitoring analysis alone is sufficient. The identification and partial repair of several superimposed failure cases such as shading, dirt, glass breakage and backsheet cracks ultimately showed that a data analysis with a determined performance ratio alone is not sufficient to draw complete conclusions about PV system and energy yield.

Keywords: Monitoring, System Performance

1 INTRODUCTION

On the way to an energy system that has a high proportion of photovoltaics, more and more people are confronted with the issues of monitoring and maintenance [1, 2]. In the course of automation and monitor solutions on the market, costs due to on-site inspections are aimed to be largely avoided [3]. Troubleshooting and solutions with the help of artificial intelligence and data evaluation is in preparation and to be seen as a business model in future [4, 5, 6].

But it is often forgotten that a professional on-site inspection is very important and facilitates the interpretation of data evaluations. On the one hand, plant visits with staff cost money, but on the other hand, this saves time in retrospect when data interpretation alone is not expedient. Further benefit is to decide if modules are to maintain, repair or recycle in case of an expired guarantee. And some failures are not seen in monitoring systems in an early stage e.g., broken glasses, back sheet cracks or even potential induced degradation (PID).

Based on an in 2012 installed PV park with 4620 crystalline silicon PV modules oriented to south direction with a tilt of 30°, both approaches, monitoring and on-site inspection were performed.

Today's monitoring systems normally have one or more temperature and irradiation sensors according to the recommendation of standard IEC 61724-1:2021 - Photovoltaic system performance - Part 1: Monitoring. With such concepts the determination of a live performance ratio is possible. But there are older PV plants existing that either have a defective monitoring system or only a feed-in counter. To make anyway trend analyses, monthly or yearly values are sufficient in combination with meteorological data. In addition, it is important to have some knowledge to the setup and what happened in the past with the PV plant. Such important information could be received by people, operating the PV plant.

2 METHODS

To check the potential of monitoring with data evaluation and inspection on site, the documentation of the PV plant has been approved, the operator was interviewed, and a short tour of the park was made (Fig. 1).

Figure 1: Partial overview of the PV park with 970 kWp

2.1 Evaluation of monitoring data

For the PV plant, there are no irradiation or temperature sensors for measurements installed. An existing inverter data collection system is installed but not proper working. Only monthly feed-in counter values of the last 9 years and since plant start are available to make analyses. To have an overview and status, only the yearly feed-in counter values were considered. Two methods to make an absolute and a relative comparison were used. This is the well-known performance ratio (PR [%]) and the double normalized yield per year (DNY [m²/MWp]).

For the PR calculation, first, the yearly yield values in kWh for the 970 kWp plant were normalized to kWh per kWp installed capacity. Since no irradiation sensor was installed, a nearby located reference weather station was used to have the sum of annual global horizontal irradiation. Afte that the horizontal irradiation value was transformed to an equivalent irradiation for the orientation and tilt of the modules. Module efficiency and module area per kWp installed power were read out of the data sheet.

The performance ratio is calculated with:

$$PR = \frac{E_{of\ the\ feed-in\ counter\ per\ year\ in\ kWh\ per\ kWp}}{Irr_{Module\ level} \times \eta_{Module} \times A_{Module\ per\ kWp}}$$

The unit check is also done:

$$PR\,[\%] = \frac{\frac{kWh}{kWp}}{\frac{kWh}{m^2} \cdot \frac{m^2}{kWp}} \cdot 100$$

For the DNY value, the annual global irradiation data were taken directly as reference and not transformed in module orientation and tilt. An additional normalization step was performed to normalize the yearly global horizontal irradiation amount to 1000 kWh/m². On this way, a comparable relative energy yield over the years is possible independent of "good" and "bad" years. The unit in the end is produced kWh per kWp installed PV power per 1000 kWh/m² global horizontal irradiation [kWh /kWp /1000 kWh/m²] and can be shortened to [m²/MWp].

Double normalized yield per year is calculated with:

$$DNY = \frac{E_{of\ the\ feed-in\ counter\ per\ year\ in\ kWh\ per\ kWp}}{G_{sum\ norm.glob.horiz.irradiation\ to\ 1000\ kWh/m^2}}$$

The unit check is also done:

$$DNY = \frac{\frac{kWh}{kWp}}{\frac{1000 \cdot kWh}{m^2}} = \left[\frac{m^2}{MWp}\right]$$

The advantage of DNY is that no module efficiency, area, orientation, or tilt is needed.

2.2 Evaluation and effect of dirt and aging

To evaluate the effect of dirt, four modules were characterized with an I-V-curve measurement device to determine the maximum power before and after cleaning. We used a calibrated HT IV 400 device with an irradiation sensor in module level and a temperature sensor fixed to the module's backside (Figure 2).

Figure 2: Determining STC power with/without dirt

With that evaluation we received two information. First, in a comparison of dirty and cleaned modules, the power loss due to dirt. Second, the aging behavior since the power of the cleaned module at STC condition is not reaching again the STC value at the manufacturer label. It is worth mentioning that for a real statistic, a minimum of samples is needed. But to have a rough result in time, four samples out of different positions are enough for an estimation and status.

2.3 Evaluation and handling of failures

Evaluating PV plants concerning failures is a challenge because good observation skills, time and background knowledge is needed. In the inspected PV park, the modules were viewed from many different perspectives. This is important because dirt deposits are better to recognize at extreme angles of view. Glass broken modules appear a little bit more matt or fringed. Such modules are often good visible by using thermal imaging systems because of cell cracks and therefore hot spots. Chalking back sheets are a sign to be in danger having back sheet cracks or coming cracks in future. If a crack track is visible e.g., behind string ribbons, a repair solution could be a possible option to prevent futural insulation problems (7).

3 RESLUTS

3.1 Results of monitoring data

In Fig. 3, the performance ratio (blue) and double normalized yield (orange) per year is shown. Both analyses approaches show the same trend over the years. Due to that fact that PR is better known, the PR was chosen for a closer look. For the purpose of a precise analysis, the PR trend was split in three sections and numbered (circled numbers). Beginning with section number 2, it can be seen that from 2015 to 2019, a continuous and linear annual PR loss of around 1.8 % is seen. In the section number 1, there is the same slope from 2012 to 2014 but from 2014 to 2015 there is a step. The answer to that break is found because there were two defective solar trackers in the PV plant that were not properly working in the first 3 years. After fixing them to a fix and same position, the PR was higher again. Therefore, the whole PV potential in the first 3 years was not used completely. In the year from 2019 to 2020 in section number 3, a stop in downwards trend is seen. This is explainable because in 2020 some improvements were done in the PV park. This was green cuttings of bushes that affected 12 strings and cleaning of about 200 modules with rainwater and a brush.

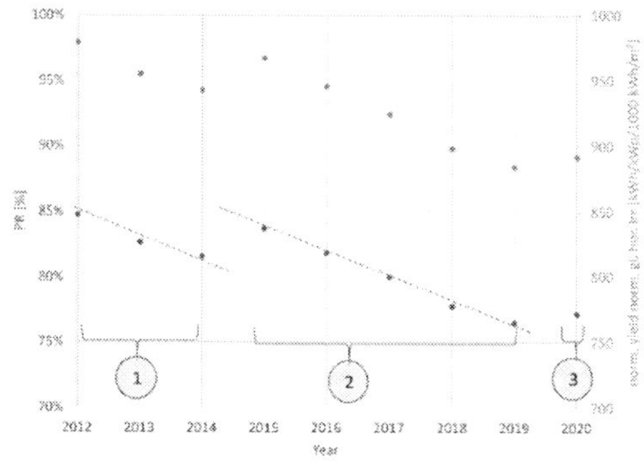

Figure 3: Trend analysis of PR (blue) and DNY (orange)

3.2 Results concerning dirt and aging

In evaluating dirt layers on PV modules, the angle of view is important to recognize the real dirt layer. Fig. 4 shows the same module in two perspectives. First from in front and second from a sharp angle. The dimension of dirt is better recognized from a sharp angle.

Figure 4: Changing angle of view detects dirt better

The exemplary I-V-curve comparison of a dirty and cleaned module in Fig. 5 shows the effect of dirt. Because of less light reaching the solar cells, current and voltage are lower. The dirt effect averaged over four measurement positions and modules was in the range of 10 %. The I-V-curve in orange colour shows the situation with dirt and the curve in blue colour without dirt after a cleaning step.

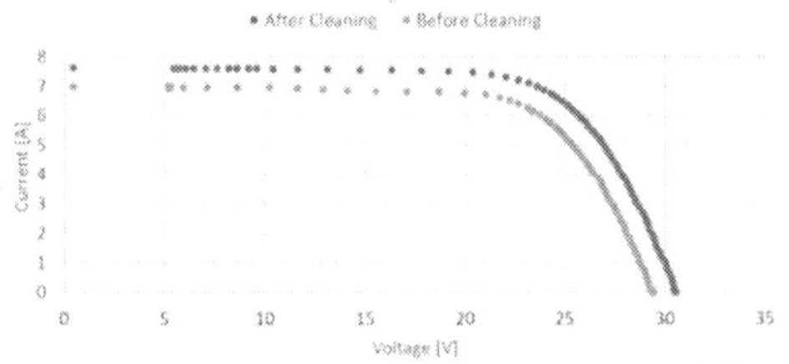

Figure 5: Measurements of dirty and cleaned module

Also, the aging behaviour of parameter STC power was checked by comparing the measurement data of cleaned modules to the labelled value. The annual degradation rate was determined in the range of 0.6 %.

3.3 Results of evaluated and handled failures

In Fig. 6, a broken module was found (right one). Compared to other not glass broken modules, the busbars are shinier, and the overall appearance is more like a mosaic view. A repair step is not possible for glass-broken modules and therefore a replacement is the only option.

Figure 6: Changing of glass-broken modules

Shaded modules because of bushes (Fig. 7) were freed by cutting them and the malfunctions in the stings were eliminated. Repair steps were not necessary because all before driven bypass didoes were in good condition.

Figure 7: Green cut for shade free operation

In a further step, some back sheets of separate PV modules were sealed with a silicone because chalking and back sheet cracks were visible. Such a repair option could avoid changing good working PV modules in case of having insulation problems.

4 DISCUSSIONS

Often analysts tend to simply average trend analyzes instead of looking at individual sections and asking "why". However, over the years some changes often occur in photovoltaic systems that must also be considered. In any case, there is one thing that is seen again and again during system checks: With each subsequent visit, one notices more and more. And every system is individual and often shows a different behavior that is not optimal for automated software evaluations.

5 CONCLUSIONS

Meanwhile a PV plant check, it was found that monitoring needs both – data analysis and on-site inspection. Two calculation approaches in comparison, performance ratio and an own idea using the DNY, lead to the same trend. By using plant insider knowledge, the interpretation of the trend was possible and explainable. Some found failure cases like glass-broken modules, dirt, shading or cracked back sheets were discussed and solved by workarounds. The annual STC power loss due to aging was around 0.6 % whereas the entire loss due to aging, dirt and shading was in the range of 1.8 %.

6 ACKNOWLEDGMENTS

This work was conducted as part of the Austrian "Energy Research Program" project PVRe² - Sustainable Photovoltaics, funded by the Austrian Climate and Energy Fund and the Austrian Research Promotion Agency (FFG), both are gratefully acknowledged.

7 REFERENCES

[1] A. Louwen, L. Koester, S. Lindig, A. Astigarraga, D. Moser, PV4.0: Combining Asset Management, PV Measurement Data and the Cost Priority Number Method in a Digital Infrastructure for PV Plant O&M, Proceedings 37th EU PVSEC, Online, 2020

[2] Grimaccia, F., Aghaei, M., Mussetta, M. et al. Planning for PV plant performance monitoring by means of unmanned aerial systems (UAS). Int J Energy Environ Eng 6, 47–54 (2015). https://doi.org/10.1007/s40095-014-0149-6

[3] International Renewable Energy Agency (IRENA) Report, Figure 3.5 Detailed breakdown of utility-scale PV total installed costs by country 2020, June 2021, ISBN: 978-92-9260-348-9, p. 74,

[4] J. M. Paredes-Parra et al., An Alternative Internet-of-Things Solution Based on LoRa for PV Power Plants: Data Monitoring and Management, Energies 2019, 12, 881; doi:10.3390/en12050881

[5] S. Samara and E. Natsheh, "Intelligent Real-Time Photovoltaic Panel Monitoring System Using Artificial Neural Networks," in IEEE Access, vol. 7, pp. 50287-50299, 2019, doi: 10.1109/ACCESS.2019.2911250.

[6] Dimitrievska, V.; Pittino, F.; Muehleisen, W.; Diewald, N.; Hilweg, M.; Montvay, A.; Hirschl, C. Statistical Methods for Degradation Estimation and Anomaly Detection in Photovoltaic Plants. Sensors 2021, 21, 3733. https://doi.org/10.3390/s21113733

[7] G. Beaucarne, E. Jadot, G.C. Eder, Y. Voronko, W. Mühleisen, Repair and Preventive Maintenance of PV Modules with Degrading Backsheets Using Flowable Silicone Sealant, Proceedings 38th EU PVSEC, Online, 2021

IMPACT OF DUST ON PV PERFORMANCE IN NIGERIA

Yusuf N. Chanchangi, Aritra Ghosh, Tapas K. Mallick, and Senthilarasu Sundaram
Environment and Sustainability Institute (ESI), University of Exeter, Penryn Campus
TR10 9FE, United Kingdom
*Corresponding author: yc486@exeter.ac.uk

ABSTRACT: This study investigates the impact of dust on polycrystalline silicon module's performance, and a 1MW solar farm developed using the same type of PV device in a city categorised as one of the most polluted cities in the region. The PV module's optimal performance was characterised using PASAN Solar Simulator prior to ambient exposure. A PV200 and survey 200R were employed for weekly performance measurement for seventeen months (date- date), and device performance was examined from sunrise to sunset at the end of the exposure. Transmission optical losses were determined to relate and validate losses by exposing low iron glass coupons considering monthly, seasonal, annual variation. The effectiveness of the employed brush cleaning mitigation approaches in the solar farm was examined against self–cleaning another approach. The findings reveal 65% Isc and 66% Impp decrement due to soiling on the modules. The outcome shows a 70.2% yield loss and a 71% efficiency drop. A wide variation of performance losses was recorded between different months and seasons in 2021, and the dry season presented the most alarming rates. The most significant power output loss recorded on the 1 MW installation was as high as 88%. The finding shows self-cleaning provided required preventive and restorative mitigation but appears to be very expensive. Although the employed brush cleaning mitigation in the studied region does not provide the required effectiveness, its less expensive property made it suitable to be applied.
Keywords: Polycrystalline Silicon, Performance, Optical Losses, Soiling, Monitoring, System Performance.

INTRODUCTION

The photovoltaic device's installed capacity is rapidly growing due to its economic and environmental merits. The developing countries allocate enormous resources for photovoltaic installation from the local budget and through international support provided by the developed world to achieve the SDG-7. However, the installations are becoming inefficient a few months after installation due to an omnipresent factor such as surface soiling that has a deleterious effect on a solar cell's performance, which is influenced by wind speed/direction, precipitation, PV surface covering material, humidity, and temperature. Dust accumulation has been reported to cause a significant performance degradation to be photovoltaic worldwide, ranging from 0 - 6% PV yield reduction in the United States (Micheli and Muller, 2017) to about 50% in Saudi Arabia (Adinoyi and Said, 2013). With the sustainable development goals deadline (2030) approaching very fast and the pledge to support developing countries, a number of PV farms have been developed in sub-Saharan countries. However, installation such as the 1 MW solar farm developed in Kano, Nigeria (11°59'02.1"N, 8°28'52.5"E) through DFID project funding faces a great challenge due to dust pollution in the region. The region has abundant solar energy potential throughout the year, with GHI (Global horizontal irradiation) of about 2148 kWh/m^2 (GSA, 2020). However, installations are not performing as required due to high dust activities in the region related to its proximity to the Saharan desert Longueville et al. (2010). In addition, it was observed that the soiling mitigation technique (cleaning with a brush) used in the solar farm is (Chanchangi et al., 2021b). Zahedi et al. (2021) provided a technical and economic review on the prominent mitigation techniques that have been previously implemented.

Areas within the global Sunbelt with abundant solar irradiance seem to suffer a high concentration of atmospheric dust (Kazem and Chaichan, 2019). The literature in the field provided less resources to determine the PV soiling rate in the region; therefore, it is imperative to investigate the impact of dust accumulation on PV performance and correlated the effectiveness of mitigation technique between the currently used and an effective method that is currently undergoing intensive study across the globe. This will provide a foundation in determining the appropriate, cost-effective mitigation technique in an optimal periodic cycle for the region.

METHOD

A p-Si module is exposed in a solar farm in Nigeria's most polluted city (IQ Air, 2021), with the highest PV investment by the Nigerian government and international bodies such as DFID (Department for International Development). PV performance was initially examined using a Pasan solar simulator with seven masks to differentiate the irradiance level. The module was then transported to a solar farm at the Bayero University Kano, Nigeria and exposed for a total duration of 17 months and 12 months without cleaning. It was once cleaned using the technique (manual brushing) used in the solar farm. On-site PV performance is measured using a PV200 and a Survey 200R every week, and at the expiration of the exposure time, a full day (sunrise to sunset) measurement was conducted. Performance data was analysed using SolarCert software, where I_{SC} outputs were compared to values generated based on the software algorithms to provide I_{SC} under STC and Nominal conditions. The PV efficiency relies on irradiation reaching the solar cells (Gupta et al., 2019; Chanchangi et al., 2021). Therefore, efficiency was determined using Eq. (1) provided by Kalogirou (2014) and Eq. (2) provided by Hachicha et al. (2019), which stated that PV system efficiency significantly depends on solar radiation getting to solar cells.

$$\eta = \frac{I_{mpp} V_{mpp}}{A G_t} \quad (1)$$

where I_{mpp} represents the current at the maximum power point, V_{mpp} is the voltage at the maximum power point, A is the active area of the PV module, and G is the incident solar irradiance. Then Eq. (2) was employed to determine the outdoor efficiency losses.

$$\Delta\eta = \frac{\eta_{Clean\,PV} - \eta_{Soiled\,PV}}{\eta_{Clean\,PV}} \times 100\% \qquad (2)$$

Fifteen pieces of 50 mm × 50 mm with thickness 0.4 mm low iron glasses were exposed to determine monthly, seasonal, and annual optical loss rates to validate the impact of dust accumulation on PV performance. Spectral transmittance measurement was conducted using Perkin Elmer Lambda 1050 UV/VIS/NIR spectrophotometer. A Maplin professional weather was installed on-site beside the module to provide accurate weather data for analysis. Fig 1 below provides a schematic presentation of on-site PV performance and weather data collection. The optical losses are calculated using Eq. (3) to determine the variation between a clean coupon and a soiled coupon based on timestamp.

$$Relative\ change\ (\Delta\tau_x) = \frac{(\tau_{clean} - \tau_x)}{\tau_{clean}}(\%) \qquad (3)$$

where Δ is the change, τ_x is transmittance data of a coupon positioned at an angle relative to a horizontal surface, τ_{clean} is the transmittance data of a clean coupon.

In addition, performance data of the solar farm was collected to examine the impact of soiling losses on the 1 MW installation. The PR (performance ratio) was calculated using Eq. (4) based on Edition 2 of IEC 61724 (cited by Chanchangi et al., 2021a). This is to highlight losses where soiling is assumed to have a more percentage.

$$PR = \left(\sum_k P_{out,k} \times \tau_k\right) / \left(\sum_k \frac{P_0 \times G_{i,k} \times \tau_k}{G_{i,ref}}\right) \qquad (4)$$

where $P_{out,}$is the measured output power (AC), τ_k is the energy recording interval time, P_0 is Array power rating (DC), $G_{i,k}$ is the measured value of solar irradiation intensity and $G_{i,ref}$ is the reference value of the solar irradiation intensity (kWh·m^{-2}).

A total of four coupons, including two acrylic plastic and two low iron glass with the dimension of 50 mm x 50 mm x 4 mm, were exposed for a month during the dry season and wet season adjacent to the experimental installation using an in-housed designed coupon holder to examine the effectiveness of two PV soiling mitigation techniques (cleaning with a brush and hydrophobic self-cleaning coating). A similar brush to the one used on-site for cleaning 1 MW was employed for cleaning the coupon, while a hybrid solution of SiO_2-TiO_2 was used as a coating to provide both anti-reflective and hydrophobicity layers. The optical transmittance variation was investigated for all the installed coupons using Eq. (3). In addition, information such as pros and cons were obtained from the literature and used for analysis. The optical efficiencies result for an uncoated coupon was calculated using the optical transmittance percentage difference between a clean coupon and the exposed one. While the optical efficiencies for the coated coupons are determined using the percentage variation between the coated but unexposed coupon and the coated and exposed coupons.

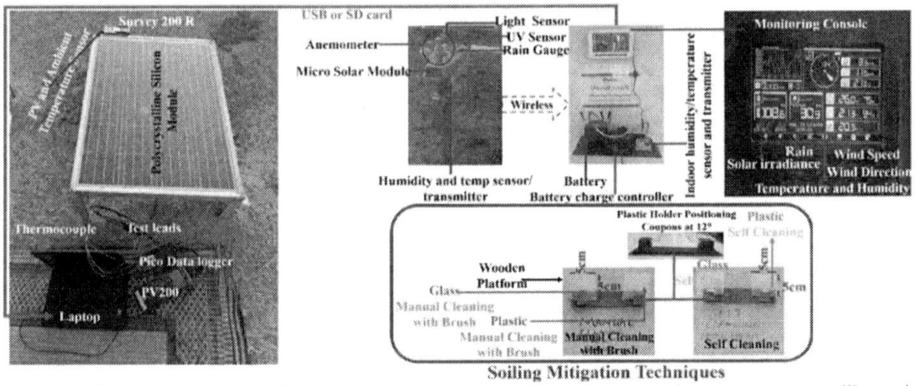

Figure 1: Schematic diagram illustrates on-site PV performance data, weather parameters, soiling mitigation technique installations, and acquisition procedure in Nigeria.

RESULTS

The 100 W pc-Si module shows efficiency reduction from the first month of exposure during the wet season. It was observed that efficiency reduced by about 21%, and after six months, the module was cleaned to improve the performance using the same technique (manual cleaning with a brush) employed for the 1 MW installation, but about 16% efficiency appears loss was not recovered. The efficiency losses elevate during the dry season and drop during the wet season, but the most significant loss was recorded during the dry season with about 71%, as illustrated in Fig. 2.

The most significant yield performance loss is presented in Fig. 3, showing a reduction of about 71.1% with a short circuit reduction of about 65% when irradiance is about 779 W/m^2. Losses are related to dust accumulation covering the entire PV surface and reducing the influx of photons reaching the solar cells to generate electricity. The optical transmittance loss findings reveal a similar result, with about 68% of losses recorded on the coupon exposed for 17 months. On the other hand, improved performances were recorded during the wet season when the rainfall provides natural cleaning.

The monthly performance output is presented in Figure 3, showing the most significant performance reduction was 70.2% in January 2021, when irradiance was 871 W/m^2. Whereas, the least performance reduction was recorded in August 2019, the beginning of the study with about 17.5% reduction when the irradiance was 1250 W/m^2. The IV curve presented shows that the PV yield recorded from August 2019 is greater than nominal due to high irradiance availability when recording the data. However, when parameters are converted to the standard test condition level, the result reveals a reduction in the Output. When considering season variation, the findings show that the worst performance was recorded during the dry season attributed to dry dust, low wind velocity, and zero precipitation. The most severe monthly optical loss was observed in January, with about 50% decline, and the least significant loss was reported in August, with a 2%.

The optical losses result for acrylic plastic coupons exposed under cleaning with brush mitigation technique demonstrates a significant loss of about 17% for the acrylic plastic during the one month of exposure in the dry season and 14% during the wet season illustrated in Fig. 5 (a). On the other hand, an optical loss of about 8% was recorded when a coated acrylic plastic was exposed under self-cleaning mitigation technique during the dry season and 3% on a similar coupon exposed during the wet season illustrated in Fig. 5 (a).

Lower optical losses were recorded considering the low iron glass coupons, with a 5% reduction on both coupons exposed under cleaning with a brush mitigation technique during the two seasons (dry and wet), as illustrated in Fig. 5 (b). On the other hand, a 5% optical reduction was recorded on a coated low iron glass coupon exposed under the self-cleaning mitigation technique during the dry season, and only about 1% reduction was recorded on a coupon exposed under the self-cleaning mitigation technique during the dry season wet season, as illustrated in Fig. 5 (b).

The 1 MW, solar farm's most significant yield performance reduction, was about 88% (117 kW/MWp) during the transition period between the wet and dry seasons. This period comes with the light rain causing cementation of windblown and wet deposited particles on PV surfaces. Despite a weekly manual cleaning with brushes, the seasonal performance variation shows about a 45% reduction during the dry season and about 35% during the wet season.

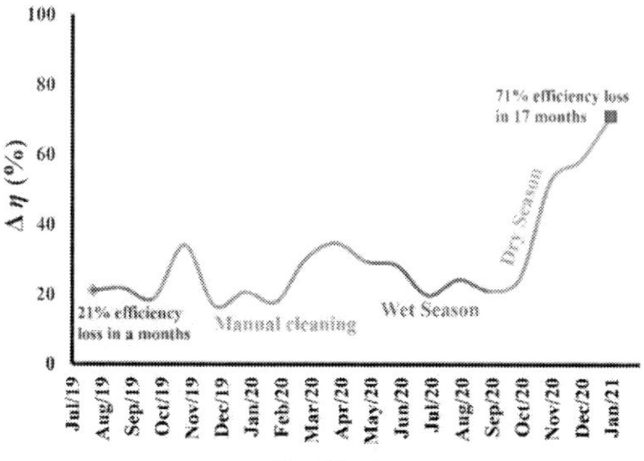

Figure 2: Monthly variation of 100 W pc-Si module efficiency losses highlighting the season where significant losses were recorded with the greatest observed during the last month of measurement (January 2021).

Performance variation table illustrating the difference between the two cleaning approaches (using a brush or the self-cleaning technique) in Table I. The table provides information such as required items, advantages, disadvantages, optical efficiency and some recommended improvement for analysis which was recorded during the study. The optical efficiencies result for an uncoated coupon was calculated using the optical transmittance percentage difference between a clean coupon and the exposed one. While the optical efficiencies for the coated coupons are determined using the percentage variation between the coated but unexposed coupon and the coated and exposed coupons.

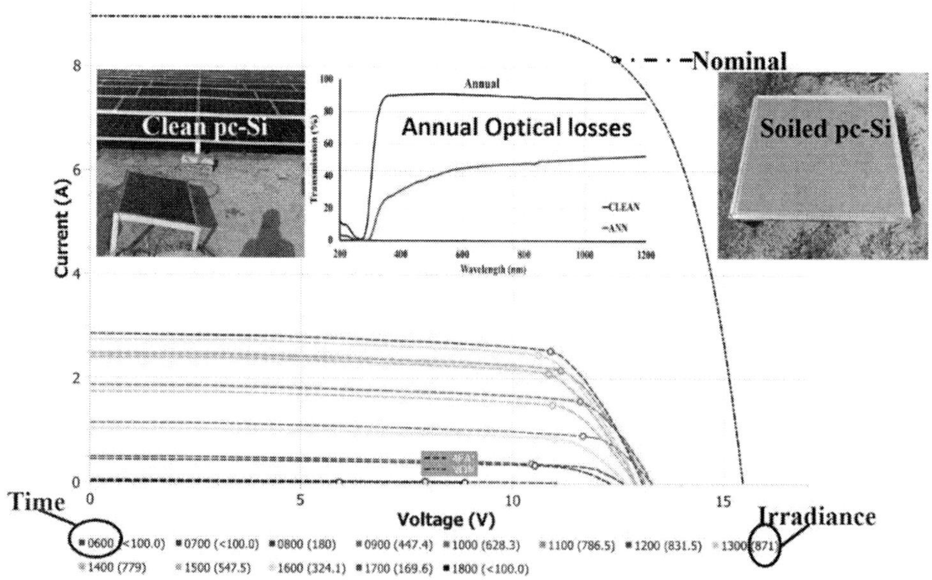

Figure 3: Hourly performance variation with corresponding optical loss plot and a digital diagram of p-Si module.

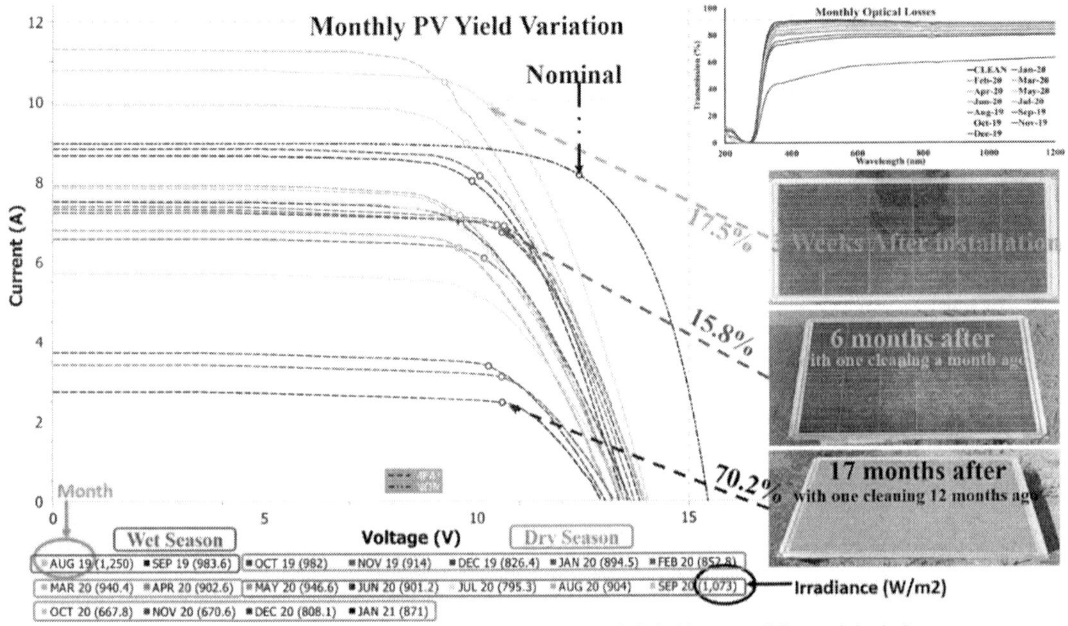

Figure 4: Monthly variation of IV curves with monthly optical losses and digital images of the module during exposure.

DISCUSSION

This section presents analyses of the experiments undertaken to characterise pc-Si performance under extreme conditions using the initial data collected from indoor measurement as the benchmark. The efficiency reductions appear to be high (21%) a month after installation and improve by about 6% in December 2019 when the module was cleaned (using manual cleaning with a brush), it continued to decrease until during the wet season when rain provides natural and improve with few percentages but did not restore efficiency to 100%. The weather parameter shows higher wind speed during the wet season (6 km/h) and heavy rain (about 131 mm/h), which is assumed to be the source of natural cleaning effect that improves the performance of the technology during the wet season. Immediately after the wet season, the efficiency reduction rapidly increases to about 71%, which appears to be highly alarming.

Similarly, after twelve months without cleaning except for the natural method (wind and rain), the yield reduced by about 70.2%, and irradiance of the day of measurement showing an I_{SC} reduction of about 65% and I_{mpp} of about 66%. If the I_{SC} and I_{mpp} reduction were the only losses related to the accumulated dust, then an average of 65% soiling losses has affected the performance of the technology. According to Kalogirou (2014), the relationship between solar radiance and I_{SC} is linear because an increase in solar radiance causes an increase in I_{SC}, and a decrease in light penetration of solar cells causes a decrease in I_{SC}. It assumed that more soiling losses could determine when the temperature excitation of some cells is be related in a synergistic approach where soiling losses and ambient temperature are combined using the temperature data of each cell. This assumption is based on an observation where dust accumulation appears not linear across the module surface (see Fig. 4).

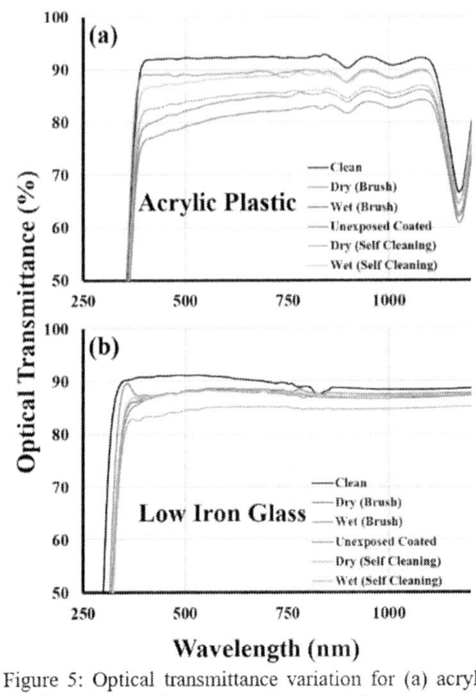

Figure 5: Optical transmittance variation for (a) acrylic plastic: clean and unexposed, coated but unexposed exposed under cleaning with a brush and under self-cleaning mitigation technique, while (b) low iron glass coupons: unexposed, coated but unexposed, exposed under cleaning with a brush and under self-cleaning mitigation technique.

Table I: Performance variation table for examined cleaning approaches.

Technique	Requirement	Advantages	Disadvantages	Optical Efficiency			Improvement
				Coupon	Season		
					Dry	Wet	
Manual cleaning	• Personnel. • Brush.	• On-demand. • Job creation.	• Unreliable. • Abrasion. • Low-performance restoration. • Expensive in developed nations.	• Plastic • Glass	83% 95%	86% 95%	• Cleaning should be conducted with water during the dry season. • Wind direction should be determined before cleaning to avoid the transfer of dust from one module to another.
Self-Cleaning (hydrophobic layer)	• Chemicals. • Determining transmittance/ wettability. • Water.	• Self-operative. • Passive approach. • Reliable in the wet season.	• Not durable on Plastic. • Water dependent. • Lifetime is very minimal.	• Plastic • Glass	92% 95%	97% 99%	• PV manufacturing companies or glass companies should accept changing covering surfaces for recycling a couple of times during the life cycle of a PV. • This technique required integrating additional techniques to support it during extreme dry, dusty weather such as Harmattan.

Optical transmittance result shows a very close finding where 68% transmittance reduction was recorded after exposing the coupon for twelve months without cleaning except the natural cleaning (by wind and rain). This finding validates the performance losses and highlights that I_{SC} and I_{mpp} losses are directly related to soiling losses. The dust density across the coupon appears to vary; therefore, multiple spectral measurements were conducted, and the point with the highest deposition was considered to provide information representing the worst-case scenario.

Performance data collected from 1 MW shows an output peak reduction of about 88% in early September 2019. The irradiance sensors reported 1028 W/m^2 for the east orientation sensor and 1094 W/m^2 for the west orientation sensor, indicating that average irradiance is more significant than 1000 W/m^2 used under standard conditions. The calculated performance ratio in decimal is 0.11387 PR, showing that the output losses is significant. Elsewhere, performance reduction was recorded at around 55% during the dry season and 65% during the wet season for particular dates. This presents alarming losses; however, it is recommended that caution be observed in generalising the solar farm's yield as some parameters such as temperature effect, shading, cable losses, and inverter losses were intentionally ignored.

In correlating weather data with the impact of soiling on PV performance, analysis illustrated performance restoration during the wet season. However, humidity promotes particles cementation towards the end of the wet season when capillary forces become active on the module, creating the foundation for greater formation all through the dry season as particles continue to build up layers. These accumulated depositions over a long time could act as a barrier to incoming photons and reduce the liberation of electrons, leading to further reduction, causing less jumping across crystals for unrestricted movement. Soiling, dust or any accumulated impurities on PV system surfaces presents a significant challenge to the performance of the PV technology (Chanchangi et al. 2020) by reducing substantial incident irradiance reaching the PV cell through scattering, attenuation, reflection and absorption (Al-Sayyab et al., 2019).

The result from examining the soiling mitigation highlights that manual cleaning is ineffective in the region of study and would be very expensive when implemented in the developed world. The lowest performance efficiency was recorded on acrylic plastic exposed under manual cleaning with a brush during the dry season, and the highest optical transmittance restoration was recorded on low iron glass exposed self-cleaning techniques during the wet season. It is assumed that the brush accumulated dust with time, and instead of only cleaning the coupon, it was cleaning and depositing dust simultaneously. In addition, it was observed that manual cleaning created abrasion on the acrylic plastic coupon, which could promote optical losses. Surprisingly, the self-cleaning method presented a better performance than manual cleaning during the dry season, which is assumed to be related to the composite coating that reduces the adhesion forces such as Van der Waal forces that are highly active between particles and particles surface. This assumption concurs with Quan et al. (2017), which stated that anti-soiling coating could reduce dust adhesion on a coated surface. However, for further studied it is recommended that SEM imaging could be used to validate this assumption. The pros and cons of the examined soiling mitigation techniques were illustrated in Table I; it also shows some necessary improvements needed to enhance each technique's performance.

The finding shows alarming soiling losses affecting performance and efficiency of pc-Si over twelve months of exposure with maintenance apart from natural cleaning. The yield of the 100 W module shows variable soiling losses with seasonal disparity; similarly, the 1 MW also follows the same pattern, revealing more significant losses during the dry season than the wet season. It was observed that manual cleaning a brush used for cleaning the 1 MW installation is less effective when compared with the self-cleaning approach; however, self-cleaning is very expensive compare to manual cleaning with a brush in the study region as labour is due to lower minimum wage. The outcome shows acrylic plastic is less durable compared to low iron glass as PV covering material.

The determining soiling losses value of a site with high solar irradiation potential and significant aerosol dust could play a decisive predictive role in identifying and optimising the maintenance schedule. Similarly, identifying an appropriate maintenance approach could prevent excessive soiling losses, restore and sustain the optimum PV performance over a period. It would prevent future installation across the region from the same challenges.

CONCLUSION

The results obtained so far show that dust accumulation has a deleterious effect on PV performance, which is significantly influenced by environmental conditions, seasonal variation, PV covering surface, and the installation pattern. The results reveal severe losses, with a 70.2% reduction in total output and 71% efficiency after 17 months of installation. The optical losses findings concur with the soiling losses, which validates the performance reductions as PV soiling losses. The 1 MW yield confirm that the soiling mitigation approach employed in the farm is ineffective, with losses highlighted across the two seasons. The finding shows self-cleaning provided required preventive and restorative mitigation but appears to be very expensive. Although the mitigation approach employed in the studied region does not provide the required effectiveness, it was less expensive.

REFERENCES

Al-Sayyab, A. K., Al Tmari, Z. Y. and Taher, M. K. (2019). Theoretical and experimental investigation of photovoltaic cell performance, with optimum tilted angle: Basra city case study. Case Studies in Thermal Engineering 14 (2019), 100421. doi:https://doi.org/10.1016/j.csite.2019.100421

Adinoyi, M. J., and Said, S. A. M. (2013). The effect of dust accumulation on the power outputs of solar photovoltaic modules. *Renewable Energy 60* (2013), 633-636. DOI: http://dx.doi.org/10.1016/j.renene.2013.06.014

Chanchangi, Y. N., Ghosh, A., Sundaram, S., & Mallick, T. K. (2021). Angular dependencies of soiling loss on photovoltaic performance in Nigeria. Solar Energy, 225, 108-121. DOI:https://doi.org/10.1016/j.solener.2021.07.001

Chanchangi, Y. N., Ghosh, A., Sundaram, S., and Mallick, T. K. (2020). An analytical indoor experimental study on the effect of soiling on PV, focusing on dust properties and PV surface material. Solar Energy, 203 (2020), 46–68. DOI:https://doi.org/10.1016/j.solener.2020.03.089

Chanchangi, Y. N., Baig, H., Ghosh, A., Sundaram, S., & Mallick, T. K. (2021a). Soiling on PV performance influenced by weather parameters in Northern Nigeria. Renewable Energy, August (2021).

Chanchangi, Y. N., Roy, A., Ghosh, A., Sundaram, S., & Mallick, T. K. (2021b). In-situ assessment of photovoltaic soiling mitigation techniques in northern Nigeria. Energy Conversion and Management, 244, 114442. DOI:https://doi.org/10.1016/j.enconman.2021.114442

GSA. (2021). Site. Retrieved from https://globalsolaratlas.info/map?c=11.894839,8.536414,11&s=11.894839,8.536414&m=site

Gupta, V., Sharma, M., Pachauri, R. K., and Babu, K. N. D. (2019). Comprehensive review on the effect of dust on the solar photovoltaic system and mitigation techniques. Solar Energy, 191(2019), 596-622. doi:https://doi.org/10.1016/j.solener.2019.08.079

IQ-Air. (2021). Nigeria. Retrieved from: https://www.iqair.com/us/Nigeria

Hachicha, A. A., Al-Sawafta, I., and Said, Z. (2019). Impact of dust on the performance of solar photovoltaic (PV) systems under United Arab Emirates weather conditions. *Renewable Energy, 141*(2019), 287-297. DOI:doi.org/10.1016/j.renene.2019.04.004

Kalogirou, S. (2014). *Solar Energy Engineering: Processes and Systems*. (Second ed.). Oxford: Academic Press, Elsevier.

Kazem, A. A., and Chaichan, M. T. (2019). The effect of dust accumulation and cleaning methods on PV panels' outcomes based on an experimental study of six locations in Northern Oman. *Solar Energy 187* (2019), 30–38. DOI: https://doi.org/10.1016/j.solener.2019.05.036

Longueville, F., Hountondji, Y., Henry, S., and Ozer, P. (2010). What do we know about the effects of desert dust on air quality and human health in West Africa compared to other regions? *Science of the Total Environment 409* (2010), 1-8. DOI: 10.1016/j.scitotenv.2010.09.025

Micheli, L., and Muller, M. (2017). An investigation of the key parameters for predicting PV soiling losses. *Prog. Photovolt: Res. Appl. 25*(2017), 291–307. DOI: 10.1002/pip. 2860

Quan, Y., and Zhang, L. (2017). Experimental investigation of the anti-dust effect of transparent hydrophobic coatings applied for solar cell covering glass. Solar Energy Materials & Solar Cells 160 (2017), 382–389. DOI:http://dx.doi.org/10.1016/j.solmat.2016.10.043

Zahedi, R., Ranjbaran, P., Gharehpetian, G. B., Mohammadi, F., & Ahmadiahangar, R. (2021). Cleaning of Floating Photovoltaic Systems: A Critical Review on Approaches from Technical and Economic Perspectives. *Energies, 14*(7), 2018. Retrieved from DOI: https://www.mdpi.com/1996-1073/14/7/2018

ACKNOWLEDGEMENT

This study is funded through a PhD research grant to Yusuf N. Chanchangi from the Petroleum Technology Development Fund, Nigeria, with additional support from the 'Joint UK Indian Renewable Energy Center (JUICE)'. The conference was funded by SuperSolar International Conference Fund. The authors gratefully acknowledge all the funding. The study was conducted in collaboration with the Bayero University, Kano (Engr. Haruna Dayyabu and Abdul Maina). The authors gratefully acknowledge the assistance of all the people involved in this research. Content is solely the authors' responsibility and may not necessarily represent funders views as they were not involved in writing the article.

CLOUDS CHARACTERIZATION AND SIMULATION TO EVALUATE THE EFFECT IN THE STABILITY OF PHOTOVOLTAIC IRRIGATION SYSTEMS

Guillén-Arenas, F.J.[1], Narvarte, L.[1], Fernández-Ramos, J.[2]
[1] Instituto de Energía Solar – Universidad Politécnica de Madrid, 28031 Madrid, España;
franciscojesus.guillen@alumnos.upm.es, Tel.: +34-95-213-4296 (F.J.G.A); navarte@ies-def.upm.es (L.N);
[2] Departamento de Electrónica, Universidad de Málaga, 29071 Málaga, España; josefer@ctima.uma.es (J.F)

ABSTRACT: One of the greatest challenges in stand-alone photovoltaic irrigation systems (PVIS) without batteries is the tuning of PID controller. Once the system is tuned, it is necessary to evaluate its performance in the most unfavorable circumstances, which occur when the passage of a cloud causes a sudden drop in available power. But there are several kinds of clouds with different effects on the system operation.
In this paper, the impact of a considerable number of clouds on the operation of a carefully tuned system has been analyzed. The clouds have been characterized based on the decrease in irradiance, the duration and the slope, and its influence on the voltage of the PV generator has been analyzed.
Subsequently, a method has been proposed to simulate the response of the system to the passage of clouds. This method can be applied on clear days, under conditions of constant irradiance, which greatly simplifies its implementation. This also greatly reduces the time required for commissioning the system.
The proposed method consists of applying a signal to the "feedforwdard" input of the PID controller that is proportional to the irradiance profile of the cloud to be simulated. Results show that the performance of the system, characterized by the voltage of the PV generator, with simulated clouds is similar to the performance with real clouds..

Keywords: Photovoltaic irrigation system, water pumping, PID tunning, cloud characterizacion, cloud simulation.

1 INTRODUCTION

The technology of stand-alone high power photovoltaic irrigation systems (PVIS) without batteries is one of the most promising to reduce the serious impact on agriculture caused by the continuous increase in prices of electricity generated by conventional energy sources [1].

A basic PVIS is made up of a photovoltaic generator, a variable frequency drive (VFD) and a centrifugal motor pump that raises a variable water flow to a water pool. The water flow depends on the irradiance in the photovoltaic generator.

The VFD performs a proportional-integral-derivative (PID) control that must keep the system operating stably on a prefixed supply voltage, that usually coincides with the maximum power point, (control variable) by varying the frequency pump operation (output variable) [2].

The key element to obtain an optimal operation of a PVIS is to perform an adequate tuning of the control algorithm, a process that presents great challenges due to the non-linear and non-time-invariant nature of the system [3]-[4]-[5]. In order to carry out the tuning process efficiently, high and stable level of PV power is necessary, which corresponds to clear days without clouds.

Once the tuning has been carried out, it is necessary to verify that the configuration of the control parameters obtained is the most suitable for the system. The most demanding test consists of checking the operation of the system in the event of sudden drops in the power supplied by the PV generator, which are normally due to the passing of clouds over it. As a result of this phenomenon there is a sudden drop in the voltage of the PV generator. If this voltage falls below the lower limit supported by the VFD, an alarm will occur that will immediately stop the system.

To carry out this test, it is necessary to wait for the appropriate meteorological conditions, which correspond to windy and sunny days "dotted" with black and compact clouds that produce sudden changes in irradiance.

Due to the different climatic conditions necessary to perform the tuning and its evaluation, these processes must be carried out in different periods of time. Considering that it is often necessary to make adjustments to the tuning parameters to improve system performance, the definitive validation of a system can take a time interval of several months or even years [6].

To solve this problem, this work presents a method that allows the evaluation of the performance of a PVIS by simulating the passage of clouds in the same climatic conditions as the tuning is carried out: on clear days with constant irradiance. This makes it possible, in the best scenario, to complete the entire control system configuration process in just one day, so that the system can operate at full capacity from initial commissioning.

The method consists of using the "feedfordward" input (present in practically all VFDs with integrated PID control) to inject a signal that produces a disturbance in the voltage of the PV generator. This disturbance is equivalent to that caused by the passage of a cloud over the generator.

To validate this method, an experimental testing system consisting of a photovoltaic generator, a VFD and a centrifugal fan has been used. This system has been very precisely tuned using the AMIGO tuning rules [7] [5].

To establish the relationship between the magnitude of the feedfordward signal and the magnitude of the cloud that it simulates, a characterization work of the most frequent clouds has been previously carried out. To do this, the system has been running for several months and all relevant variables have been recorded.

Afterwards, two of the most representative clouds have been chosen to experimentally verify the validity of the simulation method.

It has been found that the response of the PV generator voltage to the decrease in irradiance of a real cloud is similar to that which occurs on a clear day when a signal proportional to the inverted irradiance curve is applied to the feedforward input.

2 DESCRIPTION OF THE LABORATORY EXPERIMENTATION SYSTEM

The laboratory system consists of the following components:

- A PV array
- A variable frequency drive (VFD)
- An emulation system of a hydraulic pump formed by:
  - An induction motor
  - A centrifugal fan connected to motor.
- A monitoring and control system based on:
  - Arduino UNO
  - MQTT communication protocol
  - Irradiance, temperature, voltage and current sensor. Frequency is used too, but is provided by VFD
- Oscilloscope and signal generator.

The PV array consists of 4 independent strings that can be connected in parallel. Each string consists of 17 PV modules connected in series. Each PV module consists of 36 solar cells connected in series. The system also includes a calibrated photovoltaic cell, which provides the measurements of irradiance, ambient temperature and PV array temperature. The installation is fixed and oriented to the south with a tilt angle of 45°. The VFD used is the OMRON 3G3RX-A2004 model.

To measure the voltage of the PV array, a voltage sensor with 0–1000 VDC input connected to the PV array and 0–10 VDC output connected to an analog input of the VFD is used. This is the control signal used by the PID control algorithm implemented in the VFD. It is also connected to the monitoring system.

To emulate a hydraulic pump, a centrifugal fan has been used. It is inserted in an air closed circuit, simulating water flow. The fan is driven by a 275 W standard induction motor.

The monitoring system based on Arduino records data every 100 ms. It uses MQTT protocol for communication and data is saved as a csv file in the server. This allows the subsequent analysis of data and a real-time visualization.

A Digilent Analog Discovery 2 system has been used as oscilloscope and signal generator. It allows to design any kind of signal (maximum amplitude in signal generator is 5 V due to USB power supply), which is necessary to our application.

2.1 System operation

The main system consists of a PID controller which is used to maintain the voltage of PV system in the maximum power point. The PID control algorithm is performed by the VFD and it is tuned according to AMIGO rules. System works fine with that parameters and all the tests have been carried out under the same conditions.

3 METHODOLOGY OF THE TESTS

First step is to store as many data as possible in days with passing clouds. Extremely cloudy days are not interesting. It is necessary that irradiance is varying to produce disturbances in the system. For that purpose, the monitoring system allows to turn on the VFD remotely. It also can be programmed to start automatically when enough power is provided by PV system.

3.1 Analysing data and identifying clouds

As it was said before, data is stored in server every 100 ms and it also can be downloaded in real-time from server. This sample time is enough to get the irradiance and voltage variations when a cloud is passing over the PV generator causing disturbances.

Python software is used to analyse all collected data. Several programs have been developed for that purpose. Those programs are based on data analysis libraries as "pandas", "matplotlib" or "numpy". Some other libraries are necessary to filter data and build a graphic user interface.

To analyse a cloudy day, the program calculates the derivative of the irradiance curve, obtaining the same number of samples with the slope in any point of the irradiance curve. Then, a simple algorithm to identify clouds is used. It works as detailed below:

- When the value of the calculated derivative is less than a set threshold (to avoid possible noise), cloud recognition starts.
- Each of the following samples is analysed until its value is close to 0. That means irradiance is stable and cloud has finished.
- Some cloud filters can be applied, such as irradiance threshold, that allows to find clouds with an irradiance decrease greater than a set value.

Figure 1 shows some clouds recognition. It represents the irradiance and its derivative. Clouds detected are marked in different colors automatically. This is all done by the developed algorithm.

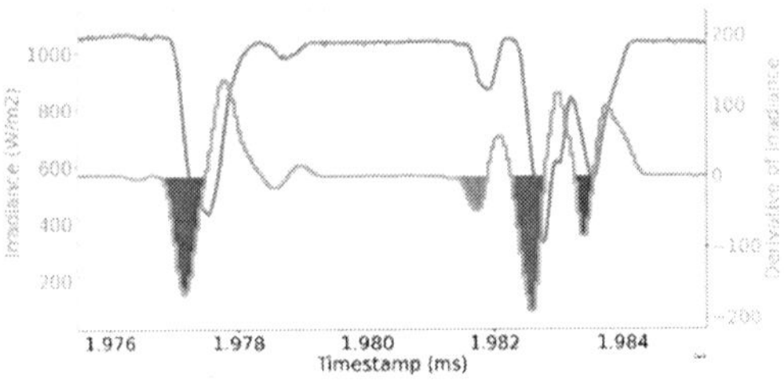

Figure 1: Cloud detection algorithm

The software also allows to export all clouds with detailed information in xlsx format.

3.2 Cloud characterization

Between 100 and 200 clouds have been collected over 6 months. To choose suitable clouds to simulate, a previous characterization is performed.

Figure 2 shows that 91% of clouds generates a 50 % or less irradiance drop. Almost all clouds (99.26 %) generate 70 % or less irradiance drop.

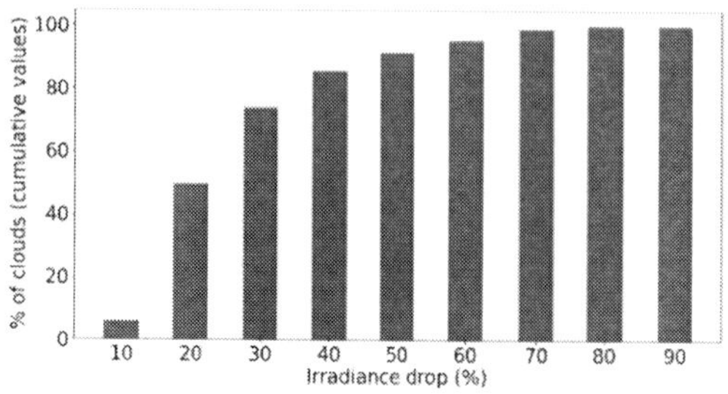

Figure 2: Cumulative cloud percentage according to irradiance drop

Considering that most clouds produce 70 % or less irradiance drop, an analysis of that clouds is done. With same irradiance drop, a cloud is more aggressive (produces a greater disturbance in the system) if duration is smaller.

Most clouds have a duration (irradiance drop time) between 4 and 6 seconds. It can be seen in Figure 3.

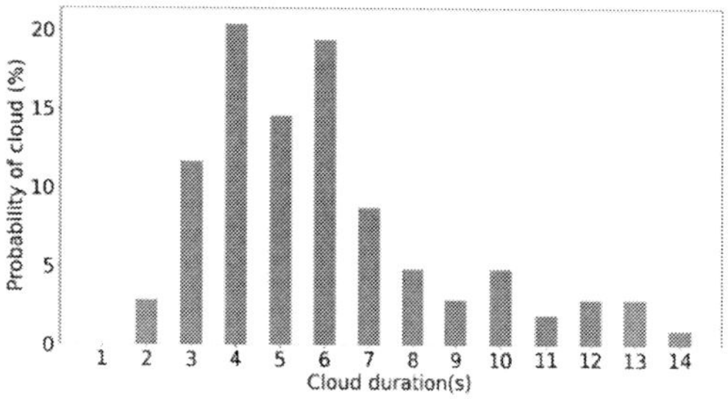

Figure 3: Probability of cloud according to cloud duration.

3.3 Clouds and system voltage disturbances

It is important to have in mind that a **cloud can be defined as a sudden decrease in irradiance**, which results in a decrease in available power. This produces a disturbance in DC voltage because PID response is not fast enough. The more aggressive a cloud is, the more voltage drop produces. The aggressiveness o the cloud depends on how fast the irradiance comes down. It can be analysed through the derivative of irradiance curve. The DC voltage decrease in a greater proportion when the slope of the derivative of the irradiance is bigger.

Two clouds have been chosen. Both have similar duration and irradiance drop (50.59 and 59.41 % respectively). The main differences are the shape produced and the irradiance starting point (that affects available power). Figure 4 and Figure 5 shows the voltage disturbance produced by the irradiance decrease of each cloud.

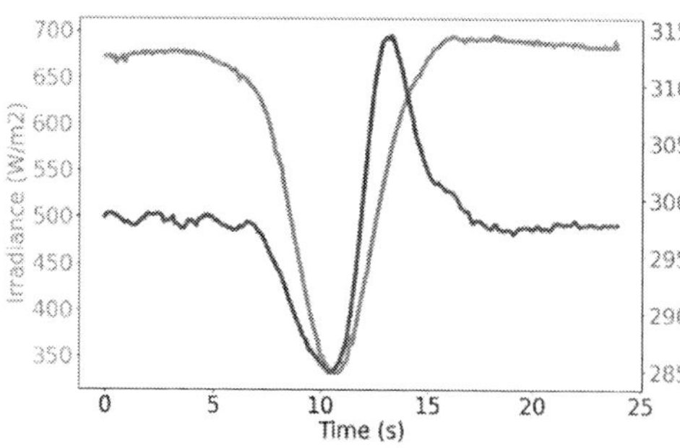

Figure 4: Cloud 1 (50.59 % irradiance drop)

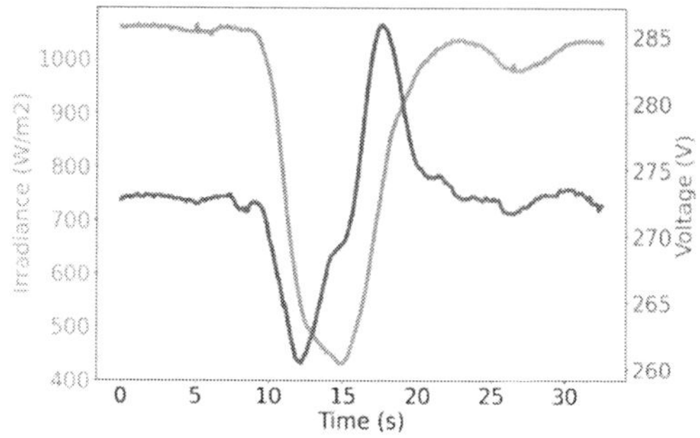

Figure 5: Cloud 2 (59.41 % irradiance drop)

The voltage variation due to irradiance drop is what characterizes how a cloud affects the system. The objective is to achieve the most similar disturbance in DC voltage with a simulated cloud.

3.4 Feedforward input

Once clouds have been characterized by its irradiance and DC voltage curves, they will be simulated in constant irradiance conditions.

The PID controller acts to keep the PV generator DC voltage close to the maximum power point by continuously adjusting the output frequency at which the motor operates. When a signal is applied to the input "feedfordward", a frequency value equivalent to the magnitude of this signal is added to this output frequency. This causes an increase in the power demanded by the motor to the PV generator.

This will cause a sudden decrease in DC voltage. The PID controller will decrease output frequency to maintain DC voltage, what is exactly what it would do when a real cloud is passing and available power drops suddenly.

A diagram of how this input works is shown in Figure 6.

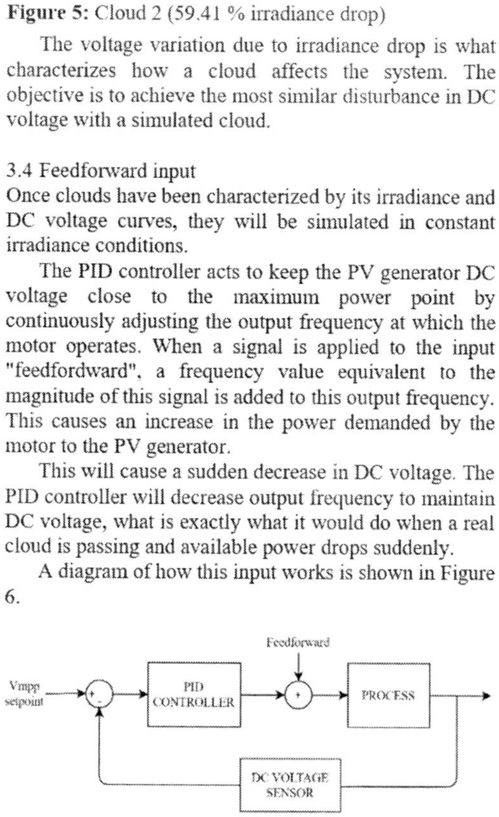

Figure 6: PID diagram with feedforward input

3.5 Creating a virtual cloud to simulate

Perfect voltage response simulation would be achieved with the change in frequency signal that is produced by cloud. That is because feedforward is an input for frequency.

Irradiance is the main characteristic of a cloud. Because of that, irradiance curve will be used to simulate the cloud. As the Figure 7 and Figure 8 show, frequency response and irradiance describe similar curves. An important point is that the system has a small PV generator. This allows to use a single irradiance sensor. In large PV systems several irradiance sensors would be necessary, obtaining irradiance as the average of all of them.

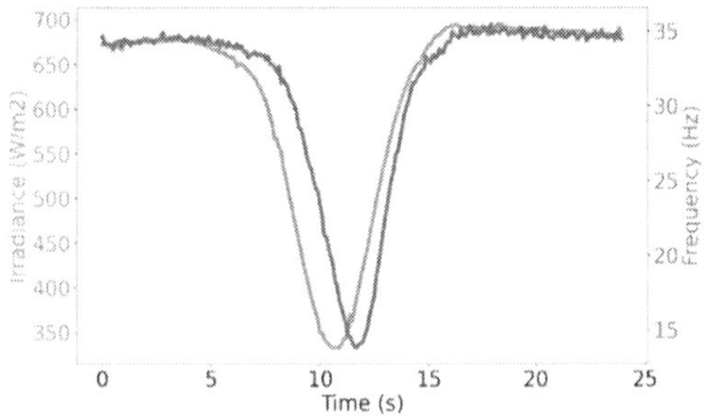

Figure 7: Irradiance and frequency in cloud 1

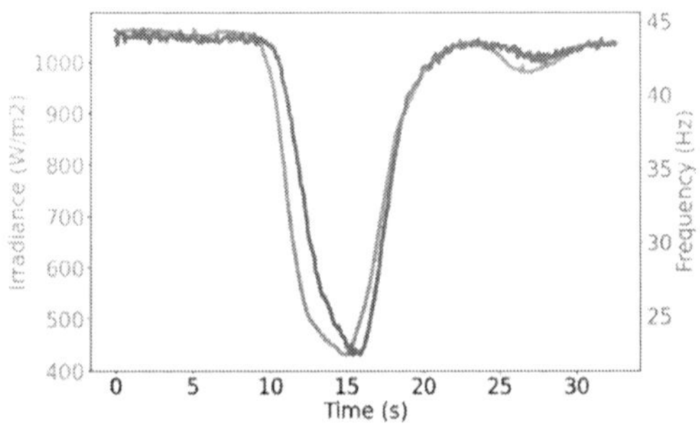

Figure 8: Irradiance and frequency in cloud 2

Considering this similarity between irradiance and frequency, a cloud to be simulated is built as follows:

- Obtain maximum value in irradiance curve of that cloud.
- Subtract each sample of the cloud to the maximum obtained previously.
- Divide by the maximum of the curve obtained in the previous step to normalize and get 1 as maximum.

The shape of this curve is the same that the inverted irradiance. The minimum point in the irradiance is now a maximum in this curve.

It is important to consider the total duration of the cloud knowing that sample time is 100 ms and we know the total number of samples. It is the only way to get the same cloud. The fact of changing this time, changes the slope in the curve, causing a different response in the system.

Cloud 2 signal to simulate is shown in Figure 9.

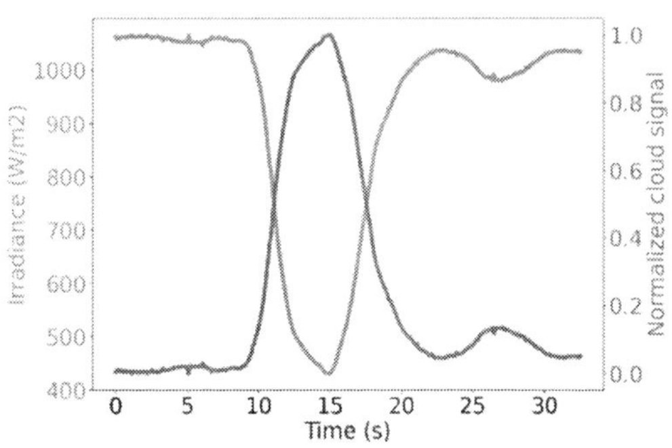

Figure 9: Signal to simulate cloud 2

Once cloud shape is obtained, is very important to consider the amplitude of the signal that will be introduced through feedforward input. Obtained curve has amplitude 1. To change it, signal must be multiplied by a constant. As results show, voltage decrease, is directly proportional to the amplitude of this signal.

4. RESULTS

Several tests have been performed to determine the relation between DC voltage and simulated clouds.

4.1 Relation between dc voltage drop and cloud signal amplitude

One of the main objectives is to achieve same DC voltage drop in the system with a simulated cloud as with a real cloud. Different amplitudes with the two clouds have been tested, obtaining the results shown in Table I.

Table I: DC voltage drop with different cloud signal amplitudes

	DC voltage drop	
Amplitude	Cloud 1	Cloud 2
1	7.19	8.73
2	14.14	17.63
3	20.84	27.63
4	27.05	34.82
5	34.56	44.1

As Figure 10 and Figure 11 show, trendline for two clouds shows a linearity between cloud signal amplitudes and DC voltage drop in the system.

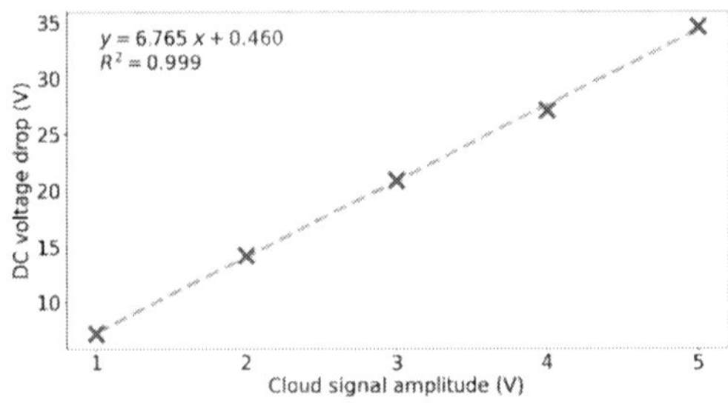

Figure 10: Relation between dc voltage drop and cloud 1 signal amplitude

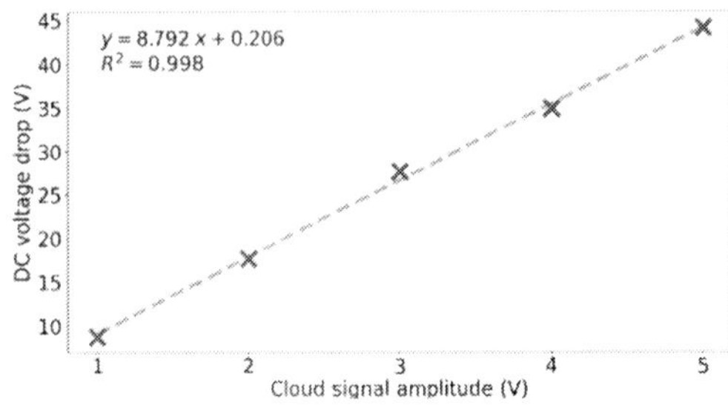

Figure 11: Relation between dc voltage drop and cloud 2 signal amplitude

4.2 DC voltage shape response similarity between real and simulated clouds

It has been possible to demonstrate that DC voltage disturbance produced by simulated clouds has always the same shape. It only depends on the shape of the cloud, not the amplitude of the simulated cloud signal. It can be seen in Figure 12 and Figure 13 for cloud 1 and 2 respectively.

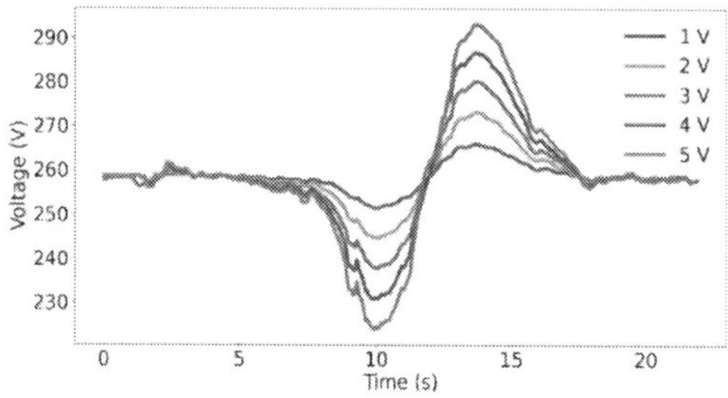

Figure 12: DC voltage disturbances produced by simulated cloud 1 (different amplitudes)

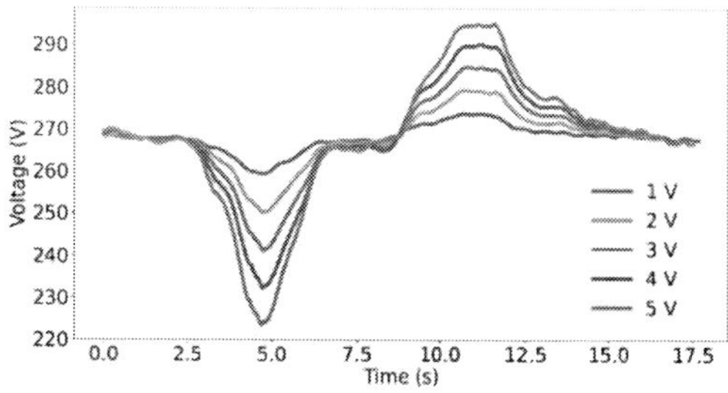

Figure 13: DC voltage disturbances produced by simulated cloud 2 (different amplitudes)

Finally, Figure 14 and Figure 15 show the DC voltage response for clouds 1 and 2, with real and simulated clouds, to compare its shape.

It can be seen that the shape of both curves is similar. The voltage scale is not the same because the samples of real and simulated clouds have been recorded in different seasons, with different temperature values in the PV cells.

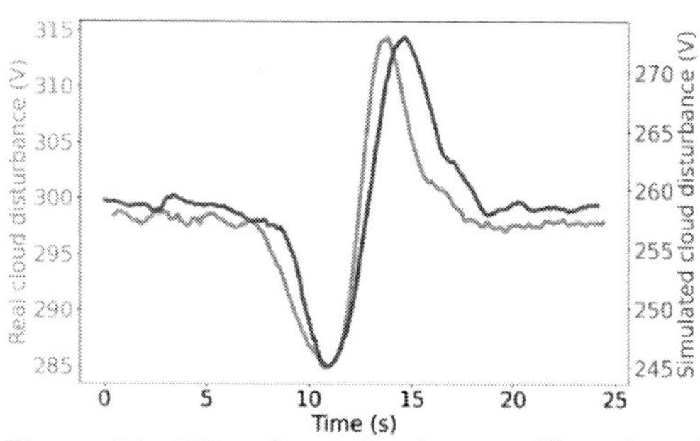

Figure 14: DC voltage disturbance with real and simulated cloud 1

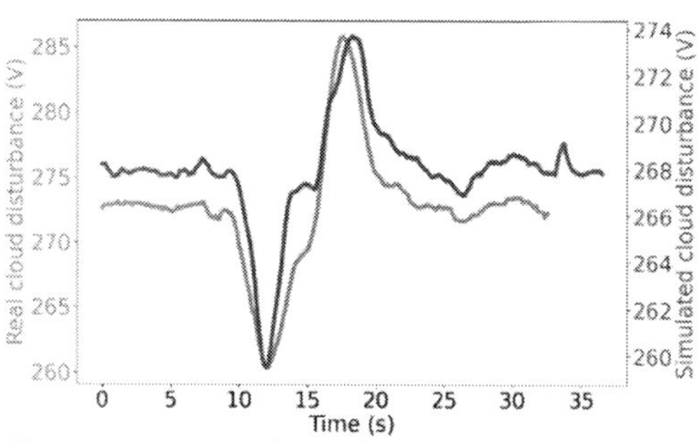

Figure 15: DC voltage disturbance with real and simulated cloud 2

5. CONCLUSIONS

Robustness is crucial in large-power stand-alones PVIS without batteries. This robustness depends fundamentally on the correct tuning of the PID control performed by the VFD.

The response of these systems is non-linear, so it is not possible to find a set of control parameters that ensure optimal operation under all conditions. Because of this, the tuning process must be completed with a performance evaluation process. The most important evaluation test consists of checking the performance of the system in sudden drops of irradiance caused by passing clouds. This can only be done in very specific weather conditions, a circumstance that can cause significant delays in the system commissioning.

This paper shows that this type of test can be performed under more favourable conditions of constant irradiance by using the "feedfordward" control input. It has been described how the shape of this signal should be designed to simulate a certain irradiance curve profile.

The algorithm designed to detect clouds and the subsequent characterization and analysis accomplished, have shown the relationship between irradiance drop and DC voltage drop. This allows to generate a disturbance in the system that causes the same effect as a real passing cloud.

It is expected that the application of this method in real PVIS will provide an important improvement in the commissioning procedure of this type of systems.

6. REFERENCES

[1] Langarita, R., Sánchez-Chóliz, J., Sarasa, C., Duarte, R., Jiménez, S., 2016. Electricity costs in irrigated agriculture: a case study for an irrigation scheme in Spain. Renew. Sustain. Energy Rev. 68, 1008–1019

[2] Alonso Abella, M., Lorenzo, E., Chenlo, F., 2003. PV water pumping systems based on standard frequency converters. Prog. Photovoltaics Res. Appl. 11, 179–191.

[3] Brito, A.U., Zilles, R., 2006. Systematized procedure for parameter characterization of a variable-speed drive used in photovoltaic pumping applications. Prog. Photovolt: Res. Appl. 14 (3), 249–260

[4] Fernández-Ramos, J., Narvarte-Fernández, L., Poza-Saura, F., 2010. Improvement of photovoltaic pumping systems based on standard frequency converters by means of programmable logic controllers. Sol. Energy 84, 101–109.

[5] Fernández-Ramos, J., Narvarte, L., López-Soria, R., Almeida, R.H., Carrêlo, I.B., 2019. An assessment of the proportional-integral control tuning rules applied to Photovoltaic Irrigation Systems based on Standard Frequency Converters. Sol. Energy 191, 468–480

[6] Herraiz J.I., Fernández-Ramos J, Almeida R.H., Báguena E.M., Castillo-Cagigal M., Narvarte L. On the tuning and performance of stand-alone large-power photovoltaic irrigation systems (under revision).

[7] Åström, K.J., Hägglund, T., 2006. Advanced PID Control. ISA – The Instrumentation, Systems, and Automation Society. Research Triangle Park, NC, USA.

Acknowledgements

This work is part of the Project MADRID-PV2 (P2018/EMT-4308) funded by the Comunidad de Madrid with the support from FEDER Funds.

DEEP LEARNING BASED IMAGE FEATURE EXTRACTION FOR PREDICTING CLIMATE RELATED DEGRADATION OF PV MODULES

L. Neumaier[1], J. Scherer[1], Ch. Schwarzlmüller[1], B. Kubicek[2], F. Mödritscher[3], Ch. Hirschl[1]

[1] Silicon Austria Labs GmbH, Europastr. 12, 9524 Villach, Austria
Corresponding Author: E-Mail: lukas.neumaier@silicon-austria.com, T: +43 664 8820014
E-Mail: juergen.scherer@silicon-austria.com, christopher.schwarzlmueller@silicon-austria.com, christina.hirschl@silicon-austria.com
[2] Austrian Institute of Technology GmbH, Giefinggasse 4, 1210 Vienna, Austria
E-Mail: bernhard.kubicek@ait.ac.at
[3] University of Applied Sciences Technikum Wien, Hoechstaedtplatz 6, 1200 Vienna
E-Mail: felix.moedritscher@technikum-wien.at

ABSTRACT: To avoid long time failures of PV power plants, reliability testing and modeling of novel photovoltaic (PV) material combinations has gained importance within the last years. In this work, conducted within the research project ADVANCE!, we present an image based deep learning approach to extract degradation-relevant information from aged PV modules. The basis of the work is a comprehensive database, established within the project INFINITY, which includes extensive time-resolved measurement and characterization data of aged PV modules, that have been subjected to precisely defined accelerated aging scenarios. Within this work, electroluminescence images serve as input for the development and application of a deep learning approach to extract degradation parameters. For defect classification and localization, we use a Mask R-CNN architecture, which allows instance-level segmentation and classification in parallel. The extracted information can be used for subsequent statistical analysis, where additional information is available, to make predictions about the operational lifetime of a module under different climatic conditions. Although the amount of data is comparatively small and the distribution of the defect classes within the dataset is not uniform, the available results show the potential of our approach.

Keywords: machine learning, deep learning, image processing, accelerated aging, lifetime prediction, electroluminescence imaging

1 INTRODUCTION

Within the last years, reliability testing of novel photovoltaic (PV) material combinations has gained importance to avoid long-time failures of PV power plants. Due to a large amount of time and equipment required for the reliability testing of photovoltaic modules, the PV community always endeavors to find possibilities for lifetime estimation based on accelerated aging (extrapolation) or the modeling of degradation mechanisms. The recently started research project ADVANCE! addresses the potential of innovative and complex statistical methods and machine learning algorithms for analysis and improved modeling of time and stress-dependent performance (reliability) of PV modules. State-of-the-art data science techniques are applied to understand specific degradation behaviors of materials used in PV modules resulting from various (climatic) stress conditions. In this work, we present an image-based deep learning approach to extract degradation-relevant information from aged PV modules. The future objective of ADVANCE! is to conjoin these image-based findings with statistical as well as chemical/physical/electrical modeling approaches and establish a comprehensive method to predict the service life of PV modules (multi-material composites) under a wide variety of operating conditions.

Several studies have already been conducted for investigating individual stress factors such as UV radiation, increased temperature/humidity stress on certain materials, or certain chemical processes, e.g. [1, 2, 3]. In contrast to that, within the project ADVANCE! we are using highly interdisciplinary research and data science approaches for opening up new paths in the analysis of the long-term and degradation behavior analysis of PV modules. The overall goal is to develop an aging model for future highly efficient PV material developments and predictive maintenance requirements for PV systems.

This paper in particular focuses on the development and application of innovative deep learning techniques for image analysis, in particular electroluminescence (EL) images. Data pre-processing includes rectification (in some images modules appear tilted and these images need to be transformed) and annotating images for machine learning. For the semantic segmentation of the EL image of a photovoltaic module the Mask R-CNN architecture (a special Deep Learning architecture for computer vision problems), that detects and classifies objects and generates bounding boxes, as well as high-resolution segmentation masks for each instance, is used [4]. In a post-processing step, relevant parameters are then extracted from the instance masks and bounding boxes.

2 APPROACH

An existing, comprehensive database established within the project INFINITY will serve as the fundamental basis for the proposed data science (hereafter referred to as INFINITY database). This database includes extensive time-resolved measurement (see Figure 1) and characterization data from 30 6-cell modules (240 samples) that have been subjected to precisely defined accelerated aging scenarios (climate-specific) [5]. In general, this dataset is not large enough to train a Deep Learning model from scratch. The PV mini-modules utilized for the current approach consist of 6 standard monocrystalline solar cells (4 busbars, 19.5 % efficiency, and 4.8 Wp @ STC, 156.75 mm x 156.75 mm), which are interconnected in series. Image data is available (i) before and after light stabilization, (ii) during aging (3-6 intermediate stages), and (iii) after accelerated aging.

Especially in the context of this work, EL images serve as input for the development and application of deep learning approaches.

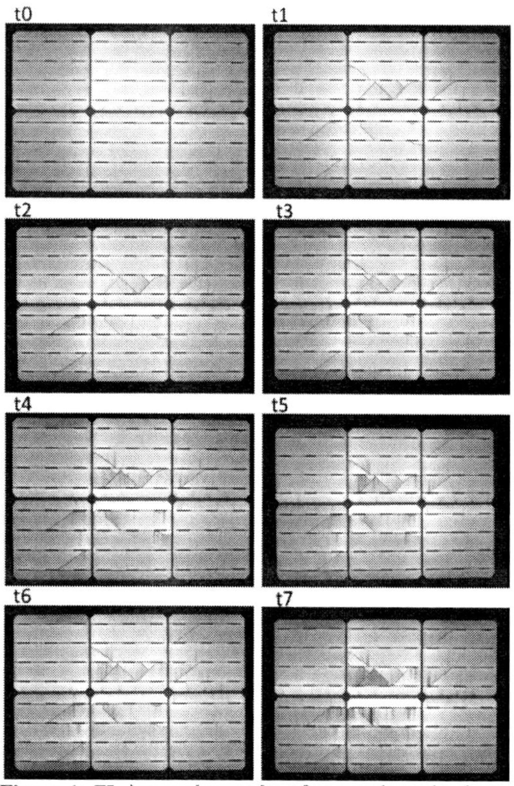

Figure 1: EL image time series of an accelerated aging procedure for a specific climatic load scenario

On the one hand, degradation effects (e.g.: lengths and positions of cracks) need to be determined, and on the other hand, a time series analysis has to be used to determine which changes a module undergoes to conclude about the respective aging procedure, i.e., the climatic zone. We use a Mask R-CNN architecture [4] for defect classification and localization (see Figure 2).

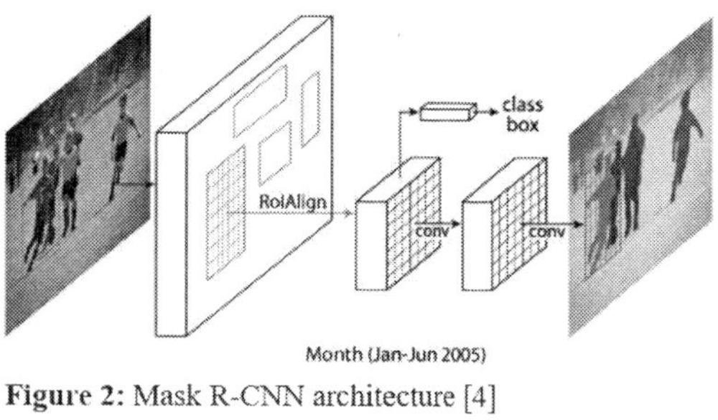

Figure 2: Mask R-CNN architecture [4]

In addition to a class prediction and an instance bounding box, the Mask R-CNN outputs a binary mask for each class. We use the ResNet-50 [6] backbone available in PyTorch [7], which is pre-trained on the COCO dataset [8] and replace the box and mask predictor for training with the appropriate number of classes.

Estimating the defect masks instead of directly estimating the metrics (e.g. crack length and finger defect area) with the model gives detailed spatial information about the defects. Additionally, it allows to calculate metrics from the masks which have not been considered at training time without retraining the model. Training the Mask R-CNN model requires ground truth masks of the defects which are not available in the dataset.

A few training samples are initially labeled (3 defect classes: finger (represented as rectangles), crack (represented as polylines), and corrosion (represented as polygons)) manually with the labeling tool VIA [9, 10] and used to train the model. To increase the number of samples and accelerate the training process, we split each module image into six images of the single cells. Each cell is then rectified and cropped to a size of 950x950 pixels. In addition, a mask underlying the same image transformations as the corresponding cell, is generated for each defect class. To further increase the number of training data we flip the existing data during the training process randomly horizontally and vertically.

For the training, 50 annotated images of modules are available, which we labeled in the course of this project. For testing our model, 13 images of modules were labeled. The training set includes 252 cells (48 cells contained no defects) with 180 cells containing defects of the class "finger", 174 cells containing defects of the class "crack" and 30 cells containing defects of the class "corrosion". Several defect classes can be contained in one cell at the same time. The test set contains 60 cells (18 cells contained no defects), with 54 cells containing defects of the class "finger", 6 cells containing defects of the class "crack" and 12 cells containing defects of the class "corrosion".

For various hyperparameters, the arithmetic means are calculated over the stratified 5-fold cross-validation runs. We weight the models by their Mean Average Precision (mAP) and Mean Average Recall (mAR) values on the segmentation task. With this approach we determine the best hyperparameters (see Table I) and train a new model with all training data, using the test data, which was not used before, to determine the final metrics.
Each model was fine-tuned for 10 epochs with different hyperparameters. We report the standard COCO challenge metrics [11] $AP_{50:5:95}$ (average precision over 10 different intersection over union (IoU) thresholds, beginning from 50% and ending with 95% with a step size of 5%), AP_{50} (average precision for an IoU threshold of 50%), AP_{75} (average precision for an IoU threshold of 75%) and AR_{100} (the average recall given 100 detections per cell) for the final model. mAR and mAP represent the mean of the corresponding AR and AP values overall defect classes.

In addition, we fully retrained (without pre-trained parameters of the ResNet-50 backbone) a model using the best hyperparameters that we identified in the stratified 5-fold cross-validation (we had to adjust the learning rate, the number of epochs, and the number of epochs after the learning rate is automatically reduced). This model was trained for 3000 epochs on the same training set and

validated on the test set. It turned out that the fully retrained model did not achieve higher performance than the pre-trained models.
Figure 3 illustrates the detection performance of one of our models on two different cells.

Table I: Best model parameters

Parameter	Value
Number of epochs	10
Size hidden layer	1024
Batch size	1
Learning Rate Finder	
gamma	0.775
Step size	1
Optimizer	
Momentum	0.9
Weight decay	0.0005
Learning rate	0.005

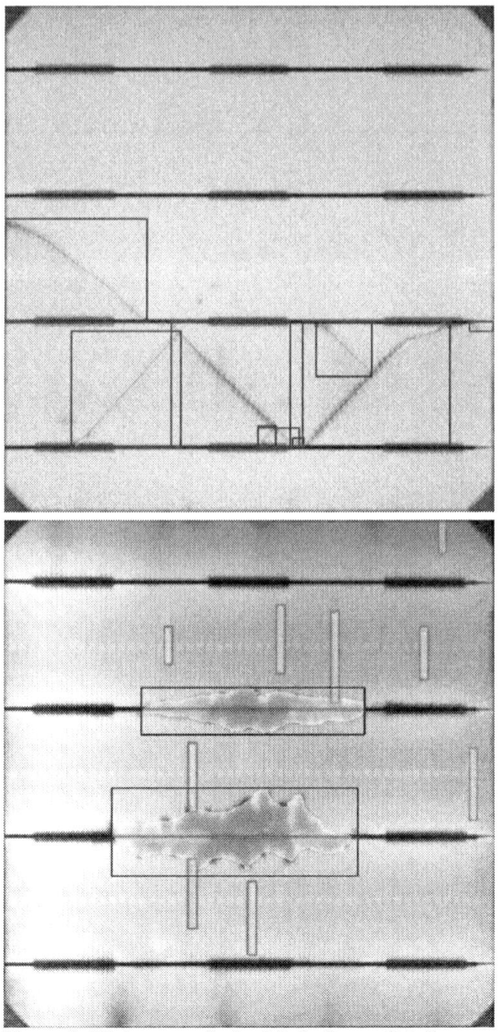

Figure 3: EL image analysis - Masks for predicted "cracks" (red), "finger defects" (green) and "corrosion" (blue) and respective bounding boxes (result obtained with the final model on two cells in the test set)

3 RESULTS

The following tables show the performance metrics of our final model. Table II contains the performance metrics averaged over all defect classes. Since the labeled defect classes are not uniformly distributed, one cannot draw a definitive conclusion from the mean average precision and the mean average recall. For this reason, it is necessary to analyze the performance for the individual defect classes separately.

Tables III, IV, and V show the performance metrics for the "finger", "crack", and "corrosion" classes, respectively. Interestingly, $AP_{50:5:95}$ is lowest for the "finger" class, although it occurs most often in the training set. Our model gives comparatively good results on the defect class "crack", although we have only a few instances in our data set. The same observation is true for the defect class "corrosion".

Table II: Performance on test data (classes "finger", "crack", and "corrosion")

Performance on test data (all defect classes)	
Segmentation	
Mean Average Precision ($mAP_{50:5:95}$)	0.450
Mean Average Precision (mAP_{50})	0.831
Mean Average Precision (mAP_{75})	0.421
Mean Average Recall (mAR_{100})	0.537
Bounding Box	
Mean Average Precision ($mAP_{50:5:95}$)	0.325
Mean Average Precision (mAP_{50})	0.833
Mean Average Precision (mAP_{75})	0.178
Mean Average Recall (mAR_{100})	0.420

Table III: Performance on test data ("finger defects" only)

Performance on test data (finger defects)	
Segmentation	
Average Precision ($AP_{50:5:95}$)	0.178
Average Precision (AP_{50})	0.697
Average Precision (AP_{75})	0.037
Average Recall (AR_{100})	0.308
Bounding Box	
Average Precision ($AP_{50:5:95}$)	0.236
Average Precision (AP_{50})	0.813
Average Precision (AP_{75})	0.059
Average Recall (AR_{100})	0.366

Table IV: Performance on test data ("crack" only)

Performance on test data (cracks)	
Segmentation	
Average Precision ($AP_{50:5:95}$)	0.779
Average Precision (AP_{50})	0.977
Average Precision (AP_{75})	0.846
Average Recall (AR_{100})	0.810
Bounding Box	
Average Precision ($AP_{50:5:95}$)	0.364
Average Precision (AP_{50})	0.844
Average Precision (AP_{75})	0.193
Average Recall (AR_{100})	0.440

Table V: Performance on test data ("corrosion" only)

Performance on test data (corrosion)	
Segmentation	
Average Precision ($AP_{50:5:95}$)	0.400
Average Precision (AP_{50})	0.835
Average Precision (AP_{75})	0.380
Average Recall (AR_{100})	0.492
Bounding Box	
Average Precision ($AP_{50:5:95}$)	0.378
Average Precision (AP_{50})	0.856
Average Precision (AP_{75})	0.280
Average Recall (AR_{100})	0.453

We assume that we can increase the performance for the defect classes "finger" and "corrosion" with additional training images. For this purpose, we will also rely on an iterative label process, where we let our best model pre-segment unknown (not annotated) cells. Afterward, the resulting segmentation will be converted into annotations, then manually checked, and corrected if necessary. The newly obtained data will be used for the next training iteration.
However, this approach will not be possible for the defect class "crack", as there are no other modules in the INFINITY dataset that have not already been used for training and validating the models.

The image dimensions of the detected and classified defects are converted into real-world dimensions since we know the dimensions of a standard module and thus of a standard cell. Thus, we can determine the length of a crack, for example. This information can be used to track the accumulated length of "cracks" in a module in an artificial aging process (see Figure 5).

The propagation of the corrosion defect class is also of interest for subsequent analyses. An example is shown in Figure 4. Figure 6 shows for a single module how the area occupied by the "corrosion" defect class changes over time.

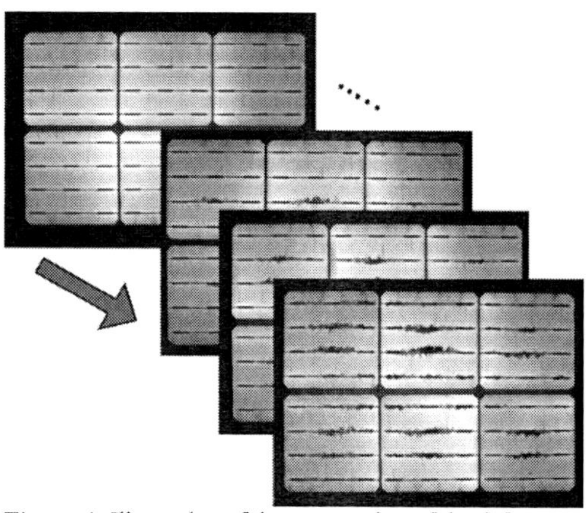

Figure 4: Illustration of the propagation of the defect "corrosion" during the artificial aging process on a single module.

This information can be used for subsequent statistical analysis, where additional information is available, to make predictions about the operational lifetime of a module under different climatic conditions.

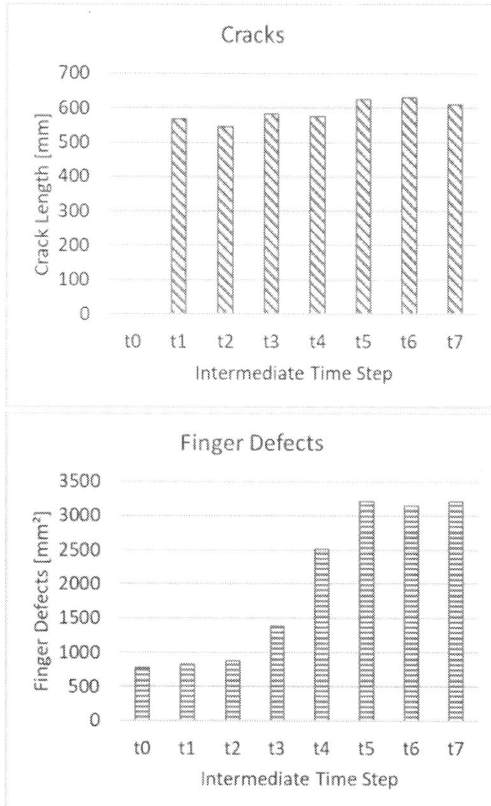

Figure 5: Accumulated area of "finger" and "crack" defects in an artificial aging process of a single module (t_0 = after 0 hours, t_1 = after 500 hours, t_2 = after 1000 hours, t_3 = after 1500 hours, t_4 = after 2000 hours, t_5 = after 2500 hours, and t_6 = after 3000 hours). The unintuitive drop-off in area measurements for the crack and finger defects is because our model reacts differently to exposure changes. During the aging process, modules were recorded with different exposures.

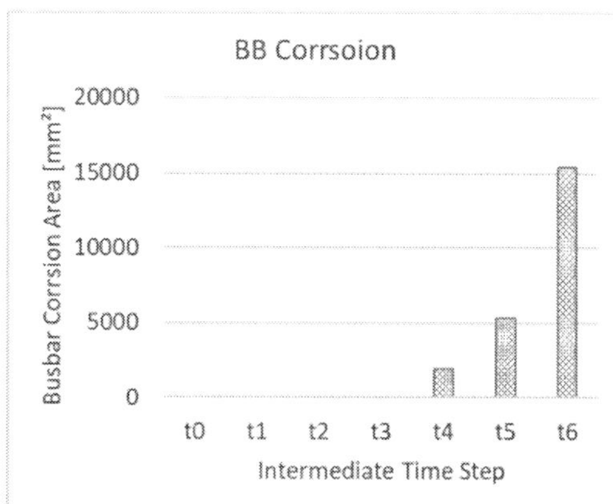

Figure 6: Accumulated area of "corrosion" defects in an artificial aging process of a single module (t_0 = after 0 hours, t_1 = after 500 hours, t_2 = after 1000 hours, t_3 = after 1500 hours, t_4 = after 2000 hours, t_5 = after 2500 hours, and t_6 = after 3000 hours).

4 CONCLUSION & OUTLOOK

Within this work, we presented a deep learning-based method for feature extraction of EL images resulting from artificially aged PV modules. In the future, the extracted data will be put into a comprehensive model to predict climate-related degradation of PV modules. This might be achieved by applying e.g., LSTMs [12] on time series data. To support and accelerate the tedious manual labeling process we are implementing an iterative semi-automatic label process. For this task, we use our best model to segment unlabeled cells. Subsequently, manual control and correction of the resulting annotations are necessary. This will allow us to increase the amount of labeled data and improve the performance of our model. As shown in the results, the performance of our model is not equally good across all classes. We assume that the performance of the defect class "corrosion" can be increased by more training data. For the defect class "crack" there is no possibility within the INFINITY dataset to increase the number of labeled cells. For this, we need to search for other datasets and incorporate them into our framework if they are compatible. We assume that the increase in training data for the defect classes "finger" and "corrosion" will improve the model performance.

5 ACKKNOWLEDGEMENT

This work was conducted as part of the Austrian "Energy Research Program" project ADVANCE! (Nr. 881133), funded by the Austrian Climate and Energy Fund and the Austrian Research Promotion Agency (FFG), both are gratefully acknowledged. We thank our colleague Federico Pittino for the fruitful discussions at the beginning of this project.

6 REFERENCES

[1] G. C. Eder, Y. Voronko, S. Dimitriadis, K. Knöbl, G. Újvári, K. A. Berger, M. Halwachs, L. Neumaier und C. Hirschl, „Climate specific accelerated ageing tests and evaluation of ageing induced electrical, physical, and chemical changes,“ *Progress in Photovoltaics: Research and Applications,* Bd. 27, pp. 934-949, 2019.

[2] C. Buerhop-Lutz, M. Hoffmann, L. Reeb, T. Pickel, J. Hauch und A. Maier, „Applying Deep Learning Algorithms to EL-images for Predicting the Module Power,“ in *Proceedings of the 36th European Photovoltaic Solar Energy Conference and Exhibition*, 2019.

[3] Y. Hu und R. H. French, „5 - Degradation and Failure Mechanisms of PV Module Interconnects,“ in *Durability and Reliability of Polymers and Other Materials in Photovoltaic Modules*, H. E. Yang, R. H. French und L. S. Bruckman, Hrsg., William Andrew Publishing, 2019, pp. 119-134.

[4] K. He, G. Gkioxari, P. Dollár und R. Girshick, „Mask R-CNN,“ in *2017 IEEE International Conference on Computer Vision (ICCV)*, 2017.

[5] G. C. Eder, Y. Voronko, K. Knöbl, S. Dimitriadis, G. Újvári, K. A. Berger und L. Neumaier, „Climate specific accelerated ageing tests,“ in *2019 IEEE 46th Photovoltaic Specialists Conference (PVSC)*, 2019.

[6] K. He, X. Zhang, S. Ren und J. Sun, „Deep Residual Learning for Image Recognition,“ in *2016 IEEE Conference on Computer Vision and Pattern Recognition (CVPR)*, 2016.

[7] A. Paszke, S. Gross, F. Massa, A. Lerer, J. Bradbury, G. Chanan, T. Killeen, Z. Lin, N. Gimelshein, L. Antiga, A. Desmaison, A. Kopf, E. Yang, Z. DeVito, M. Raison, A. Tejani, S. Chilamkurthy, B. Steiner, L. Fang, J. Bai und S. Chintala, „PyTorch: An Imperative Style, High-Performance Deep Learning Library,“ in *Advances in Neural Information Processing Systems 32*, H. Wallach, H. Larochelle, A. Beygelzimer, F. d\textquotesingle Alché-Buc, E. Fox und R. Garnett, Hrsg., Curran Associates, Inc., 2019, p. 8024–8035.

[8] T.-Y. Lin, M. Maire, S. Belongie, J. Hays, P. Perona, D. Ramanan, P. Dollár und C. L. Zitnick, „Microsoft COCO: Common Objects in Context,“ in *Computer Vision – ECCV 2014*, Cham, 2014.

[9] A. Dutta, A. Gupta und A. Zissermann, *VGG Image Annotator (VIA),* 2016. http://www.robots.ox.ac.uk/~vgg/software/via/, Version: 2.0.11, Accessed: 2021-04-13.

[10] A. Dutta und A. Zisserman, „The VIA Annotation Software for Images, Audio and Video,“ in *Proceedings of the 27th ACM International Conference on Multimedia*, New York, NY, USA, 2019.

[11] *Detection evaluation metrics used by COCO,* https://cocodataset.org/#detection-eval, Accessed: 2021-08-19.

[12] S. Hochreiter und J. Schmidhuber, „Long Short-Term Memory,“ *Neural Comput.,* Bd. 9, p. 1735–1780, 11 1997.

A self-referencing method for detecting underperforming strings in MWp-PV-generators

Claudia Buerhop, Tobias Pickel, Jens Hauch, Ian Marius Peters

HI ERN Helmholtz Institut Erlangen-Nürnberg, D-91058 Erlangen, Immerwahrstraße 2

c.buerhop-lutz@fz-juelich.de, Tel.: 0049-9131-9398-177, FAX: 0049-9131-9398-199

ABSTRACT For optimization of operation and maintenance of PV power stations the knowledge (identification and localization) of poor-performing modules, strings and inverters is important. The developed self-referencing method analyzes monitoring data, with a particular focus on energy yield. First, a well-performing unit is identified, second, the other units are set into relation to the well-preforming unit. Without weather data and prior-knowledge of the PV-system poor-performing units (strings, inverters) are identified in this fashion. Three case studies are presented. IR-images verified the failure types. Strings suffering from yield loss due to PID or bridged PV-modules where identified and the yield loss quantified. Strings in bottom rows perform less (-2.6%) than those in top rows. The noise is about 4%, which means, failures with power loss less than 4% are masked in the uncertainty level of the monitoring data, e. g. one bypassed substring, hot cells. Yield relevant module failures such as PID, bridged modules, several bypassed substrings, are identified and localized reliably using the self-referencing method.
Keywords: monitoring, PV-modules, defects, reliability

INTRODUCTION

In order to maximize or maintain the yield and lifetime of PV power stations at a high level, correct and well-defined operation and maintenance (O&M) is an important measure. Investigations with seminal information, where a problem exists, what problem exists, of what relevance it is, what further evolution can be expected, can be the key to accurate and optimum O&M planning. The fact, that poor-performing units in PV-systems may exist, is well-known from literature, e. g. IR-imaging, EL-imaging, evaluating performance ratios [1, 2].

We developed a self-referencing method for analyzing monitoring data, in order to identify faults in PV-systems, e. g. strings or inverters, address and name the location of the defective unit, classify the defect type, quantify the impact on the output (relevance of the fault).

The self-referencing method needs at least one well-performing string within the PV-system. After identifying this string, the procedure, which will be described below, of seasonal and weather related variations, especially seasonal fluctuations of the module temperature and power output of the other strings, respectively modules, will be eliminated by showing performance relative to the reference string. Afterwards, these referenced string data can be compared easily by statistical measures, e. g. over specified time intervals, module types, or array sections

Using this approach, no additional and prior-knowledge of the PV-system is necessary to practice the self-referencing approach. Furthermore, there is no need for considering weather data. For studying the uniform degradation of the system with time, the approach can be refined. Uniform degradation is not the subject of this presentation.

The suspicious, low-performing strings extracted from the monitoring data are then investigated using IR-images, which is well-known for being capable of visualizing relevant defective modules according to their temperature distribution [3-5]. IR-imaging enables also the classification of the detected failures. We give first insight into the self-referencing method, its uncertainty and the relevance of the findings.

METHODOLOGY

For this study, we used daily string yields. Because the module power in between the strings and the string lengths differs usually in PV-systems, we rated the yields to the nominal string power. The variation due to seasonality is not of primary interest, so that seasonally adjusted series can be used to highlight the variation due to the underlying changes and trends in string yields. Therefore, the reference string is the string with high overall yield and little variation over time. An algorithm calculates the regression lines of all string yields. The slope of reference string's regression line is high with the smallest possible deviation between fitted and true yield data points, in summary. Performance of all other strings is given relative to the reference string. Ideally, relative values for the string yields lower than 1 should result. Values higher than 1 can be obtained, because the reference string is the on average best-performing string but not the one with the maximum output.

EXPERIMENTAL PROCEDURE

Data of strings with a capacity of more than 40 MWp from several PV power stations located in Germany will be presented. Commissioning of the PV power stations was between 2010 and 2014. Crystalline silicon cell technologies are installed with modules powers ranging from 190 Wp up to 285 Wp. Commonly, two string inverter types were found, 15 kW inverters for three strings and 10 kW inverters for two strings. The inverters had independent mpp-tracking of the two, respectively three strings. This is of importance, since often modules of differing power classes are installed in different strings.

Monitoring data on string level was available. This study focusses on DC-data, respectively string current and string voltage. The temporal resolution ranges from 5 min up to 15 min. The data are aggregated to one data point per day, here the daily yield of a string. The calculations are carried out in Python.

For verification of the results, findings of the data analysis are compared to abnormalities found in IR-images recorded in the summer of 2020 at the sites of interest. The IR-findings are arranged in hierarchical order: module string, bridged modules, bypassed substrings, PID-affected modules, diodes, hot cells, very hot cells, interconnects, others. Multiple entries are allowed for one string.

RESULTS AND DISCUSSION

The original string yield data and the transformed relative string yield data, as described previously, are visualized in Figure 1. With the self-referencing method, the seasonal influence is eliminated. A rather small number of days are outliers, which could be removed by data cleaning and filtering. The spread in the data, respectively the increased scattering of the relative string yield, with operating time is a strong indication for poor-performing strings. Three examples/case studies will provide initial insights into possible outcomes.

For a better comparison of the performance of individual strings, average string yields are compared. Figure 2 shows case study I. Here, string yields were averaged for the period between May and August 2020. Mainly, two module types A and B are installed (visualized with different colours). Strings with modules of type A perform on a high level of 96 ± 2%. This range of roughly 4% (calculated by the standard deviation of the data) is due to various causes and includes uncertainties in voltage and current measurements, data acquisition, non-ideal mpp-tracking, and unknown true installed module power.

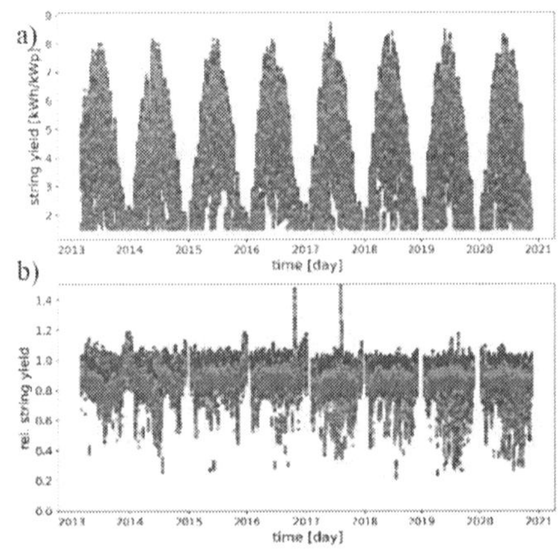

Figure 1: Monitoring data of about 450 strings of 2.5 MWp, a) original data, b) referenced relative string yield

Defects are labelled according to their appearance in the IR-images (visualized by different markers in Figure 2). If more than one defect is present in one string, the failure with the strongest impact (high in the hierarchy) is labelled for simplicity. The occurrence of the failure is symbolized by the marker size.

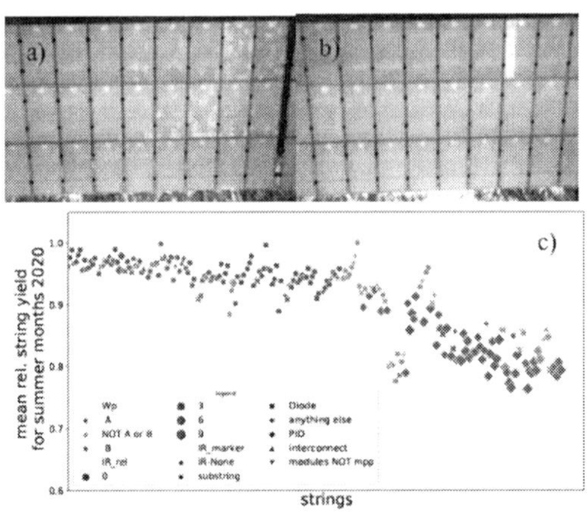

Figure 2: Averaged relative string yield for summer 2020 marked according to the findings in IR-images recorded in summer 2020, a) IR-image of PID-affected strings (upper two rows, module B), b) IR-image showing one module with a substring failure, module A, c) relative yield for 300 strings, case study I

Beside a few substring failures, PID is the dominating module failure. Especially strings with module type B are strongly affected, strings with other module types exhibit the PID-pattern rarely. With increasing number of PID-affected modules the relative yield decreases. For seven affected modules, the relative yield is about 25% below the well-performing strings. The yield loss increases with number of PID-affected modules. However, former investigations [6, 7] showed a linear correlation of the power output with number of affected PID-cells within a string.

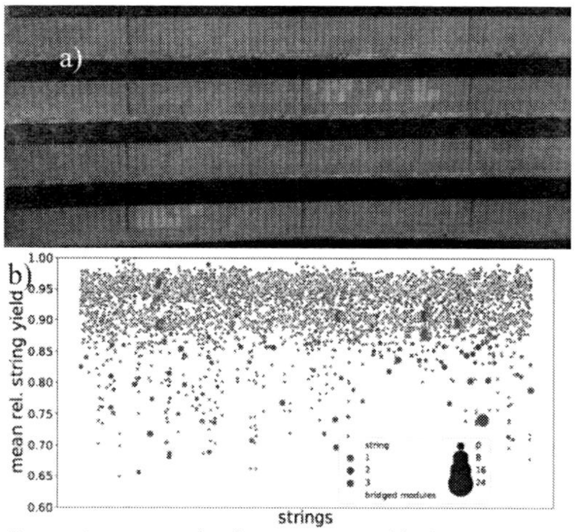

Figure 3: Averaged relative string yield a) IR-image of strings with bridged modules, , b) rel. yield for about 4500 strings, colors mark rows on a module table: 1 = lower row, 2 = center row, 3 = upper row, case study II

Figure 3 shows the distribution of the annual mean string yield of a multi-MWp PV power station, case study II. This plant suffers from hundreds of bridged modules due to insulation issues. In more than 400 strings (about 9% of the strings) at least one bridged module is detected. Yield losses amount to 1.4% for the PV-system and the year, shown in Figure 3.

Furthermore, a reduced performance of the strings in the lower row (1) is obvious. On average, row 1 produces about 2.6% less power than the top row (row 3), for the middle row 2 the difference to row 3 is only 0.5%. 56.5% of bridged modules are found in the bottom row 1, 24.5% in middle row 2, and 19% in top row 3.

In case study III, we combine the monitoring and IR-image analysis for a reliability check in more detail.

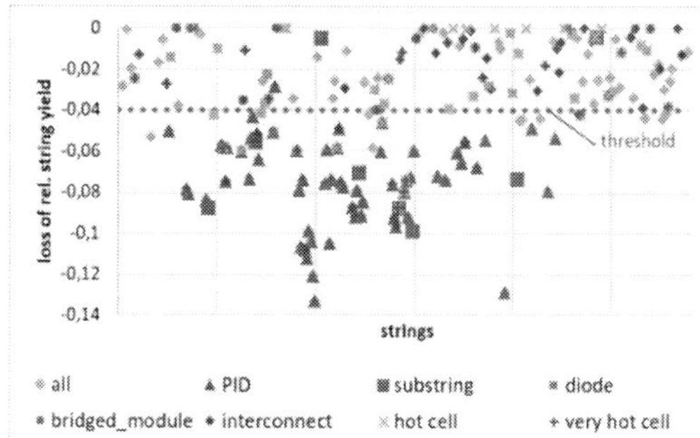

Figure 4: Comparison of failures found in IR-images with their yield loss within the inverter, 300 strings, case study III

The majority of strings perform within a range of approximately 4% yield reduction, see Figure 4. In 70% of the strings, at least one suspicious IR-feature is detected, but only 39% have a detectable impact on the yield. Failures like hot cells, interconnects, and defective diodes mainly result in yield reductions of less than 4% and are masked in the ground noise. Multiple substrings and PID cause severe and relevant yield loss larger than 4%.

Table 1: confidence matrix for IR-findings and findings using the self-referencing approach (data from Figure 4), case study III

		IR-findings	
		NO abnormalities	*abnormalities*
monitoring	$\Delta Y < -4\%$	26%	31%
	$\Delta Y > -4\%$	4%	39%

In summary, about 4% of strings are erroneously rated as poor (high yield loss) considering no abnormalities in IR-images for these strings, see also Table 1. About 44% of the distinctive features found in the IR-images have only a minor impact on yield. The corresponding module faults, which include hot cells, interconnects, diodes and single substring failures, remain undetected in the self-referencing because they are masked by background noise. This is due to the fact, that e. g. one substring failure leads theoretically to a power reduction by 1/69 = 1.4% for a string with 23 modules and a module with three substrings. Since the impact is rather small, it is almost impossible to detect such strings by this approach. For well-known, perfectly labelled strings and inverters the uncertainty range for misinterpretation is rather small with 1%, see also Figure 4.

SUMMARY

The self-referencing approach is a suitable pipeline for identifying and locating poor-performing strings/units in utility-scale MWp PV power stations without prior-knowledge, like e. g. weather data. IR-images verify that PID, bridged modules and multiple bypassed substrings are the dominating, yield-relevant failures for the PV power stations investigated. 9.5% of the strings out of 6,000 strings suffer from yield loss > 4%, which describes the method's uncertainty.

The self-referencing pipeline has the potential to be automated and applied to multi-MWp PV power stations as well as enhanced for statistical analysis and degradation studies of PV-systems.

ACKNOWLEDGEMENT

HI ERN gratefully thanks the German Federal Ministry for Economic Affairs and Energy (BMWi) for financial funding of the project COSIMA (FKZ: 0324291A)

REFERENCES

[1] T. Dierauf, et al. (2013) NREL TP-5200-57991.
[2] D. C. Jordan, et al.; Progress in Photovoltaics: Research and Applications 25 (2017) 318.
[3] J. A. Tsanakas, et al.; Renewable and Sustainable Energy Reviews 62 (2016) 695.
[4] C. Buerhop, et al., 32nd EU PVSEC (2016) 2016.
[5] C. Buerhop, et al.; Solar Energy Materials and Solar Cells 107 (2012) 154.
[6] C. Buerhop, et al., SPIE Optics and Photonics, Optics + Photonics for Sustainable Energy (2016) 99380J.
[7] C. Buerhop, et al.; IET Renewable Power Generation 11 (2017) 1253.

PERFORMANCE EVALUATION OF A HIGH-POWER PV IRRIGATION SYSTEM FOR A TWO-YEAR OPERATION PERIOD

J.I. Herraiz[1,2,*], R.H. Almeida[1], M. Castillo-Cagigal[2], L. Narvarte[1]
[1]Instituto de Energía Solar – Universidad Politécnica de Madrid, 28031 Madrid, España.
[2]Qualifying PhotoVoltaics. Centro de Empresas la Arboleda. Calle Alan Turing 1, 28031 Madrid, España.
[*]E-mail: ji.herraiz@upm.es

ABSTRACT: High-power diesel-based or grid-connected irrigation systems are giving way to high-power PV irrigation systems without batteries that reduce the energy costs by up to 80%. The design of these PV systems has not yet reached maturity and there are no experimental data available about their performance or their behavior against PV power fluctuations. This paper incorporates the definition of new indices used to assess both the robustness of the system to PV power fluctuations (the "Number of abrupt stops" and the "Passing-cloud resistance ratio") and the performance (by factoring the traditional *PR* to determine the influence of different factors external to the system) and presents their experimental values for a battery-free 213 kWp PV irrigation system, pumping to a water pool. The system has been analyzed over two years of real operation. For example, the percentage of abrupt stops was 1.3%, the passing-cloud resistance ratio was 99.8% and the performance ratio was 64.7%.
Keywords: Performance, PV irrigation system, Water pumping, Stand-alone PV system

1 INTRODUCTION

High-power diesel-based or grid-connected irrigation systems, despite being currently a majority [1], are giving way to high-power photovoltaic (PV) irrigation systems [2][3][4][5]. This process is being driven by rising energy costs in modernized agriculture [6][7][8]. Until recently, this type of system consisted of a PV generator, a frequency converter (FC) and an alternating current centrifugal motor pump [9][10][11]. PV power variability prevented ensuring the stability of the FC, and resulted in control instabilities and sudden stops of the FC, producing water hammer in the hydraulic system and overvoltages between the FC output and pump motor [12][13][14]. This problem, that led on to systems with a maximum power of 40 kWp, was solved by integrating batteries [15][16][17], but the power of the systems was not increased due to the high cost of batteries.

In the last decade, control instabilities were solved developing new control algorithms [18]. In addition, trackers with a horizontal north-south axis were incorporated to produce quasi-constant daily PV power profiles that better adjust to irrigation needs [19][20][21]. Communication protocols and irrigation controllers were developed to facilitate the control of irrigation systems. Fixed support structures with an east-west orientation (called "delta" structures) were also proposed for the PV generator to produce constant power profiles [22].

Regarding performance, unlike grid-connected PV systems, in which their performance depends basically on the quality of the system itself and for which a great experience in estimating and calculating expected performance values is available, the analysis of high-power PV irrigation systems does not have that level of expertise and there is still no information available in the literature on experimental performance data that allows us to know what the expected performance is in this type of systems. This is due to the fact that high-power PV irrigation systems present external quality factors that affect their performance: the irrigation period of the crop, the characteristics of the irrigation infrastructure to which it is connected (which does not allow to take advantage of all the available irradiation, for example, because the pump cannot work above its nominal frequency) or the behavior of the irrigator [2]. Performance indices have been proposed that factorize the traditional *performance ratio* (*PR*) into various utilization ratios that quantify the influence of these external factors on the system performance [2].

This paper provides experimental performance data and opens the door to the knowledge of the expected performance in stand-alone high-power PV irrigation systems without batteries and pumping to an elevated water pool.

This paper is structured including a methodology section that describes the PV irrigation system (PVIS) under study and the proposed new performance indices, a results section that presents the results of the performance evaluation and the discussion of these results and a conclusions section that summarizes the main conclusions of the work.

2 METHODOLOGY

2.1 Description of the PVIS

The system is located in Aldeanueva de Ebro, La Rioja, Spain (42°11'50.9''N, 1°49'58.0''W) and belongs to the Irrigator Community of "Las Planas" that brings together almost 100 farmers representing 246 ha of irrigated vineyards.

The pre-existing system consisted of two 75 kW vertical inline centrifugal pumps powered by the national grid, being able to elevate a total of 108 m^3/h of water to a 70,000 m^3 water pool located 8 km away, with a total manometric head of 225 m.

To avoid electricity consumption in high price periods, two more pumps (KSB BEV-8400/11), powered by a PV generator, where installed. The PV system is being used to pump the maximum volume of water during the day, punctually using the previous "fed by the grid" pumps at night to compensate some specific spikes in water demand. Figure 1 shows the configuration of the system.

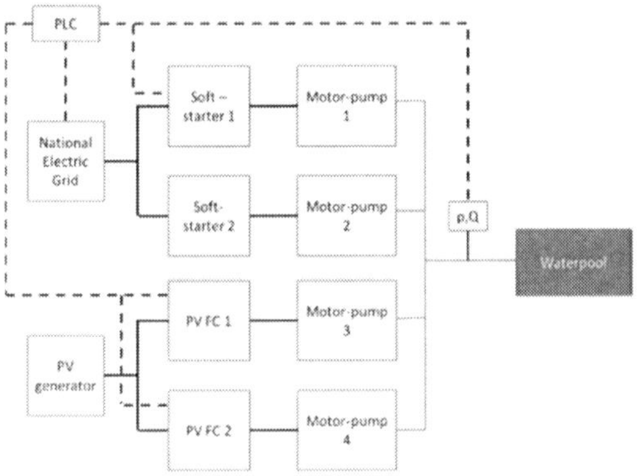

Figure 1. PVIS configuration: A PV generator, two FCs, two motor-pumps, and a PLC (all in orange) have been added to the pre-existent configuration (in green). System includes the measurement of water pressure (p) and flow (Q).

A PV generator of 213 kWp, installed in a North-South horizontal axis tracker (Nclave SP160) of ten rows with 600 PV modules (TwinPeak 2S 72 SERIES - REC Solar of 355Wp) that are connected in 40 strings of 15 modules in series, feeds both pumps through two 110 kW frequency converters (FCs) (OMRON RX 110 kW - 3G3RX-B411K-E1F). The system includes an external PLC (composed by two Siemens 6ES7131-6BH00-0BA0, a Siemens 6ES7132-6BH00-0BA0 and a Siemens 6ES7134-6GD00-0BA1) that estimates the available PV power and controls the start and stop of the two existing FCs based on four thresholds: one start threshold and one stop threshold to operate each FC. When the estimated available power is higher than the first start threshold[1] (55 kW), the PLC sends a start signal to the first FC and its pump starts operating. The FC executes the MPPT routine. A second start signal is sent to the other FC when the estimated PV power is grater than a second start threshold (95 kW). If the estimated PV power falls below any stop threshold (40 kW for the first FC and 70 kW for the second one) during a 60 second time interval, a signal is sent to the corresponding FC to stop.

The system is monitored by means of one-second records of: G, T_c, output frequency, set point frequency, V_{DC}, I_{DC}, and I_{AC} of both FCs, water flow and pressure.

The commissioning of the system was done at the end of July 2018 including a "conservative" PID tuning of the FC that ensured stable operation under "normal" operating conditions (high and constant irradiance values). This tuning was done by means of an experimental trial and error method that provided values of $K_p = 0.2$ and $T_i = 0.1$ s. At the end of September 2019 a new PID tuning of the FC was carried out to ensure an optimal operation not only at high and constant irradiance values but also at start-up (when the pump goes from standstill to the highest possible frequency in a very short time) and when sudden changes in irradiance occur. This was done by means of a systematic method that provided values of $K_p = 0.5$ and $T_i = 0.3$ s.

2.2 Indicators used to assess the PVIS in real operating conditions

In order to analyse and assess the quality of the system, new indices have been described.

2.2.1 Number of abrupt stops

An abrupt stop is a sudden and uncontrolled FC stop, and a controlled stop is that initiated by the PLC or the FC itself and that is performed gradually. Under normal operating conditions, FCs must stop in a controlled way. Nevertheless, quick PV power intermittences or malfunctions of the control system can cause high-power PVIS control instabilities that can lead on to abrupt FC stops that cause water hammer and AC overvoltages, seriously threatening the integrity of the hydraulic and electric components of the PVIS. A large number of abrupt stops will pose a greater risk of breakdowns, while controlled FC stops do not entail this type of risk.

The total number of abrupt stops and the percentage of abrupt stops with respect to the total number of stops are indicators of the good or bad functioning of the system. The absence of abrupt stops implies either no control instabilities or that the system recovers from the existing control instabilities, and a high tuning quality.

2.2.2 Passing-cloud resistance ratio

A passing-cloud is a PV power fluctuation, usually caused by a cloud passing over the PV generator and partially or totally shading it. The effects of the power fluctuation on the PVIS depend on the fluctuation itself and on the characteristics of the PVIS. There are two main characteristics of PV power fluctuations that affect their potential danger: power drop (final power minus initial power), and duration (the time it takes power to drop). These characteristics not only depend on the type of cloud (darkness, size and speed) that causes the power fluctuation, but on the system design (specially the size of the PV generator and the number and arrangement of FCs).

Not every power fluctuation is critical to the operation of the system. Deep power drops in short periods are likely to be more dangerous and cause abrupt stops of the PVIS. For that reason, it is necessary to identify which passing-clouds are potentially dangerous and specify their power drop and duration. This needs to be done considering weather conditions and PVIS design information.

A first approximation to the definition of a potentially dangerous passing-cloud considers clouds passing over the PV generator at a speed of approximately 70 km/h causing 40% to 50% PV power drops. This means that abrupt stops should be avoided for PV power ramps of duration greater than or equal to $\Delta t(\mathrm{s}) = l(\mathrm{m})/20(\mathrm{m/s})$, where l is the smallest dimension of the perimeter of the PV generator expressed in meters.

A "resisted" passing-cloud is a passing-cloud that does not lead on to a FC abrupt stop. To assess if the system resists passing-clouds a passing-cloud resistance ratio (σ_{cloud}) is defined as

$$\sigma_{cloud} = \frac{number\ of\ resisted\ passing\text{-}cloud}{number\ of\ passing\text{-}clouds}$$

in a specific period of time.

2.2.3 Performance ratio for PV irrigation systems

To analyse the global performance of PV grid-connected plants, the performance ratio (PR) [23], defined

[1] Start and stop thresholds have changed several times looking for a better tuning of the system.

as $PR = E_{PV}/((P^*/G^*)\int Gdt)$, where P^* is the peak power of the PV generator, G^* is the irradiance at standard test conditions (1000W/m^2), G is the irradiance on the plane of PV generator, and E_{PV} is the energy produced by the PV system, has been traditionally used.

High-power PVIS, due to their specificities, include *PR* losses for three different external factors not attributable to their lack of quality: the non-irrigation period, associated to water needs; the intrinsic characteristics of the PV irrigation system design; and external circumstances that may affect the *PR* like the irrigation community habits or the different rainfall over time. Taking them into consideration, the *PR* can be expressed as [2]:

$$PR = \frac{E_{PV}}{P^*/G^*} \cdot \frac{1}{\int Gdt} \cdot \frac{\int_{IP} Gdt}{\int_{IP} Gdt} \cdot \frac{\int G_{useful}dt}{\int G_{useful}dt} \cdot \frac{\int G_{used}dt}{\int G_{used}dt}$$

where:

IP is the irrigation period determined by the water needs of the crop and by the relation between water needs, pumping capacity and pumped water storage capacity (in case of pumping to a water pool).

G_{useful} is the available useful irradiance during the IP (the irradiance required to deliver the power needed to pump water, when available). It is determined by the relationship between the PV generator nominal power (P^*), the PV generator supporting structure, and the characteristics of the irrigation system. When irradiances below a minimum are available, the system cannot pump because there is not enough power, and irradiances higher than G_{useful} are partially wasted because the PV generator cannot supply more power than that consumed by the pump working at its nominal power.

G_{used} is the irradiance effectively used by the system and depends on the number and length of abrupt stops of the FCs and on other external circumstances like the availability of water or the irrigator's decisions about whether to activate the system or not.

The previous equation can be rewritten as

$$PR = PR_{PV} \times UR_{IP} \times UR_{PVIS} \times UR_{EF} \text{ [2]},$$

where:

$$PR_{PV} = \frac{E_{PV}}{P^*/G^*} \cdot \frac{1}{\int G_{used}dt}; \; UR_{IP} = \frac{\int_{IP} Gdt}{\int Gdt}$$

$$UR_{PVIS} = \frac{\int G_{useful}dt}{\int_{IP} Gdt}; \; UR_{EF} = \frac{\int G_{used}dt}{\int G_{useful}dt}$$

PR_{PV} is the *PR* considering only losses strictly related to the PV system itself; UR_{IP} is the utilization ratio related to the irrigation period; UR_{PVIS} is the utilization ratio related to the PVIS design (type of irrigation system, the ratio PV peak power - PV power required for irrigation, the tracking geometry and the accuracy of the PLC control algorithms setup); and UR_{EF} is the utilization ratio related to the irrigator's decisions.

3 RESULTS

3.1 Number of abrupt stops

Table and Table show the number of abrupt and controlled stops per month and per day, considering only days within the IP for which information is available, before and after the systematic tuning that took place in September 2019.

Table I. Number of stops before systematic tuning

Year	Month	Controlled stops	Abrupt stops	Abrupt stops (%)
2019	01	544	166	23.4
	02	270	46	14.6
	03	384	134	25.9
	04	423	259	38.0
	05	525	866	62.3
	06	489	413	45.8
	07	357	230	39.2
	08	176	92	34.3
	09	286	83	22.5
Total		3454	2289	**39.8**
Total per day		**12.8**	**8.5**	-

Table II. Number of stops after systematic tuning

Year	Month	Controlled stops	Abrupt stops	Abrupt stops (%)
2019	10	416	4	1.0
	11	336	0	0.0
	12	160	2	1.2
2020	01	75	0	0.0
	02	264	4	1.5
	03	437	17	3.7
	04	27	5	15.6
	05	4	0	0.0
	06	38	0	0.0
	07	184	5	2.6
	08	265	4	1.5
	09	369	1	0.3
	10	537	2	0.4
	11	163	0	0.0
	12	255	2	0.8
Total		3530	46	**1.3**
Total per day		**9.5**	**0.1**	-

Before applying the systematic tuning procedure, the percentage of abrupt stops was 39.8%. After the tuning change, it has been reduced to 1.3%, which means that the new tuning methodology has led on to a 98.8% reduction in the number of abrupt stops per day and most of the system stops have turned controlled stops but only a number close to one percent remain being abrupt stops. Data also show a 54.9% decrease in the total number of stops per day.

Notice that along May and June 2019 there was a high number of abrupt stops, caused by a large number of passing-clouds and system instabilities that made the passing-cloud resistance ratio lower than expected and increased the number of stops compared to other months. On the other hand, since April, May and June 2020 do not completely belong to the IP (due to weather conditions and water needs), the number of days considered for those months is very small.

3.2 Passing-cloud resistance ratio

To characterize the critical passing-clouds, a study of

the existing clouds throughout the months of March, April and May 2019 is made. The study starts from the irradiance drop information, relates it to the PV power drop and, finally, determines the critical duration of the clouds.

As an irradiance drop not always causes a PV power drop (a cloud can shade the calibrated cell that identifies the irradiance drop but not a significant part of the PV generator), the study considers just irradiance drops causing PV power drops. The percentage of clouds that cause 10% to 90% irradiance drops in a nine-seconds or less time interval is calculated. Cumulative values are represented in Figure 2.

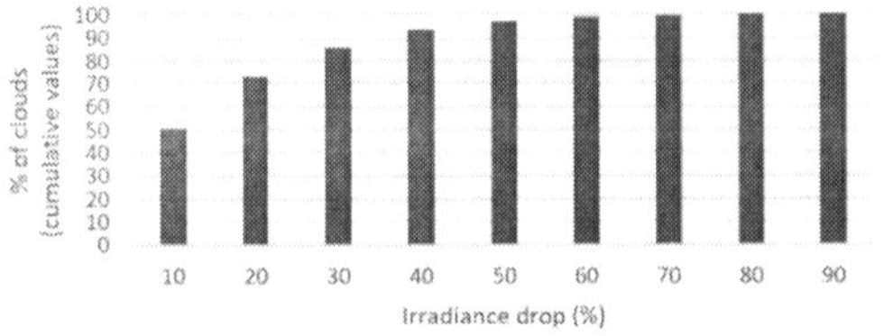

Figure 2. Cumulative distribution of detected clouds according to irradiance drop

93.7% of the clouds that produce equivalent decreases in the irradiance measured by the calibrated cell and in PV power, are associated with 40% (or less) irradiance drops, and 97.3% of the clouds are associated with 50% (or less) irradiance drops. Under the assumption that if the system resists clouds that cause a 50% irradiance drop, it will also resist clouds that cause smaller drops, it is reasonable to focus the study on this type of clouds.

Second, the irradiance drop slope is analysed for all clouds causing 40% and 50% irradiance drops in a time interval equal to or less than 9 seconds. Results are classified according to the time it takes the irradiance to drop (Figure 3).

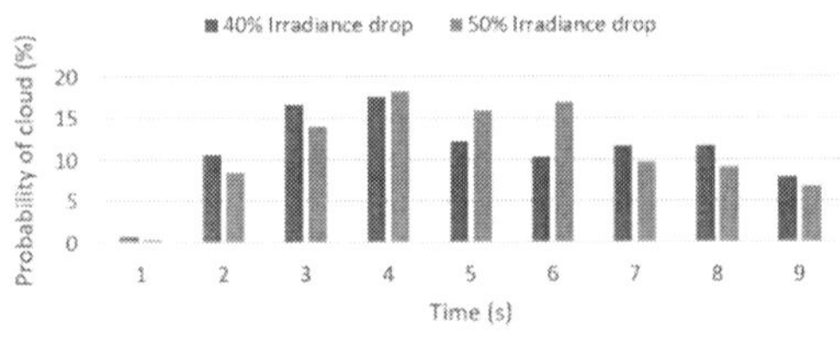

Figure 3. Probability density function of number of clouds causing 40% (blue) and 50% (orange) irradiance drops as a function of the drop time interval

The probability of a 40% irradiance drop starts to grow for clouds with a two-seconds drop interval and reaches its maximum for 3 and 4 seconds clouds. For 50% irradiance drops the maximum is reached for 4 seconds clouds. Under the assumption that if the system resists clouds with a shorter drop interval, it will also resist clouds with longer drop intervals, it is reasonable to focus the study on clouds with a four-seconds drop interval.

If the duration of the clouds is calculated from the size of the PV generator, and considering that the smallest dimension of the perimeter of the PV generator is 74 meters, it is obtained that the duration of passing-clouds (Δt) should be 3.7 s. Since the system provides logs every second, the duration needs to be rounded to either 3 or 4 seconds. This last value is the same as that obtained from the study of the existing clouds.

Cloud analysis results suggest focusing the passing-cloud resistance ratio study on passing-clouds with a 50% power drop and a four-seconds duration, associated with 50% irradiance drops. Using this criterion, the passing-clouds sample is too small, so, it is decided to expand the sample considering 40% power drop passing-clouds with a four-seconds duration, associated with 50% irradiance drops. Passing-clouds with a three-seconds duration are studied as well. Table and Table IV show the passing-cloud resistance ratio per month before and after applying the systematic tuning, respectively, for these two types of clouds. A cumulative value (considering passing-clouds with durations from 3 to 9 s) is also shown.

Table III. Monthly passing-cloud resistance ratio prior to systematic tuning (%)

Year	Month	σ_{cloud} (4 s)	σ_{cloud} (3 s)	σ_{cloud} (3 s cum.)
2019	01	95.7	96.7	93.1
	02	95.7	88.5	93.2
	03	91.2	69.8	86.0
	04	73.8	74.4	78.0
	05	50.0	48.8	52.1
	06	85.3	86.6	82.4
	07	94.1	88.9	89.0
	08	57.1	37.5	51.4
	09	57.1	36.8	64.8
Total		**80.8**	**74.1**	**79.0**

Table IV. Monthly passing-cloud resistance ratio after systematic tuning (%)

Year	Month	σ_{cloud} (4 s)	σ_{cloud} (3 s)	σ_{cloud} (3 s cum.)
2019	10	100	100	99.4
	11	100	100	100
	12	100	100	100
2020	01	100	--	100
	02	100	100	100
	03	98.6	100	99.3
	04	100	100	95.7
	05	--	--	--
	06	100	100	100
	07	100	100	100
	08	100	100	100
	09	100	100	100
	10	100	100	100
	11	100	100	100
	12	100	100	100
Total		**99.8**	**100**	**99.8**

Tables show an improvement after applying the systematic tuning method. The global passing-cloud resistance ratio before systematic tuning is 80.8% for four-seconds clouds. After systematic tuning, σ_{cloud} increases to 99.8%.

The passing-cloud resistance ratio analysis was carried out under two assumptions: i) if a PVIS resists clouds that cause deep PV power drops, it will also resist clouds that cause smaller drops; and, ii) if a PVIS resists clouds that cause a specific PV power drop in a determined period of time, it will also resist clouds for which the period is longer. Data shown in Table show that the second assumption is appropriate: the passing-cloud resistance ratio for three seconds clouds is lower than the accumulated one (74.1% vs. 79.0%) and also lower than that of clouds of a longer duration (74.1% vs. 80.8% for 4 seconds clouds). Nonetheless, the low number of abrupt stops after the systematic tuning (46 in 15 months, but only 8 of them where caused by passing-clouds) makes it impossible to corroborate this statement (nor the first one) for PVIS with a systematic tuning until a larger data sample is available.

Data also show that, after a proper tuning of the system, it is reasonable to require values greater than 95% in a general specification of the passing-cloud resistance ratio index.

3.3 Performance ratio for PV irrigation systems

Annual performance indices are shown in Table .

Table V. Annual real performance indices (%)

Year	PR_{PV}	UR_{IP}	UR_{PVIS}	UR_{EF}	PR
2019	87.1	100	70.3	98.0	60.0
2020	86.9	72.4	75.6	98.5	46.8

PR_{PV} values are similar to PR values expected for grid-connected PV systems. The decrease in the number of abrupt stops after systematic tuning does not lead on to a better PR_{PV} because abrupt stops occur when irradiance and PV power are low, so the improvement is negligible compared to the effect of days with high irradiation and PV power along the year. On the other hand, as there are two FCs, the system may react to a FC abrupt stop by starting the other FC, so the impact in PR_{PV} is minimal.

UR_{IP} is 100% in 2019 but decreases to 72.4% in 2020 because during the April-June interval the water pool was full and no irrigating was needed due to weather conditions, which makes that period not being part of the IP.

UR_{EF} values are close to 100%.

UR_{PVIS} values are the most affected by weather conditions and FC controlled stops. The decrease in the number of controlled stops after tuning and the different weather conditions during 2020 have a positive effect on this index.

The PR drops considerably from the first to the second year of operation (60.0% to 46.8%) due to the effect of UR_{IP}. If UR_{IP} had been 100% after tuning, and assuming the same values for the rest of factors, the PR would have reached 64.7%

Data show that it is reasonable to expect PR_{PV} values over 80.0%, UR_{IP} values very close to 100% if the IP is extended for most of the year, UR_{PVIS} values over 75.0% (and even above 85% with an adequate adjustment of the PV peak power to the PV power required for irrigation and the accuracy of the PLC control algorithms setup and tuning methodology) even though UR_{PVIS} is dependent on weather conditions, and UR_{EF} values very close to 100% when the irrigator uses properly the PVIS (PV irrigation centred at midday, planning of maintenance tasks in cloudy days or during the night, management of water consumption,...).

PR_{PV}, UR_{IP}, UR_{PVIS} and UR_{EF} determine PR values. Their expected values make us suggest that it is reasonable to expect PR values very close to 65% in high-power PVIS pumping to a water pool.

4 CONCLUSIONS

Stand-alone high-power PV irrigation systems without batteries is a recent innovation and there are no experimental data about the performance of this kind of systems available. This paper is a contribution to the knowledge of the experimental performance data of a PV irrigation system pumping to a water pool and, thus, to the expected values for other similar systems.

To check its quality, the experimental performance data of the system have been processed. To assess the robustness of the system to PV power fluctuations due to "passing clouds" two new indices have been proposed: the "Number of abrupt stops" and the "Passing-cloud resistance ratio". To evaluate the systems performance, the traditional PR has been factorized into four factors: PR_{PV} (assesses the performance quality of the PV system itself), UR_{IP} (measures the effect of the crop irrigation period), UR_{PVIS} (measures the effect of the hydraulic irrigation system) and UR_{EF} (considers losses due to the user behavior).

Relevant conclusions can be drawn from the results about the performance that can be expected from this type of system:

- A value of 95% can be required in the passing-cloud resistance ratio index.
- The expected PR in good quality PV irrigation systems pumping to a water pool is 65%. The expected values of the different factors are: PR_{PV}>80%, similar to what is expected in grid-connected PV systems; UR_{IP}≈100% if the needs of the crop are extended for most of the year and/or if the relative size of the pool with respect to the PV generator is appropriate; UR_{PVIS}>85% and UR_{EF}≈100% if the end user makes appropriate use of the PV irrigation system.

Nevertheless, it is necessary to confirm these expected values with experimental measurements in other high-power PV irrigation systems pumping to a water pool.

6 REFERENCES

[1] J. Carroquino, R. Dufo-López and J. L. BarnalAugustin, "Sizing of off-grid renewable energy systems for drip irrigation in Mediterranean crops," Renewable Energy, no. 76, pp. 566-574, 2015.

[2] Almeida, R.H., Carrêlo, I.B., Lorenzo, E., Narvarte, L., Fernández-Ramos, J., Martínez-Moreno, F., Carrasco, L.M., 2018a. Development and test of solutions to enlarge the power of PV irrigation and application to a 140 kW PV-diesel representative case. Energies 11, 3538-undefined. https://doi.org/10.3390/en11123538

[3] MASLOWATEN, 2016. MASLOWATEN H2020 project. [WWW Document]. URL http://maslowaten.eu/ (accessed 4.1.21).

[4] Todde, G., Murgia, L., Deligios, P.A., Hogan, R., Carrelo, I., Moreira, M., Pazzona, A., Ledda, L., Narvarte, L., 2019. Energy and environmental performances of hybrid photovoltaic irrigation systems in Mediterranean intensive and super-intensive olive orchards. Sci. Total Environ. 651, 2514–2523. https://doi.org/10.1016/j.scitotenv.2018.10.175

[5] Todde, G., Murgia, L., Carrelo, I., Hogan, R., Pazzona, A., Ledda, L., Narvarte, L., 2018. Embodied energy and environmental impact of

large-power stand-alone photovoltaic irrigation systems. Energies 11, 2110-undefined. https://doi.org/10.3390/en11082110

[6] Abu-Aligah, M., 2011. Design of photovoltaic water pumping system and compare it with diesel powered pump. Jordan J. Mech. Ind. Eng. 5, 272–280.

[7] Langarita, R., Sánchez Chóliz, J., Sarasa, C., Duarte, R., Jiménez, S., 2016. Electricity costs in irrigated agriculture: A case study for an irrigation scheme in Spain. Renew. Sustain. Energy Rev. 68, 1008–1019. https://doi.org/10.1016/j.rser.2016.05.075

[8] Lorenzo, C., Almeida, R.H., Martínez-Núñez, M., Narvarte, L., Carrasco, L.M., 2018. Economic assessment of large power photovoltaic irrigation systems in the ECOWAS region. Energy 155, 992–1003. https://doi.org/10.1016/j.energy.2018.05.066

[9] Abella, M.A., Lorenzo, E., Chenlo, F., 2003. PV water pumping systems based on standard frequency converters. Prog. Photovoltaics Res. Appl. 11, 179–191. https://doi.org/10.1002/pip.475

[10] Posorski, R., 1996. Photovoltaic water pumps, an attractive tool for rural drinking water supply, in: Solar Energy, 58. pp. 155–163. https://doi.org/10.1016/S0038-092X(96)00060-6

[11] Sontake, V.C., Kalamkar, V.R., 2016. Solar photovoltaic water pumping system - A comprehensive review. Renew. Sustain. Energy Rev. https://doi.org/10.1016/j.rser.2016.01.021

[12] Achouri, R., Nouicer, O., Mhiri, H., Bournot, P., 2013. Probable cause analysis of cracks observed on vertical centrifugal pump. Eng. Fail. Anal. 29, 1–11. https://doi.org/10.1016/j.engfailanal.2012.11.003

[13] EIP-Water, 2013. Priority area P30: water-energy nexus. European innovation partnership on water boosting opportunities – innovating water. [WWW Document]. URL https://watereurope.eu/tag/eip-water/ (accessed 4.1.21).

[14] Ismaier, A., Schlücker, E., 2009. Fluid dynamic interaction between water hammer and centrifugal pumps. Nucl. Eng. Des. 239, 3151–3154. https://doi.org/10.1016/j.nucengdes.2009.08.028

[15] Bhayo, B.A., Al-Kayiem, H.H., Gilani, S.I., 2019. Assessment of standalone solar PV-Battery system for electricity generation and utilization of excess power for water pumping. Sol. Energy 194, 766–776. https://doi.org/10.1016/j.solener.2019.11.026

[16] Das, M., Mandal, R., 2018. A comparative performance analysis of direct, with battery, supercapacitor, and battery-supercapacitor enabled photovoltaic water pumping systems using centrifugal pump. Sol. Energy 171, 302–309. https://doi.org/10.1016/j.solener.2018.06.069

[17] Khiareddine, A., Ben Salah, C., Mimouni, M.F., 2015. Power management of a photovoltaic/battery pumping system in agricultural experiment station. Sol. Energy 112, 319–338. https://doi.org/10.1016/j.solener.2014.11.020

[18] Fernández-Ramos, J., Narvarte-Fernández, L., Poza-Saura, F., 2010. Improvement of photovoltaic pumping systems based on standard frequency converters by means of programmable logic controllers. Sol. Energy 84, 101–109. https://doi.org/10.1016/j.solener.2009.10.013

[19] Bione, J., Vilela, O.C., Fraidenraich, N., 2004. Comparison of the performance of PV water pumping systems driven by fixed, tracking and V-trough generators. Sol. Energy 76, 703–711. https://doi.org/10.1016/j.solener.2004.01.003

[20] Narvarte, L., Fernández-Ramos, J., Martínez-Moreno, F., Carrasco, L.M., Almeida, R.H., Carrêlo, I.B., 2018. Solutions for adapting photovoltaics to large power irrigation systems for agriculture. Sustain. Energy Technol. Assessments 29, 119–130. https://doi.org/10.1016/j.seta.2018.07.004

[21] Narvarte, L., Lorenzo, E., 2008. Tracking and ground cover ratio. Prog. Photovoltaics Res. Appl. 16, 1–12. https://doi.org/10.1002/pip.847

[22] Almeida, R.H., Narvarte, L., Lorenzo, E., 2018b. PV arrays with delta structures for constant irradiance daily profiles. Sol. Energy 171, 23–30. https://doi.org/10.1016/j.solener.2018.06.066

[23] Luque, A., Hegedus, S., 2003. Handbook of Photovoltaic Science and Engineering, Handbook of Photovoltaic Science and Engineering. Chichester: Johm Wiley & Sons. https://doi.org/10.1002/9780470974704

A DECISION SUPPORT SYSTEM FOR COST-EFFECTIVE OPERATION AND MAINTENANCE OF PV PLANTS

S. Gallmetzer[1,2]*, A. Louwen[1], P. Ingenhoven[3,4], D. Moser[1]
[1]Institute for Renewable Energy, EURAC Research, Viale Druso 1, 39100 Bolzano, Italy,
[2]University of Trento, via Calepina 14, 38122 Trento, Italy,
[3]School of Chemical Engineering and Advanced Materials, Centre for Energy Technology, The University of Adelaide, Adelaide, SA, 5005 Australia
[4]Free University of Bolzano, Faculty of Science and Technology, piazza Università 5, 39100 Bolzano, Italy
*Phone: +39 3271325412; E-mail: sandra.gallmetzer@eurac.edu

ABSTRACT: In this paper we present a decision support system for mitigation of PV plant failures to be used in a platform supporting operation and maintenance (O&M) operators in taking the most cost-effective decision in case of performance loss. Within the Solar Bankability project risk assessment and risk management were studied. Several real maintenance tickets were collected and analyzed for typical failures during the lifetime of PV plants in order to develop a technical risk framework and to determine mitigation measures able to minimize the risk for project investments. Besides preventive and corrective mitigation measures applied before the failure happens or implemented during the operation and maintenance phase, this work introduces also direct mitigation options which can be applied in already running plants. Within PV4.0 the aim is to improve further failure data collection and to develop a fully automated decision support system that calculates the economic impact of failures and suggests a list of possible intervention strategies. The Cost Priority Number (CPN) methodology is in the core center of the PV4.0 tool since it prioritizes O&M interventions according to the economic impact of the technical risks helping so in the decision process.

Keywords: O&M, risk analysis, CPN methodology, risk mitigation measures

1 INTRODUCTION

The growing success of solar PV over other renewable technologies in the last years has many different reasons, but one of the key factors is the rapid reduction of its cost. The global weighted-average Levelized Cost of Energy (LCOE) of PV systems fell by about 82% over the last decade mainly driven by the sinking prices of PV modules [1]. Since the operational expenditures (OPEX), consisting mainly of operation and maintenance (O&M) costs, are often defined based on a fixed percentage of capital expenditures (CAPEX), with declining PV system prices the O&M market gets restricted to a limited and ever decreasing margin.

Consequently, there is a need for improvements in O&M for photovoltaic systems to reduce costs and become more effective. Depending on the requirements of the PV system and the contract for operation and maintenance, the O&M costs will vary. In any case, several improvements are being made to obtain a more standardized approach for O&M, having the potential to decrease costs and to make them more predictable. The main drivers therefore are, among others, to increase efficiency and energy delivery, to extend lifetime of the PV system, to decrease downtime due to failures and to reduce O&M costs. Thus, improvements are mainly made in the field of automation and digitalization of the workflow of technicians.

However, typical O&M contracts do not allow the O&M contractor to efficiently optimize the processes, as they are for example obliged to fix any issue as soon as possible. However, the ever-increasing availability of PV plant data, including real-time monitoring data, meteorological data, and digitally available PV plant design and component data offers many types of O&M practices by applying Industry 4.0 concepts. To make this transition towards a new type of O&M contracting, a well validated method is needed to ensure confidence in results obtained.

In the field of O&M, different improvements can be attained: automated fault detection algorithms including monitoring data in order to detect faults immediately and even predict upcoming failures, spare part management system could be included in the DSS, accurate yield and performance monitoring are obtained using digital twins of the plant. Combining all these data and a method that analyses the financial impact of failures to aid in the prioritization and decision support of O&M activities, a new approach for O&M can be achieved. Typically, high-loss failures such as a not working inverter or transformer station, need to be replaced directly after the failure occurred since their downtime causes high costs due to power loss. Soiling or hotspots, instead, are considered low-loss failures since they usually do not cause high costs because of downtime. Thus, such technical risks have a lower priority and should wait for the next ordinary maintenance to get repaired.

In this paper we present the underlying methodology and first practical implementation to this alternative approach of optimizing O&M activities.

2 METHODS

In this context, at EURAC an automated decision support system will be developed with the aim to improve operation and maintenance by quantifying the economic impact of failures in the field and suggesting a possible intervention strategy. This methodology relies on two main projects: PV4.0, which is based on industry 4.0 logics, and Solar Bankability, which initially introduced the new methodology called *CPN*.

2.1 PV4.0 Tool

Within the project PV4.0 [2] the aim is to improve further failure data collection and to develop a platform supporting operation and maintenance (O&M) operators in taking the most cost-effective decision in case of performance loss. Therefore, measurement data of PV plants are linked to a digital ticketing and asset management system and a tool including Industry 4.0 concepts shown in Figure 1 is developed. The final development of the IT platform will be done together with the ICT service provider Saidea based in the province of Trento.

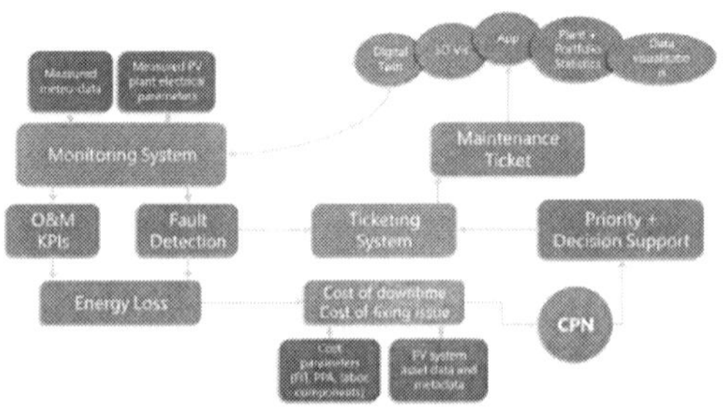

Figure 1: Layout of the PV4.0 software.

2.2 CPN Methodology

The CPN Methodology shown in Figure 2, developed within the Solar Bankability project [3], is the core center of the PV4.0 tool since it categorizes O&M interventions according to the economic impact of technical risks helping so in the decision process. This method can be described as a cost-based Failure Mode and Effect Analysis (FMEA), whereby the cost priority number (*CPN*) prioritizes risks with a higher economic impact. *CPN* is given by two contributions, the downtime costs which include all the cost coming up due to interruption of the plant and the fixing cost which consider all the different costs regarding the fixing of the failure. The unit of measurement of the *CPN* number is €/kW$_P$, making it comparable for different sizes of PV plants.

The required parameters from maintenance tickets, needed to calculate the *CPN*, are as follows [4]:

- *Failure time*: is defined as the time when the failure occurs.
- *Acknowledgement time*: is the time when the failure is detected by the O&M contractor. A new maintenance ticket will be opened.
- *Intervention time*: is defined as the time when the technician starts with the repair of the failure.
- *Repair switch-off time*: is the time when a further upper-level component is switched off to allow the repair of the broken component.
- *Repair switch-on time*: is the time when the upper-level component is again switched on after having fixed the problem.
- *Resolution time*: is defined as the time when the maintenance ticket is closed, and the issue solved.

Once the maintenance tickets present all above-described parameters, the following Key Performance Indicators (KPIs) can be calculated:

- *Detection time*: defines the time that it takes to detect a certain failure. The value depends on whether a monitoring system is present or not. With a basic or advanced monitoring system, the detection time is in the range of maximum a few minutes, but without monitoring the detection of the failure might happen only at the next scheduled maintenance intervention, that could take weeks or even months.

$$t_{dt} = acknowledgement\ time - failure\ time$$

- *Response time*: is the time needed to organize the repair or substitution of a component. In case of substitution, spare parts management is necessary and in the worst-case scenario, the spare parts must be transported from the manufacturer to the site. Consequently, the response time is strongly dependent on the availability of spare parts.

$$t_{rs} = intervention\ time - acknowledgement\ time$$

- *Repair time*: is the time needed for fixing the problem, i.e. the time the technician is on the field.

$$t_{rp} = resolution\ time - intervention\ time$$

- *Shutdown time*: considers the time period in which an upper-level component has to be switched off in order to fix the problem.

$$t_{sh} = repair\ switchoff\ time - repair\ switchon\ time$$

The *CPN* method was implemented in a python-based code and extended by the Risk Mitigation Factors (RMF) to quantify the effectiveness of mitigation measures along the PV value chain and the associated reduction in fixing costs, downtime costs and failure occurrence [5]. Solar Bankability identified two main categories of mitigation measures: preventive and corrective measures. Preventive measures are applied before the failure occurs during the product development phase and prevent the risk from happening. Mitigation measures of this category reduce the number of failures. Corrective measures, instead, are applied during the operation and maintenance phase and reduce higher losses and costs after a risk has occurred. These measures influence positively the detection time, response time and repair time.
The output from the python code provides a list of the best combinations of mitigation measures for each failure for three different cost scenarios and the corresponding cost savings.

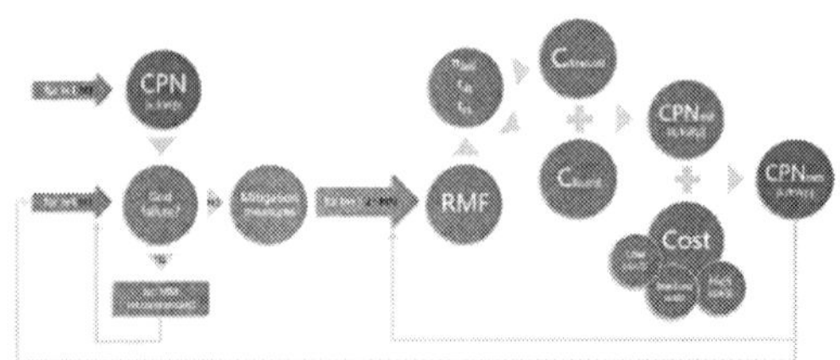

Figure 2: Overview of the CPN Methodology.

Finally, a third category of mitigation measures is introduced. The system needs to be expanded to suggest so-called direct mitigation options for application in already running plants. The three options identified are to repair the failure, to leave it unrepaired or to postpone the repair to the next periodic maintenance. Therefore, in python the code was upgraded in order take into consideration also the development of the irradiation loss in the days and months after the detection of the failure. For the repair option, the *CPN* is given by the upcoming cost for fixing the failure and thus constant. For the not-repair option, instead, the *CPN* is given by the upcoming downtime cost because of irradiation loss and depends on the hourly weather data. Finally, the third option is to repair the failure at the next periodic maintenance. Therefore, it is necessary to know first when the next ordinary maintenance is planned and second how much cost arise for the technician coming to the field for an extraordinary maintenance. Until the *CPN*, given by the costs of downtime and the cost of fixing at the next periodic maintenance, is lower than the cost of instant fixing, it is convenient to wait for the next periodic maintenance to repair the failure otherwise the failure should be repaired as soon as possible.

3 RESULTS

To provide us with a full set of measurements for calculating all relevant parameters for the CPN calculation, the maintenance tickets were assumed to be from the ABD PV plant of Bolzano airport, where EURAC is operating a test facility. Hence, the irradiation losses are calculated using ten years of irradiance data of the same plant to estimate the energy yield losses for the different Key Performance Indicators (KPIs).

3.1 Risk Assessment

This work applies for the first time the theoretical method to real maintenance tickets. The dataset analyzed presents mainly failures regarding the inverter, since they are easily visually detectable in comparison to module failures, such as hotspots or snail tracks, which usually require advanced inspection in order to be detected (for example IR or EL). It was necessary to make several adjustments to prepare the maintenance tickets for the implementation of the methodology. Since the field data are collected by different people, as it is usual, before starting to analyze and compare them, the data need to be rearranged into a common standardized format. Figure 3 shows the obtained CPN values of twenty maintenance tickets.

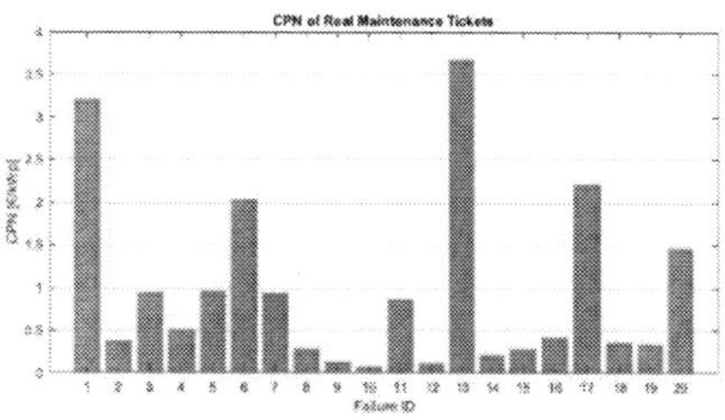

Figure 3: Cost Priority Number given in €/kW$_p$ for a dataset of real maintenance tickets.

3.2 Risk Management

The next step is to minimize the economic impact of the risks by applying a combination of mitigation measures to the various failures. Therefore, eight preventive and corrective measures and their cost considering three different scenarios were defined and their impact on the risks was calculated. Figure 4 presents the comparison between the CPN (green bars) and the best combination of the mitigated CPN considering the low-cost scenario (black bars). It is possible to identify, that by application of preventive and corrective mitigation measures, the CPN of real technical risks can be reduced only to a certain extent depending on the component affected since the KPIs are already relatively low. The best combination for most inverter failures resulted in being a basic monitoring system, which indeed reduces the time of detection of the failure, and spare part management, which significantly reduces the response time since the spare components are available immediately on the site and the time for transport can be avoided. Instead, for those maintenance tickets treating grid failures (Failure ID 2, 8-11), no measures have been suggested by the decision support system since issues with the electricity grid are outside of the scope of activities of the O&M operator.

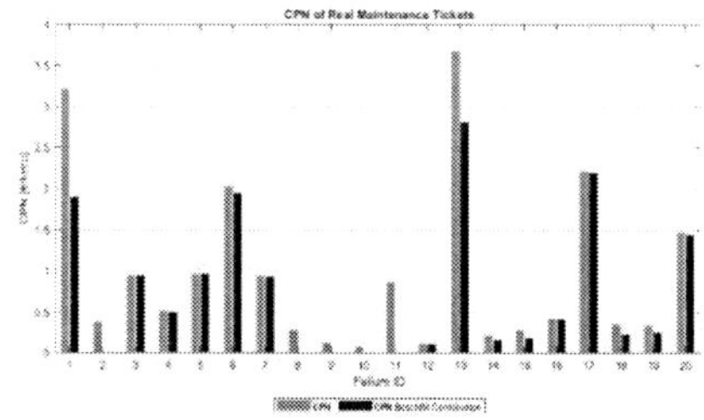

Failure ID	Component Testing	Design review	Qualification of EPC	Advanced monitoring system	Basic monitoring system	Advanced inspection	Visual inspection	Spare part management	CPNnew	Savings [€/kW$_p$]
1	0	0	0	0	1	0	0	1	1.692	1.316

Figure 4: Comparison of the CPN value with (black bars) and without (green bars) mitigation measures applied, considering low mitigation measures cost.

Besides preventive and corrective measures, direct mitigation options are introduced and applied to single maintenance tickets. Once the failure happened, the O&M operator needs to take a decision: either to repair the failure, to leave it unrepaired or to postpone the repair to the next periodic maintenance. Thus, the decision support system analyses the hourly development of two different scenarios: CPN-Repair and CPN-NotRepair. Depending on the technical risk, whether it has high impact on downtime or high fixing cost, the recommendation is to fix the failure or to leave it unrepaired.

Figures 5 and 6 show two direct mitigation options applied to a failure dealing with the theft of a module and a risk related to soiling, respectively. For both failures, at the beginning the cost related to fixing the failure, CPN-Repair, prevail over the cost related to power loss for a certain period of time after the failure happened. Since in both cases, this period is relatively small until the constantly increasing cost due to downtime overtake the cost related to fixing, the recommendation is to repair the failure.

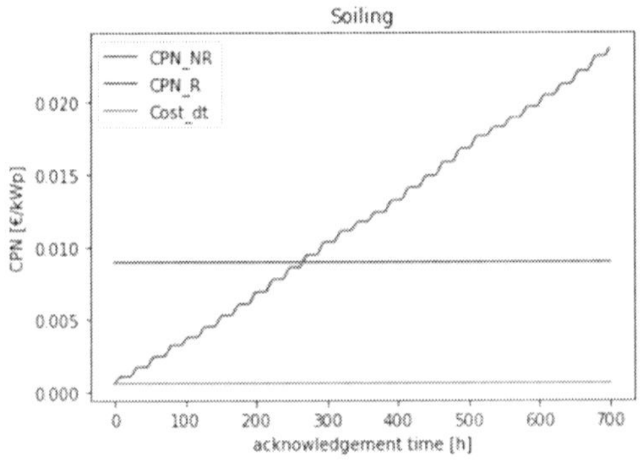

Figure 5: Comparison of CPN-Repair and CPN-NotRepair of a module with soiling risk.

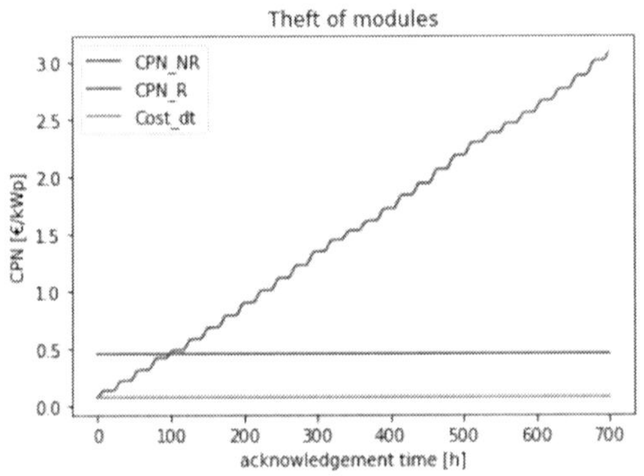

Figure 6: Comparison of CPN-Repair (in red) and CPN-NotRepair (in green) of a stolen module.

Table 1: Parameter description

CPN_NR	Cost related to downtime arising for not repairing the failure.
CPN_R	Cost if the failure is repaired immediately after detection.
Cost_dt	Cost for detection of the failure
C_inst_fix	Cost for fixing the failure instantly
CPN_inst	Sum of cost for fixing instantly and downtime cost
C_PM_fix	Cost of fixing at next periodic maintenance
CPN_PM	Sum of cost for fixing at the next PM and downtime cost

Further the tool examines the best date for fixing considering the cost of technician that arise for an extraordinary maintenance. Instant fixing requires that a technician travels to the site in order to repair the failure, which is related to high costs of travelling. Again, the two cases "Soiling" and "Theft of modules" are analysed. The first (Figure 7) presents a relatively low performance loss, thus if the next periodic maintenance is set within 600 hours from the detection of the failure, it is more convenient to wait for the next periodic maintenance. A stolen module (Figure 8), instead, causes high costs related to downtime, as it causes 100% power loss for a whole string of PV modules, thus in this case, the recommendation of the decision support system is to repair the failure instantly.

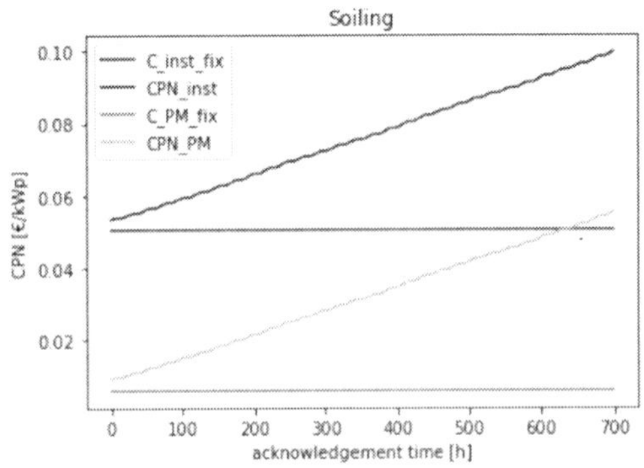

Figure 7: Comparison of the cost priority number applying instant repair of the failure (CPN_inst) or waiting for the next periodic maintenance (CPN_PM) for the technical risk "soiling".

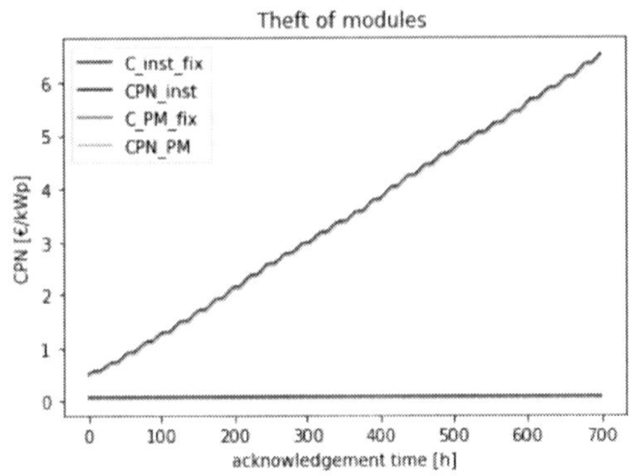

Figure 8: Comparison of the cost priority number applying instant repair of the failure (CPN_inst) or waiting for the next periodic maintenance (CPN_PM) regarding the theft of a module.

4 CONCLUSIONS

The results show that minimization of technical risks in PV projects is possible by applying several mitigation measures. Preventive and corrective measures reduce mainly the number of failures, the downtime due to detection and response and should therefore be applied during the planning phase. Direct mitigation measures, instead, suggest the best solution once the failure happened. Depending on several parameters the O&M operator should fix the failure immediately or wait for the next periodic maintenance.

A future study might also consider the uncertainties regarding weather data, since the final decision support system will calculate the irradiation loss based on actual weather data of the installed monitoring system. Finally, also safety issues should be considered in the process, which until now have been neglected.

The outcome of this research has shown that the information required for accurate CPN is indeed available to the O&M operator. The application of the CPN metric could be an aid in decision making, therefore O&M operators should set up systems where data from all stages of the PV plant lifetime, as well as from different areas of their work such as monitoring, field work and regular maintenance, are combined in order to make the best use of them. The CPN-methodology will be officially included

in the next Solar Power Europe guidelines for Operation & Maintenance and EPC. This will be a quality step ahead, meaning that the methodology that was conceived in 2017, is now ready to become mainstream.
The tool in development which should allow to evaluate the economic impact of failures with and without mitigation measures will be directly implemented in the PV4.0 project based on industry 4.0 logics, where EURAC is developing a software platform for fully digitalized O&M workflows.

5 REFERENCES

[1] International Renewable Energy Agency, "IRENA, Renewable Power Generation Costs in 2019," Abu Dhabi, 2020. [Online]. Available: https://www.irena.org/-/media/Files/IRENA/Agency/Publication/2018/Jan/IRENA_2017_Power_Costs_2018.pdf.
[2] A. Louwen, L. Koester, S. Lindig, A. Astigarraga, and D. Moser, "PV4.0: Combining Asset Management, PV Measurement Data and the Cost Priority Number Method in a Digital Infrastructure for PV Plant O&M," 2020. [Online]. Available: https://www.eupvsec-proceedings.com/proceedings?paper=49330.
[3] C. Tjengdrawira, D. Moser, U. Jahn, M. v. Amansperg, I.-T. Theologitis, and M. Heisz, "PV Investment - Technical Risk Management - Best Practice Guidelines for Risk Identification, Assessment and Mitigation.," 2017, [Online]. Available: www.solarbankability.eu.
[4] Oviedo Hernández G., Lindig S., D. Moser, and P.V. Chiantore, "Optimization of the Cost Priority Number (CPN) Methodology to the needs of a large O&M operator," 2018.
[5] U. Jahn *et al.*, "Minimizing Technical Risks in Photovoltaic Projects - Recommendations for Minimizing Technical Risks of PV Project Development and PV Plant Operation," 2016, [Online]. Available: www.solarbankability.eu.

SOLAR PHOTOVOLTAIC POWER PLANT INVENTORY AND SOLAR MODULES COUNTING BY CONVOLUTIONAL NEURAL NETWORK FROM AERIAL IMAGERY

Yuh-Ru Lin, Chao-Yang Huang , Jia-Sheng Zhang, Hao-Li Chen
Green Energy and Environment Research Laboratories, Industrial Technology Research Institute
No. 360, Gaofa 2nd Rd., Guiren Dist., Tainan City 711, Taiwan

ABSTRACT: Taiwan is an island located in the subtropics and near the Pacific Ocean, as a result, Taiwan is hit by typhoons almost every summer. Solar photovoltaic power plants suffer from storm damage and parts of solar panels could be lost. Nowadays, Taiwan is committed to promoting installations of solar photovoltaic plants. However, the government doesn't verify those existing plants are completeness or not. In this work, Aerial imagery or high-resolution satellite imagery are used to take inventory of photovoltaic systems. A training model for PV plants detection is set up by CNN (Convolutional Neural Network). After capturing the region of PV plants, a counting method, a consequent image processing techniques, is applied to count the amount of solar panels. By clarifying the amount of solar modules, those would be compared with the original information. A successful case at Liujia in Tainan verifying the counting method, the recognition rate has reach over 98%.

Keywords: Aerial imagery, computer vision, PV, Convolutional Neural Network

1 INTRODUCTION

In recent years, the exhaustion of fossil recourses and increasing energy demand driving the development of renewable energy. More and more renewable energy power plants are installed, especially in photovoltaics. Since solar energy could be easy generated by converting sunlight directly into electricity, Photovoltaics become the one of the fastest-growing renewable energy technologies.

As the amount of photovoltaic systems increase, the scale of the power plants also getting larger and systems with millions of PV modules become common. However, it becomes more difficult to find out the failure in such large area, and require a lot of maintenance as well as a large workforce even just a few of modules or system components failed. Unmanned aerial vehicles (UAVs) can be used to provide high efficiency and large-area inspections of photovoltaic system, instead of accessing places hard to arrive by human.

Periodic inspection of solar panels is very important for the effective management of photovoltaic modules. However, an island like Taiwan located in the subtropics and near the Pacific Ocean, hit by typhoons almost every summer. Solar photovoltaic power plants suffer from storm damage and parts of solar panels could be lost. Therefore, an efficient way to check whether the solar panels are damage or not may be very important for government verifying the completeness of those existing plants.

In this study, our purpose is to recognize all solar modules in a solar photovoltaic power plant. It would help make sure whether there is damage in PV plants after storm disaster or check whether still meet the registration of renewable energy equipment after several years operating.

The propose, a consequent of image processing algorithm, is used to realize aerial imagery preprocessing and combine with Convolutional Neural Network (CNN)[1][2] for image recognition to further detect solar photovoltaic system.

In this context, this study presents a helpful approach for for solar photovoltaic system detection and inventory, as well as confirming the damage of the PV power plant suffering from Typhoon. A training model for PV plants detection is set up by Convolutional Neural Network. After capturing the region of PV plants, we proposed a novel method, using several image processing techniques, applied to count the amount of solar panels.

The methods proposed in this paper could be used to not only count the number of solar modules, but also provide accurate module positioning[3][4]. Identify solar module problems is necessary for particular purpose, such as checking shading conditions, thermal abnormalities or dust soiling level. This work is useful for related works using computer vision in solar energy application to separate those region of interest (ROI) for further recognition process.

2 METHODOLOGY

In this section, a consequent of image processing techniques for separating solar panels will be described first. Then, in order to improve the accuracy rate of counting solar panels, the Convolutional Neural Network (CNN) method will be introduced.

2.1 Image Processing

Aerial images or satellite images will be used as our input images. It is recommended to use clear images with good quality.

At the first place, an image is converted to grayscale (or graylevel) image (Fig.1.a), a range of monochromatic shades from black to white. In order to count solar modules, the image is converted to binary by using thresholding (Fig.1.b). Image thresholding is a simple way of partitioning an image into foreground and background. The objects of photovoltaic system are expected to be separated to foreground. The result will make the image easier to analyze.

Then, edge detection algorithm, Canny edge detector [5], is applied to find the boundaries of objects in the image (Fig.1.c). After extracting the structure of objects in an image, the surrounding environment noise, such as grasses, trees, roof of the shed or shadow of the photovoltaic system, still included in the output boundaries. In order to extract the part of solar panels, we preliminary only left those regions that the outline is approximately close to rectangular, since a solar panel is quadrilateral. Then the amount of solar modules can be obtained by counting the number of rectangle.

Note that in this application Gaussian blur, a way to apply a low-pass filter to remove noise like "salt and pepper", should be used carefully. The detail edge features will possibly miss if the intensity of Gaussian blur filter applied too much so that solar modules cannot be cut clearly. It is also recommended to properly use morphology image processing technique, dilation and erosion, to add or remove pixel from the boundaries of the objects.

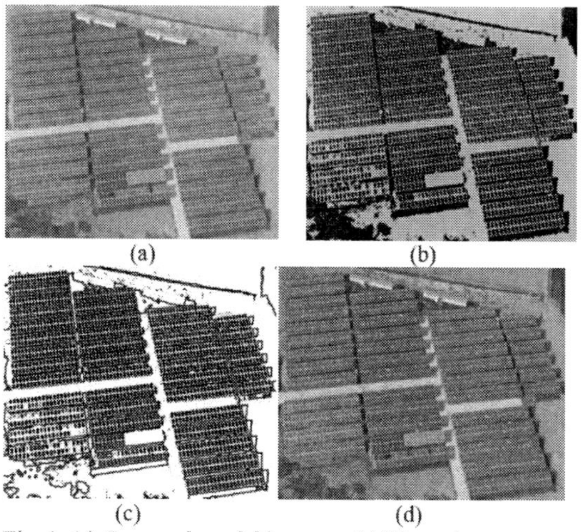

Fig. 1: (a) Gray-scale aerial imagery. (b) Binary image. (c) Contours of objects in the image after applying Edge detection algorithm. (d) Regions marked with pink determine the outline which are approximately close to rectangular.

2.2 Convolutional Neural Network (CNN)

In this work, Convolutional Neural Network (CNN) is used to recognize solar photovoltaic system to remove objects around surrounded the target PV system.

Convolution is the operation of two functions that result in a third function and a resulting output function. The features, such as edges, curves, straight lines or any geometric characteristic, from an image will be extracted different levels by applying the convolution filters.

Convolutional Neural Network (CNN) is a Deep Learning algorithm which is very similar to ordinary Neural Networks. CNN make up of neurons that have learnable weights and biases. The first layer of a Convolutional Neural Network is a Convolutional Layer, applying a convolution operation to the input, passing the result to the next layer called pooling layer. Pooling layers are used to reduce the dimensions of the feature maps. It is also useful for extracting dominant features, rotational and positional invariant, to maintain the process of effectively training of the model.

3 DISCUSSIONS AND RESULTS

With observing that if we use images that include wide area of PV plant like showed in fig.1, the counting method would filter many rectangles but not exact solar panels. Therefore, it is necessary to cut the image to extract the key area that almost include PV array, and the method mentioned to detect solar photovoltaic system with CNN could help.

First, some aerial images are preliminarily collected form PV plants and reshaped the images to the same 600*800 pixels images. Next, those images which contain PV array more than 50% of the area are labeled as "positive" data for further training. See fig 2(a). Currently, 100 training images are prepared, 50 are positive images and 50 are negative. The parameters of the CNN model are tuned show as following table 1.

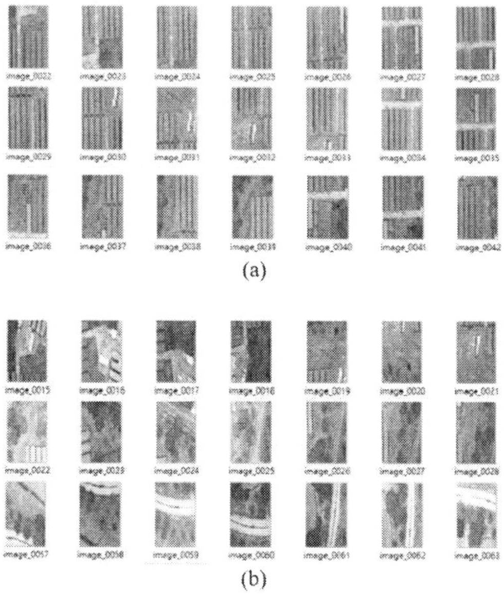

Fig 2. Training dataset. (a) "Positive" data: contain PV array more than 50%. (b) "Negative" data.

Layer (type)	Output Shape	Param #
conv2d (Conv2D)	(None, 800, 1200, 16)	800
max_pooling2d (MaxPooling2D)	(None, 400, 600, 16)	0
conv2d_1 (Conv2D)	(None, 400, 600, 16)	12560
max_pooling2d_1 (MaxPooling2	(None, 200, 300, 16)	0
conv2d_2 (Conv2D)	(None, 200, 300, 16)	12560
max_pooling2d_2 (MaxPooling2	(None, 100, 150, 16)	0
conv2d_3 (Conv2D)	(None, 100, 150, 16)	12560
max_pooling2d_3 (MaxPooling2	(None, 50, 75, 16)	0
conv2d_4 (Conv2D)	(None, 50, 75, 16)	12560
max_pooling2d_4 (MaxPooling2	(None, 25, 37, 16)	0
conv2d_5 (Conv2D)	(None, 25, 37, 16)	12560
max_pooling2d_5 (MaxPooling2	(None, 12, 18, 16)	0
dropout (Dropout)	(None, 12, 18, 16)	0
flatten (Flatten)	(None, 3456)	0
dense (Dense)	(None, 128)	442496
dropout_1 (Dropout)	(None, 128)	0
dense_1 (Dense)	(None, 2)	258

Table 1. Parameters of CNN training model.

By making good use of the PV plants recognition method to separate the image into small pieces, the accuracy of solar modules counting can be improved. At present, there is a successful case at Liujia in Tainan verifying the counting method. The recognition rate before using the separating method is 90.2%, and the recognition rate after separating reaches 98.2%.

4 CONCLUSION

In this paper, we proposed a novel method consistd of several image processing techniques for counting the amount of solar panels. In order to improve the accuracy rate of counting solar panels, a training model for PV plants detection is set up by Convolutional Neural Network. There is a successful case at Liujia in Tainan verifying the counting method and the recognition rate reaches 98.2%.

The result are encouraging and next step will be an attemp to find more PV plants to verify our results and further refine our algorithms.

5 ACKNOWLEDGEMENTS

The financial support provided by Bureau of Energy (Grant No.110-S0304) is gratefully acknowledged.

6 REFERENCES

[1] A. Krizhevsky, I. Sutskever, and G. Hinton: "ImageNet Classification with Deep Convolutional Neural Networks", In NIPS, 2012.

[2] Y. LeCun, B. Boser, J. S. Denker, D. Henderson, R. E. Howard, W. Hubbard, and L. D. Jackel: "Backpropagation applied to handwritten zip code recognition", Neural computation, 1989.

[3] F. Grimaccia , S. Leva, and A. Niccolai: "PV plant digital mapping for modules' defects detection by unmanned aerial vehicles", IET Renewable Power Generation, Volume11, Issue10, pp. 1221-1228, 2017.

[4] JA Tsanakas, LD Ha, F Al Shakarchi: "Advanced inspection of photovoltaic installations by aerial triangulation and terrestrial georeferencing of thermal/visual imagery", Renewable Energy, Volume 102, Part A, pp. 224-233, 2017.

[5] J. Canny: "A Computational Approach to Edge Detection", IEEE Trans. Pattern Anal. Mach. Intell., 6 , 679 – 697, 1986.

COMPARISON OF SOILING LOSS BETWEEN FRAMED AND FRAMELESS PHOTOVOLTAIC MODULES

*Hande KARAKAN[1], Hüseyin Anıl CEYRAN[1], Abbas YAZICI[2], Mete GÜNÖVEN[1]

1- Kalyon PV Research and Development Center, Kalyon Güneş Teknolojileri Üretim A.Ş., 06909 Ankara, Turkey
E-mail: hkarakan@kalyonpv.com; Tel: +90 5054670782
2- Kalyon Solar Energy Production Corporation, Kalyon Güneş Enerjisi Üretim A.Ş., 06909 Ankara, Turkey

ABSTRACT: Soiling rates of framed and frameless modules varies due to their design differences. To examine the soiling rate, successive outdoor tests have been conducted on extensively soiled module strings before and after cleaning. The power output of these framed and frameless module strings in horizontal or tilted positions were measured using a portable I-V tester. The horizontal irradiance has also been measured simultaneously using a reference cell. After normalizing the power measurements with respect to incident irradiance and rated module power, the loss in power due to soiling was found to be in average 11% higher for framed modules than for frameless modules. This can be explained by a longer dust and dried slurry deposition, extending from the module edge towards the first string of cells in framed modules. This comparative performance loss of framed modules due to soiling is less pronounced for tilted modules.

Keywords: solar modules, framed and frameless modules, soiling loss, dust deposition

INTRODUCTION

Throughout the photovoltaic industry, there are many different solar module technologies in use today. A solar module's current-voltage characteristic I-V is nonlinear, and temperature and irradiance fluctuations have a significant impact on energy production, as well as other environmental conditions affecting the front side transmittance of the module [1].

Solar power plants are generally built in desert areas with high amount of insolation, which also do not take up valuable agricultural or residential lands. In general, the rate of dust accumulation or soiling could be very high in such areas. When the front side of solar module is dusty, the solar module performance can decrease significantly due to soiling. [2]

Solar modules typically use aluminum frames that give structural strength that resist bending stresses [3] and provide ready attachment points for the installation. However, in double-glass bifacial modules, frames are sometimes not necessary because of the added stiffness of the double-glass structure. As such, the decision whether to use frames or not is a critical economical decision for the manufacturer during the design of photovoltaic modules, and depends mainly on mechanical loading conditions.

However, another secondary effect of the presence of the module frame is the increased dust accumulation near the edges of the module, due to the bulge created by the frame over the glass surface. Because of this, accumulated dust could extend more towards the first string of solar cells, especially after being washed towards the module edges after a light rain, as shown in Figure 1.

Figure 1: Soiled area on the framed (left) and frameless (right) modules, accumulated in a horizontal position. Blue line shows a full cell area, red line shows extent of dried slurry at the edges.

In this study, the DC power generation of two sets of module strings, each composed either of framed of frameless modules, were measured and normalized with respect to rated module power and simultaneous horizontal irradiance measurements.

The measurements were performed once on soiled modules (not cleaned by rain for at least 25 days), and once after cleaning thoroughly them. The comparison of results before and after cleaning show an 1.45% and 1.27% recovery rate respectively for horizontal and 32° tilted framed modules, compared to 1.27% and 1.18% for frameless modules.

EXPERIMENTAL METHODOLOGY

The tests were performed in the module performance and meteorology monitoring station within the 1300MWp Karapınar solar power plant, which is located in a desert climate in Konya province, Turkey. The station, shown in Figure 2, has not been fully commissioned yet. It is close to the future SCADA building of the power plant, where construction activities were ongoing before the measurements, which resulted in an accelerated yet uniform soiling over the tested modules.

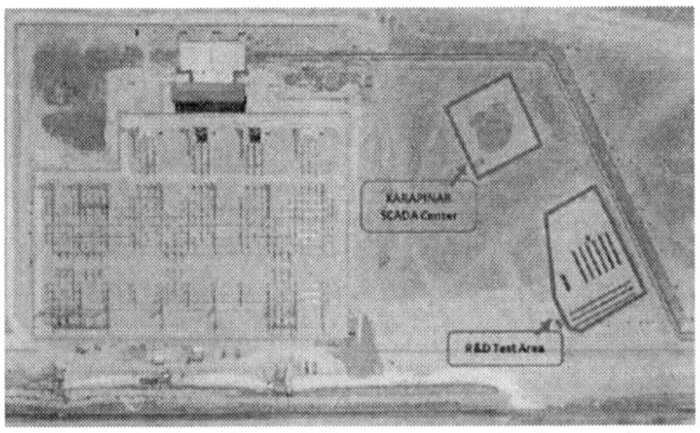

Figure 2: Top view of Karapınar monitoring station

Four module strings composed of either framed or frameless modules were used Two strings of each type are mounted on a horizontal single axis tracker but kept in a horizontal position during both soiling accumulation and the measurements, the other two strings are mounted at a fixed 32° tilt angle facing the south direction.

Strings are comprised of 13 series-connected modules each, all from KalyonPV. Framed modules consist of c-Si p-type PERC glass/backsheet modules rated at 390W. The frames result in a 3 mm high bulge above the glass surface around the edges. Frameless modules consist of c-Si p-type PERC bifacial glass/glass modules rated at 400W.

The I-V curves and generated DC maximum power points of these strings were measured using Metrel MI 3108 Eurotest portable I-V tester. The horizontal irradiance was also measured using Metrel's reference cell. These measurements were taken in approximately 30 minute-periods from morning to evening, consecutively for all four strings, over two consecutive sunny days (31.07 and 01.08). The data was then linearly interpolated to the same 30-minute intervals to enable direct comparison between measurements of different strings.

During the first day of the experiment, all modules were highly covered with dust, as they were subjected to the same increased soiling conditions for at least 25 days without a rain event. At the evening of the first day, all modules were thoroughly cleaned using only water and brush, as can be seen in Figure 3 and 4. The measurement process was repeated the next day.

Figure 3: Cleaning process of the strings.

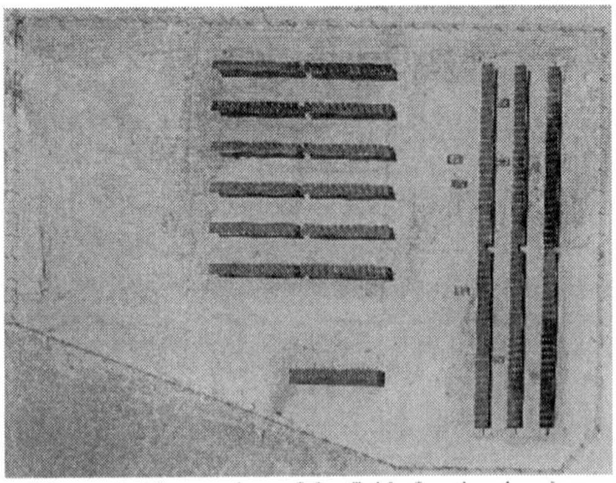

Figure 4: The up view of the field after the cleaning.

RESULTS AND DISCUSSION

Previous rain events had dislodged some of the dust over the module surface, but the water had transported the resulting slurry over the edges where it had settled. This additional soiling covered almost a full cell on framed modules and approximately a quarter of a full cell on frameless modules, as can be seen in Figure 1.

The string power generation and horizontal irradiation throughout the soiled and cleaned are shown in Figures 5 and 6 respectively.

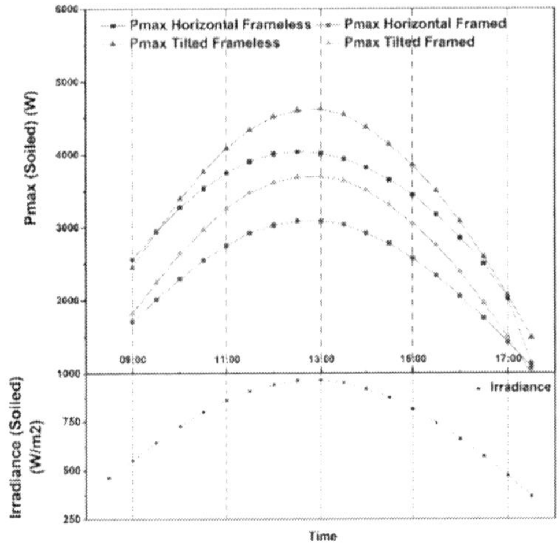

Figure 5: String maximum power and irradiation throughout the soiled day.

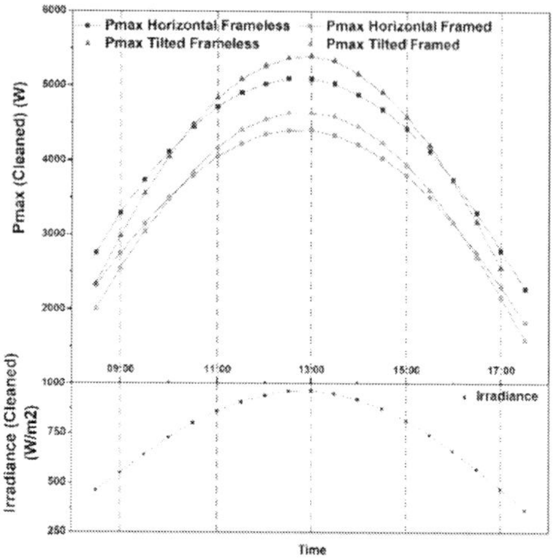

Figure 6: String maximum power and irradiation throughout the cleaned day.

As expected, during each day the power output of the tilted modules is generally higher than that of horizontal equivalent modules, except during early morning and late evening. The glass-glass bifacial modules produce higher power than glass-backsheet monofacial modules due to their higher rated power as well as rear-side bifacial gains.

In order to evaluate the performance of framed and frameless modules independently of rated power and insolation conditions, measured power for both horizontal and tilted module strings was normalized by the horizontal irradiance. The results for soiled and cleaned days are shown in Figures 7 and 8 respectively.

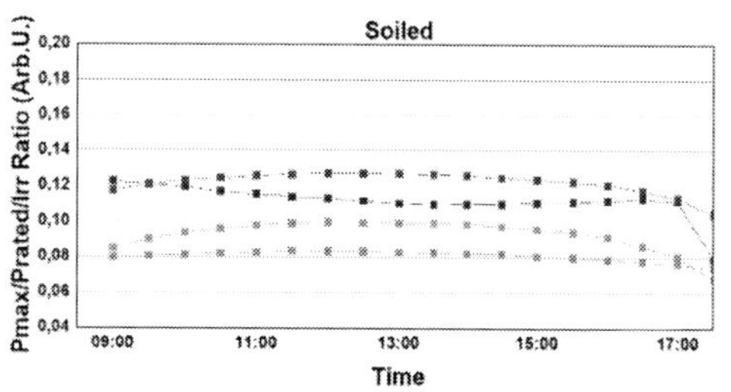

Figure 7: Maximum power and irradiation ratio versus time on soiled day.

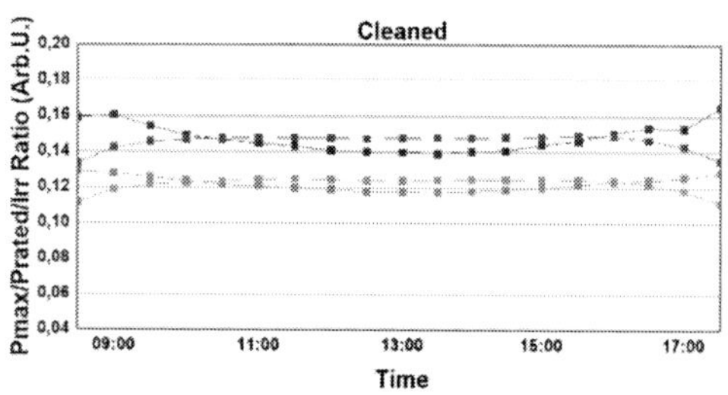

Figure 8: Maximum power and irradiation ratio versus time on cleaned day.

These P_{max}/P_{rated}/Irradiance ratios show a relatively flat evolution during the day for all module and installation types. Soiled modules have daily averaged P_{max}/P_{rated}/Irradiance ratios of 0.011 for frameless horizontal, 0.012 for frameless tilted, 0.008 for framed horizontal and 0.010 for framed tilted.

The normalized power generation for both module and installation types increased significantly after cleaning. Clean modules have daily averaged Pmax/Irradiance ratios of 0.013 for frameless horizontal, 0.014 for frameless tilted, 0.012 for framed horizontal and 0.012 for framed tilted.

Hence, the improvement rate after cleaning has been 1.27 for horizontal frameless modules, 1.18 for tilted frameless modules, 1.45 for horizontal framed modules and 1.27 for tilted framed modules. Therefore, framed modules showed, on average, 11% more improvement after cleaning, compared to frameless modules.

Similarly, the soiling loss has been calculated for these four types of modules according to Equation (1). The results have been given in Figure 9. To calculate the soiling loss, peak Pmax values has been used.

$$Soiling\ Loss = 1 - \frac{P_{max,\ soiled}}{P_{max,\ cleaned}} \quad (1)$$

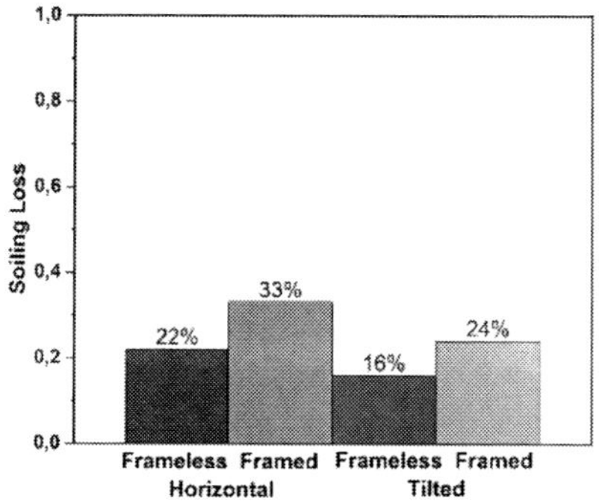

Figure 9: Soiling losses of the framed and frameless modules.

Framed modules show more soiling loss compared to their frameless counterparts, even after accounting for differences in the rated power of modules. This phenomenon could be associated with increased dust or slurry accumulation, that is transported in liquid form towards the module edges by the effect of gravity or wind, before being deposited and dried. Because of the bulge created by the aluminum frame, this slurry accumulation extends more towards the module center and over the first string of cells in a framed module.

This phenomenon is more pronounced in modules that are horizontally soiled than for tilted modules, as a larger part of the liquid slurry would flow away from the bottom edge of a tilted module with the effect of gravity.

Once this performance monitoring station will be completed and commissioned, these measurements will be done automatically over AC inverters and meteorological sensors. The effect of soiling will be investigated on module power productions, together with meteorological parameters and module temperature.

CONCLUSIONS

Based on sample of four strings of modules, and using an experiment where tested modules are extensively but homogeneously soiled before being thoroughly cleaned, we have shown that framed modules are on average 11%

more susceptible power degradation than equivalent frameless modules under equivalent irradiance. A dust deposition mechanism has been proposed that can explain the observations. These results can provide an input on the design choice of framed vs. frameless modules, or indeed on periodical cleaning intervals in order to optimize power plant performance.

REFERENCES

[1] Azouzoute, A., Hajjaj, C., Zitouni, H., El Ydrissi, M., Mertah, O., Garoum, M., & Ghennioui, A. (2021). Modeling and experimental investigation of dust effect on glass cover PV module with fixed and tracking system under semi-arid climate. Solar Energy Materials and Solar Cells, 230(October 2020), 111219. https://doi.org/10.1016/j.solmat.2021.111219

[2] E. G. Luque, F. Antonanzas-torres, and R. Escobar, "Effect of soiling in bifacial PV modules and cleaning schedule optimization," *Energy Convers. Manag.*, vol. 174, no. May, pp. 615-625, 2018, doi: 10.1016/j.enconman.2018.08.065

[3] L. Papargyri, M. Theristis, A. Livera, B. Kubicek, P. Papanastasiou and G. E. Georghiou, "NUMERICAL AND EXPERIMENTAL INVESTIGATIONS ON THE EFFECT OF DIFFERENT FRAME AND MOUNTING CONFIGURATIONS OF PV MODULES FOR CRACK PROPAGATION AND DEGRADATION", 36th European Photovoltaic Solar Energy Conference and Exhibition, Marseille, pp. 1087-1090, DOI: 10.4229/EUPVSEC20192019-4AV.1.42

[4] M. M. Fraga, B. L. de O. Campos, T. B. de Almeida, J. M. F. da Fonseca, and V. de F. C. Lins, "Analysis of the soiling effect on the performance of photovoltaic modules on a soccer stadium in Minas Gerais, Brazil," *Sol. Energy*, vol. 163, no. February, pp. 387–397, 2018, doi: 10.1016/j.solener.2018.02.025.

MODELING OF A PHOTOVOLTAIC FAULT DETECTION APPROACH CONSIDERING MACHINE LEARNING FOR THE EU PVSEC 2021

Philipp Akharath[1], Jaqueline Altkrüger[1], Harkiran Sahota[1], Volker Herbort[1], Henrik te Heesen[2]
[1] Ulm University of Applied Sciences, Prittwitzstraße 10, 89075 Ulm
[2] Trier University of Applied Sciences, Campusallee, 55768 Hoppstädten-Weiersbach

ABSTRACT: Photovoltaics play a central role in transforming the global energy system into an emission-free energy supply. For example, about 1.8 million installed photovoltaic roof systems in Germany with a nominal output of up to 30 kWp contribute to electrical power generation. Nevertheless, especially in this class of plants, it is evident that technical malfunctions can occur due to a lack of remote monitoring of electricity production and insufficient quality assurance concepts, resulting in a reduction of the yield of these photovoltaic plants [1,2,3]. So far, manual methods have been used to detect such anomalies for unlabeled data; machine learning concepts can provide an automated and adaptive alternative. For this purpose, the time and memory saving anomaly detection approach of Isolation Forests is applied. The results are compared to the ground truth provided by a manual method introduced by Leloux et al. [4]. This evaluation reveals that the approach of Isolation Forest can correctly evaluate the functionality of a plant. Nevertheless, a high rate of undetected anomalies could be caused by a misjudgment by the Ground Truth or the Isolation Forest. Therefore, it must be manually assessed until labeled data is available.

Keywords: PV System, Energy Performance, Monitoring, System Performance

1 INTRODUCTION

The global number of photovoltaic (PV) installations has increased steadily due to the global energy transition. Intending to stop climate change, rooftop PV systems contribute significantly to reducing CO_2 emissions in Germany by generating electrical energy [2]. Nearly 1.8 million rooftop systems had been installed by April 2021 [5]. The electrical energy generated by photovoltaic systems has more than doubled in the last ten years. In 2020, energy generation amounted to 50.7 TWh [6].

However, technical issues in photovoltaic systems can severely affect the energy production and thus the efficiency of a system [1, 3, 7, 8]. Different types of faults can cause this. One of the most common defects is a technical malfunction of the inverter. Interferences in the Photovoltaic modules occur less frequently but can also reduce yield in the long term. If a disturbance is not detected and corrected in time, the productivity and thus the profitability of a PV system is significantly affected, which is why a reliable fault detection procedure is essential [8]. Such faults can be identified using anomaly detection procedures, which determine abnormal data points within a data set. Such methods can be used in various application domains to detect rare but significant events [9]. An automated approach to detect such anomalies reduces the manual calibration effort and is adaptive. Alternatively, manual error detection methods have been used to profile regular data using statistical methods by identifying characteristics of standard data [10].

In the following, the functionality of the Isolation Forest is described in more detail. Then the methodology of the practical tasks with the acquisition and processing of the data is presented, as well as the use of the Isolation Forest and the implementation of the manual approach of Leloux et al. [4] for the generation of ground truth. The results of these two methodologies are then compared to evaluate the Isolation Forest.

2 ISOLATION FORESTS

The automated approach to fault identification followed in this paper is based on Liu's proposed Isolation Forest methodology [11]. Isolation Forests isolate anomalies primarily through iterative partitioning, described in more detail in the further proceeding, rather than first identifying normal data points via groupings and identifying anomalies as data points not meeting normal characteristics [12]. The approach of Isolation Forests also differs from other methods, such as the Local Outlier Factor, because it does not use calculations based on the distance or density of data points in any kind. Two major benefits of the Isolation Forest method are the reduced computational and memory requirements, and it shows a linear time complexity because there is no distance calculation [13]. Therefore, the approach of Isolation Forests can be efficiently applied to huge data sets and provide reliable results [11] when less information is at hand. The primary approach of this algorithm is to create several binary search trees, combine them into a forest and derive the necessary information from them. This requires training of the model, during which the tree structure dividing the data set into a binary tree is formed. Then, the binary search trees recursively partition the data points until the data points are isolated [12]. This isolation of the data points is done through a two-step process. First, to create these tree structures, a random attribute of the data is selected for each node. Then a random dividing value between the minimum and maximum value of this attribute is selected. Thereupon, the considered data is divided into two groups according to the value of this attribute to either the left or the right child node. This process is repeated until all values are found in one leaf node of the tree. The assumption is that anomalies are rarer than standard data and can be isolated quicker from the rest of the data set. This results in a shorter path length from the root node to the isolated data point. This process is illustrated in Figure 1 using a two-dimensional example.

Several randomly distributed binary search trees are generated to create an Isolation Forest, and their results are interpreted collectively. The data evaluation is then performed, passing all test instances to each Isolation Tree within the Isolation Forest. In this process, Anomaly Scores are calculated by using the average path length of each data point. The Anomaly Scores denote the degree of anomaly, with values close to 1 being interpreted as an anomaly with high confidence. Values less than 0.5 are considered regular. If all results in a data set show Anomaly Scores around 0.5, no anomalies have been identified by the Isolation Forest [11]. A key advantage of using Isolation Forests is that no anomalies need to be present in the training data set to identify them later. Furthermore, this method provides reliable results even for comparatively small data sets. Since both aspects apply to the data at hand, this method will be used in the further course [12].

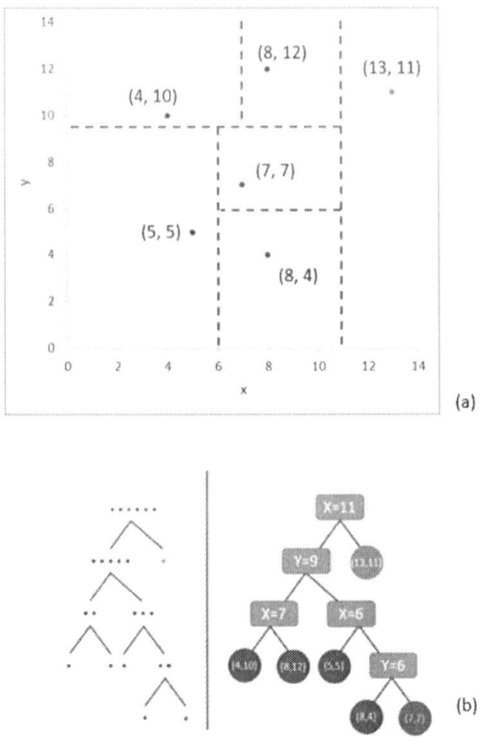

Figure 1: Schematic representation of the operation of Isolation Forest in the two-dimensional plane. Six elements are each assigned a value pair (X, Y) (a). The binary search tree (b) is used to split the instances and isolate elements (data source [12]).

3 METHODS

In the following, the implementation of the Isolation Forest used for detecting anomalies within the given data set is further described. Due to its high effectiveness and efficiency without labeling data, this publication is the method of choice. Alternative methods were only verified theoretically and are not considered nor further emphasized in this work. Finally, the results of the Isolation Forest are evaluated using a statistical approach introduced by Leloux et al. [4].

3.1 Data Acquisition and Preparation

To make a statement about anomalies that is as reliable as possible, it is necessary to compare data on as many plants as possible in the immediate vicinity. For this purpose, freely available web-based yield and master data from the remote monitoring provider Solare Datensysteme GmbH [14] were used to obtain zip code, nominal power, orientation and inclination [1]. Only data from rooftop installations that face south, have a rated power up to 30 kWp, and are in the zip code region starting with 7xxxx were considered. The database consists of over 4000 photovoltaic plants, from which the data for the exemplary comparison were obtained. For further data processing, the gathered information was normalized using the provided nominal power and aggregated from 5-minute intervals to hourly values. In this step, mismatching timestamps were adjusted. There is no information given about the functionality of a given plant, providing unlabeled data.

3.2 Data Interpretation

At this point, the prepared data is passed to the machine learning algorithm. The handling of this process depends on the tools used. For the use of the Isolation Forest process, the Python3 library Scikit-Learn is used. Since the number of trees used has a significant influence on the result, the value 100 was used, which is considered optimal according to the findings of Liu [11]. All other parameters of the model were adopted according to the default values. In Figure 2, the Isolation Forest process is represented. Here, the dataset is used for training and evaluation respectively and then assigned to the determined anomalies. After identifying the anomalous plants using the Isolation Forests, the anomalous data can be mapped to the individual systems.

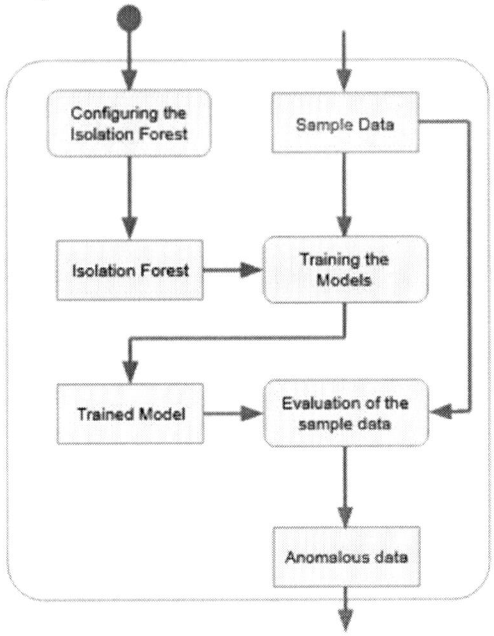

Figure 2: Activity diagram depicting the process of initializing and applying an Isolation Forest for the purpose of anomaly detection.

3.3 Ground Truth

For assessing the results of the Isolation Forest, a ground truth is required. Therefore, the approach presented by Leloux et al. [4] is followed, in which anomalies are detected by comparing normalized and weighted values of neighboring plants. This manual anomaly detection method presents a novel performance indicator - Performance to Peer (P2P) - introduced for fault detection of solar plants. This indicator is solely based on the performance comparison between the plant under consideration, referred to as the focus plant, and the surrounding systems referred to as neighboring plants. It is assumed that plants located close to each other (and thus exposed to comparable irradiation conditions), and with similar technical parameters (nominal power, orientation, inclination, etc.) produce comparable specific yields.

The P2P procedure can be divided into two phases. First, the above assumption requires assessing the neighboring plants in terms of their comparability with the plant under consideration. Therefore, the normalized yield of a neighbor is compared to the normalized yield of the focus plant. Second, this comparison is made for every neighbor within the 15 km radius proposed by Leloux and for one month to compute a weighting factor for each neighbor plant individually, according to the yield similarity.

In the second phase of the procedure, now the P2P value for evaluating the functionality of this very plant at any given time can be determined by comparing the normalized yield at that time to the yields of the neighboring plants. Therefore, a single average yield is calculated from all the eligible neighbors and the previously determined weights.

A P2P value of one indicates that the yield of that plant is comparable to its neighbors. Because of the way the P2P value is calculated, a value greater than one is possible and indicates that the yield of this plant exceeds the relative yield of its neighbors. Similarly, the P2P value can be less than one, indicating that the yield is less than the reference yield of the neighboring plants. Since this is a manual anomaly detection method, the final step is to set an appropriate threshold for the calculated P2P value to categorize the plant's performance under consideration. If the value of a plant falls below the threshold at a certain point in time, it can be assumed that a failure has occurred as the yield is not similar enough to the neighboring yields. For this purpose, Leloux presents a procedure for the dynamic determination of such a threshold value, which, however, will not be applied in the following [4].

This state-of-the-art approach is used as ground truth to verify the results of the developed machine learning approach to recognize anomalies.

4 RESULTS AND DISCUSSION

Both Leloux's [4] method and the method presented here using Isolation Forests can provide promising results to evaluate the functionality of photovoltaic systems.

For the quantitative analysis, the introduced method applying an Isolation Forest to detect anomalies in the energy production of photovoltaic plants is applied to approximately 1000 plants due to time constraints. Therefore, the plants are checked for proper operation over July 2020, resulting in roughly 31.000 values. In the course of the implementation, unique characteristics of the Isolation Forest have become apparent, which are particularly useful for determining anomalies in the available data on photovoltaic plants. Essential is a comprehensive data preparation of the raw data to exclude PV-plants with apparent misconfigurations, leading to artefacts in the data analysis. Furthermore, for comparing plants to their neighbors, the fact that Isolation Forests are well applicable on smaller data sets is relevant. This is especially relevant for regions with lower plant density.

The valuation of the Isolation Forest is compared to the assessment given by the Ground Truth. This comparison is plotted in the confusion matrix in Figure 3. In this matrix, the possible combinations of predicted values - according to the Isolation Forest - and the actual values - according to the Ground Truth by Leloux et al. - are given. In the case of categorization as True Positive or True Negative, both approaches agree for the former in the defect of the plant and for the latter in the functionality of the plant. However, in the case of categorization as False Positive, the Isolation Forest incorrectly detects an anomaly. On the other hand, in False Negative, an existing anomaly is not being recognized by the Isolation Forest.

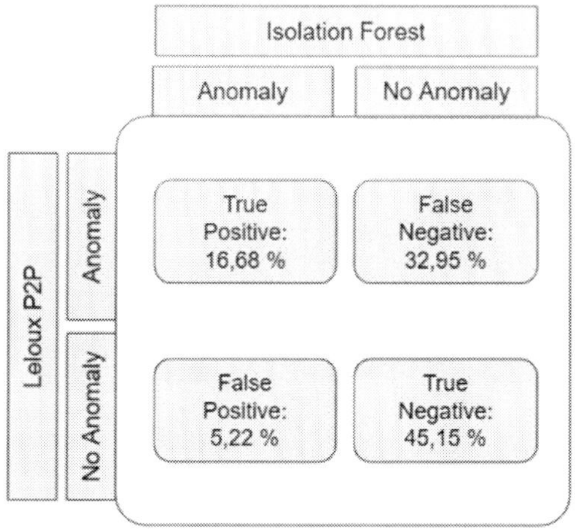

Figure 3: Confusion matrix plotting the binary classification from the comparison of the assessment by the Isolation Forest and the assessment by the Ground Truth by Leloux et al. [4].

As shown in Figure 3, the assessment of the Isolation Forest is correct for more than 60% of the plants. This includes categorizing anomalies and usual events, depicted as True Positive and True Negative in the above figure.

For a small amount of the plants - about 5 % - the Isolation Forest detects an anomaly in the plant yield while the Ground Truth is classifying it as functional, displayed as False Positive. One observed pattern of the plants categorized here was returning 0 kWh for the whole day, whereby their neighbors worked adequately. The Isolation Forest detected this as anomalous, whereas P2P categorized this behavior as non-anomalous. This is most probably due to the specific implementation of the approach by Leloux since the basic procedure aims to create a profile of standard data. Therefore, the detection of such types of errors is precisely the strength of the procedure.

Noticeable is the reasonably high number of True Negatives. In about 33% of the verifications, the Isolation Forest cannot detect an anomaly where the Ground Truth does. A first visual inspection of those plants reveals that the Isolation Forest seems to have issues categorizing strongly overperforming plants. The statistical approach by Leloux does detect these deviations as anomalous. During the manual review, it was noted that some plants overperformed by an average of 170% for most days of the month. Because the yield normalized the data basis to nominal power ratio, such performances are not realistic and must therefore be considered an anomaly. One explanation for the misinterpretation by the Isolation Forest can be the regularity of such incidents, causing it to be falsely interpreted as working ordinarily. For at least 400 plants, the interpretation of P2P can be doubted - in the manual inspection process. There was no apparent evidence for anomalous behavior.

Further, in the case of substantial underperformance of the focus plant, it is categorized as anomalous by the approach of Leloux, although most of the neighbors show comparable performance. Thus, it could be found that the assessment of the P2P can be doubted in some situations. Of course, a false-positive rate of 33% is not sustainable, but the fact that the implementation of ground truth leaves room for doubts means that the interpretation of this number can at least be developed. An evaluation of the Machine Learning approach with the help of Isolation Forests regarding the quantitative performance should therefore be done based on labeled data.

5 SUMMARY

Automated identification of technical faults in photovoltaic systems, especially roof systems with a rated power of up to 30 kWp, is used for rapid fault detection to initiate the elimination of technical faults. For this purpose, the suitability of automated anomaly identification procedures for the practical assessment of the yield quality of individual photovoltaic systems is examined. In summary, an Isolation Forest is applicable as such an automated procedure using machine learning and offers an excellent alternative to conventional manual procedures (such as Leloux's P2P procedure [4]). In contrast to the P2P approach, the application of Isolation Forests does not require prior weighting of neighboring plants and parameterization of boundaries for fault detection. Further, faster computation and less memory usage are beneficial for big data sets. Using a data set of about 1000 PV plants from southwestern Germany, it was shown that the Isolation Forest could detect technical faults and associated power production losses. Nevertheless, a misjudgment of the Isolation Forest and the implementation of ground truth create uncertainty in the assessment.

Unsupervised learning procedures do not replace labeled data since the results of these procedures have to be verified manually to allow a meaningful interpretation. Therefore, the comparison of the Isolation Forest with Leloux's statistical method can be distorted by the methodology itself and the concrete implementation, which is why such a comparison does not replace the analysis based on labeled data.

REFERENCES

[1] Schardt, Johannes; te Heesen, Henrik: Performance of roof-top PV systems in Germany from 2012 to 2019.

[2] te Heesen, Henrik; Herbort, Volker; Rumpler, Martin: Performance of roof-top PV systems in Germany from 2012 to 2018. Solar Energy, 194:128–135, 2019.

[3] te Heesen, Henrik; Herbort, Volker; Rumpler, Martin: Untersuchung des Ertrags von Photovoltaikdachanlagen bis 30 kWp in Deutschland im Zeitraum 2014 bis 2017. In: Workshops der INFORMATIK 2018-Architekturen, Prozesse, Sicherheit und Nachhaltigkeit. Köllen Druck + Verlag GmbH, 2018.

[4] Leloux, Jonathan; Narvarte, Luis; Desportes, Adrien; Trebosc, David: Performance to Peers (P2P): A benchmark approach to fault detections applied to photovoltaic system fleets. Solar Energy, 202:522–539, 2020.

[5] Marktstammdatenregister. https://www.marktstammdatenregister.de/, last access: 2021-05-14.

[6] Burger, Bruno: Nettostromerzeugnis in Deutschland im Jahr 2020. https://energy-charts.info/?l=de&c=DE, last access: 2021-04-23.

[7] te Heesen, Henrik; Herbort, Volker: Development of an algorithm to analyze the yield of photovoltaic systems. Renewable Energy, 87, 2016.

[8] Garoudja, Elyes; Harrou, Fouzi; Sun, Ying; Kara, Kamel; Chouder, Aissa; Silvestre, Santiago: Statistical fault detection in photovoltaic systems. Solar Energy, 150:485–499, 2017.

[9] Ahmed, Mohiuddin; Mahmood, Abdun Naser; Hu, Jiankun: A survey of network anomaly detection techniques. Journal of Network and Computer Applications, 60:19–31, 2016.

[10] Steenwinckel, Bram; De Paepe, Dieter; Hautte, Sander Vanden; Heyvaert, Pieter; Bentefrit, Mohamed; Moens, Pieter; Dimou, Anastasia; Van Den Bossche, Bruno; De Turck, Filip; Van Hoecke, Sofie et al.: FLAGS: A methodology for adaptive anomaly detection and root cause analysis on sensor data streams by fusing expert knowledge with machine learning. Future Generation Computer Systems, 116:30–48, 2021. Solar Energy, 217(June 2021):235–244, 2021.

[11] Liu, Fei Tony; Ting, Kai Ming; Zhou, Zhi-Hua: Isolation forest. In: 2008 eighth ieee international conference on data mining. IEEE, S. 413–422, 2008.

[12] Chen, Hansi; Ma, Hongzhan; Chu, Xuening; Xue, Deyi: Anomaly detection and critical attributes identification for products with multiple operating conditions based on isolationforest. Advanced Engineering Informatics, 46:101139, 2020.

[13] John, Hyder; Naaz, Sameena: Credit card fraud detection using local outlier factor and isolation forest. Int. J. Comput. Sci. Eng., 7(4):1060–1064, 2019.

[14] Solare Datensysteme GmbH: Energy Management System Solar-Log. https://www.solar-log.com/, last access: 2021-05-14.

EFFECTS OF THE BIGGEST SNOWFALL OF A CENTURY IN MADRID, ON THE ELECTRICITY GENERATION OF PORTRAIT VERSUS LANDSCAPE LAYOUT SOLAR PANELS

M.A. Munoz-García[1], G.P. Moreda[1], M.C. Alonso-García[2]
[1] Dep. Ingeniería Agroforestal. ETSIAAB, Universidad Politécnica de Madrid (UPM). Research group LPF TAGRALIA
Av. Puerta de Hierro, 2; 28040 Madrid, Spain
https://orcid.org/0000-0002-3670-8989
miguelangel.munoz@upm.es
[2] Dep. de Energía. Unidad de Energía Solar Fotovoltaica. CIEMAT. Madrid, Spain

ABSTRACT: On January 8th and 9th 2021, a historical snowfall occurred in Madrid. It caused severe mobility disruption in the city due to the lack of preparation for phenomena of such magnitude, and it affected the production of rooftop photovoltaic (PV) plants like the one discussed here. The fact that some of the PV panels in this PV plant are arranged vertically (portrait), caused the complete shading of their lower cells. This canceled the whole panel electricity production during snow-induced shading of those cells. When analyzing solar radiation, a drastic drop in production was observed.
Keywords: Rooftop; Shading; Small Grid-Connected PV systems

1 INTRODUCTION

In the last years, reduced cost of PV panels has fostered interest in PV power plants even in locations with limited annual irradiation like Scandinavian countries or Canada. Winter rainfall in these latitudes usually is in the form of snow, which poses a challenge to PV plant production, due to PV panel snow built-up layer-induced shading [1, 2]. The study of snow effect on PV plant power loss is also of interest in warmer latitudes like Madrid (Spain), which with a certain recurrence are subjected to extreme snowfalls, like the 40 cm one occurred on January 8-9, 2021 (Figure 1). The objective of this work was to compare how the accumulation of snow in low areas of the solar panels affects differently depending on whether the panel is placed horizontally or vertically and to determine how the snow affected the power and yield of a grid-connected rooftop PV plant. Therefore, it can be expected that when some cells in all rows of the same panel are shaded, the entire panel will stop producing. Whereas if only a few cells are shaded in a single row, the rest of the rows will continue to produce.

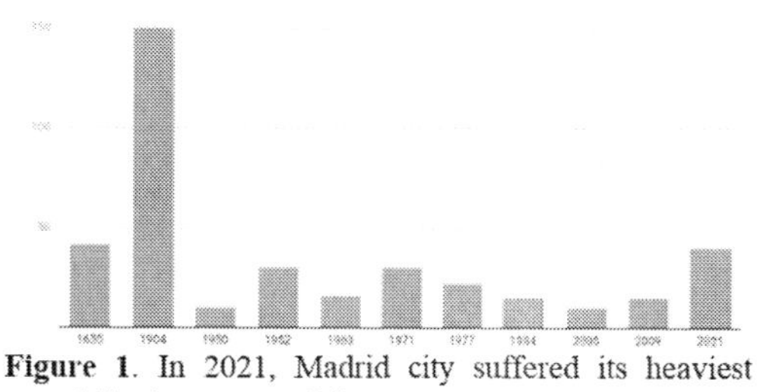

Figure 1. In 2021, Madrid city suffered its heaviest snowfall since 1904 [3]. Average accumulated snow depth in 2021 was of 40 cm, whereas in 1971 and 1952 it was of 30 cm.

2 MATERIALS AND METHODS

The School of Agricultural and Food Engineering of Madrid (ETSIAAB) hosts a 21.8 kWp grid-connected PV plant, in operation since December 2012 [4]. The plant is divided in four arrays: Three smaller (1.2 kWn each) and one larger (15 kWn), all of them with the same azimuth and tilt angle. The research reported herein refers to one of the smaller arrays and the larger array, since their modules are of the same technology (multi-crystalline Silicon) and therefore fully comparable. One layout difference between both arrays is that in the 1.2 kWn array the modules are arranged with their shorter dimension parallel to the roof plane (portrait), while the modules of the 15 kWn array are arranged with their larger dimension parallel to the roof plane (landscape), Figure 2.

Figure 2. ETSIAAB PV solar plant 5 days after the historical snowfall. Portrait layout panels still have shaded cells.

3 RESULTS AND DISCUSSION

As a preliminary conclusion, it can be determined that situations should be avoided in which all cell rows in a PV panel are shaded, even if only partially, to avoid a drastic drop in the energy produced.

Climate change could elicit snowfalls in locations where there is usually no heavy snowfall, such as Madrid city. This could block the production of electricity in photovoltaic plants. This situation will be more pronounced depending on the orientation of the panels, since depending on how many rows of cells are shaded, the production will be reduced to a greater or lesser extent.

The orientation of the panels plays a very important role in this case, depending on the connection of the cell. In our solar plant, 15 kWn are installed in landscape layout while another 1.2 kWn part is installed using a portrait layout for the panels. This is very convenient for the comparison. Obviously, the production of the whole solar plant will be totally reduced when the snow completely shades the panel. Figures 3 to 6 show how in the first days the entire plant suffers a drastic fall with the typical partial shading effect, which disappears even throughout the day.

The recovery of production was faster in panels that are oriented horizontally, since only the lower cells were affected as the snow melts and therefore the panel started producing again sooner. This can be very relevant in production and high-power plants. This also will be important if the snow load is high and may cause cracks in solar panels.

Figure 3 shows how portrait array power loss was still huge along the first sunny day after snowfall. Nonetheless, landscape array power (red bars) began to recover, with a certain delay, as the daylight hours elapsed.

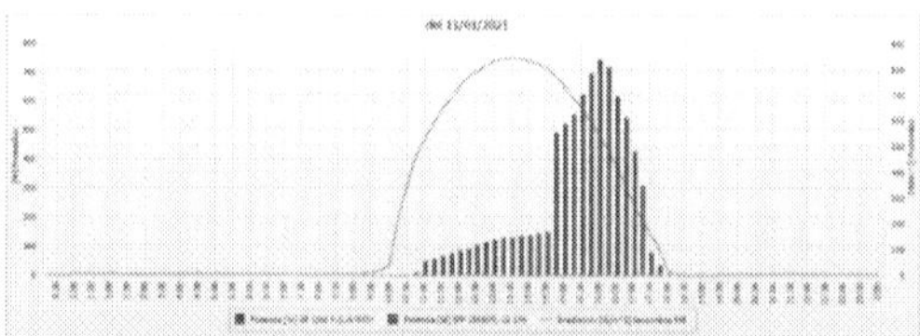

Figure 3. First sunny day after the snowfall. The portrait panels production (dark blue bars) is negligible.

Figure 3, Figure 4, Figure 5, and Figure 6, suggests that the panel arrangement influences snow-induced PV power loss due to partial shading. Like soiling, snow layer built-up on a PV panel reduces electricity production. A singularity of snow layer shading is that due to snow downward sliding, the bottom part of the panel remains shaded longer. Most PV panel manufacturers arrange panel substrings lengthwise, i.e., parallel to the panel longer side. For a 60-cell panel this entails three substrings of 20-cells each. Since an entire substring is bypassed when one of its cells is shaded [5], power loss is higher for portrait panels.

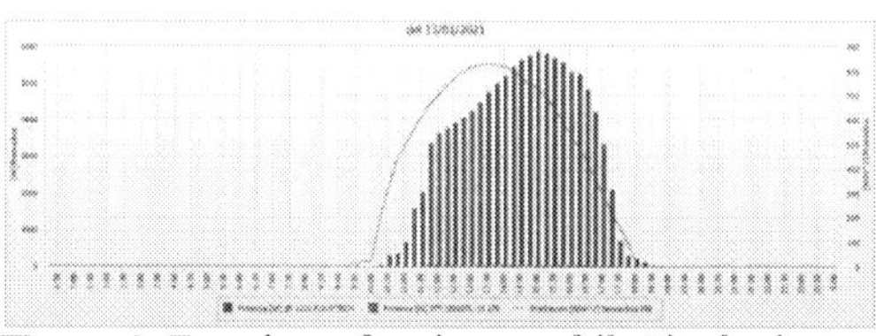

Figure 4. Two days after the snowfall: The landscape layout panels (red bars) increase their production, while portrait remains negligible. Note left-Y-axis values are higher than Figure 3 counterpart.

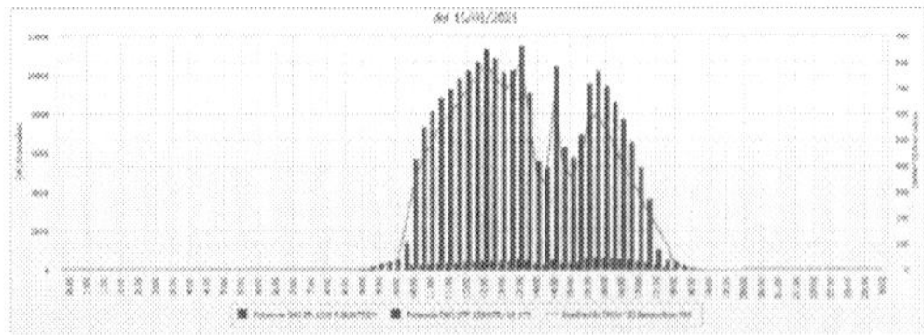

Figure 5. Five days after the snowfall. The landscape panels have practically returned to their normal winter production, while the production from the portrait panels is recovering.

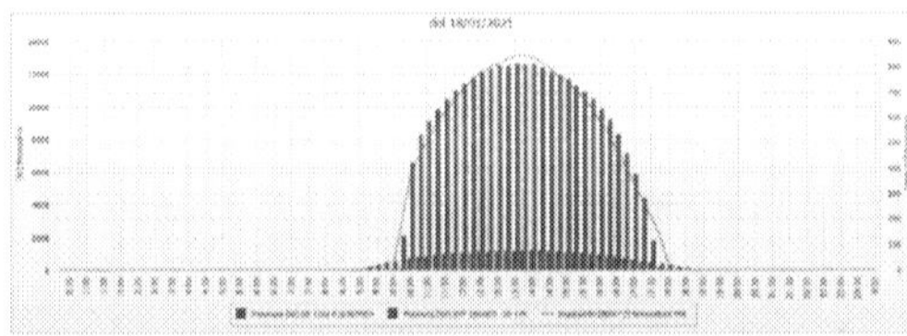

Figure 6. Eight days after the snowfall. Production of both portrait and landscape panels has returned to its normal winter level.

In a landscape PV panel, the whole top substrings become unshaded as the snow slides. The two physical mechanisms that govern snow layer clearance are sliding and melting [6]. Sliding in turn can be affected by PV panel bottom clearance to the roof surface or concrete slab ballasts. Finally, if the snow load is high enough, it could cause cracks in the shaded cells, due to hot-spot effect. This will be studied in the near future analyzing the production and probably by means of electroluminescence tests.

4 CONCLUSIONS

Given that in the future, due to climate change, extreme events such as heavy snowfall could be more likely, the study of the orientation of the panels and the effect on production will be of greater interest. This study will make it possible to determine not only aspects related to orientation but also to the elevation at which the panels should be placed based on new snow forecasts.

- Analysis of solar plant power loss due to snow-induced partial shading is of interest.

- PV panel arrangement in which one or more cell of each panel substring remains partially shaded by built-up snow entail remarkable power loss.

References

[1] Pawluk, R.E. ; Chen, Y. ; She, Y. 2019. Photovoltaic electricity generation loss due to snow- A literature review on influence factors, estimation, and mitigation. *Ren. Sust. En. Rev.*, 107 : 171-2.

[2] van Noord, M.; Landelius, T.; Andersson, S. 2021. Snow-induced PV loss modeling using production-data inferred PV system models. *Energies*, 14: 1574.

[3] Belled, S.I. 2021. De la media vara de 1655 al medio metro de 2021. Available at: https://www.diariovasco.com/sociedad/nevada-historica-madrid-20210109143215-ntrc.html (Last accessed: 8/8/2021)
[4] Muñoz-García, M.A.; Moreda, G.P.; Valle-García, A.; Novillo, J.; Alonso-García, M.C.; Chenlo, F. 2012. Grid-connected roof solar plants in educational institutions for testing and electricity generation: a case in Spain.. 27th EUPVSEC, Sept.-2012, Frankfurt, Germany.
[5] Bauwens, P.; Doutreloigne, J. 2016. NMOS-based integrated modular bypass for use in solar systems (NIMBUS): Intelligent bypass foe reducing partial shading power loss in solar panel applications. *Energies*, 9: 450.
[6] Øgaard, M.B.; Aarseth, B.L.; Skomedal, A.F.; Riise, H.N.; Sartori, S.; Selj, J.H. 2021. Identifying snow in photovoltaic monitoring data for improved snow loss modeling and snow detection. *Solar Energy*, 223: 238-7.

SOILING AND SNOW IMPACT ON A PV PLANT AT A FARM IN NORWAY

Heine Nygard Riise, Mari Øgaard and Tine Uberg Nærland
Solar Power Systems Department, Institute for Energy Technology
2007 Kjeller, Norway

Soiling of solar cell modules may impact the energy yield of a PV installation significantly. Although soiling losses are frequently reported in regions known for high soiling, the soiling impact on PV production in regions where soiling losses are expected to be limited, such as Norway, is rarely reported. On the other hand, point emission sources can increase the soiling at a site, and in this work, agricultural activity at a farm in Norway is correlated with the soiling levels experienced by a PV plant on the farmhouse roof. Soiling rates are determined from on-site measurements and estimated through analyzing the performance of the plant. Additionally, the energy loss from soiling and snow, respectively, are determined. The energy loss from soiling is quantified to 3.8 kWh/kW$_p$/yr (~0.41 % of annual production), pointing to negligible soiling loss in Norway even in the presence of a point emission source. Energy loss from snow is estimated to 20.2 kWh/kW$_p$/yr (~2.2 % of annual production), indicating that snow loss is more a more significant loss factor.

Keywords: Soiling, Snow, Performance

1 INTRODUCTION

Dirty surfaces cause the sun's rays to be blocked or scattered, so that light absorption of the solar module is entirely or partially disturbed. This is commonly termed soiling. Reports on soiling losses are frequently reported in e.g. desert areas [1], but soiling impact on PV production in the Nordic countries is rarely reported [2-4]. Relatively frequent rainfalls are typical for Eastern Norway, and thus a high number of cleaning events from precipitation is expected. Although the soiling level in Norway is expected to be low, the influence of local conditions may enhance the soiling and generate a substantial impact on energy yield [5]. Such local conditions may be pollution, industrial emission, natural phenomenons such as pollen, or, as is the case of this study, agriculture activity. Agriculture has the potential to generate high levels of soiling from e.g. plowing, harvesting, use of heavy machinery near the PV modules, or increased levels of pollen.

In this study, 57 kW$_p$ rooftop PV system located on a farmhouse in Eastern Norway is studied. The PV system, as well as the agricultural activity, was monitored over a period of 2.5 years. The Soiling Ratio (SR) measured on site by a Kipp&Zonen DustIQ is found to be high (a high SR signifies that little or no soiling is present), indicating that soiling is usually not influencing the energy production. This conclusion is supported by looking at the temperature corrected Performance Ratio (PR$_{STC}$) of the PV system. Four periods of soiling are, however, identified in both the production data and measured SR: April 2019, April 2020, May 2020, and May 2021. In addition to these four periods of soiling, September 2019 is also identified as a soiling period using the data from the DustIQ, but it does not noticeably influence the production data even though dry weather is also prevalent during this soiling period. Conversely, three periods of soiling are detected using PR$_{STC}$ which are not discernible in the DustIQ data.

Correlating the soiling estimates from the DustIQ with soiling estimates from production data is performed using a method developed for utility-scale parks [6]. The method filters and normalizes PR$_{STC}$, and soiling rates are extracted robustly using Theil Sen regression. For most of the identified soiling periods, the soiling rate estimation using PR$_{STC}$ is well correlated with the DustIQ measurements, displaying the applicability of soiling estimation from production data and increasing the confidence in the DustIQ measurements.

Further, using SR extracted from DustIQ measurements, the energy loss from soiling is quantified to 3.8 kWh/kW$_p$/yr, corresponding to 0.4 % of the average annual production. This is a remarkably low number and indicates that soiling is a negligible factor in the energy yield of PV plants in the Nordics. On the other hand, the impact of snow on the modules is more than 5 times that with 20.2 kWh/kW$_p$/yr (~2 % of annual production), showing that estimating and modelling snow losses is of higher importance for accurate energy yield assessments than soiling loss estimation in snow-rich locations.

Although snow coverage may be considered as a type of soiling, it is here considered as a separate effect from dust accumulation. Hence, in the following, the soiling term is equated to dust accumulation.

2 EXPERIMENTAL DETAILS

2.1 Description of the system

The PV-system is installed on the roof of a farmhouse belonging to the Norwegian Institute of Bioeconomy Research (NIBIO) research facility in Kapp, Norway (60°42'03''N 10°52'04''E). The system consists of 216 south-facing modules from IBC Solar (PolySol 265 W$_p$ CS4) installed in portrait orientation at 46° tilt, and 3 Fronius Symo 17.5-3-M inverters. The modules are not actively cleaned. The analysis period runs from April 2019 to July 2021. The period from July 16 2020-September 30 2020 suffers from a communication error, and has been left out of the analysis.

The horizontal and in-plane irradiance is measured by thermopile pyranometers from Kipp &Zonen. Module temperature measurements are performed using PT100-element, mounted on the backside of the modules. Soiling levels are recorded using a Kipp&Zonen DustIQ soiling monitoring system, while precipitation data are collected from a nearby meteorological station.

2.2 Performance analysis methodology

There are several ways to determine the performance of a PV plant [7]. The commonality between the methods is an attempt to normalize the energy production for a given time interval. Among the simplest ways of normalization, is calculating the specific yield of the system. Specific yield compares the energy production with the installed capacity of the system. Although specific yield enables yield comparison between systems of different size, it does not consider variations in irradiance and temperature. Irradiance and temperature greatly affect the energy production of a PV plant and will vary from plant to plant and from year to year. For a fair evaluation of PV performance, irradiance and temperature corrections are hence imperative, e.g. by specifying PR_{STC} [8]:

$$PR_{STC} = \frac{\sum_k P_k \times \tau_k}{P_0} / \frac{\sum_k C_k \times G_{i,k} \times \tau_k}{G_{i,ref}}$$

$$C_k = \left(1 + \gamma \times \left(T_{mod,k} - 25°C\right)\right)$$

where P_k is the measured AC-power of the system in time interval k, τ_k is the time interval, P_0 is installed capacity, $G_{i,k}$ is measured in-plane irradiance, $T_{mod,k}$ is module temperature, γ is the relative maximum-power temperature coefficient, and $G_{i,ref}$ is the reference irradiance (1000 W/m^2).

In this study, PR_{STC} is used to estimate the soiling level of the modules using the methodology outlined by Skomedal et al. [6]. The method assumes that short term variations and extreme values of PR_{STC} is not related to soiling, and they are consequently filtered out. In addition, periods where the modules are covered by snow are also disregarded in the analysis. Identifying snow cover is further elaborated in section 2.3. The remaining PR_{STC} data are normalized to a clean state, $PR_{STC,0}$ (defined as the 95th percentile of PR_{STC}), to generate PR^*_{STC} (here, the asterisk indicates that PR_{STC} is both normalized and filtered):

$$PR^*_{STC}(t) \cong \frac{PR_{STC}(t)}{PR_{STC,0}}$$

Henceforward, the filtered and normalized PR_{STC} is used to derive a soiling rate from the production data by first identifying cleaning events from rapid shifts in the rolling median of PR^*_{STC}, and subsequently fitting a Theil Sen-estimator to the data between cleaning events.
The soiling rate is also extracted from DustIQ measurements by manually identifying periods of soiling and fitting a Theil-Sen estimator to the SR in these periods.

Evaluating the energy loss from soiling and snow requires defining a clean and snow-free state of the system [9]. The clean and snow-free state is realized through a PVWatts model with POA and module temperature as inputs. The empirical model is optimized towards manually selected clean and snow-free production data, yielding a constant system loss of 8 % which is applied to the modeled AC power.

2.3 Filtering methodology

Data filtering is of utmost importance when analyzing the performance of PV plants, and changing the filtering procedure can have significant impact on end results and conclusions [10]. Commonly used filters in literature such as clear sky filters and irradiance thresholding was found to be too crude for this study, as the PV system rarely experiences prolonged periods of clear sky conditions and high irradiance days (hence too little data is left to robustly analyze the performance). Angle of incidence filters did not impact the results and were not used. Snow filters, however, is of high importance when considering the performance of PV plants in snow rich areas. Identifying snow in PV production data can be challenging, and several strategies are employed [9]. Typically, some sort of weather data is required. Here, we utilize both snow depths from interpolated observational data collected from senorge.no [11], and DustIQ measurements. The modules are considered covered by snow when there is snow on the ground, and the DustIQ SR is below 95 %. Using these filters, the energy loss from soiling may be separated from the energy loss from snow.

3 SOILING IMPACT ON PERFORMANCE

3.1 Correlating soiling, precipitation and farm work

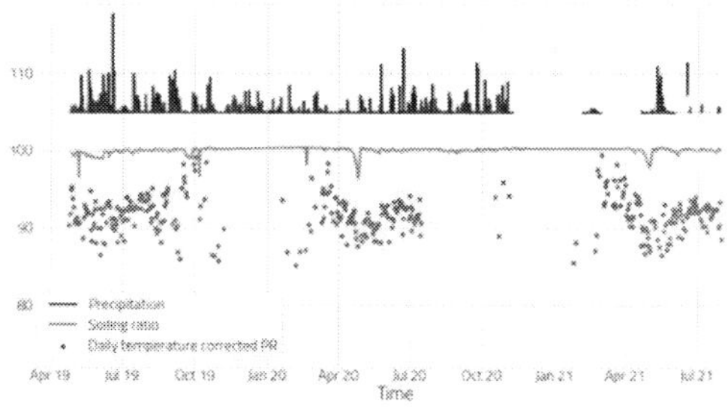

Figure 1: Measured SR, PR_{STC}, and precipitation at the site. For at least four of the increased soiling periods, low rainfalls are observed

In Figure 1 the SR for the site is seen as the golden line around 100 %. The figure also displays the filtered PR_{STC} of the system (red dots). Note that all times with snow on the modules is filtered out. At least five periods of increased levels of soiling can be seen from the SR (May 2019, September 2019, April 2020, May 2020, and May 2021). For four of these periods (September 2019, April 2020, May 2020, and May 2021), precipitation is scarce giving little or no cleaning of the modules, yielding a gradual build-up of particles on the surface of the modules. The spring periods correspond to increased outdoor agricultural activity with heavy machinery and pollen season in Norway, posing a plausible explanation of the increased soiling levels at those times. Noticeably, the soiling period in May 2019, has a significantly lower soiling rate than the spring periods in 2020 and 2021, possibly due to relatively frequent and heavy rainfall throughout May 2019. Autumn is normally also a period of increased heavy machinery activity, and evidence of this can be seen in September 2019. Unfortunately, a communication error prevents analysis of the soiling signal in autumn 2020.

3.2 Energy loss from soiling and snow

From Figure 1, it can be deduced that the soiling will impact the energy production. To estimate this impact, the daily absolute soiling loss is determined from:

$$soiling\ loss(t) = ((SR(t) - SR_0)/SR_0 \times E_{est,daily}(t)$$

Where SR_0 is defined as a clean soiling level (the 95th percentile of SR), and $E_{est,\ daily}(t)$ is the daily modeled energy production. This assumes that the optical loss that the DustIQ experiences is equal to the production loss experienced by the modules. That might not be the case at all times, but this approach is chosen in order to decouple soiling losses from other system losses, as well as eliminate artefacts due to poor data integrity in the production data. Figure 2 displays the energy loss for both soiling and snow. Especially the period in May 2020 contributes to a significant daily soiling loss since the period has both high levels of soiling and sunny days with high production. Aggregating the values yields a total energy loss from soiling of 497 kWh, corresponding to 8.7 kWh/kW$_p$. On average, the yearly soiling loss is then 3.8 kWh/kW$_p$/yr or 0.41 % of the total annual production. This minuscule value demonstrates that soiling losses in Norway are negligible, even in areas where the point emission of dust and other soiling agents might be high, such as farms.

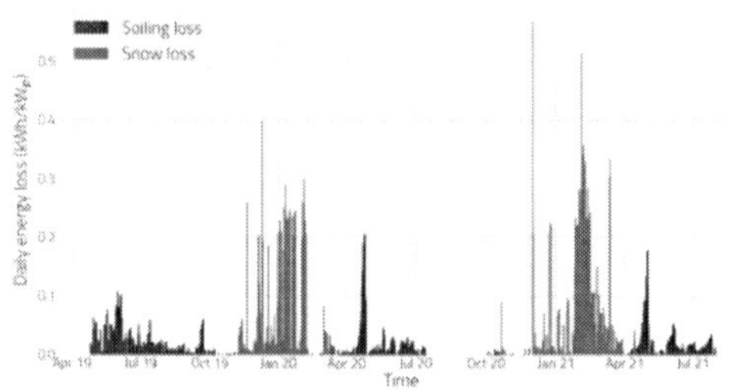

Figure 2: Estimated daily energy loss due to soiling and snow

Inspecting Figure 2 reveals that snow loss, on the other hand, has a larger impact on the energy production of the plant. In principle, the snow loss can be estimated in a similar manner as the soiling loss, but the DustIQ optical loss during snow cover is found to be erratic and not representative of the production loss of the modules. The snow loss is therefore quantified by comparing production data with modeled data during snow-covered periods:

$$snow\ loss(t) = E_{est,daily}(t) - E_{daily}(t)$$

Where and $E_{est,\ daily}(t)$ is the daily modeled energy production, and $E_{daily}(t)$ is the daily measured energy production. Applying this approach yields a snow loss of 2640 kWh, more than 5 times that of soiling. Yearly average snow loss is then 20.2 kWh/kW$_p$/yr or 2.2 % of the annual production. Adequately estimating the energy yield for such a site hence demands higher attention to the snow loss rather than the soiling loss. That is not always the case when evaluating loss mechanisms, where soiling loss generally is more investigated and understood than snow loss. This is manifested in the scientific literature where soiling related articles is 9 times more numerous than snow in PV related articles [12]. Especially in snow-rich conditions, estimating and modelling snow losses is important for accurate energy yield assessments. Estimating snow losses are, however, challenging. As an example, this PV plant experiences heavy shading during low solar elevations due to a tree in front of the farmhouse. The loss from this shading is difficult to separate from the loss from partial snow cover since they have similar impacts on the production. The snow loss of 2.2 % might hence be higher than the actual snow loss since it can't be ruled out that part of the loss originates from partial shading. Further, the modeled production may also be impacted when the pyranometers are covered in snow, giving a lower estimated loss than the actual loss.

4 SOILING RATE ESTIMATION

Quantifying the expected soiling rates in a location is important for giving an accurate estimate of the future energy yield of an installation. Direct measurements of SR using dedicated equipment can be useful, but to correlate the SR observed from the equipment to the actual soiling that the modules experience can be challenging.

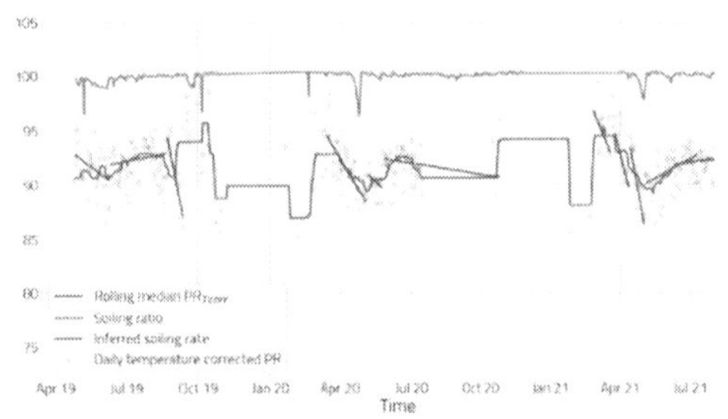

Figure 3: Measured SR, PR_{STC}, rolling median of PR_{STC}, and soiling rates inferred from PR_{STC}. A negative inferred soiling rate is found in seven separate periods, and four of these correspond to periods of declining measured SR

Figure 1 exemplifies this challenge. The filtered PR_{STC} (red dots) has large noise levels, and to see a clear trend relating to soiling is challenging. By looking at Figure 3, however, it is seen that the robust method of soiling rate determination from production data described in the paragraph on performance analysis methodology, is able to extract relevant information extracting seven periods of soiling At least four of these periods correspond to periods where the measured SR is declining, i.e. the soiling period is also identified by the DustIQ. Hence, the production data strengthens the confidence in the measured SR: It shows that the soiling that the modules experience, is correlated to the soiling measured by the DustIQ. The soiling rates are displayed in Table I. The rates extracted from the DustIQ (through simple linear regresson) is in general lower than the inferred soiling rate from production data. This indicates that measured SR, using dedicated equipment gives insight into the actual soiling that the PV system experiences, but that a calibration towards the specific system is required. Although the estimated soiling rates from PR_{STC} use actual production data, and as such, is probably closer to the true soiling rate, the filtering and normalization procedure might introduce a bias in PR^*_{STC}

which in turn alters the soiling rates extracted from production data. It is beyond the scope of this analysis to evaluate which of the soiling rates is the true soiling rate the PV system experiences.

Table I: Soiling rates estimated from DustIQ and PR_{STC}

	Estimated from Dust IQ	Estimated from PR_{STC}
May 2019	0.039 %/day	0.056 %/day
April 2020	0.32 %/day	0.13 %/day
May 2020	0.027 %/day	0.067 %/day
May 2021	0.30 %/day	0.41 %/day

5 CONCLUSIONS

This paper has investigated the impact of soiling on a BAPV plant at a farm in Kapp, Norway. The periods with high soiling levels correlate with agricultural activity with heavy machinery and prolonged periods of dry weather, except for May 2019 where rainfalls are frequent. The soiling rate of these periods are evaluated using (1) Kipp&Zonen DustIQ measurements and (2) a normalized and filtered daily PR_{STC}. It is found that the soiling rates from the two methods correlate, displaying the applicability of soiling estimation from production data and increasing the confidence of the DustIQ measurements.

The energy loss from both soiling and snow is quantified. The average annual soiling losses for the system is 3.8 kWh/kW$_p$/yr, corresponding to 0.41 % of the average annual production of the plant. For this location in Norway, soiling is hence found to have a negligible impact on the total energy yield even in the presence of a point emission source (a farm). Snow, on the other hand, gives an energy loss more than 5 times higher than that of soiling. Even for a BAPV system with high tilt, the impact of snow on the energy production is 20.2 kWh/kW$_p$/yr, corresponding to 2.2 % of the average annual production. This demonstrates that for snow-rich locations, accurately estimating and modelling snow losses is of higher importance for energy yield assessments than soiling loss estimation.

REFERENCES

[1] M. R. Maghami, H. Hizam, C. Gomes, M. A. Radzi, M. I. Rezadad, and S. Hajighorbani, "Power loss due to soiling on solar panel: A review," *Renewable and Sustainable Energy Reviews*, vol. 59, pp. 1307-1316, 2016/06/01/ 2016, doi: https://doi.org/10.1016/j.rser.2016.01.044.

[2] H. B. Pedersen, "Experimental study of soiling on photovoltaic modules in a Nordic climate," Master's thesis, Norwegian University of Life Sciences, 2015.

[3] B. R. Paudyal and A. G. Imenes, "Performance assessment of field deployed multi-crystalline PV modules in Nordic conditions," in *2019 IEEE 46th Photovoltaic Specialists Conference (PVSC)*, 16-21 June 2019 2019, pp. 1377-1383, doi: 10.1109/PVSC40753.2019.8980629.

[4] L. T. Milde, "A study of the performance of two wall attached PV systems in a Nordic climate," Master's thesis, Norwegian University of Life Sciences, 2018.

[5] L. Micheli and M. Muller, "An investigation of the key parameters for predicting PV soiling losses," *Progress in Photovoltaics: Research and Applications*, https://doi.org/10.1002/pip.2860 vol. 25, no. 4, pp. 291-307, 2017/04/01 2017, doi: https://doi.org/10.1002/pip.2860.

[6] Å. Skomedal, H. Haug, and E. S. Marstein, "Endogenous Soiling Rate Determination and Detection of Cleaning Events in Utility-Scale PV Plants," *IEEE Journal of Photovoltaics* vol. 9, p. 858, 2019.

[7] G. Belluardo, P. Ingenhoven, W. Sparber, J. Wagner, P. Weihs, and D. Moser, "Novel method for the improvement in the evaluation of outdoor performance loss rate in different PV technologies and comparison with two other methods," *Solar Energy*, vol. 117, pp. 139-152, 2015/07/01/ 2015, doi: https://doi.org/10.1016/j.solener.2015.04.030.

[8] *Photovoltaic System Performance*, International Electrotechnic Committee, 2017.

[9] M. B. Øgaard, B. L. Aarseth, Å. F. Skomedal, H. N. Riise, S. Sartori, and J. H. Selj, "Identifying snow in photovoltaic monitoring data for improved snow loss modeling and snow detection," *Solar Energy*, vol. 223, pp. 238-247, 2021/07/15/ 2021, doi: https://doi.org/10.1016/j.solener.2021.05.023.

[10] D. C. Jordan and S. R. Kurtz, "The Dark Horse of Evaluating Long-Term Field Performance—Data Filtering," *IEEE Journal of Photovoltaics*, vol. 4, p. 317, 2013.

[11] NVE. "senorge.no." (accessed 15.07.2021).

[12] "webofknowledge.com." (accessed 08.09.2021).

BIFACIAL GAIN FOR 1P AND 2P TRACKERS

P. Merodio*, F. Martínez-Moreno, E. Lorenzo

Instituto de Energía Solar – Universidad Politécnica de Madrid. Grupo de Sistemas Fotovoltaicos (IES-UPM).
Campus Sur UPM. Ctra. Valencia km. 7. ETSI Sistemas y Telecomunicación. 28031 Madrid, Spain.
*Corresponding author: pablo.merodio@upm.es

ABSTRACT: We investigate the energy gain driven by bifaciality in PV plants designed for trackers in 1P (Portrait) and 2P configurations. For that purpose, three elements have been experimentally analyzed: 1) The rear shading caused by the tracker's mounting system (torque tube and panel rails). This is characterized by the so-called Structure Shading Factor, *SSF*, defined as the share of rear irradiance that is obstructed by the structure -an input parameter required by PV software (SISIFO, PVsyst...) for energy yield estimations-. 2) The mismatch losses on the electric power delivered by the PV array due to the non-uniformity of the rear irradiance. This non-uniformity is partly due to the nature of the radiation reflected by the ground and partly to the shading caused by the tracking structure. Mismatch loses are characterized by a factor, *MML*, which quantifies either the total power reduction delivered by the PV array or just the power reduction corresponding to the rear irradiance. 3) The solar cell operation temperature. We paid especial attention to analyze the temperature differences between East and West module rows in the 2P tracker, compared to the temperature of the modules in the 1P tracker.
Keywords: Bifacial, Structure Shading Factor, Mismatch Loss Factor

1 INTRODUCTION

The rapid growth of the market share for bifacial PV modules calls for a systematic uncertainty reduction in bifacial PV energy estimations. As bifacial modules are able to harness rear side solar irradiation incident on the back surface, the resulting increase in energy yield is significantly more complex to predict. One of the main reasons is that rear irradiance is characterized by a much greater spatial non-uniformity than frontal irradiance. An important parameter related to this non-uniformity is the so-called Structure Shading Factor, *SSF*, defined as the share of rear irradiance that is obstructed by the structure where bifacial modules are fixed. It constitutes an input parameter required by PV simulation software (SISIFO, PVsyst...) for energy yield estimations. In turn, this non-uniformity translates into mismatch losses on the electric power delivered by the PV array. Mismatch loses are characterized by a factor, *MML*, which quantifies either the total power reduction delivered by the PV array or just the power reduction corresponding to the rear irradiance.

The state-of-art regarding the estimation of *SSF* and MML is mainly supported by modelling. Typically, ray-tracing software is employed for calculating *SSF* and describing the spatial distribution of the rear irradiance at the cell-level. Then electronic circuit simulation software is used for modelling the PV module, also at a cell-level, which allows translating such irradiance spatial distribution into generated current. Comparing the results of different scenarios, with and without obstructions, that is, with and without rear irradiance distribution non-uniformity, allows for calculating *MML*.

The approach followed here is however supported by experiments. Dedicated instrumentation to examine *SSF*, *MML* and cell operating temperature for 1P and 2P trackers has been, first, constructed. Then, an experimental campaign was carried out between 20/07/2020 and 28/07/2020 in the PVH testing PV plant at Cheste (Valencia, Spain). The plant includes 1P and 2P trackers equipped with 355 Wp bifacial Longi Solar LR6-72BP-355M modules composed by 72 cells. The ground is gravel-covered, with reflectivity (or albedo) close to 30%.

2 STRUCTURE SHADING FACTOR

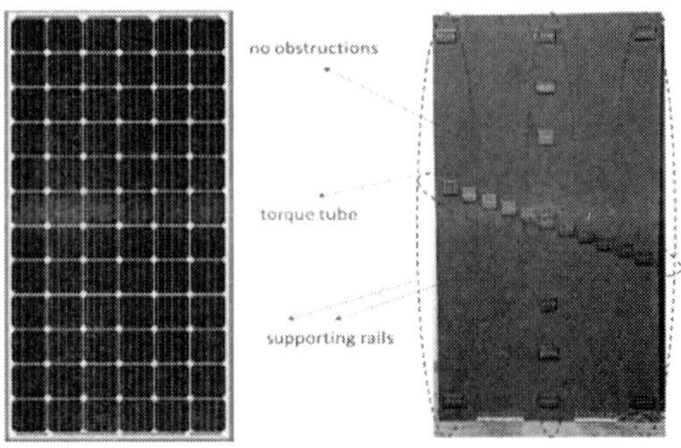

Figure 1: **(Left)** A PV module like those installed at Cheste. **(Right)** The sensor developed by IES-UPM for observing the spatial distribution of the rear irradiance

2.1 Instrumentation

Figure 1 shows the instrument developed by IES-UPM for observing the spatial distribution of the rear irradiance, G_{rear}. It consists of a wooden board with the same dimensions as a PV module (2m × 1m) on which 21 solar cells, with a surface area of 16 cm^2, have been installed in strategic positions. Each cell is individually short-circuited by means of a 0.5 or 1.5 Ω resistor, depending on the cell performance. This allows simultaneous G_{rear} measuring at all the points where the cells are located. The board is matt black painted, to avoid reflections. The 11 cells at the middle of the board are staggered to increase the resolution in the region affected by the torque tube at 1P trackers, where the variation of G_{rear} is greater. After a convenient Sun-exposure to induce Light-Induced Degradation (LID), all the 21 cells have been calibrated at the same time, directly exposing the device to the Sun, together with an external reference cell. Figure 2 shows the instrument installed in the PVH trackers at Cheste. Combining the 21 G_{rear} values measured by the cells with translational symmetry considerations inherent to the one axis horizontal trackers, allows to determine G_{rear} at all the 72 solar cell's locations on the PV module.

Figure 2: The instrument for observing the spacial distribution of irradiance installed on a 1P tracker (**left**) and 2P tracker (**right**). The main causes of shading are, respectively, the torque tube and the panel rails

2.2 *SSF* determination

For the sake of example, Figure 3 shows the G_{rear} maps deduced from measurements the 23/07/2020 at 14:12 in the middle of a 1P tracker. At the left, the map as directly deduced from measurements, i.e., including the shading from the torque tube. At the right, this shading has been (mathematically) eliminated. The surface-integration of G_{rear} (W/m^2) shown in these maps allows to calculate the corresponding radiant flux (or power, W) received by the back surface of the PV module, $\phi_E^{\text{shade}} = \iint G_{REAR}^{shade} dS$ and $\phi_E^{\text{no-shade}} = \iint G_{REAR}^{no-shade} dS$, respectively. The *SSF* is then given by:

$$SSF = \frac{\phi_E^{\text{no-shade}} - \phi_E^{\text{shade}}}{\phi_E^{\text{no-shade}}} \quad (1)$$

The resulting value for this example is *SSF* = 9.1%. It is worth noting that, as both ϕ_E^{shade} and $\phi_E^{\text{no-shade}}$, are linearly dependent on the albedo of the ground, *SSF* becomes independent of it.

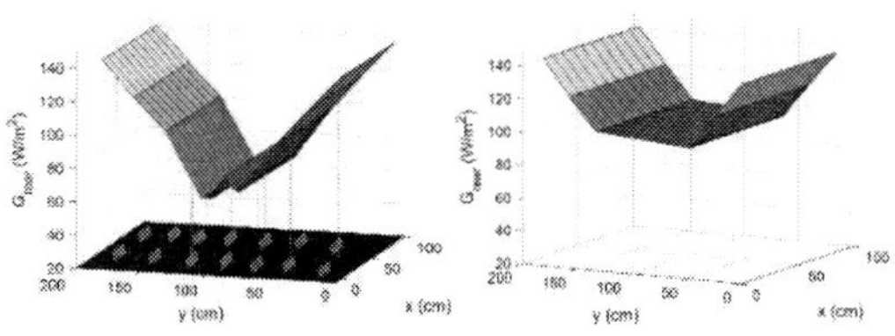

Figure 3: Rear irradiance maps with (**left**) and without (**right**) shading casted by the torque tube.

2.3 Experiments and results

G_{rear} maps and corresponding *SSF* were obtained in both 1P and 2P trackers at different times throughout the 23/07/2020 and 28/07/2020, respectively, both being very clear days.

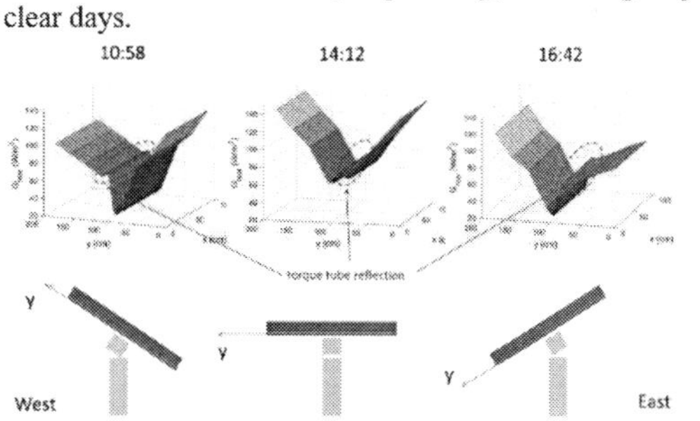

Figure 4: Rear irradiance maps (**up**) and corresponding tracker positions (**bottom**) obtained in a 1P tracker at three characteristic moments the 23/07/2020

Figure 4 shows the 1P tracker G_{rear} maps with torque tube shading obtained at three characteristic moments the 23/07/2020 (approximately two hours before noon, at noon, and two hours after noon), together with the corresponding tracker positions. The slight irradiation peaks in the vicinities of the center of the PV module (signaled with dashed-line red circles) correspond to diffuse light reflections in the torque tube, and they indicate the high sensitivity of the measuring device.

Table 1 shows the *SSF* results for nine characteristic moments throughout this day. Note that, although the region shaded by the PV array moves following the Sun's elevation, strongly modifying the ground vision from the PV module, the SSF variation is fairly small. This daily movement covers most annual positions, and that is the reason why the measured *SSF* average value is quite representative of the annual behavior of this 1P tracker. It can be argued that this value has been obtained without considering shades from pillars and other tracker components. Specific experiments to further investigate this point are already planned, but we expect negligible *SSF* variations in terms of yearly energy production.

Time	10:58	11:40	12:46	13:25	14:12
SSF (%)	10.7	10.2	8.8	8.6	9.1

14:47	15:25	16:02	16:42	**Average**
9.7	10.2	11.0	10.8	9.9

Table 1: *SSF* values measured on a 1P tracker at characteristic moments the 23/07/2020

Figure 5 shows the 2P tracker G_{rear} maps obtained at three moments the 28/07/2020. The most relevant shade is caused by the panel rails supporting the modules. The torque tube diffuse light reflection is visible again. The figure corresponds to the East row of the tracker. The G_{rear} maps for the West row can be obtained by symmetry with respect to solar noon.

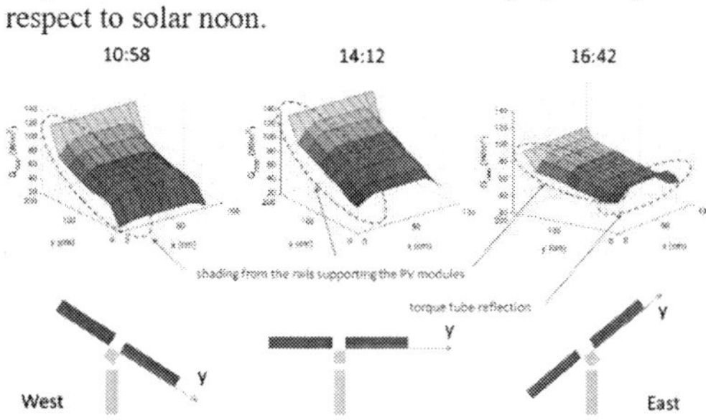

Figure 5: Rear irradiance maps (**up**) and corresponding tracker positions (**bottom**) obtained in the East row of a 2P tracker at three characteristic moments the 28/07/2020

Table 2 gives the *SSF* results for nine characteristic moments throughout this day. The aforementioned arguments supporting the representativity of these measurements regarding the annual behavior of the tracker apply also here.

Time	10:58	11:40	12:46	13:25	14:12
SSF (%)	3.58	3.62	3.62	3.62	3.68

14:47	15:25	16:02	16:42	**Average**
3.61	3.72	3.62	3.68	3.6

Table 2: *SSF* values measured at the east row of a 2 tracker at different times the 28/07/2020

3 MISMATCH LOSSES

3.1 Experimental determination

MML_T has been measured in Cheste by observing the dispersion in the instantaneous operation voltage, V_{OP}, of individual modules of a same PV array under normal operation, i.e., with the inverter at the MPP. "T" shaped quick connectors were used measure individual voltages. Figure 6 shows an illustration of the measurement procedure and a "T" connector.

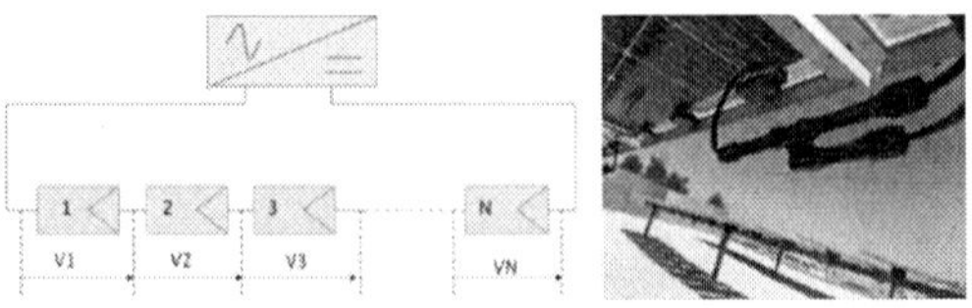

Figure 6: (**Left**) Principle of *MML* experimental observation. (**Right**) A "T" connector used for accessing the voltage of the individual modules when the PV array is injecting power to the inverter

For a particular distribution of individual voltages ${V_{OP}}^i$, the individual relative voltage deviation from the value at MPP is estimated on the assumption that the average of the distribution, $\bar{V}_{OP}$, will be the operating V_{OP} value (at MPP) of all the modules in absence of mismatch. Then, the relative voltage deviation of a module "*i*" is defined as:

$$\Delta {v_{OP}}^i = ({V_{OP}}^i - \bar{V}_{OP})/\bar{V}_{OP} \quad (2)$$

and the relative *MML* is computed as:

$$MML_T = \frac{1}{N}\sum_{i=1}^{N} abs[\Delta {v_{OP}}^i \cdot (\Delta p/\Delta v)^i] \quad (3)$$

where $\Delta p/\Delta v$ represents the ratio between the relative power loss (respect to P_{MPP}) to the relative voltage deviation (respect to V_{MPP}). That is:

$$\frac{\Delta p}{\Delta v} = \frac{(P_{V_{OP}} - P_{MPP})/P_{MPP}}{(V_{OP} - V_{MPP})/V_{MPP}} \quad (4)$$

In other words, the power loss Δp corresponding to a voltage deviation Δv from the theoretical MPP. The superscript "*i*" extends to all the N modules of the PV array. Since all the modules' I-V curves are very similar, $\Delta p/\Delta v$ is assumed to be the same for all the N modules. As an example, Figure 7 (left) shows the distribution of Δv_{OP} observed the 24/07/2020 in a PV array of 20 modules associated in series (10 in 1P, 10 in 2P, see below). The distribution is characterized by a standard deviation $\sigma_{\Delta v}$ = 1,1%. Figure 7 (right) plots Δp versus Δv, as derived from the P-V curve measured for an individual module.

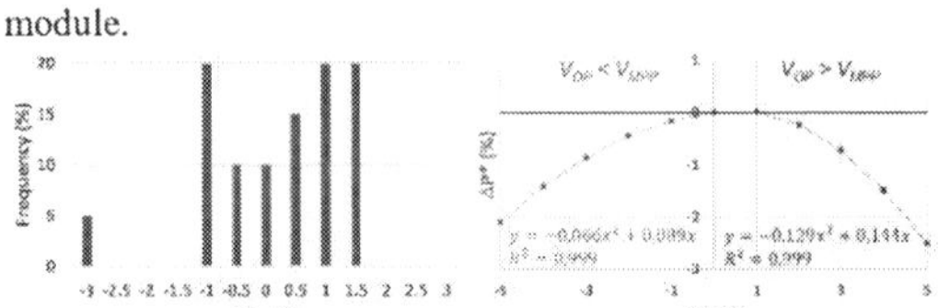

Figure 7: (**Left**) Observed distribution of the relative variations of operation voltage (with respect to the mean) in a PV array composed by 20 modules split in 1P-2P trackers (see below). (**Right**) Relative power versus voltage deviations near the MPP of the PV module

Note that, because Δp and Δv are defined in relative terms, the function Δp versus Δv is an invariant with respect to the operation conditions, insofar the Fill Factor variations can be neglected. Near the MPP, this function is properly described by a second-degree polynomial:

$$\Delta p = a\Delta v + b\Delta v^2 \quad (5)$$

where a and b are dimensionless parameters fitted to the observed curve. In our case, with Δp and Δv in %, a = 0.144(0.089) and b = -0.129(-0.06) for Δv > 0(<0). Resulting losses for this example are captured by a MML_T = 0.04%. It is worth noting that MML_T is much lower than $\sigma_{\Delta v}$.

Figure 8 (top) shows the evolution of MML_T throughout the day. Outliers are due to transitory phenomena (clouds, shadings, etc.). Apart from that, MML_T is relatively constant. The slight oscillation is likely due to a combination of module temperature oscillations and inverter MPP searching. Figure 8 (bottom) shows the frequency distribution of the voltage deviations observed throughout two different days. They are very similar, which suggest that *MML* remains constant along the year. This agrees with long term *MML* observations in existing PV generators at the IES-UPM facilities.

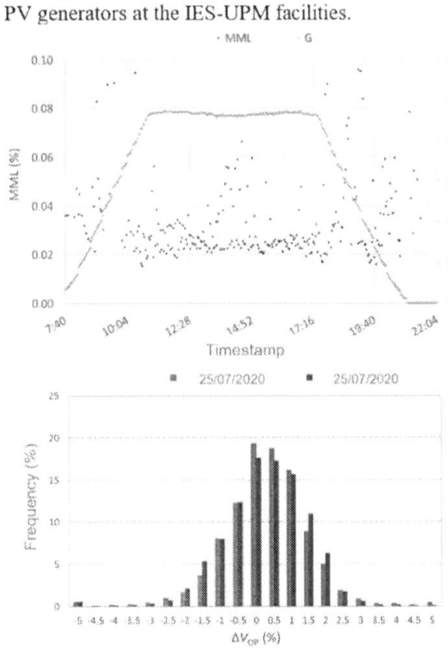

Figure 8: (**Top**) Evolution of MML_T in a 1P tracker throughout the 26/07/2020. (**Bottom**) Frequency distribution of voltage deviations with respect to average along the 24 and 25 July

Mismatch energy losses for a given period are calculated as the *MML* average weighted by the respective G_T. This way, daily losses, MML_{Td}, are given by:

$$MML_{Td} = \frac{\sum_{day} MML_T \cdot G_T}{\sum_{day} G_T} \quad (6)$$

MML_d has been measured for four PV arrays resulting from different configurations of modules (20-module strings) in 1P and 2P trackers. Figure 9 shows these configurations in the chronological order in which they were tested.

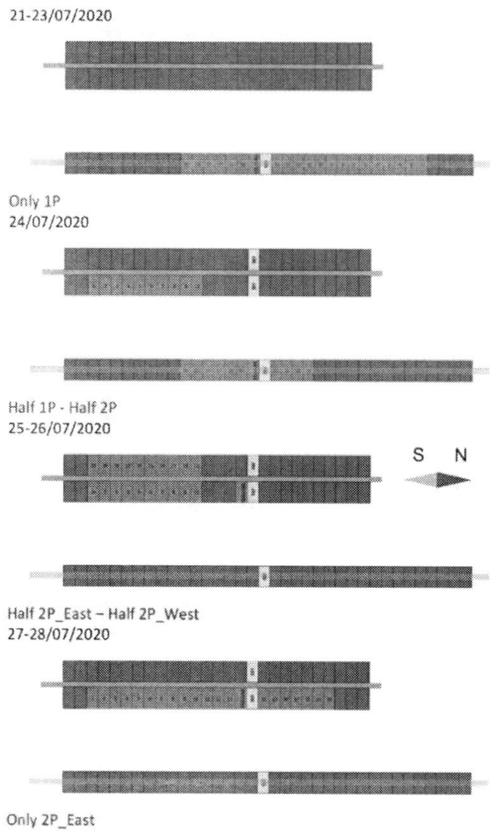

Figure 9: Different configurations of PV modules for *MML* monitoring. Modules conforming the monitored string are in light blue and numbered from 1 (positive pole) to 20 (negative pole). Reference modules for measuring the frontal and rear irradiance and cell temperature are shown in yellow (see next section). The panel for measuring spatial distribution of G_{rear} is shown in brown

3.2 Results

Table 4 shows the MML_{Td} measurements. It must be noted that this method is able to capture the impact of module-to-module mismatch resulting from manufacturing variations, different module temperatures and different rear irradiance distributions. Other effects leading to effective power reduction of the modules are not captured by this methodology when they uniformly affect all the modules. As can be seen from the tables: i) Losses in Only 1P and Only 2P East configurations must reflect the manufacturing variations and are frankly low. This is coherent with the technology of the cells (Si-mono, PERC) and with the high quality of the current PV mass production. ii) Losses in Half 2P_East - Half 2P_West configuration are of the same order than in Only 2P one. That means that the impact of the G_{rear} variability between the two rows of the tracker is negligible.

Losses in Half 1P – Half 2P configuration are around 0.1% higher than in Only 2P East configuration. This agrees with other authors [1] and with initial calculations of *MML* that exploit the results of section 2 and an empirical relation[1] between the non-uniformity of the rear irradiance and *MML*, proposed by researchers from the NREL [2]. It means that the presence of the torque tube increases MML_T in about 0.1%.

PV array	MML_{Td}(%)
Only 1P	0.05
Half 1P - Half 2P	0.11
Half 2P_East - Half 2P_West	0.03
Only 2P_East	0.02

Table 3: MML_{Td} (%) results for four different PV array configurations

4 OPERATING TEMPERATURE

Figure 10 shows a 'bifacial reference module' developed by the IES-UPM to measure the solar cell operating temperature, T_C, along with G_{front} and G_{rear} (not spatially-resolved). The module is electrically divided into three parts (this is easily done by accessing the junction boxes). The central part is kept in open circuit. The open circuit voltage is sensitive mainly to T_C and secondarily to irradiance. The lateral parts are short-circuited through calibrated resistances. A black vinyl sheet is directly sticked to the front and rear of opposite sides, and then protected with a white vinyl sheet. The black vinyl reduces internal reflection, so that irradiance is proportional to the short circuit current. The high reflectance of the white vinyl helps to avoid temperature rising due to artificial absorption by the black vinyl. This way, the corresponding short-circuit current, $I_{SC,rear}$ ($I_{SC,front}$), is only sensitive to G_{back} (G_{front}).

T_C is derived from V_{OC} through:

$$T_C = T_C^* + \frac{1}{\beta}\left[\frac{V_{OC}}{V_{OC}^*} - 1 - a.\ln\frac{G_{front}+G_{rear}}{G^*}\right] \quad (7)$$

Where T_C^* is the reference temperature for STC (25 °C), β is the thermal coefficient of voltage, V_{OC}^* is the open circuit at STC, and a is the ratio of the thermal voltage to the open circuit voltage (a = 0.05 is a reasonable approximation for crystalline silicon modules). The reference module is calibrated in β and V_{OC}^* after stabilization (Sun exposition of at least 60 kWh/m²).

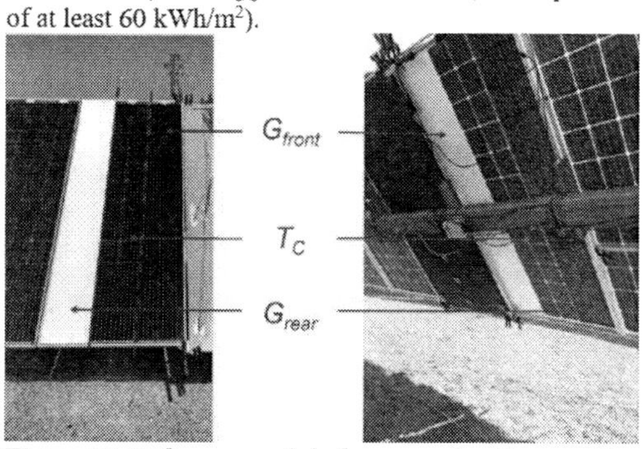

Figure 10: Reference module for measuring the operation conditions: T_C, G_{front} and G_{rear}

The key aspect of this T_C measuring procedure is that V_{OC} responds to the temperature of the entire module, instead of only one or few points, which is the case when measuring T_C by means of Pt sensors sticked to the rear of

[1] Based on extensive simulation work for a PV module similar to those mounted at Cheste (1 × 2 m²; 72 entire solar cells).

the module.

Three reference modules have been installed in Cheste. One in a 1P tracker and two in the two rows of a 2P tracker. Figure 11 shows the results of T_C measurements for a 1-minute sampling throughout 24/07/2020. Particularly interesting is the difference between the average of the T_C observed in the two rows of the 2P Tracker and the T_C observed in the 1P tracker $\Delta T_C\,(\overline{2P}-1P)$, (black spotted line). It varies in the ±2.5 °C range. The corresponding impact in daily energy production, ΔE, can be estimated as the mean of the product of such difference multiplied by the power thermal coefficient of the PV modules, γ, weighted by the irradiance:

$$\Delta E = \frac{\sum_{day} G_T \gamma \Delta T_C\,(\overline{2P}-1P)}{\sum_{day} G_T} \quad (8)$$

The result is only 0.3%. Even acknowledging that thermal behavior is likely affected by local conditions, and that it can vary with weather, this result casts doubt on the statement sometimes found in the PV field that 2P trackers entail a thermal advantage over 1P trackers.

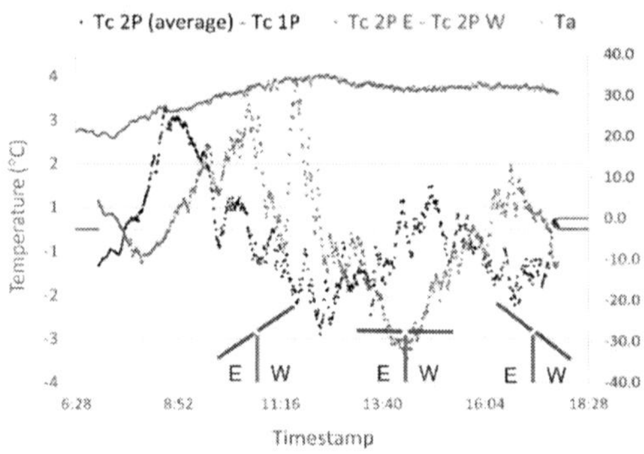

Figure 11: Differences in T_C observed in 1P and 2P trackers the 24/07/2020. In black, the difference between the average temperature of the East and West rows in the 2P tracker and the temperature of the 1P tracker, $\Delta T_C\,(\overline{2P}-1P)$; in blue, the difference between the temperatures in the East and West rows of the 2P tracker, $\Delta T_C\,(2P_{East}-2P_{West})$. Ambient temperature is given in red.

5 CONCLUSIONS

Three key aspects impacting bifacial energy gain have been experimentally investigated for the 1P and 2P trackers currently manufactured by PVH: i) The Structural Shading Factor, *SSF*, for 1P and 2P trackers. ii) The Mismatch Losses, *MML*. iii) The differences in solar cell operation temperature between 1P and 2P trackers.

Dedicated instrumentation has been designed and constructed. A subsequent experimental campaign has been carried out between the 20/07/2020 and the 28/07/2020 in the PVH testing PV plant located at Cheste (Valencia). The plant includes 1P and 2P trackers equipped with bifacial modules sized 2m × 1m and composed by 72 entire cells.

MML values reflecting module manufacturing variations are very low. The impact of the G_{rear} variability between the two rows of the 2P tracker is negligible. The torque tube in 1P trackers increases MML_T in about 0.1% compared to the 2P configuration.

The main results regarding energy yield simulations are: i) Characteristic values to be used as input for PV software: *SSF*(%) = 9.9 (1P) and 3.6 (2P). ii) Observed temperature differences between 1P and 2P trackers range between ±2.5 °C. The corresponding impact on daily energy production is low, around 0.3%.

ACKNOWLEDGEMENTS

We thank PVH for our collaboration at its PV testing plant at Cheste, where the research has been performed.

REFERENCES

[1] G. Janssen, R.S.R. Gali, K. de Groot, A.J. Carr, B.B. Van Aken, I.G. Romijin, 33rd EU PVSEC, Amsterdam (2017).

[2] C. Deline, S. Ayala Pelaez, S. MacAlpine, C. Olalla, Progress in Photovoltaics: Research and Applications, Vol. 28 (2020) 691-703.

AGEING OF TWO 5 KW PV ARRAYS AT THE IES-UPM AFTER 8 YEARS OF OPERATION

F. Martínez-Moreno(1)*, C. Rossa(1), P. Merodio(1), L. Fialho(2), N. Tuytyundzhiev(3)

(1)Instituto de Energía Solar – Universidad Politécnica de Madrid. Grupo de Sistemas Fotovoltaicos (IES-UPM).
Campus Sur UPM. Ctra. Valencia km. 7. ETSIS Telecomunicación. 28031 Madrid, Spain.
(2) Cátedra Energias Renováveis. Universidade de Évora.
Casa Cordovil, Rua D. Augusto Eduardo Nunes nº7, Évora, 7000-651, Portugal
(3) Bulgarian Academy of Sciences. Department of Solar Cells, PV Modules, Devices and Systems.
1, 15th November str., 1040, Sofia, Bulgaria

*Corresponding author: francisco.martinez@ies.upm.es

ABSTRACT: This work analyses the degradation of two 5 kW PV arrays which are located on the roof of the Campus Sur of the Universidad Politécnica de Madrid. These systems were installed in March 2013 and they have been continuously monitored while they have been injecting power into the grid, storing at the same time its DC power and the operation conditions (effective irradiance and cell temperature). These variables allow to calculate their power at Standard Test Conditions over time and, consequently, the degradation rate for each PV array in these 8 years of operation. The linear degradations obtained are similar to the ones presented by other authors, but we have found that there is not apparent degradation in the first years of operation. Besides, it seems that shading could increase the degradation rates. These figures and phenomena can be useful for a better understanding of the actual behavior of PV systems in order to obtain more accurate energy predictions.

Keywords: Ageing, PV array degradation, shading.

1 INTRODUCTION

Year by year the number of installed PV systems all around the world is increasing exponentially thanks to the maturity of PV technology, which makes PV systems more and more competitive. Just 15 years ago, the peak power tolerance of PV modules was typically ±5% [1]. Then, the uncertainty of the yearly energy predictions of PV systems was not so critical because these kind of systems was usually subsidized [2]. Nevertheless, today PV systems compete directly (even unsubsidized) with the traditional energy sources in the framework of call for tenders. Now, modules power tolerance is lower than 2% (and usually positive) and the uncertainty of the energy predictions should be low in order to obtain good conditions to get bank financing for the execution of large photovoltaic projects.

In this scenario, ageing losses begin to play an important role to obtain a more accurate prediction of the energy production. This degradation, which could be almost neglected years ago, now is critical to reduce the uncertainty of these predictions. The problem is that the degradation guaranteed by manufacturer is no more than 20% in 25 years, that is, a linear degradation of -0.8%/year [3][4]. Nowadays this figure has been reduced close to -0.5% because the previous one uses to be conservatives, as shown in several studies about the real degradation rates of PV modules after their exposition outdoors for several years [3][5][6][7]. Therefore, the initial energy reports about the production of a particular PV installation and its performance used to be underestimated.

The Instituto de Energía Solar of the Universidad Politécnica de Madrid (IES-UPM) has two PV installations in operation which have been continuously monitored since their commissioning 8 years ago. This has allowed to study their degradation rate and the actual trend of such power decrease. This will help to better understand the behaviour of these particular modules which were manufactured during the PV boom in the first decade of the 21st century [8][9].

2 THE PV INSTALLATIONS OF IES-UPM.

2.1 Description.

The IES-UPM has two PV installations located on the roof of its headquarters at Campus Sur UPM (latitude: 40.39°; longitude -3.63°) which have been injecting power into the grid from March of 2013. Both systems are made up from modules of the same manufacturer (Siliken SLK60P6L245Wp) which are mounted on a static structure tilted 30° and almost south oriented (-8°). The first one has two strings of 12 modules in series each one (5.8 kWp, Fig. 1) while the second one has three strings of 7 modules in series (5.1 kWp, Fig. 2). Each PV array is connected to an Ingecon Sun Lite 5TL inverter to inject power into the grid.

Figure 1: PV Array 1, with a peak power of 5.8 kWp.

Figure 2: PV Array 2 with a peak power of 5.1 kWp.

2.2 Monthly power at Standard Test Conditions (P^*).

These installations have been continuously monitored, storing periodically their DC power data P_{DC} (every 15 minutes). Simultaneously, the effective irradiance and cell temperature in the plane of the arrays (G_{ef} and T_C respectively) have been also recorded from a modified reference PV module [10] (the PV module on the right side of Array 1, Fig 1). These variables allow to calculate their power at Standard Test Conditions P^* (STC: $G^* = 1000 \text{W/m}^2$ and $T_C^* = 25°\text{C}$) for each single month over the 8 operating years. This can be done if the measured power P_{DC} is corrected in temperature, $P_{DC,25}$

$$P_{DC,25} = \frac{P_{DC}}{1 + \frac{\gamma}{100} \cdot (T_C - T_C^*)}$$

and is represented as a function of G_{ef}. In the previous expression, γ is the temperature coefficient of power (%/°C).

Fig. 3 shows an example of this representation related to Array 2 in March 2020. In this figure we can notice different undesired situations (shadings, inverter saturations and stops) which modify the linear behaviour of the array and, in consequence, the power at STC we are looking for.

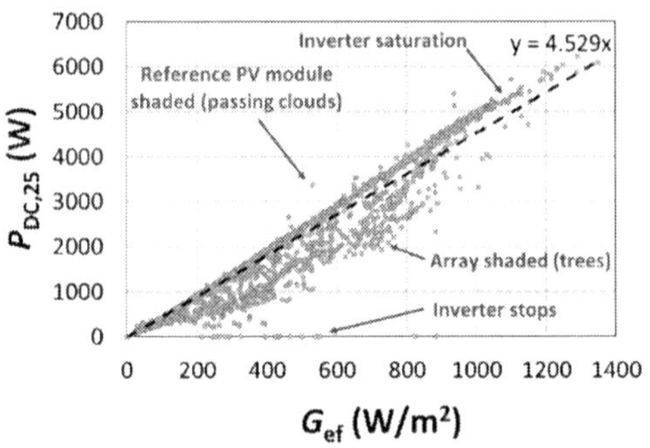

Figure 3: $P_{DC,25}$ as a function of G_{ef} (Array 2, March 2020).

When all the data related to these undesired situations (outliers) are removed, we can observe the linear relationship between $P_{DC,25}$ and G_{ef}. It allows to calculate P^*, which is obtained when $G_{ef} = G^*$ in linear fit equation. Fig. 4 shows the result for March 2020: in this month the power at STC of Array 2 is 4906 W.

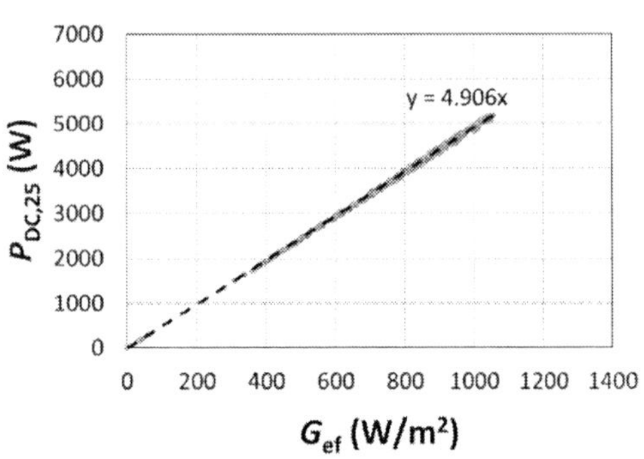

Figure 4: $P_{DC,25}$ as a function G_{ef} once outliers have been removed (Array 2, March 2020).

3 DEGRADATION OF THE ARRAYS POWER.

3.1 Linear model.

Once P^* is calculated for every single month, the degradation rate for each PV array can be analysed. Fig. 5 shows the evolution of P^* for Array 1 and Fig. 6 shows the one for Array 2. As can be seen, both arrays present some kind of oscillation along the year (it is clearer on Array 1), maybe related to spectral issues as suggested by other authors [11]. Dashed lines represent the linear least squares fit for all data for each array (excluding outliers, dark points), where the slope indicates the degradation rates: Array 1 degrades with a rate of -0.25%/year while Array 2 doubles this figure, -051%/year. Probably this difference between the arrays degradation explains why the seasonal behaviour is more evident in Array 1: the higher degradation rate of Array 2 hides the spectral phenomena. Despite the difference, the figures are in agreement with other representative studies [7].

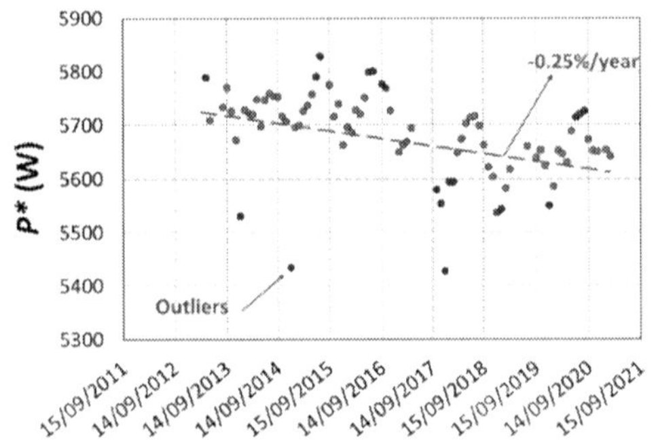

Figure 5: Linear behaviour of P^* for Array 1 from March 2013 to February 2021.

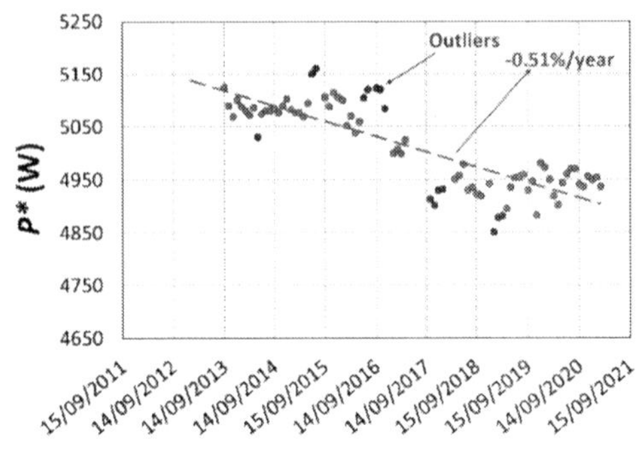

Figure 6: Linear behaviour of P^* for Array 2 from March 2013 to February 2021.

As the arrays are constituted by the same PV modules and the same inverters, they have similar interconnection configuration and they are located at the same location, we have found that this difference could be associated to their particular shading profile: the PV array which is more affected by shading from surroundings is also the PV array which has developed a higher ageing. While Array 1 is almost free of shading, the shadows from two nearby trees (one in the east, Fig. 7, another in the west, Fig. 8) reach Array 2 in the morning and in the afternoon (in the last case, shortly after solar noon with high levels of irradiance). This situation is repeated every single day and can contribute to accelerate the degradation of the modules; however although it does not have immediate harmful effects or threaten the integrity of the modules (in thermographic inspections, transient hot spots have not been detected, at least with relatively large shadows).

Figure 7: Shadow cast every day on Array 2 by a tree located at its east.

Figure 8: Shadow cast every day on Array 2 by a tree located at its west.

3.2 Two-step model.

On the other hand, once the initial light-induced degradation is overcome, we have observed that ageing is not really linear from the beginning: it could be considered that in the first years of operation there is not apparent degradation, as stated by some manufacturers [12] and as proposed by other studies [13] [14]. Later the power begins to decrease at a higher rate than the measured in the single linear model (below the manufacturer warranty) but achieving the same final power value at the end of the considered period.

In this case, it is clear that a two-step degradation profile fits much better to the actual data, as can be seen in Fig. 9 and Fig. 10: in the first 3 years the power varies in a narrow range lower than 2% (points in yellow background). Later, the degradation can be considered as linear, but now of about -0.39%/year (Array 1) and -0.81%/year (Array 2). This rates are almost double the single linear ones obtained previously.

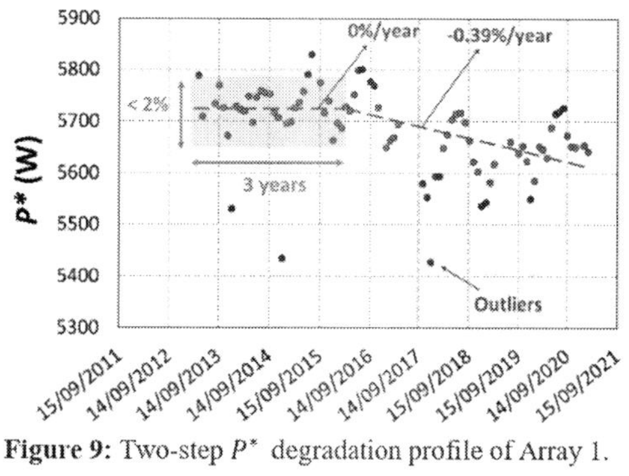

Figure 9: Two-step P^* degradation profile of Array 1.

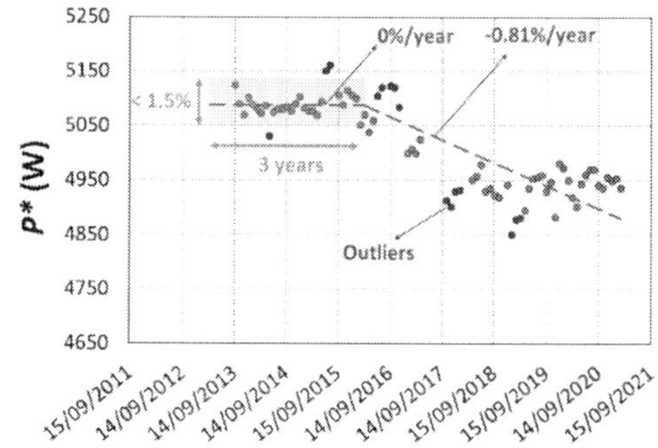

Figure 10: Two-step P^* degradation profile of Array 2.

The assumption of these models, or the linear with a reduced ageing, instead the conventional one based on the warrantied degradation (usually -0.8%/year) has a not negligible impact on the energy yield simulations of the systems.

4 IMPACT OF THE SELECTED AGEING MODEL ON THE ENERGETIC SIMULATIONS.

In order to evaluate the energetic impact of the different models of ageing we have simulated in SISIFO [15] the production of the two PV arrays under study. Our objective is to compare the energetic production for each ageing model: so we have performed these simulations assuming that the PV arrays are free of shadows.

First, we have defined the initial power for each PV array as the average power calculated in their first 3 years of operation (those inside the yellow rectangles, disregarding the outliers). So, the initial power is 5725 W for Array 1 and 5087 W for Array 2. If the linear model with the warrantied rate is used, the arrays final powers are 5358 W and 4762W respectively after 8 years. These figures turn into 5612 W and 4881 W if we consider the actual degradation measured. That is, the real power degradation of the modules in this period is between 4.7% and 2.5% lower than the warrantied.

So, as the PV modules have a better ageing, the real energy production during this period will be higher than the one obtained based on warrantied rate. Table I shows the results of the energy yield estimation for the first 8 years of operation of Array 1 for each degradation model considered:

- Manufacturer warranty degradation: linear rate of -0.8%/year.
- Actual degradation (measured): linear rate of -0.25%/year.
- Two-step degradation: without any power degradation in the first 3 years followed by a linear rate of -0.39%/year.

When a degradation rate closer to reality is used, the energy production increase is higher than 2% (2.3% just in 8 years; we must keep in mind that the warrantied lifetime of a PV installation is 25 years or higher). Moreover, if a two step model is selected, there is an additional improvement of 0.3%. Fig. 11 shows this graphically: the energy obtained from the classical model with the warrantied degradation is represented by the black area ($E_{linear\_warranty}$, the total area covers until the "x" axis), which is lower than the area covered by the actual linear degradation (the energy increase is indicated in blue, ΔE_{linear_actual}). An additional energy gain is obtained when using a two-stage model (red area, ΔE_{2_step}).

Table I: Yearly energy production of Array 1 (SISIFO simulation).

Array 1 (kWh)	Warranty	Actual	Two-step
Year 1	9797	9797	9797
Year 2	9707	9770	9797
Year 3	9617	9743	9797
Year 4	9528	9715	9758
Year 5	9437	9686	9720
Year 6	9348	9659	9681
Year 7	9259	9631	9643
Year 8	9168	9604	9604
Total	75863	77605	77799
Difference (%)	0	2.3	2.6

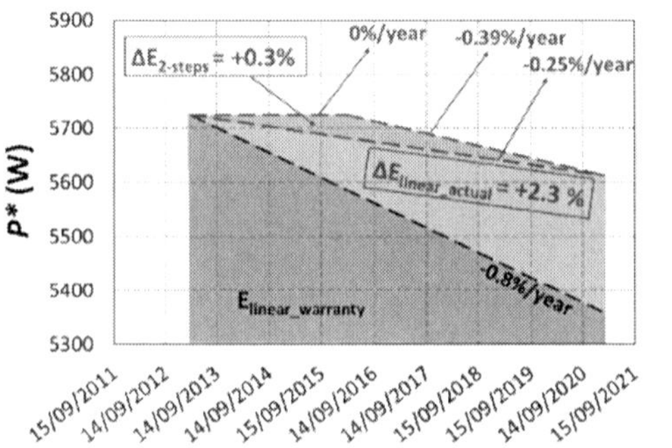

Figure 11: Energy production of Array 1 for each degradation model.

Table II shows the results obtained for Array 2. In this case the degradation models considered are:

- Manufacturer warranty degradation: linear rate of -0.8%/year.
- Actual degradation (measured): linear rate of -0.51%/year.
- Two-step degradation: without any power degradation in the first 3 years followed by a linear rate of -0.81%/year.

Table II: Yearly energy production of Array 2 (SISIFO simulation).

Array 1 (kWh)	Warranty	Actual	Two-step
Year 1	8703	8703	8703
Year 2	8624	8653	8703
Year 3	8543	8601	8703
Year 4	8464	8552	8632
Year 5	8383	8500	8562
Year 6	8304	8450	8490
Year 7	8223	8399	8419
Year 8	8144	8349	8349
Total	67388	68206	68560
Difference (%)	0	1.2	1.7

Now, the selection of a degradation rate closer to reality leads to an increase of energy production higher than 1% (again, just in 8 years). And if a two step model is selected, there is an additional improvement of 0.5%. In this case, as the actual degradation rate of Array 2 is closer to the warrantied by manufacturer, the energetic improvement with a linear model is reduced (compared to Array 1) but now the additional improvement achieved with a two-step model is almost doubled: it increases from 0.3% to 0.5%. Fig. 12 shows this graphically.

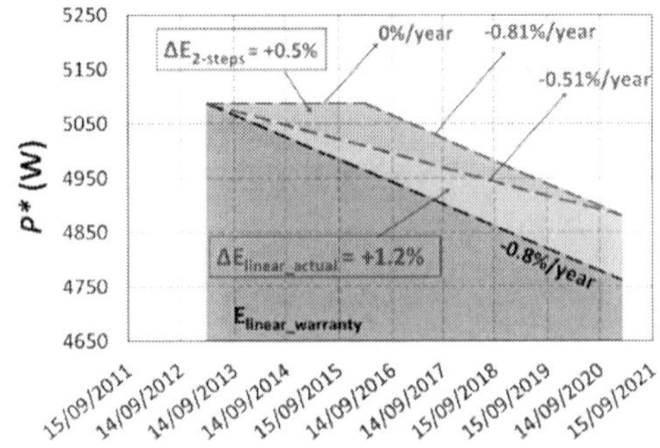

Figure 12: Energy production of Array 2 for each degradation model.

So, it is very important to select a model which best fits reality: this not only leads to more accurate predictions, but also can help to the bankability of the PV system. A higher productivity translates into in a shorter payback time and, consequently, more advantageous financing conditions. This advantages are much smaller if the strictly warrantied degradation rates are used. As we have shown, these figures used to be quite conservative at least 10 years ago. In fact, warrantied degradation rates of currently PV module are around -0.5%, figure which is closer to the actual results obtained.

5 SUMMARY

This paper reports about the actual degradation rates obtained from two PV installations of 5.8 kWp and 5.1 kWp which are operating in Madrid since 2013. Their rates have been calculated from the actual peak power measured monthly thanks to the continuous monitoring of power and operating conditions (G_{ef} and T_C) from a reference PV module.

We have observed that the manufacturer ageing warranty is too conservative: the actual degradation rates measured are lower. Besides, an ageing model which is based on a two-step degradation instead of a single linear one seems to be more appropriate to simulate the real behavior of PV modules. Energy production gains about 1%-2% are achieved in 8 years if the proper ageing model is selected. So, this increase could reach 2%-3% when the whole lifetime of the PV installation in considered (25 years).

On the other hand, we have noticed that shading over PV arrays could accelerate natural ageing. We have obtained that prolonged shading conditions throughout the whole year could lead to double the degradation rates.

6 ACKNOWLEDGEMENTS

This work has received funds from the European Union's Horizon2020 research and innovation programme under Grant Agreement 787289. This work is part of the Project MADRID-PV2 (P2018/EMT-4308) funded by the Comunidad de Madrid with the support from FEDER Funds.

7 REFERENCES

[1] TamizhMani M. G., Mikonowicz A., Kuitche J. Nameplate, datasheet and sampling requirements for PV modules. Solar ABCs, Stakeholder Webinar, 3 November 2011.

[2] Jennings C.E., Margolis R.M., Bartlett J.E. A historical analysis of investment in solar energy technologies (2000-2007). Technical Report NREL/TP-6A2-43602. December 2008.

[3] Jordan, D.C.; Kurtz, S.R. Photovoltaic Degradation Rates—an Analytical Review. Prog. Photovolt. Res. Appl., 21, 12–29. 2013

[4] Quansah D. A., Adaramola M. S., Takyi G., Preliminary assessment of degradation in field-aged multi-crystalline silicon PV modules installed in hot-humid climate of mid Ghana. 32nd EU PVSEC. 2290-2293. 2016.

[5] Dunlop E.D., Halton D. The performance of crystalline silicon photovoltaic solar modules after 22 years of continuous outdoor exposure. Prog. Photovolt: Res. Appl. 14:53–64. 2006. doi: 10.1002/pip.627

[6] X. Sun, R.V.K. Chavali, M.A. Alam. Real-time monitoring and diagnosis of photovoltaic system degradation only using maximum power point—the Suns-Vmp method. Prog Photovolt Res Appl.;27:55–66. 2019. doi: 10.1002/pip.3043.

[7] Pascual J., Martinez-Moreno F., García M., Marcos J., Marroyo L., Lorenzo E. Long-term degradation rate of crystalline silicon PV modules at commercial PV plants: An 82-MWp assessment over 10 years. Prog. Photovolt. Res. Appl., 1–9. 2021. doi: 10.1002/pip.3456

[8] SolarPower Europe (2020). Global Market Outlook for Solar Power 2020-2024. June 2020.

[9] Renewable Power Generation Costs in 2017. International Renewable Energy Agency, IRENA (2018), Abu Dhabi.

[10] Martínez-Moreno F., Lorenzo E., Muñoz J., Moreton R. On the testing of large PV arrays. Progress in Photovoltaics: Research and Applications 20, 1. 100-105. (2012).

[11] Jordan D.C., Smith R.M., Osterwald C.R., Gelak E., Kurtz S.R. Outdoor PV Degradation Comparison. 35th IEEE Photovoltaic Specialists Conference (PVSC '10) Honolulu, Hawaii, 2010.

[12] Quansah D.A., Adaramola M.S., Takyi G., Edwin I.A. Reliability and Degradation of Solar PV Modules—Case Study of 19-Year-Old Polycrystalline Modules in Ghana. Technologies 2017, 5, 22. DOI:10.3390/technologies5020022.

[13] Jordan D.C., Kurtz S.R., VanSant K., Newmiller J. Compendium of photovoltaic degradation rates. Prog. Photovolt: Res. Appl. 2016; 24:978–989

[14] T. Ishii, A. Masuda. Annual degradation rates of recent crystalline silicon photovoltaic modules. Prog. Photovolt: Res. Appl. 2017; 25:953–967. doi: 10.1002/pip.2903

[15] www.sisifo.info

THE POTENTIAL OF THE REJECTED BRINE AS A COOLANT FOR PV TEMPERATURE REDUCTION PURPOSES

Shahdad Kamfiroozi[1], Hossein Ebadi[2,*], Hossein Vahabi[3], Rezvan Talebnejad[4], Laura Savoldi[2], Ali Reza Sepaskhah[4], Barat Ghobadian[1], Shiva Gorjian[1]

1. Biosystems Engineering Department, Faculty of Agriculture, Tarbiat Modares University (TMU), Tehran, Iran
2. MAHTEP Group, Dipartimento Energia "Galileo Ferraris" (DENERG), Politecnico di Torino, Turin, Italy
3. Department of Agricultural Machinery Engineering, Faculty of Agricultural Engineering and Technology, University of Tehran, Karaj, Iran
4. Water Engineering Department and Drought Research Center, Faculty of Agriculture, Shiraz University, Shiraz, Iran

*Corresponding author
Hossein.ebadi@polito.it +39 333 1041766
Politecnico di Torino, Corso Duca degli Abruzzi 24, 10129 Torino, Italy

ABSTRACT: From the working principles of PV cells, only half of the received solar spectrum (with a wavelength of 200-800 nm) is utilized for energy conversion and the remaining portion (800-3000 nm) heats the panels, decreasing the electric efficiency. To avoid such sacrifices in performance, which could expand the economic concerns and shortening PV's lifetime, scientists have proposed various cooling mechanisms. In this study, an experimental investigation has been performed on a stand-alone PV-powered RO system to assess the electrical and thermal performance of the solar system and study the influence of the rejected brine as a coolant for PV panels. The main scope is to use brine not only for increasing the performance of solar panels by cooling effect but also for eliminating the energy needed for pumping through the utilization of the output pressure of the RO membranes. The experimental results indicated that the application of brine cooling can improve the PV efficiency from 1 to 3% depending on the feeding temperature and salinity levels.
Keywords: PV, Cooling, RO, Brine, Efficiency improvement

1 Introduction

From the working principles of PV cells, only half of the received solar spectrum (with a wavelength of 200-800 nm) is utilized for energy conversion and the remaining portion (800-3000 nm) heats the panels, decreasing the electric efficiency [1]. This is attributed to the fact that a temperature rise in solar cells could considerably decrease the semiconductor bandgap and consequently diminish the open-circuit voltage [2]. Thus, if the PV surface temperature exceeds the standard limit, efficiency decreases by a range of 0.4 – 0.5 % per °C of temperature rise [3]. To avoid such sacrifices in performance, which could expand the economic concerns and shorten PV's lifetime, scientists have proposed various cooling mechanisms. As presented by Gomaa et al. [4], a water-based cooling system could recover 10.2% energy per day compared to the 7% in an air-based cooling method. Water-based cooling systems have been documented in a large portion of studies in the literature. Tiwari and Sodha [5] proposed a numerical thermal modeling for extracting heat from a PV module by water using a series of energy balance equations. Mohsenzadeh and Hosseini [6] developed a closed-loop cooling system for a reflector assisted PV module using water as the working fluid. Results demonstrated that the temperature of the concentrated PV panel remains at 50 °C while this value could reach up to 95 °C in the absence of the cooling cycle. An active cooling system was introduced by Prudhvi and Chaitanya Sai [7] in which water is injected on the panels to absorb the excess heat and release it in a heat exchanger buried underground with a stable temperature. Thus, the system can work in a closed-cycle with a pump to circulate water where a 7.75 % rise in efficiency can be expected. In another work conducted by Moharram et al. [8], water was sprayed on the modules to cool the PV cells in a hot arid zone, Egypt. They developed a mathematical model for optimizing the set time and temperature for starting the cooling process and found that the power consumption and cooling efficiency is optimal if the system starts at a cell temperature of 45 °C. Through a numerical study, Yazdanifard et al. [9] investigated a water-based PV cooling system, known as Photovoltaic/thermal collectors (PV/T), under both laminar and turbulent conditions. Similarly, Fudholi et al. [10] studied various pipe configurations used for passing water beneath the PV panel and obtained a 13.8% electric efficiency with the spiral flow design. Bahaidarah et al. [11] also found that if the cooling is performed from the rear side of the panel, temperature could be reduced by 20% resulting in a 9% increase in electric efficiency. Moreover, Nižetić et al. [12] evaluated the performance of a PV panel cooled with sprayers from both top and bottom sides and concluded that such systems not only improve the PV efficiency by 14.1% and undermine the temperature by 30 °C, it also serve some cleaning effects for further enhancements.

Despite the benefits coming from liquid-cooling methods, the inevitable additional costs imposed by the fluid and energy used for recirculating it by a means of pumps are the main drawbacks of such systems. Thus, if new techniques could foster these problems, the application of water-based systems would be economically feasible for the deployment in the power sector. Recently with the increase in the installation of solar-powered desalination units, feed saline water has shown significant potential to be used as a coolant while getting preheated before entering the system [13]. This technique increases the feed saline water temperature, which improves not only the production of freshwater, but also the PV efficiency. However, the application of this method could still have some problems such as the energy demand for water pumping and washing operations and eventually an additional need for more pre-treatment process than before. On the other hand, due to the quality of the feed saline water used for cooling, the dissolved salts will remain on the solar panels, which requires double-washing in these systems. However, brine is a byproduct from RO desalination which is associated with negative environmental impacts due to their high concentration of

salts and other pollutant contents.
Therefore, the use of brine as an unconventional source for PV cooling purposes would be a new method for brine management. The main purpose of using brine is not only the increase of the performance of solar panels by cooling effect, but also to cut of the energy needed for pumping by the use of the output pressure of the RO membranes. Moreover, since the brine is rejected after the cooling process, the need for the additional filtration can also be avoided, bringing lesser operational costs compared to similar designs. In this study, it is aimed to investigate the applicability of brine discharged from a PVT-RO plant, as an unconventional and novel source for PV cooling purposes by conducting several experiments at different feed water salinity and temperature.

2 Materials and methods

A number of experiments were carried out at Shiraz University, Shiraz, Iran, during October and November 2020, considering various brine concentrations and temperatures rejected from a real semi-industrial scale RO plant, which was powered by two poly-crystalline PV modules with 270 Wp. As depicted in Fig. 1, the test rig consists of; (1) Saline feed water tank, (2) freshwater tank, (3) primary-pressure pump, (4) high-pressure pump, (5) RO membranes, (6) flow meter, (7) pressure gage, (8) pre-separation flirts, (9) Solar battery, (10) electric control unit, (11) solar panels, (12) brine cooling pipe.
Within the working procedures, saline feed water with a constant salinity was injected into the systems and after passing through the membranes, freshwater was accumulated inside a 100 lit storage tank. While the rejected brine with the pressure given by the system was directed toward the PV panels located on the roof of the testing facility, to induce the cooling effect and dissipate the cell temperature. 64 water sprays were formed through the hole drilling with a 50 mm distancing, which were formed on the cooling pipe mounted on the header of the PV modules as shown in Fig. 2A.

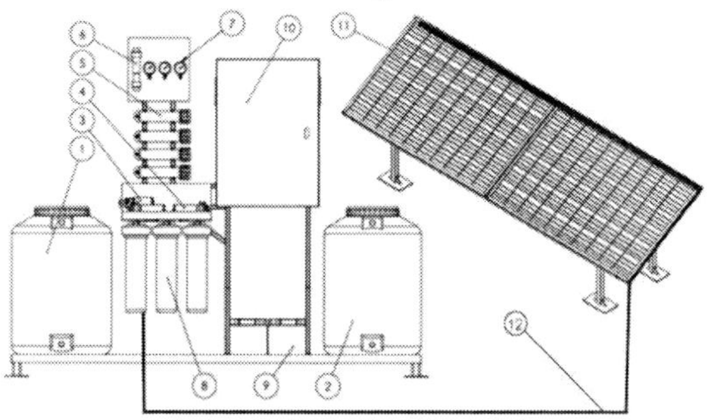

Figure 1: The experimental set-up developed for this study.

Different saline feedwater concentrations were produced with three levels of salinity (1500, 3000 and 4500 mglit^{-1}) using 50% sodium chloride (NaCl) and 50% calcium chloride ($CaCl_2$) before each test. Also, five saline feedwater temperature levels (20, 25, 30, 35, 40 °C) were achieved through the implementation of a heat exchanger with 3kW thermal power. To measure the temperature variations on the PV surface, 12 temperature sensors (NTC thermistor) were mounted on the backside of the panels (Fig.2B). A DHT11 temperature sensor was also located beside the test facility to measure the ambient temperature variations during the testing periods. The generated electricity was controlled with a Maximum Power Point Tracking (MPPT) solar controller (SON-MMPT-30 A, SUOER, China), to obtain general information of PV power generation, modify setting parameters of the controller, and protect the battery from overloading, overcharging, short circuit, reverse discharge, reverse polarity connection, and thunder lightning. Using the MMPT technique leads to ensure power generation at the maximum rate. Also, employing a rotameter, in-line with the freshwater flow allowed the measurement of freshwater flow rate, while the amount of the product water (calibration) and brine discharge flow rates were achieved through an experimental volumetric measurement at the specific time conducted at 3 replicates. Afterwards, the quality of the rejected brine and the produced freshwater were measured using an EC meter at 3 replicates. A manual data logger was also developed and utilized to record the temperature data as well as the generated voltage and current information during each test and in 2.3 s intervals. Subsequently, when the tests were done, data was logged into a PC for further processing and analysis. The implementation of 3 pressure gauges; one to measure the inlet saline feedwater pressure behind the high-pressure pumps, one to measure the pressure of the RO, and one to measure the brine outlet pressure allowed monitoring the pressures at the inlet, membranes, and outlet of the system.

Figure 2: Photographs of the PV panels used in this study; (A) cooling mechanism, (B) temperature sensors.

PV electric efficiency was calculated by comparing the standard characteristics (T=25 °C) to those obtained at operating temperatures. Since the calculation of PV efficiency (η_{PV}) is varied, the electrical assessment was carried out based on the equation recommended by Skoplaki and Palyvos [14] for evaluation consists of the (T_c) cell temperature (°C) and (G_T) total solar incident (Wm^{-2}), which is given as;

$$\eta_{PV} = \eta_{ref}\left[1 - a_1\left(T_c - T_{ref}\right) + a_2 \ln(G_T/1000)\right]$$

Where, η_{ref} is the reference electrical efficiency that is obtained based on the standard values given by the manufacturer (equal to 16.5% in this study), T_{ref} indicates the reference temperature (equal to 25 °C), a_1 and a_2 are constants as 0.005 and 0.052, respectively for silicon-based PV panels.

3 Results and discussion

3.1 PV-RO Desalination performance

Figure 3 represents the quality of the brine produced during various tests with respect to the temperature and salinity of the saline feedwater. The highest brine salt concentration has been achieved in 4500 mglit^{-1} feedwater salinity with a mean value of 6500 mglit^{-1} and the lowest bine salt concentration with a mean value of 2600 has resulted from the feedwater with 1500 mglit^{-}

[1]salinity. The effects of the temperature treatment were similar and monotonous for the salinity levels of 1500 and 4500 mglit^{-1} in the range of 20 to 40 °C and in range of 20 to 35 °C for salinity level of 3000 mglit^{-1}. However, when the temperature of 40 °C is applied, brine of salinity level of 3000 mglit^{-1} decreased, which shows the brines rejected from the salinity levels of 4500 and 1500 mglit^{-1}, undergo an increasing trend, while salinity of 3000 mglit^{-1} leads to a decreasing trend at this temperature. This implies that temperature is affecting the physical properties of the membranes such as pore size and water diffusivity, which lead to the changes in flux and permeability in different levels [13].

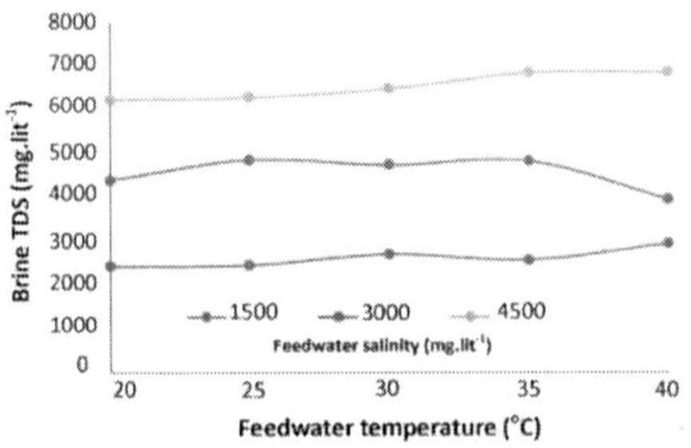

Figure 3: Quality of the brine produced under various temperatures and salinity treatment.

3.2 Thermal performance

A comparison between the mean temperatures of the PV panel under non-cooling and one of the cooling conditions with constant feedwater salinity of 1500 mg L^{-1} is shown in Fig. 4. As the system works under the conventional mode, the surface temperature of the module grows up to nearly 54 °C. However, when the cooling process is applied, the PV module takes nearly 60 time-step (138 s) to initiate the decreasing trend in temperature. As the cooling operation exceeds 200 time-step (460 s) in operation, the coolant and the cell surface reach thermal equilibrium, and the falling trend in temperature stops, leading to a constant temperature value. When the operation reaches the time-step of 500 (1150 s), cell temperatures start a very steep rise trend, which reflects the effects of the salt depositions on the PV surface and their role in reducing the contact with the coolant and consequently negative effects on the temperature reduction. The mean temperature reduction achieved after cooling for the PV panel was measured as 12.3 °C temperature reduction.

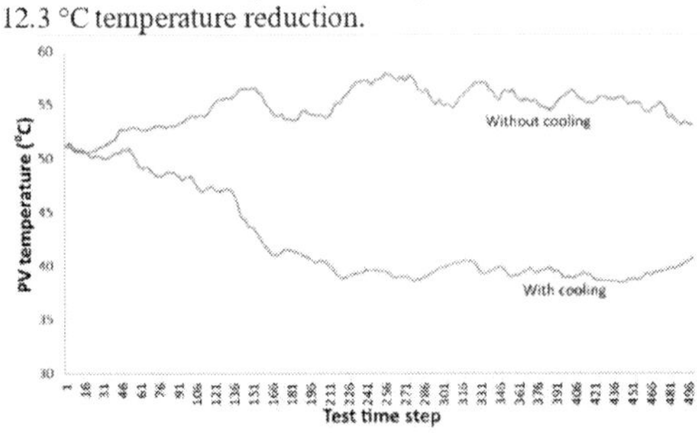

Figure 4: The impacts of the cooling process on the temperature reduction compared to non-cooling operation.

3.3 Electrical performance

Figure 5 represents the cell efficiency for the PV panel in the condition of cooling and without cooling. The cooling process has promoted the electrical PV efficiency by 1 to 3% depending on the brine temperature and feed water salinity, where the maximum increase achieved as 3 % for the panel cooled with brine produced from 25 °C and 1500 mglit^{-1} feedwater. On the other hand, the minimum improvement was 1.2 % for the panel using brine with 35 °C and 1500 mglit^{-1} feedwater. In the case of operation without cooling, the minimum efficiency was 12.9 % measured at the treatment with 30 °C feedwater temperature, and 4500 mglit^{-1} salinity and the maximum PV efficiency was 14.2% found in the treatment with 35 °C feedwater temperature, and 1500 mglit^{-1} salinity. After the cooling process, the maximum PV efficiency was calculated as 16.1% at the treatment where the panel was cooled with the brine released from the desalination of saline water with 1500 mglit^{-1} salinity and 40 °C. The reason for such PV efficiency roots in the coupling effects of high solar radiation and high brine discharge for cooling operation. Furthermore, when the mean values of the PV efficiencies were compared at each saline feedwater temperature, it was concluded that the brine with a temperature of 25 °C has the most improving effect on the electrical efficiency which is followed by temperatures of 30, 40 and 35 °C respectively. However, the cooling effects of water salinity are a complex phenomenon and depend on the solar flux, evaporation rate, wind speed, and water flow rate which need further experimental and theoretical studies.

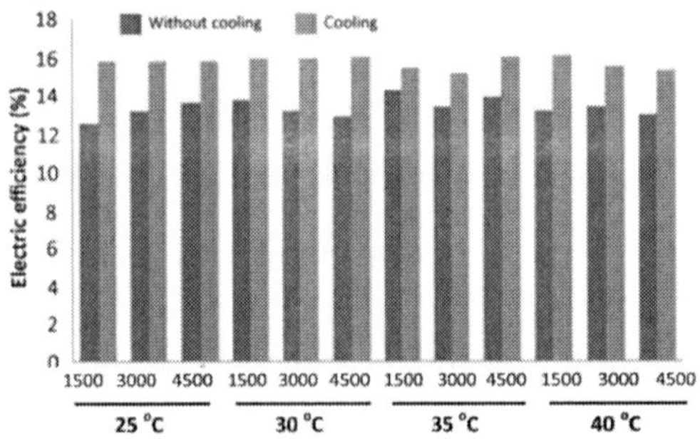

Figure 5: Electrical efficiency obtained with and without cooling.

4 Conclusion

In this study, a number of experimental tests were carried out to assess the cooling potential of brines rejected from conventional RO plants as a novel coolant for PV cooling procedures. With the simulations of various settings through five levels of temperature and three levels of salinity for saline feedwater, a range of brines was tested for a PV panel. The synergy of the cooling effects and brine management led this concept to be an appropriate alternative to conventional methods. Results showed that increasing the feedwater temperature to 40 °C could change the physical properties of the membranes, which causes fluctuations in the flux and permeability for different inputs. Moreover, the brine flow over PV panels indicated an ensured temperature reduction up to 12.3 °C. Evaluating the PV efficiency after the cooling performance revealed that the maximum increase is 3% for the panel cooled with brine rejected from 25 °C and 1500 mglit-1 feedwater, which can improve the effiecincy of the conventional PVs (without cooling) up to 20%.

5 References

[1] Elsarrag E, Pernau H, Heuer J, Roshan N, Alhorr Y, Bartholomé K. Spectrum splitting for efficient utilization of solar radiation: a novel photovoltaic–thermoelectric power generation system. Renewables Wind Water, Sol 2015;2:16. https://doi.org/10.1186/s40807-015-0016-y.

[2] Korres DN, Tzivanidis C. A symmetric and an asymmetric mini compound parabolic collector under optical investigation. Green Energy Technol 2018:649–61. https://doi.org/10.1007/978-3-319-89845-2_46.

[3] Gorjian S, Sharon H, Ebadi H, Kant K, Scavo FB, Tina GM. Recent technical advancements, economics and environmental impacts of floating photovoltaic solar energy conversion systems. J Clean Prod 2021;278:124285. https://doi.org/https://doi.org/10.1016/j.jclepro.2020.124285.

[4] Gomaa MR, Hammad W, Al-Dhaifallah M, Rezk H. Performance enhancement of grid-tied PV system through proposed design cooling techniques: An experimental study and comparative analysis. Sol Energy 2020;211:1110–27. https://doi.org/https://doi.org/10.1016/j.solener.2020.10.062.

[5] Tiwari A, Sodha MS. Performance evaluation of solar PV/T system: An experimental validation. Sol Energy 2006;80:751–9. https://doi.org/https://doi.org/10.1016/j.solener.2005.07.006.

[6] Mohsenzadeh M, Hosseini R. A photovoltaic/thermal system with a combination of a booster diffuse reflector and vacuum tube for generation of electricity and hot water production. Renew Energy 2015;78:245–52. https://doi.org/10.1016/j.renene.2015.01.010.

[7] Prudhvi P, Sai PC. Efficiency improvement of solar PV panels using active cooling. 2012 11th Int. Conf. Environ. Electr. Eng., 2012, p. 1093–7. https://doi.org/10.1109/EEEIC.2012.6221543.

[8] Moharram KA, Abd-Elhady MS, Kandil HA, El-Sherif H. Enhancing the performance of photovoltaic panels by water cooling. Ain Shams Eng J 2013;4:869–77. https://doi.org/10.1016/j.asej.2013.03.005.

[9] Yazdanifard F, Ebrahimnia-Bajestan E, Ameri M. Investigating the performance of a water-based photovoltaic/thermal (PV/T) collector in laminar and turbulent flow regime. Renew Energy 2016;99:295–306. https://doi.org/https://doi.org/10.1016/j.renene.2016.07.004.

[10] Fudholi A, Sopian K, Yazdi MH, Ruslan MH, Ibrahim A, Kazem HA. Performance analysis of photovoltaic thermal (PVT) water collectors. Energy Convers Manag 2014;78:641–51. https://doi.org/https://doi.org/10.1016/j.enconman.2013.11.017.

[11] Bahaidarah H, Subhan A, Gandhidasan P, Rehman S. Performance evaluation of a PV (photovoltaic) module by back surface water cooling for hot climatic conditions. Energy 2013;59:445–53. https://doi.org/https://doi.org/10.1016/j.energy.2013.07.050.

[12] Nižetić S, Čoko D, Yadav A, Grubišić-Čabo F. Water spray cooling technique applied on a photovoltaic panel: The performance response. Energy Convers Manag 2016;108:287–96. https://doi.org/10.1016/j.enconman.2015.10.079.

[13] Kelley LC, Dubowsky S. Thermal control to maximize photovoltaic powered reverse osmosis desalination systems productivity. Desalination 2013;314:10–9. https://doi.org/https://doi.org/10.1016/j.desal.2012.11.036.

[14] Skoplaki E, Palyvos JA. On the temperature dependence of photovoltaic module electrical performance: A review of efficiency/power correlations. Sol Energy 2009;83:614–24. https://doi.org/https://doi.org/10.1016/j.solener.2008.10.008.

Influence of shading on photovoltaic generation from winter to summer in Vanderbijlpark, South Africa

Johan Bekker

Centre for Alternative Energy, Vaal University of Technology, P/B X021, Vanderbijlpark, South Africa

bekkerj@vut.ac.za; +27 169509410

ABSTRACT: This paper presents effects and influences due to shading on solar panels installed as part of a solar system at a residential house in Vanderbijlpark, South Africa. Design considerations include the choice of inverter, solar panels as well as the arrangement and mounting of the solar panels. Results presented illustrate the effect of the shading for different scenarios from winter to summer conditions. Emphasis is placed on the difference of the solar path of the sun, resulting in different shading conditions. These results are in the form of data captured for electricity generation by the solar array, and graphical representation thereof.

Keywords: Generation, PV System, Inverter, Solar array

1 INTRODUCTION

For any photovoltaic (PV) solar installation, optimum generation of electricity from the PV panels is the key factor. This optimum performance is dependent upon many factors related to incident sunlight upon the panels. The obvious factor is the amount of shading (cloud cover, obstructions) and then also the change from summer to winter conditions. This paper represents results from an in-depth study into the figural influence of these shading factors on PV system performance, and hence electricity generation and savings. Results will be presented in the form of, among others, Power vs time graphs, representative for total energy generated under different incident sunlight conditions. Results will be discussed, and suitable conclusions will be drawn.

2 DESIGN AND INSTALLATION

The PV solar system on which this study was conducted, was a residential hybrid system, as depicted in Fig. 1 [1]. However, this study can be applicable also to off-grid as well as grid-tie systems, as it was conducted on generation of electricity by the PV panels. As was mentioned in the introduction, the aim was to investigate the difference in generation for different incident sunlight and hence shading conditions.

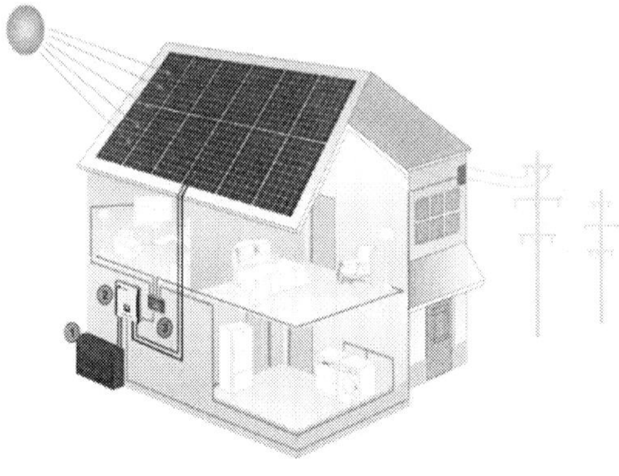

Figure 1: Example of a hybrid PV solar system. All the generated PV energy is utilized by the inverter (2) through supplying the loads through a distribution board (3), charging the batteries (1) or storing back into the utility grid. [1]

For this solar system, energy generation was done by eighteen Trina 275 W solar panels. These panels deliver a peak voltage and current of 31.8 V and 8.97 A, respectively, while the open-circuit voltage and short-circuit current are 39.3 V and 9.45 A, respectively [2]. The solar panels can be seen as an array in Fig 2. The dimensions of the solar panels are 1650 x 992 mm [2], and with sufficient spacing between them after mounting, determines the size of the north-facing roof that is needed. Panels can be installed either landscape or portrait, depending on the space available, as well as the connection needs and positions. However, it is very important to keep in mind that panels needs to be cleaned regularly, and provision must be made for space and procedures to do cleaning.

Another installation consideration is the angle at which the panels are installed. The preferred angle for Vanderbijlpark is between 16° and 36°, depending on the type of panel and load [3]. The mounting of panels at an angle is not only for better performance, but also for cleaning purposes, especially for "self-cleaning" during rainy periods. The angles allow water to not gather on the panels, but to flow away automatically. The panels in the current installation were installed at a tilt angle of around 25°. Note the trees in the background, for which the influence will be discussed under 'Results and Conclusions'.

Figure 2: The installed PV on the roof can be seen in this picture. 12 panels are installed "landscape", while 6 are installed "portrait" to optimize the incident sunlight on them. Note the trees in the background, on the neighbouring property.

For this system, a 4.6 kW Goodwe full-hybrid inverter [4] was decided upon after a load monitoring study was done over an extended period. For energy storage, two Pylontech 2000Plus Li-ion batteries with a capacity of 2.4 kWh each [5], was connected in parallel to the inverter. The installed and connected Direct Current distribution box, inverter and battery stack can be seen in Fig. 3. Ideally, the solar panels should be able to generate a total amount of electrical energy at a rate of 275 x 18 = 4950 W. This number should be kept in mind during the study, as generation will be measured against it for the different conditions, i.e., cloudy, winter/summer, and other shading influences from trees, etc.

Figure 3: The inverter (top left), DB (to the right of inverter) and battery pack (bottom) are depicted at the bottom.

In this study, the approach was to investigate and gather data for electricity generations of the specific solar array under these different conditions through voltage/current and hence power output readings. Through performing the necessary calculations, conclusions are then drawn to implement within the system for optimum performance.

3 DATA LOGGING AND ANALYSIS

Metering and data logging was done with a Goodwe Smart Energy meter [4]. This meter has Current Transformers (CT's) connected to the relevant input and output cables on the DB and has a communication cable connected to the inverter where all the data is logged. The relevant data can then be downloaded and, for this study, power generation graphs be plotted for further analysis.

4 THEORETICAL CONCEPTS

In order to analyse and understand electrical measurements and consumption, it's important to keep in mind the following basic concepts [6]:

4.1 Electric Potential Energy and Current

$$EPE = V \times q \tag{1}$$

EPE – Electric Potential Energy (Joule)
V – Electric Potential (Volts)
q – Electric charge (Coulomb)

$$I = \frac{q}{t} \tag{2}$$

I – electric current
t – time (seconds)

4.2 Electric Power

$$P = \frac{Energy}{t} = \frac{EPE}{t} = \frac{Vq}{t} = VI \tag{3}$$

P – Power (Watt)

Hence, electric potential energy can also be expressed in Watt.seconds (1 W.s = 1 J), or more commonly in electrical energy terms, in kilowatt.hours (1 kWh = 3.6 x 10^6 J). Furthermore, by measuring the current, energy can easily be calculated for a period of time, as the voltage is a constant supply.

5 RESULTS AND DISCUSSION

The performance of a solar system is, among others, determined by the efficiency of the electrical energy generation of the solar panel array. To reach optimum performance, serious consideration needs to be given to shading on the panels. The data gathered and calculations performed, will then be available for design considerations of future solar systems within the larger Gauteng / Northwest / Northern Orange Free State regions in South Africa. The installed array in the current study faces due north, which, from an orientation perspective, is ideal. The elevation is also ideal [3], so the only factor influencing the optimization of the performance is the incident sunlight, and therefore the nature of shading.

5.1 Clear vs Cloudy

The location of the solar installation in this study, i.e. Vanderbijlpark, South Africa, is known for afternoon thunderstorms (read: cloud cover) during the summer months (Nov – Febr). As a point of reference, Fig 4 is presented as an ideal electricity generation curve for the current PV panel array. This graph is representative of a day in December 2020, well into the South African summer. The amount of electrical energy generated on this day was 32.3 kWh, with a peak generation of 4.3 kW at midday. Also notable on the graph, is the fact that electricity was generated from 06h00 in the morning until around 18h00 late afternoon. It is clear that this day resulted in continuous generation represented by the near perfect symmetrical curve.

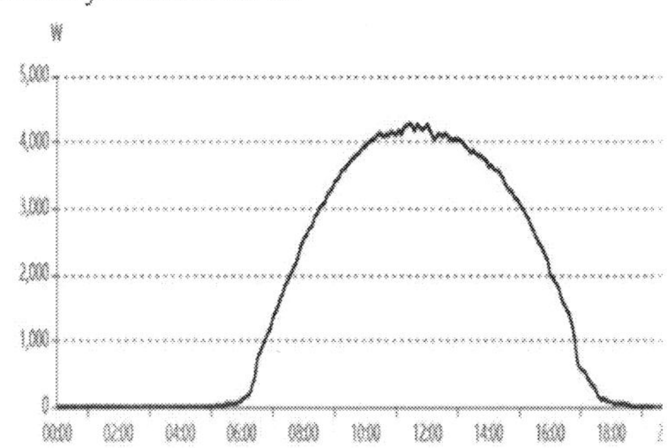

Figure 4: Near perfect curve for optimum electricity generation (Power (W) vs time) representing a summers day in December 2020. The yield was 32.3 kWh on that day, with a peak power of 4.3 kW reached at mid-day.

On the contrary, Fig 5 represents a generation graph on a summers day in Decmber 2020 that was perfectly clear in the morning, with clouds moving in at around 12 noon and resulting thunderstorms occuring for the rest of the afternoon. On this day only 14 kWh of electricity was generated as a result of the cloudcover and hence low-intensity light incident on the panels. This relation between electrical power delivered by a solar system in relation to the light intensity have been reported by various researchers [7].

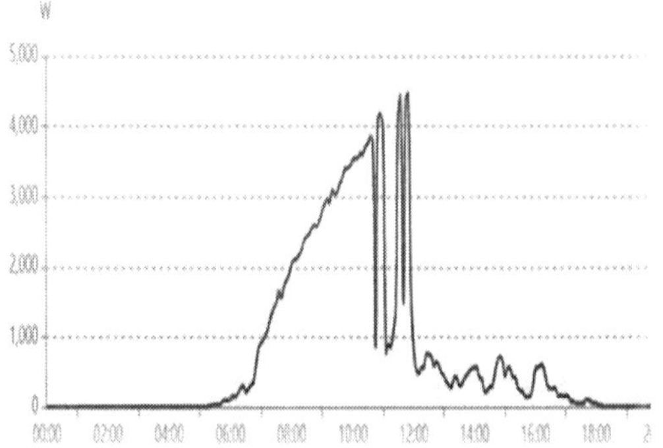

Figure 5: A summer's day when afternoon thunderstorms with rain occurred. This is quite common in this area, with the generation on this particular day only 14 kWh, with a peak output of 4.4 kW.

This graph is shown to illustrate the unpredictability of cloudcover and weather, especially during the summer months (Nov – Febr) in this region. This unpredictability makes it challenging to do a scientific study, so the current paper simply deals with awareness of summer thunderstorm unpredictability, but focus on winter shading due to the shift in the astrological path of the sun from summer to winter time.

5.2 Seasonal changes

In this section, all results presented will be for cloudless, perfectly sunny days. In Fig 6, electrical generation for a winter's day (June 2020) is represented. Firstly, note that generation on this day starts just after 07h00, and ends shortly after 16h00 in the afternoon. Another noticeable factor in this case, is the generation lost due to shading between 08h00 and around 10h30 in the morning.

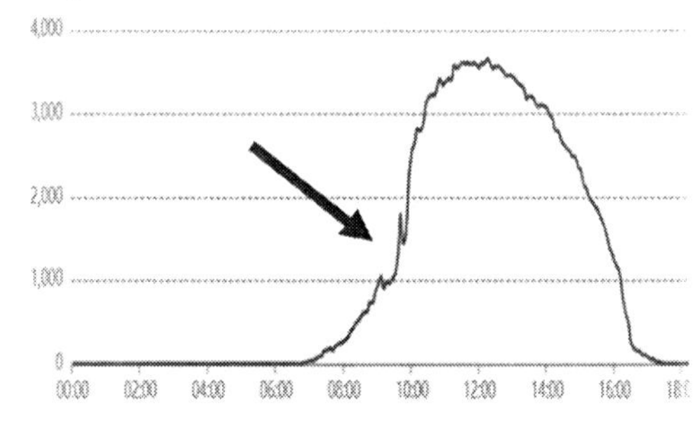

Figure 6: The PV electricity generation graph for a perfect cloudless sunshine day in May 2020. The total amount of electrical energy generated for the day was 20.20 kWh, with a peak generation of 3.6 kW. The arrow indicates the influence (loss) due to shading caused by trees.

This shading is due to the trees in the background of the photograph in Fig 2, which only has an effect during winter months (May – Aug). The effect of the shading due to the trees become more profound as the astronomical path of the sun shifts going deeper into wintertime. This is illustrated on the graph in Fig 7, with the arrow indicating the shift from the normal isometric graph without any shading. The loss of daily generation is clear to see on the graph, and also confirmed by the figures, i.e. 20.2 kWh in May, vs 19.8 kWh in June.

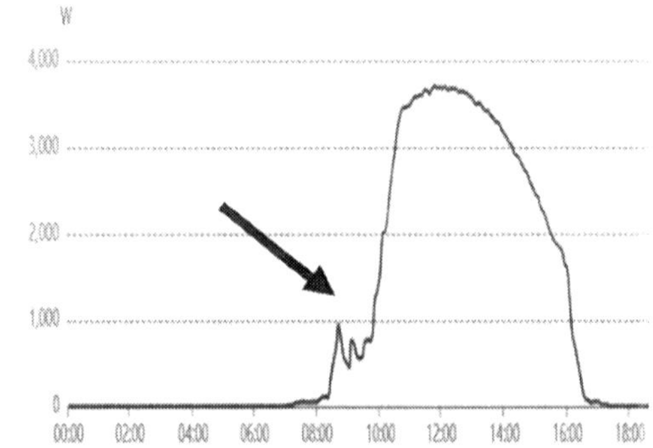

Figure 7: PV electricity generation graph for a perfect cloudless sunshine day in June 2020. The total amount of electrical energy generated for the day was 19.8 kWh, with a peak generation of 3.7 kW. The arrow again indicates the shift in generation due to the shading.

The recovery of the energy losses due to shading from the trees starts as the astronomical path of the sun starts shifting as time moves out of winter. i.e. from 21 June (shortest day) onwards. The representative PV generation graph for a perfectly cloudless day in July are shown in Fig 8. The arrow again points out the abnormality in generation due to the shading, but noticeably less than approximately a month earlier

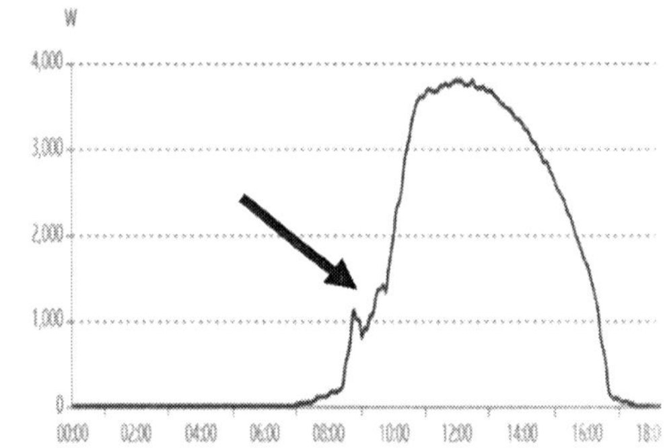

Figure 8: Generation graph for a completely cloudless sunny day in July 2020. The arrow again indicates the generation losses due to shading, which is clearly less than in June. The total generation for that day was 21.10 kWh, with a peak power of 3.8 kW.

This recovery of generation losses due to shading continues into August (Fig 9), as the generation graph shows a near-symmetrical curve and is near-nullified come September (Fig 10).

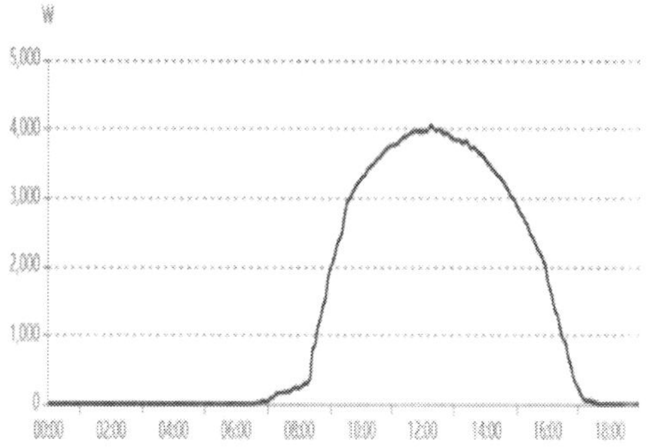

Figure 9: Electricity generation graph for a day in August 2020. The shading as illustrated in Fig 6-8 is nearly nullified, with abnormal increase present between 8h00 and 10h00, compared to the near-symmetrical graph in Fig 4. On this day, 25.10 kWh was generated, with a peak power of 4.0 kW.

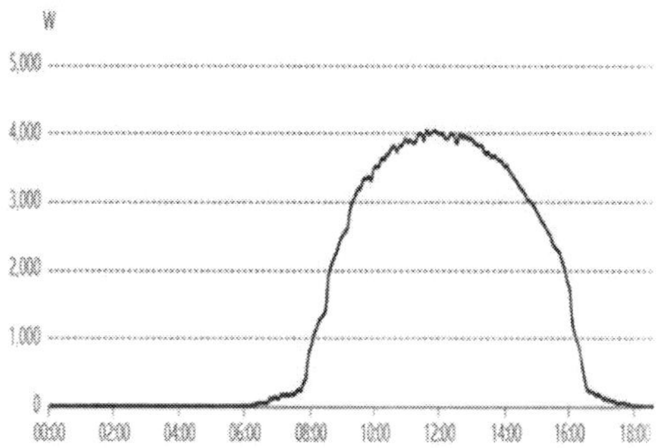

Figure 10: Generation graph on 13 September 2020 resulting in 26.3 kWh, with a 4.0 kW peak power. Note the recovery of generation between 8h00 and 10h00 to a near-symmetric curve.

The generation curves for October and November (Fig's 11 and 12 respectively) shows near symmetrical curves, confirming that shading due to the trees on a perfectly cloudless day does not influence the generation of electricity any longer.

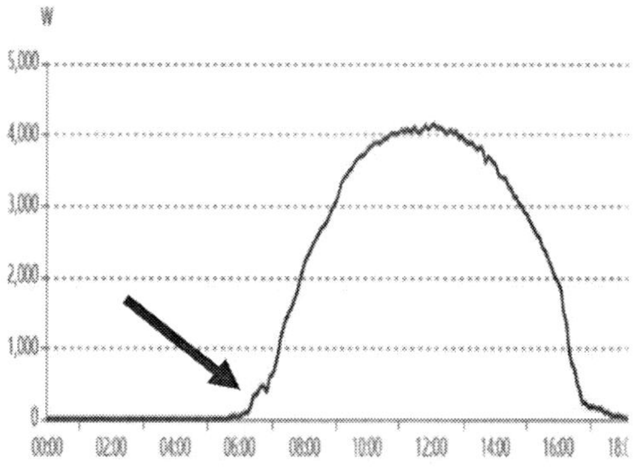

Figure 11: PV generation graph on 15 October 2020 resulting in 29.5 kWh, with a 4.14 kW peak power. Generation in the morning is now near-maximum, while the arrow also indicates a more rapid increase of PV generation between 6h30 and 8h00 due to the absence of shading.

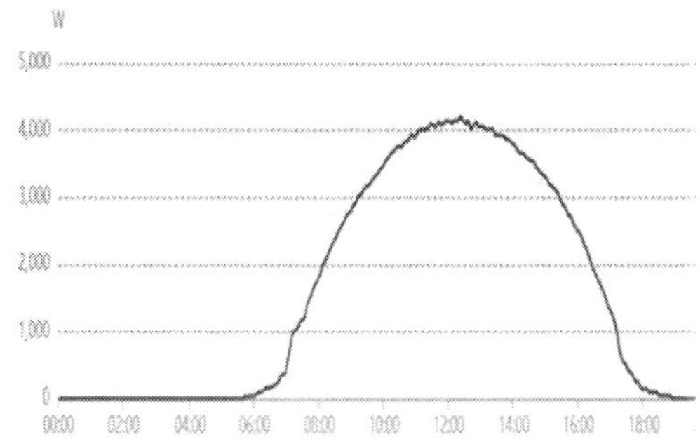

Figure 12: Near-symmetrical graph showing the PV electricity generation on 14 November 2020. Electrical energy to the amount of 31.5 kWh was generated on that day, with a peak power of 4.3 kW.

Furthermore, generation for Oct-Nov is generally at a monthly maximum for this area, despite of maximum daily generation to be higher in the months of Dec and Jan. The reason for this is that monthly rainfall in the area (a summer rainfall region) is maximum in Dec through to Feb, hence more cloud cover in these months [8]. These generation figures are represented in Table 1 and Fig 13, showing that the monthly generation for Oct/Nov is highest, with a clear decrease going into Dec due to the rainy season. This decrease is also observed into Jan/Feb (although not shown in the current graph).

Table 1: Monthly electricity generation for May-Dec 2020 during this study.

Month, 2020	Total Generation
May	610.9 kWh
June	507.3 kWh
July	603.8 kWh
August	679.4 kWh
September	685.4 kWh
October	701.4 kWh
November	697.4 kWh
December	656.5 kWh

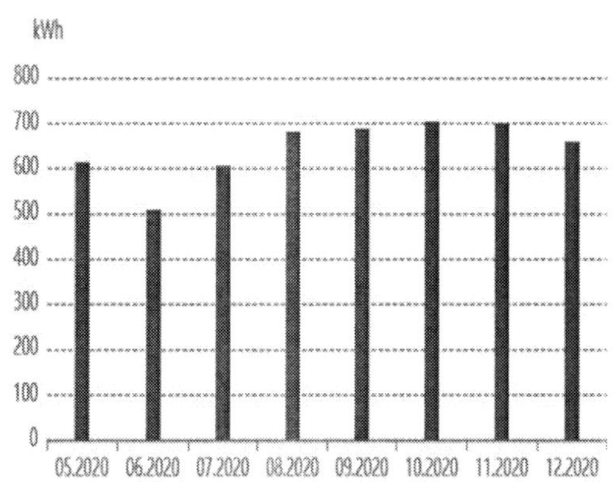

Figure13: Monthly electrical energy (kWh) generated by the PV panels in this study.

6 CONCLUSIONS AND FUTURE WORK

The generation of electricity in a solar system is done by the photovoltaic panels when sunlight is incident upon them. Therefore, shading in any form (clouds, trees, etc) is disadvantageous and the maximum possible generation will not be achieved. This paper illustrated the effect of seasonal shading due to the astronomical shift of the sun's path in winter compared to summertime. Approximately 250 kWh of electricity generation was lost from the months of May through to August due to shading that is not present in September to April. This is due to the shift in astrological path of the sun during winter months. However, because of inconsistent cloud cover during the rainy season (Dec to Feb), maximum generation during those months is also not achieved, and is on average not much more as for the winter months. It can thus be concluded that for the Southern-Gauteng region (where Vanderbijlpark is situated) and surrounding areas in South Africa, the maximum PV generation will be in the months of October and November. Therefore, design consideration for the installment of the PV panel array is important in a solar system. Apart from latitude and orientation, the amount of shading experienced due to obstructions as well as the difference between the astrological path of the sun in winter and summer should be seriously considered.

7 REFERENCES

[1] Sinetech Website. Retrieved February 2021 from http://www.sinetech.co.za/solar-house-cross-section-hybrid.html

[2] Trina website. Retrieved January 2020 from http://www.trinasolar.com/us/product/ALLMAX-M-PLUS.html

[3] Asowata, O., Swart, J., & Pienaar, C. (2012). Optimum tilt and orientation angles for photovoltaic panels in the Vaal triangle. In 2012 Asia-Pacific Power and Energy Engineering Conference, Shanghai, China.

[4] Goodwe website. Retrieved February 2020 from https://www.goodwe.com/Public/Uploads/sersups/ES%20USER%20MANUAL.pdf

[5] Pylontech website. Retrieved February 2020 from http://www.pylontech.com.cn/download.aspx?id=51

[6] J. D. Cutnell and K. W. Johnson, *Introduction to Physics*, 9th ed., John Wiley & Sons Ed. Singapore Pte. Ltd., 2013.

[7] Kumar, M.N., Saini, H.S., Anjaneyulu, K.S.R. and Singh, K, 'Solar Power Analysis Based On Light Intensity', *ICIEEE Proceedings, 2014*

[8] South African Weather Services website, Retrieved August 2021 from https://www.weathersa.co.za/home/historicalrain

This is a short version of the paper submitted to the journal EPJ Photovoltaics.
It has been selected by the Executive Committee of the EU PVSEC 2021 for submission to EPJ Photovoltaics.

EXPERIMENTAL STUDY OF THE OPERATION OF PV STRINGS AT THE MPP CLOSEST TO THE NOMINAL MPP VOLTAGE INSTEAD OF THE GLOBAL MPP

Kari Lappalainen and Seppo Valkealahti
Tampere University, Electrical Engineering
P.O. Box 692, FI-33101 Tampere, Finland
E-mail: kari.lappalainen@tuni.fi, seppo.valkealahti@tuni.fi

ABSTRACT: Under non-uniform operating conditions, photovoltaic (PV) generators can have multiple maximum power points (MPP) which may cause problems for MPP tracking. Since highly varying global MPP (GMPP) voltage causes large fluctuations in the inverter reference voltage, it would be beneficial to keep the operating point of the inverter all the time close to the nominal MPP voltage. In this way, operation of the PV system would be more predictable and straightforward. This paper presents an experimental study based on measured current–voltage curves of two PV strings of a scenario in which the MPP closest to the nominal MPP voltage is used all the time as the operating point instead of the GMPP. In total, 432000 *I–U* curves measured over 120 hours were analysed. The effects of inverter sizing on the selection of the operating point of the PV strings were also studied. The experimental results presented in this paper demonstrate that the wide operating voltage range when the GMPP is followed can be significantly reduced by operating at the MPP closest to the nominal MPP voltage at a cost of negligible energy losses.
Keywords: Voltage Fluctuation, Plant Control, Shading, PV System

1 INTRODUCTION

Photovoltaic (PV) power generators are constantly prone to variation in their operating conditions. Especially, fast irradiance fluctuations caused by overpassing cloud shadows can have negative effects on the operation of PV generators and electrical grids causing rapid fluctuations in the power fed into the grid. During uniform operating conditions, the electrical characteristic of a PV generator has only one peak, i.e., maximum power point (MPP). On the other hand, during non-uniform operating conditions, the electrical characteristics of the PV cells of the generator differ from each other, and as a result, the electrical characteristic of the entire generator may have several MPPs. The MPP where actual maximum power is achieved is called the global MPP (GMPP) while the MPPs with lower powers are called local MPPs (LMPP).

Existence of several MPPs makes MPP tracking (MPPT) more difficult and may lead to operation at an LMPP. Thus, several new algorithms for MPPT of partially shaded PV generators have been developed in the last few years [1], [2]. Moreover, the GMPP voltage can vary over a wide voltage range [3], [4]. Typically, inverters have defined allowed voltage ranges for proper operation and, accordingly, applied MPPT algorithms have defined operational voltage ranges to ensure that the GMPP is followed under varying operating conditions. Hence, the information about the applicable voltage range of the GMPP of the installed PV generator is of great value for successful choice of the inverter voltage range.

PV capacity is usually oversized with respect to the inverter meaning that the PV generator nominal DC power exceeds the inverter nominal AC power [5]. Oversizing of PV capacity restricts the output power of the PV generator to the inverter nominal power during high irradiance conditions. The inverter will operate in power limiting mode if the GMPP power of the generator rises above the inverter maximum power. In that case, the inverter operating point is forced to move to voltages higher than the GMPP voltage to reduce the power and current of the inverter. Hence, operating in power limiting mode leads to losses of available PV energy production. Moreover, it affects the operation and efficiency of the inverter: the inverter capacitor lifetime shortens [6] and the efficiency of some inverters decreases [7] with increasing DC side voltage.

In the last ten years, several studies have been presented related to the MPP characteristics of PV generators, based on simulations in [8], [9], [10], [11] and based on electrical measurements in [3], [4]. However, the effects of inverter sizing on the operating point were not considered in these studies. Moreover, fictitious irradiance values were used in [8], [11] and only the partially shaded time of the studied generators was considered in [3], [4].

Since the highly varying GMPP voltage causes large fluctuation of the inverter reference voltage, posing challenges for MPPT, it would be beneficial to keep the operating point of the inverter all the time at voltages close to the nominal MPP voltage. In this way, the operation of the PV system would be more straightforward, smoother and more predictable. This paper presents an experimental study of the scenario in which the MPP closest to the nominal MPP voltage (CMPP) is always the operating point instead of the GMPP. The study is based on measured current–voltage (*I–U*) curves of 2 PV strings located at Tampere, Finland. In total, 432000 *I–U* curves measured over 120 hours are analysed. Moreover, the effects of inverter sizing on the operating point behaviour of the PV strings are studied. An equally exhaustive study on the optimum operating point of PV strings founded on actual electrical measurements has not been presented before. The results presented in this paper are especially relevant for PV system and generator design as well as for improvement of MPPT algorithms while trying to achieve higher PV system efficiencies.

2 EXPERIMENTAL DATA AND METHODS

The experimental data consists of over 400000 measured *I–U* curves of 2 PV strings of the PV research plant of Tampere University [12]. Six consecutive days of full-time measurements were analysed for both strings. Measurement period of each day was from 8:00 to 18:00 (UTC+2). The layout of the studied PV strings, consisting of 17 and 6 series-connected PV modules, is presented in Fig. 1 and the details of the strings are compiled in Table I. The analysed measurements of

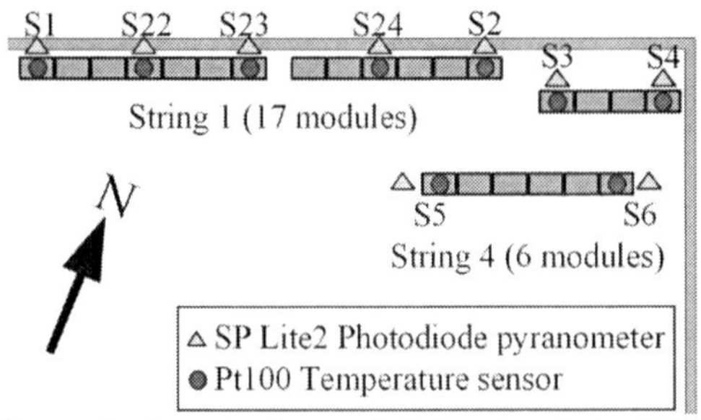

Figure 1: Layout scheme of Strings 1 and 4 of PV research plant of Tampere University.

Table I: Details of the studied PV strings.

	String 1	String 4
Number of PV modules	17	6
Nominal MPP voltage (V)	440	155
Nominal MPP power (W)	3230	1140
Length (m)	28.8	8.9

Strings 1 and 4 were performed on 14–19 and 7–12 August 2020, respectively. An *I–U* curve was measured once a second during the measurement period, using an *I–U* curve tracer where IGBTs act as a variable load. Thus, in total, 432000 measured *I–U* curves were analysed. Irradiance and temperature of seven PV modules of String 1 and two PV modules of String 4 were measured with a sampling frequency of 10 Hz. The irradiances incident on the PV modules were measured by photodiode-based SP Lite2 pyranometers, mounted at the same 45° tilt angle as the PV modules, while the PV module back-sheet temperatures were captured by Pt100 temperature sensors.

Each *I–U* curve involves 4000 measured *I–U* pairs. In order to reduce noise and other inaccuracies in the measured *I–U* curves, they were pre-processed by the following procedure. Firstly, the measurement points with identical voltage value were replaced with a single new point by averaging their current values. Thereafter, noticeably abnormal measurement points were removed. A measurement point was removed if its power differed from the power of the previous and next measurement point (to same direction) by more than 1.3 times the mean change of power between adjacent measurement points in its vicinity (previous and next 9 points). After the abnormal measurement points were removed, the measured current and voltage were smoothed separately using smooth.m function in MATLAB. An example shown in Fig. 2 illustrates the pre-processing method. A similar pre-processing method was used in [4].

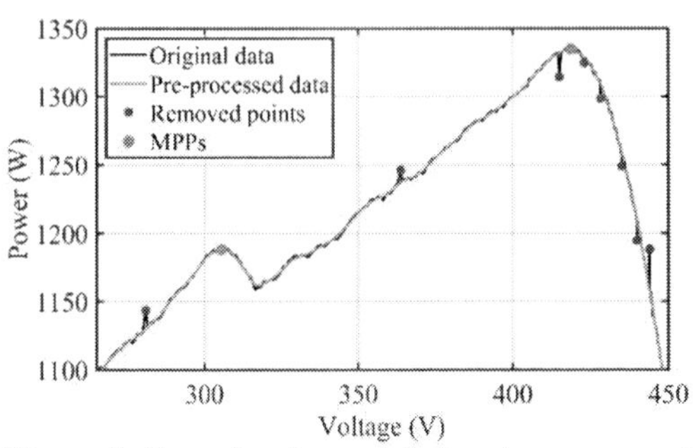

Figure 2: Example of an original and pre-processed measured *P–U* curve near the GMPP.

The effects of inverter sizing on the MPP behaviour of the PV strings were studied by altering the DC/AC ratio, i.e., the ratio of the nominal DC power of the PV string to the inverter nominal AC power. MPPT was assumed to work ideally, so that the string is operating at its GMPP (or CMPP) unless it is in power limiting mode. If the power at the GMPP (or the CMPP) exceeds the inverter nominal power, the string operates on the high voltage side of the GMPP (or the CMPP) at the lowest voltage where the inverter nominal power is not exceeded. The DC/AC ratio was altered from 0.8 to 2.0.

3 RESULTS AND DISCUSSION

The distributions of the measured voltages of the GMPP and the MPP closest to the nominal MPP voltage are presented in Fig. 3 for the studied PV strings. Both voltages were most of the time below the nominal value. This results from the typical operating conditions of the studied PV strings: during the studied periods, the irradiance was mostly lower and the cell temperature was higher than in standard test conditions, thus the MPP voltages were typically lower than the nominal MPP voltage. The GMPP voltage of String 4 was all the time below the nominal value and the GMPP voltage of String 1 was higher than the nominal value only 0.06% of time. The lowest and highest measured voltages of the two MPPs are compiled in Table II. The measured voltage ranges of the GMPP are largely in accord with the experimental results of [3] and [4] as well as with the simulation results of [9].

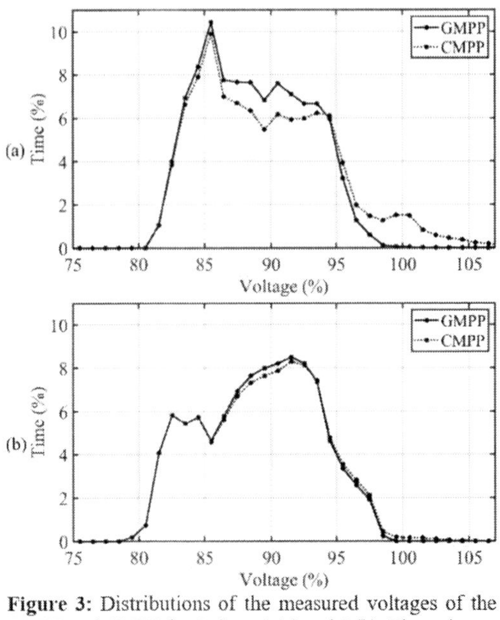

Figure 3: Distributions of the measured voltages of the GMPP and CMPP for Strings 1 (a) and 4 (b). The voltages are with respect to the nominal MPP voltages.

The differences in voltage distributions between the two MPPs are much larger for the physically longer String 1 (Fig. 3 (a)) than for the shorter String 4 (Fig. 3 (b)). The GMPP voltage range of String 1 was very wide from 38% to 104%. The voltage range can be significantly reduced by operating at the CMPP with a voltage range from 81% to 116%. This is an important finding, demonstrating that it would be beneficial for PV systems

Table II: Measured voltage ranges of the GMPP and CMPP for the studied PV strings. The voltage values are with respect to the nominal MPP voltages.

	Minimum voltage (%)	Maximum voltage (%)
String 1, GMPP	38.2	104.4
String 4, GMPP	79.2	99.6
String 1, CMPP	81.0	115.5
String 4, CMPP	79.2	108.7

to operate at the MPP closest to the nominal MPP voltage instead of the GMPP. However, for very short PV strings (String 4) corresponding advantage is not achieved. For String 4, both MPPs have the same quite high lower bound of voltage range, while the maximum CMPP voltage is higher than the maximum GMPP voltage.

The distributions of the measured powers of the GMPP and the MPP closest to the nominal MPP voltage are presented in Fig. 4 for the studied PV strings. By comparing Figs. 3 and 4, it can be seen that the differences in power between the two MPPs are smaller than in voltage. Small differences in power mean that operation at the CMPP instead of the GMPP does not cause significant energy losses. MPP powers higher than the nominal power were measured for both strings. These situations are caused by the cloud enhancement phenomenon [13]. Power values higher than the nominal value are more common for String 4 as it is physically shorter than String 1 and thus exposed to higher enhanced irradiances. One should also note in here that the clear sky irradiance in Tampere region never reaches the standard test condition irradiance of 1 kW/m^2, which causes the nominal MPP power.

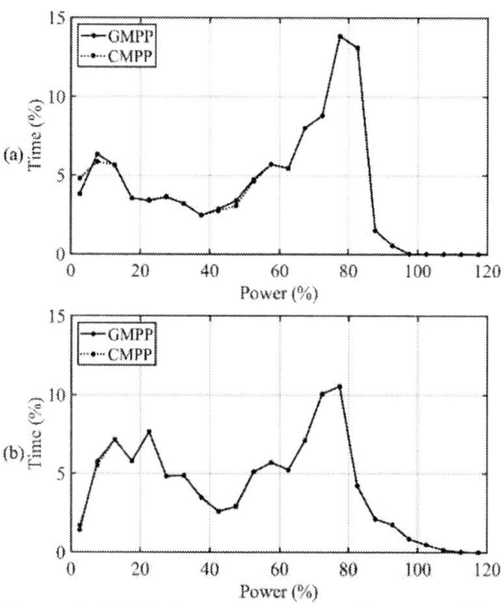

Figure 4: Distributions of the measured powers of the GMPP and CMPP for Strings 1 (a) and 4 (b). The powers are with respect to the nominal MPP powers.

Difference in length explains the differences between the studied strings in Figs. 3 and 4. String 1 is physically over three times longer than String 4 (see Table I). Thus, there are typically larger irradiance and temperature differences between the modules of String 1 than between the modules of String 4. As a result, multiple MPPs exist more often for String 1 than for String 4. Indeed, String 1 had more than one MPP 15.8% of the time while String 4 had multiple MPPs only 2.5% of the time.

In Fig. 5, the shares of time when the studied PV strings operated in power limiting mode are presented as a function of DC/AC ratio. Naturally, the shares of time when the PV strings operated in power limiting mode increased with the increasing DC/AC ratio, since the irradiance level needed to produce PV power exceeding the nominal inverter power decreases as the DC/AC ratio increases. There is a clear difference between the studied strings. With DC/AC ratios below 1.2, String 4 spends more time in power limiting mode than String 1. That is logical since with low DC/AC ratios relatively high irradiance level is needed to produce PV power exceeding the nominal inverter power. String 4 has smaller area than String 1, and thus its modules receive high enough irradiance more often. However, as the DC/AC ratios increase, the proportion of time in the power limiting mode increases sharply, and String 4 spends less time in power limiting mode than String 1. The differences between the GMPP and CMPP were very small with virtually overlapping curves but the shares were a bit larger for the GMPP.

Fig. 6 presents the highest, median and lowest operating voltages as a function of DC/AC ratio while operating at the GMPP or at the CMPP for the studied strings. The voltages increased with the increasing DC/AC ratio since the strings were operating on the high voltage side of the GMPP (or the CMPP) in power limiting mode. The maximum GMPP voltage approached the maximum CMPP voltage as the DC/AC ratio increased. Thus, the difference in voltage ranges between the MPPs increased as the DC/AC ratio increased. The maximum GMPP voltage of String 4 reached the maximum CMPP voltage already with a DC/AC ratio of 1.2 but not completely for the longer String 1.

Fig. 7 presents the relative energy losses due to operation at the CMPP instead of the GMPP and due to power curtailment. The relative energy losses due to operation at the CMPP increased with increasing DC/AC ratio for String 1 and were much larger than for String 4, which remained quite constant. However, only negligible amount of energy would be lost if the PV strings operated all the time at the CMPP instead of the GMPP. This demonstrates that the wide operating voltage range when the GMPP is followed can be significantly reduced by operating at the MPP closest to the nominal MPP voltage at a cost of negligible energy losses. The relative energy

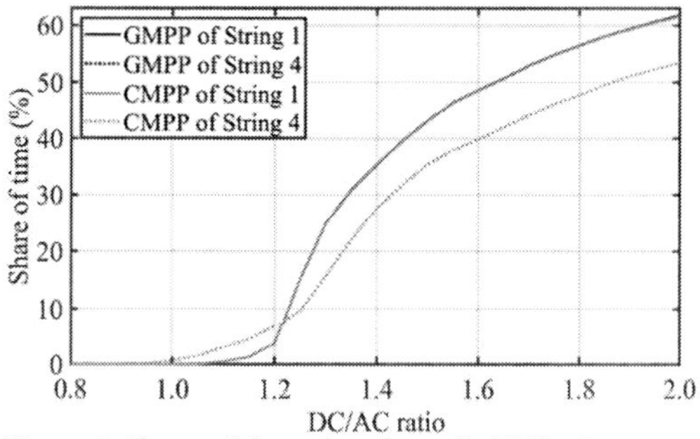

Figure 5: Shares of time when the studied PV strings were in power limiting mode when operating at the GMPP or at the CMPP as a function of DC/AC ratio.

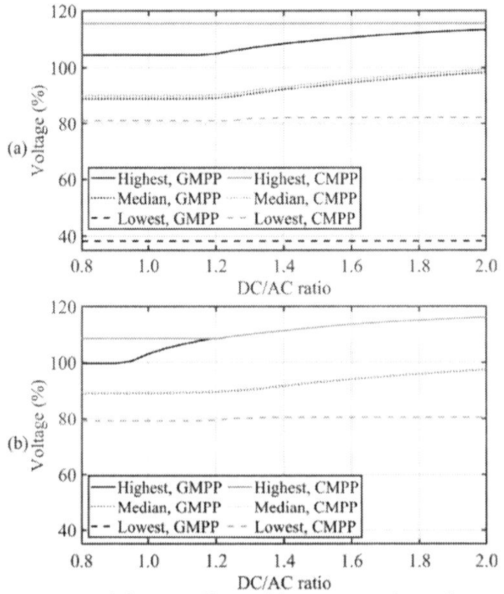

Figure 6: Highest, median and lowest operating voltages as a function of DC/AC ratio while operating at the GMPP or at the CMPP for Strings 1 (a) and 4 (b). The voltages are with respect to the nominal MPP voltages.

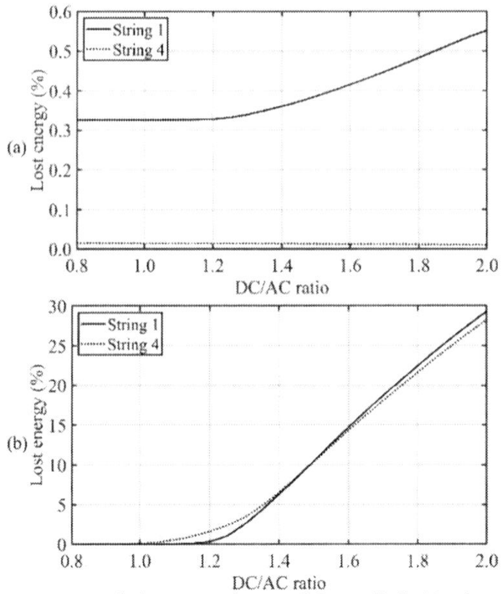

Figure 7: Relative energy losses of the studied PV strings due to operation at the CMPP instead of the GMPP (a) and due to power curtailment (b) as a function of DC/AC ratio. The relative energy losses due operation at the CMPP instead of the GMPP were calculated with respect to energy produced at the GMPP (taking power curtailment into account). The relative energy losses due to power curtailment were calculated with respect to energy produced at the GMPP without power curtailment.

losses due to power curtailment were much larger than the relative energy losses due to operation at the CMPP. They increased strongly with increasing DC/AC ratio being around 30% with a DC/AC ratio of 2.0. With small DC/AC ratios below 1.1, the strings were in power limiting mode only during cloud enhancement. As explained earlier, the share of time spent in power limiting mode increases with the increasing DC/AC ratio since the irradiance level needed to produce PV power exceeding the nominal inverter power decreases.

4 CONCLUSIONS

This paper presented an experimental study of the GMPP characteristics of PV strings and a scenario in which the MPP closest to the nominal MPP voltage is always the operating point instead of the GMPP. The effects of inverter sizing on the operating point behaviour of the PV strings were also studied. The study was based on 432000 *I–U* curves of 2 PV strings measured over 120 hours.

The experimental results demonstrate that it would be beneficial for PV systems to operate at the CMPP instead of the GMPP. For a string of 17 PV modules, a wide GMPP voltage range from 38% to 104%, with respect to the nominal MPP voltage, can be significantly reduced by operating at the CMPP with a voltage range from 81% to 116%. Moreover, the results show that only negligible amount of energy would be lost if the PV strings operated all the time at the CMPP instead of the GMPP. Thus, the wide voltage range of the GMPP can be significantly reduced by operating at the CMPP at a cost of negligible energy losses compared to operation at the GMPP. However, for very short PV strings corresponding voltage range reduction is not achieved while the energy losses due to operating in the CMPP remain small.

Energy losses due to power curtailment were found to be much larger than energy losses due to operation at the CMPP instead of the GMPP. For example, with a typical DC/AC ratio of 1.5, over 10% of available energy in both strings would be lost due to power curtailment. These losses increased strongly with increasing DC/AC ratio being close to 30% with a DC/AC ratio of 2.0. The experimental results presented in this paper confirm the findings of earlier simulation studies.

ACKNOWLEDGMENTS

We acknowledge the financial support from Business Finland for the research reported in this paper.

REFERENCES

[1] M. Mao, L. Cui, Q. Zhang, K. Guo, L. Zhou, H Huang, Energy Reports 6 (2020) 1312. https://doi.org/10.1016/j.egyr.2020.05.013.
[2] B. Yang, T. Zhu, J. Wang, H. Shu, T. Yu, X. Zhang, W. Yao, L. Sun, Journal of Cleaner Production 268 (2020) 121983. https://doi.org/10.1016/j.jclepro.2020.121983.
[3] K. Lappalainen, S. Valkealahti, Proceedings of 37th European Photovoltaic Solar Energy Conference, Vol. I (2020) 1501. https://doi.org/10.4229/EUPVSEC20202020-5CV.3.4.
[4] K. Lappalainen, S. Valkealahti, Applied Energy 301 (2021) 117436. https://doi.org/10.1016/j.apenergy.2021.117436.

[5] H.X. Wang, M.A. Munoz-García, G.P. Moreda, M.C. Alonso-García, Renewable Energy 118 (2018) 709. https://doi.org/10.1016/j.renene.2017.11.063.
[6] J.M.S. Callegari, A.F. Cupertino, V.N. Ferreira, E.M.S. Brito, V.F. Mendes, H.A. Pereira, Microelectronics Reliability 100–101 (2019) 113439. https://doi.org/10.1016/j.microrel.2019.113439.
[7] G.A. Rampinelli, A. Krenzinger, F. Chenlo Romero, Renewable and Sustainable Energy Reviews 34 (2014) 578. https://doi.org/10.1016/j.rser.2014.03.047.
[8] A.M.S. Furtado, F. Bradaschia, M.E. Cavalcanti, L.R. Limongi, IEEE Trans. Industrial Electronics 65 (2018) 3252. https://doi.org/10.1109/TIE.2017.2750623.
[9] K. Lappalainen, S. Valkealahti, IET Renewable Power Generation 13 (2019) 2864. https://doi.org/10.1049/iet-rpg.2019.0085.
[10] K. Lappalainen, S. Valkealahti, Renewable Energy 152 (2020) 812. https://doi.org/10.1016/j.renene.2020.01.119.
[11] M. Boztepe, F. Guinjoan, G. Velasco-Quesada, S. Silvestre, A. Chouder, E. Karatepe, IEEE Trans. Industrial Electronics 61 (2014) 3302. https://doi.org/10.1109/TIE.2013.2281163.
[12] D. Torres Lobera, A. Mäki, J. Huusari, K. Lappalainen, T. Suntio, S. Valkealahti, International Journal of Photoenergy 2013 (2013) 837310. https://doi.org/10.1155/2013/837310.
[13] K. Lappalainen, J. Kleissl, Journal of Renewable and Sustainable Energy 12 (2020) 043502. https://doi.org/10.1063/5.0007550.

ANALYSIS OF TEMPERATURE INERTIA OF PV MODULES USING DIFFERENT TEMPERATURE ESTIMATION MODELS

Aline Kirsten Vidal de Oliveira[1], Marília Braga[1], Soo-Young Oh[2], Sebastian Dittmann[3,4], Ralph Gottschalg[3,4], Carlos D. Rodríguez-Gallegos[5], Thomas Reindl[5], Khawla Ettalbi[6], Tom R. Betts[7], Laurie Burnham[8], Ricardo Rüther[1]
[1]Universidade Federal de Santa Catarina, Florianópolis, Brazil
E-mail: alinekvo@gmail.com, Tel.: +55 (48) 99668-9926
[2]Yeungnam University, Gyeongsan, Republic of Korea;
[3]Anhalt University of Applied Sciences, Köthen, Germany;
[4]Fraunhofer-Center for Silicon Photovoltaics CSP, Halle, Germany;
[5]Solar Energy Research Institute of Singapore (SERIS), Singapore, Singapore;
[6]Green Energy Park (Institut de Recherche en Energie Solaire et Energies Nouvelles (IRESEN)), Rabat, Morocco
[7]Centre for Renewable Energy Systems Technology (CREST), Loughborough University, United Kingdom
[8]Sandia National Laboratories, Albuquerque, USA

ABSTRACT: Module temperature is one of the main factors that directly influence photovoltaic energy production, and therefore, its precise modelling is of foremost relevance. However, most of the temperature models in the literature fail to acknowledge the transient effect due to the heat capacity of the PV module, as they instead model the temperature based on the steady-state condition. This paper addresses these modeling weaknesses by comparing two well-established steady-state methods (the so-called Ross and Faiman models) with the Oh et al. model, which includes the transient effect of the heat capacity of PV modules in its calculations. Results show that the Oh et al. model provides more accurate results for high-resolution data when compared to steady-state models. The results are even more precise when using it to calculate the module temperature using the measurement from the previous time step, which indicates that the model could be used to forecast module temperature for a time step in the future. Across all models, climate is a major factor affecting model accuracy, what demonstrates the importance of the analysis of the impact of different site conditions on PV modeling and on the selection of models that are best matched to specific climates.
Keywords: Modelling, Thermal Performance; System Performance.

1 INTRODUCTION

The accurate forecasting of PV systems generation is a major issue to optimize the dispatch of energy in the electric grid and increase their reliability. Module temperature is one of the main factors that directly influence solar generation, and therefore, its precise modelling is of foremost relevance.

Several PV module temperature models are described in the literature [1]–[5]. However, according to the Temperature Study developed by the PV CAMPER "PhotoVoltaic Collaborative to Advance Multiclimate Performance and Energy Research" initiative [6], the accuracy of these models when predicting module temperature depends on the local environmental conditions of where the PV system is installed, while the measured reference temperature might vary according to sensor type and fixation method employed.

In addition, most of the temperature models fail to acknowledge the transient effect due to the heat capacity of the PV module, as they instead model the temperature based on the steady-state condition. In order to improve the accuracy of module temperature prediction, Oh's model [7] was developed, based on the heat transfer equation. The model showed promising results. Its evaluation should, however, be expanded to other sets of data, including different climates and different time resolutions. This paper aims to address this problem presenting a comparative study of different PV module temperature (Faiman's, Ross's, and Oh's) at multiple sites worldwide.

2 MATERIALS AND METHODS

This paper aims for the assessment of the accuracy of PV module temperature estimation models and the analysis of PV module temperature data to bring greater accuracy of PV modules and PV performance prediction. To achieve these goals, data measurements of different test sites around the globe (Fig. 1) are acquired, filtered and analysed. Each test site has back-of-module (BoM) temperature measurements, among other environmental measurements, such as ambient temperature, plane-of-array irradiance, and wind speed, which are also used as inputs for the temperature models. Data is acquired with a one-minute resolution, in order to allow the evaluation of the effect of thermal inertia of the modules. The analysis of the methods' accuracy and the comparison between them is done through the Root Mean Square Error (RMSE) metric.

2.1 Test Sites and Instrumentation

This experiment is part of the Module Temperature Study held by partner institutions of PVCAMPER. The international community has ten institutional members, each of which maintains one or more outdoor field laboratories, with a total of 13 test sites (Fig.1), that span both hemispheres and most major climate zones [8]. The Module Temperature Study aims to establish the best practices for BoM temperature measurements, reduce field measurement uncertainty, assess the accuracy of PV module temperature estimation models and therefore, bring greater accuracy of PV modules and PV performance prediction.

Fig. 1 shows the location of the PV CAMPER collaborative test sites, which coincide with climatic regions of interest for PV systems. They represent diverse climates and enable the analysis of module temperature measurements in different climatic conditions. Fig. 1 also highlights, the location of the seven test sites analysed in this paper, that are also described in Table I. This paper focuses on the analyses of the temperature behaviour of multi-crystalline silicon (multi-Si) PV .

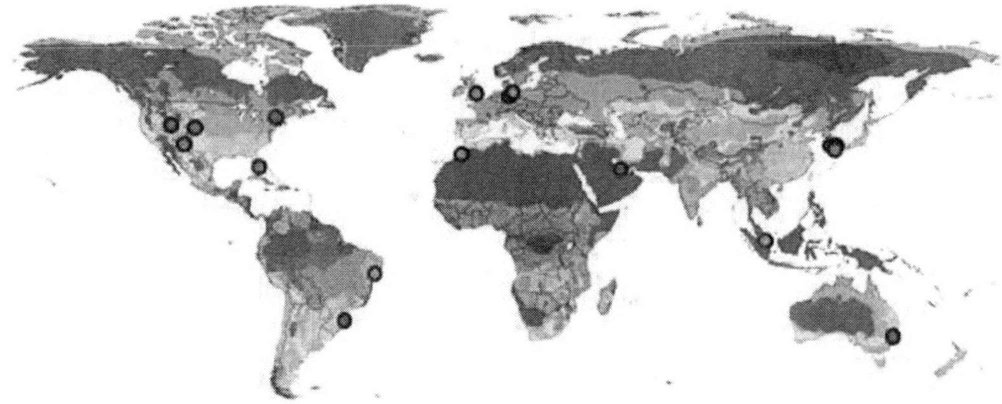

Figure 1: PV CAMPER research sites are represented by circles superimposed on a Köppen-Geiger climate map (adapted from [9]). Sites contributing data to this paper are presented as blue circles. Other sites that are part of the Temperature Study are presented as green circles. Additional PV CAMPER test sites are shown as red circles.

Table I: Description of the test sites.

Member Institution	Location	Latitude	Longitude	Global POA Irradiance	Back-of-module Temperature	Sensor Fixation
Anhalt University	Bernburg, Germany	51.77°N	11.77°E	SMP10-V	PT1000	Double-sided adhesive tape
CREST	Loughborough, United Kingdom	52.76°N	1.24°W	CMP11	PT100	Silicone adhesive sealant cocoon
IRESEN	Benguerir, Morocco	32.12°N	7.52°W	CMP21	PT100	Double-sided adhesive tape
SANDIA	Henderson, NV, USA	36.01°N	114.55°W	CMP-11	T thermocouple	Adhesive tape
SERIS	Singapore, Singapore	1.28°N	103.87°E	SMP11	PT1000	Double-sided adhesive tape
UFSC	Brotas de Macaúbas, Brazil	12.31°S	42.34°W	CMP11	PT1000	Double-sided adhesive tape
YU	Gyeongsan, South Korea	35.82°N	128.76°E	CMP10	T thermocouple	Adhesive tape

2.2 Temperature models

This study compares two well-known steady-state model temperatures, with a PV module temperature that evaluates the heat capacity of the module and the temperature inertia that it generates. The used models are presented in next sub-sections.

2.2.1 Ross's model

The most traditional method for temperature estimation in photovoltaics is the Ross's method [3], presented in Eq. (1).

$$T_{Ross} = T_{amb} + kG \quad (1)$$

Where:

- k is known as the Ross coefficient, a constant value given in °C·m²/W, that depends on module PV technology, shape, size, mounting, among other factors;
- G is the plane of array incident irradiance, given in W/m²;
- T_{amb} is the ambient temperature, in °C;
- T_{Ross} is the resultant estimated module temperature, also in °C.

In this study, k values of 0.024 °C.m2/W were used for free-standing mc-Si systems, as proposed by Segado et al., (2015).

2.2.2 Faiman's model

Another well-accepted model is the Faiman's model [1], and it shows improvements compared to Ross's model, as it includes wind speed values in its evaluation. The model is presented in Eq. (2).

$$T_{Faiman} = T_{amb} + \frac{\alpha G(1-\eta_m)}{U_c+U_v WS} \quad (2)$$

Where:

- α is the absorptance of the module (typical value is 0.9);
- η_m is the efficiency of the module (typically 0.15-0.2 for mc-Si)
- WS is the wind speed, in m/s;
- U_c is the constant heat transfer coefficient, in W/m²·°C);
- U_v is the convective heat transfer

coefficient, in W/m^2·°C / m/s);

- T_{Faiman} is the resulting estimated module temperature, in °C.

The constant and convective heat transfer coefficients are calculated through a linear regression using ambient and measured module temperature values, as described in [1]. In the absence of measured temperature values, one can find typical values for different PV technologies and mounting types in the literature. For instance, Segado et al. [4] lists typical values ranging from 23.5 to 26.5 W/m^2·°C for U_c and from 6.25 to 7.68 W/m^2·°C / m/s for U_v.

The widely used PV simulation software PVsyst® [10], employs the Faiman model in the thermal modelling in its simulations. The constant heat transfer coefficients suggested by the software range from 15 to 29 W/m^2·°C, depending on mounting type (fully insulated back, semi-integrated or free-standing systems). By default, the software sets the convective heat transfer coefficient to zero, considering the lack of quality wind speed data for simulation of PV systems and that this parameter is usually measured at higher elevation and in a free environment, which is usually not representative of the velocity at the array level. However, the user can edit these values according to the simulated system and the quality of the input data.

2.2.3 Oh's model

Oh's model is not a conventional model, it is based on the heat transfer equation including the transient effect due to the heat capacity of the PV module [7]. The model is described by Eq. (3).

$$T_{Oh(t+1)} = n\{G_{(t)} - P_{(t)} - h[T_{m(t)} - T_{amb(t)}]\} \quad (3)$$

Where:

- h is the heat transfer coefficient, found empirically, in $W/m^2 \cdot °C$;
- n is found empirically, and represents $n = \frac{\Delta T}{mC}$, in $\frac{°C}{kg \cdot J/K}$;
- P is the power output of the module, in W/m^2;
- T_m is the module temperature, in $°C$;
- T_{Oh} is the resulting estimated module temperature, in $°C$;
- ΔT is the temperature variation, in $°C$;

All variables are a function of time t and are given either in the present instant (t) or the next time step $(t + 1)$. To apply this model to complete sets of data (one full year, in this paper), two different approaches can be considered, as listed below. In this paper, both approaches are analyzed and compared among themselves, but only the first approach is compared to the other steady-state models.

1. $T_{m(1)}$, the module temperature for the first time step, is the ambient temperature. For the following calculations, $T_{m(t)}$ is $T_{Oh(t-1)}$, calculated in the previous time step $(t - 1)$.
2. $T_{m(t)}$ is the PV module temperature measured at each time step. This approach emulates a situation in which the current temperature is known, and the model intends to calculate the module temperature in the following time step. This type of situation applies to short-time forecasting of PV performance.

3 RESULTS

3.1 Time resolution

As described in the methodology section, the Oh et al. model is primarily differentiated by considering the temperature inertia of PV modules, thus allowing for a better estimation of short-term behaviour of cell temperature. To evaluate the benefits of this method, the root-mean-square error (RMSE) results are calculated for each test site and for different time steps (see Table II).

The results show that Ross and Faiman models perform better at larger time intervals in keeping with their steady-state nature. Especially for higher resolution data, the temperatures calculated with Faiman's and Ross's models vary greatly between time steps, not taking the inertia caused by the heat capacity of the module into account. The rapid changes on meteorological data decrease as the time steps get bigger, resulting in better results for the Faiman model for lower time resolution simulations.

Table II: RMSE results for the temperature models analyzed for 1-min, 5-min, 15-min and 30-min time steps. Best results for each site and method are shown in bold. Values are presented with units °C.

	Ross					Faiman					Oh**			
	1 min	**5 min**	**15 min**	**30 min**	**60 min**	**1 min**	**5 min**	**15 min**	**30 min**	**60 min**	**1 min**	**5 min**	**15 min**	**30 min**
Germany	7.4	7.3	7.1	7.0	**6.9**	3.1	3.0	2.9	2.8	**2.7**	**3.7**	3.9	4.6	-
UK	3.6	3.3	2.8	2.5	**2.2**	3.9	3.7	3.4	3.1	**2.8**	3.0	2.6	**2.4**	-
Morocco*	11.3	11.3	11.2	11.1	**10.9**	3.7	3.7	3.6	3.6	**3.5**	**3.4**	3.4	-	-
USA	4.4	4.9	3.6	3.4	**3.3**	5.0	5.4	4.3	4.1	**4.0**	**4.8**	5.2	4.5	3.9
Singapore	3.9	3.5	3.1	2.9	**2.7**	4.7	4.5	4.3	4.1	**4.0**	**2.6**	3.0	4.8	-
Brazil	6.1	5.7	5.3	**5.0**	5.3	5.9	5.6	5.1	4.8	**4.6**	**11.4**	11.4	11.5	11.7
South Korea	3.6	3.4	3.2	3.1	**3.1**	4.9	4.8	4.6	4.5	**4.3**	3.3	4.5	**3.0**	4.76

*for Morocco only 2-min data is available, thus results presented for this site have a maximum time resolution of 2-min

**Oh's model did not converge for 15-min and 30-min time steps at some test sites; thus, the corresponding spaces are left blank.

The Oh et al. model, on the other hand, performs better at shorter time intervals, as demonstrated in Fig. 2 and Fig. 3. that present examples of the comparison between the modelled and measured temperatures for a day in two different test sites. In both graphs, it is visible that the temperature predicted by Faiman's model varies greatly between time steps, not taking into consideration the inertia caused by the heat capacity of the module. Oh's model considers this factor and therefore provides more precise temperatures for high resolution predictions.

The low errors for Oh's model (case 1) also occur for the test site GER-BBG, that was already proven to be difficult for temperature modelling [6], especially when using Ross's model, that does not take the wind speed into account in the calculations. Fig. 3 shows the huge disparity between measured and modelled temperature using this model.

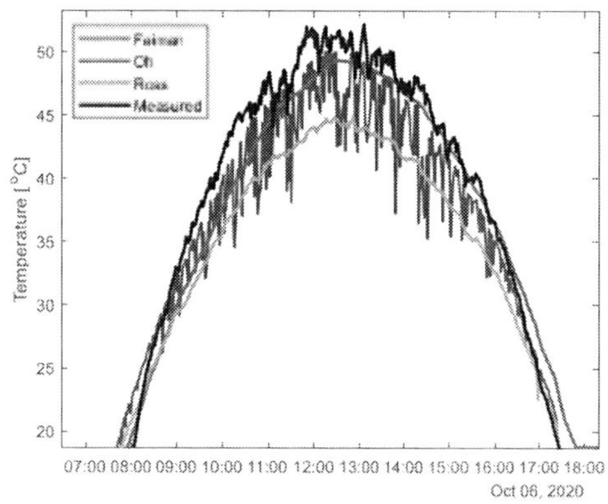

Figure 2: Measured and modelled module temperatures with 1-minute resolution for Oct 6th, 2020 for the YU-KOR test site, using different temperature models.

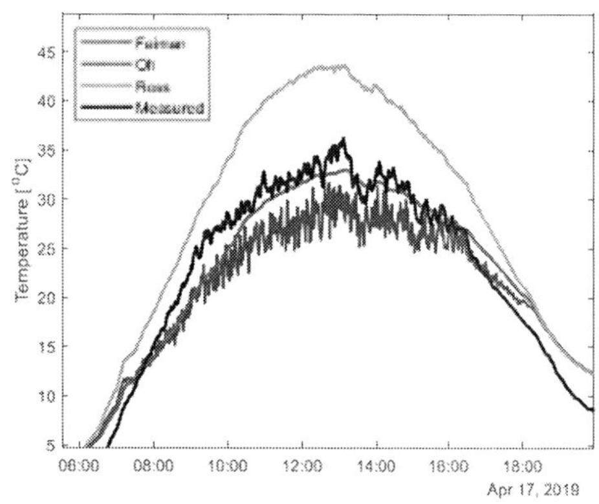

Figure 3: Measured and modelled module temperatures with 1-minute resolution for Apr 17th, 2020 for the BBG-GER test site, using different temperature models.

3.2 Climatic and environmental conditions

For 1-min estimations, the Oh et al. model outperforms the other models at four of the seven sites. Singapore produces the best results for Oh et al. model, with an RMSE of 2.8°C. This site has both low wind speed values (average 0.8 m/s and maximum 6.2 m/s) and low daily irradiation (average 4.2 kWh/m2) compared to the other sites. The Singapore site also has low estimation errors for the Ross model, which is expected considering the model does not include wind speed in its calculations. The lower RMSE values for the Singapore data could also be attributed to the site's low temperature range (13.7°C), with a minimum ambient temperature of 23.2°C and a maximum of 36.9°C (considering diurnal data only), which allows for an easier fit of data within the analyzed models. Fig. 4, Fig. 5 and Fig. 6 show measured and modeled module temperatures for this test site, where the lower error of Oh's model can be observed, as well as the improvement of the steady-state models in higher data resolutions.

The Brazilian site, which sees consistently high wind and high irradiance levels, including frequent over-irradiance events, was most challenging to model and had the highest RMSE values across the three models. The relative accuracy of the Oh et al. was not observed in this test site. This problem can be attributed to the models' inability to accurately represent high-irradiance scenarios, possibly associated to a non-linear correlation between irradiance and temperature above a certain irradiance level, especially for low frequency irradiance measurements, which may include unstable conditions of rapidly changing cloud cover leading to high irradiance events that might not be recorded if the irradiance time resolution is not high enough [10].

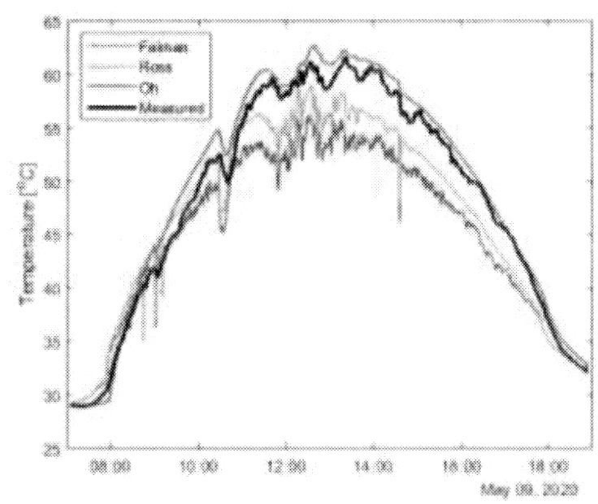

Figure 4: Measured and modelled module temperatures with 1-minute resolution for May 9th, 2020 for the SG-SG test site, using different temperature estimation models.

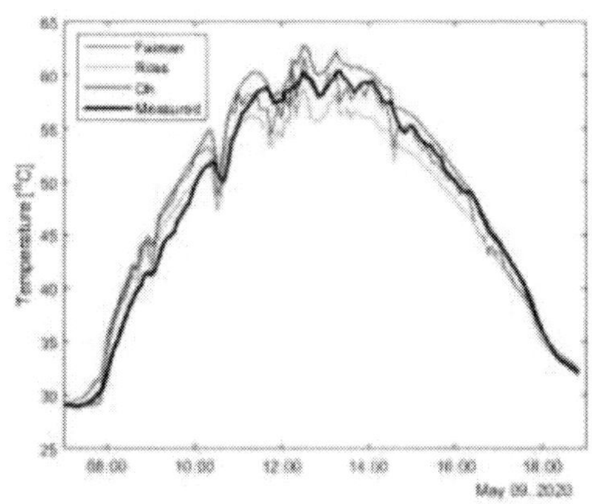

Figure 5: Measured and modelled module temperatures with 5-minute resolution for May 9th, 2020 for the SG-SG test site, using different temperature estimation models.

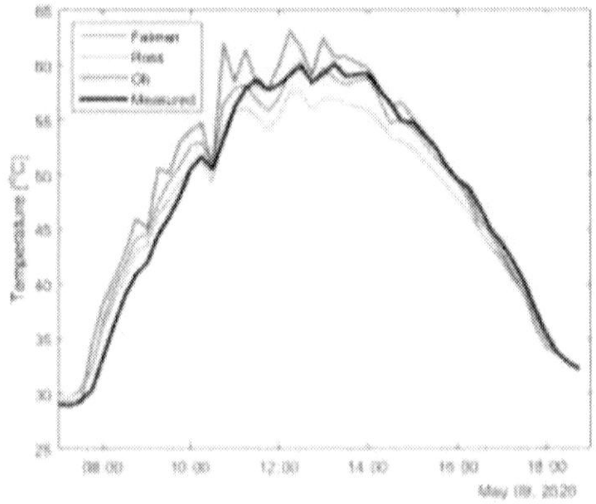

Figure 6: Measured and modelled module temperatures with 15-minute resolution for May 9th, 2020 for the SG-SG test site, using different temperature models.

The USA site presented the smallest variation among results for the three analyzed models, with a difference of 0.6°C between the highest (Faiman, 5.0°C) and lowest (Ross, 4.4°C) RMSE values. The UK site also presented a small RMSE variation between the best and worst models that can be explained by the fixation method for the module temperature sensor which is cocooned with gas-flue silicone adhesive sealant. The site is also colder than the others, with PV operational temperatures closer to ambient.

The results across all sites indicate a strong correlation between environmental conditions and the accuracy of the models under analysis.

3.3 Time step ahead prediction

Table III presents a comparison between the mean RMSE obtained for the two applications of Oh's model, as described in the methods section.

Table III: Mean RMSE results of 1-minute data for the test sites analysed.

Test Site	Mean RMSE	
	Oh 1	Oh 2
Germany	3.7	1.0
UK	3.0	0.3
Marocco	3.5	1.0
Singapore	2.6	0.4
Brazil	11.4	1.7
South Korea	3.3	0.9

Comparing the results for both approaches for Oh's model it can be observed that Oh 2 presents much better results, a consequence of the use of measured temperature from the previous time step in the calculation of the temperature in the current time step. In this case, the errors are minimal, which means that the method is very effective in predicting the PV module temperature for a time step, given that the measurement value of the previous time step is provided. Fig. 7 shows an example of one day comparison between measured and modelled temperatures for the test site SG-SG, using the case 2 of Oh's model. A part of the graph is zoomed in order to show the small differences between estimation and measurements.

This scenario applies to the forecasting of the short-term output of power plants that generally have temperature measurements available in real time through their supervisory control and data acquisition systems. It indicates that even for the most difficult site do model (Brazil), the method still presents low RMSE values for forecasting the module temperature in the following timestep.

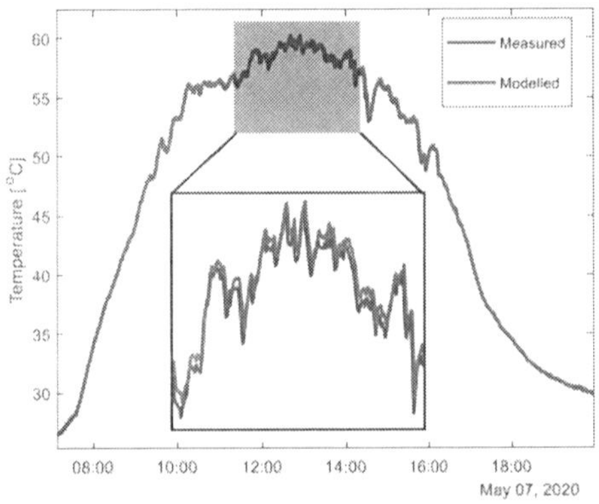

Figure 7: Measured and modelled module temperatures with 1-minute resolution for May 7th, 2020 for the SG-SG test site, using measurements for the forecasting of the next time-step module temperature (Oh 2).

4 CONCLUSIONS

This study compares two well-known steady-state temperature models (Ross's and Faiman's), with the Oh's model, that assesses the heat capacity of the module and the temperature inertia that it produces. The results show that when analysing 1-minute resolution data, Oh's model provides fairly accurate performance, compared with the two other models under study.

Especially for high resolution data, the temperatures calculated with Faiman's model vary greatly among the time steps, not taking the inertia caused by the heat capacity of the module into account. Oh's model considers this factor and therefore is the most suited model for 1-minute and 5-minute resolution simulations. These rapid changes decrease as the time steps get bigger, resulting in better results for Faiman's model for 15-minute resolution simulations. For the case Oh 2, that calculates the module temperature using the measurement from the previous time step, the errors are very small. This indicates that the method is very effective in predicting the PV module temperature for a future time step, given that the measured value of the previous time step is provided. This result is important, because it means that the model could be applied for the forecasting of the PV module temperature, what can be used to estimate module performance. Future analyses should evaluate the model's potential applications for forecasting, considering its limitations for time steps greater than 15 minutes, as described in this study.

PV CAMPER also aims to continue this module temperature study, expanding the analysis to include additional sites, and implementing a default measurement setup at all sites. Future investigations will also analyze uncertainties related to sensor fixation method, type and accuracy, to get a full understanding of the effects of these parameters on the accuracy of the reference temperature used to evaluate temperature models. In addition, PV CAMPER will assemble a set of best practices for module temperature measurements and guidelines for model choice based on climate. Future work will also expand the analysis to include other estimation models present in the literature.

5 AKNOWLEDGEMENTS

The authors are members of the PV CAMPER collaborative and wish to acknowledge with thanks their colleagues who did not participate directly in this research but who helped create the organization and are actively promoting the exchange of both ideas and data.

MB acknowledges the Brazilian Post-Graduate council CAPES for a doctoral scholarship. MB, AKVO and RR acknowledge the financial support of the Brazilian Electrical Energy Regulator ANEEL, through R&D project grants from ENGIE, ENEL, CPFL and STATKRAFT.

LB acknowledges that this work is funded in part by the U.S. Department of Energy Solar Energy Technologies Office, under Award Number 34363.

CDRG and TR acknowledge SERIS as a research institute at the National University of Singapore (NUS) supported by the National Research Foundation Singapore (NRF), the Energy Market Authority of Singapore (EMA) and the Singapore Economic Development Board (EDB).

6 REFERENCES

[1] D. Faiman, "Assessing the outdoor operating temperature of photovoltaic modules," *Prog. Photovoltaics Res. Appl.*, vol. 16, no. 4, pp. 307–315, Jun. 2008.

[2] A. Q. Jakhrani, A. K. Othman, A. R. H. Rigitand, and S. R. Samo, "Comparison of Solar Photovoltaic Module Temperature Models," *World Appl. Sci. J.*, vol. 14, no. SPL ISS 3, pp. 1–8, 2011.

[3] R. G. Ross, "Interface Design Considerations for Terrestrial Solar Cell Modules," in *Proceedings of the 12th IEEE Photovoltaic Specialists Conference*, 1976, pp. 801–806.

[4] P. Mora Segado, J. Carretero, and M. Sidrach-de-Cardona, "Models to predict the operating temperature of different photovoltaic modules in outdoor conditions," *Prog. Photovoltaics Res. Appl.*, vol. 23, no. 10, pp. 1267–1282, Oct. 2015.

[5] E. Skoplaki and J. A. Palyvos, "Operating temperature of photovoltaic modules: A survey of pertinent correlations," *Renew. Energy*, vol. 34, no. 1, pp. 23–29, 2009.

[6] A. K. V. de K. V. Oliveira *et al.*, "Comparative Analysis of Module Temperature Measurements and Estimation Methods for Various Climate Zones across the Globe," in *37th European Photovoltaic Solar Energy Conference and Exhibition*, 2019, pp. 1345–1351.

[7] S.-Y. Oh *et al.*, "Outdoor Operating Temperature Modeling of Photovoltaic Modules Including Transient Effect," in *2017 IEEE 44th Photovoltaic Specialist Conference (PVSC)*, 2017, no. 2, pp. 487–489.

[8] L. Burnham *et al.*, "Photovoltaic Collaborative to Advance Multi-Climate Performance and Energy Research (PV CAMPER)," in *36th European Photovoltaic Solar Energy Conference and Exhibition*, 2019, no. September, pp. 10–11.

[9] C. M. Peel, B. L. Finlayson, and T. A. McMahon, "Updated world map of the Köppen-Geiger climate classification," *Hydrol. Earth Syst. Sci.*, vol. 11, pp. 1633–1644, 2007.

[10] L. R. do Nascimento, T. de Souza Viana, R. A. Campos, and R. Rüther, "Extreme solar overirradiance events: Occurrence and impacts on utility-scale photovoltaic power plants in Brazil," *Sol. Energy*, vol. 186, no. January, pp. 370–381, 2019.

This is a short version of the paper submitted to the journal EPJ Photovoltaics.
It has been selected by the Executive Committee of the EU PVSEC 2021 for submission to EPJ Photovoltaics.

NEW FINDINGS ON THE PV FIRE PREVENTION – FIREFIGHTER STRATEGIES FOR IN-ROOF PV INSTALLATIONS

Urs Muntwyler, Berner Fachhochschule/ Engineering Office Muntwyler,
Hopfenrain 7, CH 3007 Bern, urs_muntwyler@gmx.ch
Eva Schüpbach, Bern University of Applied Sciences BFH
Jlcoweg 1, CH-3400 Burgdorf, schuepbach@bfh.ch

ABSTRACT: PV Installations pose a relatively low fire hazard. However, there are several reasons for some PV installations to be involved with fire. The fire risk for in-roof PV installations is 20-times greater than on-roof installations [1]. The PV Laboratory of the Berner Fachhochschule BFH and the Engineering office Muntwyler has worked for decades on the prevention of fires in PV installations [2]. In the last few years, this work has been strengthened with the support of the building insurance company (Gebäudeversicherung Bern GVB) of the state of Berne in Switzerland [3]. With a better safety strategy, several fires created by PV components could be avoided. Such a type of safety strategy was developed by the aircraft industry and is used in the gliding community. The author, an experienced glider pilot, uses this safety strategy for PV installations to improve the safety of PV installations.
Keywords: Fire risk, By-pass diode, PV-connector, Power optimiser, shading

1 SOME REASONS FOR PV FIRES

In 2014, a study on fires in PV installations (TÜV Deutschland and the PV Laboratory of the BFH-TI) examined 1,3 million PV installations (with 30 GWp) and 400 fire incidents [4].

In 180 cases the PV installation was the reason for the fire [5]. According to the technical report (IEC TR63226), in-roof installations have a 20-times higher risk for a fire than PV on-roof installations [1]. The reason for the fire is not always certain. Although, usually cross-connection of the PV connectors are found to be the cause [6].

Most of the incidents involve a fire on or in a house that destroys the PV installation on the house.

Figure 1: PV installation destroyed due to a fire in the house – most prominent case for a fire (Photo: GVB 2021 / Wolfisberg/ CH).

2 NEW IN-ROOF FIREFIGHTER STRATEGY USES FOGNAIL – TECHNOLOGY

The fire on the in-roof PV installation started at noon in July 2020 on an extremely bright and hot day. Even at the commune in Heiligenschwendi (which is above 1'000m asl), the temperatures had severely increased. The PV installation was built in 2015.

It was clear that the damages due to the water of the firefighters were remarkably severe. This was due to the Fognail technology [7] used by the firefighters of Thun, Switzerland to suppress the fire of the PV in-roof installation (see Figure 3). This method is highly effective and creates minimal water damages. More details of the effort of the fire brigade are explained in «Swissfire 118» (Ausgabe 05/2021 German and 08/ 2021 French) [8].

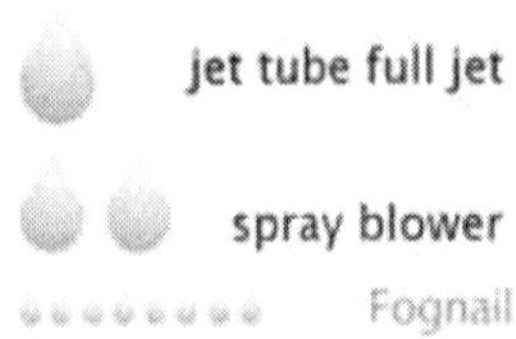

Figure 2: Fognail water spray consists of very small water vapour drops

Figure 3: Fire on a PV in-roof installation near Thun/ Switzerland summer 2020 distinguished with the fognail technology (Photo: private/ Urs Muntwyler).

Figure 4: The commander of the fire brigade of the city of Thun Roland Gfeller with an expanded Fognail to act against fires in in-roof PV installations (Photo: Urs Muntwyler).

Figure 5: Mission car with turntable fire escape and Fognail in the open compartment of the truck of the fire brigade city of Thun, Switzerland. (Photo: Urs Muntwyler).

The fire brigade of Thun has the Fognail now ready in the firefighter truck (figure 5).

3 NEW FIRE PREVENTION STRATEGIES FOR PV INSTALLATIONS

In the TÜV study [5] several reasons for a fire were mentioned. Today, the cross-connection of PV plugs is mentioned as the main reason for a fire [6].

We found several cases with destroyed by-pass diodes and a hot spot that destroyed the module and started a fire in some cases. To get a higher safety standard, this should be prevented with further layers of security.

Using a safety model from the aircraft industry [9] which is particularly common in the gliding community, we can see the problem in the safety philosophy. Like an *Emmental* cheese, with many holes, if there are not enough slices (in our case safety layers) the holes in the slice may align, so that a hazard can pass through all the slices, leading to a failure in the system.

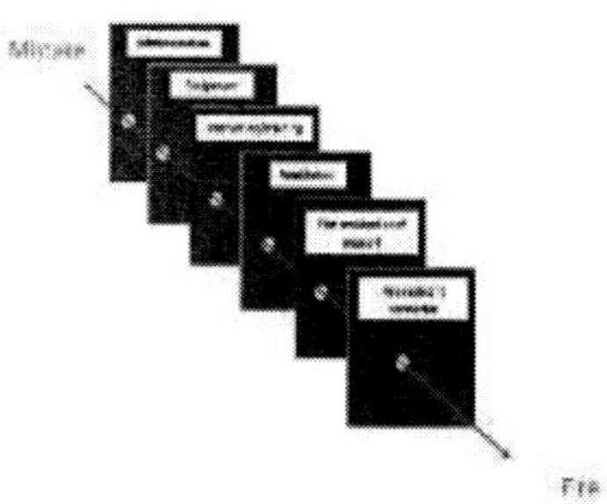

Figure 6: PV in-roof - installation does not have enough safety layers. This can lead to a fire hazard which can destroy the roof and the house.

In the case of the PV in-roof installations, we can see that even one failure of a by-pass diode can create a hot spot or a failure in a PV connector could create a fire. Therefore, we need more safety layers.

The additional safety layers could be:

- Placing a fireproof structure under the roof
- An arc detector for stopping the power flow if an arc is detected in the PV system
- An electronic converter that works below the arc voltage level

Nevertheless, the safety measure must not present new problems in the safety layer construction. For instance, installing an electronic system below the PV roof could pose a fire hazard.

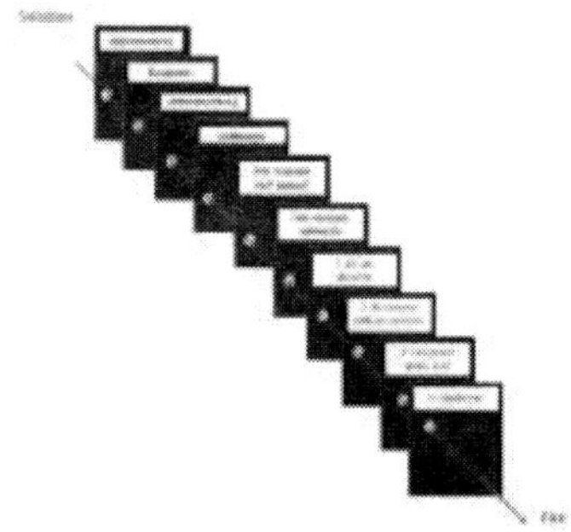

Figure 7: We propose additional safety layers in the in-roof PV installations, to avoid fire incidents.

4 CONCLUSIONS

Fires on PV plants are a serious threat to people, investments, and the perception of PV being a safe, friendly technology. However, fires on PV plants can be avoided by simple technical measures. Technical norms like the ban of cross-connections by PV connectors are an example of simple measures to increase the safety of PV plants, even if they come with additional expenses. The prevention of hot spots in PV in-roof modules is more complicated. They are either a product safety problem and/ or a PV system problem. The in-roof PV modules always fulfil all the tests in the Laboratories and fail years after due to malfunctioning of components. As a result, we must do further research to develop a PV system that is more robust and safer as it is today.

This study indicates that fire prevention, particularly for in-roof PV installations, must be improved further. Hence, it is most essential for in-roof PV installations to be subject to stronger risk assessment regarding fire safety. Albeit and thanks to innovative firefighters in Thun, Switzerland. Fires under a PV layer may effectively be addressed with comparatively low damage by applying Fognail equipment. The innovative Fognail can also be used for other fires.

5 ACKNOWLEDGMENTS

We kindly acknowledge funding from, and the collaboration with GVB (Gebäudeversicherung des Kanton Bern). The exchange and personal communications with Mr Roland Gfeller, the commander of the firefighters of the city of Thun and Alexander Dolder, chief logistics of the firefighters of the city of Thun (e.g., on 17th of December 2020) are greatly appreciated.

6 REFERENCES

[1] Managing fire risk related to Photovoltaic (PV) systems on buildings, IEC TR 63226 – Technical Report – Annex page 22, 2021
[2] NEW FINDINGS IN FIRE PREVENTION AND FIRE FIGHTING OF PV INSTALLATIONS, Urs Muntwyler, EUPVSEC 2017
[3] Merkblatt: Photovoltaikanlagen im Feuerwehreinsatz; Ausgabe vom 20.03.2015; GVB
[4] Leitfaden Bewertung des Brandrisikos in Photovoltaik-Anlagen und Erstellung von Sicherheitskonzepten zur Risikominimierung; TÜV Rheinland 2015
[5] Brandgefahr durch PV-Anlagen-Teil 1: Statistische Auswertung von Hitzeschäden und Brandfällen in PV-Anlagen; Hermann Laukamp und Georg Bopp Fraunhofer ISE; Sonnenenergie 5/2014
[6] Brandincidenten met foto voltaïsche (PV) systemen in Nederland, TNO/ ECN; 22.2.2109
[7] www.fognail.de
[8] Urs Muntwyler, Brand einer Indachanlage, Schweizerische Feuerwehr-Zeitung 2021
[9] The Field Guide to Understand Human Error, Sidney Dekker, ASHGATE, 2006

STATE OF THE ART AND PERFORMANCE OF THE PHOTOVOLTAIC (PV) SYSTEM FLEET IN BRUSSELS: AN ANALYSIS OF 8000 PV INSTALLATIONS

Babacar Sarr[1]*, Jonathan Leloux[1], Gregory Neubourg[2], Guillaume Declève[3], Renaud Tieterickx[4], Jonathan de Lathouwer[4], Régis Lambert[4]
[1]LuciSun, Sart-Dames-Avelines, Belgium
[2]Becquerel Institute, Brussels, Belgium
[3]SUNSET Energy, Brussels, Belgium
[4]BRUGEL, Brussels, Belgium
*Phone: +32 470 10 09 29; E-mail: babacar.sarr@lucisun.com

ABSTRACT: The photovoltaic sector is still relatively young in the world and in the Region of Brussels, where most installations are barely ten years old or less. The goal of this study is to provide an overall picture of the state of the art of the photovoltaic park in the Brussels Region. It includes an assessment of the installed equipment, the price of the installations and their performance. More than 8000 PV installations distributed across the Brussels Region were considered in this study with data available up to the end of the year 2019, and including residential and commercial PV systems. The analysis of installation prices highlights the downward trend in prices which continues in 2019 with a reduction ranging from 5% for large systems to 16% for small residential systems compared to 2018. Since 2013, the average decrease in the different installed power categories is almost 40%. The solar resource in Brussels has gradually increased over the last four decades, with an average annual horizontal global solar irradiation over the last forty years of about 1000 kWh/m^2. In the last ten years, the average value is estimated at 1078 kWh/m^2, about 8% above the average of the last four decades. A historical maximum was recorded in 2018, with 1172 kWh/m^2, which is about 18% higher than the average of the last four decades. The median energy yield of PV installations in the Brussels Region over the last decade was around 850 kWh/kW$_p$. The median Performance Ratio of the PV installations in Brussels Region during the last decade was about 65%, which can be considered as a typical and representative value of the PV park. In comparison, typical Performance Ratios in Europe are in the range of 70-80%. This study has also found a correlation between the performance of the installations and certain key parameters such as the installed peak power, the year of commissioning, the brand of photovoltaic panels, the power category or the type of owner of the installation. Comparisons between module brands have also shown that there is no clear correlation between the geographic origin of a module brand and its performance.
Keywords: Photovoltaic, PV, Performance, State of the art, Belgium, Brussels.

1 INTRODUCTION

The photovoltaic sector is still relatively young in the world and in the Region of Brussels, where most installations are barely ten years old or less. The overall level of knowledge and professionalization of the sector have been increasing over time, with the experience gained and the mistakes made, both on the part of panel and inverter manufacturers, as well as installers, public authorities or owners. The PV market is also fast evolving, and it is necessary to constantly investigate the return from the field experience and to incorporate the lessons learned into common knowledge and practise. Benchmarks and market analysis seek to understand how that PV sector is developing, and where it is going to. On the other hand, the assessment of the real energy productivity and performance of these photovoltaic installations seek to better understand the state of the art of the current installed PV fleet, its peculiarities, as well as the nature and frequency of possible problems that affect them. This better knowledge is an essential key for the proper development and sustainability of the photovoltaic subsidiary, and it helps to better guide public aid decisions.

The goal of this study is to provide an aggregate benchmark of the photovoltaic park in the Brussels Region and to provide an assessment of the installed equipment, the price of the installations and their performance. More than 8000 PV installations distributed across the Brussels Region were considered in this study with data available up to the end of the year 2019, and including residential and commercial PV systems.

2 INSTALLED PV CAPACITY

Figure 1 shows the evolution of the number of PV installations in Brussels, highly correlated to policy and financial incentives granted for the installation of PV in the Brussels Region between 2006 and 2019. For example, the initial peak in 2009 is correlated to the suspension of a regional bonus, equivalent to 50% of the investment for 2010. 2018 has seen a strong shift in terms of the number of newly installed capacity in all categories of installations. This was induced by a combination of factors, including the global equipment price drop and the development of new regional programs.

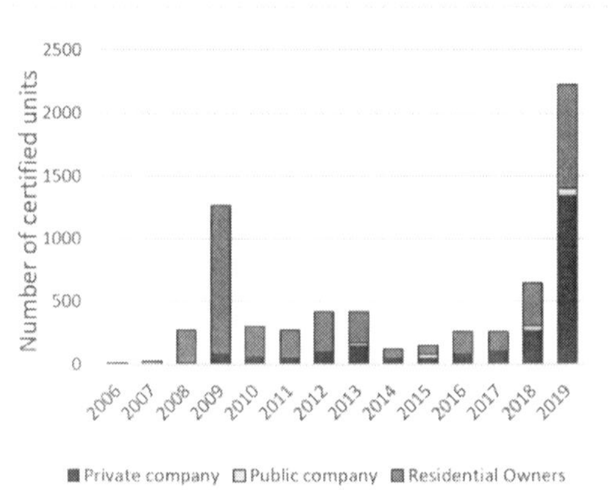

Figure 1: Evolution of number of PV installations in Brussels (2008-2019).

Figure 2 shows the evolution of the cumulative installed power of PV installations in the Brussels Region between 2008 and 2019, categorised by installed power capacity. This installed capacity has also increased over time, with a diversified size of PV installations. Small PV systems are more numerous, but large PV systems (> 250 kWp) dominate in terms of cumulative installed power since 2013 and represented in 2019 more than 50% of the total installed photovoltaic power.

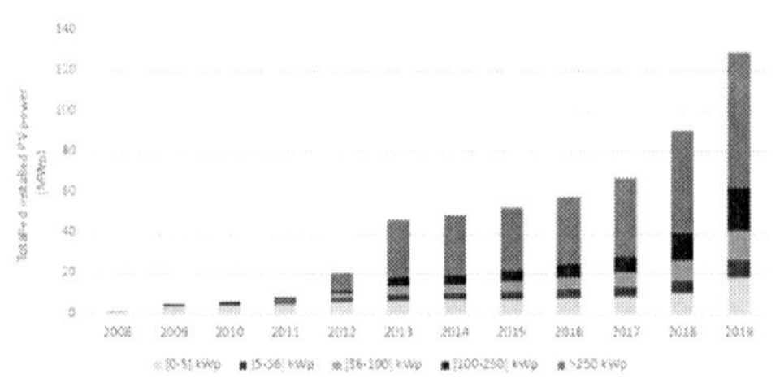

Figure 2: Evolution of the cumulative installed power of PV Installations in the Brussels Region (2008-2019) categorised by Installed power capacity.

Figure 3 shows the evolution of the installed peak power in Brussels between 2007 and 2019. A noticeable trend towards higher average installed peak powers is visible.

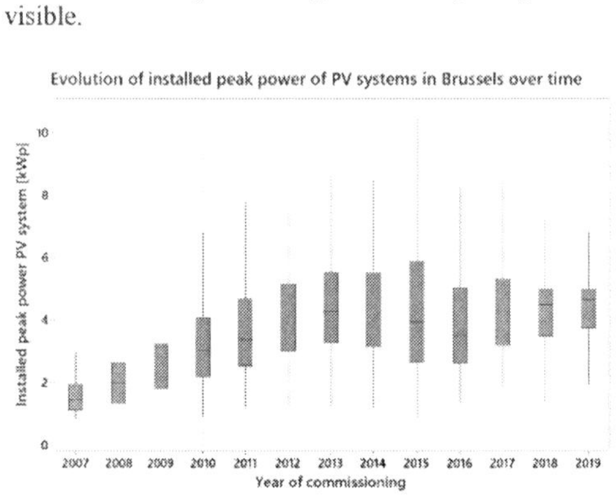

Figure 3: Evolution of the installed peak power in Brussels between 2007 and 2019.

Normalizing the installed capacity to its surface density allows for a better comparison between Brussels and other cities. Figure 4 shows that the Brussels Region has already installed more PV per km² than other cities analyzed, despite the higher population density (7500 inhabitants/km², which is twice as much as Liège or Antwerp). The Brussels Region shows an installed capacity of 105 W_p/inhabitant, which is slightly below the average of the other analyzed cities. The cities that performed better than Brussels either have large industrial areas (Gent, Antwerp, San Diego) or a completely different type of housing and population density (i.e. Honolulu with a density 50 times lower and mainly composed of suburban housing).

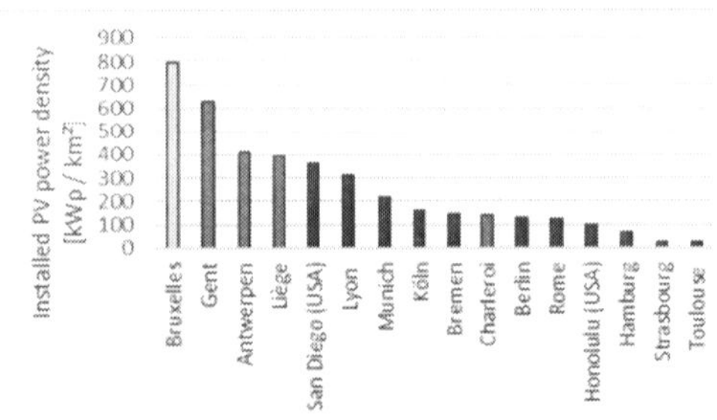

Figure 4: Comparison of the installed photovoltaic power density between Brussels and 15 other cities, in terms of kW_p/km^2 in 2019.

3 INSTALLATION COSTS

Figure 5 shows the evolution of the cost of installation per kW_p (€/kW_p) during the years with the highest availability of data. A clear trend towards the reduction of the cost of installations is visible for all categories of installations, and for all ranges of installed capacity.

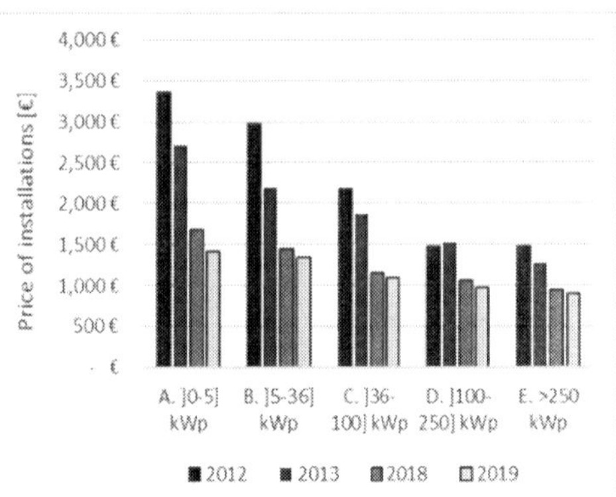

Figure 5: Evolution of average prices of the PV installations.

4 ORIENTATION OF THE PV SYSTEMS

Figure 6 shows the distribution of the orientations of the PV systems installed in Brussels. It shows that the most frequent orientations correspond to south-south-east (-22.5°) and south-south-west (+ 22.5°). These orientations are typical of the rooftops in the city center of Brussels, where the town planning is such that very few roofs face south, and many roofs have an oblique orientation between the south-east and the south-west.

This distribution of the natural orientation of the roofs probably strongly influenced the distribution of the orientations of the photovoltaic panels installed.

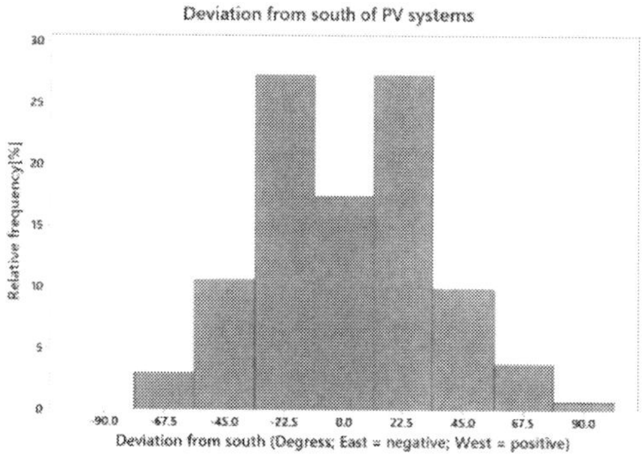

Figure 6: Distribution of the orientations of the PV systems installed in Brussels.

5 SOLAR RESOURCE

Figure 7 shows the evolution of the solar resource in Brussels from over the last seven decades, as measured by a pyranometer located in Uccle and operated by Royal Meteorological Institute of Belgium (RMI). The evolution of the solar resource is evaluated through the annual value of the global horizontal irradiation (GHI).

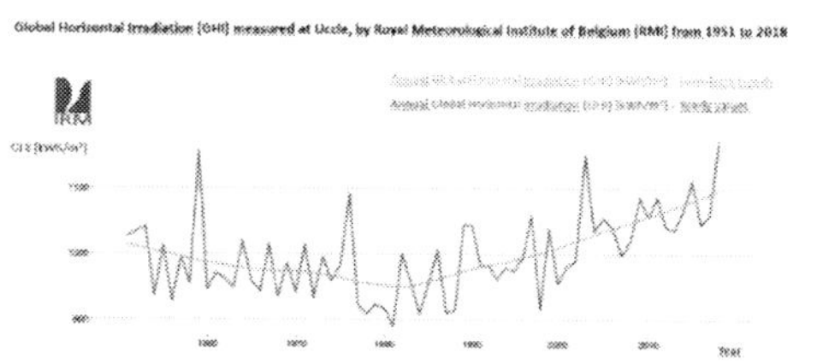

Figure 7: Evolution of the solar resource in Brussels (1951-2018), as measured by a pyranometer located in Uccle and operated by Royal Meteorological Institute of Belgium (RMI).

The annual solar irradiation has changed significantly over the past decades, and it broke new records in Belgium in 2018, with an annual global solar irradiation in the horizontal plane of 1172 kWh/m^2, measured in Uccle, the headquarters of RMI. This is the highest value measured by MRI since measurements began in the 1950s. This new record only confirms the upward trends since the early 1980s, almost 4 decades. For each decade, an increase of more than 40 kWh/m^2 has in fact been observed, or nearly 4% per decade. The historical minimum was measured in 1981, lying just below 900 kWh/m^2. The average value of irradiation between 1981 and 2018 is 997 kWh/m^2, which can be rounded to 1000 for easier retention. This historic 2018 high is about 30% higher than the 1981 low, and about 17% higher than the average value. The average irradiation over the last decade is 1,072 kWh/m2, which represents an increase of about 8% over the average of the last 40 years. As a result, most of the PV production forecasts made 10 years ago have turned out to be too pessimistic. In fact, the Belgian PV systems have produced more than expected.

The increase in solar irradiation (and therefore production) is not specific to Uccle, nor even to Belgium. Similar trends have been observed since the early 1980s in most countries in Europe and North America. The reasons for these trends are not yet fully understood. There are two main causes: the amount of aerosols in the atmosphere, and the amount of clouds, which disperse and absorb solar radiation before it reaches the ground. Recent scientific research attributes about three quarters of the trends to aerosols and one quarter to changes in cloud cover [1]. During the last decades, the Belgian sky has therefore become less loaded with aerosols, and a little less cloudy.

6 ENERGY YIELD

Figure 8 shows the distribution of the annual energy yield of the installations over the last decade. The average energy yield was 831 kWh/kW$_p$, and its median value is 844 kWh/kW$_p$, which shows a slight shift towards the lower values. The energy yields of most of the installations have values that vary between 700 and 1000 kWh/kW$_p$. Some installations have an energy yield of over 1100 kWh/kW$_p$, while others have an energy yield of less than 500 kWh/kW$_p$. There is a large dispersion between the highest and lowest energy yield, which may be due to differences in orientation, quality of the components and execution of the installation, as well as the presence of operational problems.

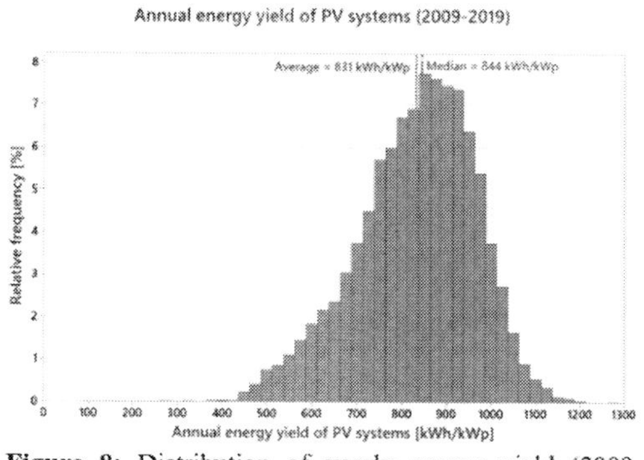

Figure 8: Distribution of yearly energy yield (2009-2019).

7 PERFORMANCE RATIO

The performance Ratio (PR) has been used to analyze the overall performance of the installations. The PR is defined as:

$$PR = \frac{Y_f}{Y_r}$$

where the final yield Y_f is defined by E_{pv}/P^{STC}, and the reference yield Y_r is defined by GTI/G^{STC}, where: E_{pv} is the net electrical energy produced by the installation during a given period, P^{STC} is the rated power of the PV generator under Standard Test Conditions (STC), G^{STC} is the global solar irradiance under STC (i.e., 1000 W/m^2), and GTI is the Global Tilted Irradiation received by the PV generator.

Figure 9 shows the distribution of the annual Performance Ratio of all the PV installations analysed in the database and for the years 2009 to 2019. It therefore provides a global picture of the typical performance of the PV installations in Brussels.

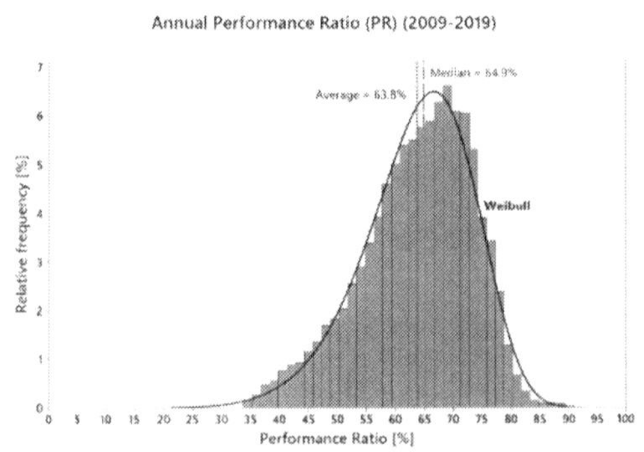

Figure 9: Distribution of the annual Performance Ratio of all the PV installations analysed in the database (2009-2019).

The distribution curve that best explains the distribution of the Performance Ratio in Brussels is a Weibull type distribution. This is an asymmetric distribution that is often observed when a population has a physical boundary at one of its extremes but not at the other. In the case of photovoltaic installations in Belgium, it is generally physically very unlikely to be able to obtain an annual PR value greater than 90%, because certain losses in the system are often present, whereas it is more frequent that installations have significantly lower PRs than average, primarily due to performance issues. The Weibull distribution suggests that the most typical value of PR to retain for a photovoltaic installation installed in Brussels during the last ten years is between 65% and 70%. A significant variation is observed around this typical value, with 90% for the best installations and 40% for the worst performers. This is a very large disparity which can be explained by a not insignificant proportion of installations suffering from undetected and/or untreated performance problems.

These typical annual PR values of 65-70% which characterize photovoltaic installations in Brussels can be compared to the PR values obtained in other studies carried out in Belgium or elsewhere in Europe. Some studies have evaluated the PR of tens of thousands of installations in Europe, whose years of commissioning are generally representative of the installed PV park in Brussels. A study of more than 31,000 photovoltaic installations in Europe located in Belgium, France, Italy, Luxembourg, Germany and the Netherlands concluded that the average PRs are between 70% and 75% [2]. Another study covering more than 31,000 photovoltaic installations in Europe mainly located in Belgium, France and the United Kingdom concluded that the PR of most installations is between 60% and 90%, with average values per country typically above 75%, and typical PR values around 80% [3]. A study of a thousand photovoltaic installations in Wallonia obtained very similar conclusions, with an average value of 78%, and a typical value of around 80% [4]. PR values in Wallonia are therefore, on average, around 5-10% higher than in BCR. Several causes can explain these differences. In Wallonia, photovoltaic panels are often installed on the roofs of houses that are relatively free of shading, while in a dense urban context such as that of Brussels, the panels are often much more affected by the shading. Differences in installation and maintenance practices may also be the cause of these observed performance differences. However, given the significant difference in the performance observed in Wallonia and in Brussels, it would be interesting to study the Brussels stock in more depth to try to better understand the reasons for these relatively low PR values and to remedy them as much as possible, for the installations of the current PV park but also and especially for the installations to come.

Figure 10 shows the distribution of the monthly Performance Ratio for the reference year 2018. It can be seen that the dispersion of the performance ratio values is larger than for the annual performance ratio. This can be explained by several factors. Firstly, the monthly variability of performance is greater than the annual variability because operating conditions (irradiation, temperature, shading, etc.) can vary greatly depending on the month of the year. Secondly, operating problems and failures can affect PV installations for one or more months before being resolved. This strong dispersion of the PR is an indication that many installations underperform strongly during at least a certain period of the year.

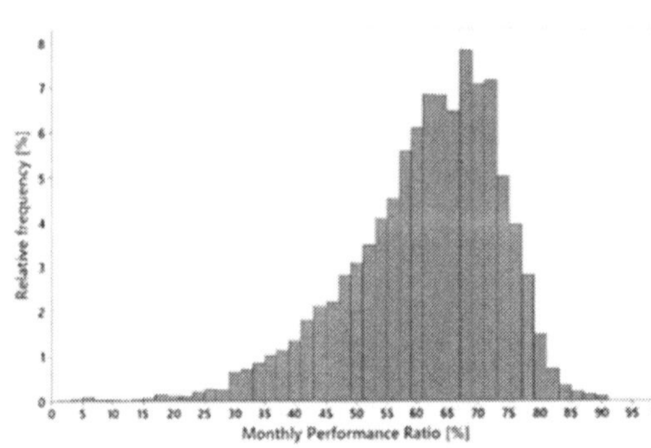

Figure 10: Distribution of monthly Performance Ratio for the reference year 2018.

Figure 11 shows the monthly evolution of the Performance Ratio during the reference year 2018. It can be seen that the PR values tends to be slightly higher during spring and summer than during autumn and winter. This means that the PR is higher when the sun is higher in the sky, which suggests that many installations are affected by strong shading during winter. These changes in PR can also be explained by other factors, including poor performance of installations under low irradiance conditions. This situation can occur when the inverters are oversized, which seems to be the case in Brussels for many of the installations. This evolution of monthly PR values with higher values in summer than in winter is characteristic of photovoltaic installations in dense urban areas. In large solar power plants, an opposite evolution is often observed during the months of the year with summer PR values slightly lower than winter values due to the thermal losses of the PV panels.

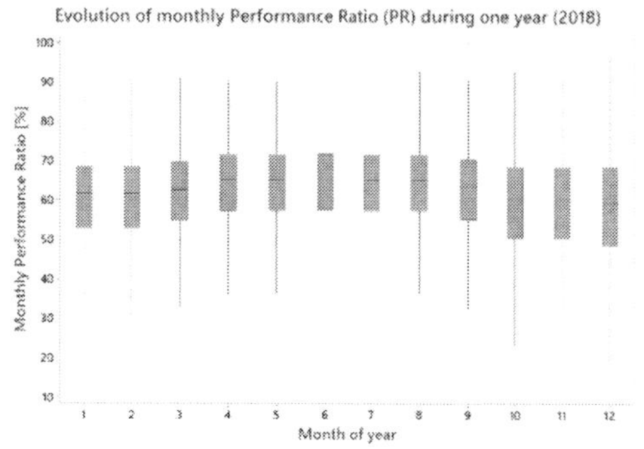

Figure 11: Evolution of monthly Performance Ratio in 2018.

Figure 12 represents the annual Performance Ratio measured for the reference year 2018, relative to the year when the installations were commissioned. The older installations tend to show a slightly lower performance. There are two basic reasons for this: ageing installations and improvements in technology and installation practices over time.

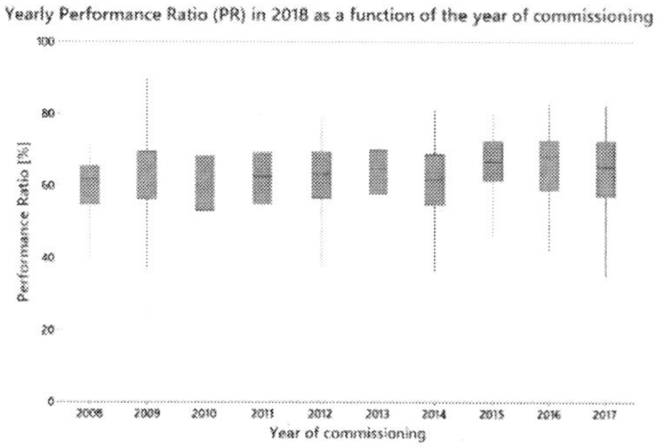

Figure 12: Performance Ratio according to the year of commissioning of the PV installations.

Figure 13 shows the typical PR range for PV installations belonging to different categories of nominal power. There is a slight tendency for the PR to increase in larger installations. This observation is in line with many previous studies and demonstrates some moderate scaling factors in the performance of PV installations. The differences in PR between categories of installed power remain modest, which shows the high modularity and reduced scaling effects of PV, making installations of very different sizes viable. PR values are higher in larger installations due to several factors, including that these installations are generally designed, installed and operated by more qualified personnel, that material quality selection is higher (tenders), and that they tend to be less affected by shading.

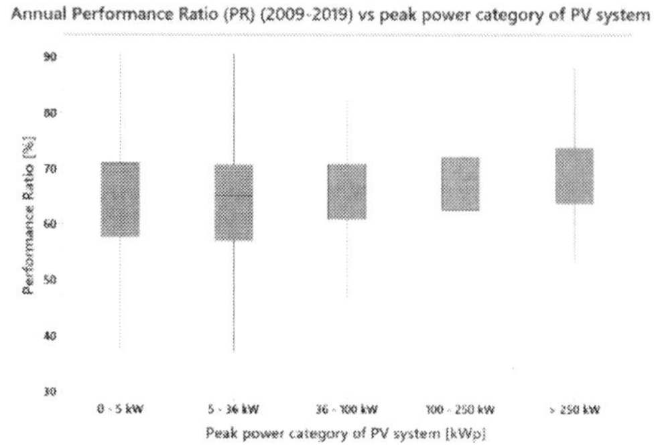

Figure 13: Performance Ratio according to PV installation category of installation size.

Figure 14 shows the annual PR according to the brand of PV panels used over the period 2009-2019. The analyses have attempted to determine whether there are statistically significant correlations between the PR of the installations and the type of PV module used. The observations obtained on the database used are in good agreement with similar observations made on other databases and elsewhere outside the Brussels Region. Most of the differences in performance between installations are not explained by the brand of panels used, but by other factors [3], since the dispersion of the PR values within each group of the same brand of panels is, in most cases, relatively large compared to the difference in PR between different groups. It can also be observed that the solar panels generally considered as "premium" show relatively good performances. This is the case for BenQ, LG, Panasonic (formerly Sanyo), or SunPower. On the other hand, some more exotic technologies that are not part of the crystalline silicon cell family tend to show poorer performances. This is notably the case for panels from Unisolar (amorphous silicon) or Solyndra (CIGS).

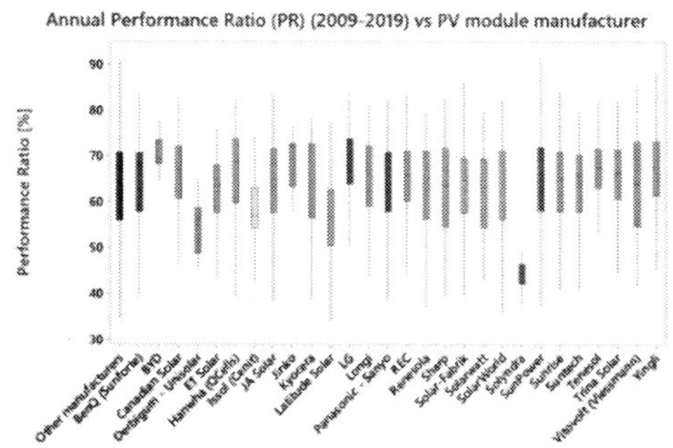

Figure 14: Annual Performance Ratio according to the PV module brands in the park of PV installations in Brussels.

Figure 15 shows the relationship between the specific power of photovoltaic panels and the Performance Ratio of the installations. No significant correlation is observed. This is a consistent result, which confirms that the notions of specific power and Performance Ratio are relatively independent. The specific power indicates the power that a panel can deliver per unit area, while the Performance Ratio indicates the amount of energy produced per unit of peak power of the installation and of solar irradiation received.

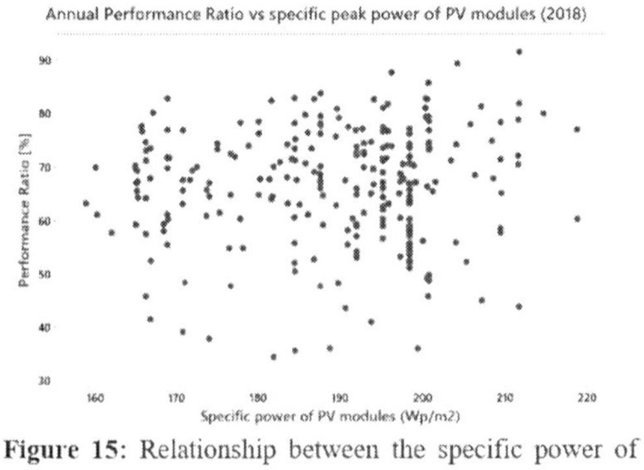

Figure 15: Relationship between the specific power of photovoltaic panels and the Performance Ratio of the installations.

Figure 16 illustrates the relationship between the Performance Ratio of a photovoltaic installation and its cost. As discussed earlier, the prices of PV installations have fallen sharply over the past ten years, and they also tend to be lower when the installation is larger. To allow for valid comparisons, the analyses were therefore limited to installations commissioned in 2018 and whose peak power is between 3 kW_p and 10 kW_p. These comparisons show that the correlation between the cost of an installation and its energy performance is generally very low. However, there is a slighly positive correlation between cost and performance, which suggests that there is some justification for the existence of price differentiation in the market. However, the low degree of general correlation shows that it is still very difficult to understand the pricing mechanisms on the current market. If the price/quality ratio does not seem to be the main factor that dictates these prices, other more important parameters could explain it. This is particularly the case with the degree of difficulty of the realization of photovoltaic installations in the urban context of Brussels.

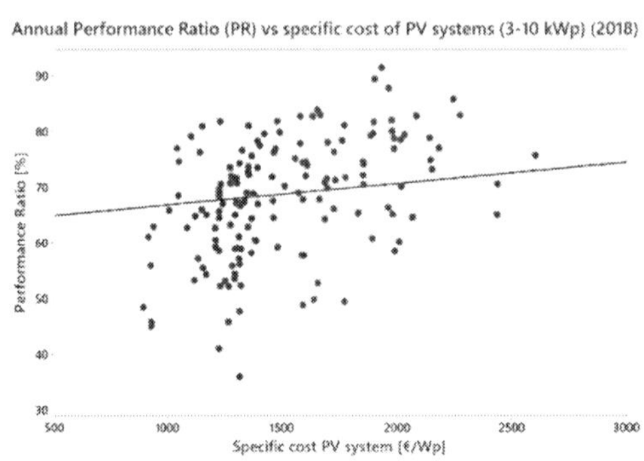

Figure 16: Relationship between the Performance Ratio of a photovoltaic installation and its cost.

8 CONCLUSION AND DISCUSSION

The analysis of installation prices highlights the downward trend in prices which continues in 2019 with a reduction ranging from 5% for large systems to 16% for small residential systems compared to 2018. Since 2013, the average decrease in the different installed power categories is almost 40%. The solar resource in Brussels has gradually increased over the last four decades, with an average annual horizontal global solar irradiation over the last forty years of about 1000 kWh/m^2. In the last ten years, the average value is estimated at 1078 kWh/m^2, about 8% above the average of the last four decades. A historical maximum was recorded in 2018, with 1172 kWh/m^2, which is about 18% higher than the average of the last four decades. The median energy yield of PV installations in the Brussels Region over the last decade was around 850 kWh/kW_p. The median Performance Ratio of the PV installations in Brussels Region during the last decade was about 65%, which can be considered as a typical and representative value of the PV park. In comparison, typical Performance Ratios in Europe are in the range of 70-80%. This study has also found a correlation between the performance of the installations and certain key parameters such as the installed peak power, the year of commissioning, the brand of photovoltaic panels, the power category or the type of owner of the installation. Comparisons between module brands have also shown that there is no clear correlation between the geographic origin of a module brand and its performance.

ACKNOWLEDGMENTS

We are grateful to Laura Rebreanu, who provided high-quality feedback and comments on a previous version of the text that was used in this article.

REFERENCES

[1] M. Wild, "Decadal changes in radiative fluxes at land and ocean surfaces and their relevance for global warming," *WIREs Clim Change*, vol. 7, no. 1, pp. 91–107, Jan. 2016, doi: 10.1002/wcc.372.

[2] J. Schardt and H. te Heesen, "Performance of roof-top PV systems in selected European countries from 2012 to 2019," *Solar Energy*, vol. 217, pp. 235–244, Mar. 2021, doi: 10.1016/j.solener.2021.02.001.

[3] J. Leloux, J. Taylor, R. Moretón, L. Narvarte, D. Trebosc, and A. Desportes, "Monitoring 30,000 PV systems in Europe: Performance, Faults, and State of the Art," p. 10, 2015.

[4] J. Leloux, L. Narvarte, and D. Trebosc, "Review of the performance of residential PV systems in Belgium," *Renewable and Sustainable Energy Reviews*, vol. 16, no. 1, pp. 178–184, Jan. 2012, doi: 10.1016/j.rser.2011.07.145.

PV Fault Detection Threshold at the Module, String, And Inverter Levels

Manjunath Matam
Florida Solar Energy Center
University of Central Florida
Cocoa, FL, USA
manjunath.matam@ucf.edu

Hubert Seigneur
Florida Solar Energy Center
University of Central Florida
Cocoa, FL, USA
seigneur@creol.ucf.edu

***Abstract*—This paper presents the low and high granular data significance in detecting and classifying the faults in a solar PV fault. In this experimental study, the low granular inverter measurements were collected every 5s and the high granular string and module I-V curves were collected every 40 min. Mainly top four PV faults were induced on the PV modules to understand their impact and signatures they leave at all the three measurement stages. Finally, statistical Analysis of Variance (ANOVA) p-values were computed with the baseline measurements. Besides, box plot analysis has been performed. The results reveal that strings data is sufficient to detect the faults, thus, no need to measure, acquire, and store a few tens of modules data.**

***Index Terms*—photovoltaic (PV), I-V curve, PV plant health monitoring, time-series, PV plant data, Machine learning.**

I. Introduction

Solar PV plants installations across the globe are being deployed at a rapid phase. The PV installations growth rate is one of the highest in the renewable energy based energy resources [1]. Increased installations would also implies the increased need for maintaining the PV plants with high reliability and trouble free, so that, PV plants achieve the desired levelized cost of energy (LCOE). With time, the PV plants start to experience different intensity and duration faults. The PV electrical faults such as the line-line fault (short-circuit among PV modules or array cables with different potential) and line-ground (ground fault) are easy to detect and there are well established protection mechanisms in place to address them in time [2], [3]. Moreover, each electrical fault with its unique signature make them easier to detect using machine learning methods [4]. However, many other PV module faults: cell cracks, partial soiling, cell interconnect failures, etc., are low power impact faults but lasts longer as the PV plant ages [5]. Moreover, they are much harder to detect.

Remote faults detection and classification is an emerging research topic in the solar PV field. With this goal, data collection is no longer limited to inverter but, expanded to string and module data as well with the advent of novel PV monitoring devices [6]. String inverter manufacturing companies also started measuring, recording and dispensing the string I-V curves for the PV consumers [7]. The fault detection algorithms proposed in [7] were effective in detecting partial shading conditions but, not other faults. Various PV degradation modes detection based on dark I-V measurements has also been proposed [8], [9]. Collecting the dark I-V in the night time requires human intervention and posses safety hazards.

Additionally, the data collected during field inspection is not limited to measuring electrical characteristics but expanded to other forms such as imaging techniques. In [10], PV faults complete theoretical review on detection and classification has been presented. In [11], infrared (IR) imaging based PV hot-spots detection method has been proposed. Requiring timely human intervention to acquire the IR images is a drawback to it. Acquiring EL/PL imaging also requires human intervention and can be expensive. Therefore, industry typically would consider field EL imaging on special occasion as in the case of hail storm damage for warranty purposes or other similar reliability concerns. Recently, EL and PL imaging is also being proposed at PV plant commissioning to capture the initial conditions of PV modules.

This paper presents the findings from an experimental study involving a 6.2 kW PV field. The objective is to investigate the inverter, string and module data significance in detecting and classifying the faults. Unlike the previous works that focused on detecting a specific fault, this paper investigates the impact that four PV faults have on the three measurements. In the process, this work identifies the unique signatures that each PV fault leave at the different measurement level.

II. PV Experimental Setup and Tests

The PV experimental setup and array in the field are shown in Figs. 1 and 2, respectively. The PV setup consists two strings, twelve modules connected in series per string. Each PV module was rated for 270 W_P. The strings are labeled S_1 (Reference), S_2 (Test). Modules in S_1 remained untouched throughout the study, whereas, faults were induced on S_2 modules. The scheme is connected to a grid-tied commercial inverter. The two strings are connected to two maximum power point tracking (MPPT) inputs separately. The inverter routinely measures voltage, current and power on the each MPPT input for every 5s. Also, a separate I-V tracing equipment is integrated into the system for measuring two strings and 24 modules I-V curves. The tracer is set to measure the I-V curves every 40 min on all the days. The inverter, string and module data were collected and stored in the database.

This material is based upon work fully supported by the U.S. Department of Energy's Solar Energy Technologies Office Award Number DE-EE0008157.

Fig. 1. PV setup to measure the module, string and inverter data in the field.

Fig. 2. Two strings, 24 modules PV array in the field.

A. Fault tests

The PV system has been in the operation from 2019 January end and working continuously since then. During this time, the PV modules in S_2 were subjected to four different fault conditions at different intervals of time for different duration while S_1 stayed the same and served as our control string. The four PV faults applied in the order are: partial soiling, cracks, cell-interconnect failures, module-interconnection failures. The partial soiling tests were performed in two cases: one module and six modules. A transparent plastic cover with 80% transmittance is used for covering the PV module(s) to represent the soiling condition. In a real world scenario, the soiling sits non-uniformly across the module with non-uniform transmittance. However, due to the difficulty in implementing and maintaining the real-soiling condition for long duration, the film approach was adopted.

The in-house developed methods have been used for creating controlled cracks, cell-interconnect failures on the PV modules. These two failures were applied on a set of four PV modules individually. A 2 Ω resistor is inserted in series in string S_2 when conducting module interconnection failure test.

III. Results Discussion

Different data filtering techniques have been employed to remove the low irradiance, dynamic weather related data that could have impact on the results. Data that did not satisfy the high irradiance filter (>600 W/m2) and one minute sky stability filter (<15 W/m2) conditions were dropped.

A. I-V curves under different fault conditions

The string and module I-V curves obtained under high irradiance conditions were shown here. All the I-V curves were corrected to STC (1000 W/m^2, 25°C). Instead of displaying 24 module I-V curves, one I-V curve from S_1, S_2 non-stressed (not faulty) and S_2 stressed (faulty) modules are only shown. This way, it is easier to distinguish the characteristics in different groups I-V curves. The string and module I-V curves obtained under soiling tests are shown in Figs. 3 and 4, respectively. The test string shows a step at 50% V_{oc} voltage. Besides, the soiled module I-V curves show less I_{sc} compared to the healthy modules in the both strings. However, soiling conditions show no impact on the test string and soiled modules V_{oc}'s.

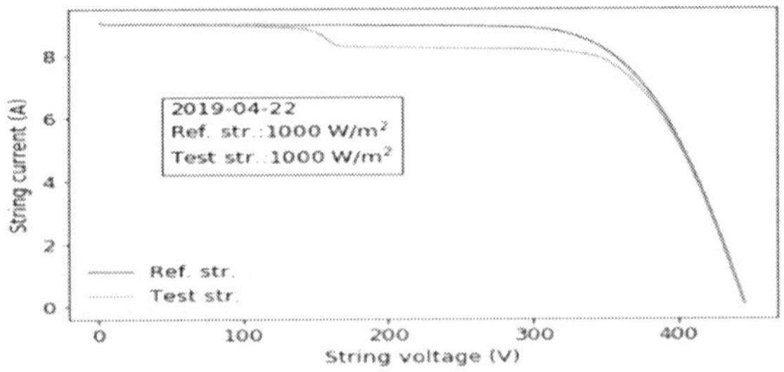

Fig. 3. S_1, S_2 I-V curves under six modules soiling test.

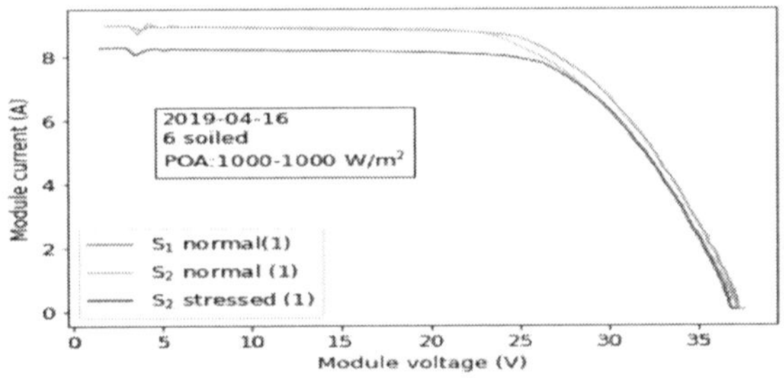

Fig. 4. Module I-V curves under six modules soiling test.

The string and module I-V curves under crack test are shown in Figs. 6 and 7, respectively. The cracks show no impact on the I-V curves except a minute difference in string and module I_{sc}'s. The string I-V curves under module interconnect failure are shown in Fig. 8. The module-interconnect failure shows a little impact on the I-V curves shape except a minute difference in string V_{oc}. However, the cell-interconnect failure test exhibits a clear difference in string I-V curves, as shown in Fig. 9, compared to the module I-V curves, as shown in Fig. 10.

B. Parameters based Boxplot analysis

Parameters based box plot analysis has been conducted to understand the pattern across various faults. After applying the filters and dropping the unreliable data, remaining data is corrected for irradiance and temperature. The box plots are drawn for string, module parameters and important parameters are only shown in Fig. 5. Each plot contains an unstressed (normal) and stressed (fault) box shown using blue and red color against a specific operating condition.

1) String parameters: String P_{mpp} box plot is shown in Fig. 5(a). In this case, string1 with no stressed modules denotes normal data shown by blue color box, string2 with stressed modules denotes stressed (fault) data shown by red color box. Besides, each box median is shown at the top. Difference between the two strings median reveals the following. Under 'normal' conditions, as anticipated, a small

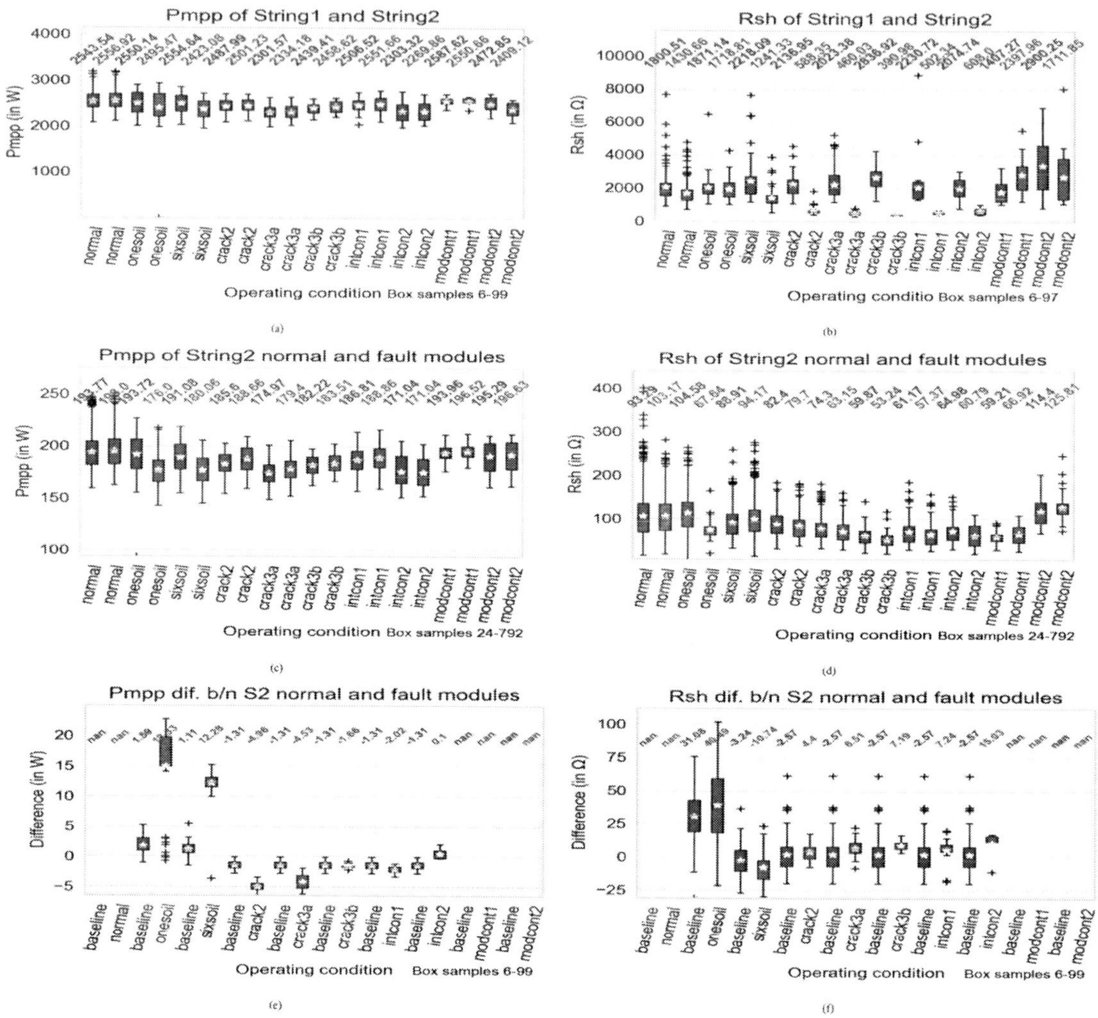

Fig. 5. Box plot of different parameters comparing unstressed (normal) and stressed (faulty) groups: (a) String Pmpp, (b) String Rsh, (c) module Pmpp, and (d) module Rsh.

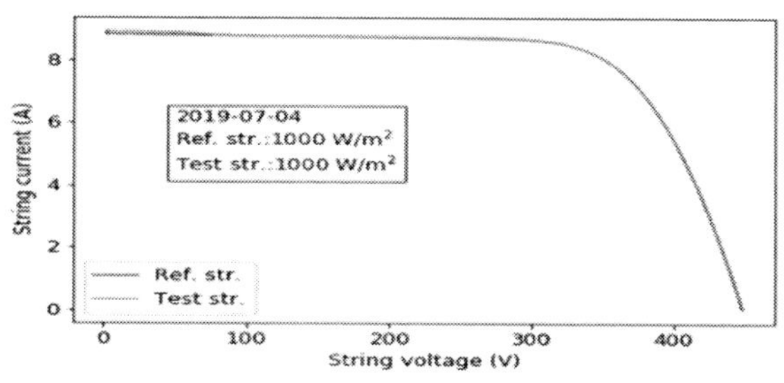

Fig. 6. S_1, S_2 I-V curves under crack test.

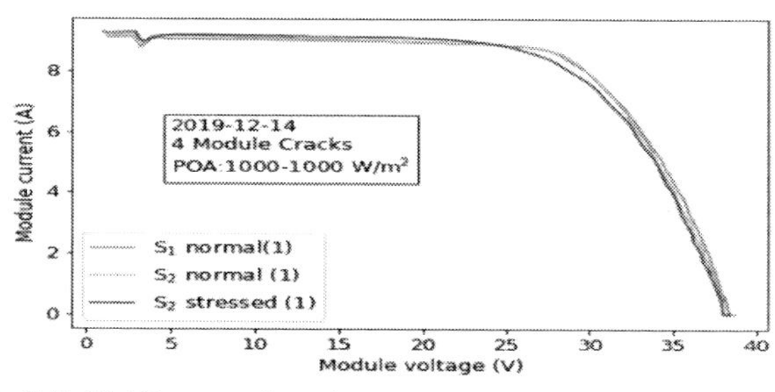

Fig. 7. Module I-V curves under crack test.

-0.52% difference is noted. This can be attributed to measurement sensors accuracy, data correction error. During soiling, an increased difference is noted with one module soiling case reporting a 2.14% and the six-module soiling reporting 5.14%. The cracks and interconnection failures report -1.8 to +1.45% losses, not significant compared to the soiling. These are low power losses causing faults. Again the module interconnection failure report greater losses as low as +1.42 to +2.57%.

String R_{sh} box plot is shown in Fig. 5(b). Unlike the P_{mpp}

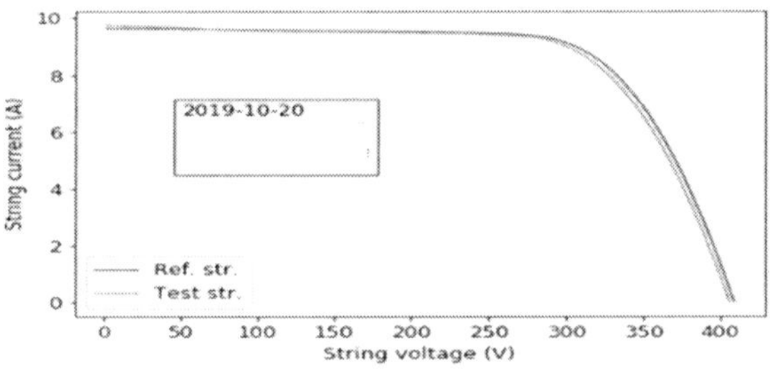

Fig. 8. S_1, S_2 I-V curves under module interconnect failure test.

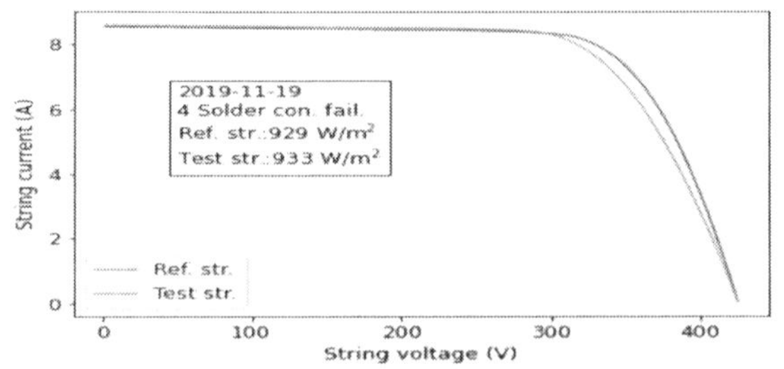

Fig. 9. S_1, S_2 I-V curves under cell-interconnect failure test.

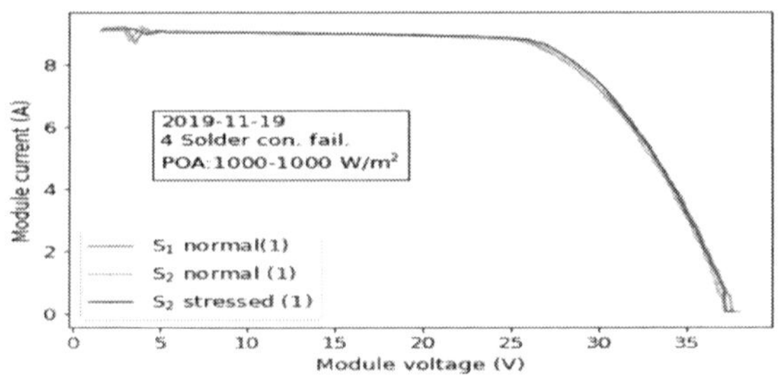

Fig. 10. Module I-V curves under cell-interconnect failure test.

box plot, R_{sh} data reports a clearly distinct pattern particular during the cracks and interconnect failures. Initial stage cracks report a straight 70% reduction and this further reduced to 86% with increased cracks.

2) Module parameters: The box plots in Figs. 5(c) and 5(d) show string2 modules data segregated into two categories: normal (unstressed) and abnormal (fault) modules. For instance, one soiling condition denotes 11 normal modules and one fault module. Unlike the strings, the modules P_{mpp} shows 5.7-9.14% losses for the module soiling, which is a significantly higher compared to the string. However, the cracks and interconnect failure show -2.5 - 0% losses, which is similar to the power losses. With increased cracks, the modules R_{sh} exhibits a clear pattern, shows a linear reduction in 3.12 - 11%. Compared to the string, the modules does not show significant difference thus minimizing the detection possibilities.

The modules data are shown again in Figs. 5(e) and 5(f) but in a different way. Unlike comparing the actual values, difference between normal and fault modules has been computed. Here, the first two months data is considered the baseline, during which no faults were induced. For instance, under one soiling condition, the blue box denotes the difference between 11 modules mean and one module computed during baseline condition. The error in median is significantly larger in this. In the P_{mpp} plot, the soiling conditions show the difference reducing in the range 1006 to 1053%. Similarly, the cracks show 26 to 277% reduction. However, the R_{sh} plot shows the difference reducing by 271 to 379% when cracks applied and 381 to 685% when interconnect failures are applied. This is significantly higher compared to the reduction observed in the strings and modules box plots. However, tracking the difference could be difficult and, sometimes, the measurement sensors accuracy, instrument calibration error, mathematical correction error could influence these results leading to the wrong conclusions.

C. *ANOVA analysis between Baseline and Fault Conditions*

After filtering for low irradiance and sky-stability condition, the data is corrected to STC conditions. Seven parameters, viz., P_{mpp}, V_{mpp}, I_{mpp}, V_{oc}, I_{sc}, R_{se}, and R_{sh} were extracted from the I-V curve. Then, the statistics based analysis of variance (ANOVA) test has been conducted on the S_2 data to identify the difference with the S_1 data under baseline and fault conditions. The baseline is when no faults were induced on S_2. On the PV plant, no tests were conducted during 01-25-2019 to 03-22-2019 and the data is labeled as 'baseline'. Different tests were conducted on the PV array starting from 03-22-2019. ANOVA test is conducted to see if the fault data set is significantly different from the baseline data set and whether this data set is generated by a chance alone. The test is conducted using the Python function as

$$p\_value = scipy.stats.f\_oneway(group1, group2) \quad (1)$$

where p_value is probability value, group1 is baseline data (in case of inverter) or reference string (in case of string and module) data and group2 is the fault data. Small p value (<0.05) denotes the highest probability that both groups are distinct, true and did not happen by any chance. It is important to note that in case of inverter, the comparison is performed between different intervals data: group1 representing baseline data and group2 representing fault data collected during different interval. However, the string and modules fault data from S_2 were readily compared with same interval S_1 data.

D. *Inverter data*

Daily performance ratio (PR) is computed for the baseline and fault data using the inverter AC power and plane of array (POA) irradiance measurements. The obtained p values under different fault conditions are shown in Fig. 11. The small p values denote that all the fault data sets are distinctly unique compared to the baseline data. However, all the baseline and fault data belong to the different periods in a year and the seasonal effects might made fault data sets look totally different from the baseline data. Any confirmation based on the inverter data is inconclusive.

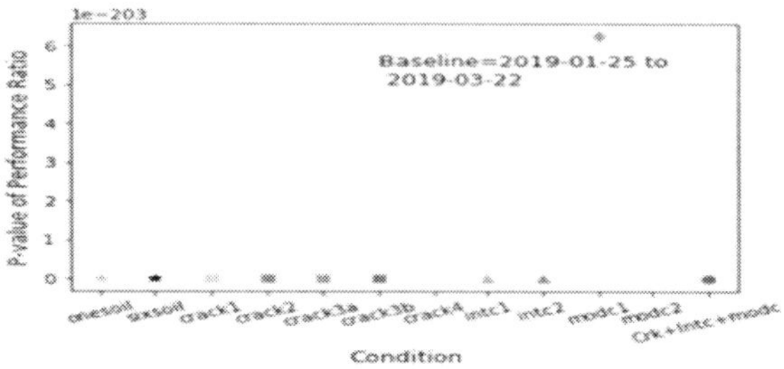

Fig. 11. ANOVA p-values of inverter PR between baseline and fault conditions.

E. *Strings data*

Seven parameters were extracted from the each string I-V curve. Unlike comparing with the baseline data, the p values are computed between the reference and test string parameters. Hence, any seasonal impacts on the fault data are removed. The p values obtained against each parameter under different fault condition are shown in Fig. 12. High p values reported against all the parameters under normal conditions confirm the similar working conditions in the both strings. Low p values were reported for P_{mpp}, V_{mpp}, and I_{mpp} under six module soiling test. It confirms the ability to detect the soiling fault using these parameters. Besides, low p values were reported for shunt resistance R_{sh} under crack faults. Also, low p values were reported for series resistance R_{se} under interconnection faults. This confirms the string level data significance in detecting four PV faults. However, this test requires a proper reference and labeled data to detect the faults.

ANOVA p values between S1, S2 strings

	normal	onesoil	sixsoil	crack2	crack3a	crack3b	crack4	intcon1	intcon2	modcont1	modcont2
Pmpp	0.6795	0.284	0.001	0.8073	0.5548	0.7351	0.7014	0.9967	0.8461	0.1758	0.0954
Vmpp	0.6151	0.9189	0	0.3661	0.0725	0.2665	0.4015	0.6243	0.6673	0	0
Impp	0.866	0.1176	0.0002	0.9402	0.8446	0.9655	0.9327	0.9425	0.9473	0.9157	0.8529
Voc	0.7927	0.2936	0.6891	0.8061	0.8096	0.4944	0.209	0.6499	0.9396	0.636	0.6149
Isc	0.5453	0.5671	0.9746	0.4764	0.1522	0.6176	0.6555	0.8704	0.9911	0.9549	0.9309
Rse	0.1148	0.9872	0.0385	0.0199	0.0001	0.0608	0.3986	0.7967	0.1701	0	0
Rsh	0.006	0.5913	0	0	0	0	0.0754	0.0035	0.0893	0.4139	0.4265

Scale: 0, 0.2, 0.4, 0.6, 0.8, 1

Fig. 12. ANOVA p-values between S_1, S_2 string parameters.

F. *Modules data*

The p values for each parameters were computed between the S_2 string normal and faulty modules data. Hence, the impact that baseline data could have on the faults data has been avoided. The obtained p values against all the parameters under different fault conditions are shown in Fig. 13. Similar to the strings, high p values are obtained under normal conditions. However, the modules report many small p values against many parameters under multiple tests. Each fault does not seem to impact a unique parameter, rather found to impact a combination of parameters. Hence, choosing a combination of parameters could help in detecting a particular fault condition.

Tracking the baseline and fault difference between the parameters over the time has revealed interesting results. The

ANOVA p values between String 2 unstressed and stressed Modules

	normal	onesoil	sixsoil	crack2	crack3a	crack3b	crack4	intcon1	intcon2	modcont1	modcont2
Pmpp	0.1762	0	0	0.0003	0	0.4178	0.3042	0.8994	0.267	0.6193	0.4584
Vmpp	0.5083	0.0141	0	0.0009	0	0.0257	0.3888	0.8111	0.0516	0.5587	0.6605
Impp	0.3366	0.0015	0	0.0741	0.0087	0.8475	0.6043	0.9539	0.9466	0.6433	0.4361
Voc	0.089	0.0008	0.0049	0	0	0	0	0	0.0048	0.001	0.0218
Isc	0.8093	0.0035	0	0.0426	0.0001	0.1779	0.1596	0.4956	0.5642	0.7658	0.5902
Rse	0.3174	0.1495	0	0.0015	0.0005	0.6484	0.2387	0.0051	0.1975	0.4693	0.7136
Rsh	0.8712	0	0.0049	0.2468	0.0004	0.0353	0.2266	0.1341	0.0737	0.0742	0.0693

Scale: 0, 0.2, 0.4, 0.6, 0.8, 1

Fig. 13. S_2 normal and faulty modules ANOVA p-values.

has been explained already in the box plots drawn for the difference in parameters in the Figs. 5(e) and 5(f). Computing the p-values between these two groups shows similar results. That is, many parameters have observed low more often under different operating conditions. However, as explained previously, tracking the difference could be extremely difficult and may sometime could lead to the wrong conclusions.

ANOVA p values for difference in parameters of S2 unstressed and stressed modules under baseline and postbaseline conditions

	normal	onesoil	sixsoil	crack2	crack3a	crack3b	crack4	intcon1	intcon2	modcont1	modcont2
Pmpp	0.9133	0	0	0	0	0.5222	0.0006	0	0.1125	0.3284	0.5763
Vmpp	0.7818	0.5817	0	0	0	0.0014	0.329	0.2463	0.0275	0.2008	0.0228
Impp	0.7155	0	0	0	0	0.0047	0.4026	0.05	0.1237	0.5104	0.1478
Voc	0	0	0.0955	0	0	0	0	0	0.0081	0.0017	0.0021
Isc	0	0	0	0	0	0	0	0	0	0	0.0182
Rse	0.4728	0	0	0.002	0.0144	0.9926	0.022	0	0.3249	0.0795	0.1009
Rsh	0.0039	0.4781	0.0001	0.4619	0.0157	0.1593	0.3623	0.1693	0.0967	0.0085	0.0912

Scale: 0, 0.2, 0.4, 0.6, 0.8, 1

Fig. 14. S_2 normal and faulty modules ANOVA p-values.

IV. Conclusion

This paper investigates the inverter, string and modules data significance in detecting the PV faults. A 6.2 kW based PV system field experiments have been conducted to study the four PV faults impact on the three data levels. The box plot analysis clearly exhibits the reduction in I-V curve R_{sh} when the PV modules cracks increase over time. Moreover, the pattern and magnitude difference is clearly visible between the normal and fault strings compared to the modules. ANOVA based p value test is conducted on the all three levels of the data. The inverter data raises the flags to indicate the PV faults but, any findings will be inconclusive. Mainly, it is difficult to differentiate the seasonal related impacts with the actual fault conditions. The string and module data have a significant advantage compared to the inverter data. The ability to compare the strings or modules data with its neighbor removes any seasonal impacts on the measurements. When compared, both the strings and modules possess a high probability to detect the four PV faults. However, in general, faults seem to impact with strong statistical significance only few parameters at the string level but multiple parameters at the module level. Nevertheless, the data indicates that string level IV can be robust enough to detect and classify faults present in the modules that form the string. Furthermore, scanning a few strings is more advantageous than scanning several tens of PV modules. Also, strings scanning significantly reduces the scanning equipment size, complexity and memory size, which makes it more cost effective and therefore attractive .

ACKNOWLEDGMENT

This material is based upon work supported by the U.S. Department of Energy's Solar Energy Technologies Office Award Number DE-EE0008157. The opinions, findings, and conclusions stated herein are those of the authors and do not necessarily reflect those of DOE.

REFERENCES

[1] P. IEA. (2021, Mar.) Renewables 2020. [Online]. Available: https://www.iea.org/reports/renewables-2020

[2] M. Falvo and S. Capparella, "Safety issues in pv systems: Design choices for a secure fault detection and for preventing fire risk," *Case Studies in Fire Safety*, vol. 3, pp. 1 – 16, 2015.

[3] Z. Yi and A. H. Etemadi, "Line-to-line fault detection for photovoltaic arrays based on multiresolution signal decomposition and two-stage support vector machine," *IEEE Trans. Ind. Electron.*, vol. 64, no. 11, pp. 8546–8556, Nov 2017.

[4] H. C. Barun Basnet and J. Bang, "An intelligent fault detection model for fault detection in photovoltaic systems," *Journal of Sensors*, vol. 2020, no. 11, pp. 1–11, 2020.

[5] J. L. Lincoln, A. M. Gabor, E. J. Schneller, H. Seigneur, J. Walters, R. Janoch, A. Anselmo, V. Huayamave, and W. Schoenfeld, "Forecasting environmental degradation power loss in solar panels with a predictive crack opening test," in *44th IEEE PVSC*, 2017, pp. 2839–2843.

[6] C. B. Jones, B. H. Ellis, J. S. Stein, and J. Walters, "Comparative review of high resolution monitoring versus standard inverter data acquisition for a single photovoltaic power plant," in *IEEE 7th WCPEC*, 2018, pp. 0715–0720.

[7] S. Spataru, D. Sera, T. Kerekes, and R. Teodorescu, "Monitoring and fault detection in photovoltaic systems based on inverter measured string i-v curves," in *Proceedings of the 31st European Photovoltaic Solar Energy Conference and Exhibition: EU PVSEC 2015*, ser. EU PVSEC Proceedings. WIP Wirtschaft und Infrastruktur GmbH and Co Planungs KG, September 2015, pp. 1667–1674.

[8] S. V. Spataru, D. Sera, P. Hacke, T. Kerekes, and R. Teodorescu, "Fault identification in crystalline silicon pv modules by complementary analysis of the light and dark current–voltage characteristics," *Progress in Photovoltaics: Research and Applications*, vol. 24, no. 4, pp. 517–532, 2016.

[9] M. Diehl. (2017, Feb.) Measuring the dark iv curves of photovoltaic systems. [Online]. Available: https://photovoltaikbuero.de/en/pv-know-how-blog-en/measuring-the-dark-iv-curves-of-photovoltaic-systems/

[10] K. AbdulMawjood, S. S. Refaat, and W. G. Morsi, "Detection and prediction of faults in photovoltaic arrays: A review," in *IEEE 12th Int. Conf. CPE-POWERENG*, 2018, pp. 1–8.

[11] C. Buerhop, F. W. Fecher, T. Pickel, A. Häring, T. Adamski, C. Camus, J. Hauch, and C. J. Brabec, "Verifying defective pv-modules by ir-imaging and controlling with module optimizers," *Progress in Photovoltaics: Research and Applications*, vol. 26, no. 8, pp. 622–630, 2018.

THE IMPACT OF TRACKING ALGORITHMS AND TIME RESOLUTION ON ENERGY YIELD MODELLING OF SINGLE AXIS TRACKER SYSTEMS

Anja Neubert[1], Mike Hamer[2], Patrick Rainey[2], Mark A. Mikofski[3]
[1]DNV Maritime Software GmbH, Emsstraße 18 c/o Glaswerk, 26135 Oldenburg, Germany
[2]Garrad Hassan & Partners Ltd, Avon Street, Temple Quay One Linear Park, BS2 0PS Bristol, UK
[3]Garrad Hassan America Inc., 155 Grand Avenue Suite 600, 94612 Oakland, USA
Corresponding author: anja.neubert@dnv.com

ABSTRACT: Estimations of the energy production of photovoltaic (PV) plants are often based on simulations with hourly resolution. For single axis tracker systems, where the tracker rotation angles change with time, this introduces systematic errors in the simulation. We discuss the effects of the simulation time resolution on the estimated energy production and the increasing need for sub-hourly simulations when using more enhanced tracking algorithms with independently rotating trackers for plants in complex terrain.
Keywords: PV Array, Tracking, Design and Simulation Software, Simulation

1 MOTIVATION

For utility-scale photovoltaic (PV) installations, tracking systems become more and more popular. For example, 88 % of all new utility-scale PV capacity added in the United States in 2019 employ tracking systems [1]. Tracker systems follow the sun's path throughout the day so that the sun's incidence angle on the PV modules is minimised to optimise the panels power output. To avoid row-to-row shading, so-called backtracking is used.

We compare power output simulations to investigate the impact on the resulting estimates of the energy production using SolarFarmer, a PV plant design and performance modelling software, developed by DNV [2] using different tracking models for a test site with single axis trackers in undulating terrain. We use standard tracking and backtracking algorithms with the same rotation assumed for all trackers and individual trackers being installed following the terrain as well as an independent tracking method, with data provided by "PV Hardware Solutions" (PVH) that are developed for tracker systems installed in complex terrain. We also simulate the independent tracking method with different time resolutions to investigate the impact of sub-hourly simulations on the energy production estimates.

2 TRACKING ALGORITHMS

2.1 Standard tracking algorithms

Trackers optimise the energy output of PV systems by following the sun, minimising the sun's incidence angle on the PV modules. However, most silicon modules are susceptible to electrical mismatch caused by uneven shade, therefore, trackers typically "backtrack" to avoid row-to-row shade occurring after sunrise and before sunset. The tracking and backtracking algorithms for flat horizontal and sloped ground are described by closed-form expressions [3], [4]. Those methods assume that all individual trackers are installed on a plane with a given slope and azimuth, so they have the same rotation angles at each time step used in an energy output simulation.
A typical daily course of the tracker rotation angle and the corresponding sun incidence angle are shown in Figure 1 based on a simulation with a time resolution of 1 minute.

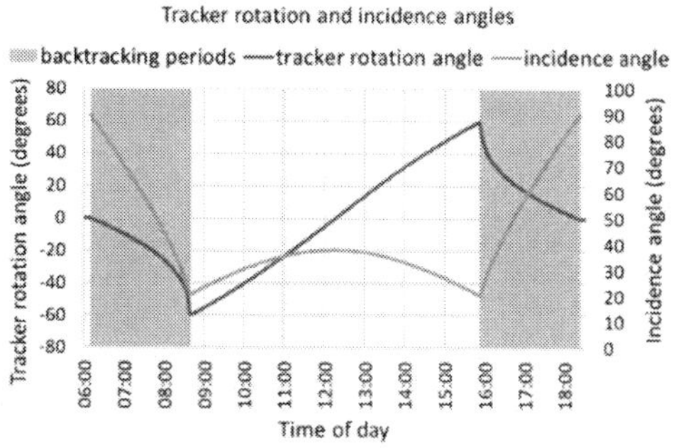

Figure 1: Example tracker rotation and corresponding incidence angle for a single day. Negative rotation means modules facing east, positive facing west.

2.2 Independent tracking algorithms

When building PV plants in complex terrain, without flattening the ground before installing the trackers, sophisticated tracking algorithms are needed to determine rotation angles for all individual trackers as they can no longer track identically without either introducing row-to-row shading or suboptimal incidence angles on the modules. Manufacturers of tracker systems have developed such algorithms that can be used in PV plant simulations. Each tracker gets its own rotation defined so that the rotation can vary significantly across the entire plant for a single time step, as indicated in Figure 2. PVH has provided tracker locations and the corresponding rotation angle time series for several PV plants and we use one specific site to model the corresponding power output.

Figure 2: Visualisation of tracker rows with individual tracker rotations

3 MODELLNG TRACKER SYSTEMS

3.1 Site description

The test site used in the simulations is located in Spain. There are terrain features within the site area as indicated in Figure 3, with terrain sloping upwards from west to east and also some height variations from north to south. The site consists of 1400 individual trackers with monofacial modules. Each tracker is 60 m long and 3 m wide. The trackers are aligned in north-south direction and the ground coverage ratio is approximately 0.4.

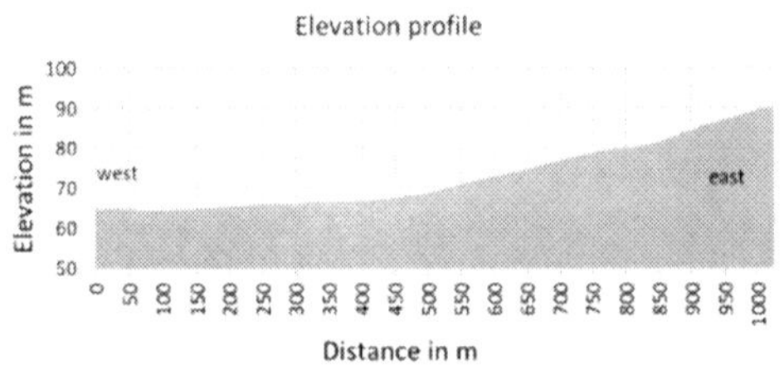

Figure 3: Elevation profile across the simulation test site

3.2 Simulation details

All simulations have been run with SolarFarmer. It allows calculations based on standard tracking algorithms where the simulation time steps are defined by the resolution of the solar resource input data. It also supports usage of custom tracker rotations defined as time series files via the C# scripting feature. The simulation time steps are then driven by the time stamps provided in the tracker rotation files. If the simulation is run with lower resolution solar resource input data, then the resource data is interpolated. A total period of one year has been modelled.

SolarFarmer uses a geometric model, based on the 3D locations of the individual trackers to determine shading effects of the direct irradiance. The shading of the diffuse irradiance uses a hemicube model to identify how much of the sky is visible from the modules and the shading of the ground that reduces the amount of reflected irradiance reaching the modules uses a simple 2D sheds model assumption. Detailed results for each simulation time stamp are available so that the individual modelling effects can be analysed in detail.

3.3 Independent and standard tracking models

PVH has provided tracker locations and the corresponding rotation angle time series to run simulations for independent tracking. In addition to the simulations with individual tracker rotations, a standard tracking model has been used that is based on the assumption that all trackers rotate the same as if they are installed on a plane. The model includes slope-aware backtracking. The plane that best fits the elevation data inside the modelled site area has a tilt of 1.7 degrees at an azimuth of 302.7 degrees. As the individual trackers are laid out to follow the terrain, it is expected that some irregular shading is introduced with the standard tracking model that causes electrical mismatch losses. Both models have been run with hourly resolution where only time stamps that define the tracker rotation at the midpoint of the hourly solar resource input data are considered.

When comparing results based on hourly resolution, the individual tracking model shows some clear advantages over the standard tracking model. Though the irradiance gain on the tilted plane relative to the input global horizontal irradiance is larger for the standard tracking model as shown in Figure 4, this model causes irregular row-to-row shading effects that do not only reduce the irradiance but also cause electrical mismatch losses in the energy output of the PV plant.

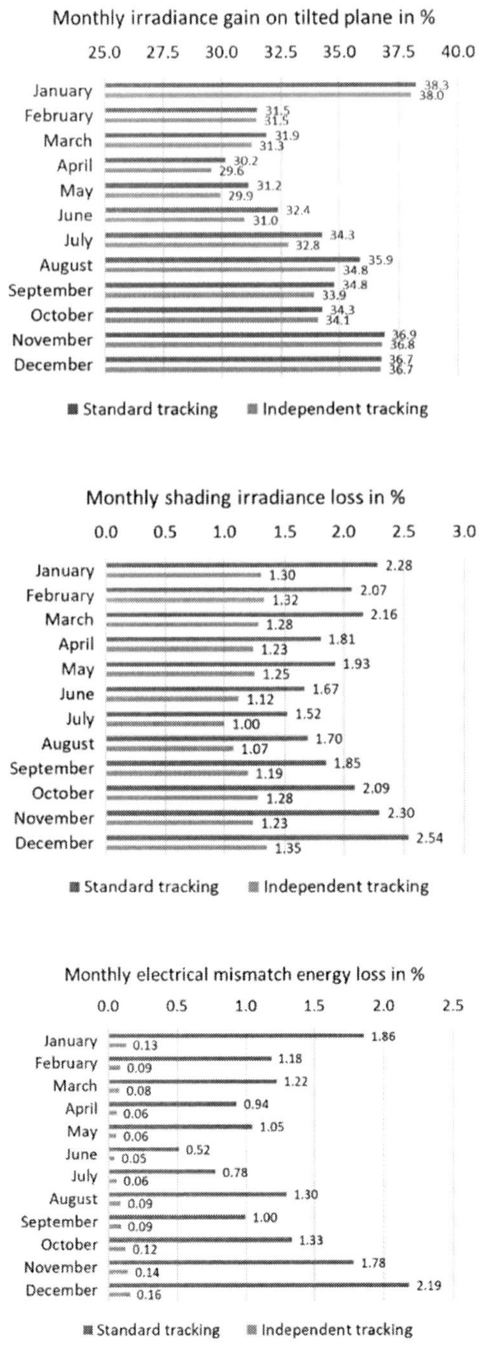

Figure 4: Monthly irradiance gain on the tilted plane (top), near shading losses (middle) and electrical mismatch losses (bottom) for the standard and independent tracking models

It should be noted that with the independent tracking model there are also shading losses, but those affect only the diffuse irradiance components and due to the more uniform effect, does not introduce irregular shading that causes electrical mismatch. A visualisation of the row-to-row shading is shown for 23rd May 07:30 AM in Figure 5 for both tracking models. It can be seen that row-to-row shading happens for the standard tracking model based on a uniform rotation angle of 21.1 degrees facing east whereas for the individual tracking model, rotations vary from 3.3 to 45.0 degrees to avoid row-to-row shading.

Figure 5: Test site visualisation for independent tracking (top) and standard tracking (bottom) models on 23rd May 07:30 AM

When looking at the PV plant AC energy output, the independent tracking model gives up to 3 % larger estimates in the winter months compared to the standard tracking model as shown in Figure 6. The overall annual energy output is about 1.1 % larger for the independent tracking model.

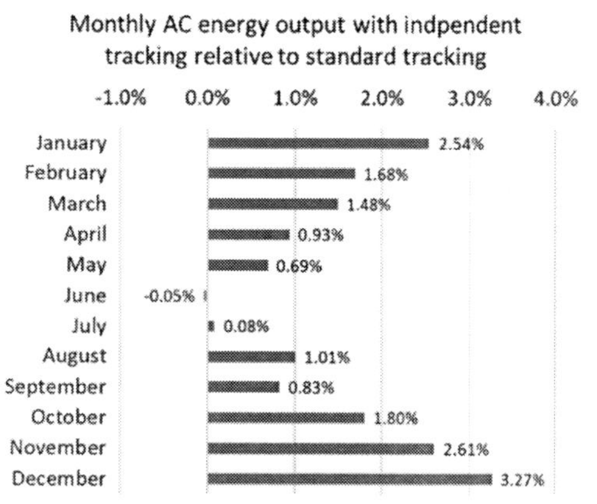

Figure 6: Monthly AC energy output with the independent tracking model relative to the standard tracking model

3.3 Impact of time resolution

The simulation for the independent tracking model has been run with a time resolution of 5 minutes to investigate the differences of sub-hourly simulations compared to hourly simulations.

Although the impact on the overall annual energy output is small with only 0.14 % higher energy when using hourly resolution compared to 5-minute resolution, there is a monthly variation, as shown in Figure 7. In the winter months, the energy output is roughly 1 % higher when using hourly resolution simulations compared to 5-minute. It should be noted that the site is not affected by inverter clipping effects as the DC/AC ratio is only 1.07 but it has been shown that inverter clipping losses can be significantly underestimated with hourly simulations [5] so running sub-hourly simulations are clearly beneficial and minimise systematic modelling bias in particular when high resolution solar resource data are available.

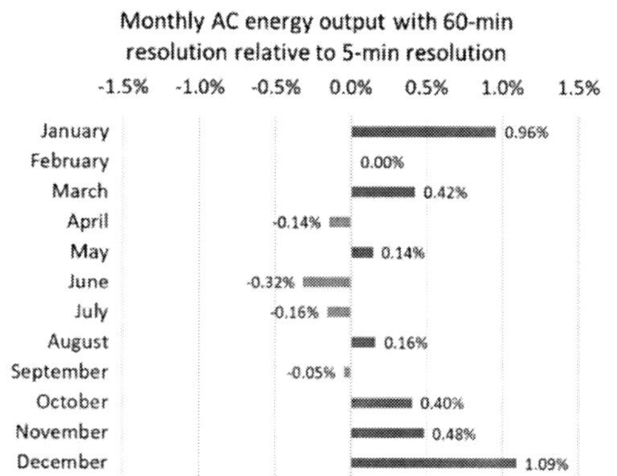

Figure 7: Monthly AC energy output with 60-min resolution relative to 5-min resolution

When looking at a specific day, the impact of backtracking becomes more evident for the precision of the modelled power output. The power output for the 5-minute and 60-minute resolution simulations together with the average tracker rotation for the entire site are shown in Figure 8 for 4th January. The hours from 10:00 to 11:00 AM and 4:00 to 5:00 PM show a systematic overprediction in the hourly simulation that is based on tracker positions at the middle of the hourly periods. For those hours, the trackers backtrack for a proportion of the time that causes systematic modelling errors when using hourly resolution.

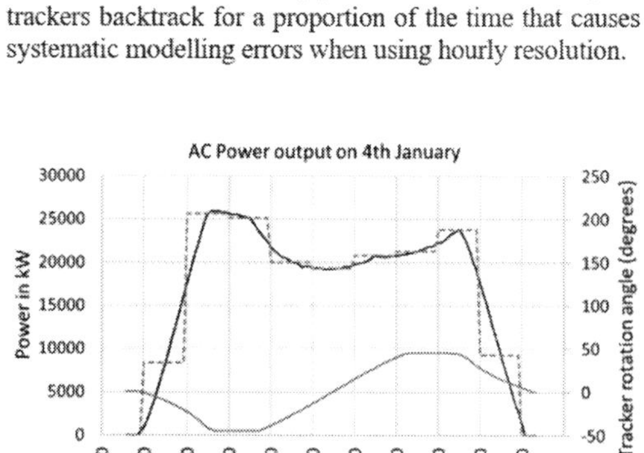

Figure 8: Power output for 5-minute and 60-minute resolution and tracker rotation angle on 4th January

The corresponding difference in the energy output when aggregated over the individual hours is shown in Figure 9. Hours including backtracking are highlighted in blue and show the largest differences of energy production based on hourly simulations relative to 5-minute resolution.

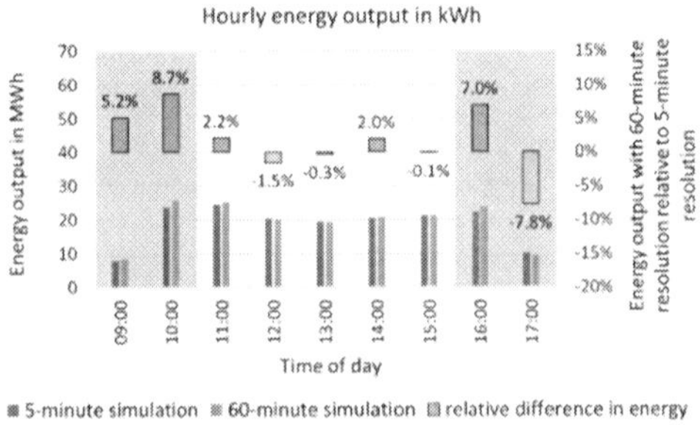

Figure 9: Hourly energy output for 5-minute and 60-minute resolution on 4th January. Periods with backtracking are highlighted in blue.

4 CONCLUSIONS

The necessity of advanced tracking algorithms for optimising the energy output of PV plants with single axis trackers installed in undulating terrain has been demonstrated in this paper. Rotating all individual trackers the same way when installed in undulating terrain can introduce shading and mismatch losses or non-optimal incidence angles and rotating trackers individually becomes beneficial when optimising the plant's energy output.

We have also shown the impact on estimated energy outputs of sub-hourly time resolution simulations compared to hourly resolution and how this can avoid systematic errors in energy assessments, in particular when seasonal or diurnal results are needed.

5 ACKNOWLEDGEMENTS

We thank PVH for providing data for individual tracker locations and their rotation angles for the simulations.

BIBLIOGRAPHY

[1] M. Bolinger, J. Seel, D. Robson, and C. Warner, "Utility-Scale Solar Data Update: 2020 Edition," 2020. Accessed: Jan. 28, 2021. [Online]. Available: https://emp.lbl.gov/sites/default/files/2020_utility-scale_solar_data_update.pdf.

[2] M. A. Mikofski, M. Lynn, J. Byrne, M. Hamer, A. Neubert, and J. Newmiller, "Accurate Performance Predictions of Large PV Systems with Shading using Submodule Mismatch Calculation," in *2018 IEEE 7th World Conference on Photovoltaic Energy Conversion (WCPEC) (A Joint Conference of 45th IEEE PVSC, 28th PVSEC & 34th EU PVSEC)*, Jun. 2018, pp. 3635–3639, doi: 10.1109/PVSC.2018.8547323.

[3] W. F. Marion and A. P. Dobos, "Rotation Angle for the Optimum Tracking of One-Axis Trackers," 2013. Accessed: Jan. 27, 2021. [Online]. Available: www.nrel.gov/publications.

[4] K. Anderson and M. Mikofski, "Slope-Aware Backtracking for Single-Axis Trackers," 2020. Accessed: Jan. 27, 2021. [Online]. Available: https://www.nrel.gov/docs/fy20osti/76626.pdf.

[5] R. Kharait, S. Raju, A. Parikh, M. A. Mikofski, and J. Newmiller, "Energy Yield and Clipping Loss Corrections for Hourly Inputs in Climates with Solar Variability," in *2020 47th IEEE Photovoltaic Specialists Conference*, 2020.

INSPECTION OF TIME SERIES CHARACTERISTICS OF IRRADIANCE GOVERNING THE SIZING OF PV SYSTEMS FOR HIGH AUTONOMY IN VIEW OF RELIABILITY OF PREDICTED SECURITY OF SUPPLY

Hans Georg Beyer
University of the Faroe Islands
Department of Science and Technology
100 Tórshavn, Faroe Islands

ABSTRACT: When designing solar power based clean energy supply systems in all scales – from autonomous devices to solar home systems, village power supply and nation-wide systems - the intended degrees of autonomy and security of supply are key figures. The identification of a reasonable system sizing – the selection of a combination of installed PV and storage capacity - is usually done by use of time step simulations based on an at minimum annual data set of the meteorological conditions. Due to the variability of the meteorological conditions on all time scales and the uncertainty of available data there is consequently an uncertainty on whether the intended performance can be reached over the live time of the systems. In this context, there is a broad discussion on the year-to-year variability of the annual irradiance sum and thus the annual PV generation. Here, effects of the the interannual and database to database variability of the fluctuation characteristics of the irradiance, governing the need for storage is inspected for the case of a simple PV-storage-load system.
Keywords: stand-alone PV Systems, sizing, solar radiation

1 INTRODUCTION

When designing solar power based clean energy supply systems in all scales – from autonomous devices to solar home systems, village power supply and nation-wide systems - the intended degrees of autonomy and supply security are key figures. Due to the variability of the meteorological conditions on all time scales there is an uncertainty on whether the intended degrees can be reached over the live time of the systems.

The identification of a reasonable system sizing – the selection of a combination of installed PV and storage capacity is usually done by use of time step simulations based on an at minimum annual data sets of the meteorological conditions. The selection of these sets – especially their length has been recognized as a critical issue (see e.g. [1,2,3]. In this context there is a broad discussion on the year-to-year or database to database variability of the annual irradiance sum and thus the annual PV generation. In contrast, the variability of the sequential characteristics of the data governing the storage sizing has received less attention.

Here, this issue – focused on storages sizes to cover the load for some days to weeks - is inspected by simulation studies for a PV-storage-load systems at location Faroe Islands (62°N, -7°E), sized to give a save supply of a constant load active during the summer months May-August. The effect of the variability of the irradiance conditions on system sizing is inspected using, on one hand, a 56-year set of daily irradiation data, derived from cloud cover observations. In addition, 3 sets – covering 11 years – derived from satellite observations and 2 reanalysis schemes are applied to inspect variations in sizing when using the different databases,

2 METHOD

In this article a scheme for storge sizing as presented in [4,5,6] is applied. In contrary to the wind/PV/storage hybrid systems inspected in articles, this work in confined to PV/storage systems. According to the geographic location at 62°N the requirement of a complete security of supply of a constant load is limited here to the summer period May-August.

The 56-year set of daily irradiation stem from a scheme as used in [6,7] for the generation of daily irradiations – and the resulting daily PV generation. The scheme for the irradiation exploits synoptic cloud cover data available for the 1958-2013 period. It is trained using satellite derived irradiances for the years 2009-2013 given by the SARAH data base [8]. The additional 3 sets are taken from the PVGIS server [9] for the location under inspection. This comprises sets derived by the CM SAF scheme (in the following labeled CM) service (the same source as used for the training of the cloud cover- based method), and two sets based on the reanalysis sets ERA5 (label E5) and COSMO-ERA6 (label CO). A basic description of these data bases is given in the PVGIS user manual [10].

For an analysis of the system performance for different combinations of PV- and storage-size, the evolution of the storage content is analyzed by a scheme that tracks the storage content according to the daily generation and load balance over the time-period of 56 years. As primary design value the PV size necessary for a mean generation in the May to August period equal to the respective consumption is determined. In the following, all PV sizes are normalized to that value. Similar, the storage content is given in multiples of the daily load [load days]. The storage size, necessary for a secure supply of the load with a given PV-size is determined by the analysis of the balance process [7], here with daily time resolution.

3 Tracing the storage content

As first example, a system with PV-size = 1 is analyzed. From the analysis of the generation to load balance process over the 56-year period, a storage of the size of 13.00 load days proves necessary to secure the continuous supply of the load (i.e. storage content always >= 0). Fig. 1 gives the respective temporal evolution of the storage content. For this example, the storage is depleted once at the end of August 1995 (day of operation Nr. 14488 after 1 of January 1958). That is, the storage size is demanded to secure the load and of that August that year. With the disconnection of the load for September to April,

1st of May 1998 finds a fully charged storage.

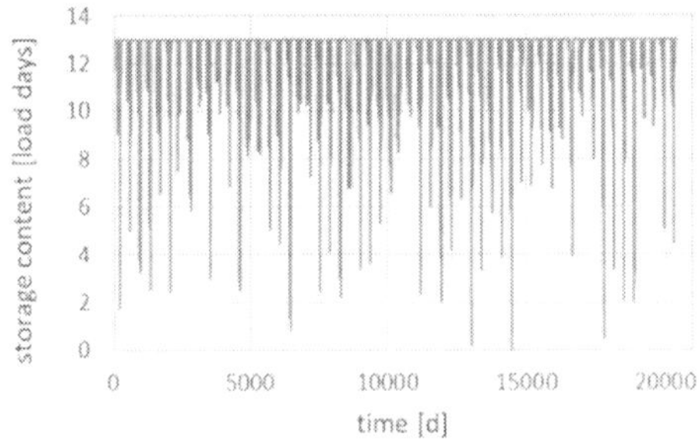

Figure 1: Daily resolved evolution of the storage content over 56 years (1958-2013) for an idealized PV/battery-storage system serving a constant load in the summer months (May-August) at Torshavn (Faroe Islands). The PV size is scaled for equal average generation and load in this period. The storage size is determined to assure the safe supply of the load. At day of operation No. 14488 (end of August 1997) this storge gets completely discharged.

An increase of the PV size by a factor of 1.2 allows for a reductio of the storage size to 7.18 load days (Fig.2). Again, the depletion of the storage happens end of August, but this time in August 1958 (day of operation Nr. 242), Increasing the PV size by 1.5 further reduces the required store size to 3.64 load days, again required in August 1958 (day of operation Nr. 242), see fig. 3.

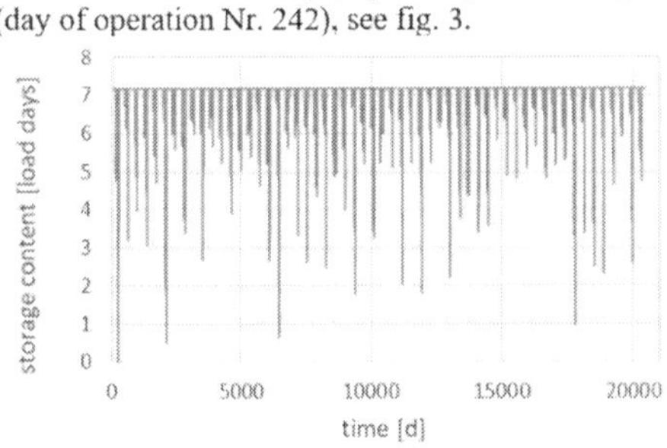

Figure 2: Same presentation as Fig. 1, but for a system with PV oversized by a factor 1.2. Storage gets depleted in the first year of operation (day 248, end of August 1958)

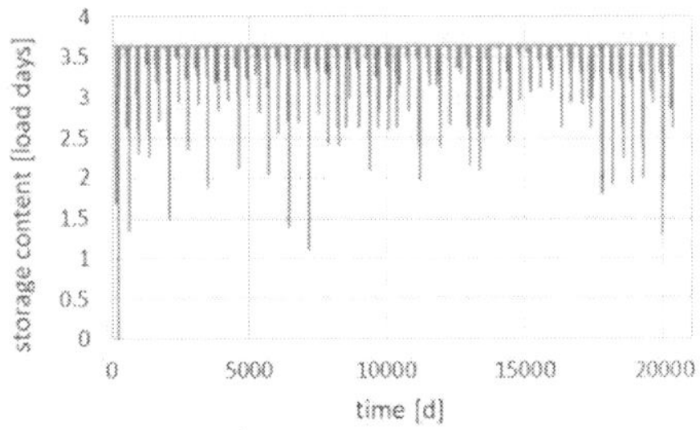

Figure 3: Same presentation as Fig. 1, but for a system with PV oversized by a factor 1.5. Storage gets depleted in the first year of operation as for the system with sizing factor 1.2

Obviously, and expectably, for the system inspected the month of August is the critical month due to the astronomical conditions. The operation in May is supported by the completely loaded storage by end of April as mentioned above.

Comparing the three cases highlights that the time of occurence of critical sitations for the storage varies with the selection of the PV size and the respective required storage size. By this, the concept of the use of a single year applied as "typical" or "worst case" year in the dimensioning process is fiutale. The system design for a reliable layout in view of the the criteriun "uninterrupted supply for a known time series of meteorological conditions is only possible given the complete knowledge of this series.

In the following, it is inspected which variability in layout results is caused by inspectimg only sections of the data base.

4 STORAGE SIZING ON THE BASIS OF SECTIONS OF THE IRRADIANCE DATA SET

The influence of the length of the dataset used for the storage sizing is inspected here by the analysis of the storage sizes stemming from the application of sections of lengths of 2, 5 and 10 years for the sizing scheme. Analyzed are all sections with the respective length, that can be formed from the basic series in a sliding mode. Basis is given here by the series of the of the storage content determined for the different PV-sizes (see Figs. 1-3). For the sections of length of 720, 1825 and 3650 days preceding a date inspected, the related storage size is determined by the difference of storage size determined from the inspection of the complete data set o the minimal storage content in the section inspected. Fig´s 4-6 show the resulting storage sizes for systems with PV-size 1, 1.2 and 1.5. It had to be stated here that the PV size relates to the sizing based on the complete set. PV sizes adapted to the individual sections would modify results.

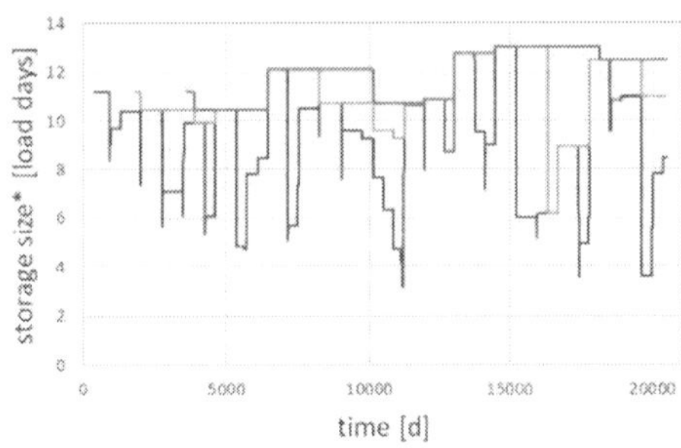

Figure 4: Storage sizes as determined from sections of the time series of lengths of 10 (red), 5 (green), and 2 (blue years). (PV-size: 1).

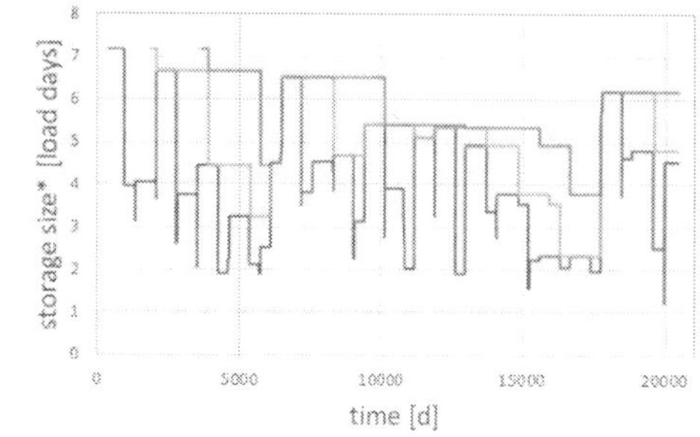

Figure 5: Same presentation as Fig.4, but for PV-size 1.2.

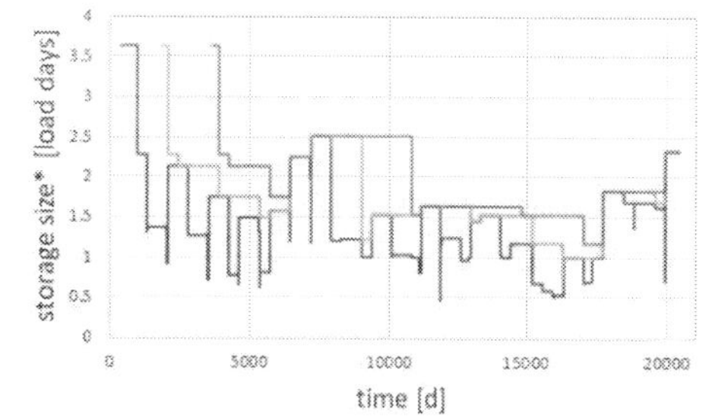

Figure 6: Same presentation as Fig.4, but for PV-size 1.5

It can be remarked, the use of an arbitrarily selected section of 10 years lengths may cause an underestimation of storage size required to secure the supply the load for the complete set by up to ~20%. For systems with increased PV size and according,ly reduced storage size this relative deviation spreads. The same pattern also appears – with augmented magnitude - when looking at the respective results for the sets with lengths of 5 and 2 years.

Fig. 7 shows the distribution of the storage sizes determined from the 10-year sets inspected for a system with PV size 1. To be sure to capture the size necessary for safe supply over the complete period 80% of all possible sets have to be inspected. Fig's. 8 and 9 give this distribution for the sets of 5- and 2-years length.

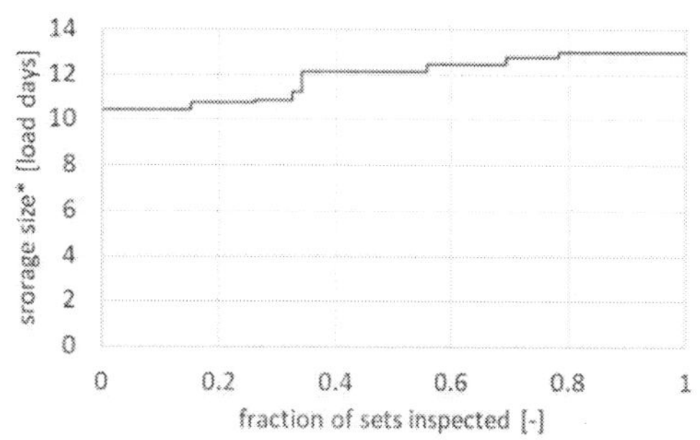

Figure 7: Distribution of the storage sizes determined from 10 years sequences for systems with PV-size 1.

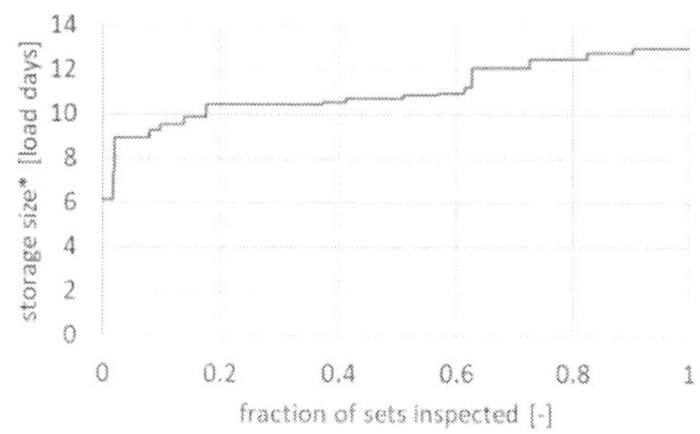

Figure 8: Same presentation as Fig. 7, but for 5 years sequences.

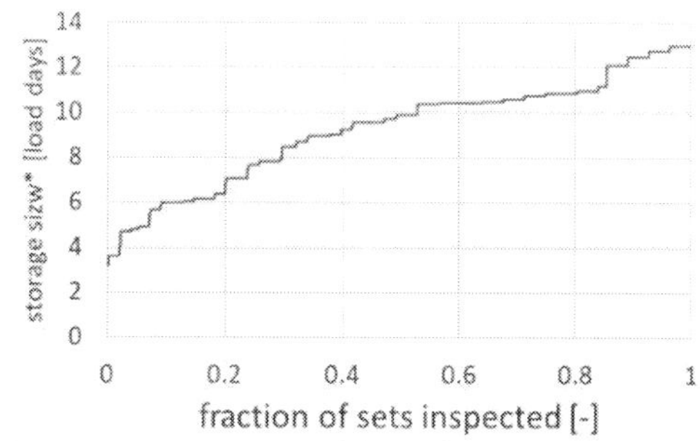

Figure 9: Same presentation as Fig. 7, but for 2 years sequences.

These finding may be used to assign an uncertainty rage to storage sizes derived from data sets with shorter durations - an information that has to accompany all sizing results.

5 VARIABILITY OF SIZING WITH SELECTED IRRADIANCE DATA BASE

As mentioned above, in addition to the uncertainties arising from the intrinsic temporal variability of irradiances, the uncertainties in measured and/or estimated radiance data add to the uncertainties in system sizing.

To inspect the resulting variability, here radiance data sets offer by the PV-GIS server [9] are used. Here, data stemming from satellite observations and derived from reanalysis schemes are used. consisting of the sets:

- PVGIS-SARAH, saellite derived, marked SC
- PVGIS-ERA5, re-analyses product, marked E5
- PVGIS-COSMO-REA6,
  re-analyses product.

These sets are commonly available for the years 2005-2015. A first view on the differences in these sets is given by the average annual irradiations (formed by average irradiances of the complete sets times 8760 hours) over the years inspected as shown in table 1.

Table 1. Average annual irradiations for the sets inspected.

SC	E5	CO
692 kWh	831 kWh	8.21 kWh

To keep the analyses scheme as used for the analysis of the multi decadal set, the data available with hourly time resolution are down-sampled to daily resolution and daily PV-generation sets are formed as described above. By the normalization to generation equal to load for May to

August, the focus is kept on the on the sensitivity of the sizing to the sequential characteristics of the sets.

For same PV-sizing, the application of these sets result in remarkable differences of the required storage sizes. Whereas the sizes resulting from the application of the SC and E5 sets differ by ~15%, the storage size from the CO set is ~50% superior to that for the SC set.

The time trace (starting from 1.1.2005) of the storage content resulting from the application of these sets are given in the figures Figs. 10-12. Differences in the patterns can be observed. Common feature is, that there is a higher depletion of the storage in the year 2006 (defining the storage size for the SA and E5 case) but the general similarity appears weak. For systems with a PV-size of 1.2 (Figs. 13-15) and 1.5 (Figs. 16-18) the findings for the resulting time traces are similar.

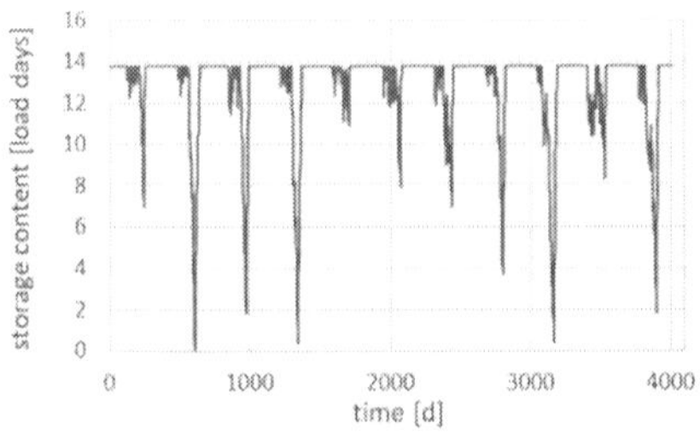

Figure 10: Daily resolved evolution of the storage content over 11 year (2005-2015), based on data from the PV-GIS server [9]. The data set used here is the satellite derived set SC (see text). PV refers to the sizing factor 1 (see section 2) defined for this 11-year set).

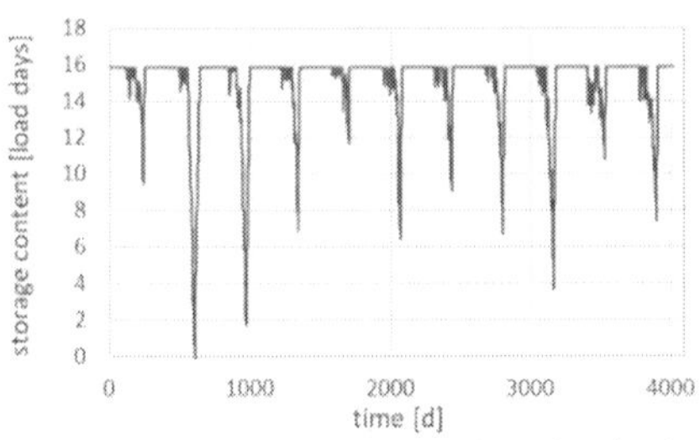

Figure 11: Same presentation as Fig. 9, but for the data set E5.

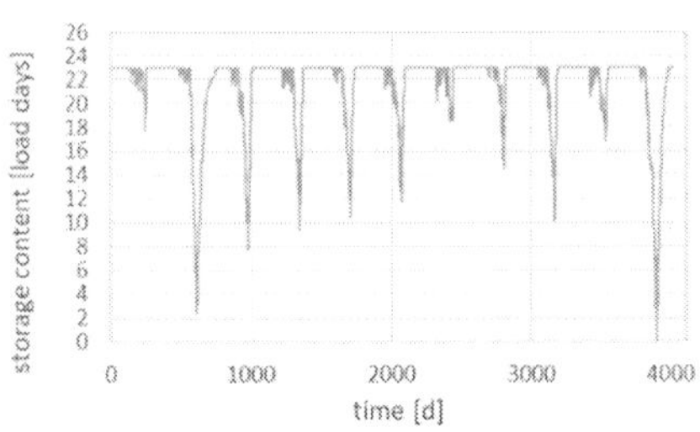

Figure 12: Same presentation as Fig. 9, but for the data set CO.

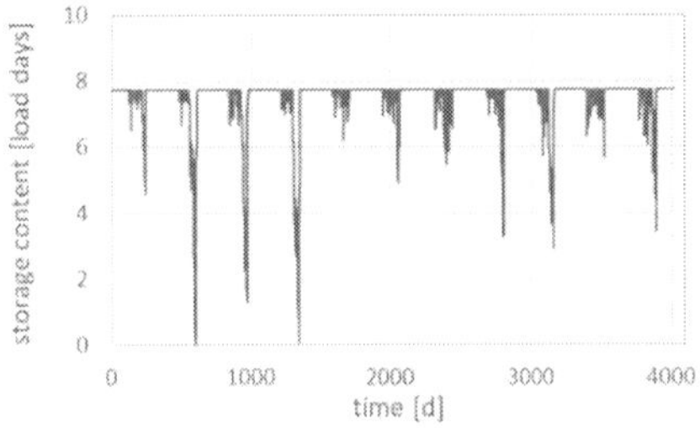

Figure 13: Same presentation as Fig. 9, but for a system with PV oversized by a factor 1.2, data aet SC,

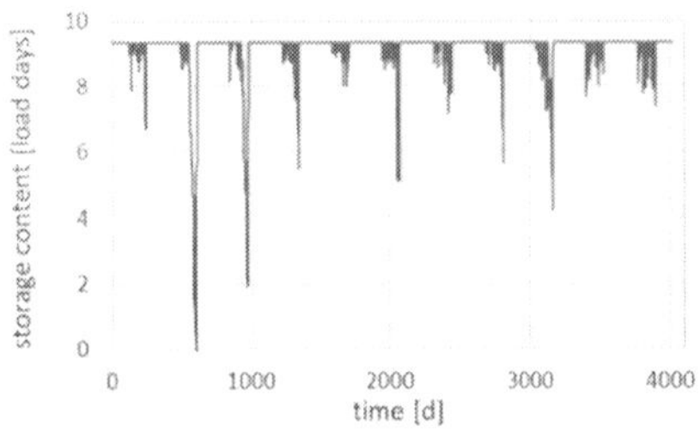

Figure 14: Same presentation as Fig. 9, but for a system with PV oversized by a factor 1.2, data set E5

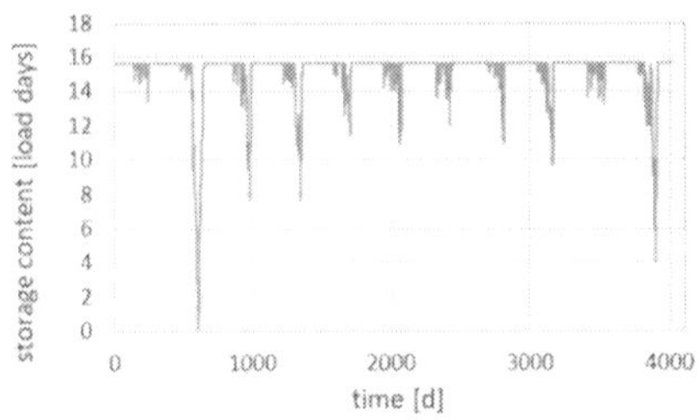

Figure 15: Same presentation as Fig. 9, but for a system with PV oversized by a factor 1.2, data set CO.

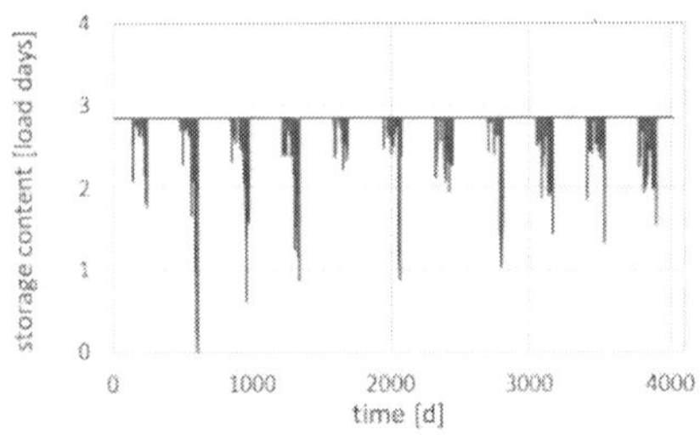

Figure 16: Same presentation as Fig. 9, but for a system with PV oversized by a factor 1.5, data set SC.

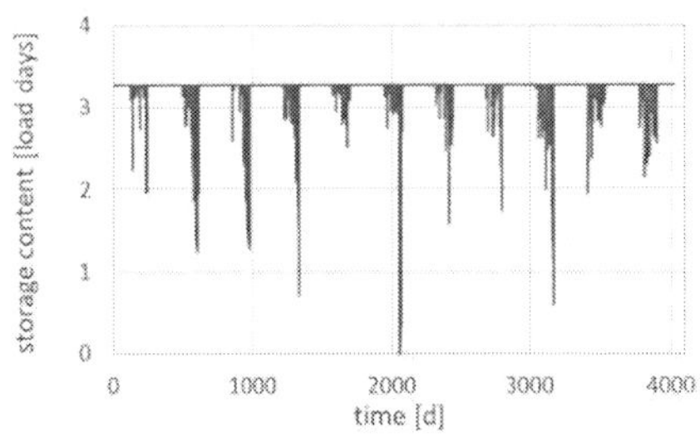

Figure 17: Same presentation as Fig. 9, but for a system with PV oversized by a factor 1.5, data set E5.

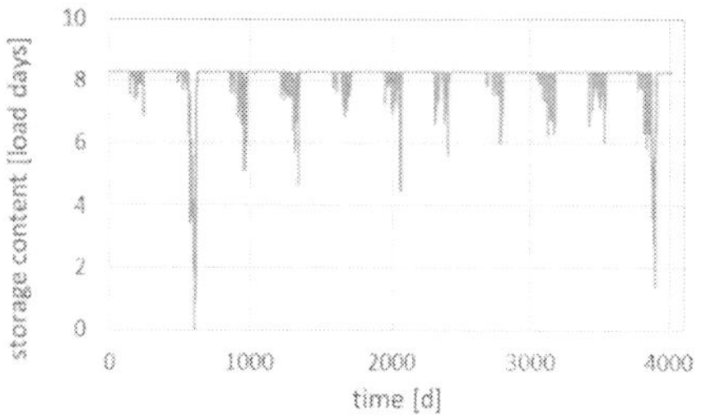

Figure 18: Same presentation as Fig. 9, but for a system with PV oversized by a factor 1.5, data set CO.

Thus, the data bases inspected obviously show considerable variability in their sequential characteristics, causing uncertainties in storage sizing that add to the uncertainties stemming from the intrinsic interannual variability of the resource itself, as discussed above. A simply bias correction, e.g. done on the bases of measured data (here the satellite-based SA set comes close) cannot overcome these discrepancies. Additional tools are needed to quantify the deviations in the sequential characteristics of modeled sets, as bases for respective model improvements - to be uses in tools for predicting future conditions as well.

6 CONCLUSIONS AND OUTLOOK

When sizing generation capacity and storage of renewable (here solar) based supply systems with intended high security of supply, the use of data sets for solar radiation (and load) covering a limited number of years only the results show a high uncertainty. This is due to the variability of yearly and monthly means and – especially for systems sensitive to thresholds and small memory (i.e. small storage size) - details of the short term sequential evolution of the sets. This casts a light on additional problems related to the use of projected evolutions of the meteorological conditions for future energy supply scenarios. The sensitivity of the system layout to details of the temporal evolution extend the requirements on the proximity to reality of projected data sets.

8 REFERENCES.

[1] J. Nitsch, T. Pregger, Y. Scholz, T.Naegler , M.Sterner, N. Gerhardt, A. von Oehsen, C. Pape, Y.M.Saint-Drenan, B. Wenzel, Langfristszenarien und Strategien für den Ausbau der erneuerbaren Energien in Deutschland bei Berücksichtigung der Entwicklung in Europa und global „Leitstudie 2010", report BMU -FKZ 03MAP146, German Federal Ministry for the Environment, Nature Conservation and Nuclear Safety (2010)
[2] S. Pfenninger, I. Staffell. Long-term patterns of European PV output using 30 years of validated hourly reanalysis and satellite data, Energy, 114, 1251-1265 (2016)
[3] R. Bryce, I. Losada, A. Kumler, B. Hodge, B., B. Roberts, C.B. Martinez-anido, Consequences of neglecting the interannual variability of the solar resource: A case study of photovoltaic power among the Hawaiian Islands. Sol. Energy 167, 61–75 (2018)
[4] H.G.Beyer, Aspects of system sizing for complete autonomy of island grids, discussed by using the example of PV-integration into the Faroe Islands grid, Solar Integration Workshop, Dublin Ireland, October (2019)
[5] H.G. Beyer, Layout of island power supply based on multi-decadal meteorological sets - discussed using the case of the Faroe Islands power system, 13th International Conference on Solar Energy for Buildings and Industry (virtual), 01-03.9 (2020)
[6] H.G. Beyer, Repercussions of interannual variability in irradiance on sizing PV for autonomous renewable energy systems, EUPVSEC 2020 (virtual), 07-11.9 (2020)
[7] A. Haas, Zur Bestimmung des Speicherbedarfs in solarelektrischen Energieversorgungsystemen aus Charakteristiken der Einstrahlung, PhD Thesis, University of Oldenburg, Germany (1995)
[8] R. Müller, R., U. Pfeifroth, C. Träger-Chatterjee R. Cremer, J. Trentmann, R. Hollmann, R., Surface solar radiation data set - Heliosat (SARAH) - Edition 1. Satellite Application Facility on Climate Monitoring (CM SAF (2015) URL: ,
[9] https://re.jrc.ec.europa.eu/pvg_tools/en/#MR
[10]https://ec.europa.eu/jrc/en/PVGIS/docs/methods

ANALYSIS OF ELECTRICAL SHADING EFFECTS IN PV SYSTEMS

Michele Oliosi, Bruno Wittmer, André Mermoud
PVsyst SA
Route de la Maison-Carrée 30, CH 1242 Satigny, Switzerland

ABSTRACT: In a PV system, partial shading of PV modules and strings leads to yield losses not only due to reduced irradiation, but also due to shaded PV cells limiting the current of a string. These so-called electrical shading losses contribute in a non-negligible way to the total losses in a PV system. The detailed calculation of these losses is non-trivial, since it is necessary to combine many IV curves from individual cell strings and determine the resulting curve which can have a much more irregular shape than the simple IV-curve of a group of unshaded strings. Hence, in-depth studies are required to understand and optimize the PV system design.
In this work, we explain the different ways in which the PVsyst software handles the electrical shadings. The detailed IV-curve calculation and three simplified models are applied to different PV system configurations, and the impact of partial shadings on the energy yield is assessed. We discuss which calculation models should be used in different situations. By comparing the results for different typical string layouts, and PV module types, we derive some strategies for minimizing electrical shading losses, that can be applied in a general way.
Keywords: Shading, PV system, Simulation, Software, Mismatch, Design

1 INTRODUCTION

Among the losses that are considered when designing a PV system, those caused by full or partial shadings are fundamental. Except for a few cases such as horizontally mounted modules, or a regular terrain allowing the use of a backtracking algorithm, at low solar elevations, shadings are usually unavoidable. Partial shadings are especially important because of the possible extra losses caused by the irradiance mismatch between strings, modules, or cells. For example, shading just a single module or even a single cell can limit the current of a whole string. This induces an additional loss on top of the irradiance deficit.

In the situation of large systems with regular rows of fixed-orientation modules, which we refer to as shed systems, mutual row shadings are the main cause of partial shading. An obvious way to reduce these losses is to increase the spacing between the rows of modules, which comes at the price of increasing the area. Since this is often not viable, we consider here only systems with a fixed ground coverage ratio (GCR).

Changing configuration of modules and strings has an impact on the electrical shading losses. Indeed, it is well known [1, 2, 3, 4], that mounting modules in landscape or portrait orientation will lead to different electrical shading losses. Other factors that influence the losses are the choice of modules (for example standard or "twin" half-cell modules), the number of modules in a string, the layout of the strings (either on a single row, or on multiple rows), the number of rows on each table/shed, etc. All these options will have an impact on the generation of electrical shading losses, and some effort has already been made to distinguish the best configurations considering some subsets of this palette (see e.g. [6, 7, 8, 9, 10, 11, 12]).

The electrical shading losses for these different options can be simulated in several ways. A complete calculation should involve the combination of the IV curves of every individual PV cell, with an accurate consideration of the respective shading conditions. However, doing this for PV systems above the MW scale, can become very time consuming. As emphasized in [10], analytical models are faster, but need several parameters which are specific to each PV module or system and are often not available. The authors of [10] argue that a simplified model using only geometrical and few module type considerations is possible in the case of simple situations, such that of regular arrays or sheds. They in fact present a simplified model, which extends earlier proposals such as [13]. As described in this work, the PVsyst software also contains such models, which can be used for a wide variety of shed systems.

In this paper we address the question of finding adequate system design choices to mitigate the electrical shading losses in the case of the mutual shadings in regular shed arrays. We also apply the models available in PVsyst to the different PV system configurations in an effort to benchmark them against each other. This allows us to present general guidelines and suggestions both for system design and modelling within PVsyst. We are able to qualitatively reproduce the field results of [10] with the detailed calculation model and present the rationale behind the simplified model.

2 METHODOLOGY

The basic idea of our study is to model and simulate a set of shed-like systems, spanning a wide range of possible module and string configurations. The simulations are run using the different types of shading models available in PVsyst: a detailed model based on the composition of IV curves, as allowed by the "module layout" definitions, and three variations of a simplified model that speed up computations for very large systems. We obtain yearly averages and hourly results for several variables of interest. This allows us to benchmark the simplified models against the detailed model and identify the configurations leading to lower electrical shading losses. This procedure is summarized in Figure 1.

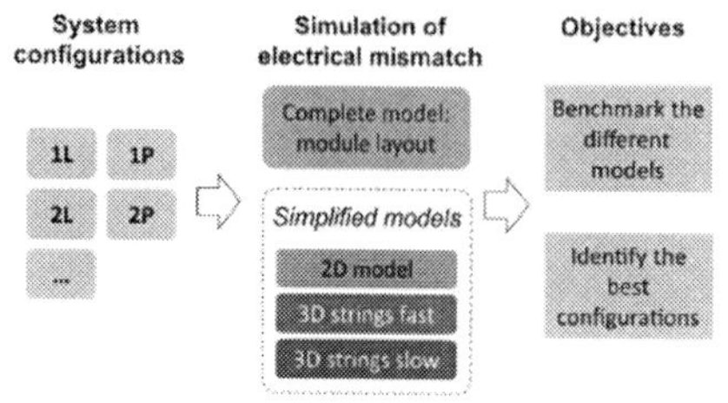

Figure 1: Procedure followed in this work. A total of 18 different configurations (listed in Table I) was studied.

Modelling with PVsyst allows one to control multiple aspects of system design. In this study we vary the configurations of PV cells, modules, and strings with the following parameters:

- Number of rows of modules in the height of the shed (1 to 4).
- Module orientation (Landscape or Portrait).
- Type of module (Standard or Twin half-cell in portrait).
- String connections layout on a single row or on multiple rows (U-shape).

The systems chosen for our study span all possibilities allowed by these four parameters. For the remainder of the work, we use the labelling described in Figure 2 for the different cases.

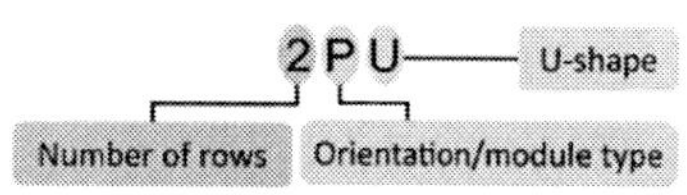

Figure 2: Configuration identifier description. The example (2PU) corresponds to 2 rows of portrait-oriented modules, with strings spanning both rows.

With this terminology, the list of simulated configurations can be found in Table I. Configurations in color are those commonly found in actual realizations; the remaining configurations are also studied to check the general character of the simplified approaches.

Table I: List of the different shed-like system configurations studied in this work. Cases in blue are those commonly used in actual realizations.

	Landscape	Portrait	Twin half-cell
1-row	1L	1P	1T
2-rows	2L/2LU	2P/2PU	2T/2TU
3-rows	3L	3P	3T
4-rows	4L/4LU	4P/4PU	4T/4TU

There are other variables which impact the losses. Non exhaustively, these are the location, the meteorological conditions, the GCR, the tilt and azimuth of the sheds. Contrary to the previous variables, this set of conditions will be fixed for all simulations. To compare yearly results and loss factors, we opt for a "realistic" setup: 20° tilt, 0° azimuth, with a GCR = 0.65, meteorological conditions synthetically generated from the Meteonorm 7.3 [14] data in Marseille (France). To study the hourly values, we opt instead for a configuration which leads to a wider range of shading conditions: 50° tilt, -90° azimuth, GCR = 0.65, and a location at 85° latitude with a manually adjusted low diffuse fraction (5-10%).

Finally special care is given to the 3D scene layout, with the goal of having the most regular shading conditions (i.e., as a horizontal band of direct sunlight deficit covering the whole shed width gradually from the bottom to the top). PVsyst allows to easily construct regular arrays; for the realistic layout we construct 2 to 6 consecutive wide sheds of 48 modules in width, with a shed-shaped rectangular shading object in front of the first shed to allow for uniform shading conditions. Instead, for the idealized setup, we also use additional shading rectangles on several rows in front and on the sides of the active sheds, to ensure that mutual shadings cover the whole horizontal width of the sheds.

Each of the electrical shading models requires further configuration. The detailed model, used as reference, requires a distribution of modules into strings, which are then distributed among the different inverter MPPT inputs. This is done through the "module layout" tool in PVsyst, and hence we will sometimes use the term "module layout model" in this work. Instead, the simplified models require either to specify the number of modules in the height of the shed for the 2D computation, or to partition the sheds into string-rectangles in the 3D scene. In some cases, for the simplified model to best match the reference, these partitions should be adjusted (i.e., not follow string boundaries anymore).

The reference and simplified models are applied to all different systems described above. The simulations run over a full year, yielding yearly average and cumulative quantities such as energies and losses, and hourly results for all the variables of interest. We are interested in the cumulative and hourly electrical shading losses, the shading fractions as well as the relative loss factors that can be derived from these.

The electrical shading factor $f_{ShdElec}$ reads

$$f_{ShdElec} = \frac{L_{ShdElec}}{P_{ArrNom} - L_{GInc} - L_{Temp}},$$

where P_{ArrNom} is the array nominal power at STC, L_{GInc} is the power loss due to the irradiance deficit with respect to the STC, L_{Temp} is the power loss due to the temperature difference with respect to the STC, and $L_{ShdElec}$ is the power loss due to electrical shadings only. The fraction of the sheds shaded from the beam component f_{ShdBm}, which we use as the main variable characterizing the shading situation, can be obtained directly as an output of the simulation. We also call it linear shading factor on beam. Finally, the electrical shading factor on beam $f_{ShdElecBm}$ requires an intermediate step: we first define the fraction of beam

$$F_{Bm} = \frac{\frac{G_{BeamEff}}{1 - f_{ShdBm}}}{G_{AlbEff} + G_{DiffEff} + \frac{G_{BeamEff}}{1 - f_{ShdBm}}},$$

where G_{AlbEff} and $G_{DiffEff}$ are the albedo and diffuse irradiances remaining after subtracting the respective shading loss, evaluated as an integral of the linear shading factor over all sky directions. $G_{BeamEff}$ is the shaded beam irradiance, obtained by discriminating the surface of shaded and unshaded active PV areas. This allows us to define the electrical shading factor on beam as

$$f_{ShdElecBm} = \frac{L_{ShdElec}}{P_{ArrRef} \cdot F_{Bm}},$$

which can be added with the linear shading factor on beam

to obtain the full shading factor on beam. Here, P_{ArrRef} is the array power at unshaded beam conditions, hence used as reference for the shading factors.

All quantities on the right-hand side of the equations can be obtained as hourly and yearly outputs of a PVsyst simulation. This study has been conducted using PVsyst 7.2.

3 MODELS USED

3.1 Irradiance components

The irradiance model in PVsyst is based on a decomposition of the global irradiance into 4 main components: 1) the beam component, 2) the circumsolar component, as well as the 3) isotropic diffuse and 4) albedo components. The beam or direct component, coming directly from the direction of the sun, leads to non-uniform irradiance conditions. Instead, the albedo and diffuse components are approximated as uniform irradiance conditions on the active PV surfaces. In the end, only the beam (and optionally the circumsolar as seen below) component can lead to electrical shading losses, which are directly caused by the electrical mismatch between different system components due to non-uniform irradiance conditions, or partial shadings.

Finally, the circumsolar component, i.e., the irradiance coming from outside a 5° cone around the sun, can be treated in PVsyst as either the beam component or included in the diffuse. Our study treats the circumsolar as the diffuse and therefore as isotropic and leading to uniform irradiance conditions.

3.2 Reference model ("module layout")

The reference model for electrical shadings used for this study is the most detailed computation available in PVsyst. It combines the IV curves of each individual submodule (i.e., a group of PV cells protected by a by-pass diode) by considering the electrical connections designed and parametrized in the "module layout" tool. This allows the computation of the shaded MPP P_{ArrShd} as well as the operating point (OP) of each inverter. The unshaded MPP P_{ArrRef} is an estimate of the array power based on the sum of the plane of array shaded albedo and diffuse components, and unshaded beam component. Comparing the power at these points with that at the unshaded MPP leads to a detailed evaluation the shading losses. The electrical shading losses are

$$L_{ShdElec} = P_{ArrRef} \cdot (1 - F_{Bm} \cdot f_{ShdBm}) - P_{ArrShd}\,.$$

The use of submodules instead of cells for the combination of IV curves can be understood as follows: whatever the number of shaded cells in a submodule, its IV curve will approximately have the same shape. This is shown in Figure 3, where the submodule IV curve is drawn for different numbers of shaded cells. Basically, up to tiny discrepancies, it is sufficient to consider submodules, instead of combining individual cells.

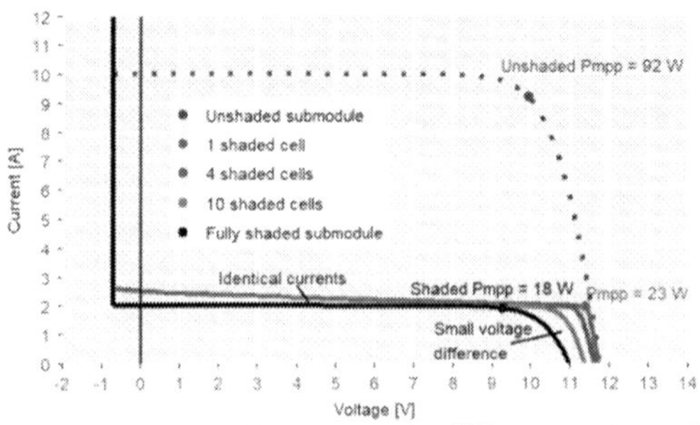

Figure 3: Submodule IV curve for different numbers of shaded cells. As long as one cell is shaded the curves are effectively the same, up to a small voltage difference.

Once the shading conditions have been determined for each submodule, their IV curve is combined with the other curves of the module, string, and finally MPPT input. This is done by adding the voltages of submodules in series to get the IV curves of the strings, and then adding the currents of the strings in parallel. The full shading losses can be split into a part proportional to the irradiance deficit part, named "linear shading losses" and the remaining "electrical shading losses".

In regular shed systems, since the shading conditions are horizontally uniform, it is possible to draw some general conclusions as to the resulting shading loss factors. One of our scenarios is for modules to be mounted in landscape orientation. In such a case, the submodules (typically 3) in each module will lie one above the other. Hence for a row of modules, 1/3, 2/3 or all submodules of the row will be shaded at once.

Figure 4 (already presented in [15]) shows the shading factor for one string, as a function of the number of submodules shaded. This is dependent on the number of strings connected to one MPPT input. We observe that for 3 strings in parallel or more, as soon as the bottom row of submodules (or cells) is shaded (i.e., 1/3 of submodules) the string becomes electrically inactive. This is due to the voltage mismatch between the strings. The losses are however lower for one (or possibly two) strings per MPPT input.

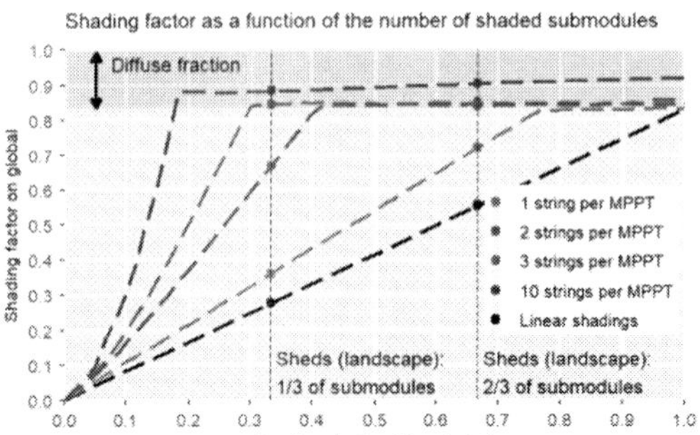

Figure 4: Electrical shading factor of a string for different numbers of strings per MPPT input. The more strings one adds per inverter input, the larger the electrical shading effect over the string. In the case of regular rows of landscape modules one shades 1/3 of the submodules at a time. For portrait oriented modules, one shades all the submodules at once.

In the case of portrait-sheds, since the submodules are arranged side-to-side, all submodules will be shaded at the same time. In this case, the shading loss factor will be maximal as soon as the bottom row of cells is shaded.

3.3 Simplified model for sheds

The simplified model takes advantage of the above observations. It is based on a partition in rectangles, each one representing a full string. As soon as the bottom of the rectangle (one cell width) is shaded, the full rectangle becomes inactive (electrically shaded). It can be applied with three different levels of sophistication in PVsyst.

The first step in the model is to determine the shading conditions on each of the string-rectangle partitions. This can be done in three ways. First, via a cross-sectional 2D unlimited sheds model (hereafter called the "2D simplified model"), allowing for analytical calculations. The partitioning in strings is configured by entering the number of modules in the height of the shed. Second, the string partitions can be defined in the 3D scene, with shading conditions interpolated from a pre-computed shading factor table. This model is called "fast 3D simplified model". Third, the same rectangles defined in the 3D scene can have their shading conditions computed hour-by-hour relative to the instantaneous sun position. We call this model the "slow 3D simplified model".

An advanced parameter in the simplified model is the fraction for electrical effect. It allows one to reduce the electrical shading losses by an overall factor. In this study we leave its default value, 100%.

Once the shading conditions of each rectangle have been computed, the simplified model bypasses the full IV-curve based calculation: the electrical shading losses are instead a simple step-like function of the shading conditions. Two factors are applied: a linear shading factor taking into account the irradiance deficit, and an electrical shading factor following the step-like function. Together these give the shading loss factor, which is applied to the beam irradiance component only. The function takes into account that if a part of the bottom cell is shaded, the current in the cell (and therefore in the submodule) is proportional to the shaded fraction of the cell. In the 2D case, the analytic calculation allows to evaluate this fraction. In the 3D case, we don't have the opportunity of applying a "partial loss" to the rectangle. Therefore, the shading on the first cell is considered null up to the middle of the cell, and complete when more than half of it is shaded. The errors should be averaged over all the hours of the simulation. The shading factors for the simplified model are represented in Figure 5.

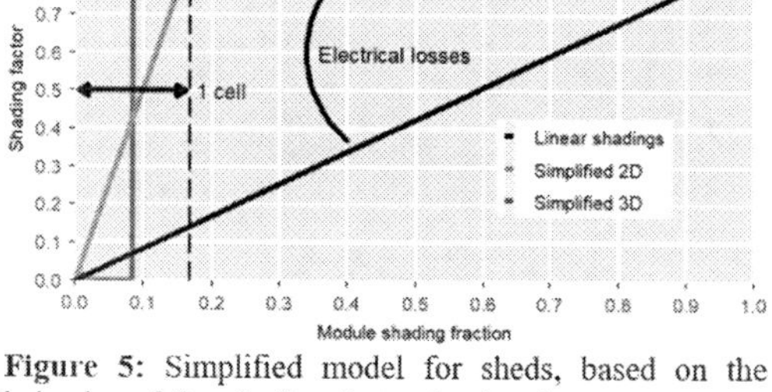

Figure 5: Simplified model for sheds, based on the behavior of the shading factor for 3 strings or more per MPPT input (see Figure 4).

In fact, the simplified model plateau behavior is based on the results for the module layout model, for landscape-oriented modules starting from 3 strings per MPPT, or for any string number of portrait-oriented modules. As a conclusion of this work, we will see that in the landscape case, for less than 3 strings per MPPT (typically for string inverters), it may be worthwhile to readjust the string-rectangle partitioning.

4 RESULTS AND DISCUSSION

Out of the one-year-long PVsyst simulations, it is possible to obtain, for a wide set of variables, both an hourly time-series and the yearly overall values. We use these two possibilities in respectively different ways: the detailed hourly results can be processed to assess the behavior of the models as a function of the shading conditions, which will vary hour by hour. Instead, we use the yearly results to benchmark the models against each other and determine which configurations should lead to lower overall electrical shading losses. As explained in Section 2, we use different setups for either hourly or yearly result evaluation. The more "realistic" setup is used to evaluate yearly results.

4.1 Detailed hourly results

The horizontally uniform shading conditions in shed systems make it possible to qualitatively characterize the shading conditions with a single variable, the beam linear shading fraction. This quantity corresponds to the shaded fraction of the active PV surface. In a perfectly regular array of sheds without module borders, this quantity is equal to the fraction of submodules heights shaded. It is thus interesting to compare the full shading factor (including the linear and the electrical shading losses) to the beam linear shading fraction. We plot these quantities for all the configurations of the study, which allows us to analyze the behavior of the models case-by-case and across the whole range of shading conditions.

In the complete model, the shading fraction always increases gradually by constant plateaus, while the simplified model exhibits slightly "curved" steps, a behavior due to the non-shaded beam fraction having been used to estimate the beam contribution to the array power. As confirmed in the next section, the simplified model is nonetheless very reliable in the case of shed systems, although it sometimes requires an adjusted rectangle partitioning.

The necessity for adjustments in the rectangle partitioning is well exemplified by the case 1L, shown alongside other cases xL in Figure 6. Instead of a single plateau, the complete model (shown in orange) increases in three well-defined constant steps corresponding to the three rows of submodules being shaded one after the other. In fact, this behavior could be anticipated already from Figure 4: in the case of 1 string per MPPT, when 1/3 of the submodules are shaded the maximum electrical shading factor has not yet been reached. To match this behavior with the simplified model (shown in blue), one should increase the number of partitions vertically.

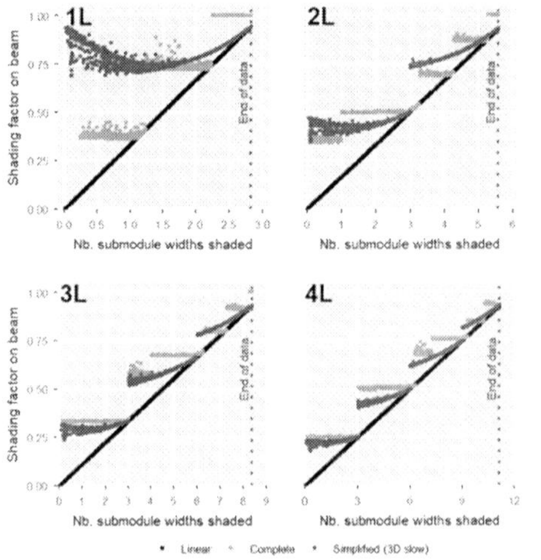

Figure 6: Full shading factor as a function of the number of submodules widths shaded, for all xL configurations. The simplified model (in blue) does not match the complete calculation (in orange) in the case 1L. In practice, the situation where only the first row of submodules is shaded has more weight on simulation results, since it occurs more frequently.

For the other cases xL, the more strings one adds to a single MPPT input, the faster the plateau of Figure 4 is reached, gradually leading to having a single plateau over the 3 first submodules shaded. This corresponds to 1 shaded row of modules—the first row is between 0 and 3 shaded submodules widths, the second between 3 and 6, and so forth. As the number of shaded rows increases beyond the first, the extra losses due to the current mismatch between strings will gradually decrease. When the last row is shaded, the behavior of the shading factor is qualitatively the same as in case 1L: there are three steps instead of a single plateau across the whole row width. Note that it is most important to accurately model electrical shadings on the first row. Indeed, shading over the first row happens more frequently, so that this situation will have more impact on simulation results.

In all cases xL except case 1L, the simplified model visibly succeeds in approximating the behavior of the detailed model. However, since the features of the detailed model are regular, it should be possible to improve the simplified model up to a much better agreement with the reference.

For the remainder of the cases, we will focus on those with 2 rows only (and always 2 strings per MPPT input), as they are representative for other number of rows (and allow us to discuss the "U-shape" connections as well). Cases 2P and 2T, presented in Figure 7, show a good agreement between the models. Both models follow the same behavior: as soon as one cell is shaded, the whole string is turned off. Still, some improvement in the form of a flatter plateau for the simplified model can be made. As discussed above, this could be done by estimating the beam fraction more accurately.

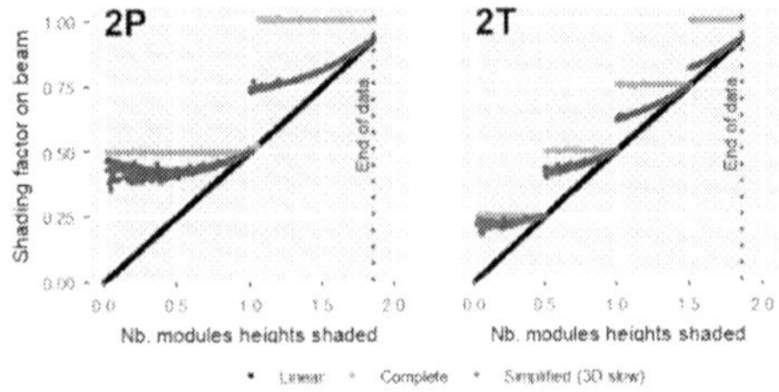

Figure 7: Full shading factor as a function of the number of submodules widths shaded, for the 2P and 2T configurations. The simplified model matches well with the complete calculation.

Instead, the U-shape cases (strings arranged vertically), exemplified in Figure 8, all need adjustments to the string-rectangle partitioning for the simplified model to match with the reference. This is because the U-shape configuration, without the constraint of a minimum inverter threshold, will simply mean that the operating point may still be chosen according to the MPP of the unshaded modules, and the electrical shading losses will be smaller. However, with a minimum voltage threshold, the inverter will more frequently select an operating point with the current limited by the shaded modules, thus leading to large electrical shading losses.

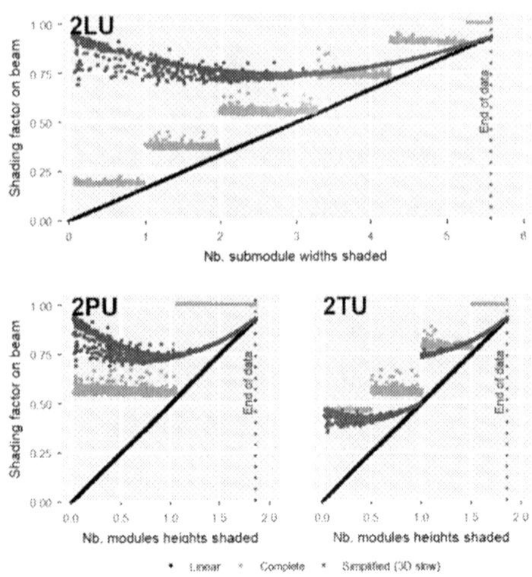

Figure 8: The case 2LU visibly needs an adjusted partitioning for the simplified model (in blue) to match with the complete calculation (in orange).

From these hourly results, it is possible to infer which configurations lead to the lower electrical shading losses. These observations will also be confirmed in the next section with the yearly results. With the understanding that the shading factor function behaves as a stepwise increasing function, having more steps, and lower ones, will mean having less electrical shading losses overall. For this reason,

a) increasing the number of rows per shed,

b) distributing the strings on the inverter inputs in terms of similar shading conditions, and

c) choosing landscape orientations for standard modules,

will ultimately decrease the electrical shading losses.

Note that throughout the section and Figs. 6-8 we have

chosen to present only the slow 3D simplified model, since the results for the other simplified models are fully comparable both qualitatively and quantitatively.

4.2 Yearly results comparison

Yearly results are used to benchmark the simplified model against the reference model. Ideally, the simplified model should yield approximately the same (yearly) average electrical shading factor than the reference model. Some cases however lead up to an order-one difference, as we will see because of inappropriate string-rectangle partitioning. Typically, a difference of up to about 1% in the shading factor may be considered satisfactory.

In Figure 9 we show the electrical shading factors for the different models in the cases that are most common in actual projects. The string-rectangle partitioning is done as intended: delimiting the surface strings, or, in the case of twin half-cell modules, half strings.

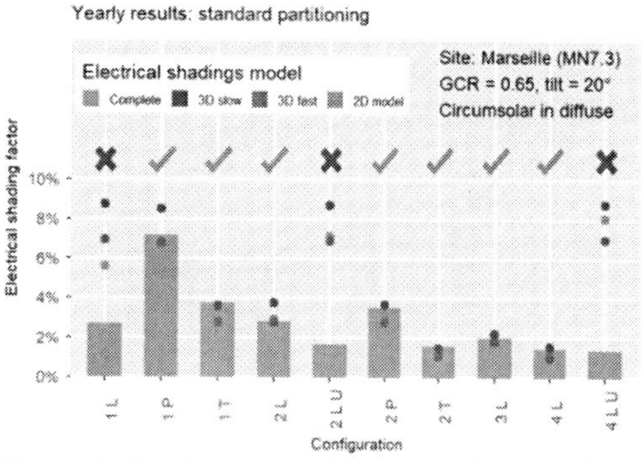

Figure 9: Yearly average electrical shading loss factor, across the most commonly used configurations. The complete calculation results (in orange) show that different configurations will lead to very different losses. The simplified models (in blue and green dots) do not always match the complete calculations, indicating that and adjusted partitioning might be needed.

For most cases, despite the small differences observed in the previous section, the yearly results show the different models matching well with each other. However, the case 1L as well as the cases with strings distributed on multiple rows (U-shape connections) do not match.

The reason behind the disparity in the cases 1L and xU can be understood from Figure 4 and from the perspective of the hourly results presented in previous section.

For the 1L case, the shading factor has a low value for 1/3 submodules shaded (around 38%, as seen in Figure 4). Since the first row of modules is more frequently shaded, this low value is the primary source of disparity. In Figure 6, this translates into a gradual increase in three steps instead of a single plateau as found by the simplified model. For the 2LU and 4LU cases, only 1/6 of the submodules are shaded at a time. In Figure 8 this translates into an increase of the shading factor over 6 steps instead of 1 for the simplified model. The most frequent shading situation is when only the lower 1/6 of the submodules are shaded, meaning that one looks at a very low value of the shading factor. Note that in Figure 4, one should look at the curve for 1 string per MPPT, because with strings in arranged vertically all of them have the same shading conditions.

To reduce the disparity for these problematic cases, we adjust the partitioning, to reflect the step-like behavior of the shading fraction. The resulting yearly shading factors are shown in Figure 10.

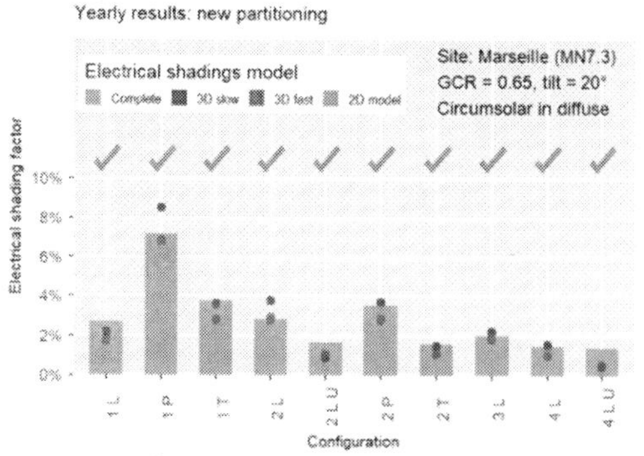

Figure 10: Idem to Figure 9, but with adjustments to the partitioning into string rectangles. The simplified model can thus match the complete calculations for all commonly used configurations.

The adjusted partitioning is shown in Table II.

Table II: Adjusted partitioning guidelines, directly applicable in the modelling in PVsyst, leading to the best match between simplified model and complete calculation. Partitionings with an (*) are the adjusted ones.

Case	Partitions in height	Case	Partitions in height
1L	2 partitions (*)	xPU	module layout (*)
xL	x partitions	xT	2x
xLU	3x partitions (*)	xTU	x (*)
xP	x partitions		

Although the most common cases can be made to match with the reference model, this is not the case for the case XPU. For example, in the 2PU case shown in Figure 8, one could think that two partitions would work best. However, since the shading factor is above 50% on the first step, the electrical shading losses end up being significatively underestimated.

Finally, it is possible to use these results to infer which configurations might produce less electrical losses. The general idea, repeated from the previous section is to:

a) increase the number of rows per shed,

b) distribute the strings on the inverter inputs in terms of similar shading conditions (alternatively reduce the number of strings per inverter input), and

c) choose landscape orientations for conventional modules.

The case xT gives approximately the same electrical shading factor as the (2x)L and the xLU case. Cases xPU and xTU are not interesting in terms of having a low electrical shading factor, and so U-shape connections are interesting only for landscape models.

4.3 Impact of inverter thresholds

A limitation for the applicability of the study happens at the level of the inverter. Indeed, inverters have upper and lower bounds on their voltage operating points. In the case of this study, a particularly important bound is the minimum inverter threshold. The different configurations lead to very different MPP voltages depending on the shading situation as shown in Figure 11 in the case of 2-row cases.

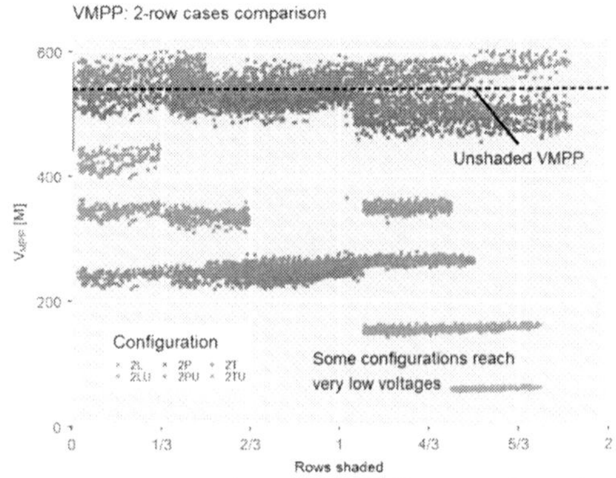

Figure 11: MPP voltage across different configurations for different shading conditions. Some situations lead to very low voltages, potentially below the inverter threshold. This will lead to extra losses.

Whenever the MPP voltage is below the inverter threshold, further losses are to be expected since the operating point of the inverter will be pushed away from the MPP. An explicit example is shown in Figure 12.

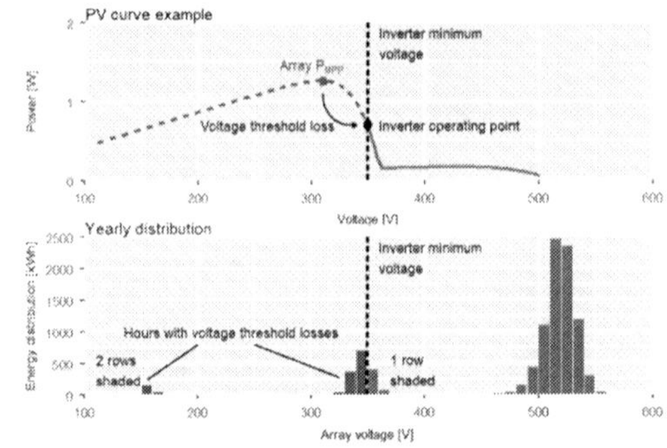

Figure 12: Example of a situation leading to extra losses due to the inverter minimum voltage threshold. Om this case hours with one or two rows shaded have an MPP voltage below the threshold.

5 CONCLUSION

To understand how the module and string arrangement in shed systems affects the electrical shading losses, we have simulated a set of different configurations in PVsyst. This also allowed us to benchmark the simplified models that are available in PVsyst in order to speed up calculations.

In general, to decrease the electrical shading losses, it is beneficial to increase the number of rows, have less strings per MPPT input, group strings by similar shading conditions, and use the landscape orientation with conventional (non-half-cell) modules. The U-shape connection may be interesting in the landscape case only.

These conclusions however hold only up to the inverter input. Once the inverter limitations are considered, some MPP with low voltages cannot be attained, and the operating point will be at a lower power, thus generating more losses. Furthermore, the losses from inverter limitations are only modellable using an IV curve approach, like the "module layout" tool of PVsyst. The losses further caused by the inverter limitations will be the subject of a future study.

We have found that in some cases, the partitioning of sheds into string-rectangles should be adjusted for the simplified model to match with the reference "module layout" model. This is the case for the 1L case that trivially has no mismatch between strings, but also for all the U-shape connection cases. We have included a table summarizing these adjusted partitioning guidelines.

A few examples of interesting extensions for this study are investigating the shading factor as a function of the GCR, as a function of the tilt, the study of tracking systems, as well as of other string or inverter layouts. A study with the circumsolar irradiance component taken into account together with the beam is expected to yield similar results.

Following this study, we plan to improve PVsyst with more options for electrical shading calculations and with a simplified model parameters ensuring a better match with the complete "module layout" calculation. In particular, we will implement a more accurate estimate of the beam contribution in the simplified model, allow more flexibility in the partitioning into rectangle-strings, as well as implement an extended "module layout" calculation, which would be applied on a subset of the system only, and then extrapolated to the whole system.

6 NOTES

This study is reproducible with PVsyst version 7.2.5. Contact the authors for additional details and results.

7 REFERENCES

[1] M. Hernandez Velasco, F. Fiedler and D. Timm, "Comparison of Partial Shading Losses in Free Field PV-Plants with Different Array Configurations," *28th European Photovoltaic Solar Energy Conference and Exhibition; 4171-4175,* 2013.

[2] N. Pearsall, The Performance of Photovoltaic (PV) Systems, Elsevier Science & Techn., 2016.

[3] P. Vanicek and S. Stein, "System-Technical Area Optimization of Large Scale Open Space PV Plants," *27th European Photovoltaic Solar Energy Conference and Exhibition; 3788-3791,* 2012.

[4] A. Mermoud, "Optimization of Row-Arrangement in PV Systems, Shading Loss Evaluations According to Module Positioning and Connexions," *27th European Photovoltaic Solar Energy Conference and Exhibition; 4009-4013,* 2012.

[5] M. Herrerías Azcué and H. Capdevila, "Detailed Calculation of Electrical Mismatch Losses for Central- and String-Inverter Configurations on Utility-Scale PV Arrays," *35th European Photovoltaic Solar Energy Conference and Exhibition; 1973-1978,* 2018.

[6] G. Knežević, D. Topić, M. Žnidarec, B. Štumberger, M. Hadžiselimović and S. Seme, "Comparison of the Shading Influence on PV Modules of Different Technologies," in *Renewable Energy Sources,* 2017.

[7] E. Paraskevadaki and S. Papathanassiou, "Estimation of PV Array Power Losses Due to Partial Shading," *25th European Photovoltaic Solar*

Energy Conference and Exhibition / 5th World Conference on Photovoltaic Energy Conversion, Vols. 6-10 September 2010, p. Spain; 4560–4564, 2010.

[8] H. Hanifi, J. Schneider and J. Bagdahn, "Reduced Shading Effect on Half-Cell Modules – Measurement and Simulation," *31st European Photovoltaic Solar Energy Conference and Exhibition; 2529-2533,* 2015.

[9] Y.-M. Saint-Drenan and T. Barbier, "Data-analysis and modelling of the effect of inter-row shading on the power production of photovoltaic plants," *Solar Energy*, vol. 184, p. 127–147, May 2019.

[10] M. Chiodetti, J. Dupuis, D. Boublil, K. Radouane and P. Dupeyrat, "Half-Cell Module Behaviour and Its Impact on the Yield of a PV Plant," *36th European Photovoltaic Solar Energy Conference and Exhibition; 1444-1448,* 2019.

[11] A. Woyte, J. Nijs and R. Belmans, "Partial shadowing of photovoltaic arrays with different system configurations: literature review and field test results," *Solar Energy*, vol. 74, p. 217–233, March 2003.

[12] N. Thakkar, D. Cormode, V. P. A. Lonij, S. Pulver and A. D. Cronin, "A simple non-linear model for the effect of partial shade on PV systems," in *2010 35th IEEE Photovoltaic Specialists Conference*, 2010.

[13] J. Remund and S. Müller, "Solar Radiation and Uncertainty Information of Meteonorm 7," in *Proceedings of the ISES Solar World Congress 2011*, 2011.

[14] A. Mermoud and T. Lejeune, "Partial shadings on PV arrays: by-pass diode benefits analysis," 2010.

DYNAMIC SIMULATION OF THE SHADING CAST BY A WIND FARM ON AN ADJACENT PHOTOVOLTAIC PLANT

Jesús Robledo[1,*], Jonathan Leloux[1], Babacar Sarr[1], Christian A. Gueymard[1,2], Patricia Darez[3]
[1] LuciSun, Rue Saint-Jean, 29, 1495 Sart-Dames-Avelines, Belgium
[2] Solar Consulting Services, P.O. Box 392, Colebrook, NH 03576, USA
[3] 350Renewables, Orinoco 90, Oficina 2101, Las Condes, Santiago, Chile

*Phone: +34 656 37 59 74; E-mail: jesus.robledo@lucisun.com

ABSTRACT: The hybridization of several renewable energy power generation sources is a viable option that can improve the use of available resources and space. However, the coexistence of different technologies, with different requirements for operation, can lead to some challenges. In the case of hybridization between a wind farm and an adjacent solar photovoltaic (PV) power plant, the wind farm can cast dynamic and static shadows on the nearby PV generators. Simple hypotheses are often chosen for their conservative character, but they often turn out to be overly pessimistic and can result in worse results in the project economics. This study evaluates the shading losses impacting a hybrid wind/PV park in a sunny region with a novel advanced 3D shading simulation tool using dynamic visualization based on the power of modern GPUs (Graphic Processor Unit. The application of this method to a case study in a sunny region shows that, as a function of the approximation that is used to model the blades of the wind turbines, the resulting estimated shading losses can be far too pessimistic, or too optimistic. The study also provides guidance on how to assess the possible existing tradeoffs between PV power installation density and shading losses, and how to apply it to the optimization of the layout of the PV plant.
Keywords: PV, simulation, shading, GPU, photovoltaic plant, wind farm

1. INTRODUCTION

The hybridization of several renewable energy power generation sources is a viable option that can improve the use of available resources and space. However, the coexistence of different technologies, with different requirements for operation, can lead to some challenges.

In the case of hybridization between a wind farm and an adjacent solar photovoltaic (PV) power plant[1], the wind farm can cast dynamic and static shadows on the nearby PV generators. Partial shading phenomena and their corresponding power losses on PV modules have been extensively studied [2]. However, these previous studies have generally dealt with fixed shading elements creating predictable shading that can be estimated over time. In the case of wind turbines, the rotation of long blades in the nearby area of the modules induces dynamic disturbances during short periods. Such disturbances can have a significant impact on the PV power generation, while being difficult to evaluate on a yearly basis.

The current state-of-the-art PV simulation tools cannot easily handle this kind of dynamic shading problems, hence simplified models are often used [3]. Depending on the simplifications chosen to evaluate the shading impacts, significant differences can be observed in the resulting dynamic shadow affected area. Simple hypotheses are often chosen for their conservative character, but they often turn out to be overly pessimistic and can result in unnecessarily reducing the energy production inputs of the project's financial model (Figure 1). Therefore, a more accurate and realistic approach is necessary.

This study evaluates the shading losses impacting a hybrid wind/PV park in a sunny region with a novel advanced 3D shading simulation tool using dynamic visualization based on the power of modern GPUs (Graphic Processor Unit).

2. POSSIBLE WIND BLADES MODELLING

Several possibilities exist to evaluate the energy losses associated with the shadows cast by the tower and blades of wind turbines on adjacent PV arrays. Wind generators are composed of two main elements: the tower and the rotating blades. Shading simulations need to account for these two shading sources. The shadows cast by the tower can easily be simulated because they are static, i.e., only affected by the sun's apparent movement, which is deterministic and relatively slow. On the contrary, the simulation of the blade-induced shading is much more complex because it requires dynamic modelling due to the nature of the movement of the blades. The blades are rotating around a horizontal axis (the hub of the turbine), and this rotation axis can change over time as a function of the dominant wind directions (the yaw mechanism within the tower). Several approaches are possible to evaluate the shading losses from the blades.

The most conservative approach is to assimilate the blades to an opaque sphere having the same diameter, thereby accounting for a large range of possible positions for these blades. This hypothesis does not require any knowledge of the dominant wind directions. The resulting shading is shown in Figure 1, where the heatmap represents the fraction of time, over one year, during which a specific area is shaded by the wind turbines. This leads to an overestimation of the shading losses that will affect the energy of the PV array because the wind turbine generators (WTGs) will not at any given time cast so much shadow on the array (the model works for all the possible wind directions, but at any given time the wind farm will experience a predominant wind).

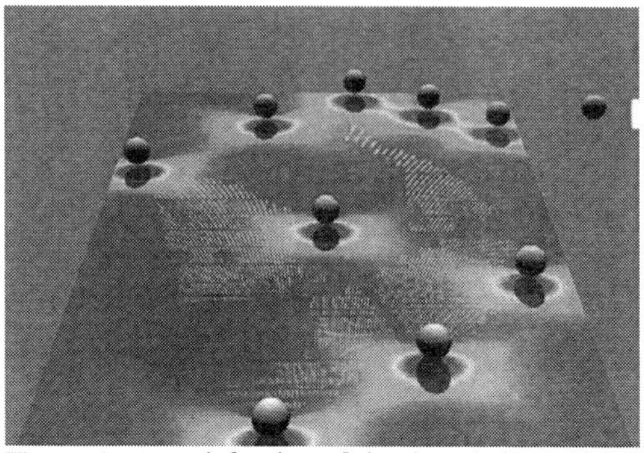

Figure 1: Annual fraction of the time during which a specific area is shaded by a wind turbine, if the blades are assumed spherical.

The most frequent approach, used in standard wind turbine shadow modeling tools, is to approximate the blades to a disk, which is facing the predominant wind direction A illustrated in Figure 2.

This is a suitable approximation for windfarm sites which have a clear predominant wind direction. As with the previous assumption, this leads to an overestimation of the shading losses.

Figure 2: Shadows projected by a wind turbine on a PV array when assimilating the blades to a disk.

The resulting shading is shown in Figure 3. The results are far less conservative than with the spheric assumptions, but they are still relatively conservative because they assume a permanent shade over the array when in reality the 3 blades are a fast-moving dynamic shadow.

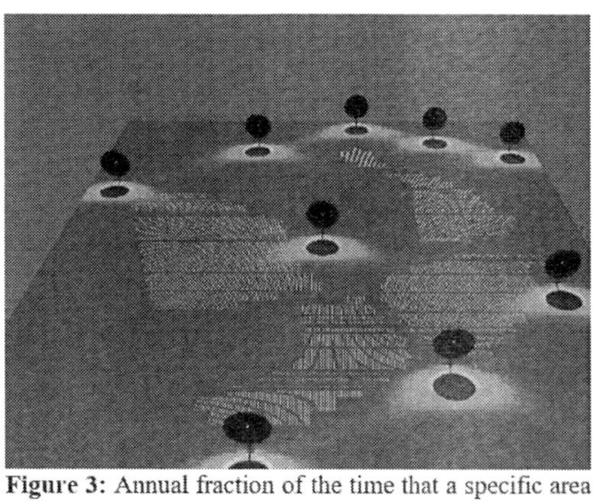

Figure 3: Annual fraction of the time that a specific area is shaded by the wind generators, when assuming that the blades are disks.

A third approach that could be considered is to only take the central tower into account, without consideration for the blades. This obviously leads to an underestimation of the shaded area, as illustrated in Figure 4.

Figure 4: Shadows projected by a wind turbine on a PV array when only accounting for the central tower.

As shown in Figure 5, the result of this assumption is unsatisfactory because it is too optimistic and does not consider the impact of the moving blades.

Figure 5: Annual fraction of the time that a specific area is shaded by the wind turbines, when simulating shadows only for the central tower without blades.

The most realistic approach is to consider both the tower and the moving blades. The shadows that are cast at a specific moment by the tower and blades are illustrated in Figure 6. Both the static (no wind or other relevant turbine downtime) and dynamic movement of the blades need to be considered, which makes this case the most complex simulation case. Obtaining a fast and accurate solution for this problem is the object of this work.

Figure 6: Shadows cast at one moment by the tower and blades.

The resulting shading is illustrated in Figure 7.

Figure 7: Annual fraction of the time that a specific area is shaded by the wind turbines, when using a dynamic simulation for the movement of the blades.

3. METHODOLOGY

With current technology, virtually all computerized graphics systems are characterized by a special-purpose Graphics Processing Units (GPU). During the last decade, the performance of GPUs has surged, mainly driven by the ever-increasing demand emanating from the video game industry. Modern GPUs are very effective at manipulating computer graphics and image processing, and their highly parallel structure makes them more efficient than general-purpose Central Processing Units (CPUs) for algorithms that are designed to process large blocks of data in parallel [4]. The availability of powerful GPUs opens very promising possibilities for the simulation of complex 3D shading scenes applied to PV systems because the impacting shadows can be evaluated with a very high spatial resolution that reaches up to the PV cell level in short calculation times. The general problem of accurately evaluating shading losses based on GPU power has already been described in a previous work [5], and since then has been applied to numerous case studies for the solar industry.

In this work, a method is developed to account for the movement of the blades during the shading simulation, making use of specialized simulation libraries incorporated in GPUs. The simulations are carried out at temporal resolutions of typically 10-min or better. For each moment, an independent position of the blades is associated to each one of the wind turbines surrounding the photovoltaic modules. The orientation of the plane of rotation of the blades is chosen so that it is representative, for each moment, of the dominant wind direction. The exact position of these blades is generated using a Monte-Carlo procedure, which provides a wide range of possible positions for the blade during the simulation period. This creates a set of blade positions that is statistically representative of reality in a one-year simulation period for a PV power plant, because a random distribution is applied over many time moments. It is worth mentioning that the wind resource at most windfarn site has high variability and the study did not focus on using wind data which was necessarily representative for the long-term conditions at the site as the purpose of this paper is to highlight the modelling approaches used.

For each moment in time, a geometric shading fraction is obtained at high spatial resolution, for each PV cell composing the PV array, as illustrated in Figure 8.

Figure 8: Visualization of the instantaneous shading caused by one wind turbine on a PV array.

This geometric shading is then converted into effective shading losses, using a diode equivalent model that allows to consider the electrical connection between the PV cells composing the PV modules and their bypass diodes [6]. The method considers the irradiance that is incident on each PV cell, as well as the electrical connection between these PV cells and the PV modules. The annual energy losses caused by shading are then estimated by integration of the shading losses calculated for each moment in time, and over the whole simulation period, typically using a Typical Meteorological Year (TMY) format [7]. This is illustrated in Figure 9.

The movements of the blade induce fast variations of the geometric factor on the PV arrays. The shadows typically appear and disappear in time fractions that are smaller than one second.

Figure 9: Integration of the shading fraction over a time period of one typical year for one wind turbine and one PV array.

4. CASE STUDY

The procedure described above is applied here to a real case study in which a PV plant with horizontal single-axis trackers (HSAT) is planned in the neighborhood of an already existing wind farm, located in a region with high irradiation. This section illustrates the main steps of the method when applied to a concrete project.

The annually integrated shading profiles of each wind turbine are obtained by weighting the geometric fraction at each time moment by the corresponding amount of solar irradiance. This produces a map of the effective shading losses on the ground and on the PV arrays, as illustrated in Figure 10 for the case of several wind turbines, and in Figure 11 for a closer view on one wind turbine, whose hub height is 150 meters and whose rotor diameter is 121 meters.

Figure 10: Cumulative shading effect over a one-year period on a large PV plant surrounded by several wind turbines.

Figure 11: Cumulative shading effect over a one-year period on a large PV plant surrounded by several wind turbines: Closer view on one wind turbine and the surrounding PV array.

A first study is carried out to determine the impact of the shadows cast by the wind turbines, taken one by one, as a function of their distance to the PV arrays. Figure 12 presents some illustrative comparison cases of effective shading losses from one wind turbine, with simulations considering only the tower (optimistic), the tower and a stationary disk (pessimistic), and the tower combined with a moving blade (realistic). The distance between the wind turbines and the PV arrays is taken along the east-west axis, and three subcases are represented: PV array in the same east-west alignment than the wind turbines, PV array 50 meters north from one wind turbine, and PV array 50m south from the wind turbine. The simulation where the moving blades are assimilated to a disk translate into shading losses that are more conservative. The shading losses estimated with the tower only or with the moving blades are much lower than with the disk assumption.

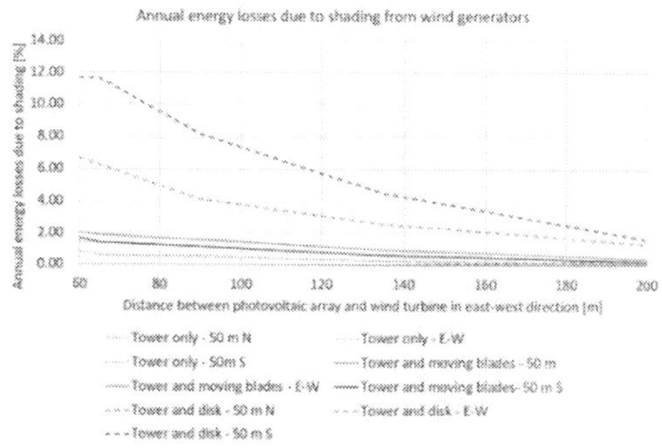

Figure 12: Illustrative comparison cases of effective shading losses, with simulations considering only the tower (optimistic), the tower and a disk (conservative), and the tower combined with a moving blade (realistic).

The difference between the tower-only assumption and the dynamic model (moving blades) is visible in Figure 13, which presents a closer view of the shading losses in a range of values between 0% and 2%. The difference is particularly significant in the direction that is facing the dominant wind direction. In the case of this test case, the shading losses from the wind turbines are significant when the distance between the wind turbines and the PV arrays is less than 150 m, i.e. approximately the hub eight of the wind turbines.

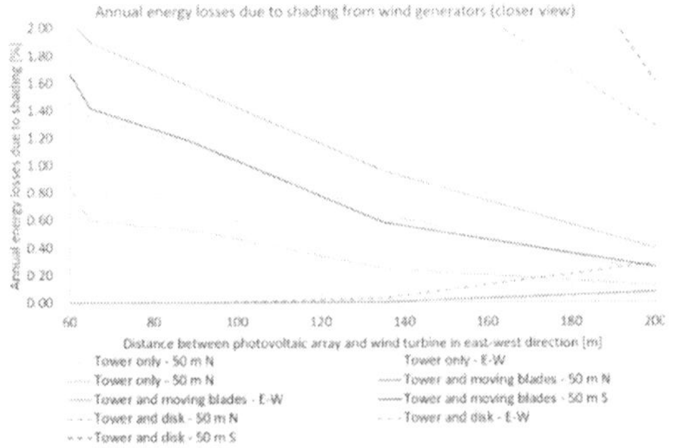

Figure 13: Same as Figure 12, but showing a closer view of the shading losses in the range 0–2%.

In practice, one relevant question for the developer of the PV plant would be to determine the trade-off between the installed power density on the available land and the resulting shading losses from the wind turbines. When the turbines are sufficiently separated from one another, and when enough land is available between them for the installation of the PV arrays, it is possible to find areas where the shading losses are negligible. That part of the land can be advantageously occupied first. The PV arrays only need to be installed closer to the areas more affected by shading when the installed capacity needs to be increased, or when other constraints are in place. Therefore, beyond a certain installed capacity, significant shading losses might appear if the PV arrays are installed too close to the wind turbines. As a result, there is also a trade-off between the irradiation losses caused by turbine-induced shading and the irradiation losses caused by mutual shading of the PV modules, or their backtracking. If the pitch distance between the trackers is reduced, then it is possible to install more PV power in areas that are relatively unaffected by the wind turbines, at the cost of more backtracking losses. On the contrary, if the pitch distance is increased, then the backtracking losses decrease, but for the same installed power, the PV arrays need to be installed in areas that are more affected by shading from the wind turbines. For the present test case, Figure 14 shows the joint evolution of the total installed capacity the pitch distance between trackers, and the turbine-induced shading losses. In this test case, the most limiting factor in the total installable PV capacity is assumed to be the grid connection rather than the available land area. Therefore, it is possible to optimize the layout so that the turbine-induced shading losses are kept relatively low, i.e., under 1% for most of the PV arrays. However, this might not be always the case, for instance if the developer wishes to install more PV capacity on the same land area.

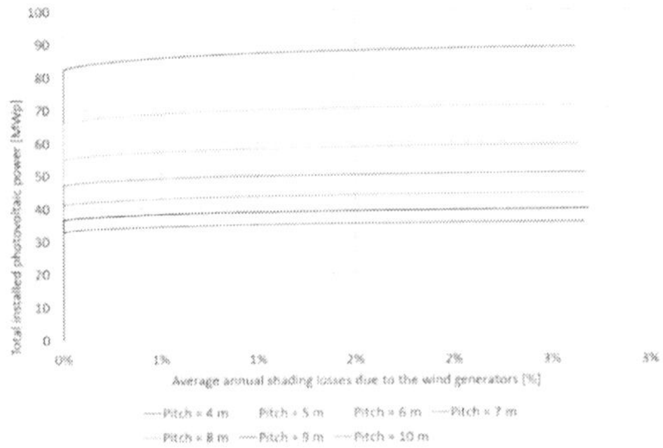

Figure 14: Joint evolution of the total installed power, the pitch distance between trackers, and the shading losses caused by the wind turbines.

In summary, for a given combination between the desired total installed capacity, the location and dimensions of the wind turbines, the nature and shape of the land, and other constraints, finding the optimum layout for the PV array requires a multi-parameter optimization. Such an optimization exercise of the layout is carried out here to exemplify the method further, and the resulting layout with the corresponding shading losses is shown in Figure 15.

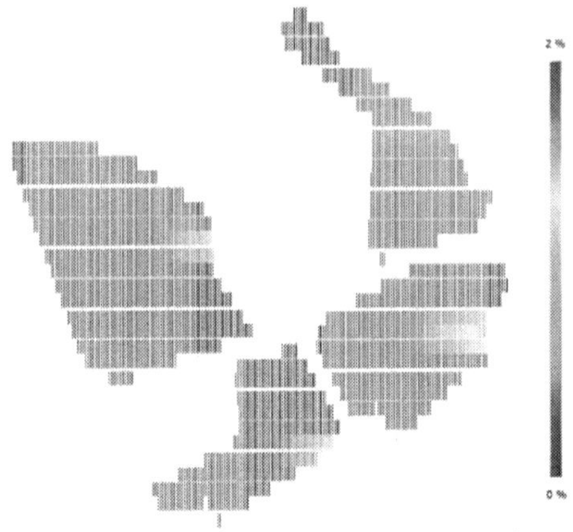

Figure 15: Layout of the PV plant that results from the optimization of the trade-offs between installed power, shading losses from wind turbines, and backtracking losses. The annual shading losses (up to 2%) are shown with the color code to the right.

5. DISCUSSION AND CONCLUSION

A novel method has been applied to the evaluation of the shading losses caused by the wind turbines on adjacent photovoltaic arrays. The application of this method to a case study in a sunny region shows that, as a function of the approximation that is used to model the blades of the wind turbines, the resulting estimated shading losses can be far too pessimistic, or too optimistic. The study also provides guidance on how to assess the possible existing tradeoffs between PV power installation density and shading losses, and how to apply it to the optimization of the layout of the PV plant.

ACKNOWLEDGMENTS

The research activities behind the simulation tool that was used in this project were partially funded by European Commission through the Horizon 2020 project SERENDI-PV (https://serendipv.eu/), which belongs to the Research and Innovation Programme, under Grant Agreement 953016. This publication is also partially based upon work from COST Action CA16235 PEARL PV (https://www.pearlpvcost.eu), supported by COST (European Cooperation in Science and Technology).

REFERENCES

[1] S. Diaf, G. Notton, M. Belhamel, M. Haddadi, A. Louche, Design and techno-economical optimization for hybrid PV/wind system under various meteorological conditions, Applied Energy, 2008, https://doi.org/10.1016/j.apenergy.2008.02.012.
[2] A. Woyte, J. Nijs, R. Belmans, Partial shadowing of photovoltaic arrays with different system configurations: literature review and field test results, Solar Energy, Volume 74, Issue 3, 2003, Pages 217-233, ISSN 0038-092X, https://doi.org/10.1016/S0038-092X(03)00155-5.
[3] PVsyst (https://www.pvsyst.com/).
[4] D. Nehab, P.V. Sander, J. Lawrence, N. Tatarchuk, J.R. Isidoro, Accelerating Real-Time Shading with Reverse Reprojection Caching, 2007, http://dx.doi.org/10.1145/1280094.1280098
[5] J. Robledo, J. Leloux, E. Lorenzo, C. A. Gueymard, From video games to solar energy: 3D shading simulation for PV using GPU. Solar Energy, 2019, https://doi.org/10.1016/j.solener.2019.09.041.
[6] F. Martínez-Moreno, J. Muñoz, E. Lorenzo, Experimental model to estimate shading losses on PV arrays, Solar Energy Materials and Solar Cells, https://doi.org/10.1016/j.solmat.2010.07.029.
[7] S. Wilcox and W. Marion, Users Manual for TMY3, Data Sets, Technical Report, p. 58, 2008.

DYNAMIC AND VISUAL SIMULATION OF BIFACIAL ENERGY GAIN FOR PHOTOVOLTAIC PLANTS

Jesús Robledo[1],*, Jonathan Leloux[1], Babacar Sarr[1], Christian A. Gueymard[1,2], Anton Driesse[3]
[1] LuciSun, Rue Saint-Jean, 29, 1495 Sart-Dames-Avelines, Belgium
[2] Solar Consulting Services, P.O. Box 392, Colebrook, NH 03576, USA
[3] PV Performance Labs, Emmy-Noether-Str. 2, 79110 Freiburg, Germany
*Phone: +34 656 37 59 74; E-mail: jesus.robledo@lucisun.com

ABSTRACT: Several simulation tools have appeared on the market to estimate the bifacial energy gain (BEG) of bifacial photovoltaic (PV) systems. These tools use two main families of methods that provide a quick estimation of the BEG or a detailed analysis of it, but not both. The current study presents an alternative methodology that provides both benefits using image generation techniques used in the video games and performing the computations in the GPU (Graphic Processor Unit) of the computer. The methodology used, as well as some examples of elaborate applications are presented, including interference with complex objects, movable elements (trackers), and intricate reflectance patterns for various reflecting materials characterized by their BRDF (Bidirectional Reflectance Distribution Function).
Keywords: Bifacial, PV, simulation, BEG, GPU, photovoltaic plants

1 INTRODUCTION

The simulation of the energy generated by bifacial photovoltaic (PV) modules is challenging because the irradiance received on their backside comes from different directions and sources. Bifacial PV has been experiencing a rapid growth in recent years. This translates into an increasing number and diversity of PV installations in the world with bifacial PV modules, and represents a large variety of scenarios, several of them with irradiance contributions involving peculiar reflectance patterns (irregular terrains, multiple reflective objects, tracking-induced dynamic movements...).

Several simulation tools have appeared on the market to estimate the bifacial energy gain (BEG) of bifacial PV systems. These tools use two main families of methods. The first family is composed of those methods that estimate the view factors at the back side of the PV modules using an analytical approach [1],[2]. This approach is relatively simple, but it can only produce a rough estimate of BEG. Another approach is to use raytracing, which makes it possible to obtain accurate simulations, but at the cost of extremely increased demands in computation time and resources [3].

This work presents a third and alternative method utilizing the processing power of the Graphics Processing Unit (GPU) of all modern computers. This avenue offers critical advantages compared to the existing approaches just mentioned. First, it is capable of reaching accuracy levels similar to raytracing. Second, it provides a dynamic and visual simulation that can be easily interpreted by non-experts. Third, it keeps the resource requirements relatively low. All this opens the door to a depth of analysis that was not achievable in practice before. The use of this method for the estimation of the bifacial energy gain is illustrated here for three typical kinds of application that are frequently encountered in the practice of solar engineering. The installation of bifacial PV modules on carports is used first to exemplify the method, because it presents several complex elements whose simulation require various steps. Two additional application cases are then presented: vertical bifacial PV structures used in agrivoltaics, and PV modules in large-scale PV plants with or without tracking systems.

2 METHODOLOGY

2.1 Method overview

The current method for the evaluation of the reflected irradiance on the backside of PV modules consists of breaking down the irradiance contribution (for the front side as well as for the rear side) incident on a single cell of each module into its diffuse and direct components. All the possible sources are grouped in a way that the overall irradiance incident on the plane of array can be obtained by transposing the direct and diffuse irradiance components that are either measured by radiometers or evaluated from spaceborne observations with appropriate radiative models. To obtain the total irradiance, I, incident on a tilted surface of any geometry, the general formulation proposed in [4] is such as:

$$I = f_b \text{ DNI} \cos\theta + f_d R_d \text{ DIF} + \text{REF} \qquad (1)$$

where DNI is the direct normal irradiance, f_b is a sun shading factor, DIF is the sky diffuse horizontal irradiance, R_d is the diffuse transposition factor from horizontal to tilt (i.e., plane of array), f_d is a sky shading factor, and REF is the irradiance reflected by the surrounding ground. For monofacial PV applications, REF is typically calculated as a simple function of the global horizontal irradiance, GHI, using various simplifying assumptions that strongly impact the accuracy of this approach, as discussed in [5]. The diffuse transposition factor is usually calculated with one of a few transposition models that include anisotropic effects. One such model is that of Hay, whose accuracy has been amply validated [6],[7]. In the present context, one important advantage of this model is that it simply decomposes DIF into two components: an isotropic component, DIF_{iso}, and a circumsolar component, DIF_{circ}.

Rearranging Eq. (1), to split REF into a beam component and a diffuse component, the following expression is obtained:

$$I = (F_1 + F_2) \cdot (\mathrm{DNI} + \mathrm{DIF}_{\mathrm{circ}}) \cos\theta + (F_3 + F_4) \cdot \mathrm{DIF}_{\mathrm{iso}} \quad (2)$$

where the F_i factors are functions that only depend on geometric, environmental, and material variables. They are time dependent but irradiance independent. The definition of those factors is as follows:

- F_1 is the fraction of direct incident sunlight incident on a cell. Consequently, F_1 is 1 if a cell is totally exposed to the sun, 0 if completely shaded, and intermediate in case the sun's disc is only partially shaded. The way to evaluate this factor is described in [8], and implies a shading computation that is based on shadow mapping techniques performed in the GPU.
- F_2 is a reflectivity factor that quantifies how the direct sunbeam are reflected over different ground elements and ultimately reach a specific solar cell, including the sun incidence angle influence. Consequently, the light source here is not constituted of parallel rays but of rays having any possible direction inside a 180° hemisphere, as seen from each cell in its specific orientation. The procedure for this computation is described in Sections 2.3 & 2.4.
- F_3 represents the ensemble of sky elements directly seen from the cell. As before, this requires a spherical view analysis, which is explained in Sections 2.3 & 2.4
- F_4 describes the reflection process over different objects that are seen from the unobscured sky. Again, this factor requires a spherical view analysis, which is explained in Sections 2.3 & 2.4.

2.2 Spherical view concept

In order to evaluate factors F_2, F_3 and F_4, the approach used here benefits from the power of GPUs in modern computers, as done in video games for fast picture generation. This provides a similar resolution as raytracing techniques, while requiring much less computing time. Essentially all objects that typically exist all around the installation's scene have simulated renderings that are included in accessible computer libraries. Hence, it is relatively easy to generate a realistic representation of any solar system and its environment.

From each side of each PV panel, a 180° spherical view over different points of the module is created, as shown in Figure 1. It is obvious that a tradeoff has to be considered between computation time and accuracy. An optimum can be found, depending on the number of viewpoints and their location over the PV plant.

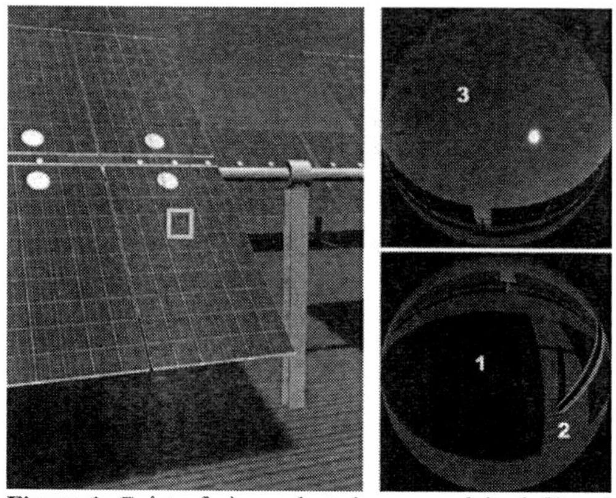

Figure 1: Point of view selected on a module (left) and spherical views generated: Front side (top right) and rear side (bottom right).

The generation of pictures is done in a way that it is possible to identify each component of the scene and assign it a specific value at pixel level, which will be eventually integrated with the rendered picture to finally obtain the desired factor F_i. As a reference, the pictures shown in Figure 1 are generated at a resolution of 512 x 512 pixels—equivalent to more than 250,000 rays with the raytracing technique—and are obtained fast (≈10–20 ms per picture).

There are three possible contributors to the source of incident radiation:

1. Objects under shade (but able to reflect irradiance from the sky);
2. Objects under direct sunlight;
3. No objects: Sky directly visible.

The procedure followed to evaluate their respective impact is detailed in the following sections.

2.3 Selection of elements for the spherical view

In order to provide a proper value for each of the sky and surface elements, three main steps are followed.

The first step is to evaluate the potential for sun visibility from a particular object. This is a pure geometric parameter, which takes any value between 0 (sky fully blocked) and 1 (sky fully visible). Figure 2 shows an example of a carport with vehicles underneath.

Figure 2: Color-coded sky visibility potential for the carport scenario, and example of upward view from two points at ground level.

The second step is to apply reflectance properties to each element of the scene. A three-parameter Bidirectional Reflectance Function (BRDF) is considered as a function of sun incidence angle and spherical coordinates of the reflected direction, as shown in Figure 3. For each reflecting element irradiated by direct sunlight, this BRDF is directly used, whereas only its average is sufficient to quantify the diffuse reflected irradiance contribution from the sky over the element.

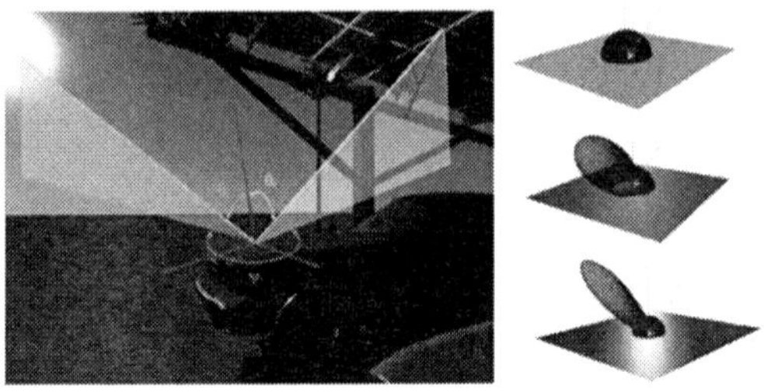

Figure 3: (Top) BRDF and angles definition; (bottom) example of different reflectance patterns in practice (Lambertian, semi-glossy, and specularly reflective).

The third step is to create a depth map reference from the sun's point of view with a parallel perspective. The goal here is to identify which elements are under shading when evaluating the irradiance coming from the sun and reflected on the surfaces. This is illustrated in Figure 4.

Figure 4: Parallel projection from the sun's point of view to identify pixels under shading. Lighter colors indicate a higher depth in the view.

2.4 Combination of elements and spherical view generation

Based on the information obtained previously, it is possible to assign a color-code value to each element in the scenario and render it from the different points of view over each PV module, and ultimately obtain pictures such as the ones shown in Figure 5.

Each of the pictures are generated using a combination of the previous elements. For the objects under direct sunlight, an evaluation of each pixel in the spheric view is done for the following elements: the reflection value based on the BRDF at that point, defined with three angles as described in Figure 3 (θ_i : sun vector with surface normal where pixel is located, θ_r : reflected vector with surface normal, Ψ_r : angle between the planes defined by sun vector and normal and the one defined by reflected vector and normal). Then, a check is done to verify if that pixel is shaded using the depth map obtained in the previous step. The integration of all the pixels of the spheric view provides the factor F_2 in Eq. (2), as shown in Figure 5a. For all the reflective objects or elements, the sky view potential is corrected by the corresponding diffuse albedo to generate the spherical view that provides the view factor F_4 in Eq. (2), shown in Figure 5b. For a direct view of the sky, which is needed to derive F_3, a detailed anisotropic sky radiance model that depends on sun position, azimuth and elevation of the pixel, such as the one described in [9], is preferable to the simplistic isotropic sky radiance model [10], although this is more involved. The resulting F_3 estimate is shown in Figure 5d. For simplicity in this example, a homogenous (isotropic) radiance distribution is used, although the effect of the varying $\cos\theta_n$ (angle between the direction of the reflected ray and the normal to the observation point, as shown in Figure 3) is shown in Figure 5d. The spherical integration of each pixel's radiance value is shown in Fig. 5 for each view. The whole process is performed in the GPU.

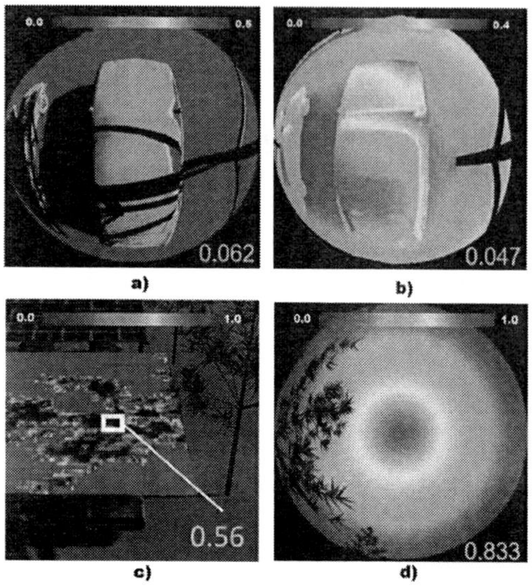

Figure 5: Views from a specific solar cell of the carport canopy: a) elements under direct sunlight; b) diffuse irradiance view-factor reflection from the sky; c) shading factor for parallel rays on the front side of the same cell; and d) sky elements direct view from the front side.

2.5 View factors implementation at cell level

From the decomposition of the incident irradiance on each panel side, as described in Eq. (2), the following factors can be identified from the values indicated in Figure 5. For the front side: $F_1 = 0.56$ (Figure 5c), F_2 is considered negligible because of the low tilt angle of the installation (it could be computed as done in Figure 5a, but it is not shown here because it would be too low to be visible), $F_3 = 0.833$ (Figure 5d), whereas F_4, just like F_2, is considered negligible because of the low tilt angle of the installation (it could be computed as done for Figure 5b but again not shown here for clarity). Similarly, for the rear side: $F_1 = 0$ because the sun is on the front side, $F_2 = 0.062$ (Figure 5a), F_3 is considered negligible here because of the low tilt angle of the installation (although it could be computed as done in Figure 5d), and $F_4 = 0.047$ (Figure 5b).

The combination of those four factors with the irradiance values of that particular moment constitute the necessary inputs to compute the irradiance incident on each specific point along each panel side. Repeating the process for all elements into which the complete PV plant can be divided makes it possible to generate a heatmap of the irradiance's spatial distribution over both surfaces, as shown in Figure 6. That heatmap is a critical step toward analyzing possible inhomogeneities in the irradiance field and evaluating their electrical impact through a known mismatch methodology. This latter step is not part of the present contribution, which rather focuses on the irradiance distribution evaluation.

Figure 6: Spatial irradiance distribution over the complete PV generator.

The total computation time is obviously dependent on the number of viewpoints where the analysis is done. Hence, different optimal solutions can be found in the trade-off between computation time and local effects identification. The latter clearly conditions the level of accuracy in the end results. As a reference, and for the present example, the computational time needed for the complete irradiance evaluation over a point at a specific time moment is ≈50 ms (depending on hardware, but with a reference GPU: NVIDIA Quadro K620 with 384 cores).

3 APPLICATIONS

As an example of the possibilities offered by this powerful technique, two additional case studies that involve complex simulation scenarios are presented here. Their results are shown and only briefly discussed, highlighting some of the particularities of the evaluation in each case. These case studies are: a large PV plant with trackers, and a system made of vertical bifacial PV modules for agrivoltaics.

3.1 Ground-mounted PV plants with 1-axis tracking

This illustrative example is based on a 1-axis tracking system, as depicted in Figure 7.

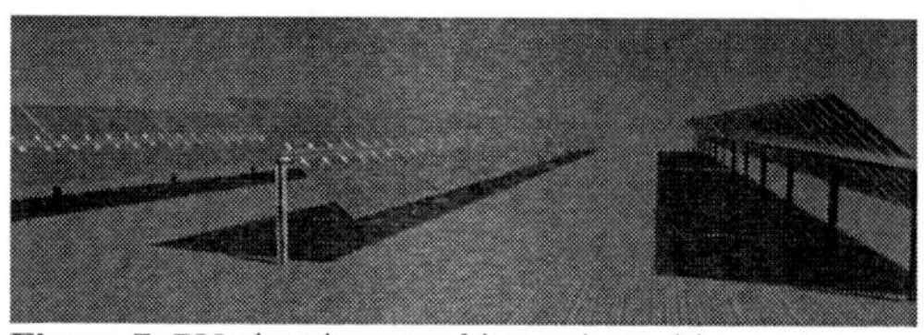

Figure 7: PV plant layout with 1-axis tracking system.

Here, the main influential characteristics are as follows:

- The module orientation is continuously variable during the day, and thus the sky-view potential is also variable.
- The rear structure creates shading and thus irradiance losses.
- The ground reflectance has specific directional properties.
- No defined "front" and "rear" side is considered with this configuration.
- The module's orientation is variable and consequently the sky view potential is variable too.

As done in the other examples above, the first step consists in evaluating the sky visibility potential—in this case that of the ground. Here, however, this potential is dependent on the tracker rotation angle, hence this variable is now time dependent. An example of this potential at a particular time step can be seen in Figure 8.

Figure 8: Sky view potential close to mid-day.

The type of ground considered here is a plowed field. Its reflectance is determined using the Roujean model from the Sasktran Python library [11]. A depth map of the installation can be seen in Figure 9 for a particular moment in the morning. In this specific case, the depth map is mainly used to evaluate the shading pattern and intensity over the ground. It also supports the evaluation of those ground elements that can be in the shade for direct sunlight reflections. For a tracking installation with backtracking implemented, it is not expected that mutual shading exists between the modules. However, when the terrain is irregular and all the trackers have the same tracking algorithm, this depth map is helpful to identify possible mutual shadows and customize the backtracking algorithm.

Figure 9: Depth map from the sun's point of view.

The spherical views for a particular instant are shown in Figure 10 for the front side and Figure 11 for the rear side. For this particular application, and by selecting a higher tilt angle than in the carport application, some of the contributions that were considered negligible in that case are now significant, hence the results are split between the front side and rear side. To better distinguish the sides, a different texture is applied to each of them (bluish for the front one and black for the rear one). No information regarding mutual shading is included because a backtracking rotation algorithm is assumed.

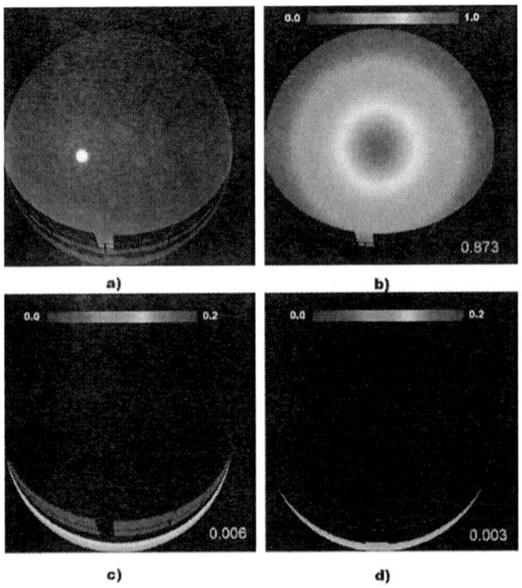

Figure 10: (a)- Realistic simulated view from the front side, for reference; (b) - sky's direct view from the front side; (c) – view factor for diffuse reflection from the sky; (d) – elements under direct sunlight.

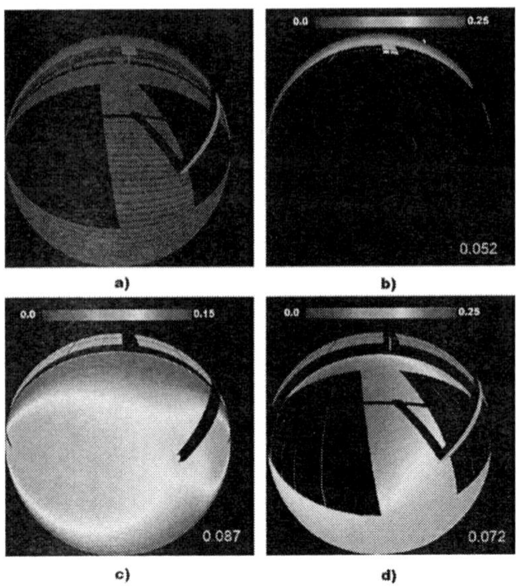

Figure 11: (a) – Realistic simulated view from the rear side, for reference; (b) - sky direct view from the rear side; (c) – view factor for diffuse reflection from the sky; (d) – elements under direct sunlight.

For the particular analysis of the irradiance incident on the rear side, two cases exemplify different module layouts: one with PV arrays formed by the installation of one row of PV modules mounted in portrait, or vertical mode (referred to as 1V), and another one with two rows of PV modules mounted one on top of the other in vertical mode (2V). As can be seen in Figure 12, the 1V configuration (left side) has the tracker bar interfering with the rear side of the modules, creating a low irradiance area. This effect is not observed in the 2V configuration. However, to properly estimate the energy losses due to this effect, a detailed mismatch analysis of the non-homogenous irradiance distribution should be done and will be the objective of future studies.

Figure 12: Irradiance distribution heat map for 1V (top) and 2V (bottom) configuration.

3.2 Vertical bifacial PV for agrivoltaics

A general overview of a typical vertically mounted installation using bifacial PV modules for agrivoltaics can be seen in Figure 13. It consists of several rows of bifacial modules installed vertically, with a significant inter-row distance.

Figure 13: Vertical modules for agrivoltaics on an uneven terrain.

In this case, the main challenges are:

- Orientation of the modules – No pre-defined "front" and "rear" side is considered with this configuration.
- The irregular terrain induces complex shading and reflection characteristics.
- Grass is known to have intricate directional reflectance properties.

As explained in Sections 2.1 to 2.5, several sequential steps must be performed to obtain the incident irradiance distribution over each side separately. The initial step consists of obtaining the sky visibility potential for the reflective elements (i.e., only the ground in this case). The result of this calculation can be seen in Figure 14. The spatial distribution shows the loss of sky visibility from the ground close to the modules, as could be expected, but increasing quickly with distance from the structures. Moreover, some differences in the visibility potential are found along the terrain due to the different tilt angles of the irregular surface, but these differences are small and cannot be detected with the contrast shown in the image.

Figure 14: Sky visibility potential from the ground.

The BRDF reflectance model implemented here is based on the Roujean model that is included in the Sasktran Python library [11]. The shape of its BRDF for a low sun angle can be seen in Figure 15. It predicts a higher reflectance for high incidence or reflected angles.

Figure 15: Directional BRDF implementation based on the Roujean model for green grass.

The depth map reference for shading analysis can be seen in Figure 16, where a low sun elevation angle is chosen to emphasize the contrast between the different elements.

Figure 16: Depth map from the sun's point of view.

An example of the different spherical views computed for a particular instant cis shown in Figure 17. For this case, all the factors can be computed for both panel sides, but only some of them are presented for conciseness. Because no "front" or "rear" side can be defined in this case, the front side is supposed to be facing the sun for the sake of this picture. As done in the other examples above, a heatmap is generated based on the view factors obtained for different points along each panel side, also considering calculated values of diffuse and direct irradiance.

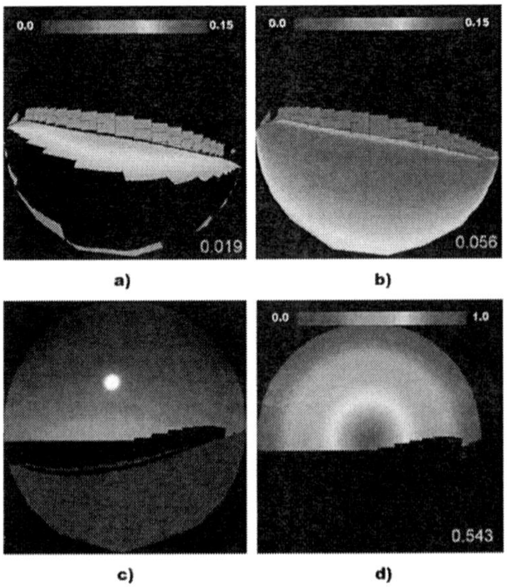

Figure 17: (a) - elements under direct sunlight for the rear side; (b) - diffuse irradiance view factor reflection from the sky; (c) – realistic simulation view from the front side, for reference; (d) – direct view of sky elements from the front side.

A result from that analysis is shown in Figure 18, showing the situation at 8:30 AM (local time) at a site located at a latitude of 45°. The effect of ground shading on the rear side can be observed by a lower irradiance incident on the bottom part of the modules, whereas the distribution is reversed for the front side (because higher irradiance is received when closer to the ground). Additionally, some effects of the terrain's slope can be observed in the top right corner of Figure 12, where the terrain is higher and freer of obstacles, thus inducing higher incident irradiance.

Figure 18: Irradiance heatmap analysis of the "front" side (top) and "rear" side (bottom).

4. CONCLUSION AND DISCUSSION

The GPU-based method presented here has already been used to assess the bifacial energy gain of many large bifacial PV plants totaling more than 1 GW since 2017, with a large diversity of applications. These include large PV plants with fixed structures or trackers, carports, canopies in rural areas, greenhouses, and vertical agrivoltaic structures. Several application cases have been illustrated in this article. The further development and scientific validation of this method is currently ongoing in the context of two European R&D projects: the H2020 project SERENDI-PV, and the COST project Pearl-PV. The method is still being improved to take more difficult situations in consideration., given the increasing diversity of bifacial PV installations, and the scarcity of reliable monitored operational data that are currently available, more effort and more collaboration will be needed in the future in order to incorporate and validate new functionalities in the tool. In this context, collaboration proposals are welcome, on the research effort as well as on the analysis of the operation data of bifacial PV installations.

ACKNOWLEDGMENTS

The research activities behind the simulation tool that was used in this project were partially funded by the European Commission through the Horizon 2020 project SERENDI-PV (https://serendipv.eu/), which belongs to the Research and Innovation Programme, under Grant Agreement 953016. This publication is also partially based upon work performed as part of COST Action CA16235 PEARL PV (https://www.pearlpvcost.eu), supported by COST (European Cooperation in Science and Technology).

REFERENCES

[1] PVsyst (https://www.pvsyst.com/help/bifacial_systems.htm)
[2] Valentin Software (https://help.valentin-software.com/pvsol/en/calculation/pv-modules/bifacial-modules/)
[3] Radiance (https://bifacial-radiance.readthedocs.io/en/latest/)
[4] C.A. Gueymard, Solar Radiation Resource: Measurement, Modeling, and Methods, https://doi.org/10.1016/B978-0-12-819727-1.00101-1.
[5] C.A. Gueymard, Surface albedo and reflectance: Review of definitions, angular and spectral effects, and intercomparison of major data sources in support of advanced solar irradiance modeling over the Americas, https://doi.org/10.1016/j.solener.2019.02.040
[6] C.A. Gueymard, Direct and indirect uncertainties in the prediction of tilted irradiance for solar engineering applications, https://doi.org/10.1016/j.solener.2008.11.004
[7] C.A. Gueymard, Solar radiation on inclined surfaces: Corrections and benchmarks, https://doi.org/10.1016/j.solener.2016.06.062
[8] J. Robledo, J. Leloux, E. Lorenzo, C. A. Gueymard, From video games to solar energy: 3D shading simulation for PV using GPU, Solar Energy, 2019, https://doi.org/10.1016/j.solener.2019.09.041.
[9] C.A. Gueymard, Progress in Sky Radiance and Luminance Modeling Using Circumsolar Radiation and Sky View Factors, 2018, https://doi.org/10.18086/eurosun2018.09.16
[10] N.R. Kamphuis, C.A. Gueymard, M.T. Holtzapple, A.T. Duggleby, K. Annamalai, Perspectives on the origin, derivation, meaning, and significance of the isotropic sky model, Solar Energy, 2020, https://doi.org/10.1016/j.solener.2020.02.067.
[10] J.-L. Roujean, M. Leroy, and P.-Y. Deschamps, A bidirectional reflectance model of the Earth's surface for the correction of remote sensing data, J. Geophys. Res., 97(D18), 1992, 20455–20468, doi:10.1029/92JD01411.

EXPERIMENTAL EVALUATION OF PERFORMANCE ENHANCEMENT IN A BIFACIAL PV SYSTEM BY A HIGHLY REFLECTIVE TEXTILE GROUND COVER

Jens D. Moschner[1,2], Georgi H. Yordanov[1,2], Louise Dupé[3]
[1] KU Leuven ESAT-ELECTA, Kasteelpark Arenberg 10, Box 2445, 3001 Leuven, Belgium
[2] EnergyVille 1, Thor Park, Poort Genk 8310, 3600 Genk, Belgium
[3] Beaulieu Technical Textiles, Boulevard Industriel 3, 7780 Comines-Warneton, Belgium

ABSTRACT: We have investigated the efficacy of a highly reflective ground cover as a reflector in combination with bifacial PV modules, using a white polymer fabric that has originally been developed for horticultural applications. This study provides an initial assessment of the potential advantage in energy yield, before testing the material on a larger scale. To test the attainable bifacial gain, we installed three identical PV modules on a south-facing rack with different undergrounds and backgrounds: the roof environment as-is, lawn grass, and Magnifield woven textile. In a first test phase, the modules were operated as solitary units in the middle of separate rack segments. In a second phase, they were surrounded by black textile covers, to simulate a realistic shading setting as it would be in a larger installation. We found that the white textile Magnifield provides a strong boost of bifacial gain relative to grass, obtaining an additional 16% in energy yield with solitary modules and 10% in a realistic rack configuration. Use of this white reflective woven textile material can aid substantially in enhancing performance and reducing kWh costs in ground-mounted PV systems of any size.
Keywords: PV system, Bifacial, Energy Performance, Optical Properties, Albedo, Reflector

1 INTRODUCTION

Bifacial PV modules have been used abundantly in ground-mounted PV systems in recent years, owing to the potential to capture additional irradiation at moderate added cost. Furthermore, the International Technology Roadmap for PV expects the market share of bifacial modules to triple in the next ten years, from 17% today to 56% in 2031 [1]. A wide range of yield gains between few percent and up to 30% has been quoted for bifacial PV systems in various different settings [2]. The gain in energy yield that can be achieved depends strongly on the system design, and in particular on the albedo of the underground. Sun et al. state that the bifacial gain is typically less than 10% for an albedo of 25%, which can be raised to 20% if the albedo reaches 50%, and even to 30% with optimized geometry, dependent on geographic location [3].

Clearly, a higher albedo can boost the performance and cost-effectiveness of such a system if it can be provided reliably and stable over time at moderate cost. One material that is excellently suited to fulfill these requirements is Magnifield, a bright white, water-permeable polymer (polypropylene) textile fabricated by Beaulieu Technical Textiles. It is already used in horticulture under the name Lumilys®, particularly in apple orchards to support fruit coloration [4]. It can be mounted with specific fixtures in a way that allows to step on it or pass over it with machinery, and that also permits temporary detachment. It reduces the growth of grass and weeds while being usable for several years, after which it can be collected and recycled.

In this study, we aim to assess the advantage in energy yield which can be obtained with this reflector material when compared to other undergrounds, in particular grass. While the albedo of grass is expected to be in the range of 0.20…0.25 [5], a very high value of 0.645 was measured in a standard laboratory setup for Magnifield in the wavelength range relevant for PV modules with silicon cells. In order to measure the possible gain in an outdoor setting, we have installed three bifacial PV modules in an open rack, which was modified to simulate the installation situation in a larger PV system in a second test phase.

2 EXPERIMENTAL

2.1 Installation

The PV test rack on the roof of the EnergyVille 1 building in Genk, Belgium, consists of south-facing segments that are 6 m wide and have an inclination of 35° relative to the horizontal. In order to test the attainable bifacial gain, we have installed three identical PV modules on a south-facing rack with different undergrounds and backgrounds. The first one, referred to as "reference", was left as-is, with the underground consisting of dark grey Sarnafil® roof cover, and some metal beams running above it. The second one was covered with a horizontal plane of Magnifield textile above the metal beams, further on called "textile". The third one was equipped with a wooden box, in which regular lawn grass was planted.

Each of three identical modules was installed in the center of one of three separate segments in landscape orientation, and the long sides of the modules were mounted on top of horizontal square metal profiles, to not shade the internal area of the module. The modules used are bifacial 60-cell LG NeON® 2 (LG340N1T-V5) with stated bifaciality coefficient of power of 0.70±0.05 and an STC power rating of 340 Watt-peak (Wp). On the rear side, the long sides of the frame have some overlap with the outer rows of cells, and a junction box is glued to the back behind two cells of the center rows, limiting the attainable bifaciality factor. The lower edge of the modules is about 95 cm above the reflector plane for the textile, while the grass is about 10 cm closer, and the roof surface is ca. 40 cm further away.

Next to each module, pairs of silicon plane-of-array irradiance sensors were installed, parallel to the front and rear surface, respectively. On the back of each module near the center, two planar Pt100 temperature sensors were attached. Electronic loads were used to conduct I-V sweeps or to keep the modules at the maximum power point, respectively, while providing high-accuracy measurements of voltage and current.

2.2 Tests

The performance of the modules was measured with the electronic loads programmed to sweep the I-V curve every 12 s, while maintaining the module at the maximum power point (MPP) otherwise. From the MPP power, we calculated the theoretical energy yield of these modules operating continuously at MPP. The performance test was conducted in four stages: in a pre-phase, the modules were operated with no special background to compare their initial performance.

Figure 1: Test setups for Phase 1 – top: reference, middle: grass, bottom: Magnifield textile.

For the test phase 1, we covered the lower frame of the rack with Magnifield for the "textile" segment, while a wooden box was installed and filled with rolled lawn grass for the "grass" segment, as Figure **1** shows. The base of the rack is approximately 3 m deep. In the background of the rack, a white wall and cable channels are present, the diffuse reflectance of which could adversely affect the experiment. Therefore, they were covered with materials similar to the respective designated reflector: black agro-textile for the "reference" segment, Magnifield for the "textile" segment, and green agro-textile for the "grass" segment.

Following these tests, we covered the upper part of the racks with black non-transparent fabric to simulate the embedment of the modules in a larger PV system. The cover encompasses the entire upper half of the rack up to a height of appr. 2.2 m and a small strip below the modules, where a cable tray is attached along the entire length, see Figure 2. Thus, the covered area measures 2.0 m from top to bottom, approximately as wide as two modules. The two test phases lasted 22 and 24 days, respectively, and were conducted subsequently in July and August of 2020. After these test phases, performance was also assessed with the rear side of the modules covered in black in order to identify any potential differences caused by the front irradiance or efficiency.

Figure 2: Setups "textile" (left) and "grass" (right) after installation of textile covers to simulate a larger system for Phase 2.

3 RESULTS AND DISCUSSION

3.1 Reflectance and albedo

The hemispheric reflectance of the Magnifield textile was measured with a spectrophotometer Perkin Elmer Lambda 950, equipped with an integrating sphere. It shows a high reflectance throughout the wavelength range relevant for PV modules, with a weighted average of 64.5% over the interval relevant for silicon solar cells (360…1100 nm), see Figure **3**.

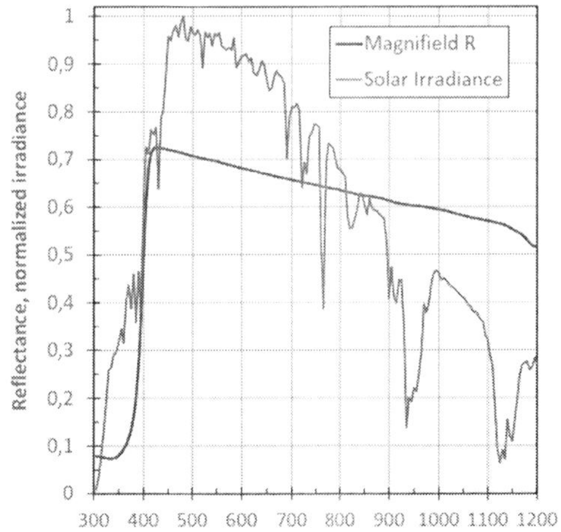

Figure 3: Global reflectance of Magnifield vs. wavelength, and solar irradiance spectrum AM1.5G according to ASTM G173-03 used for weighting.

In the pre-phase test with identical undergrounds, small differences in daily energy yield between the three modules were found. The designated "reference" module performed best, yielding 0.7% more than the "textile" module, whereas the "grass" module yielded 1.2% less. This is caused by effects of the different surroundings like shades and reflections, which affect both the front and back sides, and possibly to a small extent by different efficiencies of the modules. To distinguish effects on the front and back, we also conducted a test with the back of the modules covered, where the "reference" module yielded 0.6% more while the "textile" and "grass" ones were virtually identical in yield.

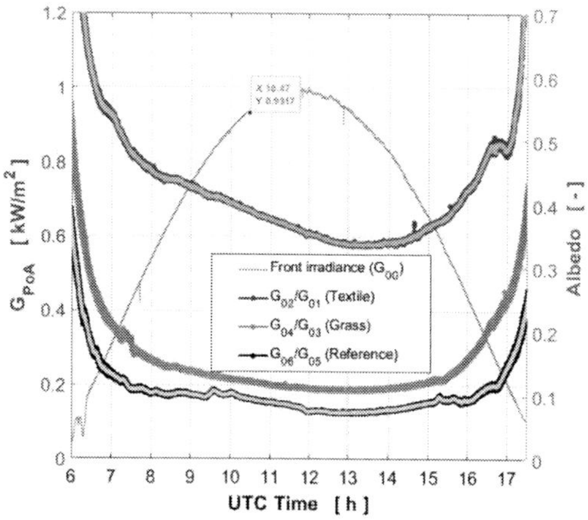

Figure 4: Daily clear-sky profile of albedos as calculated from the reference sensor measurements for the three different backgrounds (right scale) and front PoA irradiance (orange color, left scale).

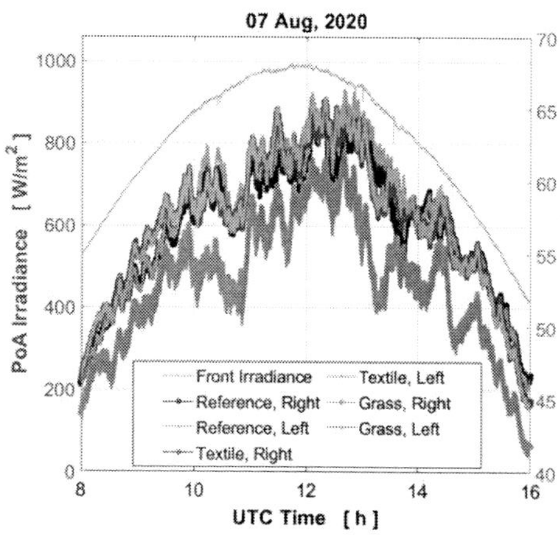

Figure 5: Daily clear-sky profiles of front PoA irradiance (left scale) and back-of-module temperatures (right scale), with black fabric surrounding each module.

For the setting with solitary modules in Phase 1, we calculated the apparent albedo as the ratio of the irradiances on the back- and front-facing sensors for each case. When comparing these ratios, the higher reflectance of the textile becomes clearly visible, as the profile for an exemplary clear-sky day in Figure **4** shows. The asymmetry in the curve is caused by the off-center position of the sensors to the east of each respective module in combination with the shading by the module on the underground.

3.2 Temperature

After installing the fabric covers around the modules for Phase 2, we compared the operating temperatures of the modules. See Figure **5** for the profiles of the same clear-sky day as in Figure **4**. While the textile and reference cases are virtually indistinguishable, the grass leads to a lower temperature by about 2…5 K, probably caused by locally increased humidity due to evaporation in combination with different infrared radiation properties. The grass was regularly irrigated during both Phases 1 and 2. Note that these experiments were conducted during summer months with unusually high ambient temperatures and a high time fraction of clear skies.

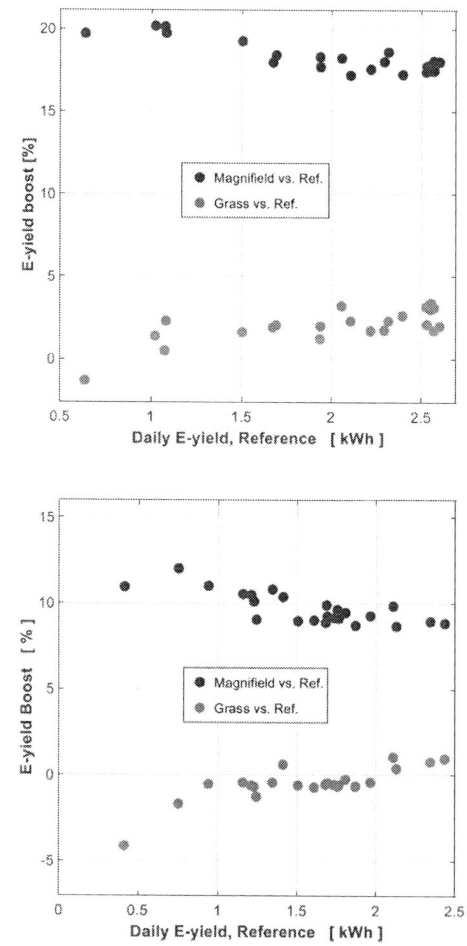

Figure 6: Increase in daily energy yield due to reflectors relative to the reference module yield, top: for solitary modules (Phase 1), bottom: for modules with surrounding covers (Phase 2).

3.3 Energy yield

The gain in energy yield compared to the reference as a function of daily yields is given in Figure **6**. On the sunnier days, the textile gains somewhat less, while with grass the gain tends to improve. We attribute this to the varying share of diffuse irradiance between sunny and overcast days. For an overview of the averaged energy yields, see Table **1**, where we differentiate between mostly clear-sky and predominantly overcast conditions.

With the solitary modules, we found a gain in energy yield of +2.2% for the grass relative to the reference, and of +18.1% for the textile. With surrounding covers, no gain was achieved with grass (-0.3%), and +9.5% for the textile. The uncertainty of the gain values is ±1.2% absolute. When differentiating overcast and sunny days, a larger gain was found in the former, with +19.7% for solitary and +11.0% for surrounded modules. Note that the bifacial gain of the reference setting (vs. monofacial) was found to be approximately +7% in previous experiments with slightly different geometry. We did not explicitly test the total bifacial gain, which is expected to be in the range of +17%with the surrounding covers.

Table 1: Energy yields per day and gains vs. reference for the sunniest and the most overcast days, and average of the respective test phases. The measurement uncertainty of the gain values is ±1.2%abs.

Seg-ment	West / ref.	East / grass		Middle / Magnifield	
Condi-tions	E yield [kWh/d]	E yield [kWh/d]	*Gain*	E yield [kWh/d]	*Gain*
Phase 1: Solitary modules, July 15...August 5, 2020					
Mostly clear sky	2.56	2.65	*+3.3%*	3.01	*+17.5%*
Mostly overcast	0.64	0.63	*-1.3%*	0.76	*+19.7%*
Average (22 days)	1.99	2.04	*+2.2%*	2.35	*+18.1%*
Phase 2: With surrounding covers, August 7...30, 2020					
Mostly clear sky	2.44	2.46	*+1.0%*	2.65	*+8.9%*
Mostly overcast	0.41	0.40	*-4.1%*	0.46	*+11.0%*
Average (24 days)	1.57	1.57	*-0.3%*	1.72	*+9.5%*

The relatively short test period was mostly sunny and has more direct irradiance than an average year, possibly leading to an underestimation of the potential gain, which tends to be higher with more diffuse irradiance and more oblique irradiance angles. Also, the modules used are not ideal given their limited bifaciality and rear angle of acceptance. On the other hand, some geometrical factors, especially the raised non-flat background of the setup, speak for some overestimation.

4 CONCLUSIONS & OUTLOOK

We conducted experiments to determine the energy yield gain by grass and white textile as reflecting materials in a bifacial PV system. From the test results, we conclude that the white Magnifield textile enhanced the energy yield of the modules by 10±2% relative to grass in a setting that is comparable to the installation in a larger ground-mounted system. Several factors related to the test set up may be influencing the result, including the non-ideal geometry, some extraneous influences on the performance, and the size constraints. Nonetheless, we found that this is the best approximation of the yield gain attainable with our setup. In any case, the gain by the reflective textile is so substantial that it would make a major difference in the performance of any bifacial PV power plant. For a large utility-scale installation, which is the application targeted ultimately with this material, the geometry of the entire system must be optimized in order to take full advantage of this in terms of costs, since the strong enhancement in albedo works in favor of a lower packaging density. In upcoming pilot-scale experiments, the installation method will be refined, and the long-term performance will be assessed.

REFERENCES

[1] International Technology Roadmap for Photovoltaic, 12th edition, April 2021, https://itrpv.vdma.org/en/
[2] Rodríguez-Gallegos et al., Global Techno-Economic Performance of Bifacial and Tracking Photovoltaic Systems, Joule 4 (2020), doi: 10.1016/j.joule.2020.05.005
[3] X. Sun et al., Optimization and Performance of Bifacial Solar Modules: A Global Perspective, Applied Energy 212, 2018, pp. 1601-1610, doi: 10.1016/j.apenergy.2017.12.041
[4] https://www.beaulieutechnicaltextiles.com/en/category/c/2/agro-textiles/s/35/fruit-coloring
[5] B. Marion, Albedo Data Sets for Bifacial PV Systems, 47th IEEE PVSC, 2020, pp. 0485-0489, doi: 10.1109/PVSC45281.2020.9300470.

LARGE SIZE FLEXIBLE AND LAMINATED SPACE PHOTOVOLTAIC ARRAYS

Thomas GUERIN*
Univ. Grenoble Alpes
CEA, LITEN, campus INES
73375, Le Bourget Du Lac, FRANCE
thomas.guerin4@cea.fr

Clément JAMIN
Univ. Grenoble Alpes
CEA, LITEN, campus INES
73375, Le Bourget Du Lac, FRANCE
clement.jamin@cea.fr

Philippe VOARINO
Univ. Grenoble Alpes
CEA, LITEN, campus INES
73375, Le Bourget Du Lac, FRANCE
philippe.voarino@cea.fr

Dominique VERGNET
AIRBUS Defence and space
Toulouse, FRANCE
dominique.vergnet@airbus.com

Paul ZEVENBERGEN
AIRBUS Defence and space
Leiden, THE NETHERLANDS
P.Zevenbergen@airbusDS.nl

Romain CARIOU*
CEA, LITEN, campus INES
73375, Le Bourget Du Lac, France
romain.cariou@cea.fr

*Corresponding authors: thomas.guerin4@cea.fr – romain.cariou@cea.fr

ABSTRACT: Space industry is facing a wave of important changes. The booming of space activities comes along with greater PV power requirements for geostationary orbit. With solar electrical propulsion and high throughput satellites, it becomes necessary to increase arrays' specific power (W/kg) and power density (W/m^3). Moving away from standard rigid panels towards foldable / flexible solar arrays is a promising way to meet this mass and volume challenges and thus address this high-power demand. The ALFAMA project [1], [2] - Advanced Lightweight and Flexible Array with Mechanical Architecture – brings innovative solutions by developing thin and foldable solar arrays with low cost fabrication processes such as lamination and printed electronics.
Keywords: Flexible photovoltaic array, Laminated space solar arrays, Thin III-V solar cells

1 GENERAL INTRODUCTION

More and more space missions require power supply greater than 25kW such as solar electrical propulsion and high throughput communication satellites. This leads to increasingly important power targets in watt per kilogram, watt per square meter and watt per cubic meter. However, the restricted volume in the spacecraft limits the number and the dimension of rigid solar panels. Therefore, the answer for this growing power demand lies neither in stacking more rigid solar arrays, nor by marginal efficiency improvements on solar cells, but rather in disruptive photovoltaic solution such as thin and flexible photovoltaic array (PVA).

Thirty years ago, European Space Agency (ESA) has developed the first flexible solar arrays for Hubble space telescope [3], [4]. Nowadays, several developments are pursued actively in the USA, with for instance the Ultraflex solar arrays from Orbital ATK's Cygnus cargo carrier in 2011 [5] or the Roll-Out Solar Array Experiment (ROSA) tested in 2017 and mounted on International Space Station (ISS) in 2021 [6, 7].

Standard space solar photovoltaic array are made with thick and rigid architecture, by using aluminum honeycomb and individual cells glued on top of this rigid substrate, with individual coverglasses on top (see Figure 1.)

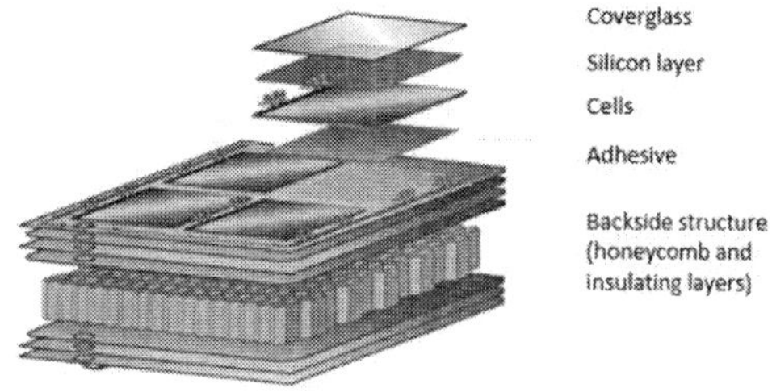

Figure 1: Schematics of standard composition of rigid solar arrays (Robert W. et. al. [8]), with individuals coverglass - cells (CIC) glued on honeycomb substrates.

Such standard space PVA have typically a > 20mm thickness that limits its W/kg and W/m^3 performances. This bottleneck can be solved by switching to thin and flexible PVA designs. The EU H2020 ALFAMA project develops such PVA solution. The work presented here focuses on large thin and flexible PVA designs and manufacturing for both vibration and deployment tests.

2 DESIGN & PROCESS DEVELOPMENTS

Before the construction of large PVA, the first steps of the project was to investigate innovative PVA architectures. This architecture study was performed on two-cells coupon and with thinned 3G30 solar cells (80µm) [9].

2.1 Designs development

Space environment constraints drive the selection of appropriate materials to design a robust architecture. Materials have to be resistant to void pressure, thermal cycles and radiations. Different design were studied in ALFAMA with various mass and thickness. As seen in Figure 2, various materials for the front and the back sides of space PVA were tested.

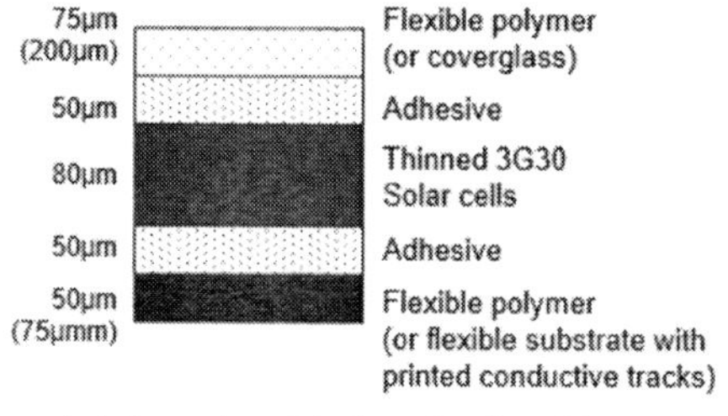

Figure 2: Schematics of designs developed in ALFAMA,

with variations on front and back of the structure.

It worth noting that each design combination result in a PVA total thickness below 500µm. Three main design variations have been selected.

Design 1 has a thin coverglass with a high optical transparency, a space-grade silicone adhesive as encapsulant with low outgassing coefficients and a flexible substrate (total thick. 450µm). Design 2 consists of symmetrical design that investigate an all-flexible polymer structure. It consists of a flexible polymer instead of coverglass and the same flexible substrate as Design 1 (total thick. 330µm). Design 3 keeps the flexible polymer in front and includes a conductive printed tracks architecture on flexible substrate at the back side of the PVA (total thick. 360µm). The cells are glued to the tracks with a space grade conductive adhesive. This allows easier processing while offering some radiation protection thanks to metal tracks beneath. These three designs compositions are compiled in Table I.

Table I: ALFAMA designs composition

	Front side	Back side	Thickness
Design 1	Coverglass	Polymer	450µm
Design 2	Polymer	Polymer	330µm
Design 3	Polymer	Printed tracks on polymer	360µm

2.2 Process developments

One of the important objective of the project was to use innovative and low-cost terrestrial solar PV processes to manufacture the flexible space PVA. In that sense, a lamination process based on liquid silicone adhesive has been developed at CEA-INES facilities for those three designs.

2.3 Performances

The two-cells coupon format allows for relative performance comparison. The design 1 is close to the standard space PVA architecture (except for its thin / flexible substrate), with the use of coverglass and classical interconnectors. Electrical performances are good with such design (+90% CTM ratio) in accordance with standard space PVA.

The use of flexible polymer in front side instead of coverglass lower the I_{sc} of Design 2 and Design 3 PV modules by approx. 12%.

In addition, the printed electronics substrate in Design 3 introduces for the time being more series resistances. One can see a summary of these conclusion with the comparative I(V) characteristics of each design in Figure 3.

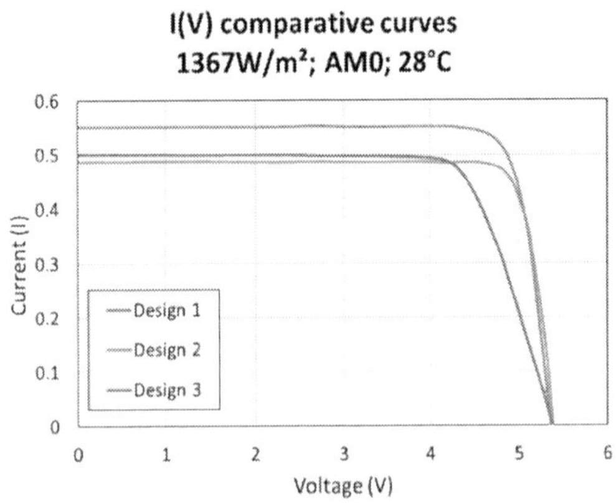

Figure 3: Comparative I(V) characteristics for each design (AM0, 1367W/m², 28°C).

Adjustments on conductive tracks and front side polymer should allow recovering some of the performance differences for design 2 and 3 with respect to design 1.

3 FABRICATION OF A FIRST LARGE PVA FOR VIBRATION TEST

The first large PVA using a combination of these designs was made to perform a vibration test that simulates the launch of a space rocket.

3.1 Realisations

The demonstrator is composed by four flexible sails and has a final surface of 3.7m². It was design with a modular approach: two laminated submodules (0.15m² each) are assembled as a first sail. They are using dummies and functional cells, large coverglass and conductive printed tracks architecture. This sail forms the first part, and three other sails with six aluminium dummies submodules each form the second part (1.11m² per sails). These two parts can be seen on Figure 4.

Figure 4: Folding step after fabrication of the large PVA made for vibration test. Top: functional and dummies laminated submodules. Bottom: Sail with aluminium mechanical dummies submodules.

3.2 Results

The vibration test was performed at ESTEC facilities (The exact parameters such as modes or vibration frequencies are confidential). The PVA was in Z-folded configuration during the test, between two pressure plates. It passed successfully the vibration test since no damage were induced during this experiment. This was confirmed by electroluminescence (EL) images before and after the vibration test (Figure 5).

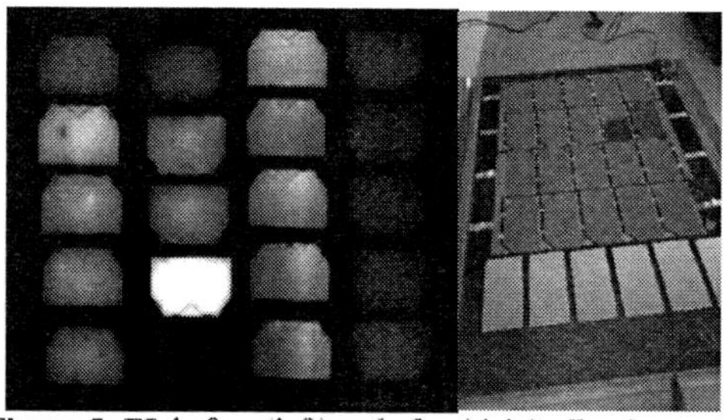

Figure 5: EL before (left) and after (right) vibration test - No cells breakage but detection of shunt instabilities since lamination and before vibration (flashy and low luminescence intensitycells visible before and after).

EL also revealed the presence of shunt instabilities at interconnector / cell interface, seen after lamination and before vibration. They are characterised by flashy and low intensity cells. These type of defects are linked to the conductive printed tracks design and are currently being investigated.

4 FABRICATION OF A SECOND LARGE PVA FOR DEPLOYMENT TEST

A second, larger PVA was made as second step in order to validate the foldability of the concept with deployment test.

4.1 Realisations

On this second large and flexible PVA, the same manufacturing processes described previously were used. Two of the previous PVA architectures were included in the sail as well: Design 1 for functional cells submodules and Design 2 for dummies cells submodules. This second demonstrator sail was manufactured with a promising yield: only one of the submodules were damaged with one cracks on a coverglass (Figure 6) and no cell breakages induced.

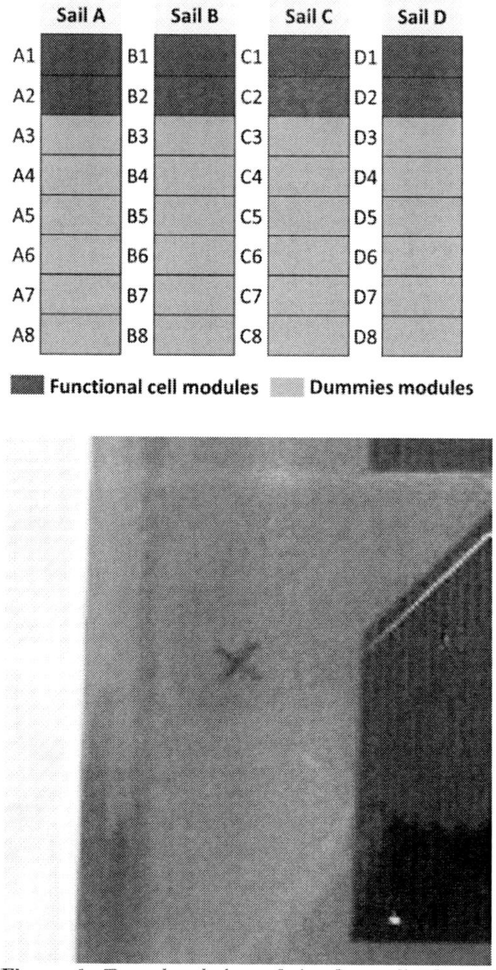

Figure 6: Top: description of the four sails for the deployment test. On 32 modules made, one had a small coverglass crack. Bottom: detail of the crack with marked location (black cross). The crack is hardly visible.

The four PVA sails, side by side in final configuration, are made with a total PVA area > 5m² (Figure 7).

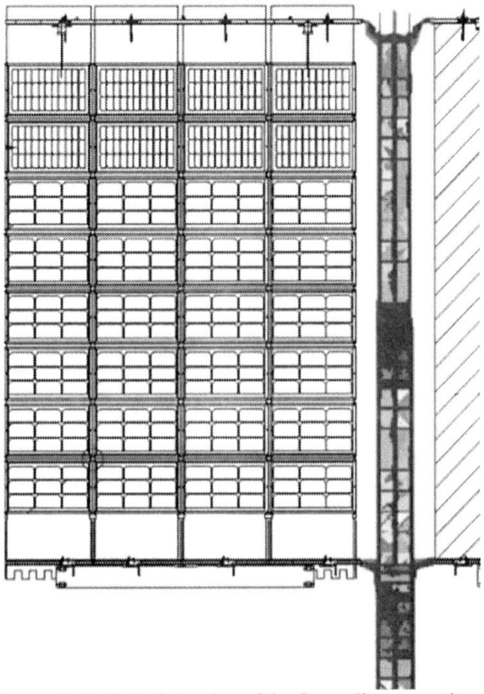

Figure 7: Technical drawing of the four sails mounted on a telescopic mast with functional cells submodules on top and dummies cells submodules in the bottom part (courtesy of ADS).

Each submodule is composed by either three strings of eleven 3G30 solar cells or nine silicon dummies cells, for a total of 32 submodules (8 functional cells and 24 dummies cells submodules). Two different cell thicknesses are tested successfully on the functional cells submodules (80µm and 50µm).

One sail is 1.30 kg, 1.69 m² and 4 cm thick once folded (Figure 8). One submodule is 140 g approx.

Figure 8: Folding step after fabrication of one of the four sails of the large PVA made for deployment test.

4.2 Results

First electrical performances measurements shown promising results. EL test proves the good integrity of large functional cells submodules without induced shunts or breaks, as the one defect one can see in the lower left part (first string) was present at cell level before lamination (Figure 9).

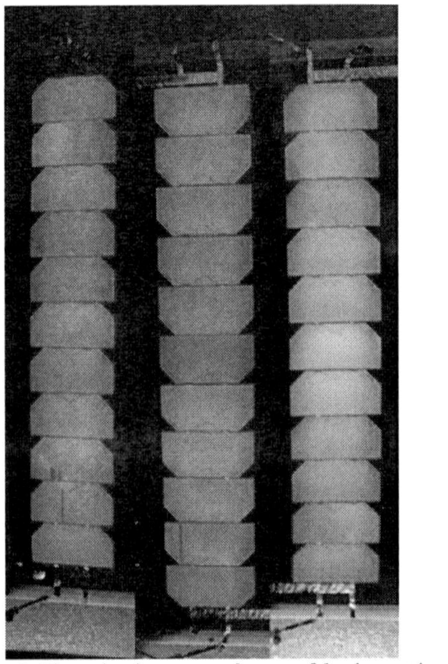

Figure 9: Electroluminescence pictures of the three strings of cells of one of the large submodule.

5 CONCLUSION AND PERSPECTIVES

This work present the design and manufacturing of large area foldable space solar photovoltaic arrays based on the low cost terrestrial lamination process. A first large demonstrator (3.7 m²) successfully passed vibration. A second larger flexible PVA (> 5 m²) was manufactured for deployment test. Electroluminescence imaging and visual inspection confirm the integrity of this innovative PVA.

Coupling the four large PVA sails with the telescopic mast mechanism developed by ADS and completing the deployment test is now the next step of the project. At module design level, the future work will focus on improving processes reliabilities for the conductive printed tracks (design 3). At the end of the project, a smaller breadboard will be developed with the presented designs but incorporating ultra-thin flexible cells developed within this ALFAMA project.

6 ACKNOWLEDGMENTS

Victor KHORENKO (AZUR SPACE Solar Power GmbH)
Sébastien NOEL, Eszter VOROSHAZI and Fabien CHABUEL (CEA).
The authors thanks the consortium members.
This project has received funding from the European Union's Horizon 2020 research and innovation Programme under grant agreement N° 821876.

7 REFERENCES

[1] http://www.alfamaproject.eu/
[2] D. Vergnet et al. in 2019 IEEE 46th Photovolt. Spec. Conf. PVSC (IEEE, Chicago, IL, USA, 2019), pp. 2808–2810
[3] Gerlach, ESA Bull. Eur. Space Agency (1990).

[4] D. Poeck, Design development, qualification and flight production of an ATOC resistant flexible solar array for Low Earth Orbit Mission, Proceedings of the European Space Power Conference, Madrid, Spain, 2-6 October 1989, ESA SP-294, August 1989.
[5] B. Spence et al., « UltraFlex-175 Solar Array Technology Maturation Achievements for NASA's New Millennium Program (NMP) Space Technology 8 (ST8) », in 2006 IEEE 4th World Conference on Photovoltaic Energy Conference, Waikoloa, HI, 2006, p. 1946-1950, doi: 10.1109/WCPEC.2006.279879..
[6] B.R. Spence et al., in 2018 IEEE 7th World Conf. Photovolt. Energy Convers. WCPEC (IEEE, Waikoloa Village, HI, 2018), pp. 3522–3529.
[7] M.K. Chamberlain, S.H. Kiefer, M. LaPointe, and P. LaCorte, Acta Astronaut. 179, 407 (2021).
[8] Robert W. Francis et al., "Thermal cycling Techniques for solar Panels", in Crosslink Fall 2005, pp. 20-23
[9] T. Guerin et al. « ALFAMA project: development of a flexible and laminated space photovoltaic array » 2020 proceedings of the EUPVSEC 37th European Photovoltaic Solar Energy Conference and Exhibition (EUPVSEC2020) 1261-1263 doi:10.4229/EUPVSEC20202020-5BO.15.1

OPTICALLY ENHANCED C-SI SOLAR CELLS FOR EPHEMERAL SPACE APPLICATIONS

David Pera, Ivo Costa, Filipe Serra, Afonso Guerra, Killian Lobato, João M. Serra, José A. Silva
Instituto Dom Luiz - Faculdade de Ciências Universidade de Lisboa
Edifício C1, Campo Grande, 1749-016 Lisboa, Portugal

ABSTRACT: The market of small-scale space satellites has been growing significantly. Concepts as CubeSats, based on open-hardware platforms became a standard, opening new opportunities for small companies such as technological start-ups. However, prototypes' associated costs are still an obstacle, mainly due to their power systems. Historically photovoltaic technology was elected as one of the most reliable solutions to power satellites. Multijunction solar cells based on III-V groups elements are the most efficient photovoltaic devices, enabling high energy densities. Reason why they became the standard for spatial applications. Nevertheless, the associated cost of such devices is a hindrance for short-term satellites' missions. For such applications, the use of more cost-effective technologies as crystalline silicon solar cells should be considered even if their efficiencies are lower. Moreover, the use of advanced light trapping concepts could further enhance the conversion efficiency while reducing solar cell's thickness and mass. A very effective way to improve light capture in crystalline silicon is surface texturization. Silicon technology has the potential to deliver 70% of energy per unit of area of the best III-V solar cells while reducing the cost by two to three orders of magnitude.
Keywords: space, cost reduction, silicon solar cell, light trapping

1 INTRODUCTION

Since the beginning of the space era, photovoltaic (PV) systems were one of the technologies of choice to power satellites. As main advantages of PV one can point the abundance of solar resource outside the atmosphere and the scalability of the technology, that allows adapting the power installed to the missions' requirements. Moreover, when discussing circum-terrestrial satellites, PV has gained ground relatively to the use of nuclear fission systems due to its safety, especially after the reentry into the atmosphere and subsequent crash of the Cosmos 954 satellite in 1978 [1].

Although, the first PV powered satellites were based on crystalline silicon technologies, due to their high efficiency, reliability in space environment and lower weight per watt produced, III-V groups' solar cells have become the dominant power technology in space applications [2].

Recently the interest in small-scale satellites has increased very significantly. Favoured by the standardization provided by the CubeSat Program [3], that opened the way to important cost reductions, the number of nanosatellite launches has exploded in the last years. Until now more than 1700 nanosatellites were launched, and it is expected that a total of 750 are launched during this year [4].

Small-size satellites such as CubeSats offer a low-priced means to develop small scientific projects or provide a proof-of-concept before large-scale prototype production [5]. This opens new opportunities for small research labs and companies such as technological start-ups.

However, the associated costs of these devices are still a hindrance to their widespread use, mainly due to the price of the power systems, mostly based on III-V solar cells.

We argue that since most of the small-size satellites are programmed for short-term missions, the use of more cost-effective photovoltaic technologies, even if less efficient such as crystalline silicon, should be considered. As is well known crystalline solar cells suffer higher degradation in space environment than the III-V solar cells, however for the life cycle of the targeted applications is less relevant.

As a mean to enable the promotion of these lower cost photovoltaic devices, advanced light trapping concepts could be applied to increase the conversion efficiency.

2 ANALISYS

Enhancing the optical performance of crystalline silicon solar cells by introducing light trapping nano structures allows to reach high conversion efficiencies while significantly reducing the cell's thickness and thus its mass [6]. For instance, reducing the thickness from the industry standard of 180μm to few tenths of μm, allows mass reductions around 80% approaching the weight/watt values of the III-V group tandem solar cells.

Due its simplicity and potential cost-efficiency, metal assisted chemical etching (MACE) techniques are being more and more considered by silicon solar cells' industrials as an effective method to improve light capture and solar cell efficiency. In this scope, we have recently presented the characterization of a maskless MACE procedure based on the use of a solution of hydrogen peroxide (H_2O_2), hydrofluoric acid (HF), and silver nitrate ($AgNO_3$). The results of the process characterization were presented elsewhere [7, 8, 9].

The surface texturizations obtained with the MACE method showed a strong reduction of the spectral reflectance in the range 350-1100 nm.

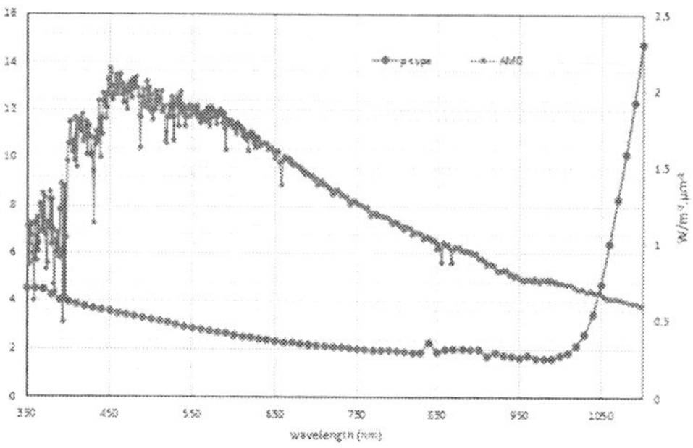

Figure 1: Reflectance curve for a p-type wafer and AM0 spectrum [10] in the range 350-1100 nm.

As shown in Figure 1, for an uncoated p-type silicon wafer the reflectance is under 4% for most of the analyzed range, leading to an effective reflectivity (eq.1) of 2.9%.

$$R_{\text{eff}} = \frac{\sum_{\lambda} R_{\lambda} n_{\lambda}}{\sum_{\lambda} n_{\lambda}} \qquad (1)$$

R_{λ} and n_{λ} are respectively the reflectance measured and the number of photons in the AM0 [10] for the wavelength λ.

Such low reflective surfaces are suitable for thinner devices as they promote high gains of photocurrent generation rates. Finite-Difference Time-Domain (FDTD) electro-magnetic simulations based on Maxwell equations were performed showing that the obtained texturizations have by itself the potential to increase the generation rate by a factor of two for 40 μm thick cells [11]. This gain could be further improved by considering optimization strategies like back reflectors.

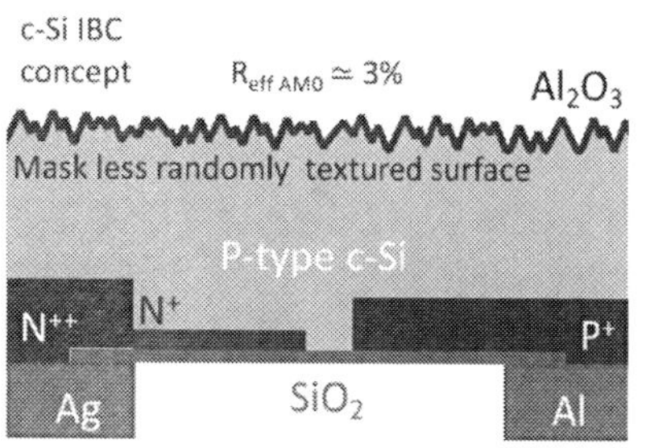

Figure 2: Concept diagram of an interdigitated solar cell. The mass reduction

An example of a solar cell structure that could benefit from this optical enhancement is the interdigitated back contact configuration (Figure 2), which is particularly adequate for MACE texturizations since both contacts and junctions are located at the back surface.

Nonetheless, the potential mass reduction resulting from the use of thinner devices' concepts is not an effective advantage for CubeSat applications as for instance, considering an 1U CubeSat configuration ($10\times10\times10 cm^3$) the difference between silicon solar cells weights of 180 μm and 40 μm is about 2 grams. As Table I, classification of satellites' systems based on their mass shows, CubeSats mainly identified as Pico, Nano, and Micro satellites have their masses ranging from 1-100 kg.

Table I: Classification of satellites according to mass [5]

	Mass (kg)
Large satellite	> 1000
Small satellite	500-1000
Mini satellite	100-500
Micro satellite	10-100
Nano satellite	1-10
Pico satellite	< 1

In terms of energy conversion, the III-V groups tandem solar cells typically used in space industry have a maximum efficiency 32 % [12], while silicon solar cells have demonstrated efficiencies ranging from 22% to 26% [13] for terrestrial STC. Considering these figures, optically enhanced crystalline silicon solar cells could assure around 70% of the energy per area unit when compared to the III-V groups solar cells. In this way the use of silicon solar cells implies larger cells areas to suppress the same energy payload. On the other hand, replacing III-V tandem solar cells by crystalline silicon ones allows a cost reduction of the satellites energy production system by two to three orders of magnitude [14], having a major impact in the CubeSat overall cost.

4 CONCLUSIONS

The use of CubeSat satellites and other small low-orbit devices for short-term missions is becoming more frequent. Reducing the production costs of these devices can further generalize their access and thus accelerate their deployment.

Optically enhanced c-Si solar cells have the potential to reduce the cost of the satellites power production systems in two to three orders of magnitude for space applications when compared to III-V solar cells, while assuring 70% of the energy yield.

5 REFERENCES

[1] A.F. Cohen, 'Cosmos 954 and the International Law of Satellite Accidents', Yale Journal of International Law 10 (1984) 78-91.
[2] R. R. King, et al. 'Advanced III-V multijunction cells for space', Proceedings 2006 IEEE 4th World Conference on Photovoltaic Energy Conversion 2 (2006) 1757-1762.
[3] 'CubeSat Design Specification Revision 13, The CubeSat Program', California Polytechnic State University (2017).
[4] www.nanosats.eu, visited on September 8, 2021.
[5] M. J Rycroft, N. Crosby, 'Smaller Satellites: Bigger Business? Concepts, Applications and Markets for Micro/Nanosatellites in a New Information World', Springer Science & Business Media (2013).
[6] Y. Liu, W. Zi, S. Liu, and B. Yan, 'Effective light trapping by hybrid nanostructure for crystalline silicon

solar cells', Solar Energy Materials and Solar Cells 140 (2015) 180-186.
[7] I. Costa, D. Pera, J. A. Silva, 'Improving light capture on crystalline silicon wafers', Materials Letters 272 (2020) 127825.
[8] D. Pera, et al., 'Advanced light-trapping structures for back-contact solar cells produced by metal-assisted chemical etching', Proceedings 37th European Photovoltaic Solar Energy Conference (2020) 354 - 357
[9] D. Pera, et al., 'Texturization of monocrystalline silicon by metal-assisted chemical etching', Proceedings 38th European Photovoltaic Solar Energy Conference (2021)
[10] ASTM, 'Reference Solar Spectral Irradiance: Air Mass 0', American Society for Testing and Materials (ASTM) Terrestrial Reference Spectra for Photovoltaic Performance Evaluation, 2000. [Online]. Available: www.nrel.gov/grid/solar-resource/spectra-astm-e490.html.
[11] D. Pera, et al., 'Computational optical analysis of 3D modelled crystalline silicon substrates randomly', Proceedings 37th European Photovoltaic Solar Energy Conference (2020) 329 - 331.
[12] www.spectrolab.com, visited on September 8, 2021.
[13] M. A. Green, et al., 'Solar cell efficiency tables (Version 55)', Progress in Photovoltaics: Research and Applications 28.1 (2020) 3-15.
[14] K. A. Horowitz, et al, 'Techno-Economic Analysis and Cost Reduction Roadmap for III-V Solar Cells', Golden, CO: National Renewable Energy Laboratory. NREL/TP-6A20-72103. https://www.nrel.gov/docs/fy19osti/72103.pdf.

6 AKNOWLEDGEMENTS

The authors acknowledge the financial support of the Portuguese Fundação para a Ciência e Tecnologia (FCT) through the projects TaCIt / PTDC/NAN-OPT / 28837 / 2017 and UIDB / 50019 / 2020 – IDL.

Performance of Photovoltaic - Thermal (PVT) and Photovoltaic (PV) Systems Under Various Weather Conditions

R.R. Vardanyan, V.K. Dallakyan, N.K. Badalyan
National Polytechnic University of Armenia
105, Teryan Str., Yerevan 0009, Republic of Armenia

ABSTRACT: Solar photovoltaic and thermal hybrid systems have a high potential for producing electricity and thermal energy with increased efficiency of the system. The hybrid system proves to be much more energy-efficient during the warm climate seasons with improved electrical efficiency due to the cooling of solar cells and gained thermal energy due to heat absorbed by the fluid from the solar cells (dual effect). Unfortunately, there is a lack of investigations of PVT systems' performance during the cold weather conditions. In this work, the performance temperatures of the PVT module and general PV module are experimentally investigated. It is shown that during cold climate seasons, the temperature of the PVT collector can become higher than the temperature of the general PV module due to the decreased convection cooling of PVT in comparison with the PV module. The ambient temperature below which the PVT module is hotter than the PV module is called the "critical". As a result of this process, the electrical efficiency of the PVT module is becoming smaller than the efficiency of the PV module. This process is called the "reverse" process. To overcome this problem the PVT system must be optimized taking into consideration the weather conditions under which the PVT system is operating.
Keywords: Photovoltaic, Solar cells, Thermal Performance, Environmental Effect

1 INTRODUCTION

The PVT hybrid system combines a solar cell, which converts solar radiation into electricity, with a solar thermal collector, which captures the remaining thermal energy and removes heat from the solar cell. In hybrid PVT systems by cooling the solar cells, electrical generation is increased and the waste heat is captured for utilization, thus maximizing the return from the available solar radiation. Thus, the PVT systems are the dual solar collector with two usable energy outputs by one collection system (two in one). In comparison with general separated PV and thermal collectors PVT systems have several advantages.

Numerous research works are conducted for the investigation of properties of PVT hybrid systems [1 - 7]. Unfortunately, there is a lack of investigations of the performance of PVT hybrid systems under various weather conditions. In this paper, the operating temperatures of the PVT collector and PV module in dependence on ambient temperature are investigated.

2 THERMAL PARAMETERS OF PVT AND PV MODULES

The temperature of solar cells in hybrid PVT collectors depends on the thermal properties of the collector and external influences as the intensity of solar radiation and ambient temperature. The analytical expression for the temperature of solar cells of hybrid PVT systems is presented as [8]:

$$T_{cell.PVT} = \frac{1}{K1+K2}[F(\tau\alpha) + K1\,Tin + K2\;Ta], \quad (1)$$

where $K1 = 2\frac{mC}{AS}$, $K2 = \frac{FU}{G}$, m is the mass of fluid in the storage tank (g), C is the heat capacity of fluid (J/g ^{0}C), A is the area of PVT solar collector receiving sun rays, S is the solar radiation energy per square meter, F is the PVT collector's heat removal factor, $(\tau\alpha)$ is the effective transmittance of the cover(τ) and absorbance of collector (α) factor, U is the overall heat loss coefficient of the collector, G is the intensity of solar radiation on the unit of the surface (W/m^2), Tin is inlet temperatures of PVT collector (^{0}C), T_a is the ambient temperature (^{0}C).

It can be seen from the (1) that the temperature of solar cells in hybrid PVT systems linearly increases with the increase of ambient temperature. The angular coefficient of this dependence is:

$$\frac{K2}{K1+K2} < 1, \quad (2)$$

which means that the inclination angle of the dependence of $T_{cell/PVT} = f(Ta)$ is smaller than 45^0.

The temperature of the solar cells of PV modules can be calculated on the base of the NOCT (Nominal Operating Cell Temperature). The NOCT is defined as the temperature reached by open-circuited cells in a module under the *standard reference environment* defined as follows: solar irradiation 800 W/m^2, ambient temperature 20 ^{0}C, wind velocity 1m/s and mounting type - open backside.

The best module operated at a NOCT of 33 °C, the worst at 58 °C and the typical module at 48 °C respectively. An approximate expression for calculating the cell temperature is given by [9]

$$T_{cell.PV} = Ta + G\,\frac{NOCT-20}{800}. \quad (3)$$

Thus the temperature of the solar cell in the PV module also increases linearly with increases in ambient temperature, and the slope angle of the line is just 45^0. It follows, that there is a point of interception of these two lines. At low temperatures, less than the point of interception $T_{cell.PVT} > T_{cell.PV}$.

The dependence of $T_{cell.PV}$ and $T_{cell.PV}$ from the ambient temperature is presented in Fig. 1. It can be seen from Fig. 1 that at high ambient temperatures, above about 0 ^{0}C, with increasing temperature the solar cells of the general PV module are heated more than the cells of the hybrid glazed PVT module. Thus, with the increase of ambient temperature, the difference between $T_{cell.PVT}$

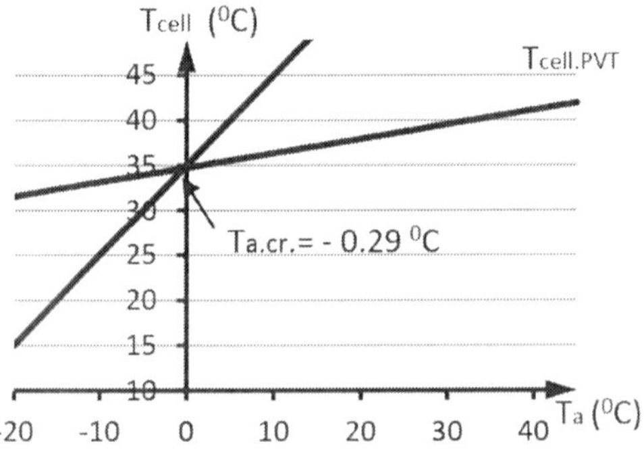

Figure 1: Dependence of the PV module and glazed PVT collector solar cells' temperatures from the ambient temperature (G = 1000 W/m^2, NOCT = 48 ^{0}C, PVT system tank capacity is 100 l, A = 2 m^2, Tin = 15 ^{0}C, $F(\tau\alpha)$ = 0.68, FU = 4.9 W/m^2 ^{0}C [10])

and $T_{cell.PV}$ is increasing. Consequently, the electrical efficiency of the PVT system is increasing also in comparison with the efficiency of the general PV module.

Under the low ambient temperatures, at $Ta < Ta.cr.$, a "reverse" the process is going on ($T_{cell.PVT} > T_{cell.PV}$) and the electrical efficiency of the PVT collector is becoming smaller in comparison with general PV module. This interception point, is called the "critical" ambient temperature [8]. Thus, at the temperatures below $Ta.cr$, due to the heating of cells, the electrical generation efficiency of the PVT system is smaller in comparison with the general PV module.

3 EXPERIMENTAL INVESTIGATIONS

To investigate experimentally the performance of PVT and PV modules under the various weather conditions, the standard PV modules with 250 W of power produced in Armenia are used. The parameters of used PV modules are similar. With the use of one PV module, the PVT module is developed by the installation of the water tubing thermal absorber under the PV module. The rare side of the module is not thermally insulated. Thus, the low thermal efficiency type PVT module was created. The thermosiphon system with a water storage tank with the capacity of 100 litres is assembled. The PVT system is a direct flow type. The prepared PVT module with the tank and the PV modules were installed and tested in the outdoor territory of the research laboratory at the National Polytechnic University of Armenia, in the Yerevan campus (Figure 3). The operating temperatures of the modules and the ambient temperature were measured for about a year. The obtained data for 11 a.m. of each day are presented below.

Figure 2: The PV module (right) and the PVT module (left) under the test

The temperatures of the PV module, PVT collector, and ambient temperature, measured during autumn 2020, for the period from 21.10.20 to 13.11.20 are presented in Figure 3.

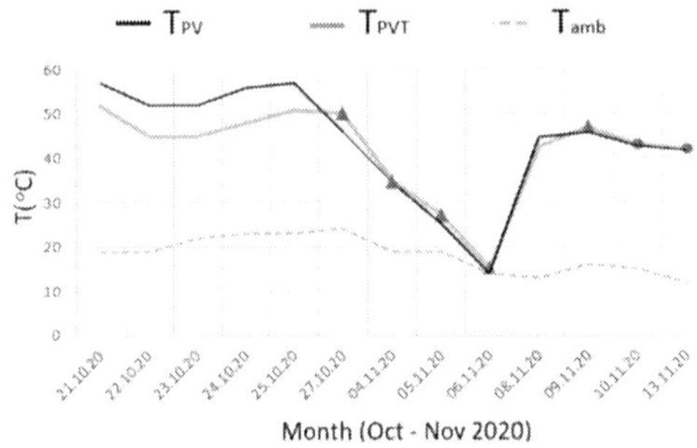

Figure 3: Temperatures of PV module, PVT collector, and the ambient temperature during the period from 21 October 2020 to 13 November 2020, measured at 11 a.m. of each day

In accordance with measured data, during the warm period, the temperature of the PVT collector is smaller than the PV module's temperature. But starting from the end of October (in Figure 3 it is the point of 27.10.20) the PVT collector becomes warmer than PV module, and the "reverse" process is going on. In Figure 3 these points are mentioned by triangle signs. This process is explained by decreased convection cooling of PVT in comparison with the PV module.

With farther decreasing of ambient temperature the "reverse" process, when the temperature of the PVT module is higher than the temperature of the PV module, is occurring more frequently. In Figure 4 the measurements conducted during January 2021 are presented. It can be seen from Figure 4 that dominantly the PVT is warmer than the PV module during almost all duration of January.

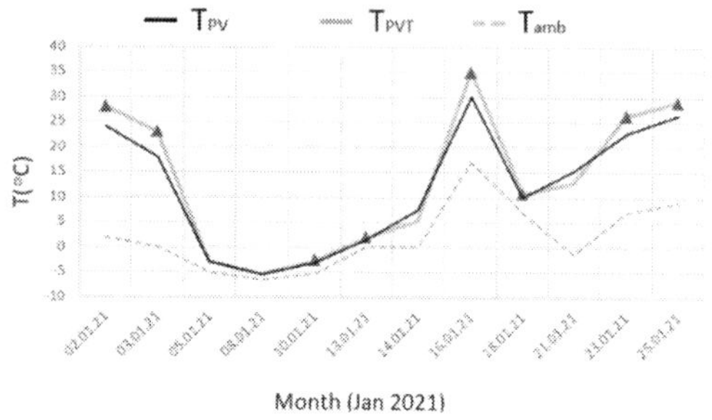

Figure 4: Temperatures of PV module, PVT collector, and the ambient temperature during January 2021

During the warmer weather climate conditions, the PVT module again is becoming cooler than the PV module. In Figure 5 the measured data for the period from the end of February to 17 of March 2021 are presented. During this period the PVT system is cooler and consequently, it is more effective than the PV module. Further, with increasing the ambient temperature the temperature of the PVT module is smaller than the temperature of the general PV module and it is operating normally with increased electrical efficiency, as expected.

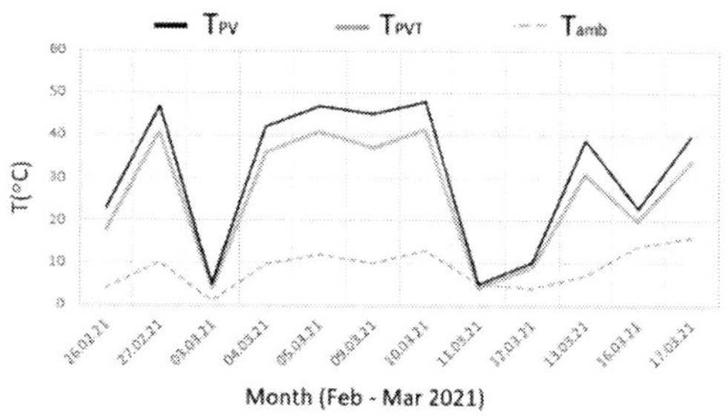

Figure 5: Temperatures of PV module, PVT collector, and the ambient temperature during the February and March 2021

Thus, the PVT hybrid modules are cooled during the warm climate conditions and the electrical efficiency of the PVT module is increased in comparison with the general PV modules. Under the cold climate conditions the PVT hybrid module the process is "reversed" and the PVT module is warmer than the PV module. The value of "critical" ambient temperature when the 'reverse" process is going on and the temperature of the PVT module becomes more than the temperature of the PV module, depends on different factors (Equations 2 and 3). To increase the electrical efficiency of the hybrid PVT collector under cold weather conditions the structure of the PVT collector must be optimized and the hybrid system operation must be properly organized.

SUMMARY

1. The temperature of solar cells in PVT hybrid system is increasing linearly with the increase of ambient temperature.
2. During warm weather conditions the temperature of the PVT hybrid module is smaller than the temperature of the general PV module, and the PVT is operating with increased electrical efficiency. During the cold climate conditions the temperature of solar cells of PVT hybrid module is higher than the temperature of the PV module.
3. There is a "critical" value of ambient temperature below which the solar cells of PVT hybrid module are warmer than the cells of a general PV module, and the efficiency of the PVT module is decreased in comparison with the efficiency of the PV module. This is called the "reverse" process.
4. To exclude the heating of solar cells in PVT module under the cold climate conditions, below the "critical" ambient temperature, the parameters of PVT module and the system operation type must be optimized.

REFERANCES

[1] T.T. Chow. A review on photovoltaic/thermal hybrid solar technology. Applied Energy, Vol. 87, (2010) 365.

[2] A. Mellora, et all, Roadmap for the next-generation of hybrid photovoltaic-thermal solar energy collectors. Solar Energy, (2018) 174, 386.

[4] H.A. Zondag Flat-plate PV-Thermal collectors and systems: A review. Elsevier, Renewable and Sustainable Energy Reviews **12**, (2008) 891.

[5] A. Mallik, M. Z. Hasan. PV/T Systems for Renewable Energy Storage: A Review. World Scientific News, 96 (2018) 83-95.

[6] S. Abdul-Ganiyu, D.A. Quansah et all. Investigation of Solar Photovoltaic-Thermal (PVT) and Solar Photovoltaic (PV) Performance: A Case Study in Ghana. Energies, Vol. 13, (2020) 1-17.

[7] S. Diwanial, S. Agrawal, A.S. Siddiqui, S. Singh. Photovoltaic–thermal (PV/T) technology: a comprehensive review on applications and its advancement. International Journal of Energy and Environmental Engineering (2020) 33–54. https://doi.org/10.1007/s40095-019-00327-y

[8] R.R.Vardanyan, V.K. Dallaqyan, M.G. Travajyan Temperature Dependent Performance of Solar Photovoltaic and Thermal Hybrid Systems. Springer, Journal of Contemporary Physics, V. 55(4), (2020) 339-344. http://link.springer.com/article/10.3103/S1068337220040180

[9] R. G. Ross. "Flat-Plate Photovoltaic Array Design Optimization", 14th IEEE Photovoltaic Specialists Conference. San Diego, CA, (1980) 1126-1132.

[10] RETScreen International. Clean energy support center. Solar water heating project analysis, Canada. www.retscreen.net

A FREE ONLINE TOOL FOR THE SIMULATION OF COLLECTIVE SELF-CONSUMPTION IN BRUSSELS

Babacar Sarr[1,*], Ziao Zhao[2], Jonathan Leloux[1], Patrick Hendrick[2], Jesús Robledo[1]
[1] LuciSun, Sart-Dames-Avelines, Belgium
[2] Aero-Thermo-Mechanics Dept. (ATM), Université Libre de Bruxelles (ULB), Belgium
*Phone: +32 470 10 09 29; E-mail: babacar.sarr@lucisun.com

ABSTRACT: In the past and until now, Belgium mainly had a mechanism in place that rewarded the PV energy production through feed-in tariffs that were attributed with a net annual balance counting. In Belgium, the mechanism of self-consumption for photovoltaic (PV) installations is projected to become mainstream. This change is going to have an important impact on the PV sector in Belgium. At the moment, there is a demand for a simple simulation tool that makes it possible to simulate the self-consumption ratio and the self-sufficiency ratio of a PV system or a group of PV systems in combination with one local consumption profile or a group of consumption profiles from the same neighborhood, in particular in the context of collective self-consumption. Indeed, combining different consumption profiles from the same neighborhood can significantly improve the self-consumption ratio to reduce the cost of a PV system for the users.
A free web application has been provided to the public for such simulation[1].
Keywords: Self-consumption, self-sufficiency, collective, Belgium, Brussels, simulation

1 INTRODUCTION

The self-consumption of the Photovoltaic (PV) systems is a strong driver in the revenue model.

Photovoltaic self-consumption is an economic model in which a PV system or a group of PV systems can be assessed to how much of PV power produced is consumed by a specific consumer or a group of consumers. In this model, the PV generated energy is consumed instantaneously as it is being produced and therefore uses on-site generated energy to cover the consumer electricity needs. It offers greater economic benefits and greater independence from the grid and future electricity rate variations. Since Belgium has had until now a mechanism in place that rewarded the PV energy production through feed in tariffs that were attributed with a net annual balance counting, the country is moving towards mechanisms of self-consumption for photovoltaic (PV) installations which are potentially projected to become mainstream. This change is going to have an important impact on the PV sector in the country.

A misalignment between PV energy generated and the local consumption whether it is for an individual prosumer or in a scenario of a collective of producers and consumers creates a lack of self-consumption, which can limit PV systems economic potential.

Assessing self-sufficiency is a parameter that combined with self-consumption form the basis of an energy community. Indeed, what individual self-consumption lacks to achieve energy independence can be provided by collective self-consumption, which can be achieved by having a community of energy producers sharing energy that they produce with a community of consumers (who also may be producers themselves and in that case called prosumers).

At the moment, there is a demand for a simple simulation tool that makes it possible to simulate the self-consumption ratio and the self-sufficiency ratio of a PV system or a group of PV systems in combination with one local consumption profile or a group of consumption profiles from the same local area, in particular in the context of collective self-consumption. Indeed, combining different consumption profiles from the same energy community can significantly improve the self-consumption ratio and thereby reduce the cost of the PV electricity generation for the users[2].

This free online tool, named Consolectiv (for CONsumption, SOLar, and collECTIVe) is the fruit of collaboration between the company Lucisun and Université Libre de Bruxelles (ULB) and can be used through a public web interface [1]. The tool is composed of an interface for calculating the self-consumption ratio, the self-sufficiency ratio and other results for the self-consumption of the energy produced by photovoltaic installations. It is aimed at the general public, policymakers, and the photovoltaic installation professionals. It has two main functionalities.

One is for the individual prosumers. It aims at helping people decide on the optimum design conditions for their photovoltaic installation, taking self-consumption into account, and the corresponding returns on investment.

Another one is for energy communities. It helps combining different types of building consumption profiles and photovoltaic producers to evaluate the global impact of these combinations on the self-consumption and self-sufficiency for a whole neighborhood of prosumers.

Therefore, both for the prosumer and for a whole energy community neighborhood, the Consolectiv tool contributes to help people make better informed decision.

2 METHODOLOGY

The tool takes inputs from the user and goes into a processing phase before generating the figures and numbers relative to the analysis of the defined systems.

Both the prosumer simulation section and the energy community simulation section take similar inputs to define the profile of energy consumer(s) and the profile of photovoltaic producer(s). The financial parameters of the PV system are opened to be modified in the prosumers section by the users. However, for the energy community, the financials parameters considered are fixed to the current context in the Brussels region area.

One of the main functionalities of the tool is to provide an energy consumption profile and the photovoltaic system generation profiles for different types of buildings that the user can select as a consumer and for different type of photovoltaic producer(s).

The self-consumption and self-sufficiency ratios of such a configured system is therefore calculated from the consumption and production profile as well as some financial metrics (only for the prosumers section) such as the payback time, and the net present value.

2.1 Energy consumption profiles

In order to have typical energy consumption profiles for different types of buildings, synthetic profiles in Typical Meteorological Year (TMY) format have been used corresponding to 17 different types of buildings, and therefore of consumption profiles. These profiles are obtained through a synthetic generation method developed by National Renewable Energy Laboratory in the USA [3] and takes into account the local climate and consumer preferences.

A set of data of commercial and residential hourly load profiles for different commercial building types and residential buildings has been used for the analyses.

The load profiles store a representative yearly consumption in an hourly format of consumed electricity, in kWh.

The tool therefore proposes a set of different profiles of consumers which when given as input a specific type of building, the electricity profile is defined. The hourly consumption data are for the different types of consumption profiles downsampled to 10 minutes granularity. This 10-min granularity is chosen to have the desired results respecting the usual time interval of metering by Distribution System Operators (DSO).

In the "prosumer" simulation section of the tool, there is a possibility to select and run an analysis with one consumption profile and in the "energy community" section different profiles can be combined.

2.2 PV generation profiles

For the photovoltaic generation part, the simulation tools use an internal simulation library from the company LuciSun which transpose the Global Horizontal Irradiance (GHI) data to the plane of array irradiance of the PV systems provided. The user can choose how the modules are mounted, by providing the system angle of inclination and orientation.

The Global horizontal irradiance data used is a Typical Meteorgical Year (TMY) dataset which is a set of meteorogical data values for every hour and representing a typical year for the region of Brussels.

The hourly irradiance data brought to the plane of array of the system(s) is downsampled to a 10-min granularity to simulate the flow of energy and the reference yield is derived from it. Considering the system peak power and an average system loss of 20% (Performance Ratio of 80%), the energy system yield of each PV system is calculated.

In the case of the prosumer simulation section, one single PV production profile is considered, and in the case of the energy community section, different profiles are combined.

2.3 Definition of self-consumption and other metrics

The 2 main principles of self-consumption and self-sufficiency will be defined below and are generally used in the world of local production. From them, ratios that allow different systems and operations can be derived and can be compared with each other to assess their efficiency and dependence on the public grid.

2.3.1 Self-consumption

Self-consumption (SC) can be defined as the share of self-consumed PV production, which is equal to the ratio of self-consumed energy from the photovoltaic system production (E_{lgc}) to the total photovoltaic energy production on the site (E_{gen}) [4].

$$SC = E_{lgc}/E_{gen}$$

The concept brought to an energy community level consists of a local exchange between neighboring consumers of renewable electricity from one or more local producers. Collective self-consumption is therefore the application of self-consumption on a collective scale.

2.3.2 Self-sufficiency

Self-sufficiency ratio indicates how independent the user or a community of consumers will be from the grid.

It can be measured by the self-sufficiency ratio (SS), which is the ratio of self-consumed energy from production (E_{lgc}) to the overall consumption of the site (E_{load})[4].

$$SS = E_{lgc}/ E_{load}$$

The higher the SS ratio, the less energy is needed from the grid.

This exchange model therefore integrates the notion of self-consumption, which consists of the possibility of consuming locally self-produced energy. The same concept of self-sufficiency can also be brought at a community level, which allows for several actors to be brought together within the same energy community.

2.3.3 Pay-back time

In the prosumers section, the payback time is calculated and given as an output when a simulation is launched for a specific prosumer.

Th payback period refers to the amount of time it takes to recover the cost of an investment. It is the equivalent time that the PV investment requires to reach a break-even point. A shorter payback is what is usually seeked, representing a more attractive investment.

This metric is used as a way to calculate the financial return of the analyzed system. It can be calculated as the cost of the investment over the whole lifetime of the system divided by the annual cash flow[5].

In our context, for a prosumer, the payback time can be modelled and obtained by analyzing the cost of an investment for a specific system without considering a PV system installed in comparison with the analyzed system considering the consumption of the same scenario but with a PV installed system.

The payback time is the intersection of the two models representing the cost of the same system considering one with a defined PV installation and another without (as in example of simulation in Figure 7).

Such an analysis is interesting, but the main problem with payback calculation period is that it ignores the time value of money in this scope.

Therefore, the Net Present Value of the investment is also considered and presented in the following section.

2.3.4 Net Present Value

The Net Present Value (NPV) is considered in order to assess better the investment profitability considering a discount rate (opportunity cost) and also the Time Value of Money. The Time value of money allocates a value to this opportunity cost.

The Net Present Value (NPV) is the difference between the present value of cash coming in and the current value of cash going out over a period of time. It is used by looking at all the money expected to be made from the investment and translating it to today actual money. The calculation of this parameter is given by the sum of present value cash flows positive or negative for each year associated with the investment, discounted so that it is expressed in today's money[5] .
For a specific prosumer simulation, the NPV is calculated for a different range of peak power values. When the price of the energy surplus is too low, the NPV reaches a peak and then decreases so there is an optimum sizing of the PV system as a function of the energy consumption.

3 WEB INTERFACE

A free web application is available to the public in order to illustrate most of the concepts explained above. The web application allows the execution of a simulation considering a prosumer scenario and also in the context of an energy community. The website is interactive and the outputs are also dynamic and can allow the users to save their simulation scenarios and plots. Users can move from one simulation profile to another and provide their scenarios inputs.

3.1 Prosumers Simulation interface

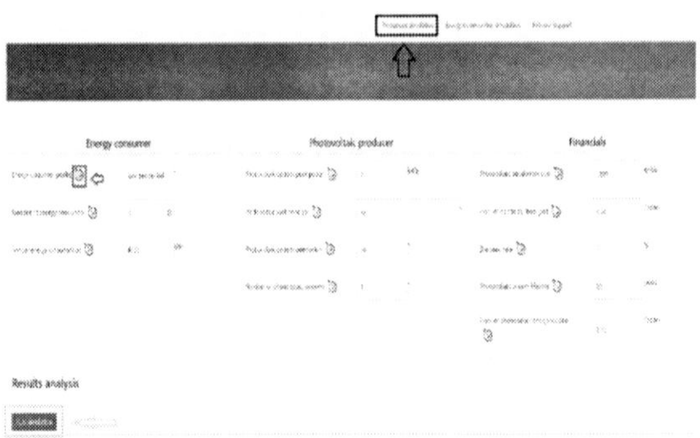

Figure 1: Web-based simple interface for prosumers self-consumption simulation.

The interface has for the different inputs provided a modal button next to the variables to be completed in the form by the users. This modal button provides information as a help element to the user when navigating the tool.
A prosumer is considered as a consumer and also a photovoltaic producer in this context. The two sections forms (energy consumer and photovoltaic producer in the "Prosumers simulation") can be therefore completed to define a prosumer type of system. There are automatic ranges of energy consumption values that are proposed as annual consumption in order to help the user get an order of magnitude of production value for such type of systems and be guided. It is actually encouraged to the user to modify the default values and test out different scenarios in order to launch a simulation. The modal buttons allow to bring some clarity when a concept is not clear and to clarify the different convention used.
Once all the inputs are included, a prosumer simulation is launched.

Figure 2 gives an example of the outputs obtained for the following input as a prosumer.

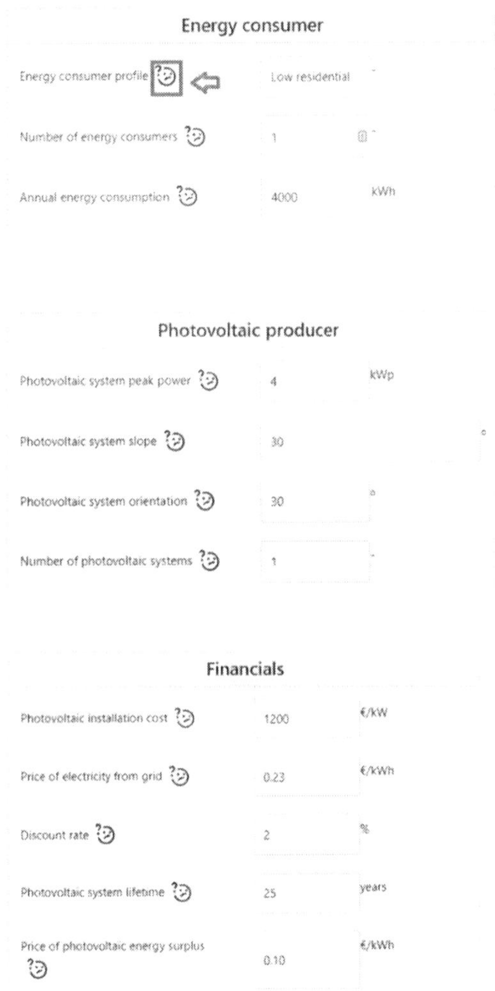

Figure 2: Prosumer analysis input example for a residential property in Brussels with a low type of consumption profile (green squared element highlighted in the first segment of the prosumers interface is an example of modal button helping the users to understand better the variables).

The outputs of the analysis provide a summary of the inputs given and other plots illustrating the analysis executed with the summary of different metrics explained in the next section where we give an example of outputs derived from the launch of simulation presented in Figure 2.

Below is an interpretation of the outputs from the simulation ran in Figure 2.

The different plots are interactive and the user can select the lines and elements they want to focus on or anlayze, hover over the curves and download the different plots.

In a first phase, the consumption profile of the system, the quantity of PV produced, the total electricity consummed , the quantity of power coming from the grid are illustrated and from these elements the PV power surplus is derived.

The illustration is provided on the highest production day and as well as on the day with the lowest PV production. For each of those two days, Consolectiv provides the profile of self consumption ratio and self sufficiency ratio.

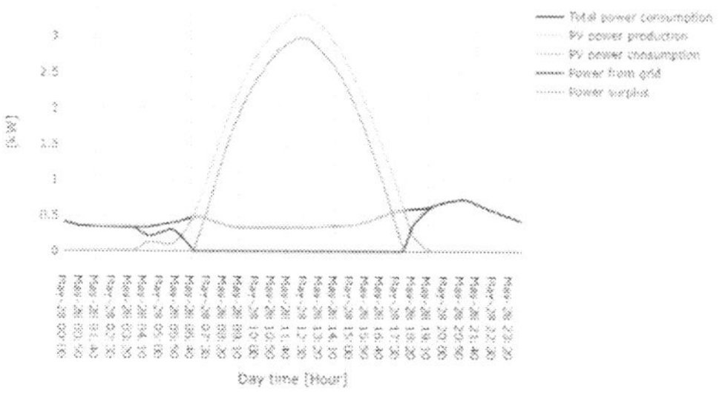

Figure 3: Electricity flow of the day with the highest PV production during the year

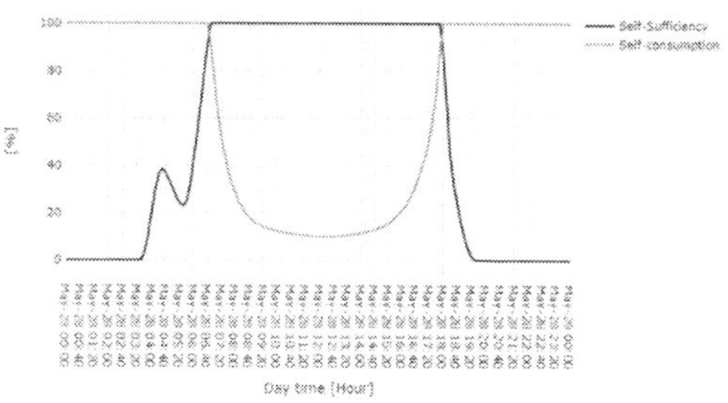

Figure 4: Time series view on self-consumption and self-sufficiency on the day with the highest PV production.

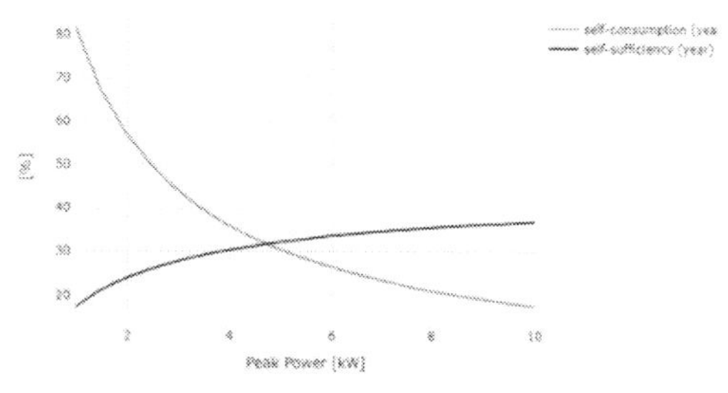

Figure 5: Yearly-integrated self-sufficiency and self-consumption vs peak power of PV installation.

Figure 5 highlights how the self-consumption and self-sufficiency ratio would evolve as a function of the PV system size. Increasing peak power with a defined consumption profile of prosumers decreases self-consumption ratio and since the peak power increases, the generated PV power of the system also tends to increase. With more increase of the peak power, the self-sufficiency ratio also does. A perfect annual balance between onsite generation and demand is the case where the self-consumption is equal to the self-sufficiency, as is shown in the intersection point in Figure 5.

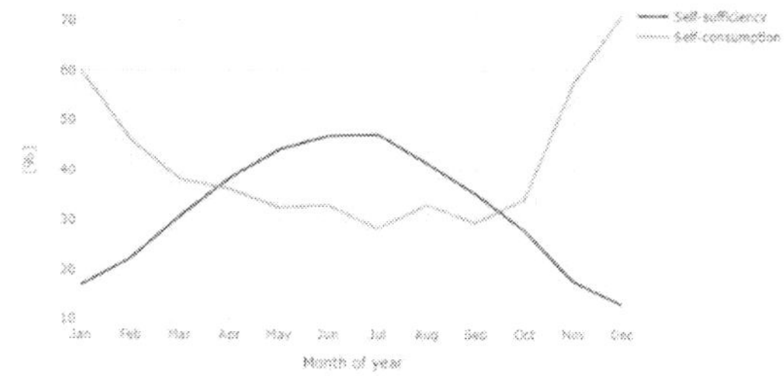

Figure 6: Monthly-integrated self-consumption and self-sufficiency for a residential PV installation in Belgium.

Figure 6 shows the monthly evolution of self-consumption and self-sufficiency ratio. A higher production in summer in Brussels for such a system translates into a self-sufficiency rate that tends to be higher during summer period and lower during winter periods. We can see the self consumption having an opposite tendency in winter when all the PV power produced for this scenario is almost consumed.

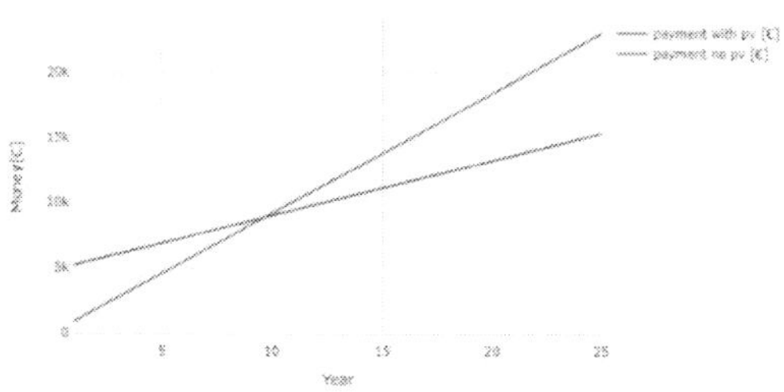

Figure 7: Illustration analysis of payback time during PV system lifetime

In the case of the prosumer simulation defined in Figure 2, and as explained in 2.3.3 , the payback time is estimated to be 10 years represented by the financial modelling for the same system with a PV system installed and another financial modelling without as seen in Figure 7.

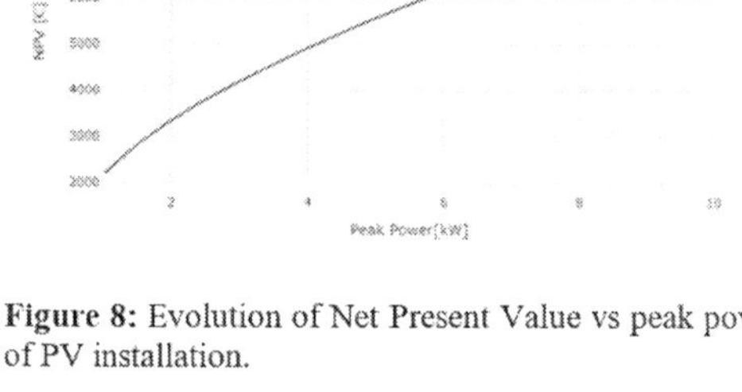

Figure 8: Evolution of Net Present Value vs peak power of PV installation.

Figure 8 shows the evolution for different ranges of

peak power values (even if the specific PV nominal power is given during the simulation) for such type of prosumer compared to the Net Present Value. The optimum peak power for such type of prosumer is considered where the NPV value is at its maximum value.

On a yearly basis, the output dashboard also provides a summary of the yearly self-consumption and self-sufficiency ratios and all the different financial metrics at the prosumer level.

Table 1: Final summary of simulation scenario presented in Figure 2.

self-consumption (year)	35.88%
self-sufficiency (year)	30.44%
Payback time	10 years
Net present value	4,915 €
Optimum peak power	10.0 kW

3.2 Energy communities simulation interface

The other interface allowing to execute a simulation mixing a community of consumers with a community of photovoltaic energy producers also provides some information on the total consumption profile and PV power produced for the community. For this section, only the self-consumption ratio and self-sufficiency ratio are provided.

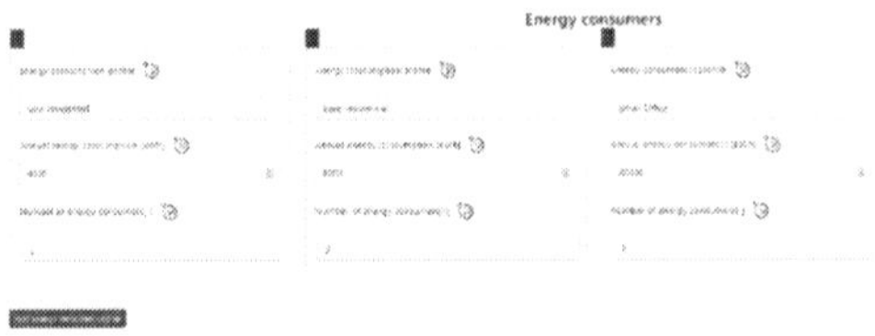

Figure 9: Example of a community of consumers entered as input in 'energy communities' simulation section of Consolectiv.

Users may add different profiles of energy consumers. In the example illustrated in Figure 9, a unit complex of two low residential consumer profiles combined with another 3 units of base residential consumers and a rental office connected to the consumer unit are created as a community of consumers.

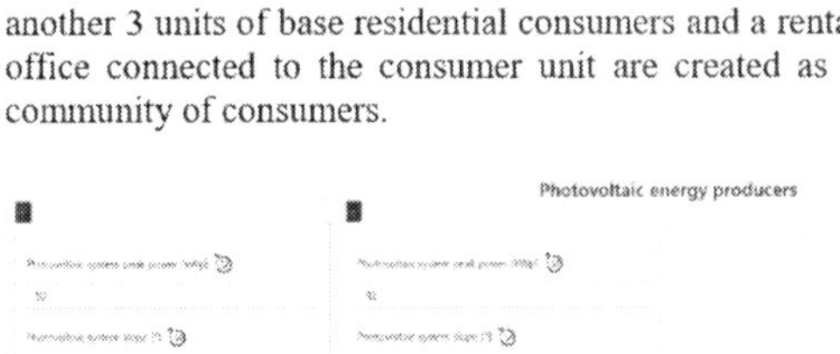

Figure 10: Selected inputs given as example for PV producers forming a community with consumers defined in Figure 9.

Below the consumers section on the interface, prior to launching a simulation, the user is invited to allocate to the community of consumers modelled in the first segment of the simulation a community of producers. An assessment of the overall scenario as a community is then exectued to provide the overall self consumption profile and self-sufficiency profile of the community.

As for the simulation for prosumers, an assesment of the highest production day profile of the overall community unit is carried out and the total consumption and PV power produced by the producers entity as well as an aggregation of the PV power coming from the grid is also provided. The same illustrations but here aggregated to the community level are given as in the prosumers section.

The self-consumption ratio and self-sufficiency ratio are derived on a 10-min basis for the whole year and also is illustrated on the user dashboard the profile for the highest production day of the year as well as the monthly aggregated form of self-consumption ratio and self-sufficiency ratio.

4 DISCUSSION

The first version of the tool has been released on the 7th of September 2021.

In future releases, some improvements to the user interface will be considered (in order to improve user experience on the application) as well as more documentation in order to help user navigate through it and also more automated suggestions will be added to it mainly for the energy community section.

A public changelog will therefore be considered in order to update the public on the changes that will be continuously done.

The next future releases will allow the possibility to consider in both types of simulation (prosumers and energy community) the possibility to integrate energy storage systems.

Also, through potential future collaborations, there may be the possibility to work with more accurate consumption profiles from Brussels if data is available, and as well replicate this tool for other regions of Belgium or in the world.

5 CONCLUSION

Consolectiv, as a tool, provides simple simulations aimed at the general public. There is an increased interest among end-users (real estate promoters, cities, regulators, etc.) and the scientific community.

Despite this interest, more research is still needed and further studies need to be done to compare the potential of the mix of different technologies. This available assessment tool may help fulfill that need to bring the concept to a broader audience.

REFERENCES

[1] *CONSOLECTIV: A free online tool for the simulation of individual or collective self-consumption in Brussels.* Lucisun, 2021. [Online]. Available: https://consolectiv.brussels/

[2] A. Pena-Bello, M. Burer, M. K. Patel, and D. Parra, "Optimizing PV and grid charging in combined applications to improve the profitability of residential batteries," *Journal of Energy Storage*, vol.

13, pp. 58–72, Oct. 2017, doi: 10.1016/j.est.2017.06.002.

[3] S. Wilcox and W. Marion, "Users Manual for TMY3 Data Sets," *Technical Report*, p. 58, 2008.

[4] R. Luthander, J. Widén, D. Nilsson, and J. Palm, "Photovoltaic self-consumption in buildings: A review," *Applied Energy*, vol. 142, pp. 80–94, Mar. 2015, doi: 10.1016/j.apenergy.2014.12.028.

[5] R. A. Brealey, S. C. Myers, and F. Allen, *Principles of corporate finance*, Thirteenth edition. New York, NY: McGraw-Hill Education, 2020.

OPTIMIZED MACHINE LEARNING METHOD FOR PV POWER PREDICTION

Horst Heck, Urs Muntwyler, Eva Schuepbach
Bern University of Applied Sciences, BFH
Jlcoweg 1, 3400 Burgdorf, Switzerland
horst.heck@bfh.ch, urs.muntwyler@bfh.ch, drs_consulting@bluewin.ch

ABSTRACT: Prediction of PV power is useful to estimate and plan power production, net stability, and own consumption. Input data for the predictions are physical parameters like solar irradiation (horizontal or inclined), temperature (of air and PV module), etc. To identify such input parameters, several methods have been proposed in the open literature. Physical models, statistical models, or a machine learning approach can be used to predict PV power. Here, we developed our own machine learning (ML) algorithm and trained it with AC-power data from our own PV monitoring network in Switzerland. Results are presented on how to optimize our algorithm in view of obtaining a precise prediction for PV power production. Such information is important for owners of PV plants to steer their own production/consumption. Especially own consumption of solar electricity in winter needs to be maximised, as PV will be enforced to successfully implement the Swiss Energy Strategy 2050.
Keywords: PV Energy Yield, Prediction, Machine Learning, Weather Data, Optimisation

1 INTRODUCTION

The Swiss Energy Strategy 2050 aims to phase-out the five nuclear power plants in Switzerland by 2050. The associated electricity gap needs to be remedied. In view of implementing the Swiss Energy Strategy 2050, production and consumption of electricity with photovoltaics (PV) is a necessity for Switzerland, especially in winter [1], [2]. At least 12 TWh (of the missing 20 TWh) will have to stem from solar energy; this amount will likely increase due to the expected lack of geothermal energy and wind energy. To achieve this production, it is suggested [2] to:

(i) generally install more PV plants in Switzerland and
(ii) introduce a seasonal splitting (summer/winter) for the PV electricity price.

According to [2], this economic promotion measure will allow the PV installers in Switzerland to foster their installations towards a higher own production in wintertime. Our study hence aims at developing a planning tool for optimising the own consumption of PV owners (especially in winter) with the help of PV energy yield predictions based on machine learning algorithms and weather data.

Predicting PV energy yield using physical models, statistical models, or a machine learning algorithm (MLA) in combination with weather data has been described in the published open literature before (see review papers [3] and [4], as an example). We developed a new ML algorithm that is fed with long term AC-power records and numerical weather predictions to predict the power output for the following day. Using this algorithm, a study was carried out [5] that illustrates the influence of different input data (like, e.g., the size of AC-power records, types of weather parameters, etc.). The results of this study are expected to help PV plant owners to choose the most suitable data sets for their prediction.

To specifically define an optimal training schedule through the year, we varied the structure of the training and test data sets to fit our ML model and simulated real-world scenarios. In a real-world scenario, the model is trained with past data and performs on new data, i.e., the training data and the test set reflect different time intervals. Due to seasonality effects, it seems to be best to train the model on data from one year and test it on data of the following year.

When we developed the new ML algorithm [5], we were, however, not in the position yet to perform such tests, as data of only 6 months were available then. Here, we present new results of our ML algorithm using data of more than 12 months. We also illustrate how our algorithm adapts to various altitudes as the seasonality at high-elevation (alpine) PV plants is different from urban PV plants located at low-elevation. These insights will enable PV plant owners to set up a training schedule detailing when and how the model should be retrained to adapt to changing weather throughout the year.

2 STRUCTURE OF THE DATA SETS

2.1 Long-term AC-power yield records

The Laboratory for Photovoltaic Systems (PV LAB) at Bern University of Applied Sciences BFH in Burgdorf, Switzerland, has monitored a network of about 30 PV plants (including local meteorological parameters) in Switzerland since the 1990s (see www.pvtest.ch). The data are quality-controlled and stored in a database at the PV LAB. Four years (6/2016-6/2020) of hourly averaged AC-power output data of six PV plants (Table I) were used to develop our own algorithm for the prediction of PV power for the next day [5]. The first three PV plants in Table I are located in urban areas close to Bern, while Jungfraujoch and Birg are alpine PV plants (Canton Bern), and Mont-Soleil represents the Jura Mountains.

Table I: Swiss PV plants included in this study

Name	kW peak	Installation	Type/ Height
Tiergarten	24.41	1994	urban (530 masl)
Schlossmatt	3.18	1995	urban (540 masl)
Worblenpark	36.4	2011	urban (493 masl)
Mont-Soleil	554.59	1992	alpine (1270 masl)
Jungfraujoch	1.15	1993	alpine (3454 masl)
Birg	2.23	1992	alpine (2677 masl)

2.2 Numerical weather predictions

To assess the influence of different numerical weather prediction models, we recorded numerical weather predictions (NWP) for global horizontal irradiation (GHI) and ambient temperature (2m above ground) from two different providers of weather predictions. Recordings from 1/2020 until 6/2021 were from MeteoSuisse (www.meteosuisse.admin.ch) and from 1/2020 until 1/2021 from Meteoblue (www.meteoblue.com).

MeteoSuisse runs 8 calculations a day for a 33-hour forecast (COSMO-1 model until 8/2020 and COSMO-1E model afterwards). Meteoblue runs prediction calculations for a 168-hour forecast two times a day.

Besides GHI and the temperature from Meteoblue, we additionally recorded predictions of the following weather parameters: relative humidity, sea level pressure, wind speed, and wind direction. The predictions were calculated using the longitude, latitude and masl (meter above sea level) of the PV plants listed in Table I.

3 METHOD: MACHINE LEARNING ALGORITHM

The machine learning (ML) algorithm we developed models the performance of a PV plant in two steps. The methods are described in detail in [5] and briefly outlined below: Step 1: Long term AC-power data is used to model the general behaviour of the PV plant; seasonality effects are averaged, and the ageing of the PV modules is modeled. Step 2: Local numerical weather predictions are added to adapt the results of Step 1 to the actual weather situation.

3.1 Step 1: Gaussian Process

The general performance of a PV plant is modelled by a so-called Gaussian Process. This method is based on the construction of a stochastic process $X(t)$, $t \in \mathbb{R}$, where every finite collection of time-points, $t_1, \ldots, t_n$ yields an n-dimensional multivariate Gaussian random variable. The process is fully described by its mean and covariance function (kernel). In machine learning, a Gaussian Process can be used to model time series. A regression on future time values yields a forecasting of the time series.

To reduce possible errors due to different daylight times, a separate Gaussian Process is trained for every hour. Furthermore, the data is disaggregated into "high performance" and "low performance" clusters. For each of these clusters, separate Gaussian Processes are performed per hour. As kernels, we use a combination of linear and periodic kernels (i.e., exponentiated sine squared kernels). Such kernels are able to model the degradation and seasonality of the data, which we both like to include into our model.

3.2 Step 2: Support Vector Regression or Random Forest

Includes weather data. The output of Step 1 (Gaussian Process) and the local numerical weather data (GHI, temperature, and even more parameters) are merged to train a Support Vector Regression (SVR) and a Random Forest (RF) algorithm. Among the machine learning algorithms we tested (SVR, RF, Gradient Boost, Artificial Neural Network, and LSTM), SVR and RF algorithms performed best on our data.

3.3 Data handling

The training data was preprocessed as follows: The AC-power data measured at the PV plants was scaled by the maximum power output measured over the four-year period used to train the Gaussian Process. The numerical weather prediction of the GHI was normalised by the so-called clear sky irradiation. The clear sky irradiation models the maximum possible irradiation for a certain date and location. We used this value since the Gaussian Process already model seasonality effects, and a GHI forecast with seasonality would disturb the modelling in Step 2.

The formula developed to normalize the GHI with the clear sky irradiance (ClearSky) is:

$$Min(\frac{GHI}{ClearSky + 1}, 1)$$

This formula avoids high values when the ClearSky is very small and restricts the maximum value to 1, which is good for the performance of the ML algorithm.

The temperature and other parameters were normalized such that the mean is 0 and the empirical standard deviation is 1 when calculated for the training set.

3.4 Hyperparameter Tuning

In order to find the best parameters for the algorithms in Step 2, a hyperparameter tuning was performed during the training and for each PV plant separately. To avoid overfitting effects of the hyperparameter tuning, a 5-fold cross validation was applied. The performance is measured (here and throughout the paper) using the mean squared error MSE.

4 TESTS WITHOUT GAUSSIAN PROCESS

In addition to the machine learning (ML) approach described in Section 3 (details see [5]), we performed tests in which Step 1 (Gaussian Process) was skipped and only SVR or RF were used to model the performance of a PV plant. In this case, GHI was normalized by scaling with the maximum value of the training data set. For these tests, we needed to keep the seasonality for the GHI value since the

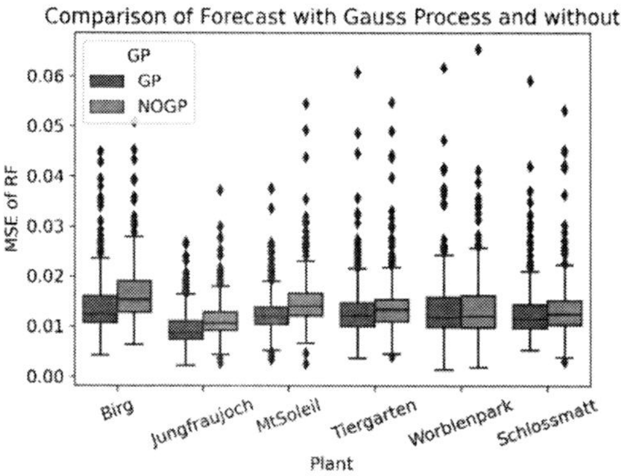

Figure 1: Gauss Process helps for modelling PV power performance1

Gaussian Process do not model these effects. Figure 1 displays the results of the two approaches for each PV plant. The performance for RF was calculated using test and training data randomly from the same time period (1/2020-12/2020). Trainings for all possible data sizes (from 1 to12 months) were performed.

Figure 1 demonstrates that the performance of our algorithm is slightly better than the approaches without Gaussian Process, and the advantage of the Gauss Process is bigger for alpine PV plants. In real-world scenarios (Figure 7 and Figure 8), including the Gaussian Process as a Step 1 seems even more beneficial.

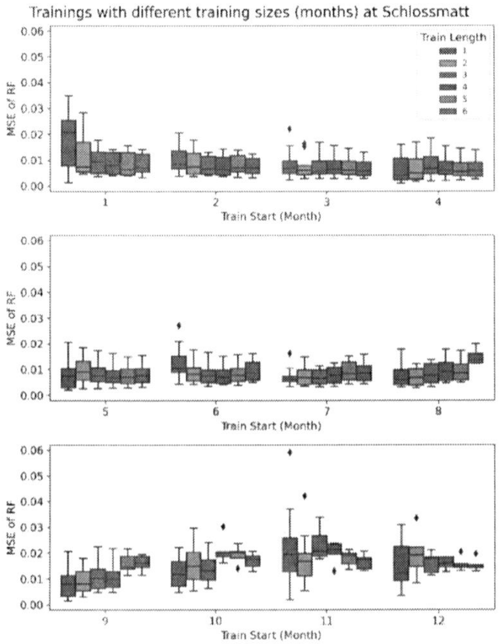

Figure 2: Tests on the training length; performance of algorithm when 1-6 months of data is used

We therefore recommend using the two-step algorithm we developed when calculating real-life PV predictions (and not approaches without Gauss Process). For the tests in Figures 7-8, we used three consecutive months from 2020 to train our algorithm. The model was tested on data from the month in the "middle", but from 2021 (e.g., train on 2-4/2020 and test on 3/2021).

5 INFLUENCE OF DIFFERENT TRAINING SETS

5.1 Tests on length of training data set

In this Section, we describe tests using different sizes of training data for fitting the model; the respective performance is illustrated in Figure 2. The x-axis in Figure 2 is the start month of the training data interval, and the different colours correspond to different training sizes (months from 1-6). It is evident that the variability is reduced when more months are included. In the first half of the year, it seems to be advantageous to use more than 3 months as a training size. In the second half of the year, 3 months of training size still seem to be a good choice, but the performance gets worse when too many months are included. This may be due to more difficult weather conditions in autumn, which yield a less precise weather prediction.

5.2 Performance of different PV plants

Figure 3 shows an exceptionally good performance of the algorithm when trained with 3 months of data from MeteoSuisse on the PV plant at Jungfraujoch. Note that, in general, the data from Meteoblue perform worse on alpine plants (see Figure 3). The underlying reason could be that MeteoSuisse operates a weather station at Jungfraujoch, and the local weather data measured by MeteoSuisse at Jungfraujoch may lead to a more precise weather forecast at this alpine site. Figure 3 also indicates that the weather forecast in spring and autumn seems more difficult than in winter and summer, when likely more stable weather conditions prevail.

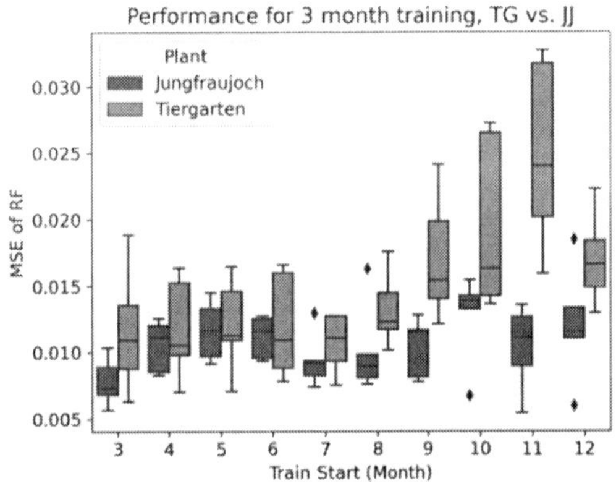

Figure 3: Comparison of performance for Tiergarten and Jungfraujoch

5.3 Comparison of two numerical weather predictions

We trained and tested our algorithm with data, and compared the results, from the two providers of numerical weather predictions (Meteoblue and MeteoSuisse). Figure 4 illustrates the comparisons and shows the performance of urban and alpine PV plants. For urban PV plants, the weather forecast from Meteoblue seems to be better for modelling the PV plant yield, while the weather forecast data from MeteoSuisse lead to better prediction results on alpine plants, especially in the winter months.

6 PERFORMANCE OF REAL-WORLD SCENARIOS

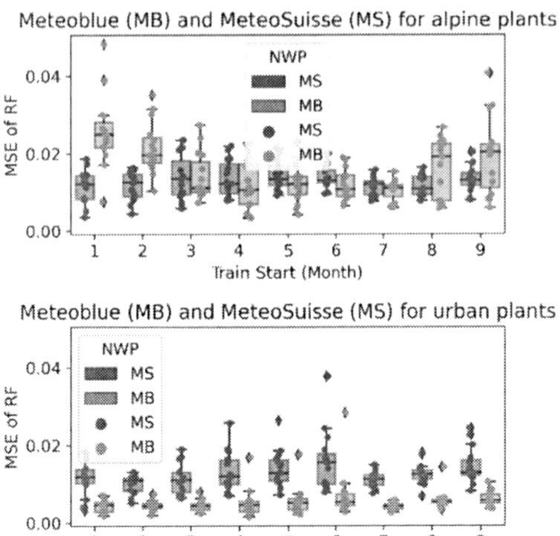

Figure 4: Performance of urban and alpine PV plants with two numerical weather predictions (see text)

Several set-ups were examined to identify optimal specifications that yield a good tradeoff for training-data-size and performance. Comparisons for two different situations were enabled due to the availability of data for 2020 (from Meteoblue) and for 2020 and 2021 (from MeteoSuisse).

One of those situations uses training data for several months and tests the performance of the algorithm on data following the training data. This simulates the situation when the algorithm is trained on past data and then yields predictions for the following weeks.

6.1 Performance for different weather prediction providers

Figure 5 presents a simulation using the two numerical weather predictions at hand with data from 2020. The

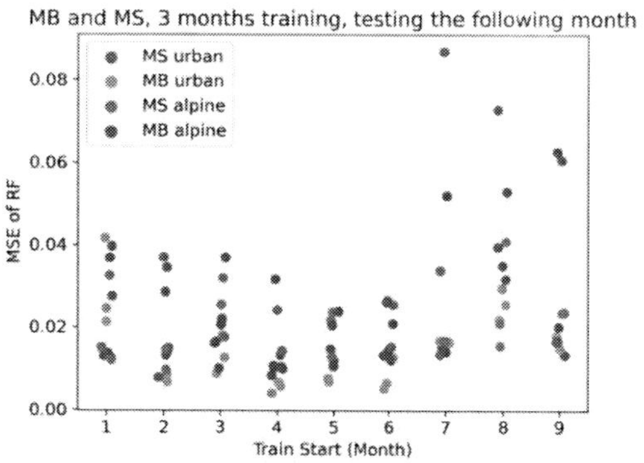

Figure 5: Direct comparison of numerical weather predictions on urban and alpine plants in a real-world scenario

graph distinguishes the numerical weather prediction (Meteoblue, MB, and MeteoSuisse, MS) as well as the type of the PV plant (urban or alpine). It appears as though the use of the predictions from Meteoblue yields a higher MSE on alpine plants than the predictions for urban plants. When using the predictions from MeteoSuisse, the difference in the performance of alpine and urban PV plants is not large. When comparing the MB and MS predictions, it seems that the algorithm performs better for urban PV plants when using Meteoblue. For alpine PV plants, the MS predictions are a better choice (see also Figure 4). Also, the advantage of the Meteoblue data is more pronounced for predictions in spring and summer.

For a real-life implementation, we recommend using several providers of numerical weather predictions as an input for our algorithm. At least, an evaluation should be performed on predictions of possible providers prior to selecting one for the actual location.

6.2 Trainings with data from last year

We calculated yet another real-world scenario. We trained the algorithm with data from 2020 and tested it on data from 2021 (the same month). MeteoSuisse predictions were used for these calculations, as data were available for 2020 and 2021. We compared three different strategies, being:

(i; blue box): Take 2 months of data from 2020 for training and tests on the same months in 2021;

(ii; orange box): Take 3 months of data from 2020 for training and test on the month "in the middle" from 2021 (e.g., train on data from 2-4/2020 and test on 3/2021);

(iii; green box): Take 2 months of training data from 2021 and test on the following month.

Figure 6 reveals that the first two strategies are comparable regarding performance. But it seems to be beneficial to train with data from the same months of the former year instead of using recent data from former months.

6.3 Effect of GP on training a year later

The test results discussed so far point out that it is better to use training data of the former year. It hence seemed interesting to analyze whether the Gauss Process

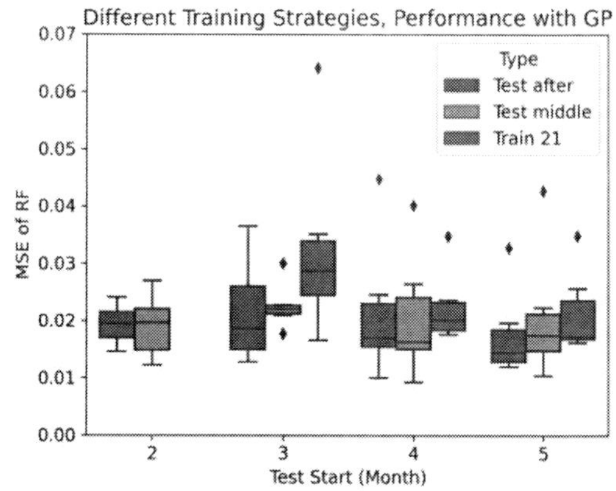

Figure 6: Comparison of different training strategies

(Step 1 in the algorithm we developed) fosters this constellation. This assumption was reasonable since the Gauss Process already models the general behavior (such as the degradation of the PV plant). Indeed, Figure 7 encourages our hypothesis and evidences that - for the strategy "train 3 months and test the middle one" (including the Gauss Process step) - a better PV power prediction is obtained than for the alternative approach (i.e., without Gauss Process).

Figure 8 highlights the performance of this training

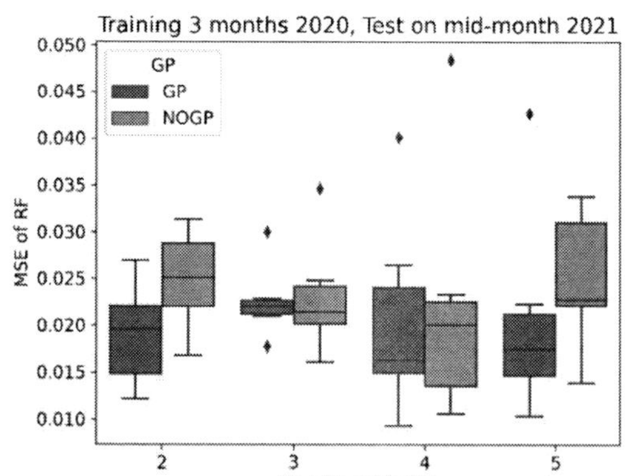

Figure 7: Performance of algorithms with and without Gauss Process in real-life situations

strategy for disaggreated PV plants (and thus allows a direct comparison with Figure 1. It is reconfirmed that alpine plants fully take advantage of the Gauss Process as compared to urban PV plants; here, the gains are even larger.

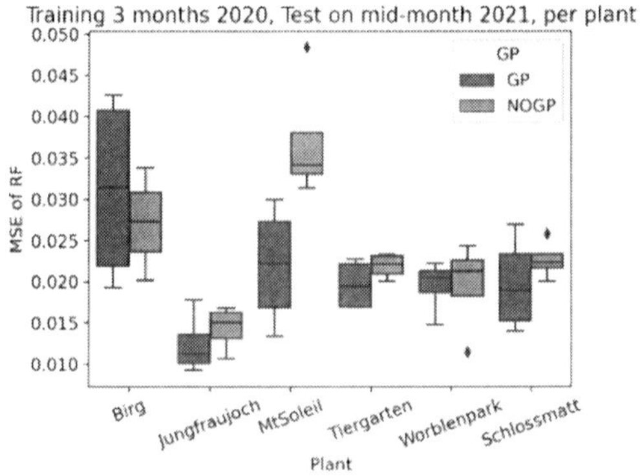

Figure 8: Performance of algorithms with and without Gauss Process in applied situation. Results per plant

CONCLUSIONS

A comprehensive analysis of the new algorithm we developed in [5] and the influence of different data structures and input parameters is presented. In this paper, we focused on different training strategies including real-world scenarios; this focus has become possible because up to 18 months of data was available in 2021. The test results evidence that it is beneficial to re-train the algorithm several times throughout the year as the re-training appears to compensate the changing seasons. The test results also make clear that the 2-step approach in our new algorithm (i.e. with Gauss Process) performs better than an alternative method using one step only (i.e. without Gauss Process). Finally, using training data from a former year instead of data from former months increases the performance of the algorithm. The findings are important for implementing our algorithm in real-world scenarios.

ACKNOWLEDGEMENTS

We are extremely thankful for the sound expertise and long-term PV measurements offered by the PV LAB staff at Bern University of Applied Sciences BFH in Burgdorf Meteoblue and MeteoSuisse are kindly acknowledged for providing the weather prediction data. The collaboration with Smart Energy Link AG Bern and Velasolaris AG Winterthur, industry partners in this project, is greatly appreciated; especially the valuable discussions and exchange of ideas with Dr. Daniel Klauser at Smart Energy Link AG (Bern). This research was part of the activities in the Swiss Centre for Competence in Energy Research on the Future Swiss Electrical Infrastructure (SCCER- FURIES), which was financially supported by the Swiss Innovation Agency (Innosuisse - SCCER program) and with Swiss industry partners. Co-financing for this project was also offered by the BFH Institute for Optimisation and Data Analysis and the BFH Institute for Energy and Mobility (IEM) Research.

REFERENCES

[1] U. Muntwyler, E. Schuepbach, Proceed. 36th European Photovoltaic Solar Energy Conference and Exhibition (EU PVSEC 2019), 9-13 September 2019, Marseille, France, 2000-2005.

[2] U. Muntwyler, F, Kuonen, E. Schüpbach, T. Schott, "Photovoltaic (PV) Winter Electricity in the Swiss Energy Strategy 2050". 36th European Photovoltaic Solar Energy Conference and Exhibition (EU PVSEC 2019), 9-13 September 2019, Marseille, France (Oral Presentation No. 5CO.13.5, unpublished).

[3] J. Antonanzas, N. Osorio, R. Escobar, R. Urraca, F.J. Martinez-de-Pison, F. Antonanzas-Torres, Solar Energy, 136 (2016) 78-111.

[4] H. Wang, Z. Lei, X. Zhang, B. Zhou, J. Peng, Energy Conversion and Management, 198 (2019) 1-16.

[5] H. Heck, A. Schmidt, E. Schuepbach, F. Kuonen, S. Bacha, U. Muntwyler, Proceed. 37th European Photovoltaic Solar Energy Conference and Exhibition (EU PVSEC2020), 9-11 September 2020 (Online), 1796–1801, doi: 10.4229/EUPVSEC20202020-6BV.5.4

ENERGY MODEL FOR A RURAL REGION IN GERMANY – RESULTS AND BALANCING OF ELECTRICITY PRODUCTION AND CONSUMPTION

Joshua Bunner, Henrik te Heesen
Trier University of Applied Sciences, Environmental Campus Birkenfeld,
Institute of Operations and Technology Management,
Campusallee, 55768 Hoppstädten-Weiersbach, Germany, h.teheesen@umwelt-campus.de

ABSTRACT
The simulation tool UCB SEnMod was developed and applied to model an energy system for a rural region in Germany (Nationalparkregion Hunsrück-Hochwald, population about 110 000), focusing on the electricity sector coupled with the transport and heat sectors. The hour-based simulations were performed for different years and possible future scenarios up to the year 2050. Various factors were considered for future energy demand, including the increasing heat pump use, the transformation to electromobility in private transport, and demographic change. The electric load for each sector, including electromobility, was modeled using standard load profiles. The model considers electricity generation via wind turbines, PV systems, and bioenergy. Also, battery storage systems are integrated into the energy model. Electricity generation by wind turbines and photovoltaic systems are simulated using long-term local weather data pro-vided by the German Weather Service (DWD). For the addition of renewable energy capacity, various future scenarios are analyzed. The baseline scenario aims to meet the electricity consumption in the region on a balance sheet basis. The energy model is based on a set of modular python scripts. For the renewable energy plants, the nominal power of wind turbines (402 MW) and PV systems (322 MWp) is revealed. A maximum storage capacity is given for the battery storage. The numbers above the arrows show the energy flows. In the analysis, both the degree of self-sufficiency and the self-consumption rate reaches about 66 %.

Keywords: Environmental Effect, Strategy, CO_2, Modeling

1 INTRODUCTION

More than 80 % of greenhouse gas emissions in Germany are energy-related and mainly attributable to the energy generation, industry, and transport sectors. To become climate neutral, more environmentally friendly technologies are therefore to be used. Furthermore, the expansion of renewable energies is intended to replace fossil fuels in the future [1].

Using the "Lappeenranta University of Technology energy system model" (LUT energy system model), Breyer et al. [2] investigated the role of photovoltaics in the global energy transition. In addition to considering a temporal resolution on an hourly basis, spatial distribution was also considered. One result of the study is that photovoltaics account for more than 40 %. Similarly, a Fraunhofer ISE study used REMod to create cost-optimized energy systems for Germany [3]. In this study, different scenarios on societal behaviors were investigated, and transformation paths were developed, considering the sector coupling of the entire energy system until the year 2050. The study results show that an energy supply based on renewable energies is safely possible for all consumption sectors despite fluctuations due to weather conditions.

Our publication shows an energy model for electricity load balancing of the national park region Hunsrück-Hochwald (see Figure 1) for the year 2050. For this purpose, the sector-coupled energy model (UCB SEnMod) developed at the Environmental Campus Birkenfeld of Trier University is used [4, 5]. The simulation modules base on the Python programming language. It allows the simulation of an energy system on an hourly basis, considering given weather data. All necessary data and suitable simulation parameters were determined for the national park region. The model's focus is on the electricity sector, coupled with the load from electromobility in private transport and the provision of electrical heat by heat pumps.

Figure 1. Overview of the Hunsrück-Hochwald National Park Region in the western part of Germany. The figure shows a map of the Hunsrück Hochwald National Park region subdivided into the associated municipalities and municipalities without associations. The background map used is from Stamen.

2 METHODS

The National Park Region power sector was modelled using Python scripts in conjunction with an SQL database. Figure 2 shows the power generators, loads, and system components that were modelled. The model is a simplified system of the UCB SEnMod tool [4, 5].

To model the electricity sector of the National Park region, wind, PV, and bioenergy electricity generators were compared to consumption. Consumption is divided into households (H0), commercial and industrial (G0), agriculture (L0), heat pumps (EW0), electric storage heating (ES0), and electric mobility. In addition, potential battery storage is considered. Finally, electricity imports and exports are calculated from the electricity load balance.

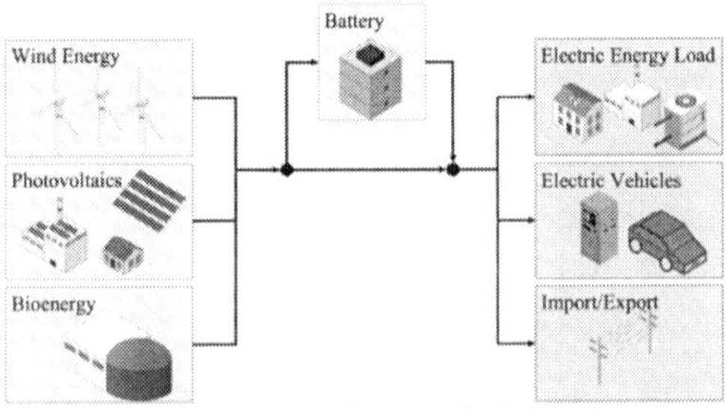

Figure 2. Structure of the modelled electricity sector based on UCB SEnMod [4, 5]. The figure shows the considered electricity variables in the modelling.

Electricity consumption was determined using data provided by Westenergie AG [6]. The data includes the electricity demand of the Rhineland-Palatinate municipalities in the national park region from 2013 to 2018 and is divided into sectors. Electricity demand from heat pumps increases as heating systems using fossil fuels are replaced. In addition, a simplifying assumption was made that electric storage heating systems will also be replaced by heat pumps starting in 2020. To model the load by heat pumps, a per capita consumption based on the assumptions of Sterchele [3] was calculated and applied to the population of the National Park region. The scenario does not assume any aggravating boundary conditions for achieving the goal of a climate-neutral energy system. The total consumption of the different sectors was distributed over the simulated year using standard load profiles. Small business, large commercial, and unique contract customers were assigned to the load profile for general commercial. The BDEW profiles were supplemented with standard load profiles for electric storage heaters (ES0) and heat pumps (EW0) from Mainzer Netze GmbH [7].

For the electricity demand of electric vehicles, several assumptions were made to model the expected annual consumption of electric vehicles in the national park region. This was done by multiplying the average energy consumption by the expected annual mileage and the assumed number of electric vehicles.

The energy consumption per 100 km of an electric vehicle depends on the vehicle, the driver's driving style, and other factors [8]. In addition, due to the novelty and small market size, the assumed consumption data for electric vehicles vary in the literature. Based on data from existing electric vehicle fleets within EU member states, consumption rates between 17.1 kWh/100 km to 21.5 kWh/100 km [9]. The results published by ADAC Ecotest show matching consumption data between 16.3 kWh/100 km and 28.1 kWh/100 km. In Ecotest, realistic consumption figures were measured, including charging losses [100]. Considering these data, a consumption of 20 kWh/100 km was assumed for the modelling. Heinz [11] modelled the electrical load due to electromobility using standard load profiles created at the Karlsruhe Institute of Technology (KIT). Due to the currently small amount of data and the deviating user behavior of electric vehicle owners. However, studies on general mobility behavior were employed. The data for the standard load profiles are based on the German Mobility Panel (MOP). The general standard load profile, according to Heinz, is used for the entire region to model the electricity load of electric mobility in the National Park region. In addition, it was initially assumed that vehicles are only charged at the residence.

In the model, the total consumption (including electromobility) of the national park region is influenced by demographic change. To estimate the demographic change, the population in 2020 was set concerning the expected population in 2040, according to the data of the Rhineland-Palatinate Statistical Office. By 2040, the population in the entire Rhineland-Palatinate part of the region will decrease by about 10 % [12]. Assuming a linear demographic change, the population in the national park region will decrease by about 5 % within ten years compared to 2020. This means that in 2030 the population will decrease to 95 % of the current level, in 2040 to 90 % and 2050 to 85 %. It was assumed that the demographic change would have the same impact on the electricity demand of all sectors. At the end of 2020, the entire National Park Region population was slightly over 132,000 residents and declines to about 112,000 by 2050.

Installed battery capacity in 2020 is just under 900 kW distributed across 172 units, according to data from the Market Master Data Registry. In the 2020 Electricity Storage Inspection [13] and Final Report [14], the average ratio of battery discharge power to capacity was 0.63 kW/kWh.

3 RESULTS

The scenario considers a future energy system in which roughly constant conditions are assumed concerning consumption and mobility behavior. The state government of Rhineland-Palatinate has already set the target of balanced coverage of the electricity demand by renewable energies for 2030. This means that as much electricity is to be fed into the grid by renewable energy systems as is consumed in total in the region. Since most of the region is in Rhineland-Palatinate, this target was considered a basis for modelling. In this case, approximately two-thirds of the electricity is to be generated by wind power and one-fourth by PV [15]. It is assumed that this target remains constant until 2050. Considering the capacities installed in 2020, this means mainly an expansion of PV capacity. Deconstruction or reduction of currently installed capacities is not assumed. Furthermore, the replacement of old plants after their operational lifetime (typically 20 years) is not explicitly considered.

To estimate the expansion of installed battery capacity, the battery capacity given in the Reference scenario of Sterchele [3] was set with the given cumulative wind and PV capacity. The resulting ratio of 0.29 GWh/GW was used to model battery capacity in the national park region as of 2050. In the scenario, a total of about 322 MWp of PV and 403 MW of wind energy are installed in 2050. Thus, the battery capacity is about 210 MWh. The expansion of battery storage from 2020 was assumed to be linear.

Figure 3 shows the electricity consumption and use in the baseline scenario for the simulated years 2020, 2030, 2040, and 2050. Despite the reduction in electricity load due to demographic change, the total annual load increases due to the expansion of heat pumps and the increase in electric vehicles. Based on the baseline data of installed capacity, the electricity load in 2020 and 2030 can already be met on balance by the RE plants that exist today. As a result, the generation in the years only changes due to different weather conditions of the reference years. Considering the goals of the state government of Rhineland-Palatinate, only the installed PV capacity was increased to cover the electricity load on balance. Thus, it reaches a share of about 26.2 % of the electricity generation in 2050. Wind energy accounts for a share of about 66.4 % and bioenergy for about 7.4 %. The shares by energy source are thus roughly in line with the target.

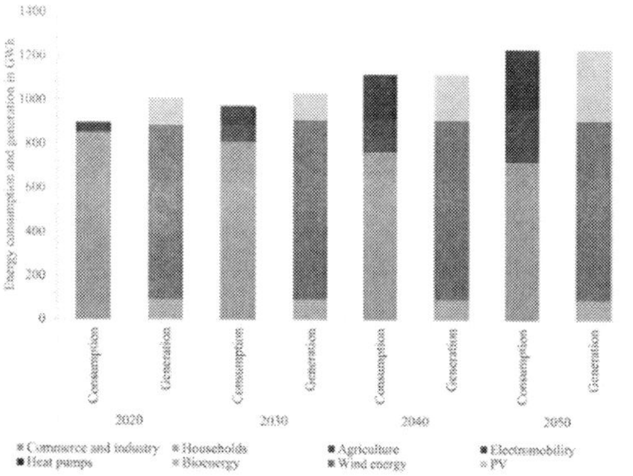

Figure 3. Electricity consumption and generation in the Reference scenario. The diagram shows the simulated annual electricity consumption and generation in the Reference scenario. Electric storage heating consumption was combined with heat pumps. Battery storage losses are not included in the graph.

In week 19, which covers 09/05/2050 to 15/05/2050, renewable energy systems have a high total electricity generation. The hourly electricity supply during this period is shown in Figure 4 top. The output of the power generators is shown here as stacked areas. From May 09, 2050, to May 10, 2050, at about 7 pm (local time), the electricity load can be fully met by renewables, mainly wind and battery storage. From about 7 pm to 9 pm, the power generation decreases, and the load is covered by the battery storage. However, from 9 pm onwards, the battery is discharged. PV power generation is zero, and wind power is at too low a level. Thus, electricity must be imported until the electricity load can be covered again by wind energy from about 12 am. Due to similar fluctuations in power generation by wind and PV, power must also be imported in the following days, especially in the evenings and at night. Particularly noticeable is the period from 05/12/2050 at 8 pm to 05/13/2050 at about 1 pm, in which almost no electricity is generated by wind power. The following days show a lower power generation by the wind power, but higher yields by the PV systems.

The power usage in week 19 is shown in Figure 4 bottom. The individual loads are shown as stacked areas. In addition, a residual load line is shown. This modelling defines this as the total electricity load minus the electricity generation from wind energy and PV. Due to the high yields of wind energy in week 19, significantly more electricity can be exported than needs to be imported. In total, about 23.4 GWh of electricity can be exported this week, while only about 1.4 GWh must be imported.

By forming the geometric mean of the result matrices for the self-consumption rate and the degree of self-sufficiency, an optimum can be depicted concerning both parameters for the installation of wind and PV. Both values (degree of self-sufficiency and self-consumption rate) deviate from the calculated mean value by the same factor. Thus, minimal values of a parameter lead to a relatively small mean value. Figure 5 shows the combination of parameters by calculating the geometric mean. The optimal PV and wind energy output ratio is below the output required to cover the load in balance. Considering this target, the capacities assumed in the 2050 scenario are already in a relatively optimal range.

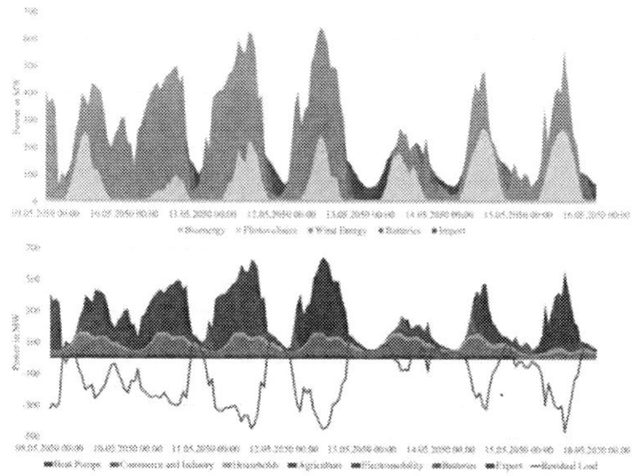

Figure 4. Electricity supply and electricity use in week 19 for the 2050 baseline scenario. The figure shows the hourly electricity supply and hourly load for week 19 in the 2050 baseline scenario. In addition, a residual load line is shown, which is defined as electricity demand minus generation from wind energy and PV. Finally, the local time is indicated.

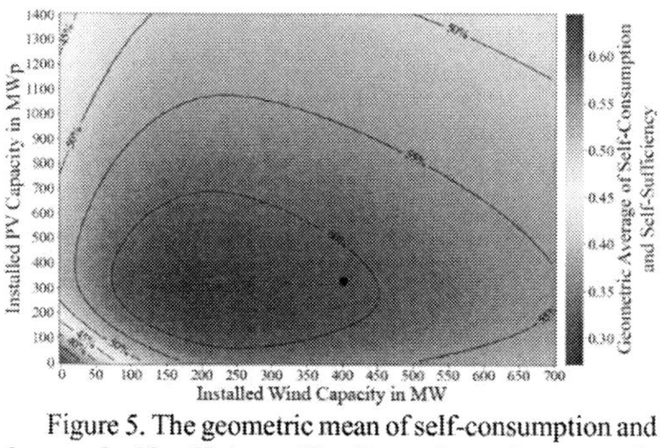

Figure 5. The geometric mean of self-consumption and degree of self-sufficiency. The figure shows the geometric mean of the result matrices for the self-consumption ratio and the degree of self-sufficiency. Thus, the optimal ratio of wind and PV power is shown considering both parameters. The highlighted point shows the installed capacities of the 2050 scenario.

4 CONCLUSION

With the UCB SEnMod energy model [4, 5], the energy system of the National Park region has been modelled. For this purpose, the currently installed renewable energy system capacities and electricity loads of different sectors were determined. Furthermore, the procedure for modelling the hourly electricity load and electricity supply and the modelling of battery storage was explained.

When looking at the results, it became apparent that the region already has sufficient renewable energy system capacity as of 2020 to meet the state government's target of covering the load in balance by 2030. However, to continue to achieve this by 2050, capacities must be further expanded, as the expansion of heat pumps and the increase in electromobility continuously increase the electricity load. In addition, the sample weeks show large, weather-related fluctuations in electricity generation that can only be partially compensated for by the battery storage systems.

The scenario was examined concerning the optimization potential for self-consumption and degree of self-sufficiency. This showed that an expansion of photovoltaics is particularly recommended to ensure the highest possible self-consumption share and degree of self-sufficiency. These can also be increased by battery storage for load balancing. In contrast, a possible load shifting through the expansion of the charging infrastructure for electromobility with assumed unregulated charging offers minimal potential for improvement.

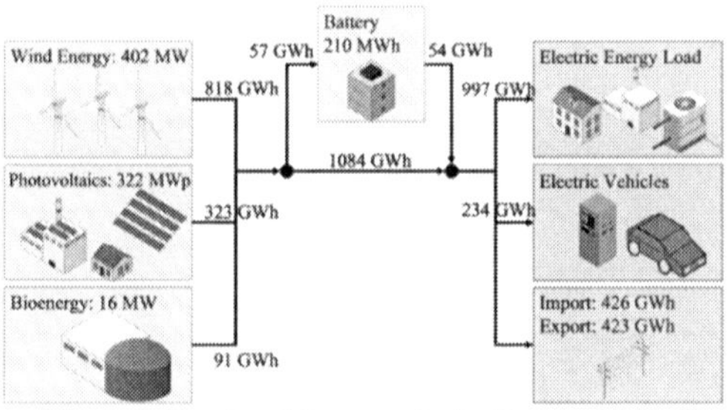

Figure 6. Results of the scenario for the year 2050. The figure shows the energy model for the scenario in 2050. For the RE plants, the nominal capacity is given. For the battery storage, the maximum storage capacity is given. The arrows represent the energy flows with the corresponding energy quantities.

An analysis of the previous expansion of renewable energy in the National Park region already shows high expansion rates. Considering the goals of the Rhineland-Palatinate state government and the results for optimizing the self-consumption rate and degree of self-sufficiency, an expansion of photovoltaic capacity of about 6.63 MW per year from 2020 is recommended. For wind energy, it is sufficient to maintain the current capacity by replacing the old plants after their operating time has expired. A smaller plant capacity could also be sufficient through repowering, provided that the same electricity production can be achieved. There is still great potential for expansion in battery storage. Although these are not necessary to cover the load, they can increase self-consumption and the degree of self-sufficiency. For these, an expansion of capacity to 210 MWh was assumed. The results for the installed capacities and electricity flows of the corresponding baseline scenario for the year 2050 are summarized in Figure 6.

REFERENCES

[1] BMU. (2020). Der Klimaschutzplan 2050. www.bmu.de. https://www.bmu.de/themen/klima-energie/klimaschutz/nationale-klimapolitik/klimaschutzplan-2050/

[2] Breyer, C., Bogdanov, D., Gulagi, A., Aghahosseini, A., Barbosa, L. S. N. S., Koskinen, O., et al. (2017). On the role of solar photovoltaics in global energy transition scenarios. *Progress in Photovoltaics: Research and Applications*, *25*(8), 727–745. https://doi.org/10.1002/pip.2885

[3] Sterchele, P., Brandes, J., Heilig, J., Wrede, D., Kost, C., Schlegl, T., et al. (2020). *Wege zu einem klimaneutralen Energiesystem: Die deutsche Energiewende im Kontext gesellschaftlicher Verhaltensweisen*. Freiburg.

[4] Leonhard, M.-C., te Heesen, H. (2020). Energy Model for a Rural Region in Germany - Methodology. *37th European Photovoltaic Solar Energy Conference and Exhibition*, 1850–1854. https://doi.org/10.4229/EUPVSEC20202020-6BV.5.22

[5] Blinn, A., & te Heesen, H. (2021). Sector Coupled Energy Model for the European Electricity, Heat and Transport Sectors – Methodology. 38th European Photovoltaic Solar Energy Conference and Exhibition.

[6] Westenergie AG. (2020). e-kommune. https://www.e-kommune.de

[7] Mainzer Netze. (2020). Synthetische Lastprofile, Mainzer Netze GmbH. https://www.mainzer-netze.de/nc/stromnetze/netzzugang/lastprofile/

[8] Helmers, E., & Weiss, M. (2017). Advances and critical aspects in the life-cycle assessment of battery electric cars. Energy and Emission Control Technologies, Volume 5, 1–18. https://doi.org/10.2147/eect.s60408

[9] Moro, A., & Lonza, L. (2018). Electricity carbon intensity in European Member States: Impacts on GHG emissions of electric vehicles. Transportation Research Part D: Transport and Environment, 64(July 2017), 5–14. https://doi.org/10.1016/j.trd.2017.07.012

[10] ADAC. (2020). Stromverbrauch Elektroautos: Aktuelle Modelle im ADAC Test. https://www.adac.de/rund-ums-fahrzeug/tests/elektromobilitaet/stromverbrauch-elektroautos-adac-test

[11] Heinz, D. (2018). Erstellung und Auswertung repräsentativer Mobilitäts- und Ladeprofile für Elektrofahrzeuge in Deutschland. Working Paper Series in Production and Energy 30. https://doi.org/10.5445/IR/100008637

[12] State Statistical Office Rhineland-Palatinate. (2020). Statistisches Landesamt Rheinland-Pfalz. https://www.statistik.rlp.de/de/startseite/

[13] Weniger, J., Maier, S., Orth, N., & Quaschning, V. (2020). Stromspeicher-Inspektion 2020. https://pvspeicher.htw-berlin.de/wp-content/uploads/Stromspeicher-Inspektion-2020.pdf

[14] Weniger, J., Orth, N., Maier, S., Böhme, N., & Quaschning, V. (2020). Bewertung und Optimierung der Energieeffizienz von Photovoltaik-Batteriesystemen (EffiBat). Berlin. https://pvspeicher.htw-berlin.de/wp-content/uploads/HTW-2020-EffiBat-Abschlussbericht.pdf

[15] Rhineland-Palatinate. (2020). Landesregierung Rheinland-Pfalz Energiewende. https://www.rlp.de/de/landesregierung/schwerpunkte/energiewende/

HYBRID PV-SYSTEMS FOR ELECTRIFICATION AND SECTOR COUPLING OF ROAD TRANSPORT SECTOR IN NORWAY

Jonathan Fagerström, Lisa Kvalbein, Janis Danebergs, Tine Uberg Nærland, Mari Øgaard, Kari Espegren
Institute for Energy Technology
Instituttveien 18, NO-2007 Kjeller, Norway

ABSTRACT: Electrification of the road transport sector will likely include both battery-electric (BEV) and hydrogen fuel cell electric vehicles (FCEV). Sector coupling of energy carriers has been described as a route forward for efficient integration of renewable energy. The objective of this work was to determine the net present cost improvements with co-localization of charging and filling stations, i.e. sector coupling of electricity and hydrogen energy carriers, and how this would impact optimal sizing of the PV production and battery storage. A grid-connected co-localized charging/filling station situated north of Oslo, Norway, was modeled in Homer Pro and Homer Grid. PV energy generation profiles were modeled using PVsyst and a snow loss model. Demand data for BEV and FCEV was synthesized based on historical traffic data (year 2015-2019). The results show that the electrification of the road transport sector passing the point of selection requires about 53 MW peak electricity demand. Co-localization, i.e., sector coupling of BEV and FCEV, indicated a marginal cost-efficiency improvement of 0.4-1.1% and increased optimal PV capacity and reduced PV-power export.
Keywords: System Performance, Charging station, PV System, Battery Storage and Control, Hydrogen

1 INTRODUCTION

Electrification of passenger cars is well established, and electrification of heavy-duty trucks is currently also gaining momentum as both battery-electric, and fuel cell models are being realized. The drive towards electrification enables the addition of locally produced renewable electricity, and it has been shown that grid-connected distributed energy systems are the preferred choice for charging stations [1]. To our knowledge, most papers focusing on charging stations with PV-battery systems have analyzed smaller stations with PV-capacities below MW-scale. However, one larger utility-scale system for fast-charging in France was analyzed, where both battery electric cars and trucks were included [2]. They showed that integration of PV successfully could increase demand coverage along major road corridors outside Paris, and the payback time for a 15-year project lifetime was found to be 7.4 years.

Sector coupling, i.e., integration of different energy carriers, is a concept used as one of the solutions for matching electricity demand with intermittent renewable electricity production [3]. Therefore, for efficient electrification of the transport sector, co-localizing hydrogen refueling stations with electric vehicle (EV) charging stations might be a means to incorporate more intermittent renewable energy to the electrification of the transport sector. Several published papers describe the optimization of both battery electric vehicles (BEV) and hydrogen fuel cell electric vehicles (FCEV), [4,5] but few of them explain the rationale for co-localization. Recently though, it was shown that co-localization of BEV and FCEV as a hybrid charging station reduced the net present cost by 2.1% [6]. The authors considered the demand for FCEV buses and BEVs and optimized local wind electricity generation in both on-grid and off-grid systems. However, their model did not cover specificities around "on-the-road chargers".

The objective of this work was, therefore, to answer the question if charging stations for BEV and filling stations for FCEV should be co-localized, and moreover, how this co-localization would impact optimal sizing of on-site PV production and battery storage. This was conducted by comparing two alternatives: 1) separate FCEV and BEV stations, and 2) co-localization of BEV and FCEV stations.

2 METHODS

2.1 System description

Three systems were included in the analyses, a BEV charging station for both battery-powered passenger cars and trucks, a single hydrogen (H_2) filling station for hydrogen fuel cell vehicles, and a co-localized charging and filling station where both battery and fuel cell vehicles charge and fill at the same station. Local electricity generation with PV and energy storage in Li-ion batteries were included in all systems for determining how renewable energy and storage contributes to energy costs for electrification of the transport sector. The co-localized station is presented in Figure 1.

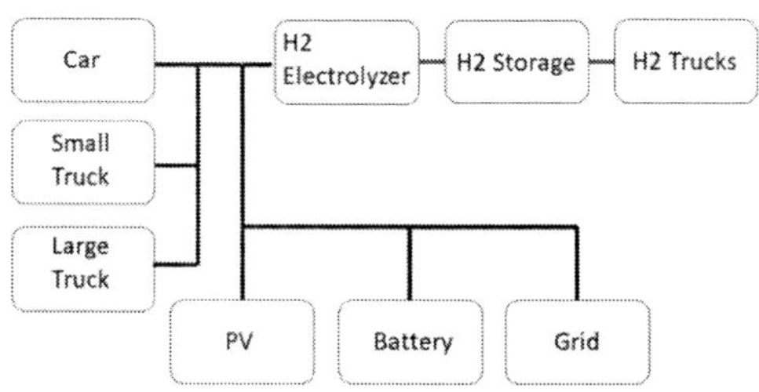

Figure 1: One-line diagram of the co-localized station.

2.2 Estimation of energy demand

In this paper, the electrification of vehicles was assumed for both light-duty (LD) and heavy-duty (HD) vehicles. The energy demand for the charging of batteries and filling of hydrogen storage tanks was estimated based on openly available traffic counts for 2015-2019 provided by the Norwegian Public Roads Administration. The location of data collection was at a major road north of Oslo, Norway. The location intersects two major transport corridors and was considered to capture most of the traffic flow between the two cities Oslo and Trondheim, hence a logical location for considering a significant "on-the-road" co-localized charging station. Data was collected as passes per hour for two different vehicle types, LD vehicles (< 5.6 m) and long HD trucks (>16 m). The truck fleet was divided into BEV and FCEV based on information from the nationally gathered data on truck type, as shown in Table I [7]. To accommodate for this division, the share of FCEV is assumed to be similar to the share of daily truck mileage above 500 km, i.e. 25%, as shown in [7]. FCEV are considered more feasible for long-haul operations [8]. The HD vehicle traffic count data was increased, as it is assumed that 27,5% of vehicles are missing in the set (vehicles with length <16 m), as shown in Table I. This additional volume has been added evenly over the day, even if such a simple approach might induce some errors. To eventually calculate hydrogen load (kg/h) throughout a typical year, the historical data on passes per hour and share of FCEV was modified with the Homer Grid EV charging module to account for uncertainties in a day-to-day and time-step variability. 50% variability were used in both cases and it was further assumed that 50% of the FCEV stop for filling at this station. The refueling amount for HD FCEV is hard to set as only one series-produced small/middle size hydrogen truck is available in Europe, with a refueling capacity of 32 kg [9]. Analysis of a tractor unit operating the route between Oslo and Trondheim concluded on an onboard storage of 46 kg, [10] while in previous hydrogen bus demo projects, the storage was 30-50 kg [11]. Based on this data an average refueling amount of 40 kg per HD vehicle was set.

Table I: Distribution of mileage for HD vehicles [7]

Vehicle		Share of total daily mileage among all HD vehicles
Truck	w/o trailer	27.5%
	with trailer	30.5%
Tractor unit with semi-trailer		42.1%
Total		100%

EV charging demand was calculated with the EV charging module in Homer Grid as hourly charging profiles (kWh/h) based on traffic count data. Charging power for LD vehicles was taken to represent the Norwegian car park of today [12]. Since the traffic count data shows passes per hour, and not actual stops at the station, it was assumed that 5% of cars and 25% of HD BEVs stopped for charging. Trucks were chosen to utilize 500 kW chargers, and the charging time was set to 40 minutes with 50% variability to represent charging of both small and large BEV trucks. Day-to-day and time-step variability was the same as for FCEV trucks, i.e. 50%.

The average number of passes per day at the charging/filling station was 556 (cars), 286 (BEV truck), and 194 (FCEV truck). To decide on the number of charging points for BEV, the peak number of passes per hour was used, and this resulted in 75 chargers for the cars and 25 for the trucks. The peak number of passes per hour for the FCEV trucks were 16. Figure 2 shows the power demand at the co-localized station.

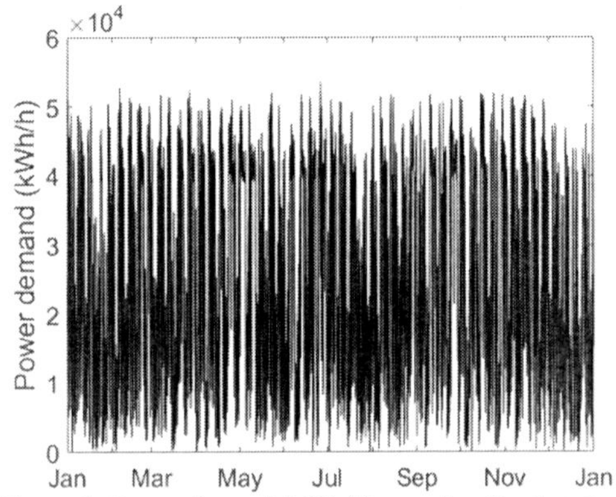

Figure 2: Power demand (kWh/h) at co-localized station for one year.

2.3 Technoeconomic analyses

The technoeconomic analysis was based on comparison of net present costs (NPC) for single BEV, single FCEV and co-localized charging/filling station. The NPC for single BEV and single FCEV stations were added and compared with NPC for the co-localized charging station where both BEVs and FCEVs are charged/filled. The commercial modelling software HOMER Pro 3.13.8 and HOMER Grid 1.7.4 were used in sequence for quantifications of NPC. The software optimizes the system based on minimizing the objective function, NPC, which is the value of all the costs the system incurs over its lifetime, minus the present value of all the revenue it earns over its lifetime [13]. Costs include capital costs, replacement costs, operation and maintenance costs, and the costs of grid electricity. Revenues include salvage value and grid sales revenue. The sequence modelling in Pro and Grid allows for a detailed simulation of electricity prices, where both energy (NOK/kWh) and power charges (NOK/kW) are considered. HOMER Pro was first used to optimize the size of the electrolyser and the hydrogen storage tank based on the estimated hydrogen load profile. The electricity consumption of the electrolyser was exported to HOMER Grid and added to the electricity demand. The total electricity profile was thereafter used to optimize the PV and battery system size. The PV-battery optimization in HOMER Grid deploys a dispatch strategy that uses a 48h perfect foresight. The project lifetime for NPC-calculations was set to 15 years, discount and inflation rate were set to 3.5% and 2% respectively.

2.4 Energy technology components

PV production data was simulated in PVsyst and exported to Homer Grid for optimization of net present cost in the three stations. The coordinates for the specific location studied in this work is 60.70°N, 11.26°E, and the PVGIS Typical Meteorological Year (TMY) database was used as input to the model. The system used in the simulation is Longi Solar LR-60 HPH 370 M G2 Si-mono, with Sungrow SG80KTL inverters, 20 modules in series, 11 strings. The nominal power was 81.4 kWp and the specific production was 834 kWh/kWp/year with a PR of 85.40%. Albedo was set to 0.20. The system is ground-mounted, and no shadow effects were taken into account. In PVsyst, snow losses are typically accounted for by increasing the monthly soiling loss value in the winter months. For the selected location, which can experience snowfalls in the winter, there is little available data on snow losses for PV systems, and reliable input for the PVsyst simulation is lacking. PV snow losses are expected to vary a lot between locations and system designs, and to get snow loss estimates for new systems, snow loss modeling is necessary. Some PV snow loss models exist, but validation is typically lacking [14]. The snow model suggested by Marion et al., [15] which is implemented in the simulation tools SAM and pvlib, has been shown to perform better than other available models [16]. Even if this model is not very well tested for ground-mounted PV systems, it is assumed to be the best available tool for giving an indication of the snow losses for our location. To estimate monthly snow losses, the implementation of the Marion et al. model described by Øgaard et al. [17] was used. Using irradiance and temperature from PVGIS [18] and snow data from senorge.no [19], the monthly snow losses for 11 years (2005-2016) was modeled for PV systems with different tilts for south facing modules. The median snow loss value for each month was then selected as monthly soiling loss value input to the PVsyst model.

The battery system was modeled as separate battery and inverter components to allow for cost-efficient design of the energy and the power capacity. The battery component was modelled as a modified advanced storage module (ASM) Li-ion module in Homer Grid to represent the lithium-manganese-cobalt-oxide (NMC) technology. Lifetime was calculated from both cycle and calendar degradation, where 3000 cycles to 90% DOD and calendar lifetime of 15 years were used [20, 21]. Battery replacement was activated when either cycle or calendar degradation reached end-of-life. The power-to-capacity-rate (C-rate) of the battery was restricted to 1, the roundtrip efficiency was 95%, and the minimum state-of-charge (SOC) was 10%. No temperature effects were included. The inverter component was modeled with an efficiency of 97%, a lifetime of 15 years, and a relative capacity of 100%, i.e. equal power for charging and discharging.

The cost of electricity was simulated according to a tariff structure that includes energy cost (NOK/kWh), power cost (NOK/kW), and a fixed monthly cost (NOK/month). Prices from Elvia, which is one of the largest utility companies in Norway, in 2021 was used as input [22]. In addition to prices from Elvia, spot prices were derived from the average of year 2017 and 2018. Spot prices have varied throughout the eight years with historic data, and the average used in this paper represents a price that is slightly above the average of the past eight years. Two price-levels were used in the analyses, one representing the case of today, and one for the year 2030. The future prices of 2030 were estimated to be about 20% higher than today, based on a yearly increase of 2%. All prices of today were scaled equally, i.e. both the energy, the power, and the fixed prices.

The electrolyzer for the FCEV case was modelled as an proton exchange membrane (PEM) module with an operation range of 0-100%. The lifetime was set to 15 years, which equals 75 000 hours with a utilization factor of 57%, and the efficiency was set to 65% and includes the electricity consumption of the compressor for the storage tank [23]. The installation cost of electrolyzers is strongly dependent on size, and a step-function according to Table II was used. The 450 bar H2-storage tank was simulated as a steel tank with a lifetime of 10 years.

Costs of technology components are summarized in Table II and 3. Since results on dimensioning and NPC-calculations are expected to be affected by installation costs and electrcity price, a sensitivity analysis was performed by using three different PV-costs (Table III) and two electricity price levels (2020 and 2030). The battery-system cost was chosen to represent a case in 2030, and is therefore about 50% lower than prices of today. Replacement costs for PV and battery were set to 50% and 75%, respectively, of installation cost. The hydrogen tank replacement cost was equal to installation cost. O&M costs for PV and battery were set to 1% of installation cost. Costs of charging and filling stations have not been included in the analysis.

Table II: Costs of electrolyzer and compressor stage, with O&M cost of 4% [23, 24]

Size (kW)	Installation cost (NOK/kW)	Replacement cost (NOK/kW)
100	40 000	22 000
1000	17 000	9 400
10000	8 200	4 300

Table III: Costs of technology components

Component	Unit	Installation cost	Ref
PV	NOK/kWp	10 000; 8 000; 5000	[25]
Battery	NOK/kWh	1700	[26]
Battery converter	NOK/kW	2200	[26]
H2-tank	NOK/kg	8100	[27]

3 RESULTS AND DISCUSSIONS

3.1 PV-production

Both the BEV and FCEV stations were initially optimized for the orientation (azimuth) and the tilt of the PV-system. Seven orientations (90, 120, 150, 180, 210, 240, 270), and six tilt angles (10, 20, 30, 40, 50, 60) were included. The results showed that south orientations with tilt angles of 30-50 were most cost-efficient. These results did not take snow shading into account, and for a proper analysis, snow coverage was modeled. The total kWh/yr including snow losses (Figure 3) shows that there is an optimum in module tilt for this location in the range of 40 – 50°. The difference in kWh/year, comparing no snow losses with snow losses, shows that the loss due to snow on the modules reduces as the tilt of the modules increases as the snow slides off the panels easier when the tilt is higher. This is taken into account in the employed model [15]. Based on these results it was concluded to perform the main system simulations with south oriented modules with a tilt of 40°. Figure 4 shows how PV production is affected by snow loss modeling for the tilt of 40° during winter months.

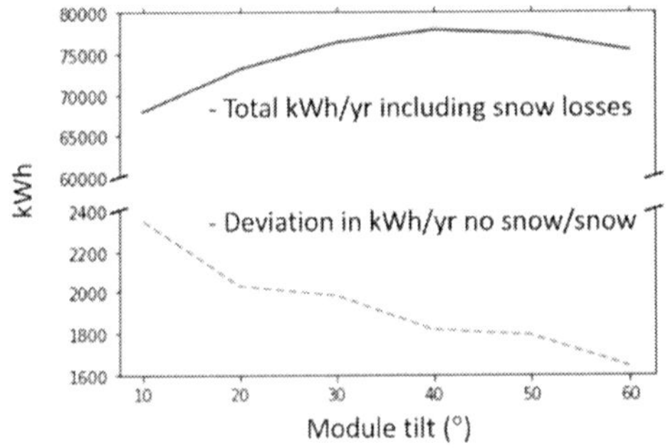

Figure 3. Total generated energy per year, including snow losses, as a function of module tilt (full line), and, the difference in the total generated energy per year with no snow losses and with snow losses (dashed line). Specific production for the system was 834 kWh/kWp/year.

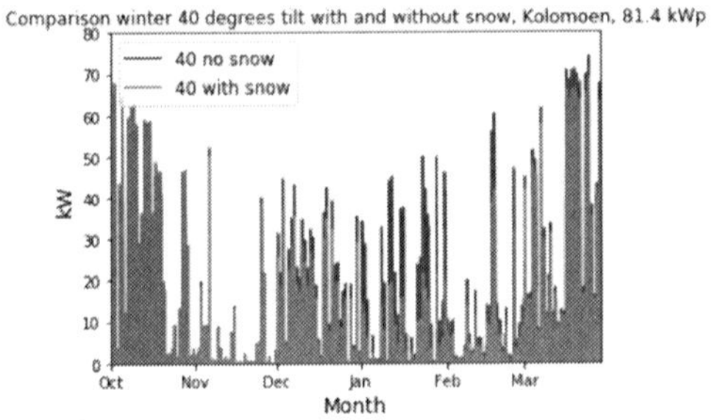

Figure 4. Modeled PV power output with and without snow loss modelling during winter months at the site north of Oslo, Norway. Specific production for the system was 834 kWh/kWp/year.

3.2 Cost-efficiency of co-localized station

The hourly peak/average demand was 14 494/4 426 kW (BEV), 39 000/19 614 kW (FCEV), and 53 494/24 041 kW (co-localized BEV and FCEV). The demand for all stations is constant throughout the year, with low demand during nights. Table IV presents the results on the NPC calculations for single (BEV and FCEV) and co-localized (BEV and FCEV) charging stations. The main objective of the paper was to determine if a co-localized station is more cost-effective than two single stations, and results show that the co-localized station has 0.4-1.1% lower NPC depending on cost assumptions of PV installation and grid electricity price. The estimated NPC is somewhat lower than previously reported, but do also represent a different system [6]. The marginal reductions in NPC are assumed to be related to more efficient utilization of produced PV-power and battery integration.

Table IV: Reduction of net present cost for the co-localized station compared to single BEV and FCEV stations.

PV cost (kNOK/kWp)	Grid cost (Year)	NPC reduction (%)
10	2020	0.4
10	2030	0.5
8	2020	0.6
8	2030	1.0
5	2020	1.0
5	2030	1.1

3.3 Cost-optimal PV capacity

Figure 5 shows results on cost-optimal PV dimensioning. For the conservative cost assumptions, optimal PV-capacity for the hybrid station is more than double the size of the two single stations, expect for the most cost-conservative case where PV was not found to be cost-optimal for the co-localized station. For more optimistic cost assumptions, optimal PV-capacity is similar to two single stations. In these optimizations, maximum installed PV-capacity was restricted to peak demand in each of the stations, i.e. 14 494 kW for BEV, 39 000 kW for FCEV, and 53 494 kW for hybrid. For reference, a PV-park at 50 MWp would require about 0.6 km^2 land area [28]. By using capacity restrictions, emphasis will be placed on covering the electricity demand at the stations rather than the export of electricity from a utility-scale PV-park. Revenues from export of electricity is however included in the NPC calculations. The optimal PV-capacity of the co-localized station depended obviously on both PV installation and electricity costs but did reach the capacity restrictions (54 MW) for the most cost-optimistic scenarios. Such a system is expected to produce about 51 GWh yearly, which is slightly more than double the size used by Mourad et al. for a BEV station without hydrogen production [2]. Removing the capacity restrictions resulted in large PV-systems for the installation cost of 5 000 NOK/kWp. At 2020 grid prices, optimal PV-capacity was about 300 MW and for 2030 grid prices, about 1 300 MW. This is however clearly questionable, and such large-scale PV-park optimizations need more details on aspects such as cost of land, grid connections, and possible conditions for power purchase agreements.

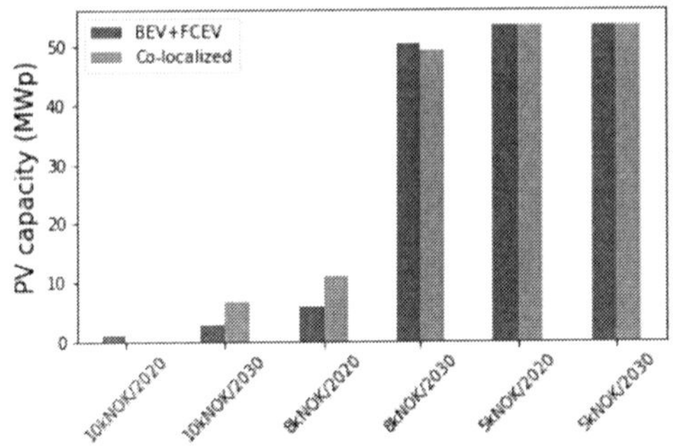

Figure 5. Comparison of cost-optimal PV capacity for two single stations and the co-localized station. X-axis represents different PV installation costs and electricity grid prices.

3.4 Reduction of PV-power export

In figure 6 it is shown how PV-power export is reduced by co-localizing the charging/filling stations. The co-localized station exports 11-68% less electricty compared to two single stations. Sector coupling of energy carriers in this manner has been proclaimed as an effective measure for increasing renewable energy in local grids, and these results support that claim. Batteries did however also contribute to reduced export.

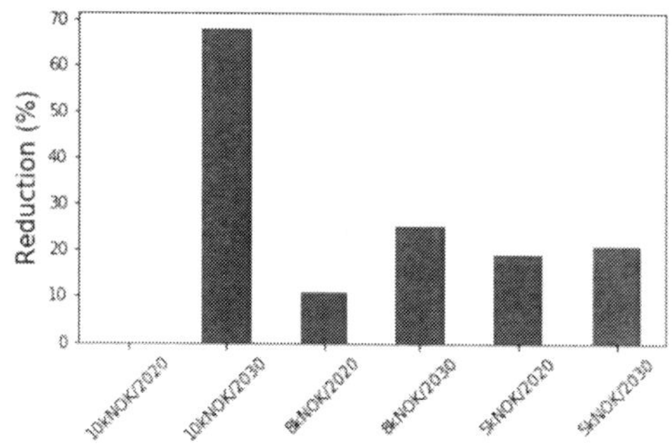

Figure 6. Reduction of PV-power export for the co-localized station compared to two single stations. X-axis represents different PV installation costs and electricity grid prices.

3.5 Impact of snow loss modeling

All results presented above are based on PV-production profiles with snow losses included, and to determine the impact of system dimensioning and LCOE, one additional optimization was performed with PV-production data without snow losses for the same orientation and tilt. The results on optimal PV-capacity and LCOE for both cases are presented in Table V and VI. Neglecting snow losses results in 0.2-0.6% lower LCOE and over-dimensioning of PV-capacity with about 40% for the cost-conservative assumptions. Although the model used to determine snow soiling was formulated based on roof-mounted systems, and more data from ground-mounted systems are needed to validate the model, these results indicate that accurate dimensioning of PV-systems in Nordic conditions should consider snow losses.

Table V: Optimal PV-capacity for co-localized station with and without snow loss modeling

PV cost (kNOK/kWp)	Grid cost (year)	PV capacity without snow loss (kWp)	PV capacity with snow loss (kWp)
10	2020	0	0
10	2030	10 695	7547
8	2020	17 372	12 075
8	2030	51 353	50 312
5	2020	53 494	53 494
5	2030	53 494	53 494

Table VI: LCOE for co-localized station with and without snow loss modeling

PV cost (kNOK/kWp)	Grid cost (year)	PV capacity without snow loss (kWp)	PV capacity with snow loss (kWp)
10	2020	-	-
10	2030	0.843	0.845
8	2020	0.703	0.706
8	2030	0.786	0.791
5	2020	0.629	0.633
5	2030	0.737	0.741

3.6 Stationary battery

The role of a stationary batteries in the charging and filling stations were also investigated. For the BEV-station, batteries were found to be cost-efficient for all cost assumptions. Energy capacities between 1 000 and 2 000 kWh were found, with C-rates ranging from 0.5-0.8. The battery capacity numbers are low in comparison to load demand and installed PV-capacity and indicate that the need for peak shaving is low due to flat net electricity consumption. For context, the demand charges (NOK/kW) used in these analyses are to the authors knowledge one of the highest in Norway, with winter months having an hourly rate of about 120 NOK/kW. Besides infrequent peak shaving operations, the dispatch strategy of the battery system was also found to perform energy arbitrage, i.e. charging/discharging based on differences in spot prices. In another study with significantly larger battery-to-PV-ratio (616 kWh battery and 445 kW PV), an energy management system was developed to benefit from energy arbitrage, and the authors concluded that although the station had excellent economic performance, benefits generated for the grid and society were larger than the economic benefits [29]. This is in line with our experience, where simulations of PV-battery systems for buildings in Norway showed that peak shaving stands for most of the savings [30].

The FCEV-station did not benefit economically from the installation of battery systems due to the flat electricity consumption profile of the electrolyzer.

Batteries in the hybrid station were found to be cost-efficient only for the electricity prices of 2030, probably as a result of the increased cost of demand charges. The assumed 2% yearly increase leads to demand charges of 144 kr/kW in 2030. Optimal battery capacities were only slightly higher than the single BEV-station, at most 3 700 kWh and 1 800 kW, despite including significantly larger PV-systems.

Besides flexibility from stationary batteries, it is believed that scheduling the operation of the electrolyzer by shifting the operation in time is a cost-efficient source of flexibility that could increase profitability further. The electricity demand profile of the FCEV station is flat around 39 MW with peak widths of one to multiple days, so by reducing electricity input to the electrolyzer for an hour during high demand at EV-chargers, the total electricity demand could be clearly lowered. This requires forecasting the consumption and state-of-charge of the hydrogen tank, and possibly also a larger hydrogen tank. Future work should therefore focus on introducing flexibility to the electrolyzer to perform peak shaving and determine profitability in relation to the battery systems.

This analysis focused on the cost-efficiency of power flows and did not include equipment costs for battery charging and hydrogen filling, nor is it an analysis of the total cost of ownership. The results presented in this paper are also specific to the distribution of battery and fuel cell vehicles, but aids as a guide towards design of single vs co-localized charging/filling stations, and the role of PV-battery systems in Nordic conditions.

4 CONCLUSIONS

Design and optimization of charging/filling stations for the electrification of light and heavy-duty vehicles in a major Norwegian road corridor were performed. The results indicate that charging stations benefit from installation of stationary battery storage and local PV-production, but due to a lack of experience of utility-scale PV-parks in Norway, further work is needed to validate the results of connecting large-scale charging stations with utility-scale PV-parks. Results from optimization of PV-system showed that production losses related to snow shading at the site in Norway are important for accurate dimensioning. Sector coupling of battery electric vehicles and hydrogen fuel cell vehicles in a co-localized station reduced net present cost by 0.4-1.1%. Based on the marginal cost reduction, it is not obvious that hybrid stations are more cost-efficient than two single stations. Despite questionable cost savings, it was shown that co-localized stations increased the cost-optimal size of local PV-system or reduced the electricity export. The future large-scale hybrid charging station was found to consume about 211 000 MWh electricity yearly with hourly peaks of about 53 MWh/h.

ACKNOWLEDGEMENTS

The authors acknowledge support from the Research Council of Norway by funding the project "Integrated Transport and Energy modelling", grant number 280989.

REFERENCES

[1] C-T. Ma, *Energies* **2019**, *12*, 4201.
[2] A. Mourad, M. Hennebel, A. Amrani, A. B. Hamida, *Energies* **2021**, *14*, 2428.
[3] G. Fridgen, R. Keller, M-F. Körner, M. Schöpf, *Energy Policy* **2020**, *147*, 111913.
[4] G. Dispenza, V. Antonucci, F. Sergi, G. Napoli, L. Andaloro, *Energy Procedia* **2017**, *143* 39-46.
[5] H. Sánches-Sáinz, C-A. García-Vázquez, F. Llorens Aborra, L. M. Fernández-Ramírez, *Sustainability* **2019**, 11, 5743.
[6] S. Bansal, Y. Zong, S. You, L. Mihet-Popa, J. Xiao, *Energies* **2020**, *13*, 2855.
[7] I. B. Hovi, D. R. Pinchasik, R. J. Thorne, E Figenbaum, (*User experiences from the early adopters of heavy-duty zero-emission vehicles in Norway: barriers and opportunities)* TØI report 1734/2019, submitted: November **2019.**
[8] Z. P. Cano, D. Banham, S. Ye, A. Hintennach, J. Lu, M. Fowler, Z. Chen, *Nature Energy* **2018,** 3(4), 279-289.
[9] Hyundai Motor Company, Hyundai XCIENT fuel cell, https://www.hyundai.news/eu/brand/hyundai-xcient-fuel-cell-heads-to-europe-for-commercial-use/, accessed: June, 2021.
[10] J. Danebergs, MSc Thesis, KTH (Stockholm), **2019.**
[11] B. Reuter, M. Faltenbacher, O. Schuller, N. Whitehouse, S. Whitehouse, (*New Bus Refuelling for European Hydrogen Bus Depots: High-Level Techno-Economic Project Summary Report) Fuel Cell and Hydrogen Joint Undertaking: Brussels, Belgium,* https://fuelcellbuses.eu/sites/default/files/documents/NewBusFuel_D4.2_High-level-techno-economic-summary-report_final.pdf, submitted March, **2017**
[12] Elbilstatistikk, www.elbilstatistikk.no. Accessed: October, **2020.**
[13] S. Bahramara, M. Parsa Moghaddam, M. R. Haghifam, *Renewable and Sustainable Energy Reviews* **2016**, *62*, 609-620.
[14] D. Ryberg, J. Freeman, *(Integration, validation, and application of a PV snow coverage model in SAM),* Technical report: NREL/TP-6A20-68705, submitted: August, **2017**.
[15] B. Marion, R. Schaefer, H. Caine, G. Sanchez, *Solar Energy* **2013**, *97*, 112–121.
[16] M.B. Øgaard, B.L. Aarseth, Å.F. Skomedal, H.N. Riise, S. Sartori, J.H. Selj, *Solar Energy **2021**, 223*, 238-247.
[17] M. B. Øgaard, H. N. Riise, J. Selj, to be published in: *EUPVSEC proceedings*, **2021**
[18] T. Huld, R. Müller, A. Gambardella, *Solar Energy* **2012**, *86*, 1803-1815.
[19] NVE, seNorge, www.senorge.no, accessed: May, **2019**
[20] P. Haidl, A. Buchroithner, B. Schweighofer, M. Bader, H. Wegleiter, *Sustainability* **2019**, *11*, 6731.
[21] W. Cole, A. Will Frazier, *(Cost Projections for Utility-Scale Battery Storage)*, Technical report: NREL/TP-6A20-73222, submitted: June, **2019.**
[22] Elvia, Nettleiepriser, https://www.elvia.no/nettleie/alt-om-nettleie/nettleiepriser-for-privatkunder-i-oslo-og-viken, accessed: May 2021.
[23] E. Rosenberg, J Danebergs, P. Seljom, L. Kvalbein, *(Documentation for IFE-TIMES-Norway v1)*, IFE/E-2020/004, submitted: September, **2020**.
[24] J. Proost, *International Journal of Hydrogen Energy* **2019**, *44*, 4406-4413.
[25] IRENA, (*Future of Solar Photovoltaic: Deployment, investment, technology, grid integration and socio-economic aspects (A Global Energy Transformation: paper))*, International Renewable Energy Agency, Abu Dhabi, **2019**
[26] I. Tsiropoulos, D. Tarvydas, N. Lebedeva, (*Li-ion batteries for mobility and stationary storage applications – Scenarios for costs and market growth)*, Publications Office of the European Union, Luxembourg, doi:10.2760/87175, submitted: **2018**.
[27] Ø. Ulleberg, R. Hancke, *International Journal of Hydrogen Energy* **2020**, *45*, 1201-1211.
[28] IFC, *(Utility-Scale Solar Photovoltaic Power Plants)*, https://www.ifc.org/wps/wcm/connect/topics_ext_content/ifc_external_corporate_site/sustainability-at-ifc/publications/publications_utility-scale+solar+photovoltaic+power+plants, submitted: June, **2015**.
[29] M. Yang, L. Zhang, Z. Zhao, L. Wang, *Journal of Cleaner Production* **2021**, *302*, 126967.
[30] J. Fagerström, K. Aamodt Espegren, J. Selj, A. Severinsen, (Forecasting and technoeconomic optimization of PV-battery systems in commercial buildings), Proceeding of ECEEE Summer Study 5-342-19, **2019**.

SECTOR COUPLED ENERGY MODEL FOR THE EUROPEAN ELECTRICITY, HEAT AND TRANSPORT SECTORS – METHODOLOGY

Alexander Blinn, a.blinn@umwelt-campus.de
Henrik te Heesen, h.teheesen@umwelt-campus.de
Trier University of Applied Sciences, Environmental Campus Birkenfeld,
Institute of Operations and Technology Management,
Campusallee, 55768 Hoppstädten-Weiersbach, Germany

ABSTRACT: Europe can ensure a reliable, economical, and environmentally compatible energy supply by moving from finite fossil fuels to regenerative and sustainable energy production. Due to the characteristics of fluctuating power generation plants, the need for storage increases but can be reduced to a minimum through energy demand reduction measures, such as general technical efficiency improvements and changes in consumer behavior, as well as through load shifting and the expansion of domestic and cross-border transmission grid capacities. Sector coupling combines the electricity, heat, and transport sectors into a holistic system. This work aims to investigate an energy system 100 % based on renewable energy by the year 2050. For this purpose, a complete sector-coupled energy model (UCB-SEnMod) of all energy sectors was modelled. The interaction of energy supply and energy consumption can be investigated with hourly resolution. The result of the present work is that a 100 % renewable energy system in Europe is technically possible. The overall system will be characterized by a high degree of photovoltaic and wind power, whose volatile power generation requires storage technologies. By displacing combustion-based technologies as far as possible and using more efficient electrical consumers, primary and final energy consumption can be significantly reduced.

Keywords: Environmental Effect, Strategy, CO_2, Modeling

1 INTRODUCTION

This work's central question is the feasibility of all sectors' energy supply by 100 % renewable energy sources. For this purpose, a complete model of all energy sectors was modelled, in which the hourly interaction of energy supply and the energy consumption is considered [1]. The installed capacity of power generation plants and the storage systems serve as the overall system's optimization parameter. The future energy system's central pillars are fluctuating renewable energy. Due to the resulting temporary high overproduction of electricity, the need for storage systems and flexible electricity use in all consumption sectors is increasing. In addition to the expansion of power generation systems, a reductionin energy consumption is essential. For example, this canhappen through insulation measures of buildings in the heating sector and the transport sector's electrification. Figure 1 shows the basic structure of the UCB-SEnMod energy model, which is presented in more detail below using Europe's energy system as an example.

2 ENERGY SUPPLY

2.1 Photovoltaic systems

A tool developed by Sandia National Laboratories is used to simulate the performance of photovoltaic power systems [2–8]. This work used the photovoltaic module CS5P 220M 2009 from Canadian Solar and the inverter SB10000TL US 208V from SMA America. The model selection of the photovoltaic module and inverter was influenced by data availability in the databases used by *pvlib*. By interconnecting ten modules in series and four in parallel, an 8.8 kWp system is modelled whose geographic location is defined by its latitude, longitude, time zone, and elevation. The orientation of the photovoltaic system corresponds to the azimuth and tilt angle of the system surface. The time- and site-specific radiation intensities GHI, DNI, and DHI required for the simulation are obtained from the Copernicus Atmospheric Monitoring Service (CAMS) from the European Centre for Medium-Range Weather Forecasts (ECMWF) [9]. The site and time associated with air temperatures and wind speeds are from the web application renewables.ninja [10].

To determine a representative orientation of a fictitious photovoltaic plant positioned in the center of the country, whose cumulative electricity yield in 2018 would be equal to that of all real plants in a country during the same period, the frequency distribution of over 30,000 photovoltaic plants positioned in Europe presented in [11, 12] is used. With these data and solar irradiance from a reference site for Germany, a suitable alignment is determined numerically. This plant's azimuth and tilt angle correspond to about 137° orientation and 27° tilt. This alignment of photovoltaic systems is assumed for all countries. This fictitious plant's power is equal to the installed power of all plants installed in the country.

For the yield simulation for 2050, a test reference year is calculated for each country, which reproduces the typical radiation and weather pattern of an entire year with hourly accuracy [13]. This calculation is based on meteorological data from 2012 to 2018. Furthermore, it is assumed that the installations existing in 2018 are distributed in all geographic regions, which may differ in their climatic conditions. Therefore, the test reference year's country-specific yield profile correlates perfectly with the predicted generated energy in 2050 [14]. That allows the yield profile to scale proportionally relative to the installed plant capacity in 2018 with the assumed plant capacity in 2050. This approach is also used for wind and hydropower simulations. The upper part of Figure 2 shows the validation of the photovoltaic yield simulations using the example for Germany in 2018 with the German Federal Network Agency data. Yield data are trend adjusted to compensate for the increase in plant capacity during the year. The cumulative electricity yield of the simulation roughly corresponds to that data from the German Federal Network Agency, but significant differences in the daily totals can be seen. That is because, as already mentioned, only one site per country is selected for the simulation.

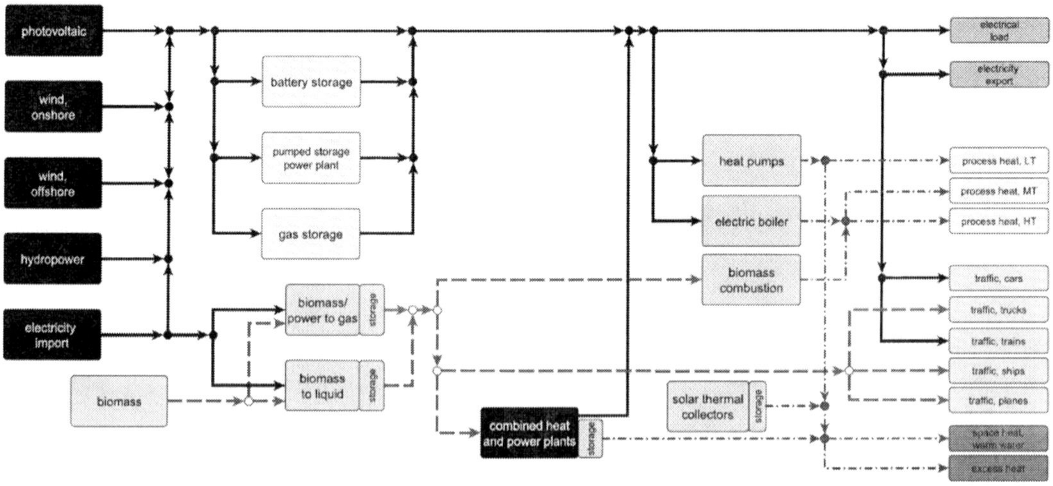

Figure 1: Topology of the UCB-SEnMod energy model.

A fairer division of the countries into clusters of different climatic conditions could reduce these differences in the total daily values. The lower part of the latter figure shows Germany's yield simulation with the same installed photovoltaic capacity but with the meteorological data of the test reference year. As expected, a substantial regression to the mean value can be seen here.

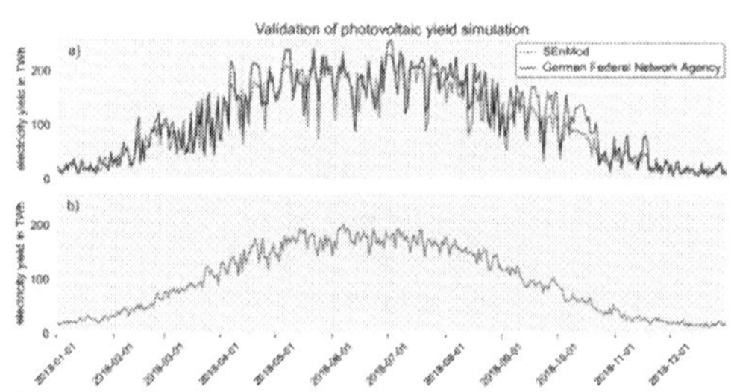

Figure 2: The upper part (a) shows the validation of the photovoltaic yield simulations using Germany's example in 2018 with the German Federal Network Agency data. The data are aggregated to daily totals. The lower part (b) shows the yield simulation with the same total capacity but calculated with the test reference year.

2.2 Onshore and offshore wind turbines

For the simulation of the net electricity yield of wind turbines in the countries, the hourly wind speeds at the hub height of the reference sites are interpolated with selected turbine coefficients and scaled with the respective installed rated power of the wind turbines in the country. For the yield calculation of onshore wind turbines, the characteristic turbine curve of the *Enercon E115-TES* with a rated power of 3 MW is used. An average hub height of 125 meters is assumed. The turbine curve of Enercon *E126-75MW* with a rated power of 7.58 MW and an assumed average hub height of 88 meters is used to determine the offshore electricity yield of countries with the necessary geographical potential for this [15]. A simple power law is used to estimate the wind speed at hub height in flat and homogeneous terrain (no boundary layer offset). When extrapolating the wind speed at heights above 100 meters, it should be noted that an exponential and logarithmic description of the height profile no longer represents the wind conditions with sufficient accuracy. Studies on the determination of profile laws valid for most atmospheric boundary layers can be found in [16, 17]. However, since insufficient data are available for the considered sites, the described effects have to be neglected in the present work.

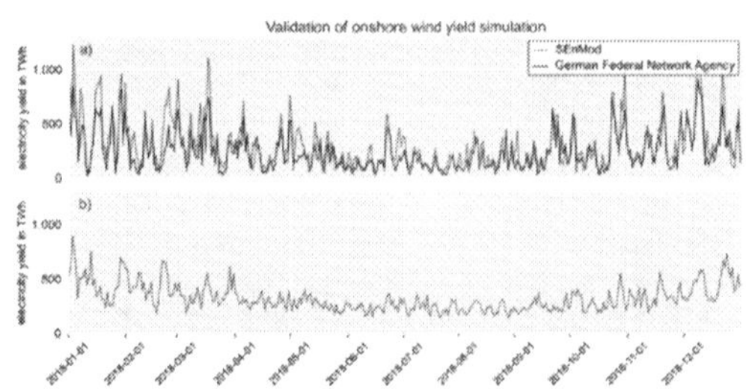

Figure 3: The upper part (a) shows the validation of onshore wind power yield simulations using Germany's example in 2018 with the German Federal Network Agency data. The lower part (b) shows the yield simulation with the same total capacity but with wind speeds averaged and offset-corrected over 2012 to 2018 for Germany. The data are aggregated to daily totals.

For onshore plants, a constant Hellmann exponent of 1/7 is assumed, and for offshore plants an exponent, due to the lower roughness length at sea, of 2/25 is assumed. The test reference year's wind speed data may be lower than the national average due to selecting a single reference site per country. The data is corrected with an offset. After adjusting the mean wind speed, the averaged wind profile for onshore and offshore turbines is extrapolated as described, and the total yield for each country is calculated. For instance, the upper part of Figure 3 and Figure 4 shows the validation of the wind power yield simulations for onshore and offshore plants for Germany in 2018 with the data from the German Federal Network Agency. As with the photovoltaic yield profile, the area of the yield profile shown integrated over the entire year 2018 roughly corresponds to that of the German Federal Network Agency, although significant differences in the daily totals can also be seen here for a reason already described. In the lower parts of Figure 3 and Figure 4, Germany's yield simulations are shown with the same installed onshore and offshore wind capacity, respectively, but with the wind speeds for Germany averaged and offset-corrected over the years 2012 to 2018. The yield profiles are trend-adjusted to compensate for the increase in turbine capacity within the year without changing the average.

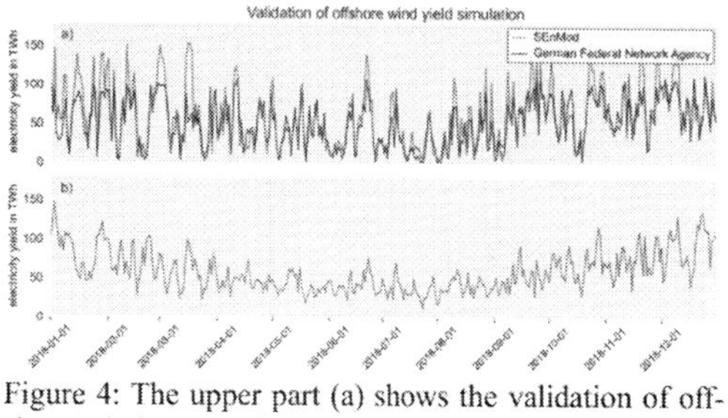

Figure 4: The upper part (a) shows the validation of offshore wind power yield simulations using Germany's example in 2018 with the German Federal Network Agency data. The lower part (b) shows the yield simulation with the same total capacity but with wind speeds averaged and offset-corrected over 2012 to 2018 for Germany. The data are aggregated to daily totals.

2.3 Hydropower plants

The basis for generating the hydropower yield profile is the electricity produced by run-of-river power plants in Germany in 2018. The profile is normalized to 1 MW of nominal capacity, aggregated to hourly values, and converted to coordinated universal time (UTC). If hydropower production has the same pattern in all countries, the profile is scaled as a function of country-specific installed rated power.

2.4 Biomass plants

In a future power system where volatile generation plants generate the most electrical power; it makes sense to take advantage of biomass plants' flexibility and planning opportunities. This model assumes that up to 50 TWh of biomass is available to regulate the power balance. The waste heat from power generation is fed into a heating network as low-temperature heat for heating purposes. The remainder of the plant's capacity produces biogenic fuels for transportation and industrial heat processes.

2.5 Solar thermal energy systems

Solar thermal systems convert solar radiation energy into thermal energy. In this model, this thermal energy is stored in a heat storage tank and provided as low-temperature heat. To determine the irradiation on the inclined surface of the solar thermal modules, the tool *pvlib* already described in section 2.1 is used. The calculation is based on the same azimuth and tilt angle as for the photovoltaic yield calculation. The meteorological data required for this are obtained from the Copernicus Atmospheric Monitoring Service (CAMS) from the European Centre for Medium-Range Weather Forecasts (ECMWF) and the web application renewables.ninja, as mentioned above, and are averaged per reference site over the years 2012 to 2018. It is assumed that the available solar thermal module area A_n corresponds to 1.5 times the population of the respective country n. For Germany, this would correspond to a roof area of about 123 m². For comparison, according to a report by Fraunhofer IWES, the potential for solar generation systems on German roofs is estimated to range from 965 to 1516 m² [18].

The method for calculating the thermal energy injected into the thermal storage $Q_{use}(t)$ at time t is described in [19] and calculated as

$$Q_{use}(t) = \eta_{coll}(t) \cdot A_n \cdot G_t. \tag{1}$$

The efficiency $\eta_{coll}(t)$ of the solar collector is given by

$$\eta_{coll}(t) = \max\left\{c_0 - c_1 \cdot \left(\frac{T_{stor}(t) - T_{amb}(t)}{G(t)}\right), 0\right\} \tag{2}$$

determined with

c_0	optical efficiency,
c_1	heat loss coefficient in W/(m² K),
$T_{stor}(t)$	temperature of the storage at time t in K,
$T_{amb}(t)$	ambient temperature at time t in K,
$G(t)$	POA irradiance at time t in Wh/m².

Here, the optical efficiency c_0 of the collector is taken to be 0.8, and the heat loss coefficient c_1 is taken to be 3 W/(m² K) [19].

2.6 Combined cycle power plants

By storing electrical energy through synthesizing gases from surplus electricity, the electricity sector can be bidirectionally coupled with the other sectors of heat, transport, and industry via the gas infrastructure. Coupling between the electricity and transport sectors is direct via electromobility or indirect via further gas processing into synthetic fuels. Regenerative power generation takes place, if required, via highly efficient gas-fired power plants. Due to these power plants' high power capacity and conversion efficiency, the heat generated during the conversion process can only be used with difficulty [19]. The model operates with an efficiency typical for converting electricity to methane, but this efficiency highly depends on the carbon dioxide source used [20]. The overall efficiency is assumed to be 0.6. That would correspond to a round-trip efficiency of 36 %.

2.7 Combined heat and power plants

In contrast to pure combined cycle power plants, combined heat and power plants serve simultaneous electricity and heat provision and are designed decentralized and monovalent in this model. The fuel used is gas, which is produced synthetically with renewable energy. The waste heat generated by the cogeneration process is used as low-temperature heat for space heating and hot water preparation and can be stored in heat storage in case of surplus.

The basis for the combined heat and power plants' design is the heat load profile of private households minus the heat energy used by solar thermal energy. The design point on the load duration curve is set at 6000 hours.
The overall efficiency with

$$\eta_{ges} = \frac{P_{el} + \dot{Q}_{th}}{\dot{B}_{gas}} = \eta_{el} + \eta_{th} \tag{3}$$

is assumed to be 0.85, and for the electrical efficiency 0.35 is assumed [21].

2.8 Heat storage

To reduce short- and long-term up to seasonal mismatches of supply and demand in the low-temperature heat sector, the heat generated by solar thermal collectors and combined heat and power plants is stored in heat storage systems. Water is used as the storage medium. The single-node model used to represent the heat storage's energy balance is described by the following differential equation [19]

$$C_{\mathrm{stor}} \frac{\mathrm{d}T_{\mathrm{stor}}}{\mathrm{d}t} = \eta_{\mathrm{charge}} \cdot P_{\mathrm{charge}} - P_{\mathrm{discharge}} - P_{\mathrm{loss}} \quad (4)$$

with

C_{stor}	total heat capacity of the storage in Wh/K,
T_{stor}	temperature of the storage medium in K,
P_{charge}	thermal power supplied to the storage in W,
$P_{\mathrm{discharge}}$	heat power removed from the storage in W,
P_{loss}	power loss of the storage tank in W,
η_{charge}	storage efficiency.

The thermal stratification in the storage medium, which positively affects the heat transfer from heat transfer fluid to the storage medium, is neglected. The following equation describes the power loss of the accumulator due to self-discharge

$$P_{\mathrm{loss}} = U_{\mathrm{A,stor}} \cdot \left(T_{\mathrm{stor}} - T_{\mathrm{amb,stor}}\right) \quad (5)$$

$$= \frac{C_{\mathrm{stor}}}{\tau_{\mathrm{stor}}} \cdot \left(T_{\mathrm{stor}} - T_{\mathrm{amb,stor}}\right) \quad (6)$$

with

$U_{\mathrm{A,stor}}$	heat loss coefficient in W/K,
$U_{\mathrm{A,stor}}$	ambient temperature of the storage tank,
$U_{\mathrm{A,stor}}$	time constant of self-discharge in 1/s.

Table 1: Parameters used for the heat storage simulation.

Parameter	Value
T_{min}	55 °C
T_{max}	95 °C
c_0	0.8
c_1	3.0 W/(m² K)
η_{charge}	0.9
c_{water}	4.184 J/(kg K)
τ_{stor}	72 h

Table 1 shows the technical and physical parameters used for the heat storage simulation. The storage volume is designed as 80 liters per installed square meter of solar thermal area. For cogeneration plants, the storage volume in liters is equivalent to 120 times the country's population. It is assumed that the temperature does not fall below a minimum value.

3 ENERGY DEMAND

3.1 Electricity

The European Network of Transmission System Operators for Electricity provides the countries' electricity load and refers to the year 2015 [22]. Those European countries that are not available in the database are estimated using countries with comparable locations and structures. Electricity demand data already includes electrical energy used to provide heat. That must be considered. It is assumed that electrical energy demand will increase about 2 % per year [23].

3.2 Heat

The heat sector's energy demand includes the provision of space heating and hot water and industrial process heat of different temperature segments. Figure 5 schematically shows the systems used to provide heat. The provision of space heating and hot water is provided by solar thermal systems, heat pumps, and combined heat and power plants. Only heat pumps are used to provide process heat in the lower temperature segment. Due to the high-temperature requirements of many industrial processes, technologies used to provide low-temperature ranges cannot be used. Electric boilers and heating furnaces provide heat for medium and high-temperature process heat.

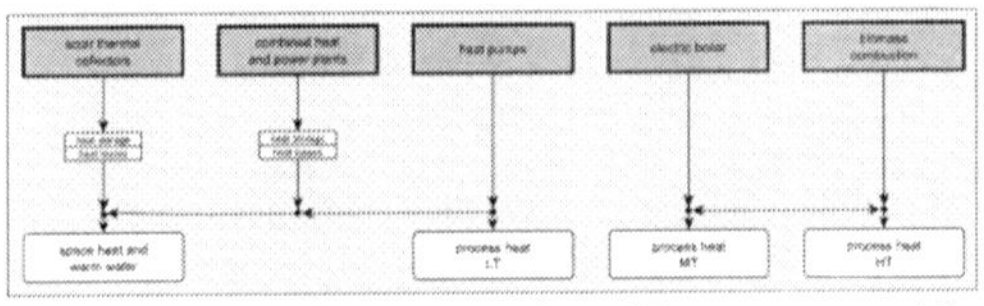

Figure 5: Schematic representation of the systems used for space heating, hot water, and process heat.

3.2.1 Space heating and water heating

The heating load profiles of the space heating and hot water supply are calculated according to the German Association for Energy and Water Management procedure and are based on standard load profiles [24]. With the standard load profile procedure, a correlation between ambient temperature and daily heat demand is mapped, and subsequently, the hourly load profile of the consumption is calculated. It is assumed that private households total space heating and water heating demand can be mapped using the standard load profile *HMF03*. Further, the total heat demand of small-scale industries, trade and services, and industry hot water consumption is mapped using the standard load profile *GHD03*. Figure 6 shows the simulated heat load profile of space heating and water heating for private households, small-scale industries, trade and services, and industry using Germany's example for the year 2018. The energy demand calculation assumes that insulation measures for existing buildings and minimum energy requirements for new buildings can reduce the energy demand for space heating by up to 60 % by 2050 [25, 26]. That also corresponds to the assumption made in the REMod-D energy model developed by Fraunhofer ISE [27].

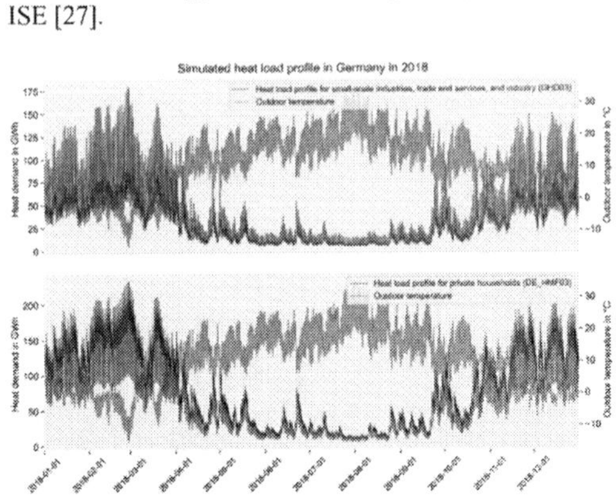

Figure 6: The figure shows the simulated heat load curve of space heating and water heating for private households, small-scale industries, trade and services, and industry and the outdoor temperature using Germany's example for the year 2018.

3.2.2 Process heat

In addition to the energy supply for water heating and

space heating, process heat consumption, which accounts for the most significant industry share, represents the third area in the heating sector [28, 29]. According to a European Commission report, 74 % of final industrial energy consumption in Europe was spent on heating, hot water, and air-conditioning cooling in 2012. With a share of 60 % of final energy consumption, processes for providing process heat of different temperature ranges took the largest share. The remaining 14 % was distributed among space heating, hot water, and air-conditioning cooling [30]. About 26 % of final energy consumption in the industry that was not used for heat-related processes was used for lighting, information and communication technologies, and mechanical energy. These industrial consumers shares are roughly consistent with [31, 32] for other years. The mean distribution of final energy consumption shares of different temperature levels for the provision of process heat at the European level for 2012 was examined in a publication by Naegler [33]. Low-temperature processes take about 8 % of industries' final energy demand, medium-temperature processes about 19 %, and the highest temperature segment processes about 33 %. The respective percentage share is multiplied by the industry's annual final energy consumption in the respective country to determine the heat demand of low-temperature, medium-temperature, and high-temperature. It is assumed that the distribution of process heat corresponds to that of the electricity load profile presented in section 3.1. For this purpose, the countries' electricity load profiles are normalized and scaled by the respective amount of process heat. As mentioned above, only heat pumps are used to provide low-temperature process heat. Medium- and high-temperature processes are supplied 80 % each by electric boilers and 20 % by biogenic fuel furnaces. Furthermore, it is assumed that efficiency measures and process changes will reduce the final energy demand for the provision of process heat by 30 % by 2050 [28]. For these purposes, studies by [34, 35] assume potential end-use energy savings of 34 to 50 by 2050.

3.3 Transportation

There are different measures to reduce final energy consumption in the transport sector. In this publication, these measures are divided into three types, which are now briefly presented.

- Type A: The amount of transportation is directly reflected in the final energy consumption. The change in kilometers travelled relative to the 2018 reference year is covered by Type C measures.
- Type B: By changing the propulsion concepts, such as, e.g., the electrification of a previously combustion-based means of transportation, are grouped under Type B. It is assumed that in 2050 the average efficiencies of the propulsion systems will be equivalent to those of today's *state of the art* propulsion systems of new vehicles of the same type [29, 36]. That would correspond to no efficiency improvements in today's powertrain designs by 2050. This assumption can be considered as a *worst-case* estimate.
- Type C: Type C measures include vehicle and infrastructure improvements that help reduce final energy consumption. Measures considered for this purpose include board energy management improvements, vehicle aerodynamics, weight and rolling resistance reduction, or changes in travel conditions, such as speed or traffic flow.

Figure 7 shows the exponential course of the final energy consumption change assumed in this model for European motorized private transport, road-bound freight transport, air transport, rail transport, and inland waterway transport by 2050 relative to 2018. Here, a constant percentage change in the respective final energy consumption of the specific transport sectors from 2018 is assumed for all European countries considered through measures A, B, and C. The country-specific development of final energy demands up to the year 2050 is calculated based on the final energy demand of the year 2018 and modelled differentiated by energy carriers.

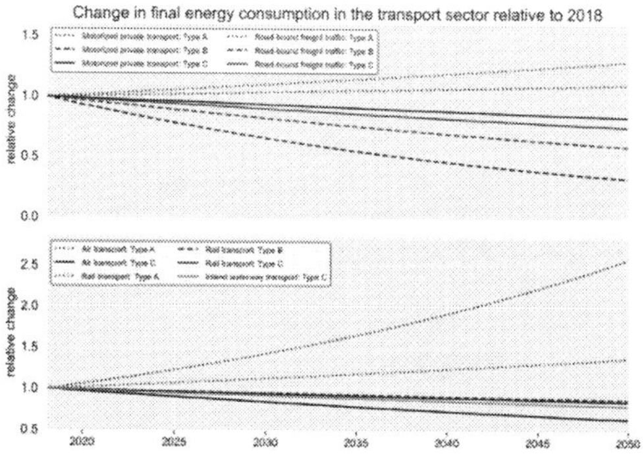

Figure 7: Change in final energy consumption due to transport performance increase and efficiency increase of the individual transport types by 2050 relative to 2018. The upper part shows the modeled final energy change of private motorized transport and road-based freight transport, and the lower section of the figure shows that of air transport, rail transport, and inland waterway transport.

It is assumed that 70 % of the final energy consumption of road transport is attributable to motorized private transport and 30 % to road-based freight transport [37]. Different methods are chosen to generate the load profiles depending on the mode of transport. The standard load profiles developed by the Karlsruhe Institute of Technology (KIT) based on long-term studies of the German Mobility Panel (MOP) [38] are used to calculate the load distribution by battery electric vehicles. The standard load profiles are differentiated by weekdays and weekends and by household types and charging strategies, which in turn depend on the availability of charging infrastructure. The basis for the load profile of road-bound freight traffic is the data on mileage on German motorways and federal roads from the Federal Office for Freight Transport. This load profile is subsequently corrected with the actual driving bans and the driving performance in the European countries. To generate the load profile of air transport, rail transport, and inland waterway transport, the required final energy of the selected fuel is distributed evenly over all hours of the year.

4 REGULATION AND BALANCING

In general, the difference between electricity production and consumption at a given time t in t: $[t, t+1\,\mathrm{h}[$ is not zero. That difference $\delta_n(t)$ of country n is defined as

$$\delta_{n(t)} = G_n^W(t) + G_n^P(t) + G_n^H(t) - L_n(t) \qquad (7)$$

with

$G_n^W(t)$ electricity yield by wind turbines in Wh,
$G_n^P(t)$ electricity yield by photovoltaic plants in Wh,
$G_n^H(t)$ electricity yield by hydropower plants in Wh,
$L_n(t)$ total electricity demand in Wh,
t timestamp in h

and must be balanced by storage processes and cross-border transit of electrical energy.

4.1 Electrical energy storage

With a high share of fluctuating power generation units, the need for storage technologies increases. If the power generation and consumption pattern repeat from year to year, the storage volume is selected so that the state of charge at the beginning of the year corresponds to that at the end of the year. The minimum storage volume required, in turn, corresponds to the maximum capacity at which the storage is fully filled for one hour per year. The systems available for electric energy storage are battery storage (s=1), pumped storage power plants (s=2), and gas storage (s=3), which are charged and discharged in the named order. The simulated storage processes are characterized by charging and discharging efficiencies and maximum charging currents. The maximum charging and discharging energy per hour were assumed to correspond to 20 % of the respective storage capacity. The bilateral efficiency for battery storage is assumed to be 0.9, pumped storage 0.8, and gas storage 0.6.

The positive part of the hourly difference $\delta_{n^+}(t)$ from electricity production and demand at time t with

$$\delta_{n^+}(t) = \max\{\delta_n(t), 0\} \tag{8}$$

is available for storage. However, the amount of energy used for storage is limited by technical and physical parameters. This total amount of energy used at time t is the sum of the subsets $C_{n,s}(t)$ $\forall s \in \{1,2,3\}$ with

$$\delta_n(t) > 0 \longrightarrow 0 \leq \sum_{s=1}^{3} C_{n,s} \leq \delta_n(t) \tag{9}$$

and

$$C_{n,s}(t) = \min \begin{cases} \delta_{n(t)} - \sum C_n'(t) \\ \dfrac{\Psi_{n,s} - \Psi_{SOC,n,s}}{\eta_{C,s}} \\ \dfrac{\Psi_{max,C,s}}{\eta_{C,s}}. \end{cases} \tag{10}$$

Here $C_n'(t)$ denotes the energy already stored in an upstream storage process at time t. The state of charge (SOC) of the storage s at the time t of the country n is finally given by

$$\Psi_{SOC,n,s}(t) = \Psi_{SOC,n,s}(t-1) + C_{n,s}(t) \cdot \eta_{C,s} \tag{11}$$

with

$\Psi_{SOC,n,s}(t)$ SOC of the storage in Wh at time t,
$\Psi_{SOC,n,s}(t-1)$ storage capacity in Wh,
$C_{n,s}(t)$ maximum charging current in Wh,
$\eta_{C,s}$ efficiency of the charging process.

Any excess energy $\Gamma_n^+(t)$ after the storage process is exported

$$\Gamma_n^+(t) = \Delta_n(t) - \textstyle\sum_{s=1}^{3} C_{n,s}. \tag{12}$$

The method can be used in a similar way for discharging operations of the storage.

4.2 International transit of electric energy

If the amount of energy $|\Gamma_n^+(t)|$ remaining in the grid after the storage processes is greater than zero, it is further reduced by importing or exporting electricity within the European grid. The index n used in the following stands for the respective country under consideration and the index m for its neighboring countries. Neighboring countries are all countries that are electrically connected to the respective country.

To simulate the energy flow from the country n to a country m, the interconnected network consisting of nodes connected by lines is assumed to behave according to Kirchhoff's laws with

$$\textstyle\sum_{n=1}^{N} \Gamma_n(t) = 0.. \tag{13}$$

Transmission losses, any transmission capacity constraints, and the actual direction of energy flow along the negative gradient of electric potential are neglected. Under the necessary condition

$$\Gamma_n^+(t) > 0 \cap \Gamma_m^-(t) < 0, \ \forall (n, m) \in \{0, 1, 2, ...\} \tag{14}$$

the energy $T_{n\rightarrow m}$ with

$$T_{n\rightarrow m} = min \begin{cases} \Gamma_n^+(t) \\ |\Gamma_m^-(t)| \end{cases} \tag{15}$$

is transferred from country n to country m. As a result, the export demand $\Gamma_n^+(t)$ of country n and the import demand $\Gamma_m^-(t)$ of country m decrease by the transferred energy $T_{n\rightarrow m}$

$$\Gamma_m^-(t) = \Gamma_m^-(t) + T_{n\rightarrow m} \tag{16}$$
$$\Gamma_n^+(t) = \Gamma_n^+(t) - T_{n\rightarrow m}. \tag{17}$$

That is repeated for all countries n and its neighboring countries m under the condition shown in (14).

In case that the electricity balance $\Gamma_{n^+}(t)$ in any hour t of a country n is still less than zero, in a subsequent process, the storage systems of the neighboring countries, provided that they do not fall below a minimum storage level, can be discharged and provide energy under the technical conditions mentioned in section 4.1. The goal is to balance the power balance $\Gamma_n(t)$ at every hour t. If the power balance cannot be balanced at every hour, the installed capacity of the generating units and, if necessary, the installed storage volume must be corrected accordingly. In real power grid systems, this would impact grid stability and, to avoid overvoltage damage and emergency shutdowns of loads, systems for power destruction or curtailment would have to be created.

5 SUMMARY AND OUTLOOK

The UCB-SEnMod energy model simulates a future energy system with a share of 100 % renewable energy. Due to the high degree of flexibility, the model can calculate different future energy system scenarios. The model consists of different components, each representing a unit of an energy system, which can be managed in the main module. The energy generation and demand analysis show that a reliable and environmentally compatible energy supply based on 100 % renewable energy is possible. The energy supply from solar and wind power form the pillars of the future energy system. The capacity of photovoltaic and wind power plants needed to cover electricity demand in 2050 is shown in Figure 8 and is estimated at nearly 4,200 GW – plus an additional 2,500 GW if the fuels demanded by the transportation sector were produced synthetically.

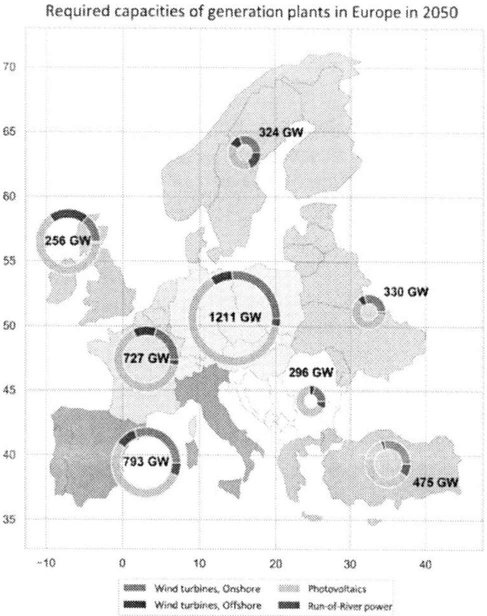

Figure 8: The figure shows the assumed installed capacity of renewable energies in the regions of Europe in 2050.

This would be equivalent to 13 to 21 times the current plant capacity. The shift from today's demand-driven energy supply – based on a few large power plants – to a highly fluctuating and decentralized energy system requires a renewal of the supply model. The storage of electrical and thermal energy and the use of flexible power generation systems (preferably with heat extraction) will play a key role and will be accompanied by further suitable measures for flexibilization and load management. The present work concludes that the discussed energy system, which is to be converted from fossil to renewable and from centralized to decentralized, is technically feasible. By displacing combustion-based loads with more efficient electric technology, primary and final energy demand can be significantly reduced, and security of supply can be increased by expanding intra- and international interconnection. In order to ensure a reliable, economical and environmentally compatible energy supply in Europe, a substantial expansion of renewable generation plants is necessary to achieve the targeted plant capacity for the year 2050. The further development of the model is currently part of the institute's work.

REFERENCES

[1] Marie-Christin Leonhard, Henrik te Heesen. Energy Model for a Rural Region in Germany - Methodology. 37th European Photovoltaic Solar Energy Conference and Exhibition 2020: 1850–1854. doi:10.4229/EUPVSEC20202020-6BV.5.22

[2] Joshua S. Stein. The photovoltaic Performance Modeling Collaborative (PVPMC). 2012 38th IEEE Photovoltaic Specialists Conference 2012. doi:10.1109/PVSC.2012.6318225

[3] Holmgren WF, Hansen CW, Mikofski MA. pvlib python: a python package for modeling solar energy systems. Journal of Open Source Software 2018; 3: 884–886. doi:10.21105/joss.00884

[4] Joshua S. Stein, William F. Holmgren, Jessica Forbess, et al. PVLIB: Open Source Photovoltaic PerformanceModeling Functions for Matlab and Python. 43rd Photovoltaic Specialists Conference 2016

[5] William F. Holmgren, Clifford W. Hansen, Mark A. Mikofski. pvlib python: a python package for modeling solar energy systems. Journal of Open Source Software 2018; 3: 884. doi:10.21105/joss.00884

[6] William F. Holmgren, Clifford W. Hansen, Mark A. Mikofski. pvlib python: a python package for modeling solar energy systems. Journal of Open Source Software 2018; 3: 884. doi:10.21105/joss.00884

[7] William F. Holmgren, Derek G. Groenendyk. An Open Source Solar Power Forecasting Tool Using PVLIB-Python. 43rd Photovoltaic Specialists Conference 2016

[8] William F. Holmgren, Robert W. Andrews, Antonio T. Lorenzo, et al. PVLIB Python 2015. 2015 IEEE 42nd Photovoltaic Specialist Conference (PVSC) 2015. doi:10.1109/PVSC.2015.7356005

[9] ECMWF - COPERNICUS Atmosphere Monitoring Service. CAMS Radiation Service. Im Internet: https://www.soda-pro.com/en/web-services/radiation/cams-radiation-service; Stand: 20.09.2020

[10] Stefan Pfenninger, Iain Staffell. Renewables.ninja. Im Internet: https://www.renewables.ninja; Stand: 05.01.2021

[11] Jonathan Leloux, Luis Narvarte, David Trebosc. Performance Analysis of 10,000 Residential PV Systems in France and Belgium. In: 26th European Photovoltaic Solar Energy Conference and Exhibition; 2011: 3939–3946. doi:10.4229/26thEUPVSEC2011-5AO.6.4

[12] Jonathan Leloux, Jamie Taylor, Rodrigo Moretón, Luis Narvarte, David Trebosc, Adrien Desportes. Monitoring 30,000 PV systems in Europe: Performance, Faults, and State of the Art. In: 31st European Photovoltaic Solar Energy Conference and Exhibition. Hamburg; 2015: 1574–1582. doi:10.4229/EUPVSEC20152015-5AO.8.1

[13] Henrik te Heesen, Volker Herbort, Martin Rumpler. Performance of roof-top PV systems in Germany from 2012 to 2018. Solar Energy 2019; 194: 128–135. doi:10.1016/j.solener.2019.10.019

[14] Detlef Ahlborn. Glättung der Windeinspeisung durch Ausbau der Windkraft? Energiewirtschaftliche Tagesfragen 2015; 65: 37–39

[15] Volker Berkhout, André Bisevic, Michael Claußner, Martin Dörenkämper, Michael Durstewitz, Stefan Faulstich, Philip Görg, Laura Große. Windenergie Report Deutschland 2018. Stuttgart: Fraunhofer Verlag; 2019

[16] Stefan Emeis. Wind Energy Meteorology: Atmospheric Physics for Wind Power Generation. 2. Aufl. Springer International Publishing; 2018. doi:10.1007/978-3-319-72859-9

[17] Stefan Emeis. Current issues in wind energy meteorology. Meteorological Applications 2014; 21: 803–819

[18] Amany von Oehsen, Yves-Marie Saint-Drenan, Thomas Stetz. Vorstudie zur Integration großer Anteile Photovoltaik in die elektrische Energieversorgung: Studie im Auftrag des BSW - Bundesverband Solarwirtschaft e.V; 2012

[19] Hans-Martin Henning, Andreas Palzer. A comprehensive model for the German electricity and heat sector in a future energy system with a dominant contribution from renewable energy technologies - Part I: Methodology. Renewable and Sustainable Energy Reviews 2014; 30: 1003–1018

[20] Michael Sterner. Bioenergy and renewable power methane in integrated 100 % renewable energy systems: Limiting global warming by transforming energy systems. Universität Kassel; 2009

[21] Bernath C, Bossmann T, Deac G, Elsland R, Fleiter T, Kühn A, Pfluger B, Ragwitz M, Rehfeldt M, Sensfuß F, Steinbach J, Cronenberg A, Ladermann A, Linke C, Maurer C, Tersteegen B, Willemsen S, Franke B, Kauertz B, Pehnt M, Rettenmaier N, Hartner M, Kranzl L, Schade W, Catenazzi G, Jakob M, Reiter U. Langfristszenarien und Strategien für den Ausbau Erneuerbarer Energien in Deutschland; 2017

[22] European Network of Transmission System Operators for Electricity. entso-e: Power Statistics. Im Internet: https://www.entsoe.eu/data/power-stats/; Stand: 02.02.2021

[23] Rolando A. Rodriguez, Sarah Becker, Gorm B. Andresen, et al. Transmission needs across a fully renewable European power system. Renewable Energy 2014; 63: 467–476. doi:10.1016/j.renene.2013.10.005

[24] Bundesverband der Energie- und Wasserwirtschaft e. V. Abwicklung von Standardlastprofilen; 2016

[25] Stephan Singer, Jean-Philippe Denruyter, Barney Jeffries. the energy report: 100 % renewable energy by 2050; 2011

[26] David Connolly, Brian Vad Mathiesen, Poul Alberg Østergaard, Bernd Möller, Steffen Nielsen, Henrik Lund Halmstad, Urban Persson, Daniel Nilsson, Sven Werner, Daniel Trier. heat roadmap europe 2050 - study for the EU27; 2012

[27] Andreas Palzer, Hans-Martin Henning. A comprehensive model for the German electricity and heat sector in a future energy system with a dominant contribution from renewable energy technologies - Part II: Results. Renewable and Sustainable Energy Reviews 2014; 30: 1019–1034

[28] Volker Quaschning. Sektorkopplung durch die Energiewende - Anforderungen an den Ausbau erneuerbarer Energien zum Erreichen der Pariser Klimaschutzziele unter Berücksichtigung der Sektorenkopplung; 2016

[29] Philip Sterchele, Julian Brandes, Judith Heilig, et al. Wege zu einem klimaneutralen Energiesystem: Die deutsche Energiewende im Kontext gesellschaftlicher Verhaltensweisen. Fraunhofer ISE 2020

[30] Tobias Fleiter, Jan Steinbach, Mario Ragwitz. Mapping and analyses of the current and future (2020 - 2030) heating/cooling fuel deployment (fossil/renewables); 2016

[31] Roland Hofer. Electricity and heat generation potentials of industrial combined heat and power plants. Brennstoff-Waerme-Kraft 1995; 47

[32] Rainer Elsland, Tobias Fleiter, Martin Jakob, Ulrich Reiter. Heating and Cooling - The transformation towards a low-carbon Heating and Cooling sector; 2017

[33] Tobias Naegler, Sonja Simon, Martin Klein, et al. Quantification of the European industrial heat demand by branch and temperature level. International Journal of Energy Research 2015; 39: 2019–2030. doi:10.1002/er.3436

[34] Norman Gerhardt, Fabian Sandau, Angela Scholz, Henning Hahn, Patrick Schumacher, Christina Sager. Interaktion EE-Strom, Wärme und Verkehr - Endbericht; 2015

[35] Umweltbundesamt. Emissionsbilanz erneuerbarer Energieträger 2018

[36] Jette Krause, Christian Thiel, Dimitrios Tsokolis, et al. EU road vehicle energy consumption and CO2 emissions by 2050 – Expert-based scenarios. Energy Policy 2020; 138. doi:10.1016/j.enpol.2019.111224

[37] Bundesministerium für Verkehr und digitale Infrastruktur. Verkehr in Zahlen 2019/2020; 2019

[38] Daniel Heinz. Erstellung und Auswertung repräsentativer Mobilitäts- und Ladeprofile für Elektrofahrzeuge in Deutschland. Karlsruhe; 2018. doi:10.5445/IR/1000086372

ENERGY MANAGEMENT SYSTEM FOR ELECTRIC BUS CHARGING HUB WITH LOCAL STORAGE AND PV ENERGY INTEGRATION

Samuli Ranta[1], Hugo Huerta[1], Dominique Roggo[1], Osmo Huhtala[1], Aleksi Heinonen[1], Ville Lavonen [1], Julien Pouget[2]

Corresponding author: Samuli Ranta
[1]Turku University of Applied Sciences, Joukahaisenkatu 7, 20520 Turku, Finland, Tel: +358 40 355 0833 [samuli.ranta@turkuamk.fi]
[2]HES-SO Valais-Wallis, Sion, Switzerland

ABSTRACT:
A modeling-based simulation and hardware-based emulator are developed to facilitate smart design of the future e-bus and e-truck charging systems and infrastructures. An existing e-bus line in the City of Turku, in Southwest Finland, was the starting point for this work. Aim of the work is to develop tools for the cities, municipalities, e-bus operators, and other decision makers to design and procure more sustainable and efficient transportation systems. Our study focuses on the accurate model of a charging hub, available PV resource and an advanced Energy Management System (EMS) to control energy flows.

Keywords: Decarbonization, Fast Charging, Modeling, Energy Management System, e-bus, Simulation

1. INTRODUCTION

The efficient and sustainable planning of e-bus charging infrastructures is a challenge many European cities and municipalities are facing in the near future. Innovative, efficient, and sustainable opportunity charging hubs integrate Renewable Energy production and energy buffers in order to reduce the impact on the distribution grid. An advanced energy management system can optimize the local use of produced energy and reduce peak power capacity of hub's components. Detailed modeling of the e-bus charging infrastructure is an important tool in solving this challenge. Scaled-down hardware emulator can be used for the model validation and development of the new control strategies and hardware. City of Turku has pledged to become carbon neutral in 2029 and as part of that strategy is planning to electrify most of the public transport bus lines by then. Electrification of an entire bus fleet is a demanding task and one of the biggest challenges is how charging can be arranged efficiently and in a sustainable way.

In the context of energy transition and sustainable mobility, it is necessary to develop energy hubs dedicated to the fast charging of the e-bus batteries with advanced distribution infrastructures on the model of micro-grids. Heavy-duty vehicles, such as e-buses and e-trucks, which can take a peak charging power of 350kW and more, could drop the voltage of the electrical distribution grid in areas where the installed capacity is not enough for an energy hub. Such power profiles will certainly have a strong impact on the Low voltage AC (LVAC) grid stability, especially if the whole bus fleet is electrified. In order to achieve the objectives of a sustainable mobility, such a hub should provide services in terms of support to the grid, efficient use of renewable energy, and energy availability at a viable cost.

The idea of the Energy Management System (EMS) of such energy hub is a perfect use case for a simulation and scale-down emulator tool. This paper describes the concept and the development of a simulation tool for e-bus charging hub and a down scale emulator [1] of a charging energy hub involving photovoltaics (PV) and local energy storage systems as well as the development and implementation of the Energy Management System.

2. METHODOLOGY

Sustainable energy for the e-bus charging requires new concepts for the energy generation. Their high peak power demand requires oversizing of distribution grid equipment in order to ensure reliable operation. These are the main challenges addressed in this project. This paper investigates sustainable solutions to utilize PV generated energy and how to reduce the peak power demand generated by fast e-bus charging stations with an optimal implementation of battery charging energy hubs through the following:

1. Measuring of the fast-charging loads from the operating fast charger at Turku pilot e-bus line.
2. Developing an e-bus fast charging hub model and simulator for predictive design of the future charging hubs.
3. Developing of a Power and Energy Management System for the energy hub.
4. Running the developed EMS against Plant Model (software)
5. Programming the developed EMS for PLC and compiling Plant Model for Real-time-target (Control Hardware in the Loop - CHIL).
6. Implementation of a hybrid e-bus charging station emulator, integrating PV production and a stationary buffer battery storage (Power Hardware in the Loop - PHIL).
7. Testing and pre-commissioning the EMS by running emulator type simulation scenarios (PHIL).

Fast charging loads and the peak power demand

Measurements from the charging stations are fundamental for the study. The total peak current demand on an actual e-bus charging station located in the city of Turku, Finland and its impact on voltage stability at a dedicated transformation sub-station is illustrated in Figure 1. Samples were completed at the Turku Airport in April 2018 and voltage, current and power were measured.

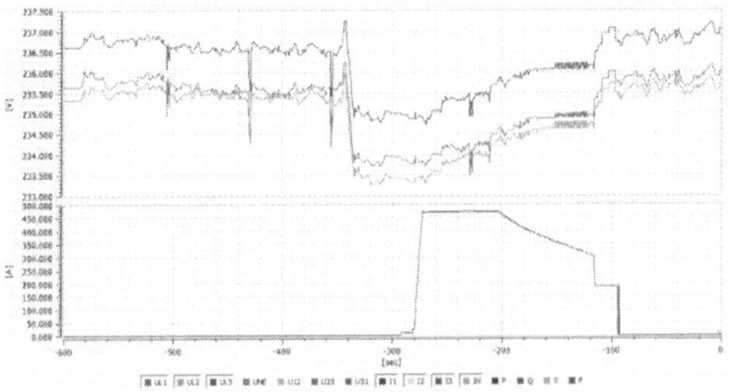

Fig. 1. e-bus power demand at the fast charging hub in Turku Airport, Finland.

The power quickly rises to 330 kW in constant current mode and when the battery approaches full charge, the charging changes to constant voltage. The problem in charging several e-buses with high power is the maximum power the local transformer can provide. A proposed solution to control high peak demands is the use of a buffer battery to reduce the high consumption peaks, especially in areas with limited grid capacity.

e-bus charging simulator

To estimate the charging hub peak power demand, including several charging stations, an e-bus charging simulator was developed. Simulator can model a system with several bus lines and estimates how many e-buses are simultaneously charged at the hub, Figure 2 shows the model concept.

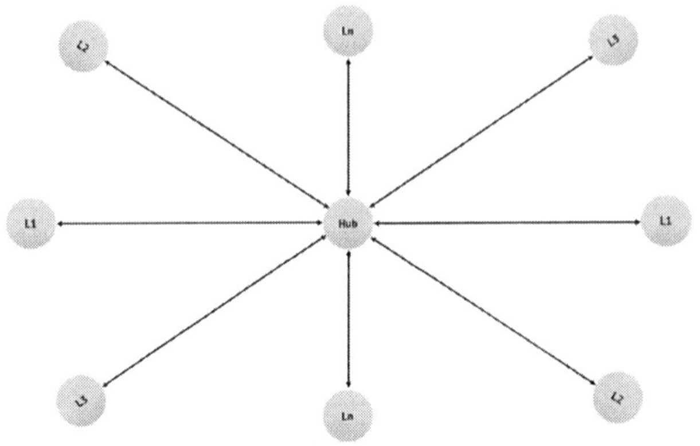

Fig. 2. Charging hub conceptualization.

Information about the bus lines – like time schedule, length of the line, number of the buses and service interval – is given as an input to the model. Additionally, the results from the charging station measurements at Turku Airport were used to model the behavior of the individual charging stations.

The model calculates energy consumption of each bus in the system and updates the state of charge (SOC) of their batteries accordingly. When the bus arrives at the charging hub, it charges the battery until it reaches the maximum limit or until the scheduled departing time. Bus arrival times to the hub are modelled by using probability distribution type of Monte Carlo simulation. Bus driving times between the hub and their start and end points are stochastic variables. Taking randomness into account makes it possible to assess charging load in a more realistic way as the deviation in driving time is relatively large compared to the charging time at the hub. As a result of the simulation, a load curve for charging hub is generated, Figure 3.

The load curve includes statistical parameters like mean and deviation of the charging power for the selected time resolution. Modelling was done for the Turku Market Square where majority of local bus lines cross each other. The simulator can be used to model any number of bus lines, in this study a maximum of five bus lines were modelled.

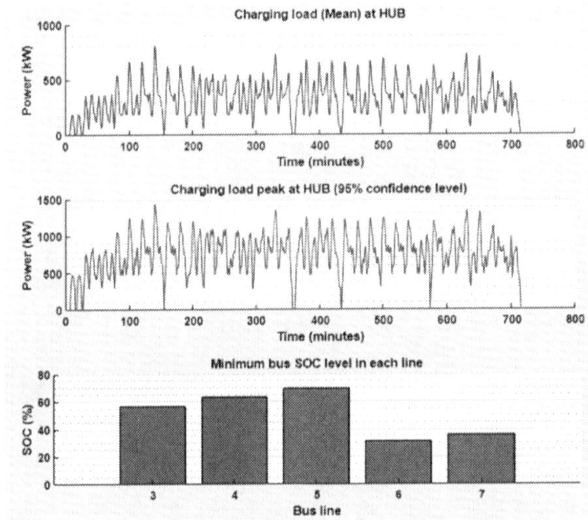

Fig.3. A simulation of a centralized charging hub at the market square in the Turku city center for 5 bus lines.

PV Generation Simulator

In order to provide PV energy data for the EMS simulations, the PV Generation Simulator [2] including a graphical user interface (GUI) developed at Turku UAS New Energy Research Center (NERC), was added to the study. It is utilized to predict the energy yield for different systems configurations. The graphical user interface (GUI) was built using MATLAB software and was developed to couple most of the PVLIB Toolbox [3] libraries and models from Sandia National Laboratories allowing an easy use for the user. The system used for the simulation was the Turku Airport 0.9 MW Carport PV plant illustrated in Figure 4.

Fig. 4. Carport PV plant (0.9 MW) simulated for the Turku Airport parking lot (Google Earth).

Energy Management System

The Energy Management System dedicated to control the energy exchanges within the LVDC microgrid [4] was developed at Haute Ecole Spécialisée de Suisse occidentale (HES-SO), Figure 5. The EMS has the objective to be implemented directly on the Hub Emulator at Turku UAS NERC.

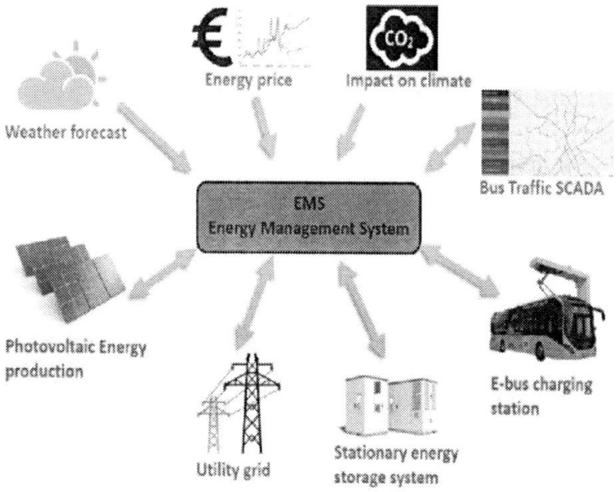

Fig.5. EMS conceptualization by HES-SO.

A strong emphasis is placed on energy services provided to the local electricity distribution grid for a more sustainable electric mobility by adding the use of PV systems for the energy storage systems. For this, a principle of the three hierarchy levels similar to the one used in the control of electrical grid was applied [5], Figure 6.

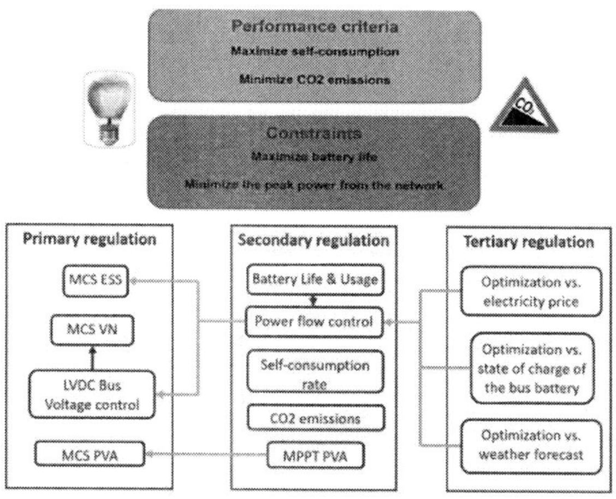

Fig.6. Performance criteria and Hierarchical Levels for the Energy Management System by HES-SO.

A primary control regulates all local power exchanges to maintain the DC bus voltage level within tolerances. The primary regulation will follow the instructions given by a higher-level secondary controller, which includes controlling energy exchanged with the battery and with the LVAC distribution network. The secondary control addresses system-level issues such as increasing battery life expectancy by limiting its use, maximizing self-consumption, minimizing CO2 emissions, and minimizing energy consumption in general. The reduction of the maximum peak power demanded from the network is also in charge of the secondary controller. A tertiary control deals with optimization, management, and overall system regulations. All predictive models of weather forecasts or price forecasts can be included in this level.

The primary and secondary controls for the EMS are implemented in Matlab/SIMULINK environment with the Macroscopic Energy Representation [6]. The EMS is tested with a real-time simulation platform (OPAL-RT) and coupled with a Programmable Logic Controller (PLC) with the down scale charging emulator.

Opportunity charging hub emulator

A down scale model in form of a LVDC microgrid based on a common energy distribution bus at 700 VDC was developed at TUAS NERC with industrial converters, Figure 7. The 1:100 down scaling applies to power and current. The relatively high voltage for the DC Bus makes the use of standard industrial components possible. A 3 kW converter configured to operate as Active Front End (AFE) between Low Voltage AC grid and DC Bus is controlling the DC bus voltage. A 51.2 V 20Ah LiFePO4 battery connected to the DC bus with 3 kW bidirectional DC/DC converter emulates the stationary buffer storage. A DC power supply emulating PV energy production is connected to the DC bus through a MPPT controller. Finally, the electric bus battery charger is achieved with a second 3 kW inverter feeding energy back to the grid. This solution provides a maximal simulation flexibility in comparison to a real e-bus battery. Most of the down scale energy hub protections are operated by the industrial converters and power supplies. A start-up sequence, protections monitoring and communication with the converters are run in PLC Power Management System (PMS). The global Energy Management System (EMS) for the whole hub is hosted in another PLC. The EMS acquires electrical values from dedicated sensors and measuring interfaces and so it calculates the power transferred by each equipment. This generates a reference current for the local energy buffer converter on the basis of the Battery State of Charge, the PV plant power and forecasted e-bus energy demand, provided by the OPAL-RT Real Time target. Scenarios and variable grid conditions can be emulated by the 50 kVA Grid simulator - amplifier. The LVDC bus emulator represents a very flexible and versatile platform for development and evaluation of control concepts for e-bus charging hubs integrating RE production.

Fig. 6. Down scale Energy hub emulator at Turku UAS NERC.

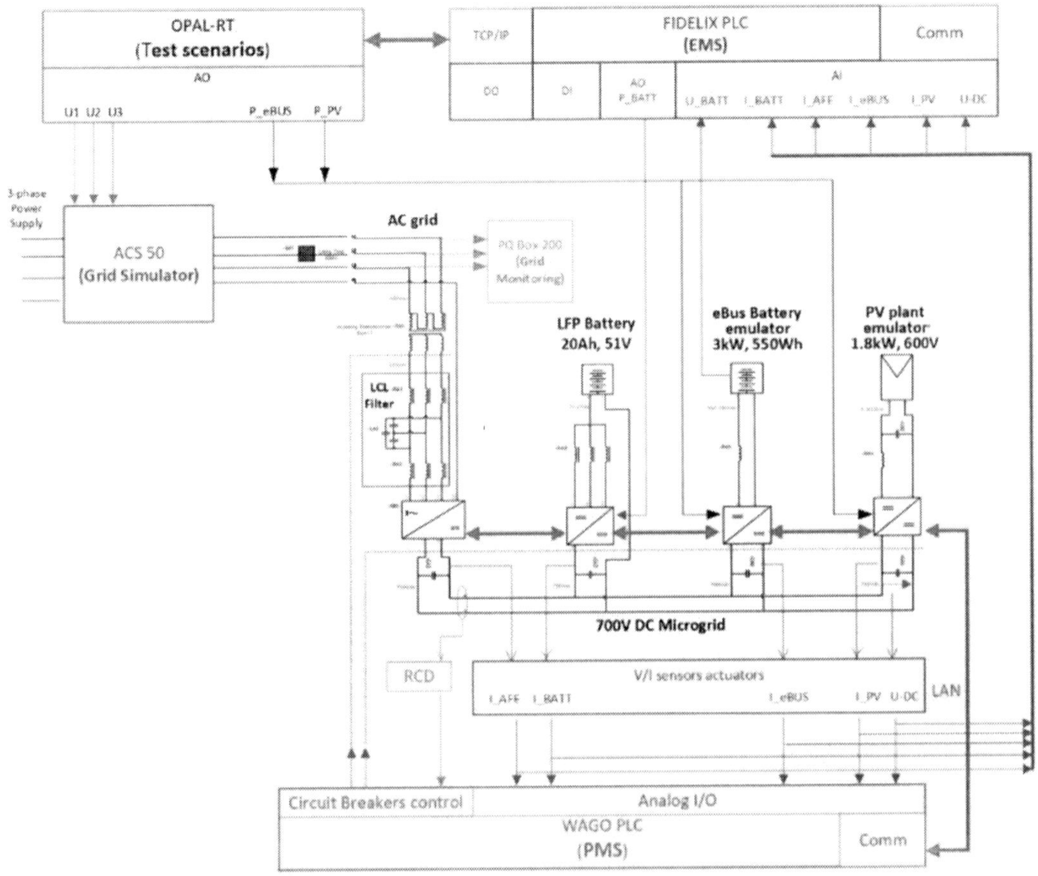

Fig. 7. Down scale Energy hub emulator schematic.

Testing EMS simulator

A typical scenario based on measured e-bus charging cycles during several days on the pilot site at Turku Airport was run to evaluate both the simulation model and the functionality of the EMS, Figure 8. The measured values for the power delivered to the E-bus battery are presented in the first plot from the left at the bottom (Green Curve). The potential renewable energy production during the corresponding time completed the EMS input for this scenario (Blue curve). The considered potential PV energy production was calculated based on measured values from both, the solar irradiance station and the PV installation at Turku UAS.

The simulation results of this scenario are presented on 3 plots at the top of Figure 8, current from the grid, current from buffer battery and battery SOC (order from left to right). Each presented curve is displaying the current fed-in to or delivered by the DC BUS for each of the corresponding equipment.

As a result of the support given by the stationary buffer battery, the power delivered to the charging hub does not come over the limit fixed in the constraints to the EMS. The SOC of the stationary buffer is maintained high, due to the relatively small energy required for one single bus trip. This is even improved with the locally produced renewable energy which can be fed directly to the E-bus, leading to maximized lifetime for the buffer battery and increased self-consumption factor.

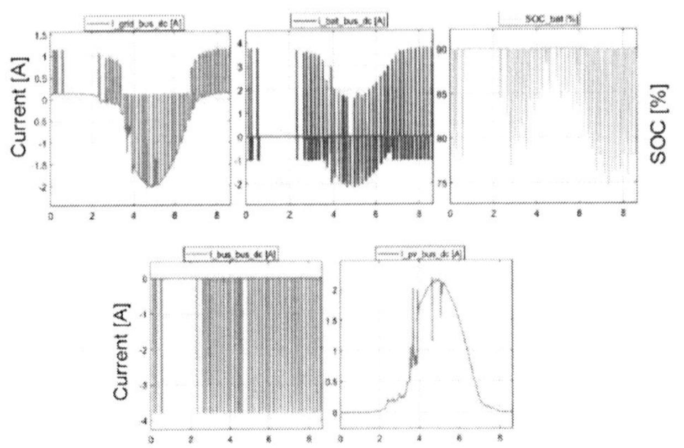

Fig.8. Simulation results for an actual charging station in Turku area, EMS by HES-SO

Simulating fast charging hub with EMS

In order to test the EMS for a charging hub with several bus lines running during the day, 2 scenarios were simulated with the e-bus charging simulator and their profiles were used as an input for the EMS.

First scenario considers 5 bus lines running for 12 hours a day, this means that a maximum of 10 buses might be charging simultaneously at some time during the day. A second scenario (optimized) considers only 3 bus lines running for 12 hours, so a maximum of 6 buses might be charging simultaneously. Irradiance data from Turku Finland for the PV simulation was selected during summertime, so they include sunny and cloudy conditions for the solar energy; Figure 9 shows the irradiance profiles used for the simulations. Results from EMS simulations are shown in Figure 10.

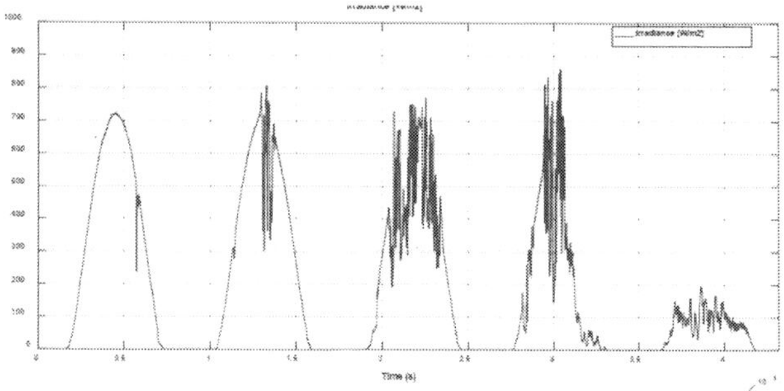

Fig. 9. Five days solar irradiance at Turku-Finland considered for the simulation.

Scenario 1

- Battery capacity 332 kWh
- Battery power 222 kW
- PV plant 920 kWp
- 10 e-bus charging at 300 kW

Scenario 2

- Battery capacity 332 kWh
- Battery power 222 kW
- PV plant 920 kWp
- 6 e-bus charging at 300 kW

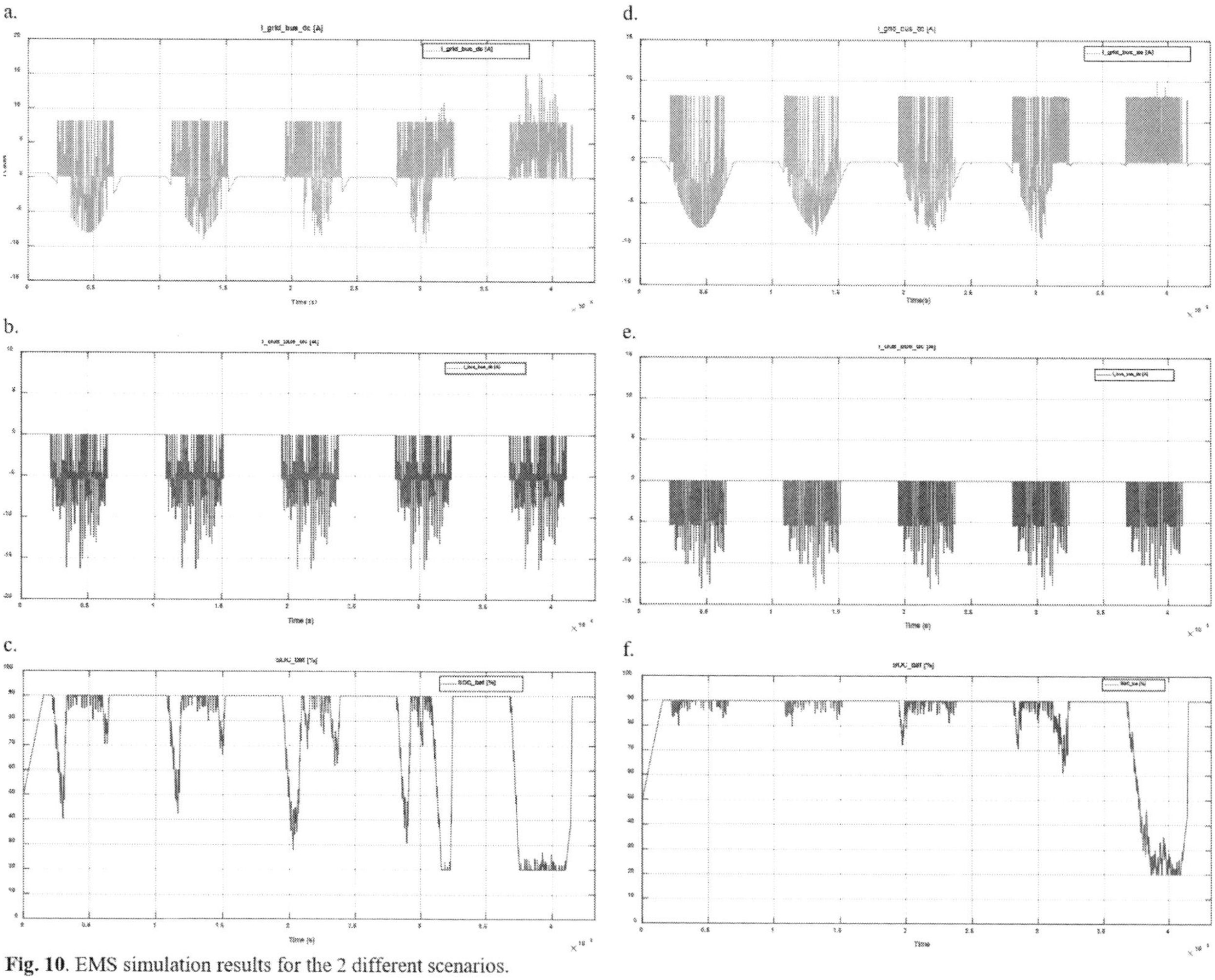

Fig. 10. EMS simulation results for the 2 different scenarios.

Results from simulations

From figure 10, results from the two scenarios can be compared.

In scenario 1, 5 bus lines running during the day, demonstrates that the EMS helps to shave the power demand from the gid, plot (a), especially during sunny days. Here, the e-buses can be charged, plot (b) by using the energy stored at the buffer battery from the PV Plant and energy from the grid, this last to avoid the complete discharge of the battery. The battery keeps the SOC above 40% (plot c) until the 3rd day, when the solar energy has some fluctuations due to cloudy conditions. During cloudy conditions, days 4 and 5, simulations show that the charging demand is oversized compared to the capacity of the buffer battery which results in very low SOC for the battery with no opportunities to be recharge it and generating high power peak demands from the grid.

Scenario 2, is presented as an optimization for the first scenario by keeping the size of the battery and by reducing the amount of e-buses running simultaneously. From the simulations it can be seen that the power from the grid was kept controlled (plot d) by reducing the load demand from the buses (plot e), the battery SOC was improved until the fifth day when cloudy conditions highly affected the production of the PV plant, and the battery stays in low SOC due to small capacity and still high demand.

3. OUTCOME

- Software simulation platform was proven to be functional
- EMS offers flexibility for different configurations and control strategies
- With this simulation tool, sizing of the PV, BESS and number of the Fast Chargers can be optimized
- Hardware CHIL setup was also commissioned and was proven to be functional

4. NEXT STEPS

- Integration of the BMS PLC for PHIL environment is next step in progress
- Future development of the EMS control strategies to include PV production and electricity price forecasting
- Improving of the e-bus energy consumption forecasting by integrating route elevation and seasonality

5. ACKNOWLEDGEMENTS

The work has been carried out under the New Energy research group at Turku UAS Finland and was partly supported by European Regional Development Fund (ERDF) funded SeBNet project (Smart Electric Bus Network Integration) Project code A73064.

References

[1] S. Hinkkanen, "A Hardware-In-The-Loop Simulator of an Electric Bus Charging Station," Thesis Work, available: http://urn.fi/URN:NBN:fi:amk-2018062114168, TUAS, 2018.

[2] S. Ranta, H. Huerta, O. Huhtala, A. Heinonen, J. S. Stein. "Modeling Tool Validation for the Yield Prediction of Bifacial East-West Vertical PV System in Nordic Conditions". EU PVSEC (2020)

[3] Sandia National Laboratories, "PVLIB_ToolBox for Matlab", https://pvpmc.sandia.gov/applications/pv_lib-toolbox/matlab/

[4] M. P. Rodriguez, "Energy management of an energy HUB for electric bus charging station", Bachelor thesis at HES-SO ValaisWallis, 2019

[5] L. Meng et al., "Review on Control of DC Microgrids and Multiple Microgrid Clusters" IEEE Journal of Emerging and selected topics in Power Electronics, pp. 928-948, 3 September 2017

[6] J. Solano-Martinez, L. Boulon, D. Hissel, M.-C. Pera, and M. Amiet, "Energetic macroscopic representation of a multiple architecture heavy duty hybrid vehicle," IEEE Vehicle Power and Propulsion Conference, 2009. VPPC '09, pp. 1322–1329, 2009.

THE BENEFITS OF COUPLING SOLAR ENERGY WITH FLEXIBLE DESALINATION AND ELECTROLYSIS

M. Ginsberg, Z. Zhang, A.A. Atia, V. Fthenakis
Columbia University
New York, NY 10027

ABSTRACT: The economics of both desalination and hydrogen production from water electrolysis are dominated by the cost of energy, so widespread availability of inexpensive solar energy incentivizes the desalination and electrolyzer industries to further advance their technologies. This study focuses on the coupling of the production of potable water, deionized water and hydrogen gas from seawater. We show that such coupling is cost-effective not only for hydrogen generation, but also for desalination, as desalination can use waste heat from electrolysis. Additionally, such coupling can be optimized when electrolyzers operate at high current density, using low-cost solar and/or wind electricity, as high density operation increases both hydrogen production and heat generation. Results of techno-economic analysis define thresholds of electricity pricing, current density, and operating temperature that make "green" electrolytic hydrogen cost-competitive with hydrogen from steam methane reforming. By using historical and projected hourly electricity prices in the US-southwest, we estimate that hydrogen can be produced from seawater in desalination-electrolyzer systems at prices near \$2/kg H_2, achieving cost parity with currently produced hydrogen. In addition, thermal energy for desalination could be supplied by the electrolyzer, creating a closed loop for continuous green H_2 production.
Keywords: photovoltaics, solar energy, desalination, electrolysis, cogeneration

1 INTRODUCTION

The photovoltaics (PV) industry has accomplished phenomenal cost reductions through improving efficiency in manufacturing and deployment. The prices of PV electricity have been reduced to parity with unsubsidized electricity rates in arid areas of high solar irradiation that need to expand the capacity of water desalination and low-cost electricity brings a new focus on electrolytic hydrogen generation. In addition, the cost of solar thermal energy is decreasing and coupling it with low-cost thermal storage enables solar energy's penetration to base-power markets. The decrease of solar energy prices over the last ten years and DOE targets for further decreases in the next ten years are shown in Figure 1.

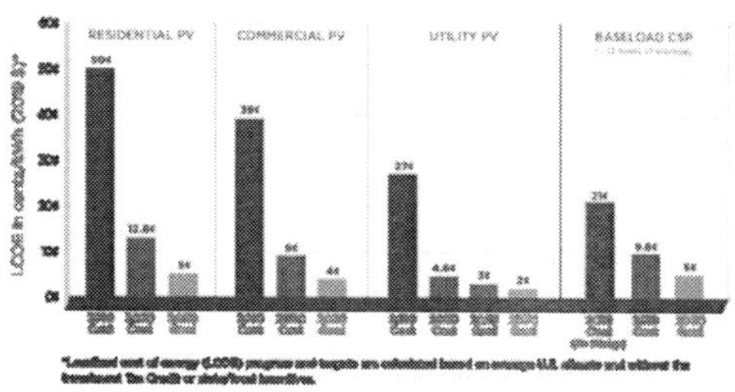

Figure 1: Solar energy prices and DOE targets; source Solar Energy Technologies Office (SETO)

The economics of both desalination and hydrogen production from water are dominated by the cost of energy and the availability of inexpensive solar energy creates markets and offers incentives to the desalination and electrolyzer industries to further advance their technologies. Co-adoption of solar and desalination technologies has a near-term potential as currently operating desalination plants can be supplemented with co-located solar plants, drastically reducing cost. The cost leverage is enhanced when grid-integrated systems also take advantage of time of use pricing in grid electricity and offer ancillary services to the grid, supplementing or displacing gas peakers. We postulate that low-cost production of hydrogen (H_2) using water electrolyzers co-located with solar plants can also be enabled when the current density of the electrolyzer is dynamically optimized based on the variability of the solar resource and energy demand. Both technologies offer ways to store, instead of curtail, solar energy. Together with co-located solar PV, we expect that an integrated desalination and electrolysis facility can benefit from multiple synergies, including but not limited to: i) desalination providing a fresh water source for water electrolysis, ii) energy storage in the form of clean H_2 fuel, iii) more effective grid services, and (iv) more flexible operation that enables higher utilization of PV-generated electricity (decreased curtailment) and lowering of the Levelized Costs of H_2O (LCOW), H_2 (LCOH_2), and electricity (LCOE). The concept of integrated technologies and bridging industry sectors that are commonly viewed individually, is shown by the schematic in Figure 2.

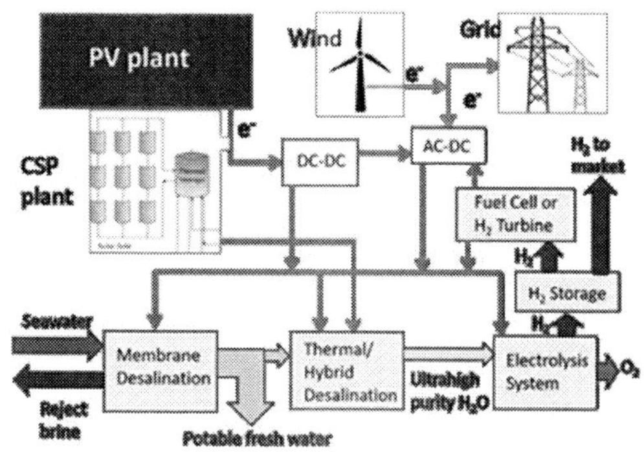

Figure 2: Integrated solar-powered plant for the production of fresh water and hydrogen from desalination and water electrolysis, respectively.

2 INTEGRATION PATHWAYS

2.1 PV_RO_Desalination

Ongoing research at Columbia University shows that PV can cost-competitively provide reverse osmosis (RO) desalination loads and ancillary services to the grid, and that emerging solar thermal desalination technologies have good potential for water with high salt content and/or zero liquid discharge (ZLD) applications. We designed integrated solar energy-desalination systems that can operate at flexible loads and provide ancillary services to the grid [1]. This concept is shown in Figure 3.

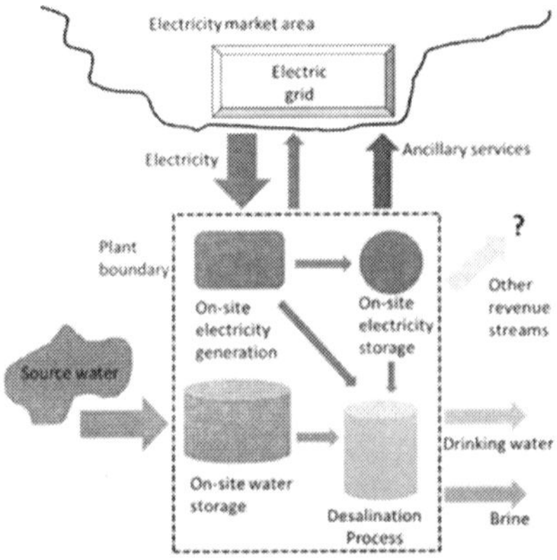

Figure 3: Overview of an Integrated PV-flexible desalination model

2.2 PV/Solar Thermal -Desalination- Electrolysis

On a parallel front, as the economics of producing hydrogen from water electrolysis are currently dominated by the cost of electricity, enabling flexibility in the operation of electrolyzers will become much more important in a renewable energy future.

Although the cost of hydrogen production from electrolysis currently costs up to 2 or 3 times more than hydrogen from SMR, as this cost is a strong function of the price of electricity, and very low-cost electricity from the sun and wind is available, clean hydrogen production is getting a new look. With a new focus on the reduction of greenhouse gases, policy is driving hydrogen research forward, particularly in Europe, Japan, Australia, and the US. The expectation is that these policies and associated incentives would have the same catalytic effect in clean hydrogen production that policy in Germany during 2001-2005 had in driving the deep reductions in solar and wind electricity production cost.

If renewable energy is used to power electrolysis, the clean hydrogen produced could be used to power automobiles and manufacture chemicals, materials, and synthetic fuels. However, in scenarios of large hydrogen production, especially in arid areas and islands, seawater may be the only sustainable source of water for electrolysis. Consequently, many R&D efforts aim at developing corrosion-resistant electrodes or new electrolyzer technologies for direct seawater electrolysis. However, our study demonstrates that coupling desalination and electrolyzer systems is a cost-effective solution for seawater electrolysis, and suggests that such coupling can be optimized to offer system cost reductions and further advance the benefits that low-cost renewable energy electricity offers to the economics of integrating desalination and electrolysis. Powering these systems with solar and/or wind electricity offers a simple pathway towards cost-parity hydrogen production.

Although intermittent high current density operation would increase the system's capacity ratio and annual production of hydrogen, there is much uncertainty about the impact of (>3 A cm^{-2}) on the lifetime of the electrolysis stack under dynamic operating conditions. Thus, we will examine the potential of adverse effects, that high current density may have on the durability of PEM electrolyzers. This work will systematically explore the influence of peak current density and operational schedule on performance degradation, and will guide the industry in two ways: a) determine optimal operational profiles for electrolyzers in the integrated solar-powered desalination/electrolysis facility that will best balance the trade-offs between efficiency, throughput, and lifetime, and b) guide the development of the next class of electrolyzers that will be optimized for variable high current density operation. This may be accomplished by varying materials of construction, catalyst loadings, membrane thickness, and cooling mechanisms in the new electrolyzers.

On the system level, the trade-off between flexibility and cost will be examined by employing TOU pricing in conjunction with solar time series and supply "duck" curves under current and future scenarios for specific ISO regions in the US. The objectives of our project is to increase the flexibility and reduce the capital cost, both via smart integration and optimized electrolyzer operation. Our techno-economic analysis examines the potential for drastic cost reductions in balance of systems and smart integration, which, in addition to very inexpensive energy, would create value by offering multiple benefits. Our research shows that H_2 from electrolysis becomes cost competitive with the current production of H_2 from steam methane reforming (SMR) when electricity prices are below $0.03/kWh and the electrolyzers operate at increasingly high current densities with declining electricity prices.

3 OVERVIEW OF SEAWATER DESALINATION TECHNOLOGIES

Desalination technologies and industry have advanced significantly during the past two decades to meet the growing freshwater demands stimulated by the compounding issues in both water quality and quantity in many regions of the world. There are more than 20,000 desalination plants operating worldwide with a total operating capacity of about 100 million m^3/d. Seawater desalination represents the largest percentage of online global capacity at 60% [2]. Once considered energy-intensive and cost-prohibitive, desalination technologies are now being embraced and are considered to be important elements of water portfolios for many communities. Desalination can be performed through thermal means by evaporation/condensation or mechanical (pressure-driven) means such as osmotic membrane processes. Thermal technologies require both heat and electricity for separation of freshwater from saline water while pressure-driven membrane technologies require only electrical energy, which is converted to mechanical energy for separating the dissolved solids from the saline water. Currently the lowest cost membrane-based technology for the desalination of seawater is reverse osmosis (RO), and the lowest cost thermal technology for seawater is multi-effect distillation (MED). The energy required per unit of freshwater product is expressed as specific Electric Energy Consumption (SEC), with units of kWh/m^3 and Specific Thermal Energy Consumption (STEC), with units of kJ/m^3 or $kWh_{(th)}/m^3$. Energy costs are around 40-60% of the cost of freshwater produced by seawater desalination. MED processes require about 45-128 $kWh_{(th)}/m^3$ of thermal energy and ~1.5 kWh/m^3 electricity for circulatory pumping. Seawater reverse osmosis process (SWRO) for producing potable water, requires about 3.5 kWh/m^3 for the Mediterranean Sea, Atlantic and Pacific oceans, and 4 kWh/m^3 for the Arabian Gulf [2].

In most applications, SWRO can produce potable water at costs lower than MED. However, the product water from RO would contain total dissolved solids (TDS) in the range of 200-500 mg/L, whereas MED produces the distilled, deionized water that can directly be fed into electrolyzers. Thus, RO units would have to be supplemented with additional processes, such as ion-exchange, or implementing multiple passes, in order to demineralize the water to the quality required by electrolysis feed of less than 0.5 mg/L TDS. These additional processes add to the cost of product water which may lessen or even eclipse the RO cost advantage when coupled with electrolysis. as discussed in the following section.

To estimate the cost of desalinating seawater and producing ultrapure water, engineering, procurement, and construction (EPC) costs were computed based on the Global Water Intelligence database [3]. Average unit EPC costs were estimated for small (<1000 m^3/day) and large (>50,000 m^3/day) plants for RO plants fed by either seawater or pure water. Furthermore, these data were restricted to online plants that serve industrial end users. O&M costs comprise labor & maintenance, chemicals, and membrane replacement (cumulatively estimated as 4% of capex), while energy costs were considered separately. For ion-exchange, water costs can range between \$0.08 - \$0.21/m^3; thus, an average cost of \$0.145/m^3 was added to the LCOW of any RO cases that included ion-exchange. The results of the comparison are shown in Figure 4.

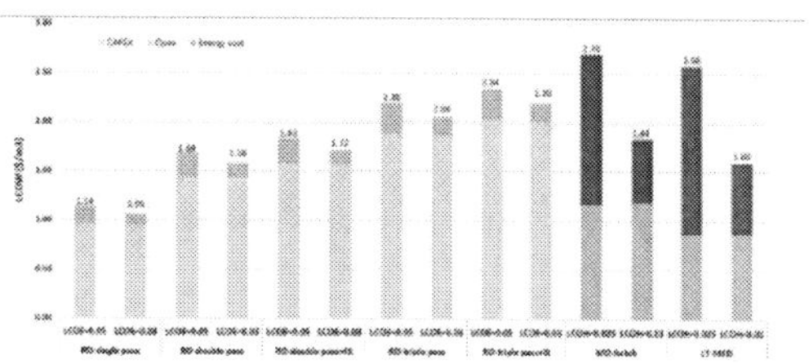

Figure 4: LCOW breakdown of different desalination technologies at a small (120 m^3/day) capacity (LCOE for MD-batch and LT-MED is 0.05 \$/m^3). LCOH of \$0.01/kWh(th) is assumed for waste heat from electrolyzers directly fed into LT-MED or MD systems. Interest rate and plant lifetime were set to 4% and 20 years, respectively.

As shown in this figure, although conventional RO can produce potable water at levelized costs of water (LCOW) much lower than that of thermal desalination plants, the addition of passes and/or ion-exchange systems to completely de-mineralize the RO permeate water, and the use of electrolyzers' waste-heat as sole thermal input into MED and MD, makes the LCOW of thermal desalination lower than that of RO-triple pass and RO+IX (assuming that waste heat will only cost \$0.01/kWh).

4 OVERVIEW OF ELECTROLYZER TECHNOLOGIES

As discussed in the introduction, currently hydrogen is produced worldwide by fossil sources employing coal gasification or natural gas reforming processes. Water electrolysis is the only method that does not emit carbon dioxide into the atmosphere, generating "zero emissions" if associated with electricity produced from renewable sources.

Commercial water electrolyzers are based on either alkaline electrolysis (AE) or polymer electrolyte membrane (PEM) electrolysis. The electrolyte in alkaline electrolyzers is usually an aqueous solution of KOH. Traditionally this technology has been the dominant technique in large-scale electrolysis and has been in use for several decades. It is a mature technology which can still compete with the much newer PEM technology. Its main disadvantages compared to PEM technology are the low current density (e.g., 0.1 to 0.4 A cm^{-2}), larger stacks, and lower flexibility in terms of load on the system. It is because of these properties that the demand for PEM technology has grown over the last few years; experts forecast that PEM technology will be dominant by 2030, but critical gaps must still be closed in the areas of materials, operating conditions, durability and economic viability [4].

The electrolyte in PEM is a solid polymer capable of conducting protons ($H+$), which results in an acidic environment. Catalysts capable of operating under these conditions are precious metals such as platinum and iridium, and the acidic nature results in short lifetimes compared to alkaline technology. The high current densities achieved by a PEM electrolyzer do however allow for very compact design. In PEM electrolyzers the electrodes are pressed against the electrolyte in a so-called zero gap configuration. These layers together are also referred to as the membrane electrode assembly (MEA) and form the heart of the cell. The MEA together with current collectors and separator plates make up the cell that, on average, is only 5-9 mm thick. This structural compactness and the high current density result in a much smaller footprint compared to alkaline technology.

As discussed earlier, our investigation entails PEM electrolyzers, because despite alkaline electrolyzers being cheaper, PEM are more amenable to high current density operation and, therefore, get more benefit from the low prices of solar and wind electricity. The installed system cost for 10 MW PEM is reported to be 755 €/kW (our best case), thus ~\$900/KW. Limited data on cost reductions beyond 2020 are available, though some projection take the cost down to 250 €/kW by 2030.

4.1 Electrolytic Hydrogen Production

Traditionally, electricity has been the single largest cost contributor to the $LCOH_2$ from electrolyzers, which are typically operated at constant current densities [5][6]. The operating current density of PEM electrolyzers is currently in the range of 1-2 A cm^{-2}, although stable operation at much higher current densities is possible with modifications to the electrolyzer power electronics. PEM electrolyzer efficiencies are commonly rated at 1.7 A cm-2. As the current density increases and more current is introduced into the unit, more hydrogen is produced although the efficiency of the electrolyzer is reduced. Figure 5a shows the current-voltage (I-V) curves and corresponding efficiencies based on experimental IV curves reported by Nel Hydrogen for PEM electrolyzers containing either 60 μm thick Nafion membranes [7]. Figure 5b presents an illustration of the tradeoff between electrolyzer current density and efficiency. In the following we investigate this trade-off by comparing modes of operation at the currently most typical current density of 1.7 A/cm^2 and at current densities of 2.6, 2.7, 3.2, 4.4 and 6 A/cm^2.

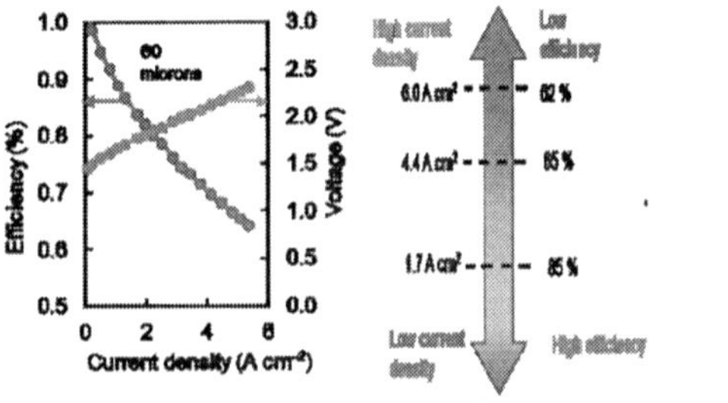

(a) (b)

Figure 5: Efficiency vs. current density in PEM electrolyzers. (a) Current-voltage (I-V) and corresponding efficiency curves based on experimental IV curves reported by Nel Hydrogen for PEM electrolyzers containing 60 µm Nafion membranes (b) Illustration of the tradeoff between electrolyzer current density and efficiency corresponding to 60 µm Nafion membranes

For all operating modes, a cumulative degradation rate (DR) of 0.15 % of cell voltage per 1000 hours was assumed, consistent with reported degradation rates for PEM electrolyzers [8]. Applying the DR as a percentage of the operating voltage had the effect of scaling DR proportionally to Jop since Vop increases with Jop (Figures 5a); thus higher Jop was assumed to entail higher degradation rates. An upper limit or threshold on Vop was set to Vmax = 2.9 V, above which electrochemical generation of ozone or other reactive oxygen species at the anode can prematurely accelerate electrolyzer degradation by damaging the membrane, ionomer, and/or other components.

The availability of abundant low-cost electricity from variable renewable energy (VRE) may justify operating electrolyzers at higher current densities, thus increasing H_2 production so that unit CAPEX (\$ kg^{-1} H_2) is decreased. Using an average electricity price of \$0.03 kWh^{-1} corresponding to CAISO 2020 data, with, we found an electrolyzer run at constant operation at various current densities within a range of 0.1 to 6 A cm^{-2}, resulted in a minimum LCOH of \$2.59 kg^{-1} at 2.7 A cm^{-2}. For the ERCOT 2019 data showing a mean price of \$0.04/kWh, $LCOH_2$ is minimized at a Jop of 2.6 A cm^{-2}, whereas for the ERCOT 2020, which has a mean electricity price of \$0.022 kWh^{-1}, we find a minimum $LCOH_2$ of \$1.93 kg^{-1} at a constant Jop of 3.2 A cm^{-2}. This shows that as electricity prices increase the optimal Jop is decreased. Also, as current density increases, the energy charge (OPEX), normalized per kg of H_2, increases and normalized CAPEX decreases because high current density operation uses more electricity and produces more water (Figure 6).

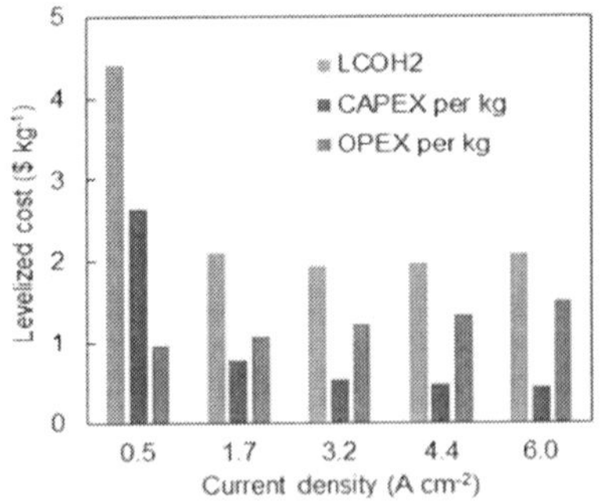

Figure 6: Electrolyzer CAPEX, OPEX and LCOH2 as a function of current density. a) LCOH2, Capex and Opex, corresponding to average LCOE of \$0.022/kWh from ERCOT 2020 wholesale prices; As current density increases, more electricity is used increasing the Capex, while the H_2 production increases, thus decreasing Capex (per kg H_2 produced)

Minimum LCOH2s were calculated based on mean electricity prices in California and Texas. By extending the electricity price range from \$0.0 to \$0.07 kWh^{-1} we model the impact of potential future electricity prices on the optimal current density and $LCOH_2$. As shown in Figure 8, hydrogen from electrolysis becomes cost competitive with the current production of H_2 from steam methane reforming (SMR) at electricity prices below \$0.03/kWh.

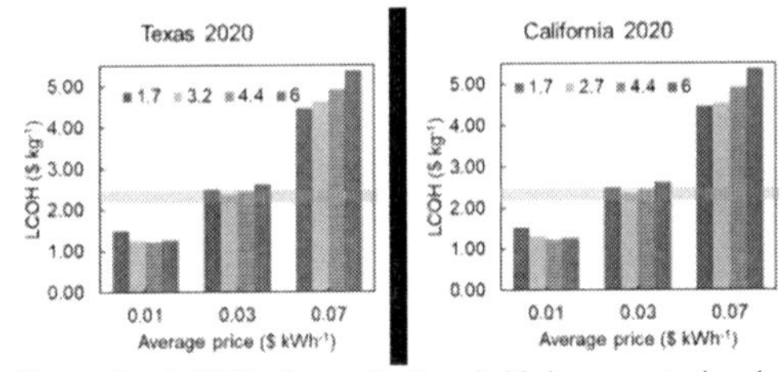

Figure 7: LCOH for typical and high current density operation as a function of average electricity price. The gray band shows the "SMR parity" range (\$2.08 kg^{-1} to \$2.27 kg^{-1}), which is the cost of producing H_2 from SMR; \$2.08 corresponds to a design capacity of 379,387 kg H_2 per day at a 90% capacity factor and a natural gas price of \$10 MMBtu^{-1}, and \$2.27 kg^{-1} corresponds to the same but with carbon capture and sequestration. Current density of 4.4 A/cm^2 reaches cost parity with SMR operation when powered by electricity costing less than \$0.03/kWh.

4.2 Electrolytic Heat Generation

Operating at cell voltages above 1.48 V generates excess heat. Heat generation occurs especially in PEM electrolyzers which usually operate at higher current densities than alkaline units; PEM electrolyzers employ a cooling system to dissipate excess heat. In this section, we are examining the use of such excess heat as an input to thermal desalination units.

The waste heat is produced at a rate directly proportional to the difference between the actual cell potential U_c and the thermoneutral voltage U_{tn} [9].

$$P_{wh} = I_{el} \cdot (U_c - U_{tn})$$

Where P_{wh} is the power dissipated as heat in a water electrolysis cell operating with an electric current I_{el} and a difference of potential Uc between anode and cathode. The estimated, according to this equation heat generation for a 10 MW PEM electrolyzer is shown in Figure 7. It is noted that heat generation is a strong function of the cell current density as increase of operating current density is associated with increasing electricity input, and as the hydrogen generation efficiency decreases (as shown in Fig. 5b), a larger fraction of the electricity input is dissipated as heat.

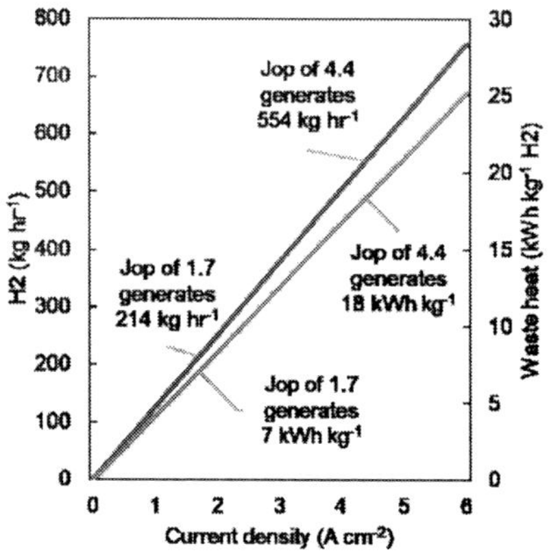

Figure 8: Hydrogen and Waste heat generation for various current densities for a PEM electrolyzer rated 10 MW at 1.7 A cm^{-2} and operating at 60 °C

When operating at a rated current density of 1.7 A cm^{-2}, a 10 MW PEM electrolyzer uses about 46 m^3 of high purity water per day to produce 5,137 kg of H_2 and 36,240 kW(th) of heat per day, assuming 100% capacity factor. If the current density of the same PEM electrolyzer is increased to 4.4 A cm^{-2}, it will consume about 120 m^3/day water to produce 13,300 kg of H_2 and 244,560 kWh(th) per day. Using a specific thermal energy consumption (STEC) of 65.5 kWh/m^3 for LT-MED and 59.2 kWh/m^3 for Batch-MD, heat input from electrolysis to desalination, assuming 50% losses, can produce 276 m^3 and 300 m^3 of deionized water from MED and Batch-MD, respectively, per day when the electrolyzer operates at Jop of 1.7 A cm^{-2} The amount of water production from MED and Batch-MD units increases to 1867 and 2065 m^3, respectively, when the electrolyzer operates at Jop of 4.4 A cm^{-2}. Thus, the amount of waste heat from the electrolyzer is sufficient for producing from thermal desalination the de-ionized water needed for electrolytic hydrogen generation, even with some heat exchange losses.

Notably, our cost comparisons assume that excess heat from the electrolyzer can be directly used into LT-MED or MD processes. This is possible when electrolyzers operate at temperatures of 65 °C or higher; operation at lower temperatures will necessitate the addition of a heat pump with increments in capital cost and heat losses. Nel H2 reports that current vintage systems operate at 60 °C; our assessment shows that operation at 70 °C is also possible with minor modifications. In fact, a theoretical investigation by Scheepers et al. (2021) [4] indicates that the optimal stack temperature for operation at 1.7 A/cm2 or higher current density is in the range of 75-90 °C.

A Siemens patent [10] has previously considered the feasibility of the coupling we suggested. According to this patent, deionized water is produced under atmospheric pressure at operating temperatures between 60° and 70° in a low-temperature distillation system using waste heat from a PEM electrolyzer. It is stated that the capacity of the low-temperature distillation system is two times higher than the deionized water flow into the electrolyzer; thus, about 50% of the produced deionized water is available for other uses. Another use for the excess heat is preheating the process water before use. However, the amount of heat required to bring the process water up to temperature is generally rather small in comparison to the heat production of the electrolyzer.

As mentioned earlier, the potential of the excess heat from electrolysis lies in the thermal desalination techniques. MED and MD both rely on an external heat source, usually steam, as well as an electrical energy source for water pumping. In the absence of an external heat source the thermal energy would have to be supplied by electric or gas steam boilers making these processes much more energy demanding than RO. As shown by the simulation results in Figure 4, if all thermal energy would be supplied by the electrolyzers these thermal desalination technologies would produce de-ionized water at lower LCOW than RO.

5 DISCUSSION

Existing water-electrolysis methods rely on highly purified water, which is a precious resource. If green hydrogen generation becomes prolific throughout the world, then seawater may be needed as a feedstock. While new technologies such as membraneless electrolyzers [11] and special electrode coatings show promise that can sustain seawater splitting without chloride corrosion, their life duration potential is uncertain and these technologies have not been commercialized yet. However, large desalination units with operation time of 20 years or longer routinely use seawater and can produce fresh water at a cost in the range of \$0.50-\$1.0/m^3 while desalination using small systems would cost two times that or more. In either case, adding desalination would readily enable the use of seawater for hydrogen production with an increment to the $LCOH_2$ of only \$0.01-0.02/kg H_2, thus just 1% of the target LCOH of \$2/kg H_2.

6 CONCLUSIONS

Deep decarbonization of sectors such as transportation, chemical and metal production will require zero-carbon hydrogen production, which currently can only be provided by splitting water molecules via electrolysis. As the economics of producing electrolytic hydrogen are dominated by the cost of electricity, enabling operation of electrolyzers to take advantage of the variability of solar and wind resources will become much more important in a renewable energy future.

We addressed the feasibility and cost benefits of electrolyzer in scenarios of high current density operation on the $LCOH_2$ and the LCOW. Our simulations show that "green" hydrogen can reach cost parity with hydrogen produced from methane at electricity prices with a mean of \$0.03 kWh^{-1}, or lower. In these scenarios, the increased H_2 output enabled by high current density operation offsets the premium on total CAPEX, and the increased OPEX costs from lower efficiencies. If all thermal energy for desalination were to be supplied by the electrolyzers thermal desalination technologies would produce deionized water at lower LCOW than RO. Similar to how we can use integrated solar energy-desalination systems, that can operate at flexible loads, as discussed in 2.1, by taking advantage of low electricity prices during high periods of VRE penetration, and using the waste heat from electrolyzers to power desalination, the $LCOH_2$ from water electrolysis can be significantly decreased to achieve cost parity with SMR-produced H_2.

7 REFERENCES

[1] Atia A., Fthenakis V.M., Active-salinity-control reverse osmosis desalination as a flexible load resource, Desalination, 468, 114062, 2019.
[2] Gude V. G., Fthenakis V., Energy Efficiency and Renewable Energy Utilization in Desalination Systems, Progress in Energy, 2 (2020) 022003 https://doi.org/10.1088/2516-1083/ab7bf6
[3] Global Water Intelligence- DesalData Database, (2019). https://www.desaldata.com/ (accessed November 14, 2018).
[4] F. F. Scheepers, M. Stahler, A. Stahler, E. Rauls, M. Müller, M. Carmo, W. Werner Lehnert, Temperature optimization for improving polymer electrolyte membrane-water electrolysis system efficiency, Applied Energy, 283, 1 February 2021, 116270.
[5] Bertuccioli, L. et al. Study on Development of Water Electrolysis in the EU (The Fuel Cells and Hydrogen Joint Undertaking, 2014).
[6] Road Map to a US Hydrogen Economy (Fuel Cell & Hydrogen Energy Association, 2020). https://www.fchea.org/us-hydrogen-study.
[7] Ayers, K. PEM Electrolysis R&D Webinar: High Current Performance, 80° C (Proton Energy Systems, 2011), https://www.energy.gov/sites/prod/files/2014/03/f11/webinarslides052311_pemelectrolysis_ayers.pdf
[8] Fuel Cells and Hydrogen Joint Undertaking. State-of-the-art and Future Targets (KPIs): Hydrogen production from renewable electricity for energy storage and grid balancing using PEM electrolysers; https://www.fch.europa.eu/soa-and-targets
[9] S. Simone Pascuzzi, A.S. Alexandros Sotirios Anifantis, I. Blanco, and G.S. Mugnozza, Electrolyzer Performance Analysis of an Integrated Hydrogen Power System for Greenhouse Heating. A Case Study, Sustainability (2016), 8, 629; doi:10.3390/su8070629
[10] METHOD FOR OPERATING AN ELECTROLYSIS SYSTEM, 28.10.2015, Patentblatt 2015/44 https://patentimages.storage.googleapis.com/e0/56/aa/d51f03ff80d9e3/EP2780491B1.pdf
[11] Esposito, V. Daniel, Membraneless electrolyzers for low-cost hydrogen production in a renewable energy future, Joule 1.4 (2017): 651-658

HIGH QUALITY SOLUTIONS OF BUILDING-INTEGRATED PHOTOVOLTAICS (BIPV) – RESULTS OF A WORLD-WIDE COMPETITION IN 2020

Gerd Becker, Fabian Flade, Roland Krippner, Bruno Schiebelsberger, Walter Weber
Solarenergieförderverein Bayern e. V. (Bavarian Association for the Promotion of Solar Energy),
Friedrich List Str. 88, 81377 München, Germany
Phone: +49 (0) 89/278134-28, info@sev-bayern.de, www.sev-bayern.de

ABSTRACT: The competition "Architectural Award Building-Integrated Solar Technology" was organized for the eighth time since its establishment in summer 2000. With 159 schemes, there has been an increase in the number of countries taking part this time, with the jury reviewing competing schemes from 26 countries. Within the competition a first prize, four second prizes, a student prize and honorable mentions were awarded – with a total prize money of € 27,000. A four-story school building in Luxemburg with PV and solar thermal systems was awarded with the first prize. There were second prizes for an apartment building in Zurich, a high rise building in Paris, a church in Germany and a solar home in India. Five honorable mentions (non-monetary prizes) were endowed: a historic building, an ice rink, a bike shelter, a solar folding roof (all in Switzerland) and an equipment barn in the USA. Many high-quality solutions indicate, that BIPV can often be applied and is able to produce electric energy and fulfill high architectural requirements.
Keywords: Building Integrated PV (BIPV), Façade, Solar Architecture

1 INTRODUCTION

The competition "Architectural Award Building-Integrated Solar Technology" was organized for the eighth time since its establishment in summer 2000. Over a period of two decades the Solarenergieförderverein Bayern (Bavarian Association for the Promotion of Solar Energy) has therefore successfully established one of the very few international competitions at the interface between architecture and solar energy.

With 159 schemes, including nine entries for the "Students Award", the number of participants has almost doubled relative to the competition 2011. There is even an increase of 20 % compared with the high number of projects in 2017.

The number of participants and the quality of the projects entering the competition are testimony to the continuing topicality of the theme [1].

There has also been an increase in the number of countries taking part this time, with the jury reviewing competing schemes from 26 countries. The German-speaking are ultimately accounts for almost 78 % of the entries.

The current competition features world-renowned firms with Renzo Piano Building Workshop (Paris), Dominique Perrault Architecte (Paris) and Shigeru Ban Architects (Tokyo). In addition, successful solar architects such as Beat Kämpfen and René Schmid from Zurich have also taken part in the competition with several projects.

The overwhelming majority of buildings are new-build-schemes. Some more characteristics of the buildings with their solar systems are indicated in Table I.

Table I: Characteristics of Building Typology

Characteristic	Percent
Pure PV systems	80 %
PV systems and solar thermal	14 %
Only solar thermal	5 %
Residential buildings	50 %
Detached and semi-detached homes	32 %
Apartment schemes/High-rise residential buildings	19 %
Buildings for transport/infrastructure	5 %
Renovation of existing building stock	16 %

During all competitions – since 2000 - the share of pure PV systems in the competition has increased continuously over the years and accounts for more than 80 % of all entries in 2020. While 14 % of the buildings use combined systems, only slightly more than 5 % exclusively use solar thermal collectors.

In terms of the proportion of different building types, residential buildings account for just over half of the 150 schemes or 50 % entered in the main competition. Detached and semi-detached homes account for 32 %, while almost 19 % are apartment schemes and, in some cases, high-rise residential buildings.

In addition to administrative buildings, commercial buildings and schools are even more widely represented. 5 % of all projects are from the transport infrastructure field. 16 % of the entries involve the renovation of existing building stock, some of which also had to observe conservation requirements.

It is the aim of this contribution to present excellent BIPV by means of the winners of the competition and to show that BIPV can be used in many cases and is able to produce electric energy and fulfill architectural requirements.

2 POTENTIAL OF BIPV

[2] indicates, that in Germany in summary the use of the building envelope as a substructure for photovoltaic systems has a very large potential. Approximately 1,000 GW of PV power could be installed on 5,000 km². PV energy generated by means of BIPV per year could amount to 671 TWh. Please note, that in the year 2020 all PV systems in Germany have produced 45.8 TWh,

3. ASSESSMENT CRITERIA

Within the competition criteria such as architectural aspects, innovativeness, energy yield/energy efficiency and communicating the system to the general public have been considered.

4 PRIZES

Within the competition a first prize, second prizes a student prize and honorable mentions were awarded – with a total prize money of € 25,000. The first prize was endowed with € 15,000, there were four second prizes each of € 2,500 and four honorable mentions.

5 FIRST PRIZE

Project: Lycée Technique pour Professions de Santé (2019), Ettelbrück (LU)
Entrant: FABECK ARCHITECTES, Koerich (LU)

. The four-story building (Fig. 1 + 2), is divided into projecting and receding structures. The primary structure is made of timber except for the escape staircases. Asymmetrical gable roofs vary in the inclination of their surfaces and are covered entirely with PV modules (320 kW). The PV generator on the roof surfaces is meticulously realized on a timber substructure with painted flashings, and continues familiar approaches from structural engineering practice.

The façade design posed greater challenges. Solar thermal collectors, to which a 91,000-liter hot water tank is connected, are installed in the southwest and west elevations at the height of the window strips. The solar thermal collectors (150 kW) are also integrated design-wise so that they align with the window openings and extend the almost square windows along their right-hand reveals. Based on a mullion/transom system, the collectors are installed flush with the surface, with concealed connection cables. The selected dimensions and proportions of the solar technical components excellently complement the façade's appearance with its vertically arranged, narrow timber battens.

Based on the energy-plus concept, the new school building vividly combines the use of innovative building services components with the use of ecological materials in a coherent building design with high functionality and design quality.

Figures 1 + 2: Two views of the award-winning solar systems (BIPV + solar thermal) of Lycée Technique pour Professions de Santé in Ettelbrück (LU) with PV in the roof and solar thermal systems in the façade
Photos: Christian Aschman | LCGDP

6 FOUR SECOND PRIZES

Four objects (see 6.1 to 6.4) were (each) endowed with a € 2,500 second prize.

6.1 Apartment Building in Zurich (CH)

Project: Wohnhaus Solaris (2017), Zürich (CH)
Entrants: huggenbergerfries Architekten, Zürich (CH); Ertex Solartechnik, Amstetten (AT)

An "aubergine-colored" apartment building (Fig. 3 + 4), with a concisely articulated structural form stands out between a railway line and a heterogeneously developed arterial road. The architects developed an ingeniously organized floor plan for the 10 residential units in the multi-faceted building in order to provide "diverse orientations with extensive views, openness and all-day sunshine". The special feature of the homogeneously formed envelope is that despite the building's complex geometry, the façade (46.5 kW) and roof (25.2 kW) are designed as solar generators and produce twice as much electricity over the year as is consumed in the building. The architects aimed to achieve a "solar house that isn't necessarily recognizable as such".

Monocrystalline modules are used with a prismatic front glass that is additionally color-coated with a special print technology. This means that the solar cells cannot be seen even at close range; the vertically textured structure gives the glass surface a matt sheen and creates a multi-shaded, colorful play of light over the course of the day in direct and diffuse light and depending on the viewing angle. The homogeneous appearance complements a façade structure with concealed fixings and unobtrusively aligned joints, so that the individual panels are barely legible from a distance.

Figures 3 + 4: "Aubergine-colored" apartment building Solaris in Zurich (CH)
Photos: Dieter Moor | Beat Bühler

6.2 High Rise Building in Paris (FR)

Project: Tribunal de Grande Instance (2017), Paris (FR)
Entrant: Renzo Piano Building Workshop, Paris (FR)

The new building for the Tribunal de Grande Instance de Paris (Fig. 5 + 6), is the largest project in the competition with a height of 160 meters and a usable floor space of more than 100,000 m². The building impresses with its staggered structure, with extensively planted terraces and the differentiated design of the double-skin façade, whose external glazing extends beyond the edges of the building on each of its longitudinal sides and thus elegantly defines the clear structure of the building.

Figures 5 + 6: Tribunal de Grande Instance de Paris
Photos: Francesca Avazinelli | Maxime Laurent

On the east and west façades of the high-rise building, horizontal PV modules cantilever outwards on steel structures to form a continuous, narrow vertical ribbon while also flanking the three uppermost storys in each of the three building sections, which respectively step back above a projecting plinth.

The PV modules are fixed back at the level of the floor slabs, out of the direct field of vision of the employees. With a depth of 1.20 m, the modules are designed with two different inclinations: the first row of modules with a 45° inclination lies directly in front of the glass façade and the second row, with a 20° inclination, is slightly offset below. The linear construction with different lengths, each staggered upwards, provides solar shading. In addition to the structural engineering requirements that had to be met for a solar power system at this height, the slender metal bands accentuate the large glass façades in a nuanced manner.

6.3 Church (2018) in Georgensgmünd (DE)

Project: St. Wunibald (2018), Georgensgmünd (DE)
Entrant: Ing. Büro Wunram, Georgensgmünd (DE)

The Catholic St. Wunibald Church in Georgensgmünd in central Franconia (Fig. 7 + 8), completed in 1966, features a large, strikingly steep gabled roof and a free-standing bell tower with its towering, pointed tent roof. At the suggestion of the church administration, locally based architect Johannes Wunram used the repair measures – in the year 2018 - as an opportunity to install a photovoltaic system on the southeast side.

Figure 7 + 8: St. Wunibald Church in Georgensgmünd with in-roof system with precisely detailed verges
Photos: Solarwatt

This measures resulted in a homogeneously integrated roof surface using an in-roof system with precisely detailed verges, which harmonizes perfectly with the color of the slate roof on the campanile.

In addition, the 78.4 kW solar power system generates around 77,000 kWh/a of electricity, whereby it is planned to increase the current self-consumption of "just under 10 %", in particular for the new electric heating system in the church interior.

6.4 Energy Plus Building in India

Project: powerHYDE (2019), Mathjalgaon (IN)
Entrant: billionBricks + Architecture BRIO, Mumbai

The non-profit organization billionBricks has presented its modular powerHYDE solar home with Architecture BRIO from Mumbai. In addition to the technical design development of this energy-plus building for the Indian subcontinent, the project is supported by a financing model that enables millions of poor and homeless people to find affordable housing solutions as needed. In India's rural regions there is an enormous demand for housing 40 million households and, consequently, a demand for sustainable energy supplies.

Figure 9 + 10: powerHYDE solar home in India
Photos: PHXINDIA | Sebastian Zachariah

Above the single-story perimeter walls, a gently inclined monopitch roof covered entirely with PV modules sits on a lightweight steel support structure (Fig. 9 + 10). This produces four times the amount of solar electricity that is actually needed, thus generating additional income opportunities for the residents by enabling them to sell the surplus energy to neighboring commercial and municipal facilities. Despite all the pragmatism, from the basic concept, choice of materials to the detailing, with respective references to regional building traditions, this has made an extremely sensible contribution to sustainable mass housing construction.

7 FIVE HONORABLE MENTIONS (NON MONETARY PRIZES)

7.1 Historic Building with new Energy Technology

Project: Schlössli (2018), Wohlen (CH)
Entrant: Megasol Energie, Deitingen (CH)

The oldest building in the competition dates back to the 15th century. The "Schlössli" in Wohlen in the Swiss canton of Aargau had led an inconspicuous existence before undergoing conversion, and after two fires only the rubble masonry walls remained reasonably intact.

Local architects Furter Eppler Partner began extensive renovation work at the end of 2016 and created an open interior space for the villagers that can be used in many different ways.

Figure 11: Historic Building in Switzerland
Photo: Megasol

The building (Fig. 11), whose new, striking gable roof rests on a surrounding, visible ring anchor made of concrete, is adorned on both sides by an inconspicuous solar power system. Monocrystalline PV modules (21.1 kW) in standard and special formats with a matt glass surface are elegantly installed on a special in-roof mounting system that is slightly separated from the roof verges. This is an unobtrusively sovereign solution that combines the historic building with new energy technology in an exemplary manner.

7.2 Project: Ice Rink Bünzmatt (2018), Wohlen (CH)

Entrants: phalt Architekten, Zürich (CH)
Planeco, Münchenstein (CH)

The ice rink at Bünzmatt Sports Park (Fig. 12) is also located in Wohlen in northern Switzerland. To cover the ice rink, phalt architects from Zurich proposed a steel structure with a segmented gable roof, with steppeddown levels at the height of the respective truss girder flanges. In seven bands, monocrystalline PV modules (378 kW) cover the upper roof surfaces, which amount to about two thirds of the total roof area.

With around 379,400 kWh/a of solar power, this enables 67 % of the sports park's total energy demand to be met. The side areas with the lower roof sections provide effective, uniform illumination with daylight, which additionally reduces electricity consumption and saves operating costs. This provides a coherent concept for the ice rink roofing, with meticulously designed verges and a concisely articulated roof silhouette.

Figure 12: Ice rink at Bünzmatt Sports Park
Photo: Roger Frei

7.3 Equipment barn in the USA

Project: Saxum Vineyard Equipment Barn (2018), Paso Robles (US)
Entrant: Clayton Korte Architects, San Antonio (US)

Saxum Vineyard is located southwest of the city of El Paso de Robles in California. Clayton Korte Architects from San Antonio designed the equipment barn to provide covered storage for farming vehicles, implements and livestock supplies. Filigree circular tube supports with diagonally arranged struts form the supporting structure and stiffen a gently inclined monopitch roof.

Photovoltaic modules are mounted in eight rows on a slim substructure, providing weather protection and the power generator (Fig. 13). The system is expected to produce about 88,700 kWh/a of solar power, which will meet the entire energy requirements for the vineyard and irrigation wells. In conjunction with the translucent wall surfaces, an excellent example of agricultural construction has been created that combines aspects such as minimal material use and recycling with aesthetic demands.

Figure 13: Saxum Vineyard Equipment Barn
Photo: Casey Dunn

7.4 Bike Shelter

Project: Bike shelter VIVA (2019), Liestal (CH)
Entrant: Antonio Stefanelli Architektur, Pratteln (CH)

A remarkable example of a solar power system in a small transport-related structure (Fig. 14) is provided by the VIVA bike shelter in the Swiss canton of Basle-Country. Situated alongside the VIVA apartment building in Liestal with 34 apartments, the elongated timber structure provides parking space and weather protection for bicycles. The translucent glass/glass solar modules (32.64 kW) used in the monopitch roof and southwest façade provide an adequate supply of daylight and deliver around 28,000 kWh of electricity per year, which is "largely used for the households' own energy supply". The clear structure of the overhead and façade glazing harmonizes well with the cladding consisting of horizontally laid, narrow timber battens.

Figure 14: Bike Shelter in the Swiss canton of Basle-Country
Photo: Stefanelli Architektur

7.5 Solar folding roof

Project: HORIZON solar folding roof (2020), Jakobsbad-Kronberg (CH)
Entrant: dhp technology, Zizers (CH)

With the HORIZON solar folding roof (Fig. 15) a roofing system has been developed and already used several times that, with its innovative lightweight construction, is able to cover large areas of commercial or infrastructural surfaces and use them for producing solar power. The movable superstructure, which enables the folding roof to be "parked" in the „garage" during poor, inclement weather conditions, has reduced the material input required for the construction by about half.

At the visitor car park for the Jakobsbad-Kronberg aerial cableway, in the canton of Appenzell Innerrhoden, the 429 kW solar power system has now been used for the first time over a roughly 4,000 m² parking area. In combination with charging stations for electric vehicles, it provides a useful link between solar power production and sustainable mobility.

Figure 15: Solar folding roof at visitor car park
Photo: dhp technology

8. CONCLUSION

The competition "Building Integrated Solar Technology" which has been held in 2020 for the eight time since 2000, shows with 159 schemes the quality of the projects and the actuality of BIPV.

A four-story building in Luxemburg from the health sector with PV and solar thermal systems was awarded with the first prize (€ 15,000). There were five second prizes (€ 2,500 each – pure PV) for an apartment building in Zurich, a high Rise Building in Paris, a church in Germany and a solar home in India. Five honorable mentions (non monetary Prizes) were endowed: a historic building, an ice rink, an equipment barn for farming vehicles, a bike shelter, and a Solar folding roof

Due to high quality solutions, BIPV can be used in many cases and is equally able to produce electric energy and fulfill architectural requirements.

9 REFERENCES

[1] Roland Krippner (Hrsg.): Building Integrated Solar Technology; Architectural design with photovoltaics and solar thermal; Detail green books. Munich: DETAIL Business Information, 2017

[2] Harry Wirth, Jan-Bleicke Eggers, Max Trommsdorff, Holger Neuhaus, Martin Heinrich, Stefan Wieland, Christian Schill; Potenziale der Integrierten Photovoltaik in Deutschland; 36. PV-Symposium /BIPV-Forum; 18. - 26. Mai 2021; Proceedings; Page 209

6.1 PV ON/IN BUILDINGS

COMPARISON OF IEC 61853-1 MATRIX EVALUATIONS BASED ON INDOOR AND OUTDOOR MEASUREMENT DATA FROM PVPS TASK 15 BIPV ROUND-ROBIN

Roland Valckenborg[1*], Karl A. Berger[2], Gusztáv Újvári [2], Gabriele C. Eder[3], Lukas Gaisberger[4], Momir Tabakovic[5], Cristina S. Polo López[6], Simon Boddaert[7], Matteo del Buono[8], Nuria Martín Chivelet[9], Asier Sanz Martinez[10], Jun-Tae Kim[11], Anne Gerd Imenes[12]

[1]*TNO, High Tech Campus 21, Eindhoven, The Netherlands*
[2]*AIT, Austrian Institute of Technology GmbH, Austria*
[3]*OFI, Austrian Research Institute for Chemistry and Technology, Austria*
[4]*University of Applied Sciences Upper Austria, Austria*
[5]*University of Applied Sciences Technikum Vienna, Austria*
[6]*SUPSI, University of Applied Sciences and Arts of Southern Switzerland, Switzerland*
[7]*CSTB, Centre Scientifique et Technique du Bâtiment, France*
[8]*EURAC Research, Italy*
[9]*CIEMAT, Madrid, Spain*
[10]*TECNALIA, Basque Research and Technology Alliance (BRTA), Spain*
[11]*Kongju National University, South Korea*
[12]*University of Agder, Norway*

*Corresponding author: roland.valckenborg@tno.nl; Tel: +31 652803729

ABSTRACT: The international standard IEC 61853-1 contains two procedures for the evaluation of PV module output (P_{mp}) over a range of irradiances (G_{POA}) and module temperatures (T_{mod}): a procedure for indoor labs and a procedure for outdoor measurements (we neglect the option of using a tracker, which is quite challenging for BIPV-systems). The data from the IEA PVPS Task 15 (Enabling Framework for the Acceleration of Building Integrated Photovoltaics, BIPV) round-robin has been used for an in-depth analysis of the differences between the indoor and outdoor determined P_{mp}-matrix. Moreover, the variety in the matrices of the eight individual test sites in various parts of the world is investigated. The approach is based on specific filtering of the outdoor data and color plot the relevant parameters in 2D matrices (of G_{POA} and T_{mod}) including the statistics of all data points that are present in each bin of that matrix. Filtering on Angle Of Incidence (AOI) and distance to center of the bin has been applied to make the comparison as precise as possible. Even after this filtering it was found impossible to decrease the standard deviation (stdev) of all points within a bin to a value lower than 5% as requested from the standard. A more realistic criterium of 10% stdev is proposed to use. The systematic fluctuations within each bin are caused by environmental factors that were not controlled or measured in our round-robin. These factors are, in random order: non-uniformity of (i) G_{POA} and (ii) T_{mod} within the module area, (iii) thermal dynamics between the T_{mod} sensor on the back of the PV-module and the actual T_{cell} that determines P_{mp}, (iv) soiling, (v) spectral effects. In conclusion, following the indoor and outdoor procedure of the IEC 61853-1 does not give the same P_{mp}-matrix for a given (BI)PV-module type. Significant differences are observed and analyzed. These results show that it might be very challenging to use the outdoor procedure without tracker for the purpose of energy rating. Leading to the recommendation to use the indoor procedure for energy rating of PV-modules and use the outdoor procedure for a well-defined quantification of energy losses caused by the aforementioned fluctuating environmental factors always present in realistic (BI)PV-systems.

Keywords: BIPV, Energy Performance, Energy Rating, Experimental Methods, Façade, PV Module, Pyranometer, Reference Cell

1 INTRODUCTION

1.1 Scope, Goal & Approach

The international standard IEC 61853-1 contains a procedure for the evaluation of PV module output over a range of irradiances and temperatures. The obtained matrix of power values allows the comparison between different PV modules in testing conditions different from the standard ones. The 61853-1 [8] gives two procedures to populate the Pmp-matrix (as a function of irradiance in the plane-of-array G_{POA} and module temperature T_{mod}): a procedure for indoor labs and a procedure for outdoor measurements (we neglect the option of using a tracker, which is quite challenging for BIPV-systems). Within the round-robin [1] of IEA PVPS Task 15 (Enabling Framework for the Acceleration of Building Integrated Photovoltaics, BIPV) both, indoor lab analysis under STC conditions, and outdoor monitoring in natural sunlight were performed. In previous work [2][3], the authors evaluated the suitability of the standard IEC 61853-1 to describe the performance of a BIPV-module in a ventilated façade structure measured outdoors in different climates and latitudes.

The current research aims to compare these matrix results with the indoor matrix, obtained under controlled irradiance and temperature conditions (STC). By analysing these data, striking differences between the two procedures were observed. Therefore, this contribution aims for an in-depth analysis of the differences in the energy rating (calculated according to IEC 61853-1) based either on indoor or outdoor measurement data. Moreover, the variety in the matrices of the eight individual test sites in various parts of the world is investigated. The approach is based on specific filtering of the outdoor data and color plotting of the relevant parameters in 2D matrices (of G_{POA} and T_{mod}) including the statistics of all data points that are present in each bin of that matrix [4]. As the indoor procedure always has the irradiation from lamps perpendicular to the module plane, the Angle-of-Incidence (AOI) of the irradiance is an important difference in the environmental conditions between the indoor and outdoor procedure. To compare indoor and outdoor data, it is necessary to consider the outdoor data points with the lowest AOI, i.e. filtering for low angles between the module normal and the direction of the direct irradiance component. Therefore, the AOI-dependence of the measured electrical parameters is calculated for each data set of the outdoor measurements, by making use of the PVLIB-library [5]. It will be shown that additional filtering of the outdoor data based on the criterium that AOI has to be below a given threshold angle eliminates the major AOI effects. Last but not least, new post-processing software allows for filtering of the data within each bin around the center point of that bin. This is especially valuable for bins in which the data is not as uniform distributed as one might expect based on the large number of data points in that bin.

1.2 Contributing sites, Experimental setup & Data handling

The round-robin has been described extensively in [1][2][3]. All details regarding the experimental setup and data post-processing can be found in those references. Since end 2020, the site of University of Agder in Norway has been added to the round-robin, which brings the amount of contributing sites to a total of eight; see Figure 1 for the locations. Unfortunately the CSTB-site had a measurement issue for a too long period, and needed to be excluded from this matrix analysis.

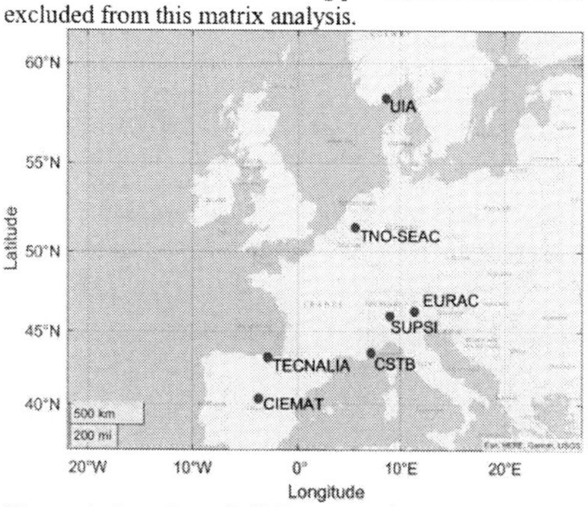

Figure 1: Location of all European sites contributing to the round-robin. Only one site is located out of Europe, which is Konju National University in South Korea.

Figure 2 gives a recap on the interpretation of the statistics of all data points that are present in each bin of a 2D matrices (of G_{POA} and T_{mod}) [4].

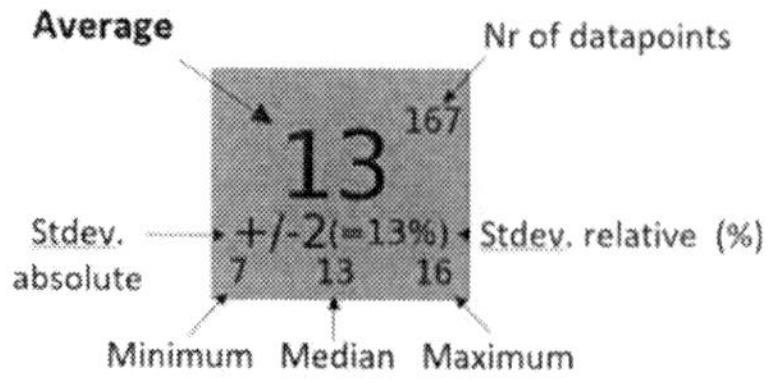

Figure 2: Description of statistics of data points in a bin.

1.3 Solar Position Calculation of PVLIB validated with Measurements

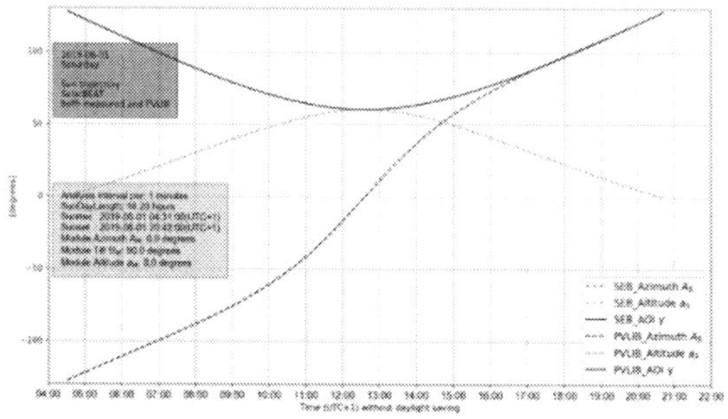

Figure 3: Azimuth (A_s) and Altitude (a_s) of the sun as function of time of the day, for an arbitrary day of the year in TNO-SEAC site in the Netherlands; dotted brown = measured values; dotted blue = PVLIB-values. AOI-calculation of measured values = dark brown line; AOI-calculation of PVLIB = dark blue line. All brown lines are hidden behind blue lines.

The calculated solar position obtained from the PVLIB library [5] has been validated with outdoor measurements with a secondary standard tracker for the site of TNO-SEAC in the Netherlands. One full year 1-minute data has been captured continuously with a data completeness of more than 99.6%. Figure 3 shows calculated and measured solar Azimuth and Altitude for a random day (1st June 2019). Additionally, the calculated AOI [5][7] based on these two different sets of angles are included. It can be seen that the correspondence is better than the line thickness in the plot. A systematic comparison shows that for all data points of the year, the RMSE of the difference between measurement and calculation is around 0.006 degrees for both Azimuth and Altitude. The difference in calculated AOI is roughly the same. These values are far beyond the measurement accuracy with a maximum occurring deviation of 0.03 degrees leading to an accuracy of better than << 0.1 degrees for the calculated AOI values.

2 INDOOR AND OUTDOOR PROCEDURE OF IEC 61853-1

2.1 Indoor Procedure

The indoor procedure to determine the P_{mp}-matrix of T_{mod} (x-axis) and G_{POA} (y-axis) is described in paragraph 8.5 of the IEC 61853-1:2011. All 23 bins of the matrix have been measured by the indoor lab facility of AIT [1]. The measurement was performed with a large A+A+A+ flasher with negligible AOI effects: using direct light only with a distance of more than 8 m between lamp and module. The resulting P_{mp}-values are visually displayed in Figure 4.

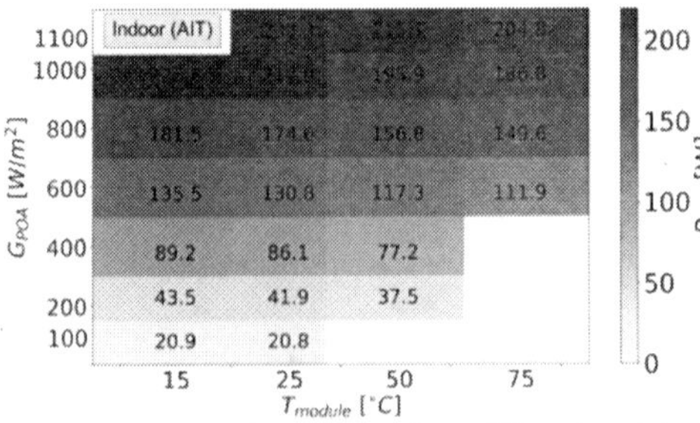

Figure 4: P_{mp} matrix as a function of T_{mod} (x-axis) and G_{POA} (y-axis) calculated according to the indoor procedure.

As this full matrix measurement is rather time consuming, it has been performed on only one module of the complete series. However, for all modules of the participating countries to the Round Robin, the Wp-rating at STC has been measured before the modules were sent to the various sites. The results [4] give an average of 219.3 +/- 0.3 Wp. Hence the relative standard deviation (stdev) is <0.2%, which is much lower than the measurement accuracy of ~1%. In conclusion, all modules have the same Wp-rating within the measurement accuracy of the normal indoor rating method of IEC 61215.

After the one-year outdoors measurements, nearly all modules were received back undamaged at AIT. On exactly the same equipment the same procedure was followed to measure the Wp-rating at STC. The modules measured before and after the outdoor installation showed P_{mp} degradation values from 0.6 to 2.1%. In all cases the electroluminescence images were without any significant changes, and free of any signs of micro-cracks. The one with the largest degradation in P_{mp} encountered a more than 50% change in series resistance. This module experienced a smooth fall during a storm, which could have caused this increased series resistance. Anyhow, the module power degradation over the outdoor measurement time is low, compared to seasonal effects, and general measurement uncertainties in outdoor monitoring.

2.2 Outdoor Procedure

The outdoor procedure to determine the P_{mp}-matrix of T_{mod} (x-axis) and G_{POA} (y-axis) without a tracker is described in paragraph 8.4 of the IEC 61853-1:2011 (denoted as P8.4 in the following). For the vertical BIPV mounted setup tested not all 23 bins have captured data. As can be seen in Figure 5 for a randomly chosen test site (TECNALIA) there is quite a large spread within each bin, because the requirements of P8.4 were not applied yet when composing this matrix. In total P8.4 refers to 6 requirements (4 req. in P8.3.2 and 3 req. in P8.3.11), which we will address in the following way:

From 8.3.2:

1. "Measurement in natural sunlight shall be performed over the range of irradiance conditions occurring during the day." → *All datapoints of complete day are included.*

2. "Short term irradiance variations caused by clouds, haze, or smoke shall be less than ±1 % of the total irradiance as measured by the reference device during the collection of each measurement point as specified in IEC 60904-1." → *This is a very difficult criterium to fulfill, because the irradiance sensors that we use (either secondary standard pyranometer or reference cell [1]) do not provide this information. Therefore, we include every measurement of irradiance as a valid measurement.*

3. "The wind speed should be less than 2 m/s." → *We did not include a local wind speed measurement next to the PV-module in the setup requirements.*

4. "To increase the range and improve the accuracy data should be taken over at least three days." → *A full year period was taken for each site.*

From 8.3.11:

1. "A minimum of three measurements shall be made at each of the test conditions on a minimum of three days" → *This is interpreted that a bin needs to be populated by at least 3 measurement points. The additional requirement of three measurement days is very difficult to implement in the code that analyses the data and composes the 2D-matrices. In general, the more datapoints in a bin, the more reliable the average will be.*

2. "Continue to collect data until the standard deviations for all V_{OC}, I_{SC} and P_{mp} values in the matrix are less than 5%." → *This criterium is very difficult to meet, even with one year of data.*

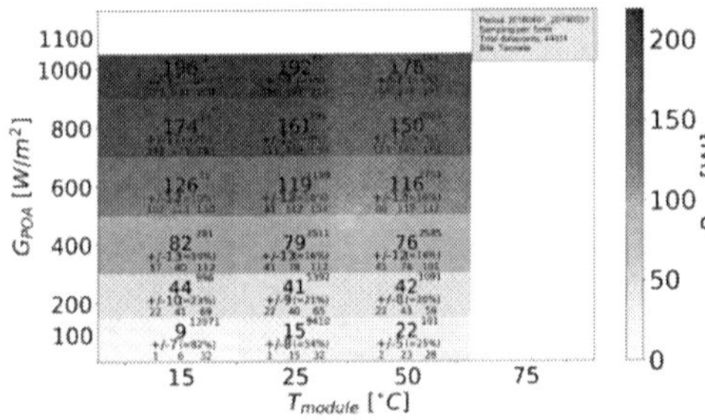

Figure 5: P_{mp} matrix as a function of T_{mod} (x-axis) and G_{POA} (y-axis) from the outdoor procedure for a random test site (TECNALIA). No data filtering.

3 CRUCIAL FILTERING FOR THE OUTDOOR PROCEDURE OF IEC 61853-1

As a best-practice the following filtering is suggested: As a first step remove the large AOI angle measurements in which the Incidence Angle Modifier (IAM) will decrease significantly the amount of light coupled into the PV-module, whereas the irradiance sensor does not measure this decreased level of irradiance (although this occurs only in those cases where a pyranometer has been used we decided to filter according to AOI in all cases). As the module temperature should be close to the center temperature of the bin, values at T_{mod} too far of the center should be filtered out in a second step. And finally, the same argument holds for irradiance: the G_{POA} should be close to the center irradiance level of the bin.

3.1 AOI-filtering

With respect to angular effects, the standard (note below P8.3.11) states: "The angular effect can be eliminated by use of a tracker". As this is outside the scope of our round-robin, we propose an additional AOI-filtering step. Indoor measurements are always performed with a solar simulator perpendicular to the panel. Outdoor measurements will have a huge spread in AOI during the day and during the season. In fact, AOI=0⁰ will never happen in vertical mockups. The AOI-calculation is included in every data set of each site. Therefore, filtering on AOI has to be applied [4] but taking into account that too stringent filtering deletes valuable data. A compromise has been found at a threshold value of AOI=60⁰. The effect on the resulting matrix is negligible for the highest irradiance bins, but rather significant for the lowest irradiance bins, see Figure 6.

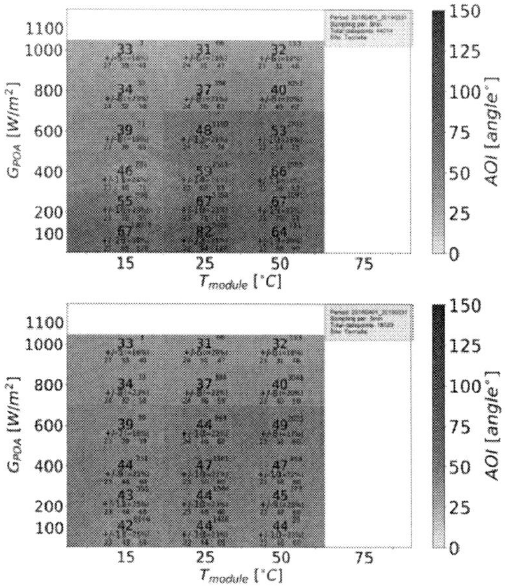

Figure 6: AOI-matrix as a function of T_{mod} (x-axis) and G_{POA} (y-axis) from the outdoor procedure for a random test site (TECNALIA). Top graph: no filtering. Bottom graph: filtering on AOI < 60⁰.

3.2 Filtering around Center Point of bin.

A very important requirement of the outdoor procedure without tracker is to collect data until the standard deviation of all bins in the V_{OC}, I_{sc} and P_{mp} matrix is less than 5%. As V_{OC} and I_{sc} were not available for all sites with a high enough time resolution, the V_{mp} and I_{mp} matrices together with the P_{mp} matrix were used. See e.g. Figure 7 for the V_{mp} matrix for random test site (TECNALIA).

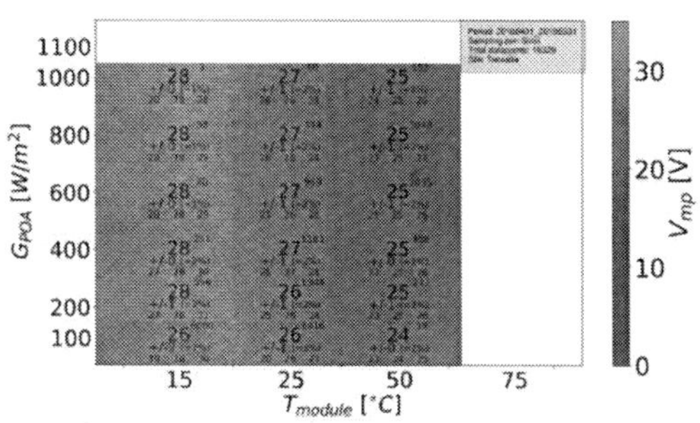

Figure 7: V_{mp} matrix as a function of T_{mod} (x-axis) and G_{POA} (y-axis) from the outdoor procedure for a random test site (TECNALIA). Filtering on AOI < 60⁰

The standard suggests that the standard deviation in these matrices will always get smaller when collecting/including more data points. However, during our analysis and post-processing of the full year data sets of all seven test sites, we found that the scatter in some bins is systematically present and caused by the environmental conditions such as non-uniformity of (i) G_{POA} and (ii) T_{mod} within the module area, (iii) thermal dynamics of the T_{mod} sensor (on the back of the PV-module) is different from the actual T_{cell} that determines P_{mp}, (iv) soiling, (v) spectral content of natural sun, which varies throughout the day and the year and depends on the air mass and the albedo. Moreover, the distorted (skewed) distribution within some bins, is an important reason for filtering around the bin-center. At least we try to be as close as possible to the center point of the bin. This can be done by filtering on a combination of maximum deviation of T_{mod} to center and maximum deviation of G_{POA} to center; a data point needs to meet both criteria. See Figure 8 for the step of filtering on T_{mod}.

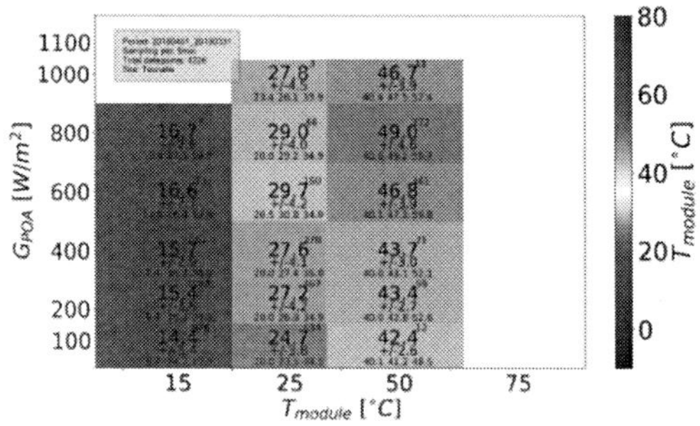

Figure 8: T_{mod} matrix as a function of T_{mod} (x-axis) and G_{POA} (y-axis) from the outdoor procedure for a random test site (TECNALIA). Filtering done on AOI < 60⁰ and T_{mod} +/- 10⁰C to center temperature of bin.

Because P_{mp} correlates in first order to G_{POA} and only in second order to T_{mod}, it was therefore decided to allow some spread in T_{mod} and be more rigorous in G_{POA}. With the first test runs of the filtering algorithm it was found that a minimum amount of residue data points in the order of 1000 is preferable, otherwise the natural spread due to environmental fluctuations is not captured enough in the result. These runs also showed that a fixed maximum deviation of T_{mod} is OK (like done in Figure 8), but a fixed maximum deviation of G_{POA} leads towards hardly any data points in the highest irradiance bins, and very many data points in the lowest irradiance bins, whereas one would prefer a more homogenous spread of data over the full matrix. Therefore, a varying criterium for maximum deviation in G_{POA} has been implemented in the filtering algorithm and is now set as a default at 5% of the center value of the bin; see Figure 9.

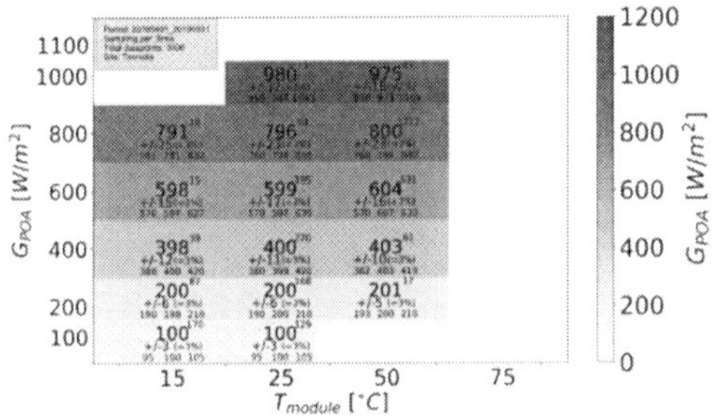

Figure 9: G_{POA} matrix as a function of T_{mod} (x-axis) and G_{POA} (y-axis) from the outdoor procedure for a random test site, TECNALIA. Filtering done on AOI < 60° and T_{mod} +/- 10°C to center temperature of bin and G_{POA}+/- 5% to center irradiance of bin.

3.3 Resulting Matrix with Filtering on AOI, T_{mod}, and G_{POA}

To demonstrate the importance of filtering, the P_{mp} matrix for the same site TECNALIA is shown in Figure 10, with exact filter settings in the caption. For this site the stdev in P_{mp} is nearly equal to the 5%-requirement of the standard. However, the matrices for other sites show less smooth behavior of the bins, with some bins having quite high (up to 25%) stdev; see e.g. Figure 10 for test site TNO-SEAC. As this spread could be caused by local albedo and hence irradiance non-uniformities within the module, which are permanently present in the measurements, it could be impossible to meet that requirement of the standard.

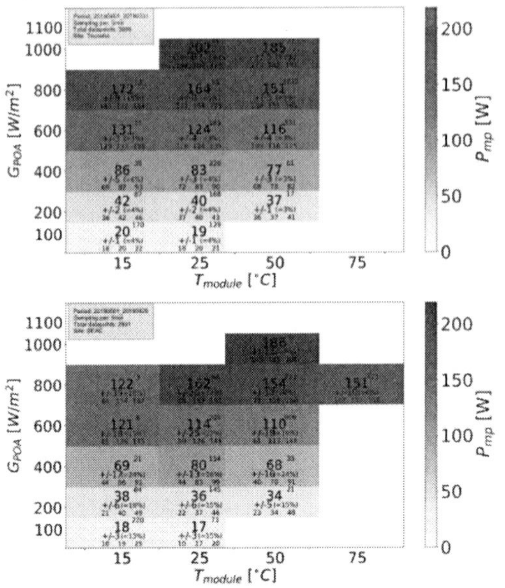

Figure 10: P_{mp} matrix as a function of T_{mod} (x-axis) and G_{POA} (y-axis) from the outdoor procedure. Filtering done on AOI < 60° and T_{mod} +/- 10°C to center temperature of bin and G_{POA} +/- 5% to center irradiance of bin. Top graph: site TECNALIA and Bottom graph: site TNO-SEAC.

4 MATRIX COMPARISONS

4.1 Average Outdoor Matrix

All test sites have followed the same well-defined procedure for this round-robin. Therefore, the P_{mp}-matrix of each site should be considered as an equally valid implementation of the standard (with the remark that not all 6 additional requirements could be implemented in practice). Moreover, the same filtering criteria as explained above are applied to the data of the test sites. From these matrices, an averaged outdoor matrix has been composed, see Figure 11.

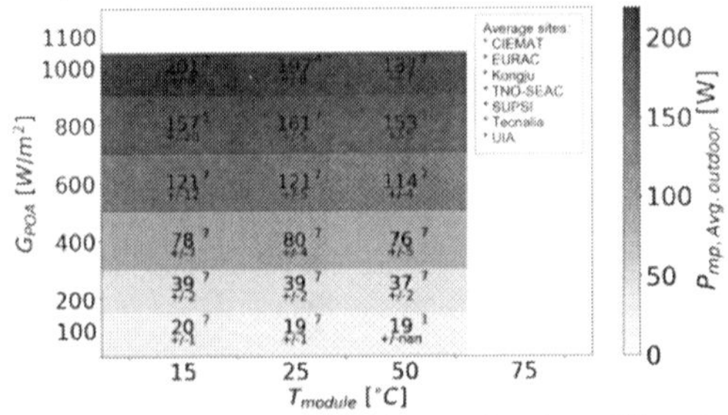

Figure 11: Average P_{mp} matrix as a function of T_{mod} (x-axis) and G_{POA} (y-axis) from the outdoor procedure. Superscript value in the bin denotes the number of sites that have the specific bin populated.

The 'quality' of each bin is determined by the quality of the underlying site P_{mp}-matrices. No manual or automatic filtering has been applied yet when composing Figure 11, but could be needed. As an example, see Figure 12 in which the deviation of a site (TECNALIA in this case) with respect to the average is shown.

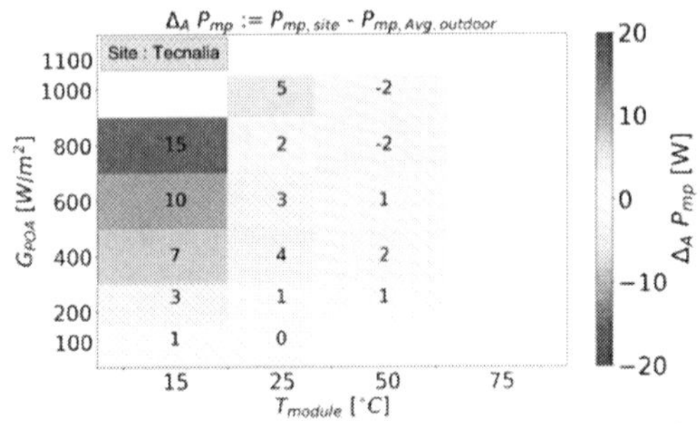

Figure 12: The deviation-matrix of a random site (TECNALIA) with respect to the average outdoor matrix.

4.2 Quality of each bin

Clearly the bin of T_{mod} = 15 °C and G_{POA} =800 W/m² needs further investigation. Browsing through all 'deviation-matrices' reveals that some kind of quality parameter for a bin would be beneficial to define. For statistical effects of a random nature, the stdev in general will decrease when taking more measurement points into account. To check this, Figure 13 was composed for all bins from all sites.

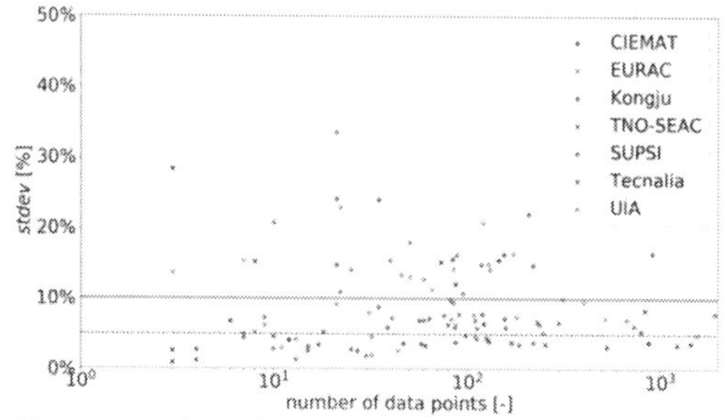

Figure 13: The stdev as a function of number of data points for each bin. Various sites are color labeled according to legend. Dotted line reflects 5%-criterium from IEC61853. Solid line shows our suggested 10%-criterium.

As can be seen in Figure 13, there is no decreasing trend. In fact, there is no dependency at all, which means that the deviations in each bin are caused by unknown (or un-measured) variations in one of the environmental parameters. When applying the 5%-criterium (dotted line in Figure 13), the amount of rejected bins is too much. Therefor we propose a 10%-criterium as a better compromise between the variations within a bin, and the number of bins available for calculating the averaged outdoor matrix. With this 10%-criterium active, the averaged outdoor matrix is plotted in Figure 14.

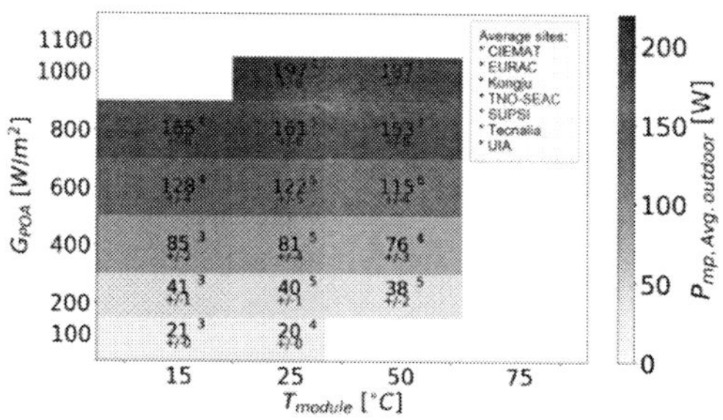

Figure 14: Same as Figure 11, but now only bins that have internally a better stdev than 10% are included in the averaging process. Moreover bins with less than 3 contributing sites have been removed from the plot.

4.3 Compare Individual Test site Outdoor Matrix to Indoor Matrix

One can also look at the deviation of the outdoor matrix of a site to the indoor matrix, as is shown in Figure 15 for TECNALIA.

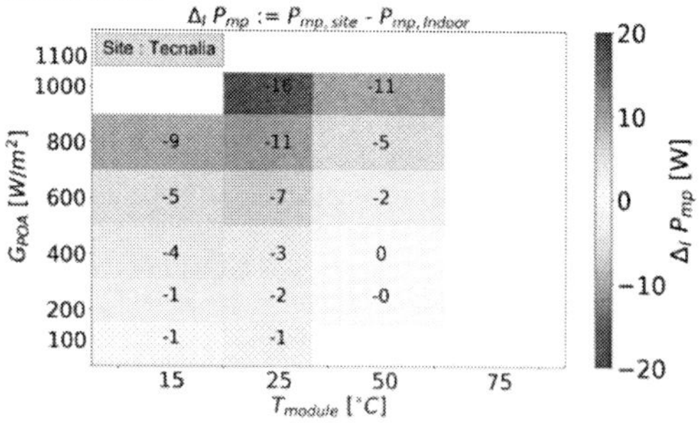

Figure 15: The deviation-matrix of site TECNALIA with respect to the indoor matrix.

Now, the shape of the deviation is quite different. As expected, the deviation is much more into negative values, which can be understood from the fact that outdoor measurements will always result in lower P_{mp} than indoor measurements, which is due to having considered all sources of losses that occur outdoors.

4.4 Compare Indoor and Outdoor Matrix

A very logical final question would be, how does the averaged P_{mp}-matrix as measured by the outdoor test sites compare with the P_{mp}-matrix as measured by the indoor lab in a solar simulator? This comparison is shown in Figure 16.

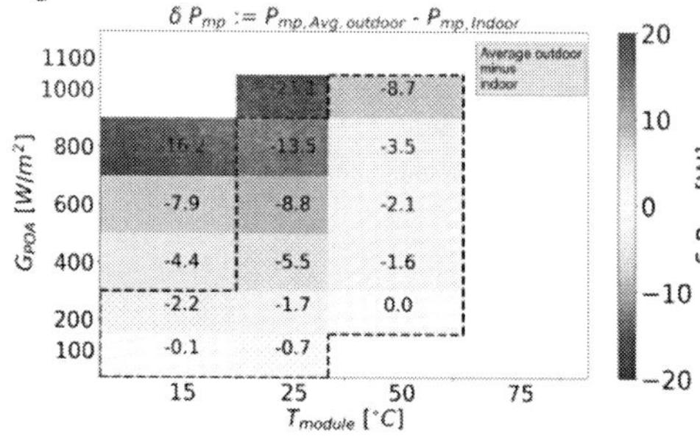

Figure 16: δPmp as defined to be the difference between the averaged outdoor P_{mp}-matrix and the indoor P_{mp}-matrix. Dotted line shows 'core' of the matrix for which specific yield is significant contributing to full year yield.

The fact that all deviations are negative values can be understood from the fact that nearly all the environmental factors will lead to less power production. Only a spectral effect could potentially lead to gains in P_{mp}. However, all the other factors (soiling, irradiance mismatch, remaining AOI-effect, etc.) will contribute to lower values in the case of an outdoor determined matrix. The specific shape of Figure 15 shows that higher differences occur at rare combinations of high irradiance and low module temperature These bins could contain groups of data with sudden variations of irradiance (fast clouds), and should not be considered, since temperature variations are much slower. This never happens indoors. To avoid this problem data during periods of high irradiance variations should not be considered. As this was impossible to implement within the scope of this paper, we leave it as a further recommendation. Please note that for a full year energy yield prediction (or energy rating according to 61853-3) the left upper part of the matrix is less of importance because it has a quite low occurrence. From the earlier analysis of this round-robin that focused on specific yield [3] we discovered which area of the matrix ('core-matrix') is significantly contributing to full year yield for the vertical BIPV mock-up. The manual drawn dotted line shows this 'core-matrix'. Nevertheless, the nature of the outdoor effects could be so complicated, that one has to conclude that outdoor measurements without a tracker would never lead to a matrix suitable for energy rating a (BI)PV-module. Unless one would find a way to better control the outdoor setup with respect to the environmental factors. Or measure and correct for the fluctuating environmental factors. This seems very difficult. Therefore, the outdoor measurements can be seen as a valuable complementary methodology to gain insight into realistic energy yields for (BI)PV-modules in realistic (BI)PV-systems.

5 DISCUSSION & CONCLUSION

To the best of our knowledge, a comparison between energy rating (IEC 61853-1) matrix evaluation based on indoor and outdoor measurement data has not been realized yet for BIPV. For a rack mounted PV-module, M. Morampudi et.al [6] reported significant differences between the two procedures, and recommended to use them complementary. In [9] outdoor data was collected for a site in the cold climate of Canada, but no explicit comparison to the indoor matrix was made.

Our experimental outdoor data originates from BIPV modules which were vertically mounted in a ventilated façade mock-up with heavy thermal insulation as representative for a BIPV-ventilated façade (south-oriented). That gives insight in the question whether one-year outdoor data in such a mock-up can be used for populating the P_{mp}-matrix according to the standard. Moreover, the individual factors that are contributing to a difference between indoor and outdoor measurements are not limited to BIPV only, but are of a more general scope applicable to any type of PV-module integration or free-standing PV systems. These factors are, amongst others, in random order: non-uniformity of (i) G_{POA} and (ii) T_{mod} within the module area, (iii) thermal dynamics because the T_{mod} sensor on the back of the PV-module will show a more integrated, smoothened behavior, than the actual T_{cell} that determines P_{mp}, (iv) soiling, (v) spectral content of natural sun, which varies throughout the day and the year and depends on the air mass and the albedo.

Our results show that filtering of AOI can be done with a threshold value of 60° that reduces the number of data points by roughly a factor of two or three. However, a subsequent second filtering step to only accept datapoints that lie some % around the center point of the bin has a more severe impact. On the one hand, the closer the data points used are concentrated around the center value of the bin, the more representative the average of the bin will be. On the other hand, more stringent data filtering reduces the number of data points drastically. A compromise seems to be that around a thousand data points should still be used for populating the matrix (starting from around 45000 data points from a full year).

In conclusion, following the indoor and outdoor procedure of the IEC 61853-1 [8] does not give the same P_{mp}-matrix for a given module-type. Small, but significant, differences are observed and analyzed. Our results show that it might be very challenging to use the outdoor procedure for the purpose of energy rating. It could turn out that the outdoor procedure should rather be used as an additional methodology closer to realistic energy yields of modules in (BI)PV-systems. That would lead to a potential recommendation: use the indoor procedure for energy rating of PV-modules and use the outdoor procedure for a well-defined quantification of energy losses caused by the aforementioned factors always present in realistic (BI)PV-systems.

6 REFERENCES

[1] P. Illich, et.al, 'Comparative Performance Measurements of Identical BIPV-Elements in Different Climatic Environments - A Round-Robin Action within the IEA PVPS Task 15 Collaboration', 6BV.1.32, 35th EU-PVSEC (2018), pp. 1794-1800

[2] L. Gaisberger, et.al, 'BIPV Round-Robin Action of IEA PVPS Task 15', 6BV.4.7, 36th EU-PVSEC (2019), pp. 1811-1816

[3] R.M.E. Valckenborg, et.al, 'IEC61853-Matrix Analysis of PVPS Task 15 BIPV Round-Robin for More Than One Year at Seven Test Sites over the World', 6DO.12.1, 37th EU-PVSEC (2020), pp. 1719-1725

[4] R.M.E. Valckenborg, B. Van Aken, 'Outdoor Performance Quantification and Understanding of Various PV Technologies using the IEC 61853 Matrix', 4CO.2.2, 36th EU-PVSEC (2019), pp. 880-886

[5]https://pvlib-python.readthedocs.io/en/stable/ accessed 2021-07-27

[6] M. Morampudi, et.al, 'Performance Comparison of PV Module Based of Temperature Coefficient in Indoor and Outdoor Conditions as per IEC 61853-1', 32nd EU-PVSEC (2015), pp. 2067-2069

[7] A. Smets et.al, 'Solar Energy, The Physics and Engineering of Photovoltaic Conversion, Technologies and Systems' (2016), EAN 9781906860325, chapter 18.2, pp. 269-271

[8] IEC 61853-1:2011, 'Photovoltaic (PV) module performance testing and energy rating - Part 1: Irradiance and temperature performance measurements and power rating', ISBN 978-2-88912-301-8

[9] Toronto and Region Conservation, 'IEC 61853-1: Photovoltaic Module Performance Testing and Energy Rating Assessment for Canada', https://sustainabletechnologies.ca/app/uploads/2016/01/IEC61853_FinalReport_Dec2015.pdf, accessed 2021-09-03

7 ACKNOWLEDGEMENTS

This work was supported by the IEA PVPS Task 15 'Enabling Framework for the Acceleration of BIPV'.

LONG-TERM PERFORMANCE OF BUILDING INTEGRATED PHOTOVOLTAIC SYSTEMS AND SHADE INDUCED DEGRADATION

Andrew Fairbrother[1]*, Hugo Quest[1,2], Ebrar Özkalay[1,3], Philipp Wälchli[2], Gabi Friesen[3], Christophe Ballif[1,4], Alessandro Virtuani[1]
1 – École Polytechnique Fédérale de Lausanne (EPFL), Institute of Electrical and Micro Engineering (IEM), Photovoltaics and Thin-Film Electronics Laboratory, Rue de la Maladière 71b, 2002 Neuchâtel (Switzerland)
2 – 3S Solar Plus AG, Schorenstrasse 39, 3645 Thun (Switzerland)
3 – University of Applied Sciences and Arts of Southern Switzerland (SUPSI-PVLab), Via Flora Ruchat-Roncati 15, 6850 Mendrisio (Switzerland)
4 – CSEM PV-Center, Rue de Jaquet-Droz 1, 2002 Neuchâtel (Switzerland)
Email: andrew.fairbrother@epfl.ch

ABSTRACT: Building integrated photovoltaic (BIPV) systems often operate under elevated temperatures and recurrent shading, which can accelerate performance loss. However, to date there are few reports on long-term performance of BIPV systems, as distinct from building applied PV (BAPV). This work reports the performance loss rates (PLR) after 5 to 10 years of operation of 55 rooftop BIPV systems in Switzerland. Median PLR was calculated to be 0.06% per year, indicating essentially no degradation, though there was a large spread of rates. Some systems were noted to be shaded at times, so a fault detection and diagnosis algorithm (FDDA) was developed to quantify shading severity. The FDDA used DC voltage and current and meteorological data as inputs and classified faults according to pre-defined thresholds. Faults caused by shading were grouped to determine the shading factor of each system. Shading factor showed a weak correlation to PLR, but delineated an upper limit, indicating it limits maximum achievable performance. Shading factor was found to increase in newer systems, and decrease with system capacity. Through innovative BIPV module and system design, the effect of shading and corresponding performance losses can be mitigated.
Keywords: Building Integrated PV (BIPV), Degradation, Monitoring, Roofing Systems, Shading

Note: this work has been published with greater detail and analysis in Solar RRL.

1 INTRODUCTION

Building integrated photovoltaics (BIPV) is a rapidly growing segment of the PV market, especially in areas where intensive land use limits deployment of utility scale PV. It is know that BIPV modules experience higher operating temperatures and near-field shading stresses compared to field mounted modules.[1–3] While there are several studies on long-term performance of utility scale PV and even building applied PV (BAPV) systems (e.g. [4,5]), there are very few for BIPV. As defined in IEC 63092-1 ("Photovoltaics in buildings"), BIPV modules are a building product intended to fulfil one or more functional requirements of the building envelope.[6] If removed from a building they must be replaced by an appropriate building product, thus they are quite distinct from BAPV modules in terms of construction, function, and operating environment.

In BIPV systems it is expected that temperature- and shaded-induced degradation will be more prevalent.[3,7–9] In fact, IEC TS 63126 addresses the first, and describes modifications to typical module qualification and safety tests (MQT and MST) to account for applications for which high temperatures are expected, such as in BIPV and hot climates.[1,8] Shade-induced degradation is less well understood. Modules typically contain two or more sub-strings of solar cells, each protected by a bypass diode (BPD) that helps mitigate the effect of partial shading conditions by allowing current to bypass the shaded or bad cell or cells in a string. For a conventional PV module and system partial shading is usually a temporary event caused by soiling or partly cloudy weather. However, in a BIPV system, shading events may be routine or continuous throughout the day because of building elements such as overhangs, dormers (windows), chimneys, or multiple roof and façade orientations, and even external elements such as nearby buildings, trees, or other structures. This extended use of bypass diodes, including regular and excessive current excursions, is not anticipated in their design. In fact, the main qualification test for module bypass diodes (MQT 18.1 in IEC 61215-2) is only conducted for 1 hour, with a possible extension to 5 hours if diode temperature does not stabilize.[10] While in operation bypass diodes dissipate heat that over time can damage the diodes and nearby components, thus reducing their functionality and ultimately the module performance.[11–13]

In this work Performance Ratio (PR) time series of 55 BIPV systems in Switzerland were calculated. All systems utilized the same type of c-Si module and were installed between 2010 and 2015, resulting in 5 to 10 years of data per system. The fleet average performance loss rate was essentially zero, but a large spread of rates existed and a closer examination of some poorly performing systems showed that shading was apparent by visual inspection or due to clear deviations from an ideal production profile. This manual method of shading evaluation was a labor intensive and subjective process, so as a next step a FDDA to identify shading and classify its severity was developed. As inputs it used meteorological conditions as well as measured and single-diode modelled DC production to classify fault cases. The main causes of these faults were linked to partial shading, so the proportion of time in a fault state was defined as the shading factor of a string/system. The shading factor was found to delineate an upper limit on the PLRs of the BIPV systems. Moreover, average shading factors were found to increase in newer systems, and decrease in larger capacity systems. Finally, improvements to BIPV module and system design are discussed that could mitigate shaded-induced performance losses.

2 EXPERIMENTAL

2.1 BIPV system characteristics

A total of 55 BIPV rooftop systems were under investigation in this work, all constructed with MegaSlate II modules manufactured by 3S Solar Plus AG and installed by Baur AG. The modules were made with monocrystalline silicon solar cells and have a glass/backsheet construction. The systems were located in central Switzerland, which has a temperate warm summer climate (Cfb), and at elevations between 500 and 600 meters above sea level. They were installed between 2010 and 2015, thereby giving 5 to 10 years of operational data for these analyses. Their cumulative capacity was 1550 kW_p, with a median value of 17 kW_p. Power production from the systems was monitored at the string level, of which there were 252 in total. Some plants contained a single string, and the largest had 18 strings. Median system tilt was 25° (range of 10° to 55°), while orientations covered all cardinal directions with the majority oriented between east-south-west.

2.2 Performance loss rate analysis

Each PV plant had a monitoring system for string and inverter yield down to 10-min. frequency. Global horizontal and diffuse irradiance were derived from satellite-based measurements (Solcast), and then translated to plane-of-array irradiance for each system.[32] Monthly PR was calculated as a function of measured yield and reference yield, using methods described by Jordan et al. and Özkalay et al.[14,15] Then PLRs were calculated using the year-on-year (YOY) methodology. This compares the PR of one month with the same month in subsequent years, and results in a distribution of PLR values for all months and years. From this degradation statistics of the system were determined. Values reported here are median PLRs for each string.

2.3 Shade detection analysis

A more quantitative analysis of the shading in the BIPV systems was achieved using a FDDA, summarized in Figure 2. In brief, it compared the measured DC inputs with simulated single-diode (performed with the pvlib-python library [16]) values calculated using satellite-derived meteorological data (from SolCast in this work). When the DC inputs passed pre-defined thresholds, the system was classified as in fault, with the fault types summarized in Table 1 by their effect. F0 is in fact not a fault, but refers to normal system operation where measured DC inputs are within expected ranges. F1 denotes operation in mostly cloudy weather, defined by a cloud opacity above 40%, where there is uniform shading across the whole system.

Faults F2 to F5 have a variety of potential causes, but can reasonably be attributed to partial shading of the systems. Potential causes of F2 faults are not obvious – BPD activation (lower voltage) should theoretically not lower the current of the string, so partial shading, connection faults, or MPPT errors are likely explanations.[17] F3 faults are clear shading faults; when the voltage decreases and the current remains within acceptable ranges, it is a clear indication of BPD activation due to shading.[18,19] For F4, when the voltage increases and current is below the defined threshold, the behavior is indicative of shading without BPD activation; the current is lower due to lower irradiance reaching the shaded modules, and the voltage increases due to its temperature dependence (lower temperatures in shade will increase the voltage) and/or to adjustments in the MPPT, which should correct to the highest MPP.[20] The F5 faults with higher voltages and no current drop are explained by the small mismatches in the current and voltage thresholds of the FDDA – these faults are often observed to be coupled with F4, and therefore have the same probable causes.

The final step of the shading analysis is the quantification of the degree of shading identified using the FDDA. This is achieved through the definition of a shading factor (SF), in this case the percentage of time during which a system experiences shaded or faulted conditions. For a given power-producing time period, e.g. a full year of data, the total amount of time while shaded is identified through the FDDA by considering the fault types F2 to F5 for partial shading. Nighttime is excluded from this computation as the DC outputs are close to zero, and this is achieved by filtering out values before the sunrise and after sunset.

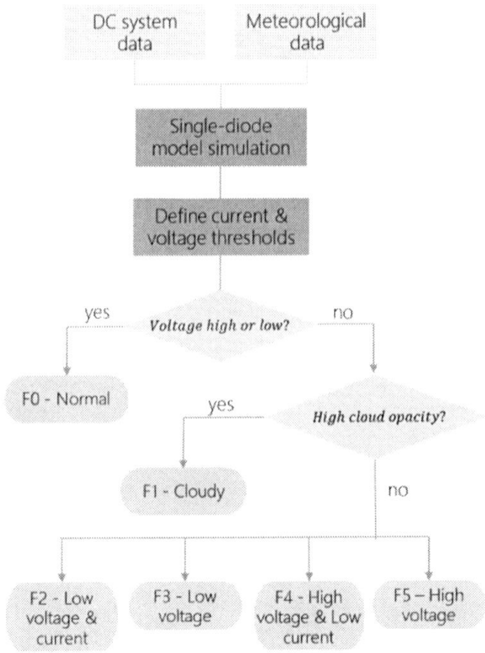

Figure 1. Shading fault detection and diagnosis algorithm (FDDA) flowchart.

Table 1. Fault classification in the shading FDDA.

Fault	Fault effect	Potential causes
F0	Normal	---
F1	Cloudy	Cloudy weather
F2	Low V, low I	Partial shading, connection fault, MPPT fault
F3	Low V	Partial shading with BPD activation, short-circuited BPD, connection fault
F4	High V, low I	Temperature drop due to shading, MPPT fault
F5	High V	Temperature drop due to shading, MPPT fault

3 RESULTS AND DISCUSSION
3.1 BIPV performance loss rates

Figure 1 show the annual PLRs for the 55 rooftop BIPV systems by installation year. Many systems had multiple strings and inverters, so in total 252 PLRs were calculated. The median fleet-wide PLR was 0.06 % per year, indicating essentially no performance change over time, though there was a large spread of positive and negative values. Note that here a negative value denotes performance loss, while a positive value signifies performance gain. In spite of the harsher operating environment for BIPV modules, many of the systems have degraded less than the average values between -0.5% and -1.0% per year reported for utility-scale and BAPV systems.[4,21,22] While there is some uncertainly inherent in PLR calculation, this could be a result of different type and quality of module materials, designs, and processes.

The spread between the best and worst performing BIPV strings was very large, 1.20% per year vs. -1.47% per year. A closer look at some poorly performing systems showed that many were characterized by several shading elements, such as nearby chimneys and trees. Moreover, the daily production profile of these systems showed several clear drops and other characteristics that were attributable to shading events (examples in next section).[3,11,20,23,24] In contrast, many well performing systems had no apparent near-field shading elements and their production profiles followed a smooth and continuous path throughout the day, mirroring the daily irradiance profile. It is suspected that module operation under routine or continuous shading accelerated degradation of the bypass diodes and cells.[11–13,23] This subjective examination of a well performing and poorly performing system highlighted the deleterious effect of shading on long-term performance, and motivated the development of a FDDA to better measure the severity of shading.

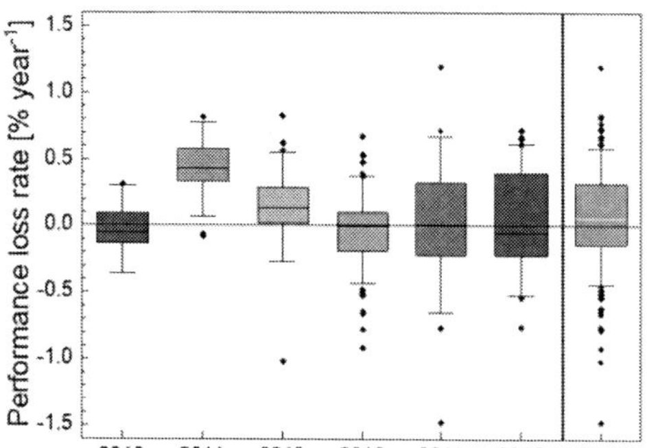

Figure 2. Performance loss rates [% year-1] for 55 BIPV systems (252 module strings) by installation year showing mean, median, confidence intervals (CI, 25-75% and 5-95%), and outliers.

3.2 Shade detection in BIPV systems

To illustrate the classification, Figure 3 shows examples of the FDDA applied to three BIPV systems, in which the DC current and voltage were compared to the single-diode simulations and their respective thresholds. The fault type heat maps on the bottom exemplifies the fault detection procedure. The unshaded BIPV system (Figure 3a) showed no faults, with both voltage and current within the simulated boundaries throughout the course of the day. The higher current in the morning and afternoon were not indicative of faults, just of higher performance than the single-diode model. The other two systems were shaded by building elements, as confirmed by on-site inspection. The first shaded system (Figure 3b) exhibited a recurring low voltage fault (F3) in the mornings and afternoons. The second shaded system (Figure 3c) and had a high voltage, low current fault (F4) occurring in the afternoons. By extending the FDDA analysis over the operational life of a system a shading factor can be computed, which is defined as the time shading fault divided by the time in operation (excluding nighttime).

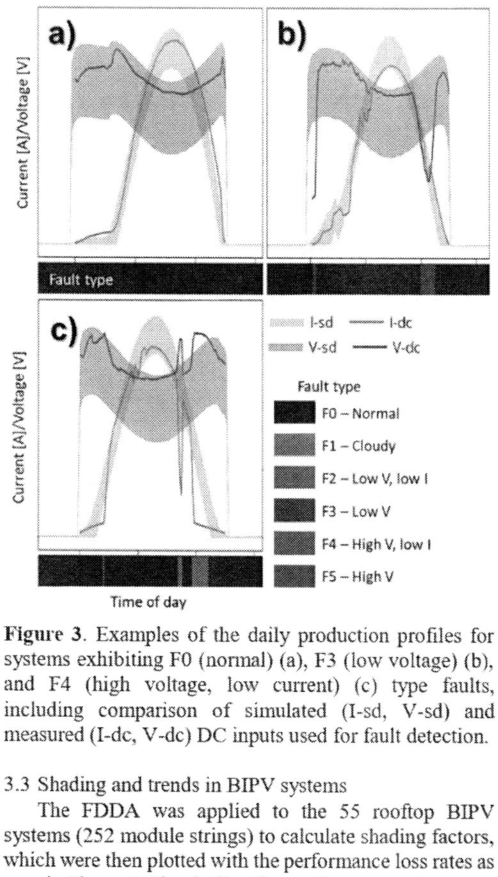

Figure 3. Examples of the daily production profiles for systems exhibiting F0 (normal) (a), F3 (low voltage) (b), and F4 (high voltage, low current) (c) type faults, including comparison of simulated (I-sd, V-sd) and measured (I-dc, V-dc) DC inputs used for fault detection.

3.3 Shading and trends in BIPV systems

The FDDA was applied to the 55 rooftop BIPV systems (252 module strings) to calculate shading factors, which were then plotted with the performance loss rates as seen in Figure 4. The shading factor alone was not a strong predictor of PLR, as in fact there are a variety of factors that contribute to system degradation besides shading.[7,25] However, shading factor did appear to present an upper limit or threshold of the PLRs, shown by the dashed line in Figure 4. This would indicate that shading limits the maximum achievable long-term performance of BIPV systems.

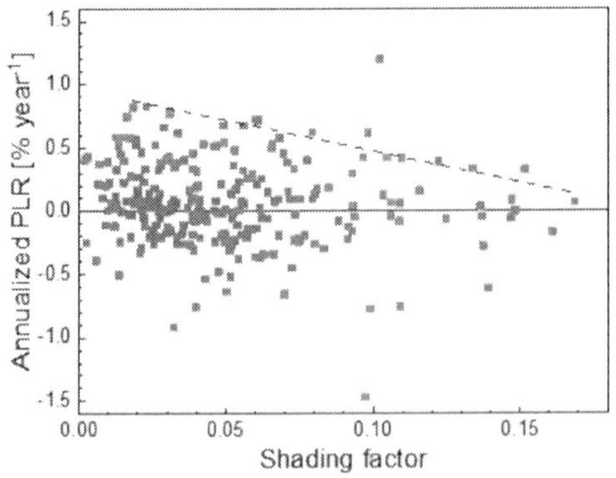

Figure 4. Performance loss rates [% year^{-1}] as a function of shading factor. The dashed line represents the upper threshold of the relationship.

Two other trends were observed for shading factor of the BIPV systems, shown in Figure 5. The first is a moderate increase in shading in newer systems. This trend is consistent from 2010 to 2014, and may be due to installation of BIPV on less ideally exposed rooftops. The best exposed roofs would offer the highest return on investment, but as systems costs have come down dramatically, even a less exposed rooftop can be viable investment. Moreover, support schemes and incentives sometimes encourage installation of PV systems without regard to local exposure (sunlight) conditions.[26] Another reason may be the increasing use of customized module sizes and shapes to fully cover roof areas, especially around shading elements.

The second trend observed is a shading factor decrease with system capacity. This is most noticeable for small capacity systems, below 30 kW_p. The reason for this is likely because larger systems tend to be on large roofs, for example warehouses, barns, or commercial buildings, which would have fewer shading elements with respect to their size and be located in larger and more open land parcels. Smaller systems on the other hand are mostly on residential buildings, including single family homes, for which the number of potential shading elements is greater compared to the roof size.

3.4 Application and mitigation of shade-induced degradation

The FDDA can be applied to any type of BIPV or even conventional PV system with intra-day monitoring data to detect shading and determine its severity. Problematic modules and strings can be identified and then removed or modified to minimize the effect of shading on the entire system. Because shading may be a contributing factor to performance loss, if a poorly performing system ultimately results in a warranty claim (e.g. $P_{max} < 80\% P_{nom}$ in the 25 year warranty period), the FDDA can assist attribution of responsibility, e.g. to the module manufacturer or system designer and installer.

Both parties can undertake action to mitigate the effect of shade-induced degradation, some of which include:

1. Design of more shade resistant modules, e.g. by installation of additional bypass diodes or diodes with higher ratings into modules.[27,28]

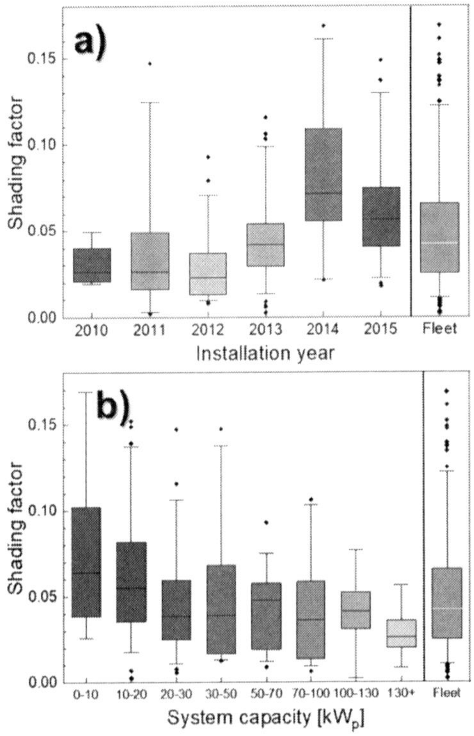

Figure 5. Shading factors as a function of installation year (a) and system capacity (b).

2. Improved education and training for system designers concerning the weaknesses and limitations of PV module design for specific applications.
3. More precise assessment of the local boundary conditions in the building envelope and surroundings to identify heavily shaded areas.
4. Optimization of the electrical layout of strings, or installation of DC/DC power optimizers and microinverters (instead of string or central inverters) so that fewer modules are affected by shading.[29,30]
5. Installation of more shade resistant modules or dummy panels in areas expected to experience heavy shading.

BIPV constitutes a small segment of the PV market, and there is still a learning curve to be climbed to optimize system performance.[9] Moreover, this work focused on rooftop BIPV systems. Façade BIPV systems are another growing segment, and have greater risk for shading than rooftops. Utility scale PV systems had a similar learning curve in the early 2000's as capacity began to grow exponentially. There were numerous examples of poor system design that resulted in non-optimal performance or early failure, such as widespread installation damage from mishandling or vertical module orientation resulting in strong losses due to inter-row shading during winter.[24,31] Experience from the early days of PV expansion have shown steadily improving performance of PV systems, and demonstrated that there is plenty of room for improvement.

4 CONCLUSIONS

The long-term performance of BIPV modules is an open question for the sector, as most reports are for field-mounted PV or BAPV systems. This lack of data is especially concerning because of the increased risk factors for BIPV systems, including elevated operating temperatures and regular shading events. In that regard, this work is one of the first studies on long-term performances of true BIPV systems as defined in IEC 63092-1. The median performance loss rate (PLR) of 55 rooftop BIPV systems (252 module strings) in Switzerland was found to be 0.06% per year, indicating an essentially stable performance over 5 to 10 years of operation. However, within the fleet a large spread of PLRs was identified, and faster degrading systems were found to be more heavily shaded. A failure detection and diagnosis algorithm (FDDA) was developed to identify deviations in DC voltage and current that were indicative of shading events. By applying the FDDA over the operational life of each string and system, a shading factor or fraction of time in shading fault was calculated. The shading factor alone was not a strong predictor of PLR, but it presented an upper boundary, effectively limiting maximum achievable performance. Shading severity was found to increase in newer systems, and decrease in larger capacity systems. These results highlight shading as a major risk factor for BIPV that has potential to accelerate system degradation. Moreover, they provide valuable insights to BIPV module manufacturers and installers, allowing them to better design products and systems to mitigate the effect of shading within a BIPV or even other integrated PV systems.

5 ACKNOWLEDGEMENTS

Authors thank Baur AG for providing access to the system data used in these analyses, and Solcast for use of their satellite-derived meteorological data.

This work has been funded in part by the European Commission (EC) under the H2020 project Be-SMART (#818009), and the Swiss National Science Foundation (SNF) under the project REBI-PV (COST IZCOZ0_182967). Additional (collaborative) support was provided by the COST Action PEARL PV (CA16235).

6 REFERENCES

[1] International Electrotechnical Commision, IEC TS 63126:2020 - Guidelines for qualifying PV modules, components and materials for operation at high temperatures, (2020).

[2] J. Oh, G. TamizhMani, E. Palomino, Temperatures of building applied photovoltaic (BAPV) modules: air gap effects, in: Reliability of Photovoltaic Cells, Modules, Components, and Systems III, SPIE, San Diego, CA, USA, 2010: p. 777305. https://doi.org/10.1117/12.861069.

[3] M. Drif, A. Mellit, J. Aguilera, P.J. Pérez, A comprehensive method for estimating energy losses due to shading of GC-BIPV systems using monitoring data, Solar Energy. 86 (2012) 2397–2404. https://doi.org/10.1016/j.solener.2012.05.008.

[4] D.C. Jordan, S.R. Kurtz, K. VanSant, J. Newmiller, Compendium of photovoltaic degradation rates: Photovoltaic degradation rates, Prog. Photovolt: Res. Appl. 24 (2016) 978–989. https://doi.org/10.1002/pip.2744.

[5] J. Leloux, J. Taylor, R. Moretón, L. Narvarte, D. Trebosc, A. Desportes, Monitoring 30,000 PV systems in Europe: Performance, Faults, and State of the Art, (2015) 9.

[6] International Electrotechnical Commision, IEC 63092-1 - Photovoltaics in buildings - Part 1: Requirements for building-integrated photovoltaic modules, (2020).

[7] M.M. Fouad, L.A. Shihata, E.I. Morgan, An integrated review of factors influencing the perfomance of photovoltaic panels, Renewable and Sustainable Energy Reviews. 80 (2017) 1499–1511. https://doi.org/10.1016/j.rser.2017.05.141.

[8] M.D. Kempe, D. Holsapple, K. Whitfield, N. Shiradkar, Standards development for modules in high temperature micro-environments, Progress in Photovoltaics: Research and Applications. 29 (2021) 445–460. https://doi.org/10.1002/pip.3389.

[9] B. Norton, P.C. Eames, T.K. Mallick, M.J. Huang, S.J. McCormack, J.D. Mondol, Y.G. Yohanis, Enhancing the performance of building integrated photovoltaics, Solar Energy. 85 (2011) 1629–1664. https://doi.org/10.1016/j.solener.2009.10.004.

[10] International Electrotechnical Commision, IEC 61215: Crystalline silicon terrestrial photovoltaic (PV) modules – Design qualification and type approval, (2020).

[11] A. Dolara, G.C. Lazaroiu, E. Ogliari, Efficiency analysis of PV power plants shaded by MV overhead lines, Int J Energy Environ Eng. 7 (2016) 115–123. https://doi.org/10.1007/s40095-016-0208-2.

[12] N.G. Dhere, N. Shiradkar, E. Schneller, V. Gade, The reliability of bypass diodes in PV modules, in: Reliability of Photovoltaic Cells, Modules, Components, and Systems VI, International Society for Optics and Photonics, 2013: p. 88250I. https://doi.org/10.1117/12.2026782.

[13] W.G. Shin, S.W. Ko, H.J. Song, Y.C. Ju, H.M. Hwang, G.H. Kang, Origin of Bypass Diode Fault in c-Si Photovoltaic Modules: Leakage Current under High Surrounding Temperature, Energies. 11 (2018) 2416. https://doi.org/10.3390/en11092416.

[14] D.C. Jordan, C. Deline, S.R. Kurtz, G.M. Kimball, M. Anderson, Robust PV Degradation Methodology and Application, IEEE J. Photovoltaics. 8 (2018) 525–531. https://doi.org/10.1109/JPHOTOV.2017.2779779.

[15] E. Özkalay, A. Virtuani, A. Fairbrother, A. Skoczek, G. Friesen, C. Ballif, Quantifying performance loss rate using ground-based vs satellite-based insolation and temperature data for small scale systems, Solar RRL. in press (2021).

[16] W.F. Holmgren, C.W. Hansen, M.A. Mikofski, pvlib python: a python package for modeling solar energy systems, Journal of Open Source Software. 3 (2018) 884. https://doi.org/10.21105/joss.00884.

[17] A. Livera, M. Theristis, G. Makrides, G.E. Georghiou, Recent advances in failure diagnosis techniques based on performance data analysis for grid-connected photovoltaic systems, Renewable Energy. 133 (2019) 126–143. https://doi.org/10.1016/j.renene.2018.09.101.

[18] S. Fadhel, C. Delpha, D. Diallo, I. Bahri, A. Migan, M. Trabelsi, M.F. Mimouni, PV shading fault detection and classification based on I-V curve using principal component analysis: Application to isolated PV system, Solar Energy. 179 (2019) 1–10. https://doi.org/10.1016/j.solener.2018.12.048.

[19] M. Seapan, Y. Hishikawa, M. Yoshita, K. Okajima, Detection of shading effect by using the current and voltage at maximum power point of crystalline silicon PV modules, Solar Energy. 211 (2020) 1365–1372. https://doi.org/10.1016/j.solener.2020.10.078.
[20] J.C. Teo, R.H.G. Tan, V.H. Mok, V.K. Ramachandaramurthy, C. Tan, Impact of Partial Shading on the P-V Characteristics and the Maximum Power of a Photovoltaic String, Energies. 11 (2018) 1860. https://doi.org/10.3390/en11071860.
[21] S. Lindig, J. Ascencio-Vásquez, J. Leloux, D. Moser, M. Topič, CLIMATE RELATED DEPENDENCE OF PERFORMANCE LOSSES OF OVER (3,500) PV SYSTEMS, (n.d.) 6.
[22] K. Kiefer, B. Farnung, B. Müller, F. Ise, K. Reinartz, I. Rauschen, C. Klünter, A. Pannhaus, Degradation in PV Power Plants: Theory and Practice, in: Proceedings of the 35th European Photovoltaic Solar Energy Conference and Exhibition, 2018: p. 5.
[23] S.W. Ko, Y.C. Ju, H.M. Hwang, J.H. So, Y.-S. Jung, H.-J. Song, H. Song, S.-H. Kim, G.H. Kang, Electric and thermal characteristics of photovoltaic modules under partial shading and with a damaged bypass diode, Energy. 128 (2017) 232–243. https://doi.org/10.1016/j.energy.2017.04.030.
[24] A. Woyte, J. Nijs, R. Belmans, Partial shadowing of photovoltaic arrays with different system configurations: literature review and field test results, Solar Energy. 74 (2003) 217–233. https://doi.org/10.1016/S0038-092X(03)00155-5.
[25] D.C. Jordan, B. Marion, C. Deline, T. Barnes, M. Bolinger, PV field reliability status—Analysis of 100 000 solar systems, Progress in Photovoltaics: Research and Applications. 28 (2020) 739–754. https://doi.org/10.1002/pip.3262.
[26] A. Virtuani, A. Fairbrother, F. Lisco, L.-E. Perret-Aebi, N. Wyrsch, C. Ballif, Do north-facing BIPV facades in Europe make sense? - Policy drivers for PV in buildings (and infrastructures), in: Proceedings of the 38th European Photovoltaic Solar Energy Conference and Exhibition, 2021.
[27] H. Hanifi, M. Pander, B. Jaeckel, J. Schneider, A. Bakhtiari, W. Maier, A novel electrical approach to protect PV modules under various partial shading situations, Solar Energy. 193 (2019) 814–819. https://doi.org/10.1016/j.solener.2019.10.035.
[28] B.B. Pannebakker, A.C. de Waal, W.G.J.H.M. van Sark, Photovoltaics in the shade: one bypass diode per solar cell revisited, Progress in Photovoltaics: Research and Applications. 25 (2017) 836–849. https://doi.org/10.1002/pip.2898.
[29] H.P. Ikkurti, S. Saha, A comprehensive techno-economic review of microinverters for Building Integrated Photovoltaics (BIPV), Renewable and Sustainable Energy Reviews. 47 (2015) 997–1006. https://doi.org/10.1016/j.rser.2015.03.081.
[30] S. Lee, J.-B. Eggers, J. Eisenlohr, K. Lee, Learning From Tetris: A New Approach for the Automated Configuration of the Interconnection Layout of BIPV Modules for Large-Scale Application, IEEE Journal of Photovoltaics. 11 (2021) 770–778. https://doi.org/10.1109/JPHOTOV.2021.3064585.
[31] N.H. Reich, B. Mueller, A. Armbruster, W.G.J.H.M. van Sark, K. Kiefer, C. Reise, Performance ratio revisited: is PR > 90% realistic?, Progress in Photovoltaics: Research and Applications. 20 (2012) 717–726. https://doi.org/10.1002/pip.1219.

EXPERIMENTAL ASSESSMENT AND DATA ANALYSIS OF COLORED PHOTOVOLTAIC IN THE FIELD OF BIPV TECHNOLOGY APPLICATION

Riccardo Roverso [1,2], Martina Pelle[1], Mattia Dallapiccola[1], Alexander Astigarraga[1], Elena Lucchi[1], Philip Ingenhoven[2,3] and Laura Maturi[1]
[1]Institute for Renewable Energies - EURAC Research, Viale Druso 1, 39100 Bolzano, Italy
[2]Free University of Bolzano, Faculty of Science and Technology, piazza Università, 5, 39100 Bolzano, Italy
[3]University of Adelaide, Faculty of Engineering, Centre of Energy Technology, Adelaide, South Australia 5005 Australia
Phone: +39 3401213455; E-mail: riccardo.roverso@eurac.edu, martina.pelle@eurac.edu, mattia.dallapiccola@eurac.edu, alexander.astigarraga@eurac.edu, elena.lucchi@eurac.edu, philip.ingenhoven@adelaide.edu.au, laura.maturi@eurac.edu

ABSTRACT: Temperature of PV modules always has been a key parameter in order to understand and predict the efficiency of PV technologies. In this work, the analysis of the temperature trend of a selected set of colored BIPV modules is performed, with the aim of correlating the irradiance recorded through a pyranometer to the temperature of the modules, recorded through Pt100 temperature sensors. A description of the different aspects of BIPV concept is provided, and a simulation strategy is introduced to understand the power loss due to the presence of colored layer on the modules' front glass. Furthermore, test facilities available at Eurac Research laboratories and the equipment used during the test campaign are described. As metric of measures, the Ross coefficient k is used to correlate the different PV systems to the temperature rise. The values found after a data calibration and filtering procedure are elaborated and compared with the outcomes of a standard not-colored BIPV installation, located in Bolzano. The modules' temperature measured through Pt100 sensors have been then compared with the results of a thermographic analysis.

Keywords: BIPV integration, data analysis, laboratory tests.

1 INTRODUCTION

The interest of understanding the effects of color on PV modules performance brought us to perform test campaigns on a set of BIPV colored installation present at the Outdoor Eurac Research PV Integration Lab, and on a standard not-colored BIPV installation at Ex-Post Building in Bolzano [1]. Both installations are located in Bolzano, Italy. The present study focuses on BIPV system energy performance in operating temperature conditions. The set of analyzed colored BIPV modules consists of roof integrated BIPV systems with different customization techniques, that mimic the appearance of standard construction materials and hide the PV cells to the view: one, hereby called "Terracotta", presents a uniform color of the front glass, similar to the classical tiles color (indicative RAL 8015); the other, hereby called "Tegola-Tiles" reproduces a typical tiles pattern, which is digitally printed on the module's front glass [2]

2 INTEGRATION OF COLORED BIPV MODULES IN BUILDINGS

The balance of technical aspect and aesthetical requirements is a key aspect of BIPV systems; from literature and standard definitions of BIPV, three integration areas are defined: aesthetical, technological, and energetic [2]. Therefore, a key aspect to consider for BIPV installations is the appearance that PV modules have on the aesthetic perception of the building elements and of the building as a whole. The purpose of integrating colored PV in buildings is to obtain feasible solutions which can be flexible, visually pleasing, colorful, and that can satisfy the requests of designers in order to create original visual effects and make the building less energy demanding.

3 COLORED BIPV MODULES PERFORMANCE: SIMULATION STRATEGY

3.1 Colored glass characteristics

Recent analysis on colored glass tried to find an effective manner to measure and quantify the efficiency due to the different technology and color which is adopted for BIPV. The work [3] gives guidelines on color efficiency and power loss based on the RAL color classification. For the purpose of this study, the spectral optical characteristics of the colored glasses have been obtained via spectrophotometry and used for calculating the energy transmitted through the front glass.

3.2 Analytical model

An analytical simulation of the "Terracotta" modules was performed, with the purpose of developing an effective methodology to calculate the performance of colored PV. The simulation strategy takes as inputs some results of the indoor tests performed over the selected set of PV modules, consisting of flash tests with a STC irradiation source, compliant with the standard IEC 60904. The aim is to obtain the modules electrical characteristics starting from standard conditions and basic information to get at first the modules I_{SC}. By using then the single diode equation [4], shunt resistance, the glass transmittance data, solar simulator irradiation spectrum, and PV cell characteristics, was possible to get the mayor features of a PV module as IV curve, I_{MPP}, V_{MPP}, P_{MPP}, FF and efficiency. At this point the efficiency was calculated, by using the formula $\eta = P_M / (Irr_{STC} A_{mod})\ 100$. in order to compare the obtained values with the literature outcomes. Different efficiency were found depending on the considered configurations. The values differ only for a 0.5% over the efficiency outcomes (Table I), with the implementation of the analytical model of Figure 1.

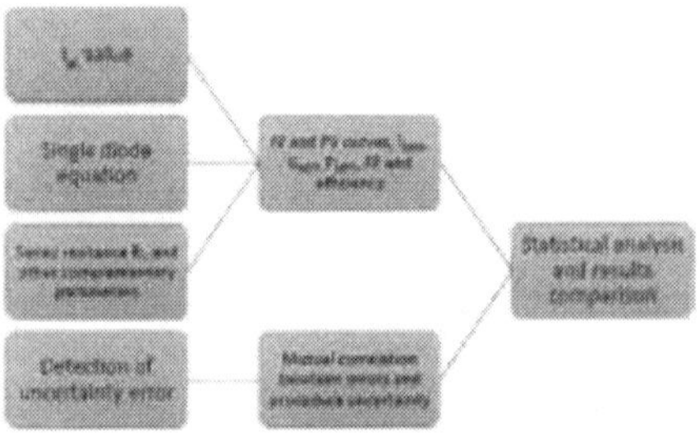

Figure 1: Developing of the analytical model. Eurac Research

Table I: Comparison of the efficiency values found. Eurac Research.

	efficiency (%)
Analytical simulated	12.67
Flash test	13.07

4 EXPERIMENTAL CAMPAIGN

4.1 Ross coefficient: colored modules at Eurac Research PV Integration Lab

By calculating the Ross coefficient [5] it is possible to correlate the temperature of the cell (and module), to the air temperature and the irradiance illuminating the module, reported by the formula $T_{mod} = T_{amb} + kG$, where k is the Ross coefficient with units of °Cm^2 W^{-1} or Km^{2}W^{-1}, G is the solar irradiance illuminating the module in Wm^{-2}, T_{mod} is the module temperature and T_{amb} is the ambient temperature in °C. The lower is the Ross coefficient, the lower is the difference between the cell temperature and the ambient temperature. Some reference values are given in Table II.

4.2 Outdoor roof mock-up

The Terracotta and Tegola-Tiles modules selected after an indoor test campaign, were installed on a testbed in the PV Integration Lab of Eurac Research (Figure 2 and Figure 3) [2]; which consists of an outdoor roof mock-up which permits to evaluate and test PV and BIPV modules in real-time conditions. From the operative point of view, it is a rotating structure of 20 square meter plane, with the possibility of tilting up to 60° from the horizontal plane and set the azimuth to any degree value, in order to reproduce any installation conditions [2].

4.3 Design of the experiment

The modules installed were previously tested at the indoor Solare PV Lab, through electroluminescence test and flash test, assuring their correct functioning. The modules were connected in series, creating two separate circuits for the Terracotta modules and the Tegola-Tiles modules. During the installation, 8 Pt100 sensors were attached to the back of the modules (Figure 4). The exact position of the sensors is shown in Figure 5.

A wooden substructure was used to install the modules trough a set of three clamps for each module. The slats position has been taken in account to understand their influence on the temperature detected by the Pt100 sensor. The electrical circuits are connected to a DC/AC inverter, with the purpose of identifying and transmitting the output performance data. The irradiance value is detected through a pyranometer installed at the side of the testbed, on the same modules' plan.

Table II: Values of Ross coefficient for various mounting type. [6]

PV Array mounting type	k (°C /W/m^2)
Free standing	0.021
Flat roof	0.026
Sloped roof: Well cooled	0.020
Sloped roof: not so well cooled	0.034
Sloped roof: highly integrated, poorly ventilated	0.056
Façade integrated: transparent PV's	0.046
Façade integrated: opaque PVs – narrow gap	0.054

Figure 2: Roof mock-up present at the outdoor PV integration Lab. Eurac Research

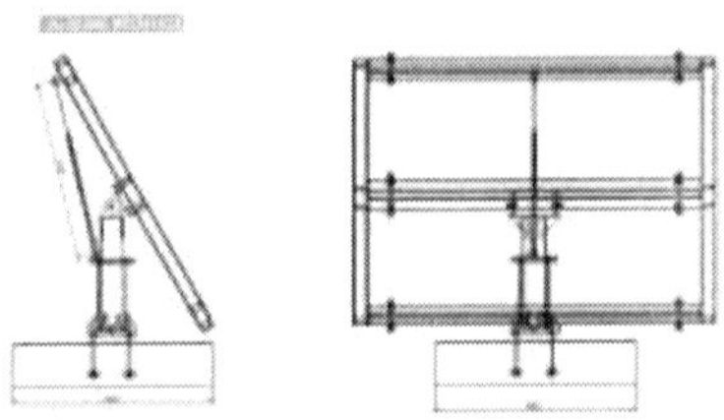

Figure 3: Frontal and lateral perspective of the roof mock-up. Eurac Research

Figure 4: Pt100 implementation on the modules back. Eurac Research

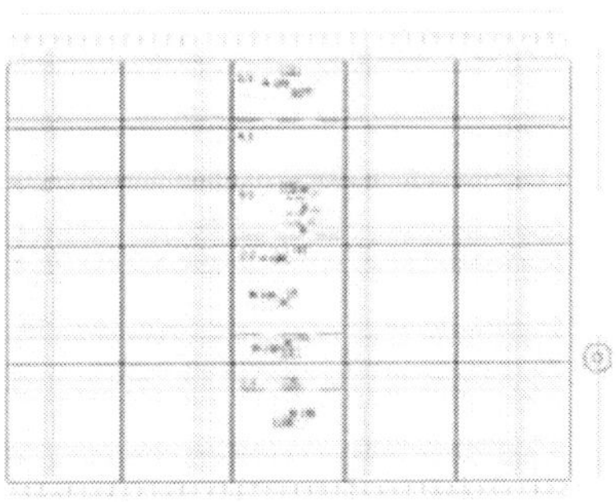

Figure 5: Localization of the Pt100 sensor on the test-bed facility. Eurac research

The instrument is a Kipp&Zonen Pyranometer with a spectral range from 285 to 2800 nm, sensitivity 7 to 14 μV/W/m^2. More specifications are present in [7]. A meteorological station is present near the rood mock-up, measuring the wind velocity, wind direction, air temperature, relative air humidity, diffuse horizontal irradiance, global horizontal irradiance, direct normal irradiance, and sun azimuth.

4.4 Modules description

The used BIPV modules are glass-glass modules, provided with monocrystalline cells, devoid of frame, where a process of mineral coating was utilized to fix the colored layer on the inner surface of the cover glass [8]. The BIPV installation at the Ex-Post consists of a not back ventilated polycrystalline system, integrated on the existing façade of the Ex-Post Building as an external cladding through a metallic structure. The modules under investigation are reported in Figure 6.

5 OUTDOOR TESTING RESULTS

5.1 Temperature calibration

Before implementing the temperature data on elaboration, calibration coefficients were applied to proper obtain values temperature with a linear regression through the formula $T_{cal} = a \cdot T_{old} + b$, where T_{cal} is the calibrated temperature, T_{old} the value to correct and a and b the calibration coefficients. In Table III, the calibration coefficient for the eight Pt100 temperature sensors are reported.

5.2 Ross coefficients calculations

The Ross coefficient can be obtained graphically by plotting the irradiance over the temperature difference between the module and the air, achieving an intercept among the scattered plots. The slope of that line represents the value of the Ross coefficient (Figure 7). Some filtering procedures are introduced as also reported in [1]. The filter excludes the points corresponding to a ratio of diffuse and global irradiance higher than 0.20 in order to consider only clear sky conditions, as the direct irradiance component is predominant on the diffuse one.

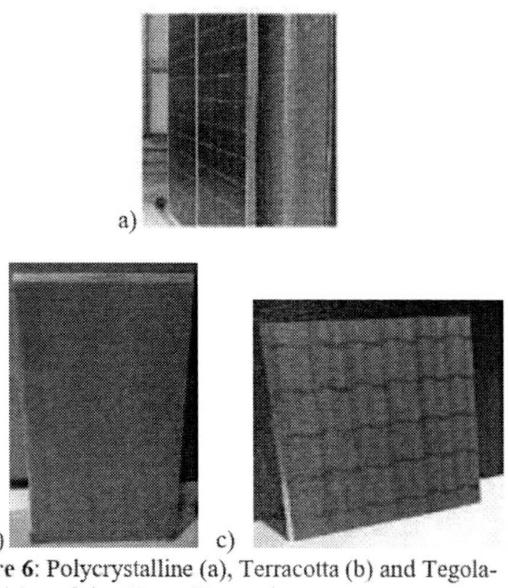

Figure 6: Polycrystalline (a), Terracotta (b) and Tegola-Tiles (c) modules. Eurac Research.

The Ross coefficient values are calculated for the modules installed at the outdoor PV Integration Lab relatively to the time period (26-02-2021 – 23-04-2021), the outcomes are shown on Figure 7.
The outcomes of Table IV are in line with the literature for a sloped roof (Table II), not so well cooled, as it is the test bed facility of the PV Integration Lab. It is possible to observe that the modules, manufactured with the same technology, but in the higher position on the testbed, register the higher value of Ross coefficient. That is due to the fact that the air behind the module gets warmer in the down-top direction of the test bed. The effect is that the modules which are positioned in the higher locations can benefit less of air cooling and can encounter a higher warming with the same irradiation conditions.

5.3 Performance Ratio (PR) filtering outcomes

A specific filter was applied only on a secondary phase, to observe if relevant changes on the graphs and Ross coefficient values are detected after the filtering implementation (Figure 8). The filter considers only the values which withstand the values of $PR_{avg} \pm \sigma$ where PR_{ave} is the average value of performance ratio of the PV array for the considered period, defined by the international standard IEC 61724 [9] $PR = (E \cdot G_{STC}) / (P_n \cdot G)$ where E is the instant power produced (W), P_n the array nominal power (W), G_{STC} the irradiance under STC conditions (1000 W/m^2), G the instant irradiance (W/m^2) σ is the standard deviation. The PR_{avg} for the two considered months is 0.785. (26-02 2021 - 23-04-2021).
Comparing these results with those without filters on PR, it is possible to notice that a significant lower amount of points are considered and that the resulting Ross coefficient slightly increases, 3.4% for the 1.3 Tegola-Tile module (from 0.0323 to 0.0334 °C W-1 m-2); 3.4% for the 2.3 Tegola-Tile module (from 0.0337 to 0.0349 °C W-1 m-2); 6% for the 3.3 Terracotta module (from 0.0332 to 0.0353 °C W^{-1} m^{-2}) and of 1.2% for 5.3 Terracotta module (from 0.0340 to 0.0343 °C W^{-1} m^{-2}).

5.4 Wind dependance

An important application of the PR filter is the wind dependance of Ross coefficient. A correlation between the Ross coefficient and wind velocity is thus evaluated plotting $k=(T_{mod,avg}-T_{amb})/G$ against wind velocity expressed in m/s (Figure 9). The filtering procedure had effect on leaving out the mayor outliers, producing a graph with a higher degree of accuracy.
It is possible to notice that the trend is linear descending, toward the maximum value of wind velocity of about 5 m/s. The wind in this case has only a slight effect due to the fact that the modules are installed over a metallic structure of the test bed and the wooden slats do not allow a greater natural ventilation under the BIPV modules.

Table III: Calibration coefficients values for the 8 Pt100 temperature sensor.

Temperature sensor	a	b
Pt100 T1	1.0056	-0.0885
Pt100 T2	1.0086	-0.1523
Pt100 T3	1.0092	-0.0882
Pt100 T4	1.0319	-0.2822
Pt100 T5	1.0089	0.0712
Pt100 T6	1.0206	-0.1319
Pt100 T7	1.0101	-0.076
Pt100 T8	1.0274	-0.137

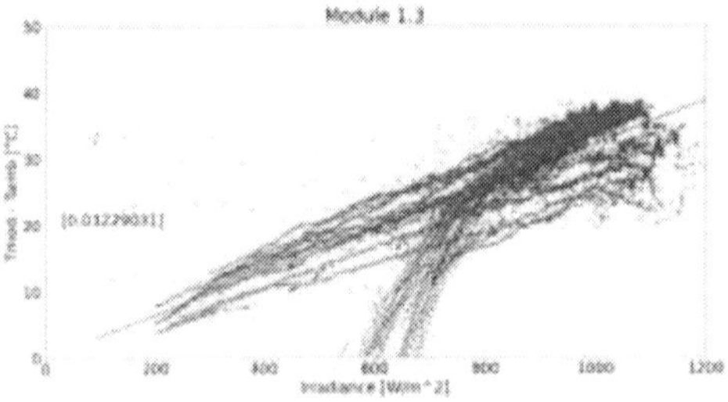

(a)

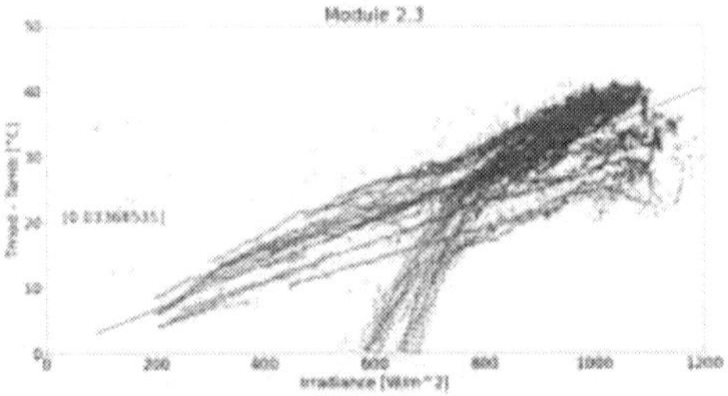

(b)

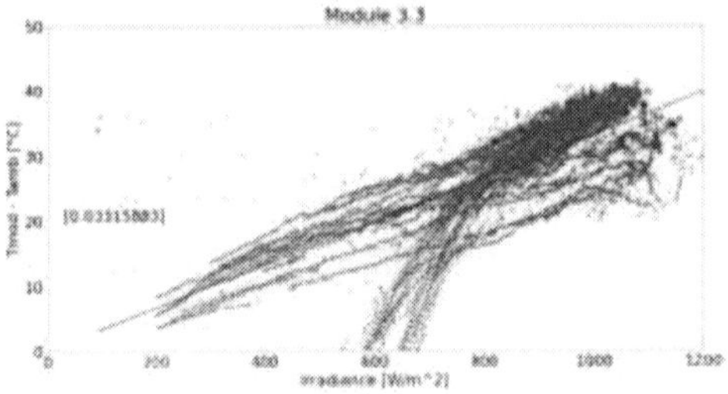

(c)

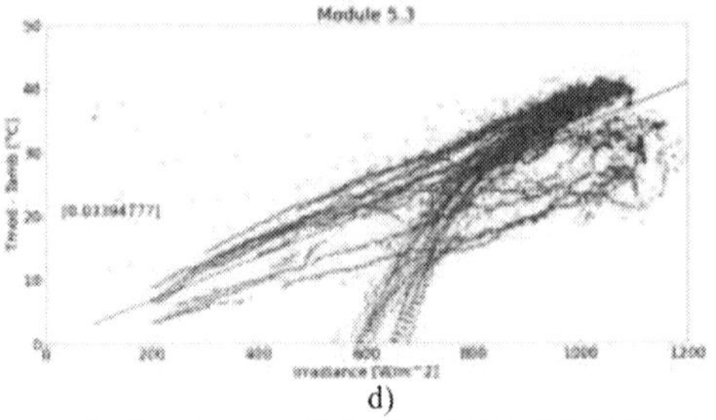

d)

Figure 7: T_{mod}-T_{amb} vs G for the 1.3(a), 2.3(b) Tegola-Tile modules and 3.3(c) and 5.3(d) Terracotta modules. Eurac Research

Table IV: Values of Ross coefficient of the installed modules. Eurac research.

Module	k (°C /W/m^2)
Tegola 1.3	0.0323
Tegola 2.3	0.0337
Terracotta 3.3	0.0331
Terracotta 5.3	0.0339

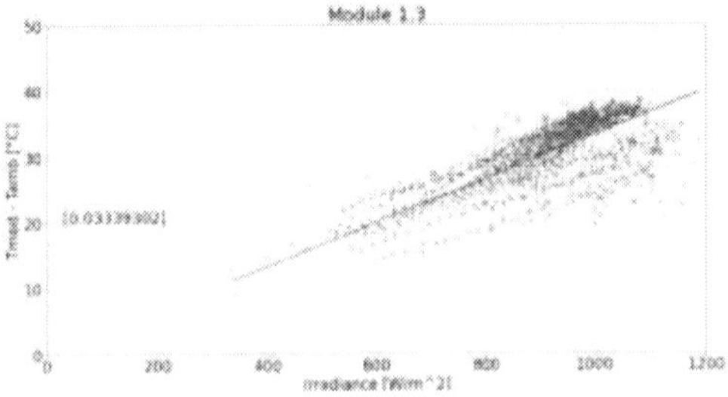

(a)

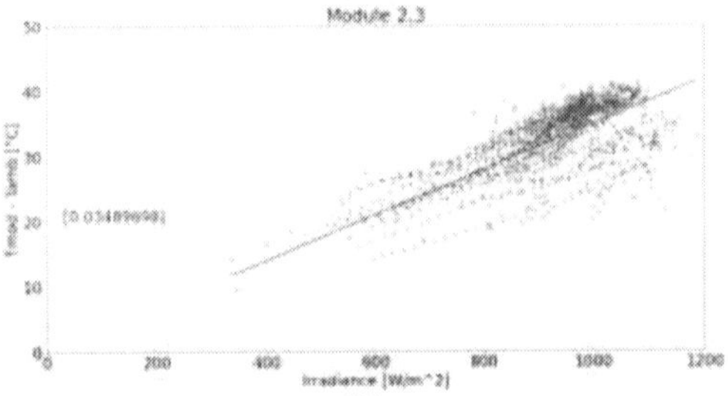

(b)

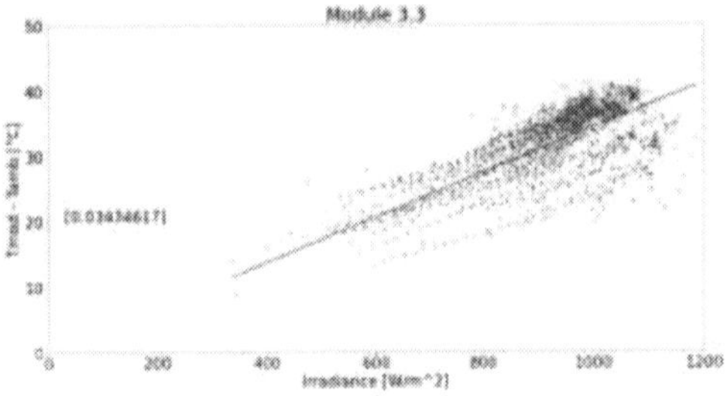

(c)

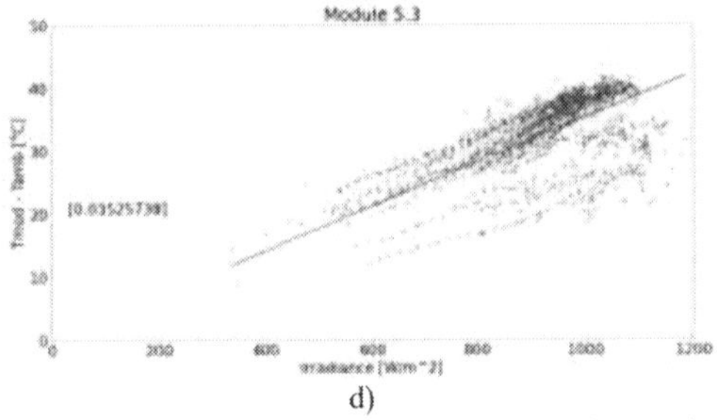

d)

Figure 8: T_{mod}-T_{amb} vs G for the 1.3(a), 2.3(b) Tegola-Tile and 3.3(c), 5.3(d) Terracotta modules after PR filtering. Eurac Research

Figure 9: Wind dependance of the Ross coefficient. Eurac Research

5.5 Thermography measurements

The thermography technique was used to visualize the temperature distribution of the modules installed on the roof mock-up, and then to correlate the temperature registered from Pt100 sensor placed on the backside module and the surface temperature registered with the infrared camera (Figure 10). The thermography measurements were performed in a sunny and cloudless day when the irradiance measured was higher than 800 W/m^2 to observe proper working thermal behavior. The IR pictures show the longitudinal temperature trend among the modules, with higher temperatures trend among the modules, with higher temperatures on the upper part, descending among the line (Figure 11). The peak temperatures can bring to an erroneous evaluation of the temperature trend, but those sharp increments are due to the metal structure which can be seen through the modules overcoming. The Pt100 sensor is positioned in the upper and lower modules, following the scheme of Figure 12. The variation between the recorded temperature through the Pt100 and the temperature recorded through the thermal camera is small (Table V). The major discrepancy on the sensor Pt100 7 could be given by the border effect which can influence the module surface temperature.

(a)

(b)

Figure 10: (a) Infrared picture of the outdoor BIPV Terracotta modules. (b) Temperature path detected with the thermo camera software. Eurac Research

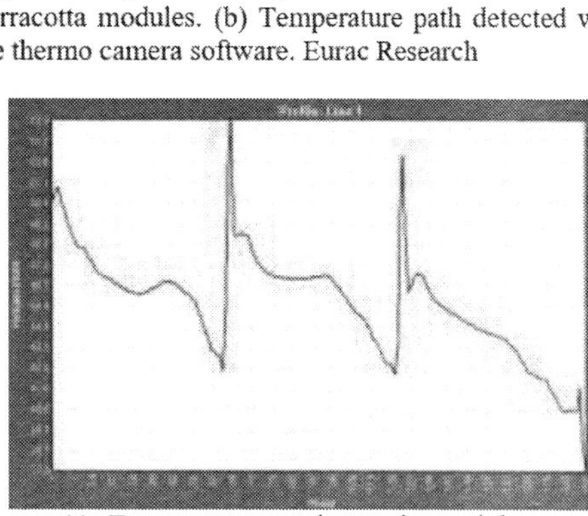

Figure 11: Temperature trend over the modules test bed (line 1 of Figure 11). Eurac Research

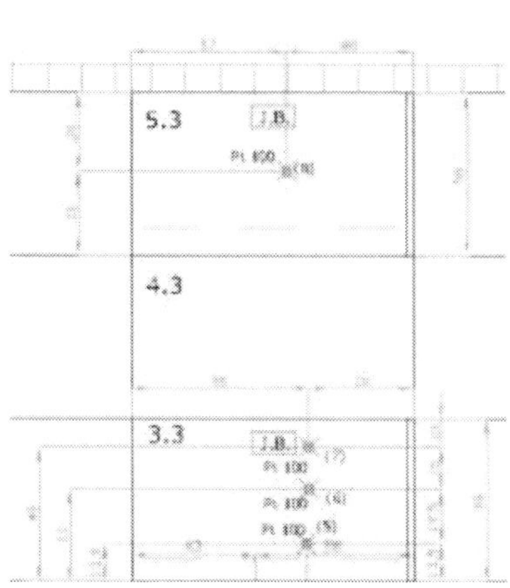

Figure 12: localization of the Pt100 sensors on the module's backside. Eurac Research

Table V: Temperature recorded with the Pt100 sensor and the thermo camera at the same instant time.

Sensor	$T_{measured}$ (°C)	T_{thermo} (°C)	ΔT(°C)	%
Pt100 T5	44.9	45.1	0.2	0.01
Pt100 T6	46.07	47.29	1.22	2.57
Pt100 T7	44.67	48.34	3.67	7.59
Pt100 T8	47.91	49.31	1.4	2.83

5.6 Ross coefficient: standard modules at Ex-Post building.

To have a meter of comparison, a previous BIPV installation has been considered, in particular the plant present at the Ex-Post building, in the Bolzano railway proximity (Figure 13) [10]. The purpose is to compare the Ross coefficient value of a standard installation against the colored BIPV one.

The data analyzed for the Ross coefficient calculations go from February 2017 to September 2017. A brief filter procedure was made over the irradiance data, excluding negative values for the irradiance.

The following graph displays the module temperature difference above ambient temperature with the increasing irradiance value (Figure 14). The fitting curve was forced to pass to zero, the slope of the blue curve is representing the Ross coefficient k. The value found is 0.0398 °C W^{-1} m^{-2} in accordance with the data presented on [12]. In this case the system is poorly ventilated with a 10 cm air gap on the back modules and the k value found confirmation on table I from literature. In a second step, the PR filtering procedure was applied as in section 5.4, leading to a Ross coefficient value of 0.0408 C W^{-1} m^{-2}.

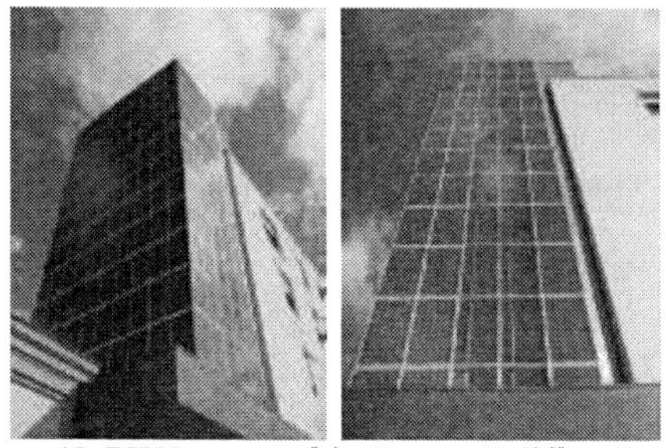

Figure 13: BIPV system of the Ex-post Building. Eurac Research

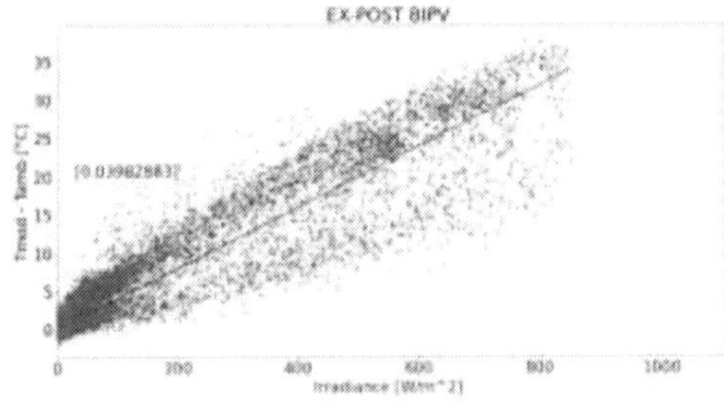

(a)

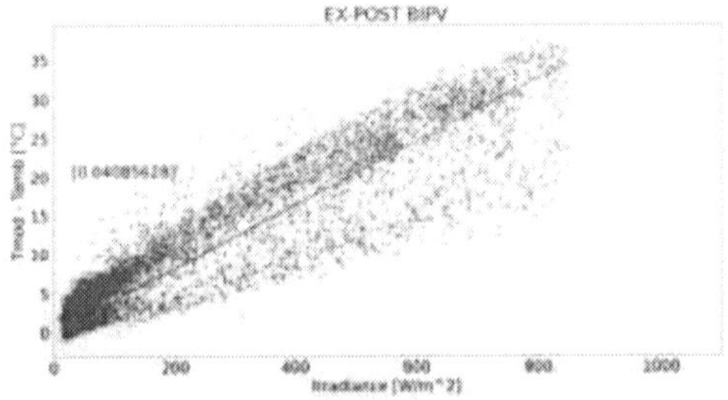

(b)

Figure 14: T_{mod}-T_{amb} vs G for the Ex-Post BIPV (a), with the implementation of the PR filter (b). Eurac Research

6 CONCLUSIONS

In this work the analysis of the Ross coefficient of a selected set of colored BIPV modules is performed, with the aim of correlating the irradiance recorded through a pyranometer to the temperature of the module. At first the analyzed BIPV technologies, the test facilities and the design of the experiment have been described. Then the Ross coefficient was introduced as metric of measurement to correlate the temperature rise of the modules to the irradiation level. The outcomes resulting from the data analysis show that values around 0.033 C W^{-1} m^{-2} are consistent with the literature [6]. The results of the analysis from the outdoor PV Integration Lab were compared with the data and the results coming from the standard not-colored BIPV installation present at Bolzano Ex-Post, with a Ross coefficient value of 0.0398 °C W^{-1} m^{-2}. This higher value it caused by the poorer ventilation conditions of the Ex-Post installation, anyway it is so possible to assert that BIPV installation color customized, with the purpose to encounter aesthetical requirements, *are not affected* by a higher heating up of the modules, which usually brings to a decline of the PV performance.

7 ACKNOWLEDGMENT

The infinite project has received funding from the European Union's Horizon 2020 research and innovation programme under grant agreement No 958397. This paper reflects only the author's view, neither the European Commission nor HADEA are responsible for any use that may be made of the information it contains.

INFINITE BUILDINGRENOVATION

The authors would like to thank the company SUNAGE "Glassfer & Sunage", for the support provided for this research.

REFERENCES

[1] L. Maturi, G. Belluardo, D. Moser, M. D. Buono. 2013. BIPV system performance and efficiency drops: overview on PV module temperature conditions of different module types.

[2] M.Pelle, E. Lucchi, L. Maturi, A. Astigarraga and F. Causone. September 2020, Energies. Coloured BIPV Technologies: Methodological and Experimental Assessment for Architecturally Sensitive Areas

[3] G. Peharz, A. Ulm. 2018, Elsevier. Quantifying the influence of colors on the performance of c-Si photovoltaic devices

[4] S. Bana, R.P. Saini. 2016, ELSEVIER, pp. 171-187. A mathematical modeling framework to evaluate the performance of single diode and double diode based SPV systems.

[5] J. E. Castillo, J. M. Russo, R. K. Kostuk, Glenn A. Rosenberg. 2011, Journal of Photonics for Energy . Thermal effects of the extended holographic regions for holographic planar concentrator.

[6] Keen-Yip Lai, Boon-Han Lim. Comparative Study for Time-specific Ross Coefficient and Overall Ross Coefficient for Estimation of Photovoltaic Module Temperature.

[7] Kipp&Zonen. [Online] Kipp&Zonen. https://www.kippzonen.com/Product/13/CMP11-Pyranometer#.YR-0aI4zbIX.

[8] A. Røyset, T. Kolås, B. P. Jelle. 2020, Energy & Buildings. Coloured building integrated photovoltaics: Influence on energy efficiency

[9] IEC. IEC 61724 Photovoltaic system performance monitoring –Guidelines for measurement, data exchange and analysis.

[10] L. Maturi, Ya B. Assoa, L. Gaillard, D. Moser, L. Kronthaler, T. G. Christophe. Thermal response of building integrated photovoltaic systems: analysis of monitoring data from installations at three sites in Europe.

OUTDOOR TEST RESULTS AND MODEL VERIFICATION OF AESTHETICS PV FAÇADE

L.H. Slooff[1], A.R. Burgers[1], K.M. De Groot[1], N.J.J. Dekker[1], T. Minderhoud[2], G. Gijzen[2], T. Sepers[3], Y. de Groot[4], W. van Strien[5], J.A.M. van Roosmalen[5].
[1]TNO Energy Transition, P.O.Box 15, 1755 ZG Petten, The Netherlands, tel: +31 650009631, fax: +31 224 56 8214, email: lenneke.slooff@tno.nl
[2] UNStudio,, PO Box 75381, 1070 AJ Amsterdam, The Netherlands
[3]TS Visuals, Schuit 14, 1724 VD Oudkarspel, The Netherlands
[4]BAM, Runnenburg 9, 3981 AZ Bunnik, The Netherlands
[5]Solar Visuals, Schuit 14, 1724 VD Oudkarspel, The Netherlands

ABSTRACT: The need for a more sustainable built environment is urgent and requires cost-efficient, smart and integrated renewable energy solutions. The surfaces of the buildings and environments we live and work in represent an enormous potential to generate sustainable energy. In order to exploit these surfaces to their full potential, it is essential to be able to predict the energy yield of a specific system. In the past years we have developed an aesthetic PV solution for façades that enables both visible and invisible integration. Recently these panels have been mounted on several façades in the Netherlands. A detailed monitoring system provides input parameters for a sophisticated model that calculates the energy yield of the system. The model results show a good overlap with the measurement data, allowing the model to be used for optimization of similar systems on other façades.
Keywords: Façade, modelling, building integration, colored PV

1 INTRODUCTION

Façades that generate electricity will become one of the solutions to reduce CO_2 emissions in the world. Large scale implementation however will require more variation in the visual appearance of the PV panels, including variations in color.
In recent years, various concepts have been studied as an alternative to the traditional blue or black c-Si modules. [1-4] and some products have become commercial available [5-13]. They differ in the technology and the color options, thereby addressing various sections of the BIPV market.
In the past years we have developed another approach for aesthetic PV solutions for façades that enables both visible and invisible integration. [14-16]. After several years of lab research and thorough testing the aesthetic solar panels have now been deployed outdoors on several test façades to gain experience in building these façades and provide a demonstration façade where people can also experience the look and feel, which is crucial for public acceptance.
The performance of the façade will strongly depend on the site where the system is installed. Geographical location, surrounding buildings, climate and system layout all have an influence on the performance. A few of the above mentioned test systems are deployed with a monitoring system to see how they perform under real outdoor conditions. The results are compared with our in-house developed energy yield model BIGEYE.
After validation of the model this enables us to calculate the energy yield of façades in various geographic regions, with various orientations and to optimize the system.

2 FAÇADE SYSTEM

2.1 Solar panels

The PV modules are fitted with customized prints that appear as the desired image to the human eye, with limited efficiency loss. The prints are engineered out of a rasterized pattern of small dots, see Fig. 1.
The pattern overlaps with the cell metallization to minimize the yield losses, also giving a better visual result compared to an equal spacing pattern.

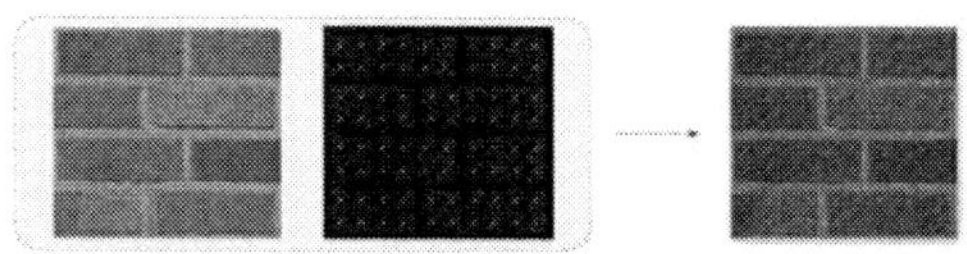

Figure 1: aesthetic PV panels are made starting with a picture with the desired image and this picture is subsequently converted into a dot pattern that overlaps with the cell metallization to minimize the losses.

Depending on the customers wishes the conversion algorithm can be chosen to optimize for yield or for aesthetics. See Fig. 2. The concept is compatible with both conventional tabbed cell and back contact foil technology.

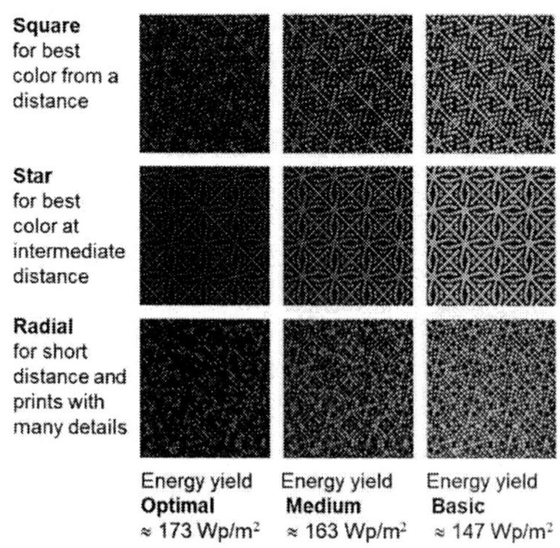

Figure 2: options for color visibility vs power density.

2.2 Façade systems

Up to now, several test fields have been built. Each with a different type of visual appearance. The first consisted of a very bright and colorful façade where the panels were optimized for appearance. See Fig. 3.

Figure 3: Aesthetic façade with bright representation of Europe and the stars of the European flag.

Fig. 4 shows a second façade that was built with a so called mimic design, where the print of the module aims to blend into the building envelope. The solar panels feature a brick pattern and have an installed capacity of 220 Wp each. The system consists of 2 arrays of 12 series connected panels facing South-East. The arrays are connected to a single GoodWe 5000 DT inverter with 2 MPP trackers, 1 for each array.

This façade was more extensively instrumented and used for the detailed performance analysis. We will refer to the left-hand array as A and the right-hand side array as B.

Figure 4: Test façade at BAM headquarters in Bunnik, the Netherlands, with mimic brick design.

Inverter data were analyzed and additional monitoring equipment was installed to monitor the AC and DC energy yield, module temperatures and global horizontal irradiance as well as in plane irradiance. The data were collected with a 1-minute sampling interval.

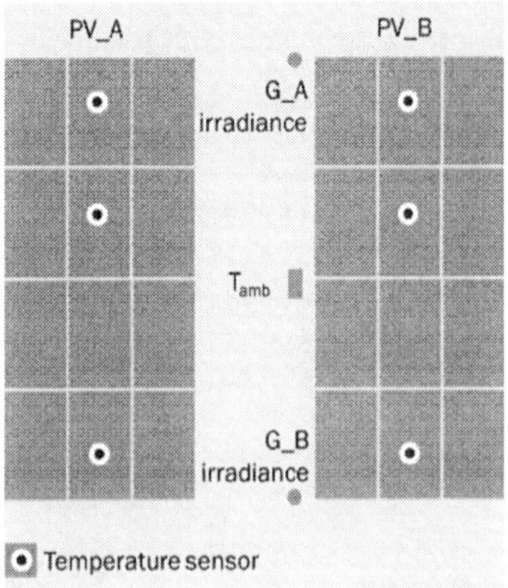

Figure 5: Schematic representation of the system and sensors.

2.3 BIGEYE model

BIGEYE is a state-of-the-art tool developed by TNO to simulate the performance of PV systems, with a focus on bifacial systems [17,18]. BIGEYE implements a 3D view factor model to accurately handle the ground and other diffuse reflectors, for instance, the irradiance from the ground to the rear side of bifacial modules. Also, in bifacial systems elements of the support structure are more likely to cast shades on a light receiving surface of the module than in conventional systems. BIGEYE has flexibility in simulating such shades, and their impact on the IV curves (mismatch) of modules and strings, taking into account the lay-out of the modules in blocks of cells protected by by-pass diodes [19].

BIGEYE deploys the Erbs model [20] to estimate diffuse horizontal irradiance (DHI) from global horizontal irradiance (GHI) when DHI is not available. BIGEYE uses the Perez model [21] to break down DHI into sky dome and circumsolar components, and to assess horizon brightening and darkening.

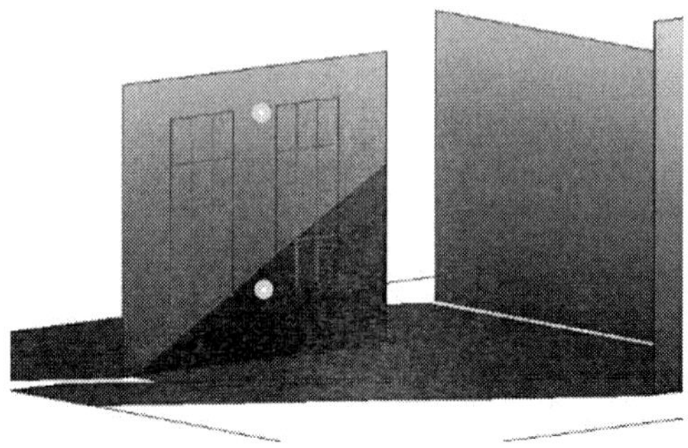

Figure 6: he scene used in the BIGEYE performance modelling.

The scene used in the BIGEYE performance modelling is shown in Fig. 6. The scene includes the PV arrays and the wall they sit on. Also the two irradiance sensor are included (yellow circles), as well as parts of the building that cast significant shades. The shade across the PV arrays in the bigeye scene matches the one seen in the photograph of Fig. 4.

The simulation of the string DC IV curve in BIGEYE is elaborate, and allows to study shading on the array of modules in fine detail. The irradiance and temperature on each cell are calculated, then the IV curve per cell. For each bypass diode protected block, the IV curve of the series connection of the cells is calculated and the characteristic of the bypass diode is applied. Next the IV curves of all the bypass diode blocks are series connected to obtain the string IV curve.

3 RESULTS AND DISCUSSION

The outdoor monitored system performs in good agreement with our modelling results. One day was analyzed in more detail. As a first step, the irradiance data of the in-plane sensors were calculated and compared with the measurements, see Fig. 7. It clearly shows the shadow of part of the building on the system in the morning, especially on the lower sensor.

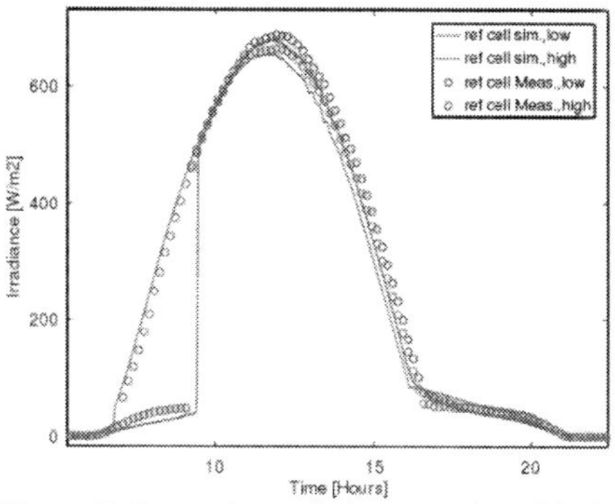

Figure 7: Comparison of BigEye results of the irradiance sensors and measured data.

The temperature measurements were compared with the measured temperature at a nearby meteorological station. The data shows that the measured ambient temperature (T_{amb}) near the façade is somewhat higher than the T_{amb} at meteorological station, especially during night. See Fig. 8 Clearly, the building itself absorbs and stores heat during daytime and radiates towards the surroundings. In our model the increase of the module temperature over T_{amb} is calculated from the in-plane irradiance.

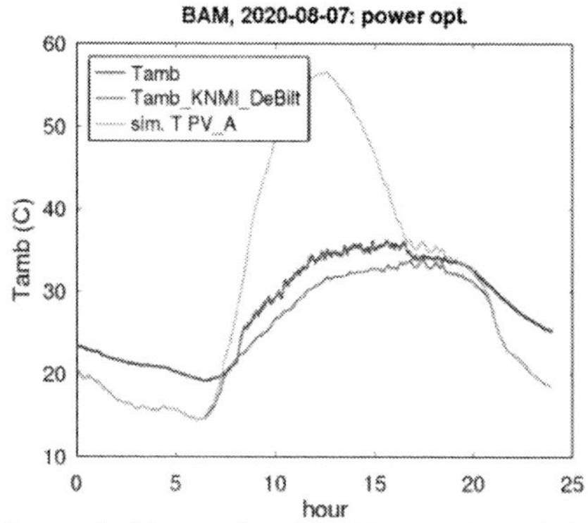

Figure 8: Measured ambient temperature at the façade and at a nearby meteorological station as well as the calculated array temperature of array A.

Fig. 9, 10 show the DC $P_{mp}/I_{mp}/V_{mp}$ (where mp is maximum power) data of the arrays and compare measured and simulated data. We see in Fig. 9 and 10 that the simulated string power starts to climb earlier than the measured one, and has a more gradual increase than the measurement. This indicates the string inverter misses out on some power in the morning.

Also the simulated DC voltage dips to lower voltage at a few points with higher irradiance, but with little impact on the power. This is a case where there are multiple local optima in the string IV curve with small difference in the DC power.

Fig. 11 shows the module IV curves (left) and the resulting array IV curve (right) for array B at three different times. At 9:05 in the morning there is quite some shade on the array and some of the panels are fully illuminated and others are shaded. In the presence of shading and bypass diodes a string IV curve will typically have multiple local optima, as observed in the blue line of Fig. 11.

In the simulations of figure 12 the P_{mp} found occurs at lower voltage with higher currents, with hence more bypass diodes active than in the highest voltage local optimum.

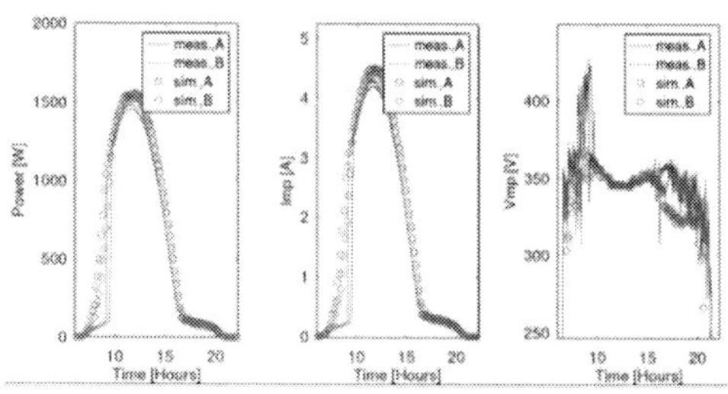

Figure 9: Comparison of measured and simulated DC string P/V/I.

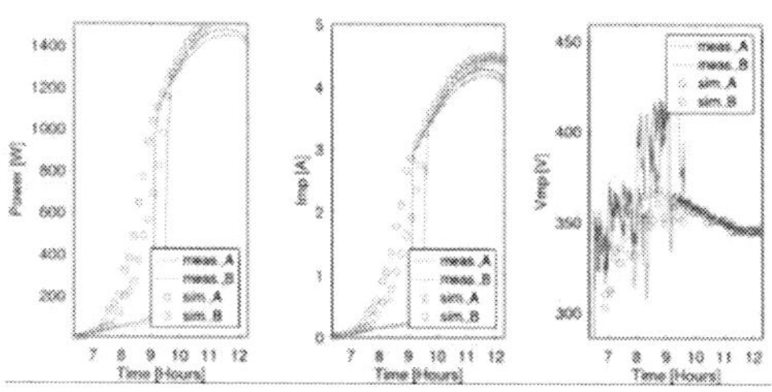

Figure 10: as figure 6, but zoom in to sun-rise.

In Fig. 10 the simulation picks the point on the IV curve with the highest power, whereas the inverter chooses the power from the local optimum with the highest voltage. From these simulation it appears that the P_{mp} tracking in the string inverters prefers high voltage local optima, which explains the abrupt changes as seen in the measured DC values.

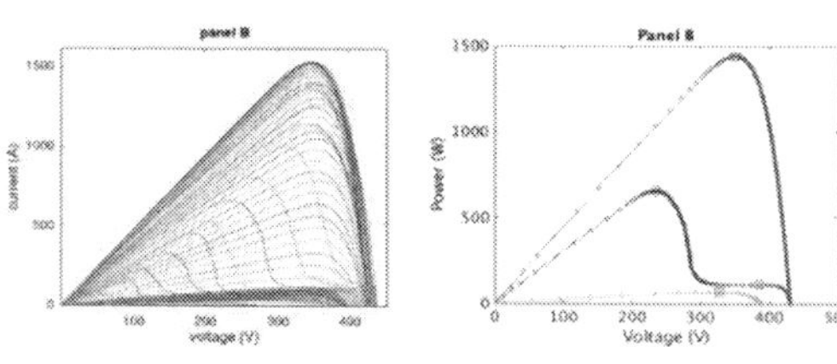

Figure 11: module IV curves (left) and the resulting array IV curve (right) for array B at three different times.

The model makes it also possible to see the effect of different system configurations on the performance. Fig. 12 shows a preliminary result for power optimizers compared to string inverters. The assumption for the string inverter is that it keeps the string at the global P_{mp} point of the IV curve.

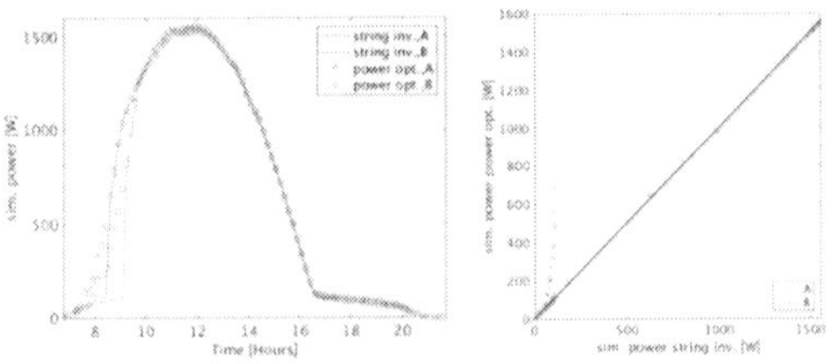

Figure 12: comparison of electrical power for string inverter (all 12 modules in series) and with power optimizer per module.

Table I: Calculated annual yield for a system with string inverter and a system with power optimizers.

kWh/KWp	String Inverter	Power Optimizer
A	674	688
B	607	631

In Table I we compared the expected PV output, for a year, with 1-hour time step in the meteorological data, for string optimizer and power optimizers. For the string optimizer we assumed that the DC side of the optimizer was kept at global maximum power point (mpp) at all times. There turned out to be negligible difference between staying at global mpp and highest voltage mpp, the situation described in figure 11 occurs infrequently, or doesn't contribute much to the overall yield.
The calculations show that the power optimizers result in a higher output power in the presence of shading. The effects of the Pmp tracking algorithm might well be larger than the effects of string inverters vs power optimizer. These aspects will be investigated later in the project in more detail.

4 CONCLUSIONS AND OUTLOOK

The aesthetic PV panels shown in this paper are an elegant option to make PV more attractive for façade implementation. They can be used to make a visual statement, branding, but can also blend in the building envelope. The performance of the panels is a trade-off between visual appearance and power, but can be tuned to some extend depending on the specific situation. Field tests were performed to test the outdoor performance in a system configuration and the results were compared with modeling data of our BIGEYE yield model for a sunny day. The good agreement between the measured and modeled data allows for optimization of the system, e.g. by looking at a different system configuration, string optimizers vs power optimizers. During shading conditions power optimizers show an improved performance. On the other hand, it was observed that the P_{mp} tracking algorithm of the inverter has a large influence on the performance as well which might have an overall larger influence. We will continue the validation of the model by looking at more days with different irradiance conditions over the year. The final goal is to use the model to design the optimal system for new façades.

5 REFERENCES

[1] C. Kutter et al, Proc. 35rd EUPVSEC, (2018) 1488-1492.
[2] E. Klampaftis, D. Ross, G. Kocher-Oberlehner and B. S. Richards, IEEE J. Photovolt., vol. 5 no. 2, (2015) 584-590.
[3] B. Bläsi, T. Kroyer, T. E. Kuhn and O. Höhn, IEEE Journal of Photovoltaics, vol. 11, no. 5 (2021) 1305-1311.
[4] J.H.Selj, T.T.Mongstad, R.Søndenå, E.S.Marstein, Solar Energy Materials and Solar Cell, 95 (2011) 2576.
[5] https://www.solarvisuals.nl/
[6] http://www.swissinso.com/
[7] http://www.lofsolar.com/
[8] www.kaleo-solar.ch/
[9] https://www.sistinesolar.com/
[10] https://www.csem.com/
[11] https://www.soluxa.solar/nl/
[12] https://kameleonsolar.com/
[13] https://www.sunergsolar.com/it/prodotti/fotovoltaico/moduli-fotovoltaici/x-color_36.html
[14] L.H. Slooff-Hoek, J.A.M. van Roosmalen, L.A.G. Okel, T. de Vries, T. Minderhoud, G. Gijzen, T. Sepers, A. Versluis, F. Frumau, M. Rietbergen, L. Polinder, E.M.B. Heller, F. de Vries, Proceedings 33rd European Photovoltaic Solar Energy Conference and Exhibition, (2017) 2110 – 2113.
[15] J.A.M. van Roosmalen, L.H. Slooff, L.A.G. Okel, M.N. van den Donker, T. de Vries, T. Minderhoud, N. Wang, G. Gijzen, T. Sepers, M. Rietbergen, L. Polinder, E.M.B. Heller, A. Versluis, F. Frumau, Proceedings Advanced Building Skin Conference, (2017) 1281-1290.
[16] L.H. Slooff-Hoek, J.A.M. van Roosmalen, L.A.G. Okel, T. Minderhoud, G. Gijzen, L.C. Polinder, F. Goethals, T. Sepers, Proceedings 35th European Photovoltaic Solar Energy Conference and Exhibition, (2018) 1483 – 1487.
[17] G.J.M. Janssen, A.R. Burgers, A. Binani, A.J. Carr, B.B. van Aken, I.G. Romijn, M. Klenk, H. Nussbaumer, and T. Baumann, Proceedings 35th European Photovoltaic Solar Energy Conference and Exhibition (2018) 1573-1577.
[18] A.R. Burgers, G.J.M. Janssen, and B.B. van Aken, BifiPV Workshop (2018).
[19] A.R. Burgers, and B.B. van Aken, BifiPV Workshop (2019).
[20] D.G.Erbs, S.A. Klein, and J.A. Duffie, Solar Energy, vol 28 (4) (1982) 293-302.
[21] R. Perez, R. Seals, P. Ineichen, R. Stewart, and D. Menicucci, Solar Energy, vol **39** (3) (1987) 221-232

DEVELOPMENT OF A NEUTRAL COLOR PHOTOVOLTAIC WINDOW BASED ON LUMINESCENT SOLAR CONCENTRATORS

Paolo Bernardoni, Giulio Mangherini, Alfredo Andreoli, Marinela Gjestila, Donato Vincenzi
University of Ferrara, Department of Physics and Earth Sciences
Via Saragat 1, building C, Ferrara, Italy

Liliana Gila, Luciano Caccianotti
Eni, Renewable Energy, Magnetic Fusion and Material Science Research Center
Via Fauser 4, Novara ,Italy

Claudio Pesenti
Eni, R&D engineer Electrification Department
Via Felice Maritano 26, San Donato Milanese (MI), Italy

ABSTRACT: This work describes a prototype of a neutral color solar window based on LSC technology that was developed by University of Ferrara and Eni and installed on a building in the Eni Donegani research center for renewable energy and the environment in Novara (Italy).
The main purpose of these devices is to generate energy from the transparent surfaces of buildings and, at the same time, to obtain both an aesthetically appealing façade and provide a good quality illumination in the rooms of the building.
Five of these prototypes with an active area of 0.61m2 were assembled and installed as the windows of a test building at the Eni Donegani site. The optical, colorimetric and electrical performances of the windows were measured both on-site and off-site at the end of the assembly process.
The prototypes showed an average transmittance of 42.2% in the standard visible range (400-700nm) and when lit by sunlight behave as a light source with a CRI of 94.4, which is comparable to halogen lamps. The power production was 2.8W when installed on a diffusive background that drops to 1.2W when installed on the test building which has a non diffusive background.
Keywords:building integration, luminescent solar concentrators, photovoltaic windows, neutral color

1 INTRODUCTION

The global need for electric energy is increasing and buildings represent a significant energy-consuming sector, indeed in Europe households represent 25.7% of the total energy consumption and 29% of the electric energy consumption [1].

On these bases, Building Integrated PhotoVoltaics (BIPV) represent a valuable set of technologies which can not only satisfy some of the electrical energy need but can also reduce the load on the electrical network moving part of the energy production closer to the location of consumption.

Hence the aim of BIPV technologies is to transform buildings from energy sinks into energy sources however, in order to achieve this goal, not all the building integration needs can be fulfilled merely adapting traditional PhotoVoltaic (PV) concepts and technologies.

In particular, the integration of PV technologies in the transparent surfaces of buildings, represents a significant challenge and the commercially available transparent PV technologies can be summarized with two main concepts [2].

The first is the use of common mono-facial or bi-facial silicon solar cells assembled in glass-to-glass panels and arranged with a spacing between the individual cells. In this case the device transparency is given by the ratio between the area not covered by the solar cells and the total area of the panel.

A second approach is to use semi-transparent thin film technologies reducing the thickness of the semiconductor film to achieve the desired transmittance level, which typically lies between 0% and 28% [3].

Both these approaches have aesthetic limitations caused by the presence of opaque areas (cells and bus bars), by the limited achievable transmittance range and by the performance penalty that occurs when colored devices are required as a consequence of the colored filters applied on the modules.

Luminescent Solar Concentrators (LSCs) represent a technology that can be effectively employed in the development of transparent PV panels. A typical LSC is represented by a slab made of a transparent material (the most common are Poly(methyl methacrylate) PMMA or PolyCarbonate (PC) colored with a fluorescent dye, these dyes can be classified in three categories: aromatic compounds, organometallic complexes and quantum dots.

When sunlight is transmitted through the slab the fluorescent dye absorbs part of the light that is later re-emitted at a longer wavelength. Since the re-emission is isotropic and the transparent material have a refractive index higher than air (i.e. n_{PMMA}=1.49 and n_{air}=1.00) a significant amount of this light is trapped inside the slab (a typical value is 75%) and transmitted to the edges of the concentrator. The trapped light can be converted into electricity by placing solar cells on the edges of the slab

The purpose of an LSC is therefore to divert on its edges part of the light incident on the slab, if the area of the edges is significantly smaller than the area of the slab the net result is a higher power density on the edges (hence the name concentrator).

Differentely from the aforementioned technologies, LSC panels are perfect to achieve a clear view through the PV device, as a continuous transparent surfaces can be obtained as well as a wide range of transmittance values and colors; which can be tuned by varying the type and concentration of the fluorescent dye [4] [5] [6].

2 DESIGN

This work describes a prototype of neutral color solar window based on LSC technology that was developed by University of Ferrara and Eni and installed on a building in the "Eni Donegani research center for renewable energy and the environment" in Novara (Italy) (Fig.1-2).

Figure 1: The prototypes installed in a test building at the Eni Donegani research center for renewable energy and the environment in Novara (Italy)

Figure 2: View from the inside of the building, despite being heavily colored when viewed from outside, the windows show only a light orange hue

The project had the following requirements:

- 5 neutral color LSC photovoltaic windows had to be built and installed on the existing windows of a building at Eni R&D Center in Novara
- The window openings had a size of 935mm x 725mm, which subtracted the frame result in 0.61m2 of active area
- Frames had to be made of aluminum with a PVC thermal break
- The window structure had to be double-glazed
- The LSC slab had to be protected from sealants and glues employed in the assembly process

Given these requirements, the prototype consist in an LSC slab of 905mm x 695mm made of PMMA with a thickness of 6mm manufactured by Arkema, a partner of Eni for LSC development. This slab uses a mix of two organic fluorescent dyes: Lumogen® F Red 305 (LR305) and Eni Yellow dye (MPDTBOP). The very low concentration (from 20 to 75ppm) of this mix allows for a wide absorption range in the visible spectrum and gives to the slab a distinctive orange color.

The photovoltaic conversion was performed by 6 arrays of monocrystalline silicon solar cells with a length of 244mm. Each array hosted 10 solar cells connected in series with a size of 24mm x 8mm. Three arrays are placed on each long side of the slab while the remaining perimeter was covered with a high reflectivity dielectric mirror film (Fig.3).

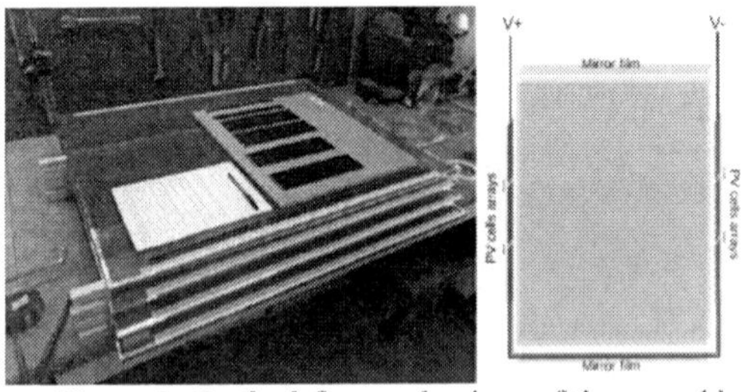

Figure 3: On the left an early phase of the assembly process in the laboratory at University of Ferrara showing the LSC slabs before the installation of the solar cell arrays, the sides with no cells are covered with a high reflectivity mirror film, on the right a representation of the prototype showing the position of the PV arrays and the edges with the mirror film.

This film had the double purpose of reflecting most of the fluorescence light to the cell arrays [7] and to protect the surface of the LSC during the sealing procedure, as the sealant would otherwise absorb or divert part of the fluorescence light.

Thermal insulation was obtained through a double glazing structure, in which the outer slab was the proper LSC device and the inner one a glass slab with a thickness of 3mm, a blue filter was applied on this glass slab (Solar Screen Bleu 40 C) whose purpose was to compensate the color shift in the visible spectrum caused by the absorption and emission of the fluorescent dyes (Fig.4).

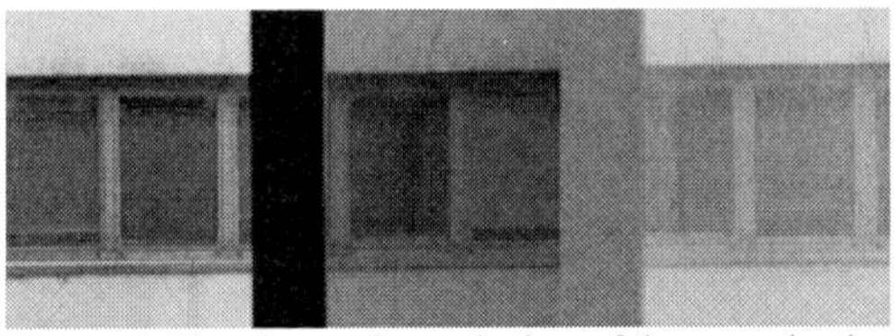

Figure 4: The picture shows the hue of the neutral color LSC (center) compared to a standard double glazing window (left), and a typical LSC dyed with LR305 at 160ppm (right). Even if a slight orange hue is present the improvement in color balance over a standard LSC is significant.

The double glazing structure was then encased in an aluminum and PVC frame with a size of 935mm x 725mm which is water proofed by mean of a silicone sealant. The electrical connections take place through two silicone-insulated wires coming out of the aluminum frame (Fig.5).

Figure 5: A prototype at the end of the assembly process and ready for the installation

This approach is particularly interesting because it allows to obtain the appeal of a colored surface on the façade without affecting heavily the quality of the lighting inside the building.

3 CHARACTERIZATION

3.1 Optical characterization

The two dyes employed in the LSC slab were at first characterized individually, two test slabs were made both with a thickness of 6mm, one with Lumogen Red 305 and one with Eni MPDTBOP.

The transmittance spectrum of the slabs was obtained using a tungsten-halogen lamp as light source and an Ocean Optics USB4000 spectrometer. This device integrates a diffraction grating and a linear CCD sensor which can measure all the spectral components between 200nm and 1050nm at the same time with a sensor resolution of 0.27nm and an optical resolution of 2nm.

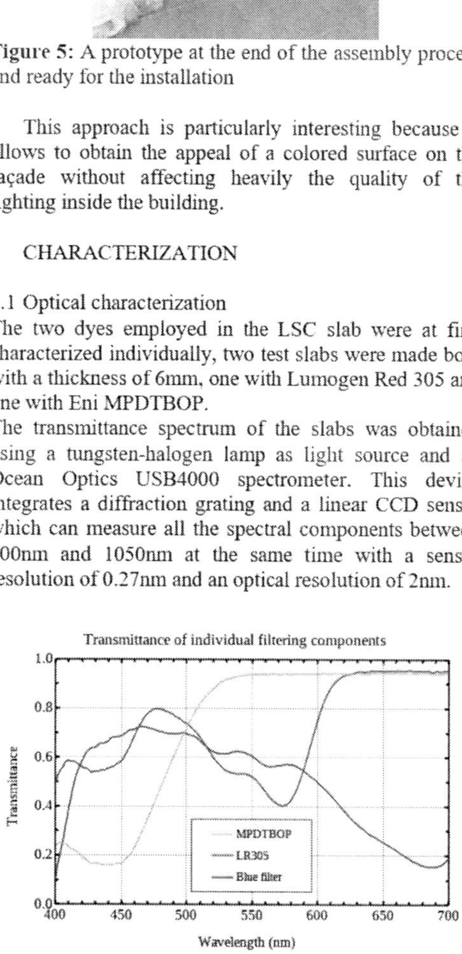

Figure 6: Transmittance spectrum of two LSCs dyed with LR305 (red) and MPDTBOP (yellow), the blue filter is added for comparison (blue).

In Fig.6 is shown the transmittance spectrum of the two slabs as well as the transmittance spectrum of the blue film used to balance chromatically the spectrum transmitted by the window. The yellow dye absorbs mainly in the blue region with an absorption peak at 430-450nm while, on the other hand, the red dye absorbs mainly in the yellow and orange regions and its absorption peak lies at 570-580nm.

The transmittance of the blue filter represents an interesting result, as expected it shows a high absorption of red and orange, but the transmittance of blue and green between 60% and 70% testifies that the absorbance is not as low as expected given the blue color.

The comparison between the transmission spectrum of the LSC device with and without the blue filter is shown in fig.7.

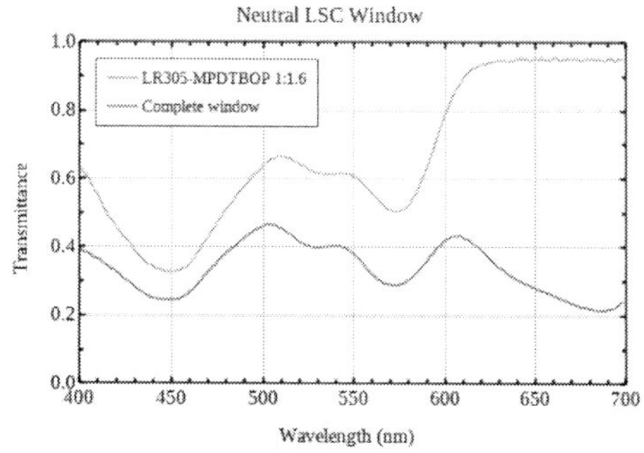

Figure 7: Transmittance spectrum of the LSC used in the prototypes (orange) and of the complete window that includes the blue filter in order to correct the color rendering of the prototype (grey).

It is evident that the spectrum is well balanced dropping only in the deep blue and deep red regions, moreover the low transmittance in the 400-460 nm range ensures a warm white color. The resulting total transmittance of the prototype in the visible range (400-700nm) is 42.2%.

3.2 Color rendering index

Colorimetric analysis has been performed evaluating the window as a light source whose emission spectrum is represented by the solar spectrum filtered using the transmittance values shown in fig.7, given that the solar spectrum at the earth surface changes depending on time and weather conditions, the CIE Standard Illuminant D65 has been used as source, D65 is also called daylight illuminant and corresponds roughly to the average midday sun spectrum in Europe taking into account both direct and diffused sunlight.

The analysis has been performed by writing a python script and using the Color-Science (v. 0.3.16) library which is an open-source Python package providing a comprehensive number of algorithms and datasets for color science and is freely available under a New BSD License [8].

Color Rendering Index (CRI) is a parameter used in colorimetry that represents a quantitative measure of how faithfully a light source can render the colors of different objects compared with a reference light source. The CRI value is the mean of the results for 8 standard samples (R1-R8). The values for 6 additional supplementary samples (R9-R14) have been obtained but these are not used by definition in the CRI calculation.

The analysis on the prototypes resulted in a CRI of 94.4 which is comparable to a tungsten-halogen lamp which typically have a CRI value between 85 and 90, a meaningful optional value R9 which represents a saturated red sample which is traditionally difficult to render for artificial lights is 86.4 which can be considered very good (R9>85). The results of the colorimetric characterization are shown in Fig.8.

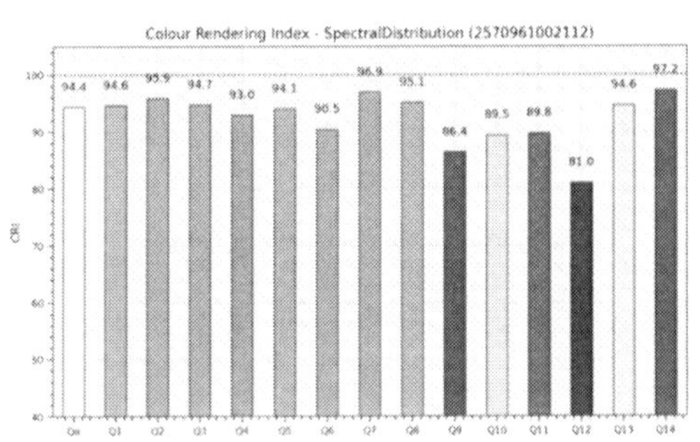

Figure 8: Color rendering index of the spectrum transmitted by the prototypes (Qa), Q1-Q8 are the values for the test samples used to compute the CRI and Q9-Q14 are the supplementary samples that do not take part in CRI calculation.

3.3 Electrical characterization

The electrical performance of the prototypes has been evaluated analyzing the IV curves obtained in various conditions, the test setup for the IV curves acquisition is represented by a high precision Keithley 2400 SourceMeter SMU unit controlled by a custom software developed in LabView and a Delta Ohm LP PYRA 02 pyranometer.

The resulting IV and PV curves are shown in fig.9. With a diffusive background the power production of a single window reaches 2.8W while once installed in the test building and without a diffusive background drops to 1.2W. This power is sufficient to power small appliances like air recycling fans, automatic blinds or LEDs lighting.

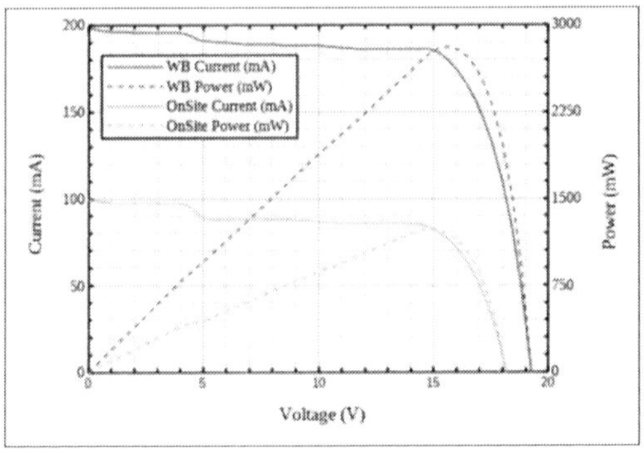

	Pmax (mW)	GNI (W/m^2)	PCE (%)
White BG	2812	990	0.460
On Site	1234	880	0.179

Figure 9: IV and PV curves of the prototype n.5 shown at the far right of Fig.1 (far left of Fig.2): the mismatch visible in the IV curves is caused by the sunlight reaching directly one side of the solar array (yellow) and IV and PV curves of the same prototype installed on a white diffusive surface (orange)

Comparing the short-circuit current of the prototypes with that of the PV arrays directly exposed to the sun it is possible to estimate the optical concentration factor of the LSC [9]. This value is 2.5x with diffusive background and 1.2x without reflecting background.

4 CONCLUSIONS

4.1 Appearance

In Fig.10 is shown one prototype both viewed from the LSC side (which is installed outwards) and from the glass side (which is installed inwards). It is evident that the color of the light that enters the building is very well balanced and shows only a slight warm tone compared to natural sunlight.

Figure 10: The picture shows the appearance of a prototype viewed from the LSC side (left) and from the filter side (right), it is evident the effect of the filter on the color rendering of the window

On the other hand, when viewed from outside of the façade, the prototypes show the original orange color of the LSC slab and, therefore, can be employed in buildings where the external design requires colored glass surfaces without compromising the quantity or the quality of the illumination inside the building.

The prototypes, for now, can be used to power small appliances and smart systems which can increase the comfort or energy efficiency such as blinds or ventilation systems, anyway it is necessary to increase their efficiency in order to have more power available and a value of 1% can be considered a reasonable target.

4.2 Future development

Three possible alternatives to increase the electrical efficiency have been identified, the simpler approach consists in increasing the concentration of the dyes but this would be at the expense of transmittance. A more reasonable one approach consists in separating the dyes in order to reduce self absorption and to create a structure with two LSC slabs, while the most promising solution is to use only one dye, with a higher concentration, and to find a better matching filter to correct the perceived color of the window.

4.3 Acknowledgments

This work has been carried out in the framework of the research project INSTM1735 OdL 4310346905 funded by Eni and would have been impossible without the support from the Eni Renewable Energy, Magnetic Fusion and Material Science Research Center.

5 BIBLIOGRAPHY

[1] European Environment Agency, Indicator assessment: final energy consumption by sector and fuel (2018). https://www.eea.europa.eu/data-and-maps/indicators/final-energy-consumption-by-sector-9/assessment-4

[2] Husain, A. A., Hasan, W. Z. W., Shafie, S., Hamidon,

M. N., & Pandey, S. S. (2018). A review of transparent solar photovoltaic technologies. Renewable and sustainable energy reviews, 94, 779-791.
[3] Onyx Solar Low-E PV Glass, https://img1.onyxsolar.com/documents/onyx-solar-asi-solutions-brochure-sqm.pdf
[4] Debije, M.G.; Verbunt, P.P.C. Thirty Years of Luminescent Solar Concentrator Research: Solar Energy for the Built Environment. Adv. Energy Mater. 2012, 2, 12–35.
[5] Meinardi F.; Bruni F.; Brovelli S. Luminescent Solar Concentrators for Building-integrated photovoltaics Natural Reviews:Materials 2017, 2, 17072

[6] Zarcone, R. Building integrated photovoltaic system for a solar infrastructure: Liv-lib'project. Energy Procedia 91 (2016): 887-896.

[7] Bernardoni, P., Mangherini, G., Gjestila, M., Andreoli, A., & Vincenzi, D. (2021). Performance Optimization of Luminescent Solar Concentrators under Several Shading Conditions. Energies, 14(4), 816.

[8] Mansencal, T, Mauderer, M. Parsons, M., Shaw, N., Wheatley, K., Cooper, S., Hill, S.. (2020, January 25). Colour 0.3.16 (Version 0.3.16). Zenodo. http://doi.org/10.5281/zenodo.3757045

[9] Inman, R.H.; Shcherbatyuk, G.V.; Medvedko, D.; Gopinathan, A.; Ghosh, S. Cylindrical Luminescent Solar Concentrators with Near-Infrared Quantum Dots. Opt. Express 2011, 19, 24308.

This is a short version of the paper submitted to the journal Solar RRL.
It has been selected by the Executive Committee of the EU PVSEC 2021 for submission to Solar RRL.

38th European Photovoltaic Solar Energy Conference and Exhibition

Development of high-efficiency solar cell modules for PV-powered vehicles

Masafumi Yamaguchi[1], Kenji Araki[2], Yasuyuki Ota[2], Kensuke Nishioka[2], Tatsuya Takamoto[3], Christian Thiel[4], Anastasios Tsakalidis[4], Arnulf Jäger-Waldau[4], Taizo Masuda[5], Kenichi Okumura[5], Takashi Nakado[6], Kazumi Yamada[6], Yusuke Zushi[7], Tsutomu Tanimoto[7], Kyotaro Nakamura[1], Ryo Ozaki[1], Nobuaki Kojima[1] and Yoshio Ohshita[1]

[1] Toyota Technological Institute, Nagoya, Japan
[2] Miyazaki University, Miyazaki, Japan
[3]Sharp Corporation, Nara, Japan
[4] European Commission-Joint Research Centre, Ispra, Italy
[5] Toyota Motor Corporation, Susono, Japan
[6]Toyota Motor Corporation, Toyota, Japan
[7]Nissan Motor Corporation, Yokosuka, Japan

Abstract - The development of photovoltaic (PV)-powered vehicles is desirable and can be very important for reducing CO_2 emissions, increasing electric range and creating new clean energy infrastructures based on PVs. According to our survey, the use of more than 30 % of high-efficiency PVs enables 30 km/day driving without external charging. With this, the majority of family cars could run by sunlight and without the need for additional fuel. Although the Toyota Prius demonstration car has shown 36.6km/day driving at solar irradiance of 6.2kWh/m^2/day, the expected driving distance of PV-powered vehicles is lower than the estimated values due to some losses, such as temperature increase of solar cell modules under sunshine conditions. This paper presents approaches on the development of high-efficiency solar cell modules. In this paper, analytical results for reduction in non-radiative recombination and resistance losses of various solar cell modules are shown. In addition, temperature coefficients of candidate solar cell modules are analyzed in order to develop highly efficient solar cell modules under sunshine conditions. The III-V compound 3-junction solar cell modules are shown to have potential efficiencies of 37% with better temperature coefficient of -0.19 %/°C compared to -0.29 %/°C for Si back contact solar cell modules and -0.26 %/°C for Si heterojunction solar cell modules. It is also shown the 3-junction solar cell modules have potential driving distance of more than 50km/day on a clear day,

1. Introduction

The development of photovoltaic (PV)-powered vehicle[1] is desirable and can be very important for reducing CO_2 emissions, increasing electric range and creating new clean energy infrastructure based on PVs. Although the Toyota Prius demonstration car[1] has shown 36.6km/day driving at solar irradiance of 6.2kWh/m^2/day, the effective driving distance of PV-powered vehicles is lower than the estimated values due to some losses such as temperature increase of solar cell modules under sunshine condition. The results suggest that loss analysis of solar cell modules and development of high-efficiency solar cell modules are fundamental.

This paper presents approaches to the development of high-efficiency solar cell modules, while analytical results for efficiencies and temperature coefficients of various solar cell modules are shown.

2. Necessity of High-efficiency Solar Cell Modules for PV-powered Vehicles

Figure 1 shows practical data regarding the driving distance of Toyota Prius 2017[1], Toyota Prius 2019 (demonstration car)[1] and Sono Motors Sion[2] as a function of module efficiency and solar irradiance in comparison with calculated results of driving distance by assuming electric mileage of 9.35km/kWh and system charging efficiency of 81.2% without cell temperature correction. Sono Motors Sion's driving distances was evaluated by using reported driving distances[2] and solar irradiation data[3] in Munich.

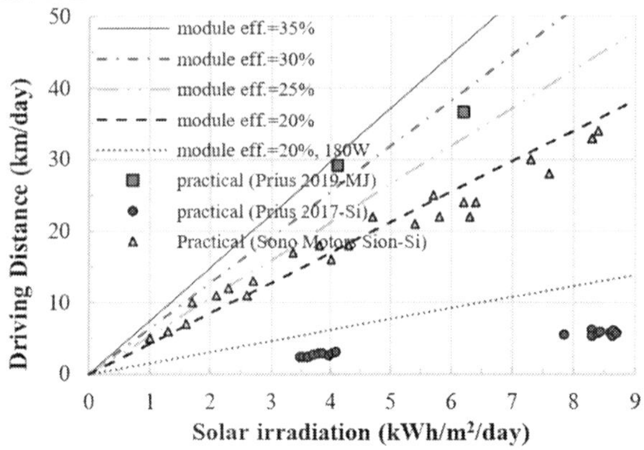

Figure 1. Practical data regarding the driving distance of Toyota Prius 2017, Toyota Prius 2019 (demonstration car) and Sono Motors Sion as a function of module efficiency and solar irradiance in comparison with calculated results of driving distance by assuming electric mileage of 9.35km/kWh and system charging efficiency of 81.2%.

The results suggest importance of high-efficiency solar cell modules, module nominal power and longer electric mileage (that is light-weight car) because Toyota demonstration car installed with III-V 3-junction solar cell module of about 30.9% efficiency has demonstrated 29.1km/day driving[1] at an average solar irradiance of 4.1kWh/m^2/day, contrary to Sono Motor Sion installed with Si back contact solar cell module of about 22%

efficiency that has shown 18-19 km/day driving[2] at the similar solar irradiance as seen in Fig. 1. The gap between calculated values and practical data suggests that there are some losses due to temperature increase rise of solar cell modules and partial shading. Therefore, the development of solar cell modules with good temperature coefficients and with effective measure to minimise the effects of partial shading is necessary.

3. Analysis of Solar Cell Module Performance towards Development of High-efficiency Solar Cell Modules

By using our analytical procedure[4], non-radiative recombination and resistance losses of various solar cell modules were analyzed. The external radiative efficiency (ERE), which is the ratio of radiatively recombined carriers against all recombined carriers, was estimated from measured open-circuit voltage V_{oc} and resistance loss was estimated from measured fill factor (FF). In this study, solar cell module performance was evaluated by using average cell performance data.

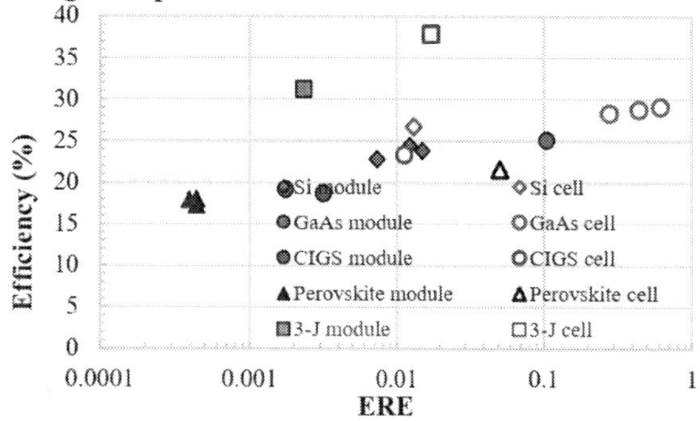

Figure 2. Changes in efficiency of various solar cell modules as a function of ERE (external radiative efficiency) in comparison with solar cells values.

Figure 2 shows changes in efficiency[5] of various solar cell modules as a function of ERE compared to those of solar cells. Although mono-crystalline Si solar cell modules have achieved the similar ERE values with solar cell values, the ERE values of the other solar cell modules are lower by 1/10 – 1/5 compared to those of solar cells. IT is important to note that ERE values of perovskite solar cell modules are about two order lower than those of perovskite solar cells. Also, the fill factor of various solar cell modules are lower than those of solar cells. The resistance losses of current perovskite solar cell modules are significantly higher than those of the other solar cell modules.

We estimated the potential conversion efficiency of various solar cell modules at different EREs. GaAs, Si, CIGS, perovskite and III-V 3-junction solar cell modules have potential efficiencies of 28%, 27%, 26%, 26% and 37%, respectively, by improving ERE into 10%.

A larger number of multi-junction solar cells is desirable for realizing higher efficiency solar cell modules after 39.2% efficiency[6] under 1-sun illumination has been demonstrated with 6-junction solar cell. Figure 3 shows realistic potential efficiencies of III-V compound multi-junction solar cells under 1-sun conditions as a function of number of junctions and ERE in comparison with ideal potential efficiencies and efficiency data reported in the literature[5]. In Fig. 3, best efficiencies[5] of III-V compound multi-junction solar cells are also plotted. 2, 3, 4, 5 and 6-junction solar cells have potential efficiencies of 36.6%, 44.0%, 48.8%, 50.4% and 51.4%, respectively, as shown in Fig. 3. However, it is thought that triple-junction is thought to be the optimal number of junction because in the case of higher number of multi-junction solar cells, i.e. of more than quad-junctions, ERE and FF decrease with an increase in number of junction.

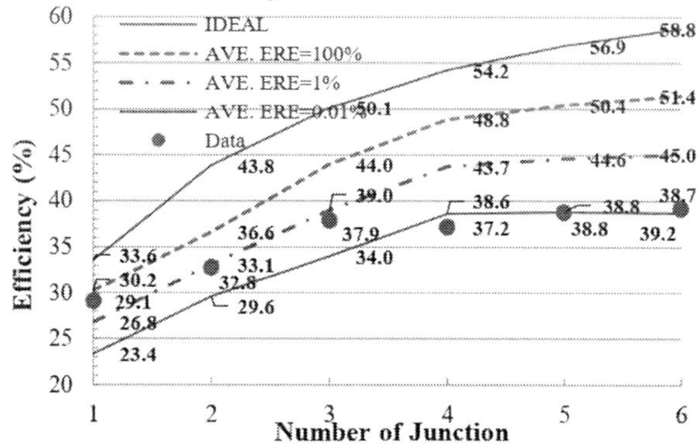

Figure 3. Realistic potential efficiencies of III-V compound multi-junction solar cells under 1-sun conditions as a function of number of junctions and external radiative efficiency (ERE) in comparison with ideal potential efficiencies and efficiency data reported in the literature.

4. Analysis of Temperature Coefficients of Solar Cell Modules towards Development of High `Performance Solar Cell Modules under High Irradiance

Conversion efficiency of solar cell modules is very sensitive to temperature and solar cell module performance decreases with increasing temperature of modules. For PV-powered vehicle applications, development of low temperature coefficient solar cell modules is very important.

Figure 4 shows the temperature rise (measured temperature above ambient temperature) of the car cabin[7], Si modules[8] and GaAs modules[8] in the case of wind speed of 0.75-1.25m/sec and measured values[9] in the case of wind speed of 3m/sec as a function of solar irradiance. Temperature rise differences from Si and GaAs modules are explained by the module size diffrence[8]. As shown in Fig. 4, module temperature depends on solar irradiance, wind speed, type and size of modules and others (wind direction, module structure, module mounting configuration, shape and operating mode of car).

As shown in Fig. 1, there is a gap between calculated values and practical data for driving distance of PV-powered vehicles. By assuming performance decrease as a result of temperature rise of solar cell modules under solar illumination condition, driving distance of PV-powered vehicles was estimated by using results for temperature rise shown in Fig. 4. In the case of Sono Mortors Sion[2], the driving distance of PV-powered vehicles is thought to be decreased with reduction in driving distance of about 10km/day under high solar irradaiance in the worst case. Therefore, development of low temperature coefficient solar cell modules is essential.

The temperature T dependence of efficiency η of solar cells and modules is approximately expressed[10,11] as a relative variation of open-circuit voltage V_{oc}, short-circuit current density J_{sc} and fill factor FF.

$$(d\eta/dT)/\eta = (d V_{oc},/dT)/ V_{oc} + (d J_{sc}/dT)/ J_{sc} + (dFF/dT)/FF = = (1 - 1.1 \Sigma E_{gn}/qV_{oc})/T + [\{J_{sc}(T) - J_{sc}(T_0)\}/(T-T_0)]/J_{sc}(T_0) + (1 - 1.02 FF_0\{(d V_{oc},/dT)/ V_{oc} - 1/T\} , (1)$$

where E_{gn} is the bandgap energy of sub-cell for n-junction solar cell, FF_0 is the ideal fill factor in the absence of resistance loss. In this study, changes in J_{sc} with temperature were calculated by considering temperature dependence of Eg.

Figure 4 shows calculated results for temperature coefficients of crystalline Si, GaAs and III-V compound 3-junction solar cell modules as a function of module efficiency in comparison with reported values[12-20] and estimated values. Better temperature coefficients can be obtained by improving ERE and Voc. Good temperature coefficients of -0.19 %/°C and -0.23 %/°C are expected to be realized for InGaP/GaAs/InGaAs 3-junction solar cell modules with module efficiencies of 37% and 31.2%, respectively although current reported values are -0.333 %/°C and -0.337 %/°C for 23.7% and 25.2% efficient InGaP/GaAs/Ge 3-junction solar cells[12].

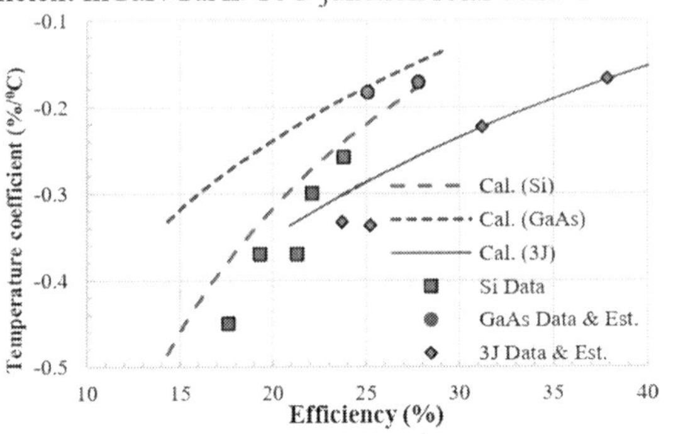

Figure 4. Calculated results for temperature coefficients of crystalline Si, GaAs and III-V compound 3-junction solar cell modules as a function of module efficiency in comparison with reported values and estimated values.

Figure 5 shows calculated results of driving distance of vehicles powered by Si, GaAs, 2-junction and 3-junction solar cell modules as a function of module efficiency and temperature coefficient. The higher efficiency solar cell modules show the better temperature coefficient and have advantageous for longer driving distance as the Toyota Prius 2019 (demonstration car) powered by 3-junction solar cell module with a module efficiency of 30.9% has demonstrated longer driving distance of 26 km/day (calibrated under the above calculation condition) driving at solar irradiance of 4 kWh/m^2/day compared to 16 km/day driving by the Sino Motors Sion powered by back contact solar cell module with module efficiency of 20-22%. The 3-junction solar cell modules have potential driving distance of 30 km/day average and more than 50 km/day on a clear day.

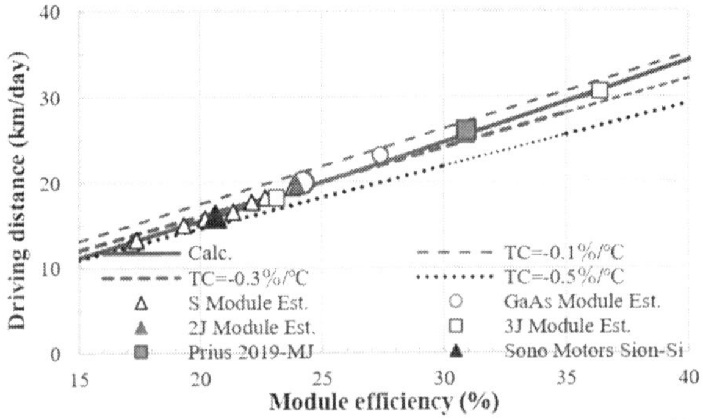

Figure 5. Calculated results for driving distance of vehicles powered by Si, GaAs, 2-junction and 3-junction solar cell modules as a function of module efficiency and temperature coefficient (TC) in comparison with estimated values of vehicles powered by Si, GaAs, 2-junction and 3-junction solar cell modules and actual driving distance calibrated of the Prius 2019[1] powered by 3-junction solar cell module and the Sono Motors Sion[2] powered by back-contact Si solar cell module.

5. Summary

The solar modules for PV-powered vehicles have a great ability to increase electric range and reduce CO_2 emissions. This paper presented the importance of developing high-efficiency solar cells modules for PV-powered vehicles. Approaches on development of further higher efficiency solar cell modules were shown. Temperature coefficients of candidate solar cell modules were also analyzed in order to develop highly efficient solar cell modules under sunshine conditions. The III-V compound 3-junction solar cell modules were shown to have high potential efficiencies of 37% with better temperature coefficient of -0.19 %/°C and were estimated to have potential driving distance of 30km/day average and more than 50km/day on a clear day.

References

1. M. Yamaguchi et al., Prog. Photovolt. DOI: 10.1002/pip.3343.
2. Sono Motors, https://sonomotors.com/
3. https://ja.weatherspark.com/70433.
4. M. Yamaguchi et al., Prog Photovol. **26**, 543 (2018).
5. M.A. Green et al., Prog Photovol. **28**, 629 (2020).
6. J.F. Geisz et al., Nature Energy, **5**, 326 (2020).
7. A. Grundstein et al., Int. J. Biometeorol. **53**, 255 (2009).
8. A. Wheeler et al., Proc. 47th IEEE PVSC (2019) p.894.
9. M. Yukawa et al., Trans. IEE Japan. **116-B**, 1101 (1996).
10. Y. Hishikawa et al., IEEE J. Photovolt.**8**, 48 (2018).
11. M.A. Green et al., Electron Lett. **18**, 97 (1982).
12. G. Siefer et al., Prog Photovol. **22**, 515 (2014).
13. Y. Hishikawa et al., AIST PV Research Symposium. (2019) T17. (in Japanese).
14. O. Dupre et al., Sol. Energy Mater. Sol. Cells **140**, 92 (2015).
15. Panasonic, https://news.panasonic.com
16. SunPower, https://www.enfsolar.com. https://www.sisolar.co.jp
17. Longi, https://www.sisolar.co.jp
18. Canadian Solar, https://www.canadiansolar.com
19. JA Solar, https://www.jasolar.com
20. Trina Solar, https://www.trinasolar.co

YIELD POTENTIAL OF VEHICLE INTEGRATED PHOTOVOLTAICS ON COMMERCIAL TRUCKS AND VANS

Christoph Kutter, Luis Eduardo Alanis, Dirk Holger Neuhaus, Martin Heinrich
Fraunhofer Institute for Solar Energy Systems ISE, Heidenhofstraße 2, 79110 Freiburg, Germany
Corresponding author: Christoph Kutter | Phone: +49 (0)761 4588 2196 | E-mail: christoph.kutter@ise.fraunhofer.de

ABSTRACT: To understand the potential of Vehicle Integrated Photovoltaics (VIPV) on commercial trucks and vans within Europe, we investigate five use cases: A) parcel delivery van, B) rural delivery truck, C) long haul, D) trailer, E) trailer with battery; we consider the European cities of Stockholm, Freiburg and Seville. For this, we first performed a vehicle geometry analysis in which we determined the potential to integrate photovoltaics on typical vehicle types with a module efficiency of 21%. Based on Global Horizontal Irradiance (GHI) datasets and location dependent conversion efficiencies we perform a yield calculation and estimate the final solar yield. A key finding is that the self-consumption during standby charging is a critical parameter in the feasibility of VIPV systems and needs to be minimized. We compute the corresponding annual solar range and find significant potential in the EU for the different use cases: A) 6637 to 11450 km, B) 3084 to 5272 km, C) 4828 to 8173 km, D) 763 to 1424 km, E) 4791 to 8134 km, depending on the city and assumed standby charging losses. Additionally, we perform a break-even analysis and find that irradiation, electricity prices, and the vehicle charging efficiency, are highly critical factors that impact the profitability of VIPV systems. Today VIPV is already profitable in Freiburg and Seville and can achieve payback times of 3,4 years for (B) and (C) in Seville and in Freiburg of 4,0 years for (B) and (C) in Stockholm within 6,9 (B) and 7,0 (C) years when mass production is achieved.

Keywords: VIPV, commercial vehicles, trucks, vans, solar energy, yield, solar range, break even, Lade-PV

1 INTRODUCTION

Vehicle Integrated Photovoltaics (VIPV) potential on trucks has been estimated to 90 GW in the EU [1] but fleet owners are still unsure about the usefulness of VIPV technology besides its ecological benefit. Previous studies have looked at fuel savings of diesel-powered refrigeration trucks using irradiance measurements in the EU, US [2] and yield calculations for Germany [3] and Australia [4]. In this study we expand on the previous findings with updated increased conversion efficiencies [5] and include a discussion of different use cases including the associated losses for power conversion and standby charging, as well as a break even analysis. We aim to provide a deeper study to answer commonly asked questions by fleet owners such as a realistic yield potential estimation or the profitability of VIPV.

Based on current available solar cell efficiencies we will discuss VIPV system sizes for typical commercial truck and van types, and calculate their solar yield for northern, central, and southern Europe. We consider the motion pattern within the yield calculation for five different scenarios: A) parcel delivery van, B) rural delivery truck, C) long haul, D) trailer, E) trailer with battery; to determine the share of solar yield harvested during standby and during driving.

The different application cases are relevant to model the energy flow within the vehicle since in standby mode the high voltage safety infrastructure currently consumes a significant amount of power in standby to allow for charging and power management. Based on typical component efficiencies (e.g. DC PR and converter efficiencies) we estimate the energy conversion flow and calculate the resulting final yield for the different use cases and the respective solar range.

Lastly, we present a break-even analysis for such VIPV systems.

2 METHOD

2.1 Vehicle Analysis

A selection of electric vehicles for commercial transportation of goods, which included different types and weight ratings, was considered for the calculations. Typical truck, van and box body dimensions have been researched and the resulting PV system installed power (P_{STC}) has been computed based on mainstream solar cell concepts. The geometrical data utilized in the study was obtained from brochures or datasheets published by the manufacturers. The roof area used for vans was calculated via geometrical analysis, as seen in Figure 1, while that of trucks and the platform cab, was calculated by using typical box dimensions for the vehicle's type and rating. The energy consumption, when not explicitly provided with reference to Worldwide Harmonised Light Vehicles Test Procedure (WLTP), was calculated with the stated range in km and battery capacity in kWh.

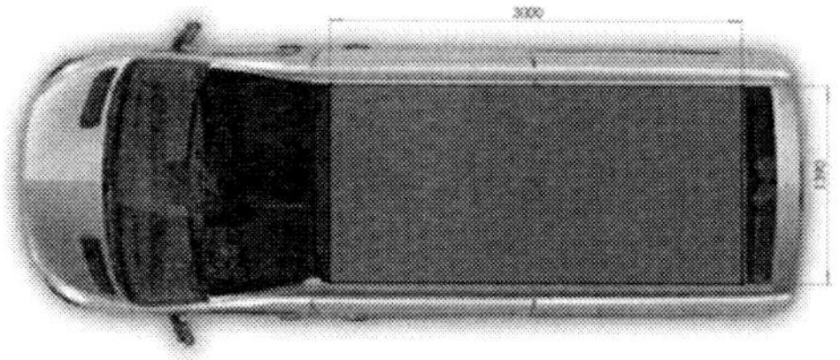

Figure 1. Mercedes-Benz eSprinter roof geometrical analysis [6]

An area usage percentage a_{usage}, which omits parts of the roof non-available for PV (such as framing, etc.), was estimated based on typical framing profiles and for each vehicle type and used to adjust the PV active area. This area was subsequently taken to calculate the potential PV power rating for each vehicle considering a PV module efficiency of 210 W_p/m^2. The vehicle model, type, consumption, roof area, area utilization factor and obtained PV power ratings are shown in Table 1. Vans are usually offered in different lengths and heights, the designation (e.g. L2H2) describes which cabin-size was considered. The highlighted vehicles were selected for the subsequent scenario-based analysis. Figure 2-5 shows illustrative images of these vehicles.

Table 1. PV power installation capacity of selected vehicles based on their roof geometry. Vehicles considered in study appear highlighted with corresponding type of scenario indicated.

Maker	Model	Type	Av. Consumption [kWh/100km]	Roof Area [m²]	Area usage	PV Power [W$_P$]	Scenario
Renault[7]	Master E-Tech	Van, 3,5t. L3	27,5	11,0	94%	2170	A
Peugeot[8]	e-Boxer	Van, 3,5t. L4H2	18,5	7,3	94%	1440	
Mercedes-Benz[6]	eSprinter	Van, 3,2t. L2H2	30,6	4,1	94%	810	
Fuso[9]	eCanter	Truck, 7,5t.	82,8	15,1	95%	3010	
Volvo[10]	FL Electric	Truck, 16t.	100,0	17,8	96%	3590	
Framo	E-165	Truck, 15t.	100,0	17,8	96%	3590	B
Mercedes-Benz[11]	eActros	Truck, 18t.	120,0	17,8	96%	3590	
MAN[12]	eTGM	Truck, 26t.	97,4	17,8	96%	3590	
Tesla[13]	Semi	Tractor unit + trailer	125,0	34,0	96%	6860	C, D
Freightliner [14]	eCascadia	Tractor unit + trailer	137,5	34,0	96%	6860	
BYD[15]	Q1R	Tractor unit + trailer	217,0	34,0	96%	6860	
ZF[16]	eTrailer	Trailer with battery	125 [see Tesla]	34,0	96%	6860	E

Figure 2. Renault Master E-Tech.
Picture: © 2021 Renault

Figure 3. Framo E-165 with a solar roof
Picture: © 2021 FhG ISE

Figure 4 Tesla Semi
Picture: © 2021 Tesla

Figure 5 ZF eTrailer
Picture: © 2021 ZF

2.2 Vehicle Operation Analysis

When electric vehicles are parked and the ignition is turned off, usually the circuit contactors at the drive train battery are opened and no charging is possible. Therefore, to allow for solar charging, the vehicle needs to be put in an operational state that allows the charging of the high voltage battery. Once the circuit contactors are closed, the safety monitoring infrastructure e.g. Battery Management System (BMS), Vehicle Management System (VMS) needs to be enabled, which leads to standby operational losses L_{CH}.

Hence, we assume that whenever the VIPV output is lower than L_{CH}, the whole system must be shut down to avoid discharging of the battery. The power consumption in fully deactivated state in this case or at night is assumed to be 0W within this study.

To show how significant the effect of the self-consumption is on the VIPV yield and to have a better understanding of how much solar energy is harvested during operation and during standby time of commercial vehicles, we perform a vehicle operation analysis. For this, we defined the hourly driving-to-standby ratio $r_{D2S,i}$ throughout a workday in each scenario and summarized them in Table 2.

We investigate five application scenarios: A) parcel delivery van, B) rural delivery truck, C) long haul, D) trailer, E) trailer with battery with its different charging efficiency.

For Scenario A the Renault Master E-Tech was considered. A 2020 study that looks at the emissions of distribution vehicles in the US found that, during operation, delivery vans can spend more time stationary than in transit when accounting for the minutes utilized on servicing customers; however, this is highly dependent on factors such as distance between customers and depot, service area, roads and parking availability [17]. In this scenario, it was decided to set the driving/standby ratio to 1:1, which would correspond to an urban environment with short transport distances. The working hours were set based on a typical parcel delivery service that operates from 7 AM to 6 PM on weekdays, from 8 AM to 6 PM on Saturdays and does not operate on Sundays. On working days, a 30-minute afternoon break for e.g. driver exchange was noted.

In Scenario B the Framo E-165 is considered. To define the motion profile, real world data from daily operation in Freiburg, Germany was analyzed.

The long-haul profile scenario C was defined according to European regulated steering times as 9 hours per day operation with a 45 min break after 4,5 h of driving [18]. On weekends the long-haul truck is parked and therefore on standby. The Tesla Semi is set as tractor unit. It is assumed that the trailer is always connected to the tractor unit.

For Scenario D and E, trailer and trailer with battery respectively, we analyzed and averaged measured motion data sets from a 2017 study in which the moving patterns of 6 reefer trailers operated in the US and in the EU were observed [2]. In scenario D, it is assumed that the trailer is only connected to the tractor unit while driving, and therefore no energy is harvested in standby mode. Within scenario D the same motion profile for all seven days per week is considered since reefer trailers mostly transport food and are therefore excluded from the driving ban on weekends.

Scenario E considers a trailer equipped with an electric battery. ZF Friedrichshafen AG recently introduced such an electric trailer system [16]. Here we assume that the PV energy can be fully utilized to charge the battery or is fed into the grid when parked over longer time frames.

Table 2. Driving-to-standby ratio per hour of the day in each scenario

Hour	Driving-to-standby ratio $r_{D2S,i}$ per hour of the day											
	A			B			C			D and E		
	Weekday	Saturday	Sunday	Weekday	Saturday	Sunday	Weekday	Saturday	Sunday	Weekday	Saturday	Sunday
0-1	0	0	0	0	0	0	0	0	0	0.07	0.07	0.07
1-2	0	0	0	0	0	0	0	0	0	0.06	0.06	0.06
2-3	0	0	0	0.01	0	0	0	0	0	0.06	0.06	0.06
3-4	0	0	0	0.40	0	0	0	0	0	0.05	0.05	0.05
4-5	0	0	0	0.55	0	0	0	0	0	0.05	0.05	0.05
5-6	0	0	0	0.52	0	0	0	0	0	0.05	0.05	0.05
6-7	0	0	0	0.47	0	0	0	0	0	0.08	0.08	0.08
7-8	0.5	0	0	0.38	0	0	1	0	0	0.1	0.1	0.1
8-9	0.5	0.5	0	0.30	0	0	1	0	0	0.13	0.13	0.13
9-10	0.5	0.5	0	0.18	0	0	1	0	0	0.13	0.13	0.13
10-11	0.5	0.5	0	0.33	0	0	1	0	0	0.14	0.14	0.14
11-12	0.25	0.25	0	0	0	0	0.5	0	0	0.15	0.15	0.15
12-13	0.5	0.5	0	0.20	0	0	0.75	0	0	0.17	0.17	0.17
13-14	0.5	0.5	0	0.47	0	0	1	0	0	0.2	0.2	0.2
14-15	0.5	0.5	0	0.35	0	0	1	0	0	0.21	0.21	0.21
15-16	0.5	0.5	0	0.17	0	0	1	0	0	0.2	0.2	0.2
16-17	0.5	0.5	0	0	0	0	1	0	0	0.19	0.19	0.19
17-18	0.5	0.5	0	0	0	0	0.5	0	0	0.16	0.16	0.16
17-19	0	0	0	0	0	0	0	0	0	0.17	0.17	0.17
19-20	0	0	0	0	0	0	0	0	0	0.16	0.16	0.16
20-21	0	0	0	0	0	0	0	0	0	0.14	0.14	0.14
21-22	0	0	0	0	0	0	0	0	0	0.11	0.11	0.11
22-23	0	0	0	0	0	0	0	0	0	0.08	0.08	0.08
23-24	0	0	0	0	0	0	0	0	0	0.08	0.08	0.08

2.3 Yield Calculation

For our estimations, the cities of Stockholm in Sweden, Freiburg im Breisgau in Germany and Seville in Spain were taken as reference for northern, central, and southern Europe, respectively. As Figure 6 illustrates, the expected Global Horizontal Irradiation is highly dependent on the latitude of each location. Plotting the expected GHI over the hours of the year in Figure 7 it becomes clear that the available radiant energy is highly variable throughout the year and, especially in Stockholm, low irradiation is predominant in winter.

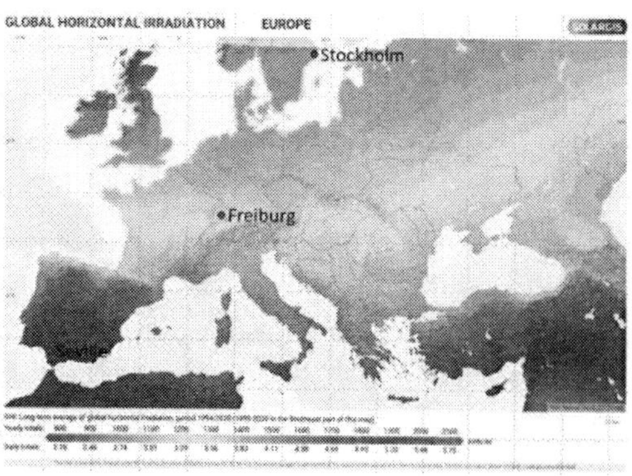

Figure 6. Geolocation of the 3 European cities taken as reference. Source: Solar resource map © 2021 Solargis

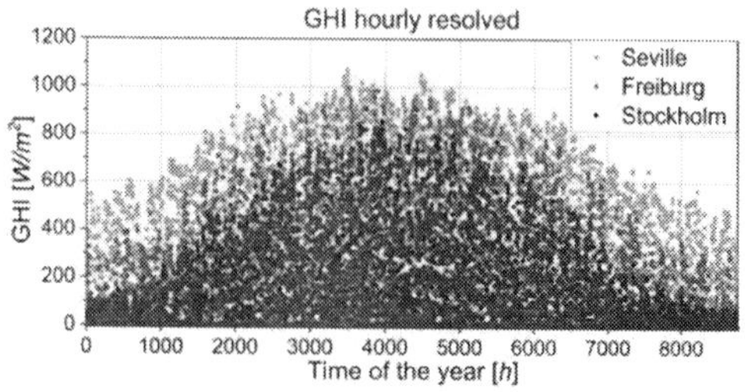

Figure 7. Hourly GHI distribution of given locations [19]

Based on hourly resolved Global Horizontal Irradiance (GHI) data sets from the Meteonorm database [19], the average annual sum of GHI per m² H_G was determined for each city and summarized in Table 3.

Table 3. Average annual horizontal irradiation per m² in given locations obtained from Meteonorm database [19].

Stockholm	Freiburg	Seville
H_G[kWh/m²]		
1,001	1,177	1,813

To assess the yield of photovoltaic systems, the following method based on data is described in [20] and extended for the selected vehicles. Shading is not considered within this study.

The used method translates H_G into a VIPV yield considering the below mentioned vehicle and location dependent parameters and the conversion efficiency within the system.

The DC Performance Ratio $PR_{DC,in}$ of PV systems is a measure for the capture efficiency of the PV array. The generator losses L_C describe the losses till the output of the PV array. For analyzing real world systems, it can be calculated by dividing Array Yield Y_A by the Reference Yield Y_R.

$$PR_{DC,in} = 1 - L_C = \frac{Y_A}{Y_R} \text{ in } \% \quad (1)$$

The Performance Ratio $PR_{DC,out}$, considers conversion losses L_s in the DC-DC converter and cabling between converter and grid, (here intermediate circuit of the vehicle) and is defined by multiplying $PR_{DC,in}$ with a conversion loss factor $(1 - L_s)$. Empirically it can be determined by diving the final yield Y_F by the Reference Yield Y_R.

$$PR_{DC,out} = PR_{DC,in} \cdot (1 - L_S) = \frac{Y_F}{Y_R} \text{ in } \% \quad (2)$$

Figure 8 shows a visualization of the different losses obtained for VIPV.

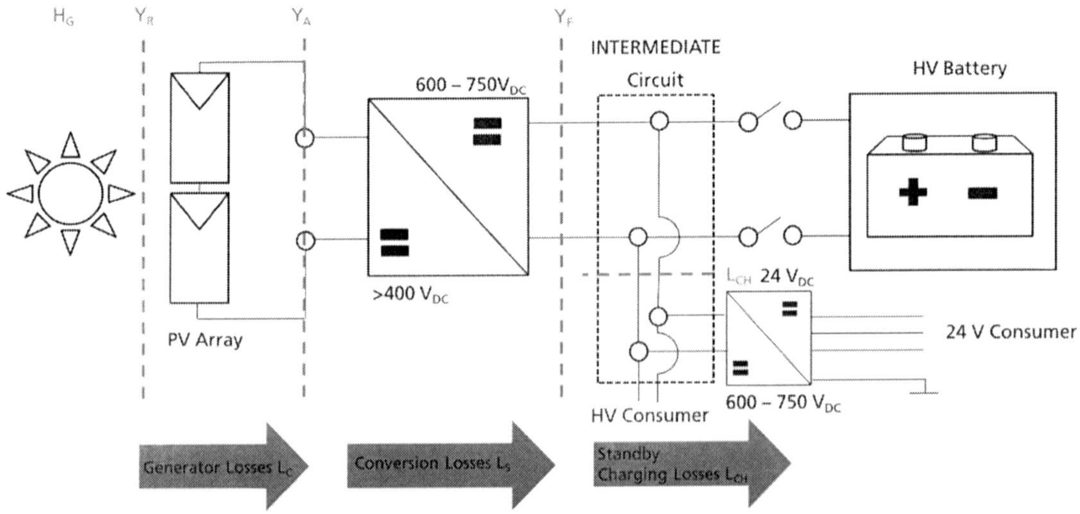

Figure 8. Flowchart showcasing the specific yield of VIPV systems. The array yield Y_A is lower than the reference yield and considers the generator losses L_C. The final yield Y_F is computed for hours driving and standby separately. It takes conversion losses L_S into account and subtracts the standby charging losses L_{CH} by the vehicle in standby charging mode, extended from [20].

Within this study we compute the final yield Y_F for every hour of the year (8760 h/year) while driving as follows:

$Y_{F,driving} = \sum_{h=1}^{8760} GHI_{location,h} \cdot A_{vehicle} \cdot \alpha_{usage} \cdot \eta_{module} \cdot PR_{DC,out} \cdot r_{D2S,h}$ in kWh/year per vehicle (3)

Meanwhile, the share of the final yield harvested in standby is computed according to equation 4. Whenever L_{CH} is higher than the VIPV power the system is assumed to be turned off and therefore yield and consumption set to 0.

$Y_{F,standby} = \sum_{h=1}^{8760} GHI_{location,h} \cdot A_{vehicle} \cdot \alpha_{usage} \cdot \eta_{module} \cdot PR_{DC,out} \cdot (1 - r_{D2S,h}) - L_{CH}$ in kWh/year per vehicle (4)

Hereby vehicle specific areas and area usage factors were taken from Table 1 and η_{module} was set to 21%. The scenario-specific hourly driving-to-standby ratio $r_{D2S,h}$ was taken from Table 2 for every day of the year.

$PR_{DC,in}$ accounts for location-specific factors of the PV array such as operating temperature, which is known to directly affect the performance of photovoltaic systems in such way that a higher PR is expected in colder climates [21].

To account for this effect, $PR_{DC,in}$ was estimated for each location based on data obtained from the Global Solar Atlas 2.0 [22]. Here the calculated specific photovoltaic power output per kWp was divided by the Global Tilted irradiation (GTI) at optimum angle and corrected for system attributed conversion losses as stated in the authors' methodology. The computed $PR_{DC,in}$ values shown in Table 4 consider the following effects:

- Soiling
- Temperature higher than STC
- Mismatch of modules
- Cabling/Ohmic losses

Table 4. DC performance ratio $PR_{DC,in}$ for each location

Stockholm	Freiburg	Seville
	[%]	
90,7	85,4	84,4

The Conversion Losses L_s consider the converter efficiency below 100%, and ohmic losses within the cabling in between the DC-DC converter and the intermediate circuit and are set to 2% referring to similar sized residential inverters [23].

For the standby operational demand L_{CH}, we performed measurements on the Framo E-165 truck, which reveal a standby consumption of up to 700 W when the ignition is on, which is in good accordance with literature [24].

On the other hand, a study evaluating the efficiency of photovoltaic home storage systems found the standby power consumption of such systems to vary between 0 and 40.8 W for a full battery and 0.1 W and 46.2 W with an empty battery [25].

Although vehicle and home storage systems concepts cannot be directly compared, the study reveals that storing PV energy can generally be done more efficiently. With this as reference, we take a self-consumption value of 25 W as a benchmark.

To illustrate the impact of the standby operational demand L_{CH} on the solar yield, we perform a sensitivity analysis considering an L_{CH} value of 0, 25, 100 and 700 W.

It is clear, however, that in future electric vehicles featuring a VIPV system, a new operation mode for "standby solar charging" will need to be implemented in the VMS, its purpose being to minimize the energy demand for charging readiness by deactivating all unnecessary consumers.

2.4 . Solar range calculation

The solar range was computed based on the average power consumption. This specification, which is vehicle-dependent, is based on the WLTP and was taken or

calculated from the advertised technical data offered by the manufacturers (see Table 1).
For electric vehicles it is typical to find this figure expressed in the units kWh/100km. With this, the solar range extension can be calculated in km per year by dividing the annual VIPV yield by the average consumption and multiplying it by a factor of 100.
It must be noted that the average consumption is highly variable from model to model and only vaguely correlates to the vehicle type, size, and weight rating. This may also imply that power efficiency of electric vehicles can still be optimized as the technology becomes more mature for trucks.

2.5 Break Even analysis for Scenario B and C

Further, we perform an Levelized Cost of Electricity (LCOE) analysis [26, 27] for a purchased VIPV system and perform a break-even analysis from the customer perspective for Scenario B and C with 25 W operational losses L_{CH} in standby and with the assumption of the parameters as shown in Table 5 taken for small scale rooftop systems from [27]. Yearly degradation was set to 1% as we expect it to be higher than within conventional modules (typically 0,25%/a) and lifetime was set to 10 years [1].

Table 5. Economic input parameters

WACC,real (Weighted Average Cost of Capital) in % [27]	2,2
Lifetime in years	10
Degradation in % per year	1

Within the break-even analysis every kWh utilized from the VIPV system replaces a kWh bought from the grid. We consider country specific household electricity prices as stated in Table 6. Charging fees at public (fast) charging stations are often even higher than household prices while non-household electricity prices e.g. at the home depot could also be lower.

Table 6. Household - electricity prices at each location [28]

Stockholm	Freiburg	Seville
	€/kWh	
0,17	0,3	0,23

The cost parameter in Table 7 indicates VIPV component prices of the prototype phase. These prices are conservative and likely to decrease with ramp up on mass production of VIPV components. Therefore, we also provide an outlook scenario with system prices of 1,0 €/Wp.

Table 7. System-related input parameters

Module price in €/Wp	0,5
Power Electronics in €/Wp	0,4
Safety Accessories in €/Wp	0,3
Installation in €/Wp	0,5
System Price in €/Wp	1,7 (or 1,0)

3 RESULTS

3.1 VIPV system size for commercial vehicles

Since the power potential of a VIPV system (or of any solar-energy-based system) depends on the surface available for catching sunlight, it was possible to estimate the maximum installed power on different vehicle types based on their roof geometry.
We find that box-bodies and trailers are usually built on dimensions that are largely standardized, featuring a typical roof area of 17.8 m² and 34 m², respectively, equivalent to approximately 3590 W_P and 6860 W_P of installed PV power considering a module performance of 210 W_P/m^2 (in line with current cell and module technologies) and an area utilization factor (α_{usage}) of 96%.
On the other hand, utility vans come in a wide range of sizes and designs, so the potential for VIPV installed power varies likewise. Due to the smaller roof surface and increased geometrical complexity, the area utilization factor (α_{usage}) used for vans was set to 94%.
We find that within the sampled utility vans, the roof areas observed range between 4,1 m² and 11 m², equivalent to a maximum installed PV power of 810 W_P and 2170 W_P, respectively.

3.2 Reference VIPV yield and solar range

The solar yield and range were calculated for the sampled vehicles in the different scenarios detailed in 2.2 considering the location-based inputs explained in 2.3.
Within the assumptions made for the expected system characteristics and taking standby operational losses L_{CH} of 25 W as reference, we find that the expected annual solar yield for the studied vehicles and scenarios would be within the range of 1825 kWh (for a Renault Master E-Tech utility van in scenario A in Stockholm) and 10216 kWh (for a Tesla Semi tractor unit with an attached trailer, operated under scenario C in Seville).
These yields would translate to an annual solar range equivalent to 6637 km and 8173 km, respectively. The overview of all studied scenarios and cities can be found in Figure 9. As Table 8 summarizes, the VIPV system can cover significant shares of the energy demand of the vehicle assuming German average annual mileage for each vehicle type: Up to 35% in Scenario A, up to 14% in Scenario B and up to 9% for long-haul trucks in C and E.

Table 8. Solar ranges, average annual mileage, and solar coverage for each scenario

	location	Solar Range for L_{CH} 25 W [km]	German average annual milage 2020 [km] [29]	Solar coverage [%]
A	Stockholm	6637	Van ≤ 3,5t: 19.038	35
	Freiburg	7397		39
	Seville	11450		60
B	Stockholm	3084	Truck >7,5t: 35.757	9
	Freiburg	3429		10
	Seville	5272		15
C	Stockholm	4828	Tractor Unit: 89.667	5
	Freiburg	5356		6
	Seville	8173		9
E	Stockholm	4791		5
	Freiburg	5317		6
	Seville	8134		9

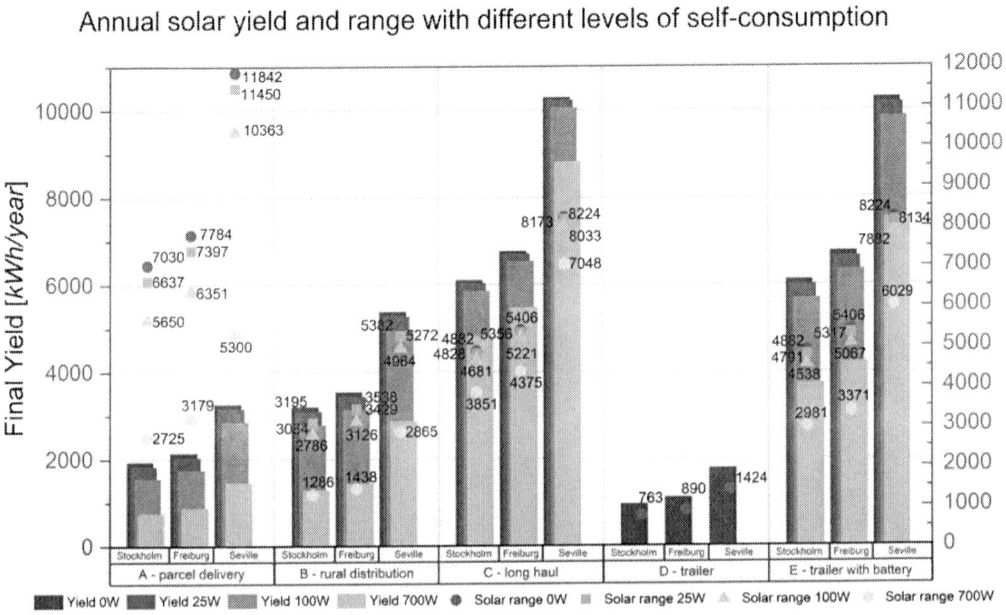

Figure 9 Results of VIPV yield and solar range calculation for Scenarios A to E and all 3 locations and varying L_{CH}

3.3 Sensitivity analysis for self-consumption

As discussed in the method section, the self-consumption of the VIPV system in standby mode plays a central role in the final solar yield and range.

For this reason, we perform a sensitivity analysis on the abovementioned calculations considering a self-consumption of 0 W (VIPV yield upper limit), of 700 W (level observed in the Framo demonstrator truck) and of 100 W; and compare them with the reference potential (25 W self-consumption) from 3.2.

Table 9 summarizes the share of the potential VIPV energy that is self-consumed in each scenario.

Since scenario D already considers that energy storage is not viable, it was excluded from the sensitivity analysis.

We find that the standby operational losses L_{CH} minimize the VIPV Yield significantly, especially for smaller vehicles and slightly more at northern locations.

Long haul trucks benefit from long operation hours and large systems so that self-consumption is lower. It becomes evident that the charging process in standby for traction batteries needs to be optimized.

Table 9. Share of VIPV energy self-consumed for varying levels of standby self-consumption

Relative standby operational losses [%]					
Scenario	L_{CH} [W] →	0	25	100	700
A	Stockholm	0	6	20	61
	Freiburg	0	5	18	59
	Seville	0	3	12	55
B	Stockholm	0	3	13	60
	Freiburg	0	3	12	59
	Seville	0	2	8	47
C	Stockholm	0	1	4	21
	Freiburg	0	1	3	19
	Seville	0	1	2	14
E	Stockholm	0	2	n/a	n/a
	Freiburg	0	2	n/a	n/a
	Seville	0	1	n/a	n/a

3.4 Break Even Analysis for VIPV Scenario B and C

In this analysis we compare the yearly yield by solar irradiation in terms of charging cost savings to the initial cost of the system as shown in Section 2.5. We find the VIPV system can be profitable in Freiburg and Seville within Scenarios B (Table 10) and C (Table 11) assuming a 10-year lifetime. In Stockholm the VIPV system is only profitable within 10 years if the system cost reaches 1,0 €/Wp and below.

The LCOE per kWh of the VIPV System are always lower than a kWh bought from the grid as a consumer in Germany and Spain. With current conservative system cost and household prices, the VIPV system reaches break-even after 6,8 years in (B) and (C) in Freiburg. The net present value (NPV) ranges from 1976 € in (B) to 3639 € (C) in Freiburg.

Table 10. Results of Break-Even-Analysis for Scenario E

	Stockholm		Freiburg		Seville	
System cost [€/Wp]	1,7	1	1,7	1	1,7	1
Break even [years]	11,8	6,9	6,8	4,0	5,8	3,4
Net present value 10 years [€]	-1478	1035	1976	4489	3320	5833
LCOE (10 years) [€/kWh]	0,22	0,13	0,20	0,12	0,13	0,08

When we consider mass production of VIPV components and prices in the range of small scale roof top systems ~1,0 €/Wp according to [5] we find much lower payback times of 3,4 years in (B) and (C) in Seville and in Freiburg of 4,0 years in (B) and (C), and VIPV to be profitable in Stockholm within 6,9 (B) and 7,0 (C) years lifetime in Scenario B and C, respectively, as shown in Table 11 and Figure 10. The LCOE drops accordingly, whereas NPV increases.

Table 11. Results of Break-Even Analysis for Scenario C

	Stockholm		Freiburg		Seville	
System cost [€/Wp]	1,7	1	1,7	1	1,7	1
Break even [years]	11,9	7,0	6,8	4,0	5,8	3,4
Net present value 10 years [€]	-2916	1882	3639	8437	6236	11033
LCOE (10 years) [€/kWh]	0,23	0,13	0,20	0,13	0,13	0,08

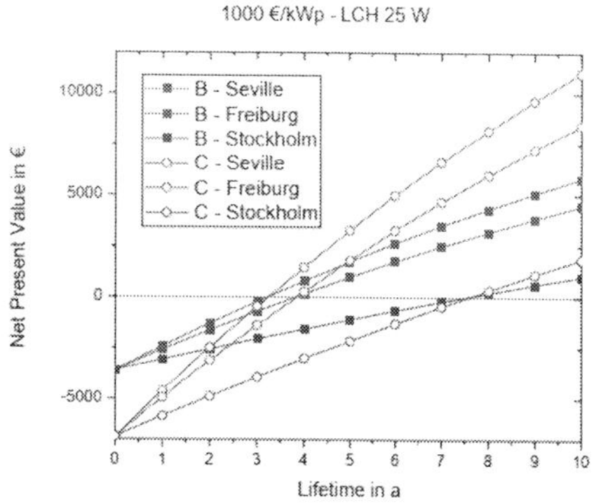

Figure 10 Break-even analysis over ten years lifetime for a Framo-E165 VIPV-System with 1,0 €/Wp system cost

The combination of low irradiation and cheap electricity is responsible for longer payback times of the VIPV system in the investigated scenarios in Sweden. For Germany, although the VIPV yield is lower on commercial vehicles, it is only slightly less profitable than in Spain due to the higher electricity price. Payback times below 5 years in Germany and Spain are realistic.

4 CONCLUSION

Based on the geometry of different electric commercial vehicles and their predicted operation patterns under specific scenarios in three European cities, we analyzed the critical variables for implementing VIPV systems that are technically and economically viable. Our main conclusions are described below.

4.1 Solar yield and range extension

VIPV installed only on the roof of the electric vehicle covers a significant part of the energy consumption of commercial vehicles within all investigated scenarios. Considering an optimized system efficiency, the following annual solar ranges can be obtained in Stockholm and Seville. The value in % states the VIPV coverage of the estimated annual energy demand of the vehicle:

- A - Parcel Delivery Vans: 6637 to 11450 km/year, solar energy coverage 35 to 60%
- B – Rural Delivery Trucks: 3084 to 5272 km/year, solar energy coverage 9 to 15%
- C – Long Haul Truck 4828 to 8173 km/year, solar energy coverage 5 to 9%
- D – Trailer (only harvesting while driving) 763 to 1424 km/year, solar energy coverage 0.9 to 1.6%
- E – Trailer with battery/grid feed-in 4791 to 8134 km/year, solar energy coverage 5 to 9%

4.2 VIPV can be economically feasible

VIPV on commercial vehicles is profitable in central and southern Europe when assuming a ten-year lifetime and current technologies and respective costs are reached. Irradiation and electricity prices and the vehicle charging efficiency have significant impact on the feasibility of VIPV systems.

We find the payback period in Scenario B and C for rural distribution and long-haul trucks to range from 3,4 years in Seville, to 4,0 years in Freiburg and 6,9 to 7,0 years in Stockholm, assuming mass production with cost of 1 €/Wp.

Additionally, VIPV features emission free energy onboard, and a gain in comfort through location independent charging and less and shorter charging stops.

4.3 Standby operational losses and the need for efficient solar charging

We found that one of the main factors to consider is the energy that the system itself must consume for tasks such as the safety monitoring of the high voltage battery charge when parked (i.e. standby) or while driving.

The analysis of the operation patterns show that the solar irradiation perceived during standby time is more significant than when driving, therefore it is critical that the standby operational losses are minimized from a system perspective.

As an example, we find that a standby self-consumption comparable to that of a residential system (25 W) would lead to operational losses of only 3% of the annual solar yield for scenario B in Stockholm, while a high level of self-consumption (700 W) would increase the losses up to 60% under the same circumstances.

To reach economic feasibility, it is necessary to review and optimize the energy flow and vehicle management system and minimize the self-consumption of the system during charging in standby.

Further, the system needs the capability to check if enough solar power is available when the vehicle is turned off. This could be achieved by using irradiation sensors or vehicle location dependent sun height calculations, to avoid waking up at night times.

5 OUTLOOK

We expect the use case VIPV to become even more feasible for the end user over time.

Firstly, with an historic efficiency increase of mainstream c-Si PV of 0,6 $\%_{abs}$ per year, the solar yield per vehicle will increase. Secondly, mass production of (VI)PV components will lead to cheaper LCOE. Every time the global, cumulative PV module production doubled in the last 40 years, the module selling price went down by 25% [5] and we expect that this trend will continue in the future. Thirdly, grid tied electricity prices tend to rise over time, which will increase the savings for each kWh produced by a VIPV system. Fourthly, the power supply on the vehicle may reduce charging times during operation of the vehicle. Finally, VIPV will benefit from more efficient electric vehicle drive trains in the future. To further boost the solar range of VIPV systems, additional areas on the commercial vehicles e.g. the sides of the box body, the cabin roof or the hood could be utilized.

6 ACKNOWLEDGEMENT
We thank Torsten Peppel, Ralf Noeske and Steffen Margenberg, from M&P Motion Control and Power Electronics GmbH, Dr. Aswin Linsenmeyer from SUNSET Energietechnik GmbH and Rolf Huhn from TBV Kühlfahrzeuge GmbH for exceptional good collaboration within the project "Lade-PV" and guidance on current component prices. Finally, we would like to thank the German Ministry of Economic Affairs and Energy (FKZ 03EE1002A "Lade-PV") for their funding.

REFERENCES

[1] C. Kutter, F. Basler, D. H. Neuhaus, M. Heinrich, J. Markert, and L. E. Alanis, "Integrated Lightweight, Glass-Free PV Module Technology for Box Bodies of Commercial Trucks," 37th European Photovoltaic Solar Energy Conference and Exhibition, pp. 1711–1718, 2020, doi: 10.4229/EUPVSEC20202020-6DO.11.6.

[2] U. Eitner et al., "Solar Potential on Commerical Trucks: Results of an Irradiance Measurement Campaign on 6 Trucks in Europe and USA," in Proceedings of the 33rd European Photovoltaic Solar Energy Conference and Exhibition (EU PVSEC), Amsterdam, Netherlands, 2017, pp. 2147–2150.

[3] M. Kühnel, B. Hanke, S. Geißendörfer, K. von Maydell, and C. Agert, "Energy Forecast for Mobile Photovoltaic Systems with Focus on Trucks for Cooling Applications," 2016, doi: 10.4229/EUPVSEC20162016-6EO.1.6.

[4] B. Elliston and M. Dennis, "Feasibility of Solar-Assisted Refrigerator Transport in Australia," Australian National University, Department of Engineering, Canberra, 2009.

[5] Fraunhofer Institute for Solar Energy Systems ISE with support of PSE Projects GmbH, Ed., "PHOTOVOLTAICS REPORT 27.06.21," Accessed: 30.08.21. [Online]. Available: https://www.ise.fraunhofer.de/content/dam/ise/de/documents/publications/studies/Photovoltaics-Report.pdf

[6] Mercedes-Benz AG, The Mercedes-Benz eSprinter. [Online]. Available: https://www.mercedes-benz.co.uk/vans/content/dam/vans/united-kingdom/brochures/december2020/https://www.mercedes-benz.co.uk/vans/en/e-sprinter-panel-van (accessed: Oct. 2 2021).

[7] Renault, New MASTER Z.E.: Dimensions. [Online]. Available: https://www.renault.co.uk/electric-vehicles/master-ze/specifications.html?bodyTypeCode=PT1261_platform_cab (accessed: Nov. 2 2021).

[8] Peugeot, Peugeot Boxer: Prices, equipment & technical specification. [Online]. Available: https://media-ct-ndp.peugeot.com/file/99/2/peugeot-boxer-spec-sheet-january-2018-web.pdf?lcdv16=2PU91DHDP670A0C0 (accessed: Nov. 2 2021).

[9] Daimler AG, FUSO eCanter. [Online]. Available: https://media.daimler.com/marsMediaSite/ko/en/45600295 (accessed: Nov. 2 2021).

[10] Volvo Truck Corporation, "Electromobility Made Easy: Electrified Urban Transport by Volvo Trucks," [Online]. Available: https://www.volvotrucks.com/en-en/trucks/alternative-fuels/electric-trucks.html

[11] Mercedes-Benz AG, Mercedes-Benz eActros: Sustainable, fully electric and quiet. [Online]. Available: https://www.mercedes-benz.com/en/vehicles/trucks/mercedes-benz-eactros-sustainable-fully-electric-and-quiet/ (accessed: Nov. 2 2021).

[12] MAN Truck & Bus SE, Fully Electric, Whisper-Quiet and Highly Efficient: The MAN eTGM. [Online]. Available: https://www.man.eu/de/en/truck/models/man-etgm/etgm.html (accessed: Nov. 2 2021).

[13] I. Tesla, Tesla Semi: Product website. [Online]. Available: https://www.tesla.com/semi (accessed: 18.08.20).

[14] Daimler AG, Electrified on the highway, Freightliner eCascadia and eM2. [Online]. Available: https://www.daimler.com/sustainability/climate/ecascadia.html (accessed: Nov. 2 2021).

[15] BYD Company Limited, Q1R. [Online]. Available: https://sg.byd.com/wp-content/uploads/2017/11/Q1R_SPEC.pdf (accessed: Nov. 2 2021).

[16] Frank Discher, Ausgezeichnet: umweltfreundlicher eTrailer von ZF erhält CLEPA-Innovationspreis 2020, 26.11.20. Accessed: 18.08.21. [Online]. Available: https://press.zf.com/press/de/releases/release_21889.html

[17] M. Figliozzi, J. Saenz, and J. Faulin, "Minimization of urban freight distribution lifecycle CO2e emissions: Results from an optimization model and a real-world case study," Transport Policy, vol. 86, pp. 60–68, 2020, doi: 10.1016/j.tranpol.2018.06.010.

[18] REGULATION (EC) No 561/2006 OF THE EUROPEAN PARLIAMENT AND OF THE COUNCIL on the harmonisation of certain social legislation relating to road transport and amending Council Regulations (EEC) No 3821/85 and (EC) No 2135/98 and repealing Council Regulation (EEC) No 3820/85. Accessed: 18.08.21. [Online]. Available: https://eur-lex.europa.eu/resource.html?uri=cellar:5cf5ebde-d494-40eb-86a7-2131294ccbd9.0005.02/DOC_1&format=PDF

[19] J. Remund, S. C. Müller, M. Schmutz, and P. Graf, "Meteonorm Version 8," 2020, doi: 10.4229/EUPVSEC20202020-5BV.3.8.

[20] K. Mertens, Photovoltaics: Fundamentals, Technology and Practice. New Jersey: Wiley, 2014.

[21] J. Leloux, J. Taylor, R. Moreton Villagrá, L. Narvarte, D. Trebosc, and A. Desportes, "Monitoring 30,000 PV Systems in Europe: Performance, Faults, and State of the Art," 2015, doi: 10.4229/EUPVSEC20152015-5AO.8.1.

[22] Solargis s.r.o, Global Solar Atlas 2.0: a free, web-based application is developed and operated by the company Solargis s.r.o. on behalf of the World Bank Group, utilizing Solargis data, with funding provided by the Energy Sector Management Assistance Program (ESMAP). [Online]. Available: https://globalsolaratlas.info./ (accessed: 18.08.21).

[23] Sungrow Power Supply Co. Ltd, Technical Datasheet SG3K/3K6/4K/4K6/5K/6K-D:

Residential String Inverter. [Online]. Available: https://en.sungrowpower.com/upload/documentFile/DS_SG3K%203K6%204K%204K6%205K%206K-D%20Datasheet(EMEA)_V12_EN.pdf.pdf (accessed: 30.08.21).

[24] C. Lodi, A. Seitsonen, E. Paffumi, M. de Gennaro, T. Huld, and S. Malfettani, "Reducing CO 2 emissions of conventional fuel cars by vehicle photovoltaic roofs," Transportation Research Part D: Transport and Environment, vol. 59, pp. 313–324, 2018, doi: 10.1016/j.trd.2018.01.020.

[25] N. Munzke, B. Schwarz, F. Büchle, and M. Hiller, "Evaluation of the efficiency and resulting electrical and economic losses of photovoltaic home storage systems," Journal of Energy Storage, vol. 33, p. 101724, 2021, doi: 10.1016/j.est.2020.101724.

[26] P. Konstantin, Praxisbuch Energiewirtschaft: Energieumwandlung, -transport und -beschaffung, Übertragungsnetzausbau und Kernenergieausstieg, 4th ed. Berlin: Springer Vieweg, 2017. [Online]. Available: http://www.springer.com/

[27] C. Kost, S. Shammugam, V. Fluri, D. Peper, A.D. Memar, T. Schlegl, "Levelized cost of electricity renewalbe energy technologies: June 2021," Jun. 2021. Accessed: 26.08.21. [Online]. Available: https://www.ise.fraunhofer.de/content/dam/ise/en/documents/publications/studies/EN2021_Fraunhofer-ISE_LCOE_Renewable_Energy_Technologies.pdf

[28] eurostat, "Electricity prices (including taxes) for household consumers, second half 2020," Accessed: 26.08.21. [Online]. Available: https://ec.europa.eu/eurostat/statistics-explained/index.php?title=Electricity_price_statistics

[29] F. Kraftfahrt-Bundesamt, Ed., "Verkehr in Kilometern 2020 - Inländerfahrleistung: Entwicklung der Fahrleistungen nach Fahrzeugarten seit 2016," Accessed: 30.08.21. [Online]. Available: https://www.kba.de/DE/Statistik/Kraftverkehr/VerkehrKilometer/vk_inlaenderfahrleistung/vk_zum_herunterladen/verkehr_in_kilometern_kurzbericht_pdf.pdf?__blob=publicationFile&v=5

ENERGY YIELD SIMULATION OF 3D CURVED VIPV MODULES

Sebastian Neven-du Mont[1,2], Hashem Alharthi[1], Christoph Kutter[1], Dirk Holger Neuhaus[1], Martin Heinrich[1]
[1] Fraunhofer Institute for Solar Energy Systems ISE, Heidenhofstr. 2, 79110 Freiburg, Germany
[2] Technical University Bergakademie Freiberg, Leipziger Str. 28, 09599 Freiberg, Germany
sebastian.neven-du.mont@ise.fraunhofer.de

ABSTRACT: The integration of photovoltaics into the body of vehicles provides many benefits for e-mobility, like inter alia the extension of the driving range, a decreased number of charging cycles per year and the reduction of CO_2-emissions of the transport sector. The greatest challenge in integrating photovoltaics in vehicles is the design-related curvature of the modules that the PV modules must follow. The impact of the curvature on the module technology and the energy yield are central research questions that therefore need to be answered. In this study we apply our previously published energy yield model for 3D curved PV modules to simulate the energy yield potential on the roof and hood of different type of cars by considering the module curvature. The curvature of the VIPV module leads to an inhomogeneous irradiance over the module surface. Thus, to an electrical mismatch within the module which can cause high yield losses. We show that the curvature of the roof decreases the yield up to 25 % when using series interconnection of cells. We present an approach to reduce losses that result though the mismatch. We propose the adaptation of the interconnection topology to the curvature by dividing the module into parallel subgroups containing cells with similar azimuth and elevation angles towards the sun. We develop interconnection topologies for roof and hood integrated PV modules based on this approach and simulate the energy yield potential of each design. We show that a reduction of only 3 % - 6 % compared to a horizontal reference can be achieved by using a reasonable low number of subgroups.
Keywords: Vehicle Integrated Photovoltaics, VIPV, Curved PV Modules, Energy Yield, Yield Simulations, Module Technology, Interconnection Topology, 3D

1 INTRODUCTION

The worldwide upscaling of PV module production led to a drastically reduction of PV manufacturing cost [1]. This opens the possibility to integrate photovoltaics onto surfaces that have not yet been used for energy production, like the outer body of cars, trains, and other vehicles [2–4].
The energy produced on the surface of these vehicles could be used directly for their electric drive. Long transportation routes of the required energy from the production sites to the customers could be avoided which obviates losses through lines, conversion and storage that go hand in hand with it. Another advantage of the spatial proximity of generation and consumption is that it reduces the load on the power grid. The worldwide deploy of electric vehicles is forecasted to increase the electricity demand of the transport sector by 35.8 % till 2050 [5].
A case study shows that 41 % of the required energy of an electric vehicle can be provided by a vehicle integrated PV system [1]. With the usage of highly efficient cell technologies and the utilization of the entire external surface of the vehicle it would even be calculative possible to achieve self-sufficiency through solar energy in Germany [6, 7].

The greatest challenge for the integration of photovoltaics in vehicles is the design-related spherical curvature of the module. First prototypes show the successful manufacturability of such PV modules [8, 9]. However, the curvature of the module surface leads to an inhomogeneous irradiance on it which needs to be considered in the design of the interconnection topology [10, 11]. The differing level of irradiance reaching the cells produce an electrical mismatch within the module which can cause severe yield losses. A solution to this problem could be the adaptation of the interconnection topology to the curvature to reduce losses resulting through this mismatch [10].

In this study we apply our previously published model [10] to simulate the energy yield potential of the curved cell pattern in VIPV modules with different interconnection topologies. Our aim is to understand the influence of different interconnection topologies and how the energy yield of cell patterns integrated in the roof and hood of cars could be maximized. We additionally simulate the energy yield of a planar module with series interconnection of cells at 0° module tilt angle to analyze the impact of the curvature on the yearly yield.

We present an approach that allows the design of efficient and yield maximizing interconnection topologies for curved PV modules. We propose the division of the module into smaller submodules whose size, form and layout are adapted to the curvature and which additionally lead to a beneficial (partial) shading behavior. The partial shading behavior is particularly crucial for PV modules operating in traffic due to the occurring shading through traffic signals, nearby buildings, and other structures.
The main influence is however the curvature since, despite the partial shading, it is always present on the module.

2 METHOD

2.1 Averaged car roof and hood designs

A survey is conveyed to analyze typical car roof and hood geometries. We analyze freely available blueprints of a variety of limousines and station wagons/ sport utility vehicles to define typical geometries (area and curvature) for each type of car. We develop car roof and hood designs based on these geometries which form the basis for our simulations. The aim is to simulate typical car roof designs by averaging the length and width (area) as well as the surface curvatures.

We investigate topologies based on the identified averaged designs. The interconnection topologies and specific cell information at STC are uploaded into our energy yield model for curved PV modules to simulate the yield for each specific design [10].

2.2 Cell Interconnection Topologies

Planar PV modules particularly use series interconnection of cells with bypass diodes. This interconnection scheme was found to lead to significant yield losses in curved PV modules due to the electrical mismatch between the cells [10]. Topologies using parallel interconnection of strings proved to be a more suitable option for curved PV modules regarding the yield maximization. However, a parallel module configuration may increase ohmic losses due to its higher currents. We therefore propose the division of the curved VIPV module into subgroups which are connected in parallel to each other to increase the overall energy yield, while maintaining low ohmic losses. To lower the ohmic losses, we use M6 half cells throughout the study instead of full cells.

The cell interconnection topologies are developed based on the maximum number of cells integrable into the averaged roof and hood surfaces. Topologies with subgroups which are connected in parallel to each other require the same number of cells in each group. Additionally, the maximum number of cells needs to be a multiple of the number of subgroups. This limits the number of possibilities of the module division in specific numbers of subgroups. Cells of one subgroup should have a similar tilt and azimuth angle towards the sun to reduce the electrical mismatch between the solar cells. We additionally consider that series interconnection of more than 50 solar cells requires the integration of bypass diodes to reduce the risk of hot spot formation. Further boundary conditions are that most car manufacturer prefer system voltages below 60 V to ensure passenger safety which can easily be obtained with this approach. Furthermore, the current of the VIPV system should not be too high since it would lead to high ohmic losses. For comparability we do not apply a boundary on the current in this study, however very high currents are achieved with certain topologies. One way to reduce those currents is to use even smaller cell size such as shingled cell interconnections [12].

Series interconnection with bypass diodes is used as reference topology to investigate if the division into subgroups leads to an increase in yield for VIPV modules. We additionally simulate the yield of a topology with parallel interconnection of all cells to determine the theoretical maximum yield that can be obtained with such modules.

2.3 Simulations

We simulate the energy yield for each design to compare the potential yield of the topologies based on our previously introduced energy yield model [10].

The energy yield simulations are done for the location Freiburg (Breisgau), Germany (48° N and 7.85° E). We use a typical meteorological year (TMY) dataset containing global horizontal irradiance (GHI) data and the sun angles (elevation α_S and azimuth γ_S) for each hour of the year. We simulate the yield for four orientations (North 0°, East 90°, South 180°, and West 270°) for each topology and define the mean of the four orientations as yield potential. We simulate M6 half-cells with the following cell IV- characteristics:

Table I: Cell IV- Parameters

M6-Format	V_{OC} [V]	I_{SC} [A]	V_{MPP} [V]	I_{MPP} [A]	P_{MPP} [W]
Half cell	0.7	5.6	0.6	5.3	3.1

In a first step we calculate the tilt and azimuth angle for each individual cell at different orientations.

Our model than calculates cell, string, and module IV- curves by considering module orientation, the interconnection topology and the level of irradiance reaching each individual cell. The module power at the maximum-power-point (MPP) is taken as reference power to calculate the yearly yield. We simulate three days prior and after the winter and summer solstice as well as autumnal and vernal equinox, thus four weeks for all topologies. We additionally simulate the yield for each hour of the year of two designs per car type and component to extrapolate the yearly energy yield of the four-week simulations and to verify that the extrapolation provides meaningful results.

The yield of all roof designs is simulated for four orientations (North, East, South, West). We analyze the influence of the curvature on the yearly yield by simulating a planar cell-pattern with series interconnection at 0° tilt angle.

We additionally simulate the yield of the hood designs with different interconnection topologies. The assembly situation (i.e., tilt angle) of the hood and roof in the vehicle are considered in this study. The simulations of the hood are more complex compared to the roof because the car itself will shade the hood at specific orientations when the elevation of the sun is low. The shading pattern on the hood will change throughout the day and year for northern, eastern, and western orientation which impedes the yield simulations for these orientations. For sake of simplicity and comparability of curvature we only simulate the yield for the southern orientation for all hood designs in this study.

The azimuth and tilt angle of the individual cells were calculated for the center of the cells. The curvature also leads to a current mismatch within a cell which will further reduce the overall yield. Optical and electrical losses were not yet considered in this study.

3 RESULTS

3.1 SUVs and Station Wagons

3.1.1 Averaged Designs

We analyzed freely available blueprints of 14 electric SUVs and station wagons to determine the average length, width, area and curvatures of the roof and hood of this type of car. The average geometries found are summarized in table II. We define the minimum width and length as usable area for PV integration and develop interconnection topologies based on the number of cells fitting in that surface. The varying radii of the surface curvature are summarized in figure 1.

Table II: Average Geometry SUV/ Station Wagons

Surface	Max. length [mm]	Min. length [mm]	Max. width [mm]	Min. width [mm]	Area [m²]
Roof	1905	1818	1206	1052	2.033
Hood	1225	996	1714	878	1.544

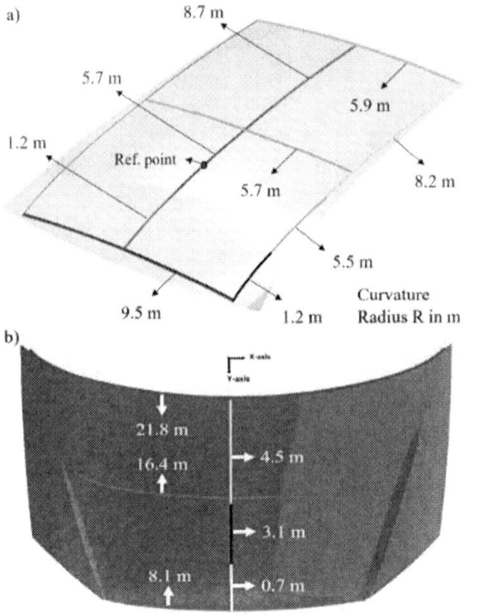

Figure 1: Surface curvature of a) the mean SUV/Station Wagon roof; b) the mean SUV/Station Wagon hood

The reference point indicates the highest point of the roof. The strongest curvature can be found at the front of the module (1.2 m). The roof surface is less curved from front to rear at the rear half of the module (R = 8.2 – 8.7 m) than from left to right (R = 5.7 - 5.9 m).

3.1.2 Cell Interconnection Topologies

With the division of the cell-matrix into subgroups the current multiplies by the number of subgroups, whereas the module voltage gets divided by it.

The maximum number of cells integrable into the averaged roof of this type of car is 120 M6 half cells (12 x 10 cells). Figure 2 displays the investigated SUV roof topologies. The reference topology (a) has 20 in series interconnected cells per string. All strings are in series and three bypass diodes are used. In topology (b) the module is divided into two subgroups (front and back) containing 60 cells in series and one bypass diode per subgroup. The two subgroups are connected in parallel. Topology (c) shows four subgroups interconnected in parallel, containing 30 cells in series each.

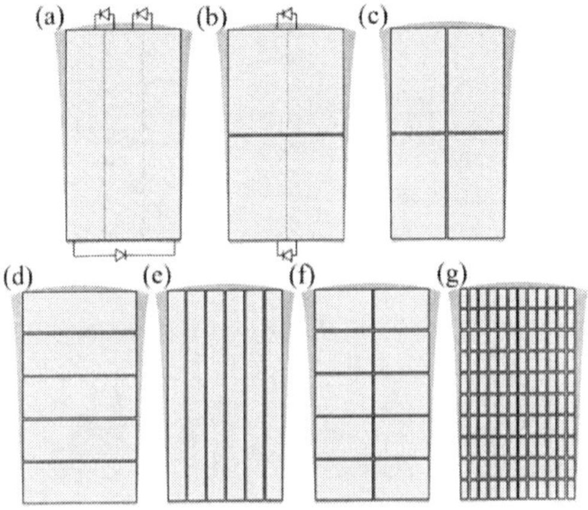

Figure 2: Investigated interconnection topologies for the averaged SUV/station wagon roof

24 half cells are connected in two rows in series along the short side of the module in each subgroup of topology (d). We use topology (e) as comparison to (d) to evaluate the influence of a horizontal and vertical division. Therefore, 20 half cells are connected in series along the long side of the module in topology (e), forming six parallel subgroups. Topology (f) shows ten parallel subgroups with 12 half-cells each. Topology (g) depicts the parallel interconnection of all 120 M6 half-cells which is used to calculate the theoretical maximum yield of this roof design.

The maximum number of cells integrable into the averaged hood of this type of car is 108 M6 half cells.

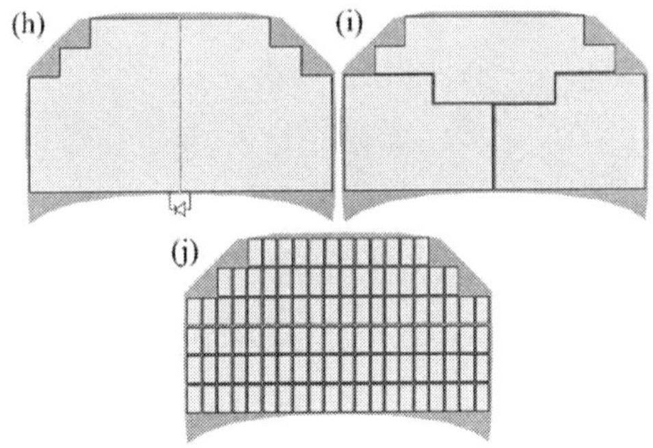

Figure 3: SUV hood topologies

The reference topology of the SUV hood (h) has 108 in series interconnected cells and one bypass diode.

Topology (i) depicts the segmentation into three parallel subgroups and topology (j) parallel interconnection of all cells.

3.1.3 IV- Characteristics

The module parameters at STC for all SUV topologies are summarized in the following table:

Table III Module IV Parameters at STC

Topology	V_{OC} [V]	I_{SC} [A]	V_{MPP} [V]	I_{MPP} [A]	P_{MPP} [W]
Roof 120 M6-half cells					
(a) Series	81.6	5.6	70.8	5.3	376.8
(b) 2 Subs	40.8	11.2	35.4	10.6	376.8
(c) 4 Subs	20.4	22.4	17.7	21.2	376.8
(d) 5 Subs	16.4	27.9	14.3	26.4	376.8
(e) 6 Subs	13.6	33.4	11.9	31.7	376.8
(f) 10 Subs	8.2	55.7	7.1	52.8	376.8
(g) Parallel	0.7	668.4	0.6	663.6	376.8
Hood 108 M6-half cells					
(h) Series	73.4	5.6	63.7	5.3	339.1
(i) 3 Subs	24.5	16.8	21.2	15.8	339.1
(j) Parallel	0.7	601.6	0.6	570.2	339.1

Standard series interconnection with three bypass diodes has the lowest module current. We simulate a topology with parallel interconnection of all cells to determine the theoretical maximum yield. Such a topology would never be used in practice due to the excessively high current. Nevertheless, it supports the evaluation of the developed topologies by defining the upper limit of the yield.

3.1.4 Energy Yield Potential

The energy yield of each roof topology is simulated for the four module orientations. The mean of the four yields is depict in Fig. 4.

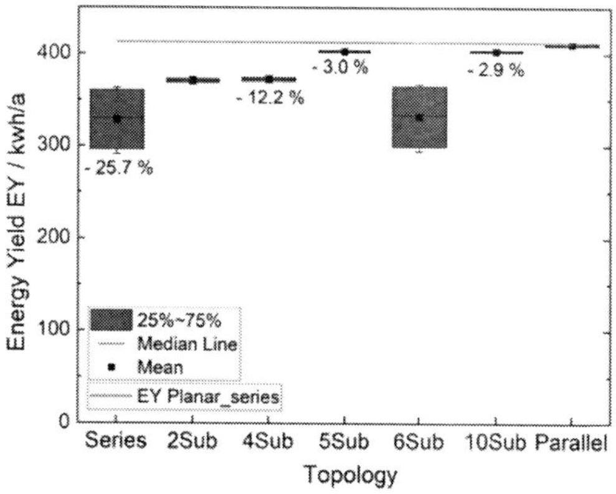

Figure 4: Energy yield potential of the SUV roof topologies compared to a planar module with series interconnection at 0° tilt angle

The graph indicates that the energy yield increases with the number of subgroups. The highest yield of 403 kWh/a is obtained with the half-cell topology with 10 Subgroups. The simulated energy yields range from 328.2 kWh/a (series) to 410.2 kWh/a (parallel). That makes up a difference of 25 %.

The curvature of the SUV roof leads to a decrease in yield of around 25.7 % when using the same topology with series interconnection (a). This loss in yield due to the curvature can be decreased to 2.6 % by adapting the topology to it, e.g., as with topology (f).

Figure 5 depicts the increase in yield compared to the reference topology with series interconnection.

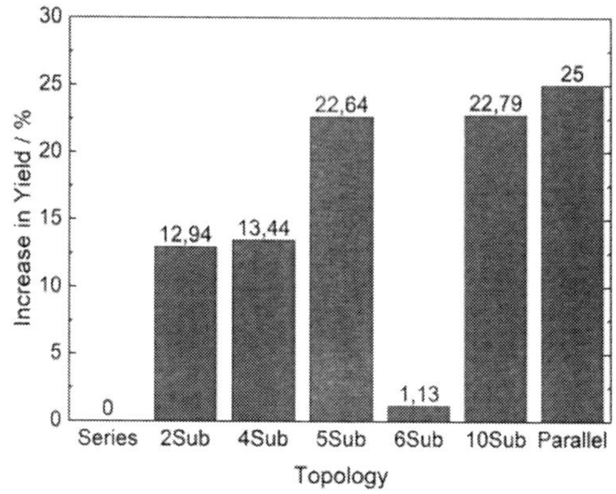

Figure 5: Increase in Yield of SUV roof topologies compared to topology (a)

The topology with four subgroups (c) has a more than 13,4 % higher yield compared to the reference topology (a). Nevertheless, its yield is only 0.5 % higher than for 2 subgroups (b). This is because the integrated bypass diodes in topology (b) have a similar effect as the division in two subgroups. Although the yield is higher for topology (c), already with topology (b) a significant higher yield can be achieved while also operating at reasonable low current leading to lower ohmic losses. Topology (d) with five subgroups increases the yield by 22,64 % in average. When comparing this topology to (e) which only leads to a total increase of 1,1 %, it clearly shows the importance of the subgroup orientation. The cells in topology (e) with six subgroups are connected from front to rear. This means that the difference of the tilt angle of the first and last cell is around 23°.

Since the cells in each subgroup of topology (d) are connected along the short side of the module, the maximum difference of the tilt angle in one subgroup is 8.2° (first subgroup at the front). The mean maximum difference of tilt angles per subgroup of all subgroups of this topology is 3.8°. This leads to the conclusion that the maximum difference of tilt angle per subgroup must be considered in the design the subgroups. The further division of the cell matrix to ten subgroups (f) leads to a yield increase of 0.15 % compared to five subgroups (d) but also to a doubling of the current. This means that the overall best option of the topologies analyzed in this study for this specific roof design is topology (d) and the interconnection should facilitate the ~30 A current.

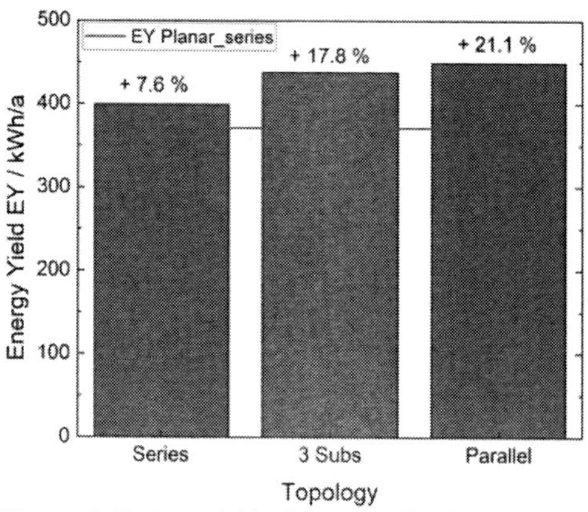

Figure 6: Energy Yield of the SUV hood topologies at southern orientation

Figure 6 indicates that our approach is also very suitable for hoods. Here the segmentation into three subgroups (i) lead to an increase in yield of 9.8 % compared to topology (h). The yields of the simulated topologies (h) to (j) are higher than for the plane cell-pattern because of their more favorable cell tilt angles when facing South.

3.1.5 Influence of Module Orientation

Figure 4 and 5 indicate that the topologies with cells connected from the front to the rear ((a) and (e)) have a higher variation in yield for different orientations.

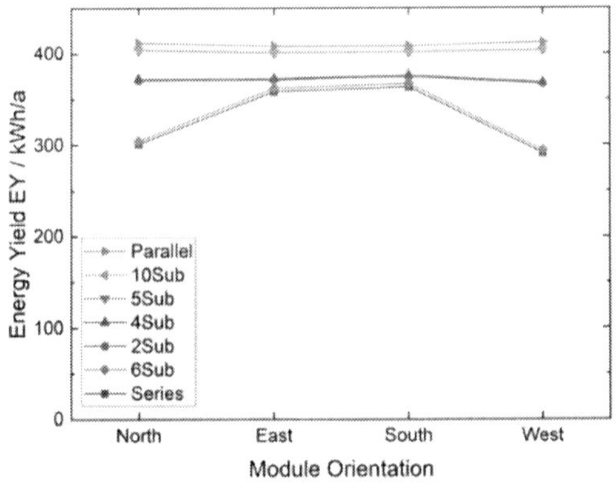

Figure 7: Energy Yield of topologies (a) to (g) with respect to module orientation

Especially the northern and western orientation lead to a reduced yield for these topologies as shown in Fig. 7. The reason of this are the cells at the front of the module. Their high tilt angle of 15.6° leads to a self-shading of the module at sun elevation angles below 15°. Due to the specific topology the self-shading and mismatch influences a large portion of the module and the shaded cells reduce the power output of the full string.

3.2 Limousines

In this part of the study, we further investigate the influence of the module orientation on the energy yield by increasing the number of simulated orientations.

Therefore, we simulate the yield of different topologies for an averaged limousine design for eight module orientations (North, North-East, East, South-East, South, South-West, West and North-West).

3.2.1 Averaged Design

We analyzed blueprints of six electric limousines to generate an averaged design for this type of car. The roof and hood geometries are summarized in table IV. Figure 8 depicts the surface curvatures of the roof.

Table IV: Average Geometry Limousines

Surface	Max. length [mm]	Min. length [mm]	Max. width [mm]	Min. width [mm]	Area [m²]
Roof	1637	1545	1180	1059	1.78
Hood	1340	1110	1643	1085	1.55

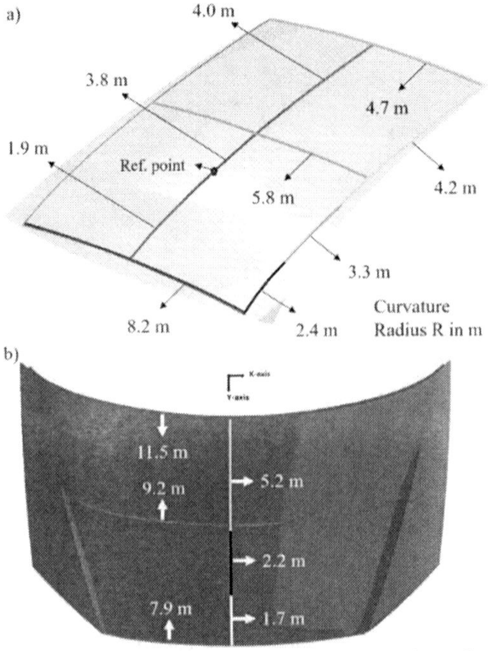

Figure 8: Surface curvature of a) the mean Limousine roof; b) the mean Limousine hood

The roof of this type of car is shorter than the roof of SUVs and station wagons which is why only 108 M6 half-cells fit in that roof. 96 M6-half cells can be integrated in averaged limousine hood.

3.2.2 Cell Interconnection Topologies

For this type of car, we compare a topology with series interconnection of 108 half-cells and three bypass diodes (k) to topologies with three parallel Subgroups containing 36 half-cells connected in series (l) and six parallel Subgroups containing 18 half-cells connected in series (m). A topology with 108 in parallel interconnected cells (n) is simulated as reference stating the maximum possible yield for this design.

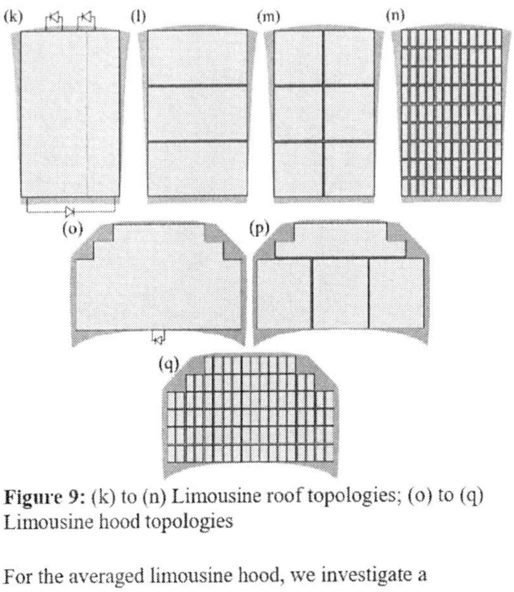

Figure 9: (k) to (n) Limousine roof topologies; (o) to (q) Limousine hood topologies

For the averaged limousine hood, we investigate a

topology with 4 Subgroups with 27 cells in series each (p) and compare the yield to the reference topologies (o) and (q).

3.2.3 IV- Characteristics
The IV-parameters of the developed topologies for the Limousine design are shown in table V.

Table V: IV- Characteristics

Topology	V_{OC} [V]	I_{SC} [A]	V_{MPP} [V]	I_{MPP} [A]	P_{MPP} [W]
Roof (108 M6 half-cells)					
(k) Series	73.4	5.6	63.7	5.3	339.1
(l) 3 Subs	24.5	17.7	21.2	15.8	339.1
(m) 6 Subs	12.2	33.4	10.6	31.6	339.1
(n) Parallel	0.7	601.5	0.6	570.2	339.1
Hood (96 M6 half-cells)					
(o) Series	65.3	5.6	56.6	5.3	301.4
(p) 3 Subs	20.8	16.7	18.9	15.9	301.4
(q) Parallel	0.7	534.7	0.6	506.9	301.4

3.2.4 Energy Yield Potential
Figure 10 shows the results of the energy yield simulations for the topologies (k) to (n). The red line indicates the energy yield of topology (k) as plane cell-pattern with a module tilt angle of 0°. The figure shows that the curvature of the limousine roof leads to a decrease in yield of 20 % when using the same topology (k).

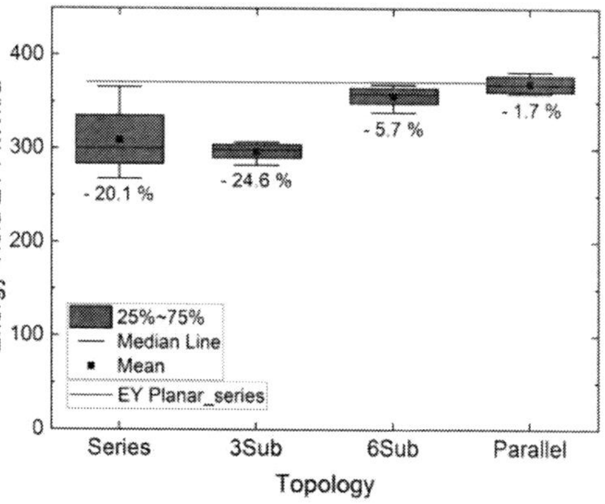

Figure 10: Energy Yield of topologies (k) to (n) for eight module orientations compared to the planar cell pattern with series interconnection

The calculated maximum yield for the limousine roof design with M6 half-cells is 369.2 kWh/a thus 18.5 % higher than for the topology with series interconnection. The results for topology (l) show that a random division of the module into subgroups will not always lead to an increase in yield. The yield is 5 % below the one with series interconnection which is 300.1 kWh/a. The reason for that is investigated below. We therefore conclude that the tilt and azimuth angles of cells within a subgroup need to be analyzed in detail to develop efficient topology designs using subgroups with parallel interconnection. The subgroups must be formed so that the maximum difference of tilt and azimuth angle are minimized for all subgroups.
However, topology (m) with six subgroups increases the yield by 14.5 % compared to the reference (k) which underpins the effectivity of this approach.
Figure 11 shows the results of the energy yield simulations for the Limousine hood designs. Here the segmentation of the cell pattern to four subgroups lead to an increase in yield of 10.4 % compared to topology (o). 377.8 kWh per year can be generated with this topology. The graph also indicates that its yield is quite close to the calculated optimum for this hood design (q), so no further segmentation would lead to any further significant yield benefit.

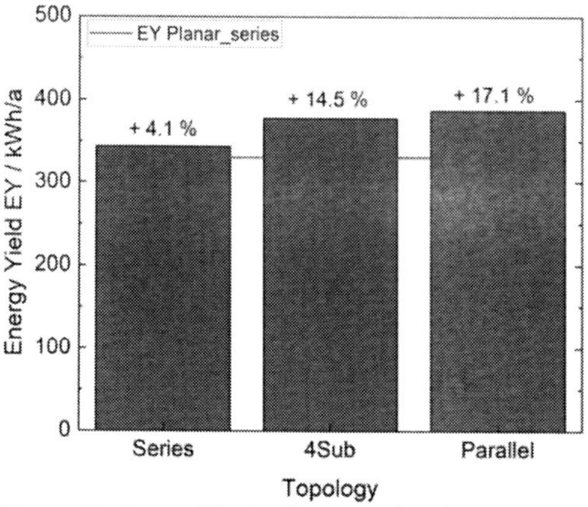

Figure 11: Energy Yield of limousine hood topologies (o) to (q) at southern orientation compared to the EY of a planar module with 0° module tilt angle

3.2.5 Influence of Module Orientation
We analyze the influence of the module orientation by additionally simulating the yield potential of the limousine roof PV module for north-eastern, south-eastern, south-western, and north-western orientation. The results are shown in figure 12.

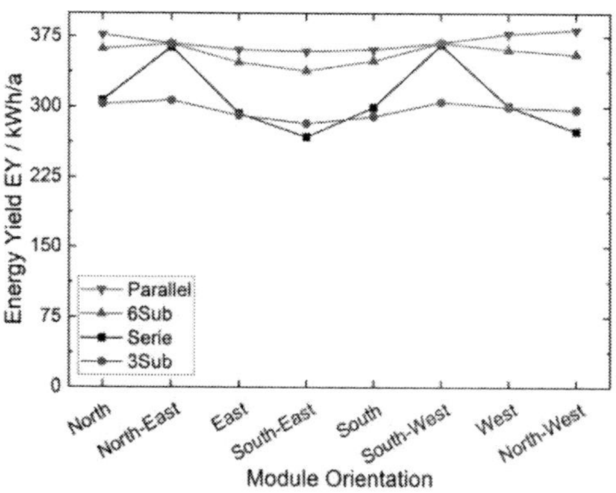

Figure 12: Energy yield of topologies (k) to (n) for eight module orientations

The resulting graph displays that for three topologies the highest yield can be achieved at north-eastern and south-western module orientation. The reason for this is the sun trajectory over the year in Freiburg. The average azimuth angle of the sun is 178° in this location, between South and South-West. When the curved VIPV module is facing North-East (315°) or South-West (135°) it will be illuminated more homogeneously over the module surface over the year which leads to an overall higher yield. This is especially relevant for the series interconnection, were a significant benefit for those respective orientations also

lead to a higher mean yield than for topology (I) with three submodules. It seems that for those two orientations the cross axis of the module is more relevant than the lengthwise axis. Therefore, the series interconnection which is less influenced by irradiation variations across the cross axis will have less mismatch than the topology (I) which is very susceptible to irradiation mismatch in the cross axis.
The yield variance between the different orientations is higher when simulating eight module orientations as shown in Fig. 12. This indicates that all orientations (360°) should be simulated for a precise yield prediction of VIPV modules. Nevertheless, the mean of all orientations is almost the same as the mean of the four main orientations. The difference of the mean EY is up to 2 %. This leads us to the conclusion that the simulation of the main orientations (North, East, South, West) allows a rough estimation of the Yield which furthermore keeps the computing time of the simulations at a reasonable length.

4 CONCLUSIONS

In this study we demonstrate the influence of the interconnection topology of curved VIPV modules on the yearly energy yield. The proposed segmentation of the VIPV module into subgroups proves to be a proper solution to reduce losses that occur due to the electric mismatch between cells in curved PV modules. We achieve a loss compared to a horizontal reference of less than 3 % for SUV or station wagon roofs and less than 6 % for limousines considering a reasonable roof layout. Table VIII displays the advantage of the interconnection topology adaptation in terms of solar range extension. The reference EY is the sum of the hood and roof EY of the reference topologies with series interconnection. The yield of the optimal roof and hood topologies display the best-case scenario. We assume no shading and a DC performance ratio of 88 %.

Table VIII: Solar Range

Car type	Average Consumption [Wh/km]	EY Ref. [kWh/a]	EY Best-case [kWh/a]	Solar Range Ref. [km/a]	Solar Range Best-case [km/a]
SUV	189.5	727.5	841.0	3378	3905
Limousine	185.5	643.6	732.7	3053	3476

The here presented approach leads to an extension of the solar driving range of 16 % for the SUV and 14 % for the Limousine design compared to series interconnection of all cells.
We show that the yield increases with the number of subgroups. High numbers of subgroups can be used when using small cell formats like half and shingle cells because of their lower currents. The presented approach is especially suitable for shingle modules.
We further demonstrate the importance of the correct design of the interconnection topology and the subgroups. The cells in one subgroup should have similar azimuth and elevation angles towards the sun. The curved module therefore should be segmented in that way that the maximum difference of the tilt in azimuth angles of cells in one subgroup are minimized for all subgroups.
The results of this study show that yield simulations should be performed for all orientations (360°) but the simulation of the four main orientations already allows a rough estimation of the yield of VIPV modules. All in all, our model supports the design, evaluation, and comparison of interconnection topologies for curved (VIPV) modules.

5 ACKNOWLEDGEMENTS

The authors thank all colleagues from Fraunhofer ISE who contributed to this work. This work was supported by the German Federal Ministry for Economic Affairs and Energy (BMWi) under the contract number 03EE1085A, acronym 3D.

6 REFERENCES

[1] ITRPV, "International Technology Roadmap for Photovoltaic (ITRPV): 2020 Results, April 2021," VDMA Photovoltaic Equipment, Frankfurt 12, Apr. 2021. Accessed: May 10 2021.

[2] Martin Heinrich, Tilmann E. Kuhn, and Frank Dimroth, "A Comparison of Different Solar Cell Technologies for Integrated Photovoltaics," 37th European Photovoltaic Solar Energy Conference and Exhibition.

[3] C. Kutter, F. Basler, M. Heinrich, and D. H. Neuhaus, "Integrated lightweight, glass-free PV module technology for box bodies of commercial trucks," in.

[4] U. Eitner et al., "Solar Potential on Commerical Trucks: Results of an Irradiance Measurement Campaign on 6 Trucks in Europe and USA," in Proceedings of the 33rd European Photovoltaic Solar Energy Conference and Exhibition (EU PVSEC), Amsterdam, Netherlands, 2017, pp. 2147–2150.

[5] Hauke Engel, Russell Hensley, Stefan Knupfer, and Shivika Sahdev, "The potential impact of electric vehicles on global energy systems," in Energy Insights by McKinsey Global Energy Perspective. [Online]. Available: https://www.mckinsey.com/industries/automotive-and-assembly/our-insights/the-potential-impact-of-electric-vehicles-on-global-energy-systems

[6] M. Heinrich et al., "Potential and challenges of vehicle integrated photovoltaics for passenger cars," in.

[7] K. Araki, L. Ji, G. Kelly, and M. Yamaguchi, "To Do List for Research and Development and International Standardization to Achieve the Goal of Running a Majority of Electric Vehicles on Solar Energy," Coatings, vol. 8, no. 7, p. 251, 2018, doi: 10.3390/coatings8070251.

[8] Tatiana Duigou, Vincent Boichon, and Fabien Chabuel, "VIPV: Process Development of Integrated Photovoltaic Cells in a Double-Curved Composite Structure for Automotive Application," in 37th European Photovoltaic Solar Energy Conference and Exhibition.

[9] Fraunhofer Institute for Solar Energy Systems, ISE, "Fraunhofer ISE präsentiert farbiges Solar-Autodach auf der IAA in Frankfurt: Presseinformation," 2019.

[10] Neven-du Mont et al., "Energy Yield Modelling of 2D and 3D Curved Photovoltaic Modules," 37th Eurpean Photovoltaic Solar Energy Conference (EUPVSEC), 2020.
[11] Javier Macías, Rebeca Herrero, Ignacio Antón, "Evaluation of the Solar Resource and Energy and Energy Generation in Vehicle Integrated Photovoltaics," 37th European Photovoltaic Solar Energy Conference and Exhibition.
[12] P. Baliozian et al., "PERC-based shingled solar cells and modules at Fraunhofer ISE," Photovoltaics International, no. 43, pp. 129–145, 2019.

STRATEGIES FOR THE ANALISIS OF SHADING EFFECTS IN VEHICLE INTEGRATED PHOTOVOLTAICS

Javier Macias, Rebeca Herrero, Rubén Núñez, Ignacio Antón
Instituto de Energía Solar, Universidad Politécnica de Madrid
Madrid, Spain, javier.macias@ies.upm.es

ABSTRACT: Vehicle Integrated Photovoltaics seeks, through solar energy, to increase the autonomy of Electric Vehicle from a renewable and clean source. To know the viability of this technology, it is necessary to develop new studies and tools. Considering the important impact that the effect of shading has on the energy production of photovoltaic surfaces, this paper aims to propose strategies to address the study and its application in VIPV technology. Due to the dynamic behaviour of an Electric Vehicle and the variability of usages it may present, it is necessary to define reference scenarios to approach the study. Thanks to this work, it will be possible to know the solar resource on the surface of an Electric Vehicle in three different shading scenarios in order to, in future studies, analyse the increase in the autonomy of an Electric Vehicle over long road distances, to know the solar resource and energy generation of a fleet of vehicles parked around an office building during a working day or to estimate optimal city routes in terms of energy generation.
Keywords: Photovoltaics, Electric Vehicle, Shading, Modelling.

1 INTRODUCTION

In Spain, around 29% of total greenhouse gas emissions are generated by the transport sector, with road transport being the largest contributor to these emissions, accounting for 95% of the total [1]. Therefore, new technologies such as renewable energies or electric vehicles have emerged with the aim of reducing these emissions and, consequently, the effects of climate change.

Vehicle Integrated Photovoltaics (VIPV) technology emerges as a point of union of both technologies, aiming to cover part or all of the energy demand required by an electric vehicle (EV) through the installation of photovoltaic (PV) cells on its surface.

The characteristics of this technology (curvature of the EV surfaces, motion or shading) require new modelling tools to study the feasibility of VIPV technology, which differ from those already conventionally used in the PV sector.

The work presented is framed within the identified series of technological challenges to address for VIPV [2], related to the modelling and energy prediction of photovoltaic surfaces in vehicles.

At this point, taking into account that the effect of shading in PV has a very relevant impact on energy generation, this paper focuses on the study of new strategies for the analysis of shading on the different surfaces of the EV along its trajectory. Through the development of these scenarios and tools, future studies aim to answer questions such as:

- What would be the increase in autonomy of an EV travelling on long road trips?
- What is the available solar resource and energy generation of a fleet of vehicles parked around an office building?
- Could optimal urban routes be estimated in terms of energy generation?

2 MODELLING ENERGY FOR VIPV

To evaluate the impact that VIPV technology would have on the propulsion of EVs, it is necessary to develop energy modeling tools to determine firstly, the solar resource on the car-roof in different places and operating points and secondly the energy generated by the PV surfaces.

A new tool that uses Photovoltaic Geographical Information System (PVGIS) as irradiation database and Python library pvlib to simulate irradiance maps on PV car-roof and its generated power, has been developed for this purpose [3, 4].

Once the solar resource available at the PV surface is known, it is possible to calculate the electrical generation considering, among other elements, the type of photovoltaic technology, the operating temperature and irradiance and the electrical connection between solar cells.

There are different PV configurations commonly used for the cells of a solar module. The electrical interconnection of PV cells (series and parallel) as well as the number of cells and bypass diodes will determine the generated power at a given instant. The effect each configuration has in the generated energy will depend on the irradiance patterns over the surface and has been analysed in previous studies [5].

Once the solar resource on the vehicle surface is known for the different scenarios including shading, it would be possible to answer the following question using the developed tool, for example:

- Estimation of the available solar resource in the different scenarios.
- Analysis of the optimal connection according to each scenario.
- Development of an algorithm to optimize the vehicle's energy production in each case.

3 STRATEGIES FOR THE ANALISIS OF SHADING EFFECTS IN VIPV

Considering the different positions that an EV may have along its trajectory, the shadows that the urban landscape casts on the PV surface of the vehicle will be different at each instant, introducing complexity in the study. Therefore, for the analysis of shading in VIPV technology, it is necessary to develop new tools that also differ from those conventionally used for the study of shading effects and to define scenarios in order to approach the study. We propose three different scenarios:

3.1 Vehicle on road:

The first scenario relates the usage of a vehicle driving on the road, where the effect of shading due to the urban landscape does not seem to be relevant. For this reason, the study of energy generation is approached following a clear sky model without shadows, as already presented in the previous section, but considering the different positions of the EV along its trajectory.

To do this, we use a Routing Machine tool for Python available in Open Street Maps (OSRM) [6]. Thanks to this tool we obtain the different positions of the vehicle in the nodes of the trajectory (Figure 1):

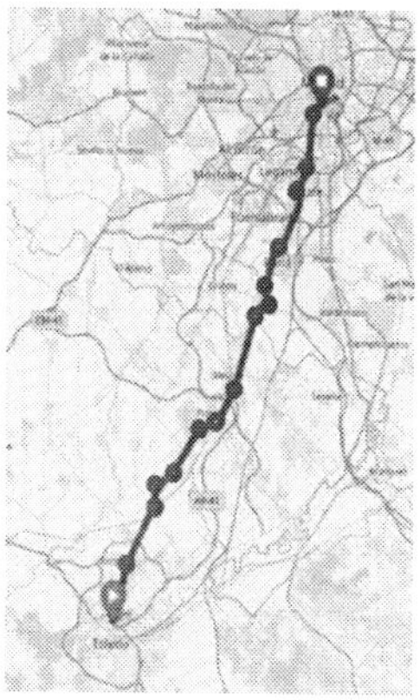

Figure 1: EV route from Toledo to Madrid. (Source: Open Street Maps Routing Machine).

Additionally, it will be necessary to calculate the orientation of the vehicle, its speed and the instant at each point along the trajectory. These parameters will affect to the irradiance map on the EV surfaces or to the thermal dissipation of the module and thus to its performance.

3.2 Vehicle parked:

The second one, would correspond to a scenario with a parked vehicle in which the shadows of the urban landscape will play a fundamental role in energy production.

This scenario proposes the study of the shadow maps cast by a typical building throughout the day, corresponding to the typical case of an EV parked around an office building throughout the working day. Following this method, we will be able to know the solar resource available on the surfaces of the EV.

In order to obtain a high density of points/cm^2 and therefore a good resolution of the projected shadow map, the reconstruction of a typical office building as shown in Figure 2 is proposed:

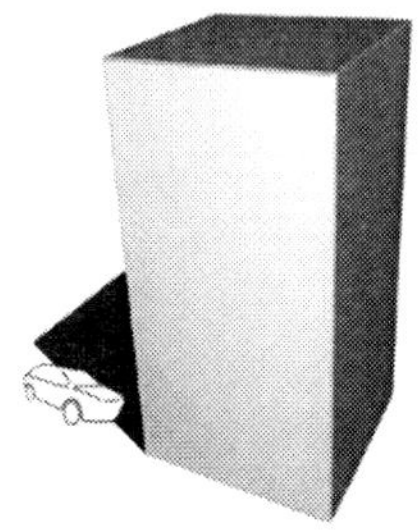

Figure 2: Simulated shadow map of an office building on 01/06/2018 at 12:00h in Madrid.

Finally, once the solar resource is known throughout the day, using the tool detailed in section 2, we can know the PV generation of an EV using VIPV technology and then extend the study to a fleet of vehicles parked in the car park. The following figure shows the irradiance map generated by the tool for the shadow map shown in Figure 2:

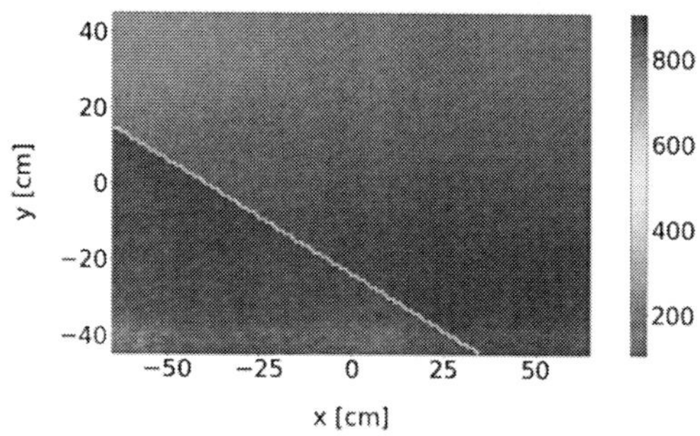

Figure 3: Irradiance map [W/m^2] for the shading conditions in Figure 2.

As we can see in the Figure 3, thanks to this methodology we can obtain partial shading simulations with a high resolution.

In this way, using the irradiance maps generated on the EV surface, we can estimate the PV generation and then extend the study to a fleet of vehicles parked around the office building during a typical working day.

3.3 Vehicle in the city:

In the third scenario, based on usage statistics (path length, peak hours, residential and working areas, etc.), different vehicle trajectories will be simulated between two points in a city for a given trip length, following the next two methodologies:

LIDAR data:

If the vehicle is moving or parked in a city, the dynamic shading casted by the urban landscape becomes relevant and, therefore, a Shadow Mapping alternative is proposed. Using the LIDAR data corresponding to the location of the vehicle and the solar resource along its trajectory, irradiance maps on the different surfaces of an EV can be simulated, including shadows.

For this purpose, an additional shading module has been developed for the tool discussed in section 2. First, the LIDAR data of the city to be studied (Madrid in this case) are extracted from the Centro Nacional de Información Geográfica [7]. Figure 4 shows a representation of the LIDAR data corresponding to the Chamberí district in Madrid:

Figure 4: LIDAR data represented using the tool for the district of Chamberí, Madrid. (Source: Centro Nacional de Información Geográfica. Ministerio de Transportes, Movilidad y Agenda Urbana - España).

Subsequently, depending on the position of the EV, a reconstruction of the adjacent buildings is carried out to obtain a higher density of points/cm^2. Figure 5 shows the before and after reconstruction:

Figure 5: Reconstruction of the LIDAR data represented using the tool for the district of Chamberí, Madrid. (Source: Centro Nacional de Información Geográfica. Ministerio de Transportes, Movilidad y Agenda Urbana - España).

Finally, knowing the position of the sun at each moment of the vehicle's trajectory, the shadows casted by the urban landscape are projected onto the plane using ray tracing methods (Figure 6):

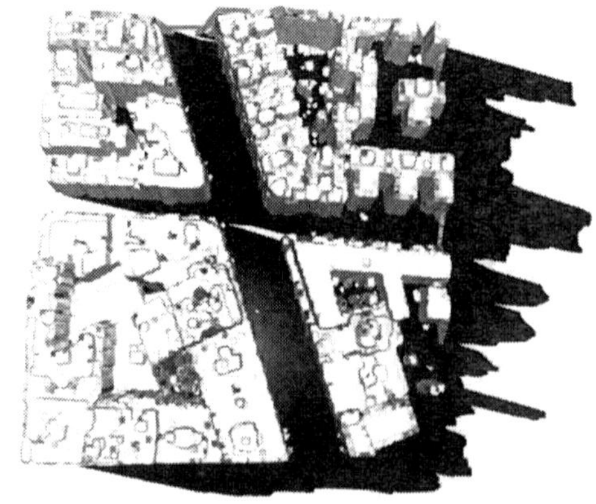

Figure 6: Shadow map of the LIDAR data represented using the tool for the district of Chamberí, Madrid.

The method presented performs ray tracing from the position of the sun to the object that casts shadows on the plane. Each point on the generated map has its relative coordinates associated, so we can relate the coordinates defining the EV trajectory to the generated shadow map. In this way, it is possible to know whether the vehicle is crossing a shaded area or not.

In positions where the vehicle surface is shaded, the direct irradiance will be zero, obtaining only diffuse irradiance as a solar resource. However, the diffuse irradiance will also depend on the refraction produced by the urban streetscape elements.

Google Open Street View:

The reverse ray tracing process can also be applied, starting from the PV surface of the vehicle to the position of the sun, so that it can be analysed whether there are intermediate objects that produce shadows or not.

For this purpose, different methodologies can be used to extract the sky images. The simulations carried out in this paper have been performed using the Google Street View API for Python. With this tool, depending on the latitude, longitude, orientation and date of the year, we can download images of the sky from the position where the vehicle is located (Figure 7):

Figure 7: Different images of the sky in Madrid. (Source: Google Street View).

Subsequently, using these images as a reference, different types of filtering are carried out in order to distinguish between sky and urban landscape objects (Figure 8):

Figure 8: Image of the sky in Madrid (left), and filtered images (center and right). (Source: Google Street View).

Thanks to this processing, it is possible to know if there are objects of the urban landscape in the line that joins the vehicle with the sun, being able to know if the EV would be receiving shading or not [8].

Finally, by extracting the different images obtained for the same position of the vehicle (from east to west at 45° intervals), we can see the complete solar path in one day (Figure 9):

Figure 9: Solar path from Plaza del Emperador Carlos V, Madrid. (Source: Google Street View).

4 CONCLUSIONS

Considering that the effect of shading on photovoltaic surfaces is a determining factor in energy production, this work proposes different strategies for its study in VIPV technology.

Taking into account the wide variety of use cases that can be found in an EV, it is necessary to consider different scenarios to approach the study. In this paper, three scenarios have been proposed to include most of the EV use cases.

Regarding the first scenario, vehicle on road, it is considered that shading effects will not significantly affect the available solar resource and therefore it has been proposed the study of the PV generation along the EV trajectory for a clear sky model without shading. Long road trips can be simulated to estimate the increase in autonomy thanks to VIPV technology.

The second scenario proposes the simulation of the map of shadows cast by a typical office building throughout the day to determine the solar resource on the surfaces of an EV parked in the adjacent car park.

Finally, in the third scenario, the study of shading for different trajectories of an EV driving in a city is proposed. At this point, the shadows cast by the urban landscape are determinant and therefore two methods based on ray tracing have been exposed. Firstly, a method based on LIDAR data is considered in which we obtain the map of shadows on the plane and secondly, thanks to Google Street View API for Python, we extract images of the sky to find out if any object is in the trajectory between the EV and the sun.

Subsequently, using the developed tool, it will be possible to know the solar resource available on the different surfaces of the EV considering shading effects, find the interconnection that maximizes energy generation or develop algorithms that allow us to optimize the energy production in each scenario. The idea is not only to develop this study for a single vehicle, but to extend it in the future for a large fleet of vehicles to shed some light on this technology.

5 ACKNOWLEDGEMENTS

This work has been supported by the project FEVEL reference RTI2018-096210-A-I00 financed by the Fondo Europeo de Desarrollo Regional (FEDER) and the Spanish Ministerio de Ciencia e Innovación and Agencia Estatal de Investigación. We acknowledge to open source libraries PVLIB, OSRM, PVGis Database and Centro Nacional de Información Geográfica for their contribution to this project.

6 REFERENCES

[1] «Avance del inventario de Emisiones GEI». Gobierno de España – Presidencia del Gobierno (2020). https://www.lamoncloa.gob.es/serviciosdeprensa/notasprensa/transicion-ecologica/Paginas/2020/080720-emisiones.aspx

[2] «To Do List for Research and Development and International Standardization to Achieve the Goal of Running a Majority of Electric Vehicles on Solar Energy». K. Araki, L. Ji, G. Kelly and M. Yamaguchi, Coatings, vol. 8, n.o 7, p. 251 (2018).

[3] «Evaluation of the Solar Resource and Energy Generation in Vehicle Integrated Photovoltaics». J. Macías, R. Herrero, I. Antón and R. Núñez, EU PVSEC Conference Proceedings (2020).

[4] «PVLIB Python: a Python package for modeling solar energy systems». W. Holmgren, C. Hansen and M. Mikofski, Journal of Open Source Software, 3(29), 884 (2018).

[5] «On the effect of cell interconnection in Vehicle Integrated Photovoltaics: modelling energy under different scenarios». J. Macías, R. Herrero, I. Antón and R. Núñez, IEEE Conference Proceedings (2021).

[6] Real-time routing with OpenStreetMap data, Luxen, Dennis and Vetter, Christian (2011).

[7] Centro Nacional de Información Geográfica. Ministerio de Transportes, Movilidad y Agenda Urbana – España. https://www.cnig.es/home

[8] «Evaluación del recurso solar en un coche eléctrico – fotovoltaico en entornos urbanos». A. Muñoz and R. Núñez (2021).

OPENING UP NEW LAND RESOURCES FOR VERTICAL BIFACIAL SOLAR MODULES USING A NEW "NATURE-CONSERVING AGRIVOLTAICS" CONCEPT

Nadine Pannicke-Prochnow[1], Jens Schneider[2], Christoph Gerhards[3], Raffael Stretz[4] and Benjamin Volz[5]
[1] Nadine Pannicke-Prochnow, Helmholtz-Centre for Environmental Research – UFZ Leipzig, Permoser Str. 15, 04318 Leipzig, Germany, phone: +49 341 2351680, email: nadine.pannicke-prochnow@ufz.de
[2] Leipzig University of Applied Science Leipzig
[3] Fraunhofer Center for International Management and Knowledge Economy IMW, Email: christoph.gerhards@imw.fraunhofer.de
[4] Helmholtz-Centre for Environmental Research - UFZ Leipzig
[5] Next2Sun GmbH

ABSTRACT: Besides climate protection, sustainable development needs increasing endeavors regarding biodiversity and nature protection as well as climate and sustainability adaptation in agriculture and concepts for integrated and multifunctional land use, addressing increasing land use competitions. To address these challenges simultaneously, we present the Agri4Power concept as a new "nature-conserving agrivoltaics" concept. It is characterized by a combination of agricultural land use (including energy plants) with sustainable electricity production through bifacial vertical photovoltaic (PV) module systems and nature and species conservation through flower strips. The concept is in line with agricultural production, may stabilize income for farmers and has the potential to contribute to the further expansion of photovoltaics at higher electricity market values. The Agri4Power concept presents socially supported opportunities to make an important contribution to energy sector transition and to improve biodiversity protection and ecosystem services in agricultural areas. Due to its considerable potentials in various applications, the Agri4Power concept may open up new land resources for renewable energy expansion.
Keywords: Bifacial, Agrivoltaics, Biodiversity, Sustainable, Environmental Effect

1 CHALLENGES FOR SUSTAINABLE DEVELOPMENT

Increasing impacts of climate change show the urgent need for a significant boost and acceleration of climate protection measures. Above all, the transition of the energy system and the rapid and sustainable expansion of renewable energy sources (RES) is an urgent need [1]. But further expansion is stagnating [2] and faces several problems. Thus, energy transition is stagnating for various reasons, for instance:

1) The land use competition for RES installations impedes the expansion of PV installations on agricultural land as this should be kept for agriculture to produce food, feed, fibre and fuels. 2) As a consequence, legal restrictions and subsidies from the German Renewable Energy Law (EEG) limit potential sites for RES installations, e. g. to marginal agricultural land or agricultural land along rail lines and motorways (200 m width). 3) A lack of societal acceptance for land use changes for RES installations, especially wind turbines, but also conventional open-field PV installations (solar parks), may hinder or delay RES projects [3, 4].

Simultaneously, sustainable development is challenged by further difficulties, particularly:

1) Besides climate change, biodiversity loss causes significant ecosystem service depletions and thus, large and increasing economic damages. A sustainable development demands the protection of biodiversity and nature in general. 2) The impacts of climate change and biodiversity loss cause yield and income insecurities for farmers. This may hinder sustainable agricultural production and hence, good nutrition (e. g due to an increased application of artificial fertilizers and pesticides). This demands for both, more climate adaptation and sustainability in agricultural practice and for an income diversification for farmers for financial hedging. 3) Increasing land use pressure, e. g. for housing and living, traffic and industry, agricultural food, feed and fuel production, renewable energies and biodiversity and nature protection cause growing land use conflicts [5]. Decreasing land use competitions demands for multifunctional land use concepts, which integrate various interests on the utilization of certain areas.

This calls for a massive expansion of RES, which must be in line with nature conservation goals, and productive and sustainable agriculture and which gains societal acceptance. In order to support this objective, we present the Agri4Power concept as a novel "nature conserving agrivoltaics" concept to address today`s major challenges climate protection and biodiversity protection as well as climate adaptation in agriculture simultaneously by integrating various land use interest in a multifunctional land use concept (fig. 1).

Figure 1: Schematic illustration of the Agri4Power concept (© Fraunhofer IMW)

The integrative Agri4Power concept combines agricultural land use (including energy plants) with sustainable electricity production through bifacial vertical photovoltaic (PV) module systems and nature and species conservation through flower strips. The division of agricultural land into various strips results from the vertical mounting of the PV modules. The strips may be used for agriculture (including energy plants) and/or as flower strips.

A detailed explanation of the Agri4power concept follows in section 2. In section 3, the advantages and in section 4, current challenges of the Agri4Power concept will be outlined. Section 5 closes with some concluding remarks.

2 ADRESSING CURRENT SUSTAINABILITY CHALLENGES USING "NATURE-PROTECTING AGRIVOLTAICS"

After a short introduction to the Agri4Power concept, the concept will be compared to similar land use approaches.

2.1 The Agri4Power concept briefly explained

The Agri4Power concept provides a modular kit consisting of the four elements 1) vertically mounted bifacial PV-modules, 2) wind turbines, 3) flower strips for biodiversity protection and 4) agriculture.

The choice of the share of each of these four elements is quite flexible. Hence, based on these elements, the concept may be compiled individually. This allows for various versions and individual modifications of the Agri4Power concept in accordance with specific site conditions and requirements of the user. Figure 2 shows a selection of versions of the Agri4Power concept, which were elaborated in workshops with farmers and other stakeholders.

Currently, version 1 is favored most as it combines greatest flexibility (agricultural and flower strips are interchangeable, crops in different rows diversifiable) with high economic gains (PV module row distance of 10 m) and biodiversity effects (flower strips > 6 m).

Version 2 may create lower economic gains as the distance between the PV module rows is quite large, when the recommendations for the single elements of the Agri4Power concept are perfectly obeyed. For instance, a distance of about 14 m PV module row distance results from a minimum width of the flower strips of about 6 m to ensure sufficient ecological effects plus a width of about 8 m for the agricultural strips to provide a working width suitable for agricultural machinery. The highest economic gains from energy production need a PV module row distance of about 8 - 12 m to prevent the modules from shading one another and to avoid large distances and cable lengths to the mains feed-in point.

In comparison, version 3 provides specific ecological valuable areas on the cropland by using the PV module rows for separating the flower strips from the agricultural area. Hence, optimal PV module row distance and flower strips widths are possible. Additionally, this may reduce pesticide inputs from agriculture to the flower strips.

In version 4, the agricultural area is divided into an ecologically valuable flowered area including the PV module rows and an area of arable land, where farmers may continue their previous agricultural practice, e. g. without the need to adapt agricultural equipment to lower widths. This version might be appropriate for the use as an agri-environmental measure to protect biodiversity, which may finance itself due to the economic gains from energy production. But for this case of application, legal security would be helpful.

Similarly, the integration of wind energy into the concept is facing some hurdles under current conditions in Germany, mainly due to bureaucratic and technical obstacles (for details see section 4.3).

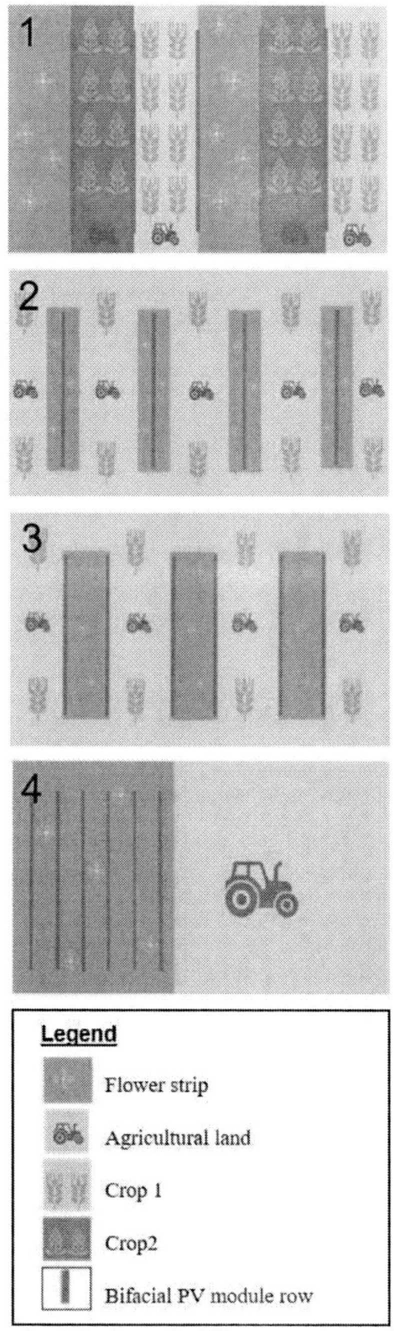

Figure 2: Selection of different versions of the Agri4Power concept elaborated in workshops with farmers and other stakeholders. Version 1 is favored most. (© DBFZ)

2.2 The Agri4Power concept in comparison to similar approaches

To address some of the sustainability challenges outlined in section 1, agrivoltaics emerged as a combination of RES and agriculture, e. g. using elevated PV modules roofing agricultural land. As a first integration of PV modules in agriculture, land use conflicts may be addressed. Ideally, synergies for agricultural production arise, e. g., a protection of fruits from heavy rain or hail due to roofing [6]. Thus, a contribution to climate adaptation in agriculture and climate protection may be done simultaneously.

As a second improvement, "Nature-conserving Agrivoltaics" concepts, such as the Agri4Power concept, additionally integrate biodiversity and nature protection into agrivoltaics. Resulting ecosystem services and impact of the vertically mounted PV modules may increase positive effects on agriculture. Multifunctionality and the integral nature of the land use concept occur as a response to increasing land use pressure and decreasing land availability for RES deployment.

Just like other PV systems or agrivoltaic concepts, the Agri4Power concept is not suitable for all sites, but rather has its niches for application (see section 4.2). As outlined before (see section 2.1), different versions and modifications of the concept create different energy revenues, agricultural yields and biodiversity effects.

A comparison of energy revenues from different energy land use options is shown in figure 3, referring to the area used for energy plants in Germany (1,55 Mio. ha of land in 2020 [7]) providing 47 TWh of bioenergy generated (dark green circle) [8]. Assuming an implementation of the Agri4Power concept on this area (light green circle), 90% of biomass generation for flexible energy production would be preserved, but an additional 488 TWh of solar energy could be generated. If 50 % of biomass production on this area are preserved, PV could generate 1473 TWh/a (@ 2 MW/ha and 950 Wh/Wp) on the other 50% of the area, when the PV area is optimized for solar energy production (dark blue circle). When the PV area is designed for additional biodiversity protection, the solar energy yield is 775 TWh/a (@ 1 MW/ha and 1000 Wh/Wp) (light blue circle). Finding the optimum mix for spatial and technical design and application of the concept, including acceptance and opportunities in the political and legal framework, especially when considering the nature protection aspects, is an important task.

The generation profile of vertical PV was implemented in a HTW energy model using the software EnergyPLAN. The model describes Germany in the sectors electricity, heat and mobility with a total of 82 % renewable energy in electricity and 66 % renewable energy in the overall system. 400 GWP PV and 150 GWp Wind is used to provide electricity, beside storage facilities of 15 TWh capacity and 80 GW peak power. Additional gas turbines using natural gas or hydrogen ensure security of supply. The beneficial generation profile of PV plants with vertical modules leads to a lower usage of natural gas and a saving of 5 Mt CO_2 emissions, half of PV capacity is build with vertical mounted bifacial modules. This corresponds to twice the amount of Germany's air traffic.

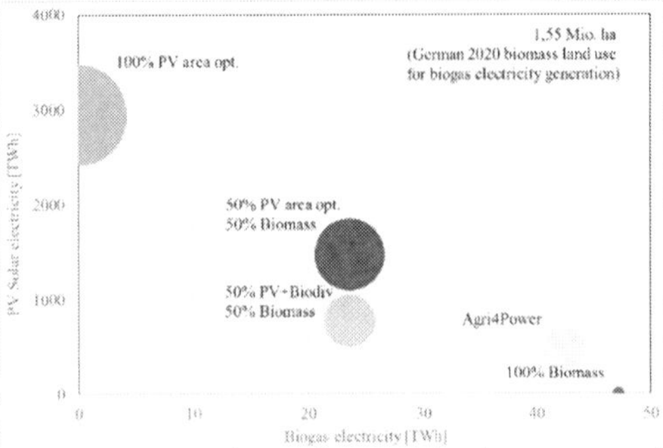

Figure 3: Comparison of potential solar energy and bioenergy yields in different land use scenarios with a share of 100 % (yellow), 50 % (blue), 10 % (light green) and 0 % PV area (dark green). The dark blue circle shows energy yield of a PV area optimized for solar energy production, the light blue circle represents the energy yield of a PV area designed for biodiversity protection. The size of all circles indicates the total energy yield from PV and bioenergy by using 1,55 Mio ha land for electricity production. (© Fraunhofer IMW).

3 POSITIVE EFFECTS OF THE AGRI4POWER CONCEPT

Firstly, the general benefits of the Agri4Power concept from a PV perspective as well from an agricultural and a biodiversity perspective will be outlined. Accordingly, benefits of the Agri4Power concept increase societal approval to the concept and may increase the acceptance of RES installations. Overall economic benefits complement the positive outcomes of the land use concept.

3.1 Possible benefits of the Agri4Power concept in general

From a PV perspective, the concept provides an application for now state of the art bifacial solar modules using both sides of the PV modules directly for energy generation. Due to vertically mounting of bifacial PV modules in east-west orientation, energy production shows peeks in morning and afternoon hours / Wp. As European electricity market value also has peaks in the morning and evening, the overall financial yield can be higher. Higher market revenues may initiate further investments in PV installations and thus, may contribute to climate protection goals.

From an agricultural perspective, the concept enables the continuation of agricultural production on an area, which is used for energy production as this is the underlying idea in all agrivoltaic concepts (for a comparison of the Agri4Power concept to similar approaches, see section 3.4). Energy production may generate additional revenues and increasing income security for farmers, especially in agricultural seasons characterized by turbulent agricultural commodity market conditions or challenging weather conditions, which may intensify as a result of progressing climate change. Benefits for agricultural production may occur due to reduced dehydration of soil and crops resulting from shadowing, slower wind speed and dew formation. Thus, the Agri4Power concept may also be used for climate adaptation in agricultural practice [9], e. g. to adapt to more intense and longer periods of drought or to support changes in crop choice and rotation.

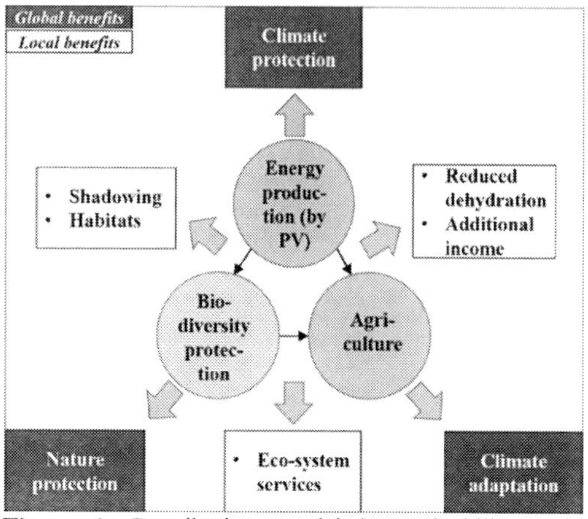

Figure 4: Contribution to global sustainability goals (grey) and synergies and local benefits (white) of "Nature-conserving Agrivoltaics" (own illustration)

From a biodiversity perspective, the flower strips embedded in the Agri4Power concept provide habitats and food supply for plants and animals, especially for insects, birds and small mammals. As well, the flora and fauna on the flower strips may benefit from the PV modules due to shadowing and reduced dehydration (for an overview, see figure 4). In addition to biodiversity protection, reduced soil working and tillage on the flower strips may also contribute to other nature protection goals, such as soil protection. Furthermore, agricultural production may benefit from the flower strips as they provide ecosystem services, e. g. pollination or natural predators reducing the need for pesticides. Positive impacts on biodiversity may be added by crop diversification on the agricultural strips, e. g. by cultivating different crops on each agricultural stripe as it is illustrated in version 1 in figure 2.

3.2 Benefits of the Agri4Power concept raise societal acceptance

As the Agri4Power concept addresses some criticism on current RES deployment, e.g. land use conflicts and trade-offs with nature and animal protection goals, it may contribute to a higher acceptance of RES expansion by integrating nature conservation and agricultural land use. Results from a national survey in Germany using a Choice Experiment and complemented by additional questions on attitudes and the Agri4Power concept itself show that more than three quarters of respondents do prefer the Agri4Power concept to the current practices of energy production via conventional south oriented monofacial PV-systems, which are spatially separated from agriculture (for more details on the survey see [10]). Nearly two thirds of the respondents assume that the combination of RE plants in general with nature conservation measures or agricultural production leads to higher acceptance of the energy sector transition. As well, nearly than two thirds of the respondents expect the Agri4Power concept to make an important contribution to energy sector transition.

A key reason for the positive societal approval of the Agri4Power concept is the positive effect of the flower strips on biodiversity and nature protection. The choice experiment showed that the majority of the respondents has a willingness to pay for increasing biodiversity in agricultural landscapes using the Agri4Power concept. The empirical results show that the variant of the concept with large flower strips, a large row spacing, a large number of PV module rows (implementing a large number of flower strips) and a low height of the PV modules receives the highest societal agreement.

3.3 Overall economic benefits

The increase of the financial revenues from energy production resulting from the shift of power supply peak to the morning and afternoon (as illustrated in section 3.1), may be an additional motivation for the implementation of further RES installations, for instance as an extension to existing south-oriented PV systems. Moreover, the shift of power supply peaks may contribute to a demand-oriented power supply and thus, to grid stability and supply reliability.

The provision of ecosystem services from the flower strips may provide additional benefits for agriculture and thus, may contribute to sustainable and more nature-compatible agricultural practice. Most of all, local communities may benefit from ecosystem services, for instance due to reduced pesticide use and drift.

Beyond that, the Agri4Power concept provides a potential for the combination with other measures contributing to sustainable development, for instance 1) reduced tillage for soil protection or even shift to organic agriculture, 2) measures for soil carbon sequestration for carbon dioxide removal as nature-based negative emission technology (NET) for climate protection on the flower strips, 3) automatisation and digitalisation in agriculture as energy demand is provided on site or 4) support of value creation in rural areas, e. g. using biorefineries by providing both, biomass and energy as necessary inputs for biorefinery processes.

4 CURRENT CHALLENGES AND FURTHER RESEARCH NEEDS

The Agri4Power concept also faces some challenges and needs further research on land use efficiency aspects, site conditions and application fields as well as on the implementation of wind energy installations into the Agri4Power concept.

4.1 Land use efficiency

Currently, insights on land use efficiency of the Agri4Power concept are not abvailable, yet. The contribution of synergies between the single components of the multifunctional land use concept may be decisive for the choice of the Agri4Power concept as they might compensate potential areal and yield-related losses for agricultural production. Areal losses result from the space requirement for the vertical mounting of the PV-modules and related infrastructure (about 10 %) as well as for the area used for biodiversity protection. This might be compensated by ecosystem services provided from these areas. Additionally, agricultural subsidies increasingly demand for areas and measures for biodiversity and nature protection resulting in a must for agricultural companies to compensate for the impacts of arable farming.

Yield-related losses might originate from the uneven shadowing of the agricultural strips. This results in uneven maturity levels of the grown crops. Due to vertically mounting in east-west orientation, during the course of the day, the centers of the strips are less shaded by the modules than the edges of the strips. The choice of the crops grown on the strips determines the effect of the shadowing for the agricultural yield. In temperate latitudes, for instance, c4 plants such as millow or maize are less shade tolerant [6] and are growing too tall, which results in a shadowing of the PV modules, what might decrease financial yields from energy production.

4.2 Suitable and unsuitable areas for the Agri4Power concept

An implementation of the Agri4Power concept is not suitable for all agricultural sites, but rather an opportunity to use niches, especially on marginal agricultural land with poor or dry soils, for organic farming (especially for the transition phase) or as an agri-environmental measure. As well, the Agri4Power concept is appropriate for pasture farming, as well as for animal husbandry and for hay production. This application field of the land use concept is already well established, for instance in Donaueschingen (Baden-Wuerttemberg) or Eppelborn (Saarland) in Germany [6].

Furthermore, post-mining landscapes provide large application opportunities for the Agri4Power concept as these areas need large periods of transition for recultivation to be used again for agricultural activities. These phases could be financially supported by energy production using the Agri4Power concept providing inter-row strips for soil recultivation activities.

Currently, crop farmers in Germany who established intensive agriculture on fertile soils using agricultural machinery with large working widths (of up to 36 m) to grow large amounts of e. g. cereals, maize or rapeseed, are less inclined to implement the Agri4Power concept on their land. From an overall economic perspective, in this field of application it is more efficient to keep land segregation in these areas with efficient agricultural production on the fertile land and to establish the Agri4Power concept as a complementation in the peripheral regions as ecologically valuable small structure or as areal division between large areas of areal land, e. g., to decrease soil erosion sensitivity [6]. For large-scale impacts of the Agri4Power concept, and especially for the targeted applications of the concept, research on different sites is needed.

As shadowing is well tolerated by some crops, e. g. celery, green leafy vegetables such as salad or special crops, such as wild garlic [6], a thoughtful crop choice and rotation is needed to make use of the shadowing and to create gains from the synergies of the agrivoltaic system. Thus, more insights from different case studies are needed to detect the optimal agricultural management to gain yield increases, but also quality increases of the grown crops.

Also a subdivision of the agricultural strips might be a good opportunity to make use of the shadowing effects: Creating a stripe pattern of 3 x 3 m wide strips within one 10 m agricultural strip allows for a diversification of crop cultivation and for precise growing of vegetables, which might also be supported by automatization in agricultural practice. The direct use of the locally generated energy from the PV modules might provide opportunities for the development of sustainable growing of vegetables, which might also be complemented by a sophisticated irrigation system. Especially for arid areas, e. g. the Mediterranean region, this might be a window of opportunity for climate adaptation of agricultural production and to increase income-earning opportunities for farmers.

4.3 Implementation of wind energy installations into the Agri4Power concept

Although it would be cost effective [11], current conditions in Germany impede the combination of wind energy and PV on one area. Firstly, areas are not designated for PV and wind simultaneously from planning authorities. Secondly, subsidies for PV installations according to the German Renewable Energy Act (EEG) are currently not possible on areas designated to wind energy. From the EEG perspective, the combination of wind and PV on one area may work out, if it is part of an innovation tender or if energy generated from PV is sold based on a long-term power purchase agreement (PPA).

Thirdly, standards for mains connection for wind and PV are different. Although the extension of mains connection incurs costs, the effort is getting more worthwhile, the longer the distance to the mains feed-in point is as cable length is a significant cost factor. To overcome these hurdles, further research on the smart linkage of different energy carriers and on possible legal adjustments would be useful. Furthermore, research on the mutual interactions regarding biodiversity effects between PV installations and wind turbines is recommendable.

5 CONCLUSIONS

All together, the concept may open up further land resources for RES deployment due to 1) the integration into agricultural land use (Agrivoltaics), 2) protecting biodiversity by flower strips ("Nature-conserving Agrivoltaics"), 3) income diversification for farmers because of additional income, 4) supporting climate adaptation in agricultural practice as contribution to sustainable agriculture and 5) enhancing societal acceptance for RES installations.

As a result, a wider audience may be reached for RES installation investments. Potential investors may be, e. g., farmers, land owners, energy communities, municipalities or regional energy providers.

The concept is not to replace other RES land use concepts, nor is it a panacea for sustainable agricultural and energy production, but it may increase opportunities for additional RES installations in various niches, such as, e. g., ecological farming, pasture farming, farming on marginal land, climate adaptation measures in agriculture, agri-environmental measures or post-mining landscape and soil recultivation.

The Agri4Power concept is not suitable for all sites, but considerable application potentials may exist especially for sites where 1) agricultural land use shall be maintained, 2) agricultural land use needs ecological (e. g. in dry areas) or financial support (e. g. on marginal sites), particularly areas in need to be adapted to the impacts of climate change, 3) (agri-)environmental measures in rural and agricultural areas need financial or spatial support to provide more area in total for biodiversity and nature protection and/or 4) local acceptance for conventional RES installations is low or hesitant.

The site-specific feasibility has to be evaluated individually, depending on local conditions and needs. In general, more research on sustainable approaches to integrate RES installations into existing land use seems useful.

In a nutshell, the Agri4Power concept presents socially supported opportunities for the contribution of RES deployment to biodiversity and nature protection and improvement of ecosystem services in agricultural areas. The concept is in line with agricultural production, may stabilize income for farmers and has the potential to contribute to the further expansion of photovoltaics at higher electricity market values.

6 ACKNOWLEDGMENTS

This work was supported by the German Federal Ministry for Economic Affairs and Energy on the basis of a decision by the German Bundestag.

7 REFERENCES

[1] IPCC: Global Warming of 1.5°C. An IPCC Special Report on the impacts of global warming of 1.5°C above pre-industrial levels and related global greenhouse gas emission pathways, in the context of strengthening the global response to the threat of climate change, sustainable development, and efforts to eradicate poverty. Full report. 2018.
[2] German Federal Ministry of Economics (BMWi): Zeitreihen zur Entwicklung der erneuerbaren Energien in Deutschland. 2020. Available online at https://www.erneuerbare-energien.de/EE/Navigation/DE/Service/Erneuerbare_Energien_in_Zahlen/Zeitreihen/zeitreihen.html (last accessed 8 September 2021).
[3] M. Sonnberger, M. Ruddat, O. Wedderhoff, A. Salup: Die gesellschaftliche Wahrnehmung der Energiewende – Ergebnisse einer deutschlandweiten Repräsentativbefragung (The societal perception of energy transition - results from a representative survey in Germany). Stuttgarter Beiträge zur Risiko- und Nachhaltigkeitsforschung Nr. 34, 2016, Stuttgart. ISSN 1614-3035, ISBN 3-938245-33-0.
[4] J. Yoo, R. C. Ready: Preference heterogeneity for renewable energy technology. Energy Economics 42 (2014) 101.
[5] German Federal Environmental Agency: Instrumente zur Reduzierung der Flächeninanspruchnahme. Aktionsplan Flächensparen. Final report. 2018. Dessau-Roßlau. 195 pp.
[6] Fraunhofer ISE: Agri-Photovoltaik: Chance für Landwirtschaft und Energiewende. Ein Leitfaden für Deutschland. 2020. Freiburg. 56 pp.
[7] Agency for Renewable Resources (FNR): Nachwachsende Rohstoffe im Überblick. 2021. https://www.fnr.de/nachwachsende-rohstoffe/nachwachsende-rohstoffe-im-ueberblick (last accessed 8 September 2021).
[8] Energy Charts: Biomasse 2020. 2021. https://energy-charts.info/charts/energy/chart.htm?l=de&c=DE&year=2020&interval=year&stacking=grouped (last accessed 8 September 2021).
[9] Jung, D., Salmon, A., Gese, P.: Agrivoltaics for farmers with shadow and electricity demand: Results of a pre-feasibility study under net billing in central Chile. AIP Conference Proceedings 2361, 030001 (2021). https://doi.org/10.1063/5.0054962.
[10] N. Pannicke-Prochnow, C. Krohn, K. Scheffler, R. Stretz: Findings from a Choice Experiment on the Acceptance for a Novel Agrivoltaic Land Use Concept with Bifacial Solar Modules. Forthcoming.
[11] Ludwig, David; Breyer, Christian; A. Solomon, A.; Seguin, Robert: Evaluation of an onsite integrated hybrid PV-Wind power plant. AIMS Energy 8 (5), 2020, S. 988–1006.

ACCURATE TESTING METHODS OF GRID-CONNECTED PV INVERTERS BY MEANS OF REAL-TIME BASED HARDWARE-IN-THE-LOOP (HIL) SIMULATION TOPOLOGIES FOR VALIDATION, TESTING, AND GRID INTEGRATION OF SOLAR PLANTS

G. Lauss, Z. Miletic, A. Banjac, C. Messner.
AIT Austrian Institute of Technology, Austria
georg.lauss@ait.ac.at

ABSTRACT: This paper presents hardware-only and real-time based methods to test and investigate PV inverters in an accurate way. Based on existing requirements for grid-connected and islanding operation modes of inverters, test procedures are well-defined and described in standardization frameworks. The increasing complexity and the rising test effort of many tests ask for alternatives to widely used hardware-only testing methods in order to be able to optimize the testing time without compromising the accuracy of results. In recent years, respective alternatives are identified as real-time based simulation topologies such as the controller hardware-in-the-loop (CHIL) or the power hardware-in-the-loop (PHIL) simulation method. Hardware-in-the-loop (HIL) simulation systems are hybrid systems consisting of one or more software system and at least one hardware system which are interconnected by means of a specific interfacing system. For both simulation topologies, the primary part of the electrical system is simulated by means of real-time capable models implemented on a digital real-time simulator (DRTS). The hardware system is represented by selected hardware components of the PV inverter itself. For CHIL simulation, the hardware system is the control board of the inverter and all other peripheral hardware of the PV inverter is modelled in software. For PHIL setups, the physical hardware device serves as the hardware system which is directly interconnected to a power interface (PI). Based on the concept of real-time based HIL setups, the accurate testing of PV inverters is possible, and results are compared to the hardware-only test setup for the purpose of validation. This work includes the explanation of the different test setups and explains different use cases in which CHIL/PHIL setups and hardware-only tests may be used for the purpose of validated investigation and accurate testing.

Keywords: Grid-connected PV Inverters, Real-time Simulation, Hardware-in-the-Loop Simulation, Inverter Control Functionality, Inverter Testing.

1 BACKGROUND AND RELEVANCE

The relevance of this work is given with the fact that different test setups are utilized for accurate testing and experimental validation of grid-connected PV inverter systems in research and industry, nowadays. These methodologies are ranging from physical laboratory setups to real-time based simulation techniques are applied for the analysis of PV components. Explaining the background of relevant test requirements for grid-connected PV inverters, existing standardization frameworks for different countries are given in [1]-[4].
Since the testing effort and the complexity of setups for respective requirements rise continuously, alternatives to existing hardware-only test setups have been developed. Testing methodologies and platforms such as real-time based simulation methods such as controller hardware-in-the-loop (CHIL) and power hardware-in-the-loop (PHIL) simulation serve as valid alternatives to hardware-only setups. Well-defined HIL topologies and representative examples for test setups can be found in [5].
For this work, different methodologies are developed, implemented and experimental results are validated for selected examples. It is the goal of this manuscript to prove that different approaches such as physical test systems and HIL setups can serve as accurate platforms for validated test results. Based on this concept for validation, test results for selected test CHIL and PHIL setups are compared to the hardware-only test setup. These real-time methods offer the ability to interface electric networks with physical device controllers in the state in which they will finally be implemented.

2 OBJECTIVES AND APPROACH

In this section, the main principles and the primary design of different testing systems are summarized and general application ranges for HIL simulation systems are introduced. It gives a brief overview of the suitability and applicability of existing computing systems commonly used for PHIL simulation. It provides an insight into the basic architecture, solver principles, operating systems, or applied models libraries eligible for HIL simulation.
Figure 1 shows methodologies applied for investigation on electromagnetic systems such as PV inverters. The usability, the controllability, and the implementation of key components are of vital importance for the setups, as highlighted in the following. Firstly, the full system hardware testing setup is the classical approach for testing PV inverters. For experimental testing purposes, the entire electrical network is built up with as a field-testing experimental setup with hardware components. Secondly the pure numerical simulation method and the single component testing setup represent useful methodologies for specific investigation tasks. For the simulation method, only software systems are used, while for the component testing, only hardware systems are coupled and interconnected. Thirdly, the HIL simulation method is a hybrid methodology which consists of at least one software and one hardware system. For the software part, the real-time capability of models, solvers, and computing systems are comprised as a necessary requirement for setting up true HIL simulation systems. The hardware part may range from a physical control board to an entire PV inverting system. Detailed information about the entire spectrum, intended application ranges, and common characteristics of the DRTS systems can be found in [6].
In general, a real-time machine consists of following systems. multiple processors operate in parallel and constitute the target platform on which the simulation runs in real-time. A host computer is used to prepare off-line models and then compile and load them on the target platform. A communication network exists for the exchange of data between multiple targets when the model is split into multiple subsystems.

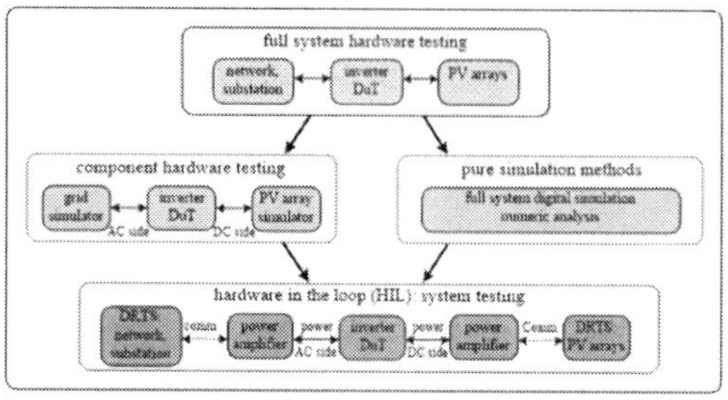

Figure 1: Overview over different methodologies for testing grid-connected PV inverters.

Data exchange between host and the target machines require a separate communication link. Specific I/O terminals are given to interface with external hardware. The CHIL method involves closed-loop simulation of two or more subsystems, at least one of which is a physical device and another software-only subsystem executed on a DRTS platform. The hardware under test (HUT) is interfaced to the DRTS using analog or digital communication, which is sampled deterministically by the DRTS. The HUT is generally an embedded controller or other device with input or output using signal-level voltages. For the PHIL simulation method, the HUT is comprised to be a physical hardware component which is sourced by power level voltage via the PI.

The real-time based CHIL simulation method is an advanced simulation method which is of major relevance for future investigations related to rapid prototyping, grid integration, pre-certification. In general, real-time computing platforms are widely used for manifold simulation techniques such as co-simulation, PHIL simulation or CHIL simulation are methods. Each of them gains increasing importance in industry and research related areas nowadays. It combines several benefits of hardware testing methods and from software simulation methods. The basic idea is comprised that a specific control board runs on its target platform, while it is tested versus a simulated model of the real-time computation system, as shown in Figure 2 (left).

The CHIL method is a real-time capable method mainly used for rapid prototyping and the specific verification of digital controllers. For the CHIL technique, at least one subsystem is a physical device and at least one is represented by a software-only subsystem executed on a DRTS platform. The interface may feature analog or digital signals as input and output signals connecting the Hardware under test (HUT) and the DRTS. This method combines many advantages of hardware testing methods and software simulation methods.

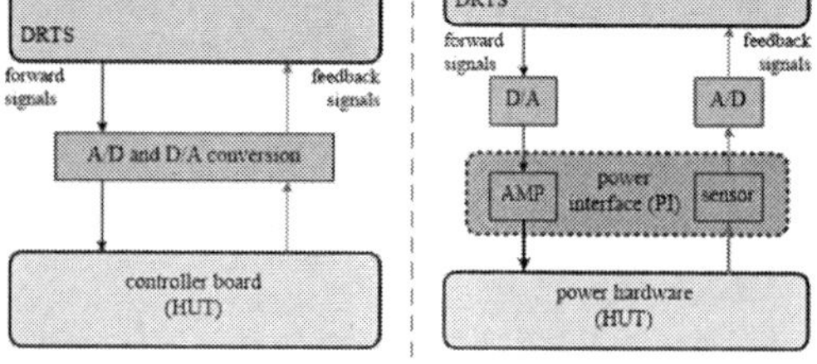

Figure 2: Principles for CHIL simulation (left) and PHIL simulation (right) concepts.

The basic idea is comprised that a specific control board runs on its target platform, while it is tested versus a simulated model of the embedded system.

Figure 2 shows the basic difference between PHIL and CHIL simulation with in a graphical way. It mainly consists of the integration of the PI which includes the amplification unit and the signal measurement. However, this structural differential implies a huge impact from a system theory point of view. Common CHIL simulation are mainly executed in real-time, however, also non-real-time simulations are possible depending on the operation and the implementation of the controller [7]. Possible communication interfaces are I/O modules, networks (TCP, UDP), or other mostly propriety interfaces that serve as a general peripheral input to the controller.

In general, a PHIL simulation system consists of multiple blocks representing structural areas connected one to each other via interacting signals. In Figure 2 (right), the basic structure of the PHIL simulation technique is highlighted in the block diagram. It highlights the causal signal flow and does not differentiating initially between high and low power scale. Three major parts are characteristic for PHIL simulation systems:

- Software Part: This component of the software side is constituted by the DRTS calculating the simulated models in real-time. Its function is clearly defined as it must always maintain the guaranteed real-time capability, collect the incoming signals, and process the outgoing signals in the given time step. The link to the interface primarily consists of the I/O ports.
- Power Interface (PI): The primary task of the PI is comprised in the bidirectional, self-contained link from the software to the hardware part. The key component of the PI is the power amplification unit and contains both forward and backward signal paths with conditioning circuits, and measurements.
- Hardware Part: It may consist of passive or active components, standard testing equipment, or custom devices. It can be assembled out of subsystems and exists in physical hardware. It is connected to the PI with cables capable of carrying power-level signals.

In what follows, two test setups are applied and run for comparison of results for the purpose of validation:

Firstly, the hardware-only test setup is shown in Figure 3. From left to right, an isolation transformer, the PV inverter, and DC and AC sources with various electric measurement devices are shown. The shown setup is entirely implemented with hardware components which requires costly and spacious laboratory equipment.

Figure 3: Image of the physical hardware testbed for experimental results based on the hardware setup.

Figure 4: Three-phase PV inverter: Picture of the real-time based CHIL testbed with the interface board and PV inverter control board.

Secondly, Figure 4 highlights the CHIL simulation setup which includes a digital real-time simulator (DRTS), a signal interface board, and the control board of the photovoltaic inverter. For this test setup, the interface board connects the DRTS placed on the left side with the ASGC control board on the right side establishing a suitable test setup for CHIL simulation system.

Based on this concept of using different test setups and real-time based simulation platforms, results are compared one to each other with respect to deviations from the hardware-only test setup indicating the resulting signal accuracy. Selected test cases according to respective standards are implemented and executed, and comparisons of measured results are presented in the following section. It is notable that the required laboratory space and related costs of the HIL setups are significantly reduced compared with the hardware-only setup. Using HIL simulation setups, physical components such as the transformer as well as connected AC and DC source can be omitted for testing and the major part of power line connections gets obsolete because cables are partly replaced by connections implemented in the software model.

3 RESULTS

Three test cases are presented in the following, demonstrating specific transition events between steady state operation modes under balanced and unbalanced operation conditions of a grid connected PV inverter. For this setup, resulting waveforms have been recorded from experimental hardware setups and HIL simulation setups for the purpose of accurate comparison and validation testing.

Figure 5 shows results for grid voltage waveforms and current phasors (top graph) as well as current waveforms (bottom graph) captured from the hardware test setup. The test demonstrates the transition from balanced operation at a nominal output power of 7 kVA to unbalanced mode by triggering a control sequence. The sequence activates the rotation of respective current phasors based on predefined phase angles, which is visualized for all tests from Figure 5 to Figure 10. For each of the tests, the selected time durations for voltages and currents are set to [-200 ms; +400 ms] and [-100 ms; +300 ms] with respect to the triggering event at $t = 0$ initialising the begin of the transition from balanced to unbalanced mode and vice versa. For the test indicated in Figure 5, the phasor of phase a is rotated by -60° and the phasor for phase b rotates by +60°. For phase c, the phasor remains identical with its settings for the balanced three-phase operation mode. The neutral phase is marked by orange colour and is individually controlled such as phases a, b, and c.

In Figure 6, the re-transition from unbalanced operation to balanced mode is depicted, and phasors a, b are re-rotated by +60° and -60° with respect to the reference setting, again. For both transition events, the voltage waveforms (top, middle graph) show the identical behaviour for time durations before, during, and after the triggered event. Therefore, the rotation of current phasors has no significant impact on grid voltages which demonstrates the proper implementation of the current control and indicates a beneficial control behaviour.

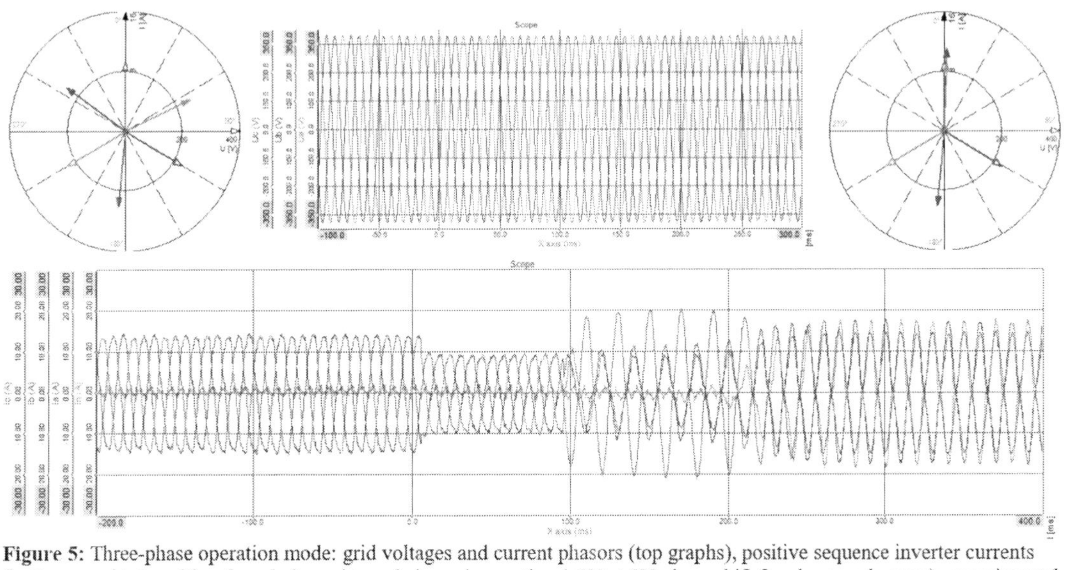

Figure 5: Three-phase operation mode: grid voltages and current phasors (top graphs), positive sequence inverter currents (bottom graph); transition from balanced to unbalanced operation (-60°, +60° phase shift for phases a, b, resp.); experimental data from the hardware setup.

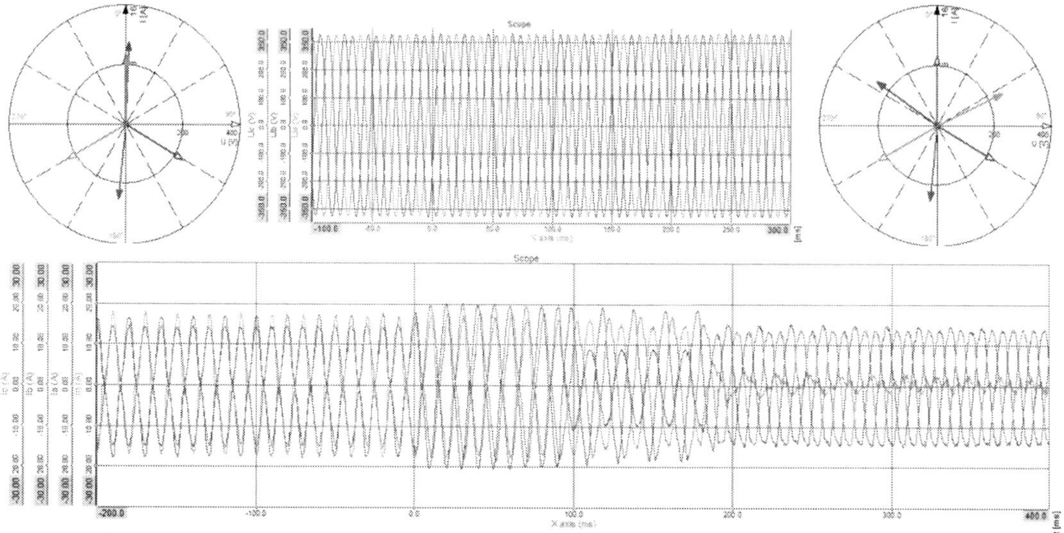

Figure 6: Three-phase operation mode: grid voltages and current phasors (top graphs), positive sequence inverter currents (bottom graph); transition from unbalanced to balanced operation mode (+60°, -60° phase shift for phases a, b, resp.); experimental data from the hardware setup.

In Figure 7, the transition from balanced three-phase operation mode to unbalanced operation mode is highlighted. In this case, the phasor of phase a is rotated by +60° and the phasor for phase b rotates by -60°. In analogy to the previous test case, the phase of phasor c remains identical with its settings for the balanced three-phase operation mode.

Figure 8 depicts waveforms and phasors of all phases for the re-transition, at which the phasor of phase a is re-rotated by -60° and the phasor for phase b re-rotates by +60°. Again, voltage waveforms for both events remain unchanged and show accurate characteristics in terms of control behaviour.

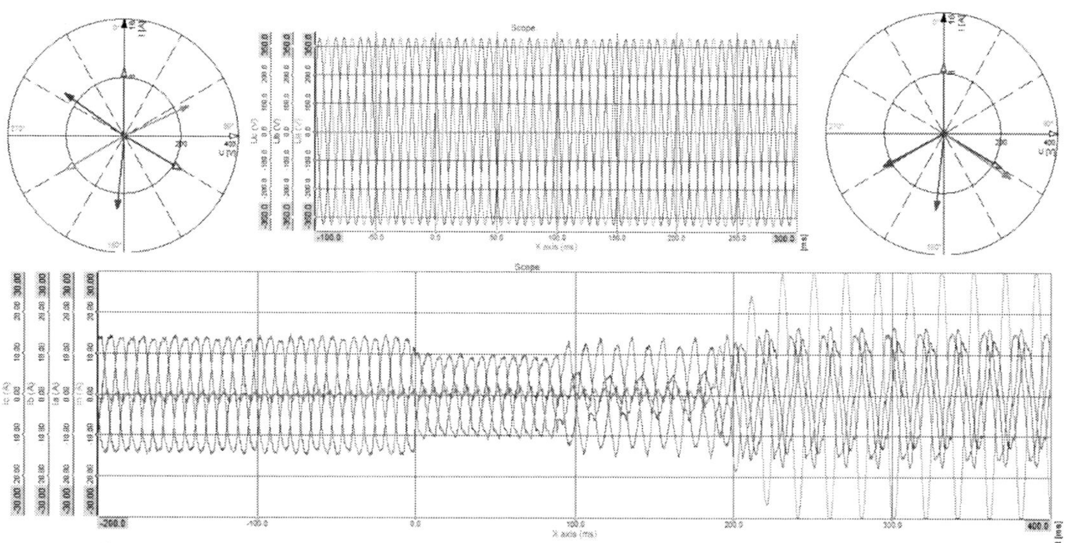

Figure 7: Three-phase operation mode: grid voltages and current phasors (top graphs), positive sequence inverter currents (bottom graph); transition from balanced to unbalanced operation (+60°, -60° phase shift for phases a, b, resp.); experimental data from the hardware setup.

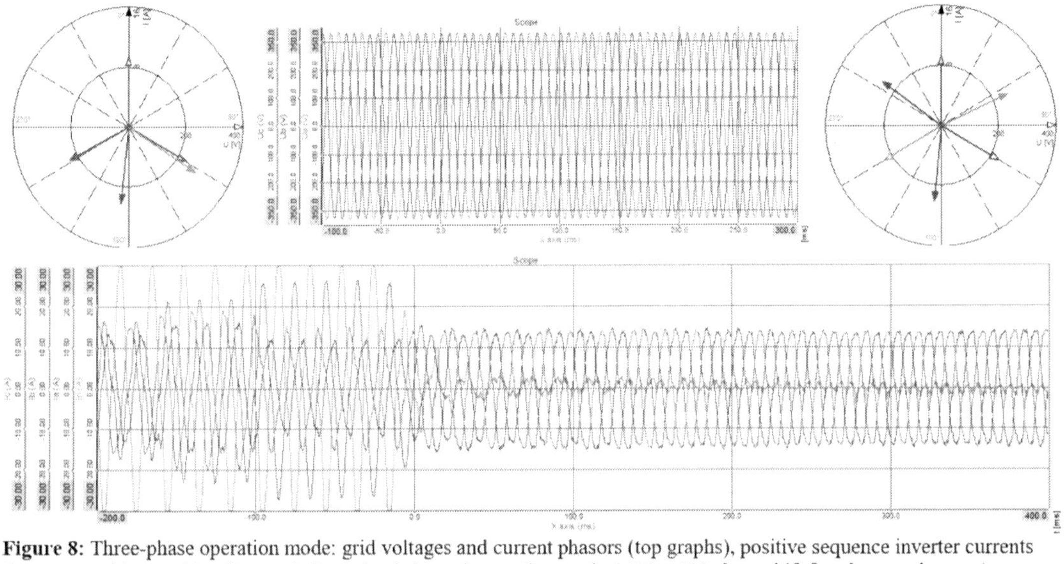

Figure 8: Three-phase operation mode: grid voltages and current phasors (top graphs), positive sequence inverter currents (bottom graph); transition from unbalanced to balanced operation mode (-60°, +60° phase shift for phases a, b, resp.); experimental data from the hardware setup.

Figure 9 presents results of the transition from balanced three-phase operation to unbalanced operation mode. For this setup, the phasor of phase a is rotated by -120°, while the phasor for phase b is rotated by +120°. In analogy to previous test cases, the phase of phasor c remains identical with its settings for the balanced three-phase operation mode.

In Figure 10, waveforms and phasors for the re-transition event in which the phasor of phase a is re-rotated by +120° and the phasor for phase b is re-rotated by -120°is shown. For the unbalanced setup, all current phasors point into the same direction which means they have the same relative angle compared to phase c. The neutral phase is the vector product of all three phases and reaches peak values of about 50 A for this asymmetrical operation mode. Results show that the transition from unbalanced to a stable steady state operation is achieved within less than 200 ms.

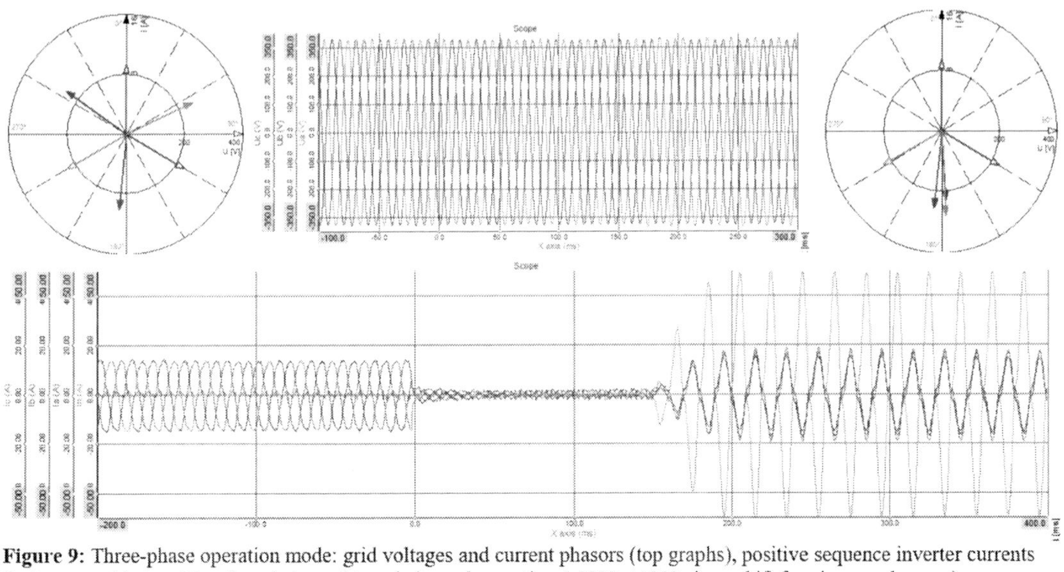

Figure 9: Three-phase operation mode: grid voltages and current phasors (top graphs), positive sequence inverter currents (bottom graph); transition from balanced to unbalanced operation (-120°, +120° phase shift for phases a, b, resp.); experimental data from the hardware setup.

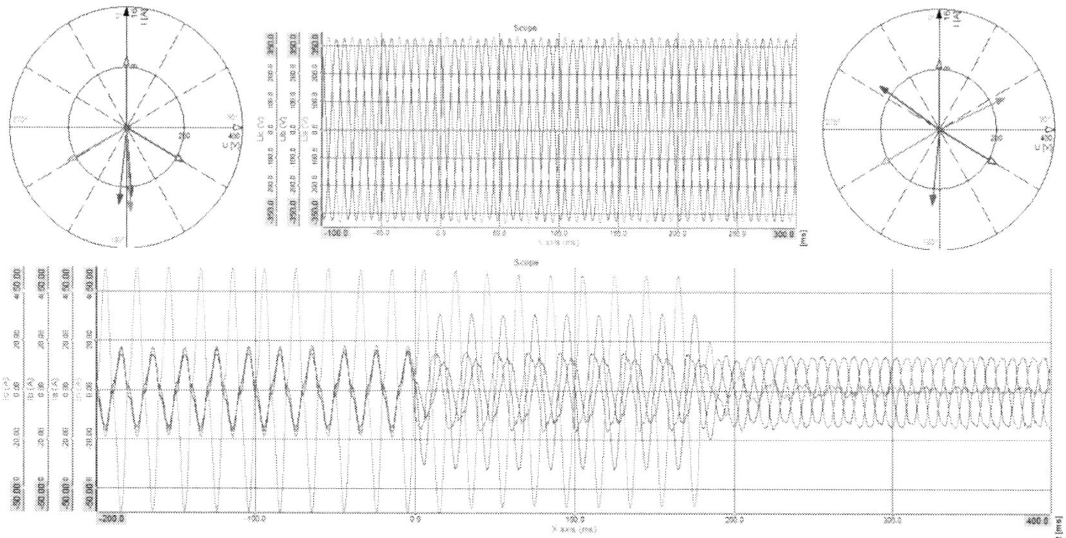

Figure 10: Three-phase operation mode: grid voltages and current phasors (top graphs), positive sequence inverter currents (bottom graph); transition from unbalanced to balanced operation mode (+120°, -120° phase shift for phases a, b, resp.); experimental data from the hardware setup.

4 CONCLUSIONS

In this work, the investigation of results for balanced and unbalanced operation modes of grid-connected PV inverters has been the major focus. Different test setups have been developed, implemented, and executed in order to obtain accurate results which can be compared one to each other for the purpose of accuracy. On the one hand, hardware-only test setups set the reference for test results. The PV inverter was physically connected to a grid emulator in a laboratory environment. On the other hand, alternative setups such as hardware-in-the-loop (HIL) simulation setups have been implemented. For the HIL setup, the control board of the PV inverter or the PV inverter itself was connected to a digital real-time simulator. By using this methodology, a complete HIL test setup has been implemented and run. Based on this concept, the validation of HIL setups could be done for steady-state operation but also for fast transients. Results confirmed the suitable applicability of HIL simulation setups as flexible platforms in order to perform valid, accurate tests.

One major contribution of this paper is given by the implementation of a grid-connected inverter which can rotate the phasors of all 3 phases in an independent way. The second novelty of the paper is the comparative validation of results based on experimental hardware setups as well as real-time based HIL setups. Results show that results are accurate and validates the given HIL methodology when being properly implemented.

Given the knowledge that an increased number of PV inverters acting as power electronic components will become an essential part of future power systems when integrating renewable energy, the proposed HIL methods will be beneficial for all manufacturers, network operators, and researchers. By using the HIL concept, the time-to-market duration time can be reduced in a n effective way and the impact of PV inverters integrated into the network can be verified and validated in an accurate way.

REFERENCES

[1] BDEW Bundesverband der Energie- und Wasserwirtschaft e. V., "Technical Directive for Medium-Voltage Generation Units," Berlin, Germany, 2008.

[2] FGW e.V. Fördergesellschaft Windenergie, "Technical Guidelines for Power Generating Units and Farms Part 3 - Determination of electrical characteristics of power generating units and systems connected to MV, HV and EHV grids," FGW e.V. Fördergesellschaft Windenergie, Berlin, Guideline 23, May 2013.

[3] UL1741: "Inverters, Converters, Controllers and Interconnection System Equipment for Use With Distributed Energy Resources", January 2010.

[4] CEI 0-21, "Reference technical rules for the connection of active and passive users to the LV electrical Utilities", November 2011.

[5] G. Lauss, M. O. Faruque, K. Schoder et al., "Characteristics and Design of Power Hardware-in-the-Loop Simulations for Electrical Power Systems," IEEE Transactions on Industrial Electronics, vol. 63, no. 1, pp. 406–417, Jan. 2016.

[6] M. O. Faruque, T. Strasser, G. Lauss, V. Jalili–Marandi, P. Forsyth, C. Dufour, V. Dinavahi, A. Monti, P. Kotsampopoulos, J. A. Martinez, K. Strunz, M. Saeedifard, X. Wang, D. Shearer, M. Paolone, "Real-Time Simulation Technologies for Power Systems Analysis, Design and Testing", *IEEE Power Energy Technol. Syst. J.*, vol. 2, no. 2, pp. 63–73, Mar. 2015.

[7] M. Steurer, F. Bogdan, W. Ren, M. Sloderbeck, S. Woodruff, "Controller and Power Hardware-In-Loop Methods for Accelerating Renewable Energy Integration", Proc. of Power Engineering Society General Meeting 2007, Tampa, USA, Jun. 24–28, 2007.

IMPACT OF MEASUREMENT DATA TIME RESOLUTION ON PREDICTED LIFETIME OF PV INVERTERS IN RESIDENTIAL SOLAR PANEL SYSTEMS

Omid Alavi*†§, Wieland Van De Sande*†§, Leander Van Cappellen*†§, Michaël Daenen*†§
*IMO-IMOMEC, Hasselt University, Wetenschapspark 1, 3590 Diepenbeek, Belgium
†imec, Kapeldreef 75, 3001 Heverlee, Belgium
§EnergyVille, Thor Park 8310, 3600 Genk, Belgium

Email: omid.alavi@uhasselt.be; Tel: +32 (0)11 26 88 26; Fax: +32 (0)11 26 82 99

ABSTRACT: The reliability of grid-tied PV inverters without integrated battery is compared with AC/DC-coupled PV-battery systems while varying the time resolution of the load and mission profiles (1 min – 60 min). Adding a battery increases self-consumption and can lower the needed grid power capacity in both DC- and AC-coupled configurations. A comparison between a battery-free PV system and a grid-tied PV-battery system shows that adding a battery to the system will not only improve the energy performance of the system but also increase the lifetime of the inverter (~15% in DC-coupled structure). The results have been extracted for two cases: (a) varying both mission profile and load profile resolutions, (b) varying the available load profile time resolution while keeping the mission profile time resolution constant at 1 min. The obtained results show that in our case, the influence of mission profile time resolution is greater than the input of load data. However, the expected PV inverter lifetime does not differ much for different time resolutions. This can highlight the importance of using measurement data with a resolution of even less than 1 minute because higher IGBT junction temperature swings can happen in a few seconds which cannot be seen in lower time resolutions.
Keywords: Lifetime estimation, Mission profile, Power devices, Time resolution, Battery.

1 INTRODUCTION

Due to the decline in PV panel prices over the recent decade, particular attention has been paid to the extensive use of solar energy. In an attempt to emerge this rising trend, researchers have made developments and improvements of PV systems a priority. However, the intermittent nature of PV power can induce major challenges on the grid when multiple residential units employ a grid-connected PV system [1]. Consequently, adding a battery to the PV systems seems to be a workable solution to ease the integration of these systems into the existing grid while PV self-consumption can be improved by using a storage system.

As the power electronic circuits are an integral part of these systems, enhancing their efficiency and reliability should be considered in the analysis. The semiconductor devices utilized in power electronics are the most failure-prone components in the system. Their lifetime is strongly influenced by environmental (mission profile) and operation (load profile) conditions [2]. Avoiding failures in the PV inverter is crucial to prevent energy losses and economic losses. On the other hand, the different system configurations will not only affect the compatibility of the entire system but also influence the efficiency, lifetime, and reliability. In Fig. 1, two different schemes have been adopted in order to link the PV panels to the battery and the grid. The first configuration is DC-coupled where the battery is connected to the DC-link directly. One and the same inverter is used as an interface between the battery and the electrical grid. Another configuration for the integration of the PV array to the battery and the grid is known as AC-coupled where the battery can be connected to the public grid at any point on the AC side. To realize this advantage, a dedicated inverter is used for the battery. However, this additional power conversion stage reduces the efficiency of the system [3].

The objective of this study is to assess the reliability of three PV system configurations: (1) without battery, (2) with a DC-coupled battery system, and (3) with an AC-coupled battery system. The reliability analysis is performed using four different mission and load profile resolutions (1 min – 60 min) which additionally provides an understanding of the sensitivity of lifetime estimations to the sampling rate. This study examines the performance of different system structures in terms of both reliability and energy waste (curtailed energy or fed from the grid). Also, in order to increase the accuracy, the power dependent efficiency of the converters is considered. The results will be compared with scenarios where an efficiency of 100% is assumed. The outcome of this study provides insights in the benefits of integrating a battery in a PV system.

2 METHODOLOGY

In this case study, a 4 kWp (SunPower SPR-200, 4 parallel – 5 series) solar array is considered. Additionally, a conventional three-phase grid-tied inverter is used for the power conversion of the PV panels and the battery. To control the charging and discharging of the battery, the approach is to maximize self-consumption [4]. Based on this control strategy, the storage system uses excess electricity when the PV production is greater than the electricity consumption. If the excess energy goes beyond the maximum battery power, the remaining energy must be injected into the grid or dissipated. On the other hand, the storage system will supply the demanded energy if the consumption exceeds the PV generation (e.g., during night). There is a limitation on the energy stored in the battery so that part of the demanded energy can be supplied by the grid. Nevertheless, the amount of energy that must be curtailed or absorbed from the grid can be specified in different scenarios.

The DC-coupled system, shown in Fig. 1(a), is comprised of three power converters: a solar DC-DC converter, a bi-directional DC-DC converter, and a three-phase inverter. Due to the internal resistance of the components inside each converter, their output power will be slightly lower than the input power, which is important to take into account in the analyses. The same procedure can also be applied on the AC-coupled system with four power converters. In both cases, the DC-link voltage is kept at a constant voltage (400 V) by using the control system. Additionally, the state-of-charge (SoC) of the battery is limited at 80 % to improve the cycle life of the battery [5]. In these cases, an LG Chem RESU3.3 battery (48 V, 63 Ah, 3 kW) is placed into the system. According to [6], the optimal battery size – combined with the PV panel with a capacity of 1 kWp/MWh – is chosen based on the optimal ratio of 0.75 kWh/MWh.

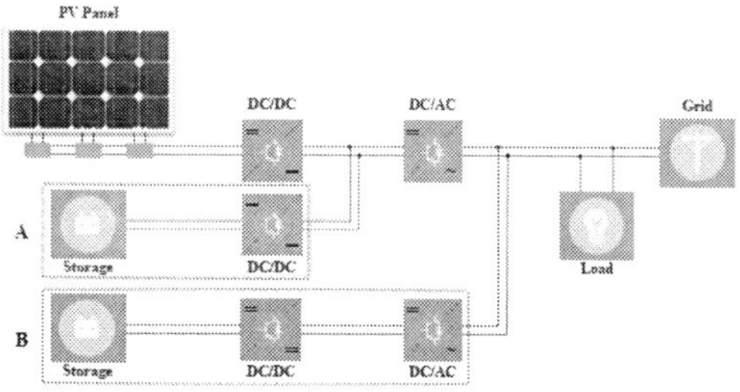

Figure 1: Typical diagram of a grid-connected PV system with (A) a DC-coupled backup battery and (B) an AC-coupled backup battery

The next step is to assess the reliability of the PV inverters. Figure 2 illustrates the used reliability assessment methodology starting from the given mission profile to the lifetime estimation of the switching devices.

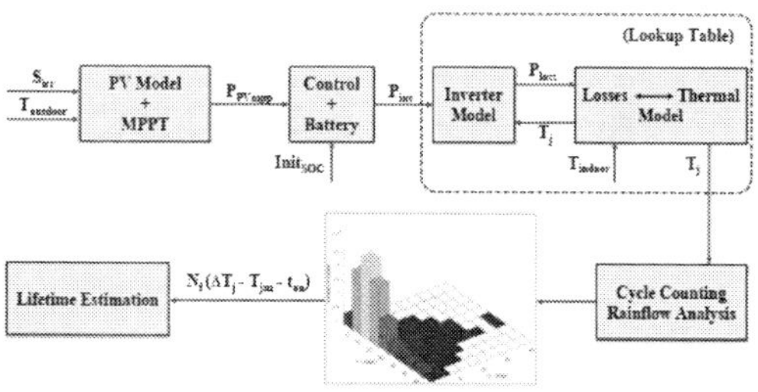

Figure 2: Lifetime and reliability assessment methodology of a PV inverter

The reliability approach used in this study has been adapted from [7]. In the inverter circuit, fast-switch insulated-gate bipolar transistors (IGBTs) rated at 600 V are used as the active switches. Based on the method displayed in Fig. 2, the one-year mission profile (solar irradiance and ambient temperature) is translated to a one-year PV generation profile. This profile is converted to the input power profile of the inverter which already includes the employed battery control strategy. In the next stage, a lookup table approach [8] has been adopted to translate the inverter input power to the power losses, then to the junction temperature of the IGBT. An important point to consider at this point is that the ambient temperature for the inverter must be assumed to be the indoor temperature T_{indoor}, since the inverter will be installed indoors in residential applications. By using the thermal model of the semiconductors and the heat sink, the input power can be translated into junction temperature profiles of the switches. The thermal model parameters are extracted from the datasheet provided by the manufacturer. Then, the rainflow cycle counting method has been used to determine the number of cycles with their corresponding amplitude (ΔT_j), average (T_{jm}), and period (t_{on}). The damage inflicted by every cycle can be calculated using the chosen lifetime model:

$$N_f = A \times (\Delta T_j)^{\alpha} \times (ar)^{\beta_1 \Delta T_j + \beta_0} \times \left[\frac{C + (t_{on})^{\gamma}}{C+1}\right] \times \exp\left(\frac{E_a}{k_b + T_{jm}}\right) \times f_d \quad (1)$$

In this model, the output N_f is the number of thermal cycles to failure. The other model parameters are provided in [9]. This equation has been validated for a certain range of ΔT_j, T_{jm} and t_{on} which means that an extrapolation is required for several cycles laying outside of this range [10]. According to Miner's rule, displayed in Eq. (2), the damage inflicted by every cycle can be linearly accumulated to calculate the total lifetime consumption (LC) of the mission profile on the IGBT [10].

$$LC = \sum_i \left(\frac{n_i}{N_{fi}}\right) \quad (2)$$

where n_i is the number of cycles obtained from the rainflow analysis with a certain combination of ΔT_j, T_{jm} and t_{on}.

3 RESULTS & DISCUSSION

The studied one-year mission and load profiles are depicted in Fig. 3. The household electric power consumption with a one-minute sampling rate has been extracted from [11]. The solar irradiance and ambient temperatures have been captured from Belgium. In this analysis, the PV panel-to-load ratio is assumed to be 100%. As the PV inverter can be installed indoors, the indoor temperature is assumed to be controlled by the heating/cooling system which can keep the inside temperature in a limited range. The approach presented in [12] has been used to produce the indoor temperature profile from the outdoor temperature. The results of this study showed that for two temperature ranges (below and above 15°C), the indoor temperature can be estimated by a linear fit.

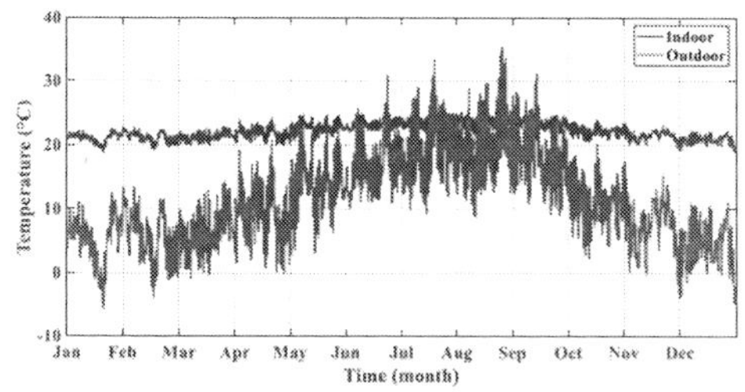

(a)

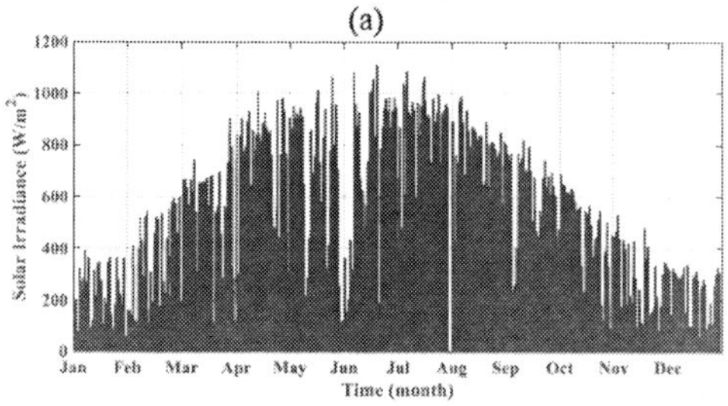

(b)

Figure 3: A one-year mission profile from Belgium with (a) the outdoor/indoor temperature and (b) the solar irradiance

The implemented control approach enables us to determine the amount of curtailed energy as well as the amount of absorbed energy fed to the load from the grid. To increase the accuracy of computations, the efficiency of converters used in the PV system is considered to be less than 100% because the input and output powers of these converters are not the equal and this should be taken into account in the calculations. Thus, as a part of analysis, three values of 95%, 97.5%, and 100% are considered as the converters' efficiency. The obtained results reveal that considering the converter's efficiency in the calculations has a great impact. As a general trend, increasing efficiency leads to a decreased thermal stress. On the other hand, the results imply that PV curtailment increases by boosting the efficiency. For instance, if the efficiency is changed from 95% to 100%, the curtailed energy value will vary from 32% to 43% of the total energy, while the absorbed energy from the grid only drops 4% from 735 to 704 kWh. The effect of the variable efficiency on the system performance has also been investigated and, since there is less than 2% difference compared with the case with a constant efficiency (95%), its use has been neglected in other stages.

In the next step, the aforesaid procedure has been applied on the given mission profile with a time resolution of 1 min. Then, the 1-minute mission profile has been converted to lower resolution data (15 min – 60 min) in order to investigate the effect of reducing time resolution on the system's reliability as well as the performance. This resolution reduction is accomplished by averaging the 1-min dataset over the chosen time resolution. This time resolution reduction has been achieved in two ways: (1) the resolution of the mission and load profiles are both reduced, (2) it is only assumed that the number of load samples is reduced, but the mission profile resolution remains the same (1 min).

For a better comparison between systems, the aforesaid approach is firstly applied on a PV system without battery storage. The results of this configuration are illustrated in Table I.

Table I: The obtained results of the PV system without battery storage

Load Profile Resolution	1 min	15 min	30 min	60 min	15 min	30 min	60 min
Mission Profile Resolution	1 min	15 min	30 min	60 min	1 min	1 min	1 min
Life Consumption LC (%)	0.0330	0.0312	0.0297	0.0287	0.0324	0.0323	0.0316
Energy Curtailment EC (kWh/y)	3112.30	3106.50	3100.70	3087.70	3102.40	3096.70	3089.20
Energy from Grid EG (kWh/y)	1284.10	1294.70	1316.70	1334.80	1291.90	1294.40	1298.50

The results can be divided into two categories: (1) both the resolutions of meteorological sensors and the meter change simultaneously, and (2) only the sampling rate of the electricity meter changes. To show the calculated lifetime sensitivity to the sampling rate, the results of these two groups are shown in Fig. 4.

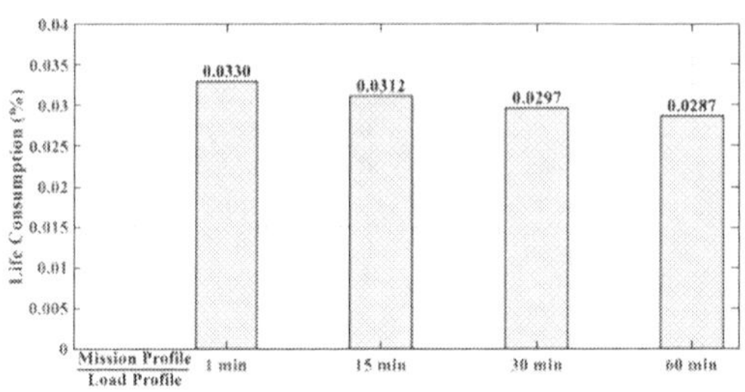

(a)

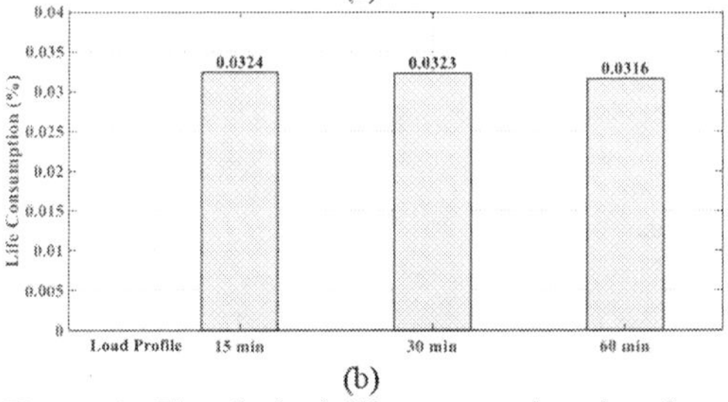

(b)

Figure 4: The obtained life consumptions based on different resolutions: (a) both mission and load profile resolutions change simultaneously, and (b) the load profile resolution changes while the mission profile resolution remains constant at 1 min

According to Fig. 4(b), by enlarging the resolution of the load profile, the calculated lifetime only experiences a slight change so that even if the time resolution is increased to 1 hour, the error is not more than 4%. On the other hand, Fig. 4(a) also shows that the sensitivity of the results to the mission profile resolution is significant as by increasing the resolution of both profiles, the error rate rises to 13%. Nevertheless, to draw a comparison between different configurations, the 1-min profiles have been selected as the reference.

Table II lists the obtained results by applying the approach on both DC and AC-coupled systems compared to the case without battery (WOBatt).

Table II: The obtained results with a 3.3 kWh battery

	WOBatt	DC	AC
Life Consumption (%)	0.0330	0.0280	0.0348
Energy Curtailment (kWh/y)	3112.3	2471.4	2398.6
Energy from Grid (kWh/y)	1284.1	735.7	701.3

As expected, by removing the battery, the amount of energy that must be curtailed as well as the absorbed energy from the grid will increase. A comparison of the battery-free system and the DC-coupled system shows that adding a battery improves the PV inverter reliability and enhances the overall system performance. The results indicate that by inserting a suitable battery into the grid-tied PV system and choosing a proper configuration, it would be feasible to increase the PV inverter reliability by 15.15%. An earlier study also confirms the improvement of the inverter lifetime by connecting a battery as a DC-coupled structure [13].

The outcome from the AC-coupled system analysis is provided in Table II. The PV inverter loading depends on the PV generation and battery operation. Hence, the main difference between the DC and AC-coupled configurations is that the excess solar generation should be exported to the battery or the grid through the PV inverter. But, in the DC-coupled design, the power of the PV inverter reduces because the battery will be charged/discharged on the DC side; thus, the PV inverter suffers less from thermal stress. Thus, the AC-coupled configuration reduces the lifespan of the PV inverter (5.45% less than the battery-free system). This is explained because of the curtailment that happens at the DC link level in the system without battery which is higher than in the case of a system with battery, so the PV inverter has to convert more energy in the system with AC coupled battery compared to a system without battery. The results indicate that the reliability of the PV inverter is 24.3% higher in a DC-coupled system than in an AC coupled system. Our findings are in line with a previous research by Sandelic et al., who found that installing PV inverters in an AC-coupled configuration leads to a shorter lifetime compared to the DC system (26.6% less) [14]. Although adding a battery in an AC-coupled configuration reduces the PV inverter reliability, this configuration effectively improves the total systems energy use, 25% less energy is curtailed compared to the battery-free system. Therefore, it can be concluded that the added battery in both AC and DC configurations has benefit to the overall performance. It is worth mentioning that a battery can also work as a filter to support the PV inverter by stabilizing its input voltage. Using a multi-port/input converter can be a suitable solution to lower the number of total components in the PV-battery system.

The effect of time resolution on the estimated lifespan is not very significant, perhaps because the highest possible resolution is considered 1 min, while within a few seconds the junction temperature can fluctuate sharply. This can lead to a severe reduction in life expectancy, especially on cloudy days. As the thermal time constant between the junction-to-heatsink ($\tau_{J\text{-}H}$) is around a few minutes, but the thermal time constant for junction-to-case layers is much lower, a further investigation on finding a marginal/optimal measurement time resolution needs to be conducted.

4 CONCLUSIONS

In this paper, a reliability assessment of the PV inverter has been performed concerning different configurations of grid-connected photovoltaic systems. Three different configurations have been chosen: a grid-tied PV system without battery, a grid-tied PV system integrated with a DC-coupled battery storage system, and a grid-tied PV system integrated with an AC-coupled battery storage system. The effect of load/mission profiles with various sampling rates (1 min – 60 min) has been determined on a PV system configuration. The results showed that changing the load and mission profile resolutions from 1 min to 60 min leads to an error of 13% in the lifetime estimation. By integrating a DC-coupled battery system into the PV system, the lifetime of the PV inverter can be increased by 15%. Although adding a battery to the system improves the overall performance in terms of energy efficiency, the AC-coupled configuration will reduce the lifetime of the PV inverter by 5% compared to the battery-free structure. For further studies, a reliability assessment of the DC-DC converters utilized in the PV battery system, and the effect of different control strategies on the lifetime estimation are suggested.

ACKNOWLEDGEMENTS

The work is (partially) supported by the energy transition funds project "BREGILAB" organized by the FPS economy, S.M.E.s, Self-employed and Energy (Met de steun van het Energietransitiefonds).

REFERENCES

[1] Sangwongwanich A. et al.: Reliability assessment of PV inverters with battery systems considering PV self-consumption and battery sizing, IEEE Energy Conversion Congress and Exposition (ECCE) 2018, pp. 7284-7291
[2] Zhang Y., Wang H., Wang Z., Yang Y., and Blaabjerg F.: The impact of mission profile models on the predicted lifetime of IGBT modules in the modular multilevel converter, 43rd Annual Conference of the IEEE Industrial Electronics Society (IECON) 2017, pp. 7980-7985
[3] Braam F., Hollinger R., Engesser M.L., Müller S., Kohrs R., and Wittwer C.: Peak shaving with photovoltaic-battery systems, IEEE PES Innovative Smart Grid Technologies 2014, pp. 1-5
[4] Linssen J., Stenzel P., and Fleer J.: Techno-economic analysis of photovoltaic battery systems and the influence of different consumer load profiles, Applied Energy Vol. 185, pp. 2019-2025
[5] Merei G., Moshövel J., Magnor D., and Sauer D.U.: Optimization of self-consumption and techno-economic analysis of PV-battery systems in commercial applications, Applied Energy Vol. 168, pp. 171-178
[6] Schram W.L., Lampropoulos I., and Van-Sark W.G.: Photovoltaic systems coupled with batteries that are optimally sized for household self-consumption: Assessment of peak shaving potential, Applied energy Vol. 223, pp. 69-81.
[7] Yang Y., Sangwongwanich A., and Blaabjerg F.: Design for reliability of power electronics for grid-connected photovoltaic systems, CPSS Transactions on Power Electronics and Applications Vol. 1, pp. 92-103
[8] Graovac D., and Pürschel M.: IGBT power losses calculation using the data-sheet parameters, Infineon Technologies AG, Application Note 2009
[9] Scheuermann U., Schmidt R., and Newman P.: Power cycling testing with different load pulse durations, 7th IET International Conference on Power Electronics, Machines and Drives (PEMD) 2014, pp. 1-6
[10] Van De Sande W. et al..: Reliability comparison of a dc-dc converter placed in building-integrated photovoltaic module frames, 7th International Conference on Renewable Energy Research and Applications (ICRERA) 2018, pp. 412-417
[11] Dua D., and Graff C.: UCI Machine Learning Repository [http://archive.ics.uci.edu/ml]. Irvine, CA: University of California, School of Information and Computer Science 2019
[12] Lee K., and Lee D.: The relationship between indoor and outdoor temperature in two types of residence, Energy Procedia Vol. 78, pp. 2851-2856
[13] Sangwongwanich A. et al.: Enhancing PV inverter reliability with battery system control strategy, CPSS Transactions on Power Electronics and Applications Vol. 3 no 2, pp. 93-101
[14] Sandelic M., Sangwongwanich A., and Blaabjerg F.: Reliability evaluation of PV systems with integrated battery energy storage systems: DC-coupled and AC-coupled configurations, Electronics Vol. 8, pp. 1-19

RELIABILITY ANALYSIS FRAMEWORK FOR A GRID-TIED PV-BATTERY SYSTEM: INFLUENCE OF PV AND BATTERY DEGRADATION ON RELIABILITY OF POWER ELECTRONIC SYSTEMS

Omid Alavi[*†§], Wieland Van De Sande[*†§], Leander Van Cappellen[*†§], Ward De Ceuninck[*†§], Michaël Daenen[*†§]
[*]IMO-IMOMEC, Hasselt University, Wetenschapspark 1, 3590 Diepenbeek, Belgium
[†]imec, Kapeldreef 75, 3001 Heverlee, Belgium
[§]EnergyVille, Thor Park 8310, 3600 Genk, Belgium

Email: omid.alavi@uhasselt.be; Tel: +32 (0)11 26 88 26; Fax: +32 (0)11 26 82 99

ABSTRACT: This paper presents a flexible framework to evaluate the reliability of power electronic systems used within a grid-connected PV-battery system. As a practical example to show the capability of this framework, the life consumption (LC) of power semiconductors (IGBTs and diodes) is determined considering both solar panel and lithium-ion battery degradation. Different panel degradation rates can be defined to the framework. The battery degradation is calculated by a closed-loop approach so that the battery degradation is adjusted based on the determined cycling and calendar aging. The outputs are obtained for three cases: (1) no degradation, (2) only PV degradation, and (3) both PV and battery degradation. The results showed that by considering the battery degradation, the battery DC-DC converter's switches will experience 14% lower LC compared to the case with both PV and battery degradation models, while the switches within the PV inverter have 2.3% higher LC when considering the battery degradation.
Keywords: Lifetime estimation, Mission profile, Power devices, Time resolution, Battery

1 INTRODUCTION

Power electronics converters are an integral part of photovoltaic (PV) systems. Lifetime (LT) of the power devices used within these converters is significantly influenced by the environmental and operating conditions of the system, also known as the mission profile. As the number of installed solar panels for household applications will increase dramatically in the coming years, special attention should be paid to the reliability of these systems. Hence, providing a flexible and comprehensive framework for reliability estimation of the PV-battery systems can be highly useful as it would be perfectly possible to modify the various parameters and inputs of the system and observe the effect of any change on the reliability.

In this paper, the proposed reliability analysis framework allows us to change all the inputs/parameters including mission profiles, PV-battery sizes, time resolution, battery characteristics/models, converters topologies, solar panel behavior, etc. Although several studies have shown the influence of different factors on the reliability of the photovoltaic systems (e.g., the mission profile [1], control methods [2], and functional modes [3]), the impact of both PV panel degradation and battery degradation on the lifetime of power converters has not been investigated. If the PV panel and battery degradation are not considered, the life consumption of subsequent years is the same as the first year which is not accurate. Here, the importance of this consideration is reflected as an example to demonstrate the capability of the adopted framework.

Ref. [4] assessed the influence of the PV panel degradation on the PV inverter lifetime. Their findings showed that considering the PV panel degradation leads to a higher estimated lifetime. A remarkable point is that many households will tend to integrate a battery storage system into their PV system in the future. As the expected lifetime of lithium-ion batteries is less than that off PV panels, the battery will be replaced multiple times over the useful life of the PV panels. Therefore, repeating the first year of damage of the power devices over the total inverter lifetime contradicts the fact that the system operation might change significantly by the battery degradation.

This paper considers a 30-year timeframe for the damage to provide more accurate results. The yearly mission profile is extracted for a location in Belgium and is duplicated for the entire timeframe. The overview of the proposed approach is illustrated in Fig. 1. Each converter is controlled by a control system, e.g., maximum power point tracking is responsible to control the buck-boost converter. The charge/discharge of the battery is determined based on maximizing the self-consumption. It should be noted that in all simulations, the converters' efficiency is variable and is considered as a function of the converter's input power.

2 BATTERY LIFETIME ESTIMATION APPROACH

As the batteries play an important role in PV applications, estimating a reliable battery lifetime is important. In general, the reduction in the battery capacity depends on the number of battery charge/discharge cycles as well as the amplitude of these cycles. Thus, the state-of-charge (SOC) of the battery over a year can be translated to cycles with different depths by a rainflow counting algorithm. Then, from the battery's cycles-to-failure (N_f) versus depth-of-discharge diagram, the number of cycles-to-failure for each cycle can be determined. Finally, Miner's rule is used to accumulate the yearly damage of the battery. There is also another aging factor reducing the battery lifetime called the calendar aging, while the battery SOC idles at a constant level. A simplified formula has been used to calculate the calendar aging based on the SOC level and the idling time. A similar procedure has been adopted in [5].

The battery lifetime depends on both the mission profile as well as the decreased PV power generation as a result of degradation. In other words, a closed-loop calculation is needed to determine the battery lifetime so that the written program detects daily data and determines the battery life consumption over that data set.

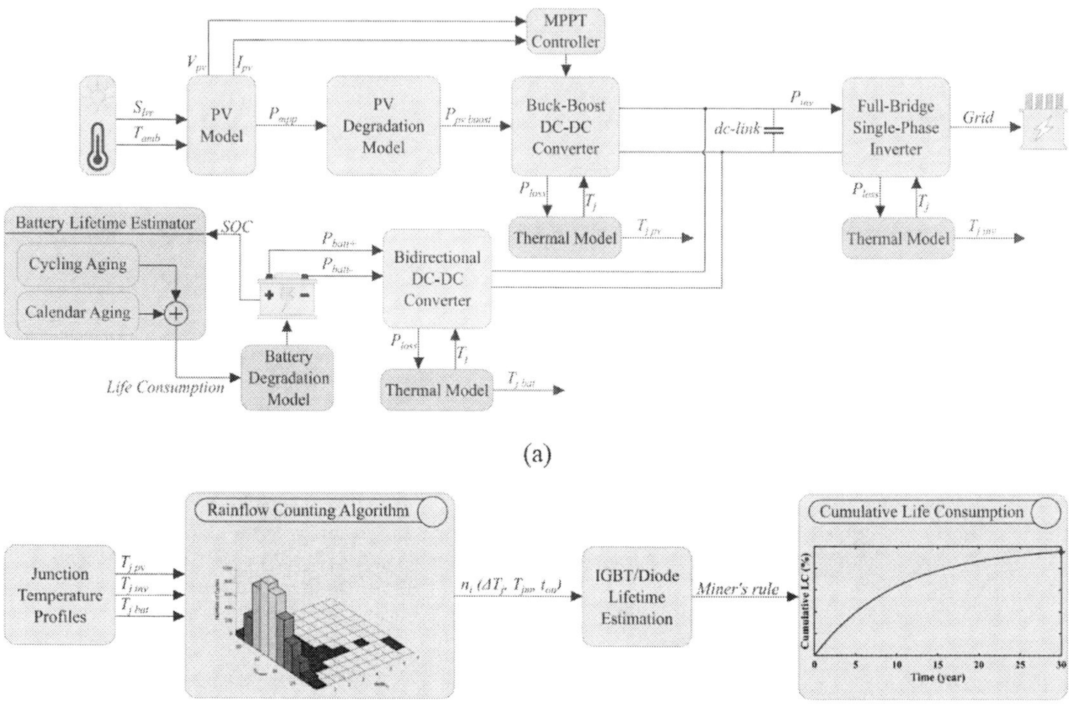

Figure 1: (a) General scheme of the proposed approach, (b) employed lifetime estimation procedure.

3 METHODOLOGY

The reliability approach used in this study has been adapted from [6]. A lookup table approach has been adopted to translate the converters' input power to the power losses, then to the junction temperature of the IGBTs/diodes. An important detail to consider at this point is that the ambient temperature of the inverter must be equal to the indoor temperature, since the inverter will be installed indoors in residential applications. Then, the rainflow counting method has been used to determine the number of thermal cycles with their corresponding amplitude (ΔT_j), average (T_{jm}), and period (t_{on}). The damage inflicted by every cycle on the bond wires can be calculated using the chosen lifetime model. This model has been explained in [7]. Finally, according to Miner's rule, the damage inflicted by every cycle can be accumulated to calculate the total life consumption (LC) of the mission profile on the IGBT/diode:

$$LC=\sum (n_i/N_{fi}) \tag{1}$$

where n_i is the number of cycles obtained from the rainflow analysis with a certain combination of ΔT_j, T_{jm} and t_{on}.

4 RESULTS AND DISCUSSION

Here, a linear degradation based on a 30-year PV lifespan is chosen as the reference case, and the analysis is performed over this timeframe. The lifetime calculations are performed for every semiconductor device used within the power converters. A part of the obtained results is shown in Fig. 2(a-b).

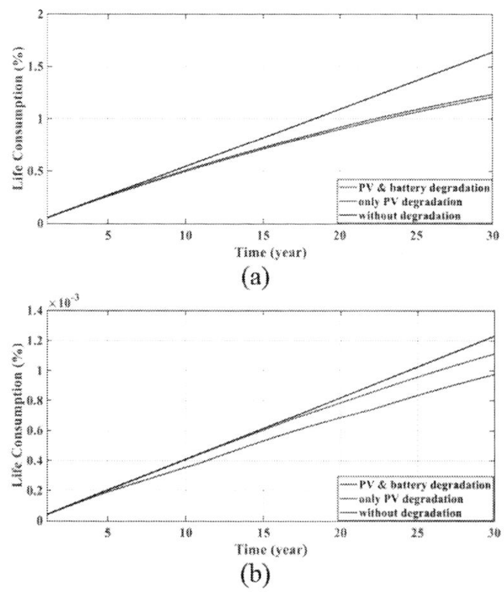

Figure 1: The estimated life consumption of the chosen IGBTs: (a) Single-phase inverter, and (b) Bidirectional DC-DC converter.

Fig. 2(a) depicts the life consumption related to the first-leg upper IGBT of the single-phase inverter. As is obvious, the estimated life consumption is much higher in the case without considering the degradation. This is expected because adding the degradation rates into the calculations decreases the input power of the converters; and as a result, this reduction relieves some of the thermal stresses in the semiconductors. By considering both the PV and battery degradation rates, the life consumption of this switch is reduced by 24.6%. Interestingly, the battery degradation has not had much impact on the inverter lifetime as it only causes approximately 2.3% difference with the case considering only PV degradation.

Fig. 2(b) illustrates the life consumption of the lower IGBT within the battery converter. The results of this converter are different from the previous case. Between the case without degradation and the case with all degradation rates included, there is an error of 26%. Considering the fading battery capacity will result in a lower life consumption as the input power of the battery converter declines. Additionally, avoiding the battery degradation into the simulations will lead to almost 14% difference compared to the case with both degradations.

The main problem with the aforesaid IGBT/diode lifetime estimation method is that the estimated life expectancy is not realistic, and the predicted life consumption can be only used to compare different scenarios. The employed lifetime model for semiconductor devices is obtained in such a way that the components are tested by a power cycling lab setup under different thermal conditions (e.g., different average temperatures T_{jm} and temperature fluctuations ΔT_j) and at the same time, different lifetime indicators such as collector-emitter voltage and junction-to-case thermal resistance are measured/monitored. When these indicators exceed their allowable value, it is assumed that the IGBT/diode is failed. These longevity tests can only be performed for high temperature fluctuations over a limited time. Due to the limitations of the empirical data on the number of cycles to failure (N_f) for low thermal fluctuations (ΔT_j), the model parameters for these small temperature swings must be extrapolated, which will lead to a non-realistic estimated lifetime.

5 CONCLUSIONS

This paper presents a flexible/comprehensive reliability analysis framework for power electronics circuits used in PV systems. The inputs and parameters of this framework can be easily modified in accordance with difference cases/scenarios. Here, an example of this framework was investigated to show the effect of PV panel degradation and battery degradation on the predicted lifetime of power electronics converters. Both calendar and cycling aging of the battery has been considered into the calculations. A linear degradation has been also added to the PV panel model.

The initial results have been provided for both the PV inverter and the DC-DC battery converter. To conclude, the primary results indicate that although the influence of the battery degradation on the inverter lifetime is negligible, it should be included for the battery converter.

The present lifetime model will lead to a non-realistic estimated lifetime more than expected lifetime, but this method can be used for comparison purposes. To overcome this lifetime prediction issue, an improved method consisting of real experimental output data on several components as well as accurate finite element method (FEM) simulations should be provided. In addition to the lifetime model related to semiconductor devices, the degradation model of the solar panel and energy storage conversion system can be replaced with a more accurate model to have a meaningful comparison not only in terms of component-level reliability but also in terms of system-level reliability.

ACKNOWLEDGEMENTS

The work is (partially) supported by the energy transition funds project "BREGILAB" organized by the FPS economy, S.M.E.s, Self-employed and Energy (Met de steun van het Energietransitiefonds).

REFERENCES

[1] De León-Aldaco SE, Calleja H, Chan F, Jiménez-Grajales HR. Effect of the mission profile on the reliability of a power converter aimed at photovoltaic applications—a case study. IEEE Transactions on Power Electronics. 2012;28(6):2998-3007.

[2] Sangwongwanich A, Angenendt G, Zurmühlen S, Yang Y, Sera D, Sauer DU, Blaabjerg F. Enhancing PV inverter reliability with battery system control strategy. CPSS Transactions on Power Electronics and Applications. 2018;3(2):93-101.

[3] de Barros RC, Brito EM, Rodrigues GG, Mendes VF, Cupertino AF, Pereira HA. Lifetime evaluation of a multifunctional PV single-phase inverter during harmonic current compensation. Microelectronics Reliability. 2018;88:1071-6.

[4] Sangwongwanich A, Yang Y, Sera D, Blaabjerg F. Lifetime evaluation of grid-connected PV inverters considering panel degradation rates and installation sites. IEEE transactions on Power Electronics. 2017;33(2):1225-36.

[5] Farinet D, Maurer M, Vacca L, Spataru SV, Stroe DI. Battery Lifetime Analysis for Residential PV-Battery System used to Optimize the Self Consumption-A Danish Scenario. In: IEEE Energy Conversion Congress and Exposition (ECCE) 2019 (pp. 6693-6698).

[6] Yang Y, Sangwongwanich A, Blaabjerg F. Design for reliability of power electronics for grid-connected photovoltaic systems. CPSS Transactions on Power Electronics and Applications. 2016;1(1):92-103.

[7] Van De Sande W, Ravyts S, Sangwongwanich A, Manganiello P, Yang Y, Blaabjerg F, Driesen J, Daenen M. A mission profile-based reliability analysis framework for photovoltaic DC-DC converters. Microelectronics Reliability. 2019;100:113383.

TARIFF DESIGN AND ECONOMIC DISPATCH IN A DC MICROGRID

Pedro Ferreira Torres[1], João Tavares Pinho[1,2], Roberto Zilles[1]
[1]Instituto de Energia e Ambiente – Universidade de São Paulo
[2]Grupo de Estudos e Desenvolvimento de Alternativas Energéticas – Universidade Federal do Pará
Avenida Professor Luciano Gualberto, 1289, bloco M. São Paulo, Brasil.

ABSTRACT: A cost allocation methodology for prosumers in a DC microgrid is proposed in this paper, based on the distribution of internal microgrid losses, energy cost and microgrid operator required income. Furthermore, an optimal economic dispatch problem is formulated to reduce overall microgrid energy acquisition costs and is solved using a meta-heuristic algorithm. The proposed approach is generic and considers different types of internal DC microgrid agents: consumers, independent producers and prosumers with behind-the-meter storage and/or generation; the microgrid connection to the main AC distribution grid is also considered in the model. The theoretical framework proposed in this paper aims to incentivize optimal energy management of flexible resources in a multi-agent DC microgrid, while is simple enough to be used as a reference in practical internal tariff designs for DC microgrids. The simulation results for a 48 V, six-node microgrid, indicate the effectiveness of both cost allocation and economic dispatch methodologies.
Keywords: DC Microgrid, Battery Storage and Control, Grid Management, Cost Allocation

1 INTRODUCTION

Microgrids are key to increase the reliability and efficiency of highly active energy systems with significant participation of distributed energy resources (DER) due to their increased capability of DER control when compared to large-scale distribution grids. DC distribution microgrids can further increase energy utilization in systems with a significant amount of DC sources and loads, as is the case of several small-scale residential, commercial, and even industrial installations. The adoption of distributed energy storage is an enabling factor that contributes to increased flexibility in the implementation of energy management strategies in microgrids, especially when associated with intermittent generation sources such as photovoltaics and wind.

In this sense, the strategic dispatch of DER is fundamental to ensure that the system will operate as close to the optimal as possible, in terms of economic value while guaranteeing technical compliance and energy security. There are several works proposing methodologies to optimize the economic dispatch of DER in microgrids [1-3]. In [1] an economic dispatch formulation is proposed to optimize energy management in a DC microgrid under real-time energy pricing, and uses a meta-heuristic algorithm to solve the problem. Meta-heuristics is also used in [2] to solve an economic dispatch optimization in a microgrid, considering uncertainties in load and generation profiles. A model predictive control is proposed in [3] to optimize the energy management in a microgrid with renewable energy resources. In addition to academic works, there is an increasing number of companies offering energy management services of DER, including demand response, for clients serviced within markets where such services are financially attractive.

In this context, two factors must be considered to enable optimal integration of DER: first, a coherent market design is important to economically incentivize microgrid users to adopt optimal management strategies; and second, the management strategy itself must be made possible through intelligent tertiary control capable devices and Information and Communications Technology (ICT) infrastructure. These two factors are correlated as the management strategy will follow market design rules and the latter must be adequate to DER management capabilities.

However, due to the increased complexity of electrical power flow calculations, technical compliance is often overlooked in optimal management methodologies. In practice, the calculated optimal economic dispatch may not be possible due to power quality constraints, for example, under-voltage that prevents proper operation of a battery power converter. This is even more noticeable in electrically weak small-scale systems in which power and voltage variations are substantial over daily operation. Such operational constraints must be considered to some extent when designing the management strategy and market rules, to ensure a viable operation. In this context, the proposed framework can be useful for microgrid owners/operators when developing the operation strategy and tariff design of a system while considering the technical requirements of the microgrid.

Fig. 1 presents the proposed framework for the integrated tariff design and economic dispatch development.

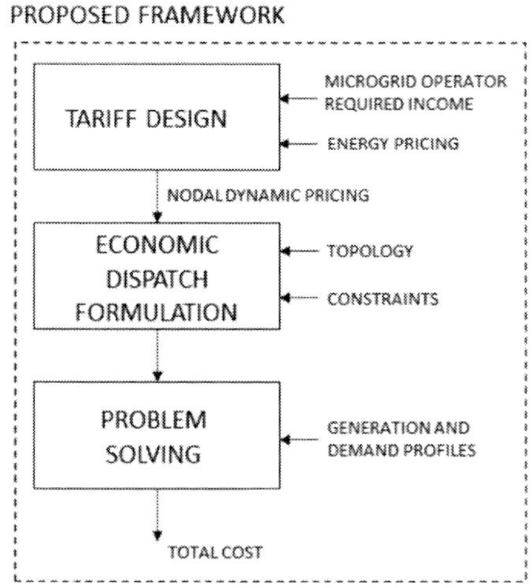

Figure 1: Structure of the proposed framework.

The process is divided in three mains blocks: (1) a

tariff scheme is developed considering that the total electricity cost is the sum of energy and distribution costs, therefore, the inputs of the tariff design block are the microgrid operator required income (distribution costs) and energy pricing (energy costs); (2) the economic dispatch is formulated considering both the DC microgrid topology and operational constraints, and, given the nodal pricing scheme obtained from the tariff design, it aims to reduce overall system electricity costs; (3) the economic dispatch problem is solved for a given set of daily generation and demand profiles of the microgrid agents.

This paper next sections are structured as follows: in section 2, the DC microgrid model is detailed, specifying individual DER models and the power flow formulation; the nodal pricing is formulated in section 3, including the loss allocation methodology for DC microgrids; section 4 describes the economic dispatch problem and in section 5 a simulation-based case study is presented to access the methodology results. The conclusions on the paper results are presented in section 6.

Notation details: single variables and constants are written in lower case italics; one-dimensional arrays are represented in lower case bold italics and matrices are in upper case bold italics.

2 DC MICROGRID MODEL

The general concept of DC microgrid approached in this paper is depicted in Fig. 2. The microgrid is composed by a set of consumers, prosumers (consumers with behind-the-meter generators and/or energy storage systems) and independent generators. For simplicity, from here on the word prosumer will be used to address all generation, storage and consumption agents connected in the microgrid. The DC microgrid can be connected to the main AC distribution grid through multiple bidirectional AC-DC converters.

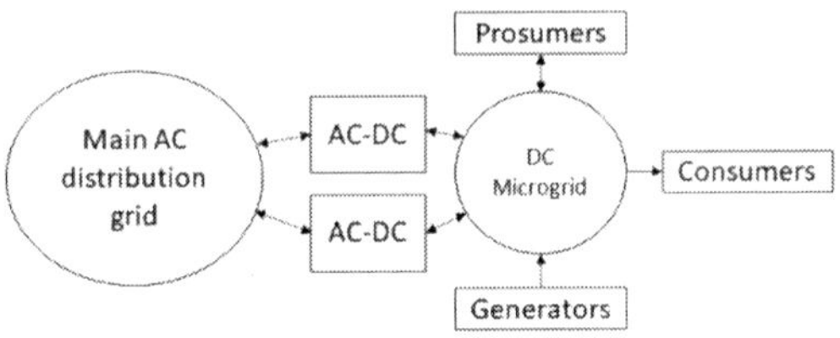

Figure 2: Generic DC microgrid concept.

2.1 DC microgrid power flow analysis

To develop a tariff scheme and further design an economic dispatch formulation that can be applied to real operation it is fundamental to model the microgrid considering the constrained operation of prosumers and the grid due to power quality (*i.e.*, admissible voltage range) and capacity conditions (*e.g.*, power ratings), as well as load characteristics. The DC power flow modeling is based on the works [1,4].

The distribution lines are modeled as resistive elements, where the conductance between nodes j and k is given as indicated in Eq. (1).

$$g_{jk} = (2\cdot 10^{6}\cdot \rho\cdot l_{jk}/a_{jk})^{-1} \quad (1)$$

where $\rho = 1.7\cdot 10^{-8}$ Ω·m is cooper resistivity at 20 °C, a_{jk} is cable cross sectional area in mm² and l_{jk} is the distance between nodes j and k, in meters. Resistance variation with temperature is not considered in this paper.

Two types of loads can be present in a given microgrid j^{th} node: constant power loads (CPL), which have fixed power ratings, $p_{CPL,j}$, given an admissible voltage range; and voltage sensitive loads modeled as constant resistance loads (CRL), with power demand according to Eq. (2).

$$p_{CRL,j} = v_j^{2}/\, r_{CRL,j} \quad (2)$$

where v_j is the node voltage and $r_{CRL,j}$ is the load resistance.

Distributed generation such as PV are modeled as non-dispatchable, constant power sources (CPS), with fixed power ratings at a given node j, $p_{CPS,j}$. Battery energy storage systems (BESS) are modeled as Thévenin-equivalent dispatchable sources (or loads), as shown in Fig. 3. j^{th} node's BESS power is given in Eq. (3).

$$p_{BESS,j} = v_{set,j}\cdot(v_{set,j} - v_j)/\, r_{th,j} \quad (3)$$

where $r_{th,j}$ is the Thévenin-equivalent resistance and $v_{set,j}$ is the setpoint voltage determined by the BESS controller. Therefore, by adjusting $v_{set,j}$, it is possible to dispatch the BESS power. Note that $r_{th,j}$ is the virtual resistance from the droop current sharing control method, not an actual physical component.

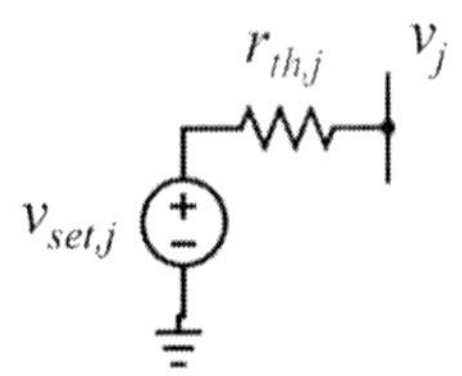

Figure 3: BESS equivalent model.

Additionally, the DC microgrid can be connected to the main AC distribution system via an AC-DC bidirectional converter. This converter is modeled as a dispatchable source similar to the BESS model, according to Eq. (4).

$$p_{AC\text{-}DC,j} = v_{conv,j}\cdot(v_{conv,j} - v_j)/\, r_{conv,j} \quad (4)$$

where $v_{conv,j}$ and $r_{conv,j}$ are the converter's voltage setpoint and Thévenin-equivalent resistance, respectively. Similarly to the BESS, the power flow from the AC main distribution grid can be controlled by adjusting $v_{conv,j}$.

In a microgrid with n nodes, the power injected in the jth node from other nodes is given by Eq. (5), where $\mathcal{N}$ is the set of microgrid nodes. Eq. (6) shows the power balance equation for the jth node.

$$p_{MG,j} = \sum v_j\cdot(v_j - v_k)\cdot g_{jk},\ \forall k \in \mathcal{N} \cap k \neq j \quad (5)$$

$$p_{BESS,j} + p_{AC\text{-}DC,j} + p_{CPS,j} + p_{MG,j} - p_{CPL,j} - p_{CRL,j} = 0 \quad (6)$$

Eq. (6) can be used to form a system of n equations for all nodes in the microgrid and solve for the set of n-node voltages, $\boldsymbol{v} = [v_1, v_2, \ldots, v_n]$. Reference [4] presents further details on the power flow analysis of a generic DC microgrid. The power demand of a prosumer connected to the jth node is given in Eq. (7).

$$p_{dem_J} = p_{CPL_J} + p_{CRL_J} - p_{BESS_J} - p_{CPS_J} \quad (7)$$

Note that p_{BESS_J} in Eq. (7) can have positive or negative values, depending whether it is discharging or charging, respectively. Also, a positive p_{dem_J} indicates that the prosumer in node j is demanding power from the microgrid, while a negative p_{dem_J} corresponds to power injection from the prosumer to the microgrid.

2.2 Microgrid distribution losses

Distribution power losses are used to verify the efficiency of the distribution system and is an important parameter to identify the microgrid use by the different prosumers, as it is related to the current flows in the lines. Furthermore, the microgrid nodal distribution-use tariff is based on the loss allocation and is detailed in section 3. Eq. (8) is used to calculate the total distribution losses as a function of nodal voltages.

$$p_{LOSS} = f_L(\mathbf{v}) = 0.5 \cdot \sum\sum g_{jk} \cdot (v_j - v_k)^2 \,,\ \forall j,k \in \mathcal{N} \quad (8)$$

3 DC MICROGRID TARIFF DESIGN

Proper tariff design is fundamental to incentivize optimal operation of the distributed energetic resources in a microgrid. Separation of energy and distribution (wire) costs and distribution cost allocation are factors that motivate strategic management of flexible resources to improve economic efficiency and thus reduce overall costs. It is important to point out that a comprehensive tariff design process must considerer several non-technical factors such as the principle of equity, affordability and other charges related to public incentives programs, but those are out of the scope of this paper.

In this paper, it is proposed that the total electricity cost for a prosumer in the microgrid is composed by the sum of two components: energy costs (c_{en}), which is based on the energy consumption; and distribution costs (c_{dist}), which is based on the demand. The tariffs correspondent to both cost components are calculated considering the distribution loss that is allocated to each prosumer.

3.1 DC Distribution Loss Allocation

The formulation for distribution loss allocation is based on the direct loss coefficient (DLC) method for AC systems presented in [5]. For the DC microgrid, an analogous method for DC distribution is developed in this work and presented as follows.

From Eq. (8), for a given initial operating state $p^0_{LOSS} = f_L(\mathbf{v}^0)$, a change in the system voltages, $\mathbf{\Delta v}$, will cause a change in total system losses. The new total system losses, p_{LOSS}, can be approximated by the first three terms of a Taylor series expansion around the initial set of node voltages as

$$p_{LOSS} \approx f_L(\mathbf{v}^0 + \mathbf{\Delta v}) \approx p^0_{LOSS} + f_L{}'(\mathbf{v}^0) \cdot \mathbf{\Delta v} + 0.5 \cdot \mathbf{\Delta v} \cdot f_L{}''(\mathbf{v}^0) \cdot \mathbf{\Delta v}^T \quad (9)$$

Given an initial operating state where all nodal voltages are $\mathbf{v}^0 = \mathbf{1}$ p.u.: $p^0_{LOSS} = 0$, $f_L{}'(\mathbf{v}^0) = 0$ and Eq. (9) can be simplified to

$$p_{LOSS} \approx 0.5 \cdot \mathbf{\Delta v} \cdot \mathbf{H} \cdot \mathbf{\Delta v}^T \quad (10)$$

where $\mathbf{H}$ is the hessian matrix of second order partial derivatives of $f_L(\mathbf{v})$ with respect to $\mathbf{v}$. Note that $\mathbf{\Delta v}$ in Eq. (10) is the voltage difference from the new operating point relative to the initial condition, *i.e.*, $\mathbf{\Delta v} = \mathbf{v} - \mathbf{v}^0$. To express losses in terms of nodal power injections, an analogy with Newton-Raphson power flow is used to equate $\mathbf{\Delta v}$ and $\mathbf{\Delta p}$ as depicted in Eq (11).

$$\mathbf{\Delta v}^T = \mathbf{J}_{avg}{}^{-1} \cdot \mathbf{\Delta p}^T \quad (11)$$

where $\mathbf{\Delta p} = \mathbf{p} - \mathbf{p}^0$ is the difference in nodal power injections between the new and the initial operating states and $\mathbf{J}_{avg}$ is an average of the Jacobian of first order derivatives of nodal power injections with respect to $\mathbf{v}$ taken at the initial and new operating states as shown in Eq. (12).

$$\mathbf{J}_{avg} = 0.5 \cdot (\mathbf{J}^0 + \mathbf{J}) \quad (12)$$

As in the initial operating state there are no nodal power injections, $\mathbf{p}^0 = \mathbf{0}$. Therefore, Eq. (10) can be rewritten as

$$p_{LOSS} \approx 0.5 \cdot \mathbf{\Delta v} \cdot \mathbf{H} \cdot \mathbf{J}_{avg}{}^{-1} \cdot \mathbf{p}^T \quad (13)$$

Eq. (13) shows how the total system losses are related to the nodal power injections, where for a given operating point, the DLC are given as shown in Eq. (14). The DLC can be used to determine the relative contribution of the power injection or demand at a given node for the overall microgrid distribution line losses.

$$\gamma = 0.5 \cdot \mathbf{\Delta v} \cdot \mathbf{H} \cdot \mathbf{J}_{avg}{}^{-1} \quad (14)$$

3.2 Distribution cost component (c_{dist})

The product of DLC given in Eq. (14) with nodal power injections should directly express the total power losses. However, approximation errors occur due to the calculation of the equivalent Jacobian. Therefore, to accurately allocate the distribution and energy costs in terms of power loss, the DLC are scaled with respect to total losses, so that the sum of individual prosumer costs equates to the total microgrid distribution and energy costs.

The distribution cost component is based on the allocation of the total required income by the microgrid operator, c_{req}, and must cover operation and maintenance costs, planned expansion costs and operator profit. There are several methodologies to define c_{req}, and its calculation is not in the scope of this paper. Here, a fixed value for c_{req} is adopted to calculate the prosumer distribution costs. It is reasonable to allocate c_{req} to the prosumers according to each prosumer grid use, which, in turn, is related to the prosumer contribution to total distribution losses. Therefore, the distribution cost component of a prosumer connected in the microgrid's j^{th} node is given by Eq. (15).

$$c_{dist_J} = p_{dem_J} \cdot \gamma_J \cdot c_{req} / p_{LOSS} \quad (15)$$

Note that the product between the prosumer demand and its respective DLC can be positive (*i.e.*, when the power demand or injection into the microgrid is increasing total system losses) or negative (*i.e.*, when the power demand or injection into the microgrid is reducing total system losses). Hence, this cost allocation method incentivizes demand-side management to reduce distribution losses and, in turn, costs. Alternatively, part of c_{req} can be evenly distributed among the prosumers, as a fixed cost of grid availability, whereas the demand-dependent cost allocation presented in Eq. (15) can be used to distribute the remainder of c_{req}. Eq. (16) describes this alternative method to calculate c_{dist}.

$$c_{dist,j} = (1-k_0)\cdot c_{req} + p_{dem,j}\cdot \gamma_j\cdot k_0\cdot c_{req}/p_{LOSS} \quad (16)$$

where k_0 is the coefficient that determines the demand-dependent parcel of $c_{dist,j}$ with respect to c_{req}.

3.2 Energy cost component (c_{en})

The energy consumption cost is also calculated considering a cost allocation method, so that the energy dissipated in the grid is allocated according to the prosumer consumption. The total microgrid energy cost $c_{en,mg}$ corresponds to the energy brought from the external AC main grid, hence

$$c_{en,mg} = \sum p_{AC\text{-}DC,j}\cdot \Delta t\cdot \alpha_j, \qquad \forall j \in N \quad (17)$$

where Δt is the calculation timestep given in hours (*e.g.*, hourly or quarter-hourly) and α_j is the AC main grid energy purchase tariff. Note that this tariff can be spatial and temporal and vary according to the time of use and grid node interconnection location. α_j can also have different values when power is exported from the microgrid to the AC main grid (in such case, generally smaller). Eq. (18) is used to allocate the total microgrid energy cost among its prosumers.

$$c_{en,j} = p_{dem,j}\cdot(1+\gamma_j)\cdot c_{en,mg}/\sum p_{dem,k}\cdot(1+\gamma_k), \forall k \in N \quad (18)$$

A negative $p_{dem,j}$, which indicates that the prosumer is injecting power into the grid during the given timestep, corresponds to a negative energy cost component for the prosumer, that is, a benefit. Such a benefit or cost is also increased or decreased according to the DLC component in Eq. (18).

The total electricity cost for a prosumer in the jth microgrid node is given as

$$c_{total,j} = c_{dist,j} + c_{en,j} \quad (19)$$

4 ECONOMIC DISPATCH FORMULATION

As shown in section 2, in the DC microgrid modeled in this paper there are two types of flexible resources that can be controlled to achieve optimal operation: the BESS and the bidirectional AC-DC main grid coupling converters. In a microgrid with np prosumers with BESS and nc nodes interconnected with the AC main grid, the dispatch of these sources is regulated by the setpoint voltages, $\mathbf{v_{set}} = [v_{set,1}, v_{set,2}, \ldots, v_{set,np}]$ and $\mathbf{v_{conv}} = [v_{conv,1}, v_{conv,2}, \ldots, v_{conv,nc}]$. Hence, for a given set of non-dispatchable sources and loads, the goal is to find a set of dispatchable sources setpoints, $\mathbf{v_{disp}} = [\mathbf{v_{set}}, \mathbf{v_{conv}}]$, that minimizes the microgrid operational and energy acquisition costs.

To determine optimal $\mathbf{v_{disp}}$, it is considered a day-ahead approach based on next-day hourly forecasts of prosumer's non-dispatchable resources (*i.e.*, CPS, CPL and CRL) and AC main grid energy purchase tariff. This is a non-linear constrained optimization problem that is solved using a genetic algorithm approach that is implemented in MATLAB's version R2021a optimization toolbox.

4.1 Objective function and constraints

Considering daily operation, the total number of time intervals is given by $i = 24/\Delta t$; therefore, $\mathbf{V_{disp}} = [\mathbf{v^1_{disp}}, \mathbf{v^2_{disp}}, \ldots, \mathbf{v^i_{disp}}]$ represents the set of dispatchable sources setpoints over time, and the total microgrid daily energy acquisition cost, $c_{en,mg,daily}$, is given by the sum of all $c_{en,mg}$ during a given day of operation. Hence, the objective function is formally described in Eq. (20).

$$f_{obj}(\mathbf{V_{disp}}) = min(c_{en,mg,daily}) \quad (20)$$

This function is subject to the following constraints:

$$soc_{min} \leq soc^t_j \leq soc_{max} \quad (21)$$

$$v_{min} \leq v^t_j \leq v_{max} \quad (22)$$

$$p_{BESS,min} \leq p^t_{BESS,j} \leq p_{BESS,max} \quad (23)$$

$$p_{AC\text{-}DC,min} \leq p^t_{AC\text{-}DC,j} \leq p_{AC\text{-}DC,max} \quad (24)$$

$$soc_{end} \leq soc^t_j \quad (25)$$

Eqs. (21) to (24) indicate the admissible operating ranges of BESS state of charge (SoC), node voltages, BESS power, and AC-DC converter power, respectively. The superscript t =1, 2, ..., i indicates the time interval. Eq. (25) indicates that the BESS must end the optimization period with a SoC no smaller than soc_{end}. SoC is calculated by the energetic balance for each BESS over time, according to Eq. (26).

$$soc^t_j = soc^{t-1}_j + p^t_{BESS,j}\cdot \Delta t/e_{BESS,j} \quad (26)$$

where soc^{t-1}_j is the SoC at the previous timestep and $e_{BESS,j}$ is the BESS rated energy capacity. For simplicity, $e_{BESS,j}$ is assumed to be constant and is not sensitive to operating temperature and charging/discharging rate.

5 CASE STUDY SIMULATION

A laboratory-scale radial unipolar 48 V DC microgrid was simulated to validate the proposed approach. It is composed of six nodes and includes constant power and constant resistance loads (CPL and CRL), and two PV-BESS prosumers (PV rated at 660 Wp and BESS rated at 500 W/ 4,800 Wh). The microgrid is connected to the AC main grid at the point of common coupling (PCC) via an AC-DC converter in node 0, rated at 1,000 W. Fig. 4 illustrates the simulated microgrid and Tab. I summarizes the cabling distances. $r_{th,3} = r_{th,5} = 0.01\ \Omega$ and $r_{conv,0} = 0.25\ \Omega$ are the virtual resistances for the BESS and AC-DC converter dispatch control. A SoC range of 0.2 to 1.0 is considered, while $soc_{end} = 0.5$. v_{min} and v_{max} are set to 0.9 and 1.1 p.u. referred to 48 V, respectively.

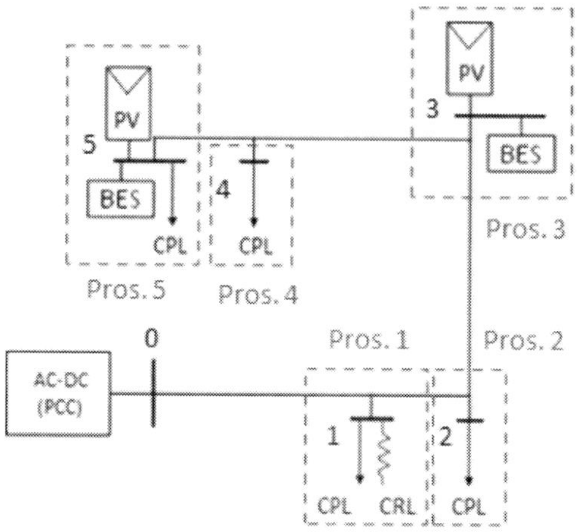

Figure 4: Configuration of the simulated DC Microgrid with PCC and prosumers indication.

Table I: DC Microgrid cabling distance.

From Node	To Node	Distance (m)	Cable Cross-Section (mm^2)
0	1	30	16
1	2	15	16
2	3	50	16
3	4	30	16
4	5	15	16

Tab. II summarizes the hourly generation and loading profiles at the microgrid nodes. The power generated by CPS (PV) and demanded by CPL as well as the resistance of CRL is considered the average within each hour. Further time discretization is possible at the expense of more computational power. However, for this case-specific energy management purposes an hourly simulation is adequate. A 3-level time of use (TOU) pricing is adopted for energy pricing from the AC main grid, with energy selling priced at 15 % of the energy buying cost, as shown in Tab. III. A daily c_{req} = \$ 2.00 is assumed.

Table II: Hourly PV generation and load profiles.

Hour	CPS Node 3 (W)	CPS Node 5 (W)	CPL Node 1 (W)	CPL Node 2 (W)	CPL Node 4 (W)	CPL Node 5 (W)	CRL Node 1 (Ω)
1	0	0	40	20	12	12	23
2	0	0	40	20	12	6	20
3	0	0	40	20	12	12	50
4	0	0	40	20	12	6	23
5	0	0	40	20	12	12	23
6	50	55	150	20	45	6	20
7	100	110	150	20	45	45	50
8	200	220	150	20	45	6	23
9	300	330	300	100	90	90	-
10	350	385	20	100	6	30	-
11	400	440	20	100	6	6	-
12	600	660	20	400	6	120	-
13	500	550	20	500	6	6	-
14	350	385	20	400	6	120	-
15	300	330	20	400	6	6	-
16	250	275	20	100	6	30	-
17	150	165	20	150	6	6	-
18	20	22	400	150	120	45	23
19	0	0	500	400	150	150	20
20	0	0	550	500	165	150	50
21	0	0	550	300	165	165	23
22	0	0	400	300	120	90	23
23	0	0	400	40	120	120	20
24	0	0	40	20	12	6	50

Table III: Energy pricing from the AC main grid.

TOU Level	$\alpha_{j,buy}$ (\$/kWh)	$\alpha_{j,sell}$ (\$/kWh)
Peak (18 h to 21 h)	1.12	0.17
Intermediate (17 h to 18 h and 21 h to 22 h)	0.73	0.11
Off-peak (other periods)	0.50	0.08

5.1 Evaluation of economic dispatch results

The economic dispatch results are presented in Figs. 5 to 7 for a single day simulation. Fig. 5 shows the hourly power injected/absorbed at the PCC and prosumer nodes. The use of batteries allows the prosumers to inject power into the microgrid even when PV generation is absent (for instance, from 19 h to 24 h). As expected, BESS dispatch is highly influenced by the peak pricing from 18 h to 21 h.

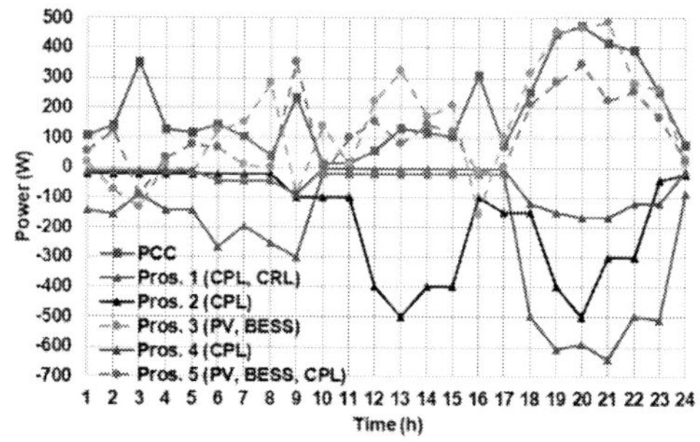

Figure 5: Hourly power demanded/injected at the PCC and prosumer nodes.

Fig. 6 shows the state of charge of the batteries at the nodes with BESS. Since the dispatch considers only a single day in advance, the batteries are discharged to the minimal SoC at the end of the day to minimize energy costs from AC main grid. Depending on the battery technology, different SoC ranges can be set to maximize the battery lifecycle and further reduce costs. It is clear from Fig. 6 that the algorithm tries to maximize the SoC during off-peak hours, so when intermediate and peak pricing begins at 18 h, BESS is ready to discharge.

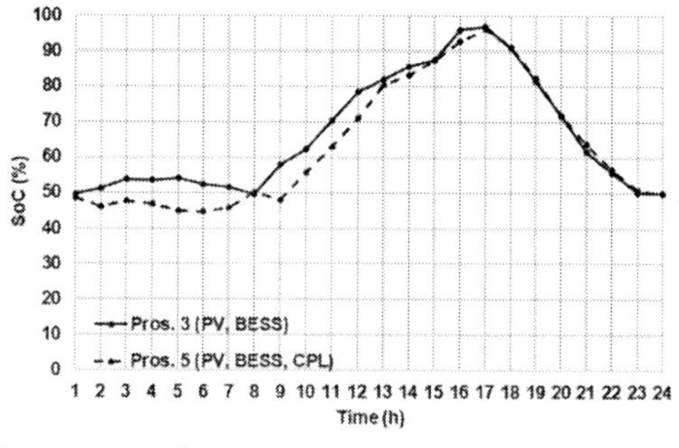

Figure 6: Hourly SoC at nodes with BESS.

Fig. 7 shows that nodal voltages remain within the admissible operating range of 0.9 to 1.1 p.u. (43.2 V and 52.8 V). It is possible to note the importance of prosumers 3 and 5 regarding voltage regulation, as voltage variations for prosumer 4 (CPL) are smaller than those for prosumers 1 and 2, even though they are closer to the PCC. This is especially evident during peak system loading, as is highlighted in Fig. 7, from 18 h to 22 h.

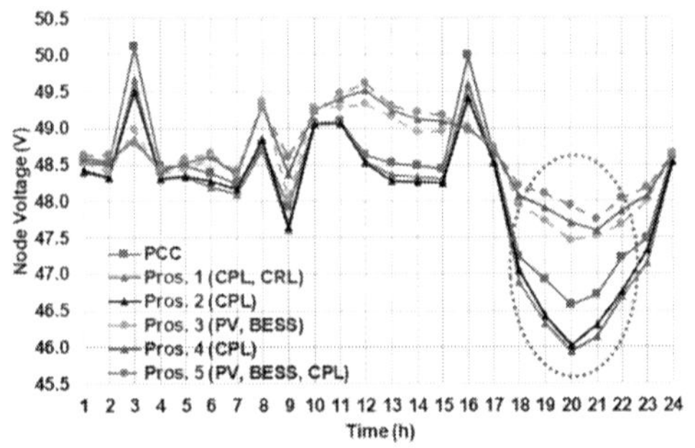

Figure 7: Hourly nodal voltages.

5.2 DLC and cost allocation result assessment

As indicated in the economic dispatch formulation in section 4, the objective function aims to minimize the overall energy acquisition cost for the DC microgrid. This cost is passed on to prosumers according to Eq. (18); thus, the minimization of $c_{en,mg,daily}$ implies in the minimization of individual prosumer energy costs, c_{en}. Even though distribution cost component is used to calculate the total electricity cost for a prosumer, it is not considered in the economic dispatch problem; however, it is important to understand how this cost is allocated to the prosumers and verify its suitability for other future economic dispatch formulations.

Figs. 8 and 9 present the hourly distribution and energy costs allocated respectively for each microgrid prosumer. Distribution costs are higher for those prosumers that contribute the most to distribution losses. Overall microgrid energy cost is $c_{en,mg,daily}$ = \$3.34, which represents the aggregated energy cost of all microgrid prosumers. However, prosumers that have generation and storage (3 and 5) presented a negative energy cost, indicating that those prosumers earned from the energy sold to the other microgrid prosumers. Tab. IV summarizes the daily electricity costs.

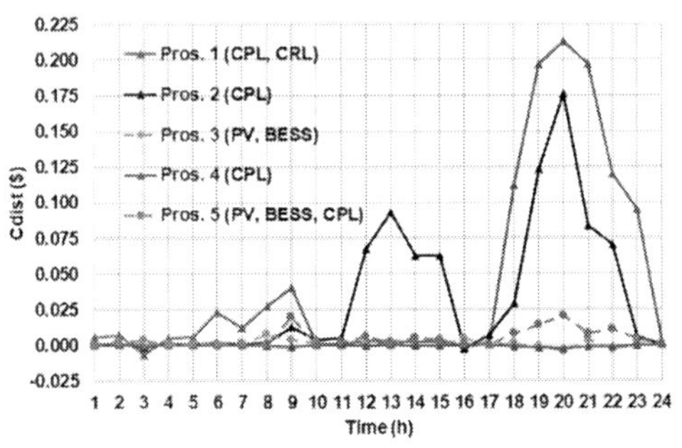

Figure 8: Hourly nodal distribution costs.

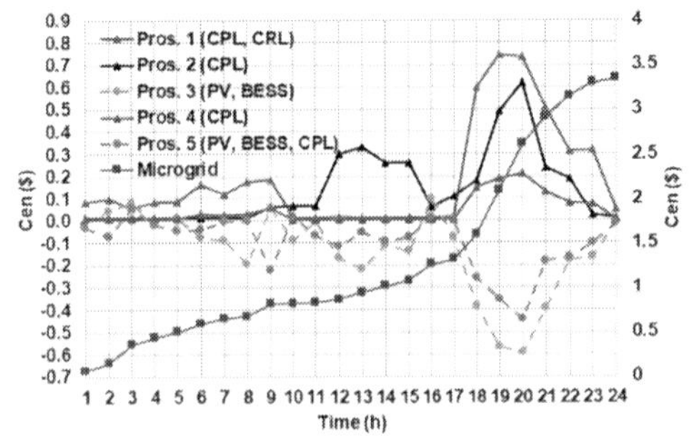

Figure 9: Hourly nodal energy costs and accumulated microgrid energy acquisition costs.

Table IV: Electricity costs for microgrid prosumers.

	C_{dist}	C_{en}	C_{total}
Pros. 1	\$ 1.07	\$ 4.42	\$ 5.49
Pros. 2	\$ 0.81	\$ 3.37	\$ 4.18
Pros. 3	\$ 0.02	\$ -3.21	\$ -3.19
Pros. 4	\$ -0.01	\$ 1.05	\$ 1.04
Pros. 5	\$ 0.11	\$ -2.29	\$ -2.18
Microgrid	\$ 2.00	\$ 3.34	

It is interesting to note that prosumer 4 has a negative, yet small, distribution cost. This indicates that this prosumer contributes to reduce total system losses. To understand this, one can refer to Fig. 4 and note that prosumer 4 is situated between prosumers 3 and 5, which have both generation and storage devices that are used to supply prosumer's 4 CPL in some extent. Without prosumer's 4 CPL, the surplus energy from the PV-storage resources in the neighboring nodes would be used to supply CPL and CRL at more distant nodes, or even be exported to the main grid, flowing through a longer distance and increasing overall system losses.

To further investigate the loss allocation methodology results, a second scenario was simulated considering that the hourly CPL of all nodes were reduced to 10 % of the base case values, so that the microgrid have significant surplus generation. Fig. 10 shows the PCC and prosumers power at the new scenario. The surplus PV generation in the microgrid is exported to the grid; hence, a negative power at the PCC appears, not only during peak PV hours, but also when PV is not present, due to BESS discharge.

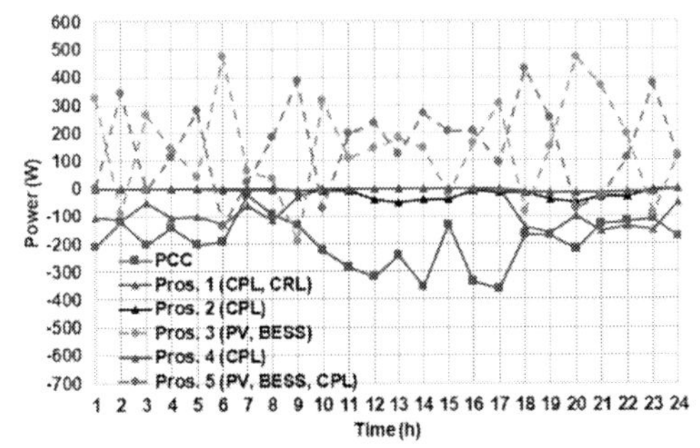

Figure 10: Hourly power demanded/injected at the PCC and prosumer nodes for the modified scenario.

Fig. 11 shows the distribution costs for the modified scenario. The surplus PV generation now has a significant participation in losses, especially the generation at prosumer 5, which is in the farthest away node from the PCC.

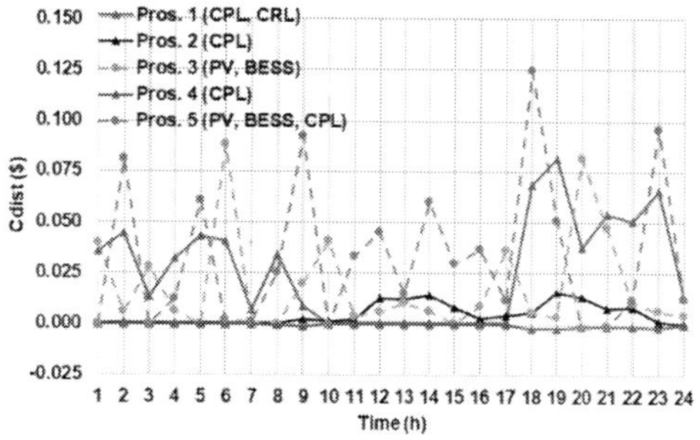

Figure 11: Hourly nodal distribution costs for the modified scenario.

6 CONCLUSIONS

The results indicate that the economic dispatch methodology can be used to optimize the energy management of DER in the DC microgrid, taking advantage of the flexibility associated with the energy storage elements, so that microgrid energy costs are reduced. For example, in the base scenario, BESS is dispatched considering the TOU energy pricing and the requirement to end the day with a minimal state of charge so that a similar day-ahead dispatch methodology can be used continuously. The dispatch also guarantees that the SoCs of the battery banks remain within optimal range to ensure a longer lifetime. The loss allocation methodology was successful in allocating energy and distribution costs to the prosumers in the microgrid according to the nodal power injections contributions to losses. As the methodology was evaluated in a small 6-node system, future works must address the following topics: (1) verify the scalability for larger microgrid topologies in terms of computational requirements and optimization performance; (2) further investigate the loss allocation methodology to correctly express loss in terms of nodal injections without the need of making approximations.

7 REFERENCES

[1] Li, C.; Bosio, F.; Chen, F.; Chaudhary, S.; Vasquez, J.; Guerrero, J. Economic Dispatch for Operating Cost Minimization Under Real-Time Pricing in Droop-Controlled DC Microgrid. IEEE Journal of Emerging and Selected Topics in Power Electronics, v. 5, n. 1, March 2017.

[2] Nikmehr, N.; Najafi-Ravadanegh, S. Optimal operation of distributed generations in micro-grids under uncertainties in load and renewable power generation using heuristic algorithm. IET Renewable Power Generation, v. 9, n. 8, November 2015.

[3] Santillán-Lemus, F.; Minor-Popocatl, H.; Aguilar-Mejía, O.; Tapia-Olvera, R. Optimal Economic Dispatch in Microgrids with Renewable Energy Sources. Energies, v. 12, n. 181, January 2019.

[4] Chen, F. Control of DC Power Distribution Systems and Low-Voltage Grid-Interface Converter Design. Doctoral Thesis in Electrical Engineering, Virginia Polytechnic Institute and State University, 2017.

[5] Mutale, J.; Strbac, C.; Curcic, S.; Jenkins, N. Allocation of losses in distribution systems with embedded generation. IET Proc. – Gen., Trans. And Dist. February 2000. DOI. 10.1049/ipgtd20000003.

EFFECT OF ENERGY STORAGE ON SELF-CONSUMPTION AND SELF-SUFFICIENCY: A FIELD STUDY IN A NORDIC CLIMATE

Patrik Ollas
Department of Energy and Resources
RISE Research Institutes of Sweden
Industrigatan 4, 501 15 Borås, Sweden
patrik.ollas@ri.se

Jon Persson
Department of Energy and Resources
RISE Research Institutes of Sweden
Industrigatan 4, 501 15 Borås, Sweden
jon.persson@ri.se

Peter Kovacs
Department of Energy and Resources
RISE Research Institutes of Sweden
Industrigatan 4, 501 15 Borås, Sweden
peter.kovacs@ri.se

ABSTRACT: This paper examines the variation in self-consumption and self-sufficiency for eight residential-size solar photovoltaic systems under Nordic climate conditions. The work uses measured photovoltaic array performances with a 30-second temporal resolution and is evaluated for a full year's operation. One of the systems is operated with battery storage to examine the effect on the system's self-consumption and self-sufficiency, where the simulated load profile is taken from a single-family residential building in Sweden. The results show a consistent negative correlation between array size and self-consumption. The relation between array size and self-sufficiency is more sensitive and is also affected by the mix of PV technologies represented in the test. Adding battery storage increases the self-consumption in the range of 0-30 percentage points weekly. Furthermore, the variation of self-consumption and self-sufficiency follow the seasonal variation in solar resource and load size and illustrate the battery's eventual redundancy in the winter when the battery's objective function is to maximize self-consumption.
Keywords: Self-consumption, Battery Storage and Control, System Performance, Self-sufficiency, Storage.

1 INTRODUCTION

The coupling of battery storage with solar photovoltaic (PV) systems is becoming more popular in several different applications, with a storage market following the path of the PV expansion. Most commonly, the battery storage is used to increase the self-consumed energy within the system's boundary by storing PV surplus via charging and later discharge during PV deficits.

The correlation between system economic profitability and system self-consumption (SC) has been examined in previous studies of solar photovoltaic [1, 2] and storage systems [3]. Luthander et al. have also conducted a comprehensive literature review of studies of self-consumed photovoltaic energy in different scenarios and measures to improve this quantity [4]. In the referred study, the relation between self-consumption, self-sufficiency, and PV array size is presented where the former has a negative correlation while the latter has a positive. A simulation study of the variation in self-consumption and self-sufficiency with varying PV array and battery storage sizes is done in [5], using close to 2000 measured consumption profiles in Sweden. Here, reported gains in SC range between 20-50 percentage points depending on the load profile.

The effect of the data set's temporal resolution on the system's self-consumption has been examined in other works, e.g., [6, 7, 8], all concluding that a coarser resolution results in a larger relative error, especially prominent for systems without battery storage. In [4], it is concluded that sub-hourly data are needed to have an accurate evaluation of the system's performance, therefore, this study uses a 30-second temporal resolution for all measured quantities.

The purpose of this work is to examine the variation in self-consumption and self-sufficiency for different residential PV array sizes (3-6 kWp) using recorded data of PV generation and energy usage from a full year's operation in Borås, Sweden. Furthermore, the effect of adding a battery to the system's self-consumption and self-sufficiency is quantified and the seasonal variation of the two factors is examined.

2 METHODS

2.1 Test Site Layout & Specifications

The measurements were done at RISE premises in Borås, Sweden. All systems are oriented due south (180° azimuth) and with a 45° tilt angle to the horizontal. The tilt was chosen to maximize the annual output at the site's latitude [9]. The technical specifications of the systems are found in Table I[1]. The selection of systems was made to reflect the residential market in Sweden and to cover its variance in buyers' choices. System "F" was equipped with battery storage and operated against a residential load profile. The load profile was taken from measurements on an electrically heated Swedish single-family house, with an annual energy demand of 16895 kWh and a peak power usage of 6 kW. Figure 1 shows the weekly energy demand and measured solar irradiance from the measurements. Here, the seasonal miscorrelation is seen between energy usage and solar irradiance. The load peaks during the colder months are caused by the heat pump's operation and reduce during the warmer periods.

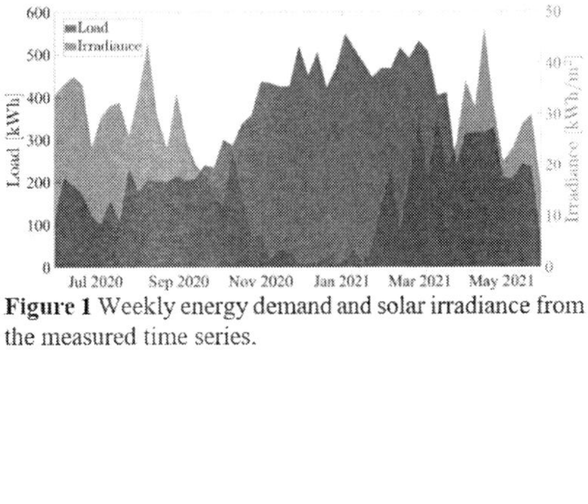

Figure 1 Weekly energy demand and solar irradiance from the measured time series.

[1] "BIPV" = Building Integrated PhotoVoltaic and "BAPV" = Building Applied PhotoVoltaic.

Table I Technical specifications of the eight systems used for the performance evaluation.

	System identification							
	A	B	C	D	E	F	G	H
Peak power [Wp]	3080	2975	3900	4080	3245	6200	4800	3289
BIPV/BAPV	BIPV/BAPV	BIPV	BAPV	BIPV	BIPV	BAPV	BAPV	BIPV
Cell technology[2]	CIGS	CdTe	c-Si	c-Si	c-Si	c-Si	c-Si	CIGS
Battery size [kWh]	-	-	-	-	-	5.7	-	-

2.2 Data Acquisition

Data were recorded in two separate sets with a 30-second temporal resolution and then resampled and synchronized to a common time step vector. The temporal resolution was selected based on findings from a related study of the effect of the temporal resolution on the usage profile and peak powers [8], [10] and to match the Class B requirements in [11]. A list of relevant measured quantities can be found in Table II. The data acquisition was continuously monitored, and obvious measuring errors and abnormalities were removed before the evaluation.

Table II Summary of measured quantities.

Data	Measurement	Unit
E_t	Generated energy for systems A–E and G–H	kWh
e_{load}	Load energy demand	kWh
e_F	Energy flows to and from system "F"	kWh
POA	Global solar irradiance in the plane of the array	W/m²

2.3 Performance Evaluation

A key performance index when evaluating the techno-economic performance of a PV array system is the share of self-consumed energy. In [4] the quantity of self-consumed energy, $m(t)$, is defined as

$$m(t) = \min\,[l(t), p(t)] \qquad (1)$$

where $l(t)$ is the load energy and $p(t)$ the PV generation in time t. When a battery is added to the system and charging and discharging is only made from excess PV generation and to the load respectively, the self-consumption is calculated as

$$m(t) = \min\,[l(t), p(t) \pm s(t)] \qquad (2)$$

where $s(t)$ is the power flow to and from the battery, and where $s(t)$ is defined negative when the battery is charged and positive when discharged. In this way, the battery's losses due to charging, storage, and discharge are all considered. In relative terms, the self-consumed energy is calculated as

$$\Phi_{SC} = \frac{\int_{t_1}^{t_2} m(t)dt}{\int_{t_1}^{t_2} p(t)dt} \qquad (3)$$

and defines the share of self-consumed energy over the period t_1-t_2. The system's self-sufficiency is defined similarly but with the load energy in the denominator as

$$\Phi_{SS} = \frac{\int_{t_1}^{t_2} m(t)dt}{\int_{t_1}^{t_2} l(t)dt} \qquad (4)$$

2.4 Battery System Operation

The battery's dispatch is done to maximize the system's self-consumption, using the algorithm described in [12]. With this dispatch, charging is done using PV surplus and discharge during PV deficit. The battery's state-of-charge (SOC) was kept within 7–97% and with maximum power for charge and discharge at 4.6 kW. In Figure 2 the physical layout of system "F" is outlined, including measurement points ("*Wh*"), load and grid interaction.

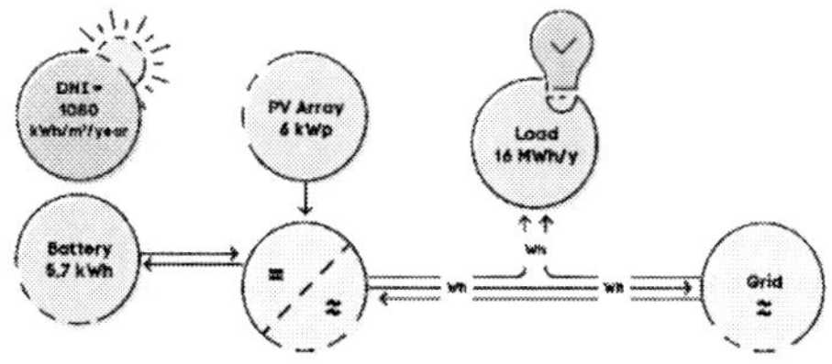

Figure 2 System "F" layout including DC coupled battery storage and measurement points "Wh".

Generated PV energy, assuming no battery charging from the grid, is then

$$p_G(t) = p_{INV}^{exp}(t) - p_{INV}^{imp}(t) \qquad (5)$$

where $p_{INV}^{exp}(t)$ and $p_{INV}^{imp}(t)$ are the power exported and imported to/from the inverter in each time step t. In the presence of a battery, the PV output will not follow the conventional output curve since surplus energy is stored in the battery. Using (5), the losses in the battery and the inverter are thus considered. With the definition of PV output from (5), the share of self-consumed energy for the PV and battery system is calculated as

$$\Phi_{SC}^F = 1 - \frac{\int_{t_1}^{t_2} p_{GRID}^{exp}(t)}{\int_{t_1}^{t_2} p_F(t)dt} \qquad (6)$$

where the ratio $p_{GRID}^{exp}(t)/p_F(t)dt$ gives the share of generated PV energy exported to the grid. Similarly, the self-sufficiency for system "F" is defined by substituting the denominator with the load usage, see (3) and (4).

In Figure 3 the operation of the PV and battery system is shown for May 29th, 2020. In the morning, the load is covered by grid import until the PV generation exceeds the load and starts charging the battery—with a resulting increase in SOC. Eventually, the battery reaches its upper SOC limit and excess PV is fed to the grid (P_{grid}^{net} becomes negative). Sometime after 6 p.m., the load is covered by discharge until just before midnight when the battery is depleted. The portion of self-consumed PV energy, SC, is represented by the shaded area.

[2] c-Si = monocrystalline silicon.

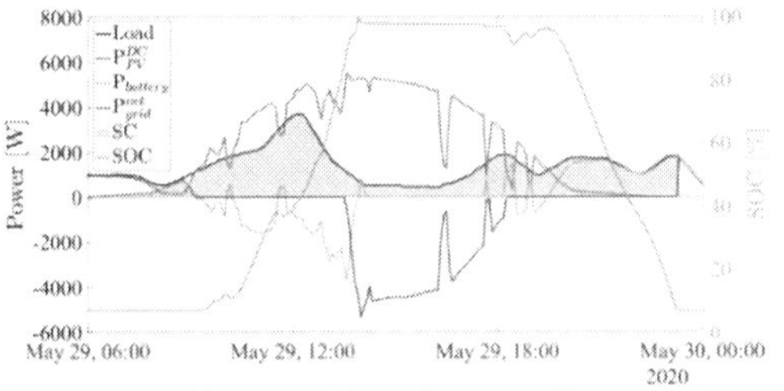

Figure 3 Working principle of the PV and battery system, with $P_{battery}$ positive for discharge and negative for charge, and P_{grid}^{net} positive for import and negative for export (from PV surplus).

As concluded in [4] one key factor to the system's self-consumption is the relative size of the PV array to the energy demand, where an increase in the PV array size decreases the self-consumption. Thus, to quantify the additional gain in self-consumption and self-sufficiency from the battery storage, the performance of system "F" is compared to a re-scaled output of system "G"—having the same cell technology—using the ratio of rated power from Table I as

$$p_G^{mod}(t) = p_G(t)\frac{P_r^F}{P_r^G} \qquad (7)$$

where $p_G(t)$ is the measured output for system "G" in each time step and P_r^i are the rated powers for systems "F" and "G". This modified output is then used to calculate the self-consumption and self-sufficiency according to (3) and (4) respectively.

3 RESULTS

As presented in [4], there is a negative correlation between self-consumption and array size. **Figure** *4* shows the resulting yearly SC as a function of array power for the seven systems without a battery. A prerequisite for the hypothesis in [4] is that the performance of each system does not differ substantially. Here, results show that despite the variation in PV cell technology and extent of roof integration, the negative trend is confirmed. For all six systems, the majority of the energy is self-consumed even without the inclusion of battery storage, given the correlation—intraday and seasonal—between the load profile and PV outputs, and the modelled sizes of the two. Currently, in Sweden, the cost savings of self-consumed energy typically exceeds the revenue for PV export [4, 13, 14]. An example of the PV array sizing in a high-latitude context is defined in [15] and demonstrated in the simulation study in [5]. Adding batteries to the smaller systems is thus questionable—giving the present battery dispatch—as the lack of PV surplus will lead to poor utilisation of the battery's capacity.

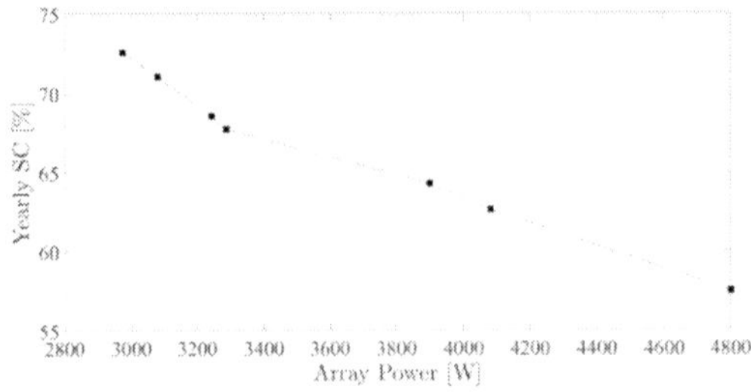

Figure 4 Yearly self-consumption (SC) as a function of array size for all tested systems.

In **Figure** *5*, the annual self-sufficiency as a function of both array power and annual energy output is presented for the seven systems without batteries. In [4], a positive correlation is claimed between self-sufficiency and array power, which is not confirmed by the results from this work. An explanation for this could be that the self-sufficiency—as defined in (4)—uses the absolute self-consumption (in kWh/a) normalised by the total load demand. Unlike the self-consumption that uses the energy yield, see (3). Since this study includes different cell technologies and roof integrations, it is more relevant to compare the yearly SS with the actual energy output. The annual yield better reflects the performance of the system and presents a more relevant index for this correlation. So, when comparing the annual SS with the energy yield, a positive correlation is identified.

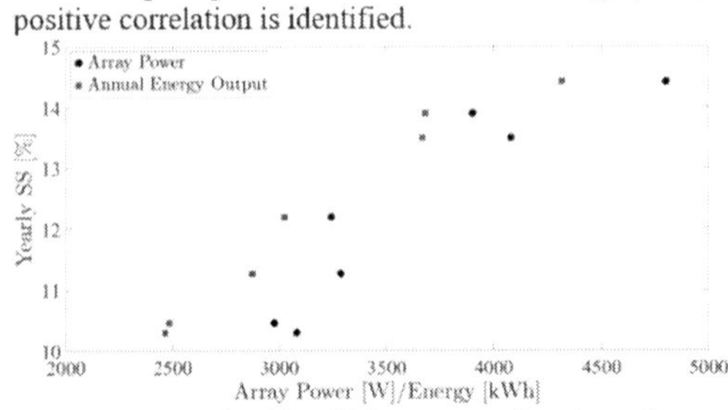

Figure 5 Annual self-sufficiency as a function of array power and energy output for all seven systems without battery storage.

The resulting self-consumption for the modified output profile defined in (7) is presented together with the calculated self-consumption for System "F" in Figure 6. Showing the weekly SC and the added gains from the inclusion of battery storage, ΔSC. There are no observed gains during the winter period where both systems give 100% SC, and this is due to the low PV outputs relative to the energy demand. Instead, the gain from the battery is observed during periods with PV surplus, with a peak SC gain of 30 percentage points. In absolute terms, the battery increases the self-consumed PV energy annually by 738 kWh.

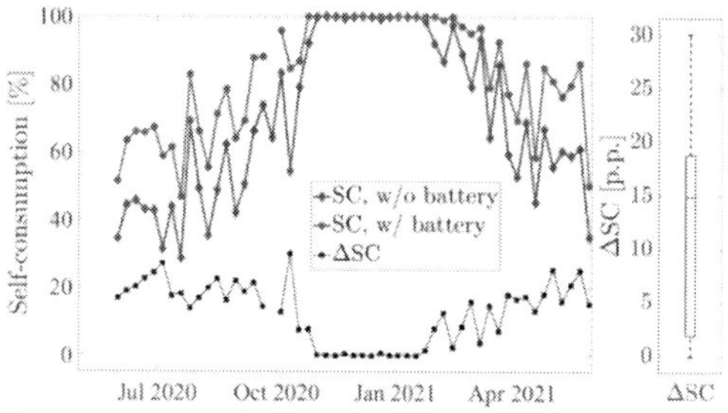

Figure 6 Quantification of weekly gains in self-consumption (SC) from the battery storage.

In Figure 7, the effect from the battery is shown for self-sufficiency using the same methodology with a modified output profile as for self-consumption. Here, peaks reaching above 40 percentage points are observed during the summer. The battery also has a negative effect on the SS during the winter—this is also noted for the SC—and is caused by the standby energy use during periods with limited PV surplus. This standby energy use is then counted as a negative output for the PV and battery system as defined by (5).

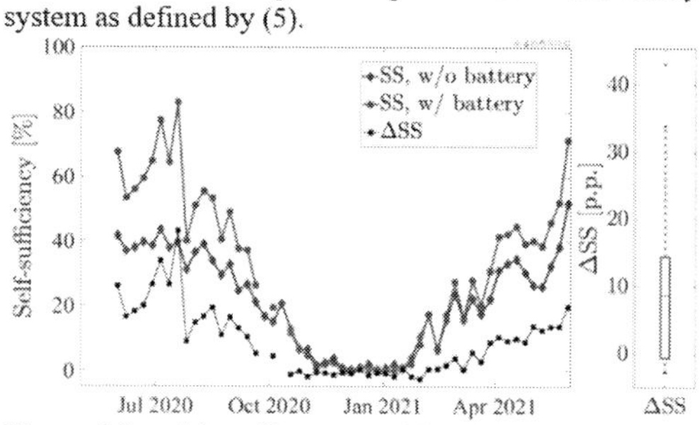

Figure 7 Resulting effect on weekly self-sufficiency (SS) from the battery storage.

Annually, the system without storage has a self-consumption of 50%, while the battery equivalent is at 70%. Comparing these results with the simulation study done in [5] at the same Array-to-Load Ratio (ALR) and Relative Battery Capacity (RBC)—3.2 and 1.05 respectively—shows good coherence with the median values of the load profiles presented in the referred study. Additionally, comparing the measured findings for the self-sufficiency—20.5 and 16.2% with and without battery respectively—also show good coherence with the results from the simulation study. The added gains in annual self-consumption (20 percentage points) are also in line with two related simulation studies from Germany using the same ratio of battery capacity to PV peak power [16, 17].

4 DISCUSSION & CONCLUSIONS

This study examines the self-consumption (SC) and self-sufficiency (SS) for eight photovoltaic systems operated against a residential building load profile in Swedish. The results are based on measurements from a full year's operation and used to quantification the effect from the inclusion of battery storage.

The results from this work are analysed against the hypotheses presented in [4], where correlations between self-consumption, self-sufficiency, and PV array size are presented. Results confirm the negative correlation between self-consumption and PV array size, despite the variation in system designs. No clear correlation is observed for self-sufficiency and array size, where the results are scattered. However, when taking into consideration the difference in system design (PV cell technology, installation method and inverter), a correlation is confirmed between self-sufficiency and annual yield. This, as the annual yield better reflects the performance when comparing various array designs.

A comparison is made for two identical PV arrays to quantify the added effect of battery storage on the system's self-consumption and self-sufficiency. Results show gains up to 30 and 40 percentage points respectively—with an annual gain of 738 kWh self-consumed PV in absolute terms. The weekly evaluation of SC and SS shows that the effect of the battery is negative during the winter with the chosen battery dispatch algorithm. This, since charging is done solely by PV surplus and with the given load demand and irradiance profiles, PV surplus is limited or non-existing during the winter. The negative effect from the battery during the winter is caused by the standby usage, resulting in negative gains in both SC and SS, as the system adds to the load demand. If the battery remains idle in the absence of PV surplus, other parallel services could be utilised, including for example peak shaving, market arbitrage and frequency regulation to increase the usage and revenues.

ACKNOWLEDGEMENT

The work is financially supported by the Swedish Energy Agency Test Lab through grant number 4057.

REFERENCES

[1] F. Cucchiella, I. D'Adamo and P. Rosa, "Industrial photovoltaic systems: An economic analysis in non-subsidized electricity markets," *Energies*, vol. 8, no. 11, pp. 12865-12880, 2015.

[2] D. Chiaroni, V. Chiesa, L. Colasanti, F. Cucchiella and I. D'Adamo, "Evaluating solar energy profitability: A focus on the role of self-consumption," *Energy Conversion and Management*, vol. 88, pp. 317-331, 2014.

[3] F. Cucchiella, I. D'Adamo, M. Gastaldi and V. Stornelli, "Solar photovoltaic panels combined with energy storage in a residential building: An economic analysis," *Sustainability*, vol. 10, no. 9, 2018.

[4] R. Luthander, J. Widén, D. Nilsson and J. Palm, "Photovoltaic self-consumption in buildings: A review," *Applied Energy*, vol. 142, pp. 80-94, 2015.

[5] E. Nyholm, J. Goop, M. Odenberger and F. Johnsson, "Solar photovoltaic-battery systems in Swedish households – Self-consumption and self-sufficiency," *Applied Energy*, vol. 183, no. 1, pp. 148-159, 2016.

[6] S. Quoilin, K. Kavvadias, A. Mercier, I. Pappone and A. Zucker, "Quantifying self-consumption linked to solar home battery systems: Statistical analysis and economic assessment," *Applied Energy*, vol. 182, pp. 58-67, 2016.

[7] S. Cao and K. Sirén, "Impact of simulation time-resolution on the matching of pv production and

household electric demand," *Applied Energy,* vol. 128, pp. 192-208, 2014.

[8] P. Ollas, J. Persson, C. Markusson and U. Alfadel, "Impact of battery sizing on self-consumption, self-sufficiency and peak power demand for a low energy single-family house with pv production in sweden," in *2018 IEEE Joint Conference of 45th IEEE PVSC, 28th PVSEC & 34th EU PVSEC), IEEE*, Waikoloa, Hawaii, 2018.

[9] M. Z. Jacobson and V. Jadhav, "World estimates of pv optimal tilt angles and ratios of sunlight incident upon tilted and tracked pv panels relative to horizontal panels," *Solar Energy,* vol. 169, pp. 55-66, 2018.

[10] M. Jaszczur, Q. Hassan, A. M. Abdulateef and J. Abdulateef, "Assessing the temporal load resolution effect on the photovoltaic energy flows and self-consumption," *Renewable Energy,* 2021.

[11] International Electrotechnical Commission, *Photovoltaic system performance - part 1: Monitoring,* IEC, 2017.

[12] R. L. Fares and M. E. Webber, "The impacts of storing solar energy in the home to reduce reliance on the utility," *Nature Energy,* vol. 2, no. 2, 2017.

[13] J. Lindahl, M. Dahlberg-Rosell and A. Oller-Westerberg, "National Survey Report of PV Power Applications in Sweden 2019," International Energy Agency Photovoltaic Power Systems Programme, 2020.

[14] I. Öhrlund, B. Stikvoort, M. Schultzberg and C. Bartusch, "Rising with the sun? Encouraging solar electricity self-consumption amongapartment owners in Sweden," *Energy Research & Social Science,* vol. 64, 2020.

[15] J. Widén, E. Wäckelgård and P. D. Lund, "Options for improving the load matching capability of distributedphotovoltaics: Methodology and application to high-latitude data," *Solar Energy,* vol. 83, pp. 1953-1966, 2009.

[16] M. Braun, K. Büdenbender, D. Magnor and A. Jossen, "Photovoltaic self-consumption in Germany: using lithium-ion storage to increase self-consumed photovoltaic energy," in *24th European photovoltaic solar energy conference (PVSEC),* 2009.

[17] E. Waffenschmidt, "Dimensioning of decentralized photovoltaic storages with limited feed-in power and their impact on the distribution grid," *Energy Procedia,* vol. 46, pp. 88-97, 2014.

MICROGRID ENERGY MANAGEMENT CONTROL WITH A VANADIUM REDOX FLOW AND A LITHIUM-ION HYBRID BATTERY SYSTEM FOR PV INTEGRATION

Ana Foles, Luís Fialho, Manuel Collares-Pereira
Renewable Energies Chair, University of Évora
Polo da Mitra da Universidade de Évora, Edifício Ário Lobo de Azevedo, 7000-083 Nossa Senhora da Tourega, Portugal

ABSTRACT: Hybrid energy storage systems combine multiple energy storage technologies, to improve the overall storage's performance and lifetime expectancy, compared to a single storage unit. This work proposes an online power/energy sharing management control of a hybrid energy storage system and a simulation tool developed by the authors. The hybridized system is composed of two PV installations (3.6 kWp and 6.7 kWp), a vanadium redox flow battery (VRFB) (5.0 kW /60 kWh) and a lithium-ion battery (LIB) (3.3 kW useful /9.8 kWh). The software simulation architecture is designed using MATLAB, where three hypothetical operation scenarios are simulated, with the aim of optimizing the energy usage according to its technical characteristics, current state operation and PV generation, with a 1-year time frame. The system's overall performance is evaluated through three technical key-performance indicators. The combination of different energy storage technologies is a feasible solution that depends on the application goals and on the assurance of the accuracy and reliability of the system and its control strategy.

Keywords: Hybrid Energy Storage System, Energy Management, Vanadium-Redox Flow Battery, Lithium-ion battery, Solar Photovoltaic.

1 INTRODUCTION

The installed capacity for decentralized renewable energy production has been increasing. Photovoltaic (PV) self-consumption maximization is one of the most popular management strategies in the solar photovoltaic sector nowadays. Renewable energy mini-grids are new grids that combine renewable energy and loads to operate on a self-sustained basis (matching discharge time within the time frame needs) [1]. Energy storage systems (ESS) help to make this match, giving additional stability to these grids. World is looking for ESS solutions that are preferentially scalable, modular, flexible, but also cheap and reliable.

Electricity storage contributes to a flexible power system. Storage can be conceived to be applied on meeting the demand and reliability in the grid peak hours; liberalized electricity markets can benefit from price arbitrage, capacity credit, ancillary resources or having customer side benefits [2]; shifting electricity from low demand to peaks, giving support to intermittent renewable energy sources, distributed generation, or smart grid initiatives, and provide alternative solutions to increase flexibility – as, demand-side management (DSM) or flexible power demand, or even smart charging strategies for Electric vehicles (EV) [3].

The energy management strategies can help in the execution of grid services such as peak shaving and curtailment avoidance, or simply balancing of loads through the power command management. Establishing the optimal technical and economic combination of different strategies and/or services is a complex task, highly dependent on the power consumption and generation profiles, market factors, applied tariffs, the existence of a grid connection to the system, general system costs, among others.

This value-stacking approach of simultaneous usage of different energy management strategies requires making trade-offs between energy needs, power needs, battery life and degradation, specific acquisition or O&M costs, etc. The hybrid energy storage system (HESS) can effectively present advantages to this multigoal task. For instance, the vanadium redox flow batteries are well suited for low power, long discharge applications, and the lithium-ion batteries in higher power applications. A Hybrid ESS can alleviate peak power stress, allowing to optimize the lifetime of these assets. Considering other fundamental technical characteristics of these systems, such as the physical limits of the technologies (State of Charge, operation temperature, etc.), allow to achieve an improved operation that, sometimes, can have competing goals.

This work presents the online operation control of a hybrid battery system VRFB-LIB, based on power-sharing techniques, for PV integration. The use of power-sharing to allocate power is studied to achieve reasonable energy distribution between the batteries and improve the utilization of energy. In the literature, several power management methods can be found as in [4]–[8]. In this work, the control strategy is proposed in three main approaches, and its effectiveness is verified through the technic key-performance indicators results, based on one-year simulation profiles. The control considers both batteries states of charge (SOC), performance, dynamics characteristics (assured by the used battery models), charge and discharge patterns. The simulation results show the improved use of energy storage devices.

The paper is organized as follows. First, the proposed hybridization of VRFB-LIB ESS is presented in Section 2. In Section 3, the methodology used to simulate the power-sharing of the hybrid solution is explained. In Section 4, the simulation results are presented. Finally, Section 5 draws the conclusions of the present work.

2 HYBRIDIZATION OF BATTERIES

2.1 Interconnection of batteries

In the case of energy storage system technologies, hybridization directly uses to the combination of more than one unit and generally is established in a single grid network. The combination is made to answer to a certain application usage case, to attend to specific needs. Hybrid battery technologies are being studied in the literature, given the distinct characteristics of each one, that can be complementary to each other. This approach can be applied for any project that combines multiple energy generation, storage, or load control technologies, co-located either physically or virtually.

Hybrid control could optimize the EMSs, allow the delivery of multiple services ("stacked"), and consider different planning horizons.

A review on Energy Storage Systems is conducted by the authors of [9], where Hybrid Energy Storage Systems (HESS) are detailed. These systems are combined considering their costs, performance, and environmental factors, emphasizing the benefits of individuals. Nevertheless, the state of the art presents gaps in the development of management strategies for these systems validated with parameters and real technical restrictions in an experimental environment.

In this work, a LIB and a VRFB with different energy and power capabilities, integrated into the microgrid of the Renewable Energies Chair of the University of Évora, are studied. Table I presents the main general characteristics of each of the energy storage technologies under study. The interconnection of different energy storage systems could result in a complementary characteristics solution, to result in an overall increase of the system's performance. The selection of a specific storage type is a result of technical and/or economic optimization.

Table I: Technical characteristics of ESS systems [3][6].

Battery/ Characteristic	VRFB	Lithium-ion
Power range (MW)	0.03-3	Up to 0.01
Discharge time	s-10h	m-h
Overall efficiency Power	0.65-0.85	0.85-0.95
Power density (W/kg)	166	50-2000
Energy density (Wh/kg)	10-35	150-350
Storage durability	h-months	Min-days
Self-discharge (per day)	Small	0.1-0.3 %
Lifetime (year)	10-20	5-15
Life cycles (cycles)	10000-13000	1500–4500
Response time	Fast	Very fast
Self-discharge	Null	Low

HESS interconnection can be achieved using different approaches. The simplest and most cost efficient is through a direct DC-connection on the battery terminals, although it lacks energy management or controllability. An alternative approach is through an active power electronics interface between the different storages to a more flexible and efficient HESS [10] across the AC grid.

3 METHODOLOGY

In this section, the methodology which allowed the one-year simulation of the proposed hybrid configurations evaluation is explained.

3.1. Microgrid constraints

The HESS is a result of the combination of the LG Chem RESU 10 lithium-ion battery (LIB) of 5.0 kW /9.8 kWp, and the redT vanadium redox flow battery (VRFB) system of 5.0 kW /60 kWh, as shown in Figure 1, and the microgrid architecture is represented in Figure 2, presented below. Since the LIB inverter has a nominal power capacity of 3.3 kW, this is the assumed maximum capacity that can be retrieved from the LIB system. These technologies were chosen to be an object for this study, due to their distinct characteristics as energy capacity, efficiency, lifetime, and response time.

Figure 1: (a) Solar photovoltaic installation 6.7 kWp; (b) Solar photovoltaic installation 3.2 kWp; (c) VRFB redT 60 kWh; (d) LIB LG Chem Resu 10, 9.8 kWh.

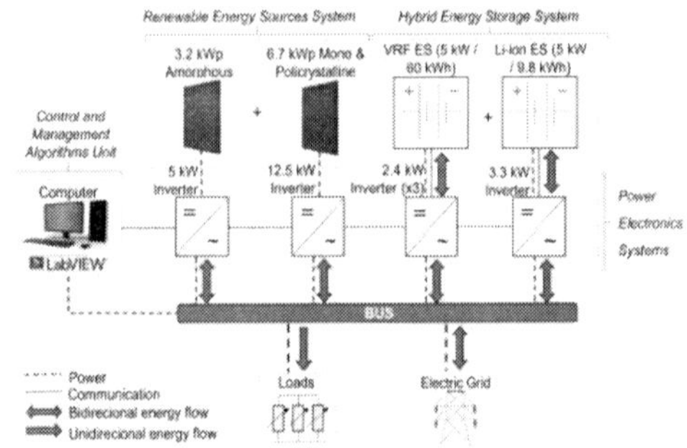

Figure 2: Scheme of the hybrid battery integration within the Renewable Energies Chair of the University of Évora's microgrid.

The power electronics efficiency of the VRFB was previously calculated and included in the developed model of the VRFB developed in [11], discussed below. In the case of the VRFB, the minimum value (limit) of the voltage is settled by the power inverter. Further development of lower voltages operation of power electronics should help the use of the battery at its fullest energy capacity. In addition, a value of 30 W total of the standby mode of the inverters was considered. The algorithms time of application is on a 1-minute basis.

2.2. PV and load consumption treatment output

One of the input variables is photovoltaic production. In this work, the actual PV generation curves were used, a complete year of data (2019) of the two installed PV systems (see Figure 3). The data from the two installations were treated separately and gathered for the results. The raw registered data is obtained with a 1- and 2-seconds interval, although, for this paper, a 1-minute average was achieved.

The load profile used is made available by the Portuguese DSO company E-REDES, with 15 minutes averaging of load data for the year 2019, for each of the Portuguese consumption sectors, in this case, for the BTN B sector, which corresponds to the overall consumption in buildings [12]. The data were treated to correspond to the 1-minute data sampled time frame for the PV.

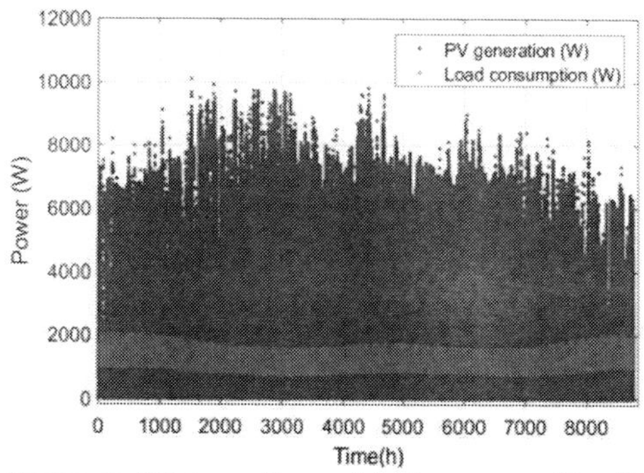

Figure 3: PV generation and load power profiles for the period of one year (2019) [12].

3.2. Modeling tool description

The model was developed with the help of the MATLAB programming tool, as a first approach to discuss the hybrid battery performance. The model was built to reproduce the operation of the batteries, describing their behaviour, also considering the inverters and general microgrid losses, computing the energy management systems' algorithms, and delivering the evaluation of the overall system, based on the selected key performance indicators (KPIs). In the studied scenarios, this tool provides information to evaluate whether the control offers a technically attractive solution to be considered for cost-effectiveness.

The developed model architecture was designed to be flexible to the addition of different battery units, data inputs, and model management of the overall system. The developed model allows the batteries to be operated together, to evaluate the overall HESS operation. For this purpose, the work main contribution is on the design algorithm to define HESS operation and control, the development of the energy routine for better energy performance, and the possibility of operating the HESS in a way to carry out multiple tasks.

3.3. Battery modelling

The operation of each BESS is achieved through its simulation model. Each battery was previously tested experimentally through characterization tests in each general operating conditions. The batteries physical constraints must be included in the model, as boundaries, such as the maximum allowable power and capacity, SOC or depth of discharge (DOD) limits, efficiencies, lifetime. Other constraints are related to the model initial definitions, being batteries-related as well. These constraints are enunciated in Table II, below.

Table II: Values of key technic parameters of the LIB and VRF batteries for modelling inputs [13].

Quantity	Value	Value reference
VRFB Maximum power (W)	5000	According to power-SOC relation and nominal power
VRFB nominal energy capacity (kWh)	60	Nominal energy capacity
VRFB SOC range (%)	10-95	According to lifetime strategy
VRFB operating voltage range (V)	50-60	Manufacturer and inverter dependent
LIB Maximum power (W)	3300	According to power-SOC relation and nominal power
LIB SOC range (%)	15-95	According to lifetime strategy
LIB nominal energy capacity (Ah)	189	Nominal energy capacity
LIB operating voltage range (V)	48-61	Manufacturer and inverter dependent
Initial SOC of both LIB and VRFB	50	Strategy starts at the middle interval
Temperature (°C)	20-35	General operating condition

Battery lifetime depends strongly on the operation conditions: temperature, state of charge and total energy throughput (electrochemical operating windows) and charge and discharge rates. Ageing depends on technology, usage conditions, storage conditions, and operation (temperature, charge/discharge rates, voltage operation limits). The knowledge currently available on these matters results from a vast combination of experimental and modelling approaches [14]. In this work, the time frame evaluation is of one year, and for that reason, no battery degradation was assumed.

3.3.1. VRFB modelling

The VRFB is integrated with the building at a real scale, in a microgrid exclusively devoted for its testing and systems operation study. The nonlinear Nernst-Plank equation is one of the modelling approaches used to describe the multiphysics of the VRFB. The model is considered an equivalent circuit that describes the major reaction occurring in real-time operation, with reduced complexity and satisfying results. The VRFB was firstly fully experimentally characterized, and the model was validated with an energy management strategy, developed in the work in [13]. The model includes thermal effects, transients, dynamic SOC, and auxiliary consumption determination.

3.3.2. Lithium-ion modelling

The LIB models currently have an emphasis on the research field, most of them developed for the automotive sector. The authors of the present work developed a model of the lithium-ion battery integrated into the microgrid. Taking the bibliographical references regarding the LIB modelling for grid applications, the simple electrical equivalent model and the modified Shepperd model were used to describe the battery of this microgrid, considering the experimental characterization results, previously achieved.

3.4. EMS and Power command – HESS Allocation Method

In this work, the energy management strategy controls the connected power-inverters and operate the units according to the pre-defined set of rules. The goal is to have a power profile smoothing to be exchanged with the grid and the maximization of self-consumption while simultaneously protecting the battery SOC to be around the defined operational limits and the power profile limits.

A coordination block is created to virtually distribute power among the operation of the system. The filter-based power provides power splitting, as shown in Figure 4. The power-sharing strategy assigns higher power ramps to the LIB and the baseload power to the VRFB. The LIB energy capacity is lower compared with the VRFB, and its SOC can reach its extremes more often, which limits battery usability. The power-sharing ratio among the energy storage system could be dynamically adjusted for SOC regulation. The systems architecture is designed to operate the VRFB as the primary energy component, and the LIB has the primary power component. The power adjustment is achieved through the inverter control power command.

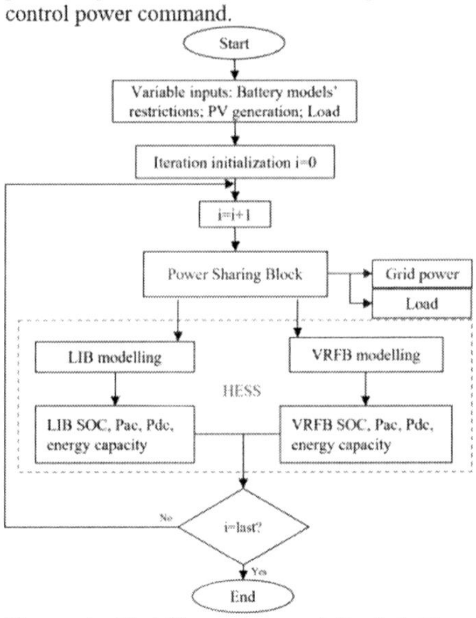

Figure 4: Modelling scheme of the hybrid energy storage system for the 3 scenarios.

The energy management strategy of self-consumption maximization is based on the HESS power allocation control. Different scenarios for the power allocation method are researched:

- **Scenario I** – The power-sharing is distributed with a fixed power ratio for each of the batteries. In this scenario, 75% of the total power is requested/retrieved from VRFB, and the rest is requested/retrieved from LIB.
- **Scenario II** – Scenario I with variable ratios. HESS is operated depending on the LIB state of charge. It obeys a power-SOC relation:

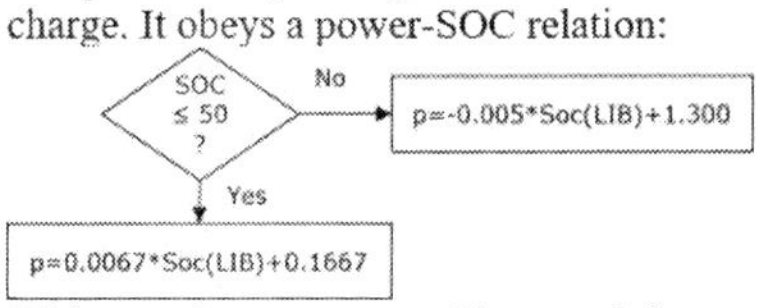

With p as the percentage. The remaining power is requested/retrieved from VRFB.

- **Scenario III** – Dynamic variability power command for LIB and fixed value for VRFB. It is the realization of two services for each battery (as if LIB executes a peak shaving strategy).

3.5. Key-Performance Indicators

The evaluation of the HESS power allocation control will be evaluated for a one-year time frame, using the key-performance indicators enunciated below.

- **Self-consumption ratio (SCR)** – Share of the PV generation consumed by the installation from the total of the PV energy generation.

$$SCR = \frac{E_{PV consumed}}{E_{PV generated}} \quad (1)$$

- **Grid-relief factor (GRF)** – The grid relief factor offers a measure of the total grid use in the overall load consumption needs.

$$GRF = \frac{E_{Grid}}{E_{Load}} \quad (2)$$

- **Overall battery use (BU)** – Share of energy of the power battery command in the overall energy consumption

$$BU = \frac{E_{fromBattery} + E_{toBattery}}{E_{Load}} \quad (3)$$

The parameters are based on the sum of the energy used throughout the days of the strategy application. The indicators are calculated for the overall strategy output results, although the BU is calculated separately for each battery.

4 SIMULATION RESULTS

The initial state of charge is 50% for both batteries, and the simulation sample time is of 1 minute, over an entire year. The simulation tool allows for obtaining the energy fluxes that occur at each moment, which, in turn, allow for the calculation of the technical performance indicators, as the example of Figure 5, shown below.

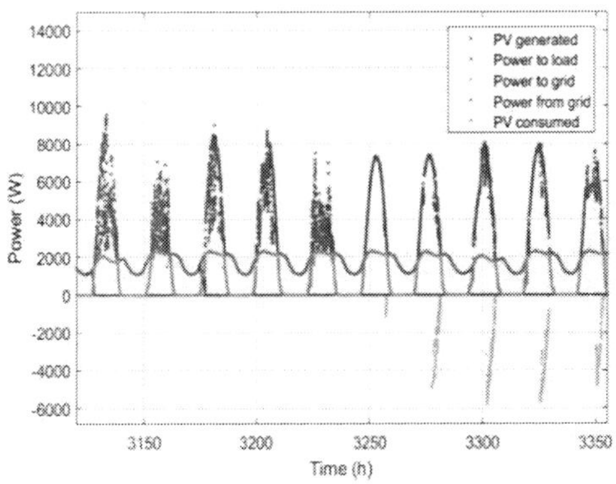

Figure 5: Output of the simulation tool, where the sum of the energy fluxes of the simulated year allows the calculation of the technical performance indicators.

The impact of the power-sharing in the studied scenarios is presented in Table III, where the calculated key-performance indicators results are shown.

Table III: Simulation results of power and energy exchange with the loads and utility grid using different power-sharing scenarios, a yearly basis.

Scenario	SCR	GRF	BU LIB	BU VRFB
I	0.53	0.35	0.55	0.67
II	0.53	0.27	0.63	0.64
III	0.53	0.36	0.54	0.64

The state of charge impact is shown in Figures 6 in (a), (b) and (c), and the AC power requested to the inverter (charge or discharge the battery) is shown in Figure 7, in (a), (b) and (c).

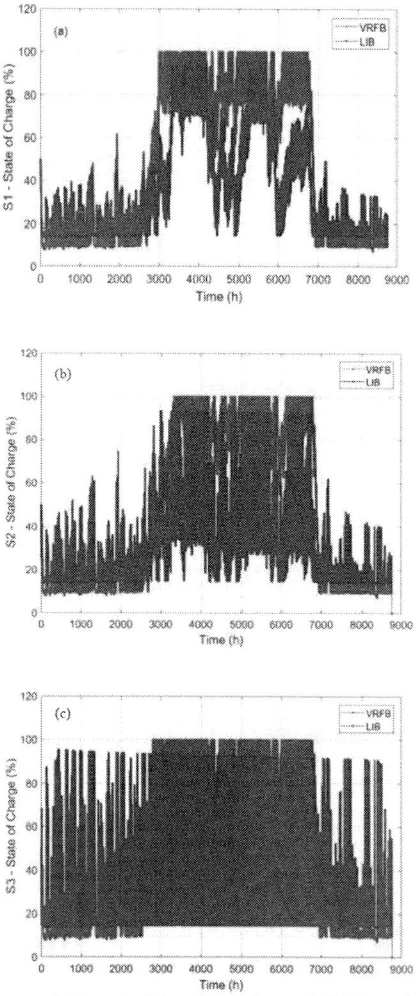

Figure 6: Output of the simulation tool: SOC evolution for both batteries for Strategy 1 (a), Strategy 2 (b) and Strategy 3 (c).

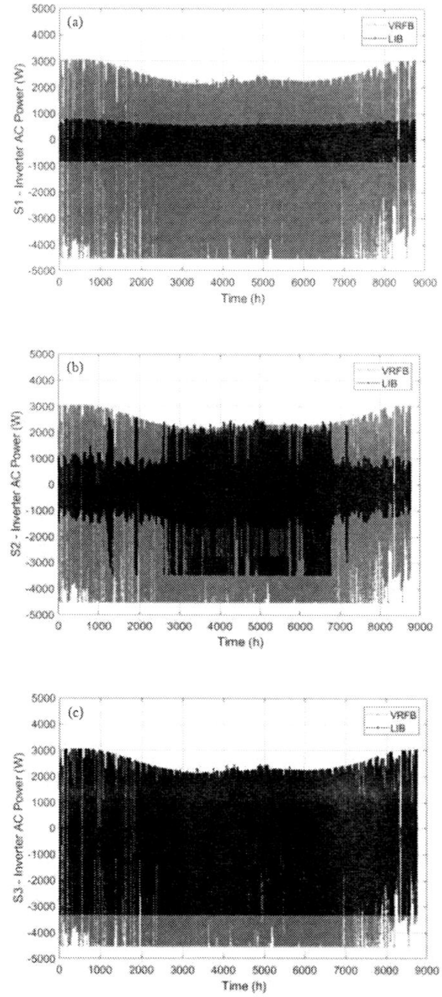

Figure 7: Output of the simulation tool: AC Power of the inverter of both batteries for Strategy 1 (a), Strategy 2 (b) and Strategy 3 (c).

5 CONCLUSIONS

A robust control can improve system dynamics, minimizing extreme SOC states and power levels, or peak temperatures caused by high C-rate operation. Scenario I is a rough approach as HESS power-sharing method. Scenario II improves the systems dynamic, by lowering extreme power SOC states. Scenario III offer the efficient realization of different tasks.

SCR is maintained in the three scenarios, the GRF is the smallest for Scenario II given the better batteries exploitation. VRFB executes the baseload in all the scenarios, shown by its BU indicator. The LIB BU indicator presents a higher use in Scenario 2 even though the LIB operation is controlled by the SOC-power profile. Summer-Winter differences are present regarding BU, and Scenario 3 presents a smaller LIB BU due to a smoothing of the grid power exchange operation.

Further conclusions should be drawn regarding a technical-economic analysis of this scenario comparison and considering battery ageing. Initial investment costs (CAPEX) should be considered for complete analysis as well as a seasonal results analysis. A robust cost analysis constitutes an important goal of future work.

ACKNOWLEDGEMENTS

The authors would like to thank the support of this work, developed under the European POCITYF project, financed by 2020 Horizon under grant agreement no. 864400. This work was also supported by the PhD. Scholarship (author Ana Foles) of FCT – Fundação para a Ciência e Tecnologia –, Portugal, with the reference SFRH/BD/147087/2019.

6 REFERENCES

[1] IRENA, *Innovation landscape for a renewable-powered future: Solutions to integrate variable renewables*. 2019.
[2] J. Liu, C. Hu, A. Kimber, and Z. Wang, "Uses, Cost-Benefit Analysis, and Markets of Energy Storage Systems for Electric Grid Applications," *J. Energy Storage*, vol. 32, no. February, p. 101731, 2020.
[3] B. Zakeri and S. Syri, "Electrical energy storage systems: A comparative life cycle cost analysis," *Renew. Sustain. Energy Rev.*, vol. 42, pp. 569–596, 2015.
[4] Y. Ye, P. Garg, and R. Sharma, "Development and demonstration of power management of hybrid energy storage for PV integration," *2013 4th IEEE/PES Innov. Smart Grid Technol. Eur. ISGT Eur. 2013*, pp. 1–5, 2013.
[5] L. Kouchachvili, W. Yaïci, and E. Entchev, "Hybrid battery/supercapacitor energy storage system for the electric vehicles," *Journal of Power Sources*, vol. 374. Elsevier B.V., pp. 237–248, 15-Jan-2018.
[6] M. U. Mutarraf, Y. Terriche, K. A. K. Niazi, F. Khan, J. C. Vasquez, and J. M. Guerrero, "Control of hybrid diesel/PV/battery/ultra-capacitor systems for future shipboard microgrids," *Energies*, vol. 12, no. 18, pp. 1–23, 2019.
[7] Z. Bai, Z. Yan, X. Wu, J. Xu, and B. Cao, "CONTROL FOR BATTERY/SUPERCAPACITOR HYBRID ENERGY STORAGE SYSTEM USED IN ELECTRIC VEHICLES," *Int. J. Automot. Technol.*, vol. 20, p. 1287?1296, 2019.
[8] Y. Zhang, Z. Yan, C. C. Zhou, T. Z. Wu, and Y. Y. Wang, "Capacity allocation of HESS in micro-grid based on ABC algorithm," *Int. J. Low-Carbon Technol.*, vol. 15, no. 4, pp. 496–505, 2020.
[9] A. Z. AL Shaqsi, K. Sopian, and A. Al-Hinai, "Review of energy storage services, applications, limitations, and benefits," *Energy Reports*, vol. 6, pp. 288–306, 2020.
[10] S. Resch and M. Luther, "Reduction of Battery-Aging of a Hybrid Lithium-Ion and Vanadium-Redox-Flow Storage System in a Microgrid Application," *Proc. - 2020 2nd IEEE Int. Conf. Ind. Electron. Sustain. Energy Syst. IESES 2020*, pp. 80–85, 2020.
[11] L. Fialho, T. Fartaria, L. Narvarte, and M. C. Pereira, "Implementation and validation of a self-consumption maximization energy management strategy in a Vanadium Redox Flow BIPV demonstrator," *Energies*, vol. 9, no. 7, 2016.
[12] D. EDP and Q-metrics, "Update of consumption, production and of self-consumption for the year 2019," p. 29, 2019.
[13] A. Foles, L. Fialho, M. Collares-Pereira, P. Horta. "Vanadium Redox Flow Battery Modelling and PV Self-Consumption Management Strategy Optimization," 2020.
[14] "Battery Lifespan | Transportation and Mobility Research | NREL." [Online]. Available: https://www.nrel.gov/transportation/battery-lifespan.html. [Accessed: 02-Jul-2021].

AGENT-BASED BEHAVIORS MODEL FOR PEER-TO-PEER ENERGY TRADING LINKED TO BLOCKCHAIN

Mazhar Sajjad, Ameni Boumaiza, Antonio Sanfilippo
Hamad-Bin Khalifa University, Qatar Environment and Energy Research Institute, Qatar
(msajjad, aboumaiza, asanfilippo}@hbku.edu.qa

ABSTRACT: The aim of this paper is to illustrate the essential components and functioning of a blockchain system and implement a generic *ABM* simulation framework for energy exchange. A major novel contribution of this work is the development of a virtual environment to test and validate the secure and equitable energy trading among participants using a simplified agent modeling approach to analyze the impacts of peer-to-peer energy trading on market operational performance. This work addresses the optimization of residential energy cost in peer-to-peer (P2P) energy trading by balancing energy supply and demand and minimizing the use of energy from the grid. The proposed model provides a constraint-based multiple-objective optimization approach where the impact of various factors is taken into account, including energy storage; rooftop PV capacity; energy from the grid; energy price, and consumers demand. The model helps policy makers assess the potential community energy trading behavior in P2P energy trading.

Keywords: Solar Energy, P2P Trading, Agents Modeling, Numerical, Behavioral Modeling.

1 INTRODUCTION

The proposed model considers various factors which have significant impact to optimize the energy cost like

- Energy storage
- Balance between the energy generation and demand
- Minimize the expensive energy transactions from the grid.

The factors that have significant impact on household energy sharing are:

- Storage capacity
- Rooftop PV solar cells
- Energy from the grid
- Energy price and consumers demand from the market.

Our model proposed some constraints on prosumer/consumer agents to

- Minimize the energy cost of each household
- Ensure the optimal usage of energy received from renewable sources (as rooftop PV)
- Balance the energy between generation and demand
- Minimize the energy transaction from the grid

The proposed model would be a useful tool for policy makers to predict the local community energy trading behavior over P2P energy trading platform.

2 APPROACH- Optimized Agent-Behavior Model

A. Agent-data Model

Each agent have certain kind of data. In our approach, the agent have data like Agent unique ID: *id*, Electricity price *P* per *KWH* in USD , Amount of electricity demanded from the market in time interval *t*, Amount of excess electricity for trading in time interval *t*, The Hyperledger fabric blockchain information, consist of private key for interaction with his blockchain node, The agent address on the Hyperledger fabric blockchain, which is used to propose buy/sell to the market, the market turn, the buy/sell energy prices, The agent refresh rate t (seconds/minutes) of each time step.

B. Agent-Behavior Model

The agent platform, allows the simulation of autonomous agents interacting with their environment, to interact in the market and trade as per defined rules, Each agent simulates one household behavior, which is defined by their already declared status either consumer, prosumer profile, Each participant agent calculate the extra or needed energy, The status of agent either consumer or prosumer would be declared from the ask/bid transaction, The offer of each buy or sell from each agent consist of bid/ask prices as well as amount of electricity in kwh to be purchased/sold over a fixed time interval respectively., . Using Random bidding strategy, the agent bid/ask prices are generated according to uniform distribution between $0.1 per kWh to $1 kWh

C. Proposed Model: Optimal Prosumer Energy Sharing Behavior

The consumption and PV generation consider 24 hours' time slot, each hour have 12 time slots with 5 min interval. The trading priority considers the total electricity consumption/generation of the prosumer over some specific time (12/24 hours).The houses with rooftop PV, the maximum power per household that could be drawn from the grid is limited to some fixed value (let say 10 kW).Each household also has a storage device with a maximum capacity of 5 kWh (chosen from the available storage devices in the market Storage Specification). The household consumption total duration are 24 hours, the maximum energy from Grid is 10KW. The household maximum storage power is 5 KW, the current storage would be calculated experimentally. The maximum storage capacity would be 5 KW, the storage efficiency is 90% and the self-discharging rate is considered 1%. The Photovoltaic (PV) maximum capacity of each hold is 200W and the PV generated power capacity is 80%. The electricity cost that buy from the grid is USD 0.03 per KWh, while the electricity price that sell to grid is USD 0.02. the daily electricity cost that buy from grid and price that sell to grid is calculated experimentally. The objective function of our model is to minimize the cost of energy of all households. The objective function is expressed as

$$Min \sum_{k \in H} C_k$$

The cost of energy of each household is calculated from energy price and amount of energy received from the grid as shown in equation.

$$C_k = \sum_{k \in K} \sum_{,t \in T} EP_{kt} \,.\, E_{g \to kt}$$

To minimize the cost of each household, the prosumer agent's behaviors needs to satisfy the following constraints.

Energy received directly from PV

If PV_generation < consumption then
PV_Energy_usage = PV_generation
Else if PV_generation >= consumption,
PV_Energy_usage = consumption

Energy Used for charging the storage devices

if PV_generation = 0 OR PV_generation < consumption
Energy_chargingStorage_devices= 0
if PV_generation > consumption then
Energy_chargingStorge_devices = PV_generation - consumption
Else Energy_chargingStorge_devices = 0

Calculate the Energy for the Current Storage

Current_Storage = (PV_generation - consumption) + Previous_StorageCharge
if CurrentStorage > Max_storage_capacity
CurrentStorage = Max_storage_capacity

Calculate the energy buying/receiving from Grid (Grid Consumption)

Grid_Consumption = consumption - (PV_generation + Current_Storage)
if (PV_generation + Current_Storage + Previous_StorageCharge) > consumption then
Grid_Consumption = 0

Calculate the amount of energy available for P2P trading (Feed-in grid)

generated_Energy = PV_generation + Current_Storage
if (generated_Energy- consumption) > 0
Energy_feedin_grid = generated_Energy – consumption

Calculate the daily electricity cost to buy from grid

*Daily-electric-cost = Grid_Consumption (calculated) * electric_cost (user defined)*

Calculate the daily electricity price that sell to Grid

*Daily-electric-price= energy_Feedin_grid (calculated) * electric_Price (user defined)*

3 RESULTS

The provided case study remains simple but it is taking into account several scenarios to demonstrate the impact of the different components of the model. In our proposed model, we are using three types of parameters: first, Electricity price per KWH, in our case the price would be defined as average of minimum and maximum. Second, the energy consumption profile of the household and third, the renewable energy generation profile. An agent represent a household with a unique id, electricity price per KWH, list of the amount of electricity consumption and demand in time interval t, list of the amount of energy generation in time interval t, amount of energy stored in storage devices. In our mode, there would be two kind of agents: consumer and prosumer. The consumer agent, wishing to buy energy from their neighbors can connect to the microgrid network. An agent who produces excess electricity can sell this electricity to neighbors connected to the microgrid. Further, our model utilization will ensure the optimal usage of energy received from renewable sources (as rooftop PV), balance the energy between generation and demand and minimize the energy transaction from the grid. The proposed model could be useful for policy makers, and the utilities to predict the household's behavior during P2P energy trading (see Figure 1).

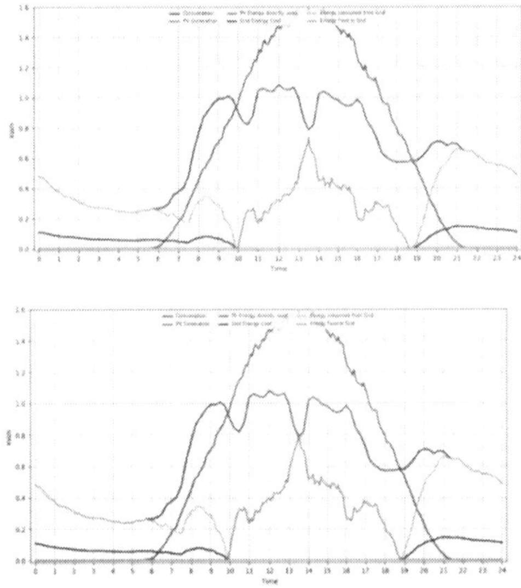

Figure 1. Energy Demand/Consumption Forecasting Results.

4 CONCLUSION

We designed the relationship among various attributes of the model that shows the optimized based agent behavior at each time slot over P2P energy trading platform. Other methods provide estimated solution through heuristic methods and do not consider it as optimization problem. The proposed model is a useful tool for policy makers to predict the local community energy trading behavior over P2P energy trading platform. The result shows that by implementing the proposed approach, we may get The minimum energy cost Maximize the balance between market supply and demand Minimize the expensive energy transactions from the grid.

5 REFERENCES

[1] Etemad RH, Lahouti F (2016) Resilient decentralized consensusbased state estimation for smart grid in presence of false data. In: International conference on acoustics, speech and signal processing (ICASSP 2016)

[2] S. Chen, B. Mulgrew, and P. M. Grant, "A clustering technique for digital communications channel equalization using radial basis function networks," IEEE Trans. on Neural Networks, vol. 4, pp. 570-578, July 1993.

[3] J. U. Duncombe, "Infrared navigation—Part I: An assessment of feasibility," IEEE Trans. Electron Devices, vol. ED-11, pp. 34-39, Jan. 1959.

VALIDATION OF A TECHNICAL SOLUTION FOR A STAND-ALONE PV HEAT PUMP SYSTEM WITHOUT BATTERIES

C. Lorenzo*, R. H. Almeida, L. Narvarte,
Instituto de Energía Solar - Universidad Politécnica de Madrid, 28031 Madrid, Spain
*Phone: +34 91 0673485, mail: celena.lorenzo@ies.upm.es

ABSTRACT: the coupling of PV generators and heat pump (HP) units permits to supply the increasing heating/cooling demand with an efficient, renewable and environmentally friendly technology. This work presents the validation results of a technical solution for a stand-alone PV-HP prototype without batteries. The lack of batteries offers economic advantages but sets the additional problem of managing the solar power fluctuations due to cloud-passing. The cloud-passing algorithm implemented in this work was able to avoid the abrupt stop of the system for 75% of the cloud-passing events registered. For the performance assessment, a combination of traditional and new Key Performance Indicators (KPIs) is proposed. The *PR* ranged from 0.16 to 0.54, the low values explained by the utilization ratios of the system (UR_{EF} ranged from 0.25 to 0.77). The PV system and HP unit good performance is reflected on the $PR_{PV,STC}$ (between 0.84 and 0.96) and on the *SPF* (between 2.49 and 3.26).

Keywords: Heat Pump, Stand-alone PV system, Key Performance Indicator

1 INTRODUCTION

In the last years, the necessity of reducing the consumption of fossil fuels has favored the use of renewable energy sources. In particular, photovoltaic (PV) energy plays a key role in any energy scenario that seeks to mitigate the climatic change [1], due to its low cost and to its versatility in terms of size and application. At the same time, the heat pump (HP) market is also growing very fast [2]: they are a technology that consume electricity, which permits to power them with PV generators (or other renewable energy sources such as wind generators), and they are very efficient in terms of electric-to-thermal energy conversion (which helps to reduce the demand of primary energy for thermal applications). Consequently, coupling these two technologies, PV generators and HPs, presents great advantages and is receiving increasing attention from the scientific community.

The objective of this work is to develop and validate a new technical solution for stand-alone PV-HP systems without batteries. The lack of batteries mainly presents economic advantages, but also sets additional technical problems. The technical solution presented here focuses on solving the problems associated to PV power fluctuations due to clouds-passing. If these fluctuations are very fast, they can derive into abrupt stops of the system that reduce its lifetime, due to electric and mechanical stress. To reach this objective, a prototype has been designed and installed at the IES-UPM (Madrid, Spain). Additionally, a combination of traditional and new Key Performance Indicators (KPIs) is proposed for the assessment of the performance of the system.

2 METHODOLOGY

2.1 Experimental set up

Figure 1 shows the schematic of the prototype installed at the IES-UPM. It consists of two thermally isolated rooms (250cm³ each) where the interior and exterior units of a commercial air-to-air heat pump have been placed. This way, one of the rooms acts as the space to be cooled and the other, as the exterior ambient. Two auxiliary heat pumps have been placed for temperature control purposes; they are powered by the conventional electric grid and are not considered for the energy balances of the PV-HP under study. The compressor of the PV-HP (located in the exterior unit) is powered by a PV generator through a Frequency Converter (FC) that is controlled by a Programmable Logic Controller (PLC). Finally, the monitoring system consists of a wattmeter (WT), a voltmeter (V) and a data logger (DL). Figure 2 shows the main components of the system: the interior (a) and exterior (b) units of the HP and the PV generator (c) with two PV modules calibrated as global irradiance (G) and cell temperature (T_c) sensors. Table I summarizes the technical specifications of the HP unit, its motor compressor and the PV generator.

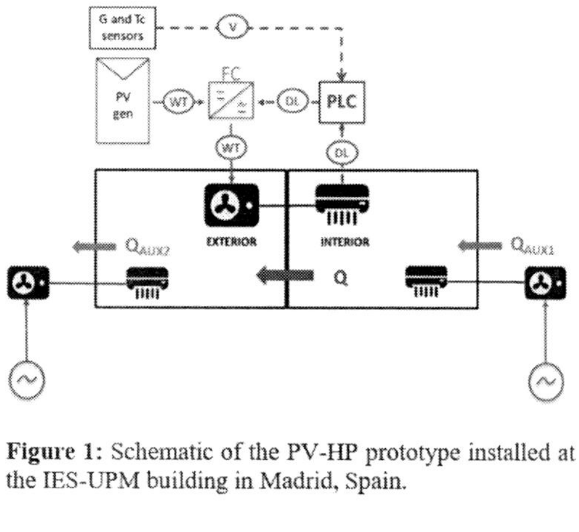

Figure 1: Schematic of the PV-HP prototype installed at the IES-UPM building in Madrid, Spain.

(a) (b)

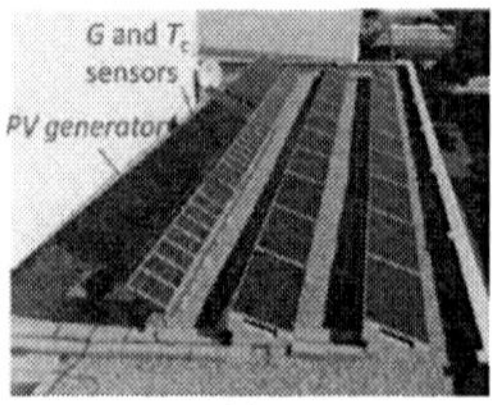

(c)

Figure 2: Images of the PV-HP prototype installed at the IES-UPM building in Madrid, Spain.

Table I: Technical specifications of the main components of the PV-HP prototype.

Heat Pump Unit	
Nominal Energy Efficiency Ratio (EER)	3.15
Nominal Cooling Capacity (kW)	3.40
Nominal AC Power Consumption (kW)	1.08
Motor compressor	
Number of poles	4
Nominal AC power (kW)	0.67
Minimum power (kW) / frequency (Hz)	0.28 / 40
PV generator	
Orientation	8ºE
Inclination	30º
Nominal DC power, P_{MPP} (kW)	0.8
Power Temperature Coefficient, γ (%/ºC)	-0.38
STC characteristics:	
Short-circuit / MPP Current (A)	2.9 / 2.73
Open-circuit / MPP Voltage (V)	363 / 293

2.2 Control algorithms

This paper presents the results of the implementation of two different control algorithms for the prototype described. The first intends to generate as much cooling power as possible by tracking the Maximum Power Point of the PV generator (therefore denominated MPPT control). The DC voltage at the MPP (V_{MPP}) at the given G and T_C is given by the following equation:

$$V_{MPP} = V^{*}_{MPP}[1 + \beta(T_C - T^{*}_C)] \quad (1)$$

Where V^{*}_{MPP} is the DC voltage at the MPP under Standard Test Conditions (STC), β is the coefficient of variation of V_{MPP} with T_C and T^{*}_C is the cell temperature at STC (25ºC).

The second control algorithm regulates the frequency of the compressor for keeping a constant temperature in the room (therefore called inverter control, which is the name given in the industry). The frequency set point ($freq_{SP}$) depends on the difference between the measured temperature in the interior room and the temperature set point (ΔT_{int}):

$$freq_{SP} = 40Hz + \frac{90Hz}{8^{o}C}\Delta T_{int} \quad \forall\, \Delta T_{int} \geq 0 \quad (2)$$

Additionally, a cloud-passing algorithm (developed by the IES-UPM for large power PV irrigation systems [3], and adapted for this PV-HP prototype) has been implemented for mitigating the effects of the solar power fluctuations. A passing cloud event can be defined as a reduction of G of at least 10% in 1 minute, which represents less than 10% of the total solar power fluctuations [4]. This reduction implies a fast DC voltage drop, with the risk of an abrupt stopping of the system due to an under-voltage alarm in the FC [5]. Additionally, if G fluctuations occur when the compressor is rotating at low frequencies, a protection valve impedes the compression and the motor starts rotating without a load. The percentage of clouds resisted (CR) -i.e. the system does not stop abruptly- is calculated as follows:

$$CR = \left(\frac{N^{o}clouds - UV - AV}{N^{o}clouds}\right)x100 \quad (3)$$

Where UV is the number under-voltage alarms and AV is the number of stops caused by the activation of the protection valve of the compressor.

2.3 Key Performance Indicators (KPIs)

a. HP unit

The performance of HP units, in terms of electric-to-thermal energy conversion, is typically given by the Seasonal Performance Factor (SPF):

$$SPF = \frac{E_{evap}}{E_{com}} \quad (4)$$

Where E_{evap} is the thermal cooling energy absorbed at the evaporator over a certain period of time and E_{com} is the electric energy consumed by the compressor over the same period of time

b. PV system

A combination of traditional and new KPIs is proposed for the assessment of the performance of the system. The Performance Ratio (PR) is typically used for PV systems and it is defined as the ratio between the energy that has been actually generated and the maximum energy that could be ideally generated over a certain period of time:

$$PR = \frac{E_{AC}}{\frac{P^{*}_{MPP}}{G^{*}}\int G(t)dt} \quad (5)$$

Where E_{AC} is the AC energy delivered by the PV generator (which in this case is equal to E_{com}), and P^{*}_{MPP} is the maximum power of the PV generator at STC and G^{*}, the global solar irradiance in the plane of the generator at STC (1,000 W/m^2).

The traditional PR is designed for grid-connected PV systems, that in theory can reach values close to 1. However, when the PV generator is coupled to a variable load (such as the compressor of the HP), this PR can be lowered by factors that do not depend on the quality of the PV system. In order to distinguish the causes of a hypothetical low PR value, this indicator has been factorized as follows [6], [7]:

$$PR = PR_{PV} \times UR_{HCp} \times UR_{PV-HP} \times UR_{EF} \quad (6)$$

Where PR_{PV} is the PR considering only losses strictly associated to the PV generator itself, UR_{HCp} is the ratio of the total irradiation throughout the heating/cooling period (HC_{p}) to the total annual irradiation, $UR_{\mathrm{PV-HP}}$ is the ratio of the irradiation that corresponds to the range of power that is useful for the compressor to the total irradiation throughout the HC_{p} and UR_{EF} is the ratio of the irradiation required by the compressor during the HC_{p} to the irradiation that could be useful for the compressor.

The dependancy of the PR_{PV} on the climatic conditions (G and T_c) can be avoided by correcting it to STC:

$$PR_{\mathrm{PV,STC}} = \frac{E_{\mathrm{AC}}}{\frac{P^*_{\mathrm{MPP}}}{G^*}\int G_{\mathrm{used}}(t)[1+\gamma(T_c(t)-T_c^*)]\frac{\eta(G)}{\eta^*}dt} \qquad (7)$$

Where $G_{\mathrm{used}}(t)$ is the irradiance used by the compressor during the HC_{p}, γ is the coefficient of variation of P_{MPP} with T_C , $\eta(G)$ and η^* are the efficiencies of the PV generator at the given G and at STC, respectively.

c. PV-HP system

For assessing the performance of a stand-alone PV-HP system, a combination of the KPIs previously defined is proposed [6], [7]:

$$SPF_{\mathrm{PV-HP}} = SPF\left(1 + PR_{\mathrm{PV}} \times UR_{\mathrm{Cp}} \times UR_{\mathrm{PV-HP}} \times UR_{\mathrm{EF}}\right) \qquad (8)$$

This way, $SPF_{\mathrm{PV-HP}}$ will be equal to the traditional SPF if there is no PV energy generation and/or use (PR=0); it will double the traditional SPF if the PV generator performs ideally and all of its electricity production is used for powering the compressor (PR=1). This KPI permits to evaluate the quality of the HP, of the PV system and of their integration.

Additionally, it is possible to obtain $SPF_{\mathrm{PV-HP,}STC}$ by substituting PR_{PV} with $PR_{\mathrm{PV,STC}}$ in Eq. (8), partially eliminating the dependency on the climatic conditions.

3 RESULTS AND DISCUSSION

3.1 Control algorithms

a. MPPT control

The test was implemented during the winter season, from 11/01/2019 to 04/02/2019. The PV-HP prototype was validated in cooling mode, so the two auxiliary HPs were used for simulating summer conditions in the two rooms.

The objective of this control algorithm is to generate as much cooling power possible by tracking the MPP of the PV generator. Over a complete sunny day (15/01/2019), the standard deviation of the MPPT error (defined as the relative error of V_{DC} respect to V_{MPP}) was less than 1%, indicating a satisfactory implementation of the control.

Figure 3 shows the daily evolution of G (in green) and P_{com} (in blue) for a very sunny day (a), 15/01/2019, and a day with some passing clouds (b), 16/01/2019. Both variables present a similar daily profile; whenever a solar power fluctuation (or G fluctuation) occurred, P_{com} fluctuated accordingly.

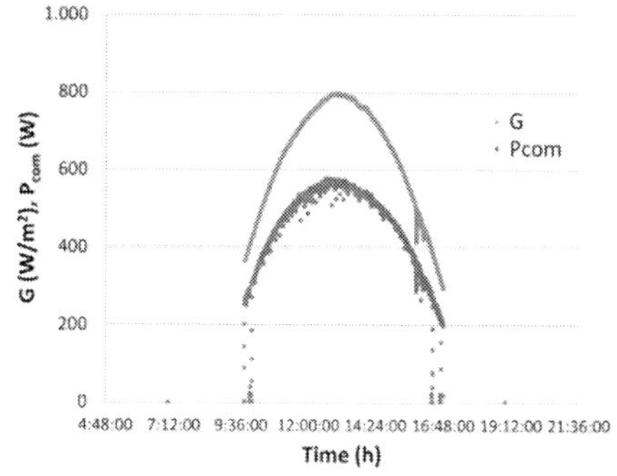

(a)

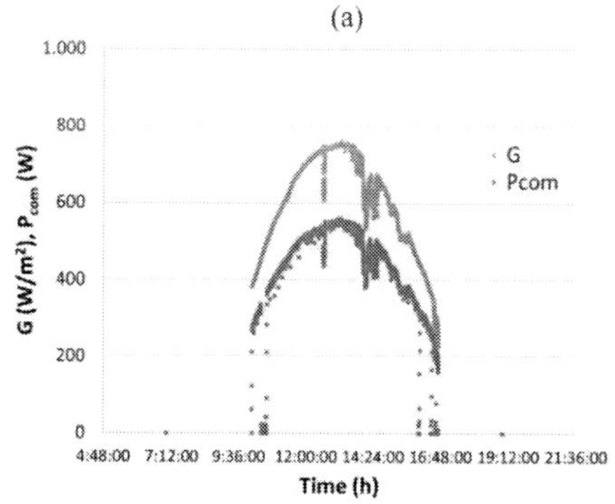

(b)

Figure 3: Effective irradiance in the plane of the PV generator (G, green line) and AC power consumed by the compressor of the PV-HP unit (P_{com}, blue line) for a sunny day (a) and a day with some passing clouds (b).

b. Inverter control

This test was implemented for two different temperature set points ($T_{\mathrm{int,SP}}$) during the summer season, from 08/07/2019 to 30/07/2019 with $T_{\mathrm{int,SP}}$=18°C and from 14/08/2019 to 04/09/2019 with $T_{\mathrm{int,SP}}$=24°C.

The objective of the inverter control is to regulate the frequency of the compressor in order to keep the room temperature as close to $T_{\mathrm{int,SP}}$ as possible. Figure 4 represents the probability distribution function of ΔT_{int} for two very sunny days, 11/07/2019 (a) and 16/08/2019 (b), with different $T_{\mathrm{int,SP}}$, 18 and 24°C respectively. It can be observed that $T_{\mathrm{int,SP}}$=18°C is never reached, so ΔT_{int} presents a symmetric distribution centered at 3.9°C. On the other hand, $T_{\mathrm{int,SP}}$=24°C was easily reached, so ΔT_{int} presents an asymmetric distribution with an average value of 0.8 °C.

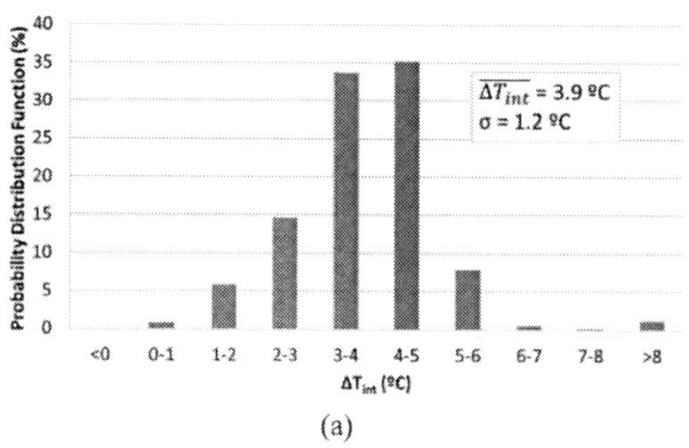

(a)

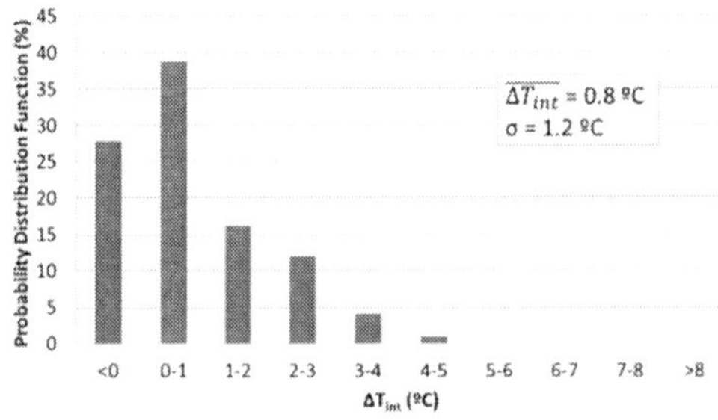

(b)

Figure 4: Probability distribution function of ΔT_{int}, for two very sunny days with different temperature set points, 18°C (a) and 24°C (b) respectively.

c. Passing-cloud algorithm

The cloud-passing algorithms was implemented for both the MPPT and inverter controls. Figure 5 shows *G* and *freq* for two consecutive clouds with the MPPT control implemented, during the day 11/01/2019. The first cloud (irradiance fluctuation of 27% in 20 seconds) was resisted, while the second one (37% in 20 seconds) provoked an abrupt stop of the compressor due to an under-voltage alarm.

Table II presents the values of the number of clouds registered, *UV*, *AV* and *CR* for the MPPT control and for the inverter control with two different temperature set points. During the MPPT test, all the abrupt stops were caused by under-voltage alarms (*AV*=0), while during the inverter control more than 90% of the stops were caused by the protection valve of the compressor. As for the *CR* values, they are very similar for the MPPT test and for the inverter test with $T_{int,SP}$=18°C (80% and 79% respectively), but significantly lower with $T_{int,SP}$=24°C (66%). The main reason is the higher *AV*, because the protection valve is more sensitive when the compressor operates at lower frequencies (which occured more frequently during the inverter test with $T_{int,SP}$=24°C).

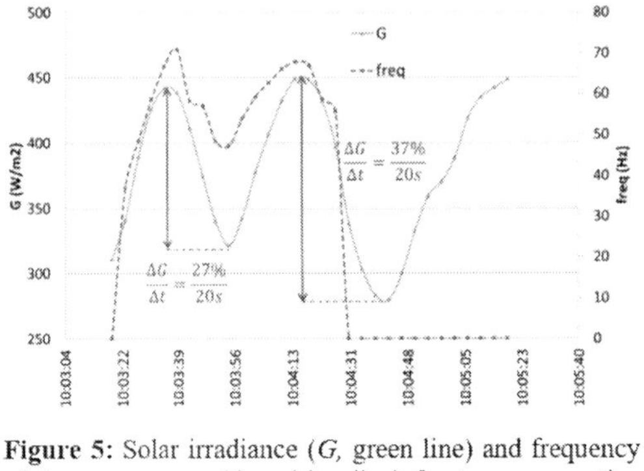

Figure 5: Solar irradiance (*G*, green line) and frequency of the compressor (*freq*, blue line) for two consecutive clouds.

Table II: number of clouds registered, *UV*, *AV* and *CR* for all the tests implemented.

	Nº clouds	*AV*	*UV*	*CR*
MPPT	156	0	31	80%
Inverter, $T_{int,SP}$=18°C	416	84	5	79%
Inverter, $T_{int,SP}$=24°C	330	110	3	66%

3.2 Key Performance Indicators (KPIs)

Table III includes the *PR* and its *URs* (UR_{PV-HP} and UR_{EF}; UR_{Cp}=1 always, so it was not presented in the table), the $PR_{PV,STC}$, the *SPF* and the $SPF_{PV-HP,STC}$ obtained for the MPPT and the inverter controls.

As regards the *PR*, it presents average values that are lower than those expected for grid-connected systems (typically higher than 0.75 [8], [9], [10]). However, this is not due to the quality of the PV system (reflected on the $PR_{PV,STC}$, which actually presents values higher than 0.8), but to the PV-HP application. The UR_{PV-HP} presented high average values (greater than 0.85) in all the cases, so it is the UR_{EF} that mostly contributed to the reduction in the *PR*. During the MPPT test, the number of starts and stops of the compressor due to the shadows affecting the PV generator in the early morning and late afternoon had to be limited (waiting for 60 minutes to start if three failed startups occurred in less than 30 minutes). This is the reason why the UR_{EF} presents a lower value than could be expected for this application. As for the inverter control, the values of UR_{EF} are lower when $T_{int,SP}$=24°C because the temperature set point is reached earlier and the compressor remains stopped for longer periods.

The *SPF* shows a good performance of the HP unit (the manufacturer indicates an *EER* of 3.15, which should be similar to the *SPF* under certain operating conditions). The differences between the three cases can be explained because the higher the P_{com}, the lower the *SPF*: as the MPPT control was implemented in winter, the available PV power permitted lower frequencies; for the inverter control, the compressor operated at low frequencies when close to the temperature set point.

Finally, the $SPF_{PV-HP,STC}$ presents the highest value for the MPPT control, mainly because of the high *SPF* and $PR_{PV,STC}$ (that indicate a good performance of the PV generator and the HP unit). For the inverter control with $T_{int,SP}$=18°C, the *PR* and its *URs* present very good values, but the *SPF* is much lower than for the two other cases. The lowest $SPF_{PV-HP,STC}$ is obtained for $T_{int,SP}$=24°C because of the low UR_{EF}, consequence of the inverter application.

The $SPF_{PV-HP,STC}$ can be used to evaluate how good these results are by comparing each test to a best-case value, calculated from the highest values of the 5 factors that make it up. For this prototype, we could achieve an $SPF_{PV-HP,STC}$ as high as 5.50 ($PR_{PV,STC}$=0.96, UR_{PV-HP}=0.93, UR_{EF}=0.77 and *SPF*=3.26,). This can be seen as an "ideal value" of this HP unit coupled to a PV system with an optimum integration and utilization.

Table III: values of the *PR* and its *URs*, the $PR_{PV,STC}$, the *SPF* and the $SPF_{PV-HP,STC}$ for all the tests implemented.

	PR	$UR_{PV\text{-}HP}$	UR_{EF}	$PR_{PV,STC}$	*SPF*	$SPF_{PV\text{-}HP,STC}$
MPPT	0.49	0.88	0.60	0.96	3.26	4.93
Inv(18°C)	0.54	0.93	0.77	0.91	2.49	4.11
Inv(24°C)	0.16	0.91	0.25	0.84	2.98	3.55

4 CONCLUSIONS

This paper presents the results of the technical validation of a stand-alone PV-HP prototype, operating in cooling mode without batteries. The following conclusions can be drawn from these results:

- The two control algorithms, MPPT and inverter, were correctly implemented: the first followed the MPP DC voltage, while the second regulated the frequency of the compressor for maintaining the temperature set point.
- The average percentage of clouds resisted was 75%. For the MPPT test, all the abrupt stops of the compressor were caused by under-voltage alarms, which could be partially solved with a better tunning of the control parameters. This is not the case of the inverter control, as 90% of the stops were brought about by the activation of a protection valve of the compressor.
- The KPIs of the system showed a good performance of its components: the low *PR* values are explained thanks to its factorization, obtaining good results for the $PR_{PV,STC}$ (performance of the PV system) and the *SPF* (performance of the HP unit).
- The $SPF_{PV-HP,STC}$ integrates the whole system performance. If combining the maximum values obtained for the 5 factors that make up this KPI (SPF,$PR_{PV,STC}$, UR_{HCp}, UR_{PV-HP} and UR_{EF}), we can establish a theoretical best-case $SPF_{PV-HP,STC}$.

5 REFERENCES

[1] "Sky Scenario," Shell International B. V., 2018. [Online]. Available: www.shell.com/skyscenario

[2] "The European Heat Pump Market and Statistics Report 2019 - Executive Summary," European Heat Pump Association, 2019.

[3] F.-R. J., L. Narvarte-Fernández, R. Hogan-Teves de Almeida, I. Barata-Carrelo, L. M. Carrasco-Moreno, and L.-P. E., "Procedimiento y dispositivo de control para sistemas de bombeo fotovoltaico," Spain Patent ES 2 607 253 A1 Patent Appl. 201730057, 2017.

[4] J. Marcos, L. Marroyo, E. Lorenzo, D. Alvira, and E. Izco, "Power output fluctuations in large scale pv plants: One year observations with one second resolution and a derived analytic model," *Progress in Photovoltaics: Research and Applications*, Article vol. 19, no. 2, pp. 218-227, 2011, doi: 10.1002/pip.1016.

[5] J. Fernández-Ramos, L. Narvarte-Fernández, and F. Poza-Saura, "Improvement of photovoltaic pumping systems based on standard frequency converters by means of programmable logic controllers," *Solar Energy*, vol. 84, no. 1, pp. 101-109, 2010, doi: 10.1016/j.solener.2009.10.013.

[6] C. Lorenzo and L. Narvarte, "Performance indicators of photovoltaic heat-pumps," *Heliyon*, vol. 5, no. 10, p. e02691, Oct 2019, doi: 10.1016/j.heliyon.2019.e02691.

[7] C. Lorenzo, L. Narvarte, R. H. Almeida, and A. B. Cristóbal, "Technical evaluation of a stand-alone photovoltaic heat pump system without batteries for cooling applications," *Solar Energy*, vol. 206, pp. 92-105, 2020, doi: 10.1016/j.solener.2020.05.097.

[8] U. G. Jahn, B.; Nasse, W, "Analysis of Photovoltaic Systems," Paris (France), 2000.

[9] N. Reich, B. Müller, A. Armbruster, W. van Sark, K. Kiefer, and C. Reise, "Performance ratio revisited: is PR > 90% realistic?," *Progress in Photovoltaics Research and Applications*, vol. 20, pp. 717-726, 09/01 2012, doi: 10.1002/pip.1219.

[10] I. Purohit and P. Purohit, "Performance assessment of grid-interactive solar photovoltaic projects under India's national solar mission," *Applied Energy*, Article vol. 222, pp. 25-41, 2018, doi: 10.1016/j.apenergy.2018.03.135.

Acknowledgements

This work was supported by the European Union's Horizon2020 research and innovation programme in the Fostering a Next GeneRation of European Photovoltaic SoCiety through Open Science project (GRECO), under grant agreement n° 787289.

This work is part of the Project MADRID-PV2 (P2018/EMT-4308) funded by the Comunidad de Madrid with the support from FEDER Funds.

REGIONAL HYDROGEN CONCEPT IN THE COMMERCIAL SECTOR FOR THE USE OF PV YIELD

Hannah Prinz, Henrik te Heesen
Trier University of Applied Sciences, Environmental Campus Birkenfeld, Institute of Operations and Technology Management, Campusallee, 55768 Hoppstädten-Weiersbach, Germany. Mail: h.teheesen@umwelt-campus.de

ABSTRACT: The electrical energy generated by photovoltaics (PV) contributes significantly to the transition to green energy production. This publication deals with a regional hydrogen storage concept for using PV yields in the commercial sector using the example of a PV carport system with battery and charging station operation at the Environmental Campus Birkenfeld, Trier University of Applied Sciences, in Germany. The PV system has a total capacity of 729.6 kWp. It is investigated whether the PV carport system with charging column operation can be realized economically. In addition, the expenses for a future without the construction and operation of the facilities is calculated, which consist only of grid purchase costs. If these are greater than the expenses of the solar carports, the investment is economically advantageous. Subsequently, the concept is complemented by a hydrogen electrolysis plant, hydrogen storage and a fuel cell to convert and store surplus PV electricity into hydrogen.

Keywords: hydrogen, renewable energy, storage, PV yield

1 INTRODUCTION

To meet the targets of the World Climate Conference in Paris, the German government plans to use 65% renewable energy for electricity consumption in Germany by 2030 [6]. In addition, it is intended that there will be 7 to 10 million electric vehicles in Germany by then to reduce emissions in the transport sector. Furthermore, solar carport systems increase the share of installed capacity of PV systems and thus the share of renewable energies in the electricity mix and offer drivers of electric vehicles the possibility to charge their cars with renewable electricity. For these reasons, they can help meet the federal government's renewable energy and electromobility goals.

Al Awadhi et al. [1] performed an economic analysis of a PV carport system over five parking spaces in the United Arab Emirates. The payback period of this south-facing system is about 13 years. Robinson et al. [12] calculated the payback time of a PV carport system with a capacity of 4 kWp in Manhattan, Kansas, the USA, between 17 and 26.5 years depending on the charging possibility. A publication by Figueiredo et al. [8] performed a profitability calculation with numerous influencing factors for a solar carport system in Almada, Portugal. Both the orientation of the system and the charging behaviour of the electric vehicles were varied. A profit was generated within the system's lifetime in the selected scenarios, with the amount of profit depending on the chosen scenario.

A publication by Lokar and Virtič [11] addresses whether a PV system in combination with a battery and hydrogen storage system can supply a residential building with energy and whether this is economically feasible.

Three of the publications mentioned deal with solar carports, also evaluating the economic feasibility of these systems. In this paper, the implementation of a solar carport concept with charging station operation for a commercial property is described using the example of the Environmental Campus Birkenfeld in Hoppstädten-Weiersbach in Germany. In addition, the system is supplemented by a hydrogen concept to store surplus electricity in the form of hydrogen.

2 METHODS

The influencing parameters of the current project are the electricity prices for grid purchase, grid feeding and electric vehicle charging. The grid purchase price was 18.44 cents/kWh in 2019 [4]. This price increased by an average of 4.42 % per year from 1999 to 2019, so it is assumed that the grid purchase cost increases by this value per year. According to § 48 EEG, the PV carport system is a system without statutory compensation, which is why the revenues from the grid feed-in are assumed to be based on the market value of solar. This averaged 4,776 cents/kWh in 2019 [2], which is why this value is considered to be constant in the calculations. The charging tariff is based on the household electricity price, which was 30.43 cents/kWh in 2019 [4]. The charging tariff was assumed to be 1 cent/kWh cheaper to create an incentive to charge at the charging stations. From 1999 to 2019, the household electricity price increased by 3.26 % per year, so the charging tariff also increased linearly by 3.26 %. The operating costs amount to 1 % of the investment costs [9]. Self-suppliers must pay the EEG levy for their electricity and consume themselves (§ 61 EEG). This amounted to 6.756 cents/kWh in 2020 [5], is reduced to 40 % for self-supply (§ 61 b EEG) and was assumed to be constant within the term. The electricity demand of the Environmental Campus Birkenfeld was 880128 kWh in 2018 and was considered to be persistent. In addition, four Fenecon Commercial series battery storage systems were planned with a battery capacity of 60 kWh each [7]. The PV carport system was designed in PV*SOL. The software allows time step simulation with hourly values and can visualize PV systems with battery systems in a three-dimensional view [14].

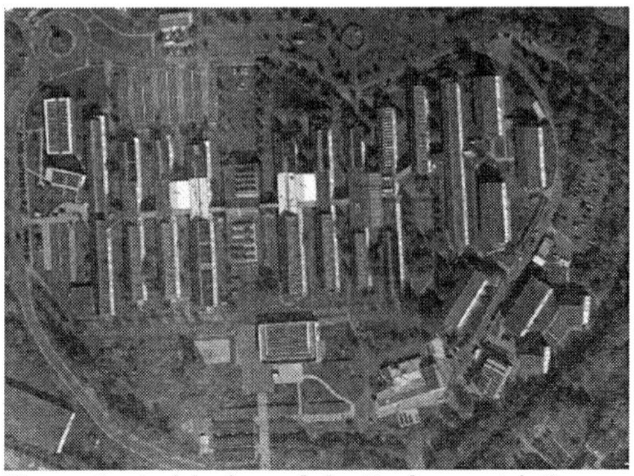

Figure 1: Aerial view of the Environmental Campus Birkenfeld with PV carports.

Three solar carport systems with a capacity of 445.4 kWp, 237.8 kWp and 46.4 kWp were installed north, south and west of the Environmental Campus Birkenfeld. The total output of the solar carports is 729.6 kWp. There can be 234 vehicles parked under the planned carports. The designed PV carport system is shown in Figure 1.

The investment costs of the plant are 2800 EUR/kWp. The investment costs of PV roof systems in 2020 amount to 976 EUR/kWp [10]. However, the current project must also include battery systems, charging infrastructure and carports. The profitability calculation includes the revenues from the charging station operation and the grid feed-in and the expenses for operating costs, costs for new charging stations, costs for direct marketing, EEG levy, grid purchase costs and the annual depreciation. The expenses were deducted from the revenues. In addition, the expenses were calculated without the construction and operation of the system, which consists only of grid purchase costs. Suppose the cumulative grid purchase costs without building the system are higher than the difference between revenues and expenses when building and operating the solar carports. In that case, it is economically advantageous to implement the carports. The net present value and payback period of the system were also calculated. For this purpose, an inflation interest rate of 1.6 % was assumed. 1.6 % is the average of the inflation interest rates from 2010 to 2019 [13].

3 RESULTS AND DISCUSSION

The number of electric vehicles is increasing based on the German government's goal of 7 to 10 million electric cars in Germany by 2030 [6]. It means that 18.05 % of passenger cars should be electrically powered vehicles in 2030. Converted to the total number of parking spaces at the Environmental Campus Birkenfeld, this means an annual increase of 10 electric cars in total.

It is assumed that the electric vehicles will be charged from Monday to Friday between 9 am and 6 pm. It was also assumed that each car drives 250 km per week since most drivers in Germany drive between 10000 km and 15000 km per year [3].

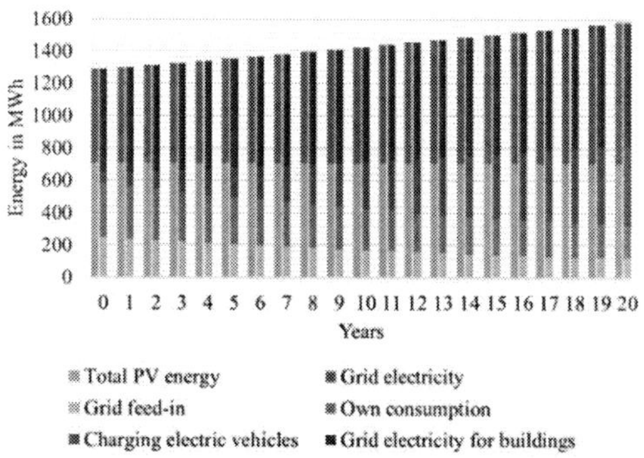

Figure 2: Simulation results for the PV carports of the Environmental Campus Birkenfeld within the lifetime of the system.

Figure 2 shows the simulation results from PV*SOL, which offers the energy flow. The year of commissioning is 2022 (year 0). The system is powered by PV electricity and electricity from the grid (left bar). The right bar shows how much electricity the electric vehicles need, how much electricity is self-consumed, how much grid electricity the buildings need and how much electricity can be fed into the public grid. Within the lifetime of the system, the PV system generates a constant 709 MWh. The electricity drawn from the grid increases from 579 MWh to 872 MWh, as more electricity is needed to charge the cars due to the increasing number of electric vehicles. Throughout the runtime, the grid feed-in and the own consumption decreases. On the other hand, electricity purchased from the grid to the buildings is increasing.

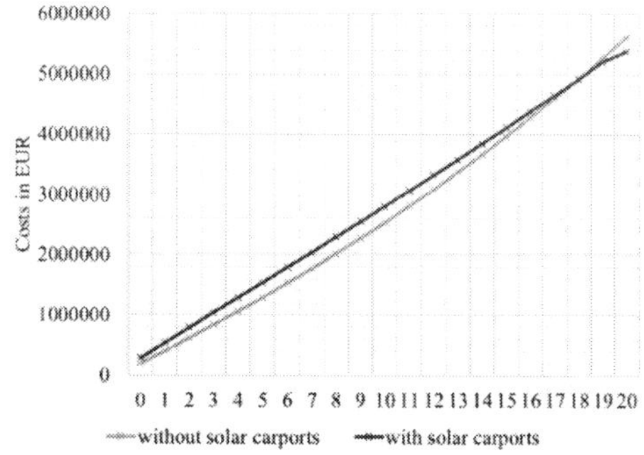

Figure 3: Cumulative total costs of the solar carports of the Environmental Campus Birkenfeld and cumulative total costs without the realization of the solar carport concept.

Figure 3 shows the cumulative total costs of the solar carport system during its lifetime. For comparison, the cumulative grid purchase costs are shown, which would arise without realizing the concept. Without the construction of the plant, the price is about EUR 5.62 million. In contrast, the realization of the project within the term only results in costs of about 5.38 million EUR. Between years 17 and 18, the graphs intersect so that the cumulative cost of the solar carports is less than the cost without implementation. During the lifetime of the system, the operator saves about 232000 EUR. Thus, the solar carport system has a net present value of 230000 EUR, and the payback period is 19 years. In the following, the influence of some parameters on the overall economic result is examined.

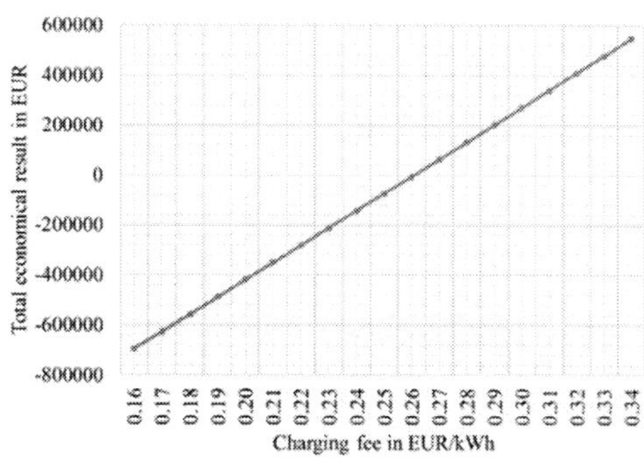

Figure 4: Overall economic result after the lifetime of the solar carport system when the charging rate is changed from 16 cents/kWh to 34 cents/kWh.

Figure 4 shows the influence of the charging rate at the charging stations on the overall economic result of the solar carport system. The charging rate shown in the diagram is the theoretical price at the charging stations in 2019. Since the commissioning of the system is not planned until

2022, the charging rates for 2022 to 2042 were calculated using the electricity price increase for the household electricity price (see 2 Methods). The charging rate of 29.43 cents/kWh is the used starting value of the electricity price at the charging stations in 2019. The larger the used starting value of the charging rate, the better the overall economic result. A value of 16 cents/kWh would mean a loss of circa 694000 EUR. A value of 34 cents/kWh would result in a saving of 548000 EUR. The initial value of the charging tariff would have to be between 26 cents/kWh and 27 cents/kWh for the system to be economically advantageous. The charging toll, therefore, plays a significant role in determining whether the plant is financially profitable.

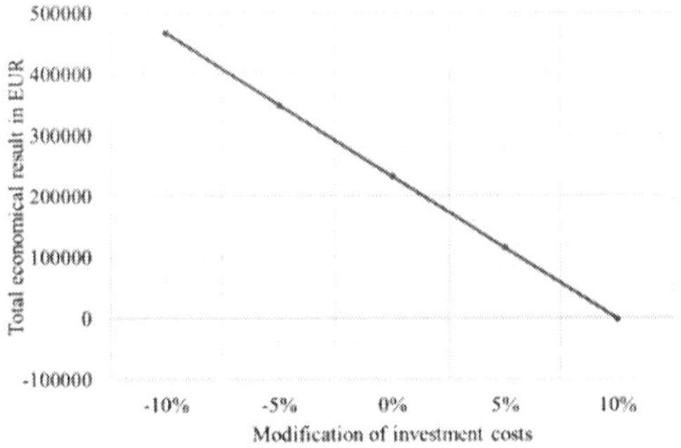

Figure 5: Overall economic result when investment costs change by -10 %, -5 %, 0 %, 5 %, and 10 %.

Figure 5 shows the effect of the change in investment costs on the overall economic result. If the investment costs are reduced by 10 %, a saving of 467000 EUR is generated within the plant's lifetime. On the other hand, if the investment costs are increased by 10 %, a loss of approximately 2000 EUR is caused.

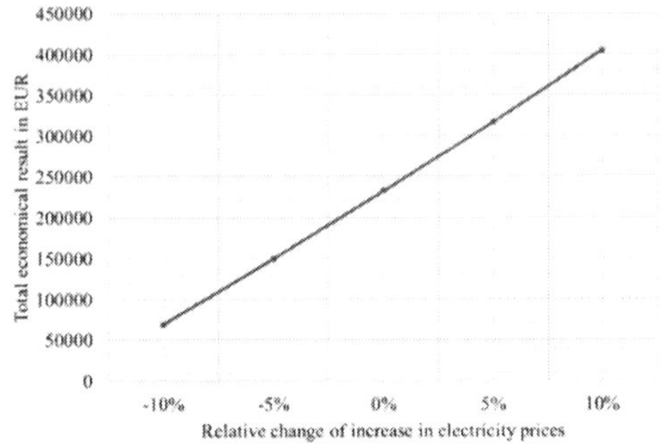

Figure 6: Overall economic result for relative change in electricity price increase of -10 %, -5 %, 0 %, 5 %, and 10 %.

Figure 6 shows the overall economic result as a function of the relative change in the electricity price increase. It is economically advantageous to realize the solar carport system both with a reduction of the electricity price increase by 10 % and an increase of the electricity price increase by 10 %. A drop of 10 % would result in a total profit of 70000 EUR. A relative rise in the electricity price increase of 10 % means a saving of 404000 EUR.
Figure 4, Figure 5, and Figure 6 show that the charging tariff, the investment costs, and the electricity price increase significantly influence the overall economic result. A change in any of these parameters can make it no longer advantageous to implement the project. The larger the charging tariff and the electricity price increase and the smaller the investment costs, the larger the savings for the plant operator compared to not realizing the project. Under the assumed conditions, the planned solar carport system is considered economically advantageous because costs can be saved compared to a future without the construction and operation of the system.

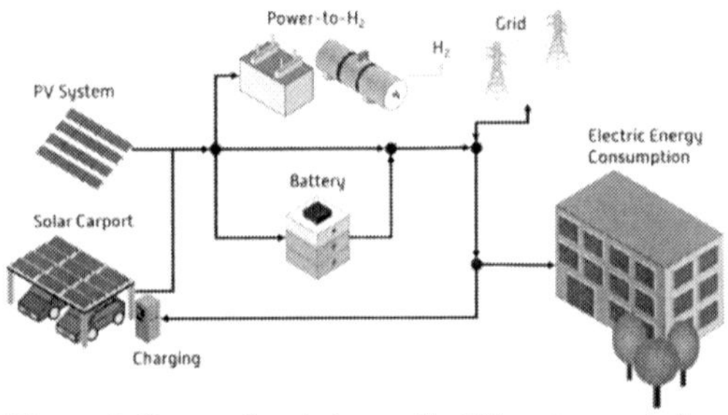

Figure 7: Energy flow between the PV system und electricity consumers.

In addition to the PV carport system with battery and charging station operation, a hydrogen electrolysis system can optimize the concept. The excess energy, described as grid feed-in Figure 2, can be used to produce hydrogen. The hydrogen can be stored in a tank and used to power hydrogen-powered vehicles on-demand or to generate electricity using a fuel cell. This could optimize the overall economic result since no statutory compensation is paid for feeding hydrogen into the grid.

When comparing the results of this work with other publications, differences but also similarities become apparent. The overall result of the present work depends strongly on the influencing parameters used, which is why these must be critically scrutinized.

In the project of Al Awadhi et al. [1] in the United Arab Emirates, a significantly simplified economic calculation of a solar carport system was carried out with constant annual yields. In addition, no battery system was included in the analysis. As a result, a payback period of 13 years was calculated, which is six years shorter than the payback period of the current project. This difference could be due to the simplified calculation methods or to a significantly higher solar radiation at the location of the PV system.

In the publication by Robinson et al. [12], the payback period of a PV carport system of 4 kWp capacity with charging capability in Manhattan, Kansas, USA, was calculated. Here, too, a constant annual yield was assumed. Operating costs and payment timing were not considered. The calculated payback period is 17 to 26.5 years, depending on the charging option. The payback time of the current project is 19 years under the assumed parameters and is thus comparable to the project in the USA.

Another scientific paper by Figueiredo et al. [8] deals with a solar carport project in Portugal, where a train station parking lot with 1045 parking spaces is equipped with solar carports and charging options. Parking fees and revenue from electricity sales to the station were included as revenue sources. Expenses included grid purchase costs, investment costs, operating costs, and management costs. Various scenarios were created in which, for example, the orientation of the PV system was varied. Depending on the scenario, the payback period was between 7 and 24 years.

The payback period of the PV carports at the Environmental Campus Birkenfeld is 19 years. The profitability calculation in this project is similar to the current project in Birkenfeld. Many influencing factors were included to make the result as accurate as possible.

Lokar and Virtič [11] published a project where a 7 kWp PV system is planned. It is combined with a battery storage system with a battery capacity of 7 kWh, a hydrogen electrolysis system, a hydrogen tank and a fuel cell. The aim is to find out whether it is possible for this system to supply the residential building with energy and whether this is economically feasible. The planned project does not include a charging station for charging electric cars. The result is that from an economic point of view, such projects are currently only possible for commercial areas. If the technology develops further and hydrogen electrolysis systems become cheaper, the project could become economical. Since the current project at the Environmental Campus Birkenfeld is a commercial plant, and the charging column operation can generate additional income, hydrogen electrolysis could improve the overall economic result here.

4 CONCLUSION AND OUTLOOK

Solar carport systems with charging station operations are suitable for advancing the federal government's renewable energy and electromobility goals. This paper deals with implementing and planning a solar carport concept with battery and charging column operation for a commercial property. In total, solar carports with a capacity of 729.6 kWp are to cover 234 parking spaces. Charging columns will also be installed, allowing drivers of electric-powered vehicles to charge their cars there from 9 am to 6 pm. The system was planned in the PV planning software PV*SOL. The simulation results and the influencing parameters were used to calculate the expenses and revenues. To calculate the total economic result, the costs were subtracted from the income. In addition, the costs without the construction and operation of the plant, which consists only of grid purchase costs, were calculated. If the cumulative grid purchase costs are higher than the total economic result of the solar carports, it is economically beneficial to implement the carports. Furthermore, under the assumed influencing parameters, the plant operator saves 232000 EUR compared to not realizing the plants. For this reason, it is economically advantageous to realize the plant.

A publication by Al Awadhi et al. states a payback period of 13 years for a solar carport system in the United Arab Emirates [1]. For a project in the USA, a payback period of 17 to 26.5 years was calculated [12]. In addition, a payback period of 7 to 24 years was calculated for solar carports in Portugal, depending on the scenario [8]. Although partially simplified calculation methods were used and the project at the Environmental Campus Birkenfeld with charging columns and batteries is extraordinary, the project results in Portugal and the USA are very similar to the present work results. In the publication of Lokar and Virtič [11], a concept of PV system, battery storage and hydrogen electrolysis system is presented. In the current project from the Environmental Campus Birkenfeld, the excess PV electricity could be used to power a hydrogen electrolysis plant. This could optimize the overall economic result.

REFERENCES

[1] Al Awadhi, N. N., Moumouni, Y., Khodary, A., 2019. Sizing of a Car Parking Photovoltaic System: An Economic Analysis. In: 2019 Advances in Science and Engineering Technology International Conferences (ASET). Dubai, United Arab Emirates, 26.03. - 10.04.2019: IEEE, 1–4.

[2] Amprion GmbH, 50Hertz Transmission GmbH, TransnetBW GmbH, TenneT TSO GmbH, 2020. Marktwertübersicht. Netztransparenz - Informationsplattform der deutschen Übertragungsnetzbetreiber. URL: https://www.netztranspa-renz.de/EEG/Marktpraemie/Marktwerte, zuletzt aktualisiert am 09.01.2020 [accessed Jan 31, 2020].

[3] Arbeitsgemeinschaft Verbrauchs- und Medienanalyse (VuMA), 2019. Autofahrer in Deutschland nach selbst gefahrenen Kilometern pro Jahr von 2015 bis 2019. Statista. URL: https://de.statista.com/statistik/daten/studie/183003/umfrage/pkw---gefahrene-kilometer-pro-jahr/ [accessed Jan 27, 2020].

[4] BDEW Bundesverband der Energie- und Wasserwirtschaft e.V., 2019. BDEW-Strompreisanalyse Juli 2019. Haushalte und Industrie. URL: https://www.bdew.de/media/documents/190723_BDEW-Strompreisanalyse_Juli-2019.pdf [accessed Jan 3, 2020]

[5] Bundesnetzagentur für Elektrizität, Gas, Telekommunikation, Post und Eisenbahnen, 2020. Preise und Rechnungen. EEG-Umlage. URL: https://www.bundesnetzagentur.de/Shared-Docs/FAQs/DE/Sachgebiete/Energie/Verbraucher/Energielexikon/EEGUmlage.html [accessed Jan 31, 2020].

[6] Die Bundesregierung, 2019. Eckpunkte für das Klimaschutzprogramm 2030. URL: https://www.bundesregierung.de/resource/blob/975202/1673502/768b67ba939c098c994b71c0b7d6e636/2019-09-20-klimaschutzprogramm-data.pdf?download=1 [accessed Jan 5, 2020].

[7] Fenecon GmbH, 2019. Installationsanleitung Fenecon Commercial 50-Serie. Technische Daten. URL: https://fenecon.de/web/content/55640?unique=962dcfd38c9835741601a5dbdaa4385a8d2c0ba6&download=true [accessed Jan 24, 2020].

[8] Figueiredo, R., Nunes, P., Brito, M. C., 2017. The feasibility of solar parking lots for electric vehicles. In: *Energy* 140, 1182–1197. DOI: 10.1016/j.energy.2017.09.024.

[9] Fraunhofer ISE, 2019. Aktuelle Fakten zur Photovoltaik in Deutschland. URL: https://www.ise.fraunhofer.de/content/dam/ise/de/documents/publications/studies/aktuelle-fakten-zur-photovoltaik-in-deutschland.pdf [accessed Jan 3, 2020].

[10] Fraunhofer ISE, 2020. Wege zu einem klimaneutralen Energiesystem. Anhang zur Studie. URL: https://www.ise.fraunhofer.de/content/dam/ise/de/documents/publications/studies/Anhang-Studie-Wege-zu-einem-klimaneutralen-Energiesystem.pdf [accessed Aug 20, 2021].

[11] Lokar, J., und Virtič, P., 2020. The potential for integration of hydrogen for complete energy self-sufficiency in residential buildings with photovoltaic and battery storage systems. In: International Journal of Hydrogen Energy, 45(60), 34566-34578. DOI: 10.1016/j.ijhydene.2020.04.170

[12] Robinson, J., Brase, G., Griswold, W., Jackson, C., Erickson, L., 2014. Business Models for Solar Powered Charging Stations to Develop Infrastructure for Electric Vehicles. In: *Sustainability* 6 (10), 7358–7387. DOI: 10.3390/su6107358.

[13] Statistisches Bundesamt, 2020. Inflationsrate in Deutschland von 1992 bis 2019 (Veränderung des Verbraucherpreisindex gegenüber Vorjahr). Statista. URL: https://de.statista.com/statistik/daten/studie/1046/umfrage/inflationsrate-veraenderung-des-verbraucherpreisindexes-zum-vorjahr/ [accessed Jan 23, 2020].

[14] Valentin Software GmbH. PV*SOL premium. Übersicht. URL: https://www.valentin-software.com/produkte/photovoltaik/57/pvsol-premium [accessed Jan 3, 2020].

SMART ENERGY MANAGEMENT FOR SME USING DIGITAL TWINS

Jesus S. da Costa Fernandes, Michael Schmidt, Rahul Khatri, Rainer Gasper, Niklas Hartmann
Hochschule Offenburg – University of Applied Sciences / INES – Institute of Energy Systems Technology
24, Badstraße, 77652, Offenburg, Germany
Phone: +49 781 205 354, Fax: +49 781 205 45354
E-mail: costa@hs-offenburg.de
Internet: www.ines.hs-offenburg.de

ABSTRACT: The twin concept is increasingly used for optimization tasks in the context of Industry 4.0 and digitization. The twin concept can also help small and medium-sized enterprises (SME) to exploit their energy flexibility potential and to achieve added value by appropriate energy marketing. At the same time, this use of flexibility helps to realize a climate-neutral energy supply with high shares of renewable energies. The digital twin reflects real production, power flows and market influences as a computer model, which makes it possible to simulate and optimize on-site interventions and interactions with the energy market without disturbing the real production processes. This paper describes the development of a generic model library that maps flexibility-relevant components and processes of SME, thus simplifying the creation of a digital twin. The paper also includes the development of an experimental twin consisting of SME hardware components and a PLC-based SCADA system. The experimental twin provides a laboratory environment in which the digital twin can be tested, further developed and demonstrated on a laboratory scale. Concrete implementations of such a digital twin and experimental twin are described as examples.

Keywords: Microgrids, Smart Grids, Energy Flexibility, Energy Management, Digital Twin, SME, Industry 4.0.

1 INTRODUCTION

On the power system side, the increasing use of intermittent renewable energy sources brings with it a greater need of energy flexibility for safe and reliable operation. Among the various other options the energy flexibility, demand-side energy flexibility and specially the industrial electricity consumption are considered to be viable options where industrial demand could be adapted (increased, decreased and shifted) in response to price or incentive signals [1,2]. This can be done by aligning with dynamic price signals or by providing and marketing flexibility in the form of balancing power. Typical measures include the shifting of operating times, the reduction or increase of the output of machines of main and auxiliary processes and the use of energy storages on site.

At the same time industrial companies are increasingly thinking about the nature of their own energy supply and are making decisions to improve energy sustainability, e.g. by installing on-site PV systems setting up local microgrids or using sector coupling. Motivations are the reduction of energy costs, the development of new sources of income on the energy market, a higher independence of external cost developments or the improvement of supply security and quality. In addition there are legal requirements for decarbonizing the mobility sector, the heating sector and industrial processes, as well as socio-technical aspects. Intelligent dynamic control of the companies' energy systems and the electricity demand can then make a major contribution to meet the need of flexibility mentioned at the beginning.

Theoretical methods for an intelligent energy management of industrial microgrids can be considered as state of the art, for example dealing with fluctuating solar energy shares or reacting on fluctuating electricity prices and other incentives. For example, the mathematical optimization and model predictive control (MPC) as approaches have proven to be effective in addressing complex control and optimization problems. These approaches have in common that they are model-based and therefore require a sufficiently accurate and comprehensive modeling of the processes with relevance to the optimization problem. This can be for example in the form of a suitable digital twin.

In order to be able to use these methods in practice the models additionally need a sufficient automation system for all relevant processes. This system continuously records the input data (system states) needed for optimization and which can automatically transfer the corresponding optimization results as control commands.

Both the effort involved in creating the digital twin and the efforts with the associated risks at the implementation in the company are hurdles to unleashing the potential for flexibility. This paper aims to contribute to significantly reduce these hurdles.

The remainder of this paper is structured as follows. In Section 2 we will describe how the twin concept can be used to leverage the energy flexibility potential of industrial companies and how the barriers to a development of a digital twin can be lowered.

In addition to a generic model library for easier generation of digital twins, the concept includes the experimental twin which provides a hardware laboratory environment for optimization approaches together with the digital twin. In Section 3 and Section 4 we present two exemplary designs of a digital twin and an experimental twin, always with the goal to harness the energy flexibility of SME based on optimization methods. The final section contains a discussion and conclusion.

2 THE FLEXIBILITY TWIN CONCEPT

The flexibility twin concept consists of a digital twin and an experimental twin, where the experimental twin supports the application of the digital twin towards the real system.

The approach of optimizing a real system by running a virtual model in parallel is a recognized tool in the context of Industry 4.0, known as digital twin. However, it has been rarely used to optimally and profitably determine the energy flexibility of SME in terms of trade activities on energy markets [3]. The digital twin makes it possible to predict and analyze effects of certain electricity purchase mechanisms on its future power consumption and to perform mathematical optimizations on this basis.

Even if the digital twin approach is very promising the hurdles for its implementation in real industrial companies are very high. With regard to the core business restrictions are expressed by owners and managers of SME who generally do not allow innovation or new impacts on production conditions if they were not sufficiently experienced or at least approved before implementation. Therefore the concept of an experimental twin presented in this work is used complementary to the digital twin.

2.1 Digital flexibility twin concept

The effort required to build up a digital twin can be very high depending on the complexity of the processes to be mapped and the desired level of accuracy. The presented work aims to represent only the components, processes, dependencies and constraints with relevance for energy flexibility by means of a digital model. In this sense one can also speak of a digital flexibility twin. The restriction to energy flexibility makes it possible to limit the level of accuracy and thus the extent of modeling efforts.

Another approach followed in this work is to reduce the modeling efforts using the commonalities of different industrial companies by appropriate abstraction. This allows numerous flexibility-relevant features of industrial companies to be represented by a generic model library. Such a generic model library has been developed under the name "IndOptFlex" at Hochschule Offenburg and is available under open source license [4].

Figure 1 shows the key features of an industrial SME that have been addressed by this generic model library. The manufacturing system (MFS) is the productive core of the SME and the model contains key components like different types of production machines as well as the interdependencies between specific machines in form of material streams, pre-products, media and energy flows. The MFS relies on technical building services (TBS) that might provide heat, cooling energy, compressed air, lighting, etc. Some industrial companies may already have their own energy system (ES) or are at least planning to build one. This includes any kind of on-site conversion with photovoltaic systems, diesel or biofuel generators or small combined heat and power plants besides electrical and chemical storages or controllable consumers such as E-Vehicle chargers. While the basic structure of components and processes in the MFS, TBS and ES can be quite similar for many companies after an appropriate abstraction, they must be parametrized for the company-specific conditions.

In addition to these generic system units, the model library must also be able to represent the flexibility measures which refer to actions such as reducing or increasing electrical consumption or shifting it to other time slots [5]. The precise characteristics of the flexibility measures that are possible in a company are again very company-specific and must be parameterized accordingly.

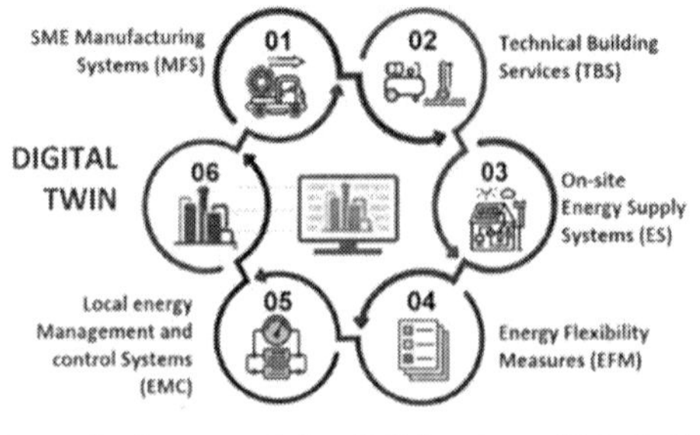

Figure 1: Concept of the digital twin as Virtual SME including local energy management and control.

The digital twin models can be used for simulation-based exploration of courses of action and their implications. This way they are helping to inform about decisions that can be taken in a traditional way. On the basis of optimization methods recommendations for control interventions can be calculated directly and implemented in a semi-automated and even fully-automated manner. For such an automated energy management mode as shown in Figure 2, it is important to consider complex optimization goals such as profiting of maximization with flexible market prices and respecting complex boundary conditions like specified minimum machine running times, constraints of production etc. For such complex optimization tasks, the formulation as a mathematical optimization problem and the use of corresponding mathematical solvers are proven as very efficient. To solve the optimization task, e.g. model-predictive control (MPC) can be applied. The MPC solves the optimization problems with continuously updated input data such as solar power prediction or price forecasts e.g. in a defined time step. Section 3 will provide more details on the digital twin and the underlying model library.

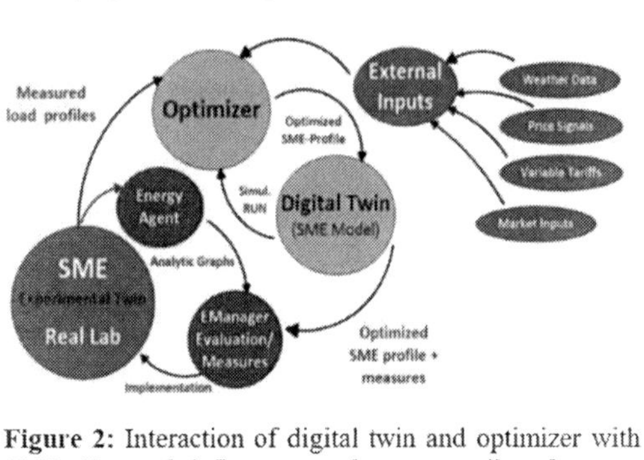

Figure 2: Interaction of digital twin and optimizer with SME. External influences and corresponding forecasts can be considered by the optimizer for predictive control. Measurements from the SME are used to update modeled system states with the optimizer and optimized schedules are sent back to the SME controls for execution.

2.2 Experimental flexibility twin concept

The experimental twin of an industrial company maps some core components of a real company as real hardware. These include e.g. motors, which are the basic motion blocks for the construction of many industrial plants and machines, power-driven heating and cooling circuits or compressors. These components of the laboratory setup are embedded to an approved automation system for monitoring, survey and data acquisition, communication and control based on programmable logic control (PLC) and mostly fieldbus solutions with the appropriate communication protocols.

In this way, it is possible to investigate how robust the optimizer operates in the face of model deviations and disturbances.

The experimental twin thus represents a laboratory environment with which energy management methods based on digital twins can be realistically tested and further developed. It also serves as a showcase for decision-makers in SME to learn about the concept and the feasibility of corresponding approaches.

The following two sections show the implementation of a digital twin with its optimizer and an experimental twin in more technical details.

3 IMPLEMENTATION OF DIGITAL TWIN

As we have seen a company can be enabled to use control interventions for energy flexibility and thereby respond to external incentive signals. To get there it is helpful to map all energy-relevant processes and components in the form of a digital system model ("digital twin").

According to the authors of [6], a general industrial enterprise can be considered as composed of various industrial systems that work together to execute the intended production by their core and auxiliary processes. In order to be able to provide digital twins with a focus on energy flexibility for a large class of industrial companies as possible, a generic model library for SME is being developed. The library contains models of the manufacturing system (MFS) in the sense of the relevant production processes and machines, as well as electrical and thermal auxiliary systems that provide heat, cooling, air or compressed air as elements of the technical building services (TBS) or modules of a classic energy system (ES) such as photovoltaic systems, combined heat and power plants, traditional generators such as backup diesel generator and on-site battery energy storage. The research methodology to develop generic models for industrial companies towards their digital twin is provided in Figure 3.

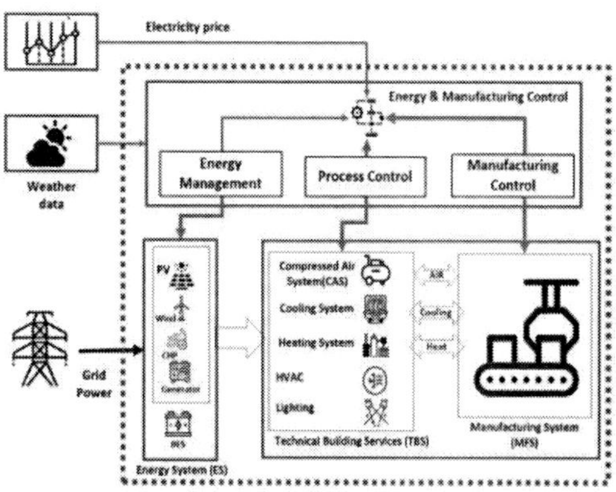

Figure 3: Schematic view of process optimization [4].

The energy and manufacturing control (EMC) block of Figure 3 serves as a central control unit equipped with model based predictive optimization. This gives the digital twin a feature to map the physical process flows and flexibilities in the form of constraints. It also contains the cost functions based on the industry's electricity purchase mechanisms or the participation rules for external platforms based on electricity prices or incentive signals. The overall goal is to adapt the load profile optimally and reduce operational costs.

Software implementation is always an important part to develop a digital twin. For this purpose, the models are designed in a most generic manner using Python programming language with its object-oriented features. An exemplary modelling diagram for the software implementation is provided in Figure 4.

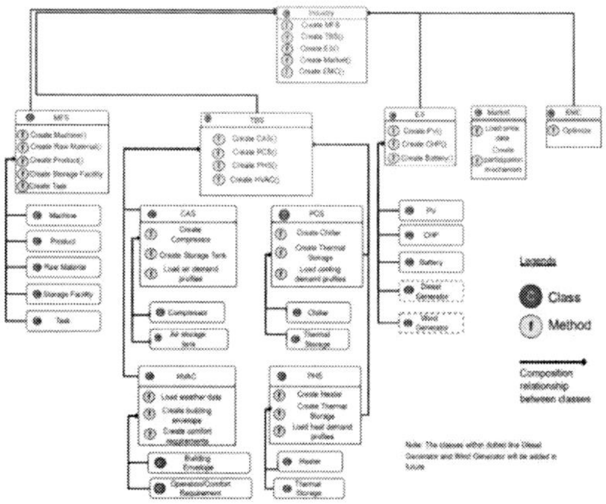

Figure 4: Modelling diagram for the generic model library of an industrial enterprise [4].

Using the generic model once an industrial company is defined and the optimization can be performed based on the generic optimization formulation written into the EMC module.

The implemented control architecture of the digital model is provided in Figure 5. The optimization model for the MFS system is a mixed integer linear programming (MILP) with scheduling optimization based on resource-task networks [7]. It results in optimal machines operation and utility demand profiles for air demand, heating or cooling demand, or lighting requirement etc. This along with external weather data for PV conversion can be fed as external parameters to a MPC based optimization for the TBS and ES systems.

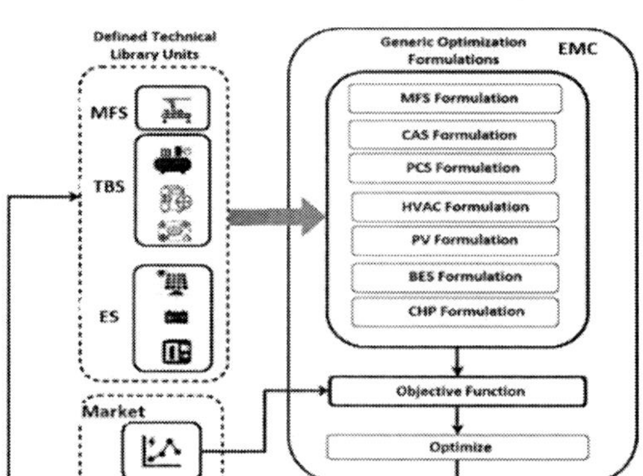

Figure 5: Optimization concept for energy and manufacturing control of the digital twin [4].

The objective or target function of both optimizations can be realized on external electricity price or incentive signals implemented to the optimizer.

4 IMPLEMENTATION OF EXPERIMENTAL TWIN

In the following, we describe the concrete experimental twin that is set up at Hochschule Offenburg, see Figure 6. As shown in Figure 7 this experimental twin is divided into subunits.

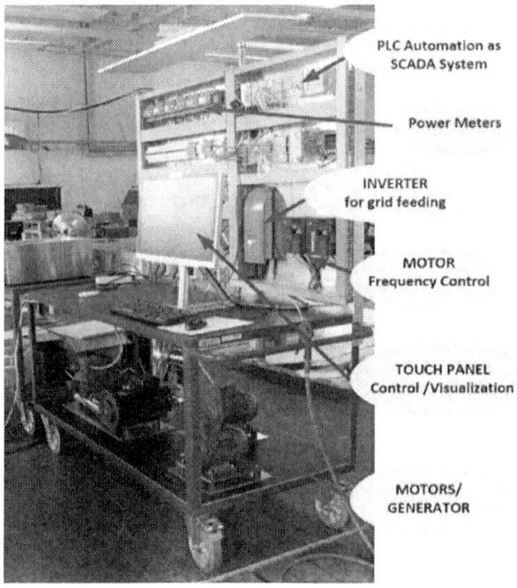

Figure 6: Setup of the experimental twin on heavy duty tables with motors, control panel, converters and switching devices.

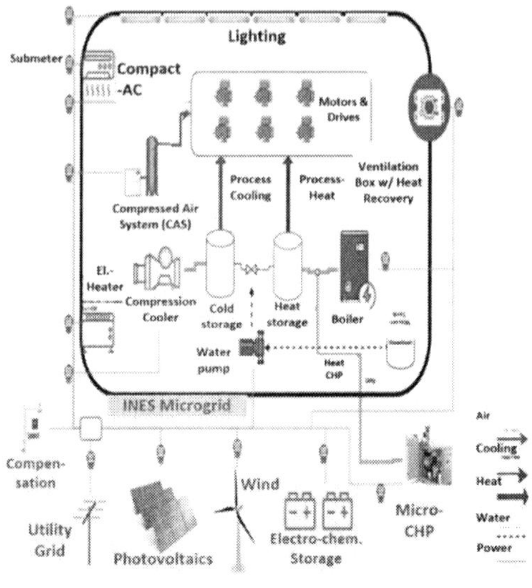

Figure 7: Schematic view of the experimental twin as lab environment with manufacturing system (MFS), technical building system (TBS) and local energy system (ES) of a SME.

4.1 Emulation of key MFS and TBS components

The hardware emulation of key SME components in the MFS is realized as follows. Rotation-based devices for transport, preparation and work of raw materials or prefabricated units of the MFS are emulated by motors which work against a controlled mechanical load which is represented by a generator unit.

The hardware emulation of key TBS components providing energy and media like heating or cooling power is actually realized by a compact cooler rated at 1.2 kW, fans and a 2 kW radiator unit dimensioned for a housing box with roundabout 30 m².

4.2 Emulation of the Energy System (ES)

The energy system (ES) in the experimental twin is built up on two contributing energy supplies. On one hand there is the conventional exchange with the utility grid mainly based on power and controlled by an energy management system. The other one is a local energy supply infrastructure made up by a bundle of energy sources like wind, solar or biomass combined with storages for energy and fuels. These form a local three-phase microgrid, see Figure 8.

Figure 8: Smart Microgrid at the Institute for Energy Systems Technology (INES) in Offenburg making up the local energy system (ES) of the experimental twin.

4.3 SCADA System for monitoring and automation

Another important component of the experimental twin is the supervisory control and data acquisition (SCADA) system. Figure 9 shows an exemplary power metering concept with a total of 7 counters (main power meter and six sub-meters). Additionally thermal counters as well as flowmeters for water, gas or other fuels and media help to capture relevant energy and media flows and to visualize them e.g. in a Sankey diagram.

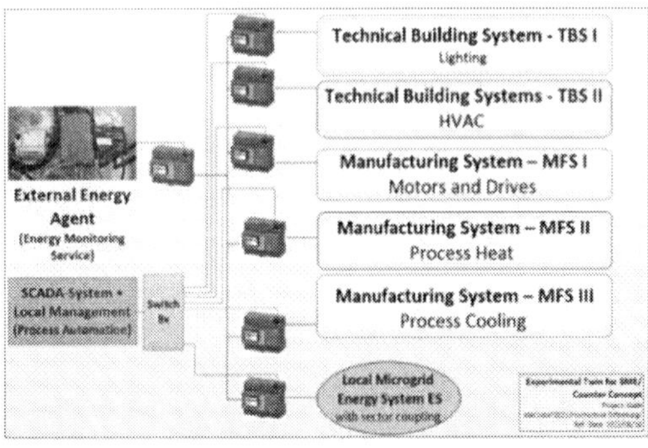

Figure 9: Counter concept to analyze energy flows and process contribution separated by the defined categories MFS and TBS and the exchange with utility gird and the local ES.

The consumption structure is organized into five groups of SME-related equipment, complemented by a local energy supply system (ES). Figure 10 shows screenshots of the visualization and control panel of the experimental twin, where the elements identify five subgroups with lighting, local heat systems, motors and drives, HVAC units, compression coolers and feed-in units like generators. Figure 11 shows an exemplary measured time series of the total power. Comparable SCADA solutions have already been used and presented in other energy-related applications [8].

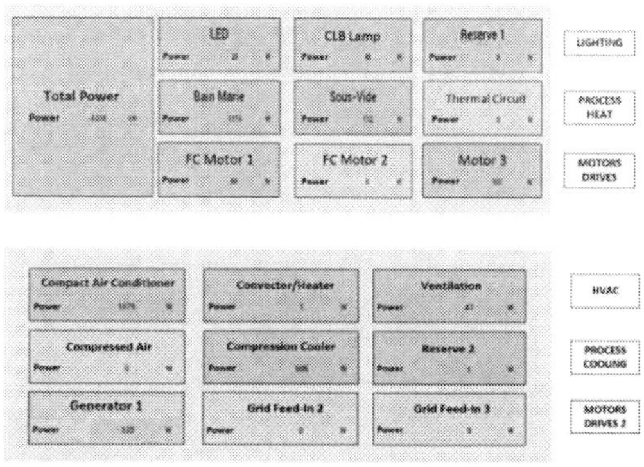

Figure 10 a)/b): The process visualization indicates the total power of the active and non-active consumers as well as feed-in units high-lighted by different colors (orange = consumption, grey = off, blue = feed-in).

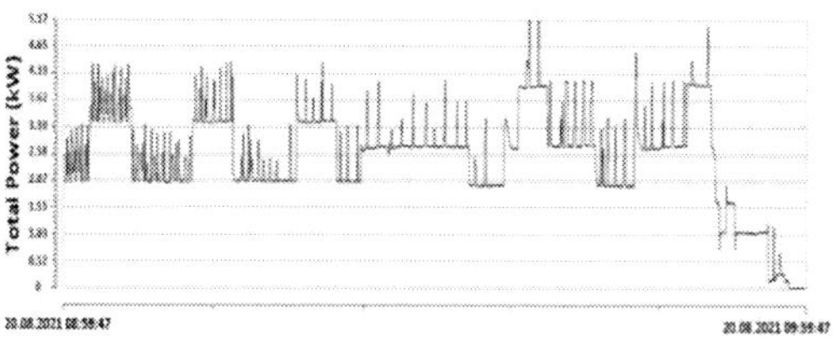

Figure 11: Exemplary total power time series as part of the PLC visualization.

4.4 Communication network and data exchange

Using established communication protocols with OPC UA variables, the data transfer runs via TCP/IP type protocol. The experimental twin data are written to a database. By using such universal communication and data exchange interfaces a high degree of transferability is ensured.

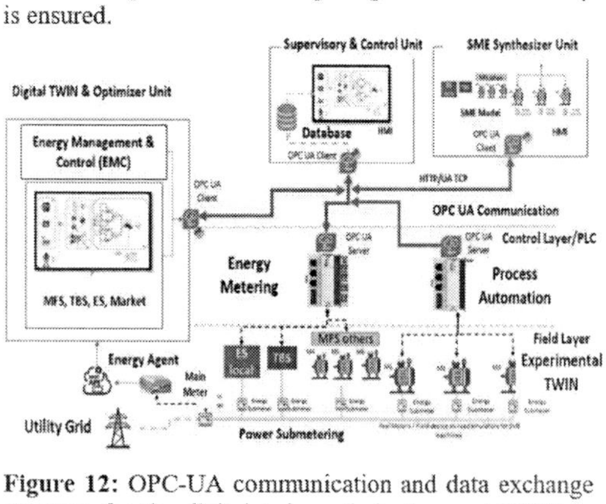

Figure 12: OPC-UA communication and data exchange structure for the digital twin optimization and SCADA-Concept.

5 DISCUSSION AND CONCLUSION

Digital twins of energy flows in an industrial enterprise including the relevant input variables and dependencies are an important prerequisite for the automated use of energy flexibility based on predictive energy management algorithms using mathematical optimization methods.

However, the entry barriers for the implementation of such methods are high. In this paper, we presented two approaches to lower the barriers of entry for the use of digital twins: on the one hand, we introduced the concept of a generic model library that simplifies the creation of a digital twin for modeling energy flexibility in SME. On the other hand, we introduced the concept of an experimental twin that replicates core components and processes of real SME at laboratory scale, allowing the testing and further development of digital twin concepts without affecting the operation of real enterprises.

These two concepts have already been implemented to a large extent at Offenburg University of Applied Sciences and the generic model library is available as open source under the name "indOptFlex" [5]. Several different SME were modeled as digital twin with respect to energy flexibility using the model library, including a plastics processing company with extruders and a steel-powder manufacturing plant. Data from real SME, which were made available within the research project "GaIN" (see acknowledgements) were also used.

The first expansion phase of the hardware part of the experimental twin including the automation was completed as shown in Section 4. The coupling work with the software part of the experimental twin and the optimizer based on the digital twin is in progress. In this context the above-mentioned plastics processing company is taken as the first use case for the complete testing and demonstration of the twin concept.

From the authors' point of view the twin concept of digital twin and experimental twin presented here can make an important contribution to develop, mature and demonstrate advanced model-based energy management algorithms for flexibility utilization for SME.

ACKNOWLEDGEMENTS

The development works on the twin approach were funded by the German Federal Ministry for Economy within the project “GaIN - Gewinnbringende Partizipation der mittelständischen Industrie am Energiemarkt der Zukunft” (BMWi Fkz 03EI6019E). Special thanks to Michael Wirwitzki for his contributions to realize and implement the process automation and PLC programming work on the experimental twin.

REFERENCES

[1] Graßl, M., & Reinhart, G. (2014). Evaluating measures for adapting the energy demand of a production system to volatile energy prices. *Procedia CIRP*, *15*, 129-134.

[2] Schott, P., Ahrens, R., Bauer, D., Hering, F., Keller, R., Pullmann, J., ... & Reinhart, G. (2018). Flexible IT platform for synchronizing energy demands with volatile markets. *it-Information Technology*, *60*(3), 155-164.

[3] L. Pérez, J. Espeche, T. Loureiro, A. Kavgić, DRIvE Project Unlocks Demand Response Potential with digital twins, *Proceedings*, vol. 65, no. 1, p. 14, 2020, doi: 10.3390/proceedings2020065014.

[4] Khatri, R., Schmidt, M., Gasper, R. (2021). Active participation of industrial enterprises in electricity markets - A generic modeling approach, Proceedings DACH+ Energy Informatics, Freiburg Sep. 2021.

[5] Tristán, A., Heuberger, F., & Sauer, A. (2020). A Methodology to Systematically Identify and Characterize Energy Flexibility Measures in Industrial Systems. *Energies*, *13*(22), 5887.

[6] Beier, J. (2017). Simulation approach towards energy flexible manufacturing systems.

[7] Castro, P. M., Sun, L., & Harjunkoski, I. (2013). Resource–task network formulations for industrial demand side management of a steel plant. *Industrial & Engineering Chemistry Research*, *52*(36), 13046-13058.

[8] da Costa Fernandes J., García-Munoz P., Sarigul G., Robert D., Keller N., Schmidt M., Bollin E., PHOTOPUR – PV powered process automation of an AOP based water decontamination, Session 6EO.1.4 at 37th EUPVSEC 2020 online, doi:10.4229/ EUPVSEC20202020-6EO.1.4, p.1791-1795,WIP Munich-DE, 2020.

[9] da Costa Fernandes J., Schmidt M., Feldmann T., Bollin E., Use of load profiles to optimize microgrids of non-residential environments, Session 6BV.3.90 at 33rd European Photovoltaic Solar Energy Conference and Exhibition, Paper DOI: 10.4229/ EUPVSEC20172017-6BV.3.90, p. 2739-2746, Amsterdam-NL, 2017.

DEGRADATION OF SUPPLY RELIABILITY IN STAND-ALONE SYSTEMS DUE TO MODELING STRATEGIES

Luis Timaná[1,2], Matheus Gemignani[1], Guido Rostegui[1], Carlos Almeida[1]
[1] Polytechnic School, Universidade de São Paulo, Av. Prof. Lúcio Martins Rodrigues 380 Travessa 4, 05508-020 São Paulo, Brazil
[2] Electric Engineering and Telecommunications Department, Universidad Católica de Colombia, Diagonal 46 A # 15 B – 10, Bogotá D.C, Colombia
lctimana@ucatolica.edu.co, Tel.: 0057-310-5303-220

ABSTRACT: This work aims to evaluate strategies to simplify the modeling of stand-alone systems, which are normally used in planning to estimate impacts on supply reliability and propose future projects. The evaluation is carried out assuming simplifications in generation and demand parameters for different scenarios, with real data and taking into account the interaction between demand, photovoltaic generation and energy storage system. The impacts on the reliability of the supply are estimated through continuity and quality indicators. The simplifications assumed in this methodology can be applied to other stand-alone systems with different renewable energy sources in combination with energy storage systems.
Keywords: planning, photovoltaic, reliability indices, standalone system, storage.

1 INTRODUCTION

Modeling of energy systems is usually based on initial premises, where conditions for simulation and data treatment are defined, among several others. In the case of energy models for the simulation of autonomous systems, some of the most common simplifications include using typical curves and reducing the analysis period by integrating the data.

These simplifications facilitate data processing, modeling, and computational effort. However, it also results in the degradation of results, leading in some cases to erroneous or imprecise conclusions. Typical consequences of simplification in planning are projections that do not meet demand or, at the other extreme, oversizing and unnecessary investments.

Planning for standalone systems is often not done in the same depth as conventional systems. This is because they have less importance and small demands compared to any national network.

With more recent technologies such as photovoltaic generation and storage systems, other simplifications are employed in planning. These characteristics applied together make planning very simple, but with unintended consequences related to system oversizing. This paper aims to evaluate some of the most used simplifications, quantitatively presenting their impacts on supply reliability and proposing a more suitable design methodology for this type of system.

Quality of the electric power service is perceived by consumers in aspects such as speed and precision of the responses to their complaints, reliability of the electricity supply and voltage levels. The quality of the service offered by the utility company is evaluated considering: continuity of supply, voltage level and commercial services [1]. Quality of the power supply is related to the number and duration of interruptions, which can be improved with greater investments in the network, such as reinforcement of lines and installation of automation devices to make the network more redundant, in addition to maintenance actions [2].

2 METHODOLOGY ANALYSIS

This paper deals with the evaluation of some reliability indicators applied to an autonomous energy system, which is made up of demand, photovoltaic generation and an energy storage system, as shown in Fig. 1. The evaluation is carried out by applying different solar irradiation scenarios, in which the effect of simplifying generation and demand parameters on demand satisfaction indicators is studied.

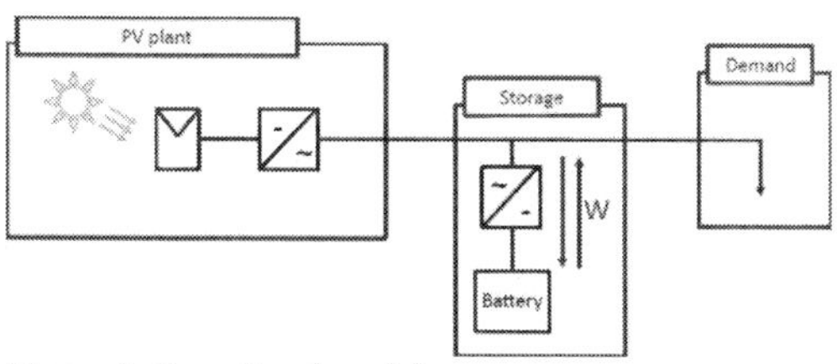

Figure 1: Operational model.

SAIDI (System Average Interruption Duration Index) and SAIFI (System Average Interruption Frequency Index) are the reliability indicators examined in this work. SAIDI measures the time of occurrence when discontinuities occur in the power supply. SAIFI represents the number of interruptions in the power supply.

In conventional applications, these quality indicators allow identifying the most critical areas of the distribution network, this helps electricity companies to maintain or improve the quality of service, so that it is within the levels required by regulatory authorities and thus avoid the payment of penalty fees [3].

UE (undistributed energy) and DE (distributed energy) indicators are also evaluated. DE is defined as the energy supplied by the photovoltaic panels and batteries. The UE is defined as energy that cannot be supplied to consumers.

Global horizontal irradiation (GHI) data used in this work were obtained from the System Advisor Model [4], a free software distributed by NREL (National Renewable Energy Laboratory). These data come from the municipality of Petrolina, Pernambuco. A typical conversion factor is used to calculate photovoltaic generation from solar irradiance, which is related to installed capacity and energy conversion efficiency. The demand curve corresponds to the average consumption of a typical family home in Brazil. The parameters of the battery energy storage system are the same for all simulated cases.

The indicators mentioned above are evaluated in four scenarios over a year, as shown in Fig. 2.

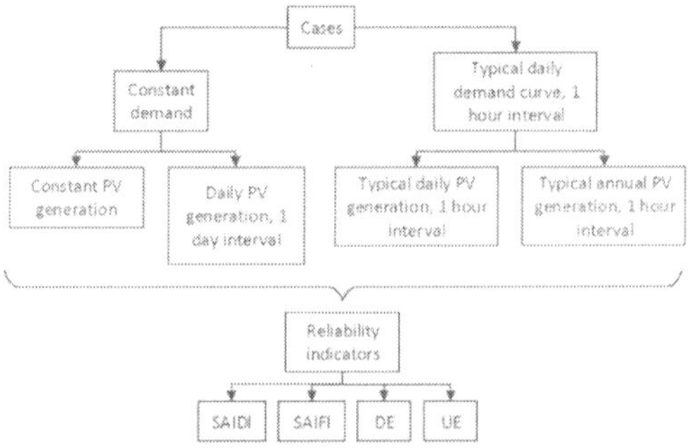

Figure 2: Case studies.

The purpose of these case studies is to compare the reliability and quality indicators of the service provided by the public utility company for different scenarios of solar irradiation and demand. As a result of this comparison, this work aims to examine the effect of simplification due to power resolution for both demand and photovoltaic generation.

In other words, SAIDI, SAIFI, UE and DE indicators are compared for the same average annual generation and demand values, but in each case, with different power curves in terms of simplification level. Constant values and typical daily and annual demand curves of hourly variation are considered.

Taking into account that a higher level of detail implies more precise results but a longer processing time is necessary, the aim is to discover how the simplifications of the problem, imposed on the demand and generation curves, can affect the results of the continuity and reliability indicators, widely used by utility companies.

The effect on the increase of photovoltaic generation between the four scenarios is also analyzed and compared. That is, each scenario is made up of sub-cases, where the same demand curve is used together with a generation curve whose mean value is different for each sub-case.

The four proposed scenarios are:

- Case 1. Constant demand and photovoltaic generation. It is the simplest possible assumption, which is often used. The calculation of quality and reliability indicators is also simple, which can lead to unrealistic results in some cases, as will be shown below. However, it can be useful if only some trends in the results need to be analyzed. The behavior of the battery bank is not taken into account, since the charging or discharging of the batteries depends on the comparison between demand and generation. As this comparison is always constant, there is no valid criterion to determine when they are loaded or unloaded.
- Case 2. Constant demand curve and daily generation curve. With this definition, it is possible to take into account the behavior of the battery bank, although with little precision, since batteries can be charged and discharged in less than a day. Thus, once batteries are fully charged or discharged, they remain inactive until a new change in demand and generation behavior occurs that allows them to be changed.
- Case 3. Typical daily demand curve and typical daily generation curve, both curves with a time interval of one hour. The demand and generation curves are periodic and show hourly variation, which allows a much more precise simulation of the battery bank, and as a consequence, more realistic values are obtained for the continuity and reliability indicators (SAIDI, SAIFI, UE and DE). However, the periodicity of the generation and demand curves might result in periodic battery behavior, which is undesirable and, in general, incorrect. In most real cases, seasonal variations are expected, especially in generation.
- Case 4. Typical daily demand curve and annual generation curve, both curves with a time interval of one hour. This is the most realistic case; an hourly variation interval allows an adequate simulation of the battery bank and the generation curve. As the generation is not periodic, it better reflects the generation changes that depend on the time of year, which can benefit or hinder quality and continuity indicators.

The following graphs show examples of generation and demand curves for a specific sub-case for each case, during a period of one month (01/01/2017 to 01/30/2017).

Fig. 3 illustrates an example of case 1, in which the demand and generation curves are constant. As the generation is greater than the demand, there is an ideal condition in which the demand is always satisfied by the photovoltaic generation and, therefore, the SAIDI and SAIFI indicators are equal to zero. It is observed that in the other evaluated sub-cases, the generation assumes different values, which are in the range of 416.7 W to 625.0 W, the minimum value being equal to the demand (which maintains the same value in all sub-cases).

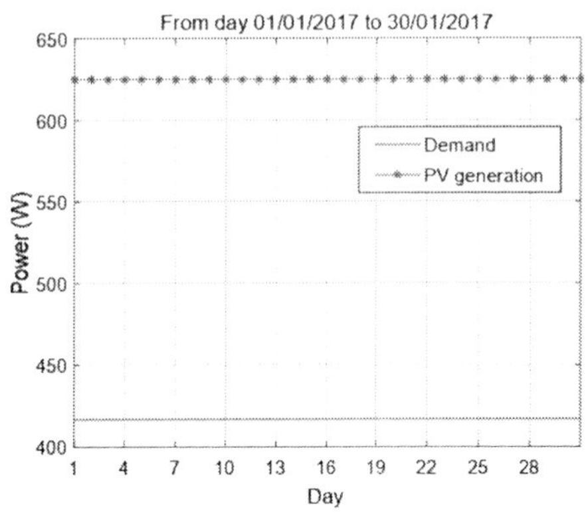

Figure 3: Case 1. Constant demand and photovoltaic generation.

Fig. 4 represents an example of case 2, in which the demand curve is constant, but the generation curve varies at intervals of one day. The average generation value is equal to 625.0 W and the demand is constant, with 416.7 W, so that in most cases the generation is higher than the demand. Taking into account a moderate size of the battery bank and a specific day with a demand greater than generation, there would be an increase in SAIDI and SAIFI, since storage could not supply the demand throughout the period.

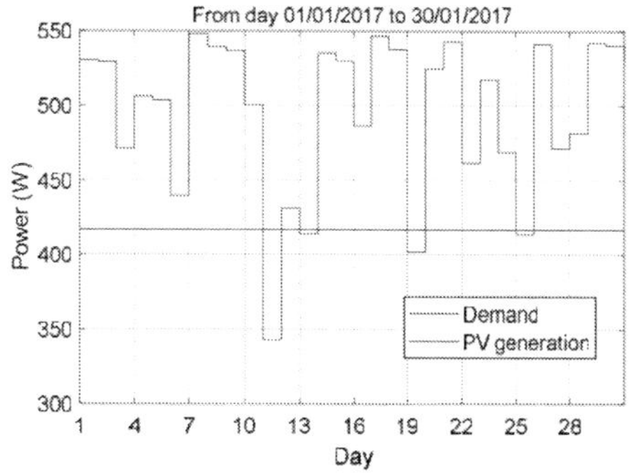

Figure 4: Case 2. Constant demand and annual PV generation curve with one day period.

A sub-case for case 3 is shown in Fig. 5, where the demand and generation curves are periodic per day. The average value of demand and generation power is 416.7 W and 625.0 W, respectively. As the sampling interval is one hour, it is possible to simulate the charging and discharging behavior of the batteries in the periods when generation exceeds demand (noon) and when there is no generation (nights). However, the behavior of the generation depending on the time of year is not taken into account, since the generation depends on the level of solar radiation.

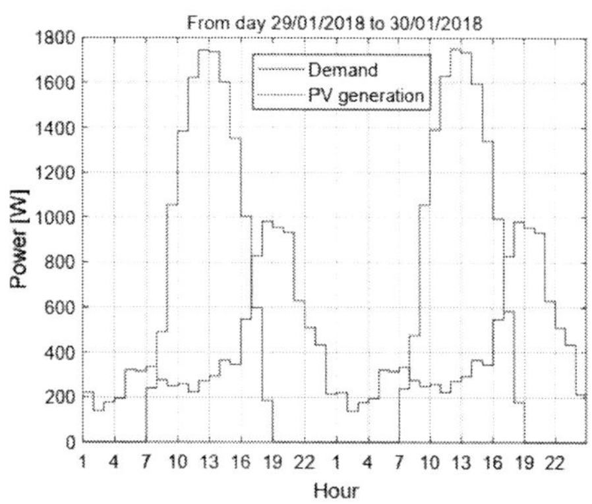

Figure 5: Case 3. Demand and photovoltaic generation curves with one day period.

Fig. 6 shows an example of case 4. The demand curve is periodic per day with an average value of 416.7 W, while the generation curve is not periodic with an average value of 625.0 W. This case constitutes a more realistic representation than the one presented above, since the generation is better modeled.

3 CASE STUDIES

The quality indicators calculated for the four cases are analyzed below.

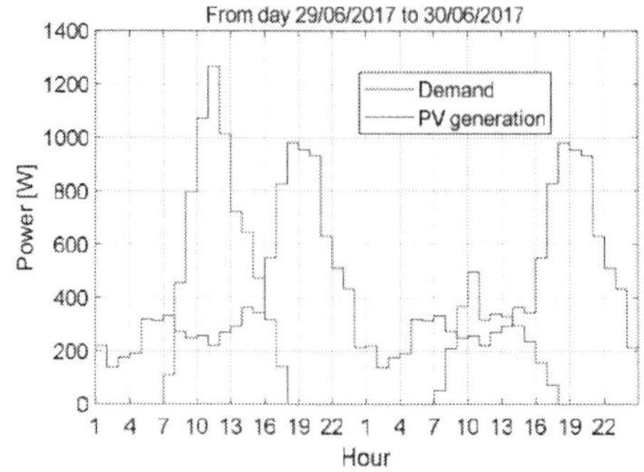

Figure 6: Case 4. Demand and annual photovoltaic generation curves. Only the demand curve is periodic (one day period).

3.1 Specification Parameters

It is intended to make a comparison of the reliability indicators, in which the same average demand value is used for all cases. This comparison is made for different average values of photovoltaic generation, as shown in the flow diagram illustrated in Fig. 7.

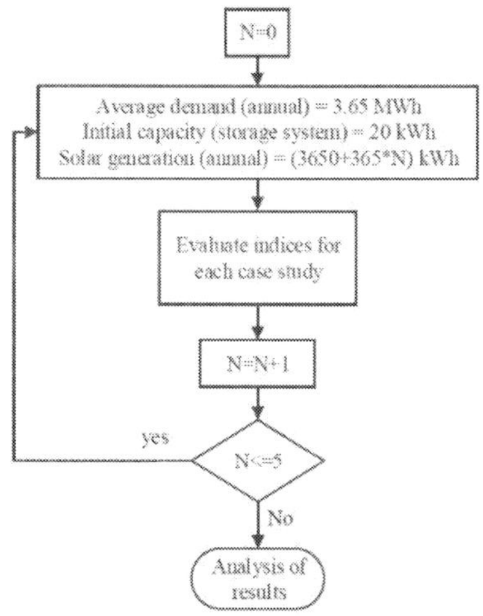

Figure 7: Simulation process of case studies.

For all cases, the following considerations are taken:

- An annual demand of 3.65 MWh is assumed, that is, the average annual power of the demand is the same in all cases.
- The effect of the increase in photovoltaic generation is evaluated, which ranges between 3.65 and 5.47 MWh per year.

Regarding batteries, it is assumed:

- The initial storage capacity of the batteries is equal to 20 kWh.
- The battery bank is fully charged at the beginning of the simulation.
- An exponential loss of battery storage capacity is assumed, which reaches 95.64% of the initial capacity at the end of the first year of use. This exponential behavior of the battery capacity is based on the periodicity of the generation and demand curves.
- In the process of charging and discharging, an

efficiency of 90% is attributed.

Taking into account that in the four scenarios the average demand is equal to 3.65 MWh and the average generation is between 3.65 and 5.47 MWh for each sub-case, it is observed:

- For case 1, this energy value corresponds to a constant power demand equal to 416.7 W. The generation is also constant with a power value defined between 416.7 W and 625.0 W, depending on the sub-case.
- For case 2, the demand is constant with a power value equal to 416.7 W and the generation curve is constant in one-day time intervals, with an average annual power value between 416.7 W and 625.0 W, depending on the sub-case.
- For case 3, a periodic demand curve per day is used, which is constant in hourly intervals, with an average value of 416.7 W. The generation curve is also periodic per day and constant in hourly intervals, with an average value between 416, 7 W and 625.0 W, depending on the sub-case.
- For case 4, a periodic demand curve per day and constant in hourly intervals is used, which has an average value of 416.7 W. The generation curve is not periodic and its average value is between 416.7 W and 625.0 W, depending on the sub-case.

All cases simulate an off-grid (off-grid) photovoltaic system, composed of solar modules, charge controller, storage system and inverter. When the generation of solar panels is greater than the demand, the surplus energy is used to charge the batteries. On the other hand, when demand is greater than generation, batteries are discharged to help supply demand. If the energy supplied by the solar panels and batteries is less than the demand, there will be energy not supplied, which implies an increase in the UE indicator. If the energy generated by the solar panels is greater than the demand and the batteries are already fully charged, so there will be wasted energy and an increase in the DE indicator.

Each case is made up of sub-cases, which differ in the average value of photovoltaic generation, although the demand curve is the same for each case (see Fig. 7). In this way, the effect of a higher generation on quality indicators can be evaluated.

3.2 Results

Fig. 8-11 show the SAIDI, SAIFI, UE and DE indicators depending on the photovoltaic generation for all cases. These indicators are calculated with a simulation time of 1 year.

In case 1, the indicators are independent of photovoltaic generation because generation and demand are considered constant over time. The generation value is always greater than or equal to that of the demand, which makes the power supply service uninterrupted, even in periods where there is no physical generation. In this case, the DE indicator is equal to the energy consumed by demand throughout the year and the UE, SAIDI and SAIFI indicators are equal to zero.

In cases 2, 3 and 4, the average energy value of photovoltaic generation is always greater than or equal to the average energy value of the demand throughout the year. However, this does not imply that the energy supplied by the photovoltaic panels will supply all the demand throughout the day, since the energy stored by the batteries during the sunny periods may not be enough to supply all the demand during the night.

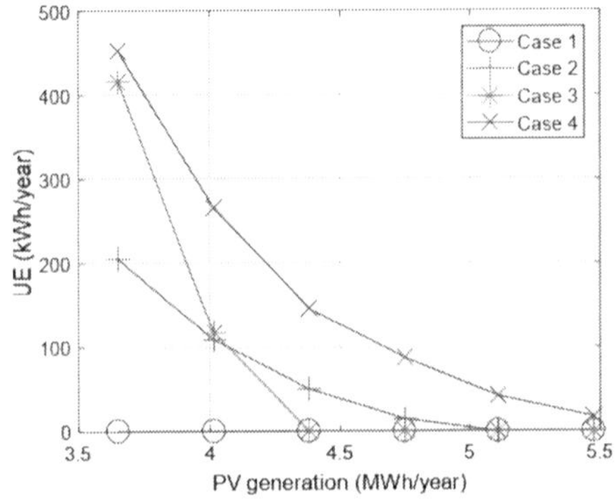

Figure 8: Undistributed Energy.

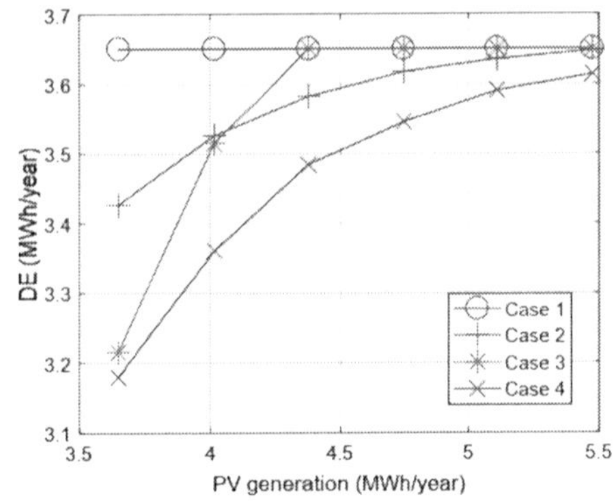

Figure 9: Distributed Energy.

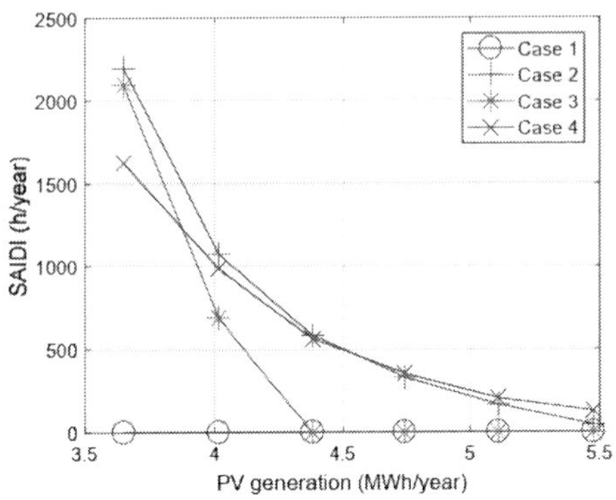

Figure 10: SAIDI (System Average Interruption Duration Index).

In addition, when the efficiency of the batteries is taken into account in both the charging and discharging process, it is necessary that the energy supplied by the panels to the batteries is greater than the demand during the night, in order to offer a full service in the power supply.

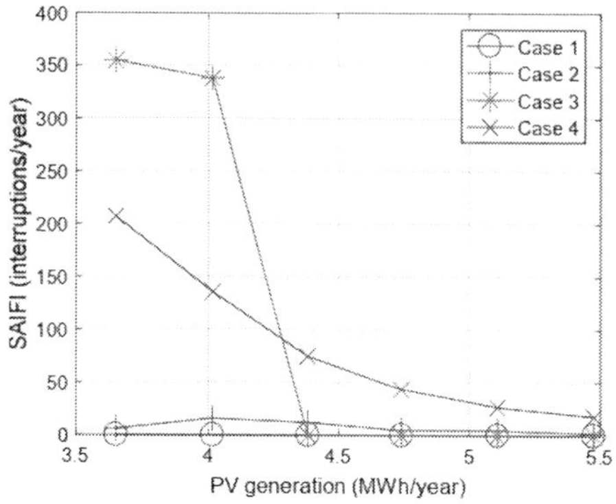

Figure 11: SAIFI (System Average Interruption Frequency Index).

The difference in results between the cases is mainly due to the precision in simulating the behavior of the batteries. When the power and generation curves are constant, the batteries are not taken into account in the calculations. When the variation of the power and generation curves is daily, the performance of the batteries can be considered, but with little precision, since they are usually fully charged or discharged in less than 24 hours, so the batteries do not operate the rest of the day, which represents unreal behavior. On the other hand, when the variation of the power and generation curves is hourly, it is possible to simulate with much more precision the behavior of the batteries when the generation is higher or lower compared to the demand.

Greater precision in simulation implies more processing time, which can become excessive or unfeasible when it comes to solving optimization problems, so there must be a balance between precision and processing speed when the solution to a problem is based on simulations.

Fig. 12 shows the simulation of case 2 for the months of January and June, when the photovoltaic generation is 3.65 MWh per year. In this case, the average power of the photovoltaic generation is equal to the average power of the demand. However, photovoltaic generation in June is lower than in January, which makes the number of interruptions more frequent in this period of less radiation.

Fig. 13 illustrates the behavior of photovoltaic generation during the last two days of January and June, for case 4, when photovoltaic generation is 3.65 MWh per year. The hours of interruption of the power supplied to the load are longer in June than in January, this is due to the difference in radiation, so this behavior is similar to the previous example.

In case 3, the photovoltaic generation is the same every month of the year because the photovoltaic generation curve is periodic per day. For this reason, Fig. 8-9 indicate that a photovoltaic generation value equal to or greater than 4.38 MWh implies that the UE and DE indicators are equal to zero.

Fig. 8-9 show that case 4 shows the highest value of UE and the lowest value of DE. The value of UE and DE, for a photovoltaic generation of 3.65 MWh, is equivalent to 12.9% and 87.1% of the total energy demand during the year. It is observed that all cases show similar results. However, it is observed that the indicators tend to approach the ideal results of case 1 as the photovoltaic generation increases.

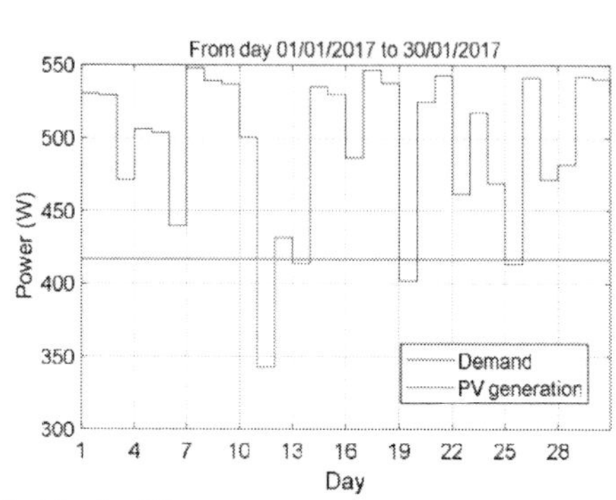

a) Month of January

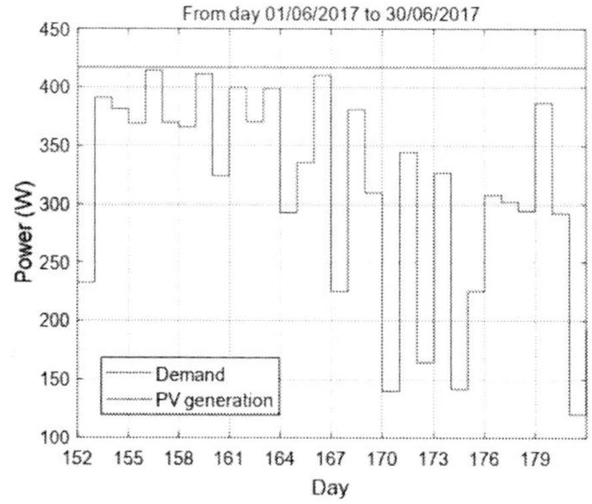

b) Month of June

Figure 12: PV generation. Case 2.

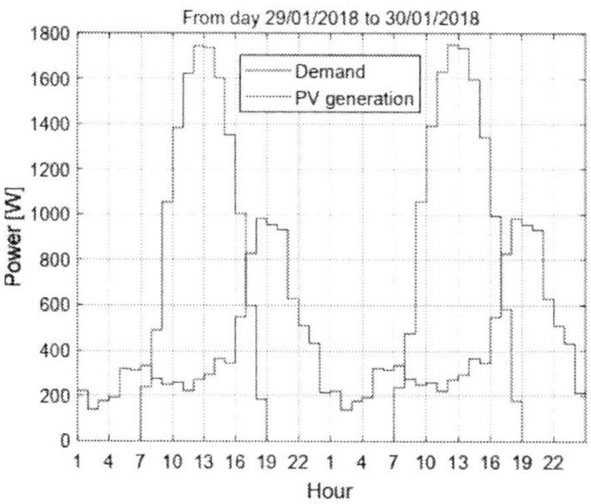

a) Last days of January

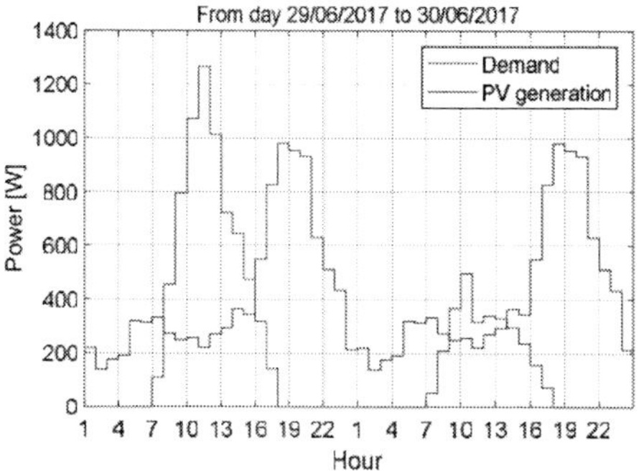

b) Last days of June

Figure 13: PV generation. Case 4.

In Fig. 11, the number of interruptions in the supply of energy on demand for case 3 is 355 W, when the photovoltaic generation is 3.65 MWh per year. The SAIFI is very high because, most days, the batteries cannot supply power to all the nightly demand; however, in the early days of the year, batteries can provide enough power as they are fully charged. In just ten days, this effect is eliminated and the batteries are completely discharged overnight.

A similar behavior is observed when photovoltaic generation is 4,015 MWh. The high photovoltaic generation reduces the number of interruptions because it allows the batteries to be charged enough to supply the nightly demand. In case 4, unlike the third, the photovoltaic generation can change a lot, since it depends on the time of year. On consecutive days of low radiation, the photovoltaic system will not be able to supply enough power to the batteries, regardless of the installed capacity.

However, in cases 3 and 4, it is observed that greater photovoltaic generation reduces the number of interruptions. Additionally, existing interruptions are not observed in the simplifications assumed in cases 1 and 2, in which it is only possible to observe daily changes in generation and demand.

4 CONCLUSIONS

For the planning of stand-alone systems, different scenarios were evaluated to compare the effect on the reliability indices as a consequence of the simplification, which is carried out on the generation and demand parameters.

ACKNOWLEDGMENT

This work was supported by the Companhia Energética de Petrolina (CEP).

REFERENCES

[1] E. Fumagalli, L. Schiavo, and F. Delestre, Service quality regulation in electricity distribution and retail. Springer Science & Business Media, 2007.

[2] M. Couto, J. Pascoal, J. D. Matos, and J. Antunes, "Approach to reduce momentary average interruption frequency index–the quality of service indicator for momentary interruptions–the experience of the portuguese dso," CIRED-Open Access Proceedings Journal, vol. 2017, no. 1, pp. 522–525, 2017.

[3] D. J. Krajnak, "Faulted circuit indicators and system reliability," in 2000 Rural Electric Power Conference. Papers Presented at the 44th Annual Conference (Cat. No. 00CH37071). IEEE, 2000, pp. A4–1.

[4] System Advisor Model (SAM) General Description (Version 2017.9.5), National Renewable Energy Laboratory, Golden, CO, 2017, release 9.5.

THE H2020 R&D PROJECT SERENDI-PV: INNOVATING TOWARDS IMPROVED RELIABILITY, HIGHER PERFORMANCE AND DISPATCHABLE GRID INTEGRATION FOR PHOTOVOLTAIC SYSTEMS

Javier del Pozo[1], Eduardo Román[1], Ricardo Alonso[1], John A. Tsanakas[2], Christian Schill[3], Monica Aleman[4], Iván Lombardero[5], Ingrid Weiss[6], Jonathan Leloux[7], Marcel Suri[8], Ismaël Lokhat[9], Joséphine Berthelot[10], Chloé Monet[11], Miguel Crespo[12], Christian Breyer[13], Basem Idlbi[14], Elias De Keyser[15], Joseph Reed[16], Marta Puente[1]7, Markus Resch[18], Íñigo Landibar[19]

[1] Tecnalia, [2] CEA – INES, [3] Fraunhofer-ISE, [4] Becquerel Institute, [5] QPV, [6] WIP, [7] LuciSun, [8] Solargis, [9] Cythelia Energy, [10] Akuo Energy, [11] CNR, [12] Grupo Cobra, [13] LUT, [14] THU, [15] Next Kraftwerke, [16] Mylight Systems, [17] Cegasa Energia. [18] Energie Gussing, [19] Ingeteam

Email and address of corresponding author/institution: javier.delpozo@tecnalia.com; T +34 946 430 850
Parque Tecnológico de Bizkaia; C/ Astondo, Edificio 700; E-48160 Derio - Bizkaia (Spain)

ABSTRACT: SERENDI-PV project is addressing some of the main concerns which would enable the increase in the penetration of PV generated power on to the European grids. On the one hand the project will address a reduction of the parameters which have the strongest influence on the risk perception for new investments in the sector in order to reduce the Levelized Cost of Energy, mainly associated with the new PV technologies such as bifacial, floating PV and building integrated PV: modelling (better energy yield assessments), faults diagnostics (increase the performance of PV systems), in-field quality control (better reliability), and forecasting. Special focus will be paid on issues that may occur to the PV plants, such as soiling (dust), snow and degradation over the years. On the other hand, the technical challenges and opportunities that will come with the increase of the penetration in the grids, such as the potential of PV to offer ancillary services, etc. will be assessed. Short-term forecasting and nowcasting fleets of distributed PV systems as a whole (aggregation) instead of PV systems alone will also play a key role in higher PV penetration and will be investigated in the project.

Keywords: PV systems, PV reliability, PV performance, grid integration, PV operations & maintenance (O&M).

1 INTRODUCTION – PROJECT RATIONALE

Photovoltaic (PV) energy is expected to play a major role in the generation of electricity in the world, including Europe, and to become one of the major sources of electrical power. However, in order for PV to reach a utility-friendly high-penetration level in the grids in the next decade, two important challenges need to be addressed: (1) To keep **reducing the Levelized Cost of Energy (LCoE) for PV** and (2) to make it possible to **integrate a rapidly increasing share of PV power into the grid**, up to high penetration levels, considering the European ambition to be the first climate neutral continent by 2050 [1].

The reduction of LCoE depends on several factors. The most sensitive parameter is the location (estimated energy yield), followed by the weighted average cost of capital (WACC) [2]. The PV industry will be able to reduce the WACC by conveying more trust to the financial sector. The WACC will be reduced by reducing the uncertainties (thus, increasing the trust) in the whole value chain.

First, SERENDI-PV will reduce the uncertainty thanks to the higher accuracy of modelling for the new PV technologies (such as bifacial PV, floating PV and BIPV) allowing better energy yield assessments.

Second, in addition to this, the better-quality controls in the field and in the lab will increase PV project's quality & lifetime and to reduce their performance uncertainty and to improve bankability of the new PV technologies. As a result, higher quality, lower uncertainty and risks in the PV projects will be achieved, what will drive to better bankability.

Third, the uncertainties in the system reliability will be addressed by applying advanced fault diagnosis in the PV plants, with special focus in the new technologies, and by using the predictive maintenance of the most complex components in PV system and with the highest impact on energy availability: PV inverter and batteries. This will be achieved by developing digital twins that will provide a better understanding of potential failures and aging processes of these components allowing to anticipate them.

Fourth, in the same way, the reduction of uncertainties by means of improving PV power forecasting (focusing on short-term and nowcasting) of new PV technologies (bifacial, floating and BIPV) and in case of specific atmospheric events (snow, dust, frog) will be addressed by SERENDI-PV. The new applications such as bifacial PV, floating PV and BIPV will enhance bankability owing to lower values and better traceability of uncertainty on PV energy yields.

In a nutshell, the uncertainty reduction achieved by SERENDI-PV will be achieved through several improvements, including better quality controls, better component and system reliability, and better energy yield assessments. Moreover, SERENDI-PV covers the whole value-chain of a PV project and therefore the results obtained are strongly linked with each other. This means the developments achieved will feedback each other, enhancing the results and increasing their value and reliability.

The second ranked challenge for PV generation is to be able to increase its grid penetration rates to very high levels without compromising the technical and financial feasibility of grids. A first step would be to ensure that PV does not introduce prohibitive costs on the grid management itself. But the actual goal is to bring cost-effective grid management solution provided by PV: grid voltage management through reactive and active power control and grid frequency management by active power control. This can be achieved by **controlling PV inverters** with reduced costs and efforts by developing **standardized smart grid** solutions based on the PV data. Another approach is hybridising with energy storage for additional ancillary services which has the potential to lead to a cost-effective reduction of the grid connection cost. In general, there will be the need for innovative monitoring and management of millions of centralized and distributed power sources to maintain system stability bringing in parallel the opportunity of additional revenues for PV. **New revenue streams** for PV will in fact be necessary to maintain profitability as further volume growth reduces costs but also exerts pressure on market prices as prices drop when large amounts of zero marginal cost PV are put on the power exchanges. Effective grid management with PV is however only possible if the behaviour of consumers and producers connected to the grid is fully understood, monitored, and forecasted with precision and the combination of PV with energy storage and flexible demand is investigated.

2 THE SERENDI-PV PROJECT: SCIENTIFIC RELEVANCE & INNOVATION GOALS

The SERENDI-PV project proposes innovations on PV systems and their grid integration to improve (1) lifetime, reliability, performance and profitability of PV generation; and (2) high-penetration of the PV generation in the grids with improved stability.

On this basis, the project's aimed leap of knowledge and advance beyond today's state-of-the-art concerns five main "innovation pathways":

1) Simulation, modelling and designs for better PV reliability, performance and profitability.

New knowledge and innovations to tackle today's challenges and R&D gaps in four aspects: modelling and mitigation of losses and performance degradation; modelling of emerging PV systems' technologies and configurations (bifacial PV, floating PV and BIPV), solar resource modelling and mitigation of uncertainties; financial risk modelling and mitigation.

2) Monitoring and image-based fault diagnosis for higher PV performance and optimized O&M.

Innovative tools and services based on six technical approaches: a) advanced fault diagnosis in large PV plants; b) fault diagnosis in new technologies (bifacial, floating); c) data analytics algorithms for diagnostics of medium/long trend processes d) fault diagnosis in medium and small (residential) size PV plants; e) digital twins for predictive maintenance of the most complex components (PV inverters and batteries) f) improved infrared thermography data analytics.

3) Quality control (QC) equipment and procedures for PV components and systems reliability.

Effective quality control throughout the whole supply chain, by innovating and validating testing equipment and procedures for:

- Laboratory facilities, providing indoor quality controls at the component level to address two issues: a) soiling and cleaning on the one hand; b) PV module degradation and failures in the floating environment on the other hand.
- Field testing toolboxes, providing outdoor quality controls at the component and system level, in order to address another two issues: c) soiling and d) PV strings ageing/degradation.
- Quality control procedures for e) PV inverters and f) energy storage, providing a coherent frame to exploit field and laboratory testing.

4) Mid-/short-term forecasting up to nowcasting for PV systems aggregations and optimized grid integration.

- enhanced algorithms for reducing uncertainties of short-term forecasting by merging data from numerical weather prediction models, satellite-based data and recent data from PV power plants for aggregation of individual PV systems to PV portfolios.
- improved algorithms for reducing uncertainties of PV power nowcasting by merging data from sky-camera, from satellite and from on-line measurements of distributed PV systems.
- improved algorithm to reduce or mitigate large forecast errors occurring in case of specific atmospheric events like snowfall, persistent fog, sudden changes of weather, and for higher depositions of dust and higher atmospheric pollution.
- extension of existing PV forecasting tools and models to address PV forecasting needs for bifacial, floating and bifacial PV systems.
- improved algorithms not only for single PV plants but also for aggregated power output of distributed PV systems (spatial averaging of multiple PV systems for different geographic scales).

5) Innovative business models for PV added revenue at high-penetration levels.

Novel business models, beyond current state-of-play, for:

- PPAs and energy market integration based on better informed trading services and reduced imbalance costs
- Promoting increased self-consumption for small and medium sized PV in particular by PV-storage combinations or energy supply and consumption matching.
- Aggregated PV in mini-smart grids by combining in locally or regionally PV production with demand and storage solutions with the goal of self-sufficiency or maximisation of collective self-consumption.
- Combined PV and storage to shift PV production to high price moments and increase of self-consumption at household level.
- Ancillary services such as control power and redispatch services to the TSO and DSO.

3 R&D ACTIVITIES

3.1 PV modelling, data analytics/diagnostics and quality control

The first step towards high PV system reliability and performance concerns the PV system components (e.g. PV modules, inverters, and batteries). Better technical risk management will address the key factors affecting PV performance. SERENDI-PV will develop solutions at various levels regarding parameters whose influence still needs to be better assessed or evaluated, such as PV module ageing, soiling and snow deposition. The innovations that are proposed will pay particular attention to ground-mounted PV plants, mainly with bifacial technology, floating PV, and rooftop residential or commercial PV systems, including BIPV.

3.1.1 PV modelling

The solar resource will be modelled through a characterisation of long-term trends and a detailed quantification of the uncertainties.

The final objective is to integrate the modelling solutions developed in the project into commercial tools. To achieve this, several key actions will be carried out. The first one will focus on the needs of the industry, and the benchmark of the current tools and models. As an example, an innovative simulation tool is currently being developed for the assessment of the solar irradiance on the back side of bifacial PV modules. This tool will make use of the recent advances in the Graphics Processing Units (GPUs) developed for the video game industry, and that will make it possible to obtain simulation with a high spatial resolution and within simulation times that are low enough to be used in commercial applications [3]. Figure 1 illustrates the application of that tool to the case of bifacial PV modules installed on carports.

Figure 1: Application for PV carports of a novel tool developed at SERENDI-PV for the accurate simulation of bifacial energy yield.

The novel simulation tools will be combined with better evaluation of technical uncertainties and their influence on financial models. These tools will build on previous pioneering developments by some partners of the project [4].

3.1.2 Data analytics/diagnostics

Advanced data analytics will be developed for the diagnosis of the situation (both monitoring-based and image-based) and predictive maintenance of the most complex components in PV system and with the highest impact on energy availability (PV inverter and batteries).

These new developments will be integrated in commercial solutions, with advanced and automated functions for data analysis, diagnosis and fault diagnosis (Fig. 2). This will be done both at the utility-large commercial scale PV plants from SCADA monitoring data, including higher granularity of condition monitoring and specific data analytics for bifacial and floating PV, as well as at medium-size commercial and small-residential scale PV plants from energy meter measurement data and through the development of a BIPV system digital twin.

Additionally, the project will develop advanced digital twins for predictive maintenance of the components with the highest complexity, uncertainty and impact on system reliability in current PV plants: inverters and batteries.

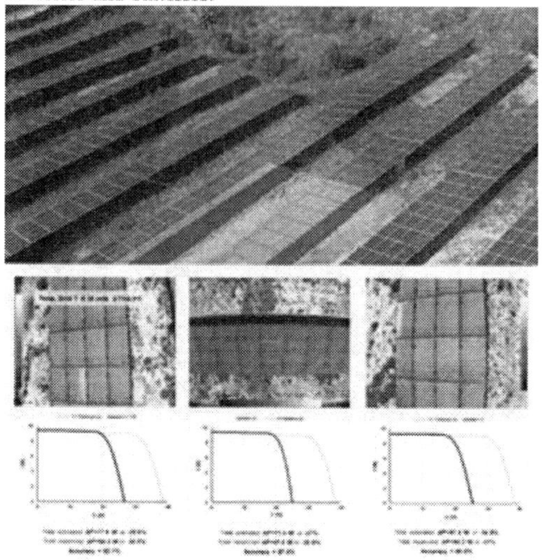

Figure 2: Example of image-based diagnostic approaches (CEA-INES) to be further developed and extended in the context of SERENDI-PV, addressing industry-relevant failure modes and new PV technologies (floating PV, bifacial PV and BIPV).

3.1.3 Quality control

Another important focus of the project will be on lab testing facility, on soiling and cleaning; field testing equipment, such as I-V curve tracer; field measurement kit for soiling quantification and improved quality control procedures (QCP) with low uncertainty included for the PV inverter. The main target in this aspect is the reduction of the uncertainty of the actual features in the field of the main devices involved in the energy generation. In this regard, the characterization methodology for PV modules (including bifacial and floating PV) together with PV inverters and batteries will be assessed. Several concrete solutions will also be developed to deal with key issues that have been identified as particularly relevant in the current state of the art of the PV sector. For example, the soiling assessment will be improved by developing equipment specifically dedicated to its assessment with both, lab and field configurations. I-V tracers will be upgraded to measure up to 1500 Vdc strings and to properly evaluate bifacial and floating PV considering the nuances involved in its performance. The degradation of PV modules in the field will be measured through low uncertainty I-V curves and through specific lab testing methodologies including the analysis of defective PV modules from the field. New characterization procedures for inverters in the field will be designed considering its transformation efficiency and the accuracy of the maximum power point tracker (MPPT) system. Finally, the characterization of batteries in the filed including battery cycle efficiency and state of health (SoH). Altogether will converge on enhanced quality control procedures and acceptance tests for bifacial and floating PV installations.

3.2 PV power forecasting, grid integration and optimization

The project will aim to reduce the level of

uncertainties of the existing services for nowcasting and short-term forecasting. The services developed will focus on a time horizon of intraday and day ahead forecast (time horizons 0 to 48 hours ahead), with adaptive temporal resolution from 1 to 60 minutes. The approach will be developed by merging models based on the use of sky cameras, satellite data and postprocessing of outputs from Numerical Weather Prediction (NWP) models. PV power production data will be assimilated with models to adapt the forecast for needs of a specific PV system or entire portfolio. The improvements will be evaluated for individual PV plants or portfolios, and in line with the most commonly used statistical characteristics in use in the PV industry.

SERENDI-PV will work towards the improvement of PV power forecasting as input for energy management, trading and grid management by TSOs and DSOs, more specifically:

- Reduction of uncertainties of mid-term forecasting (up to 9 days ahead), short-term forecasting (up to 15 min up to 1 day ahead)
- Reduction of uncertainties of nowcasting (1 to 30 minutes ahead)
- Forecasting in case of specific atmospheric events (snow, dust, frog)
- Adaptation of new simulation tools developed for new technologies (bifacial and floating PV)
- Spatial averaging effects and enhanced solutions to forecasting of aggregated PV power output at different geographical scales (neighborhoods, cities, distribution grids, grid-nodes, transmission grid, virtual power plants)
- Improving grid stability through smart grid solutions based on PV-data

The energy transition from centralized power supply in the grid through a few hundreds of power plants towards a power supply from millions of distributed energy resources (DERs) is one of the main challenges for the grid operation and a high share of PV. One of the main difficulties is the PV data collection and management as well as the integration of this huge amount of data in the operation of the grid. Therefore, the project will tackle these problems, considering the technical, economical and legal constrains. Based on this analysis, technical solutions for a reliable and standardized IT infrastructure for the PV data, considering high compatibilities for many countries, will be assessed. The discussed PV data and the corresponding IT infrastructure will be utilized to develop and test several technical solutions, including digital twin for the grid, to enable higher PV contribution and a utility friendly integration of it.

3.3 Technical/Economic Bankability, Collaborative platform and New Business Models

WACC reduction is an important pathway towards a decrease in the Levelized Cost of Energy (LCoE) of PV generation, and it needs to be addressed by showing a high level of technical risk management and a more accurate business plan thanks to a better predictive evaluation of the financial flows and financial risk modelling. Some of these points will be addressed in sections presented earlier. This requires a holistic approach in order to achieve global acceptance of 35 years lifetime in bankability, being this one of the factors that most influence the WACC.

In SERENDI-PV, we will investigate on the one hand new business models and on the other we will analyse the economic, regulatory, market design and legal constraints, which are the barriers for the concrete implementation of such models and still strongly member-state specific. For example, it is today in most countries not possible to offer PV-based ancillary services only during daytime blocks, but those are the blocks when PV is both adding stress to the system and when it can contribute to system services.

Moreover, there is a need for a common space which will enable to gather the required knowledge to lower the existing barriers to PV development. Several collaborative platforms have already been proposed in other sectors and more recently in the PV community. Some good examples are PVPMC/PVlib [5] in the USA, PVGIS [6] for solar radiation, NREL [7] for TMY data on solar radiation, PV performance and consumption data in the USA, or the open source simulation code SISIFO [8] that is the result of the EU projects PV CROPS and MASLOWATEN, or the EU COST Pearl PV project. Nevertheless, there is a need to create a holistic collaborative platform on PV based in Europe, capable of foster knowledge accumulation within the European PV community, and that focuses on the specificities of PV in Europe, while collaborating with the other existing platforms to join forces as much as possible.

SERENDI-PV will develop an online platform for simulation, design, data analytics, QC and grid integration. This platform will host physics-based and data-driven models and it will evolve to include the needs of the PV sector and new advances in technology. The best experts from the PV community, pertaining to very specialized areas, will be encouraged to bring in their knowledge to help to construct and broaden the capacities of the platform.

The platform will integrate interactive built-in learning tools that will educate the end-user in the best simulation and design practices and will make use of the newest Internet 4.0 digital technology to promote the exchange of these best practices between the PV community members. SERENDI-PV will propose a global methodology for exchanging and storing data to facilitate the integration of PV into the digital era. Sound legal framework is important to transparent and secure collection, processing and dissemination of the data and software tools. SERENDI-PV will draw the inventory of legal aspects related to data distribution and use of software tools, such as terms of use, confidentiality, modes of distribution, and will outline the best practices in the industry.

4 FIRST DELIVERABLES AND OUTLOOK

Over the first 12 months of the project, SERENDI-PV partners have already compiled and worked on the technical/preparatory reports (public and/or internal ones) on the following topics:

- KPIs on state of the art on PV reliability, performance, profitability and grid integration
- Assessment and characterization of the current PV fleet capabilities and regulatory environment for grid integration
- Definition of needs from industry, evaluation of the existing simulation tools and models available on the market
- Revision of state of the art of automatic PV performance supervision systems

- Definition of new needs and specifications on current QC procedures and equipment
- Monitoring guidelines and specific monitoring and validation plan for demonstration of modelling, diagnostics and field testing of large PV plants
- Monitoring guidelines and specific monitoring and validation plan for demonstration of modelling, diagnostics and field testing of medium-commercial and residential PV plants

These reports provide the technical groundwork and define the current regulatory/market landscape (including current industry needs), towards the core development activities on PV modelling, data analytics, diagnostics and forecasting, as well as for the optimal preparation of SERENDI-PV's collaborative platform and demonstrators. The development activities and aimed innovations have a timeline spanning towards years 2022 and 2023. Early innovation developments and key findings on the above aspects will be presented in the next EUPVSEC version.

5 ACKNOWLEDGMENTS

The authors acknowledge the valuable contribution of all SERENDI-PV project partners and involved research teams. This project has received funding from the European Union's Horizon 2020 research and innovation programme under grant agreement 953016.

6 REFERENCES

[1] European Commission, "The European Green Deal," Eur. Comm., 2019, doi: 10.1017/CBO9781107415324.004.
[2] E. Vartiainen, G. Masson, C. Breyer, D. Moser, and E. Román Medina, "Impact of weighted average cost of capital, capital expenditure, and other parameters on future utility-scale PV levelised cost of electricity," Prog. Photovoltaics Res. Appl., 2020, doi: 10.1002/pip.3189.
[3] J. Robledo, J. Leloux, B. Sarr, C.A. Gueymard, A. Driesse, Dynamic and visual simulation of the bifacial energy gain for photovoltaic plants, European photovoltaic solar conference and exhibition (EUPVSEC), 2021.
[4] B. Müller, L. Hardt, A. Armbruster, K. Kiefer, C. Reise, Yield predictions for photovoltaic power plants: empirical validation, recent advances and remaining uncertainties. Prog. Photovolt: Res. Appl., 24: 570– 583. doi: 10.1002/pip.2616, 2016.
[5] "PV Performance Modeling Collaborative | PV_LIB Toolbox." https://pvpmc.sandia.gov/applications/pv_lib-toolbox/ (accessed Feb. 26, 2021).
[6] "Photovoltaic Geographical Information System (PVGIS) | EU Science Hub." https://ec.europa.eu/jrc/en/pvgis (accessed Feb. 26, 2021).
[7] "National Renewable Energy Laboratory (NREL) Home Page | NREL." https://www.nrel.gov/ (accessed Feb. 26, 2021).
[8] "Home Page - PV Simulation." https://www.sisifo.info/en/default (accessed Feb. 26, 2021).

TECHNO-ECONOMIC ANALYSIS OF MARINE ECOSYSTEM TO ACHIEVE ZERO CARBON EMISSION

Jing Zhong Tee
University of Glasgow, School of Engineering
Glasgow G12 8QQ, UK
j.tee.1@research.gla.ac.uk

ABSTRACT: In marine industry, Flettner Rotor (FR) is being exploited as the integrated power source with conventional system on-board marine vessel as complementary resource to be introduced for fuel saving as well as reduction in carbon footprint. As such, an integrated system of Diesel Engine (DE) and FR with combined Photovoltaic (PV) and battery energy storage system (BESS) is developed to propose to meet the load demand on voyage, highly dependent on weather condition sailing in geographically. In this paper, power stability is verified with simulation (ETAP) has shown cargo ship with proposed system configuration can adapt this methodology to meet required load demand with smoothing control in load fluctuation. The return on investment will be presented to compare between proposed system and convention system for Capital Expenditure (Capex) and Operational Expenditure (Opex). As such, cost analysis on proposed system can increase the value in investment to be further developed into business model.
Keywords: Power Stability, Photovoltaic, System Performance and Economic Analysis

1 INTRODUCTION

Ever since international Marine Organisation (IMO) has been setting rule to reduce carbon emission by 40 % in 2030 [1] to prevent climate change as well as reduces global warming that causes more natural disasters [2]. Flettner rotor (FR) has been introduced and applied on ship vessel since 1918 [3]. Since flettner rotor has a size which is tall and tilting rotor that has been installed to retricted the ship from passing through the overhead bridge, tilting rotor is invented [4]. Generally, FR produces optimum power based on the wind speed and speed of ship sail towards wind direction.

However, wind speed is not constant as it changes every one second interval that causes FR to generate energy that is not constant. Moreover, diesel engine does supply inconsistent output power to propulsion system due to resistance from the water surface wave [5]. As a result, integration of DE with FR will trigger large distance on the power grid in the ship. This issue needs to be addressed and an investigation is carried out when the ship is sailing at full speed and FR with combined BESS and PV can supply electrical power to load without output power from DE. There are stochastic wind speeds which has included for the systems during simulation.

The outline of the paper is as follows. Section 2 presents system overview with control algorithms that will be tested in simulation. Section 3 shows the simulation test result. A cost analysis is evaluated and shown in section 4. Finally, the conclusion is presented in section 5.

2 SYSTEM OVERVIEW

2.1 System architecture

The detailed system as shown in Figure 1 depicts the conventional system is integrated with FR and combined BESS and PV to generate output power to load. Since the wind speed is not constant, BESS stores energy by charging from PV during sailing in daytime. In such a case, BESS will supply the remaining power loss from inconsistent power generated in FR to produce optimum power to meet the load demand. In this section, system architecture has shown individual power generation with power rating that is used in simulation as follows.

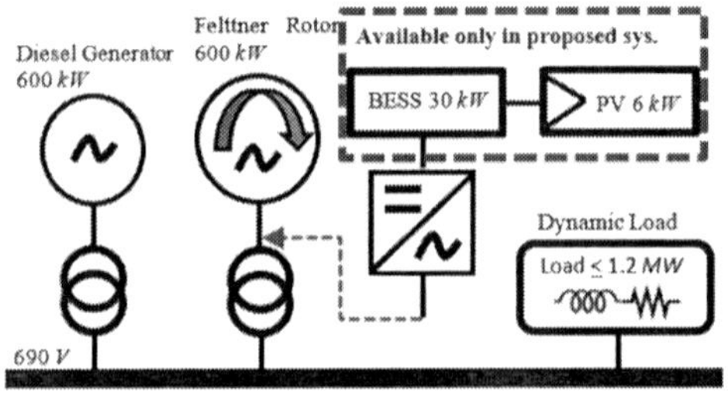

Figure 1: System, overview of the integration of DE, FR and BESS model.

2.2 Control algorithms

The process flow in the system architecture is presented in Figure 2. At the start, DE is turned on in operation to electrify the propulsion system. Once there is available wind speed, FR will be turned on to contribute more output power for propulsion system to enable ship sails at full speed. Wherever the ship is at fastest knots, DE will be switched to standby mode, and ship is sailing in this moment to achieve zero carbon emission.

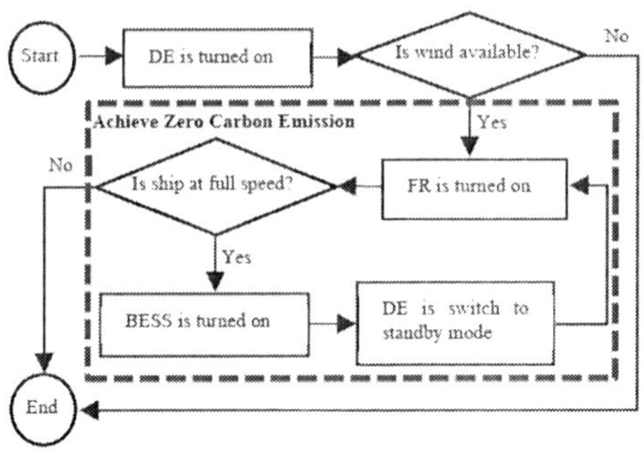

Figure 2: Simulation of the ecosystem.

In traditional system on shipboard, DE with power rating up to 4.5 MW, has voltage of 690 V in distributed power generation. In this project, the assumed power capacities of the integration of DE and FR with BESS generation system had been determined with reference from [6]. The system architecture is of micro-grid configuration requires minimum propulsion power of 600 kW with ship speed of 8 knot. The PV is of 6 kW power rating which need twenty solar panels to be installed

onboard the compass deck to generate energy in day to store into BESS. BESS is of 30 kW is used to mitigate power loss during inconsistent wind speed for smoothing control of the power consumption to the propulsion system.
The proposed system has integrated FR with combined BESS and PV to supply output power to load. FR is used in the simulated model with power rating of 600 kW is equivalent to the primary power generation DE. Since the common bus of load is of 690 V, BESS has to inverted into three-phase AC by converting into 690V in the transformer before reaching to the load. Thus, DE is of primary distributed power source and FR with combine BESS and PV as secondary distributed source.

2.3 Test scenarios

In this section, three test scenarios are described, as shown in Table I.

Table I: Three scenarios for simulation

Scenario	Event
1	DE is in starting phase to supply power to load.
2	At 5 *s*, DE is turned off and FR is turned on with stochastic wind speeds.
3	At 10 *s*, FR is turned off with continuous power from DE.

The simulation study is based on the ability of the power system, to maintain output power to load when subjected to switching of individual power generation profile. The switch event causes instability in the duration of transient power flow which will be investigated in the simulation. The simulation results are compared against load flow analysis with recovery time. For a fair comparison, the system is incorporated to have both DE and FR rated at 600 kW each and combined BESS and PV at 30 kW running onboard. FR and PV will provide electricity to the load demand according to weather condition which are not discussed in this paper as primary focus will be on power stability in output power rating to load. FR is modelled similarly from wind turbine generation (WTG) to generate the output power quality in the system.

3 TEST RESULTS

3.1 Load flow comparison

In this case, both conventional and proposed system are started in Scenario 1 where the output power is of 1 p.u.. FR is turned on at 5s with equivalent power to DE in operation mode. In this moment, DE is switched to standby mode when ship is at full speed. It can be seen from Figure 3 that the output power from the conventional system is lowered than proposed system and the power instability is of greater disturbance in conventional as compared to proposed system. The moment when Scenario 2 is switched to Scenario 3 at 10s, the FR is turned off with continuous power generation from DE. Proposed system shows constant power to load whereas conventional system has a dip to supply electrical power to load.

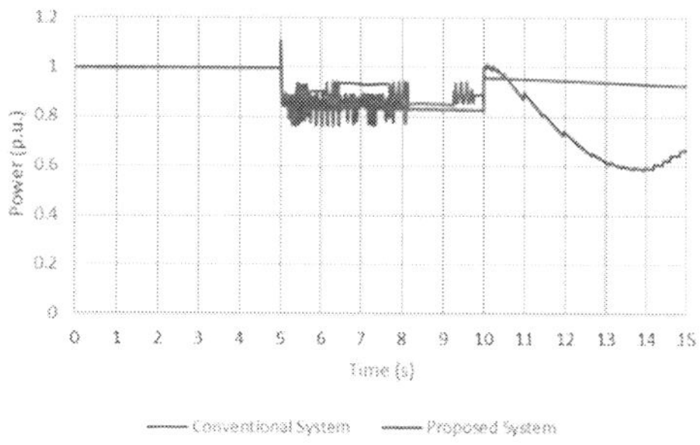

Figure 3: Both conventional and proposed system flow to load in p.u..

3.2 Voltage comparison

Conventional and proposed system start in scenario 1 at 1 p.u.. The moment scenario 1 is switched in scenario 2, conventional system oscillating to maintain between 0.8 to 1 p.u. Proposed system is oscillating between 0.9 and 1.1 p.u.. In this case, proposed system has a voltage deviation of +10/-10 % which is better than conventional of -0.2 p.u.. In scenario 3, conventional system has a large dip of -32 % as compared to proposed system which is of -10%.

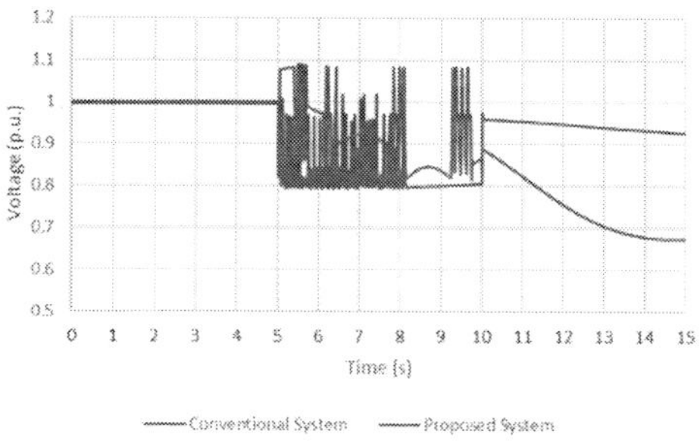

Figure 4: Voltage in p.u. of both conventional and proposed system.

3.3 Frequency comparison

The frequency in proposed system is relatively constant throughout the whole scenarios maintaining at 1 p.u. but frequency in conventional system has a huge surge followed by dip throughout simulation with a frequency transient deviation of +17/-7 %.

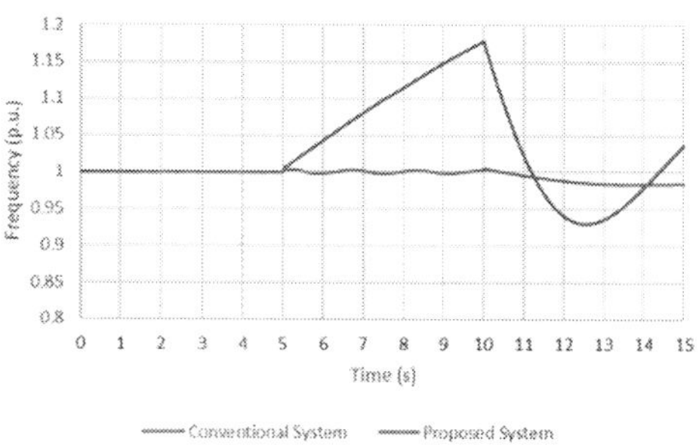

Figure 5: Frequency in p.u. of both conventional and proposed system.

4 COST ANALYSIS

The ongoing research and development FR in harnessing wind power to generate renewable energy that eliminate fuel cost which is attractive to integrate with DE onboard the ship [7]. As such, DE is a primary source which integrated with FR as a secondary source that is on standby mode. Since DE has the power factor of 0.8, the mathematical function of DE for CapEx (*\$/kW*) onboard the ship is written as follows:

$$DEc = (2610)(PR)/(PF) \qquad (1)$$

where DEc is calculated in *\$/kW*, power rating (PR) is measured in kW and power factor. (PF) is approximately valued as 0.8.

Capital cost of FR for CapEx (*\$/kW*) onboard the ship can be calculated based on reference in [8]. As such, Initial CapEx of DE, BESS and PV onboard ship can be tabulated based on the reference in [9].

The proposed system has higher Capital Expenditure (CapEx) which than the conventional system due to add-on cost from BESS and PV. However, Return On Investment (ROI) in proposed system can be approximately in three years.

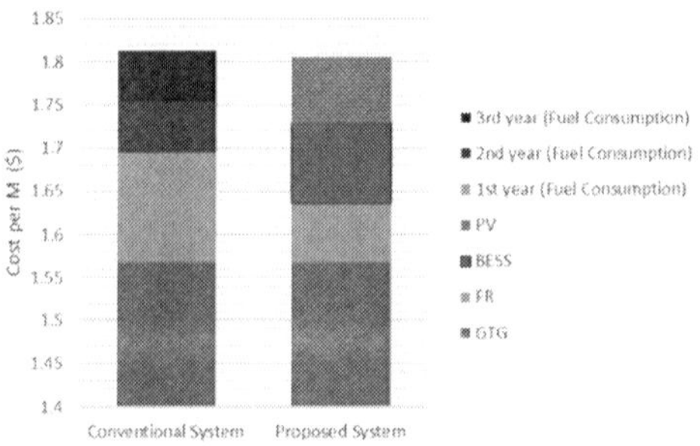

Figure 6: Cost comparison between convention system and proposed system.

5 CONCLUSIONS

In this paper, a techno-economic analysis comprising of an evaluation on decarbonisation process in ship's sailing to achieve towards zero carbon emission. It is shown in section 2 under the assumption of ship is sailing at full speed, DE can be turn to standby mode with the ecosystem that comprise of FR,BESS and PV to electrify the load. In addition, this can achieve zero carbon emission and power stability in the system due to stochastic wind speeds.

Moreover, the enhancement of quality power is better in proposed system with the add-on of combined BESS with PV which is verified in simulation to meet load demand. This is shown in simulation test result, conventional system in scenario 3 has a significant dip in output power to load. Thus, combined BESS with PV is required to integrate with DE and FR to improve in enhancement of power quality to meet dynamic load. Future work can involve integration of other RES with this existing system configuration.

6 ACKNOWLEDGMENT

This research is funded under Economic Development Board (EDB) of Singapore and Sembcorp Marine Ltd. (SCM) under Industrial Postgraduate Programme (IPP).

7 REFERENCES

[1] https://theicct.org/blog/staff/updated-imo-carbon-intensity-target-may2021
[2] Zandalinas SI, Fritschi FB, Mittler R. Global warming, climate change, and environmental pollution: Recipe for a multifactorial stress combination disaster. Trends in Plant Science. 2021 Mar 18.
[3] https://www.maritimecyprus.com/2018/11/15/flettner-rotors-applicability-in-the-maritime-industry/
[4] https://www.marineinsight.com/shipping-news/sea-cargo-ship-to-be-worlds-first-vessel-with-tilting-rotor-sails-arrives-in-rotterdam/
[5] Awwad AE, Al-Quteimat A, Al-Suod M, Ushkarenko OO, AlHawamleh A. Improving the accuracy of the active power load sharing in paralleled generators in the presence of drive motors shaft speed instability. International Journal of Electronics and Telecommunications. 2021 Aug 30;67(3):371-7.
[6] https://ecoflettner.de/ Power of Rotor and Main engine in Relation to Ships Speed.
[7] U.S. Energy Information Adminstration "Distributed Generation and Combined heat & Power System Characteristics and Costs in Buildings Sector," Indepenent Statistics & Analysis, May 27, 2017.[Online].Available:https://www.eia.gov/analysis/studies/buildings/distrigen/pdf/dg_chp.pdf.
[8] https://glomeep.imo.org/technology/flettner-rotors/#:~:text=The%20range%20of%20cost%20for,%241%2C000%2C000%20to%20%243%2C000%2C000%20(USD).
[9] J..Z. Tee, E.Z.D. Chia, K.H. Tan, L.H.I. Lim, K.Zhou, and O. Anaya-Lara, "Techno-Economic Analysis and Battery Storage Placement in grid-connected Photovoltaic (PV) system,''in Proc. 2019 EUPVSEC Conf., Sep. 2019.

DATA FORECASTING WITH APPLICATION TO BLOCKCHAIN-BASED LOCAL ENERGY MARKETS

Ameni Boumaiza, Antonio Sanfilippo
Hamad-Bin Khalifa University, Qatar Environment and Energy Research Institute, Qatar
(aboumaiza, asanfilippo}@hbku.edu.qa

ABSTRACT: The establishment of distributed energy generation through residential and commercial PV applications creates new energy markets where the traditional separation between energy producer and consumer is obliterated, giving rise to the new role of energy prosumer. Blockchain technology provides energy, consumers, prosumers and utilities with a novel, secure and cost effective energy-trading solution through the automation of direct energy transactions within a distributed database architecture rooted in cryptographic hashing and consensus-based verification. The target of this study is illustrating the forecasted households' s power profiles as well as operation of any blockchain process and apply a generic ABM simulation framework for electricity exchange (see Figure. 1). An original version of a resilient multi agent structure for a Transactive Energy (TE) type of Distributed Energy Resources (DER) within the ECCH microgrid dependent on blockchain engineering was designed & simulated. Recently proposed blockchain-based LEMs use auction designs to match future demand and supply. Thus, such blockchain-based LEMs rely on accurate short-term forecasts of individual households' energy consumption and production. Often, such accurate forecasts are simply assumed to be given. The present research tested this assumption by first evaluating the forecast accuracy achievable with state-of-the-art energy forecasting techniques for individual households and then, assessing the effect of prediction errors on market outcomes in three different supply scenarios. The evaluation showed that, although a LSTM model is capable of achieving reasonably low forecasting errors. The prediction process will be fitted to the setup of a blockchain-based LEM. Therefore, the current analysis distinguishes itself notably out of prior experiments which entirely attempt to forecast smart meter time sequence generally.

Keywords: Blockchain; local energy market; smart contract; smart meter; short-term energy forecasting; machine learning; long short-term memory (LSTM); prediction errors; market mechanism; market simulation.

1 INTRODUCTION

Although LEMs started to attract interest in academia already in the early 2000s, it is still an emerging field. Mainly driven by the widespread adoption of smart meters and Internet connected home appliances, recent work on LEMs focuses on use cases in developed and highly technologized energy grid systems [1]. While substantial work regarding LEMs in general has been done (e.g., [2, 3, 4]), there are only few examples of blockchain-based LEM designs in the existing literature. Mengelkamp et al. [5] derived seven principles for microgrid energy markets and evaluated the Brooklyn Microgrid according to those principles. Additionally, there are several industry undertakings to put blockchain-based energy trading into practice, such as Grid Singularity (gridsingularity.com) in Austria, Powerpeers (powerpeers.nl) in the Netherlands, Power Ledger (powerledger.io) in Australia, and LO3 Energy (lo3energy.com) in the United States.
Interestingly, none of the above cited works, that employ market mechanisms requiring household energy forecasts for bidding, check whether the assumed availability of such forecasts is given. However, without this assumption, trading through an auction design, as described by, e.g., Block et al. [6] or Buchmann et al. [3] and implemented in a smart contract by Mengelkamp et al. [7] is not possible. Unfortunately, this forecasting task is not trivial due to the extremely high volatility of individual households' energy patterns [8]. However, research by Arora and Taylor [9] shows that advances in the energy forecasting field also extend to household-level energy forecasting problems and serve as a promising basis for the present study.

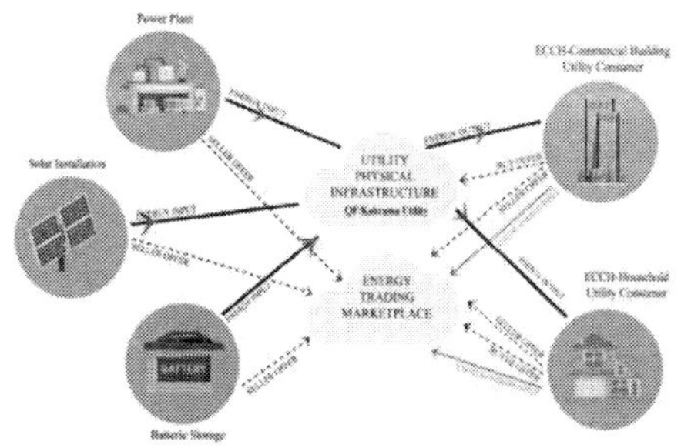

Figure 1. Overview of the architecture for a transactive energy (TE) model of Distributed Energy Resources (DER) in the ECCH microgrid based on blockchain technology.

2 Proposed Approach

2.1 Methodology

We investigated the prerequisites needed to apply blockchain based distributed community energy markets. Which means forecasting total power consumption and creation of individual people and also prosumers one-time step ahead. The prediction process will be installed to the setup of a blockchain based LEM. The following analysis questions is to determine which prediction method produces the perfect 48 ahead 15-min forward prediction for smart meter time sequence assessed in 3 min intervals using only input features produced from the historic values of the time series and calendar based features. (TE) model of Distributed Energy Resources (DER) in the ECCH microgrid based on blockchain technology has been developed.

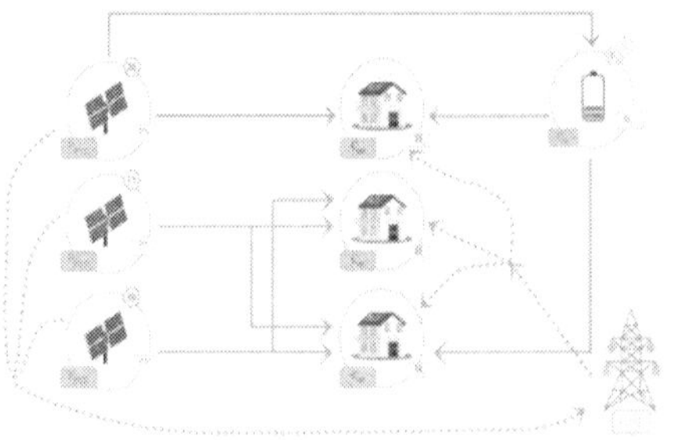

Figure 2. Overview of the architecture for a transactive energy (TE) model of Distributed Energy Resources (DER) in the ECCH microgrid based on blockchain technology.

2.2 Data

The data used in this study is for Qatar Foundation housing compound (Lot1 and Lot2). This compound contains 623 homes from different types (villas and flats). The Lot1 is adjacent to Lot2. There is a one main substation feeding 15 different substations that provide power to the ECCH homes: 6 substations for 232 homes in Lot1, and 9 substations for 391 homes in Lot2. The main substation receives electricity at 33kV and distributes it at 11kV to the secondary substations, which in turn bring down the voltage to 415V, ready for distribution to the ECCH homes. All homes connected to the same substation share the same power lines. The data used in our ABM is a combination of multiple layers of information about power transmission, buildings, generated Rooftop Photovoltaic (RPV) energy, and hourly electricity demand profiles.

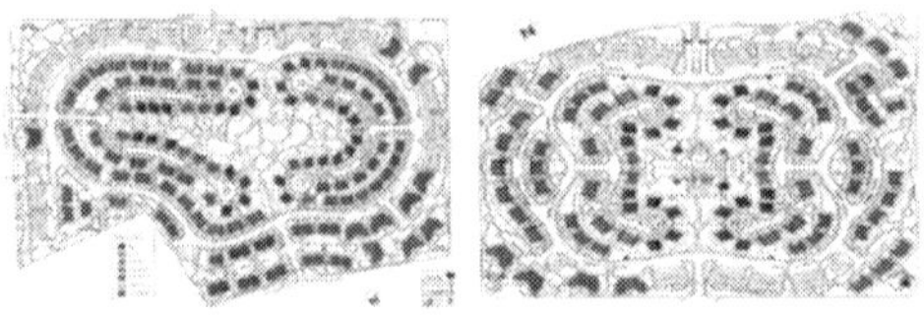

Figure 3- ECCH lots #1 (right) and #2 (left). The data used in this project is for Qatar Foundation housing compound (#lot1 and #lot2). This compound contains 623 homes from different types (villas and flats). The #lot1 is adjacent to #lot2.

3 Experimental Results

3.1 Methodology

We have developed an initial multi-agent architecture for a transactive energy (TE) model of Distributed Energy Resources (DER) in the ECCH microgrid based on blockchain technology [10]. The current model based on this initial architecture treats each building as a simple agent with a defined probability that the building owner chooses to buy or sell his rooftop solar energy excess to his neighbors. Each agent in the network would manage a set of grid-connected resources that include assets capable of producing and/or consuming electricity (e.g. fuel generators, solar PV, wind generators, electric storage and flexible electric loads) as well as sensors devices. Each agent has access to their blockchain node locally (i.e. on the same machine).

3.2 Evaluation of the Prediction Models

The LSTM prediction strategy (see Figure 5) has been employed to forecast the energy use of 10 consumers/ households fifteen min ahead. A baseline version, a LSTM RNN version, in addition to a LASSO regression will then be compared and evaluated. In this study, the LSTM model shows a higher variability within the sums of underestimation and over- errors. The LSTM method appears to be much more equally well suited for those datasets and the particular consumption patterns of theirs.

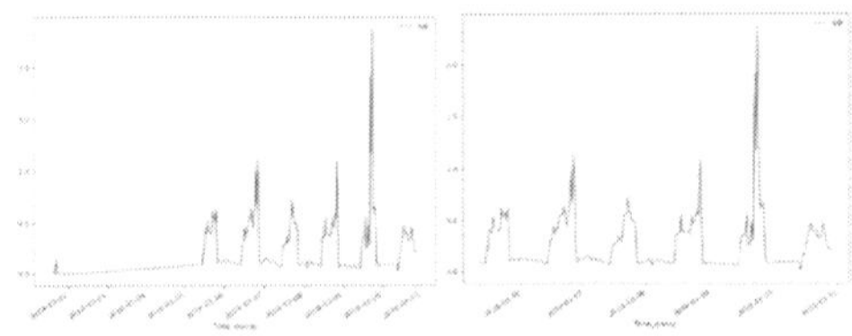

Figure 4- Overview of the architecture for a transactive energy (TE) model of Distributed Energy Resources (DER) in the ECCH microgrid based on blockchain technology.

3.2 Results

For optimization, our prediction should not be deviated more than ±5kWh. We have used a powerful type of neural network designed to handle sequence dependence. The used network for regression is time-step based framing. In this approach, we investigated various data preparation techniques to rescale the data and to make it stationary (see Figure 6). So, we started by sampling (convert 5-min data into hourly data). We observe an upward trend in the dataset over time and remark some periodicity to the dataset that probably corresponds to the vacation period. Data structure for the univariate case is described by the following figures. You may have noticed that the data preparation for the used network includes time steps. We've used a rolling average, though, and you start to see a pattern emerge, with peaks happening more and more often (see Figure 5).

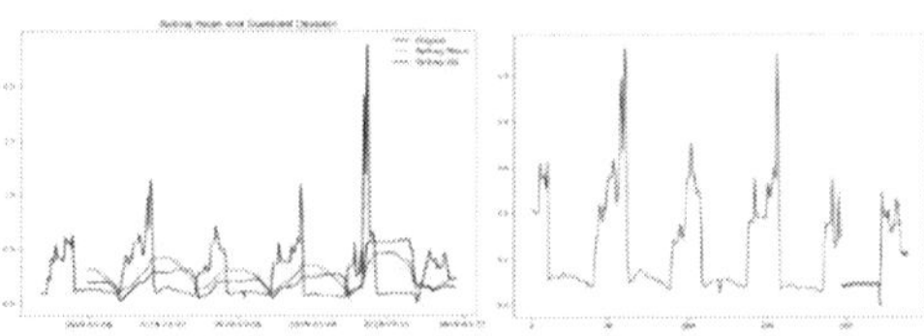

Figure 5. Short-term Energy Demand Forecasting: (a)Two variants of a LSTM network to predict the next 48 hours ahead energy demand. (b) Predictions are made at the end of each day for the maximum demand for each hour in the day ahead.

Finally, we can generate predictions using the model for both the train and test dataset to get a visual indication of the skill of the model. Running the example produces the following output (see Figure 5).

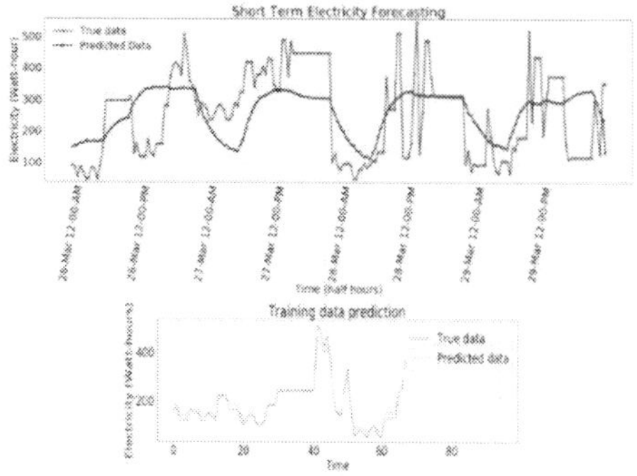

Figure 6. Short-term Energy Demand Forecasting: (a)Two variants of a LSTM network to predict the next 48 hours ahead energy demand. (b) Predictions are made at the end of each day for the maximum demand for each hour in the day ahead.

4 CONCLUSION

In this work, we demonstrated the power of forecasting energy consumption data that will be fed to a Blockchain system and evaluate the impact of the LSTM forecasting model to predict energy data in a housing community on the market trading flow. Future research will focus on implementing the whole P2P model, including Bitcoin or Ethereum information.

5 References

[1] Burger, C.; Kuhlmann, A.; Richard, P.; Weinmann, J. Blockchain in the Energy Transition. A Survey among Decision-Makers in the German Energy Endustry; Report; ESMT Berlin, Berlin, Germany, 2016.

[2] Lamparter, S.; Becher, S.; Fischer, J.G. An Agent-based Market Platform for Smart Grids. In Proceedings of the 9th International Conference on Autonomous Agents and Multiagent Systems (AAMAS): Industry Track, Toronto, ON, Canada, 10–14 May 2010; pp. 1689–1696. 8.

[3] Buchmann, E.; Kessler, S.; Jochem, P.; Böhm, K. The Costs of Privacy in Local Energy Markets. In Proceedings of the 2013 IEEE 15th Conference on Business Informatics, Vienna, Austria, 15–18 July 2013; pp. 198–207.

[4] S. Chen, B. Mulgrew, and P. M. Grant, "A clustering technique for digital communications channel equalization using radial basis function networks," IEEE Trans. on Neural Networks, vol. 4, pp. 570-578, July 1993.

[5] Ilic, D.; Silva, P.G.D.; Karnouskos, S.; Griesemer, M. An energy market for trading electricity in smart grid neighbourhoods. In Proceedings of the 2012 6th IEEE International Conference on Digital Ecosystems and Technologies (DEST), Campione d'Italia, Italy, 18–20 June 2012; pp. 1–6.

[6] Mengelkamp, E.; Notheisen, B.; Beer, C.; Dauer, D.; Weinhardt, C. A blockchain-based smart grid: Towards sustainable local energy markets. Comput. Sci. Res. Dev. 2018, 33, 207–214.

[7] Block, C.; Neumann, D.; Weinhardt, C. A Market Mechanism for Energy Allocation in Micro-CHP Grids. In Proceedings of the 41st Annual Hawaii International Conference on System Sciences (HICSS 2008), Waikoloa, HI, USA, 7–10 January 2008; pp. 172–183.

[8] Mengelkamp, E.; Gärttner, J.; Rock, K.; Kessler, S.; Orsini, L.; Weinhardt, C. Designing microgrid energy markets—A case study: The Brooklyn Microgrid. Appl. Energy 2018, 210, 870–880. [8] Wang, Y.; Chen, Q.; Hong, T.; Kang, C. Review of smart meter data analytics: Applications, methodologies, and challenges. IEEE Trans. Smart Grid 2018, 10, 1–24.

[9] Arora, S.; Taylor, J.W. Forecasting electricity smart meter data using conditional kernel density estimation. Omega 2016, 59, 47–59.

[10] Ameni Boumaiza, Mazhar Sajjad, Antonio Sanfilippo, Nassma Mohandes, "Designing an Agent-based Energy Exchange Simulation Framework for Modeling Blockchain Systems," International Conference on Sustainable Energy-Water-Environment Nexus In Desert Climate, Doha, Qatar", 2-5 Dec 2019.

LOAD DATA ACQUISITION IN RURAL EAST AFRICA FOR THE LAYOUT OF MICROGRIDS AND DEMAND–SIDE–MANAGEMENT MEASURES

Godiana Hagile Philipo[a,b], Josephine Nakato Kakande[a,c] and Stefan Krauter[a,*]
[a] Electrical Energy Technology – Sustainable Energy Concepts
Faculty of Computer Science, Electrical Engineering and Mathematics, Paderborn University.
Pohlweg 55, D–33098, Paderborn, Germany.
[b] Department of Electrical Engineering and Information Technology, Saint Augustine University of Tanzania.
P.0.Box 307, Mwanza, Tanzania
[b] Department of Electrical and Computer Engineering, CEDAT, Makerere University.
P.0.Box 7062, Kampala, Uganda
Email: philipog@campus.uni-paderborn.de
Email: jkakande@mail.uni-paderborn.de
*Corresponding author: stefan.krauter@uni-paderborn.de

ABSTRACT: With its growing population and industrialization, DREs, and solar technologies in particular, provide a sustainable means of bridging the current energy deficit in Africa, increasing supply reliability and meeting future demand. Data acquisition and data management systems allow real time monitoring and control of energy systems as well as performance analysis. However commercial data acquisition systems often have cost implications that are prohibitive for small PV systems and installations in developing countries.

In this paper, a multi-user, multi-purpose microgrid database system is designed and implemented. MAVOWATT 270 power quality analyzers by GOSSEN METRAWATT, raspberry pi modules and sensors are used for measuring, recording and storing electrical and meteorological data in East Africa. Socio-economic data is also stored in the database. The designed system employs open source software and hardware solutions which are best suited to developing regions like East Africa due to the lower cost implications.

The expected results promise a comprehensive database covering different electro-technical and socio-economic parameters useful for optimal design of microgrid systems.
Keywords: Demand side management, MySQL, Raspberry pi, Data acquisition.

1 INTRODUCTION

Microgrids are stipulated as an essential building block of the future grid [1]. They are well known alternatives to cross the electrification gap in Sub Saharan African countries. Also, microgrids help achieve the 7th sustainable development goal of "affordable, reliable, sustainable and modern energy for all by 2030" [2-4].

Despite the projected rapid increase in microgrids in rural East Africa, design and sizing of microgrids, blackouts/brownouts and sustainability are still the most significant challenges. The reason is an unpredicted increase in demand resulting in load shedding, which creates customer impatience and usage drops [5]. Commercial data acquisition systems often have cost implications that are prohibitive for small PV systems and installations in developing countries [6, 7]. Solar systems less than 25kWp often do not incorporate measurement systems [8]. Also, despite many researchers trying to solve the demand–supply imbalance, there is no comprehensive link between electrical, meteorological and social-economic data to allow proper design and planning for the layout of microgrids and demand-side management measures. Moreover, most of the African microgrids' data is not easily accessible in part due to the unavailability of proper frameworks and structures for data storage, making it difficult to find an optimal solution for the microgrid planning, layout and operation. Lack of an appropriate microgrid database makes it difficult for stakeholders such as academia, developers, government, policy makers, development partners and utilities to access pertinent data from communities and entities of interest. Limited information or data may eventually lead to increasing the cost of building and operating microgrid systems [9].

According to literature, several data acquisition systems have been developed. In [10] a real-time RDBMS was designed for the selected data requirements to avoid data mapping. Mojtaba et al. presented a monitoring and communication system for an experimental microgrid using Modbus TCP/IP communication protocol and a webpage for data monitoring [1]. Low cost open source solutions have been developed and tested [6, 7, 11]. Microcontroller based solutions such as Arduino and Raspberry Pi are able to provide low-cost intelligence and processing capacity in proximity to the sensors and/or metering equipment before data is relayed either wirelessly, via a wired medium or on a USB disk to a PC or server [7, 12, 13]. Most of the data acquisition systems already in practice have employed the high technological approach like the use of SCADA [14], CR1000 data logger [15], Modbus protocols [1], etcetera, which are expensive and do not fit the rural East African conditions.

Therefore, this work aims to design a cost-effective microgrid data acquisition and database management system for the rural East African micro grids containing data on electrical, meteorological, socio-economic and cultural parameters. The designed database will assist in the proper monitoring of progress, planning, and establishing of targets for successful microgrid operation. This work will avail data for assessing the effect of a combination of customer categories, weather conditions, photovoltaic system parameters, storage and demand to properly suggest a demand side management scheme that will cater to both blackouts and storage cost reduction. Microgrid data can also help estimate near accurate forecast of load profiles, which is essential to guarantee sustainable, reliable and affordable power supply from microgrids [16, 17].

This paper's novelty lies in the design criteria whereby the developed database employs the free and open source MYSQL software with a raspberry pi. The database integrates data from several microgrids that facilitate easy evaluation of optimal new microgrid layout requirements. Also, the scope of the stored data will be extensive, providing fodder for further research from a technology, socio-economic, didactics, culture, and policy perspective and reducing the need for disparate field visits to start collecting data that will have been provided through this work. Hence, this study's results will act as an essential push factor to guarantee a feasible microgrids operation in the developing world.

2 METHODOLOGY

The design of a MYSQL database is done for East Africa (Tanzania and Uganda). The collected data is divided into four groups; technological, socio-economic, didactics and culture. Within each group, in-depth aspects of data are considered ranging from the microgrids' electrical characteristics, household details, appliances present, storage requirements, photovoltaic supply, demand, tariff structure, billing, and other essential needs.

Data collection is to be done in three phases, the first being collection of general information on existing and planned microgrids: location, sizing, generation capacity, number of customers, load profiles etcetera, education and training arrangements followed by local data collection on factors such as demographics, consumer happiness, consumer expectations, governance, maintenance, education levels and gender-specific participation in the microgrid projects. Finally, the collection of electrical and weather data and in-depth interviews with users/customers and operators will be done.

Figure 1 illustrates the data collection framework, showing different aspects of the data collection and retrieval process.

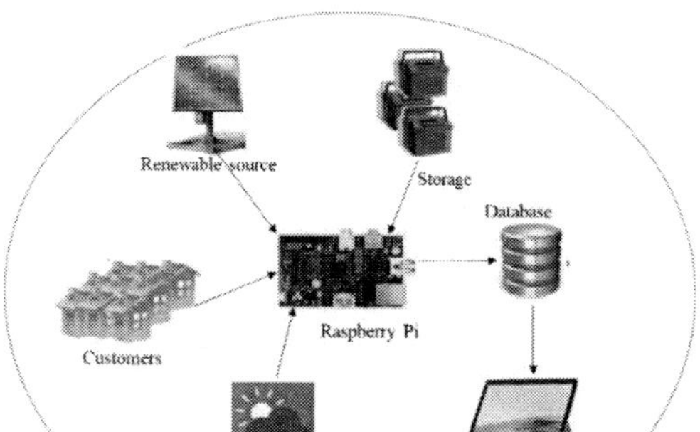

Figure 1: Data collection framework

Microgrid development, implementation and maintenance involves a very wide set of tasks and requires an interdisciplinary team to be successful and sustainable.

Important is to understand the big picture, the final goal and the links to all other involved parties. There are social, economic, technical, political, environmental aspects that correlate and for a successful operation, all tasks and results starting from project development over to technical and economic considerations need to fit together.

Therefore, these tasks require specialists as well as other stakeholders to combine information and knowledge into a sustainable solution as depicted in Figure 2.

Figure 2: Interdisciplinary approach for microgrid sustainability

3 PRELIMINARY RESULTS

3.1 The database

A local database for the microgrid systems in rural East Africa is developed. Weather data from sensors, electrical data from electricity meters and socio-economic data are received by the data acquisition component. The data is then relayed to a in a website server and logged into the MYSQL database. The database stores each type of data in different database tables and uses automated read and write SQL scripts. Figure 3 shows some of the database tables created and the internal structure of the data stored.

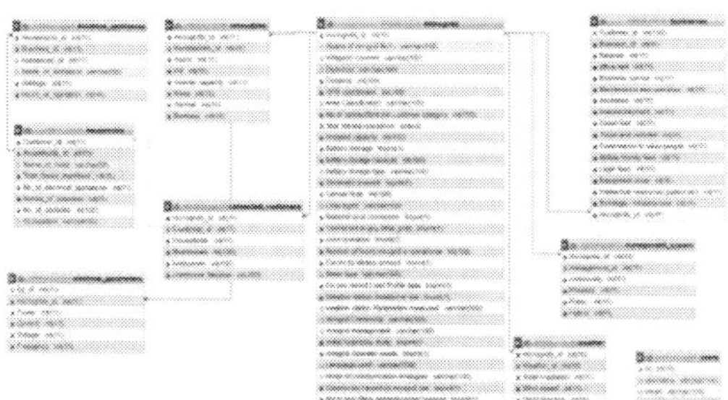

Figure 3: Microgrid database scheme for different data tables.

Results from querying the database can be viewed on a user web page designed with a free application known as Apache. This will allow new and registered users to view, update and monitor the allowed portion of data on the database. Personal computers or smart mobile phone with an installed browser and internet connection can be used to access the stored data in the database. By using a SQL script, several desired plots can be displayed.

3.2 The data visualization web application

XAMPP, an open-source web server solution that includes Apache web server, MySQL and PHP, was used to design a MySQL database that stores the microgrid-related data. The general layout of the system is depicted in Figure 4. The data acquisition system is accessed through a PHP based web application which allows two types of users (Admin and normal user). Users are able to access the database via a web browser and authorization is granted based on a valid username and password.

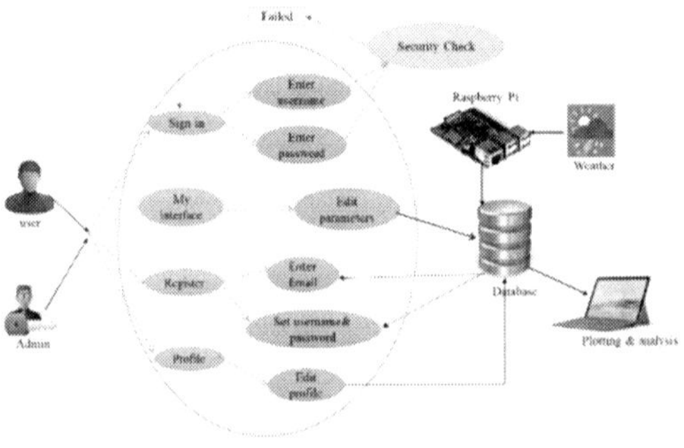

Figure 4: General layout of data acquisition system

Logged-in users can access data and accomplish several operations depending on their rights.

PHP and HTML are used to display database information and using SQL language, to query, store and modify data. Figure 5 shows the MySQL database for general microgrid data. The display of already entered data is depicted in Figure 5 and Figure 6.

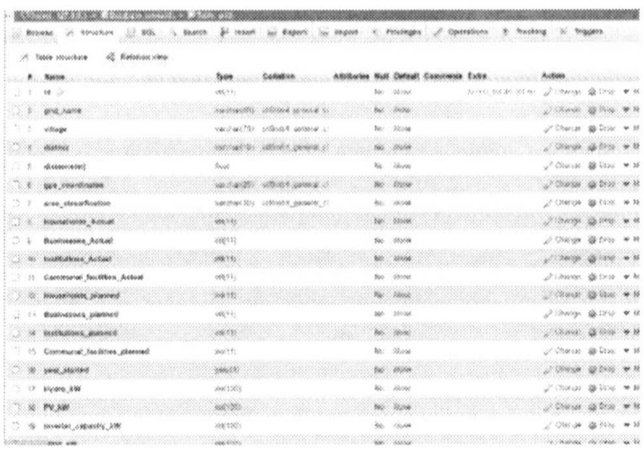

Figure 5: MySQL database with microgrid data

Microgrids General Data

id	grid_name	village	district	distance(m)
10	mwanza			0
9	Jumeme	Ukerewe	Ukara	100
8	mwanza	missungwi	masawe	200
1	Mkalama	Boma	Moshi	200

Figure 6: Web page showing microgrid data

The Raspberry Pi is a single-board embedded computer capable of carrying out computing tasks, equipped with General Purpose Input-Output (GPIO) pins. The GPIO pins enable interaction with sensors and actuators. The measurement data to be stored will ultimately be obtain from measurement systems set up in East Africa. To test the incorporation of measurement data into the database, a lab based experimental set up was done at Paderborn University in Germany. A Sense HAT and Raspberry Pi were used to obtain and relay meteorological data. The Sense HAT (Hardware Attached on Top) is an add-on board for the Raspberry Pi with integrated sensors that measure temperature (Celsius), relative humidity (%) and pressure (millibars) [18]. Because the Raspberry Pi CPU's heat raises the Sense HAT temperature measurements, an offset has to be incorporated to account for the elevated Sense HAT temperature readings.

Measurement time, temperature, humidity, and pressure values are relayed to the server which enters the data into the MSQL database. The measurement data can then be retrieved and viewed remotely via a web page (Figure 7).

Raspberry Pi Weather Log

ID	Date and Time	Temperature (Celsius)	Relative Humidity (%)	Pressure (millibars)
1276	2021-08-14 19:15:59	21.9	46.3	1024.4
1275	2021-08-14 19:14:59	22.0	46.4	1024.4
1274	2021-08-14 19:13:59	22.0	46.5	1024.4
1273	2021-08-14 19:12:59	21.9	46.5	1024.5
1272	2021-08-14 19:11:59	21.9	46.7	1024.4
1271	2021-08-14 19:10:59	22.0	46.7	1024.4
1270	2021-08-14 19:09:59	22.0	46.4	1024.4
1269	2021-08-14 19:08:59	21.8	47.5	1024.4
1268	2021-08-14 19:07:59	21.8	46.5	1024.4
1267	2021-08-14 19:06:59	22.0	46.4	1024.4
1266	2021-08-14 19:05:59	21.9	46.7	1024.4
1265	2021-08-14 19:04:59	22.0	46.5	1024.4
1264	2021-08-14 19:03:59	21.9	45.6	1024.4
1263	2021-08-14 19:02:59	22.0	46.4	1024.5
1262	2021-08-14 19:01:59	22.0	46.7	1024.4
1261	2021-08-14 19:00:59	22.0	46.5	1024.4
1260	2021-08-14 18:59:59	22.0	46.6	1024.4
1259	2021-08-14 18:58:59	21.8	46.6	1024.4
1258	2021-08-14 18:57:59	22.1	46.3	1024.4
1257	2021-08-14 18:56:59	22.1	46.1	1024.4
1256	2021-08-14 18:55:59	21.8	46.7	1024.4
1255	2021-08-14 18:54:59	22.2	46.5	1024.3
1254	2021-08-14 18:53:59	22.1	46.6	1024.4
1253	2021-08-14 18:52:59	22.1	46.6	1024.3
1252	2021-08-14 18:51:59	21.8	46.4	1024.4
1251	2021-08-14 18:50:59	22.0	45.8	1024.4

Figure 7: Weather measurement data

Time series data from the sensors are plotted using PHP and HTML and the charts can be viewed via a web browser (Figure 8).

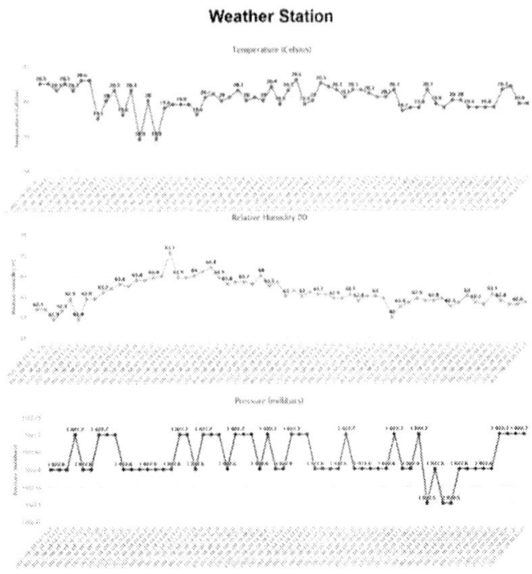

Figure 8: Real time weather data

Prediction of weather, energy generation and consumption is an important aspect of load management and ensuring reliability of RE-based microgrids. To illustrate a potential application of the weather data, prediction is done. The auto regression moving average model for next day weather prediction in a walk-forward manner is used, because of its simplicity. The recorded real time weather data is utilized as the baseline to predict and analyze the variation of weather data over time. The percentage mean square errors obtained are 0.126, 0.091 and 1.086 for temperature, pressure and relative humidity respectively. Figure 9, Figure 10 and Figure 11 show the variation between the actual data and the predicted weather data. The bottom plot in each figure is a zoom-in of the upper plot.

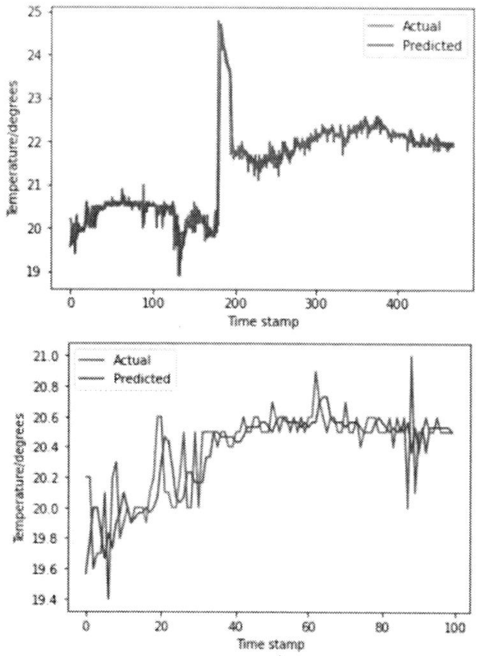

Figure 9: The actual and predicted results of temperature

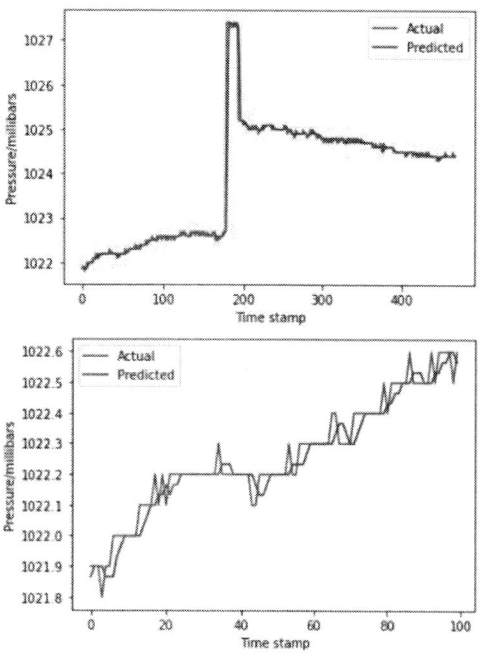

Figure 10: The actual and predicted results of pressure

Historical irradiation data in 10-minute intervals from the University of Paderborn EET-NEK (Chair of Electrical Energy Technology – Sustainable Energy Concepts) PV test setup is plotted. The data spans the period August 2013 to early October 2013. Figure 12 shows the daily variation in irradiance for all the days (irradiance profile), with the dotted line as the median irradiance.

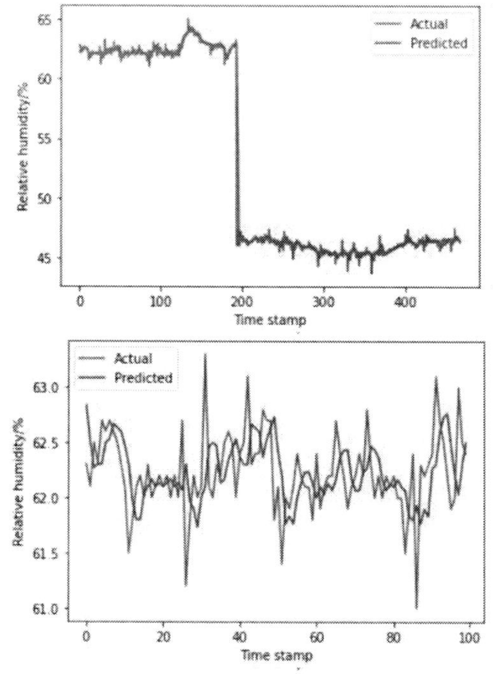

Figure 11: The actual and predicted results of humidity

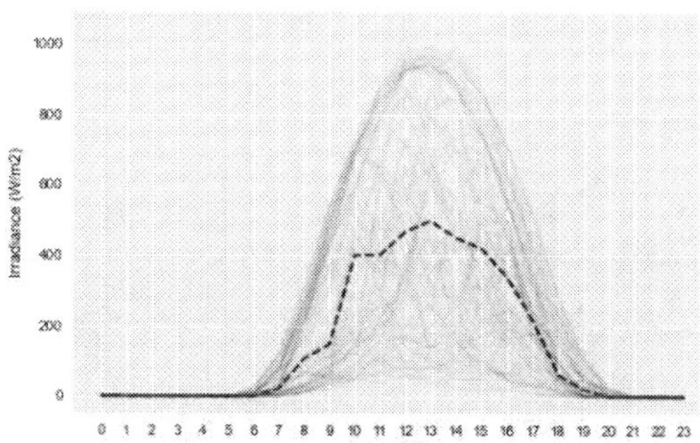

Figure 12*:* Diurnal Irradiance in W/m^2

Knowledge of customer load profiles is an important aspect of minigrid sizing and matching of generation to demand. This is particularly essential in solar PV installations where the energy generated is a factor of irradiance, and hence loads and /or storage need to be managed carefully.

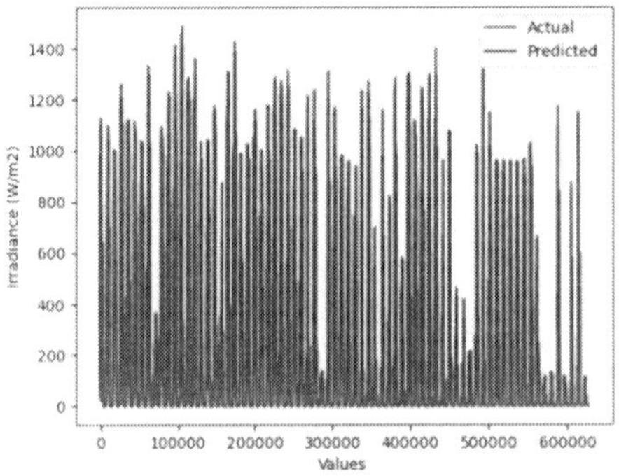

Figure 13: Actual and predicted irradiance

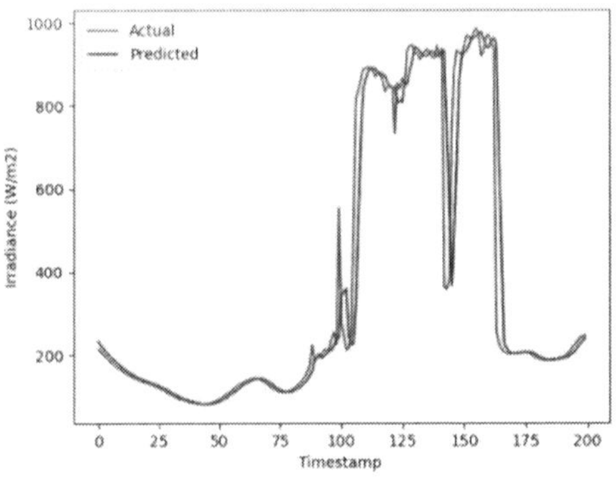

Figure 14: Actual and predicted irradiance (zoomed)

Prediction is done using auto regression moving average. Figure 13 shows the prediction results, while a zoomed-in plot is given in Figure 14. Similar and additional plotting and analysis can be done for the electrical and weather data that will be obtained in the field.

4 NEXT STEPS

Additional meteorological data will be obtained and incorporated into the database. Electrical parameters like voltage, current, active power, reactive power, power factor and energy will be measured during the field data collection phase, and the database management system will be extended to incorporate this additional data. The measurement data will also be stored locally on an SD card on the Raspberry PI to minimize potential data loss due to power outages, network connectivity issues, weather and / or voltage fluctuations [19] that can be expected in the East African microgrids.

5 CONCLUSION

This presents preliminary work on a cost-effective data acquisition and storage system. The collected data will be used for forecasting, control, remote access and monitoring. Open-source software and low-cost devices like raspberry pi modules promise cost-effective design and monitoring for rural East African microgrids. The database can interlink data from different microgrids, making it easy to assess, and plan for integration possibilities. Furthermore, aggregating technical parameters like power quality, load profiles, storage requirements, nature of customers, etcetera from a wide range of data from several microgrids can facilitate proper layout of microgrid systems and demand-side management measures. This work will trigger more researchers in different fields to assess and investigate various aspects that can lead to optimal pathways for the sustainability of microgrids.

6 ACKNOWLEDGEMENTS

The authors are thankful to the German Federal Ministry of Education and Research (BMBF) for funding this project within the compound assignment "CLIENT II – Art-D_Grids: Africa Research and Teaching Platform for Development – Sustainable Modular Grids for Grid Stability". Project: 03SF0607B.

7 REFERENCES

[1] J. L. M. Moghimi, C. Bennett, D. Leskarac, S. Stegen, "Communication Architecture and Data Acquisition for Experimental Microgrid Installations," in *IEEE PES Asia-Pacific Power and Energy ELoad and load growth models for rural microgrids, ngineering Conference (APPEEC)*, 2015, vol. 3, pp. 1–5. 2015, doi: 10.1109/APPEEC.2015.7380917

[2] R. Takalani and B. Bekker, "Load and load growth models for rural microgrids, and how to future-proof designs," *2020 Int. SAUPEC/RobMech/PRASA Conf.*, pp. 10–15, 2020, doi: 10.1109/SAUPEC/RobMech/PRASA 48453.2020.9041001.

[3] N. J. Williams, P. Jaramillo, B. Cornell, I. Lyons-Galante, and E. Wynn, "Load characteristics of East African microgrids," *Proc. - 2017 IEEE PES-IAS PowerAfrica Conf. Harnessing Energy, Inf. Commun. Technol. Afford. Electrif. Africa, PowerAfrica 2017*, pp. 236–241, 2017, doi: 10.1109/PowerAfrica.2017.7991230.

[4] N. Stevanato, F. Lombardi, E. Colmbo, S. Balderrama, and S. Quoilin, "Two-stage stochastic sizing of a rural micro-grid based on stochastic load generation," *2019 IEEE Milan PowerTech, PowerTech 2019*, pp. 1–6, 2019, doi: 10.1109/PTC.2019.8810571.

[5] S. Steyer, S. Hughes, and N. Whitney, "Predicting Low Voltage Events on Rural Micro-Grids in Tanzania," pp. 1–6, 2014, [Online]. Available: https://www.semanticscholar.org/paper/d493122e774ea6daf775e8deb7833154eb7d6aca.

[6] S. C. S. Jucá, P. C. M. Carvalho, and F. T. Brito, "A low cost concept for data acquisition systems applied to decentralized renewable energy plants," *Sensors*, vol. 11, no. 1, pp. 743–756, 2011, doi: 10.3390/s110100743.

[7] J. M. Paredes-Parra, A. Mateo-Aroca, G. Silvente-Niñirola, M. C. Bueso, and Á. Molina-García, "PV module monitoring system based on low-cost solutions: Wireless raspberry application and assessment," *Energies*, vol. 11, no. 11, 2018, doi: 10.3390/en11113051.

[8] C. Barka, H. Messaoudi-Abid, H. B. A. Setthom, A. Bennani-Ben Abdelghani, I. Slama-Belkhodja, and H. Sammoud, "A real time, wireless and low cost data acquisition system for residential PV modules," *6th IEEE Int. Energy Conf. ENERGYCon 2020*, pp. 417–422, 2020, doi: 10.1109/ENERGYCon48941.2020.9236592.

[9] V. H. Bui, A. Hussain, and H. M. Kim, "A

multiagent-based hierarchical energy management strategy for multi-microgrids considering adjustable power and demand response," *IEEE Trans. Smart Grid*, vol. 9, no. 2, pp. 1323–1333, 2018, doi: 10.1109/TSG.2016.2585671.

[10] M. Ding, T. Xie, and R. Bi, "Research on real-time database system for microgrid," *Dianwang Jishu/Power Syst. Technol.*, vol. 34, no. 11, pp. 31–37, 2010, doi: 10.1109/PEDG.2010.5545839.

[11] J. Laveyne, G. van Eetvelde, and L. Vandevelde, "Wireless Synchronisation for Low Cost Wireless Sensor Networks Using Dcf-77," *Young Res. Symp. IEEE Benelux Chapter IAS/PES/PESC*, pp. 1–2, 2014, [Online]. Available: http://hdl.handle.net/1854/LU-5638924.

[12] A. Mahjoubi, R. F. Mechlouch, and A. Ben Brahim, "A low cost wireless data acquisition system for a remote photovoltaic (PV) water pumping system," *Energies*, vol. 4, no. 1, pp. 68–89, 2011, doi: 10.3390/en4010068.

[13] A. F. Z. Abidin, M. H. Jusoh, E. James, S. A. M. Al Junid, and A. I. Mohd Yassin, "Real-Time Remote Monitoring with Data Acquisition System," *IOP Conf. Ser. Mater. Sci. Eng.*, vol. 99, no. 1, 2015, doi: 10.1088/1757-899X/99/1/012011.

[14] C. Vargas Salgado, J. Águila-León, C. Chiñas-Palacios, and D. Alfonso-Solar, "Supervisory Control And Data Acquisition system applied to a researching purpose microgrid based on Renewable Energy," in *International Conference on Innovation, Documentation and Education*, 2021, pp. 233–239, 2021, doi: 10.4995/inn2020.2020.11898.

[15] E. L. Meyer, O. O. Apeh, and O. K. Overen, "Electrical and meteorological data acquisition system of a commercial and domestic microgrid for monitoring pv parameters," *Appl. Sci.*, vol. 10, no. 24, pp. 1–18, 2020, doi: 10.3390/app10249092.

[16] E. Yoder and N. J. Williams, "Load Profile Prediction Using Customer Characteristics," *2020 IEEE PES/IAS PowerAfrica, PowerAfrica 2020*, doi: 10.1109/PowerAfrica49420.2020.9219811.

[17] P. Yazdkhasti and C. P. Diduch, "Demand side management using model-free fuzzy controller in a direct load control program," *2020 IEEE Electr. Power Energy Conf. EPEC 2020*, vol. 3, pp. 9–13, 2020, doi: 10.1109/EPEC48502.2020.9320090.

[18] M. Ambrož, "Raspberry Pi as a low-cost data acquisition system for human powered vehicles," *Meas. J. Int. Meas. Confed.*, vol. 100, pp. 7–18, 2017, doi: 10.1016/j.measurement.2016.12.037.

[19] W. Setthapun and M. Tonsing, "Big data collection procedure for on-site monitoring system of smart community with PV microgrid," *J. Renew. Energy Smart Grid Technol.*, vol. 14, no. 1, 2019, [Online]. Available:https://ph01.tcithaijo.org/index.php/RAST/article/view/185301.

SOLAR PV POWERED UNITED NATIONS HUMANITARIAN CAMPS WITH INTEGRATED DEMAND FLEXIBILITY AND TIERED RESILIENCE

Manojit Ray[1 2 *], Sampa Ray[3], Ian Dennis Miller[2], Uday Shankar[1 4]
[1]Indian Institute of Technology Kharagpur, WB 721 302, India
[2]University of Toronto, 5 King's College Road, Toronto, ON M5S 3G8, Canada
[3]Swami Vivekananda Seva Samity for Girls, Kolkata, WB 700 113, India
[4]Hidayatullah National Law University, Raipur, CG 492 002, India
*Corresponding author: manojit.ray@mail.utoronto.ca, ray.manojit@iitkgp.ac.in

ABSTRACT: About a billion people survive without electricity. Even more suffer from an unreliable, poor-quality electricity supply. Most of these people live in access challenged, isolated hamlets of the Global South. Education and healthcare suffer due to the absence of adequate electricity. Extending electricity to these settlements remains a challenge. The supply-driven electricity grid in many of these places has expanded at a rate that is below the population growth, leaving an increasingly large population in darkness. Refugee settlements suffer even more. Inadequate access to cooking fuel leads to starvation. Lack of adequate illumination after dusk gives rise to frequent security issues and intensified gender violence. However, recent technological advances allow consumer engagement. A meaningful collaboration could help in sharing available resources. Embracing advances, a favorable energy access tier may be enjoyed as compared to what the supply is capable of delivering with modest investment. This work studies how consumers can benefit with collaboration; also whether the resilience assessment can influence energy access level and cost of delivered energy in refugee camps. It may be worthwhile to note that consumers not only stand to benefit from collaborative use of essential gadgets, they can gain access and affordability with situation-responsive resilience assessment.
Keywords: energy access, refugee camps, resilience assessment, consumer engagement, responsible consumption

1 INTRODUCTION

Globally, 80 million people, a shade more than a per cent of humanity, endured forced displacement in 2020. Out of this fast-swelling group of involuntary movers, 26 million are refugees - people who are displaced away from home, across national borders. The United Nations High Commissioner for Refugees is the leading United Nations agency to aid, protect, repatriate and resettle refugees. The agency currently has more than 20.4 million refugees under its mandate across the world. Provision of shelter, sustenance, amenity and healthcare to the refugees forms an integral part of the agency's work. Multilateral and national policy advocacy by the agency, on behalf of the refugees, may favorably influence societal integration of these forcibly displaced people. Acknowledging the scale [1] and expanse of the mounting refugee population (Figure 1), the agonizing tribulation of the refugee life, and the centrality of multilateral cooperation to tackle the colossal challenge, the UN has, in 2020, adapted domain 5 of SDG indicator 10.7.2 to espouse refugee rights [2]

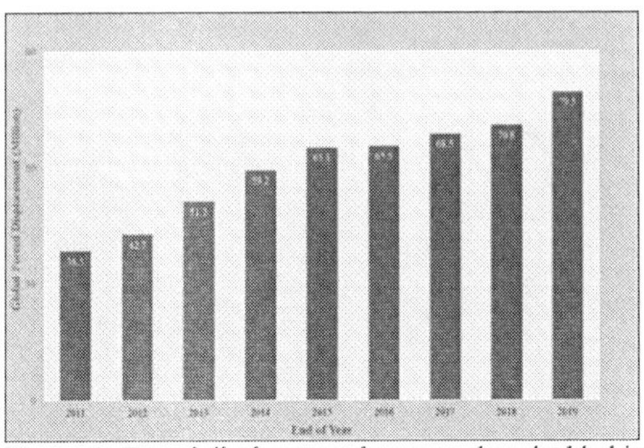

Figure 1: Forced displacement has more than doubled in less than a decade. Increasingly, people are displaced worldwide with fewer of those being able to return home.

This significant stride upholds the spirit of leaving no one behind - the central and transformative promise of the 2030 agenda for sustainable development and associated SDGs. Amongst the 17 SDGs, the seventh one is committed to 'ensure access to affordable, reliable, sustainable and modern energy for all' by 2030 [3]. The opening target of this goal is dedicated to expanding energy access and has been structured with 'proportion of population with access to electricity' as an indicator. However, sustained worldwide effort to increase energy access for all may not have impacted the displaced population much [4]. It may still be hard to imagine a life that is tougher than that of a refugee, living in a camp without essential access to energy.

Compromised access to energy in refugee camps unfavorably impacts support of essential necessities for human survival. Water supply and cooking fuel are part of bare necessities. Humanitarian operators try to address these human rights. But, water quantity and quality commonly lack in absence of access to energy. Residents require water (Figure 2) to survive, thrive and grow [5].

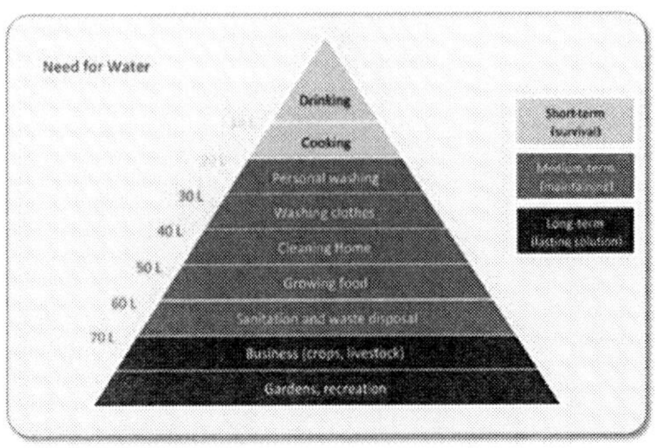

Figure 2: While 20 liters per person per day supports short-term human survival, long term community growth requires in excess of 70 liters of water per person per day.

Extending access to safe drinking water requires energy as sourcing, transportation, treatment and storage require energy - treatment being the most energy intensive of all. Moreover, access to energy remains a persistent challenge in many of the impoverished communities that commonly host refugees. Challenged access to energy unfavorably impacts access to water. Countries with a challenged energy access are also experiencing the world's worst access to safe drinking water (Fig. 3).

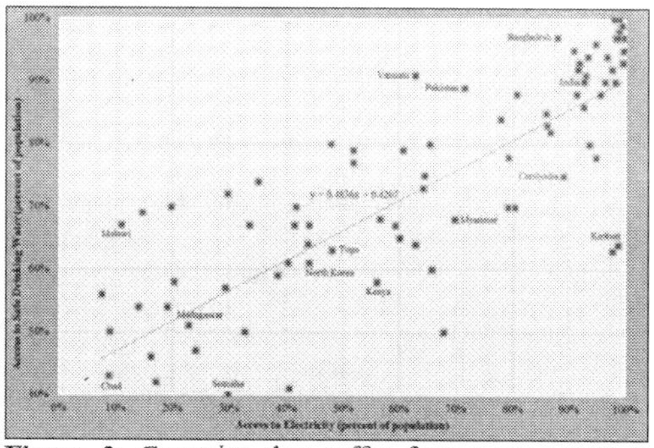

Figure 3: Countries that suffer from poor access to electricity (per cent of population) also experience poor access to safe drinking water (per cent of population)

Harnessing distributed energy resources to extend safe drinking water and clean cooking solutions, may propel humanitarian agencies to more reasonable offerings. But, currently most of the refugee camps receive insufficient water from fossil fuel supported water pumping arrangements (Figure 4). Similarly, most of the residents in refugee camps cook with the available biomass that they can collect from the nearby host community sources. Such practices are neither legal nor commonly accepted by the host communities. Conflicts, assaults and starvation are common outcomes of such unsustainable practices. Trading of humanitarian agency supplied food grains for cooking fuel is common among refugee communities. This is to ensure consumption of the remaining food supply as the humanitarian operators are commonly inadequately funded to supply the basic cooking fuel needed by the camp residents. Education, healthcare and security remain compromised as residents are required to live in darkness after sundown while using wash facilities that are far away from their home [6]. Survival and growth suffers in absence of energy access.

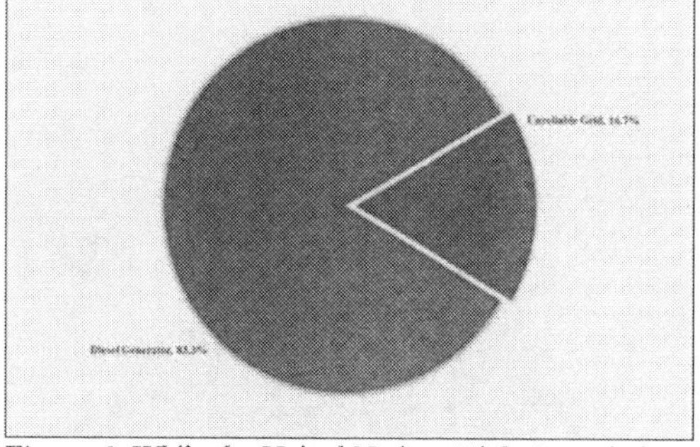

Figure 4: While the United Nations High Commissioner for Refugees supplies water to the refugee camps, most of the water pumps are oversized and fossil fuel powered.

2 METHODS AND DATA

2.1 Demand Response

The electric grid accommodates two contradictory aspects of energy supply. One aspect focuses on unfailing, quality supply even in the face of technical or other kinds of disruptions. This demands widespread redundancy, providing necessary backup support. The other aspect targets energy delivery in the most affordable way possible. It suggests optimization with focus on service reliability vis-à-vis energy price. Whatever the design philosophy of any specific grid may have been, it has always remained demand following in nature and incorporated malleable, large, and on many occasions, polluting generating stations to supply required electricity. Resulting pollution, associated with ensuing environmental unsustainability, in turn, mandated exploration of cleaner sources of generation. Participation of consumers in the electricity supply operation increasingly became essential to address the issue. Such participation can improve resource efficiency and improve affordability of access.

Demand response (Figure 5) encompasses both price based and incentive based schemes. All price based schemes and most incentive based schemes do not offer definitive load control but extend a sense of consumer choice while rewarding grid favored load movements. Such a scheme may deliver desired result when the primary aim of the utility is either raising share of intermittent sources or minimizing extensive load curtailment. Hierarchical grid control and isolation possibility backs up the grid operation in such situations. Distributed, remote, tiny settlements may not have advantage of tiered control unless large number of interconnected hamlets opt for combined demand response program [7]. In an isolated, resource deficient community settlement, it may be useful to implement and deliver results through adoption of need and preference matched predefined time sharing of available resources. Direct load control, offering assured 'supply chasing' capability, may remain the suitable choice for adoption by such a community while trying to make better use of available electricity supply on a time sharing basis to keep peak power within supply stipulated ceiling. Though the consumer autonomy may be considered as partially curtailed, the benefit of energy access may more than compensate for the same [8]. Moreover, the participation in demand response remains voluntary and the benefits may expand with the formation of a market.

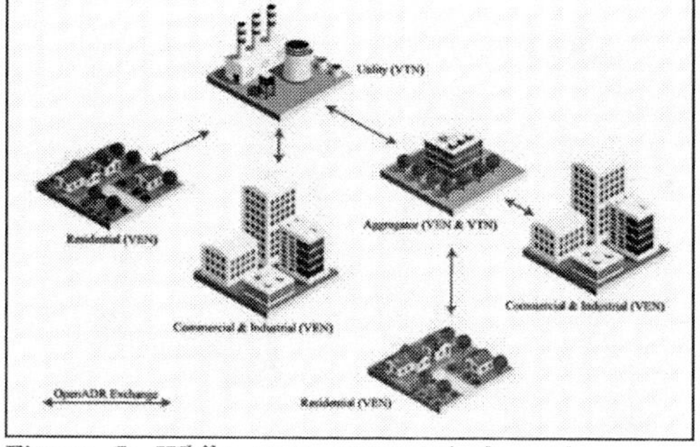

Figure 5: While open automated demand response integrates consumers in the ecosystem, it still depends on the virtual top node and the virtual end node to function.

2.2 Resilience Assessment

Resilience assessment commonly considers average load in an off-grid system. Such a practice essentially puts resilience need of all connected loads at par. However, in reality, the resilience need varies widely over a continuum between extremely critical and not critical at all. Moreover, resilience provision may not always necessarily be economically optimal if resilience provision exclusively depends on one fuel type. Fuel stacking is widely practiced worldwide for fulfilling cooking fuel necessity while large water reservoirs are built to address water need while the water pumping is infeasible in practice. Assessing resilience appropriately and integration of alternative resilience provision modalities impacts off-grid energy supply provision widely. Energy economy improves and supply disruption moves from sudden and unpredictable to gradual and predictable.

2.3 Innovative Integration

Solar photovoltaic generation provides electricity to many access-deficient communities globally. However, poor resource utilization adversely affects the cost of electricity and the financial health of the operator. To efficiently proliferate a matured technology like solar PV, an aptly designed power purchase agreement may be more appealing to all the stakeholders for a risk-mitigated price discovery. Integrating consumption flexibility and realistic resilience into the agreement crafting can usher in an era of affordable solar photovoltaic powered humanitarian settlements across the world. Such innovations at the bottom of the pyramid can profoundly impact solar PV anchored energy service operations at a higher service tier and can make a transition to demand-cantered services from supply-cantered solar photovoltaic plant designs. Improved resource utilization efficiency leads to an enhanced energy affordability.

Demand-cognizant solar PV plants can allow humanitarian agencies to benefit from harnessing solar energy with solar photovoltaic plants while avoiding the initial capital outlay. A solar PV plant developer can install a solar photovoltaic system on the land provided by the camp operator. Simultaneously, the camp operator can decide to purchase the entire power generated by the installed solar PV system. The time-specific payments to be made by the camp operator against the purchased solar power can compensate the developer over the predetermined contract period. Such an arrangement can be efficient if it considers the predominant regulatory stipulations and can embrace an anticipatory semantic to address and mitigate anticipated alterations. It can also contain necessary clauses to include mutually agreed changes in price, quantity and consumption temporality of solar photovoltaic generation and consumption.

3 RESULTS AND DISCUSSION

To optimize the solar PV plant design and minimize initial investments by the energy service operator, efficient lighting (Light-emitting diode), air circulation (Brushless Direct Current) and efficiency-certified appliances can be used by the camp operator. The energy requirement is calculated based on connected load and daily running hours of the same. The model solar PV sourcing agreement commonly provide a daily minimum guaranteed energy to the camp operator. The operator is required to pay for this daily minimum guaranteed energy, ordinarily. However, such arrangements remain cantered on the total daily energy demand of the consumer but does not attempt to take the temporal variation into cognizance. In contrast, the suggested framework of the European Federation of Energy Traders includes a time-specific detailed delivery schedule though the model maintains nonalignment with either supply focus or demand focus. Embedding pragmatic resilience into this framework with inclusion of fuel stacking possibilities can offer affordable solar energy to the humanitarian camp residents (Figure 6).

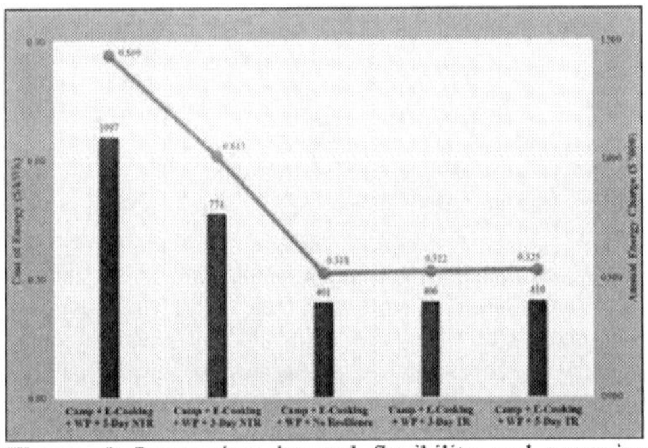

Figure 6: Integrating demand flexibility and assessing resilience by considering tiers of criticality can not only extend crucial energy access but can do so inexpensively.

The study outlines the benefit of adopting 100% solar PV generation in isolated humanitarian camps and phase out the polluting diesel generator completely. The projected near-term cost of solar PV generation and battery storage may make the attractiveness of 100% solar as compelling as solar water pumping now. Daytime use of solar PV powered water pump is cheaper and can be practiced in humanitarian settlements. E-cooking has become practical with the efficiency improvement offered by the electric pressure cooker. Energy charges for e-cooking can be lowered with the inclusion of demand flexibility and resource sharing with contemporary technological evolution. Resilience can be expensive if the traditional concept of autonomy of a solar PV system is followed. Instead, prioritizing loads and understanding specific requirements with tiered resilience can provide needed support during crisis yet keep the cost down. There is no reason not to support medicine or vaccine storage in a camp or leave community wash facilities without illumination.

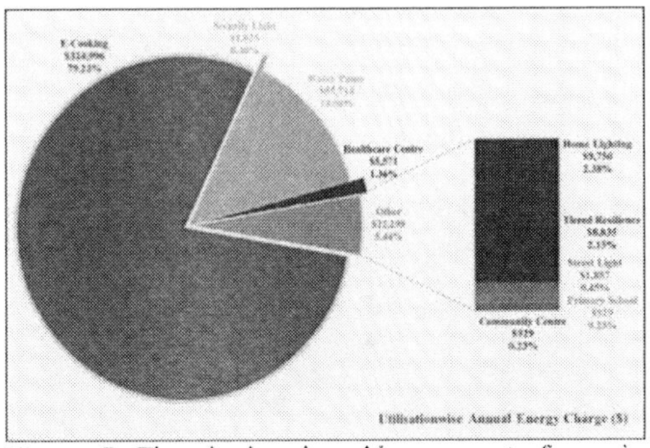

Figure 7: Though shared cooking accounts for major share of the energy charge, all camp necessities can be addressed with demand response and tiered resilience.

A proper design of a practical solar PV system can support the demand of an entire humanitarian camp while keeping expenses reasonable (Figure 7). Non-tiered resilience may require less monitoring of resilience-supporting reserves during a crisis but it can lead to rather inefficient use of resources when the share of resilience-needing critical load is low. Moreover, resilience can draw upon prevailing fuel stacking practices too. The attempt to power humanitarian camps with solar PV can benefit from technology-enabled possibilities, a clear understanding of the consumption trend and exploring possibilities to align the demand with the generation. It is also rewarding to categorize loads per resilience necessity.

4 CONCLUSION

The projected near-term cost of solar PV generation and battery storage may make the attractiveness of 100% solar as compelling as solar water pumping now. Daytime use of solar PV powered water pump is cheaper and can be practiced in humanitarian settlements. E-cooking has become practical with the efficiency improvement offered by the electric pressure cooker. Energy charges for e-cooking can be lowered with the inclusion of demand flexibility and resource sharing with contemporary technological evolution. Resilience can be expensive if the traditional concept of autonomy of a solar PV system is followed. Instead, prioritizing loads and understanding specific requirements with tiered resilience can provide needed support during crisis yet keep the cost down. There is no reason not to support medicine or vaccine storage in a camp or leave community wash facilities without illumination. A proper design of a practical solar PV system can support the demand need while keeping expenses reasonable.

5 REFERENCES

[1] The United Nations High Commissioner for Refugees. Global Trends - Forced Displacement, https://www.unhcr.org/globaltrends2019/; 2020

[2] The United Nations, Department of Economic and Social Affairs, Statistics Division. SDG Indicators, https://unstats.un.org/sdgs/metadata/?Text=&Goal=10&Target=10.7; 2020.

[3] The United Nations. Sustainable Development Goal 7, https://sustainabledevelopment.un.org/sdg7/; 2015

[4] The United Nations Institute for Training and Research. Global Plan of Action for Sustainable Energy Solutions in Situations of Displacement, https://unitar.org/sustainable-development-goals/peace/our-portfolio/global-plan-action-gpa-sustainable-energy-solutions-situations-displacement; 2020

[5] World Health Organization, Loughborough University Water Engineering and Development Centre. How much water is needed in emergencies? https://www.who.int/water_sanitation_health/publications/2011/WHO_TN_09_How_much_water_is_needed.pdf; 2013

[6] International Renewable Energy Agency. Renewables for refugee settlements: Sustainable energy access in humanitarian situations, https://www.irena.org/publications/2019/Dec/Renewable-solutions-for-refugee-settlements; 2019

[7] Ray M, Chakraborty B. Solar PV Driven Child Mortality Alleviation in the Global South. 37th European PV Solar Energy Conference and Exhibition 2020; 3-936338-73-6:1811-1814. https://doi.org/10.4229/EUPVSEC20202020-6BV.5.7.

[8] Ray M, Chakraborty B. Impact of evolving technology on collaborative energy access scaling. Renewable and Sustainable Energy Reviews 2019; 110:13-27. https://doi.org/10.1016/j.rser.2019.04.051

THE INVERSE LAMBERTIAN IRRADIATION METHOD: AN OPTICAL TOOL FOR PV SOLAR CONCENTRATORS AND URBAN HEAT ISLANDS CHARACTERIZATION

Antonio Parretta[1,2]* and Mario Tucci[3]
[1]University of Ferrara, Physics and Earth Science Department, Via Saragat 1, 44122 Ferrara (FE), Italy
Phone: +39 3401509953; E-mail: parretta@fe.infn.it
[2]Academy of Sciences of Ferrara, Via del Gregorio 13, 44121 Ferrara (FE), Italy
[3]ENEA Research Centre "Casaccia", Via Anguillarese 301, 00123 S. Maria di Galeria (RM), Italy
Phone: +39 0630484095; E-mail: mario.tucci@enea.it

ABSTRACT: The Inverse Lambertian Irradiation Method (ILIM) was developed to get the angle-resolved optical efficiency of a PV solar concentrator quickly and with a relatively simple apparatus. ILIM reverses the path of rays (reverse mode), replacing the PV receiver by a Lambertian light source and projecting the light emitted from the inlet aperture onto a far screen. When ILIM is simulated, the reverse light is projected on an absorbing hemispherical screen in order to get the map of emitted radiance. This map reproduces, apart from a constant factor, that of the optical efficiency of the concentrator operating in direct mode under collimated light. The idea behind the present work is to apply the ILIM to an urban heat island model, to get the collection efficiency of solar radiation of a receiver (a PV or thermal module, a window, etc.) in presence of a strong optical interaction with the urban environment. The receiver is then transformed into a Lambertian source and the light diffused, after the interaction with the surrounding environment (the ground, the buildings walls, any green elements, etc.), is projected onto the absorbing hemispherical dome. The absorbed irradiance, once transformed into a map of radiance, gives the map of collection efficiency. This map, when associated with the path of Sun, gives the actual power (or luminous flux) collected by the receiver, once the incident irradiance (or illuminance) as a function of the Sun position is known.
Keywords: Concentrators, Characterisation, Building Integration, Modelling, Optical Properties

1 INTRODUCTION

Solar concentrators essentially work with the direct component of solar radiation, that emitted by the solar disk, and then are operational only with clear sky. Direct solar radiation can be compared to a quasi-parallel light beam, with angular divergence $\theta_S = \pm 0.27°$, emitted by an unpolarized Lambertian source. The solar concentrators, in turn, are characterized by a certain acceptance angle, θ_{acc}, defined as the angle within which the solar radiation is collected with an efficiency higher than 50% [1-3]. Acceptance angles are almost always much greater than the angle θ_S. It is important to consider the simple and fundamental relationship between acceptance angle and optical concentration ratio, C_{opt}, when the concentrator operates in air ($n_{in} = 1$, $n_{out} = 1$), $\eta_{opt} = 1$ (ideal concentrator) and the maximum divergence for the rays at the receiver is 90°:

$$C_{opt} = 1/(\sin\theta_{acc})^2 \qquad (1)$$

Eq. (1) establishes a maximum limit to the optical solar concentration, given by the angle of acceptance of the concentrator, even if the solar angular dispersion θ_S is smaller. Recurring θ_{acc} values for a PV concentrator range from about 20° downwards to about 2.5°, corresponding to optical concentrations from about 10x up to about 500x [4]. The entry on the scene of an unpolarized Lambertian radiation, extended to an angle of 90°, occurs with the introduction of the Inverse Lambertian Irradiation Method (ILIM), which is part of an optical characterization technique of the optical unit of a PV concentrator and which has been extensively illustrated in previous works [5-17]. Here I limit myself to summarizing its functioning. As previously mentioned, the optical figure of a SC is its angle-resolved optical efficiency, $\eta_{opt}(\theta,\varphi)$, defined with respect to a parallel light beam oriented at a certain polar angle θ and azimuthal angle φ referred to the optical axis of the concentrator. For optical systems with cylindrical symmetry, like those we will consider in this paper, it is sufficient the use of only the polar angle θ. So, it comes down to finding the $\eta_{opt}(\theta)$ function. Well, it has been demonstrated [5] that the profile of this efficiency is the same as that of the radiance emitted by the concentrator from its entrance opening when it is irradiated in the opposite direction, that is on its exit opening, with an unpolarized and uniform Lambertian light with 90° divergence. By adding then the measurement of the inverse radiance profile to that of the on-axis efficiency, as described in the aforementioned literature, the absolute optical efficiency curve can be completely determined. The concept underlying the ILIM method can be summarized as follows: if the SC receives light well from a certain spatial direction (θ,φ), when operating in direct mode, it will emit light equally well from the entrance opening on the same direction (θ,φ) when operating in the inverse mode. The inverse mode is applied irradiating the exit aperture with a uniform and unpolarized Lambertian light and measuring the radiance of the emitted light. The only required condition is the validity of the principle of optical reversibility for the tested optical system [18, 19]. The ILIM method has proven to be very simple to apply and very fast, both for experimental and simulated operating conditions. It has been an important innovation in the field of optical characterization of SCs because the traditional method, called DLIM (Optical Direct Lambertian Irradiation Method) [6-7, 12, 14] is much more elaborate, and its experimental apparatus, in particular, is more expensive: it involves the measurement of the incoming flux of a collimated beam and its outgoing flux through a series of discrete measurements at different polar θ and, eventually, azimuthal angles φ, the more numerous the higher the angular resolution of the optical efficiency to be achieved. The latter is obtained, for each point, by the ratio between the outgoing and the incoming flux. Therefore, the DILM provides a series of discrete values, while the ILIM method produces, with a single measurement, a continuous map of values, whose resolution depends only on the resolution of the webcam, or CCD camera, recording the intensity map of the reverse light projected onto a planar Lambertian screen, for experimental operating conditions, or on the optical configuration of the system and on the number of incident rays, for simulated operating conditions It is useful to remember here that the ILIM method has undergone an evolution to eliminate, for experimental operating conditions, the limitations related to the distance between the inverse source and the planar screen. This is referred to as the "Parretta-Herrero method" (PH-Method) [8, 20].

The present work aims to apply the ILIM method to the determination, by optical simulations, of the solar light collection efficiency by an element of an urban heat island, like a PV or thermal module, a window, etc., here referred to as the "receiver", when the direct exposure to sunlight cannot be easily evaluated due to a strong optical interaction with the urban environment. The receiver is replaced by a Lambertian source and the light diffused all around, after interacting with the surrounding environment, is directed onto a front or back hemispherical absorber. Projecting the absorbed light on a plane orthogonal to the optical axis of the heat island, it is possible to obtain the map of radiance emitted by the Lambertian source/receiver in the front space, or on the back space, equivalent to the map of collection efficiency of the receiver under direct irradiation mode, from the front or the back space, as established by the ILIM theory. By associating this map with the path of Sun on a specific day of the year, the corresponding power, or luminous flux (depending on the type of receiver) collected by the receiver in the course of the day, can be obtained, knowing the incident irradiance, or the illuminance, as a function of the Sun position.

2 THE MEASUREMENT OF THE INVERSE RADIANCE OF A SOLAR CONCENTRATOR

For the measurement, through optical simulation, of the radiance coming out from the entrance opening of the solar concentrator, when the exit opening is irradiated by Lambertian light, the configuration shown in Fig. 1 has been adopted.

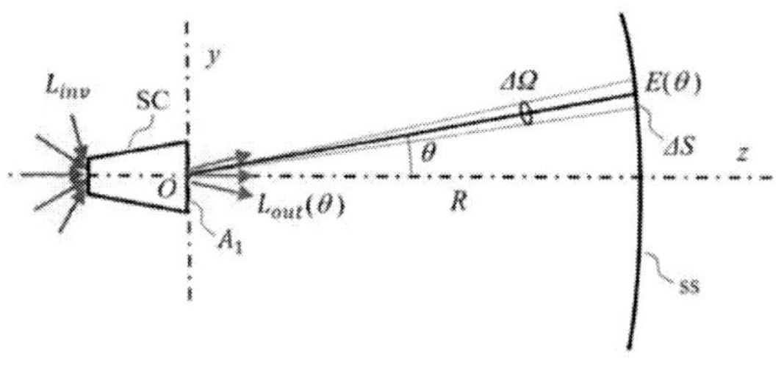

Figure 1: Scheme describing the emission of light, of radiance $L_{out}(\theta)$, from the entrance opening of the SC, irradiated in inverse way by a Lambertian source of radiance L_{inv}, collected by a spherical, absorbing screen (ss).

In this figure, it is shown the example of the irradiation of a solar concentrator SC in the inverse way. The light projected by the SC is intercepted by an absorbing spherical screen (ss), of radius R, centred on the centre O of the opening aperture of the SC. The angular aperture of the spherical screen, θ_M, is adjusted according to the SC's light emission characteristics. The radius R of the spherical screen (ss) is chosen one hundred times greater than that of the diameter of the entrance aperture of the SC, in order to have a high angular resolution on the measured radiance. The irradiance produced on the screen in the θ direction is $E(\theta)$. We then consider the surface element ΔS on the screen in the θ direction. It forms, with respect to the center O, the solid angle $\Delta\Omega$. The flux emitted by SC within this angle $\Delta\Omega$ is given by $\Delta\Phi$. The radiance emitted by SC in direction θ is then given by:

$$L_{out}(\theta) = \Delta\Phi \,/\, (A_1 \cdot \cos\theta \cdot \Delta\Omega) \quad (2)$$

where $\boldsymbol{A_1}$ is the area of the entrance opening of the SC. We have then:

$$L_{out}(\theta) = E(\theta) \cdot \Delta S \,/\, \left(A_1 \cdot \cos\theta \cdot (\Delta S/R^2)\right) = \ldots$$
$$\ldots = E(\theta) \cdot R^2 / A_1 \cdot \cos\theta \quad (3)$$

The radiance $L_{out}(\theta)$ can then be derived from the measurement of $E(\theta)$. The optical simulation code, however, does not show the function $E(\theta)$, but its projection on the $x\,/\,y$ plane, equal to $E_{pro}(\theta) = E(\theta) / \cos\theta$. Then, the program shows a circular map of radius $R \cdot \sin\theta_M$ of only the portion of $E(\theta)$ bounded by the angle θ_M. From Eq. (3) we obtain for the emitted radiance:

$$L_{out}(\theta) = E_{pro}(\theta) \cdot \cos\theta \cdot R^2 / A_1 \cdot \cos\theta = \cdots$$
$$\ldots = E_{pro}(\theta) \cdot R^2 / A_1 \quad (4)$$

Being $E_{pro}(\theta)$ proportional to $L_{out}(\theta)$, its map gives the map of the emitted radiance up to the constant R^2/A_1. Since all the elements considered have a cylindrical symmetry, the $L_{out}(\theta)$ can be symmetrized with respect to the optical axis to improve the signal-to-noise ratio. From the symmetrized map, any radial profile is then extracted which, normalized at $\theta = 0°$, provides the normalized radiance $L_{norm}(\theta)$.

3 THE INVERSE LAMBERTIAN IRRADIATION METHOD APPLIED TO SOLAR CONCENTRATORS

Figs. 2a, and 2b show, respectively, the schematic principle of the direct and inverse methods for the optical characterization of a solar concentrator. The details concerning the application of the two methods, by simulation or by laboratory experimentation, can be found in the references [5, 6, 9, 10, 13, 21], where the equivalence of the two methods is widely described. In Fig. 3 the simulation shows, as an example, the perfect superposition of the relative optical efficiency curve of direct method, and the relative emitted radiance curve of inverse method, for a nonimaging PV solar concentrator with an acceptance angle of 5°.

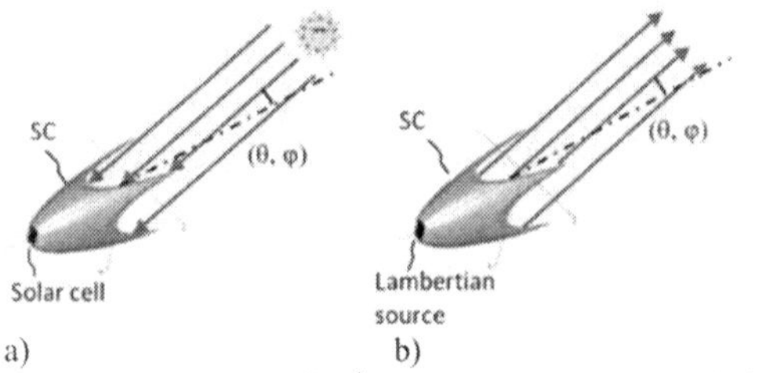

Figure 2: a) The Rondine® PV solar concentrator (SC) is irradiated in direct mode by a parallel light beam oriented at (θ, φ) direction. b) The solar concentrator (SC) is irradiated in the inverse mode and the light emitted towards (θ, φ) direction is shown.

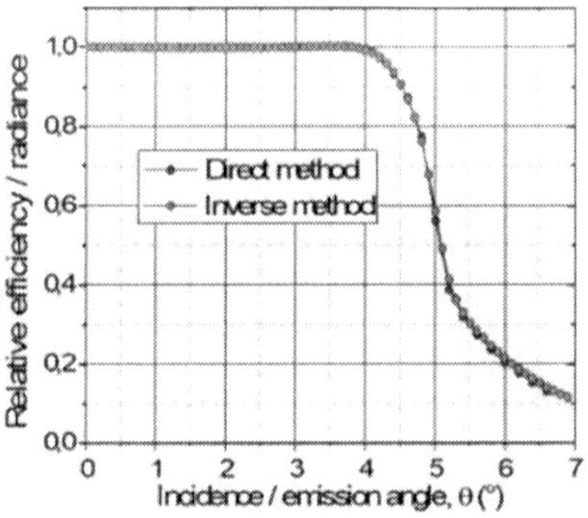

Figure 3: The optical efficiency in direct mode, relative to on-axis incidence, is compared with the emitted radiance relative to on-axis emission.

4 THE URBAN HEAT ISLAND MODEL

To apply the ILM method to an urban heat island, or urban canyon, we have chosen a simple prototype model of a business or residential center (see Fig. 4).

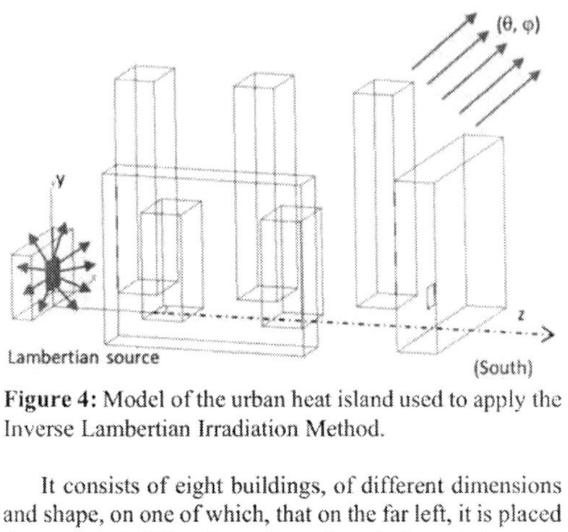

Figure 4: Model of the urban heat island used to apply the Inverse Lambertian Irradiation Method.

It consists of eight buildings, of different dimensions and shape, on one of which, that on the far left, it is placed the receiver (a PV or thermal panel, a window, etc.), south-oriented, so that it can optically interact with all the other buildings in front of it. The total size (perimeter) of the complex is: Δx = 135 mm, Δz = 220 mm (the use of dimensional units in the model is arbitrary). The urban model of Fig. 4 is characterized by an aspect ratio (average height over average distance between buildings): $\lambda_S = \bar{H} / \bar{W} = 1.7$, and classified in the Urban Climatic Zone 3 (UCZ 3) [22, 23]. Fig. 4 shows the receiver transformed into a source of Lambertian light with radiance L_{INV}. The emitted light, after interaction with the surrounding, is projected into outer space, where, in the generic direction with altazimuth coordinates (θ, φ), produces the radiance $L(\theta, \varphi)$. Now, the ILIM theory establishes the equivalence between the normalized optical efficiency of parallel light collection by the receiver from direction (θ, φ) and the normalized radiance $L(\theta, \varphi) / L_{INV}$. What remains to be done, therefore, is to derive the map of $L(\theta, \varphi) / L_{INV}$. The urban heat island is placed in the centre of a hemisphere, with radius R = 1000 mm, with the axis aligned with the North-South direction, coinciding with the normal to the receiver surface (z axis). If E (θ, φ) is the irradiance incident on the hemisphere inner surface and S_W is the receiver area, the irradiance projected on the x/y plane, $E_{PRO}(\theta, \varphi)$, is given by Eq. (5) and the radiance L (θ, φ) is given by Eq. (6). From Eqs. (5) and (6) it follows that the radiance L (θ, φ) is proportional to the projected irradiance E_{PRO} (θ, φ) (see Eq. (7)), which, finally, according to the ILIM theory, is proportional to the efficiency of light collection by the receiver from direction (θ, φ): η (θ, φ) (see Eq. (8)).

$$E_{pro}(\theta,\varphi) = \frac{E(\theta,\varphi)}{\cos\theta \cdot \cos\varphi} \quad (5)$$

$$L(\theta,\varphi) = \frac{E(\theta,\varphi) \cdot R^2}{S_w \cdot \cos\theta \cdot \cos\varphi} \quad (6)$$

$$L(\theta,\varphi) = \frac{E_{pro}(\theta,\varphi) \cdot R^2}{S_w} \quad (7)$$

$$L(\theta,\varphi) = \frac{E_{pro}(\theta,\varphi) \cdot R^2}{S_w} \quad (8)$$

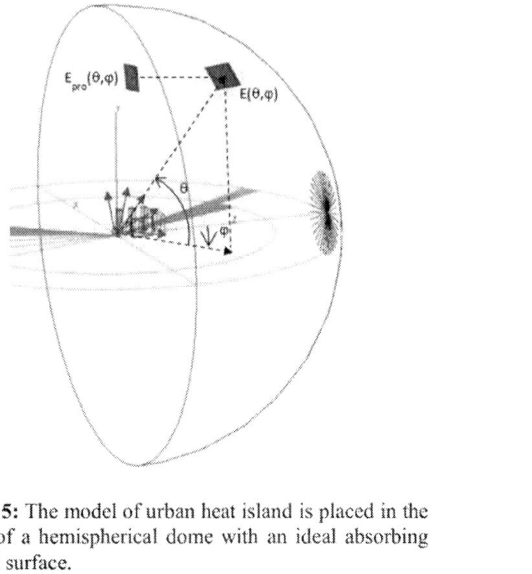

Figure 5: The model of urban heat island is placed in the centre of a hemispherical dome with an ideal absorbing internal surface.

In Fig. 5, the hemispherical dome was placed in the South direction, and then suitable to intercept the diffused light projected in that direction. However, as the light emitted by the source/receiver is also diffused backwards towards the North as an effect of the multiple reflections from the building walls, the radiance of this light will be measured making a second simulation with the hemispherical dome positioned to the North. Fig. 6 shows the front and rear projected irradiance maps obtained by setting a Lambertian diffused total reflectivity of 60% on all the model surfaces. The maps of Fig. 6 give us the optical efficiency up to a multiplicative constant (see Eq. (8)). To transform this map into an absolute efficiency map, it is sufficient to know the inverse radiance at the source/receiver, L_{INV}, or to measure the absolute optical efficiency for a particular altazimuth direction, from which a multiplicative constant for all directions can be derived.

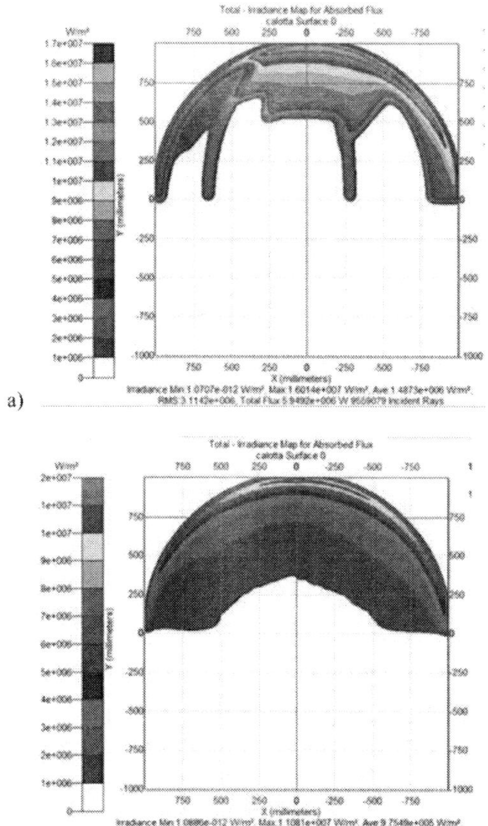

Figure 6: Maps of irradiance projected on the *x/y* plane: $E_{PRO}(x, y)$ (W/m²). a) Front hemispherical dome (South direction). b) Rear hemispherical dome (North direction).

It is useful to show the procedure followed to measure the absolute optical efficiency by the direct method (DLIM). A flat surface source is prepared, larger than the urban area, emitting parallel rays with known E_0 irradiance and oriented towards a particular direction, for example (θ, φ) = (48.6°, 0°) (see Fig. 7), corresponding to the coordinate point (0mm, 750 mm) on the *x/y* plane (see the intersection of the horizontal and vertical lines in the map of Fig. 6a) when the hemispherical dome is South oriented, and (θ, φ) = (48.6°, 180°), corresponding to the coordinate point (0mm, 750 mm) on the *x/y* plane (see the intersection of the horizontal and vertical lines in the map of Fig. 6b) when the hemispherical dome is North oriented.

The absolute collection efficiency is then the ratio between P_{REC}, the simulated power incident on the receiver in presence of the heat island, and P_{THEOR}, the calculated power incident on the receiver in absence of the heat island, as follows:

$$\eta(\theta, \varphi) = \frac{P_{rec}}{P_{theor}} = \left(\frac{1}{E_0 \cdot S_w}\right) \cdot \frac{P_{rec}}{\cos\theta \cdot \cos\varphi} \qquad (9)$$

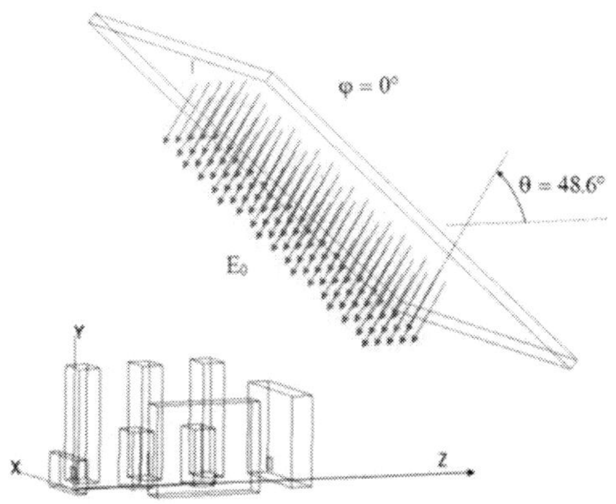

Figure 7: Irradiation of the urban heat island with parallel light, of irradiance E_0, from altazimuth direction (48.6°, 0°).

The two absolute optical efficiency maps, frontal and rear, are shown in Figs. 8a and 8b, respectively. They are plotted as a function of the altazimuth coordinates (θ, φ) which replace the coordinates (x, y) of Fig. 6. The white arrow indicates the direction used for direct light. Note the strong difference between the two maps; since the source/receiver faces South, the optical efficiency values for light from the South are higher than those for light from the North. We also note that the rear map is fainter, because the light arriving at the receiver from the back is only diffused, never direct.

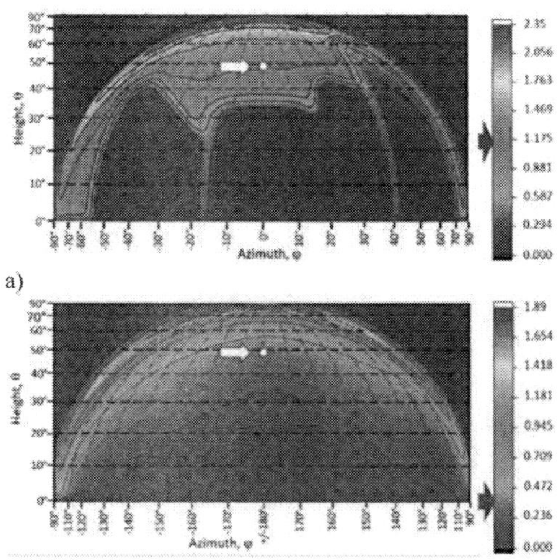

Figure 8: Maps of absolute optical efficiency, frontal (a) and rear (b). The white arrows indicate the altazimuth positions of direct light and the blue arrows the corresponding optical efficiencies. The total diffused reflectance of all the surfaces is 60%.

The absolute optical efficiency maps become very useful when associated with the path of the Sun on the celestial vault (see the example in Fig. 9). Seven Sun paths, valid for twelve days of the year, indicated in the adjacent table and including the equinoxes and solstices, are superimposed on the frontal optical efficiency map. The path of the Sun is calculated for a north latitude of 45° (near of that of Ferrara). The sun trajectories must be followed from left (East, φ negative) to right (West, φ positive). By associating the irradiance, or illuminance, data of the light source to the path of the Sun and to the map of optical efficiency, it is possible to estimate the radiant flux, or the luminous flux, depending on the type of receiver, incident on the receiver as function of time of the day. For this purpose, Fig. 10 shows, as an example, some illuminance data measured at Ferrara in the period of April-May 2020 as a function of the height of the Sun. The highest values recorded, at the same height of the Sun, allow to extract an illuminance curve at clear sky conditions. It is interesting to study the effect of ground reflectivity, R_G, on the optical efficiency maps.

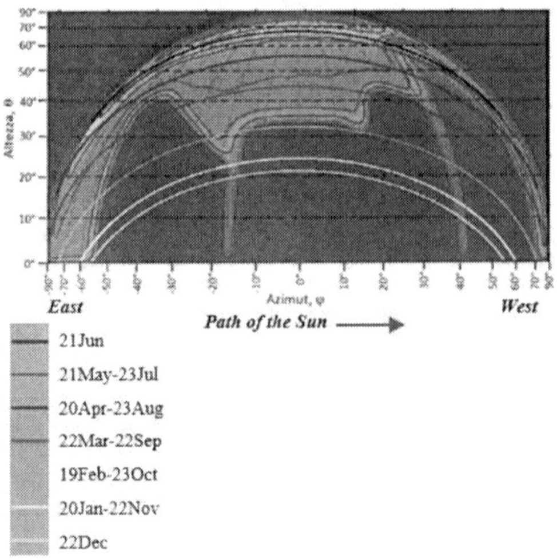

Figure 9: The paths of Sun are superimposed on the frontal optical efficiency map. Below the map, seven Sun paths, valid for twelve days of the year, are indicated in the adjacent table and include the equinoxes and the solstices.

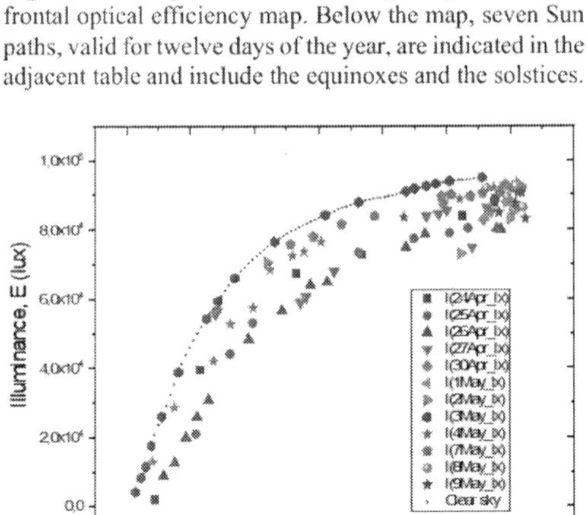

Figure 10: Experimental data of illuminance of Sun at normal direction as function of Sun height.

It must be considered, in fact, that the average temperature of the heat island strongly depends on the ground reflectivity. In Fig. 11 we compare the simulated maps for the South directions, when the ground reflectivity is reduced from 60% to 10%. In Fig. 12 we compare the simulated maps for the North directions, when the ground reflectivity is reduced from 60% to 10%. The effects are visible mainly for high values of Sun disk height, because the ground is more exposed to sunlight at these values.

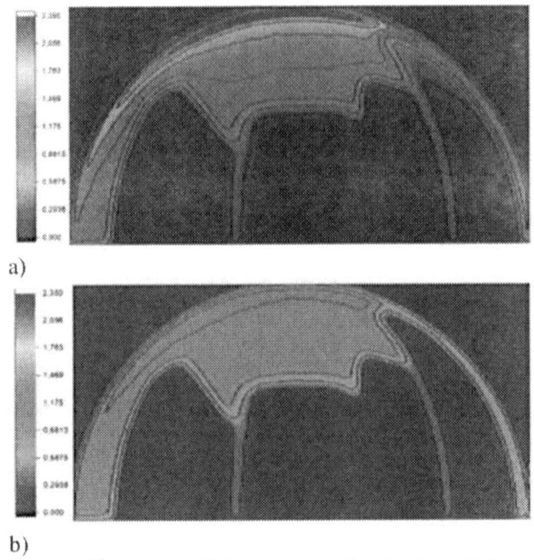

Figure 11: Front efficiency maps simulated at different ground reflectivities, R_G: a) $R_G = 60\%$; b): $R_G = 10\%$.

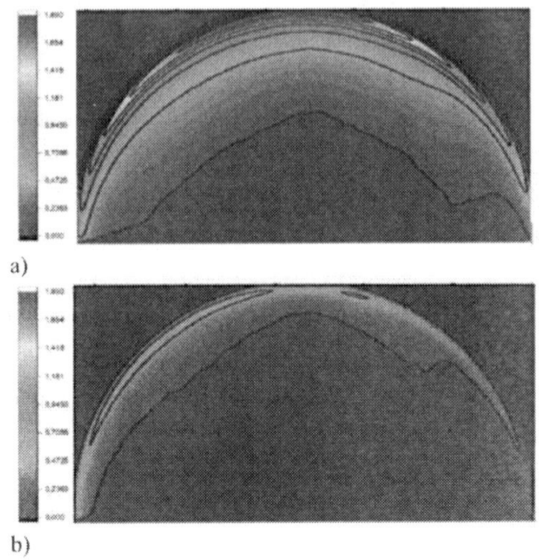

Figure 12: Back efficiency maps simulated at different ground reflectivities, R_G: a) $R_G = 60\%$; b): $R_G = 10\%$.

5 CONCLUSIONS

An optical method for the rapid and synthetic characterization of elements of an urban Heat Island is presented. The method is called Inverse Lambertian Irradiation Method (ILIM), and has been successfully applied in the past to the characterization of PV solar concentrators. In the present paper, the ILIM is applied to a model of business or residential centre (urban climate zone UCZ3) to obtain the efficiency of solar light collection by a generic receiver (PV or thermal module, window, wall, roof, etc.). The optical efficiency is presented in the form of a two-dimensional map as a function of the altazimuth coordinates of the light source (typically the direct component of solar radiation). The map is then associated to the path of the Sun, which allows to evaluate the effective power (or luminous flux) incident on the receiver once known the irradiance (or the illuminance) as a function of the height of the Sun. Some effects of ground reflectivity on the efficiency map are also shown. The ILIM will be implemented in the future by modelling a more complex urban area and by setting spectral reflectivity for the surfaces and spectral intensity for the light sources. It is potentially useful if integrated with a thermal code for the complete characterization of the Heat Island.

REFERENCES

[1] A. Luque, S. Hegedus, Handbook of Photovoltaic Science and Engineering, *John Wiley & Sons Ltd.*, Chichester, UK, 2011.

[2] A. Luque, V. M. Andreev, Concentrator Photovoltaics, *Springer-Verlag*, Berlin, D, 2007.

[3] R. Winston, J.C. Miñano, P. Benítez, 2005, Nonimaging Optics, Elsevier Academic Press.

[4] L.D. Partain, Solar cells and their applications, 1995, *Johm Wiley & Sons, Inc.*, New York (USA),

[5] A. Parretta, Optical Methods for the Characterization of PV Solar Concentrators, Chapter 13, pp. 316-416, in: *Advances in Optics: Reviews. Book Series, Vol. 3. Ed. by Sergey Y. Yurish. Published by International Frequency Sensor Association (IFSA) Publishing, S. L.*, 2018. ISBN: 978-84-697-9439-5.

[6] A. Parretta, A. Antonini, M. Butturi, E. Milan, P. Di Benedetto, D. Uderzo, P. Zurru, Optical methods for indoor characterization of small-size solar concentrators prototypes, *Advances in Science and Technology*, Vol. 74, 2010, pp. 196-204.

[7] A. Parretta, L. Zampierolo, G. Martinelli, A. Antonini, E. Milan, C. Privato, Methods of characterization of solar concentrators, in *Proceedings of the 25th European Photovoltaic Solar Energy Conference (EU PVSEC'10)*, Feria Valencia, Valencia, Spain, 6-10 Sept. 2010.

[8] A. Parretta, A. Antonini, M. A. Butturi, P. Di Benedetto, E. Milan, M. Stefancich, D. Uderzo, P. Zurru, D. Roncati, G. Martinelli, M. Armani, How to "display" the angle-resolved transmission efficiency of a solar concentrator reversing the light path, in *Proceedings of the 23rd European Photovoltaic Solar Energy Conference (EU PVSEC'08)*, Feria Valencia, Valencia, Spain, 1-5 Sept. 2008, pp. 95-98.

[9] A. Parretta, A. Antonini, M. Butturi, P. Zurru, Optical simulation of Rondine® solar concentrators by two inverse characterization methods, *Journal of Optics*, Vol. 14, 2012, 125704.

[10] A. Parretta, A. Antonini, E. Milan, M. Stefancich, G. Martinelli, M. Armani, Optical efficiency of solar concentrators by a reverse optical path method, *Optics Letters*, Vol. 33, 2008, pp. 2044-2046.

[11] A. Parretta, A. Antonini, M. Stefancich, G. Martinelli, M. Armani, Optical characterization of CPC concentrator by an inverse illumination method, in *Proceedings of the 22nd European Photovoltaic Solar Energy Conference (EU PVSEC'07)*, Fiera Milano, Milan, Italy, 3–7 Sept. 2007.

[12] A. Parretta, A. Antonini, M. Stefancich, G. Martinelli, M. Armani, Inverse illumination method for characterization of CPC concentrators, *Proceedings of SPIE*, Vol. 6652, 2007, 665205.

[13] A. Parretta, Optics of solar concentrators. Part I: Theoretical models of light collection, *International Journal of Optics and Applications*, Vol. 3, Issue 4, 2013, pp. 27-39.

[14] A. Parretta, L. Zampierolo, D. Roncati, Theoretical aspects of light collection in solar concentrators, in *Proceedings of the Conference on Optics for Solar Energy (SOLAR), Advancing the Science and Technology of Light*, The Westin La Paloma, Tucson, AZ, USA, 7-10 June 2010, p. STuE1.

[15] A. Parretta, L. Zampierolo, A. Antonini, E. Milan, D. Roncati, Theory of "inverse method" applied to characterization of solar concentrators, in *Proceedings of the 25th European Photovoltaic Solar Energy Conference (EU PVSEC'10)*, Feria Valencia, Valencia, Spain, 6-10 September 2010.

[16] A. Parretta, D. Roncati, Theory of the "inverse method" for characterization of solar concentrators, in *Proceedings of the Conference Optics for Solar Energy (SOLAR), Advancing the Science and Technology of Light*, The Westin La Paloma, Tucson, AZ, USA, 7-10 June 2010, p. STuE2

[17] A. Parretta, G. Martinelli, A. Antonini, D. Vincenzi, Direct and inverse methods of characterization of solar concentrators, in *Proceedings of the Conference Optics for Solar Energy (SOLAR), Advancing the Science and Technology of Light*, The Westin La Paloma, Tucson, AZ, USA, 7-10 June 2010, p. STuA1.

[18] R. C. Jones, On reversibility and irreversibility in optics, *Journal of Optical Society of America*, Vol. 43, 1953, pp. 138-144.

[19] Z. Knittl, The principle of reversibility and thin film optics, *Journal of Modern Optics*, Vol. 9, Issue 1, 1962, pp. 33-45.

[20] A. Parretta, F. Aldegheri, A. Antonini, M. Butturi, P. Zurru, Optical simulation of PV solar concentrators by two inverse characterization methods, *International Journal of Optics and Applications*, Vol. 2, Issue 5, 2012, pp. 62-71.

[21] A. Parretta, A. Antonini, Optics of solar concentrators. Part II: Models of light collection of 3D-CPCs under direct and collimated beams, *International Journal of Optics and Applications*, Vol. 3, Issue 5, 2013, pp. 72-102.

[22] N.H. Wong and S. K. Jusuf, *"Urban Heat Island and Mitigation Strategies at City and Building Level"*, Chap. 1 in "Advances in the Development of Cool Materials for the Built Environment", D. Kolokotsa, M. Santamouris and H. Akbari Eds. 2012, pp. 3-32.

[23] M. Mannia, G. Lobaccaro, F. Goia, A. Nicolini, F. Rossi, *"Exploiting selective angular properties of retro-reflective coatings to mitigate solar irradiation within the urban canyon"*, Solar Energy, 189 (2019) 74-85.

THE MISSING LINK: INNOVATIVE VERTICAL PV SURFACES "SOLAR WINDOWS" AND "SOLAR BLINDS"

Urs Muntwyler, Christian Renken, Eva Schüpbach, Bern University of Applied Sciences BFH, Jlcoweg 1
3400 Burgdorf-Switzerland, phone: +41 (0)34 426 68 37
urs.muntwyler@bfh.ch, christian.renken@bfh.ch

Antonin Faes, Centre Suisse d'Electronique et Microtechnique CSEM, Rue Jaquet-Droz 1,
2002 Neuchâtel-Switzerland, phone: +41 (0)32 720 56 33, antonin.faes@csem.ch

Thomas Stöckli, Schenker Storen AG, Stauwehrstrasse 34, 5012 Schönenwerd-Switzerland
phone: +41 62 858 55 11, thomas.stoeckli@storen.ch

ABSTRACT: Photovoltaic (PV) modules have traditionally been installed on roofs. In winter, fog and snow on PV modules may cause electricity losses on roof-top PV installations. Technology for snow removal is available, but there are no simple solutions. To reduce PV electricity losses on roof-top installations due to snow, PV modules should rather be mounted on vertical infrastructure (e.g., façades including balcony balustrades). While such vertical PV surfaces may offer less energy yield production over the year, they can significantly contribute to winter PV electricity in the northern (and southern) hemisphere mid-latitudes. Because of the static requirements, special PV modules are needed on façades. All façades of a building can be covered with PV modules. In contrast to PV façades, PV on windows is the one vertical element on buildings that has not been widely addressed yet by the PV industry. Yet, given the worldwide increase in glass façades, PV windows are hugely relevant. However, a window cannot be fully covered with PV, as this would obstruct its transparency. For the lengthy periods of time when transparency is not necessary, the invention of "solar blinds" are an innovative technology to harvest solar energy on vertical façades.
Keywords: Building Integration, Façade, PV Module, Rooftop, Solar Architecture, Solar Blinds

1 INTRODUCTION

1.1 Plus-house standards for buildings

Motivated by the feed-in tariff (FIT), early photovoltaic (PV) installations in Switzerland were built to produce maximum annual energy yield. PV installations were economically beneficial when they produced with lowest investment costs and, consequently, operated at the peak solar radiation input (i.e., in summer). PV modules were hence installed on roofs and at low inclinations to minimize the cost for the mounting structure, minimize wind loads, etc. and to produce high energy yields with maximum power per surface in summer (Wp/m^2).

1-story houses need about 10 kWp PV for the heating, hot water, and supply for an electric car for 4 persons. 10 kWp on a roof equals ca. 50 m^2 and hence, 1-storey single family houses with big roofs for a PV installation were among the first PV buildings in Switzerland [1].

With the introduction of new labels promoted by local states in Switzerland since the 1980s, the low-energy building standards have experienced tremendous progress in Switzerland. The new labels are "Minergie standard 2008" with 30 kWh/m^2 (Minergie = label for low energy standard created by the Swiss energy and marketing specialists Dr. Ruedi Kriesi and Heinz Übersax in 1994), Minergie-Plus (even better) and Minergie A (plus-house). Today, the official energy building standards in Switzerland for plus-houses are (i) "MuKen 2008" (MuKen = model ordonnance of the Swiss states for energy consumption of buildings) with 48 kWh/ m^2 and (ii) the new "MuKen 2014" (with 35 kWh/ m^2), see [2]. Increasingly, buildings with a PV installation in Switzerland thus tend to achieve the status of a "plus-energy house", i.e., a positive annual energy balance.

1.2 Illustration of PV buildings

This development is accelerated given the current changes in the Swiss electricity market, where extension of the PV production to seasons beyond summer is envisaged. In this context, east-west-oriented PV installations (Figures 1a, b) offer a huge advantage over the south-orientation as they cover the whole roof and, consequently, produce electricity over the whole day.

Figure 1a: East-west PV installation on a flat roof (Photo: PV LAB BFH-TI)

This is especially beneficial if the PV energy yield production needs to be maximized in view of own consumption. The lower energy/installed power ratio [kWh/ kWp] is less severe in terms of return on investment (ROI) as the price for PV modules has fallen sharply in the last decade. In standard buildings, the effect of an additional battery is less effective as compared to PV modules installed on south façades [3].

Figure 1b: East-west PV installation on an inclined roof. (Photo: PV LAB BFH-TI)

2 PV INSTALLATIONS ON VERTICAL SURFACES

With higher buildings, the ratio "roof surface/ volume" decreases. Over about three floors, it is difficult to realize a plus-house because of the small roof surface in relation to the energy consumption of the house. Yet, the area of the façades increases, which offers an ideal opportunity for PV on the façade, although this is overall less productive than the angle of mounted PV modules is 90 degrees. On a façade facing east or west, about 50% PV energy yield is produced when compared to a roof-top PV installation, and on a north side only about 20% [4]. Yet, the PV production of a south-oriented PV façade is about 70% as compared to a roof-top PV installation with ideal inclination. And in winter, the PV energy yield production is higher for a south façade than for a roof-top PV installation, as the energy production starts earlier and stops later in the day. PV façades also elegantly circumvent the issue of snow coverage of roof-top PV modules in winter. PV façades are hence a real alternative to roofs and are highly recommended [5].

Especially PV façades in alpine regions offer a very high energy yield (up to 1'500 kWh/kWp) with almost a steady summer and winter production. This was demonstrated by our PV installations at the high alpine research station of Jungfraujoch (3'500 m asl; Figure 2a).

Figure 2a: PV façade at Jungfraujoch Research Station installed by the PV LAB at BFH in 1993. (Photo: PV LAB BFH-TI)

Balcony balustrades, which are a smaller variant of PV façades, are yet another interesting option, see Figure 2a and [6]. Typical Swiss houses in alpine regions have their main façade (and often big balconies) oriented towards the south, while the roof is oriented towards east-west. PV façades thus offer a great potential and are today a very much underrated PV application. In addition, solar balconies and solar façades are highly simple solutions for powerful winter electricity production. However, they require special PV modules, e.g., fulfilling the building codes for balcony structures, or special and expensive robust PV modules and mounting structures to cope with the high wind forces encountered at high-alpine altitudes. The latter is also valid for roof-top installations in alpine regions, as high forces (up to 5'400 N / m^2 and higher) may act upon PV modules when heavy snow loads prevail during/after heavy snowfall in winter.

Figure 2b: Solar balcony in Bellwald-Switzerland. Photo: Ruedi Lehmann (Bellwald/VS)

3 SOLAR CELLS / WINDOWS, AND SOLAR BLINDS

3.1 Solar cells on façades

On very high buildings, PV façades are the preferred solution. This is illustrated (Figure 3 and Figure 4) using the example of the two high rising buildings (60 m height) in Sihlweid, Zurich, Switzerland. The two buildings constructed in the 1960s were going to be renovated in view of the "2'000W energy scenario" of the city of Zurich. An electricity production of 50'000 kWh for each building was hence needed. The roof surface of the towers is, however, small. Thus, the solution was a PV façade on all four sides of the building, realized by the PV LAB of BFH[5].

Figure 3: The two renovated buildings at Sihlweidstrasse 1 (back left) and Leimbachstrasse 215 (middle) in Zurich. The nearly identical building on the right is still in its original condition from the 1960s. Photo: Urs Muntwyler

Micro-morph tandem cells from Sharp NA 128 were installed (Figure 4). The reason was the low cost per m^2 in 2011 and 2012. In these days, high efficiency crystalline silicon PV modules could be purchased for one fourth of the price. This also enhanced the energy production by about 100% to 100'000 kWh, demonstrating the progress in PV!

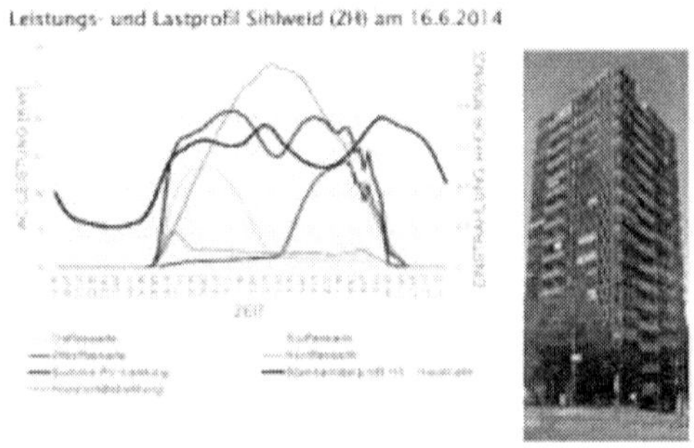

Figure 4: The energy production curve of the Sihlweid building roughly corresponds to the energy consumption curve over the day. The houses are living houses with big balconies on the south side and only few windows on the north side. (Photo: Muntwyler)

3.2 Solar blinds

With progress on glasses for windows, high rising buildings with glass façades are increasingly popular as office buildings (Figure 5). However, such buildings often suffer from overheating. In the past, often misunderstood as "good old time of energy waste", the issue of overheating was solved by powerful air conditioning installations. Today, and in a decarbonized society, sustainable energy solutions are a must, one of them being solar blinds. Even solar curtains are discussed, but these curtains are inside the house, which only partially solves the overheating problem. A better solution for energy absorption may be solar curtains when the energy is still inside the building, but only the small portion of the produced electric energy is leaving. This is a new concept and not in practical use yet.

A realistic solution are solar blinds covered with solar cells. These allow both (i) an individual regulation of the level of sunshine or shadow before the heat enters the house and (ii) energy harvesting along the façade. Several technical options are discussed, being solar cells on solar blinds outside the window and solar cells within the window.

Figure 5: Modern building with glass façade.

Thin solar cells for integration in windows (Figure 6) were invented by the Swiss company iwin (a spin-off from the world-known research institute SUPSI in Lugano). The first pilot- and demonstration installation was realized in Milano, Italy. For its innovation, iwin was awarded several prizes.

Figure 6: Illustration of an iwin solution and installation of the first iwin windows (Source: iwin)

3.3 Solar blinds outside the windows

While a solar curtain, a concept not implemented yet, exploits the energy within the building, the real innovation are solar cells on solar blinds (outside the window), see Figure 7.

Figure 7: Solar blind outside window from Solar Gaps (Photo: Solar Gaps)

3.4 Power production of an office building with and without solar blinds

We calculated the potential of the solar blind solution based on two buildings in Switzerland (Table 1). The first building is a traditional office building (Figure 8a) with a flat roof and five floors located in the Swiss Basin.

Figure 8a: The office building of the example. (Photo: Schenker Storen AG)

The façade is dominated by small structures made of

aluminum; the main surface are glass windows with solar blinds. If the solar blinds are equipped with PV cells (laminate overall efficiency of 10%), the expected yield of the solar blind surfaces on all four sides is 66'300 kWh/ year. As a comparison, the maximum roof-top production calculated based on sonnendach.ch would be 10'400 kWh/ year.

Figure 8b: The roof of the building (in the middle). Photo: Google maps

Table 1: Calculation of the solar blind energy yield production façade (preliminary results in summer 2021) assuming a PV laminate efficiency of 10%

	Total blind surface per floor (m^2)					
	EG	1. OG	2. OG	3. OG	4. OG	Total
North-east facade	68	68	68	68	68	339
South-east facade	102	102	102	102	102	509
South-west facade	68	68	68	68	68	339
North-west facade	102	102	102	102	102	509
Total	339	339	339	339	339	1695

Table 2: Installed PV power and yield per side (preliminary results in summer 2021)

Side of the facade/ PV Power installed	**Production**	
[kWp]	[kWh]	[kWh/kWp]
NE/ 34 kWp	10'200	300
SE/ 51 kWp	24'480	480
SW/ 34 kWp	16'320	480
NW/ 51 kWp	15'300	300
Total: 170 kWp	**66'300**	**390**

3.5 Potential of solar blinds for a living house with different PV applications

Solar blinds are not only an interesting option for office buildings with big glass facades. In two appartement houses in Zurich (Switzerland), the roof was covered with PV and the facades with coloured PV panels; even the balcony was covered with PV panels (Figures 9 - 10). Christian Renken (CR Energie GmbH), researcher in the PV LAB at Burgdorf and PV planer of the two appartements calculated the energy yield produced from solar blinds on all the four sides of the buildings (Figure 11). The PV solar blinds used in the calculation were specially produced products with a 13% efficiency; they are currently under development and aimed at the market in the very near future. A reduction factor of 0,56 was used to reflect the primary use of the solar blinds as windows. The results are presented in Table 3 and Table 4.

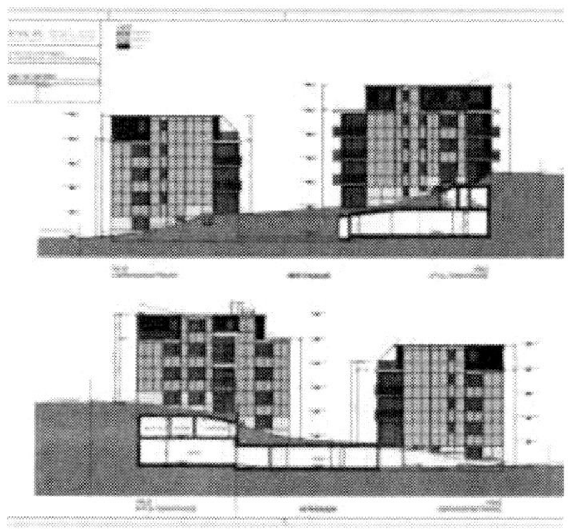

Figure 9: Schematics of the multi-story houses in Zurich-Oerlikon (Affolternstrasse 158/ Oberwiesenstrasse 67/69). Source: BKG Architekten AG / CR Energie GmbH

Figure 10: Multi-story houses at Affolternstrasse 158/ Oberwiesenstrasse 67/69, Zurich-Oerlikon. Source: BKG Architekten AG / CR Energie GmbH (CH)

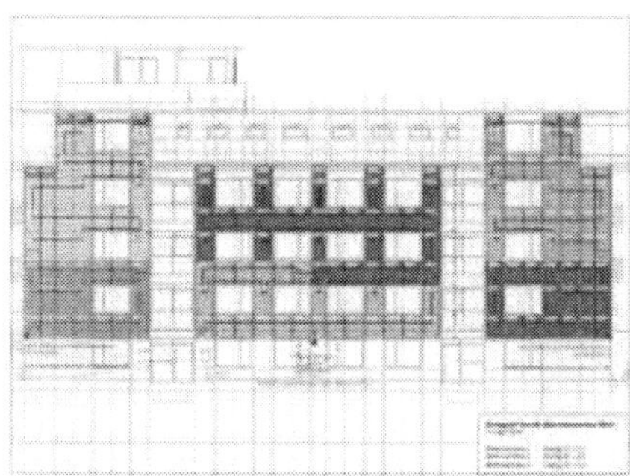

Figure 11a: Different solar blinds on the façade.

Window sizes (mm)	Nominal power (Wp)
1920x1400	350
1980x1400	350
1920x2320	575
1910x2240	550
1580x1400	280
1500x2240	440
3060x1400	630
3200x2370	1035
1700x2370	517
1700x1400	315

Figure 11b: PV solar blinds producing power from 280 - 1'035 Wp.

Table 3: PV power installed on solar blinds of the two buildings shown in Figures 9-11.

Production in kWh per window size and floor	Floor	UG	EG	1.OG	2.OG	DG			Total
AFFOLTERNSTRASSE	South facade	/	/	[illegible]	[illegible]	[illegible]			[illegible]
	East facade	[illegible]	[illegible]	[illegible]	[illegible]	[illegible]			[illegible]
	West facade	/	[illegible]	[illegible]	[illegible]	[illegible]	[illegible]	[illegible]	[illegible]
	North facade	/	/	/	/	/			/
	Total								[illegible]

Production in kWh per window size and floor	Floor	UG	EG	1.OG	2.OG	DG			Total
[illegible]	South facade	/	[illegible]	[illegible]	[illegible]	[illegible]	[illegible]	[illegible]	[illegible]
	East facade	[illegible]	[illegible]	[illegible]	[illegible]	/			[illegible]
	West facade	[illegible]	[illegible]	[illegible]	[illegible]	/			[illegible]
	North facade	[illegible]	[illegible]	[illegible]	[illegible]	[illegible]			[illegible]
	Total								[illegible]

Table 4: Detailed calculation of solar blinds in the buildings shown in Figures 9-11.

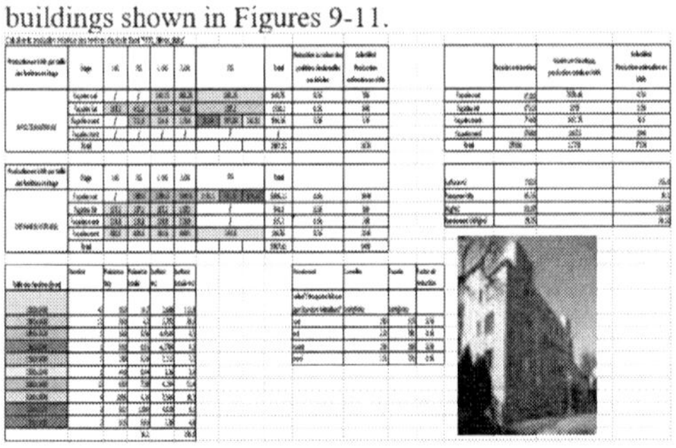

4. CONCLUSIONS

Solar blinds offer a relevant improvement of the energy yield harvested in and by buildings. As demonstrated above, a PV yield of 29'830 kWh can be produced per year by multi-story houses in Zurich with 85'500 Wp installed on solar blinds. With additional solar blinds 7'140 kWh / year could be produced. This is a considerable energy recovery as compared with a conventional roof-top PV installation. Even more so, as the energy production from solar blinds is in addition to the production of a PV installation on a roof, a balcony, and façades. In addition, the advantages of solar blinds may even more pronounced for glass office buildings. There is hence strong evidence that solar blinds are the missing link with which the gaps in PV integrated buildings can be remedied.

First companies obviously already developed solar blind products. First tests with these products as presented here indicate that some challenges must still be mastered in view of market introduction, including technology (mechanic construction, cabling, lamination of the solar cells, etc.) and user behavior. Further research in our project will also specifically address the issue below.

- What could be the price be of the PV solar blinds?
- Are the solar blinds always down?
- If the solar blinds are not always down: Over what period, are the blinds always down or in a lower position (day/month/year)?
- Who can steer the position of the solar blinds?
- Life cycle of solar blinds?

ACKNOWLEDGEMENTS

We kindly acknowledge funding from Innosuisse and Bern University of Applied Sciences BFH. Collaboration with CSEM and Schenker Storen AG is greatly appreciated.

REFERENCES

[1] U. Muntwyler, Praxis mit Solarzellen, 6th edition, Franzis Verlag (1993) page 70.

[2] MuKen, Konferenz Kantonaler Energiedirektoren, Haus der Kantone, Postfach 690, 3000 Bern (2015)

[3] L. Walther, M. Steiger, BSc Thesis (unpublished), PV LAB/BFH Burgdorf-Switzerland (2018)

[4] Ch. Renken, U. Muntwyler, E. Schüpbach, 1st SCCER-FURIES Annual Conference, ETH Lausanne (2014)

[5] D. Joss, U. Muntwyler, M. Münger, Proceed. 28th European Photovoltaic Solar Energy Conference (2013)

[6] U. Muntwyler, F. Kuonen, E. Schüpbach, Proceed. 36th European Photovoltaic Solar Energy Conference (2019)

WHEN AESTHETICS MEETS SUSTAINABILITY THROUGH SUNCOL: THE CO2-NEUTRAL WOHNUBERBAUUNG MANNEDORF PROJECT

Eni Luzi, Elena Canosci
Sunage SA
Corso San Gottardo 54B, Chiasso, 6830, Switzerland

ABSTRACT: The two residential buildings in Männedorf, Switzerland, have photovoltaic modules integrated on facades and roofs that produce renewable energy to achieve a CO2 neutral operations goal. Sunage SA, a Swiss company leader in the design, development and production of photovoltaic modules created SUNCOL, a revolutionary BIPV technology relying on fully customisable colours and materials which perfectly integrates into any facade and roof of the building. The SUNCOL BIPV can be used as traditional construction material able to generate an environmentally sound building which has an aesthetically added value. SUNCOL presents five key characteristics: colours, finishings, design, dimension, applications. The future oriented project that features the SUNCOL red-brown BIPV modules has the first fully integrated photovoltaic facade for weather protection and energy generation in monocrystalline technology with a long service life. The Männedorf project was awarded with the Watt d'Or 2021 in the Buildings and Space category by the Swiss Federal Office of Energy for the ecological and economic impact of the project.
Keywords: BIPV, SUNCOL, photovoltaic, colours

1 THE FUTURE ORIENTED APARTMENTS: THE PROJECT

The future oriented apartment buildings in Männedorf, Switzerland, combine a variety of innovative solutions to achieve a CO2 neutral operations goal. The two residential buildings have 16 apartments, photovoltaic modules integrated on facades and on roofs, and two wind turbines, these installations allow renewable electricity to be locally produced. All residents have heating and cooling energy together with the production of hot water covered.

The objective of the project is the realisation of an energy self-sufficient housing estate where inhabitants do not pay electrcity or heating costs. Residents are able to track their energy consumption via app and a display placed in the house. The building sector is responsible for over 40% of energy consumption in Switzerland. Residential buildings alone cause 11.2 million tons of CO2 emissions every year. The conditions set by the current situation are bad and make it hard to achieve the 2050 climate target. In fact, in August 2019, the Swiss Federal Council decided that Switzerland must become climate-neutral by 2050. According to this objective, greenhouse emissions must be gradually reduced by 50% until 2030, and by 100% until 2050. The Männedorf project contributes to the achievement of this green plan.

The residential development is the second flagship project of the Umwelt Arena Foundation Switzerland and the René Schmid Architects AG in cooperation with exhibition and specialist partners. It was implemented in cooperation with the Federal Materials Testing and Research Institute Empa, the HSR University of Applied Sciences Rapperswil and the Climate Foundation Switzerland.

Figure 1: Männedorf residential buildings

2 A REVOLUTIONARY BIPV PRODUCT: SUNCOL

The complete Sunage SA, a Swiss company leader in the design, development and production of photovoltaic modules that has been operating in the Swiss and EU markets for more than 12 years, created SUNCOL, a revolutionary BIPV product relying on fully customisable colours and materials which perfectly integrates into any facade and roof of the building. The brown colored structured photovoltaic facades of the two residential buildings in Männedorf are realised by Sunage SA and are used at the same time as weather protection and for energy generation. The benefits are multiple and the architect is in no way limited by colour, finishings, design or dimension of the BIPV modules anymore. The SUNCOL BIPV can be used as traditional construction material able to generate an environmentally sound building which has an aesthetically added value. Sunage wanted to guarantee the complete freedom of expression of architects, that is why id developed a wide range of modules, colours, and surface finishings to enable them to elaborate projects and designs using these elements as normal building material such as stone, metal and glass. SUNCOL presents five key characteristics: colours, finishings, design, dimension, applications.

Figure 2: Männedorf single building

2.1 Colours

The exclusive SUNCOL colouring process is the result of a long period of research and experimentation in collaboration with leading universities and research centers in order to obtain the best balance between aesthetic appearance and efficiency. The SUNCOL process is applied in the second layer of the front glass and cures up to 650 °C. Colour stability is guaranteed over time. The colour difference in 10 years is $\Delta E \leq 2$, measured on the position of the front module. SUNCOL does not only want to drive the sustainability transition of buildings but also its processes are aimed at being sustainable, in fact SUNCOL colouring processes are completely eco-friendly. There are more than 15 standard colour options to choose from and customisable colours can be created to perfectly fit with the building's requirements, the potential colour variety is limitless.

2.2 Finishings

SUNCOL finishings present 4 different options: extra clear shining glass, solar glass, satin glass, structured/tridimensional glass. The extra clear shining glass is typically used for facade or balcony photovoltaic modules; the solar glass is tipically used for roofing photovoltaic modules; the satin glass is tipically used for facade photovoltaic modules with anti-reflective finishing; the structured/tridimensional glass is tipically used for facade photovoltaic modules.

2.3 Design

SUNCOL BIPV modules can be made with uniform colour or design. The design colouring consists in the reproduction of an image or texture on the front glass of the module. This type of colouring is often used in historical and landscape contexts, since it makes possible the reproduction of the exact texture of the existing ancient material on the front glass by means of a high-resolution photograph.

2.4 Dimension

The integration of SUNCOL BIPV modules in an existing building is never easy. Sometimes, renovation projects do not allow to design all the elements of the building from scratch, therefore in order to ensure that they combine perfectly with each other, it is necessary that the modules can seamlessly adapt to the existing elements. For this reason, the SUNCOL BIPV modules are 100% customisable in shape and size. All the building surface can be covered by the modules. If there is the necessity of really small modules, Sunage has created dummy modules, non-active modules aesthetically identical to the active ones.

2.5 Applications

Being 100% customisable, the SUNCOL BIPV modules can be integrated into different parts of the buildings, such as facades, roofs, balconies. Depending on the type and area of application in the building, SUNCOL BIPV modules will have different specific physical characteristics.

Figure 3: red-brown structured SUNCOL BIPV module with vertical lines

3 ENERGY GENERATION: KEY DATA

The Männedorf project's costs, which are higher than the ones it would have had by using traditional construction materials, are amortized by giving the residents a given energy budget of heat and electricity for free. In summer there is a surplus in energy production contrasted by shortages in winter, but the powerful inhabited solar power plant uses the extra electricity that its residents do not consume in summer to compensate the lack of electricity in winter. These buildings feature the first fully integrated photovoltaic facade for weather protection and energy generation in monocrystalline technology with a long service life.

The SUNCOL red-brown colour BIPV modules manufactured by Sunage SA have an output that reaches 144 Wp/m2. The two residential buildings in the Männedorf project have a surface of 424m2 in building A and 423m2 in building B. The power installed through the SUNCOL brown-red modules is 80 kWp, the efficiency of each module is of 13.2%, the number of installed modules are 484 in building A and 480 in building B. The type of modules are Glass/Glass 4+4 mm and the finishing is front structured satin glass type BA with vertical lines. The structure used is mechanical fixing through the glass that allow the modules attachment on a wooden structure using screws.

In the underground car park there is a biogas filling station for gas vehicles and an electric charging station for electric vehicles. The electricity for the solar charging station is obtained from the building. Even the lift is energy efficient with low standby consumption.

4 AWARD & RECOGNITION: THE WATT D'OR 2021

The Männedorf project was awarded with the Watt d'Or 2021 in the Buildings and Space category by the Swiss Federal Office of Energy for the ecological and economic impact of the project. Also Sunage SA, as the solar energy partner with the highest amount of BIPV panels installed on the buildings received the Watt d'Or 2021. This prestigious recognition shows how this project completed in 2021 can only be a pioneer for a new green residential buildings generation.

Figure 4: Watt d'Or Award

MACHINE LEARNING FOR BIPV PRODUCTION

Authors: Diego Granados-López*, David González-Peña, Ana García-Rodríguez, Sol García-Rodríguez, Manuel García-Fuente.
Research Group: Solar and Wind Feasibility Technologies (SWIFT). Escuela Politécnica Superior (EPS,Universidad de Burgos). Electromechanical Engineering Department. Avda. Cantabria s/n 09006 Burgos, Spain Tel.: +34 947258853.*Corresponding author: dgranados@ubu.es

ABSTRACT: The increase in energy consumption in homes has never been as intense as in the last decade. The international alarm has resulted in various plans and strategies. This research supports the implementation of Building Integrated Photovoltaics (BIPV) which produce energy at a lower cost to the environment than traditional fuel. Also, it generates energy where it is needed, so its implementation preserves the planet's health. This research proposes a model to predict vertical PV production from vertical solar irradiation and ambient temperature measures. The predictor is an Artificial Neural Network (ANN), trained using the Levenberg-Marquardt algorithm. The research tested and compared several architectures to obtain the most accurate ANN. The one chosen was an artificial neural network with a single hidden layer. Its entry has two neurons, ambient temperature and global vertical irradiation (RaGV). This research shows that ambient Temperature and RaGV are enough to estimate BIPV production, being RaGV the most relevant variable. The introduction as input of the model of the ambient temperature increased its performance. The obtained model has good statistical results and allows to estimate the PV production of commercial panels.
Keywords: Buildinsg-integrated photovoltaics (BIPV) facilities, Machine Learning, and Neural Networks.

1 INTRODUCTION

The increasing usage of energy in factories and daily activity, combined with the extended fuel usage challenge the viability of every living thing. The international alarm is so extended that international entities propose several agreements in which the goal is to preserve the planet [1]. Renewable energies have a weaker impact on the environment; as a result, the use of solar energy is getting more and more extended.

Building Integrated Photovoltaics (BIPV) is a leading technology in the view of cost and efficiency [2]. Vertical solar panels receive less solar radiation than roofs and horizontal surfaces, particularly in the summer months, and are more affected by the density of the urban layout, despite which installations on façades can occupy large areas. Additionally, vertical PV façades will produce relatively more energy in winter than in the summer months [3] and in the early and late daylight hours, when the sun is lower in the sky. The PV building integration has good acceptance in the market. Considering their growing popularity, FOSTErinMED proposed a guide to prevent their indiscriminate usage [4]. Then it is mandatory to make a prior estimation to ensure that the power production is good enough.

The production of the PV panels depends not only on the total incoming solar radiation, commonly known as Global irradiation, but also varies with the installation design. Additionally, the surrounding area can shadow the panels and make them underused [5]. The PV modules have possibly several positions. Usually, in vertical installations, one orientation is the most profitable [6].

2 Objectives

This research follows two objectives: firstly, to describe a simple application of Supervised Machine Learning (SML) for estimating PV production; secondly, to know the seasonal performance of the model.

3 Wheather station

Experimental database is located in Astudillo (Castilla y León, Spain) (Figure 1). The experimental facility consists of four vertical pyranometers facing the cardinal orientations, and a meteorological ground measurement installation that measure ambient temperature and wind speed and direction.

Figure 1: The experimental database is located in Astudillo (Castilla y León, Spain).

The experimental campaign was carried out from 1st April to 31st December, 2016. The features that directly affect PV production are solar irradiance and ambient temperature. So that, they are the inputs of the SML model. Figure 2 shows the pyranometers that record the irradiation measures.

Figure 2: Pyranometers: Hukseflux/SR1

The experimental facility has four commercial PV panels facing the four cardinal directions.

Figure 3: Experimental facility. PV commercial panels

Figure 4 shows a boxplot of the daily PV power (W) generated by the south faced PV panel. The production was slightly bigger in August to September, but the previous months produced less than in the ending of the year, corresponding to winter season.

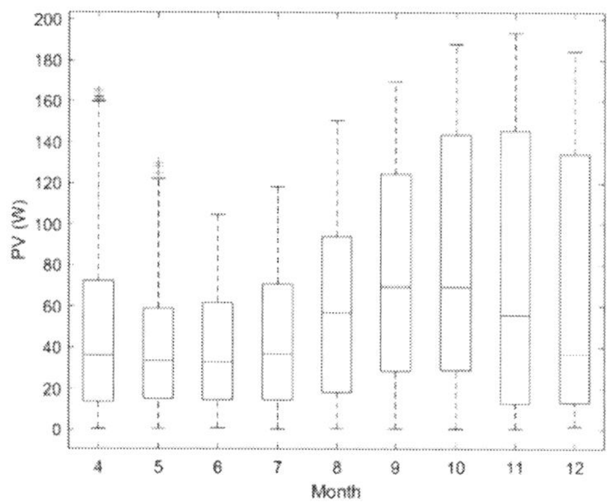

Figure 4: Boxplot of the experimental PV daily production

2 THE PROPOSED MODEL

This research aims a model to predict the instant power production of Building Integrated Photovoltaics panels (BIPV) from the measurement of irradiation and ambient temperature. The technique used is an Artificial Neural Network (ANN) in which the training process uses the algorithm Levenberg-Marquardt for fitting the weights of the network [7]. The architecture of the ANN varies a lot with the application. In this work, the one chosen has one single hidden layer and one single output, as shown Figure 5.

The meteorological variables with the most influence on PV panels production are ambient Temperature and Vertical Global irradiation (RaGV). These two variables will be the input layer of the network. All possible combinations of these two are in table 1.

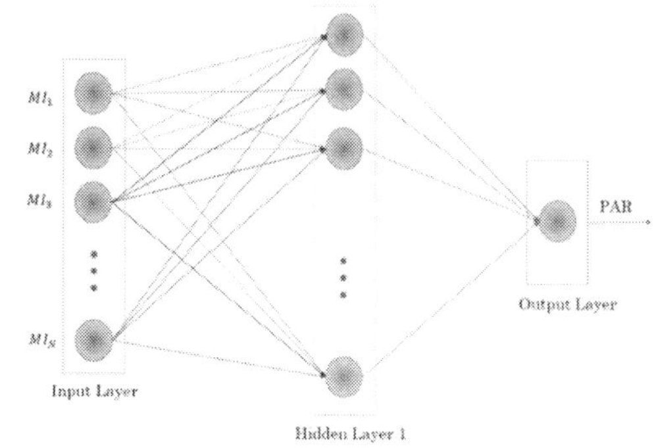

Figure 5: Architecture example of an ANN with a single hidden layer.

Table 1: Possible architectures for the ANN model:

Name	Input layer	Number of neurons in the Hidden layer 1
ANN 1	T	1
ANN 2	RaGV	1
ANN 3	T, RaGV	1
ANN 4	T, RaGV	2

Different works recommend that the number of neurons in the deep layer should not exceed the number of neurons of the previous layer [8]. So, in this case, the maximum number of neurons that is advisable to use is two. This research considers two possibilities for the input layer, [T, RaGV], as Table 1 shows.

3 ANN MODEL

As summary, this research has considered four possible architectures for the neural network that have been tested using the experimental data. ANN programming requires to divide randomly the database into three different datasets, the same for the four developed models. The original dataset splits into training set (70% of the total data) and the validation set (15%), both used during the network training, and the test set, composed by the remaining 15%,unused in the training process, and used for test the performance of the model. The number of neurons of the hidden layer 1 corresponds to the hit performance over the testing set. Table 2 lists the hit performance of the structures listed in Table 1.

Table 2: Performance of the possible architectures. R^2 (dimensionless).

Name	MBD (%)	RMSD (%)	R^2
ANN 1	-2.31	77.36	0.1539
ANN 2	-0.13	13.87	0.9728
ANN 3	0.05	12.05	0.9798
ANN 4	-0.61	11.96	0.9803

Table 2 shows that the performance of the model varies drastically depending on the input meteorological variables. On the one hand, the vertical global irradiation is just sufficient on its own to estimate the production of the PV panel. However, the temperature is incapable of predicting PV production. Finally, the model that works best is the one that considers both variables.

Figure 6 compares the performance in outdoor conditions of the ANN 4, since it has the best statistics results. Its predictions are compared in the test set because it is independent of the fitting process. The shape plot evidences a straight line with relatively low dispersion.

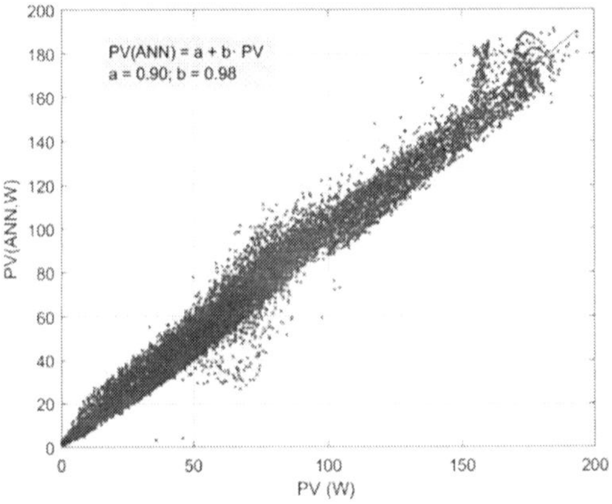

Figure 6: Modelled PV power (W) versus the experimental data.

4 RESULTS

Building Integrated PV (BIPV) is relevant in energy efficiency strategies. These facilities help to increase the independence of the building with the electricity net. However, to design an installation is necessary to know its potential, considering the location and the climate. This study proposes a method to estimate PV production from irradiation measurements, which are generally easily accessible.

The following results are remarkable:

- Temperature and the RaGV are enough for the estimation of PV power, but the most relevant one is the RaGV.
- Temperature increases the performance of the model but is unlikely to perform well on its own.
- The architecture ANN 1 has by far the worst performance.
- ANN 2 and ANN 3 have good performance
- ANN 4 has excellent results, as Figure 4 shows.
- August, September and October have more average of PV power production.

5 ACKNOWLEDGEMENTS

This research has been financial supported by projects RTI-2018-098900-B-I00 and BU021G19. One of the authors, Diego Granados-López thanks grant PIRTU Program, ORDEN EDU/556/2019, for financial support.

6 REFERENCES

[1] M. Pehl, A. Arvesen, F. Humpenöder, A. Popp, E. G. Hertwich, and G. Luderer, "Understanding future emissions from low-carbon power systems by integration of life-cycle assessment and integrated energy modelling," *Nat. Energy*, vol. 2, no. 12, pp. 939–945, 2017.

[2] H. M. Lee, J. H. Yoon, S. C. Kim, and U. C. Shin, "Operational power performance of south-facing vertical BIPV window system applied in office building," *Sol. Energy*, vol. 145, pp. 66–77, 2017.

[3] C. M. Hsieh, Y. A. Chen, H. Tan, and P. F. Lo, "Potential for installing photovoltaic systems on vertical and horizontal building surfaces in urban areas," *Sol. Energy*, vol. 93, pp. 312–321, 2013.

[4] M. Achenza and G. Desogus, "Guidelines on building integration of Photovoltaic in Mediterranean area," pp. 1–108, 2013.

[5] N. Martín-Chivelet, J. C. Gutiérrez, M. Alonso-Abella, F. Chenlo, and J. Cuenca, "Building retrofit with photovoltaics: Construction and performance of a BIPV ventilated façade," *Energies*, vol. 11, no. 7, 2018.

[6] M. Díez-Mediavilla, M. C. Rodríguez-Amigo, M. I. Dieste-Velasco, T. García-Calderón, and C. Alonso-Tristán, "The PV potential of vertical façades: A classic approach using experimental data from Burgos, Spain," *Sol. Energy*, vol. 177, no. October 2018, pp. 192–199, 2019.

[7] C. Lv *et al.*, "Levenberg-marquardt backpropagation training of multilayer neural networks for state estimation of a safety-critical cyber-physical system," *IEEE Trans. Ind. Informatics*, vol. 14, no. 8, pp. 3436–3446, 2018.

[8] J. Heaton, *Artificial Intelligence for Humans, Volume 3: Deep Learning and Neural Networks*. 2015.

SIMPLIFIED PREDICTION OF THE IMPACT OF BUILDING INTEGRATED PHOTOVOLTAIC MODULES AGEING ON THEIR PERFORMANCE

Ya Brigitte Assoa*, Anthony Rhône, Philippe Thony, Fabien Chabuel
Univ. Grenoble Alpes, CEA, LITEN, DTS, LAM, F-38000 Grenoble
* ya-brigitte.assoa@cea.fr ; philippe.thony@cea.fr ; anthony.rhone@cea.fr ; fabien.chabuel@cea.fr

ABSTRACT: A photovoltaic installation is subject to many constraints (such as weather conditions, shocks or dust) that can result in its ageing (such as delamination, discoloration or mould) and thus to performance losses. Only few measured data are available for an accurate long-term prediction of the performance of photovoltaic modules connected in series or building integrated. Some methods permitting to evaluate the impact of each kind of degradation on a photovoltaic field electrical production were proposed and compared. They were applied to a rooftop BIPV system based on its performance ratio and its degradation rate. A performance diagnosis of this installation was also proposed. Compared to the state of art which indicates a degradation rate between 0.5 and 0.8%/year for a rack mounted photovoltaic module, the BIPV system studied had a degradation rate of nearly 4%/year. The majority of the defects detected seems to be related to electrical issue, in particular at the level of the interconnection between the photovoltaic cells and of the bypass diodes in the electrical system. As further steps, a subprogram permitting to determine the evolution over time of the production of a BIPV system will be developed considering different types of installation and photovoltaic technologies.
Keywords: Building Integrated PV (BIPV), Degradation, Modelling

1 INTRODUCTION

Supported by the European directive on Near Zero Energy Buildings, the Building Integrated Photovoltaic market (BIPV) is in progressive development. However, in order to identify the relevant technologies and integration configurations, it is important to predict the thermal, electrical and mechanical behaviors of a photovoltaic system at all times and over time.

Since many years of measured data have been monitored on existing photovoltaic installations, it becomes possible to validate experimentally the prediction models of the long-term electrical performance of newer systems.

Throughout its lifetime, a photovoltaic (PV) installation is submitted to various external that can deteriorate progressively or in the short term the photovoltaic modules (delamination, breakage, moulds, discoloration, electrical failure...) and the electrical system and thus its performance. Various analyses of ageing of photovoltaic systems carried out in recent years have shown that an isolated photovoltaic module loses between 0.5 and 0.8% of production per year in average.

However, these conclusions cannot fully be applied to photovoltaic modules connected in series or building integrated due to higher operating temperatures.

The objective of this work is to establish a methodology permitting to evaluate the impact of each kind of degradation on a photovoltaic field production based on the calculation of performance indicators such as the performance ratio and the degradation rate at different time steps (day, month and annual). In this paper, three developed long-term performance prediction approaches are applied to a rooftop building integrated photovoltaic system installed in 2010 at the CEA INES site and analyzed in order to define the most relevant.

Finally, a performance diagnosis of this installation over ten years is proposed.

2 DESCRIPTION OF PREDICTION METHODS

The three performance prediction methods of PV systems presented are differentiated by the time step used (day or month) and by the consideration or not of the impact of the systems operating temperature variation due to the installation site (weather conditions, shadings...). Their application requires on the calculation of various parameters and mainly on the PV system performance ratio (PR) (see equ. 1) and the degradation rate (reduction of the system performance compared to initial data) on a defined period.

2.1 Electrical parameters considered

The main electrical parameters considered are presented, here. The performance ratio (see equ. 1) permits to take into account the impact of various operating conditions parameters (such as inverter issues, shading, wiring losses, cell mismatch or modules temperature) on a PV system electrical energy production [1] [2] [3].

$$PR = (E_{ac}/P_{max})/(E_i/G_{STC}) \quad (1)$$

With Eac, the electrical energy produced after inverter in AC (kWh), Ei the cumulated incident solar energy (kWh/m^2), $GSTC$ the incident solar radiation in standard tests conditions (STC) (of 1 kW/m^2), $Pmax$ the nominal power installed of the PV field (kWp).

A theoretical electrical efficiency equation η_{PV1} is used. It depends on $Tpvmean$, the BIPV field mean temperature [4]. (See equ. 2 and equ.3)

$$\eta_{pv1} = \eta_{ref} \cdot \left(1 - \beta_r \cdot \left(T_{pvmean} - T_{pvref}\right)\right) \quad (2)$$

$$T_{pvmean} = T_a + K_{th} \cdot G_i \quad (3)$$

With Ta, the outdoor ambient air temperature (°C), Gi, the incident solar radiation (W/m^2) and Kth, the system Ross coefficient or global heat transfer coefficient ($°C.m^2/W$).

2.2 Description of Method 1 and Method 2

Method 1 is based on monthly data and consists first in plotting the BIPV system monthly performance ratio profile according to time, then in determining the linear equation of its trend curve and finally in calculating the yearly degradation rate given by the ratio *(a/b)*, with *a* the slope of the trend curve and *b* the intercept.

Method 2 is based on selected daily data and can be time-consuming. This approach is similar to Method 1 but the PR values considered are calculated using only similar calendar days with close and satisfactory weather conditions on consecutive years (for example, each 1st of January from 2010 to 2020 could be selected for the study). Results of this method were not presented, here.

2.3 Method with temperature correction

This method could be applied to both Method 1 and Method 2. It consists in reducing the impact of the BIPV system temperature variations on its performance compared to standard test conditions (STC) in order to highlight only losses due to ageing. In this case, the performance ratio values could be calculated using PV module temperatures (see equ. 2) or electrical efficiency equations (see equation 3) and measured weather data.

3 DESCRIPTION OF THE CASE STUDY

The three methods were applied to a grid-integrated 5.04-kWp BIPV field comprising forty-two (42) polycrystalline modules and a ventilated air gap at the rear side and was inclined at 27.5°C. The 47.5 m^2 system composed of three strings of 14 photovoltaic modules (with string 1 at the upper part of the PV field) was installed in 2010.

Available measured weather data (total solar radiation in the roof plane, outdoor ambient air temperature and wind velocity and direction) and electrical data after inverter with a 1- min time step from 2010 to 2020 were filtered and used for the comparison of methods.

Figure 1: Rooftop BIPV system on an INCAS experimental building at the CEA INES site

4 ANALYSIS OF RESULTS

4.1 Methods comparison

Figures 2 and 3 present the performance ratio profiles according to time from 2010 to 2020 and their trend curves equation that permitted to obtain the PV system annual degradation rates with Method 1 and Method 1 with temperature correction.

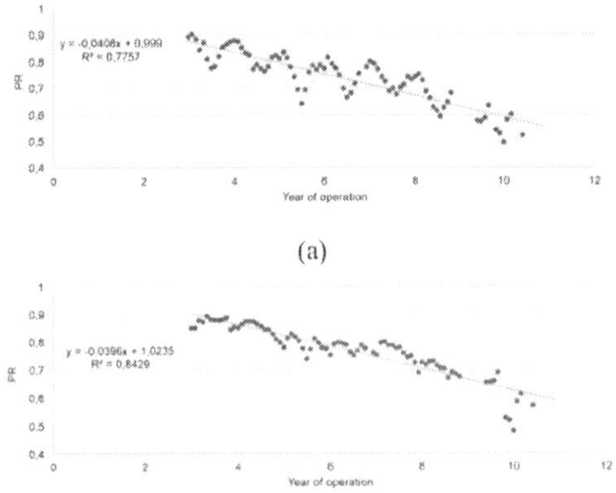

(a)

(b)

Figure 2: Performance ratio profiles of the rooftop BIPV system studied according to time obtained with Method 1 (a) and Method 1 with temperature correction (b) from 2010 to 2020 and their trend curves equations.

The comparison of these two methods has led to some observations:

- The main performance decrease is noted between years 3 and 10 of operation. (See Fig. 2a)
- Oscillations of performance ratio are mainly due to seasonal variations of the PV modules temperature.
- The temperature correction has led to a decrease of the yearly degradation rate from -4.08% to -3.86%. (See Fig. 2a and Fig. 2b)
- Method 1 with temperature correction seems the most relevant for an ageing analysis. Nevertheless, the selected correction equation should be adapted to the installation, especially, the PV module temperature equation (see equ.3).

4.2 BIPV system performance diagnosis on ten years

A performance diagnosis of the BIPV system was performed on the selected period focusing on the electrical power, the electrical current and the electrical voltage at the maximum power point. For this, after a first visual inspection and thermal images analysis, IV curves of each PV string were plotted with an IV tracer and electrical results were compared to initial flash-test results (realized in 2010 before mounting). Table 1 presents a summary of results obtained with the three strings.

Table I: Distribution of performance losses per string of the BIPV system studied from 2010 to 2020, considering the electrical power, the electrical current and the electrical voltage at the maximum power point.

From 2010 to 2020	String 1	String 2	String 3
Total electrical power (W)	35%	40%	39%
Electrical current (A)	4%	5%	4%
Electrical voltage (V)	25%	31%	31%

Important total power losses were observed on each string (between 35% and 40%) (see table I) and were coherent with degradation rates obtained with the developed prediction methods (of nearly -4%/year). The higher losses noted in string 1 (on the upper part) could be explained by most important temperatures due to stack effect in the air gap at the underside of the PV field or to possible failure on some PV modules of this string.

Moreover, most of performance losses were related to electrical voltage issues, thus PV cells seem to be disconnected in some PV modules due to interconnection or diodes failure.

Finally, losses related to electrical current were the lowest, which could indicate a reduced degradation of the PV modules front face (for example, discoloration or delamination...).

Additional studies (thermal images, further electrical parameters analysis...) should be performed in order to confirm these observations.

5 CONCLUSION

Various long-term performance prediction methods of PV systems were applied to a rooftop BIPV system based on nearly ten years of measured data.

Results has shown that Method 1 with temperature correction seems the most suitable for an ageing study of a BIPV installation. It permits to take into account a large number of data and therefore of operating conditions. Nevertheless, it is necessary to ensure that the calculation equations for module temperature and electrical efficiency are adapted to the installation. The temperature correction has led to a decrease of the yearly degradation rate from -4.08% to -3.86%. It is still relevant to avoid using this correction (Method 1), but the result will be less accurate due to the influence of temperature especially in the case of a BIPV system. Method 2 could be very accurate, but the research of similar identical days every year could be time- consuming.

The diagnosis of the installation on the considered period permitted to define the distribution of losses on PV modules strings. Most of the defects detected seems to be related to issues with interconnections within some modules and bypass diodes in the electrical system.

The degradation rate of nearly -4%/year corresponding to an observed loss of electrical production of the BIPV system of nearly 40% after ten years is higher than values presented in state of art for a rack-mounted PV module (between 0.5% and 0.8%).

As further studies, a subprogram permitting to determine the evolution over time of the production of a BIPV system will be developed considering different types of installation and photovoltaic technologies.

6 ACKNOWLEDGMENT

This work realised within INES.2S was partially supported by the French state under the Investments for the Future programme (ANR-10-IEED-0014-01).

7 REFERENCES

[1] A.M. Khalid, I. Mitra, W. Warmuth, V. Schacht, Renewable and Sustainable Energy Reviews (2016) 65.
[2] Y.B. Assoa, F. Sauzedde, B. Boillot, S. Boddaert, Energy (2017) 128.
[3] M. Cubukcu, H. Gumus, Sustainable Energy Technologies and Assessments (2020) 39.
[4] T.T. Chow, Solar Energy (2003) 75.

BE-SMART: THERMAL IMPACT OF PHOTOVOLTAIC MODULES BUILDING INTEGRATION

Ya Brigitte Assoa[1*], Denis Chavrier[1], Ioannis Tsanakas[1], Philippe Thony[1], Alexandre Mignonac[1], Anthony Blaise[1], Joao Victor Furtado Frazao de Medeiros[1], Anthony Rhône[1], Laure-Emmanuelle Perret-Aebi[2], Fabien Chabuel[1]
[1] Univ. Grenoble Alpes, CEA, LITEN, DTS, LAM, F-38000 Grenoble;
* ya-brigitte.assoa@cea.fr ; denis.chavrier@cea.fr ; ioannis.tsanakas@cea.fr ; philippe.thony@cea.fr ; alexandre.mignonac@cea.fr ; anthony.blaise@cea.fr; joao-victor.furtadofrazaodemedeiros@cea.fr ; anthony.rhone@cea.fr ; fabien.chabuel@cea.fr
[2] École Polytechnique Fédérale de Lausanne (EPFL), Institute of Microengineering (IMT), Photovoltaics and Thin-Film Electronics Laboratory, Rue de la Maladière 71b, 2002 Neuchâtel (Switzerland);
laure-emmanuelle.perret@epfl.ch

ABSTRACT: The Be-Smart project aims to contribute significantly to the strengthening of the European industrial sector in the field of solar energy and to present and demonstrate new concepts of low- cost photovoltaic (PV) products for building integration (BIPV) with a high quality architectural design (transparency, color...) named Energy positive Glasses or EpoGs. For a relevant choice of these products installation configuration, an experimental study of the impact of various parameters (see IEC TS 63126 standard) on the temperature of photovoltaic glass-glass monocrystalline silicon mini-modules is being performed for nearly one year on five different sites. A comparative analysis of first results obtained at the CEA INES site and the CEA Cadarache site is presented focusing on their maximum and their 98th percentile temperatures from December 2020 to May 2021. This most representative temperature should be considered to define EpoGs installation configuration per site. For facade integration, in cold period, the glass layers thickness seems to have an influence on photovoltaic modules temperature. As further studies, data measured in warm period will be analyzed. Then, in order to consider edges effect, similar tests will be performed on large PV modules integrated to a facade at the CEA INES site.
Keywords: Building Integrated PV (BIPV), Monitoring, Thermal behaviour

1 INTRODUCTION

The Be-Smart project aims to contribute significantly to the strengthening of the European industrial sector in the field of solar energy and to present new concepts of low- cost photovoltaic products for building integration (BIPV) with a high quality architectural design (transparency, color...). For this, colored glass-glass photovoltaic modules named Energy positive glass or EpoGs suitable especially for façade integration are designed and demonstrated in different sites in Europe. Since building integration of photovoltaic systems has a huge impact on their operating temperature [2], the knowledge of the thermal behaviour of the developed products is necessary in order to optimize the selection of the relevant installation configuration per site. Thus, an experimental study of the impact of various parameters on photovoltaic modules temperature is being performed for nearly one year on five sites with different climates (in France, in Switzerland, in Norway and in Cyprus). The influence of the mounting configuration, of the photovoltaic module glass thickness and of the climatic conditions are considered. Eighteen small size frameless monocrystalline silicon glass-glass photovoltaic modules laminated with 3.2 mm or 6 mm of glass layers were fixed on a south–oriented metal test bench according to three mounting configurations in line with the IEC TS 63126 standard (non-insulated, insulated and naturally ventilated) and three slopes (0°, 45° and 90°). This paper presents a comparative analysis of first results obtained at the CEA INES site and the CEA Cadarache site from December 2020 to May 2021 focusing on their maximum and their 98th percentile temperatures.

2 DESCRIPTION OF THE EXPERIMENTAL SETUP

2.1 Design of configurations

The photovoltaic modules mounting configuration and glass layer thickness and the climatic conditions are mainly taken into account during this study in order to optimize the choice of building integration configuration per site. The general design of the configurations to test was proposed in the framework of the Be-Smart project [2] and adapted by each partner to his site specificities.

The temperature at the 98th percentile (noted *T98*, here) considered for this analysis is the one that a PV module will reach or exceed for only 2% of the year (or 175.2h). [3]

2.2 Presentation of test benches at the CEA sites

On each site, a metal mounting structure was ground-mounted considering the site shadings and albedo effects (see Fig. 1). Eighteen (18) small size frameless photovoltaic modules comprising 3.2 mm or 6 mm glass layers were fixed on this structure:

- Six non-insulated photovoltaic modules were taken as reference.
- Then, six modules comprised a 3 cm-thick air gap insulated with 14 cm of extruded polystyrene layer.
- Finally, six modules were insulated at the rear side.

Relevant spacing and disposal of PV modules were defined in order to reduce heat transfers between configurations especially due to air heating in the configuration with the insulated air gap. Each 6 Wp PV mini-module had a dimension of 20 cm x 20 cm and was connected to two electrical resistors T10 0.1 Ohm in parallel in order to maintain operating conditions close to the maximum power point.

Figure 1: Photograph of the eighteen photovoltaic modules mounted on the test bench at CEA INES site at Le Bourget du Lac and instrumentation (weather station and thermocouples).

The instrumentation was defined in order to permit the monitoring of weather data (ambient temperature with a PT100 sensor sheltered from solar radiation, wind direction and velocity with a weather vane and an anemometer, and solar radiation in the three planes considered using pyranometers). Temperatures inside and at the backside of photovoltaic modules (with K-type thermocouples) and electrical open circuit voltages were measured.

Prior to their installation, preliminary tests were performed (flash-test and electroluminescence). The K-type thermocouples used were calibrated between -10°C and 90°C with a thermostatic bath and the consistency of the readings of pyranometers was evaluated. Moreover, a solution for the installation of a thermocouple within the photovoltaic modules permitting to avoid breakage during lamination was defined. All sensors were connected to an Agilent data logger and the monitoring was realized with a 5 min time step since March 2019.

3 ANALYSIS OF RESULTS

The impact of the site weather conditions was studied taking into account data measured at the CEA INES site and the CEA Cadarache site from December 2020 to March 2021. Moreover, most favorable mounting configurations were proposed on both sites for the month of December 2020.

3.1 Impact of the site weather conditions from December 2020 to March 2021

A comparative thermal analysis was carried out on the six first months of the test campaign focusing on their maximum (*Tmax*) and their 98th percentile temperatures from December 2020 to May 2021.

Table I summarizes the results obtained with the BIPV configurations comprising an insulated air gap on the underside, two 3.2 mm thick glass layers and tilted at 45° or at 90°. (See table I)

Table I: Intervals of maximum temperatures and 98th percentile temperatures of the tilted and the vertical ventilated BIPV mini-module with 3.2 mm thick glass layers in December 2020 at the CEA INES site (a and b, respectively) and the CEA Cadarache site (c and d, respectively).

	CEA INES (°C)	CEA Cadarache (°C)
Tilted at 45°	Tmax: [35.5 ; 54.5]	Tmax: [36.2 ; 51.2]
	T98: [24.8 ; 45.8]	T98: [31.4 ; 43.2]
Vertical	Tmax: [35.3 ; 45.8]	Tmax: [36.2 ; 46.3]
	T98: [24.4 ; 39.0]	T98: [31.5 ; 37.6]

Between December 2020 and May 2021, *Tmax* values of the tilted BIPV mini-module were between 35.5°C and 54.5°C, and its T98 values between 24.8°C and 45.8°C at the CEA INES site. Concerning the vertical BIPV mini-module, its *Tmax* values were between 35.3°C and 45.8°C and its T98 values between 24.4°C and 39.0°C. At the CEA Cadarache site, its *Tmax* values were between 36.2°C and 51.2°C and its *T98* values were between 31.4°C and 43.2°C. For the vertical system, its *Tmax* values were between 35.3°C and 45.8°C and its *T98* between 24.4°C and 39.0°C.

3.2 Impact of the site weather conditions in December 2020

Figure 2 shows the maximum and 98th percentile temperatures in December 2020 for all configurations mounted at the CEA INES (see Fig. 2a and Fig. 2b) and the CEA Cadarache site (see Fig. 2c and Fig. 2d). Solar energy values received according to PV modules slope (vertical, 45° and horizontal) were of 38.7 kWh/m², 22.27 kWh/m² and 34.09 kWh/m², respectively at the CEA INES site and of 65.50 kWh/m², 96.62 kWh/m² and 76.55 kWh/m², respectively at the CEA Cadarache site.

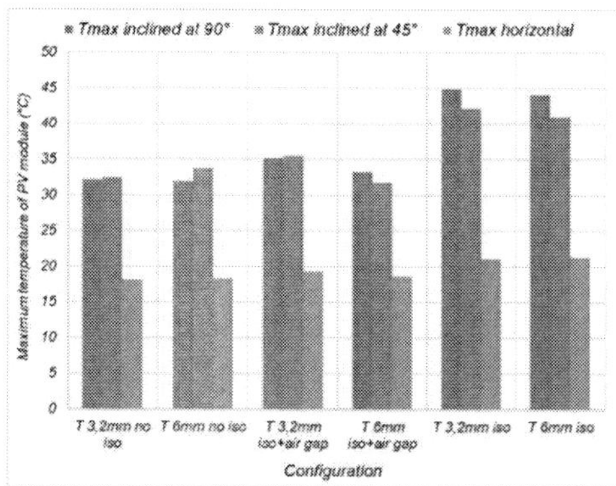

(a)

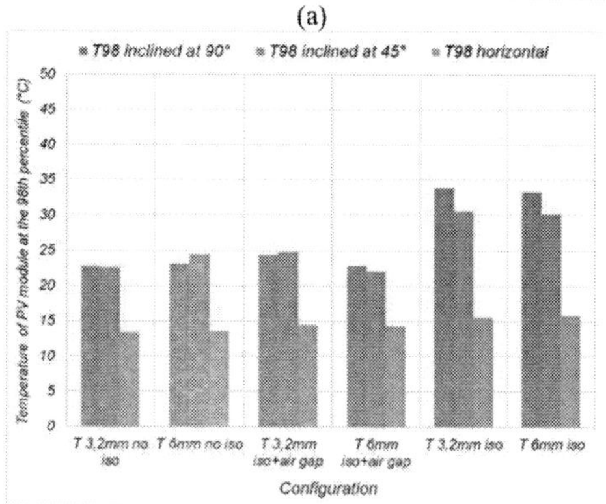

(b)

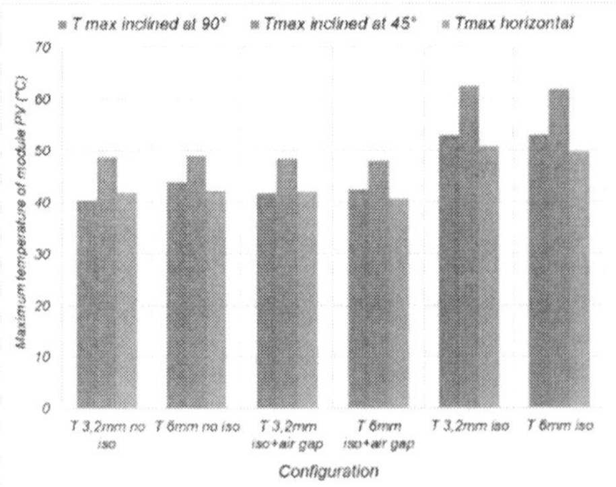

(c)

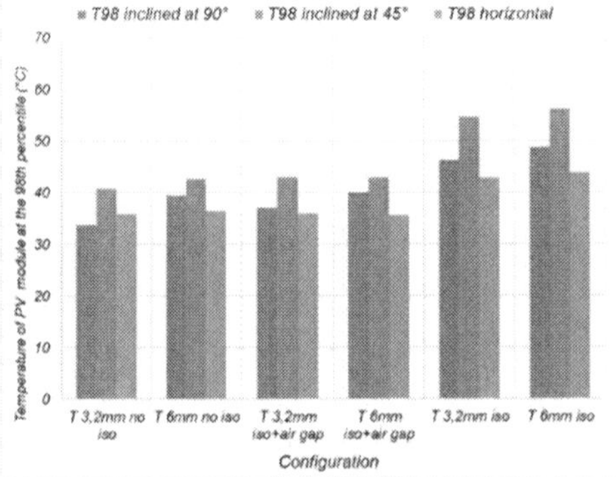

(d)

Figure 2: Maximum temperatures and 98th percentile temperatures of the PV mini-modules in December 2020 at the CEA INES site (a and b respectively) and the CEA Cadarache site (c and d respectively) (with no iso: non-insulated; iso + air gap: with an insulated air gap at the underside; iso: insulated at the underside).

Results of December 2020 have shown that the 98th percentile temperature (*T98*) is lower than the maximum temperature (*Tmax*) for all the configurations studied and *Tmax* is most the time, a peak value, as expected. Only 2% of PV module temperatures are between *T98* and *Tmax*. T98 should therefore be considered for the optimum installation of EpoGs per site.

The difference between *Tmax* and *T98* was between 4.3°C and 11.5°C at the CEA INES site and between 2.4°C and 8°C at the CEA Cadarache site. Higher temperature were observed for PV modules inclined at 45° and vertical PV modules compared to horizontal modules. This could be due to their lower angle of incidence in cold periods (lower sun height) leading to most important temperature levels.

The maximum temperatures and the *T98* values of PV modules at the CEA Cadarache site were generally higher than the ones measured at the CEA INES site, as expected, due to a higher solar radiation level. The difference between *Tmax* values of the two sites was between 6.6°C and 29.7°C. This difference was between 11°C and 28°C for *T98*, which is important.

The smaller differences observed in the case of vertical modules in both sites could be explained by a better solar exposure during the cold period, which increases their heating.

At the CEA INES site, the horizontal non-insulated and non-integrated PV configuration (with 3.2 mm- thick glass layers) was the most favorable at the thermal level (with *Tmax* value of 18.2°C and *T98* value of 13.5°C) due to an optimum natural ventilation at the underside of the PV module. For a vertical integration, ventilated BIPV mini-modules with 6 mm- thick glass layers were the most suitable (with *Tmax* value of 33.2°C and *T98* value of 22.8°C).

Then, at the CEA Cadarache site, the horizontal ventilated BIPV configuration with 6 mm- thick glass layers was the most favorable (with *Tmax* value of 40.8°C and *T98* value of 35.7°C) due, among other to higher angles of incidence of solar radiation in cold periods (low sun height). For a vertical integration, ventilated BIPV mini-modules with 3.2 mm- thick glass layers were the most suitable (with *Tmax* value of 41.8°C and *T98* value of 37.2°C).

4 CONCLUSION

Only 2% of photovoltaic module temperatures are between their maximum and their 98th percentile temperatures. Thus, the 98th percentile temperatures is a relevant parameter for the choice of the optimum installation configuration of EpoGs products per site. For facade integration, according to a cold month results, the choice of glass layers thickness seems to have an influence on the PV modules thermal behaviour. This comparison has shown that on both sites, the most favorable configuration varies per month, so a study on an annual base is required.

As further studies, a similar comparative analysis will be carried out on six months in warm period in order to validate these first observations and to define optimal configurations per site at the annual level. Finally, a similar study will also be performed on large PV modules integrated on a facade at the CEA INES site in order to take account edge effects.

5 ACKNOWLEDGMENT

This work realized within INES.2S was partially supported by the French state under the Investments for the Future programme (ANR-10-IEED-0014-01). It has received funding from the European Union's Horizon 2020 research and innovation programme under grant agreement No 818009

6 REFERENCES

[1] Y.B. Assoa, L. Mongibello, A. Carr et al., Solar Energy (2017) 155

[2] A. Fairbrother, A. Virtuani, C. Ballif, Proceedings 37th European Photovoltaic Solar Energy Conference and Exhibition (2020).

[3] M. D. Kempe, D. Holsapple, K. Whitfield, N. Shiradkar, Progress in Photovoltaic Res Appl (2021) 29:445–460.

YIELD-OPTIMIZED SMALL PHOTOVOLTAIC MODULE SOLUTIONS IN COMBINATION WITH CONCRETE FACADE ELEMENTS

Sebastian Schindler*[1], Stefan Huth[2], Jana Reise[2], Frank Hülsmeier[2], Adrian Heller[2], Paul Schenk[1], Matthias Ebert[1], Kai Wilhelm[3], Mario Butler[3], Reiner Grebe[4], Christof Erban[5]
[1]Fraunhofer Center for Silicon Photovoltaics CSP, 06120 Halle, Germany
[2]Leipzig University of Applied Science - Institute of Architecture Leipzig - ai:L, 04277 Leipzig, Germany
[3]TU Dresden - Institute of Construction Materials, 01069 Dresden
[4]Hering Bau GmbH & Co. KG, 57299 Burbach, Germany,
[5]Sunovation Produktion GmbH, 63820 Elsenfeld, Germany
*=corresponding author: phone: +49-345-5589 5523, sebastian.schindler@csp.fraunhofer.de

ABSTRACT: The building and construction sector does need to contribute in a more substantial way to worldwide endeavours to become climate-neutral in 2050. Solar integration in the building envelope still represents a niche market. Furthermore, necessary CO_2 reductions in the building sector cannot be achieved through PV roof installation alone. A recent study showed that building facades offering a so far untapped potential for the integration of photovoltaics [1]. The aim of the project is the development of an architecturally sophisticated exposed concrete facade with integrated, yield-optimized smaller sized PV modules. Design and orientation of the combined building units, concrete element and photovoltaics modules in the facade will be already optimized in the digital design process by means of Algorithms-aided Design, in which facade and architectural parameters as well as technical and economic target parameters are included and thus design variants are generated. Within this paper we will present first results of the project, the developed design and technical approaches to combine concrete and photovoltaic processes and we propose, how the building sector can contribute with modular, scalable energy facades.

Keywords: AAD, BIPV, concrete facades, PV module, simulation, solar facade

1 INTRODUCTION

Renewably generated energy is becoming increasingly important in power generation and is also an indispensable component in achieving climate goals. Photovoltaics assume a key position, but there is still great potential for expansion. Building facades in particular can be activated for energy generation. For this it is necessary to offer high-quality design solutions integrated into facades and building components.

The aim of the project is to develop an architecturally sophisticated curtain wall facade made of exposed concrete with integrated, yield-optimised small PV modules. The module alignment is optimised through algorithms-based design by including site parameters as well as technical and economic factors - already in the digital design process. For the new BIPV facade lightweight concrete modules with standardised small PV modules are developed. These can be used individually or as part of large-format facade elements.

2 METHODS AND MATERIALS

The architectural design is realized by using parametric-generative design methods. All necessary parameters are implanted in a computational model and parametrically cross-linked, see (**Figure 1**). By using Algorithms-aided Design (AAD) it is possible to describe the optimal orientation and size of the PV modules in the facade during the design process itself. Facades can be checked by AAD for their functionality with regard to solar yields and structural as well as design guiding details are worked out.

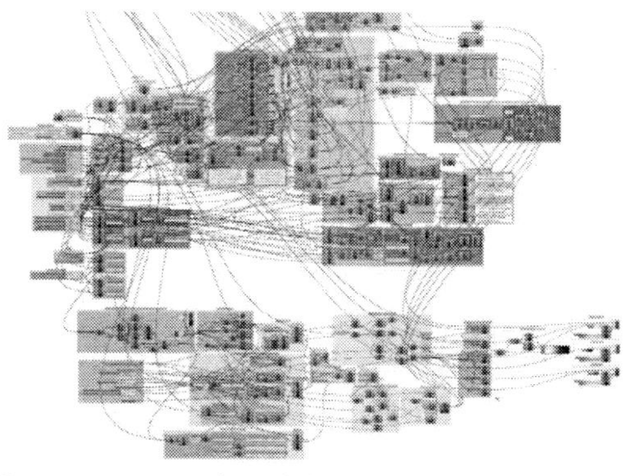

Figure 1: Screen shot of the AAD design software in which architectural and technical parameters are implemented

The focus in the concrete module development is on the production of high-quality surfaces and the color interplay between concrete and photovoltaics. For concrete fabrication conventional casting methods are used. New are the design and the utilization of a modular formwork based on the design of a base element. Multi piece arrangements of the reproducible, multiple usable base elements lead to economies of scale. For that, concrete matrix development and the use of reusable formwork play a central role.

For the photovoltaic modules, standard PV module technologies will be used. We focus on wafer-based cell technologies in combination with glass-glass or glass-backsheet laminates. By dimensioning the photovoltaic components to smaller sizes we want to emphasize alternative attachment and mounting solutions. A fanned out area of smaller PV shall also give more scope of design freedom, plus alternative interconnection layouts and irradiation orientations for yield optimization can be implemented, e.g. exclusion of shading, etc. By utilization of one standard base element, we do promote economies of scale for high volumes and a repeatable, flexibly usable PV element.

Another aspect of the overall design of such BIPV facades is the color impression, of both the concrete and the incorporated PV elements (see **Figure 2**).

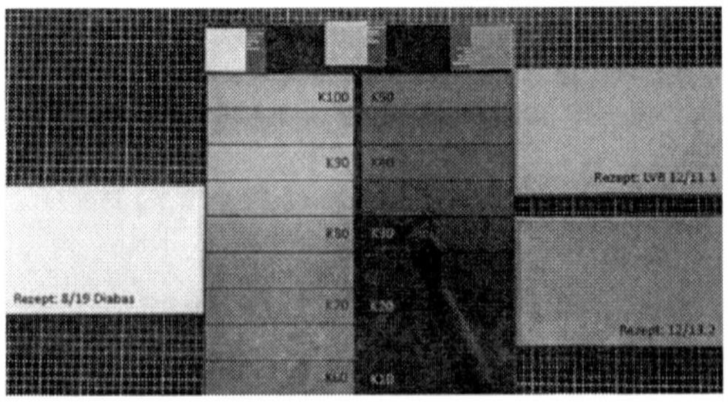

Figure 2: Colour analysis of glass and concrete surfaces

Small photovoltaic modules with a homogeneously coloured appearance create added value with regard to the individual design options of the facade. Colour studies including different angles of view are performed (see **Figure 3**), as reflection on PV elements as well as the moisture of the concrete due to weather conditions have an influence on the facade appearance.

Figure 3: Concreted working model with different glass colour patterns in sunny weather

In context of PV module technology, glass colouring are scope of the work. Also, energy yield analyses are carried out for compromise solutions of aspired colour appearance and colour-affected loss of performance.

3 RESULTS

Modularization of the basic elements is a key result of the architectural design phase. A promising design variant is a three-dimensionally shaped hexagon module that can be optimally aligned to the sun on different facades by rotating it: In the zero position, the hexagon is optimal for use on south-facing facades. On east-facing facades or as a west-facing module on south-facing facades, the BIPV hexagon module is rotated by minus 60°. For optimal use on west-facing facades or as an east-facing module on south-facing facades, the BIPV hexagon module is rotated by plus 60 ° (see **Figure 4** and **Figure 5**). In addition to the hexagonal module variant shown, other form elements have been developed in the project.

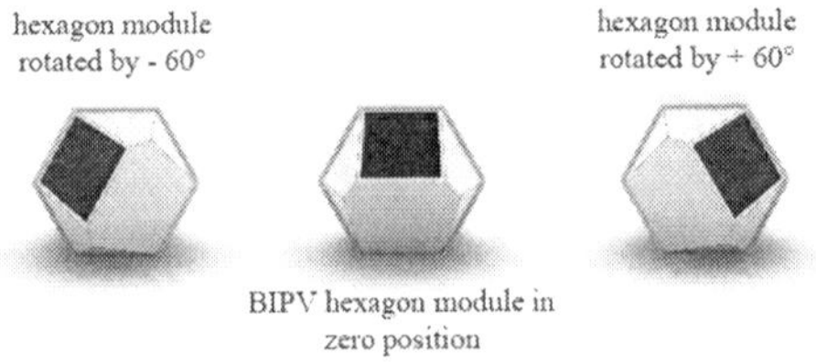

Figure 4: Alignment options for the hexagonal concrete base element

Figure 5: Rendering of a building corner with SOLARcon hexagon modules

Designed as a ventilated curtain facade, the SOLARcon approach can be used for both new and existing buildings. This results in a wide potential field of applications. By using smaller module dimensions and a direct attachment to the concrete hexagon elements, standard PV framing becomes obsolete (see **Figure 6**). Material expenditure can be significantly decreased. A standard Al profile is not necessary for handling and installation. Smaller sized edge protection and/or mounting supporting elements, e.g. for mechanical clamping or bonding, can be derived.

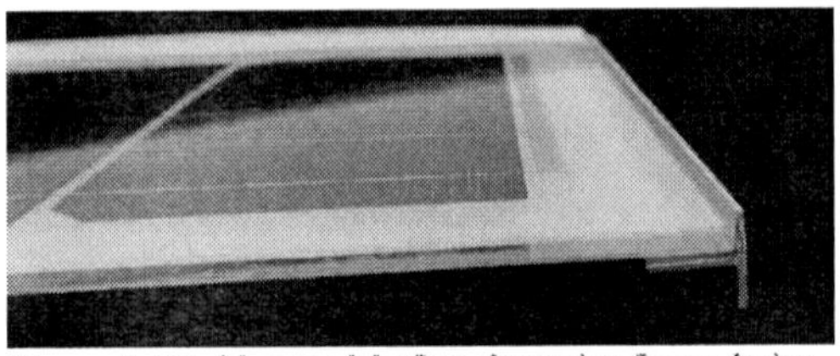

Figure 6: Working model of an alternative frame design

4 OUTLOOK

Next steps are the elaboration of further design variants, concrete casting and PV integration experiments, follow by the assessment and yield analysis of coloured PV modules. Further tasks will cover studies on overhaul solutions, life cycle and profitability assessments.

The project is supported by the Federal Ministry for Economic Affairs and Energy (grant no.: 03EE1016).

5 REFERENCES

[1] Behnisch et al. „Anwendungsszenarien von Geomassendaten zur Modellierung von Grünvolumen und Solarflächenpotenzial“, 2020

AESTHETIC EVALUATION CRITERIA FOR FAÇADE INTEGRATED PHOTOVOLTAICS IN URBAN CONTEXT

Changying Xiang[1,*], Claudia Moscoso [2], Barbara Szybinska Matusiak [1]
[1, *] Department of Architecture and Technology, Norwegian University of Science and Technology, Trondheim, Norway, Email: changying.xiang@ntnu.no
[1] Department of Architecture and Technology, Norwegian University of Science and Technology, Trondheim, Norway, Email: barbara.matusiak@ntnu.no
[2] Department of Mobility and Economics, SINTEF, Trondheim, Norway, Email: claudia.moscoso@sintef.no

ABSTRACT: There is a growing demand to utilize building façade areas for installing photovoltaics (PVs) and producing more clean energy. To support the façade integration of PVs in the urban context, this study proposed a series of evaluation criteria from an environmental aesthetic perspective. Fundamental environmental aesthetic theories were reviewed and Canter's 'sense of place' theory, Kaplan's cognitive preference model and Nasar's six positive qualities in built environment were selected as theoretical reference for this study. In addition, a literature search including aesthetic studies for building integrated PVs and solar systems was conducted. Based on the findings, relevant aesthetic factors were grouped in two levels (i.e. urban context level and building façade level). A set of new aesthetic evaluation criteria is proposed for both levels, covering both perceived physical aspects and conceptual aspects. Colour, surface texture and glossiness are suggested as key factors for further investigation.
Keywords: Façade integrated PVs; urban context; environmental aesthetics; aesthetic evaluation criteria

1 INTRODUCTION

Building Integrated Photovoltaic (BIPV) is one of the most promising strategies to produce clean energy in built environments. Architectural integrated photovoltaic is a sub-category of BIPV with special focus on the aesthetic (formal) integration [1]–[3]. In this study, **Façade Integrated Photovoltaics** (FIPVs) are defined as architectural integrated photovoltaics replacing traditional materials on building façades. PVs are viewed as a new type of façade materials like brick, wood or concrete. The aesthetic performance decides the 'aesthetic integration' quality of façade integrated PVs.

However, compared with the rapid development of advance PV technology in technical aspects, the architectural integration of PVs is still set back due to neglecting aesthetic and cultural issues of PV products [4]. Environmental psychological studies have shown that architectural aesthetic are closely related to people's well-being and health [5]. To promote the aesthetic performance of façade integrated PVs, aesthetic evaluation criteria are necessary as fundamental basis. This study aimed to add new knowledge related to aesthetics in the BIPV study area. It provides theoretical, more in-depth understanding from an urban context perspective to establish the aesthetic evaluation criteria for façade integrated PVs.

2 RESEARCH QUESTIONS AND METHODOLOGY

2.1 Research questions

The research questions of this study were: 1) **What are the most important factors influencing aesthetic perception of façade integrated PVs? 2)What aesthetic evaluation criteria are suitable for façade integrated PVs in urban context?**

2.2 Methodology

The main method of this study was literature review focusing on English-literature closely related to façade integrated PVs. A literature search within the recent 10 years (2009-2020) was conducted using *ScienceDirect*, *ResearchGate* and *Google Scholar* databases. Keywords as evaluation criteria AND BIPV, aesthetic AND BIPV, façade PV AND design were used for the first-round search. In addition, 'snow-ball' literature search was conducted in the areas of environmental aesthetic theories, aesthetic preference in built environment, aesthetic evaluation criteria for façade in urban context. Obtained data was filtered through full paper examining.

The findings of the reviewed study are synthesized and presented in a structured way. A series of relevant aesthetic factors were summarized and a set of new aesthetic evaluation criteria for façade integrated PVs were developed to answer the study's research questions.

3 THEORETICAL BACKGROUNDS

3.1 Environmental aesthetic theories and models

The theory basis of this study is environmental aesthetics. Environmental aesthetic can be defined as the interaction between human and the environment, in relation to beauty. The human-environment relationship includes both the physical interaction the psychological and physiological processes of human perception and cognition [6]. Appleton [7] investigated human's roots of aesthetic preference from an evolutionary perspective, people prefer the environment which can provide both safe feeling (refuge) and the capability to observe and understand the surroundings (prospect). Canter [8] acclaimed that the goal of environmental design is to create environments with identification and clarification. He defined the concept of 'sense of place' as a suitable exploration from interaction of three parts: physical environment, ongoing activities and concept/meaning. The human-environment interaction model by Küller [9] consists of a series of input factors (physical environment, ongoing activities, social environment and individual resource) and people's output series (control, adaptation and compensation). Sharing similarity to Appleton's theory, Kaplan [10] also claimed that people have the demand to notice, understand, evaluate and behave in an environment that could be supportive or harmful to their survival. With emphasis on visual quality of perceived environment, Kaplan's model categorizes people's preference of environmental perception into two levels: understanding and exploration. ach level has both present perception(immediate) and future perception (inferred, predicted), making in total four scenarios: Coherence, Legibility, Complexity and Mystery.

In sum, human's environmental aesthetic preferences are judged by inherited evolutionary roots and acquired knowledge (culture, social and individual experience). For a given place, both its physical environment factors and concept/meaning aspects are essential for its aesthetic evaluation. Canter's 'sense of place' theory and Kaplan's cognitive preference model as well as Küller's human-environment interaction model were selected to serve as the theoretical foundation for this study.

3.2 Environmental aesthetic preference in built environment

Research literature focusing on the concrete aesthetic preferences of built environments were also analysed and are discussed below.

The aesthetic preferences of laypersons present both differences and similarities from architects or experts. For instance, studies showed that architects tend to prefer modern style residential buildings with white colour and less materials while laypersons favour residential buildings similar to traditional styles with warm colours and different materials [11]–[13]. The difference can be explained by the dissimilar knowledge structure and aesthetic experience of observers. Cold also noticed the difference between architects and publics, she pointed out the importance of environmental aesthetic education for publics, through 'cultivating the senses' a basis for broader preference could be achieved[5]. For people's common preference in the built environment, Nasar [14] acclaimed six positive qualities in built environment, including 'order', 'moderate complexity', ' report of natural elements', ' good maintenance, good hygiene', 'openness' and 'historical significance'. Nasar's preference findings shared similarity with or were adopted by studies of many other experts [13], [15], [16].

In this study, the main 'evaluators' are mostly laypersons, but the experts play essential roles in design and creating the urban image. Therefore, to cover both evaluator groups, Nasar's six common positive qualities and 'public aesthetic senses cultivation' were chosen as adequate quality references for in the development of a basis for aesthetic evaluation criteria for BIPV.

4 AESTHETIC EVALUATION CRITERIA FOR FAÇADE INTEGRATED PHOTOVOLTAICS (FIPVS)

The logic structure of developing aesthetic evaluation criteria for façade integrated PVs is displayed in figure 1.

4.1 Aesthetic factors and evaluation criteria for BIPVs and Solar energy systems in urban context

Aesthetic factors and evaluation criteria for BIPVs and Solar energy systems in literatures were reviewed and analyzed. The study done by Farkas [1], LESO-QSV method developed by Munari-Probst and Roecker [17], as well as the AIQ-model proposed by Femenias et al. [18] were the most systematic ones with proposed aesthetic factors, criteria or design suggestions. Table I summarizes important aesthetic factors for BIPV/solar energy system from literature. Colour, shape, size, material and texture were found as the most commonly recommended aesthetic factors, while visibility, glare, identity and dummies (non-active elements with similar appearance of PVs) were also relevant.

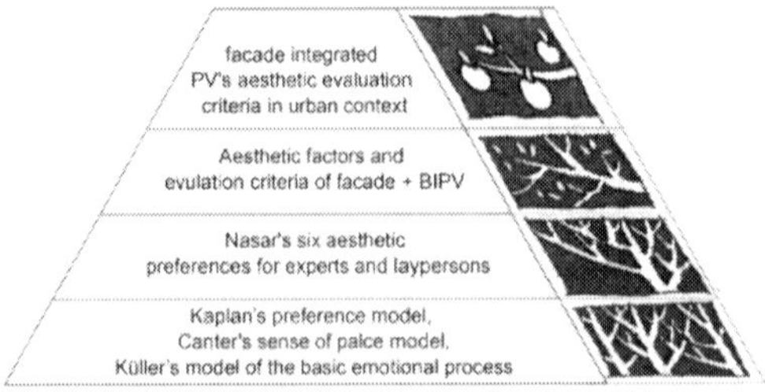

Figure 1: Logic structure of developing aesthetic evaluation criteria for façade integrated PVs. *Source: Author, inspired by drawings of Birgit Cold*

Table I: Important aesthetic factors for BIPV/solar energy system from previous literature

Important aesthetic factors	Type	Authors	Context
material, shape, colour	BIPV system	Lopez et al. 2014	European building cases
Visibility, integration degree, colour, patterns, glare	Solar energy system	Sánchez-Pantoja, Vidal, and Pastor.2018	General
Overall design, field size and position, colour and type of module framing, colour and texture of added elements, detailing of components, overall surface texture of PV module, shape and size of PV module, cell colour, cell texture, cell composition, dummies	BIPV system	Farkas.2013	Scandinavian building cases

Although evaluation criteria for BIPVs/Solar energy systems are various from different studies, many of the criteria are all very well correlated with some aspects of Nasar's six preferences: coherence with surrounding context, representing historical style, order and cleanliness etc. However, the sepecific evaluation criteria for Façade Integrated Photovoltaics (FIPVs) were rarely explored, especially from architectural perspective. Table II presents the main aesthetic criteria/design suggestions for BIPV systems found in literature.

Table II: Aesthetic criteria/suggestions for BIPVs and Solar energy system from pervious literature

Aesthetic criteria/suggestions	Type	Authors	Context
System geometry in coherency with building design logic, System materiality in coherency with building design logic, System pattern in coherency with building design logic	Solar energy system	Munari-Probst and Roecker 2019	European cities
Visibility, integration degree, colour, patterns, glare	Solar energy system	Sánchez-Pantoja, Vidal, and Pastor.2018	General
Overall design, field size and position, colour and type of module framing, colour and texture of added elements, detailing of components,	BIPV system	Farkas.2013	Scandinavian building cases

overall surface texture of PV module, shape and size of PV module, cell colour, cell texture, cell composition, dummies			

4.2 Development of aesthetic evaluation criteria for façade integrated PVs

Most of the above listed BIPV studies investigated the aesthetic factors and aesthetic evaluation criteria at the BIPV system level or building level. However, the relationship between buildings and their surrounding urban context like 'coherency' and 'moderate novelty' are essential for the environmental aesthetic quality [14]–[16], [19]. Therefore, the aesthetic of façade integrated PVs should also be addressed in a bigger scope than building level. According to the 'Open Building' concept [20], [21], a built environment can be viewed as an constant changing environment with different changing frequency (for design and decision making) at three levels: fabric (urban tissue level), base building (support level) and fit-out (interior infill level). Motivated by the concept of open building, this study analysed aesthetic evaluation in two levels: **Urban context level** and **Building façade level** (Figure 2-3)

Figure 2: Urban context level.

Figure 3: Building level.

For the first research question 'What are the most important factors influencing aesthetic perception of façade integrated PVs?', the answer is proposed as following: at urban context level, the important factors influencing aesthetic perception of façade integrated PVs consisted of ***'Dominated colour(s) of PV system perceived from distance', 'Size, shape and position of PV system' 'Dominated material of PV system', 'Dominated texture of PV system'*** and ***'Surface glossiness'.*** At building façade level, factors impacting people's aesthetic perception included ***'Colours & composition ratio of PV module', 'Shape, size and positions of PV module', 'Material of PV module', 'Texture of PV module', 'Joints'*** and ***'Dummy elements'.*** Some closely related factors could be grouped together, for instance, Colour perception was intertwined with surface material, texture and also glossiness level [22], [23] Therefore, similar to the LESO-QSV [24] and AIQ-model [18], colour was grouped together with material, texture and surface glossiness in the 'materiality' category. Size, shape and position are grouped into 'geometry' category while joints and dummy elements are categorized into 'details' group (Table III).

Table III: Aesthetic factors for façade integrated PVs

Environmental levels	Aesthetic factor groups	Aesthetic factors
Urban context level	System materiality	Dominated colour(s) of PV system perceived from distance; Dominated material of PV system; Dominated texture of PV system; Surface glossiness
	System geometry	Size, shape and position of PV system
Building level	Module materiality	Colours & composition ratio of PV module; Material of PV module; Texture of PV module;
	Module geometry	Shape, size and positions of PV module
	Details	Joints; Dummy elements

For the second research question, the formulation of aesthetic evaluation criteria of façade integrated PVs was based on previous discussed environmental aesthetic theories in section 4.2 and the above synthesized aesthetic factors in table III. Foundational aesthetic preferences were categorized into two aspects: perceived aspect and conceptual aspect. The formulated aesthetic evaluation criteria for facade integrated PVs were presented in Table IV.

Table IV: Aesthetic evaluation criteria for façade integrated PVs

		Aesthetic evaluation criteria	Related Aesthetic factor group(s)
Urban Context level	Perceived aspect	System materiality in coherence with urban context	System materiality
		System geometry in coherence with urban context	System geometry
		Moderate complexity and novelty	System materiality, system geometry
	Conceptual aspect	Presentation of positive conception	System materiality, system geometry
Building Level	Perceived aspect	Module materiality in coherence with façade design logic	Module materiality
		Module geometry in coherence with façade design logic	Module geometry
		Details in coherence with façade design logic	Details
		Moderate complexity and novelty	Module materiality, module geometry, details
	Perceived aspect	Presentation of positive conception	Module materiality, module geometry, details

In Canter's 'sense of place' theory, places are a manifestation of the physical environment, activities and conception. As a new building material, façade integrated PVs will not only alter the perceived physical components of built environment but also influence the conceptual aspect (meaning) of places in urban contexts. Conceptual characteristics of a place play important roles in people's preference to the place, people prefer places with more sense of safety or natural elements representations. Different cultures can have different symbolic preferences. Therefore, the aesthetic evaluation criteria of façade integrated PVs need to ***cover both 'perceived' and 'conceptual' aspects***. Criteria in perceived aspect were generated according to the aesthetic preferences of 'coherence', 'moderate complexity', 'novel building with context'. While the conceptual criterion named '***presentation of positive conceptions***' was generated to cover the aesthetic preferences of 'meaningful building with context' and 'public aesthetic cultivation', encouraging façade integrated PVs to present conceptions like symbolism in harmony with local culture, sense of safety, sense of nature, cultivating the public about the aesthetic of PVs etc.

In an urban context level, the geometry and materiality of façade integrated PV systems will play important roles in aesthetic performance. The coherence between façade integrated PV systems and urban context could be achieved through system materiality coherence, system geometry coherence or both (Figure 4-5).

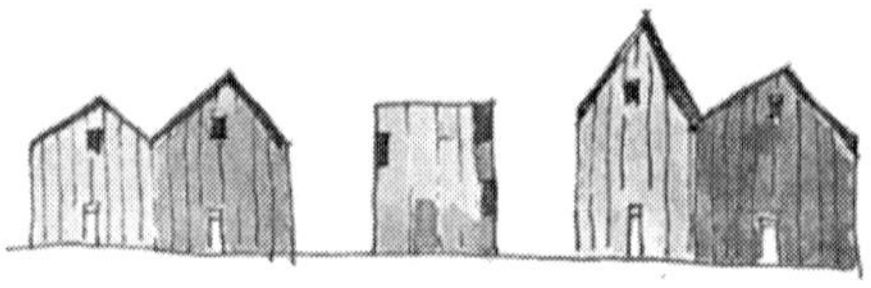

Figure 4: System materiality (colour)coherency.

Figure 5: System geometry (shape)coherency.

In a building façade level, perceived façade coherence could be achieved by integrating PV modules harmoniously with façade design logics (Figure 6). Moderate complexity or novelty at building level could be realized through moderately changing PV modules' geometry or colours.

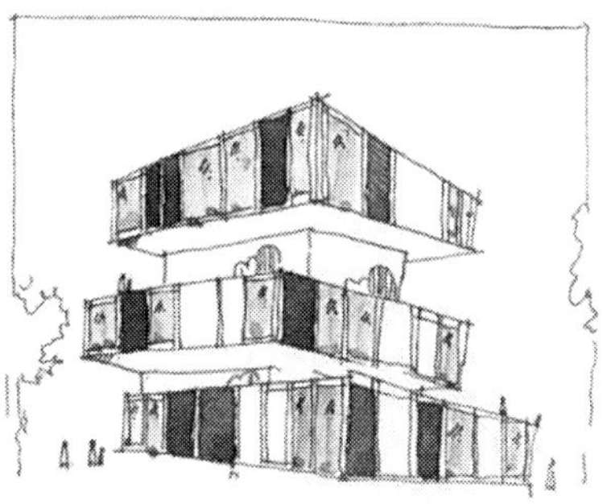

Figure 6: Façade integrated PVs in coherence with the building design logic (e.g. Dark PVs fit the façade grids).

Façade integrated PVs can present symbolic conceptions in harmony with culture and sense of nature through geometry and colour factors. Geometry and details inspiration could be learnt from façade design of Jean Nouvel's Arab World Institute in Paris. In a conceptual aspect, cultivating laypersons' aesthetic preferences can be realized through PV modules' geometry, materiality and detail factors. For example, the semi-transparent colour PV façades in the SwissTech Convention Center and the greenish metallic textured PV shadings of Kingsgate House in London are good examples to promote public's awareness of the aesthetic of PVs.

5 DISCUSSION AND CONCLUSION

Aesthetic evaluation is essential for design façade integrated PVs, especially in urban context with high density. Based on systematic literature review, this study extracted important key factors influencing the perceived aesthetic performance of façade integrated PVs, a set of evaluation criteria at both building and urban levels were also proposed from an environmental aesthetic perspective. The study findings could serve as basic reference for FIPV design strategies development for architects, urban designers or scholars in related fields.

Some key factors could be further investigated in future studies, for instance, inside the factor group materiality, **colour** is one of the most important aesthetic factors. As Kevin Lynch stated in *the image of the city* [25], colour is significant in creating the sense of place together with geometry or arrangement and plays an essential role in forming of a well-structured and identifiable mental images of city. Systematic methods combining both quantitative and qualitative colour analysis are desired other than literature review. **Texture and surface glossiness** are also important factors that deserve further attention due to limited research efforts and regulation about them. Most PV products are using glossy surfacing finishing, and many present inhomogeneous textures of PV cells, these effects all influencing the aesthetic perception of façade integrated PVs. Systematic studies exploring the intertwining relationship between colour, texture and glossiness are needed.

6 ACKNOWLEDGEMENTS

This work was performed within The Norwegian Research Center for Sustainable Solar Cell Technology (FME SUSOLTECH, project number 257639/E20). The center is co-sponsored by the Research Council of Norway and its research and industry partners. The authors also express great gratitude to Birgit Cold, who has provided inspiration of illustrations and valuable theoretical suggestions for this study.

REFERENCE

[1] K. Farkas *et al.*, "Designing photovoltaic system for architectural integration- Criteria and guidelines for product and system developers," (C.Farkas) Report T.41. A.3/2: IEA SHC Task 41 Solar Energy and Architecture, 2013.

[2] F. Frontini, M. Manfren, and L. C. Tagliabue, "A Case Study of Solar Technologies Adoption: Criteria

for BIPV Integration in Sensitive Built Environment," *Energy Procedia*, vol. 30, pp. 1006–1015, Jan. 2012.
[3] M. C. Munari Probst and C. Roecker, "Urban Acceptability of Building Integrated Solar Systems: LESO-QSV Approach," in *Proceedings of the ISES Solar World Congress 2011*, 2011, pp. 1–9.
[4] K. Farkas, I. Andersen, and A. G. Hestnes, "Architectural integration of photowoltaic cells. Overview of materials and products from an architectural point of view," in *3rd International Conference on Smart and Sustainable Built Environments (SASBE 2009)*, 2009, no. 2, p. 7.
[5] B. Cold, *Aesthetics, well-being, and health : essays within architecture and environmental aesthetics*. Aldershot : Ashgate, 2001.
[6] N. H. Perkins and R. D. Brown, "Environmental aesthetics," in *Environmental Geology. Encyclopedia of Earth Science.*, Dordrecht: Springer, Dordrecht, 1999, pp. 194–195.
[7] J. Appleton, *The experience of landscape*. Chichester: Wiley, 1975.
[8] D. Canter, "Understanding, assessing and acting in places: Is an integrative framework possible?," in *Environment, cognition, and action : an integrated approach*, 1st ed., T. Gärling and G. W. (Gary W. Evans, Eds. New York: Oxford University Press, 1991, pp. 191–209.
[9] R. Küller, "Environmental assessment from a neuro-psychological perspective," in *Environment, cognition, and action : an integrated approach*, T. Gärling and G. W. (Gary W. Evans, Eds. New York: Oxford University Press, 1991, pp. 111–147.
[10] S. Kaplan, "Aesthetics, Affect, and Cognition," *Environ. Behav.*, vol. 19, no. 1, pp. 3–32, Jan. 1987.
[11] K. Devlin and J. L. Nasar, "The beauty and the beast: Some preliminary comparisons of 'high' versus 'popular' residential architecture and public versus architect judgments of same," *J. Environ. Psychol.*, vol. 9, no. 4, pp. 333–344, Dec. 1989.
[12] K. Devlin, "an Examination of Architectural Interpretation: Architects Versus Non-Architects," *J. Archit. Plann. Res.*, vol. 7, no. 3, pp. 235–244, 1990.
[13] R. Gifford, D. W. Hine, W. Muller-Clemm, and K. . Shaw, "Why Architects and Laypersons Judge Buildings Differently," *J. Archit. Plann. Res.*, vol. 19, no. 2, pp. 131–148, 2002.
[14] J. L. Nasar, "The evaluative image of places," in *W. B. Walsh, K. H. Craik, & R. H. Price (Eds.), Person-environment psychology: new directions and perspectives*, 2nd ed., Mahwah, NJ, US: L. Erlbaum, 2000.
[15] B. Cold, "A Visual Journey into environmental aesthetics," in *Raw: Architectural Engagements with Nature.*, New York: Ashgate Publishing, 2014, pp. 47–62.
[16] H. Sotoudeh and W. M. Z. W. Abdullah, "Evaluation of fitness of design in urban historical context: From the perspectives of residents," *Front. Archit. Res.*, vol. 2, no. 1, pp. 85–93, Mar. 2013.
[17] M. C. Munari Probst and C. Roecker, "Solar energy promotion & urban context protection : LESO-QSV (Quality-Site-Visibility)method," in *31th International PLEA Conference*, 2015.
[18] P. Femenias, L. Thuvander, A. Gustafsson, S. Park, and P. Kovacs, "Improving the market up-take of energy producing solar shading: A communication model to discuss preferences for architectural integration across different professions," in *9th Nordic Conference on Construction Economics and Organization*, 2017, pp. 1–12.
[19] J.-H. Kim, M.-H. Lee, and S.-H. Han, "Development of the Evaluation Questionnaire for Facial Aesthetics of the Building Facade," *Int. J. Adv. Mech. Civ. Eng.*, no. 5, pp. 2394–2827, 2018.
[20] N. J. Habraken and J. Teicher, *The structure of the ordinary : form and control in the built environment*. M.I.T. Press, 2000.
[21] Y. Cuperus, "AN INTRODUCTION TO OPEN BUILDING," 2000.
[22] J. L. Caivano, "Cesia: its relation to color in terms of the trichromatic theory," *Farbe*, vol. 42, no. 1, pp. 51–64, 1996.
[23] J. L. Caivano, "Cesia: A system of visual signs complementing color," *Color Res. Appl.*, vol. 16, no. 4, pp. 258–268, 1991.
[24] M. C. Munari Probst and C. Roecker, "Criteria and policies to master the visual impact of solar systems in urban environments: The LESO-QSV method," *Sol. Energy*, vol. 184, pp. 672–687, May 2019.
[25] K. Lynch, *The image of the city*. MIT Press, 1960.

A STATE OF THE ART OF DESIGN CRITERIA FOR FAÇADE INTEGRATED PHOTOVOLTAICS

Changying Xiang[1,*], Barbara Szybinska Matusiak [1]
[1,*] Department of Architecture and Technology, Norwegian University of Science and Technology, Trondheim, Norway, Email: changying.xiang@ntnu.no
[1] Department of Architecture and Technology, Norwegian University of Science and Technology, Trondheim, Norway, Email: barbara.matusiak@ntnu.no

ABSTRACT: The paper presents state of the art of design criteria for façade integrated photovoltaics (FIPV). A comprehensive literature review was conducted based on a large number of English-language journal papers, conference proceedings and scientific reports, mainly from recent 10 years. The study showed that there is an obvious lack of study on design criteria specific for façade integrated photovoltaics. Current design criteria for building façade integrated photovoltaics (BIPV) were analyzed and categorized according to the scale: building level and urban context. In the building level, the AIQ-model and criteria from the project IEA SHC T.41.A.2 provide the most applicable set of criteria. In the urban scope, LESO-QVS method is one of the most advanced methods for application in current stage. It is necessary to develop systematic design criteria covering both building and urban level for façade integrated PVs. Besides, criteria should take into consideration opinions from many groups involving in developing and use of buildings, i.e. architects, authorities and also end-users. Furthermore, a corresponding application tool with rating or evaluation system should be developed for the design criteria system.
Keywords: FIPV, design criteria, urban context, architectural perspective, state of the art

1 INTRODUCTION

EU has set ambitious goals to reduce annual CO_2 emissions by 80% until 2050, which requires large scale implementation of various kind of clean energy systems. Building integrated photovoltaics (BIPVs), a promising strategy to employ clean energy in building environment, has attracted strong interest of researchers and industry since its beginning. With the rapid development of PV technology, many previous barriers like high cost, long payback time and lower efficiency have been solved or will be overcamed in foreseeable future. In last decades, most of the BIPVs related projects and research are focusing on building levels and especially on application on roofs. However, the total available roof area is limited and often not able to yield enough electricity for domestic consumption [1], which is evident in high-density urban areas. Therefore, there is an increasing demand to integrate PVs on building façades as well. Compared with roof integrated PV systems, Façade Integrated photovoltaics (FIPVs) have greater visibility and require careful architectural integration for good performance in functional, constructive, and formal (aesthetic) aspects [2], [3]. To develop advanced façade integration methods for PVs, methodical criteria for design and evaluation are necessary. This research study investigated the state of art of design criteria for façade integrated PVs. The study result could serve to develop design criteria for building façade integrated photovoltaic in urban context from an architectural point of view.

2 RESEARCH QUESTIONS AND METHODOLOGY

2.1 Research questions

This study raised three research questions for investigation: 1) What design criteria related with façade integrated PVs have been developed in the recent 10 years? (Identify the most advanced related design criteria); 2) What are their focuses and are there any common ones? (Categorize the identified criteria into different groups); 3) What evaluation methods have been employed by current design criteria? (Application methods)

2.2 Methodology

The methodology applied in this study was English-language literature review. A systematic review has been done through examining design criteria closely related to façade integration. Since the use of PVs on building facades started about ten years ago, the review is limited to literature from the recent 10 years (2009-2019). The first round of literature study was done by screening the databases *ScienceDirect*, *ResearchGate* and *Google Scholar* databases. Keywords including *design criteria BIPV, façade PV AND design* were used. The second-round literature data collection was done through 'snow-ball' search based on information obtained from the first-round search. Obtained data was filtered through abstract examining, and then selected articles were analysed and represented in a structured way.

3 DESIGN CRITERIA FOR FAÇADE INTEGRATED PVS

Studies specifically addressing design criteria for façade integrated PV are rare, therefore, literature about design criteria for BIPV which covers façade integration aspect were also analyzed. Literatures were categorized into two main groups (building level and urban level) according to their study scope.

3.1 Design criteria on the building level

Farkas [4] investigated how the form and symbolic characteristics are perceived by Norwegian architects and architectural students. Ten international public building projects with BIPV have been selected as evaluation samples, three methods are used in her study: 1-2-3 method with onsite visiting, semi-structured group discussion method and paper-based evaluation questionnaire method. The employed 1-2-3 method is an environmental aesthetic evaluation method including 3 steps: first with immediate perception then secondly with cognitive evaluation and the third step is final assessment

[5]. Based on the investigation, Farkas proposed a series of suggestions for designing BIPV covering 5 aspects: overall building design, building concept, structure, PV design and added value (Table I).

Table I: Suggestions for architectural BIPV design, adapted from [4].

Aspects	Appreciated	Characteristics
Building overall design	-silent, elegant, clear. -bold but honest	-Overall design
Building concept	-formally integrated. -compositionally integrated.	-Field size and positioning. -Composition of facade
Structure	-structurally well-integrated	-Framing. -Detailing
PV design	-present a variety of possibilities. -express material characteristics. -discrete, subtle design (PV not immediately recognized)	-Colour. -Size. -Texture of PV
Added value	-PV as ornament. -educate the public	-Artistic added value. -Symbolic added value

IEA PVPS (Photovoltaic Power Systems Programme) is one of the collaborative research and design agreements established within the IEA (International Energy Agency) and, since its establishment in 1993, the PVPS participants have been conducting a variety of joint projects in the application of photovoltaic conversion of solar energy into electricity[6]. IEA PVPS Task 7 [7] is one of the earliest tasks in IEA PVPS. The objective of Task 7 is to promote the architectural quality, the technical quality and the economic viability of PV systems in the built environment. A basic framework of good PV architecture was presented:

Naturally integrated. The PV system is a natural part of the building. Without PV, the building would be lacking something - the PV system completes the building.

Architectural pleasing. Based on a good design, the PV system add eye-catching features to the design.

Good composition. The colour and texture of the PV system should be in harmony with the other materials. Often, also a specific design of the PV system can be aimed at (e.g., frameless vs. framed modules);

Grid, harmony and composition. The sizing of the PV system matches the sizing and grid of the building.

Contextuality. The total image of a building should be in harmony with the PV system. On a historic building, tiles or slates will probably fit better than large glass modules.

Well-engineered. elegance of design details. Have details been well conceived? Has the amount of materials been minimized? Are details convincing?

Innovative new design. PV is an innovative technology, asking for innovative, creative, thinking of architects. New ideas can enhance the PV market and add value to buildings.

Similar to IEA PVPS, IEA SHC established in 1977, is one of the first programs of the International Energy Agency, to promote the use of all aspects of solar thermal energy[8]. IEA SHC Task 41 "Solar Energy and Architecture" was an international cooperative research program carried out in the period of 2009 to 2012. Aiming to promote the implementation of solar energy systems within high quality architecture and to reduce fossil energy use and GHG emission, a sub-project 'Subtask A: Criteria for Architectural Integration' of IEA SHC Task 41 was conducted by architects, researchers and educators from 14 countries [9]. The project outcomes included two publications, which were closely related with this study topic:

- *T.41.A.3/2 Designing Photovoltaic Systems for Architectural Integration*
- *T.41.A.2: Solar Energy Systems in Architecture - Integration Criteria and Guidelines.*

In the publication '*T.41.A.3/2 Designing Photovoltaic Systems for Architectural Integration*', a series of criteria and guidelines were proposed for PV product and system manufacturers/developers [10]. A definition of 'architectural integration' was also made as holistic integration including three basic architectural levels: functional, constructive, and formal. PV modules or PV products were viewed as part of building façade materials like concrete, bricks or wood other than "technically added elements". Therefore, taking into consideration functional and constructive aspects, BIPVs can function as the replacement for building envelope materials (i.e.: meet thermal and mechanical resistance demand) while producing electricity. In formal (aesthetic) aspect, key characteristics affecting integration quality were summarized: 1) Size and position of module field; 2) Module material and surface texture; 3) Absorber colour; 4) Module size and shape; 5) Type of jointing. All these key characteristics should be designed coherently with the general building design logic for formal integration. More detailly, a concept of three-levels of integration, namely basic level, medium level and advanced level was presented [10]. The publication '*T.41.A.2: Solar Energy Systems in Architecture - Integration Criteria and Guidelines*' provided architects with summarized knowledge of integrating solar thermal and photovoltaics into buildings, solving architectural integration issues and energy production requirement simultaneously. Integration criteria were given from both function/constructive aspect and formal aspect (aesthetic) aspect for a controlled and coherent integration from all points of view [3]. From function/constructive aspect, the integrated PV modules should meet compatibility function, fulfilling one or more envelop functions and meet all building construction standards. From formal aspect: 1) the position and dimension of PV systems should be coherent with the whole building's architectural composition; 2) visible materials, surface textures and colours of PV systems should be coherent with the materials, colours and textures of other building envelope parts; 3) PV module size and shape should be compatible with the building composition grid and with the dimensions of the other façade elements; 4) Jointing types should be chosen carefully.

A study addressed issues of BIPV in architectural design in China [11]showed that the function, cost, technology and aesthetics should be considered simultaneously for BIPV design. Besides, the maintenance and replacement of photovoltaic cell modules should also be taken into consideration during BIPV design. According to the authors, current BIPV products usually have around 25 years power warranty in Chinese market, which is only half of the lifetime of normal civil buildings in China. Furthermore, the rapid development of PV technology will provide more advanced BIPV products before the old products cease function. Therefore, it is necessary to emphasize the easy possibility of maintaining and replacing the BIPV products during the architectural design phase. However, no detailed criteria are presented in this study.

Yoo and Choi [12] proposed a version of ecological criteria to optimize solar architecture through an ecological convergence of a passive solar architecture and photovoltaic system. Convergence means combining passive architecture strategies with active architecture design (BIPV) together. The presented ecological design criteria of BIPV consisted of eight categories: solar radiation, energy, aesthetic view, indoor environmental quality, maintenance, practical use of structure, safety and environment friendly. However, some of the criteria could be difficult to employ during design phase, like safety or environment friendly.

Aiming to promote the deployment of PV shading in buildings markets, Femenias et al. [13] presented a statement of AIQ(Architectural Integration Qualities)-model for experts from different disciplines to discuss the aesthetic preference in architectural integration. This triangle-shaped AIQ-model presented three aesthetic characters of PV shadings: details, geometry and materiality. Detail is assessed by joints, moveable structure and attachment aspects. Geometry is evaluated in rhythm and shape while materiality is rated in aspects of texture, visual materials and colour. The overall integration of each character can be evaluated in three levels (poor, OK and good). Addition to the triangle system, visibility of PV shadings is assessed separately as 'tangibility' and rated in three levels (invisible, visible and dominant), whether high visibility is a positive value will depend on urban context or clients' preference. The research result showed that the AIQ-model serves well to support interdisciplinary discussion among professionals.

A summary of BIPV criteria on building level is generated into a table form to list the essential information from above discussed literature (Table II)

Table II: Summary of BIPV criteria in building level.

Authors and year	Integrated degree	Serving groups	Criteria aspects
Farkas 2011	General	Architect	-Building overall design. -Building concept. -Structure. -PV design. -Added value
IEA PVPS Task 7 Schoen et al. 2001	General	Architect	-Naturally integrated. -Architectural pleasing. -Good composition. -Grid, harmony and composition. -Contextuality. -Well-engineered. -Innovative new design.
IEA SHC T.41.A.3/2 Farkas et al. 2013	Basic level. Medium level. Advanced level	Manufacturer or developers	Ensure formal flexibility of the module. Provide non-active elements. Provide a complete façade system.
IEA SHC T.41.A.2 Frontini et al. 2012	General	Architect	-Position and dimension of PV systems- -Visible materials, surface textures and colours of PV systems. -PV module size and shape- -Jointing types.
Peng, Huang, and Wu 2011	General	Architect	-Function. -Cost. -Technology. -Aesthetics. -Maintance and replacement
Yoo and Choi 2017	With corresponding rating scores	Architect	-Solar radiation. -Energy. -Aesthetic view. -Indoor environmental quality. -Maintenance. -Practical use of structure. -Safety and environment friendly
Femenias et al. 2017	Three level rating	Manufacturer/ Architect / Daylight designer/ Clients	-Details. -Geometry. -Materiality. -Tangibility

Different literatures contributed design criteria in various aspects, most of them covers a wide range. However, detailed criteria from aesthetic perspective were only found from the study by Femenias et al. [13] and the project IEA SHC T.41.A.2 [3]. Other criteria are difficult to use as guideline due to their blur or undetailed aspects. For example, the criteria like 'PV design', 'architectural pleasing' or 'aesthetic view' are too abstract for users. Another common aspect of the above listed criteria set is that they are mainly aiming to serve architects or professionals, clients and users are barely considered. Additionally, many of them lack a corresponding integration rating/ranking system for the criteria.

3.1 Design criteria on the urban level

A few research have considered the integration criteria in larger urban scope, although most of them still focusing on building level, the presented criteria can be used as valuable references.

Applying BIPVs in historical urban context can be challenging due to the complex restrictions from architectural preserve regulations and the need to be coherent with historical urban image. Bonomo and Berardinis [14] proposed BIPV paradigms for the building renovation in historical contexts. The authors stated that in many cases of well-preserved historical built environment, PVs were excluded as an intervention method (e.g., city centre of Florence). BIPV intervention strategies should first concern the urban context then address to building envelope level. From semantic perspective, the authors defined BIPV design strategies according to different linguistic recognizability degrees of photovoltaics, ranging from not perceivable to the level of the "showy" integration:

Submission level. The presence of the PV system is perceived as an element deliberately alien with respect to morphology and language of building;

Evident level. The PV system is dominant in linguistically and expressive reading of the building envelope;

Integrated level. The PV language comes into equilibrium with the global perception of the building;

Moderately recognizable level. The integration is subordinated to the dominant language of the envelope;

Mimicry. BIPV system is apparently like a traditional envelope language.

In the project "Multielement II" at the Fraunhofer Institute for Wind Energy and Energy System Technology (in Kassel, Germany) a holistic evaluation system for comparison of conventional façade systems with BIPV façade systems was established[15]. The evaluation system consisted of various criteria from technical, economical, sustainability, and design aspects. The evaluation results could be used to support the certification of BIPV façade systems in the German framework of sustainable building certification systems DGNB. An evaluation method has been developed to guarantee that evaluations of different projects can be compared, both the importance of main criteria and sub-criteria can be assessed in 5-grades.

Munari Probst and Christian Roecker developed a three-step evaluation method LESO-QVS for assisting authorities to preserve the quality of pre - existing urban areas when applying solar systems [16]. All formal characteristics of the solar system (field size/position; visible materials; surface textures; colours; module shape/size; joints) were grouped into three categories (System geometry, System materiality, and System details) for coherency evaluation. Each category will be assess qualitatively using three-step scaling: fully-partially-not coherent (Figure 1-2).

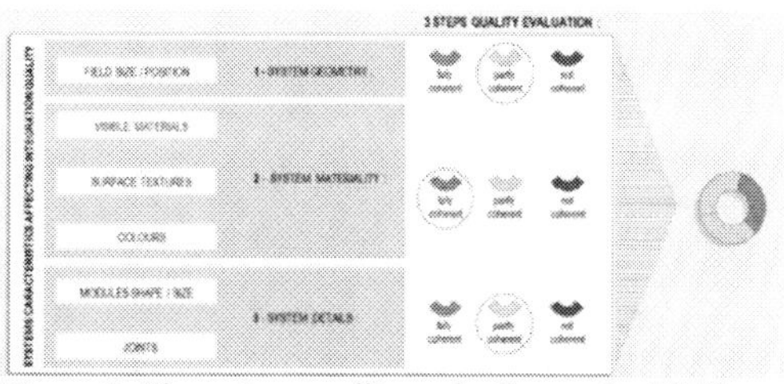

Figure 1: Three steps quality evaluation method [16].

Figure 2: Different levels of "System geometry" coherency[17]

The application of LESO-QSV method was based on the concept of 'architectural criticity'. A building sufaces' criticity level was defined by combination of sensitity level of the urban context and surface's visibility level from public domain (Figure 3). The higher the criticity level, the higher integration was required for solar energy system integration. A LESO-QSV grid software tool [18] with qualitative assessment method was further developed to support the decision making for municipalities to install solar energy systems in the urban context.

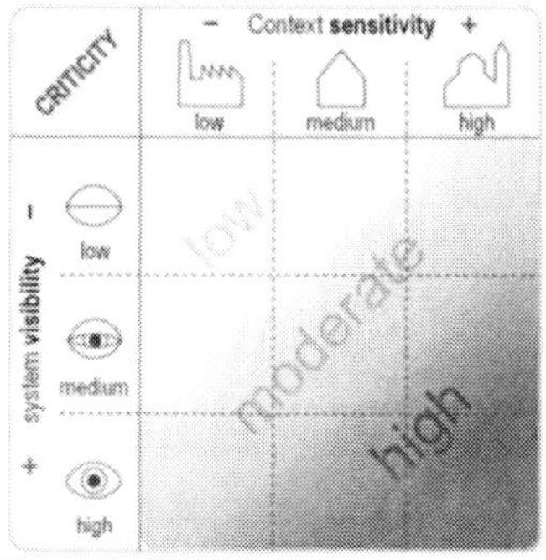

Figure 3: Concept of Criticiy[16].

A summary of BIPV criteria with consideration of urban context was generated into a table form to list the essential information from above discussed literature (see table III).

Table III: Summary of BIPV criteria include urban context consideration.

Authors and year	Integrated degree	Serving groups	Criteria aspects
Bonomo and Berardinis 2013	- Submission level - Evident level - Integrated level - Moderately recognizable level - Mimicry	Architect	-PV system is perceived as an element deliberately alien with respect to morphology and language of building; -PV system is dominant in linguistically; -PV language comes into equilibrium with the global perception of the building; -The integration is subordinated to the dominant language of the envelope; -BIPV system is apparently like a traditional envelope language.
Schuetze, Willkomm, and Roos 2015	5-grade rating	Not specified	Contemporary design; Corporate design; Functional design; Design reference to build environment; Design reference to other building components; Flexibility of dimensions; Flexibility by color variability; Flexibility by material variability; Flexibility by energy flow variability; Flexibility by functional variability
Munari Probst, Roecker and Florio 2015, 2019	3-step rating	Decision-making authority	Field size/position-coherent; Visible materials- coherent; Surface textures- coherent; Colours- coherent; Module shape/size coherent; Joints --coherent.

The three listed design criteria systems for BIPV on the urban level have an integration rating/ ranking system. In each criteria system, detailed criteria in architectural perspective are provided. The criteria system provided by Munari Probst, Roecker and Florio[16], [17] was especially practical to employ with its clear coherent suggestions on different characters of PVs.

4 DISCUSSION AND CONCLUSION

This study investigated the state of the art of design criteria for façade integrated PVs. Based on literature review of English literature in recent 10 years, it showed that current studies addressing specifically integration on facade is limited, there is a clear need to investigate design criteria for façade from architectural perspective. On the building level, the AIQ-model and criteria from project IEA SHC T.41.A.2 are among the most applicable criteria systems. On the urban level, with clear assessment 'grammar' and rating method, the LESO-QVS method is one of the most systematic methods currently.

For the further research, it is necessary to develop systematic design criteria covering both building and urban level for façade integrated PVs, as facades are not perceived isolated in urban context. Findings from the listed literature could be synthesized together and improved to serve this goal. The AIQ-model by Femenias et al. [13] provide a criteria system including 'Details', 'Geometry', 'Materiality' and 'Tangibility' on the building level while the LESO-QVS method presents characters including 'Field size', 'Visible materials', 'Surface textures', 'Colours', 'Module shape/size' and 'Joints'. These two complementary systems provide a potential basis for a holistic design criterion covering both, the building and the urban level.

Another aspect important for improving design criteria system is to involve clients' participation in defining criteria in early design stage. Most of current criteria are generated from professionals' perspective, opinions from clients could also be included to improve the criteria system. Study [19] shows that the potential clients' understanding of environmental and economic benefits of BIPV can play a significant role in the adaptation, a proper communication will promote the acceptance of employing BIPV. The AIQ-model by Femenias et al. [13] shed light in a positive communicative way for façade PV evaluation, more advanced client-participation method could be further involved in the criteria development process.

Design criteria should be combined with application tools with a rating or an evaluation system, which can work together with design criteria as a powerful tool for users. Currently there is a severe lack of suitable tool to support architects or professional for BIPV design. Kanters, Dubois and Wall[20] investigated the design process of solar-integrated buildings among Swedish architects from Stockholm, Gothenburg and Malmö. Most of interviewed architects expressed that there is a limited choice of attractive solar products on the market to apply for design and prefer to have solar products with higher degree of freedom in changing colour, size and shape. The main study result showed that architects were still using traditional design tools like hand sketches, CAAD program and physical model for designing solar-integrated buildings. In the framework of IEA SHC task 41, a web-based international survey conducted to explore architects' demands also showed that there is a commonly lack of solar application design tools to support architects' decision making in early design phases[21]. Similarly, Farkas[22] emphasizes that there was an urgent demand to provide architects with simple (compatible with already used architectural tools) computer tool supporting BIPV integration design in early design phase. Therefore, it is necessary to develop an application tool with rating or evaluation system which can work with the design criteria corporately. The rating systems of AIQ-model and LESO-QVS method are very promising to be employed for further development.

5 ACKNOWLEDGEMENTS

This work was performed within The Norwegian Research Center for Sustainable Solar Cell Technology (FME SUSOLTECH, project number 257639/E20). The center is co-sponsored by the Research Council of Norway and its research and industry partners. The authors also express great gratitude to Tore Brandstveit Haugen and Carmel Margaret Lindkvist at NTNU, who have provided valuable suggestions for this study.

REFERENCES

[1] M. C. Brito, S. Freitas, S. Guimarães, C. Catita, and P. Redweik, "The importance of facades for the solar PV potential of a Mediterranean city using LiDAR data," *Renew. Energy*, vol. 111, pp. 85–94, Oct. 2017.

[2] M. C. Munari Probst and C. Roecker, *Architectural integration and design of solar thermal systems*, 1st ed. Lausanne: EPFL Press, 2011.

[3] F. Frontini *et al.*, "Solar Energy System in Architecture-Integration Criteria and Guidelines," (MC Munari Probst, C Roecker)Report T.41.A.2: IEA SHC Task 41 Solar energy and Architecture, 2012.

[4] K. Farkas, "Perception of Formal and Symbolic Aesthetics of Photovoltaics," in *ISES Solar World Congress 2011*, 2011, pp. 1–12.

[5] B. Cold, "Metode for opplevelse og bevisstgjøring av arkitektur og steder," 1998.

[6] International Energy Agency, "IEA Photovoltaic Power Systems Programme," 2019. [Online]. Available: http://www.iea-pvps.org/index.php?id=4. [Accessed: 29-May-2019].

[7] T. Schoen, D. Prasad, D. Ruoss, P. Eiffert, and H. Sørensen, "Task 7 of the IEA PV Power Systems Program Achievements and outlook," in *17th European Photovoltaic Solar Conference*, 2001, pp. 5–8.

[8] International Energy Agency SHC, "IEA Solar Heating & Cooling Technology Collaboration Programme," 2019. [Online]. Available: https://www.iea-shc.org/. [Accessed: 02-May-2019].

[9] M. Wall *et al.*, "Achieving Solar Energy in Architecture-IEA SHC Task 41," *Energy Procedia*, vol. 30, pp. 1250–1260, Jan. 2012.

[10] K. Farkas *et al.*, "Designing photovoltaic system for architectural integration- Criteria and guidelines for product and system developers," (C.Farkas) Report T.41. A.3/2: IEA SHC Task 41 Solar Energy and Architecture, 2013.

[11] C. Peng, Y. Huang, and Z. Wu, "Building-integrated photovoltaics (BIPV) in architectural design in China," *Energy Build.*, vol. 43, no. 12, pp. 3592–3598, Dec. 2011.

[12] S. Yoo and H. Choi, "Ecological design criteria of BIPV as a shading device for solar architecture Ecological design criteria of BIPV as a shading device," in *Advanced Building Skin 2017*, 2017, no. October.

[13] P. Femenias, L. Thuvander, A. Gustafsson, S. Park, and P. Kovacs, "Improving the market up-take of energy producing solar shading: A communication model to discuss preferences for architectural

integration across different professions," in *9th Nordic Conference on Construction Economics and Organization*, 2017, pp. 1–12.
[14] P. Bonomo and P. De Berardinis, "BIPV in the Refurbishment of Minor Historical Centres: The Project of Integrability between Standard and Customized Technology," *J. Civ. Eng. Archit.*, vol. 7, no. 9, pp. 1063–1079, 2013.
[15] T. Schuetze, W. Willkomm, and M. Roos, "Development of a Holistic Evaluation System for BIPV Façades," *Energies*, vol. 8, pp. 6135–6152, 2015.
[16] M. C. Munari Probst and C. Roecker, "Solar energy promotion & urban context protection : LESO-QSV (Quality-Site-Visibility)method," in *31th International PLEA Conference*, 2015.
[17] M. C. Munari Probst and C. Roecker, "Criteria and policies to master the visual impact of solar systems in urban environments: The LESO-QSV method," *Sol. Energy*, vol. 184, pp. 672–687, May 2019.
[18] P. Florio, ; C Roecker, and ; M C Munari Probst, "Urban acceptability of solar installations: LESO-QSV grid, a software toll to support municipalities CISBAT," in *CISBAT*, 2015, pp. 981–986.
[19] D. E. Attoye, T. O. Adekunle, K. A. T. Aoul, A. Hassan, and S. O. Attoye, "A conceptual framework for a Building Integrated Photovoltaics (BIPV) Educative-communication approach," *Sustain.*, vol. 10, no. 10, pp. 1–21, 2018.
[20] J. Kanters, M.-C. Dubois, and M. Wall, "Architects' design process in solar-integrated architecture in Sweden," *Archit. Sci. Rev.*, vol. 56, no. 2, pp. 141–151, May 2013.
[21] A. Scognamiglio, K. Farkas, F. Frontini, and L. Maturi, "Architectural quality and photovoltaic products," in *27th European Photovoltaic Solar Energy Conference and Exhibition ARCHITECTURAL*, 2012, pp. 4169–76.
[22] K. Farkas, M. C. M. Probst, and M. Horvat, "Barriers and Needs for Building Integration of Solar Thermal and Photovoltaics," in *EuroSun 2010*, 2010.

FLOATING PHOTOVOLTAICS: A SWOT ANALYSIS FOR THE EU PVSEC 2021

Authors: Shahab Moghadam[1], Hesam Eldin Hayati Soloot[2], Amir Hayati Soloot[3], Ayush Jadhav[4]
Solar Edition
Oslo, Norway

ABSTRACT: Floating Photovoltaics (FPV) is a Photovoltaic (PV) based energy generation approach on water surfaces, to mitigate the solar energy high land-use footprint issue. Since 2007, the technology and industry have come a long way with a cumulative installed capacity of 1.9 GWp by 2019. This article has studied and analyzed the current status and future directions of this technology with a SWOT analysis, by reviewing scientific articles and industrial reports. Additionally, technology evolutions are addressed such as moving FPV structures from man-made still-water reservoirs to near- and offshore open seas. The technology is gaining popularity due to its potentials such as resolving green energy land usage footprint, energy security, local geographically-free energy generation, innovative business models enabler, fast deployment, and providing a solution to socio-political and community acceptance issues. Additionally, the technology has extreme potentials to meet global electricity demand by only utilizing 51.5% of the global man-made reservoirs (excluding huge potentials of near-shore and offshore applications). Finally challenges and mitigation for topics such as lack of good practices and standards for material choice due to immature technology, possible fresh water source contamination and effects on eco-system are discussed. However, these are rapidly changing as the technology and industry matures.

Keywords: Floating PV, Offshore PV, PV System and Grid Connected, Distributed PV Plant, Photovoltaics

1 INTRODUCTION

95% of net increase in global power supply is to be from renewable sources by 2025, which includes Solar, Wind and Hydropower [13]. Main enablers for this shift and increase are climate change urgency and environmentally friendly, competitive prices and last but not least distributed generation characteristics of renewables [13]. In the scope of Solar Power, one of the new emerging technologies is Floating PV or FPV.

1.1 What is FPV?

Floating Photovoltaics (FPV) is a photovoltaics (PV) based energy generation approach on the water surface. In this approach, the PV modules are installed in groups of arrays on floating structures, capable of keeping the modules floating on the water body [1], [3], [4], [5].

1.2 Why FPV?

The technology is gaining popularity due to the solutions it provides to a series of challenges with renewable energies.

- Potential gigawatt scale energy source that could meet global electricity demands, only by utilizing 51.8 % of the global potentials of FPV technology worldwide 1(offshore and near-shore not included).
- Geographical freedom, which allows to implement projects globally close by energy demanding areas. This important feature of the technology would provide Local Energy Production, Increased Energy Security and opens for new opportunities in places where energy production has been a challenge (such as islands) [1].
- Resolving green energy spatial expansion & land usage footprint [1], [4], [5], [8], [9], utilizing water space that is otherwise left unused and prevent water evaporation [1], [2]
- An approach to meet socio-political and community acceptance issues in local areas. This is by improving the aesthetic impacts of renewable energy by moving the process out of the sight, especially in case of Offshore or Near-shore Application FPV [12], [37].

1.3 Types of FPV?

FPV technology types could vary based on installation locations

- Inland Natural & Artificial Freshwater Application (lakes, reservoirs, irrigation ponds, wastewater treatment plants, lagoons, wineries, fish farms, dams, canals [5], [17])
- Offshore or Near-shore Application (Open Seas, with typically higher waves and harsher water and weather conditions)

This article's scope is to analyze the FPV technology with the SWOT analysis method. Scientific articles and industrial reports on prominent technological advancements and real-world test results are reviewed and results are presented. The review work is covered in the background section, while the SWOT table is then presented in the SWOT Analysis Method section. This is followed by detailed descriptions of the points in the table. Finally, the results are presented in the Conclusion. This method is chosen not only to explore and study the technology but also to clearly list internal strengths and weaknesses together with external factors which result in opportunities, and threats. This is to give a clear picture of the current state of the technology, industry and nevertheless the market to open discussions and cover the gaps.

2 BACKGROUND

[1] Shahab Moghadam, Project Manager at Hitachi-ABB Powergrids, Co-Founder Solar Edition, Email: sm@solaredition.com

[2] Hesam Eldin Hayati Soloot, M.Sc. Physics, University of Qom, Renewable Energy Consultant, Solar Edition

[3] Amir Hayati Soloot, Ph.D. Electrical, Norwegian University of Science and Technology, Co-Founder, Solar Edition

[4] Ayush Jadhav, Graduate student at Savitribai Phule Pune University, Renewable Energy Consultant and Research Intern, Solar Edition

2.1 History

Since 2007 that the first small-scale FPV system, 20 kW, was installed in Aichi, Japan [1], [3], the technology and industry have come a long way with a cumulative installed capacity of 1.3 gigawatt-peak (GWp) by the end of 2018 [1]. The graphs in Figure 1 shows both FPV and ground-mounted PV rate of annual and cumulative installed capacity during 2007-2018 [1], [34]. The graph illustrates the accelerated growth pace of annual installed capacity for FPV after 2014.

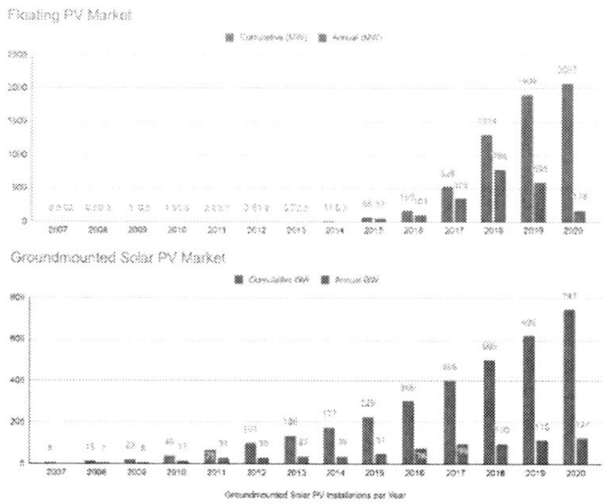

Figure 1 - Upper Graph, Global installed Floating PV capacity and annual installations [1, 50] — Lower Graph, Global installed Ground mounted PV capacity and annual installations [34], [46]

2.2 The Development Process

Despite the accelerated rise since 2014, this seems to be only the beginning of a new diversifying trend in the Solar Energy sector after rooftop PV and solar parks (solar farms), which has already entered utility-scale [34]. The global market capacity of FPV is estimated by the World Bank to be 400 GW, under conservative estimations [1], which is then only equal to 63% of world cumulative solar PV capacity by 2019 [34].

By the end of 2019, floating solar global cumulative capacity reached 1.9 GW [50] and some other market reports expect this number to reach a minimum 2.6 GW by the August 2020 [45]. On the other hand, Figure 1 shows that this growth only accounts for a small portion of annual PV market in 2019 and even less of global PV capacity. However, a closer look shows an extensive market growth[5] for FPV compared to ground mounted. Market reports also predicts an average 22% year over year growth from 2019 to 2024 for FPV market globally [2]. It is also important to mention that bad statistics for year 2020 could also reflect the COVID-19 effect on the market. Corona virus restrictions could potentially effect project executions globally.

Considering market potentials, demand and estimations, a rapid evolution is foreseen in next years, driven both by academia and industry. This evolution is necessary considering the small scale of the current FPV installations compared to ground mounted (shown in Figure 1). The small FPV portion of global solar capacity illustrates the knowledge gaps and industry maturity required which is already earned by the land-based PV industry in gigawatt scale project in recent years.

The development and research by both academia and industry should consider the FPV as a whole system, its impacts, electrical designs and suitable PV panels [10]. Additionally, it is important to consider areas such as knowledge transfer from other sectors with relevant experiences (land-based PV, Oil & Gas offshore industry, marine, water utility, etc), field experience by execution of MW / GW scale projects (ongoing in East Asia, followed by Europe), adoption of good practices (see T6) and last but not least, the development of sector specific standards (see T1). All of these are necessary for FPV to reach the same level of maturity achieved by ground-mounted PV installations over the last decade.

Examples of ongoing evolution in FPV industry are "Offshore Open-Sea test projects" [26][42], "stunning market record of daily 1 MW FPV installation" thanks to a novel floating structural design (among top 10 Intersolar Award Finalists 2020) [18] [19] and last but not least "growing projects size in the market" (Figure 2).

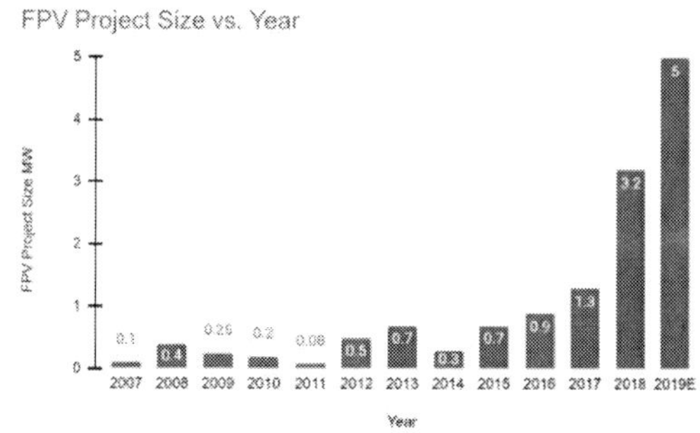

Figure 2 - Global Average of Project Size in Floating Solar PV Sector [2]

Table 1 is a walkthrough of milestones achieved by FPV technology. These milestones start from the first testbed in Aichi, Japan which was a noncommercial 20kWp plant. The last milestones on the table are the latest development in the FPV development process is the migration from manmade water reservoirs to open seas; Offshore or Near-shore Application.

[5] FPV cumulative global capacity has raised from 169 MW in 2016 [1] to estimated 2087 MW in 2020 [50]. For ground mounted this is 305 GW in 2016 and 747 GW in 2020 [34],[46]. This results in a growth rate of 12,34 times growth for FPV and 2,44 for ground mounted.

Table 1- Historical Perspective over the FPV technology achieved. Information in this table arrived from [1] and is updated with the recent technical Information and updates

Milestone	Location	Size	Date	Remarks
First FPV Installation	Aichi, Japan	20 kWp	2007	- Non-Commercial - To research on effects of water cooling on electricity output [20] - Prototype floating structure [20]
First commercial FPV Project	Far Niente Winery, USA	175 kWp	2008	- Standalone - Commercial - Save lands for agricultural use [20]
First Tracking FPV Project	Petra Winery, Italy	200 kWp	2010	- Standalone - Motorized Tracking and Rotating System - Metal Structure [20] - Use of mirrors to concentrate and reflect extra radiation on panels
First MegaWattscale FPV Project	Saitama Prefecture, Japan	1,180 kWp	2013	- Standalone - Non-metal structure made of High-density polyethylene (HDPE) [25] -Installed on a rainwater retention reservoir [25]
First FPV Project with Microinverters	Fukuoka Prefecture, Japan	300 kWp	2016	- Standalone - Installed on a farming reservoir [32] - Japan is estimated to have up to 190,000 man-made reservoirs, mostly created for farming irrigation purpose[32]
First Hybrid FPV-Hydro Power Project	Alto Rabag˜ao Dam, Portugal	218 kWp [25]	Nov. 2016 [25]	- Hybrid [25] - Installed on hydroelectric dam [25] - Challenging anchoring system designed for 60 m depth and for 30-m water level variation [25]
Largest FPV Project was connected to grid	Bengbu area of Anhui Province, China (Former Coal-mining lake)	70 005 kWp (70 MWp)	March 2019	- the largest FPV complex in the world - 194,731 PV panels of 60 cells type (335W and 360W modules) - The installation structure area covers more than 60 hectares - Bottom anchored, the solution is designed for max. 14 m depth. The system can withstand 3.9-m water level variation [36].
First Offshore solar farm in the open sea [26]	The Netherlands, North Sea	8.5 kW / 17 kW	2019/ 2020	- Offshore Standalone Plant with mooring system [26] - Designed and tested for 13m high waves [26] - Already passed the test with several storms including storm "Ciara" in Feb. 2020 (62 knots of maximum wind speeds, 5 meters maximum waves,2 knots sea currents) - The 8.5 kW platform consisted of 28 modules, which after passing initial tests was expanded to 17 kW with 56 modules in January 2020 [26]
EU funded offshore solar farm in Canary Islands	Canary Islands, Spain	250kWp	2020	European Union's Horizon 2020 research and innovation programme funded project to explore how FPV performance would be in challenging sea conditions with waves of up to 10 metres and high winds. The FPV technology for this project is Ocean Sun design with a thin polymer membrane that is used to mount PV modules, which is a water-cooled design [42][49].

2.3 Market Status and Demographic Distribution

FPV is an exciting market with high demand and great potential. Naturally, this technology is first welcomed in locations where the land is scarce, expensive or unsuitable for PV installations [1], [33]. Places such as islands and countries with high population density and limited land resources [1], [31], [2]. Currently almost 87% of the global FPV capacity is limited to Asia, with China as the leading country with South Korea and Japan following [47]. The market demand is accelerating and now is expanding this trend [33] as only Chinas Datang Power was tendering for a total capacity of 820 MW of FPV across China by December 2021 [47]. Another sign of this acceleration is also plans in South Korea to have FPV in their decarbonization plans by 2050. After the country came up in 2020 with a planned 2.1 GW FPV plant near the Saemangeum tidal flat, by the Yellow Sea [41], is now announcing yet another plan of 2.1 GW of FPV expansion by 2030 [48].

However the potential market does not end there, as the "local energy production" (geographical freedom) and "energy security" that FPV presents, open for many more market opportunities. Several European countries, such as Germany, Netherlands and France are opting for FPV projects for local energy production, achieving ambitious renewable energy targets and last but not least leading the newly emerging market [2]. Figure 3 shows the current distribution of this source of energy globally.

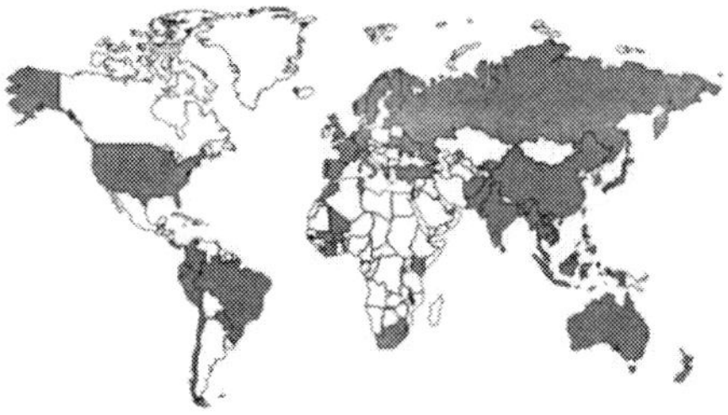

Figure 3 - Map of countries with Floating Photovoltaics Installations [47]

Currently 35 countries are hosting 350 operational floating power plants [2,47]. This is a 30% rise in the number of countries compared to 2018 when only 27 countries had FPV projects [2]. Currently market news suggests that over 60 countries are actively persuading Floating PV projects deployment [47], which is a proof for not only acceptability factor, but also acceleration in the market. It is believed that energy security and independence that solar PV technology and especially FPV presents is one of the key factors in the spread of solar power harvesting globally. The current global energy market is highly criticized, fragile, sensitive and reacts rapidly to interruptions and conflicts. Interruptions such as natural disasters, local conflicts (Hormoz bay, Red Sea pirates issue etc) or recent worldwide pandemic, COVID-19. All these turn the topic of local energy production and energy security critical and a part of the country's national security topic. Providing continuous uninterrupted energy to modern societies is one of the top objectives of governments after water and food supplies. Hence considering FPV; which can be installed in high density populated areas; would be a natural part of the critical energy mix considered at the highest level of policymaking.

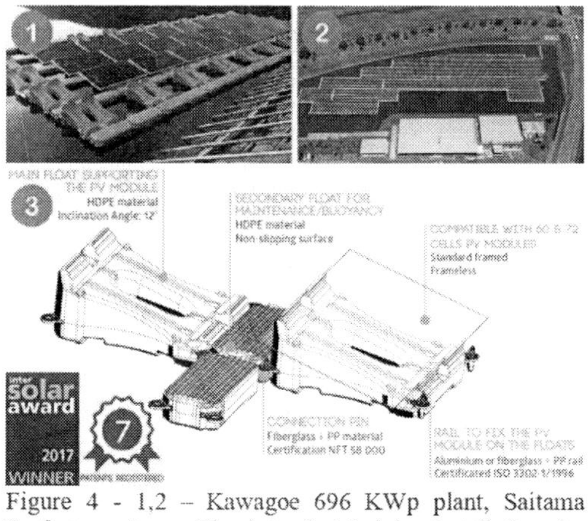

Figure 4 - 1,2 – Kawagoe 696 KWp plant, Saitama Prefecture, Japan, Fixed Angle Modular Installation, by French Ciel et Terre, 3 – Hydrelio floating structure by Ciel et Terre (images from Ciel et Terre) [51]

3 METHODOLOGY

During this section, the current status and future directions of FPV as a technology and solution is analyzed with a SWOT analysis, i.e., strengths, weaknesses, opportunities, and threats. SWOT is a useful tool to identify internal strengths and weaknesses of the topic, while exploring and identifying the opportunities and the threats caused by external factors [7], [54]. This is to illustrate the gaps in the state-of-the-art technology and start the discussion about them. For this, scientific articles and industrial reports on prominent technological advancements and real-world test results are reviewed and results are presented first in a pointwise SWOT analysis in table 2 and then later detailed out in each section. In addition, a visualized SWOT analysis illustration, Figure 9, is also prepared, which visualizes the discussions in the field.

3.1 Strengths, in Details

Waterbed Utilization:

S1. NO NEED FOR LAND

Avoiding the competition with other sectors over land [1], [4], [8], [9]; sectors such as food, residential and other industries. This results in higher chance of renewable energy project implementation (due to priorities), where land is scarce or simply wherever there is a high demand for land (islands / island countries) [33].

S2. USE OF VACANT ONWATER SPACES

Turning vacant water reservoir with high solar energy potentials into a virtual power plant producing green energy [1], [7].

S3. CONTRACTING SIMPLICITY WITH SINGLE STAKEHOLDER

Utility-scale solar power plants require vast areas which are usually owned by several stakeholders, while waterbeds are usually owned by a single stakeholder. This brings simplicity of contracting due to the single ownership of water bodies compared to lands for large-scale projects [1].

S4. LOWERING FRESH WATER EVAPORATION

Blocking direct sun radiation and providing shading for water surfaces, prevents evaporation of valuable freshwater that can be used as drinking water or for agricultural use [5], [11]. In this sector there are huge potentials to prevent around 74 billion m3 of freshwater evaporation [11].

Table 2 - Floating PV SWOT Analysis

Strength	Weakness
Waterbed Utilization S1. No Need for Land S2. Use of Vacant on-water Spaces S3. Contracting Simplicity with Single Stakeholder S4. Lowering Fresh Water Evaporation S5. Control of Algae Growth S6. Added Revenue S7. Easier Cleaning & Less Dust Soiling Effect S8. Lower Chances for Vandalism due to Access Level S9. Landscape Aesthetics Improvement (Out of Sight Strategy) **Economical** S10. Similar LCOE to Ground-Mounted Plants after certain project size (50 MW) **Energy Generation & Power Systems** S11. An energy source with potentials to meet World Electricity Demand S12. Geographical freedom & local energy production (availability) S13. Overall Higher Power Yield S14. Less Transmission Loss & Use of Current Infrastructure S15. Natural Cooling Effect (Water Cooling & Evaporative Cooling) S16. Less Shading & Surrounding Obstacles S17. Albedo Effect (although low) **Plant Design, Installation, and Maintenance** S18. Modular Design & Lack of Need for Heavy Machinery Compare to Competitors S19. Easy and Rapid Engineering, Deployment, and Installation **Environmental** S20. Less Heat Island Effect S21. Not impacting the wildlife (to our knowledge)	**Waterbed Utilization** W1. Young Immature Technology W2. CHALLENGES WITH HIGH WIND SPEED AND WAVES W3. Aesthetic Disturbance, if not Moved Out of Sight W4. Salt & Corrosion **Plant Design, Installation and Maintenance** W5. High CapEx W6. Challenging & Costly Design W7. High OpEx costs W8. High Soft Costs W9. Challenging Maintenance W10. Animal Shelter & Dirts Resulted by it W11. Wind Resistance Effect on Tilt Angle W12. Lifetime Challenges & durability **Environmental:** W14. UV corrosion
Opportunitets	Threats
Waterbed Utilization O1. LAND THIRST & SCARCITY AS MARKET BOOSTER O2. Investment & Project Scales Lowers Costs O3. INNOVATIVE FLOATING BUSINESS MODELS **Energy Generation & Power Systems** O4. Energy Demand, Local Production & Energy Security (Political aspect) O5. CO2 FOOTPRINT OPTIMIZER O6. Attractive Hybrid Solutions Coupled with Hydro Power & use of current transmission infrastructure **Plant Design, Installation and maintenance** O7. Extensive Existing Offshore Know-HOW Environmental: O8. Reuse of Current Mature Standards	**Plant Design, Installation and Maintenance** T1. Lack of Standards & Good Practices T2. Water Contamination (due to poor material choice) T3. Immature Structural Design **Environmental:** T4. Aesthetic Disturbance, Socio-Political & Social Acceptance T5. Long Term Ecological Effects T6. Lack of Environmental Standards

S5. CONTROL OF ALGAE GROWTH
Algae growth is reduced by shading the water. This could potentially improve the water quality [5], [6], [12].

S6. ADDED REVENUE
There are enormous potentials for "Hybrid Solutions", coupling already existing business models with FPV. This would result in added revenue to these sectors (less electricity bills / selling to grid). Some examples of such sectors

- Hydro-power [1], [31]
- aquaculture [1], [31]
- Farms
- Opencast lignite coal pits (e.g. China 70 MW FPV plant in a former coal-mining area of Anhui Province) [31]

S7. EASIER CLEANING & LESS DUST SOILING EFFECT
Easier access to water for cleaning is considered an advantage [1]. This could be taken to a new level by automated daily cleaning by splashing water over panels and use of innovative automatic cleaning methods with help of cleaning-robots [56]. In addition, on water beds, compared to ground-mounted plants, there is a less tendency for soiling issue from the source of dust in surrounding environment. As the panels are installed on water and mentioned in previous point, the cleaning medium is also easily available in contrast with typical solar farms in desert environments [1].

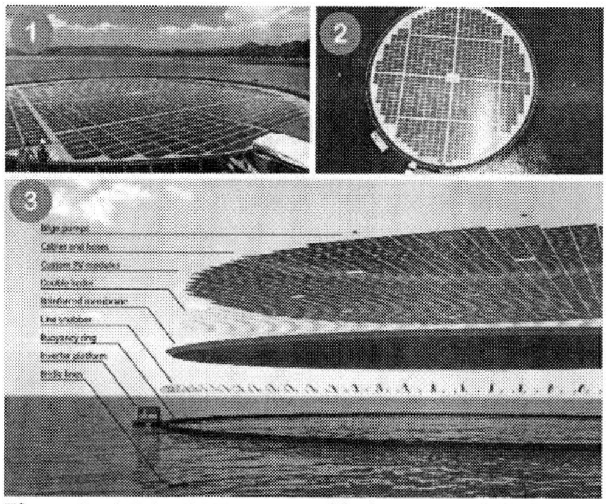

Figure 5 - 1,2,3 - DNV GL Certified Zero Angled Floating PV Structure [21], by Norwegian Ocean Sun. This solution is the candidate for South Korean 2.1 GW FPV Project [41] & ocean testbed in Canary Islands with up to 10m waves [42]

S8. LOWER CHANCES FOR VANDALISM AND SABOTAGE DUE TO ACCESS LEVEL
Due to the fact of equipment being on waterbeds, they are less accessible and therefore safer [1]. This reduces the chance of robbery and sabotage of the equipment.

S9. LANDSCAPE AESTHETICS IMPROVEMENT (OUT OF SIGHT STRATEGY)
- Violating landscape aesthetics is a challenging ongoing issue for onshore solar farms and wind parks. FPV technology is a good attempt in improving the issue on hand. The solution could vary based on distance from shore and visibility [12], [37]. As for the offshore FPV, the problem could be completely resolved if equipment is moved out of sight in the horizon and this is hopefully accessible both due to the profile characteristic of the panels and offshore technology under development.
- Blue color of the panels
- Sea based PV panels are generally easier to blend in the blue sea due to their low profile
- Shorter installations compared to offshore wind installations

These features hopefully could help resolving socio-political and community acceptance issues for FPV that renewable energies are facing in general.

Economical

S10. SIMILAR LCOE TO GROUND-MOUNTED PLANTS AFTER CERTAIN PROJECT SIZE
Based on a pretax calculation, the levelized cost of electricity (LCOE) of a 50 MW FPV plant is not significantly different from ground-mounted plants [1]. FPV plants have higher CapEx (initial invested capital), but this is blanched with higher energy yield from FPV plants [1], [17].

Energy Generation & Power Systems

S11. AN ENERGY RESOURCE WITH POTENTIALS TO MEET WORLD ELECTRICITY DEMAND
Man-made water reservoirs are found all around the world. These are built historically due to reasons such as flood prevention, seasonal water supply for drinking and irrigation, fishing, hydropower generation, energy storage, wastewater recycling etc [11], [17]. Studies suggest that there are 404,454 square km of man-made freshwater reservoirs available globally [1], a calculation and comparison with world electricity demand shows that utilizing 51.8% of this potential and turning it into FPV plants could meet world's electricity consumption.
- Utilizing 1% of potential = 521.109 TWh Energy production per year [1].
- Utilizing 51.5% of potential = 26993.44 TWh Energy production per year.
- World's annual electricity consumption in 2020 was 26823.2 TWh [16]

S12. GEOGRAPHICAL FREEDOM & LOCAL ENERGY PRODUCTION
Geographical freedom & local energy production is one of the very special features of solar energy and so FPV. Solar radiations are available source of energy globally, with different yields as it is climate specific. This also makes this point an external opportunity for FPV as well.

S13. OVERALL HIGHER POWER PRODUCTION
The characteristics of FPV system results in 5 - 10% more performance ratio (climate specific) compared to ground-mounted systems [1], [17], [27]. Field reports can go even higher, as for K-Water installations in South Korea point to 11% better yield [27] while other market reports point even up to 15% [1] higher electricity output.

S14. LESS TRANSMISSION LOSS & USE OF CURRENT INFRASTRUCTURE
Ability to have FPV power plants (energy generation) near populated areas (main consumers), opens for
- lowering the loss of energy in transmission over longer distances [1]
- utilizing the existing infrastructure to transmit the power from source to consumer [1], [2], compared to very remote areas, where it lacks transmission infrastructure or capacity.

S15. NATURAL COOLING EFFECT (WATER COOLING & EVAPORATIVE COOLING)
Depending on the design, water cooling & evaporative cooling effect on panels temperature (climate specific) [1],

[5], [10], [11], [15], [17], [27],[49] would eventually result in an increase in the power yield. However, it is important to differentiate between the air-cooled string and water-cooled string, as thermal contact directly with water would increase the yield by 5–6% [49]. Most of the current designs are not in direct contact with the water and therefore are enjoying the evaporative cooling and not direct water cooling, while some membrane based FPV systems (see Figura 4) are in direct contact with the water, where water temperature also plays an important role [49].

S16. LESS SHADING & SURROUNDING OBSTACLES

Shading is always a challenge in PV facility design and prevention of shading, forecasting / simulation of it in the design phase and its detection in live systems is always critical [28], [29], [30] no matter on land or water. PV systems installed on water have fewer surrounding obstacles around them and hence less vulnerable to shading than land-based PV installations (lack of trees, building and other obstacles on water surface). [1]

S17. WATER ALBEDO EFFECT

Although not high, albedo effect from water surface has a positive effect and the results are very design specific. Especially if bifacial modules are utilized [17], [39].

Plant Design, Installation, and Maintenance

S18. MODULAR DESIGN & LACK OF NEED FOR MACHINERY COMPARED TO COMPETITORS

Modular design of floating structures, the weight of the floats and the panels result in no need for heavy machinery for deployment compared to other technologies such as offshore wind or even oil & gas.

S19. EASY AND RAPID, ENGINEERING, DEPLOYMENT AND INSTALLATION

Simplicity of the technology and thanks to the current state of the art modular process [1], [7], [18] the engineering, deployment and installation process could be done quite rapidly. The modular small sized design would allow logistics with containers through normal shipping cargo routes compared to other renewable offshore solutions such as massive wind turbines [37] or even oil rigs. When it comes to installation speed, newest market report suggests up to 1MW installation per day, which approximately equals to 2.600 modules per day [18].

It is also worth mentioning that distance from shore is not a big concern in FPV projects, as one does not need to get very far from shore to resolve aesthetics and visibility concerns compared to offshore wind mills.

Figure 6 - 1: Pushing a modular click-in FPV system on the water (Image: FloatingSolarMounting) | 2: Installation of offshore wind turbines with heavy machinery (image from DNV) | 3: FPV is a less complicated industry when it comes to installation & maintenance (image from © Akuo Energy [1]) | 4: Enercon's new Ventum blade trailer from Goldhofer (image from Goldhofer) [55]

Environmental:

S20. LESS HEAT ISLAND EFFECT [14], [15]

Recent studies indicate that large PV power plants cause heating island effect with an ambient temperature by 3 - 4 °C higher than the wildlands at night [14]. However, in the Floating PV plants, the cooling water effect lowers or could even nullifies the heat island effect [15].

S21. NOT IMPACTING THE WILDLIFE

This is definitely an area to learn & study more FPV effects on surrounding ecosystem, but to the extent we know, wildlife is not affected (e.g., compared to birds and wind turbines issue).

Figure 7 - Upper Image: Ocean Farm from Offshore Fish Farms [57] — Lower Image: Floating Dairy Farms, Rotterdam, The Netherlands [38]

3.2 Weaknesses, in Details

W1. YOUNG TECHNOLOGY

Newly emerged technology with lack of regulation by law when utilizing water bodies, compared to ground installations using land [1], however, this is rapidly changing [1], [7] [52]. The most recent evolution in this scope has been world's first recommended practice by

DNV[6] for floating solar power plants [52].

W2. CHALLENGES WITH HIGH WIND SPEED AND WAVES

Challenges due to high wind speed, waves, typhoons and storms which could result in destructive consequences such as Figure 6. In Yamakura Dam, when the outermost array of PV structures mooring was ripped off, the ballast of wind-facing side was lost. This resulted in modules curling-up which led to an electrical fire [1], [7], [40], [44].

In addition to above mentioned, wind speed could result other types of challenges such as moving or rotating the structure. As this would change the amount of sunlight received by the system, it would affect the power output of the system [27]. However better mooring and fixed buoyant structure of the floating PV could eliminate this issue [27].

W3. AESTHETIC DISTURBANCE, IF NOT MOVED OUT OF SIGHT

The problem of Aesthetic Disturbance of the sceneries would remain, if not moved out of visual horizon. However, this is much easier challenge for FPV, as usually solar power plants are not visible from a distance due to their lower profile compared to competing renewable technologies such as wind turbines.

W4. CORROSION & SALT

Salt water in offshore or near-shore applications and corrosion over time while structure has near contact with water are challenges that industry has faced over decades. Although there are solutions to these issues due to the valuable experiences of offshore and on-water industries, this could add to CAPEX and OPEX costs in design of the system [1], [7].

Figure 8 - Destruction of Floating PV Plant by Typhoon at Yamakura Dam due to lack of torn mooring and ballast which resulted in electrical fire [40], [44]

Plant Design, Installation and Maintenance

W5. HIGH CAPEX

FPV's higher CapEx (initial investment) compared to ground-mounted systems [1], [2] could be counted as a disadvantage for this technology, but this is blanched during plant lifetime with FPV's higher energy yield for larger projects (above 50 MW) [1].

W6. CHALLENGING & COSTLY DESIGN

- Higher structural balance-of-system (SBOS) costs could stand for 34% of all-in cost in FPV projects compared to 8% in ground mounted, for a Chinese 5-10 MW project [2]. SBOS cover floating structure, mooring, and anchoring solutions.
- Fluctuations of water levels in countries with dry and wet seasons could result in challenges in design and maintenance of anchoring and mooring systems [1].
- Harder Grounding (Earching) Conditions due to being on waterbed and continuous movement of structure [1]

W7. HIGH OPEX COSTS

FPV has higher OpEx costs (operation and maintenance) compared to conventional ground-mounted systems due to the need for trained manpower and tools required to work on water [7]. However, the costs are dropping due to increase in project sizes globally since 2015 [2].

W8. HIGH SOFT COSTS

FPV has also higher soft costs which covers labor, design, engineering, supply chain and logistics [2]. It's worth mentioning that this is slowly changing by growing project sizes (figure 2, [2]) and novel designs which brings fast deployment [18].

W9. CHALLENGING MAINTENANCE

Compared to ground mounted system, maintenance and installation activities are more challenging on water (Biofouling, etc) [1]. Activities such as periodic maintenance of anchoring solutions is one of the challenging maintenance activities [1] [40].

W10. ANIMAL SHELTER & DIRTS RESULTED BY IT

Possibility for becoming an animal shelter and dirty panels due to bird droppings [1]. Here similar solutions such as the ones used in other sectors could be utilized.

W11. WIND RESISTANCE EFFECT ON TILT ANGLE

Design restrictions imposed by wind resistance calculations and designs, sets limits on the tilt angle, hence reduces power yield in high-altitude applications [1] which require steeper angles.

W12. LIFETIME CHALLENGES & DURABILITY

Floating structures have a warranty of 5 - 10 years while on-land structures main equipment has 20+ years of warranty [1]. The latter is more harmonized with panels lifetime output warranty which is commonly 20+ years for most of manufacturers.

Environmental

W13. UV CORROSION

UV corrosion effects the mechanical properties of plastic floats [24] over time. However, studies show that choice of correct plastic type, such as High-density polyethylene (HDPE), is a great improvement [24]. This is both in terms of mechanical properties change with extensive UV exposure and even water contamination [22], [23], [24].

3.3 Opportunities, in Detail

Waterbed Utilization

O1. LAND THIRST & SCARCITY AS MARKET BOOSTER

- High demand for lands in highly populated areas makes FPV an attractive alternative, which can potentially boost the market.
- Newly emerging market

[6] DNV is an international accredited registrar and classification society based in Norway.

in areas of the world where land and local energy sources are scarce (East and SouthEast Asia, Netherlands etc)[1], [7], [33] is another proof.

O2. INVESTMENT & PROJECT SCALES LOWERS COSTS

Higher investment and bigger scale projects result in lower all-in costs on a dollar-per-watt basis [2].

O3. INNOVATIVE FLOATING BUSINESS MODELS

- Electricity generation on the water is an enabler for innovative floating businesses that could be covered by solar panels on & around them (E.g. Dutch Floating Cow Facility in figure 7 [38], Fish farms, shrimp farms, etc).

Energy Generation & Power Systems:

O4. ENERGY DEMAND, LOCAL PRODUCTION & ENERGY SECURITY (POLITICAL ASPECT)

The world is in demand of energy and a solution with potentials such as FPV (meeting world electricity demand scale) is very attractive for the market, policymakers, and governments. The demand for secure and local generation of energy could be an accelerator for this young emerging technology.

O5. CO2 FOOTPRINT OPTIMIZER

FPV like Offshore Wind, provides local energy generation offshore with zero CO2 emission. This is ab attractive solution for business models that are suffering from large CO2 emission footprint and need to optimize their execution.

This opportunity and the demand in the market is an enabler for hybrid offshore business models and a booster for offshore local clean energy generation (Offshore FPV and Offshore Wind). Sectors such as Oil & Gas and Marine industry are among those under evolution. We already are witnessing emerging of business models that are based on such energy [53].

Such hybrid business models could vary from reduced reliance on gas / diesel generators in Oil & Gas industry and use of hydrogen produced offshore as fuel of marine industry (offshore fuel stations for ships) [53].

O6. ATTRACTIVE HYBRID SOLUTIONS COUPLED WITH HYDRO POWER

Hybrid solutions with hydropower plants is a very attractive and accelerator in FPV market, as many of current installations are found on hydropower plants [11,25,45,48]. Features such as following make this combination very attractive. - A potential of global 265.7 thousand square kilometers water surface [11]

- Utilizing unused available waterbed
- Utilizing the existing power transmission infrastructure [1], [31], [2]
- Optimization of day / night power output of the hybrid power plant (Source?)
- Using water in Hydropower Reservoirs as a giant Virtual Battery to pump the water to higher altitudes and produce power in absence of solar radiations [11]

Plant Design, Installation, and maintenance:

O7. EXTENSIVE OFFSHORE KNOW-HOW

Earlier experiences from offshore sector could easily boost the confidence of investors and policymakers in this technology [7].

Environmental:

O8. REUSE OF CURRENT MATURE STANDARDS

Use of already verified material and practices from drinking water standards could mitigate the concerns / risk of impacts on water quality. An Example of this approach is Ciel et Terre compliance with the BS 6920:2000 standard and use of common materials in water pipes industry such as high-density polyethylene (HDPE) plastics in their products to ensure water safety [22], [23], [24].

3.4 Threats, in Detail

Plant Design, Installation and Maintenance

T1. LACK OF STANDARDS & GOOD PRACTICES

Newly emerged technology with lack of experience, codes, standards and regulations for grid integration by local Transmission System Operators (TSO). This is rapidly changing with the growing number of projects, their size [7],[18],[21] and market demand which results in neutral certifier firms to look more into this market and develop standards and good practices [52].

T2. WATER CONTAMINATION

- Possibility for water contamination due to non-standard and poor-quality structure material
- Possibility for water contamination due to sunk & damaged solar panels (lead, cadmium, etc)

T3. IMMATURE STRUCTURAL DESIGN

Weakness against wind speed and high waves due to being on a moving platform. This is specially, if standards are not properly followed (e.g., Yamakura dam incident with 70% destruction of FPV plant) [40], [44].

Environmental:

T4. AESTHETIC DISTURBANCE, SOCIO POLITICAL & SOCIAL ACCEPTANCE

The problem of Aesthetic Disturbance, combined with socio-political and community acceptance issues could potentially become a showstopper for renewable energies. There is a historical trend for it [35] as for Norwegian national plans for expansion of wind power back in 2019 [43].

T5. LONG TERM ECOLOGICAL EFFECTS

The long-term ecological effects on the local environment, water and live ecosystems is yet to be known [1], [7], [22] and requires further studies on the subject.

T6. LACK OF ENVIRONMENTAL STANDARDS

There is a need for environmental standards and verification for the materials used in the floats [1], [7], [22]. However innovative approaches are taken to create good practices.

Good practice 1: Ciel et Terre products in the market are in compliance with the BS 6920:2000 drinking water standards of the Water Quality Association of Great Britain, which approves their installation on British drinking water reservoirs. [22], [23]

Good practice 2: Thames Water, one of the major water services in the UK, tested and verified Ciel et Terre design utilizing recyclable HDPE plastics. HDPE material is used for water pipes as per today and there should not be any more concerns about the effects of HDPE on drinking water. [22], [23], [24]

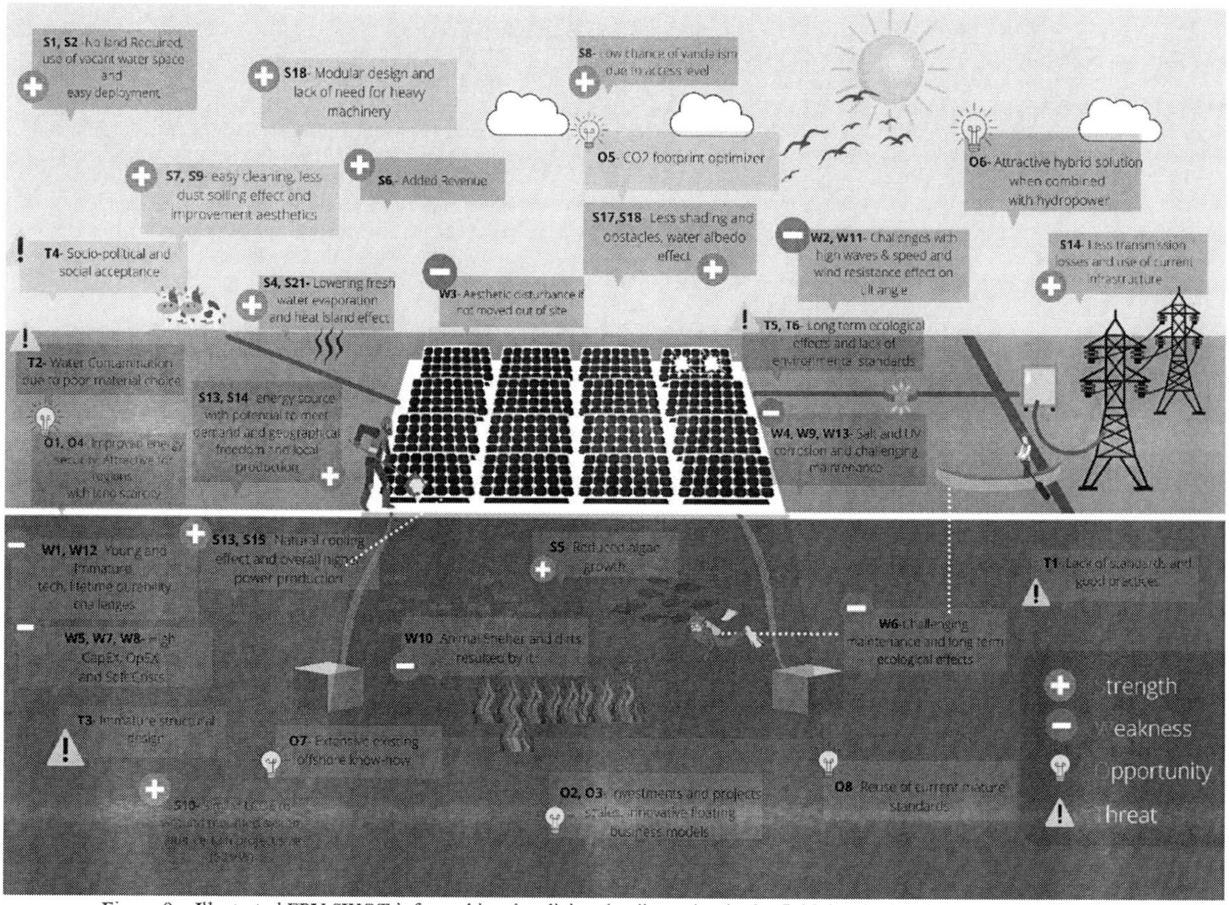

Figure 9 – Illustrated FPV SWOT infographic, visualizing the discussion in the field, inspired by illustration in [1]

4 CONCLUSIONS

Through the SWOT analysis method, facts, estimations, and market information are presented through the article. Based on this review, main drivers of FPV technology are

- Eliminating the land-usage footprint
- Potentials to meet world electricity demand
- Energy security & independence in an energy thirsty world
- Geographically-free Local energy production
- Promising potentials with available manmade reservoirs globally (404,454 square km) and vast opportunities with developing offshore and near the shore technologies.
- A strong attempt to resolve / improve the landscape aesthetics and social acceptance, by moving the renewable energy production out of sight / blending it with the water profile in horizon.
- Fast engineering, logistics, deployment through the state of the art developing modular designs of the floats structure (1 MW per day record)
- Enabler for optimization of classic business models with big carbon dioxide footprint, new innovative floating business models and hybrid models to add revenue to existing businesses (with providing local energy generation on water).

In addition to these advantages, disadvantages such as higher CAPEX and OPEX costs were discussed, which could be covered in larger project sizes by the potential higher power output. Lack of market maturity, standards, HSE concerns, and good practices has also been discussed, however examples are presented that these are rapidly changing and improving with the fast evolution of the technology, market demand for projects and the rise of projects in both numbers and sizes.

Finally, it is worth mentioning that the total electricity demand of the world could be met by only installing FPV on 51.5% of the total available man-made reservoirs globally (see S11). This shows great potential of the technology as this number does not even reflect opportunities arising by FPV plants moving to near-shore and offshore open sea locations.

5 BIBILIOGRAPHY

References

[1] (2019) "Where Sun Meets Water : Floating Solar Market Report. World Bank Report, World Bank Group; Energy Sector Management Assistance Program; Solar Energy Research Institute of Singapore. 2019. Where Sun Meets Water: Floating Solar Market Report. World Bank, Washington, DC. © World Bank - URL: https://openknowledge.worldbank.org/handle/10986/31880 , License: CCBY 3.0 IGO."

Infographic, URL:
https://www.esmap.org/where_sun_meets_water_floating_solar_market_report

[2] Wood Mackenzie Reports, by Molly Cox, 19.09.2019, "Floating solar landscape 2019", URL: https://www.woodmac.com/our-expertise/focus/Power--Renewables/floating-solar-2019/?utm_source=gtmarticle&utm_medium=web&utm_campaign=wmpr_floatingsolar ,
Green Tech Media, by Molly Cox, 19.09.2019, "The State of Floating Solar: Bigger Projects, Climbing Capacity, New Markets", URL: https://www.greentechmedia.com/articles/read/the-state-of-floating-solar-bigger-projects-and-climbing-capacity

[3] M. Rosa-Clot, G. M. Tina, 2017, "Submerged and Floating Photovoltaic Systems: Modelling, Design and Case Studies", Academic Press

[4] - Arnim Scheidel, Alevgul H. Sorman, "Energy transitions and the global land rush: Ultimate drivers and persistent consequences", Global Environmental Change, Volume 22, Issue 3,2012, Pages 588-595, ISSN 0959-3780", URL: http://www.sciencedirect.com/science/article/pii/S0959378011002068

[5] - Alok Sahu, Neha Yadav, K. Sudhakar, "Floating photovoltaic power plant: A review", Renewable and Sustainable Energy Reviews, Volume 66, 2016, Pages 815-824, ISSN 1364-0321, URL: http://www.sciencedirect.com/science/article/pii/S1364032116304841

[6] - Robert S. Spencer, Jordan Macknick, Alexandra Aznar, Adam Warren, and Matthew O. Reese, "Floating Photovoltaic Systems: Assessing the Technical Potential of Photovoltaic Systems on Man-Made Water Bodies in the Continental United States", Environmental Science & Technology 2019 53 (3), 1680-1689, DOI: 10.1021/acs.est.8b04735, URL:" https://pubs.acs.org/doi/abs/10.1021/acs.est.8b04735

[7] Sudhakar K. "SWOT analysis of floating solar plants". MOJ Solar Photoen Sys. 2019;3(1):20?22. DOI: 10.15406/mojsp.2019.03.00030 - URL: https://doi.org/10.15406/mojsp.2019.03.00030

[8] Komali Yenneti, Rosie Day, Oleg Golubchikov, "Spatial justice and the land politics of renewables: Dispossessing vulnerable communities through solar energy mega-projects", Geoforum, Volume 76, 2016, Pages 90-99, ISSN 0016-7185, URL: https://doi.org/10.1016/j.geoforum.2016.09.004

[9] Aniruddh Mohan, "Whose land is it anyway? Energy futures & land use in India", Energy Policy, Volume 110, 2017, Pages 257-262, ISSN 0301-4215, URL: https://doi.org/10.1016/j.enpol.2017.08.025

[10] Parisa Ranjbaran, Hossein Yousefi, G.B. Gharehpetian, Fatemeh Razi Astaraei, "A review on floating photovoltaic (FPV) power generation units", Renewable and Sustainable Energy Reviews, Volume 110, 2019, Pages 332-347, ISSN 1364-0321, URL: https://doi.org/10.1016/j.rser.2019.05.015

[11] Javier Farfan, Christian Breyer, "Combining Floating Solar Photovoltaic Power Plants and Hydropower Reservoirs: A Virtual Battery of Great Global Potential", Energy Procedia, Volume 155, 2018, Pages 403-411, ISSN 1876-6102, URL: https://doi.org/10.1016/j.egypro.2018.11.038

[12] Gardenio Diogo Pimentel Da Silva & David Alves Castelo Branco (2018) "Is floating photovoltaic better than conventional photovoltaic? Assessing environmental impacts", Impact Assessment and Project Appraisal, 36:5, 390-400, DOI: URL: https://doi.org/10.1080/14615517.2018.1477498

[13] Soloot, H. E. H., Agheb, E., Soloot, A. H., & Moghadam, S. (2020, December). A SWOT Analysis of Two Protection Strategies Due to the Expansion of Renewable Distributed Generation on Distribution Network. In 2020 15th International Conference on Protection and Automation of Power Systems (IPAPS) (pp. 49-52). IEEE.

[14] Barron-Gafford, G., Minor, R., Allen, N. et al. "The Photovoltaic Heat Island Effect: Larger solar power plants increase local temperatures". Sci Rep 6, 35070 (2016), URL: https://doi.org/10.1038/srep35070

[15] Anik Goswami, Paromita Sadhu, Utpal Goswami, Pradip Kumar Sadhu,"Floating solar power plant for sustainable development: A techno-economic analysis" 22 May 2019, Environmental Progress & Sustainable Energy, Volume38, Issue6, November/December 2019, e13268, URL: https://doi.org/10.1002/ep.13268

[16] "Statistical Review of World Energy 2021 — 70th edition", 2021, URL: https://www.bp.com/en/global/corporate/energy-economics/statistical-review-of-world-energy.html

[17] Marco Rosa-Clot, Giuseppe Marco Tina, Sandro Nizetic, "Floating photovoltaic plants and wastewater basins: an Australian project", Energy Procedia, Volume 134, 2017, Pages 664-674, ISSN 1876-6102, URL: https://doi.org/10.1016/j.egypro.2017.09.585

[18] PV-Tech.com News, 28.04.2020, "Project Briefing: New floating solar concepts at BayWa r.e.'s Sekdoorn", URL: https://www.pv-tech.org/project-briefing-new-floating-solar-concepts-at-baywa-r-e-s-sekdoorn/

BayWa r.e. Company Market News, 29.04.2020, "Construction of Europe's largest floating solar farm in under 8 weeks", URL: https://www.baywa-re.de/en/company/news/details/construction-of-europes-largest-floating-solar-farm-in-under-8-weeks

[19] InterSolar Website, 2020, "INTERSOLAR AWARD FINALISTS 2020", URL: https://www.intersolar.de/news/intersolar-award-2020-finalists-impress-with-innovative-solutions-for-the-solar-industry

[20] Kim Trapani, Miguel Red´on Santaf´e, 23 January 2014, "A review of floating photovoltaic installations: 2007–2013", Progress in photovoltaics, Volume23,

Issue4, April 2015, Pages 524-532, URL: https://doi.org/10.1002/pip.2466

[21] DNV, 29.04.2020, "DNV GL verifies innovative floating solar design methodology for Ocean Sun", URL: https://www.dnvgl.com/news/dnv-glverifies-innovative-floating-solardesign-methodology-for-ocean-sun-168977

[22] REC Group, White Paper, 2018, "Riding the wave of solar energy: Why floating solar installations are a positive step for energy generation"
URL 1: https://www.semanticscholar.org/paper/Riding-the-wave-of-solar-energy%3A-Why-floating-solar/e42131cdb7b19f6c915d5a07806fe4e99ff4ad61

URL 2:
https://www.recgroup.com/sites/default/files/documents/wp_-_floating_pv_rev_d_web.pdf

[23] Ciel-et-Terre Website, Product Announcement, "HYDRELIO®: THE PATENTED FLOATING PV SYSTEM", URL: https://www.ciel-et-terre.net/hydrelio-technology/our-hydrelio-solutions/

[24] Alok K. Sahu, Kumarasamy Sudhakar, "Effect of UV exposure on bimodal HDPE floats for floating solar application", Journal of Materials Research and Technology, Volume 8, Issue 1, 2019, Pages 147-156, ISSN 2238-7854, URL: https://doi.org/10.1016/j.jmrt.2017.10.002

[25] Ciel-et-Terre Website, Projects Information, Project 1 : "OKEGAWA: 1,180 KWP", URL: https://www.ciel-et-terre.net/project/okegawa-pond-1180-kwp/
Project 2 : "ALTO RABAGA˜ O: HYDRO & SOLAR", URL:
https://www.ciel-et-terre.net/project/alto-rabagao-hydro-solar/

[26] PV-Tech News Website, 2020, "Storms Ciara, Dennis no match for offshore solar trailblazer", URL: https://www.pv-tech.org/storms-ciara-dennis-no-match-for-offshore-solar-trailblazer/

- RECHARGE NEWS, 16 February 2020, "World's first offshore solar array rides out storm Ciara off Netherlands", URL:
https://www.rechargenews.com/transition/worlds-first-offshore-solar-array-rides-out-storm-ciara-off-netherlands/2-1-757022

[27] Young-Kwan Choi, 2014, "A Study on Power Generation Analysis of Floating PV System Considering Environmental Impact", International Journal of Software Engineering and Its Applications Vol.8, No.1 (2014), pp.75-84, URL:
https://www.semanticscholar.org/paper/A-Study-on-Power-Generation-Analysis-of-Floating-PV-Choi/33840dba6bb3f82ac68cde86fbbbae588a5bc180

[28] Santiago Silvestre, Sofiane Kichou, Aissa Chouder, Gustavo Nofuentes, Engin Karatepe, "Analysis of current and voltage indicators in grid connected PV (photovoltaic) systems working in faulty and partial shading conditions", Energy, Volume 86, 2015, Pages 42-50, ISSN 0360-5442, URL:
https://doi.org/10.1016/j.energy.2015.03.123

[29] Ravi Rachchh, Manoj Kumar, Brijesh Tripathi, "Solar photovoltaic system design optimization by shading analysis to maximize energy generation from limited urban area", Energy Conversion and Management, Volume 115, 2016, Pages 244-252, ISSN 0196-8904, URL:
https://doi.org/10.1016/j.enconman.2016.02.059

[30] Yang Hongxing ,Sun Yunlin, Chen Siming, Xie Liying
,Hong Ruijiang ,Shen Hui, 2014, "Investigating the Impact
of Shading Effect on the Characteristics of a Large-Scale Grid-Connected PV Power Plant in Northwest China", 763106, 2014, International Journal of Photoenergy, 1110-662X, URL:
https://doi.org/10.1155/2014/763106

[31] IRENA – International Renewable Energy Agency, 2019, "Future of Solar Photovoltaic: Deployment, investment, technology, grid integration and socio-economic aspects (A Global Energy Transformation: paper)", International Renewable Energy Agency, Abu Dhabi, November 2019, ISBN: 978-92-9260-156-0 URL: https://www.irena.org/publications/2019/Nov/Future-of-Solar-Photovoltaic

[32] Renewable Energy World News Website, Junko Movellan ,08.26.2016, "Japan's First Floating Solar System with Microinverters Installed on an Irrigation Reservoir", URL:
https://www.renewableenergyworld.com/solar/japan-s-first-floating-solar-system-with-microinverters-installed-on-an-irrigation-reservoir/

[33] International Energy Agency, 2019, "Trends in PV applications 2019", ISBN 978-3-906042-91-6: Trends in PV applications 2019, URL: https://iea-pvps.org/trends_reports/trends-in-pv-applications-2019/

[34] International Energy Agency, 2020, "Snapshot of Global PV Markets 2020", ISBN 978-3-906042-94-7: 2020 Snapshot of Global PV Markets, URL: https://iea-pvps.org/snapshot-reports/snapshot-2020/

[35] Joseph Rand, Ben Hoen, "Thirty years of North American wind energy acceptance research: What have we learned?", Energy Research & Social Science Volume 29, July 2017, Pages 135-148, URL:
https://doi.org/10.1016/j.erss.2017.05.019

[36] PV-Tech News Website, 20.03.2019 "World's largest floating solar plant connected in China", URL: https://www.pv-tech.org/news/worlds-largest-floating-solar-plant-connected-in-china

Ciel-et-Terre website, Project Information, "ANHUI CECEP: 70,005 KWP", URL: https://www.ciel-et-terre.net/project/anhui-cecep-70005-kwp/

[37] Mule' Stagno, L. (2014). "Floating Photovoltaics – technological issues, cost and practical implications". The ISE Annual conference, Qawra. 39-45.

https://www.um.edu.mt/library/oar/handle/123456789/22986

[38] Reuters News, 01.10.2019, "Moo-ving to the future? Cows try 'Floating Farm' in Rotterdam", https://www.reuters.com/article/us-netherlands-floatingfarm/moo-ving-to-the-future-cows-tryfloating-farm-in-rotterdam-idUSKBN1WG3OV

Euronews with Reuters, 05.07.2019 "Watch: World's first floating farm in operation in Rotterdam", https://www.euronews.com/2019/07/05/watch-world-s-first-floating-farm-in-operation-in-rotterdam

BBC News, by Simon Fry Technology of Business reporter, 17.08.2018, "The world's first floating farm making waves in Rotterdam", https://www.bbc.com/news/business-45130010

Floating Farm official website, URL: https://floatingfarm.nl/

[39] PV-Tech news website, By Cecilia Keating, 17.02.2020, "Enerwhere anchors offshore bifacial PV project on UAE seawater", URL: https://www.pv-tech.org/everwhere-reveals-offshore-pv-project-on-persian-gulf/

Bloomberg news agancy, By Verity Ratcliffe February 13.02.2020, "Abu Dhabi Island to Host Nation's First Floating Solar Power Plant", URL: https://www.bloomberg.com/news/articles/2020-02-13/abu-dhabi-island-to-host-nation-s-first-floating-solar-power

[40] PV Magazine, by Marian Willuhn, 22.02.2020, "The weekend read: Don't throw caution to the wind", URL: https://www.pv-magazine.com/2020/02/22/the-weekend-read-dont-throw-caution-to-the-wind/?utm_

PV Magazine, 09.09.2019, by EMILIANO BELLINI, "Japan's largest floating PV plant catches fire after Typhoon Faxai impact", URL: https://www.pv-magazine.com/2019/09/09/japans-largest-floating-pv-plant-catches-fire-after-typhoon-faxai-impact/

Ciel et Terre, Asia, Hydrelio, Other Media, Our achievements, Our partnerships, 19.08.2018, "YAMAKURA DAM WELCOMED FLOATING SOLAR PANELS ON ITS SURFACE", URL: https://www.ciel-et-terre.net/yamakura-dam-welcomed-floating-solar-panels-on-its-surface/

[41] News, pv-magazine.com (2020) "Norwegian tech for
Korea's giant floating solar complex", hyphens URL : https://www.pv-magazine.com/2020/07/16/norwegian-tech-for-koreas-giant-floating-solar-complex/

[42] News, pv-magazine.com (2021) "Floating solar pilot project near Canary Islands to face up to 10m-high waves", URL : https://www.pv-tech.org/floating-solar-pilot-project-near-canary-islands-to-face-up-to-10m-high-waves/

Ocean Sun Company Announcement, 18.01.2021, "Ocean Sun and Fred Olsen Renewables to test offshore floating solar power in the Atlantic Ocean", URL: https://oceansun.no/ocean-sun-and-fred-olsen-renewables-to-test-offshore-floating-solar-power-in-the-atlantic-ocean/

[43] News, Aftenposten (2019), "Vindkraftplanen skrotes" Published on 17. oct. 2019, https://www.aftenposten.no/norge/i/QoK09A/vindkraftplanenskrotes

News, Wind Power MOnthly (2019), "Local protests kill Norway's wind plans", published on 17 October 2019 by David Weston. https://www.windpowermonthly.com/article/1662940/localprotests-kill-norways-wind-plans

Newspaper Commentatory Article, Aftenposten (2019), "Derfor bør vindkraft p°a land stoppes — Odd Helge Gilja, Fornybar energi må spille på lag med naturen og ikke mot den." Publish in Aftenposten by Odd Helge Gilja, Professor, Universitetet i Bergen, 13. juni 2019 21:00 https://www.aftenposten.no/meninger/debatt/i/kJKlVa/derforboer-vindkraft-paa-land-stoppes-odd-helge-gilja

[44] The Asahi Shimbun National Newspaper, Youtube Channel, (2019) "Chiba, Ichihara burning dozens of solar panels on the surface of the dam, Youtube video" https://youtu.be/_EEc1MTb7DM

[45] Rocio Gonzalez Sanchez, Ioannis Kougias, Magda Moner-Girona, Fernando Fahl, Arnulf Jäger-Waldau, Assessment of floating solar photovoltaics potential in existing hydropower reservoirs in Africa, Renewable Energy, Volume 169, 2021, Pages 687-699, ISSN 0960-1481,URL: https://doi.org/10.1016/j.renene.2021.01.041

[46] IRENA (2021), Renewable capacity statistics 2021 International Renewable Energy Agency (IRENA), Abu Dhabi, ISBN 978-92-9260-342-7

[47] PV Magazine, 22.09.2020 by FRANK HAUGWITZ, SENIOR ADVISOR, APRICUM, "Floating solar PV gains global momentum", URL: https://www.pv-magazine.com/2020/09/22/floating-solar-pv-gains-global-momentum/

[48] PV Magazine, 08.03.2021, by EMILIANO BELLINI, "South Korea wants to deploy another 2.1 GW of floating PV by 2030", URL: https://www.pv-magazine.com/2021/03/08/south-korea-wants-to-deploy-another-2-1-gw-of-floating-pv-by-2030/

[49] Torunn Kjeldstad, Dag Lindholm, Erik Marstein, Josefine Selj,Cooling of floating photovoltaics and the importance of water temperature, Solar Energy, Volume 218, 2021, Pages 544-551, ISSN 0038-092X, URL: https://doi.org/10.1016/j.solener.2021.03.022

[50] Valentino Piana, Annelen Kahl, Cristina Saviozzi, René Schumann, "Floating PV in mountain artificial lakes: a checklist for site assessment", Renew. Energy Environ. Sustain. 6 4 (2021), DOI: 10.1051/rees/2021002

[51] Ciel et Terre, KAWAGOE: 696 KWP Project, Japan, URL1: http://www.floatingsolarpanels.co.uk/portfolio/

URL2: https://www.ciel-et-terre.net/project/kawagoe-696-kwp/
URL3: https://pv-magazine-usa.com/2018/09/19/floating-solar-has-a-nice-summer-going-on-252-kw-completed-5-mw-coming-for-ciel-terre/

[52] DNV Website, News, 31.03.2021 "DNV publishes world's first recommended practice for floating solar power plants", URL: https://www.dnv.com/news/dnv-publishes-world-s-first-recommended-practice-for-floating-solar-power-plants-199138

[53] Solar Edition, News Website, "a Norwegian Attempt to Reduce Oil Production CO2 Footprint, Wind + Hydrogen", URL: https://solaredition.com/deep-purple-offshore-wind-hydrogen-battery/

Solar Edition, News Website, "Deep Purple, Stable Offshore CO2 – Free Energy Production, Utilizing Wind & Hydrogen", URL: https://solaredition.com/deep-purple-stable-offshore-co2-free-energy-production-utilizing-wind-hydrogen/

Technip FMC, Company Website, "Offshore energy, powered by green hydrogen", URL: https://www.technipfmc.com/en/what-we-do/subsea/energy-transition-deep-purple/

[54] Yu Lei, Xi Lu, Mai Shi, Lei Wang, Hongyan Lv, Sirui Chen, Chengyu Hu, Qinghua Yu, Serafim Dias Henriques da Silveira, "SWOT analysis for the development of photovoltaic solar power in Africa in comparison with China", Environmental Impact Assessment Review, Volume 77, 2019, Pages 122-127, ISSN 0195-9255, URL: https://doi.org/10.1016/j.eiar.2019.04.005

[55] Images in Figure 6 are from
1: Floating Solar Mounting, URL: https://www.floatingsolarmounting.com/news/where-to-install-floating-solar-photovoltaic-projects.html
2: DNV, URL:https://www.dnv.com/expert-story/maritime-impact/Fresh-breeze-for-offshore-wind-farms.html
3; Source: World Bank Report, where suns meets water [1] © Akuo Energy, URL: https://documents1.worldbank.org/curated/en/418961572293438109/pdf/Where-Sun-Meets-Water-Floating-Solar-Handbook-for-Practitioners.pdf
4: Internationalcranes.media from Enercon & Goldhofer URL: https://www.internationalcranes.media/news/Wind-power-transport-Ride-like-the-wind/1139526.article

[56] Deb, D., & Brahmbhatt, N. L. (2018). "Review of yield increase of solar panels through soiling prevention, and a proposed water-free automated cleaning solution", Renewable and Sustainable Energy Reviews, 82, 3306–3313. doi:10.1016/j.rser.2017.10.014

[57] Zhaolong Yu, Jørgen Amdahl, David Kristiansen, Pål Takle Bore, "Numerical analysis of local and global responses of an offshore fish farm subjected to ship impacts", Ocean Engineering, Volume 194, 2019, 106653, ISSN 0029-8018, URL: https://doi.org/10.1016/j.oceaneng.2019.106653

KEY PERFORMANCE INDICES OF PHOTOVOLTAIC CARPORTS

Matthias Loup, Cyril Allenspach, Hendrik Hofmann, Roman Vogt, Fabian Carigiet, Franz Baumgartner

ZHAW, Zurich University of Applied Sciences, School of Engineering, IEFE
Technikumstrasse 9, CH-8401 Winterthur, Switzerland
Email: franz.baumgartner@zhaw.ch, Website : www.zhaw.ch/=bauf

ABSTRACT: PV carports are an ideal solution to produce electricity locally and sustainably, while taking advantage of the dual use of the space. On average, large carports have a power output of 0.17 kWp/m². The average costs amount to 2'100 CHF/kWp. The price per kWp decreases slightly when the power clearly increases. The costs of the PV modules are lower than the costs for the carport's foundation and supporting structure, which represent generally between 30 % and 60 % of the total costs. The amount of steel is estimated to 155 kg/kWp and the quantity of concrete required is approximately 650 kg/kWp. The specificities of the terrain naturally have an impact on these values. A carport parking space emits 690 kg-CO_2-eq during the construction phase and an electric vehicle (EV) 11.2 kg-Co_2-eq/100 km on average. Innovative projects include retractable roofs to offers the potential to reduce the amount of material used in the mounting structure, due to storing the PV modules in a safety box during heavy wind or snow load. The use of wood instead of metal, and screws anchor instead of concrete, as well as the addition of greenery, greatly reduces total CO_2 emissions.
Keywords: PV carport, distribution of costs and material, CO_2 footprint, e-mobility, durability.

1 INTRODUCTION

The share of newly registered electric vehicles (EV) in Switzerland doubled from 2019 to 2020 reaching eight percent, while other countries like Norway recorded more registrations for electrical cars than combustion engine passenger cars in 2020 [1]. Many countries and regions aim to reduce their greenhouse gas emissions to zero by 2050, e.g., the whole European Union. It is widely known that the total greenhouse gas emission will further rise if EVs are charged with electricity by coal fired power plants, which are predominant in countries such as Poland, China, South Africa [2]. In many countries, the total mobility sector accounts in average for about one third of total greenhouse gas emission, forcing the change to mobility system powered by renewables [3]. One solution is the local PV electricity supply by PV carports. A 2 kWp PV carport for passenger cars is able to power an EV to drive about 10'000 km annually in Switzerland [1, 4].

This publication presents the key performance indicators of today's PV carports, including a comparison of the overall costs of an installation in CHF/kWp depending on its power or the number of parking spaces. The distribution of costs between the technical study, foundations, supporting structure and PV modules is also given. The study then shows the amount of steel or concrete required for the supporting structure and the corresponding CO_2 emissions. Finally, innovative projects such as PV carports with greenery or retractable roofs are proposed.

Figure 1 : PV carport of 6.7 MWp in Courgenay, Switzerland [5].

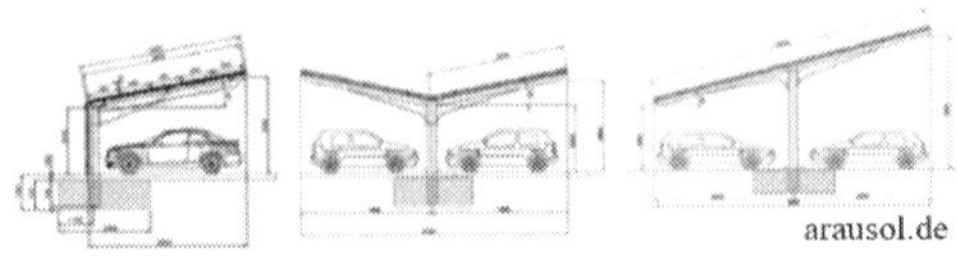

Figure 2 : Different types of PV carports, for one or two rows of cars, with variations in roof tilt. Concrete volumes are shown in light orange [6].

2 METHOD

The study determines the distribution of costs, material quantities and equivalent CO_2 emissions for the construction of small private carports with 2 places, medium ones with 10 to 20 places for blocks of flats and large ones with more than 100 places for companies and supermarkets. The results were determined by analysing already built carports, surveying engineering offices and evaluating the products offered by various suppliers in Switzerland and Europe.

3 KEY PERFORMANCE INDICES

3.1 Use of space – Costs

PV carports offer the advantage of producing electricity directly on top of infrastructure space where it is needed for EV charging. The energy can therefore be transmitted directly to the vehicles by the shortest route, minimising losses without the need of expensive storage. This technology makes it possible to take advantage of these spaces while protecting the vehicles from certain environmental influences, e.g., excessive heating by sunshine and hail. The power of large carports is on average 0.17 kWp/m² [6].

Figure 3 shows a slight downward trend in the price per kWp when the power increases significantly. This becomes visible when comparing carports with less than 2 MWp (Estavayer, Monthey, Dukovany, Aigle) with those of 6.7 MWp (Courgenay) and 9.8 MWp (Saint-Aignan). In contrast, this trend is not observed for the 4 carports with less than 2 MWp. The specificities of these and especially of the terrain have more influence at this scale.

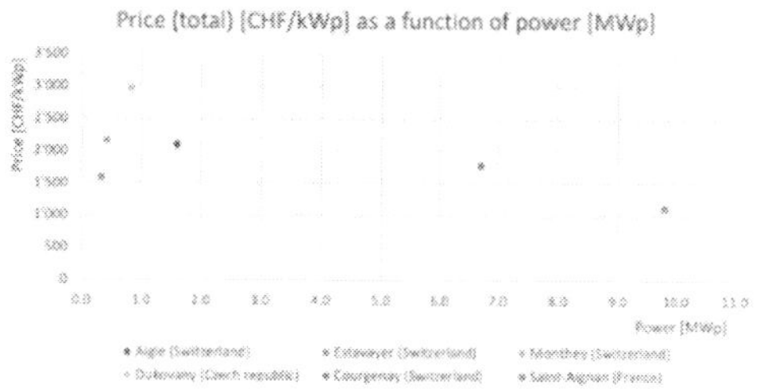

Figure 3 : Price in CHF/kWp as a function of installed power [6].

The average costs were found to be 2'100 CHF/kWp. The decreasing trend of the price per kWp as a function of the number of parking spaces is shown in Figure 4. For small carports (approx. 8 parking spaces), the price varies greatly depending on the geometry; it is generally between CHF 4'000 and 8'000/kWp.

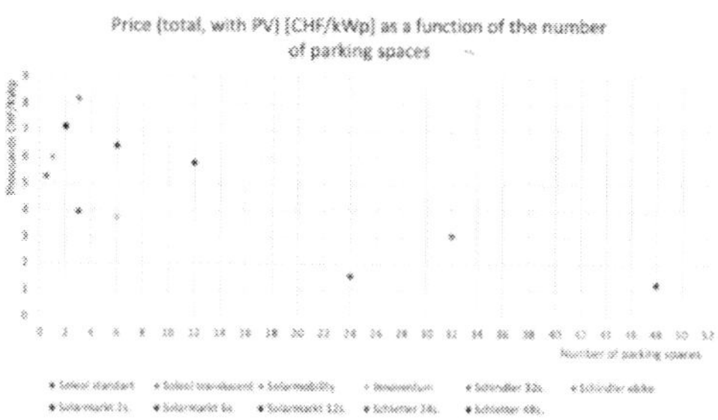

Figure 4 : Price in CHF/kWp as a function of the number of parking spaces [6].

The costs of the PV modules are lower than the costs for the carport's foundation and supporting structure, which represent generally between 30 % and 60 % of the total costs. Figure 5 shows in more detail the cost distribution for a 415 kWp carport in Aigle (Switzerland). Foundations and structure account for 50 % of the total costs, PV panels, inverter and cabling 38 % [6].

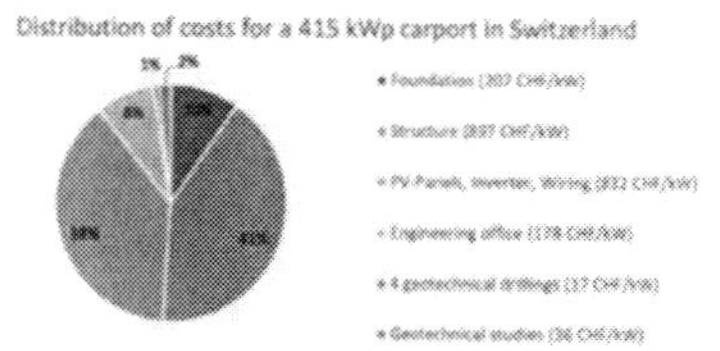

Figure 5 : Distribution of costs for a 415 kWp carport in Switzerland, in an area sometimes subject to strong winds [6].

3.2 Quantities of material used

The amount of steel is estimated to typically 155 kg/kWp. The quantity of concrete required is approximately 650 kg/kWp. The particularities of the site have a strong influence on the work to be carried out (sloping ground, landslide, strong wind, etc.).

For medium-sized carports (approx. 8 places), the price varies greatly depending on the geometry; however, it is generally between 4'000 and 8'000 CHF/kWp. The quantities of concrete required are much larger and can reach up to 1'250 kg/kWp [6].

Figure 6 : PV carport using steel structures and concrete fundaments. Performance index steel: 149 kg /kWp [6].

Table 1 : Key performance indices of amount of used steel in the structure of several PV Carports in Europe [6].

Large carports	Surface [m2]	Park spaces	Power [kWp]	Amount of steel [kg-steel]	[kg-steel/kWp]	[kg-steel/m2]
Estavayer (Switzerland)	1'802	88	327	53'000	162	29
Monthey (Switzerland)	2'500	250	415	75'000	181	30
Dukovany (Czech Republic)	8'267	330	820	115'500	141	14
Courgenay (Switzerland)	43'000		6'700	1'000'000	149	23
Saint-Aignan (France)	53'000		9'800			
Adiwatt (France)			1'350		177	30
Soleol (Switzerland)			100		120	
AVERAGE	21'714	223	2'787	310'875	155	25
MEDIAN	8'267	250	820	95'250	156	29

3.3 CO_2 content

A modern and still little used alternative to the concrete foundations are the bolted foundations. This solution is more ecological but requires more time and a geological analysis of the terrain, which is generally more expensive. The use of a wooden structure instead of metal improves architectural integration and reduces CO_2 emissions.

As shown in Figure 7, a carport parking space emits 690 kg-CO_2-eq in the construction phase and an electric vehicle 11.2 kg-Co_2-eq/100 km on average. In comparison, a gasoline-powered car emits in average 29 kg-CO_2-eq/100 km [7]. Figure 7 summarises the different emissions.

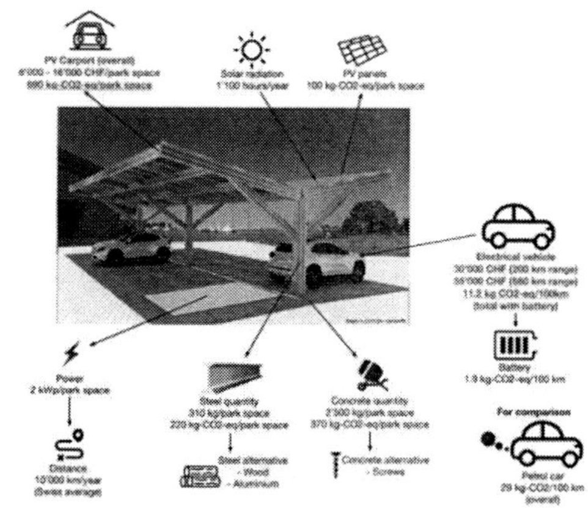

Figure 7 : Visual summary of equivalent CO_2 emissions of a PV carport [6].

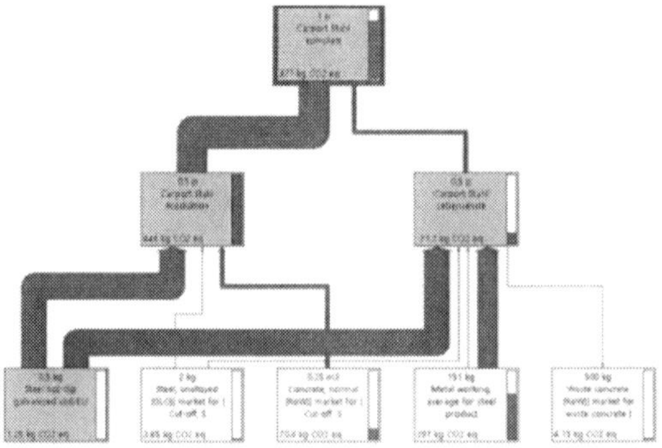

Figure 8 : LCOE of a standard steel concrete carport (foundation and structure) single parking slot results in minimum 377 kg-CO_2 for each parking space in the three phases of production, operation and disposal [8].

Table 2 : Amount of used material in the steel concrete carport (see Figure 8) [8].

Phase	Process/Material	Quantity	Unit
Production phase	Hot galvanised steel	300	kg
	Concrete	0.5	m³
	Other material	2	kg
End of life	Concrete disposal	1000	kg
	Recycling hot galvanised steel	300	kg
	Recycling metals	2	kg

Figure 9 : Wooden PV Carport from InnoVentum, a Swedish company. Architectural integration seems to be improved.

3.3 New approach – Green PV Carport

A concept of the ZHAW green PV carport was developed to get optimized key performance indices (see Figures 10 and 11). Wood is used for the main structure element together with concrete. The implementation of bifacial PV modules allows the growth of vegetation beneath the PV Generator. This would be beneficial for the ecological and sustainable perception of the carport while reducing overheating of a car without the need of a carport roof. The related CO_2 emission was estimated to be about a factor of 4 to 7 lower compared to the CO_2 emission of standard steel/concrete PV carport.

Figure 10 : Sketch of the ZHAW Green PV Carport to be using wood and vertical bifacial PV Modules, to allow growing plants beneath [8].

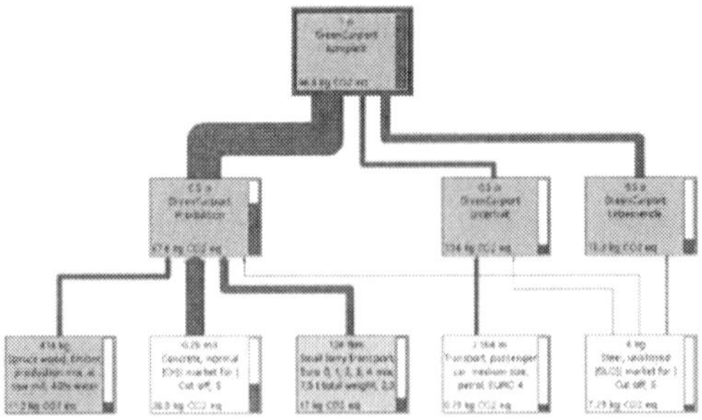

Figure 11 : LCOE of 97 kg-CO_2 (foundation and structure) for each parking space in the three phases of production, operation and disposal of the ZHAW Green Carport (see Figure 10). Thus, in comparison with Figure 8, CO_2 emissions with a wooden construction are about 3 to 4 times lower than with metal [8].

3.4 Retractable carports

Retractable carports are innovative photovoltaic systems for sustainable and local energy production. In case of unsuitable weather conditions, the PV modules are retracted and stored in a protective box. This protects them from strong weather influences (snow, hail, dew, etc.) and considerably reduces the forces impacting the supporting structure. This results in a decrease in the amount of material used and considerably reduces the costs compared to a standard installation.

The structure also offers the advantage of a double use of space and can be placed on a car park (carport), on a highway, on a water treatment plant or even above a football stadium.

For example, the URBANBOX concept is a construction with photovoltaic modules that can be moved horizontally on a fixed support structure. Thanks to the retractable roof, the forces on the structure due to wind are approximated to be 4 times smaller using less metal and concrete. In the final product, modules will have an integrated cleaning system in the protective box, which prevents yield losses due to soiling.

Figure 12 : Retractable PV carport from URBANBOX.

Another example of a retractable carport is that of DHP-Technology. In this application, the modules are suspended by a cable and can be moved in and out of a safety box. In the operating position, the PV modules have a tilt angle of 20° (see Figure 13).

Figure 13 : Retractable PV carport from DHP-Technology, Switzerland.

4 OUTLOOK

The energy transition is progressing and new technologies need to be developed and implemented in order to meet the environmental and climate targets for 2050. In Switzerland, new photovoltaic installations reached a new record of 493 MW in 2020. The final consumption of electricity in 2020 is 55,714 GWh, which means that photovoltaics currently only cover 4.7 % of the annual consumption [9].

However, some projects are in planning, such as covering 2.5 km of motorway in Zürich-Zug and 1.6 km in Wallis in Switzerland with photovoltaic panels for a total capacity of 53 MWp [10]. This enables sustainable electricity production with a double use of the land (production + road), whereas the motorway would be shaded and protected from snow. Since one third of CO_2 emissions come from the transport sector, it is relevant to envisage electricity generation by using the motorways. Since the number of electric vehicles is also increasing rapidly, charging stations could be made available with the electricity generated directly over them. The energy transport route would thus be as short as possible, which significantly minimises losses without the need of expensive storage.

Figure 14 : Motorway project in Wallis covered with PV modules [10].

5 CONCLUSION

The potential of PV carports is significant. In detail, they can be cost competitive when planned correctly. However even more important, they also tackle other relevant key performance indices beside the costs, which will be beneficial to meet customers and people's needs. The construction of a conventional steel/concrete PV carport is associated with the emissions about 480 to 690 kg-CO_2 per parking space. With the new European emission regulation limits of 95 g-CO_2 Emission in 2021 for new combustion engine cars, about 7'000 km driving distance equates to the construction phase of a single steel/concrete PV parking slot. This is close to an annual driving distance of small passenger car in Switzerland.

Taking advantage of the double use of the land, favouring self-consumption, producing locally and sustainably, choosing local wood over steel and screws over concrete, increasing the proportion of green areas and improving air circulation to minimise heating of the modules are key elements of future photovoltaic development.

6 ACKNOWLEDGMENT

We thank the companies that participated in the survey and shared the details of their PV carports.

Many thanks also to Adiwatt, Arausol, Bayo-s, InnoVentum, URBANBOX and Le Temps for their permission to use their illustrations in this publication.

7 REFERENCES

[1] Swiss Confederation, Federal Statistical Office, Mobility and transport, https://www.bfs.admin.ch/bfs/en/home/statistics/mobility-transport.html, 2021.
[2] K.M. Buresh et al. ; Energy for Sustainable Development 57 (2020) 132–140.
[3] Swiss Confederation, Federal Office for the Environment, Switzerland's greenhouse gas inventory, https://www.bafu.admin.ch/bafu/en/home/topics/climate/state/data/greenhouse-gas-inventory.html, April 2020.
[4] H.M. Neumann, D. Schär, F. Baumgartner; Prog. Photovolt: Res. Appl. 2012; 20:639–649.
[5] Adiwatt, le photovoltaïque, une histoire d'experts, adiwatt.com/ombriere-7/, 2021
[6] M. Loup, Master Theses Project Report, PV Carport, ZHAW SOE IEFE, December 2020.
[7] EnergieSchweiz, Faktenblatt Umweltauswirkungen von Personenwagen, https://www.bfe.admin.ch/bfe/de/home/news-und-medien/publikationen.exturl.html/aHR0cHM6Ly9wdWJkYi5iZmUuYWRtaW4uY2gvZGUvcHVibGljYXRpb24vZG93bmxvYWQvOTQ2MA==.html, February 2020. See also Swiss Confederation, Federal Statistical Office, https://www.bfs.admin.ch/.
[8] H. Hofmann and R. Vogt, Bachelor Project, Green PV Carport, ZHAW SOE IEFE, December 2020.
[9] Statistik Sonnenenergie, Referenzjahr 2020, Bundesamt für Energie BFE, https://www.swissolar.ch/fileadmin/user_upload/Markterhebung/2021/10539-Statistik_Sonnenenergie_Bericht_DE_def.pdf, July 2021.
[10] Zeitung LeTemps, A Fully l'autoroute se fiance avec le soleil, https://www.letemps.ch/suisse/fully-lautoroute-se-fiance-soleil, July 2021.

GEOGRAPHICAL DISTRIBUTION OF SOLAR URBAN POTENTIAL FOR VEHICLE INTEGRATED PHOTOVOLTAICS

Miguel C Brito[1], Teresa Santos[2], Rodrigo Amaro e Silva[3], David Pera[1], Filipe Moura[4], Jorge Rocha[5]
[1]Instituto Dom Luiz, Faculdade de Ciências da Universidade de Lisboa, Portugal
[2]CICS.NOVA Interdisciplinary Centre of Social Sciences, Faculty of Social Sciences and Humanities (NOVA-FCSH), Universidade NOVA de Lisboa, Portugal
[3]Centre Observation, Impacts, Énergie (OIE), MINES ParisTech, Sophia Antipolis, France
[4]CERIS, Instituto Superior Técnico, Universidade de Lisboa, Portugal
[5]Centre for Geographical Studies, Institute of Geography and Spatial Planning, Universidade de Lisboa, Portugal

ABSTRACT: Solar-powered electric vehicles can accelerate the transition to electric mobility, reducing charging needs and operating costs. In the urban environment, however, their potential is somewhat reduced by shadowing from neighbouring buildings. Here we estimate the irradiation losses and solar extended driving range for anywhere on the globe using an empirical model based on numerical simulation of solar radiation losses in simplified urban layouts. Modelling shows solar extended driving ranges above 20-30 km/day/kWp for a wide range of geographies, exceeding 40 km/day/kWp for most of the USA, China, Australia, parts of Africa and South America. The potential of parked vehicles is similar to that of driving vehicles for mid and high latitudes. These results highlight that a significant fraction of daily urban commuting can be powered by vehicle integrated photovoltaics,.
Keywords: Modelling; VIPV; Solar Radiation

1 INTRODUCTION

Solar-powered electric vehicles can significantly reduce the need for charging from the grid [1], thus reducing operating costs for the driver. The extended solar driving range brought about by onboard solar energy can also reduce the range anxiety for electric vehicle drivers, often perceived as a relevant barrier to faster adoption of electric mobility [2]. On the other hand, fewer charging events can accelerate the adoption of electric vehicles since more vehicles can be running with the same charging infrastructure, whose expansion rate is also seen as a barrier to the expansion of electric mobility.

The solar extended driving range of solar-powered electric vehicles will depend on the available solar irradiation and therefore on the severity of shadowing events, which are particularly relevant in complex urban environments. These will depend on the height of buildings and the width of the roads [3], but also the location, due to different solar geometries. The purpose of this contribution is to develop a method to assess the impact of urban shadowing as a function of latitude and solar resource.

2 METHODS

The approach has 3 main steps: the assessment of shading losses (both direct and diffuse components) for a simplified city layout at different latitudes, its fit to an empirical model and the use of the shading loss empirical model to estimate the solar extended driving range for any location in the globe.

2.1 Irradiation loss

The assessment of the impact of shadowing on the solar ViPV potential for different urban configurations and latitudes. This is carried out using the ArcGIS Solar Area Radiation tool. The input is a raster model corresponding to the different combinations of building height (5, 10, 15, and 20 m), road width (8, 12, 16, 20, 24, 28, 60, and 80 m), and road orientation (0, 45, and 90°). Then, the annual solar radiation is independently simulated for different latitudes (0, 15, 30, 45, and 60°) for direct and isotropic diffuse irradiation. After solar mapping, the radiation values were obtained by overlaying a set of points corresponding to the road central point of each urban configuration. Figure 1 represents the road layouts considered.

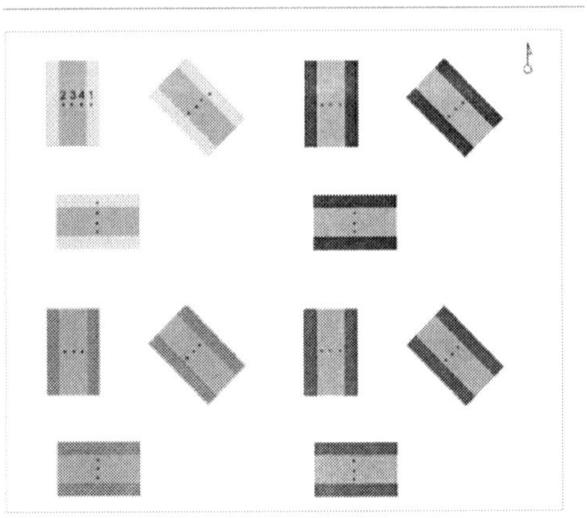

Figure 1: City layout configuration for ArcGIS irradiation modelling. Colours indicate building height.

2.2 Direct irradiation shadowing loss

The annual direct or beam irradiation shadowing loss for a driving vehicle, c_{b1}, for a driving vehicle in the middle of the street can be modelled by equation 1, where H^* is the effective height of the buildings (height of the building less the vehicle height, as shown in fig. 2), W is the street width and ϕ is the local latitude. The effect of the latitude can be interpreted as a relative shortening of the street width at higher latitudes, as the sun elevation decreases.

$$c_{b1} = 1 - \cos\left(\operatorname{atan}\left(\frac{H^*}{W\cos\phi}\right)\right) \qquad (1)$$

For parking vehicles, one ought to consider the two opposite parking sides of the street. For illustrative purposes, let us assume that the orientation of the street is north-south. During the morning, the vehicle on the left is unshaded while the vehicle on the right is shaded (fig. 3).

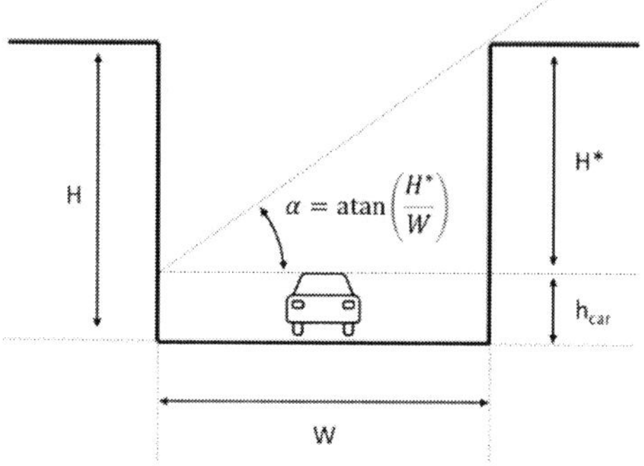

Figure 2: Definition of geometrical parameters.

Figure 3: Illustrative representation of shading of parked vehicles.

In the afternoon, the shadowing will affect the vehicle on the left but not the vehicle on the right. Thus, at any time for this day, and on average, one of the vehicles is always unshaded. The second vehicle will be shaded most of the time, except when the sun's azimuth is aligned with the street orientation, roughly estimated to be half of the time. Hence, the average shadowing loss for parking vehicles, c_{b2}, is given by equation 2.

$$c_{b2} = 1 - \frac{3}{4}\cos\left(\mathrm{atan}\left(\frac{H^*}{W\cos\phi}\right)\right) \qquad (2)$$

The empirical model can be compared to the irradiation loss assessment using the solar radiation tool, as shown in fig. 4.

One may observe that for low latitude, the model underestimates the losses for north-south streets and overestimates the losses for east-west streets. For intermediate latitude, the model is rather accurate, slightly underestimating shadowing losses for low rise buildings/wide streets while for high latitudes, the model always underestimates the shadowing losses.

It may also be noticed that data scattering is very pronounced for the parking shadowing loss, and that this scattering increases with latitude. The parking model is accurate for the north-south oriented streets. For other orientations, average values are well aligned to the model, but shadowing losses can be significantly higher or significantly lower (especially at high latitudes) depending on which side of the street the car is parked.

The overall RMSE for all height-width ratios and street orientation is as shown in Table 1. For driving vehicles, deviations from the model are less relevant at mid-latitudes. Errors for parking vehicles increase with latitude and are always larger than for driving vehicles.

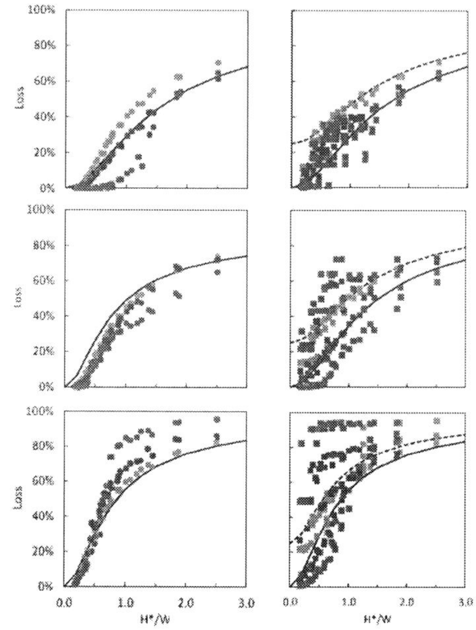

Figure 4: Direct irradiation loss due to shadowing from buildings as a function of the height-to-width ratio. Each row corresponds to different latitude (Equator – top, 30° – middle, 60° – bottom). The colour code corresponds to the orientation of the street: north-south (orange), northeast-southwest (red), east-west (blue). Circles correspond to the middle of the street (driving vehicles) and squares to parking spaces. Solid lines represent the direct irradiation model for driving vehicles (eq. 1) and dashed lines represent irradiation model for parking vehicles (eq. 2).

Table 1: Errors of the empirical fit to direct irradiation loss

Lat	RMSE driving	RMSE parking
0	9%	14.7%
30	5.7%	16.1%
60	11.8%	20.9%

2.3 Diffuse irradiation shadowing loss

If the diffuse irradiation is assumed to be isotropic, it will only depend on the sky view factor, the visible fraction of the sky, and it will not depend on the position of the sun in the sky, and thus the latitude or the time of the year (i.e., declination). For both driving and parked vehicles, the numerical simulations are well described by the shadowing loss coefficient c_d given by equation 3.

$$c_d = 1 - \cos\left(\mathrm{atan}\left(2\frac{H^*}{W}\right)\right) \qquad (3)$$

Results are shown in fig. 5. One can observe a small underestimation of the shadowing losses for wide streets (low H*/W) for parked vehicles for all street orientations. Deviations are 3.2% and 8.2% for driving and parked vehicles, respectively.

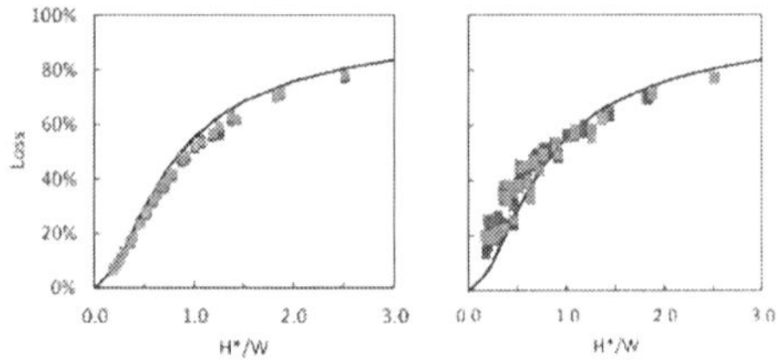

Figure 5: Diffuse irradiation loss due to shadowing from buildings as a function of the height-to-width ratio. The colour code corresponds to the orientation of the street: north-south (orange), northeast-southwest (red), east-west (blue). Circles correspond to the middle of the street (driving vehicles) and squares to parking spaces. Solid lines represent the diffuse irradiation model.

2.4 Global irradiation shadowing loss

The global irradiation shadowing loss can be constructed from its direct and diffuse components as shown in eq. 4, for driving (c_1) and parked (c_2) vehicles, where k_d is the fraction of diffuse irradiation.

$$c_1 = (1 - k_d)\, c_{b1} + k_d\, c_d$$
$$c_2 = (1 - k_d)\, c_{b2} + k_d\, c_d \qquad (4)$$

2.5 Solar extended driving range

Considering the local latitude ϕ, the average height-to-width ratio and the annual diffuse fraction, one can estimate the expected solar extended daily driving range (EDR, km/day/kWp) for each location (eq. 5), assuming an average driving efficiency (η_d) of 17 km/kWh [6].

$$EDR \;= GHI\,(1 - c)\,\eta_d\,\eta_{op} \qquad (5)$$

An extra PV operational efficiency (η_{op}) of 0.75 was introduced to account for losses associated with the curvature of the onboard PV modules or mismatch between individual solar cells [7].

3 RESULTS

To estimate the global potential of solar vehicles using equation 5, one needs annual values of global horizontal irradiation (GHI, kWh/m^2/day) and diffuse fraction. For this purpose, the ERA5 reanalysis data were used since it offers global coverage of 30 km spatial resolutions [4] (fig. 6). As it is well known, global irradiation is, in general, higher at lower latitudes with diffuse irradiation becoming dominant at higher latitudes. Nevertheless, local effects are relevant in certain regions.

Figure 7 presents the global distribution of the irradiation losses (given by eq. 4). We have also considered H=7.5m [8], h_{car}=1.5m [9] and W=8m [10]. One can observe that irradiation losses are generally higher for higher latitudes. Parked vehicles feature irradiation losses above 40-50% for most geographies, except for very high latitudes. Driving vehicles are more sensitive to latitude, with losses around 30% up to mid latitudes.

Results also show that the urban solar driving range can reach 60 to 70 km/day/kWp in the most favourable locations, at low latitudes, in parts of Africa, South America but also the Indian subcontinent, Southeast Asia, and the southern states of the USA (fig. 8).

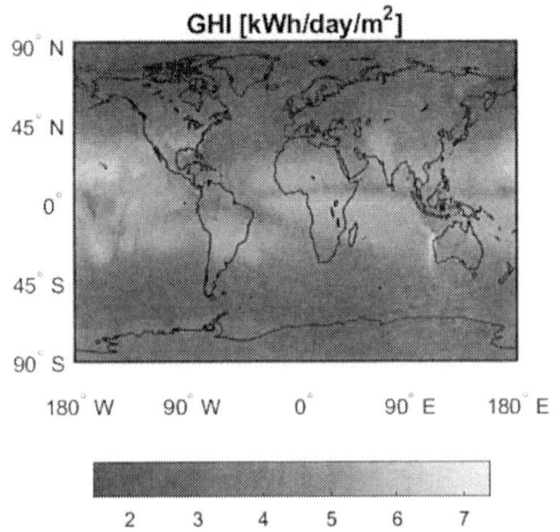

Figure 6: Global horizontal irradiation (GHI)– ERA5 reanalysis

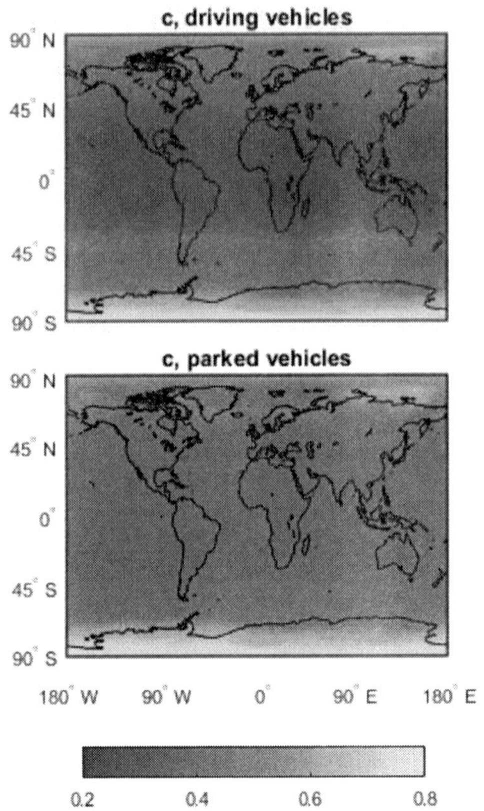

Figure 7: Irradiation losses for driving (top) and parked (bottom) vehicles.

It is also noteworthy the significant driving ranges of the order of 40 km/day/kWp in most of China, Japan, the USA, and Australia. In Europe, solar driving ranges are in the range of 20 to 30 km/day/kWp.

For parked vehicles, the solar extended driving range is significantly reduced for low latitudes, where direct irradiation losses are paramount, but quite similar to the driving range for driving vehicles for mid and high latitudes (including most of Europe).

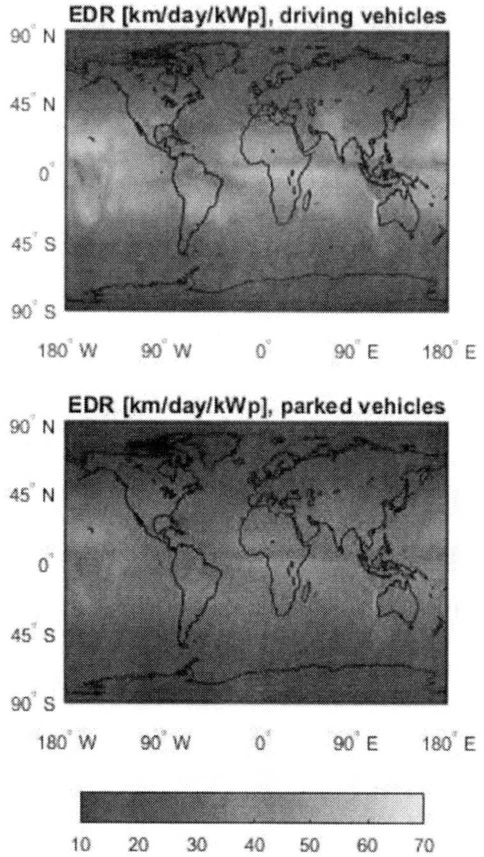

Figure 8: Solar extended driving range for driving (top) and parked (bottom) vehicles.

The ViPV potential pattern is mostly determined by the global horizontal irradiation. The impact of urban shading in Europe, in the range 0.35 to 0.55, shows that the effect of shading is mostly determined by the local latitude, the diffuse fraction having little impact on the results.

Considering median daily driving ranges of the order of 30 km/day [11] one can conclude that for most cities in the world, the annual average solar extended daily driving range exceeds the local typical needs, and therefore onboard solar power can have a significant contribution towards electric mobility, with less than 1kWp of installed photovoltaics per vehicle. These results thus underline the positive role that vehicle integrated photovoltaics may have towards the faster adoption of electric vehicles and, thus, the global reduction in CO2 emissions.

4 LIMITATIONS

It should be noted that this simplified empirical approach yields only a rough estimate of the solar potential of vehicles in urban environments. The main limitations of the present approach are associated with the urban layout configuration and the solar modelling itself:

a) Cities have complex and very different layouts, at different spatial scales, and whilst the simplified layout configurations analysed (fig.1) are representative of that variety, the modelled results (fig.4) show great sensitivity to the orientation and height-to-width ratio of the streets (in particular for parked vehicles) and therefore different urban areas in different cities will certainly yield very different solar potential.

b) For easier comparison, the solar driving range was calculated assuming a constant height-to-width ratio for all locations, which is a strong assumption since it is well known to change significantly depending on climatic, cultural, historical, etc. reasons.

c) Solar modelling was performed assuming isotropic diffuse irradiation and neglecting reflected irradiance. Furthermore, parked vehicles were assumed to be 1m away from the wall, which probably leads to an overestimation of irradiation losses for parked vehicles. On the other hand, driving vehicles were assumed to be in the middle of the street, which probably leads to an underestimation of irradiation losses for driving vehicles.

Thus, to assess the solar urban potential for vehicle integrated photovoltaics for a certain urban area it is always preferable to perform long term solar irradiance measurement on local vehicles, as reported in [12] or, at least ,detailed solar modelling, as described in [3].

5 CONCLUSIONS

We have estimated the global distribution of solar urban potential for vehicle integrated photovoltaics considering the effect of latitude and climate. Urban irradiation losses are estimates using an empirical model based on a standardized city layout simulated with ArcGIS Solar Area Radiation tool.

Results show that irradiation losses are mostly dominated by the effect of latitude, with higher losses at higher latitudes due to lower solar elevation, with minor variations introduced by the effect of climate – since diffuse irradiation losses are determined by the sky view factor and not the solar elevation (because we are assuming isotropic model for the diffuse irradiation).

The resulting solar extended driving range is of the order of 20 to 30 km/day/kWp for a wide range of geographies, highlighting the relevance of vehicle integrated photovoltaics for electric mobility. For mid and high latitudes driving and parked vehicles have similar VIPV potential.

Certain regions of the world, including most of the US, China and Australia may reach much higher VIPV potential.

6 ACKNOWLEDGEMENTS

This research was funded by national funds through FCT—Foundation for Science and Technology, I.P., within the scope of the project "UIDB/04647/2020" of CICS.NOVA—Centro Interdisciplinar de Ciências Sociais da Universidade Nova de Lisboa and project "UIDB/50019/2021" of IDL—Instituto Dom Luiz. The second author was financed by the FCT, under the Norma Transitória—DL 57/2016/CP1453/CT0004.

7 REFERENCES

[1] Carr, A.J.; van den Tillaart, E.; Burgers, A.R.; Köhler, T.; Newmanm, B.K. Vehicle Integrated Photovoltaics – Evaluation of the Energy Yield Potential Through

Monitoring and Modelling. In Proceedings of the 37th European Photovoltaic Solar Energy Conference and Exhibition; 2020; pp. 1701–1705.
[2] Berkeley, N.; Jarvis, D.; Jones, A. Analysing the take up of battery electric vehicles: An investigation of barriers amongst drivers in the UK. Transp. Res. Part D Transp. Environ. 2018, 63, 466–481, doi:10.1016/j.trd.2018.06.016.
[3] Brito et al, Urban solar potential for vehicle integrated photovoltaics, Transportation Research Part D: Transport and Environment 94, May 2021, 102810, doi:10.1016/j.trd.2021.102810
[4] Hersbach, H.; Bell, B.; Berrisford, P.; Hirahara, S.; Horányi, A.; Muñoz-Sabater, J.; Nicolas, J.; Peubey, C.; Radu, R.; Schepers, D.; et al. The ERA5 global reanalysis. Q. J. R. Meteorol. Soc. 2020, 146, 1999–2049, doi:10.1002/qj.3803.
[5] Lodi, C.; Seitsonen, A.; Paffumi, E.; De Gennaro, M.; Huld, T.; Malfettani, S. Reducing CO2 emissions of conventional fuel cars by vehicle photovoltaic roofs. Transp. Res. Part D Transp. Environ. 2018, 59, 313–324, doi:10.1016/j.trd.2018.01.020.
[6] Heinrich, M.; Kutter, C.; Basler, F.; Mittag, M.; Alanis, L.E.; Eberlein, D.; Schmid, A.; Reise, C.; Kroyer, T.; Neuhaus, D.H.; et al. Potential And Challenges Of Vehicle Integrated Photovoltaics For Passenger Cars. In Proceedings of the Presented at the 37th European PV Solar Energy Conference and Exhibition; 2020; Vol. 7, p. 11.
[7] Araki, K.; Ota, Y.; Yamaguchi, M. Measurement and Modeling of 3D Solar Irradiance for Vehicle-Integrated Photovoltaic. Appl. Sci. 2020, 10, 872, doi:10.3390/app10030872.
[8] Brzev, S., et al., 2013. GEM Building Taxonomy (Version 2.0). GEM Foundation.
[9] FIA, 2013. European Parking Garages are in Need of Improvement. Fédération Internationale de l'Automobile – Region I.
[10] PIARC, 2001. PIARC, 05.11.B – Cross Section Geometry in Unidirectional Road Tunnels. Permanent International Association of Road Congresses, 2001.
[11] Pearre, N.S.; Kempton, W.; Guensler, R.L.; Elango, V. V. Electric vehicles: How much range is required for a day's driving? Transp. Res. Part C Emerg. Technol. 2011, 19, 1171–1184, doi:10.1016/j.trc.2010.12.010.
[12] Y. Ota, K. Araki, A. Nagaoka and K. Nishioka, "Evaluating the Output of a Car-Mounted Photovoltaic Module Under Driving Conditions," in IEEE Journal of Photovoltaics, doi: 10.1109/JPHOTOV.2021.3087748.

SUSTAINABLE OPTIMIZATION OF PV VALUE CHAIN FOR ELECTROMOTIVE APPLICATIONS

Florin Popescu[1] and Stefan Wendlandt[2]
[1] Fraunhofer FOKUS, Kaiserin-Augusta-Allee 31, D-10589 Berlin, Germany,
Phone: +49 (30) 34 63 70-00, mail: florin.popescu@fokus.fraunhofer.de
[2] PI Photovoltaik-Institut Berlin AG, Wrangelstr. 100, D-10997 Berlin, Germany
Phone: +49 (30) 814 52 64-0, e-mail: wendlandt@pi-berlin.com

ABSTRACT:

While PV is poised to become the leading renewable energy source in the EU and globally, both in terms of capacity and net output, its low spatial power density poses challenges in its effective use and distribution. One of the emerging use-cases for PV power is in electromotive applications, ranging from automotive (Vehicle integrated PV – VIPV – for cars, buses, trucks), rail-transportation (urban and extra-urban rail) and marine (recreational boats and sea- and inland waterway transport). Each of these has a particular topology, both of proximity to stationary PV, energy density of storage, and available surface area for mobile PV (which integrates VIPV with fixed PV assets dedicated to vehicle fleet electricity supply). In order to maximize benefit the entire PV Value chain in these applications, we identify the individual importance of several necessary components in planning and operational intelligence, and provide relevant examples based on current and emerging technology as well as an integrative framework for the modeling of these and related value chains based on energy flow modeling.

Keywords: sustainability, mobile PV, VIPV

1 INTRODUCTION

In recent years, much attention has been paid to the widening scope of photovoltaics (PV), from traditional rooftop installations and PV farms to other use cases which involve the maximization of *value* rather than net yield. For example, vertical bifacial PV (VBIPV), provides a better load alignment to the distribution grid's daily aggregate load profile for same PV active material deployed. Marine PV or floating PV attempts to use flat areas next to the grid without competing with agricultural use. VIPV (Vehicle Integrated PV) is an attempt to increase range of battery-powered vehicles and allow bi-directional charge into the grid (when parked), with several road-bound and marine light-duty vehicles (LDVs) which integrate PV panels with battery-motor propulsion systems already on the market [1].

Cars are but a minor part of e-mobility
Major industrialized economies have at present committed to aggressive CO2 emission reductions by 2030. One of the most visible aspects of this transition from the consumer point of view is the significant and expanding market penetration of electric vehicles (ranging from microbility to to Heavy Duty commercial vehicles). While electric cars are most visible to consumers in general, and are the principal focus of governmental subsidies (now stabilizing) [2] electric single-occupancy vehicles (mainly electric scooters) provide the overwhelming share of personal-use Electric Vehicle (pEV) stock globally (i.e. service the most consumers). In fact, electric vehicles in other categories (mass transit systems of electric buses alone, along with trolleybuses light urban rail) currently provide transportation for more globally people than pEVs [2][3]. Finally, electric vehicles account for less than 5% of the world's vehicle stock, allowing plenty of room for EV market growth.

Increasing electric mobility footpring alone does not solve the climate problem.
While all EVs is perceived as reaching sustainability and climate warming mitigation goals, by itself, it merely shifts the C02 emissions topology away from cities and toward power plants. Power plants may reduce overall (non CO2, but harmful) emissions by controlled filtering, but the growth of EVs, if over-subsidized, might, absent a shift in energy production modality, even exacerbate the climate problem by contributing to the growth of personal transport-based energy consumption, currently sized at ca. 40% of aggregate crude oil demand [4][5], either by growing the number of LDEVs in circulation, or by growing their median size per passenger, which is a determinant predictor of aggregate transport-related energy consumption.

The entire chain from energy production to use must be considered.
Electric mobility is climate-neutral (or C02 emission target beneficial) only if coupled with a renewable energy source (e.g. solar, wind, hydro-electric, geothermal). This means that we must consider the energy efficiency and carbon footprint of the entire energy chain, from production to consumption. To this effect, recent authors have looked at the Energy Return on Capital Investment (EROCI), where the net energy (useful mechanical work at the wheel of a vehicle) is measured against a parity of investment in different means of energy production (fossil fuels or renewable alternatives) [4], finding a higher return for renewables than for fossil fuels, with parity among wind (terrestrial and off-shore) and PV, arguing that basic micro-conomics will eventually force a global energy production shift.

PV, sounds great, but how?
The well known problem of renewables (excepting hydro-electric power) is stochastic intermittency due to dependence on weather. In this case, a PV-heavy mix of solar and wind seems optimal, as PV matches day/night load imbalances better than wind. Therefore, assuming a large upcoming PV penetration, what can be done with all these panels?

First, we need to consider the configuration of different mobility sectors and existing stocks. For example, while electric car sales are still growing (excepting 2020), these combine plug-in hybrids, which are still fossil-based, and are only partially convertible to renewables use. Furthermore, despite recent increases in PV yield, VIPV cannot physically cover but a minor fraction of the entire energy use of the vehicle because of the inherent energy spatial density limit of solar irradiation. This covers cars, commercial vehicles, buses, boats and trucks. Then again, the bulk of e-mobility is micro-mobility: although they are far more efficient for urban transport than cars (both in energy efficiency and space usage) from a PV perspective, they have an extremely limited surface area.

Solar energy: goes anywhere because found everywhere

Among renewables, PV has two main advantages: the aforementioned generation-to-load matching, and the ability to be deployed anywhere the sun shines, i.e. the ability to reduce transmission losses over the grid. Therefore we stress that an analysis of PV-based electromotive (PVEM) applications in a wide sense, which include but are not limited to VIPV, therefore including both mobile (VIPV, batteries) and fixed assets. One example of fixed assets is that of parking lots shaded by opaque or semi-transparent PV, which allow charging points direct from PV at those points in which vehicles in a fleet are stationary, and even bi-directional charging/discharging for grid feed-in for longer pauses such as weekends.

<u>PV can cover different e-mobility market sectors</u>

In Table 1a we cover different market sectors, power, range, load characteristics of different vehicles as per commonly used vehicle categories:

Power Category	Typical Power	Subcategory	Typical Range	Electrical Supply	Typical Battery Size	Usage	Surface area
Micro-	1kW	eMM	10km	5-60 VDC	< 1kWh	Intermittent	very small
		AMR	1km			Regular	very small
Small	5-10kW	eSOV	20km	12-60VDC	1kWh-5kWh	Intermittent	very small
		eSB	2km		10kWh	Intermittent	medium
Medium	100kW	BEV				Intermittent	small
		PHEV				Intermittent	small
		eLCV	100km	up to 600VDC	100kWg	Regular	medium
		eRB	10km			Intermittent	medium
Large	500kW	eHDV	200km	Level 3 charging	500kWh	Regular	large
	0.5-5MW	eFB	2-10km	up to 10kV	0.5-5MWh	Regular	large

Table 1a. Load (approximate) and electrical distribution bus characteristics of different vehicle classes.

Term	Category / Subcategory
EV	Electric Vehicle
pEV	personal Electric Vehicle
eLDV	electric Light Duty Vehicle
PHEV	Plug-in Electric Vehicle
BEV	Battery Electric Vehicles
eLCV	Electric Light Commercial Vehicle
eSOV	electric Single Occupant Vehicle
eMM	electro micro-mobility
ePM	electro personal-mobility
eDAV	electric disability assistance vehicle
eHDV	electric Heavy Duty Vehicles
eB	electric Bus
ULR	Urban Light Rail
eHDT	electric Heavy Duty Truck
eHFT	electric Heavy Freight Truck
AMR	Autonomous Maintenance Robots
eB	Electric Boat
eSB	electric small Boat
eRB	electric recreational boat
eFB	electric Ferry boat

Table 1b. Glossary of abbreviations for vehicle classes.

Note that the vehicle categories are grouped in practice according to market characteristics rather than physical characteristics or intended use (pEV and pSOV are introduced here as to group vehicle categories which have to do with intended use: personal mobility – anything from an electric monowheel to a large car). As we can observe, there are almost as many classes of vehicle classes applicable to solar energy as there are non-airborne transportation modes. With overall EV share of less than 5% of the global vehicle stock [IEA], there is ample room for growth and for standardization across market sectors.

<u>Unmet technical challenges of PVEM</u>

As mentioned by previous authors [POSITION PAPER], the following rather self-evident challenges remain:

- The optimization of radiance collection from limited areas with complex surface geometries, often curved, while still maintaining cost-effectiveness and yield ratio of silicon.
- High mechanical, temperature and chemical resistance of PI installations. Urban and VIPV environments pose higher exposure to both mechanical stresses (impact, tension), as well as chemical loads, especially in marine/coastal environments.
- Marine applications distinguish themselves from terrestrial ones by larger surface area (both vertically and horizontally), higher tolerance for lower energy density storage and a less frequent usage schedule. Irradiation patterns are different. Therefore marine-specific electromotive energy chains are to be developed further.
- A reliable testing and norm development procedure for reliable coverage of the above challenges
- reliable accompanying models and forecasts of yield (and degradation) in complex environmental structures for maximally efficient battery and panel sizing.
- possible integration into a wider grid or micro-grid such that PV provides (at least) range boosting or reliable back-up power when off-grid.
- A high value energy-chain that leverages every Joule collected

It is the last of these challenges that we address in this article.

Managing energy flows to cover demand, profitably, carbon-neutral.

We would argue that most relevant to a circular, carbon-neutral economy is not "net" EROCI [4], since capital depends on demand and interest rate fluctuations, neither EROI (energy return on energy investment) but useful (i.e. mobile) mechanical work return on energy investment (or, given high rotary generator efficiencies, return on work investment, since work can be easily converted back into electrical energy). This means, for every Joule that we use at the wheel for a given energy chain, meaning from production, though transportation, regulation (to standard voltages, either DC or AC) and consumption, how many extra Joules remain for re-use in getting new energy? What is the net gain of this circle? This approach removes financial and market aspects from the equation (mostly). For example, in the earliest from of such a chain, coal mining to steam ship operation, one would want to know how much mechanical work it takes to mine a certain unit of coal, and how much mechanical work that provides, after it is transported to the boiler and transformed by rather inefficient (by modern standard) steam engines. We prefer the term energy *collection* rather than production, since all energy production begins not at the potentially fuel burning (steam turbine) power plant, but at the mine. We collect energy from the environment: we extract fossil and mineral fuel (by digging or drilling), we collect solar irradiation and transform it to heat and electricity, we gather geothermal energy (steam to power), we harness kinetic energy of wind (turbines) and potential energy of water (water turbines). In this source-to-destination energy balance, we'd ask: what is the overall energy efficiency of PV vs. modern "green" alternatives?

2 ENERGY CHAINS FOR MOBILE POWER

In Table 2 we outline some calculations using a few energy chains for mobility. We used hydrogen based chains such as hydrogen combustion engine, hydrogen fuel cells, green diesel (with H mixed in). We use natural gas as representative of fossil fuels since it is the cleanest and most efficiently transportable and usable, with little or no refinement needed. Finally, for PV we compare electricity provided by PV farms through the grid, and finally stored in batteries, PV from close-by fixed assets, off-grid but into batteries (hot-swappable, or fixed), and finally PV autonomous, which is VIPV with large surface areas – currently achieved only in experimental vehicles that are impracticably large, but useful as a theoretical limit.

	Natural Gas	ρ	Green diesel	ρ	H combustion	ρ	H fuel-cell	ρ	VIPV	ρ	PV fixed assets	ρ	PV Grid	ρ
Collect (EROI)	Mine (drill)	21.00	Collect	1.50	PV electrolysis	9.00	PV electrolysis	9.00	Collect	10.0	Collect	10.0	Collect	10.0
													Regulate	0.90
Transport	Pipeline	0.95	Mixed	0.99	Pipeline	0.99	Pipeline	0.99			(local)		Grid	0.45
Regulate									Regulate	0.97	Regulate	0.97	Regulate	0.90
Store	Store	0.94	Tank	0.99	Tank	0.94	Tank	0.94	Optional		Battery	0.70	Battery	0.70
							Generate	0.56						
							Regulate	0.97			Regulate	0.97	Regulate	0.97
Work	Work	0.30	Work	0.40	Work	0.30	Work	0.91	Work	0.91	Work	0.91	Work	0.91
Total chain efficiency		0.27		0.39		0.28		0.46		0.88		0.60		0.22
Total EROI (mobile)		5.63		0.59		2.51		4.13		8.81		5.95		2.25

Table 2a. Comparison of some sample "green" electromotive energy chains. Sources: [5,6-10]

Component efficiency	ρ	dρ/dx	Notes
Pipelined H transportation	0.99	< 0	← Operational in Germany
Pipelined gas transportation	0.99	< 0	since 1930s
Compressible gas storage	0.94		50% loss of LNG
Liquified gas storage	0.88		LNG
Hi-efficiency internal combustion	0.30		
Diesel engine	0.50		
Fuel cell efficiency	0.56		
DC-to-DC regulation	0.97		
Electric motor efficiency	0.91		
DC-to-HVAC regulation	0.90		inverter
Li-Ion battery	0.70		Cycle dependency
AC-to-DC regulation	0.90		converter
HV Grid transportation	0.45	< 0	

Table 2b. Efficiency values used in Table 2[a].

The information from Table 2 is best viewed graphically, see Figures 1 and 2 below.

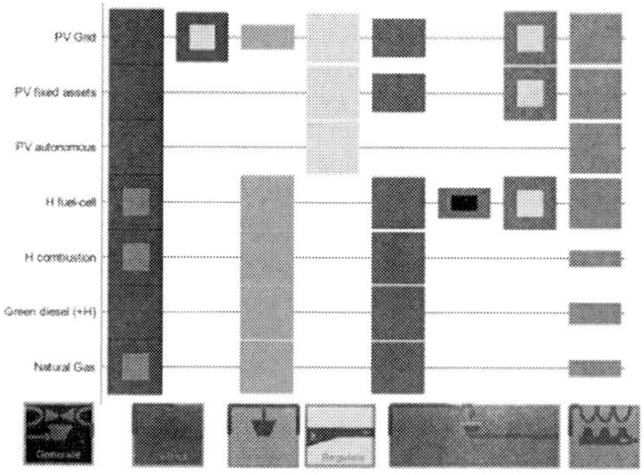

Figure 1. Energy chains of different currently available mobile transport solutions. From left to right, 1 energy unit produced undergoes several serial transformations, each with its own efficiency (height of each terminal block relative to 1 energy unit on left). "PV autonomous" is a generalized category of VIPV to all sizes.

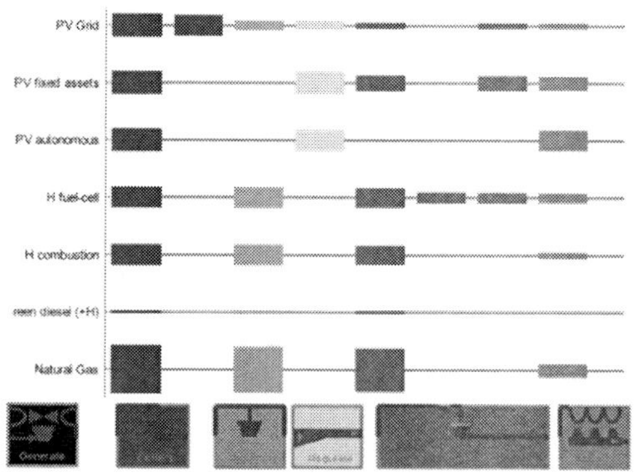

Figure 2. EROI-based value chain of mobile solutions. For each energy unit spent (height of green horizontal line), collecting a given energy types gives an energy return which is then reduced in progressive stages by the process chain in Figure 2. Note the high returns of PV in autonomous and fixed-asset configurations (off- and micro- grid), even compared to "green" fossil fuels (natural gas).

The results assembled herein have some caveats: the EROI values for PV and natural gas are subject to estimation error, and in the case of natural gas, vary by a factor of 2 above and below the cited value (the median of studies assembled in [5]). The PV value is also subject to variation according to specific assumptions. For example, we do not include in this analysis how these chains and WROI values depend on distance of collection to use, or on time, or number of cycles, we only look at initial outlays assuming no degradation. We also omit energy investments engines and batteries, which can be quite large. Unlike [4] we do not use spot price of crude as an indicator of future long-term fossil fuel prices (we do not consider demand at all), and attempt to use best-of-practice values at the time of writing, rather than future projections of technology. We also use aggregate grid loss, since in our estimation it is the correct value (rather than resistance losses based on distance), because distance independence is one of the basic functions of the grid.

Therefore, despite some variations and omissions, which can later be investigated, we agree with previous authors than PV has a WROI greater than fossil fuels, with an advantage that is bound to grow in the coming decades. The value (WROI-1) is the **gain** of the energy cycle (scaled somewhere between respective VIPV and Fixed Asset PI values). Important to understand though is that the loop time of the PV cycle (the time required to earn the WROI) is much longer than the natural gas cycle, on an order of 10-100, making fossil fuel exploration and extraction quite short-term profitable. Also to be noted that high WROI of PV depends on best practice assumptions: cost- and energy effective base technology (silicon), regular maintenance, proper installation, etc.

What of other "competitors"? Hydrogen based-chains (even if they use PV as primary energy for hydrolysis) have other blocking factors: either the need to combustion engines or loss of energy in the fuel-cell load/discharge cycle (which involves large costs of several processes, including expensive fuel-cells themselves, and need to be reduced). Green fuels (agricultural bi-products), even when mixed with hydrogen, are expensive to collect (expensive in our context meaning in terms of energy expended), and further lose energy in combustion engines.

3 MOBILE ELECTRICAL POWER TOPOLOGY

We therefore conclude that for electro-motive applications a mix of VIPV and fixed assets has the optimal WROI, with the VPIV PV to fixed PV ratio being limited by the basic geometry of the vehicle (the surface area), and the overall PV sizing dependent on the yield, with either vehicle-integrated, fixed, or hot-swappable batteries as needed, sized to load, yield and yield volatility profile. Another topic of concern related to maximizing efficiency is the need to eliminate redundant voltage regulation steps. While voltage regulation is an energy efficient process (more than 90%), it does imply added material, parts, and increased CAPEX costs.

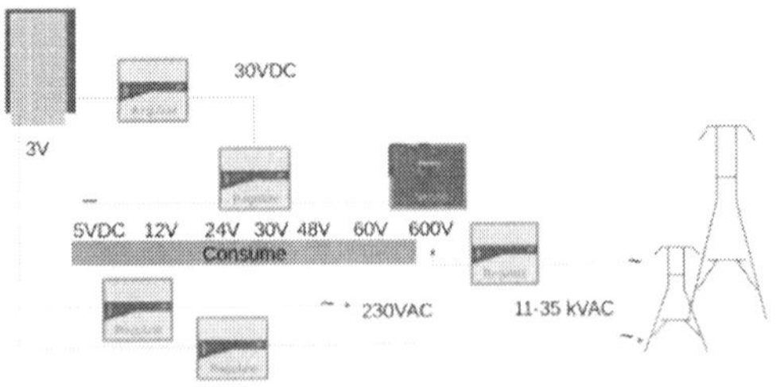

Figure 3. Energy transformations: electricity grids and buses for use of PV charge. The 600VDC Store block is a level 2 e-mobility charger. The lower VDC levels ("consume") can be used directly or to charge batteries in small power usage categories.

Therefore we propose that a basic on-panel regulator for VIPV output 30V, which can easily be then converted to other common DC bus voltages, both higher and lower, up to that required for Level 2 charging (of pEVs). A viable alternative for LDVs would be 12V, which is the standard bus voltage on cars and medium-sized vehicles. This would be used to supply power to auxiliary systems (e.g. lights, car stereo). In fact, many recreational vehicle (RVs, LCV-size) user have long used their roofs as PV installation surfaces, able to draw enough power to charge the vehicle's battery – usually enlarged in capacity with a deep-charge battery – to run auxiliary systems when parked without the use of a (noisy, fuel-guzzling) generator.

4 CHALLENGES AND SOLUTIONS

We identify several outstanding challenges toward increasing PV e-mobility footprint not addressed above:

- Mounting stability, flexibility, and chemical robustness of mobile panels affording maximal collection area at efficient yield to CAPEX ratios for VIPV.
- A need to place fixed PV assets in the nearest accessible open space allowing for traditional, efficient string layouts.
- A need to know (predict) how much energy can be collected while moving (ostensibly without connection to fixed assets, mostly stored), and how much energy is available for fixed assets.

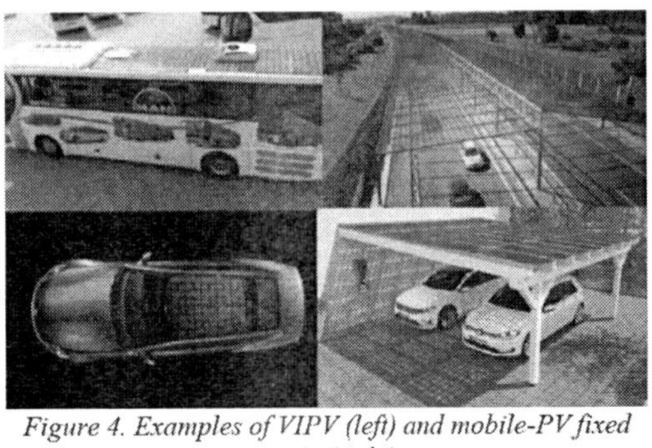

Figure 4. Examples of VIPV (left) and mobile-PV fixed assets (right).

Recommended Solutions:

- DC low-voltage bus whenever possible, avoid regulation losses.
- Fit-to-use-case battery and panel sizes (both fixed and

mobile).

- Combined Fixed PV with VIPV, mixing dispersed charging points for battery-powered mobile device when parked, using efficiently distributed, modular batteries with direct current flows. See Figure 4 for VIPV and fixed PV examples for e-mobility.
- Reliable estimates of yield and component efficiency, laboratory tested, producing models accounting for environmental factors and usage (i.e. temperature, dynamic profiles, number of cycles).

5 SUMMARY

As mobility comprises a large part of overall energy consumption, it is critical to model and design mobility systems based on sustainable energy chains. Evolution of technology has meant recent increases in Energy Return on Investment (ERoI) of PV. Accompanied by a corresponding decrease on ERoI of fossil fuels (increasing extraction costs), it points to a fixed- and mobile- PV asset as offering the highest ratio of mobile mechanical work from collection (or extraction) energy expended among common alternatives, offering win ratios on order of 6 to 1. Therefore it is critical to develop new ways of integrating high-yield PV technologies (silicone-based) into mobility networks, overcoming their low spatial density. We have identified several critical challenges and pointed to their resolution. For example, low spatial density requires reducing distances between fixed assets and urban centres, as to limit transmission losses. Both fixed urban and on- vehicle mounting of PV require new mounting and encapsulation methods to protect from mechanical stress and impact, while minimizing material cost. Electrical transformation losses may be minimized by concentrating on low-voltage DC buses for traction and control system support. Finally, minimal battery sizing (which affect CAPEX costs) can be achieved by matching accurate local forecasts of yield to expected load, both short and long-term, in mobile, fixed and mixed configurations.

8 REFERENCES

[1] Photovoltaic European Technology and Innovation Platform (2019) "Vehicle-Integrated Photovoltaics (VIPV) as a Core Source for Electricity in Road Transport." Position Paper

[2] IEA (2021) "Trends and Developments in Electric Vehicle Markets" https://www.iea.org/reports/global-ev-outlook-2021/trends-and-developments-in-electric-vehicle-markets

[3] IEA (2020) International Energy Agency (IEA), Clean Energy Ministerial, and Electric Vehicles Initiative (EVI) (June 2020). "Global EV Outlook 2020: Entering the decade of electric drive?". IEA Publications.

[4] Mark Lewis (2019) "Wells, Wires and Wheels,: EROCI and the Tough Road Ahead for Oil" BNP Paribas Asset Management Report

[5] Simon Michaux (2019) "Oil from a Critical Raw Material Perspective" Geological Survey of Finland Report 70/2019

[6]http://www1.eere.energy.gov/hydrogenandfuelcells/pdfs/accomplishments.pdf (2019)

[7] https://ec.europa.eu/energy/sites/ener/files/documents/20141211_GuidelinesElectricMotors cover.pdf (2014)

[8] https://files.sma.de/downloads/MVPS_2SC-DS-de-40.pdf (2021)

[9] Ulvestad, Marte, and Indra Overland (2012). "Natural gas and CO2 price variation: impact on the relative cost-efficiency of LNG and pipelines." The International journal of environmental studies vol. 69,3 pp 407-426.

[10] V. Michal, (2014) "Peak-Efficiency Detection and Peak-Efficiency Tracking Algorithm for Switched-Mode DC–DC Power Converters," in IEEE Transactions on Power Electronics, vol. 29, no. 12, pp. 6555-6568

MODELING AND SIMULATION OF MULTIPUMP PHOTOVOLTAIC IRRIGATION SYSTEMS

Javier R. Ledesma, Rita H. Almeida, Luis Narvarte
Grupo de Sistemas Fotovoltaicos. Instituto de Energía Solar - Universidad Politécnica de Madrid.
Campus Sur UPM, Ctra. Valencia km 7, Madrid, 28031, Spain. javier.ledesma@upm.es

ABSTRACT: In photovoltaic (PV) irrigation applications, models that allow estimating the output of a single variable speed driven (VSD) pump connected to a PV generator are available, such as the SISIFO software (www.sisifo.info). In real systems, it is common to connect several pumps in parallel to the same generator instead of a single pump. In this work, a model is generated that allows estimating the pumped water volume in a system that, sharing the same PV generator, is formed by a group of frequency-synchronized VSD pumps connected in parallel, to which a second group of frequency-synchronised VSD pumps can be optionally added. This model makes it possible to establish the number of active pumps in each group and their working frequency, depending on the PV power available, optimising some criteria. In this version, the criterion is to achieve the highest pumping volume during the irrigation period considered. This version is now freely available in the online application SISIFO.
Keywords: multipump, photovoltaic pumping, irrigation systems, real-time simulation, variable speed driven pumps.

1 INTRODUCTION

The number of photovoltaic (PV) irrigation systems is increasing worldwide. For this reason, there is the need to model and simulate multipump PV irrigation systems (PVIS) [1][2][3][4][5]. In contrast to grid-connected irrigation systems, in which the power is always available, in irrigation systems powered by a PV generator, the available power at any time is limited and not constant. Therefore, it is necessary to adapt the operation of the pumps to the available PV power. To achieve this, at least some centrifugal pumps must work at variable frequency [6][7] to adapt the power demand of the pumping group to that provided by the PV generator at each moment.

Currently, some models [8][9] allow estimating the water flow rate of a PVIS – consisting of a PV generator connected to a frequency converter, a motor and a centrifugal pump (see figure 1) –, based on meteorological data and the parameters and characteristics of the PVIS and the hydraulic system. One of these models is implemented in SISIFO [10], a freely available online software tool.

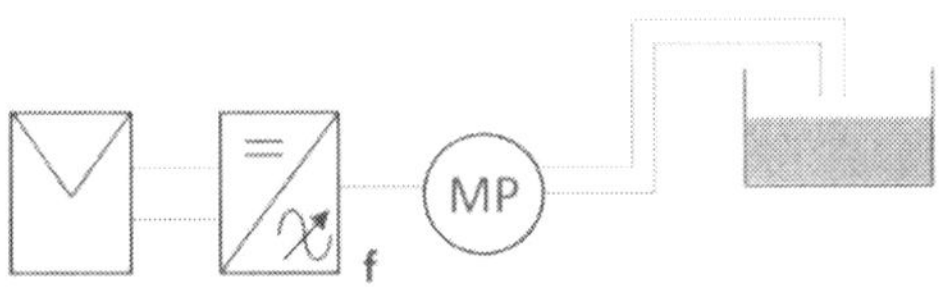

Figure 1: Components of a PV irrigation system: PV generator, frequency converter, motor, pump and, optionally, a water tank.

In real systems, it is common to find several pumps operating in parallel, but the existing simulation tools just allow estimating the water flow of a PVIS with a single pump. Usually, to estimate the total volume pumped by a group of N pumps powered by a PV generator with a peak power of P*, a single pump powered by a PV generator of P*/N is simulated and, then, the result is multiplied by N. Although, some error is assumed.

This paper describes the modeling of a multipump PVIS that meets certain conditions. The characteristic curves corresponding to the equivalent pump of a group of two or more equal centrifugal pumps, synchronized in frequency, are modeled. Then, an irrigation system is modeled – it can combine up to two groups of centrifugal pumps where:

a) each group of pumps contains only equal pumps,

b) one group of pumps may contain a different type of pump than the ones in the other group,

c) at each instant, one or more pumps in each group may be active,

d) all active pumps in the same group operate synchronized in frequency and,

e) one of the groups of pumps work at variable frequency while the second groupmay work also at variable frequency (that can be different than the frequency of the first group) or at a fixed frequency (typically the nominal one).

These conditions have been found to be common in real PVIS. This model has been implemented in the online software tool SISIFO.

2 DESCRIPTION OF THE MODEL

In this modeling, a single large PV generator is shared by several groups of pumps. The necessary balance between a group of pumps operating at fixed nominal or variable frequency with another group operating at variable frequency is modeled – this way, both groups do not demand equal instantaneous power from the PV generator. The modeling also allows the selection of the ideal combination of active-inactive pumps at each instant to maximize the water flow with the available PV power.

2.1 A single group of one or more equal VSD pumps in parallel

From the characteristic curves provided by the manufacturer for a single pump, equations (1) and (2) can be fitted

$$H = a_1 + b_1 Q_1 + c_1 Q_1^2 \quad (1)$$
$$P_2 = d_1 + e_1 Q_1 + f_1 Q_1^2 \quad (2)$$

where H is the pumping head, P_2 is the power at the pump input (or motor output), Q_1 is the water flow, and the parameters a_1, b_1, c_1, d_1, e_1 and f_1 are the parameters that fit the curves. It is straightforward to deduce that for the case of N equal pumps in parallel, then equations (3) and (4) apply

$$H = a_N + b_N Q_N + c_N Q_N^2 \quad (3)$$
$$P_{2,N} = d_N + e_N Q_N + f_N Q_N^2 \quad (4)$$

where: $Q_N = N * Q_1$, $a_N = a_1$, $b_N = \frac{b_1}{N}$, $c_N = \frac{c_1}{N^2}$, $d_N = N * d_1$, $e_N = e_1$, $f_N = \frac{f_1}{N}$

Therefore, a group of N parallel pumps working synchronised in frequency can be modelled as a single equivalent pump with the given parameters. Figure 2 shows a system consisting of a single PV generator, two frequency converters, two identical motor pumps in parallel (operating at the same frequency) and a single hydraulic system.

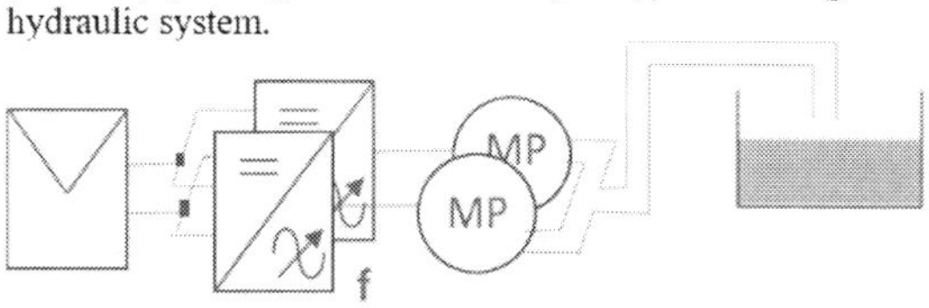

Figure 2: Components of a frequency-synchronised multipump PV irrigation system.

Figure 3 shows the characteristic curves of an individual pump and the characteristic curve of each of the equivalent pumps consisting of one or more (two in this case) equal pumps. The number of active pumps in the group at each instant determines the characteristic curve of the equivalent pump.

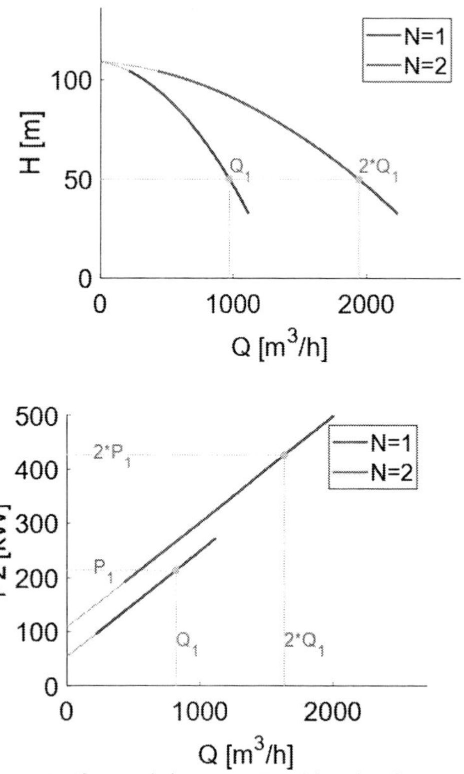

Figure 3: Characteristic curves, H-Q (above) and P2-Q (below) of a single pump and the equivalent pump made up of two equal pumps operating in parallel.

The model must therefore, depending on the available DC power of the PV generator, determine:

a) the number of active pumps, and

b) the working frequency of the pumps.

This selection is made to optimise a certain criterion, which typically corresponds to maximize the water flow with the available PV power at each instant .

2.2 Two groups of VSD pumps in parallel

In this case, as Figure 4 shows, each of the groups formed by pumps of the same pump-model can be reduced to a single equivalent pump with the characteristic curve indicated in section 2.1. In this case, the question becomes to model a system that:

1) allows to activate and deactivate pumps in each of the groups (that is, to choose different equivalent pumps resulting from combining between 0 and the number of pumps present in each group), and

2) allows to operate each equivalent pump at a different frequency f_A and f_B respectively.

In this case the DC power (P_{DC}) available in the generator is not equally distributed between the loads, because in each group the frequency converter, motor and pumps operate at different duty points with different efficiencies.

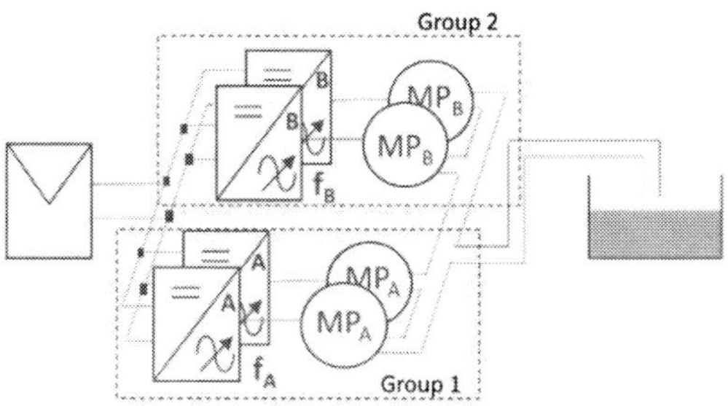

Figure 4: Components of a multipump PV irrigation system made up of two different groups of pumps.

Figure 5 shows a case consisting of a single pump in group A, and a second pump (of the same model, for simplicity) in group B. It can be seen how, depending on the available P_{DC}, the model allows to know which is the optimal combination of active pumps in each group (in green), their operating frequency (orange for the pump in the first group, and yellow for the pump in the second group), and the power sharing ratio demanded by both pumps (purple).

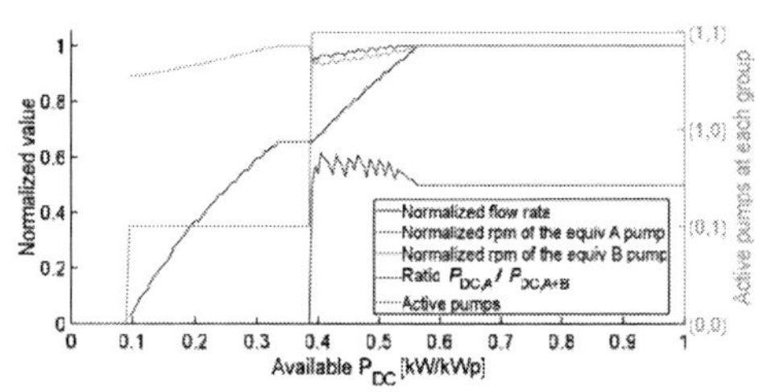

Figure 5: Optimus solution for the presented case, depending on the available DC power.

3 SIMULATION OF A REPRESENTATIVE CASE

Two pumps are installed in a real PVIS located in Najerilla, Spain (lat 42.5229, lon -2.9539). This system consists of a 500 kW PV generator (one-axis horizontal trackers) and two 200 kW equal pumps.

The results of three different configurations are

presented and analysed:

1) a single group of two identical pumps operating at the same variable frequency,

2) a group of only one pump operating at a fixed nominal frequency and a second group with the other pump operating at a variable frequency, and

3) a group of only one pump operating at a variable frequency and a group with the other pump operating at a (possibly) different variable frequency.

These combinations of two pumps of the same model make it possible to focus on the effects of different types of control (frequency-synchronized or not), and to compare the results with those usually estimated from calculations with a single pump. To understand better the importance of the modelling, a simulation of a single pump is also performed.

Table I shows the estimated annual pumped volume for these cases:

(a) G,1f_A: a single pump connected to a PV generator of half the power of the real system i.e. 250kWp, marked with (*) the case where the current hydraulic system is used and without (*) the case where the section of the hydraulic system is reduced so that the friction losses are similar,

(b) 2G,1fnom+1f_A: a system with 500kWp connected to two pumps in parallel, one of them operating at nominal fixed frequency and the second one at variable frequency,

(c) 2G,2f_A: a system with 500kWp connected to two pumps in parallel, in which the two pumps operate at the same variable frequency f_A, and

d) 2G,1f_A+1f_B: a system with 500kWp connected to two pumps in parallel, where both pumps operate at different variable frequencies.

Table I: Annual estimation for a real photovoltaic irrigation system at Najerilla (Spain), two equal-model pumps, comparing different configurations

	Volume [m3/kWp]	El [kWh/kWp]
G,1f_A(*)	4,249	1,462
G,1f_A	3,557	1,354
2G,1f_{nom}+1f_A	3,933	1,407
2G,2f_A	3,935	1,405
2G,1f_A+1f_B	3,966	1,422

Analysing these results, it can be observed that:

1) If only one pump is considered in the real system without downsizing the hydraulic system, case G,1f_A(*), an overestimation of the production is committed. If the result of a single pump is considered (case G,1f_A), adjusting the hydraulics and the PV generator, the output is underestimated. This is true regardless of the type of frequency control used. This is because, in low irradiance conditions (beginning and end of the day, and cloudy days) the PV generator (double the power of a single-pump one) can provide enough power to use one of the two pumps. Thus, figure 6 shows how the difference between one and two pumps is greater in months with lower irradiance or frequent cloudiness, as occurs in this location even in summer.
2) The difference between operating a single pump at nominal frequency and at variable frequency is small. This depends on the number of pumps to be considered, the efficiencies and other parameters. In general, it is appropriate to consider whether to simplify the control of a part of the pumps by leaving only some of them to operate at variable frequency.
3) The difference between operating one pump at nominal frequency (2G,1fnom+1fA case) or all synchronised at a variable frequency (2G,2f_A case) is small. This depends on the number of pumps to be considered, the efficiencies and other parameters but, in general, in each location it is interesting to analyse whether or not to simplify the control of a part of the pumps by fixing their frequency and leaving only some of them to operate at variable frequency.
4) Operating both pumps at an independent variable frequency (2G,1f_A+1f_B case) is generally better than the 2G,2f_A case where they operate synchronised at the same frequency. The different efficiencies of the frequency converter-motor-pump set mean that, at certain times, it is better to work at different frequencies.

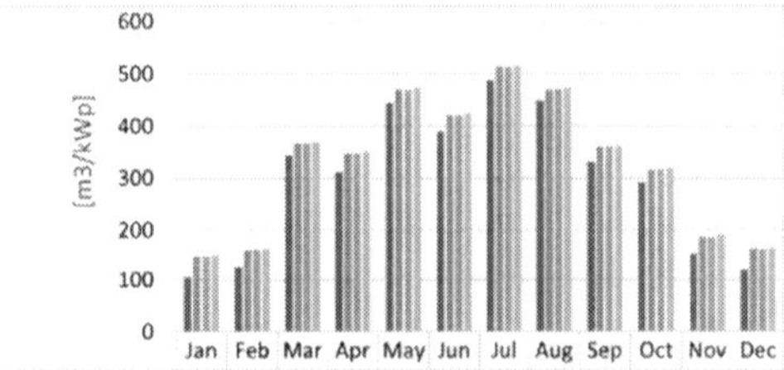

	Jan	Feb	Mar	Apr	May	Jun	Jul	Aug	Sep	Oct	Nov	Dec
1G,1fA	105	129	344	312	444	390	488	449	331	293	155	119
2G,2fA	151	163	368	348	470	422	514	470	361	315	188	164
2G,1fnom+1fA	151	163	368	348	470	421	514	470	361	316	186	164
2G,1fA+1fB	153	165	370	351	473	424	516	473	363	319	191	167

Figure 6: Monthly pumped volume for each case.

5. CONCLUSIONS

This modelling makes it possible to cover a very large number of cases in real PVIS, where it is common to find different groups of pumps working in parallel sharing a single PV generator.

This model also avoids the over- and under-estimation errors that can arise from the results estimated for a single pump.

It allows to open new lines in the design of new PVIS, to find the most suitable combination of different pumps in each location, given its meteorological data, PV generator, and hydraulic system. It also allows optimising the control of the systems (by analysing the convenience of operating the pumps combining fixed frequency, variable frequency, or two different variable frequencies).

Finally, the results obtained for a case of a simple real system have been presented, focusing on the differences in the control of its two pumps.

7. REFERENCES

[1] G. Li, Y. Jin, M. W. Akram and X. Chen, "Research and current status of the solar photovoltaic water pumping system - A review," Renewable and Sustainable Energy Reviews, vol. 79, pp. 440-458, 2017.

[2] Z. Gao, Y. Zhang, L. Gao and R. Li, "Progress on Solar Photovoltaic Pumping Irrigation Technology," Irrigation and Drainage, vol. 67, pp. 89-96, 2018.

[3] P. E. Campana, H. Li and J. Yan, "Dynamic modelling

of a PV pumping system with special consideration on water demand," Applied Energy, vol. 112, pp. 635-645, 2013.

[4] A. Ghoneim, "Design optimization of photovoltaic powered water pumping systems," Energy Conversion and Management, vol. 47, pp. 1449-1463, 2006.

[5] A.A. Hamidat and B. Benyoucef, "Mathematic models of photovoltaic motor-pump systems," Renewable Energy, vol. 33, 2008.

[6] J. F. Gülich, Centrifugal Pumps, Springer, 2010.

[7] M. A. Abella, E. Lorenzo and F. Chenlo, "PV Water Pumping Systems Based on Standard Frequency Converters," *Prog. Photovolt: Res. Appl.,* vol. 11, pp. 179-191, 2003

[8] .J. Munoz, J. Carrillo, F. Martínez-Moreno, L. Carrasco and L. Narvarte, "Modeling and simulation of large PV pumping systems," in *Proceedings of the 31th European Photovoltaic Solar Energy Conference and Exhibition*, Hamburg, 2015.

[9] R. H. Almeida, J. R. Ledesma, I. Carrêlo, L. Narvarte, G. Ferrara, L. Antipodi, "A new pump selection method for large-power PV irrigation systems at a variable frequency", in Computer Science, 2018

[10] Instituto de Energía Solar - Universidad Politécnica de Madrid, "SISIFO - Simulación de Sistemas Fotovoltaicos," [Online]. Available: https://sisifo.info/

SIMULATING THE ENERGY YIELD OF BIFACIAL PHOTOVOLTAIC MODULES INSTALLED ON CARPORTS OR CANOPIES

Jesús Robledo[1,*], Jonathan Leloux[1], Babacar Sarr[1], Christian A. Gueymard[1,2], Anton Driesse[3], Pierre-François Drouin[4]
[1] LuciSun, Rue Saint-Jean, 29, 1495 Sart-Dames-Avelines, Belgium
[2] Solar Consulting Services, P.O. Box 392, Colebrook, NH 03576, USA
[3]PV Performance Labs, Emmy-Noether-Str. 2, 79110 Freiburg, Germany
[4] Syneria, 5 Rue Taitbout, 75009 Paris, France
*Phone: +34 656 37 59 74; E-mail: jesus.robledo@lucisun.com

ABSTRACT: The installation of solar photovoltaic (PV) modules on carports or canopies is a relatively popular application of PV power generation. This advantageously combines power generation and shaded parking for vehicles in the case of carports, or power generation and shelter and shade for people, animals or goods with little waste of land. In parallel, bifacial technology has experienced a swift increase in its share of the PV market in recent years, which opens the door to various improvements and even new kinds of solutions. However, in current state-of-the-art PV simulation tools, the added complexity introduced by bifacial technology and by the complex geometry of PV installations on carports or canopies is still not fully covered and the evaluation of the bifacial energy gain (BEG) is not always accurate. In this work, a novel simulation tool is introduced, which benefits from the latest 3D evaluation libraries incorporated into the Graphic Processor Unit (GPU) of modern computers, as developed for the video game industry. This novel approach provides an evaluation of the BEG with a spatial resolution that is similar to what is achievable with backward ray-tracing techniques but within a tiny fraction of the required simulation time. With this, it is possible to simulate the energy yield of the PV plant accurately and in detail, such as for commercial projects.
Keywords: Bifacial, PV, simulation, BEG, GPU, carport, canopy, photovoltaic plants

1 INTRODUCTION

The installation of solar photovoltaic (PV) modules on carports or canopies is a relatively popular application of PV power generation. This advantageously combines power generation and shaded parking for vehicles in the case of carports, or power generation and shelter and shade for people, animals or goods with little waste of land in the case of canopies [1]. In parallel, bifacial technology has experienced a swift increase in its share of the PV market in recent years, which opens the door to various improvements and even new kinds of solutions [2]. However, in current state-of-the-art PV simulation tools, the added complexity introduced by bifacial technology is still not fully covered and the evaluation of the bifacial energy gain (BEG) is not always accurate. Some considerations are particularly important for PV installations with bifacial modules on carports or canopies, as follows:

- Arrays cannot be assimilated to "infinite" rows, as is usually done in commercial software solutions based on 2D view factors for bifacial computation [3]. Consequently, it is important to properly consider the 3D border effects on the BEG.
- Structures can be installed using different set ups or layouts depending on the location and the possible combined use of the land, hence customized distributions of dispersed elements need to be considered. This means that specific interference effects between the structures due to ground shading need to be evaluated with a case-by-case approach. This increases the number of analyses that are required, which can be very demanding in terms of computation resources.
- Ground albedo can vary for various reasons over time, so that seasonal, directional, and spectral effects need to be considered. This requires a more sophisticated approach than the simplistic Lambertian reflectance assumption for any surface that is used in conventional tools; instead, the reflectance of each material needs to be modelled with the help of a bidirectional reflectance distribution function (BRDF) that is specific to each reflecting surface. [4]

This study presents the lessons learned from the evaluation of the energy yield of several PV projects using bifacial modules on carports and canopies located in France. A novel simulation tool, using advanced computing techniques, is introduced here. This novel approach provides an evaluation of BEG with a spatial resolution that is similar to what is achievable with backward ray-tracing techniques [5], but within a tiny fraction of the required simulation time. With this, it is possible to simulate the energy yield of such small PV plants accurately and in detail, which is necessary for commercial projects.

2 METHODOLOGY

The evaluation of the bifacial energy gain is carried out with a novel simulation tool based on the use of the latest 3D evaluation libraries incorporated into the Graphic Processor Units (GPU) of modern computers. These libraries were developed for the video game industry, but they now appear useful in a variety of other demanding applications where realistic and accurate visualization is important. The achievable spatial resolution is similar to what is possible with backward ray-tracing techniques, but only necessitates a tiny fraction of the latter's required simulation time. The method behind the use of GPUs for solar energy application has been described elsewhere, e.g., for the evaluation of complex shading problems [6] and for the evaluation of bifacial irradiance [7]. Overall, the novel approach proposed here provides a detailed and accurate simulation of the energy yield of various types of PV installations, including for the small commercial projects investigated here.

3 APPLICATIONS

3.1 Carports

The evaluation of the bifacial energy gain in applications such as PV carports requires an accurate modelling of the 3D geometry on the PV array and its surrounding environment, as well as a good modelling of the reflectance of the associated materials. The approach used here benefits from the power of GPUs in modern computers, as mentioned above. The principle is simple: From each side of each PV panel, a 180° spherical view over different points of the module is created, as shown in Figure 1.

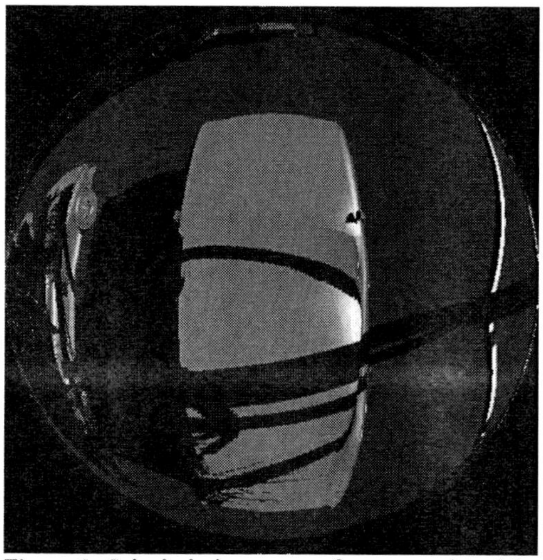

Figure 1: Spherical view as seen from one point of the PV array located on the backside of one PV module.

A tradeoff must be considered between computation time and accuracy. An optimum can be found, depending on the number of viewpoints and their location over the PV plant. As a reference, the pictures are generated here at a resolution of 512 x 512 pixels, which is equivalent to more than 250,000 rays if rather using a ray-tracing technique. These pictures (or alternative rays) are nevertheless obtained relatively fast (≈10–20 ms per picture).

There are several possible contributors to the source of incident irradiance. Some PV elements are lit by direct sunlight or by skylight, whereas others are shaded from the sun but can still receive some part of the reflected diffuse radiation that comes from the surrounding ground or objects. In the present approach, each component needs to be evaluated separately, and a correct incident irradiance magnitude must be calculated for the radiation emanating from the sun, sky, and various surface elements.

A virtual mesh is spread over the surface of all the ground-reflecting elements and any object that would be relevant, because in the immediate environment of the PV modules. For the case of carports, the ground is the flat parking lot covering (e.g., asphalt), and the relevant objects are typically the parked cars, and to a lesser extent the support structures. The mesh constitutes an ensemble of points from where the visibility to the sky is evaluated. Another important optimization must be made on the spatial resolution of this mesh to obtain accurate results without inducing exaggerated computation times. Figure 2 shows a typical mesh used on the ground and cars for the evaluation of sky-view factors.

Figure 2: Mesh on ground and car for the evaluation of sky-view factors.

Once the mesh has been defined, the evaluation of the sky-view factors between each point of this mesh and the sky is carried out, which provides a sun + sky visibility potential. This is a purely geometric parameter, which takes any value between 0 (sky fully blocked) and 1 (sky fully visible). Figure 3 shows an example of the sky view as seen from one particular point of the mesh located on the ground, as visualized in a spherical view.

Figure 3: Sky view as seen from one particular point of the mesh on the ground, visualized in a spherical view.

The result of the calculation of the sky-view factors for each point of the mesh on the ground, cars and other key objects provides a 3D map of these view factors over the whole simulation scene Figure 4 shows the spatial variation of the sky-view factors from the ground, obtained by visualization as a heatmap. Similarly, Figure 5 shows a closer view on such a heatmap for the sky-view factor originating from the front side of a car.

Figure 4: Heatmap showing the spatial variation of the sky-view factor as seen from the ground and cars.

Figure 5: Heatmap showing the spatial variation of the sky-view factor as seen from the ground and the cars.

In parallel, Figure 6 shows the spatial variation of the sky-view factors to the ground and cars as seen from one specific point located on the backside (facing down) of a PV module installed on the carport. The spatial distribution of these sky-view factors directly conditions the reflected irradiance incident on the selected element of the PV array.

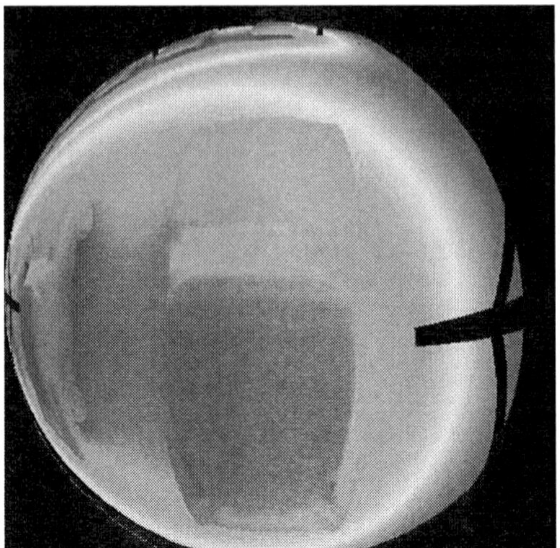

Figure 6: Spatial map of the sky-view factors to the ground and cars as seen from one specific point located on the backside of a PV module installed on a carport.

The reflectance of any material that reflects sunlight or skylight in the environment of a carport is usually complex to model. This is because (i) spatial variability in reflectance properties can be significant from one object to the other; (ii) on-site measurements are often not available for reasons of budget, personnel, or time constraints; and (iii) the reflectance of these elements can be relatively asymmetric, time varying, and sometimes spectrally variable. The exact procedures to evaluate the reflectance patterns using the BRDF approach is out of the scope of this report, but is described in other studies [2].

To implement the effect of albedo, we apply a different method for the diffuse component and for the direct component of the irradiance.

For the diffuse component, the sky-view factors obtained in Figure 6 need to be weighted by the albedo of each element seen in the spheric view (for this example, 0.1 for the ground and 0.3 for the cars). The result of the application of such a weighting procedure is illustrated in Figure 7.

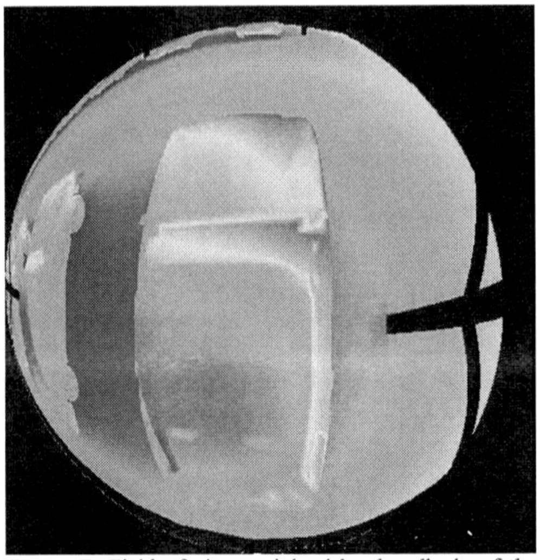

Figure 7: Field of view weighted by the albedo of the different objects seen from a point on the backside of a PV module.

The integration of all the pixels in Figure 7 provides a value (0.047 as shown in Figure 8b) that multiplied by DFI_{iso} provides the irradiance from the sky that is reflected to the node of the mesh where the spheric view is created.

For the contribution of the reflected direct irradiance from the sun, we create the view seen in Figure 8a, taking into account the sun position with respect to each object, and the shading from other elements for the directional properties of the reflectance. As done before, the integration over all the pixels of Figure 8a provides a value (0.062) that once multiplied by the direct irradiance indicates the contribution of that component to the irradiance reflected onto the node of the mesh where the spheric view is created.

On the front face, the direct and diffuse components of the irradiance are computed through other weighting factors shown in Figure 8c and 8d and that indicate the level of shading applied to the direct solar irradiance from the solar disk and to the isotropic diffuse irradiance from the rest of the sky.

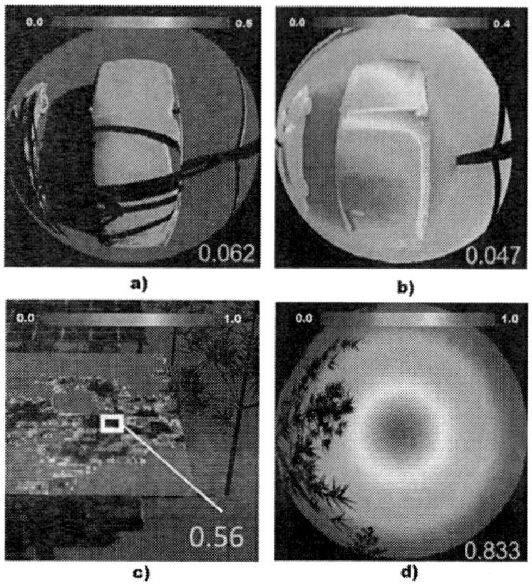

Figure 8: Views from a specific solar cell of the carport canopy: a) elements under direct sunlight; b) diffuse irradiance view-factor reflection from the sky; c) shading factor for parallel rays on the front side of the same cell; and d) sky elements direct view from the front side.

Repeating the process for all each element of the complete PV plant makes it possible to generate a heatmap of the irradiance distribution over both the front and rear module surfaces, as shown in Figure 9. That heatmap is a critical step towards analyzing inhomogeneities in the irradiance field and evaluating their mismatch-induced electrical impact through different possible methodologies.

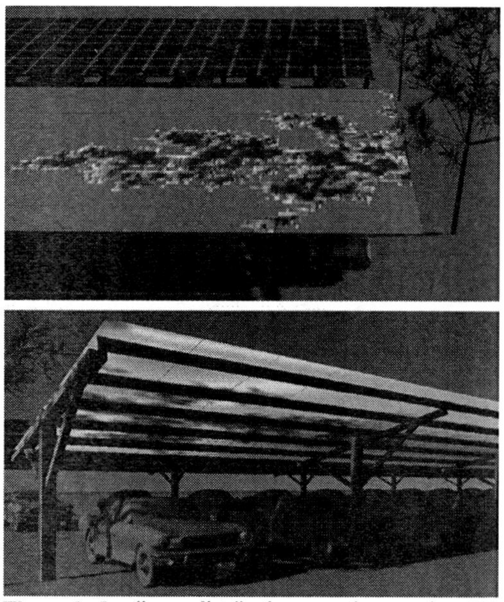

Figure 9: Irradiance distribution over the complete PV array

Once the incident irradiance profile has been obtained for each PV cell composing the PV plant, and for each 10-min time interval of the year, the irradiance is converted into PV power using a PV simulation model that accounts for all PV conversion losses in the whole system. For most of these energy losses, standard simulation routines can be used, because several of them are available and widely known [3], [8], [9], [10], [11]. The differences between these alternative models are beyond the scope of this study and are not relevant to its central topic.

Among the energy losses that need particular attention, the additional mismatch losses that are caused by the spatially inhomogeneous irradiance distribution on the PV modules need to be cautiously addressed. Several possibilities exist to simulate the impact of uneven irradiance distributions on the electrical mismatch losses. The most accurate method is to generate I-V curves for the whole PV system, considering the mapping of the irradiance down to cell level. This, however, is a very demanding exercise in terms of computing power. Other alternatives have been developed in the literature. One of them is selected here: it consists of a parametric model that estimates the mismatch losses as a function of the overall degree of inhomogeneity of the irradiance incident on the PV array [12]. This model has produced annual losses values in the range of 0.3% to 1%, depending on the specificities of each PV project.

Once the PV power is obtained for each time interval, the energy yield is calculated after summation over the time period considered.

3.2 Canopies

Similar to the installation of PV modules on carports, several others PV array configurations on canopies can benefit from bifacial energy gain. Bifacial PV modules mounted on canopies are increasingly encountered in the rural environment, in particular in agrivoltaic contexts, where the canopies are used as shelters for farm animals, some types of crops, agriculture machinery, or other farm buildings.

The geometry of these canopy installations is normally less complex than in the case of carports. For instance, Figure 10 illustrates a PV installation on canopies over a grassy field.

Figure 10: PV installation on canopies over a grassy field.

One major difficulty for the evaluation of the bifacial energy gain (BEG) in rural environments is the accurate assessment of the reflected radiation, in particular when emanating from vegetation. Predicting the reflectance properties of any object at any moment is particularly complex. Among other reasons, this complexity is caused by: (i) the lack of databases providing reliable field measurements of reflectance under well-defined conditions, (ii) the anisotropic nature of reflectance patterns, (iii) the spatial diversity and spectral sensitivity of reflectance, and (iv) its temporal variability (both diurnal and seasonal). It appears that the sensitivity of reflectance to some of these factors is generally higher in the case of natural materials such as vegetation, soil, stones, etc. than in the case of the artificial materials encountered in urban environments, such as asphalt or concrete. The exact assessment of this kind of reflectance is out of the scope of this work.

Even with a simplified geometry, significant inhomogeneities can appear in the spatial distribution of the reflected solar irradiance on the backside of PV modules, such as illustrated in Figure 13. Typically, the reflected irradiance is lower in the middle of the PV array, higher over the edges, and even higher at the corners. This is a direct consequence of the higher view factors for the points on the module backsides that are located closer to the edges and corners. When the points under scrutiny are situated closer to the center of the PV array, the surrounding PV modules contribute to decreasing the effective field of view. On the other hand, those PV modules that are situated in the center of the PV array also receive less reflected irradiance because a more important fraction of the underlying ground tends to be shaded. This inhomogeneity, if significant, can lead to additional electrical mismatch losses, as already discussed in the case of PV carports.

Figure 13: Distribution of the incident irradiance on the backside of bifacial PV arrays, showing an inhomogeneous spatial distribution.

4 CONCLUSION AND DISCUSSION

The number of installations with bifacial PV modules on carports or canopies is fast increasing. The bifaciality of the PV modules offers interesting advantages in this kind of configuration.

The accurate simulation of the energy gain for these PV installations is however challenging. This is still a rather new kind of application, and many challenges still lie ahead towards a more accurate assessment of the energy yield of these PV plants.

Among the most challenging topics are the need for accurate albedo values and reflectance profiles, including angular and spectral considerations. Another challenge is the accurate modelling of the 3D scenes that can be very complex, and the relatively high sensitivity of the bifacial energy gain to the geometric of the environment.

The GPU-based method presented here has already been used to assess the bifacial energy gain of many large bifacial PV plants totaling more than 1 GW since 2017, with a large diversity of applications, including large PV plants with fixed structures or trackers, carports, canopies in rural areas, greenhouses, or vertical agrivoltaic structures. Several application cases have been illustrated in this article. The further development and scientific validation of this method is currently ongoing in the context of two European R&D projects: the H2020 project SERENDI-PV, and the COST project Pearl-PV. However, given the increasing diversity of bifacial PV installations, and the scarcity of reliable monitored operation data that are currently available, more effort and more collaboration will be needed in the future in order to incorporate and validate new functionalities in the tool. In this context, collaboration proposals are welcome, on the research effort as well as on the analysis of the operation data of bifacial PV installations.

ACKOWLEDGMENTS

The research activities behind the simulation tool that was used in this project were partially funded by European Commission through the Horizon 2020 project SERENDI-PV (https://serendipv.eu/), which belongs to the Research and Innovation Programme, under Grant Agreement 953016. This publication is also partially based upon work from COST Action CA16235 PEARL PV (https://www.pearlpvcost.eu), supported by COST (European Cooperation in Science and Technology).

REFERENCES

[1] H.-M. Neumann, H.-M., D. Schär, D. and F. Baumgartner, F. 2012, The potential of photovoltaic carports to cover the energy demand of road passenger transport. Prog. Photovolt: Res. Appl., 20: 639-649. https://doi.org/10.1002/pip.1199.
[2] W. Gu, T. Ma, S. Ahmed, Y. Zhang, J. Peng, A comprehensive review and outlook of bifacial photovoltaic (bPV) technology, Energy Conversion and Management, 2020, https://doi.org/10.1016/j.enconman.2020.113283.
[3] https://www.pvsyst.com/.
[4] J.-L. Roujean, M. Leroy and P.Y. Deschamps, 1992, A bidirectional reflectance model of the Earth's surface for the correction of remote sensing data, J. Geophys. Res., 97(D18), 20455–20468, doi:10.1029/92JD01411.
[5] T. Whitted, 1980. An improved illumination model for shaded display. Commun. ACM 23 (6), 343–349.
[6] J. Robledo, J. Leloux, E. Lorenzo, C.A. Gueymard, From video games to solar energy: 3D shading simulation for PV using GPU, Solar Energy, 2019, https://doi.org/10.1016/j.solener.2019.09.041.
[7] J. Robledo, J. Leloux, B. Sarr, C.A. Gueymard, A. Driesse, Dynamic and visual simulation of bifacial energy gain for photovoltaic plants, European Photovoltaic Solar Energy Conference and Exhibition (EU PVSEC), 2021.
[8] https://pvpmc.sandia.gov/applications/pv_lib-toolbox/.
[9] https://sam.nrel.gov/.
[10] https://www.sisifo.info/es/default.
[11] J. Muñoz, O. Perpiñán, A simple model for the prediction of yearly energy yields for grid-connected PV systems starting from monthly meteorological data, Renewable Energy, 2016, https://doi.org/10.1016/j.renene.2016.06.023.
[12] C. Deline, S. Ayala Pelaez, S. MacAlpine, C. Olalla, Estimating and parameterizing mismatch power loss in bifacial photovoltaic systems. Prog Photovolt Res Appl. 2020; 28: 691–703. https://doi.org/10.1002/pip.3259.

LESSONS LEARNED FROM SIMULATING THE ENERGY YIELD OF AN AGRIVOLTAIC PROJECT WITH VERTICAL BIFACIAL PHOTOVOLTAIC MODULES IN FRANCE

Jesús Robledo[1,*], Jonathan Leloux[1], Babacar Sarr[1], Christian A. Gueymard[1,2], Anton Driesse[3], Pierre-François Drouin[4], Sébastien Ortega[5], Damien André[5]
[1] LuciSun, Rue Saint-Jean, 29, 1495 Sart-Dames-Avelines, Belgium
[2] Solar Consulting Services, P.O. Box 392, Colebrook, NH 03576, USA
[3] PV Performance Labs, Emmy-Noether-Str. 2, 79110 Freiburg, Germany
[4] Syneria, 5 Rue Taitbout, 75009 Paris, France
[5] Amarenco, Château de Touny Les Roses, Chemin de Touny, 81150 Lagrave, France
*Phone: +34 656 37 59 74; E-mail: jesus.robledo@lucisun.com

ABSTRACT: Agrivoltaics is a novel application of solar (PV) photovoltaic power generation where the PV modules are installed in the same field as where crops are also cultivated. This is advantageous because this combines land use while keeping attractive productions for both agricultural crops and power generation. Several configurations currently exist for these agrivoltaic installations, including rooftop PV generators, ground-mounted agrivoltaic plants, and PV installations on greenhouses. One particularly innovative kind of agrivoltaic application uses rows of vertical bifacial PV modules. In recent years, bifacial technology has experienced a swift increase in its share of the PV market, which opens the door to performance improvements and even new kinds of solutions. However, in the current state-of-the-art PV simulation tools, the added complexity introduced by bifacial technology is still not fully covered. This is even more true for agrivoltaic PV plants with vertical bifacial modules. Here, we present a summary of our most relevant findings while assessing the energy yield of an agrivoltaics plant planned with vertical bifacial PV modules in the south of France. The evaluation of the bifacial energy gain (BEG) has been carried out with a novel simulation tool based on the use of the latest 3D evaluation libraries incorporated into the Graphic Processor Units (GPU) of modern computers. These were developed for the video game industry, and they offer many interesting advantages for such bifacial PV applications. We also discuss the main challenges that still lie ahead in the path to a better design optimization and assessment of the energy yield of such vertical PV installations.

Keywords: Bifacial, Agrivoltaics, PV, simulation, BEG, GPU, photovoltaic plants

1 INTRODUCTION

Agrivoltaics is a novel application of solar photovoltaic (PV) power generation where the PV modules are installed in the same field as where crops are also cultivated. This is advantageous because this arrangement combines land use while keeping attractive productions for both agricultural crops and power generation. Several configurations currently exist for such agrivoltaic installations, including rooftop PV generators, ground-mounted agrivoltaic plants, and PV installations on greenhouses. One particularly innovative kind of agrivoltaic application uses rows of vertical bifacial PV modules. In recent years, bifacial technology has experienced a swift increase in its share of the PV market, which opens the door to performance improvements, and even to new kinds of solutions. However, in the current state-of-the-art PV simulation tools, the added complexity introduced by bifacial technology and intricate geometries is still not fully covered. This is even more true for agrivoltaic PV plants with vertical bifacial modules. Some particularities of this specific case are:

The notion of "front and rear" side of the modules is not so straightforward because the total irradiation received on both sides can be similar in various situations, such as an east/west orientation. This means that (i) the bifacial contribution cannot be seen as a simple "gain" over the main front contribution but as a comparable one, and that (ii) the sensitivity of the whole energy yield evaluation to the bifacial energy gain (BEG) may be large.

Traditionally, the geometry of agrivoltaic installations introduces complex shading effects (both for direct sunlight and for the sky and ground-reflected diffuse components), thus requiring a detailed 3D shading analysis. In that context, the total diffuse irradiance (emanating from the sky and from ground reflections) becomes a key variable. This implies that the directional reflection properties of the ground are important. Hence, a Bidirectional Reflectance Distribution Function (BRDF) should be considered instead of an ideal Lambertian response. Moreover, seasonal variations in ground albedo (caused by vegetation cycles) can be noticeable, and thus need to be accounted for.

This study presents the lessons learned from an innovative agrivoltaic project located in France with vertical bifacial modules.

The evaluation of the bifacial energy gain (BEG) has been carried out with a novel simulation tool based on the use of the latest 3D evaluation libraries incorporated into the Graphic Processor Units (GPU) of modern computers. These were developed for the video game industry, and they offer many interesting advantages for such bifacial PV applications.

2 PV INSTALLATION LOCATION AND LAYOUT

The project under scrutiny here is located in the Occitany Region (southern France), in a rural area where the land is mainly dedicated to agriculture. A PV plant was projected on a parcel (shown in Figure 1). The promoter of this project planned to install a PV plant whose nameplate peak power was slightly below 4 MW_p.

Figure 1: Satellite view of the site selected for the agrivoltaic PV plant project.

The promoter chose to use vertically mounted bifacial PV modules to keep a mixed use of the land between agriculture and photovoltaic energy generation. To facilitate mechanized agriculture, the vertical structures were projected to be installed in rows that only deviate from the north-south direction by ≈26°, following the natural orientation of the main edge of the field. As a consequence, the PV modules were oriented so that one face deviates 64° from due south towards the east, while the other face deviates 116° from due south towards the west. The PV modules have a bifaciality ratio of 70%, which is currently very common. To maximize the energy yield of this PV plant, the front (or main) side of the PV modules were installed towards the eastern direction, which receives more solar irradiation over the year, and the other side was installed towards the western orientation. The design of the mounting system is such that the ground clearance, the minimum distance between the PV modules and the ground, is exactly 1 m. This makes it possible to grow crops close to the PV modules while minimizing the impact on the crops of any potential shading caused by the installation Two rows of PV modules in landscape mode are installed on these structures. The width of the modules is of approximately 1 m. Therefore, the total height of the structure is ≈3 m. This is a good compromise between installation costs, mechanical stress induced with the wind, shelter from the wind offered to the cultures, and visual impact on the landscape. A pitch separation distance of 15 m was chosen between the rows of PV modules because it corresponds to a good tradeoff between installed PV area density and energy losses caused by shading. Figure 2 shows a 3D representation of the resulting PV plant layout that was proposed as the result of the optimization exercises.

Figure 2: 3D representation of the PV plant layout.

3 SOLAR RESOURCE AND WEATHER DATA

Several sources of solar irradiation and weather data have been consulted to assess the long-term solar resource available on the site described above. As a result of the comparison between different solar irradiation databases, an hourly TMY Global Horizontal Irradiation (GHI) time series was generated by combining 4 different sources of data (CMSAF [1], SARAH [2], Meteonorm [3] and Solargis Prospect [4]) and was eventually used to simulate the energy yield of the PV plant. The specific method used for the combination of these data sources is out of the scope of this work, but it respects relatively common practice, and in any case the exact method followed has little influence on the rest of the considerations that are presented in this article. A long-term annual value of GHI of 1,423 kWh/m^2 was retained for this project.

Several particularities arise in the case of vertically mounted bifacial PV modules when assessing the Global Tilted Irradiation (GTI) on both faces of the modules. The first critical aspect is the choice of the transposition model that has to be used to evaluate GTI on a vertical surface from the knowledge of GHI. Among other considerations, the selection of the most appropriate transposition model is of critical importance. The well-known Perez model is frequently used for typical ground-mounted PV plants, facing the equator with a low fixed tilt or with trackers. Several other models are also used by solar engineers and PV simulation software developers for such installations, with similar uncertainty levels. For vertical surfaces, however, the uncertainty associated with the transposition is generally higher, and the Perez model might not be the most suitable choice [5]. Other "anisotropic" models, such as the Hay model [6], might actually be more suited to the case of vertical surfaces [7]. One reason why the predictions of transposition models differ significantly more for vertical surfaces than for low-tilt surfaces is that, in the former case, the diffuse radiation reflected by the foreground amounts to a significant fraction (if not a preponderant fraction) of the total (sky + reflected) diffuse radiation incident on the surface. So far, transposition models have not attempted to evaluate the diffuse reflected irradiance in sufficient detail, with some exceptions, e.g. [8]. This discussion indicates that, for agrivoltaic installations (and more generally for all projects involving vertical PV modules), there is certainly room for improvement in the calculation of GTI and the overall assessment of the irradiation incident on vertical surfaces. The present study uses the Hay model for its modeling simplicity and efficiency, and because of its recognized performance in a wide variety of situations [9].

Another important challenge arises from the necessity of accurately determining the reflected component of GTI on both sides of the PV modules. As mentioned above, this contribution is very significant for vertical surfaces in general, particularly for non-equatorial orientations. This is discussed further in Section 5.

4 ALBEDO ASSESSMENT

Several methods exist to assess the albedo of the ground. For this kind of project in a rural area, several key questions arise. First, for most PV projects, reliable on-site measurements of the ground albedo are not available, for reasons of costs, manpower, available time, etc. Second, the nature of the ground under the PV plants might change after completion of the project, and even during the lifetime of the project. In the case of a vertical PV installation mixed with agriculture, the crops that are cultivated between the rows of PV modules typically changes every year, hence the albedo also changes. Moreover, the state of vegetation changes seasonally, also inducing significant albedo variations. Because of all these difficulties, some general assumptions need to be made, and the higher associated uncertainty needs to be fully acknowledged when estimating the energy yield.

For this project, the assessment of the ground albedo was carried out with the help of three widely different databases: MERRA2 [10] and GLDAS [11] (both based on reanalysis models) and Solargis Prospect [4] (based on MODIS [12] satellite observations). Their precision is however limited, in particular by the nature of their processing, as well as by their spatial resolution. The spatial resolution of MERRA2 is 0.5° x 0.625°, and that of GLDAS is 0.25° x 0.25°. Solargis offers a much better spatial resolution of 1 km x 1 km. Fortunately, it is often the case in rural regions dedicated to agriculture that the overall albedo is relatively homogeneous over large areas. Figure 3 shows a satellite view of an area of approximately 20 km x 20 km surrounding the PV project, where a typical vegetation color, varying from green to yellow, is visible. The albedo is not homogeneous on one pixel of the satellite data, but it is nevertheless relatively representative of the cultures.

Figure 3: Satellite view of the overall area of the PV project.

At the time of the study, monthly values were available from MERRA2 and GLDAS, and an average annual value was available from Solargis Prospect. It is generally accepted that MERRA2 data is of better quality, but of lower spatial resolution, than GLDAS, and that the satellite-based data from Solargis are of the highest quality. It is fortunate that, for the case of this project, the three databases provide very similar results. Figure 4 shows a comparison of the monthly albedo data from MERRA2 and GLDAS. The annual average albedo is ≈17%, with very slight seasonal variations of only ≈1%. The albedo is slightly higher in summer, which in rural areas is usually caused by the change in color of the vegetation from dark green (or dark, wet bare soil) to lighter tones. The opposite could obviously be true in cold areas impacted by snow in winter [13].

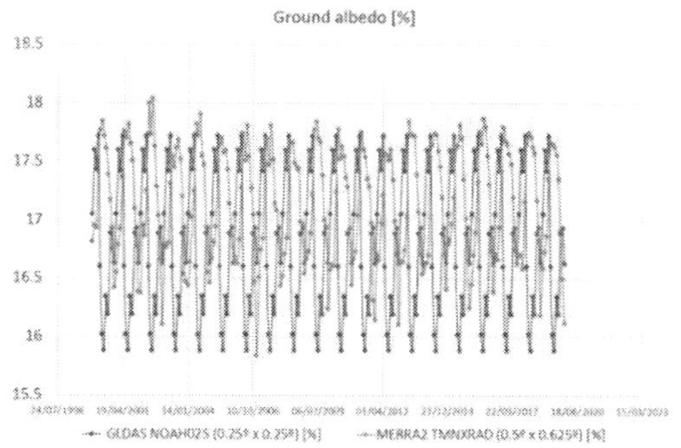

Figure 4: Monthly-average albedo during the 20 last years, from MERRA2 and GLDAS.

A detailed analysis of the most reliable albedo data obtained for homogenous parcels shows that the albedo of green grass is typically ≈14%, and that of most crops encountered in the region of southern France under scrutiny is in the range 10–25%. The albedo of the vegetation is also highly spectrally selective, mainly due to photosynthesis in plants. This translates into a very low albedo (≈0.05) in the visible and a high albedo (≈0.8) in the near infrared. This can impact the calculation of spectral effects and their impacts on PV performance, but this development is beyond the scope of this study.

In addition to spectral variations, the albedo of bare soil and vegetation has an anisotropic angular distribution. The reflectance pattern of each reflective material can be conveniently modeled by a bidirectional reflectance distribution function (BRDF). These BRDF functions are particularly important to consider for light-reflecting materials whose albedo deviates significantly from Lambertian behavior, and whose spectral reflectance is very selective, as is the case with vegetation in general.

Taking all these considerations into account led to choosing a somewhat conservative albedo value of 14%, constant over the whole year and associated with an anisotropic angular distribution represented by the BRDF of a typical green grass [14]. The BRDF used here is represented in Figure 5.

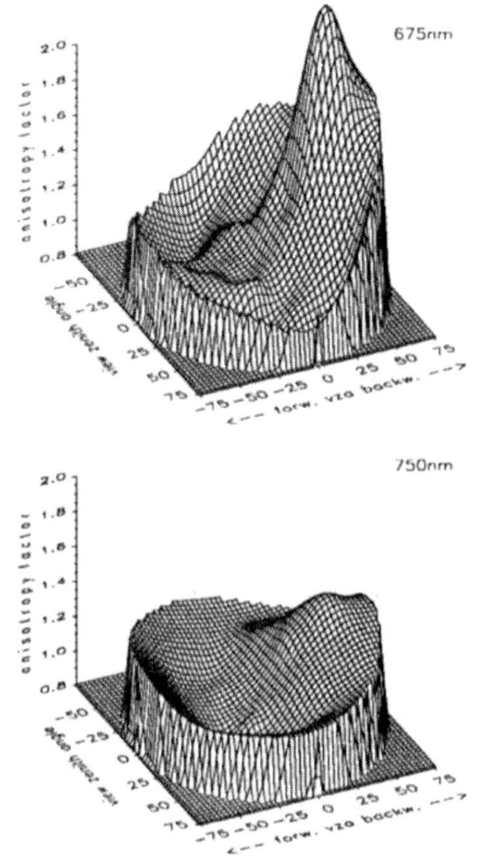

Figure 5: Directional (BRDF) of a typical green grass and its effects on the incident irradiance.

5 IRRADIANCE REFLECTION AND BIFACIAL IRRADIANCE GAIN

The evaluation of the bifacial energy gain (BEG) has been carried out with a novel simulation tool based on the use of the latest 3D evaluation libraries incorporated into the Graphic Processor Units (GPU) of modern computers. These were developed for the video game industry. The achievable spatial resolution is similar to what is possible with backward ray-tracing techniques, but only necessitates a tiny fraction of the latter's required simulation time. The method behind the use of GPUs for solar energy application has been described elsewhere, e.g., for the evaluation of complex shading problems [15] and for the evaluation of bifacial irradiance [16]. Overall, the novel approach proposed here provides a detailed and accurate simulation of the energy yield of agrivoltaic PV plants for small or large commercial projects.

In the present case, the irradiance profiles have been evaluated with high spatial resolution (i.e., at the PV cell level), and with a relatively high temporal resolution of 10 minutes. With this GPU-based approach, the simulations generate a hemispherical view of the front and back sides of each PV module, thus allowing a comprehensive representation of the environment. Figure 1 illustrates the use of this method for the evaluation of the irradiance incident on a part of the PV plant. The GPU-generated image is divided into key areas: the directly-lit part, the background sky, and the shaded areas. The contributions of the different areas are evaluated by transforming each field of view into a specific view factor. These view factors are then weighted by the amount of solar irradiance received from each component of the solar irradiance (diffuse irradiance received from the sky, reflected diffuse irradiance, and reflected direct irradiance). The total reflected irradiance incident on each PV element is simply obtained as the sum of these three components. The process is divided into five main steps.

The first step is to apply a mesh over the terrain, as illustrated in Figure 6.

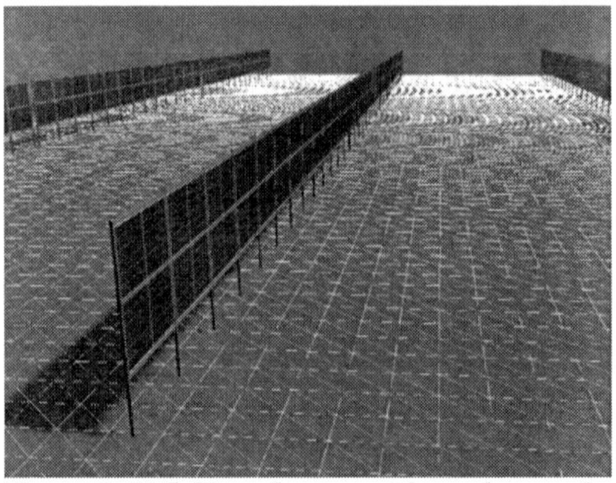

Figure 6: Definition of a geometric mesh over the terrain.

The second step is to evaluate, for each on the terrain's nodes, the visibility of the sky (sky view factor, between 0 and 1) to define the potential diffuse irradiance that could be reflected. This is illustrated in Figure 7.

Figure 7: View factor of the ground to the sky.

The third step is, from different points on the PV modules, to generate the spherical view of what is seen from it; then split the areas on the different view factors that will be applied to the irradiance components (direct, isotropic diffuse, and circumsolar diffuse; the separation of the sky diffuse irradiance into the latter two components is generated by the Hay transposition model) [9]. Some results from this step are shown in Figure 8.

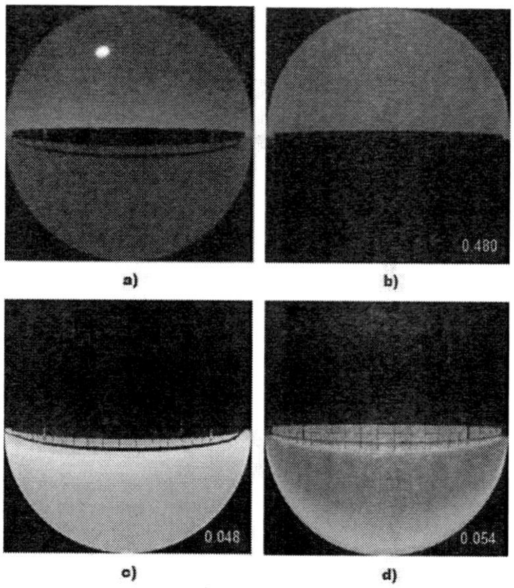

Figure 8: (a) Realistic view of the sky from one point of view on the PV array, (b) Raw sky view factor, (c) Field of view multiplied by the directional reflectance for the direct component, (d) Field of view multiplied by the effective albedo for the isotropic diffuse component.

The fourth step is, as described in [17], to multiply each view factor by the corresponding irradiance component [see Figure 8, (b) and (d) by DFIiso and (c) by $(DNI + DFI_{circ})\cos\theta$ to obtain the breakdown of each contribution.

The fifth step is to represent the result for the different points over the full PV modules to create a detailed spatial map of the magnitude of the different irradiance components.

The irradiance incident on the PV arrays can be visualized with heatmaps. They are useful to rapidly compare the effective components as a fraction of their original (unaltered) value. In summary, the value that is thus obtained quantifies the relative contribution of the incident irradiance component relative to the total irradiance.

As an illustration of this procedure, Figure 9 shows an example of heatmap quantifying the magnitude of the isotropic diffuse irradiance that is directly incident from the sky, as calculated for one specific moment. The normalized values vary here between 0.43 and 0.50. This represents the fraction of the sky that is visible by any point of the PV array. The highest values ($\approx$0.50) occur at the top of the PV array, representing half of the whole isotropic irradiance from the sky. This also corresponds to the maximum potentially visible fraction of the sky that is visible by a vertical structure. As could be expected, this value decreases when the point of observation gets closer to the ground, because of mutual shading between the rows of PV modules (one row hides parts of the sky from the surrounding PV modules).

Figure 9: Fraction of the sky that is seen by each point on the PV array, indicating the fraction of diffuse isotropic irradiance that is incident on the vertical surface.

Figure 10 represents the contribution of the reflected isotropic diffuse irradiance as a fraction of the total isotropic diffuse incident irradiance. This ratio provides an estimate of the irradiance gain created by ground reflections for that specific component. The vertical gradient of this irradiance profile follows an opposite pattern than was shown in Figure 9 because more irradiance is reflected to the points that are closer to the ground. However, the variation of this contributir is not so high (between 0.052 and 0.056) and is consequently quite uniform.

Figure 10: Contribution of the reflected isotropic diffuse irradiance to the total isotropic diffuse incident irradiance.

Figure 11 represents the contribution of the reflected direct irradiance to the total direct irradiance. The picture on the left shows the irradiance map for the side of the module that is away from sunlight at that moment, while the sun shines on the other side. For the sun-facing side, the vertical irradiance gradient indicates an increase from top to bottom, because more light is reflected to the area that is closer to the ground. For the opposite side, the lowest solar cells of the PV array receive less reflected irradiance because the shaded area on the ground occupies a greater part of their field of view. Additionally, this component is the one mainly impacted by the directionality of the BRDF function for the reflectance.

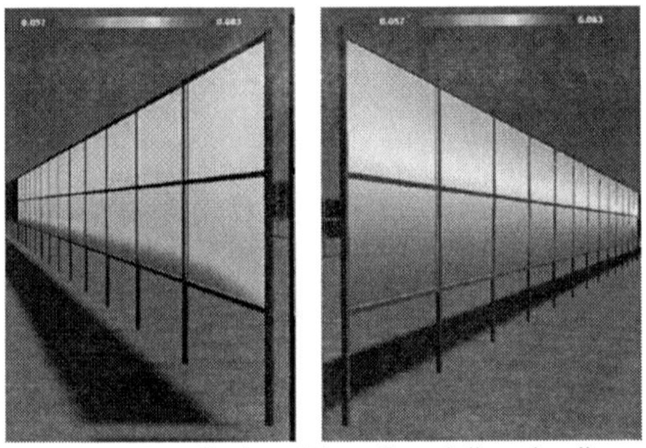

Figure 11: Contribution of the reflected direct irradiance to the total incident direct irradiance. Left: side of the module that does not receive sunlight; right: sun-facing side.

These irradiance heatmaps illustrate the relatively high degree of spatial inhomogeneity of the solar irradiance incident on the PV modules. This is an important specificity of bifacial PV module—particularly if installed vertically. This inhomeneity changes dynamically over the day becuase of the varying solar geometry and BRDF angular effects. Since spatial irradiance inhomogeneities induce mismatch losses in practice (see Section 6), all these intricate effects needs to be described in detail and taken into account in PV yield simulations, as explained in Section 6.

Figure 12 shows the resulting estimated incident irradiance for a typical clear day. The irradiance reaches higher values in the morning on the front side of this specific PV module because it is oriented towards the east (64° from south).

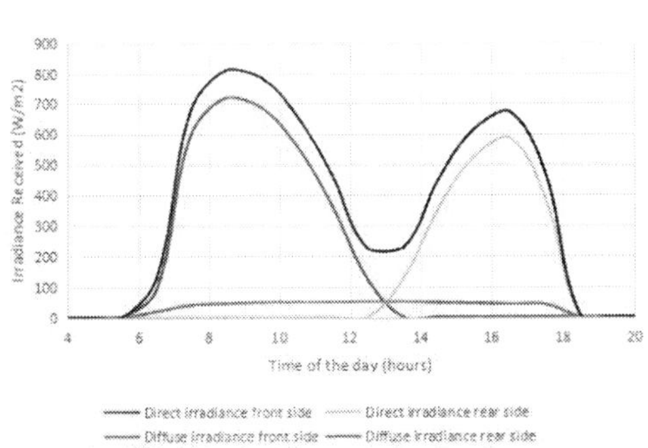

Figure 12: Estimated irradiance and its components for a typical clear day.

Finally, Figure 13 shows the corresponding estimated effective irradiance of the module, defined as the previous irradiances per side as shown in the figure 12, but with the bifaciality factor (70% for this example) applied on the rear side, as the modules are less efficient on that side.

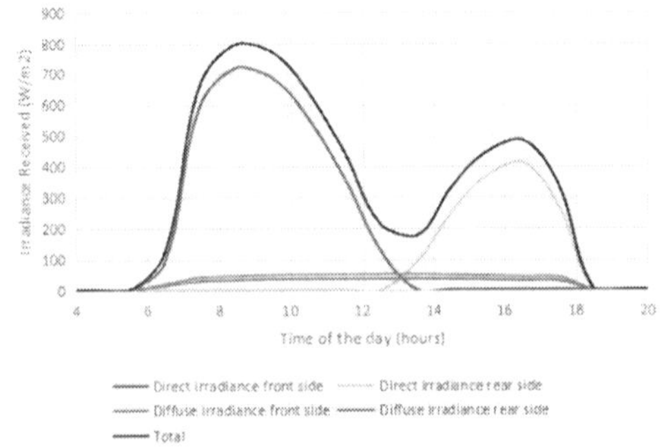

Figure 13: Direct and diffuse irradiance (from the sky and ground reflections) incident on the front and rear sides of a vertical module during a clear day and considering the bifaciality factor of the module.

6 FROM IRRADIANCE TO ENERGY

Once the incident irradiance profile has been obtained for each PV cell composing the PV plant, and for each 10-min time interval of the year, the irradiance is converted into PV power using a PV simulation model that accounts for the PV conversion losses in the whole system. For most of these energy losses, standard simulation routines can be used, because several of them are available and widely used (PVlib [17], PVsyst [18], SISIFO [19], [20], SAM [21]). The differences between these alternative models are beyond the scope of this study and are not relevant toits central topic.

Among the energy losses that need particular attention, the additional mismatch losses that are caused by the spatially inhomogeneous irradiance distribution on the PV modules need to be cautiously addressed. Several possibilities exist to simulate the impact of uneven irradiance distributions on the electrical mismatch losses. The most accurate method is to generate I-V curves for the whole PV system, considering the mapping of the irradiance down to cell level. This, however, is a very demanding exercise in terms of computing power. Other alternatives have been developed in the literature. One of them is selected here: it consists of a parametric model that estimates the mismatch losses as a function of the overall degree of inhomogeneity of the irradiance incident on the PV array [22]. This model has produced annual losses values in the range of 0.3% to 1%, depending on the specificities of each PV project.

Once the PV power is obtained for each time interval, the energy yield is calculated by summation over the time period considered.

For the test case exemplified here, the annual energy yield is estimated to be 1,093 kWh/kWp, which is approximately 10% lower than the typical PV yield of a ground-mounted monofacial PV plant facing south and with optimal tilt that would be installed at the same location. In some cases, project promoters might be willing to sacrifice a small fraction of the potential energy yield to keep an appropriate mix between agriculture and electricity generation over the available land area. The yield value just mentioned is representative of the southern France only and can greatly vary as a function of climate and latitude, with typically higher yield losses for sunnier locations at lower latitudes, and lower yield losses (or even gains) for higher latitudes or cloudier locations [23].

The Performance Ratio (PR) is often used to analyze the overall performance of PV plants. It is defined as:

$$PR = \frac{Y_f}{Y_r}$$

where the final yield Y_f is equal to the ratio E_{pv}/P^{STC}, and the reference yield Y_r is equal to the ratio GTI/G^{STC}, where: E_{pv} is the net electrical energy produced by the installation during a given period, P^{STC} is the rated power of the PV generator under Standard Test Conditions (STC), G^{STC} is the global solar irradiance under STC (i.e., 1000 W/m^2), and GTI is the Global Tilted Irradiance received by the PV generator.

There is currently a lack of agreement on the definition of the reference GTI value required to define the yield for bifacial PV modules. One part of the PV community still uses the GTI incident on the front side of the PV modules for that purpose, because the backside irradiation is more difficult to evaluate with accuracy. Nevertheless, this mechanically leads to an overestimated PR by an amount that is close to the bifacial energy gain. This definition often represents a PR increase of typically several percent for most bifacial PV plants relatively to their equivalent monofacial configuration. Nevertheless, this is particularly problematic in the case of vertical PV modules, for which the front side and backside can receive a similar irradiance. This can lead to annual PR values that are much higher than the PR values of monofacial PV modules, and also much higher than 100%, which is very confusing. For these reasons, a different reference GTI is selected here. It is defined as the sum of the GTI received on the front side and of the backside GTI multiplied by the bifaciality factor. This constitutes the *effective bifacial irradiation*. With this definition, testimated PR amounts here to a reasonable value of 82.6%.

7 DISCUSSION AND CONCLUSION

The use of bifacial PV modules for agrivoltaics presents interesting advantages, in particular regarding the combined use of the land between agriculture and photovoltaic energy generation. However, this is still a rather new kind of application, and many challenges still lie ahead towards a more accurate assessment of the energy yield of these PV plants.

Among the most challenging topics that were encountered during the course of this project, we have identified the need for accurate albedo values and reflectance profiles, including angular and spectral considerations.

The uncertainty surrounding the assessment of the inclined solar irradiance on these vertical surfaces is also particularly challenging because most of the commonly used irradiance translation models have mainly be developed and validated for south facing and optimally tilted surfaces, or for tracking surfaces. Therefore, new research is needed to improve our knowledge and develop more accurate and robust translation methods that are more suitable for vertical surfaces, in particular for vertical surfaces facing east or west.

We have also identified that, because the contribution of the reflected components of the reflected irradiance has relatively large weight in these kinds of PV installations, the energy yield of these configurations is particularly sensitive to the value and nature of the albedo, as well as to the accuracy of the simulation of the multiple light reflection processes that take plan in the PV plant. We have presented a novel simulation tool that provides a good compromise between results accuracy and computation speed. The application of that simulation tool has been illustrated on the case study of the PV plant projected in the south of France.

The use of the front side GTI for the calculation of the PR leads to important overestimations of its value. Therefore, we suggest that the PR for bifacial PV modules is calculated using the effective bifacial irradiation as a reference. This is particularly important in the case of vertically oriented PV arrays where both sides receive similar amounts of solar irradiation.

The GPU-based method presented here has already been used to assess the bifacial energy gain of many large bifacial PV plants totaling more than 1 GW since 2017, with a large diversity of applications, including large PV plants with fixed structures or trackers, carports, canopies in rural areas, greenhouses, or vertical agrivoltaic structures. Several application cases have been illustrated in this article. The further development and scientific validation of this method is currently ongoing in the context of two European R&D projects: the H2020 project SERENDI-PV, and the COST project Pearl-PV. However, given the increasing diversity of bifacial PV installations, and the scarcity of reliable monitored operation data that are currently available, more effort and more collaboration will be needed in the future in order to incorporate and validate new functionalities in the tool. In this context, collaboration proposals are welcome, on the research effort as well as on the analysis of the operation data of bifacial PV installations.

ACKNOWLEDGMENTS

The research activities behind the simulation tool that was used in this project were partially funded by European Commission through the Horizon 2020 project SERENDI-PV (https://serendipv.eu/), which belongs to the Research and Innovation Programme, under Grant Agreement 953016. This publication is also partially based upon work from COST Action CA16235 PEARL PV (https://www.pearlpvcost.eu), supported by COST (European Cooperation in Science and Technology).

REFERENCES

[1] https://www.cmsaf.eu/EN/Home/home_node.html
[2]https://ec.europa.eu/jrc/en/PVGIS/downloads/SARAH
[3] https://meteonorm.com/en/
[4] https://solargis.com/
[5] C.A. Gueymard, Direct and indirect uncertainties in the prediction of tilted irradiance for solar engineering applications, http://dx.doi.org/10.1016/j.solener.2008.11.004
[6]J.E. Hay, J.A. Davies, 1980. Calculation of the solar

irradiance incident on an inclined surface, in:
J.E. Hay, T.K. Won, (Eds.), First Canadian Solar Radiation Data Workshop, Toronto, Ontario,
Canada. pp. 59–72.
[7] C.A. Gueymard, Solar Radiation Resource: Measurement, Modeling, and Methods, https://doi.org/10.1016/B978-0-12-819727-1.00101-1, 2021, in press.
[8] C.A. Gueymard, C.A., 1987. An anisotropic solar irradiance model for tilted surfaces and its comparison with selected engineering algorithms. Solar Energy 38, 367–386, Erratum, Solar Energy 40, 175 (1988).
[9] C.A. Gueymard, 1987. An anisotropic solar irradiance model for tilted surfaces and its comparison with selected engineering algorithms. Solar Energy 38, 367-386.
[10] https://gmao.gsfc.nasa.gov/reanalysis/MERRA-2/
[11] https://ldas.gsfc.nasa.gov/gldas/.
[12] https://modis.gsfc.nasa.gov/data/.
[13] C.A. Gueymard, Surface albedo spatial variability in North America: Gridded data vs. local measurements, 2021, https://doi.org/10.1016/j.solener.2021.05.012.
[14] J.-L. Roujean, M. Leroy and P.Y. Deschamps, 1992, A bidirectional reflectance model of the Earth's surface for the correction of remote sensing data, J. Geophys. Res., 97(D18), 20455– 20468, doi:10.1029/92JD01411.
[15] J. Robledo, J. Leloux, E. Lorenzo, C.A. Gueymard,
From video games to solar energy: 3D shading simulation for PV using GPU, Solar Energy, 2019, https://doi.org/10.1016/j.solener.2019.09.041.
[16] J. Robledo, J. Leloux, B. Sarr, C.A. Gueymard, A. Driesse, Dynamic and visual simulation of bifacial energy gain for photovoltaic plants, European Photovoltaic Solar Energy Conference and Exhibition (EU PVSEC), 2021.
[17] https://pvpmc.sandia.gov/applications/pv_lib-toolbox/.
[18] https://www.pvsyst.com/.
[19] https://www.sisifo.info/es/default.
[20] J. Muñoz, O. Perpiñán, A simple model for the prediction of yearly energy yields for grid-connected PV systems starting from monthly meteorological data, Renewable Energy, 2016, https://doi.org/10.1016/j.renene.2016.06.023.
[21] https://sam.nrel.gov/.
[22] C. Deline, S. Ayala Pelaez, S. MacAlpine, C. Olalla, Estimating and parameterizing mismatch power loss in bifacial photovoltaic systems. Prog Photovolt Res Appl. 2020; 28: 691–703. https://doi.org/10.1002/pip.3259.
[23] https://pvpmc.sandia.gov/modeling-steps/5-ac-system-output/pv-performance-metrics/annual-yield/.

GROUND IRRADIANCE MODELLING: OF KEY IMPORTANCE FOR DESIGNING NATURE INCLUSIVE SOLAR PARKS AND AGRIVOLTAICS SYSTEMS

Bas B. Van Aken[1], Ashish Binani[1], Eduardo Barros[2], Antonius R. Burgers[1], Friso van der Zee[3], Alex Schotman[3] and Kay Cesar[1,*]
1 TNO Energy Transition – Solar Energy, Westerduinweg 3, Petten, The Netherlands
2 TNO Energy Transition – Applied Geosciences, Utrecht, The Netherlands
3 Wageningen Environmental Research, Wageningen University & Research, Wageningen, The Netherlands
*Corresponding author: kay.cesar@tno.nl, Phone: +31 88 866 2607

Solar electricity from solar parks in rural areas are cost effective, can be deployed fast and can therefore play an important role in the energy transition. The optimal design of a solar park is, amongst others, affected by income scheme, electricity transport capacity and land lease costs. Important design parameters for utility-scale solar parks, that may affect landscape, biodiversity and soil quality, include ground coverage ratio, size and tilt of the PV tables. Particularly, low tilt PV at high coverage reduces the amount of sunlight on the ground strongly and leads to deterioration of the soil quality over the typical 25-year lifetime. In contrast, vertical PV or an agri-PV design fairly high above the ground leads to more and homogeneous ground irradiance; these designs are favoured for pastures and croplands. In general, the amount and distribution of ground irradiance and precipitation will strongly affect which crops can grow below and between the PV tables and whether this supports the associated food chain. As agrivoltaics is the direct competition between photosynthesis and photovoltaics. Understanding when, where and how much light reaches the ground is key to relate the agri-PV solar park design to the expected agricultural and electricity yields.
Keywords: see the list of keywords

1 INTRODUCTION

Solar electricity from solar parks in rural areas are cost effective, can be deployed fast and can therefore play an important role in the energy transition. The design of a solar park, particular in case of low tilt PV at high coverage ratios, could strongly reduce the amount of sunlight on the ground. At the same time, after the 20-30 years lifetime of a utility-scale solar parks, the land should be reusable for agriculture or nature. Therefore, landowners and public authorities demand that the soil quality remains equal during this period. As we will show below, ground irradiance modelling is paramount in the design phase of the solar park to ensure that soil quality conditions are met.

For agrivoltaics, the competition of photons for photosynthesis and photovoltaics is even more obvious. In particular, when the solar park operator varies the orientation of the PV tables over the days or seasons, for example the angle in horizontal single axis-trackers, and can actively control, within the physical limits, the share of ground irradiance for photosynthesis versus the irradiance for electricity yield.

We present methodology and modelling results on the ground irradiance between and below PV tables in utility-scale and agri-PV solar parks. Line profiles of ground irradiance are compared to the open field value. We link the ground irradiance simulations with ecological observations at existing solar parks of various designs. Using the Mitscherlich equations [1], the solar park design can be evaluated for the photosynthetic potential for shade- and sun-loving species. Based on our results, we can make recommendations for park design and future research.

2 METHODOLOGY

2.1 Solar park design and ecology

Schotman *et al.* have investigated the relation between solar park design and ecology [2]. They identified two groups of south-facing designs that are regarded as 1) a safe design for soil quality, see left side of Fig. 1, and biodiversity and 2) a more "risky" design. In contrast, they also encountered other south-facing designs that had low potential to maintain soil quality. Due to the high land lease costs in the Netherlands, alternating east- and west-facing low-tilt solar parks, with very high ground coverage ratio within the active area of the solar park, are a prominent choice nowadays (Fig. 1 right). Schotman *et al.* have included two solar parks with this design in their research. Also these have little to no ecological potential.

Figure 1: (left) south-facing solar park with ample vegetation below panels and (right) east- and west-facing solar park, with nearly bare soil below the panels. Photo by Alex Schotman via [2].

In the presence of not too much nutrients and water, species-rich grasslands are favoured in the Netherlands. Enough light for vegetation growth below the panels also is a basic condition for a healthy biodiverse soil. High-quality biodiversity in solar parks in general means flower rich grasslands with many insects above and in the ground. In this work, we will investigate the light distributions for these solar park designs and look for alternative designs, that have similar light distributions as ecologically safe designs, but with industry conform coverage ratio.

2.2 Ground irradiance simulation tool

TNO has developed a software package, named BIGEYE, that simulates the electrical output of solar panels in the presence of ground reflection and nearby objects. In particular, to simulate the irradiance on the rear of the bifacial panels, it is necessary to determine the irradiance distribution on the ground below and around the solar panels. The calculation of the spatial irradiance distribution takes into account the patterns of hard shadows by the PV panels, but also the distribution of diffuse light, e.g. from clouds. The simulated output during a year, approximated in 1-hour "constant" time-steps, has been shown to be within 1% accurate compared to the actual full-year output. For more details of the BIGEYE software package and its experimental validation, including benchmarking with other simulation tools, we refer to previous ECN/TNO publications [3][4].

For the energy yield calculations, the ground irradiance, which varies point by point and with the time of the day and the seasons, is an intermediate result. However, these results can also be directly used to study the effect that objects like PV panels have on the irradiance on the ground. The amount of light that reaches the ground directly affects biodiversity, soil quality and the possibility to combine solar energy with agriculture.

The ground irradiance distribution map is calculated for each time step, e.g. 1 hour or 10 minutes, using BIGEYE's fully 3D view factor approach. For this work, the ground irradiance map for each time-step is extracted from BIGEYE and line profiles perpendicular to the PV tables are obtained. For reference purposes, the irradiance in an empty field is also calculated for the same time period. In the next step, the irradiance is used as input for the photosynthetic rate calculation, using the Mitscherlich equations [1].

3 RESULTS

3.1 Ground irradiance simulation for existing solar parks

We have generated three generalised designs based on the ecological observations and measurements of the PV table designs by Schotman [2]. Ground irradiance profiles were determined for these designs and plotted in Fig. 2.

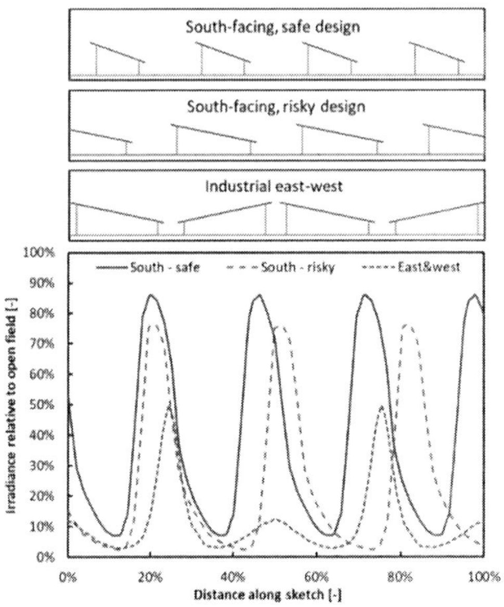

Figure 2: Cross-sections to scale and ground irradiance line profiles for three typical types of solar parks.

The industrial east- and west-facing design yields a ground irradiance profile, red dots, with two bright(er) regions. The higher bright region is quite narrow and is due to the bottom gap between two tables, where the midday sun reaches the ground through the north-south running opening. The much lower, but broader region is related to the top gap. Not only is this gap more narrow than the bottom gap, also it is much higher above the ground. This causes the direct light projection fallen through this gap to vary more from west to east due to the sun's movement during the day, see Fig. 1. The east & west profile indicates that a large fraction of the ground per pair of tables exhibits rather low ground irradiance in the range 3% to 12%.

3.2 Effect of minimum height of PV tables

Fig. 3 shows the effect of the minimum height of the PV tables on the ground irradiance distribution for the industrial east & west facing solar park design, while keeping all other parameters fixed. The profile with a ground clearance of 20 cm (red line) reveals a very sharp transition from the open field value of ~6.3 to nearly zero irradiance below the panels. In the top gap between two panels, there is a narrow region peaking at 16% of the open field irradiance level. The bottom gap, at 20 cm from the ground shows an even sharper peak with over 90% of the open field irradiance. Increasing the clearance changes the sharp step from high to near-zero irradiance to a more gradual transition but also with less high and less deep extrema.

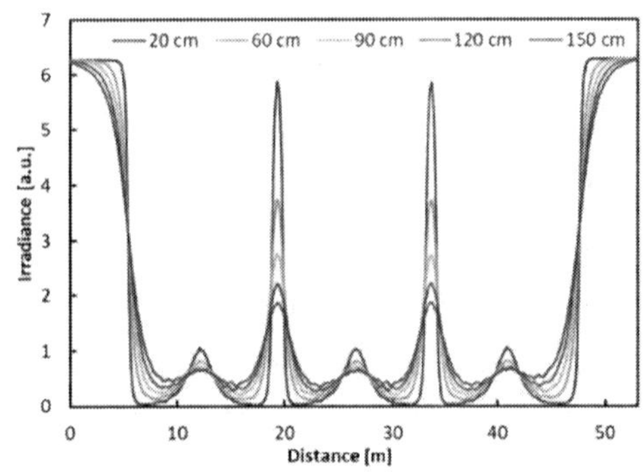

Figure 3: Variation in ground irradiance distribution as function of the minimum height for the industrial east & west design.

Increasing the minimum height of the PV tables makes the irradiance distribution more homogeneous. Although, exact boundaries between poor and good ecology cannot be drawn, it is very likely that increasing the minimum irradiance will be an important step towards nature-inclusive designs. Comparing low height (red) with high height (purple), we see that the minimum irradiance increases from near zero to about 8% of the open field irradiance. Also, the ecological observations on south-facing solar parks confirm that larger installation height leads to improved growth conditions [2].

3.3 Irradiance and photosynthesis

Fig. 4 shows the relation between the (photo-active) irradiance and the CO_2 fixation rate for several crop species [5]–[7]. These curves are generated by the Mitscherlich equations [1]. In these equations, the difference between species and their growth rate is governed by: a) the respiration rate, i.e. the metabolism of the plant without any growth; b) how fast the synthetic rate increases at low irradiance levels and c) the maximum rate of photosynthesis. For instance, clover has a high respiration rate and thus needs over 70 mmol m^{-2} s^{-1} PAR irradiance for a net-zero photosynthetic rate, in contrast with potato net-zero growth is reached at only 5 mmol m^{-2} s^{-1}. On the other hand, the rather low maximum growth rate of potato is already reached at 400 mmol m^{-2} s^{-1}, while e.g. spinach only reaches its saturation point (defined as 95% of the maximum rate) at irradiance levels above 1200 mmol m^{-2} s^{-1}. These growth rates not only determine the conditions under which crops grow better, but also what kind of plants can survive (let alone grow) in the darker regions below solar panels.

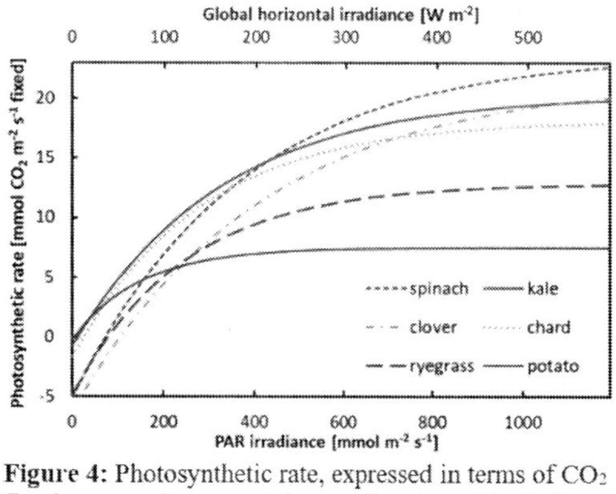

Figure 4: Photosynthetic rate, expressed in terms of CO_2 fixation per unit area and time, as function of the PAR (bottom axis) and GHI (top axis) irradiance for a selection of crops.

3.4 Increased ground irradiance designs

For ease of comparison we have made irradiance distribution diagrams that show which fraction of the land gets what amount of irradiance in Fig. 5. On the one hand, typical east- and west-facing, low tilt solar parks in the Netherlands exhibit poor potential for soil quality and biodiversity [2]. This is corroborated by the distribution diagram that shows over 80% of the active area with <20% irradiance compared to an open field and even 60% with <10%. On the other hand, south-facing systems that show good ecological potential have 40% of the active area with <20% irradiance and just under 20% at less than 10%.

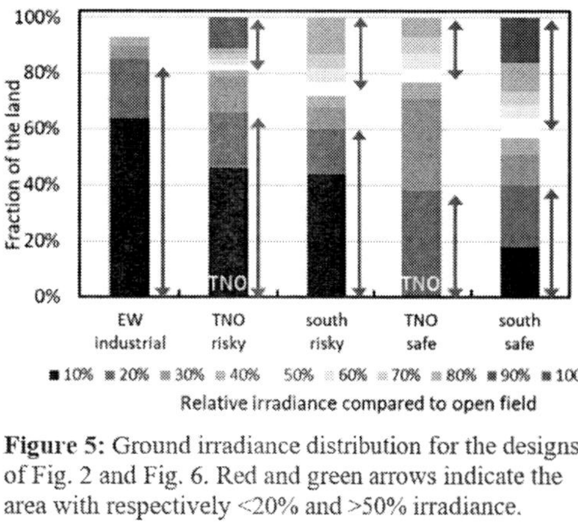

Figure 5: Ground irradiance distribution for the designs of Fig. 2 and Fig. 6. Red and green arrows indicate the area with respectively <20% and >50% irradiance.

Our results indicates that there is no sharp boundary, not even a gradual transition region. There are probably interactions between high and low irradiance regions. We deduce that as long as the darker regions are not too dark and too large and interspersed with large(r) regions with higher irradiance levels, the conditions are sufficient to ensure soil quality. In contrast, when the brighter regions are relatively small and dark and separated by larger regions with very low irradiance, the conditions are not suitable for plant growth. The first case corresponds for instance with the east- and west-facing industrial design, the latter case with the south-facing safe design. Exact conditions where the solar park design allows sufficient light on the ground to safeguard soil quality will be investigated in a four-year follow-up project EcoCertified Solar Parks, supported by the Dutch government and with participation of many project developers.

Because there is not a single value that describes the potential for soil quality, we will regard the distribution bar diagram as a fingerprint to steer the design of nature-inclusive solar parks. In particular, we will look for east- and west-facing variants that have similar or better ecological potential than the two south-facing designs. And at the same time have a ground coverage ratio that is more in line with the present industry standard.

We have varied some design parameters to increase the ground irradiance to levels comparable to the south-facing designs, see Fig. 2. The first improvement step, labelled TNO risky, is to open the gaps between the tables, both at the top and at the bottom, see Fig. 6. As a result, we obtain a bright region of about 20% of the land with over 50% of the open field irradiance. This is comparable to the irradiance distribution of the south – risky design.

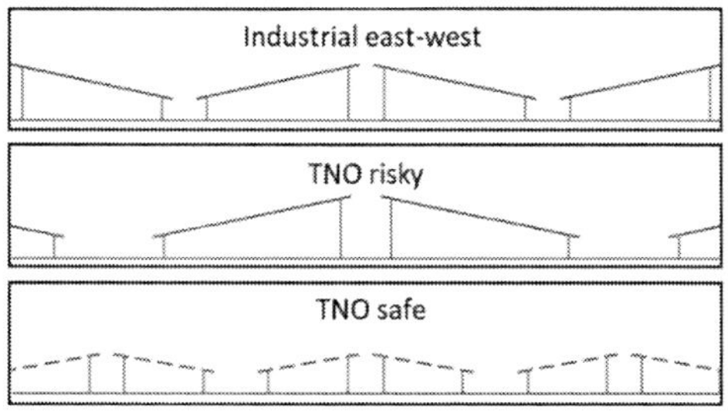

Figure 6: Cross-section to scale, comparing the industrial east-west design, Fig. 2 with two designs that lead to better ground irradiance levels and distribution.

In the next step, we also reduce the table size from 8 m to 4 m wide, decreasing all gaps correspondingly, and replace the back sheet modules with partially transparent bifacial modules. This design, labelled TNO safe, has similar bright regions as the TNO risky and south risky designs. More importantly, the fraction of dark area is strongly reduced and the minimum irradiance level is increased from around 5% to over 15% of that of the open field.

Table I: Ground coverage ratio, GCR, and area fractions with less than 10%, G < 10%, or more than 50% irradiance, G > 50%, for the five solar parks designs.

	EW industrial	TNO risky	South risky	TNO safe	South safe
GCR	89%	77%	69%	77%	53%
G < 10%	60%	45%	40%	0%	20%
G > 50%	0%	20%	25%	25%	40%

3.5 Orientation of design

One final example on the relevance of a clear understanding of the ground irradiance distribution is agrivoltaics. We have modelled a typical "high" PV design that allows tractors to drive below the structure. PV tables are 4 m across, on a 19-m pitch, and 5 m above the ground. The tables are oriented along the east-west or the north-south direction. Line profiles of the ground irradiance, taken perpendicular to the long direction of the PV tables are shown in Fig. 7. Note that the open field irradiance is approached by the irradiance at 0 m for the east-west oriented tables.

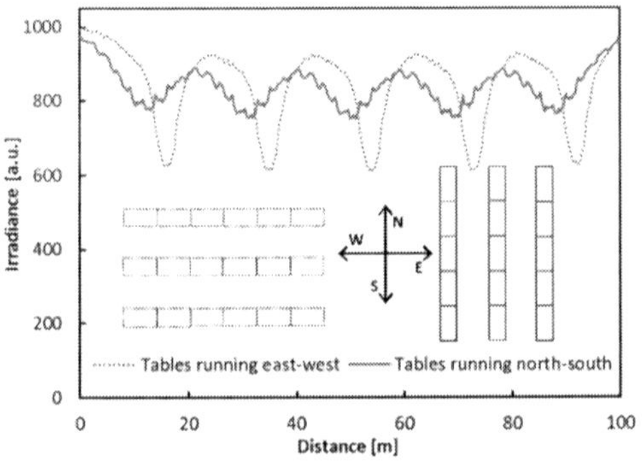

Figure 7: Ground irradiance line profiles for a simulated agri-PV system, see text, at 5 m height. The long edges of the PV tables are either running east-west (blue dots) or north-south (full green).

In the first case, the shading by the tables results in narrow east-to-west stripes with reduced irradiance, at about 60% of the open field, separated by wider stripes with irradiance up to 92%, the variation between minimum and maximum is ±19%. In the second case, the shadow and bright stripes run north-south with less pronounced minima and maxima, variation is ±8%. With this knowledge, the farmer can better decide which PV orientation she prefers, for instance how to space the rows of crops and where to plan gaps in the farming area for irrigation pipes or to drive on. Note that the average ground irradiance is, obviously, the same for both orientations.

4 DISCUSSION

Typical east- and west-facing designs of utility-scale solar parks, with a coverage ratio of 90% in the active part of the solar park, result in hardly any light on the ground underneath the PV panels, with the minimum lower than 5% of the open field irradiance. It is noted that the Dutch branch organisation Holland Solar issued a code of conduct that states a 75% maximum ground coverage. However that may be achieved by adding a green zone at the perimeter of the solar park, whilst the active part of the solar park has an effective coverage ratio often of over 90%. This green perimeter does not mitigate the soil degrading effect of low irradiance under the active areas. Note, that compared to agricultural land, green perimeters or corridors can support biodiversity development.

We have shown that distributing this additional "green" area over the active part of the solar park creates more areas with high ground irradiance. At the same time, it also decreases the land fraction with extremely low irradiance. Note that the active part coverage at 77% is still higher than both south-facing designs.

We also show the ground irradiance for a dilute PV installation, high above croplands allowing agricultural access. This contributes to a more homogeneous distribution of irradiance, compared to the utility PV park, preventing it to fall below 60% of the open field irradiance at any location. Still the orientation of the table influences the irradiation variation by over a factor two.

5 CONCLUSION

There are reasons for concern about the soil quality in utility-scale solar parks, when these have a very high coverage ratio as is often observed in east- and west-facing solar parks in the Netherlands. We have shown that over 80% of the ground receives less than 20% of the open field irradiance. In contrast, simulations of the ground irradiance of existing solar parks with sufficient potential for soil quality have only 40% of the ground receive less than 20%. Also, whereas the former has nowhere over 50% of the open field irradiance, the latter has 40% with at least 50% of the open field irradiance.

We have shown that by increasing the minimum height of the system, decreasing the size of the PV tables and decreasing the coverage ratio, the ground irradiance increases, in particular around the gaps between the tables.

The most direct way of increasing the lowest irradiance in a solar park design is to use semi-transparent PV panels, such as the commercially available bifacial glass-glass modules.

In conclusion: we have shown that we can achieve similar ground irradiance levels in an east- and west-facing design with 77% ground coverage ratio as is achieved by a south-facing design at 53% coverage.

ACKNOWLEDGEMENTS

This work was funded by Rijksvastgoedbedrijf, the Dutch central government real estate agency [8], and two projects by the Dutch ministry for Economic Affairs and Climate: SolarEcoPlus with a Topsector Energy subsidy, and the MOOI-project Sunbiose.

REFERENCES

[1] J. Goudriaan, "A Family of Saturation Type Curves, Especially in Relation to Photosynthesis," *Annals of Botany*, **43**(6), 783–785 (1979).

[2] A. Schotman, F. van der Zee, *et al.*, "Verkenning van bodem en vegetatie in 25 zonneparken in Nederland" Wageningen Environmental Research, Wageningen, 2021.

[3] G. J. M. Janssen *et al.*, "How to Maximize the kWh/kWp Ratio: Simulations of Single-Axis Tracking in Bifacial Systems," in *Proceedings of the 35th European Photovoltaic Solar Energy Conference and Exhibition*, 1573–1577 (2018).

[4] A. R. Burgers, "BIGEYE - simulation under shadow conditions," presented at the *6th workshop on bifacial PV*, Amsterdam, (2019).

[5] J. Erwin and E. Gesick, "Photosynthetic Responses of Swiss Chard, Kale, and Spinach Cultivars to Irradiance and Carbon Dioxide Concentration," *HortScience*, **52**(5), 706–712 (2017).

[6] A. Badr, P. Angers, and Y. Desjardins, "Metabolic profiling of photoautotrophic and photomixotrophic potato plantlets (Solanum tuberosum) provides new insights into acclimatization," *Plant Cell Tiss Organ Cult*, **107**(1), 13–24 (2011).
[7] S.J. Blaikie and F.M. Martin, "Effect of temperature on the photosynthesis of irrigated pasture," Melbourne, Australia, (1987).
[8] J. de Jonge and M. Ram, "Advies over een grootschalig zonnepark in de Noordoostpolder - Publicatie - College van Rijksadviseurs," (2021).

TRUE COST OF SOLAR HYDROGEN – LEVELISED COST OF HYDROGEN IN EUROPE 2021-2050

Eero Vartiainen
Fortum Growth Oy
POB 100, FI-00048 Fortum, Finland
+358 40 528 9867
eero.vartiainen@fortum.com

Christian Breyer
LUT University
Yliopistonkatu 34, 53850 Lappeenranta, Finland

David Moser
Institute for Renewable Energy, Eurac Research
Viale Druso 1, 39100, Bolzano, Italy

Eduardo Román Medina,
TECNALIA, Basque Research and Technology Alliance (BRTA)
Astondoa Kalea Edif. 700 Derio, 48170, Spain

Chiara Busto
Eni S.p.A.
Via Fauser, 4, 28100 Novara, Italy

Gaëtan Masson, Elina Bosch
Becquerel Institute
Rue Royale 146, 1000 Brussels, Belgium

Arnulf Jäger-Waldau
European Commission, Joint Research Centre (JRC)
Via E. Fermi 2749, 21027 Ispra, Italy

ABSTRACT: Green hydrogen will be an essential part of the future 100% sustainable energy and industry system. Up to one third of the required solar and wind electricity would eventually be used for water electrolysis to produce hydrogen, increasing the cumulative electrolyser capacity to about 17 TWel by 2050. Key method applied in this research are a learning curve approach for the key technologies, i.e. solar photovoltaic (PV) and water electrolysers, and levelised cost of hydrogen (LCOH). Sensitivities for the hydrogen demand and various input parameters are considered. Electrolyser capital expenditure (CAPEX) for a large utility-scale system is expected to decrease from current 400 €/kW$_{el}$ to 240 €/kW$_{el}$ by 2030 and to 80 €/kW$_{el}$ by 2050. With the continuing solar PV cost decrease, this will lead to an LCOH decrease in Europe from current 47-81 €/MWh$_{H2,LHV}$ (1.6-2.7 €/kg$_{H2}$) to 31-54 €/MWh$_{H2,LHV}$ (1.0-1.8 €/kg$_{H2}$) by 2030 and 16-27 €/MWh$_{H2,LHV}$ (0.5-0.9 €/kg$_{H2}$) by 2050, depending on the location. The share of PV electricity cost in the LCOH will increase from the current 63% to 74% by 2050.
Keywords: LCOE, solar PV, green hydrogen, LCOH

1 INTRODUCTION

In the past two years, the hydrogen sector and the hydrogen economy have experienced another wave of strong political support as enabler for a future carbon neutral energy system. This new wave is driven by an acceleration towards the energy transition and more ambitious decarbonization targets which are needed to achieve at least the ambitious 1.5°C target of the Paris Agreement. Compared to previous hypes, the current technology development, cost reduction, and a better view on the sectors which will benefit from it, clearly point to the expectation that this time the hydrogen sector may be ready to deliver and live up to their promises about hydrogen as the enabler for a future carbon neutral energy system.

There has been a long-lasting debate for what parts of an energy system hydrogen will be finally required [1–3]. Recent years have provided more and more insights that electrification is the way forward for sustainable and low-cost energy system solutions [4–7], while hard-to-abate segments are very often linked to a hydrogen route. A 100% renewable energy system in 2050 with a very high direct electrification approach and continued energy services demand growth for supporting the United Nations Sustainable Development Goals leads to 41,760 TWh of hydrogen demand [6], thereof most for e-fuels synthesis, the highest reported hydrogen demand to the knowledge of the authors.

A lot of emphasis is put on the different 'colors' of hydrogen which muddles the big picture: eventually, all energy sources need to be renewable and hydrogen will be 'green'. Solar photovoltaics (PV), together with wind power, will be the heart of the 100% renewable system since both are substantially scalable and enabled by battery storage and hydrogen [4,6]. The technology for the various conversion steps along the hydrogen value chain (electrolysis, storage, hydrogen-to-X synthesis, fuel cell, transport) is approaching market maturity, however it has not experienced yet a mass market able to achieve a significant cost reduction along a whole market segment. At the moment, the main benefits in terms of economy of scale are found in large size electrolysers where anything below 1 MW still shows very high CAPEX [8].

This paper investigates the current and future cost of utility-scale solar PV hydrogen, starting from the capital (CAPEX) and operational expenditure (OPEX) projections for solar PV and electrolysis technology. Historical learning rates (LR) are used for the future cost projections together with several volume growth scenarios. The levelised cost of hydrogen (LCOH) is calculated for five European and five non-European locations with different solar irradiation levels and with several weighted average cost of capital (WACC) rates.

A lot of reports on clean hydrogen have recently been published [3,20,23], but very few have used an up-to-date and realistic future projection of levelised cost of electricity (LCOE) for PV, as it is a major issue outside the PV community [10,11]. PV LCOE is very significant since the major part of the future LCOH is electricity cost. Another aspect is that the current electrolyser market is still very immature, most of the projects are tailor-made, site-sensitive and quite small and there is not yet real mass manufacturing. The scaling effect is very significant and for this reason it is not wise to use the past CAPEX for small projects (<1 MW_{el} on average) with truly utility-scale (>100 MW_{el}) projects. The correct estimate of LCOH is of utmost importance for the policy makers when decisions are made for a carbon neutral world between various technology options.

2 COST OF SOLAR PV

The cost of solar photovoltaic (PV) systems has decreased dramatically over the past years. Market prices of PV modules have decreased by about 90% [12] and system prices by close to 80% [13] during a decade making solar PV the cheapest form of electricity generation in many parts of the world with Power Purchase Agreements (PPAs) as low as 12 €/MWh [14] in some countries. The focus here is utility-scale PV systems with capacity of 100 MWp or more. The European Technology and Innovation Platform for Photovoltaics (ETIP PV) reported in 2020 the capital expenditure (CAPEX) for a fixed-tilt utility-scale PV system as 0.43 €/Wp [15], although in some countries like India it can be significantly lower [16], whereas in some countries like the US it is higher [17] because of soft costs and higher margins. Single-axis tracking PV is becoming increasingly more common in utility-scale systems [18], increasing the CAPEX by about 7% [19] with an increased annual yield, and further positive energy system impact [20]. Another trend is bifacial modules which also increase the yield, the price difference with corresponding mono-facial modules being currently about 0.015 €/Wp according to data collected from PVinfoLink and bifacial manufacturers [21]. Bifacial utility-scale PV system with single-axis tracking with a price of 0.47 €/Wp in 2020 is assumed here. Research insights indicate that single-axis tracking bifacial PV may become the utility-scale standard in the years to come [22].

The future development of PV CAPEX is projected with PV cumulative volume capacity growth and learning rates (LR). Fig. 1 shows three different volume growth scenarios [23]. These growth scenarios reflect the range of long-term PV deployment projections, as the International Energy Agency (IEA) and the International Renewable Energy Agency (IRENA) indicate the slow growth scenario [24,25] while Haegle et al. [26] and Bogdanov et al. [6] strongly indicate the fast growth scenario.

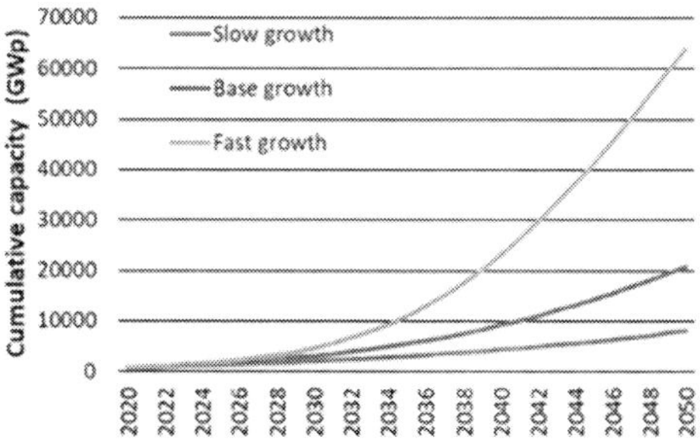

Figure 1: Global cumulative PV volume with three growth scenarios.

Historically, the average module price has decreased by an average 23.8% every time the global cumulative capacity has doubled [18]. However, from 2006 to 2020 this LR has been 40% [18]. A conservative 25% LR is assumed as base case here. For inverters, a lower LR of 20% shown by industry data [27] is assumed. For other Balance of System (BoS) components, an average 7.5% [23] LR is assumed. Since about 50% of BoS depends on the area of the modules [27], higher efficiency will in addition drive down the BoS significantly. A historical average 0.4%-point annual module efficiency increase [13] in future is assumed here. Fig. 2 shows the CAPEX development with three volume growth scenarios.

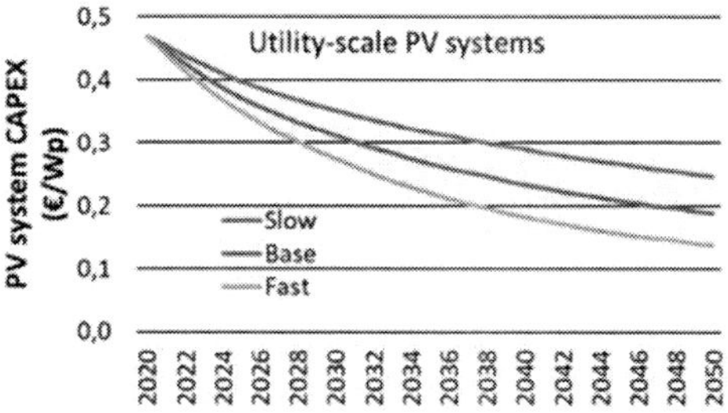

Figure 2: PV CAPEX for a system with single-axis tracking and bifacial modules with three different volume growth scenarios.

The main component of Operational Expenditure (OPEX) is usually the Operation and Maintenance (O&M). O&M cost varies greatly depending on the size of the system, scope of the O&M and location. Bloomberg New Energy Finance (BNEF) reported the average full-scope O&M contract price as 6.7 €/kWp/a in 2019, although with a large project or smaller scope it can be less than 4 €/kWp/a [28]. Here 1% of CAPEX (4.7 €/kWp/a) in 2020 is assumed as starting point. Apart from O&M, other components of OPEX include land lease, insurance, grid fees, balancing, asset management and various taxes. The amount of these cost components varies by country and by project. It is assumed that their total amount equals the O&M price, i.e., the total OPEX in 2020 was 9.4 €/kWp/a. Future OPEX is assumed to decrease with 10% LR [23].

3 COST OF ELECTROLYSIS

Production of hydrogen with Alkaline Electrolysis Cells (AEC) has been used for industrial applications since 1920 [29]. Cumulative installed water electrolysis capacity since 1956 has been estimated to be around 20 GW_{el} [30,31] although the annual market volume in 2019 was just 140 MW_{el} [32]. Future volume growth is expected to be very relevant. Here three different scenarios are used. By 2050, the installed capacity is expected to be either 1 TW_{el}, 5 TW_{el} or 17 TW_{el}. The first two are based on International Renewable Energy Agency (IRENA) scenarios [8] and the latter would be required for the 100% sustainable energy and industry system by 2050 [6,33]. For example, in a recent study [34] it was concluded that most of the examined European regions have sufficiently high technical potentials to be self-reliant using renewable energy. For the base scenario here, 5 TW_{el} global electrolyser capacity by 2050 is assumed. The different scenarios are presented in Fig. 3.

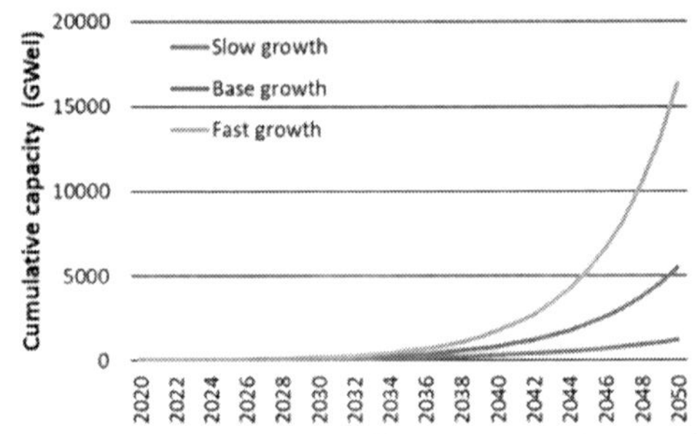

Figure 3: Global cumulative electrolyser capacity growth in three different scenarios.

Electrolyser CAPEX depends heavily on the scale. Currently a 200 kWel electrolyser has 2.3 times the unit cost of a 1 MWel electrolyser [35] and in turn a 1 MW_{el} electrolyser has 2.4 times the unit cost of a 100 MW_{el} electrolyser [8], combined this makes a factor of 5.5 between 200 kW_{el} and 100 MW_{el}. This is significant since the majority of electrolyser projects in 2018 were still below 1 MW_{el} [36], whereas 100 MW_{el} systems are now available. Scaling up is done with multi-stack systems [37], which allow modular construction and cost savings by mass manufacturing. IRENA reports [8] the investment cost for a 100 MWel electrolyser system being 450 $/$kW_{el}$ (370 €/kW_{el} with May 2021 exchange rate). This price level covers the full system cost including electrolyser stack, balance of plant (BoP), installation, civil works, grid connection and utilities.

It could be argued that in a hybrid PV and electrolyser plant the investment cost would be significantly lower since e.g. grid connection can be used by both. Another advantage is that the electrolyser plant could use directly DC generated by PV thus saving most of the power electronics cost which is about 20% of the total cost in a 100 MW_{el} system [8]. In fact, a system cost of as low as 200 $/$kW_{el}$ has been reported by offers from Chinese manufacturers [32]. However, a conservative value of 400 €/kW_{el} is used here as a base case price for a 100 MW_{el} electrolyser system in 2020.

The future electrolyser cost depends on the volume growth and the learning rate (LR). Using the volume growth scenarios shown in Fig. 3 and 18 % LR [31], the electrolyser system CAPEX would decrease according to Fig. 4. In the base scenario, system CAPEX would be 240 €/kW_{el} in 2030, 140 €/kW_{el} in 2040 and 80 €/kW_{el} in 2050.

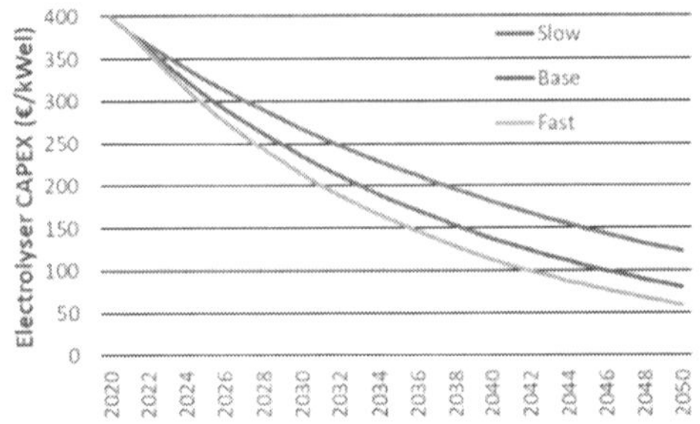

Figure 4: Electrolyser system CAPEX with three different volume growth scenarios.

Electrolyser OPEX is assumed as 1.5% of CAPEX [9] in 2020 (6 €/kW_{el}/a), decreasing with a 10% LR after that. Electrolyser efficiency is defined as lower heating value (LHV) of hydrogen divided by the electricity input. LHV for hydrogen is 33.3 kWh/kg_{H2} [38]. State-of-the-art efficiency for alkaline electrolysers is reported as 67% [8,38] which is assumed here to increase by 0.3%-points per year to 76% by 2050. Cost of catalysts are not regarded as critical since sodium hydroxide (NaOH) and sodium chloride (NaCl) are used as catalysts in alkaline water electrolysis [39]. Research on a broad variety of catalysts [40,41] may decrease the LCOH in future.

4 LEVELISED COST OF HYDROGEN (LCOH)

Levelised cost of hydrogen is the sum of electricity and electrolysis cost over the utilization rate:

$$LCOH = \frac{CAPEX_{PV} + \frac{InvRepl}{(1+WACC_{nom})^{0.5\times N}} + \sum_{t=1}^{N}\left[\frac{OPEX_{PV}(t)}{(1+WACC_{nom})^{t}}\right]}{Eff \times \sum_{t=1}^{N}\frac{Yield(0)\times(1-Degr)^{t}}{(1+WACC_{nom})^{t}}} + \frac{CAPEX_{el} + \frac{StackRepl}{(1+WACC_{nom})^{20}} + \sum_{t=1}^{N}\left[\frac{OPEX_{el}(t)}{(1+WACC_{nom})^{t}}\right]}{Eff \times \sum_{t=1}^{N}\frac{FLH(0)\times(1-Degr)^{t}}{(1+WACC_{nom})^{t}}} \quad (1)$$

where

N is economic lifetime of the system
t is year number ranging from 1 to N
$CAPEX_{PV}$ is capital expenditure of the PV system, made at $t = 0$ in €/kWp
InvRepl is the cost of inverter replacement, made at t = N/2 in €/kWp
$OPEX_{PV}(t)$ is PV system operation and maintenance expenditure in year t in €/kWp
Yield(0) is initial annual PV yield in year 0 in kWh/kWp without degradation
Degr is annual degradation of the nominal power of the system
Eff is the electrolyser efficiency
$CAPEX_{el}$ is capital expenditure of the electrolyser system, made at $t = 0$ in €/kW
StackRepl is the cost of electrolyser stack replacement, made at $t = 20$ in €/kW
OPEXel(t) is electrolyser operation and maintenance expenditure in year t in €/kW
FLH(0) is initial annual electrolyser full load hours in year 0 in h without degradation

$WACC_{nom}$ is nominal weighted average cost of capital per annum
$WACC_{real}$ is real weighted average cost of capital per annum

The relationship between $WACC_{nom}$ and $WACC_{real}$ is expressed with the formula:

$$WACC_{real} = \frac{1 + WACC_{nom}}{1 + Infl} - 1 \qquad (2)$$

where *Infl* is the annual inflation rate.

In this research, all results are given in real 2020 money. As nominal WACC rates are used here, inflation has to be taken into account in order to arrive at real values. For example, a 4% nominal WACC with 2% inflation rate corresponds to a 2% real WACC. Because the WACC rates are highly subjective and depend among other things on the country, market segment, investor type and risk appetite, a set of three different nominal WACC rates are included in the sensitivity analysis: 4%, 7% and 10%.

PV yield is calculated from Solargis database [42] for five European locations: Helsinki (Finland), Munich (Germany), Toulouse (France), Rome (Italy) and Malaga (Spain). Because the PV system seldomly generates electricity at the nominal peak power, it is wise to oversize the PV system in relation to the electrolyser input power. A ratio of 1.33 is used here which means that electrolysers have 33% more full load hours (FLH) compared to the PV yield in each location. A first-year 2% degradation and annual 0.5% degradation after that is assumed on the initial values of PV yield [18]. In a study of 11 commercial alkaline electrolyser systems with electric power inputs between 0.3 and 3.3 kW, the annual efficiency degradation was 0.10-1.50% with stack lifetimes of 50,000- 96,000 hours [43,44]. As a base case, 0.5% degradation is assumed here. Table I lists the PV yield and electrolyser FLH in each location.

Table I: PV yield and electrolyser full load hours and capacity factors in five European locations. PV yield is for a single-axis tracking system with bifacial modules.

	PV yield [kWh/(kWp·a)]	Electrolyser FLH [h/a]
Helsinki, Finland	1220	1620
Munich, Germany	1370	1820
Toulouse, France	1580	2100
Rome, Italy	1900	2530
Malaga, Spain	2110	2810

PV system lifetime is assumed to be 30 years which is becoming the industry standard for PV module production guarantees [18]. Inverter replacement is assumed to take place at half-point of the PV system lifetime. Electrolyser lifetime is also assumed as 30 years. The electrolyser stack should last up to 90,000 hours [31] which should be adequate for the 30 years expected lifetime of the PV system in the chosen European locations. However, for locations closer to the equator, the electrolyser utilization during 30 years could exceed 90,000 hours. For this reason, one electrolyser stack replacement representing 25% of the CAPEX of the time after 20 years is included in the equation. It must be noted that many reports quote much too high OPEX figures for PV electrolysers because they include the stack replacement in OPEX. This grossly exaggerates the effect of stack replacement in the LCOH because it is only made once later in the lifetime of the system and is discounted to the present time in real terms.

Fig. 5 shows the LCOH for the five European locations with the base volume growth and cost assumptions for solar PV and electrolysers. Current LCOH ranges from 47 €/MWh$_{H2,LHV}$ (1.6 €/kg$_{H2}$) in Malaga to 81 €/MWh$_{H2,LHV}$ (2.7 €/kg$_{H2}$) in Helsinki. By 2030 LCOH will decrease by about 33% and by 67% by 2050. It is notable that the cost of PV generated electricity is already about 63% of the LCOH increasing to about 74% by 2050. This suggests that electrolyser CAPEX will not play the major role in the future LCOH development. It further indicates that low-cost CAPEX electrolysers are not constrained by the solar PV yield, as it is still assumed in many publications. High FLH are only required for high CAPEX electrolysers.

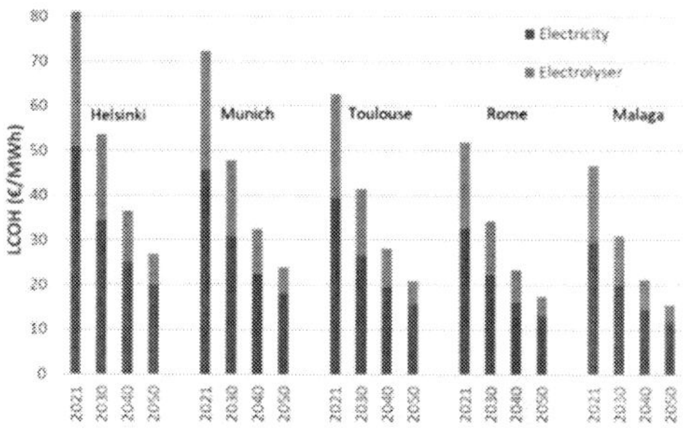

Figure 5: LCOH for five European locations with the base volume growth and cost assumptions, in €/MWh$_{H2,LHV}$.

For LCOH in non-European locations, detailed sensitivity analysis and uses cases of solar hydrogen, see the full paper published online in Solar RRL: onlinelibrary.wiley.com/doi/10.1002/solr.202100487

5 DISCUSSION

With current cost of solar hydrogen, 47-81 €/MWh$_{H2,LHV}$ (1.6-2.7 €/kg$_{H2}$) in the European locations of this study, green hydrogen is not yet competing with natural gas as a fuel. However, this will change rapidly as both solar electricity and water electrolysis cost are decreasing fast, and as soon as CO_2 emissions cost of a realistic level would be fully factored in. By 2030, LCOH of solar hydrogen in Europe will decrease to 31-54 €/MWh$_{H2,LHV}$ (1.0-1.8 €/kg$_{H2}$), making it a competitive clean fuel globally compared with hydrogen produced from natural gas with CCS. And there is an even bigger cost reduction potential when mass production of electrolysers will really kick in. So far, electrolyser projects have been quite small and often tailor-made with very high CAPEX. Currently, there is high demand of electrolysers and not enough manufacturing capacity which increases the prices. In the future, growing volumes and competition from Asian manufacturers should decrease electrolyser prices, as has happened in the PV industry.

Another factor currently increasing the LCOH in many estimations is the extremely high OPEX used in the calculations. In reality, there is not a lot of operation and maintenance to be done for an electrolysis system. And if the OPEX is calculated as a percentage of the initial CAPEX which is too high, then OPEX becomes too high too. The main component included in OPEX in most estimations is the electrolyser stack replacement. This should not be included in the OPEX every year but applied at the point of time when it is really done and then discounted to present time. With high WACC rates this makes a big difference. Furthermore, stack replacement may not be needed at all in locations with relatively low solar yield and electrolyser FLH.

Proper sizing of the system is key: the peak power of the PV system should always be oversized compared with the electrolyser input power since the PV system seldom generates electricity with its full peak power. Because of the complementary generation curves of solar and wind power, a hybrid PV-wind system would significantly increase the electrolyser FLH.

Finally, as the electrolyser technology further matures, the related risk perceived by investors should decrease, lowering the WACC rate of the investment. This is significant because as shown by the sensitivity analysis, WACC is the most important input parameter in the calculation of LCOH after location and solar yield.

Still, it needs to be clearly noted that hydrogen is part of an overall energy and industry system optimisation [6,45], which finds direct electric solutions in practically all cases lower in cost and higher in efficiency than hydrogen alternatives, in case a direct electrification and hydrogen solution would be available. However, the hard-to-abate applications cannot be directly electrified in the foreseeable future or by nature, such as long-distance marine and aviation transportation, chemicals and steel making, and these demand segments are expected to dominate green hydrogen demand in the years and decades to come.

8 CONCLUSIONS

Green hydrogen will be the main fuel as well as a fundamental energy vector for the future 100% sustainable energy and industry system. Up to one third of the required solar and wind electricity would eventually be used for water electrolysis to produce hydrogen, increasing the cumulative electrolyser capacity to about 17 TWel by 2050. There is a huge growth potential from the current 20 GW_{el} and the cost of electrolysis will decrease accordingly with mass production and volume growth, as well as with scaling up the electrolyser unit size.

Electrolyser CAPEX for a large utility-scale system is expected to decrease from the current 400 €/kW_{el} to 240 €/kW_{el} by 2030 and to 80 €/kW_{el} by 2050. Together with the continuing solar PV cost decrease, this will lead to a LCOH decrease in Europe from current 47-81 €/$MWh_{H2,LHV}$ (1.6-2.7 €/kg_{H2}) to 31-54 €/$MWh_{H2,LHV}$ (1.0-1.8 €/kg_{H2}) by 2030 and 16-27 €/$MWh_{H2,LHV}$ (0.5-0.9 €/kg_{H2}) by 2050, depending on the location. The share of PV electricity in the LCOH will increase from the current 63% to 74% by 2050. Already during this decade, solar hydrogen will be globally a less expensive fuel compared with hydrogen produced from natural gas with CCS.

ACKNOWLEDGEMENTS

This study has been made under the framework of the European Technology and Innovation Platform for Photovoltaics (ETIP PV). Open access of the full online publication in Solar RRL has been supported by the ETIP PV Secretariat which works in the framework of the ETIP PV - SEC II project. The project has received funding from the European Union's Horizon 2020 research and innovation programme under grant agreement No 825669.

DISCLAIMER

Arnulf Jäger-Waldau works at the European Commission - Joint Research Centre (JRC), Ispra, Italy. The views expressed are purely those of the author and may not in any circumstances be regarded as stating an official position of the European Commission.

REFERENCES

[1] Bockris JO. A Hydrogen Economy. Science. 1972;176(4041):1323-1323. doi:10.1126/science.176.4041.1323

[2] Barreto L, Makihira A, Riahi K. The hydrogen economy in the 21st century: a sustainable development scenario. International Journal of Hydrogen Energy. 2003;28(3):267-284. doi:10.1016/S0360-3199(02)00074-5

[3] Quarton CJ, Tlili O, Welder L, et al. The curious case of the conflicting roles of hydrogen in global energy scenarios. Sustainable Energy Fuels. 2019;4(1):80-95. doi:10.1039/C9SE00833K

[4] Jacobson MZ, Delucchi MA, Cameron MA, et al. Impacts of Green New Deal Energy Plans on Grid Stability, Costs, Jobs, Health, and Climate in 143 Countries. One Earth. 2019;1(4):449-463. doi:10.1016/j.oneear.2019.12.003

[5] Teske S, Pregger T, Simon S, et al. It Is Still Possible to Achieve the Paris Climate Agreement: Regional, Sectoral, and Land-Use Pathways. Energies. 2021;14(8):2103. doi:10.3390/en14082103

[6] Bogdanov D, Ram M, Aghahosseini A, et al. Low-cost renewable electricity as the key driver of the global energy transition towards sustainability. Energy. 2021;227:120467. doi:10.1016/j.energy.2021.120467

[7] Early decarbonisation of the European energy system pays off | Nature Communications. Accessed June 28, 2021. https://www.nature.com/articles/s41467-020-20015-4

[8] IRENA. Green Hydrogen Cost Reduction.; 2020. Accessed June 28, 2021. https://www.irena.org/publications/2020/Dec/Green-hydrogen-cost-reduction doi:10.1016/j.ijhydene.2014.12.035

[9] IEA. Data and Assumptions – The Future of Hydrogen – Analysis - Assumptions Annex.; 2019. Accessed June 28, 2021. https://www.iea.org/reports/the-future-of-hydrogen/data-and-assumptions

[10] Xiao M, Junne T, Haas J, Klein M. Plummeting costs of renewables - Are energy scenarios lagging? Energy Strategy Reviews. 2021;35:100636. doi:10.1016/j.esr.2021.100636

[11] Victoria M, Haegel N, Peters IM, et al. Solar photovoltaics is ready to power a sustainable future.

Joule. 2021;5(5):1041-1056. doi:10.1016/j.joule.2021.03.005
[12] PVinsights. Accessed June 28, 2021. http://pvinsights.com/
[13] Photovoltaics Report - Fraunhofer ISE. Fraunhofer Institute for Solar Energy Systems ISE. Accessed June 28, 2021. https://www.ise.fraunhofer.de/en/publications/studies/photovoltaics-report.html
[14] IEA PVPS. Snapshot of Global PV Markets 2021. Accessed June 28, 2021. https://iea-pvps.org/snapshot-reports/snapshot-2021/
[15] ETIP-PV. Fact Sheets about Photovoltaics - ETIP PV,PV the Cheapest Electricity Source Almost Everywhere. Accessed June 28, 2021. https://etip-pv.eu/publications/fact-sheets/
[16] Fortum to divest 500 MW of solar power plants in India – agreement with Actis for solar investments. Fortum. Accessed June 28, 2021. https://www.fortum.com/media/2021/06/fortum-divest-500-mw-solar-power-plants-india-agreement-actis-solar-investments
[17] IEA PVPS. Trends in PV Applications 2020. Accessed June 28, 2021. https://iea-pvps.org/trends_reports/trends-in-pv-applications-2020/
[18] ITRPV. ITRPV: Twelfth Edition Online. Accessed June 28, 2021. https://itrpv.vdma.org:443/viewer/-/v2article/render/67038287
[19] Feldman D, Ramasamy V, Margolis R. U.S. Solar Photovoltaic BESS System Cost Benchmark Q1 2020 Report. Published online 2021:1 files. doi:10.7799/1762492
[20] Afanasyeva S, Bogdanov D, Breyer C. Relevance of PV with single-axis tracking for energy scenarios. Solar Energy. 2018;173:173-191. doi:10.1016/j.solener.2018.07.029
[21] Bifacial modules: The challenges and advantages – pv magazine International. Accessed June 28, 2021. https://www.pv-magazine.com/2020/08/19/bifacial-modules-the-challenges-and-advantages/
[22] Rodríguez-Gallegos CD, Liu H, Gandhi O, et al. Global Techno-Economic Performance of Bifacial and Tracking Photovoltaic Systems. Joule. 2020;4(7):1514-1541. doi:10.1016/j.joule.2020.05.005
[23] Vartiainen E, Masson G, Breyer C, Moser D, Medina ER. Impact of weighted average cost of capital, capital expenditure, and other parameters on future utility-scale PV levelised cost of electricity. Progress in Photovoltaics: Research and Applications. 2020;28(6):439-453. doi:10.1002/pip.3189
[24] World Energy Outlook 2020 – Analysis. IEA. Accessed August 18, 2021. https://www.iea.org/reports/world-energy-outlook-2020
[25] Global energy transformation: A roadmap to 2050 (2019 edition). /publications/2019/Apr/Global-energy-transformation-A-roadmap-to-2050-2019Edition. Accessed August 18, 2021. https://www.irena.org/publications/2019/Apr/Global-energy-transformation-A-roadmap-to-2050-2019Edition
[26] Haegel NM, Atwater H, Barnes T, et al. Terawatt-scale photovoltaics: Transform global energy. Science. 2019;364(6443):836-838. doi:10.1126/science.aaw1845
[27] Fraunhofer ISE. Publication - Current and Future Cost of Photovoltaics.; 2015. Accessed June 28, 2021. https://www.agora-energiewende.de/en/publications/current-and-future-cost-of-photovoltaics/
[28] BNEF. PV O&M 2020: Plant Owners Get More With Less. Accessed June 28, 2021. https://about.bnef.com/
[29] Zhang X, Chan SH, Ho HK, et al. Towards a smart energy network: The roles of fuel/electrolysis cells and technological perspectives. International Journal of Hydrogen Energy. 2015;40(21):6866-6919. doi:10.1016/j.ijhydene.2015.03.133
[30] Schoots K, Ferioli F, Kramer GJ, van der Zwaan BCC. Learning curves for hydrogen production technology: An assessment of observed cost reductions. International Journal of Hydrogen Energy. 2008;33(11):2630-2645. doi:10.1016/j.ijhydene.2008.03.011
[31] Schmidt O, Gambhir A, Staffell I, Hawkes A, Nelson J, Few S. Future cost and performance of water electrolysis: An expert elicitation study. International Journal of Hydrogen Energy. 2017;42(52):30470-30492. doi:10.1016/j.ijhydene.2017.10.045
[32] Bloomberg New Energy Finance. Hydrogen Economy Outlook, Will Hydrogen Be the Molecule to Power a Clean Economy.; 2020.
[33] Ram M, Galimova T, Bogdanov D, et al. Powerfuels in a Renewable Energy World: Global Volumes, Costs, and Trading 2030 to 2050.; 2020. doi:10.13140/RG.2.2.32687.56487
[34] Kakoulaki G, Kougias I, Taylor N, Dolci F, Moya J, Jäger-Waldau A. Green hydrogen in Europe – A regional assessment: Substituting existing production with electrolysis powered by renewables. Energy Conversion and Management. 2021;228:113649. doi:10.1016/j.enconman.2020.113649
[35] Mayyas AT, Ruth MF (ORCID:0000000218380617), Pivovar BS (ORCID:0000000151815363), Bender G, Wipke KB. Manufacturing Cost Analysis for Proton Exchange Membrane Water Electrolyzers. National Renewable Energy Lab. (NREL), Golden, CO (United States); 2019. doi:10.2172/1557965
[36] Hydrogen Projects Database – Analysis. IEA. Accessed June 28, 2021. https://www.iea.org/reports/hydrogen-projects-database
[37] Proost J. Critical assessment of the production scale required for fossil parity of green electrolytic hydrogen. International Journal of Hydrogen Energy. 2020;45(35):17067-17075. doi:10.1016/j.ijhydene.2020.04.259
[38] IEA. Data and Assumptions – The Future of Hydrogen – Analysis.; 2019. Accessed June 28, 2021. https://www.iea.org/reports/the-future-of-hydrogen/data-and-assumptions
[39] Keçebaş A, Kayfeci M, Bayat M. Chapter 9 - Electrochemical hydrogen generation. In: Calise F, D'Accadia MD, Santarelli M, Lanzini A, Ferrero D, eds. Solar Hydrogen Production. Academic Press; 2019:299-317. doi:10.1016/B978-0-12-814853-2.00009-6
[40] Wang S, Lu A, Zhong C-J. Hydrogen production from water electrolysis: role of catalysts. Nano Convergence. 2021;8(1):4. doi:10.1186/s40580-021-00254-x
[41] Bernt M, Hartig-Weiß A, Tovini MF, et al. Current Challenges in Catalyst Development for PEM Water Electrolyzers. Chemie Ingenieur Technik. 2020;92(1-2):31-39. doi:10.1002/cite.201900101

[42] Solar Irradiance data. Accessed June 28, 2021. https://solargis.com/
[43] David M, Ocampo-Martínez C, Sánchez-Peña R. Advances in alkaline water electrolyzers: A review. Journal of Energy Storage. 2019;23:392-403. doi:10.1016/j.est.2019.03.001
[44] Felgenhauer M, Hamacher T. State-of-the-art of commercial electrolyzers and on-site hydrogen generation for logistic vehicles in South Carolina. International Journal of Hydrogen Energy. 2015;40(5):2084-2090. doi:10.1016/j.ijhydene.2014.12.043
[45] German bus depots clearly favoring batteries over fuel cells | S&P Global Platts. Published March 4, 2021. Accessed June 30, 2021. https://www.spglobal.com/platts/en/market-insights/latest-news/electric-power/030421-german-bus-depots-clearly-favoring-batteries-over-fuel-cells

DO NORTH-FACING BIPV FACADES IN EUROPE MAKE SENSE?

Alessandro Virtuani[1,2], Andrew Fairbrother[1], Fabiana Lisco[1], Laure-Emmanuelle Perret Aebi[1],
Nicolas Wyrsch1, and Christophe Ballif[1,3]
1 – École Polytechnique Fédérale de Lausanne (EPFL), Institute of Electrical and Micro Engineering (IEM),
Photovoltaics and Thin-Film Electronics Laboratory (PV-Lab), Rue de la Maladière 71b, 2002 Neuchâtel (Switzerland)
2 – Officina del Sole (O'Sole) srl, 20100 Milan (Italy)
3 – CSEM PV-Center, Rue de Jaquet-Droz 1, 2002 Neuchâtel (Switzerland)
Email: alessandro.virtuani@epfl.ch

ABSTRACT: In order to minimize unnecessary land exploitation and to allow PV to play a major role in the decarbonization of Europe, a massive deployment of PV in Europe should occur through the **integration of PV in buildings and infrastructures**, including the integration of PV on surfaces with a sub-optimal orientation. To assess the meaningfulness of installing PV on surfaces with sub-optimal orientations, we consider firstly the **carbon intensity balances** for PV. We show that for several cities (Milan, Frankfurt, La Valletta, …), it becomes obvious that a carbon intensity balance is largely in favor of PV not only for the optimal orientations, but also for less favorable ones: i.e. all the facades, including north-facing ones. For other cities/countries with a very low carbon footprint for the local electricity mixes (e.g. Oslo), the installation of PV may in principle not always be justifiable exclusively from a carbon balance point of view. We should however point out that in countries massively relying on **nuclear power** for their electricity supply – or planning nuclear phase-outs - other elements should be considered simultaneously. We also highlight the fact that installations in surfaces with **less optimal orientations** (e.g. north-facing facades) should possibly not be incentivized in the first place, but not expressly "prohibited" (or abandoned), as we have demonstrated that, in several countries, they are fully justifiable from the point of view of a carbon footprint balance. In addition, this may still help in promoting and creating ***"PV-awareness"*** among citizens and help architects in preserving building harmony/aesthetics.
Keywords: Building Integrated PV (BIPV), Integrated PV (I-PV), Carbon Intensity, Carbon footprint

1 INTRODUCTION

The European Commission is setting very ambitious targets for a net decarbonization of the European economy by 2050, requiring a massive electrification of the mobility and heating sectors, coupled with a major shift towards renewable energies sources, among which solar photovoltaic (PV) electricity is deemed to play a key role. By this horizon European member states may have to install from 5 to 10 GW of PV power around the Old Continent. In some countries with limited availability of land (the Netherland, Malta, Switzerland, etc.) the full deployment of PV on land may conflict with other land uses, such as agriculture, forestry, etc. Also in larger countries (e.g. Italy) the deployment of large PV parks on agricultural land is nowadays sometimes reportedly facing resistances from the local administration because of land use conflicts.

For these reasons, the adoption of PV projects leading to a double land/space use (e.g. so-called agri-PV or floating-PV) is highly welcomed.

Nevertheless, a massive deployment of PV in Europe should primarily occur through the **integration of PV in urbanized settings and into the built environment.** Including residential buildings, commercial and industrial buildings/warehouses, and more in general all the available **infrastructures**. The latter may include: noise barriers along roads and railways, car-ports, water treatment plants, bus and train stations, and many others.

In this work, we are not focusing exclusively on North-facing facades, but more in general on surfaces with sub-optimal orientations (see Fig.1), of which the former constitutes a sort of *extreme case*. To do this, we first asses the generating potential (insolation and PV energy yield) of non-optimally exposed surfaces in buildings (and elsewhere) for different European cities distributed at different latitudes (from 35° to 60°N).

To assess the meaningfulness of installing PV in surfaces with sub-optimal orientations, we do not take an economical perspective, a topic recently reviewed by other authors [1], but that of a **carbon intensity balance**. To do this, we try to assess in the first place whether – on a time horizon of 30 years – installing PV at different orientations acts as a net CO_2 sink or source, when compared to the same amount of electricity that would be generated with local electricity mixes.

Fig. 1: South- (left) and north- (right) facing BIPV (Building-Integrated PV) facades of the same building in Milan (Italy). The building has undergone a major renovation process in the years 2019 and 2020.

Thus, for the same cities, by using as a starting point updated (2021) figures for the carbon intensity (CI) of PV, we first correct these values to reflect the different generation potential of surfaces with different exposures and placed at different latitudes, coming out with some CI thresholds that would fully justify the deployment of PV. We then compare these numbers with CI values of the local (country) electricity mixes (focusing on country's electricity *consumption* rather than generation figures), finding out that for several cities/countries in Europe installing PV on north-facing facades (or other sub-optimal orientations) is fully justifiable.

2 APPROACH and METHOD

2.1 Insolation and PV energy yield of surfaces with different exposures

Insolation (H, [kWh/m²y]) and energy yield (EY, [kWh/kWp]) data for PV as a function of different orientations are obtained by JRC's (Joint Research Center – European Commission) free application tool PV-GIS, which uses satellite-derived data to estimate the availability of solar resources [2]. H and EY for South-facing facades at optimal tilt (opta) are shown in Table I for four European cities: La Valletta (Malta, MT, 35°N), Milan (Italy, IT, 45°N), Frankfurt (Germany, DE, 50°N) and Oslo (Norway, NO, 60°N). This is the orientation which maximizes the annual energy yield of a PV plant. For three different locations (Malta, Milan, Oslo), Figure 2 shows the ratio of H and EY for different orientations, normalized over the same parameters calculated for an optimal orientation (S-opta, i.e. South-facing at optimal tilt). These include values for a flat roof, for an average rooftop PV installation (*Avg roof*), and for installation facing the different cardinal points at 45° and 90° (facades) tilt, respectively. *Avg roof* values represent an average value for PV systems integrated/applied into/onto rooftops applying a constant -17% loss rate, which should account for misalignments with respect to a S-opta orientation.

As opposed to large utility-scale plants, for which it is often possible to have an optimal (or close to optimal) orientation of the PV arrays, this is generally difficult when integrating PV in roofs as the constraints will be set by the physical arrangement of the building roof/skin.

As can be observed in Fig. 2, with respect to an optimal PV energy yield, the potential of facades in Europe varies from **60% to 76%** (Malta and Oslo, respectively) for **S-facing facades**, from **46% to 49%** (*idem*) for facades with a **W and E orientation**, and from **13.1% to 17.6%** (*idem*) for **N-facing facades**. The corresponding values for Milan or Frankfurt lay in between these two extremes. Difference between E/W orientations are generally low and, for a given location, may be due to presence of different horizons or to far-shading.

2.1 Carbon intensity of solar PV electricity

LCA (Life-Cycle Analysis) figures published in literature about the carbon intensity (CI) of PV are often old or outdated, as they do not reflect the large progress made for this technology in recent years all along the value chain. Several technological improvements have in fact allowed to reduce considerably the carbon footprint of crystalline silicon (c-Si) based PV. These include, among several others: the use of thinner wafers, reduced Si losses in ingot wafering thanks to the use of diamond saws, more efficient processes for the manufacturing of mg-Si (metallurgical grade Si) and Si crystallization processes.
Two contributions have appeared in 2021 reporting updated CI figures for PV. One is a contribution from Frischknecht [3] and one from Fthenakis et al. [4]. The numbers reported by Fthenakis (for sites corresponding to three different insolation level) are slightly lower, possibly with an optimal orientation. The ones reported by Frischknecht are relative to a small-scale (3 kWp) residential rooftop PV system (with non-optimal orientation) located in Switzerland. To be a bit more conservative, and since we are focusing on PV installed in buildings rather than utility-scale systems, we have decided to use the numbers presented by Frischknecht in our analysis.

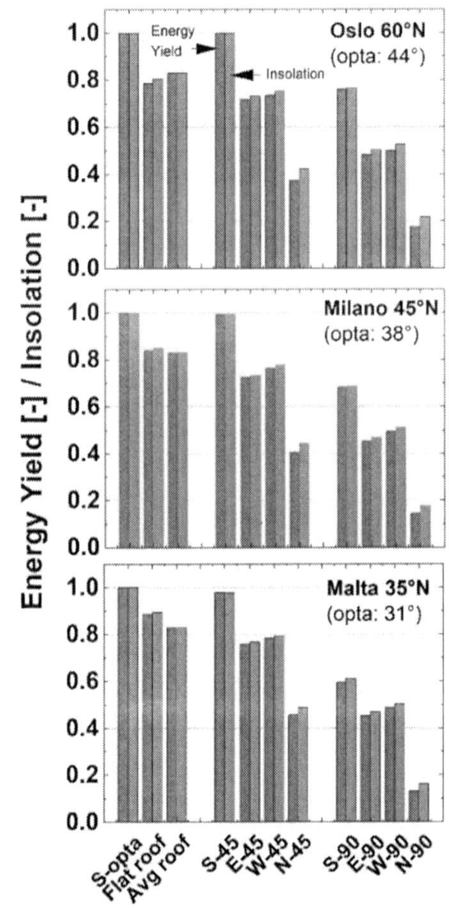

Figure 2. Ratio of the insolation H and energy yield EY for different orientations normalized over the same parameters calculated for an optimal orientation (S-opta) for three cities in Europe. *Avg roof* corresponds to an average value for PV systems integrated/applied into/onto rooftops applying a constant -17% loss rate accounting for misalignments with respect to a S-opta orientation. S-45 and S-90 correspond to a S-facing azimuth with a 45° and 90° tilt, respectively (E=East, W=West, N=North).

The assumptions (for the PV plant and energy yield) used by the author are briefly summarized in Table I (bottom). Under these assumptions, PV has a CI of **42.5 gCO_2/kWh**, assuming a 30-year-long lifetime for the PV plant and an average annual degradation rate of - 0.7%/y.

Since the energy yield (kWh/kWp) of a PV plant over its lifetime is strongly site-dependent and, for a given site, will largely be affected by the plant's orientation, we correct this number for the CI of PV (i.e. 42.5 gCO_2/kWh) to reflect the energy yield of PV plants (over their service lifetime) installed in different locations in Europe and for different orientations (see Table I). This is done by using the same assumptions for the plant service lifetime and annual degradation rates.

Table I. Yearly cumulative insolation of a south-facing surface exposed at an optimal angle (opta), corresponding yearly PV energy yield, and carbon intensity (CI) of the electricity generated by an optimally *oriented* PV system for four cities in Europe.

Location	Yearly Insolation at opta H_{opta} [kWh/m²y]	Yearly PV energy yield at opta EY_{opta} [kWh/kWpy]	PV carbon intensity at opta [gCO_2/kWh]
La Valetta MT 35° N	2'092	1'660	25
Milan, IT45°N	1'687	1'312	31.6
Frankfurt aM., DE 50°N	1'304	1'042	39.8
Oslo, NO 60°N	1'130.5	915.2	45.3

***Starting assumptions for the CI of PV (see [3]): 42.5 gCO_2/kWh** for residential rooftop PV system (3 kWp) installed Switzerland (46°N), yearly PV energy yield of 975 kWh/kW$_p$·y, corresponding to 83% of the energy yield of a S-facing system installed at the optimal angle in Bern (i.e. 1175 kWh/kW$_p$·y); lifetime of PV system 30 years with an annual degradation rate of -0.7%/y. The 42.5 gCO_2/kWh would correspond for an optimal orientation (S, 38°-tilt) in the same location to **35.2 gCO_2/kWh**, reflecting the higher energy yield. Source of the PV cell/modules: China.

2.2 Carbon intensity of European countries energy mixes

As the electricity generated by PV in buildings is generally consumed by or close to the end-user and it is injected into the low voltage (LV) or the medium voltage (MV) grid - depending on the size of the plant - to have a fairer comparison we try to make use - for CI figures of the local electricity mix – of consumption figures rather than generation ones, which are easier to retrieve and can be generally accessed through European statistical databases. In this work we used consumptions figures as published by Moro and Lonza [5], which are compensated for upstream emissions, electricity trading between countries, and for transmission and distribution losses. Other authors have recently published generation/consumption figures for the members states of the European Union and for some neighboring countries using a methodology to compensate for electricity intra-country electricity imports/exports developed by the authors [6]. These numbers are shown in Figure 4 for some representative European countries.

3 RESULTS

3.1 Carbon intensity of PV (at different orientations) vs local electricity mixes

For a PV plant located in Milan (45°N) and different orientations, Figure 3 shows the cumulative energy yield (MWh/kW$_p$) generated a PV plant under the assumption of a 30-year-long service lifetime (and an annual degradation rate of -0.7%/y); additionally the same Figure (b) shows the amount of CO2 that would be emitted by the same plant over the same temporal horizon using current PV CI figures (*PV-2021*) and under a scenario with a reduced CI (*greener-PV*). The amount of CO2 that would be emitted to generate the same amount of electricity using the CI of the local electricity mix is shown as well for a comparison, and is used to normalized the corresponding values in Figure 3 (c).

By having a closer look at Figure 3 (c), it become obvious that a carbon intensity balance in Milan is largely in favor of PV not only for the optimal orientations (opta, flat roof, avg roof and 45°-tilt), but also for the less favorable orientations: all the facades, including the N-facing one. Using 2021 values for the CI of PV a N-facing faced over 30 years is generating 50% of the CO_2 emissions that would be generated to produce the same amount of electricity using the local electricity mix. Understandably, this balance is by far more in favor of PV for all the other orientations.

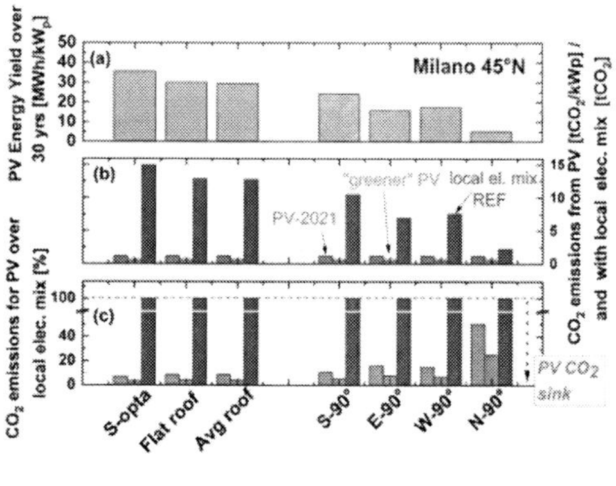

Figure 3: (a) Cumulative energy yield (kWh/kW$_p$) over 30 years – as a function of different orientations - for a PV system located in Milan (45°N); (b) shows the amount of CO_2 that would be emitted by the same plant over its service lifetime using current PV CI figures (*PV-2021*) and under a scenario with a reduced (i.e. 50%) CI (*greener-PV*). The amount of CO_2 that would be emitted to generate the same amount of electricity using the CI of the local electricity mix is shown as well for a comparison (green bars), and is used to normalized the corresponding values in Figure 3 (c).

We have performed the same analysis (not shown here) for several other cities in Europe (La Valletta, Frankfurt, etc.), for which we could drive the same conclusions as for Milan. Most notably, two extreme cases exist in Europe (see Fig. 4): (1) Malta with a very high availability of solar resources and very large CI of the local electricity mix (>1000 gCO_2eq/kWh); and (2) Norway with a low insolation (see Table I) and very low CI for its local electricity mix (~16 gCO_2eq/kWh).

Understandably, as can be inferred by Figure 4, which presents CI thresholds for PV calculated for the four cities of Table I (and for different orientations), the installation of PV in Malta (and other countries with a large carbon footprint for the local electricity mixes) makes largely sense irrespective of the orientation, including N-Facing facades. Whereas, with the current CI of PV, the installation of PV in Norway is in principle not justifiable exclusively from a carbon balance point of view (not even with an optimal orientation).

We should however point out that in countries massively relying on **nuclear power** for their electricity supply (e.g. France, Switzerland, Sweden, ...) other elements should be weighted simultaneously. Citizens or politicians in these countries may in fact oppose the use of a technology (nuclear fission), which will leave a huge burden and dangerous legacy to the coming generations in terms of disposal of nuclear fuels/infrastructures and of the costs needed for the decommissioning of the nuclear power plants (NPP). In addition, in countries planning nuclear phase-outs (Germany, Switzerland, ...) PV could be a valid alternative to other energy sources to lower or preserve a low CI budget of the local electricity mix.

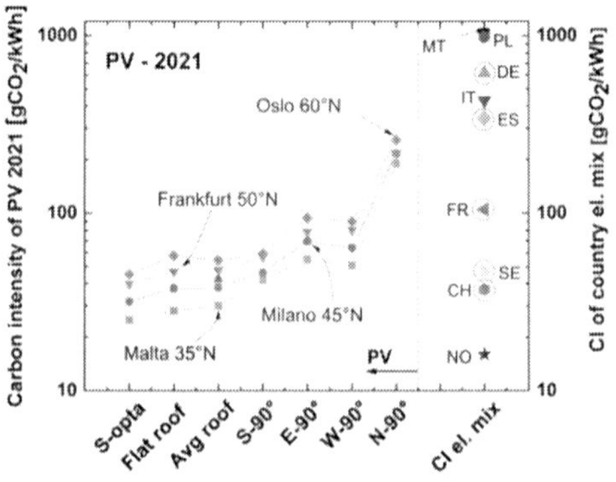

Figure 4: Carbon intensity (CI) thresholds for PV (based on 2021 figures, see Table I) corrected to compensate for different latitudes and orientations. On the right end-side, the CI of the electricity consumption mixes of several European countries are shown for a comparison (from [5]). Circles highlight countries that use nuclear as part of their national electricity mixes.

Installations in surfaces with **less optimal orientations** (e.g. N-facing facades) should possibly not be incentivized in the first place, but not expressly "prohibited" (or abandoned), as we have demonstrated that, in several countries, they are fully justifiable from the point of view of a carbon intensity balance. In addition, this may still help in promoting and creating ***"PV-awareness"*** among citizens, help architects in preserving building harmony/aesthetics, and push market deployment for BIPV and integrated PV (I-PV). Furthermore, they could still make sense from an economical point of view as – for example in the case of a new building or major renovation - they may avoid the adoption of different mounting structures and cladding elements for the different building's facades.

5 CONCLUSIONS

In order to minimize unnecessary land exploitation and to allow PV to play a major role in the decarbonization of European economies, a massive deployment of PV in Europe should primarily pass through the **integration of PV in buildings and infrastructures**, including the integration of PV in surfaces with a sub-optimal orientation.

To assess the meaningfulness of installing PV in surfaces with sub-optimal orientations, we focused on a **carbon footprint analysis**; trying to assess in the first place whether – on a temporal horizon of 30 years – installing PV at different orientations acts as a net CO_2 sink or source, when compared to the same amount of electricity that would be generated with local electricity mixes in different European countries.

For several cities (Milan, Frankfurt, La Valletta, ...) it becomes obvious that a carbon intensity balance is largely in favor of PV not only for the optimal, but also for the less favorable orientations: all the facades, including North-facing ones. For other cities/countries with a very low carbon footprint for the local electricity mixes (e.g. Norway, Switzerland, Sweden, etc.), the installation of PV may in principle not be always fully justifiable exclusively from a carbon balance point of view (not even for optimal orientations in Norway).

We should however point out that in countries massively relying on **nuclear power** for their electricity supply other elements should be weighted simultaneously. Citizens or politicians in these countries may in fact oppose the use of a technology, which will leave a huge burden and dangerous legacy to the coming generations.

We also highlight the fact that installations in surfaces with **less optimal orientations** (e.g. N-facing facades) should possibly not be incentivized in the first place, but not even abandoned, as we have demonstrated that, in several countries, they are fully justifiable from the point of view of a carbon intensity balance. In addition, this may still help in promoting and creating ***"PV-awareness"*** among citizens and help architects in preserving building harmony/aesthetics.

A follow-up of this work, we'll be that of drafting a list of recommendations that should help local authorities adopting favorable building codes and the right policies (including proper incentive schemes) to foster and maximize the diffusion of PV in buildings and infrastructures.

To summarize in brief this contribution, we make use and rephrase an older **slogan**: we should move from the approach of "*Solar PV everywhere*" to that of "**Solar PV wherever it makes sense first, and sometimes where it creates awareness**"!

6 ACKNOWLEDGEMENTS

This work has been funded in part by the European Commission (EC) under the H2020 Be-SMART (#818009) and H2020 ReCognition (#815301) projects and by the Swiss National Science Foundation (SNF) under the project REBI-PV (COST IZCOZ0_182967). Additional (collaborative) support was provided by the COST Action PEARL PV (CA16235). We gratefully acknowledge all team members at PV-lab and O'Sole.

REFERENCES

[1] H. Gholami, H.N. Rostvik, Econoomic Analyisis of BIPB systems as Building Envelope Material for Building skins in Europe, Energy 2014 (2020) 117931;

[2] PV-GIS: https://ec.europa.eu/jrc/en/pvgis;

[3] R. Frischknecht, Umweltfussabdruck Photovoltaikstrom Update 2020, presented at the 19. National Swiss Photovoltaic Symposium, July 2021;

[4] V. Fthenakis, E. Leccisi, Updated sustainability status of crystalline silicon-basedphotovoltaic systems: Life-cycle energy and environmental impact reduction trends, Prog Photovolt Res Appl. 2021;1–10;

[5] A. Moro, L. Lonza, Electricity carbon intensity in European Member States: Impacts onGHG emissions of electric vehicles, Transportation Research Part D 64 (2018) 5–14;

[6] B. Tranberg, O. Corradi, B- Lajoie, T. Gibon, I. Staffell, G.B. Andresen, Real-time carbon accounting method for the European electricity markets, Energy Strategy Reviews 26 (2019) 100367;

SimZukunft – ENERGY SCENARIOS 2050 FOR A SMALL CITY IN SWITZERLAND

Urs Muntwyler, Eva Schüpbach
Bern University of Applied Sciences BFH, Jlcoweg 1, 3400 Burgdorf, Switzerland
urs.muntwyler@bfh.ch, urs_muntwyler@gmx.ch, drs_consulting@bluewin.ch
Noah Pflugradt, Research Center Jülich, Germany, noah.pflugradt@gmail.com

ABSTRACT: SimZukunft is a P- and D (Pilot- and Demonstration) project supported by the Swiss Federal Office of Energy SFOE. The objective of the project is to define the adjustments of the grid needed in a small Swiss city when implementing the Swiss Energy Strategy 2050. Four detailed energy scenarios and associated load profiles for every single household from the present until 2050 were generated for Burgdorf, a city with 16'000 inhabitants. For each of the four energy scenarios, the impact on the grid was calculated. Impacts on the low voltage grid were analysed in detail, and a comparison was made for each of the four scenarios on how the grid would perform with or without Smart Grid functions and with significant load shifting, decentralized storage and curtailment. It is demonstrated that implementing even the most ambitious scenario "Utopia" (full decarbonization) as outlined in the official Swiss Energy Strategy 2050 until the year 2050 is not a huge challenge, when decent planning and a moderate amount of Smart Grid functionalities are introduced in Burgdorf. To achieve full carbon neutrality and achieve net-zero in Burgdorf in 2050, either large investments in the grid or in Smart Grid functionalities and decentralized storage are needed.
Keywords: Dissemination, Grid Management, Electromobility, Decarbonization

1 THE FOUR ENERGY SCENARIOS of SimZukunft

Table 1: The four energy scenarios defined in the project

Name	Description
ES-2050 POM-E	«Political Measures», Variant «Renewable Energy»: This is the path that Switzerland is currently on. The parameters describe medium renewable adoption, medium electromobility adoption, a lot of air conditioning and major renovations.
ES-2050 NEP-E	«New Energy Policy», Variant «Renewable Energy»: This is very similar to POM-E, just slightly more renewable energy. The major differences in the scenarios are not on the load side.
Utopia	Net-Zero-City, 100% electromobility, 100% heat pumps, completely carbon-neutral, major building renovations, a lot of efficiency measures.
Dystopia	Very low percentage of renewable energy, a lot of migration, climate change with high air conditioning need, low rate of renovation but high adoption of heat pumps, resulting in very high energy demands in the grid.

"Utopia" is the full decarbonization scenario with a very high contribution of PV.

2 INTRODUCTION

The project analysed the impacts upon the low- and medium-voltage grid in Burgdorf (small city in Switzerland with 16'000 inhabitants) until 2050 for different energy scenarios (shown in Table 1). Two scenarios (ES-2050 POM-E and ES-2050 NEP-E) are based on the official Swiss Energy Strategy 2050 ("Energiestrategie 2050"). The other two (Utopia and Dystopia) scenarios aim to illustrate the range of grid impacts and to identify when and under which conditions the grid in Burgdorf may fail. The Utopia scenario (full decarbonization in 2050) also is a reference for and reflects the official energy policy of Switzerland in the new "Energieperspektiven 2050+".

3 RESEARCH QUESTIONS

Original research questions in the projects were formulated both with regard to the methodology to be developed and the expected outcomes to be achieved.

The research questions on the methodology were as follows:

- Is it feasible to do grid planning by creating detailed load profiles and generation profiles for every single house?
- How far can this process be automated?
- Are there advantages to this method compared to the traditional static grid planning approach?
- Could this approach lead to lower investment costs?
- Could this approach help with better investment planning?

The research questions on the expected outcomes were as follows:

- How will the low voltage grid in a small, typical Swiss city cope with the official scenarios from the Energy Strategy 2050?
- How will it cope with a net-zero-situation where all energy is locally generated?
- What can the maximum load reduction be achieved when using smart grid technology and how much can it substitute for grid upgrades?

4 CASE STUDY BURGDORF: A SMALL SWISS CITY

Burgdorf is a district town (Figure 1) in the state of Berne with about 16'000 inhabitants in 2017 (the project start). It is situated in the entry to the "Emmental", a hilly region with a lot of agriculture, food processing enterprises and small and medium enterprises, and hosts some public administration offices from the state of Berne. There is a vast network of public transport with a fast train to Bern, the capital of Switzerland and of the state of Berne. Burgdorf is also an education hotspot and locates - besides several schools - different Departments of Bern University of Applied Sciences BFH, including the Laboratory for Photovoltaic Systems (PV LAB) of the Institute for Energy and Mobility (IEM) of BFH (Figure 2).

Figure 1: Castle of the City of Burgdorf

Figure 2: PV LAB at BFH with roof-top PV plant installed in 1993.

Burgdorf is famous for the invention of the "feed-in tariff" introduced by the local parliament in 1989. The "feed-in tariff" was proposed by a local entrepreneur (Mr Flückiger, who was a member of the local city parliament), and inspired by the Tour de Sol races.

Mr Flückiger had some engineers ("Solarmobil-Gruppe Burgdorf") participating in the Tour de Sol races. To promote the development of PV, Mr Flückiger's idea that he suggested to the local city parliament was to pay 1 Swiss Franc per kWh (about 0,9€/ kWh) to people feeding their PV electricity into the grid. His proposition was accepted by the local city parliament with all votes, and the local utility company of the city of Burgdorf, now Localnet AG, subsequently implemented the "feed-in-tariff" for PV.

As anticipated, the "feed-in tariff" created considerable actions in the field of PV application (Figure 3). For example, the PV LAB at BFH installed its roof-top PV installation in 1993 (60 kWp PV) that - besides serving as an educational infrastructure - has generated considerable benefits thereafter.

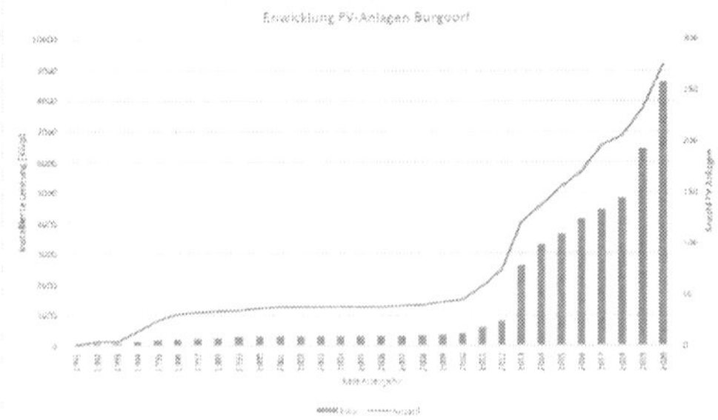

Figure 3: Development of PV plant installations in Burgdorf. After a slow start in 1991 with a first private 1 kWp plant, a strong growth after 2013 (introduction of the "national feed-in-tariff" (KEV)) is evident. Source: Localnet AG 2021.

With an installed 9 MWp (about 600 Wp/ inhabitant), Burgdorf is today still far from the 160 MWp goal (7,2 kWp/ inhabitant) as formulated by the Swiss Government in the "Energieperspektiven 2050+". The load situation in Burgdorf reflects this low level of PV (Figure 4).

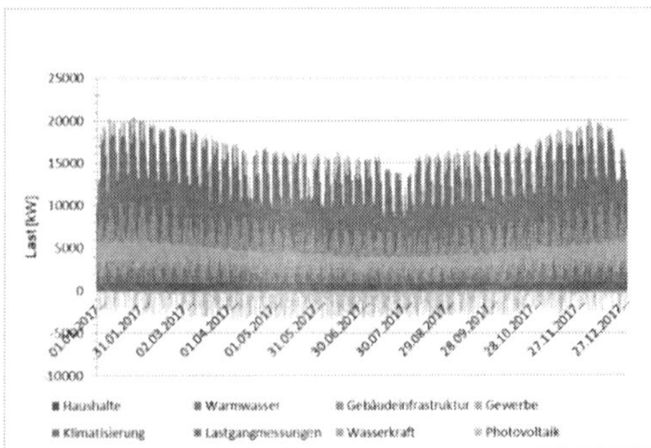

Figure 4: The maximal electric load is about 20 MW, PV in 2017 was only about 5 MWp

The detailed analysis of a winter and a summer situation for Burgdorf in 2017 indicates daily peaks ("Lastgangmessung") above 30 kVA mainly from enterprises and bigger users (Figure 5).

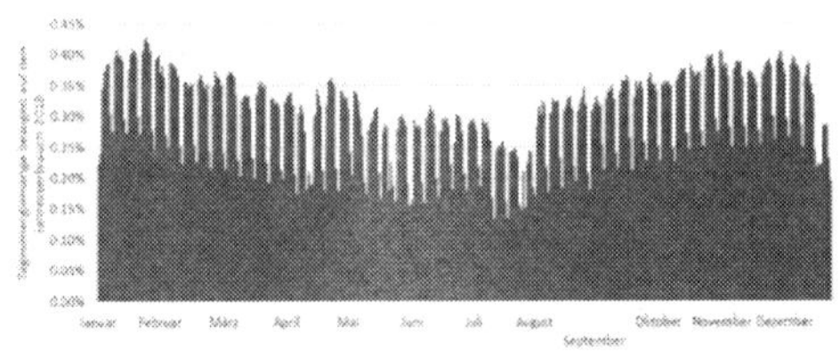

Figure 5: Daily consumed electricity in percent compared to the annual consumption in 2018. Maximum loads are **during the day in winter**

Calculations with the "Load profile generator" developed by Dr. Noah Pflugradt (the main researcher in this project) show that the measured consumption in Burgdorf (grey line in Figures 6 and 7) matches the simulated consumers quite well.

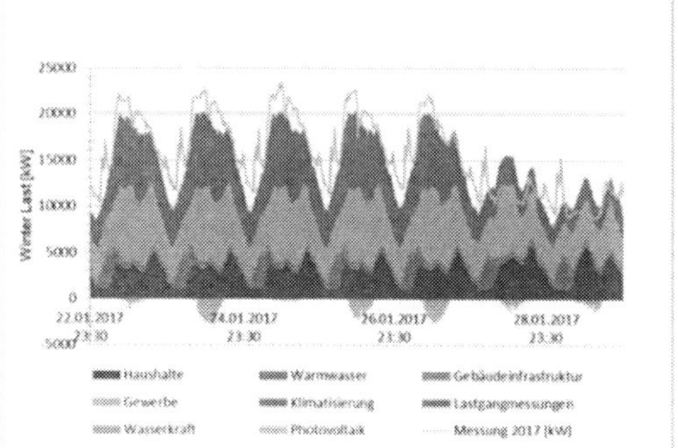

Figure 6: Consumption and PV production in wintertime in Burgdorf in 2017

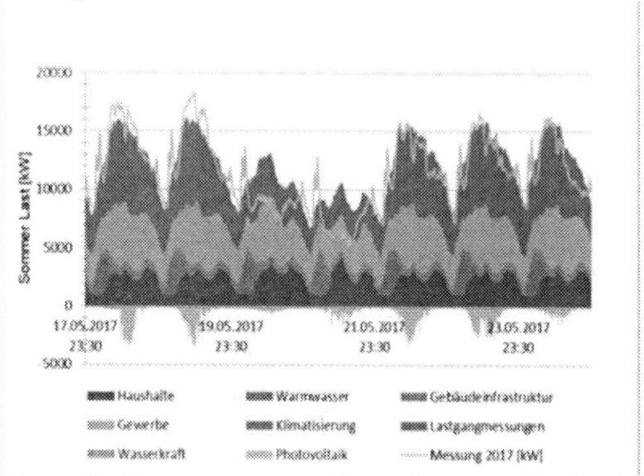

Figure 7: Summer consumption and PV production in 2017

5 DECARBONIZATION WITH 100% PV (UTOPIA)

5.1. Data sources for scenario "Utopia"

From the perspective of PV, the scenario "Utopia" is the most interesting and is described in greater detail here. Evidently, a wealth of data must be isolated, processed and merged in each scenario. Table 2 provides an overview of the data sources for the scenario "Utopia". The data are either owned by the utility company, publicly available, in the care of the local council or the Swiss Government. If the utility cooperates with the local administration, at least data ownership should offer no obstacle for the transfer of our approach to other cities in Switzerland.

Table 2: Data sources in the project

Data Source	Content	Geographic Resolution
Localnet accounting data	Consumption of every single consumer in Burgdorf.	Energy meter
Sonnen-dach.ch	Theoretical solar potential for every building in Burgdorf.	Building
Commuting data	Statistics of the Swiss Office for Statistics on the number of commuters to and from Burgdorf.	City
Population	Number of households, number of inhabitants, age pyramid in Burgdorf.	City
Building Registry	«Eidg. Gebäude- und Wohnungsregister (GWR)», This contains all buildings and appartments in Switzerland.	Building
Furnace Register	Data from the chimney sweepers (furnaces installed in the city and how old they are). Important for identifying the oil-heated buildings and estimating their energy consumption.	Building
Load measure-ments	For all large electricity customers in the project, the measured profiles from 2017 were used. The energy consumption was modified with the appropiate efficiency gain from the scenario.	Energy Meter
«Energie-bedarfsdaten Kanton Bern»	Canton Bern has a frequently updated database with the energy consumption of every single building (estimates of the energy consumption based on statistics, building data, and heuristics). This was used for the heat demand for all buildings that don't have gas heating.	Building
Car Ownership	Statistical data	City
Business Type	Manually categorized based on customer name and Google.	Customer
OpenStreet Map	OpenStreetMap was used to visualise the results.	Building
Future temperature	Used for the heat demand, air	City

and solar radiation profiles	conditioning demand and the PV profiles. Generated using Meteonorm, which uses data from the IPCC AR4 B1 scenario.	
Installed PV-and hydroplants	Lists with the technical specifications of currently installed systems.	Energy Meter

5.2. Methods

The most important steps for processing the data are described in Table 3.

Table 3: Steps for processing the data

Step	Description
Import	Imports all data in a database
Model for the present day	Merges all data into a consistent model and uses heuristics to solve conflicts. Data with a coarse geographical resolution is distributed also using heuristics.
Scenario creation	Creates the model for each scenario and each year. Uses appropriate allocation strategies for each feature, such as population, PV and electromobility to create a plausible future city.
Load profile synthesis	Generates the load profiles for each consumer and generates the input data for the LPG. Merges the profiles after calculation for each endpoint and prepares the export for Adaptricity for the grid calculations.
Grid calculations	These are performed by Adaptricity using the generated load profiles and yield detailed load results for every single cable and every substation.

6 RESULTS

The scenario "Utopia" is the official political goal in the "Energieperspective 2050+" of the Swiss Government and encompasses the phase-out of all nuclear power plants, but also a 100% decarbonization. For the case study Burgdorf in our project, the variables below are included in the scenario "Utopia" (year: 2017 / 16'000 inhabitants).

Table 4: Assumptions for the "Utopia" Scenario

UTOPIA	2020	2030	2040	2050
Target population + migration	16' 838	18' 644	20450	22257
Reduction of energy consumption of business % per 10 years	90%	90%	90%	90%
Reduction of energy consumption of consumers% per 10 years	90%	90%	90%	90%
Consumption new climatization kWh/m²	30.0	30.0	30.0	30.0
Surface with climatiz. %	0.0	0.2	0.4	0.5
Percentage [%] car user per household	44%	44%	44%	44%
Percentage [%] of EVs	5.0%	20%	70%	100%
Additional PV in GWh	3	26	50	75
House retrofit factor (0.2 = 20% from old status)	30%	30%	30%	30%
Electric boilers to heat-pumps (HP)	400	400	400	400
Gas heating to HP % of 2017	17%	28%	35%	100%
Oil heating to HP in % of 2017	17%	28%	35%	100%
Other heating to HP in % of 2017 use	17%	28%	35%	100%
Reduction electric consumption over 10 years	90%	90%	90%	90%
Percentage of building renov.[%, 0-1]	3.0%	10%	10%	10.0%

The outcomes indicate that (i) the overall grid load peak will change from winter to summer in 2050 (which is due to the needs for climatization and air conditioning) and that (2) it is the climatization of the houses that will create the new consumption peak in summer. This is the most pronounced result of the scenario "Utopia". The heat load in winter will decrease if the renovation of the existing houses is completed by 2050.

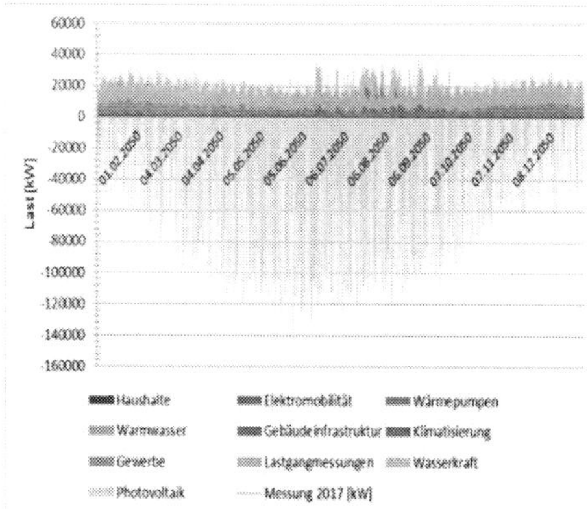

Figure 8: Annual profile in 2050; high PV injection without curtailment

Installing the 160 MWp PV needed by 2050 is not so complicated; 160 MWp is an equivalent to 10 kWp per inhabitant, or a surface of about 50m^2 (based on the population of 16'000 in Burgdorf in 2017).

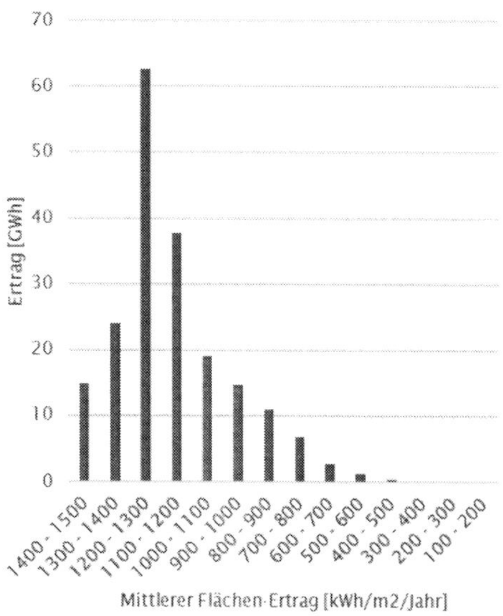

Figure 9: Size of surfaces for PV in Burgdorf

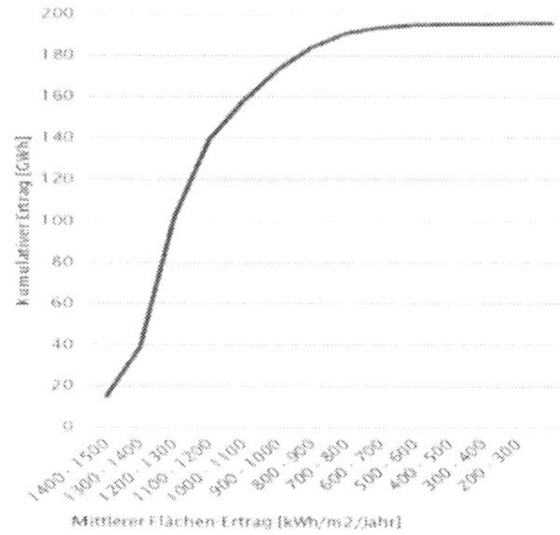

Figure 10: Cumulated yield with PV on buildings

The analysis of the surfaces on the buildings in Burgdorf shows a medium yield at 1'200 kWh/ m^2. Using the best surfaces for PV, the target of 160 GWh (corresponding to the 160 MWp) can be reached with installations above 1'000 kWh/ m^2 and year. This offers a good price/performance ratio of the PV plants. Considering the low costs of PV, the inhabitants of Burgdorf will save a substantial amount of money as they will not buy gasoline / gas and oil anymore in the "Utopia" scenario. Meaning that, in in the scenario "Utopia" in 2050, electricity will be the dominant energy carrier and will fully replace the fossil energies.

In the scenario "Utopia", the energy consumption in Burgdorf will drastically decrease by 2050 (Figure 10), even with a population of more than 22'000 inhabitants (as compared to 16'000 inhabitants in 2017). With a population of 22'000, PV per capita goes down to 7,2 kWp or 36-40m^2. The peak in 2050 will mainly be in summer (Figure 11).

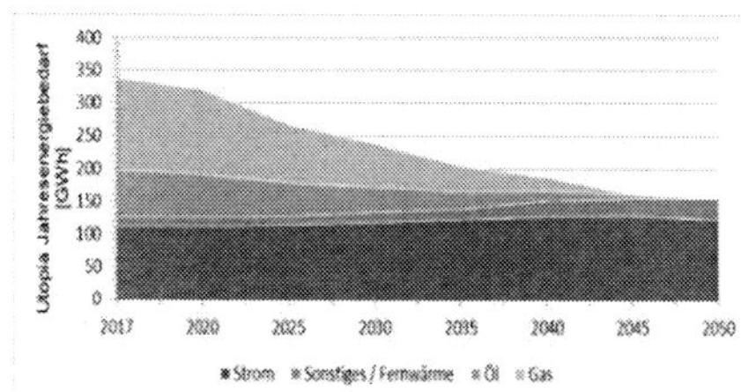

Figure 10: Yearly energy consumption will decrease in the scenario "Utopia" until 2050

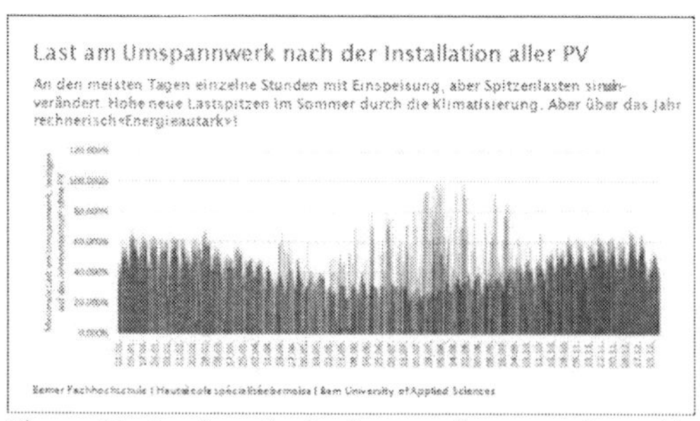

Figure 11: Load peaks in the transformer station with peaks in summertime

The grid in the transformer stations can be managed with "smart users" and supported by local batteries. With decentralized batteries of 400 MWh most of the peaks can be eliminated. But the overall peak at 100% will still occur occasionally (Figure 12).

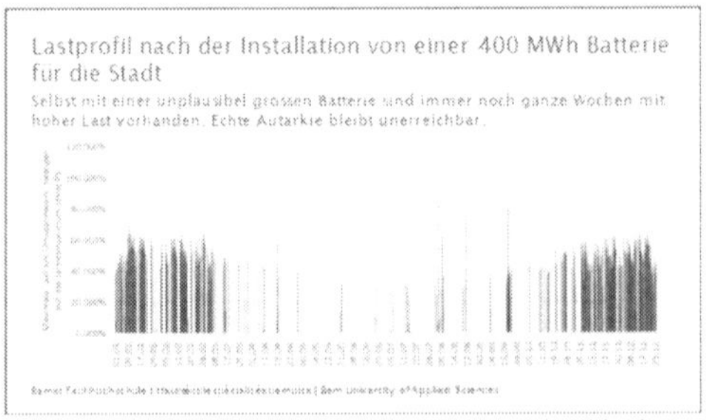

Figure 12: Most load peaks are eliminated with the 400 MWh battery (mainly in summertime), but some peaks still occur

If the number of cars per 1'000 inhabitants stays at 440/ 1'000, about 10'000 EVs will be in town. Equipped with an 80-kWh battery and using 50% of them in a bidirectional charging mode, 400 MWh will be needed. Hence, the (decentralized) 400 MWh batteries should be installed at fixed places in the city of Burgdorf. This offers a glimpse on the flexibility that the new technical solutions EV and PV offer in the future!

Nevertheless, new users such as heat pumps, electric cars and air conditioning will need some changes in the electric distribution grid in the city of Burgdorf. As illustrated in Figure 13, many substations will have to be replaced after 2030, especially in the scenario "Utopia". At the end of 2050, 70% of all substations must be replaced in Burgdorf. Policy makers in Burgdorf may hence be interested to learn that good planning will be important and will save money!

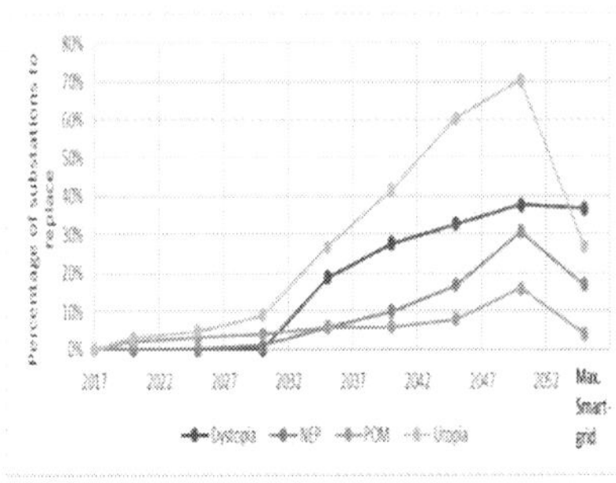

Figure 13: Percentage of substations that need to be replaced as calculated for the four energy scenarios. In the scenario "Utopia", 70 of the 101 transformers must be replaced mainly after 2030.

6 CONCLUSIONS

The project provides evidence that it is possible, in 2050, to serve a medium-size town such as Burgdorf (with a projected population of 22'000, in 2050) with PV. Smart users, PV curtailment and decentralized 400 MWh batteries will help to control the investment costs for the grid enhancement expected up to 2050.

Switzerland must produce about 80 TWh electricity (about 40TWh) in the scenario "Utopia". The existing hydro power can deliver nearly 50% of the energy, as the country produces 40 TWh hydro power. Hence, only 40 TWh PV will be needed for a full decarbonization of the country by 2050. In addition, Switzerland has 8,8 TWh storage and pump storage plants in the mountains; this helps to balance the PV peak production. Hence, the expected investments for the grid expansion as calculated in the scenario "Utopia" in this project may (in reality) be lower. In any case will the cost savings (no fossil fuel anymore) be much higher than the cost for expansion of PV, smart users, and local batteries.

The full decarbonization (i.e., scenario "Utopia") is therefore highly interesting and realistic for Switzerland, both in terms of climate protection (reduction of CO_2 emissions) and sustainability.

When "SimZukunft" started in 2017, the scenario "Utopia" was rather an extreme scenario. It has become a realistic scenario over time, as the full decarbonization of Switzerland is now the official goal of the Swiss Government for 2050 ("Energieperspektiven 2050").

The project "SimZukunft" now outlines the challenges and illustrates the avenues to be pursued for transforming Switzerland to a fully decarbonized country by 2050.

7 FURTHER INFORMATION

White Papers "SimZukunft" can be downloaded from www.pvtest.ch (publications no. 264-266).

8 ACKNOWLEDGEMENTS

This research was part of the activities of the Swiss Centre for Competence in Energy Research on the Future Swiss Electrical Infrastructure (SCCER- FURIES), which is financially supported by the Swiss Innovation Agency (Innosuisse - SCCER program) and with Swiss industry partners. We kindly acknowledge the support from the Swiss Federal Office of Energy (SFOE) and are grateful for the co-funding and collaboration with the City of Burgdorf (construction department), Localnet AG (several collaborations), Adaptricity AG, and Bern University of Applied Sciences BFH.

SOCIALRES ENERGY INNOVATION FRAMEWORK: A COMPARATIVE ANALYSIS OF EXISTING BUSINESS MODELS FOR RES COOPERATIVE, AGGREGATORS AND CROWDFUNDERS

Iban Lizarralde[1], Mikhail Hamwi[1], Silvia Caneva[2], Sonja Wilhelm[2], David van der Zande[2], Audrey Abi Akle[1], Basma Samir[1], Volker Kromrey[3], Dimitri Vedel[3], Linda Lentzen[3], Andreas Schneller[4], Jakob Hoffmann[4], Kathrin Kohl[4], Johanna Doerpinghaus[4], Elisabeth Schmid[5], Claudia Crippa[5], Eleanor Denny[6], James Carroll[6], Hangjian Wu[6], Maria Regidor[7], Sofia Mulero[7], Iban Lacoste[8], Rodrigo Ruiz[9], Santiago Campos[9], Eva Otero[9], Nuno Brito Jorge[10], Francisca Onofre[10], Karl Harder[11], Tijana Šimek[12], Daniela Leonte[13], Mihai Policarp[13]

[1]ESTIA INSTITUTE OF TECHNOLOGY, [2]WIP – RENEWABLE ENERGIES, , [3]BODENSEE-STIFTUNG / LAKE CONSTANCE FOUNDATION, [4]ADELPHI, [5]FONDAZIONE ICONS, [6]TRINITY COLLEGE DUBLIN, [7]CARTIF, [8]I-ENER, [9]ENERGETICA, [10]POWER PARITY, [11]ABUNDANCE, [12]REGEA, [13]TRACTEBEL

[1] ESTIA Institute of Technology, Université de Bordeaux
Izarbel Teknogunea
F-64210 Bidart, France
33-559-43-85-26
i.lizarralde@estia.fr

ABSTRACT

The energy sector is facing a transition phase in which the development of renewable energies and a greater social participation will be two fundamental measures of this transition. Besides, the liberalization of energy markets permits new actors to be involved and to develop original business models. Based on the ongoing research activities in the frame of the H2020 European project SocialRES, aiming to close non-technological research gaps in the energy sector, this paper contributes to identify how social and environmental standpoints impact different business model in this sector. The aim is to present a framework to analyse the business models within the energy sector and to apply this framework to the analysis of three types of businesses associated with social innovation in the renewable energy sector: Cooperatives, Aggregators and Crowdfunding platforms. The results contribute to energy entrepreneurship and identify innovative attributes applied either in the supply side or in the demand side.

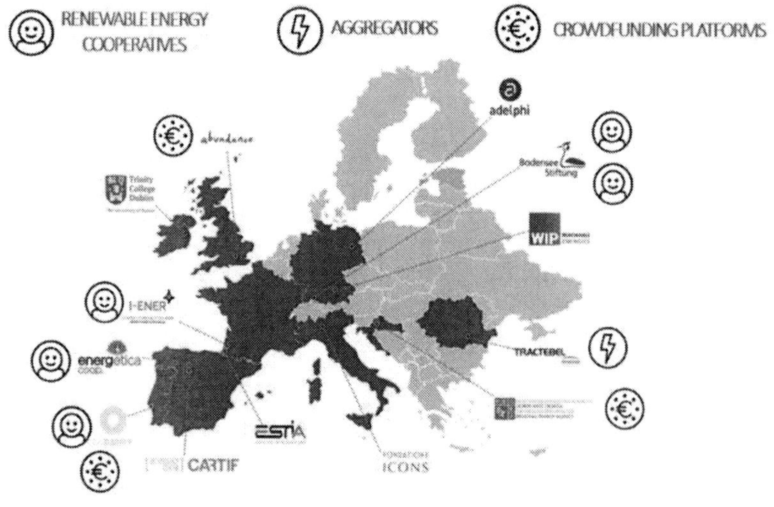

Fig. 1 SocialRES geographical and social innovation coverage

Keywords: Business Model, Social innovation, Crowdfunding, Aggregation, Cooperatives

1 AIM AND APPROACH USED

This paper to be presented at the EUPVSEC2021 aims to characterize different business model archetypes existing in the energy sector and how social and environmental standpoints impact these business models. This research work has been performed in the frame of the H2020 European project SocialRES. The aim of SocialRES is to close non-technological research gaps that impede the widespread uptake of social innovation business and service models in the European energy sector.

The SocialRES consortium brings together thirteen partners who have been chosen specifically for their expertise and experience to ensure the successful implementation of the novel research proposed in this project and to guarantee a widespread geographical coverage (Fig.1). The SocialRES project partners are briefly described in the following paragraph.

1. WIP – Renewable Energies (WIP), based in Germany, the project coordinator.
2. ESTIA Institute of Technology (ESTIA), based in France, with extensive experience in research aspect of social innovation and innovative business models.
3. CARTIF, energy expert based in Spain.
4. Bodensee-Stiftung - Lake Constance Foundation (LCF), based in Germany, as networking and innovation expert and providers of case studies for RES cooperatives.
5. Adelphi, policy think tank based in Germany.
6. Fondazione iCons (ICONS), communication, dissemination and exploitation expert based in Italy.
7. Trinity College Dublin (TCD), University based in Ireland, with extensive experience in behavioural and energy economics, and consumer engagement.
8. I-ENER, RES cooperative based in France, involved in social innovation research projects.
9. EnergEtica, RES cooperative based in Spain.
10. Power Parity, RES cooperative and crowdfunding platform based in Portugal.
11. Abundance, RES crowdfunding platform based in United Kingdom.
12. REGEA, RES crowdfunding platform based in Croatia.
13. Tractebel, RES aggregator based in Romania.

Social innovation projects address broad social issues, in this case the clean energy transition, while also driving business forward.

Three types of businesses have been considered associated with social innovation in the renewable energy sector: Cooperatives, Aggregators and Crowdfunding platforms.

On the one hand, this paper aims to present a framework to analyse the business models within the energy sector. On the other hand, and based on this framework, this paper analyses different business model archetypes existing in the energy sector.

2 SCIENTIFIC INNOVATION AND RELEVANCE

Cooperatives, Aggregators and Crowdfunding platforms are associated with social innovation in the renewable energy sector. These businesses ***facilitate an increase in energy democracy by increasing the number of local (decentralised) clean energy projects and allowing the consumer to take a more active role***.

Based on the attributes examined in the scientific literature and aiming an accurate framework to analyse the business models within the energy sector, a specific model is first presented in the paper. This conceptual framework is suitable for new entrepreneurial business model design and is being used in the frame of the SocialRES project to develop original business models merging Cooperatives, Aggregators and Crowdfunding platforms activities. Moreover, this framework can be used by researchers as a key unit of analysis in the energy transition studies.

Then, this study analyses different business model archetypes existing in the energy sector based on the scientific literature and a detailed analysis of 9 cases of social innovations implemented by crowdfunding platforms, cooperatives, and aggregators in 7 EU countries (Fig.1). Indeed, energy entrepreneurs are promoting clean energy technologies and creating innovative business models that associate commercial benefits and sustainability aspects. However, the process of finding out the appropriate business models poses challenges to new market actors, researchers, and policy makers. In this regard, ***the paper will explore how new entrepreneurs within the domains of Cooperatives, Aggregators and Crowdfunding platforms create and capture value from innovative business models based on existing technologies.***

3 PRELIMINARY RESULTS AND CONCLUSIONS

The results will be presented following the conceptual framework based on the three priorities for the future European energy system (security, affordability and sustainability) as well as other attributes applied either in the supply side or in the demand side. In the supply side, generation viewpoint is considered. This includes flexible generation means that will play an important role in the energy system. Therefore, in order to identify the value propositions related to the supply flexibility, a specific subdivision has been considered. This value proposition includes specific business models as well as special means as storage systems.

The demand side is related to the consumption part. Energy can be used to fulfil several functions which are usually classified into heat functions, mobility functions and specific electricity usages. In this case also, a specific subdivision has been considered: The demand flexibility or the Demand Response.

Lastly, a particular attribute has been considered to include the ***specific role of the Energy Communities***. These communities are created by different partners but mainly by the customers. Indeed, in this model, the role of the customer is highlighted because of the fact that customers can be members of the energy community and can realise specific activities that are not usually done in standard companies. From an energy community point of view, the customer can be an active stakeholder of the organisation in the management area or financial area (the customer is also an investor).

The results can assist new market actors and stakeholders during their early development stage to construct new business models in the energy market. The findings contribute to energy entrepreneurship and identifies innovative attributes to make the business model approach a key driver for the energy transition.

4 ACKNOWLEDGEMENTS AND DISCLAIMER

The SocialRES project receives funding from the European Union's Horizon 2020 research and innovation programme under grant agreement No. 837758. The sole responsibility for the content of this publication lies with the authors. It does not necessarily reflect the opinion of the European Commission. The European Commission is not responsible for any use that may be made of the information contained therein. Follow SocialRES (http://socialres.eu/) on the social media:

- Twitter: https://twitter.com/SocialRESeu
- LinkedIn: https://www.linkedin.com/company/socialres/
- YouTube: https://www.youtube.com/channel/UCfGxFweh8_Ba0-iIw3x_zUg

OPTIMAL ALLOCATION METHOD FOR A FAIR DISTRIBUTION OF THE BENEFITS IN AN ENERGY COMMUNITY

V. Casalicchio[1,2,*], G. Manzolini[2], M. G. Prina[1] and D. Moser[1]
[1] Institute for Renewable Energy, EURAC Research, Viale Druso 1, 39100 Bolzano, Italy
[2]Dipartimento di energia, Politecnico di Milano, Via Lambruschini, 4, 20156 Milano, Italy
*E-mail: valeria.casalicchio@eurac.edu
E-mail: giampaolo.manzolini@polimi.it, matteogiacomo.prina@eurac.edu, david.moser@eurac.edu

ABSTRACT: Energy communities have been a prominent topic since their adoption in the EU RED II Directive in 2018. However, the allocation of the economic benefit among participants and the assessment of the energy community's composition has been relatively untouched. The aim of this research is to come up with a method for distributing benefits among members in the most efficient way possible, based on their own contributions to the system. This technique will be compared to other feasible allocation strategies, and a single-objective linear programming model that considers energy exchanges and economic expenditures will be utilized to do so. The composition of energy communities has an impact on the optimization result as well as the contribution of each user: when the composition is more heterogeneous, the total effective contribution of the participants is higher, and the overall payoff in the case study increases by 12% as the heterogeneity increases from the lowest to the highest possible. Users contribute differently in this scenario, and their contribution is measured and ranges between 10% and 97%.
Keywords: energy benefit allocations, energy communities, scenario analysis, business models

1 INTRODUCTION

Energy communities (ECs) are a direct means for citizens to directly participate in energy production, consumption, and sharing and therefore help promote the use of energy from renewable sources. The 'jointly acting renewables self-consumers' and the 'renewable energy community' are the two Energy Community schemes identified by the EU Directive 2018/2001 (European Parliament and Council of 11 December 2018, 2018). This work focuses on the latter namely "[...] a legal entity based on open and voluntary participation, autonomous, effectively controlled by shareholders or members located in the proximity of the renewable energy projects [...]" (European Parliament and Council of 11 December 2018, 2018).

EC can encourage the innovative technologies diffusion, improve their energy efficiency, fight energy poverty, promote social inclusion and a sustainable lifestyle. Consumers become more empowered, and they play an active role in the energy transition and focus on optimizing self-consumption and energy sharing.

The distribution of the economic advantage among members, as well as the impact of the EC's composition, is an important component in this context that has not been analyzed and considered in detail (Barroco et al., 2020). Therefore, it is important to model a multi-energy system while also focusing on the EC's participants.

Some of the works related to EC that assess both the global EC payoff, and also its allocation among users exploit typical applications of game theory for benefit or cost sharing mechanisms, as for instance the Shapley value (Chiş & Koivunen, 2017). This mechanism assesses the average of the marginal contribution to all possible member coalitions and therefore, despite being known as one of the fairest allocation methods for shared infrastructures, it has the disadvantage of a high computational burden (Moncecchi et al., 2020).

Other approaches addressing the benefit obtained by individual users and not only by the whole system are, for example, the following:

the Shapley value (Lee et al., 2014) is used to create an electricity pricing plan that achieves a fair income split between small-scale electricity suppliers and end-users;

the Benders Decomposition algorithm (Dua et al., 2018) guarantees maximizing the self-interest through a cost allocation method which ensures a fair cost share among the members of a coalition of microgrids;

the investment is a criterion for dividing the entire economic benefit, therefore the payout is shared equally among tenants who participated in the investment to the same extent. [6].

However, the objective may not be about profit allocation, but the aim can be the minimization of the cost or the maximization of the benefit of all the prosumers. This problem can be solved, for instance, as an equilibrium problem with equilibrium constraints as in (Liu et al., 2017) or a Nash equilibrium problem as in (Celik et al., 2017) or in (Jiang et al., 2021), that nevertheless depart from the approach taken in this paper.

This study aims at proposing a method for the benefit optimal distribution among the participants by using a mathematical model that optimizes the energy flows and the expansion capacity of a EC. This method is based on the assumption that a fair distribution of the payoff among users depends on the allocation of a share which is proportional to their marginal contribution to the system.

The allocation metric is based on a heuristic mechanism and inspired by the Vickrey–Clarke–Groves auction theory mechanism (Vickrey, 1961) which evaluates the marginal contribution for all the coalitions.

This latter mechanism and the introduction of an index which measures the fairness of the adopted business model to distribute the benefit are the innovative contribution of this work. This last index allows to evaluate if and how much an applied business model for the benefit distribution among the members reflects the added value they bring to the EC.

The sections of this paper are the following: Section 2 provides an overview of the adopted model and the methodology at the base of the benefit distribution; Section 3 presents the results obtained from the application of the methodology and discusses the main aspects;

Section 4, conclusions and observations are drawn briefly.

2 METHODOLOGY

2.1 Model characteristics and structure

The EC system model (Casalicchio et al., n.d.) is based on oemof (Krien et al., 2017), and performs the operational and the expansion capacity optimization of the system by minimizing the economic expenditures. Figure 1 depicts an example EC created using the oemof components, which allow to simulate generation sources, loads, and storage systems. The outputs of the optimization are: (i) the installed capacities and the energy flows, (ii) the electricity purchased or self-produced and shared or self-consumed within the building, (iii) the overall economic benefit obtained both from the avoided electricity purchase and the refunded amounts for energy sharing.

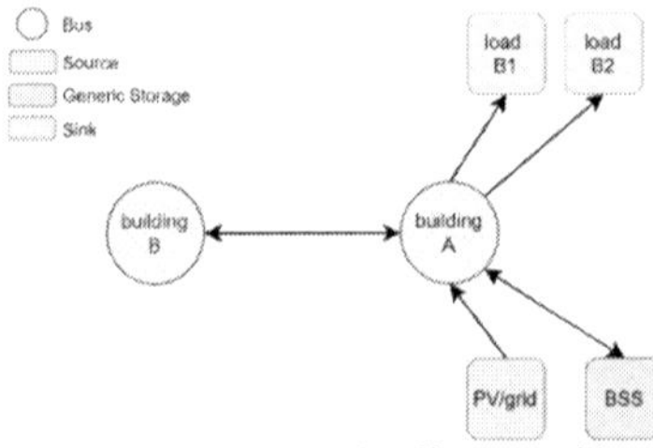

Figure 1: energy community diagram with oemof

2.2. Benefit allocation

Although the importance of benefit distribution in EC is widely acknowledged, there are few studies that examine it in depth, such as (Chiş & Koivunen, 2017), (Moncecchi et al., 2020) and (Energy@home et al., 2019). The optimal solution is one that ensures fairness, which is a relevant performance measure in allocation schemes but is also difficult to evaluate and depends on the participants involved.

In this work a distribution of the EC benefit is considered fair when it reflects the contribution of each participant. The marginal contribution represents the participant added value to the community. A heuristic approach to allocate benefits, inspired by the Vickrey–Clarke–Groves (VCG) mechanism of auction theory, is proposed. (Vickrey, 1961) The aim is to optimize global contentment, namely, to maximize the total of the bids. The price a bidder needs to pay to secure an auction item is the difference in social welfare induced by his participation. For instance, a participant offering more than others causes disadvantage - or less social welfare - to the other participants due to his presence.

This mechanism is extremely popular in the algorithmic game theory since it pursues the ensuring of a good social welfare while preserving truthfulness (e.g.(Sessa et al., 2017)). Once the allocation (participants-items) maximizing the social welfare is computed, VCG pricing ensures that each participant is incentivized to declare his actual valuation.

Similarly to auctions, in order to redistribute the payoff in energy communities, the difference in social welfare, will be considered and referred to as "contribution". Therefore in an EC system, where members make an unequal demand for resources and provide different services and production, the fairness may be based on the ratio of contribution of each member to the overall system which is calculated as in Equation 1.

Afterwards, the benefit distribution vector can be determined, where the distribution among the users is weighted by their contribution, as expressed in Equations 2 and 3.

$$MC_i(opt) = opt(EC) - opt(EC\setminus\{i\}) \quad (1)$$

$$D_{cd,i} = \frac{MC_i}{\sum_{l\in I} MC_l} \quad (2)$$

$$D_{cd} = \left(D_{cd,i}, D_{cd,i+1}, D_{cd,i+2}, \dots, D_{cd,I}\right) \quad (3)$$

Where:
MC_i = benefit marginal contribution of member i
$opt(EC)$ = savings of the EC (all members included)
$opt(EC\setminus\{i\})$ = savings of the EC (member i excluded)
i, l = indexes used for the energy community members
I = set of the energy community members
$D_{cd,i}$ = contribution distribution percentage of member i
D_{cd} = EC contribution distribution vector

This method should ensure that the benefit percentage assigned to a user does not trigger discontent among other members, who would not have equally access to that amount without that member participation.

According to the community composition, we can also have null term in the distribution vector if a member is not providing any benefit.

The best and preferable scenario is the one in which all members increase the overall common benefit.

2.3. Business Models

The distribution of benefits based on the marginal contribution of each member is feasible from a computational point of view, although the calculation time increases with the number of users of the EC. Therefore, it also useful and effective to define simpler distribution methods not involving an optimization.

In this work, some possible criteria, hereafter called Business Models (BMs), are identified for dividing the benefit among users in a more intuitive way. The main sample business models meet the following criteria for assigning the benefit among EC members:

- homogeneous distribution – including investments (BM A)
- homogeneous distribution – excluding investments (BM B)
- distribution according to the loads (BM C)
- distribution according to the number of people of each member-family (BM D)
- distribution according to the self-consumed quota – promoting the matching of demand and production (BM E)

After having selected the BM, it is possible to determine the benefit distribution vector as in Equation 4.

$$D_{BM} = \left(D_{BM,i}, D_{BM,i+1}, D_{BM,i+2}, \dots, D_{BM,I}\right) \quad (4)$$

Where:
$D_{BM,i}$ = business model distribution percentage of member i
D_{BM} = EC business model distribution vector

The business model distribution of each member can deviate from the optimal one. The deviation can be zero - the share returned to the member is perfectly in line with their contribution; the deviation can be negative and, therefore, the concerned user is receiving less than they should; the deviation can be positive and, therefore, the concerned user is receiving more than they should.

2.4. Fairness Index

based on the EC. The Gini index [15] described in Equation 5 is used to determine the EC's heterogeneity. It ranges between 0 (maximum homogeneity) and 1 (equal distribution of frequencies). The frequency is the number of occurrences of a particular member composition.

$$GI = \left(1 - \sum_{i=1}^{m_{tot}} f_i^{\,2}\right)\frac{m_{tot}}{m_{tot} - 1} \tag{4}$$

Where:
GI = Gini Index
f_i = frequency of the composition a certain member i
m_{tot} = total number of members in the EC

The Fairness Index (FI) is applicable to any EC regardless of the size and it returns a value between 0 and 1 if each individual member has a profit: the closer the value is to zero the fairer and the more suitable is the applied BM. The index also returns the number of unhappy users. The denominator of the FI involves the introduction of an additional benefit distribution called Dw (Equation 6). This distribution is the furthest from the optimal one and it maximizes the denominator of the FI. Therefore, the resulting value of the FI (Equation 7) provides a measure of how much the chosen model is "closer" to the optimal case (value goes to zero) or, on the contrary, "closer" to the worst case (value goes to one).

$$D_W = \left(D_{w,i}, D_{w,i+1}, D_{w,i+2}, \ldots, D_{w,I}\right) \tag{5}$$

$$FI(m) = \begin{cases} \dfrac{\sum_{i=1}^{m}\left|D_{BM,i} - D_{cd,i}\right|}{\sum_{i=1}^{m}\left|D_{w,i} - D_{cd,i}\right|} & if\ m = m_{tot} \\ m_{tot} - m & if\ m \neq m_{tot} \end{cases} \tag{6}$$

Where:
$D_{w,i}$ = worse distribution percentage of member i
D_w = EC worse distribution vector
m = number of members for which D>0

3 RESULTS

The EC model and the methodology introduced to allocate the payoff has been applied to case studies which have been generated ad hoc to show the potential of the introduced model. Individual users' electricity demand profiles were obtained using the LoadProfileGenerator tool [19] and tailored to the Italian case.
The regulatory framework considered for calculations and the input data is the Italian one, starting from the adoption of the RED II directive (European Parliament and Council of 11 December 2018, 2018), up to the latest documents published by ARERA [20] - [17], MiSE [18] and GSE [21]. The considered costs are the variable components of the bill applied to date by the Italian regulation, the incentives or refunded quota and the expense deductions and can be changed for different applications.

3.1. Heterogeneity impact on the case study results

Table 1 depicts the first case study involving six buildings and six equal users. It presents a first EC composed of 6 single-family housing buildings where all users are the same featuring a totally homogeneous EC (EC #1). Subsequently, not the EC configuration, but the user composition is modified, and therefore the users (EC #2, EC #3, EC #4) are gradually replaced increasing the Gini Index until the maximum heterogeneity. In all presented cases the total demand changes when the users change, only slightly from the average (10%).

A measure of the suitability of the adopted BM in relation to the previously introduced fair distribution should be provided, in order to identify which distribution, among possible ones, is more suitable to be implemented

Table 1. Case study outline

	EC #1	EC #2	EC #3	EC #4
Single housing buildings	6	6	6	6
Gini Index	0	0.33	0.8	1
Electric load [MWh]	21.6	19.5	17.6	19.4

The first results related to the energy self- consumption, the sharing, the bill savings and the investment capacity optimization of the EC in the four proposed versions are presented in Table 2.
Figure 3 shows the results related to the fairest distribution, based on the contribution of each participant. As expected for the first case, everybody contribute equally to the welfare and an uniform sharing of the benefit is fair. By increasing the heterogeneity of the system, the application of a homogeneous distribution would be not fair for 5 users out of 6 already in the second graph where only a user is different from the others. From a comparison of the BMs applied with the method based on the suggested contribution distribution, it emerges that a model is better than the others depending on the composition of the EC.
In Figure 3, the deviation from the optimal distribution to that derived from the BM considered is presented for each user: negative deviations in purple; positive deviations in light blue.
For each BM and case study introduced, the FI increases with the standard deviation as reported Table 3. Hence, it is possible to have a more accurate evidence of BM suitability. Considering a case with high heterogeneity, it will not be recommended to adopt the BM A and then to distribute the benefit equally, in fact it is more likely that the user contribution is more diversified.
shows the results for the most heterogeneous case study.

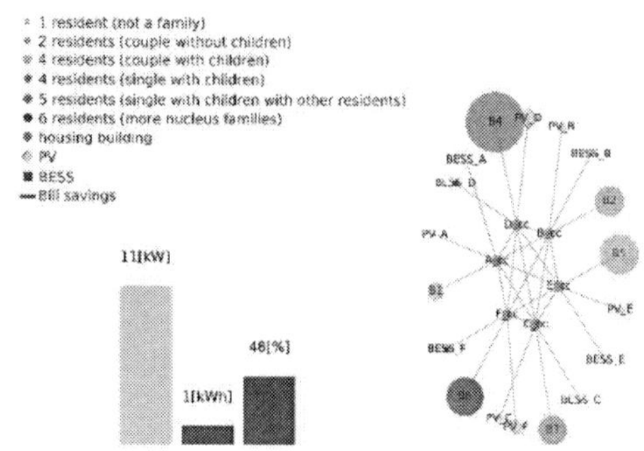

Figure 2: energy community optimization results. GI=1.0 The graphs depict the EC's composition, nodes are users and installed systems, their sizes vary according to load

and installed capacities. Bars show the global installed capacity and percentage of bill savings.

The different graphs of

Figure 3 shows the results related to the fairest distribution, based on the contribution of each participant. As expected for the first case, everybody contribute equally to the welfare and an uniform sharing of the benefit is fair. By increasing the heterogeneity of the system, the application of a homogeneous distribution would be not fair for 5 users out of 6 already in the second graph where only a user is different from the others. From a comparison of the BMs applied with the method based on the suggested contribution distribution, it emerges that a model is better than the others depending on the composition of the EC.

In Figure 3, the deviation from the optimal distribution to that derived from the BM considered is presented for each user: negative deviations in purple; positive deviations in light blue.

For each BM and case study introduced, the FI increases with the standard deviation as reported Table 3. Hence, it is possible to have a more accurate evidence of BM suitability. Considering a case with high heterogeneity, it will not be recommended to adopt the BM A and then to distribute the benefit equally, in fact it is more likely that the user contribution is more diversified.

make more comprehensible the heterogeneity or homogeneity of the EC systems. The results also show what happens with respect to investments. The following aspects can be pointed out: because of the similar demand, the installed capacity of PV does not change substantially (±10% compared to the average); unlike PV, battery energy storage systems (BESS) installed capacity decreases by about 67% as heterogeneity increases; finally, even the savings in the bill are higher in the case with higher heterogeneity, rather than in the case with higher homogeneity, likewise the shared energy.

Table 2. Optimization outputs of the case studies

	EC #1	EC #2	EC #3	EC #4
Self-consumption [%]	35.8	34.4	33.5	36.2
Energy sharing [%]	0.0	1.0	3.4	5.2
PV capacity [kW]	12.0	11.0	10.0	11.0
BESS capacity [kWh]	3.0	3.0	2.0	1.0
Bill savings [%]	43.0	42.0	43.0	48.0

Figure 3 shows the results related to the fairest distribution, based on the contribution of each participant. As expected for the first case, everybody contribute equally to the welfare and an uniform sharing of the benefit is fair. By increasing the heterogeneity of the system, the application of a homogeneous distribution would be not fair for 5 users out of 6 already in the second graph where only a user is different from the others. From a comparison of the BMs applied with the method based on the suggested contribution distribution, it emerges that a model is better than the others depending on the composition of the EC.

In Figure 3, the deviation from the optimal distribution to that derived from the BM considered is presented for each user: negative deviations in purple; positive deviations in light blue.

For each BM and case study introduced, the FI increases with the standard deviation as reported Table 3. Hence, it is possible to have a more accurate evidence of BM suitability. Considering a case with high heterogeneity, it will not be recommended to adopt the BM A and then to distribute the benefit equally, in fact it is more likely that the user contribution is more diversified.

This aspect becomes even more evident when comparing BM D and BM E: when there is maximum homogeneity, the results do not change for the suggested BMs. This is no longer valid when increasing heterogeneity.

For the case with Gini=0.33, when the BM D is adopted, the result is adequate, whereas BM E doesn't perform well in terms of fairness. For the cases with GI=0.8 and GI=1.0, both models BM D and BM E are equally less effective in terms of fairness than the others except for BM A.

Table 3. Standard deviation and Fairness Index of applied business models

	EC #1 GI = 0.0		EC #2 GI = 0.33		EC #3 GI = 0.8		EC #4 GI = 1.0	
	FI	σ	FI	σ	FI	σ	FI	σ
BM A	0.000	0.000	0.116	0.053	0.126	0.054	0.200	0.085
BM B	0.000	0.000	0.0031	0.001	0.027	0.012	**0.026**	0.011
BM C	0.000	0.000	**0.0006**	0.000	**0.025**	0.011	0.030	0.011
BM D	0.000	0.000	0.011	0.005	0.085	0.041	0.139	0.055
BM E	0.000	0.000	0.052	0.024	0.096	0.036	0.129	0.058

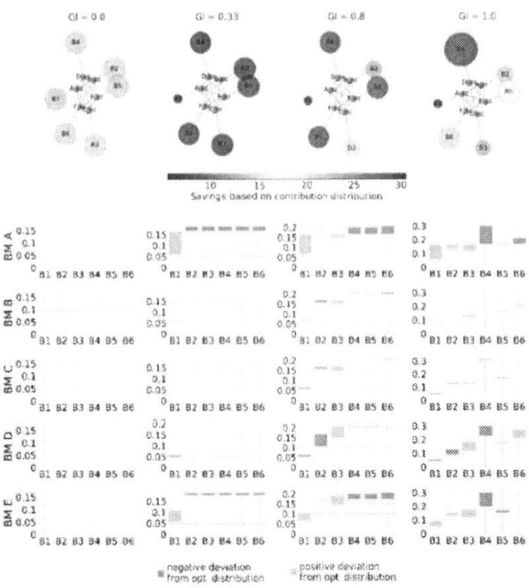

Figure 3: contribution and BMs evaluation

4 CONCLUSIONS

In this work a heuristic was developed to address the payoff distribution problem of ECs. This approach is added to those already existent, and reduces the computational time issues, which, for instance, occur with the Shapley value mentioned above. Moreover, this approach pays particular attention to the concept of fairness. Once a model has been used to evaluate the energy community's optimal expansion capacity and flows, the payoff distribution among its members must be evaluated. The optimal distribution definition assumes that each member must receive a welfare amount that reflects their contribution to the EC and to this the composition of the community must be considered. The composition determines both the result of the optimization, and also how the contribution of the user changes. It has been demonstrated that while similar users contribute equally to the community, their total effective contribution is lower than when the composition is more heterogeneous. In the case study presented, where the global demand and the installed capacity of photovoltaics are the same and we pass from a lower heterogeneity (GI=0.33) to a higher one (GI=1), the need for flexibility provided by the BESS is reduced by 67% and the benefit increases by 14%. These data demonstrate that by changing the users, each of them gives a different contribution which has to be taken into account. In the case presented, with maximum heterogeneity we have users whose presence contributes to increase the overall benefit from 10% to 97%.

The Fairness index aims at understanding how not to disadvantage anyone, since the participation is voluntary. If all users are equal, no one contributes more to the welfare than the others and any payoff uniform distribution is fair. This is no longer verified when the EC heterogeneity increases. Depending on the composition of the EC, a model will be more or less suitable. The adoption of BM A – which is based on a homogeneous distribution of payoff and investments - will not be recommended for a high heterogeneity EC because it is more likely that the users contribution is more diversified. In this case, the highest calculated FI is obtained (FI=0.200).

For the case with Gini=0.33, BM C - based on member loads - should be adopted, since it is the case in which the lowest FI is obtained (FI=0.0006). The same choice should be made if GI=0.8 even if the FI is higher (FI=0.025). By further increasing the heterogeneity to GI=1.0, the choice should fall on BM B which is based on a homogeneous distribution of the payoff (FI=0.026).

5 REFERENCES

[1] European Parliament and Council of 11 December 2018, "Directive (EU) 2018/2001," *Official Journal of the European Union*, vol. 2018, no. L 328, pp. 82–209, 2018.

[2] F. Barroco *et al.*, "LE COMUNITÀ ENERGETICHE IN ITALIA: Una guida per orientare i cittadini nel nuovo mercato dell'energia," p. 36, 2020.

[3] A. Chiş and V. Koivunen, "Coalitional game based cost optimization of energy portfolio in smart grid communities," *IEEE Transactions on Smart Grid*, vol. 10, no. 2, pp. 1960–1970, 2017.

[4] M. Moncecchi, S. Meneghello, and M. Merlo, "A game theoretic approach for energy sharing in the italian renewable energy communities," *Applied Sciences (Switzerland)*, vol. 10, no. 22, pp. 1–25, 2020, doi: 10.3390/app10228166.

[5] W. Lee, L. Xiang, R. Schober, and V. W. S. Wong, "Direct electricity trading in smart grid: A coalitional game analysis," *IEEE Journal on Selected Areas in Communications*, vol. 32, no. 7, pp. 1398–1411, 2014, doi: 10.1109/JSAC.2014.2332112.

[6] Y. Dua *et al.*, "A cooperative game approach for coordinating multi-microgrid operation within distribution systems," *IEEE Power and Energy Society General Meeting*, vol. 2018-Janua, no. December 2017, pp. 1–5, 2018, doi: 10.1109/PESGM.2017.8273876.

[7] Energy@home, Elements, RSE, and Kantar, "I prosumer condominiali," no. May, 2019.

[8] N. Liu, X. Yu, C. Wang, C. Li, L. Ma, and J. Lei, "Energy-Sharing Model with Price-Based Demand Response for Microgrids of Peer-to-Peer Prosumers," *IEEE Transactions on Power Systems*, vol. 32, no. 5, pp. 3569–3583, 2017, doi: 10.1109/TPWRS.2017.2649558.

[9] B. Celik, R. Roche, D. Bouquain, and A. Miraoui, "Coordinated neighborhood energy sharing using game theory and multi-agent systems," *2017 IEEE Manchester PowerTech, Powertech 2017*, p. 2020, 2017, doi: 10.1109/PTC.2017.7980820.

[10] A. Jiang, H. Yuan, and D. Li, "A two-stage optimization approach on the decisions for prosumers and consumers within a community in the Peer-to-peer energy sharing trading," *International Journal of Electrical Power and Energy Systems*, vol. 125, p. 106527, Feb. 2021, doi: 10.1016/j.ijepes.2020.106527.

[11] W. Vickrey, "Counterspeculation, Auctions, and Competitive Sealed Tenders," *The Journal of Finance*, vol. 16, no. 1, pp. 8–37, 1961.

[12] V. Casalicchio, G. Manzolini, M. G. Prina, and D. Moser, "From Investment Optimization to Fair Benefit Distribution in Renewable Energy Community Modelling (in review)".

[13] U. Krien, C. Kaldemeyer, and B. Schachler, "Open Energy Modelling Framework - A modular open source framework to model energy supply systems." 2017.

[14] P. G. Sessa, N. Walton, and M. Kamgarpour, "Exploring the Vickrey-Clarke-Groves Mechanism for Electricity Markets," *IFAC-PapersOnLine*, vol. 50, no. 1, pp. 189–194, 2017, doi: 10.1016/j.ifacol.2017.08.032.

[15] E. Novelli and S. Occelli, "Profili descrittivi di distribuzioni spaziali: alcune misure di diversificazione," *Cybergeo*, no. 108, pp. 1–18, 1999, doi: 10.4000/cybergeo.4930.

[16] N. Pflugradt, "LoadProfileGenerator." 2015.

[17] Arera, "Documento per la Consultazione 112/2020/R/EEL," pp. 1–31, 2020.

[18] Arera, "Deliberazione 4 agosto 318/2020/R/EEL," vol. 192, pp. 1–27, 2020.

[19] MiSe, "Gazzetta Ufficiale Della Repubblica Italiana," p. 37, 2020.

[20] GSE, "Regole tecniche per l'accesso al servizio di valorizzazione e incentivazione dell'energia elettrica condivisa." https://www.gse.it/media/comunicati/comunita-energetiche-e-autoconsumo-collettivo-online-regole-tecniche-e-portale-per-le-richieste

A Snapshot of Global PV Markets - The Latest Survey Results on PV Markets and Policies from the IEA PVPS Programme in 2020

Gaëtan Masson, Arnulf Jäger-Waldau[1], Izumi Kaizuka, Johan Lindahl, José Donoso
IEA PVPS Task 1, Belgium – g.masson@iea-pvps.org, +32 2 514 78 05
European Commission, Joint Research Centre (JRC), Via E. Fermi 2749, 21027 Ispra, Italy – Arnulf.jaeger-waldau@ec.europa.eu
RTS Corporation, Japan – kaizuka@rts-pv.com
Becquerel Sweden, Sweden – johan@becquerelsweden.se
UNEF, Spain – j.donoso@unef.es

ABSTRACT: The objective of this paper is to propose a reliable and accurate perspective on key markets and policies related to PV development in 2020 and previously. It aims at offering a clear analysis of how PV markets have developed in 2020, with updated numbers, along with analysis of the policies behind the development. In this paper are displayed and analyzed survey results for the calendar year 2020 concerning PV markets and policies, as well as other key issues. An increasing number of national markets experienced notable growth in 2020 with impacts on policy development. Figures show that over 140 GW of PV systems have been installed in the world last year. Consequently, cumulative capacity crossed the 760 GW mark in 2020.
Keywords: Photovoltaic (PV), Market, IEA PVPS

1 INTRODUCTION

This snapshot paper gives information on the development of PV power applications in the PVPS member and non-member countries and is largely based on the information provided by the 27 IEA PVPS countries plus the European Union through its European Commission. The report includes information on national market developments at the end of the former year, in this case, 2020. The International Energy Agency – Photovoltaic Power System Programme (IEA PVPS)'s Task 1 is responsible for strategy and outreach within the IEA PVPS program. This includes market and industry analysis. A key deliverable of Task 1 is the annual Snapshot of Global PV Markets publication, together with the annual flagship report Trends in PV Application.

The objective of the series of annual Snapshot and Trends reports — which have been published since 1992 (Trends) [1] and 2013 (Snapshot) [2] — is to present and interpret developments in both the PV systems and components being used in the PV power systems market and the changing applications for these products within that market. These trends are analyzed in the context of the business, policy and non-technical environment in the reporting countries.

2 THE GLOBAL PV INSTALLED CAPACITY

This paper presents the latest survey results for the calendar year 2020 concerning PV markets and policies, as well as other key issues. An increasing number of national markets experienced notable growth in 2020, and that impacted the policy development. The capacity figures given are nominal DC peak power (Wp) under standard test conditions (i.e. 1 000 W irradiance, air mass 1.5 light spectrum and 25 °C device temperature) for consistency reasons. As not all countries report DC peak power (Wp) for solar PV systems, but report AC power, these capacity figures are converted to Wp DC.

While the final figures for 2020 could slightly continue to be refined in the coming months, nearly final figures show that around 140 GW of PV systems have been installed in the world last year.

Some important trends observed are as follows:

- The global PV market grew to around 140 GW in 2020, compared to around 110 GW in 2019. This represents a year-on-year growth of more than 25%.
- Asia scored the very first place again in 2020 with China (48 GW), followed by the USA (20 GW), Vietnam (11 GW), or Japan (8.7 GW) and more appearing in the top markets.
- The market in EU grew significantly to around 20 GW. With some notable growth in Germany, the Netherlands, Poland and more.
- Preliminary numbers show that installations rose in the middle east and Egypt, while the market in India decreased to around 4 GW.
- Preliminary numbers also show some fast development in Africa and Latin America at a lower speed.
- The total installed capacity reached around 763 GW in 2020.
- Drivers for PV development include record-breaking competitiveness levels. At the same time, the industry introduced significant changes into the market, with the development of larger sized modules (based on 182 mm and 210 mm cells).
- The Covid-19 impact was not clearly observed from global PV systems installations' point of view.

[1] The views expressed are purely those of the author and may not in any circumstances be regarded as stating an official position of the European Commission

3 MARKET DEVELOPMENT

In 2020, the PV market shows a clear comeback after the 2019 slow down. Globally, the tendency is upwards by exceeding the 140 GW as expressed in Fig. 1.

Together with China, India and Japan, as well as many other countries growing as established markets in the region, Asia remains the leader of the Global PV market. Next to China (48.2 GW), Japan and India remain a relevant presence in the global market with respectively 8.7 GW and 4.4 GW installed even if in India the market decreased significantly compared to 2019.

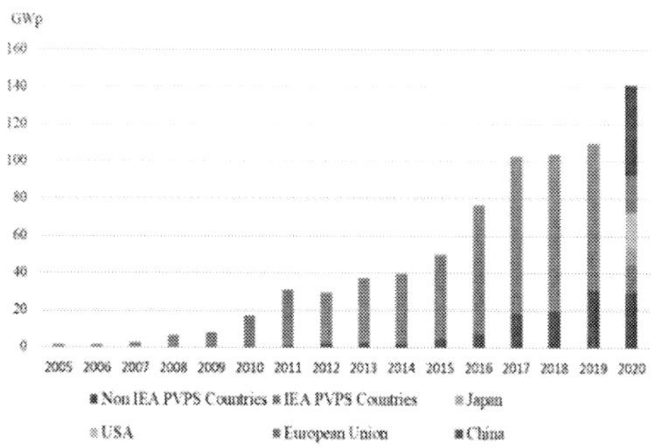

Figure 1: Evolution of annual PV installations (GW-DC)

Other Asian markets such as Vietnam (11.1 GW), Korea (4.1 GW), the Philippines (1.1 GW) and Malaysia (850 MW) have confirmed their maturity by experiencing a growth in 2020. On the contrary, the development in other markets such as Thailand, Indonesia and Singapore has been intermittent over the years.

In the Americas, the US market continued to grow (19.7 GW) similar to Brazil that installed at least 2.8 GW in 2020 in comparison to 2.2 GW in 2019. While Mexico exhibited a small decline with installation around 1,5 GW in 2020 in comparison to nearly 1.9 GW in 2019. Canada is getting roughly stable, as the market was at 200 MW in 2020 as compared to 230 MW in 2019.

In Europe, Germany took back the leading position on the continent from Spain as had been the case for already years, by installing respectively over 4.9 GW and 2.8 GW in 2020. In 2020, six European countries are above the GW mark with the Netherlands (3.0 GW), the significant increase of Poland (2.6 GW), Ukraine (1.5 GW) and Belgium (1 GW) while France is still below the GW mark with 956 MW. Other countries in Europe experienced interesting developments but at a lower level: Hungary (600 MW) and UK (545 MW). Italy (785 MW) keep progressing slowly, similarly to Switzerland (475 MW) and Austria (340 MW). In the Nordic countries, Sweden (506 MW), Denmark (264 MW), Finland (100 MW) saw a growth of their respective PV markets compared to 2019.

In the Middle East, Israel installed an additional 586 MW, continuing their progression. Several other countries in the region experienced rapid growth and hundreds of MW installations, which will mostly be connected in 2021.

In Africa, South Africa and Egypt became the first African countries to install 1 GW of PV in 2020. Projects have popped-up a bit everywhere, but few countries have installed significant amounts in 2020.

In the year 2020, nineteen countries passed the GW mark concerning the annual installed PV capacity. Eleven countries now have more than 10 GW of total capacity installed, four have more than 50 GW and China alone represented 253.6 GW. Germany, which used to lead the rankings for years, lost its leading position in 2015 and now ranks fourth (53.9 GW), the USA second (95.5 GW) and Japan third (71.8 GW). With more than 162.7 GW of total capacity, Europe is now significantly behind the Asian leader that runs at least 443 GW, and much more to come in the coming years.

4 GRID-CONNECTED CENTRALIZED AND DISTRIBUTED

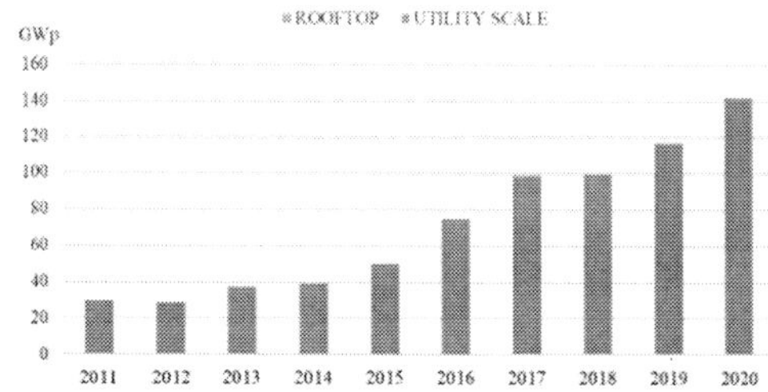

Figure 2: Segmentation of PV installation 2011-2020

Regarding the share of the grid connected and centralized installations at the global level, the trend changed a lot over the years.

During the last 6 years, the market has seen a strong development of grid connected centralized installations rather than distributed ones, as Fig. 2 illustrates. Before 2013, most PV installations used to be distributed systems, i.e. mainly installed on rooftops.

This changing trend is because centralized PV has evolved faster, e.g. in terms of cost, and most of the major PV development in emerging PV markets are coming from utility-scale PV. The success of utility-scale installations is mainly coming from the fact that the installation time and cost per Wp for utility-scale are lower than for distributed PV plants.

Moreover, as mentioned above, most of the installed capacity in emerging PV markets is the result of calls for tender. In the last years, PV electricity prices are converging towards lower prices and this is making PV electricity even more attractive in some regions. In addition, tenders offer the most competitive solution on electricity price due to the necessity to compete with low wholesale electricity price. The increasing number of tenders organized enabled the installation of utility-scale projects by stimulating competition. This has been one of the main aspects that has characterized the market in the past years, and it has strongly contributed to increasing the share of utility-scale installations on the market.

The distributed PV, after suffering from a small decline in 2018, is exhibiting an increase since 2019, reaching an even higher level than in 2017.

In the same way, the market has started to diversify, with floating PV adding to utility-scale and BIPV starts to complement BAPV in the built environment, although it remains a niche. Other emerging segments such as agricultural PV or PV integrated in vehicles (VIPV) are showing the potential for further diversification of PV components, but their current levels of development remain too low to be considered in this paper.

At the local level, the region with the highest share of grid connected centralized installations was Europe until 2011 which was holding 80% of the utility-scale system installed globally. Starting from 2012, the share of

American and Asian countries started to grow until the end of 2013, when Asia became the main region for utility-scale projects.

5 ELECTRIC VEHICLES

The electrification of transport accelerated in many countries; and almost all of which are active in the IEA PVPS program.

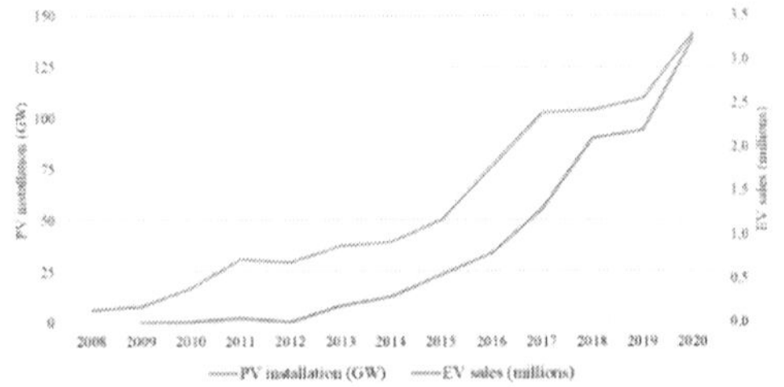

Figure 3: Comparison between PV installation and EV sales

The link between PV development and EVs is not straightforwardly understood yet, but it is simply becoming a reality. Charging electric vehicles during peak load hours implies to rethink power generation, while concepts such as virtual and collective self-consumption could rapidly provide a framework for rapid PV development.

The accelerated development of the EV market could be compared to the development of the PV market, as shown in Fig. 3. More than 3 million electric vehicles sold in 2020 (nearly half of which were sold in China), representing 4.2% of global car sales. The sales of EVs are following a trend that is even faster than the installation of PV [3].

6 ELECTRICITY PRODUCTION

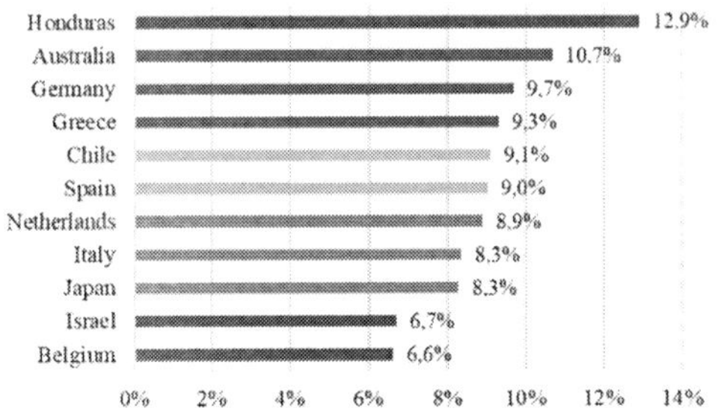

Figure 4: National PV penetration in % of the final electricity demand based on 2020 capacities

The electricity production from PV per country as shown in Fig. 4 estimates what the PV production could be based on the cumulative PV capacity at the end of 2020, close to optimum siting, orientation and average weather conditions. These numbers, which are not based on actual measurements, should therefore be considered as indicative, aiming at comparing different situations in different countries rather than official data. In several countries, the PV contribution to the final electricity demand has passed the 5% mark with Honduras in the first place with almost 13%. Australia is second with above 10%, Germany is third with an estimated 9.7%. Greece and Chile with respectively a theoretical demand coverage of 9.3% and 9.1%. Spain reached the 9% mark and China reached 4.4% in 2020. In total, PV contribution amounts to around 3.6% of the world's electricity demand in the world.

7 CONCLUSIONS

This brief overview of the situation of the PV market and its deployment in 2020 shows that the global PV market is, more than ever, dominated by a few leading countries. However, many new PV markets are developing on all continents, at different paces. Also, despite the multiplication of PPA's through competitive tenders and the decreasing costs of PV systems, which stand today at an unprecedented low level, the market remains strongly influenced and led by incentives.

8 REFERENCES

[1] IEA PVPS, TRENDS 2020 In Photovoltaic application, ISBN 978-3-907281-01-7, https://iea-pvps.org/trends_reports/trends/

[2] IEA PVPS, 2021 Snapshot of Global PV Markets, ISBN 978-3-907281-17-8, https://iea-pvps.org/snapshot-reports/snapshot-2021/

[3] IEA, Global EV Outlook 2021, https://www.iea.org/reports/global-ev-outlook-2021

POTENTIAL BIPV MARKET IN KEY EUROPEAN COUNTRIES BY 2030

Elina Bosch[1], Philippe Macé[1], Adrien Van Rechem[1]
[1]Becquerel Institute, 146 Rue Royale, 1000 Brussels, Belgium | Phone: +32 2 514 78 04

ABSTRACT:
Perspectives for BIPV market development have always been a topic of interest within the PV sector as it has often been considered as a potential growth path for PV, but this never materialized to significative deployment. The innovative methodology presented here aims at estimating a realistic market potential for BIPV by 2030, by using a two-sided and a back-casting approach for eight key European solar markets, taking into account the characteristics of their building stock. In particular, renovation rates are key inputs as they are used to define the two considered scenarios (a high scenario falling within the Renovation Wave Strategy and a Business-as-Usual scenario). Eventually, the cumulative BIPV capacity projected to be installed in the eight key European countries over the next decade could amount to 7,4 GWp to 9,5 GWp with most of the capacity being installed on residential buildings' roofs. Then, not only will a doubling of current renovation rates have clear benefits in terms of primary energy consumption reduction for the building sector, it can also be an important driver of the BIPV market representing an additional 2 GWp opportunity compared to the Business as Usual scenario.

Keywords: Building Integrated PV (BIPV), Markets, Renovation Wave

1 CONTEXT

Perspectives for BIPV (Building Integrated Photovoltaic) market development have always been a topic of interest within the PV sector. Indeed, in the last decade, BIPV applications have often been considered as a future potential growth path for PV, but this never materialized to significative deployment. Some studies have been published on the topic of BIPV market potential estimation [1] [2]. Partially building on these, as well as on a previous study by the same author as this research paper, an innovative methodology is developed, leading to estimations of BIPV market projections by 2030 in the eight considered key European countries (Belgium, France, Germany, Italy, the Netherlands, Spain and Switzerland, United Kingdom) [3].

2 OBJECTIVES

This innovative methodology presented here aims at quantifying the market potential of BIPV, by using a two-sided approach for eight key European solar markets. A better vision on perspectives for BIPV market development is a key element when it comes to defining the regulatory frameworks that shape the deployment path for renewables.

In addition, this approach will be used to develop different possible market projections scenarios for BIPV. This will permit to evaluate the impact of the stimuli caused by the Renovation Wave Strategy, published in October 2020 by the European Commission [4]. In particular, the ambitioned "doubling of the renovation rate" will be studied.

3 SCOPE AND APPROACH

First, a "supply-side" method is applied, based on the BIPV "suitability" of the existing building stock. This suitability is defined based on both architectural and solar irradiation characteristics. The second step consists in a "demand-side" approach, based on the average electricity consumption of the same existing building stock. The methodology is conducted for each country individually as well as for different segments of the building stocks (with a distinction between roof and facades and between residential and non-residential buildings). This first step of the methodology is summarised in Figure 1.

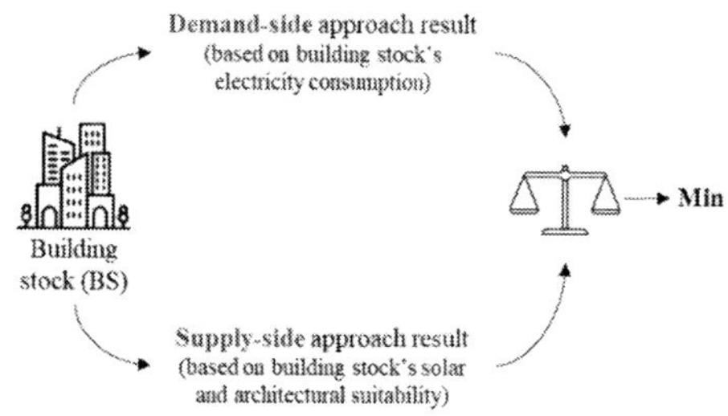

Figure 1: General overview of the supply-side and demand-side methodology

The results of both approaches are then confronted. Additional criteria such as the regulatory environments and their likely evolution, the availability and acceptance of workforce and most importantly the annual new construction rates and renovation rates are also taken into account to determine the maximal realistic cumulative BIPV potential at a long-term horizon (2100). This second step of the methodology is presented in Figure 2.

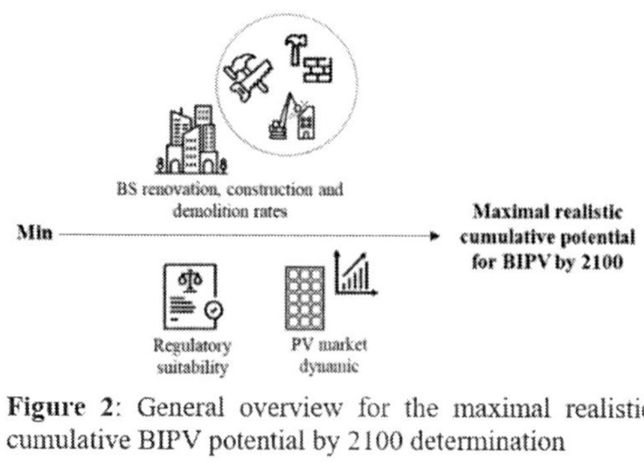

Figure 2: General overview for the maximal realistic cumulative BIPV potential by 2100 determination

Finally, a back-casting approach, presented in Figure 3, is conducted. It builds on this maximal realistic cumulative BIPV potential as well as estimated historical values.

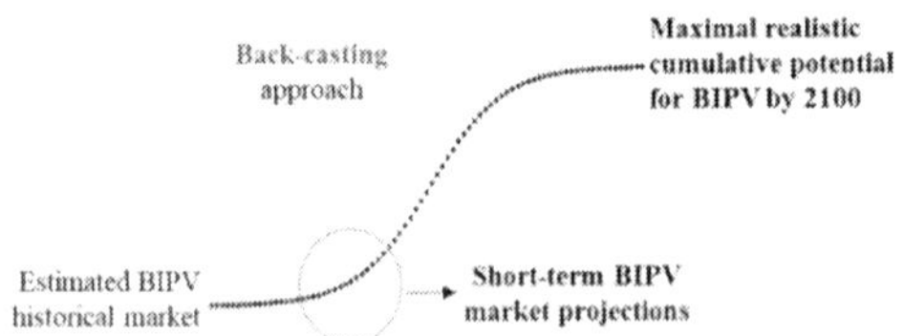

Figure 3: General overview of the back-casting methodology

3.1 Supply-side approach

Various studies that have been published on the topic of BIPV market potential estimations, focus on the "supply-side" potential [1] [2]. This approach determines the total surface that the building stock of the studied countries is able to provide for BIPV.

The starting point of this methodology is the building stock of a given country. From this building stock composed of both residential and non-residential buildings, the total building floor surface is determined based on an analysis of the available buildings' datasets [5]. Then, an analysis of average architectural characteristics for each type of building composing the building stock is necessary. The output of this "supply-side" approach's first part is the total gross roof and façade surfaces.

The second part of the "supply-side" approach is to refine these total gross roof and façade surfaces down to surfaces that are suitable for welcoming a BIPV installation. The notion of suitability refers to both the architectural and the solar suitability. Filtering by architectural suitability allows to disqualify surfaces which are already occupied by other systems such as HVAC systems, chimneys or windows for example. The solar suitability allows to consider only surfaces where the irradiation amounts to at least 4/5 of the maximum solar irradiation that would be reached at optimal tilt and azimuth (optimal azimuth only for façades). The outputs of this second part of the "supply-side" approach are the total gross roof and façade surfaces meeting the architectural and solar requirements presented above.

3.2 Demand-side approach

The "demand-side" approach looks at the average electricity consumption of the building stock. To compute the associated final electricity demand, the average electricity consumption of the residential and non-residential sector per floor surface unit is used.

3.3 Aggregation of both approaches

Eventually, the results of both the "supply-side" (annual electricity production generated by BIPV installed on all architectural and solar suitable surfaces of the building stock) and "demand-side" (annual electricity demand of the building stock) are compared. To do so, the "supply-side" approach's output, expressed in km², is converted into installed capacity GWp and eventually into energy production GWh by using a roof-specific yields and a facade-specific yield for each country.

The approach yielding the minimal output, allows to define a maximal cumulative BIPV technical market potential achievable in one point in time, considering building stock's characteristics. The underlying idea is that it cannot be installed more BIPV on more surfaces than those that are available and suitable for a given building stock, and that there is no use of installing more BIPV than would be enough to cover the electricity demand of the building stock. Indeed, current regulatory frameworks limit the possibilities to share energy production outside of buildings' perimeter, thus on-site self-consumption is naturally the most, if not the only, way to valorise PV electricity, and the oversizing of PV installations is detrimental.

3.4 Back-casting approach

In order to obtain realistic market potential figures, final constraints must be taken into account. Indeed, the BIPV market is dependent of constraints and forces at play within both the PV sector and the building and construction sector. These constraints can be of various nature (regulatory environments and their likely evolution, the availability and acceptance of workforce and most importantly the annual new construction rates and renovation rates).

To establish short-term forecasts and account for all these elements, an approach based on "back-casting" is developed. It estimates the likely long-term development of the BIPV market before evaluating the "natural" pathway from the current cumulative BIPV installed capacity leading to these figures. In our case, the maximal realistic cumulative BIPV market is estimated at the 2100 time-horizon and is calculated according to the formula presented below:

$$\begin{aligned} T_{i,x,2100,s} &= \sum_{n=2021}^{2100} M_{i,x,n} \\ &= MIN(SA_{i,x,2020}; DA_{i,x,2020}) \\ &\quad * \prod_{y=2021}^{n} (1 + NCR_{i,y} - DR_{i,y}) \\ &\quad * (RR_{i,n,s} + NCR_{i,n}) * RMI_{i,x,n} \end{aligned}$$

With:

$T_{i,x,2100,s}$ the maximal realistic cumulative BIPV market in 2100 for BIPV product x in country i, in scenario s

$M_{i,x,n}$ the annual addressable market for BIPV product x in country i, in year n

$SA_{i,x}$ the result of supply-side analysis (in m²), for BIPV product x in country i,

$DA_{i,x}$ the result of demand-side analysis (in m²), for BIPV product x in country i,

$RR_{i,n,s}$ the annual renovation rate[1] of buildings in country i (in %), in year n, in scenario s

$NCR_{i,n}$ the annual new construction rate of buildings in country i (in %), in year n

$DR_{i,n}$ the annual demolition rate of buildings in country i (in %), in year n

$RMI_{i,x,n}$ the regulatory environment and market suitability index for BIPV product x, in year n (with a value $0 \leq z \leq 1$).

[1] Renovations accounted for in the considered rates encompass various types of renovation and are not limited to only roofs and façade retrofitting.

The quantified index determining the suitability of regulatory environment and market conditions for the deployment of a BIPV product is based on a qualitative analysis.

3.5 Scenarios

Two scenarios are considered for the annual renovation rates. The first one considers a business as usual in which the renovation rates stay at their current level. The second one falls within the "Renovation wave" context and considers an annual renovation rate that increases steadily, its current level doubling by 2030. After that year, the high renovation rate is maintained. It should be noted that considering a doubling of all renovation rates for all countries and all segments is a simplified assumption. Indeed, the extent to which and the pace at which renovation rates need to be increased are country specific as they depend on the current energy performance of each Member States' building stock. It can also be added that this doubling is considered a minimal requirement (*"[The renovation] rate[s] will need at least to double to reach the EU's energy efficiency and climate objectives."*). [6]

Table I: Renovation rates per country and per building type in 2030, "Business as Usual" scenario [7]

	Residential	Non-Residential
BE	1,2%	7,0%
CH	1,5%	1,5%
DE	1,0%	1,5%
ES	2,0%	3,4%
FR	1,2%	1,6%
IT	1,8%	5,5%
NL	0,9%	2,5%
UK	1,2%	2,8%

Table II: Renovation rates per country and per building type in 2030, "Renovation wave" scenario [7]

	Residential	Non-Residential
BE	2,4%	14%
CH	3,0%	3,0%
DE	2,0%	3,0%
ES	4,0%	6,8%
FR	2,4%	3,2%
IT	3,6%	11,0%
NL	1,8%	5,0%
UK	2,4%	5,6%

Therefore, the only criteria that differentiates the two scenarios are the considered renovation rates reached by 2030. The reason why other aspects are not taken into account to define the scenarios is to really put the emphasis specifically on the impact of increasing renovation rates on the BIPV market. Thus, the underlying scenario for aspects such as distributed PV market development, technological development (e.g. power density increase) is considered to be a most probable scenario (defining the most probable value for each of these aspects for each considered year), but deviations from these trajectories are possible.

3.6 Segmentation

The whole methodology is conducted for different BIPV market segments in parallel. We considered both roofs and facades separately as well as residential and non-residential. This segmentation is a direct result of the structure of the building stock taken as initial input.

First, this allows to be able to use segment-specific values for various parameters. For example, roof-specific and façade-specific power densities are used to convert the outputs obtained in m² into Wp, and the renovation rate values differ depending on whether residential or non-residential buildings are concerned. This segmentation can also be convenient to quantify the BIPV market opportunity in terms of euros, as the different BIPV products can show very different end-user costs depending on the considered segment.

Secondly, this allows to keep track of the building stock characteristics for the whole methodological process from the initial building stock to the final market projections results. However, in addition to the initial building stock's segmentation, other parameters, more relying on a qualitative analysis rather than on quantitative statistics and data, also have a crucial impact on the final results' segmentation. These parameters include the regulatory environment and market suitability index as well as the estimated historical BIPV market segmentation.

4 RESULTS

4.1 Expected BIPV market segmentation

As shown in Figure 4, it can be expected that most of the BIPV installed capacity over the 2021-2030 period will be roof products. Indeed, around 13% of all BIPV market's installed capacity expected over that period should be installed as façade elements. Still, this percentage represents around 1 GWp installed over the next 10 years.

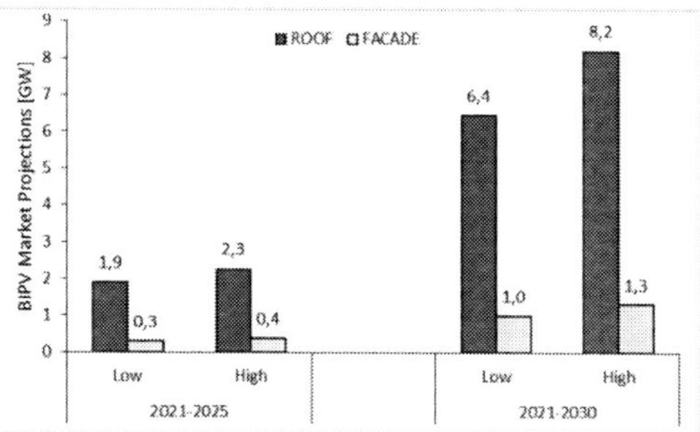

Figure 4: Expected segmentation between roof and façade systems of the cumulative installed BIPV capacity for the 2021-2025 and 2021-2030 periods, for the eight considered countries and for both scenarios.

As depicted in Figure 5, results indicate that the projected BIPV system installations should mostly take place on residential buildings. Indeed, 26% or 2,0 GWp to 2,6 GWp of the projected capacity over the 2021-2030 period should concern non-residential buildings.

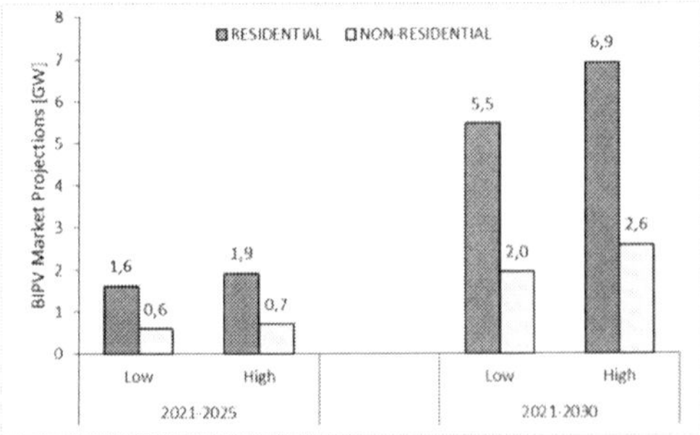

Figure 5: Expected segmentation between residential and non-residential buildings of the cumulative installed BIPV capacity for the 2021-2025 and 2021-2030 periods, for the eight considered countries and for both scenarios.

4.2 Expected influence of renovation rates doubling on BIPV market projections

As presented in Table III, reaching a doubling of current renovation rates for all types of buildings by 2030 should result in a 28% increase (or +2,1 GWp) of the installed capacity over the 2021-2030 period (almost 20% increase over the 2021-2025 period alone).

Table III: Projected cumulative installed BIPV capacity over the 2021-2025 and 2021-2030 periods for both considered scenarios expressed in GWp.

	2021-2025	2021-2030
Low	2,2	7,4
High	2,6	9,5
Delta	*+19%*	*+29%*

5 CONCLUSIONS

Eventually, the cumulative BIPV capacity projected to be installed in the eight key European countries over the next decade could amount to 7,4 GWp to 9,5 GWp with most of the capacity being installed on residential buildings' roofs.

Then, not only will a doubling of current renovation rates have clear benefits in terms of primary energy consumption reduction for the building sector, but it can also be an important driver of the BIPV market representing an additional 2 GWp opportunity compared to the Business-as-Usual scenario.

After the 2030 milestone, the effect of higher renovation rates will be even more visible.

6 REFERENCES

[1] IEA PVPS Task 7, Potential for Building Integrated Photovoltaics, 2002.

[2] P. R. Defaix and e. al., "Technical potential for photovoltaics on buildings in the EU-27," *Solar Energy (86)*, pp. 2644-2653, 2012.

[3] P. Macé, A. El Gammal and e. al., Technical Evaluation of BIPV power generation potential in EU-28, Munich (Germany): 32nd EU PVSEC, June 2016.

[4] European Commission, "A Renovation Wave for Europe - greening our buildings, creating jobs, improving lives," COM(2020) 662 fina, 14/10/2020.

[5] European Commission , "EU Building Stock Observatory," [Online]. Available: https://ec.europa.eu/energy/en/topics/energy-efficiency/energy-performance-of-buildings/eubuildings. [Accessed 28 June 2019].

[6] European Commission, "The European Green Deal," 2019.

[7] European Commission, "Comprehensive study of building energy renovation activities and the uptake," 2019.

This project has received funding from the European Union's Horizon 2020 research and innovation programme under grant agreement No 817991.

SUPER PV PROJECT INNOVATIONS - LCOE ASSESSMENT AND COMPETITIVENESS

Torstein Haarberg[1], Philippe Macé[2], Elina Bosch[2], Juras Ulbikas[3], Julius Denafas[4], Alexander Ulyashin[5]

[1]BNW-Energy AS, Trondheim, 7048, Norway, th@bnw-energy.com; [2]Becquerel Institute, Brussels, 1000, Belgium; [3]Applied Research Institute for Prospective Technologies (Protech), Vilnius, LT 10243, Lithuania; [4]SOLITEK R&D, Vilnius, 08412, Lithuania; [5]SINTEF Industry, Oslo, 0134, Norway

ABSTRACT: The performance of innovations in the ongoing Horizon 2020 project SUPER PV are assessed, both for individual innovations and at an integral level, by calculating the LCOE as the key performance variable. New prospective innovations from SUPER PV combined with price reductions and improvements elsewhere in the PV supply chain are evaluated. Such hybrid solutions can form winning combinations in the market and may point towards a flexible business model for non-vertically integrated module manufacturers.

Several SUPER PV innovations show clear commercial potential by comparing the LCOE with and without the innovation. Moreover, the *"SUPER PV best estimate"*, i.e., the combination of SUPER PV innovations and cost reductions in segments of the PV supply chain not covered by the project, leads to a total estimated LCOE reduction of 57% compared with the original benchmark at the start of the project, considerably lower than the originally targeted 37% LCOE reduction.

At this point the results from SUPER PV are overall promising. Key activities forward are validation of results at the SUPER PV demo-sites across Europe and North Africa and to prepare business models and -plans for market uptake after project end (October 2022).

Keywords: Economic Analysis, LCOE, Markets, PV System Performance

1 INTRODUCTION

The SUPER PV project started in 2018 with 26 partners from 11 countries across Europe and North Africa and targets a range of innovations at three main segments within the PV value chain: PV module innovations, power electronics enhancements and new system integration approaches, including software for design and operation of PV facilities.

The timing of SUPER PV has proven to be perfect, as key issues addressed in the project has become among the most discussed and trending topics in the solar PV community in recent years, from digitalisation of the PV sector and big data analytics, through progress in manufacturing of advanced PV modules and the introduction of bifacial solar modules, to new approaches for solar module recycling.

A common denominator for many SUPER PV innovations is enhanced energy yield. Moreover, unlike a project where solar cells are being developed as a main objective, cheaper and better solar cells in the market will add to the competitiveness of SUPER PV-based solutions. The same goes for other innovations and improvements in the solar supply chain not covered by the project.

2 SCOPE AND APPROACH

2.1 Scope

The scope of this work is to assess the performance of innovations developed in the ongoing project SUPER PV by means of Levelised Cost of Energy (LCOE) assessment as the central Key Performance Indicator (KPI), both at an individual- and an integral level. In addition to the LCOE also other KPIs were identified already from the start of the project:

- Solar module cost reduction
- BoS components costs (EUR/Wp)
- Total installation cost (EUR/Wp)
- Performance Ratio (PR) over the lifetime
- Recyclability and environmental footprint of solar modules.

The general methodology is explained below and illustrated with concrete examples in the results section.

2.2 Approach and methodology

LCOE can be thought of as the price at which electricity must be sold to the market to break even over the lifetime of the technology. It represents the present value of lifecycle cost divided by the present value of total electric power generation, in terms of, for example, €-cents per kilowatt-hour:

$$LCOE = \frac{Present\ value\ of\ lifecycle\ cost}{Present\ value\ of\ lifetime\ energy\ production}$$

The basic approach in this work is to study the effect on the LCOE by introducing changes to a PV system from new potential innovations. Individual performance assessments are done by evaluating the impact of a given innovation on the LCOE from added costs (e.g., component cost) and benefits (e.g., enhanced energy yield). If the calculated LCOE is reduced by adding an innovation, the innovation is deemed profitable, and vice versa. For most innovations this is a straightforward calculation, while for some innovations the situation is more complex. The performance of multifunctional nano-coatings is a good example of this complexity, as the integral effect of coatings requires knowledge on several variables, including coating cost and durability, changes on cleaning costs and, finally, impact on the energy yield and PR.

Integral performance assessments are done by calculating the total change in the LCOE from the combined set of innovations.

Supplemental KPIs described above are monitored continuously to achieve a best possible overview of the technical progression of the project.

3 RESULTS AND DISCUSSION

3.1 Performance of individual innovations

SUPER PV innovations cover three main segments within the PV value chain: PV module innovations, power electronics enhancements and new system integration approaches, including software for design and operation of PV facilities. Prospective innovations in SUPER PV that are subject to performance assessment are: Module front-sheet multifunctional nanocoatings, Light harvesting bifacial modules, In-laminate bypass diode, CIGS modules TPO/AlxOy barrier, PV module recycling, GaN based micro-inverter, MLPE (module level power electronics) micro-inverter, MLPE smartbox, Fault-tolerant converter, Data mining of manufacturing process, Cost-driven design and operation of PV plant software, O&M and smart monitoring with digital twin software, Bifacial modules enhanced yield with and without artificially enhanced ground albedo, Wireless communication gateways and IoT platforms, Weather monitoring, and Dust sensor.

While a thorough discussion of each individual innovation is beyond the scope of this paper, the approach and methodology are illustrated with concrete examples below. The integral effect of all prospective innovations is described in section 3.2.

Multifunctional nano-coatings will influence several defined KPIs. However, the quantitative effects from nano-coatings on the KPIs do not appear straightforward. The methodology for obtaining these relations is described below, using SUPER PV nano-coatings target values.

Target values for coating cost (€/m^2) and durability (years) are first converted into a *present value* which represents all coating costs during the anticipated lifetime of the PV facility. For consistency and cost-comparison with commercial AR-coatings, the initial coating cost is not included in this calculation, as it is considered incorporated into the solar module cost. The calculated present value of all future coating replacements is converted into a corresponding (hypothetical) annual Operational and Maintenance (O&M) cost per Wp as shown in Figure 1. As illustrated in Figure 1 *coating economics* is more favourable at increasing electric power density (W/m^2). This trivial mathematical result may nonetheless be quite interesting commercially.

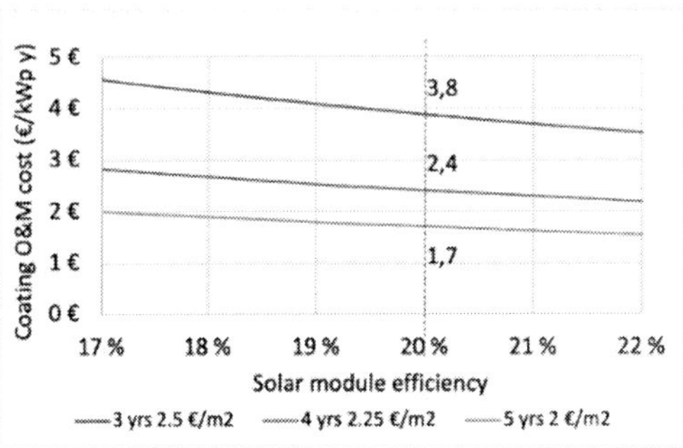

Figure 1: Present value of annual coating O&M cost at varying coating replacement rate (yrs) and -cost (€/m^2)

The overall effect of coatings should be a combination of changes in O&M cost and PR. The effect on O&M has both a negative contribution as illustrated in Figure 1, and a positive contribution with less manual cleaning and maintenance. The integral effect of all impact factors from coatings is assessed by calculating the LCOE at varying O&M and PR, shown in Figure 2.

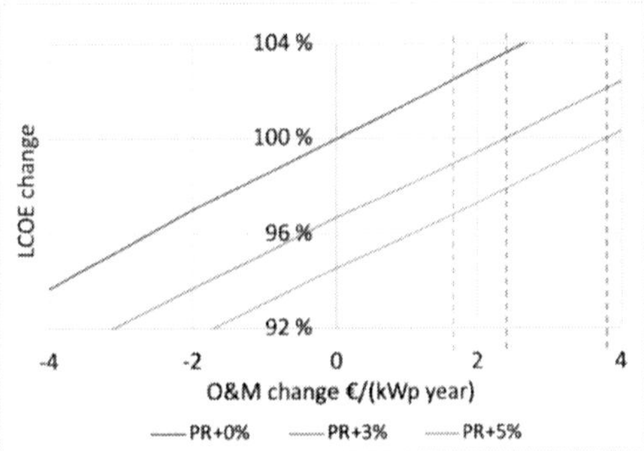

Figure 2: LCOE change vs change in PR and O&M, multifunctional nano-coatings

The LCOE is calculated at a discount rate of 5% to enable direct comparison with evaluations in [1], and the situation is profitable for LCOE change <100%. The vertical, coloured lines in Figure 2 indicate the coating cost at 20% module efficiency, from Figure 1. Ongoing field tests at the SUPER PV demo-sites will be used in conjunction with the information in Figure 1 and 2 to give a final assessment of SUPER PV multifunctional nano-coatings.

MLPE Smartbox represents a new product, and there is hence no absolute benchmark for this innovation. The overall performance of the Smartbox is assessed as the relative change in LCOE at the targeted component cost of 0.03€/Wp for varying O&M and PR, as illustrated in Figure 3. An LCOE change <100% indicates profitability. Gains in O&M and PR are to be validated in the remaining part of the project.

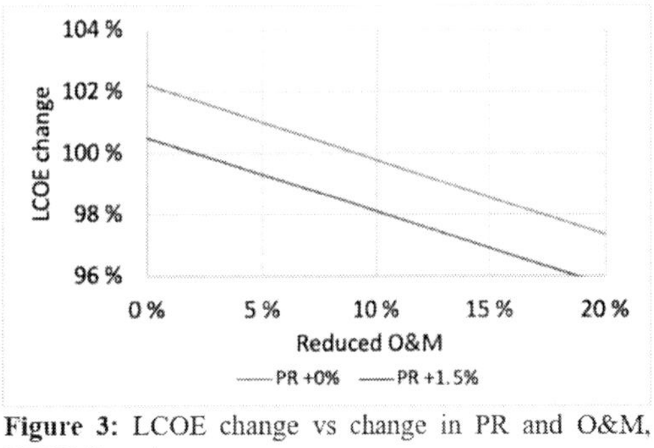

Figure 3: LCOE change vs change in PR and O&M, MLPE Smartbox

Artificially enhanced albedo may enhance the energy yield for bifacial solar modules. Surface albedo has a big effect and is also considered the largest source of uncertainty for bifacial gain. A simplified calculation of yield gain vs albedo change is given in [2], indicating a yield gain of 5-7% by enhancing the surface albedo from 30% to 70%. A material with 70% initial albedo at a cost of approximately 0.5 €ct/Wp is identified. The performance of this material is assessed coarsely, by asking: *How much extra operating* *costs can be allowed for such a material to be profitable?*

Figure 4 shows the results of the assessment, based on an LCOE calculation vs extra O&M costs and enhanced yield, where the threshold line signifies the extra O&M cost at which the high albedo material breaks even. The material is assumed to be replaced after some years, at an overall installation / replacement cost twice the material cost, i.e., 1 €ct/Wp (per replacement).

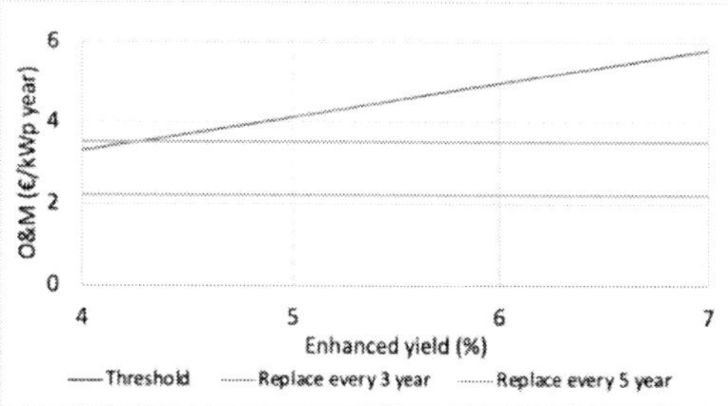

Figure 4: Estimated performance of high albedo material

Fixed costs at replacement by 3 and 5 years are plotted in the figure. It will be profitable to use the high albedo material when the total extra costs are below the blue threshold line.

3.2 Integral assessments

Three types of assessments are shown at the integral level (artificially enhanced albedo not included):

1) Comparison with the original benchmark set at the start of the SUPER PV project
2) Competitiveness in the PV market: Assessment vs defined reference cases that represent current and projected (2023) market conditions for large rooftop systems and ground mounted (utility-scale) systems, respectively
3) PV Performance in different geographies; LCOE vs long-term foreseen electricity prices in different countries in Europe and North Africa.

LCOE comparison with the original benchmark is shown in Figure 5. The original benchmark or *base case* constitutes the situation at project start, and initially a 37% LCOE reduction were targeted, as described in [3].

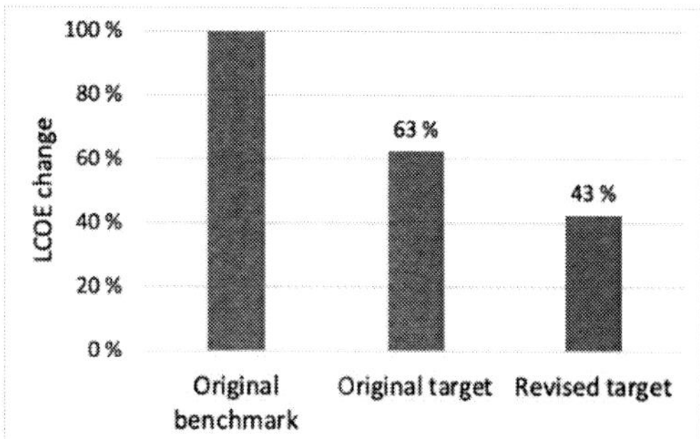

Figure 5: Calculated SUPER PV LCOE change, original and revised target

The revised target constitutes the *"SUPER PV best estimate"*, i.e., the combination of SUPER PV innovations and cost reductions in segments of the PV supply chain not covered by the project such as cheaper solar cells and reduced Balance of System costs. This combination of project innovations and external innovations leads to a total cost reduction of about 57%, as compared with the originally targeted 37%. This may point towards a flexible business model for non-vertically integrated module manufacturers which are able to capitalise on cost reductions not covered in-house.

Performance data for SUPER PV *best estimate* are shown in Table I. The targeted CAPEX cost is higher than recent system installation cost data from e.g., ITRPV [4]. This disadvantage for SUPER PV is however compensated by enhanced energy yield, giving an overall lower LCOE, as discussed below.

Table I: System performance, SUPER PV *best estimate*, ground-mounted system

Variable	Value	Unit
Installation cost, CAPEX	0.5525	€/Wp
Total OPEX	11	€/(kWp year)
Lifetime	35	year
Degradation rate	0.1	%/year
LID degradation 1st year	2	%
Initial module efficiency	20	%
PR enhancement	7	%
Bifacial enhanced yield	7.5	%

Competitiveness in the PV market is shown in Figure 6 and Figure 7. The benchmark is constituted of commercially available technologies which have been identified as being representative of current (2021) and projected (2023) market and industry reality.

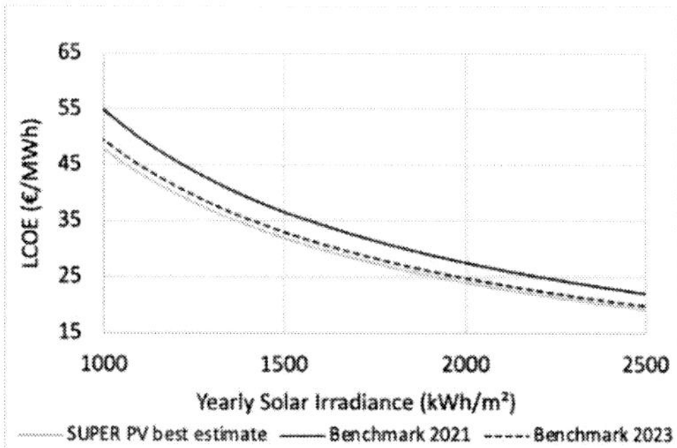

Figure 6: LCOE assessment of the SUPER PV best estimate vs defined benchmark for ground-mounted systems

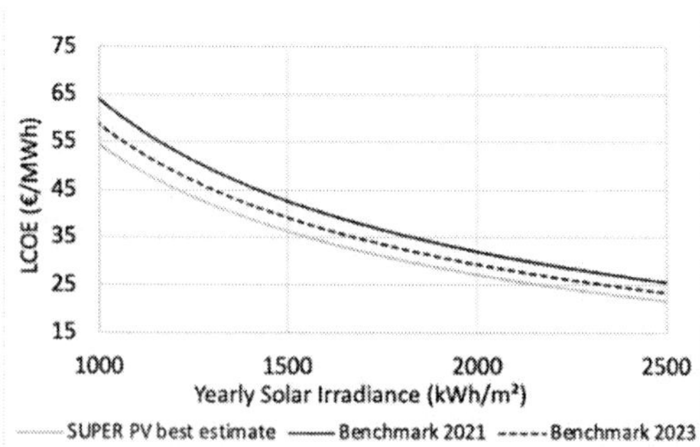

Figure 7: LCOE assessment of the SUPER PV best estimate vs defined benchmark for large rooftop systems

Even though the system CAPEX reached with the SUPER PV innovations is higher than for the benchmark, the LCOE values obtained by the "SUPER PV best estimate" are lower, both when looking at current and projected situation and at ground-mounted and large rooftop systems. This is made possible by higher yield reached by the "SUPER PV best estimate" system, thanks to significant gains in terms of Performance Ratio, which allows to harvest more solar energy at the same level of irradiance. The increased system lifetime (i.e., 35 years compared to 30 years for the benchmark) is a further explanation for these results. It should however be noted that the benchmark is constituted of monofacial modules, while the SUPER PV case assumes bifacial modules. Therefore, the Performance Ratio estimated to be reached by the SUPER PV case is heavily influenced by the bifaciality gains. Should the SUPER PV best estimate be compared to a benchmark based on bifacial modules for, it is likely that the advantage of the SUPER PV case would be substantially reduced, even though still existing.

PV performance in different geographies is evaluated; Tozeur (Tunisia), Seville (Spain,) Munich (Germany), Vilnius (Lithuania), and Oslo (Norway), applying the *SUPER PV best estimate* as described above for hypothetical utility scale PV facilities. Apart from location, the *Weighted Average Cost of Capital* (WACC) is the most important variable affecting the LCOE for solar PV [5, 6, 7]. The WACC depends on many factors such as country, market segment, and perception of risk, and hence the LCOE is calculated at varying WACC. Overall, the cost of finance for solar PV has been reduced over time in conjunction with the growing acceptance and recognition of solar power around the globe [6, 7]. It is reasonable to expect this development to continue.

The LCOE is calculated in real money at an assumed inflation rate of 2% and 0% corporate tax rate, with solar insolation data at optimum tilt angle from Solar GIS (Tozeur, Seville, Munich) and PVGIS and assumed performance ratio (PR) for *base case* shown in Table II. The PR will vary between different locations due to higher ambient temperatures in warmer climates, resulting in a reduced energy yield.

Table II: Solar insolation data and assumed PR *base case*

Location	Insolation kWh/(m^{2}day)	PR *base case* %
Tozeur	6.13	83
Seville	5.77	85
Munich	3.75	87
Vilnius	3.29	89
Oslo	3.15	89

The calculated LCOE is shown in Figure 8 as a function of the WACC. The figure illustrates clearly that local irradiation and the cost of finance are two crucial factors influencing the LCOE.

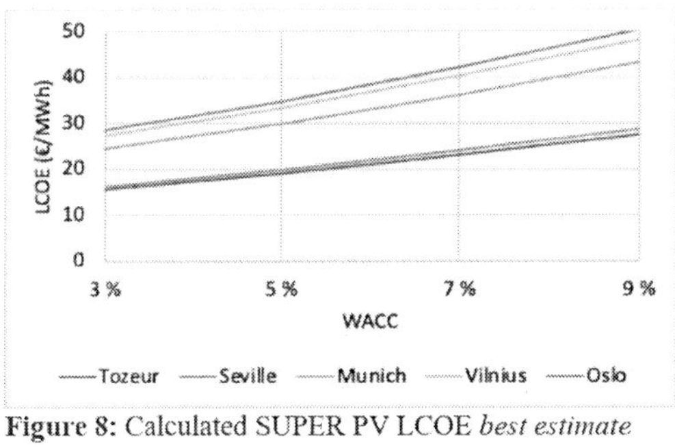

Figure 8: Calculated SUPER PV LCOE *best estimate*

The solar irradiation advantage for the southernmost locations vs the northern locations is comparable to a 6% change in WACC, i.e., the calculated LCOE for Norway at 3% WACC is close to the calculated LCOE for Tunisia at 9% WACC.

The calculated LCOE is compared with available spot price information for the relevant EU countries (Germany, Lithuania, and Spain) [8], Tunisia [9] and Norway [10], shown in Figure 9.

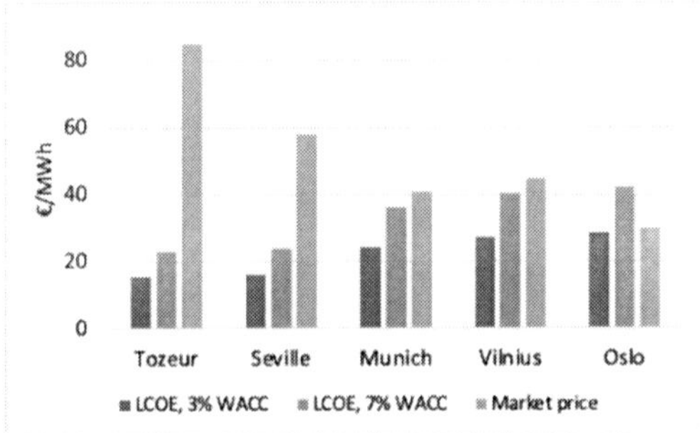

Figure 9: SUPER PV *best estimate* LCOE at 3% and 7% WACC compared with spot market prices (network/grid, taxes and levies not included)

The SUPER PV *best estimate* LCOE is lower than the market price in every country except Norway even at a relatively high WACC of 7%, and for (southern) Norway the figure indicates that solar PV is on the verge of becoming competitive.

Long-term estimates from Statnett [10] indicate a power price range of 35-60€/MWh on the European continent towards 2030-40. For southern Norway an average price in the range of 35-40€/MWh is projected after 2030.

4 CONCLUSIONS

At this point the results from SUPER PV are overall promising. Key remaining activities are to extract field data at the SUPER PV demo-sites in Lithuania, Norway, Spain, Tunisia, and Morocco to ensure final validation of performance, and to prepare business models and -plans for market uptake after project end (October 2022).

5 ACKNOWLEDGEMENTS

The SUPER PV project has received funding from the European Union's Horizon 2020 Research and Innovation Programme under Grant Agreement No. 792245. BNW-Energy and SINTEF have received additional funding for this work from the Research Council of Norway, project number 309225.

6 REFERENCES

[1] K. Ilse, L. Micheli, B. W. Figgis, K. Lange, D. Daßler, H. Hanifi, F. Wolfertstetter, V. Naumann, C. Hagendorf, R. Gottschalg, J. Bagdahn, *Techno-Economic Assessment of Soiling Losses and Mitigation Strategies for Solar Power Generation*, Joule 3, 2303–21, 2019, https://doi.org/10.1016/j.joule.2019.08.019

[2] WSP, 2019: *Technical Briefing Note - Bifacial PV Technology*, https://www.wsp.com/en-GL/insights/bifacial-pv-technology-technical-briefing-note

[3] J. Ulbikas, V. Ulbikaite, J. Denafas, R. Witteck, M. Köntges, M. Topič, F. Frontini, P. Bonomo, P. Macé, P.J. Bolt, A. Ulyashin, T. Haarberg, W. Palitzsch, B. Terheiden, I. Weiss, *Introducing the SUPER PV project - cost reduction and enhanced performance of PV systems*, 35th European Photovoltaic Solar Energy Conference and Exhibition, 2018, 2164-2168, DOI:10.4229/35thEUPVSEC20182018-7DV.2.8

[4] VDMA, *International Technology Roadmap Photovoltaic (ITRPV) 2020 Results*, 12th edition, April 2021, www.itrpv.vdma.org

[5] E. Vartiainen, G. Masson, C. Breyer, D. Moser, E. R. Medina, *Impact of weighted average cost of capital, capital expenditure, and other parameters on future utility-scale PV levelised cost of electricity*, Prog. Photovolt. Res. Appl. 2019, 1–15, https://doi.org/10.1002/pip.3189

[6] K. Chawla, M. Aggarwal, A. Dutt, *Analyzing the falling solar and wind tariffs: evidence from India*, ADBI working paper series, No. 1078, Feb. 2020, https://www.adb.org/sites/default/files/publication/566266/adbi-wp1078.pdf

[7] IRENA (2021), *Renewable Power Generation Costs in 2020*, International Renewable Energy Agency, Abu Dhabi, https://www.irena.org/-/media/Files/IRENA/Agency/Publication/2021/Jun/IRENA_Power_Generation_Costs_2020.pdf

[8] European Commission, *Energy prices and costs in Europe*, Brussels, 14.10.2020, COM (2020) 951 final, https://ec.europa.eu/energy/data-analysis/energy-prices-and-costs_en

[9] FIPA, *Factor costs in Tunisia*, Nov. 2020, http://www.investintunisia.tn/Fr/image.php?id=3181

[10] Statnett, *Long term market analysis 2020-2050*, Oct. 2020, https://www.statnett.no/globalassets/for-aktorer-i-kraftsystemet/planer-og-analyser/executive-summary_rev.pdf.

COST ASSESSMENT OF FUTURE TANDEM PEROVSKITE-SILICON MODULES AND LCOE COMPARISON TO SILICON TECHNOLOGIES IN EUROPE

Christian Moreno Castillo[1], Michael Woodhouse[2], Jean-Baptiste Puel[1,3], Arthur Julien[1], Lars Oberbeck[4], Pierre-Philippe Grand[3]
[1]Institut Photovoltaïque d'Ile-de-France, 18 Boulevard Thomas Gobert, 91120 Palaiseau, France
[2]National Renewable Energy Laboratory, 15013 Denver West Parkway, Golden, CO 80401, United States
[3]Électricité de France, R&D center, Boulevard Gaspard Monge, 91120 Palaiseau, France
[4]TotalEnergies, 2 Place Jean Millier, 92078 Paris La Défense Cedex, France

ABSTRACT: Tandem perovskite-silicon (TPS) photovoltaic (PV) technology is a promising pathway for achieving higher efficiencies than those of single-junction crystal-silicon (c-Si) technologies. In this work, a techno-economic assessment of a TPS module is performed considering a viable large-scale manufacturing process for two future scenarios in 2025 and 2030 which allowed us to identify the main cost drivers of the tandem module cost. Furthermore, to analyse the performance of tandem module-based utility-scale PV systems and understand their competitiveness in Europe, the levelized cost of electricity was compared to passivated emitter and rear cell as well as silicon heterojunction based PV systems. The aim of this work is to provide a cost and performance guidance for TPS modules which could contribute to future PV energy systems.
Keywords: Techno-economic assessment, tandem, perovskite, silicon, LCOE, levelized cost of electricity

1 INTRODUCTION

Tandem technologies combine two cells on top of each other with different bandgaps to harness the light spectrum more efficiently, enabling tandem cell efficiencies to exceed the Shockley-Queisser limit of 29.43% [1]. It has been shown that perovskite on crystalline Si (c-Si) tandem cells could achieve an efficiency of up to 39% [2]. However, the feasibility of large-scale manufacturing processes as well as tandem module reliability still need to be demonstrated.

Significant work is being done within the research community to achieve a perovskite reliability and lifetime similar to those of c-Si cells, as this is a key element for the technology to become competitive in the market. It has been reported that single-junction perovskite cells have passed damp-heat, humidity-freeze and other IEC tests [3] which bodes well for a future market introduction of the technology. First tandem modules are expected in the market in 2022, manufactured by Oxford PV [4].

Our work assumes that future tandem perovskite silicon modules will achieve annual degradation rates and lifetimes that match current c-Si technologies. The perovskite manufacturing process, used for the bottom-up cost assessment, is based on a technology that is under development at the Institut Photovoltaïque d'île de France (IPVF). We assess the costs of perovskite on SHJ tandem modules and use this data as input for the calculation of the levelized cost of electricity (LCOE) of such tandem module-based PV systems in European countries in 2025 and 2030, in comparison with passivated emitter and rear cell (PERC) module-based PV systems in 2021.

2 COST ASSESSMENT METHOD

2.1 Tandem perovskite-silicon module considerations.

A 2.17 m^2 bifacial and double-glass TPS module is considered with a 4-terminals/2 output wires voltage-matched configuration. It was also estimated a 1 GW of capacity production for a large scale manufacuring plant of TPS modules with 144 half-cut, M6 wafer-sized, bifacial silicon heterojunction (SHJ) bottom cells. The cell efficiencies expected for SHJ for 2025 and 2030 are 24,8% and 25.2%, respectively [5].

For the perovskite top cells, IPVF proposes atmospheric slot-die coating deposition for the perovskite as well as for the hole (HTL) and electron transport layer (ETL) on the front glass. The nip-type perovskite cell with a $FACsPb(IBr)_3$ absorber layer has a bandgap of 1.64 eV and an assumed stabilized efficiency of 17.4% in 2025 and 21% in 2030.

An equivalent datasheet for the TPS module was built considering the I-V curves for both top and bottom cells with module efficiencies of 26% in 2025 and 30% in 2030.

2.2 Techno-economic assessment

The cost model for the TPS module is divided into three sections. For the perovskite cells section, a bottom-up cost assessment was performed to achieve a more detailed breakdown of the costs of each manufacturing process. For the silicon cells and module assembly costs, the calculations were based on NREL data and modified to the specific requirements for the tandem configuration and adjusted to expected future values.

Once the cost for the tandem module was obtained, its price was estimated by considering reference values of 2% for the Research and Development costs (R&D), 10% for Sales General & Administrative expenses (SG&A) and 15% as specific gross margin [6].

2.3 Perovskite

The manufacturing process starts with the glass substrate entering the first sputtering chamber to deposit the Indium Tin Oxide (ITO) layer as the front Transparent Conductive Oxide (TCO), then the first laser scribing (P1) is done. It is considered that the following 3 layers are deposited by slot-die coating and have the following n-i-p configuration: tin oxide (SnO_2) for the ETL; $FACsPb(IBr)_3$ for perovskite absorber layer and copper thiocyanate (CuSCN) for HTL, each deposition is followed by a fast infrared annealing step for layer crystallization and evaporation of solvents. After the HTL deposition comes the second laser scribing (P2) and is followed by the sputtered back ITO layer. The third laser scribing (P3) is then realized and finally an alumina (Al_2O_3) layer is deposited by PECVD that acts as a pre-encapsulation to protect the cell stack against the environment during further handling before encapsulation. Figure 1 shows the final perovskite cell stack.

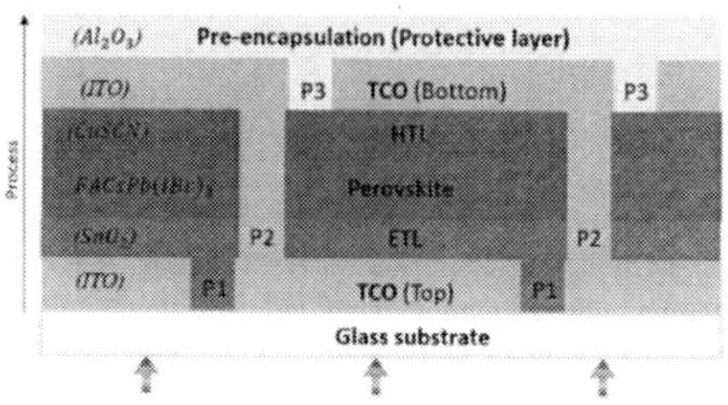

Figure 1. Schematic of the perovskite cell structure. Arrows indicate the front-side illumination.

The process, material and tool specifications for the three slot-die processes such as purity and consumption of the precursors and solvents are based on IPVF's expertise. With this information, quotations were requested from different material and tool suppliers for the 12 manufacturing steps, detailing cost, tool characteristics such as throughput, uptime, footprint, materials and consumables, utilities, labour and maintenance. This data was used as input for the bottom-up cost model that was developed at IPVF in collaboration with NREL and based on the ITRPV-VDMA Cost of Ownership (CoO) template [7].

2.4 Silicon cells

The initial silicon cell costs were obtained from NREL [8], providing cost figures for bifacial SHJ cells with M3 size with an efficiency of 24% for the year 2021. These cost figures were then adjusted to future expected values and specific considerations for the tandem application. First, the cell power is adjusted to the higher efficiencies of 24.8% and 25.2% for years 2025 and 2030, respectively, and to the larger M6 wafer size. Then we adjust the wafer cost proportionally to the amount of average poly-Si consumption per wafer, which is expected to be approx. 10% less by 2025 and 20% less by 2030 [5]. Another adjustment is the reduction of the metallization paste consumption to half of its original value since the current density in the silicon cell is expected to decrease approximately by half in a tandem configuration [9]. Finally, for the remaining direct cell conversion costs (all other process costs for converting wafer to cell except metallization pastes), a reduction of 5% for 2025 and 10% for 2030 was assumed.

2.5 Module assembly

The Balance of Module (BOM), which contains all other elements besides the PV cells such as glass, aluminium frame, junction box, back sheets, encapsulation sheets, ribbons, etc., has a strong contribution to the module cost. These materials together with the consumables for cell stringing were adjusted to both future scenarios with a cost reduction of 4.6% per year [10] except for the glass sheet costs which were obtained from PVinsights. The remaining direct module assembly costs were reduced to take the higher module efficiency into account, resulting in a lower cost per watt peak.

3 COST ASSESSMENT RESULTS

Our cost model computes the total TPS module cost per watt peak including the detailed breakdown of each perovskite manufacturing process cost. The silicon cell and module assembly costs are first added to the perovskite cell cost for the calculation of the total module cost and then divided by the module power to obtain the total TPS module cost per watt peak.

3.1 Perovskite cell cost per module

For the 2025 scenario, a cost of 25.5 \$/module was obtained for the perovskite cells. Materials and consumables represent 41% of the cost, equipment 37% and the remaining 22% consists of labour (10%), maintenance (6%), utilities (5%) and building and facilities (1%) costs.

The process with the highest materials and consumables cost is the slot-die coating of the perovskite and HTL layer followed by the TCO depositions. The processes with the highest equipment costs are both TCO depositions, followed by the slot-die processes for the perovskite layer due to the low tool throughput. The back TCO process cost is slightly higher than the front TCO process cost because of a thicker layer and the assumption of a soft deposition process to not damage the perovskite layer, resulting in increased tool size and cost. Figure 2 shows the perovskite cells manufacturing process cost per step for 2025.

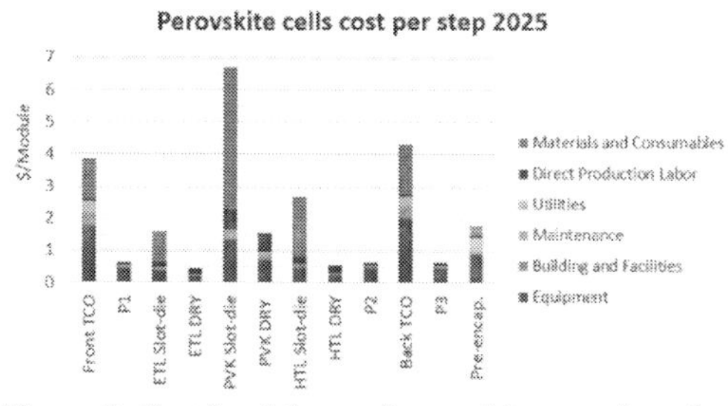

Figure 2. Cost breakdown of perovskite manufacturing process steps in 2025.

For 2030, the cost model was adjusted mainly in two aspects. The first one is that the perovskite cell efficiency increased from 17.4% to 21%, slightly reducing its cost from 25.5 \$/module to 23.9 \$/module. This decrease is explained by a lower tool count for processes such as TCO sputtering, perovskite slot-die coating and pre-encapsulation PECVD for the same plant capacity.

The second adjustment for 2030 perovskite cell costs is the consideration of a 20% per piece learning rate applied to the cumulative installed tandem capacity between 2025 and 2030. It was estimated that the cumulative Si-based tandem market will increase from ~1.1 GW in 2025 to ~11.7 GW in 2030 [5] which translates into a reduction of approx. 50% applied to each one of the manufacturing steps after the higher-efficiency adjustment. Considering both adjustments, a perovskite cell cost of 11.9 \$/module for 2030 was obtained.

Figure 3 shows the cost breakdown for perovskite cell manufacturing in 2030. The cost distribution remains similar to 2025 because the adjustments were applied evenly.

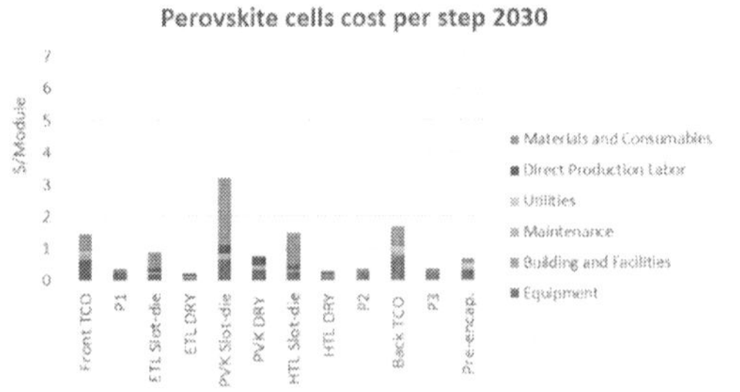

Figure 3. Cost breakdown of perovskite manufacturing process steps in 2030.

3.2 Silicon cells and module assembly cost per module

Considering the assumptions from section 2.4, the cost for the SHJ cells adjusted to the TPS technology will decrease from 58.8 $/module in 2025 to 54.0 $/module in 2030. Note that the reference SHJ cell cost is 65.3 $/module in 2021. For the module assembly costs, the considerations from section 2.5 were used, resulting in a reduction from 43.7 $/module in 2025 to 38.4 $/module in 2030, with the reference SHJ module assembly cost being 47.9 $/module in 2021.

The detailed silicon cell and module assembly costs can be seen in Figure 4 where the sum of the SHJ cells and module assembly costs are 113.1 $/module for the reference case in 2021, 102.5 $/module in 2025 and 92.4 $/module in 2030.

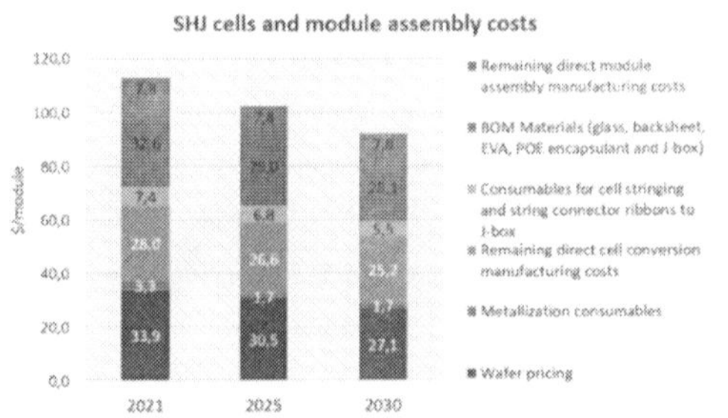

Figure 4. Silicon heterojunction cells and module assembly reference cost for 2021 [8] and cost results for 2025 and 2030.

3.3 Tandem silicon perovskite module cost

Adding up the costs for perovskite and silicon cells as well as module assembly cost, we obtain a TSP module cost of 127.9 $/module for 2025 and 104.3 $/module for 2030. Considering the TPS module efficiencies of 26% for 2025 and 30% for 2030, we obtain a power of 565 W and 652 W, respectively. The TPS module cost per watt peak is calculated to be 22.6 ¢$/$W_p$ for 2025 and 16.0 ¢$/$W_p$ for 2030.

4 LCOE COMPARISON TO SILICON TECHNOLOGIES IN EUROPE

Once the TPS module cost is calculated it is important to analyse its competitiveness at the PV system level to determine if the extra cost of adding the perovskite cell is justified by a higher energy yield and resulting lower LCOE.

We collaborated with NREL to perform LCOE analyses with their System Advisor Model (SAM) software, using the National Solar Radiation Database (NSRDB). NREL's high-performance computing allows for computations of hundreds of thousands of meteorological data points for the generation of LCOE maps for Europe.

Previous studies have been performed for the United States using PVWatts, a performance model within SAM with limited detail in the module characteristics. For this work, for the first time, the Detailed PV performance model was used to obtain more accurate results for different module technologies. Furthermore, a specific software to analyse multiple locations was included to obtain statistical data for each European country.

4.1 LCOE scenarios

PERC and SHJ are expected to be the two mainstream technologies that will dominate the market in the upcoming years, therefore, we compare the TPS to these technologies. Bifacial PERC and bifacial SHJ were used as reference scenarios for 2021. For the years of 2025 and 2030 we included two tandem scenarios: i) based on IPVF assumptions and ii) based on ITRPV data [5] which shows lower module efficiencies and therefore higher module price per Wp values. Figure 5 a) and b) show efficiencies and module prices, respectively, for the 10 different scenarios.

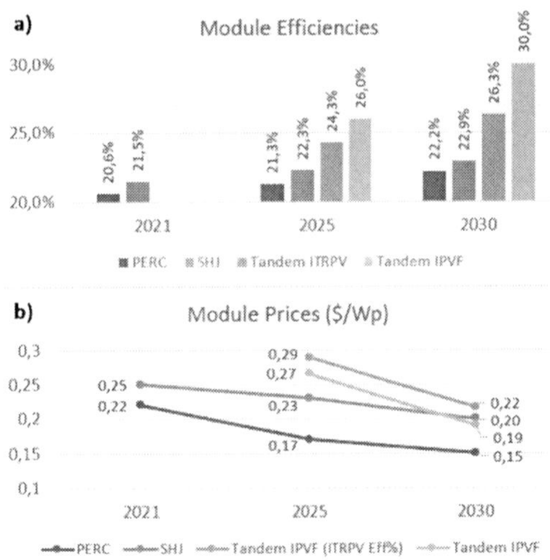

Figure 5. a) Module efficiencies and b) module prices for the 10 different LCOE scenarios considered in this study.

CAPEX and O&M costs are considered in SAM for each of the European countries [11]. The CAPEX was additionally adjusted by adding the module price difference between bifacial PERC and the specific technology analysed in each scenario. Further assumptions made in the SAM model are: lifetime degradation of 0.5%/year, 770 kW_{AC} inverter with 97% efficiency, power plant capacity of 100 MW_{DC}, DC-AC ratio of 1.23, 1-axis tracking system and ground coverage ratio of 30%.

4.2 LCOE results

Each of the 10 LCOE scenarios was computed for the European countries and statistical values were obtained for each country such as mean and median as well as the corresponding standard deviations, minima, and maxima. First results that we present here compare tandem module-based PV systems in 2025 and 2030 with PERC module-based systems in 2021 as reference.

The LCOE values for Europe in the tandem scenario of 2025 range from 36.7 $/$MWh_{AC}$ in Portugal to 100.1

$/MWh$_{AC}$ in Ireland with a median value for Europe of 65.9 $/MW$_{AC}$. For the tandem scenario in 2030, the minimum LCOE value is 34.1 $/MW$_{AC}$ in Spain and the maximum is 83.4 $/MW$_{AC}$ in Ireland with a European median of 56.7 $/MW$_{AC}$. In general, both tandem scenarios in Europe show an LCOE reduction with time and a reduction compared to the reference case of PERC-based PV systems in 2021. Table I shows the median LCOE values of the three scenarios for the European countries.

Table I. Median LCOE values for three technology scenarios in Europe.

Median LCOE ($/MWh$_{AC}$)	Reference PERC 2021	Tandem IPVF 2025	Tandem IPVF 2030
Albania	57,3	53,9	46,3
Austria	78,5	70,9	59,1
Belgium	84,8	80,0	68,8
Bosnia	69,2	64,7	55,6
Bulgaria	61,1	57,4	49,4
Croatia	68,4	64,1	55,1
Czech Rep.	73,1	70,1	60,3
Denmark	86,2	83,2	72,0
Estonia	90,5	86,1	74,2
Finland	102,1	93,9	77,8
France	66,3	63,6	52,7
Germany	72,6	72,4	61,4
Greece	48,7	46,3	39,8
Hungary	63,2	60,3	51,9
Ireland	110,5	100,1	83,4
Italy	58,0	51,2	41,8
Kosovo	61,2	57,4	49,4
Latvia	89,5	84,7	73,0
Lithuania	85,6	80,9	69,7
Luxembourg	79,9	75,2	64,7
Macedonia	58,2	54,7	47,1
Montenegro	62,6	59,0	50,8
Netherlands	74,0	75,1	64,4
Norway	99,4	78,8	75,8
Poland	74,8	70,5	59,8
Portugal	48,3	36,7	34,5
Romania	67,6	63,6	54,7
Serbia	65,4	61,4	52,8
Slovakia	70,3	67,1	57,8
Slovenia	74,2	69,5	59,7
Spain	39,2	40,1	34,1
Sweden	101,3	90,4	77,4
Switzerland	70,9	64,3	53,5
Turkey	44,6	43,8	36,9
Ukraine	64,7	61,9	53,3
United Kingdom	78,6	81,8	73,6

5 CONCLUSIONS

The bottom-up cost model for perovskite cells shows that their cost contribution to the total TPS module is low compared to the silicon cells and module assembly costs. The perovskite cells cost represents approximately 20% of the TPS module cost in 2025 and ca. 10% in 2030. While the proposed manufacturing process for the TPS module production considers some low-cost materials and process tools that require development, we believe that there is still room for further improvements of processes and materials that will enhance the perovskite cell performance.

First results of an LCOE mapping for Europe indicate that for most of the countries the median LCOE values decrease in scenarios for tandem module-based PV systems in 2025 and 2030, compared to a PERC module-based reference scenario in 2021. Note however, that in such scenarios a high reliability and lifetime of the tandem module similar to a c-Si module is needed.

However, special attention is needed in countries where system CAPEX does not decrease sufficiently over time to overcompensate the extra cost for the high-efficiency TPS module, compared to the PERC module. In such cases, the resulting LCOE from a tandem module-based system could even increase compared to the reference PV system.

REFERENCES

[1] A. Richter, M. Hermle and S. W. Glunz, "Reassessment of the limiting efficiency for crystaline silicon solar cells," *IEEE Journal of Photovoltaics*, vol. 3, pp. 1184-1191, 2013.

[2] Z. Li, Y. Zhao, X. Wang, Y. Sun, Z. Zhao, Y. Li, H. Zhou and Q. Chen, "Cost Analysis of Perovskite Tandem Photovoltaics," *Joule Cell Press,* vol. 2, pp. 1-14, 2018.

[3] N. L. Chang, J. Zheng, Y. Wu, H. Shen, F. Qi, K. Catchpole, A. Ho-Baillie and R. J. Egan, "A bottom-up cost analysis of silicon–perovskite tandem photovoltaics," *Progress in Photovoltaics,* vol. 29, pp. 401-413, 2021.

[4] Oxford PV, 23 July 2021. [Online]. Available: https://www.oxfordpv.com/news/oxford-pv-completes-build-out-its-brandenburg-factory. [Accessed 1 September 2021].

[5] VDMA, "International Technology Roadmap for Photovoltaic (ITRPV) 2020 Results," 2021.

[6] M. Woodhouse, B. Smith, A. Ramdas and R. Margolis, "Crystalline Silicon Photovoltaic Module Manufacturing Costs and Sustainable Pricing: 1H 2018 Benchmark and Cost Reduction Road Map", National Renewable Energy Laboratory, 2019.

[7] ITRPV-VDMA, "TCO-Total Cost of Ownership Calculation," ITRPV-VDMA, 2019. [Online]. Available: https://itrpv.vdma.org/web/itrpv/tco. [Accessed 1 September 2021].

[8] M. Woodhouse, D. Feldman, V. Ramasamy, T. Barnes and R. Margolis, "Technology Advances in

Solar Photovoltaic Costs, Reliability, Efficiency and Energy Yield Leading to Improved System Lifecycle Economics," *NREL Solar Futures Study in preparation for publication.*

[9] S. Pouliquen, K. Alvino, E. Drahi, S. Jutteau and L. Oberbeck, "c-Si Solar cell benchmark for tandem application," in *IPVF*, Palaiseau , 2019.

[10] L. Oberbeck, K. Alvino, B. Goraya and M. Jubault, "IPVF's Technology vision 2030," *Progress in Photovoltaics*, vol. 28, pp. 1207-1214, 2020.

[11] IHSMarkit, "Levelized Cost of Electricity for Renewables in Europe, 2020-2050," 2020.

PROJECTED COSTS FOR COMPETING PHOTOVOLTAIC TECHNOLOGIES ACHIEVING OVER 30% MODULE EFFICIENCY AT TERAWATT SCALE IN 2050

Tim Bruton
TMB Consulting
3 Sanger Drive, Send, Woking, GU23 7EB, UK
Email: tim.bruton@btinternet.com

ABSTRACT: Climate change abatement requires the achievement of net zero carbon emissions by 2050. To achieve this there has to be a shift to global energy to electricity supplied by renewable energy technologies, the majority coming from photovoltaics with 8.5TWp total global installations by 2050. This is implementable if global installation grow at 5% p.a. from a base of 773.2 GWp total installations in 2020. Learning curve studies have shown that costs as low as $0.09/Wp for silicon modules is possible but with module efficiency only at 26.4%. There is a reasonable expectation that this level can be exceed in the next 30 years. Two technologies offer efficiencies beyond 30% namely tandem cells of III-V alloys on silicon and perovskite cells on silicon. Applying learning curve theory to present best production costs for these materials shows that cost of epitaxy is a limiting factor for III-V/silicon tandems while a module efficiency of 36% is required for perovskite/silicon tandems to be cost effective against the projected cost for silicon modules.
Keywords: silicon, perovskite, III-V semiconductors

1 INTRODUCTION

It is well recognized that stabilization of climate change will require the global energy supply to be zero carbon by 2050. While the conversion of electricity generation to renewables is growing strongly and PV is seen as the lowest cost technology for new capacity[1], there has to be a transition in other sectors such as transport and heating to electricity. For transport electric vehicles with batteries are becoming more common and domestic heating using heat pumps is being given priority. Alternatively green hydrogen produced by electrolysis of water with renewable electricity has potential for decarbonisation by using fuel cells for transport and as fuel replacing fossil fuels in domestic systems and for industrial applications. The impact of these changes is a strongly growing demand for renewable electricity. IRENA has predicted that 86% of global electricity production will need to come from renewables of which photovoltaics will provide the majority and 8.5TWp of total global capacity will have to be installed by 2050 [2].773 GWp of PV was commissioned in 2020 [1]. This paper examines the potential to achieve 8.5TWp installed by 2050. It then looks at the currently emerging technologies for tandem solar cells with either III-V alloys or perovskite solar cells on silicon which have the ability to realize solar cells with efficiencies greater than 30%. The potential is evaluated for these tandems to be cost effective compared to the silicon cells alone given the reduction in cost that can be expected as global production grows.

The purpose of such a long term look ahead is not to make a case for investment now but to identify the drivers for research which will make these long term goals viable. There is a well established tradition of taking such long term views in PV technology development. In 1970 when the PV industry was producing only a few kWP p.a. for space applications Ralph predicted that cost below $1/Wp were possible given scale up to megawatt level and a range of technology improvements were implemented [3]. There was very little understanding of the how to make terrestrial solar cells at that time. When the terrestrial market developed, costs were still well above the targeted $1.0/Wp thought to be essential for a significant global PV market. The MUSIC FM study showed in 1997 that while the global PV industry was still only at 80 MWp p.a. production scaling up to 500MWp could reduce costs below $1.0/Wp [4]. Only modest technological improvements were needed and the main driver was to increase annual volume of production and subsequent growth in the industry has verified this prediction.

In this study it is shown very different technology drivers need to be deployed to achieve cost effectiveness for the III-V and perovskite silicon tandem solar cells.

2 ROADMAP TO 8.5 TWp INSTALLATION

The IRENA study [2] shows that 86% of electricity production has to come from renewable energy sources and 8.5 TWp of cumulative PV capacity. Other studies have come to similar conclusions [5]. This is an order of magnitude increase from today's installed base. If this growth were to be achieved with a linear production rate then this would be 258 GWp p.a. which is double current production. A more realistic scenario is to assume a compound growth rate and as shown in Figure 1 a 5% annual growth rate in PV installations will result in 9.2 TWp cumulative installations by 2048 meeting the IRENA target.

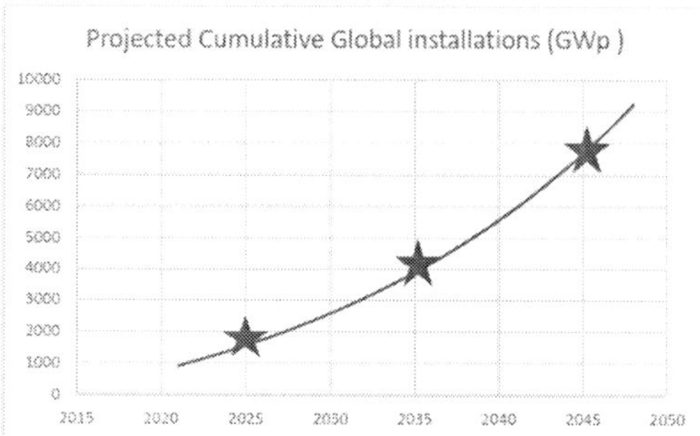

Figure 1: Projected growth in cumulative global PV installations at 5% compound growth rate to 2050 doubling of the installed base shown by stars.

The stars show when each doubling of the global installations. In total 3.2 doublings of the global deployment is projected by 2048. Annual installations will be 542 GWp p.a. in 2048. It is credible that this growth can be achieved by the PV industry. It is therefore reasonable to estimate what impact this might have on PV manufacturing costs.

3 PROJECTED MANUFACTURING COSTS

Two recent studies have looked at what manufacturing costs might be achieved by 2050 assuming manufacturing volumes are at the 10TWp level [6,7] based on the manufacturing of solar cells made with monocrystalline silicon wafers using the PERC technology. These studies applied learning curve theory where a constant factor of cost reduction is created by each doubling of installed capacity. Hoffmann [6] applied different rates for the various parts of the manufacturing process while Vartiainen applied a historic rate of 23.5% [7]. The costs generated by the application of the learning curve reduction are given in Table 1.

Table 1: Projected costs for silicon modules at TWp level of global installations.

Volume (TWp)	Cell Eff (%)	Cell type	Module cost \$/Wp	Ref.
10	26.4	Mono-PERC/L	0.09	[6]
8.6	30.0	Mono-PERC/L	0.07	[7]

The two studies project similar costs with the lower cost being due to the assumed higher efficiency in [7]. A silicon cell efficiency of 26.4% is just about the maximum achievable solar cells efficiency for a silicon cell. There is no evidence to support a 30% efficiency in a single junction silicon cell.

However it does seem improbable that in the next 30 years ways to exceed 30% module efficiency in commercial production. To achieve this technologies must be developed which allow the Schockly-Quiesser limit for a single junction cell to be exceeded. A number of possibilities exist as discussed below.

4 POTENTIAL FUTURE CELL TECHNOLOGY AT THE TERAWATT LEVEL

Silicon solar cell technology using both monocrystalline and multicrystalline wafers has been the dominant production technology for the past 40 years and maintains a share of global production at 95% and has been above 90% for the past 10 years [8]. Current global installations is only 10% of what is anticipated for 2050 and therefore there is ample time for a completely new technology to emerge and displace silicon. However as discussed in above each year incremental growth will be made at the 100GWp level and the evidence is that investments will continue to be made in the proven low risk technology that silicon represents. Many alternative solar cell technologies to silicon have been extensively researched in the past 40 years with some degree of commercialization but none has replaced crystalline silicon. The reasons for this are many and varied and are discussed at some length in [9].

It is therefore required that an add-on technology to crystalline silicon be found which can be implemented as an additional processing step in existing silicon cell manufacturing lines. Tandem cells on silicon fit this requirement and both III-V alloys on silicon and perovskite solar cells on silicon have demonstrated solar cell efficiencies at 30% or greater.

4.1 III-V Solar cells on silicon

III-V alloys offer great potential as tandem solar cells. Triple junction solar cells have been demonstrated to give very high efficiencies with a GaInP/GaAs/InGaAs tandem cell achieving a world record 37.9% efficiency [10]. In addition to the high efficiency potential the III-V solar cells have proved their durability being the space cell technology of choice since the early 1990s. The cells have al so well in terrestrial applications being deployed in concentrating systems at 500 suns and greater concentration levels with efficiencies up to 47.1% being report for 6 junction solar cells [10].

While the III-V alloys offer high efficiency potential and good stability growing tandem cells epitaxial onto silicon has proved difficult. There is a 4% lattice mismatch between silicon and GaAS and this has made growing high quality epitaxial layers difficult, a problem that has been recognized from the 1980s [11]. An alternative approach has been to fabricate the III-V cell detach it from the parent GaAs wafer and wafer bond it to a silicon cell. This approach has produced a GaInP/AlGaAS/Si tandem solar cell of 34.5% efficiency, a world record for a tandem cell on silicon. While the III-V's have the clear potential to take silicon beyond the 30% barrier the costs remain high with reliable estimates of current manufacturing costs for III-V solar cells being between \$40/Wp and \$100/Wp [12]. The potential to reduce these cost at the terawatt production level is discussed below.

4.2 Perovskite on silicon

Perovskite solar cells have emerged as promising as a promising new solar cell technology within the last decade with the lead methylammonium iodide cell demonstrating a 10% efficiency in 2012 [13]. Since than progress has been rapid with a choice of metal ion, organic molecule and halide component giving a wide range of tunable bandgaps. The current world record for a small area perovskite cell is 25.5% (0.095cm^2)[10]. However this is a still developing technology and no commercial production has taken place and there are concerns over long term stability and decrease in device efficiency with increasing area. Rather than being a standalone technology the potential for a tandem cell with a 1.8eV bandgap on a silicon cell is an ideal combination. Recent results at two research have shown that perovskite/silicon tandem cells with efficiencies close to 30% can be achieved, namely of 29.15% [14] and 29.52% [15].

The perovskite solar cells have good low cost potential being produced from solution at room temperature. Although large scale production to verify the cost potential has not yet occurred cost studies [16. 17]have been made for the perovskite silicon tandem and these are extended to terawatt production below.

5 COST AT TWp MANUFACTUTING

5.1 Silicon base case

To evaluate the potential for the tandem cells to be cost competitive the base case for silicon production at the cumulative 10TWp must be established. From [6] at that level of production a 26.4% efficiency module can be made at a cost of \$0.09/m^2 and this equates toa manufacturing cost of \$23.8/ m^2. If an additional process is used then the increase in cost must be offset by the gain in module efficiency to achieve the same cost per watt. Although there can be some increase in cost because of the reduction in BOS costs through higher efficiency but this will not be taken into account in this analysis. It is assumed that all the module costs other than the deposition of the new active layer are accounted for in the silicon base case.

5.2 III-V/Silicon tandem cost at 10TWp

As stated in [12] III-V manufacturing costs are too high given the standard MOCVD approach to epitaxy. Essig et al have looked at how costs could be reduced if MOCVD is displaced by the more recent development of Hydride Vapour Phase Epitaxy (HVPE) much higher throughput can be achieved through epitaxial reactors and costs corresponding reduced [18]. It has been projected that with full optimisation of the approach in the next few year that a 35% efficient tandem cell could be made with an cost of epitaxy of \$0.22/Wp which equates to \$77/m^2 area cost. Applying the same leaning curve reduction that has been used in the silicon case i.e. a 22.3% for each doubling [7] of the installed base and for 3.2 doublings gives a 70% reduction in cost. This then gives the cost of III-V epitaxy with 10TWp installed of \$23.1/ m^2. This added to the base silicon cost gives the cost for the tandem at \$46.9/ m^2. This a doubling of the cost and unsurprisingly therefore requires a doubling of the module efficiency to give the same module \$/Wp as silicon alone. In fact the required module efficiency becomes 52.1% . This clearly beyond any efficiency that can be achieved for a single III-V junction on silicon. Depositing multijunction cells would only further increase costs. The implication from this study is therefore for III-V on silicon tandems it is the research priority to reduce the cost of epitaxy much further rather than trying to maximise solar cell efficiency.

5.3 Perovskite /Silicon tandem cost at 10TWp

While the perovskite technology is evolving and many questions have yet to be answered the potential for low cost with existing approaches has been modelled. From [16] the current additional cost of a single junction perovskite solar cell on silicon is \$31.59/m^2. Applying the same learning curve correction reduces this cost to \$9.48/m^2 for the 10TWp cumulative installed capacity in 2050. This increases the cost for the perovskite silicon tandem to \$32.28/m^2 and this requires a module efficiency of 35.8% to be fully competitive with the single silicon solar cell case. While this is higher than current achievements which are at and early stage this appears to be a realistic target for development over the next thirty years [19]. So for perovskite silicon tandems the research emphasis has to be on maximising efficiency in the tandem module.

6 CONCLUSIONS

It has been shown that it is realistic to assume that the PV industry can expand to ensure that 8.5TWp of cumulative capacity can be installed globally by 2050. This is the level required to stabilise climate change and only a 5% annual compound growth rate is required to achieve this. Earlier studies have shown that this increase in cumulative installed capacity can reduce silicon module manufacturing costs to \$0.09/Wp but with a module efficiency of limited to 26.4%. This study has looked at the potential of both III-V/silicon and perovskite/silicon tandem cells which have achieved near 30% solar cell efficiencies to be cost effective against the single junction silicon case. The study has found that for the III-V silicon case an unrealistic 52% cell efficiency is required and future research should be directed to lowering the cost of producing the III-V active layer. The scenario is different for the perovskite where lower manufacturing costs are anticipated and a module efficiency of 35.8% is required. While this a major advance from the current achievements of near 30% efficiency it provides a realistic goal for future development.

7 REFERENCES

[1] Solar Power Europe, Global Market Outlook For Solar Power 2021-2025

[2] IRENA Energy Transformation Road Map to 2050 (2018)

[3] E L Ralph, Proc 8th IEEE PVSC (1970) 326-330

[4] T M Bruton et al , Proc. 14th EUPVSEC (1997) 672-677

[5] World Energy Council: "World Energy Insights Brief-Global scenarios Comparison Review" (2019)

[6] W Hoffmann & A Metz, Proc 36th EUPVSEC (2019) 1966-1971

[7] E Vartiainen et al, Proc 37th EUPVSEC (2020) 2046-2049

[8] FhG.ISE, Photovoltaics Report 21st July 2021

[9] T M Bruton, "Photovoltaics - from Milliwatts to Gigawatts" Pub. J Wiley (2021) 129-167

[10] M A Green et al, Progress in Photovoltaics 29 (2021) 3-15

[11] S M Vernon et al, Proc 17th IEEE PVSC (1984) 434-439

[12] K A W Horowitz et al, NREL Technical Report (2018) NREL/TP-6A20-72103

[13] M M Lee et al, Science 338 (2012) 643-647

[14] A Al Ashouri et al, Science 370 (2020) 1300-1309

[15] Oxford PV Press release 21st December 2020

[16] Z Li et al, Joule 2(2018) 1559-1572

[17] N L Chang et al, Prog. in Photovoltaics 29 (2021) 401-413

[18] E Essig et al, Nature Energy 2 (2017) 326-333

[19] H Snaith, Paper CP1.1 this conference

SMART STRATEGIES FOR THE TRANSITION IN COAL INTENSIVE REGIONS

Rita Mergner[1], Rainer Janssen[1], Dominik Rutz[1],

[2]Charalampos Malamatenios, [2]Georgia Veziryianni, [3]Dirk Knoche, [3]Anne Rademacher, [3]Rainer Schlepphorst, [4]Liliana Fonseca, [4]Rona Michie, [5]Angel Nikolaev, [6]Christian Doczekal, [7]Greg Arrowsmith, [7]Nicolas de la Vega, [8]Gloria Popescu, [8]Marian Dobrin, [9]Jasmina Mandic Lukic, [9] Miodrag Mesarovic, [10]Igor Volchyn, [10]Dmytro Bondzyk, [11]Marcin Pietrzykowski, [11]Bartłomiej Woś, [11]Marcin Chodak , [11]Justyna Likus-Cieslik, [12]Rhiannon Phillips, [12]Kenn Palmer [13]Jan Frouz, [14]Marketa Hendrychova, [15]Sabina Irimie

[1]WIP – Renewable Energies, Sylvensteinstr. 2, D - 81369 Munich, Germany
Tel. +49 89 720 12743, Fax +49 89 720 12791
E-Mail: rita.mergner@wip-munich.de; rainer.janssen@wip-munich.de
Internet: www.wip-munich.de, www.tracer-h2020.eu

[2]Centre for Renewable Energy Sources and Saving (CRES), Greece, [3]Research Institute for Post-Mining Landscapes (FIB), Germany, [4]University of Strathclyde, UK,[5] Black Sea Energy Research Centre (BSERC), Bulgaria, [6]GET, Austria, [7]EUREC, Belgium, [8]ISPE, Romania, [9]Energoprojekt ENTEL, Serbia, [10]Coal Energy Technology Institute, Ukraine, [11]University of Agriculture in Krakow, Poland, [12]Welsh Government, UK, [13]Charles University, Czech Republic, [14]Czech University of Life Sciences Prague, Czech Republic, [15]Jiu Valley Social Institute Association, Romania

ABSTRACT: The TRACER project supports a number of coal-intensive regions around Europe to design (or re-design) their Research and Innovation (R&I) strategies in order to facilitate their transition towards a sustainable energy system. The TRACER consortium consists of different target regions: South East Bulgaria, North West Bohemia - Czech Republic, Lusatian Lignite District - Germany, Western Macedonia - Greece, Upper Silesian Coalfield - Poland, West Region, Jiu Valley - Romania, Wales - UK, Kolubara - Serbia, Donetsk - Ukraine. Core activities of TRACER include the implementation of an EDP (Entrepreneurial Discovery Process) to mobilise a wide range of stakeholders in each target region to develop an appropriate governance structure and to bring regional stakeholders together to discuss and agree on a shared vision and priorities for coal transition. R&I strategies, industrial roadmaps and decision support tools will be developed jointly with key stakeholders of the TRACER target regions. Further TRACER activities include the identification and analysis of best practice examples of successful and ambitious transition processes in coal intensive regions, a detailed assessment of social, environmental and technological challenges, the elaboration of guidelines on how to mobilise investment as well as dedicated activities to stimulate R&I cooperation among coal intensive regions in Europe and beyond.

Keywords: energy transition, coal intensive regions, R&I strategies, industrial roadmaps, entrepreneurial discovery process, social challenges, re-skilling needs, SET Plan, mobilising investment.

1 INTRODUCTION

The EU aims to be climate-neutral by 2050 – an economy with net-zero greenhouse gas emissions. This objective is at the heart of the European Green Deal and in line with the EU's commitment to global climate action under the Paris Agreement. In view of these objectives, some Member States are in disadvantageous position due to the current state of the technologies used and the available local energy resources, so these specificities must be taken into account when planning and implementing the national and/or regional energy strategies.

A number of EU Member States are actively mining coal. This provides direct employment to about 185,000 people across the EU, with additional indirect jobs relying on coal production. However, the EU's decision to move to emission-free technology is not subject to discussion and it is therefore necessary to take into account the difficulties of coal regions, to anticipate and propose the energy policy that is in line with the tasks. These regions need an effective roadmap to make the necessary transition to a more diversified economic base and a more sustainable energy system, while safeguarding the social cohesion. At the same time, the production and consumption of coal in the EU has been in steady decline over the past few decades, partly due to relatively high extraction costs, which have turned Europe's coal industry less competitive, but also due to the commitment by a number of Member States to phase out coal use for power generation.

In a changing world, the EU needs to become a smart, sustainable and inclusive economy. These three mutually reinforcing priorities should help the EU and the Member States to deliver high levels of employment, productivity and social cohesion.

2 THE TRACER PROJECT

The overall objective of TRACER is to support a number of coal-intensive regions around Europe to design (or re-design) their Research and Innovation (R&I) strategies in order to facilitate their transition towards a sustainable energy system. Nine (9) European regions are involved in this procedure, 6 of them in EU Members States (Bulgaria, Czech Republic, Germany, Greece, Poland, Romania), and 3 in countries outside the EU (UK, Serbia and Ukraine).

For this purpose, regional policy makers and other key stakeholders are assisted by the TRACER partners through the provision of information, expertise and advice, as well as by decision support tools. Concrete roadmaps and blueprints on the energy transition paths will be developed in the framework of TRACER project.

In order to achieve this a number of activities were implemented. In the first year of the project an investigation of the current status in the target regions and the analysis of the best practices were made. This included an analysis of renewable energy technologies and the extent of their application in partner countries, the identification of best practices on social issues and tourism as well as the exploration of best practices of smart specialisation strategies and on financing. In the second year, regions implemented stakeholder mobilisation actions and local workshops.

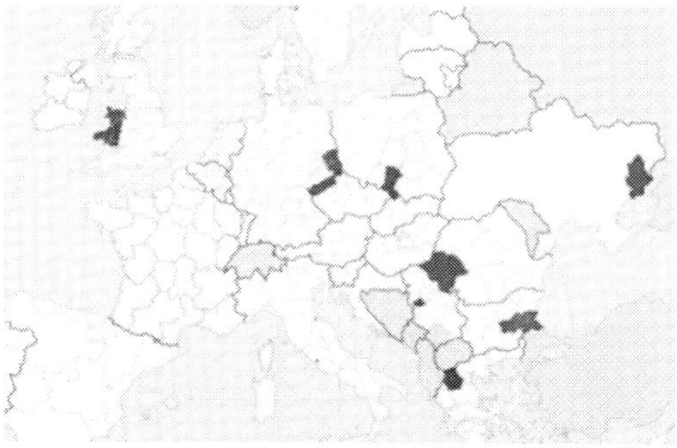

Figure 1: TRACER target regions (in red)

TRACER project is based on the cooperative approach. One of the important actions of the project is the establishment of close collaboration with other sections of Horizon 2020 and other EU funded programs, as well as the R&I Cooperation in the framework of the Implementation Plans jointly developed by European countries, as part of the EU's SET Plan and with the EU Coal Regions in Transition Platform.

3 TECHNOLOGICAL CHALLENGES AND OPPORTUNITIES

Energy generation technologies based on coal are still very widely used in the TRACER target regions. The transition can therefore be a long way, with many other aspects to be considered.

The transition process has progressed very differently in the target countries. Some are well on the way to renewable energy generation on a central (large) or decentralized (small) level. Some countries use coal not only for electricity and heat generation in power plants, but also for heating purposes at the consumer level, e.g. single stoves in households.

The path to renewable energy generation is particularly difficult, since it is progressing too slowly in many countries and some countries therefore use natural gas as an intermediate route.

Concepts are therefore required that avoid or at least minimize this intermediate step and show implementation options for renewable energy (central or decentralized).

3.1 Southeast Bulgaria (BG34)

The region of Stara Zagora is known as the "energy heart" of Bulgaria. The energy generation opportunities in the region, primarily from RES, are significant. Natural gas is currently of no interest for this purpose because there is enough cheap electricity produced from coal, but substantial new natural gas electricity capacities are expected in the period 2025-2030 to replace the coal capacities closed in that period. Fuel cell technology is not yet sufficiently used in the region, but it can be expected in the future if there is a change in the economic factors and the regulatory system. The prospect of expansion for RES is on the way.

By 2030, in parallel with the coal-fired TPP decrease, the production of renewable energy is projected to increase, according to NECP. This increase is composed mainly of solar PV, while other RES, such as wind, biomass, and biogas will grow slightly in absolute terms.

The greatest importance should be given to the future use of solar installations, but so far this process has hardly begun. The main reasons for that are:

- thermal or electric storage system to overcome the diurnal differences;
- need for connection to the distribution network to exchange energy;
- installation of protection and automation - IT systems.

Solar PV is becoming increasingly commercial on a large scale, while for small scale installations stimuli are available as well. Solar thermal is currently not supported, but subsidies are envisaged in the National Recovery and Resilience Plan (Council of Ministers, 2021).

The southern part of South East region, in which Maritsa East coal region is located, has significant solar energy potential (1450 – 1500 kWh/m^2 annually). Large-scale solar PV is a good opportunity for the post-mining development of Maritsa-East area, as it would take advantage of the existing electricity infrastructure, the available engineer's workforce, and the unusable (degraded) land.

Investment projects for the construction of new PV plants are on the way. The projects are in the solar radiation rich Stara Zagora and Haskovo regions. In Chirpan, Eneri is developing a 180 MW plant. In the industrial zone of nearby Dimitrovgrad it is working on a 65 MW project. The most significant in terms of volume is the investment plan near the village of Haskovo. Knizhovnik, which is expected to reach 400 MW. The total value of investments in these projects is estimated at over 700 million BGN (EUR 358 million).

The technological possibilities for the development of distributed RES generation in the South East Region of Bulgaria are wide enough, but the question is how much they would be implemented. Decentralized heating systems so far have not found application in Bulgaria due to a number of inconveniences – requirements for additional premises, unwillingness of households to cooperate, etc.

Given that compared to the coal sector, RES sector, particularly the solar one, being the most perspective one in the region, is not job-intensive, other opportunities should be sought, taking advantage of the well-developed infrastructure, favourable for the development of various types of business - production of electrical and electronic devices, machinery and equipment, ICT, production of second-generation biofuels.

3.3 Lusatia region, Brandenburg (DE40) and Dresden (DED2), Germany

The exit from a coal-fired power generation in Germany comes in addition to the nuclear phase out, decided already in 2011. With this, the two main power generation technologies until then, are no longer available in the future which is a complete reorganisation of the national energy industry. The historically unique transition process poses a great challenge with large effects on several areas, like power supply and labour market, which are not only related to Lusatia, but also to Germany as a whole.

Measures to improve the assured capacity from renewable sources are highly required. Batteries could be utilized in future as a fast compensation for fluctuations in electricity production by renewable sources as well as so-called vehicle-to-grid technologies which are able to return electric power from batteries from electric vehicles to the public power network. Power-to-Gas technologies can transfer surplus renewable electricity into hydrogen or methane, which could be used in several ways. Power-to-heat-technologies transfer electrical energy into heat with the help of electric boilers and heat pumps. Even though these options are already developed in a small scale and as prototypes, it will require high investments a time to set them up for a practical application.

About 4,350 MW of renewable electric power is installed in Lusatia covering 38 % of the 11,525 MW of the overall installed amount in Lusatia. Wind turbines and photovoltaics are the dominant technologies. Biomass contributes to less than 2 %.

Wind and solar energy have the greatest potential in the technological challenge and are the most cost-effective technologies to produce renewable energy in Lusatia. Other technologies, based on water, biomass or geothermal are more expensive or have less potential due to limited availabilities or are still in research stage and not ready for implementation.

Regarding PV, the law for the expansion of renewable energies (EEG, 2021) aims for an installed nominal capacity of 100 GW of photovoltaic modules in 2030. In 2020, according to initial projections, photovoltaics covered 9.3% of gross electricity consumption in Germany with electricity generation of 50 TWh (Wirth, 2021). 2 million PV modules with a nominal capacity of 53 GW were in operation. Due to this, PV is able to cover about 67% of the electricity demand at sunny days.

Brandenburg aims for an installed electric capacity from PV of 3,500 MW in 2030. In 2018, 3,673 MW have already been installed. In Saxony 35,651 photovoltaic modules with a nominal power of 1,635 MW were installed in 2016. They produced 1,463 GWh of electricity per year. 2,200 GWh per year are stated as the target value by 2022 in Saxon "Energy and Climate Programme 2012".

3.4 Western Macedonia (EL53), Greece

To ensure an easier transition to a clean energy economy, some challenges need to be overcome. They are related not only to the technology, but also to society. The region has been supplying lignite to the energy industry for more than seven decades, enough to shape a specific coal-oriented mentality of the local population. It seems to be very difficult to persuade employees, miners, engineers that this specific activity will stop. If the support consists in an early retirement bonus for the generations close to retirement age, then the younger generations will need to learn a new profession. Development of RES could be one of the options in the overall package.

Region of Western Macedonia has a huge unexploited capacity in wind and solar energy. To assess the available biomass capacity in Western Macedonia and especially in the areas of interest for the DH networks of Ptolemaida, Amynteo and Kozani, several studies have been conducted and the available data have been presented in publications. According to the Public Enterprise for District Heating in Ptolemaida (DETIP), the available biomass capacity in Western Macedonia mainly originates from forest biomass (125,000 tons/year) and agricultural residues (201,000 tons/year) with a respective total thermal content of 1,630 GWh/year.

The outputs of the World Bank's Technical Assistance Project have been presented recently and are as follows:

- land could be used for PV and wind, perhaps land repurposed from lignite mines;
- land could be used for batteries to store PV and wind and provide a service to the power system;
- lakes (or flooded land) could be used for pumped storage projects or for floating PV;
- PV and wind could produce green hydrogen to be used in power plants instead of lignite;
- PV and wind could provide electricity so cheap that it could fuel power plants instead of lignite, utilizing molten salts tanks to store energy so that the plant could continue to provide baseload power and heat post-lignite;
- biomass has the potential to provide heat and power in Western Macedonia, creating income in the agricultural sector.

The Greek government is considering a solar-based hydrogen production initiative through major-scale photovoltaic facilities planned at state-controlled coal power utility (lignite fields in Ptolemaida area) as a result of Greece's decarbonization plan.

In December 2019 cycle of auctions, the number of applications for PVs to the Regulatory Authority for Energy (RAE) for Western Macedonia increased from 3 GWp to 7.2 GWp, which shows the intense market interest. This accounts for 25% of current PV pipeline for Greece (26 GWp pending applications). The high demand is directly related to the vacuum in the electrical space that is expected to create the withdrawal of lignite plants in Western Macedonia in 2023.

Most applications have been submitted by PPC Renewables and relate to projects located in the areas of lignite mines, which are expected to utilize the connection infrastructure of the withdrawing thermal units. Among these are the 99 applications of PPC for PV parks in the open lignite mines of Ptolemaida and Amyntaio, with a capacity of 550 MWp and 350 MWp respectively, which, however, presuppose the restoration of the lignite mines. Among the rest, the PV projects at Perdikkas, Eordaia Amyntaio, with a total capacity of 204.76 MW, the PV parks Amyntaio I & II, with a capacity of 100 MWp and 75 MWp respectively, as well as the 8 PV parks at Exochi in Kozani at the Ptolemaida lignite mine stand out, with a total capacity of 173 MWp. All these projects are expected to receive soon all necessary licences.

3.5 Upper Silesia coal field (PL22), Poland

The target region for Poland is the largest industrial district in the country, i.e. the Upper Silesia Coalfield (GOP), with 19 mines operating at the moment. Energy production is Silesia (as in the whole Poland) is by far dominated by hard coal. However, there are some attempts to use also other technologies. Currently, there are large scale installations producing electricity from renewable sources in the Śląskie Voivodeship, with a total capacity of ca. 73 MW. The largest parts of installations are those using energy from biomass (including biogas), as well as hydropower. However, in terms of installation capacity, the largest share of renewable energy is installed in hydroelectric power plants, ca. 37 MW. The least installations are installations using solar radiation.

The Polskie Sieci Elektroenergetyczne electric power generation network suggests the necessity to connect PV installations to the power system due to the increasing demand for energy used by air-conditioning in the summer season. It is estimated that approximately 2000 MW of power from PV panels could be added to the system. One of the measures to achieve the necessary installed capacity is the development of prosumerism, i.e., power generated by individual users which due to technical limitations would primarily include photovoltaic projects.

3.6 Jiu Valley / West region (RO42), Romania

The transformation into a RES-based sustainable energy system with low carbon emissions, will be made gradually, with natural gas as a transition fuel, in order to provide a time buffer for upgrading and retrofitting the electricity distribution networks, as many of them are obsolete. The gradual transition of Jiu Valley micro-region to a RES-based energy system, by switching power/cogeneration units from hard-coal to natural gas is also necessary, knowing that wind and solar energy are variable resources, unable to ensure the entire energy demand, especially in extreme weather conditions generated by the increasing negative impact of climate change.

EU Just Transition Fund will provide other investment opportunities in areas such as site regeneration and decontamination, land reclamation, implementing LCT and infrastructure for clean and affordable energy supply, reducing GHG emissions, increasing energy efficiency and RES use, SMEs, including start-ups, which support economic diversification and transition, creation of new companies, including consulting services and "business incubators", R&I, including advanced technologies transfer etc.

RES potential (hydro, wind, solar, biomass, geothermal) is defined in specialized studies performed by prestigious R&I institutions. According to JRC report[1] for the West Region (RO42) the technical potential for onshore wind was assess at around 8.8 GW and for ground solar PV about 29 GW. Estimated biomass potential is app. 2,370 TJ in Hunedoara County (78% agricultural, 22% forestry), however there is no data available for Jiu Valley.

National projections for 2030 reflect an increase of up to 5,255 MW in the wind capacities and to 5,054 MW in the PV capacities, but 60% of the lignite-based power plants will be still operational. The natural gas-fired capacities include both new highly efficient and flexible units and retrofitted ones, considering the age of the current natural gas-fired plants, estimating that the decrease due to decommissioning will exceed the increase foreseen through the new capacities.

For Jiu Valley micro-region it was estimated that the evolution of RES use will be slow until 2025, with an increase only in PV of about 50% compared to 2020. The growth rate becomes more pronounced after 2025 and until 2040, stagnating after that.

3.7 Kolubara region (RS11 and RS21), Serbia

There are no other than coal fired technologies for utility scale electricity generation except a very small number of mini hydro and rooftop PV installations. These small installations are connected to the national power grid. That is because of an attractive high subsidized price under the existing feed-in tariffs scheme.

The most frequently used energy generation technologies for small applications other than those burning coal are stoves and other facilities burning wood (chips and pellets), and water heating boilers burning natural gas and fuel oil. Relatively small number of users outside the capital city of Belgrade have gas installations. However, the use of electricity for space heating and cooking is very significant. Besides conventional electrical heaters and stoves, there may also be found hybrid installations for heating and cooling, based on air or (very rarely) ground-based heat pumps.

Under the present Action plan to increase the share of renewable energy sources (RES) in gross final energy consumption, Serbia was expected to reach 27% target in 2020, but it fails to do so, although subsidies are provided by law. New target to increase its use of RES in electricity production by 2030 in Serbia is yet to be set. Some estimates of the expected new target for Serbia under the Energy Community (EnC) Treaty translates to at least 5,037 MW generation capacity on RES, of which 3,409 MW on hydro, 869 MW on solar PV, 723 MW on wind and at least 36 MW on biomass. A considerable portion of these new capacities, particularly those based on biomass and solar potentials, may be constructed within the Kolubara target region.

The Energy Utility of Serbia (EPS) is preparing a comprehensive study with the main objective to evaluate the possibilities of solar energy used to generate electricity and heat within its facilities, both as mounted on the existing structures (buildings, etc.) and put on the open space properties (closed mines areas, ash and overburden disposal areas, etc.).

The efficiency of conversion of solar radiation to electricity in the Kolubara region spans over a wide range, depending on both technology (fixed or moving panels) and the age of solar cells. On an annual basis, the energy efficiency of fixed PV panels is slightly above 11% for a number of older installations such as the one on the roof of Energoprojekt's head office building in Belgrade. Several experimental PV solar installations built in the Kolubara region with the moving panels technology have demonstrated about twice higher efficiency. A wide range of capacities may be found when renewable energy and new energy technologies in the Kolubara region are considered. While the existing roof-top PV installations are of the order of several tens of kilowatts, the expected utility-size PV solar power plants may reach several hundred megawatts and, if associated with adequate energy storage capacities, may play an important role in the power generation mix in the region.

It is necessary to remove administrative barriers, increase permitting capacities at the level of relevant authorities (local and governmental level) and provide long-term investment environment to achieve needed increase of RES capacity.

[1] Kapetaki, Z., R. e. (2020). Clean energy technologies in coal regions: Opportunities for jobs and growth: Deployment potential and impacts. Luxembourg: EUR 29895 EN, Publications Office of the European Union. doi:10.2760/063496

3.8 Donetsk region, Ukraine

In general, the potential of Donetsk region allows efficient use of renewable energy sources. Thus, according to experts, the technically achievable potential of the region is 25.4 billion kWh per year. One of the largest shares in its structure belongs to biomass (32.9%).

Energy strategy of Ukraine until 2035 is aimed at innovative development of the energy sector and investment in new generation capacity to replace decommissioning one. The choice of the type of generation will depend on the forecast price situation for fuel and the intensity of development of each type of generation, which will increase the level of competition between them, from the introduction of smart technologies to equalize consumption peaks.

It is envisaged to constantly expand the use of all types of renewable energy, which will become one of the tools to ensure the energy security of the state. In the short and medium term (until 2025), the share of renewable energy in Ukraine is projected to increase to 12% of primary energy and at least to 25% by 2035. At the same time, the share of energy produced from coal is planned to be reduced to 12% by 2035.

3.9 Wales (UKL1, UKL2), United Kingdom

Besides wind and solar power, other renewable sources are in early stages of deployment or only operate at a very local level. The capacity of 'smaller' energy generation sources, such as anaerobic digestion, heat pumps, sewage gas and waste energy, has sharply increased since 2012. For example, a single (municipal) waste energy project started in 2015 in Cardiff, Trident Park, generating 30 MW in electricity and 25 MW of heat. Using a 12 km long insulated pipeline from the newly built incinerator, an annual 425,000 tonnes of non-recyclable waste is converted in heat energy for c. 20,000 homes There are also pilot project relating to district heating and peer-to-peer energy generation.

An important aspect of renewable energy projects, particularly in comparison with a fossil fuel-based system, is their degree of local ownership. This covers projects owned by Welsh households, communities, local authorities, housing associations, charities (including religious organisations), education establishments, and Wales-based businesses, farms and estates. Data sources usually underestimate their numbers, but locally owned renewable energy project accumulated to an amount of over 64,000 in 2018, with a total capacity of 540 MWel and 243 MWth. This is respectively 17% and 37% of the total renewable electrical and heat energy capacity. In comparison, 19% of gas and coal electricity generation capacity is locally owned. One of the key principles of the Welsh energy strategy is the support of local ownership, in order to reduce costs, climate change effects and reliance on import. In addition, community ownership delivers key opportunities for changing local attitudes towards new renewable energy developments, such as wind farms. Concrete local ownership targets include:

- all new energy projects in Wales should have an element of local ownership from 2020;
- at least 1 GW of renewable energy capacity should be locally owned by 2030;
- concrete investment in public sector organisations, using interest-free loans, to finance renewable energy installations (in particular solar PV).

5 NEXT STEPS

TRACER will continue working on a key challenge to mobilise a wide range of stakeholders in 9 coal regions across Europe and prepare the R&I strategies. Active involvement of the different stakeholder groups is being implemented by a large set of soft support measures, such as guidelines, specific reports, working group meetings, matchmaking meetings, workshops etc.

ACKNOWLEDGEMENTS

The authors would like to thank the European Commission and the INEA for the support of the TRACER project. TRACER has received funding from the European Union's Horizon 2020 research and innovation programme under grant agreement No. 836819.

Disclaimer: The sole responsibility for the content of this paper lies with the authors. It does not necessarily reflect the opinion of the European Union. Neither the INEA nor the European Commission are responsible for any use that may be made of the information contained therein.

TRACER
TRANSITION IN COAL INTENSIVE REGIONS

SOLAR-ERA.NET
EUROPEAN COOPERATION IN PV RTDI AND BEYOND
HIGHLIGHTS, LESSONS LEARNED AND FUTURE PERSPECTIVES

Stefan Nowak, NET Nowak Energy & Technology, Waldweg 8, 1717 St. Ursen, Switzerland,
T: +41 26 494 00 30, stefan.nowak@netenergy.ch
Marcel Gutschner, NET Nowak Energy & Technology, marcel.gutschner@netenergy.ch
Thomas Biel, NET Nowak Energy & Technology, thomas.biel@netenergy.ch
Stefan Oberholzer, Swiss Federal Department for Environment, Transports, Energy and Communication,
stefan.oberholzer@bfe.admin.ch
Christoph Hünnekes, Forschungszentrum Jülich – PtJ, ch.huennekes@fz-juelich.de
Kambulakwao Chakanga, Forschungszentrum Jülich – PtJ, k.chakanga@fz-juelich.de
Renate Horbelt, Forschungszentrum Jülich – PtJ, r.horbelt@fz-juelich.de
Melanie Dürr, Project Management Organisation Energy, Technology, Sustainability (ETN), me.duerr@fz-juelich.de
Estrella Fernandez, Ministry of Science, Universities and Innovation, estrella.fernandez@aei.gob.es
Beatriz Gómez, Ministry of Science, Universities and Innovation, beatriz.gomez@aei.gob.es
Irene Carlos, Fundación Española para la Ciencia y la Tecnología (FECYT), irene.carlos@fecyt.es
Julio Marchamalo, Fundación Española para la Ciencia y la Tecnología (FECYT), julio.marchamalo@fecyt.es
Marina Sopeña, Centro para el Desarrollo Tecnológico Industrial (CDTI), marina.sopena@cdti.es
Pierre-Jean Rigole, Swedish Energy Agency, pierre-jean.rigole@energimyndigheten.se
Otto Bernsen, Rijksdienst voor Ondernemend Nederland (RVO), otto.bernsen@rvo.nl
Laurence Polain, Service Public de Wallonie (SPW), laurence.polain@spw.wallonie.be
Geert Carchon, Hermesfonds / Agentschap Innoveren & Ondernemen (VLAIO), geert.carchon@vlaio.be
Evi Afentaki, Geniki Grammatia Erevnas Kai Technologias (GSRT), pafe@gsrt.gr
Ioanna Sergidou Loizou, Research and Innovation Foundation (RIF), iloizou@research.org.cy
Pierre Rale, Agence de l'Environnement et de la Maîtrise de l'Energie (ADEME), pierre.rale@ademe.fr
Pascal Bain, Agence nationale de la recherche (ANR), pascal.bain@agencerecherche.fr
Kaan Karaösz, Turkiye Bilimsel ve Teknolojik Arastirma Kurumu (TUBITAK), kaan.karaoz@tubitak.gov.tr
Maria Bianco, Ministero dell'Istruzione, dell'Università e della Ricerca (MIUR), maria.bianco@miur.it
Elvira Lutter, Klima- und Energiefonds, elvira.lutter@klimafonds.gv.at
Anita Hipfinger, Österreichische Forschungsförderungsgesellschaft (FFG), anita.hipfinger@ffg.at
Gideon Friedmann, Ministry of National Infrastructure, Energy and Water Resources, gideonf@energy.gov.il

ABSTRACT: SOLAR-ERA.NET is a network that brings together more than 20 RTD and innovation programmes in the field of solar electricity technologies in the European Research Area. The network of national and regional funding organisations has been established in order to increase transnational cooperation between RTD and innovation programmes and to contribute to achieving the objectives of the Strategic Energy Technology Plan (SET-Plan) through dedicated transnational activities (especially transnational calls). Through 7 joint calls, 88 research and innovation projects are supported, involving 460 organisations and mobilising more than 110 MEUR. This contribution focusses mainly on results from the transnational SOLAR-ERA.NET calls.

Keywords: R&D and Demonstration Programmes, Photovoltaic, Funding and Incentives, Strategy

1 INTRODUCTION AND CONTEXT

SOLAR-ERA.NET is a European network of national and regional funding organisations and RTD and innovation programmes in the field of solar electricity generation, i.e. photovoltaics (PV) and solar thermal electricity (STE) / concentrating solar power (CSP). SOLAR-ERA.NET shall contribute to reach the objectives of the European SET-Plan, by carrying out the coordination and support actions between national and regional research and innovation programmes.

The SET-Plan aims to increase, coordinate and focus EU support on key low-carbon energy technologies in order to achieve Europe's 2020 and 2030 energy and climate objectives in the future.

2 NETWORK AND MISSION

2.1 Network

SOLAR-ERA.NET involves more than 20 national and regional RTD and innovation programmes dealing with PV and CSP. This high level of involvement of most relevant stakeholders provides excellent outreach and allows for a solid coordination needed for an efficient and coherent approach in the highly diverse RTD landscape.

The network has been composed of some 20 organisations being programme owners and managers. Most countries and regions belonging to the SOLAR-ERA.NET consortium also participate in most calls.

2.2 Missions and goals

SOLAR-ERA.NET has two fundamental missions. As a network supported by the EC within the ERA-NET (2012-2016) respectively ERA-NET Cofund (since 2016) scheme, the mission is to improve the coordination and cooperation between national and regional programmes. In the context of the SET-Plan, the mission of the network is to implement central parts of the SET-Plan on a transnational level and thus contribute to achieving the goals defined in the SET-Plan.

The more specific and essential goals and activities of SOLAR-ERA.NET are i) to launch joint calls for RTD proposals by national and regional RTD and innovation programmes and ii) to define and support the best joint activities, strategic information exchange and use of implementation tools.

By identifying and choosing priority topics based on the Implementation Plans for PV respectively CSP for transnational calls, the SOLAR-ERA.NET network selects and funds industrially relevant transnational RTD and innovation projects in the field of solar electricity technologies.

In quantitative, financial terms, SOLAR-ERA.NET shall result in a total funding volume by the participating national and regional programmes of some 10 to 15 MEUR per call for innovative projects.

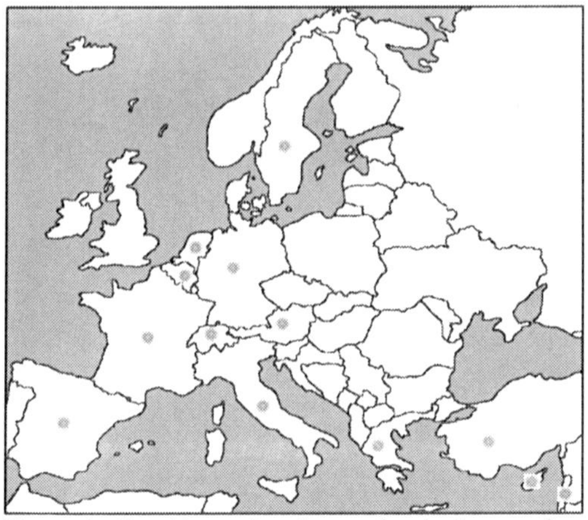

Figure 1: Countries and regions active in the ongoing SOLAR-ERA.NET Cofund 2.

3 ACTIVITIES, FACTS AND FIGURES

3.1 Set up of transnational calls

SOLAR-ERA.NET started in November 2012. Seven joint calls have been launched so far, with the participation of up to 17 countries / regions covering PV and CSP topics. Call topics in photovoltaics (PV) and concentrating solar power (CSP) are commonly defined by the SOLAR-ERA.NET consortium and adopted by the participating programmes and countries according to their national / regional priorities. The recent topics can be found in Figure 2.

The transnational call is based on a 2-step / stage-procedure with i) a preproposal and ii) a full proposal stage. Preproposals are checked according to the national / regional regulations and then discussed in the SOLAR-ERA.NET consortium setting up lists of preproposals recommended respectively not recommended for stepping on to the full proposal phase.

Once the full proposals are evaluated by independent international experts as well as by national experts, a list of projects suggested for funding is established.

The first four transnational SOLAR-ERA.NET joint calls were each allocated with a total public funding budget of approximately 10 to 12 MEUR provided by the participating national / regional programmes and agencies.

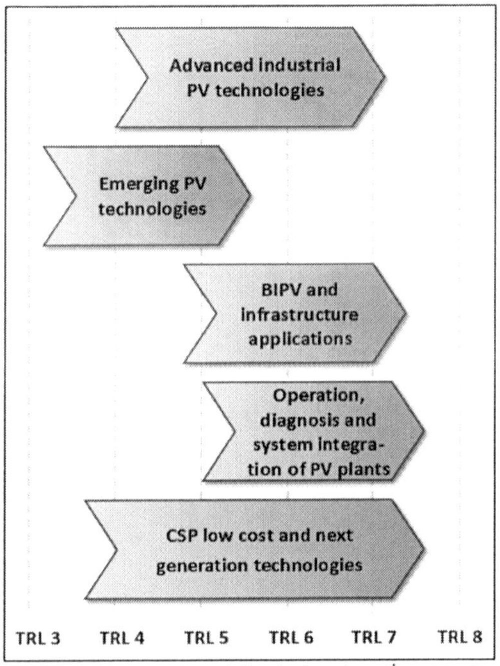

Figure 2: Topics defined for the 7th SOLAR-ERA.NET joint call carried out in 2019/2020

Within the ERA-NET Cofund scheme (since 2016), SOLAR-ERA.NET calls (5th and 6th call) come along with an even higher funding budget as the EC provides the so-called Top-Up presenting 33% of the total funding effectively used. The selection procedure includes a ranking list, established by independent international experts, which has to be strictly followed from top downwards. Finally, SOLAR-ERA.NET carried out an additional joint call (7th call, 2019/2020) and expects to launch another additional joint call (8th call, 2021/2022) in autumn 2021 – both additional joint calls support transnational projects with funding from the participating national and regional agencies.

Applicants may also come from other countries but they do have to provide their own funding for their participation in the projects.

3.2 Participation in transnational calls

The joint calls found good interest in the solar power industry sector and research community. 310 preproposals were submitted with a total budget of around 410 MEUR (of which 287 MEUR funding requested) involving 1'467 partners (multiple counting – some partners have been involved in several preproposals). About half of the preproposals were selected for the second stage of the call procedure, resulting in 166 full proposals.

88 transnational projects are supported. The total public funding amounts to approx. 76 MEUR for a total project volume of around 112 MEUR. 460 organisations are involved in these transnational projects funded (multiple counting). A slight majority of the organisations

involved belong to the categories of small and medium sized enterprises and large enterprises.

Figure 3 displays the thematic split of the 88 projects.

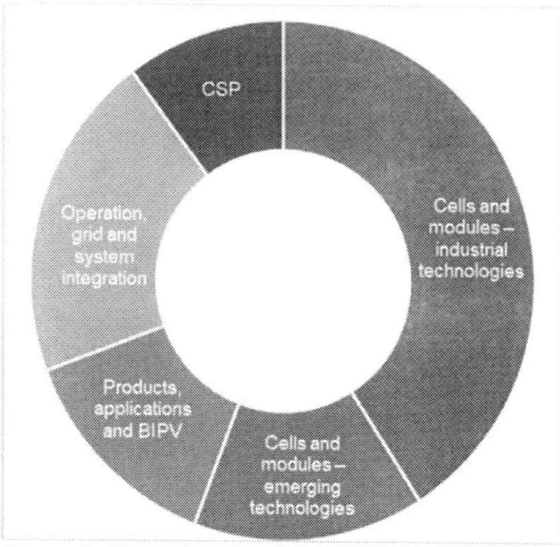

Figure 3: Thematic split of the 88 SOLAR-ERA.NET projects.

Figures 4 and 5 show the number of projects and the funding share by country / region resulting from the seven SOLAR-ERA.NET calls comprising 88 projects.

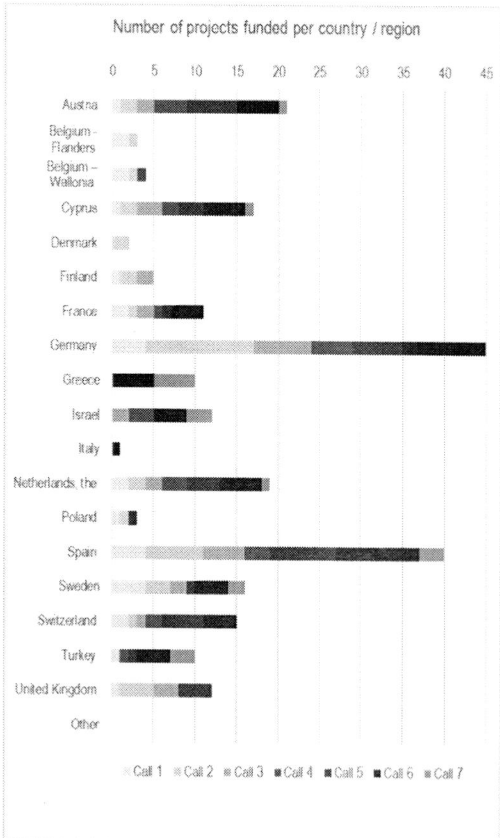

Figure 4: Number of projects by country / region as well as by call resulting from the seven SOLAR-ERA.NET calls.

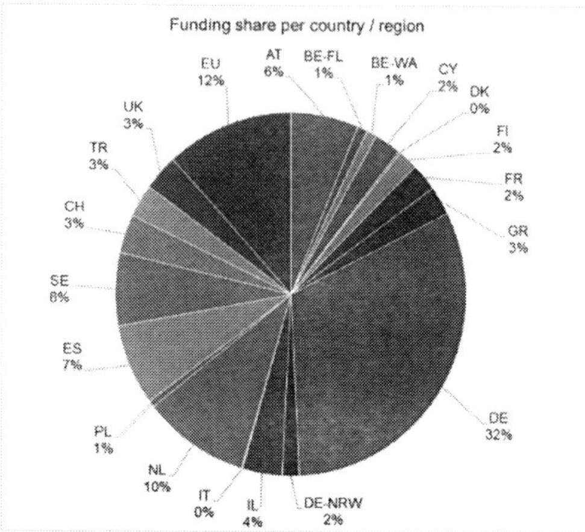

Figure 5: Funding share per country / region for the seven joint calls carried out by SOLAR-ERA.NET.

4 RESULTS AND OUTLOOK

4.1 Selected results

On the level of research, development and innovation, SOLAR-ERA.NET has enabled the support of 88 transnational PV and CSP projects mobilising some 112 million euros and involving some 460 organisations from the scientific community and the industry sector.

44 projects were funded within the first four SOLAR-ERA.NET calls. They have been meanwhile all finalized and can be summarised as follows:

- HESiTSC: exploited the innovative direct III-V/Si heterojunction concept for improving the efficiency and paving the way for an aggressive cost reduction of c-Si based tandem solar cells (SiTSC) at industrial scale.
- Monoscribe: integrated the selective laser ablation processes and printing processes allowing for producing roll-to-roll CIGS PV modules with varying voltages, sizes and shapes.
- CNT-PV: developed press-transferred, semi-transparent, extremely flexible single-wall carbon nanotube films (CNT) functioning as a hybrid hole-transporting and collecting layer in the perovskite solar cell and avoiding expensive hole-transporting materials and silver or gold contacts.
- InnoModu: demonstrated that it is possible to build PV modules with reduced silver content and without using lead.
- NovaZOlar: eliminated Cd-containing buffer layer in CIGS solar cells and achieved a record efficiency of 21,0%.
- NOVACOST: developed eco-friendly metallic salt-based inks suitable for CZTS deposition and building devices with an efficiency of 6,6%, paving the way to 10% efficiencies.
- LIMES: worked on the glass composition and adding optically active components which absorb harmful UV light (at least 4% less) and simultaneously convert those UV photons into visible light (related efficiency

increase of 4%). The optimisation of the glass composition allowed for thinner glass and improvements of the crack resp. chemical resistance by a factor of 3 respectively 4.

- PV me: integrated organic PV modules into steel resp. glass based façades, developed a hard coat that can be roll-to-roll coated on top barrier foils and realised a micro-inverter specifically suitable for organic PV modules. The project ended with one of the largest organic-based BIPV façade installation in Benelux.
- BLACK: demonstrated that high quality black silicon surface is applicable in industrial-scale solar cell processes, namely in passivated and emitter rear cells. Additionally, the nanostructured surfaces eliminated the light-induced power conversion degradation.
- AER II: developed, industrialised and brought to market the aesthetic energy roof (AER) concept. A project realised with this concept won the prize "Most beautiful solar project of the Netherlands".
- PV4FACADES: developed low-cost, high-efficiency c-Si back contact cell technologies (metal wrap through and interdigitated back contact) with improved aesthetics and higher energy yield for industrial take-up.
- InGrid: developed a new module technology for high efficiency rear contact and bifacial solar cells by interconnecting cells with conductive tracks embedded in the encapsulation material, replacing the ribbon typically used.
- PV2GRID: designed a grid side converter improving performance ensuring the proper grid integration of PV systems under any grid conditions.
- INTESEM: implemented a virtual power plant (VPP) to optimise intermittent generation and storage assets and to facilitate efficient distribution of energy. The simulations showed a 10-20% better revenue on the day-ahead market compared to a normal PV site.
- ACCESS-CIGS: explored atmospheric cost competitive elemental Sulpho-Selenisation for CIGS in order to reduce the CIGSSe processing cost and improve the conversion efficiency.
- U-light: developed new light weight, high efficiency and long life-time modules by using thin, strong, low-cost glass and mineral or organic compositions like glass-fibre reinforced plastic.
- HighCast: focused on research and development of advanced casting and wafering techniques for producing silicon wafers suitable for higher efficiency solar cells, using alternative casting techniques comprising "High Performance Multicrystalline Silicon" (HPmcSi) and "Mono-Casting".
- APPI: developed high-efficiency Passivated Emitter and Rear Cell (PERC) solar cells manufactured by using low-cost atmospheric pressure processing, namely advanced texturing, high-efficiency emitter and atmospheric pressure passivation.
- FunGlass: developed chemically functionalised surfaces for glass enabling optimised (anti-reflective, light trapping and light diffusion) glass properties without deposition of additional coatings.
- SPRINTCELL: proposed an innovative thin-film photovoltaic fabrication route offering lower manufacturing costs, using earth abundant photovoltaic materials, namely Cu_2ZnSnS_4, in terms of natural reserve and annual ore production, and selecting technologies for large-scale commercial powder synthesis, advanced powder milling, novel solar ink formulation and solution-based non-vacuum film processing.
- HVolt-PV: designed and developed High Voltage IBC Photovoltaic i-Cells and Modules by integrating several smaller cells onto a single silicon wafer and eliminating line losses, achieving efficiency of >20%.
- PROOF: worked on an industrialised concept for BIPV/BAPV roofs visualising several added values and minimising additional costs.
- EDITOR: evaluated and tested the dispatchability of a parabolic trough collector system with concrete storage, successfully covering solar thermal needs in an industrial context and application.
- SolFieOpt: developed optimization tools, which deliver very cost-efficient heliostat layouts for solar tower power plants.
- DINAMIC: worked on dilute Nitride based concentrator multijunction solar cells, aiming at efficiencies over 47%.
- ALCHEMI: demonstrated a new type of low cost, high concentration photovoltaic (HCPV) module using small III-V multi-junction solar cells, with a DC module efficiency (η) >37%, operating at a concentration factor of ~1000x.
- IPERMON: developed an innovative performance monitoring system including formulation of data acquisition guidelines, development of performance loss, failure and degradation rate detection algorithms, improved forecasting and, finally, a reliable solution for PV monitoring.
- Bifalo: fabricated pPERT cells and respective full size (60 cells) modules with a bifacial factor of 0,9, resulting in a relative increase of bifacial gain – compared to monofacial modules – by 28% and 44% demonstrated for two different Bifalo module designs.
- HESTPV: developed a pure tin / lead-free and mixed tin-lead perovskite based solar cells reaching power conversion efficiency as high as 14 % and a novel measurement platform for lifetime evaluation of perovskite solar cells.
- INFORPV: developed a PV power production forecasting solution with improved accuracy for point and aggregated forecasts at day- and hour-ahead forecasting horizons, with interoperable data-flow modules using an interface with numerical weather predictions and PV system monitoring systems for automated active grid management (alerting and signalling to mitigate power quality issues).
- FrontCIGS: demonstrated a corrosion-resistant transparent contact for flexible CIGS modules, which should enable a less expensive frontsheet.
- HIPER: realized a record four terminal tandem device by combining a 22,7% c-Si solar cell with a 18,0% semi-transparent perovskite solar cell, yielding in 28,0% for the hybrid tandem solar cell, and tested industrially scalable processes resulting in a 22,4% 6 inch hybrid tandem module.
- SITEF: demonstrated the loop scale functionality and applicability of a new silicone heat transfer fluid and associated parabolic trough collector components at temperatures up to 450°C.
- SNOOPI: set up an autonomous and transferable Smart Network Control with Coordinated PV Infeed Box controlling PV and battery inverters so that all

inverters along a feeder provide reactive power in a coordinated way.
- THESEUS worked on the development of novel, very high efficiency tandem flat plate solar cells utilising III-V semiconductors grown onto silicon substrates, using advanced epitaxial techniques to grow SiGe and SiGeSn structures, and their integration into high efficiency modules.
- Refined PV: demonstrated the ability of printing high aspect ratio ultra-fine finger lines and of connecting effectively down to ≤20 mg of Ag on the cell front side.
- BIPVpod: focused on the industrial research and development of improved methods for the production of BIPV modules and optimised the production of customer-specific "Panels on Demand" of a fixed PV size and colouring for small production batches at low additional costs.
- HIPPO: developed large area silicon solar cells with a full area passivated rear contact and wire inter-connected solar modules with bifacial properties.
- Liquid Si 2.0: focused on solution-based silicon deposition and processing starting from hydrogenated polysilanes as a cost efficient alternative.
- SIMON: demonstrated the applicability of new products which have improved viscosity and lower vapor pressure, thus facilitating a substitution scenario for the heat transfer fluid currently used in solar power plants with parabolic troughs in operation
- DURACIS: explored new alternative encapsulation and optical glue materials and concepts, compatible with their implementation into already existing industrial CIGS pilot lines and allowing a significant extension of the lifetime while substantially reducing costs.
- ENHANCE: worked on a novel PV system architecture equipped with kinetic battery storage system, PV inverter with advanced capabilities, distribution network monitoring tool and management system.
- PEARL TF-PV: improved the applicability and understanding of electroluminescence imaging methods to large-scale measurements on thin-film CIGS and CZTS solar cells and modules.

SOLAR-ERA.NET Cofund 1

16 projects are funded within the SOLAR-ERA.NET Cofund 1 call carried out in 2017, namely:

- PANELPV: Sandwich panels with integrated PV with freedom of size and color
- MASTERPV: Innovative manufacturing solutions for cost-efficient semitransparent BIPV
- Pearl: PERC meets self-aligned selective emitter technologies based on inkjet printing and silver less plating
- BI-FACE: High-efficiency bifacial PV Modules and Systems for flat roof applications
- Cover Power: Smart Glass Coatings for Innovative BiPV Solutions
- CEFRABID: Clean energy from road acoustic barriers infrastructure development
- NEXT-FOIL: Next generation conductive solar foil for flexible photovoltaics
- HEAVENLY: High-efficiency PERT and IBC cell development focussing on paste and CVD optimization for long-term stability
- Erigeneia: Enabling rising penetration and added value of photovoltaic generation by implementation of advanced storage systems
- Hyconsys: Hydrogen control in solar thermal parabolic trough heat transfer fluid systems
- ENMESH: ENabling Micro-ConcEntrator PhotovoltaicS with Novel Interconnection MetHods
- PVtool: Development of tools for effective control of large PV power plants
- RHINO: Realization of High efficiency Industrial N-type sOlar cells
- PROGNOSIS: Intra-hour prediction of solar electricity generation from Photovoltaics
- 1500-SiC: Develop a new photovoltaic Inverter with SiC full power operation at 1500V
- NELL: Novel encapsulant for long lifetime high voltage PID-resistant PV modules

These projects are finalised latest by end of 2021 and will be presented by the respective project coordinators during a public webinar on 6 October 2021.

Within the framework of the SOLAR-ERA.NET Cofund 2 cofunded joint call carried out in 2018/2019, 19 projects are running, namely:

- UNIQUE: Carbon Based Perovskite Solar Cells with UNI-Directional Electron Bulk Transport: in the QUEst of a Short Time to Market
- ECOSun: Economic COgeneration by Efficiently COncentrated SUNlight
- PERDRY: Dry production routes for large-area benign metal halide perovskite solar cells
- 1C4PV: One intelligent cloud for PV Assets Diagnosis and Maintenance
- CHEER-UP: Low Cost High EfficiEnt and Reliable Umg Pv cells
- SCALEUP: Large scale molecular simulation of perovskite solar cells
- BOBTANDEM: Band Offset selective Barrier Three Terminal perovskite on silicon high efficiency Tandem Solar Cell
- SUCCESS: Sequential, high Uniformity, Cost Competitive Elemental Selenization and Sulphurization for CIGSSe2
- In4CIS: New in-line optical methodologies for advanced assessment of high efficiency CIGS industrial processes
- TubeMon: New online flux density and temperature measuring systems for MONitoring and optimized operation of external TUBE receivers
- CUSTCO: Cost efficient, upscalable and stable transparent conductive oxides for silicon solar cells based on passivated contacts.
- FUN: Sputtered and otherwise deposited a-Si for Fabricating passivated screen- printed contacts for an indUstrially feasible productioN
- PANAMA: Prescriptive analytics and advanced workforce management for optimized O&M of solar power plants
- AmBiPV: Adapted Modules for Bifacial Photovoltaics
- NFA4R2ROPV: Industrial roll-to-roll (R2R) printing of highly efficient non-fullerene acceptor (NFA)-based organic photovoltaics (OPV)

- PV-ANALYTIC: Advanced photovoltaic system monitoring and analytics solution enhanced with intelligent interoperable data-driven features for efficient big data real-time analysis, failure diagnosis, automated management and integrated micro-grid control
- ROM-PV: Reducing the photovoltaic operation and maintenance (O&M) costs through an advanced online platform
- Prognosis: Diagnosis, maintenance and operation of PV plants
- Nano4CSP: Nanomaterials for reduced maintenance costs in CSP plants

Finally, 9 projects were selected for funding within the SOLAR-ERA.NET Cofund 2 additional joint call (7th call). 5 projects are still in the negotiation and formalisation phase. 4 projects are formalised and running:

- PrOperPhotoMiLe: Towards Prediction of Operational Lifetime of Perovskite Photovoltaics: Acceleration Factors in Stability Study through Machine Learning
- PV40+: PV module with an enhanced lifetime of more than 40 years and reduced environmental impact
- Bussard: Busbarless solar cells with passivated contacts for Next Gen Module Integration
- HIPER XL: HIgh-Efficiency Si PERovskite Tandem Solar Cells eXtra Large

4.2 Conclusions and Outlook

The successful projects demonstrate i) potential commercial impact / relevance to industrial and market needs / contribution to the SET-Plan and added transnational value, ii) scientific and technological excellence and iii) high quality and efficiency of the implementation and the management.

SOLAR-ERA.NET provides added-value and opportunities for the research and industry sector, namely:

- Complementarity between national and EU funded programmes
- Building a more robust and consistent research landscape
- Increasing strengths and reducing weaknesses
- Stepping stone for international cooperation
- Flexibility and bottom-up nature, no "one size fits all" approach
- New routes and innovative transnational research for medium-sized projects

Considering the experience made in seven joint calls, the following success factors can be identified for potential applicants:

- Start with a good idea and have a vision how to realize it
- Build strong, complementary and trustful partnerships
- Communicate with partners and funding agencies
- Respect formal requirements / conditions
- Avoid redundancies with existing projects / proposals
- Identify innovation – excellence
- Head for industrial orientation – impact
- Provide convincing project plan – implementation
- Secure a high transnational value
- SMART Goals (Specific, Measurable, Ambitious, Realistic, Time Bound) / Key Performance Indicators

SOLAR-ERA.NET not only allows for supporting innovative research projects on a transnational level but also provides added value in terms of networking and programming, namely:

- New industry-led innovation projects and partnerships in a strongly competitive environment
- More transnational cooperation bottom-up and "cut to measure"
- Larger and more diverse portfolio
- Common initiatives and implementation activities
- Networking and collaboration between countries across Europe
- Overview on European research and insight into trends
- Best practice for supporting projects on the transnational level

Yet, there is a number of challenges respectively opportunities for further improvements related to public-public-partnerships in the future Horizon Europe programme.

- Adequacy and adaptability of European, national and regional funding budgets (Funding needs identified in the SET-Plan are considerably higher than funding budgets actually made available.)
- Flexibility to overcome the "funding gap" by following the ranking list with individual funding budgets running out already further up in the ranking list
- Consideration and inclusion of less strong ERA countries
- Visibility of impact through transnational projects and related projects
- Targeted and fair redistribution / use of Top-Up (Ideally, all funding agencies and transnational projects get the same share of the Top-Up.)
- Minimum extra administrative workload for agencies and project partners on transnational level
- Synchronisation of the timeline with all funding agencies (especially in the negotiation and contractual phase)
- Pragmatic handling (e.g. making reserve list possible)

Summing up, SOLAR-ERA.NET has become a key instrument for cooperation among research and innovation agencies in Europe.

The coordinator of a finalised transnational project summarised the role of SOLAR-ERA.NET as follows: "SOLAR-ERA.NET allows for real added value, namely transnational cooperation with complementary skills, equipment and facilities, along the value chain from research to industry and better reaching out to the different markets in Europe and beyond."

Next upcoming SOLAR-ERA.NET activities planned are i) the launch of the 8th joint call and ii) the webinar "Exchange of experiences" (with presentations on the outcomes of transnational projects supported within the SOLAR-ERA.NET framework) on 6 October 2021. In the context of the future Horizon Europe, new options are discussed – notably the Clean Energy Transition Partnership (CETP) – for public-public partnerships bringing together national and regional funding organisations and RTD and innovation programmes in the thematically larger field of renewable energy including photovoltaics.

PV is in the pole position to become a backbone of the future clean energy system. Yet new bold initiatives are needed, taking advantage of the wide experience and good practice gained over the past decade, keeping the momentum and accelerating the efforts.

REFERENCES

Further information on running projects from the transnational joint calls as well as news on ongoing and future SOLAR-ERA.NET activities are channelled through the www.solar-era.net website and the websites of the funding agencies involved.

ACKNOWLEDGEMENTS

The support of the SOLAR-ERA.NET joint activities by the EC under grant agreement N° 321571 within the 7th Framework Programme, SOLAR-ERA.NET Cofund under grant agreement N° 691664 and SOLAR-ERA.NET Cofund 2 under grant agreement N° 786483 within Horizon 2020 is gratefully acknowledged.

IMPACT OF PUBLIC AND PRIVATE FUNDING ON THE DEVELOPMENT OF THE PHOTOVOLTAIC SECTOR AND THE ACHIEVEMENT OF 2030 ENERGY TRANSITION TARGETS

D. Moser[1]*, F. De Nigri[1], S. Pezzutto[1], S. Gantioler[1], M. Aleman[2], G. Masson[2]
[1]Institute for Renewable Energy, EURAC Research, Viale Druso 1, 39100 Bolzano, Italy
[2]Becquerel Institute, Rue Royale 146, 1000 Brussels, Belgium
*Phone: +39 0471 055 627; E-mail: david.moser@eurac.edu,

ABSTRACT: This paper is part of the PV IMPACT project funded by the EU H2020 program. The main objective is to shed light on the research and development funding for photovoltaics (PV) by analysing funding from public (European Commission and national budgets) and private sources so to get detailed understanding of past R&I expenditures for the PV sector and related research topics, current level and potential future trends. The countries under scope are the EU27+UK, Norway and Turkey. Climate and energy related targets should be met by 2030, so keeping track of investments in a field that enables technology innovation is an important effort as PV will play a major role in the coming decade. The CORDIS database provides ample data on H2020 projects, allowing the compilation of relevant information. As of December 2020, a total of ~360 M EUR have been distributed for PV R&D since 2014. A large amount of these funds (~194 M EUR, equivalent to 54% of the whole funds) are destined for the sole purpose of improving PV module efficiency. At Member States level, leveraging on the availability of national databases, it is possible to define on which areas of the SET-PLAN each Member State is investing most.
Keywords: see the list of keywords

1 INTRODUCTION

Public funding is crucial in bridging the gap from theory to reality and in attracting private investments by de-risking low TRL technologies. It took PV a long time before becoming commercially viable. The cumulative production of solar PV modules started to become relevant in volume only in the 2000s, so until that point the price drop per Watt peak (Wp) had been mainly the result of R&D breakthroughs. Once mass manufacturing began, economic factors (like economy of scale) accelerated that price drop. The importance of R&D towards the effective deployment of RES in the European Union (EU) is highlighted in multiple publications like the Energy Roadmap 2050, the European Strategic Energy Plan, or the EU Energy Security Strategy. In general, many studies can be found that support R&D as a crucial element towards generating innovation and productivity in any type of field. In this work we have answered to the following relevant questions:

- What is the amount of funding that has been allocated through the years to R&D for PV in Europe?
- What is the incidence of private investments versus public investments in Europe?
- What is the relationship between the knowledge stock and technical KPIs?
- Which are the main areas of interest for the allocation of funding?
- How are the different Member States investing in PV R&D?

2 METHODOLOGY

2.1 Methodology top down

It was found that from all the available current funding schemes and programs from the EC (Figure 1), approximately 80-90% of the financing came from the Horizon 2020 (H2020) framework program (FP). To obtain the necessary inputs, the Community Research and Development Information Service (CORDIS[1]) database has been utilized as the main source. This funding source

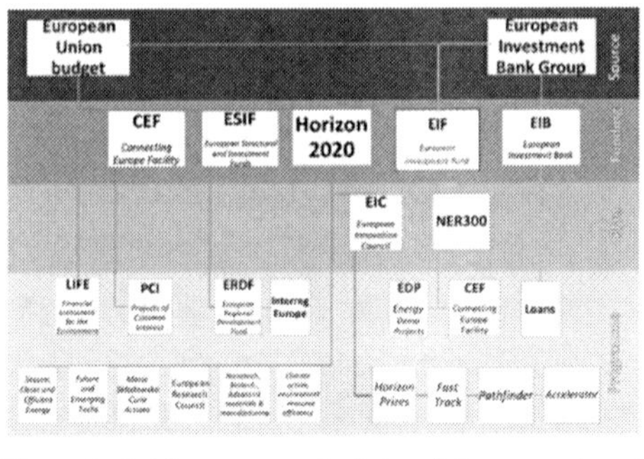

Figure 1: Public sources of photovoltaics research and development funding in the European Union (2019). The last level "Programme", indicates the name of the fund providing the financing resources which can be in the form of grants or loans

is not the only one coming out of the EC, but it covers the highest share in resources. Since these values were given in nominal terms, they were adjusted to real terms to show consistency throughout the timeframe. To do so, time-based EU27+UK annual inflation rates determined by the International Monetary Fund has been utilized, using 2018 as the reference year. The CORDIS database does not provide information by type of technology involved, so a selection of relevant projects was done. After a first screening that considered specific keywords from the project's abstracts (which described their activities and goals), a selection of PV related projects was produced. Following this, a secondary screening was performed to divide these projects into seven groups according to the type of activity carried out and based upon the classification by the Strategic Energy Technology (SET)-Plan Implementation Plan[2]. For information regarding funding at national level (MS, Norway and Turkey), the International Energy Agency's (IEA) data service was used[3].

Private PV R&D global data is scarce. Consequently, a different approach to the public sector's one had to be

[1] https://cordis.europa.eu/

[2] https://www.iwg-pv.eu/

[3]https://www.iea.org/data-and-statistics/data-product/energy-technology-rd-and-d-budget-database-2

taken. The Organization for Economic Co-operation and Development (OECD) defines three levels for the PV supply chain: upstream/raw materials processing, midstream/manufacturing, and downstream/power generation. An analysis of the activities from PV projects under H2020 shows that most funding goes towards improving module efficiency. The most probable source of PV R&D might thus come from PV panels manufacturing companies in the midstream stage. Thus, a list of the key European market players was composed and examined. The fact that the absolute majority of these companies are in private hands is very important, since private companies (in contrast to the publicly listed) have no legal requirements to provide financial reports to government agencies and shareholders. But even though some companies provide R&D information, this is either incomplete or does not go into specifics. Therefore, a combination of top-down and bottom-up approach was chosen. For the top-down, EurObserv'ER information (a project that provides a barometer that monitors the total turnover of PV companies in Europe) was combined with an average of R&D/turnover ratio from the few publicly listed companies available (bottom-up). Finally, a cross-check was undertaken with regard to studies looking into private R&D investments according to different renewable energy technologies, including PV, for a rough validation of the produced numbers.
Finally, in De Negri et al [1] the knowledge stock method was used as a way to reach an approximate economic value of the accumulated know-how of PV. This method provides a solution to the high fluctuating nature of R&D funding and takes into consideration the depreciating value of knowledge across time as it tends to be replaced by new one

$$KS_{R\&D}=(1+\rho)(1-\delta)\times KS_{(1-t)}+R\&D_{(t-x)} \quad [EUR]\ (1)$$

ρ: Spillover, δ: Depreciation, t: Time period, X: Time lag. The first term, spillover, means there is a positive external effect from indirect sources of research. The second term, depreciation, means knowledge tends to become useless through time. Finally, the third variable is the time lag, (X). This determines the time needed for a research project to produce a scientific result.

2.2 Methodology bottom-up

The database for the "bottom-up" data collection on R&I projects was created on the PV Impact website (www.pvimpact.eu[4]). A request for reporting projects was disseminated throughout the network of PV Impact partners, as a result several stakeholders did report their projects into the database. Furthermore, some PV Impact partners have also directly reported information on nationally funded projects in their countries based on information retrieved from accessible existing national databases (e.g., Photowatt for France and EURAC for Italy). Transnational projects from Solar ERA Net website were also added to the database by partners (Photowatt, BI). In total, more than 350 nationally funded projects for the period 2014-2020 were collected through this approach in close collaboration with the SET-PLAN PV-IWG.

[4]https://pvimpact.eu/towards-set-plan-targets-ip-progress/pv-projects-database/

3 RESULTS AND DISCUSSION

The CORDIS database provides ample data on H2020 projects, allowing the compilation of relevant information. As of December 2020, a total of ~360 M EUR have been distributed for PV R&D since 2014. These funds were assigned to seven categories (Building-integrated PV, cross-sectoral research at lower technology readiness level, development of PV power plants and diagnostics, manufacturing technologies, new technologies and materials, others, and technologies for silicon solar cells and modules with higher quality) according to the classification proposed by the SET-Plan PV Implementation Plan defined in 2017 (Figure 2). A large amount of these funds (~194 M EUR, equivalent to 54% of the whole funds) are destined for the sole purpose of improving PV module efficiency.

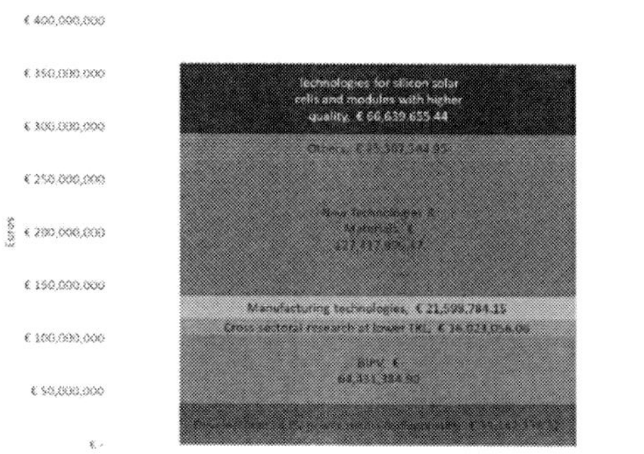

Figure 2: Accumulated photovoltaics research and development funding from Horizon 2020 (2014-2020) grouped into seven categories by activity type as per the PV Implementation Plan (TRL stands for technology readiness level).

To answer one of the research questions, a comparison between total public and private funding was done. The results, which indicate a predominance of the private sector funding, ranging roughly between 60–70% (Figure 3). This is in line with what was presented by other sources, which provided similar ranges concerning RES in general, or for PV but at a world scale. Finally, large corporations, as well as small and medium businesses, depend on contributions made in the form of public grants or loans for PV R&D. Any contributions that come at a low cost will be most welcome as these companies need to keep producing innovative goods and services but at the same time spend the lowest possible in doing so. Smaller companies will especially seek funding from a diversity of sources. This normally also involves private capital investors that take on substantial financial risks. Public offerings and bank loans are more uncommon because they are in general more costly. The way companies finance their operations is varied and might give further indications for R&D investing performance.

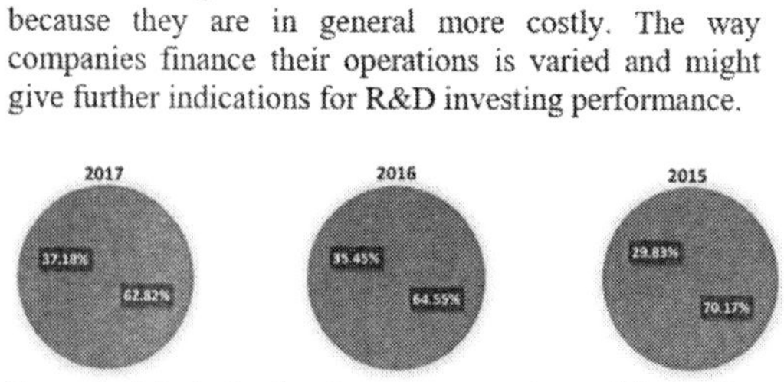

Figure 3: Private funding (green) compared to public funding (blue) by year

Based on Equation (1), only data of public PV R&D funding was used (from CORDIS and IEA), because it has proven difficult to construct an equivalent dataset for the entire EU private sector. Compiling all the data, the result is influenced by the predominance of national funding. By 2030 the total projected value exceeds 4000 M EUR. A comparison between the Knowledge Stock (KS) (both sources combined) and the production price of solar PV modules is proposed (Figure 4). A very high negative correlation of −0.84 was the result, indicating a very close relationship between investment in R&D and a drop in PV costs per Watt-peak (Wp). This does not necessarily imply that PV R&D is the main reason for cheaper production costs, rather, it provides an argument for increased levels of funding to sustain this trend, given also constant innovation can be assumed to be beneficial for any technological deployment.

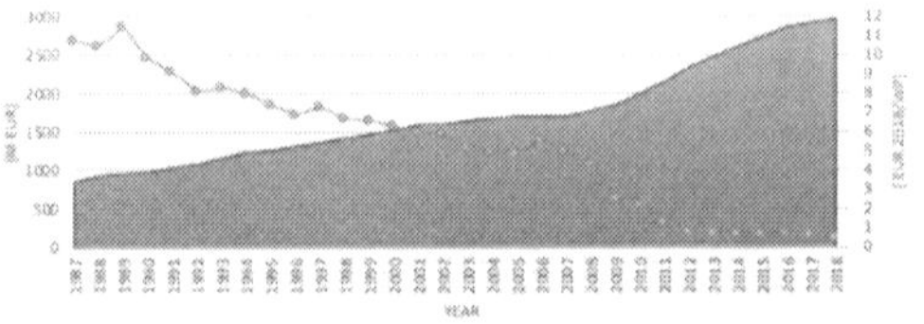

Figure 4: Left axis, blue area: Knowledge stock calculations for photovoltaics research and development originated from investment by the European Union member states, Norway, Turkey, and the European Union in M EUR (2018 prices and exchange rates). Right axis, orange line: solar module price, inflation-adjusted EUR/Watt-peak (Wp) (2018 prices and exchange rates)

Regarding the bottom-up approach in project collection using the PV Impact database and the collaboration with the IWG, this activity has led so far to a total of 350 projects. H2020 projects were not included as the intention is to focus on Member States funded projects (e.g., via internal funding or ERDF funding) and transnational projects (e.g., CO-FUND actions and INTERREG). The projects were included in a common excel database where the most important information used for the analysis are:

- Funding type
- Member state
- Public funding
- Private funding
- Activity related to PV IP
- PV share of R&D activities
- TRL

The following issues have been identified in the database:

i) For some projects, the R&D activities are not entirely focus on PV R&D and defining a % of share of overall budget is not always easy. We recommend at least in CORDIS for H2020 projects and in the future for Horizon Europe projects to ask the coordinators to provide a rough division for technologies
ii) Share of private funding is not always available
iii) Some projects cover various activities of the PV Implementation Plan and in case of lack of specific information we have assumed that the budget is equally spread between activities.

Based on the available information from the database is possible to start the mapping activity of funding at Member State level.

Analysis that can be carried out (depending on data availability) are:

- Mapping TOPIC of preference from the Implementation Plan at member state level (for total, public and private funding)
- Mapping overall budget per country for R&D&I PV sector (normalised pro capita)
- Mapping % capture (Defined as funding in the database compared to overall budget for R&D&I for PV in a specific country)

The overall budget for some Member States was taken from the IEA public spending database.

Table 1 shows the status of the database as for 31 August 2021 with the sum of projects budget at country level and assigned to the various activities of the Implementation Plan. For each country we have identified which activity has the most prominent budget level and the share over the overall budget. For example, for Germany the overall budget is 442.2 M€, 5.3 €/person. The country with the highest captured budget per capita is Germany with 5.3 €/person, followed by Cyprus and Norway with 4.6 and 4 €/person, respectively.

The first check to be carried out is to try to understand the share of the overall R&D&I which is captured by the database so far. For information regarding funding at national level (MS, Norway and Turkey), the International Energy Agency's (IEA) data service was used. From their information service hub, the source used was the "Detailed country RD&D budgets" report that provides R&D expenditures by country at 2019 prices and exchange rates (meaning no adjustments needed to be made for inflation). The data is provided individually, therefore it had to be summed up to reach a regional total segmented by year. In doing so, the only category considered was "312 Photovoltaics".
Table 2 shows the results in terms of % of captured budget by PV IMPACT compared to the overall R&D&I budget for the PV. Countries where information is available per year along the period 2014-2019 are highlighted in green. The other countries included in the IEA database do not present sufficient data (highlighted in yellow and red); some countries in Table 1 are not included in the IEA database. For green labelled countries we have for example Germany with a % of captured budget of 70%, followed by Norway with around 17%, and France with 7%. This is not surprisingly as for Germany, the database ENARGUS[5] was used to gather information about projects funded by Germany. Similar activity was done for the Netherlands using the database https://projecten.topsectorenergie.nl/projecten. Unfortunately for the Netherlands, the information available in the IEA database is limited and no considerations can be done on the % of captured budget.

Figure 5, and Figure 6 map the values included in Table 1. Figure 5 shows the per capita expenditure per country, Figure 6 show the TOP TOPIC of the Implementation Plan per Country where the shade of colours indicates the share of the TOPIC.

[5] https://www.enargus.de/

4 CONCLUSIONS

The analysis of this work built on the results presented in [1] by updating the figures to 2020. The analysis also includes the first analysis from the bottom up approach which allows the collection of information at Member States level to map the PV R&D focus for the seven considered categories of the PV Implementation Plan using the PV IMPACT project database. We acknowledge that the analysis of the overall budgets and potentially also their distribution, is influenced by the differences in the accessibility of R&D spending data between the Member States. Efforts are still ongoing to gather as much information about national R&D funding as possible to provide a more precise picture about the status.

ACKNOWLEDGEMENTS

The Authors acknowledge the funding received by the PV IMPACT. This project has received funding from the European Union's Horizon 2020 research and innovation programme under grant agreement No 842547. David Moser and Monica Aleman would like to thank the successful collaboration with the Secretation and the Chairs of the IWG.

REFERENCES

[1] J. F. De Negri, S. Pezzutto, S. Gantioler, D. Moser, and W. Sparber, "A Comprehensive Analysis of Public and Private Funding for Photovoltaics Research and Development in the European Union, Norway, and Turkey," *Energies*, vol. 13, no. 11, Art. no. 11, Jan. 2020, doi: 10.3390/en13112743.

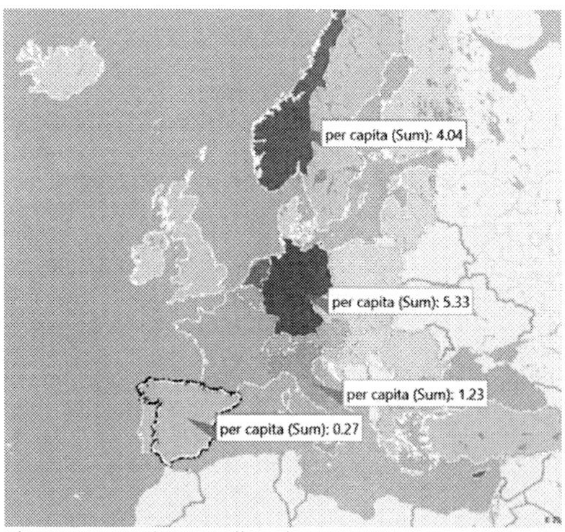

Figure 5: Per capita expenditure in PV R&D

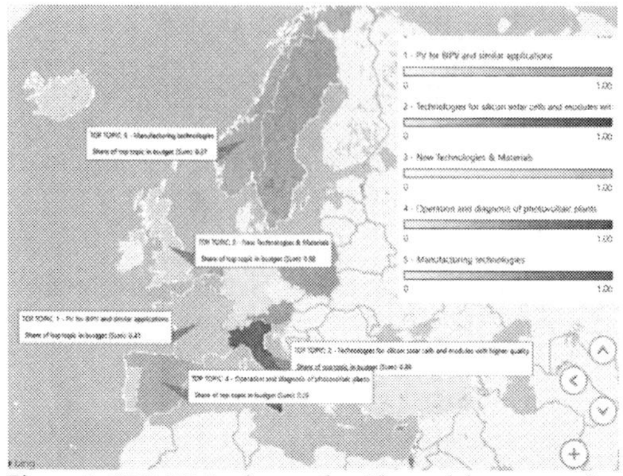

Figure 6: TOP PV Implementation Plan activity per member state. The colour intensity shows the share of the specific topic of the overall PV R&D budget

Table 1: Summary of total PV R&D budget from the PV IMPACT database and contribution towards each topic of the SET-PLAN PV Implementation Plan

Country	TOTAL	Pro capita	Population	PUBLIC	PRIVATE	IP activity 1	2	3	4	5	6	SUM	MAX	TOP TOPIC	Share in budget
	Meuros	Euros/person	Million people	Meuros	Meuros	Meuros	Meuros	Meuros	Meuros	Meuros	Meuros	Meuros	Meuros		%
Austria	4.6	0.5	8.9	3.1	1.6	0.5	0.0	1.7	2.2	0.3	0.0	4.6	2.2	4	47%
Belgium	8.3	0.7	11.5	5.8	0.5	4.2	0.0	1.6	1.9	0.4	0.0	8.3	4.2	1	50%
Bulgaria	0.2	0.0	7.0	0.2	0.0	0.2	0.0	0.0	0.0	0.0	0.0	0.2	0.2	1	100%
Croatia															
Cyprus	4.1	4.6	0.9	3.9	0.2	0.6	0.0	1.3	2.1	0.0	0.0	4.1	2.1	4	51%
Czech Republic															
Denmark															
Estonia															
Finland															
France	51.2	0.8	67.0	28.2	26.0	20.8	1.7	5.6	7.0	15.5	3.6	51.2	20.8	1	41%
Germany	442.2	5.5	83.0	321.8	120.6	66.6	120.8	166.4	55.0	29.5	6.6	442.2	166.4	3	38%
Greece	0.5	0.0	10.7	0.5	0.0	0.5	0.0	0.0	0.0	0.0	0.0	0.5	0.5	1	100%
Hungary															
Ireland															
Italy	74.5	1.2	60.4	23.0	51.5	3.1	66.0	3.3	1.1	0.5	0.6	74.5	66.0	2	89%
Latvia															
Lithuania															
Luxembourg															
Malta															
Netherlands	56.0	3.2	17.3	42.2	1.9	10.7	17.2	16.3	7.1	4.3	0.0	55.6	17.2	2	31%
Poland	0.2	0.0	38.0	0.1	0.0	0.0	0.0	0.0	0.0	0.0	0.0	0.2	0.2	1	100%
Portugal	0.5	0.0	10.3	0.5	0.0	0.0	0.1	0.3	0.0	0.1	0.0	0.5	0.3	3	64%
Romania															
Slovakia															
Slovenia	0.1	0.0	2.1	0.1	0.0	0.0	0.0	0.1	0.0	0.0	0.0	0.1	0.1	3	100%
Spain	12.4	0.3	46.9	9.7	2.6	1.7	2.7	3.5	3.7	1.7	0.2	12.4	3.7	4	29%
Sweden	1.4	0.1	10.2	1.3	0.2	0.0	0.0	0.5	0.3	0.6	0.0	1.4	0.6	5	44%
UK	1.9	0.0	66.7	1.0	0.9	0.0	0.2	1.1	0.1	0.0	0.0	1.9	1.1	3	58%
Iceland															
Liechtenstein															
Norway	21.0	4.0	5.3	9.6	11.5	0.8	6.4	5.8	4.2	7.7	0.0	21.0	7.7	5	37%
Turkey	7.6	0.1	82.0	7.3	0.3	0.5	2.5	2.8	2.6	0.3	0.1	7.6	2.8	3	37%
Switzerland	4.7	0.5	8.5	1.8	2.9	1.2	1.6	1.9	0.0	0.0	0.0	4.7	1.9	3	41%
SUM	691.8	1.5	546.8	406.3	221.8	89.5	227.6	173.7	89.0	109.9	7.5	691.0	227.6	2	33%
Israel	2.1	0.2	9.1	1.3	1.0	0.0	0.5	0.3	1.5	0.0	0.0	2.1	1.5	4	69%
North Macedonia	0.1	0.0	2.1	0.1	0.0	0.1	0.0	0.0	0.0	0.0	0.0	0.1	0.1	1	100%
SUM	693.9	1.2	558.0	406.6	222.8	89.7	228.1	173.9	90.5	109.3	7.6	693.1	228.1	2	33%

Table 2: % of captured budget by the PV Impact database compared to the IEA total public PV R&D. In green countries with data available for 2014-2019, in yellow where 1-3 years are missing, in red where 4-6 years are missing.

IEA 2014-2019	% captured
Austria	5.9%
Belgium	26.1%
Czech Republic	
Denmark	
Estonia	
Finland	
France	7.3%
Germany	70.3%
Greece	
Hungary	
Ireland	
Italy	
Luxembourg	
Netherlands	207.4%
Poland	
Portugal	65.5%
Slovak Republic	
Spain	489.6%
Sweden	3.4%
United Kingdon	8.0%
Norway	16.9%
Turkey	272.9%
Switzerland	0.9%

EVALUATING THE IMPACT OF CLIMATE CHANGE ON THE TECHNO-ECONOMIC PERFORMANCE OF BUILDING INTEGRATED PHOTOVOLTAIC SYSTEMS IN PROTECTED CONTEXTS

Martina Pelle[1,2,*], Mattia Dallapiccola[1], Anita Tatti[2], Francesco Causone[2], Laura Maturi[1], David Moser[1]
[1] Institute for Renewable Energy, EURAC Research, Viale Druso 1, 39100 Bolzano, Italy
[2]Dipartimento di energia, Politecnico di Milano, Via Lambruschini, 4, 20156 Milano, Italy
* Phone: +39 0471 055 730; E-mail: martina.pelle@eurac.edu
E-mail: mattia.dallapiccola@eurac.edu, anita.tatti@polimi.it, francesco.causone@polimi.it, laura.maturi@eurac.edu e david.moser@eurac.edu

ABSTRACT: BIPV (building-integrated photovoltaics) technologies may play a key role in the transition towards low-carbon energy communities. A relevant aspect to consider to correctly assessing the techno-economic and environmental performance of BIPV applications is the dynamic variation of the climate boundaries conditions during the entire systems' lifespan. This study provides an innovative approach to optimize BIPV applications in the early design phase. By using a real case study, that consists of four residential buildings, located in Milan, Italy, including about 142 residential units, simulations have been run with two different weather datasets, one representative of the current weather conditions and one representative of their projection in 2050, to investigate the possible impacts of climate change on some relevant key performance indicators (KPIs), with a particular focus on the minimization of the levelized cost of electricity (LCOE). This approach is meant to foster a more accurate design of BIPV systems in urban contexts. Results show that the LCOE decreases by almost 50% from 2020 to 2050 optimizations, and the self-sufficiency target is guaranteed with a smaller installed PV capacity.
Keywords: Design, architectural and landscape aspects, colored BIPV, climate change, heritages.

1 INTRODUCTION

In light of the latest European Directives on renewable energy source (RES) [1], BIPV (building-integrated photovoltaics) technologies may play a key role in the transition towards low-carbon energy communities, since they incorporate RES in buildings serving at the same time as envelope components and active power generation elements, fostering a better matching between the buildings' energy demand and on-site energy production. The correct assessment of the techno-economic performance of BIPV systems is crucial to foster their optimal penetration in the built environment. In this context, evaluating the economic profitability of their installation is crucial. Nevertheless, BIPV simulation and optimization tools usually do not assess the techno-economic performance of the system for different climate boundary conditions during its entire lifespan, and this could lead to significant errors in the plant design and incorrect evaluation of the system's economic profitability. Therefore, this study provides an innovative approach to optimize BIPV applications in the early design phase by considering different climate conditions, evaluating the impact of climate change on some relevant key performance indicators (KPIs) for BIPV installation, to foster a more accurate design of these systems in urban contexts, where heritage or protected landscape are located, as in the case of the proposed case study. In this work, the chosen optimization function has been minimization of the levelized cost of electricity (LCOE, €/kWh) constrained to self-sufficiency (SS) greater than 55%. This choice allows assessing the economic performance of the BIPV systems while targeting environmental goals set by the European Directives. The output of the optimization is the configuration of the system that guarantees the minimum LCOE and respects the constraint related to self-sufficiency. In the following paragraphs, we present the case study, the methodological approach, the input required to run the optimizations, and the results. Furthermore, two additional simulations have been conducted to understand how a system designed for operation in current weather would perform under different weather conditions. These simulations took as input the optimal BIPV configuration resulted for the optimizations with 2020 weather file and simulated by using the weather projection in 2050. This approach would enable more accurate design and, consequently, more cost-effective BIPV applications. Furthermore, it fosters the design of more resilient integrated building-plant systems, better tackling European energy efficiency and renewable energy share targets [2, 3].

2 CASE STUDY

The case study is a large public residential estate located in the south-eastern suburban area of Milan, Italy. It is composed of four-stories buildings, each built in the '80s, with a gross floor area of about 10300 m^2, accounting for 142 residential units and about 450 inhabitants. Since 2018 the buildings have undergone a deep energy renovation focused on the substantial reduction of energy demand. The buildings' renovation project has been designed to comply with current national and regional regulations on energy performance and with those related to heritages or protected landscapes since the buildings are located in a heritage-protected district, where architectural constraints exist due to the presence of a medieval Abbey. For this reason, design criteria have included limitations and restrictions on the use of specific colors for the envelope material/elements. These constraints hinder the installation of standard high-performance PV modules, whose usual dark blue appearance cannot comply with the heritage's preservation criteria in terms of defined acceptable colors and materials. Therefore, this analysis has been conducted by using specific techno-economic parameters of hidden colored BIPV modules, that hide completely the PV cell to the human eye, under a uniform-colored layer. The chosen colors result in compliance with the specific regulation on heritages or protected landscapes in the analyzed area.

3 METHODOLOGY

The use of future weather projections to generate data sets to support the design of a more resilient building envelope is extensively been reported in the literature [4, 5]. This approach helps to tackle the energy efficiency target set by the EU Directives, by minimizing the effect of climate change on the buildings' energy efficiency. On the contrary, few studies use this methodology to address an effective penetration of RES in buildings, enhancing a more accurate design of the integrated building-plants system. The core of the calculations of the current work is an optimization tool described in detail in [6] which is able, from simplified inputs, to optimize the capacity of the photovoltaic system, module position, and battery capacity. The optimization tool takes as inputs the weather file, the hourly electric demand profiles, the 3D model of the buildings and the surrounding context and the available surfaces eligible for installation of photovoltaic modules and runs a ray-tracing simulation based on RADIANCE [7] to obtain the hourly irradiance profile for all the possible module position. Then, the system is optimized iteratively changing the number and the position of modules and the battery capacity. At every iteration, the performance of the system is evaluated in terms of a selected objective function. At the end of the optimization process, the tool provides the final configuration of the system and a set of indicators useful to evaluate the analysis. Different target functions can be chosen for the system optimization: in this work, the chosen optimization function has been minimization of the levelized cost of electricity (LCOE), as calculated in [8], expressed as €/kWh constrained to self-sufficiency (SS) greater than 55% . The output of the optimization is the configuration of the system that guarantees the minimum LCOE and respects the constraint related to self-sufficiency. The threshold for self-sufficiency has been set at 55% after the prescription of the Piano Nazionale Integrato per l'Energia ed il Clima (PNIEC) [9], the national energy plan that considers 55% as the 2030 target percentage for electricity generation from renewable energy sources in Italy. Assuming that the objective is to generate the electricity that is consumed, imposing a minimum of 55% of self-sufficiency to buildings using BIPV systems, it is equivalent to say that at least the 55% of the power generation of the building is generated from RES, thus being compliant to the PNIEC prescription. The optimizations have been performed by using two different weather scenarios, one representative of the current weather conditions (year 2020) and one corresponding to a future projection to 2050. More details about the two weather datasets are provided in section 3.1. Two different configurations for the electric demand profiles have been considered and optimized in both the weather scenarios: (i) scenario EL: the BIPV system serves the building only for the hourly electric demand of lights and electric appliances (ii) scenario EL + HP: the BIPV system serves the building for both the hourly electric load of lights and electric appliances and the hourly electric consumption of the heat pump used for space heating (SH), cooling (SC) and domestic hot water (DHW) production. This led to four different optimizations, summarized in Table 1. The results of these optimizations have been presented in section 4. Furthermore, to understand how a system designed for operation in current weather would perform under different weather conditions, two additional simulations

Table 1. Optimizations and simulations details: ID, load profile for lights and appliances (EL) and the heat pump (HP) for space heating (SH) cooling (SC) and domestic hot water (DHW), and optimization target.

ID	Load profile	Optimization target
2020 EL	Lights and appl.	min LCOE, min SS=55%
2050 EL	Lights and appl.	min LCOE, min SS=55%
2020 EL+HP	Lights and appl. + heat pump (SH, SC, DHW)	min LCOE, min SS=55%
2050 EL+HP	Lights and appl. + heat pump (SH, SC, DHW)	min LCOE, min SS=55%
2050 EL (PVS conf 2020)	Lights and appl.	-
2050 EL+HP (PVS conf 2020)	Lights and appl. + heat pump (SH, SC, DHW)	-

have been conducted. These simulations took as input the optimal BIPV configuration resulted from the optimizations done with 2020 weather conditions and simulated them by using the weather projection in 2050. This procedure has been applied to both the load profile scenario, EL and EL+HP. The results of the simulations have been compared to the result of the optimizations in 2050, and they are presented in section 4.

3.1 Inputs and objective functions

The following inputs are required to run the optimization of the system:

(i) Measuring geometry is the surface that is considered available for the installation of modules. The area is then automatically divided into modules and for all modules, a sensor is placed in the centroid for irradiance calculation with the software RADIANCE.

(ii) Context geometry represents the buildings under analysis, the ground, and all the surrounding objects that could shade the surfaces under analysis. The material assigned to the surrounding objects has an impact on the reflectance and transmittance of the surfaces.

(iii) Weather data sets. To simulate the effect of climate change on the analyzed BIPV plant, the software CCWorldWeatherGen has been used [10]. Starting from historical weather data from the PVGIS database [11], climate projections have been made through morphing and stretching methods, based on IPCC3 [12], to generate future weather data sets for the years 2050 (Figure 1). These weather data set have been used to optimize the BIPV plants, to highlights how the optimal techno-economic configuration could change by changing boundary conditions during the system lifespan. The weather file available on the PVGIS TMY tool, for the city of Milan, has been used as a representative of the current weather scenario.

(iv) Demand profiles. In this work, the profile of occupant-related electric energy used by appliances and lights has been generated through stochastic schedules, which have been created using a data-driven procedure, developed in [13]. The schedules have been set up starting from metered data of the residential buildings under analysis, using machine learning techniques (i.e., Self-organizing map coupled with k-means algorithm). The procedure results in the identification of five clusters of the hourly probability of the electric energy used by lights and appliances. A cluster consists of a group of users with similar daily hourly profiles. The five clusters have been then multiplied for each apartment by an

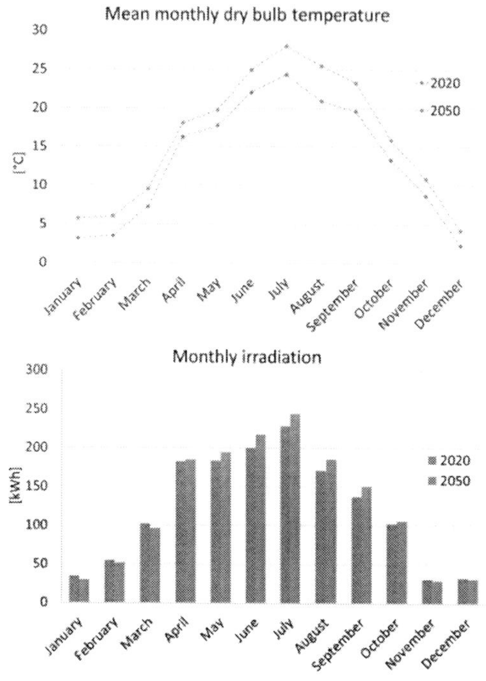

Figure 1. From the top: mean monthly dry bulb temperatures and monthly irradiation for 2020 and 2050 weather files.

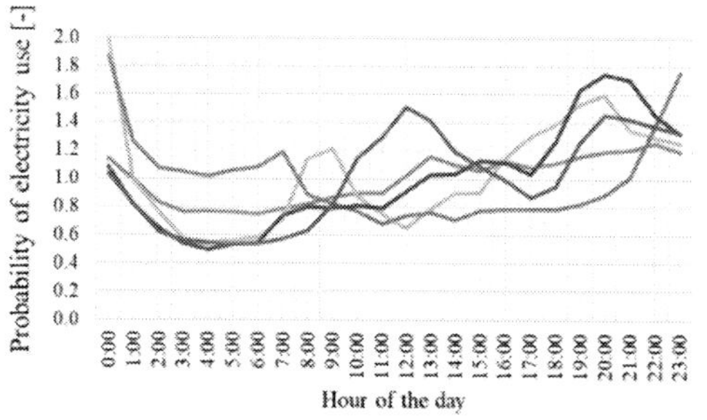

Figure 2. The five clusters of hourly probability of the electric energy used by lights and appliances.

average value of electric energy use derived from the same database used for the generation of the profile.

These profiles have been randomly combined to build up the yearly hourly schedule for the district, by considering different user typologies (low or medium electric energy users, corresponding to the first and second clusters' quartiles). For the electric energy use for space heating, cooling, and DHW, following the approach adopted in [8] and considering the present and foreseen future diffusion of heat pumps, it has been assumed that the thermal demand of the buildings is completely covered by heat pumps, both for the heating and the cooling season. For the calculation of the electric consumption profile of the heat pumps, the EDGE-HP (Electric Demand GEnerator for Heat Pumps) tool explained in detail in [8] has been used. The tool takes as input the yearly cumulative thermal demand for heating, cooling, and DHW preparation of the district (sum of the demand of the single buildings). The yearly cumulative demand has been calculated as the energy needs of the four buildings after deep retrofit design. Secondly, the tool calculates the dimensionless hourly thermal profiles which depend on the hour of the day, the day of the week, and the outdoor temperature according to the relationships obtained from synthetic load profiles calculated in a previous study. For the details of the calculations, we refer to the project report [14]. Then, the dimensionless thermal demand is rescaled for matching the yearly cumulative thermal demands previously calculated and finally, the profiles have been divided by an external temperature-dependent coefficient of performance - COP (heating) and energy efficiency ratio - EER (cooling) obtained from the performance map of commercially available air to water heat pumps. The total electric consumption profile is finally obtained summing the contribution related to space heating, cooling, and DHW. To obtain the two consumption profiles required for the optimization and simulation of the 2020 and 2050 conditions, the procedure was repeated using both the 2020 and 2050 weather files and yearly cumulative demand data.

(v) Techno-economic parameters. The parameters used in the analysis include components costs, cost of electricity, maintenance costs, degradation of module efficiency, the performance ratio of the system, etc. The system is then optimized to maximize or minimize a selected objective function. Hereby, the results of the minimization of LCOE constrained to self-sufficiency greater than 55 % are reported. In other words, at each iteration, the system is simulated over the entire lifetime and the LCOE is calculated. The optimal configuration selected is the one that guarantees the minimum LCOE and respects the constraint related to self-sufficiency. The cost for the BIPV system has been set as 400 €/m^2 for 2020 [15], and it refers to the cost of the BIPV envelope solution as a whole, including BIPV modules, mounting system, electrical components, and installation. The chosen modules are of the type hidden colored, with front glass colored with terracotta dyes, with modules' efficiency of 13%. This allows the multilevel integration of the BIPV system in the buildings, concerning the energy, functional and aesthetic integration, allowing to hide the PV cells from the view and thus making possible the installation of the BIPV system in the protected context where the four buildings are located [16]. The cost of the storage system has been set at 563.4 €/kWh for 2020 and the annual maintenance cost has been calculated as a percentage of about 2% on the BIPV system' initial cost. The system' cost in 2020 and the cost projections in 2050 have been issued from data available at [17, 18]. Concerning the price of electricity, it will likely grow in the future years, but it is difficult to have reliable projections for future scenarios. Therefore, we chose a conservative approach for the simulations, considering this parameter constant during the years and equal to the current price for the consumer. All the economic parameters are summarized in Table 2.

Table 2. Economic parameters for the optimizations and simulations in the years 2020 and 2050.

	2020	2050
cost of the finished BIPV system [€/kWp]	3000	1647.1
cost of the storage system [€/kWh]	563.4	126
annual maintenance costs [€/kWp year]	62.7	34.42
price of electricity for the consumer [€/kWh]	0.2226	0.2226

4 RESULTS

In Table 3 the results of the minimization of LCOE have been reported. It is possible to notice that optimizations in the 2050 scenario provide a noticeable impact on the LCOE, which decreased by about 50% (Figure 3). When considering only the electric load for lights and appliances (2050 EL) the LCOE decreased by 49%, compared to 2020 EL optimization. This result is due to the combined effect of the decreasing costs of the BIPV system components and the increase of the available irradiation per square meter over the year. The 55% target of self-sufficiency is guaranteed with a smaller installed capacity (-9% compared to 2020 EL optimization) and a higher capacity of the installed electric storage (+14% compared to 2020 EL optimization, Figure 4). Furthermore, while the annual cumulative balance between production and consumption decreases, due to the lower installed capacity, the optimal configuration in 2050 allows increasing the self-consumption of about 9%. This influences the self-consumed-LCOE, i.e., the LCOE calculated by considering the sole self-consumed fraction of the produced electricity. On the other hand, the decarbonization of the residential sector requires the shift from traditional natural gas/oil-powered thermal generators to electric heat pumps. This will have a great impact on the electric demand of buildings and additionally, the heat pumps load will be directly influenced by the outdoor temperatures. For this reason, the effect of the different climate scenarios on the electric consumption of the heat pumps (increased cooling load, decreased heating demand during winter) has been analyzed in detail and its effect on the optimal configuration of the BIPV system evaluated (2020 EL+HP and 2050 EL+HP). Based on the reported results, it is possible to conclude that the 2050 scenario provides a noticeable impact on the LCOE, which decreased by 51%. In this case, the 55% target of self-sufficiency is guaranteed with a smaller installed capacity of 10% and a smaller capacity of the electric storage of 2%, because of a smaller electric load if compared to the same case in 2020 (2020 EL+HP). On the other hand, the self-consumption decreases for both 2020 and 2050 optimizations, if compared with the cases without the heat pump load This is could be due to a higher mismatch between hourly production and consumption. This overproduction could be (i) lost, if not fed into the grid, (ii) made available for a district renewable energy community, or (iii) if the load were more flexible, there would be the possibility to increase the SS and the SC.

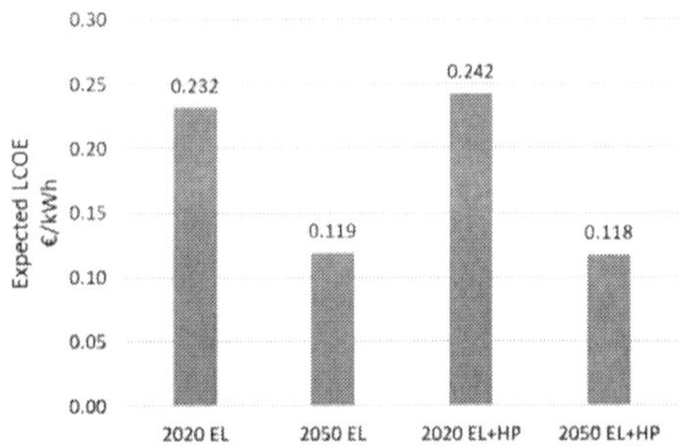

Figure 3. Results of the optimization: minimum LCOE.

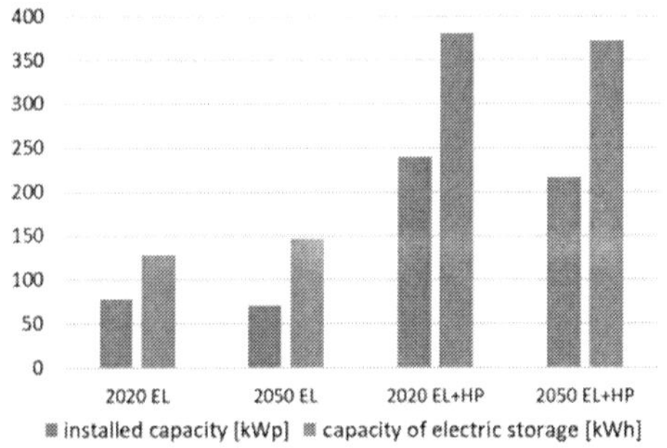

Figure 4. Results of the optimization: installed PV capacity and capacity of the electric storage.

Table 3. Results of the minimization of LCOE for the four scenarios summarized in Table 1.

	2020 EL	2050 EL	2020 EL+HP	2050 EL+HP
installed area [m²]	600	546	1840	1664
installed capacity [kWp]	78.00	70.98	239.20	216.32
variation vs. 2020		- 9 %		- 10 %
capacity of electric storage [kWh]	128	146	381	372
variation vs. 2020		+ 14 %		- 2 %
ratio battery/PV capacity	1.64	2.06	1.59	1.72
system cost [€/kWp]	3924	1907	3897	1864
system cost [€/m²]	510	248	507	242
self-sufficiency [%]	55.0	55.0	55.0	55.0
expected LCOE [€/kWh]	0.2316	0.1187	0.2423	0.1177
variation vs. 2020		- 49 %		- 51 %
self-consumption [%]	73.3	79.6	63.7	63.4
expected self-consumed-LCOE [€/kWh]	0.3054	0.1443	0.3649	0.1768
annual cumulative production [kWh]	93773	86332	272607	254058
cumulative consumption [kWh]	124988.4	124988.4	315911.7	292905.4
variation vs. 2020		-		- 7 %
annual cumulative balance production/consumption	0.7503	0.6907	0.8629	0.8674
specific emissions of the whole produced electricity [kg CO_2-eq/MWh]	110.48	111.06	116.34	113.37

To understand how a system designed for operation in current weather would perform under different weather conditions, two additional simulations have been conducted. These simulations took as input the optimal BIPV configuration resulted for the optimizations with 2020 weather file and simulated by using the weather projection in 2050, for both the load profile scenarios, EL and EL+HP. The results of this comparison have been reported in Table 4. The LCOE slightly decreases for the case of the sole EL, but in this case, the 55% target for the SS is not reached (Figure 5). This is probably since the system in 2050 produces less energy in winter and more energy in summer compared to 2020, because of the difference in the irradiation levels in the two seasons between the two weather scenarios (see Figure 6). As a consequence, the SC and SS decrease in the cold season, when the ratio between production and consumption is already low, while in summer remains constant, since the BIPV system matches better the load. If we compare the EL+HP case, it is possible to notice that the LCOE slightly increases with the weather scenario 2050, with a variation of +2% (Figure 5). The configuration of the system, optimized for the 2020 weather scenario, led to lower energy production (-2%) due to the combined and opposite effects of the increasing temperatures and irradiation in 2050. On the other hand, the variations of the climate conditions are responsible for a lower electric load of the heat pump, since the changing weather has, proportionally, more beneficial effects on the energy need for space heating than cooling. As a consequence, the SS target is met and exceeded, while the SC slightly decreases.

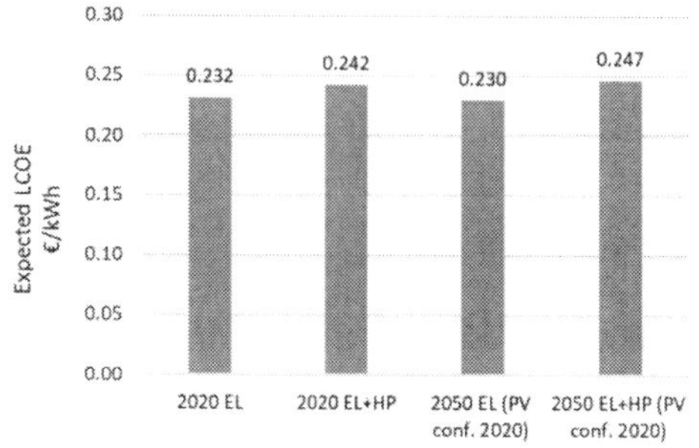

Figure 5. Results of the simulations: LCOE.

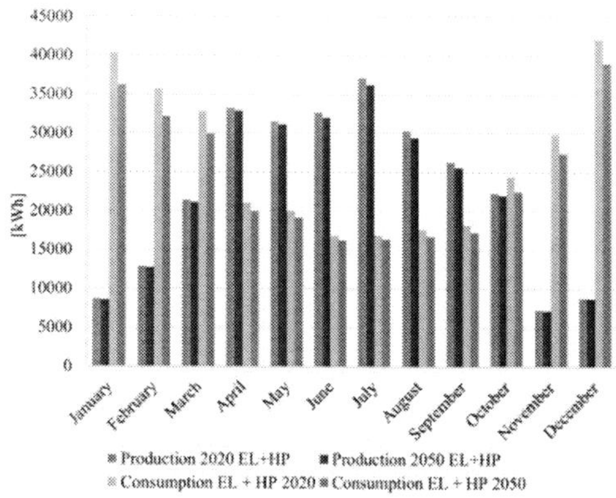

Figure 6. Monthly production Vs consumption of the optimal configuration for 2020, simulated with the 2020 and the 2050 weather file.

Table 4. Results of the simulations for the two scenarios summarized in Table 1.

	2020 EL	2050 EL (PV conf. 2020)	2020 EL+HP	2050 EL+HP (PV conf. 2020)
installed area [m²]	600		1840	
installed capacity [kWp]	78.00		239.20	
capacity of electric storage [kWh]	128		381	
ratio battery/PV capacity	1.64		1.59	
system cost [€/kWp]	3924		3897	
system cost [€/m²]	510		507	
self-sufficiency [%]	55.0	54.3	55.0	57.5
expected LCOE [€/kWh]	0.2316	0.2304	0.2423	0.2467
variation %		-1%		2%
self-consumption [%]	73.3	72.0	63.7	62.9
expected self-consumed-LCOE [€/kWh]	0.3054	0.3084	0.3649	0.3740
annual cumulative production [kWh]	93773	94233	272607	267827
variation %		0.5%		-2%
cumulative consumption [kWh]	124988.4	124988.41	315911.7	292905.4
annual cumulative balance production/consumption	0.7503	0.7539	0.8629	0.9144
specific emissions of the whole produced electricity [kg CO2-eq/MWh]	110.48	109.88	116.34	118.47

5 CONCLUSIONS

This study provides an innovative approach to optimize BIPV applications in the early design phase by considering different climate conditions (years 2020 and 2050). BIPV technologies represent an effective way to incorporate RES in the building envelope, serving at the same time as envelope components and active power generation elements. Nonetheless, the correct assessment of the techno-economic performance of BIPV systems is crucial to foster their optimal penetration in the built environment. This is particularly important in architectural contexts where heritages or protected landscapes are present. These areas have specific design limitations on the envelope material/elements and constraints that hinder the installation of standard high-performance not-integrated PV modules. Therefore, the presented analysis has been conducted by using techno-economic parameters of hidden colored BIPV modules, compliant with the specific regulation on heritages or protected landscapes in the analyzed area. Firstly, the configuration of the BIPV system has been optimized to minimize LCOE with constrained self-sufficiency (SS) greater than 55%. The optimization has been conducted by considering the two weather scenarios and two load profiles, one including the electric use for lights and appliances and one including the electric use of a heat pump, covering the building's energy needs for heating, cooling and DHW. The optimizations, having the same load scenario for 2020 and 2050, have been compared. Results show that the LCOE has been halved from 2020 and 2050, due to the decreasing costs of the BIPV system components in future years and the increase of the available irradiation per square meter over the year. The SS target is guaranteed with a smaller installed PV capacity in 2050 for both cases (EL and EL+HP). Then, to understand how a system designed for operating in current weather scenarios would perform under different weather conditions, we took as input the optimal configurations of the BIPV system in 2020 and simulated them with the weather projection in 2050, for both the load profile scenarios, EL and EL+HP. It is possible to notice that the BIPV systems produce almost the same or slightly less energy in 2050, because of the combined and opposite effects of the increasing temperatures and irradiation from 2020 to 2050. As a consequence, the SS target is not met in the case of EL 2050, while for the case 2050 EL+HP, the system performs better in terms of SS, while the SC slightly decreases, because the changing weather conditions have, proportionally, more beneficial effects on the energy need for space heating than cooling, resulting in a lower electric load of the heat pump. Future improvement of the present work would include the dynamic variation of the weather conditions during the system lifespan for the optimization, including the effect that changing climate conditions have on the energy use of an electric heat pump that serves the building energy needs for space heating, cooling and DHW. A sensitivity analysis on the most important factors having an impact in the long term (climate change, inhabitants' behaviour, change of demand profile) would be also needed to assess separately the influence of each varying parameter on the techno-economic performance of the BIPV systems.

AKNOWLEDGMENT

Operation co-financed by the European Union, European Regional Development Fund, the Italian Government, the Swiss Confederation and Cantons, as part of the Interreg V-A Italy-Switzerland Cooperation Program, within the context of the BIPV meets History project (grant n. 603882).

REFERENCES

[1] European Parlament, *DIRECTIVE (EU) 2018/2001 on the promotion of the use of energy from renewable sources*. 2018.

[2] European Parlament, *DIRECTIVE (EU) 2010/31. Energy performance of buildings*. 2010.

[3] European Parlament, *DIRECTIVE (EU) 2018/2002 on energy efficiency*. 2018.

[4] L. Pagliano, S. Carlucci, F. Causone, A. Moazami, and G. Cattarin, "Energy retrofit for a climate resilient child care centre," *Energy Build.*, vol. 127, pp. 1117–1132, 2016.

[5] M. F. Jentsch, P. A. B. James, L. Bourikas, and A. B. S. Bahaj, "Transforming existing weather data for worldwide locations to enable energy and building performance simulation under future climates," *Renew. Energy*, vol. 55, pp. 514–524, 2013.

[6] M. Lovati, G. Salvalai, G. Fratus, L. Maturi, R. Albatici, and D. Moser, "New method for the early design of BIPV with electric storage: A case study in northern Italy," *Sustain. Cities Soc.*, vol. 48, no. December 2018, p. 101400, Jul. 2019.

[7] "Lawrence Berkeley National Laboratory, RADIANCE." [Online]. Available: http://radsite.lbl.gov/.

[8] M. Dallapiccola, G. Barchi, J. Adami, and D. Moser, "The Role of Flexibility in Photovoltaic and Battery Optimal Sizing towards a Decarbonized Residential Sector," *Energies*, vol. 14, no. 8, p. 2326, Apr. 2021.

[9] Italian Ministry of Economic Development, Italian Ministry of the Environment and Protection of Natural Resources and the Sea, and Italian Ministry of Infrastructure and Transport, "Integrated national Energy and climate Plan (PNIEC)." 2019.

[10] "University of Southampton. CCWorldWeatherGen." [Online]. Available: https://energy.soton.ac.uk/ccworldweathergen/.

[11] "Photovoltaic Geographical Information System (PVGIS) | EU Science Hub." [Online]. Available: https://ec.europa.eu/jrc/en/pvgis. [Accessed: 08-Sep-2021].

[12] "IPCC — Intergovernmental Panel on Climate Change." [Online]. Available: https://www.ipcc.ch/. [Accessed: 08-Sep-2021].

[13] F. Causone, S. Carlucci, M. Ferrando, A. Marchenko, and S. Erba, "A data-driven procedure to model occupancy and occupant-related electric load profiles in residential buildings for energy simulation," *Energy Build.*, vol. 202, p. 109342, Nov. 2019.

[14] S. Pezzutto *et al.*, "Hotmaps Project, D2.3 WP2 Report – Open Data Set for the EU28," *www.hotmaps-project.eu*, 2018. [Online]. Available: https://www.hotmaps-project.eu/wp-content/uploads/2018/03/D2.3-Hotmaps_for-upload_revised-final_.pdf. [Accessed: 08-Sep-2021].

[15] Scuola universitaria professionale della Svizzera italiana (SUPSI) and Becquerel Iinstitute, "Building Integrated Photovoltaics : A practical handbook for solar buildings' stakeholders Status Report," 2020.

[16] M. Pelle, E. Lucchi, L. Maturi, A. Astigarraga, and F. Causone, "Coloured BIPV Technologies: Methodological and Experimental Assessment for Architecturally Sensitive Areas," *Energies*, vol. 13, no. 17, p. 4506, Sep. 2020.

[17] D. Moser, C. Breyer, G. Masson, and E. Vartiainen, "Improving the Competitiveness of Solar PV with Electricity Storage," *33rd Eur. Photovolt. Sol. Energy Conf. Exhib.*, pp. 2783–2789, Nov. 2017.

[18] "Electricity storage and renewables: Costs and markets to 2030." [Online]. Available: https://www.irena.org/publications/2017/Oct/Electricity-storage-and-renewables-costs-and-markets. [Accessed: 08-Sep-2021].

DEMAND-AWARE ELECTRICITY PRICE PREDICTION BASED ON LSTM AND WAVELET TRANSFORM

Koki Iwabuchi*[1], Kenshiro Kato[1], Daichi Watari[1], Ittetsu Taniguchi[1],
Franchy Catthoor[2], Elham Shirazi[2], Takao Onoye[1]
[1] Graduate School of Information Science and Technology, Osaka University
1-5 Yamadaoka, Suita, Osaka 565-0871, Japan
[2] IMEC, Kapeldreef 75, 3001 Heverlee, Belgium / ESAT KU Leuven, Kasteelpark Arenberg 10, 3001 Heverlee, Belgium
*Corresponding author: iwabuchi.koki@ise.eng.osaka-u.ac.jp, Tel: +81-6-6879-4528, Fax: +81-6-6879-4529

ABSTRACT: This paper proposes a novel electricity price prediction method using past electricity demand. The previous method predicts the electricity price using LSTM after wavelet transform of the past electricity price series. Wavelet transform decomposes the series into more smooth and stable series, and accurate prediction is performed by LSTM with these series. The proposed method also uses the electricity demand series, and the demand is also applied wavelet transform for effective LSTM step. Experimental results show the effectiveness of the proposed method, and the accuracy is drastically improved over the previous method.
Keywords: Dynamic pricing, Electricity price forecast, Electricity demand

1 INTRODUCTION

Demand response is changing electricity demand based on a change in the price of electricity, so the electricity price is capable indirectly to control the electricity demand. Dynamic pricing is one of the promising ways to realize demand response, and various countries introduce this scheme. The electricity price is set by the intersection of the generation and demand curves. Price prediction is required for any demand control strategy in real-time electricity price for their decision-making.

Various researches on electricity price prediction have been done [1-4]. Zhao, et al. proposed a short-term electricity price forecasting method based on ARIMA (Autoregressive Integrated Moving Average) model [1]. ARIMA is quite popular for the time series data analysis, and it is also often used for electricity price prediction. Lago, et al. proposed a deep learning approach for the electricity price prediction [2]. Lago, et al. demonstrated the effectiveness of deep learning approaches, which include 27 common approaches, under the same benchmark. Lago, et al. also showed that the load or generation in a neighboring market brings better performance. Xu, et al. also proposed deep learning approaches for day-ahead electricity price prediction [3]. The input of Xu's model includes previous price and load/wind forecast data. Xu, et al. demonstrated the effectiveness of LSTM and GRU.

Chang, et al. proposed a hybrid model based on wavelet transform and Adam optimized LSTM neural network for accurate electricity price forecasting [4]. Chang's method first applies the wavelet transform for the time series of electricity price, and the nonlinear sequence can be decomposed, to obtain a stable series. For these data, Chang's method inputs them to the LSTM model, which is optimized by Adam, to obtain the appropriate behavior, and forecast the precise electricity price. Chang's method is quite accurate, but this method only uses the price information. Thus, further improvement of accuracy is expected using various data such as demand and generation.

This paper proposes demand aware electricity price prediction based on LSTM and wavelet transform. The idea of this paper is to improve Chang's model [4], and the electricity demand profile are also utilized for the prediction. Figure 1 shows the measured electricity price and demand profile in New South Wales in May 2013. We can observe that both series are synchronized, and the peak and valleys happen at the same time. This means the electricity prices and demand are correlated, and this becomes a big motivation to predict the electricity price using both past price and demand series. The contribution of this paper is novel electricity price prediction model starting from Chang's model [4]. The proposed model applies wavelet transform for both electricity price and demand patterns, and these decomposed patterns are given to the LSTM model. To the best of our knowledge, this is the first paper which applies wavelet transform decomposition for the electricity demand to help in predict the future electricity price.

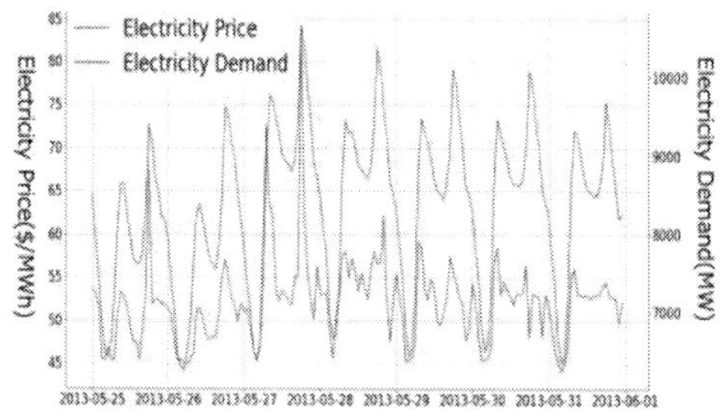

Figure 1: Relation between Electricity Price and Demand in New South Wales

2 PROPOSED METHOD

This section describes the overview of the proposed method. The proposed method is based on Chang's method [4], and an overview of Chang's method takes the following steps.

(1) Data collection
(2) Data preprocessing
(3) Electricity price decomposition using wavelet transform
(4) Adam-optimized LSTM only with price series

After data is collected, preprocessing is performed to remove abnormal values. Then electricity price decomposition is performed by wavelet transform. Because the original price series usually includes various noise, wavelet transform is applied to decompose the series into stable and smooth series. In this step, Daubechies 5 is used as a mother wavelet, and five levels of decomposition are considered. For decomposition series, a one-step-ahead price is predicted. Figure 2 shows the overview of Chang's Adam-LSTM model, and for each decomposed series, a two-column data-set is given to the model. After that, the model is evaluated.

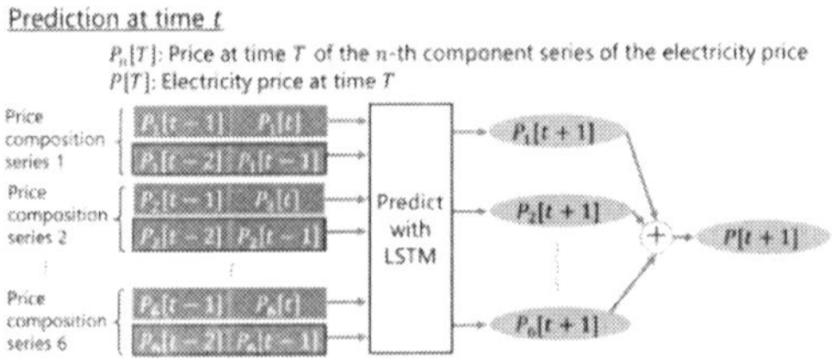

Figure 2: Chang's Adam-LSTM model [4]

The proposed method basically follows the same steps, but step (3) and (4) are largely extended to utilize the electricity demand series.

(1) Data collection
(2) Data preprocessing
(3) Electricity price and demand decomposition using wavelet transform
(4) Adam-optimized LSTM with decomposed price and demand series

Step (3) in the proposed method decomposes electricity price and demand series, respectively. We utilized Daubechies 5 as mother wavelet, and five level of decomposition are considered. After that, the decomposed series are given to the model shown in Figure 3. In the same as Chang's model, a two-column data-set is given to the model, and the predicted price is obtained.

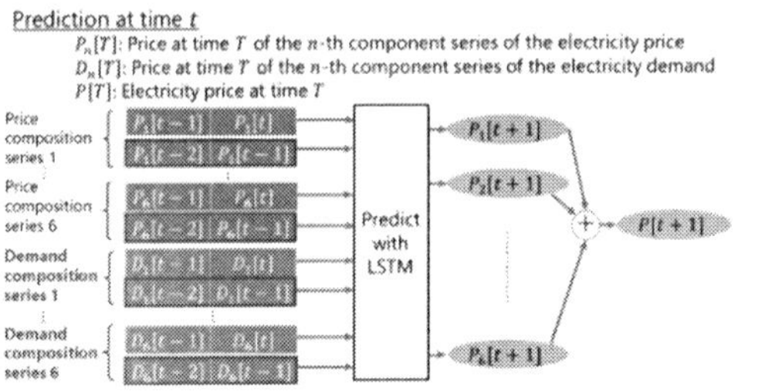

Figure 3: Proposed Model

3 EXPERIMENTAL RESULTS

In order to demonstrate the efficiency of the proposed method, hourly electricity price and demand data from New South Wales in May 2013 are utilized. MSE and MAE are introduced as performance metrics. Let x_t and $\widehat{x_t}$ be the real value and the predicted value of t-th forecasting point, respectively. MSE and MAE are defined as follows:

$$MSE = \frac{1}{T}\sum_{t=1}^{T}(\widehat{x_t} - x_t)^2$$

$$MAE = \frac{1}{T}\sum_{t=1}^{T}|\widehat{x_t} - x_t|$$

where T is the number of test samples. We compared the proposed method with Chang's method under the same condition described in the reference [4].

Table I shows the comparison of one-step-ahead results, in which the data of one day (24 points) are set as a test set. From the reference [4], the MSE and MAE of Chang's model were 2.80 and 1.45, respectively. We preformed the same evaluation of Chang's model and replicated MSE and MAE were 2.92 and 1.42, respectively. Under this condition, we also implemented the proposed model, and evaluated MSE and MAE. As shown in Table I, MSE and MAE are drastically improved for one-step-ahead prediction. This shows that the historical price and demand series are both containing different relevant information for the prediction. By properly decomposing them with the wavelet transform, they can be combined as input to the LSTM engine.

Figure 4 shows the comparison of predicted price and actual price on the test-day. The green curve is the actual price, and the blue curve is the predicted price obtained by Chang's method. The orange curve is the predicted price obtained by the proposed method. We can see the orange curve follows the peak and bottom of the green curve at the same timing. So the general trend is very accurately followed with small errors.

Table I: Comparison of One-step-ahead in NSW

	Chang's model [4]		Proposed model
	Described	Replicated	
MSE	2.80	2.92	1.35
MAE	1.45	1.42	0.84

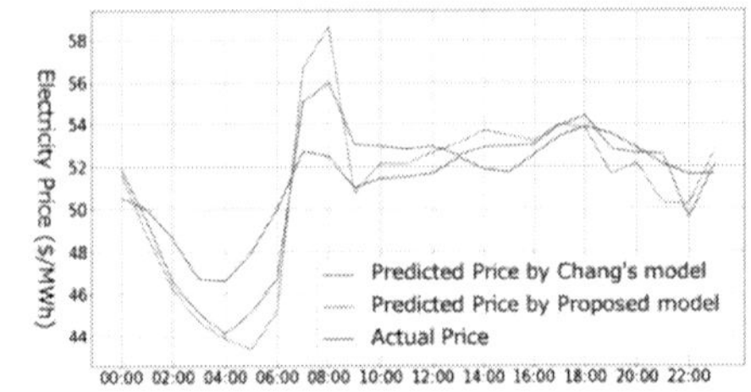

Figure 4: Comparison of predicted price and actual price on May 31, 2013

4 CONCLUSION

This paper proposed a novel electricity price prediction method using past electricity demand series. The previous method predicts the electricity price using LSTM after wavelet transform of the past electricity price series. The proposed method includes the electricity demand series as input, and the price prediction is performed by LSTM after wavelet transform of both electricity price and demand series. Experimental results show the effectiveness of the proposed method. Future work includes detailed analysis and includes the other data such as the amount of electricity generation.

REFERENCES

[1] Z. Zhao, C. Wang, M. Nokleby, C. J. Miller, "Improving short-term electricity price forecasting

using day-ahead LMP with ARIMA models," IEEE Power & Energy Society General Meeting, 2017.
[2] J. Lago, F. De Ridder, B. De Schutter, "Forecasting spot electricity prices: Deep learning approaches and empirical comparison of traditional algorithms," Elsevier Applied Energy, vol.221, pp.386—405, 2018.
[3] J. Xu, R. Baldick, "Day-Ahead Price Forecasting in ERCOT Market Using Neural Network Approaches," Proc. ACM e-Energy, pp.486—491, 2019.
[4] Z. Chang, Y. Zhang, W. Chen, "Electricity price prediction based on hybrid model of Adam optimized LSTM neural network and wavelet transform," Elsevier Energy, vol.187, Nov. 2019.

THE ASSET OF MEASUREMENTS FOR PROJECT FINANCE AND PROJECT VALUE

Weber, Simon, Granados, Raúl, Meyer, Richard and Westphal, Boris
Suntrace GmbH
Grosse Elbstrasse 145c, Hamburg (Germany)
+49 40 809 0354-0
meteo@suntrace.de

ABSTRACT: During the development of a photovoltaic plant project, the expected energy yield plays a key role for financial viability. Up to now, the uncertainty of the potential yield is dominated by the solar resource risk. The evaluation of solar resource for an average geographical situation still shows an uncertainty of approximately 5%. This risk can be significantly reduced by adaption to the specific site using qualified ground-based measurements. By measurements of highest quality and multiple years of overlap with satellite timeseries, the uncertainty of the solar resource may be reduced to a minimum of 2%. Such high accuracy of solar resource data sharpens forecasted project yields and improves financing terms. This study shows the impacts of on-site measurements to leveraged project finance and project value. Results show that costs for a one-year measurement campaign already pay off for plant sizes in the order of 10 MW. Moreover, the direct impact by high-quality on-site solar measurements translates into an increase of up to 200 kUSD each 10 MW.

Keywords: De-risking renewable resources, Cost reduction, Risk reduction, Techno-economical optimization, Solar radiation

1 LOWERING YIELD UNCERTAINTY BY MEASUREMENTS ENABLES FINANCING

Solar resources account for a decisive share of uncertainty when determining the return on investment of large-scale photovoltaic power plants [1]. For instance, yield predictions and yield certificates mainly comprise of three components: Technical plant specifications, solar resource evaluation and estimation of changes in energy yield over the lifetime of the system [2]. It is proven, that irradiance conversion models and uncertainty of incoming irradiance have the highest influence. Subsequently, meteorological data predominantly affect plant performance [3]. The required degree of solar data granularity can be adapted according to project progress, maturity, size and budget. To get an idea of the cost-benefit relationship, it is important to understand some basic mechanisms of uncertainty reduction in solar resource assessments.

1.1. Financing of utility-scale PV projects

For yield and return calculations, investors commonly consider the average, most probable solar resource conditions. These are typically described by the P50 value, which is defined as the value being exceeded with 50% probability. Lenders rather rely on P90 values for their debt sizing i.e. the solar irradiance value that is expected to be exceeded to 90%. Given that the P50 or mean value does not change substantially, more precise measurements will increase the P90 value as illustrated in Figure 1. Through this *P90-Effect*, the investor can demonstrate a higher cash flow to lenders, who, in turn, would allow a higher share of debt to the project - generally resulting in lower capital cost. Lender's may further reduce the Debt Service Coverage Ratio (DSCR) threshold or offer lower interest rates. All these yield a higher real return on investment.

A second effect of more accurate solar resource estimates is that the investor can price the bid in a Power Purchase Agreement (PPA) negotiation more accurately with lower risk deductions. Accordingly, the investor can price the tariff with a higher certainty on the solar fuel factor.

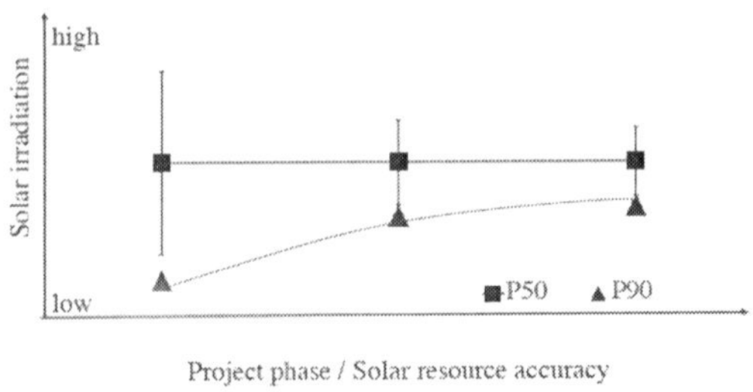

Figure 1: Progressing project phase requests de-risked assumptions. Reduced uncertainty in solar resource will increase the P90 value.

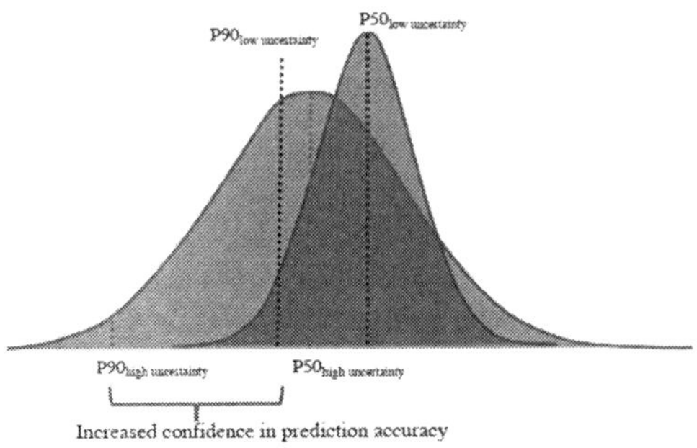

Figure 2: The P90 Effect leading to higher predicted yield due to reduced uncertainty.

The depicted P90-Effect assumes that the general estimate of a resource distribution is fairly accurate, indicated by an overall good overlap of resource distribution. Yet, the higher uncertainty associated with satellite-only data does not allow for more optimistic yield predictions. Opposed to that, the risk deduction effect assumes that without multi-source analysis and on-site ground-based measurement data you cannot even be sure about the accurateness of the position of your resource distribution.

A thorough assessment in combination with high-quality on-site measurements may reduce an initially quite high resource uncertainty of 8% down to 2% - and in turn will increase the P90-Best Estimate by roughly 2.5% as illustrated in Figure 2. Such an improvement may increase Net Present Value (NPV) by up to 200 kUSD each 10 MW.

1.2 On-site measurements during project development

For the development of utility-scale solar parks there is an economic need for solar resource assessments for all phases of project development: At Pre-Feasibility Stage, cost-efficient resources are exploited to estimate preliminary ground-truth conditions. Since both private and public networks for ground-based measurement stations may provide data of higher accuracy, even initial phases benefit from existing ground-truth data. To achieve reliable meteorological data, solar resource assessments rely on qualified satellite-derived solar radiation data validated by high-level ground-based measurements [4], [5]. To ensure highest cost-efficiency even at Pre-Feasibility Phase, knowledge on accessible information is a vital tool. At this conceptual phase, consultancy services may reduce uncertainty by analysing and comparing multiple available sources [6], [7] and thereby provide higher reliability to the found resource distribution. Yet, one major source for uncertainty derives from the spaceborne data's coarse resolution: Data of low temporal and spatial resolution holds great potential for uncertainty. Consequently, the project's Feasibility Phase foresees the procurement of high-quality data after having gained first precaution.

The expansion of solar projects around the globe has increased the necessity of well-proven project development strategies. With these enhanced requirements, Due Diligence has shifted to the centre of attention of most lenders and project developers. To successfully pass Due Diligence for projects of this size, project management must seek a bankable expert opinion [8]. The expert's detailed assessment of the quality and corrections of satellite data as well as measurements is required for risk analysis. For bankable expert opinions on meteorological conditions, data of high accuracy is required to satisfy conservative approaches of banks and lenders. Thus, solar measurement campaigns are often carried out in this phase.

The solar measurement campaign provides data of high quality to a detailed profitability study and the final plant design. Since even solar measurement equipment of highest standards may contain biased or distorted data, it is recommended to quality check all recordings. At best, post-processing of solar measurements is implemented as standardised procedure throughout the measurement campaign to enhance data quality and ensure representativeness of on-site conditions.

For balancing out seasonal affects at satellite data correction, it is highly recommended to conduct measurement for one year. Thus, in this study, expenditures for a 1-year on-site measurement campaigns, including post-processing of data, are considered.

Having passed Due Diligence, the project reaches its operational phase. Best operational strategies rely on continuous monitoring of meteorological conditions and forecasts determine plant control. For the needs of plant control, both solar and auxiliary parameters are monitored. For instance, times of solar-field safety shutdowns are determined from wind speed monitoring and forecasting. Therefore, monitoring meteorological and auxiliary parameters helps to reduce the risk of malfunction and downtime [9]. Utility-size solar parks commonly have two to three solar measurement stations evenly distributed across the plant to enable the plant's full potential and allow redundant surveillance. Assumptions stated in this paper consider the financial benefit of solar measurement stations during plant development. Equipment being utilized beyond Due Diligence further increases the value of expenditures for measurement campaigns during project development. Sensitive sensors such as pyranometers will require re-calibration after approx. 2 years but may be considered for > 5 years of measurements. Therefore, it is worth investing in robust and durable equipment.

1.3 Guidelines for ground-based solar measurements

Together with other organisations, the World Bank's Energy Sector Management Assistance Program (ESMAP) first introduced the Multi-Tier Framework for measuring energy access. In accordance, solar measurement equipment has been assimilated to meet tiered standards. For CSP, Tier 1 and Tier 2 standards become applicable[1]. For PV projects, all three tiers may be considered. For Floating PV, the offshore measurement of irradiance can be challenging. A practical solution is commonly to measure irradiance at the shore and auxiliary parameters offshore. For such onshore measurements, again all 3 tiers can be considered. However, the most feasible tier is moreover defined by the project site: High maintenance requirements of Tier-1 type favour selection of Tier-2 and Tier-3 stations for remote areas.

Table 1: Estimated 1σ-uncertainty of the long-term average of GHI

Station type	**Main sensor**	**Uncertainty**		
		Difficult	Average	easy situations
no station	Satellite only	8.0	5.0	4.0
Tier 3	High-quality pyranometer	4.0	3.5	3.0
Tier 2	High-quality pyranometer + RSI	3.5	3.0	2.5
Tier 1	Sun Tracker, high-quality pyranometers + pyrheliometer	3.0	2.5	2.0

For all technologies and tiers, it must be ensured that the stations are well-maintained and equipped with high-quality pyranometers [2]. Furthermore, it is advisable to consider low-cost photodiodes for redundancy and regular quality checks. To capture the full meteorological year,

[1] Classification of Solar Measurements for CSP according to World Bank [10].

[2] Class A-C according to ISO9060:2018 and IEC 61724-1:2017, International standards for measurement of solar radiation and photovoltaic system performance.

ground-based measurements are recommended to continue for at least 1 year (12 months). Given project development often faces very limited time, shorter periods can be considered: According to [11], ground data should have at least 6 months of overlap with satellite data to assure measurement-corrected satellite data usually lead to improved data quality. With improved site adaption techniques already 3 months of overlap may lead to improvement. Nonetheless, 12 months of overlap continue to give substantial improvement [5].

Depending on the basic quality of the satellite algorithm, the satellite viewing geometry of the site, and the specific type and heterogeneity of land cover, its topography and typical micro-climate, the basic uncertainty of satellite-derived solar resource data can vary strongly. Also, the quality of station location of the site on the and spatial context Based on the evaluation of 100 satellite-derived solar resource data sets through ground-based measurements of at least Tier-3 standard, the expert guess for the uncertainty is given in Table 1. The given numbers refer to stations, which are properly installed, maintained, and well documented. If the measurement stations cannot proof sufficient quality, they are not of value. In such situations it is often better ignoring unqualified measurements to avoid erroneous measurements would misdirect good satellite-derived data sets.

2 QUANTIFIYING THE VALUE OF MEASUREMENTS

The evaluation Key Performance Indicators (KPIs) such as Internal Rate of Return (IRR) and project's NPV enables reliable cost-benefit analysis of CAPEX investments such as for solar measurement campaigns. Among various projects, a clear upward trend in NPV due to advanced resource and risk assessments can be observed. By following a stepwise approach, the increase in NPV results in up to 40%. To quantify this increase in accordance to both technical and financial project-specific assumptions, this study presents a financial model that allows to analyse the return on investment by taking solar resource uncertainty into account.

4.1 Financial modelling of project value as a function of solar resource accuracy

To evaluate the effect of solar resource uncertainty on NPV, our model simulates a PV project's cash flow as shown in Figure 3. Costs follow today's market values [12] and leverage is based on review of financing conditions for global renewable energy projects [13]. According to [12], CAPEX expenditures for utility scale PV plants range in the order of 1.300 USD/kW related to AC-power. According to our experience in 2021, CAPEX prices can drop down to 900 USD/kW$_p$ (DC) and below. Electricity tariffs in this study are set according to 20-year Power Purchase Agreements.

In this study, three 50 MW PV plants are investigated: A site with low GHI levels of 1000 kWh/(m^{2}a) in Germany (DE), a site with intermediate GHI of 1500 kWh/(m^2 a) in Vietnam (VN) and a site with more favorable solar resource of 2000 kWh/(m^2 a) in Egypt (EG). We are aware that in all countries - especially in Egypt - much more favorable GHI values occur. The assumptions were set to provide conservative, yet representative findings.

Expenditures for a 1-year measurement campaign range from approx. 45 to 80 kUSD. These considerable costs include expenses for professional station maintenance and data post-processing i.e. data quality assurance and documentation. These quality assurance tasks are seen as essential to reach the intended improvement of uncertainty and subsequent increase in project finance.

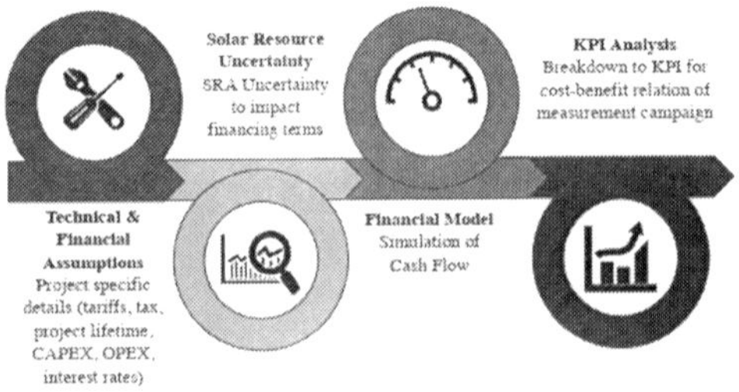

Figure 3: Schematic of financial model applied in this study.

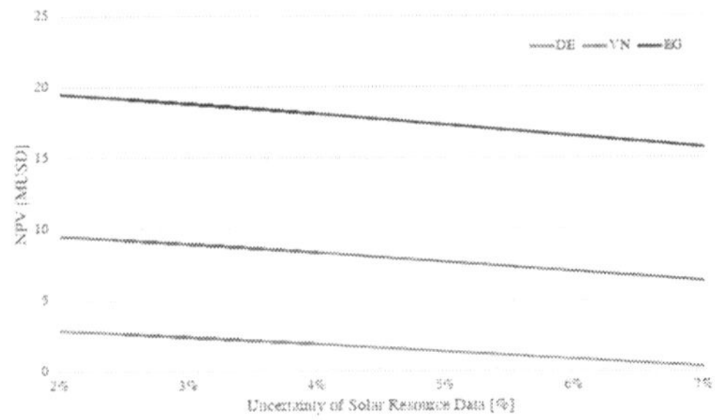

Figure 4: Dependency of project's NPV on solar resource accuracy.

The impact by solar resource accuracy on NPV is shown in Figure 4. Overall, it can be said that a higher tier of measurement equipment directly translates into higher accuracy and subsequently into a higher increase in NPV. Whilst NPV of a solar PV project is dominated by its overall solar resource (intercomparison of DE, VN and EG), a substantial financial gain is listed due to an increase in solar resource data. The increase in NPV with lower solar resource uncertainty can be observed across all geographical situations and markets. For sites with low solar resource, the case of DE demonstrates that project financing under high uncertainty would hardly lead to positive NPVs while the ultimate accuracy of 2% would still make a business case.

To illustrate the return on investment for an on-site measurement campaign, Figure 5 shows the increase in NPV as a function of installed power. The graph illustrates the results derived for measurement campaigns with Tier-1 and Tier-3 standard systems. Results for Tier-2 solutions lie in between and are not shown here for better overview. To its first order, the widespread ΔNPV is driven by the magnitude of solar resource and the accompanied accuracy i.e. tier of measurement equipment. Deducted by the expenditures for a 1-year campaign (solid and dashed grey line), solar measurements already pay off for plant sizes above 10 MW. For regions with high solar resource, this threshold can move to even smaller plant sizes. With the efforts of such high-level risk assessments, developers ease their financial negotiations by providing testified, bankable evidence on ground conditions, providing comfort to lenders and investors.

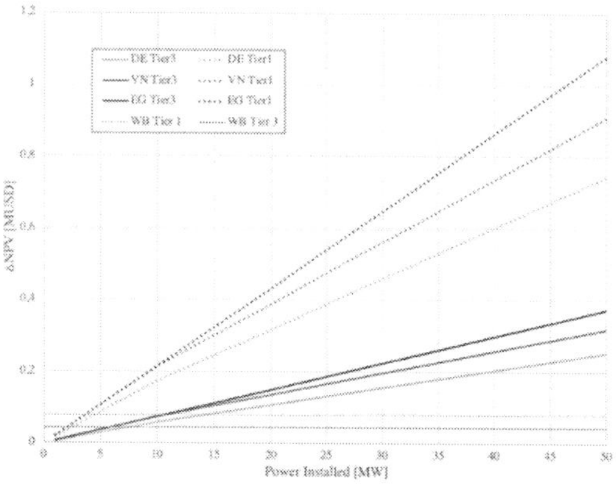

Figure 5: NPV shows return on investment to be reached for project's above 10 MW plant size.

3 VENTURING MEASUREMENT CAMPAIGNS TO ITS FULLEST

Whilst irradiance is the predominant driver in energy yield, it has already become standard to extend campaigns by the measurement for auxiliary parameters. Compared to the effort of commissioning and maintaining high-quality meteorological instruments, the additional benefit soiling-, wind- and corrosion sensors bring to plant design come at low cost.

In the field of service life and damage analysis, the assessment of corrosivity investigates the quality and aging behaviour of PV modules as well as their materials and individual components. Here, service providers draw on an extensive wealth of experience from the analysis and evaluation of numerous cases of damage from practical field application [3]. With high-level corrosion assessments, plant design and subsequently plant lifetime are optimized, CAPEX requirements are sharpened during project development and the risk of technical failure during operation is reduced.

Soiling directly affects energy yield by reducing insolation. In regions with strong soiling rates and poor cleaning schedules soiling rates for GHI may reach above 10% within one month [14], [15]. Even under optimized cleaning of PV plants, losses in solar power production are estimated to be at least 3 to 4% on global average. By 2023, this figure is expected to rise to 4 to 7% [16]. Especially arid and desert areas suffer from severe soiling. To keep losses at a minimum, frequent cleaning of solar panels is necessary. At utilize size, cleaning efforts are a main factor to OPEX and thereby stress the need for optimizing cleaning practices. Soiling sensors enable the assessment of expected soiling impact to any plant size and design during project development and enable the optimization of cleaning cycles during operation.

Addressing both plant lifetime and OPEX, high-level solar resource assessments accompanied with auxiliary measurements hold great potential to enhance project revenue. The combined uncertainty of lifetime energy yield amounts to approx. 8%. Within this uncertainty, solar resource assessment and changes in system performance provide the highest share [17]. In conclusion, solar measurement campaigns offer improvement to a solar project's most vulnerable sources for risk and therefore must be seen by project developers as valuable tool.

4 CONCLUSIONS

Both our experience and the financial model show optimization of utility-sized solar power plants using meteorological data of highest possible accuracy to be indispensable. Often, many cost-relevant decisions are taken based on rough guesses [1]. On-site measurements enable objective decisions, quantified by key financial parameters leading to both highest success during project development and maximum revenue during operation. Without high-level solar resource assessment, the possible range of deviations of economic key figures is too large and may lead to severe losses or unfavourable financing conditions, especially in a competitive market environment. Compared to risk reduction and their impact on economic key figures, investments in meteorological on-site measurements are small and should be considered for any utility scale solar power plant. The additional CAPEX for such measurement campaigns proves to pay-off for plant sizes as small as 10 MW.

Nevertheless, the initial costs of a solar measurement station seem to be a relatively high burden for project developers. For mitigating these costs in the risky phase of early project development, it is recommended to consider purchasing measurement campaigns as a service. If such campaigns are conducted by independent experts instead of the project developer, the creditability of the measurements is higher. In addition, independent experts often provide know-how to the project to identify necessary action to fulfil requirements for risk assessments. In case a project site turns out not to be feasible, the campaign costs are much lower than purchasing a station. This approach is providing higher success rates while lowering potential stranded costs for the developer.

Figures provided in this study considered standardised Tier-configuration for the measurement of solar irradiance. The financial benefit of an assessment of corrosivity or the measurement of soiling bring to project finance is not reflected in the presented model. Accordingly, stated potential for increase in NPV is conservative and suggests further investigations.

---

[3] The assessment shall comply at least with ISO 9226:2012 and ISO 8407:2009.

REFERENCES

[1] C. Reise, B. Müller, D. Moser, G. Belluardo, and P. Ingenhoven, 'Uncertainties in PV System Yield Predictions and Assessments', Internationa Energy Agency - Photovoltaic Power Systems Programme, Report IEA-PVPS T13-12:2018, Apr. 2018. Accessed: Oct. 12, 2018. [Online]. Available: http://www.iea-pvps.org/index.php?id=477

[2] B. Müller, P. Bostock, B. Farnung, and C. Reise, 'A Framework to Calculate Uncertainties for Lifetime Energy Yield Predictions of PV Systems', in *2017 IEEE 44th Photovoltaic Specialist Conference (PVSC)*, Washington, DC, Jun. 2017, pp. 2267–2271. doi: 10.1109/PVSC.2017.8366185.

[3] B. Müller, C. Reise, W. Heydenreich, and K. Kiefer, 'Are Yield Certificates Reliable? A Comparison to Monitored Real World Results Björn Müller, Christian Reise, Wolfgang Heydenreich, Klaus Kiefer', *Conference Paper*, vol. 22nd European Photovoltaic Solar Energy Conference and Exhibition, 2007.

[4] P. Ineichen, 'Long term satellite global, beam and diffuse irradiance validation', *Energy Procedia*, vol. 48, no. 0, pp. 1586–1596, 2014, doi: http://dx.doi.org/10.1016/j.egypro.2014.02.179.

[5] J. Polo *et al.*, 'Preliminary survey on site-adaptation techniques for satellite-derived and reanalysis solar radiation datasets', *Solar Energy*, vol. 132, pp. 25–37, Jul. 2016, doi: 10.1016/j.solener.2016.03.001.

[6] R. Meyer, 'Recommendations for bankable meteorological site assessments for solar thermal power plants', in *Proceedings of SolarPACES 2010*, Perpignan, France, 2010, p. 8. [Online]. Available: http://www.solarpaces2010.org/

[7] R. Meyer *et al.*, *Combining solar irradiance measurements and various satellite-derived products to a site-specific best estimate.* Las Vegas, NV, USA, 2008.

[8] A. C. McMahan, C. N. Grover, and F. E. Vignola, 'Chapter 4 - Evaluation of Resource Risk in Solar-Project Financing', in *Solar Energy Forecasting and Resource Assessment*, J. Kleissl, Ed. Boston: Academic Press, 2013, pp. 81–95. [Online]. Available: http://www.sciencedirect.com/science/article/pii/B9780123971777000048

[9] K. Chhatbar, 'The influence of meteorological parameters on the energy yield of solar thermal power plants'. 2011.

[10] M. Sengupta *et al.*, *Best Practices Handbook for the Collection and Use of Solar Resource Data for Solar Energy Applications.* NREL/TP-5D00-63112, 2015.

[11] K. Schumann, H. G. Beyer, K. Chhatbar, and R. Meyer, 'Improving satellite-derived solar resource analysis with parallel ground-based measurements.', in *ISES Solar World Congress*, Kassel, Germany, Sep. 2011, p. 12. [Online]. Available: http://proceedings.ises.org/paper/swc2011/swc2011-0328-Schumann.pdf

[12] NREL, '2021 Electricity ATB Technologies and Data'. 2021. [Online]. Available: https://atb.nrel.gov/electricity/2021

[13] NREL, 'Electricity Annual Technology Baseline (ATB) Data Download', 2021, [Online]. Available: https://atb.nrel.gov/electricity/2021/data

[14] R. Meyer, R. Granados, M. Schwandt, H. Dorff, and A. Petersen, 'Measuring soiling impact on solar power plants & their application for monitoring CSP and PV plants', presented at the EU-COST-Action inDust: Workshop at Intersolar Europe, Munich, Germany, May 16, 2019.

[15] A. Heimsath *et al.*, 'Monitoring of Soiling with the AVUS Instrument – Technical and Economic Assessment', presented at the The SolarPACES Conference October 2018, Casablanca, Morocco, 2018.

[16] K. Ilse *et al.*, 'Techno-Economic Assessment of Soiling Losses and Mitigation Strategies for Solar Power Generation', *Joule*, vol. 3, no. 10, pp. 2303–2321, Oct. 2019, doi: 10.1016/j.joule.2019.08.019.

[17] B. Müller, L. Hardt, A. Armbruster, K. Kiefer, and C. Reise, 'Yield predictions for photovoltaic power plants: empirical validation, Recent advances and remaining uncertainties', in *Proc. of the 29th EU PVSEC Conference*, Amsterdam, Netherlands, 2014, pp. 2591–2599. doi: 10.4229/EUPVSEC20142014-5BV.1.7.

Development of Conceptual Framework for Time of Generation feed-in -tariff for a rooftop photovoltaic system

Rahul R Urs, Mousa Marzband
Dep. of Math, Physics and Electrical Engineering, Northumbria University, Newcastle, UK

Ameena Al-Sumaiti
Dep. of Electrical and Computer Engineering, Khalifa University, Abu Dhabi, UAE

Abdullah M. Abusorah
Dep. of Electrical and Computer Engineering, King Abdulaziz University, Saudi Arabia

ABSTRACT: In the deployment of photovoltaic systems in the residential and commercial sectors, rooftop solar scheme such as feed-in-tariffs (FIT) have had a significant impact. Although this impact benefits the prosumer, the regulation of feed-in-tariffs based on time of generation (ToG) is an important factor for further growth of its integration. The methodology framework for setting feed-in-tariffs based on time of generation is proposed and evaluated in this paper. In contrast with the conventional feed-in-tariffs and the proposed feed-in-tariffs for a simulation model, the return on investment is observed to be more due to maximized total savings upon integration of grid connected photovoltaic (PV) system. Upon investigation of the proposed framework in this paper, it was found that the total savings would be 6.87% more than the current policy adopted worldwide.

Keywords: photovoltaic system, rooftop systems, economic analysis, feed-in-tariff, funding and incentives

Nomenclature

Acronyms	
ToG	Time of generation
FIT	Feed-in-Tariff
FIT_N	Feed-in-Tariff for energy net metered in-dependent of ToG
PV	Photovoltaic
P_1	Peak Period
P_2	Off- peak period
P_0	Moderate period
FiT_1	Feed-in-Tariff for peak period
FiT_2	Feed-in-Tariff for off-peak period
FiT_0	Feed-in-Tariff for moderate period
Rs.	Currency in Indian Rupee (INR)
Indexes	
p	Period of electricity demand
d	Day number of the month
Variables	
E^c, E^g	Energy consumption/ generation in current policy (kWh)
$E^c_{p,d}$, $E^g_{p,d}$	Energy consumption/ generation for a particular period p in a particular day d in the proposed policy (kWh)
E^N	Energy net metered (kWh)
E^c_d, E^c_d	Energy Consumption/ generation in a particular day d (kWh)
E^c_m, E^c_m	Energy Consumption/ generation in a particular month m (kWh)
Functions	
$FiT_{N,p}$	Feed-in-Tariff for period (p) net metered

1 Introduction

The world has gradually implemented photovoltaic (PV) systems to produce electricity in order to minimize carbon emissions from generating electricity. Owners of residential and industrial buildings also known as prosumer's around the world interested to contribute to minimize carbon emissions set by their respective governments, have shown considerable interest in installing PV panels on their rooftops. A PV rooftop system is a form of PV system in which the conversion panels of solar energy are installed on the rooftop of the building and the energy produced is consumed within the building or exported to the local electricity utility. In [1] , the international renewable energy agency has estimated that a rooftop PV system will be about 40 percent of the global total solar capacity. Due to supporting policies such as net metering and fiscal incentives that make PV more appealing than purchasing local electricity power, this rooftop PV system implementation has been observed to increase significantly.

However, countries that have not witnessed significant rooftop PV penetration need to re-visit their net metering policy or implement cost-based methods of compensation [2]. While net metering is the most adapted system in the world, time-variant rates are more likely to reduce the savings gained than time-invariant rates. [3]. The authors in [4] have shown that, in formulating the net metering scheme, consideration should not only be given to the geographical divergence of solar radiation, but also to the different levels of electricity demand. Also, the authors in [5] found that there will be sufficient incentives for residential solar PV investors to continue in the current mode of solar rooftop installations if overage electricity (i.e. feed-in-tariff (FIT)) is credited at or above the levelized cost of electricity (LCOE) rate, although the profitability of these investments may be significantly reduced.

In [6] , when the super-premium feed in the tariff scheme was replaced with the less premium feed in the tariff in Australia and California, the authors reported the fall of growth in rooftop PV installations. For all these limitations shown in [2]–[6], this paper is intended to contribute as follows:-

- A new framework is developed to maximise the benefit of both utility and the prosumer and thereby reduce period of return on investment for prosumer.
- Mathematical model in formulated to the existing net metering scheme for enforcing time of generation FIT.
- Comparison of the present net-metering FIT policy with the proposed net-metering FIT policy.

In this paper, three periods such as peak period (P1), off peak period (P2) and moderate period (P3) are considered based on the prosumer's electricity demand. The peak period is during which the energy usage of the rooftop premises is higher than at any other time and the load demand cannot be reduced at all because it is mandatory to use such appliances. Off peak duration is those where it is possible to compensate for the consumption and is lower than moderate period. However, in the moderate period, it is a balancing duration in which consumption is comparatively neither peak nor off peak and is within the range of electricity load demand forecast.

The rest of paper is organized as follows. In Section 2, the specification of the green building considered in this work is shown, the framework of the proposed system is explained in section 3. The mathematical modelling is discussed in Section 4. Finally, results and discussions and conclusions are presented in Sections 5 and 6, respectively.

2 Specification of green building

In this paper, a green building is considered to be located in the region Bangalore, south of India. It is assumed that the owner of this building has installed a 3 kW grid connected PV system with current utility and FIT of the local utility company as shown in table for simulation of the mathematical model formulated in this paper.

Table 1 Specifications of green building

Site	*Residential Building*
Location	Bangalore
Elevation	920 m a.s.l
Latitude	12° 59' 9.852" N
Longitude	77° 33' 17.6688" E
Annual GHI	2073 kWh/m2
Size of the rooftop plant	3 kW
Current electricity tariff	Rs. 5/kWh
Current FIT	Rs. 3.56/kWh
Monthly electricity consumption	360 kWh
Monthly electricity generation	450 kWh

3 Methodology

The current worldwide net metering scheme states that electricity exports are adjusted against imports for each billing cycle, so that electricity generated is deducted from the total electricity consumed over a given period of time [8], [9]. If the electricity generated exceeds the electricity consumed during the billing period, the electric utility shall pay the tariff agreed in the power purchase agreement (PPA) for any excess electricity injected into the grid. This tariff is known as the feed-in-

Figure 1 Global Horizontal Irradiation map obtained from [7] for India

tariff (FIT) as shown in Figure 2 Current framework adopted worldwide for net metering scheme. Furthermore, provided that if the electricity supplied by the utility exceeds the electricity generated during any billing period by the prosumers of the solar rooftop PV system, the utility shall charge at the rate applicable to that eligible user, due to increase in net electricity consumption. While this has so far encouraged building owners to install PV systems on their premises, in order to further promote PV integration, a major change in FIT is required. Also, this system would not benefit the utility by providing same FIT for the energy purchased from prosumer during off peak period.

Here a conceptual framework is proposed where the net metering FIT would vary based on energy generated during different electricity demands of the prosumer as shown in Figure 3. These electricity demands can be classified into three periods such as peak period (P_1), off peak period (P_2) and moderate period (P_0) and their feed-in-tariffs as FIT_1, FIT_2 and FIT_0 respectively. The methodology framework developed in this paper is as shown in the figure. The steps followed in the proposed methodology are as follows:-

Step 1 : The difference between power consumption and generation data by the prosumer's building is obtained to find the power net metered.

Step 2 : The period during which electricity is net-metering is identified.

Step 3 : With the respective period, the incentive for power net metered based on time of generation FIT in terms of FIT_m is obtained.

Step 4 : The costs incurred during all three periods is summed up for a month to obtain total $FIT_{N,m}$.

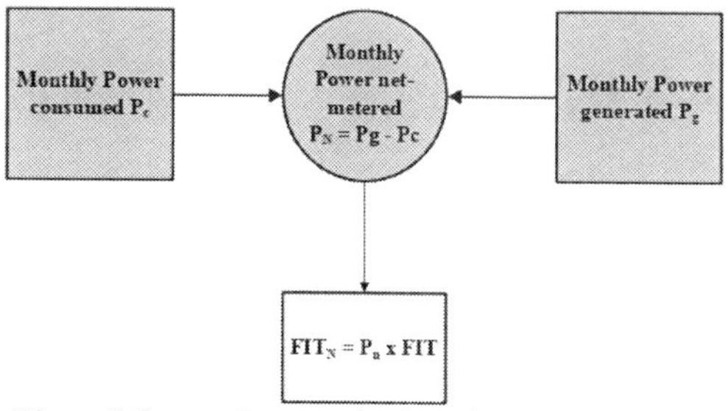

Figure 2 Current framework adopted worldwide for net metering scheme

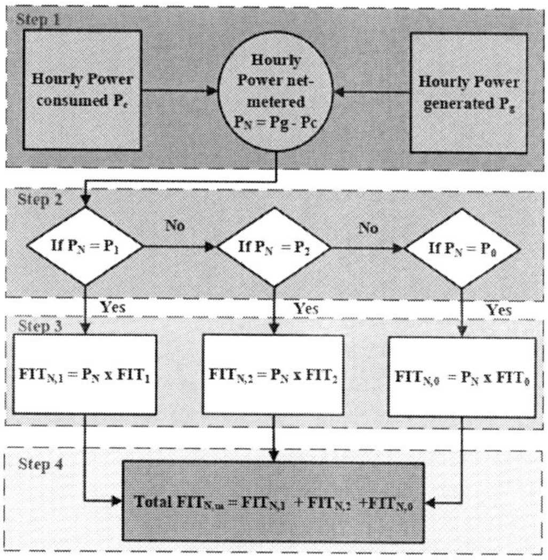

Figure 3 Proposed Framework

4 Mathematical Modelling

The monthly energy consumed from a building is the summation of the daily electricity consumption for that particular month as given by

$P_{c,M} = \sum_{n=1}^{30} P_{c,n}$ where n is days of a month (1)

Similarly, the monthly energy generated from solar photovoltaic system of a building is the summation of the daily generation for that particular month as given by

$P_{g,M} = \sum_{n=1}^{30} P_{g,n}$ where n is days of a month (2)

Also, the current method of net metering policy adapted globally is given by eqn (3). The net metering incentive is the product of difference of energy generation($P_{g,M}$) and consumption($P_{c,M}$) with agreed FIT from power purchase agreement (PPA).

$FIT_N = (P_N) \times (FIT)$ where P_N is $P_{g,M} - P_{c,M}$ (3).

The total savings from this policy is the summation of electricity consumption bill from utility compensated and the net metering incentive (FIT_N) obtained by eqn.

Total savings = Electricity consumption bill + FIT_N.....(4)

The proposed method considers all three periods of the electricity demand and is formulated in below equations. The energy consumed and generated in all three periods P_0,P_1 and P_2 of a month is considered in the equations and is modified as given in eqn (5) and eqn (6)

$$P_{c,M} = \sum_{n=1}^{30} \sum_{m=0}^{2} (P_c)_{m,n} \quad (5)$$

$$P_{g,M} = \sum_{n=1}^{30} \sum_{m=0}^{2} (P_g)_{m,n} \quad (6)$$

The proposed method of net metering policy is to define different FIT for all three different periods. The net metering bill throughout the month is equal to the product of difference in power generation and consumption during a particular period with $FIT_{N,m}$ agreed in power purchase agreement for the respective period (m) as shown in eqn.

$$FIT_{N,m} = \sum_{n=1}^{30} \sum_{m=0}^{2} \{(FIT)_m \times [(P_g)_{m,n} - (P_c)_{m,n}]\}$$

.... (7)

Also, the total savings from this proposed policy is the summation of electricity consumption bill from utility compensated and the net metering incentive ($FIT_{N,m}$) obtained by eqn.

Total savings = Electricity consumption bill + $FIT_{N,m}$ (8)

5 Result and Discussions

Taking into account the necessary changes in the existing policy, the proposed model was formulated. The current policy has neglected to consider the time of day for the generation of electricity, while the proposed policy accounts for the time of generation and categorizes it according to energy demand in the region of interest. Moreover, the proposed policy supports different FiT for different periods of generation based on peakperiod, off-peak period and moderate period.

The proposed model was applied for a green building in Bangalore, India considering a 3-kW simulation model. It is assumed that power generation being more than the power consumption of the building with rooftop PV plant throughout the year. The region considered in this paper is blessed with annual global horizontal irradiance (GHI) of 2073 kWh/m^2 [7], [10], has an average GHI of 5-6 kWh/day. The case study of the building considered here has a monthly energy consumption of 180 kWh, 120 kWh and 60 kWh during the peak period, off-peak period and moderate period respectively, as shown in Table II and III. Also, the PVsystem in this building generates 210 kWh, 150 kWh and 90 kWh during the peak period, off-peak period and moderate period respectively of a month as shown in Table 2 and 3.

Table 2 Total savings from the current policy

Power Parameters				Current Policy				
	Pg in kWh	**Pc in kWh**	**Pn in kWh**	**Utility Tariff**	**Electricity Bill**	**FiT**	**Net Metering incentive**	**Total savings**
Peak Period	210	180	30	5	900	3.56	106.8	1006.8
Off Peak	150	120	30	5	600	3.56	106.8	706.8
Moderate	90	60	30	5	300	3.56	106.8	406.8
Total	**450**	**360**	**90**		**1800**		**320.4**	**2120.4**

Table 3 Total savings from proposed policy

Power Parameters				Proposed Policy				
	Pg in kWh	**Pc in kWh**	**Pn in kWh**	**Utility Tariff**	**Electricity Bill**	**FiT**	**Net Metering incentive**	**Total savings**
Peak Period	210	180	30	6.25	1125	4.45	133.5	1258.5
Off Peak	150	120	30	4.25	510	3.026	90.78	600.78
Moderate	90	60	30	5	300	3.56	106.8	406.8
Total	**450**	**360**	**90**		**1335**		**331.08**	**2266.08**

From the current net-metering policy, the total power consumption of a month $P_{c.M}$ from eqn (1) is 360 kWh and $P_{g.M}$ from eqn (2) is 450 kWh. Also, the average utility tariff is considered to be Rs.5/kWh and FiT is considered to be Rs.3.56/kWh. Hence the power net metered (P_N) i.e., difference of power consumed and generated is 90 kWh and the net metering bill from eqn (3) is Rs.320.4. Thereby the total savings from this policy is the summation of electricity bill and net metering bill from eqn (4) is Rs.2120.4.

In the proposed net-metering policy, both the utility tariff and FIT is considered to be 25% surcharged than the current tariffs during peak period, 15% rebated than the current tariffs during off peak period and during moderate period it remains same as current tariff. The monthly net metering bill from eqn (7) is Rs.331.08 and the total savings from eqn (8) is Rs.2266.08. Upon comparison of models of both the policies in table, it was found that the net metering incentive to be 3.33% more from the proposed policy than the current policy. Also, the total savings from proposed policy was 6.87% more than the current policy.

6 Conclusion

A framework was developed to further promote the global integration of PV rooftop systems in the residential and commercial sectors by proposing a time-based (ToG) feed-in - tariff (FIT). During the peak time, this established framework would support the prosumer by selling the power at more realistic costs and the utility would benefit by buying the power from the prosumer at lower reasonable costs during the off-peak period. Thus, during the moderate time, the established system would not deceive both the prosumer and the utility. In addition, a case study was carried out considering a 3 kW plant installed in Bangalore, south of India, and local policies

were considered in the model operation. When comparing the model results, it was found that the return on investment will be more from the proposed policy than the current policy implemented in this region because of the 6.87% increase in prosumer total savings.

7 Future Work

The suggested approach in this paper will be tested in real-time systems in regions where the time of day (ToD) is the basis for the electricity consumption tariff. The established structure would also be optimised to further enhance the advantages of both the prosumer and the utility and thus influence the worldwide user of electricity to be a prosumer.

8 References

[1] I. . R. E. Agency, *FUTURE OF SOLAR PHOTOVOLTAIC Deployment, investment, technology, grid integration and socio-economic aspects A Global Energy Transformation paper About IRENA*, 2019. [Online]. Available: www.irena.org/publications.

[2] "Quantifying net energy metering subsidies," *Electricity Journal*, vol. 32,Oct 2019.

[3] J. Thakur and B. Chakraborty, "Impact of increased solar penetration on bill savings of net metered residential consumers in india," *Energy*, vol. 162, pp. 776–786, Nov 2018.

[4] X. Jia, H. Du, H. Zou, and G. He, "Assessing the effectiveness of china's net-metering subsidies for household distributed photovoltaic systems," *Journal of Cleaner Production*, vol. 262, July 2020.

[5] S. Comello and S. Reichelstein, "Cost competitiveness of residential solar PV: The impact of net metering restrictions," *Renewable and Sustainable Energy*

Reviews, vol. 75, pp. 46–57, Aug 2017.

[6] "Net energy metering laws yield fewer rooftop solar PVs," *The Elec- tricity Journal*, vol. 30, p. 88, April 2017.

[7] "Global solar atlas 2.0, a free, web-based application is developed and operated by the company solargis s.r.o. on behalf of the world bank group, utilizing solargis data, with funding provided by the energy sector management assistance program (esmap)." 2019. [Online].
Available: https://globalsolaratlas.info/download/india

[8] "Difference between gross metering and net me- tering, difference between gross." [Online]. Available:https://cef.ceew.in/masterclass/explains/difference-between-gross
-metering-and-net-metering

[9] "Net metering, seia." [Online]. Available: https://www.seia.org/initiatives
/net-metering

[10] "Jrc photovoltaic geographical information system (PVGIS)- european commission." [Online]. Available: https://re.jrc.ec.europa.eu/pvg*ools/ en/#MR*

HOW INNOVATIVE CITIZEN FINANCING SCHEMES ENABLE LARGE-SCALE ENERGY EFFICIENCY MEASURES IN THE BUILDING SECTOR.

Sonja Wilhelm[1], Silvia Caneva[1], David van der Zande[1], Jean-Francois Marchand[2], Miguel Casas[2], Fabian Pause[3], Maximilian Wimmer[3], Johanna Kamm[3], Luís Couto[4], Barbara De Kezel [5], Velimir Šegon[6], Tijana Šimek[6], Renata Adomavičienė[7], Kristina Vaskelienė[7], Asta Gladkauskienė[7], Dalia Juškevičienė[7], Deimantė Banytė[7], Žilvinas Kačiuška[7], Conny Weber[8]

[1]WIP – RENEWABLE ENERGIES, [2]ENERGINVEST, [3]STIFTUNG UMWELTENERGIERECHT, [4]GOPARITY [5]VEB, [6]REGEA, [7]VIPA, [8]ECN

[1] WIP Renewable Energies
D-64210 Munich, Germany
Sonja.wilhelm@wip-munich.de
Silvia.caneva@wip-munich.de

ABSTRACT

Buildings play an important role in everyone's life: they serve as homes, workplaces, schools, shopping centres, hospitals or other public buildings. Nevertheless, about 75% of the EU building stock is not energy efficient, leading to a large amount of the energy produced being wasted. To minimize energy loss, existing buildings need to undergo energy efficient renovations. However, there is a particular deficit in finding suitable financing schemes for the implementation of energy efficiency measures. Innovative financing solutions are needed to enable energy-efficient retrofits on a large scale and to strengthen collaboration between all stakeholders (project promoters, public/private financial institutions, end-users) and especially with citizens. Based on the ongoing research activities in the frame of the H2020 European project CitizEE, this paper presents the latest advancements of the innovative Citizen Financing Scheme of the Pilot VIPA based in Lithuania. The VIPA scheme enables citizens to become prosumers (people who both consume and produce electricity) through the combination and leveraging of different funding sources.

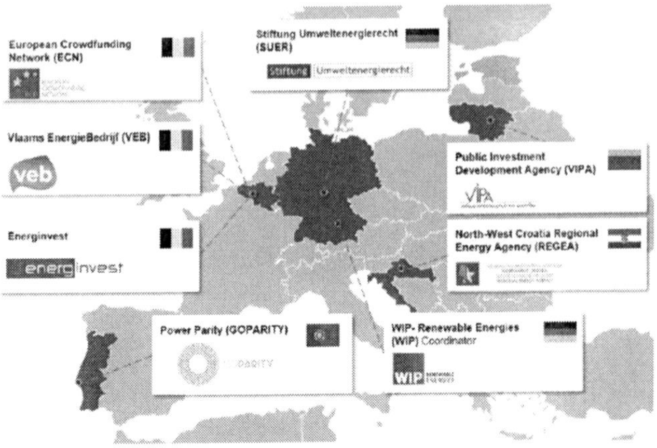

Fig. 1 CitizEE geographical and social innovation coverage

Keywords: Citizen Financing Schemes, Investment Platform, Energy Efficiency, Local Authorities

1 AIM AND APPROACH USED

This paper, to be presented at the 38th EU PVSEC, aims to introduce innovative Citizen Financing models leveraging different financing sources for energy efficiency in the building sector. This research work has been performed in the frame of the H2020 European project CitizEE. The aim of the CitizEE project is to support European local authorities to increase investments for energy efficiency in the building sector by attracting private investments from citizens.

The CitizEE consortium brings together eight partners who have been chosen specifically for their expertise and experience to ensure the successful implementation of the novel research proposed in this project and to guarantee a widespread geographical coverage (Fig. 1). The CitizEE project partners are briefly described in the following paragraph.

1. WIP – Renewable Energies (WIP), the project coordinator, communication and dissemination expert based in Germany.
2. ENERGINVEST, based in Belgium, with extensive experience in operational innovative financing models.
3. SUER, renewables and efficiency law expert based in Germany.
4. GOPARITY, RES cooperative and crowdfunding platform based in Portugal.

5. VEB, energy efficiency project aggregator and service provider based in Belgium.
6. REGEA, energy efficiency project aggregator and service provider based in Croatia.
7. VIPA, agency for financial services and instruments for public sector based in Lithuania.
8. ECN, expert in crowdfunding policy and industry based in Belgium.

European innovation strategy projects are conceived to support public entities, in this case public authorities, to drive the necessary systemic changes to reach climate neutrality (under the Green Deal call) while also ensuring an inclusive ecological and economical transition.

2 SCIENTIFIC INNOVATION AND RELEVANCE

Citizen Financing Schemes are a relatively new concept still far from having reached its full potential. They combine under their roof different financing sources and have proven to be very successful in financing projects in the field of renewable energy. Examples of such innovative citizen financing schemes are the combination of citizen financing, crowdfunding or cooperative financing, which are then complemented with public financing instruments. However, citizen access to such alternative financing mechanisms remains largely unexploited.

The CitizEE project counteracts on this by establishing ***national*** and/or ***regional Investment Platforms (IPs)***, which are backed by the European Fund for Strategic Investment (and its successor the InvestEU programme). The IPs serve as investment facility that channels public and private financing from several co-investors (citizens, public banks and institutions etc.). In that sense, IPs bring together all market players: citizens, associations and the economic sector. The additional integration of innovative ***Citizen Financing Schemes*** into those ***Investment Platforms*** enables the ***set-up of large-scale community energy efficiency programs*** and attract ***private investments*** while also ***strengthening know-how of regional/national key stakeholders***.

The conceptual framework of the proposed project is suitable to be replicated by other European public authorities. Moreover, this framework can be used by researchers as a key unit of analysis for leveraging energy efficiency measures. Further scientific objectives are the evaluation of the competitiveness and the demonstration of the viability and market potential of the established Citizen Financing Schemes. In addition, recommendations for improving the market and regulatory framework as well as the enhancement of the in-house capacity of all relevant stakeholders are being conducted.

As innovative financing models for energy efficiency retrofits are still far from having reached its full potential, cooperation between different actors (public authorities, financial institutions, and project developers) and their interactions towards innovative citizen financing models for energy efficiency are required. In this regard, ***the paper will introduce the advancements of the Pilot VIPA, one of the four CitizEE pilots, which enables citizens to become prosumers (people who both consume and produce electricity) through the innovative combination and leveraging of different funding sources.***

3 PRELIMINARY RESULTS AND CONCLUSIONS

The results will be presented following the conceptual framework. The VIPA ***Citizen Financing Scheme*** enables citizens to become prosumers (people who both consume and produce electricity) through loans (see Figure 2 below). The final loan amount may be constituted by:

- Investments from the Investment Platform
- Investments from citizens
- Investments from P2P platform operators.

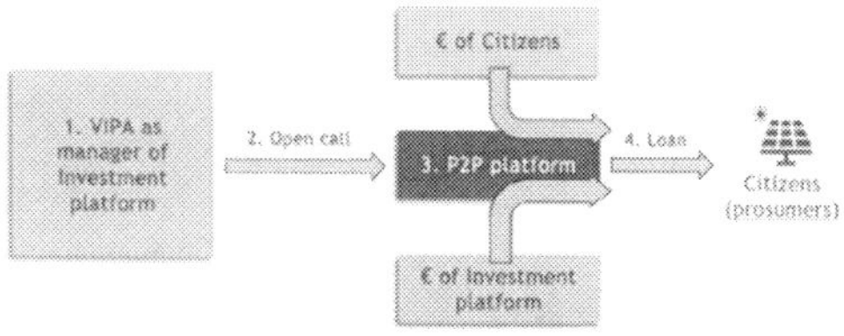

Fig. 2 Overview of the CFs4EE financing scheme of VIPA

VIPA is acting as the overall manager of the Investment platform (Nr.1 in Figure 2), responsible for the management and overall implementation of the platform, including risk management, compliance, and assurance. VIPA also provides the staff and IT infrastructure to run the platform.

One essential key aspect of the citizen financing model are P2P platforms (Peer-to-Peer services are a decentralized platform whereby individuals interact with each other on a platform, which is administrated by a P2P platform operator). The two main reasons for using a P2P platform in the financing scheme are:

- leveraging funds
- complying with consumer financing laws in Lithuania (IP do not have the license to provide consumer financing).

The P2P platform operators will be selected through an open call and act as financial intermediaries and may act as private co-investors (Nr.2 & 3 in Figure 2). Currently, VIPA is developing financial products and is negotiating with potential P2P operators (an open call for P2P platform operators will be held at the end of Q3 2021). The financing scheme also offers the possibility for prosumers to get subsidy from the state of around 320 EUR per 1kW. ***The innovative Citizen Financing Scheme leverages different financing sources allowing citizens to become prosumers.*** The findings contribute to set up large scale community energy efficiency programs, attract private investments and strengthens the know-how of European key stakeholder.

4 ACKNOWLEDGEMENTS AND DISCLAIMER

The CitizEE project receives funding from the European Union's Horizon 2020 coordination and support action under grant agreement No. 847147. The sole responsibility for the content of this publication lies with the authors. It does not necessarily reflect the opinion of the European Commission. The European Commission is not responsible for any use that may be made of the information contained therein. Follow CitizEE (http://citizee.eu/) on the social media:

- LinkedIn: #CitizEE
- Twitter: https://twitter.com/CitizEE_project

EDUCATION AND AWARENESS CAN BRAND PV TECHNOLOGY BEAUTIFUL BEFORE TURNING IT INVISIBLE

Manojit Ray[1 2] *, Indranil Majumdar[3], Tulsi Ram[4], Basab Chakraborty[1]
[1]Indian Institute of Technology Kharagpur, WB 721 302, India
[2]University of Toronto, 5 King's College Road, Toronto, ON M5S 3G8, Canada
[3]Toshani Industries Private Limited, Kolkata, WB 700 128, India
[4]Government of Himachal Pradesh, Mandi, HP 175 019, India
*Corresponding author: manojit.ray@mail.utoronto.ca, ray.manojit@iitkgp.ac.in

ABSTRACT: Worldwide, three-quarter of a billion people live without electricity. Besides, hundreds of million use a solar lantern a few hours a day. Solar photovoltaic powered mini-grids are increasingly extending improved energy access to deprived regions. However, poor load-factor and expensive storage adversely impact reach and growth. Moreover, these mini-grids commonly do not support intermittent large loads to avoid drooping resource utilization. Though electric cooking is very efficient and non- polluting, feeble mini-grids struggle to power these. Comparable supply constraints affect water treatment infrastructure that can potentially save millions from contaminant and pathogen impacted water. In cold climates, space heating remains essential but may not be supported by the electric supply system. Washing machines are efficient in terms of water usage but need appropriate supply to operate. Consumers can collaborate to avoid some of these challenges but education and awareness of the prevailing situation and the potential can help in solar PV proliferation. Though solar PV is still perceived by many communities as suitable for tiny loads only, education, awareness and suitable ecosystem can change that. Increasingly, solar PV can integrate well into the supply infrastructure and eventually become the backbone of distributed 100% renewable sourced communities across nations and geographies.
Keywords: energy access, engaged consumer, consumer awareness, consumer education, consumer collaboration

1 INTRODUCTION

The era of energy access enhancement through grid-connected systems is gradually giving way to an age of decentralized solutions. Decentralized solutions are easier to scale but more challenging to operate. Regions around the world are shifting from fossil fuel based generation to greener sources of energy to meet the increasing electricity demand. Historically, harvesting wind energy remained the more popular clean source of energy after hydro. However, with falling cost and associated surging popularity of solar photovoltaic, it has taken lead in contributing to expanding electricity generating capacities. It integrates well with complex grid networks, powers up isolated rural homes and floats over water bodies while generating. It is also the most widely owned electricity generating source of electricity globally [1].

Increased urbanization has propelled adoption of the rooftop mounted variety of photovoltaic electricity generation as it avoids eating into shrinking unoccupied ground space. Falling prices of solar photovoltaic systems, improved prediction support of renewable resources, advancement of energy storage technology, and policy supported proliferation of energy efficient gadgets are expected to accelerate acceptance of diverse, clean energy source secured, decentralized energy access provision solutions. However, increasing renewable integration poses diverse challenges. Renewable integration and associated challenges may be deliberated upon by dividing progressive integration into three distinct stages [2].

Regions in the first stage of renewable integration are in the primary phase of renewable adoption and frequently witness renewable source promotion through inducements. These inducements take care of the innate risk associated with early adoption of cleaner sources and can take the form of capital, finance, fiscal, or revenue support. In the first stage, total generation from renewable sources remain rather limited and often does not complicate electricity supply management in a major way. Electricity supply system has been traditionally designed for one way power flow, carrying energy generated at large power stations to distributed consumption centers. Power balance must always be maintained, requiring need-based flow control. In the first stage of renewable adoption, grid management does not necessitate substantial additional balancing intervention [3].

The second stage of renewable adoption experiences a larger share of energy, sourced from intermittent renewable energy generating stations connected to the electric supply network. These generating facilities are often modest in size as compared to large central generating stations but are numerous in number and are also scattered across a renewable resource-rich region. Energy source within the same grid control region may vary across generating stations and energy contribution of different stations in an area would likely be temporally diverse. In many renewable generating facilities, abrupt generation swing is difficult to anticipate prior to the event. Generation becomes largely intermittent in nature and marks the departure from the load following nature of central facilities. This renewable adoption stage demands cautious energy resource planning and intensified system balancing effort to maintain supply reliability. This second stage of renewable integration also draws not-so-positive outlook of renewable adoption as the major share of total energy still comes from traditional sources but the complexity of management increases exponentially [4].

The third and final stage of renewable adoption offers to change the characteristics of the grid, shifting it from monopolistic service to distributed, prosumer focused, and adaptable service. Progression of renewable adoption may nudge regulators to graduate to benefactors while passive consumers may morph into technology-aided, supply chasing, active agents with adequate education and essential awareness. Challenges accompanying this transition are at least as splendid as the opportunities. Remarkably, regions lacking an appropriate supply system may more easily attempt to integrate evolving technical advances and socio-economic opportunities to circumvent the trying penultimate stage. These regions may embrace the final stage of clean energy source integration with access to affordable electricity through responsible engagement. Awareness building, cultural reorientation, risk mitigation, trust development, regulatory agility, financing innovation, and responsible agent participation are some of the pivotal factors for the transition [5].

Interestingly, the pace and expanse of the transition depends critically on the consumer engagement. Consumers engage effectively only when the education, awareness and policy making nudge supports such engagement. Fear is an important motivator and prevents many consumers to engage and benefit from the opportunity in absence of proper education and awareness.

2 METHODS AND DATA

2.1 Evolution of Energy Access

Binary measurement of household electrification indicated 'presence' or 'absence' of energy supply, leaving no trace about the real consumption. Often 'presence' meant an electrified source of illumination in the house, an unembellished household electricity connection or even the existence of an electricity supply structure outside the house. This arrangement of household energy access assessment was devoid of representativeness to make access measurement meaningful [6].

Recent efforts to make energy access measurement more meaningful by deepening focus on service usability resulted in the development of the multi-tier access measurement framework. Care has been taken to make this new framework as technology agnostic as possible. Energy supply attributes change across tiers. Utilities may upgrade supply infrastructure and move across access tiers. Some of the access tier improvement may also occur because of efficiency improvement or the enhanced ability of equipment to handle lower quality power. Accurate and reliable information about energy delivery system may encourage coordination among different stakeholders to avoid abrupt supply failure. Evolving technology, consumer behavior and the drifting power-service-energy relationship may make some of the transition levels between tiers of access less pertinent over time. This may require periodic revision of defined access tier targets and associated thresholds, keeping the underlying energy access measurement framework intact [7].

Energy access index can be calculated taking indices of household energy, productive engagement and community facilities into account. Household energy access index, in turn, integrates household electricity, household heating and household cooking (Figure 1) indices. This article centers on household electricity and household heating indices while exploring impact of consumers' approach.

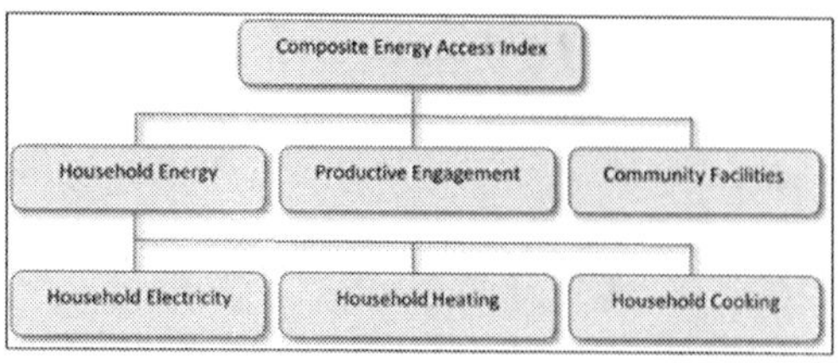

Figure 1: Composite energy access index, like household energy access index, integrates constituent components.

2.2 Communities with Feeble Supply

Towering Himalayan mountain ranges surrounding the Spiti Valley in India deprives the valley of the advantage of monsoon compared to many other parts of the country. The Spiti Valley receives snowfall over a prolonged winter period and experiences very low temperatures spanning an extended period (Figure 2).

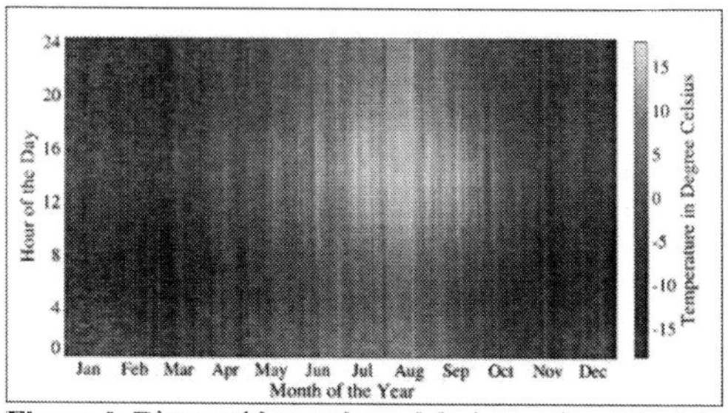

Figure 2: Bitter cold experienced during prolonged winter months necessitates space and water heating in the valley.

In such low temperatures, space heating becomes indispensable. Inhabitants of the valley have been using a firewood burning contraption, locally called 'Bukhari', for space heating. It has a wide cylindrical fire chamber attached to a narrow top cylinder. The base of the wider cylinder accommodates burning firewood. Narrow cylinder acts as a chimney and spreads heat in the room. This space heating solution is inefficient in energy conversion and billows toxic smoke. In practice, chimney may not always be sufficiently long to exhaust smoke outside the heated unventilated room. Local administration encourages individuals not to use traditional heating methods. Noxious inhalation, acute child burn cases and complete fatality are not uncommon. To address concerns, people prefer electric space heating solution but feeble electricity supply prohibits the use of space heaters owing to inadequate infrastructure.

Climate change has impacted the valley, resulting in less water being available during the brief farming period. In a typical year, at present, even tiny parcels of available farmland cannot all be cultivated because of severe water scarcity. Bushes no longer grow in enough quantity during the very short and mild summer period, leading to a shortage of firewood in winter. Usage of firewood for space heating grievously impacts health. Firewood provisioning has gradually proceeded from a traditional household activity to comprehensive commercial business. This has resulted in firewood price rise to a level where most of the households have been forced to limit the use of the traditional space heater to an hour in the late evening. Residents of the hamlet still use traditional space heating in a close environment ignoring administrative appeals and spend the night in toxic fume-filled rooms. Moreover, the electricity supply ceases to provide power occasionally (Figure 3).

In absence of permission to connect comparatively larger loads to the electricity supplying grid, households consume electricity for task lighting, small electronic device including cell phone charging and entertainment device powering. Mini television sets are used along with a Direct to Home satellite set-top box. Average households use up to four energy efficient light sources and two small electronic gadget chargers. An electric room heater and a washing machine have remained aspirational gadgets for the impoverished consumers of the valley for long.

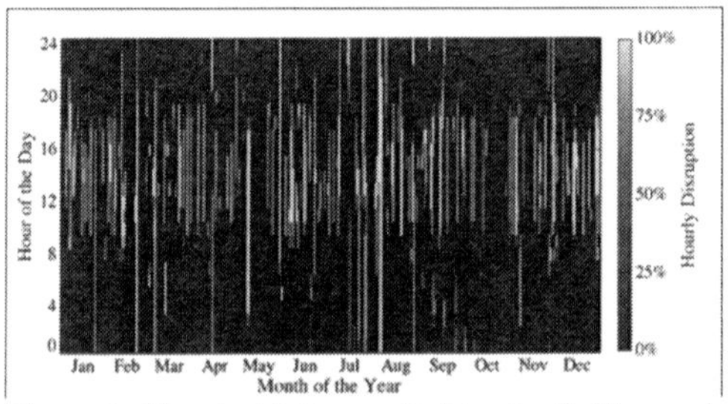

Figure 3: The electricity supplied by the feeble supply system of the valley remains unreliable through the year

2.3 Communities with No Supply

Communities in the Global South often experience insufficient to no energy access together with economic, educational, household, and health challenges. One of the challenges, child mortality, has seen remarkable progress with a reduction in incidents over the past decades. Yet, an estimated 5.3 million children were lost in 2018. Global South experiences the highest child mortality rate in the world. In 2018, the region had an average under-five mortality rate of 1 in 13 children dying before his or her fifth birthday - several times higher than the average fatality of 1 in 200 children in high-income economies. Pneumonia and diarrhea are the two primary catalysts of fatality among under-five children. UNICEF and WHO have proposed a plan to address the needs of a growing child. Safe drinking water and clean indoor environment are marked as essentials for the survival and growth of a child (Figure 4).

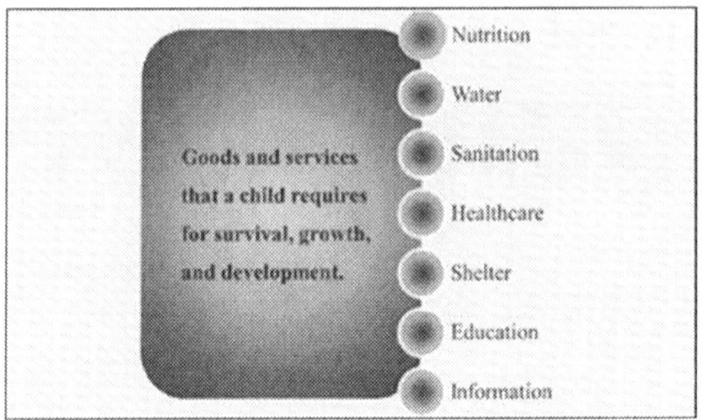

Figure 4: A young child requires safe drinking water and clean shelter among others to survive, grow and develop

But, the provision of safe drinking water requires water treatment and it necessitates a high access tier electricity supply. Befitting access to electricity may extend an added benefit of gender equity by reducing the time and stress of fetching water from faraway places. Similarly, provision of clean indoor environment requires the adoption of clean cooking. But, required high access tier electricity supply remains a persistent challenge in many of the impoverished communities. Cost-effective and suitable electricity access may mitigate the challenge.

Consumer collaboration can support energy sharing and device sharing, if required. Such collaboration can enable a consumer enjoy a higher access tier than the service provision (Figure 5). A device requiring a high access tier supply may be powered with a more modest supply provision with integration of evolving technologies and resulting adoption of collaborative opportunities. Moreover, it impacts nations differently as the income equality and per capita income influences the benefit that can be realized in a community (Figure 6)

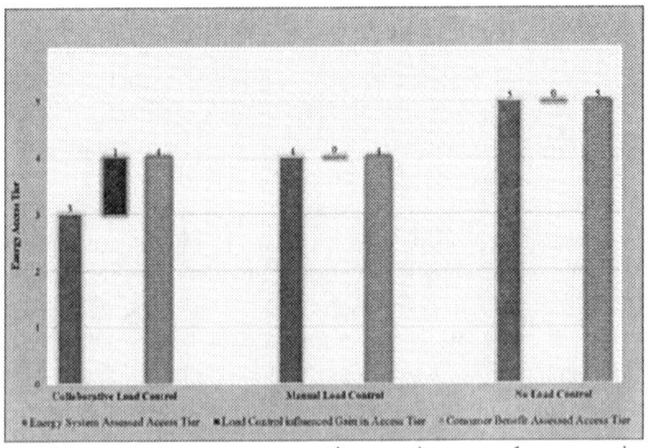

Figure 5: Consumers can enjoy an improved access tier compared to supply with informed effective collaboration

The Gini coefficient indicates the distributional equality of income and the same impacts the gain that can be realized through collaborative consumption of electricity while attempting to scale access tier enjoyed by educated and aware consumers.

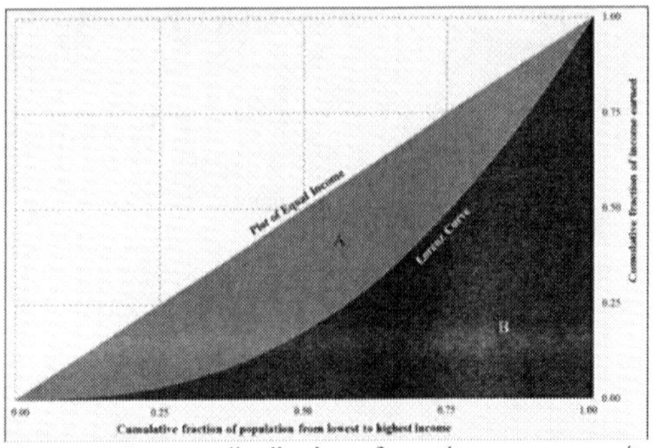

Figure 6: Income distribution of a nation or community along with the per capita income influences shared gain

2.4 Demand Response

The electric grid accommodates two contradictory aspects of energy supply. One aspect focuses on unfailing, quality supply even in the face of technical or other kinds of disruptions. This demands widespread redundancy, providing necessary backup support. The other aspect targets energy delivery in the most affordable way possible. It suggests optimization with focus on service reliability vis-à-vis energy price. Whatever the design philosophy of any specific grid may have been, it has always remained demand following in nature and incorporated malleable, large, and on many occasions, polluting generating stations to supply required electricity. Resulting pollution, associated with ensuing environmental unsustainability, in turn, mandated exploration of cleaner sources of generation. Participation of consumers in the electricity supply operation increasingly became essential to address the issue. Such participation can improve resource efficiency and improve affordability of access.

Demand response (Figure 7) encompasses both price based and incentive based schemes. All price based schemes and most incentive based schemes do not offer definitive load control but extend a sense of consumer choice while rewarding grid favored load movements. Such a scheme may deliver desired result when the primary aim of the utility is either raising share of intermittent sources or minimizing extensive load curtailment. Hierarchical grid control and isolation possibility backs up the grid operation in such situations. Distributed, remote, tiny settlements may not have advantage of tiered control unless large number of interconnected hamlets opt for combined demand response program [8]. In an isolated, resource deficient community settlement, it may be useful to implement and deliver results through adoption of need and preference matched predefined time sharing of available resources. Direct load control, offering assured 'supply chasing' capability, may remain the suitable choice for adoption by such a community while trying to make better use of available electricity supply on a time sharing basis to keep peak power within supply stipulated ceiling. Though the consumer autonomy may be considered as partially curtailed, the benefit of energy access may more than compensate for the same. Moreover, the participation in demand response remains voluntary and the benefits may expand with the formation of a market.

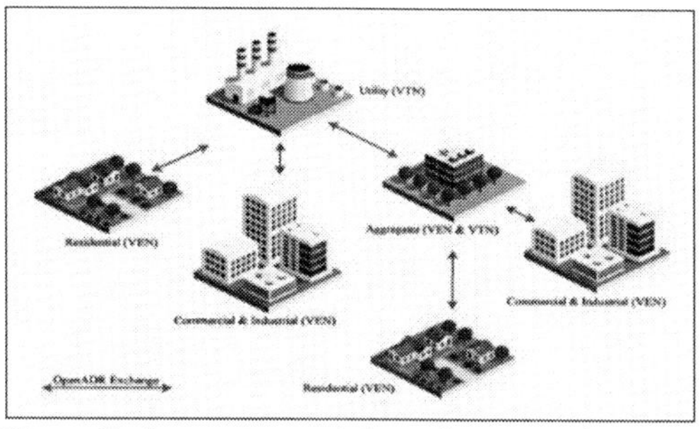

Figure 7: Open automated demand response integrates consumers in the ecosystem through tiered nodes

2.5 Multi Criteria Decision

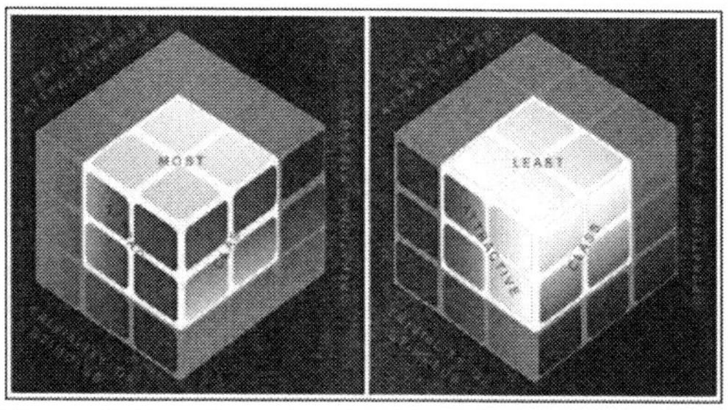

Figure 8: Operation, attractiveness and stimulus may form the basis for multi-criteria decision making framework

In the multi-criteria decision making framework (Figure 8), triviality of stimulus, operational strength and sectoral attractiveness remains integrated.

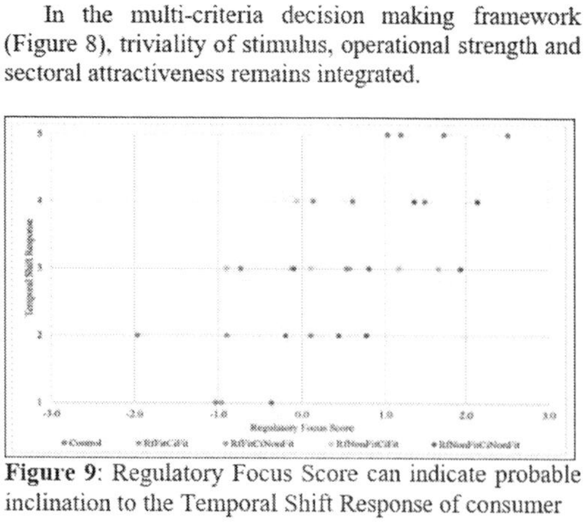

Figure 9: Regulatory Focus Score can indicate probable inclination to the Temporal Shift Response of consumer

Consumer education, consumer awareness and consumer goal pursuit motivation (Figure 9) influences outcome.

3 RESULTS AND DISCUSSION

The consumers can share available energy resources effectively to enjoy access or to improve access level realized. To optimize the solar PV plant design and minimize initial investments by the energy service operator, efficient lighting (Light-emitting diode), air circulation (Brushless Direct Current) and efficiency-certified appliances can be used by the consumers. All essential household gadgets can be powered while keeping the cost of energy low (Figure 10).

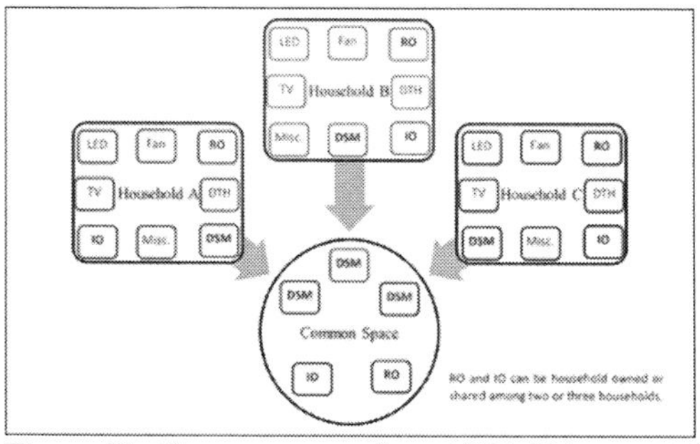

Figure 10: Consumers can share 100% PV generated energy resources as well as essential gadgets of household

All countries of the global south with energy access deficiency can benefit with collaborative sharing through education and awareness. However, the benefit gets influenced by the per capita income as well as income distribution (Figure 11).

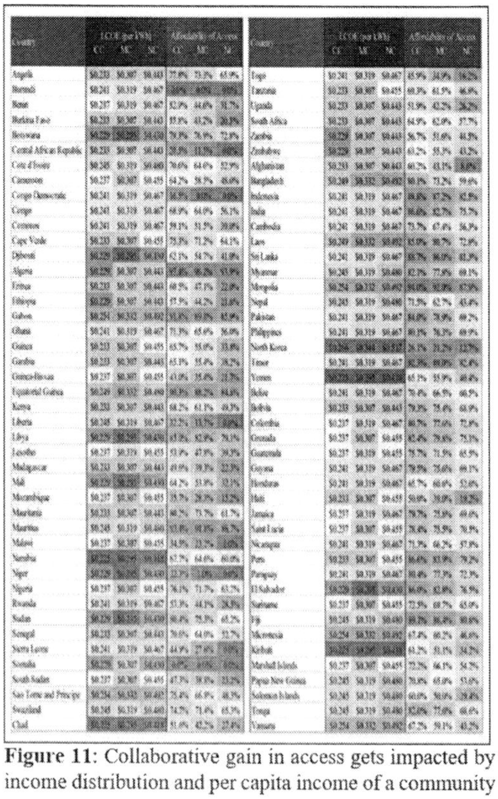

Figure 11: Collaborative gain in access gets impacted by income distribution and per capita income of a community

A feeble supply provision can be collaboratively used for all essentials including powering space heating and washing machine of a household (Figure 12).

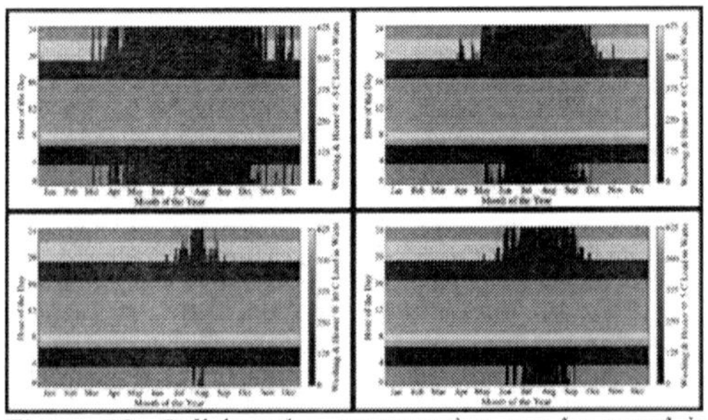

Figure 12: Collaborative consumption can be tuned in response to energy availability and economic affordability

The unprecedented growth of solar PV has received support from policy makers and dwindling cost but the growth can be much more impressive with active role played by the educated and aware consumers (Figure 13).

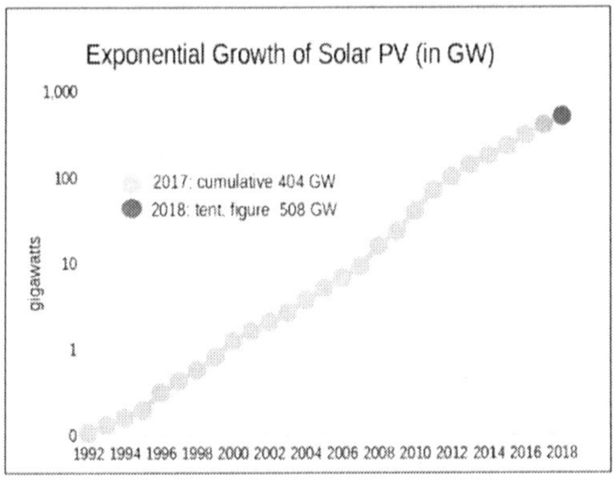

Figure 13: Unprecedented growth of solar PV adoption can result into widespread energy access expansion

Consumer orientation deeply impacts participation and effective gain realized with collaboration (Figure 14).

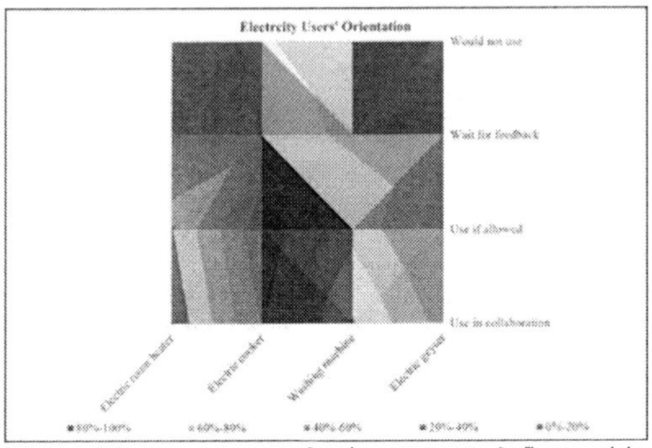

Figure 14: Consumer motivation may get influenced by diverse sets of influences and the same may impact access

4 CONCLUSION

Consumer education turns out to be the key for successful policy implementation. Awareness about own need and photovoltaic capabilities becomes mandatory for appreciating real ability of solar photovoltaic to fulfil consumer need. Surprisingly, people who are less involved in direct economic activity are more interested to learn. This actually opens an opportunity to tap economically unused or less-used human resources to propagate possibility into target society. Unsurprisingly, settlements with higher level of difficulty and less resources at their disposal are more inclined to participate in the collaborative use cases. This approach may make them fast learners. People want heater in cold climate but not cooking support or even geyser. School children are familiar with solar photovoltaic but thinks these are good for illumination, phone charging and occasional television viewing. Same is with religious organizations, settlements and even commercial establishments. Religious organizations, though revered in society, are largely averse to exploration. Commercial establishments are more malleable but less collaboration seeking. Remarkably, large load support capability awareness is just missing. While electric room heater and washing machine has higher share of people expressing interest in collaborative use, electric cooker does not find many takers without feedback from other users. Electric geyser has even lower traction because consumers believe a large amount of heat gets wasted when only a small amount of hot water is used, leaving balance water to cool.

5 REFERENCES

[1] The United Nations High Commissioner for Refugees. Global Trends - Forced Displacement, https://www.unhcr.org/globaltrends2019/; 2020

[2] The United Nations, Department of Economic and Social Affairs, Statistics Division. SDG Indicators, https://unstats.un.org/sdgs/metadata/?Text=&Goal=10&Target=10.7; 2020.

[3] The United Nations. Sustainable Development Goal 7, https://sustainabledevelopment.un.org/sdg7/; 2015

[4] The United Nations Institute for Training and Research. Global Plan of Action for Sustainable Energy Solutions in Situations of Displacement, https://unitar.org/sustainable-development-goals/peace/our-portfolio/global-plan-action-gpa-sustainable-energy-solutions-situations-displacement; 2020

[5] World Health Organization, Loughborough University Water Engineering and Development Centre. How much water is needed in emergencies? https://www.who.int/water_sanitation_health/publications/2011/WHO_TN_09_How_much_water_is_needed.pdf; 2013

[6] International Renewable Energy Agency. Renewables for refugee settlements: Sustainable energy access in humanitarian situations, https://www.irena.org/publications/2019/Dec/Renewable-solutions-for-refugee-settlements; 2019

[7] Ray M, Chakraborty B. Solar PV Driven Child Mortality Alleviation in the Global South. 37th European PV Solar Energy Conference and Exhibition 2020; 3-936338-73-6:1811-1814. https://doi.org/10.4229/EUPVSEC20202020-6BV.5.7.

[8] Ray M, Chakraborty B. Impact of evolving technology on collaborative energy access scaling. Renewable and Sustainable Energy Reviews 2019; 110:13-27. https://doi.org/10.1016/j.rser.2019.04.051

A BRAND NEW TRAINING PLATFORM AIMED AT UPSCALING SOLAR CAPACITY BUILDING

Gaiddon, B.[1], De l'Epine, M.[1], Enjolras, P.[1], Verdeil, O.[2], Anquetin, S.[2], Gagnaire, D.[2]
[1]Hespul, 14 place Jules Ferry, 69006 Lyon, France
email: bruno.gaiddon@hespul.org , melodie.delepine@hespul.org , paul.enjolras@hespul.org
[2]INES PFE, 60 avenue Lac Léman, 73370 Le Bourget-du-Lac, France
email: olivier.verdeil@ines-solaire.org , solenn.anquetin@ines-solaire.org , dimitri.gagnaire@ines-solaire.org

ABSTRACT: The Training and Evaluation Platform of the French National Institute of Solar Energy (INES PFE) is a training center dedicated to solar energy. Previously, most training sessions organized by INES PFE were based on an inconvenient and outdated training platform initially designed for the residential PV market. In order to drive capacity building for stakeholders involved in the planning, design, operation and maintenance of PV systems, INES PFE decided in 2019 to design a new training platform, equipping it with cutting edge systems and components used worldwide. The aim of this paper is to describe the technical features of the training platform and in particular two models, one dedicated to PV systems with storage for non-interconnected grids, another dedicated to the connection of large-scale PV systems to the Medium Voltage grid, developed in partnership with Hespul, a French not-for-profit organization specialized in the promotion of PV.
Keywords: Education and Training, Large Grid-connected PV systems, Storage, Developing Countries

1 PRESENTATION OF INES PFE

INES PFE [1] is the Training and Evaluation Platform of the French National Institute of Solar Energy. It is a training center dedicated to solar energy located directly on the INES campus in France, near the Swiss and Italian borders. Its role is to accelerate capacity building of all stakeholders involved in the planning, design, operation and maintenance of PV systems whilst insuring a high level of quality in the work they will undertake. Most training sessions organized by INES PFE used to take place in an outdated training platform initially designed for the residential PV market and located 10 km away from the INES campus, inadequate to reconcile practical and theoretical training sessions. Following an increased demand for capacity building of decision makers, designers and installers of PV systems as well as PV operation and maintenance companies, INES PFE decided in 2019 to design a new training platform, equipping it with cutting edge systems and components used worldwide.

The old training platform was dismounted in the summer of 2020 and the new training platform came into operation at the end of 2020. The first phase is now fully completed with a 500m2 outdoor area dedicated to ground mounted PV, a 950 m2 outdoor area with 3 tiny houses dedicated to roof-top-mounted PV and a 500 m2 indoor area with inverters and other electrical components (see Figure 1).

Figure 1: Overview of the INES PFE training platform

The new platform is designed not only to support the European professionals on a wider range of skills and application developments, but also to respond to the fast-increasing demands from African decision-makers and operators. In order to develop new training sessions, INES PFE signed several partnership agreements with international or national organizations such as the International Solar Alliance (ISA) [2] in India and Hespul [3] in France, a French not-for-profit organization specialized in the promotion of PV. The aim of this paper is to describe the technical features of two specific models on this new training platform used for training sessions developed and organized in partnership with Hespul, one dedicated to PV systems with storage for non-interconnected grids and another dedicated to the connection of large-scale PV systems to the Medium Voltage grid.

2 REDESIGN OF THE MINI-GRID MODEL

2.1 Background

In March 2019, the Agency for the Safety of Air Navigation in Africa and Madagascar (Asecna) [4] based in Senegal came to INES PFE for a dedicated training session related to large-scale PV systems. Asecna was looking for technical information to manage the construction of a 1MWp PV system equipped with storage in order to reduce the use of Diesel generators of the International Airport of Nouakchott in Mauritania. Also, following the successful implementation by SMA of the PV system with storage on Statia island in the Caribbean in 2017 [5], many PV systems with storage have been deployed on small islands to reduce the use of, or shut down Diesel generators several hours a day. This new application of PV systems led Hespul and INES PFE to create a new training session dedicated to PV systems with storage for non-interconnected grids in order to reduce the consumption of Diesel [6], bringing to redesign an old Mini-Grid model.

2.2 Features of the Mini-Grid model

The objective of the Mini-grid model is to show several operating modes of a PV system connected to a non-interconnected grid such as:

- Mode 1 (Grid connection): the grid is stable and the PV system is connected with the production either self-consumed or injected to the grid,

- Mode 2 (Parallel operation): a diesel generator powers the grid and the frequency is of low quality. The power delivered by the PV system is limited due to the parallel operation of the Diesel generator,
- Mode 3 (Diesel off): the diesel generator is off and a PV grid forming inverter with storage powers the loads.

Based on the single line diagram of the International Airport of Nouakchott, a single line diagram of the Mini-Grid model has been designed to demonstrate these 3 operating modes (See figure 2).

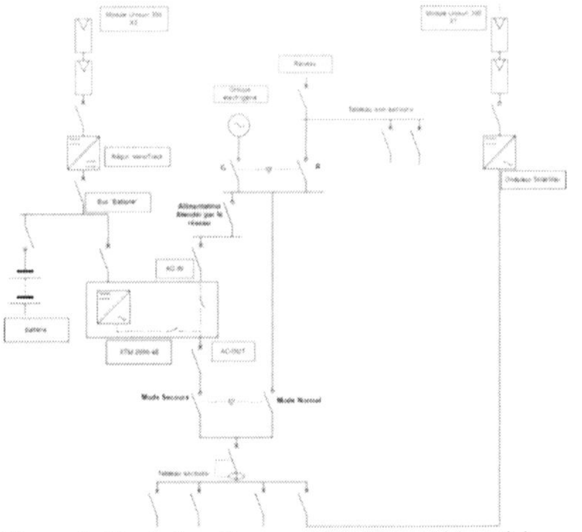

Figure 2: Single line diagram of the Mini-Grid model

The Mini-Grid model is installed within one of the tiny houses of INES PFE and is composed of the following devices:

- a 6 kW single phase SDMO Diesel generator,
- a reversing switch to select the type of grid with 2 choices: stable interconnected grid or weaker grid powered by a Diesel generator,
- a general switchboard to connect loads such as lights, electric heaters and plugs to connect any other devices,
- a 2,1 kWp grid connected PV system connected to the general electric board (7x300Wp Unisun PV modules and 1x1800W Sputnik Solarmax inverter),
- a reversing switch to select the operating mode of the general switchboard with 2 choices: normal operation mode with connection to the mains or back-up mode if case of grid failure,
- and a Studer Innotec Xtender grid forming inverter connected to a 0,6 kWp PV system and a 48Vdc battery set of 460 Ah.

The construction work of this Mini-Grid model occurred from May to July 2021 and is now completed (see Figure 3). From September 2021 onwards it will be used for new training sessions of INES PFE dedicated to PV systems with storage for non-interconnected grids in order to reduce the consumption of Diesel.

Figure 3: The new Mini-Grid model of INES PFE

3 DESIGN OF A MEDIUM VOLTAGE MODEL

3.1 Legal framework and market size

The technical requirements for the connection of PV systems to the grid are set by the Ministry Order of June 2020 [7]. Article 24 of this text gives the grid voltage as a function of the nominal power of the PV system: above 250 kW, a PV system must be connected to the medium-voltage grid (MV) and up-to a power of 12 or 17 MW (see table I).

Table I: Grid voltage as a function of the nominal power of the PV system

Grid voltage	Maximum power
Low voltage LV (230-400V)	250 kVA
Medium Voltage MV (20.000V)	12 or 17 MW

According to Enedis, the largest French distribution grid operator, there were 1560 PV systems connected to the MV by the end of March 2021 for a total power of 4,9 GW, i.e. approximately 50% of the installed capacity of PV in France [8]. This range of PV systems has therefore a significant impact on the French PV market and, as the number of PV systems in operation is limited, this generates an increased need for capacity building related to the connection of PV systems to the MV grid.

3.2 Features of the MV model

The objective of the MV model is to show to trainees of INES PFE typical devices used to connect large-scale PV systems to the MV grid. The connection of PV systems to the MV grid requires the use of specific MV devices called switchgears, given that each switchgear has a specific electrical function: switching, metering, protection, ... Once installed, switchgears are generally switched on and safety measures require that only skilled staff members with a protective equipment can approach them. Thus, it is not common to have the opportunity to approach such equipment, to see what is behind the front panel or to operate it.

The MV model of INES PFE replicates the grid connection point of a PV system with a power smaller than 1,25 MVA (see Figure 4). It is composed of the following devices:

- a MV Switch used to physically connect the power cables of the grid with the PV system,
- a MV Switchgear dedicated to voltage measurement equipped with 3 MV fuses and 3 MV/LV

transformers used to connect the external automatic disconnection device,
- a MV circuit breaker to protect the MV/LV transformer,
- a MV/LV transformer,
- a main LV circuit breaker.

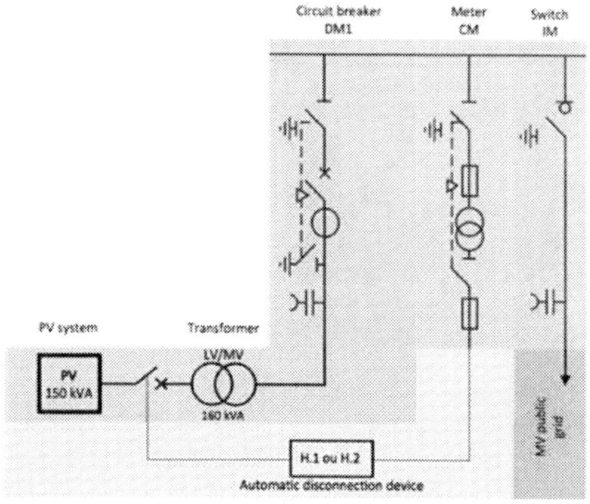

Figure 4: Single line diagram of the MV model

These MV switchgears have been provided by 2 different local grid operators, Soréa [9] in the French Alps and GreenAlp [10] in Grenoble, and are not connected to a power source, with the side panels removed to better understand how they operate. The MV model is in place at INES since October 2020 (see Figure 5).

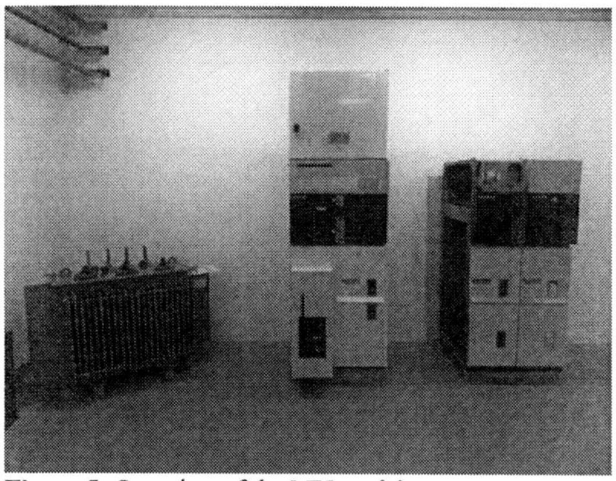

Figure 5: Overview of the MV model

It is used during training sessions dedicated to the development of large-scale PV systems targeting project developers and engineering consultants [11]. During these training sessions, trainees learn how to design the connection point of a large-scale PV system to a MV grid and how to operate MV switchgears to switch on, switch off the PV systems or to disconnect the MV switchgears to the MV grid (see Figure 6).

INES PFE and Hespul are now considering other potential cooperations to further jointly develop new training sessions and new PV models.

Figure 6: Use of the MV model during a training session dedicated to the connection to the grid of large-scale PV systems

4 ACKNOWLEDGEMENTS

The training platform of INES PFE has received funding from the French State within INES.2S [12], an Energy Transition Institute dedicated to developing the French photovoltaic industry, and from the Auvergne-Rhône-Alpes regional government [13]. The authors therefore wish to thank them for the funding received. They also wish to thank all companies that donated technical material such as Soréa and GreenAlp.

5 REFERENCES

[1] www.ines-solaire.org
[2] https://isolaralliance.org
[3] www.hespul.org
[4] https://asecnaonline.asecna.aero/index.php/en/
[5] E. Garralaga Rojas, H. Sadri, W. Krueger, '*Experience on MW-Sized Hybrid PV, Battery Storage and Genset System; Case Study of St. Eustatius Island*', proceedings of the 36th European Photovoltaic Solar Energy Conference and Exhibition, Marseille, 2019
[6] https://www.ines-solaire.org/renforcer-capacites/formation/systemes-photovoltaiques-stockage-et-reseaux-pour-pays-emergents-et-zones-insulaires/
[7] Arrêté du 9 juin 2020 relatif aux prescriptions techniques de conception et de fonctionnement pour le raccordement aux réseaux d'électricité www.legifrance.gouv.fr/eli/arrete/2020/6/9/TRER2007744A/jo/texte
[8] https://data.enedis.fr
[9] www.sorea-maurienne.fr/
[10] https://greenalp.fr
[11] https://www.ines-solaire.org/renforcer-capacites/formation/conception-exploitation-et-maintenance-des-grandes-centrales-photovoltaiques-250-kwc/
[12] https://www.ines-solaire.org/en/ines-2s/
[13] www.auvergnerhonealpes.fr

STRATEGIES AND TECHNOLOGIES TO ACHIEVE A EUROPEAN FOSSIL-ENERGY-FREE AGRICULTURE: THE AGROFOSSILFREE PROJECT

Dominik Rutz[1], Veronika Hofmeier[1], Chuan Ma[1], Rainer Janssen[1],
Athanasios Balafoutis[2] (project coordinator), Konstantinos Vaiopoulos[2], Claus Aage Grøn Sørensen [3], Dimitris Manolakos[4], Alex Koutsouris[4], George Papadakis[4], Magdalena Borzecka[5], Vanja Bisevac[6], Daan Creupelandt[7], Julio Román[8], Frank Oudshoorn[9], Daniele Rossi[10], Martyna Próchniak[11], Zisis Tsiropoulos[12], Harm Brinks[13], Barry Caslin[14], Jorge Sneij[15], Mirentxu Asín[16]

[1]WIP – Renewable Energies, Sylvensteinstr. 2, D - 81369 Munich, Germany
Tel. +49 89 720 12743, Fax +49 89 720 12791
E-Mail: dominik.rutz@wip-munich.de
Internet: www.wip-munich.de, www.agrofossilfree.eu

[2]Centre for Research and Technology Hellas (coordinator), [3]Aarhus University, [4]Agricultural University of Athens, [5]Institute of Soil Science and Plant Cultivation, [6]European Agricultural Machinery Association, [7]REScoop, [8]European Conservation Agriculture Federation, [9]Landbrug & Fodevarer, [10]Confagricoltura, [11]Lublin Agricultural Advisory Center in Konskowola, [12]AGENSO, [13]DELPHY BV, [14]TEAGASC, [15]Trama TecnoAmbiental, [16]Iniciativas Innovadoras

ABSTRACT: The AgroFossilFree project is supporting Fossil-Energy-Free Technologies and Strategies (FEFTS) in agriculture by coordinating several stakeholder actions and by providing information on FEFTS in agriculture. Thereby, special emphasis is placed on applications of photovoltaics (PV) in agriculture. The core objective of the project is to set-up an online platform for farmers, advisors, researchers, industrial partners and other actors on which they can find useful information on FEFTS in agriculture. The AgroFossilFree project contributes to the High Level EU Strategies (i.e. EU Green Deal and Farm to Fork strategy) as it aims to decrease the use of fossil energy in any farming process. It also contributes to close the gap between the available FEFTS and the everyday EU agricultural practices by promoting effective exchange of novel ideas and by exchanging information between research, industry, extension services and the farming community. If this gap closes, European agriculture could become more sustainable with minimum energy, environmental and socioeconomic impacts. With the knowledge that is collected, classified and presented to actors in the agricultural field, the European agricultural sector is able to promote and comply with the fossil energy use reduction policies, the related legal and regulatory frameworks and sustainable food production practices. This paper presents the current status of the AgroFossilFree project which is funded by the Horizon 2020 Programme of the European Commission.

Keywords: Photovoltaic, Dissemination, Agriculture, Strategy, Agri-PV, Defossilisation

1 INTRODUCTION

Emissions from global agriculture are, among emissions from other sectors, a significant driver for climate change. The agricultural sector was responsible for about 10% of the EU's total GHG emissions in 2015, excluding land use, land use change, and net forestry removals [1]. These emissions are dominantly associated with livestock production (both enteric fermentation and manure management) and the application of fertilisers to agricultural soils. A recent study [2] found that the emissions related to the production of energy in agriculture (except for the horticultural crop type) were found to be of minor importance compared to other emission sources.

Nevertheless, in absolute numbers, large amounts of fossil energy are used in agriculture for fertilizer and pesticide production as well as for heating, cooling, transport and electricity. The AgroFossilFree project aims to support any fossil fuel saving measures and replacements by renewable energies, also named as Fossil-Energy-Free Technologies and Strategies (FEFTS).

FEFTS refer to the tools that are required to address cleaner and more efficient energy production and use in agriculture. FEFTS are categorised based on the associated agricultural technology application. The two main categories are about Clean Energy Production and Energy Efficiency Improvement, while an auxiliary category refers to Carbon Sequestration.

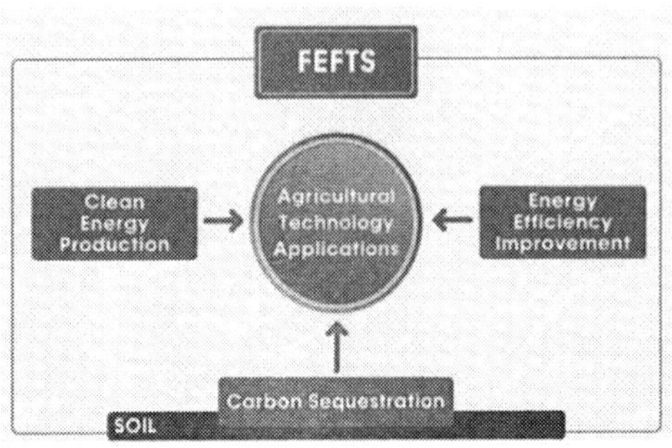

Figure 1: Overview on Fossil-Energy-Free Technologies and Strategies (FEFTS)

2 THE AGROFOSSILFREE PROJECT

The AgroFossilFree project (Figure 1) is supporting integration of FEFTS in agriculture by coordinating several stakeholder actions and by providing information on FEFTS in agriculture. Thereby, special emphasis is placed on applications of photovoltaics (PV) in agriculture.

Figure 2: Logo of the AgroFossilFree project

The core objective of the project is to set-up an online platform (AgEnergy Platform) for farmers, advisors, researchers, industrial partners and other actors on which they can find useful information on FEFTS in agriculture.

The AgroFossilFree project contributes to the High Level EU Strategies (i.e. EU Green Deal and Farm to Fork strategy) as it aims to decrease the use of fossil energy in any farming process. It also contributes to close the gap between the available FEFTS and the everyday EU agricultural practices by promoting effective exchange of novel ideas and by exchanging information between research, industry, extension services and the farming community. If this gap closes, European agriculture could become more sustainable with minimum energy, environmental and socioeconomic impacts. With the knowledge that is collected, classified and presented to actors in the agricultural field, the European agricultural sector is able to promote and comply with the fossil energy use reduction policies, the related legal and regulatory frameworks and sustainable food production practices.

The expected results of the project are:

- Assessment and evaluation of the current energy use status in EU agriculture and the requirements to achieve a European fossil-energy-free agriculture.
- Identification and presentation of available and directly applicable FEFTS from applied research results to market solutions and explore existing financing tools for de-fossilising activities.
- Collaboration with all relevant stakeholders in thematic groups using interactive physical and online methodologies to produce community-based ideas for FEFTS integration in agricultural systems at regional and EU levels.
- Creation of an online platform with all available FEFTS for stakeholders and a Decision Support Toolkit to propose interventions and financing tools.
- Creation of policy guidelines on EU, regional and national levels to be communicated and to promote FEFTS in real agricultural activities in the near future.

The AgroFossilFree project is funded by the Horizon 2020 Programme of the European Commission (Grant Agreement Number 101000496), started in October 2020 and has a duration of 3 years. An overview on the status of the project is presented in the following chapters.

3 RESULTS

3.1 Current energy use status in EU agriculture

The current energy status in EU agriculture has been evaluated and the main findings indicate that energy use throughout EU agriculture is significant and fossil fuel dependent. According to Eurostat [3], agriculture accounts for 3.2% of total energy consumption, 56% which is derived directly from crude oil and petroleum products, 17% from electricity, 14% from gas and 9% from renewables and biofuels. However, AgroFossileFree results suggest that if indirect energy use associated with the production and transport of fertilizers and pesticides is included, the proportion of energy use in the EU-27 would be 62% higher overall.

The results also show that energy use, its concentrations and breakdown, vary significantly per production system (open-field, livestock, greenhouses). According to AgroFossilFree estimates of the crops and production systems, the annual energy inputs for arable agriculture are 1227 PJ, for orchards and vineyards 208 PJ, for meat production systems 501 PJ, and for dairy production systems 543 PJ.

For open-field agriculture, it is found that the use of fertilizer is the largest energy consuming activity in EU agriculture, accounting for around 50% of all energy inputs. On-farm diesel use accounts for 30%, while other uses are mainly dedicated to irrigation, storage and drying which accounts for 8%. Pesticides and seeds each account for 5% of total energy inputs. In all livestock systems, except for beef production systems, animal feed is the main energy input accounting for around three quarters of all energy requirements. The production of animal feed consumes around 60% of the cereal production in the EU and requires significant high-protein imports. On-farm electricity use, which currently mainly comes from fossil sources, is also significant but varies considerably depending on the production system. In high yielding and high-energy intensive greenhouses energy demand is dominated for heating and cooling. By contrast, lower yielding and less energy-intensive systems use little to no heating/cooling and instead energy use is mainly associated with fertilizers, diesel use for machinery, irrigation and other activities.

The above illustrate that for the EU to achieve the goals outlined in the Green Deal and Farm to Fork strategy, it is likely that the adoption of new technologies, and improvements in energy efficiency and the further development and adoption of non-fossil energy sources for agriculture is required. In addition to this, a transition to more sustainable agricultural practices and farming systems is required. For instance, conservation agriculture can significantly reduce on farm energy use and carbon emissions as well as sequester considerable amounts of carbon (up to 190M tonnes per year) and improve the overall climate resilience of the agricultural sector.

Providing an overview of energy use in EU agriculture is a challenging topic due to the plethora of the available, and in most cases inconsistent data. However, there are multiple areas that would benefit from further research. The understanding and estimates of energy use in EU agriculture would benefit from additional studies on energy use in all three main production systems, especially for greenhouse agriculture. In addition, further research into the energy use of non-conventional systems and crops that are cultivated on a minor scale is required as well as further research into the correlations between farm indicators (size, location, specialization) and energy use.

3.2 Overview on farmers' needs, innovative ideas and

interests

In the AgroFossilFree project farmers' needs, innovative ideas and interests were assessed by social research methods. 50-60 interviews with farmers were conducted in each of the 8 target countries of the AgroFossilFree project.

In addition, 4-5 interviews with experts in the field of energy in agriculture were conducted in each target country. Currently, the data are being processed and analysed and results will be available in autumn 2021.

Finally, a series of online and physical workshops will be organized to collect input from stakeholders. The preparation of the workshops has already started and the events will be implemented in winter 2021/22.

3.3 Inventories of FEFTS

In order to feed the AgroFossilFree online information platform, which will be launched in autumn 2021, a methodology for an inventory for data collection was elaborated. The methodology includes various categories which will facilitate the search function on the online platform. The project consortium is currently including information on FEFTS, to be fed into the online platform:

- Scientific papers
- Research projects
- Commercial technologies
- Training and advising material
- Financial support mechanisms

A database was created and in total more than 400 entries already registered. Once, the AgroFossilFree online information platform will be published, also external users will have the opportunity to register new entries.

3.4 AgEnergy Platform

AgroFossilFree will create the AgEnergy Platform to be its main online tool of promoting FEFTS worldwide. The first step of developing the website of the project (www.agrofossilfree.eu) that will interconnect interested stakeholders to the Platform has been made. The website is very informative and explains the objectives of the project in detail. The AgEnergy Platform is under construction and the wireframes of its contents are already in place. It is expected that the Platform will be launched in autumn 2021.

4 CONCLUSION

Agriculture, besides other important sectors, has to actively take various measures to mitigate climate change and thus to contribute to the target of climate neutrality in Europe as fast as possible.

These measures include innovations on livestock production, the application of fertilisers to agricultural soils, the application of pesticides, and the energy consumption in agriculture. The AgroFossilFree project contributes, with the creation of the AgEnergy platform, to knowledge transfer and easy access to information on fossil free technologies in agriculture.

4 REFERENCES

[1] European Parliament (2020) Briefing: EU agricultural policy and climate change. - https://www.europarl.europa.eu/RegData/etudes/BRIE/2020/651922/EPRS_BRI(2020)651922_EN.pdf

[2] Gołasa P. et al. (2021) Sources of Greenhouse Gas Emissions in Agriculture, with Particular Emphasis on Emissions from Energy Used. - Energies 2021, 14, 3784. https://doi.org/10.3390/en14133784 file:///C:/Users/rutz/AppData/Local/Temp/energies-14-03784-v2.pdf

[3] Eurostat (2021) Agri-environmental indicator - energy use. - https://ec.europa.eu/eurostat/statistics-explained/index.php?title=Agri-environmental_indicator_-_energy_use [06.09.2021]

5 ACKNOWLEDGEMENT & DISCLAIMER

This project has received funding from the European Union's Horizon 2020 research and innovation programme under grant agreement No 101000496. The contents of this publication are the sole responsibility of the authors and do not necessarily reflect the opinion of the European Union.

OPEN-SCHOOLING ON SOLAR ENERGY AND GREEN MOBILITY - ACTION TARGETING ON INCREASING THE STUDENTS' INTEREST IN SCIENCE AND THEIR CONFIDENCE IN A SUSTAINABLE FUTURE

G. Mantescu, G. Gorghiu, M. Bizoi
Valahia University of Targoviste, Aleea Sinaia 13, 130004, Targoviste, Romania,
phone +40 245 206 101, fax +40 245 217 692, www.valahia.ro
gabriela.mantescu@valahia.ro, gabriel.gorghiu@valahia.ro, mihai.bizoi@valahia.ro

ABSTRACT: As stated by the *European Green Deal*, the transition to low-emission or near-zero energy systems, equipped with modern and efficient structures, represents a priority addressed at the European level, from the technological, organizational, and educational point of view. One of the challenges that support the demand for technical specialists for a sustainable future in the field of renewable energy and green mobility is to keep up with the evolution of scientific results, equipment, and developed strategies. In this respect, the European educational systems are interested to develop specific competencies, starting from primary and secondary levels. In order to support the skills acquired in the classrooms, the interventions of academic teachers and scientific researchers are relevant for the dissemination of scientific developments at the international level. In this context, in the frame of the CONNECT project, a particular science action has been developed, oriented on "Sustainable future: solar energy and green mobility". The volume of information brings to the attention of young people theoretical elements in the field of solar energy, applications with photovoltaic systems, and possibilities of storing photovoltaic energy in hydrogen production. Targeting also to stimulate the orientation of youngers toward science and technology, the most efficient application relied on photovoltaic systems in Romania are presented, as well as those developed in Valahia University of Targoviste. In addition, an important emphasis is placed on the possibilities for developing ecological mobility.
Keywords: Education, Photovoltaic (PV), Sustainable Energy, Green Mobility, CONNECT Project

1. INTRODUCTION

Renewable Energy Systems and Hydrogen are set to become the targets of future sustainability and innovation in the world. An important role for the development of the global economy is played by the solar equipment. Hydrogen, is also considered one of the most important energy carriers dedicated to mobility sector as fuel and, to the energy sector as a mean to handle the intermittent nature of renewable energy systems [1]. The Green Deal stated the need of transition to low-emission or near-zero energy systems, equipped with modern and efficient structures, a priority addressed at the European level with technological, organizational and educational targets [2]. Therefore, a large number of specialists will be required for these sectors in the coming decades to make possible the transition to a sustainable economy.

The declining interest of youth in Romania for engineering specializations or for technical colleague profiles, is due to a higher volume of work required to obtain a diploma, compared to other specialization of education. Also, in the context of the rapid evolution of the technological equipment's on the energy market as well as the versatility of the labor market, the training and skills of youth after obtaining a diploma in engineering is not always appreciated by the labor market at its true values.

In the context of the increasingly complex volume of information that appears in the field of renewable energies, with a tradition of over 25 years of scientific research in solar systems, engineering education at Valahia University of Targoviste (UVT) has adopted a flexible curriculum specific to training specialists, creatively and suitable for the development and use of new facilities and equipment.

Following the recent evolution and level of training of young students in European countries, the Valahia University of Targoviste was involved in different educational actions, carrying out in the recent years specifically activities to reduce youth migration to other specialization and attracting them to the top energy sectors, such as building integrated photovoltaic (PV) systems and distributed energy PV systems. Despite all these efforts, at national and institutional level, the situation is still not balanced, that was highlighted in the evaluation made by the Romanian-German Chamber of Commerce and Industry, AHK Romania, regarding the college graduate in electrical engineering and in different study fields [3].

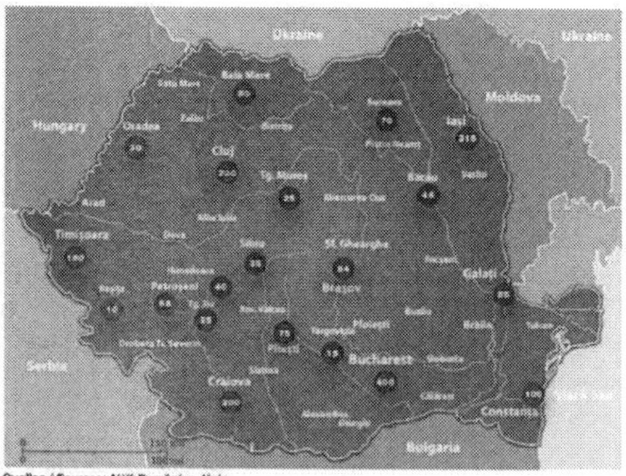

Figure 1: Number of college graduate in electrical engineering, in Romania, in 2017. Source: AHK Romania

Through their educational, technical and innovative capabilities, UVT specialists have joined in the recent years to the international work teams within the European Projects IRRESISTIBLE [4] and ENGAGE [5] to develop methods and training modules of formal and non-formal education in solar energy and nanotechnologies in order to attracting the students from secondary education to the technical specializations of industrial chemistry and electrical engineering.

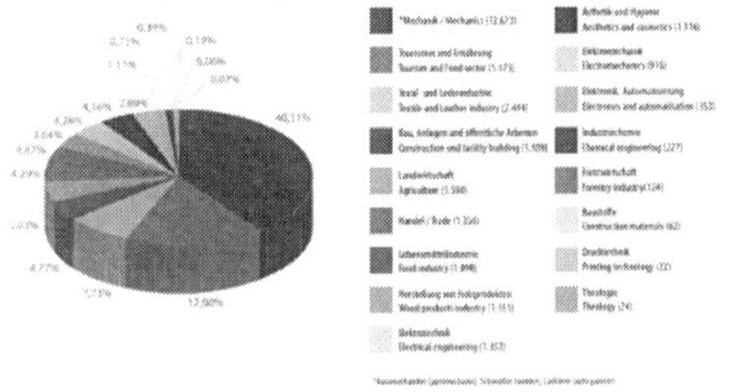

Figure 2: Number of apprentices by study field from Romania, in 2017. Source, AHK Romania

2. APROACH, INNOVATION AND RELEVANCE

Receptive to the initiatives promoted by the GREEN DEAL, a team of specialists, researchers and professors in the field of energy, chemistry, IT and education from UVT, approached a new method of communication and transmission of information to the secondary schools, "The open schooling on solar energy and electrical mobility" within the European Project CONNECT, "*Inclusive open schooling through engaging and future-oriented science* ". The main goal of the educational module is to remove the non-technological barriers, raising awareness among young students about the benefits of using photovoltaic energy and electric mobility, stimulating the choosing of a professional career in accordance with their aspirations, increasing the students' interest in science and their confidence in a sustainable future. To make open schooling inclusive and widespread in science education, the authors of this paper had organized within Project CONNECT, various online meetings with secondary school teachers, in Romania.

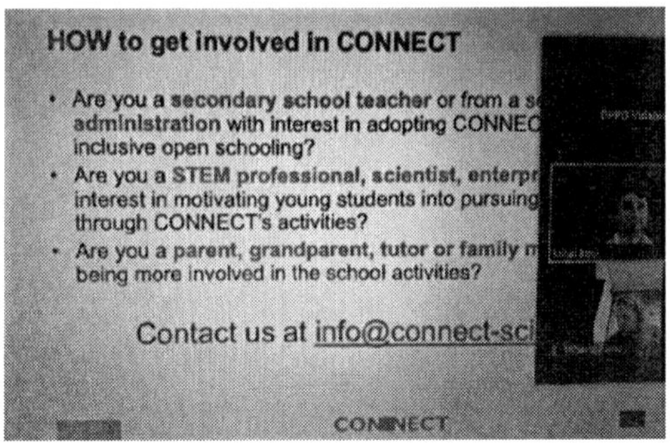

Figure 3: The online meeting of the Project CONNECT

The thematical discussions were based on the following questions:

Q1. How can open schooling increase the value and priority of partnership activities and reduce the barriers to entry so more schools are willing and able to adopt them?

Q2. How can open schooling facilitate the interaction between all partners (science professionals, teachers, students and their families) to result in effective learning experiences linked to R&I?

Q3. How can open schooling make science careers more inclusive by giving disadvantaged students more experience and insight into professional STEM jobs, supported by professionals and their families?

Following the analysis of the results obtained during the online meetings and thanks to a selection of specialized information found in [6], [7], [8], [9], [10], [11], [12] the authors of this paper adapted the innovative support for learning "CARE-DO-MODEL" developed within the Project CONNECT, in the thematic of photovoltaic conversion and its applications, Hydrogen and electric mobility, with the issues:

- developing essential skills for students interested in solar PV and electric mobility applications;
- developing practical skills that support the transfer of knowledge activities;
- creating an information transfer plan, using typical tools and methods for practical demonstrations.

With CARE-DO-MODEL will be possibly making the transfer of knowledge in these areas, in two aspects:

- We care: informal learning with professionals;
- We know: formal learning focuses on the student scientific understanding and skills.

In accordance with the school curricula agreed by schools or technical colleges, didactic materials with technical educational character in the field of photovoltaic energy, hydrogen storage and electrical mobility were created presentation sheets for the most representative applications in the field made in the country and in the world as well as presentation sheets of the great personalities in these sectors which have influenced the development of specific technologies. To support the skills acquired in the classrooms, the interventions and explanations of the academic teachers and scientific researchers from UVT are relevant for the dissemination of educational action *"Sustainable future: solar energy and green mobility"*, together with the specific equipment of solar energy, located in the Laboratories of the Institute of Science and Technology for Multidisciplinary Research (ICSTM) from UVT, and presented in figures 4-8.

Figure 4: The PV façade of ICSTM, Targoviste, Romania Source: Photos gallery of the paper author

Figure 5: PV traker, in front of the ICSTM building. Source: Photos gallery of the paper authors. Source: Photos gallery of the paper author

Figure 6: The inverters of the PV grid-connected system from ICSTM. Source: Photos gallery of the paper authors

Figure 7: Visit on the grid-connected PV system located at the terrace of ICSTM building. Source: Photos gallery of the paper authors

3. EXPLORATIVE METHODS ADAPTED WITHIN THE CARE-DO-MODEL

The informal learning method promoted in this paper was based on 4 chapters of the educational module *"Sustainable future: solar energy and green mobility"* dedicated to open schooling, in the aim of the Project CONNECT, to investigate the use of renewable energy, in particular the use of photovoltaics for hydrogen production and electrical mobility aplications, in accord with the content of the physics and chemistry curricula for the secondary level classes, as follow:
(1) theoretical explanatory of photovoltaic conversion;
(2) applications with PV systems;
(3) an overall fuel cell and hydrogen explanatory content;
(4) applications of green mobility.

Figure 8: Photovoltaic PV Panel Prototyping Laboratory. Source: Photos gallery of the paper authors

Figure 9: Presentation of different types of PV panels. Source: Photos gallery of the paper authors

In the framework of the ICSTM Laboratories, it was presented to the students from the secondary classes the state of art of equipment and new directions of research development in the sector of PV technologies.

The experimental demonstrations of photovoltaic conversion was performed with an experimental solar energy kit. The splitting of water into molecular hydrogen and molecular oxygen using solar radiation, conceptually divided into four fundamental processes as charge carrier generation, charge carrier separation, charge carrier transport, and charge carrier transfer at the electrolyte interface was demonstrate using an Hydrogen fuell cel kit and the Solar energy kit, figures10.
Within the action "We do" and with the experimental demontration students gain the knowledge of how the gases are fed into a fuel cell, where they are converted into electricity and how the current drives an electric motor, turning propeller to produce motion.

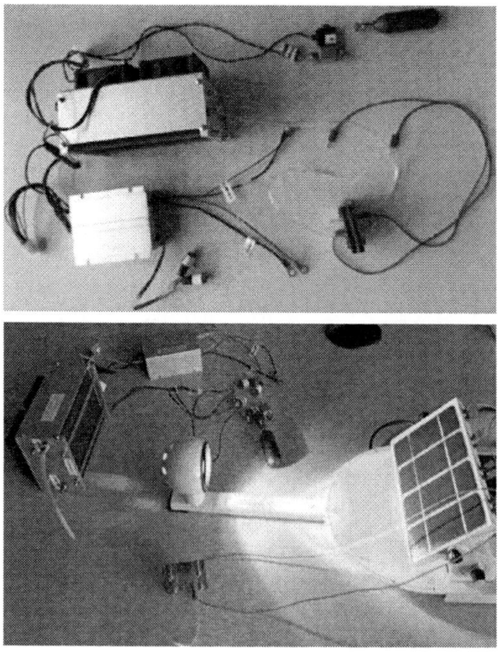

Figures 10: Kits for demonstration of the Hydrogen production, as a mean to handle the intermittent nature of PV systems. Source: Photos gallery of the paper authors

Within both solar kits demonstration, students can test and undersand a wide variety of demonstration:

- measuring the solar radiation;
- determining the solar I-V characteristic curves of PV cells;
- investigate the advantage and the disadvantage of fuel cells over conventional means of power production and analyse the difference between car with combustion engine, electric car with battery, cars with hydrogen fuel cells.

4. CONCLUSIONS

A major role for the development of a global economy in the transition to low-emission or near-zero energy systems is played by renewable energy sources and Hydrogen.

Education will play, also, a major role in the future in providing specialists for renewable energy technologies and for hydrogen sector. Through his technical ability to innovate and thanks to his organizational spirit to develop and apply new technological processes, the engineer will surely become the central character of future development, being the engine that will dynamically influence other sectors of life, ensuring the balance offered by a clean environment.

The aim of this paper was performed with in the thematic demonstratin of use of photovoltaics for hydrogen production, in accord with the content of the physics and chemistry curricula in secondary level classes. An extensive material was necessary for the dissemination of specialized information on solar-hydrogen technology to the students from the secondary classes

Evaluating the suitability and success of activities the authors of this paper suggesting further action dedicated to improve the curricula of Master classes from UVT.

ACKNOWLEDGEMENTS

This work was funded through the Project "*CONNECT - Inclusive open schooling through engaging and future-oriented science*", in the frame of the HORIZON 2020 - EU.5.d. Programme - "Encourage citizens to engage in science through formal and informal science education, and promote the diffusion of science-based activities, namely in science centers and through other appropriate channels", Topic: "SwafS-01-2018-2019-2020 - Open schooling and collaboration on science education", Grant agreement ID: 872814.

REFERENCES

[1] Bölkow, L. *Romania - Opportunities for Hydrogen Energy Technologies Considering the National Energy & Climate Plans*, Fuel Cells and Hydrogen 2 Joint Undertaking (FCH 2 JU), Study on Opportunities arising from the inclusion of Hydrogen Energy Technologies in the National Energy & Climate Plans FCH2 JU, 2020, www.fch.europa.eu

[2] European Commission. The European Green Deal. vol. Brussels, Belgium: Communicate; 2019

[3] Romania, a Growing Market, Brosure of important economic indicators and data on Romania, edited by the Romanian-German Chamber of Commerce and Industry, AHK Romania

[4] IRRESISTIBLE, "Including Responsible Research and Innovationin Cutting Edge Science and Inquiry based Science Education to Improve Teacher's, Ability of Bridging Learning Environments", Grant no. 612367, European Programme FP7-SCIENCE-IN-SOCIETY-2013-1 ACTIVITY5.2.2 Young People and Science; Topic SiS.2013.2.2.1-1: Raising youth awareness to Responsible Research and Innovation through Inquiry Based Science Education, 2013-2016;

[5] ENGAGE, "Equipping the next generation for Responsible Research and Innovation", Grant no. 612269, European Union's Seventh Framework Programme for research, technological development and demonstration

[6] Pan, A.C., Golembieski Machado, A., Menna de Oliveira, L.P., Alves Schmitt, L., Antunes De Paulo, A., Ribeiro Gomes, L.F., Souza da Silva, R. School Sustainable: *Ecological Farming, Solar Energy and Rainwater Capture as Element Educator in Country Gravatai/RS-Brazil*, Proceedings of the 33rd European Photovoltaic Solar Energy Conference and Exhibition, DOI: 10.4229/EUPVSEC20172017-7DV.1.50

[7] Hartmann, U., Haselhuhn, R., Vanicek, P. *Education and Training, Fundamentals, Grid-Connected, Stand-Alone PV System, Photovoltaic (PV)*, Proceedings of the 28th European Photovoltaic Solar Energy Conference and Exhibition, DOI: 10.4229/28thEUPVSEC2013-6BV.8.22

[8] W.G.J.H.M. van Sark, A.C. de Waal, J. Uithol, N. Dols, F. Houben, R. Kuepers, M. Scherrenburg, B. van Lith, F. Benjamin, *Energy Performance of a 1.2 MWp Photovoltaic System Distributed over Eight Buildings at Utrecht University Campus*, Proceedings of the 33rd European Photovoltaic Solar Energy Conference and Exhibition, DOI: 10.4229/EUPVSEC20172017-6BV.1.31

[9] Visa, I., Perniu, D., Jaliu, C., Visa, M., Duta, A. *European Training In-Service Tool on Sustainable Energy*, Proceedings of the 23rd European Photovoltaic Solar Energy Conference and Exhibition, DOI: 10.4229/23rdEUPVSEC2008-6DV.5.22

[10] Mateescu, C., Popa, I. *European best practices and policies in promoting green mobility*, Conference: The 12th Electric Vehicles Conference & Show, ICPE Solar Park, Electrotehnica, Electronica, Automatica (EEA), 2017, vol. 65, no. 4, pp. 12-16, ISSN 1582-5175

[11] Kakoulakia, G., Kougiasa, I., Taylora, N., Dolcib, F., Moyab, J., Jäger-Waldaua, J. *Green hydrogen in Europe – A regional assessment: Substituting existing production with electrolysis powered by renewables*, Energy Conversion and Management, Volume228 (2021) 113649

[12] C. S. Polo López, F. Frontini, P. Bonomo, Enhancing solar research by using ICT and explorative web-based methods for communication, education and training, 35th European Photovoltaic Solar Energy Conference and Exhibition, pp 2199 - 2204

AUTHOR INDEX

WIP – Renewable Energies
Sylvensteinstr. 2
81369 Munchen
Germany

ISBN 978-1-7138-5213-1

CURRAN ASSOCIATES INC.
proceedings.com